W9-AQF-198

Brooker | Widmaier | Graham | Stiling

BIOLOGY

THIRD EDITION

Robert J. Brooker
University of Minnesota - Twin Cities

Eric P. Widmaier
Boston University

Linda E. Graham
University of Wisconsin - Madison

Peter D. Stiling
University of South Florida

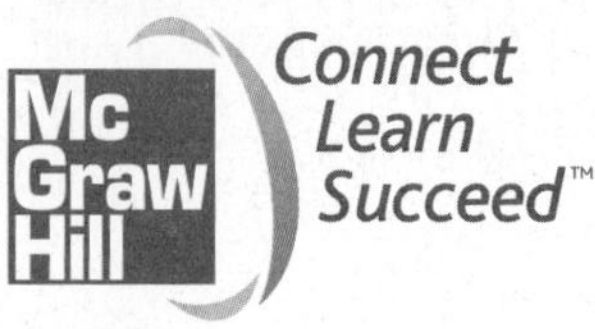

BIOLOGY, THIRD EDITION

Published by McGraw-Hill, a business unit of The McGraw-Hill Companies, Inc., 1221 Avenue of the Americas, New York, NY 10020.

This book is printed on acid-free paper.

1 2 3 4 5 6 7 8 9 0 DOW/DOW 1 0 9 8 7 6 5 4 3

ISBN 978-0-07-353224-0
MHID 0-07-353224-X

Senior Vice President, Products & Markets: *Kurt L. Strand*
Vice President, General Manager, Products & Markets: *Marty Lange*
Vice President, Content Production & Technology Services: *Kimberly Meriwether David*
Managing Director: *Michael S. Hackett*
Director: *Lynn Breithaupt*
Brand Manager: *Rebecca Olson*
Director of Development–Biology: *Elizabeth M. Sievers*
Director of Digital Content Development: *Tod Duncan*
Executive Marketing Manager: *Patrick E. Reidy*
Content Project Manager: *Sandy Wille*
Senior Buyer: *Laura Fuller*
Senior Media Project Manager: *Tammy Juran*
Senior Designer: *David W. Hash*
Cover/Interior Designer: *John Joran*
Cover Image: *Stichodactyla haddoni* (saddle carpet anemone) and *Amphiprion ocellaris* (clownfish) © Pedro GoniO
Content Licensing Specialist: *John Leland*
Photo Research: *Danny Meldung / Photo Affairs*
Compositor: *Lachina Publishing Services*
Typeface: *9.5/12 Minion Pro*
Printer: *R. R. Donnelley*

Library of Congress Cataloging-in-Publication Data

Biology / Robert J. Brooker . . . [et al.]. – 3rd ed.
p. cm.
Includes index.
ISBN 978-0-07-353224-0 — ISBN 0-07-353224-X (hard copy : alk. paper) 1. Biology–Textbooks.
I. Brooker, Robert J.
QH308.2.B564445 2014
570–dc23

2012005419

www.mhhe.com

Brief Contents

About the Authors iv
Improving Biology Education v
Guided Tour x
Acknowledgments xviii
Contents xxv

1 An Introduction to Biology 1

Unit I Chemistry

2 The Chemical Basis of Life I: Atoms, Molecules, and Water 21
3 The Chemical Basis of Life II: Organic Molecules 42

Unit II Cell

4 General Features of Cells 66
5 Membrane Structure, Synthesis, and Transport 97
6 An Introduction to Energy, Enzymes, and Metabolism 118
7 Cellular Respiration and Fermentation 136
8 Photosynthesis 154
9 Cell Communication 174
10 Multicellularity 193

Unit III Genetics

11 Nucleic Acid Structure, DNA Replication, and Chromosome Structure 212
12 Gene Expression at the Molecular Level 235
13 Gene Regulation 257
14 Mutation, DNA Repair, and Cancer 279
15 The Eukaryotic Cell Cycle, Mitosis, and Meiosis 297
16 Simple Patterns of Inheritance 321
17 Complex Patterns of Inheritance 343
18 Genetics of Viruses and Bacteria 359
19 Developmental Genetics 380
20 Genetic Technology 400
21 Genomes, Proteomes, and Bioinformatics 420

Unit IV Evolution

22 The Origin and History of Life on Earth 438
23 An Introduction to Evolution 459
24 Population Genetics 478
25 Origin of Species and Macroevolution 495
26 Taxonomy and Systematics 514

Unit V Diversity

27 Archaea and Bacteria 533
28 Protists 553
29 Plants and the Conquest of Land 575
30 The Evolution and Diversity of Modern Gymnosperms and Angiosperms 597
31 Fungi 618
32 An Introduction to Animal Diversity 638
33 The Invertebrates 652
34 The Vertebrates 684

Unit VI Flowering Plants

35 An Introduction to Flowering Plant Form and Function 716
36 Flowering Plants: Behavior 737
37 Flowering Plants: Nutrition 756
38 Flowering Plants: Transport 772
39 Flowering Plants: Reproduction 793

Unit VII Animals

40 Animal Bodies and Homeostasis 813
41 Neuroscience I: Cells of the Nervous System 831
42 Neuroscience II: Evolution, Structure, and Function of the Nervous System 854
43 Neuroscience III: Sensory Systems 875
44 Muscular-Skeletal Systems and Locomotion 901
45 Nutrition and Animal Digestive Systems 920
46 Control of Energy Balance, Metabolic Rate, and Body Temperature 944
47 Circulatory Systems 964
48 Respiratory Systems 986
49 Excretory Systems and Salt and Water Balance 1008
50 Endocrine Systems 1030
51 Animal Reproduction 1056
52 Animal Development 1076
53 Immune Systems 1094

Unit VIII Ecology

54 An Introduction to Ecology and Biomes 1117
55 Behavioral Ecology 1146
56 Population Ecology 1168
57 Species Interactions 1187
58 Community Ecology 1207
59 Ecosystem Ecology 1225
60 Biodiversity and Conservation Biology 1244

Appendix A: Periodic Table of the Elements A-1
Appendix B: Answer Key A-2
Glossary G-1
Photo Credits C-1
Index I-1

About the Authors

Left to right: Eric Widmaier, Linda Graham, Peter Stiling, and Rob Brooker

Robert J. Brooker

Rob Brooker (Ph.D., Yale University) received his B.A. in biology at Wittenberg University, Springfield, Ohio, in 1978. At Harvard, he studied lactose permease, the product of the *lacY* gene of the *lac* operon. He continues working on transporters at the University of Minnesota, where he is a Professor in the Department of Genetics, Cell Biology, and Development and has an active research laboratory. At the University of Minnesota, Dr. Brooker teaches undergraduate courses in biology, genetics, and cell biology. In addition to many other publications, he has written two undergraduate genetics texts published by McGraw-Hill: *Genetics: Analysis & Principles,* 4th edition, copyright 2012, and *Concepts of Genetics*, copyright 2012.

Eric P. Widmaier

Eric Widmaier received his Ph.D. in 1984 in endocrinology from the University of California at San Francisco. His research focuses on the control of body mass and metabolism in mammals, the hormonal correlates of obesity, and the effects of high-fat diets on intestinal cell function. Dr. Widmaier is currently Professor of Biology at Boston University, where he teaches undergraduate human physiology and recently received the university's highest honor for excellence in teaching. Among other publications, he is a coauthor of *Vander's Human Physiology: The Mechanisms of Body Function*, 13th edition, published by McGraw-Hill, copyright 2014.

Linda E. Graham

Linda Graham received her Ph.D. in botany from the University of Michigan, Ann Arbor. Her research explores the evolutionary origin of land-adapted plants, focusing on their cell and molecular biology as well as ecological interactions. Dr. Graham is now Professor of Botany at the University of Wisconsin-Madison. She teaches undergraduate courses in biology and plant biology. She is the coauthor of, among other publications, *Algae*, 2nd edition, copyright 2008, a major's textbook on algal biology, and *Plant Biology*, 2nd edition, copyright 2006, both published by Prentice Hall/Pearson.

Peter D. Stiling

Peter Stiling obtained his Ph.D. from University College, Cardiff, Wales, in 1979. Subsequently, he became a postdoctoral fellow at Florida State University and later spent two years as a lecturer at the University of the West Indies, Trinidad. During this time, he began photographing and writing about butterflies and other insects, which led to publication of several books on local insects. Dr. Stiling is currently a Professor of Biology at the University of South Florida at Tampa. His research interests include plant-insect relationships, parasite-host relationships, biological control, restoration ecology, and the effects of elevated carbon dioxide levels on plant herbivore interactions. He teaches graduate and undergraduate courses in ecology and environmental science as well as introductory biology. He has published many scientific papers and is the author of *Ecology: Global Insights and Investigations*, published by McGraw-Hill, copyright 2012.

The authors are grateful for the help, support,
and patience of their families, friends, and students,
Deb, Dan, Nate, and Sarah Brooker,
Maria, Rick, and Carrie Widmaier,
Jim, Michael, and Melissa Graham, and
Jacqui, Zoe, Leah, and Jenna Stiling.

The Learning Continues in the Digital Environment

The digital offerings for the study of biology have become a key component of both instructional and learning environments. In response to this, the author team welcomes Dr. Ian Quitadamo as Lead Digital Author for *Biology*, 3rd edition. As Lead Digital Author, Ian oversaw the development of digital assessment tools in Connect®. Ian's background makes him a uniquely valuable addition to the third edition of *Biology*.

Ian Quitadamo

Ian Quitadamo is an Associate Professor with a dual appointment in Biological Sciences and Science Education at Central Washington University in Ellensburg, Washington. He teaches introductory and majors biology courses and cell biology, genetics, and biotechnology, as well as science teaching methods courses for future science teachers and interdisciplinary content courses in alternative energy and sustainability. Dr. Quitadamo was educated at Washington State University and holds a BA in biology, Masters degree in genetics and cell biology, and an interdisciplinary Ph.D. in science, education, and technology. Previously a researcher of tumor angiogenesis, he now investigates critical thinking and has published numerous studies of factors that affect student critical thinking performance. He has received the Crystal Apple award for teaching excellence, led various initiatives in critical thinking and assessment, and is active in training future and currently practicing science teachers. He served as a coauthor on *Biology*, 11th edition, by Mader and Windelspecht, copyright 2013 and is the lead digital author for *Biology*, 10th edition by Raven, copyright 2014, both published by McGraw-Hill.

Improving Biology Education: We Listened to You

A New Vision for Learning

A New Vision for Learning describes what we set out to accomplish with this third edition of our textbook. As authors and educators, we know your goal is to ensure your students are prepared for the future—their future course work, lab experiences, and careers in the sciences. Building a strong foundation in biology for your students required a new vision for how they learn.

Through our classroom experiences and research work, we became inspired by the prospect that the first and second editions of *Biology* could move biology education forward. We are confident that this new edition of *Biology* is another step in the right direction because we listened to you. Based on our own experience and our discussions with educators and students, we continue to concentrate our efforts on these crucial areas:

- Experimentation and the process of science
- Modern content
- Evolutionary perspective
- Emphasis on visuals
- Accuracy and consistency
- Critical thinking
- Media—Active teaching and learning with technology

Continued feedback from instructors using this textbook and other educators in the field of biology has been extremely valuable in refining the presentation of the material. Likewise, we have used the textbook in our own classrooms. This hands-on experience has provided much insight for areas for improvement. Our textbook continues to be comprehensive and cutting-edge, featuring an evolutionary focus and an emphasis on scientific inquiry.

The first edition of *Biology* was truly innovative in its visual program, and the third edition continues to emphasize this highly instructional visual program. In watching students study as well as in extensive interviews, it is clear that students rely heavily on the artwork as their primary study tool. As you will see when you scan through our book, the illustrations have been crafted with the student's perspective in mind. They are very easy to follow, particularly those that have multiple steps, and have very complete explanations of key concepts. We have taken the approach that students should be able to look at the figures and understand the key concepts, without having to glance back and forth between the text and art. Many figures contain text boxes that explain what the illustration is showing. In those figures with multiple steps, the boxes are numbered and thereby guide the students through biological processes.

The figures are of excellent quality and one of the reasons we switched to Brooker.

Michael Cullen, University of Evansville

The visuals are definitely a strength for this chapter. They are easy to interpret and illustrate the basic themes in genetic engineering in a way that students can comprehend.

Kim Risley, University of Mount Union

A New Vision Serving Teachers and Learners

To accurately and thoroughly cover a course as wide-ranging as biology, we felt it was essential that our team reflect the diversity of the field. We saw an opportunity to reach students at an early stage in their education and provide their biology training with a solid and up-to-date foundation. We have worked to balance coverage of classic research with recent discoveries that extend biological concepts in surprising new directions or that forge new concepts. Some new discoveries were selected because they highlight scientific controversies, showing students that we don't have all the answers yet. There is still a lot of work for new generations of biologists. With this in mind, we've also spotlighted discoveries made by diverse people doing research in different countries to illustrate the global nature of modern biological science.

This is an excellent textbook for biology majors, and the students should keep the book as a future reference. The thoughts flow very well from one topic to the next.

Gary Walker, Youngstown State University

As active teachers and writers, one of the great joys of this process for us is that we have been able to meet many more educators and students during the creation of this textbook. It is humbling to see the level of dedication our peers bring to their teaching. Likewise, it is encouraging to see the energy and enthusiasm so many students bring to their studies. We hope this book and its media package will serve to aid both faculty and students in meeting the challenges of this dynamic and exciting course. For us, this remains a work in progress, and we encourage you to let us know what you think of our efforts and what we can do to serve you better.

Rob Brooker, Eric Widmaier, Linda Graham, Peter Stiling

CHANGES TO THIS EDITION

The author team is dedicated to producing the most engaging and current text that is available for undergraduate students who are majoring in biology. We have listened to educators and reviewed documents, such as *Vision and Change, A Call to Action*, which includes a summary of recommendations made at a national conference organized by the American Association for the Advancement of Science (see www.visionandchange.org). We want our textbook to reflect core competencies and provide a more learner-centered approach. To achieve these goals, we have made the following innovations to *Biology*, third edition.

- **Principles of Biology:** Based on educational literature and feedback from biology educators, we have listed 12 Principles of Biology in Chapter 1. These 12 principles align with the overarching core concepts described in *Vision and Change*, with 1 to 3 principles included per core concept.
- The Principles of Biology are threaded throughout the entire textbook. This is achieved in two ways. First, the principles are explicitly stated in selected figure legends in every chapter. Such legends are given a Principle of Biology icon. In addition, a question at the end of each chapter is directly aimed at a particular principle.
- **Unit openers**: We have added Unit openers to the third edition. These openers serve two purposes. They allow the student to see the "big picture" of the unit. In addition, the unit openers draw attention to the principles of biology that will be emphasized in that unit.
- **BioConnections**: We have also added a new feature called BioConnections, which are found in selected figure legends in each chapter. The BioConnections inform students of how a topic in one chapter is connected to a topic in another.
- **Learning Outcomes**: As advocated in *Vision and Change*, educational materials should have well-defined learning goals. In this third edition of *Biology*, we begin each section of every chapter with a set of Learning Outcomes. These outcomes inform students of the skills they will acquire when mastering the material and provide a tangible understanding of how such skills may be assessed. The assessment in Connect was developed using these Learning Outcomes as a guide in formulating online questions, thereby linking the learning goals of the text with the assessment in Connect.

With regard to the scientific content in the textbook, the author team has worked with hundreds of faculty reviewers to refine this new edition and to update the content so that our students are exposed to the most current material. Some of the key changes that have occurred are summarized below.

- **Chapter 1. An Introduction to Biology:** As mentioned earlier, we have explicitly stated 12 Principles of Biology in Figure 1.4 and described them on pages 2 through 5.

Chemistry Unit

- **Chapter 2. The Chemical Basis of Life I: Atoms, Molecules, and Water:** Figure 2.1 (diagram of simple atoms) is now introduced before the Rutherford experiment, so that students are exposed immediately to the basic nature of atoms. In addition, Figure 2.16 has been revised (spatially) to better represent partial charges on water molecules to make it clear how attractive forces occur between water molecules.
- **Chapter 3. The Chemical Basis of Life II: Organic Molecules:** We expanded the discussion of protein domains in the context of evolution and adjusted Figure 3.21 to more accurately reflect domains of STAT, including the addition of the Linker Domain. We also improved color coding of pentoses and hexoses as well as amino acid side chains, and other chemical groups, throughout the chapter to maintain consistency.

Cell Unit

- **Chapter 4. General Features of Cells:** A new figure has been added that illustrates the relationship between cell size and the surface area/volume ratio (see Figure 4.8). Several BioConnections are made between figures in this chapter and topics in other units, thereby bridging the gap between cell biology and life at the organismal level.

- **Chapter 5. Membrane Structure, Synthesis, and Transport:** The section on Membrane Structure is now divided into two sections: one that emphasizes structure and another that emphasizes fluidity. Similarly, the section on Membrane Transport is now divided into two sections: an overview of membrane transport and then a section that describes how transport proteins function.
- **Chapter 6. An Introduction to Energy, Enzymes, and Metabolism:** The added learning outcomes to this chapter inform the student of the key concepts and how they relate to the laws of chemistry and physics. The discussion of reversible and irreversible inhibitors of enzymes has been expanded.
- **Chapter 7. Cellular Respiration and Fermentation:** The discussion of cellular respiration from the previous edition is now separated into six sections that begin with an overview and then emphasize different processes of cell respiration, such as glycolysis, citric acid cycle, and oxidative phosphorylation. A greater emphasis is placed on how these processes are regulated. Also, a new figure has been added that describes the mechanism of ATP synthesis via ATP synthase (see Figure 7.12).
- **Chapter 8. Photosynthesis:** BioConnections in this chapter connect photosynthesis at the cellular level to topics described in the Plant Biology unit.
- **Chapter 9. Cell Communication:** The initial discussion of cellular receptors and their activation has been subdivided into two topics: how signaling molecules bind to receptors and how receptors undergo conformational changes. A brief description of the intrinsic pathway of apoptosis has also been added.
- **Chapter 10. Multicellularity:** BioConnections in figure legends have been added to connect this cell biology chapter to topics in the Animal and Plant Biology units. Biology Principle legends remind students that "Structure determines function" and that "New properties emerge from complex interactions."

Genetics Unit

- **Chapter 11. Nucleic Acid Structure, DNA Replication, and Chromosome Structure:** A new illustration has been added that shows how two replication forks emanate from an origin of replication (see Figure 11.19b).
- **Chapter 12. Gene Expression at the Molecular Level:** Several illustrations have been revised for student clarity. BioConnections in figure legends relate gene transcription to other topics such as DNA replication.
- **Chapter 13. Gene Regulation:** The section on eukaryotic gene regulation has been divided into two sections: one section emphasizes regulatory transcription factors and the other emphasizes changes in chromatin structure. Three new figures have been added to this chapter that pertain to chromatin remodeling (see Figure 13.17), nucleosome arrangements in the vicinity of a structural gene (see Figure 13.19), and the most current model for how eukaryotic genes are activated (see Figure 13.20).
- **Chapter 14. Mutation, DNA Repair, and Cancer:** Several illustrations have been revised to match presentation styles in other chapters. BioConnections relate information in this genetics chapter to topics in previous cell biology chapters.
- **Chapter 15. The Eukaryotic Cell Cycle, Mitosis, and Meiosis:** At the request of reviewers, a new illustration has been added that shows how nondisjunction can occur during meiosis (see Figure 15.18).
- **Chapter 16. Simple Patterns of Inheritance:** The section on Variations in Inheritance Patterns and Their Molecular Basis has been streamlined to have fewer types of inheritance patterns.
- **Chapter 17. Complex Patterns of Inheritance:** The topic of maternal effect has been moved to Chapter 19, where it is discussed at the molecular level. The last section of this chapter focuses on epigenetics.
- **Chapter 18. Genetics of Viruses and Bacteria:** BioConnections relate the genetics of viruses and bacteria to the structure and function of cells.
- **Chapter 19. Developmental Genetics:** New insets describing the radial pattern of plant growth have been added to Figures 19.2 and 19.22. Biology principle legends remind students that developmental biology is an experimental science that uses model organisms, and that new properties of life arise by complex interactions.
- **Chapter 20. Genetic Technology:** The technique of DNA sequencing now emphasizes the more modern approach of using fluorescently labeled nucleotides (see Figure 20.9). BioConnections relate the techniques described in this chapter to topics found in other chapters, such as DNA replication.
- **Chapter 21. Genomes, Proteomes, and Bioinformatics:** Biology principle legends help students understand how bioinformatics techniques illuminate principles of evolution.

Evolution Unit

- **Chapter 22. The Origin and History of Life:** New information regarding the experiments of Urey and Miller has been added. Figure 22.12 has been revised to reflect the importance of endocytosis for the emergence of the first eukaryotic cells.
- **Chapter 23. An Introduction to Evolution:** The historical development of Darwin's theory of evolution has been set off in its own subsection. BioConnections relate the study of evolution to bioinformatics techniques.
- **Chapter 24. Population Genetics:** The introductory material on natural selection has been revised to better emphasize that natural selection is typically related to two aspects of reproductive success: traits that are directly associated with reproduction, such as gamete viability, and the ability to survive to reproductive age.
- **Chapter 25. Origin of Species and Macroevolution:** The topic of species concepts has been highlighted in its own separate subsection. Biology principle legends remind students that populations of organisms evolve from one generation to the next.
- **Chapter 26. Taxonomy and Systematics:** Phylogenetic trees have been revised to include more plant examples and to include changes in the environment (see Figures 26.3 and 26.7). The use of DNA sequence changes in primates to hypothesize a phylogenetic tree has been moved to the molecular clock section of the this chapter, which reflects how the data relate to neutral changes in DNA sequences (see Figure 26.13).

Diversity Unit

- **Chapter 27. Bacteria and Archaea:** A new Feature Investigation has been added that highlights the recent discovery that soil bacteria from diverse habitats are able to break down and consume a wide variety of commonly used antibiotics and in the process become resistant to them. This new feature reinforces the general concept that heterotrophic organisms require an organic carbon food source and reiterates the roles of natural selection and horizontal gene transfer in the evolution and spread of traits such as antibiotic resistance. A new illustration has been added that compares and contrasts the structure of bacterial and archaeal membranes, thereby building student knowledge about cell membrane structure and function.

- **Chapter 28. Protists:** This chapter includes an updated concept of protist diversification based on recent research findings. New information about malarial disease development reinforces earlier chapter content on the process and significance of phagocytosis, important both in the lives of protists and immune function in humans.
- **Chapter 29. Plants and the Conquest of Land:** Art specially produced for this edition illustrates a new concept of vegetation spacing in landscapes of the Carboniferous (coal age), a time period when plants dramatically changed Earth's atmospheric chemistry and climate, thereby setting the stage for modern ecosystems. The new illustration helps to emphasize human need to understand past interactions of biology and physical environment as one way to predict future change. A new Genomes and Proteomes feature focuses on the evolutionary and potential pharmaceutical value of comparative genomic analyses leveraged by the availability of increasing numbers of plant genomes. This new material reinforces the concept that plants produce many types of secondary chemical compounds that influence humans and other organisms.
- **Chapter 30. The Evolution and Diversity of Modern Gymnosperms and Angiosperms:** The discussion has been expanded to emphasize the roles of whole-genome duplication and polyploidy in the diversification of the flowering plants on which humans depend.
- **Chapter 31. Fungi:** This chapter includes updated concepts of fungal diversification based on recent research, accompanied by new illustrations of cryptomycota and microsporidia and descriptions of their evolutionary and ecological significance.
- **Chapter 32. An Introduction to Animal Diversity:** The chapter now presents a single unified animal phylogeny rather than two alternate versions based on body plans or molecular data. In addition, the section on animal characteristics has been rewritten, as has the section on specific features of embryonic development. New conceptual questions were added.
- **Chapter 33. The Invertebrates:** The section on Lophotrochozoa has been reorganized, and new photographs have been added. The section on annelids has been completely modernized and rewritten.
- **Chapter 34. The Vertebrates:** The treatment of early vertebrates, the hagfish and lamprey has undergone a major revision, and new phylogenies have been constructed and used throughout. Mammal phylogeny has been rewritten to include information from new molecular studies, and a new figure has been added (Figure 34.26). Bird taxonomy has also been updated in the text and tables. New conceptual questions have been added.

Plant Unit

- **Chapter 36. Flowering Plants: Behavior:** The chapter features new material on the function of circadian rhythms, displayed not only in plants, but also in animals and microorganisms.
- **Chapter 38. Flowering Plants: Transport:** The Feature Investigation on xylem transport has been updated to reflect recent research results.
- **Chapter 39. Flowering Plants: Reproduction:** A new Feature Investigation has been added that focuses on how physics and mathematical modeling can be used to explain how flowers bloom. This new material is a response to recent pedagogical motivation to more effectively integrate physics and mathematics into biology training at an early stage.

Animal Unit

Key changes to the Animal Unit include updating statistics on human disease to reflect current numbers. Cross references have been modified to give specific figure and table numbers rather than simply "see Chapter XX." In many chapters, the end-of-chapter Conceptual Questions were altered to make them more challenging and thought-provoking.

- **Chapter 40. Animal Bodies and Homeostasis:** The discussion of *Hox* genes and their role in organ development was simplified. The chapter-opening introduction was revised to more closely relate to the opening photo and to better tie in with a main theme of the chapter (Homeostasis).
- **Chapter 41. Neuroscience I: Cells of the Nervous System:** The chapter was reorganized to cover resting potentials, action potentials, and synapses in separate sections. The description of microelectrodes and membrane potential recording in squid giant axons was revised to improve understanding. In Figure 41.5 a preimpaled tracing of membrane potential was added to emphasize the negative resting potential in this experiment, and the experimental tracing was adjusted to better reflect what is actually seen in such an experiment. The five keyed steps associated with Figure 41.10 now have headers to help emphasize what is happening at each of those steps. Figures 41.11 and 41.12, showing movement of positive charges along an axon in action potential propagation, were improved. In Figure 41.14, voltage-gated calcium channels were added to the active zone of the axon terminal to reflect recent advances in our understanding of their location on presynaptic cells. In Figure 41.17 trimeric G protein was added to its metabotropic receptor to emphasize its mechanism of action. Table 41.3 was altered such that drugs are organized into subcategories according to therapeutic value, or as illicit or recreational drugs.
- **Chapter 42. Neuroscience II: Evolution and Function of the Brain and Nervous System:** Many of the figures were improved. Figure 42.5 now shows a side-by-side comparison of human and frog nervous systems, emphasizing their similarities. In Figure 42.7, meninges layers were enhanced for visual clarity. Figure 42.9 was redrawn and relabeled to more clearly identify the parts of the human brain. Figure 42.12, showing the lobes of the cortex, is a new figure (formerly part of another figure) and now includes the major functions of each lobe. Figure 42.14a includes an improved photo of *Aplysia*. In Figure 42.15, the presynaptic cell changes that occur with learning are now identified using color and line coding. In Figure 42.18, a new line art illustration of plaques and tangles in the brain of a person with Alzheimer's disease now accompanies the light-micrograph.

- **Chapter 43. Neuroscience III: Sensory Systems:** The Genomes and Proteomes Connection was expanded to describe the evolution of color vision (including the genetics of color blindness). The descriptions of the evolution of eyes and vision was expanded. In Figure 43.5 and elsewhere, the middle ear bones are now color-coded to help distinguish them. Improved color matching in small and large (blow-ups) images of fly ommatidium has been incorporated into Figure 43.14. In Figure 43.15, the macula is now labeled on the illustration of the eye/retina. Figures 43.21, 43.25 and 43.27 were improved with better sizing, labeling, color, and detail for realism.
- **Chapter 44. The Muscular-Skeletal System and Locomotion:** In Figure 44.5, an SEM of a sarcomere (relaxed vs. contracted) was added to the line art to provide a real-life image of sliding filaments. The labeling of Figure 44.6 was improved to help clarify this complex figure. Figure 44.8 was improved for clarity and accuracy of cross-bridge cycling. In Figure 44.10, the sarcoplasmic reticulum and T-tubules of muscle were redrawn to include lateral sacs, and to improve clarity, accuracy, and detail. In Figure 44.15b, better comparison photos were provided to illustrate the differences between normal and osteoporotic bone. The Genomes and Proteomes Connection was updated with more recent data.
- **Chapter 45. Nutrition and Animal Digestive Systems:** The chapter has been reorganized so that it begins with a new section that provides an overview of nutrition and ingestion. The layout for Figure 45.3 was revised for easier flow and clarity. Figure 45.4b now includes a very clear and dramatic new photo of leech mouthparts. Figure 45.8, the digestive system of a ruminant, was simplified for clarity. In Figures 45.9, 45.10, 45.11, and 45.15, the labeling was improved for greater detail.
- **Chapter 46. Control of Energy Balance, Metabolic Rate, and Body Temperature:** Figure 46.16 is a new CDC-derived figure showing rates of obesity in U.S. counties across the country. The discussion of brown adipose tissue has been simplified.
- **Chapter 47. Circulatory Systems:** Several figures were enlarged and simplified to improve clarity. In many cases, additional labels and leader lines have been added or modified to improve accuracy and clarity. The distinguishing anatomic and physiologic features of cardiac muscle are now reviewed early in the chapter (with reference back to Chapter 44). The Genomes and Proteomes Connection was changed from the genetics of hemophilia (some of this information was retained in the main text) to a new feature: A Four-Chambered Heart Evolved from Simple Contractile Tubes. This new Genomes and Proteomes Connection also includes a new Figure 47.4).
- **Chapter 48. Respiratory Systems:** The chapter was reorganized so that the Mechanisms of Oxygen Transport section precedes the section on Control of Ventilation. This is consistent now with other chapters in this unit in which mechanisms are described before control. A new SEM photomicrograph was added to Figure 48.7, showing a spiracle on the body surface of an insect.
- **Chapter 49. Excretory Systems and Salt and Water Balance:** In Figure 49.14, tight junctions were added to cells in the epithelia. Figure 49.9 was improved by making the Malpighian tubules more visible. Figure 49.10c was improved by more accurate and visible labeling, by extending the collecting duct, and by removing the background coloring so that this complex image is much more easily viewed. Figures 49.11 and 49.12 were improved with better labels and leaders. New Figure 49.15 was added, showing the mechanism of action of antidiuretic hormone on medullary collecting duct cells, including aquaporin migration.
- **Chapter 50. Endocrine Systems:** Figure 50.5b (thyroid hormone synthesis) was simplified to improve clarity. A new photo of a flounder was added to Figure 50.14b, to show thyroid-induced effects on eye development. The Genomes and Proteomes Connection was moved to later in the chapter after a discussion of steroid hormones.
- **Chapter 51. Animal Reproduction:** Several discussions were clarified and enhanced with more specific detail, including various morphologic and functional aspects of ovarian cycle, follicular development, and parturition. Figure 51.7 (flow chart of endocrine control of male reproduction) was improved by a new drawing and now includes the hormone inhibin.
- **Chapter 53. Immune Systems:** A new section and Feature Investigation on Toll-Like Receptors and pathogen-associated molecular patterns (PAMPs) were added. Figure 53.2 was enhanced with additional detail to include the presence of nitric oxide and macrophages. Figure 53.7, the structure of immunoglobulin, was simplified for clarity. In Figure 53.8, the number of "variable" segments of an immunoglobulin was changed to reflect more recent understanding of structure. Former Figure 53.9 was split into two figures (Figures 53.10 and 53.11) to help walk the reader through this complex material. A new figure was added to the Public Health section, showing the number of people living with HIV infection over the past 20 years.

Ecology Unit

- **Chapter 54. An Introduction to Ecology and Biomes:** A new section on Continental Drift and Biogeography was added, which addresses the importance of evolution and dispersal on the distribution of life on Earth.
- **Chapter 55. Behavioral Ecology:** The discussion on mating systems has been reorganized and a new figure added to better explain the concepts. The Genomes and Proteomes Connection was updated. Two new conceptual questions have been included.
- **Chapter 56. Population Ecology:** Section 56.3, How Populations Grow, has been expanded slightly. In addition, the data on human population growth has been updated. Two new conceptual questions have been included.
- **Chapter 57. Species Interactions:** The information on enslaver parasites and the section on Parasitism have been expanded to include more detail. The Genomes and Proteomes Connection was updated. Two new conceptual questions have been included.
- **Chapter 58. Community Ecology:** The Genomes and Proteomes Connection was updated. Two new conceptual questions have been included.
- **Chapter 59. Ecosystem Ecology:** The section on biogeochemical cycles was reorganized, and some figures have been updated. The Genomes and Proteomes Connection was replaced with a newer, relevant topic.
- **Chapter 60. Biodiversity and Conservation Biology:** The chapter opens with new introductory material and the section "Why Conserve Biodiversity?" has been updated. New information on conservation strategies has been added.

A NEW VISION FOR LEARNING: PREPARING STUDENTS FOR THE FUTURE

MAKING CONNECTIONS

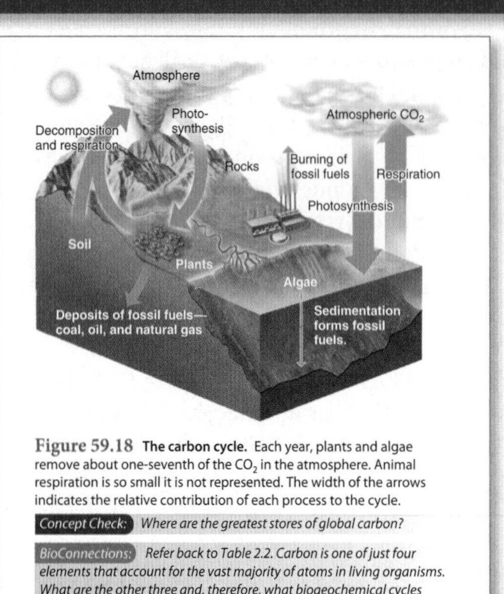

Figure 59.18 **The carbon cycle.** Each year, plants and algae remove about one-seventh of the CO_2 in the atmosphere. Animal respiration is so small it is not represented. The width of the arrows indicates the relative contribution of each process to the cycle.

Concept Check: *Where are the greatest stores of global carbon?*

BioConnections: *Refer back to Table 2.2. Carbon is one of just four elements that account for the vast majority of atoms in living organisms. What are the other three and, therefore, what biogeochemical cycles might be the most important to us?*

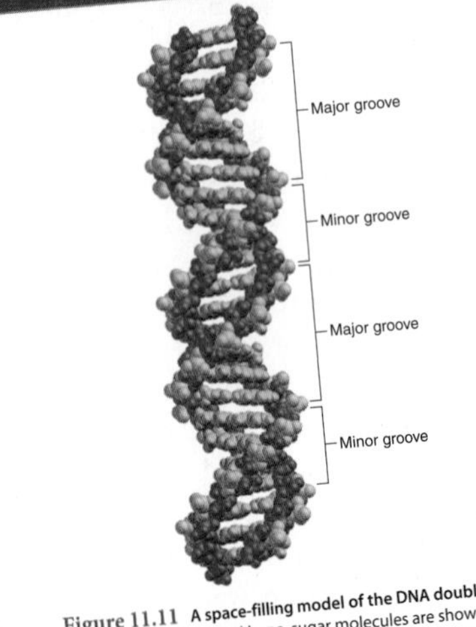

Figure 11.11 **A space-filling model of the DNA double helix.** In the sugar-phosphate backbone, sugar molecules are shown in blue, and phosphate groups are yellow. The backbone is on the outermost surface of the double helix. The atoms of the bases, shown in green, are more internally located within the double-stranded structure. Notice the major and minor grooves that are formed by this arrangement.

BIOLOGY PRINCIPLE Structure determines function. The major groove provides a binding site for proteins that control the expression of genes.

Principles of Biology are introduced in Chapter 1 and are then threaded throughout the entire textbook. This is achieved in two ways. First, the principles are explicitly stated in selected figure legends for figures in which the specific principle is illustrated. The legends that relate to a Principle of Biology are highlighted with an icon.

In addition, a Conceptual Question at the end of each chapter is directly aimed at exploring a particular principle related to the content of the chapter.

BioConnections New in the third edition, BioConnections are questions found in selected figure legends in each chapter that help students make connections between biological concepts. BioConnections help students understand that their study of biology involves linking concepts together and building on previously learned information. Answers to the BioConnections are found in Appendix B.

Conceptual Questions

1. What are the four key characteristics of the genetic material? What was Frederick Griffith's contribution to the study of DNA, and why was it so important?
2. The Hershey and Chase experiment used radioactive isotopes to track the DNA and protein of phages as they infected bacterial cells. Explain how this procedure allowed them to determine that DNA is the genetic material of this particular virus.
3. A principle of biology is that *structure determines function.* Discuss how the structure of DNA underlies different aspects of its function.

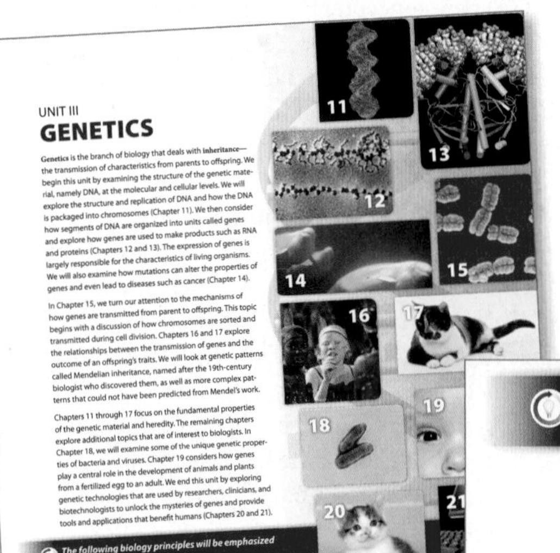

UNIT III

GENETICS

Genetics is the branch of biology that deals with **inheritance**—the transmission of characteristics from parents to offspring. We begin this unit by examining the structure of the genetic material, namely DNA, at the molecular and cellular levels. We will explore the structure and replication of DNA and how the DNA is packaged into chromosomes (Chapter 11). We then consider how segments of DNA are organized into units called genes and explore how genes are used to make products such as RNA and proteins (Chapters 12 and 13). The expression of genes is largely responsible for the characteristics of living organisms. We will also examine how mutations can alter the properties of genes and even lead to diseases such as cancer (Chapter 14).

In Chapter 15, we turn our attention to the mechanisms of how genes are transmitted from parent to offspring. This topic begins with a discussion of how chromosomes are sorted and transmitted during cell division. Chapters 16 and 17 explore the relationships between the transmission of genes and the outcome of an offspring's traits. We will look at genetic patterns called Mendelian inheritance, named after the 19th-century biologist who discovered them, as well as more complex patterns that could not have been predicted from Mendel's work.

Chapters 11 through 17 focus on the fundamental properties of the genetic material and heredity. The remaining chapters explore additional topics that are of interest to biologists. In Chapter 18, we will examine some of the unique genetic properties of bacteria and viruses. Chapter 19 considers how genes play a central role in the development of animals and plants from a fertilized egg to an adult. We end this unit by exploring genetic technologies that are used by researchers, clinicians, and biotechnologists to unlock the mysteries of genes and provide tools and applications that benefit humans (Chapters 20 and 21).

Unit openers have been added in the third edition and serve two purposes. They allow the student to see the "big picture" of the unit. In addition, the unit openers draw attention to the principles of biology that will be emphasized in that unit.

The following biology principles will be emphasized in this unit:

- ***The genetic material provides a blueprint for reproduction:*** *Throughout this unit, we will see how the genetic material carries the information to sustain life.*
- ***Structure determines function:*** *In Chapters 11 through 15, we will examine how the structures of DNA, RNA, genes, and chromosomes underlie their functions.*
- ***Living organisms interact with their environment:*** *In Chapters 16 and 17, we will explore the interactions between an organism's genes and its environment.*
- ***Living organisms grow and develop:*** *In Chapter 19, we will consider how a genetic program is involved in the developmental stages of animals and plants.*
- ***Biology affects our society:*** *In Chapters 20 and 21, we will examine genetic technologies that have many applications in our society.*
- ***Biology is an experimental science:*** *Every chapter in this unit has a Feature Investigation that describes a pivotal experiment that provided insights into our understanding of genetics.*

A GUIDED LEARNING SYSTEM AND CRITICAL THINKING

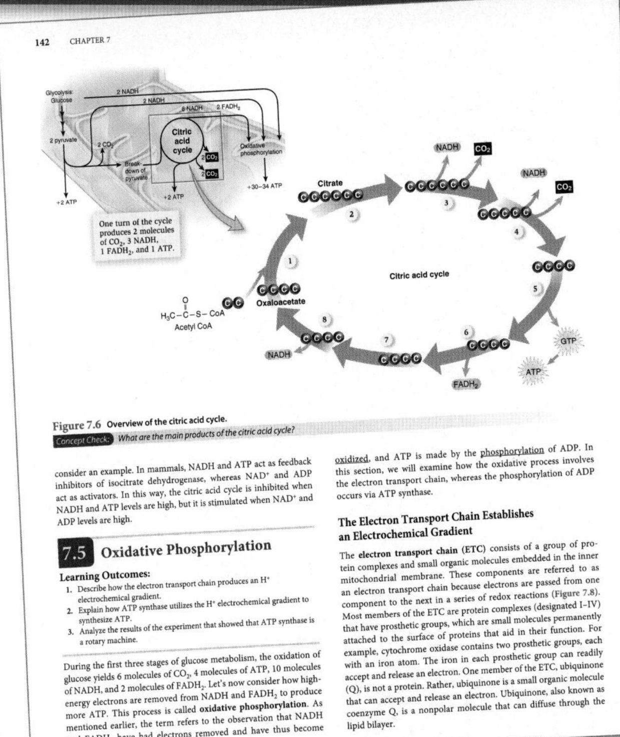

142 CHAPTER 7

Figure 7.6 Overview of the citric acid cycle.

Concept Check: *What are the main products of the citric acid cycle?*

consider an example. In mammals, NADH and ATP act as feedback
inhibitors of isocitrate dehydrogenase, whereas NAD^+ and ADP
act as activators. In this way, the citric acid cycle is inhibited when
NADH and ATP levels are high, but it is stimulated when NAD^+ and
ADP levels are high.

7.5 Oxidative Phosphorylation

Learning Outcomes:

1. Describe how the electron transport chain produces an H^+ electrochemical gradient.
2. Explain how ATP synthase utilizes the H^+ electrochemical gradient to synthesize ATP.
3. Analyze the results of the experiment that showed that ATP synthase is a rotary machine.

During the first three stages of glucose metabolism, the oxidation of
glucose yields 6 molecules of CO_2, 4 molecules of ATP, 10 molecules
of NADH, and 2 molecules of $FADH_2$. Let's now consider how high-
energy electrons are removed from NADH and $FADH_2$ to produce
more ATP. This process is called **oxidative phosphorylation**. As
mentioned earlier, the term refers to the observation that NADH
and $FADH_2$ have had electrons removed and have thus become
oxidized, and ATP is made by the phosphorylation of ADP. In
this section, we will examine how the oxidative process involves
the electron transport chain, whereas the phosphorylation of ADP
occurs via ATP synthase.

The Electron Transport Chain Establishes an Electrochemical Gradient

The **electron transport chain** (ETC) consists of a group of protein complexes and small organic molecules embedded in the inner
mitochondrial membrane. These components are referred to as
an electron transport chain because electrons are passed from one
component to the next in a series of redox reactions (Figure 7.8).
Most members of the ETC are protein complexes (designated I–IV)
that have prosthetic groups, which are small molecules permanently
attached to the surface of proteins that aid in their function. For
example, cytochrome oxidase contains two prosthetic groups, each
with an iron atom. The iron in each prosthetic group can readily
accept and release an electron. One member of the ETC, ubiquinone
(Q), is not a protein. Rather, ubiquinone is a small organic molecule
that can accept and release an electron. Ubiquinone, also known as
coenzyme Q, is a nonpolar molecule that can diffuse through the
lipid bilayer.

At the beginning of each section, **Learning Outcomes** have been added in the third edition that inform students of the skills they will acquire when mastering the material and provide a specific understanding of how such skills may be assessed. The assessments in Connect use these Learning Outcomes as a guide to developing online questions.

Critical Thinking—in the text . . .

- **ConceptChecks** are questions that go beyond simple recall of information and ask students to apply or interpret information presented in the illustrations.
- Questions with the **Feature Investigations** continually ask the student to check their understanding and push a bit further.

. . . continued online

- ***NEW! Quantitative Question Bank in Connect®*** Developing quantitative reasoning skills is important to the success of today's students. In addition to the Question Bank and Test Bank in Connect® a separate bank of quantitative questions is readily available for seamless use in homework/practice assignments, quizzes, and exams. These algorithmic-style questions provide an opportunity for students to more deeply explore quantitative concepts and to experience repeated practice that enables quantitative skill building over time.

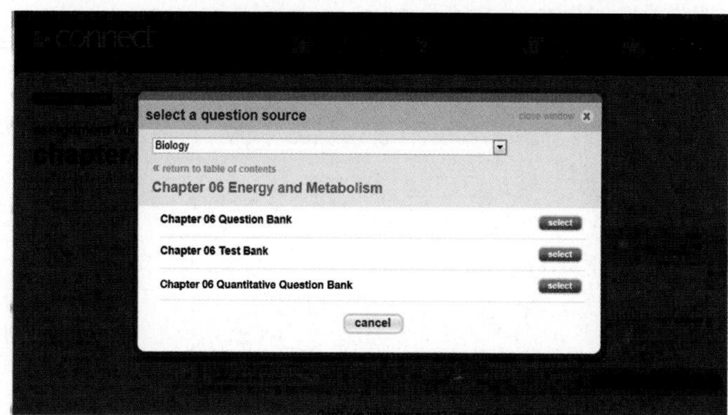

EXPERIMENTAL APPROACH

Feature Investigations provide a complete description of experiments, including data analysis, so students can understand how experimentation leads to an understanding of biological concepts. There are two types of ***Feature Investigations***. Most describe experiments according to the scientific method. They begin with observations and then progress through the hypothesis, experiment, data, and the interpretation of the data (conclusion). Some ***Feature Investigations*** involve discovery-based science, which does not rely on a preconceived hypothesis. The illustrations of the Feature Investigations are particularly innovative by having parallel drawings at the experimental and conceptual levels. By comparing the two levels, students will be able to understand how the researchers were able to interpret the data and arrive at their conclusions.

I really like the Feature Investigation so students can begin to grasp how scientists come to the conclusions that are simply presented as facts in these introductory texts.

Richard Murray, Hendrix College

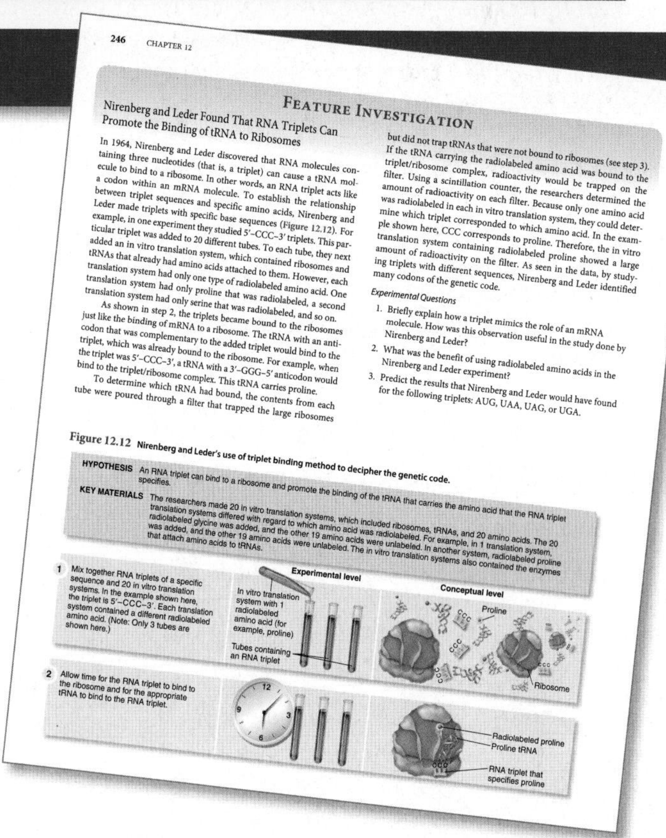

246 CHAPTER 12

FEATURE INVESTIGATION

Nirenberg and Leder Found That RNA Triplets Can Promote the Binding of tRNA to Ribosomes

In 1964, Nirenberg and Leder discovered that RNA molecules containing three nucleotides (that is, a triplet) can cause a tRNA molecule to bind to a ribosome. In other words, an RNA triplet acts like
a codon within an mRNA molecule. To establish the relationship
between triplet sequences and specific amino acids, Nirenberg and
Leder made triplets with specific base sequences (Figure 12.12). For
example, in one experiment they studied 5′–CCC–3′ triplets. This particular triplet was added to 20 different tubes. To each tube, they next
added an in vitro translation system, which contained ribosomes and
tRNAs that already had amino acids attached to them. However, each
translation system had only one type of radiolabeled amino acid. One
translation system had only proline that was radiolabeled, a second
translation system had only serine that was radiolabeled, and so on.

As shown in step 2, the triplets became bound to the ribosomes
just like the binding of mRNA to a ribosome. The tRNA with an anticodon that was complementary to the added triplet would bind to the
triplet, which was already bound to the ribosome. For example, when
the triplet was 5′–CCC–3′, a tRNA with a 3′–GGG–5′ anticodon would
bind to the triplet/ribosome complex. This tRNA carries proline.

To determine which tRNA had bound, the contents from each
tube were poured through a filter that trapped the large ribosomes
but did not trap tRNAs that were not bound to ribosomes (see step 3).
If the tRNA carrying the radiolabeled amino acid was bound to the
triplet/ribosome complex, radioactivity would be trapped on the
filter. Using a scintillation counter, the researchers determined the
amount of radioactivity on each filter. Because only one amino acid
was radiolabeled in each in vitro translation system, they could determine which triplet corresponded to which amino acid. In the example shown here, CCC corresponds to proline. Therefore, the in vitro
translation system containing radiolabeled proline showed a large
amount of radioactivity on the filter. As seen in the data, by studying triplets with different sequences, Nirenberg and Leder identified
many codons of the genetic code.

Experimental Questions

1. Briefly explain how a triplet mimics the role of an mRNA molecule. How was this observation useful in the study done by Nirenberg and Leder?
2. What was the benefit of using radiolabeled amino acids in the Nirenberg and Leder experiment?
3. Predict the results that Nirenberg and Leder would have found for the following triplets: AUG, UAA, UAG, or UGA.

Figure 12.12 Nirenberg and Leder's use of triplet binding method to decipher the genetic code.

HYPOTHESIS An RNA triplet can bind to a ribosome and promote the binding of the tRNA that carries the amino acid that the RNA triplet specifies.

KEY MATERIALS The researchers made 20 in vitro translation systems, which included ribosomes, tRNAs, and 20 amino acids. The 20 translation systems differed with regard to which amino acid was radiolabeled. For example, in 1 translation system, radiolabeled glycine was added, and the other 19 amino acids were unlabeled. In another system, radiolabeled proline was added, and the other 19 amino acids were unlabeled. The in vitro translation systems also contained the enzymes that attach amino acids to tRNAs.

EVOLUTIONARY PERSPECTIVE

Modern techniques have enabled researchers to study many genes simultaneously, allowing them to explore genomes (all the genes an organism has) and proteomes (all the proteins encoded by those genes). This allows us to understand biology in a broader way. Beginning in Chapter 3, each chapter contains a topic called the ***Genomes & Proteomes Connection*** that provides an understanding of how genomes and proteomes underlie the inner workings of cells and explains how evolution works at the molecular level. The topics that are covered in the *Genomes & Proteomes Connection* are very useful in preparing students for future careers in biology. The study of genomes and proteomes has revolutionized many careers in biology, including those in medicine, research, biotechnology, to name a few.

This is one of the best features of these chapters. It is absolutely important to emphasize evolution themes at the molecular level in undergraduate biology courses.

Jorge Busciglio, University of California—Irvine

ANIMAL BODIES AND HOMEOSTASIS 819

GENOMES & PROTEOMES CONNECTION

Organ Development and Function Are Controlled by *Hox* Genes

In previous chapters, you have learned about a family of ancient, highly conserved genes called *Hox* genes that are found in all animals. *Hox* genes determine the timing and spatial patterning of the anteroposte- rior body axis during development. For example, we saw in Chapter 19 how these genes determine the number and position of legs and wings in *Drosophila*. *Hox* genes play a similar role in determining the spatial patterning of the vertebrate body and appendages. Recently, scientists have begun exploring the role of *Hox* genes in the development and spatial patterning of the organs that make up animals' organ systems.

By generating mutant mice that fail to express one or more *Hox* genes (the genes are said to have been knocked out; see Chapter 20), researchers have discovered the important role these genes play in determining where within the vertebrate body particular organs form. Recall from Chapter 19 (refer back to Figures 19.16 and 19.17) that mouse *Hox* genes are arranged in four clusters, designated A–D, with multiple genes per cluster. Because these homologous genes (for example, *HoxA-3*, *HoxB-3*, and *HoxD-3*) are found within a single species, they are considered paralogous genes. Such paralogous genes typically act in concert to regulate similar developmental processes. For example, *HoxA-3* is important for development of anterior parts of the body, including the neck. When this gene is knocked out, mouse embryos show defects in neck structure, such as abnormal blood vessels. Also, the organs within the neck—including the thymus, thyroid, and parathyroid glands—do not develop normally. When two or more paralogs of this group are knocked out, certain neck organs fail to form at all. Experimental deletion of *Hox* genes associated with other body segments does not affect the development and function of neck glands and organs. Likewise, investigators have uncovered vital roles of different *Hox* genes in lung development within the thorax and the proper positioning and development of the vertebrate kidneys in the abdomen.

Of particular interest is the discovery that *Hox* genes are important not only for spatial patterning of organs but also for their growth, development, and function. Genes in the *Hox* 1 and 3 groups, for instance, help determine the final branching patterns of the airways of the lungs, the final size of the lungs, and the ability of the lungs to produce secretions that are important for breathing air after birth. Other *Hox* genes have been shown to control cell proliferation, shape changes, apoptosis, cell migration, and cell-cell adhesion within various organs. This is also true in invertebrates, such as the leech, where *Hox* genes are first expressed during organ formation, and in *Drosophila*, where the final shape and size of the heart are partly controlled by *Hox* genes.

Body Fluids Are Distributed into Compartments

All animal bodies are composed primarily of water. Most of the water in an animal's body is contained inside its cells and therefore is called **intracellular fluid** (from the Latin *intra*, meaning inside of). The rest of the water in its body exists outside of the cells and is therefore called

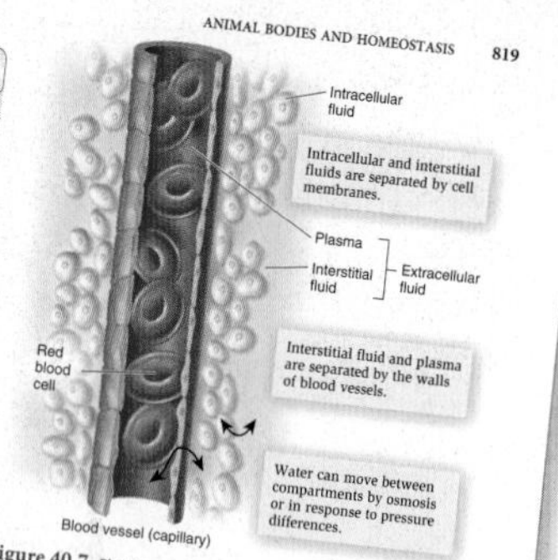

Figure 40.7 **Fluid compartments in a typical vertebrate.** Most of the fluid within an animal's body exists within cells (intracellular fluid). Extracellular fluid is that portion of the body's fluid that lies outside cells (interstitial fluid) and within blood vessels (plasma), such as the capillary shown here. Arrows indicate directions of water movement between adjacent compartments.

Concept Check: *What would happen to the distribution of water in the fluid compartments of an animal's body if a blood vessel were damaged such that it leaked its contents?*

extracellular fluid (from the Latin *extra*, meaning outside of). Extracellular fluid is composed of the fluid part of blood, called **plasma**, and the fluid-filled spaces that surround cells, called **interstitial fluid** (from the Latin *inter*, meaning between) (**Figure 40.7**). In vertebrates and some invertebrates, plasma and interstitial fluid are kept separate within a closed circulatory system. In many invertebrates with open circulatory systems (see Chapter 47), however, plasma and interstitial fluid are intermingled into a single fluid called hemolymph.

The fluids depicted in Figure 40.7 are said to be enclosed in compartments. In a typical vertebrate, the total water volume in the three compartments (intracellular fluid, plasma, and interstitial fluid) accounts for about two-thirds of body weight, with solids comprising the rest. Of the total body water, up to two-thirds is intracellular and one-third extracellular, the majority of which is located in the interstitial compartment.

Plasma membranes separate the intracellular fluid from the extracellular fluid. The two components of extracellular fluid in vertebrates and some invertebrates—the interstitial fluid and the plasma—are separated by the walls of the blood vessels (for example, arteries, capillaries, and veins).

The solute composition of the extracellular fluid is very different from that of the intracellular fluid. Maintaining differences in solute composition across the plasma membrane is an important way in which

A VISUAL OUTLINE

Because students rely on the art as a primary study tool, the authors worked with a team of editors, scientific illustrators, educators, and students to create an accurate, up-to-date, realistic, and visually appealing illustration program that is also easy to follow and instructive. The artwork and photos serve as a visual outline and guide students through complex processes.

The illustrations were very effective in detailing the processes. The drawings were more detailed than our current book, which allowed for a better idea of what the proteins (or whatever the object) structure was.

Amy Weber, student, Ohio University

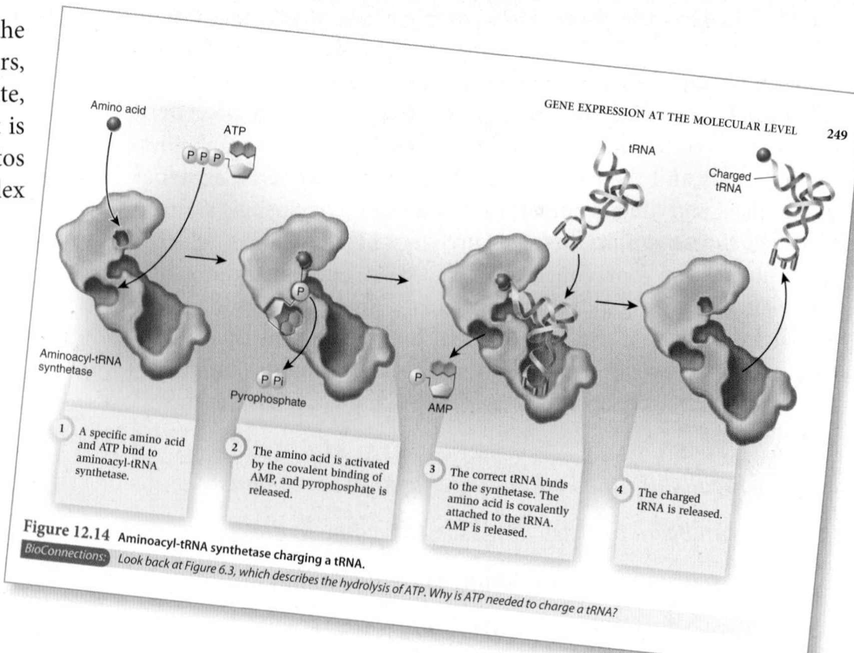

Figure 12.14 **Aminoacyl-tRNA synthetase charging a tRNA.**

BioConnections: *Look back at Figure 6.3, which describes the hydrolysis of ATP. Why is ATP needed to charge a tRNA?*

THE DIGITAL STORY...

Digital assessment is a major focus in higher education. Online tools promise anywhere, anytime access combined with the possibility of learning tailored to individual student needs. Digital assessments should span the spectrum of Bloom's taxonomy within the context of best-practice pedagogy. The increased challenge at higher Bloom's levels will help students grow intellectually and be better prepared to contribute to society.

Significant faculty demand for content at higher Bloom's levels led us to examine assessment quality and consistency of our Connect® content and to develop a scientific approach to systematically increase Bloom's levels and develop internally consistent and balanced digital assessments that promote student learning.

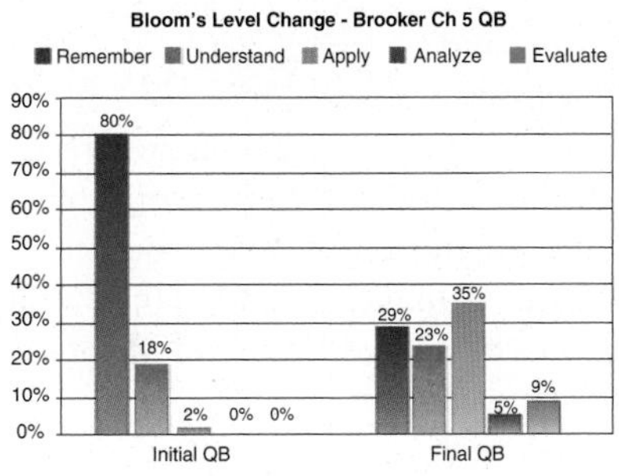

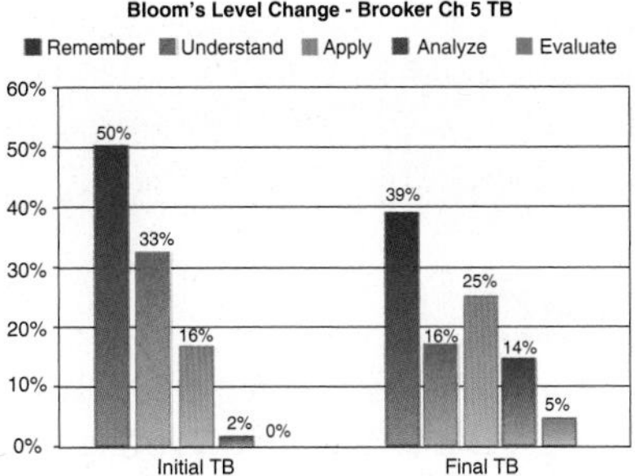

Our goal was to increase assessment quality of our Connect content to meet faculty and student needs. Our objective was to have 30% of all digital assessment questions in Connect at the Apply, Analyze, or Evaluate levels of Bloom's taxonomy. With thousands of existing questions, that is no small task. Consistent with best-practices research on how students learn, we took a comprehensive look at our existing digital assessments to determine Bloom's levels across our assignable content. Because this project was too extensive for a single person to accomplish, we assembled a team of faculty from research, comprehensive, liberal arts universities, and community colleges. Digital team members were selected based on commitment to student learning, biology content expertise, openness to a new vision for digital assessment and professional development, and question-writing skills.

Under the direction of lead digital author, Ian Quitadamo, team members were calibrated to a common perception of Bloom's taxonomy. The team then evaluated our existing Question Bank, Test Bank, Animation Quizzes, and Video Quizzes for appropriate level of Bloom's and compiled the results into a comprehensive database that was statistically analyzed. Results showed adequate coverage at the lower level of Bloom's taxonomy but less so at the higher levels of Bloom's. Knowing that assessment drives learning quality, we focused our efforts on "Blooming up" existing content and developing new assessments that examine students' problem-solving skills. The end result of our team's scientific approach to developing digital content is a collection of engaging, diagnostic assessments that strengthen student ability to think critically, build connections across biology concepts, and develop quantitative reasoning skills that ultimately underlie student academic success and ability to contribute to society.

We would like to acknowledge our digital team and thank them for their tireless efforts:

Kerry Bohl, *University of South Florida*
David Bos, *Purdue University*
Scott Bowling, *Auburn University*
Scott Cooper, *University of Wisconsin, La Crosse*
Cynthia Dadmun, *Freelance content expert*
Jenny Dechaine, *Central Washington University*
Elizabeth Drumm, *Oakland Community College-Orchard Ridge Campus*
Susan Edwards, *Appalachian State University*
Julie Emerson, *Amherst College*
Brent Ewers, *University of Wyoming*
Chris Himes, *Massachusetts College of Liberal Arts*
Cintia Hongay, *Clarkson University*
Heather Jezorek, *University of South Florida*
Kristy Kappenman, *Central Washington University*
Jamie Kneitel, *California State University, Sacramento*
Marcy Lowenstein, *Florida International University*
Carolyn Martineau, *DePaul University*
Christin Munkittrick, *Freelance content expert*
Chris Osovitz, *University of South Florida*
Anneke Padolina, *Virginia Commonwealth University*
Marius Pfeiffer, *Tarrant County College*
Marceau Ratard, *Delgado Community College*
Nicolle Romero, *Freelance content expert*
Amanda Rosenzweig, *Delgado Community College*
Kathryn Spilios, *Boston University*
Jen Stanford, *Drexel University*
Martin St. Maurice, *Marquette University*
Salvatore Tavormina, *Austin Community College*
Sharon Thoma, *University of Wisconsin, Madison*
Gloriana Trujillo, *University of New Mexico*
Jennifer Wiatrowski, *Pasco-Hernando Community College*

Quantitative question bank

David Bos, *Purdue University*
Chris Osovitz, *University of South Florida*
Martin St. Maurice, *Marquette University*

A NEW VISION IN PREPARING YOUR COURSE

MCGRAW-HILL HIGHER EDUCATION AND BLACKBOARD HAVE TEAMED UP

Blackboard®, the Web-based course management system, has partnered with McGraw-Hill to better allow students and faculty to use online materials and activities to complement face-to-face teaching. Blackboard features exciting social learning and teaching tools that foster more logical, visually impactful, and active learning opportunities for students. You'll transform your closed-door classrooms into communities where students remain connected to their educational experience 24 hours a day.

This partnership allows you and your students access to McGraw-Hill's Connect® and McGraw-Hill Create™ right from within your Blackboard course—all with one single sign-on. Not only do you get single sign-on with Connect and Create, you also get deep integration of McGraw-Hill content and content engines right in Blackboard. Whether you're choosing a book for your course or building Connect assignments, all the tools you need are right where you want them—inside of Blackboard.

Gradebooks are now seamless. When a student completes an integrated Connect assignment, the grade for that assignment automatically (and instantly) feeds your Blackboard grade center.

McGraw-Hill and Blackboard can now offer you easy access to industry leading technology and content, whether your campus hosts it or we do. Be sure to ask your local McGraw-Hill representative for details.

MCGRAW-HILL CONNECT® BIOLOGY

McGraw-Hill Connect® Biology provides online presentation, assignment, and assessment solutions. It connects your students with the tools and resources they'll need to achieve success. With Connect Biology you can deliver assignments, quizzes, and tests online. A robust set of questions and activities are presented and aligned with the textbook's learning outcomes. As an instructor, you can edit existing questions and author entirely new problems. Track individual student performance—by question, assignment, or in relation to the class overall—with detailed grade reports. Integrate grade reports easily with Learning Management Systems (LMS), such as WebCT and Blackboard—and much more. ConnectPlus Biology provides students with all the advantages of Connect Biology plus 24/7 online access to an eBook. This media-rich version of the book is available through the McGraw-Hill Connect platform and allows seamless integration of text, media, and assessments.

To learn more, visit www.mcgrawhillconnect.com

MY LECTURES—TEGRITY®

Mc Graw Hill tegrity.

McGraw-Hill Tegrity® records and distributes your class lecture with just a click of a button. Students can view them anytime/anywhere via computer, iPod, or mobile device. It indexes as it records your PowerPoint® presentations and anything shown on your computer so students can use keywords to find exactly what they want to study. Tegrity is available as an integrated feature of McGraw-Hill Connect Biology and as a standalone.

PERSONALIZED AND ADAPTIVE LEARNING

Mc Graw Hill LearnSmart™

McGraw-Hill LearnSmart™ is available as an integrated feature of McGraw-Hill Connect® Biology. It is an adaptive learning system designed to help students learn faster, study more efficiently, and retain more knowledge for greater success. LearnSmart assesses a student's knowledge of course content through a series of adaptive questions. It pinpoints concepts the student does not understand and maps out a personalized study plan for success. This innovative study tool also has features that allow instructors to see exactly what students have accomplished and a built-in assessment tool for grading assignments. Visit the following site for a demonstration. www.mhlearnsmart.com

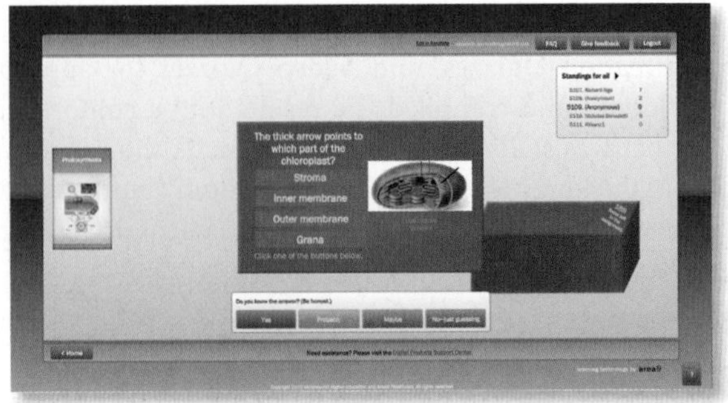

LabSmart™

Based on the same world-class super-adaptive technology as LearnSmart, McGraw-Hill LabSmart is a must-see, outcomes-based lab simulation. It assesses a student's knowledge and adaptively corrects deficiencies, allowing the student to learn faster and retain more knowledge with greater success.

First, a student's knowledge is adaptively leveled on core learning outcomes: Questioning reveals knowledge deficiencies that are corrected by the delivery of content that is conditional on a student's response. Then, a simulated lab experience requires the student to think and act like a scientist: Recording, interpreting, and analyzing data using simulated equipment found in labs and clinics. The student is allowed to make mistakes—a powerful part of the learning experience! A virtual coach provides subtle hints when needed; asks questions about the student's choices; and allows the student to reflect on and correct those mistakes. Whether your need is to overcome the logistical challenges of a traditional lab, provide better lab prep, improve student performance, or make your online experience one that rivals the real world, LabSmart accomplishes it all.

Learn more at www.mhlabsmart.com

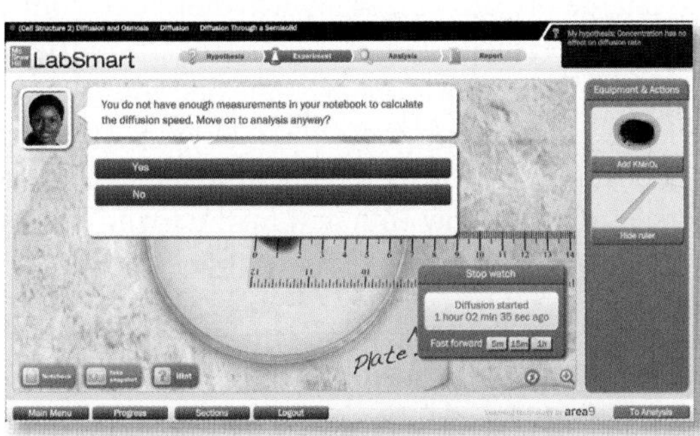

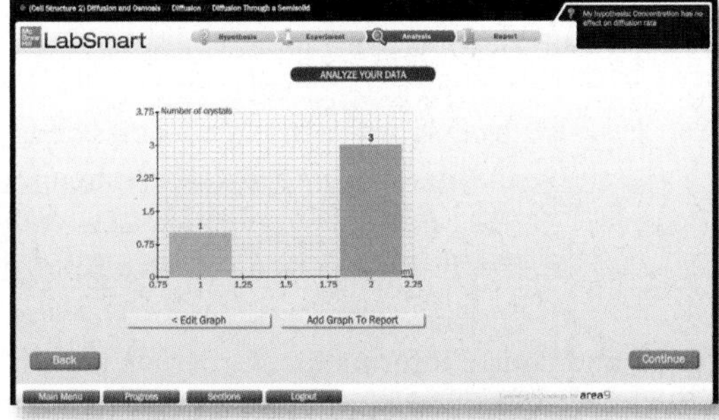

Preparing for Majors Biology

Do your Majors Biology students struggle the first few weeks of class, trying to get up to speed? McGraw-Hill can help.

McGraw-Hill has developed an adaptive learning tool designed to increase student success and aid retention through the first few weeks of class. Using this digital tool Majors Biology students can master some of the most fundamental and challenging principles of biology before they might begin to struggle in the first few weeks of class.

An initial diagnostic establishes a student's baseline comprehension and knowledge; then the program generates a learning plan tailored to the student's academic needs and schedule. As the student works through the learning plan, the program tracks the student's progress, delivering appropriate assessment and learning resources (e.g., tutorials, figures, animations, etc.) as needed. If students incorrectly answer questions around a particular learning objective, they are asked to review learning resources around that objective before re-assessing their mastery of the objective.

Using this program, students can identify the content they don't understand, focus their time on content they need to know but don't, and therefore improve their chances of success in the Majors Biology course.

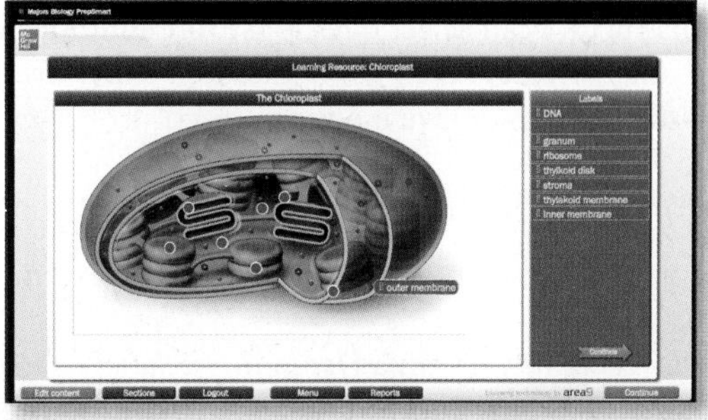

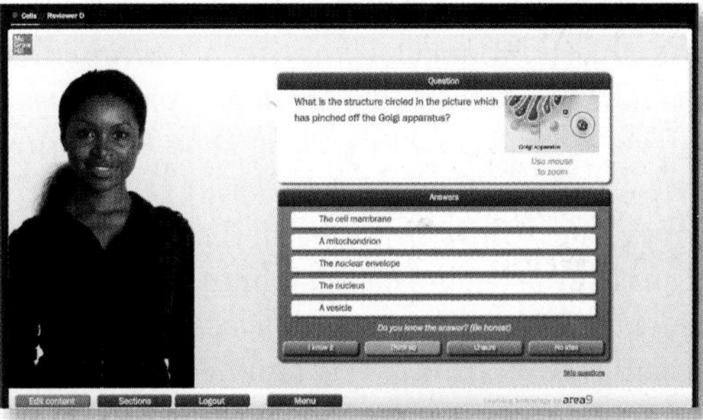

POWERFUL PRESENTATION TOOLS IN CONNECT BIOLOGY

Everything you need for outstanding presentation in one place!

- FlexArt Image PowerPoints—including every piece of art that has been sized and cropped specifically for superior presentations as well as labels that you can edit, flexible art that can be picked up and moved, tables, and photographs.
- Animation PowerPoints—Numerous full-color animations illustrating important processes. Harness the visual impact of concepts in motion by importing these slides into classroom presentations or online course materials.
- Lecture PowerPoints with animations fully embedded.
- Labeled and unlabeled JPEG images—Full-color digital files of all illustrations that can be readily incorporated into presentations, exams, or custom-made classroom materials.

FULLY DEVELOPED TEST BANK

The Digital Team revised the Test Bank to fully align with the learning outcomes and complement questions written for the Question Bank intended for homework assignments. A thorough review process has been implemented to ensure accuracy. Provided within a computerized test bank powered by McGraw-Hill's flexible electronic testing program EZ Test Online, instructors can create paper and online tests or quizzes in this easy-to-use program! A new tagging scheme allows you to sort questions by Learning Outcome, Bloom's level, topic, and section. Imagine being able to create and access your test or quiz anywhere, at any time, without installing the testing software. Now, with EZ Test Online, instructors can select questions from multiple McGraw-Hill test banks or create their own, and then either print the test for paper distribution or give it online.

Contributors for other digital assets:

FlexArt Image PowerPoints—Sharon Thoma, *University of Wisconsin, Madison*
Lecture PowerPoints—Cynthia Dadmun, *freelance content expert*
eBook Quizzes—Nancy Boury, *Iowa State University* and Lisa Bonneau, *Mount Marty College*
Website—Kathleen Broomall, *University of Cincinnati, Clermont College* and Carla Reinstadtler, *freelance content expert*
LearnSmart™—Lead: Laurie Russell, *St. Louis University*, Authors and reviewers: Isaac Barjis, *New York City College of Technology*; Tonya Bates, *University of North Carolina, Charlotte*; Kerry Bohl, *University of South Florida*; Johnny El-Rady, *University of South Florida*; Elizabeth Harris, *Appalachian State University*; Shelley Jansky, *University of Wisconsin, Madison*; Teresa McElhinny, *Michigan State University*; Murad Odeh, *South Texas College*; Nilo Marin, *Broward College*

Flexible Delivery Options

Brooker et al. *Biology* is available in many formats in addition to the traditional textbook to give instructors and students more choices when deciding on the format of their biology text.

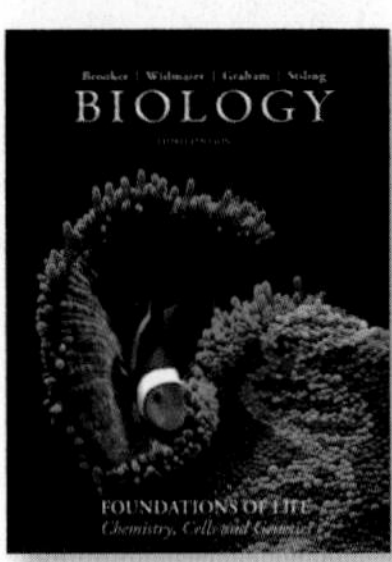

Foundations of Life—Chemistry, Cells, and Genetics
ISBN: 007777583X
Units 1, 2, and 3

Evolution, Diversity, and Ecology
ISBN: 0077775848
Units 4, 5, and 8

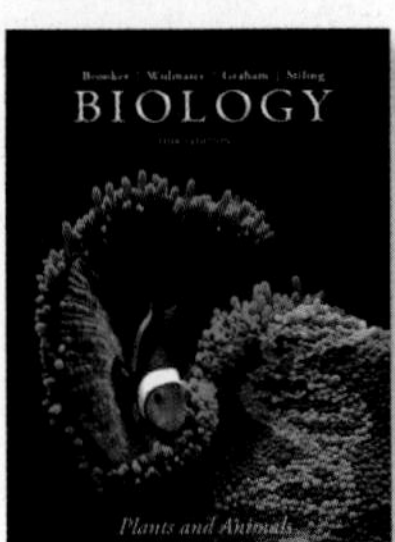

Plants and Animals
ISBN: 0077775856
Units 6 and 7

Also available, customized versions for all of your course needs. You're in charge of your course, so why not be in control of the content of your textbook? At McGraw-Hill Custom Publishing, we can help you create the ideal text—the one you've always imagined. Quickly. Easily. With more than 20 years of experience in custom publishing, we're experts. But at McGraw-Hill we're also innovators, leading the way with new methods and means for creating simplified value-added custom textbooks.

The options are never-ending when you work with McGraw-Hill. You already know what will work best for you and your students. And here, you can choose it.

MCGRAW-HILL CREATE™

With ***McGraw-Hill Create***™, you can easily rearrange chapters, combine material from other content sources, and quickly upload content you have written, like your course syllabus or teaching notes. Find the content you need in Create by searching through thousands of leading McGraw-Hill textbooks. Arrange your book to fit your teaching style. Create even allows you to personalize your book's appearance by selecting the cover and adding your name, school, and course information. Order a Create book and you'll receive a complimentary print review copy in 3–5 business days or a complimentary electronic review copy (eComp) via e-mail in minutes. Go to www.mcgrawhillcreate.com today and register to experience how McGraw-Hill Create empowers you to teach *your* students *your* way. **www.mcgrawhillcreate.com**

LABORATORY MANUALS

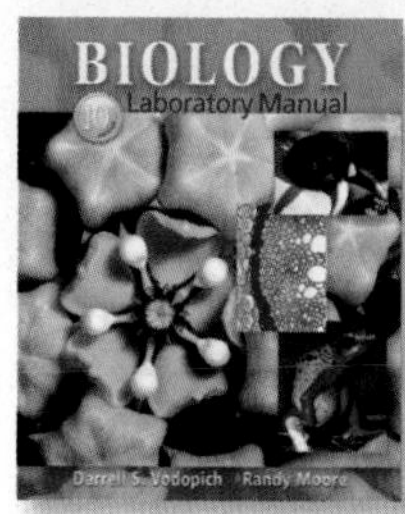

Biology Laboratory Manual, Tenth Edition
Vodopich/Moore
ISBN: 0-07-353225-8

This laboratory manual is designed for an introductory majors-level biology course with a broad survey of basic laboratory techniques. The experiments and procedures are simple, safe, easy to perform, and especially appropriate for large classes. Few experiments require a second class meeting to complete the procedure. Each exercise includes many photographs, traditional topics, and experiments that help students learn about life. Procedures within each exercise are numerous and discrete so that an exercise can be tailored to the needs of the students, the style of the instructor, and the facilities available.

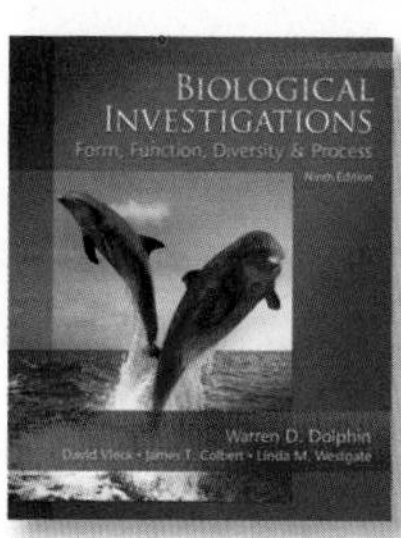

Biological Investigations Lab Manual, Ninth Edition
Dolphin
ISBN: 0-07-338305-8

This independent lab manual can be used for a one- or two-semester majors-level general biology lab and can be used with any majors-level general biology textbook. The labs are investigative and ask students to use more critical thinking and hands-on learning. The author emphasizes investigative, quantitative, and comparative approaches to studying the life sciences.

Mc Graw Hill LabSmart™

Based on the same world-class super-adaptive technology as LearnSmart, McGraw-Hill LabSmart is a must-see, outcomes-based lab simulation. It assesses a student's knowledge and adaptively corrects deficiencies, allowing the student to learn faster and retain more knowledge with greater success. Whether your need is to overcome the logistical challenges of a traditional lab, provide better lab prep, improve student performance, or create an online experience that rivals the real world, LabSmart accomplishes it all.

A STEP AHEAD IN QUALITY

360° DEVELOPMENT PROCESS

McGraw-Hill's 360° Development Process is an ongoing, never-ending, education-oriented approach to building accurate and innovative print and digital products. It is dedicated to continual large-scale and incremental improvement, driven by multiple user feedback loops and checkpoints. This is initiated during the early planning stages of our new products, and intensifies during the development and production stages, then begins again upon publication in anticipation of the next edition.

This process is designed to provide a broad, comprehensive spectrum of feedback for refinement and innovation of our learning tools, for both student and instructor. The 360° Development Process includes market research, content reviews, course- and product-specific symposia, accuracy checks, and art reviews. We appreciate the expertise of the many individuals involved in this process.

General Biology Symposia

Every year McGraw-Hill conducts several General Biology Symposia, which are attended by instructors from across the country. These events are an opportunity for editors from McGraw-Hill to gather information about the needs and challenges of instructors teaching the major's biology course. It also offers a forum for the attendees to exchange ideas and experiences with colleagues they might not have otherwise met. The feedback we have received has been invaluable, and has contributed to the development of Biology and its supplements. A special thank you to recent attendees:

Thomas Abbott, *University of Connecticut*
Sylvester Allred, *Northern Arizona University*
Julie Anderson, *University of Wisconsin–Eau Claire*
Kim Baker, *University of Wisconsin–Green Bay*
Michael Bell, *Richland College*
Brian Berthelsen, *Iowa Western Community College*
Joe Beuchel, *Triton College*
Arlene Billock, *University of Louisiana–Lafayette*
Stephane Boissinot, *Queens College, the City University of New York*
David Bos, *Purdue University*
Scott Bowling, *Auburn University*
Jacqueline Bowman, *Arkansas Technical University*
Randy Brooks, *Florida Atlantic University*
Arthur Buikema, *Virginia Polytechnic Institute*
Anne Bullerjahn, *Owens Community College*
Helaine Burstein, *Ohio University*
Raymond Burton, *Germanna Community College*
Peter Busher, *Boston University*
Ruth Buskirk, *University of Texas–Austin*
Richard Cardullo, *University of California–Riverside*
Frank Cantelmo, *St. Johns University*
Jennifer Ciaccio, *Dixie State College*
Anne Barrett Clark, *Binghamton University*
Allison Cleveland, *University of South Florida–Tampa*
Clark Coffman, *Iowa State University*
Jennifer Coleman, *University of Massachusetts–Amherst*
Sehoya Cotner, *University of Minnesota*
Mitch Cruzan, *Portland State University*
Karen A. Curto, *University of Pittsburgh*
Rona Delay, *University of Vermont*
Mary Dettman, *Seminole State College of Florida*
Laura DiCaprio, *Ohio University*
Kathyrn Dickson, *California State College–Fullerton*
Cathy Donald-Whitney, *Collin County Community College*
Moon Draper, *University of Texas–Austin*
Tod Duncan, *University of Colorado–Denver*
Brent Ewers, *University of Wyoming*
Stanley Faeth, *Arizona State University*
Michael Ferrari, *University of Missouri–Kansas City*
David Fitch, *New York University*
Donald French, *Oklahoma State University*
Douglas Gaffin, *University of Oklahoma*
John Geiser, *Western Michigan University*
Karen Gerhart, *University of California–Davis*
Julie Gibbs, *College of DuPage*
Cynthia Giffen, *University of Wisconsin–Madison*
Sharon Gill, *Western Michigan University*
William Glider, *University of Nebraska–Lincoln*
Steven Gorsich, *Central Michigan University*
Christopher Gregg, *Louisiana State University*
Stan Guffey, *The University of Tennessee*
Sally Harmych, *University of Toledo*
Bernard Hauser, *University of Florida–Gainesville*
Jean Heitz, *University of Wisconsin–Madison*
Mark Hens, *University of North Carolina–Greensboro*
Albert Herrera, *University of Southern California*
Ralph James Hickey, *Miami University of Ohio–Oxford*
Jodi Huggenvik, *Southern Illinois University–Carbondale*
Brad Hyman, *University of California–Riverside*
Rick Jellen, *Brigham Young University*
Michael Kempf, *University of Tennessee–Martin*
Kyoungtae Kim, *Missouri State University*
Sherry Krayesky, *University of Louisiana–Lafayette*
Jerry Kudenov, *University of Alaska–Anchorage*
Josephine Kurdziel, *University of Michigan*
Ellen Lamb, *University of North Carolina–Greensboro*
Brenda Leady, *University of Toledo*
Graeme Lindbeck, *Valencia Community College*
David Longstreth, *Louisiana State University*
Lucile McCook, *University of Mississippi*
Susan Meiers, *Western Illinois University*
Michael Meighan, *University of California–Berkeley*
John Merrill, *Michigan State University*
John Mersfelder, *Sinclair Community College*
Melissa Michael, *University of Illinois–Urbana-Champaign*
Michelle Mynlieff, *Marquette University*
Leonore Neary, *Joliet Junior College*
Shawn Nordell, *Saint Louis University*
John Osterman, *University of Nebraska–Lincoln*
Stephanie Pandolfi, *Michigan State University*
Anneke Padolina, *Virginia Commonwealth University*
C.O. Patterson, *Texas A&M University*
Nancy Pencoe, *University of West Georgia*
Roger Persell, *Hunter College*
Marius Pfeiffer, *Tarrant County College NE*
Steve Phelps, *University of Florida*
Debra Pires, *University of California–Los Angeles*
Thomas Pitzer, *Florida International University*
Steven Pomarico, *Louisiana State University*
Jo Anne Powell-Coffman, *Iowa State University*
Lynn Preston, *Tarrant County College*
Ian Quitadamo, *Central Washington University*
Rajinder Ranu, *Colorado State University*
Marceau Ratard, *Delgado Community College–City Park*
Melanie Rathburn, *Boston University*
Robin Richardson, *Winona State University*
Mike Robinson, *University of Miami*
Amanda Rosenzweig, *Delgado Community College–City Park*
Connie Russell, *Angelo State University*
Laurie Russell, *St. Louis University*
David Scicchitano, *New York University*
Timothy Shannon, *Francis Marion University*
Brian Shmaefsky, *Lone Star College–Kingwood*
Richard Showman, *University of South Carolina*
Allison Silveus, *Tarrant County College–Trinity River Campus*
Robert Simons, *University of California–Los Angeles*
Steve Skarda, *Linn Benton Community College*
Steven D. Skopik, *University of Delaware*
Phillip Sokolove, *University of Maryland–Baltimore County*
Martin St. Maurice, *Marquette University*
Brad Swanson, *Cental Michigan University*
David Thompson, *Northern Kentucky University*
Maureen Tubbiola, *St. Cloud State University*
Ashok Upadhyaya, *University of South Florida–Tampa*
Anthony Uzwiak, *Rutgers University*

Rani Vajravelu, *University of Central Florida*
Gary Walker, *Appalachian State University*
Pat Walsh, *University of Delaware*
Elizabeth Weiss-Kuziel, *University of Texas–Austin*
Clay White, *Lone Star College–CyFair*
Leslie Whiteman, *Virginia State University*
Jennifer Wiatrowski, *Pasco-Hernando Community College*
David Williams, *Valencia Community College, East Campus*
Holly Williams, *Seminole Community College*
Michael Windelspecht, *Appalachian State University*
Robert Winning, *Eastern Michigan University*
Mary Wisgirda, *San Jacinto College, South Campus*
Michelle Withers, *West Virginia University*
Kevin Wolbach, *University of the Sciences in Philadelphia*
Jay Zimmerman, *St. John's University*

Third Edition Reviewers

John Alcock, *Arizona State University*
Brian Ashburner, *University of Toledo*
Elizabeth A. Bailey, *Alamance Community College*
Arlene Billock, *University of Louisiana–Lafayette*
Bronwyn Bleakley, *Stonehill College*
Randy Brewton, *University of Tennessee–Knoxville*
Becky Brown, *College of Marin*
Carolyn J. W. Bunde, *Idaho State University*
Joseph Bundy, *University of North Carolina–Greensboro*
Romi Burks, *Southwestern University*
Genevieve Chung, *Broward College*
Joe Coelho, *Quincy University*
Michael Cullen, *University of Evansville*
Cynthia Doffitt, *Mississippi State University*
David W. Eldridge, *Baylor University*
Teresa Fischer, *Indian River State College*
Greg Fox, *College of the Canyons*
Arundhati Ghosh, *University of Pittsburgh*
Leonard Ginsberg, *Western Michigan University*
Elizabeth Godrick, *Boston University*
Robert Greene, *Niagara University*
Timothy Grogan, *Valencia Community College*
Theresa Grove, *Valdosta State University*
Jane Henry, *Baton Rouge Community College*
Margaret Horton, *The University of North Carolina–Greensboro*
Jerry Kaster, *University of Wisconsin–Milwaukee*
Bryan Krall, *Parkland College*
Craig Longtine, *North Hennepin Community College*
Kathryn Nette, *Cuyamaca College*
Deb Pires, *University of California–Los Angeles*
Nicola Plowes, *Arizona State University*
Kim Risley, *University of Mount Union*
Laurel Roberts, *University of Pittsburgh*
Rebecca Sheller, *Southwestern University*
Mark A. Shoop, *Tennessee Wesleyan College*
Om Singh, *University of Pittsburgh–Bradford*
Keith Snyder, *Southern Adventist University*
Hattie Spencer, *Mississippi Valley State University*
Ken Spitze, *University of West Georgia*
Joyce Stamm, *University of Evansville*
Ivan Still, *Arkansas Tech University*
Mark Sturtevant, *Oakland University*
Scott Tiegs, *Oakland University*
D. Alexander Wait, *Missouri State University*
Delon Washo-Krupps, *Arizona State University*
Chad Wayne, *University of Houston*
Mollie Wegner, *University of California–Davis*
Emily Williamson, *Mississippi State University*
Erica B. Young, *University of Wisconsin–Milwaukee*

First and Second Edition Reviewers and Contributors

Eyualem Abebe, *Elizabeth City State University*
James K. Adams, *Dalton State College*
Nihal Ahmad, *University of Wisconsin–Madison*
John Alcock, *Arizona State University*
Myriam Alhadefl-Feldman, *Lake Washington Technical College*
Sylvester Allred, *Northern Arizona University*
Jonathan W. Armbruster, *Auburn University*
Joseph E. Armstrong, *Illinois State University*
Dennis Arvidson, *Michigan State University*
David K. Asch, *Youngstown State University*
Tami Asplin, *North Dakota State University*
Amir Assadi-Rad, *Delta College*
Karl Aufderheide, *Texas A&M University*
Idelisa Ayala, *Broward Community College*
Lisa M. Baird, *University of San Diego*
Adebiyi Banjoko, *Chandler-Gilbert Community College*
Gerry Barclay, *Highline Community College*
Susan Barrett, *Massasoit Community College*
Diane Bassham, *Iowa State University*
Donald Reon Baud, *University of Memphis*
Vernon Bauer, *Francis Marion University*
Chris Bazinet, *St. John's University*
Ruth E. Beattie, *University of Kentucky*
Michael C. Bell, *Richland College*
Steve Berg, *Winona State University*
Giacomo Bernardi, *University of California–Santa Cruz*
Deborah Bielser, *University of Illinois–Urbana-Champaign*
Arlene G. Billock, *University of Louisiana–Lafayette*
Eric Blackwell, *Delta State University*
Andrew R. Blaustein, *Oregon State University*
Kristopher A. Blee, *California State University–Chico*
Steve Blumenshine, *California State University–Fresno*
Jason Bond, *East Carolina University*
Heidi B. Borgeas, *University of Tampa*
Russell Borski, *North Carolina State University*
James Bottesch, *Brevard Community College/Cocoa Campus*
Scott Bowling, *Auburn University*
Robert S. Boyd, *Auburn University*
Eldon J. Braun, *University of Arizona*
Michael Breed, *University of Colorado–Boulder*
Robert Brewer, *Cleveland State Community College*
Randy Brewton, *University of Tennessee*
Peggy Brickman, *University of Georgia*
Cheryl Briggs, *University of California–Berkeley*
George Briggs, *State University College–Geneseo*
Mirjana M. Brockett, *Georgia Institute of Technology*
W. Randy Brooks, *Florida Atlantic University*
Jack Brown, *Paris Junior College*
Peter S. Brown, *Mesa Community College*
Mark Browning, *Purdue University*
Cedric O. Buckley, *Jackson State University*
Don Buckley, *Quinnipiac University*
Arthur L. Buikema, Jr., *Virginia Tech University*
Rodolfo Buiser, *University of Wisconsin–Eau Claire*
Anne Bullerjahn, *Owens Community College*
Carolyn J.W. Bunde, *Idaho State University*
Ray D. Burkett, *Southeast Tennessee Community College*
Scott Burt, *Truman State University*
Stephen R. Burton, *Grand Valley State University*
Jorge Busciglio, *University of California*
Stephen P. Bush, *Coastal Carolina University*
Thomas Bushart, *University of Texas–Austin*
Peter E. Busher, *Boston University*
Malcolm Butler, *North Dakota State University*
David Byres, *Florida Community College South Campus*
Jennifer Campbell, *North Carolina State University*
Jeff Carmichael, *University of North Dakota*
Clint E. Carter, *Vanderbilt University*
Patrick A. Carter, *Washington State University*
Timothy H. Carter, *St. John's University*
Merri Lynn Casem, *California State University–Fullerton*
Domenic Castignetti, *Loyola University of Chicago*
Deborah A. Cato, *Wheaton College*
Maria V. Cattell, *University of Colorado*
David T. Champlin, *University of Southern Maine*
Tien-Hsien Chang, *Ohio State University*
Estella Chen, *Kennesaw State University*
Sixue Chen, *University of Florida*
Brenda Chinnery-Allgeier, *University of Texas–Austin*
Young Cho, *Eastern New Mexico University*
Jung H. Choi, *Georgia Institute of Technology*
Genevieve Chung, *Broward Community College–Central*
Philip Clampitt, *Oakland University*
Curtis Clark, *Cal Poly–Pomona*
T. Denise Clark, *Mesa Community College*
Allison Cleveland Roberts, *University of South Florida–Tampa*
Janice J. Clymer, *San Diego Mesa College*
Randy W. Cohen, *California State University–Northridge*
Patricia Colberg, *University of Wyoming*
Craig Coleman, *Brigham Young University–Provo*
Linda T. Collins, *University of Tennessee–Chattanooga*
William Collins, *Stony Brook University*
Jay L. Comeaux, *Louisiana State University*
Bob Connor II, *Owens Community College*
Joanne Conover, *University of Connecticut*
John Cooley, *Yale University*
Ronald H. Cooper, *University of California–Los Angeles*
Vicki Corbin, *University of Kansas–Lawrence*
Anthony Cornett, *Valencia Community College*
Daniel Costa, *University of California–Santa Cruz*
Sehoya Cotner, *University of Minnesota*
Will Crampton, *University of Central Florida*
Mack E. Crayton III, *Xavier University of Louisiana*
Louis Crescitelli, *Bergen Community College*
Charles Creutz, *University of Toledo*
Karen Curto, *University of Pittsburgh*
Kenneth A. Cutler, *North Carolina Central University*
Anita Davelos Baines, *University of Texas–Pan American*
Cara Davies, *Ohio Northern University*
Mark A. Davis, *Macalester College*
Donald H. Dean, *The Ohio State University*
James Dearworth, *Lafayette College*
Mark D. Decker, *University of Minnesota*
Jeffery P. Demuth, *Indiana University*
Phil Denette, *Delgado Community College*
John Dennehy, *Queens College*
William Dentler, *University of Kansas*
Smruti A. Desai, *Lonestar College–Cy Fair*
Donald Deters, *Bowling Green State University*

Hudson R. DeYoe, *University of Texas–Pan American*
Laura DiCaprio, *Ohio University*
Randy DiDomenico, *University of Colorado–Boulder*
Robert S. Dill, *Bergen Community College*
Kevin Dixon, *University of Illinois–Urbana–Champaign*
John S. Doctor, *Duquesne University*
Warren D. Dolphin, *Iowa State University*
David S. Domozych, *Skidmore College*
Robert P. Donaldson, *George Washington University*
Cathy A. Donald-Whitney, *Collin County Community College*
Kristiann M. Dougherty, *Valencia Community College*
Kari M.H. Doyle, *San Jacinto College*
Marjorie Doyle, *University of Wisconsin–Madison*
John Drummond, *Lafayette College*
Ernest Dubrul, *University of Toledo*
Jeffry L. Dudycha, *William Patterson University of New Jersey*
Charles Duggins, Jr., *University of South Carolina*
Richard Duhrkopf, *Baylor University*
James N. Dumond, *Texas Southern University*
Tod Duncan, *University of Colorado–Denver*
Susan Dunford, *University of Cincinnati*
Roland Dute, *Auburn University*
Ralph P. Eckerlin, *Northern Virginia Community College*
Jose L. Egremy, *Northwest Vista College*
William D. Eldred, *Boston University*
David W. Eldridge, *Baylor University*
Inge Eley, *Hudson Valley Community College*
Lisa K. Elfring, *University of Arizona*
Kurt J. Elliot, *Northwest Vista College*
Johnny El-Rady, *University of South Florida*
Seema Endley, *Blinn College*
Bill Ensign, *Kennesaw State University*
David S. Epstein, *J. Sergeant Reynolds Community College*
Shannon Erickson Lee, *California State University–Northridge*
Gary N. Ervin, *Mississippi State University*
Frederick Essig, *University of Southern Florida*
Sharon Eversman, *Montana State University*
Brent E. Ewers, *University of Wyoming*
Stan Faeth, *Arizona State University*
Susan Fahrbach, *Wake Forest University*
Peter Fajer, *Florida State University*
Paul Farnsworth, *University of Texas–San Antonio*
Zen Faulkes, *University of Texas–Pan American*
Paul D. Ferguson, *University of Illinois–Urbana–Champaign*
Fleur Ferro, *Community College of Denver*
Miriam Ferzli, *North Carolina State University*
Margaret F. Field, *Saint Mary's College of California*
Jose Fierro, *Florida State College–Jacksonville*
Melanie Fierro, *Florida State College–Jacksonville*
Teresa G. Fischer, *Indian River College*
David Fitch, *New York University*
Jorge A. Flores, *West Virginia University*
Irwin Forseth, *University of Maryland*
David Foster, *North Idaho College*
Paul Fox, *Danville Community College*
Sandra Fraley, *Dutchess Community College*
Pete Franco, *University of Minnesota*
Steven N. Francoeur, *Eastern Michigan University*
Wayne D. Frasch, *Arizona State University*
Barbara Frase, *Bradley University*
Robert Friedman, *University of South Carolina*
Adam J. Fry, *University of Connecticut*
Bernard L. Frye, *University of Texas–Arlington*
Caitlin Gabor, *Texas State University–San Marcos*
Anne M. Galbraith, *University of Wisconsin–La Crosse*
Mike Ganger, *Gannon University*
Deborah Garrity, *Colorado State University*
John R. Geiser, *Western Michigan University*
Nicholas R. Geist, *Sonoma State University*
Patricia A. Geppert, *University of Texas–San Antonio*
Shannon Gerry, *Wellesley College*
Cindee Giffen, *University of Wisconsin–Madison*
Frank S. Gilliam, *Marshall University*
Chris Gissendanner, *University of Louisiana at Monroe*
Jon Glase, *Cornell University*
Florence K. Gleason, *University of Minnesota*
Elmer Godeny, *Baton Rouge Community College*
Elizabeth Godrick, *Boston University*
Robert Gorham, *Northern Virginia Community College*
James M. Grady, *University of New Orleans*
Brian Grafton, *Kent State University*
John Graham, *Bowling Green State University*
Barbara E. Graham-Evans, *Jackson State University*
Christine E. Gray, *Blinn College*
Christopher Gregg, *Louisiana State University*
John Griffis, *Joliet Junior College*
LeeAnn Griggs, *Massasoit Community College*
Tim Grogan, *Valencia Community College–Osceola*
Richard S. Groover, *J. Sergeant Reynolds Community College*
Gretel Guest, *Durham Technical Community College*
Stan Guffey, *University of Tennessee*
Cameron Gundersen, *University of California*
Rodney D. Hagley, *University of North Carolina–Wilmington*
George Hale, *University of West Georgia*
Patricia Halpin, *University of California–Los Angeles*
William Hanna, *Massasoit Community College*
Gary L. Hannan, *Eastern Michigan University*
David T. Hanson, *University of New Mexico*
Christopher J. Harendza, *Montgomery County Community College*
Kyle E. Harms, *Louisiana State University*
Sally E. Harmych, *University of Toledo*
Betsy Harris, *Appalachian State University*
M.C. Hart, *Minnesota State University–Mankato*
Barbara Harvey, *Kirkwood Community College*
Carla Ann Hass, *The Pennsylvania State University*
Mary Beth Hawkins, *North Carolina State University*
Brian T. Hazlett, *Briar Cliff University*
Harold Heatwole, *North Carolina State University*
Cheryl Heinz, *Benedictine University*
Jutta B. Heller, *Loyola University–Chicago*
Susan Hengeveld, *Indiana University–Bloomington*
Mark Hens, *University of North Carolina–Greensboro*
Steven K. Herbert, *University of Wyoming–Laramie*
Edgar Javier Hernandez, *University of Missouri–St. Louis*
Albert A. Herrera, *University of Southern California*
David L. Herrin, *University of Texas–Austin*
Helen Hess, *College of the Atlantic*
David S. Hibbert, *Clark University*
R. James Hickey, *Miami University of Ohio–Oxford*
Tracey E. Hickox, *University of Illinois–Urbana–Champaign*
Terri Hildebrand, *Southern Utah University*
Juliana Hinton, *McNeese State University*
Anne Hitt, *Oakland University*
Mark A. Holbrook, *University of Iowa*
Robert D. Hollister, *Grand Valley State University*
Richard G. Holloway, *Northern Arizona University*
Harriette Howard-Lee Block, *Prairie View A&M University*
Dianella Howarth, *St. John's University*
Kelly Howe, *University of New Mexico*
Carrie Hughes, *San Jacinto College*
Barbara Hunnicutt, *Seminole Community College*
Bradley Hyman, *University of California–Riverside*
Ella Ingram, *Rose-Hulman Institute of Technology*
Vicki J. Isola, *Hope College*
Jeffrey Jack, *University of Louisville*
Desirée Jackson, *Texas Southern University*
Joseph J. Jacquot, *Grand Valley State University*
John Jaenike, *University of Rochester*
Ashok Jain, *Albany State University*
Eric Jellen, *Brigham Young University*
Judy Jernstedt, *University of California–Davis*
Lee Johnson, *Ohio State University*
Elizabeth A. Jordan, *Moorpark College*
Robyn Jordan, *University of Louisiana at Monroe*
Susan Jorstad, *University of Arizona*
Walter S. Judd, *University of Florida*
David Julian, *University of Florida*
Nick Kaplinsky, *Swarthmore College*
Vesna Karaman, *University of Texas at El Paso*
Istvan Karsai, *East Tennessee State University*
Nancy Kaufmann, *University of Pittsburgh*
Stephen R. Kelso, *University of Illinois–Chicago*
Heather R. Ketchum, *Blinn College*
Eunsoo Kim, *University of Wisconsin–Madison*
Denice D. King, *Cleveland State Community College*
Stephen J. King, *University of Missouri–Kansas City*
Bridgette Kirkpatrick, *Collin County Community College*
John Z. Kiss, *Miami University*
Ted Klenk, *Valencia Community College–West*
David M. Kohl, *University of California–Santa Barbara*
Anna Koshy, *Houston Community College–NW*
David Krauss, *Borough of Manhattan Community College*
Sherry Krayesky, *University of Louisiana–Lafayette*
John Krenetsky, *Metropolitan State College–Denver*
Karin E. Krieger, *University of Wisconsin–Green Bay*
William Kroll, *Loyola University–Chicago*
Paul Kugrens, *Colorado State University*
Pramod Kumar, *University of Texas–San Antonio*
Josephine Kurdziel, *University of Michigan*
David T. Kurjiaka, *Ohio University*
Allen Kurta, *Eastern Michigan University*
Paul K. Lago, *University of Mississippi*
William Lamberts, *College of St. Benedict/Saint John's University*
David Lampe, *Duquesne University*
Pamela Lanford, *University of Maryland*
Marianne M. Laporte, *Eastern Michigan University*
Arlen T. Larson, *University of Colorado–Denver*
John Latto, *University of California–Berkeley*
John C. Law, *Community College of Allegheny County*
Jonathan N. Lawson, *Collin County Community College*
Brenda Leady, *University of Toledo*
Tali D. Lee, *University of Wisconsin–Eau Claire*
Hugh Lefcort, *Gonzaga University*
Michael Lentz, *University of North Florida*
John Lepri, *University of North Carolina–Greensboro*
Army Lester, *Kennesaw State University*
Jennifer J. Lewis, *San Juan College*

Q Quinn Li, *Miami University, Ohio*
Nardos Lijam, *Columbus State Community College*
Yusheng Liu, *East Tennessee State University*
Pauline A. Lizotte, *Valencia Community College*
Jason L. Locklin, *Temple College*
Robert Locy, *Auburn University*
Albert R. Loeblich III, *University of Houston*
Thomas A. Lonergan, *University of New Orleans*
James A. Long, *Boise State University*
Craig Longtine, *North Hennepin Community College*
David Lonzarich, *University of Wisconsin–Eau Claire*
Donald Lovett, *The College of New Jersey*
James B. Ludden, *College of DuPage*
Albert MacKrell, *Bradley University*
Paul T. Magee, *University of Minnesota–Minneapolis*
Jay Mager, *Ohio Northern University*
Christi Magrath, *Troy University*
Richard Malkin, *University of California–Berkeley*
Charles H. Mallery, *University of Miami*
Nilo Marin, *Broward College*
Kathleen A. Marrs, *IUPUI–Indianapolis*
Diane L. Marshall, *University of New Mexico*
Peter J. Martinat, *Xavier University of Louisiana*
Cindy Martinez Wedig, *University of Texas–Pan American*
Joel Maruniak, *University of Missouri*
Joe Matanoski, *Stevenson University*
Patricia Matthews, *Grand Valley State University*
Barbara May, *College of St. Benedict/St. John's University*
Kamau Mbuthia, *Bowling Green State University*
Norah McCabe, *Washington State University*
Chuck McClaugherty, *Mount Union College*
Regina S. McClinton, *Grand Valley State University*
Greg McCormac, *American River College*
Andrew McCubbin, *Washington State University*
David L. McCulloch, *Collin County Community College*
Mark A. McGinley, *Texas Tech University*
Kerry McKenna, *Lord Fairfax College*
Tanya K. McKinney, *Xavier University of Louisiana*
Carrie McMahon Hughes, *San Jacinto College–Central Campus*
Joseph McPhee, *LaGuardia Community College*
Judith Megaw, *Indian River Community College*
Mona C. Mehdy, *University of Texas–Austin*
Brad Mehrtens, *University of Illinois–Urbana-Champaign*
Susan Meiers, *Western Illinois University*
Michael Meighan, *University of California–Berkeley*
Douglas Meikle, *Miami University*
Allen F. Mensinger, *University of Minnesota–Duluth*
Catherine Merovich, *West Virginia University*
John Merrill, *Michigan State University*
Richard Merritt, *Houston Community College*
Jennifer Metzler, *Ball State University*
Melissa Michael, *University of Illinois–Urbana-Champaign*
James Mickle, *North Carolina State University*
Brian T. Miller, *Middle Tennessee State University*
Hugh A. Miller III, *East Tennessee State University*
Thomas E. Miller, *Florida State University*
Sarah L. Milton, *Florida Atlantic University*
Dennis J. Minchella, *Purdue University*
Subhash C. Minocha, *University of New Hampshire*
Manuel Miranda-Arango, *University of Texas at El Paso*
Patricia Mire, *University of Louisiana–Lafayette*
Michael Misamore, *Texas Christian University*
Jasleen Mishra, *Houston Community College–Southwest*
Alan Molumby, *University of Illinois, Chicago*
Daniela S. Monk, *Washington State University*
W. Linn Montgomery, *Northern Arizona University*
Daniel Moon, *University of North Florida*
Jennifer Moon, *University of Texas–Austin*
Janice Moore, *Colorado State University*
Richard C. Moore, *Miami University*
Mathew D. Moran, *Hendrix College*
Jorge A. Moreno, *University of Colorado–Boulder*
David Morgan, *University of West Georgia*
Roderick M. Morgan, *Grand Valley State University*
James V. Moroney, *Louisiana State University*
Ann C. Morris, *Florida State University*
Molly R. Morris, *Ohio University*
Christa P.H. Mulder, *University of Alaska–Fairbanks*
Mike Muller, *University of Illinois–Chicago*
Darrel C. Murray, *University of Illinois–Chicago*
Richard J. Murray, *Hendrix College*
Melissa Murray Reedy, *University of Illinois–Urbana-Champaign*
Michelle Mynlieff, *Marquette University*
Jennifer Nauen, *University of Delaware*
Allan D. Nelson, *Tarleton State University*
Raymond Neubauer, *University of Texas–Austin*
Jacalyn Newman, *University of Pittsburgh*
Robert Newman, *University of North Dakota*
Laila Nimri, *Seminole Community College*
Colleen J. Nolan, *St. Mary's University*
Shawn E. Nordell, *St. Louis University*
Margaret Nsofor, *Southern Illinois University–Carbondale*
Dennis W. Nyberg, *University of Illinois–Chicago*
Nicole S. Obert, *University of Illinois–Urbana-Champaign*
Olumide Ogunmosin, *Texas Southern University*
Wan Ooi, *Houston Community College–Central*
David G. Oppenheimer, *University of Florida*
John C. Osterman, *University of Nebraska–Lincoln*
Brian Palestis, *Wagner College*
Ravishankar Palanivelu, *University of Arizona*
Julie M. Palmer, *University of Texas–Austin*
Peter Pappas, *Community College of Morris*
Lisa Parks, *North Carolina State University*
C. O. Patterson, *Texas A&M University*
Ronald J. Patterson, *Michigan State University*
Linda M. Peck, *University of Findlay*
David Pennock, *Miami University*
Shelley W. Penrod, *North Harris College*
Beverly Perry, *Houston Community College*
John S. Peters, *College of Charleston*
Chris Petersen, *College of the Atlantic*
David K. Peyton, *Morehead State University*
Marius Pfeiffer, *Tarrant County College NE*
Jay Phelan, *University of California–Los Angeles*
Jerry Phillips, *University of Colorado–Colorado Springs*
Susan Phillips, *Brevard Community College*
Randall Phillis, *University of Massachusetts–Amherst*
Eric R. Pianka, *The University of Texas–Austin*
Paul Pilliterri, *Southern Utah University*
Debra B. Pires, *University of California–Los Angeles*
Thomas Pitzer, *Florida International University*
Terry Platt, *University of Rochester*
Peggy E. Pollak, *Northern Arizona University*
Uwe Pott, *University of Wisconsin–Green Bay*
Linda F. Potts, *University of North Carolina–Wilmington*
Jessica Poulin, *University at Buffalo, SUNY*
Kumkum Prabhakar, *Nassau Community College*
Joelle Presson, *University of Maryland*
Mitch Price, *Pennsylvania State University*
Richard B. Primack, *Boston University*
Gregory Pryor, *Francis Marion University*
Penny L. Ragland, *Auburn University*
Lynda Randa, *College of Dupage*
Rajinder S. Ranu, *Colorado State University*
Marceau Ratard, *Delgado Community College*
Melanie K. Rathburn, *Boston University*
Robert S. Rawding, *Gannon University*
Flona Redway, *Barry University*
Jennifer Regan, *University of Southern Mississippi*
Stuart Reichler, *University of Texas–Austin*
Jill D. Reid, *Virginia Commonwealth University*
Anne E. Reilly, *Florida Atlantic University*
Linda R. Richardson, *Blinn College*
Kim Risley, *Mount Union College*
Elisa Rivera-Boyles, *Valencia Community College*
Laurel B. Roberts, *University of Pittsburgh*
James V. Robinson, *University of Texas–Arlington*
Kenneth R. Robinson, *Purdue University*
Luis A. Rodriguez, *San Antonio College*
Chris Romero, *Front Range Community College–Larimer Campus*
Chris Ross, *Kansas State University*
Anthony M. Rossi, *University of North Florida*
Doug Rouse, *University of Wisconsin–Madison*
Kenneth H. Roux, *Florida State University*
Ann E. Rushing, *Baylor University*
Laurie K. Russell, *St. Louis University*
Scott Russell, *University of Oklahoma*
Christina T. Russin, *Northwestern University*
Charles L. Rutherford, *Virginia Tech University*
Margaret Saha, *College of William and Mary*
Sheridan Samano, *Community College of Aurora*
Hildegarde Sanders, *Stevenson University*
Kanagasabapathi Sathasivan, *University of Texas–Austin*
David K. Saunders, *Augusta State University*
Stephen G. Saupe, *College of St. Benedict*
Jon B. Scales, *Midwestern State University*
Daniel C. Scheirer, *Northeastern University*
H. Jochen Schenk, *California State University–Fullerton*
John Schiefelbein, *University of Michigan*
Deemah Schirf, *University of Texas–San Antonio*
Mark Schlueter, *College of Saint Mary*
Chris Schneider, *Boston University*
Susan Schreier, *Towson University*
Scott Schuette, *Southern Illinois University–Carbondale*
David Schwartz, *Houston Community College–Southwest*
Dean D. Schwartz, *Auburn University*
David A. Scicchitano, *New York University*
Erik Scully, *Towson University*
Robin Searles-Adenegan, *University of Maryland*
Pat Selelyo, *College of Southern Idaho*
Pramila Sen, *Houston Community College–Central*
Tim Shannon, *Francis Marion University*
Jonathan Shaver, *North Hennepin Community College*
Brandon Sheafor, *Mount Union College*
Ellen Shepherd Lamb, *University of North Carolina–Greensboro*
Mark Sheridan, *North Dakota State University*
Dennis Shevlin, *The College of New Jersey*
Patty Shields, *University of Maryland*

Cara Shillington, *Eastern Michigan University*
Richard M. Showman, *University of South Carolina*
Michele Shuster, *New Mexico State University*
Scott Siechen, *University of Illinois–Urbana-Champaign*
Martin Silberberg, *McGraw-Hill chemistry author*
Anne Simon, *University of Maryland*
Sue Simon Westendorf, *Ohio University–Athens*
Robert Simons, *University of California–Los Angeles*
John B. Skillman, *California State University–San Bernadino*
J. Henry Slone, *Francis Marion University*
Lee Smee, *Texas A&M University*
Phillip Snider, *Jr., Gadsden State Community College*
Dianne Snyder, *Augusta State University*
Nancy Solomon, *Miami University*
Sally Sommers Smith, *Boston University*
Punnee Soonthornpoct, *Blinn College*
Vladimir Spiegelman, *University of Wisconsin–Madison*
Bryan Spohn, *Florida State College at Jacksonville*
Lekha Sreedhar, *University of Missouri–Kansas City*
Bruce Stallsmith, *University of Alabama–Huntsville*
Richard Stalter, *St. John's University*
Susan J. Stamler, *College of Dupage*
Mark P. Staves, *Grand Valley State University*
William Stein, *Binghamton University*
Mark E. Stephansky, *Massasoit Community College*
Philip J. Stephens, *Villanova University*
Dean Stetler, *University of Kansas–Lawrence*
Brian Stout, *Northwest Vista College*
Kevin Strang, *University of Wisconsin–Madison*
Antony Stretton, *University of Wisconsin–Madison*
Gregory W. Stunz, *Texas A&M University–Corpus Christi*
Mark Sturtevant, *Oakland University*
C.B. Subrahmanvam, *Florida A&M University*
Julie Sutherland, *College of Dupage*
Mark Sutherland, *Hendrix College*
Brook Swanson, *Gonzaga University*
Debbie Swarthout, *Hope College*
David Tam, *University of North Texas*
Roy A. Tassava, *Ohio State University*
Judy Taylor, *Motlow State Community College*
Randall G. Terry, *Lamar University*
Sharon Thoma, *University of Wisconsin*
Shawn A. Thomas, *College of St. Benedict/St. John's University*
Carol Thornber, *University of Rhode Island*
Patrick A. Thorpe, *Grand Valley State University*
Scott Tiegs, *Oakland University*
Kristina Timmerman, *St. John's University*
Daniel B. Tinker, *University of Wyoming*
Marty Tracey, *Florida International University*
Paul Trombley, *Florida State University*
John R. True, *Stony Brook University*
Encarni Trueba, *Community College of Baltimore County Essex*
Cathy Tugmon, *Augusta State University*
J. M. Turbeville, *Virginia Commonwealth University*
Marshall Turell, *Houston Community College*
Ashok Upadhyaya, *University of South Florida–Tampa*
Anthony J. Uzwiak, *Rutgers University*
Rani Vajravelu, *University of Central Florida*
William Velhagen, *New York University*
Wendy Vermillion, *Columbus State Community College*
Sara Via, *University of Maryland*
Neal J. Voelz, *St. Cloud State University*
Thomas V. Vogel, *Western Illinois University*
Samuel E. Wages, *South Plains College*
Jyoti R. Wagle, *Houston Community College System–Central*
R. Steven Wagner, *Central Washington University–Ellensburg*
Charles Walcott, *Cornell University*
John Waldman, *Queens College–CUNY*
Randall Walikonis, *University of Connecticut*
Gary R. Walker, *Youngstown State University*
Jeffrey A. Walker, *University of Southern Maine*
Sean E. Walker, *California State University–Fullerton*
Delon E. Washo-Krupps, *Arizona State University*
Fred Wasserman, *Boston University*
Steven A. Wasserman, *University of California–San Diego*
R. Douglas Watson, *University of Alabama–Birmingham*
Arthur E. Weis, *University of California–Irvine*
Doug Wendell, *Oakland University*
Howard Whiteman, *Murray State University*
Susan Whittemore, *Keene State College*
Jennifer Wiatrowski, *Pasco-Hernando Community College*
Sheila Wicks, *Malcolm X College*
Donna Wiersema, *Houston Community College*
Regina Wiggins-Speights, *Houston Community College–Northeast*
David H. Williams, *Valencia Community College*
Lawrence R. Williams, *University of Houston*
Ned Williams, *Minnesota State University–Mankato*
E. Gay Williamson, *Mississippi State University*
David L. Wilson, *University of Miami*
Mark S. Wilson, *Humboldt State University*
Bob Winning, *Eastern Michigan University*
Jane E. Wissinger, *University of Minnesota*
Michelle D. Withers, *Louisiana State University*
Clarence C. Wolfe, *Northern Virginia Community College*
Gene K. Wong, *Quinnipiac University*
David Wood, *California State University–Chico*
Bruce Wunder, *Colorado State University*
Richard P. Wunderlin, *University of South Florida*
Mark Wygoda, *McNeese State University*
Joanna Wysocka-Diller, *Auburn University*
H. Randall Yoder, *Lamar University*
Marilyn Yoder, *University of Missouri–Kansas City*
Marlena Yost, *Mississippi State University*
Robert Yost, *Indiana University–Purdue*
Kelly Young, *California State University–Long Beach*
Linda Young, *Ohio Northern University*
Ted Zerucha, *Appalachian State University*
Scott D. Zimmerman, *Southwest Missouri State University*

International Reviewers

Dr. Alyaa Ragaei, *Future University, Cairo*
Heather Addy, *University of Calgary*
Mari L. Acevedo, *University of Puerto Rico at Arecibo*
Heather E. Allison, *University of Liverpool, UK*
David Backhouse, *University of New England*
Andrew Bendall, *University of Guelph*
Marinda Bloom, *Stellenbosch University, South Africa*
Tony Bradshaw, *Oxford-Brookes University, UK*
Alison Campbell, *University of Waikato*
Bruce Campbell, *Okanagan College*
Clara E. Carrasco, Ph.D., *University of Puerto Rico–Ponce Campus*
Keith Charnley, *University of Bath, UK*
Ian Cock, *Griffith University*
Margaret Cooley, *University of NSW*
R. S. Currah, *University of Alberta*
Logan Donaldson, *York University*
Theo Elzenga, *Rijks Universiteit Groningen, Netherlands*
Neil C. Haave, *University of Alberta*
Tom Haffie, *University of Western Ontario*
Louise M. Hafner, *Queensland University of Technology*
Annika F. M. Haywood, *Memorial University of Newfoundland*
William Huddleston, *University of Calgary*
Shin-Sung Kang, *KyungBuk University*
Wendy J. Keenleyside, *University of Guelph*
Christopher J. Kennedy, *Simon Fraser University*
Bob Lauder, *Lancaster University*
Richard C. Leegood, *Sheffield University, UK*
Thomas H. MacRae, *Dalhousie University*
R. Ian Menz, *Flinders University*
Kirsten Poling, *University of Windsor*
Jim Provan, *Queens University, Belfast, UK*
Richard Roy, *McGill University*
Han A.B. Wösten, *Utrecht University, Netherlands*

A NOTE FROM THE AUTHORS

The lives of most science-textbook authors do not revolve around an analysis of writing techniques. Instead, we are people who understand science and are inspired by it, and we want to communicate that information to our students. Simply put, we need a lot of help to get it right.

Editors are a key component that help the authors modify the content of their book so it is logical, easy to read, and inspiring. The editorial team for this *Biology* textbook has been a catalyst that kept this project rolling. The members played various roles in the editorial process. Rebecca Olsen, Sponsoring Editor (Major Biology) did an outstanding job of overseeing the third edition. Her insights with regard to pedagogy, content, and organization have been invaluable. Elizabeth Sievers, Director of Development-Biology, has been the master organizer. Liz's success at keeping us on schedule is greatly appreciated.

Our Freelance Developmental Editor, Joni Frasier, worked directly with the authors to greatly improve the presentation of the textbook's content. She did a great job of editing chapters and advising the authors on improvements for the third edition. We would also like to acknowledge our copy editor, Linda Davoli, for keeping our grammar on track.

Another important aspect of the editorial process is the actual design, presentation, and layout of materials. It's confusing if the text and art aren't on the same page, or if a figure is too large or too small. We are indebted to the tireless efforts of Sandy Wille, Content Project Manager; and David Hash, Senior Designer at McGraw-Hill. Likewise, our production company, Lachina Publishing Services, did an excellent job with the paging, revision of existing art, and the creation of new art for the third edition. Their artistic talents, ability to size and arrange figures, and attention to the consistency of the figures have been remarkable.

We would like to acknowledge the ongoing efforts of the superb marketing staff at McGraw-Hill. Special thanks to Patrick Reidy, Executive Marketing Manager-Life Sciences, for his ideas and enthusiasm for this book.

Finally, other staff members at McGraw-Hill Higher Education have ensured that the authors and editors were provided with adequate resources to achieve the goal of producing a superior textbook. These include Kurt Strand, Senior Vice President, Products & Markets, and Marty Lange, Vice President, General Manager, Products & Markets, and Michael Hackett, Director for Life Sciences.

Contents

CONTENTS

Chapter 1

An Introduction to Biology 1

1.1 Principles of Biology and the Levels of Biological Organization 2

1.2 Unity and Diversity of Life 6

Genomes & Proteomes Connection: The Study of Genomes and Proteomes Provides an Evolutionary Foundation for Our Understanding of Biology 9

1.3 Biology as a Scientific Discipline 13

Feature Investigation: Observation and Experimentation Form the Core of Biology 17

UNIT I Chemistry

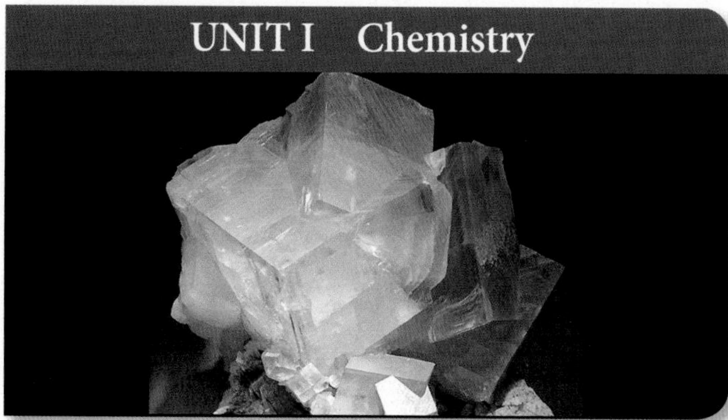

Chapter 2

The Chemical Basis of Life I: Atoms, Molecules, and Water 21

2.1 Atoms 21

Feature Investigation: Rutherford Determined the Modern Model of the Atom 22

2.2 Chemical Bonds and Molecules 28

2.3 Properties of Water 33

Chapter 3

The Chemical Basis of Life II: Organic Molecules 42

3.1 The Carbon Atom and the Study of Organic Molecules 42

3.2 Formation of Organic Molecules and Macromolecules 45

3.3 Carbohydrates 46

3.4 Lipids 49

3.5 Proteins 52

Feature Investigation: Anfinsen Showed That the Primary Structure of Ribonuclease Determines Its Three-Dimensional Structure 58

Genomes & Proteomes Connection: Proteins Contain Functional Domains Within Their Structures 60

3.6 Nucleic Acids 60

UNIT II Cell

Chapter 4

General Features of Cells 66

4.1 Microscopy 66

4.2 Overview of Cell Structure 69

Genomes & Proteomes Connection: The Proteome Largely Determines the Characteristics of a Cell 70

4.3 The Cytosol 73

4.4 The Nucleus and Endomembrane System 78

Feature Investigation: Palade Demonstrated That Secreted Proteins Move Sequentially Through Organelles of the Endomembrane System 81

4.5 Semiautonomous Organelles 86

4.6 Protein Sorting to Organelles 89

4.7 Systems Biology of Cells: A Summary 92

Chapter 5

Membrane Structure, Synthesis, and Transport 97

5.1 Membrane Structure 98

Genomes & Proteomes Connection: Approximately 25% of All Genes Encode Transmembrane Proteins 99

5.2 Fluidity of Membranes 101

5.3 Synthesis of Membrane Components in Eukaryotic Cells 103

5.4 Overview of Membrane Transport 105

5.5 Transport Proteins 108

Feature Investigation: Agre Discovered That Osmosis Occurs More Quickly in Cells with a Channel That Allows the Facilitated Diffusion of Water 109

5.6 Exocytosis and Endocytosis 113

Chapter 6

An Introduction to Energy, Enzymes, and Metabolism 118

6.1 Energy and Chemical Reactions 118

Genomes & Proteomes Connection: Many Proteins Bind ATP and Use That ATP as a Source of Energy 121

6.2 Enzymes and Ribozymes 122

Feature Investigation: The Discovery of Ribozymes by Sidney Altman Revealed That RNA Molecules May Also Function as Catalysts 126

6.3 Overview of Metabolism 128

6.4 Recycling of Organic Molecules 131

Chapter 7

Cellular Respiration and Fermentation 136

7.1 Overview of Cellular Respiration 136

7.2 Glycolysis 138

Genomes & Proteomes Connection: Cancer Cells Usually Exhibit High Levels of Glycolysis 139

7.3 Breakdown of Pyruvate 139

7.4 Citric Acid Cycle 141

7.5 Oxidative Phosphorylation 142

Feature Investigation: Yoshida and Kinosita Demonstrated That the γ Subunit of the ATP Synthase Spins 147

7.6 Connections Among Carbohydrate, Protein, and Fat Metabolism 149

7.7 Anaerobic Respiration and Fermentation 150

Chapter 8

Photosynthesis 154

8.1 Overview of Photosynthesis 154

8.2 Reactions That Harness Light Energy 157

Genomes & Proteomes Connection: The Cytochrome Complexes of Mitochondria and Chloroplasts Contain Evolutionarily Related Proteins 161

8.3 Molecular Features of Photosystems 161

8.4 Synthesizing Carbohydrates via the Calvin Cycle 165

Feature Investigation: The Calvin Cycle Was Determined by Isotope-Labeling Methods 167

8.5 Variations in Photosynthesis 169

Chapter 9

Cell Communication 174

9.1 General Features of Cell Communication 174

9.2 Cellular Receptors and Their Activation 178

9.3 Signal Transduction and the Cellular Response 181

9.4 Hormonal Signaling in Multicellular Organisms 187

Genomes & Proteomes Connection: A Cell's Response to Hormones and Other Signaling Molecules Depends on the Proteins It Makes 187

9.5 Apoptosis: Programmed Cell Death 188

Feature Investigation: Kerr, Wyllie, and Currie Found That Hormones May Control Apoptosis 189

Chapter 10

Multicellularity 193

10.1 Extracellular Matrix and Cell Walls 194

Genomes & Proteomes Connection: Collagens Are a Family of Proteins That Give Animals a Variety of ECM Properties 195

10.2 Cell Junctions 198

Feature Investigation: Loewenstein and Colleagues Followed the Transfer of Fluorescent Dyes to Determine the Size of Gap-Junction Channels 202

10.3 Tissues 204

UNIT III Genetics

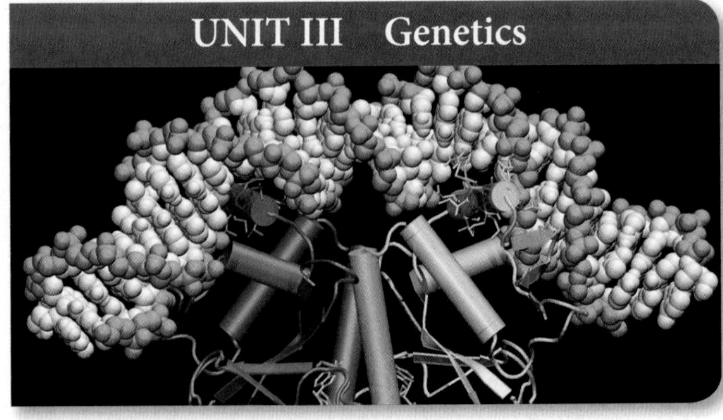

Chapter 11

Nucleic Acid Structure, DNA Replication, and Chromosome Structure 212

11.1 Biochemical Identification of the Genetic Material 212

Feature Investigation: Avery, MacLeod, and McCarty Used Purification Methods to Reveal That DNA Is the Genetic Material 214

11.2 Nucleic Acid Structure 217

11.3 An Overview of DNA Replication 221

11.4 Molecular Mechanism of DNA Replication 224
Genomes & Proteomes Connection: DNA Polymerases Are a Family of Enzymes with Specialized Functions 228
11.5 Molecular Structure of Eukaryotic Chromosomes 230

Chapter 12

Gene Expression at the Molecular Level 235

12.1 Overview of Gene Expression 236
12.2 Transcription 239
12.3 RNA Processing in Eukaryotes 241
12.4 Translation and the Genetic Code 243
Feature Investigation: Nirenberg and Leder Found That RNA Triplets Can Promote the Binding of tRNA to Ribosomes 246
12.5 The Machinery of Translation 247
Genomes & Proteomes Connection: Comparisons of Small Subunit rRNAs Among Different Species Provide a Basis for Establishing Evolutionary Relationships 250
12.6 The Stages of Translation 251

Chapter 13

Gene Regulation 257

13.1 Overview of Gene Regulation 257
13.2 Regulation of Transcription in Bacteria 260
Feature Investigation: Jacob, Monod, and Pardee Studied a Constitutive Mutant to Determine the Function of the Lac Repressor 264
13.3 Regulation of Transcription in Eukaryotes: Roles of Transcription Factors and Mediator 267
13.4 Regulation of Transcription in Eukaryotes: Changes in Chromatin Structure and DNA Methylation 271
13.5 Regulation of RNA Processing and Translation in Eukaryotes 274
Genomes & Proteomes Connection: Alternative Splicing Tends to Be More Prevalent in Complex Eukaryotic Species 275

Chapter 14

Mutation, DNA Repair, and Cancer 279

14.1 Mutation 279
Feature Investigation: The Lederbergs Used Replica Plating to Show That Mutations Are Random Events 282
14.2 DNA Repair 286
14.3 Cancer 288
Genomes & Proteomes Connection: Mutations in Approximately 300 Human Genes May Promote Cancer 295

Chapter 15

The Eukaryotic Cell Cycle, Mitosis, and Meiosis 297

15.1 The Eukaryotic Cell Cycle 297
Feature Investigation: Masui and Markert's Study of Oocyte Maturation Led to the Identification of Cyclins and Cyclin-Dependent Kinases 302
15.2 Mitotic Cell Division 303
Genomes & Proteomes Connection: The Genomes of Diverse Animal Species Encode Approximately 20 Proteins Involved in Cytokinesis 307
15.3 Meiosis and Sexual Reproduction 308
15.4 Variation in Chromosome Structure and Number 314

Chapter 16

Simple Patterns of Inheritance 321

16.1 Mendel's Laws of Inheritance 322
16.2 The Chromosome Theory of Inheritance 327
16.3 Pedigree Analysis of Human Traits 330
16.4 Sex Chromosomes and X-Linked Inheritance Patterns 331
Feature Investigation: Morgan's Experiments Showed a Correlation Between a Genetic Trait and the Inheritance of a Sex Chromosome in *Drosophila* 333
16.5 Variations in Inheritance Patterns and Their Molecular Basis 335
Genomes & Proteomes Connection: Recessive Alleles That Cause Diseases May Have Multiple Effects on Penotype 336
16.6 Genetics and Probability 339

Chapter 17

Complex Patterns of Inheritance 343

17.1 Gene Interaction 344
17.2 Genes on the Same Chromosome: Linkage, Recombination, and Mapping 346
Feature Investigation: Bateson and Punnett's Crosses of Sweet Peas Showed That Genes Do Not Always Assort Independently 346
17.3 Extranuclear Inheritance: Organelle Genomes 350
Genomes & Proteomes Connection: Chloroplast and Mitochondrial Genomes Are Relatively Small, but Contain Genes That Encode Important Proteins 350
17.4 Epigenetic Inheritance 353

Chapter 18

Genetics of Viruses and Bacteria 359

18.1 Genetic Properties of Viruses 360
18.2 Viroids and Prions 368
18.3 Genetic Properties of Bacteria 370
18.4 Gene Transfer Between Bacteria 373
Feature Investigation: Lederberg and Tatum's Work with *E. coli* Demonstrated Gene Transfer Between Bacteria and Led to the Discovery of Conjugation 374
Genomes & Proteomes Connection: Horizontal Gene Transfer Is the Transfer of Genes Between the Same or Different Species 377

Chapter 19

Developmental Genetics 380

19.1 General Themes in Development 380
19.2 Development in Animals 385
Genomes & Proteomes Connection: A Homologous Group of Homeotic Genes Is Found in Nearly All Animals 390
Feature Investigation: Davis, Weintraub, and Lassar Identified Genes That Promote Muscle Cell Differentiation 393
19.3 Development in Plants 395

Chapter 20

Genetic Technology 400

20.1 Gene Cloning 400
20.2 Genomics: Techniques for Studying Genomes 406
Genomes & Proteomes Connection: A Microarray Can Identify Which Genes Are Transcribed by a Cell 409
20.3 Biotechnology 410
Feature Investigation: Blaese and Colleagues Performed the First Gene Therapy to Treat ADA Deficiency 416

Chapter 21

Genomes, Proteomes, and Bioinformatics 420

21.1 Bacterial and Archaeal Genomes 420
Feature Investigation: Venter, Smith, and Colleagues Sequenced the First Genome in 1995 421
21.2 Eukaryotic Genomes 423
21.3 Proteomes 429
21.4 Bioinformatics 431

UNIT IV Evolution

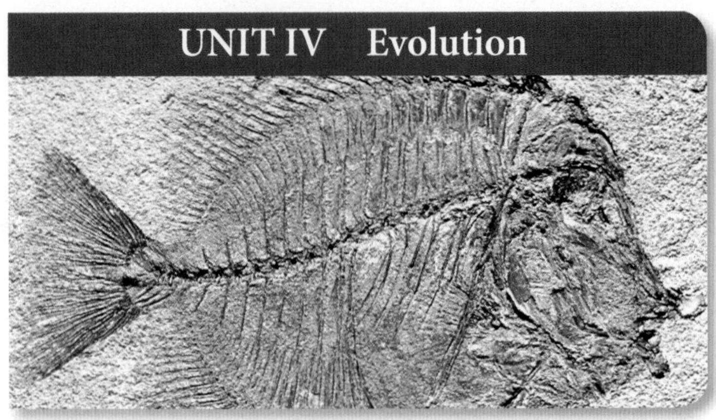

Chapter 22

The Origin and History of Life on Earth 438

22.1 Origin of Life on Earth 439
Feature Investigation: Bartel and Szostak Demonstrated Chemical Evolution in the Laboratory 443
22.2 The Fossil Record 446
22.3 History of Life on Earth 448
Genomes & Proteomes Connection: The Origin of Eukaryotic Cells Involved a Union Between Bacterial and Archaeal Cells 451

Chapter 23

An Introduction to Evolution 459

23.1 The Theory of Evolution 460
Feature Investigation: The Grants Observed Natural Selection in Galápagos Finches 463
23.2 Evidence of Evolutionary Change 465
23.3 The Molecular Processes That Underlie Evolution 472
Genomes & Proteomes Connection: New Genes in Eukaryotes Have Evolved via Exon Shuffling 473

Chapter 24

Population Genetics 478

24.1 Genes in Populations 478
Genomes & Proteomes Connection: Genes Are Usually Polymorphic 479
24.2 Natural Selection 482
24.3 Sexual Selection 486
Feature Investigation: Seehausen and van Alphen Found That Male Coloration in African Cichlids Is Subject to Female Choice 487
24.4 Genetic Drift 489
24.5 Migration and Nonrandom Mating 491

Chapter 25

Origin of Species and Macroevolution 495

25.1 Identification of Species 495

25.2 Mechanisms of Speciation 500

Feature Investigation: Podos Found That an Adaptation to Feeding May Have Promoted Reproductive Isolation in Finches 502

25.3 The Pace of Speciation 506

25.4 Evo-Devo: Evolutionary Developmental Biology 507

Genomes & Proteomes Connection: The Study of the *Pax6* Gene Indicates That Different Types of Eyes Evolved from a Simpler Form 511

Chapter 26

Taxonomy and Systematics 514

26.1 Taxonomy 514

26.2 Phylogenetic Trees 517

26.3 Cladistics 521

Feature Investigation: Cooper and Colleagues Compared DNA from Extinct Flightless Birds and Existing Species to Propose a New Phylogenetic Tree 524

26.4 Molecular Clocks 527

26.5 Horizontal Gene Transfer 528

Genomes & Proteomes Connection: Due to Horizontal Gene Transfer, the "Tree of Life" Is Really a "Web of Life" 529

UNIT V Diversity

Chapter 27

Archaea and Bacteria 533

27.1 Diversity and Evolution 533

27.2 Structure and Movement 538

27.3 Reproduction 542

27.4 Nutrition and Metabolism 544

27.5 Ecological Roles and Biotechnology Applications 545

Feature Investigation: Dantas and Colleagues Found That Many Bacteria Can Break Down and Consume Antibiotics as a Sole Carbon Source 545

Genomes & Proteomes Connection: The Evolution of Bacterial Pathogens 549

Chapter 28

Protists 553

28.1 An Introduction to Protists 553

28.2 Evolution and Relationships 556

Genomes & Proteomes Connection: Genome Sequences Reveal the Different Evolutionary Pathways of *Trichomonas vaginalis* and *Giardia intestinalis* 557

28.3 Nutritional and Defensive Adaptations 563

Feature Investigation: Burkholder and Colleagues Demonstrated That Strains of the Dinoflagellate Genus *Pfiesteria* Are Toxic to Mammalian Cells 565

28.4 Reproductive Adaptations 567

Chapter 29

Plants and the Conquest of Land 575

29.1 Ancestry and Diversity of Modern Plants 575

Genomes & Proteomes Connection: Comparison of Plant Genomes Reveals Genetic Changes That Occured During Plant Evolution 583

29.2 An Evolutionary History of Land Plants 586

29.3 The Origin and Evolutionary Importance of the Plant Embryo 588

Feature Investigation: Browning and Gunning Demonstrated That Placental Transfer Tissues Facilitate the Movement of Organic Molecules from Gametophytes to Sporophytes 589

29.4 The Origin and Evolutionary Importance of Leaves and Seeds 591

Chapter 30

The Evolution and Diversity of Modern Gymnosperms and Angiosperms 597

30.1 The Evolution and Diversity of Modern Gymnosperms 597

30.2 The Evolution and Diversity of Modern Angiosperms 604

Genomes & Proteomes Connection: Whole-Genome Duplications Influenced the Evolution of Flowering Plants 607

Feature Investigation: Hillig and Mahlberg Analyzed Secondary Metabolites to Explore Species Diversification in the Genus *Cannabis* 612

30.3 The Role of Coevolution in Angiosperm Diversification 613
30.4 Human Influences on Angiosperm Diversification 615

Chapter 31

Fungi 618

31.1 Evolution and Distinctive Features 618
31.2 Fungal Asexual and Sexual Reproduction 622
31.3 Diversity of Fungi 624
31.4 Fungal Ecology and Biotechnology 630
Genomes & Proteomes Connection: Genomic Comparison of Forest Fungi Reveals Coevolution with Plants 633
Feature Investigation: Márquez and Associates Discovered That a Three-Partner Symbiosis Allows Plants to Cope with Heat Stress 633

Chapter 32

An Introduction to Animal Diversity 638

32.1 Characteristics of Animals 638
32.2 Animal Classification 640
Genomes & Proteomes Connection: Changes in *Hox* Gene Expression Control Body Segment Specialization 645
32.3 Molecular Views of Animal Diversity 646
Feature Investigation: Aguinaldo and Colleagues Used SSU rRNA to Analyze the Taxonomic Relationships of Arthropods to Other Taxa 649

Chapter 33

The Invertebrates 652

33.1 Parazoa: Sponges, the First Multicellular Animals 653
33.2 Radiata: Jellyfish and Other Radially Symmetrical Animals 654
33.3 Lophotrochozoa: The Flatworms, Rotifers, Bryozoans, Brachiopods, Mollusks, and Annelids 656
Feature Investigation: Fiorito and Scotto's Experiments Showed Invertebrates Can Exhibit Sophisticated Observational Learning Behavior 663
33.4 Ecdysozoa: The Nematodes and Arthropods 667
Genomes & Proteomes Connection: Barcoding: A New Tool for Classification 675
33.5 Deuterostomia: The Echinoderms and Chordates 677

Chapter 34

The Vertebrates 684

34.1 Vertebrates: Chordates with a Backbone 684
34.2 Gnathostomes: Jawed Vertebrates 687
34.3 Tetrapods: Gnathostomes with Four Limbs 692
Feature Investigation: Davis and Colleagues Provide a Genetic-Developmental Explanation for Limb Length in Tetrapods 693
34.4 Amniotes: Tetrapods with a Desiccation-Resistant Egg 696
34.5 Mammals: Milk-Producing Amniotes 702
Genomes & Proteomes Connection: Comparing the Human and Chimpanzee Genetic Codes 709

UNIT VI Flowering Plants

Chapter 35

An Introduction to Flowering Plant Form and Function 716

35.1 From Seed to Seed—The Life of a Flowering Plant 716
35.2 How Plants Grow and Develop 720
35.3 The Shoot System: Stem and Leaf Adaptations 724
Feature Investigation: Lawren Sack and Colleagues Showed That Palmate Venation Confers Tolerance of Leaf Vein Breakage 726
Genomes & Proteomes Connection: Genetic Control of Guard-Cell Development 729
35.4 Root System Adaptations 732

Chapter 36

Flowering Plants: Behavior 737

36.1 Overview of Plant Behavioral Responses 737
36.2 Plant Hormones 740
Feature Investigation: An Experiment Performed by Briggs Revealed the Role of Auxin in Phototropism 742
Genomes & Proteomes Connection: Gibberellin Function Arose in a Series of Stages During Plant Evolution 744
36.3 Plant Responses to Environmental Stimuli 747

Chapter 37

Flowering Plants: Nutrition 756

37.1 Plant Nutritional Requirements 756

37.2 The Role of Soil in Plant Nutrition 760

Feature Investigation: Hammond and Colleagues Engineered Smart Plants That Can Communicate Their Phosphate Needs 764

37.3 Biological Sources of Plant Nutrients 766

Genomes & Proteomes Connection: Development of Legume–Rhizobia Symbioses 767

Chapter 38

Flowering Plants: Transport 772

38.1 Overview of Plant Transport 772

38.2 Uptake and Movement of Materials at the Cellular Level 773

38.3 Tissue-Level Transport 776

38.4 Long-Distance Transport 778

Feature Investigation: Holbrook and Associates Revealed the Dynamic Role of Xylem in Transport 781

Genomes & Proteomes Connection: Microarray Studies of Gene Transcription Reveal Xylem- and Phloem-Specific Genes 789

Chapter 39

Flowering Plants: Reproduction 793

39.1 An Overview of Flowering Plant Reproduction 793

39.2 Flower Production, Structure, and Development 797

Feature Investigation: Liang and Mahadevan Used Time-Lapse Video and Mathematical Modeling to Explain How Flowers Bloom 800

39.3 Male and Female Gametophytes and Double Fertilization 802

39.4 Embryo, Seed, Fruit, and Seedling Development 805

39.5 Asexual Reproduction in Flowering Plants 808

Genomes & Proteomes Connection: The Evolution of Plantlet Production in *Kalanchoë* 809

UNIT VII Animals

Chapter 40

Animal Bodies and Homeostasis 813

40.1 Organization of Animal Bodies 813

Genomes & Proteomes Connection: Organ Development and Function Are Controlled by *Hox* Genes 819

40.2 The Relationship Between Form and Function 821

40.3 Homeostasis 822

Feature Investigation: Pavlov Demonstrated the Relationship Between Learning and Feedforward Processes 826

Chapter 41

Neuroscience I: Cells of the Nervous System 831

41.1 Cellular Components of Nervous Systems 831

41.2 Electrical Properties of Neurons and the Resting Membrane Potential 834

41.3 Generation and Transmission of Electrical Signals Along Neurons 838

41.4 Neurons Communicate Electrically or Chemically at Synapses 842

Feature Investigation: Otto Loewi Discovered Acetylcholine 846

Genomes & Proteomes Connection: Varied Subunit Compositions of Neurotransmitter Receptors Allow Precise Control of Neuronal Regulation 848

41.5 Impact on Public Health 850

Chapter 42

Neuroscience II: Evolution, Structure, and Function of the Nervous System 854

42.1 The Evolution and Development of Nervous Systems 854

42.2 Structure and Function of the Nervous Systems of Humans and Other Vertebrates 858

Genomes & Proteomes Connection: Many Genes Have Been Important in the Evolution and Development of the Cerebral Cortex 866

42.3 Cellular Basis of Learning and Memory 867

Feature Investigation: Gaser and Schlaug Showed That the Sizes of Certain Brain Structures Differ Between Musicians and Nonmusicians 870

42.4 Impact on Public Health 871

Chapter 43

Neuroscience III: Sensory Systems 875

43.1 An Introduction to Sensory Receptors 875

43.2 Mechanoreception 877

43.3 Thermoreception and Nociception 884

43.4 Electromagnetic Reception 884

43.5 Photoreception 885

Genomes & Proteomes Connection: Color Vision Is an Ancient Adaptation in Animals 889

43.6 Chemoreception 893

Feature Investigation: Buck and Axel Discovered a Family of Olfactory Receptor Proteins That Bind Specific Odor Molecules 895

43.7 Impact on Public Health 897

Chapter 44

Muscular-Skeletal Systems and Locomotion 901

44.1 Types of Animal Skeletons 901

44.2 The Vertebrate Skeleton 903

44.3 Skeletal Muscle Structure and the Mechanism of Force Generation 904

Genomes & Proteomes Connection: Did an Ancient Mutation in Myosin Play a Role in the Development of the Human Brain? 907

44.4 Skeletal Muscle Function 911

Feature Investigation: Evans and Colleagues Activated a Gene to Produce "Marathon Mice" 913

44.5 Animal Locomotion 915

44.6 Impact on Public Health 916

Chapter 45

Nutrition and Animal Digestive Systems 920

45.1 Overview of Animal Nutrition and Ingestion 920

45.2 Principles of Digestion and Absorption of Food 927

45.3 Vertebrate Digestive Systems 928

45.4 Mechanisms of Digestion and Absorption in Vertebrates 933

Genomes & Proteomes Connection: Genetics Explains Lactose Intolerance 934

45.5 Regulation of Digestion 936

Feature Investigation: Bayliss and Starling Discovered a Mechanism by Which the Small Intestine Communicates with the Pancreas 937

45.6 Impact on Public Health 940

Chapter 46

Control of Energy Balance, Metabolic Rate, and Body Temperature 944

46.1 Nutrient Use and Storage 944

46.2 Regulation of the Absorptive and Postabsorptive States 947

Genomes & Proteomes Connection: GLUT Proteins Transport Glucose in Animal Cells 948

46.3 Energy Balance 950

Feature Investigation: Coleman Revealed a Satiety Factor in Mammals 953

46.4 Regulation of Body Temperature 955

46.5 Impact on Public Health 960

Chapter 47

Circulatory Systems 964

47.1 Types of Circulatory Systems 964

Genomes & Proteomes Connection: A Four-Chambered Heart Evolved from Simple Contractile Tubes 968

47.2 Blood and Blood Components 969

47.3 The Vertebrate Heart and Its Function 971

47.4 Blood Vessels 974

47.5 Relationships Among Blood Pressure, Blood Flow, and Resistance 977

Feature Investigation: Furchgott Discovered a Vasodilatory Factor Produced by Endothelial Cells 978

47.6 Adaptive Functions of Closed Circulatory Systems 981

47.7 Impact on Public Health 983

Chapter 48

Respiratory Systems 986

48.1 Physical Properties of Gases 986

48.2 Types of Respiratory Systems 988

48.3 Structure and Function of the Mammalian and Avian Respiratory Systems 992

Feature Investigation: Schmidt-Nielsen Mapped Airflow in the Avian Respiratory System 997

48.4 Mechanisms of Oxygen Transport in Blood 999

Genomes & Proteomes Connection: Hemoglobin Evolved Over 500 Million Years Ago 1001

48.5 Control of Ventilation in Mammalian Lungs 1002

48.6 Adaptations to Extreme Conditions 1003

48.7 Impact on Public Health 1004

Chapter 49

Excretory Systems and Salt and Water Balance 1008

49.1 Principles of Homeostasis of Internal Fluids 1008

Feature Investigation: Cade and Colleagues Discovered Why Athletes' Performances Wane on Hot Days 1013

49.2 Comparative Excretory Systems 1015

49.3 Structure and Function of the Mammalian Kidney 1018

Genomes & Proteomes Connection: Aquaporins Comprise a Large Family of Proteins That Are Ubiquitous in Nature 1025

49.4 Impact on Public Health 1026

Chapter 50

Endocrine Systems 1030

50.1 Types of Hormones and Their Mechanisms of Action 1031

50.2 Links Between the Endocrine and Nervous Systems 1035

50.3 Hormonal Control of Metabolism and Energy Balance 1037

Feature Investigation: Banting, Best, MacLeod, and Collip Were the First to Isolate Active Insulin 1042

50.4 Hormonal Control of Mineral Balance 1044

50.5 Hormonal Control of Growth and Development 1047

50.6 Hormonal Control of Reproduction 1049

50.7 Hormonal Responses to Stress 1050

Genomes & Proteomes Connection: Hormones and Receptors Evolved as Tightly Integrated Molecular Systems 1052

50.8 Impact on Public Health 1052

Chapter 51

Animal Reproduction 1056

51.1 Asexual and Sexual Reproduction 1056

Feature Investigation: Paland and Lynch Provided Evidence That Sexual Reproduction May Promote the Elimination of Harmful Mutations in Populations 1058

51.2 Gametogenesis and Fertilization 1059

51.3 Mammalian Reproductive Structure and Function 1062

51.4 Pregnancy and Birth in Mammals 1068

Genomes & Proteomes Connection: The Evolution of the Globin Gene Family Has Been Important for Internal Gestation in Mammals 1070

51.5 Impact on Public Health 1072

Chapter 52

Animal Development 1076

52.1 Principles of Embryonic Development 1076

52.2 General Events of Embryonic Development 1077

52.3 Control of Cell Differentiation and Morphogenesis During Animal Development 1086

Genomes & Proteomes Connection: Groups of Embryonic Cells Can Produce Specific Body Structures Even When Transplanted into Different Animals 1088

Feature Investigation: Richard Harland and Coworkers Identified Genes Expressed Specifically in the Organizer 1089

52.4 Impact on Public Health 1091

Chapter 53

Immune Systems 1094

53.1 Types of Pathogens 1094

53.2 Innate Immunity 1095

Feature Investigation: Lemaitre and Colleagues Identify an Immune Function for Toll Protein in *Drosophila* 1098

53.3 Acquired Immunity 1100

Genomes & Proteomes Connection: Recombination and Hypermutation Produce an Enormous Number of Different Immunoglobulin Proteins 1105

53.4 Impact on Public Health 1111

UNIT VIII Ecology

Chapter 54

An Introduction to Ecology and Biomes 1117

54.1 The Scale of Ecology 1118

Feature Investigation: Callaway and Aschehoug's Experiments Showed That the Secretion of Chemicals Gives Invasive Plants a Competitive Edge Over Native Species 1118

54.2 Ecological Methods 1120

54.3 The Environment's Effect on the Distribution of Organisms 1122

Genomes & Proteomes Connection: Temperature Tolerance May Be Manipulated by Genetic Engineering 1126

54.4 Climate and Its Relationship to Biological Communities 1129

54.5 Major Biomes 1132

54.6 Continental Drift and Biogeography 1142

Chapter 55

Behavioral Ecology 1146

55.1 The Influence of Genetics and Learning on Behavior 1146

Genomes & Proteomes Connection: Some Behavior Results from Simple Genetic Influences 1147

55.2 Local Movement and Long-Range Migration 1150

Feature Investigation: Tinbergen's Experiments Show That Digger Wasps Use Landmarks to Find Their Nests 1151

55.3 Foraging Behavior 1154

55.4 Communication 1155

55.5 Living in Groups 1157

55.6 Altruism 1158

55.7 Mating Systems 1162

Chapter 56

Population Ecology 1168

56.1 Understanding Populations 1168

56.2 Demography 1172

Feature Investigation: Murie's Collections of Dall Mountain Sheep Skulls Permitted the Accurate Construction of Life Tables 1174

56.3 How Populations Grow 1176

Genomes & Proteomes Connection: Hexaploidy Increases the Growth of Coastal Redwood Trees 1181

56.4 Human Population Growth 1182

Chapter 57

Species Interactions 1187

57.1 Competition 1188

Feature Investigation: Connell's Experiments with Barnacle Species Revealed Each Species' Fundamental and Realized Niches 1191

57.2 Predation, Herbivory, and Parasitism 1194

Genomes & Proteomes Connection: Transgenic Plants May Be Used in the Fight Against Plant Diseases 1201

57.3 Mutualism and Commensalism 1202

57.4 Bottom-Up and Top-Down Control 1204

Chapter 58

Community Ecology 1207

58.1 Differing Views of Communities 1207

58.2 Patterns of Species Richness 1209

58.3 Calculating Species Diversity 1212

Genomes & Proteomes Connection: Metagenomics May Be Used to Measure Species Diversity 1212

58.4 Species Diversity and Community Stability 1214

58.5 Succession: Community Change 1215

58.6 Island Biogeography 1219

Feature Investigation: Simberloff and Wilson's Experiments Tested the Predictions of the Equilibrium Model of Island Biogeography 1221

Chapter 59

Ecosystem Ecology 1225

59.1 Food Webs and Energy Flow 1226

Genomes & Proteomes Connection: Using Genetically Engineered Plants to Remove Pollutants 1232

59.2 Biomass Production in Ecosystems 1232

59.3 Biogeochemical Cycles 1235

Feature Investigation: Stiling and Drake's Experiments with Elevated CO_2 Showed an Increase in Plant Growth but a Decrease in Herbivory 1238

Chapter 60

Biodiversity and Conservation Biology 1244

60.1 What Is Biodiversity? 1245

60.2 Why Conserve Biodiversity? 1245

Feature Investigation: Ecotron Experiments Showed the Relationship Between Biodiversity and Ecosystem Function 1247

60.3 The Causes of Extinction and Loss of Biodiversity 1250

60.4 Conservation Strategies 1254

Genomes & Proteomes Connection: Can Cloning Save Endangered Species? 1261

Appendix A Periodic Table of the Elements A-1

Appendix B Answers to End-of-Chapter and Concept Check Questions A-2

Glossary G-1

Credits C-1

Index I-1

Chapter Outline

1.1 Principles of Biology and the Levels of Biological Organization
1.2 Unity and Diversity of Life
1.3 Biology as a Scientific Discipline
Summary of Key Concepts
Assess and Discuss

1 An Introduction to Biology

Clownfish and anemones. The habits of clownfish and anemones present biologists with interesting mysteries to solve. Why doesn't the anemone sting the clownfish? How does a male clownfish transform its body into a female clownfish?

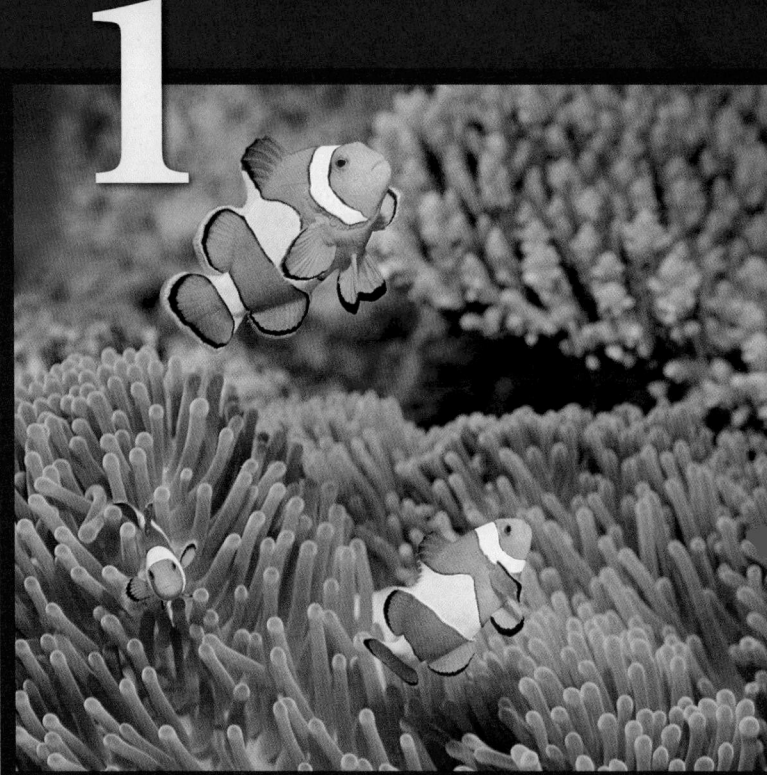

Biology is the study of life. The diverse forms of life found on Earth provide biologists with an amazing array of organisms to study. In many cases, the investigation of living things leads to discoveries that no one would have imagined. For example, researchers determined that the venom from certain poisonous snakes contains a chemical that lowers blood pressure in humans. By analyzing that chemical, scientists developed drugs to treat high blood pressure (**Figure 1.1**).

Certain ancient civilizations, such as the Greeks, Romans, and Egyptians, discovered that the bark of the white willow tree can be used to fight fever. Modern chemists determined that willow bark contains a substance called salicylic acid, which led to the development of the related compound acetylsalicylic acid, more commonly known as aspirin (**Figure 1.2**).

In the 20th century, biologists studied soil bacteria that naturally produce "chemical weapons" to kill competing bacteria in their native environment. These chemicals have been characterized and used to develop antibiotics such as streptomycin to treat bacterial infections (**Figure 1.3**).

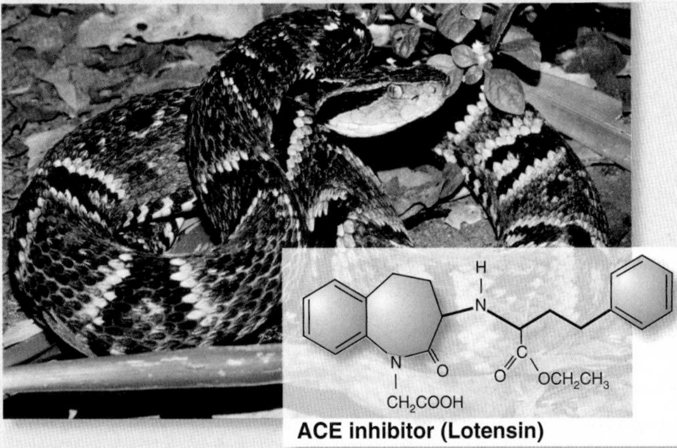

Figure 1.1 The Brazilian arrowhead viper and inhibitors of high blood pressure. Derivatives of a chemical, called an angiotensin-converting enzyme (ACE) inhibitor, are found in the venom of the Brazilian arrowhead viper and are commonly used to treat high blood pressure.

Figure 1.2 The white willow and aspirin. Modern aspirin, acetylsalicylic acid, was developed after analyzing a chemical found in the bark of the white willow tree.

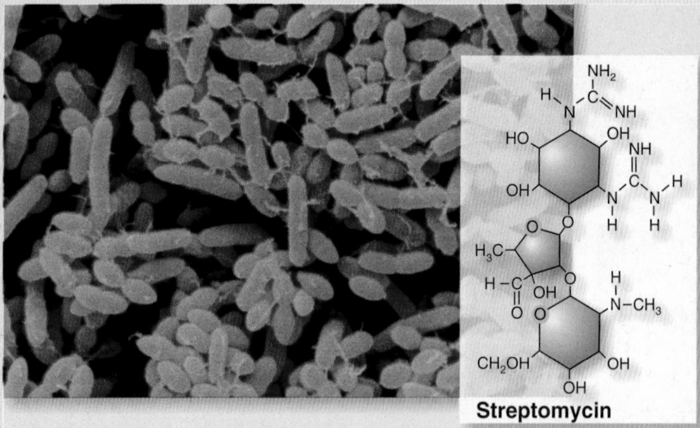

Figure 1.3 The soil bacterium *Streptomyces griseus*, which naturally produces the antibiotic streptomycin, kills competing bacteria in the soil. Doctors administer streptomycin to treat bacterial infections.

These are but a few of the many discoveries that make biology an intriguing discipline. The study of life not only reveals the fascinating characteristics of living species but also leads to the development of medicines and research tools that benefit the lives of people.

To make new discoveries, biologists view life from many different perspectives. What is the composition of living things? How is life organized? How do organisms reproduce? Sometimes the questions posed by biologists are fundamental and even philosophical in nature. How did living organisms originate? Can we live forever? What is the physical basis for memory? Can we save endangered species? Can we understand intriguing aspects of body form and function, such as the ability of clownfish to change from a male to a female?

Future biologists will continue to make important advances. Biologists are scientific explorers looking for answers to some of life's most enduring mysteries. Unraveling these mysteries presents an exciting challenge to the best and brightest minds. The rewards of a career in biology include the excitement of forging into uncharted territory, the thrill of making discoveries that can improve the health and lives of people, and the impact of trying to preserve the environment and protect endangered species. For these and many other compelling reasons, students seeking challenging and rewarding careers may wish to choose biology as a lifelong pursuit.

In this chapter, we will begin our survey of biology by examining the basic principles that underlie the characteristics of all living organisms and the fields of biology. We will then take a deeper look at the process of evolution and how it has led to the development of genomes—the entire genetic compositions of living organisms—which explains the unity and diversity that we observe among living species. Finally, we will explore the general approaches that scientists follow when making new discoveries.

1.1 Principles of Biology and the Levels of Biological Organization

Learning Outcomes:

1. Describe the principles of biology.
2. Explain how life can be viewed at different levels of biological complexity.

Since biology is the study of life, a good way to begin a biology textbook is to distinguish living organisms from nonliving objects. At first, the distinction might seem obvious. A person is alive, but a rock is not. However, the distinction between living and nonliving may seem less obvious when we consider microscopic entities. Is a bacterium alive? What about a virus or a chromosome? In this section, we will examine the principles that underlie the characteristics of all forms of life and explore other broad principles in biology. We will then consider the levels of organization that biologists study, ranging from atoms and small molecules to very large geographical areas.

The Study of Life Has Revealed a Set of Unifying Principles

Biologists have studied many different species and learned that a set of principles applies to all fields of biology. Twelve broad principles are described in Figure 1.4. The first eight principles are often used as criteria to define the basic features of life. You will see these principles at many points as you progress through this textbook and they are indicated with the icon. In particular, we will draw your attention to them at the beginning of each unit, and we will refer to them within particular figures in Chapters 3 through 60. It should be noted that the principles of biology are also governed by the laws of chemistry and physics, which are discussed in Chapters 2, 3, and 6.

Principle 1: Cells are the simplest units of life. The term **organism** can be applied to all living things. Organisms maintain an internal order that is separated from the environment (Figure 1.4a). The simplest unit of such organization is the **cell**, which we will examine in Unit II. One of the foundations of biology is the **cell theory**, which states that (1) all organisms are composed of cells, (2) cells are the smallest units of life, and (3) new cells come from pre-existing cells via cell division. Unicellular organisms are composed of one cell, whereas multicellular organisms such as plants and animals contain many cells. In plants and animals, each cell has an internal order, and the cells within the body have specific arrangements and functions.

Principle 2: Living organisms use energy. The maintenance of organization requires energy. Therefore, all living organisms acquire energy from the environment and use that energy to maintain their internal order. Cells carry out a variety of chemical reactions that are responsible for the breakdown of nutrients. Such reactions often release energy in a process called **respiration**. The energy may be used to synthesize the components that make up individual cells and living organisms. Chemical reactions involved with the breakdown and synthesis of cellular molecules are collectively known as **metabolism**. Plants, algae, and certain bacteria can directly harness light energy to produce their own nutrients in the process of **photosynthesis**

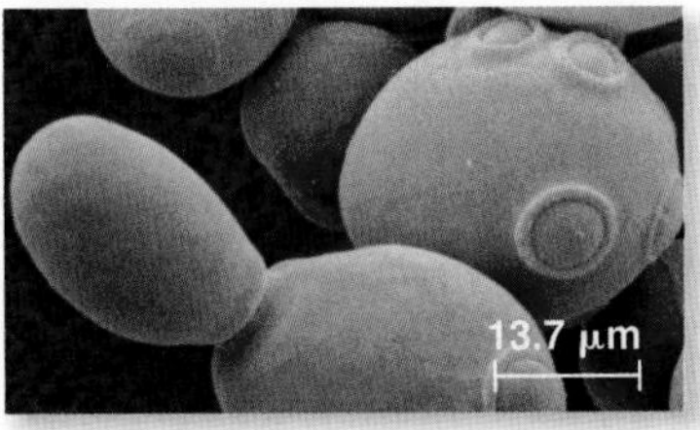

(a) Cells are the simplest unit of life:
Organisms maintain an internal order. The simplest unit of organization is the cell. Yeast cells are shown here.

(b) Living organisms use energy:
Organisms need energy to maintain internal order. These algae harness light energy via photosynthesis. Energy is used in chemical reactions collectively known as metabolism.

(c) Living organisms interact with their environment:
Organisms respond to environmental changes. These plants are growing toward the light.

(d) Living organisms maintain homeostasis:
Organisms regulate their cells and bodies, maintaining relatively stable internal conditions, a process called homeostasis.

(e) Living organisms grow and develop:
Growth produces more or larger cells, whereas development produces organisms with a defined set of characteristics.

(f) The genetic material provides a blueprint for reproduction:
To sustain life over many generations, organisms must reproduce. Due to the transmission of genetic material, offspring tend to have traits like their parents.

(g) Populations of organisms evolve from one generation to the next:
Populations of organisms change over the course of many generations. Evolution results in traits that promote survival and reproductive success.

(h) All species (past and present) are related by an evolutionary history:
The three mammal species shown here share a common ancestor, which was also a mammal.

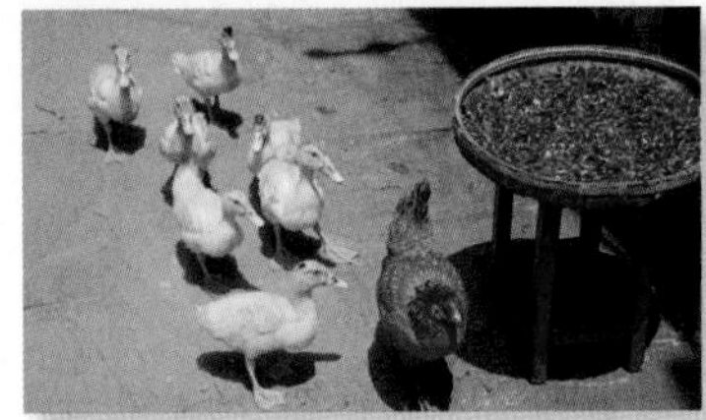

(i) Structure determines function:
In the example seen here, webbed feet (on ducks) function as paddles for swimming. Nonwebbed feet (on chickens) function better for walking on the ground.

(j) New properties of life emerge from complex interactions:
Our ability to see is an emergent property due to interactions among many types of cells in the eye and neurons that send signals to the brain.

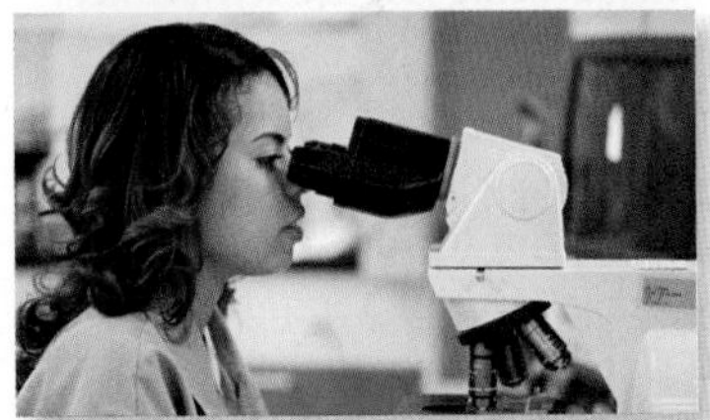

(k) Biology is an experimental science:
The discoveries of biology are made via experimentation, which leads to theories and biological principles.

(l) Biology affects our society:
Many discoveries in biology have had major effects on our society. For example, biologists developed Bt-corn, which is resistant to insect pests and widely planted by farmers.

Figure 1.4 Twelve principles of biology. The first eight principles are often used as criteria for defining the basic features of life. Note: The principles described here are considered very broad. Biologists have determined many others that have a more defined focus.

BioConnections: *Look ahead to Figure 52.11. Which of these principles is this figure emphasizing?*

(Figure 1.4b). They are the primary producers of food on Earth. In contrast, some organisms, such as animals and fungi, are consumers—they must use other organisms as food to obtain energy.

Principle 3: Living organisms interact with their environment. To survive, living organisms must be able to interact with their environment, which includes other organisms they may encounter. All organisms must respond to environmental changes. For example, bacterial cells have mechanisms to detect that certain nutrients in the environment are in short supply, whereas others are readily available. In the winter, many species of mammals develop a thicker coat of fur that protects them from the cold temperatures. Plants respond to changes in the angle of the sun. If you place a plant in a window, it will grow toward the light (Figure 1.4c).

Principle 4: Living organisms maintain homeostasis. As we have just seen, one way that organisms respond to environmental variation is to change their characteristics. Although life is a dynamic process, living cells and organisms regulate their cells and bodies to maintain relatively stable internal conditions, a process called **homeostasis** (from the Greek, meaning to stay the same). The degree to which homeostasis is achieved varies among different organisms. For example, most mammals and birds maintain a relatively stable body temperature in spite of changing environmental temperatures (Figure 1.4d), whereas reptiles and amphibians tolerate a wider fluctuation in body temperature. By comparison, all organisms continually regulate their cellular metabolism so nutrient molecules are used at an appropriate rate and new cellular components are synthesized when they are needed.

Principle 5: Living organisms grow and develop. All living organisms have an ability to grow and develop. **Growth** produces more or larger cells. In plants and animals, a fertilized egg undergoes multiple cell divisions to develop into a mature organism with many cells. Among unicellular organisms such as bacteria, new cells are relatively small, and they increase in volume by the synthesis of additional cellular components. **Development** is a series of changes in the state of a cell, tissue, organ, or organism, eventually resulting in organisms with a defined set of characteristics (Figure 1.4e).

Principle 6: The genetic material provides a blueprint for reproduction. All living organisms have a finite life span. To sustain life, organisms must **reproduce**, or generate offspring (Figure 1.4f). A key feature of reproduction is that offspring tend to have characteristics that greatly resemble those of their parent(s). How is this possible? All living organisms contain genetic material composed of **DNA (deoxyribonucleic acid)**, which provides a blueprint for the organization, development, and function of living things. During reproduction, a copy of this blueprint is transmitted from parent to offspring. DNA is **heritable**, which means that offspring inherit DNA from their parents.

As discussed in Unit III, **genes**, which are segments of DNA, govern the characteristics, or traits, of organisms. Most genes are transcribed into a type of **RNA (ribonucleic acid)** molecule called messenger RNA (mRNA) that is then translated into a **polypeptide** with a specific amino acid sequence. A **protein** is composed of one or more polypeptides. The structures and functions of proteins are largely responsible for the traits of living organisms.

Principle 7: Populations of organisms evolve from one generation to the next. The first six characteristics of life, which we have just considered, apply to individual organisms over the short run. Over the long run, another universal characteristic of life is **biological evolution**, or simply **evolution**, which refers to a heritable change in a population of organisms from generation to generation. As a result of evolution, populations become better adapted to the environment in which they live. For example, the long snout of an anteater is an adaptation that enhances its ability to obtain food, namely ants, from hard-to-reach places (Figure 1.4g). Over the course of many generations, the fossil record suggests that the long snout occurred via biological evolution in which modern anteaters evolved from populations of organisms with shorter snouts.

Principle 8: All species (past and present) are related by an evolutionary history. Principle 7 considers evolution as an ongoing process that happens from one generation to the next. Evidence from a variety of sources, including the fossil record and DNA sequences, also indicates that all organisms on Earth share a common ancestry. For example, the three species of mammals shown in Figure 1.4h shared a common ancestor in the past, which was also a mammal. We will discuss evolutionary relationships later in Section 1.2 and more thoroughly in Chapter 26.

As described later in this chapter, biologists often view evolution within the context of genomes and proteomes. The term **genome** refers to the complete genetic composition of an organism or species. Because most genes encode proteins, these genetic changes are often associated with changes in the **proteome**, which is the complete protein composition of a cell or organism. By studying how evolution affects genomes and proteomes, biologists can better understand how the changes that occur during evolution affect the characteristics of species. Because evolution is a core unifying principle in biology, we will draw your attention to it in Chapters 4 through 60 by having a brief topic that we call *Genomes and Proteomes Connection.* This topic connects the principle of evolution to the subject matter in that chapter.

Principle 9: Structure determines function. In addition to the preceding eight characteristics of life, biologists have identified other principles that are important in all fields of biology. The principle "structure determines function" pertains to very tiny biological molecules and to very large biological structures. For example, at the microscopic level, a cellular protein called actin naturally assembles into structures that are long filaments. The function of these filaments is to provide support and shape to cells. At the macroscopic level, let's consider the feet of different birds (Figure 1.4i). Aquatic birds have webbed feet that function as paddles for swimming. By comparison, the feet of nonaquatic birds are not webbed and are better adapted for grasping food, perching on branches, and running along the ground. In this case, the structure of a bird's feet, webbed versus nonwebbed, is a critical feature that affects their function.

Principle 10: New properties of life emerge from complex interactions. In biology, when individual components in an organism interact with each other or with the external environment to create novel structures and functions, the resulting characteristics are called **emergent properties**. For example, the human eye is composed of many different types of cells that are organized to sense incoming light and transmit signals to the brain (Figure 1.4j). Our ability to see is an emergent property of this complex arrangement of different cell types. As discussed later in this chapter, biologists use the term **systems biology** to describe the study of how new properties of life arise by complex interactions of its components.

Principle 11: Biology is as an experimental science. Biology is an inquiry process. Biologists are curious about the characteristics of living organisms and ask questions about those characteristics. For example, a cell biologist may wonder why a cell produces a specific protein when it is confronted with high temperature. An ecologist may ask herself why a particular bird eats insects in the summer

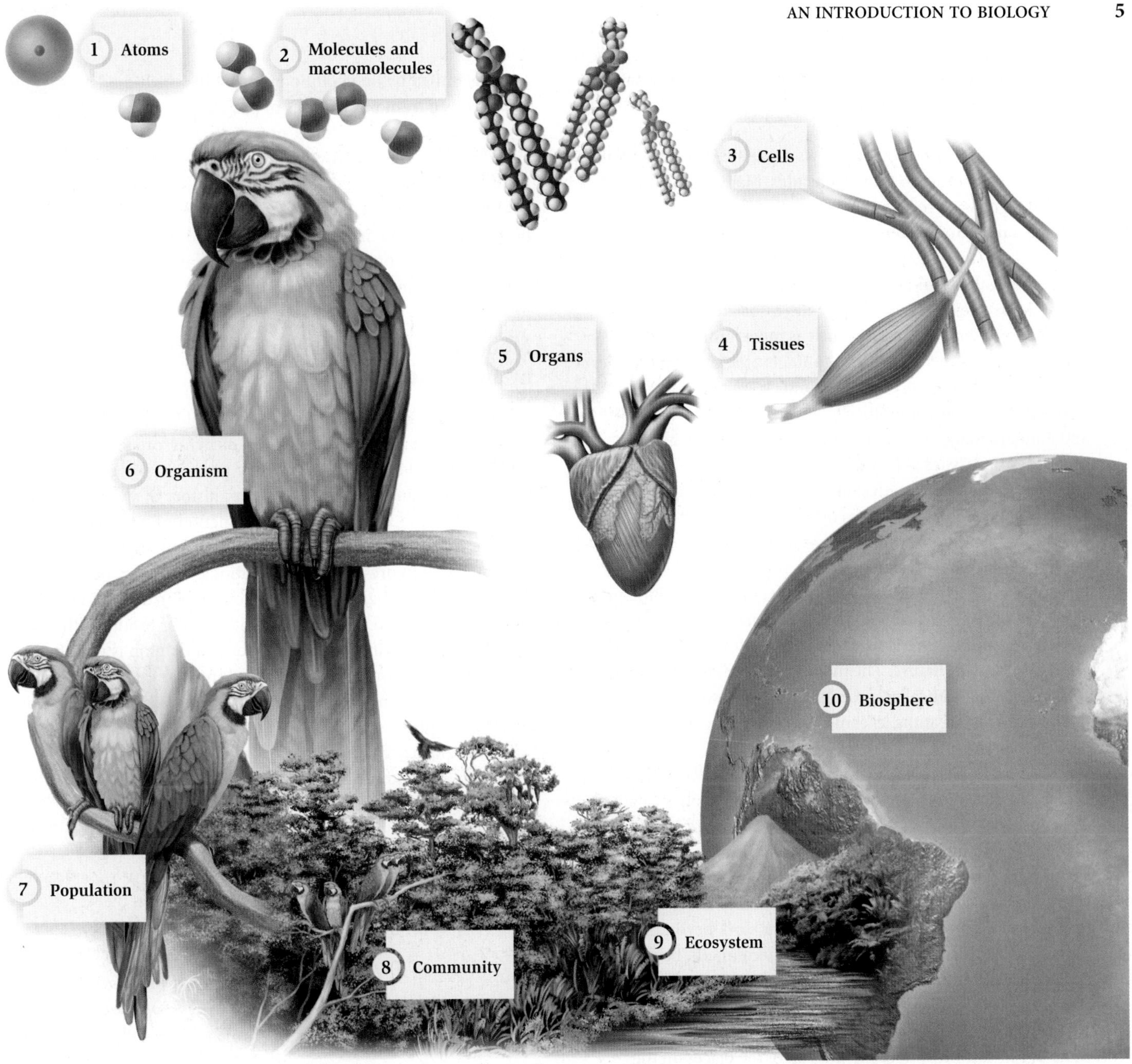

Figure 1.5 The levels of biological organization.

Concept Check: *At which level of biological organization would you place a herd of buffalo?*

and seeds in the winter. To answer such questions, biologists typically gather additional information and ultimately form a hypothesis, which is a proposed explanation for a natural phenomenon. The next stage is to design one or more experiments to test the validity of a hypothesis (Figure 1.4k).

Like evolution, experimentation is such a key aspect of biology that the authors of this textbook have emphasized it in every chapter. As discussed later, a consistent element in Chapters 2 through 60 is a "Feature Investigation"—an actual study by current or past researchers that showcases the experimental approach.

Principle 12: Biology affects our society. The influence of biology is not confined to textbooks and classrooms. The work of biologists has far-reaching effects in our society. For example, biologists have discovered drugs that are used to treat many different human diseases. Likewise, biologists have created technologies that have many uses. Examples include the use of microorganisms to make medical products, such as human insulin, and the genetic engineering of crops to make them resistant to particular types of insect pests (Figure 1.4l).

Living Organisms Can Be Studied at Different Levels of Organization

The organization of living organisms can be analyzed at different levels of biological complexity, starting with the smallest level of organization and progressing to levels that are physically much larger and more complex. Figure 1.5 depicts a scientist's view of the levels of biological organization.

1. **Atoms:** An **atom** is the smallest unit of an element that has the chemical properties of the element. All matter is composed of atoms.
2. **Molecules and macromolecules:** As discussed in Unit I, atoms bond with each other to form **molecules**. Many molecules bonded together to form a polymer such as a polypeptide is called a **macromolecule**. Carbohydrates, proteins, and nucleic acids (for example, DNA and RNA) are important macromolecules found in living organisms.
3. **Cells:** Molecules and macromolecules associate with each other to form larger structures such as membranes. A **cell** is surrounded by a membrane and contains a variety of molecules and macromolecules. A cell is the simplest unit of life.
4. **Tissues:** In the case of multicellular organisms such as plants and animals, many cells of the same type associate with each other to form **tissues**. An example is muscle tissue.
5. **Organs:** In complex multicellular organisms, an **organ** is composed of two or more types of tissue. For example, the heart is composed of several types of tissues, including muscle, nervous, and connective tissue.
6. **Organism:** All living things can be called **organisms**. Biologists classify organisms as belonging to a particular **species**, which is a related group of organisms that share a distinctive form and set of attributes in nature. The members of the same species are closely related genetically. In Units VI and VII, we will examine plants and animals at the level of cells, tissues, organs, and complete organisms.
7. **Population:** A group of organisms of the same species that occupy the same environment is called a **population**.
8. **Community:** A biological **community** is an assemblage of populations of different species. The types of species found in a community are determined by the environment and by the interactions of species with each other.
9. **Ecosystem:** Researchers may extend their work beyond living organisms and also study the physical environment. Ecologists analyze **ecosystems**, which are formed by interactions of a community of organisms with their physical environment. Unit VIII considers biology from populations to ecosystems.
10. **Biosphere:** The **biosphere** includes all of the places on the Earth where living organisms exist. Life is found in the air, in bodies of water, on the land, and in the soil.

1.2 Unity and Diversity of Life

Learning Outcomes:

1. Explain the two basic mechanisms by which evolutionary change occurs: vertical descent with mutation and horizontal gene transfer.
2. Outline how organisms are classified (taxonomy).
3. Describe how changes in genomes and proteomes underlie evolutionary changes.

Unity and diversity are two words that often are used to describe the living world. As we have seen, all modern forms of life display a common set of characteristics that distinguish them from nonliving objects. In this section, we will explore how this unity of common traits is rooted in the phenomenon of biological evolution. Life on Earth is united by an evolutionary past in which modern organisms have evolved from populations of pre-existing organisms.

Evolutionary unity does not mean that organisms are exactly alike. The Earth has many different types of environments, ranging from tropical rain forests to salty oceans, hot and dry deserts, and cold mountaintops. Diverse forms of life have evolved in ways that help them prosper in the different environments the Earth has to offer. In this section, we will begin to examine the unity and diversity that exists within the biological world.

Modern Forms of Life Are Connected by an Evolutionary History

Life began on Earth as primitive cells about 3.5–4 billion years ago (bya). Since that time, populations of living organisms underwent evolutionary changes that ultimately gave rise to the species we see today. Understanding the evolutionary history of species can provide key insights into the structure and function of an organism's body, because evolutionary change frequently involves modifications of characteristics in pre-existing populations. Over long periods of time, populations may change so that structures with a particular function become modified to serve a new function. For example, the wing of a bat is used for flying, and the flipper of a dolphin is used for swimming. Evidence from the fossil record indicates that both structures were modified from a limb that was used for walking in a pre-existing ancestor (**Figure 1.6**).

Evolutionary change occurs by two mechanisms: vertical descent with mutation and horizontal gene transfer. Let's take a brief look at each of these mechanisms.

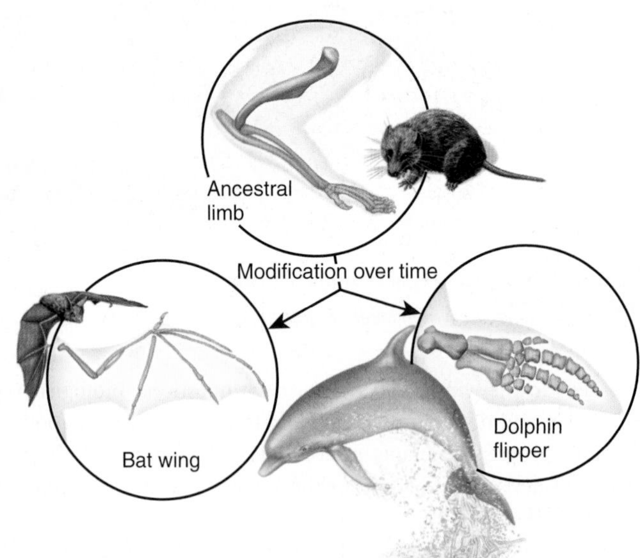

Figure 1.6 An example showing a modification that has occurred as a result of biological evolution. The wing of a bat and the flipper of a dolphin are modifications of a limb that was used for walking in a pre-existing ancestor.

Concept Check: *Among mammals, give two examples of how the tail has been modified and has different purposes.*

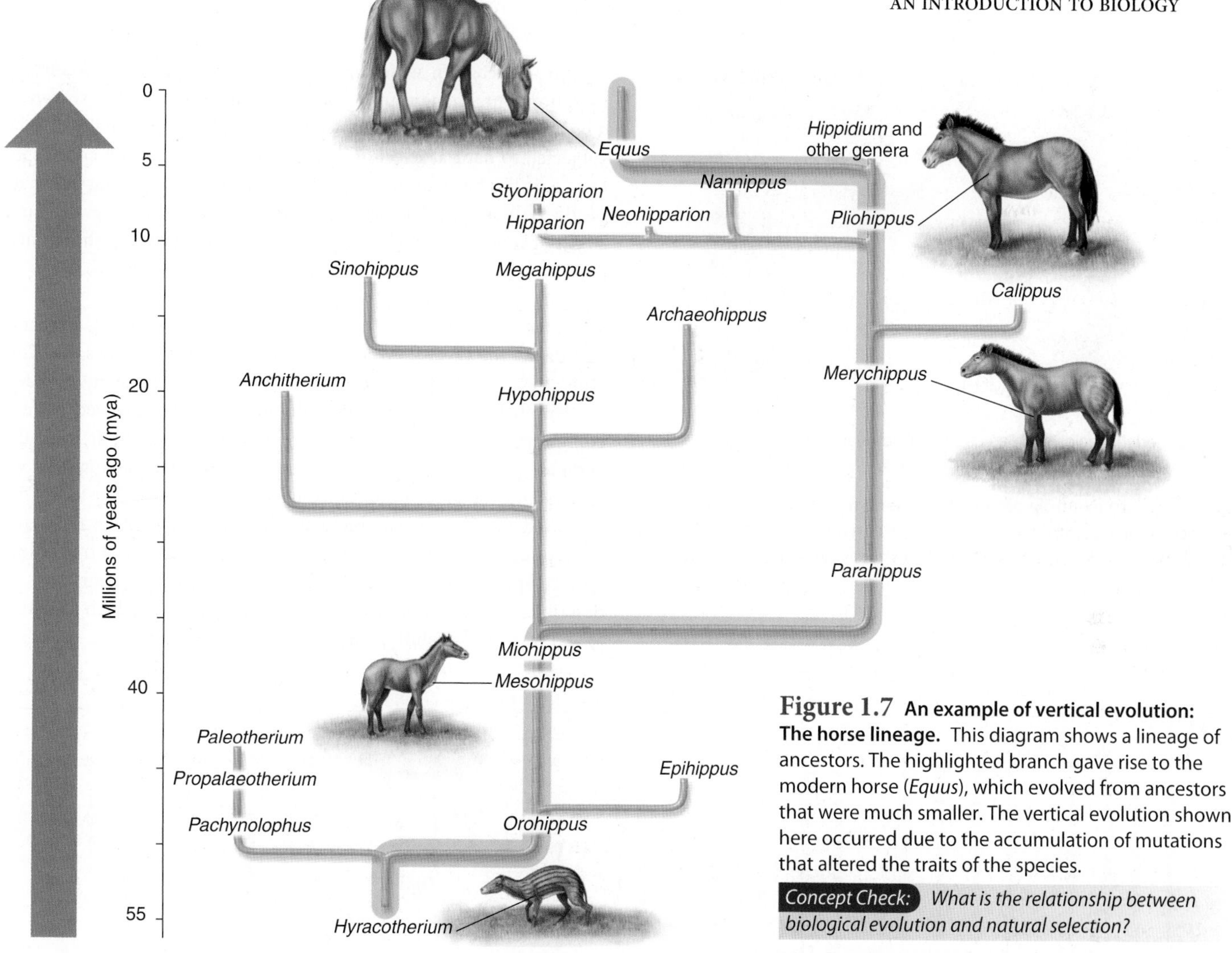

Figure 1.7 An example of vertical evolution: The horse lineage. This diagram shows a lineage of ancestors. The highlighted branch gave rise to the modern horse (*Equus*), which evolved from ancestors that were much smaller. The vertical evolution shown here occurred due to the accumulation of mutations that altered the traits of the species.

Concept Check: What is the relationship between biological evolution and natural selection?

Vertical Descent with Mutation The traditional way to study evolution is to examine a progression of changes in a series of ancestors. Such a series is called a **lineage**. Figure 1.7 shows a portion of the lineage that gave rise to modern horses. This type of evolution is called **vertical evolution** because it occurs in a lineage. Biologists have traditionally depicted such evolutionary change in a diagram like the one shown in Figure 1.7. In this mechanism of evolution, new species evolve from pre-existing ones by the accumulation of **mutations**, which are random changes in the genetic material of organisms. But why would some mutations accumulate in a population and eventually change the characteristics of an entire species? One reason is that a mutation may alter the traits of organisms in a way that increases their chances of survival and reproduction. When a mutation causes such a beneficial change, the frequency of the mutation may increase in a population from one generation to the next, a process called **natural selection**. This topic is discussed in Units IV and V. Evolution also involves the accumulation of neutral changes that do not benefit or harm a species, and evolution sometimes involves rare changes that may be harmful.

With regard to the horses shown in Figure 1.7, the fossil record has revealed adaptive changes in various traits such as size and tooth morphology. The first horses were the size of dogs, whereas modern horses typically weigh more than a half ton. The teeth of *Hyracotherium* were relatively small compared with those of modern horses. Over the course of millions of years, horse teeth have increased in size, and a complex pattern of ridges has developed on the molars. How do evolutionary biologists explain these changes in horse characteristics? They can be attributed to natural selection producing adaptations to changing global climates. Over North America, where much of horse evolution occurred, large areas changed from dense forests to grasslands. The horses' increase in size allowed them to escape predators and travel greater distances in search of food. The changes seen in horses' teeth are consistent with a dietary shift from eating tender leaves to eating grasses and other vegetation that are more abrasive and require more chewing.

Horizontal Gene Transfer The most common way for genes to be transferred is in a vertical manner. This can involve the transfer of genetic material from a mother cell to daughter cells, or it can occur via gametes—sperm and egg—that unite to form a new organism. However, as discussed in later chapters, genes are sometimes transferred between organisms by other mechanisms. These other mechanisms are collectively known as **horizontal gene transfer**. In some cases, horizontal gene transfer can occur between members of different species. For example, you may have heard in the news media that resistance to antibiotics among bacteria is a growing medical problem. As discussed in Chapter 18, genes that confer antibiotic resistance are sometimes transferred between different bacterial species (**Figure 1.8**).

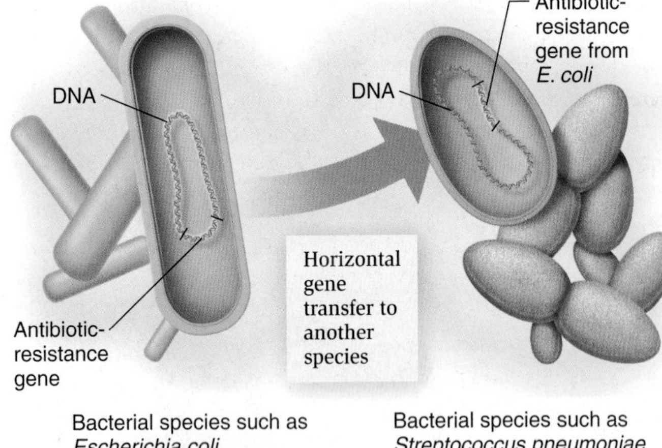

Figure 1.8 An example of horizontal gene transfer: Antibiotic resistance. One bacterial species may transfer a gene to a different bacterial species, such as a gene that confers resistance to an antibiotic.

In a lineage in which the time scale is depicted on a vertical axis, horizontal gene transfer between different species is shown as a horizontal line (**Figure 1.9**). Genes transferred horizontally may be subjected to natural selection and promote changes in an entire species. This has been an important mechanism of evolutionary change, particularly among bacterial species. In addition, during the early stages of evolution, which occurred a few billion years ago, horizontal gene transfer was an important part of the process that gave rise to all modern species.

Traditionally, biologists have described evolution using diagrams that depict the vertical evolution of species on a long time scale. This type of evolutionary tree is shown in Figure 1.7. For many decades, the simplistic view held that all living organisms evolved from a common ancestor, resulting in a "tree of life" that could describe the vertical evolution that gave rise to all modern species. Now that we understand the great importance of horizontal gene transfer in the evolution of life on Earth, biologists have needed to re-evaluate the concept of evolution as it occurs over time. Rather than a tree of life, a more appropriate way to view the unity of living organisms is to describe it as a "web of life," as shown in Figure 1.9, which accounts for both vertical evolution and horizontal gene transfer.

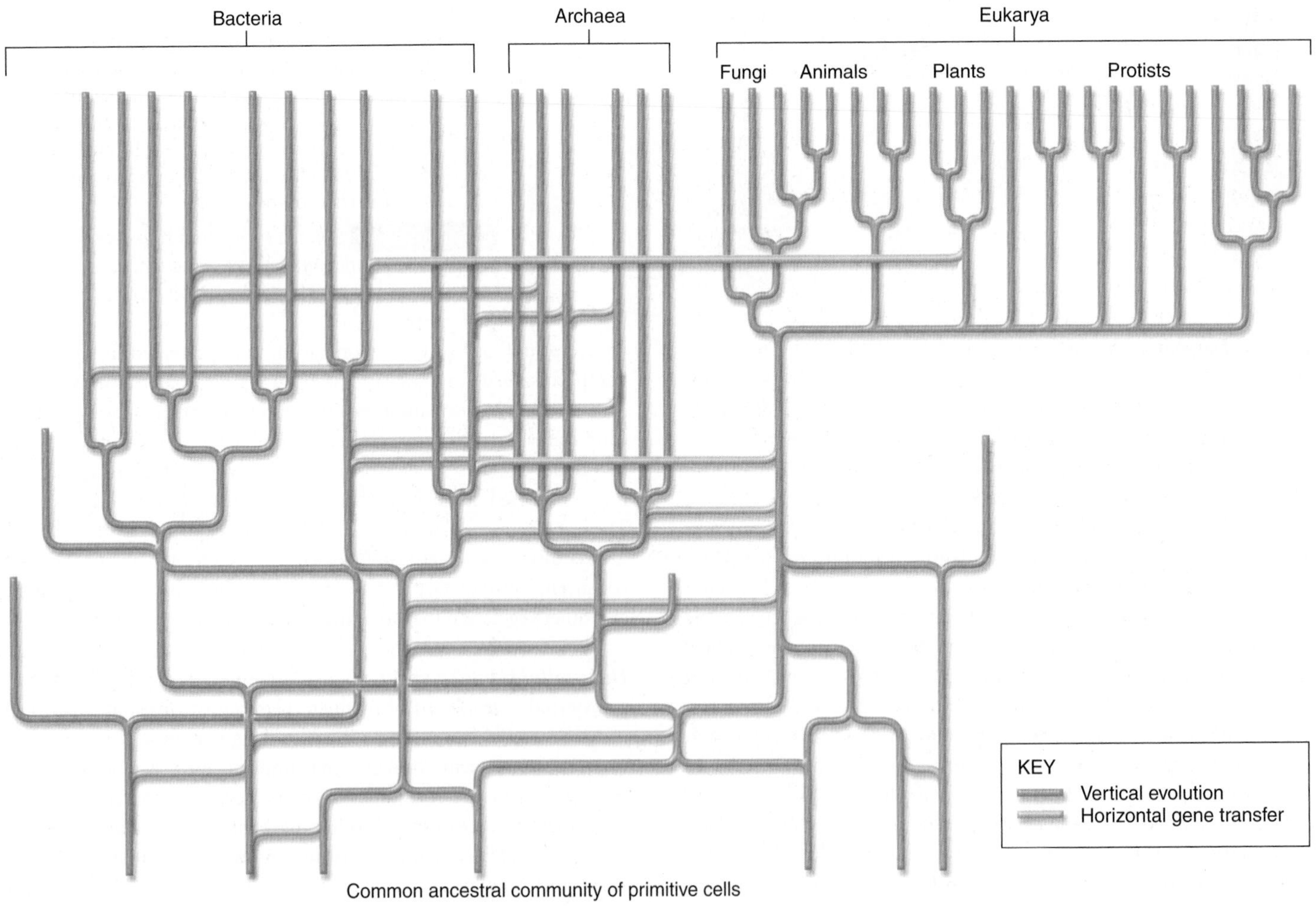

Figure 1.9 The web of life, showing both vertical evolution and horizontal gene transfer. This diagram of evolution includes both of these important mechanisms in the evolution of life on Earth. Note: Archaea are unicellular species that are similar in cell structure to bacteria.

Concept Check: How does the concept of a tree of life differ from that of a web of life?

The Classification of Living Organisms Allows Biologists to Appreciate the Unity and Diversity of Life

As biologists discover new species, they try to place them in groups based on their evolutionary history. This is a difficult task because researchers estimate that the Earth has between 10 and 100 million different species! The rationale for categorization is based on vertical descent. Species with a recent common ancestor are grouped together, whereas species whose common ancestor was in the very distant past are placed into different groups. The grouping of species is termed **taxonomy**.

Let's first consider taxonomy on a broad scale. You may have noticed that Figure 1.9 showed three main groups of organisms. From an evolutionary perspective, all forms of life can be placed into those three large categories, or domains, called **Bacteria**, **Archaea**, and **Eukarya** (Figure 1.10). Bacteria and archaea are microorganisms that are also termed **prokaryotic** because their cell structure is relatively simple. At the molecular level, bacterial and archaeal cells show significant differences in their compositions. By comparison, organisms in domain Eukarya are **eukaryotic** and have larger cells with internal compartments that serve various functions. A defining distinction between prokaryotic and eukaryotic cells is that eukaryotic cells have a **cell nucleus** in which the genetic material is surrounded by a membrane. The organisms in domain Eukarya were once subdivided into four major categories, or kingdoms, called Protista (protists), Plantae (plants), Fungi, and Animalia (animals). However, as discussed in Chapter 26 and Unit V, this traditional view became invalid as biologists gathered new information regarding the evolutionary relationships of these organisms. We now know the protists do not form a single kingdom but instead are divided into several broad categories called supergroups.

Taxonomy involves multiple levels in which particular species are placed into progressively smaller and smaller groups of organisms that are more closely related to each other evolutionarily. Such an approach emphasizes the unity and diversity of different species. As an example, let's consider the clownfish, which is shown on the textbook cover and examined in Figure 1.11. Several species of clownfish have been identified. One species of clownfish, which is orange with white stripes, has several common names, including Ocellaris clownfish. The broadest grouping for this clownfish is the domain, namely, Eukarya, followed by progressively smaller divisions, from kingdom (Animalia) to species. In the animal kingdom, clownfish are part of a phylum, Chordata, the chordates, which is subdivided into classes. Clownfish are in a class called Actinopterygii, which includes all ray-finned fishes. The common ancestor that gave rise to ray-finned fishes arose about 420 million years ago (mya). Actinopterygii is subdivided into several smaller orders. The clownfish are in the order Perciformes (bony fish). The order is, in turn, divided into families; the clownfish belong to the family of marine fish called Pomacentridae, which are often brightly colored. Families are divided into genera (singular, genus). The genus *Amphiprion* is composed of 28 different species; these are various types of clownfish. Therefore, the genus contains species that are very similar to each other in form and have evolved from a common (extinct) ancestor that lived relatively recently on an evolutionary time scale.

Biologists use a two-part description, called **binomial nomenclature**, to provide each species with a unique scientific name. The scientific name of the Ocellaris clownfish is *Amphiprion ocellaris*. The first part is the genus, and the second part is the specific epithet, or species descriptor. By convention, the genus name is capitalized, whereas the specific epithet is not. Both names are italicized. Scientific names are usually Latinized, which means they are made similar in appearance to Latin words. The origins of scientific names are typically Latin or Greek, but they can come from a variety of sources, including a person's name.

Genomes & Proteomes Connection

The Study of Genomes and Proteomes Provides an Evolutionary Foundation for Our Understanding of Biology

The unifying concept in biology is evolution. We can understand the unity of modern organisms by realizing that all living species evolved from an interrelated group of ancestors. However, from an experimental perspective, this realization presents a dilemma—we cannot take a time machine back over the course of 4 billion years to carefully study the characteristics of extinct organisms and fully appreciate the series of changes that have led to modern species. Fortunately, though, evolution has given biologists some wonderful puzzles to study, including the fossil record and the genomes of modern species. As mentioned, the term genome refers to the complete genetic composition of an organism or species (**Figure 1.12a**). The genomes of bacteria and archaea usually contain a few thousand genes, whereas those of eukaryotes may contain tens of thousands. The genome is critical to life because it performs these functions:

- *Stores information in a stable form:* The genome of every organism stores information that provides a blueprint for producing its characteristics.
- *Provides continuity from generation to generation:* The genome is copied and transmitted from generation to generation.
- *Acts as an instrument of evolutionary change:* Every now and then, the genome undergoes a mutation that may alter the characteristics of an organism. In addition, a genome may acquire new genes by horizontal gene transfer. The accumulation of such changes from generation to generation produces the evolutionary changes that alter species and produce new species.

An exciting advance in biology over the past couple of decades has been the ability to analyze the DNA sequence of genomes, a technology called **genomics**. For instance, we can compare the genomes of a frog, a giraffe, and a petunia and discover intriguing similarities and differences. These comparisons help us to understand how new traits evolved. For example, all three types of organisms have the same kinds of genes needed for the breakdown of nutrients such as sugars. In contrast, only the petunia has genes that allow it to carry out photosynthesis.

An extension of genome analysis is the study of **proteomes**, which refers to all of the proteins that a cell or organism makes. The function of most genes is to encode polypeptides that become units in proteins. As shown in **Figure 1.12b**, these include proteins that form

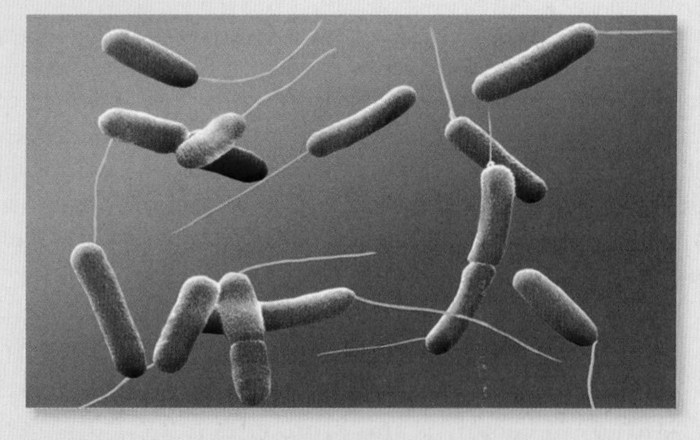

(a) Domain Bacteria: Mostly unicellular prokaryotes that inhabit many diverse environments on Earth.

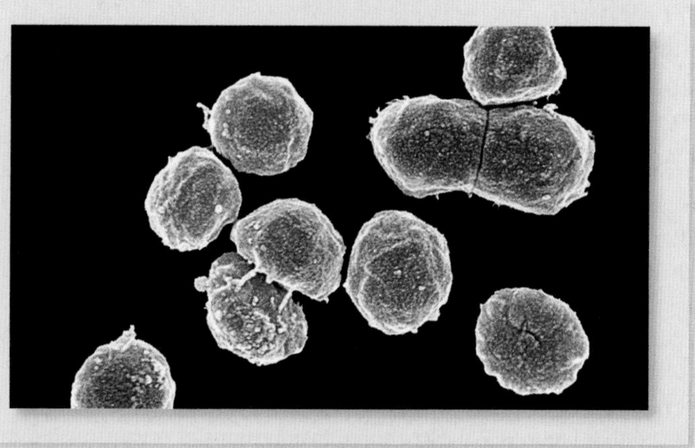

(b) Domain Archaea: Unicellular prokaryotes that often live in extreme environments, such as hot springs.

Protists: Unicellular and small multicellular organisms that are now subdivided into seven broad groups based on their evolutionary relationships.

Plants: Multicellular organisms that can carry out photosynthesis.

Fungi: Unicellular and multicellular organisms that have a cell wall but cannot carry out photosynthesis. Fungi usually survive on decaying organic material.

Animals: Multicellular organisms that usually have a nervous system and are capable of locomotion. They must eat other organisms or the products of other organisms to live.

(c) Domain Eukarya: Unicellular and multicellular organisms having cells with internal compartments that serve various functions.

Figure 1.10 The three domains of life. Two of these domains, **(a)** Bacteria and **(b)** Archaea, are prokaryotic cells. The third domain, **(c)** Eukarya, comprises species that are eukaryotes.

BioConnections: Look ahead to Figure 26.1. Are fungi more closely related to plants or animals?

Taxonomic group	Clown anemonefish is found in	Approximate time when the common ancestor for this group arose	Approximate number of modern species in this group	Examples
Domain	Eukarya	2,000 mya	> 5,000,000	
Kingdom	Animalia	600 mya	> 1,000,000	
Phylum	Chordata	525 mya	50,000	
Class	Actinopterygii	420 mya	30,000	
Order	Perciformes	80 mya	7,000	
Family	Pomacentridae	~ 40 mya	360	
Genus	*Amphiprion*	~ 9 mya	28	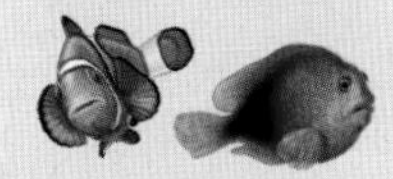
Species	*ocellaris*	< 3 mya	1	

Figure 1.11 **Taxonomic classification of the clownfish.** This simplified classification does not include a taxonomic group, called a supergroup, which is described in Chapter 26.

Concept Check: *Why is it useful to place organisms into taxonomic groupings?*

a cytoskeleton and proteins that function in cell organization and as enzymes, transport proteins, cell-signaling proteins, and extracellular proteins. The genome of each species carries the information to make its proteome—the hundreds or thousands of proteins that each cell of that species makes. Proteins are largely responsible for the structures and functions of cells and organisms. The technical approach called **proteomics** involves the analysis of the proteome of a single species and the comparison of the proteomes of different species. Proteomics helps us understand how the various levels of biology are related to one another, from the molecular level—at the level of protein molecules—to the higher levels, such as how the functioning of proteins produces the characteristics of cells and organisms and affects the ability of populations of organisms to survive in their natural environments.

In the chapters that follow, we use a recurring theme in a brief topic called "Genomes & Proteomes Connection" as a concrete way of understanding the unifying concept of evolution in biology. These brief topics will allow you to appreciate how evolution produced the characteristics of modern species. These topics explore how the genomes of different species are similar to each other and how they are different. You will learn how genome changes affect the proteome and thereby control the traits of modern species. Ultimately, these concepts provide you with a way to relate information at the molecular level to the traits of organisms and their survival within ecosystems.

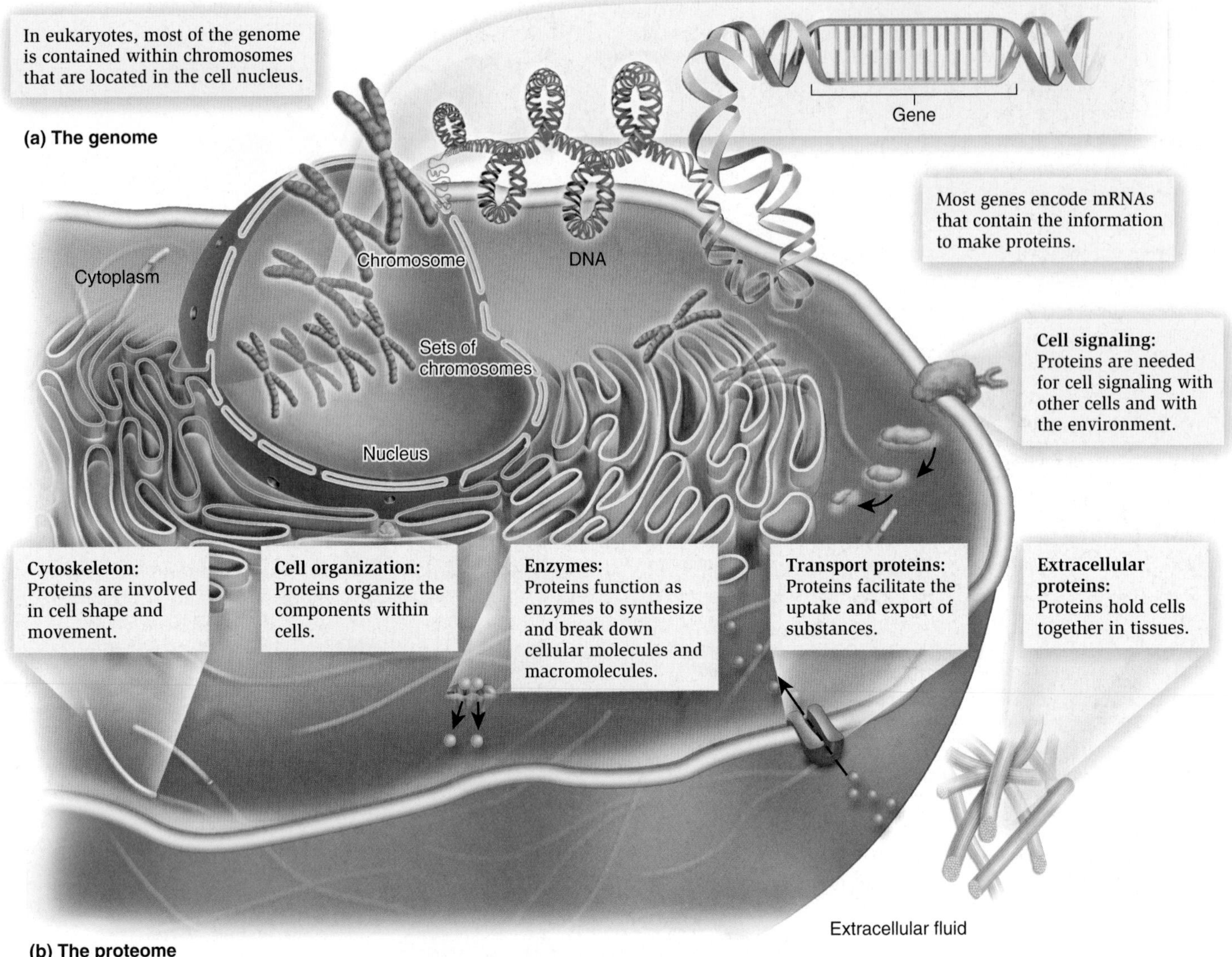

In eukaryotes, most of the genome is contained within chromosomes that are located in the cell nucleus.
(a) The genome
Gene
Most genes encode mRNAs that contain the information to make proteins.
Cytoplasm
Chromosome
DNA
Sets of chromosomes
Nucleus
Cell signaling: Proteins are needed for cell signaling with other cells and with the environment.
Cytoskeleton: Proteins are involved in cell shape and movement.
Cell organization: Proteins organize the components within cells.
Enzymes: Proteins function as enzymes to synthesize and break down cellular molecules and macromolecules.
Transport proteins: Proteins facilitate the uptake and export of substances.
Extracellular proteins: Proteins hold cells together in tissues.
Extracellular fluid
(b) The proteome

Figure 1.12 **Genomes and proteomes.** **(a)** The genome, which is composed of DNA, is the entire genetic composition of an organism. Most of the genetic material in eukaryotic cells is found in the cell nucleus. The primary function of the genome is to encode the proteome **(b)**, which is the entire protein complement of a cell or organism. Six general categories of proteins are illustrated. Proteins are largely responsible for the structure and function of cells and complete organisms.

Concept Check: Biologists sometimes say the genome is a storage unit of life, whereas the proteome is largely the functional unit of life. Explain this statement.

Genomes and Proteomes Are Fundamental to an Organism's Characteristics

Clownfish (*Amphiprion ocellaris*), which we considered earlier in Figure 1.11, are coral reef fish that live among anemones on the ocean floor (see textbook cover). Normally, the nematocysts (stinging cells) of an anemone's tentacles discharge when fish brush against the tentacles, paralyzing the fish. However, clownfish are not routinely stung by the anemone's tentacles. The mechanism that makes this possible is not clearly established, but one widely held view is that the biochemical makeup of the clownfishes' mucus layer provides protection from the stinging cells. The relationship between the clownfish and anemone is symbiotic—it is an advantage to both species. Anemones provide clownfish with a safe refuge. The clownfish drive away butterfly fish that eat anemones, and they also may clean away debris and parasites from the anemone.

One anemone is typically surrounded by a harem of clownfish consisting of a large female, a smaller reproductive male, and even smaller nonreproductive juveniles. Clownfish are protandrous hermaphrodites—that is, they can switch from male to female! When the female of a harem dies, the reproductive male changes sex to become a female and the largest of the juveniles matures into a reproductive male. How does this occur? Unlike male and female humans that differ in their sex chromosomes (XY males and XX females), the genomes of male and female clownfish don't contain different sex chromosomes. In contrast to humans, the opposite sexes of clownfish are not determined by genome differences. Instead, sex differences in clownfish arise solely as a result of proteome differences (**Figure 1.13**).

A juvenile clownfish has both male and female immature sex organs. Hormone levels, particularly those of testosterone and 17-estradiol, control particular genes in the genome. High levels of testosterone promote the expression of genes in the genome that encode proteins that cause the male organs to mature. Alternatively, if the 17-estradiol level becomes high and testosterone is decreased, gene expression is altered in a way that leads to the synthesis of some new types of proteins and prevents the synthesis of others. In other words, changes in hormones alter the composition of the proteome. When this occurs, the female organs grow, and the male reproductive system degenerates; the male fish becomes female. These sex changes, which are irreversible, are due to sequential changes in the proteomes of clownfish.

What factor determines the hormone levels in clownfish? Females seem to control the other clownfish in the harem through aggressive behavior, thereby preventing the formation of other females. This aggressive behavior suppresses an area of the brain in the other clownfish that is responsible for the production of certain hormones that are needed to promote female development. If a clownfish is left by itself in an aquarium, it will automatically develop into a female because this suppression does not occur.

Is there a survival or reproductive advantage of protandrous hermaphroditism? Though various hypotheses have been proposed, the answer is not well understood. Perhaps someone reading this textbook will one day conduct research that will answer this question. When this question is answered, biologists may understand how natural selection played a role in the phenomenon of protandrous hermaphroditism observed in clownfish.

A harem of clownfish contains a single large female. The large female prevents other members of the harem from developing into females through aggressive dominance, which affects their hormone levels.

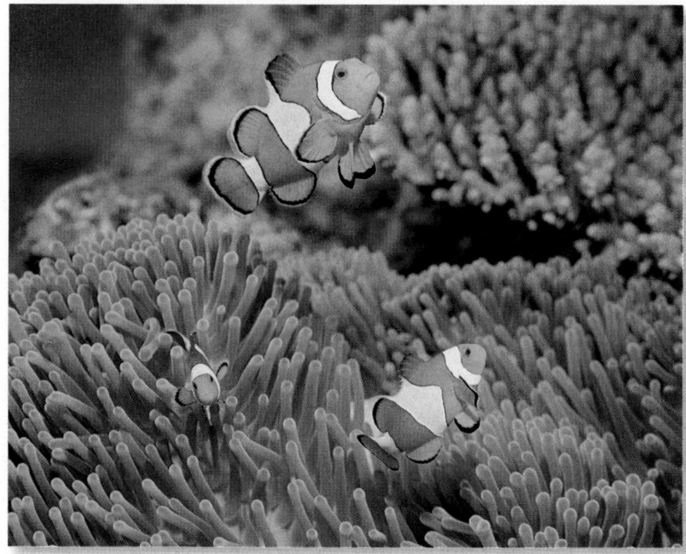

The other members of the harem are sexually immature juveniles.

The second largest fish is a mature male. When the female dies, this male will develop into a female.

Figure 1.13 **How an understanding of genomes and proteomes provides us with a foundation to understand the characteristics of living organisms.**

Concept Check: *What determines the morphological differences between male and female clownfish? Is it a difference between their genomes, their proteomes, or both?*

1.3 Biology as a Scientific Discipline

Learning Outcomes:

1. Explain how researchers study biology at different levels, ranging from molecules to ecosystems.
2. Distinguish between discovery-based science and hypothesis testing, and describe the steps of the scientific method.

What is science? Surprisingly, the definition of science is not easy to state. Most people have an idea of what science is, but actually articulating that idea proves difficult. In biology, we can define **science** as the observation, identification, experimental investigation, and theoretical explanation of natural phenomena.

Science is conducted in different ways and at different levels. Some biologists study the molecules that compose life, and others try to understand how organisms survive in their natural environments. Experimentally, researchers often focus their efforts on **model organisms**—organisms studied by many different researchers so they can compare their results and determine scientific principles that apply more broadly to other species. Examples include *Escherichia coli* (a bacterium), *Saccharomyces cerevisiae* (a yeast), *Drosophila melanogaster* (fruit fly), *Caenorhabditis elegans* (a nematode worm), *Mus musculus* (mouse), and *Arabidopsis thaliana* (a flowering plant). Model organisms offer experimental advantages over other species. For example, *E. coli* is a very simple organism that can be easily grown in the laboratory. By limiting their work to a few such model organisms, researchers can gain a deeper understanding of these species. Importantly, the discoveries made using model organisms help us to understand how biological processes work in other species, including humans.

In this section, we will examine how biologists follow a standard approach, called the **scientific method**, to test their ideas. We will explore how scientific knowledge makes predictions that can be experimentally tested. Even so, not all discoveries are the result of researchers following the scientific method. Some discoveries are simply made by gathering new information. For example, as described earlier in Figures 1.1 to 1.3, the characterization of many plants and animals has led to the development of important medicines. In this section, we will also consider how researchers often set out on "fact-finding missions" aimed at uncovering new information that may eventually lead to modern discoveries in biology.

Biologists Investigate Life at Different Levels of Organization

In Figure 1.5, we examined the various levels of biological organization. The study of these different levels depends not only on the scientific interests of biologists but also on the tools available to them. The study of organisms in their natural environments is a branch of biology called **ecology**, which considers populations, communities, and ecosystems (**Figure 1.14a**). In addition, researchers examine the structures and functions of plants and animals, which are disciplines called **anatomy** and **physiology** (**Figure 1.14b**). With the advent of microscopy, **cell biology**, which is the study of cells, became an important branch of biology in the early 1900s and remains so today (**Figure 1.14c**). In the 1970s, genetic tools became available for studying single genes and the proteins they encode. This genetic technology enabled researchers to study individual molecules, such as proteins, in living cells and thereby spawned the field of **molecular biology**. Together with chemists and biochemists, molecular biologists focus their efforts on the structure and function of the molecules of life (**Figure 1.14d**). Such researchers want to understand how biology works at the molecular and even atomic levels. Overall, the 20th century saw a progressive increase in the number of biologists who used an approach to understanding biology called **reductionism**—reducing complex systems to simpler components as a way to understand how the system works. In biology, reductionists study the parts of a cell or organism as individual units.

In the 1980s, the pendulum began to swing in the other direction. Scientists have invented new tools that allow them to study groups of genes (genomic techniques) and groups of proteins (proteomic techniques). As mentioned earlier, biologists now use the term **systems biology** to describe research aimed at understanding how emergent properties arise. This term is often applied to the study of cells. In this context, systems biology may involve the investigation of groups of genes that encode proteins with a common purpose (**Figure 1.14e**). For example, a systems biologist may conduct experiments that try to characterize an entire cellular process, which is driven by dozens of different proteins. However, systems biology is not new. Animal and plant physiologists have been studying the functions of complex organ systems for centuries. Likewise, ecologists have been characterizing ecosystems for a very long time. The novelty and excitement of systems biology in recent years have been the result of new experimental tools that allow us to study complex interactions at the molecular level. As described throughout this textbook, the investigation of genomes and proteomes has provided important insights into many interesting topics in systems biology.

A Hypothesis Is a Proposed Idea, Whereas a Theory Is a Broad Explanation Backed by Extensive Evidence

Let's now consider the process of science. In biology, a **hypothesis** is a proposed explanation for a natural phenomenon. It is a proposition based on previous observations or experimental studies. For example, with knowledge of seasonal changes, you might hypothesize that maple trees drop their leaves in the autumn because of the shortened amount of daylight. An alternative hypothesis might be that the trees drop their leaves because of colder temperatures. In biology, a hypothesis requires more work by researchers to evaluate its validity.

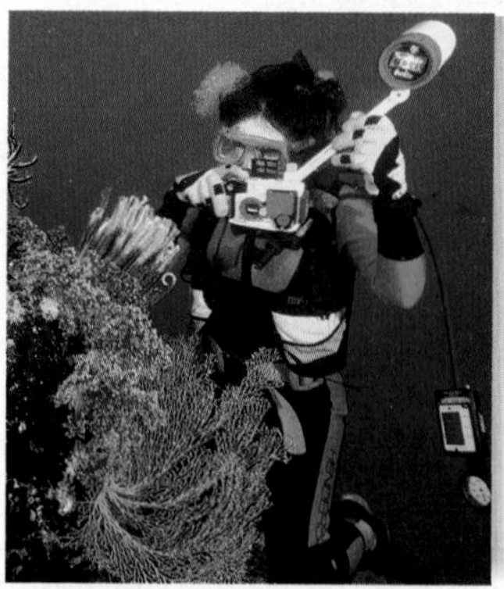

Ecologists study species in their native environments.

(a) Ecology—population/community/ecosystem levels

Anatomists and physiologists study how the structures of organisms are related to their functions.

(b) Anatomy and physiology—tissue/organ/organism levels

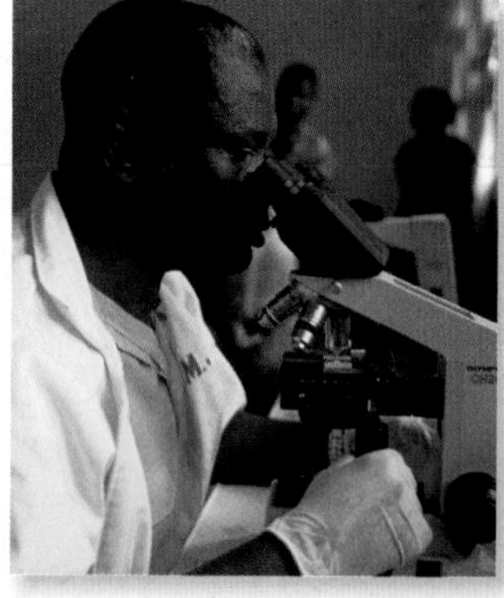

Cell biologists often use microscopes to learn how cells function.

(c) Cell biology—cellular levels

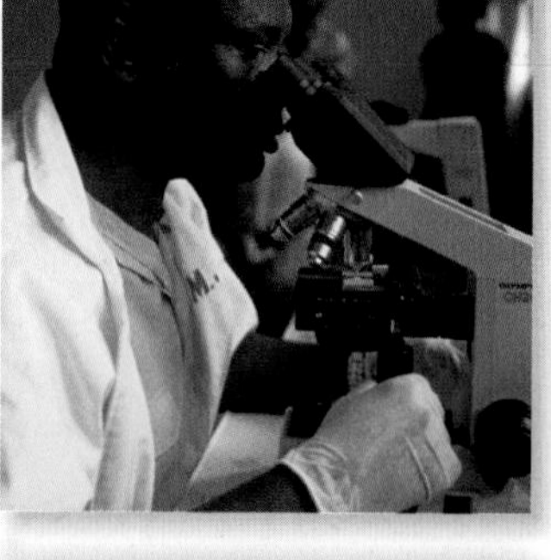

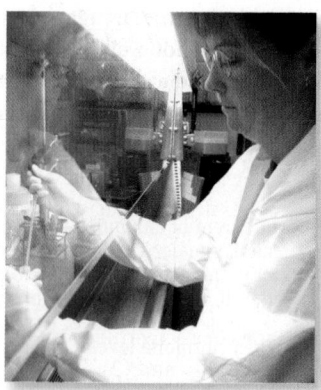

Molecular biologists and biochemists study the molecules and macromolecules that make up cells.

(d) Molecular biology—atomic/molecular levels

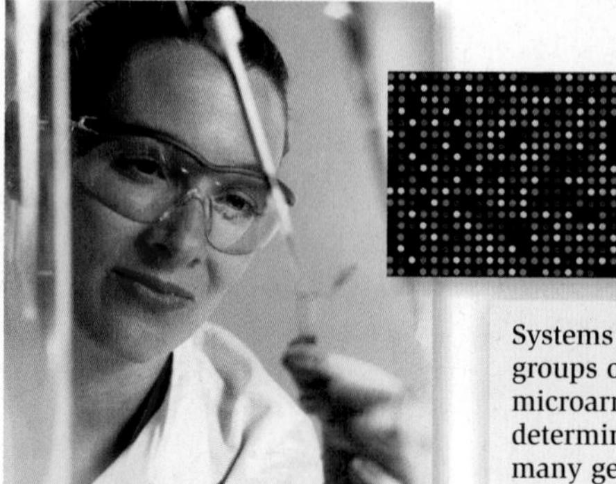

Systems biologists may study groups of molecules. The microarray shown in the inset determines the expression of many genes simultaneously.

(e) Systems biology—all levels, shown here at the molecular level

Figure 1.14 Biological investigation at different levels.

A useful hypothesis must make **predictions**—expected outcomes that can be shown to be correct or incorrect. In other words, a useful hypothesis is testable. If a hypothesis is incorrect, it should be **falsifiable**, which means that it can be shown to be incorrect by additional observations or experimentation. Alternatively, a hypothesis may be correct, so further work will not disprove it. In such cases, we would say that the researcher(s) has failed to reject the hypothesis. Even so, in science, a hypothesis is never really proven but rather always remains provisional. Researchers accept the possibility that perhaps they have not yet conceived of the correct hypothesis. After many experiments, biologists may conclude their hypothesis is consistent with known data, but they should never say the hypothesis is proven.

By comparison, the term **theory**, as it is used in biology, is a broad explanation of some aspect of the natural world that is substantiated by a large body of evidence. Biological theories incorporate observations, hypothesis testing, and the laws of other disciplines such as chemistry and physics. Theories are powerful because they allow us to make many predictions about the properties of living organisms. As an example, let's consider the theory that DNA is the genetic material and that it is organized into units called genes. An overwhelming body of evidence has substantiated this theory. Thousands of living species have been analyzed at the molecular level. All of them have been found to use DNA as their genetic material and to express genes that produce the proteins that lead to their characteristics. This theory makes many valid predictions. For example, certain types of mutations in genes are expected to affect the traits of organisms. This prediction has been confirmed experimentally. Similarly, this theory predicts that genetic material is copied and transmitted from parents to offspring. By comparing the DNA of parents and offspring, this prediction has also been confirmed. Furthermore, the theory explains the observation that offspring resemble their parents. Overall, two key attributes of a scientific theory are (1) consistency with a vast amount of known data and (2) the ability to make many correct predictions. Two other important biological theories we touched on in this chapter are the cell theory and the theory of evolution by natural selection.

The meaning of the term "theory" is sometimes muddled because it is used in different situations. In everyday language, a theory is often viewed as little more than a guess. For example, a person might say, "My theory is that Professor Simpson did not come to class today because he went to the beach." However, in biology, a theory is much more than a guess. A theory is an established set of ideas that explains a vast amount of data and offers valid predictions that can be tested. Like a hypothesis, a theory can never be proven to be true. Scientists acknowledge that they do not know everything. Even so, biologists would say that theories are extremely likely to be true, based on all known information. In this regard, theories are viewed as **knowledge**, which is the awareness and understanding of information.

Discovery-Based Science and Hypothesis Testing Are Scientific Approaches That Help Us Understand Biology

The path that leads to an important discovery is rarely a straight line. Rather, scientists ask questions, make observations, ask modified questions, and may eventually conduct experiments to test their hypotheses. The first attempts at experimentation may fail, and new experimental approaches may be needed. To suggest that scientists follow a rigid scientific method is an oversimplification of the process of science. Scientific advances often occur as scientists dig deeper and deeper into a topic that interests them. Curiosity is the key phenomenon that sparks scientific inquiry. How is biology actually conducted? As discussed next, researchers typically follow two general types of approaches: discovery-based science and hypothesis testing.

Discovery-Based Science The collection and analysis of data without the need for a preconceived hypothesis is called **discovery-based science**, or simply **discovery science**. Why is discovery-based science carried out? The information gained from discovery-based science may lead to the formation of new hypotheses and, in the long run, may have practical applications that benefit people. Researchers, for example, have identified and begun to investigate newly discovered genes within the human genome without already knowing the function of the gene they are studying. The goal is to gather additional clues that may eventually allow them to propose a hypothesis that explains the gene's function. Discovery-based science often leads to hypothesis testing.

Hypothesis Testing In biological science, the scientific method, also known as **hypothesis testing**, is usually followed to formulate and test the validity of a hypothesis. This strategy may be described as a five-step method:

1. Observations are made regarding natural phenomena.
2. These observations lead to a hypothesis that tries to explain the phenomena. A useful hypothesis is one that is testable because it makes specific predictions.
3. Experimentation is conducted to determine if the predictions are correct.
4. The data from the experiment are analyzed.
5. The hypothesis is considered to be consistent with the data, or it is rejected.

The scientific method is intended to be an objective way to gather knowledge. As an example, let's return to our scenario of maple trees dropping their leaves in autumn. By observing the length of daylight throughout the year and comparing that data with the time of the year when leaves fall, one hypothesis might be that leaves fall in response to a shorter amount of daylight (Figure 1.15). This hypothesis makes a prediction—exposure of maple trees to shorter daylight will cause their leaves to fall. To test this prediction, researchers would design and conduct an experiment.

How is hypothesis testing conducted? Although hypothesis testing may follow many paths, certain experimental features are common to this approach. First, data are often collected in two parallel manners. One set of experiments is done on the **control group**, while another set is conducted on the **experimental group**. In an ideal experiment, the control and experimental groups differ by only one factor. For example, an experiment could be conducted in which two groups of trees are observed, and the only difference between their environments is the length of light each day. To conduct such an experiment, researchers would grow small trees in a greenhouse where they could keep other factors such as temperature, water, and nutrients the same between the control and experimental groups, while providing them with different amounts of daylight. In the control group, the number

1 OBSERVATIONS The leaves on maple trees fall in autumn when the days get colder and shorter.

2 HYPOTHESIS The shorter amount of daylight causes the leaves to fall.

3 EXPERIMENTATION
Small maple trees are grown in 2 greenhouses where the only variable is the length of light.

Control group:
Amount of daily light remains constant for 180 days.

Experimental group:
Amount of daily light becomes progressively shorter for 180 days.

4 THE DATA

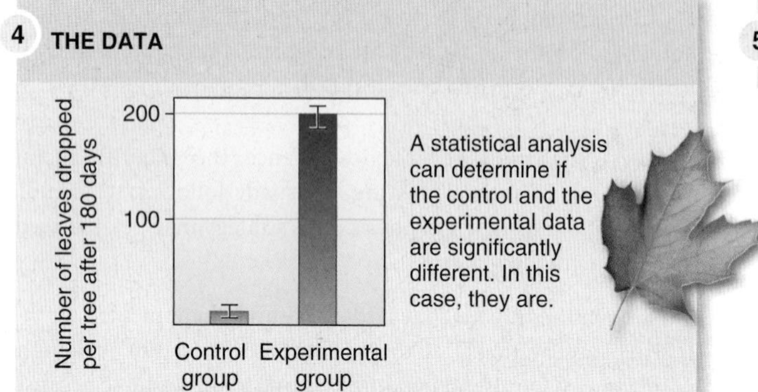
Number of leaves dropped per tree after 180 days
200
100
Control group
Experimental group
A statistical analysis can determine if the control and the experimental data are significantly different. In this case, they are.

5 CONCLUSION The hypothesis cannot be rejected.

Figure 1.15 The steps of the scientific method, also known as hypothesis testing. In this example, the goal is to test the hypothesis that maple trees drop their leaves in the autumn due to shortening length of daylight.

Concept Check: *What is the purpose of a control group in hypothesis testing?*

of hours of light provided by lightbulbs is kept constant each day, whereas in the experimental group, the amount of light each day becomes progressively shorter to mimic seasonal light changes. The researchers would then record the number of leaves dropped by the two groups of trees over a certain period of time.

Another key feature of hypothesis testing is data analysis. The result of experimentation is a set of data from which a biologist tries to draw conclusions. Biology is a quantitative science. When experimentation involves control and experimental groups, a common form of analysis is to determine if the data collected from the two groups are truly different from each other. Biologists apply statistical analyses to their data to determine if the control and experimental groups are likely to be different from each other because of the single variable that is different between the two groups. When they are statistically significant, the differences between the control and experimental data are not likely to have occurred as a matter of random chance.

In our tree example shown in Figure 1.15, the trees in the control group dropped far fewer leaves than did those in the experimental group. A statistical analysis could determine if the data collected from the two greenhouses are significantly different from each other. If the two sets of data are found not to be significantly different, the hypothesis would be rejected. Alternatively, if the differences between the two sets of data are significant, as shown in Figure 1.15, biologists would conclude that the hypothesis is consistent with the data, though it is not proven. A hallmark of science is that valid experiments are **repeatable**, which means that similar results are obtained when the experiment is conducted on multiple occasions. For our example in Figure 1.15, the data would be valid only if the experiment was repeatable.

As described next, discovery-based science and hypothesis testing are often used together to learn more about a particular scientific topic. As an example, let's look at how both approaches have led to successes in the study of the disease called cystic fibrosis.

The Study of Cystic Fibrosis Provides Examples of Discovery-Based Science and Hypothesis Testing

Let's consider how biologists made discoveries related to the disease cystic fibrosis (CF), which affects about 1 in every 3,500 Americans. Persons with CF produce abnormally thick and sticky mucus that obstructs the lungs and leads to life-threatening lung infections. The thick mucus also blocks ducts in the pancreas, which prevents the digestive enzymes this organ produces from reaching the intestine. Without these enzymes, the intestine cannot fully absorb proteins and fats, which can cause malnutrition. Persons with this disease may also experience liver damage because the thick mucus can obstruct the liver. On average, people with CF in the United States currently live into their late 30s. Fortunately, as more advances have been made in treatment, this number has steadily increased.

Because of its medical significance, many scientists are interested in CF and are conducting studies aimed at gaining greater information regarding its underlying cause. The hope is that knowing more about the disease may lead to improved treatment options, and perhaps even a cure. As described next, discovery-based science and hypothesis testing have been critical to gaining a better understanding of this disease.

The CFTR Gene and Discovery-Based Science In 1935, Dorothy Andersen determined that cystic fibrosis is a genetic disorder. Persons with CF have inherited two faulty *CFTR* genes, one from each parent. (We now know this gene encodes a protein named the cystic fibrosis transmembrane regulator, abbreviated CFTR.) In the 1980s, researchers used discovery-based science to identify this gene. Their search for the *CFTR* gene did not require any preconceived hypothesis regarding the function of the gene. Rather, they used genetic strategies similar to those described in Chapter 20. Research groups headed by Lap-Chee Tsui, Francis Collins, and John Riordan identified the *CFTR* gene in 1989.

The discovery of the gene made it possible to devise diagnostic testing methods to determine if a person carries a faulty *CFTR* gene. In addition, the characterization of the *CFTR* gene provided important clues about its function. Researchers observed striking similarities between the *CFTR* gene and other genes that were already known to encode proteins that function in the transport of substances across membranes. Based on this observation, as well as other kinds of data, the scientists hypothesized that the function of the normal *CFTR* gene is to encode a transport protein. In this way, the identification of the *CFTR* gene led them to conduct experiments aimed at testing a hypothesis of its function.

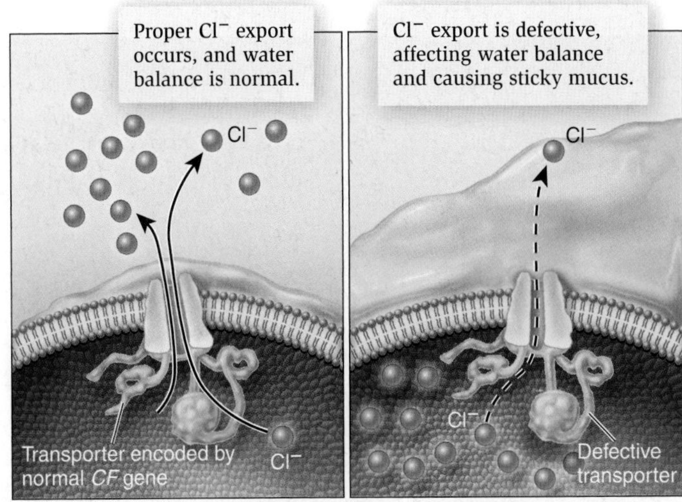

Figure 1.16 A hypothesis suggesting an explanation for the function of a gene defective in patients with cystic fibrosis. The normal *CFTR* gene, which does not carry a mutation, encodes a protein that transports chloride ions (Cl^-) across the plasma membrane to the outside of the cell. In persons with CF, this protein is defective due to a mutation in the *CFTR* gene.

Concept Check: *Explain how discovery-based science helped researchers to hypothesize that the CFTR gene encodes a transport protein.*

The CFTR Gene and Hypothesis Testing Researchers considered the characterization of the *CFTR* gene along with other studies showing that patients with the disorder have an abnormal regulation of salt balance across their plasma membranes. They hypothesized that the normal *CFTR* gene encodes a protein that functions in the transport of chloride ions (Cl^-) across the membranes of cells (Figure 1.16). This hypothesis led to experimentation in which they tested normal cells and cells from CF patients for their ability to transport Cl^-. The CF cells were found to be defective in chloride transport. In 1990, scientists successfully transferred the normal *CFTR* gene into CF cells in the laboratory. The introduction of the normal gene corrected the cells' defect in chloride transport. Overall, the results showed that the *CFTR* gene encodes a protein that transports Cl^- across the plasma membrane. A mutation in this gene causes it to encode a defective protein, leading to a salt imbalance that affects water levels outside the cell, which explains the thick and sticky mucus in CF patients. In this example, hypothesis testing has provided a way to evaluate a hypothesis about how a disease is caused by a genetic change.

Feature Investigation

Observation and Experimentation Form the Core of Biology

Biology is largely about the process of discovery. Therefore, a recurring theme of this textbook is how scientists design experiments, analyze data, and draw conclusions. Although each chapter contains many examples of data collection and experiments, a consistent element is a "Feature Investigation"—an actual study by current or past researchers. Some of these involve discovery-based science, in which biologists collect and interpret data in an attempt to make discoveries that are not hypothesis driven. Most Feature Investigations, however, involve hypothesis testing in which a hypothesis is stated and the experiment and resulting data are presented. Figure 1.15 shows the general form of Feature Investigations.

The Feature Investigations allow you to appreciate the connection between science and scientific theories. As you read a Feature Investigation, you may find yourself thinking about different approaches and alternative hypotheses. Different people can view the same data and arrive at very different conclusions. As you progress through the experiments in this textbook, we hope you will try to develop your own skills at formulating hypotheses, designing experiments, and interpreting data.

Experimental Questions

1. Discuss the difference between discovery-based science and hypothesis testing.
2. What are the steps in the scientific method, also called hypothesis testing?
3. When conducting an experiment, explain how a control group and an experimental group differ from each other.

Figure 1.17 **The social aspects of science.** At scientific meetings, researchers gather together to discuss new data and discoveries. Research that is conducted by professors, students, lab technicians, and industrial participants is sometimes hotly debated.

Science as a Social Discipline

Finally, it is worthwhile to point out that biology is a social as well as a scientific discipline. Different laboratories often collaborate on scientific projects. After performing observations and experiments, biologists communicate their results in different ways. Most importantly, papers are submitted to scientific journals. Following submission, papers usually undergo a **peer-review process** in which other scientists, who are experts in the area, evaluate the paper and make suggestions regarding its quality. As a result of peer review, a paper is either accepted for publication, rejected, or the authors of the paper may be given suggestions for how to revise the work or conduct additional experiments before it is acceptable for publication.

Another social aspect of research is that biologists often attend meetings where they report their most recent work to the scientific community (Figure 1.17). They comment on each other's ideas and work, eventually shaping together the information that builds into scientific theories over many years. As you develop your skills at scrutinizing experiments, it is helpful to discuss your ideas with other people, including fellow students and faculty members. Importantly, you do not need to "know all the answers" before you enter into a scientific discussion. Instead, a more realistic way to view science is as an ongoing and never-ending series of questions.

Summary of Key Concepts

Biology is the study of life. Discoveries in biology help us understand how life exists, and they also have many practical applications, such as the development of drugs to treat human diseases. (Figures 1.1, 1.2, 1.3)

1.1 Principles of Biology and the Levels of Biological Organization

- Eight principles underlie the characteristics that are common to all forms of life. All living things (1) are composed of cells as their simplest unit; (2) use energy; (3) interact with their environment; (4) maintain homeostasis; (5) grow and develop; and (6) have genetic material for reproduction. Also, (7) populations of organisms evolve from one generation to the next, and (8) are connected by an evolutionary history (Figure 1.4).
- Additional principles include (9) structure determines function; (10) new properties of life emerge from complex interactions; (11) biology is an experimental science; and (12) biology influences our society.
- Living organisms can be viewed at different levels of biological organization: atoms, molecules and macromolecules, cells, tissues, organs, organisms, populations, communities, ecosystems, and the biosphere (Figure 1.5).

1.2 Unity and Diversity of Life

- Changes in species often occur as a result of modification of pre-existing structures (Figure 1.6).
- During vertical evolution, mutations in a lineage alter the characteristics of species from one generation to the next. Individuals with greater reproductive success are more likely to contribute to future generations, a process known as natural selection. Over the long run, this process alters species and may produce new species (Figure 1.7).
- Horizontal gene transfer may involve the transfer of genes between different species. Along with vertical evolution, it is an important process in biological evolution, producing a web of life (Figures 1.8, 1.9).
- Taxonomy is the grouping of species according to their evolutionary relatedness to other species. Going from broad to narrow groups, each species is placed into a domain, kingdom, phylum, class, order, family, and genus (Figures 1.10, 1.11).
- The genome is the genetic composition of a species. It provides a blueprint for the traits of an organism, is transmitted from parents to offspring, and acts as an instrument for evolutionary change. The proteome is the collection of proteins that a cell or organism makes (Figure 1.12).
- An analysis of genomes and proteomes helps us to understand how information at the molecular level relates to the characteristics of individuals and how they survive in their native environments (Figure 1.13).

1.3 Biology as a Scientific Discipline

- Biological science is the observation, identification, experimental investigation, and theoretical explanation of natural phenomena.
- Biologists study life at different levels, ranging from ecosystems to the molecular components in cells (Figure 1.14).
- A hypothesis is a proposal to explain a natural phenomenon. A useful hypothesis makes a testable prediction. A biological theory is a broad explanation that is substantiated by a large body of evidence.
- Discovery-based science is an approach in which researchers conduct experiments and analyze data without a preconceived hypothesis.
- The scientific method, also called hypothesis testing, is a series of steps to formulate and test the validity of a hypothesis. The experimentation often involves a comparison between control and experimental groups (Figure 1.15).
- The study of cystic fibrosis provides an example in which both discovery-based science and hypothesis testing led to key insights regarding the nature of the disease (Figure 1.16).

- Each chapter in this textbook has a "Feature Investigation," an actual study by current or past researchers that highlights the experimental approach and helps you appreciate how science has led to key discoveries in biology.
- Biology is a social discipline in which scientists often work in teams. To be published, a scientific paper is usually subjected to a peer-review process in which other scientists evaluate the paper and make suggestions regarding its quality. Advances in science often occur when scientists gather and discuss their data (Figure 1.17).

Assess and Discuss

Test Yourself

1. The process in which living organisms maintain a relatively stable internal condition is
   a. adaptation.
   b. evolution.
   c. metabolism.
   d. homeostasis.
   e. development.
2. Populations of organisms change over the course of many generations. Many of these changes result in increased survival and reproduction. This phenomenon is
   a. evolution.
   b. homeostasis.
   c. development.
   d. genetics.
   e. metabolism.
3. All of the places on Earth where living organisms are found is termed
   a. the ecosystem.
   b. a community.
   c. the biosphere.
   d. a viable land mass.
   e. a population.
4. Which of the following is an example of horizontal gene transfer?
   a. the transmission of an eye color gene from father to daughter
   b. the transmission of a mutant gene causing cystic fibrosis from father to daughter
   c. the transmission of a gene conferring pathogenicity (the ability to cause disease) from one bacterial species to another
   d. the transmission of a gene conferring antibiotic resistance from a mother cell to its two daughter cells
   e. all of the above
5. The scientific name for humans is *Homo sapiens*. The name *Homo* is the _____ to which humans are classified.
   a. kingdom
   b. phylum
   c. order
   d. genus
   e. species
6. The complete genetic makeup of an organism is called
   a. the genus.
   b. the genome.
   c. the proteome.
   d. the genotype.
   e. the phenotype.
7. A proposed explanation for a natural phenomenon is
   a. a theory.
   b. a law.
   c. a prediction.
   d. a hypothesis.
   e. an assay.
8. In science, a theory should
   a. be equated with knowledge.
   b. be supported by a substantial body of evidence.
   c. provide the ability to make many correct predictions.
   d. do all of the above.
   e. b and c only
9. Conducting research without a preconceived hypothesis is called
   a. discovery-based science.
   b. the scientific method.
   c. hypothesis testing.
   d. a control experiment.
   e. none of the above.
10. What is the purpose of using a control group in scientific experiments?
    a. A control group allows the researcher to practice the experiment first before actually conducting it.
    b. A researcher can compare the results in the experimental group and control group to determine if a single variable is causing a particular outcome in the experimental group.
    c. A control group provides the framework for the entire experiment so the researcher can recall the procedures that should be conducted.
    d. A control group allows the researcher to conduct other experimental changes without disturbing the original experiment.
    e. All of the above are correct.

Conceptual Questions

1. Of the first eight principles of biology described in Figure 1.4, which apply to individuals and which apply to populations?
2. Explain how it is possible for evolution to result in unity among different species yet also produce amazing diversity.
3. In your own words, describe the twelve principles of biology that are detailed at the beginning of this chapter.

Collaborative Questions

1. Discuss whether or not you think that theories in biology are true. Outside of biology, how do you decide if something is true?
2. In certain animals, such as alligators, sex is determined by temperature. When alligator eggs are exposed to low temperatures, most alligator embryos develop into females. Discuss how this phenomenon is related to genomes and proteomes.

Online Resource

www.brookerbiology.com

Stay a step ahead in your studies with animations that bring concepts to life and practice tests to assess your understanding. Your instructor may also recommend the interactive eBook, individualized learning tools, and more.

UNIT I

CHEMISTRY

Living organisms can be thought of as collections of discrete packages—one or more cells—within which occur a wide array of chemical reactions. These reactions occur between atoms and molecules and may require, or in some cases release, energy. Chemical reactions and interactions between molecules play a role in virtually all aspects of a cell's activities. In order to understand how living organisms function, grow, develop, behave, and interact with their environments, therefore, we first need to understand some basic principles of atomic and molecular structure and the forces that allow them to interact with each other. We begin this unit with an overview of **inorganic chemistry**—that is, the nature of atoms and molecules, with the exception of those that contain rings or chains of carbon. Such carbon-containing molecules form the basis of **organic chemistry** and are covered in Chapter 3.

The following biology principles will be emphasized in this unit:

- ***Living organisms use energy:*** *We will see how the chemical energy stored in certain of the bonds of molecules such as sugars and fats can be released and used by living organisms to perform numerous functions that support life, including growth, digestion, and locomotion, among many others.*
- ***Structure determines function:*** *As described in Chapter 3, the three-dimensional structure of molecules is critical in enabling them to specifically interact with each other.*
- ***The genetic material provides a blueprint for reproduction, and all species (past and present) are related by an evolutionary history:*** *Nucleic acids, the basis of inherited genetic material, are first introduced in Chapter 3.*
- ***New properties emerge from complex interactions:*** *You will learn in this unit how various types of forces create complex molecules from simple atoms and impart precise shapes on these molecules. The newly created molecule has properties that are different from those of its component atoms.*
- ***Biology affects our society:*** *In Chapter 2, we will see how an understanding of chemistry has transformed the ability of physicians to diagnose disease in humans. One example of an application of chemistry to medicine discussed in Chapter 2 is the PET scan.*

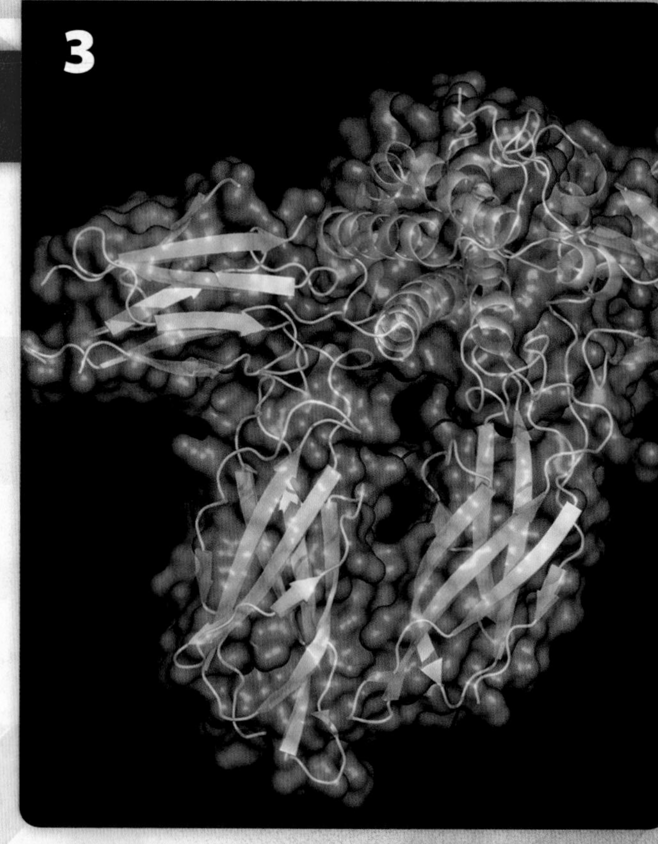

Chapter Outline

2.1 Atoms

2.2 Chemical Bonds and Molecules

2.3 Properties of Water

Summary of Key Concepts

Assess and Discuss

The Chemical Basis of Life I: Atoms, Molecules, and Water

2

Crystals of sodium chloride (NaCl), a molecule composed of two elements.

Biology—the study of life—is founded on the principles of chemistry and physics. All living organisms are a collection of atoms and molecules bound together and interacting with each other through the forces of nature. Throughout this textbook, we will see how chemistry can be applied to living organisms as we discuss the components of cells, the functions of proteins, the flow of nutrients in plants and animals, and the evolution of new genes. This chapter lays the groundwork for understanding these and other concepts. We begin with an overview of the nature of atoms and molecules, focusing on the structure of the atom and how it was discovered. We next explore the various ways that atoms combine with other atoms to create molecules, looking at the different types of chemical bonds between atoms, how these bonds form, and how they determine the structures of molecules. We end with an examination of the water molecule and the properties that make it a crucial component of living organisms and their environment.

2.1 Atoms

Learning Outcomes:

1. Describe the general structure of atoms and their constituent particles.
2. Discuss the way electrons orbit the nucleus of an atom within discrete energy levels.
3. Relate atomic structure to the periodic table of the elements.
4. Quantify atomic mass using units such as daltons and moles.
5. Explain how a single element may exist in more than one form, called isotopes, and how certain isotopes have importance in human medicine.
6. List the elements that make up most of the mass of all living organisms.

All life-forms are composed of **matter**, which is defined as anything that contains mass and occupies space. In living organisms, matter may exist in any of three states: solid, liquid, or gas. All matter is composed of **atoms**, which are the smallest functional units of matter that form all chemical substances and ultimately all organisms; they cannot be further broken down into other substances by ordinary chemical or physical means. Atoms, in turn, are composed of smaller, subatomic components, collectively referred to as particles. Chemists study the properties of atoms and **molecules**, which are two or more atoms bonded together. A major interest of the physicist, by contrast, is to uncover the properties of subatomic particles. Chemistry and physics merge when one attempts to understand the mechanisms by which atoms and molecules interact. When atoms and molecules are studied in the context of a living organism, the science of biochemistry emerges. In this section, we explore the physical properties of atoms so we can understand how atoms combine to form molecules of biological importance.

Atoms Are Composed of Subatomic Particles

There are many types of atoms in living organisms. The simplest atom, hydrogen, is approximately 0.1 nanometers (nm, 1×10^{-10} meters) in diameter, roughly one-millionth the diameter of a human hair. Each specific type of atom—nitrogen, hydrogen, oxygen, and so on—is called an **element** (or chemical element), which is defined as a pure substance of only one kind of atom.

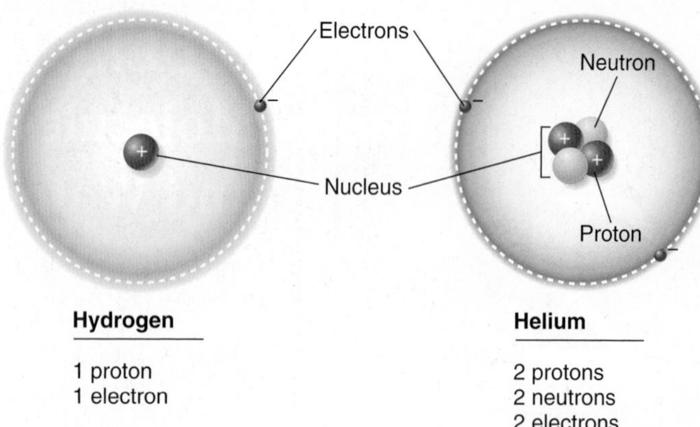

Figure 2.1 Diagrams of two simple atoms. This is a model of the two simplest atoms, hydrogen and helium. The nucleus consists of protons and neutrons, and electrons are found outside the nucleus. Note: In all figures of atoms, the sizes and distances are not to scale.

Three subatomic particles—**protons** (p^+), **neutrons** (n^0), and **electrons** (e^-)—are found within atoms (**Figure 2.1**). The protons and neutrons are confined to a very small volume at the center of an atom, the **atomic nucleus**, whereas the electrons are found in regions at various distances from the nucleus. With the exception of ions—atoms that have gained or lost one or more electrons (described later in this chapter)—the numbers of protons and electrons in a given type of atom are identical, but the number of neutrons may vary. Each of the subatomic particles has a different electric charge. Protons have one unit of positive charge, electrons have one unit of negative charge, and neutrons are electrically neutral. Like charges always repel each other, and opposite charges always attract each other. It is the opposite charges of the protons and electrons that create an atom—the positive charges in the nucleus attract the negatively charged electrons.

Because the protons are located in the atomic nucleus, the nucleus has a net positive charge equal to the number of protons it contains. The entire atom has no net electric charge, however, because the number of negatively charged electrons around the nucleus is equal to the number of positively charged protons in the nucleus.

This basic concept of the structure of the atom was not established until a landmark experiment conducted by Ernest Rutherford during the years 1909–1911, as described next.

FEATURE INVESTIGATION

Rutherford Determined the Modern Model of the Atom

Nobel laureate Ernest Rutherford was born in 1871 in New Zealand, but he did his greatest work at McGill University in Montreal, Canada, and later at the University of Manchester in England. At that time, scientists knew that atoms contained charged particles but had no idea how those particles were distributed. Neutrons had not yet been discovered, and many scientists, including Rutherford, hypothesized that the positive charge and the mass of an atom were evenly dispersed throughout the atom.

In a now-classic experiment, Rutherford aimed a fine beam of positively charged α (alpha) particles at an extremely thin sheet of gold foil only 400 atoms thick (**Figure 2.2**). Alpha particles are the two protons and two neutrons that comprise the nuclei of helium atoms; you can think of them as helium atoms without their electrons (see Figure 2.1). Surrounding the gold foil were zinc sulfide screens that registered any α particles passing through or bouncing off the foil, much like film in a camera detects light. Rutherford hypothesized that if the positive charges of the gold atoms were uniformly distributed, many of the positively charged α particles would be slightly deflected, because one of the most important features of electric charge is that like charges repel each other. Due to their much smaller mass, he did not expect electrons in the gold atoms to have any effect on the ability of an α particle to move through the metal foil.

Surprisingly, Rutherford discovered that more than 98% of the α particles passed right through as if the foil was not there and only

Figure 2.2 Rutherford's gold foil experiment, demonstrating that most of the volume of an atom is empty space.

HYPOTHESIS Atoms in gold foil are composed of diffuse, evenly distributed positive charges that should usually cause α particles to be slightly deflected as they pass through.

KEY MATERIALS Thin sheet of gold foil, α particle emitter, zinc sulfide detection screen.

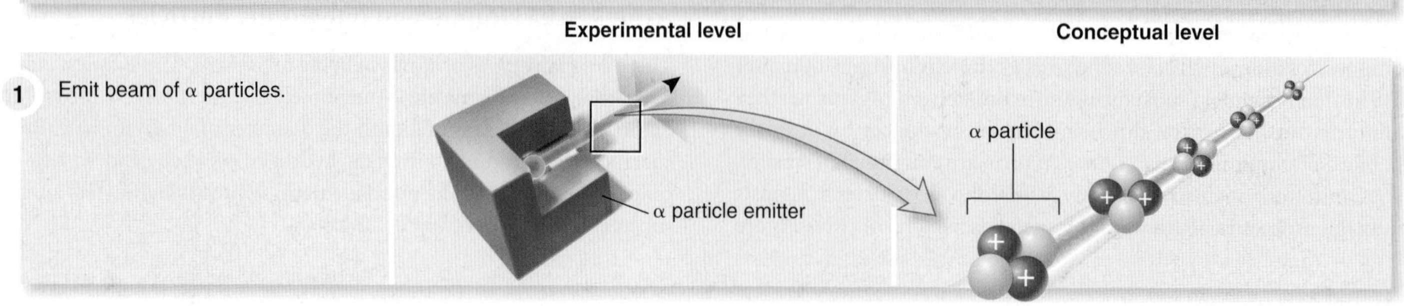

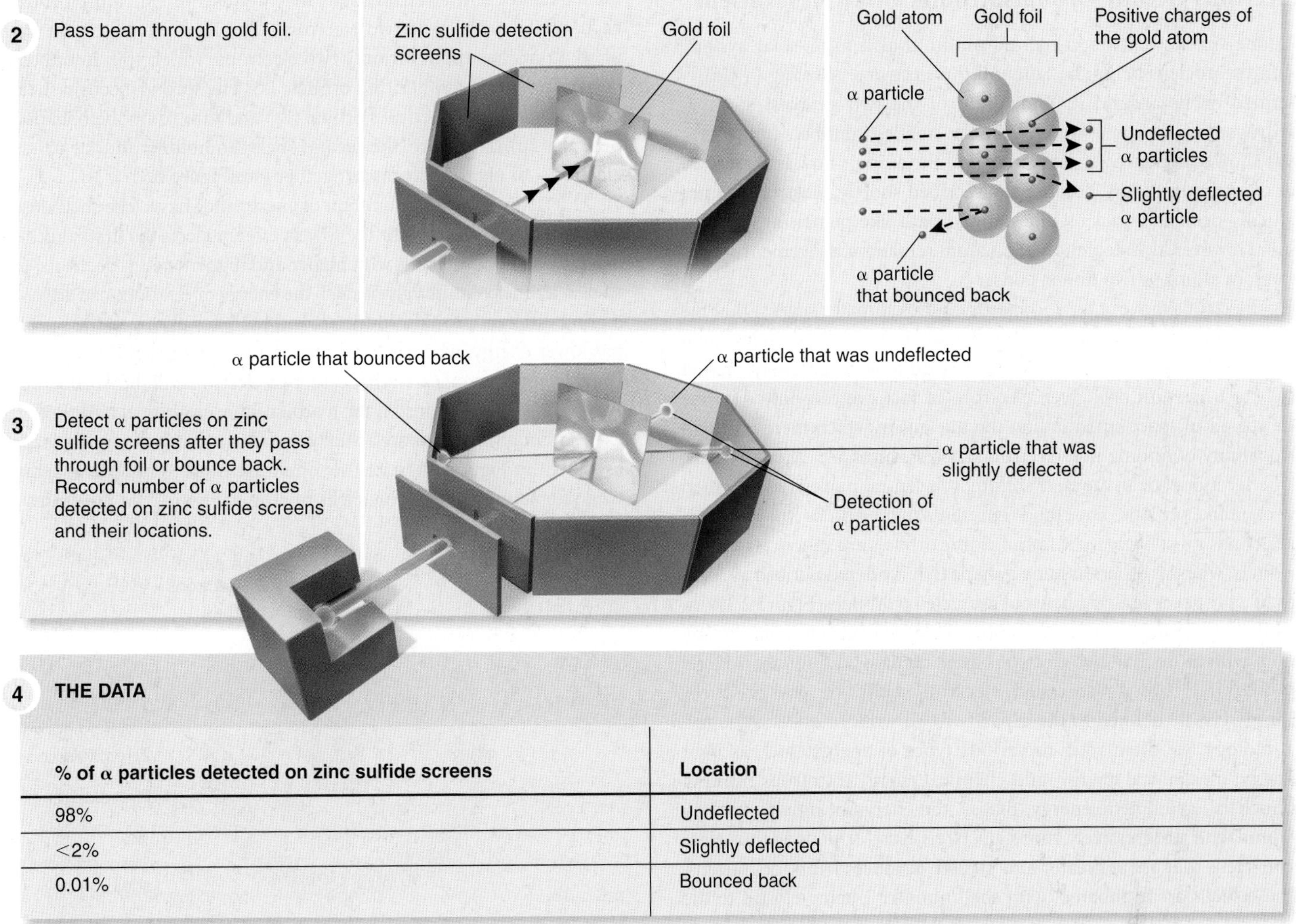

% of α particles detected on zinc sulfide screens	Location
98%	Undeflected
<2%	Slightly deflected
0.01%	Bounced back

5 CONCLUSION Most of the volume of an atom is empty space, with the positive charges concentrated in a small volume.

6 SOURCE Rutherford, E. 1911. The scattering of α and β particles by matter and the structure of the atom. *Philosophical Magazine* 21:669–688.

a small percentage was slightly deflected; a few even bounced back at a sharp angle! To explain the 98% that passed right through, Rutherford concluded that most of the volume of an atom is empty space. To explain the few α particles that bounced back at a sharp angle, he postulated that most of the atom's positive charge was localized in a highly compact area at the center of the atom. The existence of this small, dense region of highly concentrated positive charge—which today we call the atomic nucleus—explains how some α particles could be so strongly deflected by the gold foil. The α particles would bounce back on the rare occasion when they directly collided with an atomic nucleus. Therefore, based on these results, Rutherford rejected his original hypothesis that atoms are composed of diffuse, evenly distributed positive charges.

From this experiment, without being able to actually visualize an atom, Rutherford proposed a transitional model of an atom, with its small, positively charged nucleus surrounded at relatively great distances by negatively charged electrons. Today we know that more than 99.99% of an atom's volume is outside the nucleus. Indeed, the nucleus accounts for only about 1/10,000 of an atom's diameter—most of an atom is empty space!

Experimental Questions

1. Before the experiment conducted by Ernest Rutherford, how did many scientists envision the structure of an atom?
2. What was the hypothesis tested by Rutherford?
3. What were the results of the experiment? How did Rutherford interpret the results?

Electrons Occupy Orbitals Around an Atom's Nucleus

At one time, scientists visualized an atom as a mini-solar system, with the nucleus being the Sun and the electrons traveling in clearly defined orbits around it. A diagram of the two simplest atoms—hydrogen and helium—which have the smallest numbers of protons, was shown in Figure 2.1. This model of the atom is now known to be an oversimplification, because as described shortly, electrons do not actually orbit the nucleus in a defined path like planets around the Sun. However, this depiction of an atom remains a convenient way to diagram atoms in two dimensions.

Electrons move at terrific speeds. Some estimates suggest that the electron in a typical hydrogen atom could circle the Earth in less than 20 seconds! Partly for this reason, it is difficult to precisely predict the exact location of a given electron. In fact, we can only describe the region of space surrounding the nucleus in which there is a high probability of finding that electron. These regions are called **orbitals**. A better model of an atom, therefore, is a central nucleus surrounded by cloudlike orbitals. The cloud represents the region in which a given electron is most likely to be found. Some orbitals are spherical, called *s* orbitals, whereas others assume a shape that is often described as similar to a propeller or dumbbell and are called *p* orbitals (Figure 2.3). An orbital can contain a maximum of two electrons. Consequently, any atom with more than two electrons must contain additional orbitals.

Orbitals occupy so-called **electron shells**, or energy levels. **Energy** can be defined as the capacity to do work or effect a change. In biology, we often refer to various types of energy, such as light energy, mechanical energy, and chemical energy. Electrons orbiting a nucleus have kinetic energy, that is, the energy of moving matter. Atoms with progressively more electrons have orbitals within electron shells that are at greater and greater distances from the nucleus. These shells are numbered, with shell number 1 being closest to the nucleus. Different electron shells may contain one or more orbitals, each orbital with up to two electrons. The innermost electron shell of all atoms has room for only two electrons, which spin in opposite directions within a spherical *s* orbital (1*s*). The second electron shell is composed of one spherical *s* orbital (2*s*) and three dumbbell-shaped *p* orbitals (2*p*). Therefore, the second shell can hold up to four pairs of electrons, or eight electrons altogether (see Figure 2.3).

Electrons vary in the amount of energy they have. The shell closest to the nucleus fills up with the lowest energy electrons first, and then each subsequent shell fills with higher and higher energy electrons, one shell at a time. Within a given shell, the energy of electrons can also vary among different orbitals. In the second shell, for example, the *s* orbital has lower energy, whereas the three *p* orbitals have slightly higher and roughly equal energies. In that case, two electrons fill the *s* orbital first. Any additional electrons fill the *p* orbitals one electron at a time.

Although electrons are actually found in orbitals of varying shapes, as shown in Figure 2.3, scientists often use more simplified diagrams when depicting the electron shells of atoms. Figure 2.4a illustrates an

Orbital name	1*s*	2*s*	2*p*
	Nucleus		
Number of electrons per electron shell	2	2 per orbital; 8 total	
Orbital shape	Spherical	First orbital: spherical	Second to fourth orbital: dumbbell-shaped

Figure 2.3 Diagrams of individual electron orbitals. Electrons are found outside the nucleus in orbitals that may resemble spherical or dumbbell-shaped clouds. The orbital cloud represents a region in which the probability is high of locating a particular electron. For this illustration, only two shells are shown; the heaviest elements contain a total of seven shells.

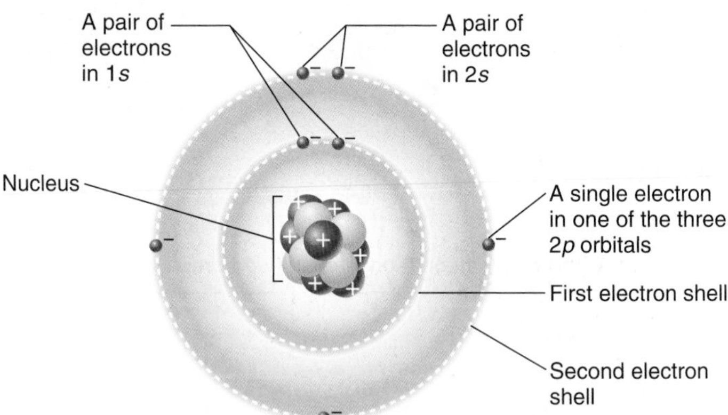

(a) Simplified depiction of a nitrogen atom (7 electrons; 2 electrons in first electron shell, 5 in second electron shell)

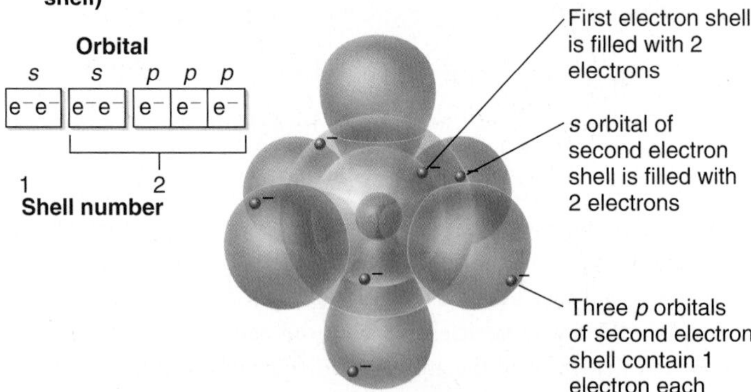

(b) Nitrogen atom showing electrons in orbitals

Figure 2.4 Diagrams showing the multiple electron shells and orbitals of a nitrogen atom. The nitrogen atom is shown **(a)** simplified and **(b)** with all of its orbitals and shells. An atom's shells fill up one by one. In shells containing more than one orbital, the orbital with lowest energy fills first. Subsequent orbitals gain one electron at a time, shown schematically in boxes, where e^- represents an electron. Heavier elements contain additional shells and orbitals.

Concept Check: *Explain the difference between an electron shell and an orbital.*

example involving the element nitrogen. An atom of this element has seven protons and seven electrons. Two electrons fill the first shell, and five electrons are found in the outer shell. Two of these fill the 2*s* orbital and are shown as a pair of electrons in the second shell. The other three electrons in the second shell are found singly in each of the three *p* orbitals. The diagram in Figure 2.4a makes it easy to see whether electrons are paired within the same orbital and whether the outer shell is full. Figure 2.4b shows a more scientifically accurate depiction of a nitrogen atom, showing how the electrons occupy orbitals with different shapes.

Most atoms have outer shells that are not completely filled with electrons. Nitrogen, as we just saw, has a first shell filled with two electrons and a second shell with five electrons (see Figure 2.4a). Because the second shell can actually hold eight electrons, the outer shell of a nitrogen atom is not full. As discussed later in this chapter, atoms that have unfilled electron shells tend to share, release, or obtain electrons to fill their outer shell. Those electrons in the outermost shell are called the **valence electrons**. As you will learn shortly, in certain cases such electrons allow atoms to form chemical bonds with each other, in which two or more atoms become joined together to create a new substance.

Each Element Has a Unique Number of Protons

Each chemical element has a specific and unique number of protons in its nucleus that distinguishes it from another element. The number of protons in an atom is its **atomic number**. For example, hydrogen, the simplest atom, has an atomic number of 1, corresponding to its single proton. Magnesium has an atomic number of 12, corresponding to its 12 protons. Recall that with the exception of ions, the number of protons and electrons in a given atom are identical. Therefore, the atomic number is also equal to the number of electrons in the atom, resulting in a net electric charge of zero.

Figure 2.5 shows the first three rows of the periodic table of the elements, which arranges the known elements according to their atomic number and electron shells (see Appendix A for the complete periodic table). A one- or two-letter symbol is used as an abbreviation for each element. The rows (known as "periods") indicate the number of electron shells. For example, hydrogen (H) has one shell, lithium (Li) has two shells, and sodium (Na) has three shells. The columns (called "groups"), from left to right, indicate the numbers of electrons in the outer shell. The outer shell of lithium (Li) has one electron, beryllium (Be) has two, boron (B) has three, and so forth. This

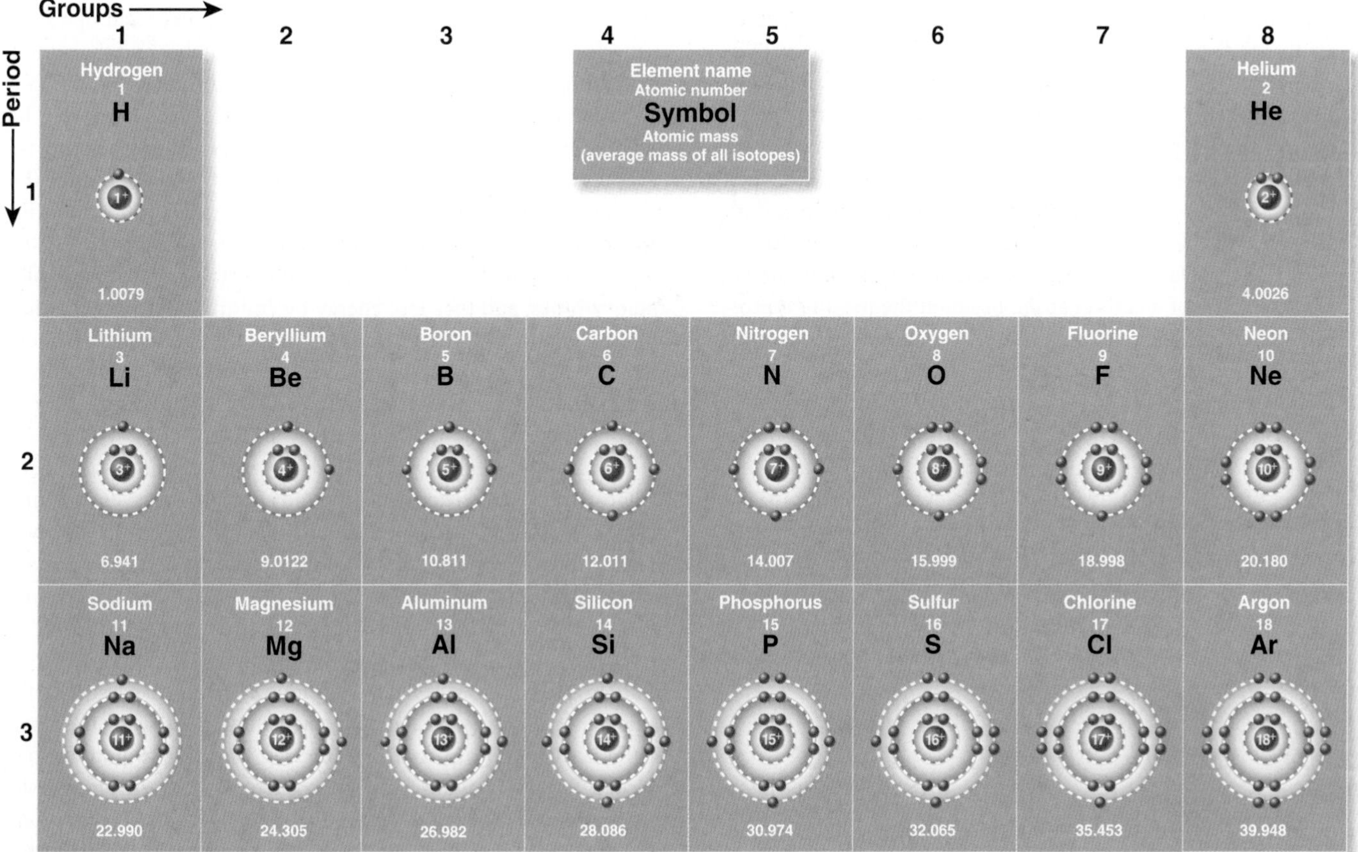

Figure 2.5 The first three rows of the periodic table of the elements. The elements are shown in models that depict the electron shells in different colors and the total number of electrons in each shell. The occupancy of orbitals is that of the elements in their pure state. The red sphere represents the nucleus of the atom, and the numerical value with the $^+$ designation represents the number of protons and, therefore, the positive charge of the nucleus. Elements are arranged in groups (columns) and periods (rows). For the complete periodic table, see Appendix A.

organization of the periodic table tends to arrange elements based on similar chemical properties. The similarities of elements within a group occur because they have the same number of valence electrons, and therefore, they have similar chemical bonding properties.

Atoms Have a Small but Measurable Mass

Atoms are extremely small and thus have very little mass. A single hydrogen atom, for example, has a mass of about 1.67×10^{-24} g (grams). Protons and neutrons are nearly equal in mass, and each are more than 1,800 times the mass of an electron (**Table 2.1**). Because of their tiny size relative to protons and neutrons, the mass of the electrons in an atom is ignored in calculations of atomic mass.

The **atomic mass** scale indicates an atom's mass relative to the mass of other atoms. By convention, the most common type of carbon atom, which has six protons and six neutrons, is assigned an atomic mass of exactly 12. On this scale, a hydrogen atom has an atomic mass of 1, indicating that it has 1/12 the mass of a carbon atom. A magnesium atom, with an atomic mass of 24, has twice the mass of a carbon atom.

The term mass is sometimes confused with weight, but these two terms refer to different features of matter. Weight is derived from the gravitational pull on a given mass. For example, a man who weighs 154 pounds on Earth would weigh only 25 pounds if he were standing on the Moon, and he would weigh 21 trillion pounds if he could stand on a neutron star. However, his mass is the same in all locations because he has the same amount of matter. Because we are discussing mass on Earth, we can assume that the gravitational tug on all matter is roughly equivalent, and thus the terms become essentially interchangeable for our purpose.

Atomic mass is measured in units called daltons, after the English chemist John Dalton, who, in postulating that matter is composed of tiny indivisible units he called atoms, laid the groundwork for atomic theory. One **dalton (Da)**, also known as an atomic mass unit (amu), equals 1/12 the mass of a carbon atom, or about the mass of a proton or a hydrogen atom. Therefore, the most common type of carbon atom has an atomic mass of 12 Da.

Because atoms such as hydrogen have a small mass, but atoms such as carbon have a larger mass, 1 g of hydrogen contains more atoms than 1 g of carbon. A mole (mol) of any substance contains the same number of particles as there are atoms in exactly 12 g of carbon. Twelve grams of carbon equals 1 mol of carbon, and 1 g of hydrogen equals 1 mol of hydrogen. As first described by Italian physicist Amedeo Avogadro, 1 mol of any element contains the same number of atoms—6.022×10^{23}. For example, 12 g of carbon contain 6.022×10^{23} atoms, and 1 g of hydrogen, whose atoms have 1/12 the mass of a carbon atom, also contains 6.022×10^{23} atoms. This number, which is known today as Avogadro's number, is large enough to be somewhat mind-boggling, and thus gives us an idea of just how small atoms really are. To visualize the enormity of this number, imagine that people could move through a turnstile at a rate of 1 million people per second. Even at that incredible rate, it would require almost 20 billion years for 6.022×10^{23} people to move through that turnstile!

Table 2.1 Characteristics of Major Subatomic Particles

Particle	Location	Charge	Mass relative to electron
Proton	Nucleus	+1	1,836
Neutron	Nucleus	0	1,839
Electron	Around the nucleus	−1	1

Isotopes Vary in Their Number of Neutrons

Although the number of neutrons in most biologically relevant atoms is often equal to the number of protons, many elements can exist in multiple forms, called **isotopes**, that differ in the number of neutrons they contain. For example, the most abundant form of the carbon atom, ^{12}C, contains six protons and six neutrons, and thus has an atomic number of 6 and an atomic mass of 12 Da, as described earlier. The superscript placed to the left of ^{12}C is the sum of the protons and neutrons. The rare carbon isotope ^{14}C, however, contains six protons and eight neutrons. Although ^{14}C has an atomic number of 6, it has an atomic mass of 14 Da. Nearly 99% of the carbon in living organisms is ^{12}C. Consequently, the *average* atomic mass of carbon is very close to, but actually slightly greater than, 12 Da because of the existence of a small amount of heavier isotopes. This explains why the atomic masses given in the periodic table do not add up exactly to the predicted masses based on the atomic number and the number of neutrons of a given atom (for example, see carbon in Figure 2.5).

Isotopes of an atom often have similar chemical properties but may have very different physical properties. For example, many isotopes found in nature are inherently unstable; the length of time they persist is measured in half-lives—the time it takes for 50% of the isotope to decay. Some persist for very long times; for example, ^{14}C has a half-life of more than 5,000 years. Such unstable isotopes are called **radioisotopes**, and they lose energy by emitting subatomic particles and/or radiation. At the very low amounts found in nature, radioisotopes usually pose no serious threat to life, but exposure of living organisms to high amounts of radioactivity can result in the disruption of cellular function, cancer, and even death.

Modern medical treatment and diagnosis make use of the special properties of radioactive compounds in many ways. For example, beams of high-energy radiation can be directed onto cancerous parts of the body to kill cancer cells. In another example, one or more atoms in a metabolically important molecule, such as the sugar glucose, can be chemically replaced with a radioactive isotope of fluorine (^{18}F) to create a molecule called fluorodeoxyglucose (FDG). ^{18}F has a half-life of about 110 minutes. When a solution containing such a modified radioactive glucose is injected into a person's bloodstream, the organs of the body take it up from the blood just as they would ordinary glucose. Special imaging techniques, such as the positron-emission tomography (PET) scan shown in **Figure 2.6**, can detect the amount of the radioactive FDG in the body's organs. In this way, it is possible to visualize whether organs such as the heart or brain are functioning normally, or at an increased or decreased rate. A PET scan of the heart that showed reduced uptake of glucose from the blood might indicate the blood vessels of the heart were damaged and are

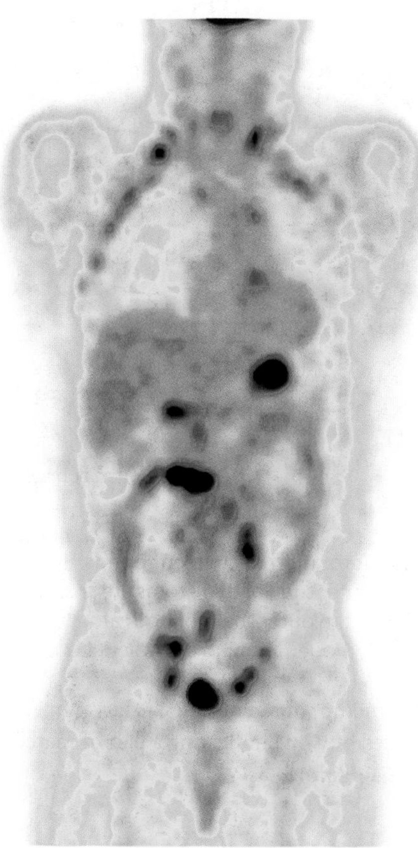

Figure 2.6 Diagnostic image of the human body using radioisotopes. A procedure called positron-emission tomography (PET) scanning highlights regions of the body that are actively using glucose, the body's major energy source. Radioactivity in this image shows up as a color. The dark patches are regions of extremely intense activity, which were later determined to be cancer in this patient.

BIOLOGY PRINCIPLE Biology affects our society. Applying an understanding of chemistry to biology has transformed the ability of physicians to diagnose disease in humans. In the United States alone, between 1 and 2 million PET scans such as this one are performed each year, helping to localize the sites and extent of diseased structures, greatly facilitating subsequent drug or surgical treatments.

Table 2.2 Chemical Elements Essential for Life in Many Organisms*

Element	Symbol	% Human body mass	% All atoms in human body
Most abundant in living organisms (approximately 95% of total mass)			
Oxygen	O	65	25.5
Carbon	C	18	9.5
Hydrogen	H	9	63.0
Nitrogen	N	3	1.4
Mineral elements (less than 1% of total mass)			
Calcium	Ca		
Chlorine	Cl		
Magnesium	Mg		
Phosphorus	P		
Potassium	K		
Sodium	Na		
Sulfur	S		
Trace elements (less than 0.01% of total mass)			
Boron	B		
Chromium	Ch		
Cobalt	Co		
Copper	Cu		
Fluorine	F		
Iodine	I		
Iron	Fe		
Manganese	Mn		
Molybdenum	Mo		
Selenium	Se		
Silicon	Si		
Tin	Sn		
Vanadium	V		
Zinc	Zn		

*Although these are the most common elements in living organisms, many other trace and mineral elements have reported functions. For example, aluminum is believed to be a cofactor for certain chemical reactions in animals, but it is generally toxic to plants.

therefore depriving the heart of nutrients. PET scans can also reveal the presence of cancer—a disease characterized by uncontrolled cell growth. The scan of the individual shown in Figure 2.6, for example, identified numerous regions of high activity, suggestive of cancer.

The Mass of All Living Organisms Is Largely Composed of Four Elements

Just four elements—oxygen, carbon, hydrogen, and nitrogen—account for the vast majority of atoms in living organisms (**Table 2.2**). These elements typically make up about 95% of the mass of living organisms. Much of the oxygen and hydrogen occur in the form of water, which accounts for approximately 60% of the mass of most animals and up to 95% or more in some plants. Carbon is a major building block of all living matter, and nitrogen is a vital element in all proteins. Note in Table 2.2 that although hydrogen accounts for about 63% of all the atoms in the body, it makes up only a small percentage of the mass of the human body. That is because the atomic mass of hydrogen is so much smaller than that of heavier elements such as oxygen.

Other important elements in living organisms include the mineral elements. Calcium and phosphorus, for example, are important constituents of the skeletons and shells of animals. Minerals such as potassium and sodium are key regulators of water movement and electric currents that occur across the surfaces of many cells.

In addition, all living organisms require **trace elements**. These elements are present in extremely small quantities but still are essential for normal growth and function. For example, iron plays an important role in how vertebrates transport oxygen in their blood, and copper serves a similar role in some invertebrates.

2.2 Chemical Bonds and Molecules

Learning Outcomes:

1. Compare and contrast the types of atomic interactions that lead to the formation of molecules.
2. Explain the concept of electronegativity and how it contributes to the formation of polar and nonpolar covalent bonds.
3. Describe how a molecule's shape is important for its ability to interact with other molecules.
4. Relate the concepts of a chemical reaction and chemical equilibrium.

The linkage of atoms with other atoms serves as the basis for life and also gives life its great diversity. Two or more atoms bonded together make up a molecule. Atoms can combine with each other in several ways. For example, two oxygen atoms can combine to form one oxygen molecule, represented as O_2. This representation is called a **molecular formula**. It consists of the chemical symbols for all of the atoms that are present (here, O for oxygen) and a subscript that tells you how many of those atoms are present in the molecule (in this case, two). The term **compound** refers to a molecule composed of two or more different elements. Examples include water (H_2O), with two hydrogen atoms and one oxygen atom, and the sugar glucose ($C_6H_{12}O_6$), which has 6 carbon atoms, 12 hydrogen atoms, and 6 oxygen atoms.

One of the most important features of compounds is their emergent physical properties. This means that the properties of a compound differ greatly from those of its elements. Let's consider sodium as an example. Pure sodium (Na) is a soft, silvery white metal that can be cut with a knife. When sodium forms a compound with chlorine (Cl), table salt (NaCl) results. NaCl is a white, relatively hard crystal (as seen in the chapter opening photo) that dissolves in water. Thus, the properties of sodium in a compound can be dramatically different from its properties as a pure element.

The atoms in molecules are held together by chemical bonds. In this section, we will examine the different types of chemical bonds, how these bonds form, and how they determine the structures of molecules.

Covalent Bonds Join Atoms Through the Sharing of Electrons

Covalent bonds, in which atoms share a pair of electrons, can occur between atoms whose outer shells are not full. A fundamental principle of chemistry is that *atoms tend to be most stable when their outer shells are filled with electrons.* Figure 2.7 shows this principle as it applies to the formation of hydrogen fluoride (HF), a molecule with many important industrial and medical applications such as petroleum refining, fluorocarbon formation, and pharmaceutical production. The outer shell of a hydrogen atom is full when it contains two electrons, though a hydrogen atom has only one electron. The outer shell of a fluorine atom is full when it contains eight electrons, though a fluorine atom has only seven electrons in its outer shell. When HF is made, the two atoms share a pair of electrons, which spend time orbiting both nuclei. This allows both of the outer shells of those atoms to be full. Covalent bonds are strong chemical bonds, because the shared electrons behave as if they belong to each atom.

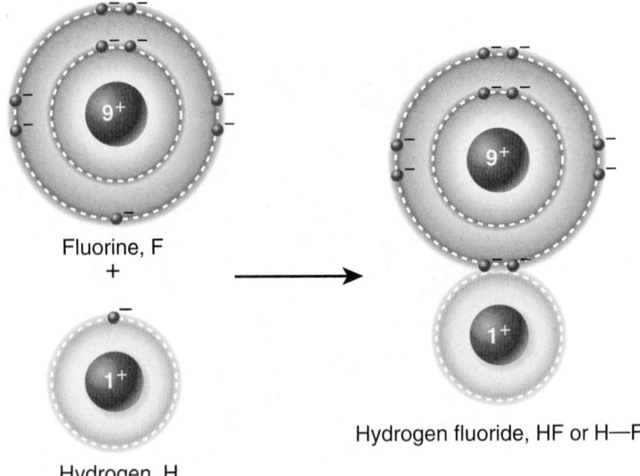

Figure 2.7 The formation of covalent bonds. In covalent bonds, electrons from the outer shell of two atoms are shared with each other in order to complete the outer shells of both atoms. This simplified illustration shows hydrogen forming a covalent bond with fluorine.

When the structure of a molecule is diagrammed, each covalent bond is represented by a line indicating a pair of shared electrons. For example, HF is diagrammed as

H—F

A molecule of water (H_2O) can be diagrammed as

H—O—H

The structural formula of water indicates that the oxygen atom is covalently bound to two hydrogen atoms.

Each atom forms a characteristic number of covalent bonds, which depends on the number of electrons required to fill the outer shell. The atoms of some elements important for life, notably carbon, form more than one covalent bond and become linked simultaneously to two or more other atoms. Figure 2.8 shows the number of covalent bonds formed by several atoms commonly found in the molecules of living cells—hydrogen, oxygen, nitrogen, and carbon.

For many types of atoms, their outermost shell is full when they contain eight electrons, an octet. The **octet rule** states that many atoms are most stable when they have eight electrons in their outermost electron shell. This rule applies to most atoms found in living organisms, including oxygen, nitrogen, carbon, phosphorus, and sulfur. These atoms form a characteristic number of covalent bonds to make an octet in their outermost shell (see Figure 2.8). However, the octet rule does not always apply. For example, hydrogen has an outermost shell that can contain only two electrons, not eight.

In some molecules, a **double bond** occurs when atoms share two pairs of electrons (four electrons) rather than one pair. As shown in Figure 2.9, this is the case for an oxygen molecule (O_2), which can be diagrammed as

O=O

Another common example occurs when two carbon atoms form bonds in compounds. They may share one pair of electrons (single bond) or two pairs (double bond), depending on how many other covalent bonds each carbon forms with other atoms. In rare cases,

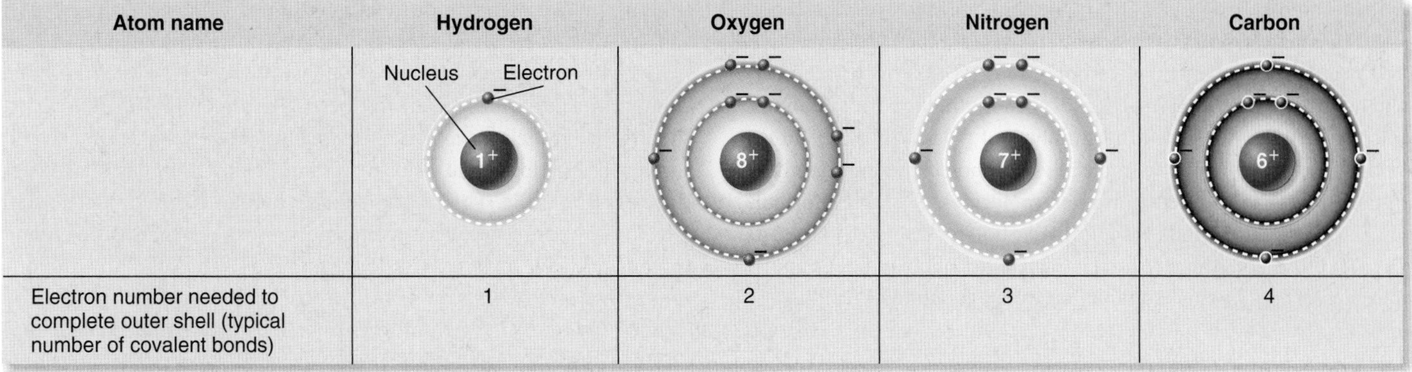

Atom name
Hydrogen
Oxygen
Nitrogen
Carbon
Nucleus
Electron
1+
8+
7+
6+
Electron number needed to complete outer shell (typical number of covalent bonds)
1
2
3
4

Figure 2.8 The number of covalent bonds formed by common essential elements found in living organisms. These elements form different numbers of covalent bonds due to the electron configurations in their outer shells.

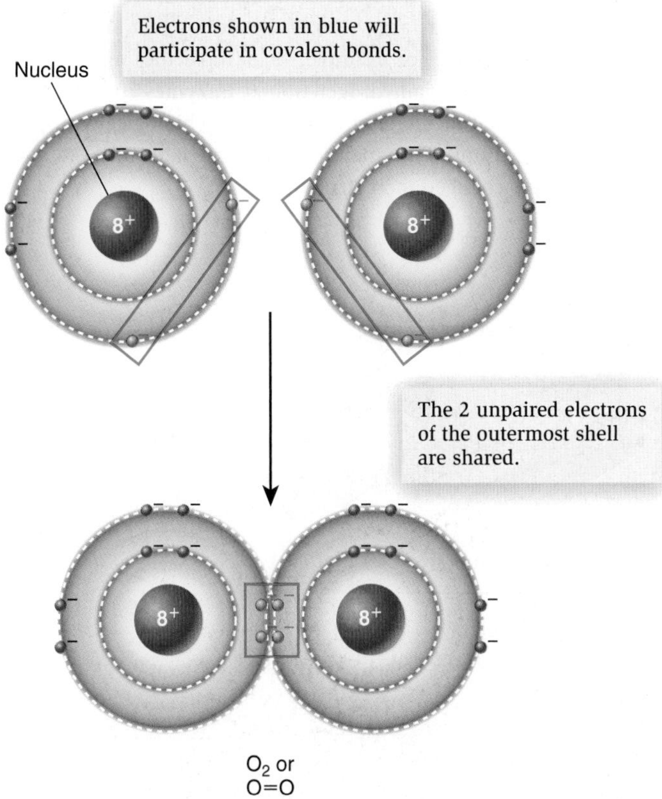

Figure 2.9 A double bond between two oxygen atoms.

Concept Check: Explain how an oxygen molecule obeys the octet rule.

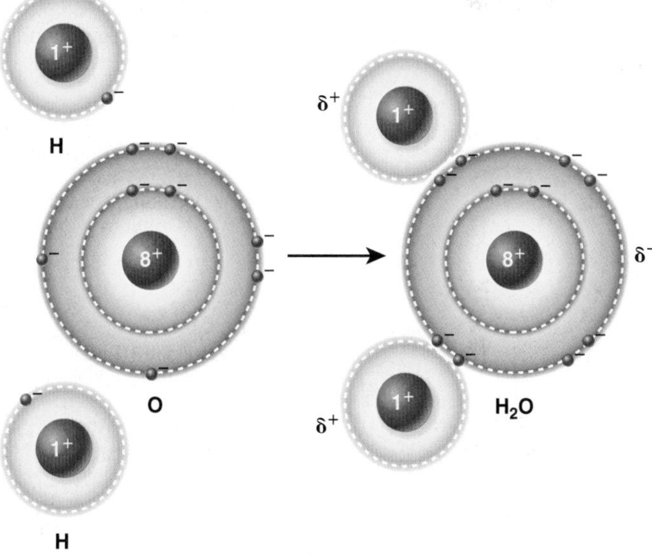

Figure 2.10 Polar covalent bonds in water molecules. In a water molecule, two hydrogen atoms share electrons with an oxygen atom. Because oxygen has a higher electronegativity, the shared electrons spend more time closer to oxygen. This gives oxygen a partial negative charge, designated δ^-, and each hydrogen a partial positive charge, designated δ^+.

carbon can even form triple bonds, in which three pairs of electrons are shared between two atoms.

Some atoms attract shared electrons more readily than do other atoms. The **electronegativity** of an atom is a measure of its ability to attract electrons in a bond with another atom. When two atoms with different electronegativities form a covalent bond, the shared electrons are more likely to be closer to the nucleus of the atom of higher electronegativity than to the atom of lower electronegativity. Such bonds are called **polar covalent bonds**, because the distribution of electrons around the nuclei creates a polarity, or difference in electric charge, across the molecule. Water is a classic example of a molecule containing polar covalent bonds. Since oxygen is much more electronegative than hydrogen, the shared electrons tend to be pulled closer to the oxygen nucleus than to either of the hydrogens. This unequal sharing of electrons gives the molecule a region of partial negative charge (indicated by the Greek letter δ and a minus sign, δ^-) and two regions of partial positive charge (δ^+) (**Figure 2.10**).

Atoms with high electronegativity, such as oxygen and nitrogen, have a relatively strong attraction for electrons. These atoms form polar covalent bonds with hydrogen atoms, which have low electronegativity. Examples of polar covalent bonds include O—H and N—H. In contrast, bonds between atoms with similar electronegativities, for example between two carbon atoms (C—C) or between carbon and hydrogen atoms (C—H), are called **nonpolar covalent bonds**. Molecules containing significant numbers of polar bonds are known as **polar molecules**, whereas molecules composed predominantly of nonpolar bonds are called **nonpolar molecules**. A single molecule may have different regions with nonpolar bonds and polar bonds. As we will explore later, the physical characteristics of polar and nonpolar molecules, especially their solubility in water, are quite different.

Hydrogen Bonds Allow Interactions Between and Within Molecules

An important result of certain polar covalent bonds is the ability of one molecule to loosely associate with another molecule through a weak interaction called a **hydrogen bond**. A hydrogen bond forms when a hydrogen atom from one polar molecule becomes electrically attracted to an electronegative atom, such as an oxygen or nitrogen atom, in another polar molecule. Hydrogen bonds, like those between water molecules, are represented in diagrams by dashed or dotted lines to distinguish them from covalent bonds (Figure 2.11a). A single hydrogen bond is very weak. The strength of a hydrogen bond is only a small percentage of the strength of the polar covalent bonds linking the hydrogen and oxygen within a water molecule.

Hydrogen bonds can also occur within a single large molecule. Many large molecules may have dozens, hundreds, or more hydrogen bonds within their structure. Collectively, many hydrogen bonds add up to a strong force that helps maintain the three-dimensional structure of a molecule. This is particularly true in deoxyribonucleic acid (DNA)—the molecule that makes up the genetic material of living organisms. DNA exists as two long, twisting strands of many millions of atoms. The two strands are held together along their length by hydrogen bonds between different portions of the molecule (Figure 2.11b). Due to the large number of hydrogen bonds, it takes considerable energy to separate the strands of DNA.

In contrast to the cumulative strength of many hydrogen bonds, the weakness of individual bonds is also important. When an interaction between two molecules involves relatively few hydrogen bonds, such interactions tend to be weak and readily broken. The reversible nature of hydrogen bonds allows molecules to interact and then to become separated again. For example, small molecules may bind to proteins called enzymes via hydrogen bonds. **Enzymes** are molecules found in all cells that facilitate or catalyze many biologically important chemical reactions. The small molecules are later released after the enzymes have changed their structure.

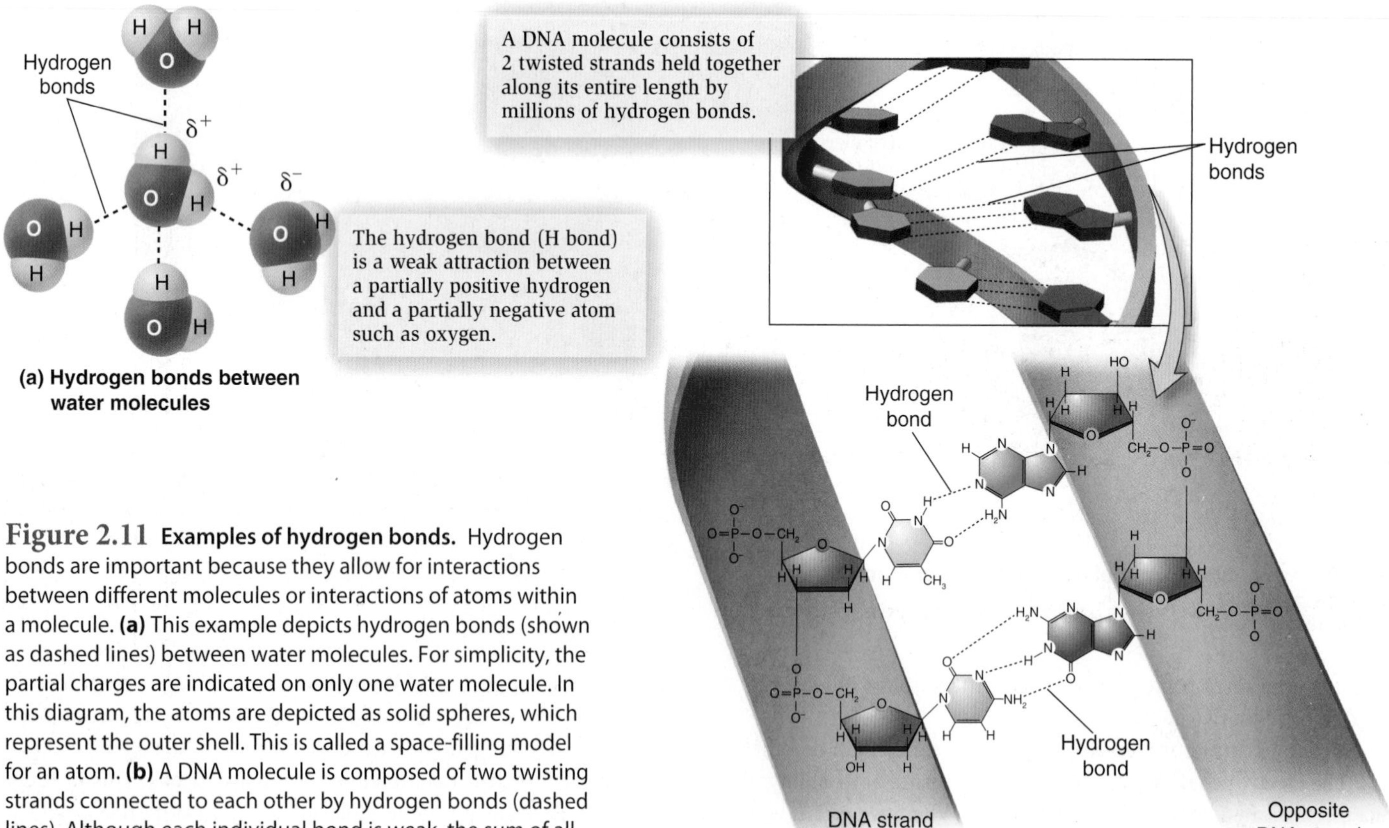
Hydrogen bonds
δ+
δ+
δ−
The hydrogen bond (H bond) is a weak attraction between a partially positive hydrogen and a partially negative atom such as oxygen.
(a) Hydrogen bonds between water molecules
A DNA molecule consists of 2 twisted strands held together along its entire length by millions of hydrogen bonds.
Hydrogen bonds
Hydrogen bond
Hydrogen bond
DNA strand
Opposite DNA strand
(b) Hydrogen bonds within a DNA molecule

Figure 2.11 **Examples of hydrogen bonds.** Hydrogen bonds are important because they allow for interactions between different molecules or interactions of atoms within a molecule. **(a)** This example depicts hydrogen bonds (shown as dashed lines) between water molecules. For simplicity, the partial charges are indicated on only one water molecule. In this diagram, the atoms are depicted as solid spheres, which represent the outer shell. This is called a space-filling model for an atom. **(b)** A DNA molecule is composed of two twisting strands connected to each other by hydrogen bonds (dashed lines). Although each individual bond is weak, the sum of all the hydrogen bonds in a large molecule like DNA imparts considerable stability to the molecule.

Concept Check: *In Chapter 11, you will learn that the two DNA strands must first separate into two single strands for DNA to be replicated. Do you think the process of strand separation requires energy, or do you think the strands can separate spontaneously?*

Hydrogen bonds are similar to a special class of bonds that are collectively known as **van der Waals forces**. In some cases, temporary attractive forces that are even weaker than hydrogen bonds form between molecules. These van der Waals forces arise because electrons orbit atomic nuclei in a random, probabilistic way, as described previously. At any moment, the electrons in the outer shells of the atoms in a nonpolar molecule may be evenly distributed or unevenly distributed. In the latter case, a fleeting electrical attraction may arise with other nearby molecules. Like hydrogen bonds, the collective strength of these temporary attractive forces between molecules can be quite strong.

Ionic Bonds Involve an Attraction Between Positive and Negative Ions

Atoms are electrically neutral because they contain equal numbers of negative electrons and positive protons. If an atom or molecule gains or loses one or more electrons, it acquires a net electric charge and becomes an **ion** (Figure 2.12a). For example, when a sodium atom (Na), which has 11 electrons, loses one electron, it becomes a sodium ion (Na^+) with a net positive charge. Ions that have a net positive charge are called **cations**. A sodium ion still has 11 protons, but only 10 electrons. Ions such as Na^+ are depicted with a superscript that indicates the net charge of the ion. A chlorine atom (Cl), which has 17 electrons, can gain an electron and become a chloride ion (Cl^-) with a net negative charge—it still has 17 protons but now has 18 electrons. Ions with a net negative charge are called **anions**.

Table 2.3 lists the ionic forms of several elements. Hydrogen atoms and most mineral and trace elements readily form ions. The ions listed in this table are relatively stable because the outer electron shells of the ions are full. For example, a sodium atom has one electron in its third (outermost) shell. If it loses this electron to become Na^+, it no longer has a third shell, and the second shell, which is full, becomes its outermost shell (see Figure 2.12).

Table 2.3 Ionic Forms of Some Common Elements in Living Organisms

Atom	Chemical symbol	Ion	Ion symbol	Electrons gained or lost
Calcium	Ca	Calcium ion	Ca^{2+}	2 lost
Chlorine	Cl	Chloride ion	Cl^-	1 gained
Hydrogen	H	Hydrogen ion	H^+	1 lost
Magnesium	Mg	Magnesium ion	Mg^{2+}	2 lost
Potassium	K	Potassium ion	K^+	1 lost
Sodium	Na	Sodium ion	Na^+	1 lost

Alternatively, a Cl atom has seven electrons in its outermost shell. If it gains an electron to become a chloride ion (Cl^-), its outer shell becomes full with eight electrons. Some atoms can gain or lose more than one electron. For instance, a calcium atom, which has 20 electrons, loses 2 electrons to become a calcium ion, depicted as Ca^{2+}.

An **ionic bond** occurs when a cation binds to an anion. Figure 2.12a shows an ionic bond between Na^+ and Cl^- to form NaCl. Salt is the general name given to compounds formed from an attraction between a positively charged ion (a cation) and negatively charged ion (an anion). Examples of salts include NaCl, KCl, and $CaCl_2$. Salts may form crystals in which the cations and anions form a regular array. Figure 2.12b shows a NaCl crystal, in which the sodium and chloride ions are held together by ionic bonds. Ionic bonds are easily broken in water—the environment of the cell—releasing the individual ions.

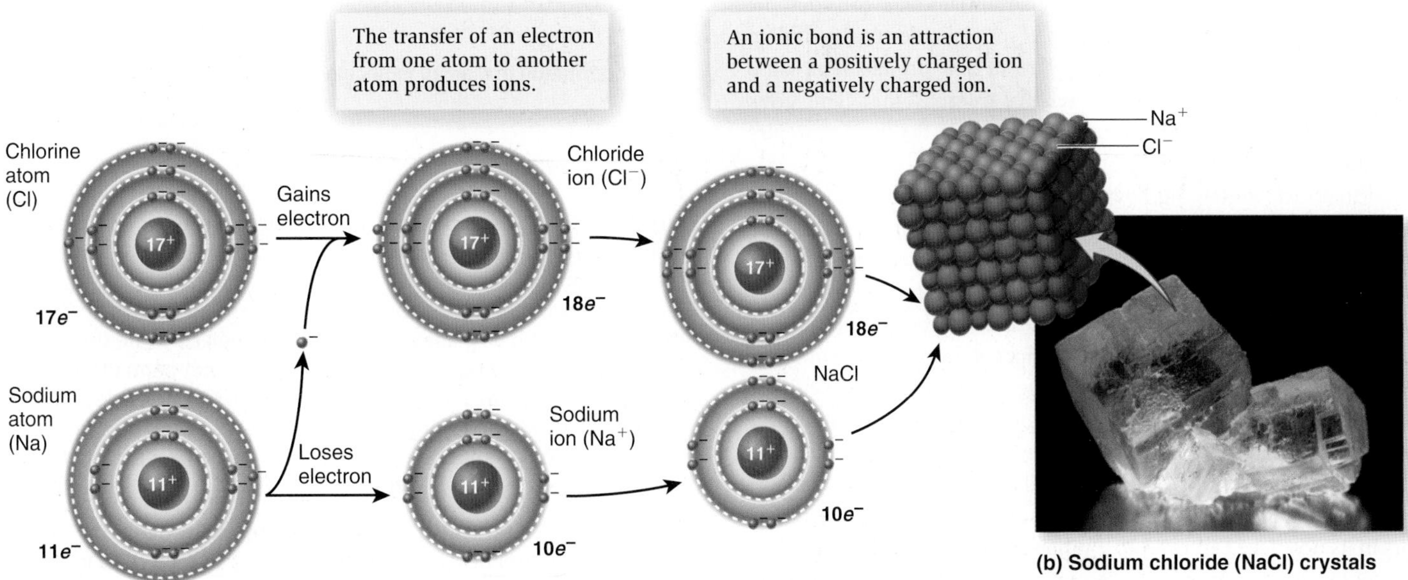

Figure 2.12 Ionic bonding in table salt (NaCl). (a) When an electron is transferred from a sodium atom to a chlorine atom, the resulting ions are attracted to each other via an ionic bond. **(b)** In a salt crystal, a lattice is formed in which the positively charged sodium ions (Na^+) are attracted to negatively charged chloride ions (Cl^-).

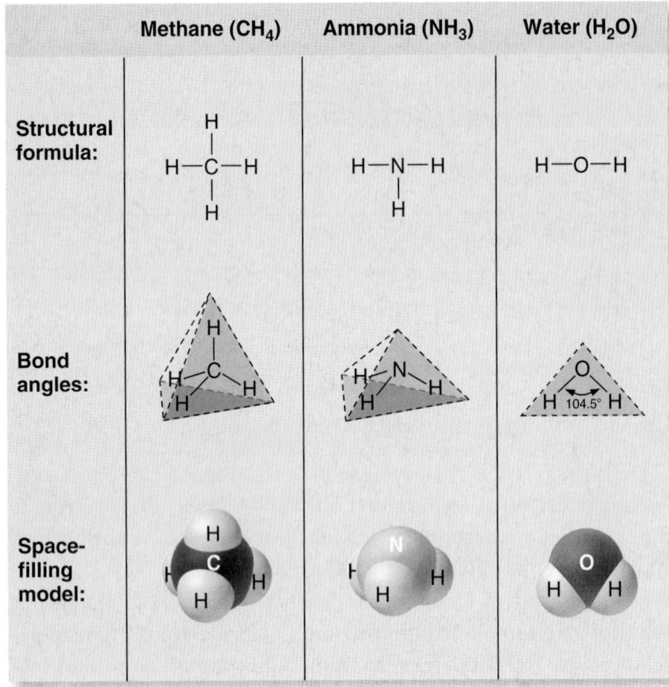

Figure 2.13 Shapes of molecules. Molecules may assume different shapes, depending on the types of bonds between their atoms. The angles between groups of atoms are well defined. For example, in liquid water at room temperature, the angle formed by the covalent bonds between the two hydrogen atoms and the oxygen atom is approximately 104.5°. This bond angle can vary slightly, depending on the temperature and degree of hydrogen bonding between adjacent water molecules.

Molecules May Change Their Shapes

When atoms combine, they can form molecules with various three-dimensional shapes, depending on the arrangements and numbers of bonds between their atoms. As an example, let's consider the arrangements of covalent bonds in a few simple molecules, including water (Figure 2.13). These molecules form new orbitals that cause the atoms to have defined angles relative to each other. This gives groups of atoms very specific shapes, as shown in the three examples of Figure 2.13.

Molecules containing covalent bonds are not rigid, inflexible structures. Think of a single covalent bond, for example, as an axle around which the joined atoms can rotate. Within certain limits, the shape of a molecule can change without breaking its covalent bonds. As illustrated in Figure 2.14a, a molecule of six carbon atoms bonded together can assume a number of shapes as a result of rotations around various covalent bonds. The three-dimensional, flexible shape of molecules contributes to their biological properties. As shown in Figure 2.14b, the binding of one molecule to another may affect the shape of one of the molecules. An animal can smell food, for instance, because odor molecules interact with special proteins called receptors in its nose (see Figure 43.25). When an odor molecule encounters a receptor, the two molecules recognize each other by their unique shapes, somewhat like a key fitting into a lock. As molecules in the food interact with the receptor, the shape of the receptor changes. When we look at how an animal's brain receives information from

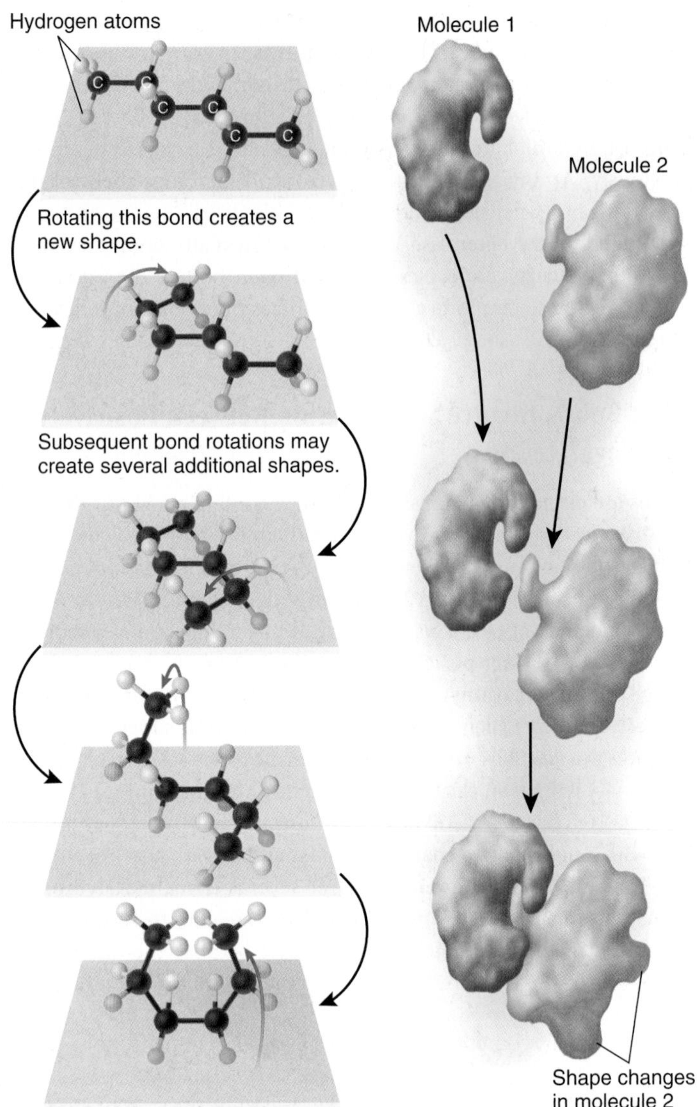

Figure 2.14 Shape changes in molecules. A single molecule may assume different three-dimensional shapes without breaking any of the covalent bonds between its atoms, as shown in **(a)** for a six-carbon molecule. Hydrogen atoms above the blue plane are shown in white; those below the blue plane are blue. **(b)** Two molecules are shown schematically as having complementary shapes that permit them to interact. Upon interacting, the flexible nature of the molecules causes molecule 2 to twist sufficiently to assume a new shape. This change in shape is often an important mechanism by which one molecule influences the activity of another.

BIOLOGY PRINCIPLE Structure determines function. The three-dimensional structure of the two molecules in (b) is critical in ensuring that they are capable of specifically interacting with each other and not with other molecules. Thus, the function of the molecules is defined by their abilities to bind to each other, which in turn depends on their unique structures.

other parts of the body, we will see that the altered shape of the receptor initiates a signal that communicates information about the smell of the food to the animal's brain (look ahead to Chapter 43).

Free Radicals Are a Special Class of Highly Reactive Molecules

Recall that an atom or an ion is most stable when each of its orbitals is occupied by its full complement of electrons. A molecule containing an atom with a single, unpaired electron in its outer shell is known as a **free radical**. Free radicals can react with other molecules to "steal" an electron from one of their atoms, thereby filling the orbital in the free radical. In the process, this may create a new free radical in the donor molecule, setting off a chain reaction.

Free radicals can be formed in several ways, including exposure of cells to radiation and toxins. Free radicals can do considerable harm to living cells—for example, by causing a cell to rupture or by damaging the genetic material. Surprisingly, the lethal effect of free radicals is sometimes put to good use. Some cells in animals' bodies create free radicals and use them to kill invading cells such as bacteria. Likewise, for many years people have used weak solutions of hydrogen peroxide to kill bacteria, as in a dirty skin wound. When applied to the wound, hydrogen peroxide can break down to create free radicals, which can then attack bacteria (this practice is no longer recommended because of the possibility of damage to skin cells).

Despite the exceptional case of fighting off bacteria, though, most free radicals that arise in an organism need to be inactivated so they do not kill healthy cells. Protection from free radicals is afforded by molecules that can donate electrons to the free radicals without becoming highly reactive themselves. Examples of such protective compounds are certain vitamins known as antioxidants (for example, vitamins C and E), which are found in fruits and vegetables, and the numerous plant compounds known as flavonoids. This is one reason why a diet rich in fruits and vegetables is beneficial to our health.

Free radicals are diagrammed with a dot next to the atomic symbol. Examples of biologically important free radicals are superoxide anion, $O_2{\cdot}^-$; hydroxyl radical, ·OH; and nitric oxide, NO·. Note that free radicals can be either charged or neutral.

Chemical Reactions Change Elements or Compounds into Different Compounds

A **chemical reaction** occurs when one or more substances are changed into other substances by the making or breaking of chemical bonds. This can happen when two or more elements or compounds combine to form a new compound, when one compound breaks down into two or more molecules, or when electrons are added to or removed from an atom.

Chemical reactions share many similar properties. First, they all require a source of energy for molecules to encounter each other. The energy required for atoms and molecules to interact is provided partly by heat, or thermal, energy. In the complete absence of any heat (a temperature called absolute zero), atoms and molecules would be totally stationary and unable to interact. Heat energy causes atoms and molecules to vibrate and move, a phenomenon known as Brownian motion. Second, chemical reactions that occur in living organisms often require more than just Brownian motion to proceed at a reasonable rate. Such reactions need to be catalyzed. As discussed in Chapter 6, a **catalyst** is a substance that speeds up the rate of a chemical reaction. Enzymes are proteins found in all cells that catalyze important chemical reactions. Third, chemical reactions tend to proceed in a particular direction but eventually reach a state of equilibrium.

To understand what we mean by "direction" and "equilibrium" in this context, let's consider a chemical reaction between methane (a component of natural gas) and oxygen. When a single molecule of methane reacts with two molecules of oxygen, one molecule of carbon dioxide and two molecules of water are produced:

$$\underset{\text{(methane)}}{CH_4} + \underset{\text{(oxygen)}}{2\,O_2} \rightleftharpoons \underset{\text{(carbon dioxide)}}{CO_2} + \underset{\text{(water)}}{2\,H_2O}$$

As it is written here, methane and oxygen are the starting materials, or **reactants**, and carbon dioxide and water are the **products**. The bidirectional arrows indicate that this reaction can proceed in both directions. Whether a chemical reaction is likely to proceed in a forward ("left to right") or reverse ("right to left") direction depends on changes in free energy, which you will learn about in Chapter 6. If we began with only methane and oxygen, the forward reaction would be very favorable. The reaction would produce a large amount of carbon dioxide and water, as well as heat. This is why natural gas is used as a fuel to heat homes. However, all chemical reactions eventually reach **chemical equilibrium**, in which the rate of the formation of products equals the rate of the formation of reactants; in other words, there is no longer a change in the concentrations of products and reactants. In the case of the reaction involving methane and oxygen, this equilibrium occurs when nearly all of the reactants have been converted to products. In biological systems, however, many reactions do not have a chance to reach chemical equilibrium. For example, the products of a reaction may immediately be converted within a cell to a different product through a second reaction, or used by a cell to carry out some function. When a product is removed from a reaction as fast as it is formed, the reactants continue to form new product until all the reactants are used up.

A final feature common to chemical reactions in living organisms is that many reactions occur in watery environments. Such chemical reactions involve reactants and products that are dissolved in water. Next, we will examine the properties of this amazing liquid and its importance to biology.

2.3 Properties of Water

Learning Outcomes:

1. Describe how hydrogen bonding determines many properties of water.
2. List the properties of water that make it a valuable solvent, and distinguish between hydrophilic and hydrophobic substances.
3. Explain how the molarity of a solution—the number of moles of a solute per liter of solution—is used to measure the concentration of solutes in solution.
4. Discuss the properties of water that are critical for the survival of living organisms.

5. Explain how water has the ability to ionize into hydroxide ions (OH^-) and into hydrogen ions (H^+), and how the H^+ concentration is expressed as a solution's pH.
6. Give examples of how buffers maintain a stable environment in an animal's body fluids.

It would be difficult to imagine life without water. People can survive for a month or more without food but usually die in less than a week without water. The bodies of all organisms are composed largely of water; most of the cells in an organism's body are filled with water and surrounded by it. Up to 95% of the weight of certain plants comes from water. In humans, typically 60–70% of body weight is from water. The brain is roughly 70% water, blood is about 80% water, and the lungs are nearly 90% water. Even our bones are about 20% water! In addition, water is an important liquid in the surrounding environments of living organisms. For example, vast numbers of species are aquatic organisms that live in watery environments.

Thus far in this chapter, we have considered the features of atoms and molecules and the nature of bonds and chemical reactions between atoms and molecules. In this section, we will turn our attention to issues related to the liquid properties of living organisms and the environment in which they live. Most of the chemical reactions that occur in nature involve molecules that are dissolved in water, including those reactions that happen inside the cells of living organisms and in the spaces that surround them (**Figure 2.15**).

However, not all molecules dissolve in water. In this section, we will examine the properties of chemicals that influence whether they dissolve in water, and we consider how biologists measure the amounts of dissolved substances. In addition, we will examine some of the other special properties of water that make it a vital component of living organisms and their environments.

Ions and Polar Molecules Readily Dissolve in Water

Substances dissolved in a liquid are known as **solutes**, and the liquid in which they are dissolved is the **solvent**. In all living organisms, the solvent for chemical reactions is water, which is the most abundant solvent in nature. Solutes dissolve in a solvent to form a **solution**. Solutions made with water are called **aqueous solutions**. To understand why a substance dissolves in water, we need to consider the chemical bonds in the solute molecule and those in water. As discussed earlier, the covalent bonds linking the two hydrogen atoms to the oxygen atom in a water molecule are polar. Therefore, the oxygen in water has a slight negative charge, and each hydrogen has a slight positive charge. To dissolve in water, a substance must be electrically attracted to water molecules. For example, table salt (NaCl) is a solid crystalline substance because of the strong ionic bonds between positive sodium ions (Na^+) and negative chloride ions (Cl^-). When a crystal of sodium chloride is placed in water, the partially negatively charged oxygens of water molecules are attracted to the Na^+, and the partially positively charged hydrogens are attracted to the Cl^- (**Figure 2.16**). Clusters of water molecules surround the ions, allowing the Na^+ and Cl^- to separate from each other and enter the water—that is, to dissolve.

Generally, molecules that contain ionic and/or polar covalent bonds dissolve in water. Such molecules are said to be **hydrophilic**, which literally means "water-loving." In contrast, molecules composed predominantly of carbon and hydrogen are relatively insoluble in water, because carbon-carbon and carbon-hydrogen bonds are nonpolar. These molecules do not have partial positive and negative charges and, therefore, are not attracted to water molecules. Such molecules are **hydrophobic**, or "water-fearing." Oils are a familiar example of hydrophobic molecules. Try mixing vegetable oil with water and observe the result. The two liquids separate into an oil layer and water layer, with very little oil dissolving in the water.

Although hydrophobic molecules dissolve poorly in water, they normally dissolve readily in nonpolar solvents. For example, cholesterol is a compound found in the blood and cells of animals. It is a hydrophobic molecule that is barely soluble in water but easily dissolves in nonpolar solvents used in chemical and biological

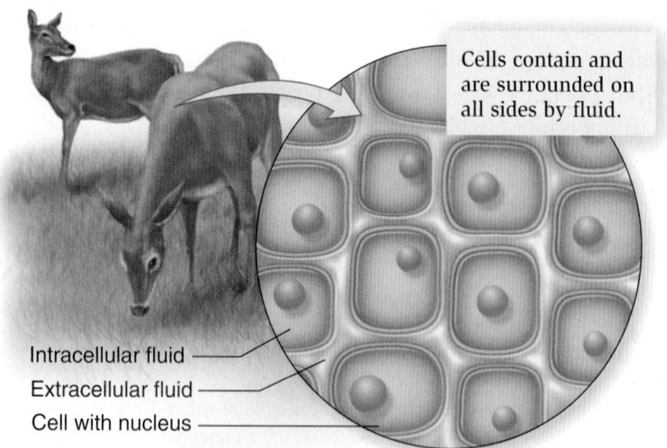

Figure 2.15 Fluids inside and outside of cells. Aqueous solutions exist in the intracellular fluid and in the extracellular fluid. Chemical reactions are always ongoing in both fluids.

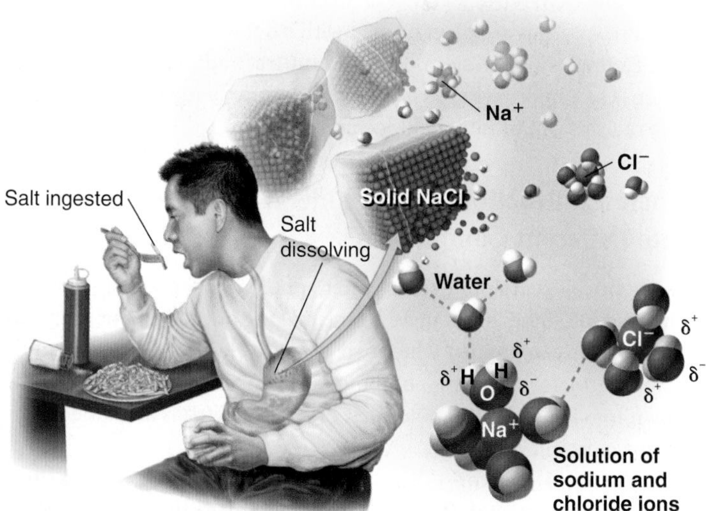

Figure 2.16 Table salt (NaCl crystals) dissolving in water. The ability of water to dissolve sodium chloride crystals depends on the electrical attraction between the polar water molecules and the charged sodium (Na^+) and chloride ions (Cl^-). Water molecules surround each ion as it becomes dissolved. For simplicity, the partial charges are indicated for only two water molecules.

laboratories, such as acetone or chloroform. Biological membranes like those that encase cells are made up in large part of nonpolar compounds. Because of this, cholesterol also inserts itself into biological membranes, where it helps to maintain the membrane structure.

Molecules that have both polar or ionized regions at one or more sites and nonpolar regions at other sites are called **amphipathic** (or amphiphilic, from the Greek for "both loves"). When mixed with water, long amphipathic molecules may aggregate into spheres called **micelles**, with their polar (hydrophilic) regions at the surface of the micelle, where they are attracted to the surrounding water molecules. The nonpolar (hydrophobic) ends are oriented toward the interior of the micelle (**Figure 2.17**). Such an arrangement minimizes the interaction between water molecules and the nonpolar ends of the amphipathic molecules, which face inward. Nonpolar molecules can dissolve in the central nonpolar regions of these clusters and thus exist in an aqueous environment in far higher amounts than would otherwise be possible based on their low solubility in water. Familiar examples of amphipathic molecules are those in detergents, which can form micelles that help to dissolve oils and nonpolar molecules found in dirt. The detergent molecules found in soap have polar and nonpolar ends. Oils on your skin dissolve in the nonpolar regions of the detergent, and the polar ends help the detergent rinse off in water, taking the oil with it.

In addition to micelles, amphipathic molecules may form structures consisting of double layers of molecules called bilayers. Such bilayers have two hydrophilic surfaces facing outside, in contact with water, and a hydrophobic interior facing away from water. As you will learn in Chapter 5, bilayers play a key role in cellular membrane structure (look ahead to Figure 5.1).

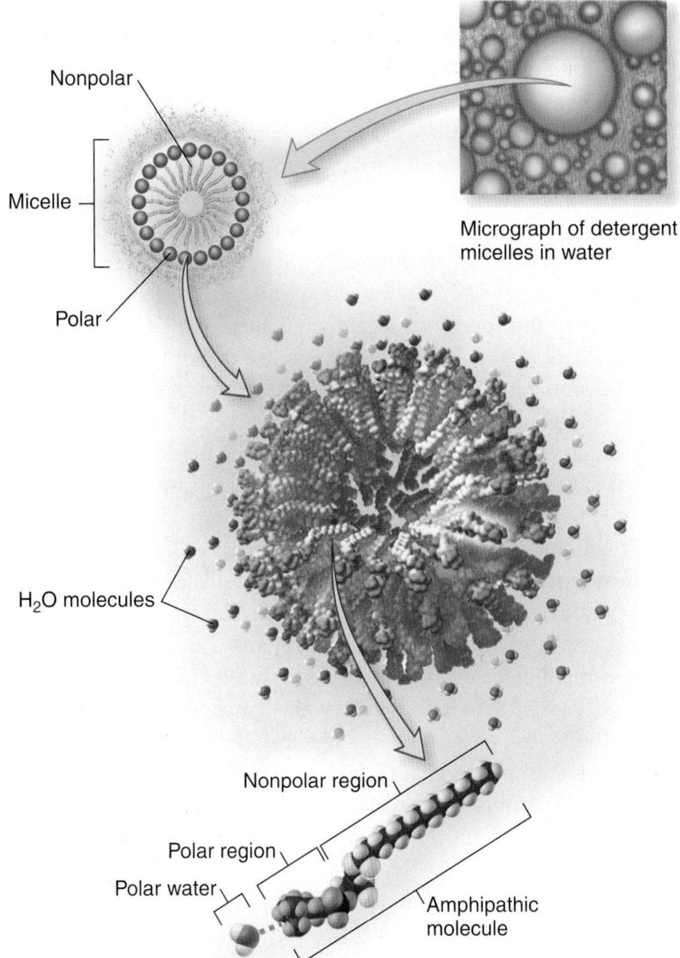

Figure 2.17 The formation of micelles by amphipathic molecules. In water, amphipathic molecules tend to arrange themselves so their nonpolar regions are directed away from water molecules and the polar regions are directed toward the water and can form hydrogen bonds with it.

Concept Check: *When oil dissolves in soap, where is the oil found?*

The Amount of a Dissolved Solute per Unit Volume of Liquid Is Its Concentration

Solute **concentration** is defined as the amount of a solute dissolved in a unit volume of solution. For example, if 1 gram (g) of NaCl were dissolved in enough water to make 1 liter (L) of solution, we would say that its solute concentration is 1 g/L.

A comparison of the concentrations of two different substances on the basis of the number of grams per liter of solution does not directly indicate how many molecules of each substance are present. For example, let's compare 10 g each of glucose ($C_6H_{12}O_6$) and sodium chloride (NaCl). Because the individual molecules of glucose have more mass than those of NaCl, 10 g of glucose contains fewer molecules than 10 g of NaCl. Therefore, another way to describe solute concentration is according to the moles of dissolved solute per volume of solution. To make this calculation, we must know three things: the amount of dissolved solute, the molecular mass of the dissolved solute, and the volume of the solution.

The **molecular mass** of a molecule is equal to the sum of the atomic masses of all the atoms in the molecule. For example, glucose ($C_6H_{12}O_6$) has a molecular mass of 180 ($[6 \times 12] + [12 \times 1] + [6 \times 16] = 180$).

As mentioned earlier, 1 mole (abbreviated mol) of a substance is the amount of the substance in grams equal to its atomic or molecular mass. The **molarity** of a solution is defined as the number of moles of a solute dissolved in 1 L of solution. A solution containing 180 g of glucose (1 mol) dissolved in enough water to make 1 L is a 1 **molar** solution of glucose (1 mol/L). By convention, a 1 mol/L solution is usually written as 1 M, where the capital M stands for molar and is defined as mol/L. If 90 g of glucose (half its molecular mass) were dissolved in enough water to make 1 L, the solution would have a solute concentration of 0.5 mol/L, or 0.5 M.

The concentrations of solutes dissolved in the fluids of living organisms are usually much less than 1 M. Many have concentrations in the range of millimoles per liter (1 mM = 0.001 M = 10^{-3} M), and others are present in even smaller concentrations—micromoles per liter (1 μM = 0.000001 M = 10^{-6} M) or nanomoles per liter (1 nM = 0.000000001 M = 10^{-9} M), or even less.

Water Exists in Three States

Let's now consider some general features of water and how dissolved solutes affect its properties. Water is an abundant compound on Earth that exists in all three states of matter—solid (ice), liquid (water), and gas (water vapor). At the temperatures found over most regions of the planet, water is found primarily as a liquid in which the weak hydrogen bonds between molecules are continuously being formed, broken, and formed again. If the temperature rises, the rate at which hydrogen bonds break increases, and molecules of water escape into the gaseous state, becoming water vapor. If the temperature falls, hydrogen bonds are broken less frequently, so larger and larger clusters of water molecules are formed, until at 0°C water freezes into a crystalline matrix—ice. The water molecules in ice tend to lie in a more orderly and "open" arrangement, that is, with greater intermolecular distances, which makes ice less dense than water. This is why ice floats on water (**Figure 2.18**). Compared with water, ice is also less likely to participate in most types of chemical reactions.

Changes in state, such as changes between the solid, liquid, and gaseous states of water, involve an input or a release of energy. For example, when energy is supplied to make water boil, it changes from the liquid to the gaseous state—a process called vaporization. The heat required to vaporize 1 mole of any substance at its boiling point is called the substance's **heat of vaporization**. For water, this value is very high, because of the high number of hydrogen bonds between the molecules. It takes more than five times as much heat to vaporize water than it does to raise the temperature of water from 0°C to 100°C. In contrast, energy is released when water freezes to form ice. A substance's **heat of fusion** is the amount of heat energy that must be withdrawn or released from a substance to cause it to change from the liquid to the solid state. For water, this value is also high.

Another important feature for living organisms is that water has a very high **specific heat**, defined as the amount of heat energy required to raise the temperature of 1 gram of a substance by 1°C (or conversely, the amount of heat energy that must be lost to lower the temperature by 1°C). A high specific heat means that it takes considerable heat to raise the temperature of water. A related concept is heat capacity, which refers to the amount of heat energy required to raise the temperature of an entire object or substance. A lake has a greater heat capacity than does a bathtub filled with water, but both have the same specific heat because both are the same substance (ignoring for the moment that neither are pure H_2O). These properties of water contribute to the relatively stable temperatures of large bodies of water compared with inland temperatures. Large bodies of water tend to have a moderating effect on the temperature of nearby land masses. These three features, the high heats of vaporization and fusion, and the high specific heat of water, mean that water is extremely stable as a liquid. Not surprisingly, therefore, living organisms have evolved to function best within a range of temperatures consistent with the liquid phase of water.

The temperature at which a solution freezes or vaporizes is influenced by the amounts of dissolved solutes. These are examples of a solution's **colligative properties**, defined as those properties that depend strictly on the total number of dissolved solutes, not on the specific type of solute. Pure water freezes at 0°C and vaporizes at 100°C. Addition of solutes to water lowers its freezing point below 0°C and raises its boiling point to above 100°C. Adding a small amount of the compound ethylene glycol—antifreeze—to the water in a car's radiator, for instance, lowers the freezing point of the water and consequently prevents it from freezing in cold weather. Similarly, the presence of large amounts of solutes partly explains why the oceans do not freeze when the temperature falls below 0°C.

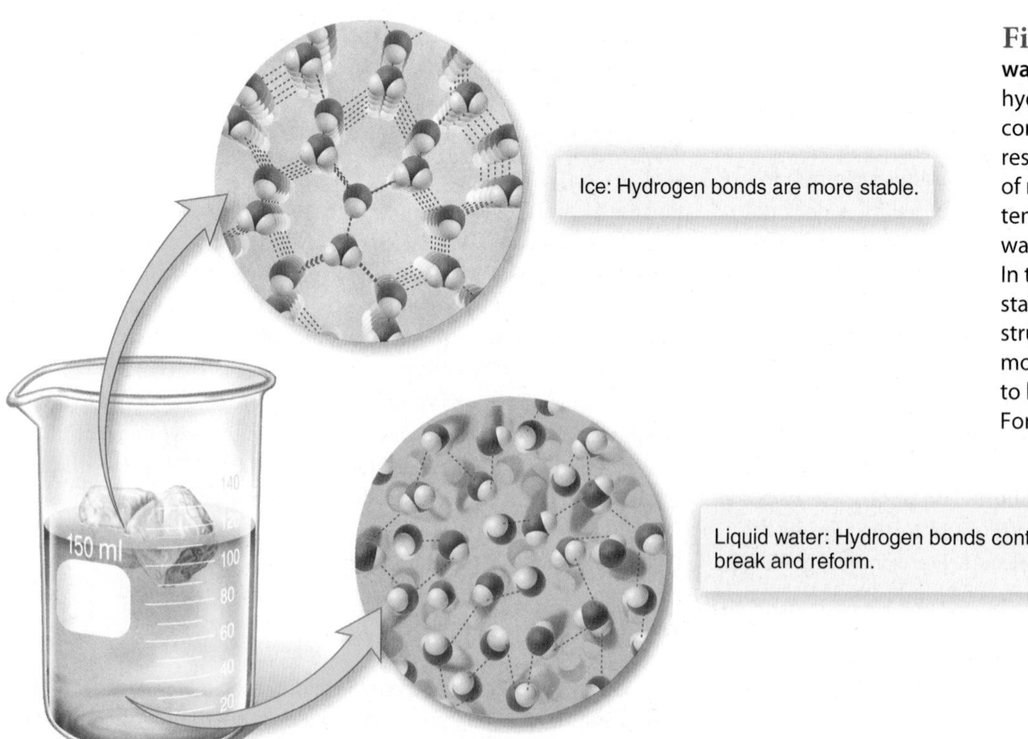
Ice: Hydrogen bonds are more stable.
Liquid water: Hydrogen bonds continually break and reform.
150 ml
140
120
100
80
60
40
20

Figure 2.18 Structure of liquid water and ice. In its liquid form, the hydrogen bonds between water molecules continually form, break, and re-form, resulting in a changing arrangement of molecules from instant to instant. At temperatures at or below its freezing point, water forms a crystalline matrix called ice. In this solid form, hydrogen bonds are more stable. Ice has a hexagonally shaped crystal structure. The greater space between H_2O molecules in this crystal structure causes ice to have a lower density than liquid water. For this reason, ice floats on water.

Water Performs Many Important Tasks in Living Organisms

As discussed previously, water is the primary solvent in the fluids of all living organisms, from unicellular bacteria to the largest sequoia tree. Water permits atoms and molecules to interact in ways that would be impossible in their nondissolved states. In Unit II, we will consider many ions and molecules that are solutes in living cells.

Even so, it is important to recognize that in addition to acting as a solvent, water serves many other remarkable functions that are critical for the survival of living organisms. For example, water molecules participate in many chemical reactions of this general type:

$$R_1—R_2 + H—O—H \rightarrow R_1—OH + H—R_2$$

R is a general symbol used in this case to represent a group of atoms. In this equation, R_1 and R_2 are distinct groups of atoms. On the left side, $R_1—R_2$ is a compound in which the groups of atoms are connected by a covalent bond. To be converted to products, a covalent bond is broken in each reactant, $R_1—R_2$ and H—O—H, and OH and H (from water) form covalent bonds with R_1 and R_2, respectively. Reactions of this type are known as **hydrolysis reactions** (from the Greek *hydro*, meaning water, and *lysis*, meaning to break apart), because water is used to break apart another molecule (Figure 2.19a). As discussed in Chapter 3 and later chapters, many large molecules are broken down into smaller, biologically important units by hydrolysis reactions.

Alternatively, other chemical reactions in living organisms involve the removal of a water molecule so that a covalent bond can be formed between two separate molecules. For example, let's consider a chemical reaction that is the reverse of our previous hydrolysis reaction:

$$R_1—OH + H—R_2 \rightarrow R_1—R_2 + H—O—H$$

Such a reaction involves the formation of a covalent bond between two molecules. Two or more molecules combining to form one larger molecule with the loss of a small molecule is called a **condensation reaction**. In the example shown here, a molecule of water is lost during the reaction; this is a specific type of condensation reaction called a **dehydration reaction**. As discussed in later chapters, this is a common reaction used to build larger molecules in living organisms.

Another feature of water is that it is incompressible—its volume does not significantly decrease when subjected to high pressure. This has biological importance for many organisms that use water to provide force or support. For example, water supports the bodies of worms and some other invertebrates, in a structure called a hydrostatic skeleton, and it provides turgidity (stiffness) and support for plants (Figure 2.19b).

Water is also the means by which unneeded and potentially toxic waste compounds are eliminated from an animal's body (Figure 2.19c). In mammals, for example, the kidneys filter out soluble waste products derived from the breakdown of proteins and other compounds. The filtered products remain in solution in a watery fluid, which eventually becomes urine and is excreted.

Recall from our discussion of water's properties that it takes considerable energy in the form of heat to convert water from a liquid to a gas. This feature has great biological significance. Although everyone

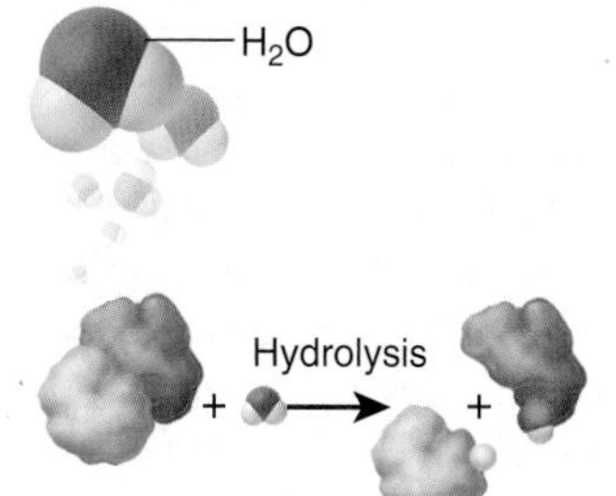

(a) Water participates in chemical reactions.

(b) Water provides support. The plant on the right is wilting due to lack of water.

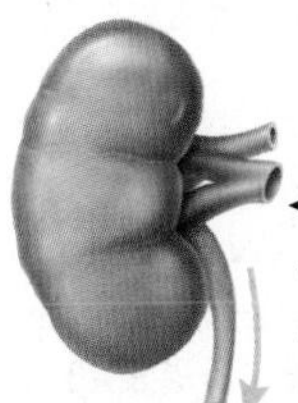

Blood enters and is purified by kidney cells.
Waste products are carried away in the watery urine.

(c) Water is used to eliminate soluble wastes.

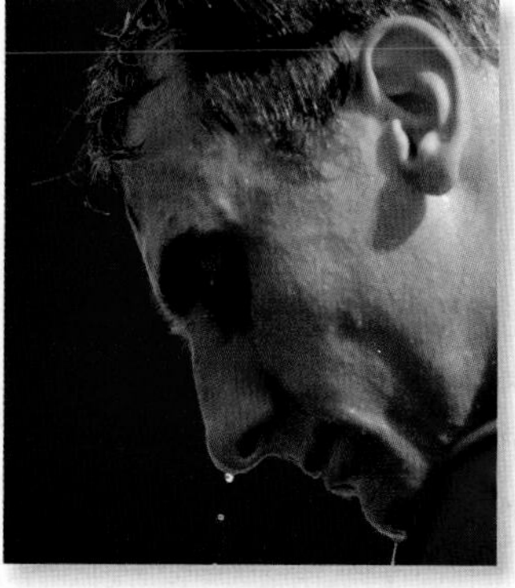

(d) Evaporation helps some animals dissipate body heat.

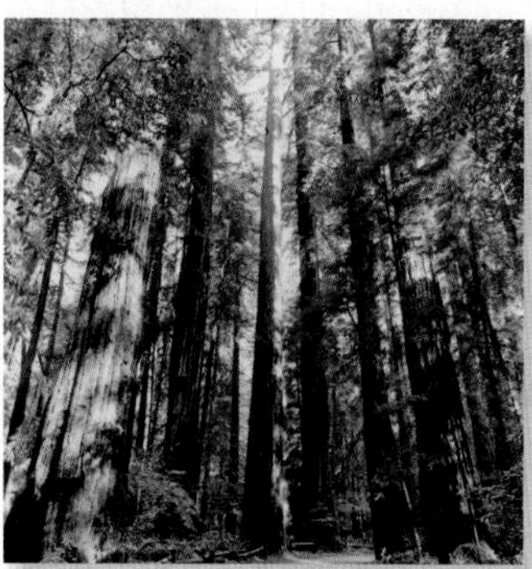

(e) The cohesive force of water molecules aids in the movement of fluid through vessels in plants.

(f) Water in saliva serves as a lubricant during—or as shown here, in anticipation of—feeding.

(g) The surface tension of water explains why this water strider doesn't sink.

Figure 2.19 **Some of the amazing functions of water in biology.** In addition to acting as a solvent, water serves many crucial functions in nature.

BIOLOGY PRINCIPLE Living organisms maintain homeostasis. Healthy organisms have a normal supply of water that maintains such homeostatic functions as body temperature, salt balance, and waste levels.

is familiar with the fact that boiling water is converted to water vapor, water can vaporize into the gaseous state even at ordinary temperatures. This process is known as **evaporation**. The simplest way to understand this is to imagine that in any volume of water at any temperature, some vibrating water molecules have higher energy than others. Those with the highest energy break their hydrogen bonds and escape into the gaseous state. The important point, however, is that even at ordinary temperatures, it still requires the same energy to change water from liquid to gas. Therefore, the evaporation of sweat from an animal's skin requires considerable energy in the form of body heat, which is then lost to the environment. Evaporation is an important mechanism by which many animals cool themselves on hot days (Figure 2.19d).

The hydrogen-bonding properties of water affect its ability to form droplets and to adhere to surfaces. The phenomenon of water molecules attracting each other is called **cohesion**. Water exhibits strong cohesion due to hydrogen bonding. Cohesion aids in the movement of water through the vessels of plants (Figure 2.19e). A property similar to cohesion is **adhesion**, which refers to the ability of water to be attracted to, and thus adhere to, a surface that is not electrically neutral. Water tends to cling to surfaces to which it can hydrogen bond, such as a paper towel. In organisms, the adhesive properties of water allow it, for example, to coat the surfaces of the digestive tract of animals and act as a lubricant for the passage of food (Figure 2.19f).

Surface tension is a measure of the attraction between molecules at the surface of a liquid. In the case of water, the attractive force between hydrogen-bonded water molecules at the interface between water and air is what causes water to form droplets. The surface water molecules attract each other into a configuration (roughly that of a sphere) that reduces the number of water molecules in contact with air. You can see this by slightly overfilling a glass with water; the water forms an oval shape above the rim. Likewise, surface tension allows certain insects, such as water striders, to walk on the surface of a pond without sinking (Figure 2.19g).

Hydrogen Ion Concentrations Are Changed by Acids and Bases

Pure water has the ability to ionize to a very small extent into **hydroxide ions (OH^-)** and into hydrogen ions that exist as single protons (H^+). (In nature or in laboratory conditions, hydrogen atoms may exist as any of several rare types of positively or negatively charged ions; in this text, we will use the term hydrogen ion to refer to the common H^+ form.) In pure water, the concentrations of H^+ and OH^- are both 10^{-7} mol/L, or 10^{-7} M. An inherent property of water is that the product of the concentrations of H^+ and OH^- is always 10^{-14} M at 25°C. Therefore, in pure water, $[H^+][OH^-] = [10^{-7}\ M][10^{-7}\ M] = 10^{-14}$ M. (The brackets around the symbols for the hydrogen and hydroxide ions indicate concentration.)

When certain substances are dissolved in water, they may release or absorb H^+ or OH^-, thereby altering the relative concentrations of these ions. Substances that release hydrogen ions in solution are called **acids**. Two examples are hydrochloric acid and carbonic acid:

$$\underset{\text{(hydrochloric acid)}}{HCl} \rightarrow H^+ + \underset{\text{(chloride ion)}}{Cl^-}$$

$$\underset{\text{(carbonic acid)}}{H_2CO_3} \rightleftharpoons H^+ + \underset{\text{(bicarbonate ion)}}{HCO_3^-}$$

Hydrochloric acid is called a **strong acid** because it almost completely dissociates into H^+ and Cl^- when added to water (which is why the arrow is not bidirectional in this reaction). By comparison, carbonic acid is a **weak acid** because some of it remains in the H_2CO_3 state when dissolved in water (note the bidirectional arrows $\rightleftharpoons$).

Compared with an acid, a **base** has the opposite effect when dissolved in water—it absorbs hydrogen ions in solution. This can occur in different ways. Some bases, such as sodium hydroxide (NaOH), release OH^- when dissolved in water:

$$\underset{\text{(sodium hydroxide)}}{NaOH} \rightarrow \underset{\text{(sodium ion)}}{Na^+} + OH^-$$

Recall that the product of $[H^+]$ and $[OH^-]$ is always 10^{-14} M. When a base such as NaOH raises the OH^- concentration, some of the hydrogen ions bind to these hydroxide ions to form water. Therefore, increasing the OH^- concentration lowers the H^+ concentration. In another example, ammonia reacts with water to produce ammonium ion:

$$\underset{\text{(ammonia)}}{NH_3} + H_2O \rightleftharpoons \underset{\text{(ammonium ion)}}{NH_4^+} + OH^-$$

Both NaOH and ammonia have the same effect—they lower the concentration of H^+. NaOH achieves this by directly increasing the OH^- concentration, whereas NH_3 reacts with water to produce OH^-.

The H^+ Concentration of a Solution Determines the Solution's pH

The addition of acids and bases to water can greatly change the $H^+$and OH^- concentrations over a very broad range. Therefore, scientists use a log scale to describe the concentrations of these ions. The H^+ concentration is expressed as the solution's **pH**, which is defined as the negative logarithm to the base 10 of the H^+ concentration (a logarithmic scale is used because the concentrations of hydrogen ions can vary over a very wide range).

$$pH = -\log_{10} [H^+]$$

To understand what this equation means, let's consider a few examples. A solution with a H^+ concentration of 10^{-7} M has a pH of 7. A concentration of 10^{-7} M is the same as 0.1 μM. A solution in which $[H^+] = 10^{-6}$ M has a pH of 6. 10^{-6} M is the same as 1.0 μM. A solution at pH 6 is said to be **acidic**, because it contains more H^+ ions than OH^- ions. Note that as the acidity increases, the pH decreases. A solution in which the pH is 7 is said to be neutral because $[H^+]$ and $[OH^-]$ are equal. A solution with a pH above 7 is considered to be **alkaline**. Figure 2.20 considers the pH values of some familiar fluids. Keep in mind that each change of one pH unit represents a 10-fold difference in H^+ concentration.

Why is pH of importance to biologists? The answer lies in the observation that H^+ and OH^- can readily bind to many kinds of ions and molecules. For this reason, the pH of a solution can affect:

- the shapes and functions of molecules
- the rates of many chemical reactions
- the ability of two molecules to bind to each other
- the ability of ions or molecules to dissolve in water

Due to the various effects of pH, many biological processes function best within very narrow ranges of pH, and even small shifts can have a negative effect. In living cells, the pH ranges from about 6.5 to 7.8 and is carefully regulated to avoid major shifts in pH. The blood of the human body has a normal range of about pH 7.35 to 7.45 and is therefore slightly alkaline. Certain diseases, such as kidney disease, can decrease or increase blood pH by a few tenths of a unit. When this happens, the enzymes in the body that are required for normal metabolism can no longer function optimally, leading to additional illness. As described next, living organisms have molecules called buffers to help prevent such changes in pH.

Buffers Minimize Fluctuations in the pH of Fluids

What factors might alter the pH of an organism's fluids? In plants, external factors such as acid rain and other forms of pollution can reduce the pH of water entering the roots. In animals, exercise generates lactic acid, and certain disease states can raise or lower the pH of blood.

Organisms have several ways to cope with changes in pH. In vertebrate animals such as mammals, for example, structures like the kidney secrete acidic or alkaline compounds into the bloodstream when the blood pH becomes imbalanced. Similarly, the kidneys can transfer hydrogen ions from the fluids of the body into the urine and adjust the pH of the body's fluids in that way. Another mechanism by which pH balance is regulated in diverse organisms involves the actions of acid-base buffers. A **buffer** is a compound that minimizes pH fluctuations in the fluids of living organisms. A buffer solution is composed of a weak acid and its related base. One such buffer solution contains carbonic acid (H_2CO_3) and bicarbonate ions (HCO_3^-), called the bicarbonate pathway, which functions to keep the pH of an animal's body fluids within a narrow range:

$$CO_2 + H_2O \rightleftharpoons H_2CO_3 \rightleftharpoons H^+ + HCO_3^-$$

(carbonic acid) (bicarbonate)

This buffer system can work in both directions. For example, if the pH of an animal's body fluids were to increase (that is, the H^+ concentration decreased), the bicarbonate pathway would proceed from left to right. Carbon dioxide would combine with water to make carbonic acid, and then the carbonic acid would dissociate into H^+ and HCO_3^-. This would increase the H^+ concentration and thereby decrease the pH. Alternatively, when the pH of an animal's blood decreases (that is, the H^+ concentration increases), this pathway runs in reverse. Bicarbonate combines with H^+ to make H_2CO_3, which then dissociates to CO_2 and H_2O. This process removes H^+ from the blood, restoring it to its normal pH, and the CO_2 is exhaled from the lungs. Many buffers exist in nature. Buffers found in living organisms function most efficiently at the normal range of pH values found in that organism.

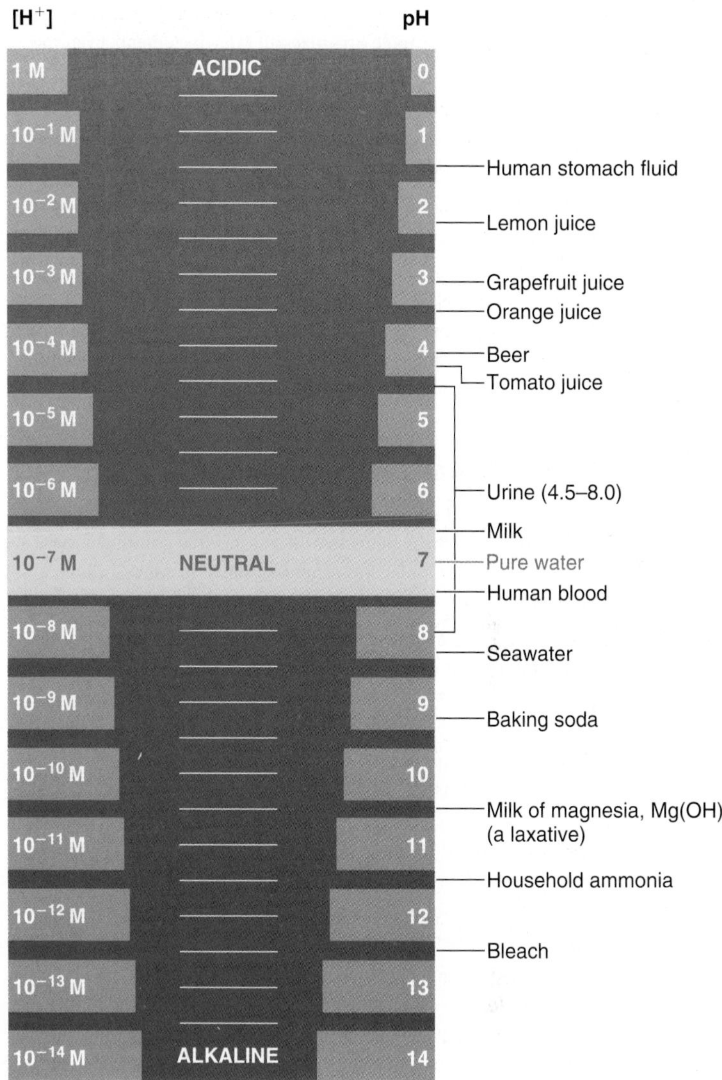

Figure 2.20 **The pH scale and the relative acidities of common substances.**

Concept Check: *What is the OH^- concentration at pH 8?*

BioConnections: *Look ahead to Figure 54.16. The plant life shown growing in part (b) of that figure has been exposed to rain that was acidified due to contaminants arising from the burning of fossil fuels. If the pH of the soil were 5.0, what would the H^+ concentration be?*

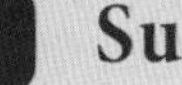

Summary of Key Concepts

2.1 Atoms

- Atoms are the smallest functional units of matter that form all chemical elements and cannot be further broken down into other substances by ordinary chemical or physical means. Atoms are composed of protons (p^+, positive charge), electrons (e^-, negative charge), and (except for hydrogen) neutrons (n^0, electrically neutral). Electrons are found in orbitals around the nucleus (Table 2.1, Figures 2.1, 2.2, 2.3, 2.4).

- Each element contains a unique number of protons—its atomic number. The periodic table organizes all known elements by atomic number and electron shells (Figure 2.5).
- Each atom has a small but measurable mass, measured in daltons (Da). The atomic mass scale indicates an atom's mass relative to the mass of other atoms.
- Many atoms exist as isotopes, which differ in the number of neutrons they contain. Some isotopes are unstable radioisotopes and emit radiation (Figure 2.6).
- Four elements—oxygen, carbon, hydrogen, and nitrogen—account for the vast majority of atoms in living organisms. In addition, living organisms require mineral and trace elements that are essential for growth and function (Table 2.2).

2.2 Chemical Bonds and Molecules

- The properties of a molecule—two or more atoms bonded together—are different from the properties of the atoms that combined to form it. A compound is a molecule composed of two or more different elements.
- Atoms tend to form bonds that fill their outer shell with electrons. Covalent bonds, in which atoms share electrons, are strong chemical bonds. Atoms form two covalent bonds—a double bond—when they share two pairs of electrons (Figures 2.7, 2.8, 2.9).
- The electronegativity of an atom is a measure of its ability to attract bonded electrons. When two atoms with different electronegativities combine, they form a polar covalent bond because the distribution of electrons around the atoms creates a difference in electric charge across the molecule. Polar molecules, such as water, are largely composed of polar bonds, and nonpolar molecules are composed predominantly of nonpolar bonds (Figure 2.10).
- An important result of polar covalent bonds is the ability of one molecule to loosely associate with another molecule through weak interactions called hydrogen bonds. The van der Waals forces are weak electrical attractions that arise between molecules due to the probabilistic orbiting of electrons in atoms (Figure 2.11).
- If an atom or molecule gains or loses one or more electrons, it acquires a net electric charge and becomes an ion. The strong attraction between two oppositely charged ions forms an ionic bond (Table 2.3, Figure 2.12).
- The three-dimensional, flexible shape of molecules allows them to interact and contributes to their biological properties (Figures 2.13, 2.14).
- A free radical is an unstable molecule that interacts with other molecules by taking away electrons from their atoms.
- A chemical reaction occurs when one or more substances are changed into different substances. All chemical reactions eventually reach an equilibrium, unless the products of the reaction are continually removed.

2.3 Properties of Water

- Water is the solvent for most chemical reactions in all living organisms, both inside and outside of cells. Atoms and molecules dissolved in water interact in ways that would be impossible in their nondissolved states (Figure 2.15).
- Solutes dissolve in a solvent to form a solution. Solute concentration refers to the amount of a solute dissolved in a unit volume of solution. The molarity of a solution is defined as the number of moles of a solute dissolved in 1 L of solution (Figure 2.16).
- Molecules with ionic and polar covalent bonds generally are hydrophilic, whereas nonpolar molecules, composed predominantly of carbon and hydrogen, are hydrophobic. Amphipathic molecules, such as detergents, have polar and nonpolar regions (Figure 2.17).
- H_2O exists as ice, liquid water, and water vapor (gas) (Figure 2.18).
- The colligative properties of water depend on the number of dissolved solutes and allow it to function as an antifreeze in certain organisms.
- Water's high heat of vaporization and high heat of fusion make it very stable in liquid form.
- Water molecules participate in many chemical reactions in living organisms. Hydrolysis breaks down large molecules into smaller units, and dehydration reactions combine two smaller molecules into one larger one. In living organisms, water provides support, is used to eliminate wastes, dissipates body heat, aids in the movement of liquid through vessels, serves as a lubricant, and its surface tension allows certain insects to walk on water (Figure 2.19).
- The pH of a solution refers to its hydrogen ion concentration. The pH of pure water is 7 (a neutral solution). Alkaline solutions have a pH higher than 7, and acidic solutions have a pH lower than 7 (Figure 2.20).
- Buffers are compounds that minimize pH fluctuations in the fluids of living organisms. Buffer systems in animals can raise or lower pH to keep the pH of body fluids within a narrow range.

Assess and Discuss

Test Yourself

1. ____________________ make(s) up the nucleus of an atom.
   a. Protons and electrons
   b. Protons and neutrons
   c. DNA and RNA
   d. Neutrons and electrons
   e. DNA only
2. Living organisms are composed mainly of which atoms?
   a. calcium, hydrogen, nitrogen, and oxygen
   b. carbon, hydrogen, nitrogen, and oxygen
   c. hydrogen, nitrogen, oxygen, and helium
   d. carbon, helium, nitrogen, and oxygen
   e. carbon, calcium, hydrogen, and oxygen
3. The ability of an atom to attract electrons in a bond with another atom is termed its
   a. hydrophobicity.
   b. electronegativity.
   c. solubility.
   d. valence.
   e. both a and b.

4. Hydrogen bonds differ from covalent bonds in that
   a. covalent bonds can form between any type of atom, and hydrogen bonds form only between H and O.
   b. covalent bonds involve sharing of electrons, and hydrogen bonds involve the complete transfer of electrons.
   c. covalent bonds result from equal sharing of electrons, but hydrogen bonds involve unequal sharing of electrons.
   d. covalent bonds involve sharing of electrons between atoms, but hydrogen bonds are the result of weak attractions between a hydrogen atom of a polar molecule and an electronegative atom of another polar molecule.
   e. covalent bonds are weak bonds that break easily, but hydrogen bonds are strong links between atoms that are not easily broken.
5. A free radical
   a. is a positively charged ion.
   b. is an atom with one unpaired electron in its outer shell.
   c. is a stable atom that is not bonded to another atom.
   d. can cause considerable cellular damage.
   e. both b and d.
6. Chemical reactions in living organisms
   a. require energy to begin.
   b. usually require a catalyst to speed up the process.
   c. are usually reversible.
   d. occur in liquid environments, such as water.
   e. all of the above.
7. Solutes that easily dissolve in water are said to be
   a. hydrophobic.
   b. hydrophilic.
   c. polar molecules.
   d. all of the above.
   e. b and c only.
8. The sum of the atomic masses of all the atoms of a molecule is its
   a. atomic weight.
   b. molarity.
   c. molecular mass.
   d. concentration.
   e. polarity.
9. Reactions that involve water in the breaking apart of other molecules are known as __________ reactions.
   a. hydrophilic
   b. hydrophobic
   c. dehydration
   d. anabolic
   e. hydrolytic
10. A difference between a strong acid and a weak acid is
   a. strong acids have a higher molecular mass than weak acids.
   b. strong acids completely (or almost completely) ionize in solution, but weak acids do not completely ionize in solution.
   c. strong acids give off two hydrogen ions per molecule, but weak acids give off only one hydrogen ion per molecule.
   d. strong acids are water-soluble, but weak acids are not.
   e. strong acids give off hydrogen ions, and weak acids give off hydroxyl groups.

Conceptual Questions

1. Distinguish between the types of bonds commonly found in biological molecules.
2. What is the significance of molecular shape, and what may change the shape of molecules?
3. A principle of biology is that *new properties emerge from complex interactions*. How is this principle evident even at the molecular level? What examples of emergent properties of molecules can you find in this chapter, in which atoms with one type of property combine to form molecules with completely different properties?

Collaborative Questions

1. Discuss the properties of the three subatomic particles of atoms.
2. Discuss several properties of water that make it possible for life to exist.

Online Resource

www.brookerbiology.com

Stay a step ahead in your studies with animations that bring concepts to life and practice tests to assess your understanding. Your instructor may also recommend the interactive eBook, individualized learning tools, and more.

The Chemical Basis of Life II: Organic Molecules

3

Chapter Outline

3.1 The Carbon Atom and the Study of Organic Molecules
3.2 Formation of Organic Molecules and Macromolecules
3.3 Carbohydrates
3.4 Lipids
3.5 Proteins
3.6 Nucleic Acids
Summary of Key Concepts
Assess and Discuss

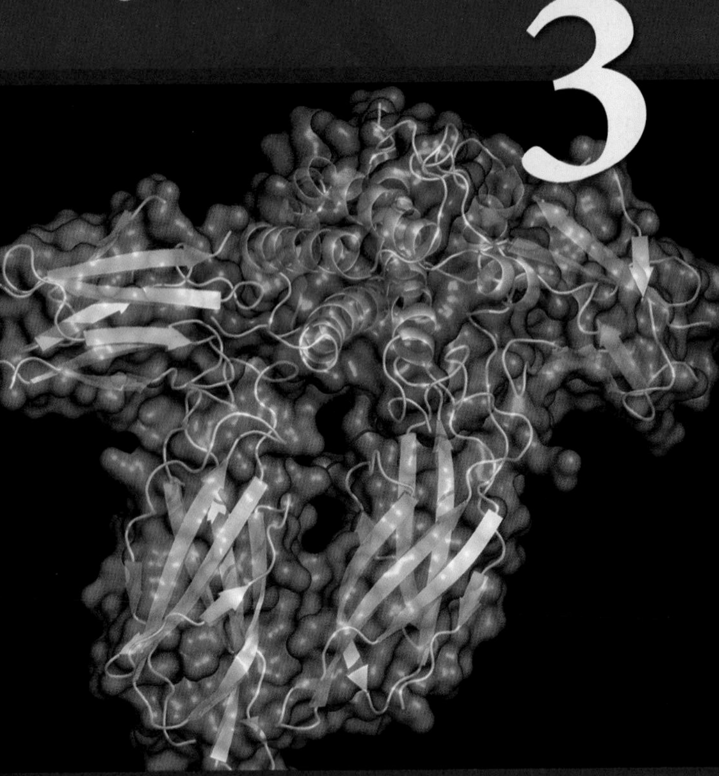

A model showing the structure of a protein—a type of organic macromolecule.

In Chapter 2, we learned that all life is composed of subatomic particles that form atoms, which, in turn, combine to form molecules. Molecules may be simple in atomic composition, as in water (H_2O) or hydrogen gas (H_2), or may bind with other molecules to form larger ones. Of the countless possible molecules that can be produced from the known elements in nature, certain types contain carbon and are found in all forms of life. These carbon-containing molecules are collectively referred to as **organic molecules**, so named because they were first discovered in living organisms. Among these are lipids and large, complex compounds called **macromolecules**, which include carbohydrates, proteins, and nucleic acids. In this chapter, we will survey the structures of these molecules and examine their chief functions. We begin with the element whose chemical properties are fundamental to the formation of biologically important molecules: carbon. This element provides the atomic scaffold on which life is built.

3.1 The Carbon Atom and the Study of Organic Molecules

Learning Outcomes:

1. Explain the properties of carbon that make it the chemical basis of all life.
2. Describe the variety and chemical characteristics of common functional groups of organic compounds.
3. Compare and contrast the different types of isomeric compounds.

The science of carbon-containing molecules is known as **organic chemistry**. In this section, we will examine the bonding properties of carbon that create molecules with distinct functions and shapes.

Interestingly, the study of organic molecules was long considered a fruitless endeavor because of a concept called vitalism, which persisted into the 19th century. Vitalism held that organic molecules were created by, and therefore imparted with, a vital life force that was contained within a plant or an animal's body. Supporters of vitalism argued there was no point in trying to synthesize an organic compound, because such molecules could arise only through the intervention of mysterious qualities associated with life. As described next, this would all change due to the pioneering experiments of Friedrich Wöhler in 1828.

Wöhler's Synthesis of an Organic Compound Transformed Misconceptions About the Molecules of Life

Friedrich Wöhler was a German physician and chemist interested in the properties of inorganic and organic compounds. He spent some time studying urea ($(NH_2)_2CO$), a natural organic product formed from the breakdown of proteins in an animal's body. In mammals, urea accumulates in the urine formed by the kidneys, and then is excreted from the body. During the course of his studies, Wöhler purified urea from the urine of mammals. He noted the color, size, shape, and other characteristics of the crystals that formed when urea was isolated. This experience would serve him well in later years when he quite accidentally helped to put the concept of vitalism to rest.

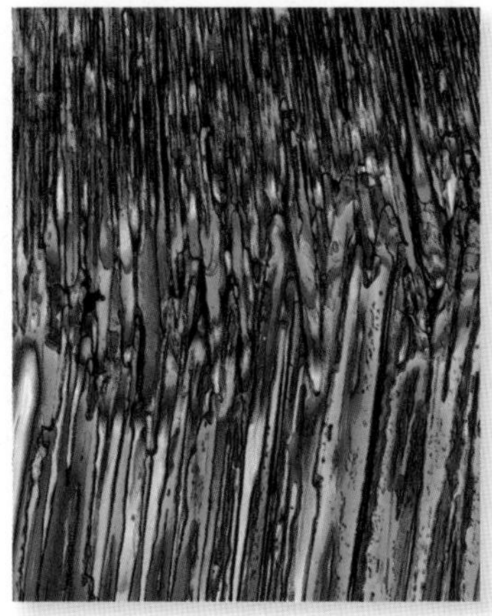

Figure 3.1 Crystals of urea as viewed with a polarizing microscope (approximately 80× magnification).

Concept Check: How did prior knowledge of urea allow Wöhler to realize he had synthesized urea from ammonia and cyanic acid?

In 1828, while exploring the reactive properties of ammonia and cyanic acid, Wöhler attempted to synthesize an inorganic molecule, ammonium cyanate (NH_4OCN), which is not found in living organisms. Instead, to his surprise, Wöhler discovered that ammonia and cyanic acid reacted to produce a third compound, which, when heated, formed familiar-looking crystals (**Figure 3.1**). After careful analysis, he concluded that these crystals were in fact urea. In short, no mysterious life force was required to create this organic molecule. Other scientists, such as Hermann Kolbe, would soon demonstrate that organic compounds such as acetic acid (CH_3COOH) could be synthesized directly from simpler molecules. These studies were a major breakthrough in the way in which scientists viewed life, and so began the field of science now called organic chemistry. Since that time, the fields of chemistry and biology have been understood to be intricately related.

Central to Wöhler's and Kolbe's reactions is the carbon atom. Urea and acetic acid, like all organic compounds, contain carbon atoms bound to other atoms. Let's now consider the chemical features of carbon that make it such an important element in living organisms.

Carbon Forms Four Covalent Bonds with Other Atoms

One of the properties of the carbon atom that makes life possible is its ability to form four covalent bonds with other atoms, including other carbon atoms. This occurs because carbon has four electrons in its outer (second) shell, and it requires eight electrons, or four additional electrons, to fill its second shell (**Figure 3.2**). In living organisms, carbon atoms most commonly form covalent bonds with other carbon atoms and with hydrogen, oxygen, nitrogen, and sulfur atoms. Bonds between two carbon atoms, between carbon and oxygen, or between carbon and nitrogen can be single or double, or in the case of certain C≡C and C≡N bonds, triple. The variation in bonding of carbon with other carbon atoms and with different elements allows a vast number of organic compounds to be formed from only a few chemical elements. This is made all the more impressive because carbon bonds may occur in configurations that are linear, ringlike, or highly branched. Such molecular shapes can produce molecules with a variety of functions.

Carbon and hydrogen have similar electronegativities (see Chapter 2); therefore, carbon-carbon and carbon-hydrogen bonds are nonpolar. As a consequence, molecules with predominantly or entirely hydrogen-carbon bonds, called **hydrocarbons**, are hydrophobic and poorly soluble in water. In contrast, when carbon forms polar covalent bonds with more electronegative atoms, such as oxygen or nitrogen, the molecule is much more soluble in water due to the electrical attraction of polar water molecules. The ability of carbon to form

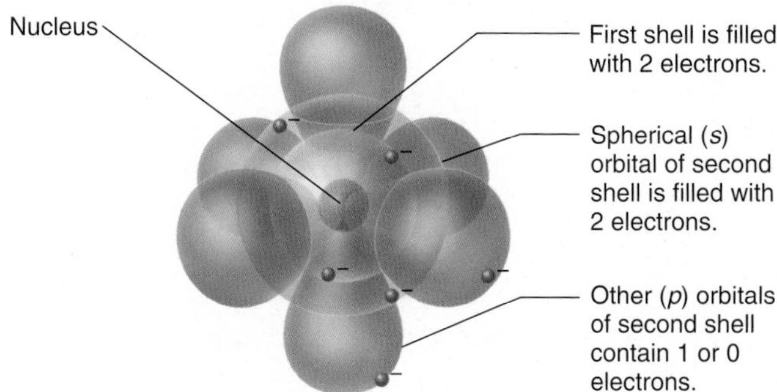
Nucleus
First shell is filled with 2 electrons.
Spherical (s) orbital of second shell is filled with 2 electrons.
Other (p) orbitals of second shell contain 1 or 0 electrons.

(a) Electron orbitals in carbon

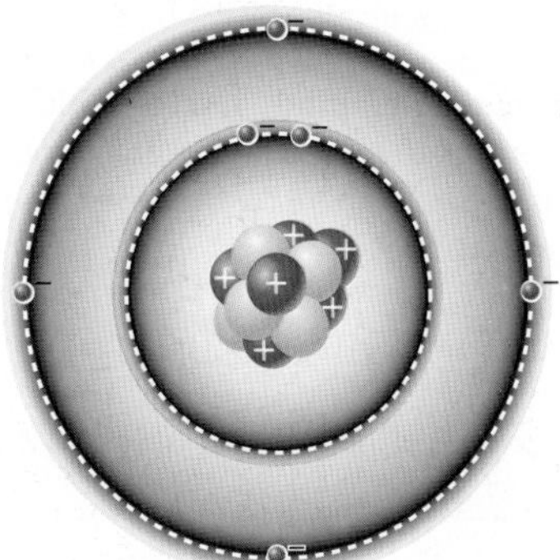

(b) Simplified depiction of carbon's electron shells

Figure 3.2 Models for the electron orbitals and shells of carbon. Carbon atoms have only four electrons in their outer (second) electron shell, which allows carbon to form four covalent bonds. When carbon forms four covalent bonds, the result is four hybrid orbitals of equal energy.

both polar and nonpolar bonds (**Figure 3.3**) contributes to its ability to serve as the backbone for an astonishing variety of biologically important molecules.

One last feature of carbon that is important to biology is that carbon bonds are stable within the large range of temperatures associated with life. This property arises in part because the carbon atom is very small relative to most other atoms; therefore, the distance between carbon atoms forming a carbon-carbon bond is quite short. Shorter bonds tend to be stronger and more stable than longer bonds between two large atoms. Thus, carbon bonds are compatible with what we observe about life-forms today; namely, living organisms can inhabit environments with a range of temperatures, from the Earth's frigid icy poles to the superheated water of deep-sea vents.

Carbon Atoms Can Bond to Several Biologically Important Functional Groups

Aside from the simplest hydrocarbons, most organic molecules and macromolecules contain **functional groups**—groups of atoms with characteristic chemical features and properties. Each type of functional group exhibits similar chemical properties in all molecules in which it occurs. For example, the amino group (NH_2) acts like a base. At the pH found in living organisms, amino groups readily bind H^+ to become NH_3^+, thereby removing H^+ from an aqueous solution and raising the pH. As discussed later in this chapter, amino groups are widely found in proteins and also in other types of organic molecules.

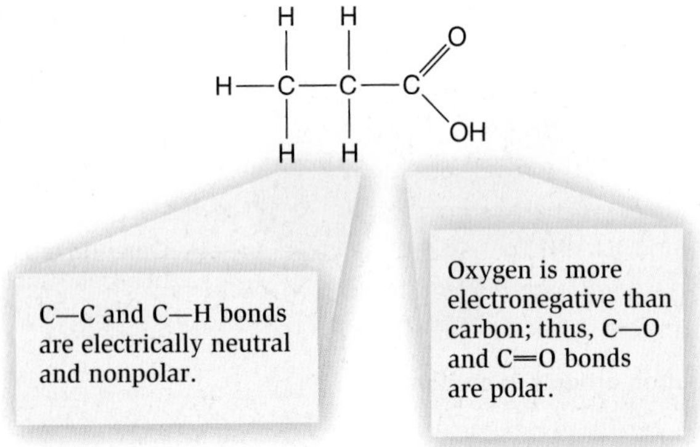

Propionic acid

C—C and C—H bonds are electrically neutral and nonpolar.

Oxygen is more electronegative than carbon; thus, C—O and C═O bonds are polar.

Figure 3.3 Nonpolar and polar bonds in an organic molecule. Carbon can form both nonpolar and polar bonds, and both single and double bonds, as shown here for the molecule propionic acid, a common food preservative.

Table 3.1 describes examples of functional groups found in many different types of organic molecules. We will discuss each of these groups at numerous points throughout this textbook.

Carbon-Containing Molecules May Exist in Multiple Forms Called Isomers

When Wöhler did his now-famous experiment, he was surprised to discover that urea, ($(NH_2)_2CO$), and ammonium cyanate, (NH_4OCN), apparently contained the exact same ratio of carbon, nitrogen, hydrogen, and oxygen atoms, yet they were different molecules with distinct chemical and biological properties. Two molecules with an identical chemical formula but different structures and characteristics are called **isomers**.

Figure 3.4 depicts three ways in which isomers may occur. **Structural isomers** contain the same atoms but in different bonding relationships. Urea and ammonium cyanate fall into this category; a simpler example of structural isomers (isopropyl alcohol and propyl alcohol) is illustrated in Figure 3.4a.

Table 3.1 Some Biologically Important Functional Groups That Bond to Carbon

Functional group* (with shorthand notation)	Formula	Examples of where they are found	Properties
Amino $-NH_2$	R—N(H)H	Amino acids (proteins)	Weakly basic (can accept H^+); polar; forms part of peptide bonds
†Carbonyl (–CO) Ketone	R—C(═O)—R'	Steroids, waxes, and proteins	Polar; highly chemically reactive; forms hydrogen bonds
Aldehyde	R—C(═O)—H	Linear forms of sugars and some odor molecules	Polar; highly chemically reactive, forms hydrogen bonds
Carboxyl (–COOH)	R—C(═O)OH	Amino acids, fatty acids	Acidic (gives up H^+ in water); forms part of peptide bonds
Hydroxyl (–OH)	R—OH	Steroids, alcohol, carbohydrates, some amino acids	Polar; forms hydrogen bonds with water
Methyl ($-CH_3$)	R—C(H)(H)—H	May be attached to DNA, proteins, and carbohydrates	Nonpolar
Phosphate ($-PO_4^{2-}$)	R—O—P(═O)(O⁻)—O⁻	Nucleic acids, ATP, phospholipids	Polar; weakly acidic and thus negatively charged at typical pH of living organisms
Sulfate ($-SO_4^-$)	R—O—S(═O)(═O)—O⁻	May be attached to carbohydrates, proteins, and lipids	Polar; negatively charged at typical pH
Sulfhydryl (–SH)	R—SH	Proteins that contain the amino acid cysteine	Polar; forms disulfide bridges in many proteins

†A carbonyl group is C═O. In a ketone, the carbon forms covalent bonds with two other carbon atoms. In an aldehyde, the carbon is linked to a hydrogen atom.
*This list contains many of the functional groups that are important in biology. However, many more functional groups have been identified by biochemists. R and R' represent the remainder of the molecule.

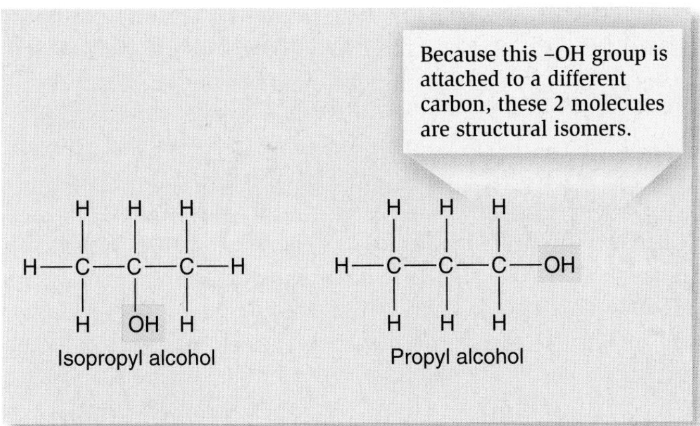
Because this -OH group is attached to a different carbon, these 2 molecules are structural isomers.
Isopropyl alcohol
Propyl alcohol

(a) Structural isomers

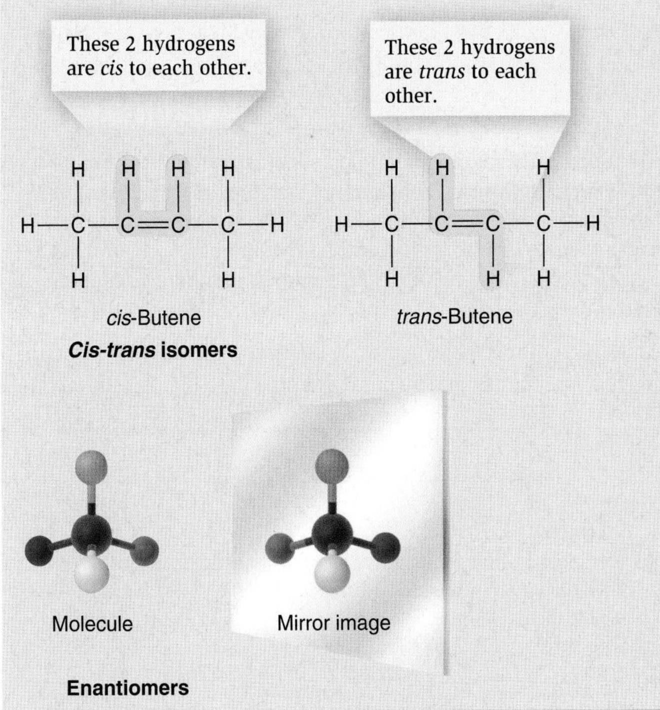
These 2 hydrogens are *cis* to each other.
These 2 hydrogens are *trans* to each other.
cis-Butene
trans-Butene
Cis-trans isomers
Molecule
Mirror image
Enantiomers

(b) Two types of stereoisomers

Figure 3.4 Types of isomers. Isomers are molecules with the same chemical formula but different structures. The differences in structure, though small, are sufficient to result in very different biological properties. Isomers can be grouped into **(a)** structural isomers and **(b)** stereoisomers.

Stereoisomers have identical bonding relationships, but the spatial positioning of their atoms differs. Two types of stereoisomers are *cis-trans* isomers and enantiomers. In ***cis-trans*** **isomers**, like those shown in Figure 3.4b, the two hydrogen atoms linked to the two carbons of a C=C double bond may be on the same side of the carbons, in which case the C=C bond is called a *cis* double bond. If the hydrogens are on opposite sides, it is a *trans* double bond. *Cis-trans* isomers may have very different chemical properties from each other, most notably their stability and sensitivity to heat and light. For instance, the light-sensitive region of your eye contains a molecule called retinal, which may exist in either a *cis* or *trans* form because of a pair of double-bonded carbons in its string of carbon atoms. In darkness, the *cis*-retinal form predominates. The energy of sunlight, however, causes retinal to isomerize to the *trans* form. The *trans*-retinal activates the light-capturing cells in the eye.

A second type of stereoisomer, called an **enantiomer**, exists as a pair of molecules that are mirror images. Four different atoms can bind to a single carbon atom in two possible ways, designated a left-handed and a right-handed structure. The resulting structures are not identical, but instead are mirror images of each other (Figure 3.4b). A convenient way to visualize the mirror-image properties of enantiomers is to look at a pair of gloves. No matter which way you turn or hold a left-hand glove, for example, it cannot fit properly on your right hand. Any given pair of enantiomers shares identical chemical properties, such as solubility and melting point. However, due to the different orientation of atoms in space, their ability to noncovalently bind to other molecules can be strikingly different. For example, as you learned in Chapter 2, **enzymes** are molecules that catalyze (speed up) the rates of many biologically important chemical reactions. Typically, a given enzyme is very specific in its action, and an enzyme that recognizes one enantiomer of a pair often does not recognize the other. That is because the actions of enzymes depend on the spatial arrangements of the particular atoms in a molecule.

3.2 Formation of Organic Molecules and Macromolecules

Learning Outcomes:

1. Diagram how small molecules may be assembled into larger ones by dehydration reactions and how hydrolysis reactions can reverse this process.
2. List the four major classes of organic molecules and macromolecules found in living organisms.

As we have seen, organic molecules have various shapes due to the bonding properties of carbon. During the past two centuries, biochemists have studied many organic molecules found in living organisms and determined their structures at the molecular level. Many of these compounds are relatively small molecules, containing a few or a few dozen atoms. However, some organic molecules are extremely large macromolecules composed of thousands or even millions of atoms. Such large molecules are formed by linking together many smaller molecules called **monomers** (meaning one part) and are thus also known as **polymers** (meaning many parts). The structure of macromolecules depends on the structure of their monomers, the number of monomers linked together, and the three-dimensional way in which the monomers are linked.

As introduced in Chapter 2, the process by which two or more molecules combine into a larger one, with the loss of a small molecule, is called a **condensation reaction**. When an organic macromolecule is formed, two smaller molecules combine by condensation, producing a larger molecule along with a molecule of water. This specific type of condensation reaction is called a **dehydration reaction**, because a molecule of water is removed when the monomers combine.

Monomers
HO- -H + HO- -H → HO- - -H → HO- - - -H → HO- - - - -H
H_2O
H_2O
H_2O
HO- -H
HO- -H
A polymer begins as two monomers combine in a dehydration reaction.
Elongation of the polymer continues with additional dehydration reactions.
The final length of a polymer may consist of thousands of monomers.

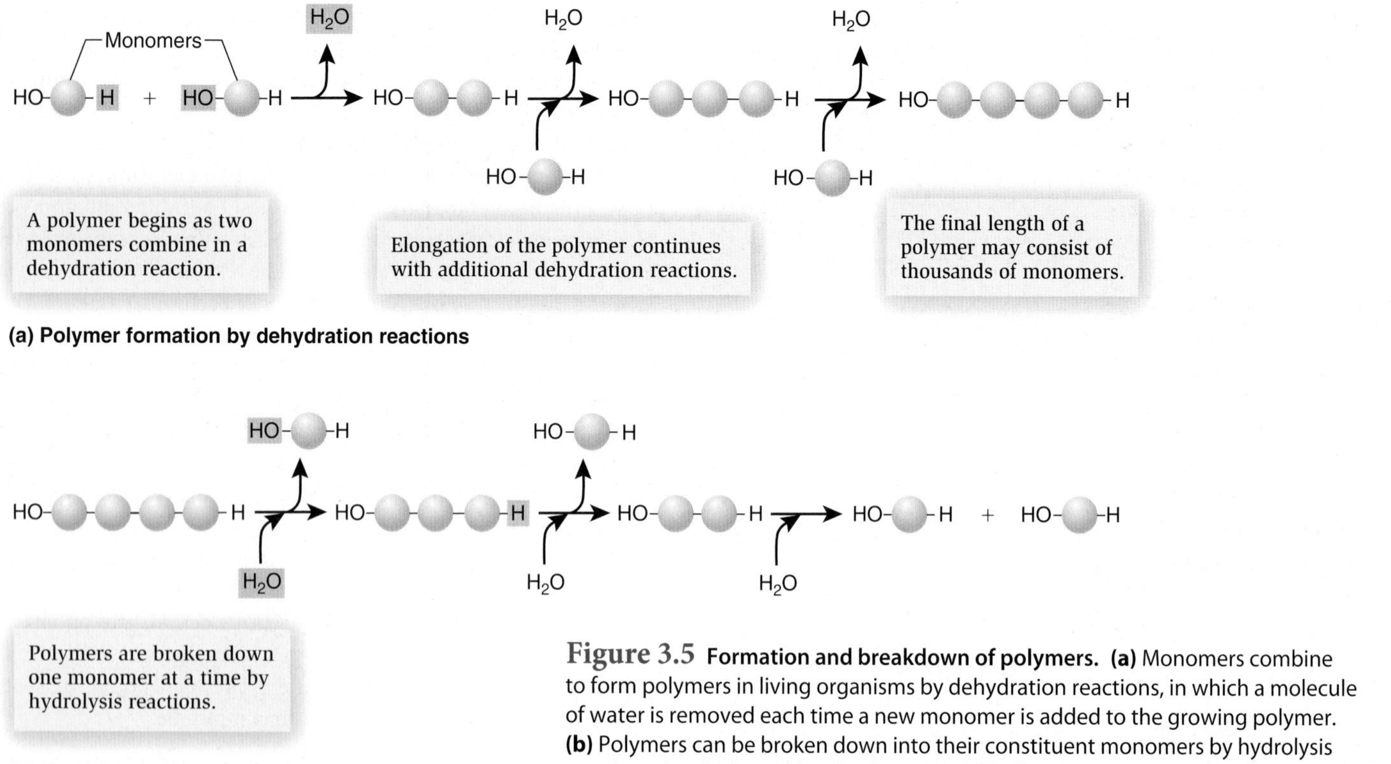

(a) Polymer formation by dehydration reactions

HO- - - - - -H → HO- - - - -H → HO- - - -H → HO- -H + HO- -H
HO- -H
HO- -H
H_2O
H_2O
H_2O
Polymers are broken down one monomer at a time by hydrolysis reactions.

(b) Breakdown of a polymer by hydrolysis reactions

Figure 3.5 Formation and breakdown of polymers. **(a)** Monomers combine to form polymers in living organisms by dehydration reactions, in which a molecule of water is removed each time a new monomer is added to the growing polymer. **(b)** Polymers can be broken down into their constituent monomers by hydrolysis reactions, in which a molecule of water is added each time a monomer is released.

An idealized dehydration reaction is illustrated in Figure 3.5a. Notice that the length of a polymer may be extended again and again with additional dehydration reactions. Some polymers can reach great lengths by this mechanism. For example, as you will learn in Chapter 46, nutrients in an animal's food are transported out of the digestive tract into the body fluids as monomers. If more energy-yielding nutrients are consumed than are required for the animal's activities, the excess nutrients may be processed by certain organs into extremely long polymers consisting of tens of thousands of monomers. The polymers are then stored in this convenient form to provide a source of energy when food is not available, such as during sleep, when the animal is not eating but nevertheless still requires energy to carry out all the various activities required to maintain cellular function.

Polymers, however, are not recognized by the cellular machinery that functions to release the chemical energy stored in the bonds of molecules. Consequently, polymers must first be broken down into their constituent monomers, which then, under the right conditions, can release some of the energy stored in their bonds. As you learned in Chapter 2, the process by which a polymer is broken down into monomers is called a **hydrolysis reaction**, because a molecule of water is added back each time a monomer is released (Figure 3.5b). Therefore, the formation of polymers in organisms is generally reversible; once formed, a polymer can later be broken down. These processes may repeat themselves over and over again as dictated by changes in the various cellular activities of an organism. Dehydration and hydrolysis reactions are both catalyzed by enzymes.

By analyzing the cells of many different species, researchers have determined that all forms of life have organic molecules and macromolecules that fall into four broad categories, based on their chemical and biological properties: carbohydrates, lipids, proteins, and nucleic acids. In the next sections, we will survey the structures of these organic compounds and begin to examine their biological functions.

3.3 Carbohydrates

Learning Outcomes:

1. Distinguish among different forms of carbohydrate molecules, including monosaccharides, disaccharides, and polysaccharides.
2. Relate the functions of plant and animal polysaccharides to their structure.

Carbohydrates are composed of carbon, hydrogen, and oxygen atoms in or close to the proportions represented by the general formula $C_n(H_2O)_n$, where n is a whole number. This formula gives carbohydrates their name—carbon-containing compounds that are hydrated, that is, contain water. Most of the carbon atoms in a carbohydrate are linked to a hydrogen atom and a hydroxyl functional group. However, other functional groups, such as amino and carboxyl groups, are also found in certain carbohydrates. As discussed next, sugars are relatively small carbohydrates, whereas polysaccharides are large macromolecules.

Sugars Are Small Carbohydrates That May Taste Sweet

Sugars are small carbohydrates that in many, but not all, cases taste sweet. The simplest sugars are the monomers known as **monosaccharides** (from the Greek, meaning single sugars). The most common types are molecules with five carbons, called pentoses, and those with six carbons, called hexoses. Important pentoses are ribose ($C_5H_{10}O_5$) and the closely related deoxyribose ($C_5H_{10}O_4$), which are part of RNA and DNA molecules, respectively, and are described later in this chapter. The most common hexose is glucose ($C_6H_{12}O_6$). Like other monosaccharides, glucose is very water-soluble and thus circulates in the blood or fluids of animals, where it can be transported across cell membranes. Once inside a cell, enzymes can break down glucose into smaller molecules, releasing energy that was stored in the chemical bonds of glucose. This energy is then stored in the bonds of another molecule, called adenosine triphosphate, or ATP (see Chapter 7), which, in turn, powers a variety of cellular processes. In this way, sugar is often used as a source of energy by living organisms.

Figure 3.6a depicts the bonds between atoms in a monosaccharide in both linear and ring forms. The ring structure is a better approximation of the true shape of the molecule as it mostly exists in solution, with the carbon atoms numbered by convention, as shown. The ring is made from the linear structure by an oxygen atom, which forms a bond that bridges two carbons in which carbon 1 binds to the oxygen of carbon 5. The hydrogen atoms and the hydroxyl groups may lie above or below the plane of the ring structure.

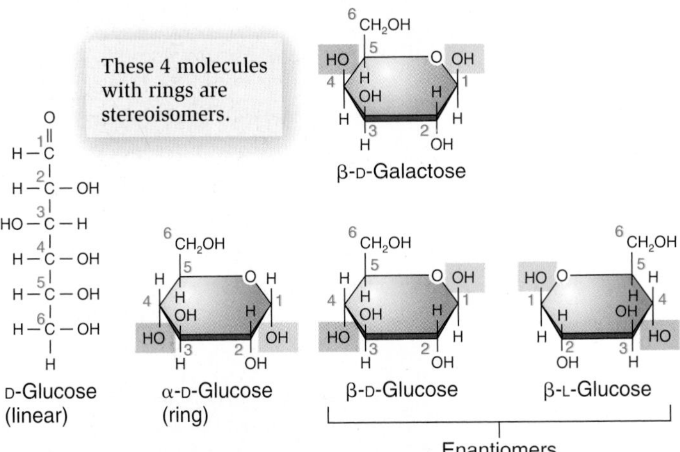

(a) Linear and ring structures of D-glucose **(b) Isomers of glucose**

Figure 3.6 Monosaccharide structure. **(a)** A comparison of the linear and ring structures of glucose. In solution, such as the fluids of organisms, nearly all glucose is in the ring form. **(b)** Isomers of glucose. The locations of the C-1 and C-4 hydroxyl groups are emphasized with green and orange boxes, respectively. Glucose exists as stereoisomers designated *α*- and *β*-glucose, which differ in the position of the —OH group attached to carbon atom number 1. Glucose and galactose differ in the position of the —OH group attached to carbon atom number 4. Enantiomers of glucose, called D-glucose and L-glucose, are mirror images of each other. D-Glucose is the form used by living cells.

BIOLOGY PRINCIPLE Living organisms use energy. The chemical energy stored in certain of the bonds of glucose molecules can be harnessed by living organisms. This energy can be used to perform numerous functions that support life, including the synthesis of new molecules, growth, digestion, locomotion, and many others.

Concept Check: *With regard to their binding to enzymes, why do enantiomers such as D- and L-glucose have different biological properties?*

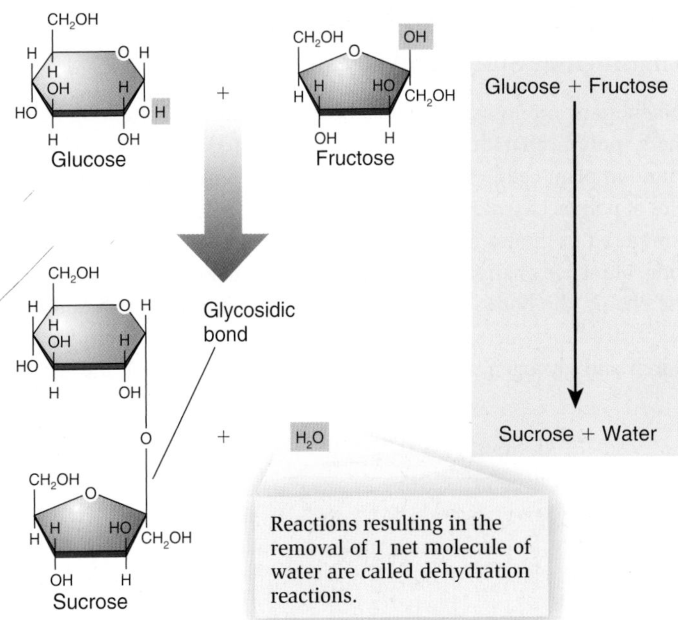

Figure 3.7 Formation of a disaccharide. Two monosaccharides can bond to each other to form a disaccharide, such as sucrose, maltose, or lactose, by a dehydration reaction.

Concept Check: *What type of reaction is the reverse of the one shown here, in which a disaccharide is broken down into two monosaccharides?*

Figure 3.6b compares different types of isomers of glucose. Glucose can exist as D- and L-glucose, which are mirror images of each other, or enantiomers. Other types of isomers are formed by changing the relative positions of the hydrogens and hydroxyl groups along the sugar ring. For example, glucose exists in two interconvertible forms, with the hydroxyl group attached to the number 1 carbon atom lying either above (the *β* form of glucose, Figure 3.6b) or below (the α form, Figure 3.6a) the plane of the ring. As discussed later, the isomers of glucose have different biological properties. In another example, if the hydroxyl group on carbon atom number 4 of glucose is above the plane of the ring, the sugar called galactose is created (Figure 3.6b).

Monosaccharides can join together by dehydration to form larger carbohydrates. **Disaccharides** (meaning two sugars) are carbohydrates composed of two monosaccharides. A familiar disaccharide is sucrose, or table sugar, which is composed of the monomers glucose and fructose (Figure 3.7). Sucrose is the major transport form of sugar in plants. The linking together of most monosaccharides involves the removal of a hydroxyl group from one monosaccharide and a hydrogen atom from the other, giving rise to a molecule of water and covalently bonding the two sugars together through an oxygen atom. The bond formed between two sugar molecules by such a dehydration reaction is called a **glycosidic bond**. Conversely, hydrolysis of a glycosidic bond in a disaccharide breaks the bond by adding back the water, thereby uncoupling the two monosaccharides. Other disaccharides frequently found in nature are maltose, formed in animals during the digestion of large carbohydrates in the intestinal tract, and lactose, present in the milk of mammals. Maltose is α-D-glucose linked to α-D-glucose, and lactose is *β*-D-galactose linked to *β*-D-glucose.

Polysaccharides Are Carbohydrate Polymers That Include Starch and Glycogen

When many monosaccharides are linked together to form long polymers, **polysaccharides** (meaning many sugars) are made. **Starch**, found in plant cells, and **glycogen**, present in animal cells, are examples of polysaccharides (Figure 3.8). Both of these polysaccharides are composed of thousands of α-D-glucose molecules linked together in long, branched chains, differing only in the extent of branching along the chain. The bonds that form in polysaccharides are not random but instead form between specific carbon atoms of each molecule. In starch and glycogen, the bonds form between carbons 1 and 4 and between carbons 1 and 6. The high degree of branching in glycogen contributes to its solubility in animal tissues, such as muscle tissue. This is because the extensive branching creates a more open structure, in which many hydrophilic hydroxyl (—OH) side groups have access to water and can hydrogen-bond with it. Starch, because it is less branched, is less soluble, which contributes to the properties of plant structures (think of a potato or a kernel of corn).

Some polysaccharides, such as starch and glycogen, are used to store energy in cells. Like disaccharides, polysaccharides can be hydrolyzed in the presence of water to yield monosaccharides, which are broken down to provide the energy that can be stored in ATP. Starch and glycogen, the polymers of α-glucose, provide an efficient

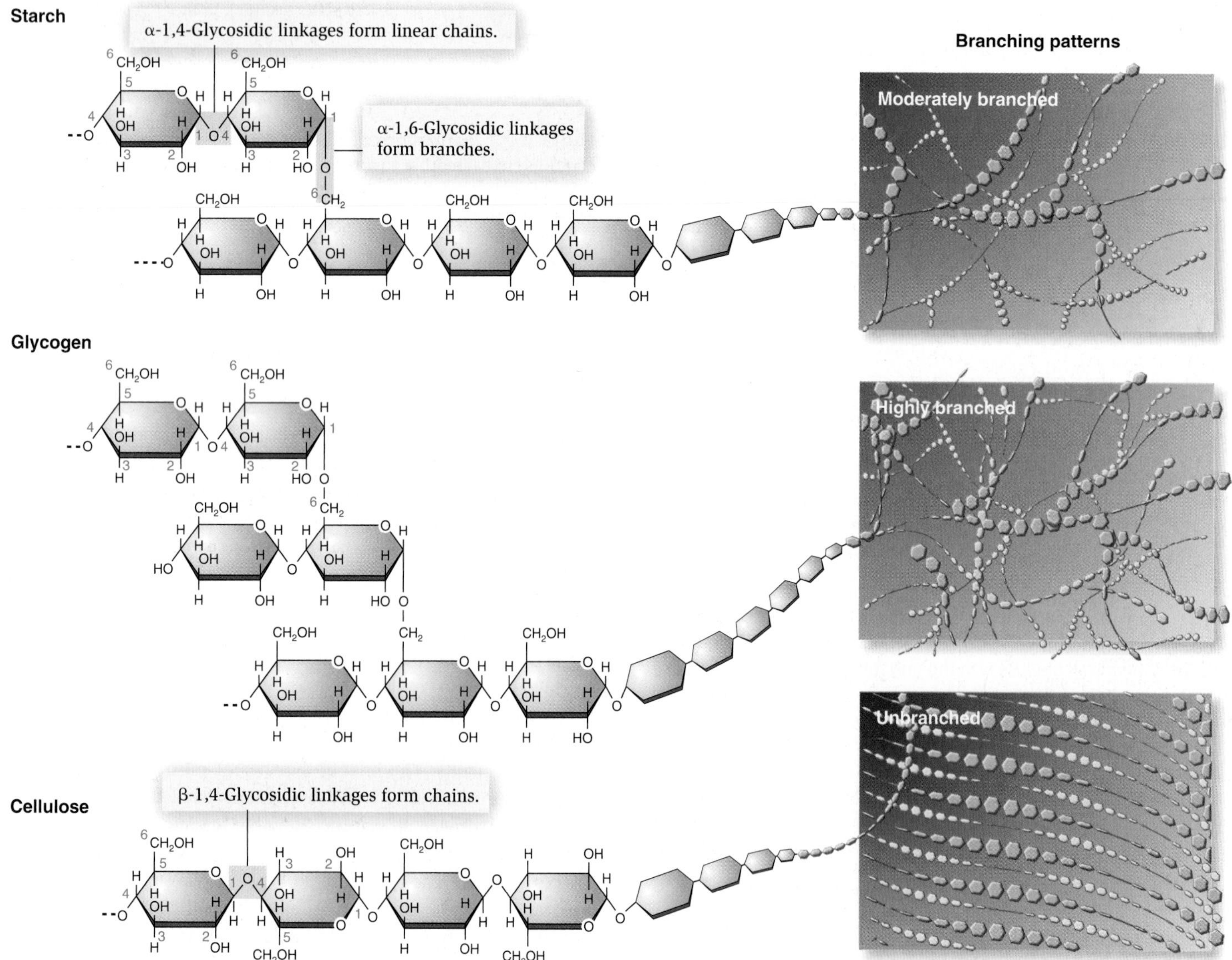

Figure 3.8 Polysaccharides that are polymers of glucose. These polysaccharides differ in their arrangement, extent of branching, and type of glucose isomer. Note: In cellulose, the bonding arrangements cause every other glucose to be upside down with respect to its neighbors.

BioConnections: *Look ahead to Figures 10.5 and 10.6 for its role in plant structure and to Figures 35.10 and 35.11 for the role of cellulose in plant growth. Considering the amount of plant life on Earth, what might you conclude about the abundance of cellulose on the planet?*

means of storing energy for those times when a plant or animal cannot obtain sufficient energy from its environment or diet for its metabolic requirements.

Other polysaccharides provide a structural role, rather than storing energy. The plant polysaccharide **cellulose** is a polymer of β-D-glucose, with a linear arrangement of carbon-carbon bonds and no branching (see Figure 3.8). Each glucose monomer in cellulose is in an opposite orientation from its adjacent monomers ("flipped over"), forming long chains of several thousand glucose monomers. The bond orientations in β-D-glucose prevent cellulose from being hydrolyzed for ATP production in most types of organisms. As noted earlier, this is because many enzymes are highly specific for one type of molecule. The enzymes that break the bonds between monomers of α-D-glucose in starch do not recognize the shape of the polymer made by the bonds between β-D-glucose monomers in cellulose. Therefore, plant cells can break down starch without breaking down cellulose. In this way, cellulose can be used for other functions, notably in the formation of the rigid cell-wall structure characteristic of plants. The linear arrangement of bonds in cellulose provides opportunities for vast numbers of hydrogen bonds between cellulose molecules, which stack together in sheets and provide great strength to structures like plant cell walls. The hydrogen bonds form between hydroxyl groups on carbons 3 and 6.

Unlike most animals and plants, some organisms do have an enzyme capable of breaking down cellulose. For example, certain bacteria present in the gastrointestinal tracts of grass and wood eaters, such as cows and termites, respectively, can digest cellulose into usable monosaccharides because they contain an enzyme that can hydrolyze the bonds between β-D-glucose monomers. Humans lack this enzyme; therefore, we eliminate in the feces most of the cellulose ingested in our diet. Undigestible plant matter we consume is commonly referred to as fiber.

Other polysaccharides also play structural roles. **Chitin**, a tough, structural polysaccharide, forms the external skeleton of insects and crustaceans (shrimp and lobsters) as well as the cell walls of fungi. The sugar monomers within chitin have nitrogen-containing groups attached to them. **Glycosaminoglycans** are large polysaccharides that play a structural role in animals. For example, they are abundantly found in cartilage, the tough, fibrous material found in bone and certain other animal structures. Glycosaminoglycans are also abundant in the extracellular matrix that provides a structural framework surrounding many of the cells in an animal's body (see Figure 10.4).

3.4 Lipids

Learning Outcomes:

1. List the several different classes of lipid molecules important in living organisms.
2. Diagram the structure of a triglyceride and explain how it is affected by the presence of saturated and unsaturated fatty acids.
3. Explain why some fats are solid at room temperature, and others are liquid.
4. Discuss how fats function as energy-storage molecules.
5. Apply knowledge of the structure of phospholipids to the formation of cellular membranes.
6. Describe the chemical nature of steroids and give an example of their biological importance.

Lipids are hydrophobic molecules composed mainly of hydrogen and carbon atoms, and some oxygen. The defining feature of lipids is that they are nonpolar and therefore insoluble in water. Lipids account for about 40% of the organic matter in the average human body and include fats, phospholipids, steroids, and waxes.

Triglycerides Are Made from Glycerol and Fatty Acids

Triglycerides or triacylglycerols (often simply called "fats"), are formed by bonding glycerol to three fatty acids (Figure 3.9). Glycerol is a three-carbon molecule with one hydroxyl group (—OH) bonded to each carbon. A fatty acid is a chain of carbon and hydrogen atoms with a carboxyl group (—COOH) at one end. Each of the hydroxyl groups in glycerol is linked to the carboxyl group of a fatty acid by the removal of a molecule of water by a dehydration reaction. The resulting bond is an example of a type of chemical bond called an ester bond.

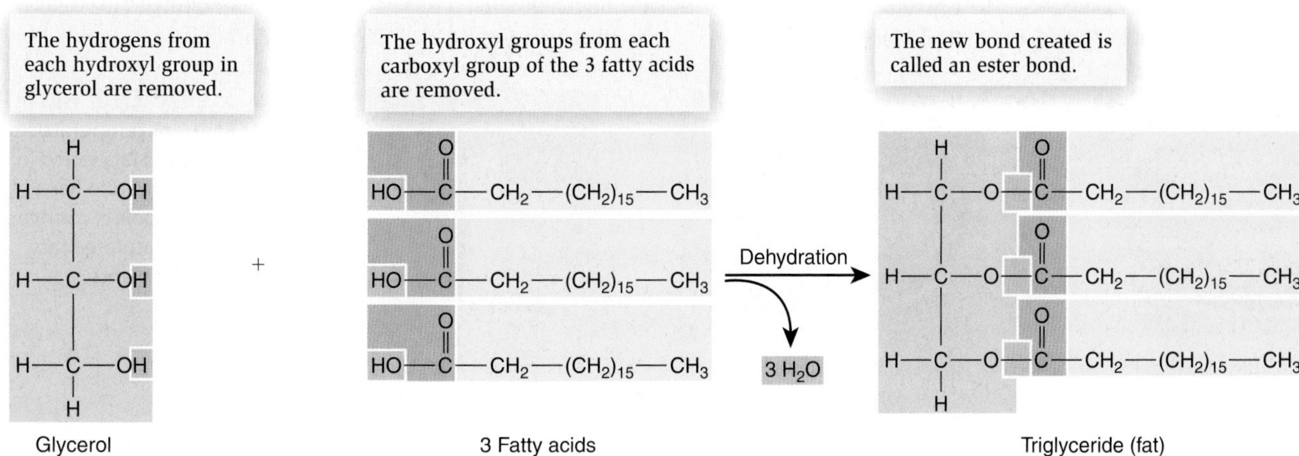

Figure 3.9 The formation of a triglyceride. The formation of a triglyceride requires three dehydration reactions in which fatty acids are bonded to glycerol. Note in this figure and in Figure 3.10, a common shorthand notation is used for depicting fatty acid chains, in which a portion of the CH_2 groups are illustrated as $(CH_2)_n$, where n may be 2 or greater.

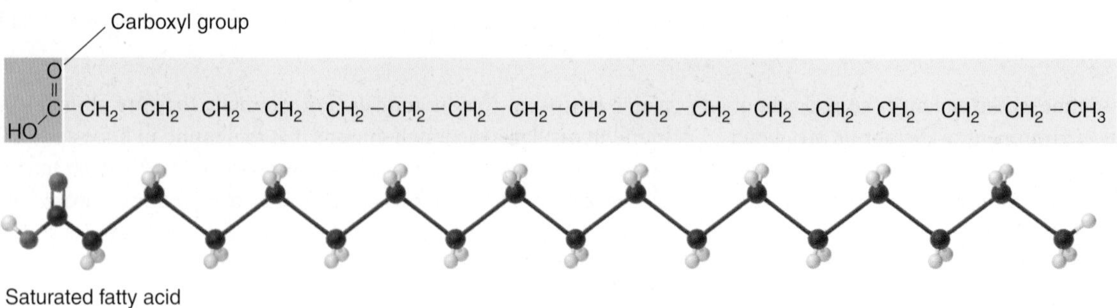

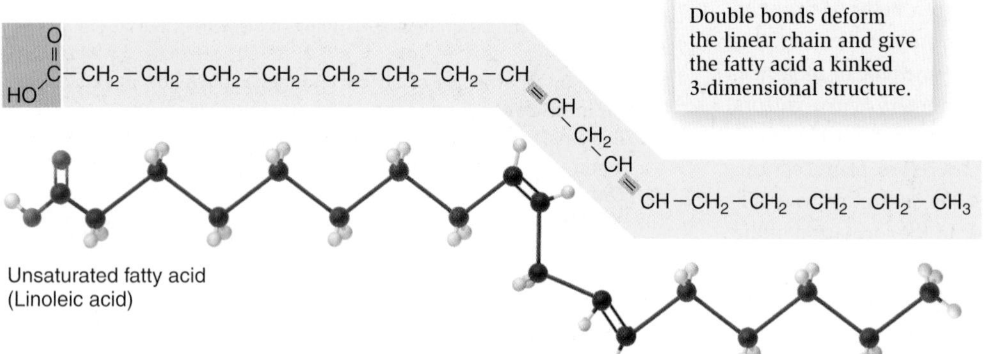

Figure 3.10 **Examples of fatty acids.** Fatty acids are hydrocarbon chains with a carboxyl functional group at one end and either no double-bonded carbons (saturated) or one or more double bonds (unsaturated). Stearic acid, for example, is an abundant saturated fatty acid in animals, whereas linoleic acid is an unsaturated fatty acid found in plants. Note that the presence of two C=C double bonds introduces two kinks into the shape of linoleic acid. As a consequence, unsaturated fatty acids are not able to pack together as tightly as saturated fatty acids.

The fatty acids found in fats and other lipids may differ with regard to their lengths and the presence of double bonds (**Figure 3.10**). Most fatty acids in nature have an even number of carbon atoms, with 16- and 18-carbon fatty acids being the most common in the cells of plants and animals. Fatty acids also differ with regard to the presence of double bonds. When all the carbons in a fatty acid are linked by single covalent bonds, the fatty acid is said to be a **saturated fatty acid**, because all the carbons are saturated with covalently bound hydrogen. Alternatively, some fatty acids contain one or more C=C double bonds and consequently fewer C—H bonds; these fatty acids are known as **unsaturated fatty acids**. The C=C double bond introduces a kink into the linear shape of a fatty acid. A fatty acid with one C=C bond is a monounsaturated fatty acid, whereas a fatty acid with two or more C=C bonds constitutes a polyunsaturated fatty acid. In organisms such as mammals, certain fatty acids are necessary for good health but cannot be synthesized by the body. Such fatty acids are called essential fatty acids, because they must be obtained in the diet; one example is linoleic acid (see Figure 3.10).

Fats (triglycerides) that contain high amounts of saturated fatty acids can pack together tightly, resulting in numerous intermolecular interactions that stabilize the fat. Saturated fats have a high melting point and tend to be solid at room temperature. Animal fats generally contain a high proportion of saturated fatty acids. For example, beef fat contains high amounts of stearic acid, a saturated fatty acid with a melting point of 70°C (see Figure 3.10). When you heat a hamburger on the stove, the stearic acid and other saturated animal fats melt, and liquid grease appears in the frying pan (**Figure 3.11**). When allowed to cool to room temperature, however, the liquid grease in the pan returns to its solid form.

Because of kinks in their chains, unsaturated fatty acids cannot stack together as tightly as saturated fatty acids. Fats high in unsaturated fatty acids usually have low melting points and are liquids at room temperature. Such fats are called oils. Fats derived from plants generally contain unsaturated fatty acids. For example, olive oil contains high amounts of oleic acid, a monounsaturated fatty acid with a melting point of 16°C. Fatty acids with additional double bonds

High temperature converts solid, saturated fats to liquid.

After cooling, saturated fats return to their solid form.

(a) Animal fats at high and low temperatures

Unsaturated fats have low melting points and are liquid at room temperature.

(b) Vegetable fats at low temperature

Figure 3.11 **Fats at different temperatures.** Saturated fats found in animals tend to have high melting points compared with unsaturated fats found in plants.

Concept Check: Certain types of fats used in baking are called shortenings. Shortenings are often made from vegetable oils by a process called hydrogenation, in which hydrogen causes double bonds to become single bonds. What do you think happens to such oils when they are hydrogenated?

have even lower melting points; linoleic acid (see Figure 3.10) has two double bonds and melts at –5°C. Safflower and sunflower oils contain high amounts of linoleic acid.

Most unsaturated fatty acids, including linoleic acid, exist in nature in the *cis* form (see Figure 3.4). Of particular importance to human health, however, are *trans* fatty acids, which are formed by an artificial process in which the natural *cis* form is altered to a *trans* configuration. This gives the fats that contain such fatty acids a more compact, linear structure and, therefore, a higher melting point. Although this process has been used for many years to produce fats with a longer shelf-life and with better characteristics for baking, it is now understood that *trans* fats are linked with human disease. Notable among these is coronary artery disease, caused by a narrowing of the blood vessels that supply the heart with blood.

Like starch and glycogen, fats are important for storing energy. The hydrolysis of triglycerides releases the fatty acids from glycerol, and these products can then be metabolized to provide energy to make ATP. Certain organisms, most notably mammals, have the ability to store large amounts of energy by accumulating fats. The number of C—H bonds in a molecule of fat or carbohydrate determines in part how much energy the molecule can yield. Fats are primarily long chains of C—H bonds, whereas glucose and other carbohydrates have numerous C—OH bonds. Consequently, 1 gram of fat stores more energy than does 1 gram of starch or glycogen. Fat is therefore an efficient means of energy storage for mobile organisms in which excess body mass may be a disadvantage. In animals, fats can also play a structural role by forming cushions that support organs. In addition, fats provide insulation under the skin that helps protect many terrestrial animals during cold weather and marine mammals in cold water.

Phospholipids Are Amphipathic Lipids

Phospholipids, another class of lipids, are similar in structure to triglycerides but with one important difference. The third hydroxyl group of glycerol is linked to a phosphate group instead of a fatty acid. In most phospholipids, a small polar or charged nitrogen-containing molecule is attached to this phosphate (**Figure 3.12a**). The glycerol backbone, phosphate group, and charged molecule constitute a polar hydrophilic region at one end of the phospholipid, whereas the fatty acid chains provide a nonpolar hydrophobic region at the opposite end. Recall from Chapter 2 that molecules with polar and nonpolar regions are called amphipathic molecules.

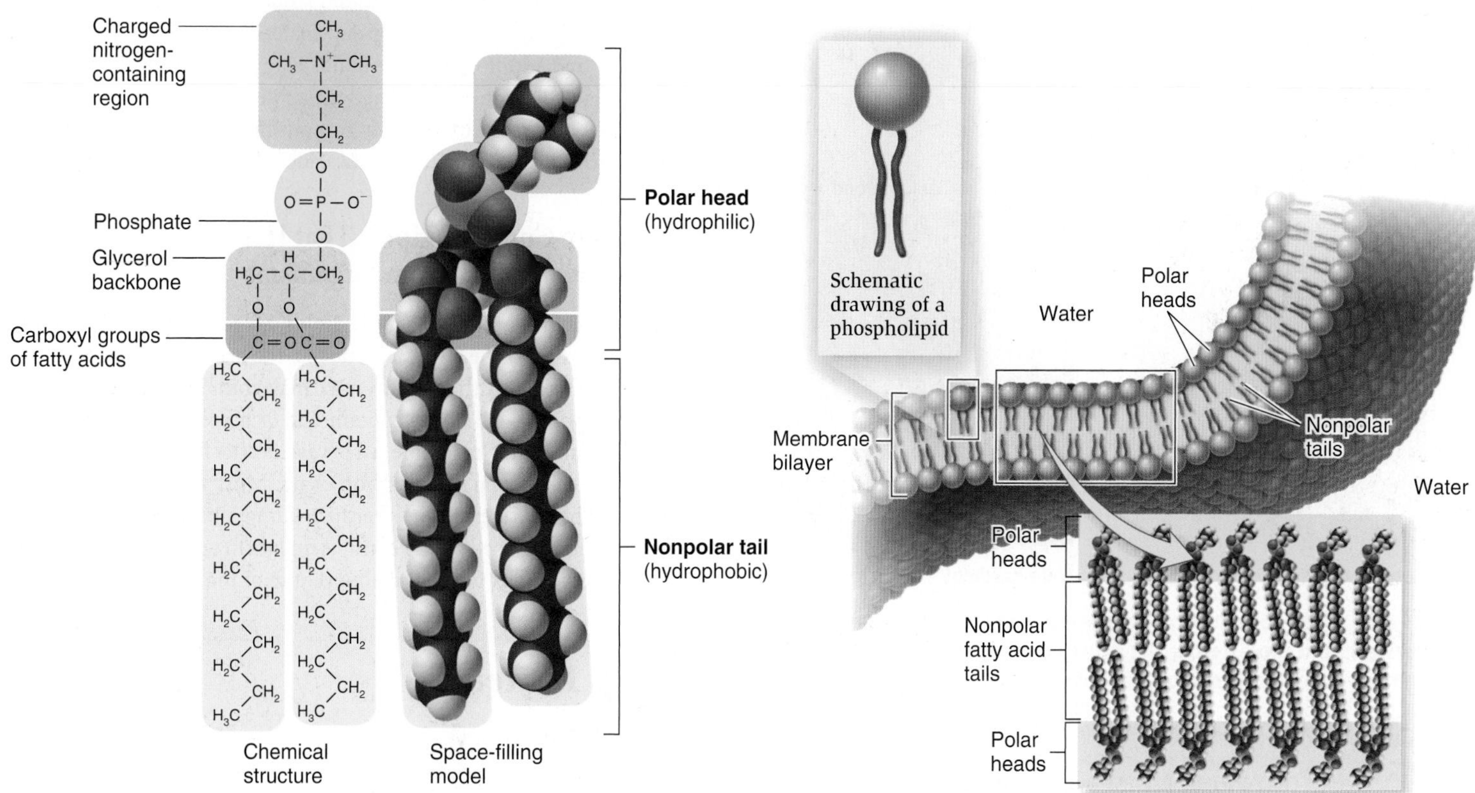

Figure 3.12 **Structure of phospholipids.** **(a)** Chemical structure and space-filling model of phosphatidylcholine, a common phospholipid found in living organisms. Phospholipids contain both polar and nonpolar regions, making them amphipathic. The fatty-acid tails are the nonpolar region. The rest of the molecule is polar. **(b)** Arrangement of phospholipids in a biological membrane, such as the plasma membrane that encloses cells. The hydrophilic regions of the phospholipid face the watery environments on either side of the membrane, whereas the hydrophobic regions associate with each other in the interior of the membrane, forming a bilayer.

Concept Check: *When water and oil are added to a test tube, the two liquids form two separate layers (think of oil and vinegar in a bottle of salad dressing). If a solution of phospholipids was added to a mixture of water and oil, where would the phospholipids dissolve?*

BioConnections: *For a more detailed view of the components of cellular membranes, including the phospholipid bilayer, look ahead to Figure 5.1.*

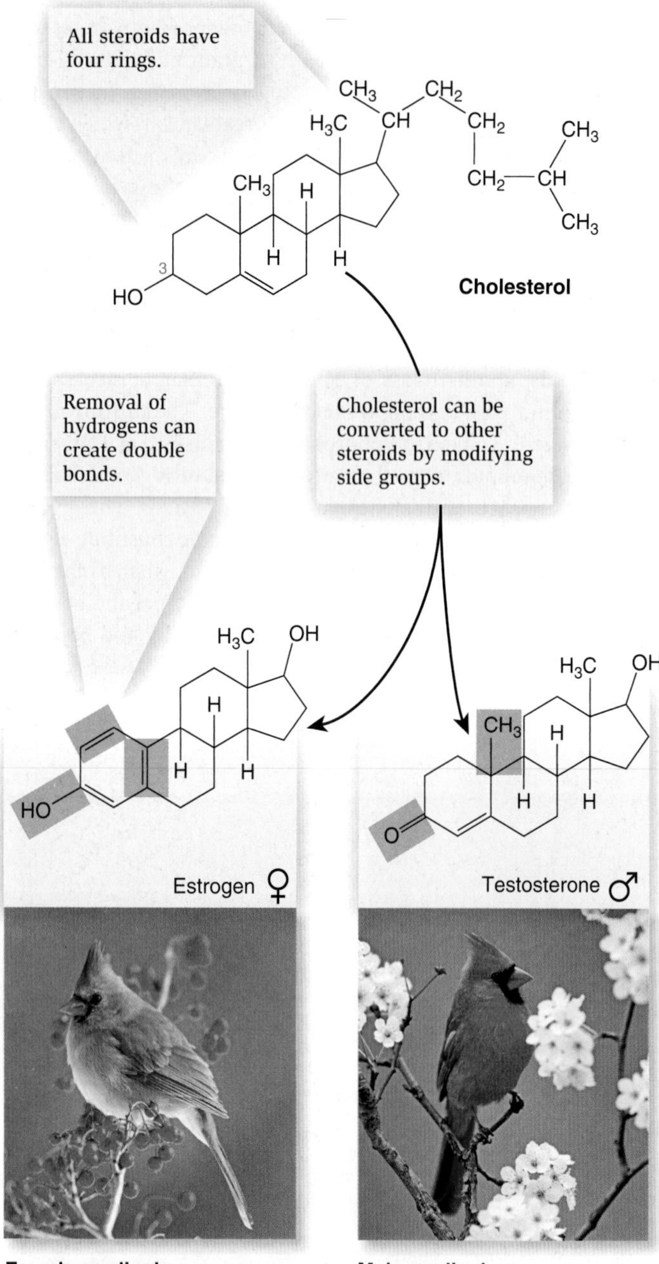

Figure 3.13 Structure of cholesterol and steroid hormones derived from cholesterol. The structure of a steroid has four rings. Steroids include cholesterol and molecules derived from cholesterol, such as steroid hormones. These include the reproductive hormones estrogen and testosterone.

BIOLOGY PRINCIPLE Structure determines function. The few seemingly minor chemical (structural) differences between estrogen and testosterone are sufficient to completely alter their biological functions. The striking differences in appearance of these two cardinals are just one example of sex-dependent differences in form and function in the animal world that are due to these two molecules.

In water, phospholipids become organized into bilayers, with their hydrophilic polar ends interacting with the water molecules and their hydrophobic nonpolar ends facing the interior, where they are shielded from water. As you will learn in detail in Chapter 5, this bilayer arrangement of phospholipids is critical for determining the structure of cellular membranes, as shown in Figure 3.12b.

Steroids Contain Ring Structures

Steroids have a distinctly different chemical structure from that of the other types of lipid molecules discussed thus far. Four fused rings of carbon atoms form the skeleton of all steroids. One or more polar hydroxyl groups are attached to this ring structure, but they are not numerous enough to make a steroid highly water-soluble. For example, steroids with a hydroxyl group are known as sterols—one of the most well known being cholesterol (Figure 3.13, top). Cholesterol is found in the blood and cellular membranes of animals. Due to its low solubility in water, at high concentrations cholesterol can contribute to the formation of blockages in major blood vessels.

In steroids, tiny differences in chemical structure can lead to profoundly different biological properties. For example, all steroid hormones are derived from cholesterol and share similarities in structure, but with some important differences. Estrogen is a steroid hormone found in high amounts in female vertebrates. Estrogen differs from testosterone, a steroid hormone found largely in males, by having one less methyl group, a hydroxyl group instead of a ketone group (see Table 3.1), and additional double bonds in one of its rings (Figure 3.13, bottom). However, these seemingly small differences are sufficient to make these two molecules largely responsible for whether an animal exhibits male or female characteristics, including feather color in birds.

Waxes Are Complex Lipids That Help Prevent Water Loss from Organisms

Many plants and animals produce lipids called waxes that are typically secreted onto their surface, such as the leaves of plants and the cuticles of insects. Although any wax may contain hundreds of different compounds, all waxes contain one or more hydrocarbons and long structures that resemble a fatty acid attached by its carboxyl group to another long hydrocarbon chain. Most waxes are very nonpolar and therefore exclude water, providing a barrier to water loss. They may also be used as structural elements in colonies like those of bees, where beeswax forms the honeycomb of the hive.

3.5 Proteins

Learning Outcomes:

1. Give examples of the general types of functions that are carried out in cells by different types of proteins.
2. Explain how amino acids are joined to form a polypeptide, and distinguish between a polypeptide and protein.
3. Describe the levels of protein structure and the factors that determine them.
4. Outline the bonding forces important in determining protein shape and function.
5. Explain what domains are and their importance in proteins.

Proteins are polymers found in all cells and play critical roles in nearly all life processes (Table 3.2). The word protein comes from the

Table 3.2 Major Categories and Functions of Proteins

Category	Functions	Examples
Proteins involved in gene expression and regulation	Make mRNA from a DNA template; synthesize polypeptides from mRNA; regulate genes	RNA polymerase assists in synthesizing RNA from DNA.
Motor proteins	Initiate movement	Myosin provides the contractile force of muscles.
Defense proteins	Protect organisms against disease	Antibodies help destroy bacteria or viruses.
Metabolic enzymes	Increase rates of chemical reactions	Hexokinase is an enzyme involved in sugar metabolism.
Cell signaling proteins	Enable cells to communicate with each other and with the environment	Taste receptors in the tongue allow animals to taste molecules in food.
Structural proteins	Support and strengthen structures	Actin provides shape to the cytoplasm of plant and animal cells. Collagen gives strength to tendons.
Transporters	Promote movement of solutes across plasma membranes	Glucose transporters move glucose from outside cells to inside cells, where it can be used for energy.

Greek *proteios* (meaning of the first rank), which aptly describes their importance. Proteins account for about 50% of the organic material in a typical animal's body.

Proteins Are Made Up of Amino Acid Monomers

Proteins are composed of carbon, hydrogen, oxygen, nitrogen, and small amounts of other elements, notably sulfur. The building blocks of proteins are **amino acids**, compounds with a structure in which a carbon atom, called the α-carbon, is linked to an amino group (NH_2) and a carboxyl group (COOH). The α-carbon also is linked to a hydrogen atom and a side chain, which is given a general designation R. Proteins are polymers of amino acids.

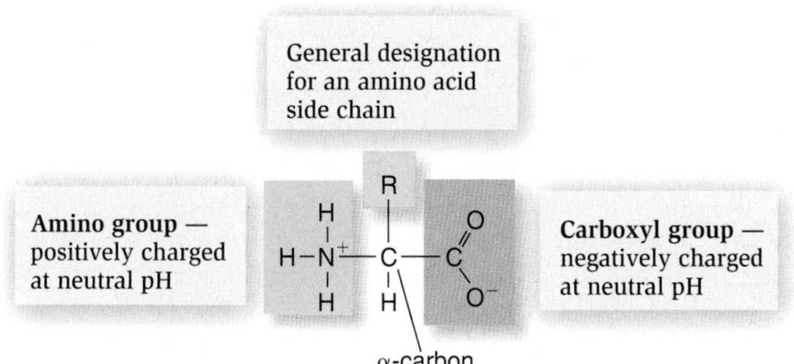

When an amino acid is dissolved in water at neutral pH, the amino group accepts a hydrogen ion and is positively charged, whereas the carboxyl group loses a hydrogen ion and is negatively charged. The term amino acid is the name given to such molecules because they have an amino group and also a carboxyl group that acts as an acid.

All amino acids except glycine may exist in more than one isomeric form, called the D and L forms, which are enantiomers. Note that glycine cannot exist in D and L forms because there are two hydrogens bound to its α-carbon, making it symmetric. Only L-amino acids and glycine are found in proteins. D-isomers are found in the cell walls of certain bacteria, where they may play a protective role against molecules secreted by the host organism in which the bacteria live.

The 20 amino acids found in proteins are distinguished by their side chains (Figure 3.14). The amino acids are categorized by the properties of their side chains, whether nonpolar, polar and uncharged, or polar and charged. The varying structures of the side chains are critical features of protein structure and function. The arrangement and chemical features of the side chains cause proteins to fold and adopt their three-dimensional shapes. In addition, certain amino acids may be critical in protein function. For example, amino acid side chains found within the active sites of enzymes are important in catalyzing chemical reactions.

Amino acids are joined together by a dehydration reaction that links the carboxyl group of one amino acid to the amino group of another (Figure 3.15a). The covalent bond formed between a carboxyl and amino group is called a **peptide bond**. When multiple amino acids are joined by peptide bonds, the resulting molecule is called a **polypeptide** (Figure 3.15b). The backbone of the polypeptide in Figure 3.15 is highlighted in yellow. The amino acid side chains project from the backbone. When two or more amino acids are linked together, one end of the resulting molecule has a free amino group. This is the amino end, or **N-terminus**. The other end of the polypeptide, called the carboxyl end, or **C-terminus**, has a free carboxyl group. As shown in Figure 3.15c, amino acids within a polypeptide are numbered from the N-terminus to the C-terminus.

The term polypeptide refers to a structural unit composed of a linear sequence of amino acids and does not imply anything about functionality. In other words, an artificial molecule with a random sequence of amino acids could be synthesized experimentally in the laboratory; this molecule would be a polypeptide but would have no function. In contrast, a **protein** is a molecule composed of one or more polypeptides that have been folded and twisted into a precise three-dimensional shape that carries out a particular function or functions. Many proteins also have carbohydrates (glycoproteins) or lipids (lipoproteins) attached at various points along their amino acid chain; these modifications impart unique functions to such proteins.

Proteins Have a Hierarchy of Structure

Scientists view protein structure at four progressive levels: primary, secondary, tertiary, and quaternary, shown schematically in Figure 3.16. Each higher level of structure depends on the preceding levels. For example, changing the primary structure may affect the secondary, tertiary, and quaternary structures. Let's now consider each level separately.

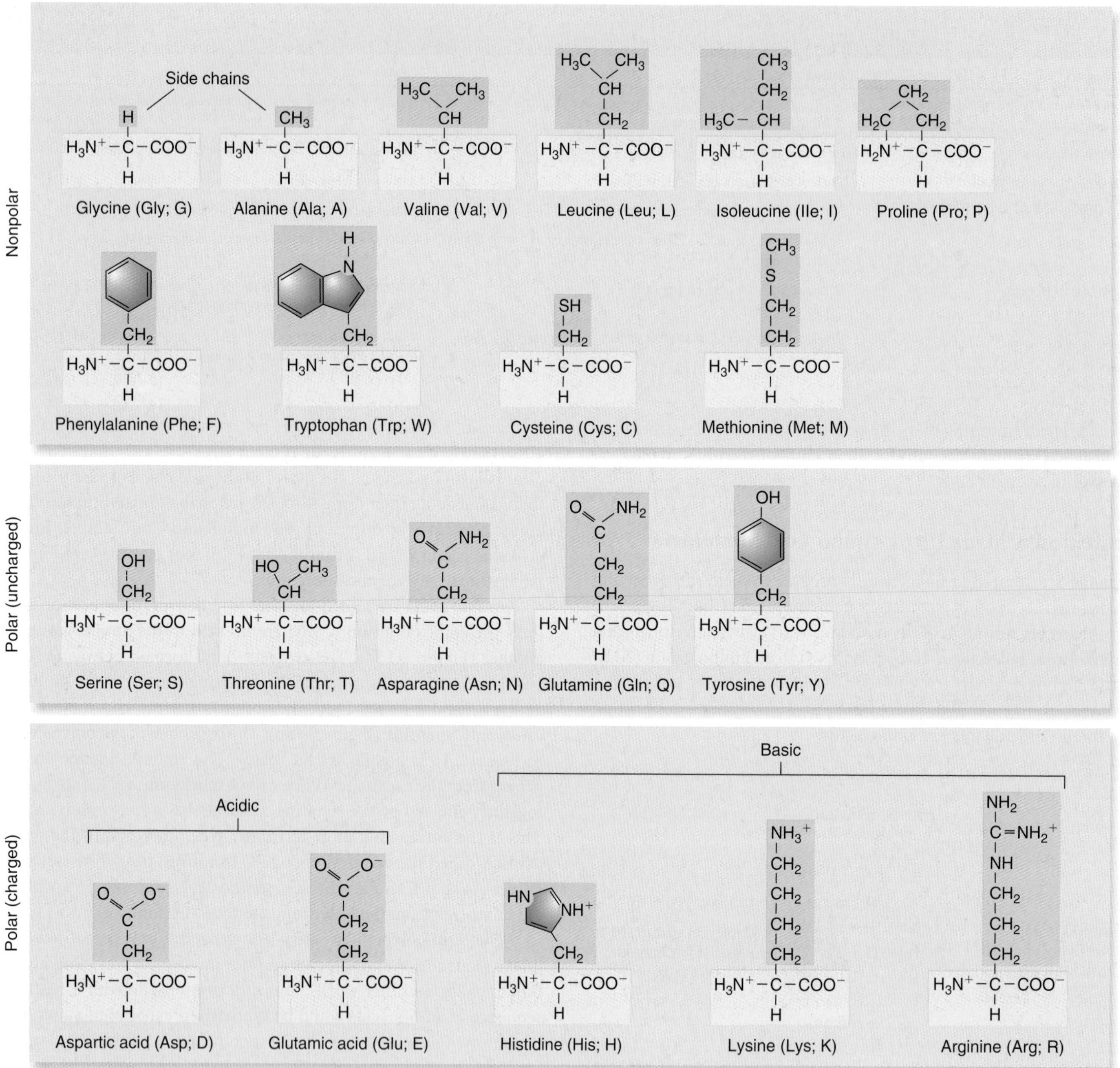

Figure 3.14 The 20 amino acids found in living organisms. The various amino acids have different chemical properties (for example, nonpolar versus polar) due to the nature of their different side chains. These properties contribute to the differences in the three-dimensional shapes and chemical properties of proteins, which, in turn, influence their biological functions. Tyrosine has both polar and nonpolar characteristics and is listed in just one category for simplicity. The common three-letter and one-letter abbreviations for each amino acid are shown in parentheses.

Primary Structure The **primary structure** (see Figure 3.16) of a protein is its amino acid sequence, from beginning to end. The primary structures of proteins are determined by genes. As we will explore in Chapter 12, genes carry the information for the production of polypeptides with specific amino acid sequences.

Figure 3.17 shows the primary structure of ribonuclease, which functions as an enzyme to degrade ribonucleic acid (RNA) molecules after they are no longer required by a cell. As described later and in Unit III of this textbook, RNA is a key part of the mechanism of protein synthesis. Ribonuclease is composed of a relatively short polypeptide consisting of 124 amino acids. An average polypeptide is about 300 to 500 amino acids in length, but some polypeptides form proteins that are a few thousand amino acids long.

Secondary Structure The amino acid sequence of a polypeptide, together with the fundamental constraints of chemistry and physics,

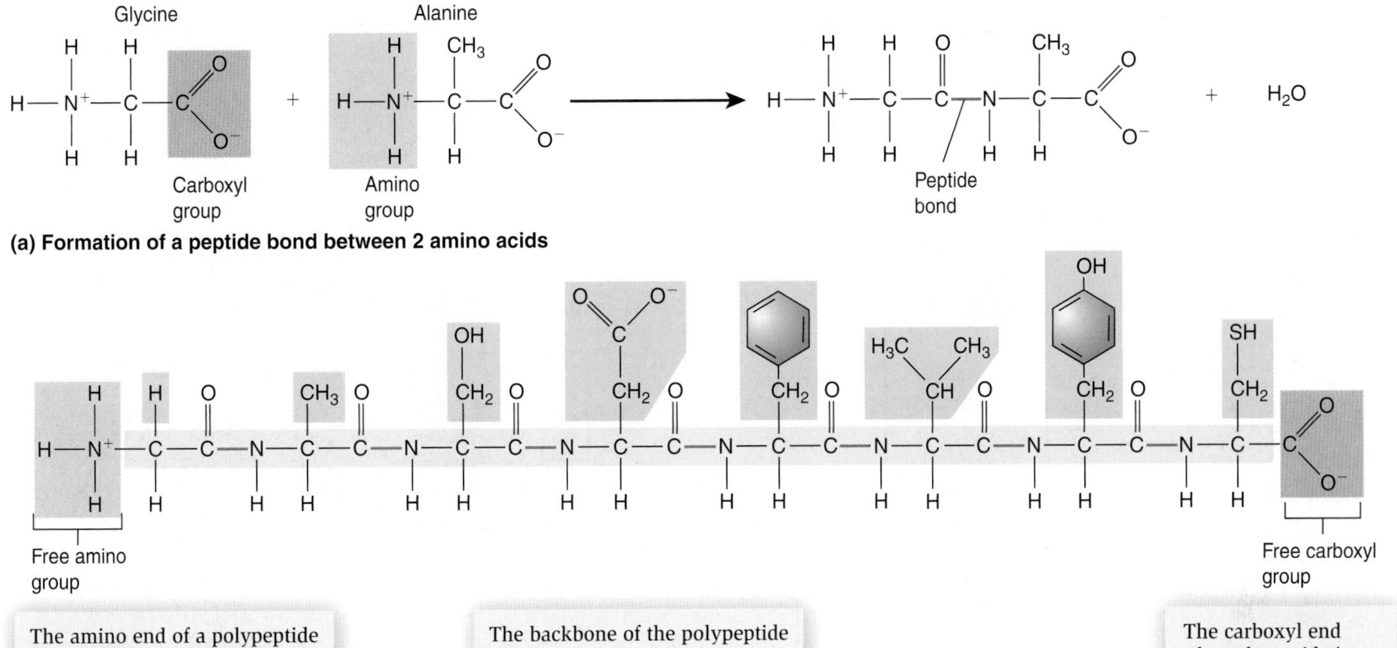
Glycine
Alanine
Carboxyl group
Amino group
Peptide bond
H_2O
(a) Formation of a peptide bond between 2 amino acids
Free amino group
Free carboxyl group
The amino end of a polypeptide is called the N-terminus.
The backbone of the polypeptide is highlighted in yellow.
The carboxyl end of a polypeptide is called the C-terminus.
(b) Polypeptide—a linear chain of amino acids

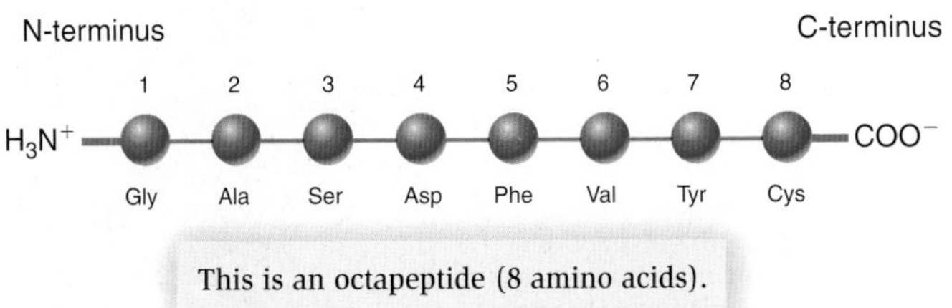
N-terminus
C-terminus
1 2 3 4 5 6 7 8
H_3N^+
COO^-
Gly Ala Ser Asp Phe Val Tyr Cys
This is an octapeptide (8 amino acids).

(c) Numbering system of amino acids in a polypeptide

Figure 3.15 The chemistry of polypeptide formation. Polypeptides are polymers of amino acids. They are formed by linking amino acids via dehydration reactions to make peptide bonds. Every polypeptide has an amino end, or N-terminus, and a carboxyl end, or C-terminus.

Concept Check: How many water molecules would be produced in making a polypeptide that is 72 amino acids long by dehydration reactions?

cause a protein to fold into a more compact structure. Amino acids can rotate around bonds within a polypeptide. Consequently, proteins are flexible and can fold into a number of shapes, just as a string of beads can be twisted into many configurations. Folding can be irregular, or certain regions can have a repeating folding pattern. Such repeating patterns are called **secondary structure**. The two basic types of a protein's secondary structure are the α helix and the β pleated sheet.

In an α helix, the polypeptide backbone forms a repeating helical structure that is stabilized by hydrogen bonds along the length of the backbone. As shown in Figure 3.16, the hydrogen linked to a nitrogen atom forms a hydrogen bond with an oxygen atom that is double-bonded to a carbon atom. These hydrogen bonds occur at regular intervals within the polypeptide backbone and cause the backbone to twist into a helix.

In a β pleated sheet, regions of the polypeptide backbone come to lie parallel to each other. Hydrogen bonds between a hydrogen linked to a nitrogen atom and a double-bonded oxygen form between these adjacent, parallel regions. When this occurs, the polypeptide backbone adopts a repeating zigzag, or pleated, shape.

The α helices and β pleated sheets are key determinants of a protein's characteristics. For example, α helices in certain proteins are composed primarily of nonpolar amino acids. Proteins containing many such regions with an α helix structure tend to anchor themselves into a lipid-rich environment, such as a cell's plasma membrane. In this way, a protein whose function is required in a specific location such as a plasma membrane can be retained there. Secondary structure also contributes to the great strength of certain proteins, including the keratins found in hair and hooves; the proteins that make up the silk webs of spiders; and collagen, the chief component of cartilage in vertebrate animals.

Some regions along a polypeptide chain do not assume an α helix or β pleated sheet conformation and consequently do not have a secondary structure. These regions are sometimes called random coiled regions. However, this term is somewhat misleading because the shapes of random coiled regions are usually very specific and important for the protein's function.

Tertiary Structure As the secondary structure of a polypeptide becomes established due to the particular primary structure, side chains of amino acids interact with each other. The polypeptide folds and refolds upon itself to assume a complex three-dimensional shape—its **tertiary structure** (see Figure 3.16). The tertiary structure is the three-dimensional shape of a single polypeptide. Tertiary

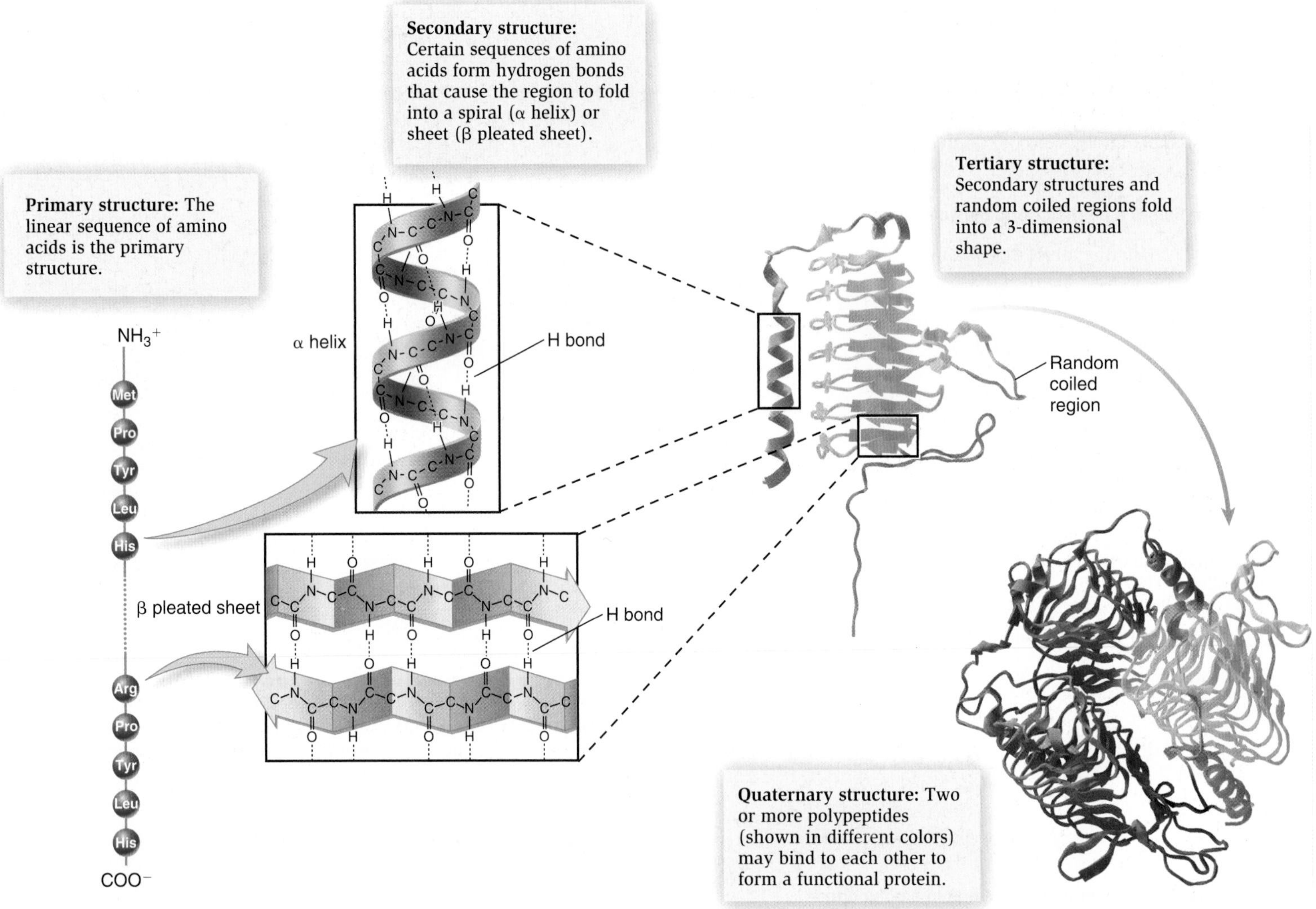

Secondary structure: Certain sequences of amino acids form hydrogen bonds that cause the region to fold into a spiral (α helix) or sheet (β pleated sheet).

Primary structure: The linear sequence of amino acids is the primary structure.

Tertiary structure: Secondary structures and random coiled regions fold into a 3-dimensional shape.

NH_3^+

Met Pro Tyr Leu His

α helix

H bond

β pleated sheet

H bond

Arg Pro Tyr Leu His

COO^-

Random coiled region

Quaternary structure: Two or more polypeptides (shown in different colors) may bind to each other to form a functional protein.

Figure 3.16 The hierarchy of protein structure. The R groups are omitted for simplicity.

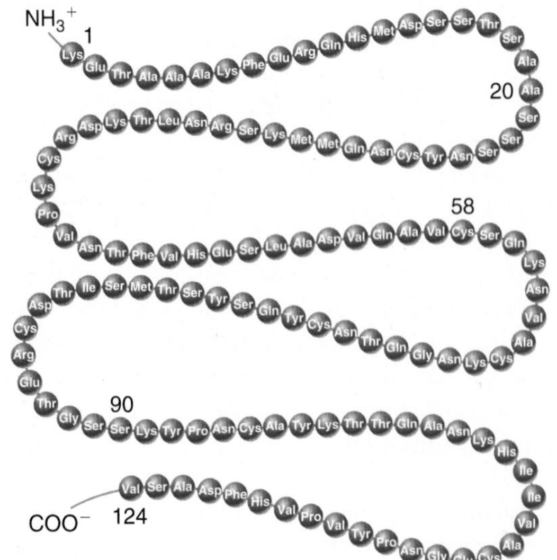

Figure 3.17 The primary structure of ribonuclease. The example shown here is ribonuclease from cows, which contains 124 amino acids.

structure includes all secondary structures plus any interactions involving amino acid side chains. For some proteins, such as ribonuclease, the tertiary structure is the final structure of a functional protein. However, as described next, other proteins are composed of two or more polypeptides and adopt a quaternary structure.

Quaternary Structure Many functional proteins are composed of two or more polypeptide chains that each adopt a tertiary structure and then assemble with each other (see Figure 3.16). The individual polypeptide chains are called **protein subunits**. Subunits may be identical polypeptides or they may be different. When proteins consist of more than one polypeptide chain, they are said to have **quaternary structure** and are also known as **multimeric proteins** (meaning multiple parts). Multimeric proteins are widespread in organisms. A common example is the oxygen-binding protein called hemoglobin, found in the red blood cells of vertebrate animals. Four protein subunits combine to form one hemoglobin protein. Each subunit can bind a single molecule of oxygen; therefore, each hemoglobin protein can carry four molecules of oxygen in the blood.

Protein Structure Is Influenced by Several Factors

The amino acid sequences of polypeptides are the features that distinguish the structure of one protein from another. As polypeptides are synthesized in a cell, they fold into secondary and tertiary structures, which assemble into quaternary structures for many proteins. Several factors determine the way proteins adopt their secondary, tertiary, and quaternary structures. As mentioned, the laws of chemistry and physics, together with the amino acid sequence, govern this process. As shown in Figure 3.18, five factors are critical for protein folding and stability:

1. *Hydrogen bonds*—The large number of weak hydrogen bonds within a polypeptide and between polypeptide chains adds up to a collectively strong force that promotes protein folding and stability. As we already learned, hydrogen bonding is a critical determinant of protein secondary structure and also is important in tertiary and quaternary structure.
2. *Ionic bonds and other polar interactions*—Some amino acid side chains are positively or negatively charged. Positively charged side chains may bind to negatively charged side chains via ionic bonds. Similarly, uncharged polar side chains in a protein may bind to ionic amino acids. Ionic bonds and polar interactions are particularly important in tertiary and quaternary structure.
3. *Hydrophobic effect*—Some amino acid side chains are nonpolar. These amino acids tend to exclude water. As a protein folds, the hydrophobic amino acids are likely to be found in the center of the protein, minimizing contact with water. As mentioned, some proteins have stretches of nonpolar amino acids that anchor them in the hydrophobic portion of membranes. The hydrophobic effect plays a major role in tertiary and quaternary structures.
4. *van der Waals forces*—Atoms within molecules have temporary weak attractions for each other if they are an optimal distance apart. This weak attraction is termed the van der Waals force (see Chapter 2). If two atoms are very close together, their electron clouds will repel each other. If they are far apart, the van der Waals force will diminish. The van der Waals forces are particularly important in determining tertiary structures.

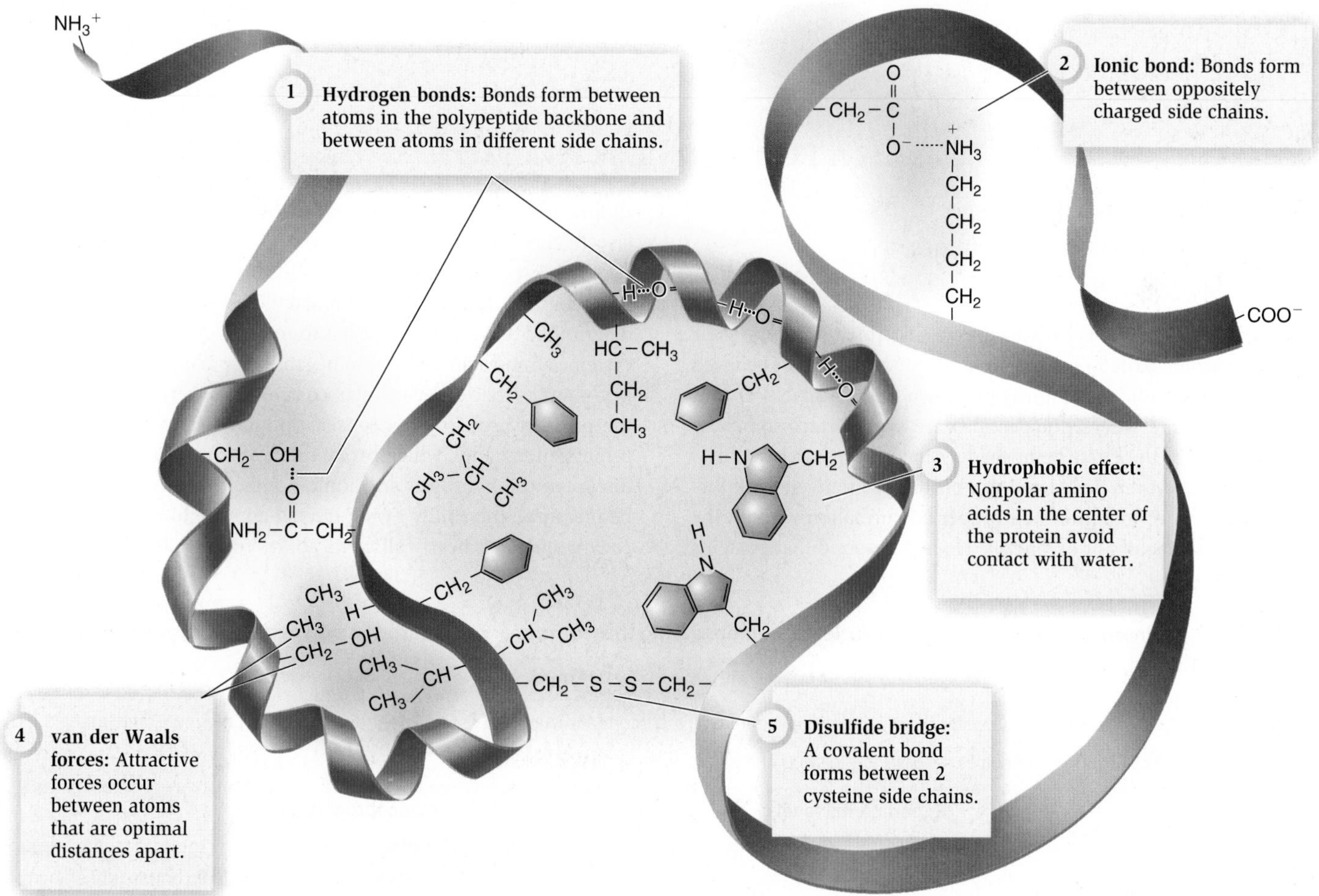

Figure 3.18 **Factors that influence protein folding and stability.**

BIOLOGY PRINCIPLE New properties emerge from complex interactions. This key principle of biology is apparent even at the molecular level. Note in this figure how several distinct types of chemical interactions can produce a protein with a complex shape. Compare this with the chapter-opening photo of a real protein, and the several intermediate levels of protein structure shown in Figure 3.16. It is the three-dimensional shape of different proteins that determines their ability to interact with other molecules, including other proteins.

5. *Disulfide bridges*—The side chain of the amino acid cysteine contains a sulfhydryl group (—SH), which can react with a sulfhydryl group in another cysteine side chain (see Figure 3.14). The result is a **disulfide bridge** or disulfide bond, which links the two amino acid side chains together (—S—S—). Disulfide bridges are covalent bonds that can occur within a polypeptide or between different polypeptides. Though other forces are usually more important in protein folding, the covalent nature of disulfide bridges can help to stabilize the tertiary structure of a protein.

The first four factors just described are also important in the ability of different proteins to interact with each other. As discussed throughout Unit II and other parts of this textbook, many cellular processes involve steps in which two or more different proteins interact with each other. For this to occur, one protein must recognize and bind to the surface of the other. Such binding is usually very specific. The surface of one protein precisely fits into the surface of another (Figure 3.19). Such **protein-protein interactions** are critically important so that cellular processes can occur in a series of defined steps. In addition, protein-protein interactions are important in building complicated cellular structures that provide shape and organization to cells.

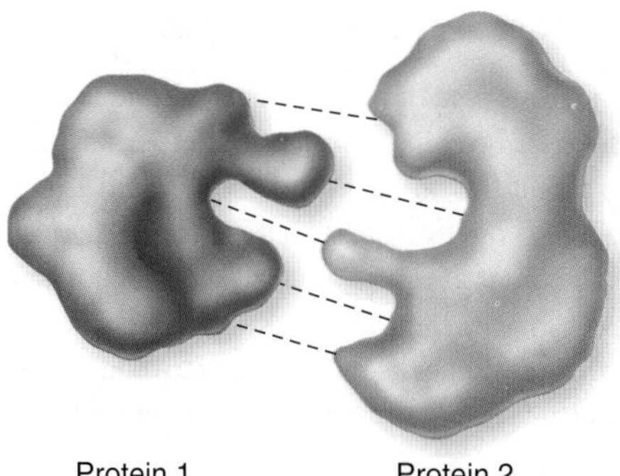

Figure 3.19 Protein-protein interaction. Two different proteins may interact with each other due to hydrogen bonding, ionic bonding, the hydrophobic effect, and van der Waals forces.

Concept Check: *If the primary structure of protein 1 in this figure were experimentally altered by the substitution of several incorrect amino acids for the correct ones, would protein 1 still be able to interact with protein 2?*

FEATURE INVESTIGATION

Anfinsen Showed That the Primary Structure of Ribonuclease Determines Its Three-Dimensional Structure

Prior to the 1960s, the mechanisms by which proteins assume their three-dimensional structures were not understood. Scientists believed either that correct folding required unknown cellular factors or that ribosomes, the site where polypeptides are synthesized, somehow shaped proteins as they were being made. American researcher Christian Anfinsen, however, postulated that proteins contain all the information necessary to fold into their proper conformation without the need for cellular factors or organelles. He hypothesized that proteins spontaneously assume their most stable conformation based on the laws of chemistry and physics (Figure 3.20).

To test this hypothesis, Anfinsen studied ribonuclease, an enzyme that degrades RNA molecules. Biochemists had already determined that ribonuclease has four disulfide bridges between eight cysteine amino acids. Anfinsen began with purified ribonuclease. The key point is that other cellular components were not present, only the purified protein. He exposed ribonuclease to a chemical called β-mercaptoethanol, which broke the S—S bonds, and to urea, which disrupted the hydrogen and ionic bonds. Following this treatment, he measured the ability of the treated enzyme to degrade RNA. The enzyme had lost nearly all of its ability to degrade RNA. Therefore,

Figure 3.20 Anfinsen's experiments with ribonuclease, demonstrating that the primary structure of a polypeptide plays a key role in protein folding.

HYPOTHESIS Within their amino acid sequence, proteins contain all the information needed to fold into their correct, three-dimensional shapes.

KEY MATERIALS Purified ribonuclease, RNA, denaturing chemicals, size-exclusion columns.

	Experimental level	Conceptual level	
1 Incubate purified ribonuclease in test tube with RNA, and measure its ability to degrade RNA.		Purified ribonuclease; S-S	Numerous H bonds and ionic bonds (not shown) and 4 S—S bonds. Protein is properly folded.

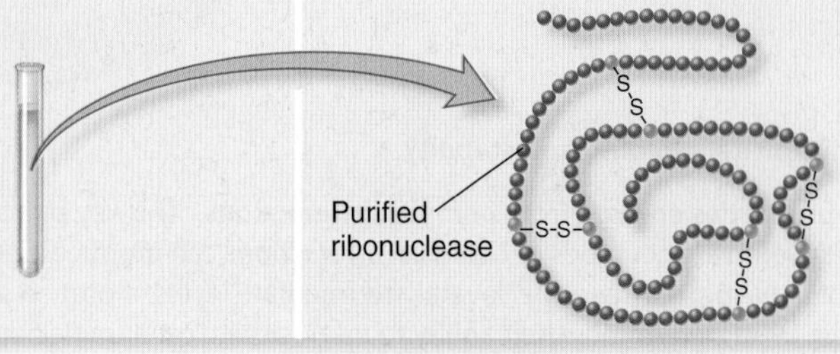

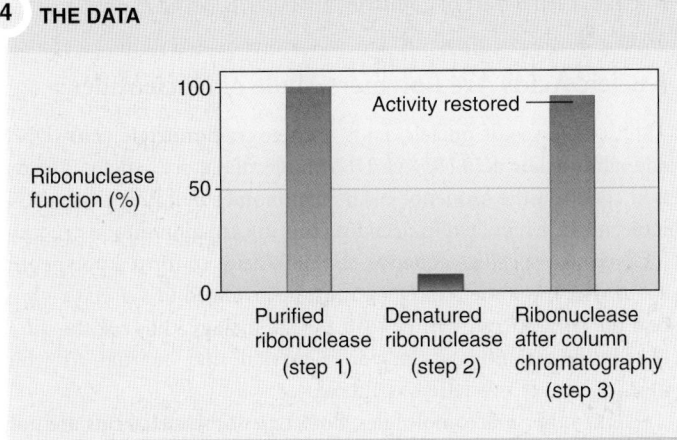

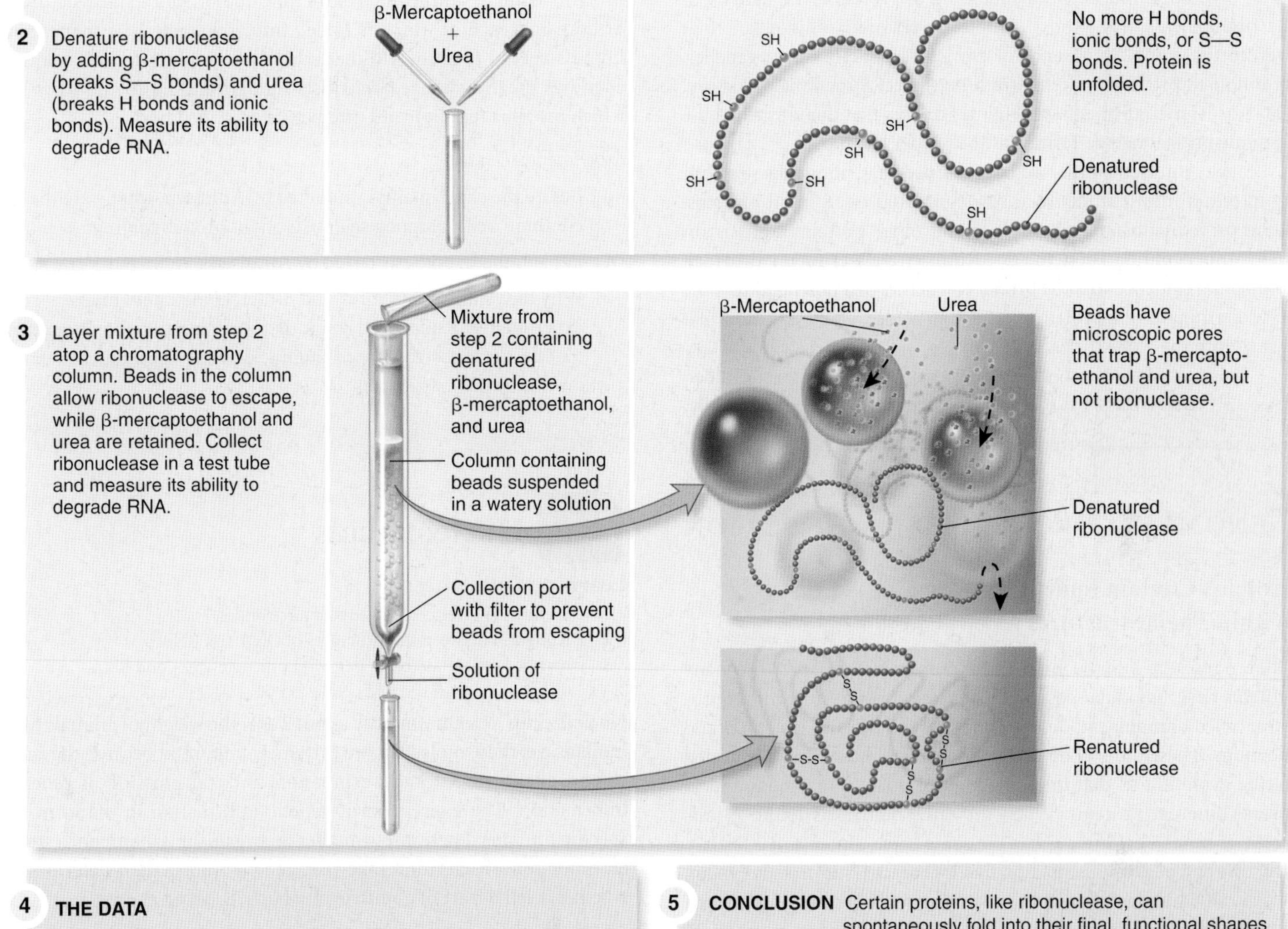

2 Denature ribonuclease by adding β-mercaptoethanol (breaks S—S bonds) and urea (breaks H bonds and ionic bonds). Measure its ability to degrade RNA.
β-Mercaptoethanol + Urea
SH
No more H bonds, ionic bonds, or S—S bonds. Protein is unfolded.
Denatured ribonuclease
3 Layer mixture from step 2 atop a chromatography column. Beads in the column allow ribonuclease to escape, while β-mercaptoethanol and urea are retained. Collect ribonuclease in a test tube and measure its ability to degrade RNA.
Mixture from step 2 containing denatured ribonuclease, β-mercaptoethanol, and urea
Column containing beads suspended in a watery solution
Collection port with filter to prevent beads from escaping
Solution of ribonuclease
β-Mercaptoethanol
Urea
Beads have microscopic pores that trap β-mercapto-ethanol and urea, but not ribonuclease.
Denatured ribonuclease
-S-S-
Renatured ribonuclease

4 **THE DATA**

Ribonuclease function (%)

100

50

0

Activity restored

Purified ribonuclease (step 1)

Denatured ribonuclease (step 2)

Ribonuclease after column chromatography (step 3)

5 **CONCLUSION** Certain proteins, like ribonuclease, can spontaneously fold into their final, functional shapes without assistance from other cellular structures or factors. (However, as described in the text, this is not true of many other proteins.)

6 **SOURCE** Haber, E., and Anfinsen, C.B. 1961. Regeneration of enzyme activity by air oxidation of reduced subtilisin-modified ribonuclease. *Journal of Biological Chemistry* 236:422–424.

Anfinsen concluded that when ribonuclease was unfolded or denatured, it was no longer functional.

The key step in this experiment came when Anfinsen removed the urea and β-mercaptoethanol from the solution. Because these molecules are much smaller than ribonuclease, removing them from the solution was accomplished with a technique called size-exclusion chromatography. In size-exclusion chromatography, solutions are layered atop a glass column of beadlike particles and allowed to filter down through the column to an open collection port at the bottom. The particles in the column have microscopic pores that trap small molecules like urea and mercaptoethanol but that permit large molecules such as ribonuclease to pass down the length of the column and out the collection port.

Using size-exclusion chromatography, Anfinsen was able to purify the ribonuclease out of the original solution. He then allowed the purified enzyme to sit in water for up to 20 hours, after which he retested the ribonuclease for its ability to degrade RNA. The result revolutionized our understanding of proteins. The activity of the ribonuclease was almost completely restored! This meant that even in the complete absence of any cellular factors or organelles, an unfolded protein can refold into its correct, functional structure. This was later confirmed by chemical analyses that demonstrated the disulfide bridges had re-formed at the proper locations.

Since Anfinsen's time, we have learned that ribonuclease's ability to refold into its functional structure is not seen in all proteins. Some proteins do require enzymes and other proteins to assist in their proper folding. Nonetheless, Anfinsen's experiments provided compelling evidence that the primary structure of a polypeptide is the key determinant of a protein's tertiary structure, an observation that earned him the Nobel Prize in Chemistry for 1972.

As investigations into the properties of proteins have continued since Anfinsen's classic experiments, it has become clear that most proteins contain within their structure one or more substructures, or domains, each of which is folded into a characteristic shape that imparts special functions to that region of the protein. This knowledge has greatly changed scientists' understanding of the ways in which proteins function and interact, as described next.

Experimental Questions

1. Before the experiments conducted by Anfinsen, what were the common beliefs among scientists about protein folding?
2. Explain the hypothesis tested by Anfinsen.
3. Why did Anfinsen use urea and β-mercaptoethanol in his experiments? Explain the result that was crucial to the discovery that the tertiary structure of ribonuclease may depend entirely on the primary structure.

GENOMES & PROTEOMES CONNECTION

Proteins Contain Functional Domains Within Their Structures

Modern research into the functions of proteins has revealed that many proteins have a modular design. This means that portions within proteins, called **domains**, modules, or motifs, have distinct structures and functions. These units of amino acid sequences have been duplicated during evolution so that the same kind of domain may be found in several different proteins. When the same domain is found in different proteins, the domain has the same three-dimensional tertiary structure and performs a function that is characteristic of that domain.

As an example, Figure 3.21 shows a member of a family of related proteins that are known to play critical roles in regulating how certain genes are turned on and off in living cells. This protein is called a signal transducer and activator of transcription (STAT) protein.

Each domain of this protein is involved in a distinct biological function, a common occurrence in proteins with multiple domains. For example, one of the domains is labeled the SH2 domain (see Figure 3.21). Many different proteins contain this domain. It allows such proteins to recognize other proteins in a very specific way. The function of SH2 domains is to bind to tyrosine amino acids to which phosphate groups have been added by cellular enzymes. When an amino acid receives a phosphate group in this way, it is said to be phosphorylated. As might be predicted, proteins that contain SH2 domains all bind to phosphorylated tyrosines in the proteins they recognize. As a second example, a STAT protein has another domain called a DNA-binding domain. This portion of the protein has a structure that specifically binds to DNA.

Overall, the domain structure of proteins enables them to have multiple, discrete regions, each with its own structure and purpose in the functioning of the protein. Individual domain sequences and structures are often highly conserved across all life and can be thought of as evolutionary modules; they may have originated as discrete proteins of their own that were later combined and rearranged in multiple, different ways to produce new proteins.

3.6 Nucleic Acids

Learning Outcomes:

1. Describe the three components of nucleotides.
2. Distinguish between the structures of DNA and RNA.
3. Describe how certain bases pair with others in DNA and RNA.

Nucleic acids account for only about 2% of the weight of animals like humans, yet these molecules are extremely important because they are responsible for the storage, expression, and transmission of genetic information. The expression of genetic information in the form of specific proteins determines whether an organism is a human, a frog, an onion, or a bacterium. Likewise, genetic information determines whether a cell is part of a muscle or a bone, a leaf or a root.

Nucleic Acids Are Polymers Made of Nucleotides

The two classes of nucleic acids are **deoxyribonucleic acid (DNA)** and **ribonucleic acid (RNA)**. DNA molecules store genetic information coded in the sequence of their monomer building blocks. RNA molecules are involved in decoding this information into instructions for linking a specific sequence of amino acids to form a polypeptide chain. The monomers in DNA must be arranged in a precise way so that the correct code can be read. As an analogy, think of the difference in the meanings of the words "marital" and "martial," in which the sequence of two letters is altered.

Like other macromolecules, both types of nucleic acids are polymers consisting of linear sequences of repeating monomers. Each monomer, known as a **nucleotide**, has three components: (1) a phosphate group, (2) a pentose (five-carbon) sugar (either ribose or deoxyribose), and (3) a single or a double ring of carbon and nitrogen atoms known as a base (Figure 3.22). A nucleotide of DNA is called a deoxyribonucleotide; that of RNA is a ribonucleotide. The phosphates and sugar molecules form the backbone of a DNA or RNA strand, with the bases projecting from the backbone. The carbon atoms of the sugar are numbered 1′–5′ (**Figure 3.23**).The phosphate groups link the 3′ carbon of one nucleotide to the 5′ carbon of the next (see Figure 3.23).

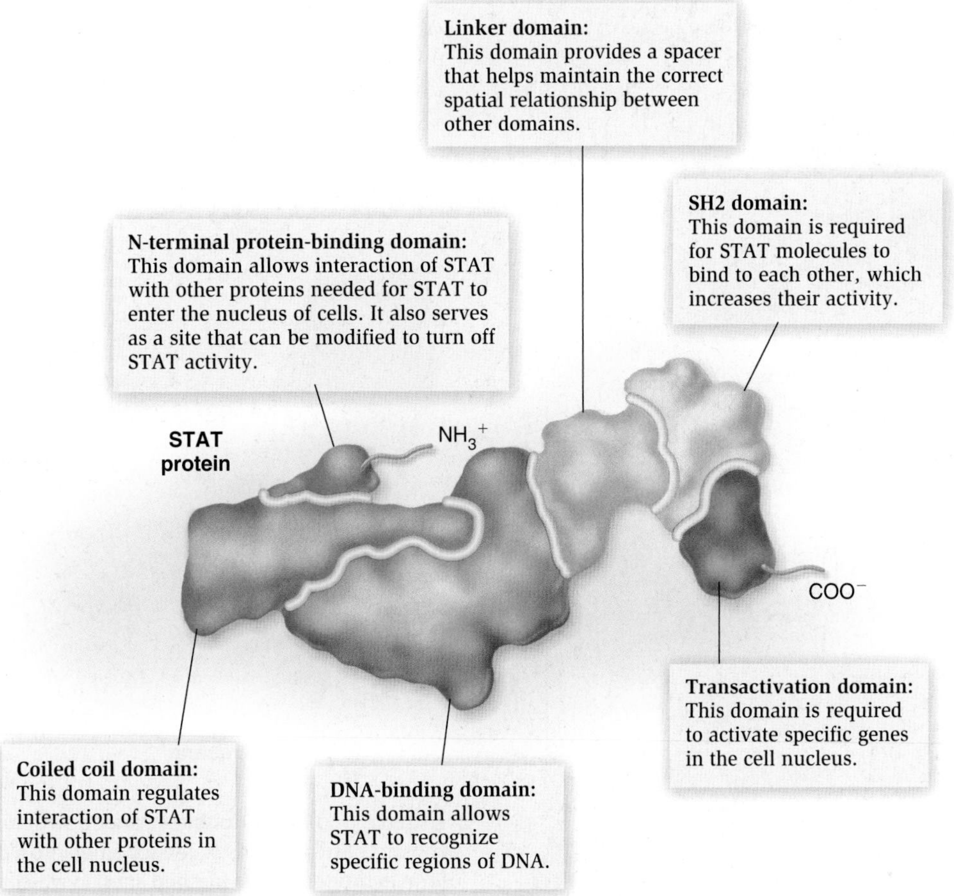

Linker domain:
This domain provides a spacer that helps maintain the correct spatial relationship between other domains.
SH2 domain:
This domain is required for STAT molecules to bind to each other, which increases their activity.
N-terminal protein-binding domain:
This domain allows interaction of STAT with other proteins needed for STAT to enter the nucleus of cells. It also serves as a site that can be modified to turn off STAT activity.
STAT protein
NH_3^+
COO^-
Transactivation domain:
This domain is required to activate specific genes in the cell nucleus.
Coiled coil domain:
This domain regulates interaction of STAT with other proteins in the cell nucleus.
DNA-binding domain:
This domain allows STAT to recognize specific regions of DNA.

Figure 3.21 The domain structure of a signal transducer and activator of transcription (STAT) protein.

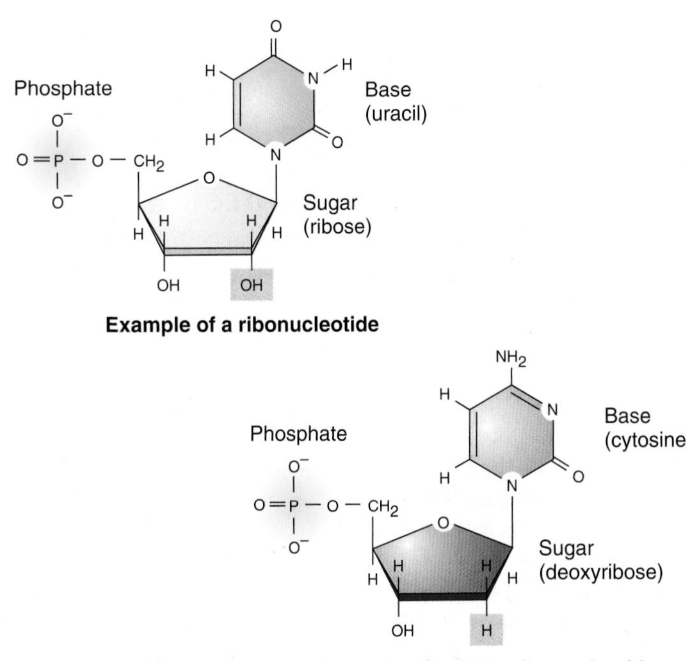

Figure 3.22 Examples of two nucleotides. A nucleotide has a phosphate group, a five-carbon sugar, and a nitrogenous base.

DNA Is Composed of Purines and Pyrimidines

The nucleotides in DNA contain the five-carbon sugar **deoxyribose**. Four different nucleotides are present in DNA, corresponding to the four different bases that can be linked to deoxyribose. The **purine** bases, **adenine (A)** and **guanine (G)**, have fused double rings of carbon and nitrogen atoms, and the **pyrimidine** bases, **cytosine (C)** and **thymine (T)**, have a single-ring structure (see Figure 3.23).

A DNA molecule consists of two strands of nucleotides coiled around each other to form a double helix (Figure 3.24). Purines and pyrimidines occur in both strands. The two strands are held together by hydrogen bonds between a purine base in one strand and a pyrimidine base in the opposite strand. The ring structure of each base lies in a flat plane perpendicular to the sugar-phosphate backbone, somewhat like steps on a spiral staircase.

As we will see in Chapter 11, only certain bases can pair with others, due to the location of the hydrogen-bonding groups in the four bases (see Figure 3.24). Two hydrogen bonds can be formed between adenine and thymine (A-T pairing), but three hydrogen bonds are formed between guanine and cytosine (G-C pairing). In a DNA molecule, A on one strand is always paired with T on another strand, and G with C. This base pairing maintains a constant distance between the sugar-phosphate backbones of the two strands as they coil around each other. If we know the amount of one type of base

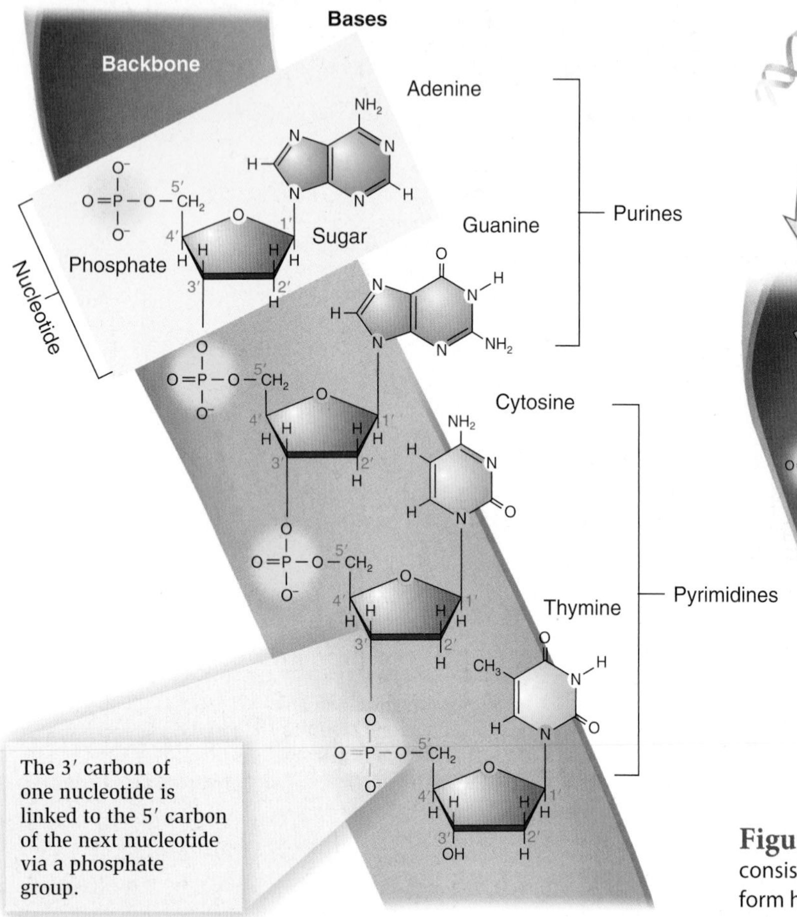

The 3′ carbon of one nucleotide is linked to the 5′ carbon of the next nucleotide via a phosphate group.

Figure 3.23 Structure of a DNA strand. Nucleotides are linked to each other to form a strand of DNA. The four bases found in DNA are shown. A strand of RNA is similar except the sugar is ribose, and uracil is substituted for thymine.

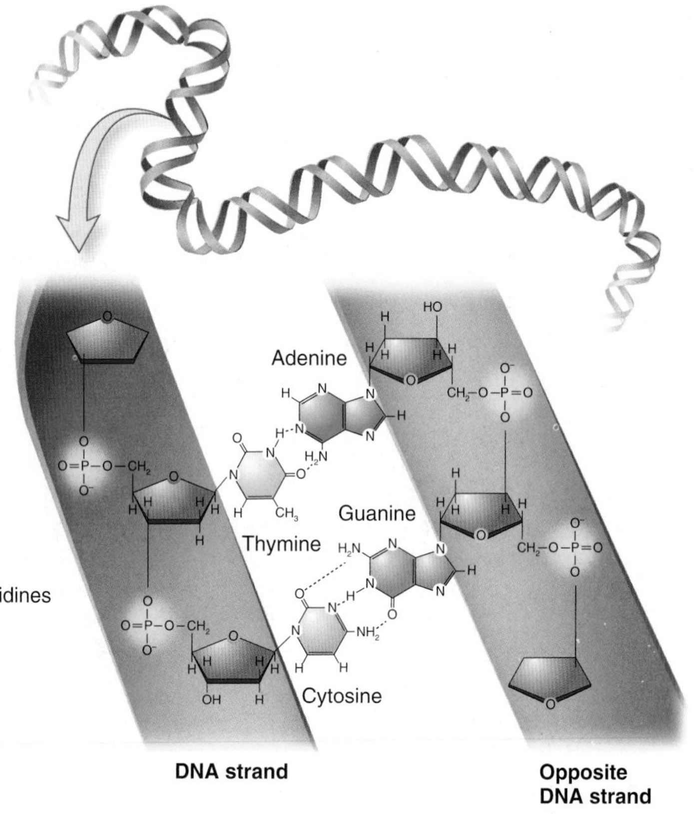

Figure 3.24 The double-stranded structure of DNA. DNA consists of two strands coiled together into a double helix. The bases form hydrogen bonds (dashed lines) in which A pairs with T, and G pairs with C.

Concept Check: If the sequence of bases in one strand of a DNA double helix is known, can the base sequence of the opposite strand be predicted?

in a DNA molecule, we can predict the relative amounts of each of the other three bases. For example, if a DNA molecule were composed of 20% A bases, then there must also be 20% T bases. That leaves 60% of the bases that must be G and C combined. Because the amounts of G and C must be equal, this particular DNA molecule must be composed of 30% each of G and C. This specificity provides the mechanism for duplicating and transferring genetic information (see Chapter 11).

RNA Is Usually Single Stranded and Comes in Several Forms

RNA molecules differ in only a few respects from DNA. Except in some viruses, RNA consists of a single rather than double strand of nucleotides. In RNA, the sugar in each nucleotide is **ribose** rather than deoxyribose. Also, the pyrimidine base thymine in DNA is replaced in RNA with the pyrimidine base **uracil (U)** (see Figure 3.22). The other three bases—adenine, guanine, and cytosine—are found in both DNA and RNA. Certain forms of RNA called messenger RNA (mRNA), ribosomal RNA (rRNA), and transfer RNA (tRNA) are responsible for converting the information contained in DNA into the formation of a new polypeptide. This topic will be discussed in Chapter 12.

Summary of Key Concepts

3.1 The Carbon Atom and the Study of Organic Molecules

- Organic chemistry is the science of studying carbon-containing molecules, which are compounds found in living organisms (Figure 3.1).
- One property of the carbon atom that makes life possible is its ability to form four covalent bonds (polar or nonpolar) with other atoms. The combination of different elements and different types of bonds allows a vast number of organic compounds to be formed from a relatively few chemical elements (Figures 3.2, 3.3).
- Carbon bonds are stable at the different temperatures associated with life.
- Organic compounds may contain functional groups (Table 3.1).
- Carbon-containing molecules can exist as isomers, which have identical molecular composition but different structures and

characteristics. Structural isomers contain the same atoms but in different bonding relationships. Stereoisomers have identical bonding relationships but different spatial positioning of their atoms. Two types of stereoisomers are *cis-trans* isomers and enantiomers (Figure 3.4).

3.2 Formation of Organic Molecules and Macromolecules

- The four major classes of organic molecules are carbohydrates, lipids, proteins, and nucleic acids. Organic molecules exist as monomers or polymers. Polymers are large macromolecules built up by a type of condensation reaction called a dehydration reaction, in which individual monomers combine with each other. Polymers are broken down into monomers by hydrolysis reactions (Figure 3.5).

3.3 Carbohydrates

- Carbohydrates are composed of carbon, hydrogen, and oxygen atoms. Cells can break down glucose, an important carbohydrate, releasing energy which is then stored in the bonds of ATP.
- Carbohydrates include monosaccharides (the simplest sugars), disaccharides, and polysaccharides. The polysaccharides starch (in plant cells) and glycogen (in animal cells) store energy in cells. Some polysaccharides, notably cellulose, serve a support or structural function (Figures 3.6, 3.7, 3.8).

3.4 Lipids

- Lipids, composed predominantly of hydrogen and carbon atoms, are nonpolar and very insoluble in water. Major classes of lipids include fats, phospholipids, steroids, and waxes.
- Fats, also called triglycerides, are formed by bonding glycerol with three fatty acids. In a saturated fatty acid, all the carbons are linked by single covalent bonds. Unsaturated fatty acids contain one or more C=C double bonds. Animal fats generally contain a high proportion of saturated fatty acids, and vegetable fats contain more unsaturated fatty acids (Figures 3.9, 3.10, 3.11).
- Phospholipids are similar in structure to triglycerides, except that one glycerol is linked to a phosphate group instead of a fatty acid. They contain both polar and nonpolar regions, making them amphipathic (Figure 3.12).
- Steroids are constructed of four fused rings of carbon atoms. Small differences in steroid structure can lead to profoundly different biological properties, such as the differences between estrogen and testosterone (Figure 3.13).
- Waxes, another class of lipids, are nonpolar and repel water, and are often found as protective coatings on the leaves of plants and the outer surfaces of animals' bodies.

3.5 Proteins

- Proteins are composed of carbon, hydrogen, oxygen, nitrogen, and small amounts of other elements such as sulfur. Proteins are macromolecules that play critical roles in almost all life processes. The proteins of all living organisms are composed of the same set of 20 amino acids, corresponding to 20 different side chains (Figure 3.14, Table 3.2).
- Amino acids are joined by linking the carboxyl group of one amino acid to the amino group of another, forming a peptide bond. A polypeptide is a structural unit composed of amino acids. A protein is a functional unit composed of one or more polypeptides that have been folded and twisted into precise three-dimensional shapes (Figure 3.15).
- The four levels of protein structure are primary (its amino acid sequence), secondary (α helices or β pleated sheets), tertiary (folding to assume a three-dimensional shape), and quaternary (multimeric proteins that consist of more than one polypeptide chain). The three-dimensional structure of a protein determines its function—for example, by creating binding sites for other molecules (Figures 3.16, 3.17, 3.18, 3.19, 3.20, 3.21).

3.6 Nucleic Acids

- Nucleic acids are responsible for the storage, expression, and transmission of genetic information. The two types of nucleic acids are deoxyribonucleic acid (DNA) and ribonucleic acid (RNA). Both are molecules consisting of repeating monomers known as nucleotides. Each nucleotide is composed of a phosphate group, a five-carbon sugar (either ribose or deoxyribose), and a single or double ring of carbon and nitrogen atoms called a base (Figures 3.22, 3.23).
- A DNA molecule consists of two strands of nucleotides coiled around each other to form a double helix, held together by hydrogen bonds between a purine base (adenine or guanine) on one strand and a pyrimidine base (cytosine or thymine) on the opposite strand. DNA molecules store genetic information coded in the sequence of their monomers (Figure 3.24).
- RNA consists of a single strand of nucleotides. The sugar in each nucleotide is ribose rather than deoxyribose, and the base uracil replaces thymine. RNA molecules are involved in decoding this information into instructions for linking amino acids in a specific sequence to form a polypeptide chain.

Assess and Discuss

Test Yourself

1. Molecules that contain the element __________ are considered organic molecules.
   a. hydrogen
   b. carbon
   c. oxygen
   d. nitrogen
   e. calcium
2. __________ was the first scientist to synthesize an organic molecule. The organic molecule synthesized was _________.
   a. Kolbe, urea
   b. Wöhler, urea
   c. Wöhler, acetic acid
   d. Kolbe, acetic acid
   e. Wöhler, glucose

3. The versatility of carbon to serve as the backbone for a variety of different molecules is due to
   a. the ability of carbon atoms to form four covalent bonds.
   b. the fact that carbon usually forms ionic bonds with many different atoms.
   c. the abundance of carbon in the environment.
   d. the ability of carbon to form covalent bonds with many different types of atoms.
   e. both a and d.
4. _________ are molecules that have the same molecular composition but differ in structure and/or bonding association.
   a. Isotopes
   b. Isomers
   c. Free radicals
   d. Analogues
   e. Ions
5. ____________ is a storage polysaccharide commonly found in the cells of animals.
   a. Glucose
   b. Sucrose
   c. Glycogen
   d. Starch
   e. Cellulose
6. In contrast to other fatty acids, essential fatty acids
   a. are always saturated fats.
   b. cannot be synthesized by the organism and are necessary for survival.
   c. can act as building blocks for large, more complex macromolecules.
   d. are the simplest form of lipids found in plant cells.
   e. are structural components of plasma membranes.
7. Phospholipids are amphipathic, which means they
   a. are partially hydrolyzed during cellular metabolism.
   b. are composed of a hydrophilic portion and a hydrophobic portion.
   c. may be poisonous to organisms if in combination with certain other molecules.
   d. are molecules composed of lipids and proteins.
   e. are all of the above.
8. The monomers of proteins are __________, and these are linked by polar covalent bonds commonly referred to as ______________ bonds.
   a. nucleotides, peptide
   b. amino acids, ester
   c. hydroxyl groups, ester
   d. amino acids, peptide
   e. monosaccharides, glycosidic
9. A ________ is a portion of protein with a particular structure and function.
   a. peptide bond
   b. domain
   c. phospholipids
   d. wax
   e. monosaccharide
10. The ________ of a nucleotide determines whether it is a component of DNA or a component of RNA.
   a. phosphate group
   b. five-carbon sugar
   c. side chain
   d. fatty acid
   e. Both b and d are correct.

Conceptual Questions

1. Distinguish between different types of isomers.
2. Explain the difference between saturated and unsaturated fatty acids, and how the structural differences between them contribute to differences in their properties.
3. A principle of biology is that *structure determines function.* What does this mean for organic molecules such as carbohydrates, lipids, and proteins?

Collaborative Questions

1. Discuss the differences between different types of carbohydrates.
2. Discuss some of the roles that proteins play in organisms.

Online Resource

www.brookerbiology.com

Stay a step ahead in your studies with animations that bring concepts to life and practice tests to assess your understanding. Your instructor may also recommend the interactive eBook, individualized learning tools, and more.

UNIT II

CELL

Cell biology is the study of life at the cellular level. Although cells are the simplest units of life, biologists have come to realize that they are wonderfully complex and interesting, providing information about all living things. In this unit, Chapter 4 begins with an overview of cell structure and function and explores the technique of microscopy. In Chapter 5, we will examine the structure and synthesis of cell membranes and the transport of substances in and out of the cell. Chapters 6, 7, and 8 are largely devoted to metabolism, the sum of the chemical reactions in a cell or organism. Chapter 6 explores the topic of energy and considers how enzymes facilitate chemical reactions in a cell. Chapter 7 examines the pathways for carbohydrate breakdown and how those pathways are used to make an energy intermediate called ATP. In Chapter 8, we will explore the process of photosynthesis in which the energy of sunlight drives the synthesis of carbohydrates. Finally, Chapters 9 and 10 consider the ways that cells interact with their environment and with each other. In Chapter 9, we will examine how cells respond to signals, either those that come directly from their environment or those that are made by other cells. In Chapter 10, we will explore how cells interact with each other to produce a multicellular organism.

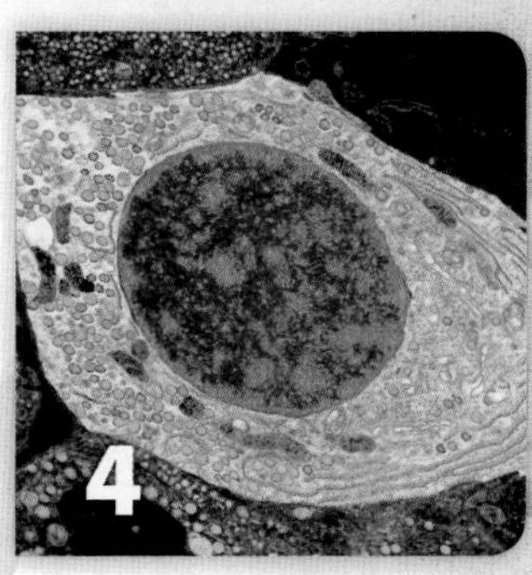

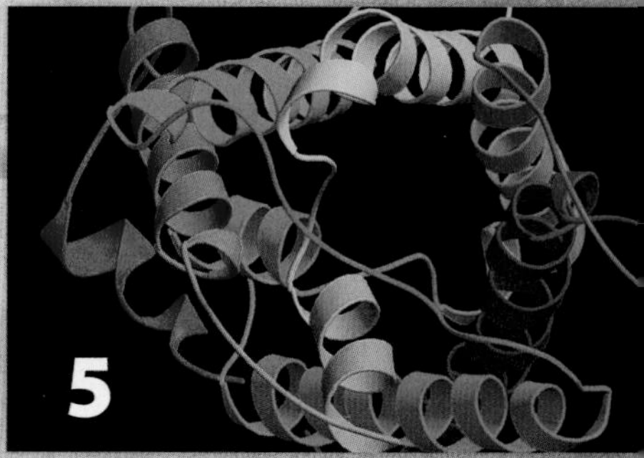

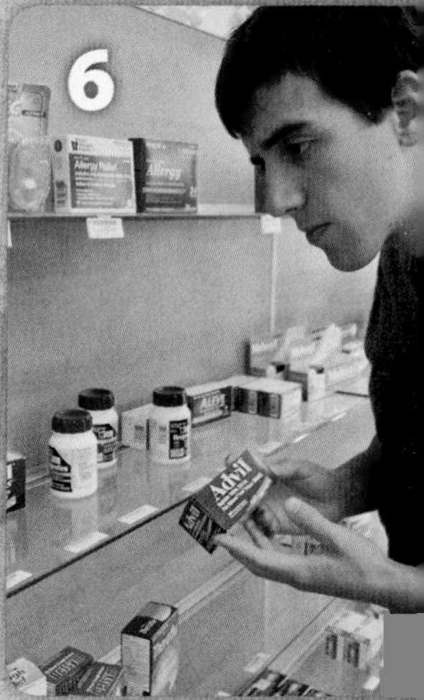

The following biology principles will be emphasized in this unit:

- ***Cells are the simplest units of life:*** *Throughout this unit, we will examine how life exists at the cellular level.*
- ***Living organisms use energy:*** *Chapters 6, 7, and 8 examine how cells utilize and store energy contained within organic molecules such as glucose.*
- ***Living organisms interact with their environment:*** *Chapters 5 and 9 largely focus on the plasma membrane of the cell and how it provides an interface between the cell and its environment.*
- ***Structure determines function:*** *Throughout this unit, we will see many examples in which the structure of proteins determines their cellular functions.*
- ***Biology is an experimental science:*** *Every chapter has a Feature Investigation that describes a pivotal experiment that provided insights into the workings of cells.*

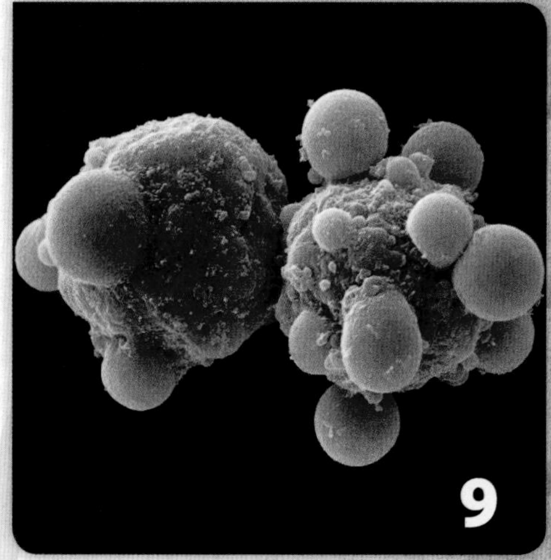

4

5

6

7

8

9

10

General Features of Cells

4

Chapter Outline

4.1 Microscopy
4.2 Overview of Cell Structure
4.3 The Cytosol
4.4 The Nucleus and Endomembrane System
4.5 Semiautonomous Organelles
4.6 Protein Sorting to Organelles
4.7 Systems Biology of Cells: A Summary
Summary of Key Concepts
Assess and Discuss

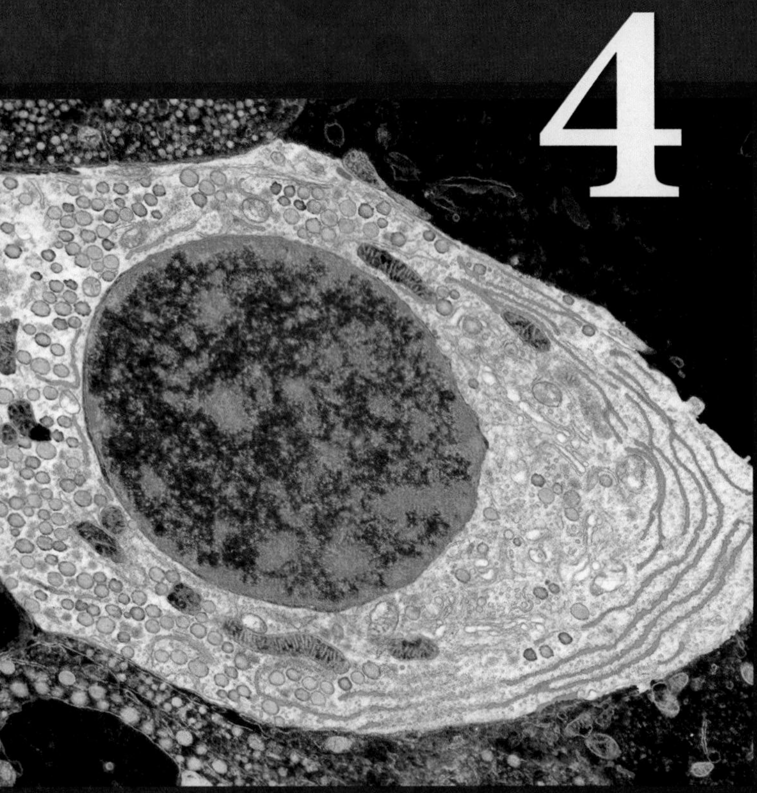

A cell from the pituitary gland. The cell in this micrograph was viewed by a technique called transmission electron microscopy, which is described in this chapter. The micrograph was artificially colored using a computer to enhance the visualization of certain cell structures.

Emily had a persistent cough ever since she started smoking cigarettes in college. However, at age 35, it seemed to be getting worse, and she was alarmed by the occasional pain in her chest. When she began to lose weight and noticed that she became easily fatigued, Emily decided to see a doctor. The diagnosis was lung cancer. Despite aggressive treatment of the disease with chemotherapy and radiation therapy, she succumbed to lung cancer 14 months after the initial diagnosis. Emily was 36.

Topics such as cancer are within the field of **cell biology**—the study of individual cells and their interactions with each other. Researchers in this field want to understand the basic features of cells and apply their knowledge in the treatment of diseases such as cystic fibrosis, sickle-cell disease, and lung cancer.

The idea that organisms are composed of cells originated in the mid-1800s. German botanist Matthias Schleiden studied plant material under the microscope and was struck by the presence of many similar-looking compartments, each of which contained a dark area. Today we call those compartments cells and the dark area the nucleus. In 1838, Schleiden speculated that cells are living entities and that plants are aggregates of cells arranged according to definite laws.

Schleiden was a good friend of the German physiologist Theodor Schwann. Over dinner one evening, their conversation turned to the nuclei of plant cells, and Schwann remembered having seen similar structures in animal tissue. Schwann conducted additional studies that showed large numbers of nuclei in animal tissue at regular intervals and also located in cell-like compartments. In 1839, Schwann extended Schleiden's hypothesis to animals. About two decades later, German biologist Rudolf Virchow proposed that *omnis cellula e cellula*, or "every cell originates from another cell." This idea arose from his research, which showed that diseased cells divide to produce more diseased cells.

The **cell theory**, which is credited to both Schleiden and Schwann with contributions from Virchow, has three parts.

1. All living organisms are composed of one or more cells.
2. Cells are the smallest units of life.
3. New cells come only from pre-existing cells by cell division.

Most cells are so small they cannot be seen with the unaided eye. However, as cell biologists have begun to unravel cell structure and function at the molecular level, the cell has emerged as a unit of incredible complexity and adaptability. In this chapter, we will begin our examination of cells with an overview of their structures and functions. Later chapters in this unit will explore certain aspects of cell biology in greater detail. But first, let's look at the tools and techniques that allow us to observe cells.

4.1 Microscopy

Learning Outcomes:

1. Explain the three important parameters in microscopy: resolution, contrast, and magnification.
2. Compare and contrast the different types of light and electron microscopes and their uses.

The **microscope** is a magnification tool that enables researchers to study the structure and function of cells. A **micrograph** is an image taken with the aid of a microscope. The first compound microscope—a

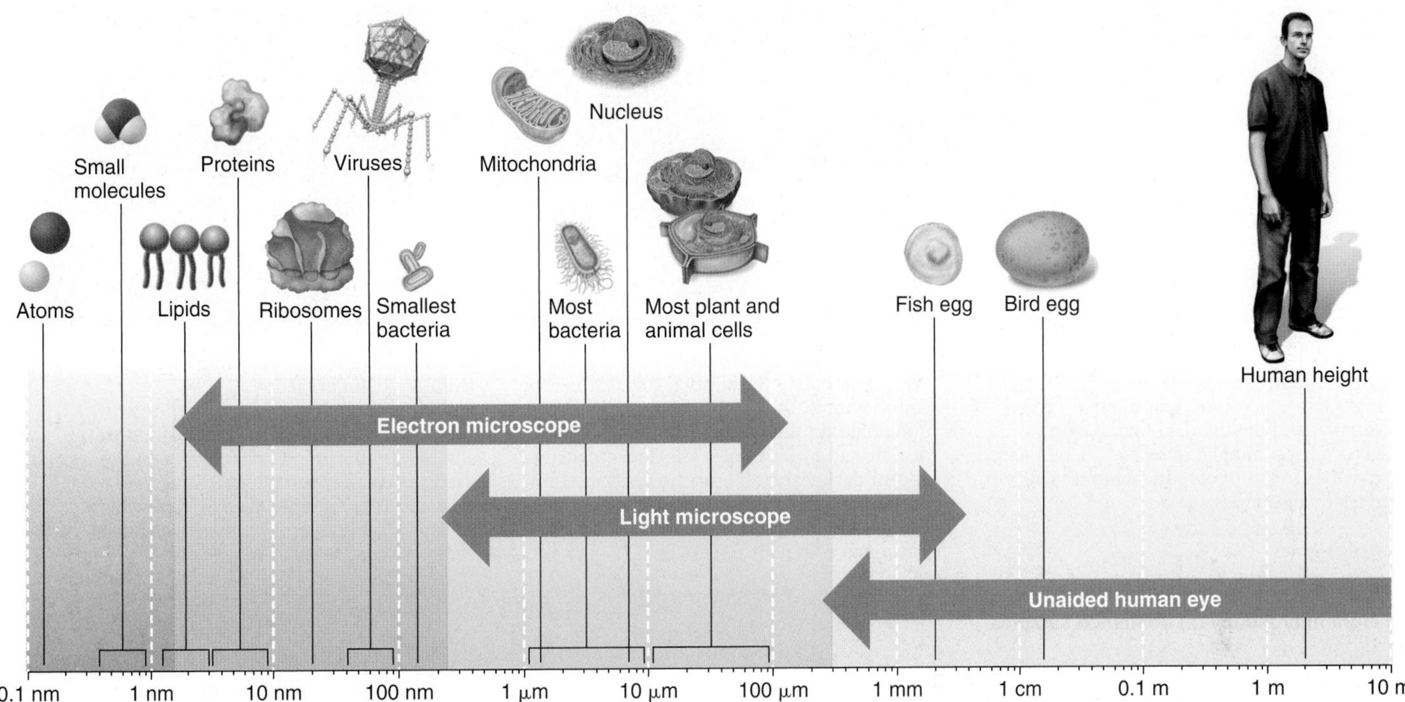

Figure 4.1 A comparison of the sizes of various chemical and biological structures, and the resolving power of the unaided eye, light microscope, and electron microscope. The scale at the bottom is logarithmic to accommodate the wide range of sizes in this drawing.

Concept Check: *Which type of microscopy would you use to observe a virus?*

microscope with more than one lens—was invented in 1595 by Zacharias Jansen of Holland. In 1665, an English biologist, Robert Hooke, studied cork under a primitive compound microscope he had made. He actually observed cell walls because cork cells are dead and have lost their internal components. Hooke coined the word *cell*, derived from the Latin word *cellula*, meaning small compartment, to describe the structures he observed. Ten years later, the Dutch merchant Anton van Leeuwenhoek refined techniques of making lenses and was able to observe single-celled microorganisms such as bacteria.

Three important parameters in microscopy are resolution, contrast, and magnification. **Resolution**, a measure of the clarity of an image, is the ability to observe two adjacent objects as distinct from one another. For example, a microscope with good resolution enables a researcher to distinguish two adjacent chromosomes as separate objects, which would appear as a single object under a microscope with poor resolution. The second important parameter in microscopy is **contrast**. The ability to visualize a particular cell structure may depend on how different it looks from an adjacent structure. Staining the cellular structure of interest with a dye can make viewing much easier. The application of stains, which selectively label individual components of the cell, greatly improves contrast. As described later, fluorescent molecules are often used to selectively stain cellular components. However, staining should not be confused with colorization. Many of the micrographs shown in this textbook are colorized, or artificially colored, to emphasize certain cellular structures, such as different parts of a cell (see the chapter opener, for example). In colorization, particular colors are added to micrographs with the aid of a computer. Finally, **magnification** is the ratio between the size of an image produced by a microscope and its actual size. For example, if the image size is 100 times larger than its actual size, the magnification is designated 100×. Depending on the quality of the lens and illumination source, every microscope has an optimal range of magnification before objects appear too blurry to be readily observed.

Microscopes are categorized into two groups based on the source of illumination. A **light microscope** utilizes light for illumination, whereas an **electron microscope** uses a beam of electrons for illumination. Very good light microscopes resolve structures that are as close as 0.2 μm (micron, or micrometer) from each other. The resolving power of a microscope depends on several factors, including the wavelength of the source of illumination. Resolution is improved when the illumination source has a shorter wavelength. A major advance in microscopy occurred in 1931 when Max Knoll and Ernst Ruska invented the first electron microscope. Because the wavelength of an electron beam is much shorter than visible light, the resolution of the electron microscope is far better than any light microscope. For biological samples, the resolution limit is typically around 2 nm (nanometers), which is about 100 times better than the light microscope. **Figure 4.1** shows the range of resolving powers of the electron microscope, light microscope, and unaided eye and compares them with various cells and cell structures.

Over the past several decades, enormous technological advances have made light microscopy a powerful research tool. Improvements in lens technology, microscope organization, sample preparation, sample illumination, and computerized image processing have enabled researchers to invent different types of light microscopes, each with its own advantages and disadvantages (**Figure 4.2**).

Similarly, improvements in electron microscopy occurred during the 1930s and 1940s, and by the 1950s, the electron microscope

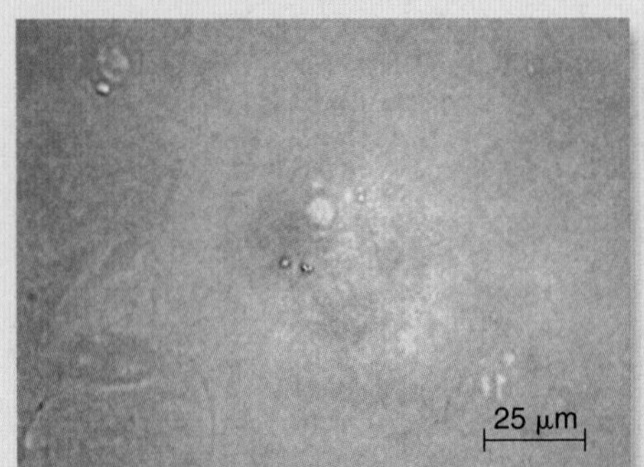

Standard light microscopy (bright field, unstained sample). Light is passed directly through a sample, and the light is focused using glass lenses. Simple, inexpensive, and easy to use but offers little contrast with unstained samples.

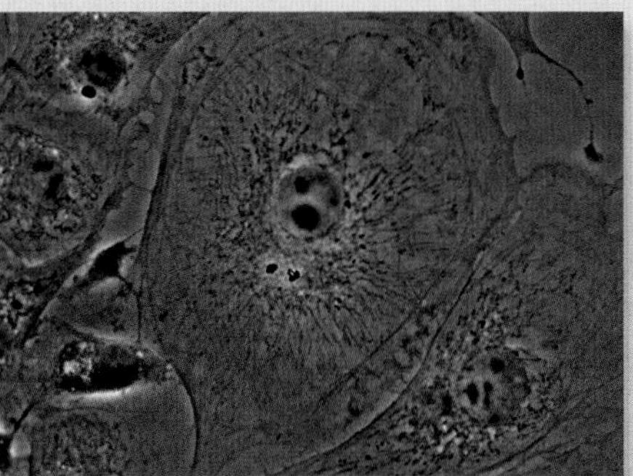

Phase contrast microscopy. As an alternative to staining, this microscope controls the path of light and amplifies differences in the phase of light transmitted or reflected by a sample. The dense structures appear darker than the background, thereby improving the contrast in different parts of the specimen. Can be used to view living, unstained cells.

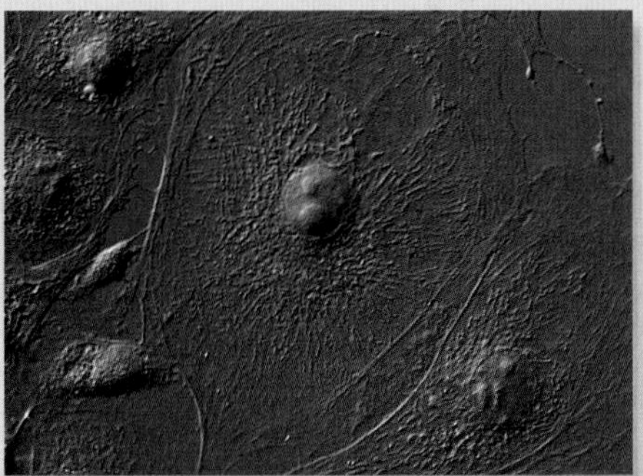

Differential interference contrast (Nomarski) microscopy. Similar to a phase contrast microscope in that it uses optical modifications to improve contrast in unstained specimens. Can be used to visualize the internal structures of cells and is commonly used to view whole cells or large cell structures such as nuclei.

(a) Light microscopy on unstained samples

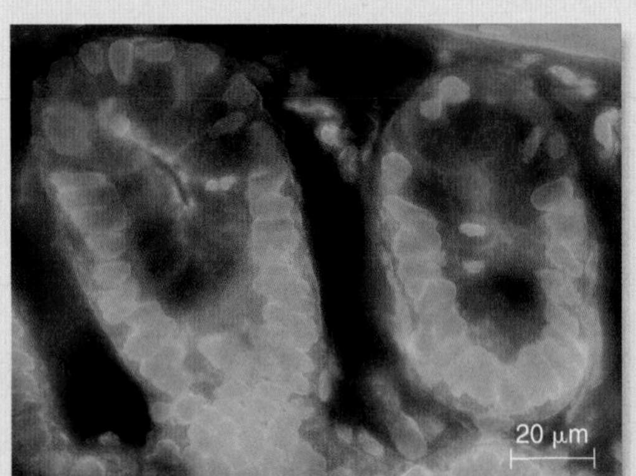

Standard (wide-field) fluorescence microscopy. Fluorescent molecules specifically label a particular type of cellular protein or organelle. A fluorescent molecule absorbs light at a particular wavelength and emits light at a longer wavelength. This microscope has filters that illuminate the sample with the wavelength of light that a fluorescent molecule absorbs, and then only the light that is emitted by the fluorescent molecules is allowed to reach the observer. To detect their cellular location, researchers often label specific cellular proteins using fluorescent antibodies that bind specifically to a particular protein.

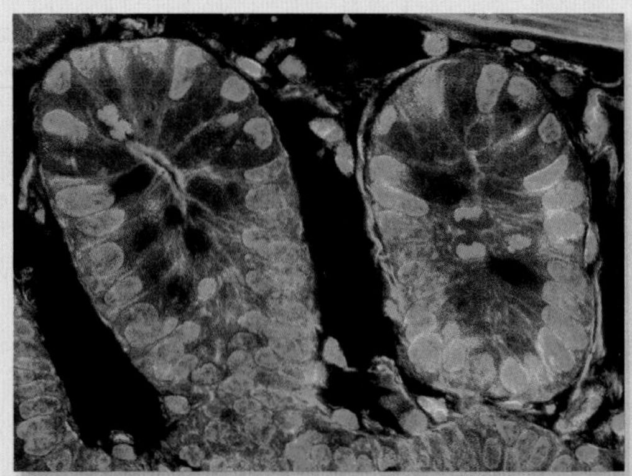

Confocal fluorescence microscopy. Uses lasers that illuminate various points in the sample. These points are processed by a computer to give a very sharp focal plane. In this example, this microscope technique is used in conjunction with fluorescence microscopy to view fluorescent molecules within a cell.

(b) Fluorescence microscopy

Figure 4.2 Examples of light microscopy. (a) These micrographs compare three microscopic techniques on the same unstained sample of cells. These are endothelial cells that line the interior surface of arteries in the lungs. **(b)** These two micrographs compare standard (wide-field) fluorescence microscopy with confocal fluorescence microscopy. The sample is a section through a mouse intestine, showing two villi, projections from the small intestine that are described in Chapter 45. In this sample, the nuclei are stained green, and the actin filaments (discussed later in this chapter) are stained red.

was playing a major role in advancing our understanding of cell structure. Two general types of electron microscopy have been developed: transmission electron microscopy and scanning electron microscopy. In **transmission electron microscopy (TEM)**, a beam of electrons is transmitted through a biological sample. To provide contrast, the sample is stained with a heavy metal, which binds to certain cellular structures such as membranes. The sample is then adhered to a copper grid and placed in a transmission electron microscope. When the beam of electrons strikes the sample, some of them hit the heavy metal and are scattered, while those that pass through without being

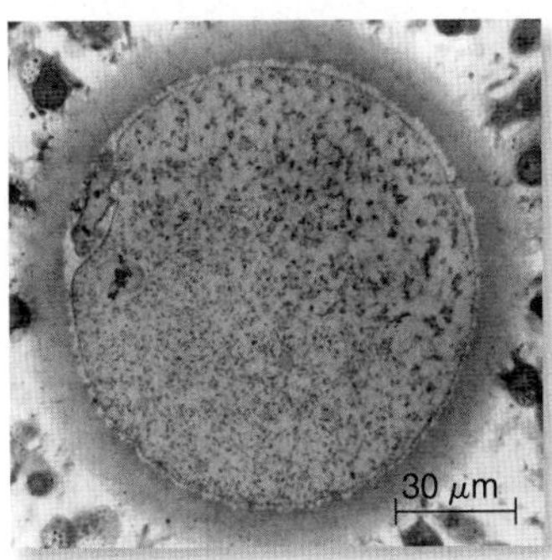

(a) Transmission electron micrograph (TEM)

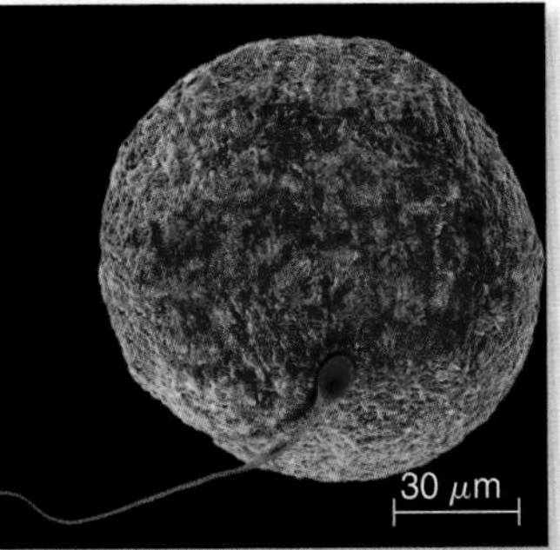

(b) Scanning electron micrograph (SEM)

Figure 4.3 **A comparison of transmission and scanning electron microscopy.** **(a)** Section through a developing human egg cell, observed by TEM, shortly before it was released from an ovary. **(b)** An egg cell, with an attached sperm, was coated with heavy metal and observed via SEM. This SEM is colorized.

Concept Check: *What is the primary advantage of SEM?*

scattered are focused to form an image on a photographic plate or screen (Figure 4.3a). The metal-stained regions of the sample that scatter electrons appear as darker areas, due to reduced electron penetration. TEM provides a cross-sectional view of a cell and its organelles and gives the best resolution compared with other forms of microscopy. However, such microscopes are expensive and cannot be used to view living cells.

Scanning electron microscopy (SEM) is used to view the surface of a sample. A biological sample is coated with a thin layer of heavy metal, such as gold or palladium, and then is exposed to an electron beam that scans its surface. Secondary electrons are emitted from the sample, which are detected and create an image of the three-dimensional surface of the sample (Figure 4.3b).

4.2 Overview of Cell Structure

Learning Outcomes:

1. Compare and contrast the general structural features of prokaryotic and eukaryotic cells.
2. Explain how the proteome underlies the structure and function of cells.
3. Analyze how cell size and shape affect the surface area/volume ratio.

Cell structure is primarily determined by four factors: (1) matter, (2) energy, (3) organization, and (4) information. In Chapters 2 and 3, we considered the first factor. The matter found in living organisms is composed of atoms, molecules, and macromolecules. Each type of cell synthesizes a unique set of molecules and macromolecules that contribute to cell structure. We will discuss the second factor, energy, throughout this unit, particularly in Chapters 6 through 8. Energy is needed to produce molecules and macromolecules and to carry out many cellular functions.

The third phenomenon that underlies cell structure is organization. A cell is not a haphazard bag of components. The molecules and macromolecules that constitute cells are found at specific sites. For instance, if we compare the structure of a muscle cell in two different humans, or two muscle cells within the same individual, we would see striking similarities in their overall structures. All living cells have the ability to build and maintain their internal organization. Proteins often bind to each other in much the same way that building blocks snap together. These types of **protein-protein interactions** create intricate cell structures and also facilitate processes in which proteins interact in a consistent series of steps.

The fourth critical factor of cell structure is information. Cell structure requires instructions. These instructions are found in the blueprint of life, namely, the genetic material, which is discussed in Unit III. Every species has a distinctive **genome**, the entire complement of its genetic material. Likewise, each living cell has a copy of the genome. The **genes** within each species' genome contain the information to produce cells with particular structures and functions. This information is passed from cell to cell and from parent to offspring to yield new generations of cells and new generations of offspring. In this section, we will explore the general structure of cells and examine how the genome contributes to cell structure and function.

Prokaryotic Cells Have a Simple Structure

Based on cell structure, all forms of life can be placed into two categories called prokaryotes and eukaryotes. We will first consider **prokaryotic cells**, which have a relatively simple structure. The term comes from the Greek *pro* and *karyon*, which means before a kernel—a reference to the kernel-like appearance of what would later be named the cell nucleus. Prokaryotic cells lack a membrane-enclosed nucleus.

From an evolutionary perspective, the two categories of organisms that have prokaryotic cells are **bacteria** and **archaea**. Both types are microorganisms that are relatively small with cell sizes that usually range between 1 micrometer (μm) and 10 μm in diameter. Bacteria are abundant throughout the world, being found in soil, water, and even our digestive tracts. Most bacterial species are not harmful to humans, and they play vital roles in ecology. However, some species are pathogenic—they cause disease. Examples of pathogenic bacteria include *Vibrio cholerae*, the source of cholera, and *Bacillus anthracis*, which causes anthrax. Archaea are also widely found throughout the world, though they are less common than bacteria and often occupy extreme environments such as hot springs and deep-sea vents.

Figure 4.4 shows a typical bacterial cell. The **plasma membrane**, which is a double layer of phospholipids and embedded proteins, forms an important barrier between the cell and its external environment. The **cytoplasm** is the region of the cell contained within the plasma membrane. Certain structures in the bacterial cytoplasm are visible via microscopy. These include the **nucleoid region**, where the genetic material (DNA) is located, and **ribosomes**, which are involved in polypeptide synthesis.

Some bacterial structures are located outside the plasma membrane. Nearly all species of bacteria and archaea have a relatively rigid **cell wall** that supports and protects the plasma membrane and cytoplasm. The cell-wall composition varies widely among prokaryotic cells but commonly contains peptides and carbohydrates. The cell wall, which is relatively porous, allows most nutrients in the environment to reach the plasma membrane. Many bacteria also secrete a **glycocalyx**, an outer viscous covering surrounding the bacterium. The glycocalyx traps water and helps protect bacteria from drying out. Certain strains of bacteria that invade animals' bodies produce a very thick, gelatinous glycocalyx called a **capsule** that may help them

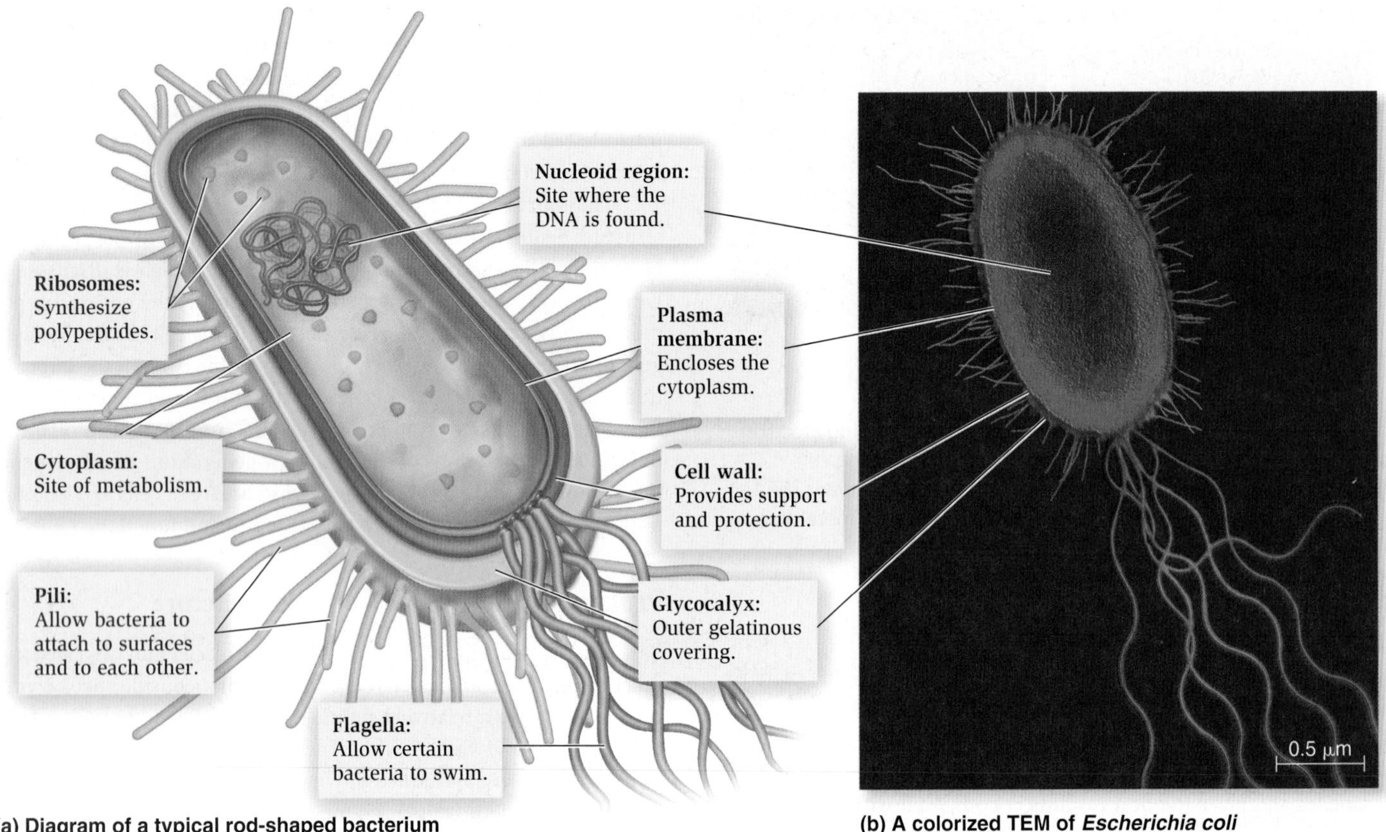
Nucleoid region:
Site where the DNA is found.
Ribosomes:
Synthesize polypeptides.
Plasma membrane:
Encloses the cytoplasm.
Cytoplasm:
Site of metabolism.
Cell wall:
Provides support and protection.
Pili:
Allow bacteria to attach to surfaces and to each other.
Glycocalyx:
Outer gelatinous covering.
Flagella:
Allow certain bacteria to swim.
0.5 μm
(a) Diagram of a typical rod-shaped bacterium
(b) A colorized TEM of *Escherichia coli*

Figure 4.4 Structure of a typical bacterial cell. Prokaryotic cells, which include bacteria and archaea, lack internal compartmentalization.

avoid being destroyed by the animal's immune (defense) system or may aid in the attachment to cell surfaces. Finally, many prokaryotic cells have appendages such as pili and flagella. **Pili** allow cells to attach to surfaces and to each other. **Flagella** provide prokaryotic cells with a way to move, also called motility.

Eukaryotic Cells Are Compartmentalized by Internal Membranes to Create Organelles

Aside from bacteria and archaea, all other species are **eukaryotes** (from the Greek, meaning true nucleus), which include protists, fungi, plants, and animals. Paramecia and algae are types of protists; yeasts and molds are types of fungi. Figure 4.5 illustrates the morphology of a typical animal cell. Eukaryotic cells possess a true nucleus, where most of the DNA is housed. A nucleus is a type of **organelle**—a membrane-bound compartment with its own unique structure and function. Like prokaryotic cells, all eukaryotic cells have a plasma membrane. In contrast to prokaryotic cells, eukaryotic cells exhibit extensive **compartmentalization**, which means they have many membrane-bound organelles that separate the cell into different regions. Cellular compartmentalization allows a cell to carry out specialized chemical reactions in different places.

Some general features of cell organization, such as a nucleus, are found in nearly all eukaryotic cells. Even so, the shape, size, and organization of cells vary considerably among different species and even among different cell types of the same species. For example, micrographs of a human skin cell and a human neuron show that, although these cells contain the same types of organelles, their overall morphologies are quite different (Figure 4.6).

Plant cells possess a collection of organelles similar to animal cells (Figure 4.7). Additional structures found in plant cells but not animal cells include chloroplasts, a central vacuole, and a cell wall.

GENOMES & PROTEOMES CONNECTION

The Proteome Largely Determines the Characteristics of a Cell

Many organisms, such as animals and plants, are multicellular, meaning that a single organism is composed of many cells. However, the cells of most multicellular organisms are not all identical. For example, your body contains skin cells, neurons, muscle cells, and many other types. An intriguing question, therefore, is how does a single organism produce different types of cells?

To answer this question, we need to consider the distinction between an individual's genome and proteome. Recall that the genome constitutes all types of genetic material, namely DNA, that an organism has. Most genes encode the production of polypeptides, which assemble into functional proteins. The **proteome** is defined as the complete protein composition of a cell or organism. A recurring biological principle is *structure determines function*. As discussed in this unit, the structures and functions of proteins are primarily responsible for the structures and functions of cells.

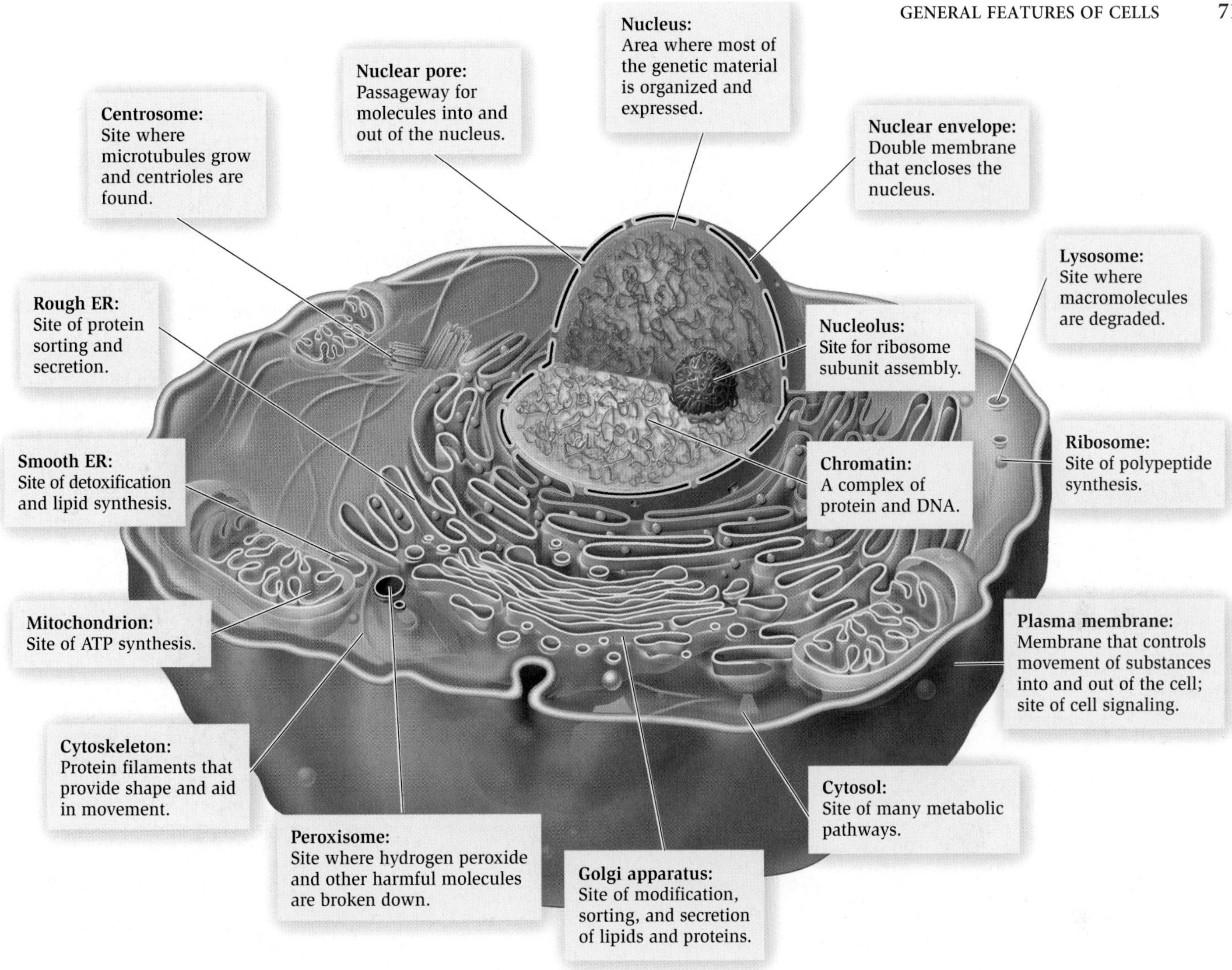

Centrosome: Site where microtubules grow and centrioles are found.

Nuclear pore: Passageway for molecules into and out of the nucleus.

Nucleus: Area where most of the genetic material is organized and expressed.

Nuclear envelope: Double membrane that encloses the nucleus.

Lysosome: Site where macromolecules are degraded.

Rough ER: Site of protein sorting and secretion.

Nucleolus: Site for ribosome subunit assembly.

Smooth ER: Site of detoxification and lipid synthesis.

Chromatin: A complex of protein and DNA.

Ribosome: Site of polypeptide synthesis.

Mitochondrion: Site of ATP synthesis.

Plasma membrane: Membrane that controls movement of substances into and out of the cell; site of cell signaling.

Cytoskeleton: Protein filaments that provide shape and aid in movement.

Cytosol: Site of many metabolic pathways.

Peroxisome: Site where hydrogen peroxide and other harmful molecules are broken down.

Golgi apparatus: Site of modification, sorting, and secretion of lipids and proteins.

Figure 4.5 General structure of an animal cell.

BIOLOGY PRINCIPLE Cells are the simplest units of life. A cell, such as the animal cell illustrated here, is the smallest unit that satisfies all of the characteristics of living organisms. These characteristics are discussed in Chapter 1 (see Figure 1.4).

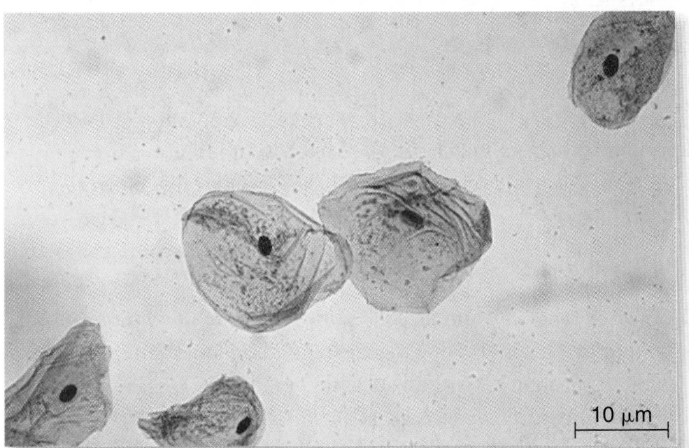

(a) Human skin cell

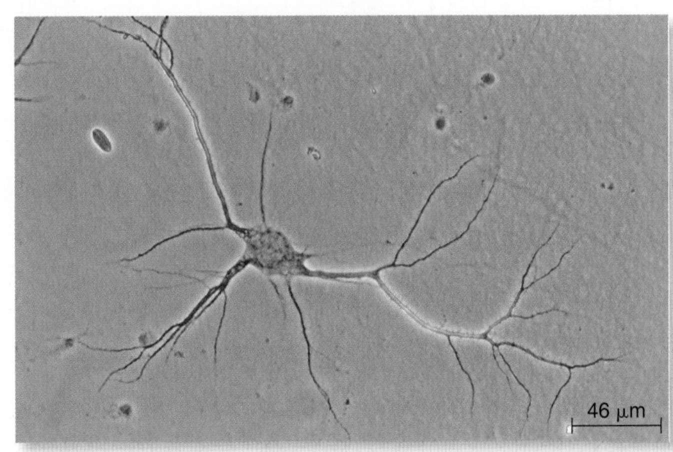

(b) Human nerve cell

Figure 4.6 Variation in morphology of eukaryotic cells. Light micrographs of **(a)** a human skin cell and **(b)** a human neuron (a cell of the nervous system). Although these cells have the same genome and same types of organelles, their general morphologies are quite different.

BioConnections: *Look ahead to Figure 13.21. How does alternative splicing affect protein structure and function?*

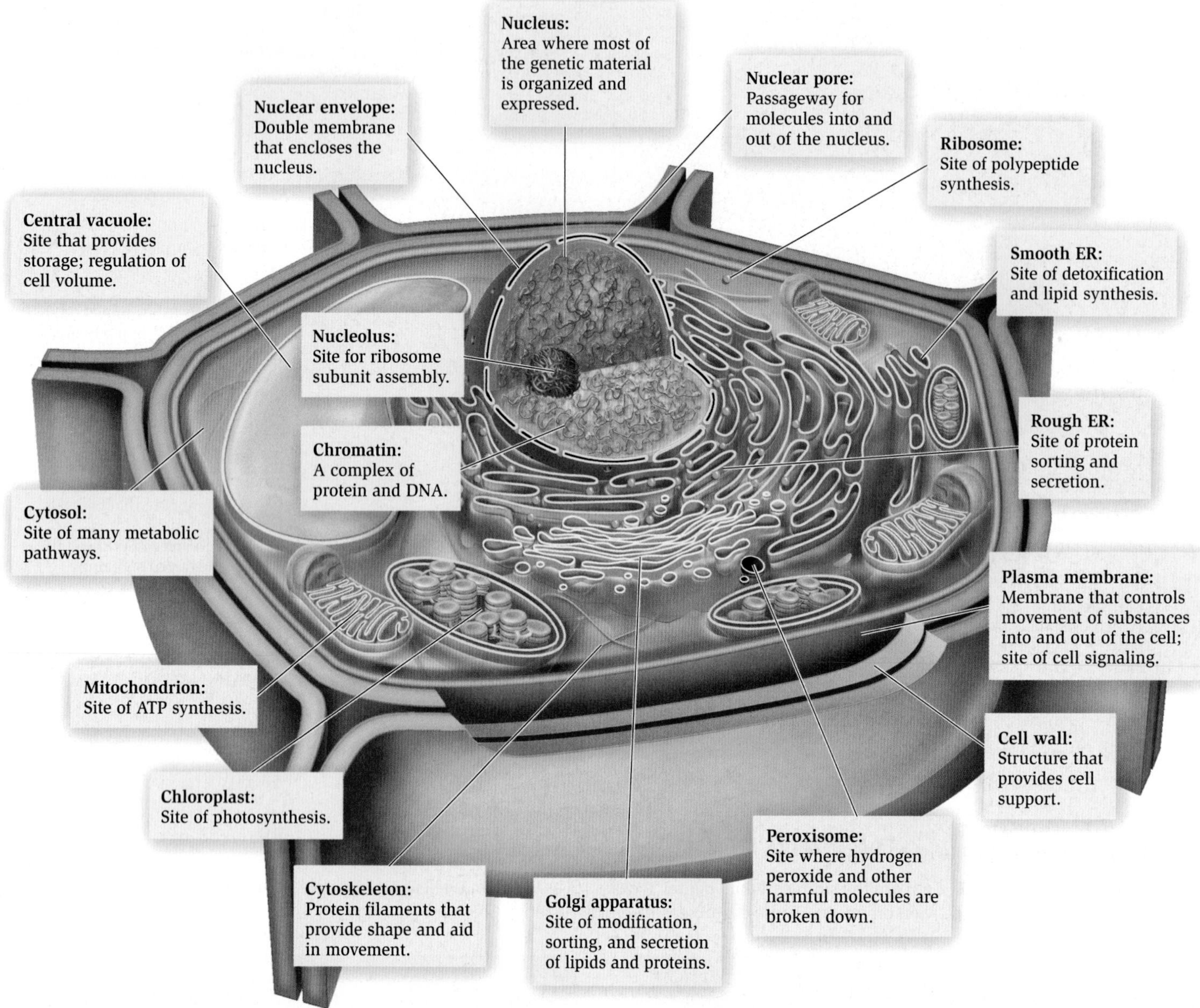

Nucleus:
Area where most of the genetic material is organized and expressed.
Nuclear envelope:
Double membrane that encloses the nucleus.
Nuclear pore:
Passageway for molecules into and out of the nucleus.
Ribosome:
Site of polypeptide synthesis.
Central vacuole:
Site that provides storage; regulation of cell volume.
Smooth ER:
Site of detoxification and lipid synthesis.
Nucleolus:
Site for ribosome subunit assembly.
Rough ER:
Site of protein sorting and secretion.
Chromatin:
A complex of protein and DNA.
Cytosol:
Site of many metabolic pathways.
Plasma membrane:
Membrane that controls movement of substances into and out of the cell; site of cell signaling.
Mitochondrion:
Site of ATP synthesis.
Cell wall:
Structure that provides cell support.
Chloroplast:
Site of photosynthesis.
Peroxisome:
Site where hydrogen peroxide and other harmful molecules are broken down.
Cytoskeleton:
Protein filaments that provide shape and aid in movement.
Golgi apparatus:
Site of modification, sorting, and secretion of lipids and proteins.

Figure 4.7 General structure of a plant cell. Plant cells lack lysosomes and centrioles. Unlike animal cells, plant cells have an outer cell wall; a large central vacuole that functions in storage and the regulation of cell volume; and chloroplasts, which carry out photosynthesis.

Concept Check: What are the functions of the cell structures and organelles that are (1) found in animal cells but not plant cells and (2) found in plant cells but not animal cells?

As an example, let's consider skin cells and neurons—two cell types that have dramatically different organization and structure (see Figure 4.6). In any particular individual, the genes in a human skin cell are identical to those in a human neuron. However, their proteomes are different. The proteome of a cell largely determines its structure and function. Several phenomena underlie the differences observed in the proteomes of different cell types.

1. *Certain proteins found in one cell type may not be produced in another cell type.* This phenomenon is due to differential gene regulation, discussed in Chapter 13.
2. *Two cell types may produce the same protein but in different amounts.* This is also due to gene regulation and to the rates at which a protein is synthesized and degraded.
3. *The amino acid sequences of particular proteins can vary in different cell types.* As discussed in Chapter 13, the mRNA from a single gene can produce two or more polypeptides with different amino acid sequences via a process called alternative splicing.
4. *Two cell types may alter their proteins in different ways.* After a protein is made, its structure may be changed in a variety of ways. These include the covalent attachment of molecules, such as

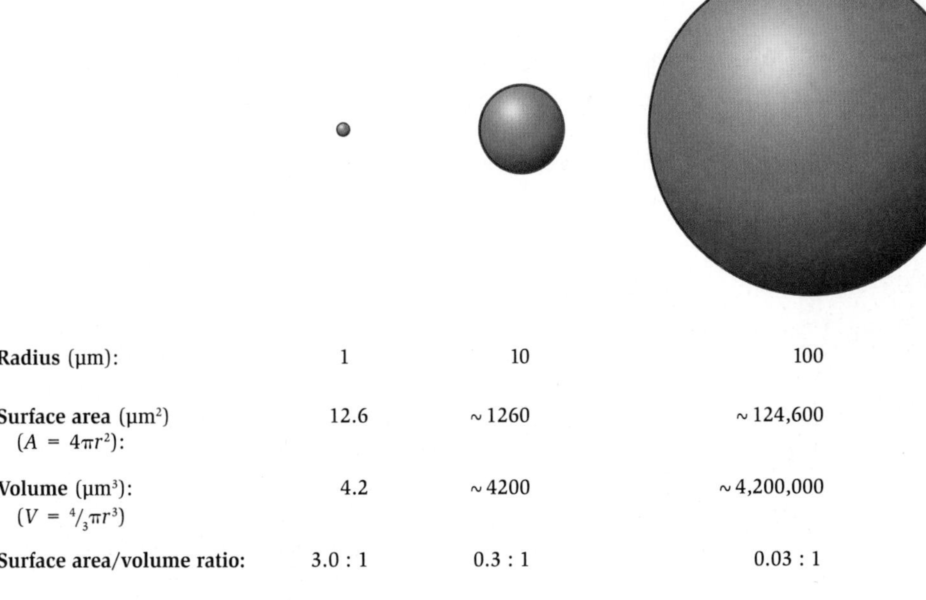

Radius (μm): 1 10 100
Surface area (μm²) (A = 4πr²): 12.6 ~1260 ~124,600
Volume (μm³) (V = 4/3πr³): 4.2 ~4200 ~4,200,000
Surface area/volume ratio: 3.0 : 1 0.3 : 1 0.03 : 1

Figure 4.8 Relationship between cell size and the surface area/volume ratio. As cells get larger, the surface area/volume ratio gets smaller. Note: The three spheres shown here are not drawn precisely to scale.

BioConnections: Look ahead to Figure 40.9. How does the surface area/volume ratio affect the shapes of structures involved with gas exchange?

phosphate and carbohydrates, and the cleavage of a protein to a smaller size, which are discussed in Chapter 21.

These four phenomena enable skin cells and neurons to produce different proteomes and therefore different structures with different functions. Likewise, the proteomes of skin cells and neurons differ from those of other cell types such as muscle and liver cells. Ultimately, the proteomes of cells are largely responsible for producing the traits of organisms, such as the color of a person's eyes.

During the last few decades, researchers have also discovered an association between proteome changes and disease. For example, the proteomes of healthy lung cells are different from the proteomes of lung cancer cells. Furthermore, the proteomes of cancer cells change as the disease progresses. A key challenge for biologists is to understand the synthesis and function of proteomes in different cell types and how proteome changes may lead to disease conditions such as cancer.

Surface Area and Volume Are Critical Parameters That Affect Cell Sizes and Shapes

As we have seen, a common feature of most cells is their small size. For example, most bacterial cells are about 1–10 μm in diameter, and a typical eukaryotic cell is 10–100 μm in diameter (see Figure 4.1). Though some exceptions are known, such as an ostrich egg, small size is a nearly universal characteristic of cells. In general, large organisms attain their large sizes by having more cells, not by having larger cells. For example, the various types of cells found in an elephant and a mouse are roughly the same sizes. However, an elephant has many more cells than a mouse.

Why are cells usually small? One key factor is the interface between a cell and its extracellular environment, which is the plasma membrane. For cells to survive, they must import substances across their plasma membranes and export waste products. If the internal volume of a cell is large, it will require a greater amount of nutrient uptake and waste export. The rate of transport of substances across the plasma membrane, however, is limited by its surface area. Therefore, a critical issue for sustaining a cell is the surface area/volume ratio. This concept is illustrated in Figure 4.8, which considers a simplified case in which cells are spherical. As cells get larger, the surface area of their plasma membrane increases with the square of the radius ($A = 4\pi r^2$), whereas the volume increases with the cube of the radius ($V = 4/3\pi r^3$). Therefore, as the radius of the cell gets larger, the surface area-to-volume ratio gets smaller. Biologists hypothesize that most cells are small because a high surface area/volume ratio is needed for cells to sustain an adequate level of nutrient uptake and waste export.

One way for cells to partially overcome the dilemma posed by the surface area-to-volume ratio is to have elongated and irregularly shaped surfaces. For example, look back at Figure 4.6. The skin cell is roughly spherical, whereas the neuron is very elongated and has an irregularly shaped surface. If a skin cell and a neuron cell had the same internal volume, the neuron would have a much higher surface area/volume ratio.

4.3 The Cytosol

Learning Outcomes:

1. Identify the location of the cytosol in a eukaryotic cell and list its general functions.
2. Describe the three types of protein filaments that make up the cytoskeleton.
3. Explain how motor proteins interact with microtubules or actin filaments to promote cellular movement.

Thus far, we have focused on the general features of prokaryotic and eukaryotic cells. In the rest of this chapter, we will survey the various compartments of eukaryotic cells with a greater emphasis on structure and function. Figure 4.9 highlights an animal and plant cell according to four different regions. We will start with the **cytosol** (shown in yellow), the region of a eukaryotic cell that is outside the membrane-bound organelles but inside the plasma membrane. The other regions of the cell, which we will examine later in this chapter, include the

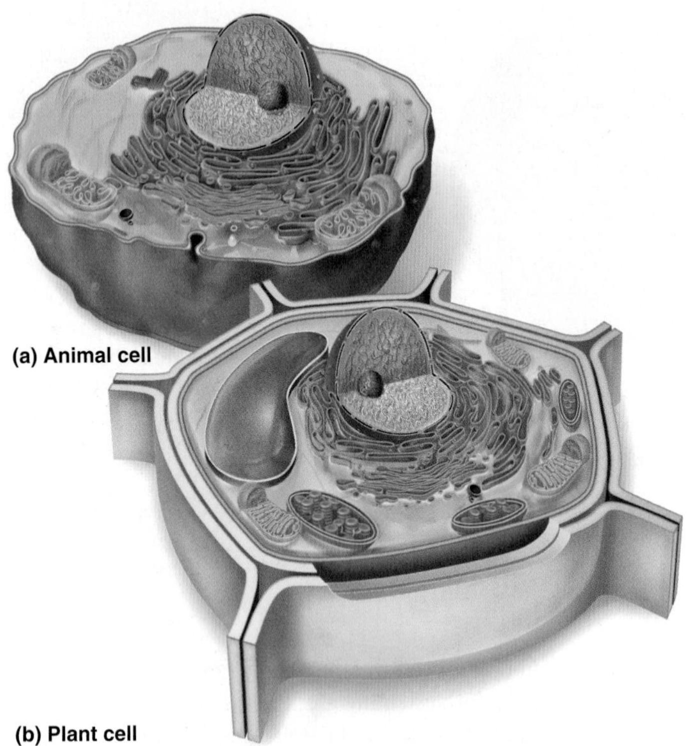

Figure 4.9 Compartments within (a) animal and (b) plant cells. The cytosol, which is outside the organelles but inside the plasma membrane, is shown in yellow. The membranes of the endomembrane system are shown in purple, and the fluid-filled interiors are pink. The peroxisome is dark purple. The interior of the nucleus is blue. Semiautonomous organelles are shown in orange (mitochondria) and green (chloroplasts).

interior of the nucleus (blue), the endomembrane system (purple and pink), and the semiautonomous organelles (orange and green). As in prokaryotic cells, the term cytoplasm refers to the region enclosed by the plasma membrane. This includes the cytosol and the organelles.

Synthesis and Breakdown of Molecules Occur in the Cytosol

Metabolism is defined as the sum of the chemical reactions by which cells produce the materials and utilize the energy necessary to sustain life. Although specific steps of metabolism also occur in cell organelles, the cytosol is a central coordinating region for many metabolic activities of eukaryotic cells. Metabolism often involves a series of steps called a metabolic pathway. Each step in a metabolic pathway is catalyzed by a specific **enzyme**—a protein that accelerates the rate of a chemical reaction. In Chapters 6 and 7, we will examine the functional properties of enzymes and consider a few metabolic pathways that occur in the cytosol and cell organelles.

Some pathways involve the breakdown of a molecule into smaller components, a process termed **catabolism**. Such pathways are needed by the cell to utilize energy and also to generate molecules that provide the building blocks to construct macromolecules. Conversely, other pathways are involved in **anabolism**, the synthesis of molecules and macromolecules. For example, polysaccharides are made by linking sugar molecules. To make proteins, amino acids are covalently connected to form a polypeptide, using the information within an mRNA (see Chapter 12). Translation occurs on ribosomes, which are found in various locations in the cell. Some ribosomes may float freely in the cytosol, others are attached to the outer membrane of the nuclear envelope and endoplasmic reticulum membrane, and still others are found within the mitochondria or chloroplasts.

The Cytoskeleton Provides Cell Shape, Organization, and Movement

The **cytoskeleton** is a network of three different types of protein filaments: **microtubules**, **intermediate filaments**, and **actin filaments** (Table 4.1). Each type is constructed from many protein monomers. The cytoskeleton is a striking example of protein-protein interactions. The cytoskeleton is found primarily in the cytosol and also in the nucleus along the inner nuclear membrane. Let's first consider the structure of cytoskeletal filaments and their roles in the construction and organization of cells. Later, we will examine how they are involved in cell movement.

Microtubules Microtubules are long, hollow, cylindrical structures about 25 nm in diameter composed of protein subunits called α and β tubulin. The assembly of tubulin to form a microtubule results in a polar structure with a plus end and a minus end (see Table 4.1). Microtubules grow only at the plus end, but can shorten at either the plus or minus end. A single microtubule can oscillate between growing and shortening phases, a phenomenon termed **dynamic instability**. Dynamic instability is important in many cellular activities, including the sorting of chromosomes during cell division.

The sites where microtubules form within a cell vary among different types of organisms. Nondividing animal cells contain a single structure near their nucleus called the **centrosome**, also called the microtubule-organizing center (see Table 4.1). Within the centrosome are the **centrioles**, a conspicuous pair of structures arranged perpendicular to each other. In animal cells, microtubule growth typically starts at the centrosome in such a way that the minus end is anchored there. In contrast, most plant cells and many protists lack centrosomes and centrioles. Microtubules are created at many sites that are scattered throughout a plant cell. In plants, the nuclear membrane appears to function as a microtubule-organizing center.

Microtubules are important for cell shape and organization. Organelles such as the Golgi apparatus are attached to microtubules. In addition, microtubules are involved in the organization and movement of chromosomes during mitosis and in the orientation of cells during cell division, events we will examine in Chapter 15.

Intermediate Filaments Intermediate filaments are another class of cytoskeletal filament found in the cells of many but not all animal species. Their name is derived from the observation that they are intermediate in diameter between actin filaments and microtubules. Intermediate filament proteins bind to each other in a staggered array to form a twisted, ropelike structure with a diameter of approximately 10 nm (see Table 4.1). They function as tension-bearing fibers that help maintain cell shape and rigidity. Intermediate filaments tend to

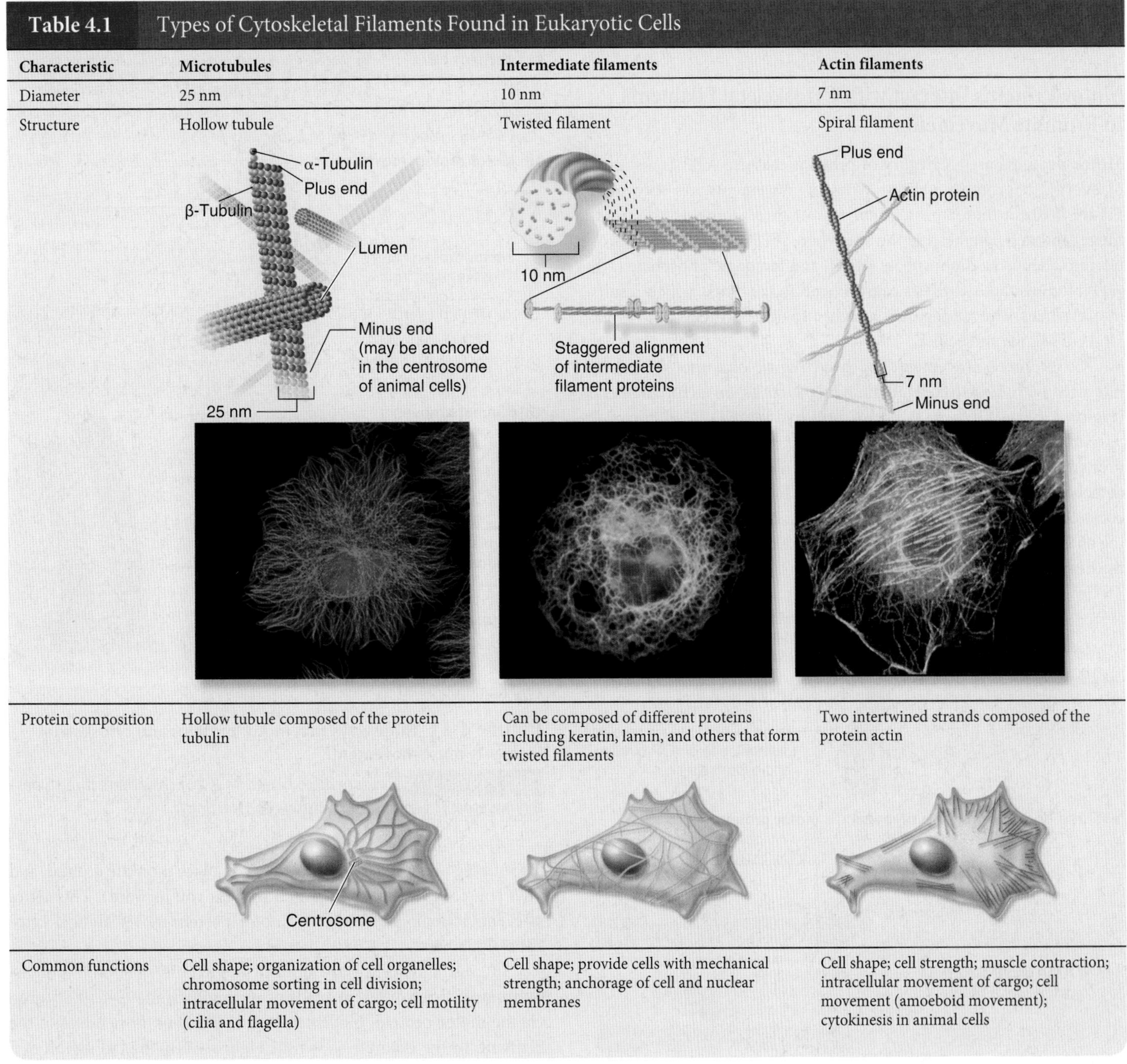

Table 4.1 Types of Cytoskeletal Filaments Found in Eukaryotic Cells

Characteristic	Microtubules	Intermediate filaments	Actin filaments
Diameter	25 nm	10 nm	7 nm
Structure	Hollow tubule	Twisted filament	Spiral filament
Protein composition	Hollow tubule composed of the protein tubulin	Can be composed of different proteins including keratin, lamin, and others that form twisted filaments	Two intertwined strands composed of the protein actin
Common functions	Cell shape; organization of cell organelles; chromosome sorting in cell division; intracellular movement of cargo; cell motility (cilia and flagella)	Cell shape; provide cells with mechanical strength; anchorage of cell and nuclear membranes	Cell shape; cell strength; muscle contraction; intracellular movement of cargo; cell movement (amoeboid movement); cytokinesis in animal cells

be relatively permanent. By comparison, microtubules and actin filaments readily lengthen and shorten in cells.

Several types of proteins assemble into intermediate filaments. Keratins form intermediate filaments in skin, intestinal, and kidney cells, where they are important for cell shape and mechanical strength. They are also a major constituent of hair and nails. In addition, intermediate filaments are found inside the cell nucleus. As discussed later in this chapter, nuclear lamins form a network of intermediate filaments that line the inner nuclear membrane and provide anchorage points for the nuclear pores.

Actin Filaments Actin filaments are also known as **microfilaments**, because they are the thinnest cytoskeletal filaments. They are long, thin fibers approximately 7 nm in diameter (see Table 4.1). Like microtubules, actin filaments have plus and minus ends, and they are very dynamic structures in which each strand grows at the plus end by the addition of actin monomers. This assembly process produces a fiber composed of two strands of actin monomers that spiral around each other.

Despite their thinness, actin filaments play a key role in cell shape and strength. Although actin filaments are dispersed throughout the cytosol, they tend to be highly concentrated near the plasma membrane. In many types of cells, actin filaments support the plasma membrane and provide shape and strength to the cell. The sides of actin filaments are often anchored to other proteins near the plasma membrane, which explains why actin filaments are typically found there. The plus ends grow toward the plasma membrane and play a key role in cell shape and movement.

Motor Proteins Interact with Cytoskeletal Filaments to Promote Movements

Motor proteins are a category of proteins that use ATP as a source of energy to promote various types of movements. As shown in Figure 4.10a, a motor protein consists of three domains: the head, hinge, and tail. The head is the site where ATP binds and is hydrolyzed to adenosine diphosphate (ADP) and inorganic phosphate (P_i). ATP binding and hydrolysis cause a bend in the hinge, which results in movement. The tail region is attached to other proteins or to other kinds of cellular molecules.

To promote movement, the head region of a motor protein interacts with a cytoskeletal filament, such as an actin filament (Figure 4.10b). When ATP binds and is hydrolyzed, the motor protein attempts to "walk" along the filament. The head of the motor protein is initially attached to a filament. To move forward, the head detaches from the filament, cocks forward, binds to the filament, and cocks backward. To picture how this works, consider the act of walking and imagine that the ground is a cytoskeletal filament, your leg is the head of the motor protein, and your hip is the hinge. To walk, you lift your leg up, you move it forward, you place it on the ground, and then you cock it backward (which propels you forward). This series of events is analogous to how a motor protein moves along a cytoskeletal filament.

Interestingly, cells have utilized the actions of motor proteins to promote three different kinds of movements: movement of cargo via the motor protein, movement of the filament, or bending of the filament. In the example shown in Figure 4.11a, the tail region of a motor protein called kinesin is attached to a cargo, so the motor protein moves the cargo from one location to another. Alternatively, a motor protein called myosin can remain in place and cause the filament to move (Figure 4.11b). This occurs during muscle contraction, which is described in Chapter 44.

A third possibility is that both the motor protein and filament are restricted in their movement due to the presence of linking proteins. In this case, when motor proteins called dynein attempt to walk toward the minus end, they exert a force that causes the microtubules to bend (Figure 4.11c). In certain kinds of cells, microtubules and motor proteins facilitate movement involving cell appendages called **flagella** and **cilia** (singular, flagellum and cilium). The difference between the two is that flagella are usually longer than cilia and are typically found singly or in pairs. Both flagella and cilia cause movement by a bending motion. In flagella, movement occurs by a

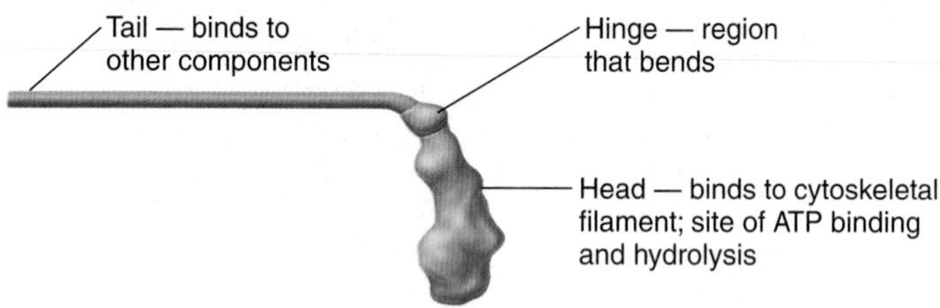

Hinge — region that bends

(a) Three-domain structure of myosin, a motor protein

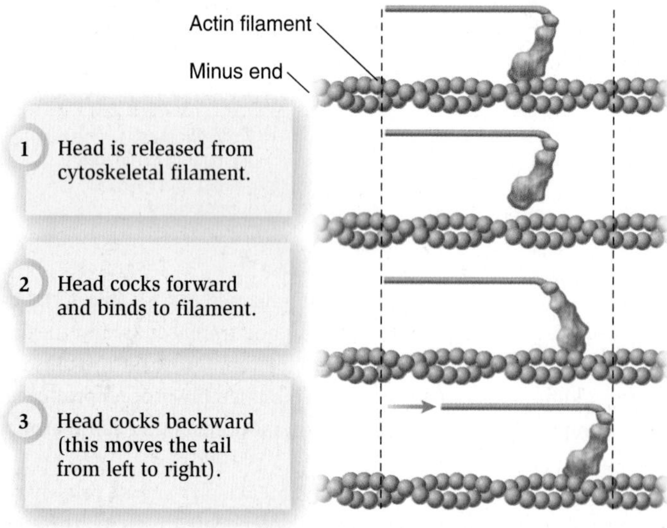

3 Head cocks backward (this moves the tail from left to right).

(b) Movement of a motor protein along a cytoskeletal filament

Figure 4.10 Motor proteins and their interactions with cytoskeletal filaments. The example illustrated here is the motor protein myosin (discussed in Chapter 44), which interacts with actin filaments. **(a)** Three-domain structure of myosin. **(b)** Conformational changes in a motor protein that allow it to "walk" along a cytoskeletal filament.

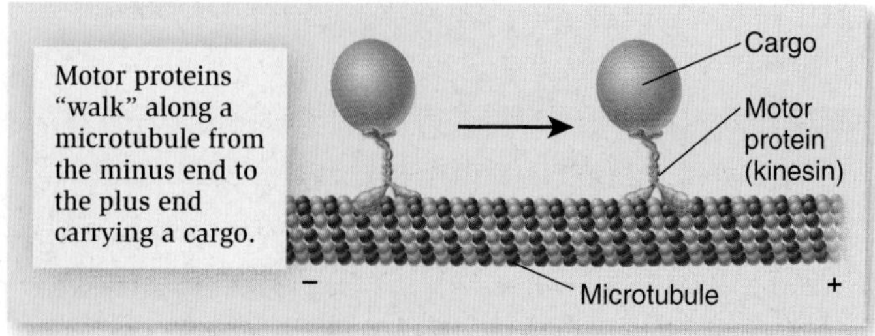

Motor proteins "walk" along a microtubule from the minus end to the plus end carrying a cargo.
Cargo
Motor protein (kinesin)
Microtubule
–
+

(a) Motor protein moves

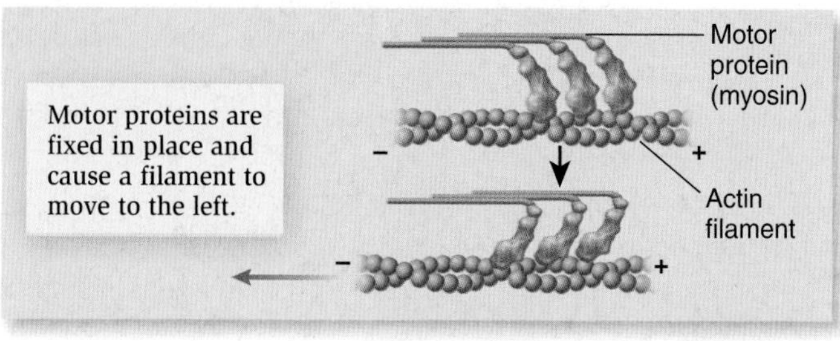

(b) Filament moves

(c) Filaments bend

Figure 4.11 Three ways that motor proteins and cytoskeletal filaments cause movement.

BioConnections: Look ahead to Figure 44.6. Which of these three types of movements occurs during muscle contraction?

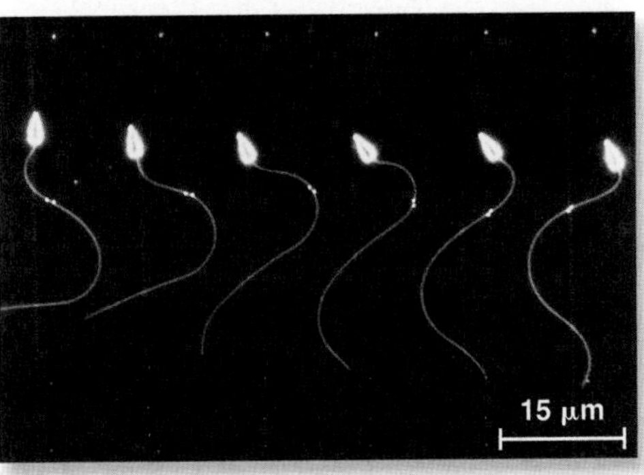

(a) Time-lapse photography of a human sperm moving its flagellum

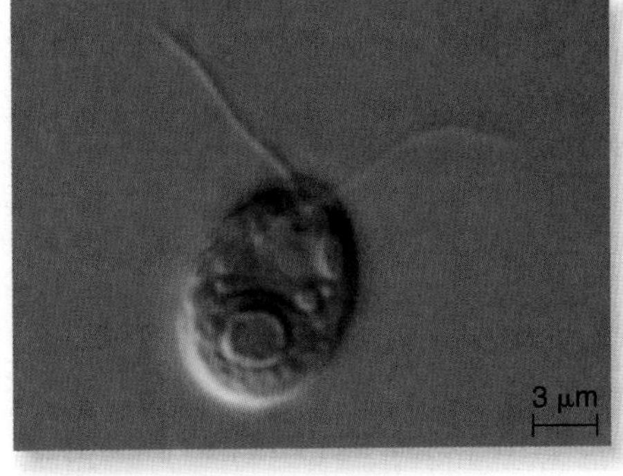

(b) *Chlamydomonas* with 2 flagella

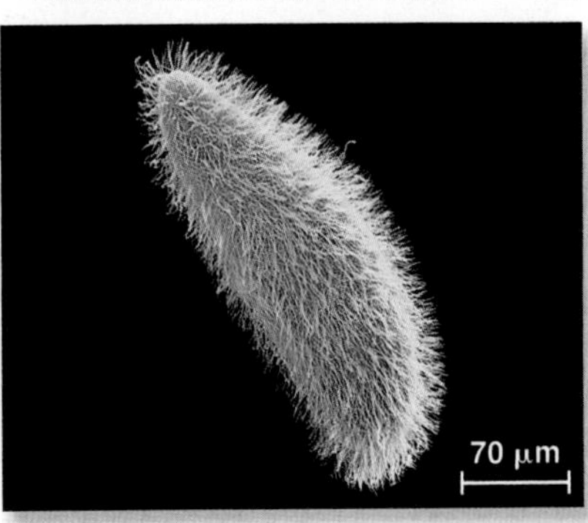

(c) *Paramecium* with many cilia

Figure 4.12 Cellular movements due to the actions of flagella and cilia. (a) Sperm swim by means of a single, long flagellum that moves in a whiplike motion, as shown by this human sperm. **(b)** The swimming of *Chlamydomonas reinhardtii*, a unicellular green algae, also involves a whiplike motion at the base, but the motion is precisely coordinated between two flagella. This results in swimming behavior that resembles a breaststroke. **(c)** Ciliated protozoa such as this *Paramecium* swim via many shorter cilia.

Concept Check: During the movement of a cilium or flagellum, describe the type of movements that occur between the motor proteins and microtubules.

whiplike motion that is due to the propagation of a bend from the base to the tip. A single flagellum may propel a cell such as a sperm cell with a whiplike motion (Figure 4.12a). Alternatively, a pair of flagella may move in a synchronized manner to pull a microorganism through the water (think of a human swimmer doing the breaststroke). Certain unicellular algae swim in this manner (Figure 4.12b).

By comparison, cilia are often shorter than flagella and tend to cover all or part of the surface of a cell. Protists such as paramecia may have hundreds of adjacent cilia that beat in a coordinated fashion to propel the organism through the water (Figure 4.12c).

Despite their differences in length, flagella and cilia share the same internal structure called the **axoneme**. The axoneme contains microtubules, the motor protein dynein, and linking proteins (Figure 4.13). In the cilia and flagella of most eukaryotic organisms, the microtubules form an arrangement called a 9 + 2 array. The outer nine are doublet microtubules, which are composed of a partial

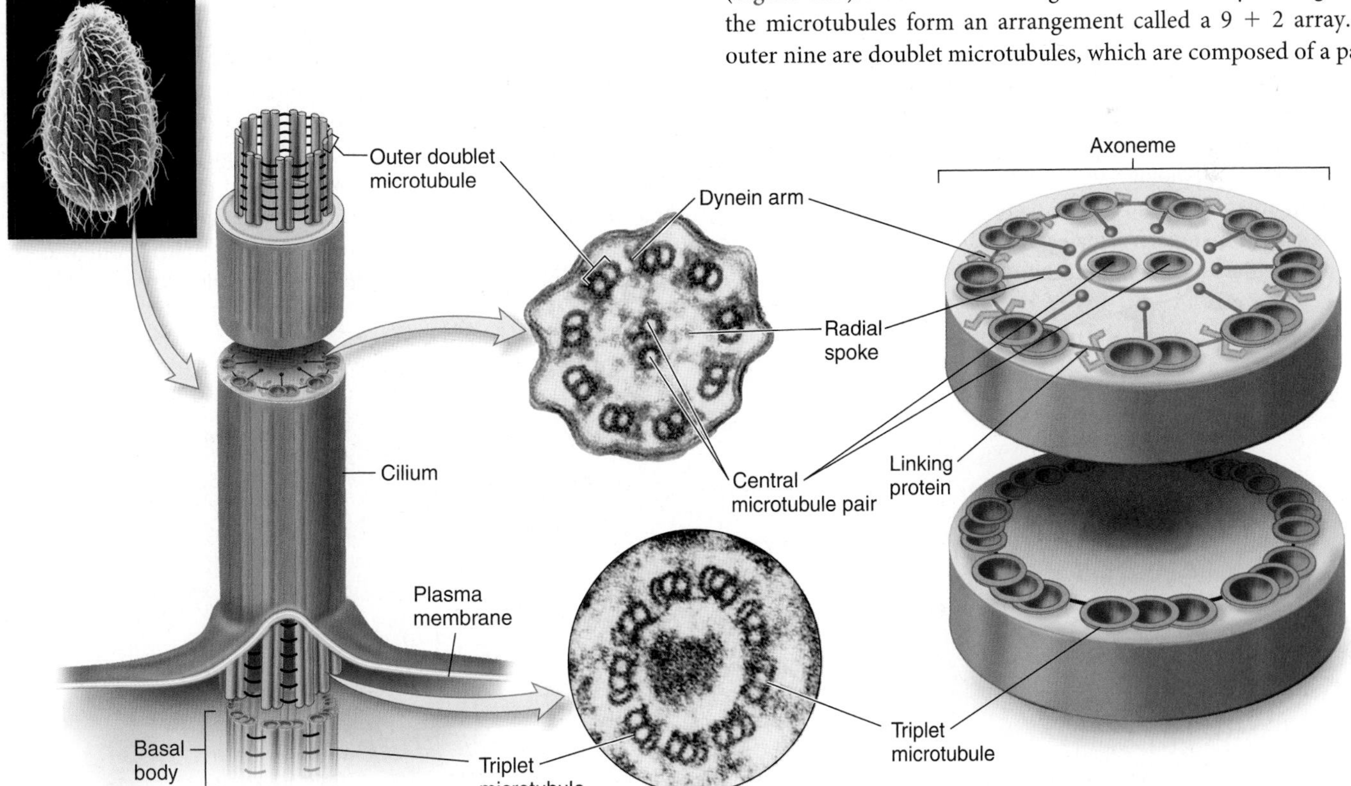

Figure 4.13 Structure of a eukaryotic cilium or flagellum. The structure of a cilium of a protist, *Tetrahymena thermophila* (see inset), consists of a 9 + 2 arrangement of nine outer doublet microtubules and two central microtubules. This structure is anchored to the basal body, which has nine triplet microtubules, in which three microtubules are fused together. Note: The structure of the basal body is very similar to centrioles in animal cells.

BioConnections: Look ahead to Figures 28.4, 28.7, and 28.8. What are some different uses of cilia and flagella among protists?

microtubule attached to a complete microtubule. Each of the two central microtubules consists of a single microtubule. Radial spokes project from the outer doublet microtubules toward the central pair. The microtubules in flagella and cilia emanate from **basal bodies**, which are anchored to the cytoplasmic side of the plasma membrane. At the basal body, the microtubules form a triplet structure. Much like the centrosome of animal cells, the basal bodies provide a site for microtubules to grow.

The movement of both flagella and cilia involves the propagation of a bend, which begins at the base of the structure and proceeds toward the tip (see Figure 4.12a). The bending occurs because dynein is activated to walk toward the minus end of the microtubules. ATP hydrolysis is required for this process. However, the microtubules and dynein are not free to move relative to each other because of linking proteins. Therefore, instead of dyneins freely walking along the microtubules, they exert a force that bends the microtubules (see Figure 4.11c). The dyneins at the base of the flagellum or cilium are activated first, followed by dyneins that are progressively closer to the tip, and the resulting movement propels the organism.

4.4 The Nucleus and Endomembrane System

Learning Outcomes:

1. Describe the structure and organization of the cell nucleus.
2. Outline the structures and general functions of the components of the endomembrane system.
3. Distinguish between the rough endoplasmic reticulum and the smooth endoplasmic reticulum.
4. Identify three important functions of the plasma membrane.

In Chapter 2, we learned that the nucleus of an atom contains protons and neutrons. In cell biology, the term **nucleus** has a different meaning. It is an organelle found in eukaryotic cells that contains most of the cell's genetic material. A small amount of genetic material is also found outside the nucleus, in mitochondria and chloroplasts.

The membranes that enclose the nucleus are part of a larger network of membranes called the **endomembrane system**. This system includes not only the nuclear envelope, which encloses the nucleus, but also the endoplasmic reticulum, Golgi apparatus, lysosomes, vacuoles, and peroxisomes. The prefix *endo* (from the Greek, meaning inside) originally referred only to these organelles and internal membranes. However, we now know that the plasma membrane is also part of this integrated membrane system (**Figure 4.14**). Some of these membranes, such as the outer membrane of the nuclear envelope and the membrane of the endoplasmic reticulum, have direct connections to one another. Other organelles of the endomembrane system pass materials to each other via **vesicles**—small membrane-enclosed spheres. In this section, we will examine the nucleus and survey the structures and functions of the organelles and membranes of the endomembrane system.

The Eukaryotic Nucleus Contains Chromosomes

The nucleus is the internal compartment that is enclosed by a double-membrane structure termed the **nuclear envelope** (Figure 4.15). In most cells, the nucleus is a relatively large organelle that typically occupies 10–20% of the total cell volume. **Nuclear pores** are formed where the inner and outer nuclear membranes make contact with each other. The pores provide a passageway for the movement of molecules and macromolecules into and out of the nucleus. Although cell biologists view the nuclear envelope as part of the endomembrane system, the materials within the nucleus are not (see Figure 4.15).

Inside the nucleus are the chromosomes and a filamentous network of proteins called the nuclear matrix. Each **chromosome** is composed of genetic material, namely DNA, and many types of proteins that help to compact the chromosome to fit inside the nucleus. The complex formed between DNA and such proteins is termed **chromatin**. The **nuclear matrix** consists of two parts: the nuclear lamina, which is composed of intermediate filaments that line the inner nuclear membrane, and an internal nuclear matrix, which is connected to the lamina and fills the interior of the nucleus. The nuclear

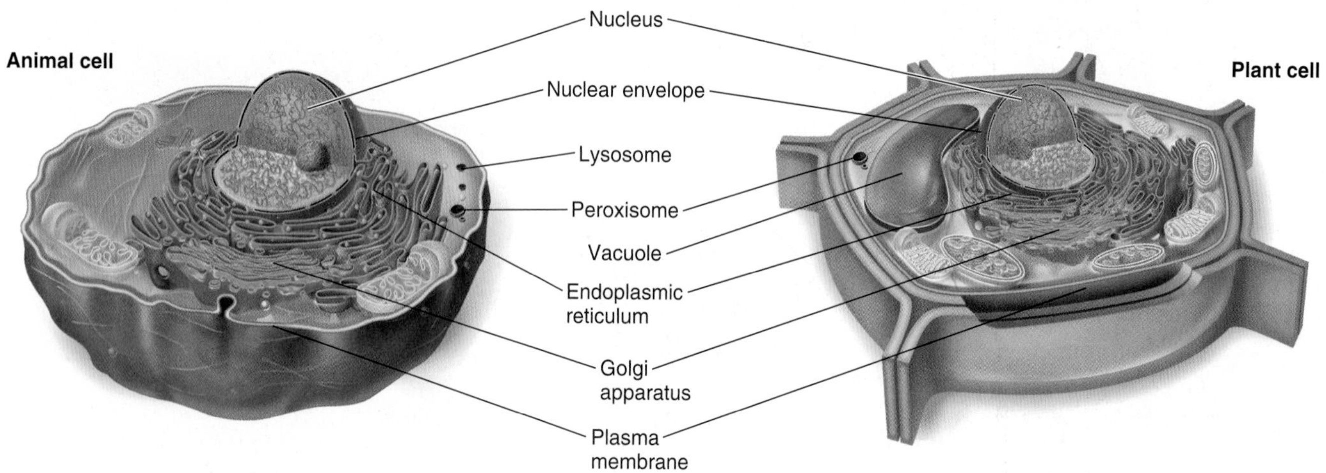

Figure 4.14 **The nucleus and endomembrane system.** This figure highlights the internal compartment of the nucleus (blue), the membranes of the endomembrane system (purple), and the fluid-filled interiors of the endomembrane system (pink). The nuclear envelope is part of the endomembrane system, but the interior of the nucleus is not.

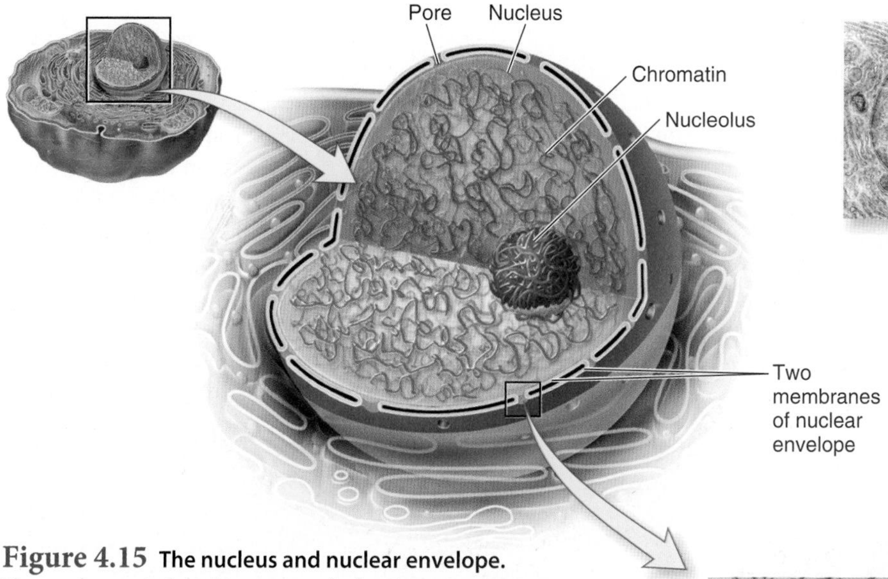

Pore
Nucleus
Chromatin
Nucleolus
Two membranes of nuclear envelope
Nucleolus
Chromatin
Nuclear lamina
Nuclear envelope
Pore in nuclear envelope
5.4 μm
Pore complexes
0.4 μm
Nucleus
Nuclear pore complex
Chromatin in nucleus
Internal nuclear matrix
Inner membrane
Outer membrane
Nuclear envelope
Nuclear lamina
Cytosol

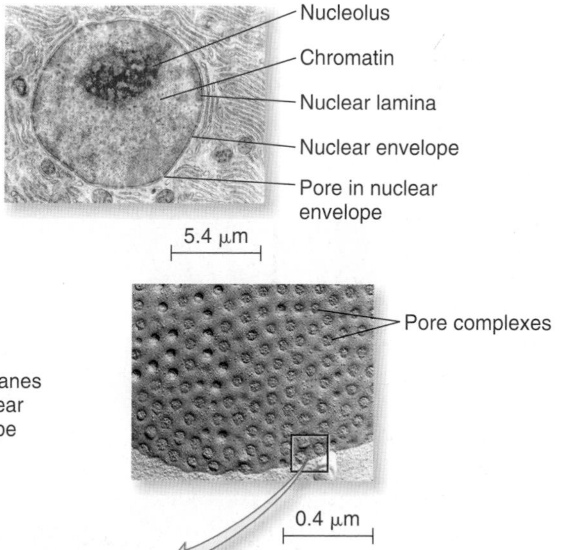

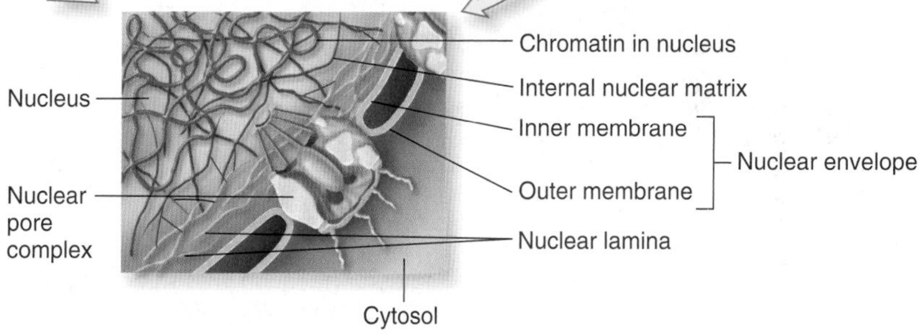

Figure 4.15 The nucleus and nuclear envelope. The nuclear envelope is composed of an inner membrane and an outer membrane that meet at the nuclear pores. The inner nuclear membrane is lined with lamin proteins to form the nuclear lamina. The interior of the nucleus contains chromatin, which is attached to the nuclear matrix, and a nucleolus, where ribosome subunits are assembled. Middle right: © Dr. Richard Kessel & Gene Shih/ Visuals Unlimited.

Concept Check: *What is the function of the nuclear lamina and the internal nuclear matrix?*

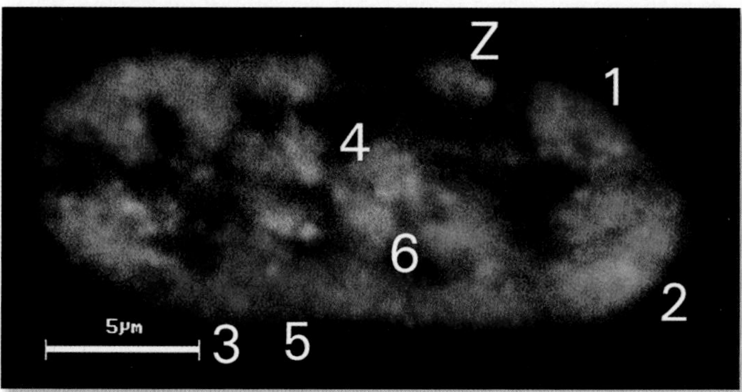

Figure 4.16 Chromosome territories in the cell nucleus. Chromosomes from a chicken were labeled with chromosome-specific probes. Seven types of chicken chromosomes are stained with a different dye. Each chromosome occupies its own distinct, nonoverlapping territory within the cell nucleus. Reprinted by permission from Macmillan Publishers Ltd. Cremer, T., and Cremer, C. Chromosome territories, nuclear architecture and gene regulation in mammalian cells. *Nature Reviews/Genetics*, Vol. 2(4), Figure 2, 292–301, 2001.

BioConnections: *Look ahead to Figure 15.8. What happens to chromosome territories during cell division?*

matrix serves to organize the chromosomes within the nucleus. Each chromosome is located in a distinct, nonoverlapping **chromosome territory**, which is visible when cells are exposed to dyes that label specific types of chromosomes (**Figure 4.16**).

The primary function of the nucleus is the protection, organization, replication, and expression of the genetic material. These topics are discussed in Unit III. Another important function is the assembly of ribosome subunits—cellular structures involved in producing polypeptides during the process of translation. The assembly of ribosome subunits occurs in the **nucleolus** (plural, nucleoli), a prominent region in the nucleus of nondividing cells. A ribosome is composed of two subunits: one small and one large (see Chapter 12, Table 12.3). Each subunit contains one or more RNA molecules and several types of proteins. Most of the RNA molecules that are components of ribosomes are made in the vicinity of the nucleolus. By comparison, the ribosomal proteins are produced in the cytosol and then imported into the nucleus through the nuclear pores. The ribosomal proteins and RNA molecules then assemble in the nucleolus to form the ribosomal subunits. Finally, the subunits exit through the nuclear pores into the cytosol, where they are needed for protein synthesis.

The Endoplasmic Reticulum Initiates Protein Sorting and Carries Out Metabolic Functions

The **endoplasmic reticulum (ER)** is a network of membranes that form flattened, fluid-filled tubules, or **cisternae** (**Figure 4.17**). The terms endoplasmic (Greek, for in the cytoplasm) and reticulum (Latin, for little net) refer to the location and shape of this organelle when viewed under a microscope. The term **lumen** describes the internal space of an organelle. The ER membrane encloses a single compartment called the **ER lumen**. There are two distinct, but continuous types of ER: rough ER and smooth ER.

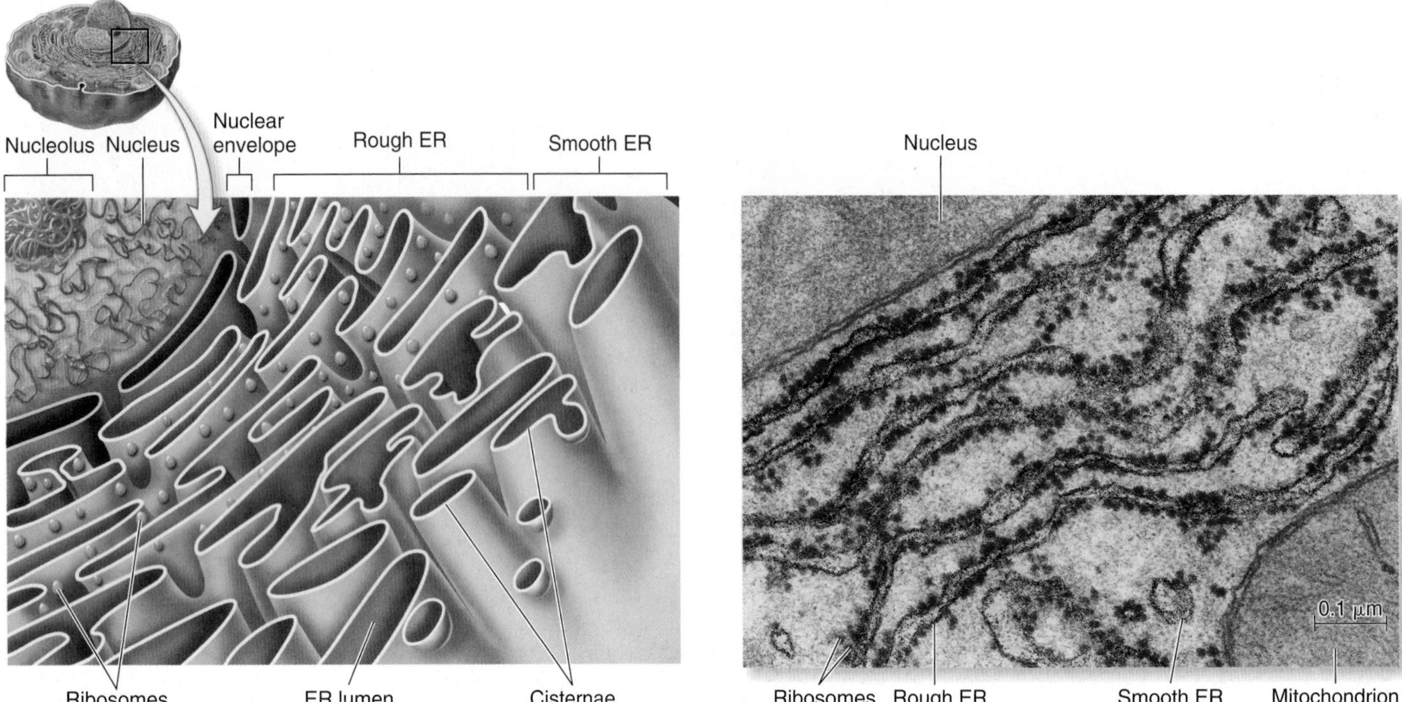

Figure 4.17 Structure of the endoplasmic reticulum. (Left side) The endoplasmic reticulum (ER) is composed of a network of flattened tubules called cisternae that enclose a continuous ER lumen. The rough ER is studded with ribosomes, whereas the smooth ER lacks ribosomes. The rough ER is continuous with the outer nuclear membrane. (Right side) A colorized TEM of the ER. The lumen of the ER is colored yellow and the ribosomes are red.

Rough ER The outer surface of the **rough endoplasmic reticulum (rough ER)** is studded with ribosomes, giving it a bumpy appearance. Rough ER plays a key role in the sorting of proteins that are destined for the ER, Golgi apparatus, lysosomes, vacuoles, plasma membrane, or outside of the cell. This topic is described later in Section 4.6. In conjunction with protein sorting, a second function of the rough ER is the insertion of certain newly made proteins into the ER membrane. A third important function of the rough ER is the attachment of carbohydrates to proteins and lipids. This process is called **glycosylation**. The topics of membrane protein insertion and protein glycosylation will be discussed in Chapter 5, because they are important features of cell membranes.

Smooth ER The **smooth endoplasmic reticulum (smooth ER)**, which lacks ribosomes, functions in diverse metabolic processes. The extensive network of smooth ER membranes provides an increased surface area for key enzymes that play important metabolic roles. In liver cells, enzymes in the smooth ER detoxify many potentially harmful organic molecules, including barbiturate drugs and ethanol. These enzymes convert hydrophobic toxic molecules into more hydrophilic molecules, which are easily excreted from the body. Chronic alcohol consumption, as in alcoholics, leads to a greater amount of smooth ER in liver cells, which increases the rate of alcohol breakdown. This explains why people who consume alcohol regularly must ingest more alcohol to experience its effects. It also explains why alcoholics often have enlarged livers.

The smooth ER of liver cells also plays a role in carbohydrate metabolism. The liver cells of animals store energy in the form of glycogen, which is a polymer of glucose. Glycogen granules, which are in the cytosol, sit very close to the smooth ER membrane. When chemical energy is needed, enzymes are activated that break down the glycogen to glucose-6-phosphate. Then, an enzyme in the smooth ER called glucose-6-phosphatase removes the phosphate group, and glucose is released into the bloodstream.

Another important function of the smooth ER in all eukaryotes is the accumulation of calcium ions (Ca^{2+}). The smooth ER contains calcium pumps that transport Ca^{2+} into the ER lumen. The regulated release of Ca^{2+} into the cytosol is involved in many vital cellular processes, including muscle contraction in animals.

Finally, enzymes in the smooth ER are critical in the synthesis and modification of lipids. For example, the smooth ER is the primary site for the synthesis of phospholipids, which are the main lipid component of eukaryotic cell membranes. This topic is described in Chapter 5. In addition, enzymes in the smooth ER are necessary for certain modifications of the lipid cholesterol that are needed to produce steroid hormones such as estrogen and testosterone.

The Golgi Apparatus Directs the Processing, Sorting, and Secretion of Cellular Molecules

The **Golgi apparatus** (also called the Golgi body, Golgi complex, or simply Golgi) was discovered by the Italian microscopist Camillo Golgi in 1898. It consists of a stack of flattened membranes, with each flattened membrane enclosing a single compartment. The Golgi stacks are named according to their orientation in the cell. The *cis* Golgi is near the ER membrane, the *trans* Golgi is closest to the plasma membrane, and the medial Golgi is found in the middle. Materials are transported between the Golgi stacks via membrane vesicles that bud

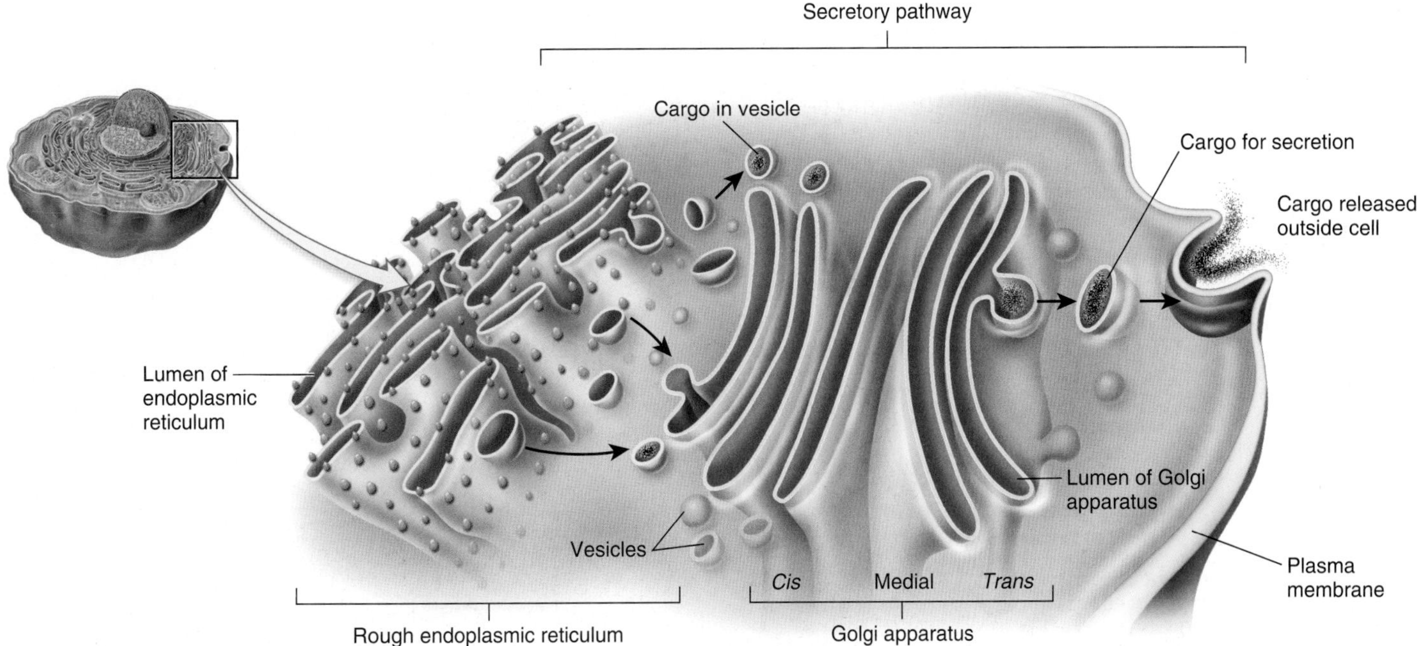

Figure 4.18 The Golgi apparatus and secretory pathway. The Golgi is composed of stacks of membranes that enclose separate compartments. Transport to and from the Golgi compartments occurs via membrane vesicles. Vesicles bud from the ER and go to the Golgi, and vesicles from the Golgi fuse with the plasma membrane to release cargo to the outside. The pathway from the ER to the Golgi to the plasma membrane is termed the secretory pathway.

Concept Check: *If we consider the Golgi apparatus as three compartments (cis, medial, and trans), describe the compartments that a protein travels through to be secreted.*

from one compartment in the Golgi (for example, the *cis* Golgi) and fuse with another compartment (for example, the medial Golgi).

The Golgi apparatus performs three overlapping functions: (1) processing, (2) protein sorting, and (3) secretion. We will discuss protein sorting in Section 4.6. Enzymes in the Golgi apparatus process, or modify, certain proteins and lipids. As mentioned earlier, carbohydrates can be attached to proteins and lipids in the endoplasmic reticulum. Glycosylation continues in the Golgi. For this to occur, a protein or lipid is transported via vesicles from the ER to the *cis* Golgi. Most of the glycosylation occurs in the medial Golgi.

A second type of processing event is **proteolysis**, whereby enzymes called **proteases** make cuts in polypeptides. For example, the hormone insulin is first made as a large precursor termed proinsulin. In the Golgi apparatus, proinsulin is packaged with proteases into vesicles. The proteases cut out a portion of the proinsulin to create a smaller insulin polypeptide that is a functional hormone. This happens just prior to secretion, which is described next.

The Golgi apparatus packages different types of materials (cargo) into **secretory vesicles** that fuse with the plasma membrane, thereby releasing their contents outside the cell. Proteins destined for secretion are synthesized into the ER, travel to the Golgi, and then are transported by vesicles to the plasma membrane. The vesicles then fuse with the plasma membrane, and the proteins are secreted to the outside of the cell. The entire route is called the **secretory pathway** (Figure 4.18). In addition to secretory vesicles, the Golgi also produces vesicles that travel to other parts of the cell, such as the lysosomes.

FEATURE INVESTIGATION

Palade Demonstrated That Secreted Proteins Move Sequentially Through Organelles of the Endomembrane System

As we have seen, one of the key functions of the endomembrane system is protein secretion. The identification of the secretory pathway came from studies of George Palade and his colleagues in the 1960s. He hypothesized that proteins follow a particular intracellular pathway to be secreted. Palade's team conducted pulse-chase experiments, in which the researchers administered a pulse of radioactive amino acids to cells so they made radioactive proteins. A few minutes later, the cells were given a large amount of nonradioactive amino acids. This is called a "chase" because it chases away the ability of the cells to make any more radioactive proteins. In this way, radioactive proteins were produced only briefly. Because they were labeled with radioactivity, the fate of these proteins could be monitored over time. The goal of a pulse-chase experiment is to determine where the radioactive proteins are produced and the pathway they take as they travel through a cell.

Palade chose to study the cells of the pancreas. This organ secretes enzymes and protein hormones that play a role in digestion and metabolism. Therefore, these cells were chosen because their primary activity is protein secretion. To study the pathway for protein secretion, Palade and colleagues injected a radioactive version of the amino acid leucine into the bloodstream of male guinea pigs.

Figure 4.19 Palade's use of the pulse-chase method to study protein secretion.

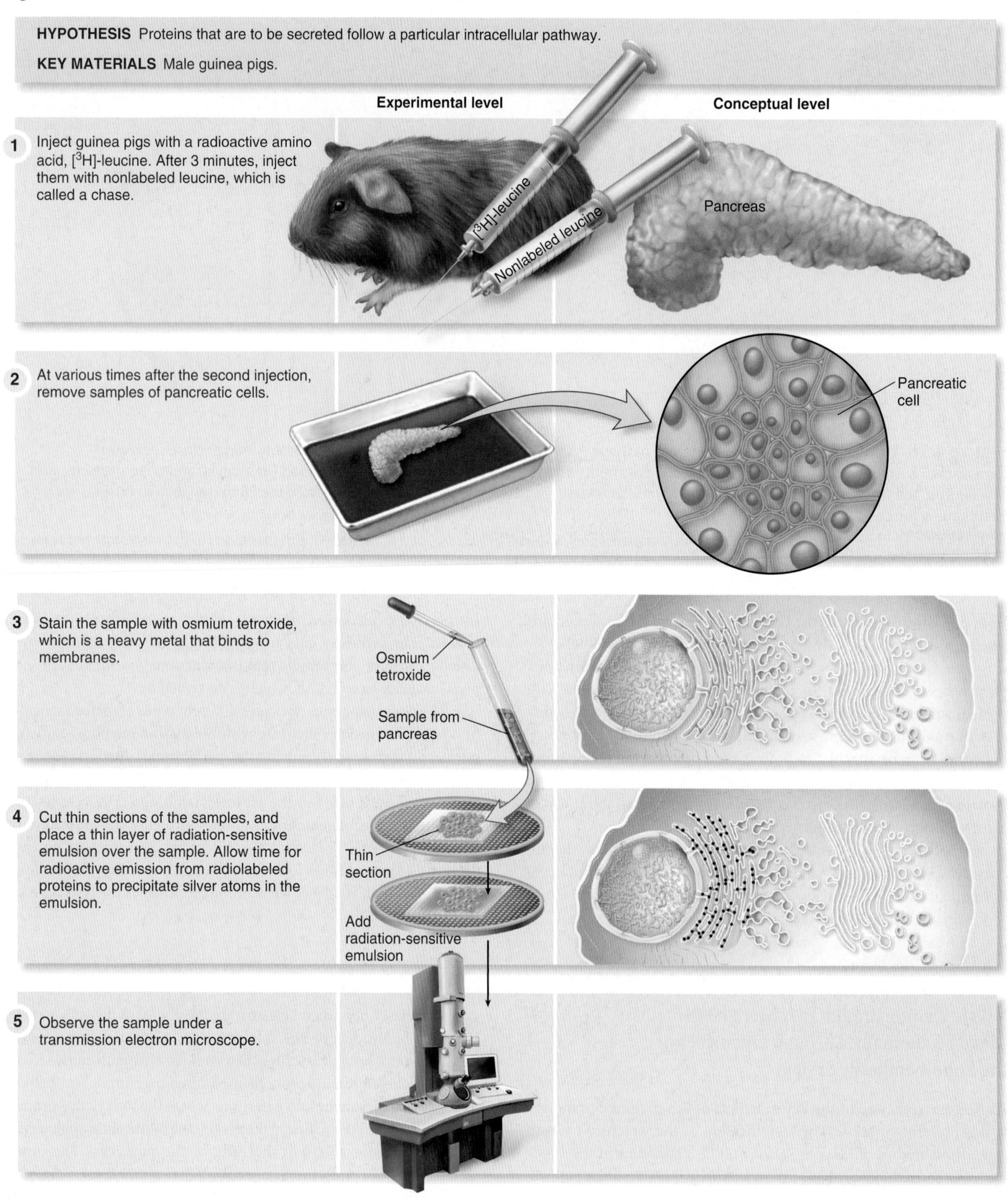

HYPOTHESIS Proteins that are to be secreted follow a particular intracellular pathway.
KEY MATERIALS Male guinea pigs.
Experimental level
Conceptual level
1 Inject guinea pigs with a radioactive amino acid, [³H]-leucine. After 3 minutes, inject them with nonlabeled leucine, which is called a chase.
[³H]-leucine
Nonlabeled leucine
Pancreas
2 At various times after the second injection, remove samples of pancreatic cells.
Pancreatic cell
3 Stain the sample with osmium tetroxide, which is a heavy metal that binds to membranes.
Osmium tetroxide
Sample from pancreas
4 Cut thin sections of the samples, and place a thin layer of radiation-sensitive emulsion over the sample. Allow time for radioactive emission from radiolabeled proteins to precipitate silver atoms in the emulsion.
Thin section
Add radiation-sensitive emulsion
5 Observe the sample under a transmission electron microscope.

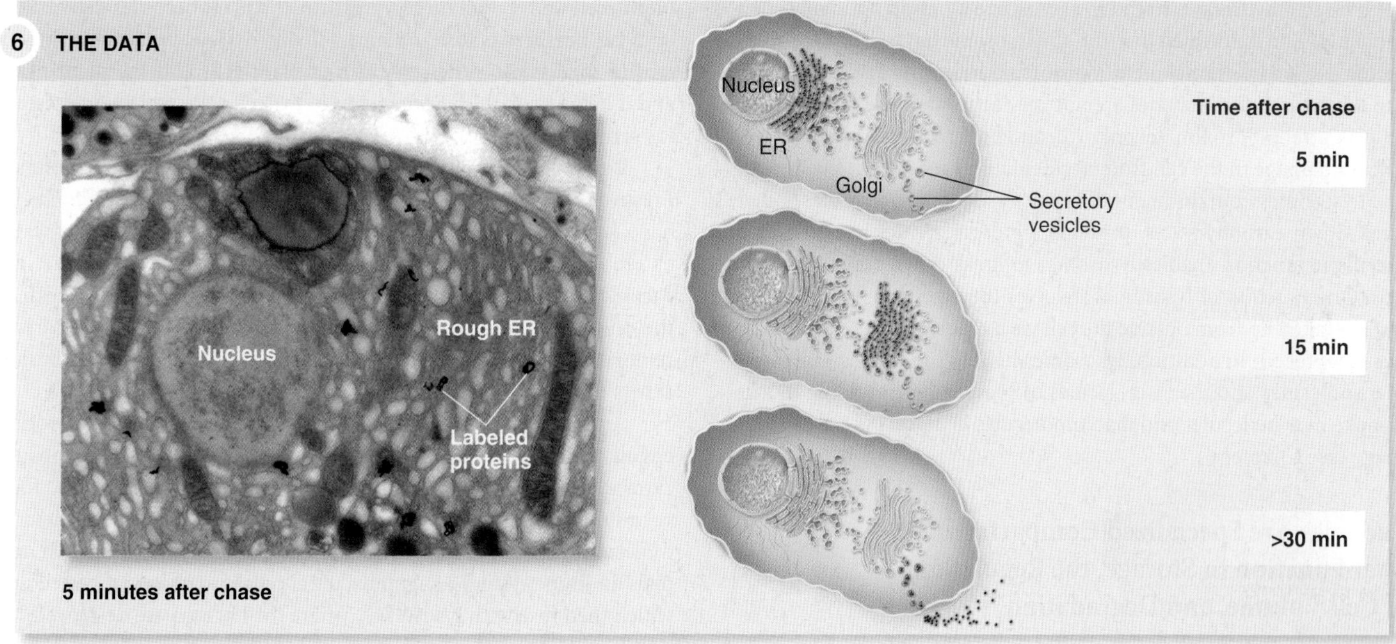

7 CONCLUSION To be secreted, proteins move from the ER to the Golgi to secretory vesicles and then to the plasma membrane, where they are released to the outside of the cell.

8 SOURCE Caro, L.G., and Palade, G.E. 1964. Protein synthesis, storage, and discharge in the pancreatic exocrine cell. An autoradiographic study. *Journal of Cell Biology* 20:473–495.

The radiolabeled leucine traveled in the bloodstream and was quickly taken up by cells of the body, including those in the pancreas. Three minutes later, they injected nonradioactive leucine (**Figure 4.19**). At various times after the second injection, samples of pancreatic cells were removed from the animals. The cells were then prepared for transmission electron microscopy (TEM). The sample was stained with osmium tetroxide, a heavy metal that became bound to membranes and showed the locations of the cell organelles. In addition, the sample was coated with a radiation-sensitive emulsion containing silver. When radiation was emitted from radioactive proteins, it interacted with the emulsion in a way that caused the precipitation of silver, which became tightly bound to the sample. In this way, the precipitated silver marked the location of the radiolabeled proteins. Unprecipitated silver in the emulsion was later washed away. Because silver atoms are electron dense, they produce dark spots in a TEM. Therefore, dark spots revealed the locations of radioactive proteins.

The micrograph in the data of Figure 4.19 illustrates the results that were observed 5 minutes after the completion of the pulse-chase injections. Very dark objects, namely radioactive proteins, were observed in the rough ER. As shown schematically to the right of the actual data, later time points indicated that the radioactive proteins moved from the ER to the Golgi, and then to secretory vesicles near the plasma membrane. In this way, Palade followed the intracellular pathway of protein movement. His experiments provided the first evidence that secreted proteins are synthesized into the rough ER and move through a series of cellular compartments before they are secreted. These findings also caused researchers to wonder how proteins are targeted to particular organelles and how they move from one compartment to another, topics that are described in Section 4.6.

Experimental Questions

1. Explain the procedure of a pulse-chase experiment. What is the pulse, and what is the chase? What was the purpose of the approach?
2. Why were pancreatic cells used for this investigation?
3. What were the key results of the experiment of Figure 4.19? What did the researchers conclude?

Lysosomes Are Involved in the Intracellular Digestion of Macromolecules

We now turn to another organelle of the endomembrane system, **lysosomes**, which are small organelles found in animal cells that break down macromolecules. Lysosomes contain many **acid hydrolases**, which are hydrolytic enzymes that use a molecule of water to break a covalent bond. This type of chemical reaction is called hydrolysis:

$$R_1\text{—}R_2 + H_2O \xrightarrow{\text{Acid hydrolase}} R_1\text{—}OH + R_2\text{—}H$$

The acid hydrolases found in a lysosome function optimally at an acidic pH. The fluid-filled interior of a lysosome has a pH of approximately 4.8. If a lysosomal membrane breaks, releasing acid hydrolases into the cytosol, the enzymes are not very active because the cytosolic pH is neutral (approximately pH 7.2) and buffered. This prevents significant damage to the cell from accidental leakage.

Lysosomes contain many different types of acid hydrolases that break down carbohydrates, proteins, lipids, and nucleic acids. This enzymatic function enables lysosomes to break down complex materials. One function of lysosomes involves the digestion of substances that are taken up from outside the cell via a process called endocytosis (see Chapter 5). In addition, lysosomes help to break down intracellular molecules and macromolecules to recycle their building blocks to make new molecules and macromolecules in a process called autophagy (see Chapter 6).

Vacuoles Are Specialized Compartments That Function in Storage, the Regulation of Cell Volume, and Degradation

Vacuoles are prominent organelles in plant cells, fungal cells, and certain protists. In animal cells, vacuoles tend to be smaller and are more commonly used to temporarily store materials or transport substances. In animals, such vacuoles are sometimes called storage vesicles. The term vacuole (Latin, for empty space) came from early microscopic observations of these compartments. We now know that vacuoles are not empty but instead contain fluid and sometimes even solid substances. Most vacuoles are made from the fusion of many smaller membrane vesicles.

The functions of vacuoles are extremely varied, and they differ among cell types and even environmental conditions. The best way to appreciate vacuole function is to consider a few examples. Mature plant cells often have a large **central vacuole** that occupies 80% or more of the cell volume (**Figure 4.20a**). The central vacuole serves two important purposes. First, it stores a large amount of water, enzymes, and inorganic ions such as calcium. It also stores other materials including proteins and pigments. Second, it performs a space-filling function. The large size of the vacuole exerts a pressure on the cell wall, called turgor pressure. If a plant becomes dehydrated and this pressure is lost, a plant will wilt. Turgor pressure is important in maintaining the structure of plant cells and the plant itself, and it helps to drive the expansion of the cell wall, which is necessary for growth.

Certain species of protists use vacuoles to maintain cell volume. Freshwater organisms such as the alga *Chlamydomonas reinhardtii* have small, water-filled **contractile vacuoles** that expand as water enters the cell (**Figure 4.20b**). Once they reach a certain size, the vacuoles suddenly contract, expelling their contents to the exterior of the cell (look ahead to Figure 5.16). This mechanism is necessary to remove the excess water that continually enters the cell by diffusion across the plasma membrane.

Another function of vacuoles is degradation. Some protists engulf their food into large food vacuoles in the process of phagocytosis (**Figure 4.20c**). As in the lysosomes of animal cells, food vacuoles contain hydrolytic enzymes that break down the macromolecules within the food. Macrophages, a type of cell found in animals' immune systems, engulf bacterial cells into phagocytic vacuoles, which then fuse with lysosomes, where the bacteria are destroyed.

Peroxisomes Catalyze Detoxifying Reactions

Peroxisomes, discovered by Christian de Duve in 1965, are relatively small organelles found in all eukaryotic cells. Peroxisomes consist of a single membrane that encloses a fluid-filled lumen. A typical eukaryotic cell contains several hundred of them.

Peroxisomes catalyze a variety of chemical reactions, including some reactions that break down organic molecules and others that are biosynthetic. In mammals, for example, large numbers of peroxisomes are found in liver cells, where toxic molecules accumulate and are broken down. A common by-product of the breakdown of toxins is hydrogen peroxide, H_2O_2:

$$\underset{\text{(toxin)}}{RH_2} + O_2 \longrightarrow R + H_2O_2$$

Hydrogen peroxide has the potential to damage cellular components. In the presence of metals such as iron (Fe^{2+}), which are found

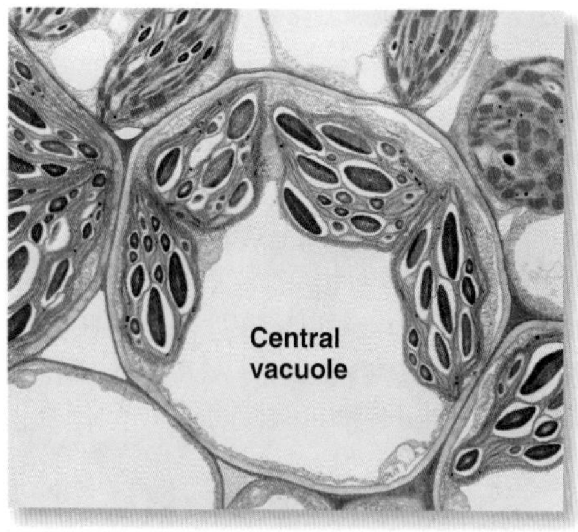

(a) Central vacuole in a plant cell

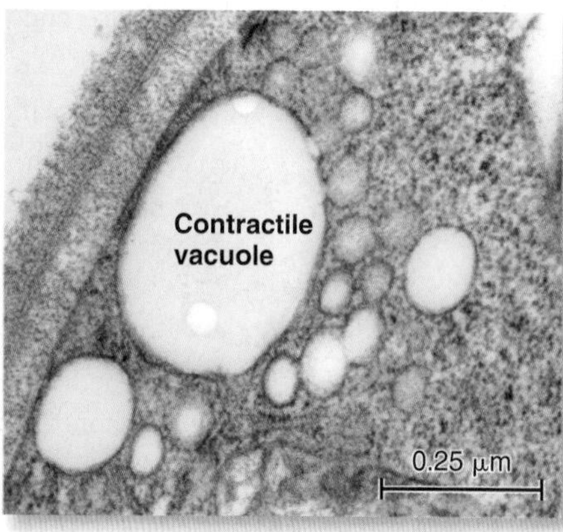

(b) Contractile vacuoles in an algal cell

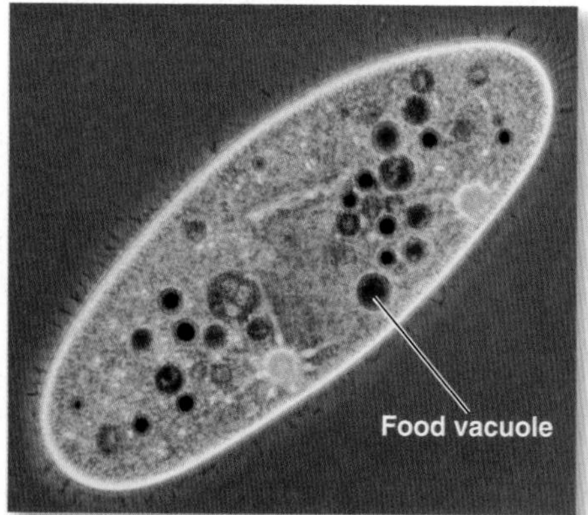

(c) Food vacuoles in a paramecium

Figure 4.20 **Examples of vacuoles.** These are TEMs. Part (c) is colorized.

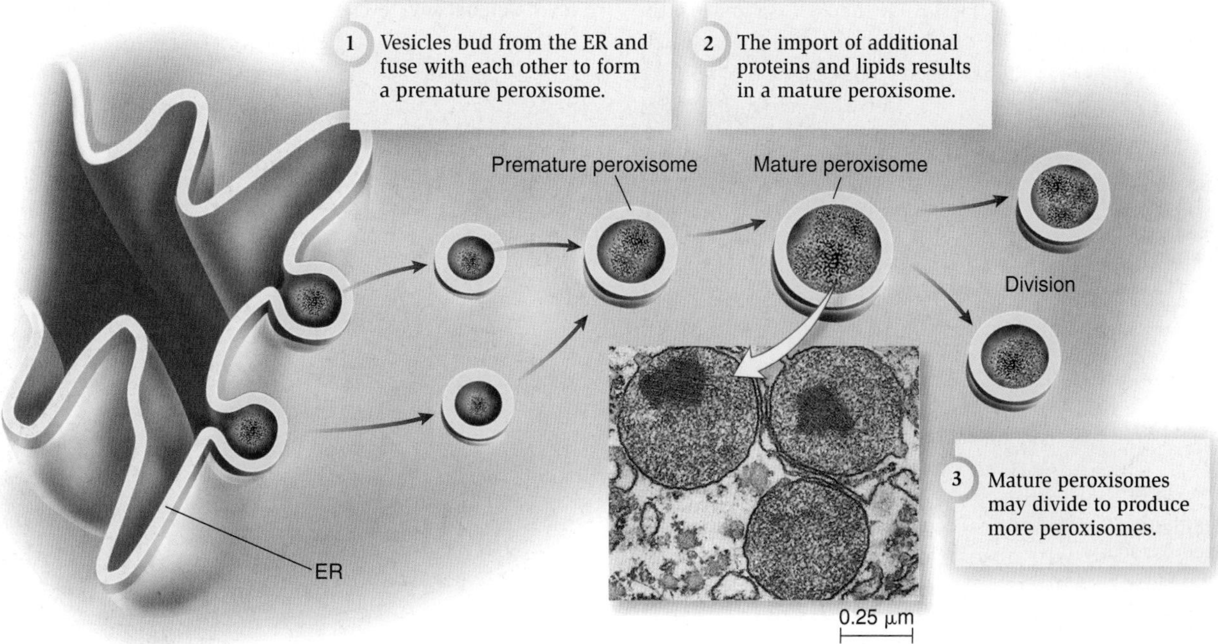

1 Vesicles bud from the ER and fuse with each other to form a premature peroxisome.
2 The import of additional proteins and lipids results in a mature peroxisome.
Premature peroxisome
Mature peroxisome
Division
3 Mature peroxisomes may divide to produce more peroxisomes.
ER
0.25 μm

Figure 4.21 **Formation of peroxisomes.** The inset is a TEM of mature peroxisomes.

naturally in living cells, H_2O_2 is broken down to form a hydroxide ion (OH-) and a molecule called a hydroxide free-radical (·OH):

$$Fe^{2+} + H_2O_2 \rightarrow Fe^{3+} + OH^- + \cdot OH \text{ (hydroxide free-radical)}$$

The ·OH is highly reactive and can damage proteins, lipids, and DNA. Therefore, it is beneficial for cells to break down H_2O_2 in an alternative manner that does not form a ·OH. Peroxisomes contain an enzyme called **catalase** that breaks down H_2O_2 to make water and oxygen gas (hence the name peroxisome):

$$2\,H_2O_2 \xrightarrow{\text{Catalase}} 2\,H_2O + O_2$$

Aside from detoxification, peroxisomes usually contain enzymes involved in the metabolism of fats and amino acids. For example, plant seeds contain specialized organelles called **glyoxysomes**, which are similar to peroxisomes. Seeds often store fats instead of carbohydrates. Because fats have higher energy per unit mass, a plant can make seeds that are smaller and less heavy. Glyoxysomes contain enzymes that are needed to convert fats to sugars. These enzymes become active when a seed germinates and the seedling begins to grow.

Peroxisomes were once viewed as semiautonomous organelles like mitochondria and chloroplasts, because they appeared to be self-replicating, that is, produced by the division of pre-existing peroxisomes. However, recent research indicates that peroxisomes are derived from the endomembrane system. A general model for peroxisome formation is shown in Figure 4.21, though the details may differ among animal, plant, and fungal cells. To initiate peroxisome formation, vesicles bud from the ER membrane and form a premature peroxisome. Following the import of additional proteins, the premature peroxisome becomes a mature peroxisome. Once the mature peroxisome has formed, it may then divide to further increase the number of peroxisomes in the cell.

The Plasma Membrane Is the Interface Between a Cell and Its Environment

The cytoplasm of eukaryotic cells is surrounded by a plasma membrane, which is part of the endomembrane system and provides a boundary between a cell and the extracellular environment. Proteins in the plasma membrane perform many important functions that affect the activities inside the cell (Figure 4.22). First, many plasma membrane proteins are involved in **membrane transport**. Some of these proteins function to transport essential nutrients or ions into the cell, and others are involved in the export of substances. Due to the functioning of these protein transporters, the plasma membrane is selectively permeable; it allows only certain substances in and out. We will examine the structure and function of the plasma membrane, as well as a variety of transporters, in Chapter 5.

A second vital function of the plasma membrane is **cell signaling**. To survive and adapt to changing conditions, cells must be able to sense changes in their environment. In addition, the cells of a multicellular organism need to communicate with each other to coordinate their activities. The plasma membrane of all cells contains receptors that recognize signals—either environmental agents or molecules secreted by other cells. When a signaling molecule binds to a receptor, it activates a signal transduction pathway—a series of steps that cause the cell to respond to the signal. For example, when you eat a meal, the hormone insulin is secreted into your bloodstream. This hormone binds to receptors in the plasma membrane of your cells, which results in a cellular response that allows your cells to increase their uptake of certain molecules found in food, such as glucose. We will explore the details of cell signaling in Chapter 9.

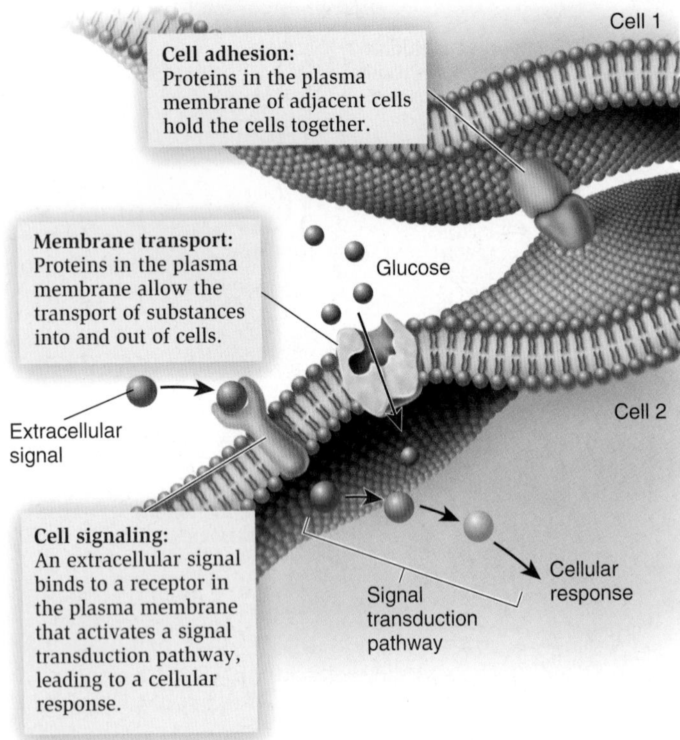
Cell 1
Cell adhesion:
Proteins in the plasma membrane of adjacent cells hold the cells together.
Membrane transport:
Proteins in the plasma membrane allow the transport of substances into and out of cells.
Glucose
Extracellular signal
Cell 2
Cell signaling:
An extracellular signal binds to a receptor in the plasma membrane that activates a signal transduction pathway, leading to a cellular response.
Signal transduction pathway
Cellular response

Figure 4.22 **Major functions of the plasma membrane.** Three important roles are membrane transport, cell signaling, and cell adhesion.

Concept Check: *Which of these three functions do you think is the most important for cell metabolism?*

A third important role of the plasma membrane in animal cells is **cell adhesion**. Protein-protein interactions among proteins in the plasma membranes of adjacent cells promote cell-to-cell adhesion. This phenomenon is critical for animal cells to properly interact to form a multicellular organism and allows cells to recognize each other. The structures and functions of proteins involved in cell adhesion will be examined in Chapter 10.

4.5 Semiautonomous Organelles

Learning Outcomes:

1. Outline the structures and general functions of mitochondria and chloroplasts.
2. Discuss the evidence for the endosymbiosis theory.

We now turn to those organelles in eukaryotic cells that are considered semiautonomous: mitochondria and chloroplasts. These organelles grow and divide, but they are not completely autonomous because they depend on other parts of the cell for their internal components (**Figure 4.23**). For example, most of the proteins found in mitochondria are imported from the cytosol. In this section, we will survey the structures and functions of the semiautonomous organelles in eukaryotic cells and consider their evolutionary origins. In Chapters 7 and 8, we will explore the functions of mitochondria and chloroplasts in greater depth.

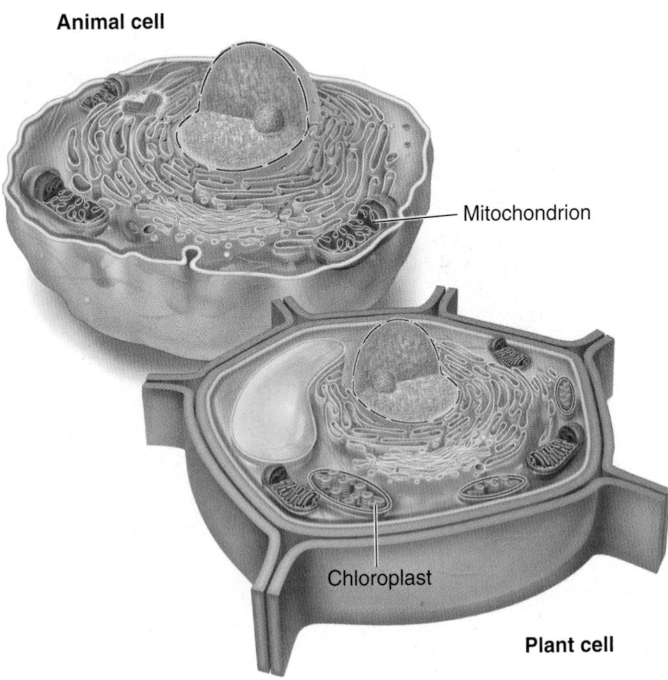
Animal cell
Mitochondrion
Chloroplast
Plant cell

Figure 4.23 **Semiautonomous organelles.** These are the mitochondria and chloroplasts.

BIOLOGY PRINCIPLE **Living organisms use energy.** Chloroplasts capture light energy and synthesize organic molecules. Mitochondria break down organic molecules and make ATP that is used as an energy source to drive many different cellular processes.

Mitochondria Supply Cells with Most of Their ATP

Mitochondrion (plural, mitochondria) literally means thread granule, which is what mitochondria look like under a light microscope—either threadlike or granular-shaped. They are similar in size to bacteria. A typical cell may contain a few hundred to a few thousand mitochondria. Cells with particularly heavy energy demands, such as muscle cells, have more mitochondria than other cells. Research has shown that regular exercise increases the number and size of mitochondria in human muscle cells to meet the expanded demand for energy.

A mitochondrion has an outer membrane and an inner membrane separated by a region called the intermembrane space (**Figure 4.24**). The inner membrane is highly invaginated (folded) to form projections called **cristae**. The cristae greatly increase the surface area of the inner membrane, which is the site where ATP is made. The compartment enclosed by the inner membrane is the **mitochondrial matrix**.

The primary role of mitochondria is to make ATP. Even though mitochondria produce most of a cell's ATP, mitochondria do not create energy. Rather, their primary function is to convert chemical energy that is stored within the covalent bonds of organic molecules into a form that can be readily used by cells. Covalent bonds in sugars, fats, and amino acids store a large amount of energy. The breakdown of these molecules into simpler molecules releases energy that is used to make ATP. Many proteins in living cells utilize ATP to carry out

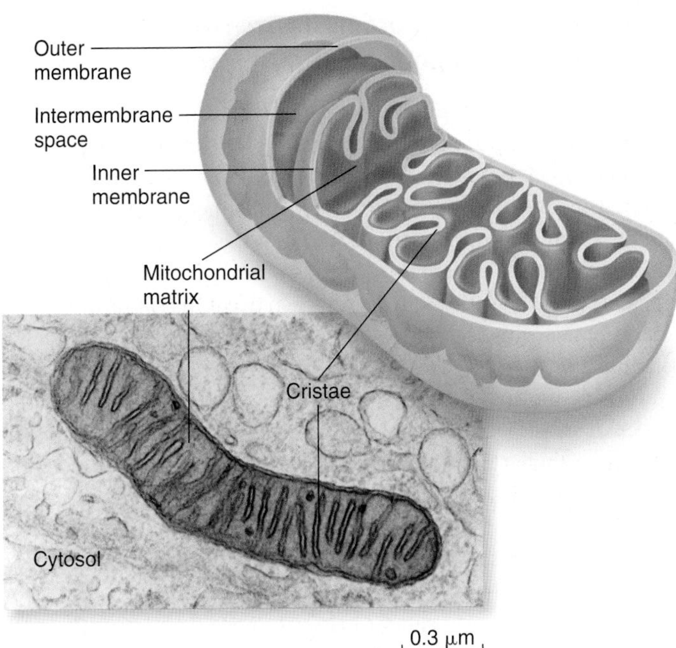

Figure 4.24 **Structure of a mitochondrion.** This figure emphasizes the membrane organization of a mitochondrion, which has an outer and inner membrane. The invaginations of the inner membrane are called cristae. The mitochondrial matrix lies inside the inner membrane. The micrograph is a colorized TEM.

Concept Check: *What is the advantage of having a highly invaginated inner membrane?*

their functions, such as muscle contraction, uptake of nutrients, cell division, and many other cellular processes.

Mitochondria perform other functions as well. They are involved in the synthesis, modification, and breakdown of several types of cellular molecules. For example, the synthesis of certain hormones requires enzymes that are found in mitochondria. Another interesting role of mitochondria is to generate heat in specialized fat cells known as brown fat cells. Groups of brown fat cells serve as "heating pads" that help to revive hibernating animals and protect sensitive areas of young animals from the cold.

Chloroplasts Carry Out Photosynthesis

Chloroplasts are organelles that capture light energy and use some of that energy to synthesize organic molecules such as glucose. This process, called **photosynthesis**, is described in Chapter 8. Chloroplasts are found in nearly all species of plants and algae. Figure 4.25 shows the structure of a typical chloroplast. Like the mitochondrion, a chloroplast contains an outer and inner membrane. An intermembrane space lies between these two membranes. A third system of membranes, the **thylakoid membrane**, forms many flattened, fluid-filled tubules that enclose a single, convoluted compartment called the thylakoid lumen. These tubules tend to stack on top of each other to form a structure called a **granum** (plural, grana). The **stroma** is the compartment of the chloroplast that is enclosed by the inner membrane but outside the thylakoid membrane.

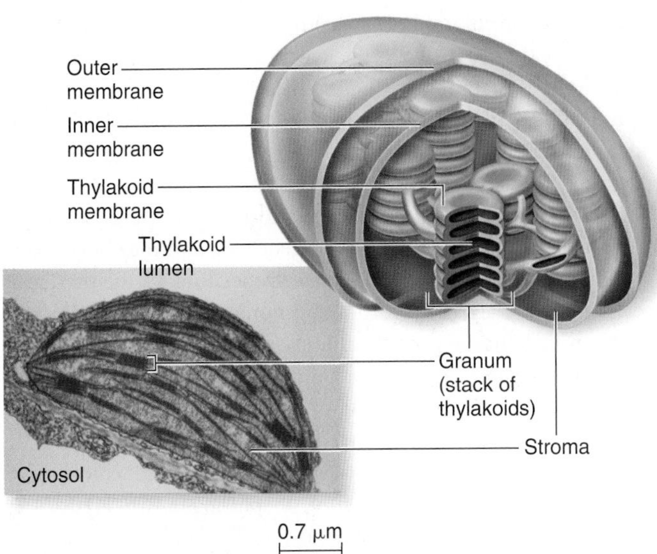

Figure 4.25 **Structure of a chloroplast.** Like a mitochondrion, a chloroplast is enclosed in a double membrane. In addition, it has an internal thylakoid membrane system that forms flattened compartments. These compartments stack on each other to form grana. The stroma is located inside the inner membrane but outside the thylakoid membrane. This micrograph is a colorized TEM.

Chloroplasts are a specialized version of plant organelles that are more generally known as **plastids**. All plastids are derived from unspecialized **proplastids**. The various types of plastids are distinguished by their synthetic abilities and the types of pigments they contain. Chloroplasts, which carry out photosynthesis, contain the green pigment chlorophyll. The abundant number of chloroplasts in the leaves of plants gives them their green color. Chromoplasts, a second type of plastid, function in synthesizing and storing the yellow, orange, and red pigments known as carotenoids. Chromoplasts give many fruits and flowers their colors. In autumn, the chromoplasts also give many leaves their yellow, orange, and red colors. A third type of plastid, leucoplasts, typically lacks pigment molecules. An amyloplast is a leucoplast that synthesizes and stores starch. Amyloplasts are common in underground structures such as roots and tubers.

Mitochondria and Chloroplasts Contain Their Own Genetic Material and Divide by Binary Fission

To fully appreciate the structure and organization of mitochondria and chloroplasts, we also need to briefly examine their genetic properties. In 1951, Yasutane Chiba exposed plant cells to Feulgen, a DNA-specific dye, and discovered that the chloroplasts became stained. Based on this observation, he was the first to suggest that chloroplasts contain their own DNA. Researchers in the 1970s and 1980s isolated DNA from both chloroplasts and mitochondria. These studies revealed that the DNA of these organelles resembled smaller versions of bacterial chromosomes.

The chromosomes found in mitochondria and chloroplasts are referred to as the **mitochondrial genome** and **chloroplast genome**, respectively, and the chromosomes found in the nucleus of the cell constitute the **nuclear genome**. Like bacteria, the genomes of most mitochondria and chloroplasts are composed of a single circular chromosome. Compared with the nuclear genome, they are very small. For example, the amount of DNA in the human nuclear genome (about 3 billion base pairs) is about 200,000 times greater than the mitochondrial genome. In terms of genes, the human genome has approximately 20,000–25,000 different genes, whereas the human mitochondrial genome has only a few dozen. Chloroplast genomes tend to be larger than mitochondrial genomes, and they have a correspondingly greater number of genes. Depending on the particular species of plant or algae, a chloroplast genome is about 10 times larger than the mitochondrial genome of human cells.

Just as the genomes of mitochondria and chloroplasts resemble bacterial genomes, the production of new mitochondria and chloroplasts bears a striking resemblance to the division of bacterial cells. Like their bacterial counterparts, mitochondria and chloroplasts increase in number via **binary fission**, or splitting in two. Figure 4.26 illustrates the process for a mitochondrion. The mitochondrial chromosome, which is found in a region called the nucleoid, is duplicated, and the organelle divides into two separate organelles. Mitochondrial and chloroplast division are needed to maintain a full complement of these organelles when cell growth occurs following cell division. In addition, environmental conditions may influence the sizes and numbers of these organelles. For example, when plants are exposed to more sunlight, the number of chloroplasts in leaf cells increases.

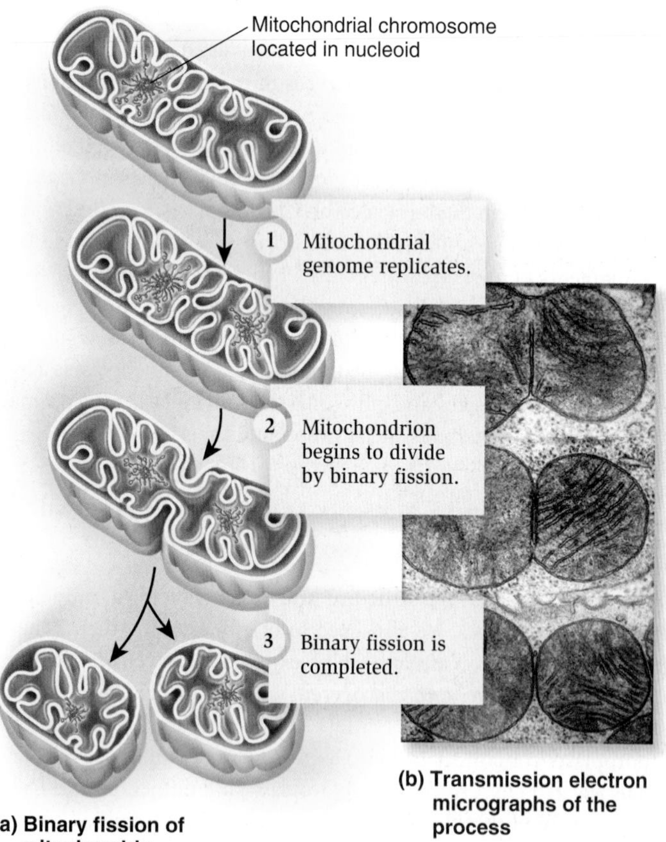
Mitochondrial chromosome located in nucleoid
1 Mitochondrial genome replicates.
2 Mitochondrion begins to divide by binary fission.
3 Binary fission is completed.
(a) Binary fission of mitochondria
(b) Transmission electron micrographs of the process

Figure 4.26 Division of mitochondria by binary fission.

BioConnections: Look ahead to Figure 18.14. How is this process similar to bacterial cell division, and how is it different?

Mitochondria and Chloroplasts Are Derived from Ancient Symbiotic Relationships

The observation that mitochondria and chloroplasts contain their own genetic material may seem puzzling. Perhaps you might think that it would be simpler for a eukaryotic cell to have all of its genetic material in one place—the nucleus. The distinct genomes of mitochondria and chloroplasts can be traced to their evolutionary origin, which involved an ancient symbiotic association.

A symbiotic relationship occurs when two different species live in direct contact with each other. **Endosymbiosis** describes a symbiotic relationship in which the smaller species—the symbiont—actually lives inside the larger species. In 1883, Andreas Schimper proposed that chloroplasts were descended from an endosymbiotic relationship between cyanobacteria (a bacterium capable of photosynthesis) and eukaryotic cells. In 1922, Ivan Wallin also hypothesized an endosymbiotic origin for mitochondria.

In spite of these interesting ideas, the question of endosymbiosis was largely ignored until the discovery that mitochondria and chloroplasts contain their own genetic material. In 1970, the issue of endosymbiosis as the origin of mitochondria and chloroplasts was revived by Lynn Margulis in her book *Origin of Eukaryotic Cells*. During the 1970s and 1980s, the advent of molecular genetic techniques allowed researchers to analyze genes from mitochondria, chloroplasts, bacteria, and eukaryotic nuclear genomes. Researchers discovered that genes in mitochondria and chloroplasts are very similar to bacterial genes. Likewise, mitochondria and chloroplasts are strikingly similar in size and shape to certain bacterial species. These observations provided strong support for the **endosymbiosis theory**, which proposes that mitochondria and chloroplasts originated from bacteria that took up residence within a primordial eukaryotic cell (**Figure 4.27**). Over the next 2 billion years, the characteristics of these intracellular bacterial cells gradually changed to those of a mitochondrion or chloroplast. The origin of eukaryotic cells is discussed in more detail in Chapter 22.

Symbiosis occurs because the relationship is beneficial to one or both species. According to the endosymbiosis theory, this relationship provided eukaryotic cells with useful cellular characteristics. Chloroplasts, which were derived from cyanobacteria, have the ability to carry out photosynthesis. This benefits plant cells by giving them the ability to use the energy from sunlight. By comparison, mitochondria are thought to have been derived from a different type of bacteria known as purple bacteria or α-proteobacteria. In this case, the endosymbiotic relationship enabled eukaryotic cells to synthesize greater amounts of ATP. How the relationship would have been beneficial to a cyanobacterium or purple bacterium is less clear, though the cytosol of a eukaryotic cell may have provided a stable environment with an adequate supply of nutrients.

During the evolution of eukaryotic species, many genes that were originally found in the genome of the primordial purple bacteria and cyanobacteria have been transferred from the organelles to the nucleus. This has occurred many times throughout evolution, so modern mitochondria and chloroplasts have lost most of the genes that still exist in present-day purple bacteria and cyanobacteria. Some researchers speculate that the movement of genes into the nucleus makes it easier for the cell to control the structure, function, and

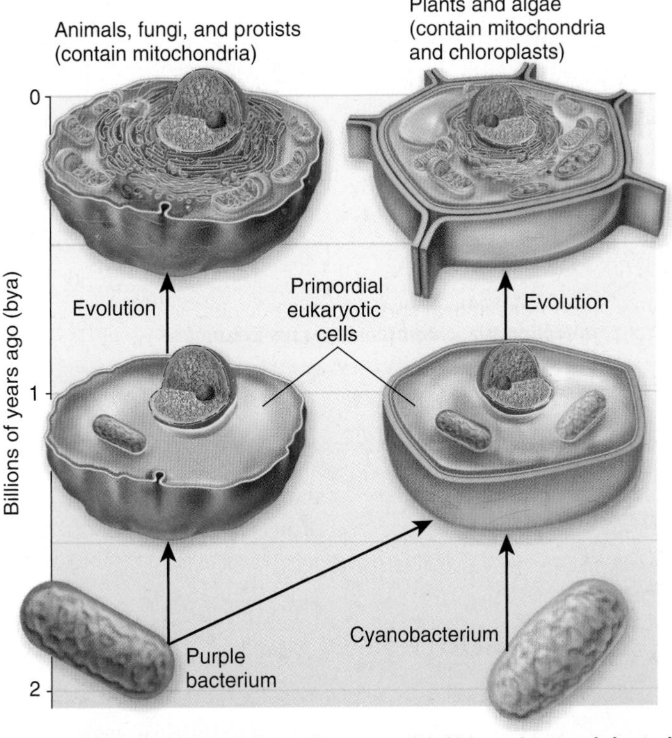

Figure 4.27 A simplified view of the endosymbiosis theory. **(a)** According to this concept, modern mitochondria were derived from purple bacteria, also called α-proteobacteria. Over the course of evolution, their characteristics changed into those found in mitochondria today. **(b)** A similar phenomenon occurred for chloroplasts, which were derived from cyanobacteria (blue-green bacteria), a bacterium that is capable of photosynthesis.

Concept Check: Discuss the similarities and differences between modern bacteria and mitochondria.

division of mitochondria and chloroplasts. In modern cells, hundreds of different proteins that make up these organelles are encoded by genes that have been transferred to the nucleus. These proteins are made in the cytosol and then taken up into mitochondria or chloroplasts. We will discuss this topic next.

4.6 Protein Sorting to Organelles

Learning Outcomes:

1. List which categories of proteins are sorted cotranslationally and which are sorted post-translationally.
2. Describe the steps that occur during the cotranslational sorting of proteins to the endoplasmic reticulum.
3. Explain how proteins are moved via vesicles through the endomembrane system.
4. Outline the steps of post-translational sorting of proteins to mitochondria.

As we have seen, eukaryotic cells contain a variety of membrane-bound organelles. Each protein that a cell makes usually functions within one cellular compartment or is secreted from the cell. How does each protein reach its appropriate destination? For example, how does a mitochondrial protein get sent to the mitochondrion rather than to a different organelle such as a lysosome? In eukaryotes, most proteins contain short stretches of amino acid sequences that direct them to their correct cellular location. These sequences are called **sorting signals**, or **traffic signals**. Each sorting signal is recognized by specific cellular components that facilitate the proper routing of that protein to its correct location.

Most eukaryotic proteins begin their synthesis on ribosomes in the cytosol, using messenger RNA (mRNA) that contains the information for polypeptide synthesis (Figure 4.28). The cytosol provides amino acids, which are used as building blocks to make these proteins during translation. Cytosolic proteins lack any sorting signal, so they remain there. By comparison, the synthesis of proteins destined for the ER, Golgi, lysosomes, vacuoles, or secretory vesicles begins in the cytosol and then halts temporarily until the ribosome has become bound to the ER membrane. After this occurs, translation resumes and the polypeptide is synthesized into the ER. Proteins that are destined for the ER, Golgi, lysosome, vacuole, plasma membrane, or secretion are first directed to the ER. This is called **cotranslational sorting** because the first step in the sorting process begins while translation is occurring. Finally, the uptake of most proteins into the nucleus, mitochondria, chloroplasts, and peroxisomes occurs after the protein is completely made (that is, completely translated) in the cytosol. This is called **post-translational sorting** because sorting does not happen until translation is finished. In this section, we will consider how cells carry out cotranslational and post-translational sorting.

The Cotranslational Sorting of Some Proteins Occurs at the Endoplasmic Reticulum Membrane

The concept of sorting signals in proteins was first proposed by Günter Blobel in the 1970s. Blobel and colleagues discovered a sorting signal in proteins that sends them to the ER membrane, which is the first step in cotranslational sorting (Figure 4.29). To be directed to the rough ER membrane, a polypeptide must contain a sorting signal called an **ER signal sequence**, which is a sequence of about 6–12 amino acids that are predominantly hydrophobic and usually located near the N-terminus. As the ribosome is making the polypeptide in the cytosol, the ER signal sequence emerges from the ribosome and is recognized by a protein-RNA complex called **signal recognition particle (SRP)**. SRP has two functions. First, it recognizes the ER signal sequence and pauses translation. Second, SRP binds to an SRP receptor in the ER membrane, which docks the ribosome over a channel. At this stage, SRP is released and translation resumes. The growing polypeptide is threaded through the channel to cross the ER membrane. If the protein is not a membrane protein, it will be released into the lumen of the ER. In most cases, the ER signal sequence is removed by an enzyme, signal peptidase. In 1999, Blobel won the Nobel Prize for Physiology or Medicine for his discovery of sorting signals in proteins. The process shown in Figure 4.29 illustrates another important role of protein-protein interactions—a series of interactions causes the steps of a process to occur in a specific order.

Some proteins are meant to function in the ER. Such proteins contain ER retention signals in addition to the ER signal sequence.

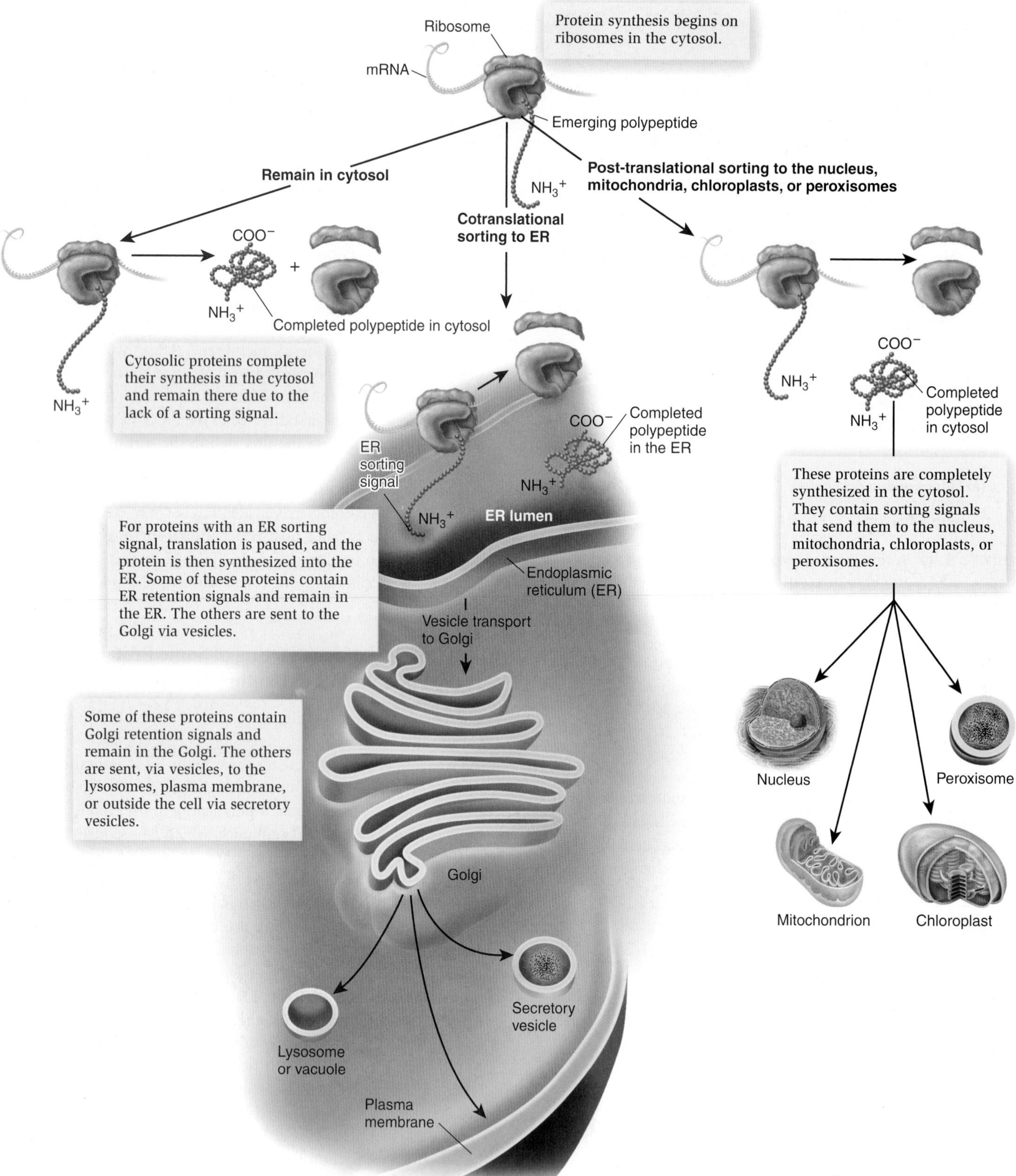

Protein synthesis begins on ribosomes in the cytosol.
Ribosome
mRNA
Emerging polypeptide
Remain in cytosol
NH_3^+
Cotranslational sorting to ER
Post-translational sorting to the nucleus, mitochondria, chloroplasts, or peroxisomes
COO^-
NH_3^+
Completed polypeptide in cytosol
Cytosolic proteins complete their synthesis in the cytosol and remain there due to the lack of a sorting signal.
NH_3^+
ER sorting signal
COO^-
Completed polypeptide in the ER
NH_3^+
ER lumen
For proteins with an ER sorting signal, translation is paused, and the protein is then synthesized into the ER. Some of these proteins contain ER retention signals and remain in the ER. The others are sent to the Golgi via vesicles.
Endoplasmic reticulum (ER)
Vesicle transport to Golgi
Some of these proteins contain Golgi retention signals and remain in the Golgi. The others are sent, via vesicles, to the lysosomes, plasma membrane, or outside the cell via secretory vesicles.
Golgi
Lysosome or vacuole
Secretory vesicle
Plasma membrane
COO^-
Completed polypeptide in cytosol
These proteins are completely synthesized in the cytosol. They contain sorting signals that send them to the nucleus, mitochondria, chloroplasts, or peroxisomes.
Nucleus
Peroxisome
Mitochondrion
Chloroplast

Figure 4.28 **Three pathways for protein sorting in a eukaryotic cell.** Proteins either remain in the cytosol, or they are sorted to the ER (cotranslational sorting), or sorted after the protein is completely made (post-translational sorting).

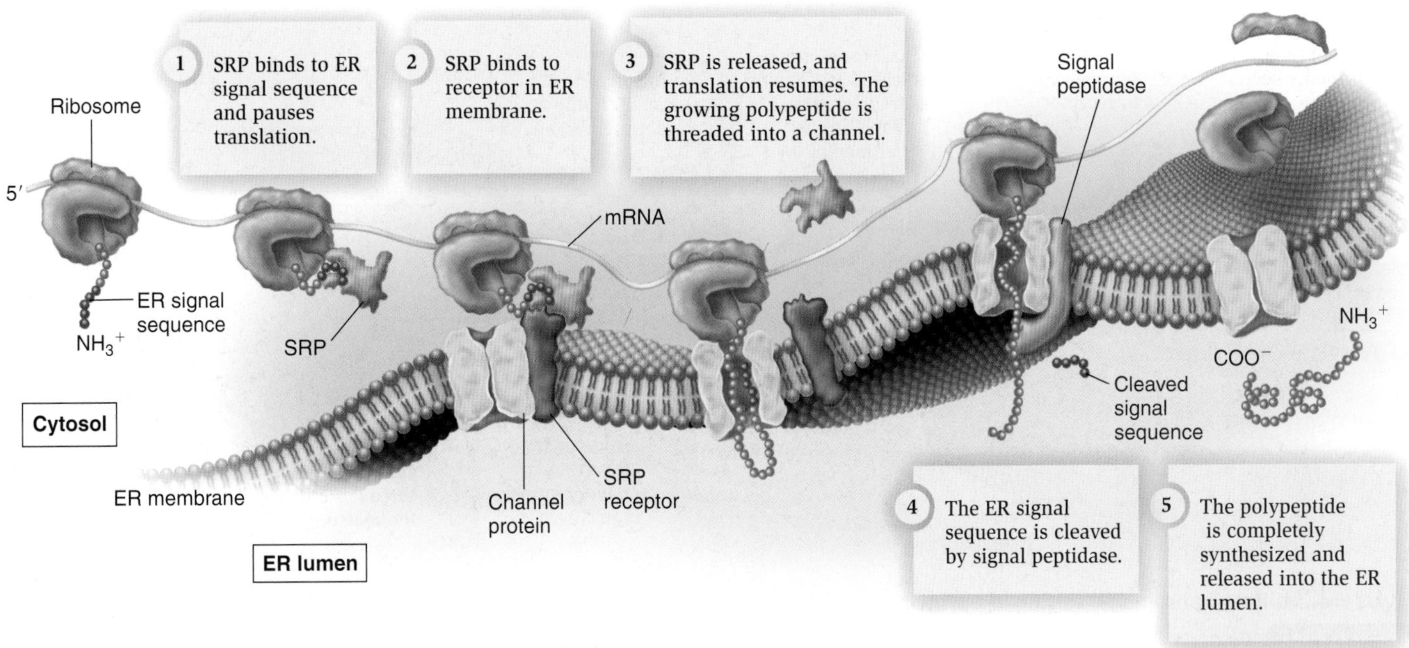

Figure 4.29 **First step in cotranslational sorting: sending proteins to the ER.**

Concept Check: *What prevents an ER protein from being completely synthesized in the cytosol?*

Alternatively, other proteins that are destined for the Golgi, lysosomes, vacuoles, plasma membrane, or secretion leave the ER and are transported to their correct location (see Figure 4.28). This transport process occurs via vesicles that are formed from one compartment and then move through the cytosol and fuse with another compartment. Vesicles from the ER may go to the Golgi, and then vesicles from the Golgi may go to the lysosomes, vacuoles, or plasma membrane. Sorting signals within proteins' amino acid sequences are responsible for directing them to the correct location.

Figure 4.30 describes the second step in cotranslational sorting, vesicle transport from the ER to the Golgi. A cargo, such as protein molecules, is loaded into a developing vesicle by binding to cargo receptors in the ER membrane. Vesicle formation is facilitated by the binding of coat proteins, which, by shaping the surrounding membrane into a sphere, helps a vesicle to bud from a given membrane. As a vesicle forms, other proteins called V-snares are incorporated into the vesicle membrane (hence the name V-snare). Many types of V-snares are known to exist. The particular V-snare that is found in

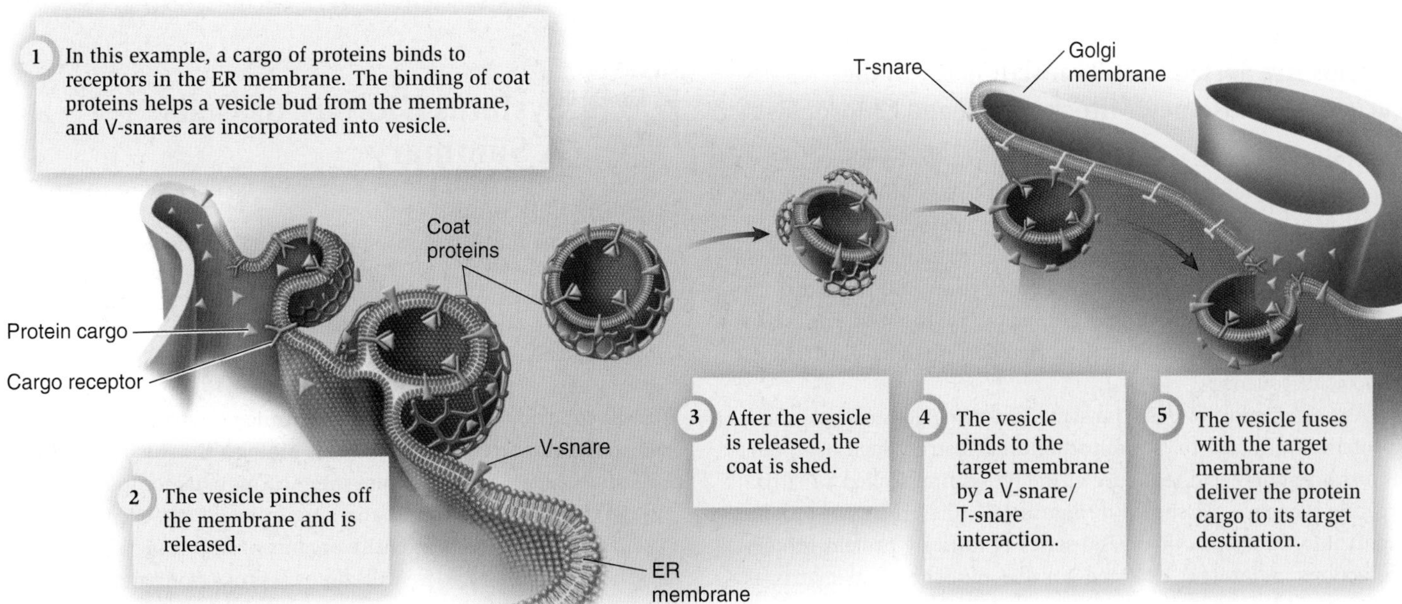

Figure 4.30 **Second step in cotranslational sorting: vesicle transport from the ER to the Golgi.**

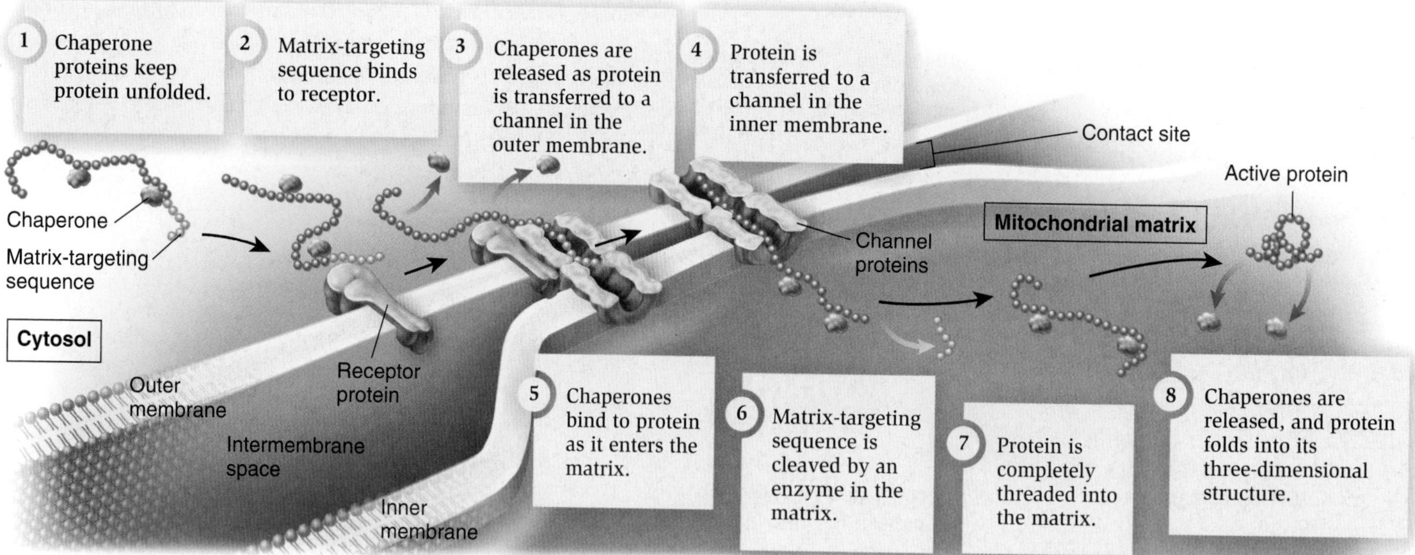

Figure 4.31 **Post-translational sorting of a protein to the mitochondrial matrix.**

Concept Check: *What do you think would happen if chaperone proteins did not bind to a mitochondrial matrix protein before it was imported into the mitochondrion?*

a vesicle membrane depends on the type of cargo it carries. After a vesicle is released from one compartment, such as the ER, the coat is shed. The vesicle then travels through the cytosol. But how does the vesicle know where to go? The answer is that the V-snares in the vesicle membrane are recognized by T-snares in a target membrane. After V-snares recognize T-snares, the vesicle fuses with the membrane containing the T-snares. The recognition between V-snares and T-snares ensures that a vesicle carrying a specific cargo moves to the correct target membrane in the cell. Like the sorting of proteins to the ER membrane, the formation and sorting of vesicles also involves a series of protein-protein interactions that cause the steps to occur in a defined manner.

Proteins Are Sorted Post-Translationally to the Nucleus, Peroxisomes, Mitochondria, and Chloroplasts

The organization and function of the nucleus, peroxisomes, mitochondria, and chloroplasts depend on the uptake of proteins from the cytosol. Most of their proteins are synthesized in the cytosol and then taken up into their respective organelles. For example, most proteins involved in ATP synthesis are made in the cytosol and taken up into mitochondria after they have been completely synthesized. For this to occur, a protein must have the appropriate sorting signal as part of its amino acid sequence.

As one example of post-translational sorting, let's consider how a protein is directed to the mitochondrial matrix. Such a protein has a short amino acid sequence at the N-terminus called a matrix-targeting sequence. As shown in Figure 4.31, the process of protein import into the matrix involves a series of intricate protein-protein interactions. A protein destined for the mitochondrial matrix is first made in the cytosol, where proteins called **chaperones** keep it in an unfolded state. A receptor protein in the outer mitochondrial membrane recognizes the matrix-targeting sequence. The protein is released from the chaperone as it is transferred to a channel in the outer mitochondrial membrane. Because it is in an unfolded state, the mitochondrial protein can be threaded through this channel, and then through another channel in the inner mitochondrial membrane. These channels lie close to each other at contact sites between the outer and inner membranes. As the protein emerges in the matrix, other chaperone proteins already in the matrix continue to keep it unfolded. Eventually, the matrix-targeting sequence is cleaved, and the entire protein is threaded into the matrix. At this stage, the chaperone proteins are released, and the protein can fold into its three-dimensional active structure.

4.7 Systems Biology of Cells: A Summary

Learning Outcomes:

1. Outline the differences in complexity among bacteria, animal, and plant cells.
2. Describe how a eukaryotic cell can be viewed as four interacting systems: the nucleus, cytosol, endomembrane system, and semiautonomous organelles.

We will conclude this chapter by reviewing cell structure and function from a perspective called **systems biology**, the study of how new properties of life arise by complex interactions of its components. The "system" being studied can be anything from a metabolic pathway to a cell, an organ, or even an entire organism. In this section, we view the cell as a system. First, we will compare prokaryotic and eukaryotic cells as systems, and then examine the four interconnected parts that make up the system that is the eukaryotic cell.

Bacterial Cells Are Relatively Simple Systems Compared to Eukaryotic Cells

Bacterial cells are relatively small and lack the extensive internal compartmentalization characteristic of eukaryotic cells (**Table 4.2**). On the outside, bacterial cells are surrounded by a cell wall, and many species have flagella. Animal cells lack a cell wall, and only certain cell types have flagella or cilia. Like bacteria, plant cells also have cell walls but their chemical composition is different from that of bacterial cells. Plant cells rarely have flagella.

As mentioned earlier in this chapter, the cytoplasm is the region of the cell enclosed by the plasma membrane. Ribosomes are found in the cytoplasm of all cell types. In bacteria, the cytoplasm is a single compartment. The bacterial genetic material, usually a single chromosome, is found in the nucleoid region, which is not surrounded by a membrane. By comparison, the cytoplasm of eukaryotic cells is highly compartmentalized. The cytosol is the area that surrounds many different types of membrane-bound organelles. For example, eukaryotic chromosomes are found in the nucleus that is surrounded by a double membrane. In addition, all eukaryotic cells have an endomembrane system and mitochondria, and plant cells also have chloroplasts.

A Eukaryotic Cell Is a System with Four Interacting Parts

We can view a eukaryotic cell as a system with four interacting parts: the interior of the nucleus, the cytosol, the endomembrane system, and the semiautonomous organelles (**Figure 4.32**). These four regions have their own structure and organization, while also playing a role in the structure and organization of the entire cell.

Nucleus The nucleus houses the genome. Earlier in this chapter, we learned how the genome plays a key role in producing the proteome through the process of gene expression. The collection of proteins that a cell makes is primarily responsible for the structure and function of the entire cell. Gene regulation, which occurs in the cell nucleus, is very important in producing specific cell types and enabling cells to respond to environmental changes. The collection of filamentous proteins called the nuclear matrix serves to organize and protect the chromosomes within the nucleus.

Cytosol The cytosol is the region that is enclosed by the plasma membrane but outside the organelles. It is an important coordination center for cell function and organization. Along with the plasma membrane, the cytosol coordinates responses to the environment. Factors in the environment may stimulate signaling pathways in the cytosol that affect the functions of cellular proteins and the regulation of genes in the cell nucleus.

The cytosol also has a large influence on cell structure because it is the compartment where many small molecules are metabolized in the cell. This region receives molecules that are taken up from the environment. In addition, many metabolic pathways for the synthesis and breakdown of cellular molecules are found in the cytosol, and pathways in organelles are often regulated by events there. Most of the proteins that constitute the proteome are made in the cytosol.

A particularly important component of the eukaryotic cell is the cytoskeleton, which provides organization to the cell and facilitates cellular movements. The formation and function of the cytoskeleton is caused by an amazing series of protein-protein interactions. In most cells, the cytoskeleton is a dynamic structure, enabling its composition to respond to environmental and developmental changes.

Endomembrane System The endomembrane system can be viewed as a smaller system within the confines of the eukaryotic cell. The endomembrane system includes the nuclear envelope, endoplasmic reticulum (ER), Golgi apparatus, lysosomes, vacuoles, peroxisomes,

Table 4.2 A Comparison of Cell Complexity Among Bacterial, Animal, and Plant Cells

Structures	Bacteria	Animal cells	Plant cells
Extracellular structures			
Cell wall*	Present	Absent	Present
Flagella/cilia	Flagella sometimes present	Cilia or flagella present on certain cell types	Rarely present†
Plasma membrane	Present	Present	Present
Interior structures			
Cytoplasm	Usually a single compartment inside the plasma membrane	Composed of membrane-bound organelles that are surrounded by the cytosol	Composed of membrane-bound organelles that are surrounded by the cytosol
Ribosomes	Present	Present	Present
Chromosomes and their location	Typically one circular chromosome per nucleoid region; a nucleoid region is not a separate compartment.	Multiple linear chromosomes in the nucleus; nucleus is surrounded by a double membrane. Mitochondria also have chromosomes.	Multiple linear chromosomes in the nucleus; nucleus is surrounded by a double membrane. Mitochondria and chloroplasts also have chromosomes.
Endomembrane system	Absent	Present	Present
Mitochondria	Absent	Present	Present
Chloroplasts	Absent	Absent	Present

* Note that the biochemical composition of bacterial cell walls is very different from plant cell walls.
†Some plant species produce sperm cells with flagella, but flowering plants produce sperm within pollen grains that lack flagella.

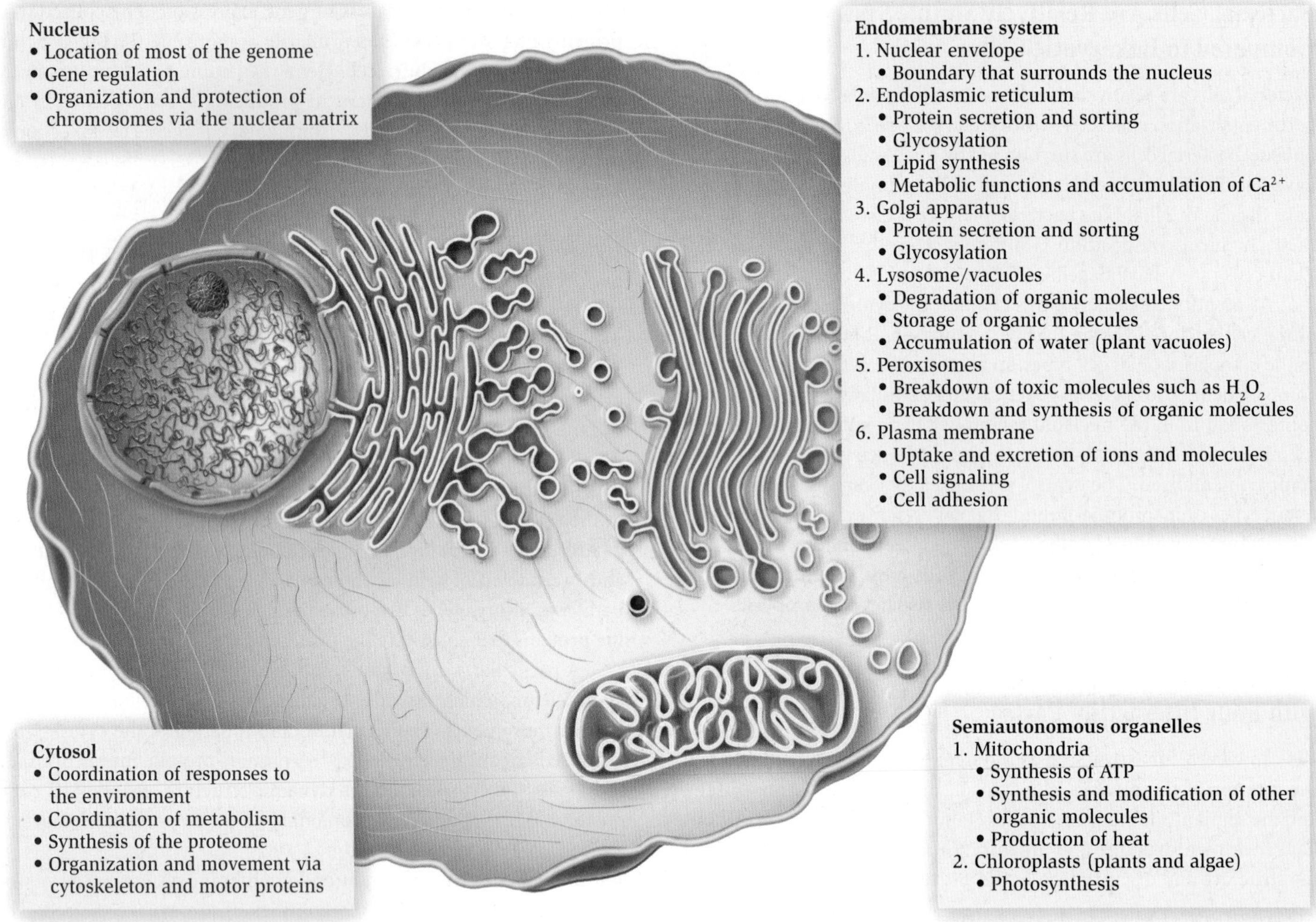

Figure 4.32 **The four interacting parts of eukaryotic cells: nucleus, cytosol, endomembrane system, and semiautonomous organelles.** An animal cell is illustrated here.

and plasma membrane. This system forms a secretory pathway that is crucial in the movement of larger substances, such as carbohydrates and proteins, out of the cell. The export of carbohydrates and proteins plays a key role in the organization of materials that surround cells.

The endomembrane system also contributes to the overall structure and organization of eukaryotic cells in other ways. The ER and Golgi are involved in protein sorting and in the attachment of carbohydrates to lipids and proteins. In addition, most of a cell's lipids are made in the smooth ER membrane and distributed to other parts of the cell. The smooth ER also plays a role in certain metabolic functions, such as the elimination of alcohol, and is important in the accumulation of Ca^{2+}.

Another important function of the endomembrane system that serves the needs of the entire cell is the breakdown and storage of organic molecules. Lysosomes in animal cells and vacuoles in the cells of other organisms assist in breaking down various types of macromolecules. The building blocks are then recycled back to the cytosol and used to construct new macromolecules. Vacuoles often play a role in the storage of organic molecules such as carbohydrates, proteins, and fats. In plants, vacuoles may store large amounts of water. Peroxisomes are involved in the breakdown and synthesis of organic molecules and degrade toxic molecules such as hydrogen peroxide.

The plasma membrane is also considered a part of the endomembrane system. It plays an important role as a selective barrier that allows the uptake of nutrients and the excretion of waste products. Proteins in the plasma membrane facilitate the transport of substances into and out of cells. The plasma membrane contains different types of receptors that provide a way for a cell to sense changes in its environment and communicate with other cells. Finally, in animals, proteins in the plasma membrane promote the adhesion of adjacent cells.

Semiautonomous Organelles The semiautonomous organelles include the mitochondria and chloroplasts. Regarding organization, these organelles tend to be rather independent. They exist in the cytosol much like a bacterium would grow in a laboratory medium. Whereas a bacterium takes up essential nutrients from the growth medium, the semiautonomous organelles take up molecules from the cytosol. The organelles use these molecules to carry out their functions and maintain their organization. Like bacteria, the semiautonomous organelles divide by binary fission to produce more of themselves.

Although the semiautonomous organelles rely on the rest of the cell for many of their key components, they also give back to the cell in ways that are vital to maintaining cell organization. Mitochondria take up organic molecules from the cytosol and give back ATP, which is used throughout the cell to drive processes that are energetically unfavorable. This energy is vital for cell structure and function. Mitochondria also modify certain organic molecules and may produce heat. By comparison, the chloroplasts capture light energy and synthesize organic molecules. These organic molecules store energy and are broken down when energy is needed. In addition, organic molecules, such as sugars and amino acids, are used as building blocks to synthesize many different types of cellular molecules, such as carbohydrate polymers and proteins.

Summary of Key Concepts

4.1 Microscopy

- Three important parameters in microscopy are magnification, resolution, and contrast. A light microscope utilizes light for illumination, whereas an electron microscope uses an electron beam. Transmission electron microscopy (TEM) provides the best resolution of any form of microscopy, and scanning electron microscopy (SEM) produces an image of a three-dimensional surface (Figures 4.1, 4.2, 4.3).

4.2 Overview of Cell Structure

- Cell structure is determined by four factors: matter, energy, organization, and information. Every living organism has a genome. The genes within the genome contain the information to produce cells with particular structures and functions.
- We can classify all forms of life into two categories based on cell structure: prokaryotic and eukaryotic cells.
- Bacteria and archaea have prokaryotic cells with a relatively simple structure that lacks a membrane-enclosed nucleus. Structures in prokaryotic cells include the plasma membrane, cytoplasm, nucleoid region, and ribosomes. Prokaryotic cells also have a cell wall and many have a glycocalyx (Figure 4.4).
- Eukaryotic cells are compartmentalized into organelles and contain a nucleus that houses most of the DNA. The surface area/volume ratio is thought to limit cell size (Figures 4.5, 4.6, 4.7, 4.8).
- The proteome of a cell determines its structure and function.

4.3 The Cytosol

- The cytosol is a central coordinating region for many metabolic activities of eukaryotic cells, including polypeptide synthesis (Figure 4.9).
- The cytoskeleton is a network of three different types of protein filaments: microtubules, intermediate filaments, and actin filaments. Microtubules are important for cell shape, organization, and movement. Intermediate filaments help maintain cell shape, rigidity, and strength. Actin filaments support the plasma membrane and play a key role in cell strength, shape, and movement (Table 4.1, Figures 4.10, 4.11, 4.12, 4.13).

4.4 The Nucleus and Endomembrane System

- The primary function of the nucleus is the organization and expression of the genetic material. A second important function is the assembly of ribosomes in the nucleolus (Figures 4.14, 4.15, 4.16).
- The endomembrane system includes the nuclear envelope, endoplasmic reticulum (ER), Golgi apparatus, lysosomes, vacuoles, peroxisomes, and plasma membrane. The rough endoplasmic reticulum (rough ER) plays a key role in the initial sorting of proteins. The smooth endoplasmic reticulum (smooth ER) functions in metabolic processes such as detoxification, carbohydrate metabolism, accumulation of calcium ions, and synthesis and modification of lipids. The Golgi apparatus performs three overlapping functions: processing, protein sorting, and secretion. Lysosomes degrade macromolecules and help digest substances taken up from outside the cell (endocytosis) and inside the cell (autophagy) (Figures 4.17, 4.18).
- Palade's pulse-chase experiments demonstrated that secreted proteins move sequentially through the ER and Golgi apparatus (Figure 4.19).
- Types and functions of vacuoles include central vacuoles; contractile vacuoles; and phagocytic, or food, vacuoles (Figure 4.20).
- Peroxisomes catalyze a variety of chemical reactions, including those involved with the breakdown of toxic molecules such as hydrogen peroxide, and also typically contain enzymes involved in the metabolism of fats and amino acids. Peroxisomes are made via budding from the ER, followed by maturation and division (Figure 4.21).
- Proteins in the plasma membrane perform many important roles that affect activities inside the cell, including membrane transport, cell signaling, and cell adhesion (Figure 4.22).

4.5 Semiautonomous Organelles

- Mitochondria and chloroplasts are considered semiautonomous because they grow and divide, but they still depend on other parts of the cell for their internal components (Figure 4.23).
- Mitochondria produce most of a cell's ATP, which is utilized by many proteins to carry out their functions. Other mitochondrial functions include the synthesis, modification, and breakdown of cellular molecules and the generation of heat in specialized fat cells (Figure 4.24).
- Chloroplasts, which are found in nearly all species of plants and algae, carry out photosynthesis (Figure 4.25).
- Mitochondria and chloroplasts contain their own genetic material and divide by binary fission (Figure 4.26).
- According to the endosymbiosis theory, mitochondria and chloroplasts originated from bacteria that took up residence in early eukaryotic cells (Figure 4.27).

4.6 Protein Sorting to Organelles

- Eukaryotic proteins are sorted to their correct cellular destination (Figure 4.28).
- The cotranslational sorting of ER, Golgi, lysosomal, vacuolar, plasma membrane, and secreted proteins begins in the cytosol, while translation is occurring, and involves sorting signals and vesicle transport (Figures 4.29, 4.30).
- Most proteins of the nucleus, mitochondria, chloroplasts, and peroxisomes are synthesized in the cytosol and taken up after the protein is completely made; this is called post-translational sorting (Figure 4.31).

4.7 Systems Biology of Cells: A Summary

- Systems biology is the study of how new properties of life arise by complex interactions of its components. In systems biology, the cell is viewed in terms of its structural and functional connections, rather than its individual molecular components.
- Prokaryotic and eukaryotic cells differ in their levels of organization. In eukaryotic cells, four regions—the nucleus, cytosol, endomembrane system, and semiautonomous organelles—work together to produce dynamic organization (Table 4.2, Figure 4.32).

Assess & Discuss

Test Yourself

1. The cell doctrine states that
   a. all living things are composed of cells.
   b. cells are the smallest units of living organisms.
   c. new cells come from pre-existing cells by cell division.
   d. all of the above.
   e. a and b only.
2. When using microscopes, the resolution refers to
   a. the ratio between the size of the image produced by the microscope and the actual size of the object.
   b. the degree to which a particular structure looks different from other structures around it.
   c. how well a structure takes up certain dyes.
   d. the ability to observe two adjacent objects as being distinct from each other.
   e. the degree to which the image is magnified.
3. If a motor protein was held in place and a cytoskeletal filament was free to move, what type of motion would occur when the motor protein was active?
   a. The motor protein would "walk" along the filament.
   b. The filament would move.
   c. The filament would bend.
   d. All of the above would happen.
   e. Only b and c would happen.
4. The process of polypeptide synthesis is called
   a. metabolism.
   b. transcription.
   c. translation.
   d. hydrolysis.
   e. both c and d.
5. Each of the following is part of the endomembrane system except
   a. the nuclear envelope.
   b. the endoplasmic reticulum.
   c. the Golgi apparatus.
   d. lysosomes.
   e. mitochondria.
6. Vesicle transport occurs between the ER and the Golgi in both directions. Let's suppose a researcher added a drug to cells that inhibited vesicle transport from the Golgi to the ER but did not affect vesicle transport from the ER to the Golgi. If you observed cells microscopically after the drug was added, what would you expect to see happen over the course of 1 hour?
   a. The ER would get smaller, and the Golgi would get larger.
   b. The ER would get larger, and the Golgi would get smaller.
   c. The ER and Golgi would stay the same size.
   d. Both the ER and Golgi would get larger.
   e. Both the ER and Golgi would get smaller.
7. Functions of the smooth endoplasmic reticulum include
   a. detoxification of harmful organic molecules.
   b. metabolism of carbohydrates.
   c. protein sorting.
   d. all of the above.
   e. a and b only.
8. The central vacuole in many plant cells is important for
   a. storage.
   b. photosynthesis.
   c. structural support.
   d. all of the above.
   e. a and c only.
9. Let's suppose an abnormal protein contains three targeting sequences: an ER signal sequence, an ER retention sequence, and a mitochondrial matrix-targeting sequence. The ER retention sequence is supposed to keep proteins within the ER. Where would you expect this abnormal protein to go? Note: Think carefully about the timing of events in protein sorting and which events occur cotranslationally and which occur post-translationally.
   a. It would go to the ER.
   b. It would go the mitochondria.
   c. It would go to both the ER and mitochondria equally.
   d. It would remain in the cytosol.
   e. It would be secreted.
10. Which of the following observations would not be considered evidence for the endosymbiosis theory?
    a. Mitochondria and chloroplasts have genomes that resemble smaller versions of bacterial genomes.
    b. Mitochondria, chloroplasts, and bacteria all divide by binary fission.
    c. Mitochondria, chloroplasts, and bacteria all have ribosomes.
    d. Mitochondria, chloroplasts, and bacteria all have similar sizes and shapes.
    e. all of the above

Conceptual Questions

1. Describe two specific ways that protein-protein interactions are involved with cell structure or cell function.
2. Explain how motor proteins and cytoskeletal filaments interact to promote three different types of movements: movement of a cargo, movement of a filament, and bending of a filament.
3.  A principle of biology is that *structure determines function*. Explain how the invaginations of the inner mitochondrial membrane are related to mitochondrial function.

Collaborative Questions

1. Discuss the roles of the genome and proteome in determining cell structure and function.
2. Discuss and draw the structural relationship between the nucleus, the rough endoplasmic reticulum, and the Golgi apparatus.

Online Resource

www.brookerbiology.com

Stay a step ahead in your studies with animations that bring concepts to life and practice tests to assess your understanding. Your instructor may also recommend the interactive eBook, individualized learning tools, and more.

Chapter Outline

5.1 Membrane Structure
5.2 Fluidity of Membranes
5.3 Synthesis of Membrane Components in Eukaryotic Cells
5.4 Overview of Membrane Transport
5.5 Transport Proteins
5.6 Exocytosis and Endocytosis
Summary of Key Concepts
Assess and Discuss

Membrane Structure, Synthesis, and Transport

5

A model for the structure of aquaporin. This protein, found in the plasma membrane of many cell types, such as red blood cells and plant cells, allows the rapid movement of water molecules across the membrane.

When he was 28, Andrew began to develop a combination of symptoms that included fatigue, joint pain, abdominal pain, and a loss of sex drive. His doctor conducted some tests and discovered that Andrew had abnormally high levels of iron in his body. Iron is a mineral found in many foods. Andrew was diagnosed with a genetic disease called hemochromatosis, which caused him to absorb more iron than he needed. This was due to an overactive protein involved in the transport of iron through the membranes of intestinal cells and into the body. Unfortunately, when the human body takes up too much iron, it is stored in body tissues, especially the liver, heart, pancreas, and joints. The extra iron can damage a person's organs. In Andrew's case, the disease was caught relatively early, and treatment—which includes a modification in diet along with medication that inhibits the absorption of iron—prevented more severe symptoms. Without treatment, however, hemochromatosis can cause organ failure. Later signs and symptoms include skin discoloration, arthritis, liver disease, diabetes mellitus, and heart failure.

The disease hemochromatosis illustrates the importance of membranes in regulating the traffic of ions and molecules into and out of cells. Cellular membranes, also known as biological membranes, are an essential characteristic of all living cells. The **plasma membrane** separates the internal contents of a cell from its external environment. With such a role, you might imagine that the plasma membrane would be thick and rigid. Remarkably, the opposite is true. All cellular membranes, including the plasma membrane, are thin (typically 5–10 nm) and somewhat fluid. It would take 5,000–10,000 membranes stacked on top of each other to equal the thickness of the page you are reading! Despite their thinness, membranes are impressively dynamic structures that effectively maintain the separation between a cell and its surroundings. Membranes provide an interface to carry out many vital cellular activities (**Table 5.1**).

In this chapter, we will begin by considering the components that provide the structure and fluid properties of membranes and then explore how they are synthesized. Finally, we will examine one of a membrane's primary functions—membrane transport. Biological membranes regulate the traffic of substances into and out of the cell and its organelles. As you will learn, this occurs via transport proteins and via exocytosis and endocytosis.

Table 5.1 Important Functions of Cellular Membranes

Function
Selective uptake and export of ions and molecules
Cell compartmentalization
Protein sorting
Anchoring of the cytoskeleton
Production of energy intermediates such as ATP and NADPH
Cell signaling
Cell and nuclear division
Adhesion of cells to each other and to the extracellular matrix

5.1 Membrane Structure

Learning Outcomes:

1. Describe the fluid-mosaic model of membrane structure.
2. Identify the three different types of membrane proteins.
3. Explain the technique of freeze-fracture electron microscopy.

An important biological principle is that *structure determines function*. Throughout this chapter, we will see how the structure of cellular membranes enables them to compartmentalize the cell while selectively importing and exporting vital substances. The two primary components of membranes are phospholipids, which form the basic matrix of a membrane, and proteins, which are embedded in the membrane or loosely attached to its surface. A third component is carbohydrate, which may be attached to membrane lipids and proteins. In this section, we will be mainly concerned with the organization of these components to form a biological membrane and how they are important in the overall function of membranes.

Biological Membranes Are a Mosaic of Lipids, Proteins, and Carbohydrates

Figure 5.1 shows the biochemical organization of a membrane, which is similar in composition among all living organisms. The framework of the membrane is the **phospholipid bilayer**, which consists of two layers of phospholipids. Recall from Chapter 3 that phospholipids are **amphipathic** molecules. They have a hydrophobic (water-fearing) or nonpolar region, and also a hydrophilic (water-loving) or polar region. The hydrophobic tails of the lipids, referred to as fatty acyl tails, are found in the interior of the membrane, and the hydrophilic heads are on the surface. Biological membranes also contain proteins, and most membranes have carbohydrates attached to lipids and proteins. Overall, the membrane is considered a mosaic of lipid, protein, and carbohydrate molecules. The membrane structure illustrated in Figure 5.1 is referred to as the **fluid-mosaic model**, originally proposed by S. Jonathan Singer and Garth Nicolson in 1972. As discussed later, the membrane exhibits properties that resemble a fluid because lipids and proteins can move relative to each other within the membrane. Table 5.2 summarizes some of the historical experiments that led to the formulation of the fluid-mosaic model.

Half of a phospholipid bilayer is termed a **leaflet**. Each leaflet faces a different region. For example, the plasma membrane contains a cytosolic leaflet and an extracellular leaflet (see Figure 5.1). With regard to lipid composition, the two leaflets of cellular membranes are highly asymmetrical. Certain types of lipids may be more abundant in one leaflet compared to the other. A striking asymmetry occurs with glycolipids—lipids with carbohydrate attached. These are found primarily in the extracellular leaflet. The carbohydrate portion of a glycolipid protrudes into the extracellular medium.

Proteins Associate with Membranes in Three Different Ways

Although the phospholipid bilayer forms the basic foundation of cellular membranes, the protein component carries out many key functions. Some of these functions were considered in Chapter 4. For example, we saw how membrane proteins in the smooth ER membrane function as enzymes that break down glycogen. Later in this chapter, we will explore how membrane proteins are involved in transporting ions and molecules across membranes. In later chapters, we will examine how membrane proteins are responsible for other functions,

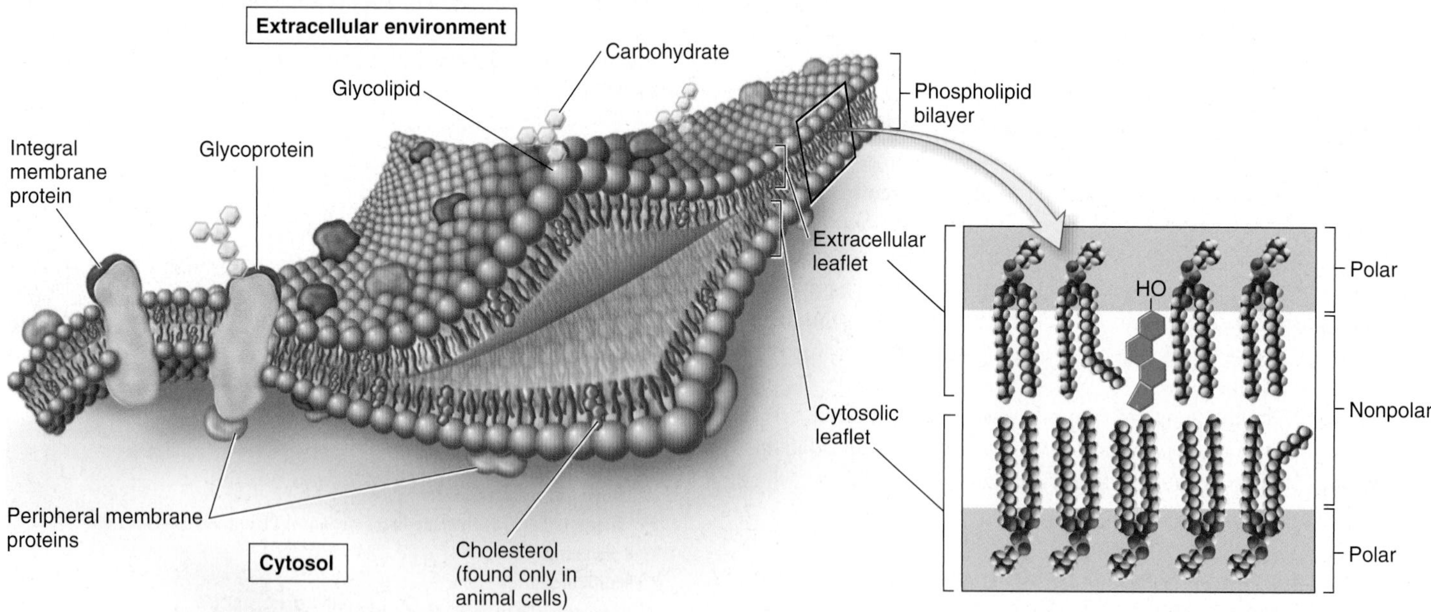

Figure 5.1 **Fluid-mosaic model of membrane structure.** The basic framework of a plasma membrane is a phospholipid bilayer. Proteins may span the membrane and may be bound on the surface to other proteins or to lipids. Proteins and lipids, which have covalently bound carbohydrates, are called glycoproteins and glycolipids, respectively. The inset shows nine phospholipids and one cholesterol molecule in a bilayer, and it emphasizes the two leaflets and the polar and nonpolar regions of the bilayer.

Table 5.2 Historical Developments That Led to the Formulation of the Fluid-Mosaic Model

Date	Description
1917	Irving Langmuir made artificial membranes experimentally by creating a monolayer of lipids on the surface of water. The polar heads interacted with water, and nonpolar tails projected into the air.
1925	Evert Gorter and F. Grendel proposed that lipids form bilayers around cells. This was based on measurements of lipid content enclosing red blood cells that showed there was just enough lipid to surround the cell with two layers.
1950s	Electron microscopy studies carried out by J.D. Robertson and others revealed that membranes look like a train track—two dark lines separated by a light space (look ahead to Figure 5.3a). Researchers determined that the dark lines in these experiments are the phospholipid heads, which were heavily stained, whereas the light region between them is their phospholipid tails.
1966	Using freeze fracture electron microscopy (look ahead to Figure 5.3b), Daniel Branton concluded that membranes are bilayers, because the freeze fracture procedure splits membranes in half, thus revealing proteins in the two membrane leaflets.
1972	S. Jonathan Singer and Garth Nicolson proposed the fluid-mosaic model described in Figure 5.1. Their model was consistent with the observation that membrane proteins are globular, and some are known to span the phospholipid bilayer and project from both sides.

including ATP synthesis (Chapter 7), photosynthesis (Chapter 8), cell signaling (Chapter 9), and cell-to-cell adhesion (Chapter 10).

Membrane proteins have three different ways of associating with a membrane (Figure 5.2). An **integral membrane protein** cannot be released from the membrane unless the membrane is dissolved with an organic solvent or detergent—in other words, you have to disrupt the integrity of the membrane to remove it. An integral membrane protein can associate with a membrane in two ways. **Transmembrane proteins**, the most common type, have one or more regions that are physically inserted into the hydrophobic interior of the phospholipid bilayer. These regions, the transmembrane segments, are stretches of nonpolar amino acids that span or traverse the membrane from one leaflet to the other. In most transmembrane proteins, each transmembrane segment is folded into an α helix structure. Such a segment is stable in a membrane because the nonpolar amino acids interact favorably with the hydrophobic fatty acyl tails of the lipid molecules.

A second type of integral membrane protein is a **lipid-anchored protein**. This type of protein associates with a membrane because it has a lipid molecule that is covalently attached to an amino acid side chain within the protein. The fatty acyl tails are inserted into the hydrophobic portion of the membrane and thereby keep the protein firmly attached to the membrane.

Peripheral membrane proteins, also called extrinsic proteins, represent a third way that proteins can associate with membranes. They do not interact with the hydrophobic interior of the phospholipid bilayer. Instead, they are noncovalently bound to regions of integral membrane proteins that project out from the membrane (see Figure 5.2), or they are bound to the polar head groups of phospholipids. Peripheral membrane proteins are typically bound to the membrane by hydrogen and/or ionic bonds.

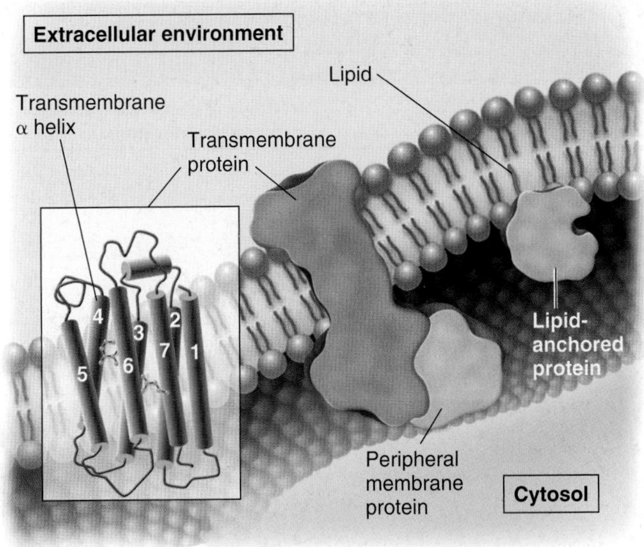

Figure 5.2 Types of membrane proteins. Integral membrane proteins are of two types: transmembrane proteins and lipid-anchored proteins. Peripheral membrane proteins are noncovalently bound to the hydrophilic regions of integral membrane proteins or to the polar head groups of lipids. Inset: The protein bacteriorhodopsin contains seven transmembrane segments, depicted as cylinders, in an α helix structure. Bacteriorhodopsin is found in halophilic (salt-loving) archaea.

GENOMES & PROTEOMES CONNECTION

Approximately 25% of All Genes Encode Transmembrane Proteins

Membrane proteins participate in some of the most important cellular processes, including transport, energy transduction, cell signaling, secretion, cell recognition, metabolism, and cell-to-cell contact. Research studies have revealed that cells devote a sizeable fraction of their energy and metabolic machinery to the synthesis of membrane proteins. These proteins are particularly important in human medicine—approximately 70% of all medications exert their effects by binding to membrane proteins. Examples include the drugs aspirin, ibuprofen, and acetaminophen, which are widely used to relieve pain and inflammatory conditions such as arthritis. These drugs bind to cyclooxygenase, a protein in the ER membrane that is necessary for the synthesis of chemicals that play a role in pain sensation and inflammation.

Because membrane proteins are so important biologically and medically, researchers have analyzed the genomes of many species and asked the question, "What percentage of genes encodes transmembrane proteins?" To answer this question, they have developed tools to predict the likelihood that a gene encodes a transmembrane protein. For example, the occurrence of transmembrane α helices can be predicted from the amino acid sequence of a protein. All 20 amino acids can be ranked according to their tendency to enter a hydrophobic or hydrophilic environment. With these values, the amino acid sequence of a protein can be analyzed using computer software to determine the average hydrophobicity of short amino acid sequences within the protein. A stretch of 18 to 20 amino acids in an α helix is long enough to span the membrane. If such a stretch contains a high percentage of hydrophobic amino acids, it is predicted to be a

Table 5.3 Estimated Percentage of Genes That Encode Transmembrane Proteins*

Organism	Percentage of genes that encode transmembrane proteins
Archaea	
Archaeoglobus fulgidus	24.2
Methanococcus jannaschii	20.4
Pyrococcus horikoshii	29.9
Bacteria	
Escherichia coli	29.9
Bacillus subtilis	29.2
Haemophilus influenzae	25.3
Eukaryotes	
Homo sapiens	29.7
Drosophila melanogaster	24.9
Arabidopsis thaliana	30.5
Saccharomyces cerevisiae	28.2

* Data from Stevens and Arkin (2000) Do more complex organisms have a greater proportion of membrane proteins in their genomes? *Proteins* 39: 417–420. Although the numbers may vary due to different computer programs and estimation techniques, the same general trends have been observed in other similar studies.

transmembrane α helix. However, such computer predictions must eventually be verified by experimentation.

Using a computer approach, many research groups have attempted to calculate the percentage of genes that encode transmembrane proteins in various species. **Table 5.3** shows the results of one such study. The estimated percentage of transmembrane proteins is substantial: 20–30% of all genes may encode transmembrane proteins. This trend is found throughout all domains of life, including archaea, bacteria, and eukaryotes. For example, about 30% of human genes encode transmembrane proteins. With a genome size of 20,000 to 25,000 different genes, the total number of genes that encode different transmembrane proteins is estimated to be 6,000 to 7,500. The functions of many of them have yet to be determined. Identifying their functions will help researchers gain a better understanding of human biology. Likewise, medical researchers and pharmaceutical companies are interested in the identification of new transmembrane proteins that could be targets for effective new medications.

Membrane Structure Can Be Viewed with an Electron Microscope

Electron microscopy, discussed in Chapter 4, is a valuable tool to probe membrane structure and function. In transmission electron microscopy (TEM), a biological sample is thin sectioned and stained with heavy-metal dyes such as osmium tetroxide. This compound binds tightly to the polar head groups of phospholipids, but it does not bind well to the fatty acyl tails. As shown in **Figure 5.3a**, membranes stained with osmium tetroxide resemble a railroad track. Two thin dark lines, which are the stained polar head groups, are separated by a

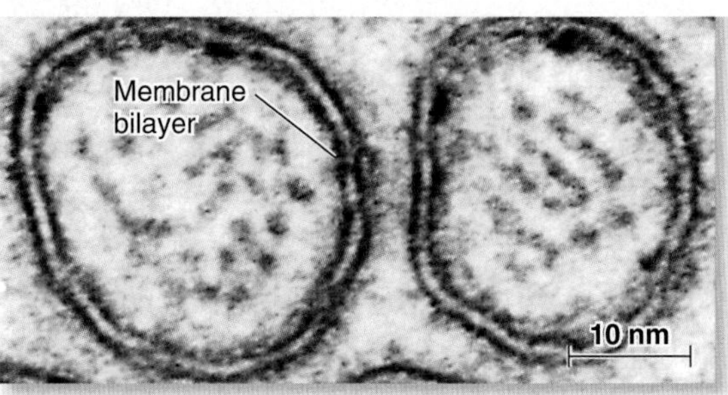

(a) Transmission electron microscopy (TEM)

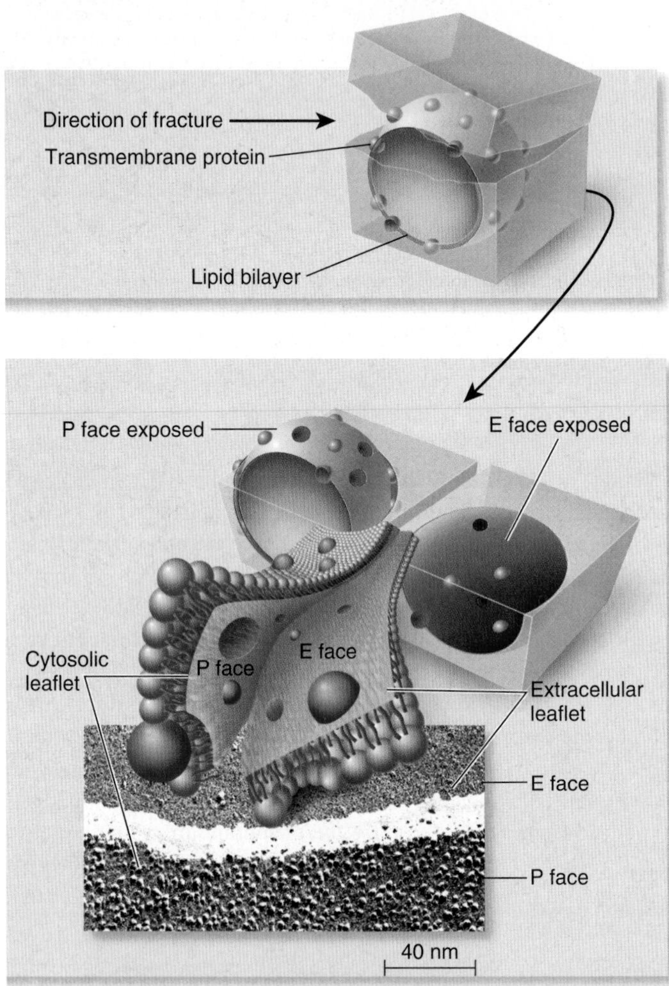

(b) Freeze fracture electron microscopy (FFEM)

Figure 5.3 Electron micrographs of a biological membrane. **(a)** In the standard form of TEM, a membrane appears as two dark parallel lines. These lines are the lipid head groups, which stain darkly with osmium tetroxide. The fatty acyl tails do not stain well and appear as a light region sandwiched between the dark lines. **(b)** In the technique of freeze fracture electron microscopy (FFEM), a sample is frozen in liquid nitrogen and fractured. The sample is then coated with metal and viewed under the electron microscope.

Concept Check: If a heavy metal labeled the hydrophobic tails rather than the polar head groups (as osmium tetroxide does), do you think you would see a bilayer (that is, a railroad track) under TEM?

uniform light space about 2 nm thick. This railroad track morphology is seen when cell membranes are subjected to electron microscopy.

A specialized form of TEM, freeze fracture electron microscopy (FFEM), is used to analyze the interiors of phospholipid bilayers. Russell Steere invented this method in 1957. In FFEM, a sample is frozen in liquid nitrogen and split with a knife (Figure 5.3b). The knife does not actually cut through the bilayer, but it fractures the frozen sample. Due to the weakness of the central membrane region, the leaflets separate into a P face (the protoplasmic face that was next to the cytosol) and the E face (the extracellular face). Most transmembrane proteins do not break in half. They remain embedded within one of the leaflets, usually in the P face. The samples, which are under a vacuum, are then sprayed with a heavy metal such as platinum, which coats the sample and reveals architectural features within each leaflet. When viewed with an electron microscope, membrane proteins are visible as bumps that provide significant three-dimensional detail about their form and shape.

5.2 Fluidity of Membranes

Learning Outcomes:

1. Describe the fluidity of membranes.
2. Analyze how membrane fluidity is affected by lipid composition.

Let's now turn our attention to the dynamic properties of membranes. Although a membrane provides a critical interface between a cell and its environment, it is not a solid, rigid structure. Rather, biological membranes exhibit properties of **fluidity**, which means that individual molecules remain in close association yet have the ability to readily move within the membrane. In this section, we will examine the fluid properties of biological membranes.

Membranes Are Semifluid

Though membranes are often described as fluid, it is more appropriate to say they are **semifluid**. In a fluid substance, molecules can move in three dimensions. By comparison, most phospholipids can rotate freely around their long axes and move laterally within the membrane leaflet (Figure 5.4a). This type of motion is considered two-dimensional, which means it occurs within the plane of the membrane. Because rotational and lateral movements keep the fatty acyl tails within the hydrophobic interior, such movements are energetically favorable. At 37°C, a typical lipid molecule exchanges places with its neighbors about 10^7 times per second, and it can move several micrometers per second. At this rate, a lipid can traverse the length of a bacterial cell (approximately 1 μm) in only 1 second and the length of a typical animal cell in 10 to 20 seconds.

In contrast to rotational and lateral movements, the "flip-flop" of lipids from one leaflet to the opposite leaflet does not occur spontaneously. Flip-flop is energetically unfavorable because the hydrophilic polar head of a phospholipid would have to travel through the hydrophobic interior of the membrane. How are lipids moved from one leaflet to the other? The transport of lipids between leaflets requires the action of the enzyme flippase, which requires energy input in the form of ATP (Figure 5.4b).

Although most lipids diffuse rotationally and laterally within the plane of the lipid bilayer, researchers have discovered that certain types of lipids in animal cells tend to strongly associate with each other to form structures called lipid rafts. As the term raft suggests, a **lipid raft** is a group of lipids that float together as a unit within a larger sea of lipids. Lipid rafts have a lipid composition that differs from the surrounding membrane. For example, they usually have a high amount of cholesterol. In addition, lipid rafts may contain unique sets of lipid-anchored proteins and transmembrane proteins.

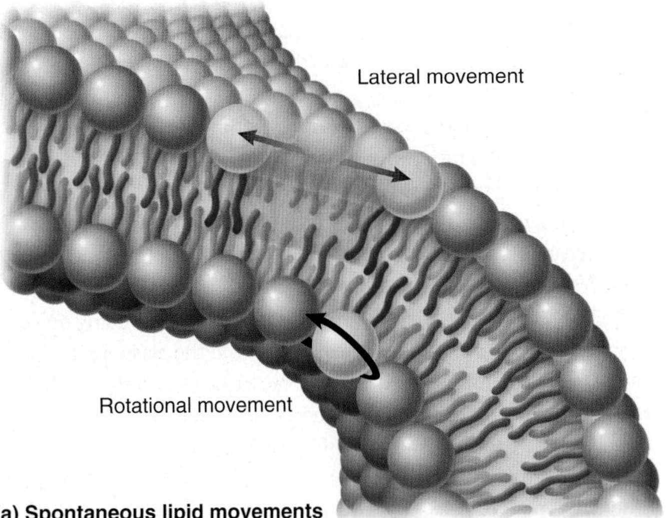

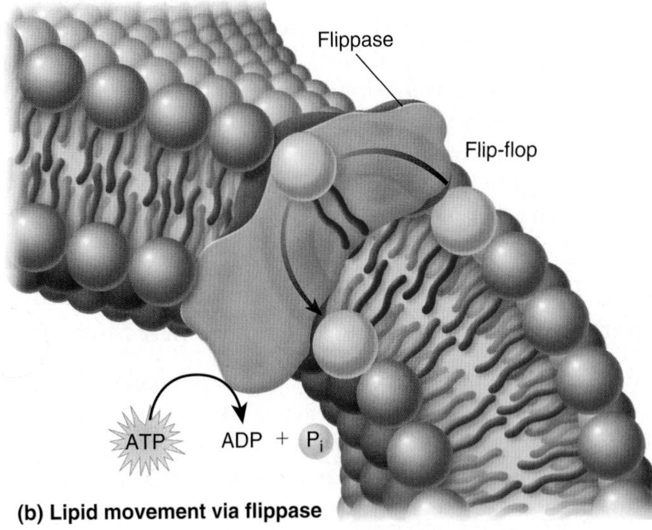

Figure 5.4 Semifluidity of the lipid bilayer. **(a)** Spontaneous movements in the bilayer. Lipids can rotate (that is, move 360°) and move laterally (for example, from left to right in the plane of the bilayer). **(b)** Flip-flop does not happen spontaneously, because the polar head group would have to pass through the hydrophobic region of the bilayer. Instead, the enzyme flippase uses ATP to flip phospholipids from one leaflet to the other.

Concept Check: *In an animal cell, how can changes in lipid composition affect membrane fluidity?*

The functional importance of lipid rafts is the subject of a large amount of current research. Lipid rafts may play an important role in endocytosis (discussed later in this chapter) and cell signaling.

Lipid Composition Affects Membrane Fluidity

The biochemical properties of phospholipids affect the fluidity of the phospholipid bilayer. One key factor is the length of fatty acyl tails, which range from 14 to 24 carbon atoms, with 18 to 20 carbons being the most common. Shorter acyl tails are less likely to interact with each other, which makes the membrane more fluid.

A second important factor is the presence of double bonds in the acyl tails. When a double bond is present, the lipid is said to be **unsaturated** with respect to the number of hydrogens that are bound to the carbon atoms (refer back to Figure 3.10). A double bond creates a kink in the fatty acyl tail (see inset to Figure 5.1), making it more difficult for neighboring tails to interact and making the bilayer more fluid. As described in Chapter 3, unsaturated lipids tend to be more liquid than saturated lipids, which often form solids at room temperature (refer back to Figure 3.11).

A third factor affecting fluidity is the presence of cholesterol, a short and rigid molecule produced by animal cells (see inset to Figure 5.1). Plant cell membranes contain phytosterols that resemble cholesterol in their chemical structure. Cholesterol tends to stabilize membranes; its effects depend on temperature. At higher temperatures, such as those observed in mammals that maintain a constant body temperature, cholesterol makes the membrane less fluid. At lower temperatures, such as icy water, cholesterol has the opposite effect. It makes the membrane more fluid and prevents it from freezing.

An optimal level of bilayer fluidity is essential for normal cell function, growth, and division. If a membrane is too fluid, which may occur at higher temperatures, it can become leaky. However, if a membrane becomes too solid, which may occur at lower temperatures, the functioning of membrane proteins will be inhibited. How can organisms cope with changes in temperature? The cells of many species adapt to changes in temperature by altering the lipid composition of their membranes. For example, when the water temperature drops, the cells of certain fish will incorporate more cholesterol in their membranes, making the membrane more fluid. If a plant cell is exposed to high temperatures for many hours or days, it will alter its lipid composition to have longer fatty acyl tails and fewer double bonds, which will make the membrane less fluid.

Many Transmembrane Proteins Can Rotate and Move Laterally, But Some Are Restricted in Their Movement

Like lipids, many transmembrane proteins may rotate and laterally move throughout the plane of a membrane. Because transmembrane proteins are larger than lipids, they move within the membrane at a much slower rate. Flip-flop of transmembrane proteins does not occur, because the proteins also contain hydrophilic regions that project out from the phospholipid bilayer, and it would be energetically unfavorable for the hydrophilic regions of membrane proteins to pass through the hydrophobic portion of the phospholipid bilayer.

In 1970, Larry Frye and Michael Edidin conducted an experiment that verified the lateral movement of transmembrane proteins (**Figure 5.5**). Mouse and human cells were mixed together and exposed to agents that caused them to fuse with each other to produce mouse-human cell hybrids. Some cells were cooled to 0°C, while others were incubated at 37°C before being cooled. Both sets of cells were then exposed to fluorescently labeled antibodies that became specifically bound to a mouse transmembrane protein called H-2. The fluorescent label was observed with a fluorescence microscope. If the cells were maintained at 0°C, a temperature that greatly inhibits lateral movement, the fluorescence was seen on only one side of the fused cell. However, if the cells were incubated for several hours at 37°C and then cooled to 0°C, the fluorescence was distributed throughout the plasma membrane of the fused cell. This occurred because the higher temperature allowed the lateral movement of the H-2 protein throughout the fused cell.

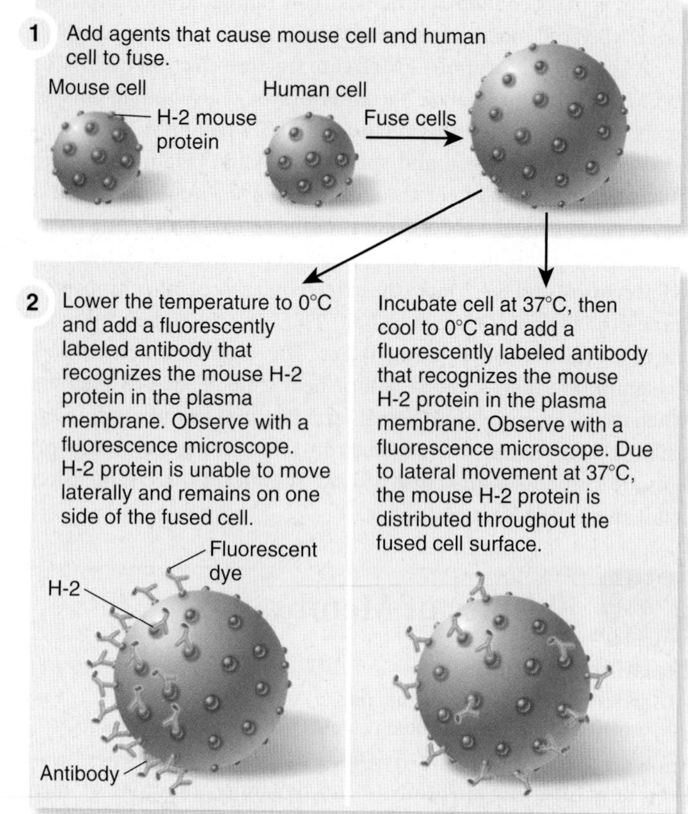

1 Add agents that cause mouse cell and human cell to fuse.
Mouse cell
Human cell
H-2 mouse protein
Fuse cells
2 Lower the temperature to 0°C and add a fluorescently labeled antibody that recognizes the mouse H-2 protein in the plasma membrane. Observe with a fluorescence microscope. H-2 protein is unable to move laterally and remains on one side of the fused cell.
Incubate cell at 37°C, then cool to 0°C and add a fluorescently labeled antibody that recognizes the mouse H-2 protein in the plasma membrane. Observe with a fluorescence microscope. Due to lateral movement at 37°C, the mouse H-2 protein is distributed throughout the fused cell surface.
Fluorescent dye
H-2
Antibody

Figure 5.5 **A method to measure the lateral movement of membrane proteins.**

BIOLOGY PRINCIPLE Biology is an experimental science. This experiment verified that membrane proteins can diffuse laterally within the plane of the lipid bilayer.

Concept Check: *Explain why the H-2 proteins are found on only one side of the cell when the cells were incubated at 0°C.*

Unlike the example shown in Figure 5.5, not all transmembrane proteins are capable of rotational and lateral movement. Depending on the cell type, 10–70% of membrane proteins may be restricted in their movement. Transmembrane proteins may be bound to components of the cytoskeleton, which restricts the proteins from moving (**Figure 5.6**), or may be attached to molecules that are outside the cell, such as the interconnected network of proteins that forms the extracellular matrix of animal cells (see Chapter 10).

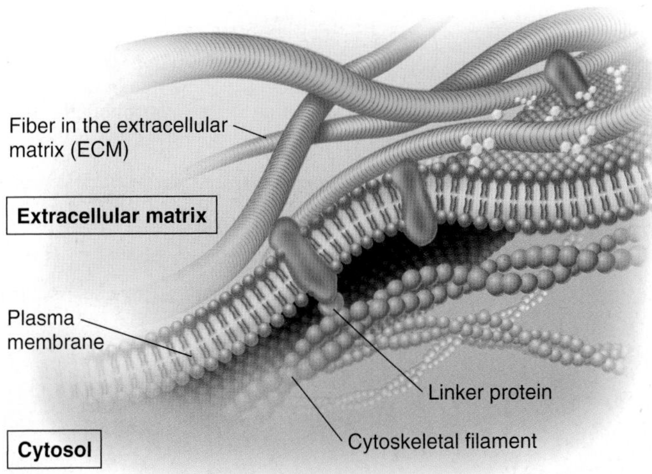

Figure 5.6 Attachment of transmembrane proteins to the cytoskeleton and extracellular matrix of an animal cell. Some transmembrane proteins have regions that extend into the cytosol and are anchored to large cytoskeletal filaments via linker proteins. Being bound to these large filaments restricts the movement of these proteins. Similarly, some transmembrane proteins are bound to large, immobile fibers in the extracellular matrix, which restricts their movements.

BioConnections: Look ahead to Figure 10.8. Discuss how transmembrane proteins are important in the binding of cells to each other and the binding of cells to the extracellular matrix.

5.3 Synthesis of Membrane Components in Eukaryotic Cells

Learning Outcomes:

1. Outline the synthesis of lipids at the ER membrane.
2. Explain how transmembrane proteins are inserted into the ER membrane.
3. Describe the process of glycosylation and its functional consequences.

As we have seen, cellular membranes are composed of lipids, proteins, and carbohydrates. Most of the membrane components of eukaryotic cells are made at the endoplasmic reticulum (ER). In this section, we will begin by considering how phospholipids are synthesized at the ER membrane. We will then examine the process by which transmembrane proteins are inserted into the ER membrane and explore how some proteins are glycosylated.

Lipid Synthesis Occurs at the ER Membrane

In eukaryotic cells, the cytosol and endomembrane system work together to synthesize most lipids. This process occurs at the cytosolic leaflet of the smooth ER membrane. Figure 5.7 shows a simplified pathway for the synthesis of phospholipids. The building blocks for a phospholipid are two fatty acids, each with an acyl tail, one glycerol molecule, one phosphate, and a polar head group. These building blocks are made via enzymes in the cytosol, or they are taken into cells from food. To begin the process of phospholipid synthesis, the fatty acids are activated by attachment to an organic molecule called coenzyme A (CoA). This activation promotes the bonding of the two fatty acids to a glycerol-phosphate molecule, and the resulting molecule is inserted into the cytosolic leaflet of the ER membrane. The phosphate is removed from glycerol, and then a polar molecule already linked to phosphate is attached to glycerol. In the example shown in Figure 5.7, the polar head group contains choline, but many other types are possible. Phospholipids are initially inserted into the cytosolic leaflet. Flippases in the ER membrane transfer some of the newly made lipids to the other leaflet so similar amounts of lipids are found in both leaflets.

The lipids made in the ER membrane are transferred to other membranes in the cell by a variety of mechanisms. Phospholipids in the ER can diffuse laterally to the nuclear envelope. In addition, lipids are transported via vesicles to the Golgi, lysosomes, vacuoles, or plasma membrane. A third mode of lipid transfer involves **lipid exchange proteins**, which extract a lipid from one membrane, diffuse through the cell, and insert the lipid into another membrane. Such transfer can occur between any two membranes, even between the endomembrane system and semiautonomous organelles. For example, lipid exchange proteins transfer lipids between the ER and mitochondria. In addition, chloroplasts and mitochondria synthesize certain types of lipids that are transferred from these organelles to other cellular membranes via lipid exchange proteins.

Most Transmembrane Proteins Are First Inserted into the ER Membrane

In Chapter 4, we learned that eukaryotic proteins contain sorting signals that direct them to their proper destination (see Figure 4.28). With the exception of proteins destined for semiautonomous organelles, most transmembrane proteins contain an ER signal sequence that directs them to the ER membrane. If a polypeptide also contains a stretch of 20 amino acids that are mostly hydrophobic and form an α helix, this region will become a transmembrane segment. In the example shown in Figure 5.8, the polypeptide contains one such sequence. After the ER signal sequence is removed by signal peptidase (refer back to Figure 4.29), a membrane protein with a single transmembrane segment is the result. Other polypeptides may contain more than one transmembrane segment. Each time a polypeptide sequence contains a stretch of 20 hydrophobic amino acids that forms an α helix, an additional transmembrane segment is synthesized into the membrane. From the ER, membrane proteins can be transferred via vesicles to other regions of the cell, such as the Golgi, lysosomes, vacuoles, or plasma membrane.

Glycosylation of Proteins Occurs in the ER and Golgi Apparatus

Glycosylation refers to the process of covalently attaching a carbohydrate to a lipid or protein. When a carbohydrate is attached to a lipid, a **glycolipid** is created, whereas attachment of a carbohydrate to a protein produces a **glycoprotein**.

What is the function of glycosylation? Though the roles of carbohydrate in cell structure and function are not entirely understood, some functional consequences of glycosylation have emerged. Glycolipids and glycoproteins often play a role in cell surface recognition.

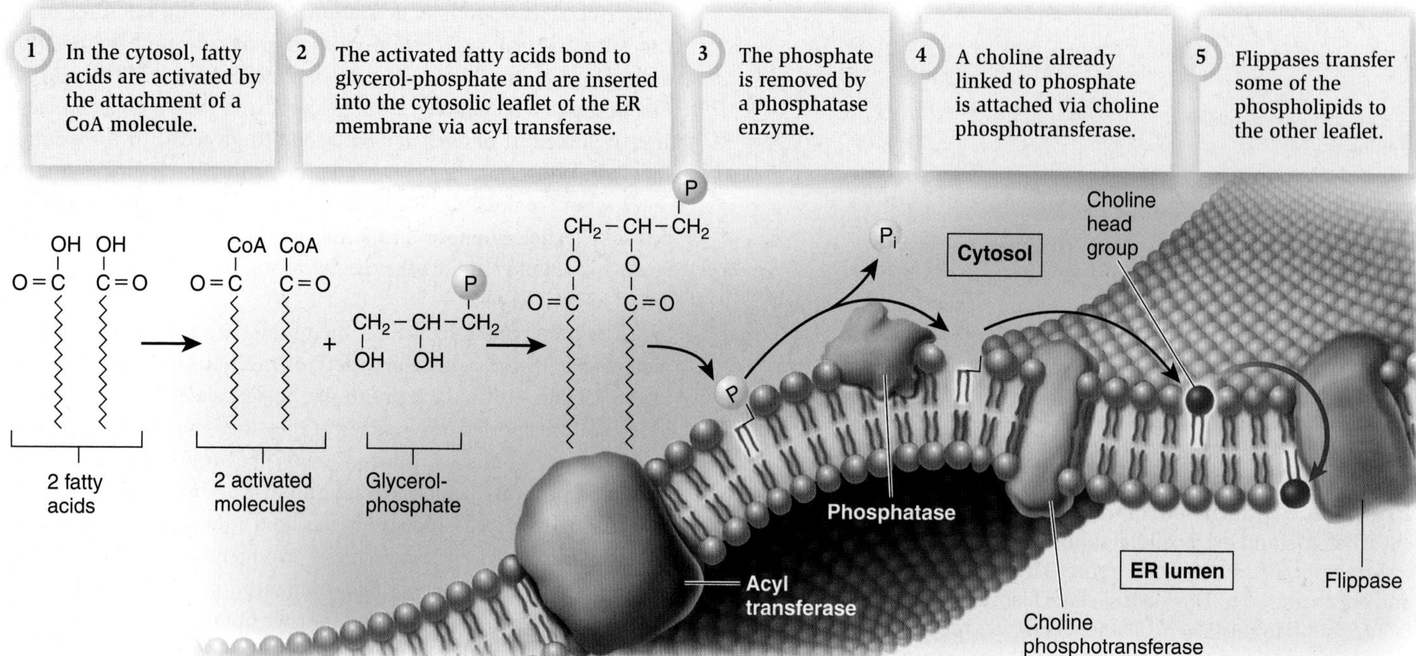
1 In the cytosol, fatty acids are activated by the attachment of a CoA molecule.
2 The activated fatty acids bond to glycerol-phosphate and are inserted into the cytosolic leaflet of the ER membrane via acyl transferase.
3 The phosphate is removed by a phosphatase enzyme.
4 A choline already linked to phosphate is attached via choline phosphotransferase.
5 Flippases transfer some of the phospholipids to the other leaflet.
2 fatty acids
2 activated molecules
Glycerol-phosphate
Cytosol
Choline head group
Phosphatase
Acyl transferase
ER lumen
Choline phosphotransferase
Flippase

Figure 5.7 A simplified pathway for the synthesis of membrane phospholipids at the ER membrane. Note: Phosphate is abbreviated P when it is attached to an organic molecule and P_i when it is unattached. The subscript i refers to the inorganic form of phosphate.

Concept Check: How are phospholipids transferred to the leaflet of the ER membrane that faces the ER lumen?

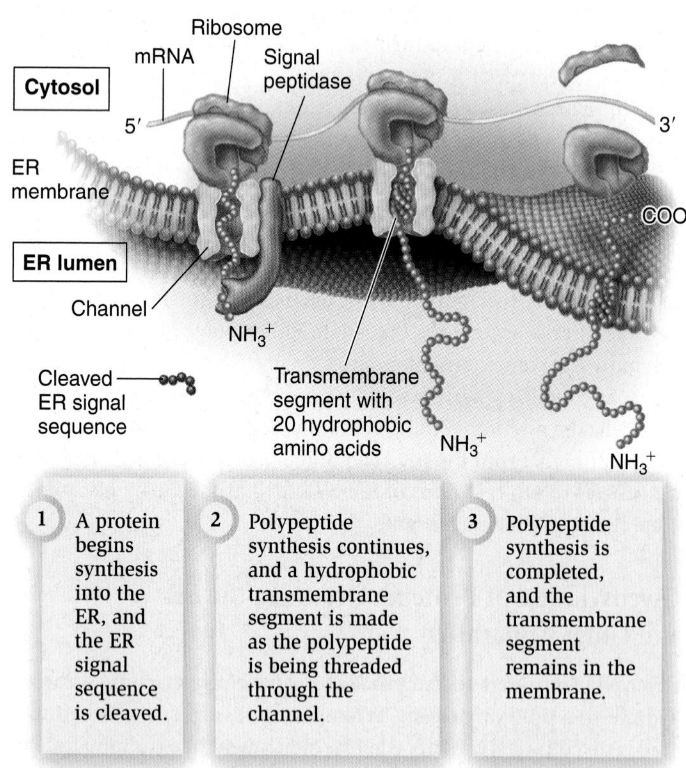
Ribosome
mRNA
Signal peptidase
Cytosol
5′
3′
ER membrane
ER lumen
Channel
NH_3^+
COO^-
Cleaved ER signal sequence
Transmembrane segment with 20 hydrophobic amino acids
NH_3^+
NH_3^+
1 A protein begins synthesis into the ER, and the ER signal sequence is cleaved.
2 Polypeptide synthesis continues, and a hydrophobic transmembrane segment is made as the polypeptide is being threaded through the channel.
3 Polypeptide synthesis is completed, and the transmembrane segment remains in the membrane.

Figure 5.8 Insertion of membrane proteins into the ER membrane.

Concept Check: What structural feature of a protein causes a region to form a transmembrane segment?

When glycolipids and glycoproteins are found in the plasma membrane, the carbohydrate portion is located in the extracellular region. During embryonic development in animals, significant cell movement occurs. Layers of cells slide over each other to create body structures such as the spinal cord and internal organs. The proper migration of individual cells and cell layers relies on the recognition of cell types via the carbohydrates on their cell surfaces.

Carbohydrates often have a protective effect. The carbohydrate-rich zone on the surface of certain animal cells shields the cell from mechanical and physical damage. Similarly, the carbohydrate portion of glycosylated proteins protects them from the harsh conditions of the extracellular environment and degradation by extracellular proteases, which are enzymes that digest proteins.

Two forms of protein glycosylation occur in eukaryotes: N-linked and O-linked. N-linked glycosylation, which also occurs in archaea, involves the attachment of a carbohydrate to the amino acid asparagine in a polypeptide chain. It is called N-linked because the carbohydrate attaches to a nitrogen atom of the asparagine side chain. For this to occur, a group of 14 sugar molecules are built onto a lipid called dolichol, which is found in the ER membrane. This carbohydrate tree is then transferred to an asparagine as a polypeptide is synthesized into the ER lumen through a channel protein (Figure 5.9). It attaches only to asparagines occurring in the sequence asparagine–X–threonine or asparagine–X–serine, where X can be any amino acid except proline. An enzyme in the ER, oligosaccharide transferase, recognizes this sequence and transfers the carbohydrate tree from dolichol to the asparagine. Following this initial glycosylation step, the carbohydrate tree is further modified as other enzymes in the ER attach additional sugars or remove sugars. After a glycosylated protein is transferred to the Golgi by vesicle transport, enzymes in the

Golgi usually modify the carbohydrate tree as well. N-linked glycosylation commonly occurs on membrane proteins that are transported to the cell surface.

The second form of glycosylation, O-linked glycosylation, occurs only in the Golgi apparatus. This form involves the addition of a string of sugars to the oxygen atom of serine or threonine side chains in polypeptides. In animals, O-linked glycosylation is important for the production of proteoglycans, which are highly glycosylated proteins that are secreted from cells and help to organize the extracellular matrix that surrounds cells. Proteoglycans are also a component of mucus, a slimy material that coats many cell surfaces and is secreted into fluids such as saliva. High concentrations of carbohydrates give mucus its slimy texture.

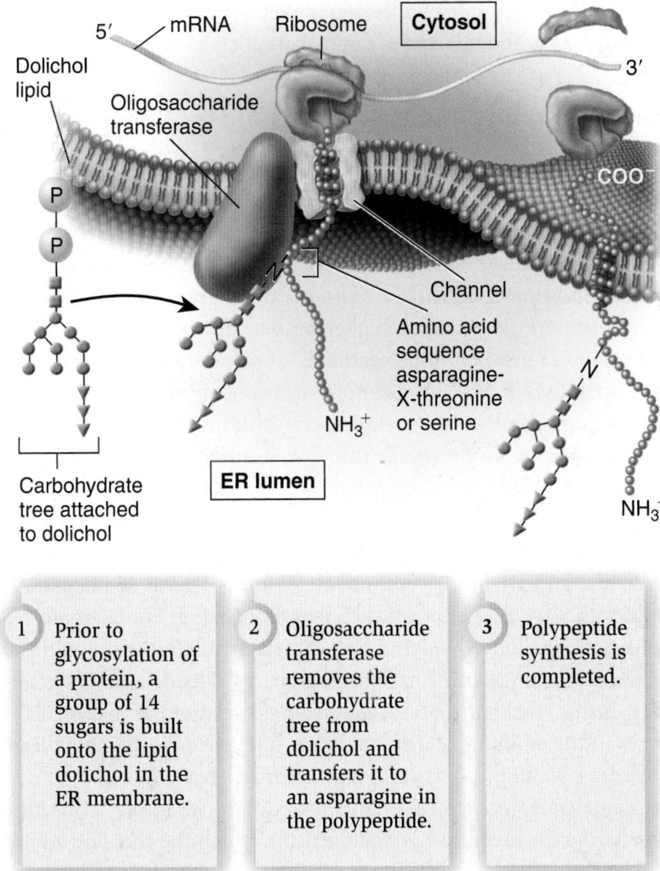

Figure 5.9 N-linked glycosylation in the endoplasmic reticulum.

5.4 Overview of Membrane Transport

Learning Outcomes:

1. Compare and contrast diffusion, facilitated diffusion, passive transport, and active transport.
2. Explain the process of osmosis and how it affects cell structure.

We now turn to one of the key functions of membranes, **membrane transport**—the movement of ions and molecules across biological membranes. All cells contain a plasma membrane that exhibits **selective permeability**, allowing the passage of some ions and molecules but not others. As a protective envelope, its structure ensures that essential molecules such as glucose and amino acids enter the cell, metabolic intermediates remain in the cell, and waste products exit. The selective permeability of the plasma membrane allows the cell to maintain a favorable internal environment.

Substances can move directly across a membrane in three general ways (Figure 5.10). **Diffusion** occurs when a substance moves from a region of high concentration to a region of lower concentration. Some substances can move directly through a biological membrane via diffusion. In **facilitated diffusion**, a transport protein provides a passageway for a substance to diffuse across a membrane. Diffusion and facilitated diffusion are examples of **passive transport**—the transport of a substance across a membrane that does not require an input of energy. In contrast, a third mode of transport, called **active transport**, moves a substance from an area of low concentration to one of high concentration with the aid of a transport protein. Active transport requires an input of energy from a source such as ATP.

In this section, we will begin with a discussion of how the phospholipid bilayer presents a barrier to the movement of ions and molecules across membranes. We will then consider the concept of gradients across membranes and how such gradients affect the movement of water.

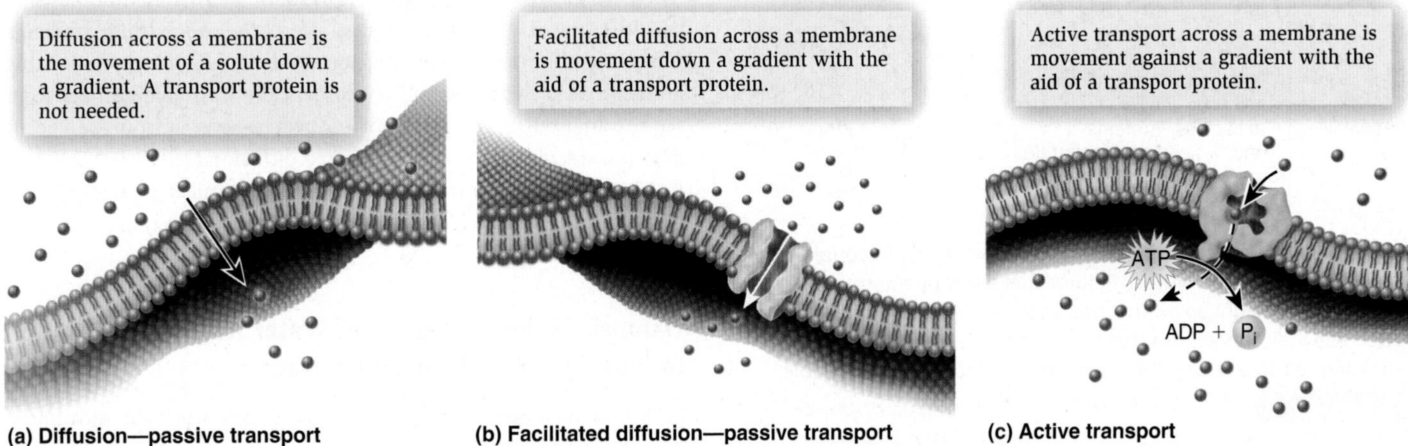

Diffusion across a membrane is the movement of a solute down a gradient. A transport protein is not needed.
Facilitated diffusion across a membrane is movement down a gradient with the aid of a transport protein.
Active transport across a membrane is movement against a gradient with the aid of a transport protein.
ATP
ADP + Pᵢ
(a) Diffusion—passive transport
(b) Facilitated diffusion—passive transport
(c) Active transport

Figure 5.10 Three general types of membrane transport.

The Phospholipid Bilayer Is a Barrier to the Diffusion of Hydrophilic Solutes

Because of their hydrophobic interiors, phospholipid bilayers are a barrier to the movement of ions and hydrophilic molecules. Such ions and molecules are called **solutes**; they are dissolved in water, which is a solvent. The rate of diffusion across a phospholipid bilayer depends on the chemistry of the solute and its concentration. Figure 5.11 compares the relative permeabilities of various solutes through an artificial phospholipid bilayer that does not contain any proteins or carbohydrates. Gases and a few small, uncharged molecules can readily diffuse across the bilayer. However, the permeability of ions and larger polar molecules, such as sugars, is relatively low, and the permeability of macromolecules, such as proteins and polysaccharides, is even lower.

When we consider the steps of diffusion among different solutes, the greatest variation occurs in the ability of solutes to enter the hydrophobic interior of the bilayer. As an example, let's compare urea and diethylurea. Diethylurea is much more hydrophobic because it contains two nonpolar ethyl groups ($-CH_2CH_3$) (Figure 5.12). For this reason, it can pass more quickly through the hydrophobic region of the bilayer. The rate of diffusion of diethylurea through a phospholipid bilayer is about 50 times faster than urea.

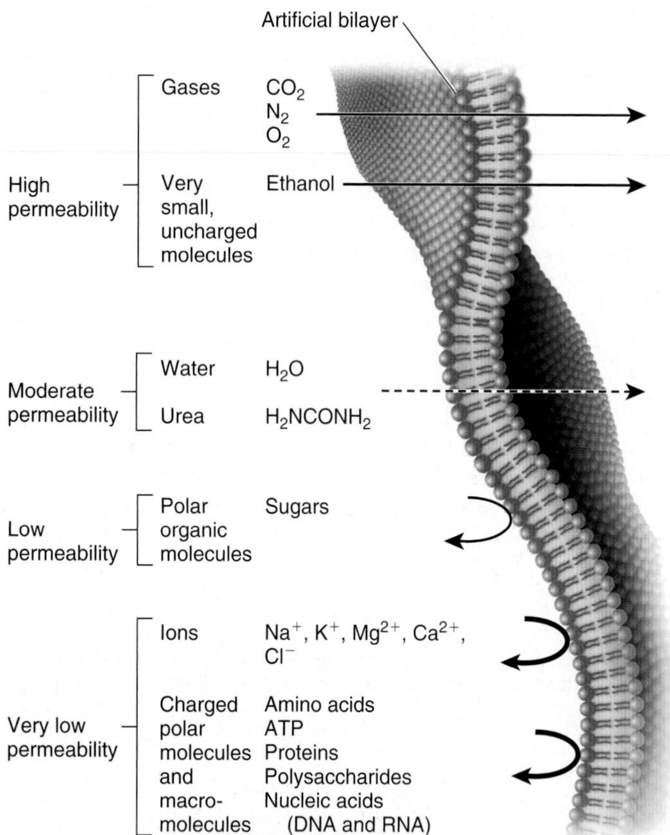

Figure 5.11 Relative permeability of an artificial phospholipid bilayer to a variety of solutes. Solutes that easily penetrate are shown with a straight arrow that passes through the bilayer. The dotted line indicates solutes that have moderate permeability. The remaining solutes shown at the bottom are relatively impermeable.

BioConnections: Which amino acid, described in Chapter 3 (see Figure 3.14), would you expect to cross an artificial membrane more quickly, leucine or lysine?

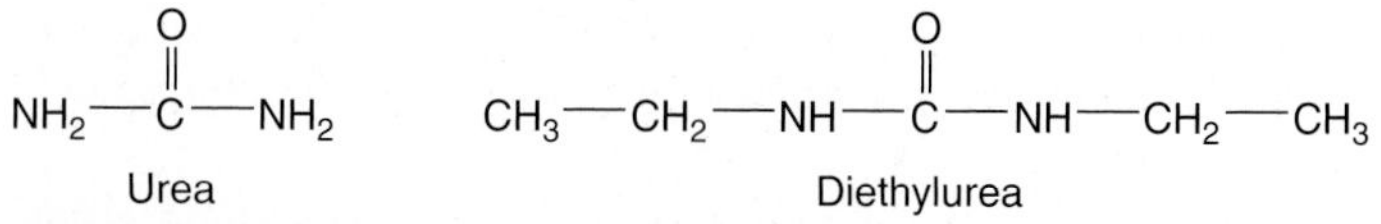

$NH_2-C(=O)-NH_2$ (Urea)

$CH_3-CH_2-NH-C(=O)-NH-CH_2-CH_3$ (Diethylurea)

Figure 5.12 Structures of urea and diethylurea.

Concept Check: Which molecule would you expect to pass through a phospholipid bilayer more quickly, methanol (CH_3OH) or methane (CH_4)?

Cells Maintain Gradients Across Their Membranes

A hallmark of living cells is their ability to maintain a relatively constant internal environment that is distinctively different from their external environment. Solute gradients are formed across the plasma membrane and across organellar membranes. When we speak of a **transmembrane gradient** or **concentration gradient**, we mean the concentration of a solute is higher on one side of a membrane than the other. Transmembrane gradients of solutes are a universal feature of all living cells. For example, immediately after you eat a meal containing carbohydrates, a higher concentration of glucose is found outside your cells than inside; this is an example of a chemical gradient (Figure 5.13a).

Gradients involving ions have two components—electrical and chemical. An **electrochemical gradient** is a dual gradient with both electrical and chemical components (Figure 5.13b). It occurs with solutes that have a net positive or negative charge. For example, let's consider a gradient involving Na^+. An electrical gradient could exist in which the amount of net positive charge outside a cell is greater than inside. In Figure 5.13b, an electrical gradient is due to differences in the amounts of different types of ions across the membrane, including sodium, potassium, and chlorine (Na^+, K^+, and Cl^-). At the same time, a chemical gradient—a difference in Na^+ concentration across the membrane—could exist in which the concentration of Na^+ outside is greater than inside. The Na^+ electrochemical gradient is composed of both an electrical gradient due to charge differences across the membrane along with a chemical gradient for Na^+.

One way to view the transport of solutes across membranes is to consider how the transport process affects the pre-existing gradients across membranes. Passive transport tends to dissipate a pre-existing gradient. It is a process that is energetically favorable and does not require an input of energy. As mentioned, passive transport can occur in two ways, via diffusion or facilitated diffusion (see Figure 5.10a,b). By comparison, active transport produces a chemical gradient or electrochemical gradient. The formation of a gradient requires an input of energy.

Osmosis Is the Movement of Water Across Membranes to Balance Solute Concentrations

Let's now turn our attention to how gradients affect the movement of water across membranes. When the concentrations of dissolved particles (solutes) on both sides of the plasma membrane are equal,

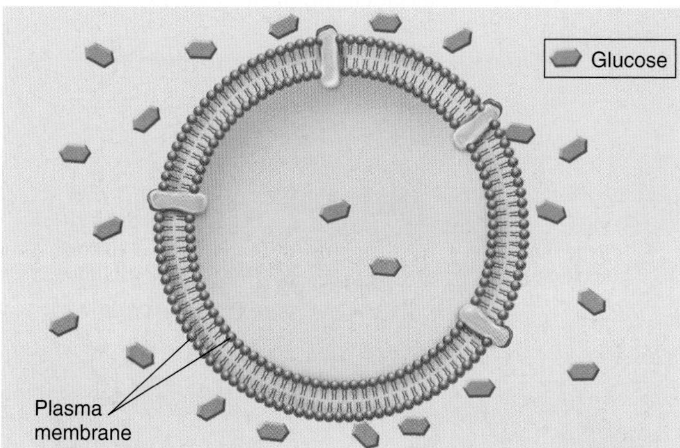

(a) Chemical gradient for glucose—a higher glucose concentration outside the cell

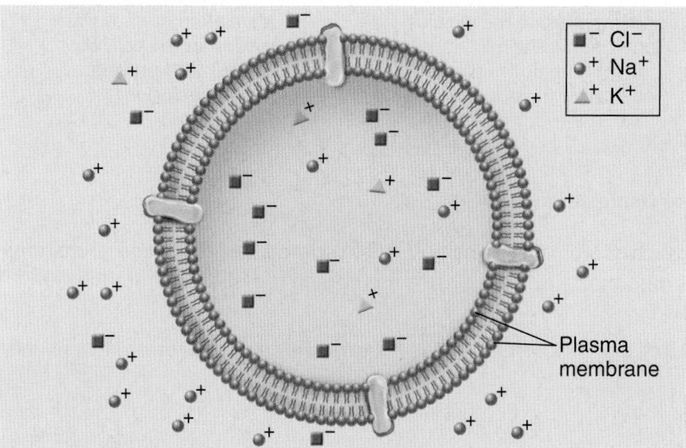

(b) Electrochemical gradient for Na^+—more positive charges outside the cell and a higher Na^+ concentration outside the cell

Figure 5.13 Gradients across cell membranes.

BioConnections: Look ahead to Figure 41.10. What types of ion gradients are important for the conduction of an action potential across the plasma membrane of a neuron?

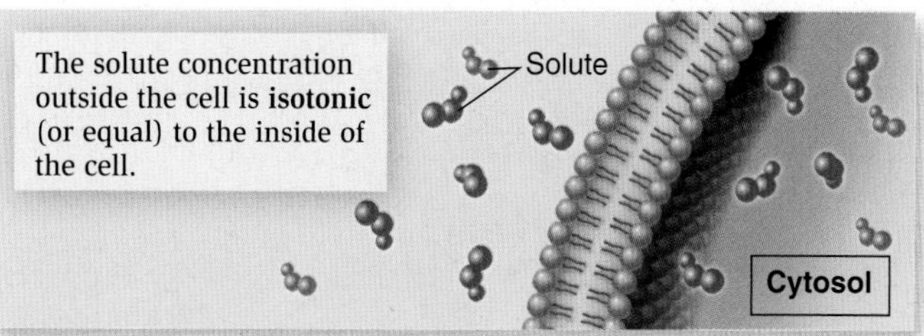

(a) Outside isotonic

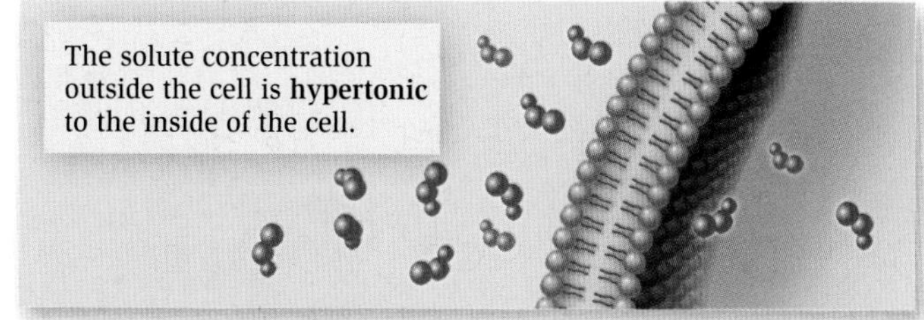

(b) Outside hypertonic

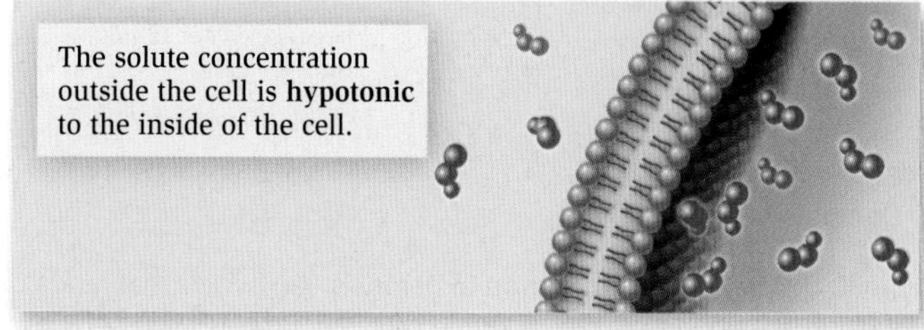

(c) Outside hypotonic

Figure 5.14 Relative solute concentrations outside and inside cells.

the two solutions are said to be **isotonic** (Figure 5.14a). However, we have also seen that transmembrane gradients commonly exist across membranes. When the concentration of solutes outside the cell is higher, it is said to be **hypertonic** relative to the inside of the cell (Figure 5.14b). Alternatively, the outside of the cell could be **hypotonic**—have a lower concentration of solutes relative to the inside (Figure 5.14c).

If solutes cannot readily move across the membrane, water will move and tend to balance the solute concentrations. In this process, called **osmosis**, water diffuses across a membrane from the hypotonic compartment (a lower concentration) into the hypertonic compartment (a higher concentration). Animal cells, which are not surrounded by a rigid cell wall, must maintain a balance between the extracellular and intracellular solute concentrations; they are isotonic. Animal cells contain a variety of transport proteins that sense changes in cell volume and allow the necessary movements of solutes across the membrane to prevent osmotic changes and maintain normal cell shape. However, if animal cells are placed in a hypotonic medium, water will diffuse into them to equalize solute concentrations on both sides of the membrane. In extreme cases, a cell may take up so much water that it ruptures, a phenomenon called osmotic lysis (Figure 5.15a). Alternatively, if animal cells are placed in a hypertonic medium, water will exit the cells via osmosis and equalize solute concentrations on both sides of the membrane, causing them to shrink in a process called crenation.

How does osmosis affect cells with a rigid cell wall, such as bacteria, fungi, algae, and plant cells? If the extracellular fluid is hypotonic, a plant cell will take up a small amount of water, but the cell wall prevents osmotic lysis from occurring (Figure 5.15b). Alternatively, if the extracellular fluid surrounding a plant cell is hypertonic, water will exit the cell and the plasma membrane will pull away from the cell wall, a process called **plasmolysis**.

Some freshwater microorganisms, such as amoebae and paramecia, are found in extremely hypotonic environments where the external solute concentration is always much lower than the concentration of solutes in their cytosol. Because of the great tendency for water to move into the cell by osmosis, such organisms contain one or more contractile vacuoles to prevent osmotic lysis. A contractile vacuole takes up water from the cytosol and periodically discharges it by fusing the vacuole with the plasma membrane (Figure 5.16).

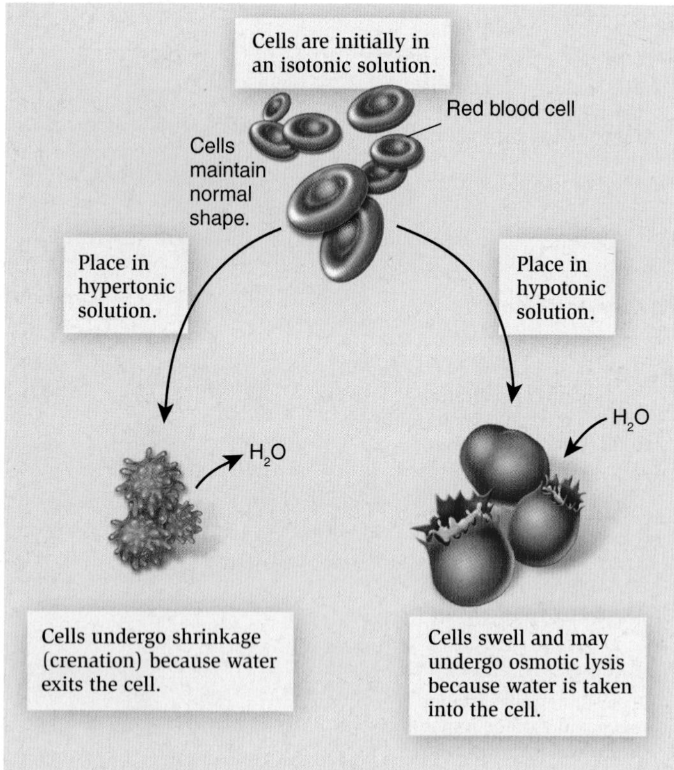

(a) Osmosis in animal cells

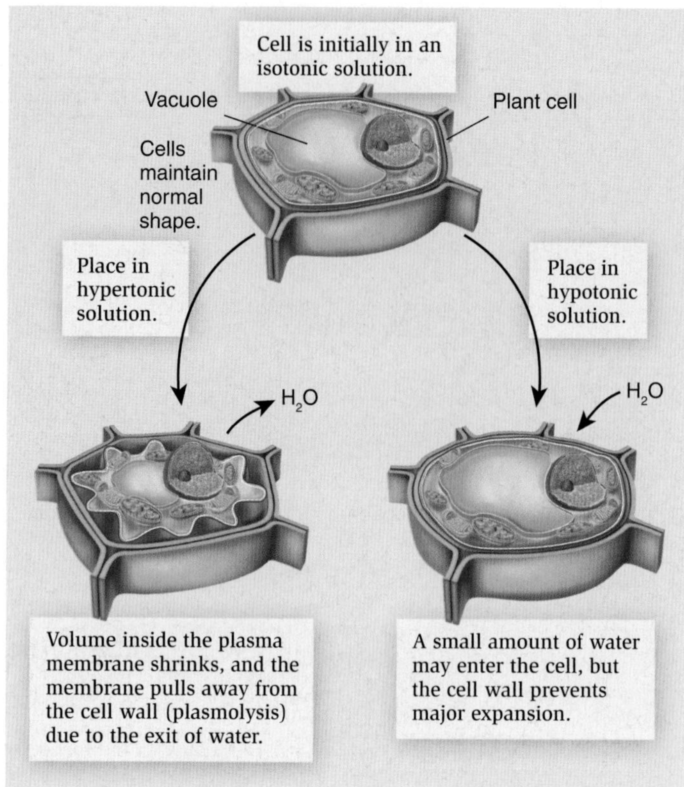

(b) Osmosis in plant cells

Figure 5.15 The phenomenon of osmosis. (a) In cells that lack a cell wall, such as animal cells, osmosis may promote cell shrinkage (crenation) or swelling. **(b)** In cells that have a rigid cell wall, such as plant cells, hypertonic medium causes the plasma membrane to pull away from the cell wall, whereas a hypotonic medium causes only a minor amount of expansion.

Concept Check: Let's suppose the inside of a cell has a solute concentration of 0.3 M, and the outside is 0.2 M. If the membrane is impermeable to solutes, which direction will water move?

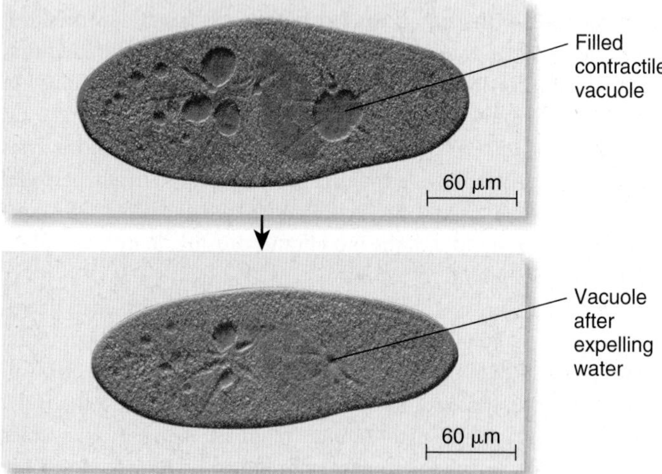

Figure 5.16 The contractile vacuole in *Paramecium caudatum*. In the upper photo, a contractile vacuole is filled with water from radiating canals that collect fluid from the cytosol. The lower photo shows the cell after the contractile vacuole has fused with the plasma membrane (which would be above the plane of this page) and released the water from the cell.

BIOLOGY PRINCIPLE Living organisms maintain homeostasis. In this example, the paramecium maintains a relatively constant internal volume by using contractile vacuoles to remove excess water.

5.5 Transport Proteins

Learning Outcomes:

1. Outline the functional differences between channels and transporters.
2. Compare and contrast uniporters, symporters, and antiporters.
3. Explain the difference between primary active transport and secondary active transport.
4. Describe the structure and function of pumps.

Because the phospholipid bilayer is a physical barrier to the diffusion of most hydrophilic molecules and ions, cells can separate their internal contents from the external environment. However, this barrier also poses a potential problem because cells must take up nutrients from the environment and export waste products. How do cells overcome this dilemma? Over the course of millions of years, species have evolved a multitude of **transport proteins**—transmembrane proteins that provide a passageway for the movement of ions and hydrophilic molecules across the phospholipid bilayer. Transport proteins play a central role in the selective permeability of biological membranes. In this section, we will examine the two categories of transport proteins—channels and transporters—based on the manner in which they move solutes across the membrane.

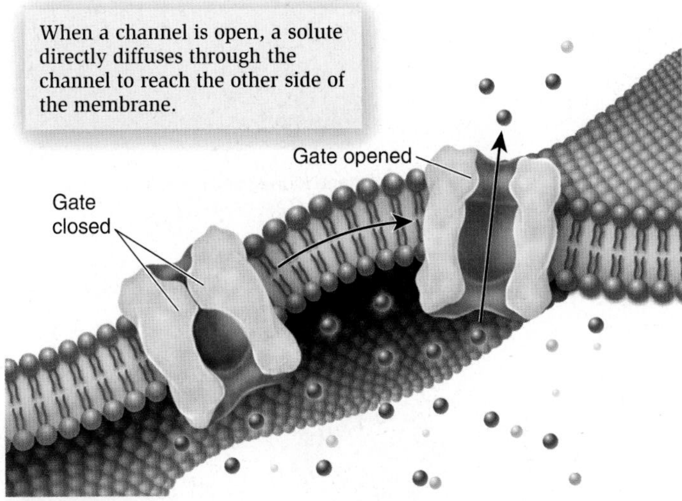

When a channel is open, a solute directly diffuses through the channel to reach the other side of the membrane.
Gate opened
Gate closed

Figure 5.17 Mechanism of transport by a channel protein.

Concept Check: *What is the purpose of gating?*

Channels Provide an Open Passageway for Solute Movement

A **channel** is a transmembrane protein that forms an open passageway for the facilitated diffusion of ions or molecules across the membrane (Figure 5.17). Solutes move directly through a channel to get to the other side. When a channel is open, the transmembrane movement of solutes can be extremely rapid, up to 100 million ions or molecules per second!

Most channels are **gated**, which means they open to allow the diffusion of solutes and close to prohibit diffusion. The phenomenon of gating allows cells to regulate the movement of solutes. For example, gating may involve the direct binding of a molecule to the channel protein itself. These gated channels are controlled by the noncovalent binding of small molecules—called ligands—such as hormones or neurotransmitters. The ligands are often important in the transmission of signals between neurons and muscle cells or between two neurons.

FEATURE INVESTIGATION

Agre Discovered That Osmosis Occurs More Quickly in Cells with a Channel That Allows the Facilitated Diffusion of Water

As discussed earlier in this chapter, osmosis is the flow of water to balance solute concentrations. Water can slowly cross biological membranes by diffusion through the phospholipid bilayer. However, in the 1980s, researchers discovered that certain cell types allow water to move across the plasma membrane at a much faster rate than would be predicted by diffusion. For example, water moves very quickly across the membrane of red blood cells, which causes them to shrink and swell in response to changes in extracellular solute concentrations. Likewise, bladder and kidney cells, which play a key role in regulating water balance in the bodies of vertebrates, allow the rapid movement of water across their membranes. Based on these observations, researchers speculated that certain cell types might have channel proteins in their plasma membranes that enable the rapid movement of water.

One approach to characterizing a new protein is to first identify a protein based on its relative abundance in a particular cell type and then attempt to determine the protein's function. This rationale was applied to the discovery of proteins that allow the rapid movement of water across membranes. Peter Agre and his colleagues first identified a protein that was abundant in red blood cells and kidney cells but not found in high amounts in many other cell types. Though they initially did not know the function of the protein, its physical structure was similar to other proteins that were already known to function as channels. They named this protein CHIP28, which stands for channel-forming integral membrane protein with a molecular mass of 28,000 daltons (Da). During the course of their studies, they also identified and isolated the gene that encodes CHIP28.

In 1992, Agre and his colleagues conducted experiments to determine if CHIP28 functions in the transport of water across membranes (Figure 5.18). Because they already had isolated the gene that encodes CHIP28, they could make many copies of this gene in a test tube (in vitro) using gene cloning techniques (see Chapter 20). Starting with many copies of the gene in vitro, they added an enzyme to transcribe the gene into mRNA that encodes the CHIP28 protein. This mRNA was then injected into frog oocytes, chosen because frog oocytes are large, easy to inject, and lack pre-existing proteins in their plasma membranes that allow the rapid movement of water. Following injection, the mRNA was translated into CHIP28 proteins that were inserted into the plasma membrane of the oocytes. After allowing sufficient time for this to occur, the oocytes were placed in a hypotonic medium. As a control, oocytes that had not been injected with CHIP28 mRNA were also exposed to a hypotonic medium.

As you can see in the data, a striking difference was observed between oocytes that expressed CHIP28 versus the control. Within minutes, oocytes that contained the CHIP28 protein were seen to swell due to the rapid uptake of water. Three to five minutes after being placed in a hypotonic medium, they actually lysed! By comparison, the control oocytes did not swell as rapidly, and they did not rupture even after 1 hour. Taken together, these results are consistent with the hypothesis that CHIP28 functions as a channel that allows the facilitated diffusion of water across the membrane. Many subsequent studies confirmed this observation. Later, CHIP28 was renamed **aquaporin** to indicate its newly identified function of allowing water to diffuse through a channel in the membrane. More recently, the three-dimensional structure of aquaporin was determined (see chapter opening figure). In 2003, Agre was awarded the Nobel Prize in Chemistry for this work.

Experimental Questions

1. What observations about particular cell types in the human body led to the experimental strategy of Figure 5.18?
2. What were the characteristics of CHIP28 that made Agre and associates speculate that it may transport water? In your own words, briefly explain how they tested the hypothesis that CHIP28 has this function.
3. Explain how the results of the experiment of Figure 5.18 support the proposed hypothesis.

Figure 5.18 **The discovery of water channels (aquaporins) by Agre.**

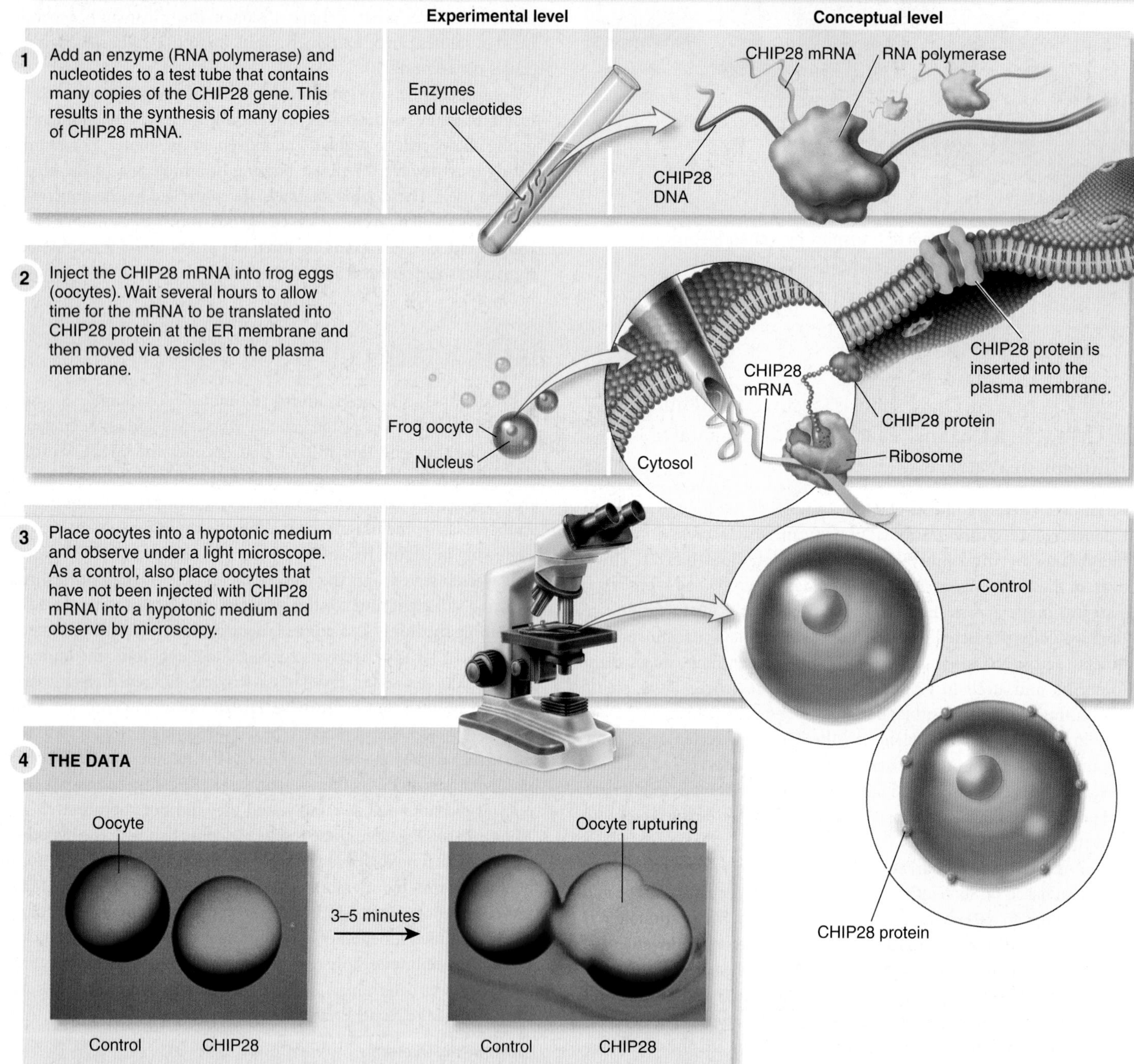

HYPOTHESIS CHIP28 may function as a water channel.
KEY MATERIALS Prior to this work, a protein called CHIP28 was identified that is abundant in red blood cells and kidney cells. The gene that encodes this protein was cloned, which means that many copies of the gene were made in a test tube.
Experimental level
Conceptual level
1 Add an enzyme (RNA polymerase) and nucleotides to a test tube that contains many copies of the CHIP28 gene. This results in the synthesis of many copies of CHIP28 mRNA.
Enzymes and nucleotides
CHIP28 mRNA
RNA polymerase
CHIP28 DNA
2 Inject the CHIP28 mRNA into frog eggs (oocytes). Wait several hours to allow time for the mRNA to be translated into CHIP28 protein at the ER membrane and then moved via vesicles to the plasma membrane.
Frog oocyte
Nucleus
Cytosol
CHIP28 mRNA
CHIP28 protein is inserted into the plasma membrane.
CHIP28 protein
Ribosome
3 Place oocytes into a hypotonic medium and observe under a light microscope. As a control, also place oocytes that have not been injected with CHIP28 mRNA into a hypotonic medium and observe by microscopy.
Control
CHIP28 protein
4 THE DATA
Oocyte
Oocyte rupturing
3–5 minutes
Control
CHIP28
Control
CHIP28

5 **CONCLUSION** The CHIP28 protein, now called aquaporin, allows the rapid movement of water across the membrane.

6 **SOURCE** Preston, G.M., Carroll, T.P., Guggino, W.B., and Agre, P. 1992. Appearance of water channels in *Xenopus* oocytes expressing red cell CHIP28 protein. *Science* 256:385–387.

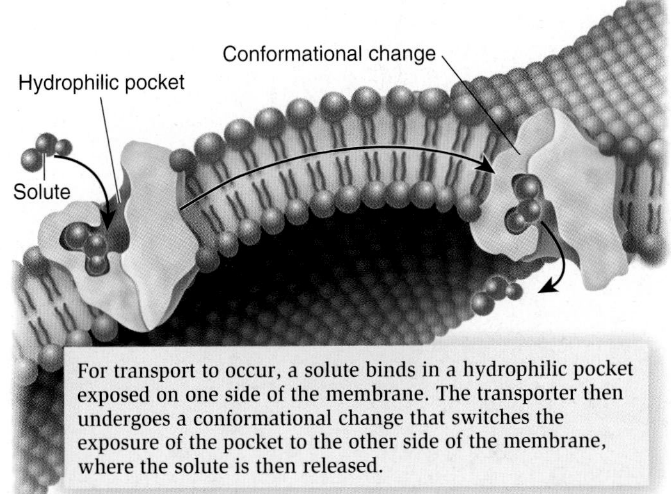

Conformational change
Hydrophilic pocket
Solute
For transport to occur, a solute binds in a hydrophilic pocket exposed on one side of the membrane. The transporter then undergoes a conformational change that switches the exposure of the pocket to the other side of the membrane, where the solute is then released.

Figure 5.19 **Mechanism of transport by a transporter, also called a carrier.**

BIOLOGY PRINCIPLE **Structure determines function.** Two structural features—a hydrophilic pocket and the ability to wobble back and forth between two conformations—allow transporters to move ions and molecules across the membrane.

Transporters Bind Their Solutes and Undergo Conformational Changes

Let's now turn our attention to a second category of transport proteins known as **transporters.*** These transmembrane proteins bind their solutes in a hydrophilic pocket and undergo a conformational change that switches the exposure of the pocket from one side of the membrane to the other side (Figure 5.19). For example, in 1995, American biologist Robert Brooker and colleagues proposed that a transporter called lactose permease, which is found in the bacterium *E. coli*, has a hydrophilic pocket that binds lactose. They further proposed that the two halves of the transporter protein come together at an interface that moves in such a way that the lactose-binding site alternates between an outwardly accessible pocket and an inwardly accessible pocket, as shown in Figure 5.19. This idea was later confirmed by studies that determined the structure of the lactose permease and related transporters.

Transporters provide the principal pathway for the uptake of organic molecules, such as sugars, amino acids, and nucleotides. In animals, they also allow cells to take up certain hormones and neurotransmitters. In addition, many transporters play a key role in export. Waste products of cellular metabolism must be released from cells before they reach toxic levels. For example, a transporter removes lactic acid, a by-product of muscle cells during exercise. Other transporters, which are involved with ion transport, play an important role in regulating internal pH and controlling cell volume. Transporters tend to be much slower than channels. Their rate of transport is typically 100 to 1,000 ions or molecules per second.

Transporters are named according to the number of solutes they bind and the direction in which they transport those solutes (Figure 5.20). **Uniporters** bind a single ion or molecule and transport it across the membrane. **Symporters** bind two or more ions or molecules and transport them in the same direction. **Antiporters** bind two or more ions or molecules and transport them in opposite directions.

* Transporters are also called carriers. However, this term is misleading because transporters do not physically carry their solutes across the membrane.

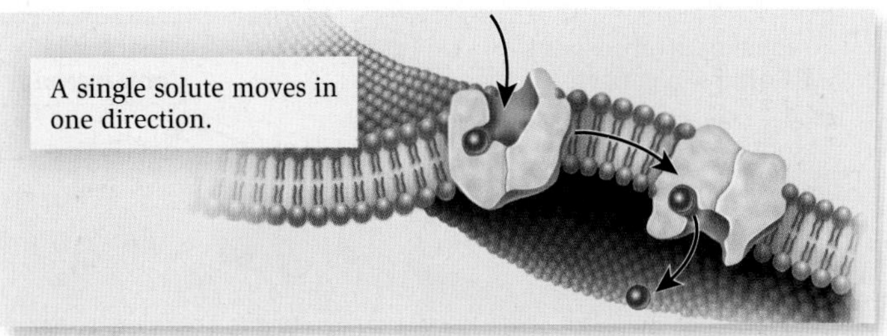

(a) Uniporter

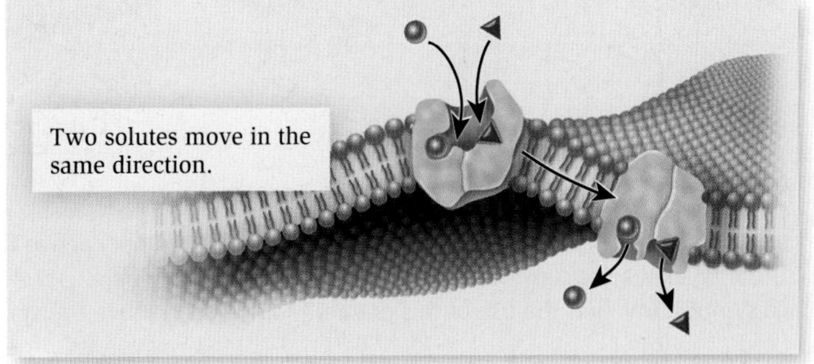

(b) Symporter

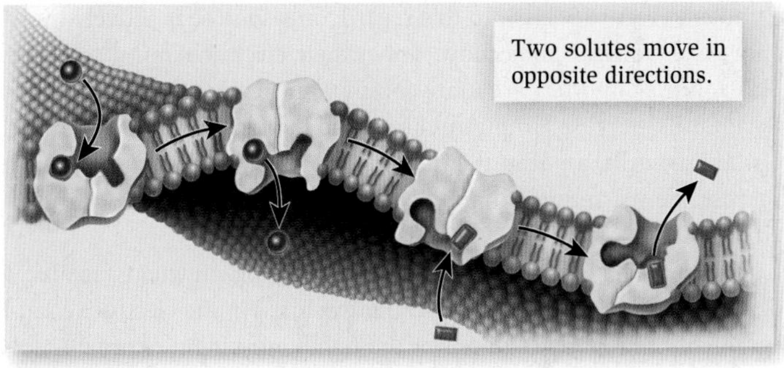

(c) Antiporter

Figure 5.20 **Types of transporters based on the direction of transport.**

Active Transport Is the Movement of Solutes Against a Gradient

As mentioned, active transport is the movement of a solute across a membrane against its concentration gradient—that is, from a region of low concentration to higher concentration. Active transport is energetically unfavorable and requires an input of energy. **Primary active transport** involves the functioning of a **pump**—a type of transporter that directly uses energy to transport a solute against a concentration gradient. Figure 5.21a shows a pump that uses ATP to transport H^+ against a gradient. Such a pump can establish an H^+ electrochemical gradient across a membrane.

Secondary active transport involves the use of a pre-existing gradient to drive the active transport of another solute. For example, an H^+/sucrose symporter uses an H^+ electrochemical gradient,

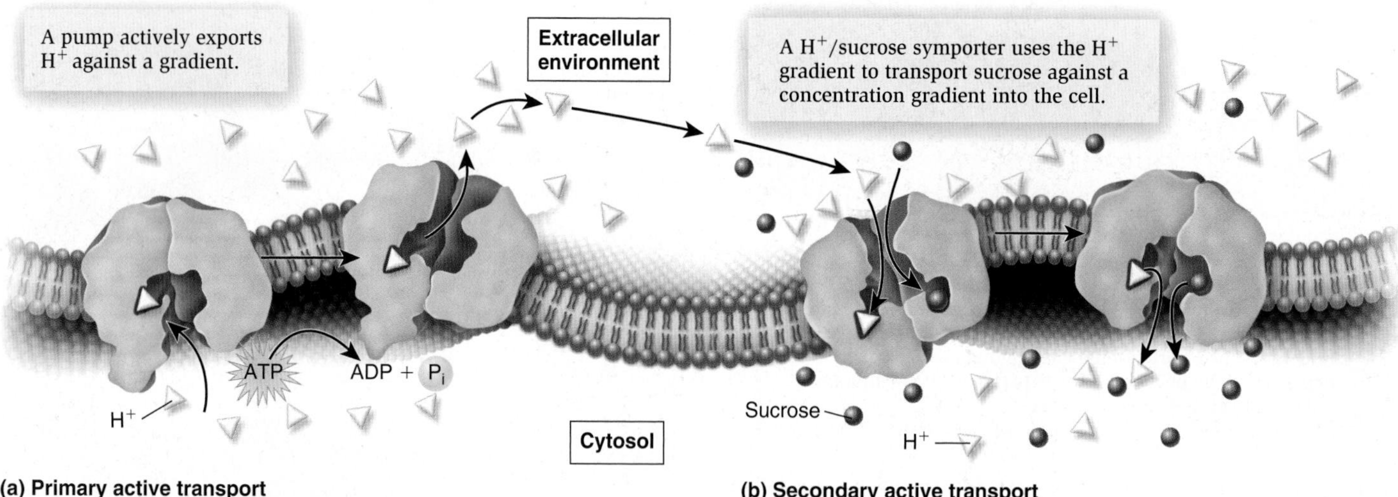
A pump actively exports H^+ against a gradient.
Extracellular environment
A H^+/sucrose symporter uses the H^+ gradient to transport sucrose against a concentration gradient into the cell.
ATP
ADP + P_i
H^+
Cytosol
Sucrose
H^+
(a) Primary active transport
(b) Secondary active transport

Figure 5.21 **Types of active transport.** **(a)** During primary active transport, a pump directly uses energy, in this case from ATP, to transport a solute against a concentration gradient. The pump shown here uses ATP to establish an H^+ electrochemical gradient. **(b)** Secondary active transport via symport involves the use of this gradient to drive the active transport of a solute, such as sucrose.

established by a pump, to move sucrose against its concentration gradient (Figure 5.21b). In this regard, only sucrose is actively transported. Hydrogen ions move down their electrochemical gradient. H^+/solute symporters are more common in bacteria, fungi, algae, and plant cells, because H^+ pumps are found in their plasma membranes. In animal cells, a pump that exports Na^+ maintains a Na^+ gradient across the plasma membrane. Na^+/solute symporters are prevalent in animal cells.

Symporters enable cells to actively import nutrients against a gradient. These proteins use the energy stored in the electrochemical gradient of H^+ or Na^+ to power the uphill movement of organic solutes such as sugars, amino acids, and other needed molecules. Therefore, with symporters in their plasma membrane, cells can scavenge nutrients from the extracellular environment and accumulate them to high levels within the cytoplasm.

ATP-Driven Ion Pumps Generate Ion Electrochemical Gradients

The phenomenon of active transport was discovered in the 1940s based on the study of the transport of sodium ions (Na^+) and potassium ions (K^+). In animal cells, the concentration of Na^+ is lower inside the cell than outside, whereas the concentration of K^+ is higher inside the cell than outside. After analyzing the movement of these ions across the plasma membrane of muscle cells, neurons, and red blood cells, researchers determined that the export of Na^+ is coupled to the import of K^+. In the late 1950s, Danish biochemist Jens Skou proposed that a single transporter is responsible for this phenomenon. He was the first to describe an ATP-driven ion pump, which was later named the Na^+/K^+-ATPase. This pump actively transports Na^+ and K^+ against their gradients by using the energy from ATP hydrolysis. The plasma membrane of a typical animal cell contains thousands of Na^+/K^+-ATPase pumps that maintain large concentration gradients in which the concentration of Na^+ is higher outside the cell and the concentration of K^+ is higher inside the cell.

Let's take a closer look at the Na^+/K^+-ATPase that Skou discovered. Every time one ATP is hydrolyzed, the Na^+/K^+-ATPase functions as an antiporter that pumps three Na^+ into the extracellular environment and two K^+ into the cytosol (Figure 5.22a). Because one cycle of pumping results in the net export of one positive charge, the Na^+/K^+-ATPase also produces an electrical gradient across the membrane. For this reason, it is called an **electrogenic pump**, because it generates an electrical gradient.

By studying the interactions of Na^+, K^+, and ATP with the Na^+/K^+-ATPase, researchers have pieced together a molecular road map of the steps that direct the pumping of ions across the membrane (Figure 5.22b). The Na^+/K^+-ATPase alternates between two conformations, designated E1 and E2. In E1, the ion-binding sites are accessible from the cytosol—Na^+ binds tightly to this conformation, whereas K^+ has a low affinity. In E2, the ion-binding sites are accessible from the extracellular environment—Na^+ has a low affinity, and K^+ binds tightly.

To examine the pumping mechanism of the Na^+/K^+-ATPase, let's begin with the E1 conformation. Three Na^+ bind to the Na^+/K^+-ATPase from the cytosol (Figure 5.22b). When this occurs, ATP is hydrolyzed to ADP and phosphate. Temporarily, the phosphate is covalently bound to the pump, an event called phosphorylation. The pump then switches to the E2 conformation. The three Na^+ are released into the extracellular environment, because they have a lower affinity for the E2 conformation. In this conformation, two K^+ bind from the outside. The binding of two K^+ causes the release of phosphate, which, in turn, causes a switch to E1. Because the E1 conformation has a low affinity for K^+ the two K^+ are released into the cytosol. The Na^+/K^+-ATPase is now ready for another round of pumping.

The Na^+/K^+-ATPase is a critical ion pump in animal cells because it maintains Na^+ and K^+ gradients across the plasma

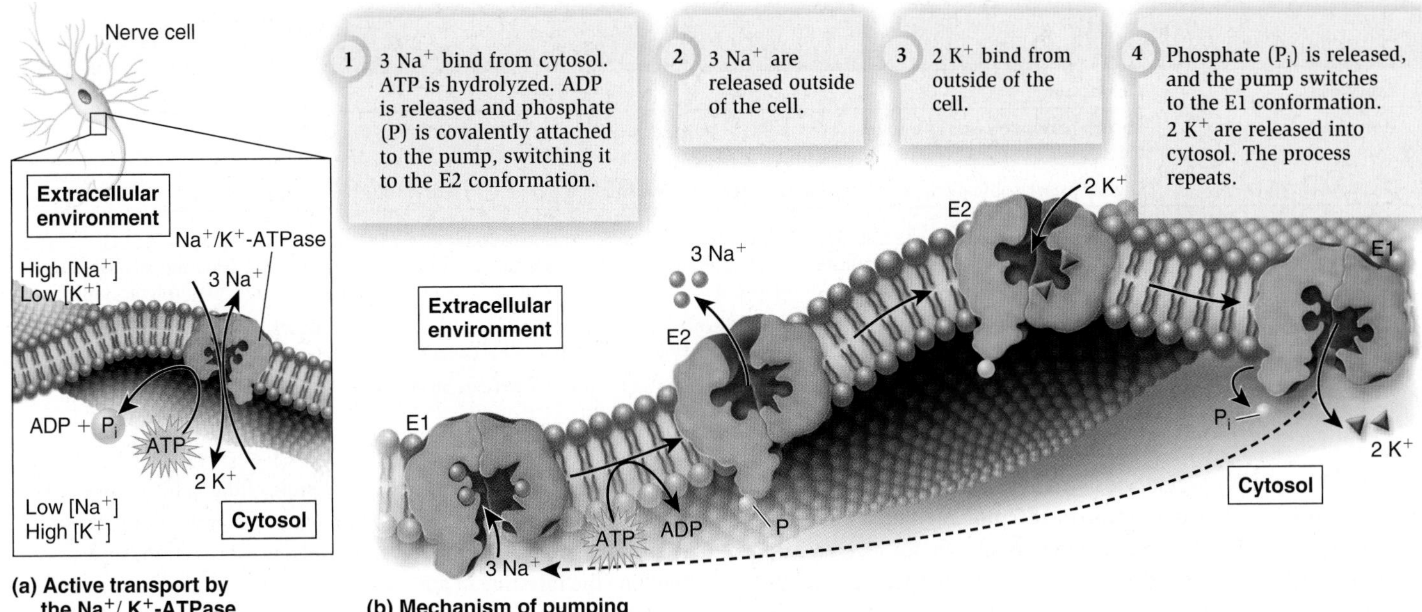
Nerve cell
Extracellular environment
Na+/K+-ATPase
High [Na+]
Low [K+]
3 Na+
ADP + Pi
ATP
2 K+
Low [Na+]
High [K+]
Cytosol
1 3 Na+ bind from cytosol. ATP is hydrolyzed. ADP is released and phosphate (P) is covalently attached to the pump, switching it to the E2 conformation.
2 3 Na+ are released outside of the cell.
3 2 K+ bind from outside of the cell.
4 Phosphate (Pi) is released, and the pump switches to the E1 conformation. 2 K+ are released into cytosol. The process repeats.
Extracellular environment
E1
E2
3 Na+
E2
2 K+
E1
Pi
2 K+
Cytosol
ATP
ADP
P
3 Na+
(a) Active transport by the Na+/ K+-ATPase
(b) Mechanism of pumping

Figure 5.22 **Structure and function of the Na$^+$/K$^+$-ATPase.** **(a)** Active transport by the Na$^+$/K$^+$-ATPase. Each time this protein hydrolyzes one ATP molecule, it pumps out three Na$^+$ and pumps in two K$^+$. **(b)** Pumping mechanism. This figure illustrates the protein conformational changes between E1 and E2. As this occurs, ATP is hydrolyzed to ADP and phosphate. During the process, phosphate is covalently attached to the protein but is released after two K$^+$ bind.

Concept Check: If a cell contained ATP and Na$^+$, but K$^+$ were missing from the extracellular medium, how far through these steps could the Na$^+$/K$^+$-ATPase proceed?

membrane. Many other types of ion pumps are also found in the plasma membrane and in organellar membranes. Ion pumps play the primary role in the formation and maintenance of ion gradients that drive many important cellular processes (**Table 5.4**). ATP is commonly the source of energy to drive ion pumps, and cells typically use a substantial portion of their ATP to keep them working. For example, neurons use up to 70% of their ATP just to operate ion pumps!

Table 5.4 Important Functions of Ion Electrochemical Gradients

Function	Description
Transport of ions and molecules	Symporters and antiporters use H$^+$ and Na$^+$ gradients to take up nutrients and export waste products (see Figure 5.21).
Production of energy intermediates	In the mitochondrion and chloroplast, H$^+$ gradients are used to synthesize ATP.
Osmotic regulation	Animal cells control their internal volume by regulating ion gradients between the cytosol and extracellular fluid.
Neuronal signaling	Na$^+$ and K$^+$ gradients are involved in conducting action potentials, the signals transmitted by neurons.
Muscle contraction	Ca^{2+} gradients regulate the ability of muscle fibers to contract.
Bacterial swimming	H$^+$ gradients drive the rotation of bacterial flagella.

5.6 Exocytosis and Endocytosis

Learning Outcome:

1. Describe the steps in exocytosis and endocytosis.

We have seen that most small substances are transported via membrane proteins such as channels and transporters, which provide a passageway for the movement of ions and molecules directly across the membrane. Eukaryotic cells have two other mechanisms, exocytosis and endocytosis, to transport larger molecules such as proteins and polysaccharides, and even very large particles. Both mechanisms involve the packaging of the transported substance, sometimes called the cargo, into a membrane vesicle or vacuole. **Table 5.5** describes some examples.

Exocytosis During **exocytosis**, material inside the cell is packaged into vesicles and then excreted into the extracellular environment (**Figure 5.23**). These vesicles are usually derived from the Golgi apparatus. As the vesicles form, a specific cargo is loaded into their interior. The budding process involves the formation of a protein coat around the emerging vesicle. The assembly of coat proteins on the surface of the Golgi membrane causes the bud to form. Eventually, the bud separates from the membrane to form a vesicle. After the vesicle is released, the coat is shed. Finally, the vesicle fuses with the plasma membrane and releases the cargo into the extracellular environment.

Endocytosis During **endocytosis**, the plasma membrane invaginates, or folds inward, to form a vesicle that brings substances into

Table 5.5 Examples of Exocytosis and Endocytosis

Exocytosis	Description
Hormones	Certain hormones, such as insulin, are composed of polypeptides. To exert its effect, insulin is secreted via exocytosis into the bloodstream from beta cells of the pancreas.
Digestive enzymes	Digestive enzymes that function in the lumen of the small intestine are secreted via exocytosis from exocrine cells of the pancreas.
Endocytosis	**Description**
Uptake of vital nutrients	Many important nutrients are very insoluble in the bloodstream. Therefore, they are bound to proteins in the blood and then taken into cells via endocytosis. Examples include the uptake of lipids (bound to low-density lipoprotein) and iron (bound to transferrin protein).
Root nodules	Nitrogen-fixing root nodules found in certain species of plants, such as legumes, are formed by the endocytosis of bacteria. After endocytosis, the bacterial cells are contained within a membrane-enclosed compartment in the nitrogen-fixing tissue of root nodules.
Immune system	Cells of the immune system, known as macrophages, engulf and destroy bacteria via phagocytosis.

the cell. Three types of endocytosis are receptor-mediated endocytosis, pinocytosis, and phagocytosis. In **receptor-mediated endocytosis**, a receptor in the plasma membrane is specific for a given cargo (**Figure 5.24**). Cargo molecules binding to their specific receptors stimulate many receptors to aggregate, and then coat proteins bind to the membrane. The protein coat causes the membrane to invaginate and form a vesicle. Once it is released into the cell, the vesicle sheds its coat. In most cases, the vesicle fuses with an internal membrane organelle, such as a lysosome, and the receptor releases its cargo. Depending on the cargo, the lysosome may release it directly into the cytosol or digest it into simpler building blocks before releasing it.

Other specialized forms of endocytosis occur in certain types of cells. **Pinocytosis** (from the Greek, meaning cell-drinking) involves the formation of membrane vesicles from the plasma membrane as a way for cells to internalize the extracellular fluid. This allows cells to sample the extracellular solutes. Pinocytosis is particularly important in cells that are actively involved in nutrient absorption, such as cells that line the intestine in animals.

Phagocytosis (from the Greek, meaning cell-eating) involves the formation of an enormous membrane vesicle called a phagosome, or phagocytic vacuole, which engulfs a large particle such as a bacterium. Only certain kinds of cells can carry out phagocytosis. For example, macrophages, which are cells of the immune system in mammals, kill bacteria via phagocytosis. Macrophages engulf bacterial cells into phagosomes. Once inside the cell, the phagosome fuses with a lysosome, and the digestive enzymes within the lysosome destroy the bacterium.

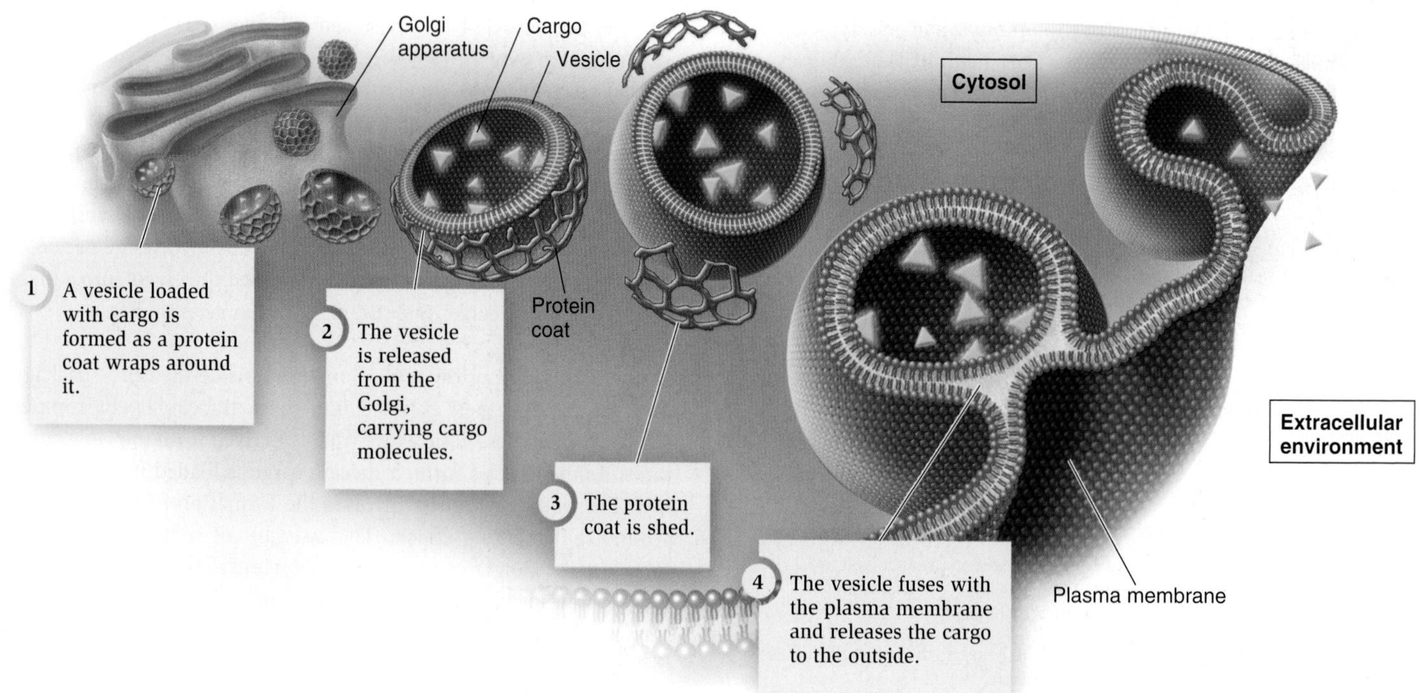
Golgi apparatus
Cargo
Vesicle
Cytosol
1 A vesicle loaded with cargo is formed as a protein coat wraps around it.
2 The vesicle is released from the Golgi, carrying cargo molecules.
Protein coat
3 The protein coat is shed.
4 The vesicle fuses with the plasma membrane and releases the cargo to the outside.
Plasma membrane
Extracellular environment

Figure 5.23 **Exocytosis.**

Concept Check: *What is the function of the protein coat?*

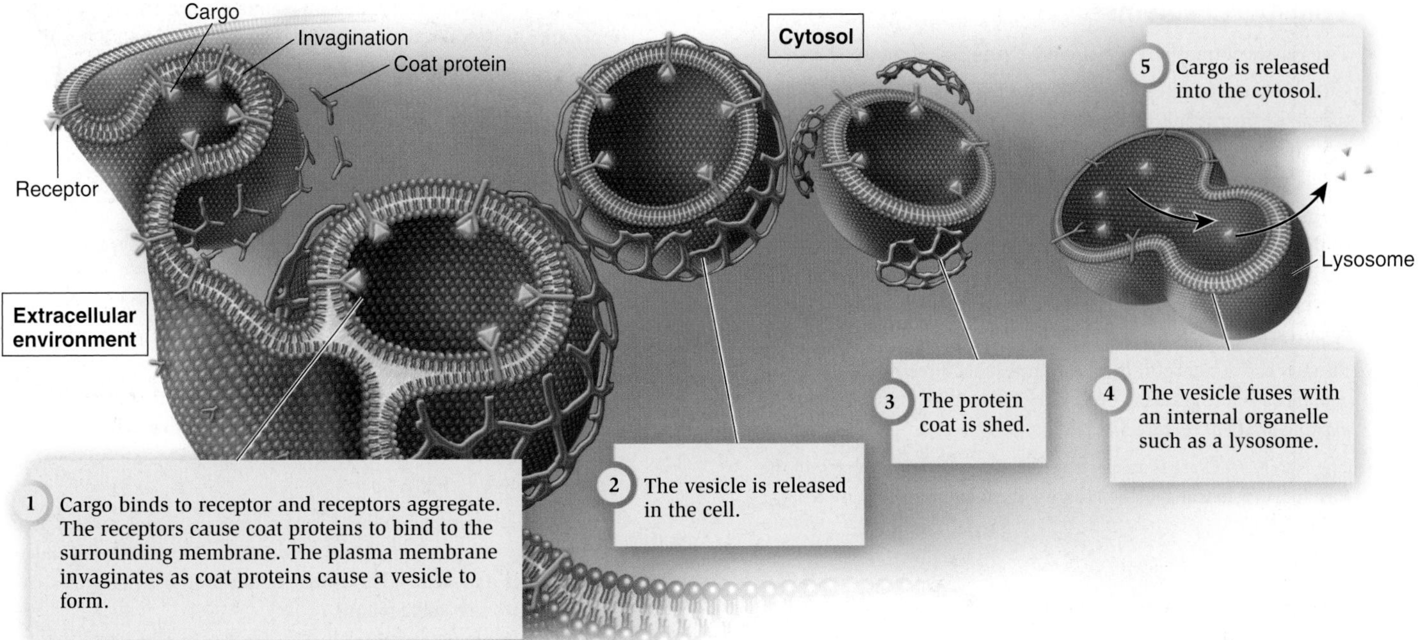

Cargo
Invagination
Coat protein
Receptor
Extracellular environment
Cytosol
1 Cargo binds to receptor and receptors aggregate. The receptors cause coat proteins to bind to the surrounding membrane. The plasma membrane invaginates as coat proteins cause a vesicle to form.
2 The vesicle is released in the cell.
3 The protein coat is shed.
4 The vesicle fuses with an internal organelle such as a lysosome.
5 Cargo is released into the cytosol.
Lysosome

Figure 5.24 Receptor-mediated endocytosis.

Summary of Key Concepts

5.1 Membrane Structure

- A plasma membrane separates a cell from its surroundings. The plasma membrane and organellar membranes provide interfaces for carrying out vital cellular activities (Table 5.1).
- The accepted model of membranes is the fluid-mosaic model, and its basic framework is the phospholipid bilayer. Cellular membranes also contain proteins, and most membranes have attached carbohydrates (Figure 5.1, Table 5.2).
- The three main types of membrane proteins are transmembrane proteins, lipid-anchored proteins, and peripheral membrane proteins. Transmembrane proteins and lipid-anchored proteins are classified as integral membrane proteins. Researchers are working to identify new membrane proteins and their functions because these proteins are important biologically and medically (Figure 5.2, Table 5.3).
- Electron microscopy is a valuable tool for studying membrane structure and function. Freeze fracture electron microscopy (FFEM) is used to analyze the interiors of phospholipid bilayers (Figure 5.3).

5.2 Fluidity of Membranes

- Bilayer semifluidity is essential for normal cell function, growth, and division. Lipids and many proteins can move rotationally and laterally, but the flip-flop of lipids from one leaflet to the opposite does not occur spontaneously. Some membrane proteins are restricted in their movements (Figures 5.4, 5.5, 5.6).
- The chemical properties of phospholipids—such as tail length and the presence of double bonds—and the amount of cholesterol affect the fluidity of membranes.

5.3 Synthesis of Membrane Components in Eukaryotic Cells

- In eukaryotic cells, most membrane phospholipids are synthesized at the cytosolic leaflet of the smooth ER membrane. Flippases move some phospholipids to the other leaflet (Figure 5.7).
- Most transmembrane proteins are first inserted into the ER membrane (Figure 5.8).
- Glycosylation of proteins occurs in the ER and Golgi apparatus (Figure 5.9).

5.4 Overview of Membrane Transport

- Biological membranes exhibit selective permeability. Diffusion occurs when a solute moves from a region of high concentration to a region of lower concentration. Passive transport of a solute across a membrane can occur via diffusion or facilitated diffusion. Active transport is the movement of a substance against a gradient (Figure 5.10).
- The phospholipid bilayer is relatively impermeable to many substances (Figures 5.11, 5.12).
- Living cells maintain an internal environment that is separated from their external environment. Transmembrane gradients are established across the plasma membrane and across organellar membranes (Figure 5.13, Table 5.4).
- In the process of osmosis, water diffuses through a membrane from a solution that is hypotonic (lower concentration of dissolved particles) into a solution that is hypertonic (higher concentration of dissolved particles). Solutions with identical concentrations are isotonic. Some cells have contractile vacuoles to eliminate excess water (Figures 5.14, 5.15, 5.16).

5.5 Transport Proteins

- Two classes of transport proteins are channels and transporters.
- Channels form an open passageway for the direct diffusion of solutes across the membrane. One example is aquaporin, which allows the movement of water. Most channels are gated, which allows cells to regulate the movement of solutes (Figures 5.17, 5.18).
- Transporters, which tend to be slower than channels, bind their solutes in a hydrophilic pocket and undergo a conformational change that switches the exposure of the pocket to the other side of the membrane. They can be uniporters, symporters, or antiporters (Figures 5.19, 5.20).
- Primary active transport involves pumps that directly use energy to generate a solute gradient. Secondary active transport uses a pre-existing gradient (Figure 5.21).
- The Na^+/K^+-ATPase is an electrogenic pump that uses energy in the form of ATP to transport ions across the membrane (Figure 5.22, Table 5.4).

5.6 Exocytosis and Endocytosis

- In eukaryotes, exocytosis and endocytosis are used to transport large molecules and particles. Exocytosis is a process in which material inside the cell is packaged into vesicles and excreted into the extracellular environment. During endocytosis, the plasma membrane folds inward to form a vesicle that brings substances into the cell. Receptor-mediated endocytosis, pinocytosis, and phagocytosis are forms of endocytosis (Figures 5.23, 5.24, Table 5.5).

Assess and Discuss

Test Yourself

1. Which of the following statements best describes the chemical composition of biological membranes?
   a. Biological membranes are bilayers of proteins with associated lipids and carbohydrates.
   b. Biological membranes are composed of two layers—one layer of phospholipids and one layer of proteins.
   c. Biological membranes are bilayers of phospholipids with associated proteins and carbohydrates.
   d. Biological membranes are composed of equal numbers of phospholipids, proteins, and carbohydrates.
   e. Biological membranes are composed of lipids with proteins attached to the outer surface.
2. Which of the following events in a biological membrane would not be energetically favorable and therefore not occur spontaneously?
   a. the rotation of phospholipids
   b. the lateral movement of phospholipids
   c. the flip-flop of phospholipids to the opposite leaflet
   d. the rotation of membrane proteins
   e. the lateral movement of membrane proteins
3. Let's suppose an insect, which doesn't maintain a constant body temperature, was exposed to a shift in temperature from 60°F to 80°F. Which of the following types of membrane changes would be the most beneficial to help this animal cope with the temperature shift?
   a. increase the number of double bonds in the fatty acyl tails of phospholipids
   b. increase the length of the fatty acyl tails of phospholipids
   c. decrease the amount of cholesterol in the membrane
   d. decrease the amount of carbohydrate attached to membrane proteins
   e. decrease the amount of carbohydrate attached to phospholipids
4. Carbohydrates of the plasma membrane
   a. are bonded to a protein or lipid.
   b. are located on the outer surface of the plasma membrane.
   c. can function as cell markers for recognition by other cells.
   d. all of the above
   e. a and c only
5. A transmembrane protein in the plasma membrane is glycosylated at two sites in the polypeptide sequence. One site is Asn—Val—Ser and the other site is Asn—Gly—Thr. Where in this protein would you expect these two sites to be found?
   a. in transmembrane segments
   b. in hydrophilic regions that project into the extracellular environment
   c. in hydrophilic regions that project into the cytosol
   d. could be anywhere
   e. b and c only
6. The tendency for Na^+ to move into the cell could be due to
   a. the higher numbers of Na^+ outside the cell, resulting in a chemical concentration gradient.
   b. the net negative charge inside the cell attracting the positively charged Na^+.
   c. the attractive force of K^+ inside the cell pulling Na^+ into the cell.
   d. all of the above.
   e. a and b only.
7. Let's suppose the solute concentration inside the cells of a plant is 0.3 M and outside is 0.2 M. If we assume that the solutes do not readily cross the membrane, which of the following statements best describes what will happen?
   a. The plant cells will lose water, and the plant will wilt.
   b. The plant cells will lose water, which will result in a higher turgor pressure.
   c. The plant cells will take up a lot of water and undergo osmotic lysis.
   d. The plant cells will take up a little water and have a higher turgor pressure.
   e. Both a and b are correct.
8. What features of a membrane are major contributors to its selective permeability?
   a. phospholipid bilayer
   b. transport proteins
   c. glycolipids on the outer surface of the membrane
   d. peripheral membrane proteins
   e. both a and b

9. What is the name given to the process in which solutes are moved across a membrane against their concentration gradient?
   a. diffusion
   b. facilitated diffusion
   c. osmosis
   d. passive diffusion
   e. active transport
10. Large particles or large volumes of fluid can be brought into the cell by
    a. facilitated diffusion.
    b. active transport.
    c. endocytosis.
    d. exocytosis.
    e. all of the above.

Conceptual Questions

1. With your textbook closed, draw and describe the fluid-mosaic model of membrane structure.
2. Describe two different ways that integral membrane proteins associate with a membrane. How do peripheral membrane proteins associate with a membrane?
3. A principle of biology is that *living organisms interact with their environment*. Discuss how lipid bilayers, channels, and transporters influence the ability of cells to interact with their environment.

Collaborative Questions

1. Proteins in the plasma membrane are often the target of medicines. Discuss why you think this is the case. How would you determine experimentally that a specific membrane protein was the target of a drug?
2. With regard to bringing solutes into the cell across the plasma membrane, discuss the advantages and disadvantages of diffusion, facilitated diffusion, active transport, and endocytosis.

Online Resource

www.brookerbiology.com

Stay a step ahead in your studies with animations that bring concepts to life and practice tests to assess your understanding. Your instructor may also recommend the interactive eBook, individualized learning tools, and more.

An Introduction to Energy, Enzymes, and Metabolism

6

Chapter Outline

6.1 Energy and Chemical Reactions
6.2 Enzymes and Ribozymes
6.3 Overview of Metabolism
6.4 Recycling of Organic Molecules
Summary of Key Concepts
Assess and Discuss

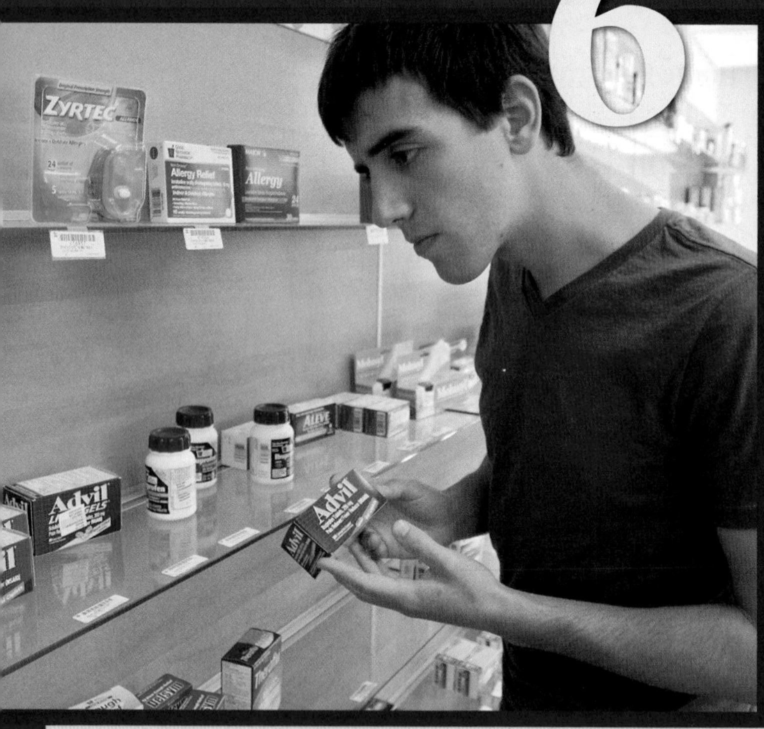

Common drugs that act as enzyme inhibitors. **Drugs such as aspirin and ibuprofen exert their effects by inhibiting an enzyme that speeds up a chemical reaction in the cell.**

Have you ever taken aspirin or ibuprofen to relieve a headache or reduce a fever? Do you know how these medications work? If you answered "no" to the second question, you're not alone. Over 2,000 years ago, humans began treating pain with powder from the bark and leaves of the willow tree, which contains a compound called salicylic acid. Modern aspirin is composed of a derivative of salicylic acid called acetylsalicylic acid, which is gentler to the stomach. Only recently, however, have we learned how such drugs work. Aspirin and ibuprofen are examples of drugs that inhibit a specific enzyme called cyclooxygenase. This enzyme is needed to synthesize molecules called prostaglandins, which play a role in inflammation, pain, and fever. Aspirin and ibuprofen exert their effects by inhibiting cyclooxygenase, thereby blocking the production of prostaglandins and in turn relieving pain and fever.

Enzymes are proteins that act as critical catalysts to speed up thousands of different reactions in cells. As discussed in Chapter 2, a **chemical reaction** is a process in which one or more substances are changed into other substances. Such reactions may involve molecules attaching to each other to form larger molecules, molecules breaking apart to form two or more smaller molecules, rearrangements of atoms within molecules, or the transfer of electrons from one atom to another. Every living cell continuously performs thousands of such chemical reactions to sustain life. **Metabolism** is the sum total of all chemical reactions that occur within an organism. Metabolism also refers to a specific set of chemical reactions occurring at the cellular level. For example, biologists may speak of sugar metabolism or fat metabolism. Most types of metabolism involve the breakdown or synthesis of organic molecules. Cells maintain their structure by using organic molecules. Such molecules provide the building blocks for constructing cells, and the chemical bonds within organic molecules store energy that is used to drive cellular processes.

In this chapter, we begin with a general discussion of energy and chemical reactions. We will examine what factors control the direction of a chemical reaction and what determines its rate, paying particular attention to the role of enzymes. We then consider metabolism at the cellular level, examining how chemical reactions are often coordinated with each other in metabolic pathways. We will also explore the variety of ways in which metabolic pathways are regulated and how nucleotides and amino acids are recycled.

6.1 Energy and Chemical Reactions

Learning Outcomes:

1. Define energy and distinguish between potential and kinetic energy.
2. State the first and second laws of thermodynamics and discuss how they relate to living things.
3. Explain how the change in free energy determines the direction of a chemical reaction and how chemical reactions eventually reach a state of equilibrium.
4. Distinguish between exergonic and endergonic reactions in terms of the energy of the reactants and products and the free energy change.
5. Describe how cells use the energy released by ATP hydrolysis to drive endergonic reactions.

Two general factors govern the fate of a given chemical reaction in a living cell—its direction and rate. To illustrate this point, let's consider this generalized chemical reaction:

$$aA + bB \rightleftharpoons cC + dD$$

where A and B are the reactants, C and D are the products, and *a*, *b*, *c*, and *d* are the number of moles of reactants and products. This reaction is reversible, which means that A + B could be converted to C + D, or C + D could be converted to A + B. The direction of the reaction, whether C + D are made (the forward direction) or A + B are made (the reverse direction), depends on energy and on the concentrations of A, B, C, and D. In this section, we will begin by examining the interplay of energy and the concentration of reactants as they govern the direction of a chemical reaction. You will learn that cells use energy intermediate molecules, such as ATP, to drive chemical reactions in a desired direction.

Energy Exists in Different Forms

To understand why a chemical reaction occurs, we first need to consider **energy**, which is the ability to promote change or do work. Physicists often consider energy in two general forms: kinetic energy and potential energy (Figure 6.1). **Kinetic energy** is energy associated with movement, such as the movement of a baseball bat from one location to another. By comparison, **potential energy** is the energy that a substance or object possesses due to its structure or location. The energy contained within covalent bonds in molecules is also a type of potential energy called **chemical energy**. The breakage of those bonds is one way that living cells harness this energy to perform cellular functions. Table 6.1 summarizes chemical and other forms of energy that are common in biological systems.

An important issue in biology is the ability of energy to be converted from one form to another. The study of energy interconversions is called **thermodynamics**. Physicists have determined that two laws govern energy interconversions:

1. **The first law of thermodynamics**—The first law of thermodynamics, also called the law of conservation of energy, states that energy cannot be created or destroyed. However, energy can be transferred from one place to another and can be transformed from one type to another (as when, for example, chemical energy is transformed into heat).

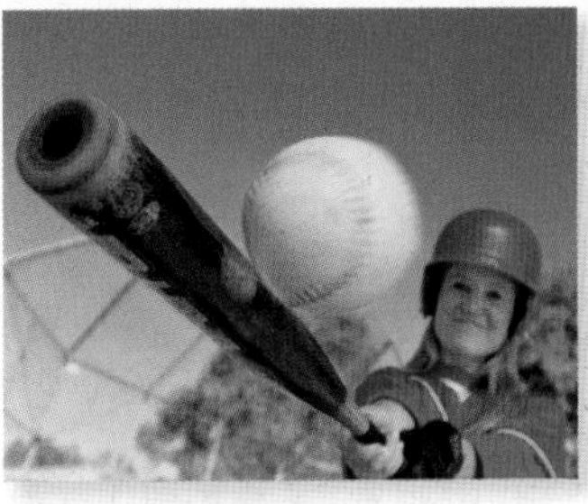

(a) Kinetic energy

(b) Potential energy

Figure 6.1 **Examples of energy.** **(a)** Kinetic energy, such as swinging a bat, is energy associated with motion. **(b)** Potential energy is stored energy, as in a bow that is ready to fire an arrow.

2. **The second law of thermodynamics**—The second law states that any energy transfer or transformation from one form to another increases the degree of disorder of a system, called **entropy** (Figure 6.2). Entropy is a measure of the randomness of molecules in a system. When a physical system becomes more disordered, the entropy increases. As the energy becomes more evenly distributed, that energy is less able to promote change or do work. When energy is converted from one form to another, some energy may become unusable by living organisms. For example, a chemical reaction may release unusable heat. Next, we will see how the two laws of thermodynamics place limits on the ways that living cells use energy for their own needs.

The Change in Free Energy Determines the Direction of a Chemical Reaction

Energy is necessary for organisms to exist. Energy is required for many cellular processes, including chemical reactions, cellular movements such as those occurring in muscle contraction, and the maintenance of cell organization. To understand how organisms use energy, we need to distinguish between the energy that can be used to

Table 6.1 Types of Energy That Are Important in Biology

Energy type	Description	Biological example
Light	Light is a form of electromagnetic radiation that is visible to the eye. The energy of light is packaged in photons.	During photosynthesis, light energy is captured by pigments in chloroplasts (described in Chapter 8). Ultimately, this energy is used to reduce carbon, thus producing organic molecules.
Heat	Heat is the transfer of kinetic energy from one object to another or from an energy source to an object. In biology, heat is often viewed as kinetic energy that can be transferred due to a difference in temperature between two objects or locations.	Many organisms, including humans, maintain their bodies at a constant temperature. This is achieved, in part, by chemical reactions that generate heat.
Mechanical	Mechanical energy is the energy possessed by an object due to its motion or its position relative to other objects.	In animals, mechanical energy is associated with movements due to muscle contraction such as walking.
Chemical	Chemical energy is potential energy stored in the chemical bonds of molecules. When the bonds are broken and rearranged, large amounts of energy may be released.	The covalent bonds in organic molecules, such as glucose and ATP, store large amounts of energy. When bonds are broken in larger molecules to form smaller molecules, the chemical energy that is released can be used to drive cellular processes.
Electrical/Ion gradient	The movement of charge or the separation of charge can provide energy. Also, a difference in ion concentration across a membrane constitutes an electrochemical gradient, which is a source of potential energy.	During a stage of cellular respiration called oxidative phosphorylation (described in Chapter 7), an H^+ gradient provides the energy to drive ATP synthesis.

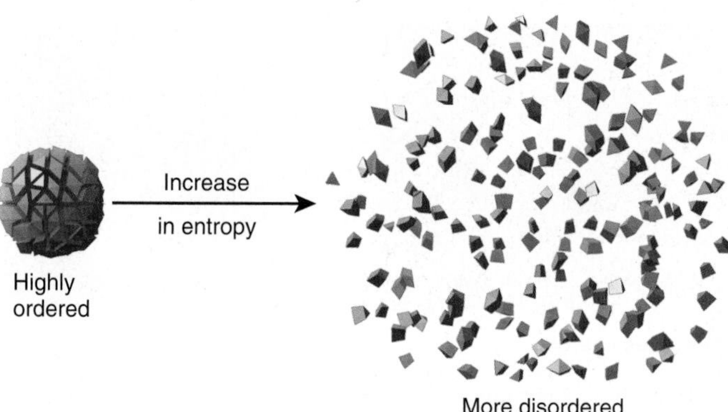

Figure 6.2 **Entropy, a measure of the disorder of a system.** An increase in entropy means an increase in disorder.

Concept Check: Which do you think has more entropy, a NaCl crystal at the bottom of a beaker of water or the same beaker of water after the Na^+ and Cl^- in the crystal have dissolved in the water?

promote change or do work (usable energy) and the energy that cannot (unusable energy).

$$\text{Total energy} = \text{Usable energy} + \text{Unusable energy}$$

Why is some energy unusable? The main culprit is entropy. As stated by the second law of thermodynamics, energy transfers or transformations involve an increase in entropy, a degree of disorder that cannot be harnessed in a useful way. The total energy is termed **enthalpy (*H*)**, and the usable energy—the amount of available energy that can be used to promote change or do work—is called the **free energy (*G*)**. The letter *G* is in recognition of the American physicist J. Willard Gibbs, who proposed the concept of free energy in 1878. The unusable energy is the system's entropy (*S*). Gibbs proposed that these three factors are related to each other in the following way:

$$H = G + TS$$

where *T* is the absolute temperature in kelvins (K). Because our focus is on free energy, we can rearrange this equation as

$$G = H - TS$$

A critical issue in biology is whether a process does or does not occur spontaneously. For example, will glucose be broken down into carbon dioxide and water? Another way of framing this question is to ask: "Is the breakdown of glucose a spontaneous reaction?" A spontaneous reaction or process is one that occurs without being driven by an input of energy. However, a spontaneous reaction does not necessarily proceed quickly. In some cases, the rate of a spontaneous reaction can be quite slow. For example, the breakdown of sugar is a spontaneous reaction, but the rate at which sugar in a sugar bowl breaks down into CO_2 and H_2O is very slow.

The key way to evaluate if a chemical reaction is spontaneous is to determine the free-energy change that occurs as a result of the reaction:

$$\Delta G = \Delta H - T\Delta S$$

where the Δ sign (the Greek letter delta) indicates a change, such as before and after a chemical reaction. If a chemical reaction has a negative free-energy change ($\Delta G < 0$), the products have less free energy than the reactants, and, therefore, free energy is released during product formation. Such a reaction is said to be **exergonic**. Exergonic reactions are spontaneous. Alternatively, if a reaction has a positive free-energy change ($\Delta G > 0$), requiring the addition of free energy from the environment, it is termed **endergonic**. An endergonic reaction is not a spontaneous reaction.

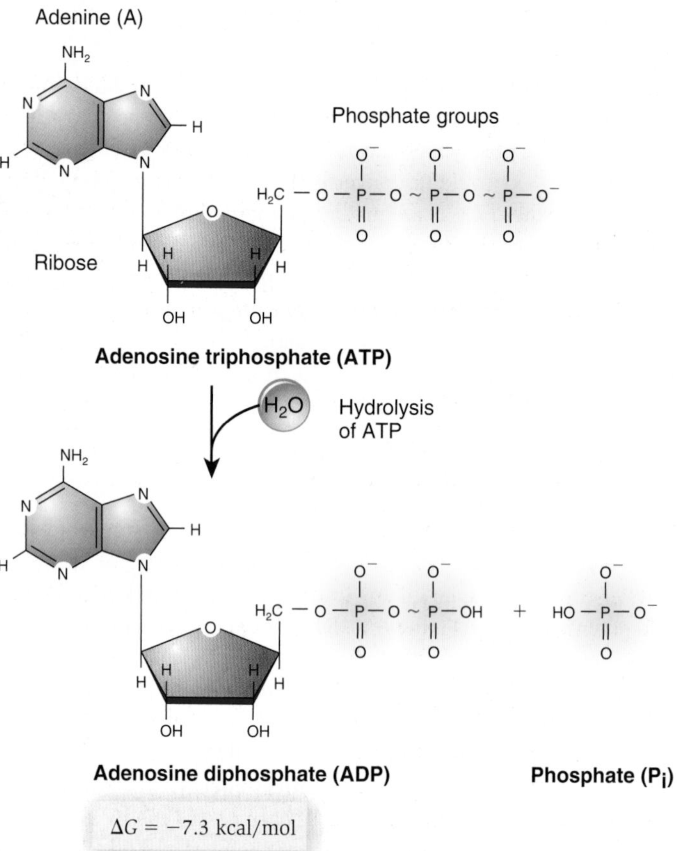

Figure 6.3 **The hydrolysis of ATP to ADP and P_i.** As shown in this figure, ATP has a net charge of –4, while ADP and P_i are shown with net charges of −2 each. When these compounds are shown in chemical reactions with other molecules, the net charges are also indicated. Otherwise, these compounds are simply designated ATP, ADP, and P_i. At neutral pH, ADP^{2-} dissociates to ADP^{3-} and H^+.

BIOLOGY PRINCIPLE Living organisms use energy. The hydrolysis of ATP is an exergonic reaction that is used to drive many cellular reactions.

If ΔG for a chemical reaction is negative, the reaction favors the conversion of reactants to products, whereas a reaction with a positive ΔG favors the formation of reactants. Chemists have determined free-energy changes for a variety of chemical reactions, which allows them to predict their direction. As an example, let's consider **adenosine triphosphate (ATP)**, which is a molecule that is a common energy source for all cells. ATP is broken down to adenosine diphosphate (ADP) and inorganic phosphate (HPO_4^{2-}, abbreviated to P_i). Because water is used to remove a phosphate group, chemists refer to this as the hydrolysis of ATP (**Figure 6.3**). In the reaction of converting 1 mole of ATP to 1 mole of ADP and P_i, ΔG equals –7.3 kcal/mol. Because this is a negative value, the reaction strongly favors the formation of products. As discussed later, the energy liberated by the hydrolysis of ATP is used to drive a variety of cellular processes.

Chemical Reactions Eventually Reach a State of Equilibrium

Even when a chemical reaction is associated with a negative free-energy change, not all of the reactants are converted to products. The reaction reaches a state of **chemical equilibrium** in which the rate of formation of products equals the rate of formation of reactants. Let's consider the generalized reaction

$$aA + bB \rightleftharpoons cC + dD$$

where again A and B are the reactants, C and D are the products, and *a*, *b*, *c*, and *d* are the number of moles of reactants and products. An equilibrium occurs, such that

$$K_{eq} = \frac{[C]^c[D]^d}{[A]^a[B]^b}$$

where K_{eq} is the equilibrium constant. Each type of chemical reaction has a specific value for K_{eq}.

Biologists make two simplifying assumptions when determining values for equilibrium constants. First, the concentration of water does not change during the reaction, and the pH remains constant at pH 7. The equilibrium constant under these conditions is designated with a prime symbol, K_{eq}'. If water is one of the reactants, as in a hydrolysis reaction, it is not included in the chemical equilibrium equation. As an example, let's consider the chemical equilibrium for the hydrolysis of ATP.

$$ATP^{4-} + H_2O \rightleftharpoons ADP^{2-} + P_i^{2-}$$

$$K_{eq}' = \frac{[ADP][P_i]}{[ATP]}$$

Experimentally, the value for K_{eq}' for this reaction has been determined to be approximately 1,650,000 M. Such a large value indicates that the equilibrium greatly favors the formation of products—ADP and P_i.

Cells Use ATP to Drive Endergonic Reactions

In living organisms, many vital processes require the addition of free energy; that is, they are endergonic and do not occur spontaneously. Fortunately, organisms have a way to overcome this problem. Cells often couple exergonic reactions with endergonic reactions. If an exergonic reaction is coupled with an endergonic reaction, the endergonic reaction will proceed spontaneously if the net free-energy change for both processes combined is negative. For example, consider the following reactions:

$$\text{Glucose} + \text{Phosphate}^{2-} \rightarrow \text{Glucose-6-phosphate}^{2-} + H_2O \qquad \Delta G = +3.3 \text{ kcal/mol}$$

$$ATP^{4-} + H_2O \rightarrow ADP^{2-} + P_i^{2-} \qquad \Delta G = -7.3 \text{ kcal/mol}$$

Coupled reaction:

$$\text{Glucose} + ATP^{4-} \rightarrow \text{Glucose-6-phosphate}^{2-} + ADP^{2-} \qquad \Delta G = -4.0 \text{ kcal/mol}$$

The first reaction, in which phosphate is covalently attached to glucose, is endergonic, and by itself is not spontaneous. The second, the hydrolysis of ATP, is exergonic. If the two reactions are coupled, however, the net free-energy change for both reactions combined is exergonic ($\Delta G = -4.0$ kcal/mol). In the coupled reaction, a phosphate is directly transferred from ATP to glucose, in a process called **phosphorylation**. This coupled reaction proceeds spontaneously because the net free-energy change is negative. Exergonic reactions, such as the breakdown of ATP, are commonly coupled to chemical reactions and other cellular processes that would otherwise be endergonic.

Genomes & Proteomes Connection

Many Proteins Bind ATP and Use That ATP as a Source of Energy

Over the past several decades, researchers have studied the functions of many types of proteins and discovered numerous examples in which a protein uses the hydrolysis of ATP to drive a chemical reaction or cellular process (**Table 6.2**). In humans, a typical cell uses millions of ATP molecules per second. At the same time, the breakdown of food molecules to form smaller molecules (an exergonic reaction) releases energy that allows cells to make more ATP from the phosphorylation of ADP (an endergonic reaction). The recycling of ATP occurs at a remarkable pace. An average person hydrolyzes about 100 pounds of ATP per day, yet at any given time we do not have 100 pounds of ATP in our bodies. For this to happen, each molecule of

Table 6.2 Examples of Proteins That Use ATP for Energy

Type	Description
Metabolic enzymes	Many enzymes use ATP to catalyze endergonic reactions. For example, hexokinase uses ATP to attach phosphate to glucose, producing glucose-6-phosphate.
Transporters	Ion pumps, such as the Na^+/K^+-ATPase, use ATP to pump ions against a gradient (see Chapter 5).
Motor proteins	Motor proteins such as myosin use ATP to facilitate cellular movement, as in muscle contraction (see Chapter 44).
Chaperones	Chaperones are proteins that use ATP to aid in the folding and unfolding of cellular proteins (see Chapter 4).
DNA-modifying enzymes	Many proteins such as helicases and topoisomerases use ATP to modify the conformation of DNA (see Chapter 11).
Aminoacyl-tRNA synthetases	These enzymes use ATP to attach amino acids to tRNAs (see Chapter 12).
Protein kinases	Protein kinases are regulatory proteins that use ATP to attach a phosphate to proteins, thereby phosphorylating the protein and affecting its function (see Chapter 9).

ATP, adenosine triphosphate; tRNA, transfer RNA.

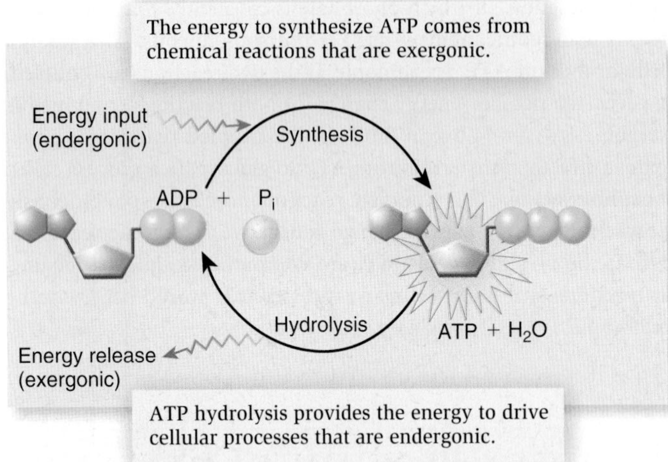
The energy to synthesize ATP comes from chemical reactions that are exergonic.
Energy input (endergonic)
Synthesis
ADP + P_i
Hydrolysis
ATP + H_2O
Energy release (exergonic)
ATP hydrolysis provides the energy to drive cellular processes that are endergonic.

Figure 6.4 The ATP cycle. Living cells continuously recycle ATP. The energy released from the breakdown of food molecules into smaller molecules is used to synthesize ATP from ADP and P_i. The hydrolysis of ATP to ADP and P_i is used to drive many different endergonic reactions and processes that occur in cells.

Concept Check: *If a large amount of ADP was broken down in the cell, how would this affect the ATP cycle?*

ATP undergoes about 10,000 cycles of hydrolysis and regeneration during an ordinary day (**Figure 6.4**).

By studying the structures of many proteins that use ATP, biochemists have discovered that particular amino acid sequences within proteins function as ATP-binding sites. This information has allowed researchers to predict whether a newly discovered protein uses ATP or not. When an entire genome sequence of a species has been determined, the genes that encode proteins can be analyzed to find out if the encoded proteins have ATP-binding sites in their amino acid sequences. Using this approach, researchers have been able to analyze **proteomes**—all of the proteins that a given cell makes—and estimate the percentage of proteins that are able to bind ATP. This approach has been applied to the proteomes of bacteria, archaea, and eukaryotes.

Researchers have discovered that, on average, over 20% of all proteins bind ATP. However, this number is likely to underestimate the total percentage of ATP-utilizing proteins because we may not have identified all of the types of ATP-binding sites in proteins. In humans, who have an estimated genome size of 20,000 to 25,000 different genes, a minimum of 4,000 to 5,000 of those genes encode proteins that use ATP. From these numbers, we can see the enormous importance of ATP as a source of energy for living cells.

6.2 Enzymes and Ribozymes

Learning Outcomes:

1. Explain how enzymes increase the rates of chemical reactions by lowering the activation energy.
2. Describe how enzymes bind their substrates with high specificity and undergo induced fit.
3. Analyze the velocity of chemical reactions and evaluate the effects of competitive and noncompetitive inhibitors.
4. Explain how additional factors, such as nonprotein molecules or ions, temperature, and pH, influence enzyme activity.
5. Identify the unique feature of ribozymes.

For most chemical reactions in cells to proceed at a rapid pace, such as the breakdown of glucose, a catalyst is needed. A **catalyst** is an agent that speeds up the rate of a chemical reaction without being permanently changed or consumed by it. In living cells, the most common catalysts are **enzymes**, which are proteins. The term was coined in 1876 by a German physiologist, Wilhelm Kühne, who discovered trypsin, an enzyme in pancreatic juice that is needed for the digestion of food proteins. In this section, we will explore how enzymes increase the rates of chemical reactions. Interestingly, some biological catalysts are RNA molecules called ribozymes. We will examine a few examples in which RNA molecules carry out catalytic functions.

Enzymes Increase the Rates of Chemical Reactions

Thus far, we have examined aspects of energy and considered how the laws of thermodynamics are related to the direction of chemical reactions. If a chemical reaction has a negative free-energy change, the reaction will be spontaneous; it will tend to proceed in the direction of reactants to products. Although thermodynamics governs the direction of an energy transformation, it does not control the rate of a chemical reaction. For example, the breakdown of the molecules in gasoline to smaller molecules is an exergonic reaction. Even so, we could place gasoline and oxygen in a container and nothing much would happen (provided it wasn't near a flame). If we came back several days later, we would expect to see the gasoline still sitting there. Perhaps if we came back in a few million years, the gasoline would have been broken down. On a timescale of months or a few years, however, the chemical reaction would proceed very slowly.

In living cells, the rates of enzyme-catalyzed reactions typically occur millions of times faster than the corresponding uncatalyzed reactions. An extreme example is the enzyme catalase, which is found in peroxisomes, organelles responsible for the breakdown of toxic molecules such as hydrogen peroxide (see Chapter 4). This enzyme catalyzes the breakdown of hydrogen peroxide (H_2O_2) into water and oxygen. Catalase speeds up this reaction 10^{15}-fold faster than the uncatalyzed reaction!

Why are catalysts necessary to speed up a chemical reaction? Chemical reactions between molecules involve bond breaking and bond forming. When a covalent bond is broken or formed, this process initially involves the straining or stretching of one or more bonds in the starting molecule(s), and/or it may involve the positioning of two molecules so they interact with each other properly. Let's consider the reaction in which ATP is used to phosphorylate glucose:

$$\text{Glucose} + \text{ATP}^{4-} \rightarrow \text{Glucose-6-phosphate}^{2-} + \text{ADP}^{2-}$$

For a reaction to occur between glucose and ATP, the molecules must collide in the correct orientation and possess enough energy so the chemical bonds can be changed. As glucose and ATP approach each other, their electron clouds cause repulsion. To overcome this repulsion, an initial input of energy, called the **activation energy**, is

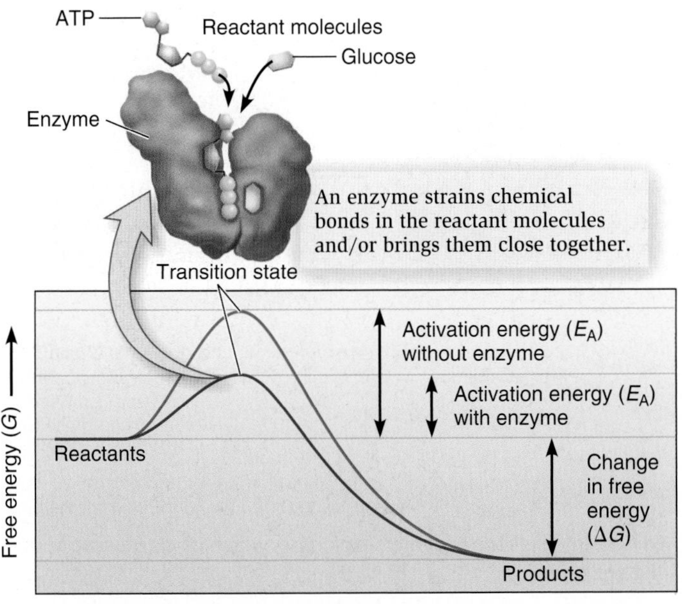

Figure 6.5 **Activation energy of a chemical reaction.** This figure depicts an exergonic reaction. The activation energy (E_A) is needed for molecules to achieve a transition state. One way that enzymes lower the activation energy is by straining chemical bonds in the reactants so less energy is required to attain the transition state. A second way is by binding two reactants so they are close to each other and in a favorable orientation.

Concept Check: How does lowering the activation energy affect the rate of a chemical reaction? How does it affect the direction?

required (**Figure 6.5**). Activation energy (E_A) allows the molecules to get close enough to cause a rearrangement of bonds. With the input of activation energy, glucose and ATP can achieve a **transition state** in which the original bonds have stretched to their limit. Once the reactants have reached the transition state, the chemical reaction can readily proceed to the formation of products, which in this case is glucose-6-phosphate and ADP.

The activation energy required to achieve the transition state is a barrier to the formation of products. This barrier is the reason why the rate of many chemical reactions is very slow. Two common mechanisms can overcome this barrier and thereby accelerate a chemical reaction. First, the reactants could be exposed to a large amount of heat. For example, as we noted previously, if gasoline is sitting at room temperature, nothing much happens. However, if the gasoline is exposed to a flame or spark, it breaks down rapidly, perhaps at an explosive rate! Alternatively, a second strategy is to lower the activation energy barrier. Enzymes lower the activation energy to a point where a small amount of available heat can push the reactants to a transition state.

How do enzymes lower the activation energy barrier of chemical reactions? Enzymes are generally large proteins that bind relatively small reactants. When bound to an enzyme, the bonds in the reactants can be strained, thereby making it easier for them to achieve the transition state (see Figure 6.5). This is one way that enzymes lower the activation energy. In addition, when a chemical reaction involves two or more reactants, the enzyme provides a site in which the reactants are positioned very close to each other in an orientation that facilitates the formation of new covalent bonds. This also lowers the necessary activation energy for a chemical reaction.

Straining the reactants and bringing them close together are two common ways that enzymes lower the activation energy barrier. In addition, enzymes may facilitate a chemical reaction by changing the local environment of the reactants. For example, amino acids in an enzyme may have charges that affect the chemistry of the reactants. In some cases, enzymes lower the activation energy by directly participating in the chemical reaction. For example, certain enzymes that hydrolyze ATP form a covalent bond between phosphate and an amino acid in the enzyme. However, this is a temporary condition. The covalent bond between phosphate and the amino acid is quickly broken, releasing the phosphate and returning the amino acid back to its original condition. An example of such an enzyme is Na^+/K^+-ATPase, described in Chapter 5.

Enzymes Recognize Their Substrates with High Specificity and Undergo Conformational Changes

Thus far, we have considered how enzymes lower the activation energy of a chemical reaction, and thereby increase its rate. Let's consider some other features of enzymes that enable them to serve as effective catalysts in chemical reactions. The **active site** is the location in an enzyme where the chemical reaction takes place. The **substrates** for an enzyme are the reactant molecules that bind to an enzyme at the active site and participate in the chemical reaction. For example, hexokinase is an enzyme whose substrates are glucose and ATP (**Figure 6.6**). The binding between an enzyme and substrate produces an **enzyme-substrate complex**.

A key feature of nearly all enzymes is their ability to bind their substrates with a high degree of specificity. For example, hexokinase recognizes glucose but does not recognize other similar sugars, such as fructose and galactose, very well. In 1894, the German chemist Emil Fischer proposed that the recognition of a substrate by an enzyme resembles the interaction between a lock and key: only the right-sized key (the substrate) will fit into the keyhole (active site) of the lock (the enzyme). Further research revealed that the interaction between an enzyme and its substrates also involves movements or conformational changes in the enzyme itself. As shown in Figure 6.6 step 2, these conformational changes cause the substrates to bind more tightly to the enzyme, a phenomenon called **induced fit**, which was proposed by American biochemist Daniel Koshland in 1958. Only after this induced fit takes place does the enzyme catalyze the conversion of reactants to products. Induced fit is a key phenomenon that lowers the activation energy.

Enzyme Function Is Influenced by the Substrate Concentration and by Inhibitors

The degree of attraction between an enzyme and its substrate(s) is called the **affinity** of the enzyme for its substrate(s). Some enzymes recognize their substrates with very high affinity, which means they have a strong attraction for their substrates. Such enzymes bind their

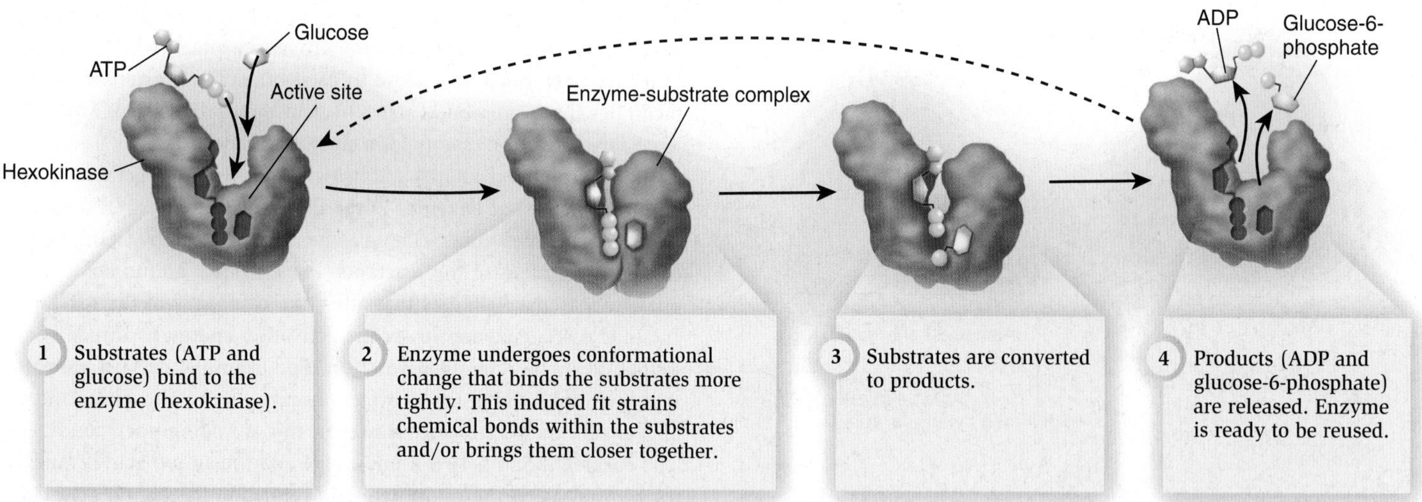

Glucose
ATP
Active site
Hexokinase
Enzyme-substrate complex
ADP
Glucose-6-phosphate
1 Substrates (ATP and glucose) bind to the enzyme (hexokinase).
2 Enzyme undergoes conformational change that binds the substrates more tightly. This induced fit strains chemical bonds within the substrates and/or brings them closer together.
3 Substrates are converted to products.
4 Products (ADP and glucose-6-phosphate) are released. Enzyme is ready to be reused.

Figure 6.6 The steps of an enzyme-catalyzed reaction. The example shown here involves the enzyme hexokinase, which binds glucose and ATP. The products are glucose-6-phosphate and ADP, which are released from the enzyme.

BIOLOGY PRINCIPLE Structure determines function. A key function of enzymes is their ability to bind their substrates with high specificity. This is due to the structure of the enzyme's active site.

Concept Check: During which step is the activation energy lowered?

substrates even when the substrate concentration is relatively low. Other enzymes recognize their substrates with lower affinity; the enzyme-substrate complex is likely to form when the substrate concentration is higher.

Let's consider how biologists analyze the relationship between substrate concentration and enzyme function. In the experiment of Figure 6.7a, tubes labeled A, B, C, and D each contained 1 μg of enzyme but they varied in the amount of substrate that was added. This enzyme recognizes a single type of substrate and converts it to a product. The samples were incubated for 60 seconds, and then the amount of product in each tube was measured. In this example, the velocity or rate of the chemical reaction is expressed as the amount of product produced per second. As we see in Figure 6.7a, the velocity increases as the substrate concentration increases, but eventually reaches a plateau. Why does the plateau occur? At high substrate concentrations, nearly all of the active sites of the enzyme are occupied with substrate, so increasing the substrate concentration further has a negligible effect. At this point, the enzyme is saturated with substrate, and the velocity of the chemical reaction is near its maximal rate, called its $\boldsymbol{V_{max}}$.

Figure 6.7a also helps us understand the relationship between substrate concentration and velocity. The $\boldsymbol{K_M}$ is the substrate concentration at which the velocity is half its maximal value. The K_M is also called the Michaelis constant in honor of the German biochemist Leonor Michaelis, who carried out pioneering work with the Canadian biochemist Maud Menten on the study of enzymes. The K_M is a measure of the substrate concentration required for a chemical reaction to occur. An enzyme with a high K_M requires a higher substrate concentration to achieve a particular reaction velocity compared to an enzyme with a lower K_M.

For an enzyme-catalyzed reaction, we can view the formation of product as occurring in two steps: (1) binding or release of substrate and (2) formation of product:

$$E + S \rightleftharpoons ES \rightarrow E + P$$

where E is the enzyme, S is the substrate, ES is the enzyme-substrate complex, and P is the product.

If the second step—the rate of product formation—is much slower than the rate of substrate release, the K_M is inversely related to the affinity between the enzyme and substrate. For example, let's consider an enzyme that breaks down ATP into ADP and P_i. If the rate of formation of ADP and P_i is much slower than the rate of ATP release, the K_M and affinity show an inverse relationship. Enzymes with a high K_M have a low affinity for their substrates—they bind them more weakly. By comparison, enzymes with a low K_M have a high affinity for their substrates—they bind them more strongly.

Now that we understand the relationship between substrate concentration and the velocity of an enzyme-catalyzed reaction, we can explore how inhibitors may affect enzyme function. These can be categorized as reversible inhibitors that bind noncovalently to an enzyme or irreversible inhibitors that usually bind covalently to an enzyme and permanently inactivate its function.

Reversible Inhibitors Cells often use reversible inhibitors to modulate enzyme function. We will consider examples in Chapter 7 in our discussion of glucose metabolism. **Competitive inhibitors** are molecules that bind noncovalently to the active site of an enzyme and inhibit the ability of the substrate to bind. Such inhibitors compete with the substrate for the ability to bind to the enzyme. Competitive inhibitors usually have a structure or a portion of their structure that mimics the structure of the enzyme's substrate. As seen in

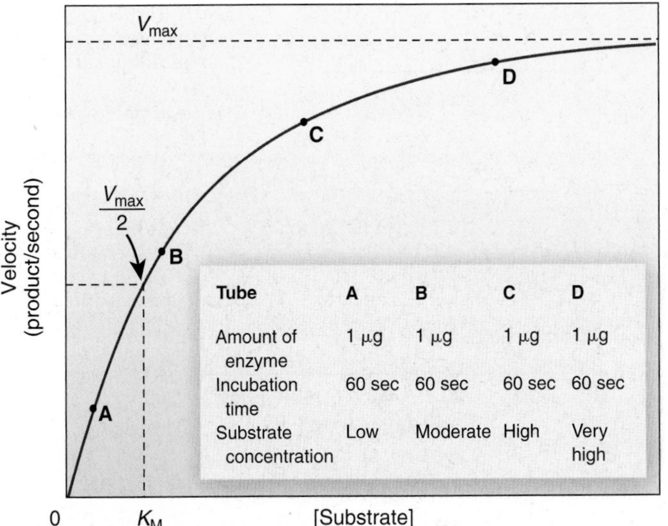

Tube	A	B	C	D
Amount of enzyme	1 μg	1 μg	1 μg	1 μg
Incubation time	60 sec	60 sec	60 sec	60 sec
Substrate concentration	Low	Moderate	High	Very high

(a) Reaction velocity in the absence of inhibitors

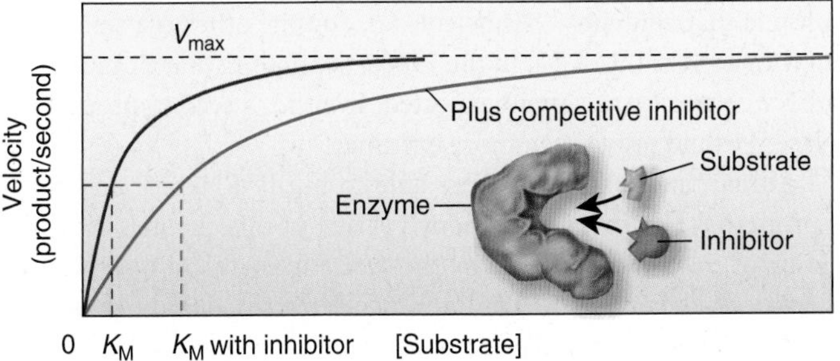

(b) Competitive inhibition

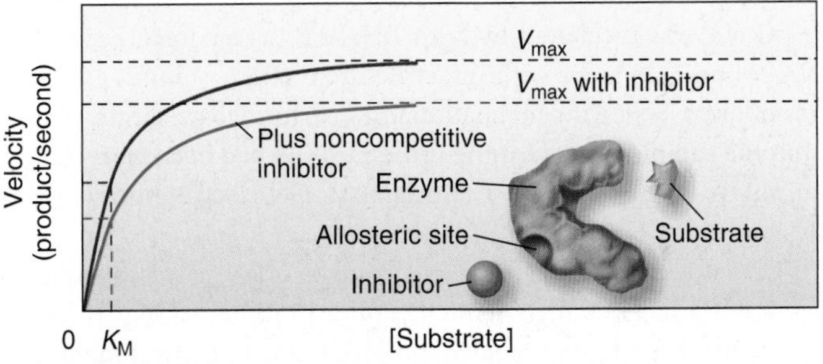

(c) Noncompetitive inhibition

Figure 6.7 **The relationship between velocity and substrate concentration in an enzyme-catalyzed reaction, and the effects of inhibitors.** **(a)** In the absence of an inhibitor, the maximal velocity (V_{max}) is achieved when the substrate concentration is high enough to be saturating. The K_M value is the substrate concentration at which the velocity is half the maximal velocity. **(b)** A competitive inhibitor binds to the active site of an enzyme and raises the K_M for the substrate. **(c)** A noncompetitive inhibitor binds to an allosteric site outside the active site and lowers the V_{max} for the reaction.

Concept Check: Enzyme A has a K_M of 0.1 mM, whereas enzyme B has a K_M of 1.0 mM. They both have the same V_{max}. If the substrate concentration was 0.5 mM, which reaction—the one catalyzed by enzyme A or B—would have the higher velocity?

Figure 6.7b, when competitive inhibitors are present, the apparent K_M for the substrate increases—a higher concentration of substrate is needed to achieve the same rate of the chemical reaction. In this case, the effects of the competitive inhibitor can be overcome by increasing the concentration of the substrate.

By comparison, Figure 6.7c illustrates the effects of a **noncompetitive inhibitor**. As seen here, this type of inhibitor lowers the V_{max} for the reaction without affecting the K_M. A noncompetitive inhibitor binds noncovalently to an enzyme at a location outside the active site, called an **allosteric site**, and inhibits the enzyme's function.

Irreversible Inhibitors Irreversible inhibitors usually bind covalently to an enzyme to inhibit its function. For example, some irreversible inhibitors bind covalently to an amino acid at the active site of an enzyme, thereby preventing it from catalyzing a chemical reaction. An example of an irreversible inhibitor is diisopropyl phosphorofluoridate (DIFP). DIFP is a type of nerve gas that was developed as a chemical weapon. This molecule covalently reacts with the enzyme, acetylcholinesterase, which is important for the proper functioning of neurons.

Irreversible inhibition is not a common way for cells to control enzyme function. Why do cells usually control enzymes via reversible inhibitors? The answer is that a reversible inhibitor allows an enzyme to be used again, when the inhibitor concentration becomes lowered. Being able to reuse an enzyme is energy-efficient. In contrast, irreversible inhibitors permanently inactivate an enzyme, thereby preventing its further use.

Additional Factors Influence Enzyme Function

Enzymes, which are proteins, sometimes require nonprotein molecules or ions to carry out their functions. **Prosthetic groups** are small molecules that are permanently attached to the surface of an enzyme and aid in enzyme function. **Cofactors** are usually inorganic ions, such as Fe^{3+} or Zn^{2+}, that temporarily bind to the surface of an enzyme and promote a chemical reaction. Finally, some enzymes use **coenzymes**, organic molecules that temporarily bind to an enzyme and participate in the chemical reaction but are left unchanged after the reaction is completed. Some of these coenzymes are synthesized by cells, but many of them are taken in as dietary vitamins by animal cells.

The ability of enzymes to increase the rate of a chemical reaction is also affected by the surrounding environment. In particular, the temperature, pH, and ionic conditions play an important role in the proper functioning of enzymes. Most enzymes function maximally in a narrow range of temperature and pH. For example, many human enzymes work best at 37°C (98.6°F), which is the normal body temperature. If the temperature is several degrees above or below an enzyme's optimum temperature due to infection or environmental causes, the function of many enzymes is greatly inhibited (Figure 6.8). Very high temperatures may denature a protein, causing it to unfold and lose its three-dimensional shape, thereby inhibiting its function.

Enzyme function is also sensitive to pH. Certain enzymes in the stomach function best at the acidic pH found in this organ. For example, pepsin is a protease—an enzyme that digests proteins into peptides—that is released into the stomach. The optimal pH for pepsin function is around pH 2.0, which is extremely acidic. By comparison, many cytosolic enzymes function optimally at a more neutral pH, such as pH 7.2, which is the pH normally found in the cytosol of human cells. If the pH was significantly above or below this value, enzyme function would be decreased for cytosolic enzymes.

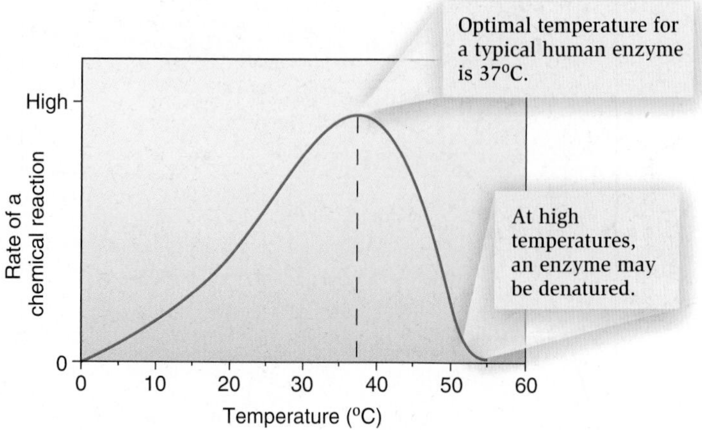

Figure 6.8 Effects of temperature on a typical human enzyme. Most enzymes function optimally within a narrow range of temperature. Many human enzymes function best at 37°C, which is body temperature.

Feature Investigation

The Discovery of Ribozymes by Sidney Altman Revealed That RNA Molecules May Also Function as Catalysts

Until the 1980s, scientists thought that all biological catalysts are proteins. One avenue of study that dramatically changed this view came from the analysis of ribonuclease P (RNase P), a catalyst initially found in the bacterium *Escherichia coli* and later identified in all species examined. RNase P is involved in the processing of tRNA molecules—a type of molecule required for protein synthesis. Such tRNA molecules are synthesized as longer precursor molecules called ptRNAs, which have 5′ and 3′ ends. (The 5′ and 3′ directionality of RNA molecules is described in Chapter 11.) RNase P breaks a covalent bond at a specific site in precursor tRNAs, which releases a fragment at the 5′ end and makes them shorter (Figure 6.9).

Sidney Altman and his colleagues became interested in the processing of tRNA molecules and turned their attention to RNase P in *E. coli*. During the course of their studies, they purified this enzyme and, to their surprise, discovered it has two subunits—one is an RNA molecule that contains 377 nucleotides, and the other is a small protein with a mass of 14 kDa. In the 1980s, the finding that a catalyst has an RNA subunit was very unexpected. Even so, a second property of RNase P would prove even more exciting.

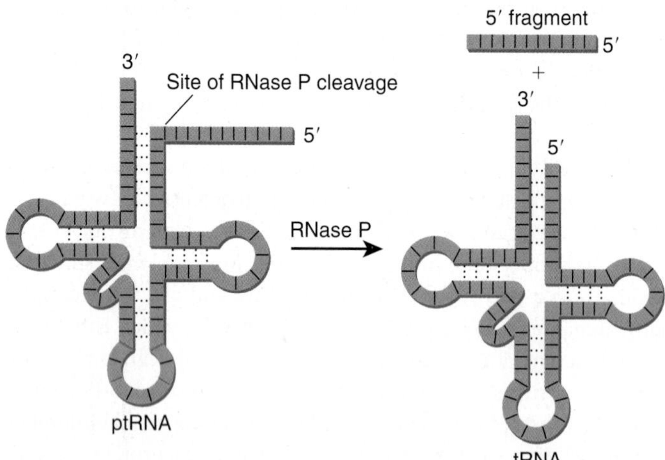

Figure 6.9 The function of RNase P. A specific bond in a precursor tRNA (ptRNA) is cleaved by RNase P, which releases a small fragment at the 5′ end. This results in the formation of a mature tRNA.

BioConnections: Look ahead to Figure 12.19. How do you imagine translation would be affected if RNase P did not work properly?

Altman and colleagues were able to purify RNase P and study its properties in vitro. As mentioned earlier in this chapter, the functioning of enzymes is affected by the surrounding conditions. Cecilia Guerrier-Takada in Altman's laboratory determined that magnesium ion (Mg^{2+}) had a stimulatory effect on RNase P function. In the experiment described in Figure 6.10, the effects of Mg^{2+} were studied in greater detail. The researchers analyzed the effects of low (10 mM $MgCl_2$) and high (100 mM $MgCl_2$) magnesium concentrations on the processing of a ptRNA. At low or high magnesium concentrations, the ptRNA was incubated without RNase P (as a control); with the RNA subunit alone; or with intact RNase P (RNA subunit and protein subunit). Following incubation, they performed gel electrophoresis on the samples to determine if the ptRNAs had been cleaved into two pieces—the tRNA and a 5′ fragment. Gel electrophoresis separates molecules on the basis of their masses.

Let's now look at the data. As a control, ptRNAs were incubated with low (lane 1) or high (lane 4) concentrations of $MgCl_2$ in the absence of RNase P. As expected, no processing to a lower molecular mass tRNA was observed. When the RNA subunit alone was incubated with ptRNA molecules in the presence of low $MgCl_2$ (lane 2), no processing occurred, but it did occur if the protein subunit was also included (lane 3).

The surprising result is shown in lane 5, in which the ptRNA was incubated with the RNA subunit alone in the presence of a high concentration of $MgCl_2$. The RNA subunit by itself was sufficient to cleave the ptRNA to a smaller tRNA and a 5′ fragment! Presumably, the high $MgCl_2$ concentration helps to keep the RNA subunit in a conformation that is catalytically active. Alternatively, the protein subunit plays a similar role in a living cell.

Subsequent work confirmed these observations and showed that the RNA subunit of RNase P is a true catalyst—it accelerates the rate of a chemical reaction and is not permanently altered by it. Around

Figure 6.10 The discovery that the RNA subunit of RNase P is a catalyst.

HYPOTHESIS The catalytic function of RNase P could be carried out by its RNA subunit or by its protein subunit.

KEY MATERIALS Purified precursor tRNA (ptRNA) and purified RNA and protein subunits of RNase P from *E. coli.*

Experimental level | Conceptual level

1 Into each of five tubes, add ptRNA.

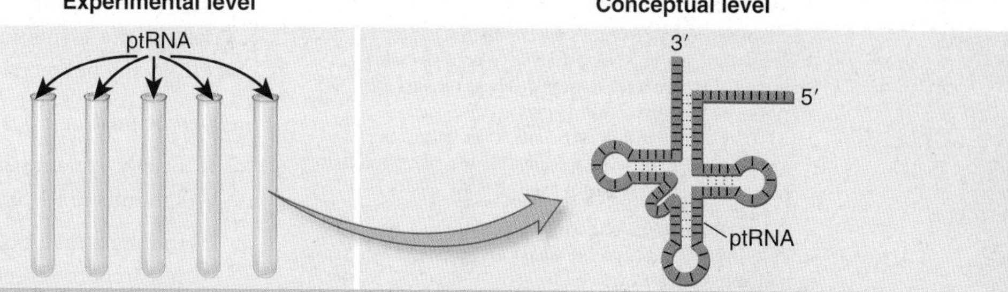

2 In tubes 1–3, add a low concentration of $MgCl_2$; in tubes 4 and 5, add a high $MgCl_2$ concentration.

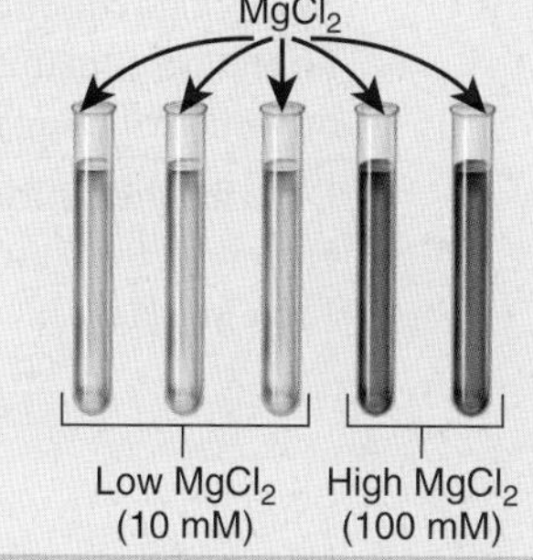

3 Into tubes 2 and 5, add the RNA subunit of RNase P alone; into tube 3, add both the RNA subunit and the protein subunit of RNase P. Incubate to allow digestion to occur. Note: Tubes 1 and 4 are controls that have no added subunits of RNase P.

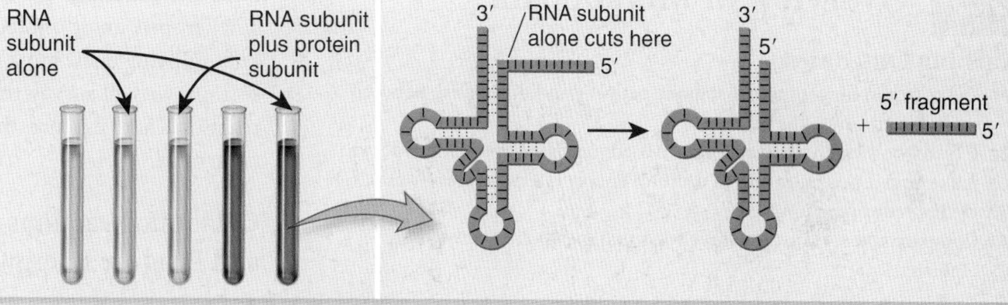

4 Carry out gel electrophoresis on each sample. In this technique, samples are loaded into a well on a gel and exposed to an electric field as described in Chapter 20. The molecules move toward the bottom of the gel and are separated according to their masses: Molecules with higher masses are closer to the top of the gel. The gel is exposed to ethidium bromide, which stains RNA.

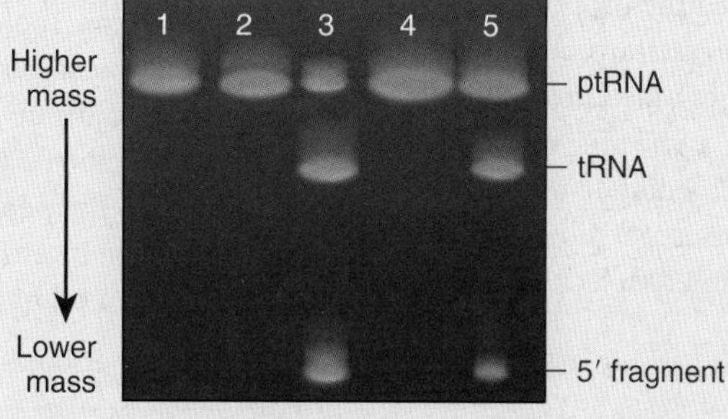

Catalytic function will result in the digestion of ptRNA into tRNA and a smaller 5′ fragment.

5 **THE DATA**

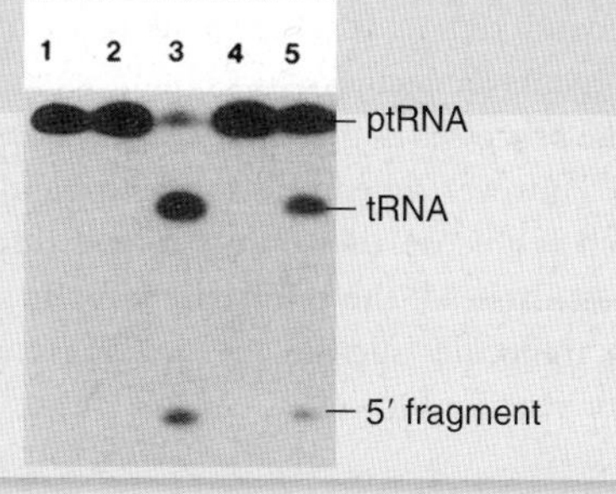

6 **CONCLUSION** The RNA subunit alone can catalyze the breakage of a covalent bond in ptRNA at high Mg concentrations. It is a ribozyme.

7 **SOURCE** Altman, S. 1990. Enzymatic cleavage of RNA by RNA. *Bioscience Reports* 10:317–337.

Table 6.3 Types of Ribozymes

General function	Biological examples
Processing of RNA molecules	1. RNase P: As described in this chapter, RNase P cleaves precursor tRNA molecules (ptRNAs) to a mature tRNAs.
	2. Spliceosomal RNA: As described in Chapter 12, eukaryotic pre-mRNAs often have regions called introns that are later removed. These introns are removed by a spliceosome that is composed of RNA and protein subunits. The RNA within the spliceosome is believed to function as a ribozyme that removes the introns from pre-mRNA.
	3. Certain introns found in mitochondrial, chloroplast, and bacterial RNAs are removed by a self-splicing mechanism.
Synthesis of polypeptides	The ribosome has an RNA component that catalyzes the formation of covalent bonds between adjacent amino acids during polypeptide synthesis.

the same time, Thomas Cech and colleagues determined that a different RNA molecule found in the protist *Tetrahymena thermophila* also has catalytic activity. The term **ribozyme** is now used to describe an RNA molecule that catalyzes a chemical reaction. In 1989, Altman and Cech received the Nobel Prize in Chemistry for their discovery of ribozymes.

Since the pioneering work of Altman and Cech, researchers have discovered that ribozymes play key catalytic roles in cells (**Table 6.3**). They are primarily involved in the processing of RNA molecules from precursor to mature forms. In addition, a ribozyme in the ribosome catalyzes the formation of covalent bonds between adjacent amino acids during polypeptide synthesis.

Experimental Questions

1. Briefly explain why it was necessary to purify the individual subunits of RNase P to show that it is a ribozyme.
2. In the Altman experiment involving RNase P, explain how the researchers experimentally determined if RNase P or subunits of RNase P were catalytically active or not. Explain why the researchers conducted experiments in which they measured the formation of mature tRNA without adding the protein subunit or without adding the RNA subunit.
3. Describe the critical results that showed RNase P is a ribozyme. How does the concentration of Mg^{2+} affect the function of the RNA in RNase P?

6.3 Overview of Metabolism

Learning Outcomes:

1. Describe the concept of a metabolic pathway and distinguish between catabolic and anabolic reactions.
2. Explain how catabolic reactions are used to generate building blocks to make larger molecules and to produce energy intermediates.
3. Define redox reaction.
4. Compare and contrast the three ways that metabolic pathways are regulated.

In the previous sections, we examined the underlying factors that govern individual chemical reactions and explored the properties of enzymes and ribozymes. In living cells, chemical reactions are often coordinated with each other and occur in sequences called **metabolic pathways**, each step of which is catalyzed by a specific enzyme (**Figure 6.11**). These pathways are categorized according to whether the reactions lead to the breakdown or synthesis of substances. **Catabolic reactions** result in the breakdown of larger molecules into smaller ones. Such reactions are often exergonic. By comparison, **anabolic reactions** involve the synthesis of larger molecules from smaller precursor molecules. This process usually is endergonic and, in living cells, must be coupled to an exergonic reaction. In this section, we will survey the general features of catabolic and anabolic reactions and explore the ways in which metabolic pathways are controlled.

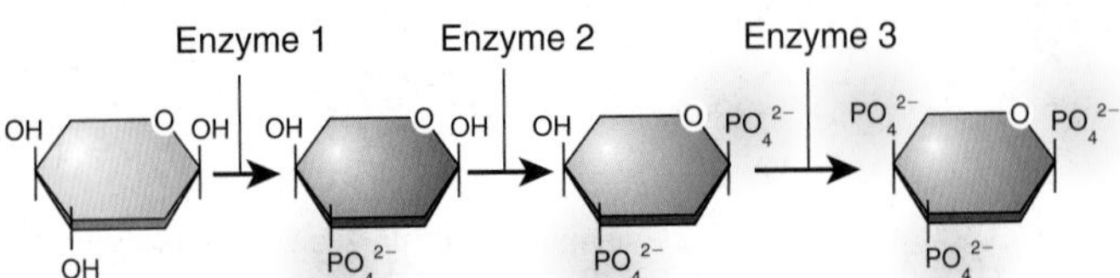

Figure 6.11 A metabolic pathway. In this metabolic pathway, a series of different enzymes catalyze the attachment of phosphate groups to various sugars, beginning with a starting substrate and ending with a final product.

Catabolic Reactions Recycle Organic Building Blocks and Produce Energy Intermediates Such as ATP

Catabolic reactions result in the breakdown of larger molecules into smaller ones. One reason for the breakdown of macromolecules is to recycle their organic molecules, which are used as building blocks to construct new molecules and macromolecules. For example, polypeptides, which make up proteins, are composed of a linear sequence of amino acids. When a protein is improperly folded or is no longer needed by a cell, the peptide bonds between the amino acids in the protein are broken by enzymes called proteases. This generates amino acids that can be used in the construction of new proteins.

$$\text{Protein} \xrightarrow{\text{Proteases}} \rightarrow \rightarrow \rightarrow \rightarrow \rightarrow \rightarrow \rightarrow \rightarrow \rightarrow \text{Many individual amino acids}$$

We will consider the mechanisms of recycling later in Section 6.4.

A second reason for the breakdown of macromolecules and smaller organic molecules is to obtain energy that is used to drive endergonic processes in the cell. Covalent bonds store a large amount of energy. However, when cells break covalent bonds in organic molecules such as glucose, they do not directly use the energy released in this process. Instead, the released energy is stored in **energy intermediates**, molecules such as ATP, which are directly used to drive endergonic reactions in cells.

As an example, let's consider the breakdown of glucose into two molecules of pyruvate. As discussed in Chapter 7, the breakdown of glucose to pyruvate involves a catabolic pathway called glycolysis. Some of the energy released during the breakage of covalent bonds in glucose is harnessed to synthesize ATP. However, this does not occur in a single step. Rather, glycolysis involves a series of steps in which covalent bonds are broken and rearranged. This process produces molecules that can readily donate a phosphate group to ADP, thereby producing ATP. For example, phosphoenolpyruvate has a phosphate group attached to pyruvate. Due to the arrangement of bonds in phosphoenolpyruvate, this phosphate bond is unstable and easily broken. Therefore, the phosphate can be readily transferred from phosphoenolpyruvate to ADP:

Unstable phosphate bond

O^-–C=O–CO–P=CH_2 (Phosphoenolpyruvate) + 2 ADP → (Pyruvate kinase) → 2 ATP + O^-–C=O–C=O–CH_3 (Pyruvate) $\Delta G = -7.5$ kcal/mol

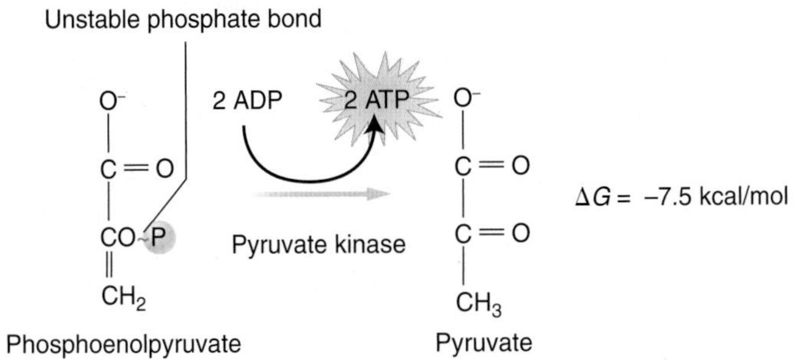

This is an exergonic reaction and therefore favors the formation of products. In this step of glycolysis, the breakdown of an organic molecule, namely phosphoenolpyruvate, results in the formation of pyruvate and the synthesis of an energy intermediate molecule, ATP, which can then be used by a cell to drive endergonic reactions. This way of synthesizing ATP, termed **substrate-level phosphorylation**, occurs when an enzyme directly transfers a phosphate from an organic molecule to ADP, thereby making ATP. In this case, the enzyme pyruvate kinase transfers a phosphate from phosphoenolpyruvate to ADP. Another way to make ATP is via **chemiosmosis**. In this process, energy stored in an ion electrochemical gradient is used to make ATP from ADP and P_i. We will consider this other mechanism in Chapter 7.

Redox Reactions Are Important in the Metabolism of Small Organic Molecules

During the breakdown of small organic molecules, **oxidation**—the removal of one or more electrons from an atom or molecule—may occur. This process is called oxidation because oxygen is frequently involved in chemical reactions that remove electrons from other molecules. By comparison, **reduction** is the addition of electrons to an atom or molecule. Reduction is so named because the addition of a negatively charged electron reduces the net charge of a molecule.

Electrons do not exist freely in solution. When an atom or molecule is oxidized, the electron that is removed must be transferred to another atom or molecule, which becomes reduced. This type of reaction is termed a **redox reaction**, which is short for a reduction-oxidation reaction. As a generalized equation, an electron may be transferred from molecule A to molecule B as follows:

$$Ae^- + B \rightarrow A + Be^-$$
(oxidized) (reduced)

As shown in the right side of this reaction, A has been oxidized (that is, had an electron removed), and B has been reduced (that is, had an electron added). In general, a substance that has been oxidized has less energy, whereas a substance that has been reduced has more energy.

During the oxidation of organic molecules such as glucose, the electrons are used to produce energy intermediates such as NADH (Figure 6.12). In this process, an organic molecule has been oxidized, and **NAD^+ (nicotinamide adenine dinucleotide)** has been reduced to NADH. Cells use NADH in two common ways. First, as we will see in Chapter 7, the oxidation of NADH is a highly exergonic reaction that can be used to make ATP. Second, NADH can donate electrons to other organic molecules and thereby energize them. Such energized molecules can more readily form covalent bonds. Therefore, as described next, NADH is often needed in reactions that involve the synthesis of larger molecules through the formation of covalent bonds between smaller molecules.

Anabolic Reactions Require an Input of Energy to Make Larger Molecules

Anabolic reactions are also called **biosynthetic reactions**, because they are necessary to make larger molecules and macromolecules. We will examine the synthesis of macromolecules in several chapters of this textbook. For example, RNA and protein biosynthesis are described in Chapter 12. Cells also need to synthesize small organic molecules, such as amino acids and fats, if they are not readily available from food sources. Such molecules are made by the formation of covalent linkages between precursor molecules. For example, glutamate (an amino acid) is made by the covalent linkage between α-ketoglutarate (a product of sugar metabolism) and ammonium (NH_4^+).

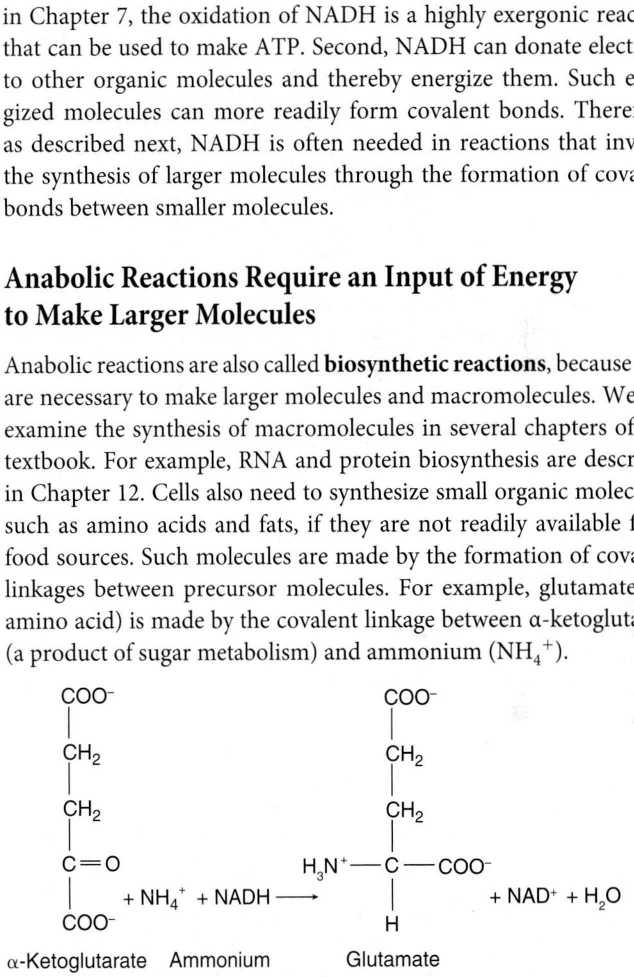

COO^-–CH_2–CH_2–C=O–COO^- (α-Ketoglutarate) + NH_4^+ (Ammonium) + NADH → COO^-–CH_2–CH_2–H_3N^+–C(H)–COO^- (Glutamate) + NAD^+ + H_2O

Subsequently, another amino acid, glutamine, is made from glutamate and ammonium.

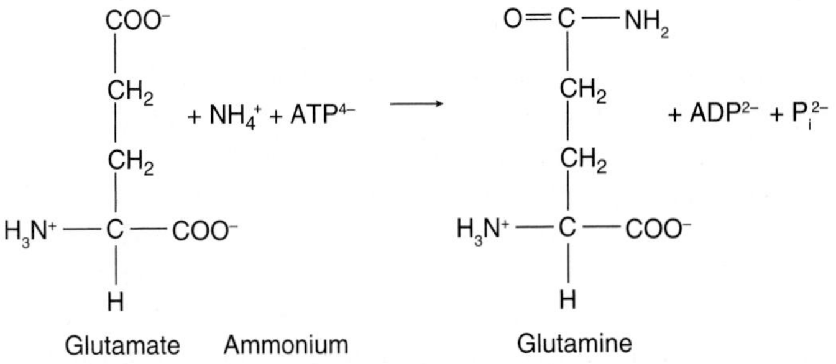

COO^-–CH_2–CH_2–H_3N^+–C(H)–COO^- (Glutamate) + NH_4^+ (Ammonium) + ATP^{4-} → O=C–NH_2–CH_2–CH_2–H_3N^+–C(H)–COO^- (Glutamine) + ADP^{2-} + P_i^{2-}

In both reactions, an energy intermediate molecule such as NADH or ATP is needed to drive the reaction forward.

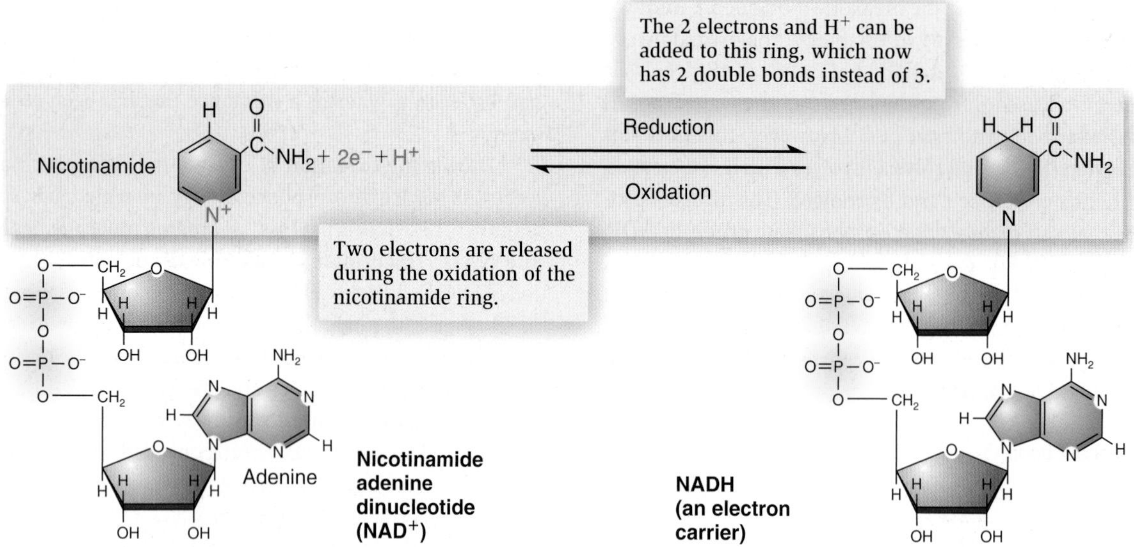
The 2 electrons and H+ can be added to this ring, which now has 2 double bonds instead of 3.
Nicotinamide
Reduction
Oxidation
Two electrons are released during the oxidation of the nicotinamide ring.
Adenine
Nicotinamide adenine dinucleotide (NAD+)
NADH (an electron carrier)

Figure 6.12 **The reduction of NAD+ to produce NADH.** NAD+ is composed of two nucleotides, one with an adenine base and one with a nicotinamide base. The oxidation of organic molecules releases electrons that bind to NAD+ (and along with a hydrogen ion) result in the formation of NADH. The two electrons and H+ are incorporated into the nicotinamide ring. Note: The actual net charges of NAD+ and NADH are minus one and minus two, respectively. They are designated NAD+ and NADH to emphasize the net charge of the nicotinamide ring, which is involved in oxidation-reduction reactions.

Concept Check: *Which is the oxidized form, NAD+ or NADH?*

Metabolic Pathways Are Regulated in Three General Ways

The regulation of metabolic pathways is important for a variety of reasons. Catabolic pathways are regulated so organic molecules are broken down only when they are no longer needed or when the cell requires energy. During anabolic reactions, regulation ensures that a cell synthesizes molecules only when they are needed. The regulation of catabolic and anabolic pathways occurs at the genetic, cellular, and biochemical levels.

Gene Regulation Because enzymes in every metabolic pathway are encoded by genes, one way that cells control chemical reactions is via gene regulation. For example, if a bacterial cell is not exposed to a particular sugar in its environment, it will turn off the genes that encode the enzymes that are needed to break down that sugar. Alternatively, if the sugar becomes available, the genes are switched on. Chapter 13 examines the steps of gene regulation in detail.

Cellular Regulation Metabolism is also coordinated at the cellular level. Cells integrate signals from their environment and adjust their chemical reactions to adapt to those signals. As discussed in Chapter 9, cell-signaling pathways often lead to the activation of protein kinases—enzymes that covalently attach a phosphate group to target proteins. For example, when people are frightened, they secrete a hormone called epinephrine into their bloodstream. This hormone binds to the surface of muscle cells and stimulates an intracellular pathway that leads to the phosphorylation of several intracellular proteins, including enzymes involved in carbohydrate metabolism. These activated enzymes promote the breakdown of carbohydrates, an event that supplies the frightened individual with more energy. Epinephrine is sometimes called the "fight-or-flight" hormone because the added energy prepares an individual to either stay and fight or run away quickly. After a person is no longer frightened, hormone levels drop, and other enzymes called phosphatases remove the phosphate groups from enzymes, thereby restoring the original level of carbohydrate metabolism.

Another way that cells control metabolic pathways is via compartmentalization. The membrane-bound organelles of eukaryotic cells, such as the endoplasmic reticulum and mitochondria, serve to compartmentalize the cell. As discussed in Chapter 7, this allows specific metabolic pathways to occur in one compartment in the cell but not in others.

Biochemical Regulation A third and very prominent way that metabolic pathways are controlled is at the biochemical level. In this case, the noncovalent binding of a molecule to an enzyme directly regulates its function. As discussed earlier, one form of biochemical regulation involves the binding of molecules such as competitive or noncompetitive inhibitors (see Figure 6.7). An example of noncompetitive inhibition is a type of regulation called **feedback inhibition**, in which the product of a metabolic pathway inhibits an enzyme that acts early in the pathway, thus preventing the overaccumulation of the product (**Figure 6.13**).

Many metabolic pathways use feedback inhibition as a form of biochemical regulation. In such cases, the inhibited enzyme has two binding sites. One site is the active site, where the reactants are converted to products. In addition, enzymes controlled by feedback inhibition also have an allosteric site, where a molecule can bind noncovalently and affect the function of the active site. The binding of a

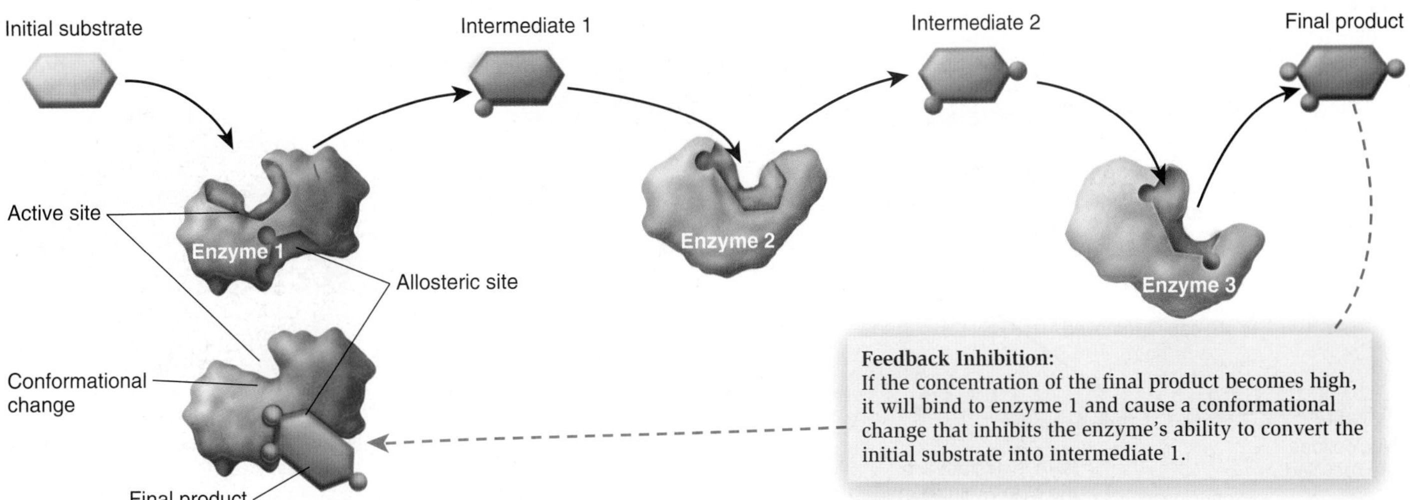
Initial substrate
Intermediate 1
Intermediate 2
Final product
Active site
Enzyme 1
Allosteric site
Enzyme 2
Enzyme 3
Conformational change
Final product
Feedback Inhibition:
If the concentration of the final product becomes high, it will bind to enzyme 1 and cause a conformational change that inhibits the enzyme's ability to convert the initial substrate into intermediate 1.

Figure 6.13 Feedback inhibition. In this process, the final product of a metabolic pathway inhibits an enzyme that functions in the pathway, thereby preventing the overaccumulation of the product.

BIOLOGY PRINCIPLE Living organisms maintain homeostasis. In this example, feedback inhibition prevents the formation of too much product of a metabolic pathway.

BioConnections: Look ahead to Figure 7.3, which describes a metabolic pathway called glycolysis. Feedback inhibition occurs such that high levels of ATP inhibit phosphofructokinase, an enzyme that catalyzes the conversion of fructose-6-phosphate and ATP to fructose 1,6-bisphosphate and ADP. How is this beneficial to the cell?

molecule to an allosteric site causes a conformational change in the enzyme that inhibits its catalytic function. Allosteric sites are often found in the enzymes that catalyze the early steps in a metabolic pathway. Such allosteric sites typically bind molecules that are the products of the metabolic pathway. When the products bind to these sites, they inhibit the function of these enzymes, thereby preventing the formation of too much product.

Cellular and biochemical regulation are important and rapid ways to control chemical reactions in a cell. For a metabolic pathway composed of several enzymes, which enzyme in a pathway should be controlled? In many cases, a metabolic pathway has a **rate-limiting step**, which is the slowest step in a pathway. If the rate-limiting step is inhibited or enhanced, such changes will have the greatest influence on the formation of the product of the metabolic pathway. Rather than affecting all of the enzymes in a metabolic pathway, cellular and biochemical regulation are often directed at the enzyme that catalyzes the rate-limiting step. This is an efficient and rapid way to control the amount of product of a pathway.

6.4 Recycling of Organic Molecules

Learning Outcomes:

1. Evaluate the relationship between the recycling of organic molecules and cellular efficiency.
2. Outline how the building blocks of mRNAs and proteins are recycled.
3. Describe how the components of cellular organelles are recycled via autophagy.

As mentioned earlier in this chapter, another important feature of metabolism is the recycling of organic molecules, such as nucleotides and amino acids, which are the building blocks of larger molecules. Except for DNA, which is stably maintained and inherited from cell to cell, other large molecules such as RNA, proteins, lipids, and polysaccharides typically exist for a relatively short period of time. Biologists often speak of the **half-life** of molecules, which is the time it takes for 50% of the molecules to be broken down and recycled. For example, a population of messenger RNA molecules in bacteria has an average half-life of about 5 minutes, whereas mRNAs in eukaryotes tend to exist for longer periods of time, on the order of 30 minutes to 24 hours or even several days.

Why is recycling important? To compete effectively in their native environments, all living organisms must efficiently use and recycle the organic molecules that are needed as building blocks to construct larger molecules and macromolecules. Otherwise, they would waste a great deal of energy making such building blocks. For example, organisms conserve an enormous amount of energy by reusing the amino acids that are needed to construct proteins. In this section, we will explore how nucleotides and amino acids are recycled and consider a mechanism for the recycling of materials found in an entire organelle.

The Nucleotides Within Messenger RNA Molecules in Eukaryotes Are Recycled by Two Mechanisms

The genome of every cell contains many genes that are transcribed into RNA. Most of these RNA molecules, called messenger RNA, or mRNA, encode proteins that ultimately determine the structure and function of cells. As described in Chapter 12, eukaryotic mRNAs contain a cap at their 5′ end. A tail is found at their 3′ end consisting of many adenine bases (look ahead to Figure 12.9). In most cases, degradation of mRNA begins with the removal of nucleotides in the poly A tail at the 3′ end (**Figure 6.14**). After the tail gets shorter, two mechanisms of degradation may occur.

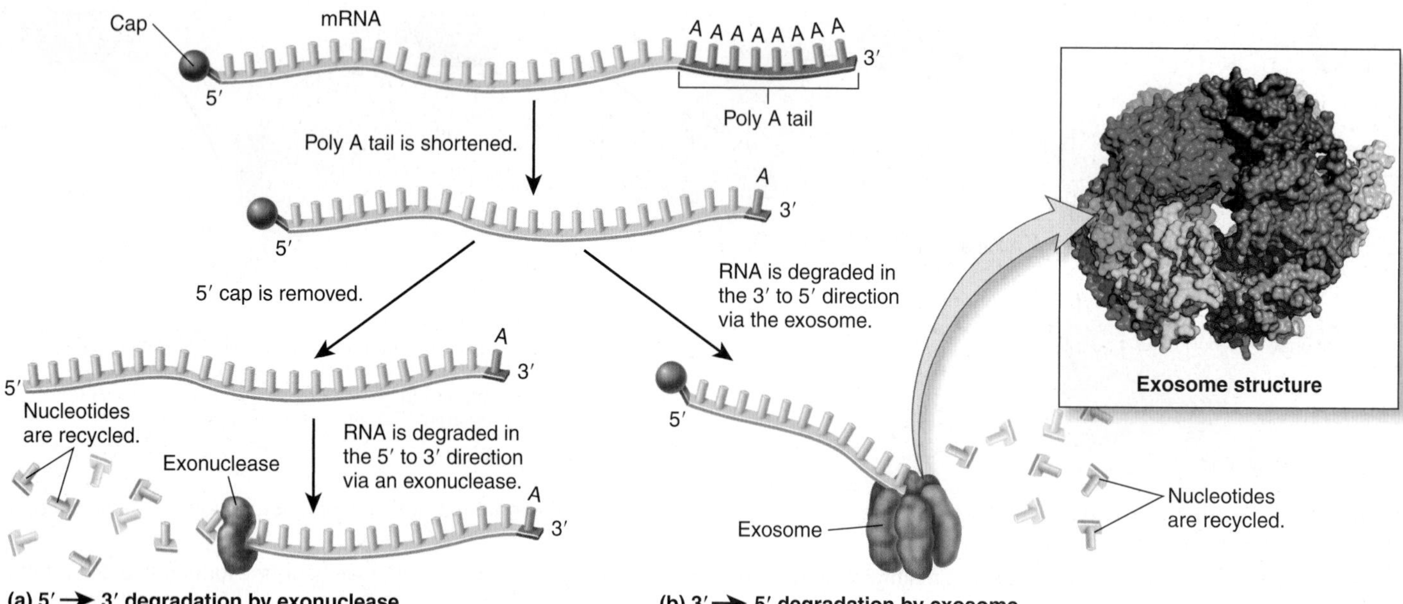

Figure 6.14 **Two pathways for mRNA degradation in eukaryotic cells.** Degradation usually begins with a shortening of the poly A tail. After tail shortening, either **(a)** the 5′ cap is removed and the RNA is degraded in a 5′ to 3′ direction by an exonuclease, or **(b)** the mRNA is degraded in the 3′ to 5′ direction via an exosome. The reason why cells have two different mechanisms for RNA degradation is not well understood.

In one mechanism, the 5′ cap is removed, and the mRNA is degraded by an **exonuclease**—an enzyme that cleaves off nucleotides, one at a time. In this case, the exonuclease removes nucleotides starting at the 5′ end and moving toward the 3′. The nucleotides can then be reused to make new RNA molecules.

The other mechanism involves mRNA being degraded by an **exosome**, a multiprotein complex discovered in 1997. Exosomes are found in eukaryotic cells and some archaeal cells, whereas in bacterial cells a simpler complex called the degradosome carries out a similar function. The core of the exosome is a six-membered protein ring to which other proteins are attached (see inset to Figure 6.14). Certain proteins within the exosome are exonucleases that degrade the mRNA starting at the 3′ end and moving toward the 5′ end, thereby releasing nucleotides that can be recycled.

Proteins in Eukaryotes and Archaea Are Broken Down in the Proteasome

Cells continually degrade proteins that are faulty or no longer needed. To be degraded, proteins are recognized by **proteases**—enzymes that cleave the bonds between adjacent amino acids. The primary pathway for protein degradation in archaea and eukaryotic cells is via a protein complex called a **proteasome**. Similar to the exosome, which has a central cavity surrounded by a ring of proteins, the core of the proteasome consists of four stacked rings, each composed of seven protein subunits (Figure 6.15a). The proteasomes of eukaryotic cells also contain caps at each end that control the entry of proteins into the proteasome.

Figure 6.15b describes the steps of protein degradation via eukaryotic proteasomes. A string of proteins called **ubiquitin** is covalently attached to the target protein. This event directs the target protein to a proteasome cap, which has binding sites for ubiquitin. The cap also has enzymes that unfold the protein and inject it into the internal cavity of the proteasome core. The ubiquitin proteins are removed during entry and released to the cytosol for reuse. Inside the proteasome, proteases degrade the protein into small peptides and amino acids. The process is completed when the peptides and amino acids are recycled back into the cytosol. The amino acids can be reused to make new proteins.

Ubiquitin targeting has two functions. First, the enzymes that attach ubiquitin to its target recognize improperly folded proteins, allowing cells to identify and degrade nonfunctional proteins. Second, changes in cellular conditions may warrant the rapid breakdown of particular proteins. For example, cell division requires a series of stages called the cell cycle, which depends on the degradation of specific proteins. After these proteins perform their functions in the cycle, ubiquitin targeting directs them to the proteasome for degradation.

Autophagy Recycles the Contents of Entire Organelles

As described in Chapter 4, lysosomes contain many different types of acid hydrolases that break down proteins, carbohydrates, nucleic acids, and lipids. This enzymatic function enables lysosomes to break down complex materials. One function of lysosomes involves the digestion of substances that are taken up from outside the cell. This process, called endocytosis, is described in Chapter 5. In addition, lysosomes help digest intracellular materials. In a process known as **autophagy** (from the Greek, meaning eating one's self), cellular material, such as a worn-out organelle, becomes enclosed in a double

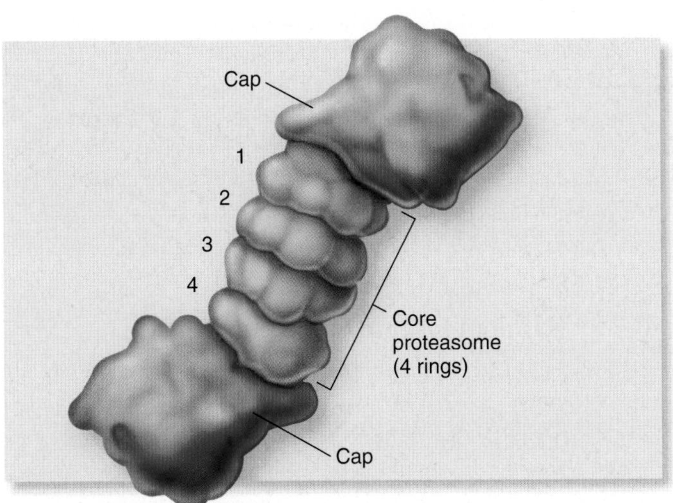

(a) Structure of the eukaryotic proteasome

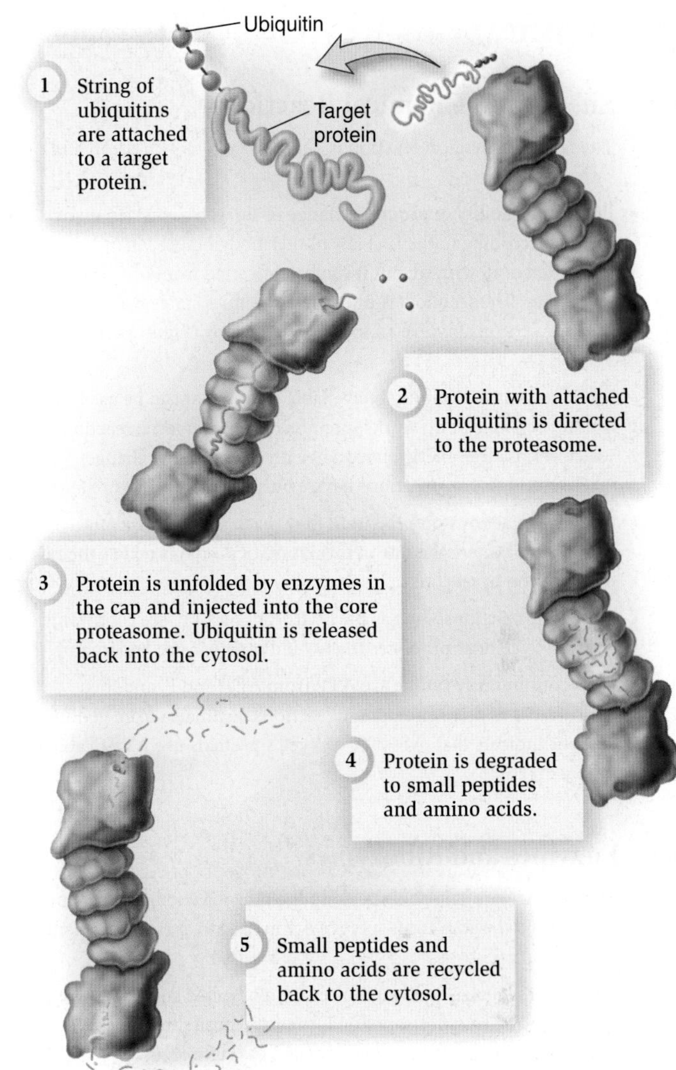

(b) Steps of protein degradation in eukaryotic cells

Figure 6.15 Protein degradation via the proteasome.

Concept Check: *What are advantages of protein degradation?*

membrane (Figure 6.16). This double membrane is formed from a tubule that elongates and eventually wraps around the organelle to form an **autophagosome**. The autophagosome then fuses with a lysosome, and the material inside the autophagosome is digested. The small molecules released from this digestion are recycled back into the cytosol.

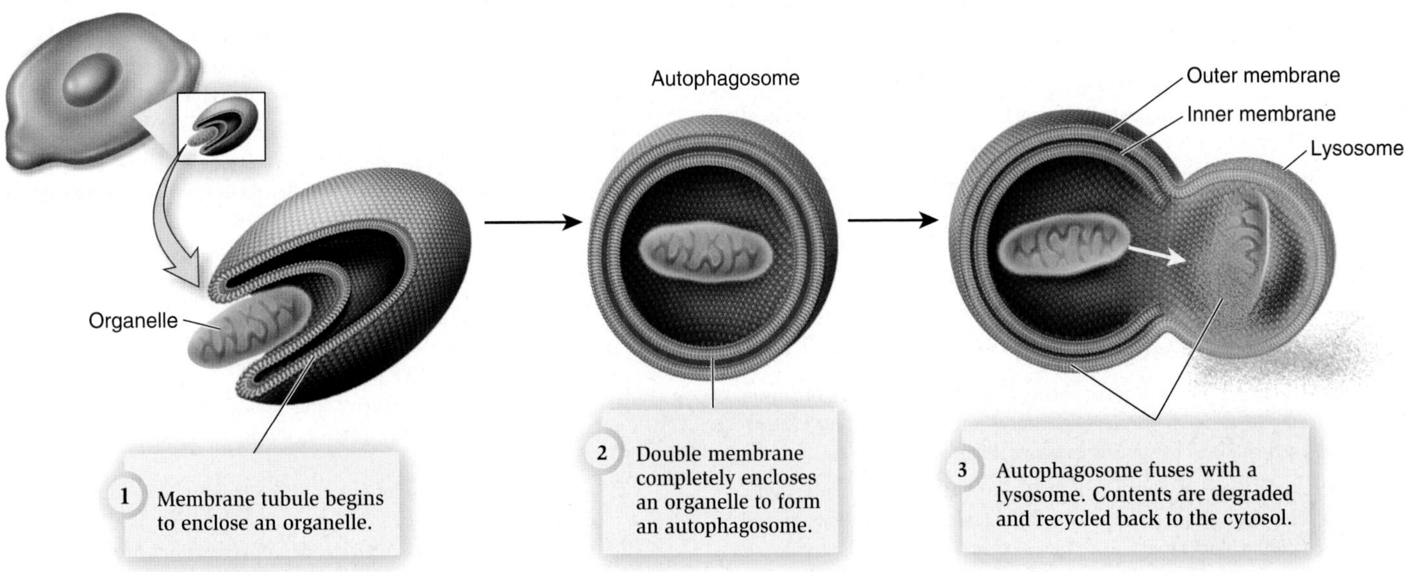

Organelle
1 Membrane tubule begins to enclose an organelle.
Autophagosome
2 Double membrane completely encloses an organelle to form an autophagosome.
Outer membrane
Inner membrane
Lysosome
3 Autophagosome fuses with a lysosome. Contents are degraded and recycled back to the cytosol.

Figure 6.16 Autophagy.

Summary of Key Concepts

6.1 Energy and Chemical Reactions

- The fate of a chemical reaction is determined by its direction and rate.
- Energy, the ability to promote change or do work, exists in many forms. According to the first law of thermodynamics, energy cannot be created or destroyed, but it can be converted from one form to another. The second law of thermodynamics states that energy interconversions involve an increase in entropy (Figures 6.1, 6.2, Table 6.1).
- Free energy is the amount of available energy that can be used to promote change or do work. Spontaneous or exergonic reactions, which release free energy, have a negative free-energy change, whereas endergonic reactions have a positive change (Figure 6.3).
- Chemical reactions proceed until they reach a state of chemical equilibrium, where the rate of formation of products equals the rate of formation of reactants.
- Exergonic reactions, such as the hydrolysis of ATP, are commonly coupled to cellular processes that would otherwise be endergonic.
- Cells continuously synthesize ATP from ADP and P_i and then hydrolyze it to drive endergonic reactions. Estimates from genome analysis indicate that over 20% of a cell's proteins use ATP (Table 6.2, Figure 6.4).

6.2 Enzymes and Ribozymes

- Enzymes are proteins that speed up the rate of a chemical reaction by lowering the activation energy (E_A) needed to achieve a transition state (Figure 6.5).
- Enzymes recognize reactant molecules, also called substrates, with a high specificity. Conformational changes cause substrates to bind more tightly to enzymes, called induced fit (Figure 6.6).
- Each enzyme-catalyzed reaction exhibits a maximal velocity (V_{max}). The K_M is the substrate concentration at which the velocity of the chemical reaction is half of the V_{max} value. Competitive inhibitors raise the apparent K_M for the substrate, whereas noncompetitive inhibitors lower the V_{max} (Figure 6.7).
- Enzyme function may be affected by a variety of other factors, including prosthetic groups, cofactors, coenzymes, temperature, and pH (Figure 6.8).
- Altman and colleagues discovered that the RNA subunit within RNase P is a ribozyme, an RNA molecule that catalyzes a chemical reaction. Other ribozymes play key roles in the cell (Figures 6.9, 6.10, Table 6.3).

6.3 Overview of Metabolism

- Metabolism is the sum of the chemical reactions in a living organism. Metabolic pathways consist of chemical reactions that occur in steps and are catalyzed by specific enzymes (Figure 6.11).
- Catabolic reactions involve the breakdown of larger molecules into smaller ones. These reactions recycle organic molecules that are used as building blocks to make new molecules. The organic molecules are also broken down to make energy intermediates such as ATP.
- Some chemical reactions are redox reactions in which electrons are transferred from one molecule to another. These can be used to make energy intermediates such as NADH (Figure 6.12).
- Anabolic reactions require an input of energy to synthesize larger molecules and macromolecules.
- Metabolic pathways are controlled by gene regulation, cellular regulation, and biochemical regulation. An example of biochemical regulation is feedback inhibition (Figure 6.13).

6.4 Recycling of Organic Molecules

- Recycling of organic molecules saves a great deal of energy for living organisms.
- Messenger RNAs in eukaryotes are degraded by 5′ to 3′ exonucleases or by exosomes (Figure 6.14).
- Proteins in eukaryotes and archaea are degraded by proteasomes (Figure 6.15).
- Lysosomes digest intracellular material through the process of autophagy (Figure 6.16).

Assess and Discuss

Test Yourself

1. According to the second law of thermodynamics,
   a. energy cannot be created or destroyed.
   b. each energy transfer decreases the disorder of a system.
   c. energy is constant in the universe.
   d. each energy transfer increases the disorder of a system.
   e. chemical energy is a form of potential energy.
2. Reactions that release free energy are
   a. exergonic.
   b. spontaneous.
   c. endergonic.
   d. endothermic.
   e. both a and b.
3. Enzymes speed up reactions by
   a. providing chemical energy to fuel a reaction.
   b. lowering the activation energy necessary to initiate the reaction.
   c. causing an endergonic reaction to become an exergonic reaction.
   d. substituting for one of the reactants necessary for the reaction.
   e. none of the above.
4. Which of the following factors may alter the function of an enzyme?
   a. pH
   b. temperature
   c. cofactors
   d. all of the above
   e. b and c only
5. In biological systems, ATP functions by
   a. providing the energy to drive endergonic reactions.
   b. acting as an enzyme and lowering the activation energy of certain reactions.
   c. adjusting the pH of solutions to maintain optimal conditions for enzyme activity.
   d. regulating the speed at which endergonic reactions proceed.
   e. interacting with enzymes as a cofactor to stimulate chemical reactions.

6. In a chemical reaction, NADH is converted to $NAD^+ + H^+$. We would say that NADH has been
   a. reduced.
   b. phosphorylated.
   c. oxidized.
   d. decarboxylated.
   e. methylated.
7. Currently, scientists are identifying proteins that use ATP as an energy source by
   a. determining whether those proteins function in anabolic or catabolic reactions.
   b. determining if the protein has a known ATP-binding site.
   c. predicting the free energy necessary for the protein to function.
   d. determining if the protein has an ATP synthase subunit.
   e. all of the above.
8. With regard to its effects on an enzyme-catalyzed reaction, a competitive inhibitor
   a. lowers the K_M only.
   b. lowers the K_M and lowers the V_{max}.
   c. raises the K_M only.
   d. raises the K_M and lowers the V_{max}.
   e. raises the K_M and raises the V_{max}.
9. In eukaryotes, mRNAs may be degraded by
   a. a 5′ to 3′ exonuclease.
   b. the exosome.
   c. the proteasome.
   d. all of the above.
   e. a and b only.
10. Autophagy provides a way for cells to
    a. degrade entire organelles and recycle their components.
    b. control the level of ATP.
    c. engulf bacterial cells.
    d. export unwanted organelles out of the cell.
    e. inhibit the first enzyme in a metabolic pathway.

Conceptual Questions

1. With regard to rate and direction, discuss the differences between endergonic and exergonic reactions.
2. Describe the mechanism and purpose of feedback inhibition in a metabolic pathway.
3. 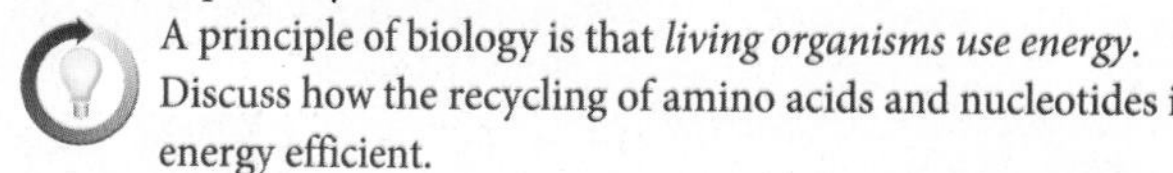
   A principle of biology is that *living organisms use energy.* Discuss how the recycling of amino acids and nucleotides is energy efficient.

Collaborative Questions

1. Living cells are highly ordered units, yet the universe is heading toward higher entropy. Discuss how life can maintain its order in spite of the second law of thermodynamics. Are we defying this law?
2. What is the advantage of using ATP as a common energy source? Another way of asking this question is, "Why does ATP provide an advantage over using a bunch of different food molecules?" For example, instead of just having a Na^+/K^+-ATPase in a cell, why not have many different ion pumps, each driven by a different food molecule, like a Na^+/K^+-glucosase (a pump that uses glucose), a Na^+/K^+-sucrase (a pump that uses sucrose), a Na^+/K^+-fatty acidase (a pump that uses fatty acids), and so on?

Online Resource

www.brookerbiology.com

Stay a step ahead in your studies with animations that bring concepts to life and practice tests to assess your understanding. Your instructor may also recommend the interactive eBook, individualized learning tools, and more.

Cellular Respiration and Fermentation

7

Chapter Outline

7.1 Overview of Cellular Respiration
7.2 Glycolysis
7.3 Breakdown of Pyruvate
7.4 Citric Acid Cycle
7.5 Oxidative Phosphorylation
7.6 Connections Among Carbohydrate, Protein, and Fat Metabolism
7.7 Anaerobic Respiration and Fermentation
Summary of Key Concepts
Assess and Discuss

Physical endurance. Conditioned athletes, like these marathon runners, have very efficient metabolism of organic molecules such as glucose.

Carmen became inspired while watching the 2008 Summer Olympics and set a personal goal to run a marathon—a distance of 42.2 kilometers, or 26.2 miles. Although she was active in volleyball and downhill skiing in high school, she had never attempted distance running. At first, running an entire mile was pure torture. She was out of breath, overheated, and unhappy, to say the least. However, she became committed to endurance training and within a few weeks discovered that running a mile was a "piece of cake." Two years later, Carmen participated in her first marathon and finished with a time of 4 hours and 11 minutes—not bad for someone who had previously struggled to run a single mile!

How had Carmen's training allowed her to achieve this goal? Perhaps the biggest factor is that the training had altered the metabolism in her leg muscles. For example, the network of small blood vessels supplying oxygen to her leg muscles became more extensive, providing more efficient delivery of oxygen and removal of wastes. Second, her muscle cells developed more mitochondria. Recall from Chapter 4 that the primary role of mitochondria is to make ATP, which is used as a source of energy. With these changes, Carmen's leg muscles were better able to break down organic molecules in her food and utilize them to make ATP.

The cells in Carmen's leg muscles had become more efficient at **cellular respiration**, which comprises the metabolic reactions that a cell uses to get energy from food molecules and release waste products. When we eat food, we are using much of that food for energy. People often speak of "burning calories." Although metabolism does generate some heat, the chemical reactions that take place in the cells of living organisms are uniquely different from those that occur, say, in a fire. When wood is burned, the reaction produces enormous amounts of heat in a short period of time—the reaction lacks control. In contrast, the metabolism that occurs in living cells is extremely controlled. The food molecules from which we harvest energy give up that energy in a very restrained manner rather than all at once, as in a fire. An underlying theme in metabolism is the remarkable control that cells possess when they coordinate chemical reactions. A key emphasis of this chapter is how cells use the energy stored within the chemical bonds of organic molecules.

We will begin by surveying a group of chemical reactions that involves the breakdown of carbohydrates, namely, the sugar glucose. As you will learn, cells carry out an intricate series of reactions so that glucose can be "burned" in a very controlled fashion when oxygen is available. We will then examine how cells use organic molecules in the absence of oxygen via processes known as anaerobic respiration and fermentation.

7.1 Overview of Cellular Respiration

Learning Outcome:

1. Describe the four metabolic pathways that are needed to break down glucose to CO_2.

Cellular respiration is a process by which living cells obtain energy from organic molecules and release waste products. A primary aim of cellular respiration is to make adenosine triphosphate, or ATP. When oxygen (O_2) is used, this process is termed **aerobic respiration**. During aerobic respiration, O_2 is consumed, and CO_2 is released via the oxidation of organic molecules. When we breathe, we inhale the oxygen needed for aerobic respiration and exhale the CO_2, a by-product of the process. For this reason, the term respiration has a second meaning, which is the act of breathing.

Different types of organic molecules, such as carbohydrates, proteins, and fats, can be used as energy sources to drive aerobic respiration. In this chapter, we will largely focus on the use of glucose as an energy source for cellular respiration.

$$C_6H_{12}O_6 + 6\ O_2 \rightarrow 6\ CO_2 + 6\ H_2O + \text{Energy intermediates} + \text{Heat}$$

(Glucose)

$$\Delta G = -686 \text{ kcal/mol}$$

We will focus on the breakdown of glucose in a eukaryotic cell in the presence of oxygen. Certain covalent bonds within glucose store a large amount of chemical bond energy. When glucose is broken down via oxidation, ultimately to CO_2 and water, the energy within those bonds is released and used to make three energy intermediates: ATP, NADH, and $FADH_2$. This process involves four metabolic pathways: (1) glycolysis, (2) the breakdown of pyruvate, (3) the citric acid cycle, and (4) oxidative phosphorylation (Figure 7.1):

1. **Glycolysis:** In glycolysis, glucose (a compound with six carbon atoms) is broken down to two pyruvate molecules (with three carbons each), producing a net energy yield of two ATP molecules and two NADH molecules. The two ATP are synthesized via **substrate-level phosphorylation**, which occurs when an enzyme directly transfers a phosphate from an organic molecule to ADP. In eukaryotes, glycolysis occurs in the cytosol.
2. **Breakdown of pyruvate:** The two pyruvate molecules enter the mitochondrial matrix, where each one is broken down to an acetyl group (with two carbons each) and one CO_2 molecule. For each pyruvate broken down via oxidation, one NADH molecule is made by the reduction of NAD^+.
3. **Citric acid cycle:** Each acetyl group is incorporated into an organic molecule, which is later oxidized to liberate two CO_2 molecules. One ATP, three NADH, and one $FADH_2$ are made in

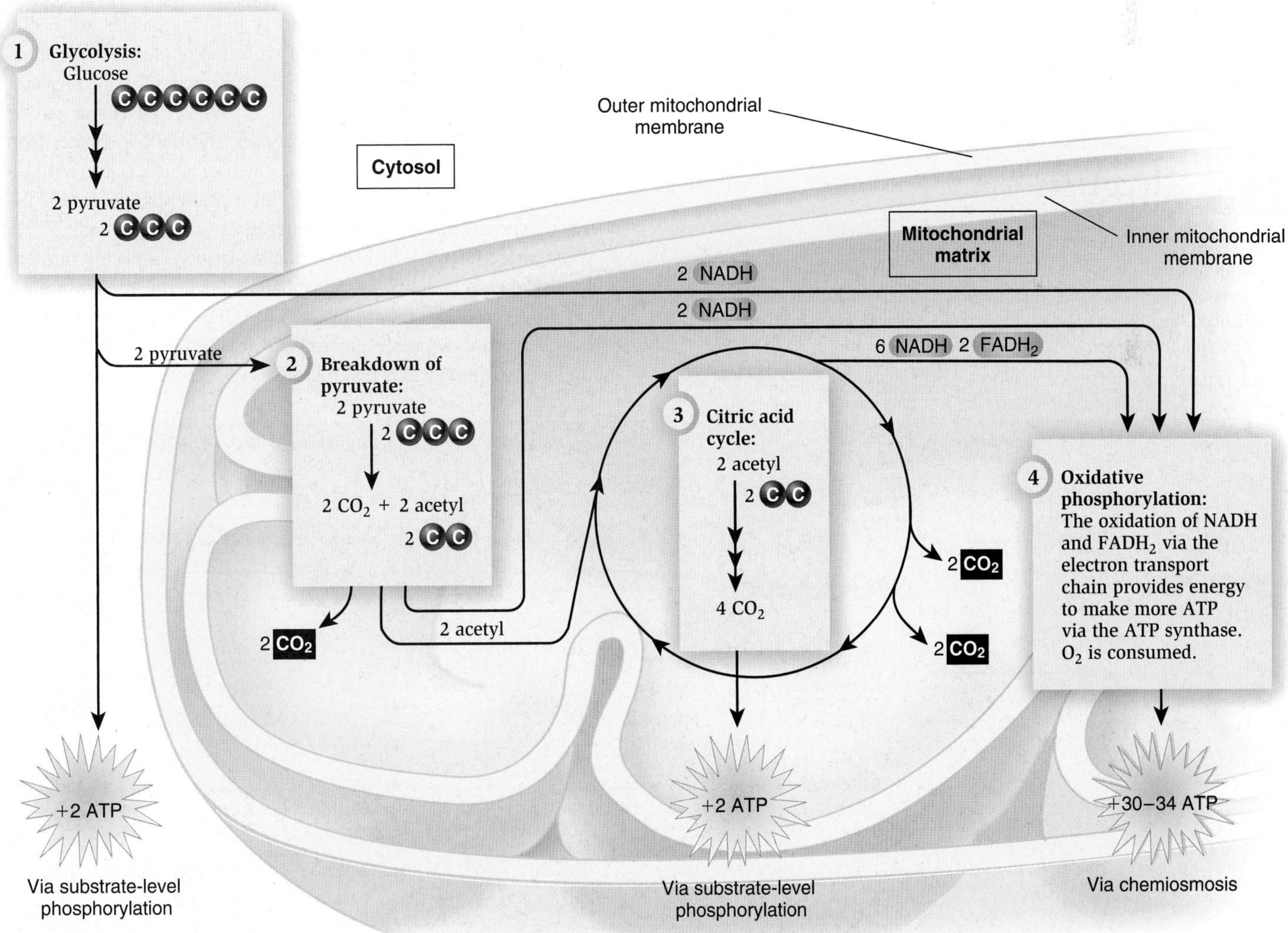

Figure 7.1 An overview of cellular respiration.

BIOLOGY PRINCIPLE Living organisms use energy. Molecules such as glucose store a large amount of energy. The breakdown of glucose is used to make energy intermediates, such as ATP molecules, which drive many types of cellular processes.

this process. Because there are two acetyl groups (one from each pyruvate), the total yield is four CO_2, two ATP via substrate-level phosphorylation, six NADH, and two $FADH_2$. This process occurs in the mitochondrial matrix.

4. **Oxidative phosphorylation:** The NADH and $FADH_2$ made in the three previous stages contain high-energy electrons that can be readily transferred in a redox reaction to other molecules. Once removed from NADH or $FADH_2$, these high-energy electrons release some energy, and through an electron transport chain, that energy is harnessed to produce an H^+ electrochemical gradient. In **chemiosmosis**, energy stored in the H^+ electrochemical gradient is used to synthesize ATP from ADP and P_i. The overall process of electron transport and ATP synthesis is called oxidative phosphorylation because NADH or $FADH_2$ have been oxidized and ADP has become phosphorylated to make ATP. Approximately 30 to 34 ATP molecules are made via oxidative phosphorylation.

In eukaryotes, oxidation phosphorylation occurs along the **cristae**, which are invaginations of the inner mitochondrial membrane. The invaginations greatly increase the surface area of the inner membrane and thereby increase the amount of ATP that can be made. In bacteria and archaea, oxidative phosphorylation occurs along the plasma membrane.

7.2 Glycolysis

Learning Outcomes:

1. Outline the three phases of glycolysis and the net products.
2. Describe the series of enzymatic reactions that constitute glycolysis.
3. Explain the underlying basis for detecting tumors using positron-emission tomography.

Thus far, we have examined the general features of the four metabolic pathways that are involved with the breakdown of glucose. We will now turn our attention to a more detailed understanding of the pathways for glucose metabolism, beginning with glycolysis.

Glycolysis Is a Metabolic Pathway That Breaks Down Glucose to Pyruvate

Glycolysis (from the Greek *glykos*, meaning sweet, and *lysis*, meaning splitting) involves the breakdown of glucose, a simple sugar, into two molecules of a compound called pyruvate. This process can occur in the presence of oxygen, that is, under aerobic conditions, and it can also occur in the absence of oxygen. During the 1930s, the efforts of several German biochemists, including Gustav Embden, Otto Meyerhof, and Jacob Parnas, determined that glycolysis involves 10 steps, each one catalyzed by a different enzyme. The elucidation of these steps was a major achievement in the field of **biochemistry**—the study of the chemistry of living organisms. Researchers have since discovered that glycolysis is the common pathway for glucose breakdown in bacteria, archaea, and eukaryotes. Remarkably, the steps of glycolysis are virtually identical in nearly all living species, suggesting that glycolysis arose very early in the evolution of life on our planet.

The 10 steps of glycolysis can be grouped into three phases (**Figure 7.2**). The first phase (steps 1–3) involves an energy investment. Two ATP molecules are hydrolyzed, and the phosphates from those ATP molecules are attached to glucose, which is converted to fructose-1,6-bisphosphate. The energy investment phase raises the free energy of glucose, thereby allowing later reactions to be exergonic. The cleavage phase (steps 4–5) breaks this six-carbon molecule into two molecules of glyceraldehyde-3-phosphate. The energy liberation phase (steps 6–10) produces four ATP, two NADH, and two molecules of pyruvate. Because two molecules of ATP are used in the energy investment phase, the net yield is two molecules of ATP.

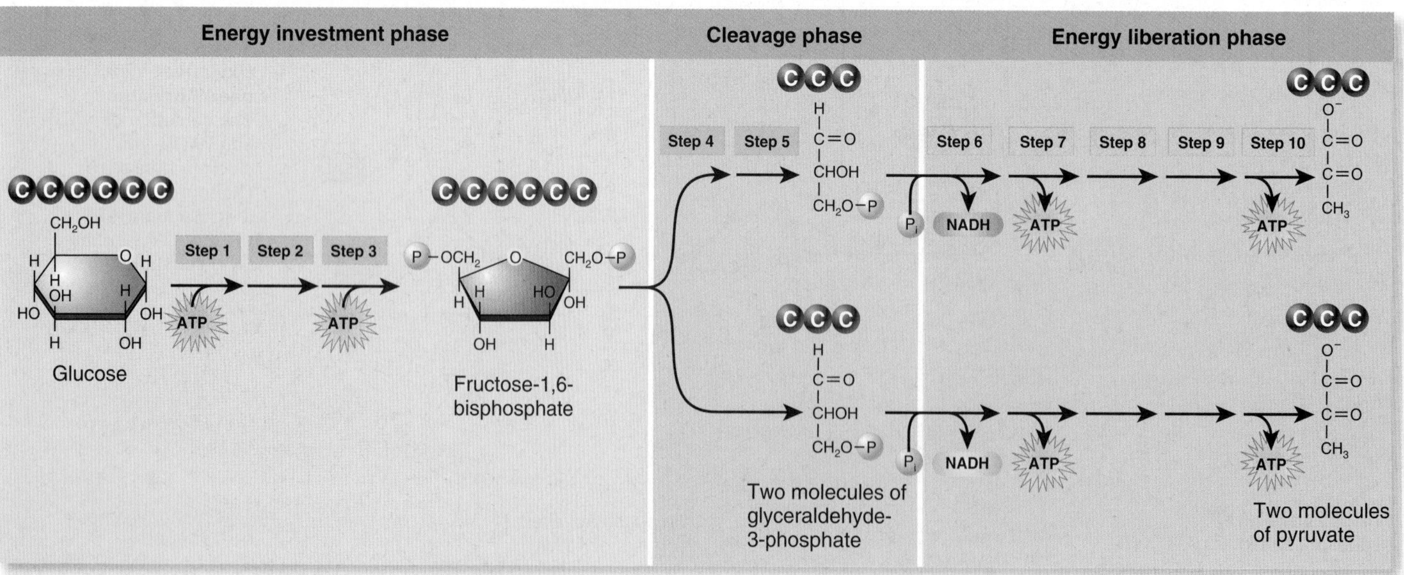

Figure 7.2 Overview of glycolysis.

Concept Check: *Explain why the three phases are named the energy investment phase, the cleavage phase, and the energy liberation phase.*

BioConnections: *Look ahead to Table 44.1. With regard to oxygen needs, what is an advantage of glycolytic muscle fibers?*

Figure 7.3 (shown across the top of pages 140 and 141) describes the details of the 10 reactions of glycolysis. The net reaction of glycolysis is as follows:

$$\underset{\text{Glucose}}{C_6H_{12}O_6} + 2\,NAD^+ + 2\,ADP^{2-} + 2\,P_i^{2-} \rightarrow$$

$$\underset{\text{Pyruvate}}{2\,CH_3(C{=}O)COO^-} + 2\,H^+ + 2\,NADH + 2\,ATP^{4-} + 2\,H_2O$$

Regulation of Glycolysis How do cells control glycolysis? The rate of glycolysis is largely regulated by the availability of substrates, such as glucose, and by feedback inhibition. A key control point involves the enzyme phosphofructokinase, which catalyzes the third step in glycolysis, the step believed to be the slowest, or rate-limiting, step. When a cell has a sufficient amount of ATP, feedback inhibition occurs. At high concentrations, ATP binds to an allosteric site in phosphofructokinase, causing a conformational change that renders the enzyme functionally inactive. This prevents the further breakdown of glucose and thereby inhibits the overproduction of ATP. (Allosteric sites and rate-limiting steps are discussed in Chapter 6.) Alternatively, if ADP binds to this enzyme, its function is stimulated, thereby increasing the rate of glycolysis.

Genomes & Proteomes Connection

Cancer Cells Usually Exhibit High Levels of Glycolysis

In 1931, the German physiologist Otto Warburg discovered that certain cancer cells preferentially use glycolysis for ATP production in contrast to healthy cells, which mainly generate ATP from oxidative phosphorylation. This phenomenon, termed the Warburg effect, is very common among different types of tumors. The Warburg effect is the foundation for the detection of cancer via a procedure called positron-emission tomography (PET, see Figure 2.6). In this technique, patients are injected with a radioactive glucose analogue called $[^{18}F]$-fluorodeoxyglucose (FDG). As a glucose analogue, FDG is taken up by high-glucose-using cells such as cancer cells. The scanner detects regions of the body that metabolize FDG rapidly, which are visualized as bright spots on the PET scan.

Figure 7.4 shows a PET scan of a patient with lung cancer. The bright regions next to the arrows are tumors that show abnormally high levels of glycolysis. This result occurred because the genome found in cancer cells exhibits an increased expression of genes that encode enzymes involved with glycolysis. In other words, cancer cells exhibit changes in their proteomes compared to their normal counterparts. Research has shown that the enzymes of glycolysis are overexpressed in approximately 80% of all types of cancer, including lung, skin, colon, liver, pancreatic, breast, ovarian, and prostate cancers. The three enzymes of glycolysis whose overexpression is most commonly associated with cancer are glyceraldehyde-3-phosphate dehydrogenase, enolase, and pyruvate kinase (shown in Figure 7.3). In many cancers, all 10 glycolytic enzymes are overexpressed!

How does the overexpression of glycolytic enzymes affect tumor growth? While the genetic changes associated with tumor growth are complex, researchers have speculated that an increase in glycolysis may favor the growth of the tumor due to changes in oxygen levels. As a tumor grows, the internal regions of the tumor tend to become hypoxic, or deficient in oxygen. The hypoxic state inside a tumor may contribute to the overexpression of glycolytic genes and lead to a higher level of glycolytic enzymes within the cancer cells. This favors glycolysis as a means of making ATP, because it does not require oxygen. Making ATP via glycolysis is an advantage to cancer cells, because such cells would have trouble making ATP via oxidative phosphorylation, which requires oxygen. Based on these findings, some current research is aimed at discovering drugs that inhibit glycolysis in cancer cells as a way to prevent their growth.

7.3 Breakdown of Pyruvate

Learning Outcome:

1. Describe how pyruvate is broken down and acetyl CoA is made.

In eukaryotes, glycolysis produces pyruvate in the cytosol, which is then transported into the mitochondrion. Once in the mitochondrial matrix, pyruvate molecules are broken down (oxidized) by

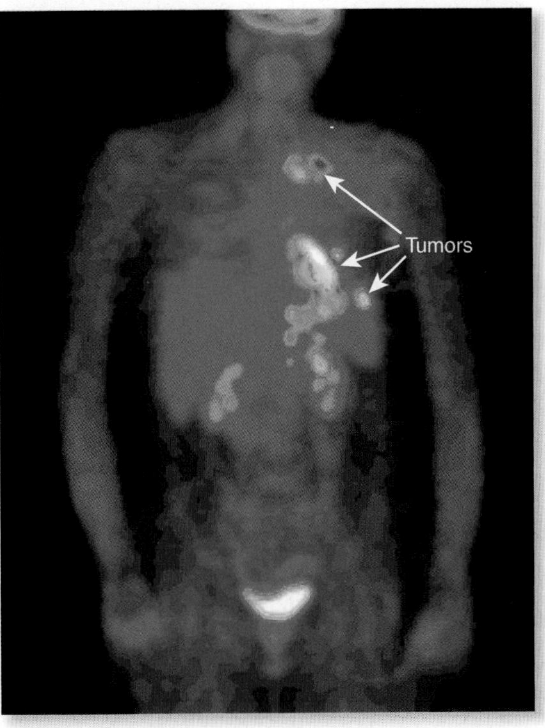

Figure 7.4 A PET scan of a patient with lung cancer. The bright regions in the lungs are tumors (see arrows). Organs such as the brain, which are not cancerous, appear bright because they perform high levels of glucose metabolism. Also, the kidneys and bladder appear bright because they filter and accumulate FDG. (Note: FDG is taken up by cells and converted to FDG-phosphate by hexokinase, the first enzyme in glycolysis. However, because FDG lacks an —OH group, it is not metabolized further. Therefore, FDG-phosphate accumulates in metabolically active cells.)

BioConnections: *Look back at Figure 2.6. Why is FDG radiolabeled?*

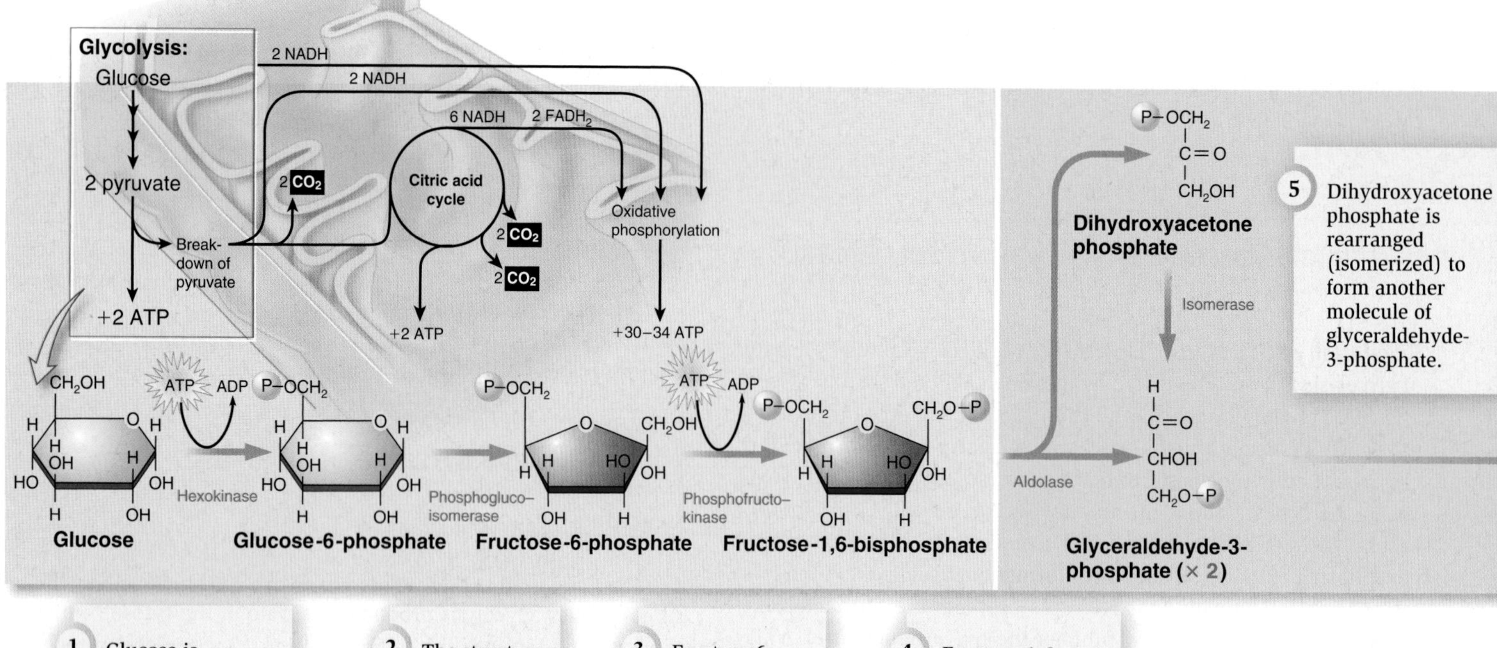

Figure 7.3 A detailed look at the steps of glycolysis. The pathway begins with a 6-carbon molecule (glucose) that is eventually broken down into 2 molecules that contain 3 carbons each. The notation **x 2** in the figure indicates that 2 of these 3-carbon molecules are produced from each glucose molecule.

BIOLOGY PRINCIPLE Living organisms maintain homeostasis. To maintain a relatively constant level of ATP in the cytosol, the activity of phosphofructokinase is controlled by feedback inhibition. A high level of ATP inhibits its function.

Concept Check: Which of these organic molecules donate a phosphate group to ADP during substrate-level phosphorylation?

an enzyme complex called pyruvate dehydrogenase (Figure 7.5). A molecule of CO_2 is removed from each pyruvate, and the remaining acetyl group is attached to an organic molecule called coenzyme A (CoA) to produce acetyl CoA. (In chemical equations, CoA is depicted as CoA—SH to emphasize how the SH group participates in the chemical reaction.) During this process, two high-energy electrons are removed from pyruvate and transferred to NAD^+ and together with H^+ produce a molecule of NADH. For each pyruvate, the net reaction is as follows:

$$\underset{\text{Pyruvate}}{{}^-O-\overset{\overset{O}{\|}}{C}-\overset{\overset{O}{\|}}{C}-CH_3} + \underset{\text{CoA}}{CoA-SH} + NAD^+ \longrightarrow$$

$$\underset{\text{Acetyl CoA}}{CoA-S-\overset{\overset{O}{\|}}{C}-CH_3} + CO_2 + NADH$$

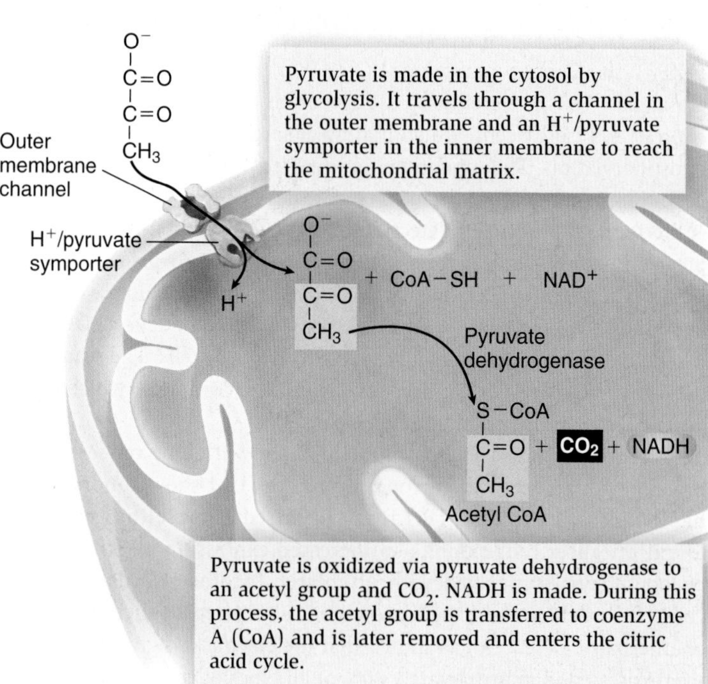

Figure 7.5 Breakdown of pyruvate and the attachment of an acetyl group to CoA.

The acetyl group is attached to CoA via a covalent bond to a sulfur atom. The hydrolysis of this bond releases a large amount of free energy, making it possible for the acetyl group to be transferred to other organic molecules. As described next, the acetyl group is removed from CoA and enters the citric acid cycle.

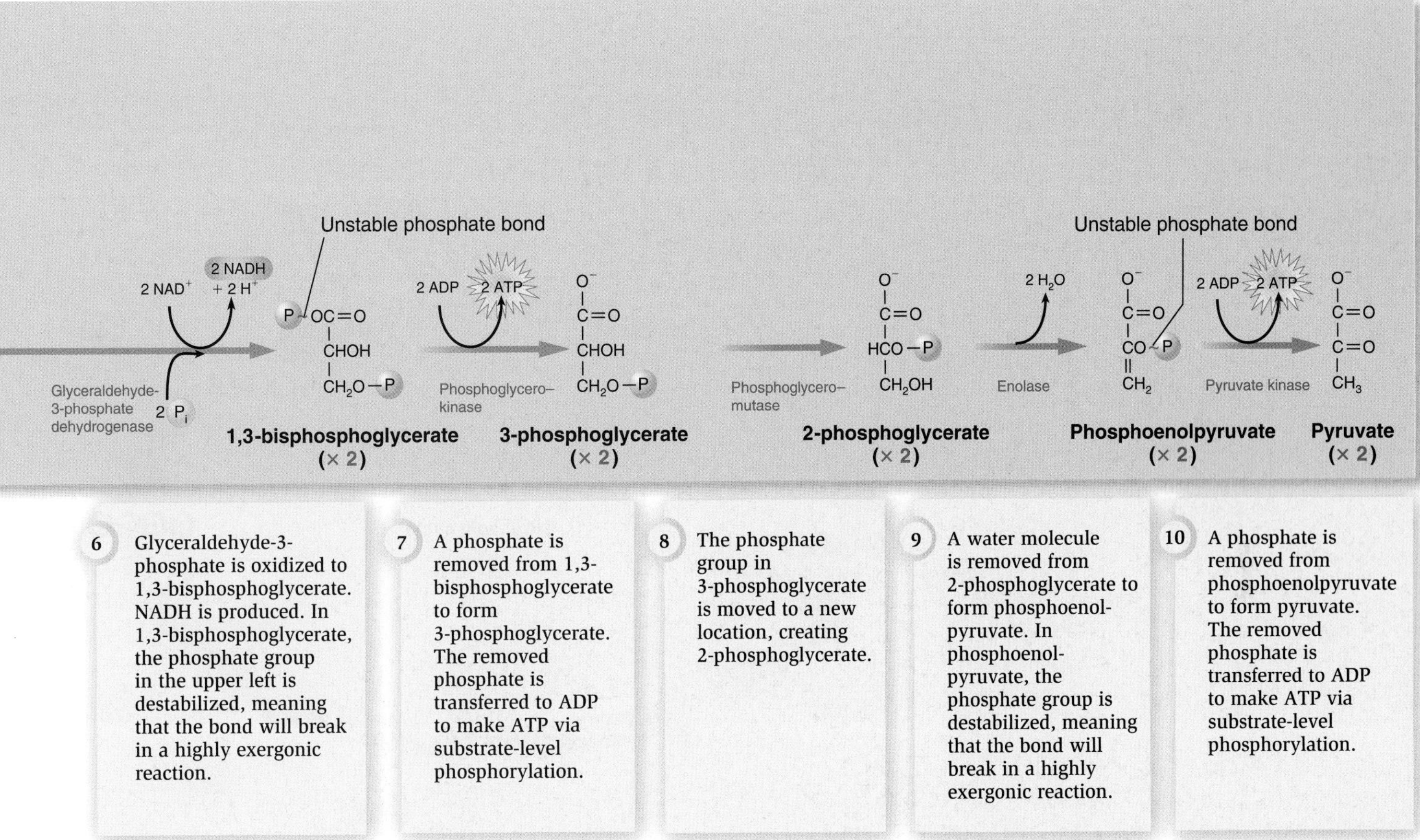

Unstable phosphate bond
2 NAD+
2 NADH + 2 H+
Glyceraldehyde-3-phosphate dehydrogenase
2 Pi
1,3-bisphosphoglycerate (× 2)
2 ADP
2 ATP
Phosphoglycero-kinase
3-phosphoglycerate (× 2)
Phosphoglycero-mutase
2-phosphoglycerate (× 2)
2 H2O
Enolase
Unstable phosphate bond
Phosphoenolpyruvate (× 2)
2 ADP
2 ATP
Pyruvate kinase
Pyruvate (× 2)
6 Glyceraldehyde-3-phosphate is oxidized to 1,3-bisphosphoglycerate. NADH is produced. In 1,3-bisphosphoglycerate, the phosphate group in the upper left is destabilized, meaning that the bond will break in a highly exergonic reaction.
7 A phosphate is removed from 1,3-bisphosphoglycerate to form 3-phosphoglycerate. The removed phosphate is transferred to ADP to make ATP via substrate-level phosphorylation.
8 The phosphate group in 3-phosphoglycerate is moved to a new location, creating 2-phosphoglycerate.
9 A water molecule is removed from 2-phosphoglycerate to form phosphoenol-pyruvate. In phosphoenol-pyruvate, the phosphate group is destabilized, meaning that the bond will break in a highly exergonic reaction.
10 A phosphate is removed from phosphoenolpyruvate to form pyruvate. The removed phosphate is transferred to ADP to make ATP via substrate-level phosphorylation.

7.4 Citric Acid Cycle

Learning Outcomes:

1. Discuss the concept of a metabolic cycle.
2. Describe how an acetyl group enters the citric acid cycle and identify the net products of the cycle.

The third stage of glucose metabolism introduces a new concept, that of a **metabolic cycle**. During a metabolic cycle, particular molecules enter the cycle while others leave. The process is cyclical because it involves a series of organic molecules that are regenerated with each turn of the cycle. The idea of a metabolic cycle was first proposed in the early 1930s by the German biochemist Hans Krebs. While studying carbohydrate metabolism in England, he analyzed cell extracts from pigeon muscle and determined that citric acid and other organic molecules participated in a cycle that resulted in the breakdown of carbohydrates to carbon dioxide. This cycle is called the **citric acid cycle**, or the Krebs cycle, in honor of Krebs, who was awarded the Nobel Prize in Physiology or Medicine in 1953.

An overview of the citric acid cycle is shown in Figure 7.6. In the first step of the cycle, the acetyl group (with two carbons) is removed from acetyl CoA and attached to oxaloacetate (with four carbons) to form citrate (with six carbons), also called citric acid. Then in a series of several steps, two CO_2 molecules are released. As this occurs, a total of three molecules of NADH, one molecule of $FADH_2$, and one molecule of guanine triphosphate (GTP) are made. The GTP, which is made via substrate-level phosphorylation, is used to make ATP. After a total of eight steps, oxaloacetate is regenerated so the cycle can begin again, provided acetyl CoA is available. Figure 7.7 shows a more detailed view of the citric acid cycle. For each acetyl group attached to CoA, the net reaction of the citric acid cycle is as follows:

$$\text{Acetyl-CoA} + 2\,H_2O + 3\,NAD^+ + FAD + GDP^{2-} + P_i^{2-} \rightarrow$$

$$\text{CoA—SH} + 2\,CO_2 + 3\,NADH + FADH_2 + GTP^{4-} + 3\,H^+$$

Regulation of the Citric Acid Cycle How is the citric acid cycle controlled? The rate of the cycle is largely regulated by the availability of substrates, such as acetyl-CoA and NAD^+, and by feedback inhibition. The three steps in the cycle that are highly exergonic are those catalyzed by citrate synthase, isocitrate dehydrogenase, and α-ketoglutarate dehydrogenase (see Figure 7.7). Each of these steps can become rate-limiting under certain circumstances, and the way that each enzyme is regulated varies among different species. Let's

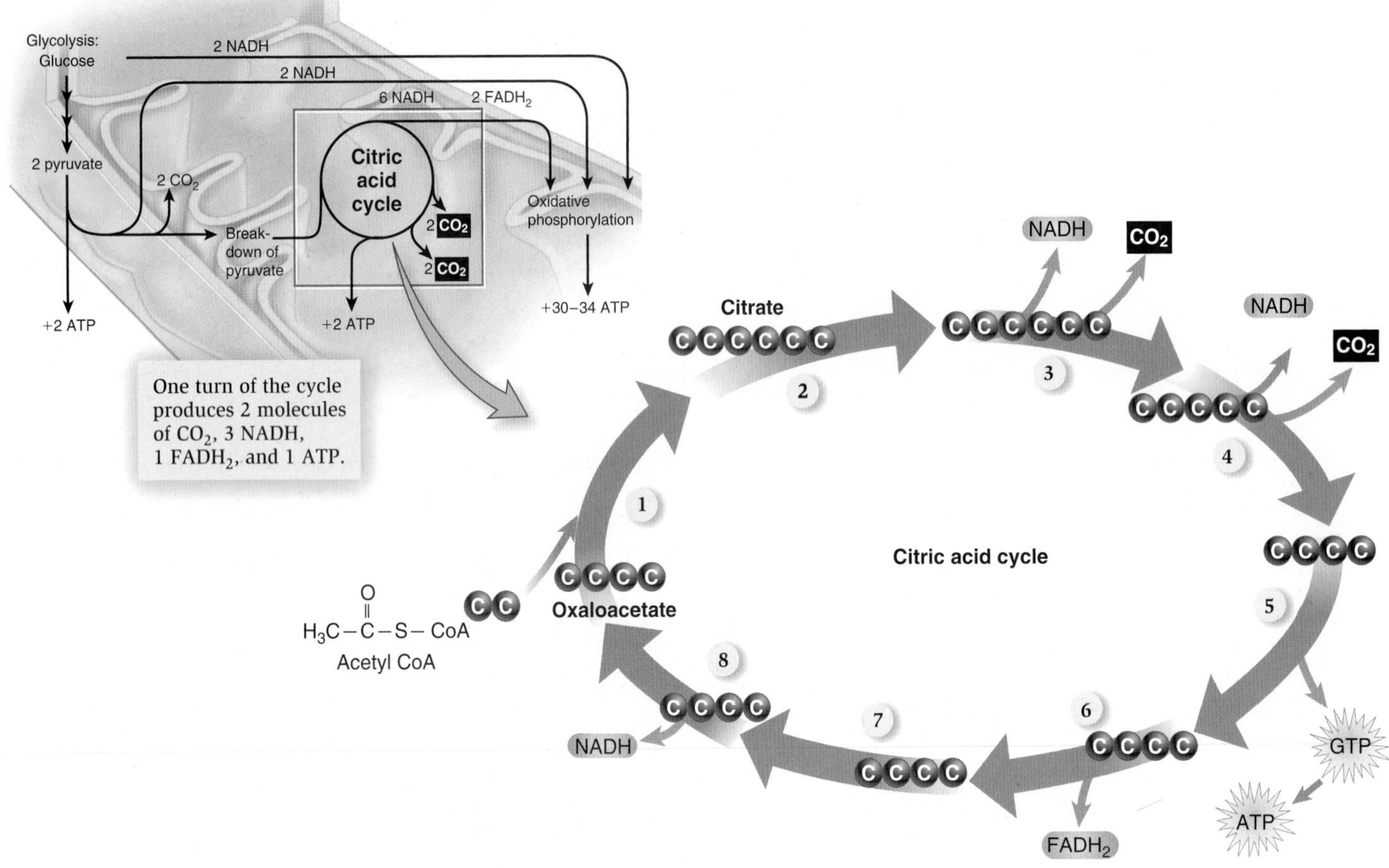

Figure 7.6 **Overview of the citric acid cycle.**

Concept Check: *What are the main products of the citric acid cycle?*

consider an example. In mammals, NADH and ATP act as feedback inhibitors of isocitrate dehydrogenase, whereas NAD^+ and ADP act as activators. In this way, the citric acid cycle is inhibited when NADH and ATP levels are high, but it is stimulated when NAD^+ and ADP levels are high.

7.5 Oxidative Phosphorylation

Learning Outcomes:

1. Describe how the electron transport chain produces an H^+ electrochemical gradient.
2. Explain how ATP synthase utilizes the H^+ electrochemical gradient to synthesize ATP.
3. Analyze the results of the experiment that showed that ATP synthase is a rotary machine.

During the first three stages of glucose metabolism, the oxidation of glucose yields 6 molecules of CO_2, 4 molecules of ATP, 10 molecules of NADH, and 2 molecules of $FADH_2$. Let's now consider how high-energy electrons are removed from NADH and $FADH_2$ to produce more ATP. This process is called **oxidative phosphorylation**. As mentioned earlier, the term refers to the observation that NADH and $FADH_2$ have had electrons removed and have thus become oxidized, and ATP is made by the phosphorylation of ADP. In this section, we will examine how the oxidative process involves the electron transport chain, whereas the phosphorylation of ADP occurs via ATP synthase.

The Electron Transport Chain Establishes an Electrochemical Gradient

The **electron transport chain (ETC)** consists of a group of protein complexes and small organic molecules embedded in the inner mitochondrial membrane. These components are referred to as an electron transport chain because electrons are passed from one component to the next in a series of redox reactions (**Figure 7.8**). Most members of the ETC are protein complexes (designated I–IV) that have prosthetic groups, which are small molecules permanently attached to the surface of proteins that aid in their function. For example, cytochrome oxidase contains two prosthetic groups, each with an iron atom. The iron in each prosthetic group can readily accept and release an electron. One member of the ETC, ubiquinone (Q), is not a protein. Rather, ubiquinone is a small organic molecule that can accept and release an electron. Ubiquinone, also known as coenzyme Q, is a nonpolar molecule that can diffuse through the lipid bilayer.

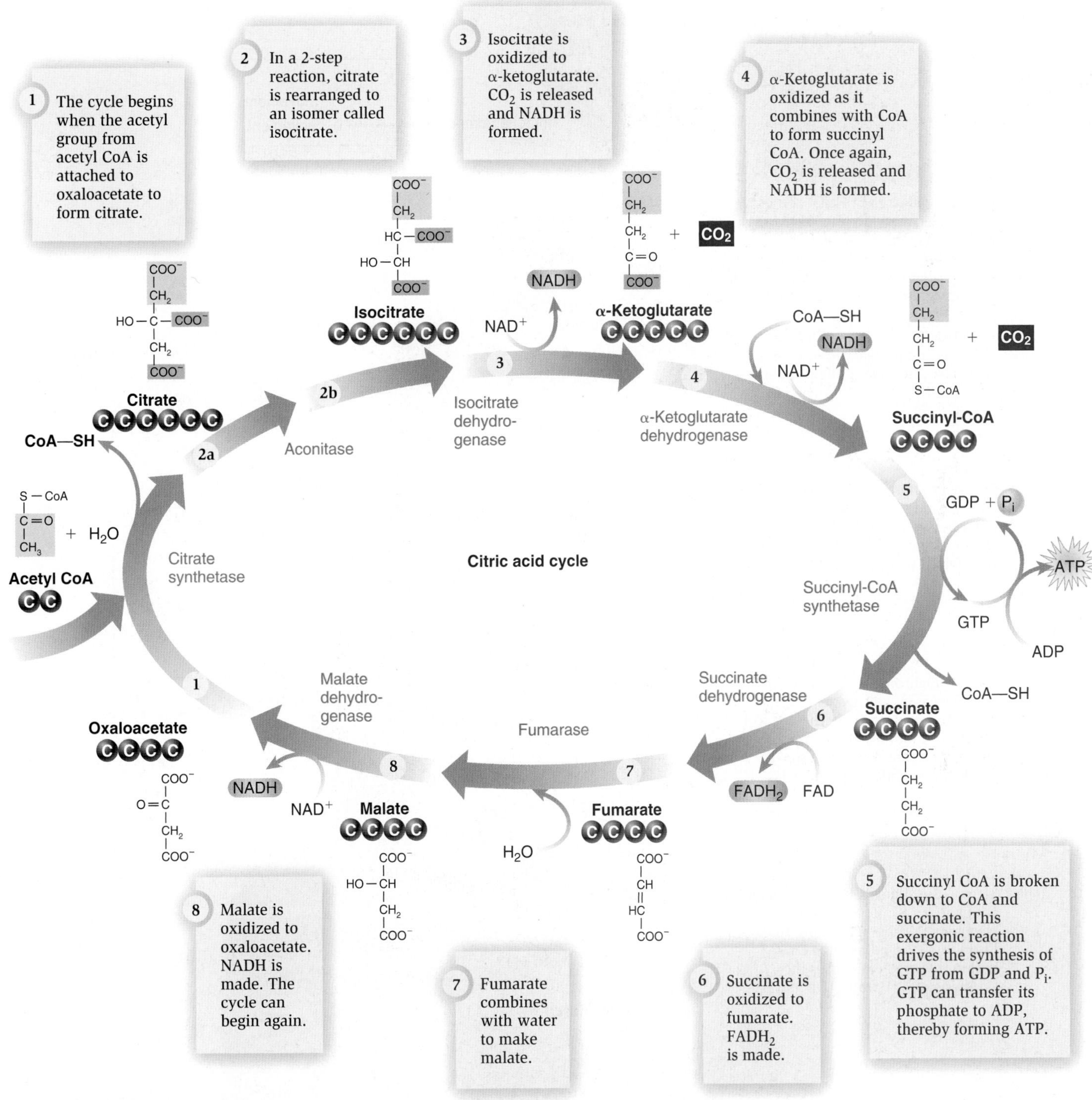

Figure 7.7 A detailed look at the steps of the citric acid cycle. The blue boxes indicate the location of the acetyl group, which is oxidized at step 6. (It is oxidized again in step 8.) The green boxes indicate the locations where CO_2 molecules are removed.

The red line in Figure 7.8 shows the path of electron flow. The electrons, which are originally located on NADH or $FADH_2$, are transferred to components of the ETC. The electron path is a series of redox reactions in which electrons are transferred to components with increasingly higher electronegativity. As discussed in Chapter 2, electronegativity is a measure of an atom's ability to attract electrons. At the end of the chain is oxygen, which is the most electronegative and the final electron acceptor. The ETC is also called the **respiratory chain** because the oxygen we breathe is used in this process.

NADH and $FADH_2$ donate their electrons at different points in the ETC. Two high-energy electrons from NADH are first transferred one at a time to NADH dehydrogenase (complex I). They are then transferred to ubiquinone (Q), cytochrome *b-c₁* (complex III), cytochrome *c*, and cytochrome oxidase (complex IV). The final electron

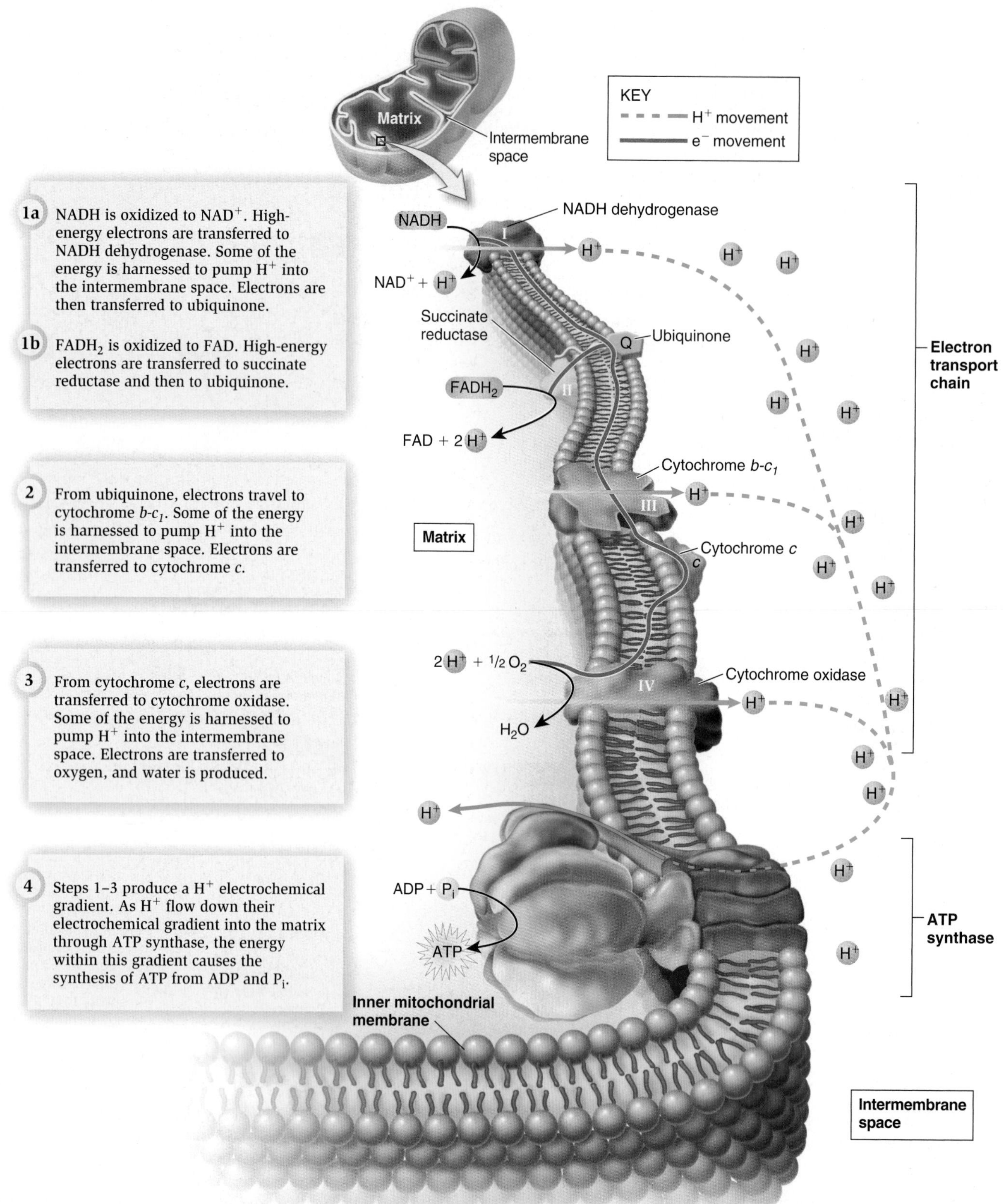

Matrix
Intermembrane space
KEY
H^+ movement
e^- movement
1a NADH is oxidized to NAD^+. High-energy electrons are transferred to NADH dehydrogenase. Some of the energy is harnessed to pump H^+ into the intermembrane space. Electrons are then transferred to ubiquinone.
1b $FADH_2$ is oxidized to FAD. High-energy electrons are transferred to succinate reductase and then to ubiquinone.
2 From ubiquinone, electrons travel to cytochrome b-c_1. Some of the energy is harnessed to pump H^+ into the intermembrane space. Electrons are transferred to cytochrome c.
3 From cytochrome c, electrons are transferred to cytochrome oxidase. Some of the energy is harnessed to pump H^+ into the intermembrane space. Electrons are transferred to oxygen, and water is produced.
4 Steps 1–3 produce a H^+ electrochemical gradient. As H^+ flow down their electrochemical gradient into the matrix through ATP synthase, the energy within this gradient causes the synthesis of ATP from ADP and P_i.
NADH
$NAD^+ + H^+$
NADH dehydrogenase
I
Succinate reductase
Q
Ubiquinone
$FADH_2$
II
$FAD + 2H^+$
Cytochrome b-c_1
III
Matrix
Cytochrome c
$2H^+ + 1/2\,O_2$
Cytochrome oxidase
IV
H_2O
H^+
Electron transport chain
ADP + P_i
ATP
ATP synthase
Inner mitochondrial membrane
Intermembrane space

Figure 7.8 Oxidative phosphorylation. This process consists of two distinct events: the electron transport chain (ETC) and ATP synthesis. The ETC oxidizes, or removes electrons from, NADH or $FADH_2$ and pumps H^+ across the inner mitochondrial membrane. In chemiosmosis, ATP synthase uses the energy in this H^+ electrochemical gradient to phosphorylate ADP, thereby synthesizing ATP. In this figure, an O_2 atom is represented as ½ O_2 to emphasize that the ETC reduces oxygen when it is in its molecular (O_2) form.

Concept Check: *Can you explain the name of cytochrome oxidase? Can you think of another appropriate name?*

acceptor is O_2. By comparison, $FADH_2$ transfers electrons to succinate reductase (complex II), then to ubiquinone, and the rest of the chain.

As shown in Figure 7.8, some of the energy that is released during the movement of electrons is used to pump H^+ across the inner mitochondrial membrane, from the matrix and into the intermembrane space. This active transport establishes a large **H^+ electrochemical gradient**, in which the concentration of H^+ is higher outside of the matrix than inside and an excess of positive charge exists outside the matrix. Because hydrogen ions consist of protons, the H^+ electrochemical gradient is also called the **proton-motive force**. NADH dehydrogenase, cytochrome *b-c*$_1$, and cytochrome oxidase are H^+ pumps. While traveling along the electron transport chain, electrons release free energy, and some of this energy is captured by these proteins to actively transport H^+ out of the matrix into the intermembrane space against the H^+ electrochemical gradient. Because the electrons from $FADH_2$ enter the chain at an intermediate step, they release less energy and so result in fewer hydrogen ions being pumped out of the matrix than do electrons from NADH.

Why do electrons travel from NADH or $FADH_2$ to the ETC and then to O_2? As you might expect, the answer lies in free-energy changes. The electrons found on the energy intermediates have a high amount of energy. As they travel along the ETC, free energy is released (**Figure 7.9**). The movement of one electron from NADH to O_2 results in a very negative free-energy change of approximately −25 kcal/mol. That is why the process is spontaneous and proceeds in the forward direction. Because it is a highly exergonic reaction, some of the free energy can be harnessed to do cellular work. In this case, some energy is used to pump H^+ across the inner mitochondrial membrane and establish an H^+ electrochemical gradient that is then used to power ATP synthesis.

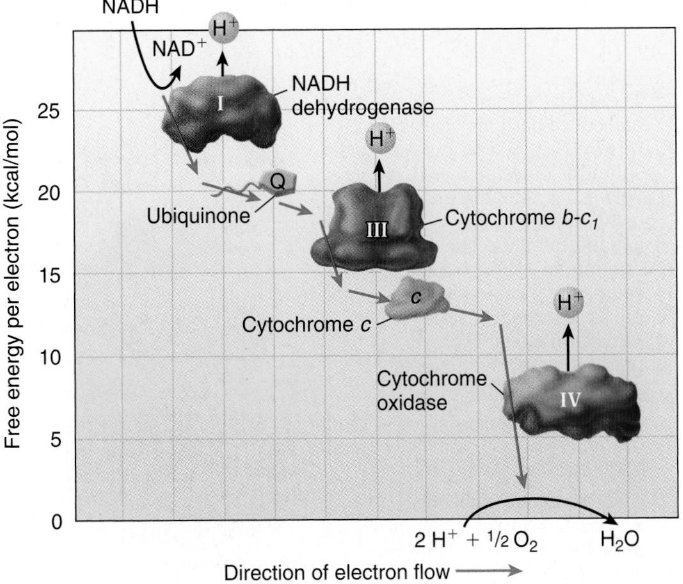

Figure 7.9 The relationship between free energy and electron movement along the electron transport chain. As electrons hop from one site to another along the electron transport chain, they release energy. Some of this energy is harnessed to pump H^+ across the inner mitochondrial membrane. The total energy released by a single electron is approximately −25 kcal/mol.

Chemicals that inhibit the flow of electrons along the ETC can have lethal effects. For example, one component of the ETC, cytochrome oxidase (complex IV), is inhibited by cyanide. The deadly effects of cyanide ingestion occur because the ETC is shut down, preventing cells from making enough ATP for survival.

ATP Synthase Makes ATP via Chemiosmosis

The second event of oxidative phosphorylation is the synthesis of ATP by an enzyme called **ATP synthase**. The H^+ electrochemical gradient across the inner mitochondrial membrane is a source of potential energy. How is this energy used? The passive flow of H^+ back into the matrix is an exergonic process. The lipid bilayer is relatively impermeable to H^+. However, H^+ can pass through the membrane-embedded portion of ATP synthase. This enzyme harnesses some of the free energy that is released as the ions flow through its membrane-embedded region to synthesize ATP from ADP and P_i (see bottom of Figure 7.8). This is an example of an energy conversion: Energy in the form of an H^+ gradient is converted to chemical bond energy in ATP. The synthesis of ATP that occurs as a result of pushing H^+ across a membrane is called chemiosmosis (from the Greek *osmos*, meaning to push). The theory behind it was proposed by Peter Mitchell, a British biochemist who was awarded the Nobel Prize in Chemistry in 1978.

Regulation of Oxidative Phosphorylation How is oxidative phosphorylation controlled? This process is regulated by a variety of factors, including the availability of ETC substrates, such as NADH and O_2, and by the ATP/ADP ratio. When ATP levels are high, ATP binds to a subunit of cytochrome oxidase (complex IV), thereby inhibiting the ETC and oxidative phosphorylation. By comparison, when ADP levels are high, oxidative phosphorylation is stimulated for two reasons: (1) ADP stimulates cytochrome oxidase, and (2) ADP is a substrate that is used (with P_i) to make ATP.

NADH Oxidation Makes a Large Proportion of a Cell's ATP

For each molecule of NADH that is oxidized and each molecule of ATP that is made, the two chemical reactions of oxidative phosphorylation can be represented as follows:

$$NADH + H^+ + 1/2\, O_2 \rightarrow NAD^+ + H_2O$$

$$ADP^{2-} + P_i^{2-} \rightarrow ATP^{4-} + H_2O$$

The oxidation of NADH to NAD^+ results in an H^+ electrochemical gradient in which more hydrogen ions are in the intermembrane space than are in the matrix. The synthesis of one ATP molecule is thought to require the movement of three to four ions down their H^+ electrochemical gradient into the matrix.

When we add up the maximal amount of ATP that can be made by oxidative phosphorylation, most researchers agree it is in the range of 30 to 34 ATP molecules for each glucose molecule that is broken down to CO_2 and H_2O. However, the maximum amount of ATP is rarely achieved, for two reasons. First, although 10 NADH and 2 $FADH_2$ are available to make the H^+ electrochemical gradient across the inner mitochondrial membrane, a cell uses some of these molecules for anabolic pathways. For example, NADH is used in the synthesis of organic molecules such as glycerol (a component of phospholipids). It is also used to make lactic acid, which is secreted from muscle cells during strenuous exercise (described later in Figure 7.16a). Second, the mitochondrion may use some of the H^+ electrochemical gradient for other purposes. For example, the gradient is used for the uptake of pyruvate into the matrix via an H^+/pyruvate symporter (see Figure 7.5). Therefore, the actual amount of ATP synthesis is usually a little less than the maximum number of 30 to 34. Even so, when we compare the amount of ATP that is made by glycolysis (2), the citric acid cycle (2), and oxidative phosphorylation (30–34), we see that oxidative phosphorylation provides a cell with a much greater capacity to make ATP.

Experiments with Purified Proteins in Membrane Vesicles Verified Chemiosmosis

To show experimentally that ATP synthase directly uses an H^+ electrochemical gradient to make ATP, researchers needed to purify the enzyme and study its function in vitro. In 1974, Efraim Racker and Walther Stoeckenius purified ATP synthase and another protein called bacteriorhodopsin, which is found in certain species of archaea. Previous research had shown that bacteriorhodopsin is a light-driven H^+ pump. Racker and Stoeckenius took both purified proteins and inserted them into membrane vesicles (**Figure 7.10**). ATP synthase was oriented so its ATP-synthesizing region was on the outside of the vesicles. Bacteriorhodopsin was oriented so it would pump H^+ into the vesicles. They added ADP and P_i on the outside of the vesicles. In the dark, no ATP was made. However, when they shone light on the vesicles, a substantial amount of ATP was synthesized. Because bacteriorhodopsin was already known to be a light-driven H^+ pump, these results convinced researchers that ATP synthase uses an H^+ electrochemical gradient as an energy source to make ATP.

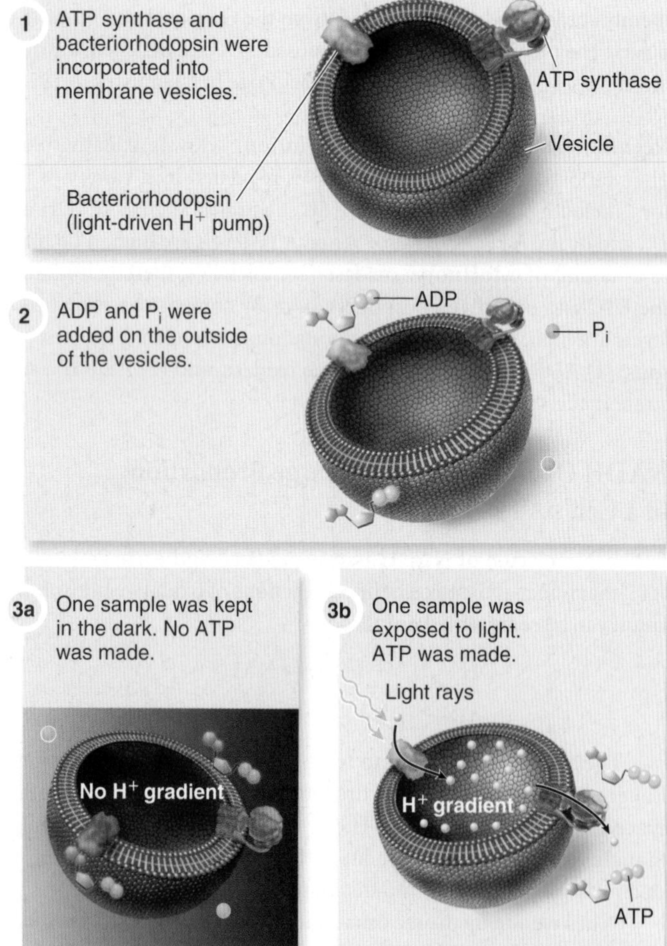

1 ATP synthase and bacteriorhodopsin were incorporated into membrane vesicles.
ATP synthase
Vesicle
Bacteriorhodopsin (light-driven H+ pump)
2 ADP and Pi were added on the outside of the vesicles.
ADP
Pi
3a One sample was kept in the dark. No ATP was made.
No H+ gradient
3b One sample was exposed to light. ATP was made.
Light rays
H+ gradient
ATP

Figure 7.10 The Racker and Stoeckenius experiment. In this experiment, bacteriorhodopsin pumped H^+ into vesicles, and the resulting H^+ electrochemical gradient was sufficient to drive ATP synthesis via ATP synthase.

Concept Check: Is the functioning of the electron transport chain always needed to make ATP via ATP synthase?

ATP Synthase Is a Rotary Machine That Makes ATP as It Spins

The structure and function of ATP synthase are particularly intriguing and have received much attention over the past few decades. ATP synthase is a rotary machine (**Figure 7.11**). The region embedded in the membrane is composed of three types of subunits called *a*, *b*, and *c*. Approximately 10 to 14 *c* subunits form a ring in the membrane. One *a* subunit is bound to this ring, and two *b* subunits are attached to the *a* subunit and protrude from the membrane. The nonmembrane-embedded subunits are designated with Greek letters. One ε and one γ subunit bind to the ring of *c* subunits. The γ subunit forms a long stalk that pokes into the center of another ring of three α and three β subunits. Each β subunit contains a catalytic site where ATP is made. Finally, the δ subunit forms a connection between the ring of α and β subunits and the two *b* subunits.

When hydrogen ions pass through a narrow channel at the contact site between a *c* subunit and the *a* subunit, a conformational

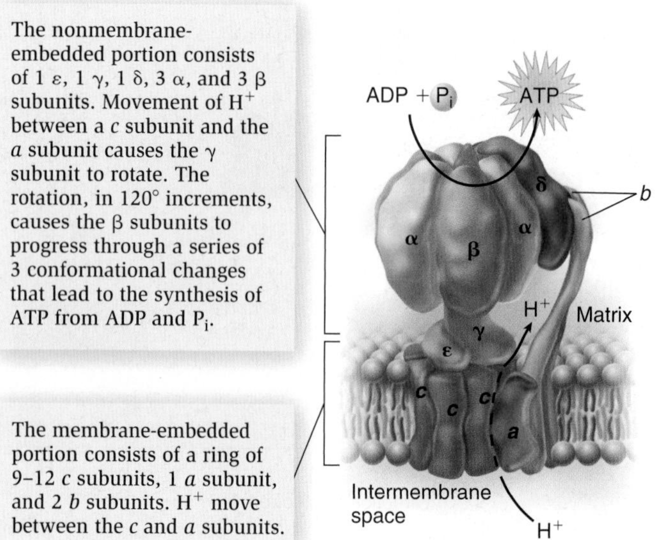

The nonmembrane-embedded portion consists of 1 ε, 1 γ, 1 δ, 3 α, and 3 β subunits. Movement of H+ between a c subunit and the a subunit causes the γ subunit to rotate. The rotation, in 120° increments, causes the β subunits to progress through a series of 3 conformational changes that lead to the synthesis of ATP from ADP and Pi.
The membrane-embedded portion consists of a ring of 9–12 c subunits, 1 a subunit, and 2 b subunits. H+ move between the c and a subunits.
ADP + Pi
ATP
δ
b
α
β
α
H+
Matrix
γ
ε
c
c
c
a
Intermembrane space
H+

Figure 7.11 The subunit structure and function of ATP synthase.

Concept Check: If the β subunit in the front center of this figure is in conformation 2, what are the conformations of the β subunit on the left and the β subunit on the back right?

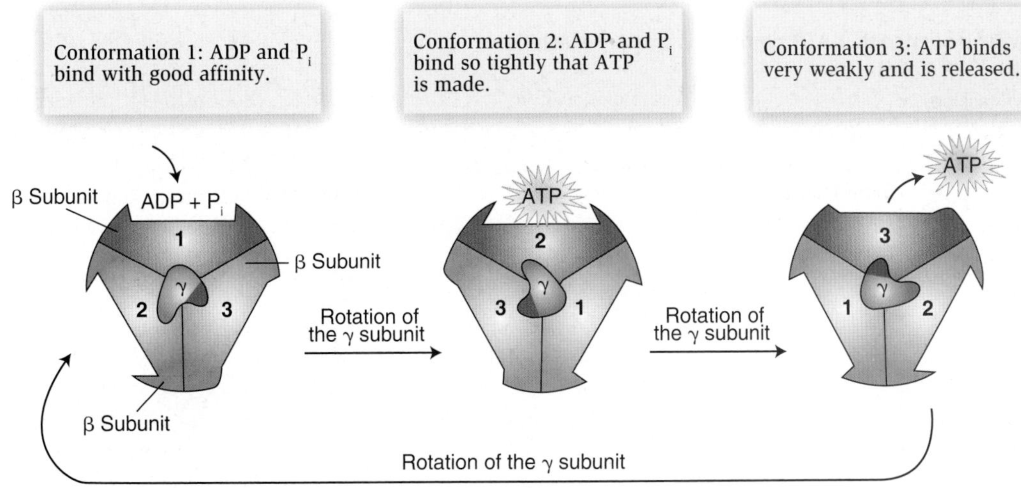

Figure 7.12 Conformational changes that result in ATP synthesis. For simplicity, the α subunits are not shown. This drawing emphasizes the conformational changes in the β subunit shown at the top. The other two β subunits also make ATP. All three β subunits alternate between three conformational states due to their interactions with the γ subunit.

change causes the γ subunit to turn clockwise (when viewed from the intermembrane space). Each time the γ subunit turns 120°, it changes its contacts with the three β subunits, which, in turn, causes the β subunits to change their conformations. How do these conformational changes promote ATP synthesis? The answer is that the conformational changes occur in a way that favors ATP synthesis and release. As shown in **Figure 7.12**, the conformational changes in the β subunits happen in the following order:

- Conformation 1: ADP and P_i bind with good affinity.
- Conformation 2: ADP and P_i bind so tightly that ATP is made.
- Conformation 3: ATP (and ADP and P_i) bind very weakly, and ATP is released.

Each time the γ subunit turns 120°, it causes a β subunit to change to the next conformation. After conformation 3, a 120° turn by the γ subunit returns a β subunit back to conformation 1, and the cycle of ATP synthesis can begin again. Because ATP synthase has three β subunits, each subunit is in a different conformation at any given time.

American biochemist Paul Boyer proposed the concept of a rotary machine in the late 1970s. In his model, the three β subunits alternate between three conformations, as described previously. Boyer's original idea was met with great skepticism, because the concept that part of an enzyme could spin was very novel, to say the least. In 1994, British biochemist John Walker and his colleagues determined the three-dimensional structure of the nonmembrane-embedded portion of the ATP synthase. The structure revealed that each of the three β subunits had a different conformation—one with ADP bound, one with ATP bound, and one without any nucleotide bound. This result supported Boyer's model. In 1997, Boyer and Walker shared the Nobel Prize in Chemistry for their work on ATP synthase. As described next in the Feature Investigation, other researchers subsequently visualized the rotation of the γ subunit.

FEATURE INVESTIGATION

Yoshida and Kinosita Demonstrated That the γ Subunit of ATP Synthase Spins

In 1997, Japanese biochemist Masasuke Yoshida, physicist Kazuhiko Kinosita, and colleagues set out to experimentally visualize the rotary nature of ATP synthase (**Figure 7.13**). The membrane-embedded region of ATP synthase can be separated from the rest of the protein by treatment of mitochondrial membranes with a high concentration of salt, releasing the portion of the protein containing one γ, three α, and three β subunits. The researchers adhered the $\gamma\alpha_3\beta_3$ complex to a glass slide so the γ subunit was protruding upward. Because the γ subunit is too small to be seen with a light microscope, the rotation of the γ subunit cannot be visualized directly. To circumvent this problem, the researchers attached a large, fluorescently labeled actin filament to the γ subunit via linker proteins. The fluorescently labeled actin filament is very long compared with the γ subunit and can be readily seen with a fluorescence microscope.

Because the membrane-embedded portion of the protein is missing, you may be wondering how the researchers could get the γ subunit to rotate. The answer is they added ATP. Although the normal function of the ATP synthase is to make ATP, it can also hydrolyze ATP. In other words, ATP synthase can run backwards. As shown in the data for Figure 7.13, when the researchers added ATP, they observed that the fluorescently labeled actin filament rotated in a counterclockwise direction, which is opposite to the direction that the γ subunit rotates when ATP is synthesized. Actin filaments were observed to rotate for more than 100 revolutions in the presence of ATP. These results convinced the scientific community that ATP synthase is a rotary machine.

Experimental Questions

1. The components of ATP synthase are too small to be visualized by light microscopy. For the experiment of Figure 7.13, how did the researchers observe the movement of ATP synthase?
2. In the experiment of Figure 7.13, what observation did the researchers make that indicated ATP synthase is a rotary machine? What was the control of this experiment? What did it indicate?
3. Were the rotations seen by the researchers in the data of Figure 7.13 in the same direction as expected in the mitochondria during ATP synthesis? Why or why not?

Figure 7.13 **Yoshida and Kinosita provide evidence that ATP synthase is a rotary machine.**

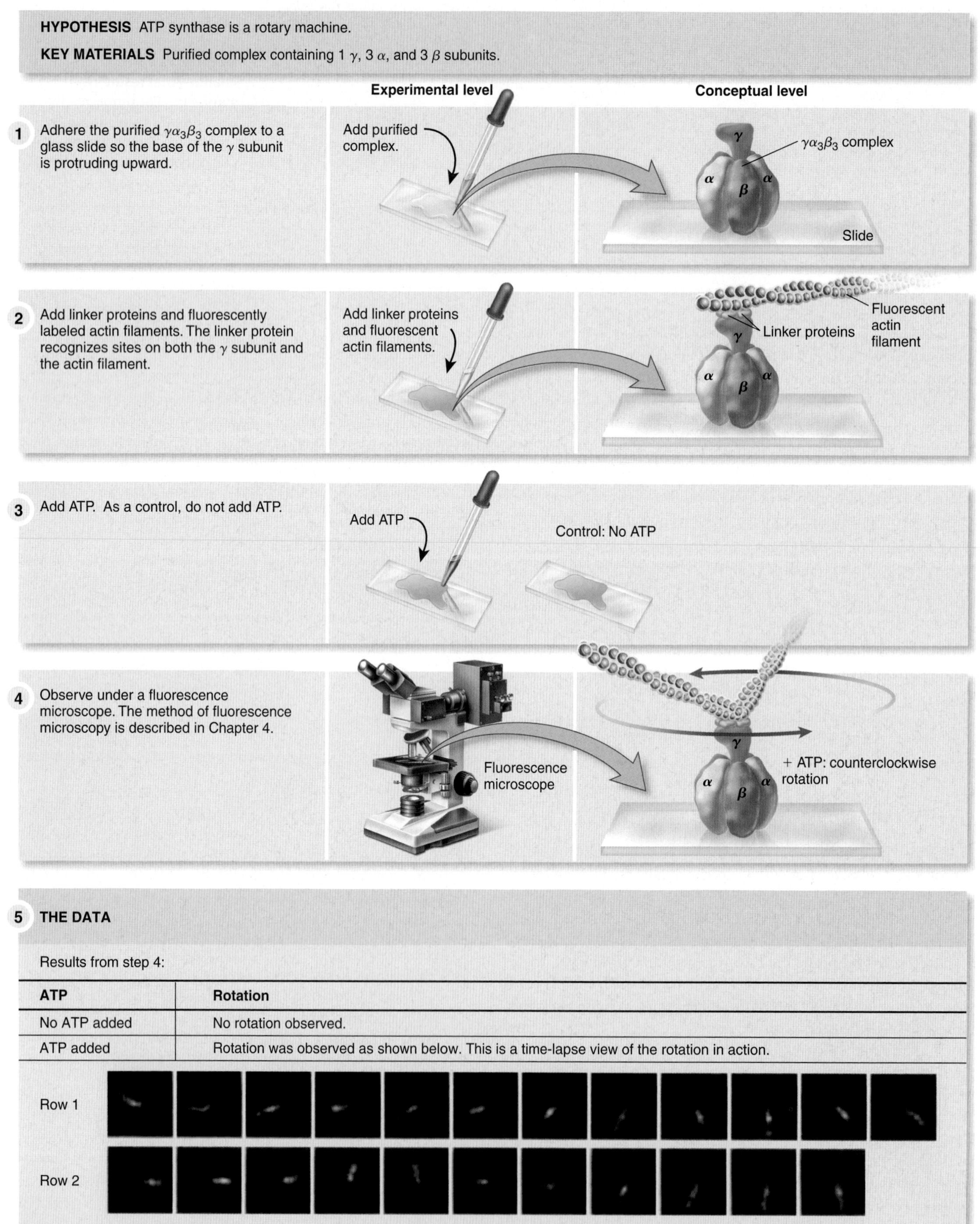

HYPOTHESIS ATP synthase is a rotary machine.
KEY MATERIALS Purified complex containing 1 γ, 3 α, and 3 β subunits.
Experimental level
Conceptual level
1 Adhere the purified $\gamma\alpha_3\beta_3$ complex to a glass slide so the base of the γ subunit is protruding upward.
Add purified complex.
$\gamma\alpha_3\beta_3$ complex
γ
α
β
α
Slide
2 Add linker proteins and fluorescently labeled actin filaments. The linker protein recognizes sites on both the γ subunit and the actin filament.
Add linker proteins and fluorescent actin filaments.
Fluorescent actin filament
Linker proteins
3 Add ATP. As a control, do not add ATP.
Add ATP
Control: No ATP
4 Observe under a fluorescence microscope. The method of fluorescence microscopy is described in Chapter 4.
Fluorescence microscope
+ ATP: counterclockwise rotation
5 THE DATA
Results from step 4:

ATP	Rotation
No ATP added	No rotation observed.
ATP added	Rotation was observed as shown below. This is a time-lapse view of the rotation in action.

Row 1
Row 2

6 **CONCLUSION** The γ subunit rotates counterclockwise when ATP is hydrolyzed. It would be expected to rotate clockwise when ATP is synthesized.

7 **SOURCE** Reprinted by permission from Macmillan Publishers Ltd. Noji, H., Yasuda, R., Yoshida, M., and Kinosita, K. 1997. Direct observation of the rotation of F_1-ATPase. *Nature* 386:299–303.

7.6 Connections Among Carbohydrate, Protein, and Fat Metabolism

Learning Outcome:

1. Explain how carbohydrate, protein, and fat metabolism are interconnected.

When you eat a meal, it usually contains not only carbohydrates (including glucose) but also proteins and fats. These molecules are broken down by some of the same enzymes involved with glucose metabolism.

As shown in Figure 7.14, proteins and fats can enter into glycolysis or the citric acid cycle at different points. Proteins are first acted on by enzymes, either in digestive juices or within cells, that cleave the bonds connecting individual amino acids. Because the 20 amino acids differ in their side chains, amino acids and their breakdown products can enter at different points in the pathway. Breakdown products of some amino acids can enter at later steps of glycolysis, or an acetyl group can be removed from certain amino acids and become attached to CoA and then enter the citric acid cycle (see Figure 7.14). Other amino acids are modified and enter the citric acid cycle.

Fats are typically broken down to glycerol and fatty acids. Glycerol can be modified to glyceraldehyde-3-phosphate and enter glycolysis. Fatty acyl tails can have two carbon acetyl units removed, which bind to CoA and enter the citric acid cycle. By using the same pathways for the breakdown of sugars, amino acids, and fats, cellular metabolism is more efficient because the same enzymes are used for the breakdown of different starting molecules.

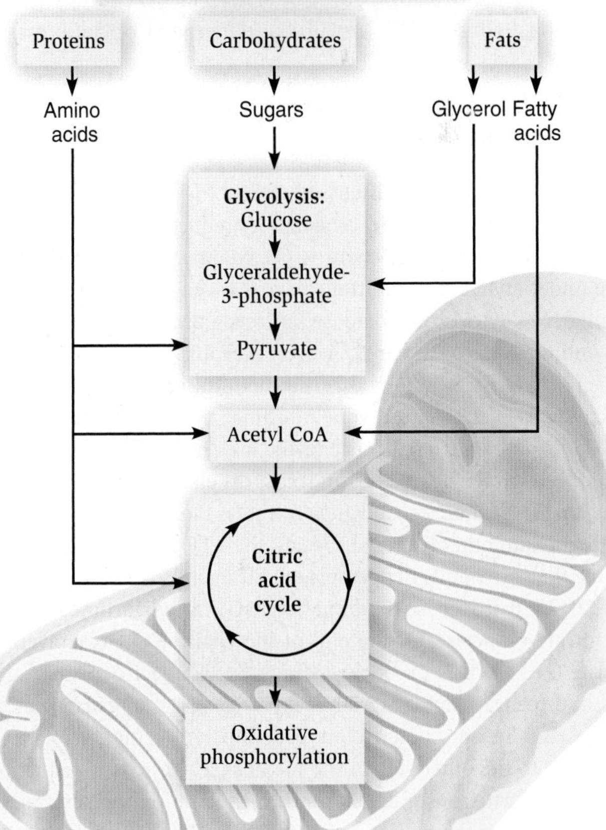

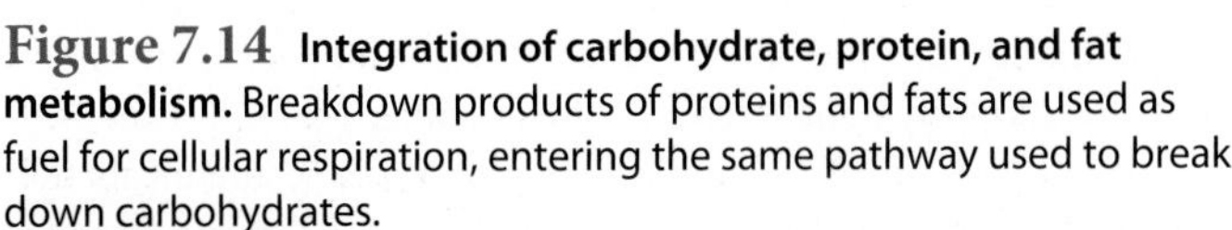

Figure 7.14 Integration of carbohydrate, protein, and fat metabolism. Breakdown products of proteins and fats are used as fuel for cellular respiration, entering the same pathway used to break down carbohydrates.

Concept Check: What is a cellular advantage of integrating protein, carbohydrate, and fat metabolism?

7.7 Anaerobic Respiration and Fermentation

Learning Outcomes:

1. Describe how certain microorganisms make ATP using a final electron acceptor of their electron transport chain that is not oxygen.
2. Explain how muscle and yeast cells use fermentation to synthesize ATP under anaerobic conditions.

Thus far, we have surveyed catabolic pathways that result in the complete breakdown of glucose in the presence of oxygen. Cells also commonly metabolize organic molecules in the absence of oxygen. The term **anaerobic** is used to describe an environment that lacks oxygen. Many bacteria and archaea and some fungi exist in anaerobic environments but still have to oxidize organic molecules to obtain sufficient amounts of energy. Examples include microbes living in your intestinal tract and those living deep in the soil. Similarly, when a person exercises strenuously, the rate of oxygen consumption by muscle cells may greatly exceed the rate of oxygen delivery—particularly at the start of strenuous exercise. Under these conditions, muscle cells become anaerobic and must obtain sufficient energy in the absence of oxygen to maintain their level of activity.

Organisms have evolved two different strategies to metabolize organic molecules in the absence of oxygen. One mechanism is to use a substance other than O_2 as the final electron acceptor of an electron transport chain, a process called **anaerobic respiration**. A second approach is to produce ATP only via substrate-level phosphorylation without any net oxidation of organic molecules, a process called fermentation. In this section, we will consider examples of both strategies.

Some Microorganisms Carry Out Anaerobic Respiration

At the end of the ETC discussed earlier in Figure 7.8, cytochrome oxidase recognizes O_2 and catalyzes its reduction to H_2O. The final electron acceptor of the chain is O_2. Many species of bacteria that live under anaerobic conditions have evolved enzymes that function similarly to cytochrome oxidase but recognize molecules other than O_2 and use them as the final electron acceptor.

For example, under anaerobic conditions *Escherichia coli*, a bacterial species found in your intestinal tract, produces an enzyme called nitrate reductase. This enzyme uses nitrate (NO_3^-) as the final electron acceptor of an electron transport chain. **Figure 7.15** shows a simplified ETC in *E. coli* in which nitrate is the final electron acceptor. In *E. coli* and other bacterial species, the ETC is in the plasma membrane that surrounds the cytoplasm. Electrons travel from NADH to NADH dehydrogenase to ubiquinone (Q) to cytochrome *b* and then to nitrate reductase. At the end of the chain, NO_3^- is converted to nitrite (NO_2^-). This process generates an H^+ electrochemical gradient in three ways. First, NADH dehydrogenase pumps H^+ out of the cytoplasm. Second, ubiquinone picks up H^+ in the cytoplasm and carries it to the other side of the membrane. Third, the reduction of nitrate to nitrite consumes H^+ in the cytoplasm. The generation of an H^+ gradient via these three processes allows *E. coli* cells to make ATP via chemiosmosis under anaerobic conditions.

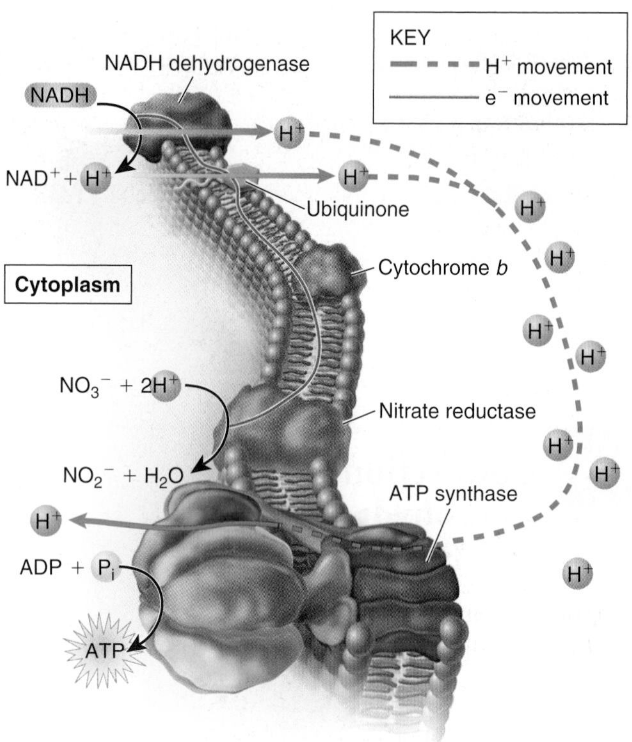

Figure 7.15 **An example of anaerobic respiration in *E. coli*.** When oxygen is absent, *E. coli* can use nitrate instead of oxygen as the final electron acceptor in an electron transport chain. This generates an H^+ electrochemical gradient that is used to make ATP via chemiosmosis. Note: As shown in this figure, ubiquinone (Q) picks up H^+ on one side of the membrane and deposits it on the other side. A similar event happens during aerobic respiration in mitochondria (see Figure 7.8), except that ubiquinone transfers H^+ to cytochrome *b-c*$_1$, which pumps it into the intermembrane space.

Fermentation Is the Breakdown of Organic Molecules Without Net Oxidation

Many organisms, including animals and yeast, use only O_2 as the final electron acceptor of their ETCs. When confronted with anaerobic conditions, these organisms must have a different way of producing sufficient ATP. One strategy is to make ATP via glycolysis, which can occur under both anaerobic or aerobic conditions. Under anaerobic conditions, cells do not use the citric acid cycle or the ETC, but make ATP only via glycolysis.

A key issue is that glycolysis requires NAD^+ and generates NADH. Under aerobic conditions, oxygen acts as a final electron acceptor, and the high-energy electrons from NADH can be used to make more ATP. To make ATP, NADH is oxidized to NAD^+. However, this cannot occur under anaerobic conditions in yeast and animals, and, as a result, NADH builds up and NAD^+ decreases. This is a potential problem for two reasons. First, at high concentrations, NADH haphazardly donates its electrons to other molecules and promotes the formation of free radicals, highly reactive chemicals that damage DNA and cellular proteins. For this reason, yeast and animal cells exposed to anaerobic conditions must have a way to remove the excess NADH generated from the breakdown of glucose. The second problem is the decrease in NAD$^+$. Cells need to regenerate NAD$^+$ to keep glycolysis running and make ATP via substrate-level phosphorylation.

How do muscle cells cope with the buildup of NADH and decrease in NAD$^+$? When a muscle is working strenuously and becomes anaerobic, as in high-intensity exercise, the pyruvate from glycolysis is reduced to make lactate. (The uncharged, or protonated, form is called lactic acid.) The electrons to reduce pyruvate are derived from NADH, which is oxidized to NAD$^+$ (**Figure 7.16a**). Therefore, this process decreases NADH and reduces its potentially harmful effects. It also increases the level of NAD$^+$, thereby allowing glycolysis to continue. The lactate is secreted from muscle cells. Once sufficient oxygen is restored, the lactate produced during strenuous exercise can be taken up by cells, converted back to pyruvate, and used for energy, or it may be used to make glucose by the liver and other tissues.

Yeast cells cope with anaerobic conditions differently. During wine making, a yeast cell metabolizes sugar under anaerobic conditions. The pyruvate is broken down to CO_2 and a two-carbon molecule called acetaldehyde. The acetaldehyde is then reduced by NADH to make ethanol, while NADH is oxidized to NAD$^+$ (**Figure 7.16b**). Similar to lactate production in muscle cells, this decreases NADH and increases NAD$^+$, thereby preventing the harmful effects of NADH and allowing glycolysis to continue.

The term **fermentation** is used to describe the breakdown of organic molecules to harness energy without any net oxidation (that is, without any removal of electrons). The pathways of breaking down glucose to lactate or ethanol are examples of fermentation. Although electrons are removed from an organic molecule such as glucose to make pyruvate and NADH, the electrons are donated back to an organic molecule in the production of lactate or ethanol. Therefore, there is no net removal of electrons from an organic molecule. Compared with oxidative phosphorylation, fermentation produces far less ATP, for two reasons. First, glucose is not oxidized completely to CO_2 and H_2O. Second, the NADH made during glycolysis cannot be used to make more ATP. Overall, the complete breakdown of glucose in the presence of oxygen yields 34 to 38 ATP molecules. By comparison, the anaerobic breakdown of glucose to lactate or ethanol yields only 2 ATP molecules.

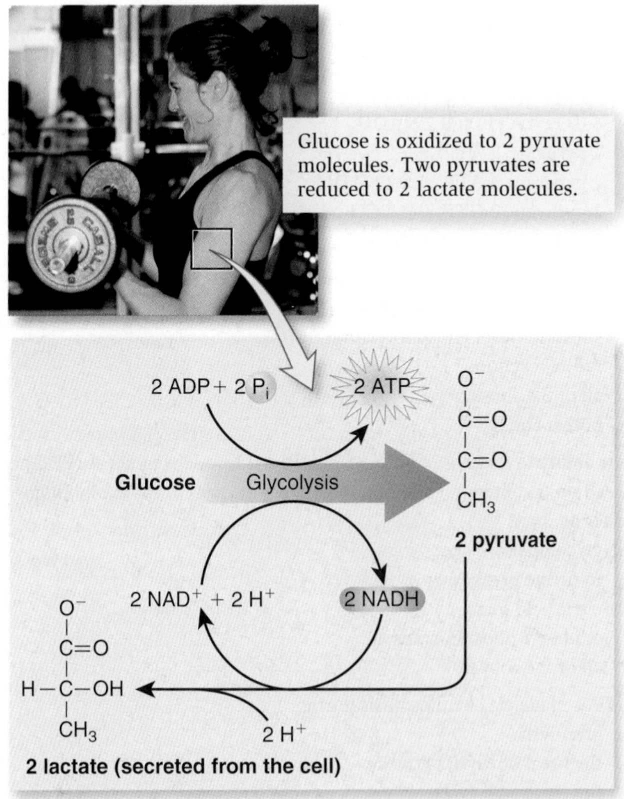

(a) Production of lactic acid

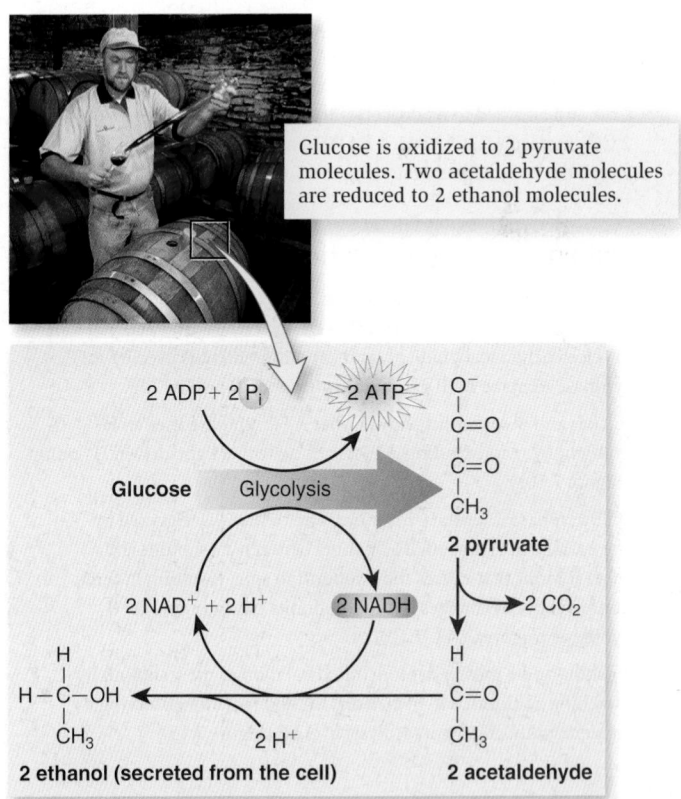

Glucose is oxidized to 2 pyruvate molecules. Two acetaldehyde molecules are reduced to 2 ethanol molecules.
2 ADP + 2 P_i
2 ATP
Glucose
Glycolysis
2 pyruvate
2 CO_2
2 NAD$^+$ + 2 H$^+$
2 NADH
2 H$^+$
2 ethanol (secreted from the cell)
2 acetaldehyde

(b) Production of ethanol

Figure 7.16 **Examples of fermentation.** In these examples, NADH is produced by the oxidation of an organic molecule, and then the NADH is converted back to NAD$^+$ by donating electrons to a different organic molecule such as pyruvate **(a)** or acetaldehyde **(b)**.

BIOLOGY PRINCIPLE Biology affects our society. Fermentation by microorganisms is used in wine making, beer brewing, and bread making.

Summary of Key Concepts

7.1 Overview of Cellular Respiration

- Cells obtain energy via cellular respiration, which involves the breakdown of organic molecules and the export of waste products.
- The breakdown of glucose occurs in four stages: glycolysis, pyruvate breakdown, citric acid cycle, and oxidative phosphorylation (Figure 7.1).

7.2 Glycolysis

- During glycolysis, which occurs in the cytosol, glucose is split into two molecules of pyruvate, with a net yield of 2 molecules of ATP and 2 of NADH. ATP is made by substrate-level phosphorylation (Figures 7.2, 7.3).
- Cancer cells preferentially carry out glycolysis, which enables the detection of tumors via a procedure called positron-emission tomography (PET) (Figure 7.4).

7.3 Breakdown of Pyruvate

- Pyruvate is broken down to CO_2 and an acetyl group that becomes attached to CoA. NADH is made during this process (Figure 7.5).

7.4 Citric Acid Cycle

- During the citric acid cycle, an acetyl group is removed from acetyl CoA and attached to oxaloacetate to make citrate. In a series of steps, 2 CO_2 molecules, 3 NADH, 1 $FADH_2$, and 1 ATP are made, after which the cycle begins again (Figures 7.6, 7.7).

7.5 Oxidative Phosphorylation

- Oxidative phosphorylation involves two events: (1) The electron transport chain (ETC) oxidizes NADH or $FADH_2$ and generates an H^+ electrochemical gradient, and (2) this gradient is used by ATP synthase to make ATP via chemiosmosis (Figures 7.8, 7.9).
- Racker and Stoeckenius showed that ATP synthase uses an H^+ gradient by reconstituting ATP synthase with a light-driven H^+ pump (Figure 7.10).
- ATP synthase is a rotary machine. The rotation is triggered by the passage of H^+ through a channel between a *c* subunit and the *a* subunit that causes the γ subunit to spin, resulting in three conformational changes in the β subunits that promote ATP synthesis (Figures 7.11, 7.12).
- Yoshida and Kinosita demonstrated rotation of the γ subunit by attaching a fluorescently labeled actin filament and observing its movement during the hydrolysis of ATP (Figure 7.13).

7.6 Connections Among Carbohydrate, Protein, and Fat Metabolism

- Proteins and fats can enter into glycolysis or the citric acid cycle at different points (Figure 7.14).

7.7 Anaerobic Respiration and Fermentation

- Anaerobic respiration occurs in the absence of oxygen. Certain microorganisms carry out anaerobic respiration in which the final electron acceptor of the ETC is a substance other than oxygen, such as nitrate (Figure 7.15).
- During fermentation, organic molecules are broken down without any net oxidation (that is, without any net removal of electrons). Examples include lactic acid production in muscle cells and ethanol production in yeast cells (Figure 7.16).

Assess and Discuss

Test Yourself

1. Which of the following pathways occurs in the cytosol?
   a. glycolysis
   b. breakdown of pyruvate to an acetyl group
   c. citric acid cycle
   d. oxidative phosphorylation
   e. all of the above
2. To break down glucose to CO_2 and H_2O, which of the following metabolic pathways is *not* involved?
   a. glycolysis
   b. breakdown of pyruvate to an acetyl group
   c. citric acid cycle
   d. photosynthesis
   e. c and d only
3. The net products of glycolysis are
   a. 6 CO_2, 4 ATP, and 2 NADH.
   b. 2 pyruvate, 2 ATP, and 2 NADH.
   c. 2 pyruvate, 4 ATP, and 2 NADH.
   d. 2 pyruvate, 2 GTP, and 2 CO_2.
   e. 2 CO_2, 2 ATP, and glucose.
4. During glycolysis, ATP is produced by
   a. oxidative phosphorylation.
   b. substrate-level phosphorylation.
   c. redox reactions.
   d. all of the above.
   e. both a and b.
5. The ability to diagnose tumors using [^{18}F]-fluorodeoxyglucose (FDG) is based on the phenomenon that most types of cancer cells exhibit higher levels of
   a. glycolysis.
   b. pyruvate breakdown.
   c. citric acid metabolism.
   d. oxidative phosphorylation.
   e. all of the above.
6. ATP is made via chemiosmosis during
   a. glycolysis.
   b. the breakdown of pyruvate.
   c. the citric acid cycle.
   d. oxidative phosphorylation.
   e. all of the above.

7. Certain drugs act as ionophores that cause the mitochondrial membrane to be highly permeable to H^+. How would such drugs affect oxidative phosphorylation?
   a. Movement of electrons down the ETC would be inhibited.
   b. ATP synthesis would be inhibited.
   c. ATP synthesis would be unaffected.
   d. ATP synthesis would be stimulated.
   e. Both a and b are correct.
8. The source of energy that *directly* drives the synthesis of ATP during oxidative phosphorylation is
   a. the oxidation of NADH.
   b. the oxidation of glucose.
   c. the oxidation of pyruvate.
   d. the H^+ gradient.
   e. the reduction of O_2.
9. Compared with oxidative phosphorylation in mitochondria, a key difference of anaerobic respiration in bacteria is that
   a. more ATP is made.
   b. ATP is made only via substrate-level phosphorylation.
   c. O_2 is converted to H_2O_2 rather than H_2O.
   d. something other than O_2 acts as a final electron acceptor of the ETC.
   e. b and d.
10. When a muscle becomes anaerobic during strenuous exercise, why is it necessary to convert pyruvate to lactate?
   a. to decrease NAD^+ and increase NADH
   b. to decrease NADH and increase NAD^+
   c. to increase NADH and increase NAD^+
   d. to decrease NADH and decrease NAD^+
   e. to keep oxidative phosphorylation running

Conceptual Questions

1. The electron transport chain is so named because electrons are transported from one component to another. Describe the purpose of the ETC.
2. What causes the rotation of the γ subunit of the ATP synthase? How does this rotation promote ATP synthesis?
3. A principle of biology is that *living organisms maintain homeostasis.* How is glucose breakdown regulated to maintain homeostasis? What would be some potentially harmful consequences if glucose metabolism was not regulated properly?

Collaborative Questions

1. Discuss the advantages and disadvantages of aerobic respiration, anaerobic respiration, and fermentation.
2. Read more about PET scans from other sources. Which types of cancers are most easily detected by this procedure and which types are not? Is the ability to detect cancer via a PET scan related to the state of hypoxia?

Online Resource

www.brookerbiology.com

Stay a step ahead in your studies with animations that bring concepts to life and practice tests to assess your understanding. Your instructor may also recommend the interactive eBook, individualized learning tools, and more.

Photosynthesis

Chapter Outline

8.1 Overview of Photosynthesis
8.2 Reactions That Harness Light Energy
8.3 Molecular Features of Photosystems
8.4 Synthesizing Carbohydrates via the Calvin Cycle
8.5 Variations in Photosynthesis
Summary of Key Concepts
Assess and Discuss

8

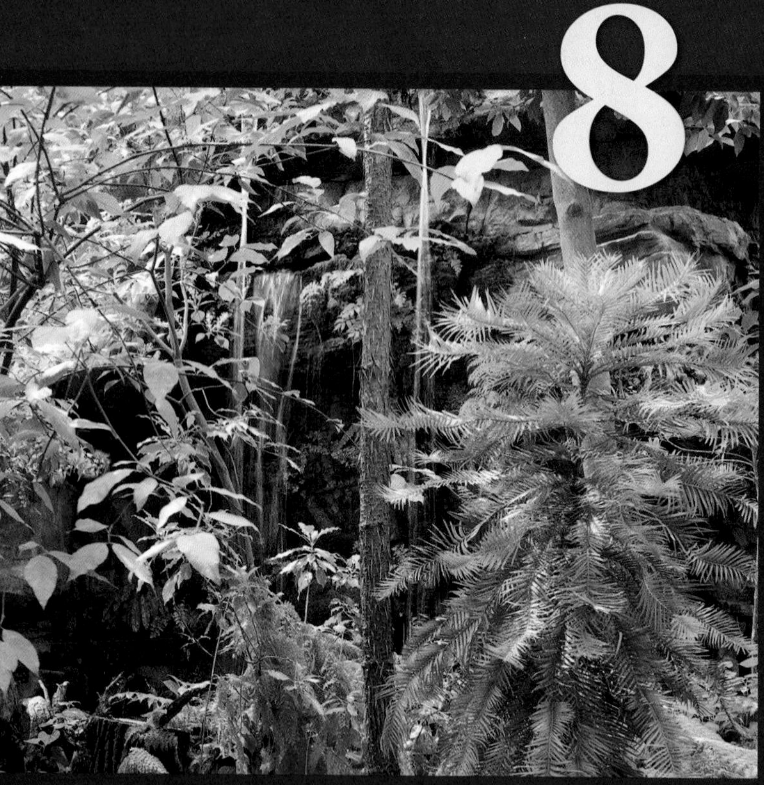

A tropical rain forest in the Amazon. Plant life in tropical rain forests carries out a large amount of the world's photosynthesis and supplies the atmosphere with a sizeable fraction of its oxygen.

Take a deep breath. Nearly all of the oxygen in every breath you take is made by the abundant plant life, algae, and cyanobacteria on Earth. More than 20% of the world's oxygen is produced in the Amazon rain forest in South America alone (see chapter opening photo). Biologists are alarmed about the rate at which such forests are being destroyed by human activities such as logging, mining, and oil extraction. Rain forests once covered 14% of the Earth's land surface, but they now occupy less than 6%. At their current rate of destruction, rain forests may be nearly eliminated in less than 40 years. Such an event may lower the level of oxygen in the atmosphere and thereby have a harmful effect on living organisms on a global scale.

In rain forests and across all of the Earth, the most visible color on land is green. The green color of plants is due to a pigment called chlorophyll. This pigment provides the starting point for the process of **photosynthesis**, in which the energy from light is captured and used to synthesize glucose and other organic molecules. Nearly all living organisms ultimately rely on photosynthesis for their nourishment, either directly or indirectly. Photosynthesis is also responsible for producing the oxygen that makes up a large portion of the Earth's atmosphere. Therefore, all aerobic organisms rely on photosynthesis for cellular respiration.

We begin this chapter with an overview of photosynthesis as it occurs in green plants and algae. We will then explore the two stages of photosynthesis in more detail. In the first stage, called the **light reactions**, light energy is absorbed by chlorophyll and converted to chemical energy in the form of two energy intermediates: ATP and NADPH. During the second stage, known as the **Calvin cycle**, ATP and NADPH are used to drive the synthesis of carbohydrates. We conclude with a consideration of the variations in photosynthesis that occur in plants existing in hot and dry conditions.

8.1 Overview of Photosynthesis

Learning Outcomes:

1. Write the general equations that represent the process of photosynthesis.
2. Explain how photosynthesis powers the biosphere.
3. Describe the general structure of chloroplasts.
4. Compare and contrast the two phases of photosynthesis: the light reactions and carbon fixation.

In the mid-1600s, a Flemish physician, Jan Baptista Van Helmont, conducted an experiment in which he transplanted the shoot of a young willow tree into a bucket of soil and allowed it to grow for 5 years. After this time, the willow tree had added 164 pounds to its original weight, but the soil had lost only 2 ounces. Van Helmont correctly concluded that the willow tree did not get most of its nutrients from the soil. He also hypothesized that the mass of the tree came from the water he had added over the 5 years. This hypothesis was partially correct, but we now know that CO_2 from the air is also a major contributor to the growth and mass of plants.

In the 1770s, Jan Ingenhousz, a Dutch physician, immersed green plants under water and discovered they released bubbles of oxygen. Ingenhousz determined that sunlight was necessary for oxygen production. During this same period, Jean Senebier, a Swiss botanist, found that CO_2 is required for plant growth. With this accumulating information, Julius von Mayer, a German physicist, proposed in 1845 that plants convert light energy from the Sun into chemical energy.

For the next several decades, plant biologists studied photosynthesis in plants, algae, and bacteria. Researchers discovered that some photosynthetic bacteria use hydrogen sulfide (H_2S) instead of water (H_2O) for photosynthesis, and these organisms release sulfur instead of oxygen. In the 1930s, based on this information, Dutch-American microbiologist Cornelis van Niel proposed a general equation for photosynthesis that applies to plants, algae, and photosynthetic bacteria alike.

$$CO_2 + 2\,H_2A + \text{Light energy} \rightarrow CH_2O + A_2 + H_2O$$

where A is oxygen (O) or sulfur (S) and CH_2O is the general formula for a carbohydrate. This is a redox reaction in which CO_2 is reduced and H_2A is oxidized.

In green plants, A is oxygen and 2 A is a molecule of oxygen that is designated O_2. Therefore, this equation becomes

$$CO_2 + 2\,H_2O + \text{Light energy} \rightarrow CH_2O + O_2 + H_2O$$

When the carbohydrate produced is glucose ($C_6H_{12}O_6$), we multiply each side of the equation by 6 to obtain:

$$6\,CO_2 + 12\,H_2O + \text{Light energy} \rightarrow C_6H_{12}O_6 + 6\,O_2 + 6\,H_2O$$

$$\Delta G = +685 \text{ kcal/mol}$$

In this redox reaction, CO_2 is reduced during the formation of glucose, and H_2O is oxidized during the formation of O_2. Notice that the free-energy change required for the production of 1 mole of glucose from carbon dioxide and water is a whopping +685 kcal/mol! As we learned in Chapter 6, endergonic reactions are driven forward by coupling the reaction with an exergonic process that releases free energy. In this case, the energy from sunlight ultimately drives the synthesis of glucose.

In this section, we will survey the general features of photosynthesis as it occurs in green plants and algae. Later sections will examine the various steps in this process.

Photosynthesis Powers the Biosphere

The term **biosphere** describes the regions on the surface of the Earth and in the atmosphere where living organisms exist. Organisms can be categorized as heterotrophs and autotrophs. **Heterotrophs** must consume food—organic molecules from their environment—to sustain life. Most species of bacteria and protists, as well as all species of fungi and animals, are heterotrophs. By comparison, **autotrophs** sustain themselves by producing organic molecules from inorganic sources such as CO_2 and H_2O. **Photoautotrophs** are autotrophs that use light as a source of energy to make organic molecules. These include green plants, algae, and some bacterial species such as cyanobacteria.

Life in the biosphere is largely driven by the photosynthetic power of green plants and algae. The existence of most species relies on a key energy cycle that involves the interplay between organic molecules (such as glucose) and inorganic molecules, namely, O_2, CO_2, and H_2O (**Figure 8.1**). Photoautotrophs make a large proportion of the Earth's organic molecules via photosynthesis, using light energy, CO_2, and H_2O. During this process, they also produce O_2. To supply their energy needs, both photoautotrophs and heterotrophs metabolize organic molecules via cellular respiration. As described in

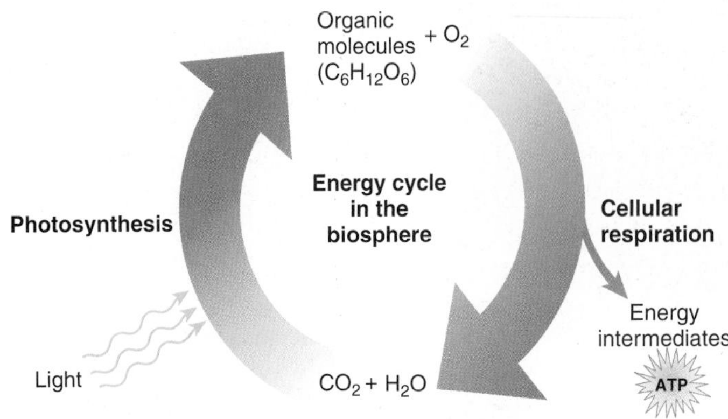

Figure 8.1 An important energy cycle between photosynthesis and cellular respiration. Photosynthesis uses light, CO_2 and H_2O to produce O_2 and organic molecules. The organic molecules can be broken down to CO_2 and H_2O via cellular respiration to supply energy in the form of ATP; O_2 is reduced to H_2O.

BIOLOGY PRINCIPLE Living organisms use energy. Photosynthetic species capture light energy and store it in organic molecules, which are used by photosynthetic and nonphotosynthetic species as sources of energy.

Concept Check: Which types of organisms carry out cellular respiration? Is it heterotrophs, autotrophs, or both?

Chapter 7, cellular respiration generates CO_2 and H_2O and is used to make ATP. The CO_2 is released into the atmosphere and can be reused by photoautotrophs to make more organic molecules such as glucose. In this way, an energy cycle between photosynthesis and cellular respiration sustains life on our planet.

In Plants and Algae, Photosynthesis Occurs in the Chloroplast

Chloroplasts are organelles found in plant and algal cells that carry out photosynthesis. These organelles contain large quantities of **chlorophyll**, which is a pigment that gives plants their green color. All green parts of a plant contain chloroplasts and can perform photosynthesis, although the majority of photosynthesis occurs in the leaves (**Figure 8.2**). The tissue in the internal part of the leaf, called the **mesophyll**, contains cells with chloroplasts. For photosynthesis to occur, the mesophyll cells must obtain water and carbon dioxide. The water is taken up by the roots of the plant and is transported to the leaves by small veins. Carbon dioxide gas enters the leaf, and oxygen exits via pores called stomata (singular, stoma or stomate; from the Greek, meaning mouth).

Like the mitochondrion, a chloroplast contains an outer and inner membrane, with an intermembrane space lying between the two. A third membrane, called the **thylakoid membrane**, contains pigment molecules, including chlorophyll. The thylakoid membrane forms many flattened, fluid-filled tubules called **thylakoids**, which enclose a single, convoluted compartment known as the **thylakoid lumen**. Thylakoids stack on top of each other to form a structure called a **granum** (plural, grana). The **stroma** is the fluid-filled region of the chloroplast between the thylakoid membrane and the inner membrane (see Figure 8.2).

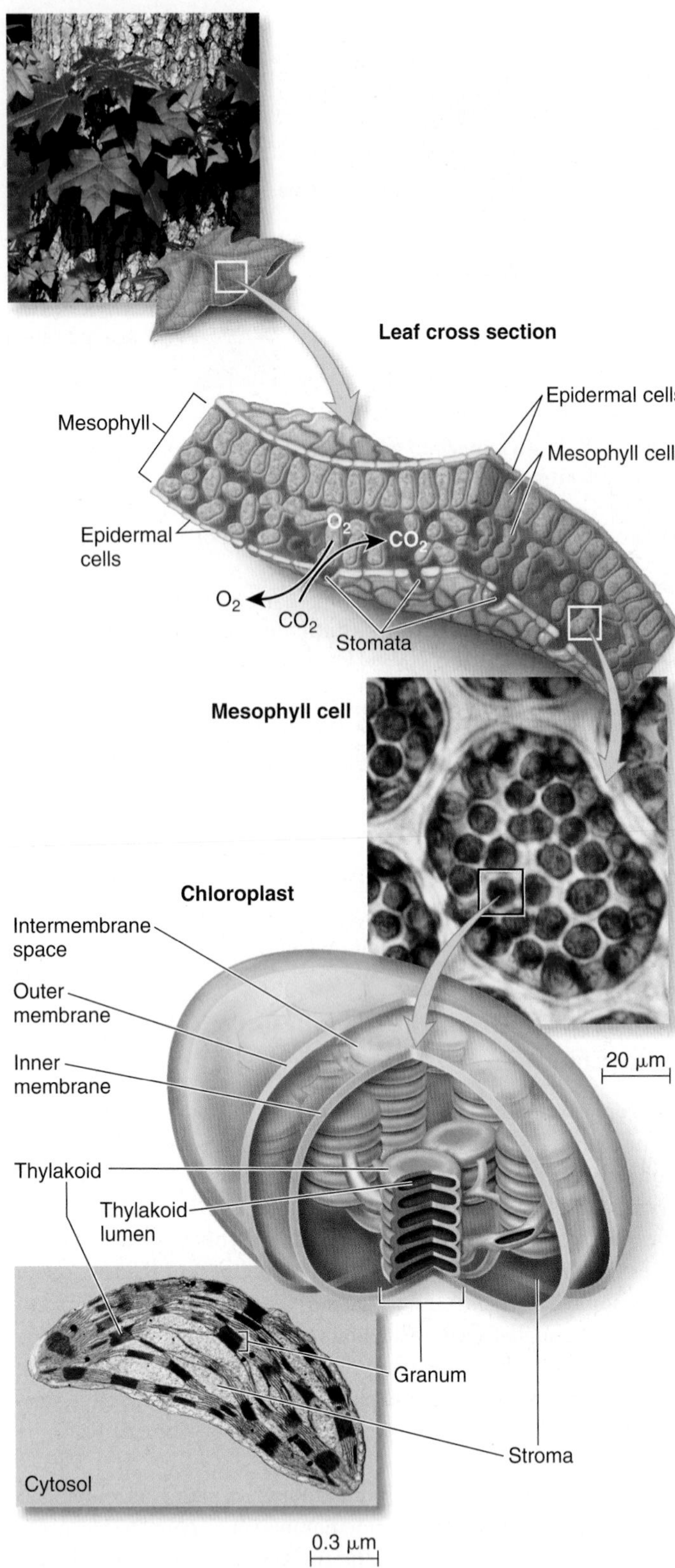

Figure 8.2 **Leaf organization.** Leaves are composed of layers of cells. The epidermal cells are on the outer surface, both top and bottom, with mesophyll cells sandwiched in the middle. The mesophyll cells contain chloroplasts and are the primary sites of photosynthesis in most plants.

BioConnections: *Look ahead to Figure 38.17. How many guard cells make up a stoma (plural, stomata)?*

Photosynthesis Occurs in Two Stages: Light Reactions and the Calvin Cycle

How does photosynthesis take place? As mentioned, the process of photosynthesis occurs in two stages called the light reactions and the Calvin cycle. The term photosynthesis is derived from the association between these two stages: The prefix photo refers to the light reactions that capture the energy from sunlight needed for the synthesis of carbohydrates that occurs in the Calvin cycle. The light reactions take place at the thylakoid membrane, and the Calvin cycle occurs in the stroma (Figure 8.3).

The light reactions involve an amazing series of energy conversions, starting with light energy and ending with chemical energy that is stored in the form of covalent bonds. The light reactions produce three chemical products: ATP, NADPH, and O_2. ATP and NADPH are energy intermediates that provide the needed energy and electrons to drive the Calvin cycle. Like NADH, **NADPH (nicotinamide adenine dinucleotide phosphate)** is an electron carrier that can accept two electrons. Its structure differs from NADH by the presence of an additional phosphate group. The structure of NADH is described in Chapter 6 (see Figure 6.12).

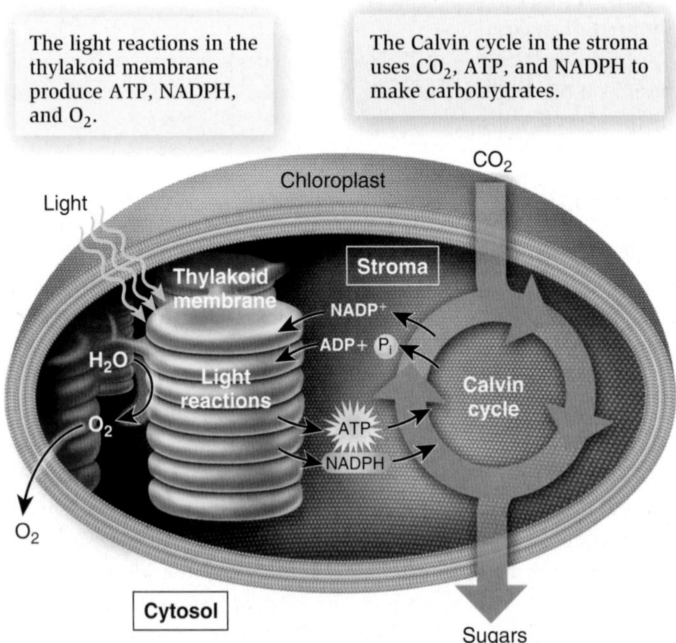

Figure 8.3 **An overview of the two stages of photosynthesis: light reactions and the Calvin cycle.** The light reactions, through which ATP, NADPH, and O_2 are made, occur at the thylakoid membrane. The Calvin cycle, in which enzymes use ATP and NADPH to incorporate CO_2 into carbohydrate, occurs in the stroma.

Concept Check: *Can the Calvin cycle occur in the dark?*

8.2 Reactions That Harness Light Energy

Learning Outcomes:

1. Define the general properties of light.
2. Describe how pigments absorb light energy and the types of pigments found in plants and green algae.
3. Outline the steps in which photosystem II and I capture light energy and produce O_2, ATP, and NADPH.
4. Explain the process of cyclic photophosphorylation in which only ATP is made.

According to the first law of thermodynamics discussed in Chapter 6, energy cannot be created or destroyed, but it can be transferred from one place to another and transformed from one form to another. During photosynthesis, energy in the form of light is transferred from the Sun, some 92 million miles away, to a pigment molecule in a photosynthetic organism such as a plant. What follows is an interesting series of energy transformations in which light energy is transformed into electrochemical energy and then into energy stored within chemical bonds.

In this section, we will explore this series of transformations, collectively called the light reactions of photosynthesis. We begin by examining the properties of light and then consider the features of chloroplasts that allow them to capture light energy. The remainder of this section focuses on how the light reactions of photosynthesis generate three important products: ATP, NADPH, and O_2.

Light Energy Is a Form of Electromagnetic Radiation

Light is essential to support life on Earth. Light is a type of electromagnetic radiation, so named because it consists of energy in the form of electric and magnetic fields. Electromagnetic radiation travels as waves caused by the oscillation of the electric and magnetic fields. The **wavelength** is the distance between the peaks in a wave pattern. The **electromagnetic spectrum** encompasses all possible wavelengths of electromagnetic radiation, from relatively short wavelengths (gamma rays) to much longer wavelengths (radio waves) (Figure 8.4). Visible light is the range of wavelengths detected by the human eye, commonly between 380–740 nm. As discussed later, visible light provides the energy to drive photosynthesis.

Physicists have also discovered that light has properties that are characteristic of particles. Albert Einstein formulated the photon theory of light in which he proposed that light is composed of discrete particles called **photons**—massless particles traveling in a wavelike pattern and moving at the speed of light (about 300 million m/sec). Each photon contains a specific amount of energy. An important difference between the various types of electromagnetic radiation, shown in Figure 8.4, is the amount of energy found in the photons. Shorter wavelength radiation carries more energy per unit of time than longer wavelength radiation. For example, the photons of gamma rays carry more energy than those of radio waves.

The Sun radiates the entire spectrum of electromagnetic radiation, but the atmosphere prevents much of this radiation from reaching the Earth's surface. For example, the ozone layer forms a thin shield in the upper atmosphere, protecting life on Earth from much

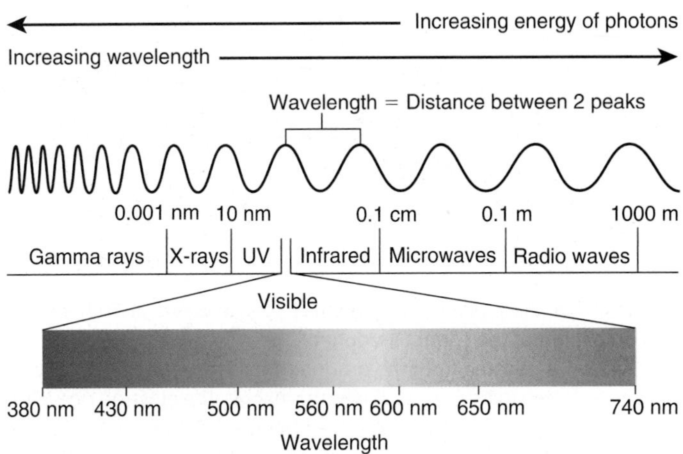

Figure 8.4 The electromagnetic spectrum. The bottom portion of this figure emphasizes visible light—the wavelengths of electromagnetic radiation visible to the human eye. Light in the visible portion of the electromagnetic spectrum drives photosynthesis.

Concept Check: Which has higher energy, gamma rays or radio waves?

of the Sun's ultraviolet (UV) rays. Even so, a substantial amount of electromagnetic radiation does reach the Earth's surface. The effect of light on living organisms is critically dependent on the energy of the photons that reach them. The photons found in gamma rays, X-rays, and UV rays have very high energy. When molecules in cells absorb such energy, the effects can be devastating. Such radiation can cause mutations in DNA and even lead to cancer. By comparison, the energy of photons found in visible light is much milder. Molecules can absorb this energy in a way that does not cause damage. Next, we will consider how molecules in living cells absorb the energy within visible light.

Pigments Absorb Light Energy

When light strikes an object, one of three things happens. First, light may simply pass through the object. Second, the object may change the path of light toward a different direction. A third possibility is that the object may absorb the light. The term **pigment** is used to describe a molecule that can absorb light energy. When light strikes a pigment, some of the wavelengths of light energy are absorbed, while others are reflected. For example, leaves look green to us because they reflect radiant energy of the green wavelength. Various pigments in the leaves absorb the energy of other wavelengths. At the extremes of color reflection are white and black. A white object reflects nearly all of the visible light energy falling on it, whereas a black object absorbs nearly all of the light energy. This is why it is coolest to wear white clothes on a sunny, hot day.

What do we mean when we say that light energy is absorbed? In the visible spectrum, light energy may be absorbed by boosting electrons to higher energy levels (Figure 8.5). Recall from Chapter 2 that electrons are located around the nucleus of an atom (refer back to Figure 2.4). The location in which an electron is likely to be found is called its orbital. Electrons in different orbitals possess different amounts of energy. For an electron to absorb light energy and be boosted to an orbital with a higher energy, it must overcome

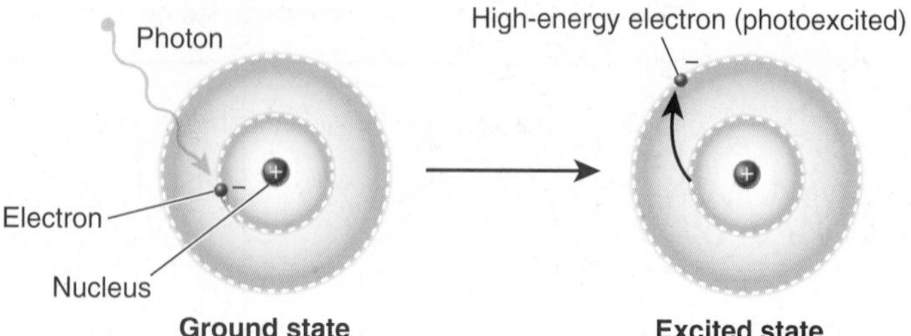

Figure 8.5 Absorption of light energy by an electron. When a photon of light of the correct amount of energy strikes an electron, the electron is boosted from the ground (unexcited) state to a higher energy level (an excited state). When this occurs, the electron occupies an orbital that is farther away from the nucleus of the atom. At this farther distance, the electron is held less firmly and is considered unstable.

Concept Check: *Describe the three things that could happen to enable a photoexcited electron to become more stable.*

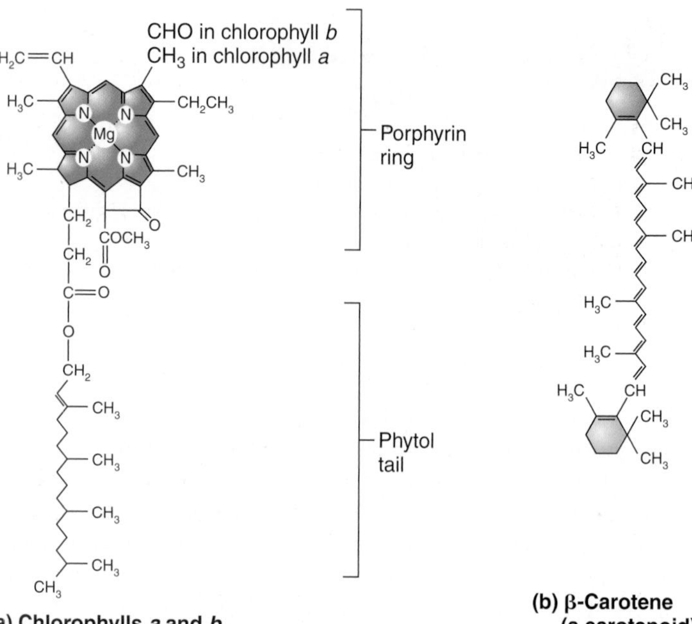

Figure 8.6 Structures of pigment molecules. (a) The structure of chlorophylls *a* and *b*. As indicated, chlorophylls *a* and *b* differ only at a single site, at which chlorophyll *a* has a —CH_3 group and chlorophyll *b* has a —CHO group. **(b)** The structure of *β*-carotene, an example of a carotenoid. The dark green and light green areas in parts (a) and (b) are the regions where a delocalized electron can hop from one atom to another.

the difference in energy between the orbital it is in and the orbital to which it is going. For this to happen, an electron must absorb a photon that contains precisely that amount of energy. Different pigment molecules contain a variety of electrons that can be shifted to different energy levels. Therefore, the wavelength of light that a pigment absorbs depends on the amount of energy needed to boost an electron to a higher orbital.

After an electron absorbs energy, it is said to be in an excited state. Usually, this is an unstable condition. The electron may release the energy in different ways. First, when an excited electron drops back down to a lower energy level, it may release heat. For example, on a sunny day, the sidewalk heats up because it absorbs light energy that is released as heat. A second way that an electron can release energy is in the form of light. Certain organisms, such as jellyfish, possess molecules that make them glow. This glow is due to the release of light when electrons drop down to lower energy levels, a phenomenon called fluorescence.

In the case of photosynthetic pigments, however, a different event happens that is critical for the process of photosynthesis. Rather than releasing energy, an excited electron in a photosynthetic pigment is removed from that molecule and transferred to another molecule where the electron is more stable. When this occurs, the energy in the electron is said to be "captured," because the electron does not readily drop down to a lower energy level and release heat or light.

Plants Contain Different Types of Photosynthetic Pigments

In plants, different pigment molecules absorb the light energy used to drive photosynthesis. Two types of chlorophyll pigments, termed **chlorophyll *a*** and **chlorophyll *b***, are found in green plants and green algae. Their structure was determined in the 1930s by German chemist Hans Fischer (**Figure 8.6a**). In the chloroplast, both chlorophylls *a* and *b* are bound to integral membrane proteins in the thylakoid membrane.

The chlorophylls contain a porphyrin ring and a phytol tail. A magnesium ion (Mg^{2+}) is bound to the porphyrin ring. An electron in the porphyrin ring follows a path in which it spends some of its time around several different atoms. Because this electron isn't restricted to a single atom, it is called a delocalized electron. The delocalized electron can absorb light energy. The phytol tail in chlorophyll is a long hydrocarbon structure that is hydrophobic. Its function is to anchor the pigment to the surface of hydrophobic proteins within the thylakoid membrane of chloroplasts.

Carotenoids are another type of pigment found in chloroplasts (**Figure 8.6b**). These pigments impart a color that ranges from yellow to orange to red. Carotenoids are often the major pigments in flowers and fruits. In leaves, the more abundant chlorophylls usually mask the colors of carotenoids. In temperate climates where the leaves change colors, the quantity of chlorophyll in the leaf declines during autumn. The carotenoids become readily visible and produce the yellows and oranges of autumn foliage.

An **absorption** spectrum is a graph that plots a pigment's light absorption as a function of wavelength. Each of the photosynthetic pigments shown in **Figure 8.7a** absorbs light in different regions of the visible spectrum. The absorption spectra of chlorophylls *a* and *b* are slightly different, though both chlorophylls absorb light most strongly in the red and violet parts of the visible spectrum and absorb green light poorly. Green light is reflected, which is why leaves appear green during the growing season. Carotenoids absorb light in the blue and blue-green regions of the visible spectrum, reflecting yellow and red.

Why do plants have different pigments? Having different pigments allows plants to absorb light at many different wavelengths. In this way, plants are more efficient at capturing the energy in sunlight.

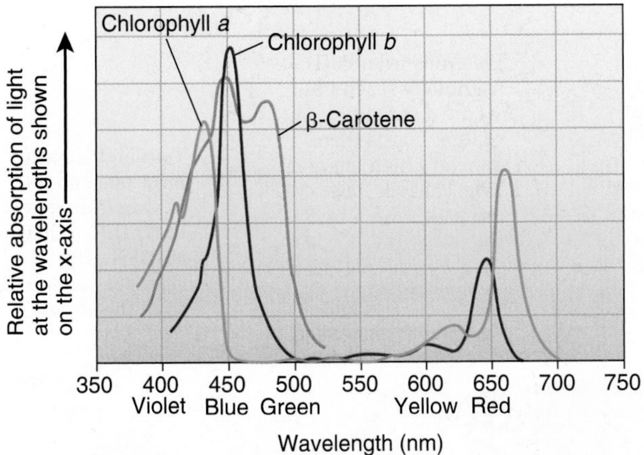

(a) Absorption spectra

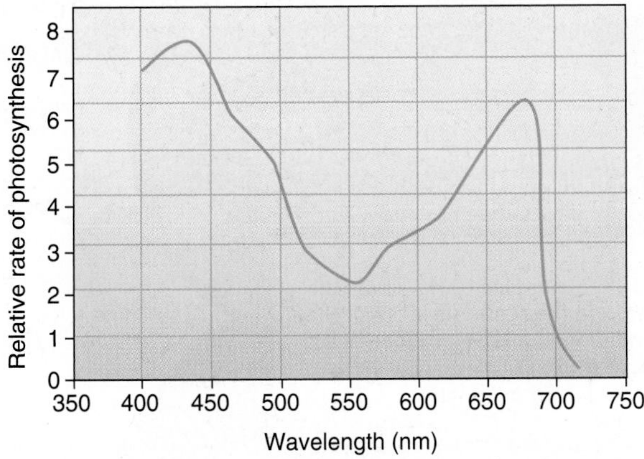

(b) Action spectrum

Figure 8.7 **Properties of pigment function: absorption and action spectra.** **(a)** These absorption spectra show the absorption of light by chlorophyll *a*, chlorophyll *b*, and *β*-carotene. **(b)** An action spectrum of photosynthesis depicting the relative rate of photosynthesis in green plants at different wavelengths of light.

Concept Check: *What is the advantage of having different pigment molecules?*

This phenomenon is highlighted in an **action spectrum**, which plots the rate of photosynthesis as a function of wavelength (Figure 8.7b). The highest rates of photosynthesis in green plants correlate with the wavelengths that are strongly absorbed by the chlorophylls and carotenoids. Photosynthesis is poor in the green region of the spectrum, because these pigments do not readily absorb this wavelength of light.

Photosystems II and I Work Together to Produce ATP and NADPH

As noted previously, photosynthetic organisms have the unique ability not only to absorb light energy but also to capture that energy in a stable way. Many organic molecules can absorb light energy. For example, on a sunny day, molecules in your skin absorb light energy and release the energy as heat. The heat that is released, however, cannot be harnessed to do useful work. A key feature of photosynthesis is the ability of pigments to capture light energy and transfer it to other molecules that can hold on to the energy in a stable fashion and ultimately produce energy-intermediate molecules that can do cellular work.

Let's now consider how chloroplasts capture light energy. The thylakoid membranes of the chloroplast contain two distinct complexes of proteins and pigment molecules called **photosystem I (PSI)** and **photosystem II (PSII)** (Figure 8.8). Photosystem I was discovered before photosystem II, but photosystem II is the initial step in photosynthesis. We will consider the structure and function of PSII in greater detail later in this chapter.

As described in steps 1 and 2 of Figure 8.8, light excites electrons in pigment molecules, such as chlorophylls, which are located in regions of PSII and PSI called light-harvesting complexes. Rather than releasing their energy in the form of heat, the excited electrons follow a path shown by the red arrow. The combined action of photosystem II and photosystem I is termed **noncyclic electron flow** because the electrons move linearly from PSII to PSI and ultimately reduce $NADP^+$ to NADPH.

Initially, the excited electrons move from a pigment molecule called P680 in PSII to other electron carriers called pheophytin (Pp), Q_A, and Q_B. The excited electrons are moved out of PSII by Q_B. PSII also oxidizes water, which generates O_2 and adds H^+ into the thylakoid lumen. The electrons released from the oxidized water molecules are used to replenish the electrons that leave PSII via Q_B.

After a pair of electrons reaches Q_B, each one enters an electron transport chain (ETC)—a series of electron carriers—located in the thylakoid membrane. The ETC functions similarly to the one found in mitochondria. From Q_B, an electron goes to a cytochrome complex; then to plastocyanin (Pc), a small protein; and then to photosystem I. Along its journey from photosystem II to photosystem I, the electron releases some of its energy at particular steps and is transferred to the next component that has a higher electronegativity. The energy released is harnessed to pump H^+ into the thylakoid lumen.

A key role of photosystem I is to make NADPH (see Figure 8.8, step 3). When light strikes the light-harvesting complex of photosystem I, this energy is also transferred to a reaction center, where a high-energy electron is removed from a pigment molecule, designated P700, and transferred to a primary electron acceptor. A protein called ferredoxin (Fd) can accept two high-energy electrons, one at a time, from the primary electron acceptor. Fd then transfers the two electrons to the enzyme $NADP^+$ reductase. This enzyme transfers the two electrons to $NADP^+$ and together with an H^+ produces NADPH. The formation of NADPH results in fewer H^+ in the stroma.

The synthesis of ATP in chloroplasts is achieved by a chemiosmotic mechanism called **photophosphorylation**, which is similar to that used to make ATP in mitochondria. In chloroplasts, ATP synthesis is driven by the flow of H^+ from the thylakoid lumen into the stroma via ATP synthase (Figure 8.8, step 4). An H^+ electrochemical gradient is generated in three ways: (1) the splitting of water, which places H^+ in the thylakoid lumen; (2) the movement of high-energy electrons from photosystem II to photosystem I, which pumps H^+ into the thylakoid lumen; and (3) the formation of NADPH, which consumes H^+ in the stroma.

A key difference between photosystem II and photosystem I is how the pigment molecules receive electrons. As discussed in more detail later, $P680^+$ receives an electron from water. By comparison,

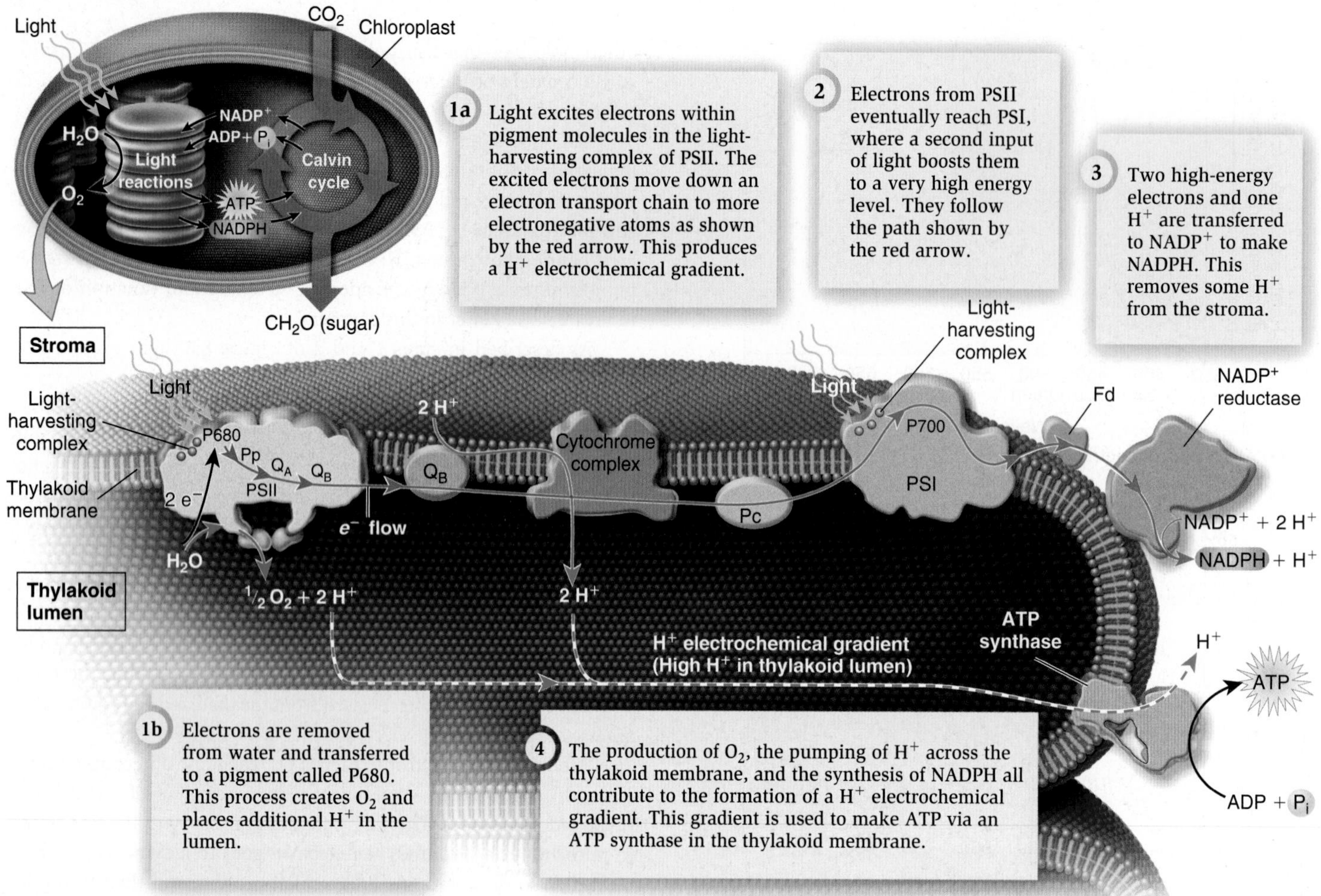

Figure 8.8 The synthesis of ATP, NADPH, and O_2 by the concerted actions of photosystems II and I. The linear process of electron movement from photosystem II to photosystem I to NADPH is called noncyclic electron flow.

Concept Check: *Are ATP, NADPH, and O_2 produced in the stroma or in the thylakoid lumen?*

$P700^+$—the oxidized form of P700—receives an electron from Pc. Therefore, photosystem I does not need to split water to reduce $P700^+$ and does not generate oxygen.

In summary, the steps of the light reactions of photosynthesis produce three chemical products: O_2, NADPH, and ATP:

1. O_2 is produced in the thylakoid lumen by the oxidation of water by photosystem II. Two electrons are removed from water, which produces 2 H^+ and $^1/_2\,O_2$. The two electrons are transferred to $P680^+$ molecules.
2. NADPH is produced in the stroma from high-energy electrons that start in photosystem II and are boosted a second time in photosystem I. Two high-energy electrons and one H^+ are transferred to $NADP^+$ to produce NADPH.
3. ATP is produced in the stroma via ATP synthase that uses an H^+ electrochemical gradient.

Cyclic Electron Flow Produces Only ATP

As mentioned, the mechanism of harvesting light energy described in Figure 8.8 is called noncyclic electron flow because it is a linear process. This electron flow produces ATP and NADPH in roughly equal amounts. However, as we will see later, the Calvin cycle uses more ATP than NADPH. How can plant cells avoid making too much NADPH and not enough ATP? In 1959, Daniel Arnon discovered a pattern of electron flow that is cyclic and generates only ATP (**Figure 8.9**). Arnon termed the process **cyclic photophosphorylation** because (1) the path of electrons is cyclic, (2) light energizes the electrons, and (3) ATP is made via the phosphorylation of ADP. Due to the path of electrons, the mechanism is also called **cyclic electron flow.**

When light strikes photosystem I, high-energy electrons are sent to the primary electron acceptor and then to ferredoxin (Fd). The key difference in cyclic photophosphorylation is that the high-energy electrons are transferred from Fd to Q_B. From Q_B, the electrons then go to the cytochrome complex, then to plastocyanin (Pc), and back to photosystem I. As the electrons travel along this cyclic route, they release energy, and some of this energy is used to transport H^+ into the thylakoid lumen. The resulting H^+ gradient drives the synthesis of ATP via ATP synthase.

Cyclic electron flow is favored when the level of $NADP^+$ is low and NADPH is high. Under these conditions, there is sufficient NADPH to run the Calvin cycle, which is described later. Alternatively, when $NADP^+$ is high and NADPH is low, noncyclic electron flow is favored, so more NADPH can be made. Cyclic electron flow is also favored when ATP levels are low.

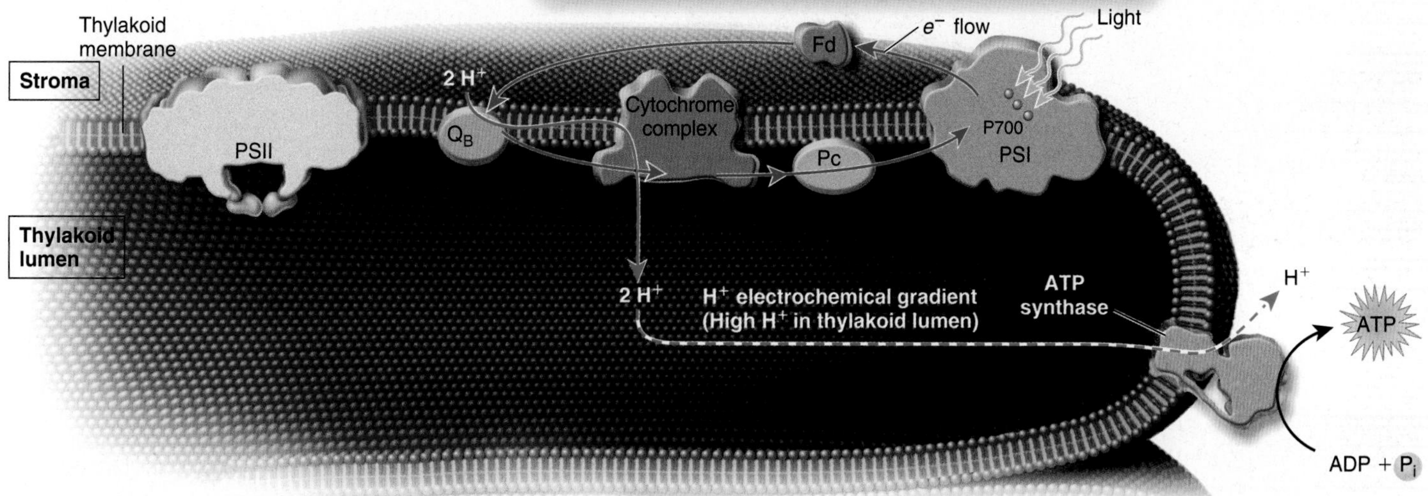
When light strikes photosystem I, electrons are excited and sent to ferredoxin (Fd). From Fd, the electrons are then transferred to Q_B, to the cytochrome complex, to plastocyanin (Pc), and back to photosystem I. This produces a H^+ electrochemical gradient, which is used to make ATP via ATP synthase.
Thylakoid membrane
Stroma
Thylakoid lumen
PSII
2 H+
Q_B
Cytochrome complex
Fd
e− flow
Light
P700
PSI
Pc
2 H+
H+ electrochemical gradient (High H+ in thylakoid lumen)
ATP synthase
H+
ATP
ADP + Pi

Figure 8.9 **Cyclic photophosphorylation.** In this process, an electron follows a cyclic path that is powered by photosystem I (PSI). This contributes to the formation of an H^+ electrochemical gradient, which is then used to make ATP by ATP synthase.

Concept Check: *Why is having cyclic photophosphorylation an advantage to a plant over having only noncyclic electron flow?*

GENOMES & PROTEOMES CONNECTION

The Cytochrome Complexes of Mitochondria and Chloroplasts Contain Evolutionarily Related Proteins

A recurring theme in cell biology is that evolution has resulted in groups of genes that encode proteins that play similar but specialized roles in cells—an example of descent with modification. When two or more genes are similar because they are derived from the same ancestral gene, they are called **homologous genes**. As discussed in Chapter 23, homologous genes encode proteins that have similar amino acid sequences and often perform similar functions.

A comparison of the electron transport chains of mitochondria and chloroplasts reveals homologous genes. In particular, let's consider the cytochrome complex found in the thylakoid membrane of plants and algae, called cytochrome b_6-f (**Figure 8.10a**) and cytochrome b-c_1, which is found in the ETC of mitochondria (**Figure 8.10b**; also refer back to Figure 7.9). Both cytochromes b_6-f and b-c_1 are composed of several protein subunits. One of those proteins is called cytochrome b_6 in cytochrome b_6-f and cytochrome b in cytochrome b-c_1.

By analyzing the sequences of the genes that encode these proteins, researchers discovered that cytochrome b_6 and cytochrome b are homologous proteins. These proteins carry out similar functions: Both of them accept electrons from a quinone (Q_B or ubiquinone) and both donate an electron to another protein within their respective complexes (cytochrome f or cytochrome c_1). Likewise, both proteins function as H^+ pumps that capture some of the energy that is released from electrons to transport H^+ across the membrane. In this way, evolution has produced a family of cytochrome b-type proteins that play similar but specialized roles.

8.3 Molecular Features of Photosystems

Learning Outcomes:

1. Describe how the light-harvesting complex absorbs light energy and how it is transferred via resonance energy transfer.
2. Diagram the path of electron flow through photosystem II.
3. Explain how O_2 is produced by photosystem II.

The previous section provided an overview of how chloroplasts absorb light energy and produce ATP, NADPH, and O_2. As you have learned, two photosystems—PSI and PSII—play critical roles in two aspects of photosynthesis. First, both PSI and PSII absorb light energy and capture that energy in the form of excited electrons. Second, PSII oxidizes water, thereby producing O_2. In this section, we will take a closer look at how these events occur at the molecular level.

Photosystem II Captures Light Energy and Produces O_2

PSI and PSII have two main components: a light-harvesting complex and a reaction center. **Figure 8.11** shows how these components function in PSII. In 1932, American biologist Robert Emerson and an undergraduate student, William Arnold, originally discovered the **light-harvesting complex** in the thylakoid membrane. It is composed of several dozen pigment molecules that are anchored to transmembrane proteins. The role of the complex is to directly absorb photons of light. When a pigment molecule absorbs a photon, an electron is boosted to a higher energy level. As shown in Figure 8.11, the energy (not the electron itself) is transferred to adjacent pigment molecules by a process called **resonance energy transfer**. The energy may be

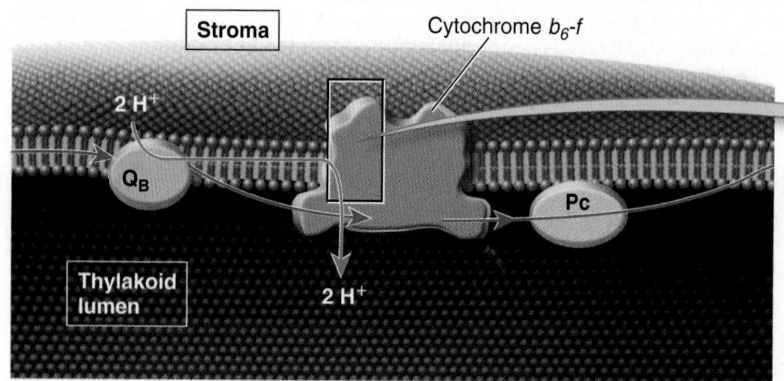

(a) Cytochrome b_6-f in the chloroplast

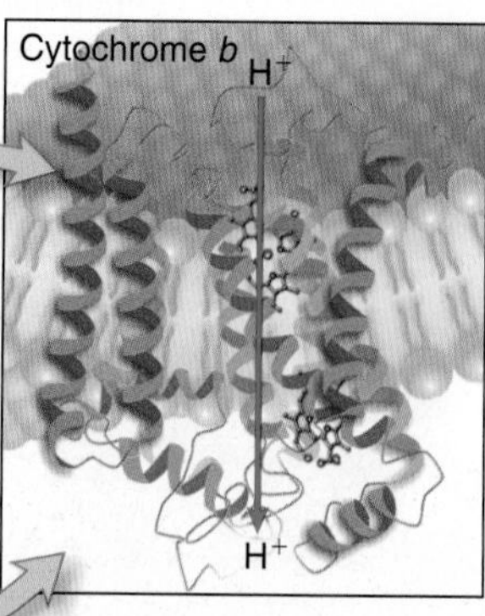

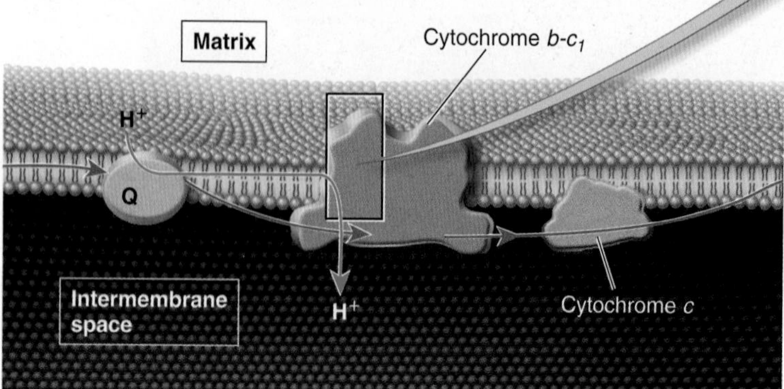

(b) Cytochrome b-c_1 in the mitochondrion

Figure 8.10 Homologous proteins in the electron transport chains of chloroplasts and mitochondria. **(a)** Cytochrome b_6-f is a complex involved in electron and H^+ transport in chloroplasts, and **(b)** cytochrome b-c_1 is a complex involved in electron and H^+ transport in mitochondria. These complexes contain homologous proteins designated cytochrome b_6 in chloroplasts and cytochrome b in mitochondria. The inset shows the three-dimensional structure of cytochrome b, which was determined by X-ray crystallography. It is an integral membrane protein with several transmembrane helices and two heme groups, which are prosthetic groups involved in electron transfer. The structure of cytochrome b_6 has also been determined and found to be very similar.

BIOLOGY PRINCIPLE Populations of organisms evolve from one generation to the next. Evolution has resulted in proteins with similar but specialized functions.

Concept Check: *Explain why the three-dimensional structures of cytochrome b and cytochrome b_6 are very similar.*

transferred among multiple pigment molecules until it is eventually transferred to a special pigment molecule designated P680, which is located within the reaction center of PSII. The P680 pigment is so named because it can directly absorb light at a wavelength of 680 nm. However, P680 is more commonly excited by resonance energy transfer from another chlorophyll pigment. In either case, when an electron in P680 is excited, it is designated P680*. The light-harvesting complex is also called the antenna complex because it acts like an antenna that absorbs energy from light and funnels that energy to P680 in the reaction center.

A high-energy (photoexcited) electron in a pigment molecule is relatively unstable. It may abruptly release its energy by giving off heat or light. Unlike the pigments in the light-harvesting complex that undergo resonance energy transfer, P680* can actually release its high-energy electron and become $P680^+$.

$$P680^* \rightarrow P680^+ + e^-$$

The role of the reaction center is to quickly remove the high-energy electron from P680* and transfer it to another molecule, where the electron is more stable. This molecule is called the **primary electron acceptor** (see Figure 8.11). The transfer of the electron from P680* to the primary electron acceptor is remarkably fast. It occurs in less than a few picoseconds! (One picosecond equals one-trillionth of a second, also noted as 10^{-12} sec.) Because this occurs so quickly, the excited electron does not have much time to release its energy in the form of heat or light.

After the primary electron acceptor has received this high-energy electron, the light energy has been captured and can be used to perform cellular work. As discussed earlier, the work it performs is to synthesize the energy intermediates ATP and NADPH.

Let's now consider what happens to $P680^+$, which has given up its high-energy electron. After $P680^+$ is formed, it is necessary to replace the electron so that P680 can function again. Therefore, another role of the reaction center is to replace the electron that is removed when P680* becomes $P680^+$. This missing electron of $P680^+$ is replaced with a low-energy electron from water (see Figure 8.11).

$$H_2O \rightarrow 2\,H^+ + {}^1\!/_2\,O_2 + 2\,e^-$$

$$2\,P680^+ + \underset{\text{(from water)}}{2\,e^-} \rightarrow 2\,P680$$

The oxidation of water results in the formation of oxygen gas (O_2), which is used by many organisms for cellular respiration. Photosystem II is the only known protein complex that can oxidize water, resulting in the release of O_2 into the atmosphere.

Photosystem II Is an Amazing Redox Machine

All cells rely on redox reactions to store and utilize energy and to form covalent bonds in organic molecules. Photosystem II is a particularly remarkable example of a redox machine. As we have learned, this complex of proteins removes high-energy electrons from a pigment

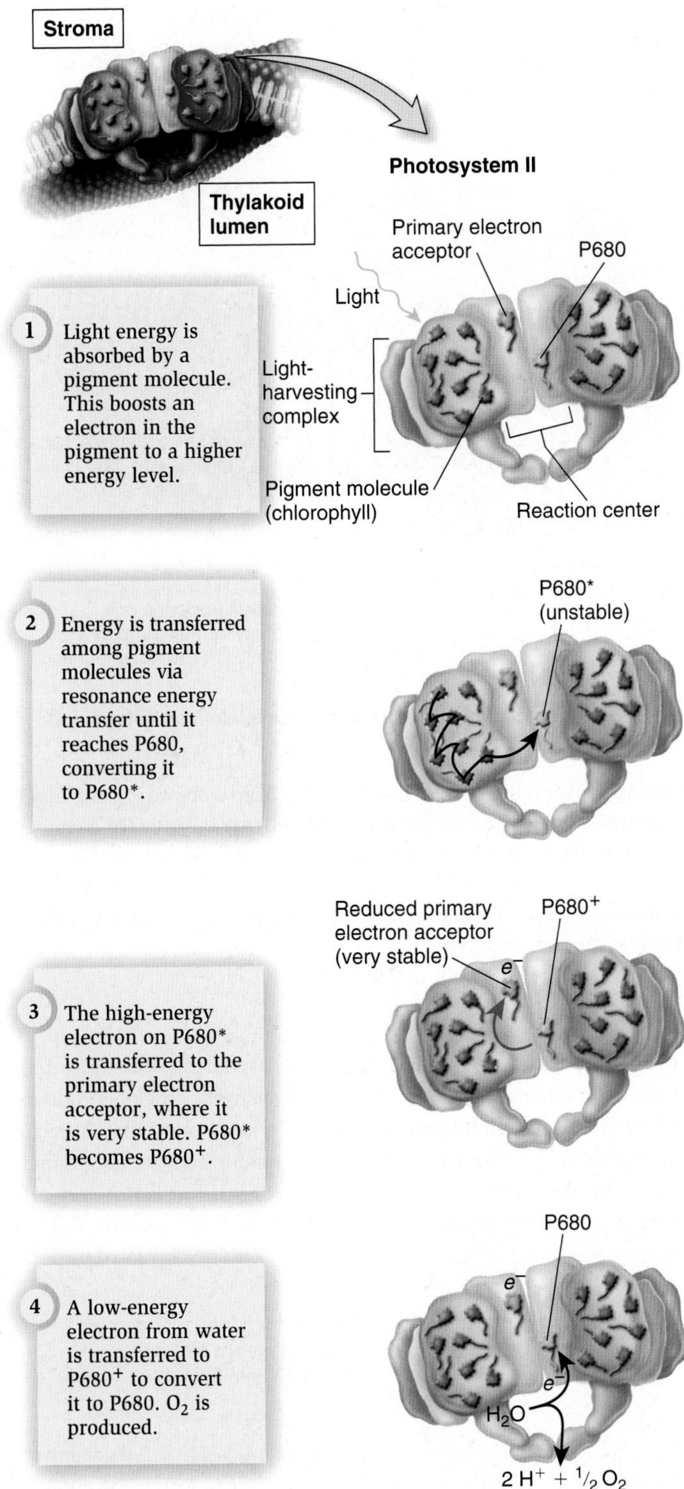

Figure 8.11 **A closer look at how photosystem II harvests light energy and oxidizes water.** Note: Two electrons are released during the oxidation of water, but they are transferred one at a time to $P680^+$.

molecule and transfers them to a primary electron acceptor. Perhaps even more remarkable is that photosystem II can remove low-energy electrons from water—a very stable molecule that holds onto its electrons tightly. The removal of electrons from water results in the formation of oxygen (O_2).

Many approaches have been used to study how photosystem II works. In recent years, much effort has been aimed at determining the biochemical composition of the protein complex and the roles of its individual components. The number of protein subunits varies somewhat from species to species and may vary due to environmental changes. Typically, photosystem II is composed of approximately 19 different protein subunits. Two subunits, designated D1 and D2, contain the reaction center, which carries out the redox reactions (**Figure 8.12a**). Two other subunits, called CP43 and CP47, bind the pigment molecules that form the light-harvesting complex. Many additional subunits regulate the function of photosystem II and provide structural support.

Figure 8.12a illustrates the pathway of electron movement through photosystem II. The red arrows indicate the movement of a high-energy electron, and the black arrows show the path of a low-energy electron. Let's begin with a high-energy electron. When the electron on P680 becomes boosted to a higher energy level, usually by resonance energy transfer, this high-energy electron then moves to the primary electron acceptor, which is a chlorophyll molecule lacking Mg^{2+}, called pheophytin (Pp). Pheophytin is permanently bound to photosystem II and transfers the electron to a plastoquinone molecule, designated Q_A, which is also permanently bound to photosystem II. Next, the electron is transferred to another plastoquinone molecule designated Q_B, which can accept two high-energy electrons and bind two H^+. As shown earlier in Figure 8.8, Q_B can diffuse away from the reaction center.

Let's now consider the path of a low-energy electron. The oxidation of water occurs in a region called the **manganese cluster**. This site is located on the side of D1 that faces the thylakoid lumen. The manganese cluster has four Mn^{2+}, one Ca^{2+}, and one Cl^-. Two water molecules bind to this site. D1 catalyzes the removal of four low-energy electrons from the two H_2O molecules to produce O_2 and four H^+. Each low-energy electron is transferred, one at a time, to an amino acid in D1 (a tyrosine, Tyr) and then to $P680^+$ to produce P680.

In 2004, So Iwata, James Barber, and colleagues determined the three-dimensional structure of photosystem II using a technique called **X-ray crystallography**. In this method, researchers must purify a protein or protein complex and expose it to conditions that cause the proteins to associate with each other in an ordered array. In other words, the proteins form a crystal. When a crystal is exposed to X-rays, the resulting pattern can be analyzed mathematically to determine the three-dimensional structure of the crystal's components. Major advances in this technique over the last couple of decades have enabled researchers to determine the structures of relatively large macromolecular complexes such as photosystem II (**Figure 8.12b**). The structure shown there is a dimer: it has two PSII complexes, each with 19 protein subunits. As seen in Figure 8.12b, the intricacy of the structure of photosystem II rivals the complexity of its function.

The Use of Light Flashes of Specific Wavelengths Provided Experimental Evidence for the Existence of PSII and PSI

An experimental technique that uses light flashes at particular wavelengths has been important in helping researchers to understand the function of photosystems. In this method, pioneered by Robert Emerson,

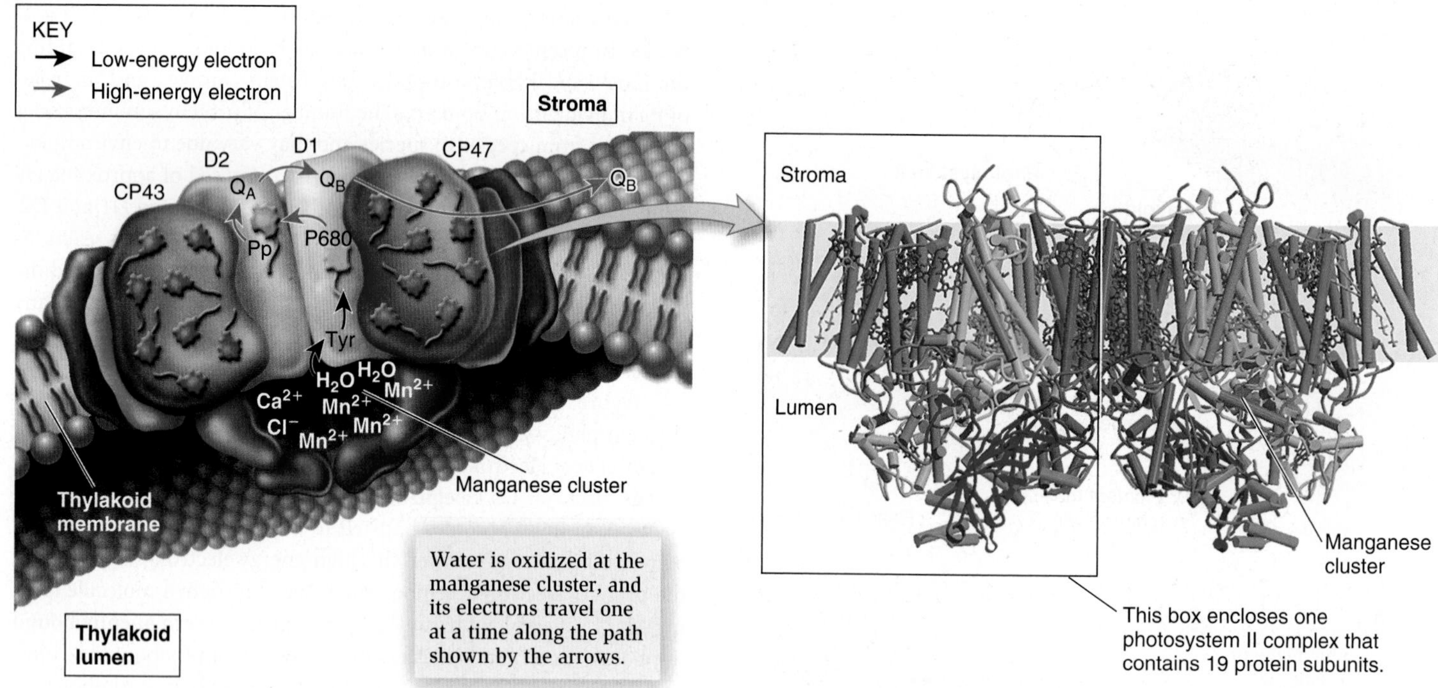

(a) The path of electron flow through photosystem II

(b) Three-dimensional structure of photosystem II as determined by X-ray crystallography

Figure 8.12 **The molecular structure of photosystem II. (a)** Schematic drawing showing the path of electron flow from water to Q_B. The CP43 and CP47 protein subunits wrap around D1 and D2. Pigments in CP43 and CP47 transfer energy to P680 by resonance energy transfer. **(b)** The three-dimensional structure of photosystem II as determined by X-ray crystallography. In the crystal structure, the colors are CP43 (green), D2 (orange), D1 (yellow), and CP47 (red).

BIOLOGY PRINCIPLE Structure determines function. The structural arrangement of the manganese cluster, P680, pheophytin, Q_A, and Q_B provides a pathway of electron movement from photosystem II to the electron transport chain.

a photosynthetic organism is exposed to a particular wavelength of light, after which the rate of photosynthesis is measured by the amount of CO_2 consumed or the amount of O_2 produced. In the 1950s, Emerson performed a particularly intriguing experiment that greatly stimulated photosynthesis research (**Figure 8.13**). He subjected algae to light flashes of different wavelengths and obtained a mysterious result. When he exposed algae to a wavelength of 680 nm, he observed a low rate of photosynthesis. A similarly low rate of photosynthesis occurred when he exposed algae to a wavelength of 700 nm. However, when he exposed the algae to both wavelengths of light simultaneously, the rate of photosynthesis was more than double the rate observed at only one wavelength. This phenomenon was termed the **enhancement effect**. We know now that it occurs because light of 680-nm wavelength can readily activate the pigment (P680) in the reaction center in photosystem II but is not very efficient at activating pigments in photosystem I. In contrast, light of

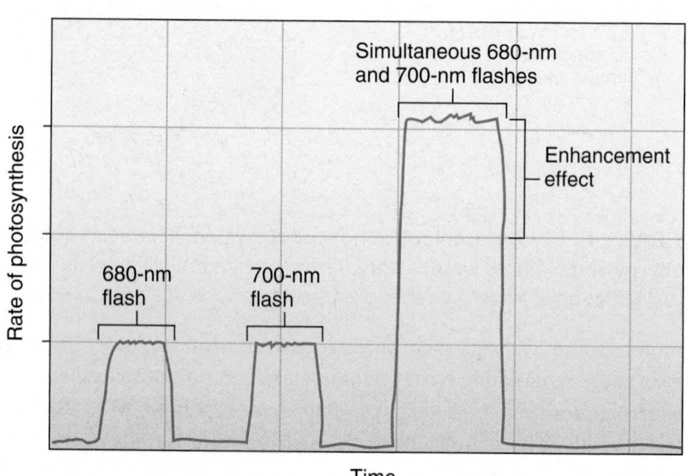

Figure 8.13 **The enhancement effect observed by Emerson.** When photosynthetic organisms such as green plants and algae are exposed to 680-nm and 700-nm light simultaneously, the resulting rate of photosynthesis is much more than double the rate produced by each wavelength individually.

BIOLOGY PRINCIPLE Biology is as an experimental science. This experiment provided key evidence for the existence of two photosystems.

Concept Check: Would the enhancement effect be observed if two consecutive flashes of light occurred at 680 nm?

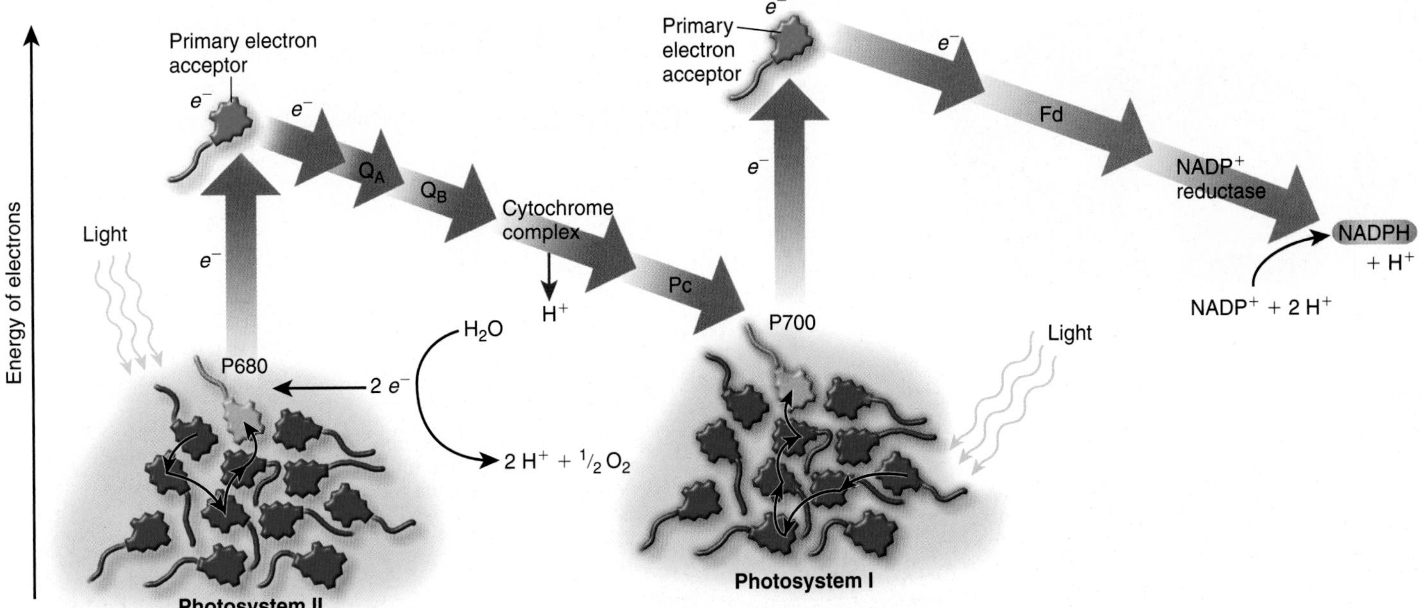

Figure 8.14 The Z scheme, showing the energy of an electron moving from photosystem II to NADP⁺. The oxidation of water releases two electrons that travel one at a time from photosystem II to NADP⁺. As seen here, the input of light boosts the energy of the electron twice. At the end of the pathway, two electrons are used to make NADPH.

Concept Check: *During its journey from photosystem II to NADP⁺, at what point does an electron have the highest amount of energy?*

700-nm wavelength is optimal at activating the pigments in photosystem I but not those in photosystem II. When algae are exposed to both wavelengths, however, the pigments in both photosystems are maximally activated.

When researchers began to understand that photosynthesis results in the production of both ATP and NADPH, Robin Hill and Fay Bendall also proposed that photosynthesis involves two photoactivation events. According to their model, known as the **Z scheme**, an electron proceeds through a series of energy changes during photosynthesis (**Figure 8.14**). The Z refers to the zigzag shape of this energy curve. Based on our modern understanding of photosynthesis, we now know these events involve increases and decreases in the energy of an electron as it moves from photosystem II through photosystem I to $NADP^+$ during noncyclic electron flow. An electron on a nonexcited pigment molecule in photosystem II has the lowest energy. In photosystem II, light boosts an electron to a much higher energy level. As the electron travels from photosystem II to photosystem I, some of the energy is released. The input of light in photosystem I boosts the electron to an even higher energy than it attained in photosystem II. The electron releases a little energy before it is eventually transferred to $NADP^+$.

8.4 Synthesizing Carbohydrates via the Calvin Cycle

Learning Outcomes:

1. Outline the three phases of the Calvin cycle.
2. Explain how Calvin and Benson identified the components of the Calvin cycle.

In the previous sections, we learned how the light reactions of photosynthesis produce ATP, NADPH, and O_2. We will now turn our attention to the second phase of photosynthesis, the Calvin cycle, in which ATP and NADPH are used to make carbohydrates. The Calvin cycle consists of a series of steps that occur in a metabolic cycle. In plants and algae, it occurs in the stroma of chloroplasts. In cyanobacteria, the Calvin cycle occurs in the cytoplasm of the bacterial cells.

The Calvin cycle takes CO_2 from the atmosphere and incorporates the carbon into organic molecules, namely, carbohydrates. As mentioned earlier, carbohydrates are critical for two reasons. First, they provide the precursors to make the organic molecules and macromolecules of nearly all living cells. The second key reason is the storage of energy. The Calvin cycle produces carbohydrates, which store energy. These carbohydrates are accumulated inside plant cells. When a plant is in the dark and not carrying out photosynthesis, the stored carbohydrates are used as a source of energy. Similarly, when an animal consumes a plant, it uses the carbohydrates as an energy source.

In this section, we will examine the three phases of the Calvin cycle. We will also explore the experimental approach of Melvin Calvin and his colleagues that enabled them to elucidate the steps of this cycle.

The Calvin Cycle Incorporates CO_2 into Carbohydrate

The Calvin cycle, also called the Calvin-Benson cycle, was determined by chemists Melvin Calvin and Andrew Adam Benson and their colleagues in the 1940s and 1950s. This cycle requires a massive input of energy. For every 6 carbon dioxide molecules that are incorporated into a carbohydrate such as glucose ($C_6H_{12}O_6$), 18 ATP molecules are hydrolyzed and 12 NADPH molecules are oxidized:

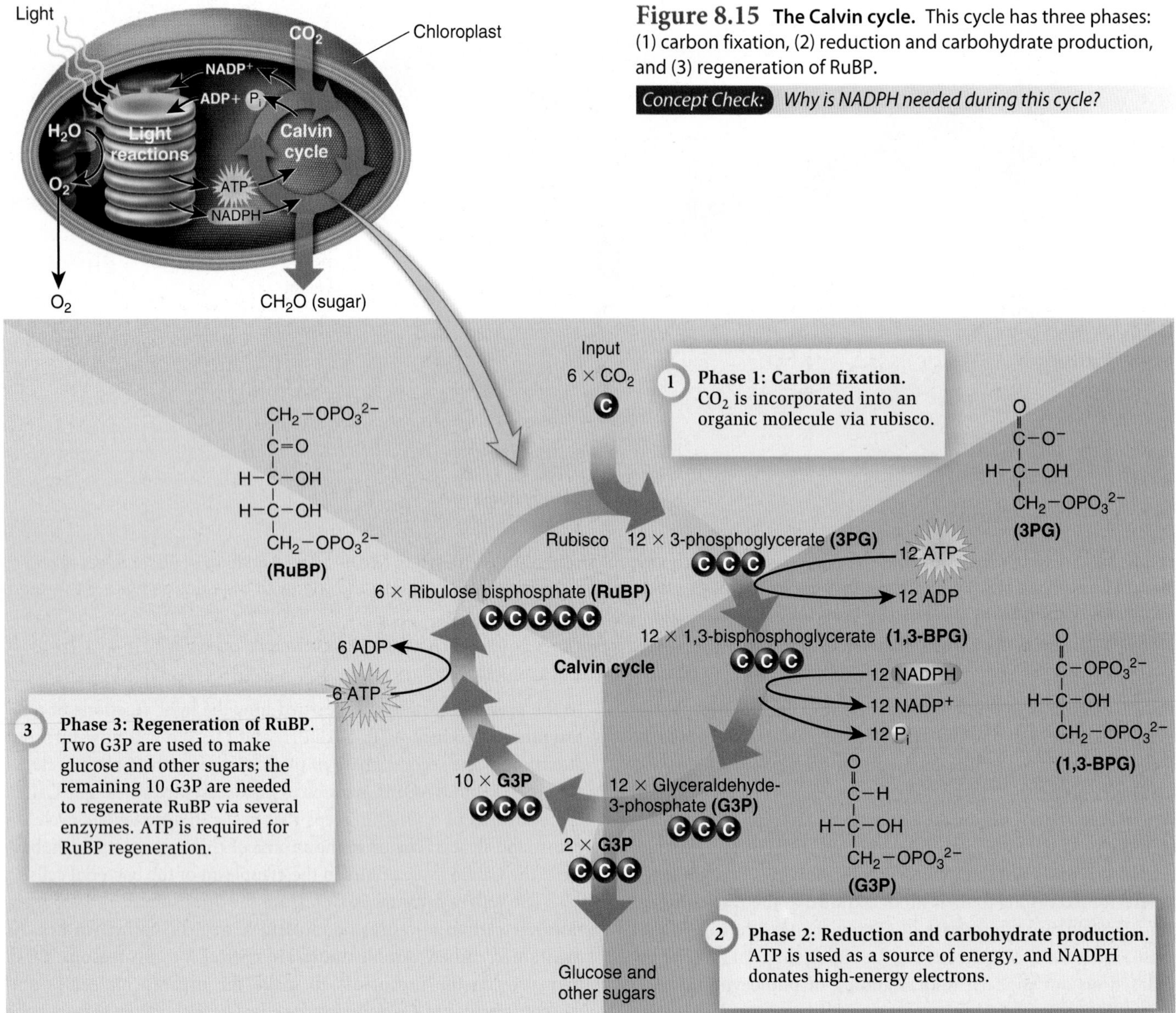

Figure 8.15 **The Calvin cycle.** This cycle has three phases: (1) carbon fixation, (2) reduction and carbohydrate production, and (3) regeneration of RuBP.

Concept Check: *Why is NADPH needed during this cycle?*

$$6\ CO_2 + 12\ H_2O \rightarrow C_6H_{12}O_6 + 6\ O_2 + 6\ H_2O$$

$$18\ ATP + 18\ H_2O \rightarrow 18\ ADP + 18\ P_i$$

$$12\ NADPH \rightarrow 12\ NADP^+ + 12\ H^+ + 24\ e^-$$

Although biologists commonly describe glucose as a product of photosynthesis, glucose is not directly made by the Calvin cycle. Instead, molecules of glyceraldehyde-3-phosphate, which are products of the Calvin cycle, are used as starting materials for the synthesis of glucose and other molecules, including sucrose. After glucose molecules are made, they may be linked together to form a polymer of glucose called starch, which is stored in the chloroplast for later use. Alternatively, the disaccharide sucrose may be made and transported out of the leaf to other parts of the plant.

The Calvin cycle can be divided into three phases. These phases are carbon fixation, reduction and carbohydrate production, and regeneration of ribulose bisphosphate (RuBP) (**Figure 8.15**).

Carbon Fixation (Phase 1) During **carbon fixation**, CO_2 is incorporated into RuBP, a five-carbon sugar. The term *fixation* means that the carbon has been removed from the atmosphere and fixed into an organic molecule that is not a gas. More specifically, the product of the reaction is a six-carbon intermediate that immediately splits in half to form two molecules of 3-phosphoglycerate (3PG). The enzyme that catalyzes this step is named RuBP carboxylase/oxygenase, or **rubisco**. It is the most abundant protein in chloroplasts and perhaps the most abundant protein on Earth! This observation underscores the massive amount of carbon fixation that happens in the biosphere.

Reduction and Carbohydrate Production (Phase 2) In the second phase, ATP is used to convert 3PG to 1,3-bisphosphoglycerate (1,3-BPG). Next, electrons from NADPH reduce 1,3-BPG to glyceraldehyde-3-phosphate (G3P). G3P is a carbohydrate with three carbon atoms. The key difference between 3PG and G3P is that 3PG has a C—O bond, whereas the analogous carbon in G3P is a C—H bond (see Figure 8.15). The C—H bond occurs because the G3P molecule has been reduced by the addition of two electrons from NADPH. Compared with 3PG, the bonds in G3P store more energy and enable G3P to readily form larger organic molecules such as glucose.

As shown in Figure 8.15, only some of the G3P molecules are used to make glucose or other carbohydrates. Phase 1 begins with 6 RuBP molecules and 6 CO_2 molecules. Twelve G3P molecules are made at the end of phase 2, and only two of these G3P molecules are used in carbohydrate production. As described next, the other 10 G3P molecules are needed to keep the Calvin cycle turning by regenerating RuBP.

Regeneration of RuBP (Phase 3) In the last phase of the Calvin cycle, a series of enzymatic steps converts the 10 G3P molecules into 6 RuBP molecules, using 6 molecules of ATP. After the RuBP molecules are regenerated, they serve as acceptors for CO_2, thereby allowing the cycle to continue.

As we have just seen, the Calvin cycle begins by using carbon from an inorganic source, that is, CO_2, and ends with organic molecules that will be used by the plant to make other molecules. You may be wondering why CO_2 molecules cannot be directly linked to form these larger molecules. The answer lies in the number of electrons that orbit carbon atoms. In CO_2, the carbon atom is considered electron poor. Oxygen is a very electronegative atom that monopolizes the electrons it shares with other atoms. In a covalent bond between carbon and oxygen, the shared electrons are closer to the oxygen atom.

By comparison, in an organic molecule, the carbon atom is electron rich. During the Calvin cycle, ATP provides energy and NADPH donates high-energy electrons, so the carbon originally in CO_2 has been reduced. The Calvin cycle combines less electronegative atoms with carbon atoms so that C—H and C—C bonds are formed. This allows the eventual synthesis of larger organic molecules, including glucose, amino acids, and so on. In addition, the covalent bonds within these molecules are capable of storing large amounts of energy.

FEATURE INVESTIGATION

The Calvin Cycle Was Determined by Isotope-Labeling Methods

The steps in the Calvin cycle involve the conversion of one type of molecule to another, eventually regenerating the starting material, RuBP. In the 1940s and 1950s, Calvin and his colleagues used ^{14}C, a radioisotope of carbon, to label and trace molecules produced during the cycle (**Figure 8.16**). They injected ^{14}C-labeled CO_2 into cultures of the green algae *Chlorella pyrenoidosa* grown in an apparatus called a "lollipop" (because of its shape). The *Chlorella* cells were given different lengths of time to incorporate the ^{14}C-labeled carbon, ranging from fractions of a second to many minutes. After this incubation period, the cells were abruptly placed into a solution of alcohol to inhibit enzymatic reactions and thereby stop the cycle.

The researchers separated the newly made radiolabeled molecules by a variety of methods. The most commonly used method was two-dimensional paper chromatography. In this approach, a sample containing radiolabeled molecules was spotted onto a corner of the paper at a location called the origin. The edge of the paper was placed in a solvent, such as phenol-water. As the solvent rose through the paper, so did the radiolabeled molecules. The rate at which they rose depended on their structures, which determined how strongly they interacted with the paper. This step separated the mixture of molecules spotted onto the paper at the origin.

The paper was then dried, turned 90°, and then the edge was placed in a different solvent, such as butanol-propionic acid-water. Again, the solvent rose through the paper (in a second dimension), thereby separating molecules that may not have been adequately separated during the first separation step. After this second separation step, the paper was dried and exposed to X-ray film, a procedure called autoradiography. Radioactive emission from the ^{14}C-labeled molecules caused dark spots to appear on the film.

The pattern of spots changed depending on the length of time the cells were incubated with ^{14}C-labeled CO_2. When the incubation period was short, only molecules that were made in the first steps of the Calvin cycle were seen—3-phosphoglycerate (3PG) and 1,3-bisphosphoglycerate (1,3-BPG). Longer incubations revealed molecules synthesized in later steps—glyceraldehyde-3-phosphate (G3P) and ribulose bisphosphate (RuBP).

A challenge for Calvin and his colleagues was to identify the chemical nature of each spot. They achieved this by a variety of chemical methods. For example, a spot could be cut out of the paper, the molecule within the paper could be washed out or eluted, and then the eluted molecule could be subjected to the same procedure that included a radiolabeled molecule whose structure was already known. If the unknown molecule and known molecule migrated to the same spot in the paper, this indicated they were likely to be the same molecule. During the late 1940s and 1950s, Calvin and his coworkers identified all of the ^{14}C-labeled spots and the order in which they appeared. In this way, they determined the series of reactions of what we now know as the Calvin cycle. For this work, Calvin was awarded the Nobel Prize in Chemistry in 1961.

Experimental Questions

1. What was the purpose of the study conducted by Calvin and his colleagues?
2. In Calvin's experiments shown in Figure 8.15, why did the researchers use ^{14}C? Why did they examine samples at several different time periods? How were the different molecules in the samples identified?
3. What were the results of Calvin's study?

Figure 8.16 **The determination of the Calvin cycle using ^{14}C-labeled CO_2 and paper chromatography.**

GOAL The incorporation of CO_2 into carbohydrate involves a biosynthetic pathway. The aim of this experiment was to identify the steps.

KEY MATERIALS The green alga *Chlorella pyrenoidosa* and ^{14}C-labeled CO_2.

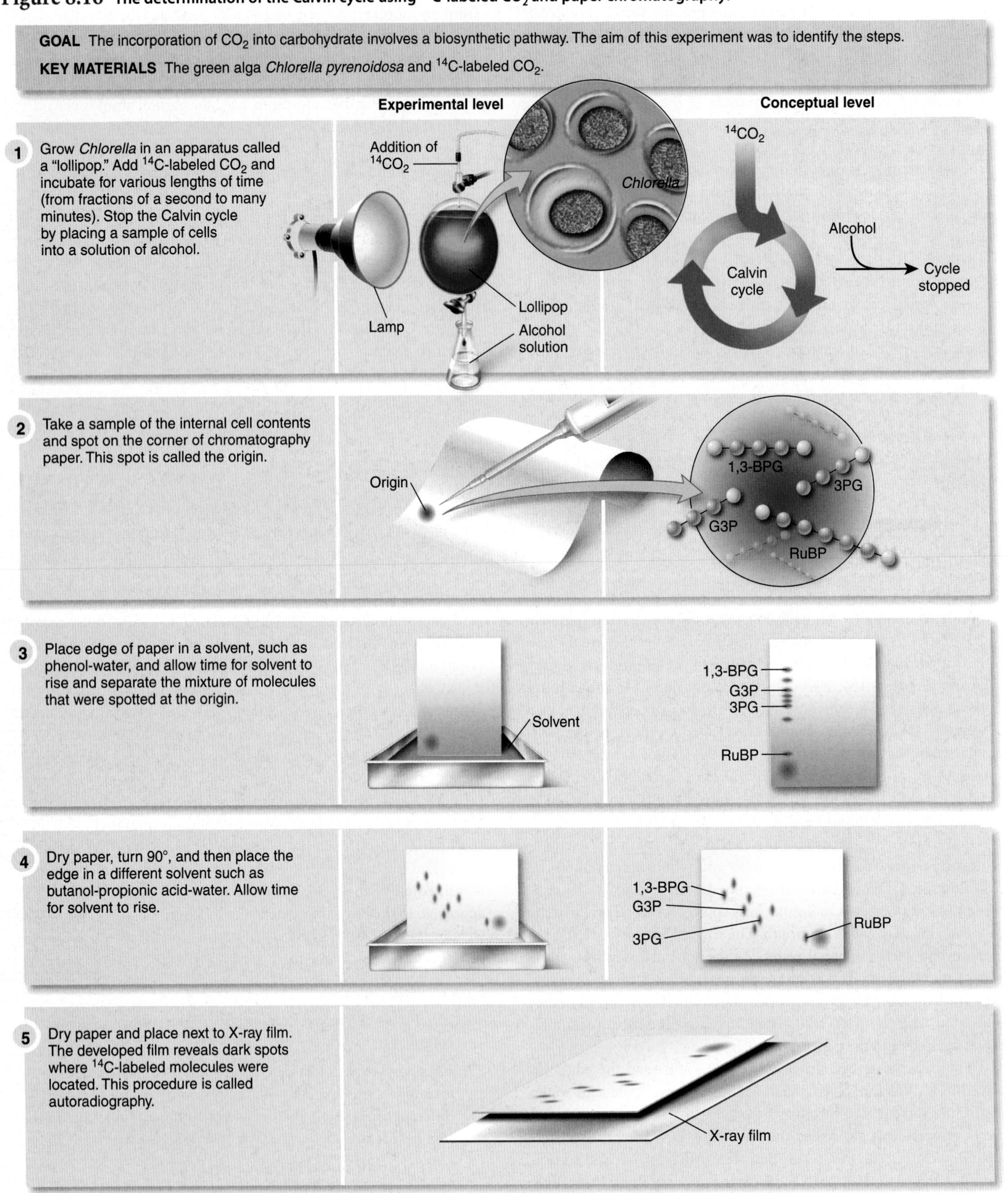

Experimental level
Conceptual level
1
Grow *Chlorella* in an apparatus called a "lollipop." Add ^{14}C-labeled CO_2 and incubate for various lengths of time (from fractions of a second to many minutes). Stop the Calvin cycle by placing a sample of cells into a solution of alcohol.
Addition of $^{14}CO_2$
Chlorella
Lamp
Lollipop
Alcohol solution
$^{14}CO_2$
Calvin cycle
Alcohol
Cycle stopped
2
Take a sample of the internal cell contents and spot on the corner of chromatography paper. This spot is called the origin.
Origin
1,3-BPG
3PG
G3P
RuBP
3
Place edge of paper in a solvent, such as phenol-water, and allow time for solvent to rise and separate the mixture of molecules that were spotted at the origin.
Solvent
1,3-BPG
G3P
3PG
RuBP
4
Dry paper, turn 90°, and then place the edge in a different solvent such as butanol-propionic acid-water. Allow time for solvent to rise.
1,3-BPG
G3P
3PG
RuBP
5
Dry paper and place next to X-ray film. The developed film reveals dark spots where ^{14}C-labeled molecules were located. This procedure is called autoradiography.
X-ray film

6 THE DATA*

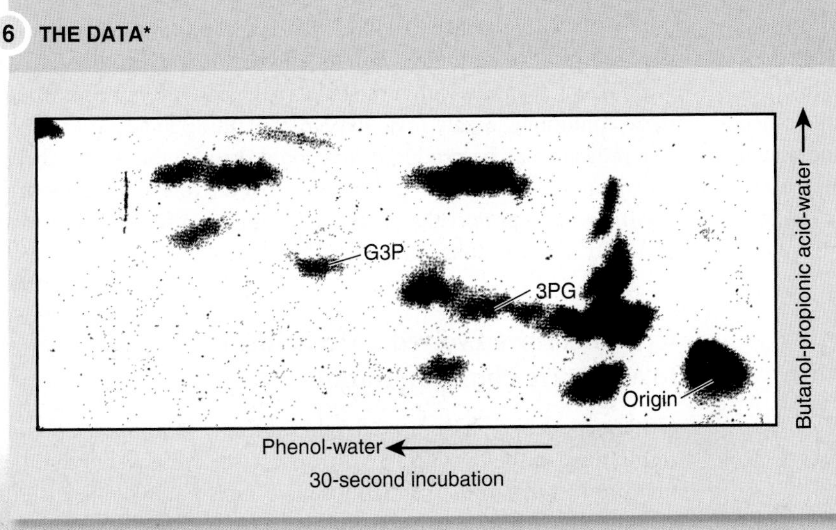

*An autoradiograph from one of Calvin's experiments.

7 **CONCLUSION** The identification of the molecules in each spot elucidated the steps of the Calvin cycle.

8 **SOURCE** Calvin, M. December 11, 1961. The path of carbon in photosynthesis. Nobel Lecture.

8.5 Variations in Photosynthesis

Learning Outcomes:

1. Explain the concept of photorespiration.
2. Compare and contrast how C_4 and CAM plants avoid photorespiration and conserve water.

Thus far, we have considered photosynthesis as a two-stage process in which the light reactions produce ATP, NADPH, and O_2, and the Calvin cycle uses the ATP and NADPH in the synthesis of carbohydrates. This two-stage process is a universal feature of photosynthesis in all green plants, algae, and cyanobacteria. However, certain environmental conditions such as temperature, water availability, and light intensity alter the way in which the Calvin cycle operates. In this section, we begin by examining how hot and dry conditions may reduce the output of photosynthesis. We then explore two adaptations that certain plant species have evolved that conserve water and help to maximize photosynthetic efficiency in such environments.

Photorespiration Decreases the Efficiency of Photosynthesis

In the previous section, we learned that rubisco adds a CO_2 molecule to an organic molecule, RuBP, to produce two molecules of 3-phosphoglycerate (3PG):

$$RuBP + CO_2 \rightarrow 2\ 3PG$$

For most species of plants, the incorporation of CO_2 into RuBP is the only way for carbon fixation to occur. Because 3PG is a three-carbon molecule, these plants are called **C_3 plants**. Examples of C_3 plants include wheat and oak trees (Figure 8.17). About 90% of the plant species on Earth are C_3 plants.

(a) Wheat plants

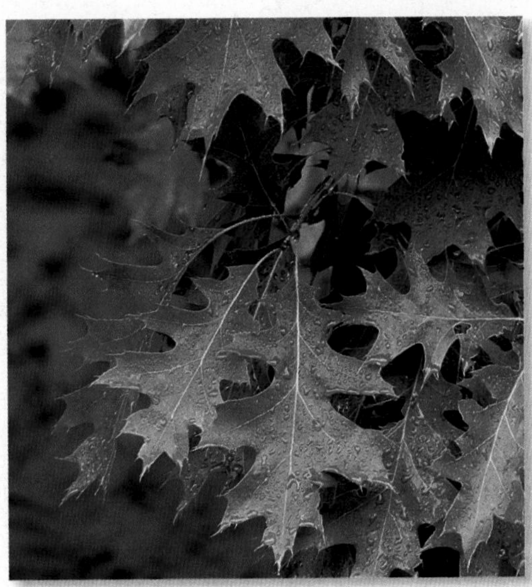

(b) Oak leaves

Figure 8.17 Examples of C_3 plants. The structures of **(a)** wheat and **(b)** white oak leaves are similar to that shown in Figure 8.2.

Researchers have discovered that the active site of rubisco can also add O_2 to RuBP, although its affinity for CO_2 is over 10-fold better than that for O_2. Even so, when CO_2 levels are low and O_2 levels are high, rubisco adds an O_2 molecule to RuBP. This produces only one molecule of 3PG and a two-carbon molecule called phosphoglycolate. The phosphoglycolate is then dephosphorylated to glycolate, which is released from the chloroplast. In a series of several steps, the two-carbon glycolate is eventually oxidized in peroxisomes and mitochondria to produce an organic molecule plus a molecule of CO_2:

$$\text{RuBP} + O_2 \rightarrow \text{3PG} + \text{Phosphoglycolate}$$

$$\text{Phosphoglycolate} \rightarrow \text{Glycolate} \rightarrow \rightarrow \text{Organic molecule} + CO_2$$

This process, called **photorespiration**, uses O_2 and liberates CO_2. Photorespiration is considered wasteful because it releases CO_2, thereby limiting plant growth.

Photorespiration is more likely to occur when plants are exposed to a hot and dry environment. To conserve water, the stomata of the leaves close, inhibiting the uptake of CO_2 from the air and trapping the O_2 that is produced by photosynthesis. When the level of CO_2 is low and O_2 is high, photorespiration is favored. If C_3 plants are subjected to hot and dry environmental conditions, as much as 25–50% of their photosynthetic work is reversed by the process of photorespiration.

Why do plants carry out photorespiration? The answer is not entirely clear. One common view is that photorespiration is an evolutionary relic. When rubisco first evolved some 3 billion years ago, the atmospheric oxygen level was low, so photorespiration would not have been a problem. Another view is that photorespiration may have a protective advantage. On hot and dry days when the stomata are closed, CO_2 levels within the leaves fall, and O_2 levels rise. Under these conditions, highly toxic oxygen-containing molecules such as free radicals may be produced that could damage the plant. Therefore, plant biologists have hypothesized that the role of photorespiration may be to protect the plant against the harmful effects of such toxic molecules by consuming O_2 and releasing CO_2.

C_4 Plants Have Evolved a Mechanism to Minimize Photorespiration

Certain species of plants have developed a way to minimize photorespiration. In the early 1960s, Hugo Kortschak discovered that the first product of carbon fixation in sugarcane is not 3GP but instead is a molecule with four carbon atoms. Species such as sugarcane are called **C_4 plants** because of this four-carbon molecule. Later, Marshall Hatch and Roger Slack confirmed this result and identified the molecule as oxaloacetate. For this reason, the pathway is sometimes called the Hatch-Slack pathway.

Some C_4 plants have a unique leaf anatomy that allows them to avoid photorespiration (**Figure 8.18**). An interior layer in the leaves of many C_4 plants has a two-cell organization composed of mesophyll cells and bundle-sheath cells. CO_2 from the atmosphere enters the mesophyll cells via stomata. Once inside, the enzyme PEP carboxylase adds CO_2 to phosphoenolpyruvate (PEP), a three-carbon molecule,

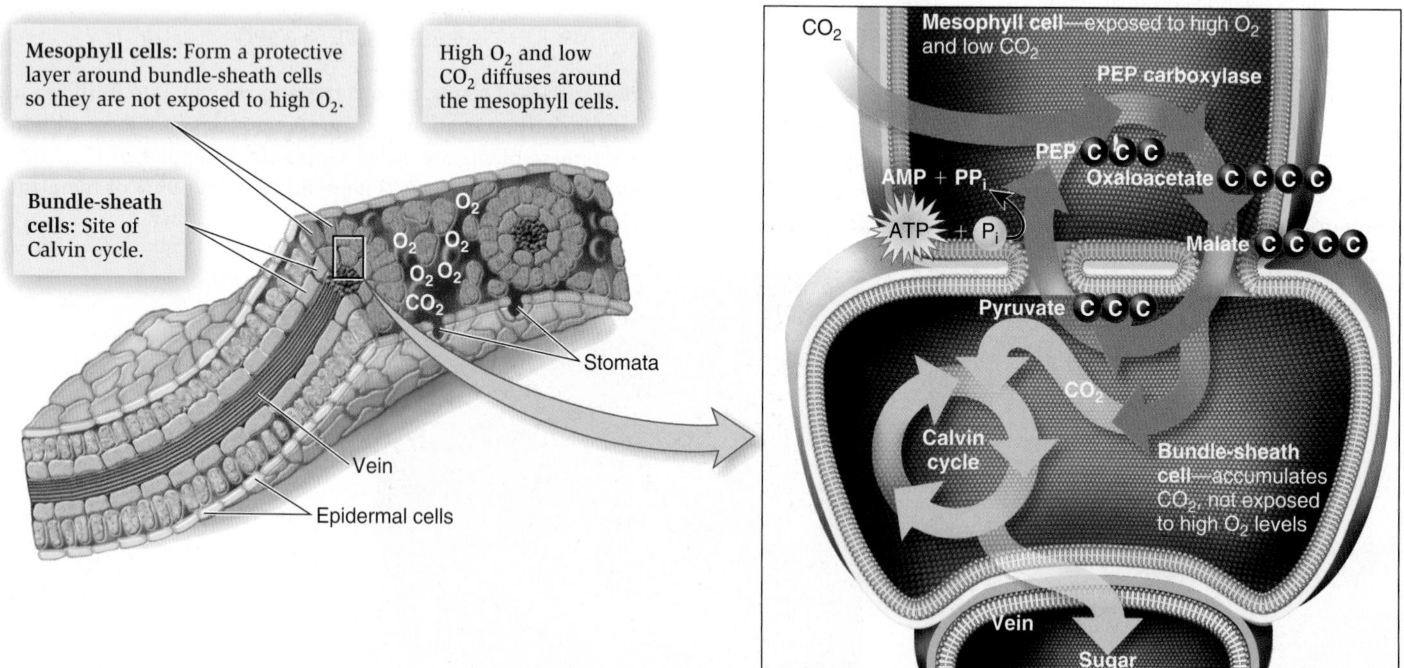

Figure 8.18 **Leaf structure and its relationship to the C_4 cycle.** C_4 plants have mesophyll cells, which initially take up CO_2, and bundle-sheath cells, where much of the carbohydrate synthesis occurs. Compare this leaf structure with the structure of C_3 leaves shown in Figure 8.2.

Concept Check: *How does this cellular arrangement minimize photorespiration?*

BioConnections: *Look ahead to Figure 38.9. How do plants get water needed for photosynthesis into their leaves?*

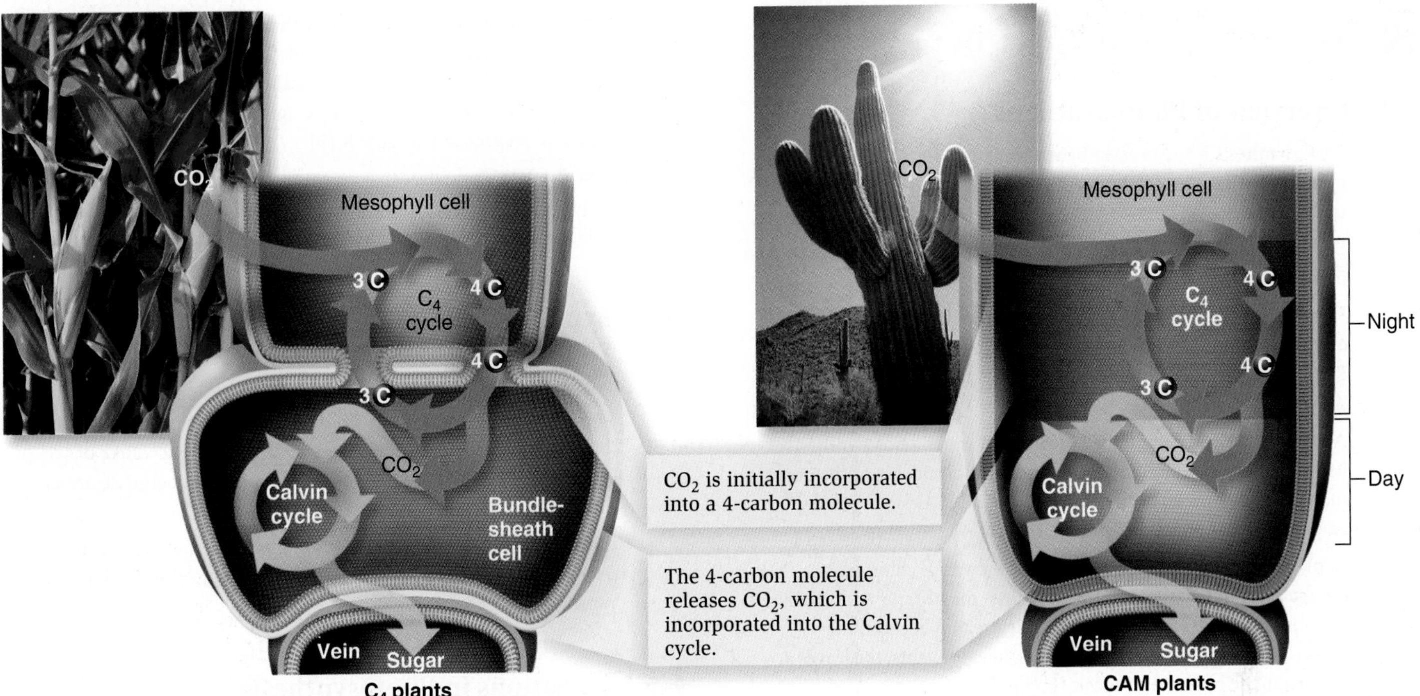

Figure 8.19 **A comparison of C_4 and CAM plants.** The name C_4 plant describes those plants in which the first organic product of carbon fixation is a four-carbon molecule. Using this definition, CAM plants are a type of C_4 plant. CAM plants, however, do not separate the functions of making a four-carbon molecule and the Calvin cycle into different types of cells. Instead, they make a four-carbon molecule at night and break down that molecule during the day so the CO_2 can be incorporated into the Calvin cycle.

Concept Check: *What are the advantages and disadvantages among C_3, C_4, and CAM plants?*

to produce oxaloacetate, a four-carbon molecule. PEP carboxylase does not recognize O_2. Therefore, unlike rubisco, PEP carboxylase does not promote photorespiration when CO_2 is low and O_2 is high. Instead, PEP carboxylase continues to fix CO_2.

As shown in Figure 8.18, oxaloacetate is converted to the four-carbon molecule malate, which is transported into the bundle-sheath cell. Malate is then broken down into pyruvate and CO_2. The pyruvate returns to the mesophyll cell, where it is converted to PEP via ATP, and the cycle in the mesophyll cell can begin again. The CO_2 enters the Calvin cycle in the chloroplasts of the bundle-sheath cells. Because the mesophyll cell supplies the bundle-sheath cell with a steady supply of CO_2, the concentration of CO_2 remains high in the bundle-sheath cell. Also, the mesophyll cells shield the bundle-sheath cells from high levels of O_2. This strategy minimizes photorespiration, which requires low CO_2 and high O_2 levels to proceed.

Which is better—being a C_3 or a C_4 plant? The answer is that it depends on the environment. In warm and dry climates, C_4 plants have an advantage. During the day, they can keep their stomata partially closed to conserve water. Furthermore, they minimize photorespiration. Examples of C_4 plants are sugarcane, crabgrass, and corn. In cooler climates, C_3 plants have the edge because they use less energy to fix CO_2. The process of carbon fixation that occurs in C_4 plants uses ATP to regenerate PEP from pyruvate (see Figure 8.18), which C_3 plants do not have to expend.

CAM Plants Are C_4 Plants That Take Up CO_2 at Night

We have just learned that certain C_4 plants prevent photorespiration by providing CO_2 to the bundle-sheath cells, where the Calvin cycle occurs. This mechanism spatially separates the processes of carbon fixation and the Calvin cycle. Another strategy followed by other C_4 plants, called **CAM plants**, separates these processes in time. CAM stands for crassulacean acid metabolism, because the process was first studied in members of the plant family Crassulaceae. CAM plants are water-storing succulents such as cacti, bromeliads (including pineapple), and sedums. To avoid water loss, CAM plants keep their stomata closed during the day and open them at night, when it is cooler and the relative humidity is higher.

How, then, do CAM plants carry out photosynthesis? **Figure 8.19** compares CAM plants with the other type of C_4 plants we considered in Figure 8.18. Photosynthesis in CAM plants occur entirely within mesophyll cells, but the synthesis of a C_4 molecule and the Calvin cycle occur at different times. During the night, the stomata of CAM plants open, thereby allowing the entry of CO_2 into mesophyll cells. CO_2 is joined with PEP to form the four-carbon molecule oxaloacetate. This is then converted to malate, which accumulates during the night in the central vacuoles of the cells. In the morning, the stomata close to conserve moisture. The accumulated malate in the mesophyll cells leaves the vacuole and is broken down to release CO_2, which then drives the Calvin cycle during the daytime.

Summary of Key Concepts

8.1 Overview of Photosynthesis

- Photosynthesis is the process by which plants, algae, and cyanobacteria capture light energy that is used to synthesize carbohydrates.
- During photosynthesis, carbon dioxide, water, and energy are used to make carbohydrates and oxygen.
- Heterotrophs must obtain organic molecules in their food, whereas autotrophs make organic molecules from inorganic sources. Photoautotrophs use the energy from light to make organic molecules.
- An energy cycle occurs in the biosphere in which photosynthesis uses light, CO_2, and H_2O to make organic molecules, and the organic molecules are broken back down to CO_2 and H_2O via cellular respiration to supply energy in the form of ATP (Figure 8.1).
- In plants and algae, photosynthesis occurs within chloroplasts, organelles with an outer membrane, inner membrane, and thylakoid membrane. The stroma is the fluid-filled region between the thylakoid membrane and inner membrane. In plants, the leaves are the major site of photosynthesis (Figure 8.2).
- The light reactions of photosynthesis capture light energy to make ATP, NADPH, and O_2. These reactions occur at the thylakoid membrane. Carbohydrate synthesis via the Calvin cycle uses ATP and NADPH from the light reactions and happens in the stroma (Figure 8.3).

8.2 Reactions That Harness Light Energy

- Light is a form of electromagnetic radiation that travels in waves and is composed of photons with discrete amounts of energy (Figure 8.4).
- Electrons can absorb light energy and be boosted to a higher energy level—an excited state (Figure 8.5).
- Photosynthetic pigments include chlorophylls *a* and *b* and carotenoids. These pigments absorb light energy in the visible spectrum to drive photosynthesis (Figures 8.6, 8.7).
- During noncyclic electron flow, electrons from photosystem II follow a pathway along an electron transport chain (ETC) in the thylakoid membrane. This pathway generates an H^+ gradient that is used to make ATP. In addition, light energy striking photosystem I (PSI) boosts electrons to a very high energy level that allows the synthesis of NADPH (Figure 8.8).
- During cyclic photophosphorylation, electrons are activated in PSI and flow through the ETC back to PSI. This cyclic electron route produces an H^+ gradient that is used to make ATP (Figure 8.9).
- Cytochrome b_6 in chloroplasts and cytochrome b in mitochondria are homologous proteins involved in electron transport and H^+ pumping (Figure 8.10).

8.3 Molecular Features of Photosystems

- In the light-harvesting complex of photosystem II (PSII), pigment molecules absorb light energy that is transferred to the reaction center via resonance energy transfer. A high-energy electron from P680* is transferred to a primary electron acceptor. An electron from water is then used to replenish the electron lost from P680* (Figures 8.11, 8.12).
- Emerson showed that compared with single light flashes at 680 nm and 700 nm, light flashes at both wavelengths more than doubled the amount of photosynthesis, a result called the enhancement effect. This occurred because these wavelengths activate pigments in PSII and PSI, respectively (Figure 8.13).
- The Z scheme proposes that an electron absorbs light energy twice, at both PSII and PSI, losing some of that energy as it flows along the ETC in the thylakoid membrane (Figure 8.14).

8.4 Synthesizing Carbohydrates via the Calvin Cycle

- The Calvin cycle is composed of three phases: carbon fixation, reduction and carbohydrate production, and regeneration of ribulose bisphosphate (RuBP). In the cycle, ATP is used as a source of energy, and NADPH is used as a source of high-energy electrons to incorporate CO_2 into carbohydrate (Figure 8.15).
- Calvin and Benson determined the steps in the Calvin cycle by isotope-labeling methods in which the products of the Calvin cycle were separated by chromatography (Figure 8.16).

8.5 Variations in Photosynthesis

- C_3 plants incorporate CO_2 only into RuBP to make 3PG, a three-carbon molecule (Figure 8.17).
- Photorespiration occurs when the level of O_2 is high and CO_2 is low, which happens under hot and dry conditions. During this process, some O_2 is used and CO_2 is liberated. Photorespiration is inefficient because it reverses the incorporation of CO_2 into an organic molecule.
- Some C_4 plants avoid photorespiration because the CO_2 is first incorporated, via PEP carboxylase, into a four-carbon molecule, which is pumped from mesophyll cells into bundle-sheath cells. This maintains a high concentration of CO_2 in the bundle-sheath cells, where the Calvin cycle occurs. The high CO_2 concentration minimizes photorespiration (Figure 8.18).
- CAM plants, a type of C_4 plant, prevent photorespiration by fixing CO_2 into a four-carbon molecule at night and then running the Calvin cycle during the day with their stomata closed to reduce water loss (Figure 8.19).

Assess and Discuss

Test Yourself

1. The water necessary for photosynthesis
   a. is split into H_2 and O_2.
   b. is directly involved in the synthesis of carbohydrate.
   c. provides the electrons to replace lost electrons in photosystem II.
   d. provides H^+ needed to synthesize G3P.
   e. does none of the above.
2. The reaction center pigment differs from the other pigment molecules of the light-harvesting complex in that
   a. the reaction center pigment is a carotenoid.
   b. the reaction center pigment absorbs light energy and transfers that energy to other molecules without the transfer of electrons.
   c. the reaction center pigment transfers excited electrons to the primary electron acceptor.

d. the reaction center pigment does not transfer excited electrons to the primary electron acceptor.
e. the reaction center acts as an ATP synthase to produce ATP.

3. The cyclic electron flow that occurs via photosystem I produces
a. NADPH.
b. oxygen.
c. ATP.
d. all of the above.
e. a and c only.

4. During noncyclic electron flow, the high-energy electron from P680*
a. eventually moves to $NADP^+$.
b. becomes incorporated in water molecules.
c. is pumped into the thylakoid space to drive ATP production.
d. provides the energy necessary to split water molecules.
e. falls back to the low-energy state in photosystem II.

5. During the first phase of the Calvin cycle, carbon dioxide is incorporated into ribulose bisphosphate (RuBP) by
a. oxaloacetate.
b. rubisco.
c. RuBP.
d. quinone.
e. G3P.

6. The NADPH produced during the light reactions is necessary for
a. the carbon fixation phase, which incorporates carbon dioxide into an organic molecule of the Calvin cycle.
b. the reduction phase, which produces carbohydrates in the Calvin cycle.
c. the regeneration of RuBP of the Calvin cycle.
d. all of the above.
e. a and b only.

7. The majority of the G3P produced during the reduction and carbohydrate production phase is used to produce
a. glucose.
b. ATP.
c. RuBP to continue the cycle.
d. rubisco.
e. all of the above.

8. Photorespiration
a. is the process where plants use sunlight to make ATP.
b. is an inefficient way plants can produce organic molecules and in the process use O_2 and release CO_2.
c. is a process that plants use to convert light energy to NADPH.
d. occurs in the thylakoid lumen.
e. is the normal process of carbohydrate production in cool, moist environments.

9. Photorespiration is avoided in C_4 plants because
a. these plants separate the formation of a four-carbon molecule from the rest of the Calvin cycle in different cells.
b. these plants carry out only anaerobic respiration.
c. the enzyme PEP functions to maintain high CO_2 concentrations in the bundle-sheath cells.
d. all of the above.
e. a and c only.

10. Plants commonly found in hot and dry environments that carry out carbon fixation at night are
a. oak trees.
b. C_3 plants.
c. CAM plants.
d. all of the above.
e. a and b only.

Conceptual Questions

1. What are the two stages of photosynthesis? What are the key products of each stage?
2. What is the function of NADPH in the Calvin cycle?
3. A principle of biology is that *living organisms use energy.* At the level of the biosphere, what is the role of photosynthesis in the utilization of energy by living organisms?

Collaborative Questions

1. Discuss the advantages and disadvantages of being a heterotroph or a photoautotroph.
2. Biotechnologists are trying to genetically modify C_3 plants to convert them to C_4 or CAM plants. Why would this be useful? What genes might you introduce into C_3 plants to convert them to C_4 or CAM plants?

Online Resource

www.brookerbiology.com

Stay a step ahead in your studies with animations that bring concepts to life and practice tests to assess your understanding. Your instructor may also recommend the interactive eBook, individualized learning tools, and more.

Cell Communication

9

Chapter Outline

9.1 General Features of Cell Communication

9.2 Cellular Receptors and Their Activation

9.3 Signal Transduction and the Cellular Response

9.4 Hormonal Signaling in Multicellular Organisms

9.5 Apoptosis: Programmed Cell Death

Summary of Key Concepts

Assess and Discuss

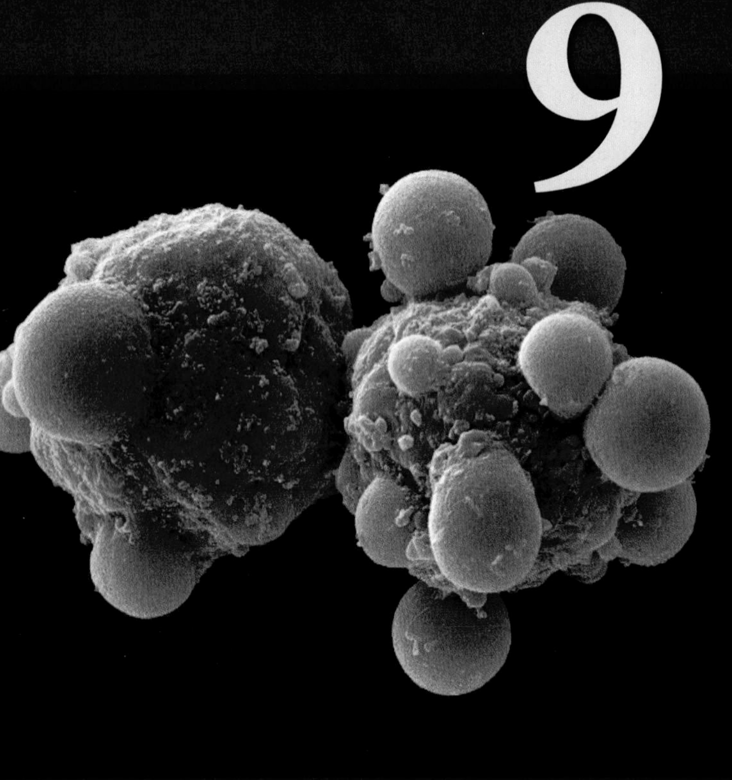

Programmed cell death. The two cells shown here are breaking apart due to signaling molecules that initiated a pathway that programmed their death.

Over 2 billion cells will die in your body during the next hour. In an adult human body, approximately 50–70 billion cells die each day due to programmed cell death—the process in which a cell breaks apart into small fragments (see chapter opening photo). In a year, your body produces and purposely destroys a mass of cells that is equal to your own body's weight! Though this may seem like a scary process, it's actually keeping you healthy. Programmed cell death, also called apoptosis, ensures that your body maintains a proper number of cells. It also eliminates cells that are worn out or potentially harmful, such as cancer cells. Programmed cell death can occur via signals that intentionally cause particular cells to die, or it can result from a failure of proper cell communication. It may also happen when environmental agents cause damage to a cell. Programmed cell death is one example of a response that involves **cell communication**—the process through which cells can detect, interpret, and respond to signals in their environment. A **signal** is an agent that can influence the properties of cells.

In this chapter, we will examine how cells detect environmental signals and also how they produce signals that enable them to communicate with other cells. Communication at the cellular level involves not only receiving and sending signals but also their interpretation. For this to occur, a signal must be recognized by a cellular protein called a **receptor**. When a signal and receptor interact, the receptor changes shape, or conformation, thereby changing the way the receptor interacts with cellular factors. These interactions eventually lead to some type of response in the cell. We begin the chapter with the general features of cell communication, and then discuss the main ways in which cells receive, process, and respond to signals sent by other cells. As you will learn, cell communication involves an amazing diversity of signaling molecules and cellular proteins that are devoted to this process. We conclude by looking at the role of cell communication in apoptosis in which a cell becomes programmed to die.

9.1 General Features of Cell Communication

Learning Outcomes:

1. Describe the two general reasons for cell signaling: responding to environmental changes and cell-to-cell communication.
2. Compare and contrast the five ways that cells communicate with each other based on the distance between cells.
3. Outline the three-stage process of cell signaling.

All living cells, including bacteria, archaea, protists, fungi, and plant and animal cells, conduct and require cell communication to survive. Cell communication, also known as cell signaling, involves both incoming and outgoing signals. For example, on a sunny day, cells can sense their exposure to ultraviolet (UV) light—a physical signal—and respond accordingly. In humans, UV light acts as an incoming signal to promote the synthesis of melanin, a protective pigment that helps to prevent the harmful effects of UV radiation. In addition, cells produce outgoing signals that influence the behavior of neighboring cells. Plant cells, for example, produce hormones that influence the pattern of cell elongation so the plant grows toward light. Cells of all living organisms both respond to incoming signals and produce outgoing signals. Cell communication is a two-way street.

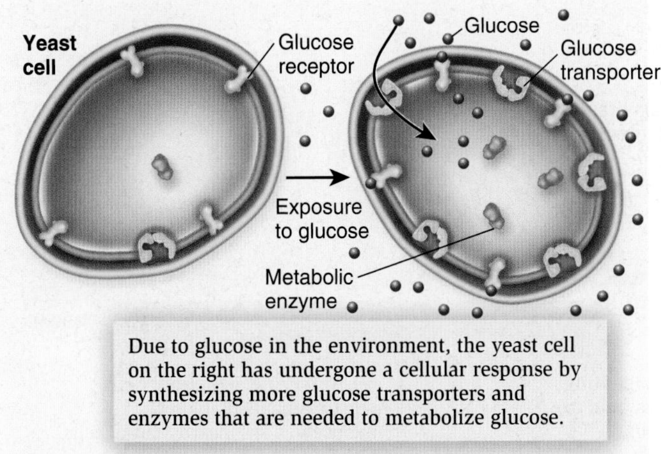

Yeast cell
Glucose receptor
Glucose
Glucose transporter
Exposure to glucose
Metabolic enzyme
Due to glucose in the environment, the yeast cell on the right has undergone a cellular response by synthesizing more glucose transporters and enzymes that are needed to metabolize glucose.

Figure 9.1 **Response of a yeast cell to glucose.** When glucose is absent from the extracellular environment, the cell is not well prepared to take up and metabolize this sugar. However, when glucose is present, some of that glucose binds to receptors in the membrane, which leads to changes in the amounts and properties of intracellular and membrane proteins so the cell can readily use glucose.

BIOLOGY PRINCIPLE Living organisms interact with their environment. In this example, cell signaling allows yeast cells to respond to the presence of glucose in their environment.

Concept Check: What is the signaling molecule in this example?

In this section, we begin by considering why cells need to respond to signals. We will then examine various forms of signaling that are based on the distance between the cells that communicate with each other. Finally, we will examine the main steps that occur when a cell is exposed to a signal and elicits a response to it.

Cells Detect and Respond to Signals from Their Environment and from Other Cells

Before getting into the details of cell communication, let's take a general look at why cell communication is necessary. The first reason is that cells need to respond to a changing environment. Changes in the environment are a persistent feature of life, and living cells are continually faced with alterations in temperature and availability of nutrients, water, and light. A cell may even be exposed to a toxic chemical in its environment. Being able to respond to change at the cellular level is called a **cellular response**.

As an example, let's consider the response of a yeast cell to glucose in its environment (**Figure 9.1**). Some of the glucose acts as a signaling molecule that binds to a receptor and causes a cellular response. In this case, the cell responds by increasing the number of glucose transporters needed to take glucose into the cell and also by increasing the number of metabolic enzymes required to utilize glucose once it is inside. The cellular response allows the cell to use glucose efficiently.

A second reason for cell signaling is the need for cells to communicate with each other—a type of cell communication also called **cell-to-cell communication**. In one of the earliest experiments demonstrating cell-to-cell communication, Charles Darwin and his son Francis Darwin studied phototropism, the phenomenon in which plants grow toward light (**Figure 9.2**). The Darwins observed that the actual bending occurs in a zone below the growing shoot tip. They concluded that a signal must be transmitted from the growing tip to lower parts of the shoot. Later research revealed that the signal is a molecule called auxin, which is transmitted from cell to cell. A higher amount of auxin accumulates on the nonilluminated side of the shoot and promotes cell elongation on that side of the shoot only, thereby causing the shoot to bend toward the light source.

Cell-to-Cell Communication Can Occur Between Adjacent Cells and Between Cells That Are Long Distances Apart

Researchers have determined that organisms have a variety of different mechanisms to achieve cell-to-cell communication. The mode of communication depends, in part, on the distance between the cells that need to communicate with each other. Let's first examine the various ways in which signals are transferred between cells. Later in this chapter, we will learn how such signals elicit a cellular response.

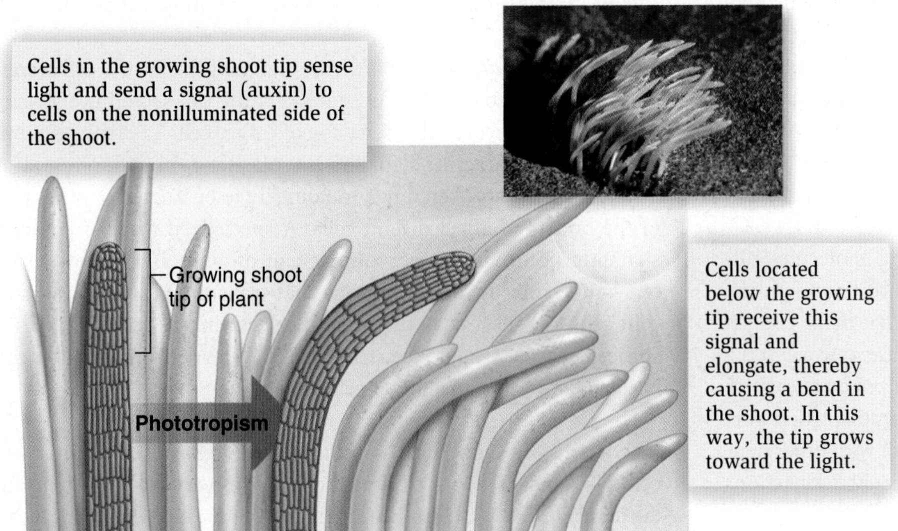

Cells in the growing shoot tip sense light and send a signal (auxin) to cells on the nonilluminated side of the shoot.
Growing shoot tip of plant
Phototropism
Cells located below the growing tip receive this signal and elongate, thereby causing a bend in the shoot. In this way, the tip grows toward the light.

Figure 9.2 **Phototropism in plants.** This process involves cell-to-cell communication that leads to a shoot bending toward light just beneath its actively growing tip.

BioConnections: Look ahead to Figure 36.5. How does light affect the distribution of auxin produced by a plant's growing shoot tip?

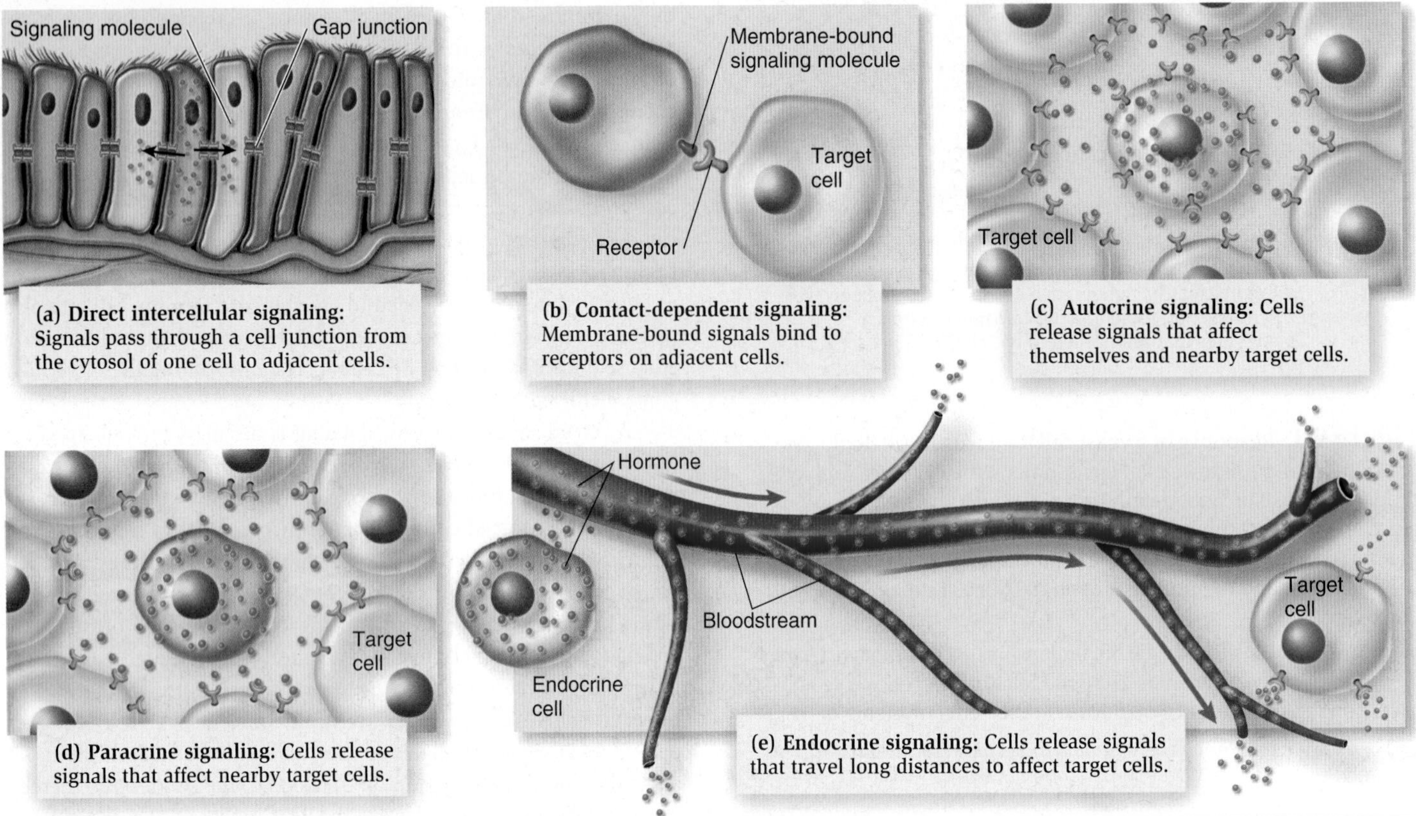

Figure 9.3 Types of cell-to-cell communication based on the distance between cells.

Concept Check: *Which type of signal, paracrine or endocrine, is likely to exist for a longer period of time? Explain why this is necessary.*

One way to categorize cell signaling is by the manner in which the signal is transmitted from one cell to another. Signals are relayed between cells in five common ways, all of which involve a cell that produces a signal and a target cell that receives the signal (**Figure 9.3**).

Direct Intercellular Signaling In a multicellular organism, cells adjacent to each other may have contacts, called cell junctions, that enable them to pass ions, signaling molecules, and other materials between the cytosol of one cell and the cytosol of another (Figure 9.3a). For example, cardiac muscle cells, which cause your heart to beat, have intercellular connections called gap junctions that allow the passage of ions needed for the coordinated contraction of cardiac muscle cells. We will examine how gap junctions work in Chapter 10.

Contact-Dependent Signaling Not all signaling molecules diffuse from one cell to another. Some molecules are bound to the surface of cells and provide a signal to other cells that make contact with the surface of that cell (Figure 9.3b). In this case, one cell has a membrane-bound signaling molecule that is recognized by a receptor on the surface of another cell. This occurs, for example, when portions of neurons (nerve cells) grow and make contact with other neurons. This is important for the formation of the proper connections between neurons.

Autocrine Signaling In autocrine signaling, a cell secretes signaling molecules that bind to receptors on its own cell surface or on neighboring cells of the same cell type, stimulating a response (Figure 9.3c). What is the purpose of autocrine signaling? It is often important for groups of cells to sense cell density. When cell density is high, the concentration of autocrine signals is also high. In some cases, such signals inhibit further cell growth, thereby limiting cell density.

Paracrine Signaling In paracrine signaling, a specific cell secretes a signaling molecule that does not affect the cell secreting the signal but instead influences the behavior of target cells in close proximity (Figure 9.3d). Paracrine signaling is typically of short duration. Usually, the signal is broken down too quickly to be carried to other parts of the body and affect distant cells. A specialized form of paracrine signaling occurs in the nervous system of animals. Neurotransmitters—molecules made in neurons that transmit a signal to an adjacent cell—are released at the end of the neuron and traverse a narrow space called the synapse (see Chapter 41). The neurotransmitter then binds to a receptor in a target cell.

Endocrine Signaling In contrast to the previous mechanisms of cell signaling, endocrine signaling occurs over relatively long distances (Figure 9.3e). In both animals and plants, molecules involved in

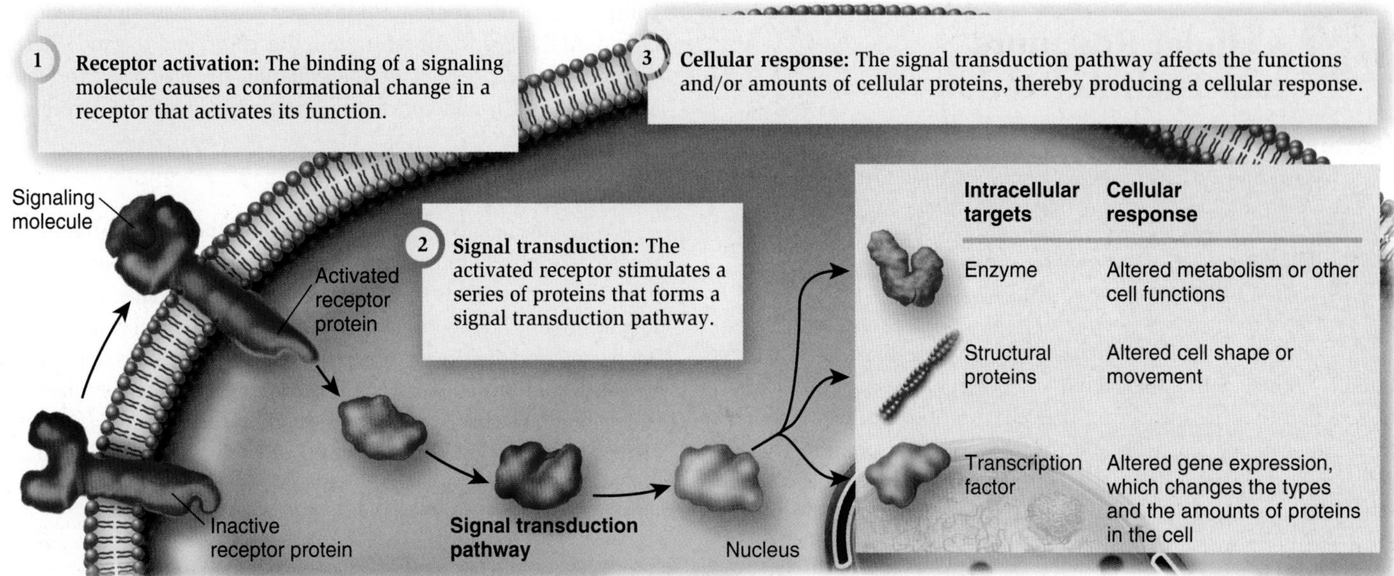

1 Receptor activation: The binding of a signaling molecule causes a conformational change in a receptor that activates its function.
3 Cellular response: The signal transduction pathway affects the functions and/or amounts of cellular proteins, thereby producing a cellular response.
Signaling molecule
Activated receptor protein
2 Signal transduction: The activated receptor stimulates a series of proteins that forms a signal transduction pathway.
Intracellular targets
Cellular response
Enzyme
Altered metabolism or other cell functions
Structural proteins
Altered cell shape or movement
Transcription factor
Altered gene expression, which changes the types and the amounts of proteins in the cell
Inactive receptor protein
Signal transduction pathway
Nucleus

Figure 9.4 **The three stages of cell signaling: receptor activation, signal transduction, and a cellular response.**

Concept Check: *For most signaling molecules, explain why a signal transduction pathway is necessary.*

long-distance signaling are called **hormones**. They usually last longer than signaling molecules involved in autocrine and paracrine signaling. In mammals, endocrine signaling involves the secretion of hormones into the bloodstream, which may affect virtually all cells of the body, including those that are far from the cells that secrete the signaling molecules. In flowering plants, hormones move through the plant vascular system and also move through adjacent cells. Some hormones are even gases that diffuse into the air. Ethylene, a gas given off by plants, plays a variety of roles, such as accelerating the ripening of fruit.

Cells Usually Respond to Signals by a Three-Stage Process

Up to this point, we have learned that signals influence the behavior of cells in close proximity or at long distances, interacting with receptors to elicit a cellular response. What events occur when a cell encounters a signal? In most cases, the binding of a signaling molecule to a receptor causes the receptor to activate a signal transduction pathway, which then leads to a cellular response. **Figure 9.4** diagrams the three common stages of cell signaling: receptor activation, signal transduction, and a cellular response.

Stage 1: Receptor Activation In the initial stage, a signaling molecule binds to a receptor in the target cell, causing a conformational change in the receptor that activates its function. In most cases, the activated receptor initiates a response by causing changes in a series of proteins that collectively forms a signal transduction pathway, as described next.

Stage 2: Signal Transduction During signal transduction, the initial signal is converted—or transduced—to a different signal inside the cell. This process is carried out by a group of proteins that form a **signal transduction pathway**. These proteins undergo a series of changes that may result in the production of an intracellular signaling molecule. However, some receptors are intracellular and do not activate a signal transduction pathway. As discussed later, certain types of intracellular receptors directly cause a cellular response.

Stage 3: Cellular Response Cells respond to signals in several different ways. Figure 9.4 shows three common categories of proteins that are controlled by cell signaling: enzymes, structural proteins, and transcription factors.

Many signaling molecules exert their effects by altering the activity of one or more enzymes. For example, certain hormones provide a signal that the body needs energy. These hormones activate enzymes that are required for the breakdown of molecules such as carbohydrates.

Cells also respond to signals by altering the functions of structural proteins in the cell. For example, when animal cells move during embryonic development or when an amoeba moves toward food, signals play a role in the rearrangement of actin filaments, which are components of the cytoskeleton. The coordination of signaling and changes in the cytoskeleton enable cells to move in the correct direction.

Cells may also respond to signals by affecting the function of **transcription factors**—proteins that regulate the transcription of genes. Some transcription factors activate gene expression. For example, when cells are exposed to sex hormones, transcription factors activate genes that change the properties of cells, which can lead to changes in the sexual characteristics of entire organisms. As discussed in Chapter 51 (see Section 51.3), estrogens and androgens are responsible for the development of secondary sex characteristics in humans, including breast development in females and beard growth in males, respectively.

9.2 Cellular Receptors and Their Activation

Learning Outcomes:

1. Calculate how a signaling molecule, or ligand, binds to its receptor with an affinity measured as a dissociation constant.
2. Explain how a signaling molecule activates a receptor.
3. Identify the three general types of cell surface receptors.
4. Describe intracellular receptors, using estrogen receptors as an example.

In this section, we will take a closer look at receptors and how they interact with signaling molecules. We will compare receptors based on whether they are located on the cell surface or inside the cell. In this chapter, our focus will be on receptors that respond to chemical signaling molecules. Other receptors discussed in Units VI and VII respond to mechanical motion (mechanoreceptors), temperature changes (thermoreceptors), and light (photoreceptors).

Signaling Molecules Bind to Receptors

The ability of cells to respond to a signal usually requires precise recognition between a signal and its receptor. In many cases, the signal is a molecule, such as a steroid or a protein, that binds to the receptor. A signaling molecule binds to a receptor in much the same way that a substrate binds to the active site of an enzyme, as described in Chapter 6. The signaling molecule, which is called a **ligand**, binds noncovalently to the receptor with a high degree of specificity. The binding occurs when the ligand and receptor happen to collide in the correct orientation with enough energy to form a **ligand • receptor complex**.

$$[\text{Ligand}] + [\text{Receptor}] \underset{k_{off}}{\overset{k_{on}}{\rightleftharpoons}} [\text{Ligand} \cdot \text{Receptor complex}]$$

Brackets [ ] refer to concentration. The value k_{on} is the rate at which binding occurs. After a complex forms between the ligand and its receptor, the noncovalent interaction between a ligand and receptor remains stable for a finite period of time. The term k_{off} is the rate at which the ligand • receptor complex falls apart or dissociates.

In general, the binding and release between a ligand and its receptor are relatively rapid, and therefore an equilibrium is reached when the rate of formation of new ligand • receptor complexes equals the rate at which existing ligand • receptor complexes dissociate:

$$k_{on}[\text{Ligand}][\text{Receptor}] = k_{off}[\text{Ligand} \cdot \text{Receptor complex}]$$

Rearranging,

$$\frac{[\text{Ligand}][\text{Receptor}]}{[\text{Ligand} \cdot \text{Receptor complex}]} = \frac{k_{off}}{k_{on}} = K_d$$

K_d is called the **dissociation constant** between a ligand and its receptor. The K_d value is inversely related to the affinity between the ligand and receptor. A low K_d value indicates that a receptor has a high affinity for its ligand.

Let's look carefully at the left side of this equation and consider what it means. At a ligand concentration at which half of the receptors are bound to a ligand, the concentration of the ligand • receptor complex equals the concentration of receptor that doesn't have ligand bound. At this ligand concentration, [Receptor] and [Ligand • Receptor complex] cancel out of the equation because they are equal. Therefore, at a ligand concentration at which half of the receptors have bound ligand:

$$K_d = [\text{Ligand}]$$

When the ligand concentration is above the K_d value, most of the receptors are likely to have ligand bound to them. In contrast, if the ligand concentration is substantially below the K_d value, most receptors will not be bound by their ligand. The K_d values for many different ligands and their receptors have been experimentally determined. How is this information useful? It allows researchers to predict when a signaling molecule is likely to cause a cellular response. If the concentration of a signaling molecule is far below the K_d value, a cellular response is not likely because relatively few receptors form a complex with the signaling molecule.

Receptors Undergo Conformational Changes

Unlike enzymes, which convert their substrates into products, receptors do not usually alter the structure of their ligands. Instead, the ligands alter the structure of their receptors, causing a conformational change (**Figure 9.5**). In this case, the binding of the ligand to its receptor changes the receptor in a way that activates its ability to initiate a cellular response.

Because the binding of a ligand to its receptor is a reversible process, the ligand and receptor also dissociate. Once the ligand is released, the receptor is no longer activated.

Cells Contain a Variety of Cell Surface Receptors That Respond to Extracellular Signals

Most signaling molecules are either small hydrophilic molecules or large molecules that do not readily pass through the plasma membrane of cells. Such extracellular signals bind to **cell surface**

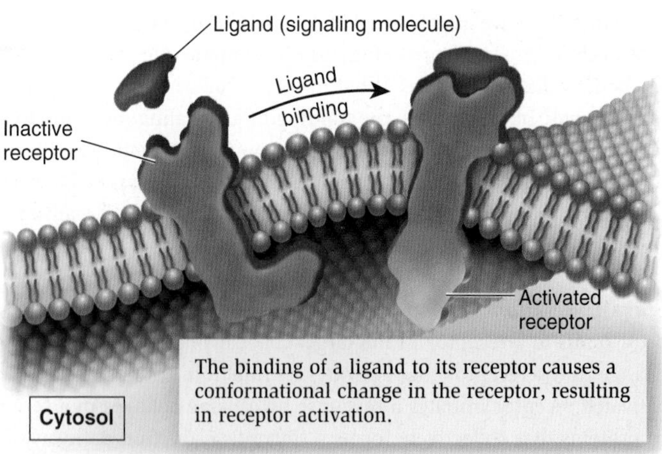
Ligand (signaling molecule)
Ligand binding
Inactive receptor
Activated receptor
Cytosol
The binding of a ligand to its receptor causes a conformational change in the receptor, resulting in receptor activation.

Figure 9.5 **Receptor activation.**

BioConnections: *Look back at Figure 6.6. How is the binding of a ligand to its receptor similar to the binding of a substrate to an enzyme? How are they different?*

receptors—receptors found in the plasma membrane. A typical cell is expected to contain dozens or even hundreds of different cell surface receptors that enable the cell to respond to different kinds of extracellular signaling molecules. By analyzing the functions of cell surface receptors from many different organisms, researchers have determined that most fall into one of three categories: enzyme-linked receptors, G-protein-coupled receptors, and ligand-gated ion channels, which are described next.

Enzyme-Linked Receptors Receptors known as **enzyme-linked receptors** are found in all living species. Many human hormones bind to this type of receptor. For example, when insulin binds to an enzyme-linked receptor in muscle cells, it enhances the ability of those cells to use glucose. Enzyme-linked receptors typically have two important domains: an extracellular domain, which binds a signaling molecule, and an intracellular domain, which has a catalytic function (Figure 9.6a). When a signaling molecule binds to the extracellular domain, a conformational change is transmitted through the membrane-embedded portion of the protein that affects the conformation of the intracellular catalytic domain. In most cases, this conformational change causes the intracellular catalytic domain to become functionally active.

Most types of enzyme-linked receptors function as **protein kinases**, enzymes that transfer a phosphate group from ATP to specific amino acids in a protein (Figure 9.6b). For example, tyrosine kinases attach phosphate to the amino acid tyrosine, whereas serine/threonine kinases attach phosphate to the amino acids serine and threonine. In the absence of a signaling molecule, the catalytic domain of the receptor remains inactive. However, when a signal binds to the extracellular domain, the catalytic domain is activated. Under these conditions, the receptor may phosphorylate itself or it may phosphorylate intracellular proteins. The attachment of a negatively charged phosphate changes the structure of a protein and thereby can alter its function. Later in this chapter, we will explore how this event leads to a cellular response, such as the activation of enzymes that affect cell function.

G-Protein-Coupled Receptors Receptors called **G-protein-coupled receptors (GPCRs)** are found in the cells of all eukaryotic species and are particularly common in animals. GPCRs typically contain seven transmembrane segments that wind back and forth through the plasma membrane. The receptors interact with intracellular proteins called **G proteins**, which are so named because of their ability to bind guanosine triphosphate (GTP) and guanosine diphosphate (GDP). GTP is similar in structure to ATP except it has guanine as a base instead of adenine. In the 1970s, the existence of G proteins was first proposed by Martin Rodbell and colleagues, who found that GTP is needed for certain hormone receptors to cause an intracellular response. Later, Alfred Gilman and coworkers used genetic and biochemical techniques to identify and purify a G protein. In 1994, Rodbell and Gilman won the Nobel Prize in Physiology or Medicine for their pioneering work.

Figure 9.7 shows how a GPCR and a G protein interact. At the cell surface, a signaling molecule binds to a GPCR, causing a conformational change that activates the receptor enabling it to bind to a G protein. The G protein, which is a lipid-anchored protein, releases GDP and binds GTP instead. GTP binding changes the conformation of the G protein, causing it to dissociate into an α subunit and a β/γ dimer. Later in this chapter, we will examine how the α subunit interacts with other proteins in a signal transduction pathway to elicit a cellular response. The β/γ dimer also plays a role in signal transduction. For example, it can regulate the function of ion channels in the plasma membrane.

When a signaling molecule and GPCR dissociate, the GPCR is no longer activated, and the cellular response is reversed. For the G protein to return to the inactive state, the α subunit first hydrolyzes its

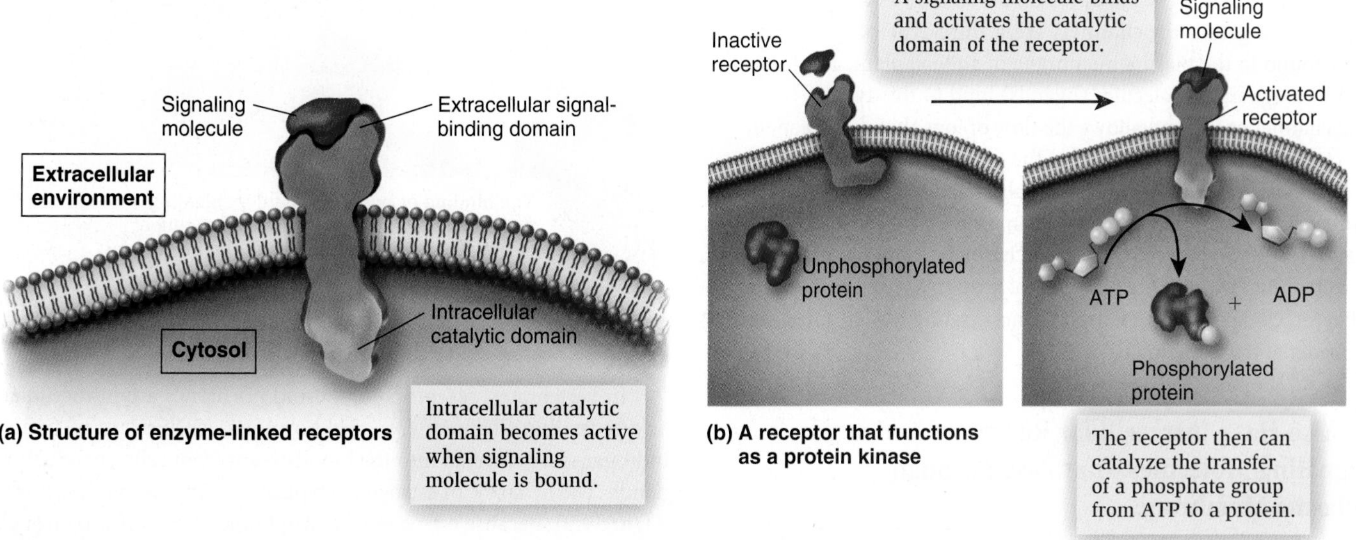
Signaling molecule
Extracellular signal-binding domain
Extracellular environment
Cytosol
Intracellular catalytic domain
(a) Structure of enzyme-linked receptors
Intracellular catalytic domain becomes active when signaling molecule is bound.
Inactive receptor
A signaling molecule binds and activates the catalytic domain of the receptor.
Signaling molecule
Activated receptor
Unphosphorylated protein
ATP
+
ADP
Phosphorylated protein
(b) A receptor that functions as a protein kinase
The receptor then can catalyze the transfer of a phosphate group from ATP to a protein.

Figure 9.6 **Enzyme-linked receptors.**

Concept Check: Based on your understanding of ATP as an energy intermediate, is the phosphorylation of a protein via a protein kinase an exergonic or endergonic reaction? How is the energy of protein phosphorylation used—what does it accomplish?

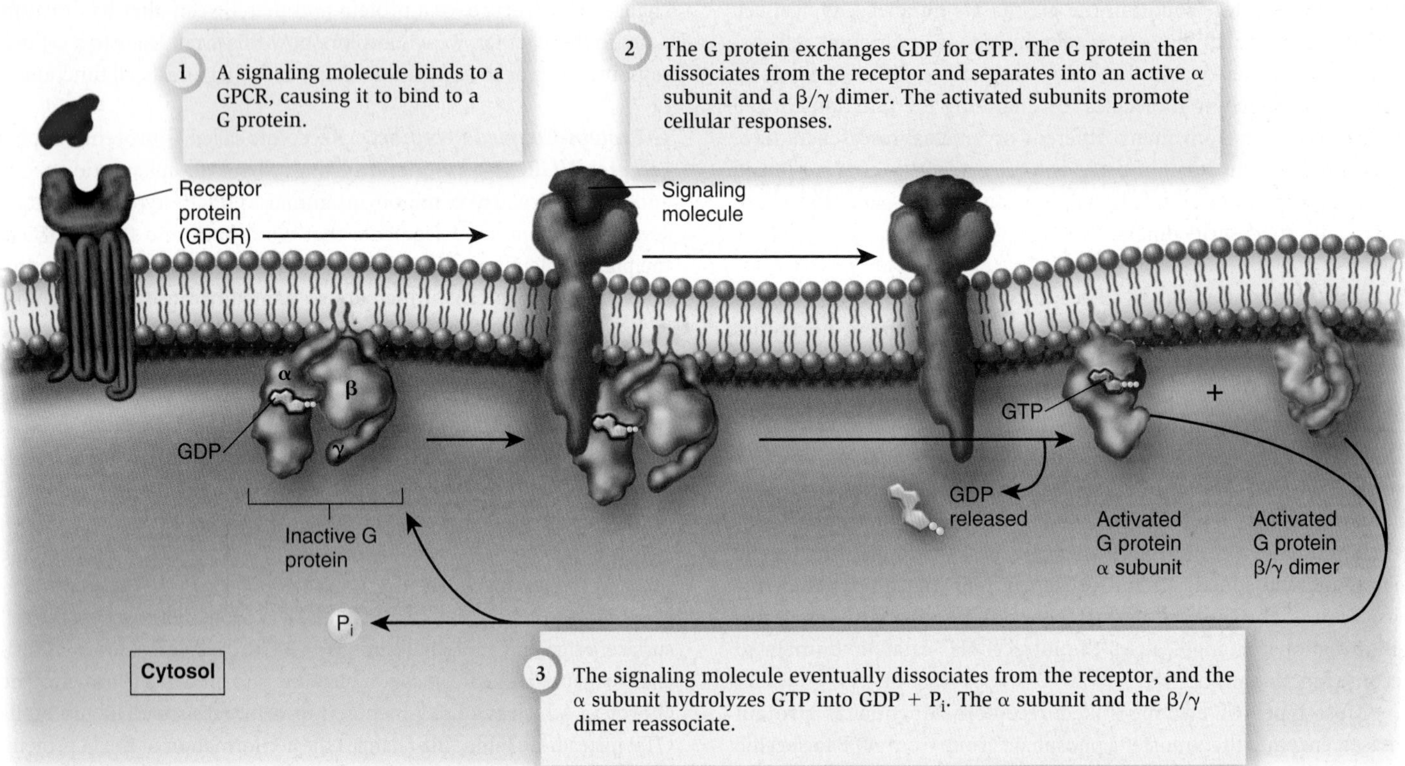

Figure 9.7 The activation of G-protein-coupled receptors (GPCRs) and G proteins. Note: All three receptors shown in this figure are meant to be the same receptor, but the one on the left is drawn with greater detail to emphasize that it has seven transmembrane segments.

Concept Check: *What has to happen for the α and β/γ subunits of the G protein to reassociate with each other?*

bound GTP to GDP and P_i. After this occurs, the α and *β/γ* subunits reassociate with each other to form an inactive G protein complex.

Ligand-Gated Ion Channels As described in Chapter 5, ion channels are proteins that allow the diffusion of ions across cellular membranes. **Ligand-gated ion channels** are a third type of cell surface receptor found in the plasma membrane of animal, plant, and fungal cells. When signaling molecules (ligands) bind to this type of receptor, the channel opens and allows the flow of ions through the membrane, changing the concentration of the ions in the cell (**Figure 9.8**).

In animals, ligand-gated ion channels are important in the transmission of signals between neurons and muscle cells and between two neurons. In addition, ligand-gated ion channels in the plasma membrane allow the influx of Ca^{2+} into the cytosol. As discussed later in this chapter, changes in the cytosolic concentration of Ca^{2+} often play a role in signal transduction.

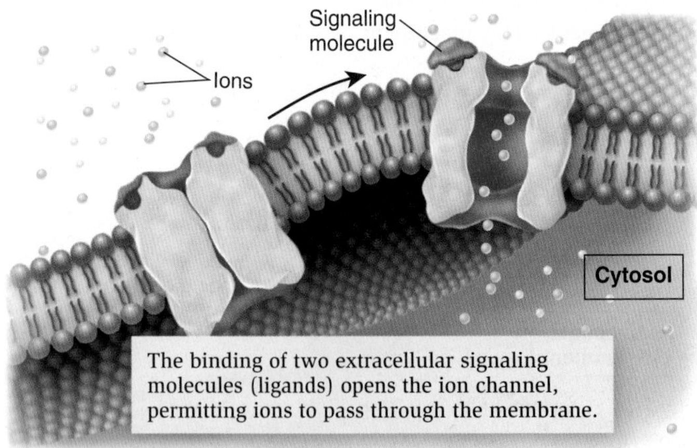

Signaling molecule
Ions
Cytosol
The binding of two extracellular signaling molecules (ligands) opens the ion channel, permitting ions to pass through the membrane.

Figure 9.8 The function of a ligand-gated ion channel.

Cells Also Have Intracellular Receptors Activated by Signaling Molecules That Pass Through the Plasma Membrane

Although most receptors for signaling molecules are located in the plasma membrane, some are found inside the cell. In these cases, an extracellular signaling molecule must diffuse through the plasma membrane to gain access to its receptor.

In vertebrates, receptors for steroid hormones are intracellular. As discussed in Chapter 51, steroid hormones, such as estrogens and androgens, are secreted into the bloodstream from cells of endocrine glands. The behavior of estrogen is typical of many steroid hormones (**Figure 9.9**). Because estrogen is hydrophobic, it can diffuse through the plasma membrane of a target cell and bind to a receptor inside the cell. Some steroids bind to receptors in the cytosol, which then travel into the nucleus. Other steroid hormones, such as estrogen, bind to receptors in the nucleus. After binding, the estrogen • receptor

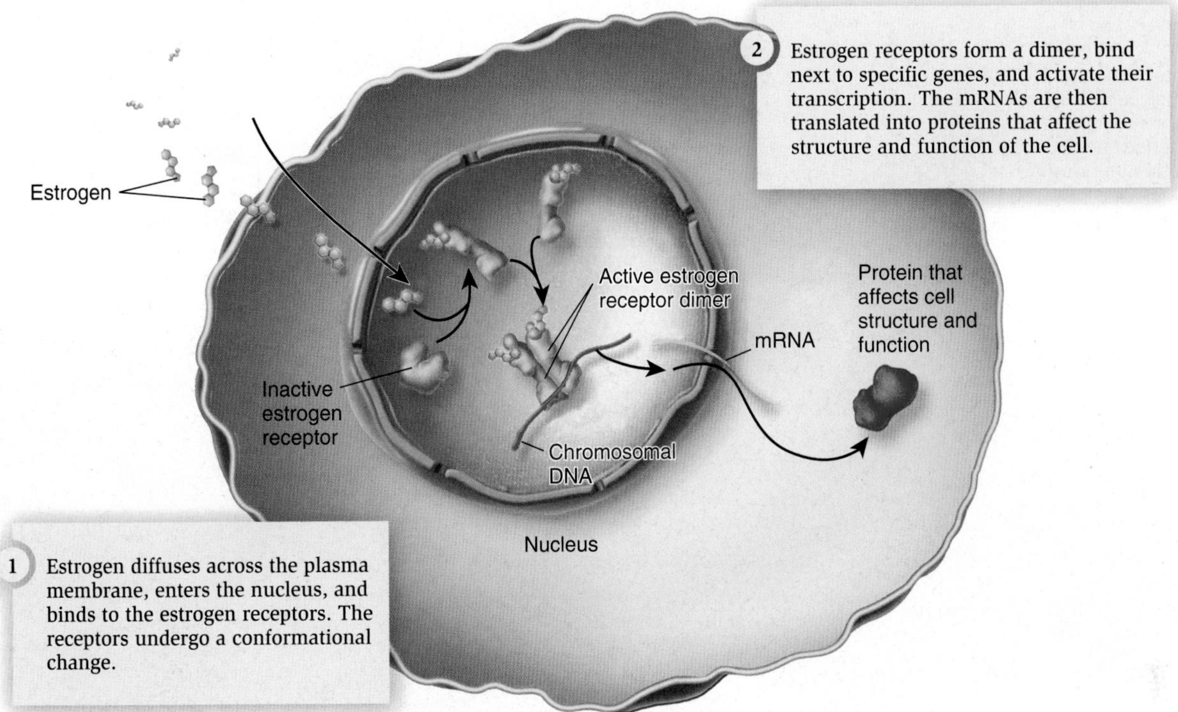

Estrogen
2 Estrogen receptors form a dimer, bind next to specific genes, and activate their transcription. The mRNAs are then translated into proteins that affect the structure and function of the cell.
Active estrogen receptor dimer
Protein that affects cell structure and function
mRNA
Inactive estrogen receptor
Chromosomal DNA
Nucleus
1 Estrogen diffuses across the plasma membrane, enters the nucleus, and binds to the estrogen receptors. The receptors undergo a conformational change.

Figure 9.9 **Estrogen receptor in mammalian cells.** This is an example of an intracellular receptor.

complex undergoes a conformational change that enables it to form a dimer with another estrogen • receptor complex. The dimer then binds to the DNA and activates the transcription of specific genes. The estrogen receptor is an example of a transcription factor—a protein that regulates the transcription of genes. The expression of specific genes changes cell structure and function in a way that results in a cellular response.

9.3 Signal Transduction and the Cellular Response

Learning Outcomes:

1. For signaling molecules that bind to receptor tyrosine kinases or G-protein-coupled receptors, describe the signal transduction pathways and how those pathways lead to a cellular response.
2. Relate the function of second messengers to signal transduction pathways.
3. List examples of second messengers and explain how they exert their effects.

We now turn our attention to the intracellular events that enable a cell to respond to a signaling molecule that binds to a cell surface receptor: signal transduction and a cellular response. In most cases, the binding of a signaling molecule to its receptor stimulates a signal transduction pathway. We begin by examining a pathway that is controlled by an enzyme-linked receptor. We will then examine pathways and cellular responses that are controlled by G-protein-coupled receptors. As you will learn, these pathways sometimes involve the production of intracellular signals called second messengers.

Receptor Tyrosine Kinases Activate Signal Transduction Pathways Involving a Protein Kinase Cascade That Alters Gene Transcription

Receptor tyrosine kinases are a category of enzyme-linked receptors that are found in all animals and also in choanoflagellates, which are the protists that are most closely related to animals (see Chapter 32). However, they are not found in bacteria, archaea, or other eukaryotic species. (Bacteria do have receptor histidine kinases, and all eukaryotes have receptor serine/threonine kinases.) The human genome contains about 60 different genes that encode receptor tyrosine kinases that recognize various types of signaling molecules such as hormones.

Figure 9.10 describes a simplified signal transduction pathway for epidermal growth factor (EGF). A **growth factor** is a signaling molecule that promotes cell division. Multicellular organisms, such as plants and animals, produce a variety of different growth factors to coordinate cell division throughout the body. In vertebrate animals, EGF is secreted from endocrine cells, travels through the bloodstream, and binds to a receptor tyrosine kinase called the EGF receptor. EGF is responsible for stimulating epidermal cells, such as skin cells, to divide. Following receptor activation, the three general parts of the signal transduction pathway are (1) relay proteins (also called adaptor proteins) activate a protein kinase cascade; (2) the protein kinase cascade phosphorylates proteins in the cell such as transcription factors; and (3) the phosphorylated transcription factors stimulate gene transcription. Next, we will consider the details of this pathway.

EGF Receptor Activation For receptor activation to occur, two EGF receptor subunits each bind a molecule of EGF. The binding of

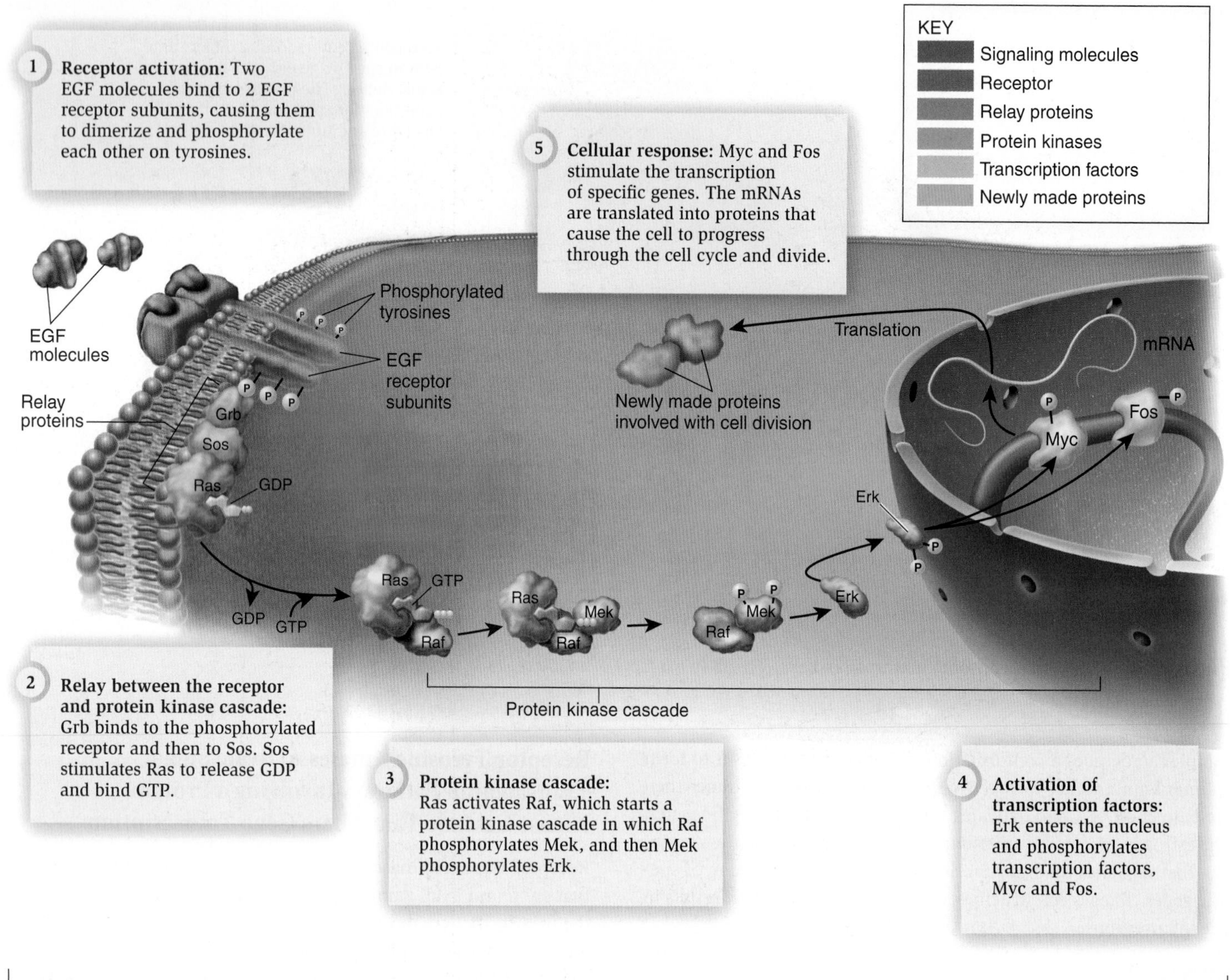
KEY
Signaling molecules
Receptor
Relay proteins
Protein kinases
Transcription factors
Newly made proteins
1 Receptor activation: Two EGF molecules bind to 2 EGF receptor subunits, causing them to dimerize and phosphorylate each other on tyrosines.
5 Cellular response: Myc and Fos stimulate the transcription of specific genes. The mRNAs are translated into proteins that cause the cell to progress through the cell cycle and divide.
EGF molecules
Relay proteins
Phosphorylated tyrosines
EGF receptor subunits
Grb
Sos
Ras
GDP
GTP
Raf
Mek
Erk
P
Translation
mRNA
Newly made proteins involved with cell division
Myc
Fos
Protein kinase cascade
2 Relay between the receptor and protein kinase cascade: Grb binds to the phosphorylated receptor and then to Sos. Sos stimulates Ras to release GDP and bind GTP.
3 Protein kinase cascade: Ras activates Raf, which starts a protein kinase cascade in which Raf phosphorylates Mek, and then Mek phosphorylates Erk.
4 Activation of transcription factors: Erk enters the nucleus and phosphorylates transcription factors, Myc and Fos.
Signal transduction (steps 2–4)

Figure 9.10 **The epidermal growth factor (EGF) pathway that promotes cell division.**

BioConnections: Look ahead to Figures 14.11 and, in particular, 14.12. Certain mutations alter the structure of the Ras protein so it does not hydrolyze GTP. Such mutations cause cancer. Explain why.

EGF causes the subunits to dimerize and phosphorylate each other on tyrosines within the receptors, which is why they are named receptor tyrosine kinases. Because the receptor phosphorylates itself, this event is called **autophosphorylation**. Next comes the signal transduction pathway.

Relay Proteins The phosphorylated form of the EGF receptor is first recognized by a relay protein of the signal transduction pathway called Grb. This interaction changes the conformation of Grb, causing it to bind another relay protein in the signal transduction pathway termed Sos, thereby changing the conformation of Sos. The activation of Sos causes a third relay protein called Ras to release GDP and bind GTP. The GTP form of Ras is the active form.

Protein Kinase Cascade The function of Grb, Sos, and Ras is to relay a cellular signal to additional proteins in the signal transduction pathway that form a **protein kinase cascade**. This cascade involves the sequential activation of multiple protein kinases. Activated Ras binds to Raf, the first protein kinase in the cascade. Raf then phosphorylates Mek, which becomes active and, in turn, phosphorylates Erk. Raf, Mek, and Erk, the protein kinase cascade, are all examples of **mitogen-activated protein kinases (MAP-kinases)**. This type of protein kinase was first discovered because it is activated in the presence of mitogens—agents that cause a cell to divide. EGF is an example of a mitogen.

Activation of Transcription Factors and the Cellular Response The phosphorylated form of Erk enters the nucleus and phosphorylates transcription factors such as Myc and Fos. What is the cellular response? Once these transcription factors are phosphorylated, they stimulate the transcription of genes that encode proteins that promote cell division. After these proteins are made, the cell is stimulated to divide.

Growth factors such as EGF cause a rapid increase in the expression of many genes in mammals, perhaps as many as 100. As discussed in Chapter 14, growth factor signaling pathways are often involved in cancer. Mutations that cause proteins in these pathways to become hyperactive result in cells that divide uncontrollably!

Second Messengers Such as Cyclic AMP Are Key Components of Many Signal Transduction Pathways

Let's now turn to examples of signal transduction pathways and cellular responses that involve G-protein-coupled receptors (GPCRs). Extracellular signaling molecules that bind to cell surface receptors are sometimes referred to as first messengers. After first messengers bind to receptors such as GPCRs, many signal transduction pathways lead to the production of **second messengers**—small molecules or ions that relay signals inside the cell. The signals that result in second messenger production often act quickly, in a matter of seconds or minutes, but their duration is usually short. Therefore, such signaling typically occurs when a cell needs a quick and short cellular response.

Production of cAMP Mammalian and plant cells make several different types of G protein α subunits. One type of α subunit binds to **adenylyl cyclase**, an enzyme in the plasma membrane. This interaction stimulates adenylyl cyclase to synthesize **cyclic adenosine monophosphate (cyclic AMP, or cAMP)** from ATP (**Figure 9.11**). cAMP is an example of a second messenger.

Signal Transduction Pathway Involving cAMP Let's explore a signal transduction pathway in which the GPCR recognizes the hormone epinephrine (also called adrenaline). This hormone is sometimes called the "fight or flight" hormone. Epinephrine is produced when an individual is confronted with a stressful situation and helps the individual deal with a perceived threat or danger.

First, epinephrine binds to its receptor and activates a G protein (**Figure 9.12**). The α subunit then activates adenylyl cyclase, which catalyzes the production of cAMP from ATP. One effect of cAMP is to activate protein kinase A (PKA), which is composed of four subunits: two catalytic subunits that phosphorylate specific cellular proteins, and two regulatory subunits that inhibit the catalytic subunits when they are bound to each other. cAMP binds to the regulatory subunits of PKA. The binding of cAMP separates the regulatory and catalytic subunits, which allows each catalytic subunit to be active.

Cellular Response via PKA How does PKA activation lead to a cellular response? The catalytic subunit of PKA phosphorylates specific cellular proteins such as enzymes, structural proteins, and transcription factors. The phosphorylation of enzymes and structural proteins influences the structure and function of the cell. Likewise, the phosphorylation of transcription factors leads to the synthesis of new proteins that affect cell structure and function.

As a specific example of a cellular response, **Figure 9.13** shows how a skeletal muscle cell responds to elevated levels of epinephrine. When PKA becomes active, it phosphorylates two enzymes—phosphorylase kinase and glycogen synthase. Both of these enzymes are involved with the metabolism of glycogen, which is a polymer of glucose used to store energy. When phosphorylase kinase is phosphorylated, it becomes activated. The function of phosphorylase kinase is to phosphorylate another enzyme in the cell called glycogen phosphorylase, which then becomes activated. This enzyme causes glycogen breakdown by phosphorylating glucose units at the ends of a glycogen polymer, which releases individual glucose-phosphate molecules from glycogen:

$$\text{Glycogen}_n + \text{P}_\text{i} \xrightarrow{\text{Glycogen phosphorylase}} \text{Glycogen}_{n-1} + \text{Glucose-phosphate}$$

where n is the number of glucose units in glycogen.

When PKA phosphorylates glycogen synthase, the function of this enzyme is inhibited rather than activated (see Figure 9.13). The function of glycogen synthase is to make glycogen. Therefore, the effect of cAMP is to prevent glycogen synthesis.

Taken together, the effects of epinephrine in skeletal muscle cells are to stimulate glycogen breakdown and inhibit glycogen synthesis. This provides these cells with more glucose molecules, which they can use for the energy needed for muscle contraction. In this way, the individual is better prepared to fight or flee.

Reversal of the Cellular Response As mentioned, signaling that involves second messengers is typically of short duration. When the signaling molecule is no longer produced and its level falls, a larger

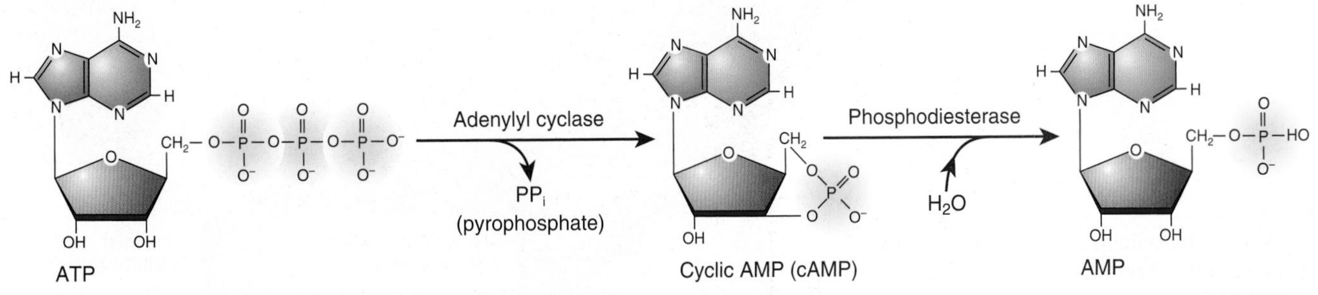

Figure 9.11 The synthesis and breakdown of cyclic AMP. Cyclic AMP (cAMP) is a second messenger formed from ATP by adenylyl cyclase, an enzyme in the plasma membrane. cAMP is inactivated by the action of an enzyme called phosphodiesterase, which converts cAMP to AMP.

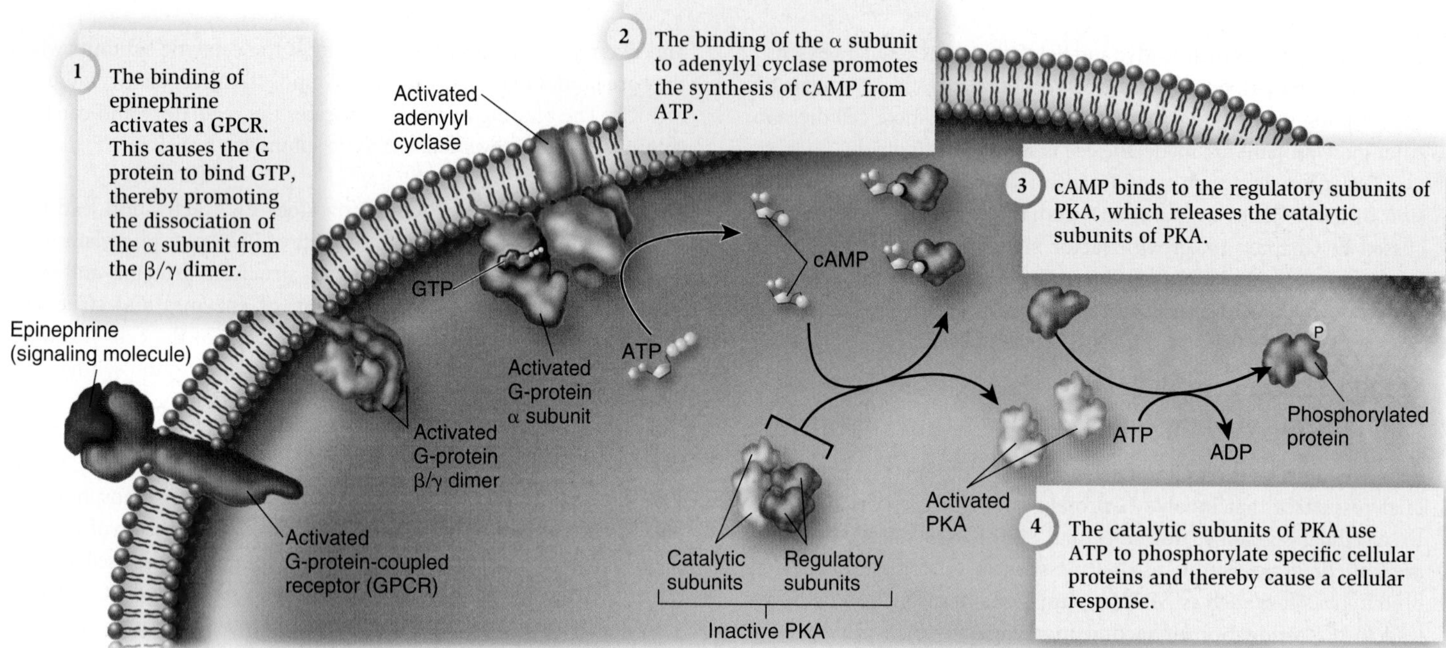

1 The binding of epinephrine activates a GPCR. This causes the G protein to bind GTP, thereby promoting the dissociation of the α subunit from the β/γ dimer.
Activated adenylyl cyclase
2 The binding of the α subunit to adenylyl cyclase promotes the synthesis of cAMP from ATP.
3 cAMP binds to the regulatory subunits of PKA, which releases the catalytic subunits of PKA.
cAMP
GTP
ATP
Epinephrine (signaling molecule)
Activated G-protein α subunit
Activated G-protein β/γ dimer
Activated G-protein-coupled receptor (GPCR)
Catalytic subunits
Regulatory subunits
Inactive PKA
Activated PKA
ATP
ADP
P
Phosphorylated protein
4 The catalytic subunits of PKA use ATP to phosphorylate specific cellular proteins and thereby cause a cellular response.

Figure 9.12 **A signal transduction pathway involving cAMP.** The pathway leading to the formation of cAMP and subsequent activation of protein kinase A (PKA), which is mediated by a G-protein-coupled receptor (GPCR).

Concept Check: *In this figure, which part is the signal transduction pathway, and which is the cellular response?*

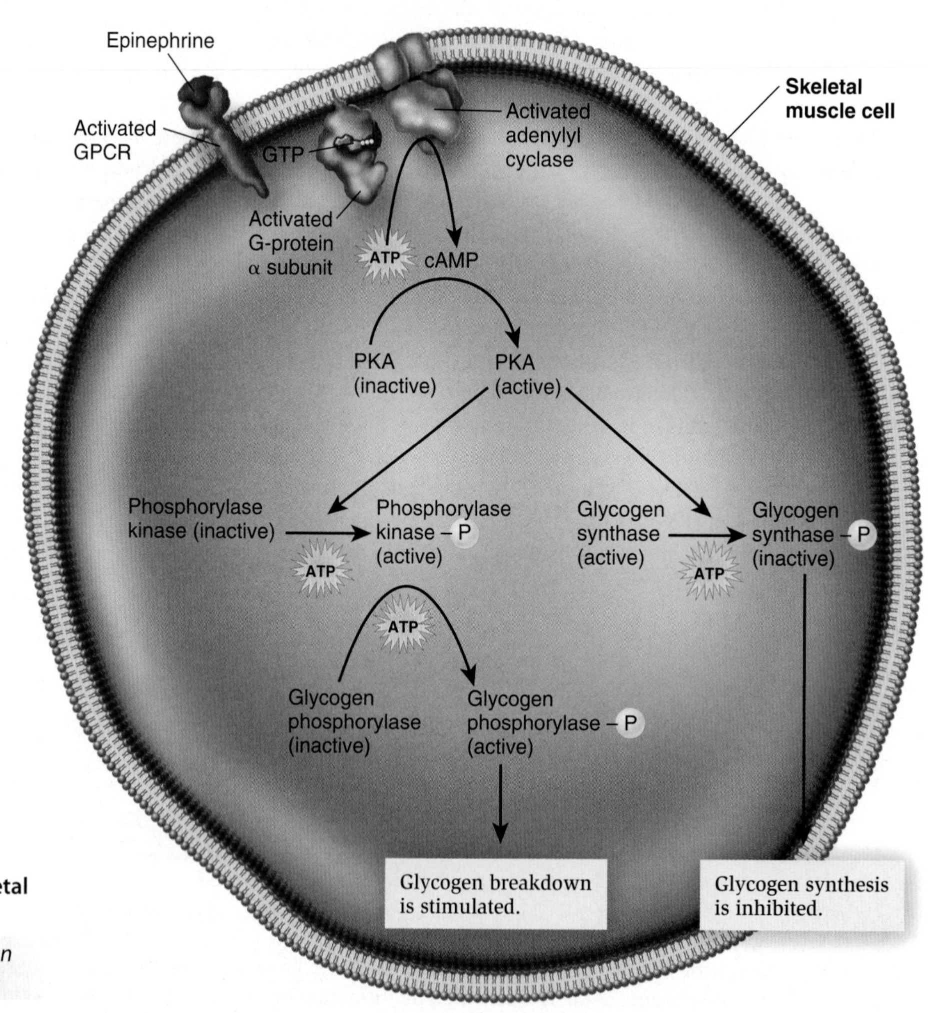

Figure 9.13 **The cellular response of a skeletal muscle cell to epinephrine.**

Concept Check: *Explain whether phosphorylation activates or inhibits enzyme function.*

percentage of the receptors are not bound by their ligands. When a ligand dissociates from the GPCR, the GPCR becomes deactivated. Intracellularly, the α subunit hydrolyzes its GTP to GDP, and the α subunit and β/γ dimer reassociate to form an inactive G protein (see step 3, Figure 9.7). The level of cAMP decreases due to the action of an enzyme called **phosphodiesterase**, which converts cAMP to AMP:

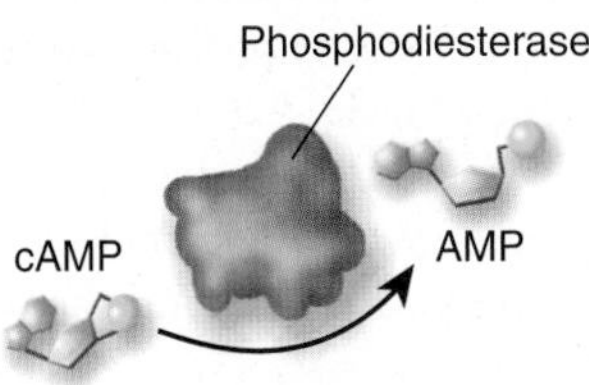

As the cAMP level falls, the regulatory subunits of PKA release cAMP, and the regulatory and catalytic subunits reassociate, thereby inhibiting PKA. Finally, enzymes called **protein phosphatases** are responsible for removing phosphate groups from proteins, which reverses the effects of PKA:

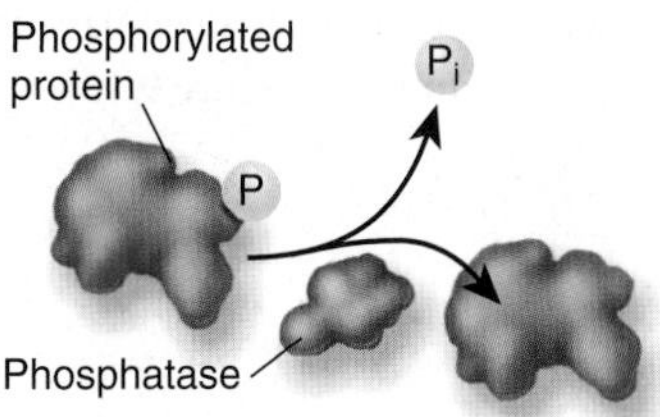

The Main Advantages of Second Messengers Are Amplification and Speed

In the 1950s, Earl Sutherland determined that many different hormones cause the formation of cAMP in a variety of cell types. This observation, for which he won the Nobel Prize in Physiology or Medicine in 1971, stimulated great interest in the study of signal transduction pathways. Since Sutherland's discovery, the production of second messengers such as cAMP has been found to have two important advantages: amplification and speed.

Amplification of the signal involves the synthesis of many cAMP molecules, which, in turn, activate many PKA proteins (**Figure 9.14**). Likewise, each PKA protein phosphorylates many target proteins in the cell to promote a cellular response.

A second advantage of second messengers such as cAMP is speed. Because second messengers are relatively small and water-soluble, they can diffuse rapidly through the cytosol. For example, Brian Bacskai and colleagues studied the response of neurons to a signaling molecule called serotonin, which is a neurotransmitter that binds to a GPCR. In humans, low serotonin is believed to play a role in depression, anxiety, and other behavioral disorders. To monitor cAMP levels, neurons grown in a laboratory were injected with a fluorescent protein that changes its fluorescence when cAMP is made. As shown in the right micrograph in **Figure 9.15**, such cells made a substantial amount of cAMP within 20 seconds after the addition of serotonin.

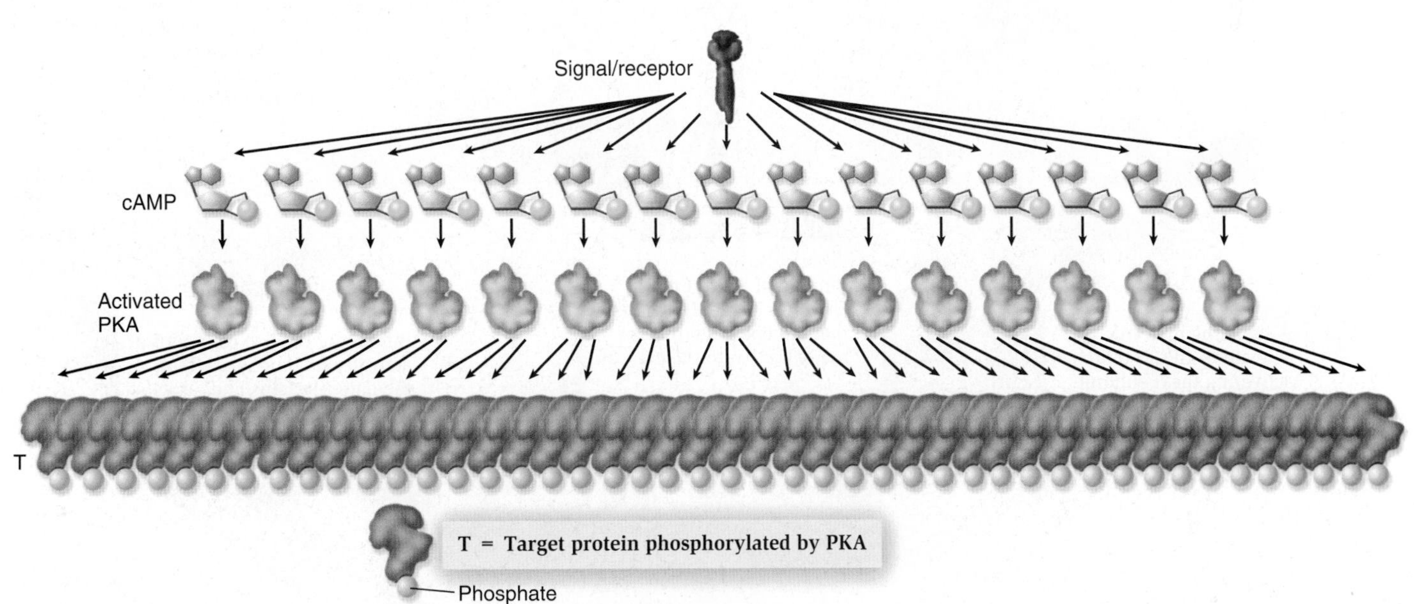

Figure 9.14 **Signal amplification.** An advantage of a signal transduction pathway is the amplification of a signal. In this case, a single signaling molecule leads to the phosphorylation of many, perhaps hundreds or thousands of, target proteins.

Concept Check: *In the case of signaling pathways involving hormones, why is signal amplification an advantage?*

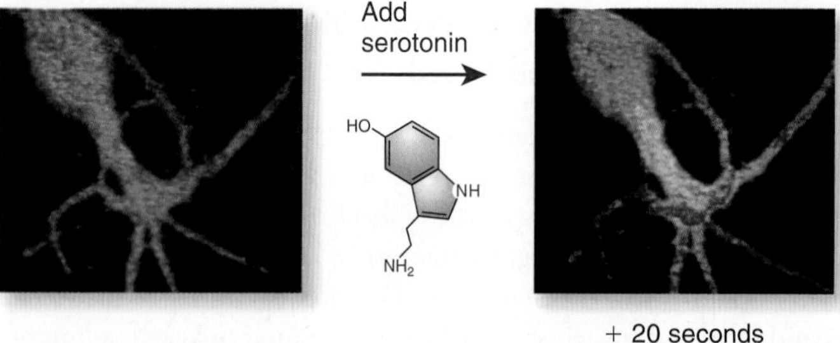

Figure 9.15 The rapid speed of cAMP production. The micrograph on the left shows a neuron prior to its exposure to serotonin, a signaling molecule; the micrograph on the right shows the same cell 20 seconds after exposure. Blue indicates a low level of cAMP, yellow is an intermediate level, and red/purple is a high level.

Signal Transduction Pathways May Produce Other Second Messengers, Such as Diacylglycerol, Inositol Trisphosphate, and Ca^{2+}

Cells use several different types of second messengers, and more than one type may be used at the same time. Let's now consider a second way that an activated G protein influences a signal transduction pathway that produces second messengers. This pathway produces the second messengers diacylglycerol (DAG) and inositol trisphosphate (IP_3) and also causes cellular effects by altering the levels of Ca^{2+} in the cell.

Production of DAG and IP_3 To start this pathway, a signaling molecule binds to its GPCR, which, in turn, activates a G protein. However, rather than activating adenylyl cyclase, as described in Figure 9.12, the α subunit of this G protein activates an enzyme called phospholipase C, which breaks a covalent bond in a plasma membrane phospholipid called PIP_2, producing the two second messengers DAG and IP_3 (**Figure 9.16**). Many molecules of DAG and IP_3 are produced when phospholipase C becomes active, thereby leading to signal amplification.

Release of Ca^{2+} into the Cytosol Due to active transport via a Ca^{2+}-ATPase, the lumen of the ER contains a very high concentration of Ca^{2+} compared with the cytosol. After IP_3 is released into the cytosol, it binds to a ligand-gated ion channel in the ER membrane. The binding of IP_3 causes the ion channel to open, releasing many Ca^{2+} ions into the cytosol, which further amplifies the signal. Calcium ions act as second messengers to elicit a cellular response in a variety of ways, two of which are shown in Figure 9.16 and described next.

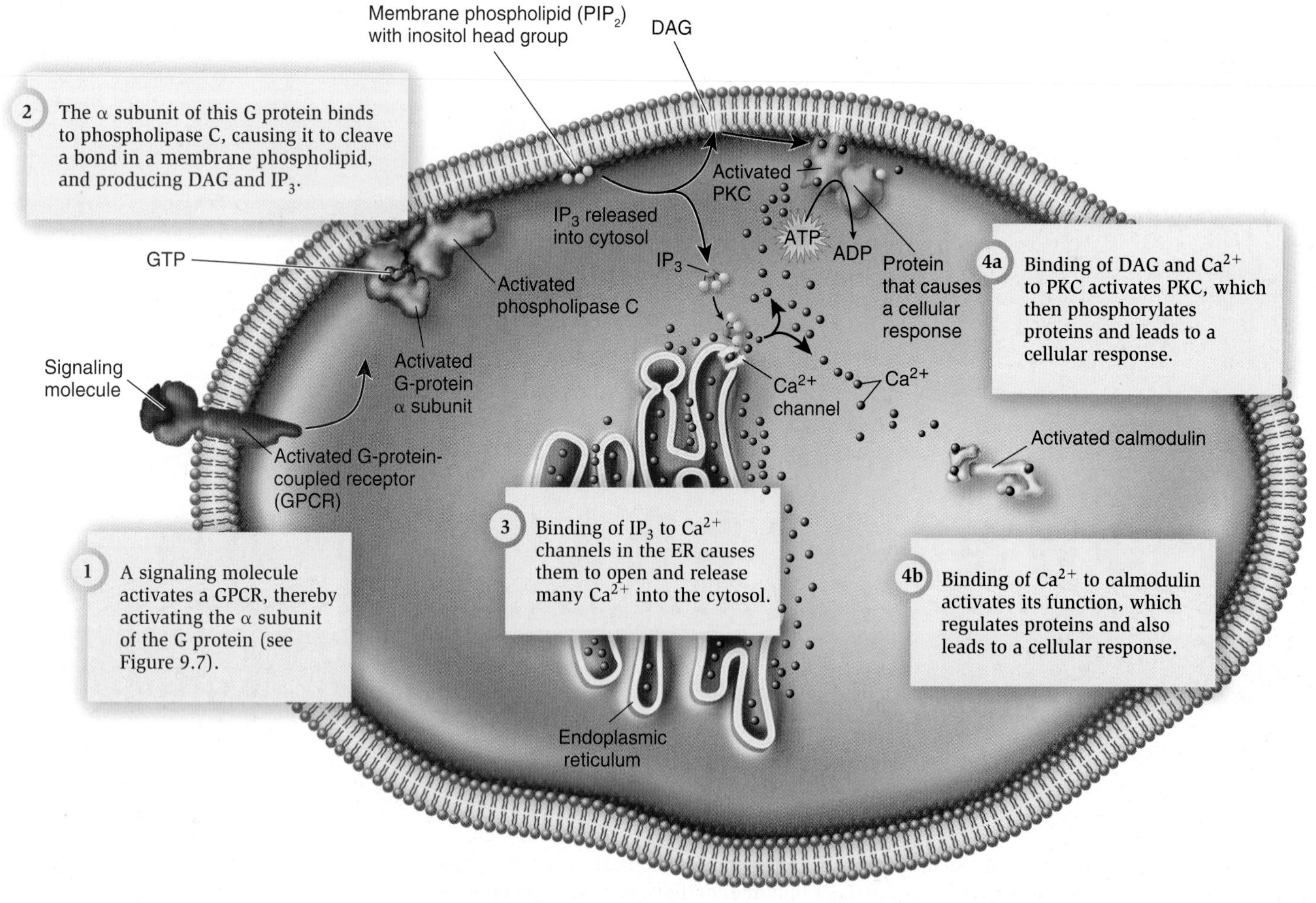

Membrane phospholipid (PIP_2) with inositol head group
DAG
2 The α subunit of this G protein binds to phospholipase C, causing it to cleave a bond in a membrane phospholipid, and producing DAG and IP_3.
Activated PKC
IP_3 released into cytosol
IP_3
ATP
ADP
GTP
Activated phospholipase C
Protein that causes a cellular response
4a Binding of DAG and Ca^{2+} to PKC activates PKC, which then phosphorylates proteins and leads to a cellular response.
Signaling molecule
Activated G-protein α subunit
Ca^{2+} channel
Ca^{2+}
Activated calmodulin
Activated G-protein-coupled receptor (GPCR)
3 Binding of IP_3 to Ca^{2+} channels in the ER causes them to open and release many Ca^{2+} into the cytosol.
1 A signaling molecule activates a GPCR, thereby activating the α subunit of the G protein (see Figure 9.7).
4b Binding of Ca^{2+} to calmodulin activates its function, which regulates proteins and also leads to a cellular response.
Endoplasmic reticulum

Figure 9.16 A signal transduction pathway involving diacylglycerol (DAG), inositol trisphosphate (IP_3), and changing Ca^{2+} levels.

Cellular Response via Protein Kinase C Ca^{2+} and DAG bind to protein kinase C (PKC), which activates the kinase. Once activated, PKC phosphorylates specific cellular proteins, thereby altering their function and leading to a cellular response. In smooth muscle cells, for example, PKC phosphorylates proteins that are involved with muscle contraction.

Cellular Response via Calmodulin Ca^{2+} also binds to a protein called calmodulin, which is a calcium-modulated protein. The Ca^{2+}-calmodulin complex then interacts with specific cellular proteins and alters their functions. For example, calmodulin regulates proteins involved in carbohydrate breakdown in liver cells.

9.4 Hormonal Signaling in Multicellular Organisms

Learning Outcomes:

1. Explain how the cellular response to a particular hormone can vary among different cell types.
2. Describe how a cell's response to a hormone depends on the types of proteins it makes.

Thus far, we have considered how signaling molecules bind to particular types of receptors, thereby activating a signal transduction pathway that leads to a cellular response. In this section, we will consider how hormones in multicellular organisms exert a variety of responses. As you will learn, the type of cellular response that is caused by a given hormone depends on the type of cell. Each cell type responds to a particular hormone in its own unique way. The variation in a cellular response is determined by the types of proteins, such as receptors and signal transduction proteins, that each cell type makes.

The Cellular Response to a Given Hormone Varies Among Different Cell Types

As we have seen, signaling molecules usually exert their effects on cells via signal transduction pathways that control the functions and/or synthesis of specific proteins. In multicellular organisms, one of the amazing effects of hormones is their ability to coordinate cellular activities. One example is epinephrine, which is secreted from endocrine cells. As mentioned, epinephrine is also called the fight-or-flight hormone because it quickly prepares the body for strenuous physical activity in response to a perceived danger. Epinephrine is also secreted into the bloodstream when someone is exercising vigorously.

Epinephrine has different effects throughout the body (Figure 9.17). We have already discussed how it promotes the breakdown of glycogen in skeletal muscle cells (see Figure 9.13). In the lungs, it relaxes the airways, allowing a person to take in more oxygen. In the heart, epinephrine stimulates heart muscle cells so the heart beats faster. Interestingly, one of the effects of caffeine can be explained by this mechanism. Caffeine inhibits phosphodiesterase, which converts cAMP to AMP. Phosphodiesterase functions to remove cAMP once a signaling molecule, such as epinephrine, is no longer present. When phosphodiesterase is inhibited by caffeine, cAMP persists for a longer period of time and prolongs the effects

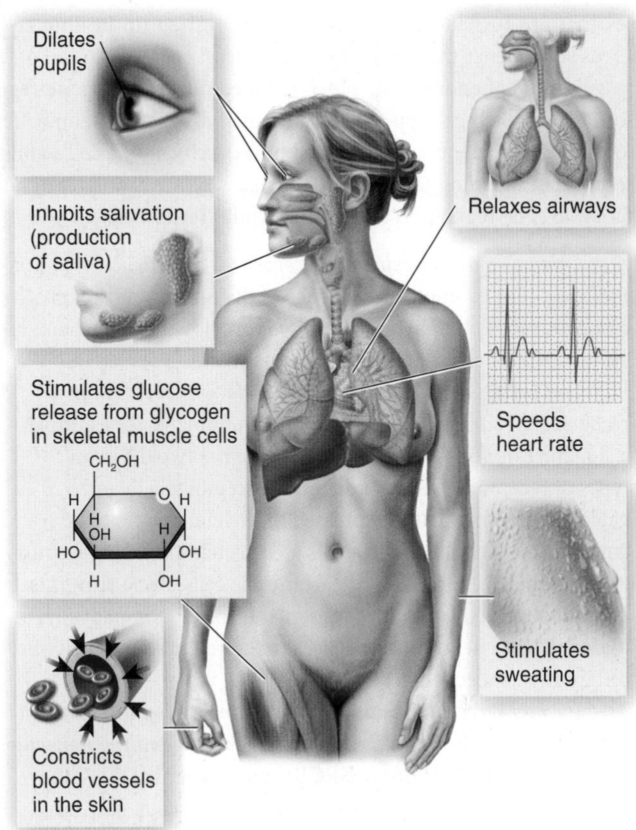

Figure 9.17 **The effects of epinephrine in humans.** This hormone prepares the body for fight or flight.

BIOLOGY PRINCIPLE New properties emerge from complex interactions. Because different cell types respond to epinephrine in different ways, an emergent property of epinephrine production is that a multicellular organism can respond to this hormone in a very complex way, preparing it to fight or flee from a perceived threat.

of signaling molecules like epinephrine. Therefore, even low levels of epinephrine have a greater effect. This is one of the reasons why drinks containing caffeine, including coffee and many energy drinks, provide a feeling of vitality and energy.

Genomes & Proteomes Connection

A Cell's Response to Hormones and Other Signaling Molecules Depends on the Proteins It Makes

As Figure 9.17 shows, a hormone such as epinephrine produces diverse responses throughout the body. How do we explain the observation that various cell types respond so differently to the same hormone? As a multicellular organism develops from a fertilized egg, the cells of the body become differentiated into particular types, such as heart and lung cells. The mechanisms that underlie this differentiation process are described in Chapter 19. Although different cell types, such as heart and lung cells, contain the same set of genes—the same genome—they are not expressed in the same pattern. Certain genes that are turned off in heart cells are turned on in lung cells, whereas some genes that are turned on in heart cells are turned off in lung cells. This phenomenon, which is called **differential gene regulation**, causes each cell type to have its own distinct proteome. The set of proteins made in any given cell type is critical to a cell's ability to respond to signaling molecules. The following are examples of how differential gene regulation affects the cellular response:

1. *A cell may or may not express a receptor for a particular signaling molecule.* For example, not all cells of the human body express a receptor for epinephrine. These cells are not affected when epinephrine is released into the bloodstream.
2. *Different cell types have different cell surface receptors that recognize the same signaling molecule.* In humans, for example, a signaling molecule called acetylcholine has two different types of receptors. One acetylcholine receptor is a ligand-gated ion channel that is expressed in skeletal muscle cells. Another acetylcholine receptor is a G-protein-coupled receptor (GPCR) that is expressed in heart muscle cells. Because of this, acetylcholine activates different signal transduction pathways in skeletal and heart muscle cells. Therefore, these cells respond differently to acetylcholine.
3. *Two (or more) receptors may work the same way in different cell types but have different affinities for the same signaling molecule.* For example, two different GPCRs may recognize the same hormone, but the receptor expressed in liver cells may have a higher affinity (that is, a lower K_d) for the hormone than does a receptor expressed in muscle cells. In this case, liver cells respond to a lower hormone concentration than muscle cells do.
4. *The expression of proteins involved in intracellular signal transduction pathways may vary in different cell types.* For example, one cell type may express the proteins that are needed to activate PKA, but another cell type may not.
5. *The expression of proteins that are controlled by signal transduction pathways may vary in different cell types.* For example, the presence of epinephrine in skeletal muscle cells leads to the activation of glycogen phosphorylase, an enzyme involved in glycogen breakdown. However, this enzyme is not expressed in all cells of the body. Glycogen breakdown is only stimulated by epinephrine if glycogen phosphorylase is expressed in that cell.

9.5 Apoptosis: Programmed Cell Death

Learning Outcomes:

1. Define and describe apoptosis.
2. Analyze the results of experiments indicating that certain hormones control apoptosis.
3. Outline the extrinsic pathway of apoptosis.

We will end our discussion of cell communication by considering one of the most dramatic responses that eukaryotic cells exhibit—**apoptosis**, or programmed cell death. During this process, a cell orchestrates its own destruction! The cell first shrinks and forms a rounder shape due to the internal destruction of its nucleus and cytoskeleton (**Figure 9.18**). The plasma membrane then forms irregular extensions that eventually become blebs—small cell fragments that break away from the cell as it destroys itself (also see chapter opening photo).

Cell biologists have discovered that apoptosis plays many important roles. During embryonic development in animals, it is needed to sculpt the tissues and organs. For example, the fingers on a human hand are initially webbed, but become separated during embryonic development when the cells between the fingers undergo apoptosis (see Chapter 19, Figure 19.4). Apoptosis is also necessary in adult organisms to maintain the proper cell number in tissues and organs, and it eliminates cells that have become worn out or infected by viruses, or have the potential to cause cancer. In this section, we will examine the pioneering work that led to the discovery of apoptosis and explore its molecular mechanism.

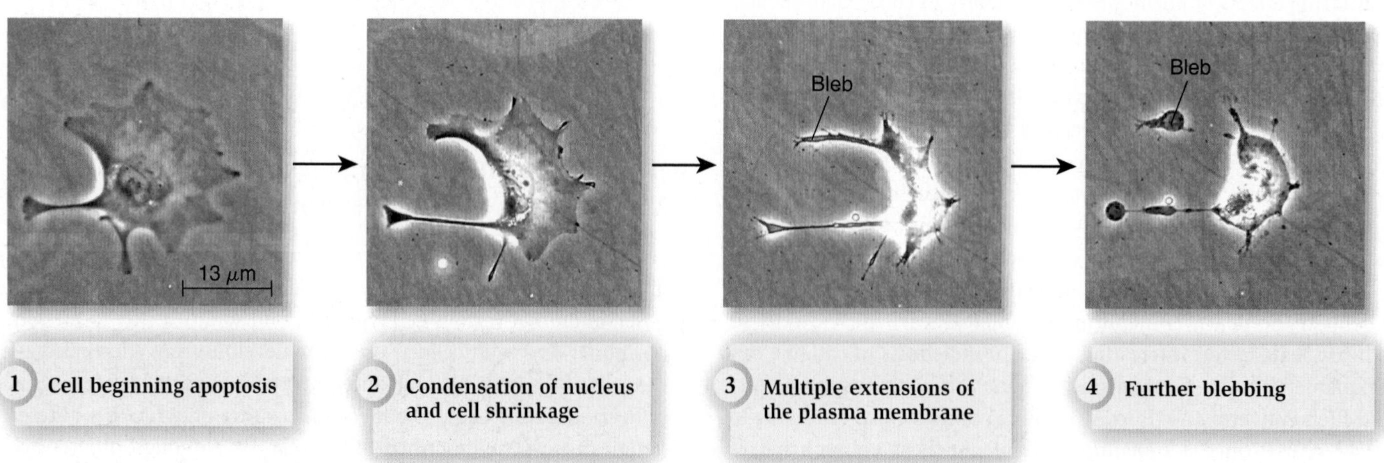

Figure 9.18 Stages of apoptosis.

FEATURE INVESTIGATION

Kerr, Wyllie, and Currie Found That Hormones May Control Apoptosis

How was apoptosis discovered? One line of evidence involved the microscopic examination of tissues in mammals. In the 1960s, British pathologist John Kerr microscopically examined liver tissue that was deprived of oxygen. Within hours of oxygen deprivation, he observed that some cells underwent a process that involved cell shrinkage. Around this time, similar results had been noted by other researchers, such as Scottish pathologists Andrew Wyllie and Alastair Currie, who had studied cell death in the adrenal glands. In 1973, Kerr, Wyllie, and Currie joined forces to study this process further.

Prior to their collaboration, other researchers had already established that certain hormones affect the growth of the adrenal glands, which sit atop the kidneys. Adrenocorticotropic hormone (ACTH) was known to increase the number of cells in the adrenal cortex, which is the outer layer of the adrenal glands. By contrast, prednisolone was shown to suppress the synthesis of ACTH and cause a decrease in the number of cells in the cortex. In the experiment described in **Figure 9.19**, Kerr, Wyllie, and Currie wanted to understand how ACTH and prednisolone exert their effects. They subjected rats to four types of treatments. The control rats were injected with saline (salt water). Other rats were injected with prednisolone alone, prednisolone plus ACTH, or ACTH alone. After 2 days, samples of adrenal cortex were obtained from the rats and observed by light microscopy. Even in control samples, the researchers occasionally observed cell death via apoptosis (see micrograph under The Data). However, in prednisolone-treated rats, the cells in the adrenal cortex were found to undergo a dramatically higher rate of apoptosis. Multiple cells undergoing apoptosis were found in 9 out of every 10 samples observed under the light microscope. Such a high level of apoptosis was not observed in control samples or in samples obtained from rats treated with both prednisolone and ACTH or ACTH alone. Therefore, ACTH appears to prevent apoptosis.

The results of Kerr, Wyllie, and Currie are important for two reasons. First, their results indicated that tissues decrease their cell number

Figure 9.19 Discovery of apoptosis in the adrenal cortex by Kerr, Wyllie, and Currie.

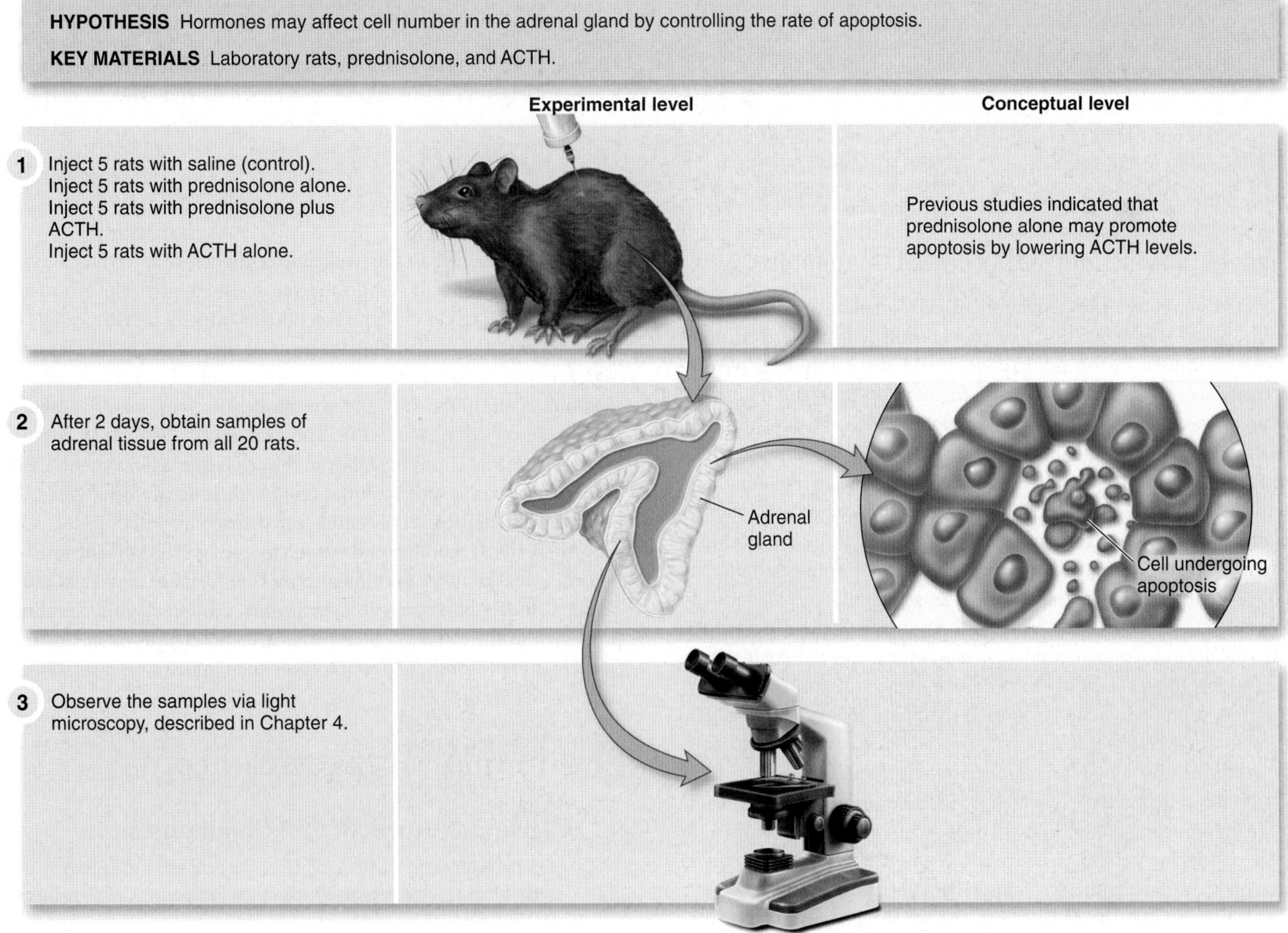

HYPOTHESIS Hormones may affect cell number in the adrenal gland by controlling the rate of apoptosis.
KEY MATERIALS Laboratory rats, prednisolone, and ACTH.
Experimental level
Conceptual level
1 Inject 5 rats with saline (control). Inject 5 rats with prednisolone alone. Inject 5 rats with prednisolone plus ACTH. Inject 5 rats with ACTH alone.
Previous studies indicated that prednisolone alone may promote apoptosis by lowering ACTH levels.
2 After 2 days, obtain samples of adrenal tissue from all 20 rats.
Adrenal gland
Cell undergoing apoptosis
3 Observe the samples via light microscopy, described in Chapter 4.

4 THE DATA

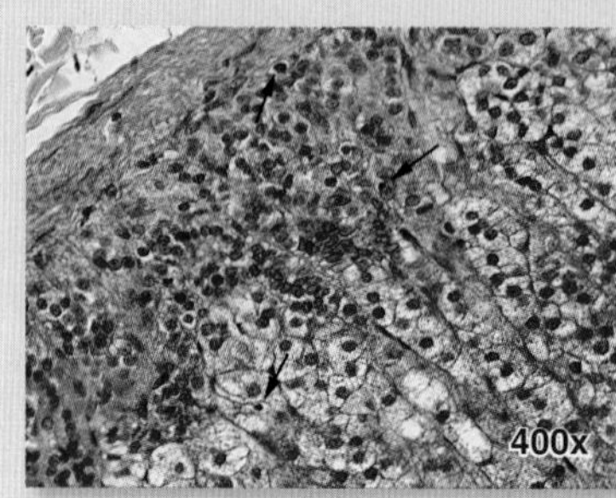

Micrograph of adrenal tissue showing occasional cells undergoing apoptosis (see arrows)

Treatment	Number of animals	Glands with enhanced apoptosis*/ Total number of animals
Saline	5	0/10
Prednisolone	5	9/10
Prednisolone + ACTH	5	0/10
ACTH	5	0/10

*Samples from two adrenal glands were removed from each animal. Enhanced apoptosis means that cells undergoing apoptosis were observed in every sample under the light microscope.

5 **CONCLUSION** Prednisolone alone, which lowers ACTH levels, causes some cells to undergo apoptosis. During this process, the cells shrink and form blebs as they kill themselves. Apoptosis is controlled by hormones.

6 **SOURCE** Wyllie, A.H., Kerr, J.F.R., Macaskill, I.A.M., and Currie, A.R. 1973. Adrenocortical cell deletion: the role of ACTH. *Journal of Pathology* 111:85–94.

via a mechanism that involves cell shrinkage and eventually blebbing. Second, they showed that cell death followed a program that, in this case, was induced by the presence of prednisolone (which decreases ACTH). They coined the term apoptosis to describe this process.

Experimental Questions

1. In the experiment of Figure 9.19, explain the effects on apoptosis in the control rats (saline injected) versus those injected with prednisolone alone, predinisolone + ACTH, or ACTH alone.
2. Prednisolone inhibits the production of ACTH in rats. Do you think it inhibited the ability of rats to make their own ACTH when they were injected with both prednisolone and ACTH? Explain.
3. Of the four groups—control, prednisolone alone, prednisolone + ACTH, and ACTH alone—which would you expect to have the lowest level of apoptosis? Explain.

Signal Transduction Pathways Lead to Apoptosis

Apoptosis involves the activation of cell-signaling pathways. One pathway, called the extrinsic pathway, begins with the activation of **death receptors** on the cell surface. When death receptors bind to extracellular signaling molecules, a pathway is stimulated that leads to apoptosis. **Figure 9.20** shows a simplified pathway for this process. In this example, the signaling molecule is a protein composed of three identical subunits—a trimeric protein. Such trimeric signaling molecules are typically produced by cells of the immune system that recognize abnormal cells and target them for destruction. For example, when a cell is infected with a virus, cells of the immune system may target the infected cell for apoptosis. The signaling molecule binds to three death receptors, which causes them to aggregate into a trimer. This results in a conformational change that exposes a domain on the death receptors called the death domain. Once the death domain is exposed, it binds to adaptors, which then bind to a procaspase. The complex between the death receptors, adaptors, and procaspase is called the death-inducing signaling complex (DISC).

Once the procaspase, which is inactive, is part of the death-inducing signaling complex, it is converted by proteolytic cleavage to caspase, which is active. An active **caspase** functions as a protease—an enzyme that digests other proteins. After it is activated, the caspase is then released from the DISC. This caspase is called an initiator caspase because it initiates the activation of many other caspases in the cell. These other caspases are called executioner, or effector, caspases because they are directly responsible for digesting intracellular proteins and causing the cell to die. The executioner caspases digest a variety of intracellular proteins, including the proteins that constitute the cytoskeleton and nuclear lamina as well as proteins involved with DNA replication and repair. In this way, the executioner caspases cause the cellular changes described in Figure 9.18. The caspases also activate an enzyme called DNase that chops the DNA in the cell into small fragments. This event may be particularly important for eliminating virally infected cells because it also destroys viral genomes that are composed of DNA.

Alternatively, another pathway of apoptosis, called the intrinsic or mitochondrial pathway, is stimulated by DNA damage that could cause cancer. Mitochondria release cytochrome c (a small mitochondrial protein) into the cytosol, which forms a complex with other proteins called an **apoptosome**. The apoptosome then initiates the activation of caspases.

Summary of Key Concepts

9.1 General Features of Cell Communication

- A signal is an agent that can influence the properties of cells. A signal binds to a receptor to elicit a cellular response. Cell signaling enables cells to sense and respond to environmental changes and to communicate with each other (Figures 9.1, 9.2).

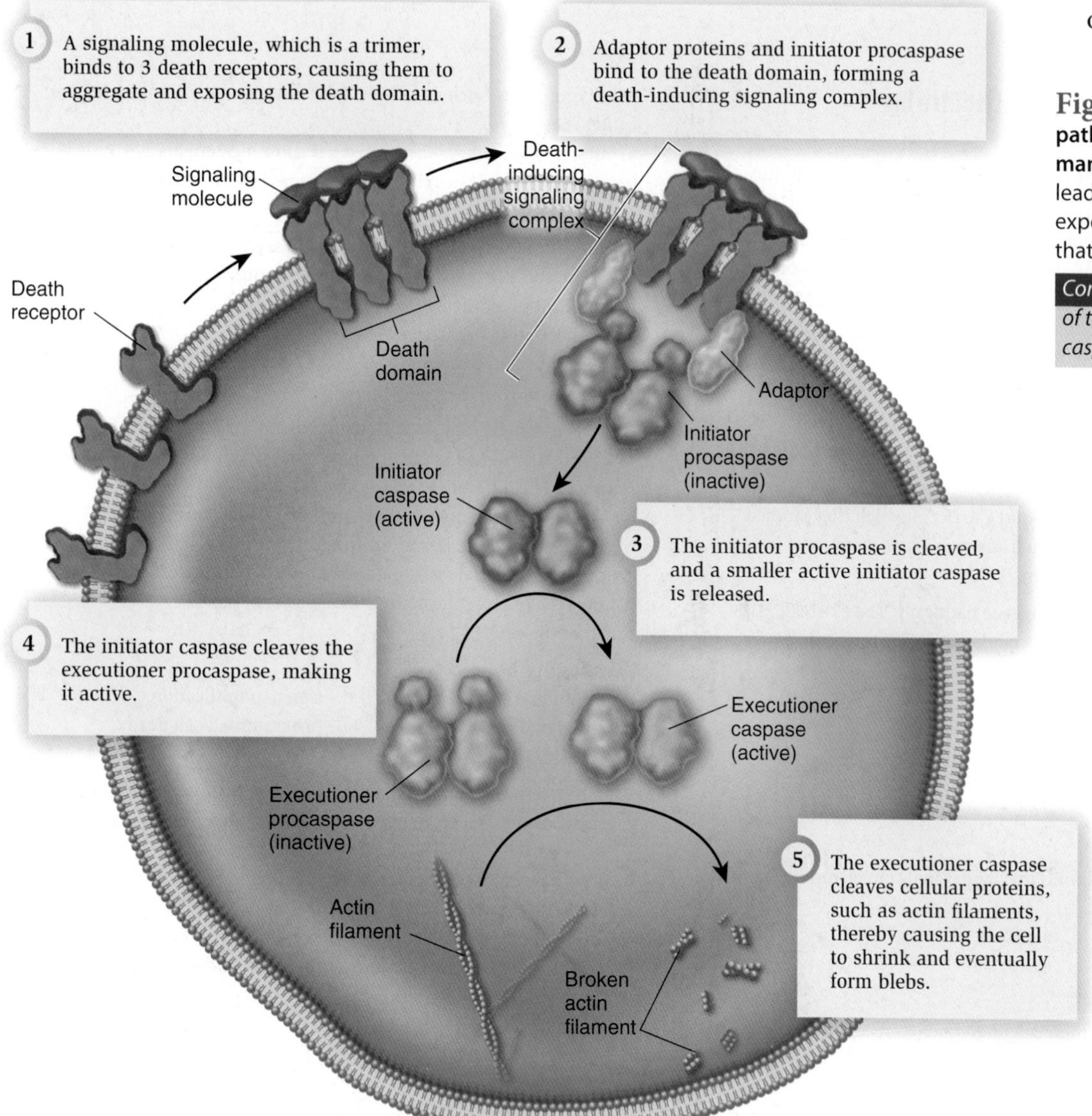

1 A signaling molecule, which is a trimer, binds to 3 death receptors, causing them to aggregate and exposing the death domain.
2 Adaptor proteins and initiator procaspase bind to the death domain, forming a death-inducing signaling complex.
Signaling molecule
Death-inducing signaling complex
Death receptor
Death domain
Adaptor
Initiator procaspase (inactive)
Initiator caspase (active)
3 The initiator procaspase is cleaved, and a smaller active initiator caspase is released.
4 The initiator caspase cleaves the executioner procaspase, making it active.
Executioner caspase (active)
Executioner procaspase (inactive)
5 The executioner caspase cleaves cellular proteins, such as actin filaments, thereby causing the cell to shrink and eventually form blebs.
Actin filament
Broken actin filament

Figure 9.20 The extrinsic pathway for apoptosis in mammals. This simplified pathway leads to apoptosis when cells are exposed to an extracellular signal that causes cell death.

Concept Check: *How are the roles of the initiator and the executioner caspases different in this process?*

- Cell-to-cell communication varies in the mechanism and distance that a signal travels. Signals are relayed between cells in five common ways: direct intercellular, contact-dependent, autocrine, paracrine, and endocrine signaling (Figure 9.3).
- Cell communication is usually a three-stage process involving receptor activation, signal transduction, and a cellular response. A signal transduction pathway is a group of proteins that convert an initial signal to a different signal inside the cell (Figure 9.4).

9.2 Cellular Receptors and Their Activation

- A signaling molecule, also called a ligand, binds to a receptor with an affinity that is measured as a dissociation constant or K_d value. The binding of a ligand to a receptor is usually very specific and alters the conformation of the receptor (Figure 9.5).
- Enzyme-linked receptors have some type of catalytic function. Many of them are protein kinases that phosphorylate proteins (Figure 9.6).
- G-protein-coupled receptors (GPCRs) interact with G proteins to initiate a cellular response (Figure 9.7).
- Ligand-gated ion channels are receptors that allow the flow of ions across cellular membranes (Figure 9.8).
- Although most receptors involved in cell signaling are found on the cell surface, some receptors, such as the estrogen receptor, are intracellular receptors (Figure 9.9).

9.3 Signal Transduction and the Cellular Response

- Signaling pathways influence whether or not a cell divides. An example is the pathway that is stimulated by epidermal growth factor, which binds to a receptor tyrosine kinase (Figure 9.10).
- Second messengers, such as cAMP, play a key role in signal transduction pathways, such as those that occur via GPCRs. These pathways are reversible once the signal is degraded (Figures 9.11, 9.12).
- An example of a pathway that uses cAMP is found in skeletal muscle cells responding to elevated levels of epinephrine, the "fight-or-flight" hormone. Epinephrine enhances the function of enzymes that increase glycogen breakdown and inhibits enzymes that cause glycogen synthesis (Figure 9.13).
- Second messenger pathways amplify the signal and occur with great speed (Figures 9.14, 9.15).
- Diacylglycerol (DAG), inositol trisphosphate (IP_3), and Ca^{2+} are other second messengers involved in signal transduction (Figure 9.16).

9.4 Hormonal Signaling in Multicellular Organisms

- Hormones such as epinephrine exert different effects throughout the body (Figure 9.17).
- The way in which any particular cell type responds to a signaling molecule depends on the set of proteins it makes. The amounts of these proteins are controlled by differential gene regulation.

9.5 Apoptosis: Programmed Cell Death

- Apoptosis is the process of programmed cell death in which the nucleus and cytoskeleton break down and eventually the cell breaks apart into blebs (Figure 9.18).
- Microscopy studies of Kerr, Wyllie, and Currie, in which they studied the effects of hormones on the adrenal cortex, were instrumental in the identification of apoptosis (Figure 9.19).
- Apoptosis occurs via extrinsic or intrinsic pathways. The extrinsic pathway is stimulated when an extracellular signaling molecule binds to death receptors (Figure 9.20).

Assess and Discuss

Test Yourself

1. An agent that allows a cell to respond to changes in its environment is termed
   a. a cell surface receptor.
   b. an intracellular receptor.
   c. a structural protein.
   d. a signal.
   e. apoptosis.
2. When a cell secretes a signaling molecule that binds to receptors on neighboring cells as well as the same cell, this is called ____________ signaling.
   a. direct intercellular
   b. contact-dependent
   c. autocrine
   d. paracrine
   e. endocrine
3. Which of the following does *not* describe a typical cellular response to signaling molecules?
   a. activation of enzymes within the cell
   b. change in the function of structural proteins, which determine cell shape
   c. alteration of levels of certain proteins in the cell by changing the level of gene expression
   d. change in a gene sequence that encodes a particular protein
   e. All of the above are examples of cellular responses.
4. A receptor has a K_d for its ligand of 50 nM. This receptor
   a. has a higher affinity for its ligand than does a receptor with a K_d of 100 nM.
   b. has a higher affinity for its ligand than does a receptor with a K_d of 10 nM.
   c. is mostly bound by its ligand when the ligand concentration is 100 nM.
   d. must be an intracellular receptor.
   e. both a and c
5. ____________ binds to receptors inside cells.
   a. Estrogen
   b. Epinephrine
   c. Epidermal growth factor
   d. All of the above
   e. None of the above
6. Small molecules, such as cAMP, that relay signals within the cell are called
   a. first messengers.
   b. ligands.
   c. G proteins.
   d. second messengers.
   e. transcription factors.
7. The benefit of second messengers in signal transduction pathways is
   a. an increase in the speed of a cellular response.
   b. duplication of the ligands in the system.
   c. amplification of the signal.
   d. all of the above.
   e. a and c only.
8. All cells of a multicellular organism may not respond in the same way to a particular ligand (signaling molecule) that binds to a cell surface receptor. The difference in response may be due to
   a. the type of receptor for the ligand that the cell expresses.
   b the affinity of the ligand for the receptor in a given cell type.
   c. the type of signal transduction pathways that the cell expresses.
   d. the type of target proteins that the cell expresses.
   e. all of the above.
9. Apoptosis is the process of
   a. cell migration.
   b. cell signaling.
   c. signal transduction.
   d. signal amplification.
   e. programmed cell death.
10. Which statement best describes the extrinsic pathway for apoptosis?
   a. Caspases recognize an environmental signal and expose their death domain.
   b. Death receptors recognize an environmental signal, which then leads to the activation of caspases.
   c. Initiator caspases digest the nuclear lamina and cytoskeleton.
   d. Executioner caspases are part of the death-inducing signaling complex (DISC).
   e. all of the above

Conceptual Questions

1. What are the two general reasons that cells need to communicate?
2. What are the three stages of cell signaling? What stage does not occur when the estrogen receptor is activated?
3.  A principle of biology is that *living organisms interact with their environment.* Discuss how cell signaling helps organisms to interact with their environment.

Collaborative Questions

1. Discuss and compare several different types of cell-to-cell communication. What are some advantages and disadvantages of each type?
2. How does differential gene regulation enable various cell types to respond differently to the same signaling molecule? Why is this useful to multicellular organisms?

Online Resource

www.brookerbiology.com

Stay a step ahead in your studies with animations that bring concepts to life and practice tests to assess your understanding. Your instructor may also recommend the interactive eBook, individualized learning tools, and more.

Chapter Outline

10.1 Extracellular Matrix and Cell Walls

10.2 Cell Junctions

10.3 Tissues

Summary of Key Concepts

Assess and Discuss

Multicellularity

10

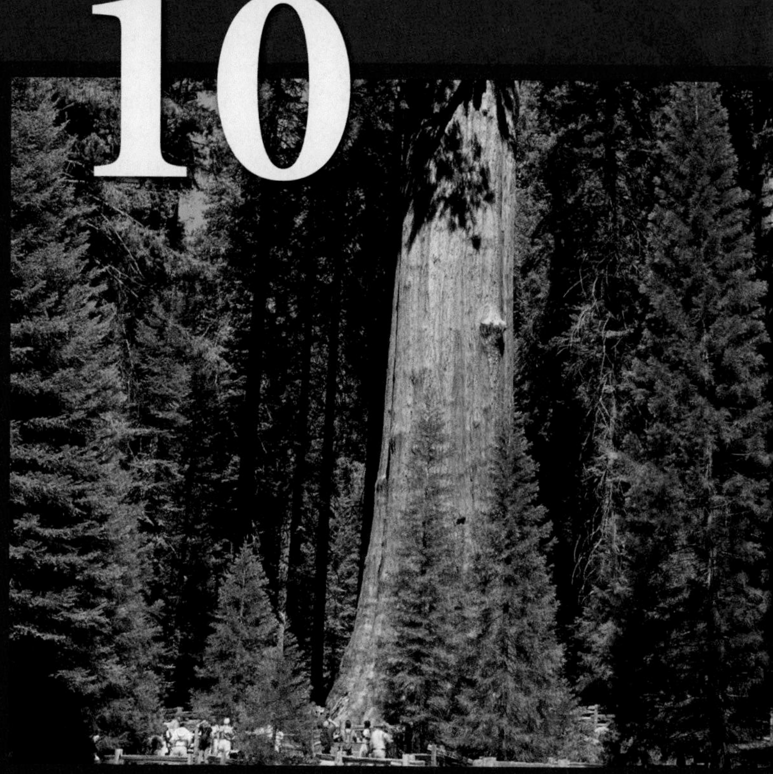

The General Sherman tree in Sequoia National Park, a striking example of the size that multicellular organisms can reach. This tree is thought to be the largest organism (by mass) in the world.

BIOLOGY PRINCIPLE Living organisms grow and develop. In multicellular organisms, growth involves the production of many new cells.

What is the largest living organism on Earth? The size of an organism can be defined by its volume, mass, height, length, or the area it occupies. A giant fungus (*Armillaria ostoyae*), growing in the soil in the Malheur National Forest in Oregon, spans 8.9 km^2, or 2,200 acres, which makes it the largest single organism by area. Most of the organism lies below ground, so it is not visible from the surface. In the Mediterranean Sea, marine biologists discovered a giant aquatic plant (*Posidonia oceanica*) that is 8 km, or 4.3 miles, in length, making it the world's longest organism. With regard to mass, the largest organism is probably a tree named the General Sherman tree, which is 83.8 meters tall (275 feet), nearly the length of a football field (see chapter opening photo). This giant sequoia tree (*Sequoiadendron giganteum*) is estimated to weigh nearly 2 million kg (over 2,000 tons)—equivalent to a herd of 400 elephants!

An organism composed of more than one cell is said to be **multicellular**. The preceding examples illustrate the amazing sizes that certain multicellular organisms have attained. As we will discuss in Chapter 22, multicellular organisms came into being approximately 1 billion years ago. Some species of protists are multicellular, as are most species of fungi. In this chapter, we will focus on plants and animals, which are always multicellular organisms.

The main benefit of multicellularity arises from the division of labor between different types of cells in an organism. For example, the intestinal cells of animals and the root cells of plants have become specialized for nutrient uptake. Other types of cells in a multicellular organism perform different roles, such as reproduction. In animals, most of the cells of the body—somatic cells—are devoted to the growth, development, and survival of the organism, whereas specialized cells—gametes—function in sexual reproduction.

Multicellular species usually have much larger genomes than unicellular species. The increase in genome size is associated with an increase in proteome size—multicellular organisms produce a larger array of proteins than do unicellular species. The additional proteins play a role in three general phenomena. First, in a multicellular organism, cell communication is vital for the proper organization and functioning of cells. Many more proteins involved in cell communication are made in multicellular species. Second, both the arrangement of cells within the body and the attachment of cells to each other require a greater variety of proteins in multicellular species than in unicellular species. Finally, additional proteins play a role in cell specialization because proteins that are needed for the structure and function of one cell type may not be needed in a different cell type, and vice versa. Likewise, additional proteins are needed to regulate the expression of genes so these proteins are expressed in the proper cell types.

In this chapter, we consider characteristics specific to the cell biology of multicellular organisms. We will begin by exploring the material that is produced by animal and plant cells to form an extracellular matrix or cell wall, respectively. This material plays many important roles in the structure, organization, and functioning of cells within multicellular organisms. We will then turn our attention to cell junctions, specialized structures that enable cells to make physical contact with one another. Cells within multicellular organisms form junctions that help to make a cohesive and well-organized body. Finally, we examine the organization and function of tissues, groups of cells that have a similar structure and function. In this chapter, we will survey the general features of tissues from a cellular perspective. Units VI and VII will explore the characteristics of particular plant and animal tissues in greater detail.

10.1 Extracellular Matrix and Cell Walls

Learning Outcomes:

1. Explain the functional roles of the extracellular matrix in animals.
2. Outline the major structural components of the ECM of animals.
3. Describe the structure and function of plant cell walls.

Organisms are not composed solely of cells. A large portion of an animal or plant consists of a network of material that is secreted from cells and forms a complex meshwork outside of cells. In animals, this is called the **extracellular matrix (ECM)**, whereas plant cells are surrounded by a **cell wall**. The ECM and cell walls are a major component of certain parts of animals and plants, respectively. For example, bones and cartilage in animals are composed largely of ECM, and the woody portions of plants are composed mostly of cell walls. Although the cells within wood eventually die, the cell walls they have produced provide a rigid structure that supports the plant for years or even centuries.

Over the past few decades, cell biologists have examined the synthesis, composition, and function of the ECM in animals and the cell walls in plants. In this section, we will begin by examining the structure and role of the ECM in animals, focusing on the functions of the major ECM components: proteins and polysaccharides. We will then explore the cell wall of plant cells and consider how it differs in structure and function from the ECM of animal cells.

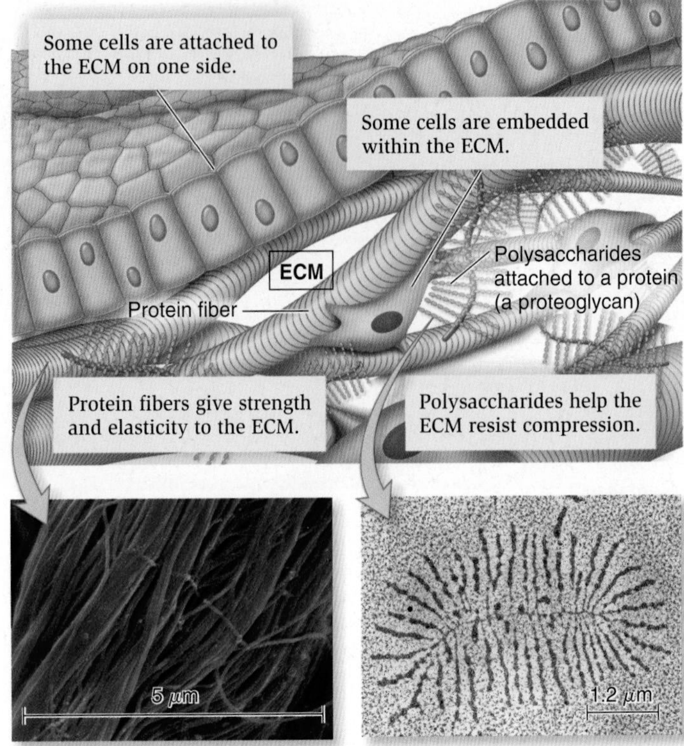
Some cells are attached to the ECM on one side.
Some cells are embedded within the ECM.
ECM
Protein fiber
Polysaccharides attached to a protein (a proteoglycan)
Protein fibers give strength and elasticity to the ECM.
Polysaccharides help the ECM resist compression.
5 μm
1.2 μm

Figure 10.1 **The extracellular matrix (ECM) of animal cells.** The micrograph (SEM) at the bottom left shows collagen fibers, a type of protein fiber found in the ECM. The micrograph (TEM) at the bottom right shows a proteoglycan, which consists of polysaccharides attached to a protein.

Concept Check: *What are the four functions of the ECM in animals?*

The Extracellular Matrix in Animals Supports and Organizes Cells and Plays a Role in Cell Signaling

Unlike the cells of bacteria, archaea, fungi, and plants, the cells of animals are not surrounded by a rigid cell wall that provides structure and support. However, animal cells secrete materials that form an ECM that also provides support and helps to organize cells. Certain animal cells are completely embedded within an extensive ECM, whereas other cells may adhere to the ECM on only one side. **Figure 10.1** illustrates the general features of the ECM and its relationship to cells. The major macromolecules of the ECM are proteins and polysaccharides. The most abundant proteins are those that form large fibers. The polysaccharides give the ECM a gel-like character.

As we will see, the ECM found in animals performs many important roles, including strength, structural support, organization, and cell signaling.

- **Strength:** The ECM is the "tough stuff" of animals' bodies. In the skin of mammals, the strength of the ECM prevents tearing. The ECM found in cartilage resists compression and provides protection to the joints. Similarly, the ECM protects the soft parts of the body, such as the internal organs.
- **Structural support:** The bones of many animals are composed primarily of ECM. Skeletons not only provide structural support but also facilitate movement via the functioning of attached muscles.
- **Organization:** The attachment of cells to the ECM plays a key role in the proper arrangement of cells throughout the body. In addition, the ECM binds many body parts together, such as tendons to bones.
- **Cell signaling:** A newly discovered role of the ECM is cell signaling. One way that cells in multicellular organisms sense their environment is via changes in the ECM.

Let's now consider the synthesis and structure of ECM components found in animals.

Adhesive and Structural Proteins Are Major Components of the ECM of Animals

In the 1850s, German biologist Rudolf Virchow suggested that all extracellular materials are made and secreted by cells. Around the same time, biologists realized that gelatin and glue, which are produced by the boiling of animal tissues, must contain a common fibrous substance. This substance was named **collagen** (from the Greek, meaning glue-producing). Since that time, the advent of experimental techniques in chemistry, microscopy, and biophysics has enabled scientists to probe the structure of the ECM. We now understand that the ECM contains a mixture of several different components, including proteins such as collagen, which form fibers.

The proteins found in the ECM are grouped into adhesive proteins, such as fibronectin and laminin, and structural proteins, such

Table 10.1 Proteins in the ECM of Animals

General type	Example	Function
Adhesive	Fibronectin	Connects cells to the ECM and helps to organize components in the ECM.
	Laminin	Connects cells to the ECM and helps to organize components in the basal lamina, a specialized ECM found next to epithelial cells (described in Section 10.3).
Structural	Collagen	Forms large fibers and interconnected fibrous networks in the ECM. Provides tensile strength.
	Elastin	Forms elastic fibers in the ECM that can stretch and recoil.

as collagen and elastin (**Table 10.1**). How do adhesive proteins work? Fibronectin and laminin have multiple binding sites that bind to other components in the ECM, such as protein fibers and polysaccharides. These same proteins also have binding sites for receptors on the surfaces of cells. Therefore, adhesive proteins are so named because they adhere ECM components together and to the cell surface. They provide organization to the ECM and facilitate the attachment of cells to the ECM.

Structural proteins, such as collagen and elastin, form large fibers that give the ECM its strength and elasticity. A key function of collagen is to impart tensile strength, which is a measure of how much stretching force a material can bear without tearing apart. Collagen provides high tensile strength to many parts of an animal's body. It is the main protein found in bones, cartilage, tendons, skin, and the lining of blood vessels and internal organs. In the bodies of mammals, more than 25% of the total protein mass consists of collagen, much more than any other protein. Approximately 75% of the protein in mammalian skin is composed of collagen. Leather is largely a pickled and tanned form of collagen.

Figure 10.2 depicts the synthesis and assembly of collagen. As described in Chapter 4 (see Figure 4.28), proteins, such as collagen, that are secreted from eukaryotic cells are first directed from the cytosol to the endoplasmic reticulum (ER), then to the Golgi apparatus, and subsequently are secreted from the cell via vesicles that fuse with the plasma membrane. Individual procollagen polypeptides (called α chains) are synthesized into the lumen of the ER. Three procollagen polypeptides then associate with each other to form a procollagen triple helix. The amino acid sequences at both ends of the polypeptides, termed extension sequences, promote the formation of procollagen and prevent the formation of a much larger fiber. After procollagen is secreted from the cell, extracellular enzymes remove the extension sequences. Once this occurs, the protein, now called collagen, can form larger structures. Collagen proteins assemble in a staggered way to form relatively thin collagen fibrils, which then align and produce large collagen fibers. The many layers of these proteins give collagen fibers their tensile strength.

In addition to tensile strength, elasticity is needed in regions of the body such as the lungs and blood vessels, which regularly expand and return to their original shape. In these places, the ECM contains an abundance of elastic fibers composed primarily of the protein **elastin** (**Figure 10.3**). Elastin proteins form many covalent crosslinks to make a fiber with remarkable elastic properties. In the absence of a

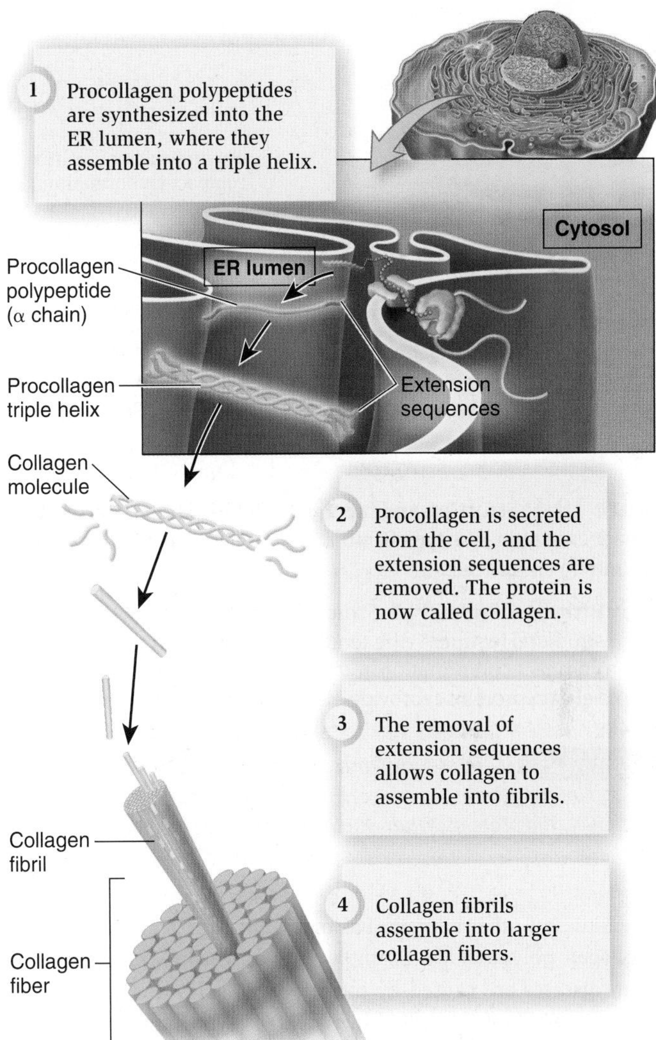

1 Procollagen polypeptides are synthesized into the ER lumen, where they assemble into a triple helix.
Cytosol
ER lumen
Procollagen polypeptide (α chain)
Procollagen triple helix
Extension sequences
Collagen molecule
2 Procollagen is secreted from the cell, and the extension sequences are removed. The protein is now called collagen.
3 The removal of extension sequences allows collagen to assemble into fibrils.
Collagen fibril
4 Collagen fibrils assemble into larger collagen fibers.
Collagen fiber

Figure 10.2 Formation of collagen fibers. Collagen is one type of structural protein found in the ECM of animal cells.

Concept Check: What prevents large collagen fibers from forming intracellularly?

stretching force, each protein tends to adopt a compact conformation. When subjected to a stretching force, however, the compact proteins become more linear, with the covalent crosslinks holding the fiber together. When the stretching force has ended, the proteins naturally return to their compact conformation. In this way, elastic fibers behave much like a rubber band, stretching under tension and snapping back when the tension is released.

GENOMES & PROTEOMES CONNECTION

Collagens Are a Family of Proteins That Give Animals a Variety of ECM Properties

Researchers have determined that animals make many different types of collagen fibers. These are designated as type I, type II, and so on. At least 27 different types of collagens have been identified in humans. To make different types of collagens, the human genome, as well as

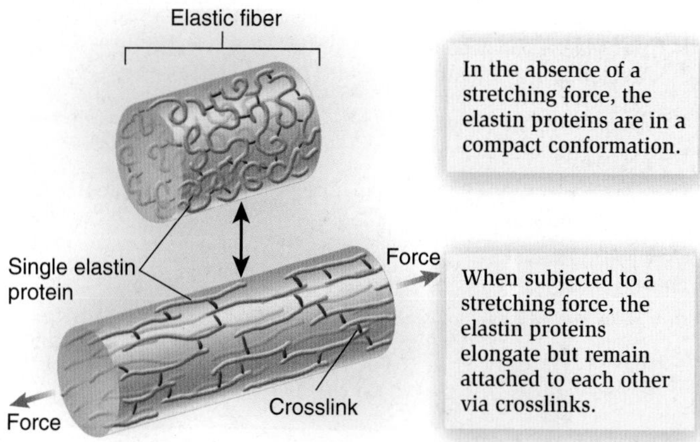
Elastic fiber
In the absence of a stretching force, the elastin proteins are in a compact conformation.
Single elastin protein
Force
When subjected to a stretching force, the elastin proteins elongate but remain attached to each other via crosslinks.
Crosslink
Force

Figure 10.3 **Structure and function of elastic fibers.** Elastic fibers are made of elastin, one type of structural protein found in the ECM surrounding animal cells.

BIOLOGY PRINCIPLE **Structure determines function.** Two structural features—the tendency of the elastin protein to adopt a compact conformation and the covalent crosslinking of multiple elastin proteins—provide elastin fibers with the function of elasticity.

Concept Check: Suppose you started with an unstretched elastic fiber and treated it with a chemical that breaks the crosslinks between adjacent elastin proteins. What would happen when the fiber is stretched?

the genomes of other animals, has many different genes that encode procollagen polypeptides. Examples are known in which inherited human diseases are caused by mutations in genes that encode collagen proteins. For example, Ehlers-Danlos syndrome is caused by mutations in one of several different collagen genes. Characteristic symptoms are extremely elastic skin and hyperflexible joints.

Collagens have a common structure, in which three polypeptides wind around each other to form a triple helix (see Figure 10.2). Each polypeptide is an α chain. In some collagens, all three α chains are identical, whereas in others, the α chains may be encoded by different collagen genes. Nevertheless, the triple helix structure is common to all collagen proteins.

Why are different collagens made? Each of the many different types of collagen polypeptides has a similar yet distinctive amino acid sequence that affects the structure of not only individual collagen proteins but also the resulting collagen fibers. For example, the amino acid sequence may cause the α chains within each collagen protein to bind to each other very tightly, thereby creating rigid proteins that form a relatively stiff fiber. Such collagen fibers are found in bone and cartilage.

The amino acid sequence of the α chains also influences the interactions between the collagen proteins within a fiber. For example, the amino acid sequences of certain α chains promote a looser interaction that produces a more bendable or thin fiber. More flexible collagen fibers support the lining of your lungs and intestines. In addition, domains within the collagen polypeptide affect the spatial arrangement of collagen proteins. The collagen shown in Figure 10.2 forms fibers in which collagen proteins align themselves in parallel arrays. However, not all collagen proteins form long fibers. For example, type IV collagen proteins interact with each other in a meshwork pattern. This meshwork acts as a filter around capillaries.

Table 10.2 Examples of Collagen Types in Humans

Type	Sites of synthesis*	Structure and function
I	Tendons, ligaments, bones, and skin	Forms a relatively rigid and thick fiber. Very abundant, provides most of the tensile strength to the ECM.
II	Cartilage, discs between vertebrae	Forms a fairly rigid and thick fiber but is more flexible than type I. Permits smooth movements of joints.
III	Arteries, skin, internal organs, and around muscles	Forms thin fibers, often arranged in a meshwork pattern. Allows for greater elasticity in tissues.
IV	Skin, intestine, and kidneys; also found around capillaries	Does not form long fibers. Instead, the proteins are arranged in a meshwork pattern that provides organization and support to cell layers. Functions as a filter around capillaries.

* The sites of synthesis denote where a large amount of the collagen type is made.

Differential gene regulation, in which the expression of genes differs under various environmental conditions and in specialized cell types, controls which types of collagens are made throughout the body and in what amounts they are made. Of the 27 types of collagens identified in humans, Table 10.2 considers types I to IV, each of which varies in where it is primarily synthesized and its structure and function. Collagen genes are regulated, so the required type of collagen is made in the correct sites in your body. In skin cells, for example, the genes that encode the polypeptides that make up collagen types I, III, and IV are turned on, but the synthesis of type II collagen is minimal.

The regulation of collagen synthesis has received a great deal of attention due to the phenomenon of wrinkling. As we age, the amount of collagen that is synthesized in our skin significantly decreases. The underlying network of collagen fibers, which provides scaffolding for the surface of our skin, loosens and unravels. This is one of the factors that cause the skin of older people to sink, sag, and form wrinkles. Various therapeutic and cosmetic agents have been developed to prevent or reverse the appearance of wrinkles, most with limited benefits. For example, many face and skin creams contain collagen as an ingredient. Another approach is collagen injections, in which small amounts of collagen (from cows) are injected into areas where the body's collagen has weakened, filling the depressions to the level of the surrounding skin. Because collagen is naturally broken down in the skin, the injections are not permanent and last only about 3 to 6 months.

Animal Cells Also Secrete Polysaccharides into the ECM

Polysaccharides are the second major component of the ECM of animals. As discussed in Chapter 3, polysaccharides are polymers of simple sugars. Among vertebrates, the most abundant types of polysaccharides in the ECM are **glycosaminoglycans (GAGs)**. These molecules are long, unbranched polysaccharides containing a repeating disaccharide unit (Figure 10.4a). GAGs are highly negatively charged molecules that tend to attract positively charged ions and water. The

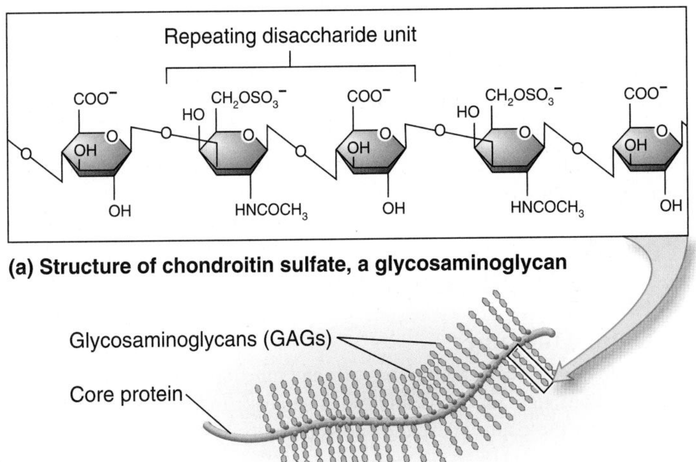

(a) Structure of chondroitin sulfate, a glycosaminoglycan

(b) General structure of a proteoglycan

Figure 10.4 Structures of glycosaminoglycans and proteoglycans. These large molecules are found in the ECM of animal cells. **(a)** Glycosaminoglycans (GAGs) are composed of repeating disaccharide units. They range in length from several dozen to 25,000 disaccharide units. The GAG shown here is chondroitin sulfate, which is a component of cartilage. **(b)** Proteoglycans are composed of a long, linear core protein with many GAGs attached. Note that each GAG is typically 80 disaccharide units long but only a short chain of sugars is shown in this illustration.

Concept Check: What structural feature of GAGs and proteoglycans give them a gel-like character?

majority of GAGs in the ECM are linked to core proteins, forming **proteoglycans** (Figure 10.4b).

Providing resistance to compression is the primary function of GAGs and proteoglycans. Once secreted from cells, these macromolecules form a gel-like component in the ECM. How is this gel-like property important? Due to its high water content, the ECM is difficult to compress and thereby serves to protect cells. GAGs and proteoglycans are found abundantly in regions of the body that are subjected to harsh mechanical forces, such as the joints of the human body. Two examples of GAGs are chondroitin sulfate, which is a major component of cartilage, and hyaluronic acid, which is found in the skin, eyes, and joint fluid. Hyaluronic acid is also used to treat wrinkles and give skin fullness.

Among many invertebrates, an important ECM component is **chitin**, a nitrogen-containing polysaccharide. Chitin forms the hard protective outer covering (called an exoskeleton) of insects, such as crickets and grasshoppers, and crustaceans, such as lobsters and shrimp. As these animals grow, they must periodically shed this rigid outer layer and secrete a new, larger one—a process called molting (look ahead to Figure 32.12).

The Cell Wall of Plants Provides Strength and Resistance to Compression

Let's now turn our attention to the cell walls of plants. Plants cells are surrounded by a cell wall, a protective layer that forms outside of the plasma membrane. Like animal cells, the cells of plants are surrounded by material that provides tensile strength and resistance to compression. The cell walls of plants, however, are usually thicker, stronger, and more rigid than the ECM found in animals. Plant cell walls provide rigidity for mechanical support and also play a role in the maintenance of cell shape and the direction of cell growth. As we learned in Chapter 5, the cell wall also prevents expansion when water enters the cell, thereby preventing osmotic lysis.

The cell walls of plants are composed of a primary cell wall and a secondary cell wall (Figure 10.5). These walls are named according to the timing of their synthesis—the primary cell wall is made before the secondary cell wall. During cell division, the **primary cell wall** develops between two newly made daughter cells. It is usually very flexible and allows new cells to increase in size. The main macromolecule of the plant cell wall is **cellulose**, a polysaccharide made of repeating molecules of glucose attached end to end. These glucose polymers associate with each other via hydrogen bonding to form microfibrils that provide great tensile strength (Figure 10.6).

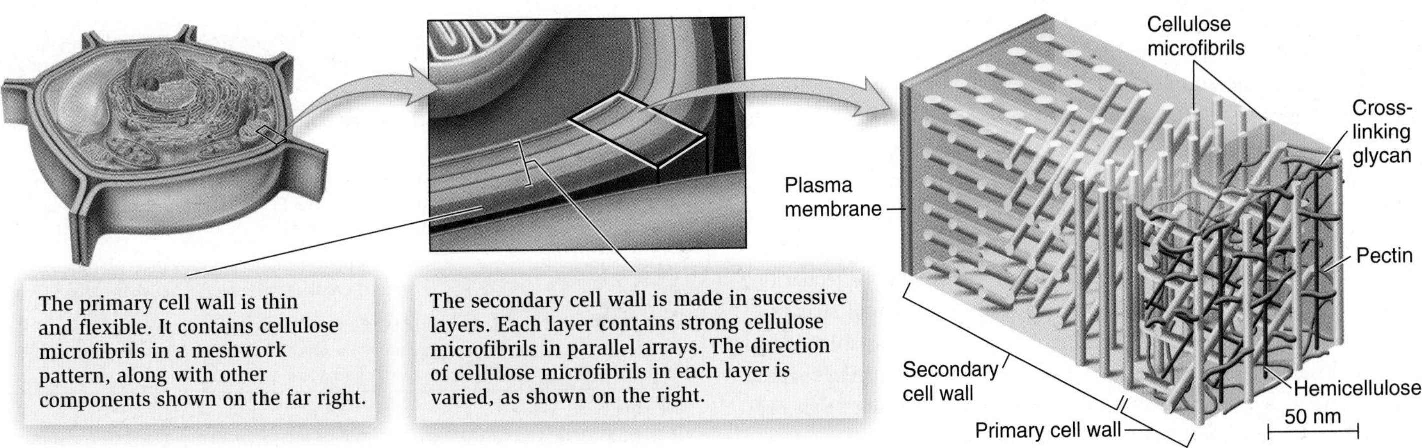

The primary cell wall is thin and flexible. It contains cellulose microfibrils in a meshwork pattern, along with other components shown on the far right.
The secondary cell wall is made in successive layers. Each layer contains strong cellulose microfibrils in parallel arrays. The direction of cellulose microfibrils in each layer is varied, as shown on the right.
Plasma membrane
Cellulose microfibrils
Cross-linking glycan
Pectin
Hemicellulose
50 nm
Secondary cell wall
Primary cell wall

Figure 10.5 Structure of the cell wall of plant cells. The primary cell wall is relatively thin and flexible. It contains cellulose (tan), hemicellulose (red), crosslinking glycans (blue), and pectin (green). The secondary cell wall, which is produced only by certain plant cells, is made after the primary cell wall and is synthesized in successive layers.

Concept Check: With regard to cell growth, what would happen if the secondary cell wall was made too soon?

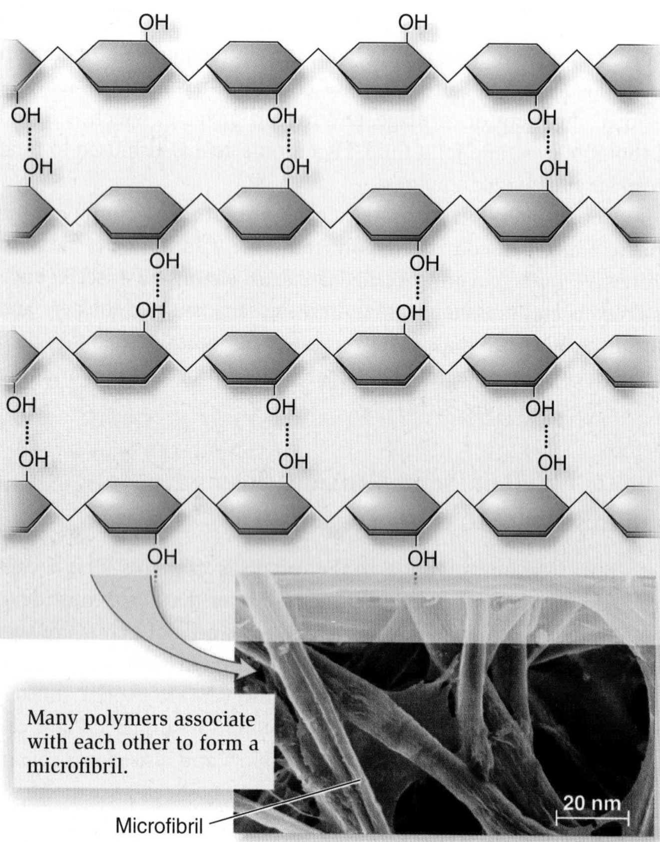

Figure 10.6 **Structure of cellulose, the main macromolecule of the primary cell wall.** Cellulose is made of repeating glucose units linked end to end that hydrogen-bond to each other to form microfibrils (SEM).

Cellulose was discovered in 1838 by the French chemist Anselme Payen, who was the first scientist to attempt to separate wood into its component parts. After treating different types of wood with nitric acid, Payen obtained a fibrous substance that was also found in cotton and other plants. His chemical analysis revealed that the fibers were made of the carbohydrate glucose. Payen called this substance cellulose (from the Latin, meaning consisting of cells). Cellulose is probably the single most abundant organic molecule on Earth. Wood consists mostly of cellulose, and cotton and paper are almost pure cellulose. The mechanism of cellulose synthesis is described in Chapter 35.

In addition to cellulose, other components found in the primary cell wall include hemicellulose, glycans, and pectins (see Figure 10.5). Hemicellulose is another linear polysaccharide, with a structure similar to that of cellulose, but it contains sugars other than glucose in its structure and usually forms thinner microfibrils. Glycans, polysaccharides with branching structures, are also important in cell wall structure. The crosslinking glycans bind to cellulose and provide organization to the cellulose microfibrils. Pectins, which are highly negatively charged polysaccharides, attract water and have a gel-like character that provides the cell wall with the ability to resist compression.

The **secondary cell wall** is synthesized and deposited between the plasma membrane and the primary cell wall (see Figure 10.5) after a plant cell matures and has stopped increasing in size. It is made in layers by the successive deposition of cellulose microfibrils and other components. Whereas the primary wall structure is relatively similar in nearly all cell types and species, the structure of the secondary cell wall is more variable. The secondary cell wall often contains components in addition to those found in the primary cell wall. For example, phenolic compounds called lignins, which are found in the woody parts of plants, are very hard and impart considerable strength to the secondary wall structure.

10.2 Cell Junctions

Learning Outcomes:

1. Compare and contrast the structure and function of anchoring junctions, tight junctions, and gap junctions found between animal cells.
2. Analyze the results of experiments that determined the size of gap junction channels.
3. Describe the structure and function of the middle lamella and plasmodesmata that connect adjacent plant cells.

Thus far, we have learned that the cells of animals and plants produce an ECM or cell wall that provides strength, support, and organization. To become a multicellular organism, cells within the body must also be linked to each other. In animals and plants, this is accomplished by specialized structures called **cell junctions** (**Table 10.3**). In this section, we will examine different types of cell junctions in animals and plants.

Animal cells, which lack the structural support provided by the cell wall, have a more varied group of cell junctions than plant cells. In animals, three types of junctions are found between cells: anchoring junctions play a role in anchoring cells to each other or to the ECM; tight junctions seal cells together to prevent small molecules from leaking across a layer of cells; and gap junctions allow cells to communicate directly with each other.

In plants, cellular organization is somewhat different because plant cells are surrounded by a rigid cell wall. Plant cells are connected to each other by a component called the middle lamella, which cements their cell walls together. They also have junctions termed plasmodesmata that allow adjacent cells to communicate with each other. In this section, we will examine these various types of junctions found between the cells of animals and plants.

Table 10.3 Common Types of Cell Junctions

Type	Description
Animals	
Anchoring junctions	Cell junctions that hold adjacent cells together or attach cells to the ECM. Anchoring junctions are mechanically strong.
Tight junctions	Junctions between adjacent cells in a layer that prevent the leakage of material between cells.
Gap junctions	A collection of channels that permit the direct exchange of ions and small molecules between the cytosol of adjacent cells.
Plants	
Middle lamella	A polysaccharide layer that cements together the cell walls of adjacent cells.
Plasmodesmata	Passageways between the cell walls of adjacent cells that can be opened or closed. When open, they permit the direct diffusion of ions and molecules between the cytosol of cells.

Anchoring Junctions Link Animal Cells to Each Other and to the ECM

The advent of electron microscopy allowed researchers to explore the types of junctions that occur between cells and within the ECM. In the 1960s, Marilyn Farquhar, George Palade, and colleagues conducted several studies showing that various types of cellular junctions connect cells to each other. Over the past few decades, researchers have begun to unravel the functions and molecular structures of these types of junctions, collectively called **anchoring junctions**, which attach cells to each other and to the ECM. Anchoring junctions are particularly common in parts of the body where the cells are tightly connected and form linings. An example is the layer of cells that line the small intestine. Anchoring junctions keep intestinal cells tightly adhered to one another, thereby forming a strong barrier between the lumen of the intestine and the blood. A key component of anchoring junctions that form the actual connections are integral membrane proteins called **cell adhesion molecules (CAMs)**. Two types of CAMs are cadherins and integrins.

Anchoring junctions are grouped into four main categories, according to their functional roles and their connections to cellular components. Figure 10.7 shows these junctions between cells of the mammalian small intestine.

1. **Adherens junctions** connect cells to each other via cadherins. In many cases, these junctions are organized into bands around cells. In the cytosol, adherens junctions bind to cytoskeletal filaments called actin filaments.
2. **Desmosomes** also connect cells to each other via cadherins. They are spotlike points of intercellular contact that rivet cells together. Desmosomes are connected to cytoskeletal filaments called intermediate filaments.
3. **Hemidesmosomes** connect cells to the extracellular matrix via integrins. Like desmosomes, they interact with intermediate filaments.
4. **Focal adhesions** also connect cells to the ECM via integrins. In the cytosol, focal adhesions bind to actin filaments.

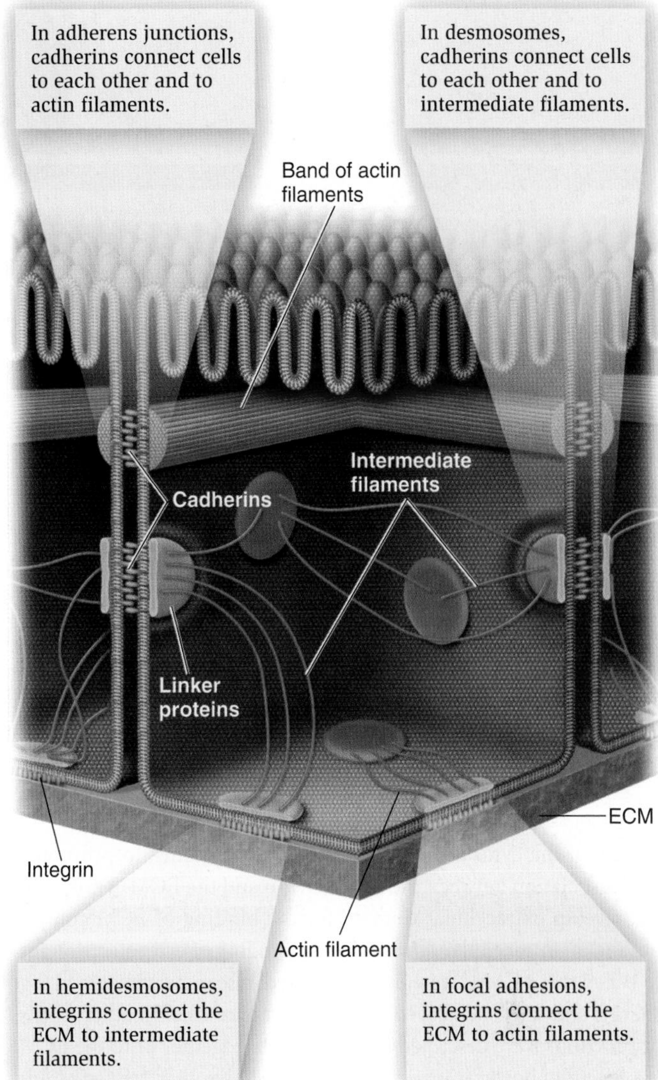

In adherens junctions, cadherins connect cells to each other and to actin filaments.
In desmosomes, cadherins connect cells to each other and to intermediate filaments.
Band of actin filaments
Intermediate filaments
Cadherins
Linker proteins
ECM
Integrin
Actin filament
In hemidesmosomes, integrins connect the ECM to intermediate filaments.
In focal adhesions, integrins connect the ECM to actin filaments.

Figure 10.7 Types of anchoring junctions. This figure shows these junctions in three adjacent intestinal cells.

Concept Check: Which junctions are cell-to-cell junctions and which are cell-to-ECM junctions?

Let's now consider the molecular components of anchoring junctions. As noted, **cadherins** are CAMs that create cell-to-cell junctions (Figure 10.8a). The extracellular domains of two cadherin proteins, each in adjacent cells, bind to each other to promote cell-to-cell adhesion. This binding requires the presence of calcium ions (Ca^{2+}), which change the conformation of the cadherin protein such that cadherins in adjacent cells bind to each other. (This calcium dependence is where cadherin gets its name—Ca^{2+}-dependent adhering molecule.) On the interior of the cell, linker proteins connect cadherins to actin or intermediate filaments of the cytoskeleton. This promotes a more stable interaction between two cells because their strong cytoskeletons are connected to each other.

The genomes of vertebrates and invertebrates contain multiple cadherin genes, which encode slightly different cadherin proteins. The expression of particular cadherins allows cells to recognize each other. Dimer formation follows a homophilic, or like-to-like, binding mechanism. To understand the concept of homophilic binding, let's consider an example. One type of cadherin is called E-cadherin, and another is N-cadherin. E-cadherin in one cell binds to E-cadherin in an adjacent cell to form a homodimer. However, E-cadherin in one cell does not bind to N-cadherin in an adjacent cell to form a heterodimer. Similarly, N-cadherin binds to N-cadherin but not to E-cadherin in an adjacent cell. Why is such homophilic binding important? By expressing only certain types of cadherins, each cell binds only to other cells that express the same cadherin types. This phenomenon plays a key role in the proper arrangement of cells throughout the body, particularly during embryonic development.

Integrins, a group of cell-surface receptor proteins, are a second type of CAM, one that creates connections between cells and the ECM. Integrins do not require Ca^{2+} to function. Each integrin protein is composed of two nonidentical subunits. In the example shown in Figure 10.8b, an integrin is bound to fibronectin, an adhesive protein in the ECM that binds to other ECM components such as collagen fibers. Like cadherins, integrins also bind to actin or intermediate

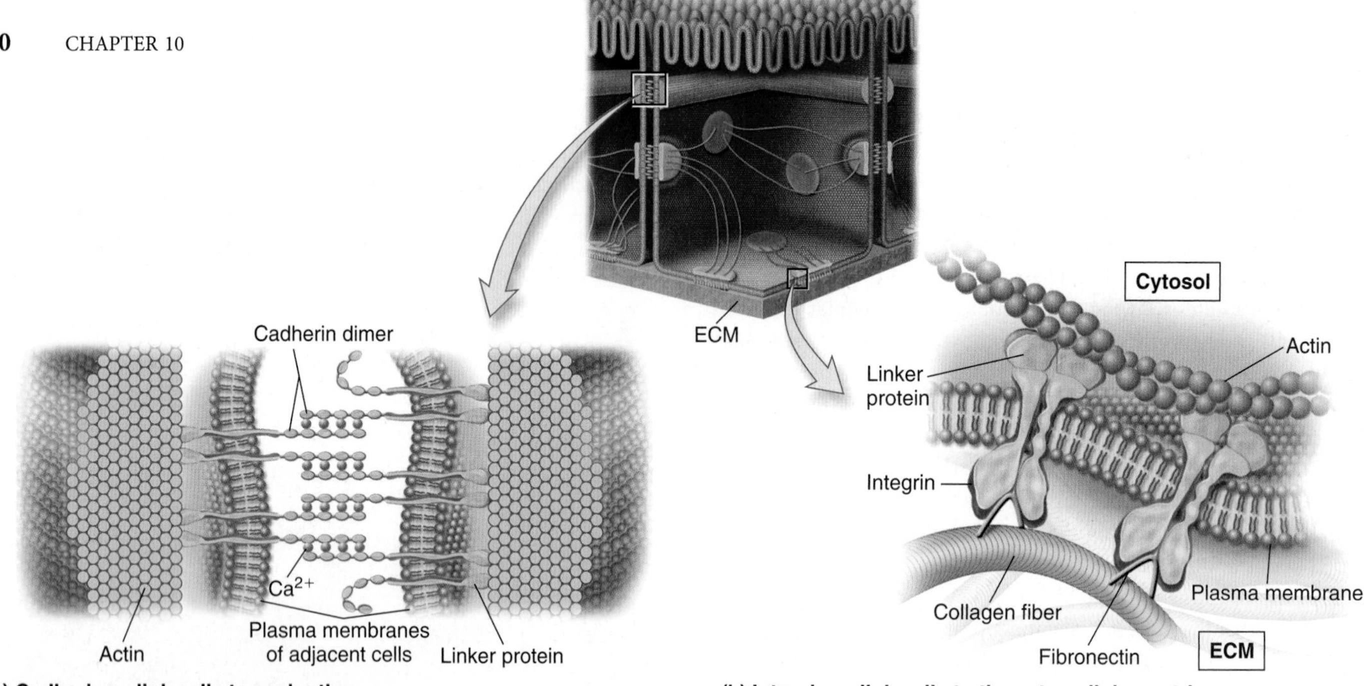

Figure 10.8 Types of cell adhesion molecules (CAMs). Cadherins and integrins are CAMs that form connections in anchoring junctions. **(a)** A cadherin in one cell binds to a cadherin of an identical type in an adjacent cell. This binding requires Ca^{2+}. In the cytosol, cadherins bind to actin or intermediate filaments of the cytoskeleton via linker proteins. **(b)** Integrins link cells to the ECM and form intracellular connections to actin or intermediate filaments. Each integrin protein is composed of two nonidentical subunits, a heterodimer.

filaments in the cytosol of the cell, via linker proteins, to promote a strong association between the cytoskeleton and the ECM. Thus, integrins have an extracellular domain for the binding of ECM components and an intracellular domain for the binding of cytosolic proteins.

When these CAMs were first discovered, researchers imagined that cadherins and integrins played only a mechanical role. In other words, their functions were described as holding cells together or to the ECM. More recently, however, experiments have shown that cadherins and integrins are also important in cell communication. The formation or breaking of cell-to-cell and cell-to-ECM anchoring junctions affects signal transduction pathways within the cell. Similarly, intracellular signal transduction pathways affect cadherins and integrins in ways that alter intercellular junctions and the binding of cells to ECM components.

Abnormalities in CAMs such as integrins are often associated with the ability of cancer cells to metastasize, that is, to move to other parts of the body. CAMs are critical for keeping cells in their correct locations. When they become defective due to cancer-causing mutations, cells lose their proper connections with the ECM and adjacent cells and may spread to other parts of the body.

Tight Junctions Prevent the Leakage of Materials Across Animal Cell Layers

In animals, **tight junctions**, or occluding junctions, are a second type of junction, one that forms a tight seal between adjacent cells, thereby preventing material from leaking between cells. As an example, let's consider the intestine. The cells that line the intestine form a sheet that is one cell thick. One side faces the intestinal lumen, and the other faces the ECM and blood vessels (Figure 10.9). Tight junctions between these cells prevent the leakage of materials from the lumen of the intestine into the blood and vice versa.

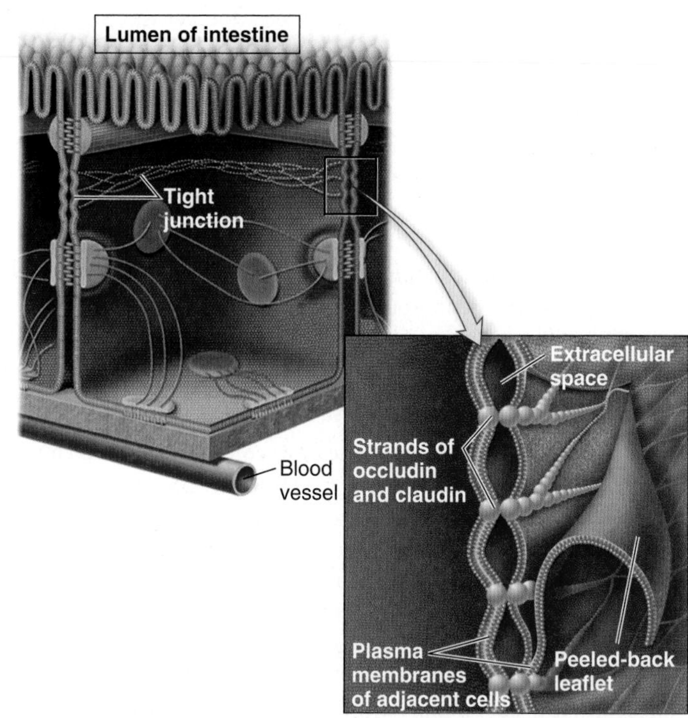

Figure 10.9 Tight junctions between adjacent intestinal cells. In this example, tight junctions form a seal that prevents the movement of material between cells, from the intestinal lumen into the blood, and vice versa. The inset shows the interconnected network of occludin and claudin that forms the tight junction.

Concept Check: What do you think is the role of tight junctions in the epidermal layers of your skin?

BioConnections: Look ahead to Figure 45.11. What problems might arise if tight junctions did not connect the cells that line your small intestine?

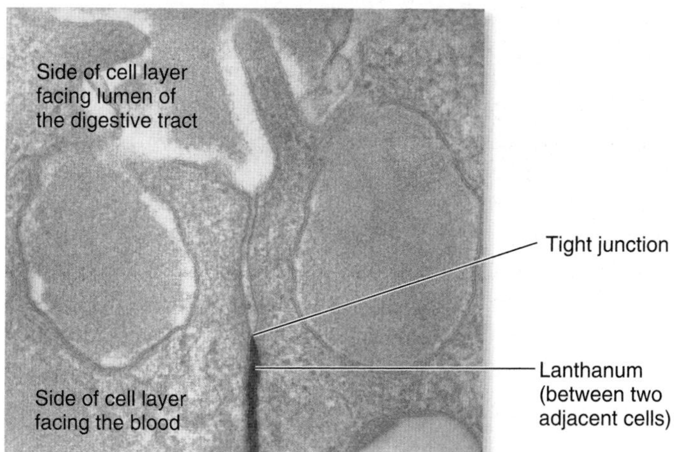

Figure 10.10 An experiment demonstrating the function of a tight junction. When lanthanum was injected into the bloodstream of a rat, it diffused between the cells in the region up to a tight junction but could not diffuse past the junction to the other side of the cell layer.

Concept Check: What results would you expect if a rat was fed lanthanum and then a sample of intestinal cells was observed under the electron microscope?

Tight junctions are made by membrane proteins, called occludin and claudin, that form interlaced strands in the plasma membrane (see Figure 10.9 inset). These strands of proteins, each in adjacent cells, bind to each other, thereby forming a tight seal between cells. Tight junctions are not mechanically strong like anchoring junctions, because they do not have strong connections with the cytoskeleton. Therefore, adjacent cells that have tight junctions also have anchoring junctions to hold the cells in place.

The amazing ability of tight junctions to prevent the leakage of material across cell layers has been demonstrated by dye-injection studies. In 1972, Daniel Friend and Norton Gilula injected lanthanum into the bloodstream of a rat. Lanthanum is an electron-dense metallic element that can be visualized under the electron microscope. A few minutes later, a sample of a cell layer in the digestive tract was removed and visualized by electron microscopy. As seen in the micrograph in Figure 10.10, lanthanum diffused into the region between the cells that faces the blood, but it could not move past the tight junction to the side of the cell layer facing the lumen of the digestive tract.

Gap Junctions in Animal Cells Provide a Passageway for Intercellular Transport

A third type of junction found in animals is called a **gap junction**, because a small gap occurs between the plasma membranes of cells connected by these junctions (Figure 10.11). Gap junctions are abundant in tissues and organs where the cells need to communicate with each other. For example, cardiac muscle cells, which cause your heart to beat, are interconnected by many gap junctions. Because gap junctions allow the passage of ions, electrical changes in one cardiac muscle cell are easily transmitted to an adjacent cell that is connected via gap junctions. This is needed for the coordinated contraction of cardiac muscle cells.

In vertebrates, gap junctions are composed of an integral membrane protein called connexin. Invertebrates have a structurally similar protein called innexin. Six connexin proteins in one vertebrate cell form a channel called a **connexon**. A connexon in one cell aligns with a connexon in an adjacent cell to form an intercellular channel (see inset to Figure 10.11). The term *gap junction* refers to an organized association in which many connexons are close to each other in the plasma membrane and form many intercellular channels.

The connexons allow the passage of ions and small molecules, including amino acids, sugars, and signaling molecules such as Ca^{2+}, cAMP, and IP_3. In this way, gap junctions allow adjacent cells to share metabolites and directly signal each other. At the same time, gap junction channels are too small to allow the passage of RNA, proteins, or polysaccharides. Therefore, cells that communicate via gap junctions still maintain their own distinctive set of macromolecules.

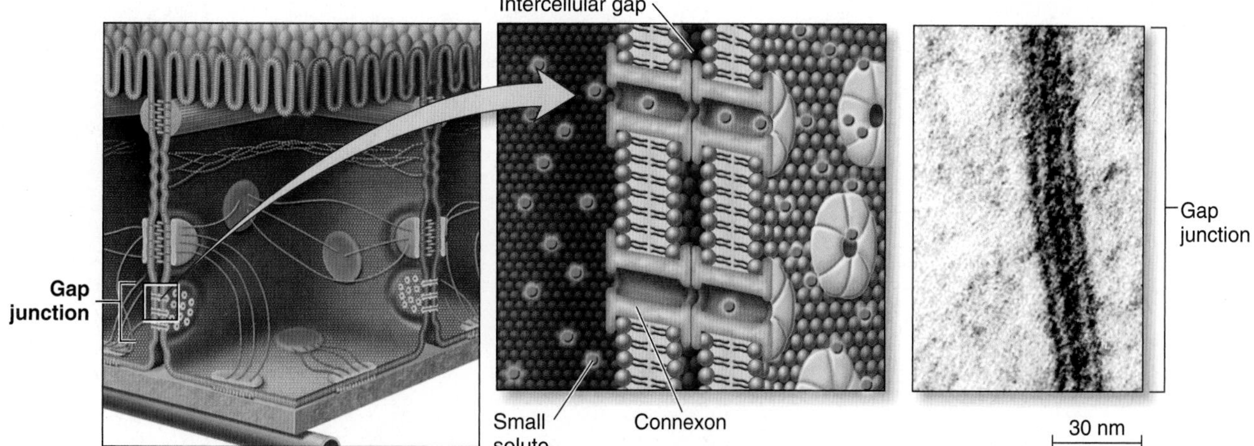

Figure 10.11 Gap junctions between adjacent cells. Gap junctions form intercellular channels that allow the passage of small solutes with masses less than 1,000 Da. A connexon consists of six proteins called connexins. Two connexons align to form an intercellular channel. The micrograph shows a gap junction, which is composed of many connexons, between intestinal cells.

FEATURE INVESTIGATION

Loewenstein and Colleagues Followed the Transfer of Fluorescent Dyes to Determine the Size of Gap-Junction Channels

As mentioned, gap junctions allow the passage of ions and small molecules, those with a mass up to about 1,000 Da. This property of gap junctions was determined in experiments involving the transfer of fluorescent dyes. During the 1960s, several research groups began using fluorescent dyes to study cell morphology and function. As discussed in Chapter 4, the location of fluorescent dyes within cells can be seen via fluorescence microscopy. In 1964, Werner Loewenstein and colleagues observed that a fluorescent dye could move from one cell to an adjacent cell, which prompted them to investigate this phenomenon further.

In the experiment shown in **Figure 10.12**, Loewenstein and colleagues grew rat liver cells in the laboratory, where they formed a single layer. The adjacent cells formed gap junctions. Single cells were injected with various dyes composed of fluorescently labeled amino acids or peptide molecules with different masses, and then the cell layers were observed via fluorescence microscopy. As shown in The Data, the researchers observed that dyes with a molecular mass up to 901 Da passed from cell to cell. Dyes with a larger mass, however, did not move intercellularly. Loewenstein and other researchers subsequently investigated dye transfer in other cell types and species. Though some variation is found when comparing different cell types and species, the researchers generally observed that molecules with a mass greater than 1,000 Da do not pass through gap junctions.

Experimental Questions

1. What was the purpose of the study conducted by Loewenstein and colleagues?
2. Explain the experimental procedure used by Loewenstein to determine the size of molecules that can pass through gap-junction channels.
3. What did the results of Figure 10.12 indicate about the size of gap-junction channels?

Figure 10.12 **Use of fluorescent molecules by Loewenstein and colleagues to determine the size of gap-junction channels.**

HYPOTHESIS Gap-junction channels allow the passage of ions and molecules, but there is a limit to how large the molecules can be.

KEY MATERIALS Rat liver cells grown in the laboratory, a collection of fluorescent dyes.

1 Grow rat liver cells in a laboratory on solid growth medium until they become a single layer. At this point, adjacent cells have formed gap junctions.

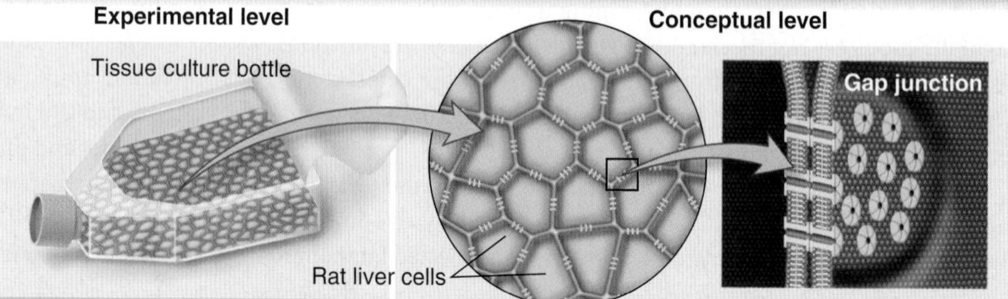

2 Inject 1 cell in the layer with fluorescently labeled amino acids or peptides. Note: Several dyes with different molecular masses were tested.

3 Incubate for various lengths of time (for example, 40–45 minutes). Observe cell layer under the fluorescence microscope to determine if the dye has moved to adjacent cells.

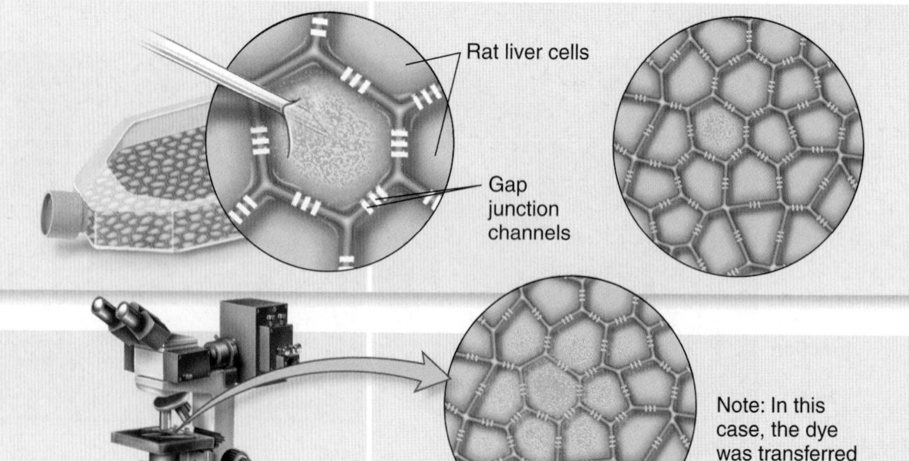

4 THE DATA

Mass of dye (in daltons)	Transfer to adjacent cells*	Mass of dye	Transfer to adjacent cells*
376	++++	851**	–
464	++++	901	+++
536	+++	946	–
559	++++	1004	–
665	+	1158	–
688	++++	1678	–
817	+++	1830	–

*The number of pluses indicates the relative speed of transfer. Four pluses denotes fast transfer, whereas one plus is slow transfer. A minus indicates that transfer between cells did not occur. **In some cases, molecules with less mass did not pass between cells compared with molecules with a higher mass. This may be due to differences in their structures (for example, charges) that influence whether or not they can easily penetrate the channel.

5 **CONCLUSION** Gap junctions allow the intercellular movement of molecules that have a mass of approximately 900 Da or less.

6 **SOURCE** Flagg-Newton, J., Simpson, I., and Loewenstein, W.R. 1979. Permeability of the Cell-to-Cell Membrane Channels in Mammalian Cell Junction. *Science* 205:404–407.

The Middle Lamella Cements Adjacent Plant Cell Walls Together

In animals, we have seen that cell-to-cell contact, via anchoring junctions, tight junctions, and gap junctions, involves interactions between membrane proteins in adjacent cells. In plants, cell junctions are quite different. Rather than using membrane proteins to form cell-to-cell connections, plant cells make an additional component called the **middle lamella** (plural, lamellae), which is found between most adjacent plant cells (**Figure 10.13**). When plant cells are dividing, the middle lamella is the first layer formed. The primary cell wall is then made. The middle lamella is rich in pectins, negatively charged polysaccharides that are also found in the primary cell wall (see Figure 10.5). Pectins attract water and make a hydrated gel. Ca^{2+} and Mg^{2+} interact with the negative charges in the pectins and cement the cell walls of adjacent cells together.

The process of fruit ripening illustrates the importance of pectins in holding plant cells together. An unripened fruit, such as a green tomato, is very firm because the rigid cell walls of adjacent cells are firmly attached to each other. During ripening, the cells secrete a group of enzymes called pectinases, which digest pectins in the middle lamella as well as those in the primary cell wall. As this process continues, the attachments between cells are broken, and the cell walls become less rigid. For this reason, a red ripe tomato is much less firm than an unripe tomato.

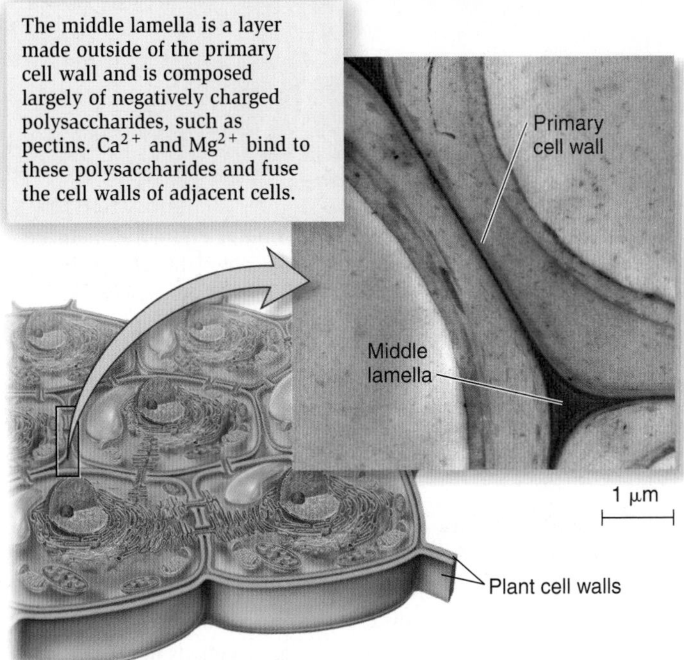

The middle lamella is a layer made outside of the primary cell wall and is composed largely of negatively charged polysaccharides, such as pectins. Ca2+ and Mg2+ bind to these polysaccharides and fuse the cell walls of adjacent cells.
Primary cell wall
Middle lamella
1 μm
Plant cell walls

Figure 10.13 **Plant cell-to-cell junctions known as middle lamellae.**

Concept Check: How are middle lamellae similar to the anchoring junctions and desmosomes found between animal cells? How are they different?

Plasmodesmata Are Channels Connecting the Cytoplasm of Adjacent Plant Cells

In 1879, Eduard Tangl, a Russian botanist, observed intercellular connections in the seeds of the strychnine tree and hypothesized that the cytoplasm of adjacent cells is connected by ducts in the cell walls. He was the first to propose that direct cell-to-cell communication integrates the functioning of plant cells. The ducts or intercellular channels that Tangl observed are now known as **plasmodesmata** (singular, plasmodesma).

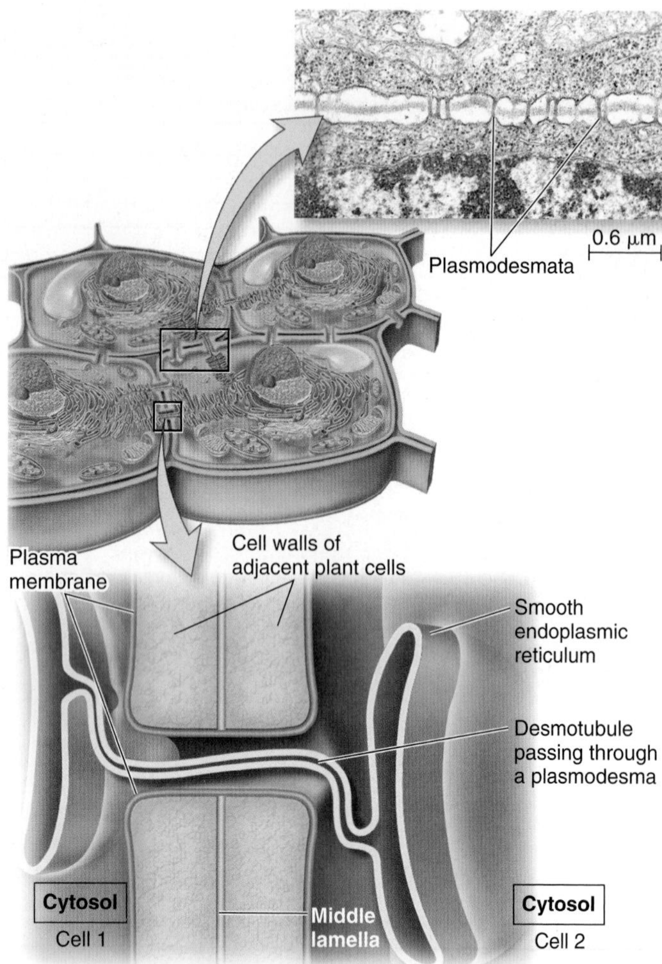

Figure 10.14 Structure of plasmodesmata. Plasmodesmata are cell junctions connecting the cytosol of adjacent plant cells, allowing water, ions, and molecules to pass from cell to cell. At these sites, the plasma membrane of one cell is continuous with the plasma membrane of an adjacent cell. In addition, the smooth ER from one cell is connected to that of the adjacent cell via a desmotubule.

BioConnections: Look ahead to Figure 38.8. How do plasmodesmata play a role in the movement of nutrients through a plant root?

Plasmodesmata are functionally similar to gap junctions in animal cells because they allow the passage of ions and molecules between the cytosol of adjacent plant cells. However, the structure of plasmodesmata is quite different from that of gap junctions. As shown in Figure 10.14, plasmodesmata are channels in the cell walls of adjacent cells. At these sites, the plasma membrane of one cell is continuous with the plasma membrane of the other cell, which permits the diffusion of molecules from the cytosol of one cell to the cytosol of the other. In addition to a cytosolic connection, plasmodesmata also have a central tubule, called a desmotubule, connecting the smooth ER membranes of adjacent cells.

Plasmodesmata can change the size of their opening between closed, open, and dilated states. In the open state, they allow the passage of ions and small molecules, such as sugars and cAMP. In this state, plasmodesmata play a similar role to gap junctions between animal cells. Plasmodesmata tend to close when a large pressure difference occurs between adjacent cells. Why does this happen? One reason is related to cell damage. When a plant is wounded, damaged cells lose their turgor pressure. (Turgor pressure is described in Chapter 38, Figure 38.4.) The closure of plasmodesmata between adjacent cells helps to prevent the loss of water and nutrients from the wound site.

Unlike gap junctions between animal cells, researchers have recently discovered that plasmodesmata can dilate to also allow the passage of macromolecules and even viruses between adjacent plant cells. Though the mechanism of dilation is not well understood, the wider opening of plasmodesmata is important for the passage of proteins and mRNA during plant development. It also provides a key mechanism whereby viruses can move from cell to cell.

10.3 Tissues

Learning Outcomes:

1. List the six basic cell processes that produce tissues and organs.
2. Outline the structures and functions of the four types of animal tissues: epithelial, connective, nervous, and muscle tissues.
3. Identify the structures and functions of the three types of plant tissues: dermal, ground, and vascular tissues.
4. Compare and contrast the similarities and differences between animal and plant tissues.

A **tissue** is a part of an animal or plant consisting of a group of cells having a similar structure and function. In this section, we will view tissues from the perspective of cell biology. Animals and plants contain many different types of cells. Humans, for example, have over 200 different cell types, each with a specific structure and function. Even so, these cells can be grouped into a few general categories. For example, muscle cells found in your heart (cardiac muscle cells), in your biceps (skeletal muscle cells), and around your arteries (smooth muscle cells) look somewhat different under the microscope and have unique roles in the body. Yet due to structural and functional similarities, all three types are categorized as muscle tissue. In this section, we will begin by surveying the basic processes that cells undergo to make tissues. Next, we will examine the main categories of animal and plant tissues. We will conclude by taking a more in-depth look at some differences and similarities between selected animal and plant tissues, focusing in particular on the functions of the ECM and cell junctions.

Six Basic Cell Processes Produce Tissues and Organs

A multicellular organism such as a plant or animal contains many cells. For example, an adult human has somewhere between 10 and 100 trillion cells in her or his body. Cells are organized into tissues, and tissues are organized into organs. An **organ** is a collection of two or more tissues that performs a specific function or set of functions. The heart is an organ found in the bodies of complex animals, and a leaf is an organ found in plants. We will examine the structures and functions of organs in Units VI and VII.

How are tissues and organs formed? To form tissues and organs, cells undergo six basic processes that influence their morphology, arrangement, and number: cell division, cell growth, differentiation, migration, apoptosis, and the formation of cell connections.

1. *Cell division:* As discussed in Chapter 15, eukaryotic cells progress through a cell cycle that leads to cell division.
2. *Cell growth:* Following cell division, cells take up nutrients and usually expand in volume. Cell division and cell growth are the primary mechanisms for increasing the size of tissues, organs, and organisms.
3. *Differentiation:* Due to gene regulation, cells differentiate into specialized types of cells. Cell differentiation is described in Chapter 19.
4. *Migration:* During embryonic development in animals, cells migrate to their appropriate positions within the body. Also, adults have cells that can move into regions that have become damaged. Cell migration does not occur during plant development.
5. *Apoptosis:* Programmed cell death, also known as apoptosis (discussed in Chapter 9), is necessary to produce certain morphological features of the body. For example, during development in mammals, the formation of individual fingers and toes requires the removal, by apoptosis, of the skin cells between them.
6. *Cell connections:* In the first section of this chapter, we learned that cells produce an extracellular matrix or cell wall that provides strength and support. In animals, the ECM serves to organize cells within tissues and organs. In plants, the cell wall is largely responsible for the shapes of plant tissues. Different types of cell junctions in both animal and plant cells enable cells to make physical contact and communicate with one another.

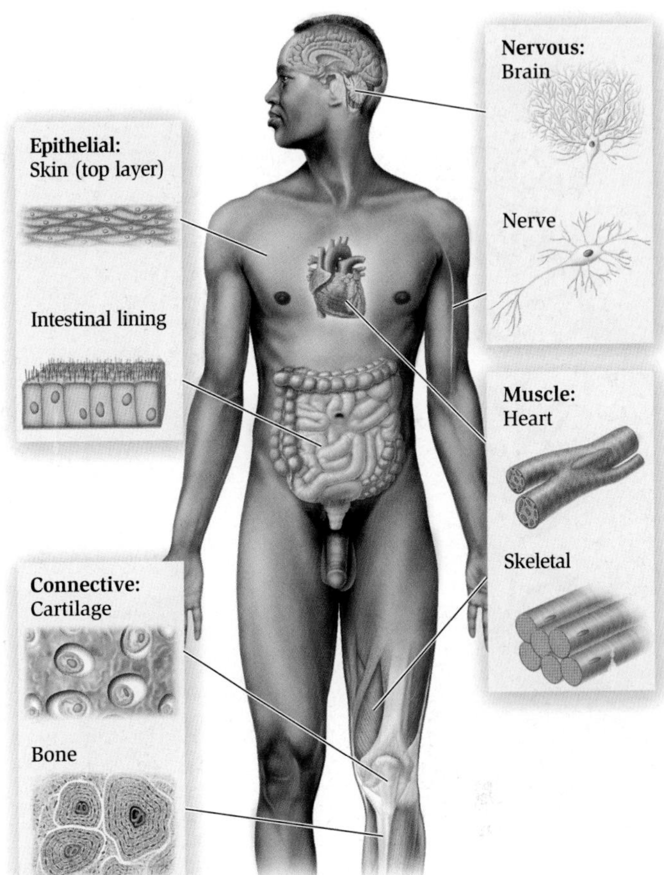

Figure 10.15 **Examples of the four general types of tissues—epithelial, connective, nervous, and muscle—found in animals.**

Concept Check: *Which of these four types of tissues has the most extensive ECM?*

Animals Are Composed of Epithelial, Connective, Nervous, and Muscle Tissues

The body of an animal contains four general types of tissue—epithelial, connective, nervous, and muscle—that serve very different purposes (Figure 10.15).

Epithelial Tissue **Epithelial tissue** is composed of cells that are joined together via tight junctions and form continuous sheets. Epithelial tissue covers or forms the lining of all internal and external body surfaces. For example, epithelial tissue lines organs such as the lungs and digestive tract. In addition, epithelial tissue forms skin, a protective surface that shields the body from the outside environment.

Connective Tissue Most **connective tissue** provides support to the body and/or helps to connect different tissues to each other. Connective tissue is rich in ECM. In some cases, the tissue contains only a sparse population of cells that are embedded in the ECM. Examples of connective tissue include cartilage, tendons, bone, fat tissue, and the inner layers of the skin. Blood is also considered a form of connective tissue because it provides liquid connections to various regions of the body.

Nervous Tissue **Nervous tissue** receives, generates, and conducts electrical signals throughout the body. In vertebrates, these electrical signals are integrated by nervous tissue in the brain and transmitted down the spinal cord to the rest of the body. Chapter 41 considers the cellular basis of nerve signals, and Chapters 42 and 43 examine the organization of nervous systems in animals.

Muscle Tissue **Muscle tissue** can generate a force that facilitates movement. Muscle contraction is needed for bodily movements such as walking and running and also plays a role in the movement of materials throughout the body. For example, contraction of heart muscle propels blood through your body, and smooth muscle contractions move food through the digestive system. The properties of muscle tissue in animals are examined in Chapter 44.

Plants Contain Dermal, Ground, and Vascular Tissues

Plant biologists usually classify tissues as simple or complex. Simple tissues are composed of one or possibly two cell types. Complex tissues are composed of two or more cell types but lack an organization that would qualify them as organs. The bodies of most plants contain three general types of tissue—dermal, ground, and vascular—each with a different structure suited to its functions (Figure 10.16).

Dermal Tissue A **dermal tissue** is a complex tissue that forms a covering on various parts of the plant. The term **epidermis** refers to

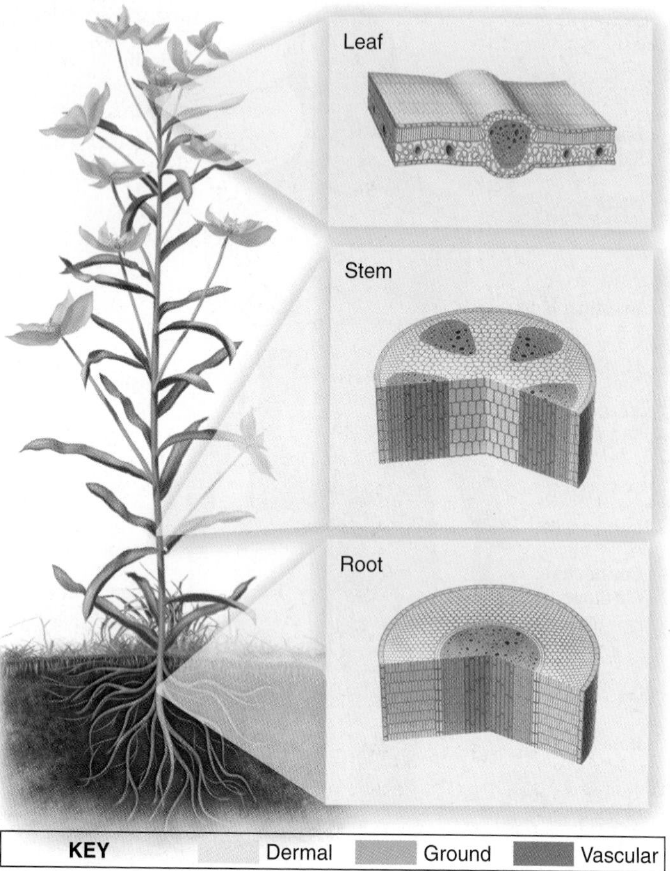

Figure 10.16 Locations of the three general types of tissues—dermal, ground, and vascular—found in plants.

Concept Check: *Which of these three types of tissues is found on the surfaces of leaves, stems, and roots?*

the newly made tissue on the surfaces of leaves, stems, and roots. Surfaces of leaves are usually coated with a waxy cuticle to prevent water loss. In addition, leaf epidermis often has hairs, or trichomes, which are specialized types of epidermal cells. Trichomes have diverse functions, including the secretion of oils and leaf protection. In leaves, epidermal cells called guard cells form pores known as stomata, which permit gas exchange. The function of the root epidermis is the absorption of water and nutrients. The root epidermis does not have a waxy cuticle because such a cuticle would inhibit water and nutrient absorption.

Ground Tissue Most of a plant's body is made of **ground tissue**, which has a variety of functions, including photosynthesis, storage of carbohydrates, and support. Ground tissue is subdivided into three types of simple tissues: parenchyma, collenchyma, and sclerenchyma. Let's look briefly at each of these types of ground tissue (also see Figure 35.7).

1. Parenchyma tissue is very active metabolically. The mesophyll, the central part of the leaf that carries out the bulk of photosynthesis, is parenchyma tissue. Parenchyma tissue also functions in the storage of carbohydrates. The cells of parenchyma tissue usually lack a secondary cell wall.
2. Collenchyma tissue provides structural support to the plant body, particularly to growing regions such as the periphery of the stems and leaves. Collenchyma cells tend to have thick, secondary cell walls but do not contain much lignin. Therefore, they provide support but are also able to stretch.
3. Sclerenchyma tissue also provides structural support to the plant body, particularly to those parts that are no longer growing, such as the dense, woody parts of stems. The secondary cell walls of sclerenchyma cells tend to have large amounts of lignin, which thereby provide rigid support. In many cases, sclerenchyma cells are dead at maturity, but their cell walls continue to provide structural support during the life of the plant.

Vascular Tissue Most plants living today are vascular plants. In these species, which include ferns and seed plants, the **vascular tissue** is a complex tissue composed of cells that are interconnected and form conducting vessels for water and nutrients. The two types of vascular tissue are called xylem and phloem. The xylem transports water and mineral ions from the root to the rest of the plant, and the phloem distributes the products of photosynthesis and a variety of other nutrients throughout the plant. Some types of modern plants, such as mosses, are nonvascular plants that lack conducting vessels. These plants tend to be small and live in damp, shady places.

Animal and Plant Tissues Have Striking Differences and Similarities

Because plants and animals appear strikingly different, it is not too surprising that their cells and tissues show conspicuous differences. For example, the vascular tissue of plants (which transports water and nutrients) does not resemble any one tissue in animals. The blood vessels of animals (which transport blood carrying oxygen and nutrients throughout the body) are hollow tubes that contain both connective and muscle tissue. In addition, animals have two tissue types that are not found in plants: muscle and nervous tissue. Even so, plants are capable of movement and the transmission of signals via action potentials. For example, the Venus flytrap (*Dionaea muscipula*) has two modified leaves that resemble a clamshell. When an insect touches the trichomes on the surface of these leaves, an electrical signal is triggered that causes the leaves to move closer to each other, thereby trapping the unsuspecting insect.

Although cellular differences are prominent between plants and animals, certain tissues show intriguing similarities. Why are there similarities between some animal and plant tissues? The answer is that certain types of cell organization and structure provide functions that are needed by both animals and plants. For example, the epithelial tissue of animals and the dermal tissue of plants both form a protective covering over the organism. Also, the connective tissue of animals and the ground tissue of plants both play a role in structural support. Let's take a closer look at the similarities between these tissues in animals and plants.

A Comparison of Epithelial and Dermal Tissues Both the epithelial tissue in animals (also called an epithelium) and dermal tissue in plants form layers of cells. Let's begin with epithelial tissue. An

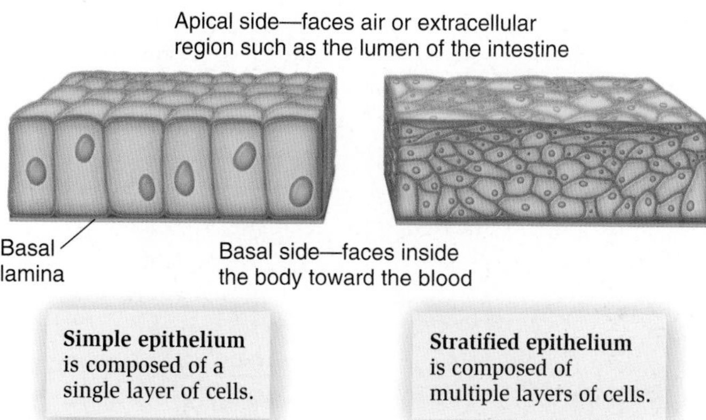
Apical side—faces air or extracellular region such as the lumen of the intestine
Basal lamina
Basal side—faces inside the body toward the blood
Simple epithelium is composed of a single layer of cells.
Stratified epithelium is composed of multiple layers of cells.

Figure 10.17 **Simple and stratified epithelia in animals.**

epithelium can be classified according to its number of layers. Simple epithelium is one cell layer thick, whereas stratified epithelium has several layers (**Figure 10.17**). In both cases, the epithelium has a polarity, which is due to an asymmetry in its organization. The outer, or apical, side of an epithelium is exposed to air or to a watery fluid such as the lumen of the intestine. The inner, or basal, side faces the blood. The basal side of the epithelium rests on some type of support, such as another type of tissue or on a form of ECM called the basal lamina.

A hallmark of epithelial cells is that they form many connections with each other. For example, in the simple epithelium that lines the intestine (**Figure 10.18**), adjacent cells form anchoring junctions with each other and with the basal lamina. These anchoring junctions hold the cells firmly in place. Tight junctions, found near the apical surface, prevent the leakage of materials from the lumen of the intestine into the blood. Instead, nutrients are selectively transported from the intestinal lumen into the cytosol of the epithelial cell and then are exported across the basal side of the cell into the blood. This phenomenon, called transepithelial transport, allows the body to take up the nutrients it needs while preventing unwanted materials from getting into the bloodstream. Epithelial cells are connected via gap junctions, which allow the exchange of nutrients and signaling molecules throughout the epithelium.

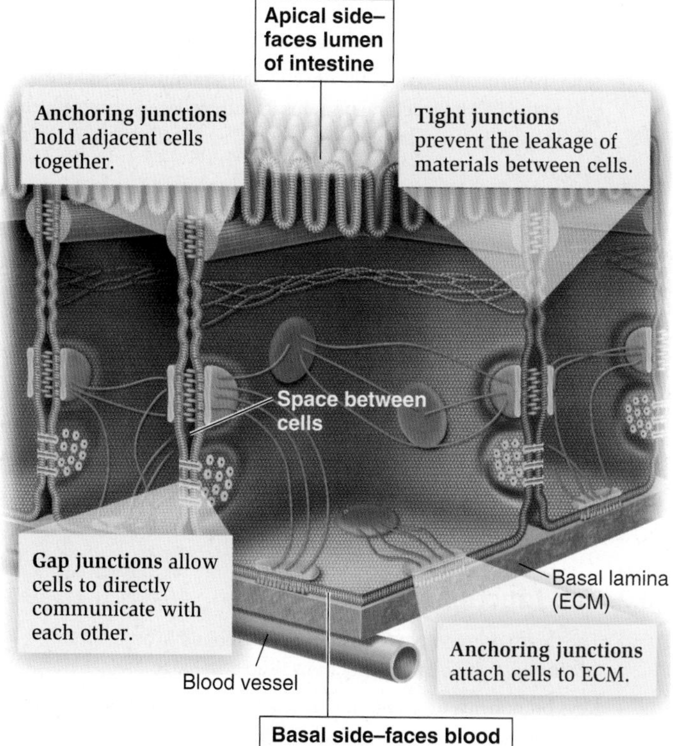
Apical side–faces lumen of intestine
Anchoring junctions hold adjacent cells together.
Tight junctions prevent the leakage of materials between cells.
Space between cells
Gap junctions allow cells to directly communicate with each other.
Basal lamina (ECM)
Blood vessel
Anchoring junctions attach cells to ECM.
Basal side–faces blood

Figure 10.18 **Connections between cells of a simple epithelium that lines the intestine.** This figure emphasizes the three major types of cell junctions that are common in epithelial tissue.

In flowering plants, the epidermis covers all of the newly made parts of a plant. For example, the upper and lower sides of leaves are covered by epidermis, which is usually a single layer of closely packed cells (**Figure 10.19a**). Epidermal cells have a thick primary cell wall and are tightly interlocked by their middle lamella. As a consequence, plant epidermal cells are tightly woven together, much like epithelial cell layers in animals.

As a plant ages, the epidermis may be replaced by another dermal tissue called the periderm (**Figure 10.19b**). In woody plants, an

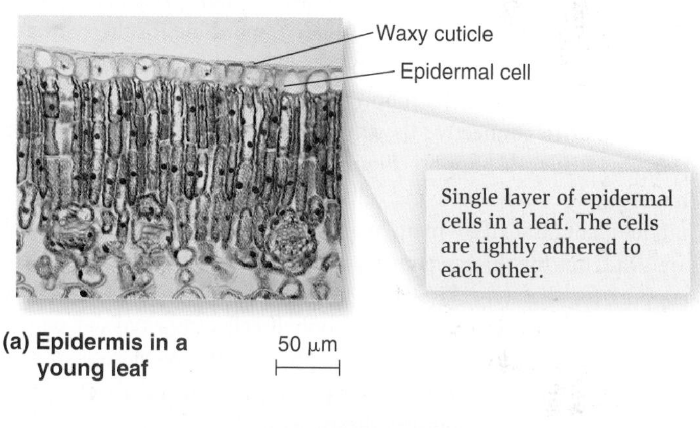
Waxy cuticle
Epidermal cell
Single layer of epidermal cells in a leaf. The cells are tightly adhered to each other.
(a) Epidermis in a young leaf
50 μm

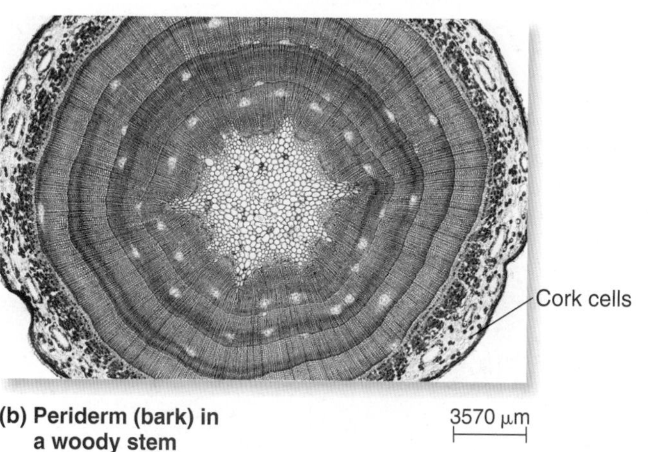
Cork cells
(b) Periderm (bark) in a woody stem
3570 μm

Figure 10.19 **Dermal tissues in plants.**

BIOLOGY PRINCIPLE New properties emerge from complex interactions. In this case, the interactions of dermal cells result in the new property that dermal tissue provides a protective covering.

Concept Check: Do you notice any parallels between simple epithelium and epidermis, and between stratified epithelium and periderm?

example of periderm is the bark on trees. Periderm protects a plant from pathogens, prevents excessive water loss, and provides insulation. The periderm consists of interconnected cork cells, which may be several layers thick. The cork cells have extremely thick cell walls. When cork cells reach maturity, they die, but their cell walls continue to provide support.

A Comparison of Connective and Ground Tissue In contrast to epithelial tissue, which is mostly composed of cells, the connective tissue of animals is largely composed of ECM and has relatively few cells. In animal connective tissue, cell-to-cell contact is somewhat infrequent. Instead, cells are usually adhered to the ECM via integrins, as shown earlier in Figure 10.8b. In some cases, the primary function of cells within connective tissue is to synthesize the components of the ECM. For example, let's consider cartilage, a connective tissue found in joints such as your knees. The cells that synthesize cartilage, known as chondrocytes, actually represent a small proportion of the total volume of cartilage. As shown in Figure 10.20, the chondrocytes are found in small cavities within the cartilage called lacunae (singular, lacuna). In some types of cartilage, the chondrocytes represent only 1–2% of the total volume of the tissue! Chondrocytes are the only cells found in cartilage. They are solely responsible for the synthesis of protein fibers, such as collagen, as well as the glycosaminoglycans and proteoglycans that are found in cartilage.

Similar to connective tissue in animals, ground tissue in plants provides structural support. Figure 10.21 shows a scanning electron micrograph of sclerenchyma cells found in *Arabidopsis thaliana*, a model plant studied by many plant biologists. At maturity, the cells are dead, but the thick secondary cell walls continue to provide rigid support for the stem. However, not all cells in ground tissue have thick cell walls. For example, mesophyll cells are a type of parenchyma cell in the leaf that carry out photosynthesis. Mesophyll cells have relatively thin cell walls that permit the transmission of light, which is required for photosynthesis.

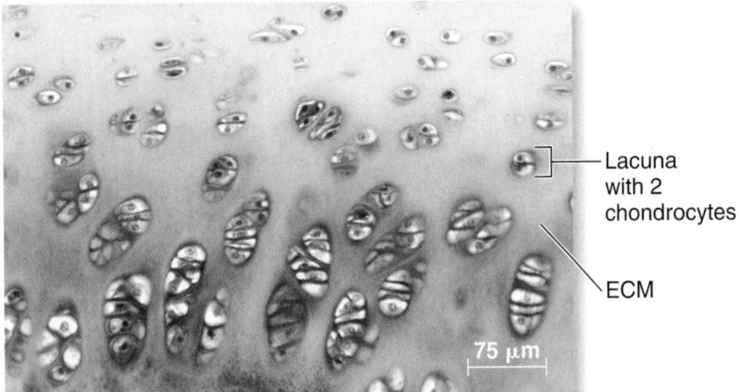

Figure 10.20 An example of connective tissue in animals that is rich in extracellular matrix. This micrograph of cartilage shows chondrocytes in the ECM. The chondrocytes, which are responsible for making the components of cartilage, are found in cavities called lacunae.

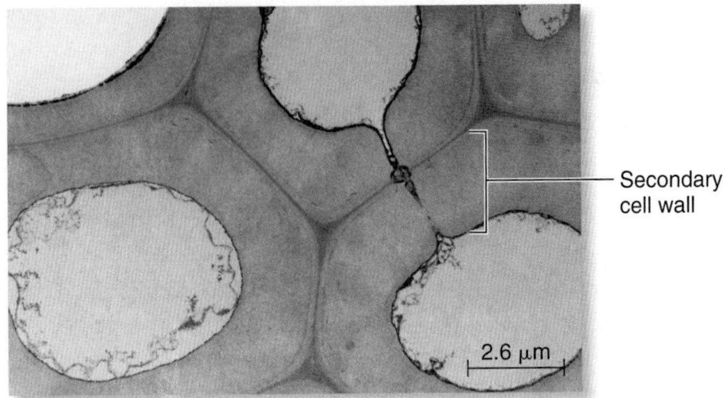

Figure 10.21 An example of ground tissue in plants. This scanning electron micrograph shows sclerenchyma cells from *Arabidopsis thaliana*. The cells themselves are dead; only their thick secondary cell walls remain. These cell walls provide structural support to the plant.

Concept Check: *Do you notice any parallels between ground tissue in plants and connective tissue in animals?*

Summary of Key Concepts

10.1 Extracellular Matrix and Cell Walls

- The extracellular matrix (ECM) is a network of material that forms a complex meshwork outside of animal cells; the cell wall is a similar component of plant cells.
- Proteins and polysaccharides are the major constituents of the ECM in animals. These materials are involved in strength, structural support, organization, and cell signaling (Figure 10.1).
- Adhesive proteins, such as fibronectin and laminin, help adhere cells to the ECM. Structural proteins include collagen fibers, which provide tensile strength, and elastic fibers, which allow regions of the body to stretch (Table 10.1, Figures 10.2, 10.3).
- Animals make many different types of collagen fibers, and differential gene regulation controls the place in the body where they are made (Table 10.2).
- Glycosaminoglycans (GAGs) are polysaccharides of repeating disaccharide units that give a gel-like character to the ECM of animals. Proteoglycans consist of a long, linear core protein with attached GAGs (Figure 10.4).
- Plant cells are surrounded by a cell wall composed of primary and secondary cell walls. The primary cell wall is composed largely of cellulose. The secondary cell wall is made after the primary cell wall and is often thick and rigid (Figures 10.5, 10.6).

10.2 Cell Junctions

- The three common types of cell junctions found in animals are anchoring, tight, and gap junctions (Table 10.3).
- Key components of anchoring junctions are cell adhesion molecules (CAMs), which bind cells to each other or to the ECM. The four types of anchoring junctions are adherens junctions, desmosomes, hemidesmosomes, and focal adhesions (Figure 10.7).

- Cadherins and integrins are two types of CAMs. Cadherins link cells to each other, whereas integrins link cells to the ECM. In the cytosol, CAMs bind to actin or intermediate filaments (Figure 10.8).
- Tight junctions between cells, composed of occludin and claudin, prevent the leakage of materials across a layer of cells (Figures 10.9, 10.10).
- Gap junctions consist of channels composed of connexons, which permit the direct passage of ions and small molecules between adjacent cells (Figure 10.11).
- Experiments of Loewenstein and colleagues showed that gap junctions permit the passage of substances with a molecular mass of less than about 1,000 Da (Figure 10.12).
- Plant junctions include middle lamella and plasmodesmata. The cell walls of adjacent plant cells are cemented together via middle lamellae, which are rich in pectins—negatively charged polysaccharides (Figure 10.13).
- The plasma membranes and endoplasmic reticula of adjacent plant cells are connected via plasmodesmata that allow the passage of ions and molecules between the cytosol of adjacent cells (Figure 10.14).

10.3 Tissues

- Cells are organized into tissues, and tissues are organized into organs. A tissue is a group of cells that have a similar structure and function, and an organ is composed of two or more tissues that carry out a particular function or set of functions.
- Six processes—cell division, cell growth, differentiation, migration, apoptosis, and the formation of cell connections—produce tissues and organs.
- The four general kinds of tissues found in animals are epithelial, connective, nervous, and muscle tissues (Figure 10.15).
- The three general kinds of tissues found in plants are dermal, ground, and vascular tissues (Figure 10.16).
- Epithelial and dermal tissues form layers of cells that are highly interconnected. These layers can be one cell thick or several cells thick, and they serve as protective coverings for various parts of animal and plant bodies (Figures 10.17, 10.18, 10.19).
- Connective and ground tissues provide structural support in animals and plants (Figures 10.20, 10.21).

Assess and Discuss

Test Yourself

1. The function of the extracellular matrix (ECM) in animals is
   a. to provide strength.
   b. to provide structural support.
   c. to organize cells and other body parts.
   d. cell signaling.
   e. all of the above.
2. The protein found in the ECM of animals that provides strength and resistance to tearing when stretched is
   a. elastin.
   b. cellulose.
   c. collagen.
   d. laminin.
   e. fibronectin.
3. The polysaccharide that forms the hard outer covering of many invertebrates is
   a. collagen.
   b. chitin.
   c. chondroitin sulfate.
   d. pectin.
   e. cellulose.
4. The extension sequence found in procollagen polypeptides
   a. causes procollagen to be synthesized into the ER lumen.
   b. causes procollagen to form a triple helix.
   c. prevents procollagen from forming large collagen fibers.
   d. causes procollagen to be secreted from the cell.
   e. Both b and c are correct.
5. The dilated state of plasmodesmata allows the passage of
   a. water.
   b. ions.
   c. small molecules.
   d. macromolecules and viruses.
   e. all of the above.
6. The gap junctions of animal cells differ from the plasmodesmata of plant cells in that
   a. gap junctions serve as communicating junctions and plasmodesmata serve as anchoring junctions.
   b. gap junctions prevent extracellular material from moving between adjacent cells but plasmodesmata do not.
   c. gap junctions allow for direct exchange of cellular material between cells but plasmodesmata cannot allow the same type of exchange.
   d. gap junctions are formed by specialized proteins that form channels through the membranes of adjacent cells but plasmodesmata are not formed by specialized proteins.
   e. All of the above are correct.
7. Which of the following is involved in the process of tissue and organ formation in multicellular organisms?
   a. cell division
   b. cell growth
   c. cell differentiation
   d. cell connections
   e. all of the above
8. The tissue type common to animals that functions in the conduction of electrical signals is
   a. epithelial.
   b. dermal.
   c. muscle.
   d. nervous.
   e. ground.
9. A type of tissue that is rich in ECM or has cells with a thick cell wall would be
   a. dermal tissue in plants.
   b. ground tissue in plants.
   c. nervous tissue in animals.
   d. connective tissue in animals.
   e. both b and d.

10. Which of the following is *not* a correct statement when comparing plant tissues and animal tissues?
    a. Nervous tissue of animals plays the same role as vascular tissue in plants.
    b. The dermal tissue of plants is similar to epithelial tissue of animals in that both provide a covering for the organism.
    c. The epithelial tissue of animals and the dermal tissue of plants have special characteristics that limit the movement of material between cell layers.
    d. The ground tissue of plants and the connective tissue of animals provide structural support for the organism.
    e. The statements in b, c, and d are correct comparisons between animal and plant tissues.

Conceptual Questions

1. What are key differences between the primary cell wall and the secondary cell wall of plant cells?
2. What are similarities and differences in the structures and functions of cadherins and integrins found in animal cells?
3. A principle of biology is that *organisms grow and develop.* Discuss how cell junctions play a key role in this process.

Collaborative Questions

1. Discuss the similarities and differences between the ECM of animals and the cell walls of plants.
2. Cell junctions in animals are important in preventing cancer cells from metastasizing—moving to other parts of the body. Certain drugs bind to CAMs and influence their structure and function. Some of these drugs may help to prevent the spread of cancer cells. What would you hypothesize to be the mechanism by which such drugs work? What might be some harmful side effects?

Online Resource

www.brookerbiology.com

Stay a step ahead in your studies with animations that bring concepts to life and practice tests to assess your understanding. Your instructor may also recommend the interactive eBook, individualized learning tools, and more.

UNIT III

GENETICS

Genetics is the branch of biology that deals with **inheritance**—the transmission of characteristics from parents to offspring. We begin this unit by examining the structure of the genetic material, namely DNA, at the molecular and cellular levels. We will explore the structure and replication of DNA and how the DNA is packaged into chromosomes (Chapter 11). We then consider how segments of DNA are organized into units called genes and explore how genes are used to make products such as RNA and proteins (Chapters 12 and 13). The expression of genes is largely responsible for the characteristics of living organisms. We will also examine how mutations can alter the properties of genes and even lead to diseases such as cancer (Chapter 14).

In Chapter 15, we turn our attention to the mechanisms of how genes are transmitted from parent to offspring. This topic begins with a discussion of how chromosomes are sorted and transmitted during cell division. Chapters 16 and 17 explore the relationships between the transmission of genes and the outcome of an offspring's traits. We will look at genetic patterns called Mendelian inheritance, named after the 19th-century biologist who discovered them, as well as more complex patterns that could not have been predicted from Mendel's work.

Chapters 11 through 17 focus on the fundamental properties of the genetic material and heredity. The remaining chapters explore additional topics that are of interest to biologists. In Chapter 18, we will examine some of the unique genetic properties of bacteria and viruses. Chapter 19 considers how genes play a central role in the development of animals and plants from a fertilized egg to an adult. We end this unit by exploring genetic technologies that are used by researchers, clinicians, and biotechnologists to unlock the mysteries of genes and provide tools and applications that benefit humans (Chapters 20 and 21).

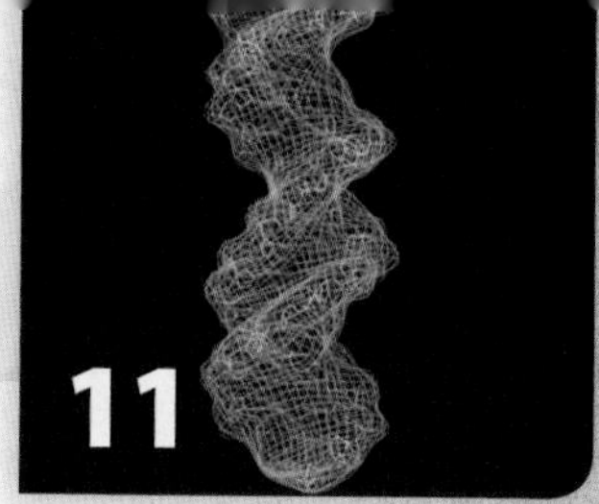

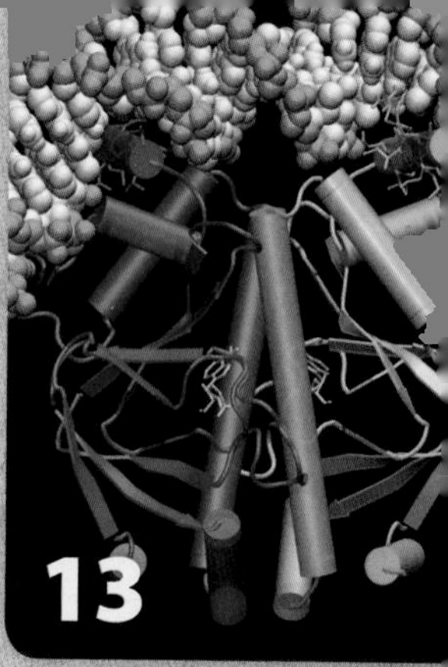

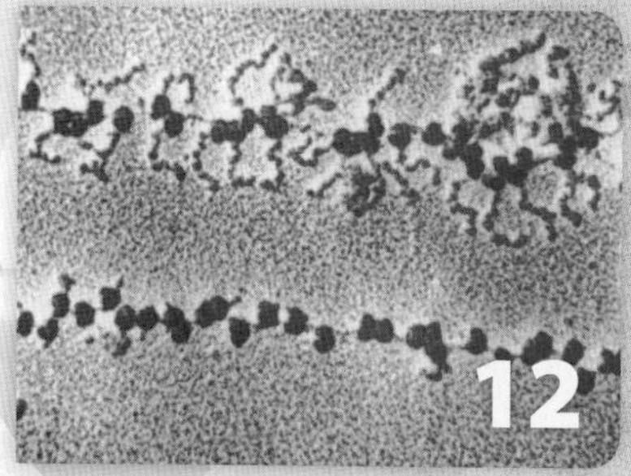

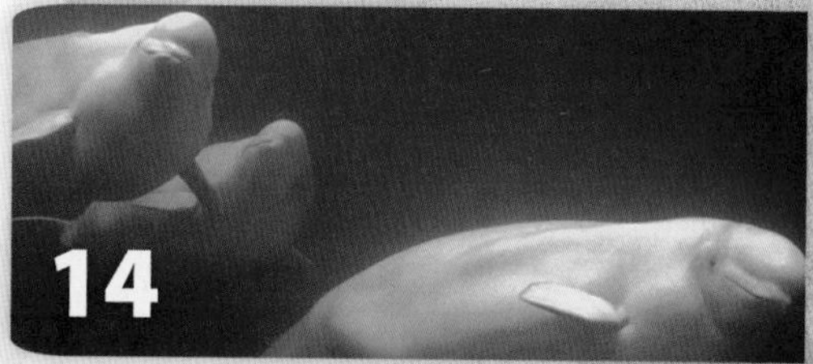

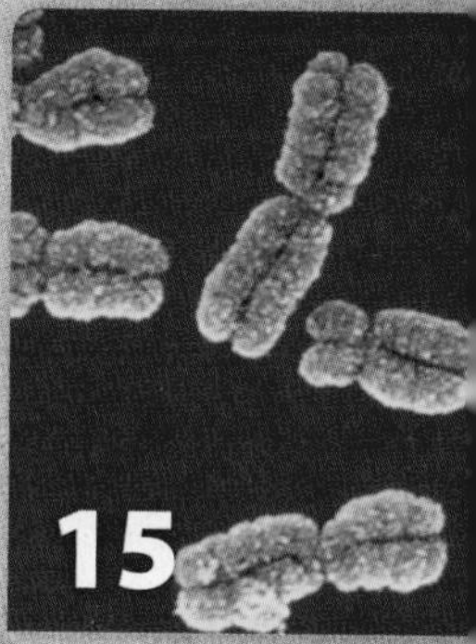

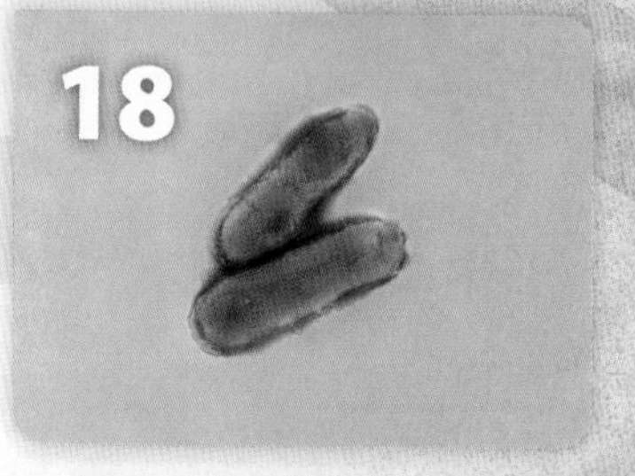

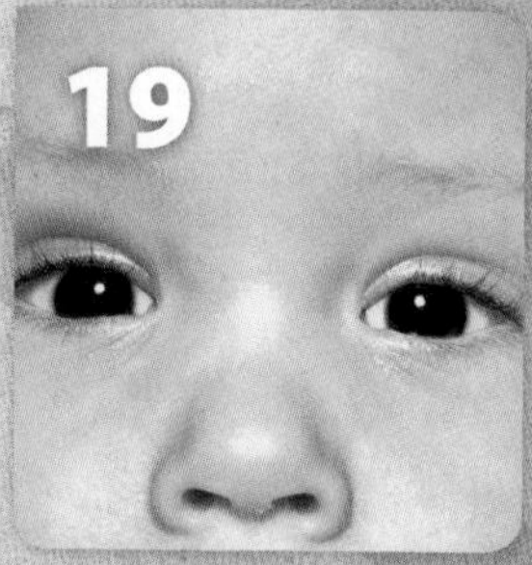

11 12 13 14 15 16 17 18 19 20 21

The following biology principles will be emphasized in this unit:

- ***The genetic material provides a blueprint for reproduction:*** *Throughout this unit, we will see how the genetic material carries the information to sustain life.*
- ***Structure determines function:*** *In Chapters 11 through 15, we will examine how the structures of DNA, RNA, genes, and chromosomes underlie their functions.*
- ***Living organisms interact with their environment:*** *In Chapters 16 and 17, we will explore the interactions between an organism's genes and its environment.*
- ***Living organisms grow and develop:*** *In Chapter 19, we will consider how a genetic program is involved in the developmental stages of animals and plants.*
- ***Biology affects our society:*** *In Chapters 20 and 21, we will examine genetic technologies that have many applications in our society.*
- ***Biology is an experimental science:*** *Every chapter in this unit has a Feature Investigation that describes a pivotal experiment that provided insights into our understanding of genetics.*

Nucleic Acid Structure, DNA Replication, and Chromosome Structure

11

Chapter Outline

11.1 Biochemical Identification of the Genetic Material

11.2 Nucleic Acid Structure

11.3 An Overview of DNA Replication

11.4 Molecular Mechanism of DNA Replication

11.5 Molecular Structure of Eukaryotic Chromosomes

Summary of Key Concepts

Assess and Discuss

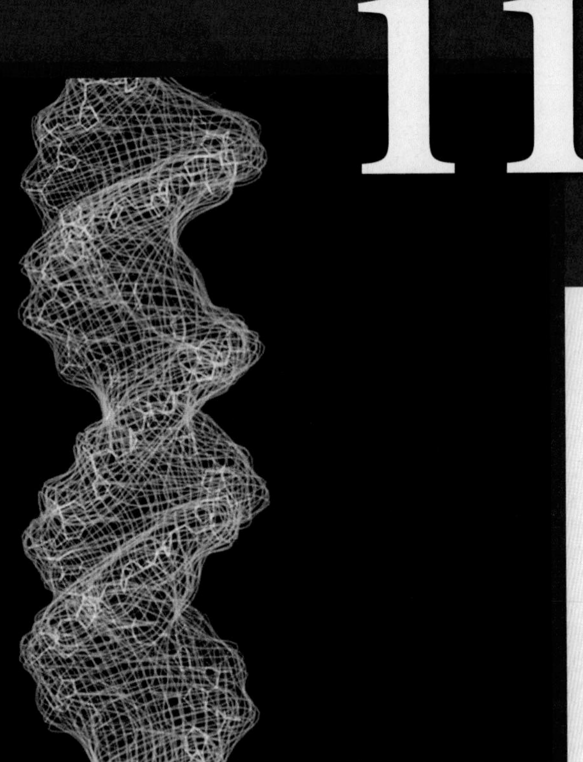

A molecular model for the structure of a DNA double helix.

On October 17, 2001, Mario K. was set free after serving 16 years in prison. He had been convicted of a sexual assault and murder that occurred in 1985. The charges were dropped because investigators discovered that another person, Edwin M., had actually committed the crime. How was Edwin M. identified as the real murderer? In 2001, he committed another crime, and his DNA was entered into a computer database. Edwin's DNA matched the DNA that had been collected from the victim in 1985, and other evidence was then gathered indicating that Edwin M. was the true murderer. Like Mario K., over 200 other inmates have been exonerated when DNA tests have shown that a different person was responsible for the crime.

Deoxyribonucleic acid, or **DNA**, is the genetic material that provides the blueprint to produce an individual's traits. Each person's DNA is distinct and unique. Even identical twins show minor differences in their DNA sequences. We begin our survey of genetics by examining DNA at the molecular level. Once we understand how DNA works at this level, it becomes easier to see how the function of DNA controls the properties of cells and ultimately the characteristics of unicellular and multicellular organisms. The past several decades have seen exciting advances in techniques and approaches to investigating and even to altering the genetic material. Not only have these advances greatly expanded our understanding of molecular genetics, such technologies are also widely used in related disciplines such as biochemistry, cell biology, and microbiology. Likewise, genetic techniques have many important applications in biotechnology and are used in criminal justice, including forensics, to provide evidence of guilt or innocence.

To a large extent, our understanding of genetics comes from our knowledge of the molecular structure of DNA. In this chapter, we begin by considering some classic experiments that provided evidence that DNA is the genetic material. We will then survey the molecular features of DNA, which allows us to appreciate how DNA can store information and be accurately copied. Though this chapter is largely concerned with DNA, we will also consider the components of ribonucleic acid (RNA), which show striking similarities to DNA. Lastly, we will examine the molecular composition of chromosomes where the DNA is found.

11.1 Biochemical Identification of the Genetic Material

Learning Outcomes:

1. List the four key criteria that the genetic material must fulfill.
2. Analyze the results of the experiments that identified DNA as the genetic material.

DNA carries the genetic instructions for the characteristics of living organisms. In the case of multicellular organisms such as plants and animals, the information stored in the genetic material enables a fertilized egg to develop into an embryo and eventually into an adult organism. In addition, the genetic material allows organisms to survive in their native environments. For example, an individual's DNA provides the blueprint to produce enzymes that are needed to metabolize nutrients in food. To fulfill its role, the genetic material must meet the following key criteria:

1. **Information:** The genetic material must contain the information necessary to construct an entire organism.
2. **Replication:** The genetic material must be accurately copied.

3. **Transmission:** After it is replicated, the genetic material can be passed from parent to offspring. It also must be passed from cell to cell during the process of cell division.
4. **Variation:** Differences in the genetic material must account for the known variation within each species and among different species.

How was the genetic material discovered? The quest to identify the genetic material began in the late 1800s, when a few scientists postulated that living organisms possess a blueprint that has a biochemical basis. In 1883, German biologist August Weismann and his Swiss colleague Karl Nägeli championed the idea that a chemical substance exists within living cells that is responsible for the transmission of traits from parents to offspring. During the next 30 years, experimentation along these lines centered on the behavior of **chromosomes**, the cellular structures that we now know contain the genetic material. Taken literally, chromosome is from the Greek words *chromo* and *soma*, meaning colored body, which refers to the observation of early microscopists that chromosomes are easily stained by colored dyes. By studying the transmission patterns of chromosomes from cell to cell and from parent to offspring, researchers were convinced that chromosomes carry the determinants that control the outcome of traits.

Ironically, the study of chromosomes initially misled researchers regarding the biochemical identity of the genetic material. Chromosomes contain two classes of macromolecules: proteins and DNA. Scientists of this era viewed proteins as being more biochemically complex because they are made from 20 different amino acids. Furthermore, biochemists already knew that proteins perform an amazingly wide range of functions, and complexity seemed an important prerequisite for the blueprint of an organism. By comparison, DNA seemed less complex, because it contains only four types of repeating units, called nucleotides, which will be described later in this chapter. In addition, the functional role of DNA in the nucleus had not been extensively investigated prior to the 1920s. Therefore, from the 1920s to the 1940s, most scientists were expecting research studies to reveal that proteins are the genetic material. Contrary to this expectation, however, the experiments described in this section were pivotal in showing that DNA carries out this critical role.

Griffith's Bacterial Transformation Experiments Indicated the Existence of a Genetic Material

Studies in microbiology were important in developing an experimental strategy for identifying the genetic material. In the late 1920s, an English microbiologist, Frederick Griffith, studied a type of bacterium known then as pneumococci and now classified as *Streptococcus pneumoniae*. Some strains of *S. pneumoniae* secrete a polysaccharide capsule, but other strains do not. When streaked on petri plates containing solid growth media, capsule-secreting strains have a smooth colony morphology. Those strains unable to secrete a capsule have a colony morphology that looks rough. In mammals, smooth strains of *S. pneumoniae* may cause pneumonia and other symptoms. In mice, such infections are usually fatal.

As shown in Figure 11.1, Griffith injected live and/or heat-killed bacteria into mice and then observed whether or not the bacteria caused them to die. He investigated the effects of two strains of *S. pneumoniae*:

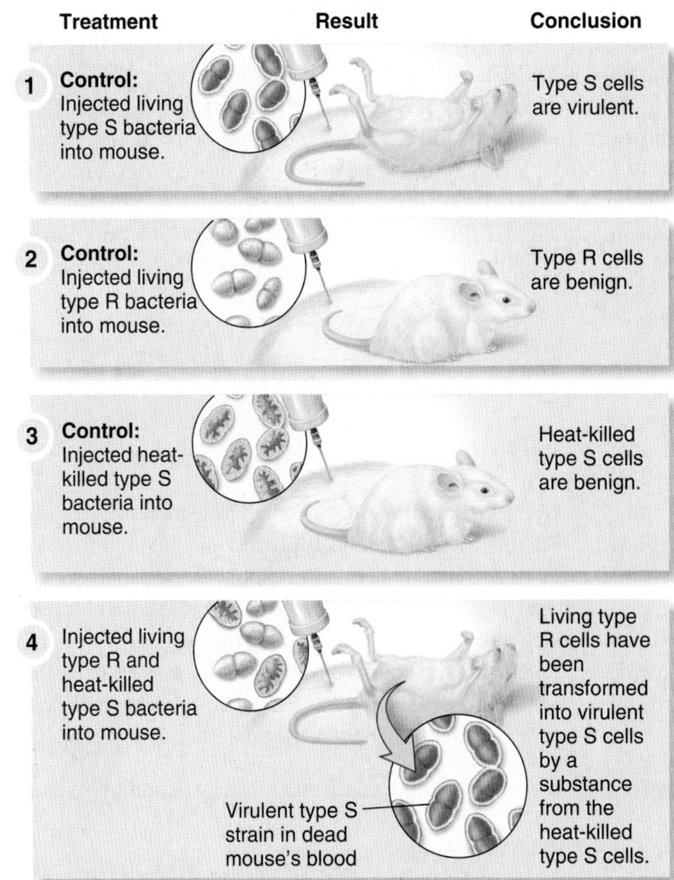

Treatment
Result
Conclusion
1 Control: Injected living type S bacteria into mouse.
Type S cells are virulent.
2 Control: Injected living type R bacteria into mouse.
Type R cells are benign.
3 Control: Injected heat-killed type S bacteria into mouse.
Heat-killed type S cells are benign.
4 Injected living type R and heat-killed type S bacteria into mouse.
Virulent type S strain in dead mouse's blood
Living type R cells have been transformed into virulent type S cells by a substance from the heat-killed type S cells.

Figure 11.1 Griffith's experiments that showed the transformation of bacteria by a "transformation principle." Note: To determine if a mouse's blood contained live bacteria, a sample of blood was applied to solid growth media to determine if smooth or rough bacterial colonies would form.

BioConnections: Look ahead to Figure 18.18. How does bacterial transformation play a role in the transfer of genes, such as antibiotic resistance genes, from one bacterial species to another?

type S for smooth and type R for rough. When injected into a live mouse, the type S strain killed the mouse (Figure 11.1, step 1). The capsule made by type S strains prevents the mouse's immune system from killing the bacterial cells. Following the death of the mouse, many type S bacteria were found in the mouse's blood. By comparison, when type R bacteria were injected into a mouse, the mouse survived, and after several days, living bacteria were not found in the live mouse's blood (Figure 11.1, step 2). In a follow-up to these results, Griffith also heat-killed the smooth bacteria and then injected them into a mouse. As expected, the mouse survived (Figure 11.1, step 3).

A surprising result occurred when Griffith mixed live type R bacteria with heat-killed type S bacteria and then injected them into a mouse—the mouse died (Figure 11.1, step 4). The blood from the dead mouse contained living type S bacteria! How did Griffith explain these results? He postulated that a substance from dead type S bacteria transformed the type R bacteria into type S bacteria. Griffith called this process **transformation**, and he termed the unidentified material responsible for this phenomenon the "transformation principle."

Let's consider what these observations mean with regard to the four criteria of the genetic material: information, replication, transmission, and variation. According to Griffith's results, the transformed bacteria had acquired the information (criterion 1) to make a capsule from the heat-killed cells. For the transformed bacteria to proliferate and thereby kill the mouse, the substance conferring the ability to make a capsule must be replicated (criterion 2) and then transmitted (criterion 3) from mother to daughter cells during cell division. Finally, Griffith already knew that variation (criterion 4) existed in the ability of his strains to produce a capsule (S strain) or not produce a capsule (R strain). Taken together, these observations are consistent with the idea that the formation of a capsule is governed by genetic material. The experiment of Figure 11.1, step 4, was consistent with the idea that some genetic material from the heat-killed type S bacteria had been transferred to the living type R bacteria and provided those bacteria with a new trait. At the time of his studies, however, Griffith could not determine the biochemical composition of the transforming substance.

FEATURE INVESTIGATION

Avery, MacLeod, and McCarty Used Purification Methods to Reveal That DNA Is the Genetic Material

Exciting discoveries sometimes occur when researchers recognize that another scientist's experimental approach may be modified and then used to dig deeper into a scientific question. In the 1940s, American physician Oswald Avery and American biologists Colin MacLeod and Maclyn McCarty were also interested in the process of bacterial transformation. During the course of their studies, they realized that Griffith's observations could be used as part of an experimental strategy to biochemically identify the genetic material. They asked the question, "What substance is being transferred from the dead type S bacteria to the live type R bacteria?"

To answer this question, Avery, MacLeod, and McCarty needed to purify the general categories of substances found in living cells. They used established biochemical procedures to purify classes of macromolecules, such as proteins, DNA, and RNA, from the type S streptococcal strain. Initially, they discovered that only the purified DNA could convert type R bacteria into type S. To further verify that DNA is the genetic material, they performed the investigation outlined in **Figure 11.2**. They purified DNA from the type S bacteria and mixed it with type R bacteria. After allowing time for DNA uptake,

Figure 11.2 **The Avery, MacLeod, and McCarty experiments that identified DNA as Griffith's transformation principle—the genetic material.**

HYPOTHESIS A purified macromolecule from type S bacteria, which functions as the genetic material, will be able to convert type R bacteria into type S.

KEY MATERIALS Type R and type S strains of *Streptococcus pneumoniae.*

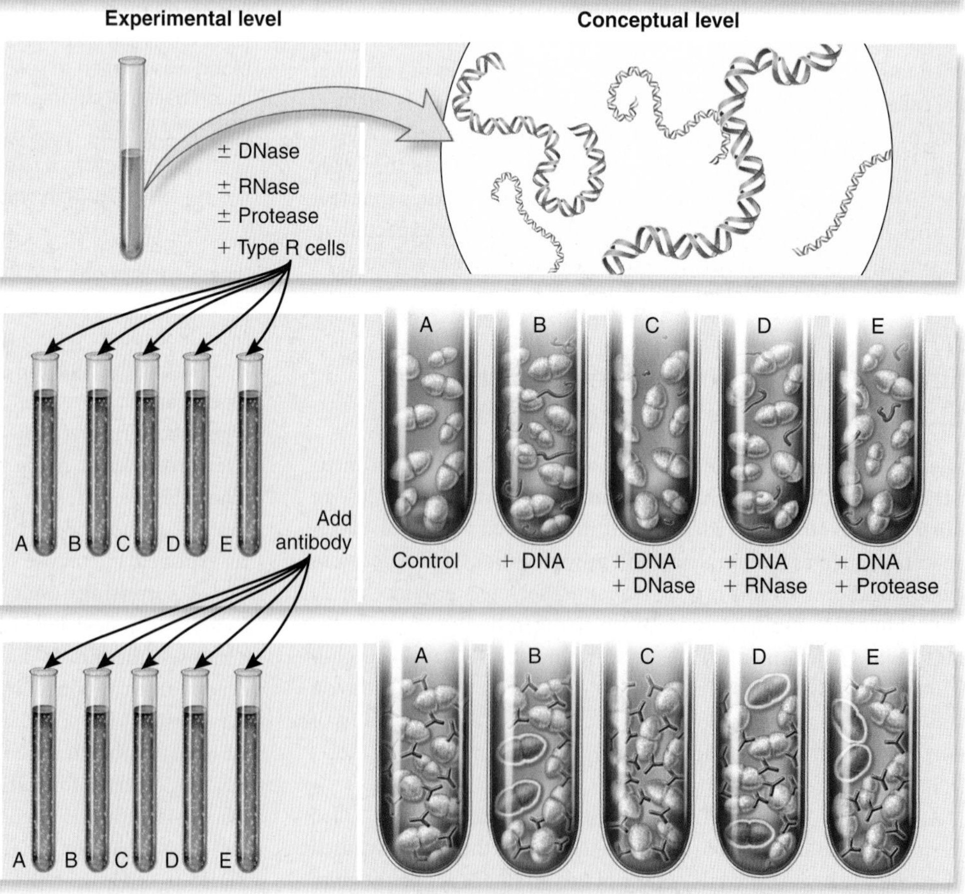

Experimental level
Conceptual level
1 Purify DNA from a type S strain. This involves breaking open cells and separating the DNA away from other components by centrifugation.
± DNase
± RNase
± Protease
+ Type R cells
2 Mix the DNA extract with type R bacteria. Allow time for the DNA to be taken up by the type R cells, converting a few of them to type S. Also, carry out the same steps but add the enzymes DNase, RNase, or protease to the DNA extract, which digest DNA, RNA, and proteins, respectively. As a control, don't add any DNA extract to some type R cells.
A B C D E
Add antibody
A B C D E
Control
+ DNA
+ DNA + DNase
+ DNA + RNase
+ DNA + Protease
3 Add an antibody, a protein made by the immune system of mammals, that specifically recognizes type R cells that haven't been transformed. The binding of the antibody causes the type R cells to aggregate.
A B C D E
A B C D E

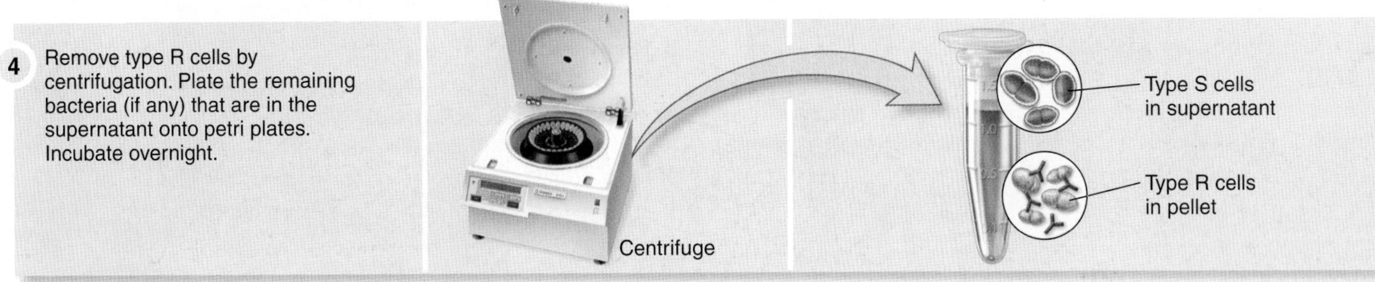

4
Remove type R cells by centrifugation. Plate the remaining bacteria (if any) that are in the supernatant onto petri plates. Incubate overnight.
Centrifuge
Type S cells in supernatant
Type R cells in pellet

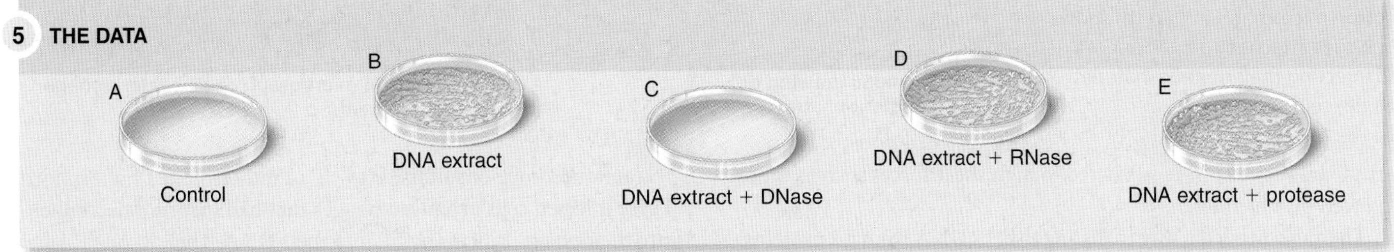

6 **CONCLUSION** DNA is responsible for transforming type R cells into type S cells.

7 **SOURCE** Avery, O.T., MacLeod, C.M., and McCarty, M. 1944. Studies on the Chemical Nature of the Substance Inducing Transformation of Pneumococcal Types. *Journal of Experimental Medicine* 79:137–158.

they added an antibody that aggregated any nontransformed type R bacteria, which were then removed by centrifugation. The remaining bacteria were incubated overnight on petri plates.

As a control, if no DNA extract was added, no type S bacterial colonies were observed on the petri plates (see plate A in step 5). When they mixed their S strain DNA extract with type R bacteria, some of the bacteria were converted to type S bacteria (see plate B in step 5 of Figure 11.2). This result was consistent with the idea that DNA is the genetic material. Even so, a careful biochemist could argue that the DNA extract might not have been 100% pure. For this reason, the researchers realized that a small amount of contaminating material in the DNA extract could actually be the genetic material. The most likely contaminating substances in this case would be RNA or protein. To address this possibility, Avery, MacLeod, and McCarty treated the DNA extract with enzymes that digest DNA (called **DNase**), RNA **(RNase)**, or protein **(protease)** (see step 2). When the DNA extracts were treated with RNase or protease, the type R bacteria were still converted into type S bacteria, suggesting that contaminating RNA or protein in the extract was not acting as the genetic material (see step 5, plates D and E). Moreover, when the extract was treated with DNase, it lost the ability to convert type R bacteria into type S bacteria (see plate C). Taken together, these results were consistent with the idea that DNA is the genetic material.

Experimental Questions

1. Avery, MacLeod, and McCarty worked with two strains of *Streptococcus pneumoniae* to determine the biochemical identity of the genetic material. Explain the characteristics of the *S. pneumoniae* strains that made them particularly well suited for such an experiment.
2. What is a DNA extract?
3. In the experiment of Avery, MacLeod, and McCarty, what was the purpose of using the protease, RNase, and DNase if only the DNA extract caused transformation?

Hershey and Chase Determined That DNA Is the Genetic Material of T2 Bacteriophage

In a second avenue of research conducted in 1952, the efforts of Alfred Hershey and Martha Chase centered on the study of a virus named T2. This virus infects bacterial cells, in this case *Escherichia coli*, and is therefore known as a **bacteriophage**, or simply a **phage**. A T2 phage has an outer covering called the phage coat that contains a head (capsid), sheath, tail fibers, and base plate (**Figure 11.3**). We now know the phage coat is composed entirely of protein. DNA is found inside the head of T2. From a biochemical perspective, T2 is very simple because it is composed of only proteins and DNA.

The genetic material of T2 provides a blueprint to make new phages. To replicate, all viruses must introduce their genetic material into the cytoplasm of a living host cell. In the case of T2, this involves attaching its tail fibers to the bacterial cell wall and injecting its genetic material into the cytoplasm (**Figure 11.3b**). However, at the time of Hershey and Chase's work, it was not known if the phage was injecting DNA or protein.

To determine whether DNA or protein is the genetic material of T2, Hershey and Chase devised a method for separating the phage coat, which is attached to the outside of the bacterium, from the genetic material, which is injected into the cytoplasm. They reasoned that the attachment of T2 on the surface of the bacterium could be

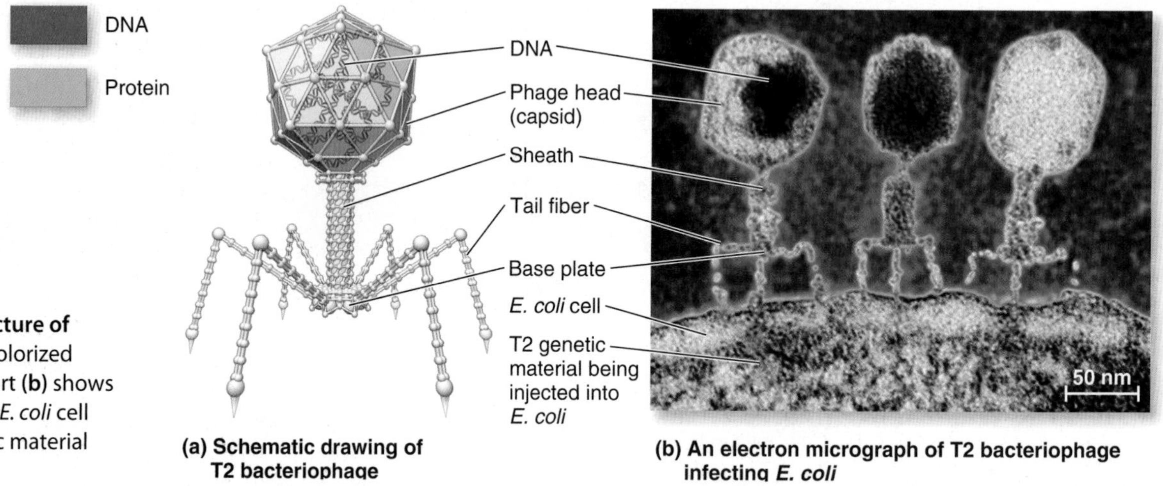

Figure 11.3 The structure of T2 bacteriophage. The colorized electron micrograph in part **(b)** shows T2 phages attached to an *E. coli* cell and injecting their genetic material into the cell.

(a) Schematic drawing of T2 bacteriophage

(b) An electron micrograph of T2 bacteriophage infecting *E. coli*

disrupted if the cells were subjected to high shear forces such as those produced by a blender.

They also needed a way to distinguish T2 DNA from T2 proteins. Hershey and Chase used radioisotopes (described in Chapter 2) as a way to label these molecules. Sulfur atoms are found in phage proteins but not in DNA, whereas phosphorus atoms are found in DNA but not in phage proteins. They exposed T2-infected bacterial cells to ^{35}S (a radioisotope of sulfur) or to ^{32}P (a radioisotope of phosphorus). These infected cells produced phages that had incorporated either ^{35}S into their phage proteins or ^{32}P into their DNA. The ^{35}S- or ^{32}P-labeled phages were then used in the experiment shown in Figure 11.4.

Let's now consider the steps in this experiment. In separate tubes, they took samples of T2 phage, one in which the proteins were labeled with ^{35}S and the other in which the DNA was labeled with ^{32}P, and mixed

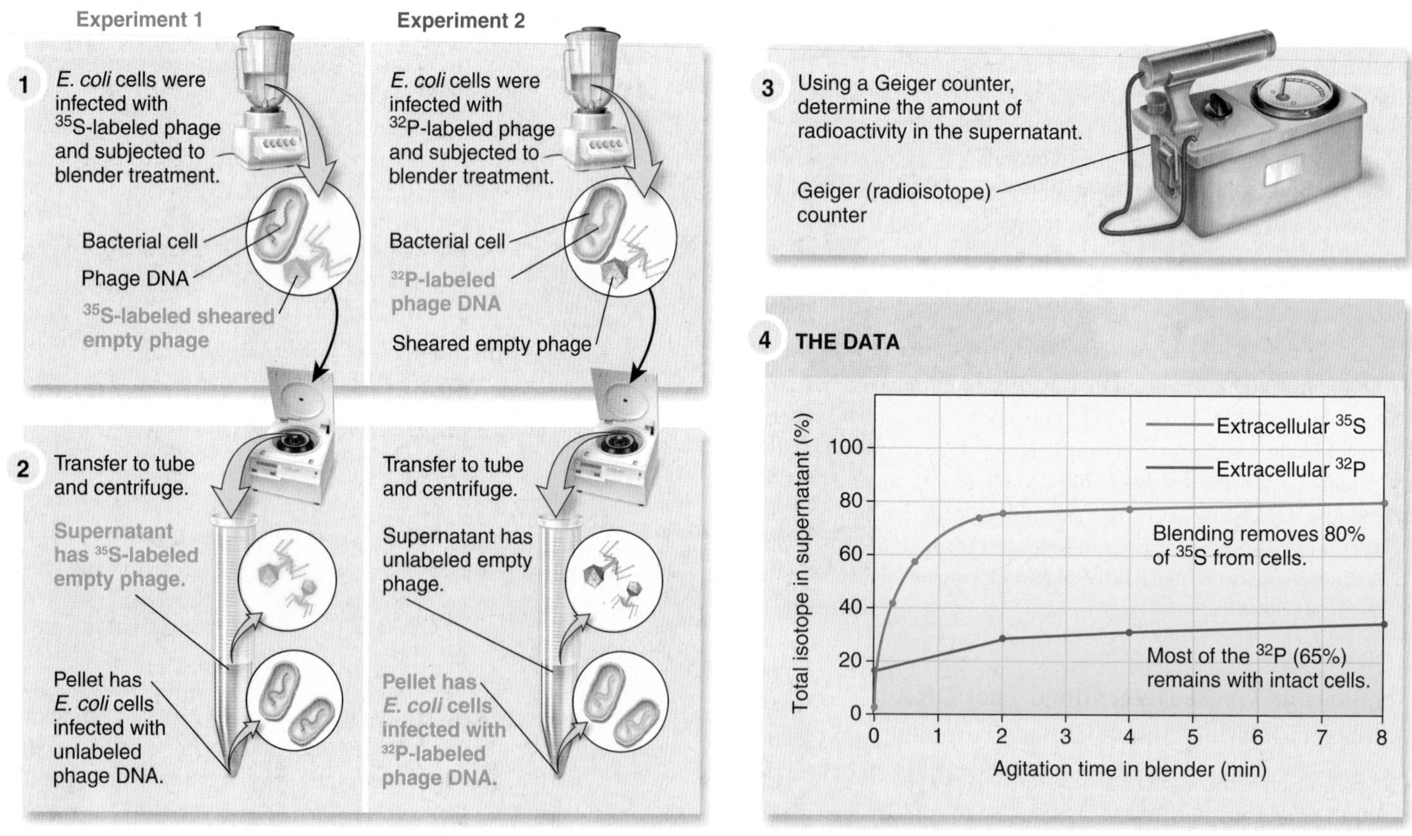
Experiment 1
Experiment 2
1 E. coli cells were infected with 35S-labeled phage and subjected to blender treatment.
E. coli cells were infected with 32P-labeled phage and subjected to blender treatment.
Bacterial cell
Phage DNA
35S-labeled sheared empty phage
Bacterial cell
32P-labeled phage DNA
Sheared empty phage
2 Transfer to tube and centrifuge.
Supernatant has 35S-labeled empty phage.
Pellet has E. coli cells infected with unlabeled phage DNA.
Transfer to tube and centrifuge.
Supernatant has unlabeled empty phage.
Pellet has E. coli cells infected with 32P-labeled phage DNA.
3 Using a Geiger counter, determine the amount of radioactivity in the supernatant.
Geiger (radioisotope) counter
4 THE DATA
Total isotope in supernatant (%)
100
80
60
40
20
0
Extracellular 35S
Extracellular 32P
Blending removes 80% of 35S from cells.
Most of the 32P (65%) remains with intact cells.
0 1 2 3 4 5 6 7 8
Agitation time in blender (min)

Figure 11.4 Hershey and Chase experiment showing that the genetic material of T2 phage is DNA. Green coloring indicates radiolabeling with either ^{35}S or ^{32}P. Note: The phages and bacteria are not drawn to scale; bacteria are much larger.

BIOLOGY PRINCIPLE Biology is an experimental science. This experiment, along with the previous one by Avery, MacLeod, and McCarty, was consistent with the idea that DNA is the genetic material.

Concept Check: *In these experiments, what was the purpose of using two different isotopes, ^{35}S and ^{32}P?*

them with *E. coli* cells for a short period of time. This allowed the phages enough time to inject their genetic material into the bacterial cells. The samples were then subjected to a shearing force, using a blender for up to 8 minutes. This treatment removed the phage coats from the surface of the bacterial cells without causing cell lysis. Each sample was then subjected to centrifugation at a speed that caused the heavier bacterial cells to form a pellet at the bottom of the tube, while the lighter phage coats remained in the supernatant, the liquid above the pellet. The amount of radioactivity in the supernatant (emitted from either ^{35}S or ^{32}P) was measured using an instrument called a Geiger counter.

As you can see in the data of Figure 11.4, most of the ^{35}S isotope (80%) was found in the supernatant. This represents the phage coats. In contrast, only about 35% of the ^{32}P was found in the supernatant following shearing. This indicates that most of the phage DNA (65%) was injected into the bacterial cells, which formed the pellet. This is the expected outcome if DNA is the genetic material of T2.

11.2 Nucleic Acid Structure

Learning Outcomes:

1. Outline the structural features of DNA at five levels of complexity.
2. Describe the structure of nucleotides, a DNA strand, and the DNA double helix.
3. Discuss and interpret the work of Franklin; Chargaff; and Watson and Crick.

A unifying principle in biology is that structure determines function. When biologists want to understand the function of a material at the molecular and cellular level, they focus some of their efforts on determining its biochemical structure. In this regard, an understanding of DNA's structure has proven to be particularly exciting because the structure makes it easier for us to understand how DNA can store information, how it is replicated and then transmitted from cell to cell, and how variation in its structure can occur.

DNA and its molecular cousin, RNA, are known as **nucleic acids**, polymers consisting of nucleotides, which are responsible for the storage, expression, and transmission of genetic information. This term is derived from the discovery of DNA by Swiss physician Friedrich Miescher in 1869. He identified a novel phosphorus-containing substance from the nuclei of white blood cells found in waste surgical bandages. He named this substance nuclein. As the structure of DNA and RNA became better understood, they were found to be acidic molecules, which means they release hydrogen ions (H^+) in solution and have a net negative charge at neutral pH. Thus, the name nucleic acid was coined.

DNA is a very large macromolecule composed of smaller building blocks. We can consider the structural features of DNA at different levels of complexity (Figure 11.5):

1. **Nucleotides** are the building blocks of DNA.
2. A **strand** of DNA is formed by the covalent linkage of nucleotides in a linear manner.
3. Two strands of DNA hydrogen-bond with each other to form a **double helix**. In a DNA double helix, two DNA strands are twisted together to form a structure that resembles a spiral staircase.
4. In living cells, DNA is associated with an array of different proteins to form chromosomes. The association of proteins with DNA organizes the long strands into a compact structure.

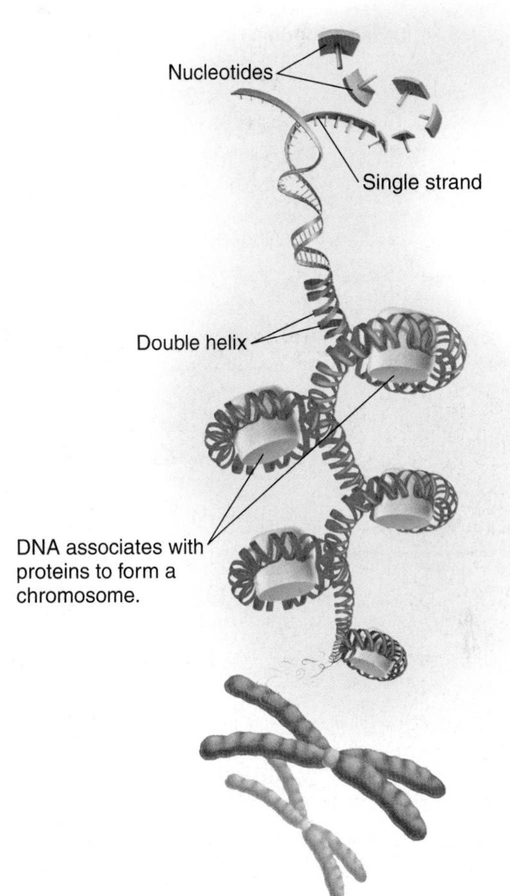
Nucleotides
Single strand
Double helix
DNA associates with proteins to form a chromosome.

Figure 11.5 **Levels of DNA structure to create a chromosome.**

5. A **genome** is the complete complement of an organism's genetic material. For example, the genome of most bacteria is a single circular chromosome, whereas eukaryotic cells have DNA in their nucleus, mitochondria, and chloroplasts.

The first three levels of complexity will be the focus of this section. Level 4 will be discussed in Section 11.5, and level 5 is examined in Chapter 21.

Nucleotides Contain a Phosphate, a Sugar, and a Base

A nucleotide has three components: a phosphate group, a pentose (five-carbon) sugar, and a nitrogenous base. The nucleotides in DNA and RNA contain different sugars. Deoxyribose is found in DNA, and ribose is found in RNA. The base and phosphate group are attached to the sugar molecule at different sites (Figure 11.6).

Five different bases are found in nucleotides, although any given nucleotide contains only one base. The five bases are subdivided into two categories, the **purines** and the **pyrimidines**, due to differences in their structures (see Figure 11.6). The purine bases, **adenine (A)** and **guanine (G)**, have a double-ring structure, whereas the pyrimidine bases, **thymine (T)**, **cytosine (C)**, and **uracil (U)**, have a single-ring structure. Adenine, guanine, and cytosine are found in both DNA and RNA. Thymine is found only in DNA, whereas uracil is found only in RNA.

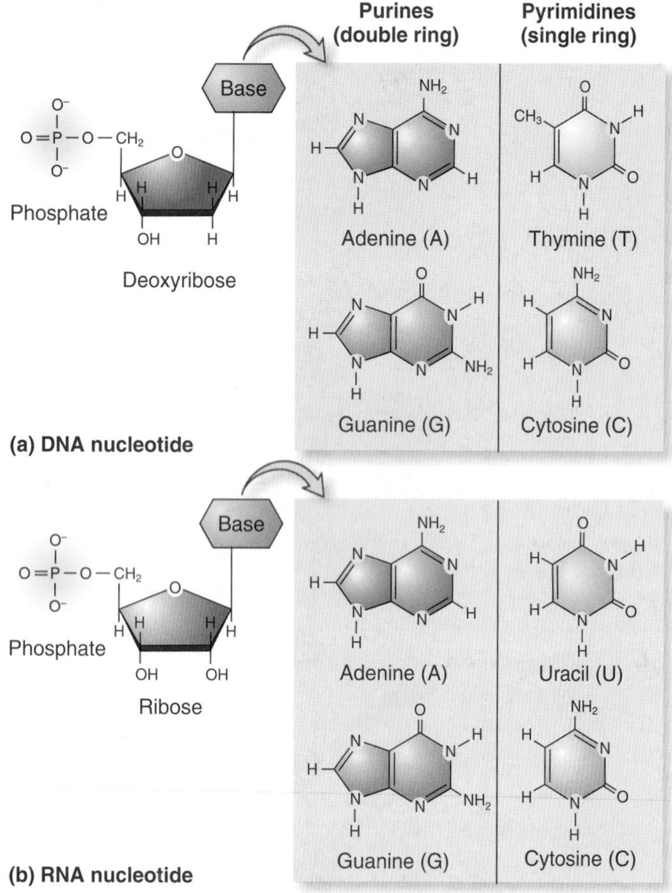

Figure 11.6 Nucleotides and their components. For simplicity, the carbon atoms in the ring structures are not shown.

Concept Check: Which pyrimidine(s) is/are found in both DNA and RNA?

A conventional numbering system describes the locations of carbon and nitrogen atoms in the sugars and bases (**Figure 11.7**). The prime symbol (′) is used to distinguish the numbering of carbons in the sugar. The atoms in the ring structures of the bases are not given the prime designation. The sugar carbons are designated 1′ (that is, "one prime"), 2′, 3′, 4′, and 5,′ with the carbon atoms numbered in a clockwise direction starting with the carbon atom to the right of the ring oxygen atom. The fifth carbon is outside the ring. A base is attached to the 1′ carbon atom, and a phosphate group is attached at the 5′ position. Compared with ribose (see Figure 11.6), deoxyribose lacks a single oxygen atom at the 2′ position; the prefix deoxy- (meaning without oxygen) refers to this missing atom.

A Strand Is a Linear Linkage of Nucleotides with Directionality

The next level of nucleotide structure is the formation of a strand of DNA or RNA in which nucleotides are covalently attached to each other in a linear fashion. **Figure 11.8** depicts a short strand of DNA with four nucleotides. The linkage is a phosphoester bond (a covalent bond between phosphorus and oxygen) involving a sugar molecule in one nucleotide and a phosphate group in the next nucleotide. Another

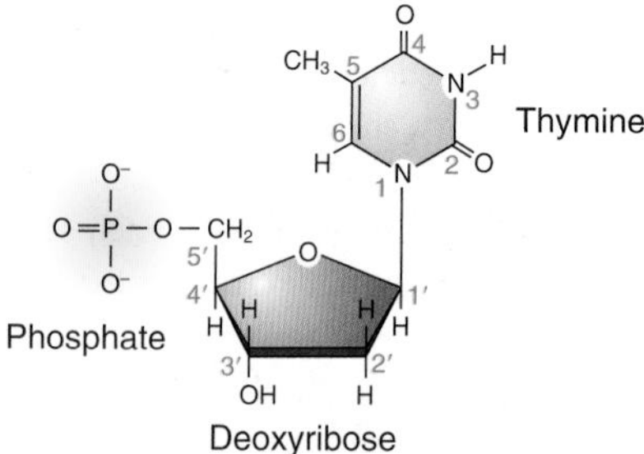

Figure 11.7 Conventional numbering in a DNA nucleotide. The carbons in the sugar are given a prime designation, whereas those in the base are not.

Concept Check: What is the numbering designation of the carbon atom to which the phosphate is attached?

way of viewing this linkage is to notice that a phosphate group connects two sugar molecules. From this perspective, the linkage in DNA and RNA strands is called a **phosphodiester linkage**, which has two phosphoester bonds. The phosphates and sugar molecules form the **backbone** of a DNA or RNA strand, and the bases project from the backbone. The backbone is negatively charged due to the negative charges of the phosphate groups.

An important structural feature of a nucleic acid strand is the orientation of the nucleotides. Each phosphate in a phosphodiester linkage is covalently bonded to the 5′ carbon in one nucleotide and to the 3′ carbon in the other. In a strand, all sugar molecules are oriented in the same direction. For example, in the strand shown in Figure 11.8, all of the 5′ carbons in every sugar molecule are above the 3′ carbons. A strand has a **directionality** based on the orientation of the sugar molecules within that strand. In Figure 11.8, the direction of the strand is said to be 5′ to 3′ when going from top to bottom. The 5′ end of a DNA strand has a phosphate group, whereas the 3′ end has an —OH group.

From the perspective of function, a key feature of DNA and RNA structure is that a strand contains a specific sequence of bases. In Figure 11.8, the sequence of bases is thymine—adenine—cytosine—guanine, or TACG. To indicate its directionality, the strand is abbreviated 5′–TACG–3′. Because the nucleotides within a strand are attached to each other by stable covalent bonds, the sequence of bases in a DNA strand remains the same over time, except in rare cases when mutations occur. The sequence of bases in DNA and RNA is the critical feature that allows them to store and transmit information.

A Few Key Experiments Paved the Way to Solving the Structure of DNA

What experimental approaches were used to analyze DNA structure? X-ray diffraction was a key experimental tool that led to the discovery of the DNA double helix. When a substance is exposed to X-rays, the atoms in the substance cause the X-rays to be scattered (**Figure 11.9**). If the substance has a repeating structure, the pattern of scattering, known as the diffraction pattern, is mathematically related to the structural arrangement of the atoms causing the scattering. The diffraction pattern is analyzed using mathematical theory to provide information regarding the three-dimensional structure

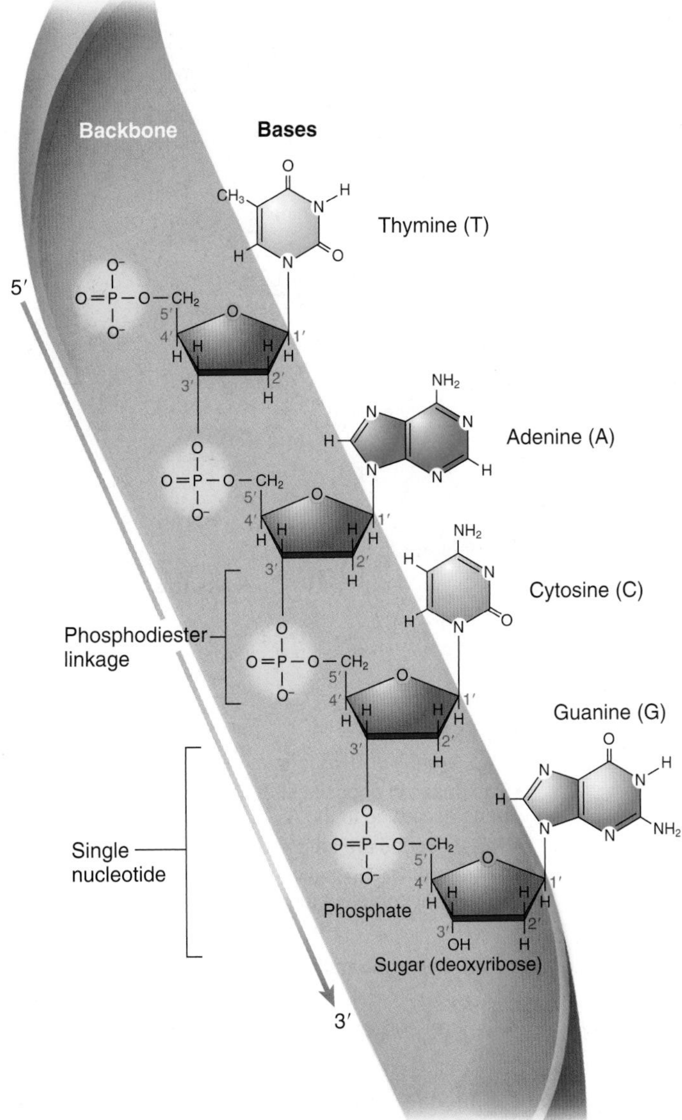

Figure 11.8 The structure of a DNA strand. Nucleotides are covalently bonded to each other in a linear manner. Notice the directionality of the strand and that it carries a particular sequence of bases. An RNA strand has a very similar structure, except the sugar is ribose rather than deoxyribose, and uracil is substituted for thymine.

BIOLOGY PRINCIPLE The genetic material provides a blueprint for reproduction. The covalent linkage of a sequence of bases allows DNA to store information.

Concept Check: What is the difference between a phosphoester bond and a phosphodiester linkage?

of the molecule. British biophysicist Rosalind Franklin, working in the 1950s in the same laboratory as Maurice Wilkins, was a gifted experimentalist who made marked advances in X-ray diffraction techniques involving DNA. The diffraction pattern of DNA fibers produced by Franklin suggested a helical structure with a diameter that is relatively uniform and too wide to be a single-stranded helix. In addition, the pattern provided information regarding the number of nucleotides per turn and was consistent with a 2-nm (nanometers)

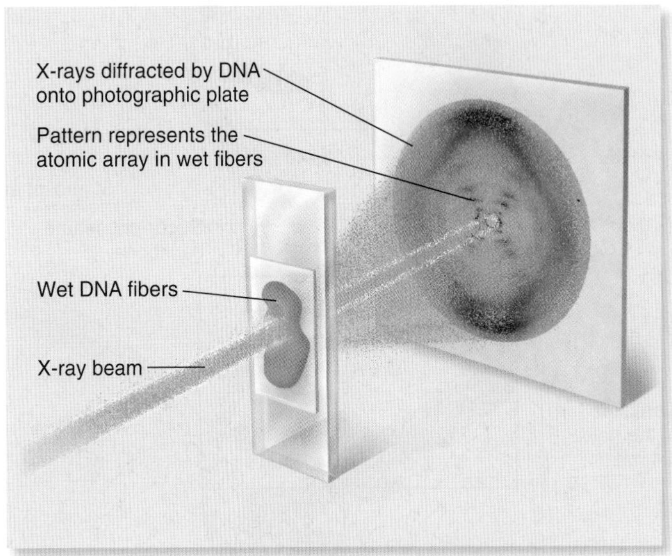

Figure 11.9 Rosalind Franklin's X-ray diffraction of DNA fibers. The exposure of DNA wet fibers to X-rays causes the X-rays to be scattered and the pattern of scattering is related to the position of the atoms in the DNA fibers.

spacing between the strands in which a purine (A or G) bonds with a pyrimidine (T or C).

Another piece of information that proved to be critical for the determination of the double helix structure came from the studies of Austrian-born American biochemist Erwin Chargaff. In 1950, Chargaff analyzed the base composition of DNA that was isolated from many different species. His experiments consistently showed that the amount of adenine in each sample was similar to the amount of thymine, and the amount of cytosine was similar to the amount of guanine (**Table 11.1**).

In the early 1950s, more information was known about the structure of proteins than that of nucleic acids. American biochemist Linus Pauling correctly proposed that some regions of proteins fold into a structure known as an α helix. To determine the structure of the α helix, Pauling built large models by linking together simple ball-and-stick

Table 11.1 Base Content in the DNA from a Variety of Organisms as Determined by Chargaff

	Percentage of bases (%)			
Organism	**Adenine**	**Thymine**	**Guanine**	**Cytosine**
Escherichia coli (bacterium)	26.0	23.9	24.9	25.2
Streptococcus pneumoniae (bacterium)	29.8	31.6	20.5	18.0
Saccharomyces cerevisiae (yeast)	31.7	32.6	18.3	17.4
Turtle	28.7	27.9	22.0	21.3
Salmon	29.7	29.1	20.8	20.4
Chicken	28.0	28.4	22.0	21.6
Human	30.3	30.3	19.5	19.9

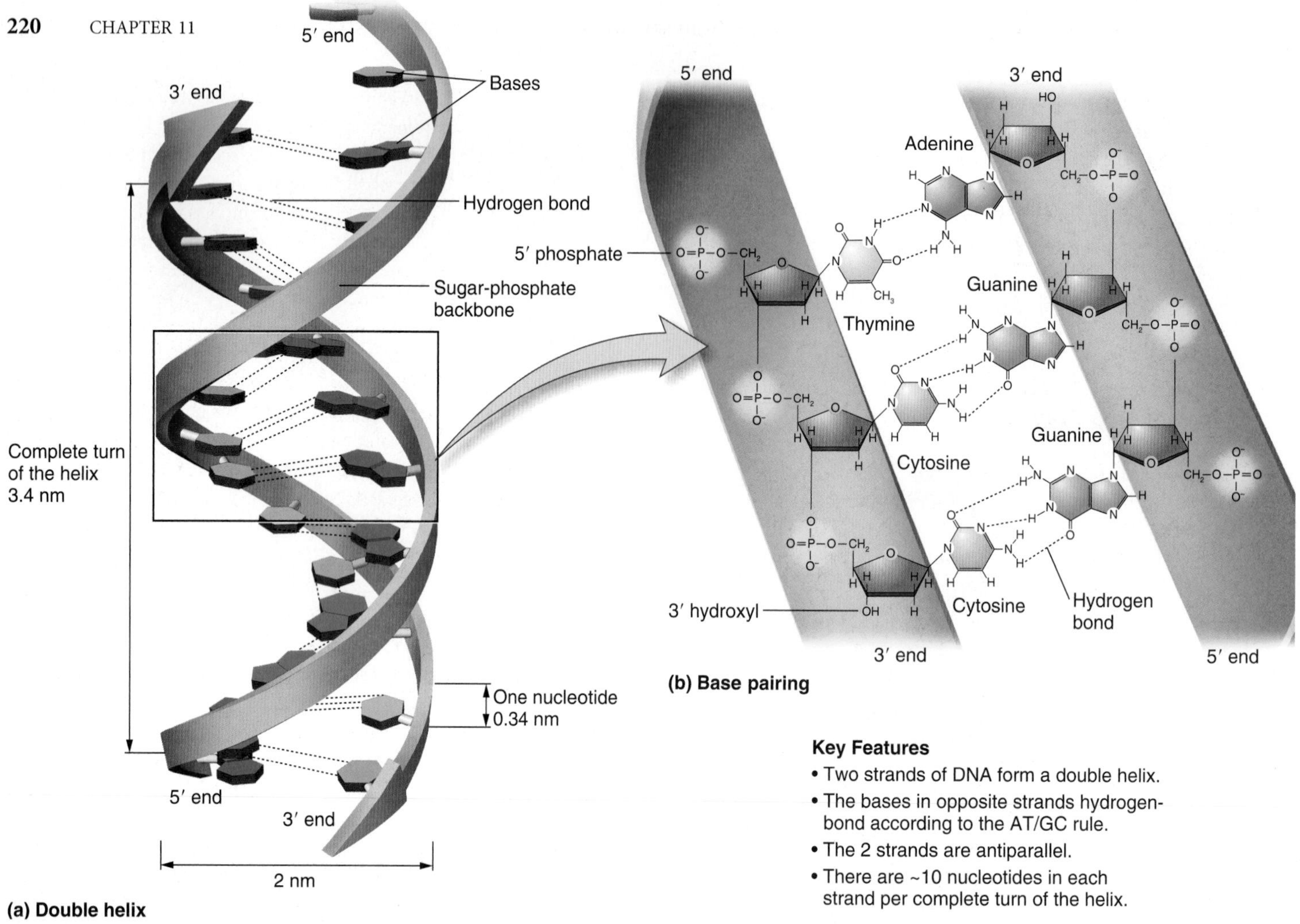

(a) Double helix

(b) Base pairing

Key Features

- Two strands of DNA form a double helix.
- The bases in opposite strands hydrogen-bond according to the AT/GC rule.
- The 2 strands are antiparallel.
- There are ~10 nucleotides in each strand per complete turn of the helix.

Figure 11.10 **Structure of the DNA double helix.** As seen in part **(a)**, DNA is a helix composed of two antiparallel strands. Part **(b)** shows the AT/GC base pairing that holds the strands together via hydrogen bonds.

Concept Check: *If one DNA strand is 5′–GATTCGTTC–3′, what is the complementary strand?*

units. In this way, he could see if atoms fit together properly in a complicated three-dimensional structure. This approach is still widely used today, except that now researchers construct three-dimensional models using computers. Use of the ball-and-stick approach was instrumental in solving the structure of the DNA double helix.

Watson and Crick Deduced the Double Helix Structure of DNA

Thus far, we have considered the experimental studies that led to the determination of the DNA double helix. American biologist James Watson and English biologist Francis Crick, working together at Cambridge University, assumed that nucleotides are linked together in a linear fashion and that the chemical linkage between two nucleotides is always the same. In collaboration with Wilkins, they then set out to build ball-and-stick models that incorporated all of the known experimental observations.

Modeling of chemical structures involves trial and error. Watson and Crick initially considered several incorrect models. One model was a double helix in which the bases were on the outside of the helix. In another model, each base formed hydrogen bonds with the identical base in the opposite strand (A to A, T to T, G to G, and C to C). However, model-building revealed that purine-purine pairs were too wide and pyrimidine-pyrimidine pairs were too narrow to fit the uniform diameter of DNA revealed from Franklin's work. Eventually, they realized that the hydrogen bonding of adenine to thymine was structurally similar to that of guanine to cytosine. In both cases, a purine base (A or G) bonds with a pyrimidine base (T or C). With an interaction between A and T and between G and C, the ball-and-stick models showed that the two strands would form a double helix structure in which all atoms would fit together properly.

Watson and Crick proposed the structure of DNA, which was published in the journal *Nature* in 1953. In 1962, Watson, Crick, and Wilkins were awarded the Nobel Prize in Physiology or Medicine. Unfortunately, Rosalind Franklin had died before this time, and the Nobel Prize is awarded only to living recipients.

DNA Has a Repeating, Antiparallel Helical Structure Formed by the Complementary Base Pairing of Nucleotides

The structure that Watson and Crick proposed is a double-stranded, helical structure with the sugar-phosphate backbone on the outside and the bases on the inside (Figure 11.10a). This structure is stabilized by hydrogen bonding between the bases in opposite strands to form **base pairs**. A distinguishing feature of base pairing is its specificity. An adenine (A) base in one strand forms two hydrogen bonds with a thymine (T) base in the opposite strand, or a guanine (G) base forms three hydrogen bonds with a cytosine (C) (**Figure 11.10b**). This **AT/GC rule** is consistent with Chargaff's observation that DNA contains equal amounts of A and T, and equal amounts of G and C. According to the AT/GC rule, purines (A and G) always bond with pyrimidines (T and C) (recall that purines have a double-ring structure, whereas pyrimidines have single rings). This keeps the width of the double helix relatively constant. One complete turn of the double helix is 3.4 nm in length and comprises 10 base pairs.

Due to the AT/GC rule, the base sequences of two DNA strands are **complementary** to each other. That is, you can predict the sequence in one DNA strand if you know the sequence in the opposite strand. For example, if one DNA strand has the sequence of 5′–GCGGATTT–3′, the opposite strand must be 3′–CGCCTAAA–5′. With regard to their 5′ and 3′ directionality, the two strands of a DNA double helix are **antiparallel**. If you look at Figure 11.10, one strand runs in the 5′ to 3′ direction from top to bottom, while the other strand is oriented 3′ to 5′ from top to bottom.

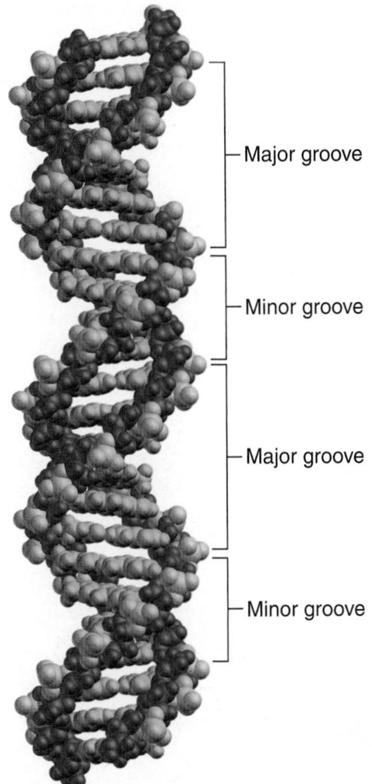

Figure 11.11 A space-filling model of the DNA double helix. In the sugar-phosphate backbone, sugar molecules are shown in blue, and phosphate groups are yellow. The backbone is on the outermost surface of the double helix. The atoms of the bases, shown in green, are more internally located within the double-stranded structure. Notice the major and minor grooves that are formed by this arrangement.

BIOLOGY PRINCIPLE Structure determines function. The major groove provides a binding site for proteins that control the expression of genes.

The DNA model in Figure 11.10a is called a ribbon model, which clearly shows the components of the DNA molecule. However, other models are also used to visualize DNA. The model for the DNA double helix shown in **Figure 11.11** is a space-filling model in which the atoms are depicted as spheres. Why is this model useful? This type of structural model emphasizes the surface of DNA. As you can see in this model, the sugar-phosphate backbone is on the outermost surface of the double helix; the backbone has the most direct contact with water. The atoms of the bases are more internally located within the double-stranded structure. The indentations where the atoms of the bases make contact with the surrounding water are termed grooves. Two grooves, called the **major groove** and the **minor groove**, spiral around the double helix. As discussed in later chapters, the major groove provides a location where a protein can bind to a particular sequence of bases and affect the expression of a gene (for example, look ahead to Figure 13.10).

11.3 An Overview of DNA Replication

Learning Outcomes:

1. Discuss and interpret the experiments of Meselson and Stahl.
2. Describe the double-stranded structure of DNA and explain how the AT/GC rule underlies the ability of DNA to be replicated semiconservatively.

The structure of DNA immediately suggested to Watson and Crick a mechanism by which DNA can be copied. They proposed that during this process, known as **DNA replication**, the original DNA strands are used as templates for the synthesis of new DNA strands. In this section, we will look at an early experiment that helped to determine the mechanism of DNA replication and then examine the structural characteristics that enable a double helix to be faithfully copied.

Meselson and Stahl Investigated Three Proposed Mechanisms of DNA Replication

Researchers in the late 1950s considered three different models for the mechanism of DNA replication (**Figure 11.12**). In all of these models, the two newly made strands are called the **daughter strands**, and the original strands are the **parental strands**. The first model is a **semiconservative mechanism** (Figure 11.12a). In this model, the double-stranded DNA is half conserved following the replication process so the new double-stranded DNA contains one parental strand and one daughter strand. This model is consistent with the proposal of Watson and Crick. However, other models were possible and had to be ruled out. According to a second model, called a **conservative mechanism**, both parental strands of DNA remain together following DNA replication (Figure 11.12b). The original arrangement of parental strands is completely conserved, and the two newly made daughter strands are also together following replication. Finally, a third possibility, called a **dispersive mechanism**, proposed that segments of parental DNA and newly made DNA are interspersed in both strands following the replication process (Figure 11.12c).

In 1958, American biologists Matthew Meselson and Franklin Stahl devised an experimental approach to distinguish among these

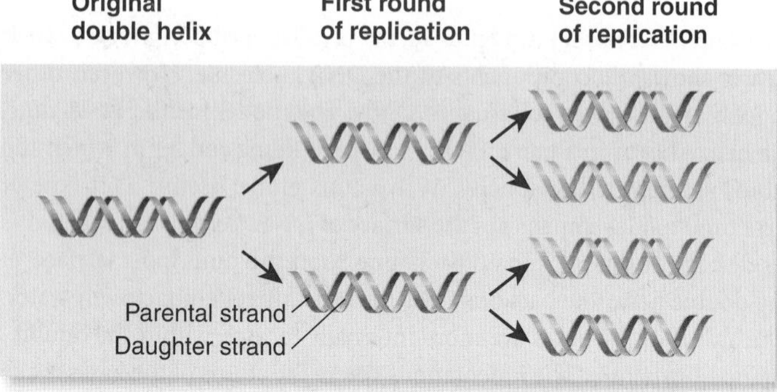

(a) Semiconservative mechanism. DNA replication produces DNA molecules with 1 parental strand and 1 newly made daughter strand.

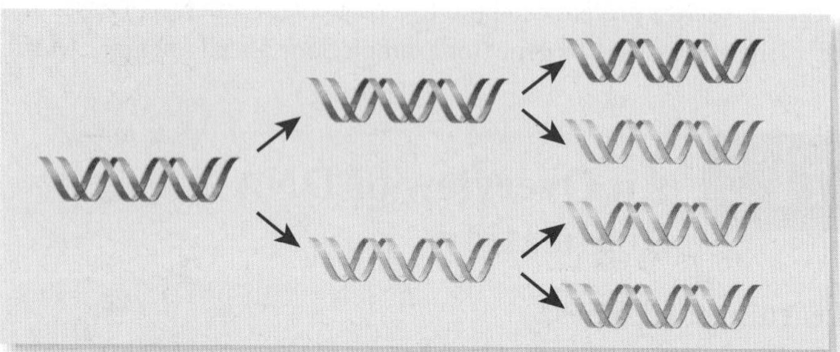

(b) Conservative mechanism. DNA replication produces 1 double helix with both parental strands and the other with 2 new daughter strands.

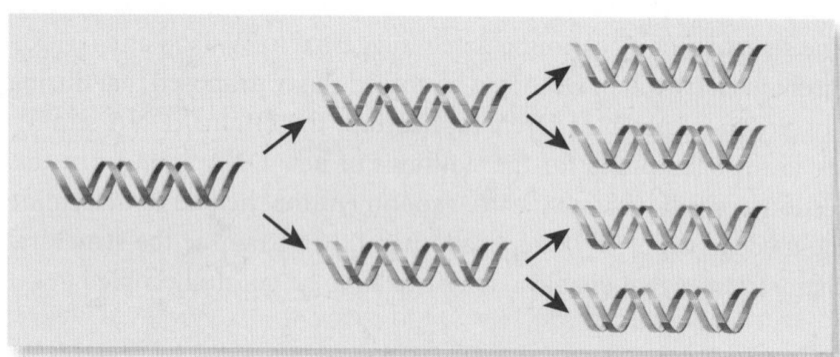

(c) Dispersive mechanism. DNA replication produces DNA strands in which segments of new DNA are interspersed with the parental DNA.

Figure 11.12 Three proposed mechanisms for DNA replication. The strands of the original (parental) double helix are shown in red. Two rounds of replication are illustrated with the daughter strands shown in blue.

three mechanisms. An important feature of their research was the use of isotope labeling. Nitrogen, which is found in DNA, occurs in a common light (^{14}N) form and a rare heavy (^{15}N) form. Meselson and Stahl studied DNA replication in the bacterium *Escherichia coli*. They grew *E. coli* cells for many generations in a medium that contained only the ^{15}N form of nitrogen (**Figure 11.13**). This produced a population of bacterial cells in which all of the DNA was heavy labeled. Then they switched the bacteria to a medium that contained only ^{14}N as its nitrogen source. The cells were allowed to divide, and samples were collected after one generation (that is, one round of DNA replication), two generations, and so on. Because the bacteria were doubling in a medium that contained only ^{14}N, all of the newly made DNA strands were labeled with light nitrogen, but the original strands would remain labeled with the heavy form.

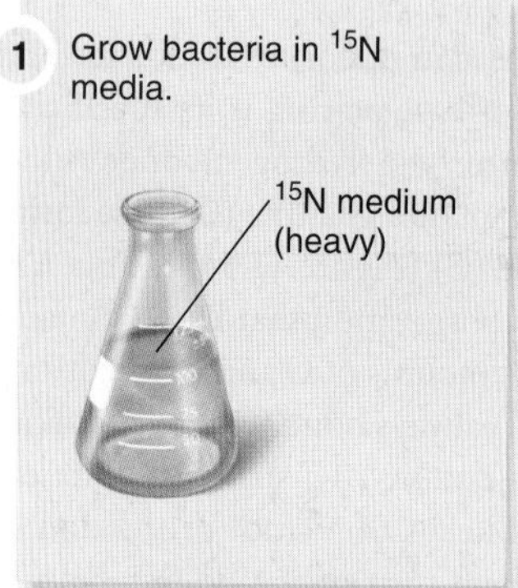

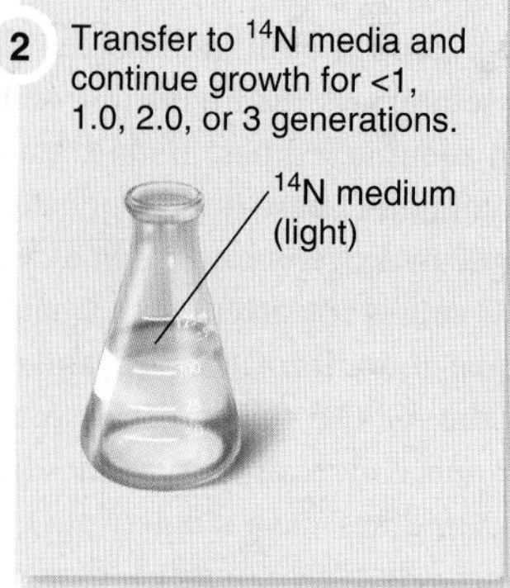

1 Grow bacteria in ^{15}N media.

2 Transfer to ^{14}N media and continue growth for <1, 1.0, 2.0, or 3 generations.
^{14}N medium (light)

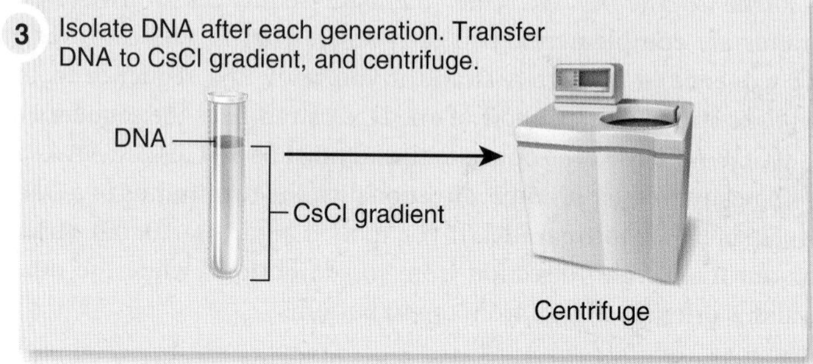

CsCl gradient

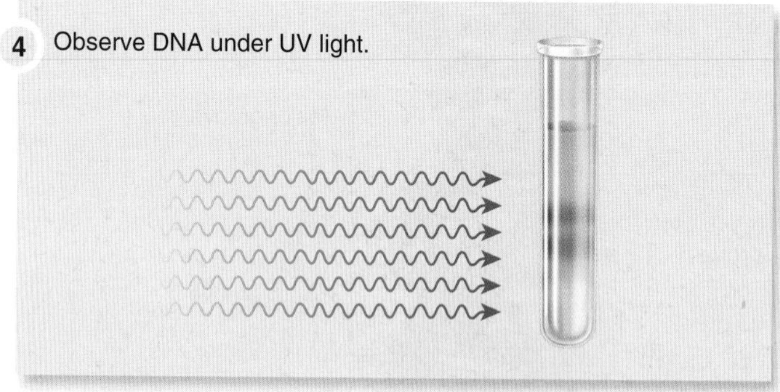

4 Observe DNA under UV light.

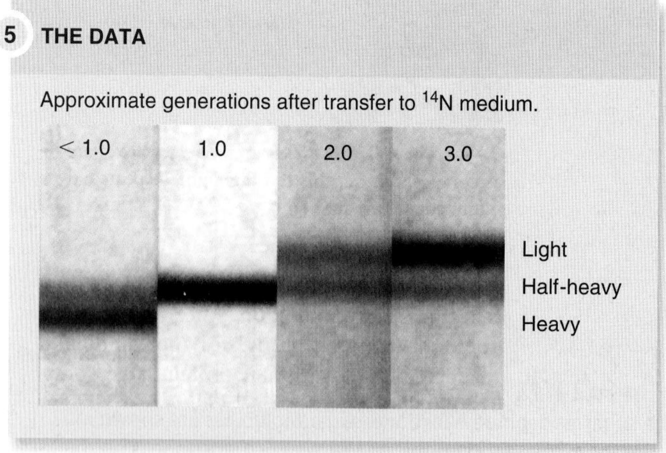

3 Isolate DNA after each generation. Transfer DNA to CsCl gradient, and centrifuge.

Figure 11.13 The Meselson and Stahl experiment showing that DNA replication is semiconservative.

Concept Check: *If this experiment was conducted for four rounds of DNA replication (that is, four generations), what would be the expected fractions of light DNA and half-heavy DNA according to the semiconservative model?*

How were the DNA molecules analyzed? Meselson and Stahl used centrifugation to separate DNA molecules based on differences in density. Samples were placed on the top of a solution that contained a salt gradient, in this case, cesium chloride (CsCl). A double helix containing all heavy nitrogen has a higher density and therefore travels closer to the bottom of the gradient. By comparison, if both DNA strands contained ^{14}N, the DNA would have a low density and remain closer to the top of the gradient. If one strand contained ^{14}N and the other strand contained ^{15}N, the DNA would be half-heavy and have an intermediate density, ending up near the middle of the gradient.

After one cell doubling (that is, one round of DNA replication), all of the DNA exhibited a density that was half-heavy or intermediate density (Figure 11.13, step 5). These results are consistent with both the semiconservative and dispersive models. In contrast, the conservative mechanism predicts two different DNA bands: one of high density and one of low density. Because the DNA was found in a single half-heavy band after one doubling, the conservative model was disproved.

After two cell doublings, both light DNA and half-heavy DNA bands were observed. This result was also predicted by the semiconservative mechanism of DNA replication, because half of the DNA molecules should contain all light DNA, while the other molecules should be half-heavy (see Figure 11.12a). However, in the dispersive mechanism, all of the DNA strands would have been 1/4 heavy after two generations. This mechanism predicts that the heavy nitrogen would be evenly dispersed among four double helices, each strand containing 1/4 heavy nitrogen and 3/4 light nitrogen (see Figure 11.12c). This prediction did not agree with the data. Taken together, the results of the Meselson and Stahl experiment are consistent only with a semiconservative mechanism for DNA replication.

DNA Replication Proceeds According to the AT/GC Rule

As originally proposed by Watson and Crick, DNA replication relies on the complementarity of DNA strands according to the AT/GC rule. During the replication process, the two complementary strands of DNA separate and serve as **template strands**, also called parental strands, for the synthesis of daughter strands of DNA (**Figure 11.14a**). After the double helix has separated, individual nucleotides have access to the template strands in a region called the replication fork. First, individual nucleotides hydrogen-bond to the template strands according to the AT/GC rule: an adenine (A) base in one strand bonds with a thymine (T) base in the opposite strand, or a guanine (G) base bonds with a cytosine (C). Next, a covalent bond is formed between the phosphate of one nucleotide and the sugar of the previous nucleotide. The end result is that two double helices are made that have the same base sequence as the original DNA molecule (**Figure 11.14b**). This is a critical feature of DNA replication, because it enables the

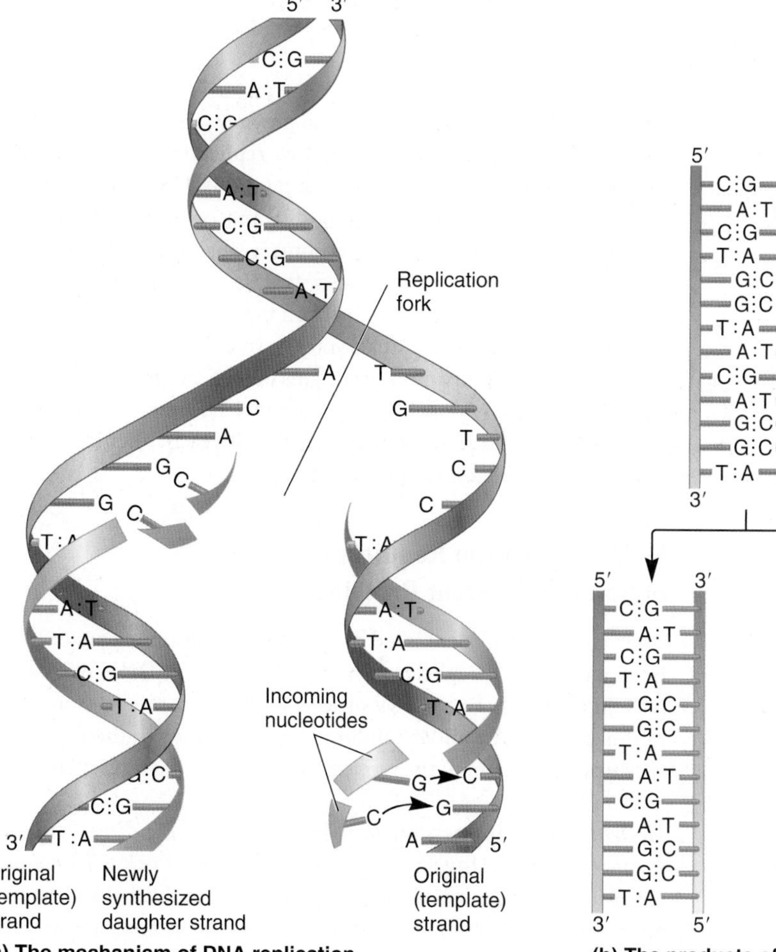

Figure 11.14 DNA replication according to the AT/GC rule. **(a)** The mechanism of DNA replication as originally proposed by Watson and Crick. As we will see in Section 11.4, the synthesis of one newly made strand (the leading strand on the left side) occurs in the direction toward the replication fork, whereas the synthesis of the other newly made strand (the lagging strand on the right side) occurs in small segments away from the fork. **(b)** DNA replication produces two copies of DNA with the same sequence as the original DNA molecule.

BIOLOGY PRINCIPLE Structure determines function. A double-stranded structure that obeys the AT/GC rule underlies the function of DNA replication.

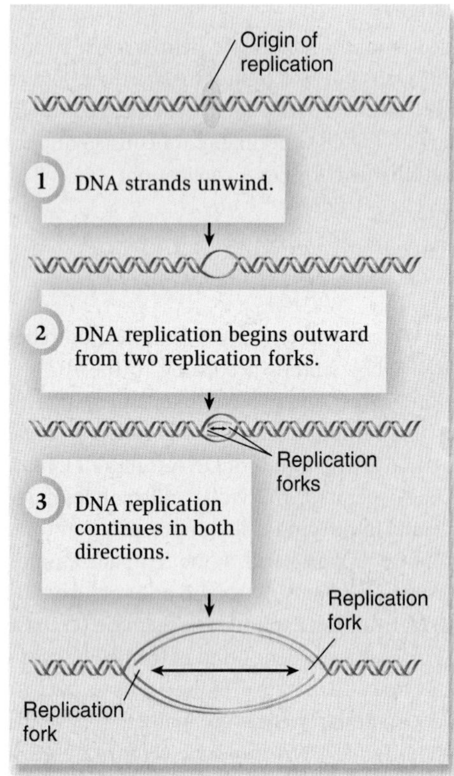

(a) Bidirectional replication

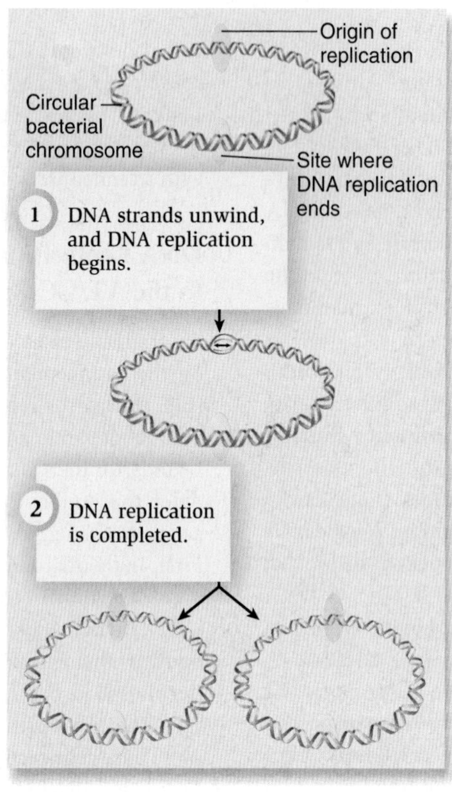

(b) Single origin of replication in bacteria

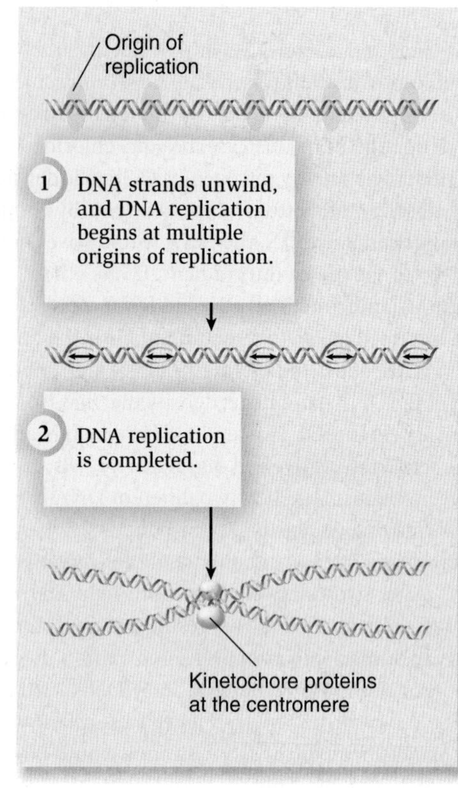

(c) Multiple origins of replication in eukaryotes

Figure 11.15 **The bidirectional replication of DNA.** **(a)** DNA replication proceeds in both directions from an origin of replication. **(b)** Bacterial chromosomes have a single origin of replication, whereas **(c)** eukaryotes have multiple origins. Following DNA replication in eukaryotes, the two copies remain attached to each other at the centromere via kinetochore proteins.

replicated DNA molecules to retain the same information (that is, the same base sequence) as the original molecule. In this way, DNA has the remarkable ability to direct its own duplication.

11.4 Molecular Mechanism of DNA Replication

Learning Outcomes:

1. Describe how the synthesis of new DNA strands begins at an origin of replication.
2. List the functions of helicase, topoisomerase, single-strand binding protein, primase, and DNA polymerase at the replication fork.
3. Outline the key differences between the synthesis of the leading and lagging strands.
4. List three reasons why DNA replication is very accurate.
5. Explain how DNA replication occurs at telomeres in eukaryotic chromosomes.

Thus far, we have examined the general mechanism of DNA replication, known as semiconservative replication, and we've seen how DNA synthesis obeys the AT/GC rule. In this section, we will examine the details of DNA replication as it occurs inside living cells. As you will learn, several different proteins are needed to initiate DNA replication and allow it to proceed quickly and accurately.

DNA Replication Begins at an Origin of Replication

Where does DNA replication begin? An **origin of replication** is a site within a chromosome that serves as a starting point for DNA replication. At the origin, the two DNA strands unwind (Figure 11.15a). DNA replication proceeds outward from two **replication forks**, a process termed **bidirectional replication** (see Figure 11.15a). The number of origins of replication varies among different organisms. In bacteria, which have a small circular chromosome, a single origin of replication is found. Bidirectional replication starts at the origin of replication and proceeds until the new strands meet on the opposite side of the chromosome (Figure 11.15b). Eukaryotes have larger chromosomes that are linear. They require multiple origins of replication so the DNA can be replicated in a reasonable length of time. The newly made strands from each origin eventually make contact with each other to complete the replication process (Figure 11.15c).

DNA Replication Requires the Action of Several Different Proteins

Thus far, we have considered how DNA replication occurs outward from an origin of replication in a region called a DNA replication fork. In all living species, a set of several different proteins is involved in this process. An understanding of the functions of these proteins is critical to explaining the replication process at the molecular level.

Helicase, Topoisomerase, and Single-Strand Binding Proteins: Formation and Movement of the Replication Fork To act as a template for DNA replication, the strands of a double helix must separate, and the resulting fork must move. As mentioned, an origin of replication serves as a site where this separation initially occurs. The strand separation at each fork then moves outward from the origin via the action of an enzyme called **DNA helicase**. At each fork, DNA helicase binds to one of the DNA strands and travels in the 5′ to 3′ direction toward the fork (Figure 11.16). It uses energy from ATP to separate the DNA strands and keeps the fork moving forward. The action of DNA helicase generates additional coiling just ahead of the replication fork that is alleviated by another enzyme called **DNA topoisomerase**, which helps to untwist the strands.

After the two template DNA strands have separated, they must remain that way until the complementary daughter strands have been made. The function of **single-strand binding proteins** is to coat both of the single strands of template DNA and prevent them from re-forming a double helix. In this way, the bases within the template strands are kept exposed so they can act as templates for the synthesis of complementary strands.

DNA Polymerase and Primase: Synthesis of DNA Strands The enzyme **DNA polymerase** is responsible for covalently linking nucleotides together to form DNA strands. American biochemist Arthur Kornberg originally identified this enzyme in the 1950s. The structure of DNA polymerase resembles a human hand with the DNA threaded through it (Figure 11.17a). As DNA polymerase slides along the

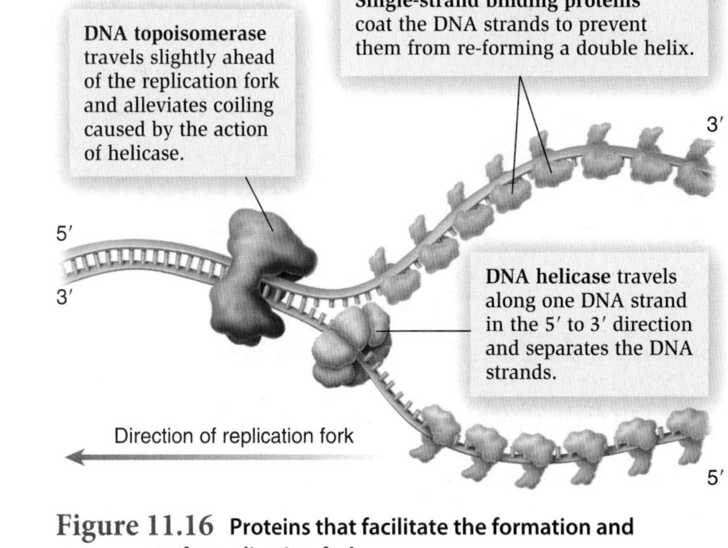
DNA topoisomerase travels slightly ahead of the replication fork and alleviates coiling caused by the action of helicase.
Single-strand binding proteins coat the DNA strands to prevent them from re-forming a double helix.
DNA helicase travels along one DNA strand in the 5′ to 3′ direction and separates the DNA strands.
Direction of replication fork
5′
3′
3′
5′

Figure 11.16 Proteins that facilitate the formation and movement of a replication fork.

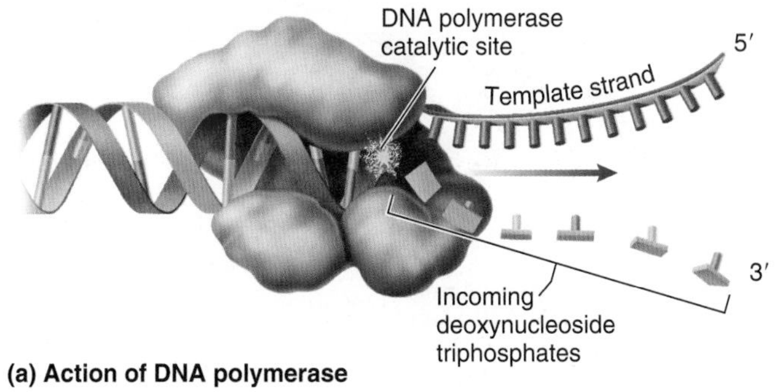

(a) Action of DNA polymerase

(b) Chemistry of DNA replication

Figure 11.17 Enzymatic synthesis of DNA. **(a)** Incoming deoxynucleoside triphosphates first hydrogen-bond to the template strand according to the AT/GC rule. DNA polymerase recognizes these deoxynucleoside triphosphates and attaches a deoxynucleoside monophosphate to the 3′ end of a growing strand. **(b)** DNA polymerase breaks the bond between the first and second phosphate in a deoxynucleoside triphosphate, causing the release of pyrophosphate. This provides the energy to form a covalent bond between the resulting deoxynucleoside monophosphate and the previous nucleotide in the growing strand. The pyrophosphate is broken down to two phosphates.

Concept Check: Does the oxygen in a new phosphoester bond come from the sugar or from the phosphate?

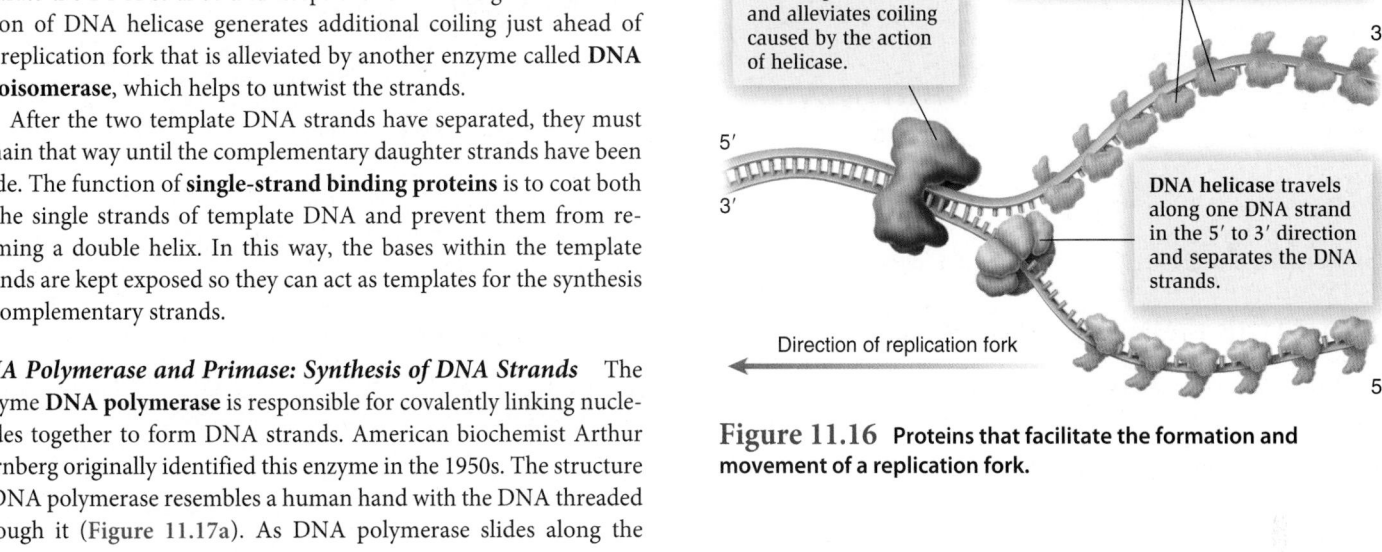

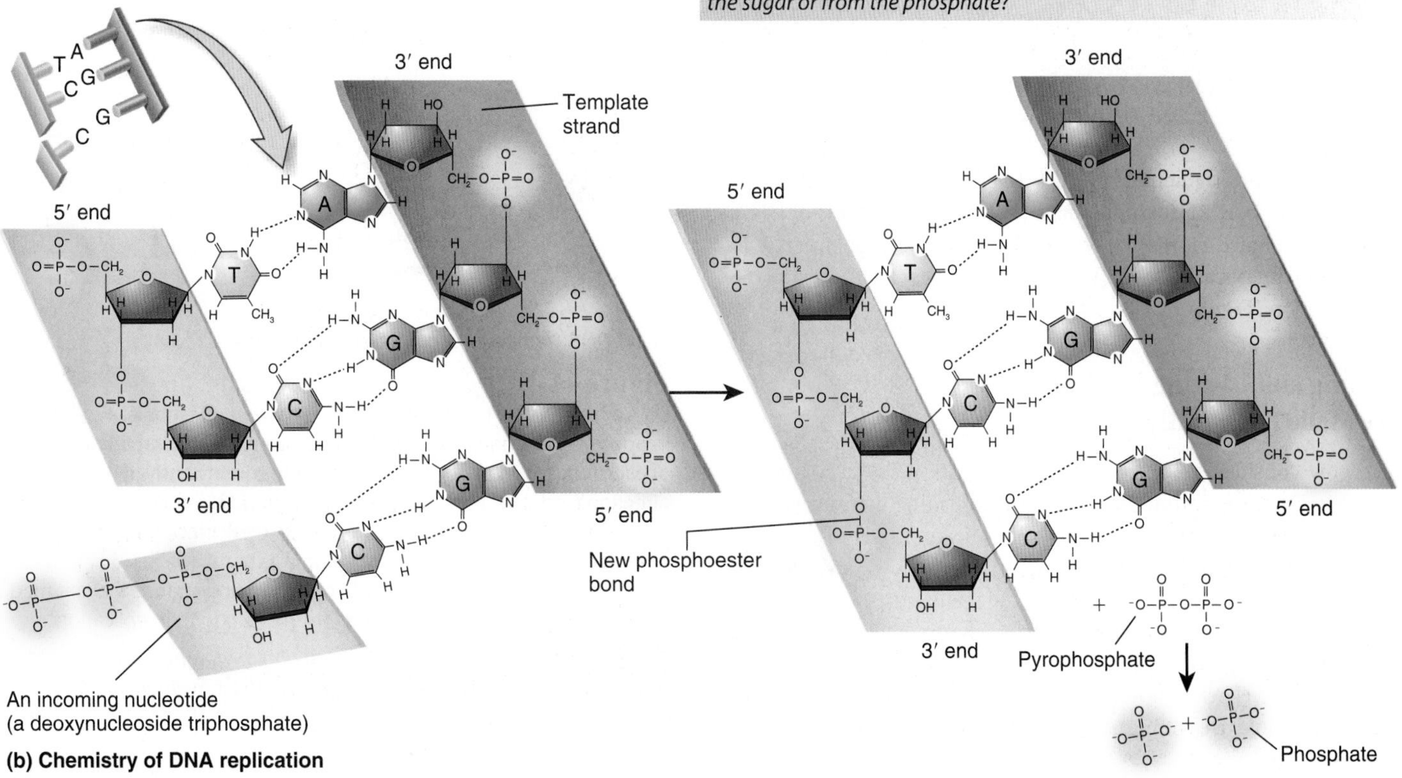

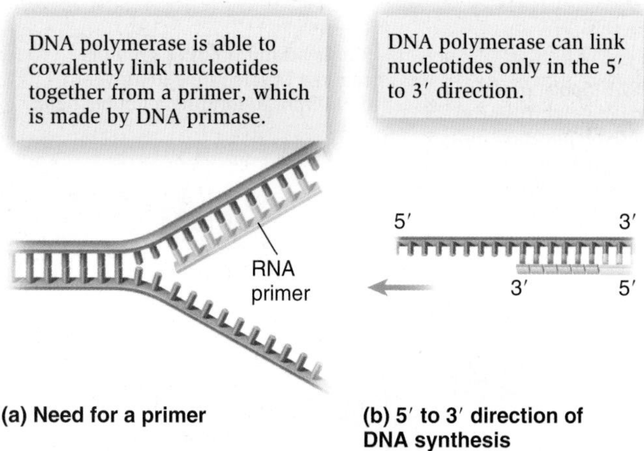

Figure 11.18 **Enzymatic features of DNA polymerase.** **(a)** DNA polymerase needs a primer to begin DNA synthesis, and **(b)** it can synthesize DNA only in the 5′ to 3′ direction.

DNA, free nucleotides with three phosphate groups, called **deoxynucleoside triphosphates**, hydrogen-bond to the exposed bases in the template strand according to the AT/GC rule. At the catalytic site, DNA polymerase breaks a bond between the first and second phosphate and then attaches the resulting nucleotide with one phosphate group (a deoxynucleoside monophosphate) to the 3′ end of a growing strand via a phosphoester bond. The breakage of the covalent bond that releases pyrophosphate is an exergonic reaction that provides the energy to covalently connect adjacent nucleotides (Figure 11.17b). The pyrophosphate is broken down to two phosphates. The rate of synthesis is truly remarkable. In bacteria, DNA polymerase synthesizes DNA at a rate of 500 nucleotides per second, whereas eukaryotic species make DNA at a rate of about 50 nucleotides per second.

DNA polymerase has two additional enzymatic features that affect how DNA strands are made. First, if a DNA or RNA strand is already attached to a template strand, DNA polymerase can elongate such a pre-existing strand by making DNA. However, DNA polymerase is unable to begin DNA synthesis on a bare template strand. A different enzyme called **DNA primase** is required if the template strand is bare. DNA primase makes a complementary primer that is actually a short segment of RNA, typically 10 to 12 nucleotides in length. These short RNA strands start, or prime, the process of DNA replication (**Figure 11.18a**). A second feature of DNA polymerase is that once synthesis has begun, it can synthesize new DNA only in a 5′ to 3′ direction (**Figure 11.18b**).

Leading and Lagging DNA Strands Are Made Differently

Let's now consider how new DNA strands are made at a replication fork. DNA replication occurs near the opening that forms each replication fork (**Figure 11.19a**, step 1). The synthesis of a strand always begins with an RNA primer (depicted in yellow), and the new DNA is made in the 5′ to 3′ direction. The manner in which the two daughter strands are synthesized is strikingly different. One strand, called the **leading strand**, is made in the same direction that the fork is moving. The leading strand is synthesized as one long continuous molecule. By comparison, the other daughter strand, termed

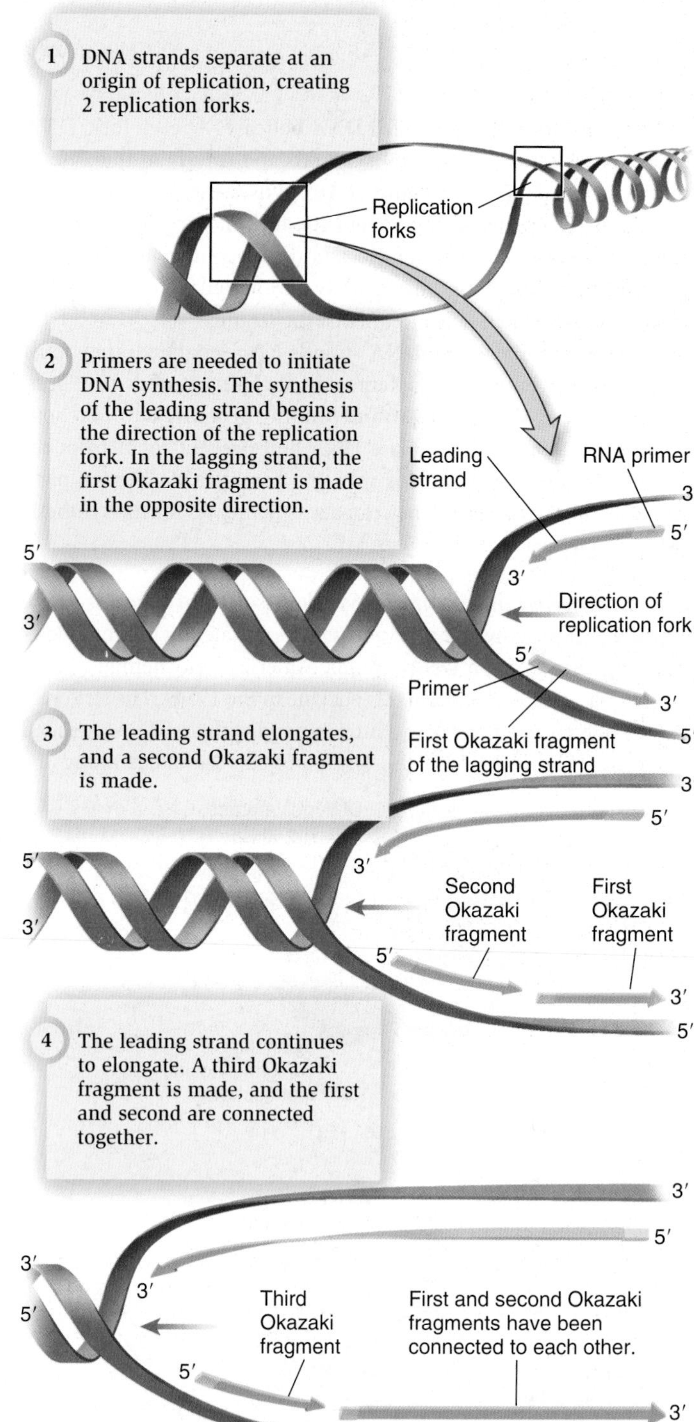

Figure 11.19 **Synthesis of new DNA strands.** The separation of DNA at the origin of replication produces two replication forks that move in opposite directions. New DNA strands are made near the opening of each fork. **(a)** The leading strand is made continuously in the same direction the fork is moving. The lagging strand is made as small pieces in the opposite direction. Eventually, these small pieces are connected to each other to form a continuous lagging strand. **(b)** This diagram illustrates the locations of the leading and lagging strands that are made during bidirectional DNA replication from one origin of replication.

Concept Check: *Which strand, the leading or lagging strand, is made discontinuously in the direction opposite to the movement of the replication fork?*

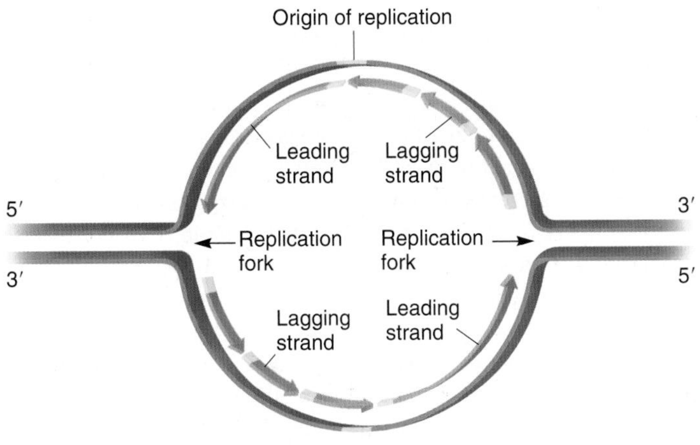

(b) Replication from an origin

the **lagging strand**, is made as a series of small fragments that are subsequently connected to each other to form a continuous strand. These DNA fragments are known as **Okazaki fragments**, after Japanese biologists Reiji and Tsuneko Okazaki, who initially discovered them in the late 1960s. The synthesis of Okazaki fragments occurs in the opposite direction of the replication fork. For example, the lower fragment seen in Figure 11.19a, steps 2 and 3, is synthesized from left to right. As shown in Figure 11.19a, step 4, the RNA primer is eventually removed, and adjacent Okazaki fragments are connected to each other to form a continuous strand of DNA. Figure 11.19b shows the leading and lagging strands that are made during bidirectional DNA replication from a single origin of replication.

Figure 11.19 was meant to emphasize the synthesis of new DNA strands. Figure 11.20 shows the proteins involved with the synthesis of the leading and lagging strands in *E. coli.* In this bacterium, two different DNA polymerases, called DNA polymerase I and III, are primarily responsible for DNA replication. In the leading strand, DNA primase makes one RNA primer at the origin, and then DNA polymerase III attaches nucleotides in a 5′ to 3′ direction as it slides toward the opening of the replication fork. DNA polymerase III has a subunit called the clamp protein that allows the enzyme to slide along the template strand without falling off, a characteristic called **processivity**.

In the lagging strand, DNA is also synthesized in a 5′ to 3′ direction, but this synthesis occurs in the direction away from the replication fork. In the lagging strand, short segments of DNA are made discontinuously as a series of Okazaki fragments, each of which requires its own primer. DNA polymerase III synthesizes the remainder of the fragment (Figure 11.20, step 2). To complete the synthesis of Okazaki fragments within the lagging strand, three additional events occur: the removal of the RNA primers, the synthesis of DNA in the

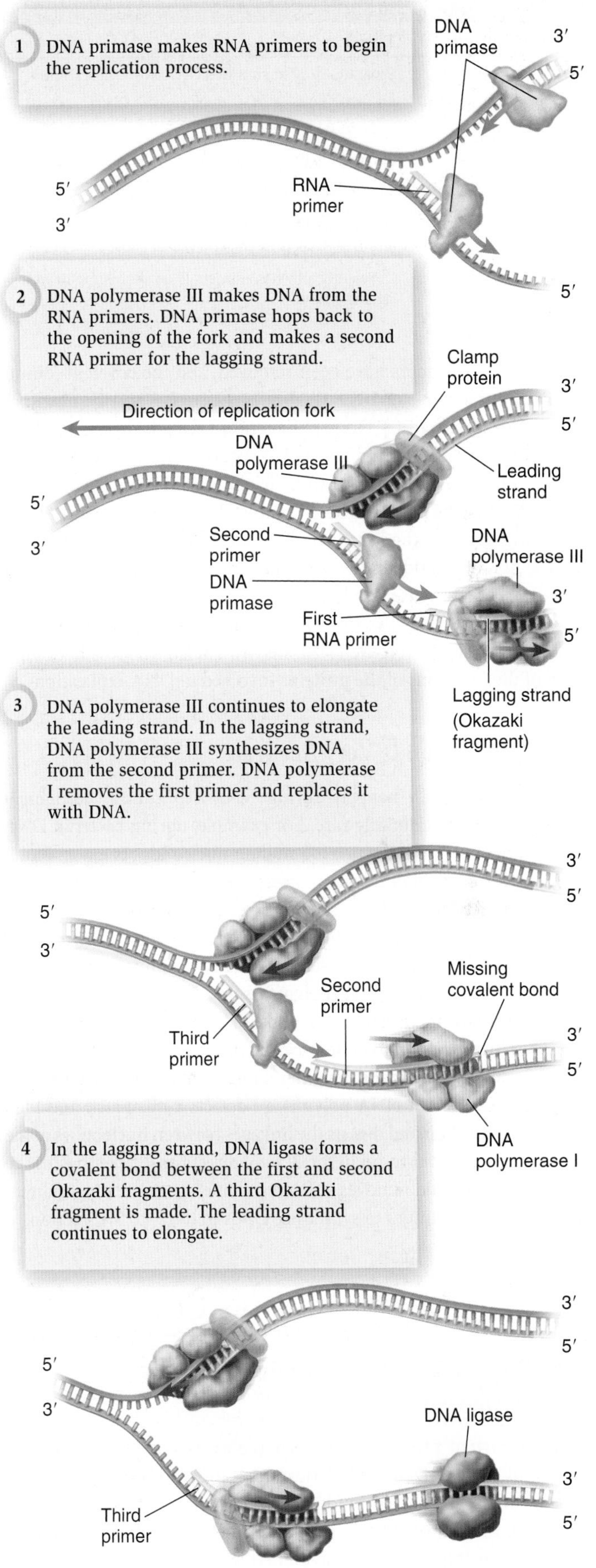

Figure 11.20 Proteins involved with the synthesis of the leading and lagging strands in *E. coli.*

Concept Check: Briefly describe the movement of primase in the lagging strand in this figure. In which direction does it move when it is making a primer, from left to right or right to left? Describe how it must move after it is done making a primer and has to start making the next primer at a new location. Does it have to hop from left to right or from right to left?

1 DNA primase makes RNA primers to begin the replication process.

DNA primase
3′
5′
5′
3′
RNA primer
5′

2 DNA polymerase III makes DNA from the RNA primers. DNA primase hops back to the opening of the fork and makes a second RNA primer for the lagging strand.

Direction of replication fork
Clamp protein
3′
5′
DNA polymerase III
Leading strand
5′
3′
Second primer
DNA primase
First RNA primer
DNA polymerase III
3′
5′
Lagging strand (Okazaki fragment)

3 DNA polymerase III continues to elongate the leading strand. In the lagging strand, DNA polymerase III synthesizes DNA from the second primer. DNA polymerase I removes the first primer and replaces it with DNA.

3′
5′
5′
3′
Second primer
Missing covalent bond
Third primer
3′
5′
DNA polymerase I

4 In the lagging strand, DNA ligase forms a covalent bond between the first and second Okazaki fragments. A third Okazaki fragment is made. The leading strand continues to elongate.

3′
5′
5′
3′
DNA ligase
3′
Third primer
5′

Table 11.2 Proteins Involved in DNA Replication

Common name	Function
DNA helicase	Separates double-stranded DNA into single strands
Single-strand binding protein	Binds to single-stranded DNA and prevents it from re-forming a double helix
Topoisomerase	Removes tightened coils ahead of the replication fork
DNA primase	Synthesizes short RNA primers
DNA polymerase	Synthesizes DNA in the leading and lagging strands, removes RNA primers, and fills in gaps
DNA ligase	Covalently attaches adjacent Okazaki fragments in the lagging strand

area where the primers have been removed, and the covalent joining of adjacent fragments of DNA (Figure 11.20, steps 3 and 4). The RNA primers are removed by DNA polymerase I, which digests the linkages between nucleotides in a 5′ to 3′ direction. After the RNA primer is removed, DNA polymerase I fills in the vacant region with DNA. However, once the DNA has been completely filled in, a covalent bond is missing between the last nucleotide added by DNA polymerase I and the first nucleotide in the adjacent Okazaki fragment. An enzyme known as **DNA ligase** catalyzes the formation of a covalent bond between these two DNA fragments to complete the replication process in the lagging strand (Figure 11.20, step 4). **Table 11.2** provides a summary of the functions of the proteins involved in DNA replication.

DNA Replication Is Very Accurate

Although errors can happen during DNA replication, permanent mistakes are extraordinarily rare. For example, during bacterial DNA replication, only 1 mistake per 100 million nucleotides is made. Biologists use the term "high fidelity" to refer to a process that occurs with relatively few mistakes. How can we explain such a remarkably high fidelity for DNA replication? First, hydrogen bonding between A and T or between G and C is more stable than between mismatched pairs. Second, the active site of DNA polymerase is unlikely to catalyze bond formation between adjacent nucleotides if a mismatched base pair is formed. Third, DNA polymerase can identify a mismatched nucleotide and remove it from the daughter strand. This event, called **proofreading**, occurs when DNA polymerase detects a mismatch and then reverses its direction and digests the linkages between nucleotides at the end of a newly made strand in the 3′ to 5′ direction. Once it passes the mismatched base and removes it, DNA polymerase then changes direction again and continues to synthesize DNA in the 5′ to 3′ direction.

GENOMES & PROTEOMES CONNECTION

DNA Polymerases Are a Family of Enzymes with Specialized Functions

Three important properties of DNA replication are speed, fidelity, and completeness. DNA replication must proceed quickly and with great accuracy, and gaps should not be left in the newly made strands. To ensure these three requirements are met, living species produce more than one type of DNA polymerase, each of which may differ in the rate and accuracy of DNA replication and/or the ability to prevent the formation of DNA gaps.

The genomes of living species have multiple DNA polymerase genes, which were produced by random gene duplication events. During evolution, mutations have altered each gene to produce a family of DNA polymerase enzymes with more specialized functions. Natural selection has favored certain mutations that result in DNA polymerase properties that are suited to the organism in which they are found. For comparison, let's consider the families of DNA polymerases found in the bacterium *E. coli* and humans (**Table 11.3**). Why does *E. coli* produce 5 DNA polymerases and humans produce 12 or more? The answer lies in specialization and the functional requirements of each species.

In *E. coli*, DNA polymerase III is responsible for most DNA replication. It synthesizes DNA very rapidly and with high fidelity. By comparison, the role of DNA polymerase I is to remove the RNA primers and fill in the short vacant regions with DNA. DNA polymerases II, IV, and V are involved in repairing DNA and in replicating DNA that has been damaged. DNA polymerases I and III become stalled when they encounter DNA damage and may be unable to make a complementary strand at such a site. By comparison, DNA polymerases II, IV, and V do not stall. Although their rate of synthesis is not as rapid as DNA polymerases I and III, they ensure that DNA replication is complete.

In human cells, DNA polymerase α (alpha) has its own "built-in" primase subunit. It synthesizes RNA primers followed by short DNA regions. Two other DNA polymerases, δ (delta) and ε (epsilon), then extend the DNA at a faster rate. DNA polymerase γ (gamma) functions in the mitochondria to replicate mitochondrial DNA.

When DNA replication occurs, the general DNA polymerases (α, δ, or γ) may be unable to replicate over an abnormality in DNA structure (a lesion). If this happens, lesion-replicating polymerases are attracted to the damaged DNA. These polymerases have special properties that enable them to synthesize a complementary strand

Table 11.3 DNA Polymerases in *E. coli* and Humans

Polymerase types*	Functions
E. coli	
III	Replicates most of the DNA during cell division
I	Removes RNA primers and fills in the gaps
II, IV, and V	Repairs damaged DNA and replicates over DNA damage
Humans	
α (alpha)	Makes RNA primers and synthesizes short DNA strands
δ (delta), ε (epsilon)	Displaces DNA polymerase α and then replicates DNA at a rapid rate
γ (gamma)	Replicates the mitochondrial DNA
η (eta), κ (kappa), ι (iota), ζ (zeta)	Replicates over damaged DNA
α, β (beta), δ, ε, σ (sigma), λ (lambda), μ (mu), φ (phi), θ (theta)	Repairs DNA or has other functions

*Certain DNA polymerases have more than one function.

over the lesion. Each type of lesion-replicating polymerase may be able to replicate over different kinds of DNA damage, thereby ensuring that DNA replication is complete.

Other human DNA polymerases play an important role in DNA repair. The need for multiple repair enzymes is rooted in the various ways that DNA can be damaged, as described in Chapter 14. Multicellular organisms must be particularly vigilant about repairing DNA, because unrepaired DNA can lead to cancer.

Telomerase Attaches DNA Sequences at the Ends of Eukaryotic Chromosomes

We will end our discussion of DNA replication by considering a specialized form of DNA replication that happens at the ends of eukaryotic chromosomes. This region, called the **telomere**, has a short nucleotide sequence that is repeated a few dozen to several hundred times in a row (**Figure 11.21**). The repeat sequence shown here, 5′–GGGTTA–3′, is the sequence found in human telomeres. Other organisms have different repeat sequences. For example, the sequence found in the telomeres of maize is 5′–GGGTTTA–3′. A telomere has a region at the 3′ end that is termed a 3′ overhang, because it does not have a complementary strand.

As discussed previously, DNA polymerase synthesizes DNA only in a 5′ to 3′ direction and requires a primer. For these reasons, DNA polymerase cannot copy the tip of a DNA strand with a 3′ end. Therefore, if this replication problem was not overcome, a linear chromosome would become progressively shorter with each round of DNA replication. In 1984, American molecular biologist Carol Greider and Australian-born American molecular biologist Elizabeth Blackburn discovered an enzyme called **telomerase** that prevents chromosome shortening by attaching many copies of a DNA repeat sequence to the ends of chromosomes (**Figure 11.22**). Telomerase contains both protein and RNA. The RNA part of telomerase has a sequence that is complementary to the DNA repeat sequence. This allows telomerase to bind to the 3′ overhang region of the telomere. Following binding, the RNA sequence beyond the binding site functions as a template, allowing telomerase to synthesize a 6-nucleotide sequence at the end of the DNA strand. The enzyme then moves to the new end of this DNA strand and attaches another 6 nucleotides to the end. This occurs many times, thereby greatly lengthening the 3′ end of the DNA

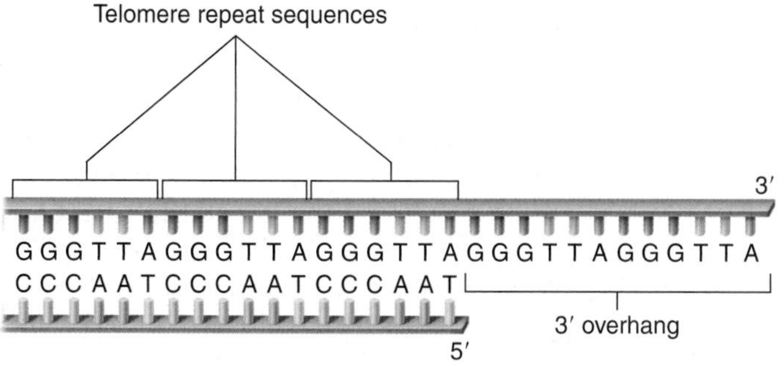

Telomere repeat sequences
3′
GGGTTAGGGTTAGGGTTAGGGTTAGGGTTA
CCCAATCCCAATCCCAAT
3′ overhang
5′

Figure 11.21 Telomere sequences at the end of a human chromosome. The telomere sequence shown here is found in humans and other mammals. The length of the telomere and the 3′ overhang varies among different species and cell types.

Telomere
Eukaryotic chromosome
Telomere
1 Telomerase binds to a DNA repeat sequence.
Repeat sequence
RNA in telomerase
Telomerase
2 Telomerase synthesizes a 6-nucleotide repeat sequence.
3 Telomerase moves 6 nucleotides to the right and begins to make another repeat.
4 Primase makes an RNA primer near the end of the telomere, and DNA polymerase synthesizes a complementary strand in the 5′ to 3′ direction. The RNA primer is eventually removed.
TAGGGTTAGGGTTAGGGTTAGGGTTAGGGTTA
ATCCCAATCCCAATCCCAAUCCCAAUCCC
RNA primer that is eventually removed

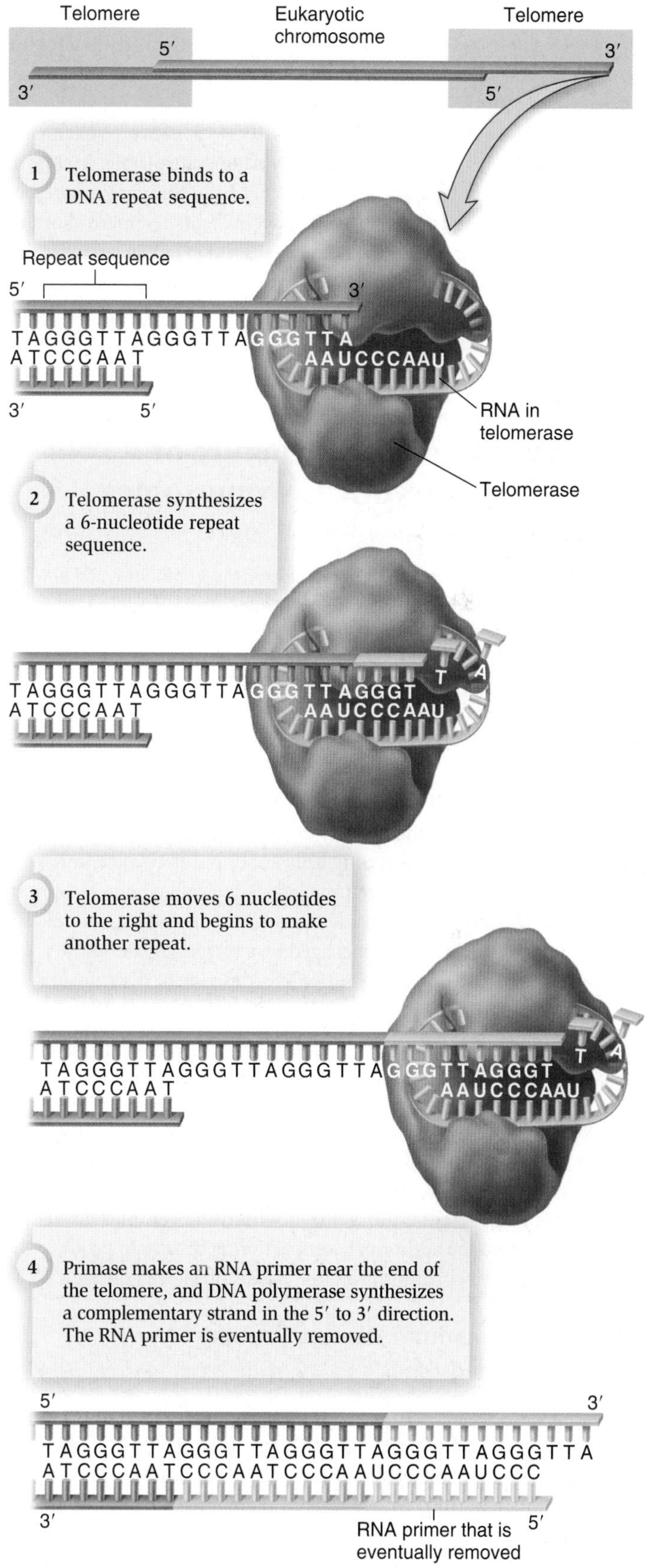

Figure 11.22 Mechanism of DNA replication by telomerase.

Concept Check: *What does telomerase use as a template to make DNA?*

in the telomeric region. This lengthening provides an upstream site for an RNA primer to be made. DNA polymerase then synthesizes the complementary DNA strand. In this way, the progressive shortening of eukaryotic chromosomes is prevented.

Telomerase function is also associated with cancer. When cells become cancerous, they continue to divide uncontrollably. In 90% of all types of human cancers, telomerase has been found to be present at high levels in the cancerous cells. This prevents telomere shortening and may play a role in the continued growth of cancer cells. The mechanism whereby cancer cells are able to increase the function of telomerase is not well understood and is a topic of active research.

Greider and Blackburn shared the 2009 Nobel Prize in Physiology or Medicine with Jack Szostak for their work on telomeres.

11.5 Molecular Structure of Eukaryotic Chromosomes

Learning Outcomes:

1. Describe the structure of nucleosomes and the 30-nm fiber, and how the 30-nm fiber forms radial loop domains.
2. Outline the various levels of compaction that lead to a metaphase chromosome.

We now turn our attention to the structure of eukaryotic chromosomes. A typical eukaryotic chromosome contains a single, linear, double-stranded DNA molecule that may be hundreds of millions of base pairs in length. If the DNA from a single set of human chromosomes were stretched end to end, the length would be over 1 meter! By comparison, most eukaryotic cells are only 10–100 μm (micrometers) in diameter, and the cell nucleus is typically about 2–4 μm in diameter. Therefore, to fit inside the nucleus, the DNA in a eukaryotic cell must be folded and packaged by a staggering amount.

The term chromosome is used to describe a discrete unit of genetic material. For example, a human somatic cell contains 46 chromosomes. By comparison, the term chromatin has a biochemical meaning. **Chromatin** is used to describe the complex of DNA and protein that makes up eukaryotic chromosomes. Chromosomes are very dynamic structures that alternate between tight and loose compaction states. In this section, we will focus our attention on two issues of chromosome structure. First, we will consider how chromosomes are compacted and organized within the cell nucleus. Then, we will examine the additional compaction necessary to produce the highly condensed chromosomes that occur during cell division.

DNA Wraps Around Histone Proteins to Form Nucleosomes

The first way DNA is compacted is by wrapping itself around a group of proteins called **histones**. As shown in Figure 11.23, a repeating structural unit of eukaryotic chromatin is the **nucleosome**, which is 11 nanometers (nm) in diameter and composed of 146 or 147 bp (base pairs) of DNA wrapped around an octamer of histone proteins. An octamer contains two molecules each of four types of histone proteins: H2A, H2B, H3, and H4. Histone proteins are very basic proteins because they contain a large number of positively charged

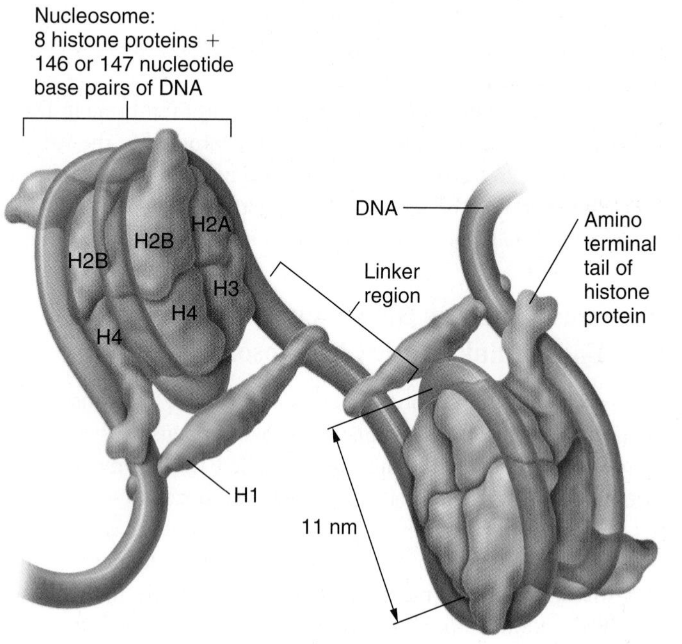

Figure 11.23 **Structure of a nucleosome.** A nucleosome is composed of double-stranded DNA wrapped around an octamer of histone proteins. A linker region connects two adjacent nucleosomes. Histone H1 is bound to the linker region, as are other proteins not shown in this figure.

lysine and arginine amino acids. The negative charges found in the phosphates of DNA are attracted to the positive charges on histone proteins. The amino terminal tail of each histone protein protrudes from the histone octamer. As discussed in Chapter 13, these tails can be covalently modified and play a key role in gene regulation.

The nucleosomes are connected by linker regions of DNA that vary in length from 20 to 100 bp, depending on the species and cell type. A particular histone named histone H1 is bound to the linker region, as are other types of proteins. The overall structure of connected nucleosomes resembles beads on a string. This structure shortens the length of the DNA molecule about sevenfold.

Nucleosomes Form a 30-nm Fiber

Nucleosome units are organized into a more compact structure that is 30 nm in diameter, known as the **30-nm fiber** (Figure 11.24a). Histone H1 and other proteins are important in the formation of the 30-nm fiber, which shortens the nucleosome structure another sevenfold. The structure of the 30-nm fiber has proven difficult to determine because the conformation of the DNA may be substantially altered when extracted from living cells. A current model for the 30-nm fiber was proposed by Rachel Horowitz-Scherer and Christopher Woodcock in the 1990s (Figure 11.24b). According to their model, linker regions in the 30-nm structure are variably bent and twisted, with little direct contact observed between nucleosomes. The 30-nm fiber forms an asymmetric, three-dimensional zigzag of nucleosomes. At this level of compaction, the overall picture of chromatin that emerges is an irregular, fluctuating structure with stable nucleosome units connected by bendable linker regions.

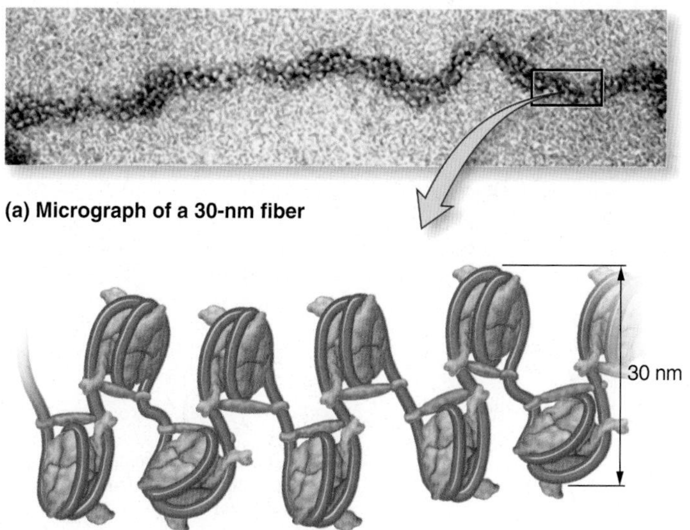

Figure 11.24 **The 30-nm fiber.** **(a)** A photomicrograph of the 30-nm fiber. **(b)** In this three-dimensional zigzag model, the linker DNA forms a bendable structure with little contact between adjacent nucleosomes.

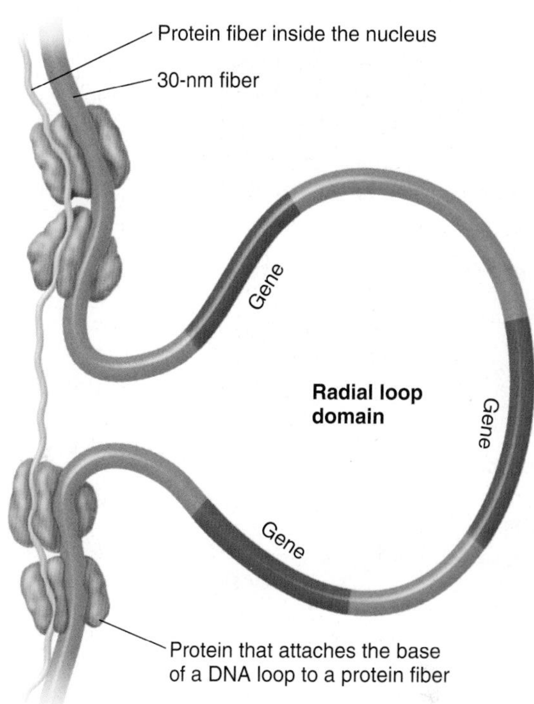

Figure 11.25 Attachment of the 30-nm fiber to a protein fiber to form a radial loop domain.

Concept Check: *What holds the bottoms of the loops in place?*

Chromatin Loops Are Anchored to the Nuclear Matrix

Thus far, we have examined two mechanisms that compact eukaryotic DNA: the formation of nucleosomes and their arrangement into a 30-nm fiber. Taken together, these two events shorten the folded DNA about 49-fold. A third level of compaction involves interactions between the 30-nm fibers and a filamentous network of proteins in the nucleus called the **nuclear matrix**. This matrix consists of the **nuclear lamina**, which is composed of protein fibers that line the inner nuclear membrane (see Chapter 4, Figure 4.15), and an internal nuclear matrix that is connected to the lamina and fills the interior of the nucleus. The internal nuclear matrix is an intricate network of irregular protein fibers plus many other proteins that bind to these fibers. The nuclear matrix is involved in the compaction of the 30-nm fiber by participating in the formation of **radial loop domains**. These loops, often 25,000–200,000 bp in size, are anchored to the nuclear matrix (**Figure 11.25**).

How are chromosomes organized within the cell nucleus? In nondividing cells, each chromosome occupies its own discrete region in the cell nucleus that usually does not overlap with the territory of adjacent chromosomes (see Chapter 4, Figure 4.16). In other words, different chromosomes are not substantially intertwined with each other even when they are in a noncompacted condition.

The compaction level of chromosomes in the cell nucleus is not completely uniform. This variability can be seen with a light microscope and was first observed by German cytologist Emil Heitz in 1928. He used the term **heterochromatin** to describe the highly compacted regions of chromosomes. By comparison, the less condensed regions are known as **euchromatin**. Euchromatin is the form of chromatin in which the 30-nm fiber forms radial loop domains. In heterochromatin, these radial loop domains are compacted even further. In nondividing cells, most chromosomal regions are euchromatic, and some localized regions are heterochromatic.

During Cell Division, Chromosomes Undergo Maximum Compaction

When cells prepare to divide, the chromosomes become even more compacted or condensed. This aids in their proper alignment during metaphase, which is a stage of eukaryotic cell division described in Chapter 15. **Figure 11.26** illustrates the levels of compaction that contribute to the formation of a metaphase chromosome. DNA in the nucleus is always compacted by forming nucleosomes and condensing into a 30-nm fiber (Figure 11.26a,b,c). In euchromatin, the 30-nm fibers are arranged in radial loop domains that are relatively loose, meaning that a fair amount of space is between the 30-nm fibers (Figure 11.26d). The average width of such loops is about 300 nm.

By comparison, heterochromatin involves a much tighter packing of the loops, so little space is between the 30-nm fibers (Figure 11.26e). Heterochromatic regions tend to be wider, about 700 nm. When cells prepare to divide, all of the euchromatin becomes highly compacted. The compaction of euchromatin greatly shortens the chromosomes. In a metaphase chromosome, which contains two copies of the DNA (Figure 11.26f), the width averages about 1,400 nm, but the length of a metaphase chromosome is much shorter than the same chromosome in the nucleus of a nondividing cell.

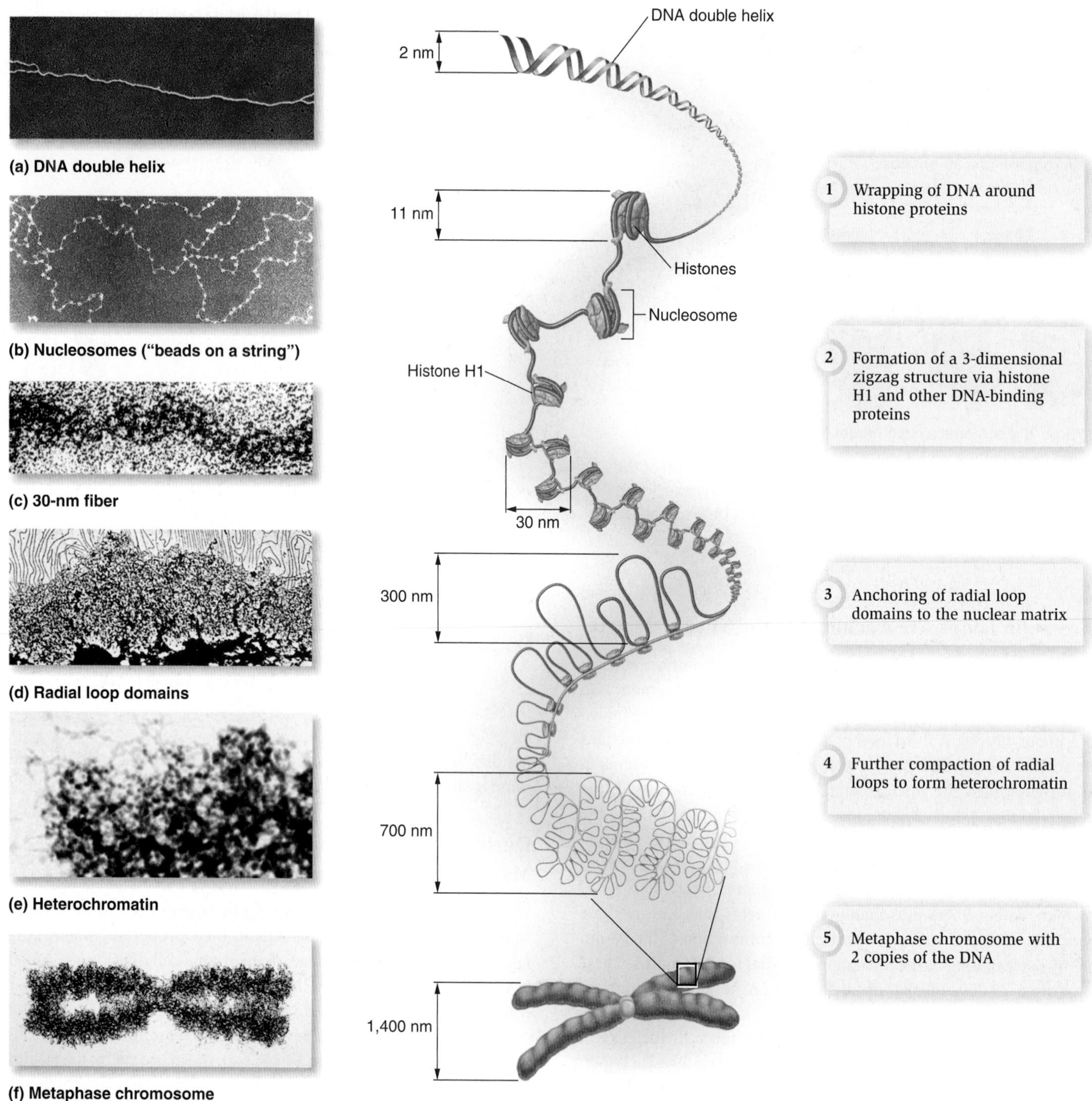

Figure 11.26 The steps in eukaryotic chromosomal compaction leading to the metaphase chromosome.

Concept Check: *After they have replicated and become compacted in preparation for cell division, chromosomes are often shaped like an X, as in part (f) of this figure. Which proteins are primarily responsible for this X shape?*

BioConnections: *Look ahead to Figure 15.8. Why do you think it is necessary for the chromosomes to become compact in preparation for cell division?*

Summary of Key Concepts

11.1 Biochemical Identification of the Genetic Material

- The genetic material carries information to produce the traits of organisms. It is accurately replicated and transmitted from cell to cell and parent to offspring. The genetic material carries differences that explain the variation among different organisms.
- Griffith's work with type S and type R bacteria indicated the existence of a genetic material, which he called the transformation principle (Figure 11.1).
- Avery, MacLeod, and McCarty used biochemical methods to show that DNA is the genetic material (Figure 11.2).
- Hershey and Chase labeled T2 phage with ^{35}S and ^{32}P and determined that DNA is the genetic material of this phage (Figures 11.3, 11.4).

11.2 Nucleic Acid Structure

- DNA is composed of nucleotides, which are covalently linked to form DNA strands. Two DNA strands are held together by hydrogen bonds between the bases to form a double helix. DNA associates with various proteins to form a chromosome (Figure 11.5).
- Nucleotides are composed of a phosphate, sugar, and nitrogenous base. The sugar can be deoxyribose (DNA) or ribose (RNA). The purine bases are adenine and guanine, and the pyrimidine bases are thymine (DNA only), cytosine, and uracil (RNA only). The atoms in a nucleotide are numbered in a conventional way (Figures 11.6, 11.7).
- In a strand of DNA (or RNA), the sugars are connected by covalent bonds in a 5′ to 3′ direction (Figure 11.8).
- The X-ray diffraction data of Franklin, the biochemical data of Chargaff (that is, the amount of A = T and the amount of G = C), and the ball-and-stick model approach of Pauling helped reveal the structure of DNA (Figure 11.9, Table 11.1).
- Watson and Crick determined that DNA is a double helix in which the DNA strands are antiparallel and obey the AT/GC rule (Figures 11.10, 11.11).

11.3 An Overview of DNA Replication

- Meselson and Stahl used ^{15}N- and ^{14}N-isotope labeling methods to show that DNA is replicated by a semiconservative mechanism in which the product of DNA replication is one original strand and one new strand (Figures 11.12, 11.13).
- During DNA replication, the parental double-stranded DNA separates and each strand serves as a template for the synthesis of daughter strands. Nucleotides hydrogen-bond to the template strands according to the AT/GC rule: an adenine (A) base in one strand bonds with a thymine (T) base in the opposite strand, or a guanine (G) base bonds with a cytosine. The result of DNA replication is two double helices with the same base sequence as the parental DNA (Figure 11.14).

11.4 Molecular Mechanism of DNA Replication

- DNA synthesis occurs bidirectionally from an origin of replication. The synthesis of new DNA strands happens near each replication fork (Figure 11.15).
- DNA helicase separates DNA strands, single-strand binding proteins keep them separated, and DNA topoisomerase alleviates coiling ahead of the fork (Figure 11.16).
- Deoxynucleoside triphosphates bind to the template strands according to the AT/GC rule. DNA polymerase recognizes these deoxynucleoside triphosphates and attaches a deoxynucleoside monophosphate to the 3′ end of a growing strand (Figure 11.17).
- DNA polymerase requires a primer and synthesizes new DNA strands only in the 5′ to 3′ direction (Figure 11.18).
- The leading strand is made continuously, in the same direction the fork is moving. The lagging strand is made in the opposite direction of the replication fork as short Okazaki fragments that are synthesized and connected together (Figure 11.19).
- DNA primase makes one RNA primer in the leading strand and multiple RNA primers in the lagging strand. In *E. coli*, DNA polymerase III extends these primers with DNA, and DNA polymerase I removes the primers when they are no longer needed and fills in with DNA. DNA ligase connects adjacent Okazaki fragments in the lagging strand (Figure 11.20, Table 11.2).
- DNA replication is very accurate because (1) hydrogen bonding according to the AT/CG rule is more stable; (2) DNA polymerase is unlikely to catalyze bond formation if a mismatched base pair is formed; and (3) DNA polymerase carries out proofreading.
- Living organisms have several different types of DNA polymerases with specialized functions (Table 11.3).
- The ends of linear, eukaryotic chromosomes have telomeres composed of repeat sequences. Telomerase binds to the telomere repeat sequence and synthesizes a 6-nucleotide repeat. This happens many times in a row to lengthen one DNA strand of the telomere. DNA primase, DNA polymerase, and DNA ligase are needed to synthesize the complementary DNA strand (Figures 11.21, 11.22).

11.5 Molecular Structure of Eukaryotic Chromosomes

- Chromosomes are structures in living cells that carry the genetic material. Chromatin is the name given to the complex of DNA and protein that makes up chromosomes.
- In eukaryotic chromosomes, the DNA is wrapped around histone proteins to form nucleosomes. Nucleosomes are further compacted into 30-nm fibers. The linker regions are variably twisted and bent into a zigzag pattern (Figures 11.23, 11.24).
- A third level of compaction of eukaryotic chromosomes involves the formation of radial loop domains in which the bases of 30-nm fibers are anchored to a network of proteins called the nuclear matrix. This level of compaction is called euchromatin. In heterochromatin, the loops are even more closely packed together (Figure 11.25).
- During cell division, chromosomes become even more condensed (Figure 11.26).

Assess and Discuss

Test Yourself

1. Why did researchers initially believe the genetic material was protein?
   a. Proteins are more biochemically complex than DNA.
   b. Proteins are found only in the nucleus, but DNA is found in many areas of the cell.
   c. Proteins are much larger molecules and can store more information than DNA.
   d. all of the above
   e. both a and c
2. Considering the components of a nucleotide, what component is always different when comparing nucleotides in a DNA strand or an RNA strand?
   a. phosphate group
   b. pentose sugar
   c. nitrogenous base
   d. both b and c
   e. a, b, and c
3. Which of the following equations is appropriate when considering DNA base composition?
   a. %A + %T = %G + %C
   b. %A = %G
   c. %A = %G = %T = %C
   d. %A + %G = %T + %C
4. If the sequence of a segment of DNA is 5′–CGCAACTAC–3′, what is the appropriate sequence for the opposite strand?
   a. 5′–GCGTTGATG–3′
   b. 3′–ATACCAGCA–5′
   c. 5′–ATACCAGCA–3′
   d. 3′–GCGTTGATG–5′
5. Of the following statements, which is correct when considering the process of DNA replication?
   a. New DNA molecules are composed of two completely new strands.
   b. New DNA molecules are composed of one strand from the old molecule and one new strand.
   c. New DNA molecules are composed of strands that are a mixture of sections from the old molecule and sections that are new.
   d. none of the above
6. Meselson and Stahl were able to demonstrate semiconservative replication in *E. coli* by
   a. using radioactive isotopes of phosphorus to label the old strand and visually determining the relationship of old and new DNA strands.
   b. using different enzymes to eliminate old strands from DNA.
   c. using isotopes of nitrogen to label the DNA and determining the relationship of old and new DNA strands by density differences of the new molecules.
   d. labeling viral DNA before it was incorporated into a bacterial cell and visually determining the location of the DNA after centrifugation.
7. During replication of a DNA molecule, the daughter strands are not produced in exactly the same manner. One strand, the leading strand, is made toward the replication fork, while the lagging strand is made in fragments in the opposite direction. This difference in the synthesis of the two strands is the result of which of the following?
   a. DNA polymerase is not efficient enough to make two "good" strands of DNA.
   b. The two template strands are antiparallel, and DNA polymerase makes DNA only in the 5′ to 3′ direction.
   c. The lagging strand is the result of DNA breakage due to UV light.
   d. The cell does not contain enough nucleotides to make two complete strands.
8. In eukaryotic cells, chromosomes consist of
   a. DNA and RNA.
   b. DNA only.
   c. RNA and proteins.
   d. DNA and proteins.
   e. RNA only.
9. A nucleosome is
   a. a dark-staining body composed of RNA and proteins found in the nucleus.
   b. a protein that helps organize the structure of chromosomes.
   c. another word for a chromosome.
   d. a structure composed of DNA wrapped around eight histones.
   e. the short arm of a chromosome.
10. The conversion of euchromatin into heterochromatin involves
    a. the formation of more nucleosomes.
    b. the formation of less nucleosomes.
    c. a greater compaction of loop domains.
    d. a lesser compaction of loop domains.
    e. both a and c.

Conceptual Questions

1. What are the four key characteristics of the genetic material? What was Frederick Griffith's contribution to the study of DNA, and why was it so important?
2. The Hershey and Chase experiment used radioactive isotopes to track the DNA and protein of phages as they infected bacterial cells. Explain how this procedure allowed them to determine that DNA is the genetic material of this particular virus.
3. A principle of biology is that *structure determines function.* Discuss how the structure of DNA underlies different aspects of its function.

Collaborative Questions

1. A trait that some bacterial strains exhibit is resistance to killing by antibiotics. For example, certain strains of bacteria are resistant to the drug tetracycline, whereas other strains are sensitive to this antibiotic. Describe an experiment you would carry out to demonstrate that tetracycline resistance is an inherited trait carried in the DNA of the resistant strain.
2. How might you provide evidence that DNA is the genetic material in mice?

Online Resource

www.brookerbiology.com

Stay a step ahead in your studies with animations that bring concepts to life and practice tests to assess your understanding. Your instructor may also recommend the interactive eBook, individualized learning tools, and more.

Chapter Outline

12.1 Overview of Gene Expression

12.2 Transcription

12.3 RNA Processing in Eukaryotes

12.4 Translation and the Genetic Code

12.5 The Machinery of Translation

12.6 The Stages of Translation

Summary of Key Concepts

Assess and Discuss

Gene Expression at the Molecular Level

12

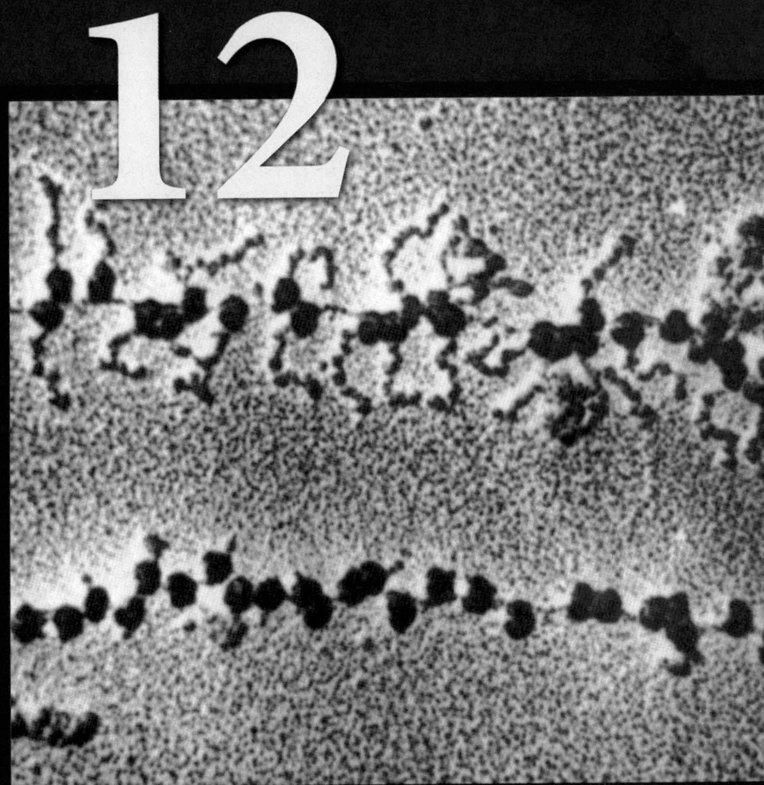

An electron micrograph of many ribosomes in the act of translating two mRNA molecules into many polypeptides. The complex of one mRNA and many ribosomes is called a polysome. Short polypeptides are seen emerging from the ribosomes.

Mina, age 21, works part-time in an ice-cream shop and particularly enjoys the double-dark chocolate and chocolate fudge brownie flavors on her breaks. She exercises little and spends most of her time studying or watching television. Mina is effortlessly thin. She never worries about what or how much she eats. By comparison, her close friend, Rezzy, has struggled with her weight as long as she can remember. Compared with Mina, she feels like she must constantly deprive herself of food just to maintain her current weight—a weight she would describe as 30 pounds too much.

How do we explain the differences between Mina and Rezzy? Two fundamental factors are involved. Our weight is strongly influenced by the environment, especially our diet, as well as by social and behavioral factors. The amount and types of food we eat are correlated with weight gain. However, there is little doubt that our weight is also influenced by variation in our genes. Obesity, the condition of having too much body fat, runs in families. The degree of obesity is often similar between genetically identical twins who have been raised apart. Why has genetic variation resulted in some genes that cause certain people to gain weight? A popular hypothesis is that some people have inherited "thrifty genes" as hand-me-downs from their ancestors, who periodically faced famines and food scarcity. Such thrifty genes would be advantageous in allowing people to store body fat more easily and to use food resources more efficiently when times are lean. The negative side is that when food is abundant, unwanted weight gain—and associated diseases such as diabetes and heart disease—can constitute a serious health problem.

Why do we care about our genes? Let's consider this question with regard to obesity. Researchers have identified several key genes that influence a person's predisposition to becoming obese. Dozens more are likely to play a minor role. By identifying those genes and studying the proteins specified by those genes, researchers may gain a better understanding of how genetic variation causes certain people to gain weight more easily than others. In addition, this knowledge has led to the development of drugs that are used to combat obesity.

We can broadly define a gene as a unit of heredity. Geneticists view gene function at different biological levels. In Chapter 16, we will examine how genes affect the traits or characteristics of individuals. For example, we will consider how the transmission of genes from parents to offspring affects the color of the offspring's eyes and their likelihood of becoming color blind. In this chapter, we will explore how genes work at the molecular level. You will learn how DNA sequences are organized to form genes and how those genes are used as a template to make RNA copies, ultimately leading to the synthesis of a functional protein. The term **gene expression** can refer to gene function either at the level of traits or at the molecular level. In reality, the two phenomena are intricately woven together. The expression of genes at the molecular level affects the structure and function of cells, which, in turn, determine the traits that an organism expresses.

We begin this chapter by considering how researchers came to realize that most genes store the information to make proteins. We then explore the steps of gene expression as they occur at the molecular level. These steps include the use of a gene as a template to make an RNA molecule, the processing of the RNA into a functional molecule (in eukaryotes), and the use of RNA to direct the formation of a protein.

12.1 Overview of Gene Expression

Learning Outcomes:

1. Analyze the results of the experiments of Garrod and of Beadle and Tatum.
2. Outline the general steps of gene expression at the molecular level, which together constitute the central dogma.
3. Explain how proteins are largely responsible for determining an organism's characteristics.

Even before DNA was known to be the genetic material, scientists had asked the question, "How does the functioning of genes produce the traits of living organisms?" At the molecular level, a similar question can be asked. "How do genes affect the composition and/or function of molecules found within living cells?" An approach that was successful in answering these questions involved the study of **mutations**, which are changes in the genetic material that can be inherited. Mutations may affect the genetic blueprint by altering gene function. For this reason, research that focused on the effects of mutations proved instrumental in determining the molecular function of genes.

In this section, we will consider two early experiments in which researchers studied the effects of mutations in humans and in a bread mold. Both studies led to the conclusion that the role of some genes is to carry the information to produce enzymes, which are a type of protein. Then we will examine the general features of gene expression at the molecular level.

The Study of Inborn Errors of Metabolism Suggested That Some Genes Carry the Information to Make Enzymes

In 1908, Archibald Garrod, a British physician, proposed a relationship between genes and the production of enzymes. Prior to his work, biochemists had studied many metabolic pathways that consist of a series of conversions of one molecule to another, each step catalyzed by an enzyme. Figure 12.1 illustrates part of the metabolic pathway for the breakdown of phenylalanine, an amino acid commonly found in human diets. The enzyme phenylalanine hydroxylase catalyzes the conversion of phenylalanine to tyrosine, another amino acid. A different enzyme, tyrosine aminotransferase, converts tyrosine into the next molecule, called *p*-hydroxyphenylpyruvic acid. In each case, a specific enzyme catalyzes a single chemical reaction.

Much of Garrod's early work centered on the inherited disease alkaptonuria, in which the patient's body accumulates abnormal levels of homogentisic acid (also called alkapton). This compound, which is bluish black, results in discoloration of the skin and cartilage and causes the urine to appear black. Garrod hypothesized that the accumulation of homogentisic acid in these patients is due to a defect in an enzyme, namely, homogentisic acid oxidase (see Figure 12.1). Furthermore, he already knew that alkaptonuria is an inherited condition that follows a recessive pattern of inheritance. As discussed in Chapter 16, if a disorder is recessive, an individual with the disease has inherited the mutant (defective) gene that causes the disorder from both parents.

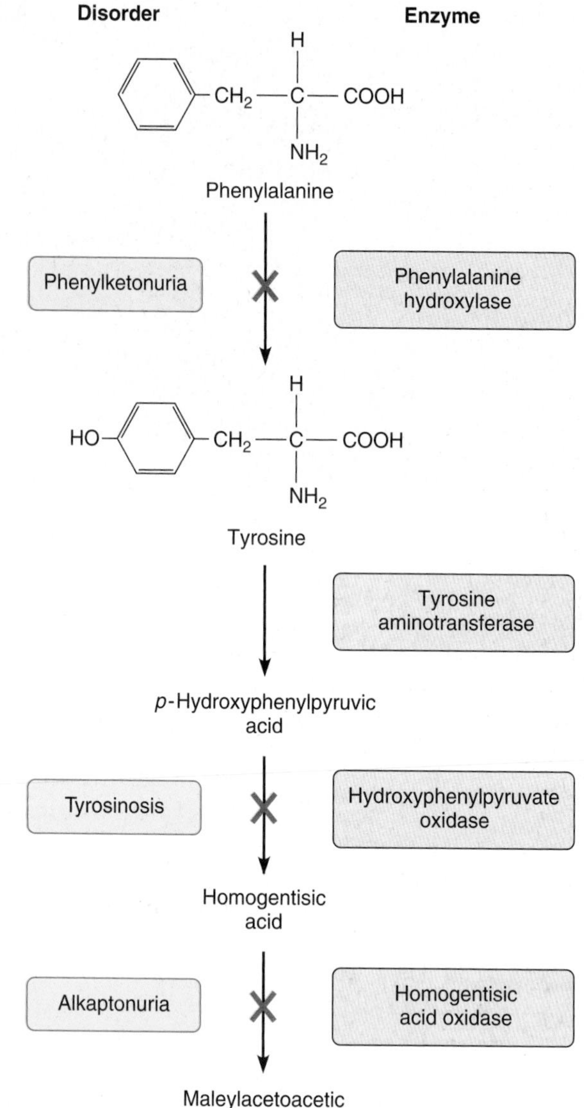

Figure 12.1 The metabolic pathway that breaks down phenylalanine and its relationship to certain genetic diseases. Each step in the pathway is catalyzed by a different enzyme, shown in the boxes on the right. If one of the enzymes is not functioning, the previous compound builds up, causing the disorders named in the boxes on the left.

Concept Check: What disease would occur if a person had inherited two defective copies of the gene that encodes phenylalanine hydroxylase?

How did Garrod explain these observations? In 1908, he proposed a relationship between the inheritance of a mutant gene and a defect in metabolism. In the case of alkaptonuria, if an individual inherited the mutant gene from both parents, she or he would not produce any normal enzyme and would be unable to metabolize homogentisic acid. Garrod described alkaptonuria as an **inborn error of metabolism**. An inborn error refers to a mutation in a gene that is inherited from one or both parents. At the turn of the last century,

During **RNA processing**, which is described later in this chapter, the RNA transcript, termed **pre-mRNA**, is modified in ways that make it a functionally active mRNA (**Figure 12.3b**).

Another difference between bacteria and eukaryotes is the cellular location of transcription and translation. In bacteria, both events occur in the same location, namely, the cytoplasm. In eukaryotes, transcription occurs in the nucleus. The mRNA then exits the nucleus through a nuclear pore, and translation occurs in the cytosol.

Though the direction of information flow, that is, from DNA to RNA to protein, is the most common pathway, exceptions do occur. For example, certain viruses use RNA as a template to synthesize DNA. Such viruses are described in Chapter 18.

The Protein Products of Genes Determine an Organism's Characteristics

The genes that constitute the genetic material provide a blueprint for the characteristics of every organism. They contain the information necessary to produce an organism and allow it to favorably interact with its environment. Each structural gene stores the information for the production of a polypeptide, which then becomes a unit within a functional protein. The activities of proteins determine the structure and function of cells. Furthermore, the characteristics of an organism are rooted in the activities of cellular proteins.

The main purpose of the genetic material is to encode the production of proteins in the correct cell, at the proper time, and in suitable amounts. This is an intricate task, because living cells make thousands of different kinds of proteins. Genetic analyses have shown that a typical bacterium can make a few thousand different proteins, and estimates for eukaryotes range from several thousand in simpler eukaryotes to tens of thousands in more complex eukaryotes like humans.

12.2 Transcription

Learning Outcomes:

1. Describe how a gene is an organization of DNA sequences that can be transcribed into RNA.
2. Outline the three stages of transcription and the role of RNA polymerase in this process.
3. Explain how genes within the same chromosome vary in their direction of transcription.
4. Compare and contrast transcription in bacteria and eukaryotes.

DNA is an information storage unit. For genes to be expressed, the information in them must be accessed at the molecular level. Rather than accessing the information directly, however, a working copy of the DNA, composed of RNA, is made. This occurs by the process of transcription, in which a DNA sequence is copied into an RNA sequence. Importantly, transcription does not permanently alter the structure of DNA. Therefore, the same DNA can continue to store information even after an RNA copy has been made. In this section, we will examine the steps necessary for genes to act as transcriptional units. We will also consider some differences in these steps between bacteria and eukaryotes.

At the Molecular Level, a Gene Is Transcribed and Produces a Functional Product

What is a gene? At the molecular level, a **gene** is defined in the following way:

> *A gene is an organized unit of DNA sequences that enables a segment of DNA to be transcribed into RNA and ultimately results in the formation of a functional product.*

When a structural gene is transcribed, an mRNA is made that specifies the amino acid sequence of a polypeptide. After it is made, the polypeptide becomes a functional product. The mRNA is an intermediary in polypeptide synthesis. Among all species, most genes are structural genes. However, for some genes, the functional product is the RNA itself. The RNA from a nonstructural gene is never translated. Two important products of nonstructural genes are transfer RNA and ribosomal RNA. **Transfer RNA (tRNA)** translates the language of mRNA into that of amino acids. **Ribosomal RNA (rRNA)** forms part of ribosomes, which provide the site where translation occurs. We'll learn more about these two types of RNA later in this chapter.

A gene is composed of specific base sequences organized in a way that allows the DNA to be transcribed into RNA. **Figure 12.4** shows the general organization of sequences in a structural gene. Transcription begins next to a site in the DNA called the **promoter**, whereas the **terminator** specifies the end of transcription. Therefore, transcription occurs between these two boundaries. As shown in Figure 12.4, the DNA is transcribed into mRNA from the end of the

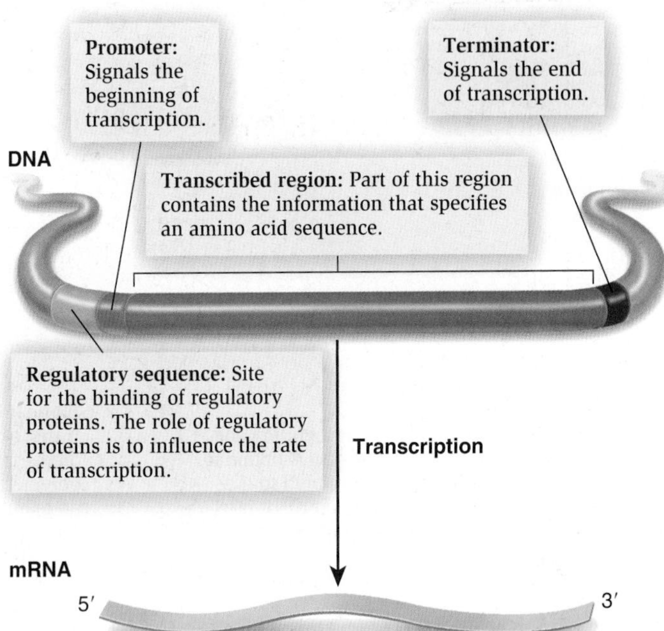
Promoter: Signals the beginning of transcription.
Terminator: Signals the end of transcription.
DNA
Transcribed region: Part of this region contains the information that specifies an amino acid sequence.
Regulatory sequence: Site for the binding of regulatory proteins. The role of regulatory proteins is to influence the rate of transcription.
Transcription
mRNA
5′
3′

Figure 12.4 A structural gene as a transcriptional unit.

BIOLOGY PRINCIPLE The genetic material provides a blueprint for reproduction. Transcription is the first step in accessing the information that is stored in DNA.

Concept Check: *How would removing a terminator from a gene affect transcription? Where would transcription end?*

promoter through the coding sequence to the terminator. Within this transcribed region is the information that will specify the amino acid sequence of a polypeptide when the mRNA is translated.

Other DNA sequences are involved in the regulation of transcription. These **regulatory sequences** function as sites for the binding of regulatory proteins, which are discussed in Chapter 13. When a regulatory protein binds to a regulatory sequence, the rate of transcription is affected. Some regulatory proteins enhance the rate of transcription, whereas others inhibit it.

During Transcription, RNA Polymerase Uses a DNA Template to Make RNA

Transcription occurs in three stages, called initiation, elongation, and termination, during which various proteins interact with DNA sequences (Figure 12.5). **Initiation** is a recognition step. In bacteria such as *E. coli*, a protein called **sigma factor** binds to **RNA polymerase**, the enzyme that synthesizes strands of RNA. Sigma factor also recognizes the base sequence of a promoter and binds there. An example of a promoter sequence is described in the legend to Figure 12.5. The role of sigma factor is to cause RNA polymerase to bind to the promoter. The initiation stage is completed when the DNA strands are separated near the promoter to form an **open complex** that is approximately 10–15 bp long.

During **elongation**, RNA polymerase synthesizes the RNA transcript. For this to occur, sigma factor is released and RNA polymerase slides along the DNA in a way that maintains an open complex as it goes. The DNA strand that is used as a template for RNA synthesis is called the **template strand**. The opposite DNA strand is called the **coding strand**. The coding strand has the same sequence of bases as the resulting mRNA, except that thymine in the DNA is substituted for uracil in the RNA. The coding strand is so named because, like mRNA, it carries the information that codes for a polypeptide.

During the elongation stage of transcription, nucleotides bind to the template strand and are covalently connected in the 5′ to 3′ direction (see inset of step 2, Figure 12.5). The complementarity rule used in this process is similar to the AT/GC rule of DNA replication, except that uracil (U) in RNA substitutes for thymine (T) in DNA. For example, a DNA template with a sequence of 3′–TACAAT-GTAGCC–5′ will be transcribed into an RNA sequence reading

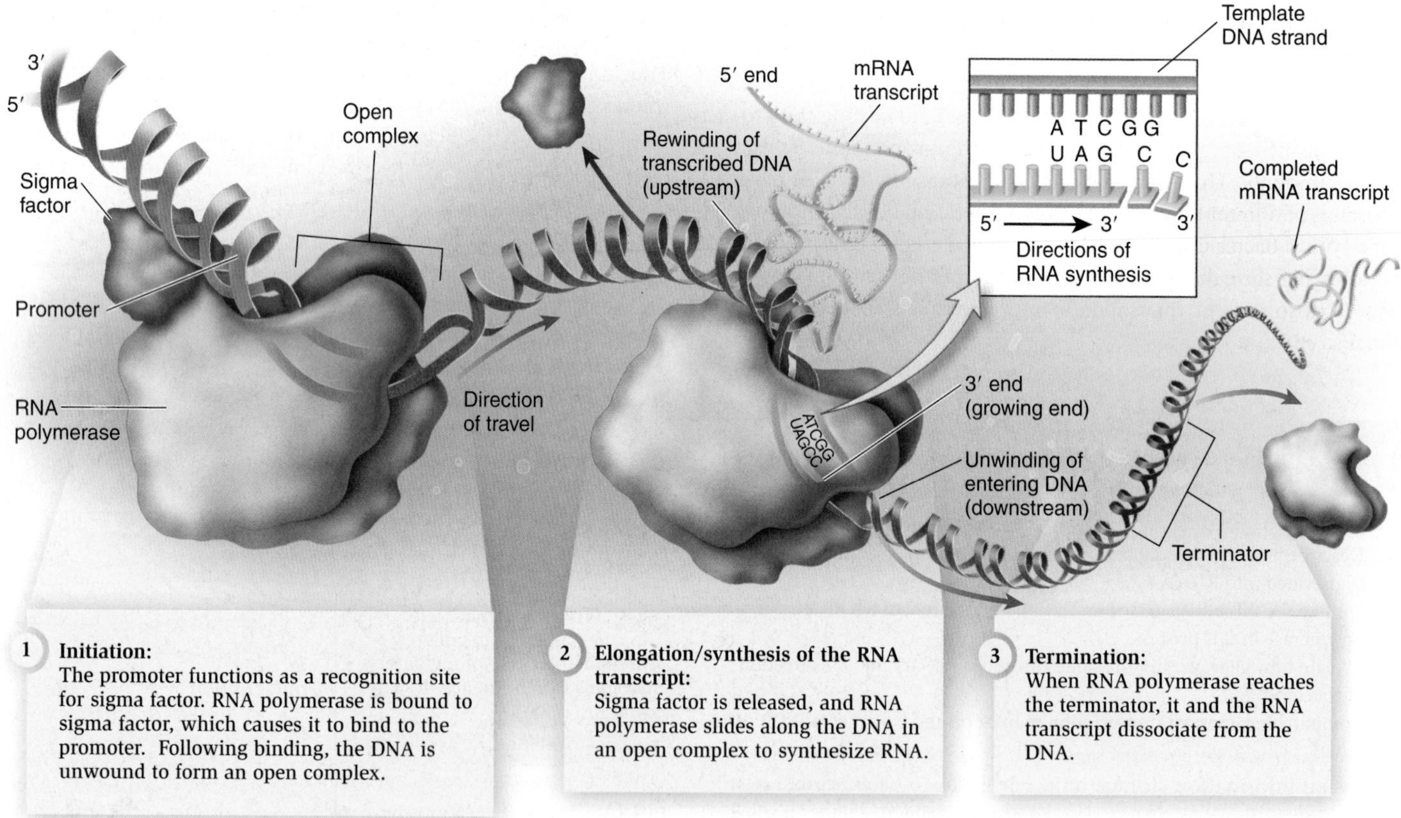
3′
5′
Sigma factor
Promoter
RNA polymerase
Open complex
Direction of travel
5′ end
mRNA transcript
Rewinding of transcribed DNA (upstream)
Template DNA strand
A T C G G
U A G C C
5′ 3′ 3′
Directions of RNA synthesis
Completed mRNA transcript
3′ end (growing end)
Unwinding of entering DNA (downstream)
Terminator
ATCGG
UAGCC
1 Initiation:
The promoter functions as a recognition site for sigma factor. RNA polymerase is bound to sigma factor, which causes it to bind to the promoter. Following binding, the DNA is unwound to form an open complex.
2 Elongation/synthesis of the RNA transcript:
Sigma factor is released, and RNA polymerase slides along the DNA in an open complex to synthesize RNA.
3 Termination:
When RNA polymerase reaches the terminator, it and the RNA transcript dissociate from the DNA.

Figure 12.5 Stages of transcription. Transcription can be divided into initiation, elongation, and termination. The inset emphasizes the direction of RNA synthesis and base pairing between the DNA template strand and RNA. An example of a promoter sequence in *E. coli* is:

5′-TTGACATGATAGAAGCACTCTACTATATT-3′
3′-AACTGTACTATCTTCGTGAGATGATATAA-5′

This region is 29 bp long. The bases that are specifically recognized by sigma factor are shown in red. The sequences of promoters for different genes are fairly diverse, particularly in eukaryotic species.

BioConnections: Look back at DNA polymerase described in Figure 11.17. What are similarities and differences between the function of DNA polymerase and that of RNA polymerase?

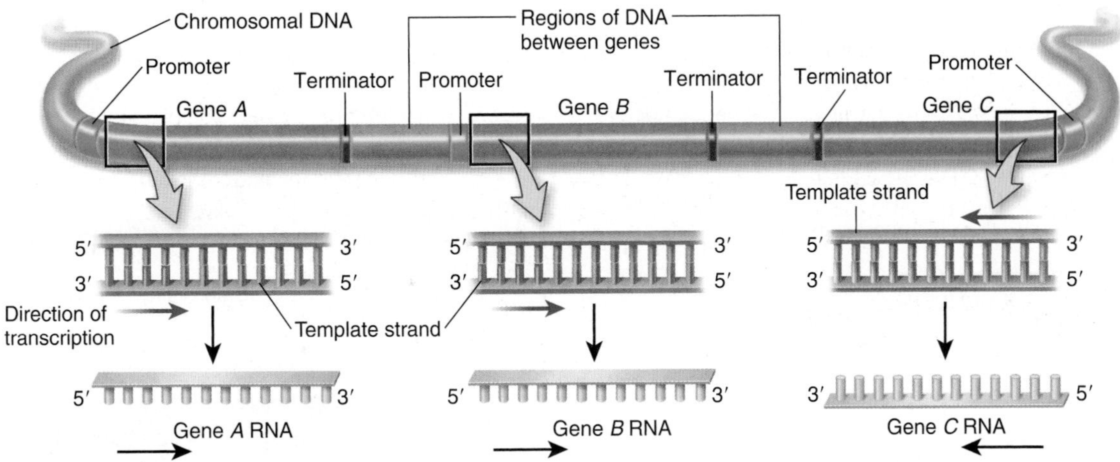

Figure 12.6 The transcription of three different genes found in the same chromosome. RNA polymerase synthesizes each RNA transcript in a 5′ to 3′ direction, sliding along a DNA template strand in a 3′ to 5′ direction. However, the use of the template strand can vary from gene to gene. For example, genes *A* and *B* use the bottom strand, while gene *C* uses the top strand.

5′–AUGUUACAUCGG–3′. In bacteria, the rate of RNA synthesis is about 40 nucleotides per second! Behind the open complex, the DNA rewinds back into a double helix. Eventually, RNA polymerase reaches a terminator, which causes it and the newly made RNA transcript to dissociate from the DNA. This event constitutes the **termination** of transcription.

When considering the transcription of multiple genes within a chromosome, the DNA strand that is used as the template strand varies among different genes. Figure 12.6 shows three genes adjacent to each other within a chromosome. Genes *A* and *B* are transcribed from left to right, using the bottom DNA strand as the template strand. By comparison, gene *C* is transcribed from right to left, using the top DNA strand as a template strand. In all three cases, however, the synthesis of the RNA transcript begins at a promoter and always occurs in a 5′ to 3′ direction. The template strand is read in the 3′ to 5′ direction.

Transcription in Eukaryotes Involves More Proteins

The basic features of transcription are similar among all organisms. The genes of all species have promoters, and the transcription process occurs in the stages of initiation, elongation, and termination. However, compared with bacteria, the transcription of eukaryotic genes tends to involve a greater complexity of protein components. For example, three forms of RNA polymerase, designated I, II, and III, are found in eukaryotes. The catalytic portion of RNA polymerase responsible for the synthesis of RNA has a similar protein structure in all species. RNA polymerase II is responsible for transcribing the mRNA from eukaryotic structural genes, whereas RNA polymerases I and III transcribe nonstructural genes such as the genes that encode tRNAs and rRNAs. By comparison, bacteria have a single type of RNA polymerase that transcribes all genes, though many bacterial species have more than one type of sigma factor that can recognize different promoters.

The initiation stage of transcription in eukaryotes is also more complex. Recall that in bacteria such as *E. coli*, sigma factor recognizes the promoter of genes. By comparison, RNA polymerase II of eukaryotes always requires five general transcription factors to initiate transcription. **Transcription factors** are proteins that influence the ability of RNA polymerase to transcribe genes. In addition, the regulation of gene transcription in eukaryotes typically involves the function of several different proteins. The roles of eukaryotic transcription factors are considered in Chapter 13.

12.3 RNA Processing in Eukaryotes

Learning Outcomes:

1. Explain the process of splicing that produces mature eukaryotic mRNA.
2. Describe the addition of the 5′ cap and 3′ poly A tail to eukaryotic mRNA.

As noted previously, eukaryotic mRNA transcripts undergo modification, or RNA processing, to produce a functional mRNA. Transcription initially produces a longer RNA, called **pre-mRNA**, which undergoes certain processing events before it exits the nucleus. The final product is called a **mature mRNA**, or simply mRNA (Figure 12.7).

In the late 1970s, when the experimental tools became available to study eukaryotic genes at the molecular level, the scientific community was astonished by the discovery that the coding sequences within many eukaryotic structural genes are separated by DNA sequences that are transcribed but not translated into protein. These intervening sequences that are not translated are called **introns**, whereas sequences contained in the mature mRNA are termed **exons**. Exons are expressed regions, whereas introns are intervening regions that are not expressed because they are removed from the mRNA.

To become a functional mRNA, the pre-mRNA undergoes a process known as **RNA splicing**, or simply splicing, in which introns are removed and the remaining exons are connected to each other (see Figure 12.7). In addition to splicing, eukaryotic pre-mRNA transcripts are modified in other ways, including the addition of caps and tails to the ends of the mRNA. After these modifications have been completed, the mRNA leaves the nucleus and enters the cytosol, where translation occurs. In this section, we will examine the molecular mechanisms that account for RNA processing events and consider why they are functionally important.

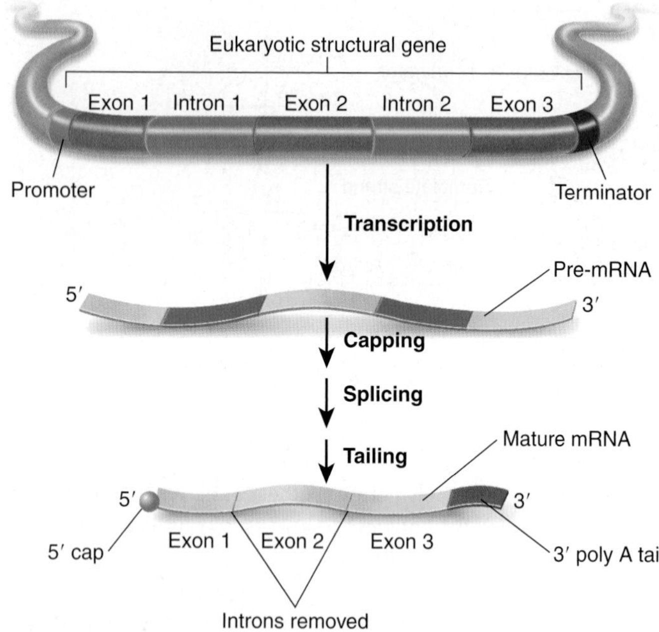

Figure 12.7 **Modifications to eukaryotic pre-mRNA that are needed to produce a mature mRNA molecule.**

Splicing Involves the Removal of Introns and the Linkage of Exons

Introns are found in many but not all eukaryotic genes. Splicing is less frequent among unicellular eukaryotic species, such as yeast, but is a widespread phenomenon among more complex eukaryotes. In animals and flowering plants, most structural genes have one or more introns. For example, an average human gene has about nine introns. The sizes of introns vary from a few dozen nucleotides to over 100,000! A few bacterial genes have been found to have introns, but they are rare among bacterial and archaeal species.

Introns are precisely removed from eukaryotic pre-mRNA by a large complex called a **spliceosome** that is composed of several different subunits known as snRNPs (pronounced "snurps"). Each snRNP contains small nuclear RNA and a set of proteins. This small nuclear RNA is the product of a nonstructural gene. Intron RNA is defined by a particular sequence within the intron termed the branch site and by two intron-exon boundaries, called the 5′ splice site and the 3′ splice site (**Figure 12.8**). Spliceosome subunits bind to specific sequences at these three locations. This binding causes the intron to loop outward, which brings the two exons close together. The 5′ splice site is then cut, and the 5′ end of the intron becomes covalently attached to the branch site. In the final step, the 3′ splice site is cut, and the two exons are covalently attached to each other. The intron is released and eventually degraded.

In some cases, the function of the spliceosome is regulated so the splicing of exons for a given mRNA can occur in two or more ways. This phenomenon, called **alternative splicing**, allows a single gene to encode two or more polypeptides with differences in their amino acid sequences. As described in Chapter 13, alternative splicing allows complex eukaryotic species to use the same gene to make different proteins at different stages of development or in different cell types. This increases the size of the proteome while minimizing the size of the genome.

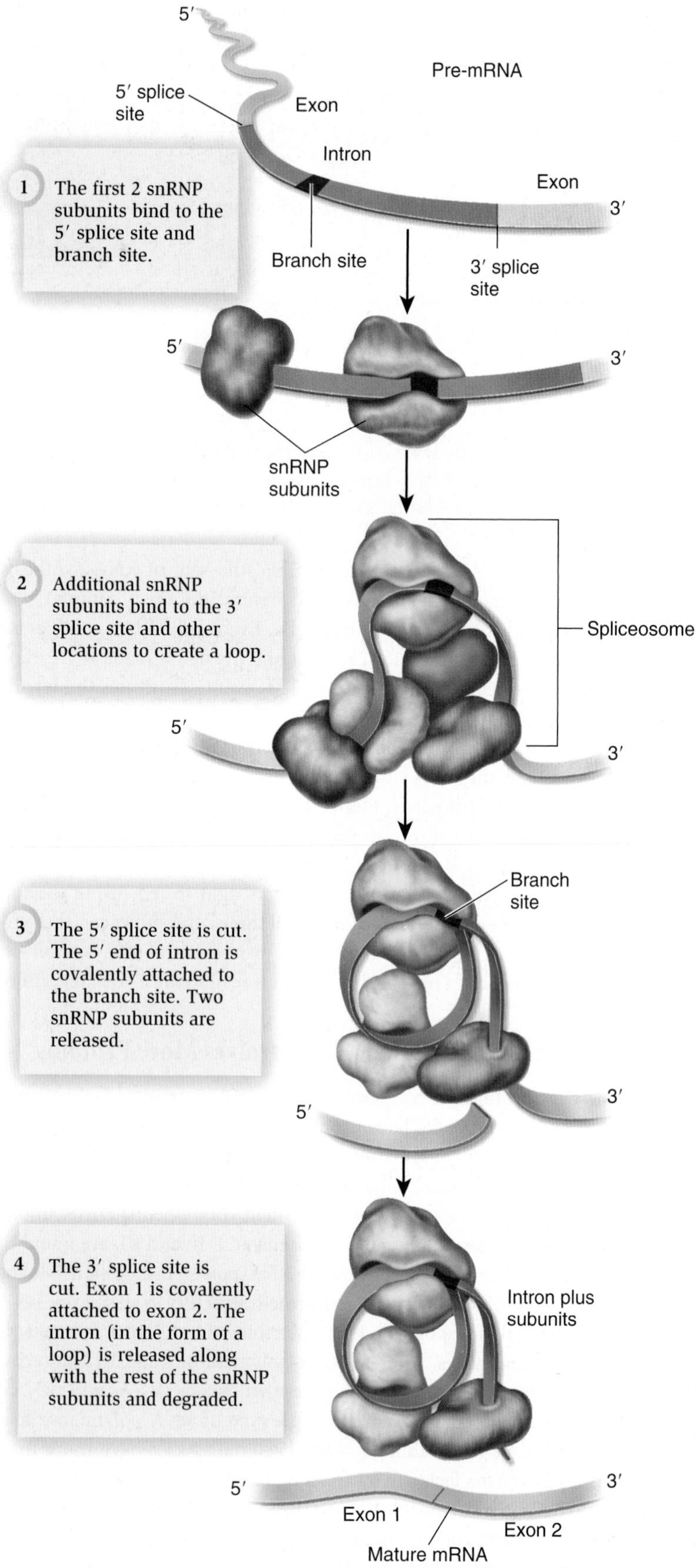

5′
5′ splice site
Exon
Pre-mRNA
Intron
Exon
3′
Branch site
3′ splice site

1 The first 2 snRNP subunits bind to the 5′ splice site and branch site.

snRNP subunits

2 Additional snRNP subunits bind to the 3′ splice site and other locations to create a loop.

Spliceosome

3 The 5′ splice site is cut. The 5′ end of intron is covalently attached to the branch site. Two snRNP subunits are released.

Branch site

4 The 3′ splice site is cut. Exon 1 is covalently attached to exon 2. The intron (in the form of a loop) is released along with the rest of the snRNP subunits and degraded.

Intron plus subunits

Exon 1
Exon 2
Mature mRNA

Figure 12.8 **The splicing of a eukaryotic pre-mRNA by a spliceosome.**

Although primarily found in mRNAs, introns occasionally occur in rRNA and tRNA molecules of certain species. These introns, however, are not removed by the action of a spliceosome. Instead, such rRNAs and tRNAs are **self-splicing**, which means the RNA itself can catalyze the removal of its own intron. Portions of the RNA act like an enzyme to cleave the covalent bonds at the intron-exon boundaries and connect the exons together. An RNA molecule that catalyzes a chemical reaction is termed a **ribozyme**.

RNA Processing Also Involves the Addition of a 5′ Cap and a 3′ Poly A Tail to Eukaryotic mRNAs

Mature mRNAs of eukaryotes have a modified form of guanine covalently attached at the 5′ end, an event known as **capping** (Figure 12.9a). Capping occurs while a pre-mRNA is being made by RNA polymerase, usually when the transcript is only 20 to 25 nucleotides in length. What are the functions of the cap? The 7-methylguanosine structure, called a **5′ cap**, is recognized by cap-binding proteins, which are needed for the proper exit of mRNAs from the nucleus. After an mRNA is in the cytosol, the cap structure helps to prevent mRNA degradation and is recognized by other cap-binding proteins that enable the mRNA to bind to a ribosome for translation.

At the 3′ end, most mature eukaryotic mRNAs have a string of adenine nucleotides, typically 100 to 200 nucleotides in length, referred to as a **poly A tail** (Figure 12.9b). The poly A tail is not encoded in the gene sequence. Instead, the tail is added enzymatically after a pre-mRNA has been completely transcribed. A long poly A tail aids in the export of mRNAs from the nucleus. It also causes a eukaryotic mRNA to be more stable and thereby exist for a longer period of time in the cytosol. Interestingly, new research has shown that some bacterial mRNAs also have poly A tails attached to them. However, the poly A tail has an opposite effect in bacteria, where it causes the mRNA to be rapidly degraded. The importance of a poly A tail in bacterial mRNAs is not well understood.

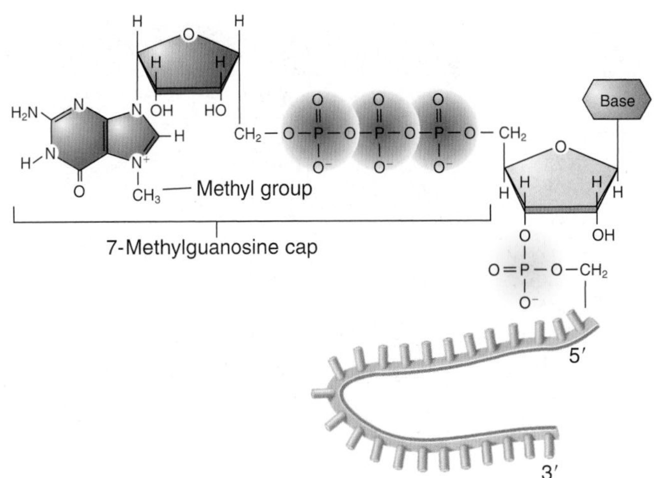

(a) Cap structure at the 5′ end of eukaryotic mRNA

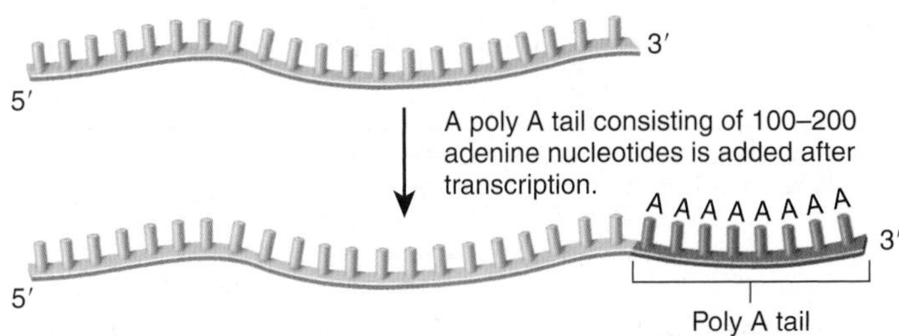

(b) Addition of a poly A tail at the 3′ end of eukaryotic mRNA

Figure 12.9 **Modifications that occur at the ends of mRNA in eukaryotic cells.** **(a)** A guanosine cap is attached to the 5′ end. This guanine base is modified by the attachment of a methyl group. The linkage between the cap and the mRNA is a 5′ to 5′ linkage. **(b)** A poly A tail is added to the 3′ end.

Concept Check: Do the ends of structural genes have a poly T region that provides a template for the synthesis of a poly A tail in mRNA? Explain.

12.4 Translation and the Genetic Code

Learning Outcomes:

1. Explain how the genetic code specifies the relationship between the sequence of codons in mRNA and the amino acid sequence of a polypeptide.
2. Analyze the experiments of Nirenberg and Leder that led to the deciphering of the genetic code.

In the two previous sections, we considered the first stages of the central dogma—how an RNA transcript is made from DNA and how eukaryotes process that transcript. Recall that this type of RNA is called messenger RNA (mRNA), because its function is to transmit information from DNA to cellular components called ribosomes, where polypeptide synthesis occurs. In this section, we will consider the next stage, translation, which is the synthesis of polypeptides using information from the mRNA. To understand the process of translation, we will first examine the **genetic code**, which specifies the relationship between the sequence of nucleotides in the mRNA and the sequence of amino acids in a polypeptide.

The Genetic Code Specifies the Amino Acids within a Polypeptide

The ability of mRNA to be translated into an amino acid sequence of a polypeptide relies on the genetic code. The code is read in groups of three nucleotide bases known as **codons**. The genetic code consists of 64 different codons (**Table 12.1**). The sequence of three bases in most codons specifies a particular amino acid. For example, the codon CCC specifies the amino acid proline, whereas the codon GGC encodes the amino acid glycine. From the analysis of many different species, including bacteria, archaea, protists, fungi, plants, and animals, researchers have found that the genetic code is nearly universal. Only a few rare exceptions to the genetic code have been discovered.

Why are there 64 codons, as shown in Table 12.1? Because there are 20 types of amino acids, at least 20 different codons are needed so each amino acid can be specified by a codon. With four types of bases in mRNA (U, C, A, and G), a genetic code containing two bases in a codon would not be sufficient, because only 4^2, or 16, different codons would be possible. A three-base system can specify 4^3, or 64, different codons, which is far more than the number of amino acids. The genetic code is said to be **degenerate** because more than one

Table 12.1 The Genetic Code*

First position	Second position: U		C		A		G		Third position
U	UUU	Phe	UCU	Ser	UAU	Tyr	UGU	Cys	U
	UUC		UCC		UAC		UGC		C
	UUA	Leu	UCA		UAA	Stop	UGA	Stop	A
	UUG		UCG		UAG	Stop	UGG	Trp	G
C	CUU	Leu	CCU	Pro	CAU	His	CGU	Arg	U
	CUC		CCC		CAC		CGC		C
	CUA		CCA		CAA	Gln	CGA		A
	CUG		CCG		CAG		CGG		G
A	AUU	Ile	ACU	Thr	AAU	Asn	AGU	Ser	U
	AUC		ACC		AAC		AGC		C
	AUA		ACA		AAA	Lys	AGA	Arg	A
	AUG	Met/start	ACG		AAG		AGG		G
G	GUU	Val	GCU	Ala	GAU	Asp	GGU	Gly	U
	GUC		GCC		GAC		GGC		C
	GUA		GCA		GAA	Glu	GGA		A
	GUG		GCG		GAG		GGG		G

*Exceptions to the genetic code are sporadically found among various species. For example, AUA encodes methionine in yeast and mammalian mitochondria.

codon can specify the same amino acid (see Table 12.1). For example, the codons GGU, GGC, GGA, and GGG all code for the amino acid glycine. In most instances, the third base in the codon is the degenerate or variable base.

During Translation, mRNA Is Used to Make a Polypeptide with a Specific Amino Acid Sequence

Let's look at the organization of a bacterial mRNA to see how translation occurs (Figure 12.10). A ribosomal-binding site is located near the 5′ end of the mRNA. The **start codon**, which specifies the amino acid methionine, is only a few nucleotides from the ribosomal-binding site. Beyond this, a large portion of an mRNA functions as a **coding sequence**—a region that specifies the linear amino acid sequence of a polypeptide. A typical polypeptide is a few hundred amino acids in length. The coding sequence consists of a series of codons. Finally, one of three **stop codons** signals the end of translation. These codons, also known as **termination codons**, are UAA, UAG, and UGA.

The start codon also defines the **reading frame** of an mRNA, which refers to the order in which codons are read during translation. Beginning at the start codon, each adjacent codon is read as a group of three bases, also called a **triplet**, in the 5′ to 3′ direction. For example, look at the following two mRNA sequences and their corresponding amino acid sequences.

Ribosomal-binding site — Start codon

mRNA 5′–AUAAGGAGGUUACG(AUG)(CAG)(CAG)(GGC)(UUU)(ACC)–3′

Polypeptide Met - Gln - Gln - Gly - Phe - Thr

Ribosomal-binding site — Start codon

mRNA 5′–AUAAGGAGGUUACG(AUG)(UCA)(GCA)(GGG)(CUU)(UAC)C–3′

Polypeptide Met - Ser - Ala - Gly - Leu - Tyr

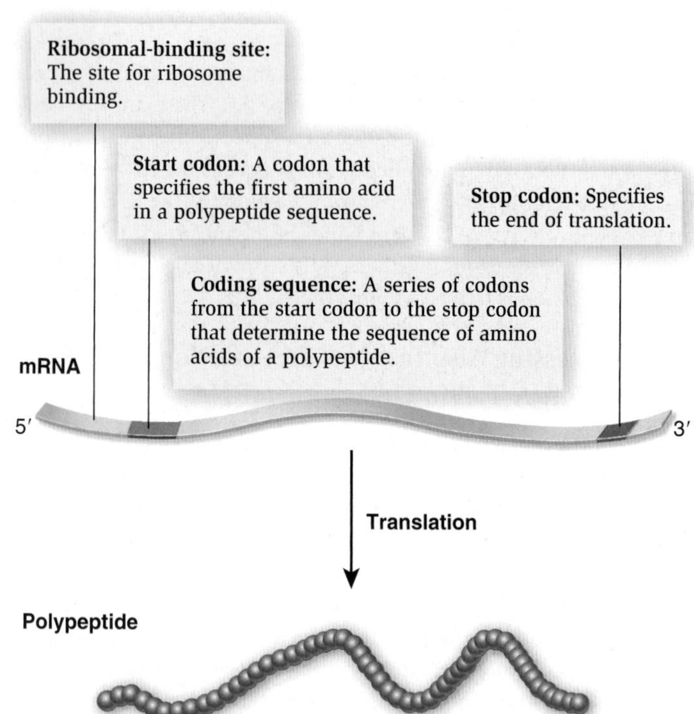

Figure 12.10 The organization of a bacterial mRNA as a translational unit. The string of blue balls represents a sequence of amino acids in a polypeptide. During and following translation, a sequence of amino acids folds into a more compact structure as described in Chapter 3 (see Figure 3.16).

Concept Check: If a mutation eliminated the start codon from a gene, how would the mutation affect transcription, and how would it affect translation?

The first sequence shows how the mRNA codons would be correctly translated into amino acids. In the second sequence, an additional U has been added to the same sequence after the start codon. This shifts the reading frame, thereby changing the codons as they occur in the 5′ to 3′ direction. The polypeptide produced from this series of codons has a very different sequence of amino acids. From this comparison, we can also see that the reading frame is not overlapping, which means that each base functions within a single codon.

DNA Stores Information, Whereas mRNA and tRNA Access That Information to Make a Polypeptide

The relationships among the DNA sequence of a gene, the mRNA transcribed from the gene, and the polypeptide sequence are shown schematically in Figure 12.11. Recall that the template strand is used to make mRNA. The resulting mRNA strand corresponds to the coding strand of DNA, except that U in the mRNA substitutes for T in the DNA. The 5′ end of the mRNA contains an untranslated region (5′ UTR) as does the 3′ end (3′ UTR). The middle portion contains a series of codons that specify the amino acid sequence of a polypeptide.

To translate a nucleotide sequence of mRNA into an amino acid sequence, recognition occurs between mRNA and transfer RNA (tRNA) molecules. Transfer RNA, which is described in Section 12.5, functions as the "translator" or intermediary between an mRNA

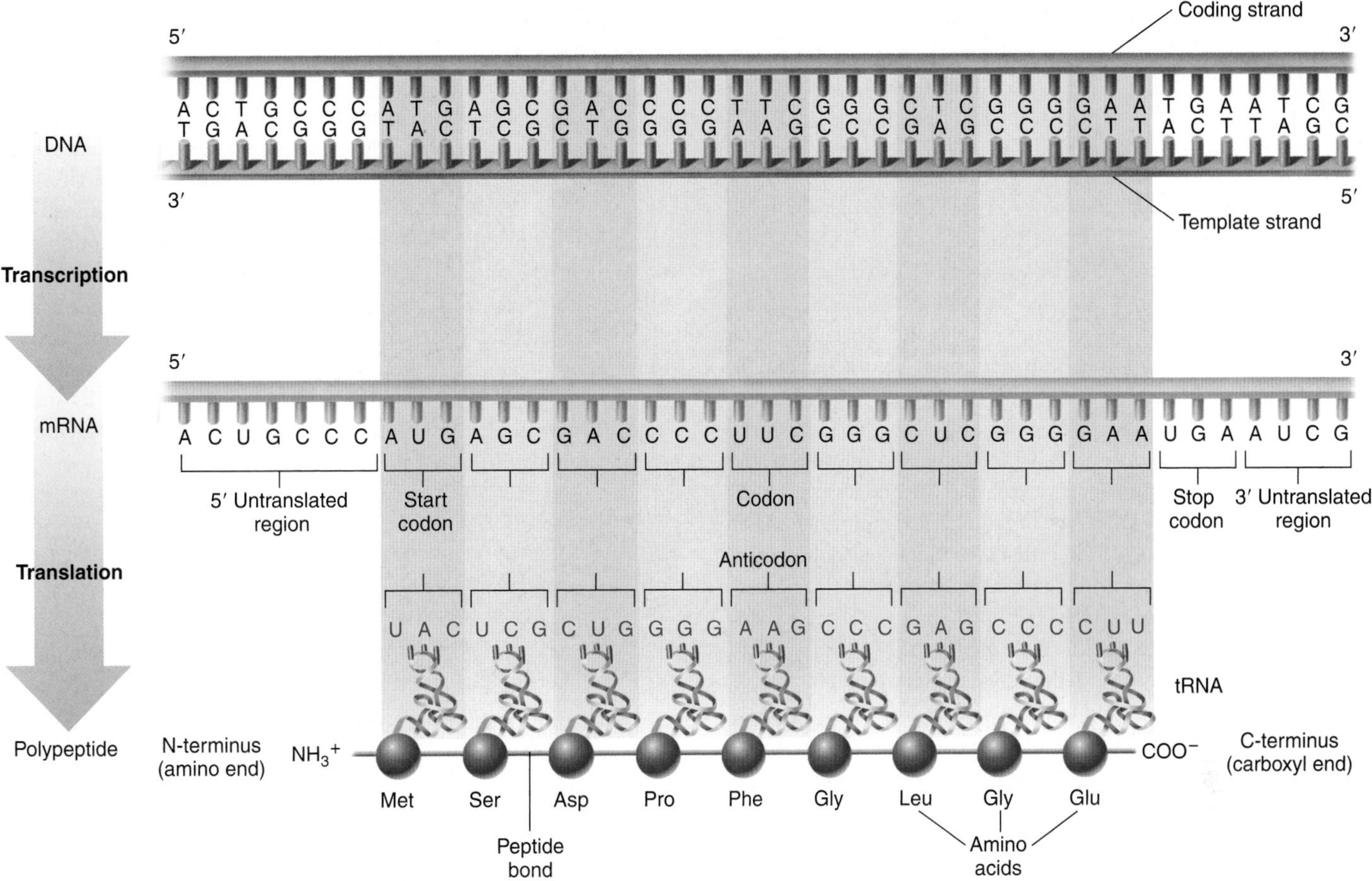

Figure 12.11 **Relationships among the coding sequence of a gene, the codon sequence of an mRNA, the anticodons of tRNA, and the amino acid sequence of a polypeptide.**

Concept Check: *If an anticodon in a tRNA molecule has the sequence 3′–ACC–5′, which amino acid does it carry?*

codon and an amino acid. The **anticodon** is a three-base sequence in a tRNA molecule that is complementary to a codon in mRNA. Due to this complementarity, the anticodon in the tRNA and a codon in an mRNA bind to each other. Furthermore, the anticodon in a tRNA corresponds to the amino acid that it carries. For example, if the anticodon in a tRNA is 3′–AAG–5′, it is complementary to a 5′–UUC–3′ codon. According to the genetic code, a UUC codon specifies phenylalanine (Phe). Therefore, a tRNA with a 3′–AAG–5′ anticodon must carry phenylalanine. As another example, a tRNA with a 3′–GGG–5′ anticodon is complementary to a 5′–CCC–3′ codon, which specifies proline. This tRNA must carry proline (Pro).

As seen at the bottom of Figure 12.11, the direction of polypeptide synthesis parallels the 5′ to 3′ orientation of mRNA. The first amino acid is said to be at the amino end, or **N-terminus**, of the polypeptide. The term N-terminus refers to the presence of a nitrogen atom (N) at this end, whereas amino end indicates the presence of an amino group (NH_2). **Peptide bonds** connect the amino acids together. These covalent bonds form between the carboxyl group (COOH) of the previous amino acid and the amino group of the next amino acid. The last amino acid in a completed polypeptide does not have another amino acid attached to its carboxyl group. This last amino acid is said to be located at the carboxyl end, or **C-terminus**. A carboxyl group is always found at this end of the polypeptide. Note that at neutral pH, the amino group is positively charged (NH_3^+), whereas the carboxyl group is negatively charged (COO^-).

Synthetic RNA Helped to Decipher the Genetic Code

Now let's look at some early experiments that allowed scientists to decipher the genetic code. During the early 1960s, the genetic code was determined by the collective efforts of several researchers, including American biochemist Marshall Nirenberg, Spanish-American biochemist Severo Ochoa, and American geneticist Philip Leder. Prior to their studies, other scientists had discovered that bacterial cells can be broken open and components from the cytoplasm can synthesize polypeptides. This is termed an in vitro or cell-free translation system. Nirenberg and Ochoa made synthetic RNA molecules using an enzyme that covalently connects nucleotides together. Using this synthetic mRNA, they then determined which amino acids were incorporated into polypeptides. For example, if an RNA molecule had only adenine-containing nucleotides (for example, 5′–AAAAAAAAAAAAAAAAAAA–3′), a polypeptide was produced that contained only lysine. This result indicated that the AAA codon specifies lysine.

Another method used to decipher the genetic code involved the chemical synthesis of short RNA molecules, as described next in the Feature Investigation.

FEATURE INVESTIGATION

Nirenberg and Leder Found That RNA Triplets Can Promote the Binding of tRNA to Ribosomes

In 1964, Nirenberg and Leder discovered that RNA molecules containing three nucleotides (that is, a triplet) can cause a tRNA molecule to bind to a ribosome. In other words, an RNA triplet acts like a codon within an mRNA molecule. To establish the relationship between triplet sequences and specific amino acids, Nirenberg and Leder made triplets with specific base sequences (**Figure 12.12**). For example, in one experiment they studied 5′–CCC–3′ triplets. This particular triplet was added to 20 different tubes. To each tube, they next added an in vitro translation system, which contained ribosomes and tRNAs that already had amino acids attached to them. However, each translation system had only one type of radiolabeled amino acid. One translation system had only proline that was radiolabeled, a second translation system had only serine that was radiolabeled, and so on.

As shown in step 2, the triplets became bound to the ribosomes just like the binding of mRNA to a ribosome. The tRNA with an anticodon that was complementary to the added triplet would bind to the triplet, which was already bound to the ribosome. For example, when the triplet was 5′–CCC–3′, a tRNA with a 3′–GGG–5′ anticodon would bind to the triplet/ribosome complex. This tRNA carries proline.

To determine which tRNA had bound, the contents from each tube were poured through a filter that trapped the large ribosomes but did not trap tRNAs that were not bound to ribosomes (see step 3). If the tRNA carrying the radiolabeled amino acid was bound to the triplet/ribosome complex, radioactivity would be trapped on the filter. Using a scintillation counter, the researchers determined the amount of radioactivity on each filter. Because only one amino acid was radiolabeled in each in vitro translation system, they could determine which triplet corresponded to which amino acid. In the example shown here, CCC corresponds to proline. Therefore, the in vitro translation system containing radiolabeled proline showed a large amount of radioactivity on the filter. As seen in the data, by studying triplets with different sequences, Nirenberg and Leder identified many codons of the genetic code.

Experimental Questions

1. Briefly explain how a triplet mimics the role of an mRNA molecule. How was this observation useful in the study done by Nirenberg and Leder?
2. What was the benefit of using radiolabeled amino acids in the Nirenberg and Leder experiment?
3. Predict the results that Nirenberg and Leder would have found for the following triplets: AUG, UAA, UAG, or UGA.

Figure 12.12 **Nirenberg and Leder's use of triplet binding method to decipher the genetic code.**

HYPOTHESIS An RNA triplet can bind to a ribosome and promote the binding of the tRNA that carries the amino acid that the RNA triplet specifies.

KEY MATERIALS The researchers made 20 in vitro translation systems, which included ribosomes, tRNAs, and 20 amino acids. The 20 translation systems differed with regard to which amino acid was radiolabeled. For example, in 1 translation system, radiolabeled glycine was added, and the other 19 amino acids were unlabeled. In another system, radiolabeled proline was added, and the other 19 amino acids were unlabeled. The in vitro translation systems also contained the enzymes that attach amino acids to tRNAs.

1 Mix together RNA triplets of a specific sequence and 20 in vitro translation systems. In the example shown here, the triplet is 5′–CCC–3′. Each translation system contained a different radiolabeled amino acid. (Note: Only 3 tubes are shown here.)

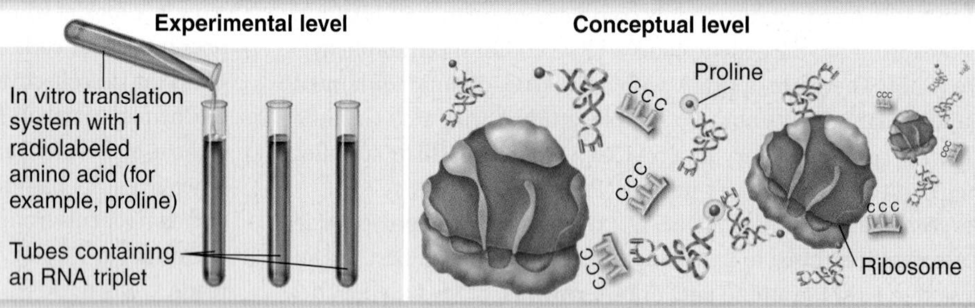

2 Allow time for the RNA triplet to bind to the ribosome and for the appropriate tRNA to bind to the RNA triplet.

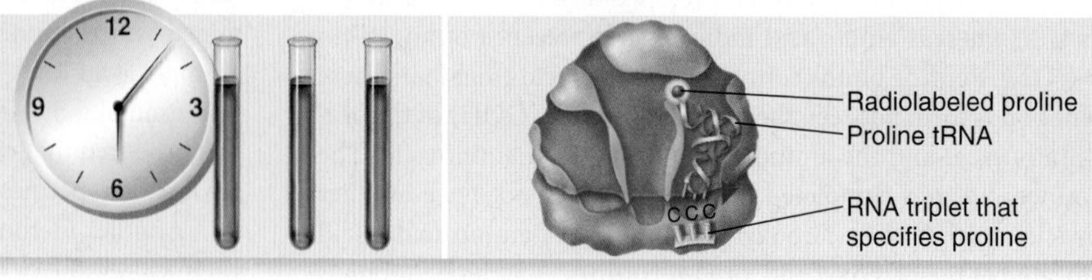

3 Pour each mixture through a filter that allows the passage of unbound tRNA but does not allow the passage of ribosomes.

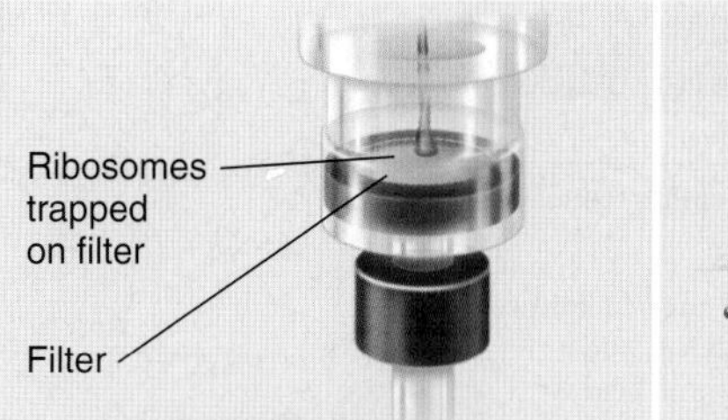

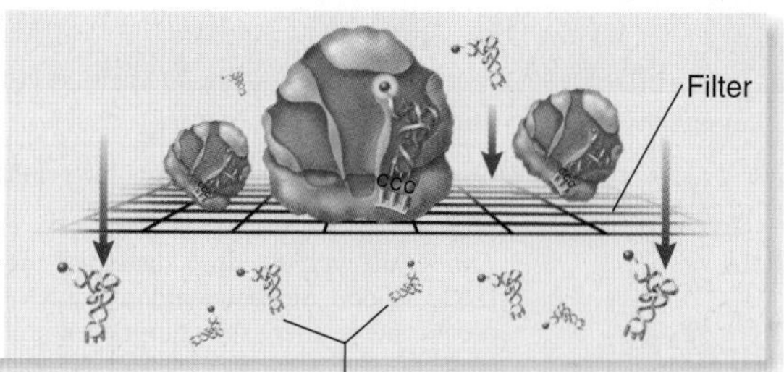

4 Count radioactivity on the filter.

5 **THE DATA**

Triplet	Radiolabeled amino acid trapped on the filter	Triplet	Radiolabeled amino acid trapped on the filter
5′ – AAA – 3′	Lysine	5′ – GAC – 3′	Aspartic acid
5′ – ACA – 3′	Threonine	5′ – GCC – 3′	Alanine
5′ – ACC – 3′	Threonine	5′ – GGU – 3′	Glycine
5′ – AGA – 3′	Arginine	5′ – GGC – 3′	Glycine
5′ – AUA – 3′	Isoleucine	5′ – GUU – 3′	Valine
5′ – AUU – 3′	Isoleucine	5′ – UAU – 3′	Tyrosine
5′ – CCC – 3′	Proline	5′ – UGU – 3′	Cysteine
5′ – CGC – 3′	Arginine	5′ – UUG – 3′	Leucine
5′ – GAA – 3′	Glutamic acid		

6 **CONCLUSION** This method enabled the researchers to identify many of the codons of the genetic code.

7 **SOURCE** Leder, Philip, and Nirenberg, Marshall W. 1964. RNA Codewords and Protein Synthesis, III. On the nucleotide sequence of a cysteine and a leucine RNA codeword. *Proceedings of the National Academy of Sciences* 52:1521–1529.

12.5 The Machinery of Translation

Learning Outcomes:

1. Describe the structure and function of tRNA.
2. Explain how aminoacyl-tRNA synthases attach amino acids to tRNAs.
3. Outline the structural features of bacterial and eukaryotic ribosomes.
4. Analyze how ribosomal RNA (rRNA) is used to evaluate evolutionary relationships among different species.

Let's now turn our attention to the components found in living cells that are needed to use the genetic code and translate mRNA into polypeptides. Earlier in this chapter, we considered transcription, the first step in gene expression. To transcribe an RNA molecule, a preexisting DNA template strand is used to make a complementary RNA strand. A single enzyme, RNA polymerase, catalyzes this reaction. By comparison, translation requires more components because the sequence of codons in an mRNA molecule must be translated into a sequence of amino acids according to the genetic code. A single protein cannot accomplish such a task. Instead, many different proteins and RNA molecules interact in an intricate series of steps to achieve the synthesis of a polypeptide. A cell must make many different components, including mRNAs, tRNAs, ribosomes, and translation factors, so polypeptides can be made (**Table 12.2**).

Though the estimates vary from cell to cell and from species to species, most cells use a substantial amount of their energy to translate mRNA into polypeptides. In *E. coli*, for example, approximately 90% of the cellular energy is used for this process. This value underscores the complexity and importance of translation in living organisms. In this section, we will focus on the components of the translation machinery. The last section of the chapter will describe the stages of translation as they occur in living cells.

Table 12.2 Components of the Translation Machinery

Component	Function
mRNA	Contains the information for a polypeptide sequence according to the genetic code.
tRNA	A molecule with two functional sites: one site, termed the anticodon, binds to a codon in mRNA, and a second site is where an appropriate amino acid is attached.
Ribosome	Composed of many proteins and rRNA molecules, the ribosome provides a location where mRNA and tRNA molecules can properly interact with each other. The ribosome also catalyzes the formation of covalent bonds between adjacent amino acids so that a polypeptide can be made.
Translation factors	Proteins needed for the three stages of translation. Initiation factors are required for the assembly of mRNA, the first tRNA, and ribosomal subunits. Elongation factors are needed to synthesize the polypeptide. Release factors are needed to recognize the stop codon and disassemble the translation machinery. Several translation factors use GTP as an energy source to carry out their functions.

Transfer RNAs Share Common Structural Features

To understand how tRNAs act as carriers of the correct amino acids during translation, researchers have examined their structural characteristics. The tRNAs of bacteria, archaea, and eukaryotes share common features. As originally proposed in 1965 by American biochemist Robert Holley, the two-dimensional structure of a tRNA exhibits a cloverleaf pattern. The structure has three stem-loops and a fourth stem with a 3′ single-stranded region (Figure 12.13a). The stem in a stem-loop is a region where the RNA is double-stranded due to complementary base pairing via hydrogen binding, whereas the loop is a region without base pairing. The anticodon is located in the loop of the second stem-loop region. The 3′ single-stranded region is the amino acid attachment site. The three-dimensional structure of tRNA molecules involves additional folding of the secondary structure (Figure 12.13b).

The cells of every organism make many different tRNA molecules, each encoded by a different gene. A tRNA is named according to the amino acid it carries. For example, $tRNA^{Ser}$ carries a serine. Because the genetic code contains six different serine codons, as shown in Table 12.1, a cell produces more than one type of $tRNA^{Ser}$.

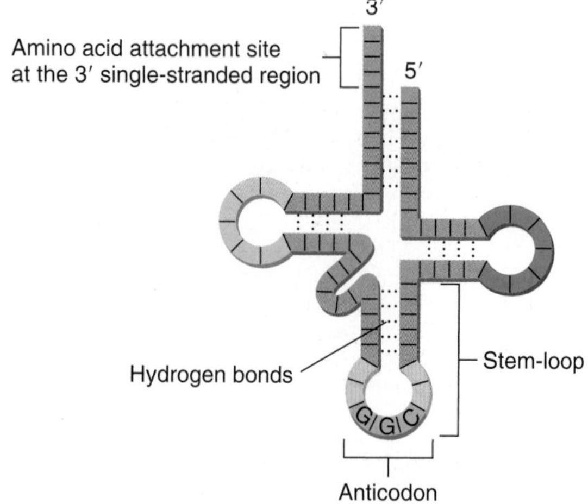

(a) Two-dimensional structure of tRNA

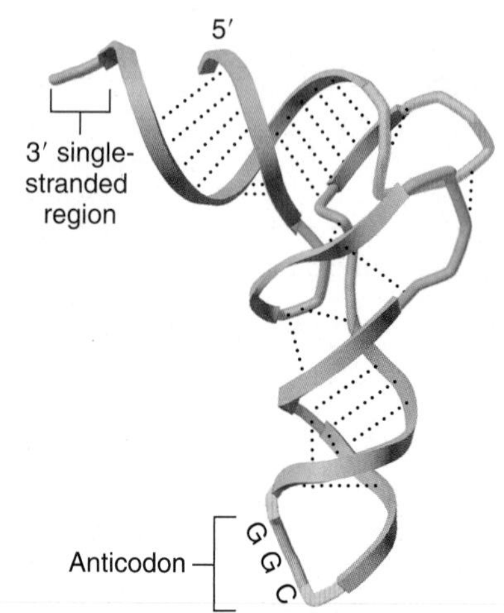

(b) Three-dimensional structure of tRNA

Figure 12.13 **Structure of tRNA.** **(a)** The two-dimensional or secondary structure of tRNA is that of a cloverleaf, with the anticodon within the middle stem-loop structure. The 3′ single-stranded region (acceptor stem) is where an amino acid can attach. **(b)** The actual three-dimensional structure folds in on itself.

BIOLOGY PRINCIPLE Structure determines function. The structure of tRNA has two functional sites: a 3′ single-stranded region where an amino acid is attached and an anticodon that binds to a codon on mRNA.

Concept Check: What is the function of the anticodon?

Aminoacyl-tRNA Synthetases Charge tRNAs by Attaching an Appropriate Amino Acid

To perform its role during translation, a tRNA must have the appropriate amino acid attached to its 3′ end. The enzymes that catalyze the attachment of amino acids to tRNA molecules are known as **aminoacyl-tRNA synthetases**. Cells make 20 distinct types of aminoacyl-tRNA synthetase enzymes, with each type recognizing just one of the 20 different amino acids. Each aminoacyl-tRNA synthetase is named for the specific amino acid it attaches to tRNA. For example, alanyl-tRNA synthetase recognizes alanine and attaches this amino acid to all tRNAs with alanine anticodons.

Aminoacyl-tRNA synthetases catalyze chemical reactions involving an amino acid, a tRNA molecule, and ATP (Figure 12.14). First, a specific amino acid and ATP bind to the enzyme. Next, the amino acid is activated by the covalent attachment of an AMP molecule, and pyrophosphate is released. In a third step, the activated amino acid is covalently attached to the 3′ end of a tRNA molecule, and AMP is released. Finally, the tRNA with its attached amino acid, called a **charged tRNA** or an **aminoacyl tRNA**, is released from the enzyme.

The ability of each aminoacyl-tRNA synthetase to recognize an appropriate tRNA has been called the second genetic code. A precise recognition process is necessary to maintain the fidelity of genetic information. If the wrong amino acid was attached to a tRNA, the amino acid sequence of the translated polypeptide would be incorrect. However, aminoacyl-tRNA synthetases are amazingly accurate enzymes. The wrong amino acid is attached to a tRNA less than once in 100,000 times! The anticodon region of the tRNA is usually

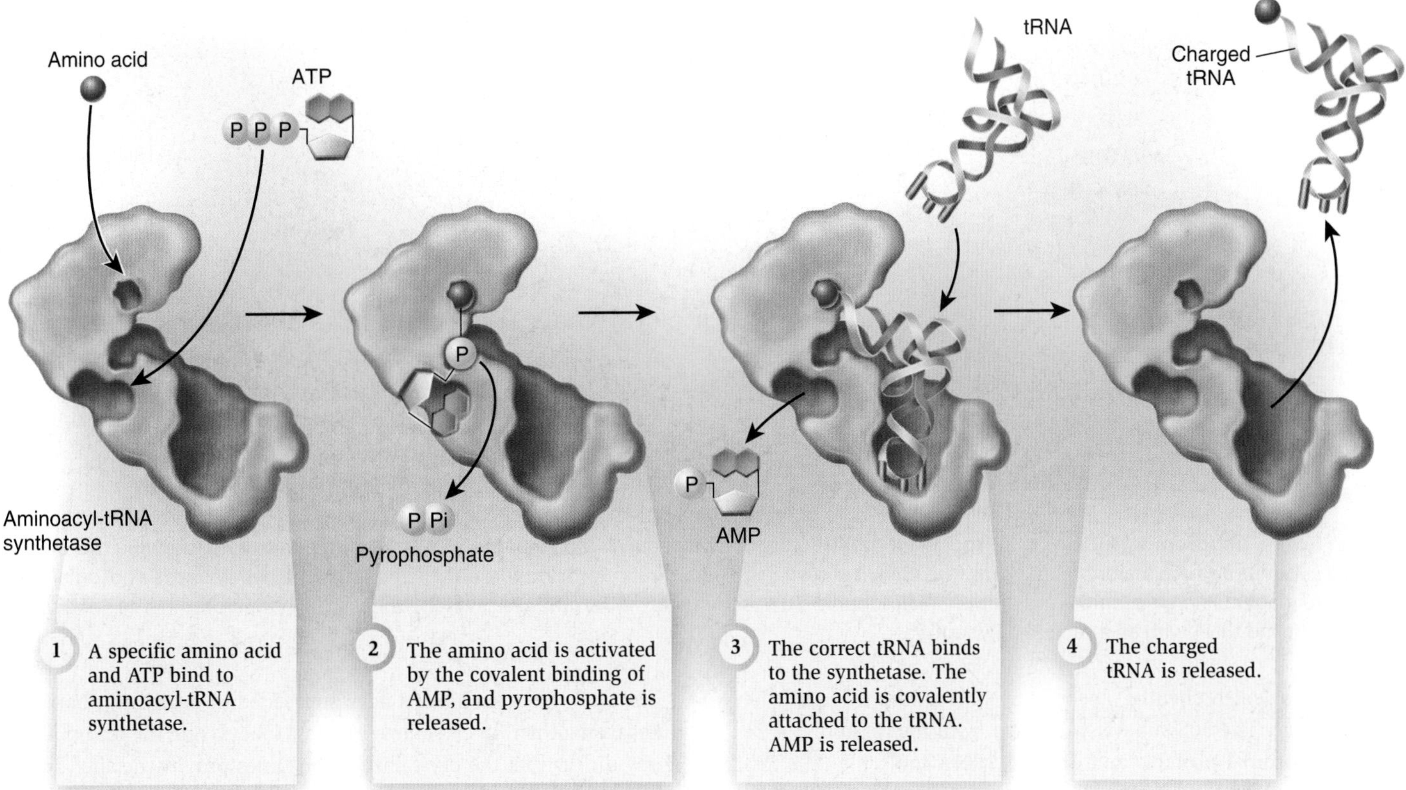
Amino acid
ATP
P P P
Aminoacyl-tRNA synthetase
P
P Pi
Pyrophosphate
P
AMP
tRNA
Charged tRNA
1 A specific amino acid and ATP bind to aminoacyl-tRNA synthetase.
2 The amino acid is activated by the covalent binding of AMP, and pyrophosphate is released.
3 The correct tRNA binds to the synthetase. The amino acid is covalently attached to the tRNA. AMP is released.
4 The charged tRNA is released.

Figure 12.14 **Aminoacyl-tRNA synthetase charging a tRNA.**

BioConnections: *Look back at Figure 6.3, which describes the hydrolysis of ATP. Why is ATP needed to charge a tRNA?*

important for recognition by the correct aminoacyl-tRNA synthetase. In addition, the base sequences in other regions may facilitate binding to an aminoacyl-tRNA synthetase.

Ribosomes Are Assembled from rRNA and Proteins

Let's now turn our attention to the **ribosome**, the site where translation takes place. The ribosome is often described as a molecular machine. Bacterial cells have one type of ribosome, which translates all mRNAs in the cytoplasm. Because eukaryotic cells are compartmentalized into cellular organelles bounded by membranes, biochemically distinct ribosomes are found in different cellular compartments. The most abundant type of eukaryotic ribosome functions in the cytosol. In addition, mitochondria have ribosomes, and plant and algal cells have ribosomes in their chloroplasts. As you might expect from the endosymbiosis theory described in Chapter 4 (see Figure 4.27), the compositions of mitochondrial and chloroplast ribosomes are more similar to bacterial ribosomes than they are to eukaryotic cytosolic ribosomes. Unless otherwise noted, the term eukaryotic ribosome refers to ribosomes in the cytosol, not to those found in organelles.

A ribosome is a large complex composed of structures called the large and small subunits. The term subunit is perhaps misleading, because each ribosomal subunit is itself assembled from many different proteins and one or more RNA molecules. In the bacterium *E. coli*, the small ribosomal subunit is called 30S, and the large subunit is 50S (**Table 12.3**). The designations 30S and 50S refer to the rate at which these subunits sediment when subjected to a centrifugal force. This rate is described as a sedimentation coefficient in Svedberg units (S)

Table 12.3 Composition of Bacterial and Eukaryotic Ribosomes

	Small subunit	Large subunit	Assembled ribosome
Bacterial			
Sedimentation coefficient	30S	50S	70S
Number of proteins	21	34	55
rRNA	16S rRNA	5S rRNA, 23S rRNA	16S rRNA, 5S rRNA, 23S rRNA
Eukaryotic			
Sedimentation coefficient	40S	60S	80S
Number of proteins	33	49	82
rRNA	18S rRNA	5S rRNA, 5.8S rRNA, 28S rRNA	18S rRNA, 5S rRNA, 5.8S rRNA, 28S rRNA

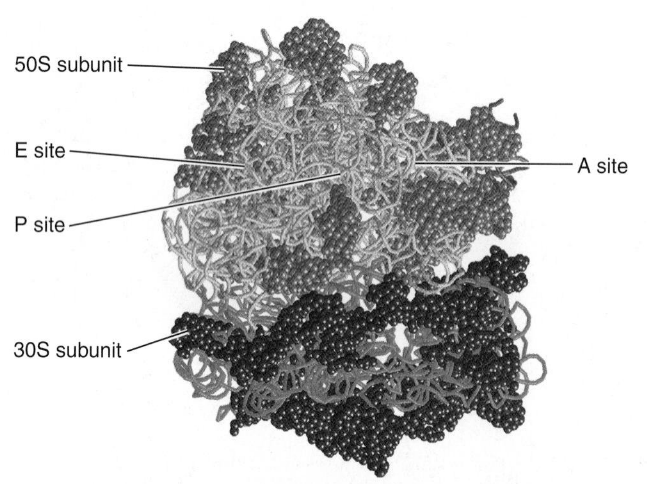

(a) Bacterial ribosome model based on X-ray diffraction studies

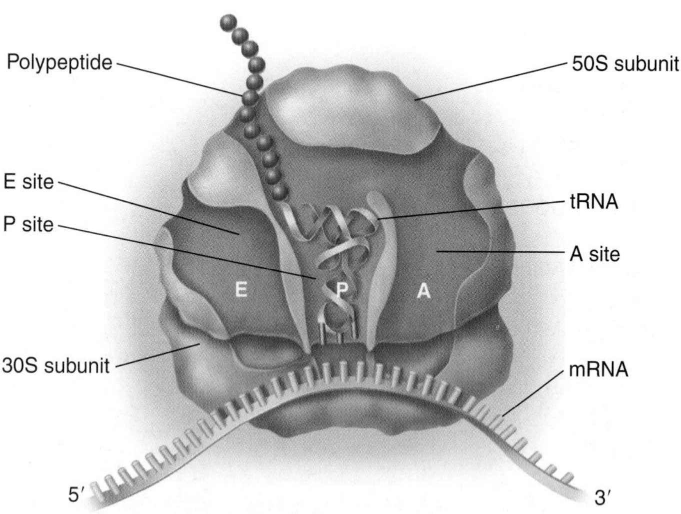

(b) Schematic model for ribosome structure

Figure 12.15 **Ribosome structure.** **(a)** A model for the structure of a bacterial ribosome based on X-ray diffraction studies, showing the large and small subunits and the major binding sites. The rRNA is shown in gray (large subunit) and turquoise (small subunit), whereas the ribosomal proteins are magenta (large subunit) and purple (small subunit). **(b)** A schematic model emphasizing functional sites in the ribosome, and showing bound mRNA and tRNA with an attached polypeptide.

in honor of Swedish chemist Theodor Svedberg, who invented the ultracentrifuge. The 30S subunit is formed from the assembly of 21 different ribosomal proteins and one 16S rRNA molecule. The 50S subunit contains 34 different proteins and two different rRNA molecules, called 5S and 23S. Together, the 30S and 50S subunits form a 70S ribosome. (Svedberg units don't add up linearly, because the sedimentation coefficient is a function of both size and shape.) In bacteria, ribosomal proteins and rRNA molecules are synthesized in the cytoplasm, and the ribosomal subunits are assembled there as well.

Eukaryotic ribosomes consist of subunits that are slightly larger than their bacterial counterparts (Table 12.3). In eukaryotes, 40S and 60S subunits combine to form an 80S ribosome. The 40S subunit is composed of 33 proteins and an 18S rRNA, and the 60S subunit has 49 proteins and 5S, 5.8S, and 28S rRNAs. The synthesis of eukaryotic rRNA occurs in the nucleolus, a region of the nucleus that is specialized for that purpose. The ribosomal proteins are made in the cytosol and imported into the nucleus. The rRNAs and ribosomal proteins are then assembled within the nucleolus to make the 40S and 60S subunits. The 40S and 60S subunits are exported into the cytosol, where they associate to form an 80S ribosome during translation.

Due to structural differences between bacterial and eukaryotic ribosomes, certain chemicals may bind to bacterial ribosomes but not to eukaryotic ribosomes, and vice versa. Some **antibiotics**, which are chemicals that inhibit the growth of certain microorganisms, bind only to bacterial ribosomes and inhibit translation. Examples include erythromycin and chloramphenicol. Because these chemicals do not inhibit eukaryotic ribosomes, they have been effective drugs for the treatment of bacterial infections in humans and domesticated animals.

Components of Ribosomal Subunits Form Functional Sites for Translation

To understand the structure and function of the ribosome at the molecular level, researchers have determined the locations and functional roles of individual ribosomal proteins and rRNAs. In recent years, a few research groups have succeeded in purifying ribosomes and causing them to crystallize in a test tube. Using the technique of X-ray diffraction, the crystallized ribosomes provide detailed information about ribosome structure. Figure 12.15a shows a model of a bacterial ribosome.

During bacterial translation, the mRNA lies on the surface of the 30S subunit, within a space between the 30S and 50S subunits (**Figure 12.15b**). As a polypeptide is synthesized, it exits through a hole within the 50S subunit. Ribosomes contain discrete sites where tRNAs bind and the polypeptide is synthesized. In 1964, James Watson proposed a two-site model for tRNA binding to the ribosome. These sites are known as the **peptidyl site (P site)** and **aminoacyl site (A site)**. In 1981, German geneticists Knud Nierhaus and Hans-Jörg Rheinberger expanded this to a three-site model (Figure 12.15b). The third site is known as the exit site (E site). In Section 12.6, we will examine the roles of these sites in the synthesis of a polypeptide.

GENOMES & PROTEOMES CONNECTION

Comparisons of Small Subunit rRNAs Among Different Species Provide a Basis for Establishing Evolutionary Relationships

Translation is a fundamental process that is vital for the existence of all living species. Research indicates that the components needed for translation arose very early in the evolution of life on our planet in ancestors that gave rise to all known living species. For this reason, all organisms have translational components that are evolutionarily related to each other. For example, the rRNA found in the small subunit of ribosomes is similar in all forms of life, though it is slightly larger in eukaryotic species (18S) than in bacterial species (16S). In other words, the gene for the small subunit rRNA (SSU rRNA) is found in the genomes of all organisms.

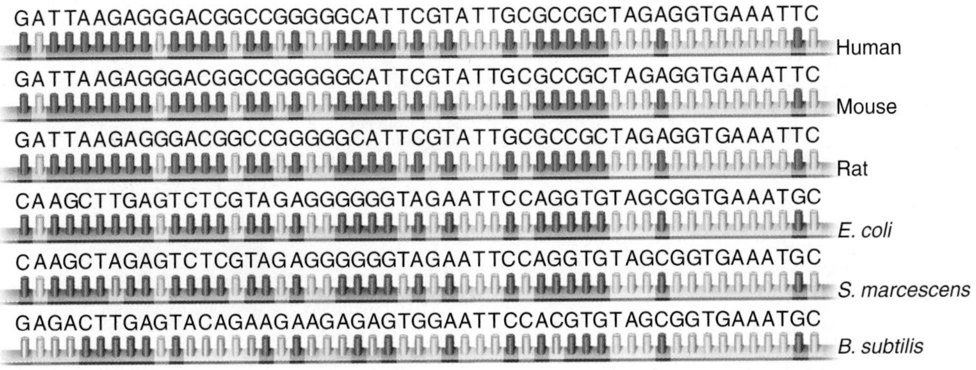

GATTAAGAGGGACGGCCGGGGGCATTCGTATTGCGCCGCTAGAGGTGAAATTC Human
GATTAAGAGGGACGGCCGGGGGCATTCGTATTGCGCCGCTAGAGGTGAAATTC Mouse
GATTAAGAGGGACGGCCGGGGGCATTCGTATTGCGCCGCTAGAGGTGAAATTC Rat
CAAGCTTGAGTCTCGTAGAGGGGGGTAGAATTCCAGGTGTAGCGGTGAAATGC *E. coli*
CAAGCTAGAGTCTCGTAGAGGGGGGTAGAATTCCAGGTGTAGCGGTGAAATGC *S. marcescens*
GAGACTTGAGTACAGAAGAAGAGAGTGGAATTCCACGTGTAGCGGTGAAATGC *B. subtilis*

Figure 12.16 Comparison of small subunit rRNA gene sequences from three mammalian and three bacterial species. Note the many similarities (yellow) and differences (green and red) among the sequences. The gray color indicates differences among the three bacterial species.

Concept Check: Based on the gene sequences shown here, pick two species that are closely related evolutionarily and two that are distantly related.

How is this observation useful? One way that geneticists explore evolutionary relationships is to compare the sequences of evolutionarily related genes. At the molecular level, gene evolution involves changes in DNA sequences. After two different species have diverged from each other during evolution, the genes of each species have an opportunity to accumulate changes, or mutations, that alter the sequences of those genes. After many generations, evolutionarily related species contain genes that are similar but not identical to each other, because each species accumulates different mutations. In general, if a very long time has elapsed since two species diverged evolutionarily, their genes tend to be quite different. In contrast, if two species diverged relatively recently on an evolutionary time scale, their genes tend to be more similar.

Figure 12.16 compares a portion of the sequence of the SSU rRNA gene from three mammalian and three bacterial species. The colors highlight different types of comparisons. The bases shaded in yellow are identical in five or six species. Sequences of bases that are identical or very similar in different species are said to be **evolutionarily conserved**. Presumably, these sequences were found in the primordial gene that gave rise to modern species. Perhaps because these sequences may have some critical function, they have not changed over evolutionary time. Those sequences shaded in green are identical in all three mammals, but differ compared with one or more bacterial species. Actually, if you scan the mammalian species, you may notice that all three sequences are identical in this region. The sequences shaded in red are identical or very similar in the bacterial species, but differ compared with the mammalian SSU rRNA genes. The sequences from *Escherichia coli* and *Serratia marcescens* are more similar to each other than the sequence from *Bacillus subtilis* is to either of them. This observation suggests that *E. coli* and *S. marcescens* are more closely related evolutionarily than either of them is to *B. subtilis*.

12.6 The Stages of Translation

Learning Outcomes:

1. Describe the three stages of translation.
2. Summarize the similarities and differences between translation in bacteria and eukaryotes.

As in transcription, the process of translation occurs in three stages called initiation, elongation, and termination. Figure 12.17 provides an overview of the process. During initiation, an mRNA, the first tRNA, and the ribosomal subunits assemble into a complex. Next, in the elongation stage, the ribosome moves in the 5′ to 3′ direction from the start codon in the mRNA toward the stop codon, synthesizing a polypeptide according to the sequence of codons in the mRNA. Finally, the process is terminated when the ribosome reaches a stop codon and the complex disassembles, releasing the completed polypeptide. In this section, we will examine the steps in this process as they occur in living cells.

Translation Is Initiated with the Assembly of mRNA, tRNA, and the Ribosomal Subunits

During **initiation**, a complex is formed between an mRNA molecule, the first tRNA, and the ribosomal subunits. In all species, the assembly of this complex requires the help of proteins called **initiation factors** that facilitate the interactions between these components (see Table 12.2). The assembly also requires an input of energy. Guanosine triphosphate (GTP) is hydrolyzed by certain initiation factors to provide the necessary energy.

In the absence of translation, the small and large ribosomal subunits exist separately. To begin assembly in bacteria, mRNA binds to the small ribosomal subunit (Figure 12.18). The binding of mRNA to this subunit is facilitated by a short ribosomal-binding site near the 5′ end of the mRNA. The ribosomal binding site is a sequence of bases that is complementary to a portion of the 16S rRNA within the small ribosomal subunit. The mRNA becomes bound to the ribosome because the ribosomal-binding site and rRNA hydrogen-bond to each other by base-pairing. The start codon is usually just a few nucleotides downstream (that is, toward the 3′ end) from the ribosomal-binding site. A specific tRNA, which functions as the initiator tRNA, recognizes the start codon in mRNA (AUG) and binds to it. In eukaryotes, this tRNA carries a methionine, whereas in bacteria it carries a methionine that has been modified by the attachment of a formyl group. To complete the initiation stage, the large ribosomal subunit associates with the small subunit. At the end of this stage, the initiator tRNA is located in the P site of the ribosome.

In eukaryotic species, the initiation phase of translation differs in two ways from the process in bacteria. First, instead of a RNA sequence that functions as a ribosomal-binding site, eukaryotic mRNAs have a guanosine cap (5′ cap) at their 5′ end. This 5′ cap is recognized by cap-binding proteins that promote the binding of the mRNA to the small ribosomal subunit. Also, unlike bacteria, in which the start codon is very close to a ribosomal-binding site, the location

Figure 12.17 **An overview of the stages of translation.**

BIOLOGY PRINCIPLE Living organisms use energy. The process of translation uses a sizeable amount of a cell's energy.

Completed polypeptide
Large subunit
Initiator tRNA
Polypeptide
Small subunit
5′
mRNA
Start codon
3′
Stop codon
1 Initiation: mRNA, tRNA, and the ribosomal subunits form a complex.
2 Elongation: The ribosome travels in the 5′ to 3′ direction and synthesizes a polypeptide.
3 Termination: The ribosome reaches a stop codon, and all of the components disassemble, releasing a completed polypeptide.

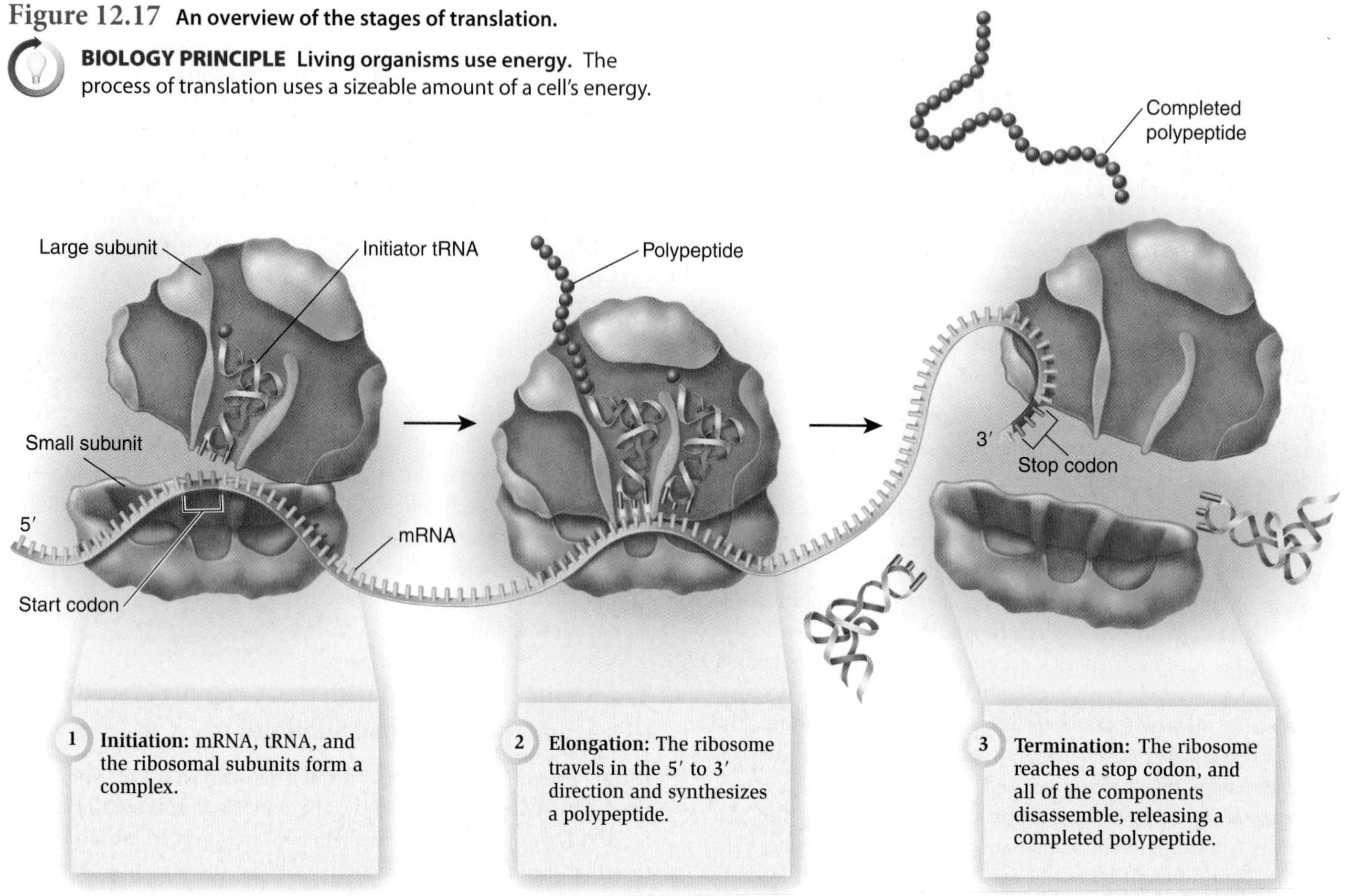

of start codons in eukaryotes is more variable. In 1978, American biochemist Marilyn Kozak proposed that the small ribosomal subunit identifies a start codon by beginning at the 5′ end and then scanning along the mRNA in the 3′ direction in search of an AUG sequence. In many, but not all, cases the first AUG codon is used as a start codon. By analyzing the sequences of many eukaryotic mRNAs, Kozak and her colleagues discovered that the sequence around an AUG codon is important for it to be used as a start codon. The sequence for optimal start codon recognition is shown here:

Upstream of start codon	Start codon	Downstream coding region

. . . G C C (A or G) C C (**A U G**) G . . . .

Aside from an AUG codon itself, a guanine just past the start codon and the sequence of six bases directly upstream from the start codon are important for start codon selection. If the first AUG codon is within a site that deviates markedly from this optimal sequence, the small subunit may skip this codon and instead use another AUG codon farther downstream. Once the small subunit selects a start codon, an initiator tRNA binds to the start codon, and then the large ribosomal subunit associates with the small subunit to complete the assembly process.

Polypeptide Synthesis Occurs During the Elongation Stage

As its name suggests, the stage of translation called **elongation** involves the covalent bonding of amino acids to each other, one at a time, to produce a polypeptide. Even though this process involves several different components, translation occurs at a remarkable rate. Under normal cellular conditions, the translation machinery can elongate a polypeptide at a rate of 15 to 18 amino acids per second in bacteria and 6 amino acids per second in eukaryotes!

To elongate a polypeptide by one amino acid, a tRNA brings a new amino acid to the ribosome, where it is attached to the end of a growing polypeptide. In step 1 of Figure 12.19, translation has already proceeded to a point where a short polypeptide is attached to the tRNA located in the P site of the ribosome. This is called peptidyl tRNA. In the first step of elongation, a charged tRNA carrying a single amino acid binds to the A site. This binding occurs because the anticodon in the tRNA is complementary to the codon in the mRNA. The hydrolysis of GTP by proteins that function as **elongation factors** provides the energy for the binding of the tRNA to the A site (see Table 12.2). At this stage of translation, a peptidyl tRNA is in the P site and a charged tRNA (an aminoacyl tRNA) is in the A site. This is how the P and A sites came to be named.

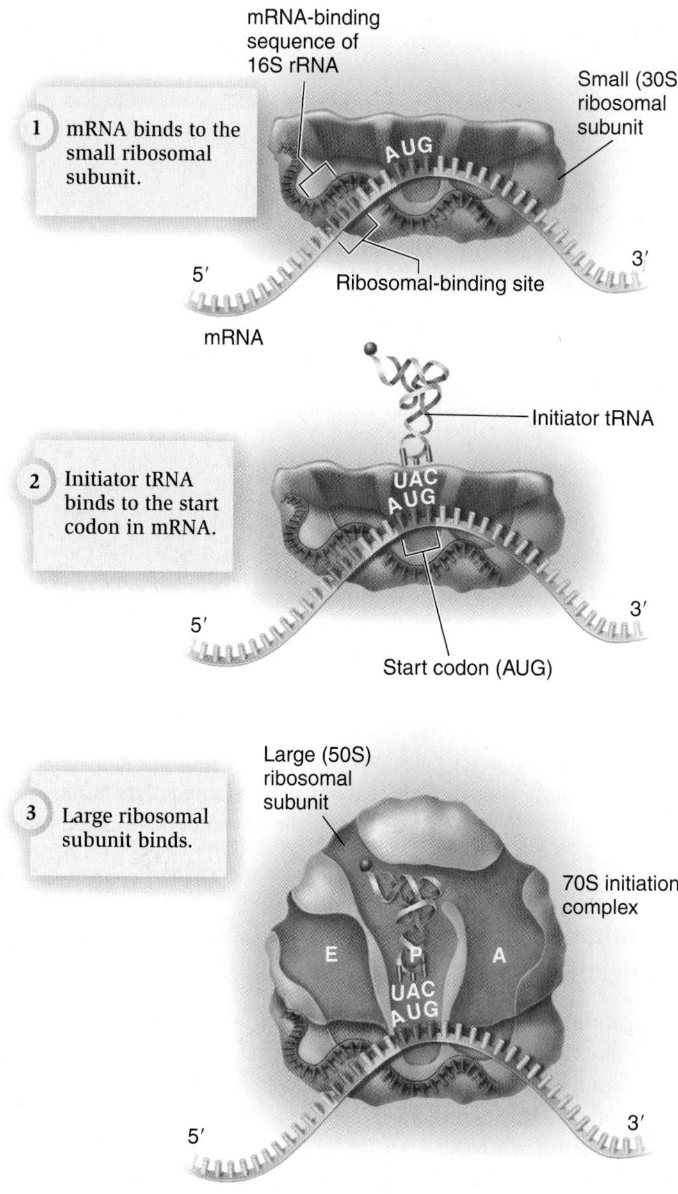

Figure 12.18 **Initiation of translation in bacteria.**

Concept Check: *What promotes the binding between the mRNA and the small ribosomal subunit?*

In the second step, a peptide bond is formed between the amino acid at the A site and the growing polypeptide, thereby lengthening the polypeptide by one amino acid. As this occurs, the polypeptide is removed from the tRNA in the P site and transferred to the amino acid at the A site, an event termed a **peptidyl transfer reaction**. This reaction is catalyzed by a region of the 50S subunit known as the peptidyltransferase center, which is composed of several proteins and rRNA. In 2000, American biochemist Thomas Steitz, American biophysicist Peter Moore, and their colleagues proposed that the rRNA is responsible for catalyzing the peptide bond formation between adjacent amino acids. It is now accepted that the ribosome is a ribozyme!

After the peptidyl transfer reaction is complete, the third step involves the movement or translocation of the ribosome toward the 3′ end of the mRNA by exactly one codon. This shifts the tRNAs in the

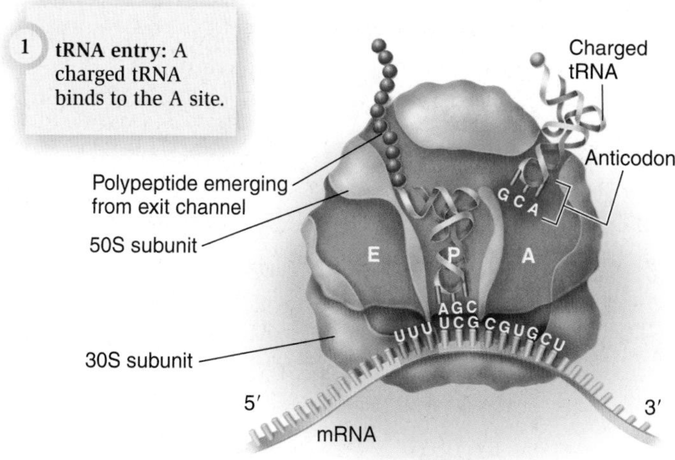

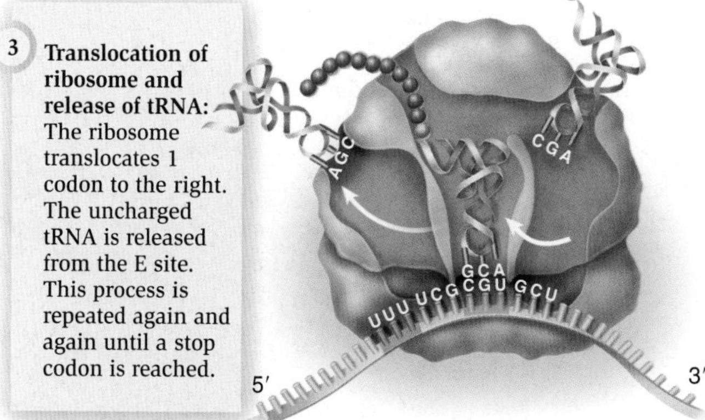

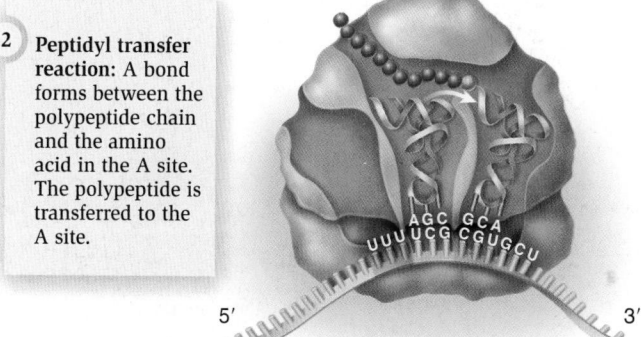

Figure 12.19 **Elongation stage of translation in bacteria.** In this drawing, the amino acids that were already contained within a polypeptide are shown in blue. The amino acids attached to incoming tRNAs in this figure are shown in yellow to distinguish them from the amino acids that were already in the polypeptide.

P and A sites to the E and P sites, respectively. The uncharged tRNA exits the E site. Notice that the next codon in the mRNA (GCU in Figure 12.19) is now exposed at the unoccupied A site. At this point, a charged tRNA can enter the A site, and the same series of steps will add the next amino acid to the polypeptide.

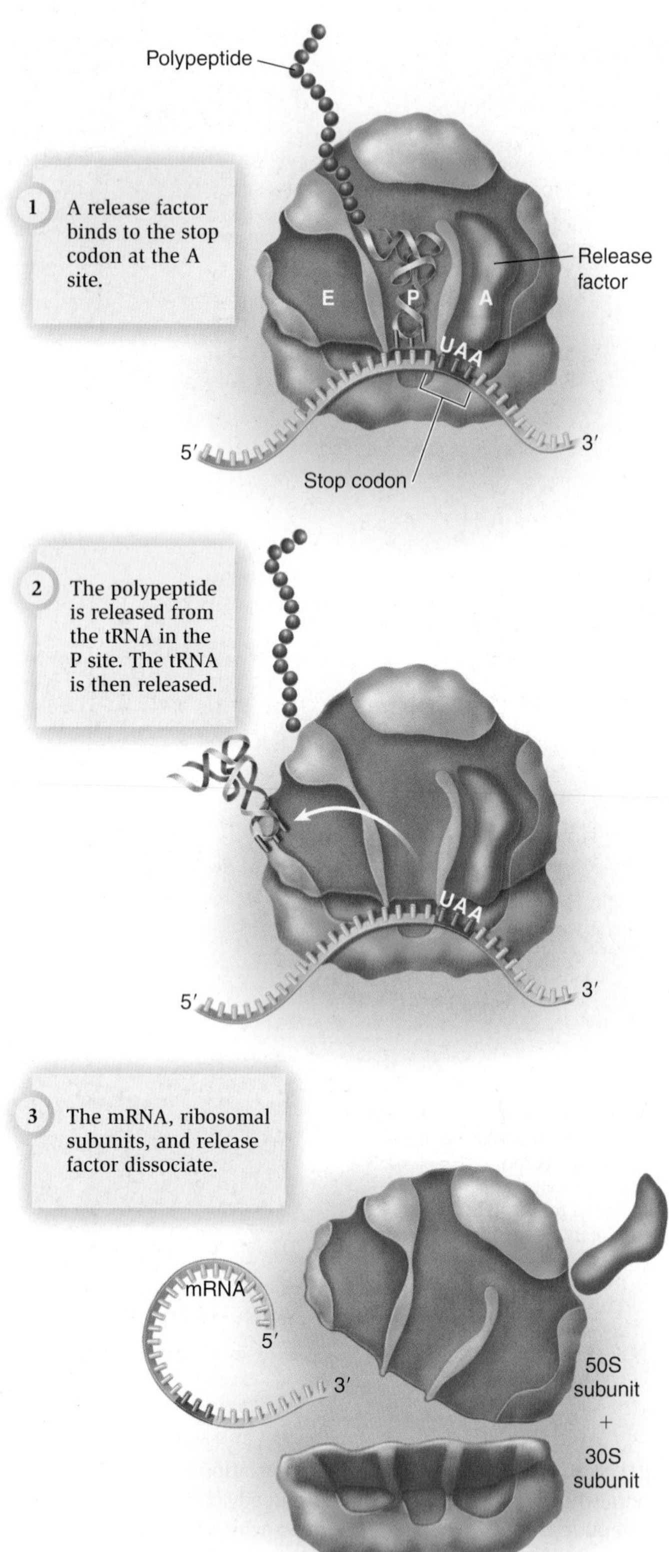
Polypeptide
1 A release factor binds to the stop codon at the A site.
E
P
A
Release factor
UAA
5′
3′
Stop codon
2 The polypeptide is released from the tRNA in the P site. The tRNA is then released.
UAA
5′
3′
3 The mRNA, ribosomal subunits, and release factor dissociate.
mRNA
5′
3′
50S subunit
+
30S subunit

Figure 12.20 Termination of translation in bacteria.

Table 12.4 Comparison of Bacterial and Eukaryotic Translation

	Bacterial	Eukaryotic
Cellular location	Cytoplasm	Cytosol*
Ribosome composition	70S ribosomes:	80S ribosomes:
	30S subunit: 21 proteins + 1 rRNA	40S subunit: 33 proteins + 1 rRNA
	50S subunit: 34 proteins + 2 rRNAs	60S subunit: 49 proteins + 3 rRNAs
Initiator tRNA	$tRNA^{Formyl\text{-}methionine}$	$tRNA^{Methionine}$
Initial binding of mRNA	Requires a ribosomal-binding site	Requires a 7-methylguanosine cap
Selection of a start codon	Just downstream from the ribosomal-binding site	According to Kozak's sequences
Termination factors	Two factors: RF1 and RF2	One factor: eRF

*The components for eukaryotic translation described in this table refer to those that are used in the cytosol. Different types of ribosomes are used for translation inside mitochondria and chloroplasts.

Figure 12.19 considers a single ribosome in the act of translating an mRNA. In living cells, it is common for multiple ribosomes to be gliding along the same mRNA and synthesizing polypeptides. The complex of a single mRNA and multiple ribosomes is called a **polysome** (see Chapter opening micrograph).

Termination Occurs When a Stop Codon Is Reached in the mRNA

Elongation continues until a stop codon moves into the A site of a ribosome. The three stop codons, UAA, UAG, and UGA, are not recognized by a tRNA with a complementary sequence. Instead, they are recognized by a protein known as a **release factor**. The three-dimensional structure of a release factor protein mimics the structure of tRNAs, which allows it to fit into the A site.

Figure 12.20 illustrates the termination of translation. In step 1 of this figure, a release factor binds to the stop codon at the A site. The completed polypeptide is attached to a tRNA in the P site. In the second step, the bond between the polypeptide and the tRNA is hydrolyzed, causing the polypeptide and tRNA to be released from the ribosome. In the third step, the mRNA, ribosomal subunits, and release factor dissociate. The termination stage of translation is similar in bacteria and eukaryotes except that bacteria have two different termination factors that recognize stop codons (RF1 and RF2), whereas eukaryotes have only one (eRF). **Table 12.4** compares some of the key differences between bacterial and eukaryotic translation.

Summary of Key Concepts

12.1 Overview of Gene Expression

- Based on his studies of inborn errors of metabolism, Garrod hypothesized that certain genetic diseases are caused by a defect in a gene encoding an enzyme (Figure 12.1).
- By studying the nutritional requirements of bread mold, Beadle and Tatum proposed the one-gene/one-enzyme hypothesis in which a single gene controls the synthesis of a single enzyme (Figure 12.2).
- A polypeptide is a unit of structure. A protein, composed of one or more polypeptides, is a unit of function.
- At the molecular level, the central dogma states that most genes are transcribed into mRNA, and then the mRNA is translated into polypeptides. Eukaryotes modify their RNA transcripts to make them functional (Figure 12.3).
- The molecular expression of genes is an underlying factor that determines an organism's characteristics.

12.2 Transcription

- A site in a gene called a promoter specifies where transcription begins. A terminator specifies where transcription will end (Figure 12.4).
- In bacteria, the initiation of transcription begins when sigma factor binds to RNA polymerase and to a promoter. During elongation, an RNA transcript is synthesized due to base pairing of nucleotides to the template strand of DNA as RNA polymerase slides along the DNA. RNA polymerase and the RNA transcript dissociate from the DNA at the terminator (Figure 12.5).
- The genes along a chromosome are transcribed in different directions using either DNA strand as a template. RNA is always synthesized in a 5′ to 3′ direction (Figure 12.6).

12.3 RNA Processing in Eukaryotes

- In eukaryotes, transcription produces a pre-mRNA that is capped, spliced, and given a poly A tail (Figure 12.7).
- During RNA splicing, intervening sequences called introns are removed from eukaryotic pre-mRNA by a spliceosome (Figure 12.8).
- The 5′ cap is a methylated guanine base at the 5′ end of the mRNA. The poly A tail is a string of adenine nucleotides that is added to the 3′ end (Figure 12.9).

12.4 Translation and the Genetic Code

- Based on the genetic code, each of the 64 codons specifies a start codon (methionine), other amino acids, or a stop codon (Table 12.1, Figure 12.10).
- The template strand of DNA is used to make mRNA with a series of codons. Recognition between mRNA and many tRNA molecules determines the amino acid sequence of a polypeptide. A polypeptide has a directionality in which the first amino acid is at the N-terminus, or amino terminus, whereas the last amino acid is at the C-terminus, or carboxyl terminus (Figure 12.11).
- Nirenberg and Leder used the ability of RNA triplets to promote the binding of tRNA to ribosomes as a way to determine many of the codons of the genetic code (Figure 12.12).

12.5 The Machinery of Translation

- Translation requires mRNA, charged tRNAs, ribosomes, and many translation factors (Table 12.2).
- tRNA molecules have a cloverleaf structure. Two important sites are the amino acid attachment site at the 3′ end and the anticodon, which forms base-pairs with a codon in mRNA (Figure 12.13).
- The enzyme aminoacyl-tRNA synthetase attaches the correct amino acid to a tRNA molecule, producing a charged tRNA (Figure 12.14).
- Ribosomes are composed of a small and large subunit, each consisting of rRNA molecules and many proteins. Bacterial and eukaryotic ribosomes differ in their molecular composition (Table 12.3).
- Ribosomes have three sites, termed the A, P, and E sites, which are locations for the binding and release of tRNA molecules (Figure 12.15).
- The gene that encodes the small subunit rRNA (SSU rRNA) has been used to determine evolutionary relationships among different species (Figure 12.16).

12.6 The Stages of Translation

- Translation occurs in three stages called initiation, elongation, and termination (Figure 12.17).
- During initiation of translation, the mRNA assembles with the ribosomal subunits and the initiator tRNA molecule, which carries methionine, the initial amino acid (Figure 12.18).
- During elongation, amino acids are added one at a time to a growing polypeptide (Figure 12.19).
- Termination of translation occurs when the binding of a release factor to a stop codon causes the release of the completed polypeptide from the tRNA and the disassembly of the mRNA, ribosomal subunits, and release factor (Figure 12.20).
- Though translation in bacteria and eukaryotes is strikingly similar, some key differences have been observed (Table 12.4).

Assess and Discuss

Test Yourself

1. Which of the following best represents the central dogma of gene expression?
   a. During transcription, DNA codes for polypeptides.
   b. During transcription, DNA codes for mRNA, which codes for polypeptides during translation.
   c. During translation, DNA codes for mRNA, which codes for polypeptides during transcription.
   d. none of the above
2. Transcription of a gene begins at a site on DNA called ___ and ends at a site on DNA known as ___.
   a. the start codon, the stop codon
   b. a promoter, the stop codon
   c. the start codon, the terminator
   d. a promoter, the terminator
   e. none of the above

3. The functional product of a structural gene is
   a. tRNA.
   b. mRNA.
   c. rRNA.
   d. a polypeptide.
   e. a, b, and c.
4. During eukaryotic RNA processing, the nontranslated sequences that are removed are called
   a. exons.
   b. introns.
   c. promoters.
   d. codons.
   e. ribozymes.
5. The ___ is the site where the translation process takes place.
   a. mitochondria
   b. nucleus
   c. ribosome
   d. lysosome
   e. ribozyme
6. The small subunit of a ribosome is composed of
   a. a protein.
   b. an rRNA molecule.
   c. many proteins.
   d. many rRNA molecules.
   e. many proteins and one rRNA molecule.
7. The region of the tRNA that is complementary to a codon in mRNA is
   a. the acceptor stem.
   b. the codon.
   c. the peptidyl site.
   d. the anticodon.
   e. the adaptor loop.
8. During the initiation of translation, the first codon, ____, enters the ________ and associates with the initiator tRNA.
   a. UAG, A site
   b. AUG, A site
   c. UAG, P site
   d. AUG, P site
   e. AUG, E site
9. The movement of the polypeptide from the tRNA in the P site to the tRNA in the A site is referred to as
   a. peptide bonding.
   b. aminoacyl binding.
   c. translation.
   d. the peptidyl transfer reaction.
   e. elongation.
10. The synthesis of a polypeptide occurs during which stage of translation?
    a. initiation
    b. elongation
    c. termination
    d. splicing

Conceptual Questions

1. Describe the one-gene/one-enzyme hypothesis and the more modern modifications of this hypothesis. Briefly explain how studying the pathway that leads to arginine synthesis allowed Beadle and Tatum to conclude that one gene sometimes encodes one enzyme.
2. What is the function of an aminoacyl-tRNA synthetase?
3. A principle of biology is that *the genetic material provides a blueprint for reproduction.* Explain how the information within the blueprint is accessed at the molecular level.

Collaborative Questions

1. Why do you think some complexes, such as spliceosomes and ribosomes, have both protein and RNA components?
2. Discuss and make a list of the similarities and differences in the events that occur during the initiation, elongation, and termination stages of transcription and translation.

Online Resource

www.brookerbiology3e.com

Stay a step ahead in your studies with animations that bring concepts to life and practice tests to assess your understanding. Your instructor may also recommend the interactive eBook, individualized learning tools, and more.

Chapter Outline

13.1 Overview of Gene Regulation

13.2 Regulation of Transcription in Bacteria

13.3 Regulation of Transcription in Eukaryotes: Roles of Transcription Factors and Mediator

13.4 Regulation of Transcription in Eukaryotes: Changes in Chromatin Structure and DNA Methylation

13.5 Regulation of RNA Processing and Translation in Eukaryotes

Summary of Key Concepts

Assess and Discuss

13 Gene Regulation

A model for a protein that binds to DNA and regulates genes. The catabolite activator protein (CAP), shown in dark and light blue, is binding to the DNA double helix, shown in orange and white. The CAP, shown again in Figure 13.10, activates gene transcription.

Emilio took a weight-lifting class in college and was surprised by the results. Within a few weeks, he was able to lift substantially more weight. He was inspired by this progress and continued lifting weights after the semester-long course ended. A year later, he was not only much stronger, but he could see physical changes in his body. Certain muscles, such as the biceps and triceps in his upper arms, were noticeably larger. How can we explain the increase in mass of Emilio's muscles? Unknowingly, when he was lifting weights, Emilio was affecting the regulation of his genes. Certain genes in his muscle cells were being "turned on" during his workouts, which then led to the synthesis of proteins that increased the mass of Emilio's muscles.

At the molecular level, **gene expression** is the process by which the information within a gene is made into a functional product, such as an RNA molecule or a protein. Most genes in all species are regulated so the proteins they specify are produced at appropriate times and in specific amounts. The term **gene regulation** refers to the ability of cells to control the expression of their genes. By comparison, some genes have relatively constant levels of expression in all conditions over time. These are called **constitutive genes**. Frequently, constitutive genes encode proteins that are always required for the survival of an organism, such as certain metabolic enzymes.

The importance of gene regulation is underscored by the number of genes devoted to this process in an organism. For example, in *Arabidopsis thaliana*, a plant that is studied by many plant geneticists, over 5% of its genome is involved with regulating gene transcription. This species has more than 1,500 different genes that encode proteins that regulate the transcription of other genes.

In this chapter, we will begin with an overview that emphasizes the benefits of gene regulation and the general mechanisms that achieve such regulation in bacteria and in eukaryotes. Later sections will describe how bacteria regulate gene expression in the face of environmental change and the more complex nature of gene regulation in eukaryotes.

13.1 Overview of Gene Regulation

Learning Outcomes:

1. Discuss the various ways that organisms benefit from gene regulation.
2. Identify where gene regulation can occur in the pathway of gene expression for bacteria and eukaryotes.

How do living organisms benefit from gene regulation? One reason is that it conserves energy. Proteins that are encoded by genes will be produced only when they are needed. In multicellular organisms, gene regulation also ensures that genes are expressed in the appropriate cell types and at the correct stage of development. In this section, we will examine a few examples that illustrate the important consequences of gene regulation. We will also survey the major points in the gene expression process at which genes are regulated in bacterial and eukaryotic cells.

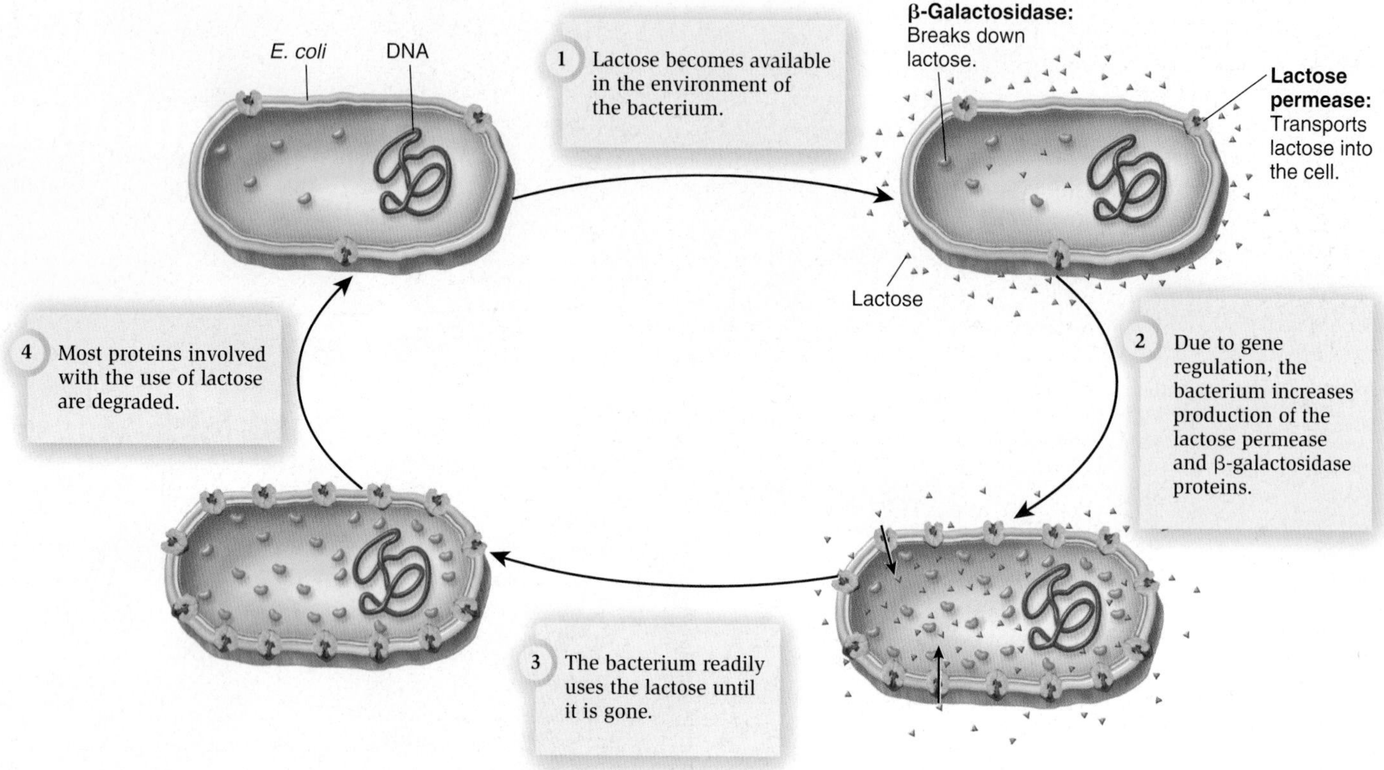

Figure 13.1 Gene regulation of lactose utilization in *E. coli*.

BIOLOGY PRINCIPLE Living organisms use energy. Gene regulation provides a way for organisms to avoid wasting energy. They make proteins only when they are needed.

Bacteria Regulate Genes to Respond to Changes in Their Environment

The bacterium *Escherichia coli* can use many types of sugars as food sources, thereby increasing its chances of survival. With regard to gene regulation, we will focus on how it uses lactose, which is the sugar found in milk. *E. coli* carries genes that code for proteins that enable it to take up lactose from the environment and metabolize it.

Figure 13.1 illustrates the effects of lactose on the regulation of those genes. In order to utilize lactose, an *E. coli* cell requires a transporter, called lactose permease, that facilitates the uptake of lactose into the cell, and an enzyme, called β-galactosidase, that catalyzes the breakdown of lactose. When lactose is not present in the environment, an *E. coli* cell makes very little of these proteins. However, when lactose becomes available, the bacterium produces many more copies of these proteins, enabling it to readily use lactose from its environment. Eventually, all of the lactose in the environment is used up. At this point, the genes encoding these proteins will be shut off, and most of the proteins will be degraded. Overall, gene regulation conserves energy because it ensures that the proteins needed for lactose utilization are made only when lactose is present in the environment.

Eukaryotic Gene Regulation Produces Different Cell Types in a Single Organism

One of the most amazing examples of gene regulation is the phenomenon of **cell differentiation**, the process by which cells become specialized into particular types. In humans, for example, cells may differentiate into muscle cells, neurons, skin cells, or other types. **Figure 13.2** shows micrographs of three types of cells found in humans. As seen here, their morphologies are strikingly different. Likewise, their functions within the body are also quite different. Muscle cells are important in body movements, neurons function in cell signaling, and skin cells form a protective outer surface to the body.

Gene regulation is responsible for producing different types of cells within a multicellular organism. The three cell types shown in Figure 13.2 contain the same **genome**, meaning they carry the same set of genes. However, their **proteomes**—the collection of proteins they make—are quite different. Certain proteins are found in particular cell types but not in others. Alternatively, a protein may be present in all three cell types, but the relative amounts of the protein may be different. The amount of a given protein depends on many factors, including how strongly the corresponding gene is turned on and how much protein is synthesized from mRNA. Gene regulation plays a major role in determining the proteome of each cell type.

Eukaryotic Gene Regulation Enables Multicellular Organisms to Proceed Through Developmental Stages

In multicellular organisms that progress through developmental stages, certain genes are expressed at particular stages of development but not others. Let's consider an example of such gene regulation in mammals. Early stages of development occur in the uterus of female mammals. Following fertilization, an embryo develops inside the uterus. In humans, the embryonic stage lasts from fertilization to 8 weeks. During this stage, major developmental changes produce

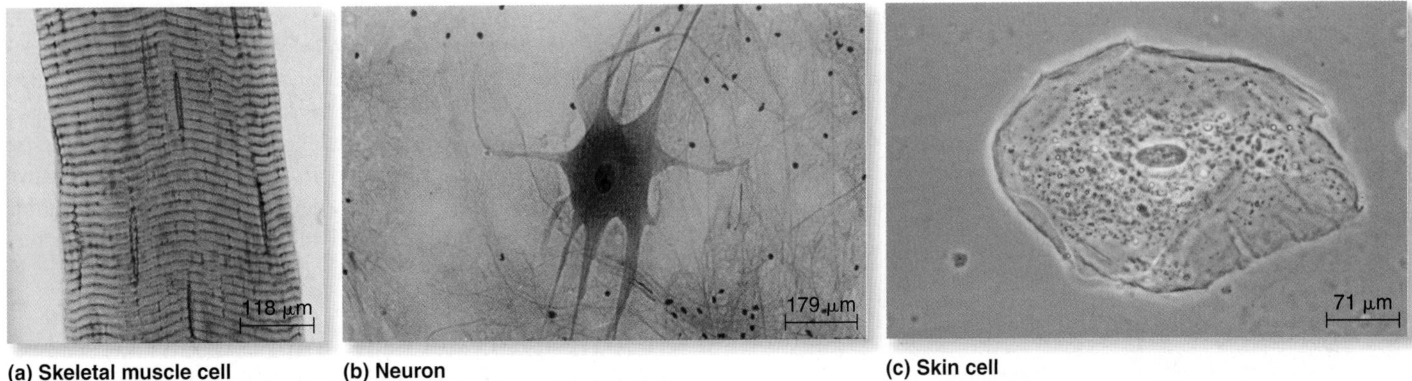

Figure 13.2 **Examples of different cell types in humans.** These cells have the same genetic composition. Their unique morphologies are due to differences in the proteins they make.

Concept Check: *How does gene regulation underlie the different morphologies of these cells?*

the various body parts. The fetal stage occurs from 8 weeks to birth (41 weeks). This stage is characterized by continued refinement of body parts and a large increase in size.

The oxygen demands of a rapidly growing embryo and fetus are quite different from the needs of the mother. Gene regulation plays a vital role in ensuring that an embryo and fetus get the proper amount of oxygen. Hemoglobin is a protein that delivers oxygen to the cells of a mammal's body. A hemoglobin protein is composed of four globin polypeptides, two encoded by one globin gene and two encoded by another globin gene (**Figure 13.3**). The genomes of mammals carry several genes (designated with Greek letters) that encode slightly different globin polypeptides. During the embryonic stage of development, the epsilon (ε)- and zeta (ζ)-globin genes are turned on. At the fetal stage, these genes are turned off, and the alpha (α)- and gamma (γ)-globin genes are turned on. Finally, at birth, the γ-globin gene is turned off, and the beta (β)-globin gene is turned on.

Fertilization
8 weeks
41 weeks
Embryo → Fetus → Adult

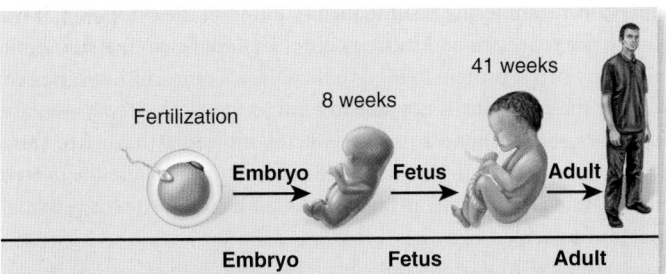

	Embryo	Fetus	Adult
Hemoglobin protein	2 ζ-globins 2 ε-globins	2 α-globins 2 γ-globins	2 α-globins 2 β-globins
Oxygen affinity	Highest	High	Moderate
Gene expression			
α-globin gene	**Off**	**On**	**On**
β-globin gene	**Off**	**Off**	**On**
γ-globin gene	**Off**	**On**	**Off**
ζ-globin gene	**On**	**Off**	**Off**
ε-globin gene	**On**	**Off**	**Off**

Figure 13.3 **Regulation of human globin genes at different stages of development.**

BIOLOGY PRINCIPLE **Living organisms grow and develop.** Gene regulation is an important process that allows organisms to proceed through developmental stages.

How do the embryo and fetus acquire oxygen from their mother's bloodstream? The hemoglobin produced during the embryonic and fetal stages has a higher binding affinity for oxygen than does the hemoglobin produced after birth. Therefore, the embryo and fetus can remove oxygen from the mother's bloodstream and use that oxygen for their own needs. This occurs across the placenta, where the mother's bloodstream is adjacent to the bloodstream of the embryo or fetus. In this way, gene regulation enables mammals to develop internally, even though the embryo and fetus are not breathing on their own. Gene regulation ensures that the correct hemoglobin protein is produced at the right time in development. We'll discuss how gene expression controls the process of development in greater detail in Chapter 19.

Gene Regulation Occurs at Different Points in the Process from DNA to Protein

Thus far, we have learned that gene regulation has a dramatic influence on the ability of organisms to respond to environmental changes, produce different types of cells, and progress through developmental stages. For genes that encode proteins, the regulation of gene expression can occur at any of the steps that are needed to produce a functional protein.

In bacteria, gene regulation most commonly occurs at the level of transcription, which means that bacteria regulate how much mRNA is made from genes (**Figure 13.4a**). When geneticists say a gene is "turned off," they mean that very little or no mRNA is made from that gene, whereas a gene that is "turned on" is transcribed into mRNA. Because transcription is the first step in gene expression, transcriptional regulation is a particularly efficient way to regulate genes because cells avoid wasting energy when the product of the gene is not needed. A second way for bacteria to regulate gene expression is to control the ability of an mRNA to be translated into protein. This form of gene regulation is less common in bacteria. Last, gene expression can be regulated at the protein or post-translational level.

In eukaryotes, gene regulation occurs at many levels, including transcription, RNA processing, translation, and after translation is completed (**Figure 13.4b**). As in their bacterial counterparts,

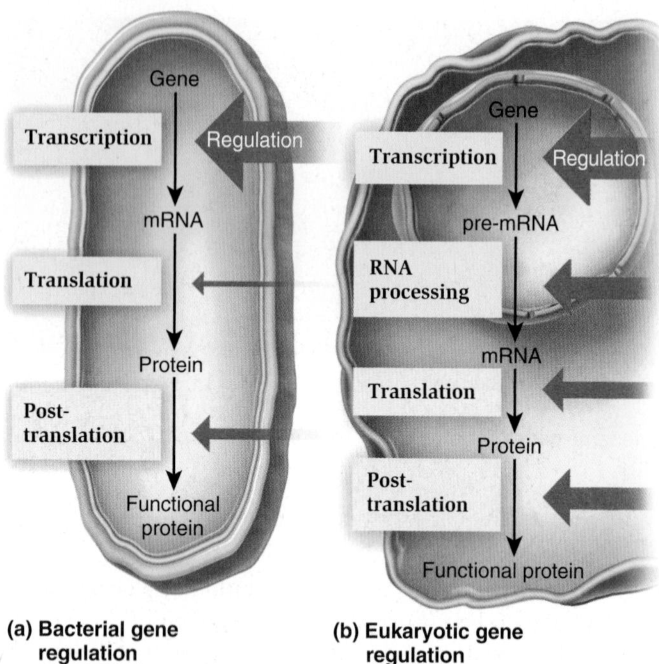

Figure 13.4 Overview of gene regulation in (a) bacteria and (b) eukaryotes. The relative width of the red arrows indicates the prominence with which gene regulation is used to control the production of functional proteins.

transcriptional regulation is a prominent form of gene regulation for eukaryotes. As discussed later in this chapter, eukaryotic genes are transcriptionally regulated in several different ways, some of which are not found in bacteria. As discussed in Chapter 12, eukaryotes process their mRNA transcripts in ways that do not commonly occur in bacteria (refer back to Figure 12.3). For example, RNA splicing is a widespread phenomenon in eukaryotes. Later in this chapter, we will examine how this process is regulated to produce two or more different types of mRNA from a single gene. Eukaryotes can also regulate an mRNA after its modification. The translation of mRNA may be regulated by small, inhibitory RNA molecules or by RNA-binding proteins that prevent translation from occurring.

As in bacteria, eukaryotic proteins can be regulated in a variety of ways other than gene regulation, including cellular regulation and biochemical regulation (such as feedback inhibition). These various types of regulation are best understood within the context of cell biology, so they were primarily discussed in Unit II (see Chapter 6 and Figure 6.13). Post-translational modifications are also summarized in Chapter 21 (look ahead to Figure 21.10b).

13.2 Regulation of Transcription in Bacteria

Learning Outcomes:

1. Explain how regulatory transcription factors and small effector molecules are involved in the regulation of transcription.
2. Describe the organization of the *lac* operon and how it is under negative and positive control.
3. Analyze the results of the experiments of Jacob, Monod, and Pardee.
4. Describe how the *trp* operon is under negative control.

As we have seen, when a bacterium is exposed to a particular nutrient in its environment, such as a sugar, the genes are expressed that encode proteins needed for the uptake and metabolism of that sugar. In addition, bacteria have genes that encode enzymes that synthesize molecules such as particular amino acids. In such cases, the control of gene expression often occurs at the level of transcription. In this section, we will examine the underlying molecular mechanisms that bring about transcriptional regulation in bacteria.

Transcriptional Regulation Involves Regulatory Transcription Factors and Small Effector Molecules

In most cases, regulation of transcription involves the actions of **regulatory transcription factors**—proteins that bind to **regulatory sequences** in the DNA in the vicinity of a promoter and affect the rate of transcription of one or more nearby genes. These transcription factors either decrease or increase the rate of transcription of a gene. **Repressors** are regulatory transcription factors that bind to the DNA and decrease the rate of transcription. This is a form of regulation called **negative control**. **Activators** bind to the DNA and increase the rate of transcription, a form of regulation termed **positive control** (Figure 13.5a).

In conjunction with regulatory transcription factors, molecules called **small effector molecules** often play a critical role in transcriptional regulation. A small effector molecule exerts its effects by binding to a regulatory transcription factor and causing a conformational change in the protein. In many cases, the effect of the conformational change determines whether or not the protein can bind to the DNA. **Figure 13.5b** illustrates an example involving a repressor. When the small effector molecule is not present in the cytoplasm, the repressor binds to the DNA and inhibits transcription. However, when the small effector molecule is subsequently found in the cytoplasm, it will bind to the repressor and cause a conformational change that inhibits the ability of the protein to bind to the DNA. Transcription can occur because the repressor is not able to bind to the DNA. Repressors and activators that respond to small effector molecules have two functional regions called domains. One domain is a site where the protein binds to the DNA, whereas the other is the binding site for the small effector molecule.

The *lac* Operon Contains Genes That Encode Proteins Involved in Lactose Metabolism

In bacteria, structural genes are sometimes clustered together and under the transcriptional control of a single promoter. This arrangement is known as an **operon**. The transcription of the genes occurs as a single unit and results in the production of a **polycistronic mRNA**, an mRNA that encodes more than one protein. What advantage is this arrangement? An operon organization allows a bacterium to coordinately regulate a group of genes that encode proteins whose functions are used in a common pathway.

The genome of *E. coli* carries an operon, called the ***lac* operon**, that contains the genes for the proteins that allow it to metabolize lactose (see Figure 13.1). **Figure 13.6a** shows the organization of this operon as it is found in the *E. coli* chromosome, as well as the polycistronic mRNA that is transcribed from it. The *lac* operon

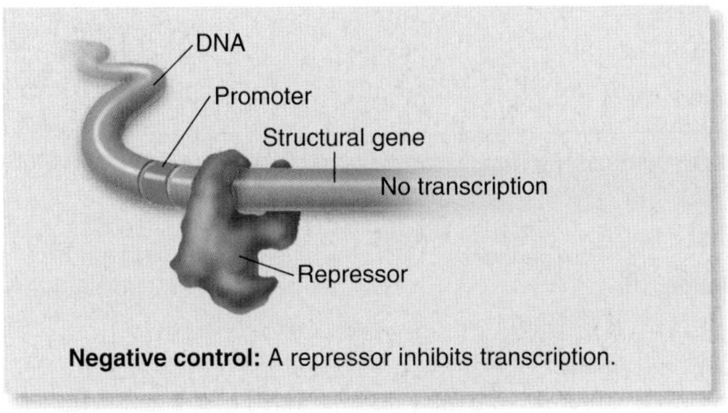
DNA
Promoter
Structural gene
No transcription
Repressor
Negative control: A repressor inhibits transcription.

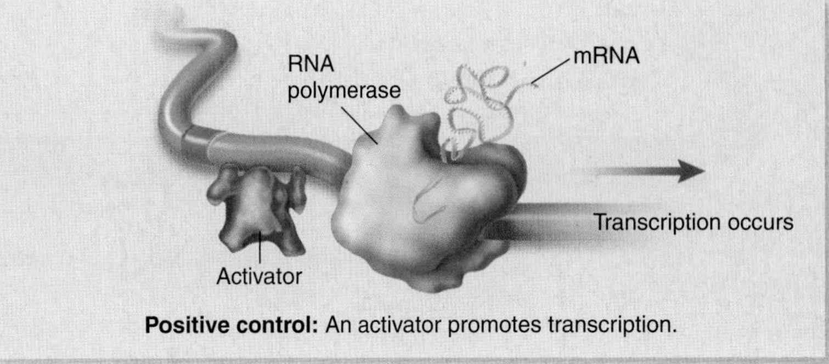
RNA polymerase
mRNA
Transcription occurs
Activator
Positive control: An activator promotes transcription.

(a) Actions of regulatory transcription factors

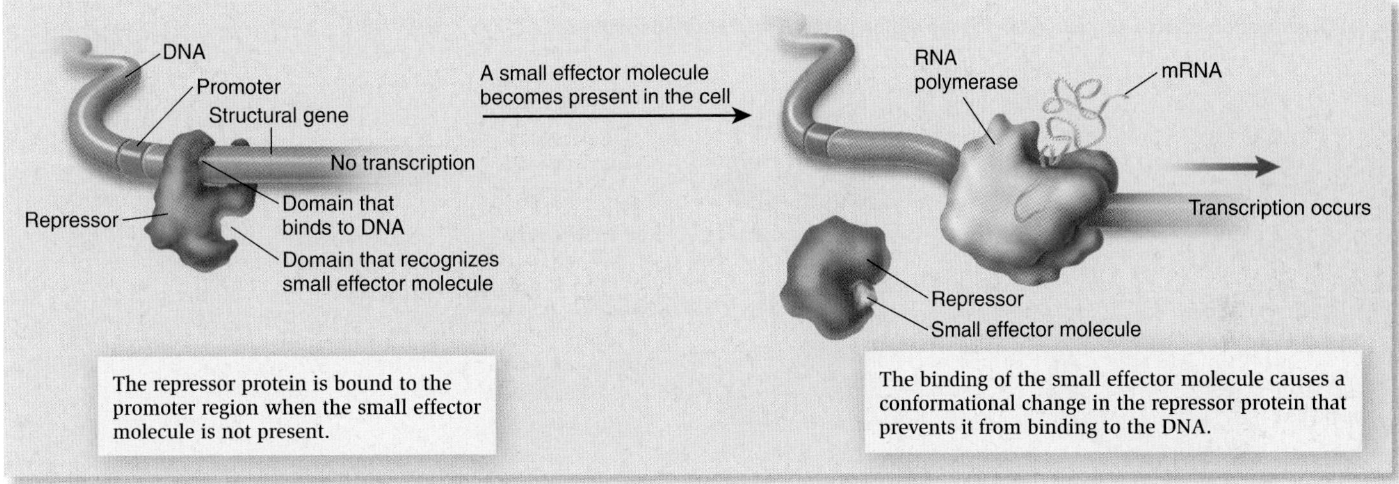
DNA
Promoter
Structural gene
A small effector molecule becomes present in the cell
No transcription
Repressor
Domain that binds to DNA
Domain that recognizes small effector molecule
The repressor protein is bound to the promoter region when the small effector molecule is not present.
RNA polymerase
mRNA
Transcription occurs
Repressor
Small effector molecule
The binding of the small effector molecule causes a conformational change in the repressor protein that prevents it from binding to the DNA.

(b) Action of a small effector molecule on a repressor

Figure 13.5 Actions of regulatory transcription factors and small effector molecules. **(a)** Regulatory transcription factors are proteins that exert negative or positive control. **(b)** One way that a small effector molecule may exert its effects is by preventing a repressor protein from binding to the DNA.

contains a promoter, *lacP*, that is used to transcribe three structural genes: *lacZ*, *lacY*, and *lacA*. *LacZ* encodes β-galactosidase, which is an enzyme that breaks down lactose (**Figure 13.6b**). As a side reaction, β-galactosidase also converts a small percentage of lactose into allolactose, a structurally similar sugar or lactose analogue. As described later, allolactose is important in the regulation of the *lac* operon. The *lacY* gene encodes lactose permease, which is a membrane protein required for the transport of lactose into the cytoplasm of the bacterium. The *lacA* gene encodes galactoside transacetylase, which covalently modifies lactose and lactose analogues by attaching an acetyl group ($—COCH_3$). Although the functional necessity of this enzyme remains unclear, the attachment of acetyl groups to nonmetabolizable lactose analogues may prevent their toxic buildup in the cytoplasm.

Near the *lac* promoter are two regulatory sequences designated the operator and the CAP site (see Figure 13.6a). The **operator**, or *lacO* site, is a regulatory sequence in the DNA. The sequence of bases at the *lacO* site provides a binding site for a repressor protein. The **CAP site** is a regulatory sequence recognized by an activator protein.

Adjacent to the *lac* operon is the *lacI* gene, which encodes the **lac repressor**. This repressor protein is important for the regulation of the *lac* operon. The *lacI* gene, which is constitutively expressed at fairly low levels, has its own promoter called the *i* promoter. The *lacI* gene is not considered a part of the *lac* operon. Let's now take a look at how the *lac* operon is regulated by the lac repressor.

The *lac* Operon Is Under Negative Control by a Repressor Protein

In the late 1950s, the first researchers to investigate gene regulation were French biologists François Jacob and Jacques Monod at the Pasteur Institute in Paris, France. Their focus on gene regulation stemmed from an interest in the phenomenon known as enzyme adaptation, which had been identified early in the 20th century. Enzyme adaptation occurs when a particular enzyme appears within a living cell only after the cell has been exposed to the substrate for that enzyme. Jacob and Monod studied lactose metabolism in *E. coli* to investigate this phenomenon. When they exposed bacteria to lactose, the levels of lactose-using enzymes in the cells increased by 1,000- to 10,000-fold. After lactose was removed, the synthesis of the enzymes abruptly stopped.

The first mechanism of regulation that Jacob and Monod discovered involved the lac repressor, which binds to the sequence of nucleotides found at the *lac* operator site. Once bound, the lac repressor prevents RNA polymerase from transcribing the *lacZ*, *lacY*, and

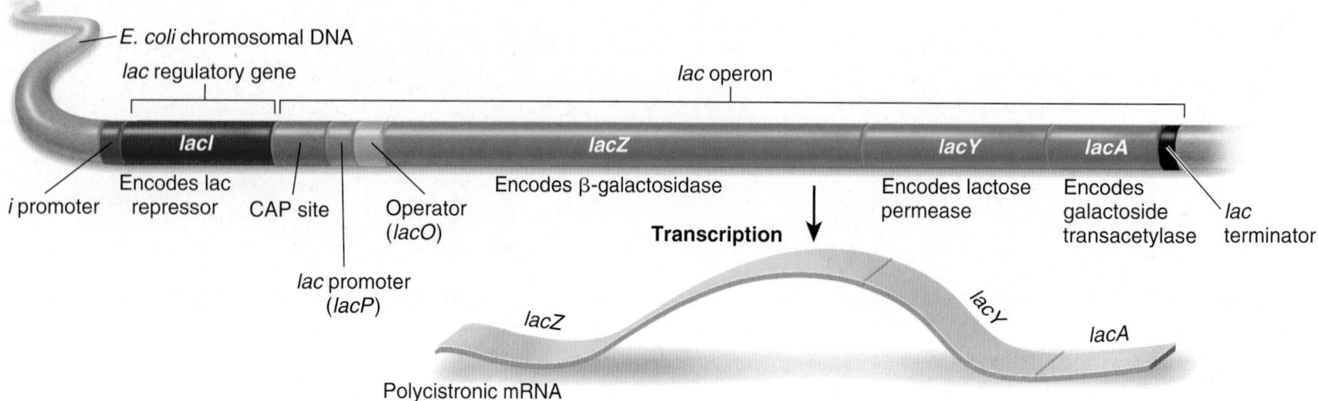

(a) Organization of DNA sequences in the *lac* region of the *E. coli* chromosome

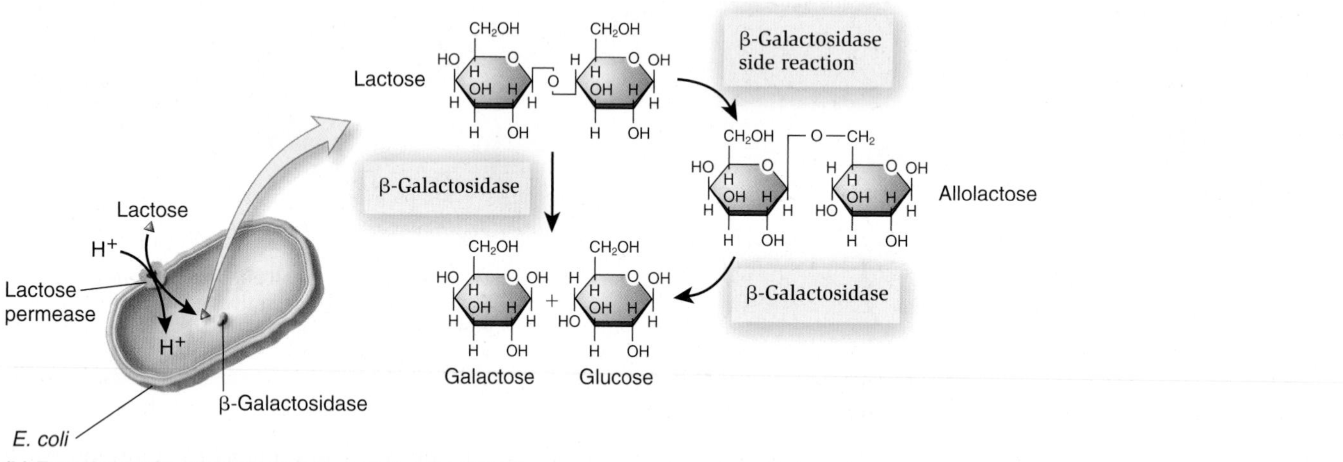

(b) Functions of lactose permease and β-galactosidase

Figure 13.6 **The *lac* operon.** **(a)** This diagram depicts a region of the *E. coli* chromosome that contains the *lacI* gene and the adjacent *lac* operon, as well as the polycistronic mRNA transcribed from the operon. The mRNA is translated into three proteins: lactose permease, β-galactosidase, and galactoside transacetylase. **(b)** Lactose permease cotransports H^+ with lactose. Bacteria maintain an H^+ gradient across their cytoplasmic membrane that drives the active transport of lactose into the cytoplasm. β-Galactosidase cleaves lactose into galactose and glucose. As a side reaction, it can also convert lactose into allolactose.

Concept Check: *Which genes are under the control of the lac promoter?*

lacA genes (**Figure 13.7a**). RNA polymerase can bind to the promoter when the lac repressor is bound to the operator site, but cannot move past the operator to transcribe the *lacZ*, *lacY*, and *lacA* genes.

Whether or not the lac repressor binds to the operator site depends on allolactose, which is the previously mentioned side product of the β-galactosidase enzyme (see Figure 13.6b). How does allolactose control the lac repressor? Allolactose is an example of a small effector molecule. The lac repressor protein contains four identical subunits, each one recognizing a single allolactose molecule. When four allolactose molecules bind to the lac repressor, a conformational change occurs that prevents the repressor from binding to the operator. Under these conditions, RNA polymerase is free to transcribe the operon (**Figure 13.7b**).

The regulation of the *lac* operon enables *E. coli* to conserve energy because lactose-utilizing proteins are made only when lactose is present in the environment. Allolactose is an **inducer**, a small effector molecule that increases the rate of transcription, and the *lac* operon is said to be an **inducible operon**. When the bacterium is not exposed to lactose, no allolactose is available to bind to the lac repressor. Therefore, the lac repressor binds to the operator site and inhibits its transcription. In reality, the repressor does not completely inhibit transcription, so very small amounts of β-galactosidase, lactose permease, and galactoside transacetylase are made. Even so, the levels are far too low for the bacterium to readily use lactose. When the bacterium is exposed to lactose, a small amount can be transported into the cytoplasm via lactose permease, and β-galactosidase converts some of it to allolactose (see Figure 13.6b). The cytoplasmic level of allolactose gradually rises until allolactose binds to the lac repressor, which induces the *lac* operon and promotes a high rate of transcription of the *lacZ*, *lacY*, and *lacA* genes. Translation of the encoded polypeptides produces the proteins needed for lactose uptake and metabolism as described previously in Figure 13.1.

When lactose is absent from the environment, allolactose is not made and the lac repressor is free to bind to the *lac* operator. The transcription of the operon is inhibited.
lac regulatory gene
lac operon
RNA polymerase
lac promoter
Operator
lacI
lacP
lacO
lacZ
lacY
lacA
mRNA
Lac repressor binds to the operator and inhibits transcription.
Lac repressor (active)

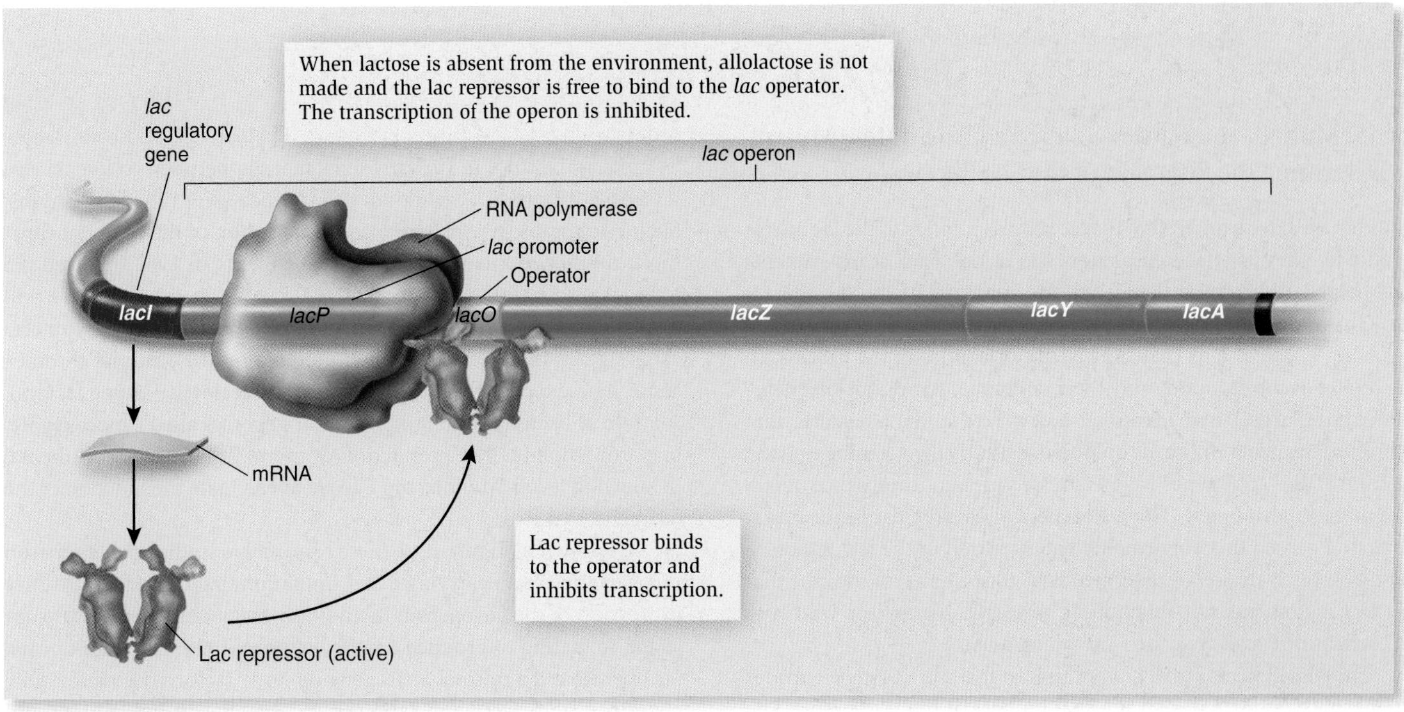

(a) Lactose absent from the environment

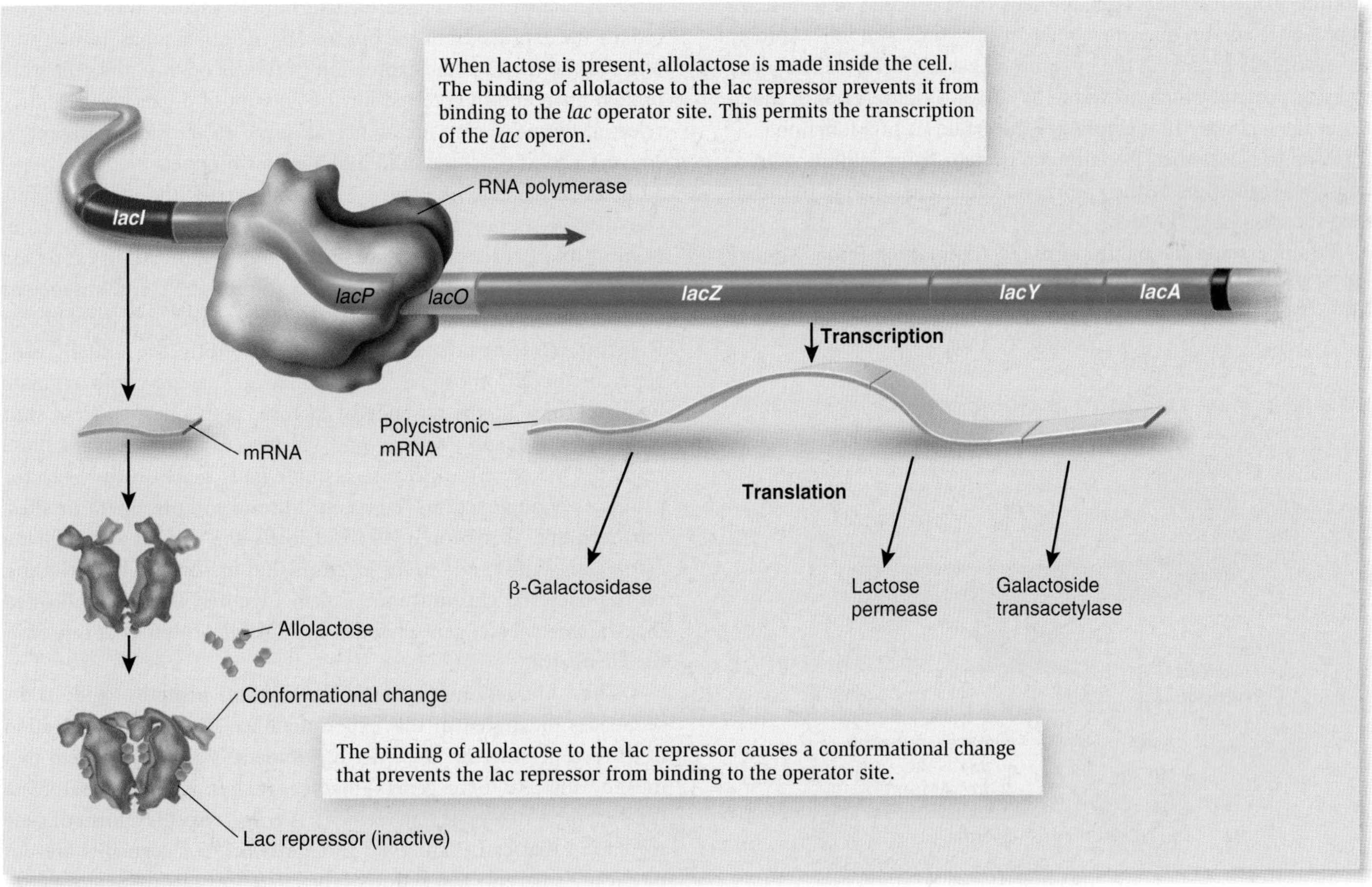

When lactose is present, allolactose is made inside the cell. The binding of allolactose to the lac repressor prevents it from binding to the *lac* operator site. This permits the transcription of the *lac* operon.
RNA polymerase
lacI
lacP
lacO
lacZ
lacY
lacA
Transcription
Polycistronic mRNA
mRNA
Translation
β-Galactosidase
Lactose permease
Galactoside transacetylase
Allolactose
Conformational change
The binding of allolactose to the lac repressor causes a conformational change that prevents the lac repressor from binding to the operator site.
Lac repressor (inactive)

(b) Lactose present

Figure 13.7 **Negative control of an inducible set of genes: function of the lac repressor in regulating the *lac* operon.**

Concept Check: *With regard to regulatory proteins and small effector molecules, explain the meaning of the terms "negative control" and "inducible."*

FEATURE INVESTIGATION

Jacob, Monod, and Pardee Studied a Constitutive Mutant to Determine the Function of the Lac Repressor

Thus far, we have learned that the lac repressor binds to the *lac* operator site to exert its effects. Let's now take a look back at experiments that helped researchers determine the function of the lac repressor. Our understanding of *lac* operon regulation came from studies involving *E. coli* strains that showed abnormalities in the process. In the 1950s, Jacob, Monod, and their colleague, American biochemist Arthur Pardee, had identified a few rare mutant bacteria that expressed the genes of the *lac* operon constitutively, meaning that the *lacZ*, *lacY*, and *lacA* genes were expressed even in the absence of lactose in the environment. The researchers discovered that some mutations that caused this abnormality had occurred in the *lacI* region of the DNA. Such strains were termed *lacI*$^-$ (*lacI* minus) to indicate that the *lacI* region was not functioning properly. Normal or wild-type *lacI* strains of *E. coli* are called *lacI*$^+$ (*lacI* plus).

The researchers initially hypothesized that the *lacI* gene encodes an enzyme that degrades an internal inducer of the *lac* operon. The *lacI*$^-$ mutation was thought to inhibit this enzyme, thereby allowing the internal inducer to always be synthesized. In this way, the *lacI*$^-$ mutation made it unnecessary for cells to be exposed to lactose for induction. However, over the course of this study and other studies, they eventually arrived at the hypothesis that the *lacI* gene encodes a repressor protein, which proved to be correct (**Figure 13.8**). A mutation in the *lacI* gene that eliminates the synthesis of a functional lac repressor prevents the lac repressor protein from inhibiting transcription. At the time of their work, however, the function of the lac repressor was not yet known.

To understand the nature of the *lacI*$^-$ mutation, Jacob, Monod, and Pardee applied a genetic approach. Although bacterial conjugation is described in Chapter 18, let's briefly examine this process in order to understand this experiment. The earliest studies of Jacob, Monod, and Pardee in 1959 involved matings between recipient cells, termed F$^-$ (F minus), and donor cells, which were called Hfr strains. Such Hfr strains were able to transfer a portion of the bacterial chromosome to a recipient cell. Later experiments in 1961 involved the transfer of circular segments of DNA known as F factors. We will consider this later type of experiment here. Sometimes an F factor also carries genes that were originally found within the bacterial chromosome. These types of F factors are called F′ factors (F prime factors). A strain of bacteria containing F′ factor genes is called a **merozygote**, or partial diploid. The production of merozygotes was instrumental in allowing Jacob, Monod, and Pardee to elucidate the function of the *lacI* gene.

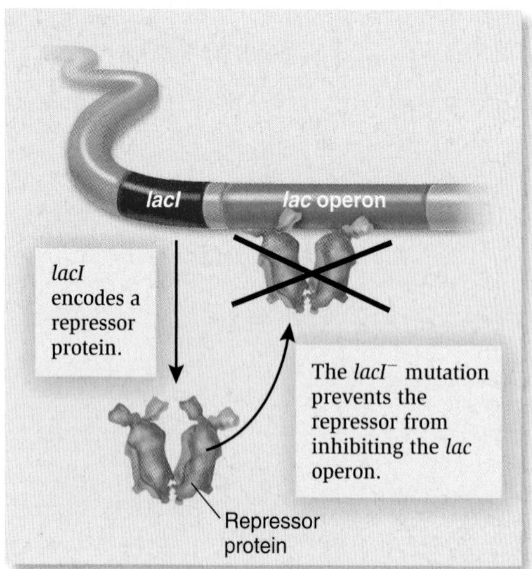
lacI
lac operon
lacI encodes a repressor protein.
The lacI⁻ mutation prevents the repressor from inhibiting the lac operon.
Repressor protein

Figure 13.8 **A hypothesis for the function of the *lacI* gene.**

As shown in **Figure 13.9**, these researchers studied the *lac* operon in a bacterial strain carrying a *lacI*$^-$ mutation that caused constitutive expression of the *lac* operon. In addition, the mutant strain was subjected to mating to produce a merozygote that also carried a normal *lac* operon and a normal *lacI*$^+$ gene on an F′ factor. The merozygote contained both *lacI*$^+$ and *lacI*$^-$ genes. The constitutive mutant and corresponding merozygote were grown separately in liquid media and then divided into two tubes each. In half of the tubes, the cells were incubated with lactose to determine if lactose was needed to induce the expression of the operon. In the other tubes, lactose was omitted. To monitor the expression of the *lac* operon, the cells were broken open and then tested for the amount of β-galactosidase they released by measuring the ability of any β-galactosidase present to convert a colorless compound into a yellow product.

The data table of Figure 13.9 summarizes the effects of this constitutive mutation and its analysis in a merozygote. As Jacob, Monod, and Pardee already knew, the *lacI*$^-$ mutant strain expressed the *lac* operon constitutively, in both the presence and absence of lactose. However, when a normal *lac* operon and *lacI*$^+$ gene on an F′ factor were introduced into a cell harboring the mutant *lacI*$^-$ gene on the chromosome, the normal *lacI*$^+$ gene could regulate both operons. In the absence of lactose, both operons were shut off. How did Jacob, Monod, and Pardee eventually explain these results? This occurred because a single *lacI*$^+$ gene on the F′ factor produces enough repressor protein to bind to both operator sites. Furthermore, this protein is diffusible—can spread through the cytoplasm—and binds to *lac* operons that are on the F′ factor and on the bacterial chromosome. Taken together, the data indicated that the normal *lacI* gene encodes a diffusible protein that represses the *lac* operon.

The interactions between regulatory proteins and DNA sequences illustrated in this experiment have led to the definition of three genetic terms. A ***cis*-acting element** is a DNA segment that must be adjacent to the gene(s) that it regulates. The *lac* operator site is an example of a *cis*-acting element. A ***trans*-effect** is a form of gene regulation that can occur even though two DNA segments are not physically adjacent. The action of the lac repressor on the *lac* operon is a *trans*-effect. A ***cis*-effect** is mediated by a *cis*-acting element that binds regulatory proteins, whereas a *trans*-effect is mediated by genes that encode diffusible regulatory proteins.

Experimental Questions

1. What were the key observations made by Jacob, Monod, and Pardee that led to the development of their hypothesis regarding the *lacI* gene and the regulation of the *lac* operon?
2. What was the eventual hypothesis proposed by the researchers to explain the function of the *lacI* gene and the regulation of the *lac* operon?
3. How did Jacob, Monod, and Pardee test the hypothesis? What were the results of the experiment? How do these results support the idea that the *lacI* gene produces a repressor protein?

Figure 13.9 **The experiment performed by Jacob, Monod, and Pardee to study a constitutive *lacI⁻* mutant.**

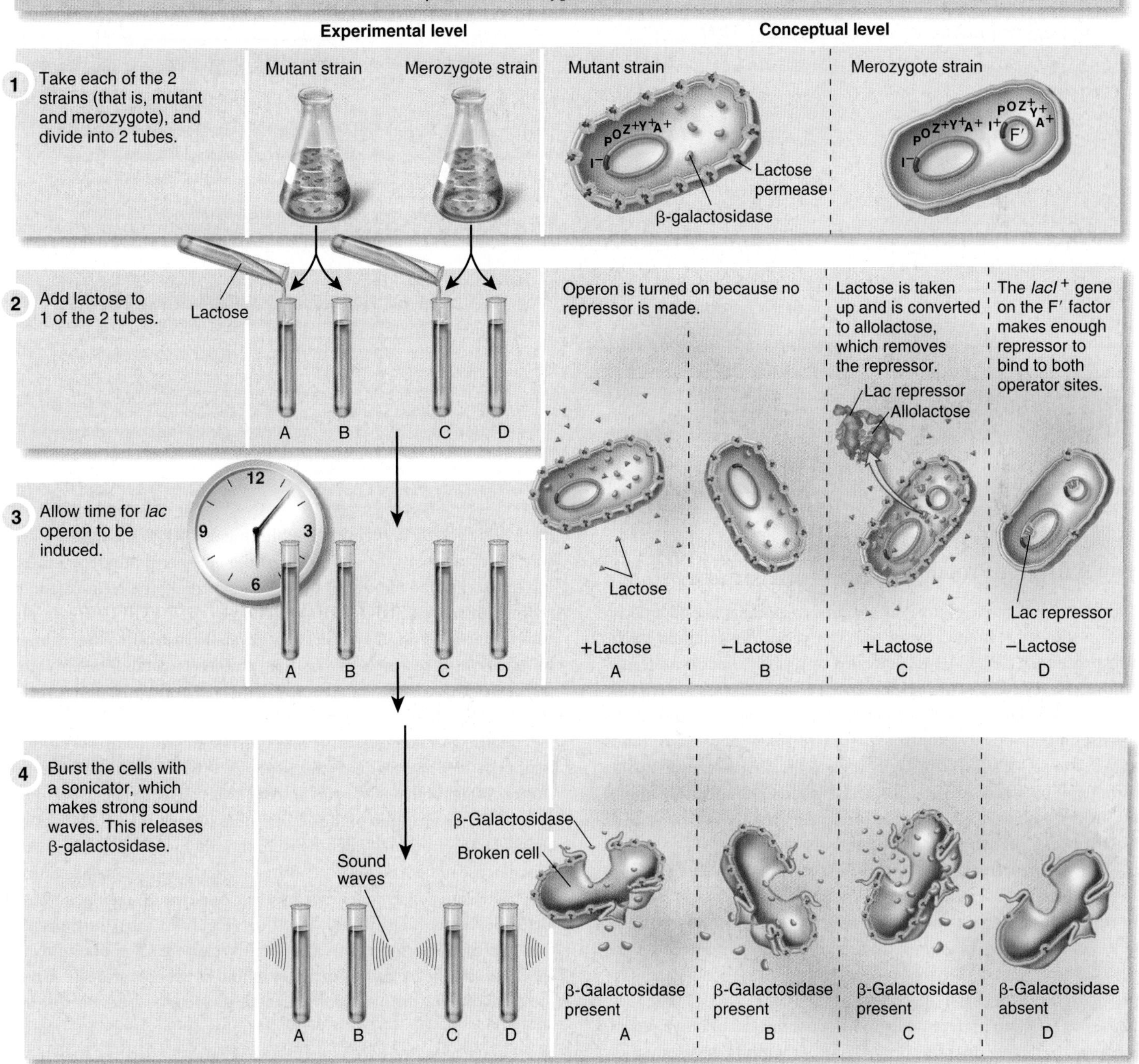

HYPOTHESIS The *lacI*⁻ mutation inhibits the lac repressor and thereby allows the constitutive expression of the *lac* operon. Note: This correct hypothesis actually arose from the results of this study and other studies.
KEY MATERIALS A constitutive *lacI*⁻ mutant strain was already characterized. An F′ factor carrying a normal *lacI*⁺ gene and *lac* operon was introduced into this strain to produce a merozygote strain. Note: POZ⁺Y⁺A⁺ refers to a normal *lac* operon.
Experimental level
Conceptual level
1 Take each of the 2 strains (that is, mutant and merozygote), and divide into 2 tubes.
Mutant strain
Merozygote strain
Mutant strain
Merozygote strain
POZ⁺Y⁺A⁺
I⁻
Lactose permease
β-galactosidase
POZ⁺Y⁺A⁺ I⁺
F′
2 Add lactose to 1 of the 2 tubes.
Lactose
A B C D
Operon is turned on because no repressor is made.
Lactose is taken up and is converted to allolactose, which removes the repressor.
Lac repressor
Allolactose
The *lacI*⁺ gene on the F′ factor makes enough repressor to bind to both operator sites.
3 Allow time for *lac* operon to be induced.
12
9
3
6
A B C D
Lactose
Lac repressor
+Lactose A
−Lactose B
+Lactose C
−Lactose D
4 Burst the cells with a sonicator, which makes strong sound waves. This releases β-galactosidase.
Sound waves
β-Galactosidase
Broken cell
A B C D
β-Galactosidase present A
β-Galactosidase present B
β-Galactosidase present C
β-Galactosidase absent D

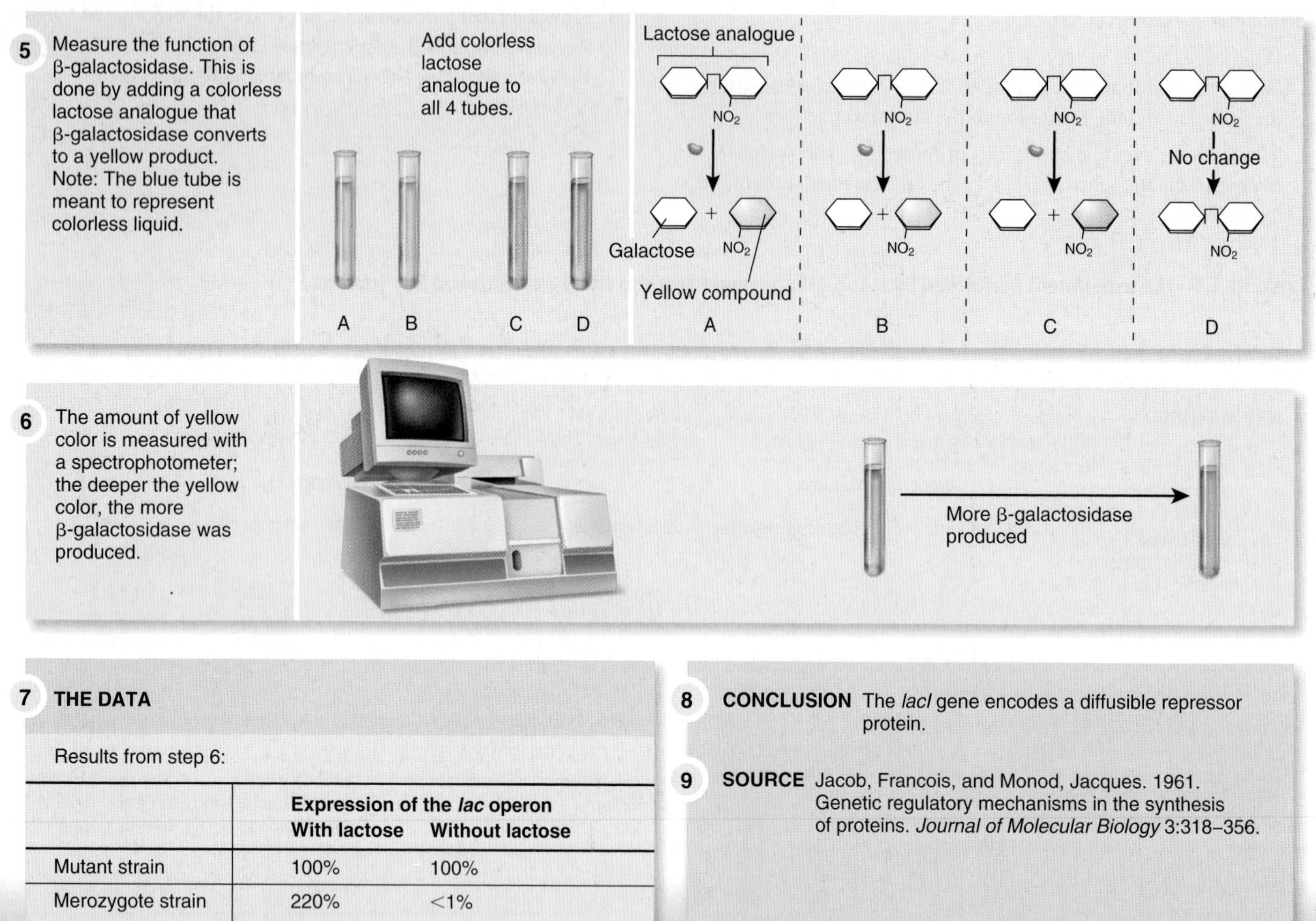

5 Measure the function of β-galactosidase. This is done by adding a colorless lactose analogue that β-galactosidase converts to a yellow product. Note: The blue tube is meant to represent colorless liquid.
Add colorless lactose analogue to all 4 tubes.
A B C D
Lactose analogue
NO2
Galactose
Yellow compound
No change
A B C D
6 The amount of yellow color is measured with a spectrophotometer; the deeper the yellow color, the more β-galactosidase was produced.
More β-galactosidase produced

7 **THE DATA**

Results from step 6:

	Expression of the *lac* operon With lactose	Without lactose
Mutant strain	100%	100%
Merozygote strain	220%	<1%

8 **CONCLUSION** The *lacI* gene encodes a diffusible repressor protein.

9 **SOURCE** Jacob, Francois, and Monod, Jacques. 1961. Genetic regulatory mechanisms in the synthesis of proteins. *Journal of Molecular Biology* 3:318–356.

The *lac* Operon Is Also Under Positive Control by an Activator Protein

In addition to negative control by a repressor protein, the *lac* operon is also positively regulated by an activator called the **catabolite activator protein (CAP)**. CAP is controlled by a small effector molecule, **cyclic AMP (cAMP)**, which is produced from ATP via an enzyme known as adenylyl cyclase. Gene regulation involving CAP and cAMP is an example of positive control (**Figure 13.10**). When cAMP binds to CAP, the cAMP-CAP complex binds to the CAP site near the *lac* promoter. This causes a bend in the DNA that enhances the ability of RNA polymerase to bind to the promoter. In this way, the rate of transcription is increased.

The key functional role of CAP is to allow *E. coli* to choose between different sources of sugar. In a process known as **catabolite repression**, the presence of a preferred energy source inhibits the use of other energy sources. In this case, transcription of the *lac* operon is inhibited by the presence of glucose, which is a catabolite (it is broken down—catabolized—inside the cell). This gene regulation allows *E. coli* to preferentially use glucose instead of other sugars, such as lactose. How does this occur? Glucose inhibits the production of cAMP, thereby preventing the binding of CAP to the DNA. In this way, glucose blocks the activation of the *lac* operon, thereby inhibiting transcription. Though it may seem puzzling, the term catabolite repression was coined before the action of the cAMP-CAP complex was understood at the molecular level. Historically, the primary observation of researchers was that glucose (a catabolite) inhibited (repressed) lactose metabolism. Further experimentation revealed that CAP is actually an activator protein.

Figure 13.11 considers the four possible environmental conditions that an *E. coli* bacterium might experience with regard to these two sugars. When both lactose and glucose levels are high (Figure 13.11a), the rate of transcription of the *lac* operon is low, because CAP does not activate transcription. Under these conditions, the bacterium primarily uses glucose rather than lactose. Why is this a benefit to the bacterium? The bacterium conserves energy by using one type of sugar at a time. If lactose levels are high and glucose is low (Figure 13.11b), the transcription rate of the *lac* operon is very high because CAP is bound to the CAP site and the lac repressor is not bound to the operator site. Under these conditions, the bacterium metabolizes lactose. When lactose levels are low, the lac repressor prevents transcription of the *lac* operon, whether glucose levels are high or low (Figure 13.11c,d).

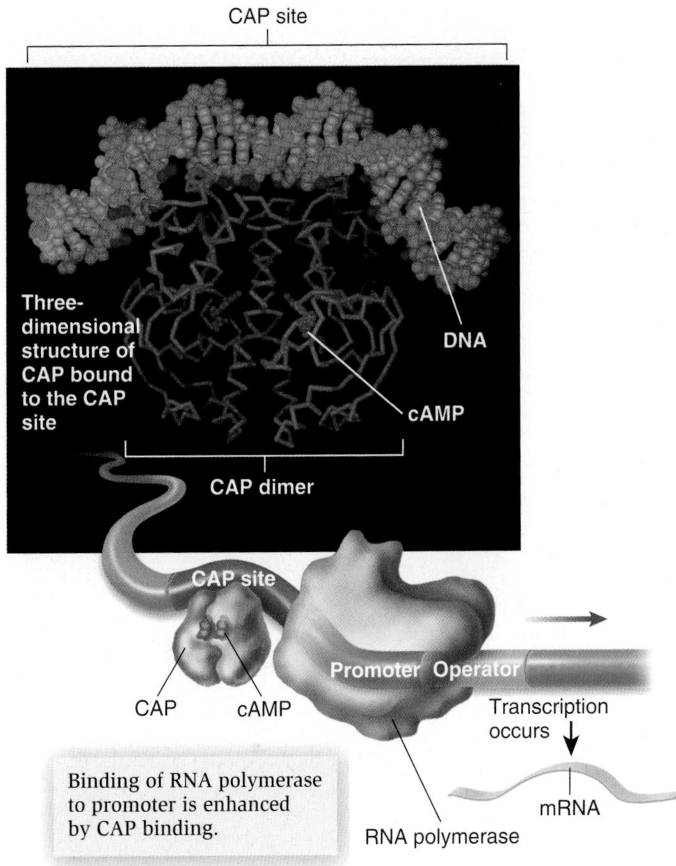

Figure 13.10 **Positive control of the *lac* operon by the catabolite activator protein (CAP).** When cAMP is bound to CAP, CAP binds to the DNA and causes it to bend. This bend facilitates the binding of RNA polymerase.

BioConnections: Look back at Figure 9.12. What is the function of cAMP in eukaryotic cells?

The *trp* Operon Is Under Negative Control by a Repressor Protein

So far in this section, we have examined the regulation of the *lac* operon. Let's now consider an example of an operon that encodes enzymes involved in biosynthesis rather than breakdown. Our example is the ***trp* operon** of *E. coli*, which encodes enzymes that are required to make the amino acid tryptophan, a building block of cellular proteins. More specifically, the *trpE, trpD, trpC, trpB,* and *trpA* genes encode enzymes that are involved in a pathway that leads to tryptophan synthesis.

The *trp* operon is regulated by a repressor protein that is encoded by the *trpR* gene. The binding of the repressor to the *trp* operator site inhibits transcription. The ability of the trp repressor to bind to the *trp* operator is controlled by tryptophan, which is the product of the enzymes that are encoded by the operon. When tryptophan levels within the cell are very low, the trp repressor cannot bind to the operator site. Under these conditions, RNA polymerase readily transcribes the operon (**Figure 13.12a**). In this way, the cell expresses the genes that encode enzymes that result in the synthesis of tryptophan, which is in short supply. Alternatively, when the tryptophan levels within the cell are high, tryptophan turns off the *trp* operon. Tryptophan acts as a small effector molecule, or **corepressor**, by binding to the trp repressor protein. This causes a conformational change in the repressor that allows it to bind to the *trp* operator site, inhibiting the ability of RNA polymerase to transcribe the operon (**Figure 13.12b**). Therefore, the bacterium does not waste energy making tryptophan when it is abundant.

When comparing the *lac* and *trp* operons, the actions of their small effector molecules are quite different. The lac repressor binds to its operator in the absence of its small effector molecule, whereas the trp repressor binds to its operator only in the presence of its small effector molecule. The *lac* operon is categorized as an inducible operon because allolactose, its small effector molecule, induces transcription. By comparison, the *trp* operon is considered to be a **repressible operon** because its small effector molecule, namely tryptophan, represses transcription.

13.3 Regulation of Transcription in Eukaryotes: Roles of Transcription Factors and Mediator

Learning Outcomes:

1. Explain the concept of combinatorial control.
2. Describe how RNA polymerase and general transcription factors initiate transcription at the core promoter.
3. Compare and contrast how activators, coactivators, repressors, TFIID, and mediator play a role in gene regulation.

Regulation of transcription in eukaryotes follows some of the same principles as those found in bacteria. For example, activator and repressor proteins are involved in regulating genes by influencing the ability of RNA polymerase to initiate transcription. In addition, many eukaryotic genes are regulated by small effector molecules. However, some important differences also occur. In eukaryotic species, genes are almost always organized individually, not in operons. In addition, eukaryotic gene regulation tends to be more intricate, because eukaryotes are faced with complexities that differ from their bacterial counterparts. For example, eukaryotes have more complicated cell structures that contain many more proteins and a variety of cell organelles. Many eukaryotes, such as animals and plants, are multicellular and contain different cell types. As discussed earlier in this chapter, animal cells may differentiate into neurons, muscle cells, and skin cells, and so on. Furthermore, animals and plants progress through developmental stages that require changes in gene expression.

By studying transcriptional regulation, researchers have discovered that most eukaryotic genes, particularly those found in multicellular species, are regulated by many factors. This phenomenon is called **combinatorial control** because the combination of many factors determines the expression of any given gene. At the level of

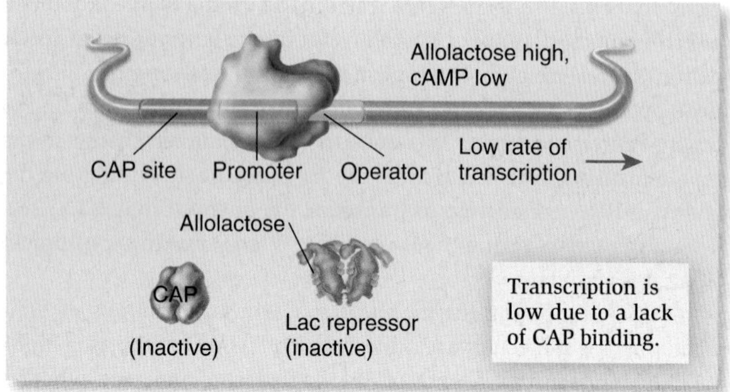

(a) Lactose high, glucose high

CAP site
Allolactose high,
cAMP high
High rate of
transcription
Promoter
Operator
CAP
cAMP
Allolactose
Lac repressor
(inactive)
Binding of RNA polymerase
to promoter is enhanced by
CAP binding. Transcription
rate is high.

(b) Lactose high, glucose low

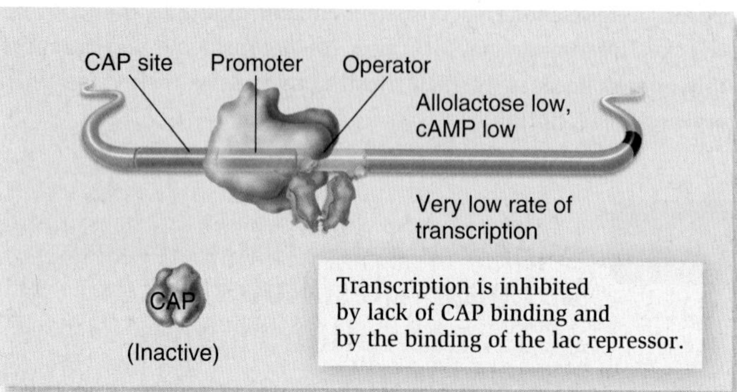

(c) Lactose low, glucose high

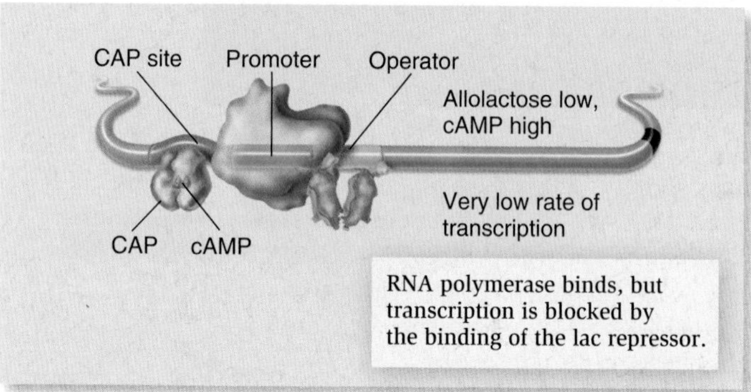

(d) Lactose low, glucose low

Figure 13.11 **Effects of lactose and glucose on the expression of the *lac* operon.**

Concept Check: *What are the advantages of having both an activator and a repressor protein?*

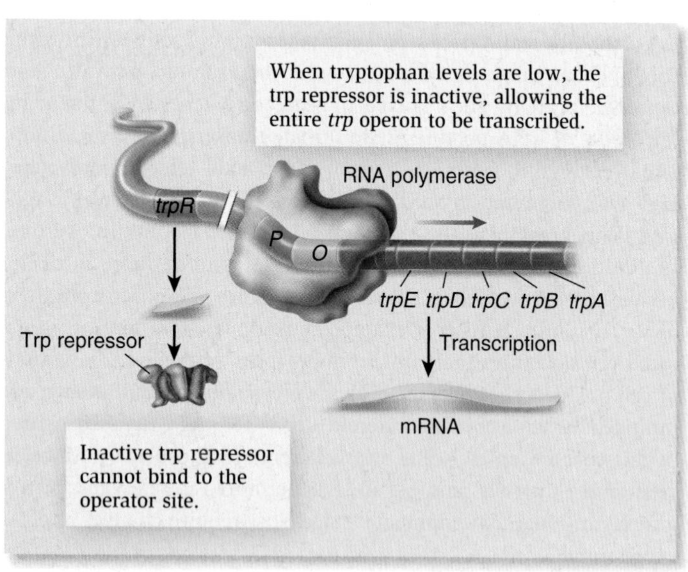

(a) Low tryptophan

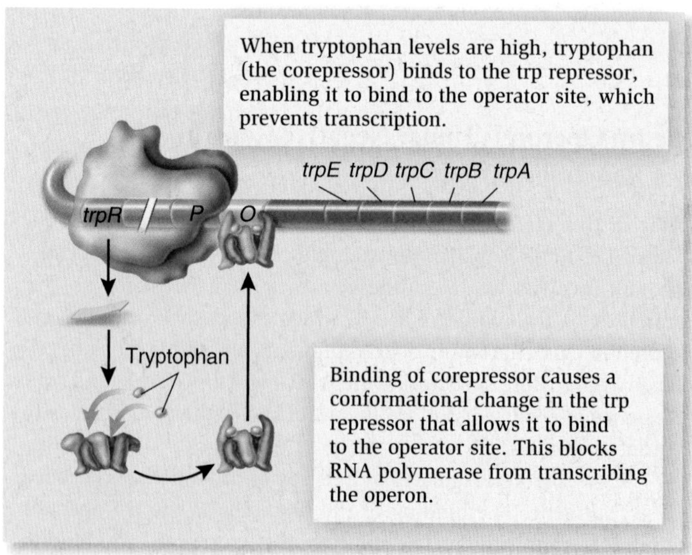

When tryptophan levels are high, tryptophan
(the corepressor) binds to the trp repressor,
enabling it to bind to the operator site, which
prevents transcription.
trpE trpD trpC trpB trpA
trpR
P
O
Tryptophan
Binding of corepressor causes a
conformational change in the trp
repressor that allows it to bind
to the operator site. This blocks
RNA polymerase from transcribing
the operon.

(b) High tryptophan

Figure 13.12 **Negative control of a repressible set of genes: function of the trp repressor and corepressor (tryptophan) in regulating the *trp* operon.**

Concept Check: *How are the functions of the lac repressor and trp repressor similar to each other, and how are they different?*

transcription, common factors that contribute to combinatorial control include the following:

1. One or more activators may stimulate the ability of RNA polymerase to initiate transcription.
2. One or more repressors may inhibit the ability of RNA polymerase to initiate transcription.
3. The function of activators and repressors may be modulated in several ways, which include the binding of small effector molecules, protein-protein interactions, and covalent modifications.
4. Activators are necessary to alter chromatin structure in the region where a gene is located, thereby making it easier for the gene to be recognized and transcribed by RNA polymerase.
5. DNA methylation usually inhibits transcription, either by preventing the binding of an activator or by recruiting proteins that inhibit transcription.

All five of these factors may contribute to the regulation of a single gene, or possibly only three or four will play a role. In most cases, transcriptional regulation is aimed at controlling the initiation of transcription at the promoter. In this section and the following section, we will survey these basic types of gene regulation in eukaryotic species.

Eukaryotic Structural Genes Have a Core Promoter and Regulatory Elements

To understand gene regulation in eukaryotes, we first need to consider the DNA sequences that are needed to initiate transcription. For eukaryotic structural genes that encode proteins, three features are common among most promoters: **regulatory elements**, a **TATA box**, and a **transcriptional start site** (Figure 13.13).

The TATA box and transcriptional start site form the **core promoter**. The transcriptional start site is the place in the DNA where transcription actually begins. The TATA box, which is a

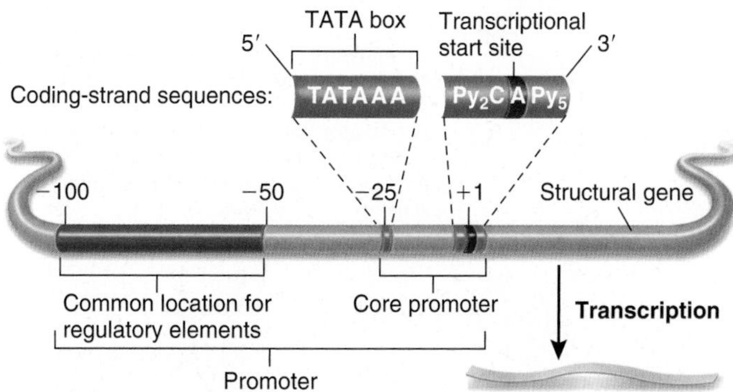

Figure 13.13 **A common organization of sequences for the promoter of a eukaryotic structural gene.** The core promoter has a TATA box and a transcriptional start site. The TATA box sequence is 5′–TATAAA–3′. However, not all structural genes in eukaryotes have a TATA box. The A highlighted in dark blue is the transcriptional start site. This A marks the site of the first A in the RNA transcript. The sequence that flanks the A of the transcriptional start site is two pyrimidines, then C, then five pyrimidines. Py refers to pyrimidine—cytosine or thymine. Regulatory elements, such as enhancers and silencers, are usually found upstream from the core promoter.

5′–TATAAA–3′ sequence, is usually about 25 bp upstream from a transcriptional start site. The TATA box is important in determining the precise starting point for transcription. If it is missing from the core promoter, transcription may start at a variety of different locations. The core promoter, by itself, results in a low level of transcription that is termed **basal transcription.**

Regulatory elements (or regulatory sequences) are DNA segments that regulate eukaryotic genes. As described later, regulatory elements are recognized by regulatory transcription factors that control the ability of RNA polymerase to initiate transcription at the core promoter. Some regulatory elements, known as **enhancers**, play a role in the ability of RNA polymerase to begin transcription, thereby enhancing the rate of transcription. When enhancers are not functioning, most eukaryotic genes have very low levels of transcription. Other regulatory elements, known as **silencers**, prevent transcription of a given gene when its expression is not needed. When these sequences function, the rate of transcription is decreased.

A common location for regulatory elements is the region that is 50–100 bp upstream from the transcriptional start site (see Figure 13.13). However, the locations of regulatory elements vary greatly among different eukaryotic genes. Regulatory elements can be quite distant from the promoter, even 100,000 bp away, yet exert strong effects on the ability of RNA polymerase to initiate transcription at the core promoter! Regulatory elements were first discovered by Japanese molecular biologist Susumu Tonegawa and coworkers in the 1980s. While studying genes that play a role in immunity, they identified a region that was far away from the core promoter but was needed for high levels of transcription to take place.

RNA Polymerase II, General Transcription Factors, and Mediator Are Needed to Transcribe Eukaryotic Structural Genes

As discussed in Chapter 12, three forms of RNA polymerases, designated I, II, and III, are found in eukaryotes. RNA polymerase II transcribes structural genes that encode proteins. By studying transcription in a variety of eukaryotic species, researchers have identified three types of proteins that play a role in initiating transcription at the core promoter of structural genes. These are RNA polymerase II, five different proteins called **general transcription factors (GTFs)**, and a large protein complex called mediator.

RNA polymerase II and GTFs must come together at the TATA box of the core promoter so transcription can be initiated. A series of interactions occurs between these proteins so RNA polymerase II can bind to the DNA. The completed assembly of RNA polymerase II and GTFs at the TATA box is known as the **preinitiation complex** (Figure 13.14).

Another component needed for transcription in eukaryotes is the mediator protein complex. **Mediator** is composed of many proteins that bind to each other to form an elliptically shaped complex that partially wraps around RNA polymerase II and the GTFs. Mediator derives its name from the observation that it mediates interactions between the preinitiation complex and regulatory transcription factors such as activators or repressors that bind to enhancers or silencers. The function of mediator is to control the rate at which RNA polymerase can begin to transcribe RNA at the transcriptional start site.

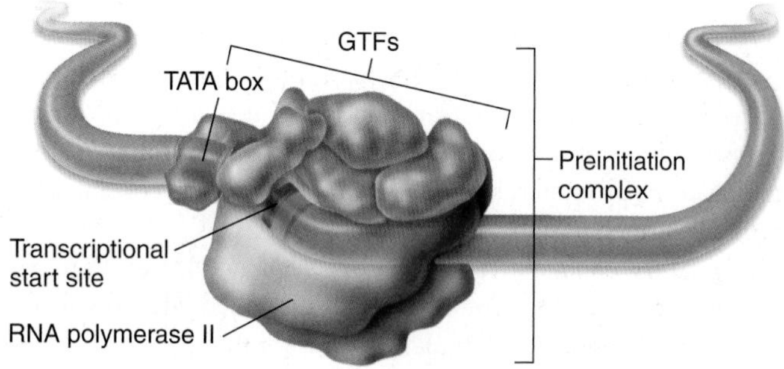

Figure 13.14 **The preinitiation complex.** General transcription factors (GTFs) and RNA polymerase II assemble into the preinitiation complex at the core promoter in eukaryotic structural genes.

Activators and Repressors May Influence the Function of GTFs or Mediator

In eukaryotes, regulatory transcription factors called activators and repressors bind to enhancers or silencers, respectively, and regulate the rate of transcription of genes. Activators and repressors commonly regulate the function of RNA polymerase II by binding to GTFs or mediator. As shown in **Figure 13.15**, some activators bind to an enhancer and then influence the function of GTFs. For example, an activator may improve the ability of a GTF called transcription factor II D (TFIID) to initiate transcription. The function of TFIID is to recognize the TATA box and begin the assembly process. An activator may recruit TFIID to the TATA box, thereby promoting the assembly of GTFs and RNA polymerase II into the preinitiation complex. In contrast, repressors may bind to a silencer and inhibit the function of TFIID. Certain repressors exert their effects by preventing the binding of TFIID to the TATA box or by inhibiting the ability of TFIID to assemble other GTFs and RNA polymerase II at the core promoter.

In addition to affecting general transcription factors, a second way that regulatory transcription factors control RNA polymerase II is via mediator (**Figure 13.16**). In this example, an activator also interacts with a **coactivator**—a protein that increases the rate of transcription but does not directly bind to the DNA itself. The activator-coactivator complex stimulates the function of mediator, thereby causing RNA polymerase II to proceed to the elongation phase of transcription more quickly. Alternatively, repressors have the opposite effect to those seen in Figure 13.16. When a repressor inhibits mediator, RNA polymerase II cannot progress to the elongation stage.

A third way that regulatory transcription factors influence transcription is by recruiting proteins that affect chromatin structure in the promoter region, as described next.

1 An activator binds to an enhancer.

Activator

Enhancer

2 The activator enhances the ability of a GTF called TFIID to bind to the TATA box.

TFIID

TATA box

3 TFIID promotes the assembly of the preinitiation complex.

TFIID

TATA box

Preinitiation complex

Figure 13.15 **Effect of an activator via TFIID, a general transcription factor.**

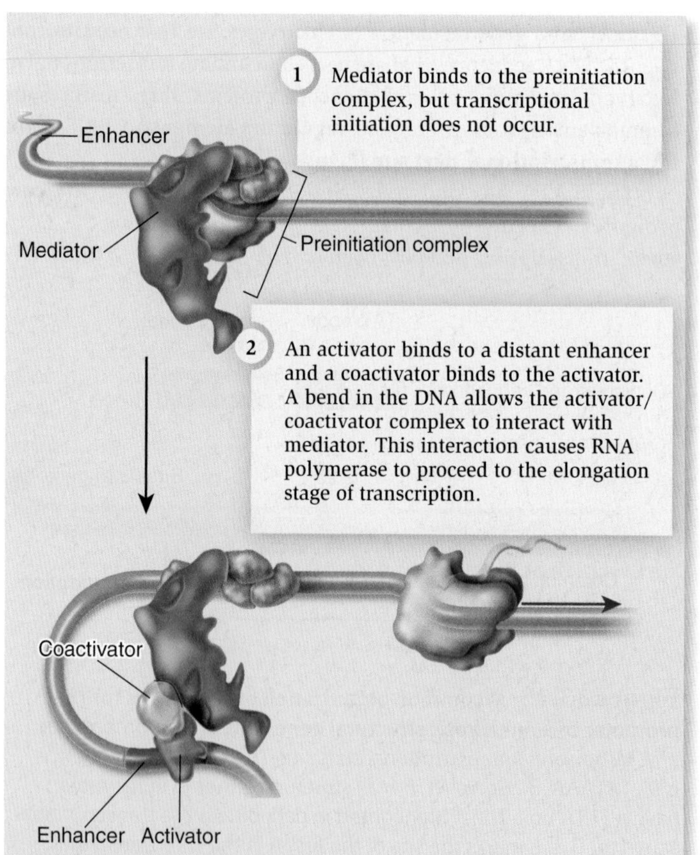

1 Mediator binds to the preinitiation complex, but transcriptional initiation does not occur.
Enhancer
Mediator
Preinitiation complex
2 An activator binds to a distant enhancer and a coactivator binds to the activator. A bend in the DNA allows the activator/coactivator complex to interact with mediator. This interaction causes RNA polymerase to proceed to the elongation stage of transcription.
Coactivator
Enhancer
Activator

Figure 13.16 **Effect of an activator via mediator.**

Concept Check: *When an activator interacts with mediator, how does this affect the function of RNA polymerase?*

13.4 Regulation of Transcription in Eukaryotes: Changes in Chromatin Structure and DNA Methylation

Learning Outcomes:

1. Describe how eukaryotic genes are flanked by nucleosome-free regions and how nucleosomes are altered during gene transcription.
2. Explain how DNA methylation affects transcription.

In eukaryotes, DNA is associated with proteins to form a structure called **chromatin**—the complex of DNA and proteins that makes up eukaryotic chromosomes (see Chapter 11, Figures 11.23 and 11.26). How does the structure of chromatin affect gene transcription? Recall from Chapter 11 that nucleosomes are composed of DNA wrapped around an octamer of histone proteins. Depending on the locations and arrangements of nucleosomes, a region containing a gene may be in a **closed conformation**, and transcription may be difficult or impossible. Transcription requires changes in chromatin structure that allow transcription factors to gain access to and bind to the DNA in the promoter region. Such chromatin, said to be in an **open conformation**, is accessible to GTFs and RNA polymerase II so transcription can take place. In this section, we will examine how chromatin is converted from a closed to an open conformation. We will also explore how **DNA methylation**—the attachment of methyl groups to the base cytosine—affects chromatin conformation and gene expression.

Transcription Is Controlled by Changes in Chromatin Structure

In recent years, geneticists have been trying to identify the steps that promote the interconversion between the closed and open conformations of chromatin. One way to change chromatin structure is through **ATP-dependent chromatin-remodeling complexes**, which are a group of proteins that alter chromatin structure. Such complexes use energy from ATP hydrolysis to drive a change in the locations and/or compositions of nucleosomes, thereby making the DNA more or less amenable to transcription. Therefore, chromatin remodeling is important for both the activation and repression of transcription.

How do ATP-dependent chromatin-remodeling complexes change chromatin structure? Three effects are possible. One result is that these complexes may bind to chromatin and change the locations of nucleosomes (**Figure 13.17a**). This may involve a shift of the relative positions of a few nucleosomes or a change in the relative spacing of nucleosomes over a long stretch of DNA. A second effect is that remodeling complexes may evict histone octamers from the DNA, thereby creating gaps where nucleosomes are not found (**Figure 13.17b**). A third possibility is that chromatin-remodeling complexes may change the composition of nucleosomes by removing standard histone proteins from an octamer and replacing them with histone variants (**Figure 13.17c**). A **histone variant** is a histone protein that has a slightly different amino acid sequence from the standard histone proteins described in Chapter 11. Some histone variants promote gene transcription, whereas others inhibit it.

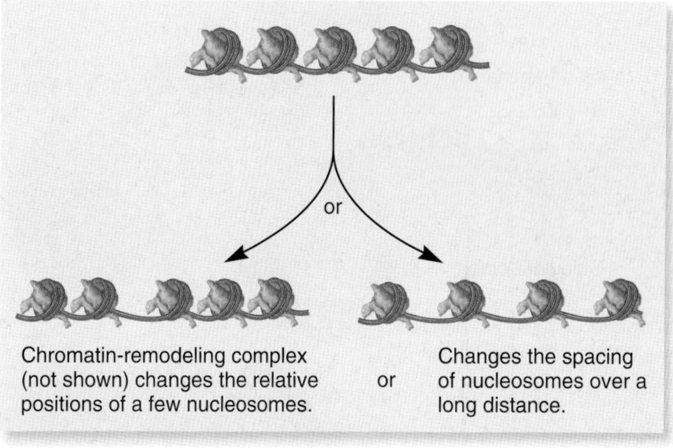

(a) Change in nucleosome position

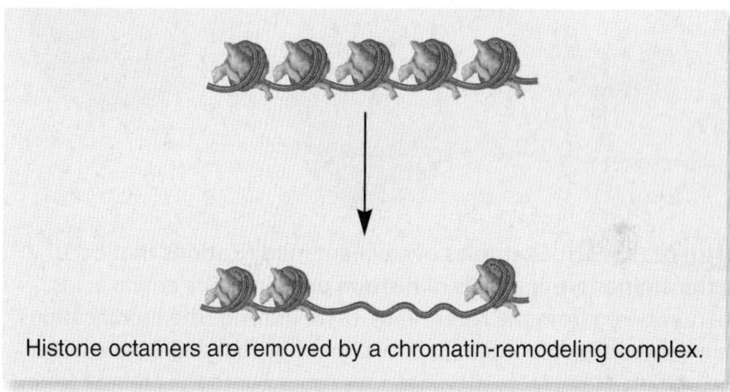

(b) Histone eviction

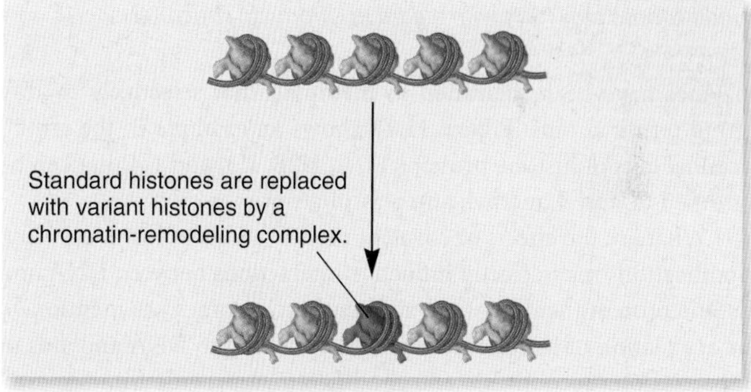

(c) Replacement with variant histones

Figure 13.17 **ATP-dependent chromatin remodeling.** Chromatin-remodeling complexes may **(a)** change the locations of nucleosomes, **(b)** remove histones from the DNA, or **(c)** replace standard histones with variant histones. The chromatin-remodeling complex, which is composed of a group of proteins, is not shown in this figure.

Histone Modifications Affect Gene Transcription

In recent years, researchers have learned that the amino terminal tails of histone proteins are subject to several types of covalent modifications. For example, an enzyme called **histone acetyltransferase** attaches acetyl groups (—$COCH_3$) to the amino terminal tails of histone proteins. When acetylated, histone proteins do not bind as tightly to the DNA, which aids in transcription. Over 50 different

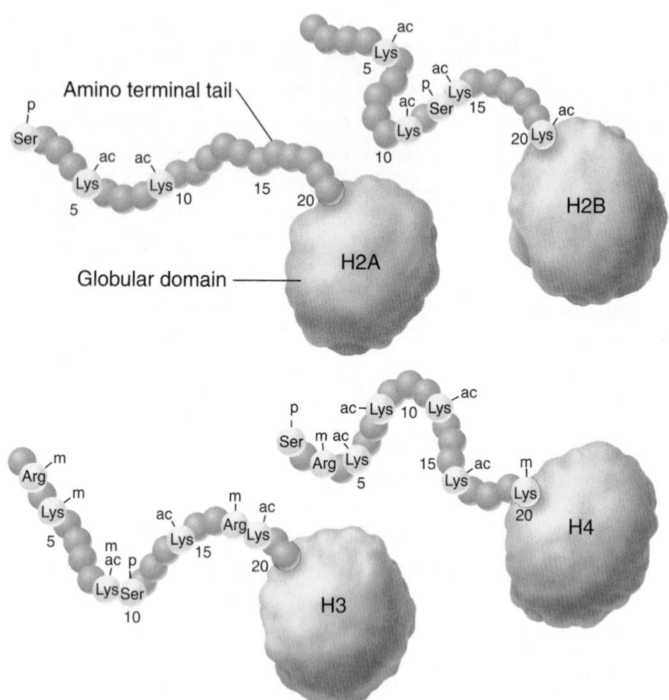
Amino terminal tail
Globular domain
H2A
H2B
H3
H4
Ser
Lys
Arg
ac
m
p

Figure 13.18 Examples of covalent modifications that occur to the amino terminal tails of histone proteins. The amino acids are numbered from the N-terminus, or amino end. The modifications shown here are m for methylation, p for phosphorylation, and ac for acetylation. Many more modifications can occur to the amino terminal tails. These modifications are reversible.

Concept Check: What are the two opposing effects that histone modifications may have with regard to chromatin structure?

enzymes have been identified in mammals that selectively modify amino terminal tails. Figure 13.18 shows an example in the amino terminal tails of histone proteins H2A, H2B, H3, and H4 that can be modified by acetyl, methyl, and phosphate groups.

What are the effects of covalent modifications of histones? First, modifications may directly influence interactions between DNA and histone proteins, and between adjacent nucleosomes. As mentioned, the acetylation of histones loosens their binding to DNA and aids in transcription. Second, histone modifications provide binding sites that are recognized by other proteins. According to the **histone code hypothesis**, proposed by American biologists Brian Strahl and David Allis in 2000, the pattern of histone modification is recognized by proteins much like a language or code. One pattern of histone modification may attract proteins that inhibit transcription. Alternatively, a different combination of histone modifications may attract proteins, such as ATP-dependent chromatin-remodeling complexes, that promote gene transcription. In this way, the histone code plays a key role in accessing the information within the genomes of eukaryotic species.

Eukaryotic Genes Are Flanked by Nucleosome-Free Regions

Studies over the last 10 years or so have revealed that many eukaryotic genes show a common pattern of nucleosome organization (Figure 13.19). For active genes or those genes that can be activated, the core promoter is found at a **nucleosome-free region (NFR)**, which is a site that is missing nucleosomes. The NFR is typically 150 bp in length. Although the NFR may be required for transcription, it is not, by itself, sufficient for gene activation. At any given time in the life of a eukaryotic cell, many genes that contain an NFR are not being actively transcribed. The NFR is flanked by two nucleosomes that are termed the −1 and +1 nucleosomes. These nucleosomes often contain histone variants that promote transcription. The nucleosomes downstream from the +1 nucleosome tend to be evenly spaced near the beginning of a eukaryotic gene, but their spacing becomes less regular farther downstream. The end of many eukaryotic genes is followed by another NFR. This arrangement at the end of genes may be important for transcriptional termination.

Transcriptional Activation Involves Changes in Nucleosome Locations, Composition, and Histone Modifications

A key role of certain activators is to recruit ATP-dependent chromatin-remodeling complexes and histone-modifying enzymes to the promoter region of eukaryotic genes. Though the order of recruitment may differ among specific activators, this appears to be critical for transcriptional initiation and elongation. In the scenario shown in Figure 13.20, an activator binds to an enhancer in the NFR. The activator then recruits chromatin-remodeling complexes and

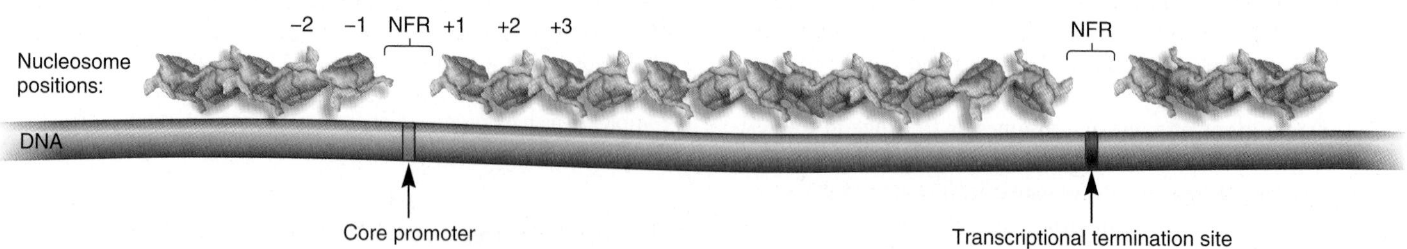
−2 −1 NFR +1 +2 +3
NFR
Nucleosome positions:
DNA
Core promoter
Transcriptional termination site
A nucleosome-free region (NFR) is found at the beginning and end of many genes. Nucleosomes tend to be precisely positioned near the beginning and end of a gene, but are less regularly distributed elsewhere.

Figure 13.19 Nucleosome arrangements in the vicinity of a eukaryotic structural gene.

BioConnections: Look back at Figure 11.23. What is the composition of a nucleosome?

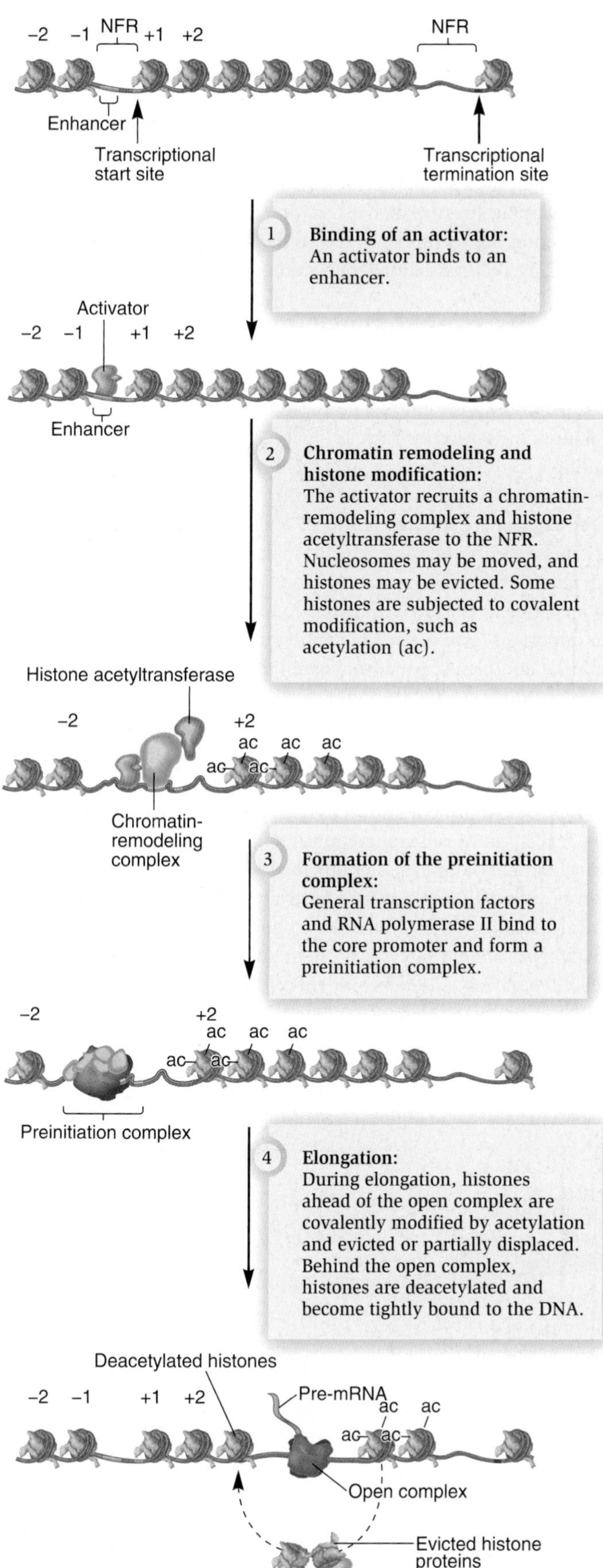
Many genes are flanked by nucleosome-free regions (NFR) and well-positioned nucleosomes.
−2 −1 NFR +1 +2
NFR
Enhancer
Transcriptional start site
Transcriptional termination site
1 Binding of an activator: An activator binds to an enhancer.
Activator
−2 −1 +1 +2
Enhancer
2 Chromatin remodeling and histone modification: The activator recruits a chromatin-remodeling complex and histone acetyltransferase to the NFR. Nucleosomes may be moved, and histones may be evicted. Some histones are subjected to covalent modification, such as acetylation (ac).
Histone acetyltransferase
−2
+2
ac ac ac
ac ac
Chromatin-remodeling complex
3 Formation of the preinitiation complex: General transcription factors and RNA polymerase II bind to the core promoter and form a preinitiation complex.
−2
+2
ac ac ac
ac ac
Preinitiation complex
4 Elongation: During elongation, histones ahead of the open complex are covalently modified by acetylation and evicted or partially displaced. Behind the open complex, histones are deacetylated and become tightly bound to the DNA.
Deacetylated histones
−2 −1 +1 +2
Pre-mRNA
ac ac
ac ac
Open complex
Evicted histone proteins
Chaperone

Figure 13.20 **A simplified model for the transcriptional activation of a eukaryotic structural gene.**

histone-modifying enzymes to this region. The chromatin-remodeling complex may shift nucleosomes or temporarily evict nucleosomes from the promoter region. Nucleosomes containing certain histone variants are thought to be more easily removed from the DNA than those containing the standard histones. Histone-modifying enzymes, such as histone acetyltransferase, covalently modify histone proteins and may affect nucleosome contact with the DNA. The actions of chromatin-remodeling complexes and histone-modifying enzymes facilitate the binding of general transcription factors and RNA polymerase II to the core promoter, thereby allowing the formation of a preinitiation complex (see Figure 13.20, step 2).

Further changes in chromatin structure are necessary for elongation to occur. RNA polymerase II cannot transcribe DNA that is tightly wrapped in nucleosomes. For transcription to occur, histones are evicted, partially displaced, or destabilized so RNA polymerase II can pass. Evicted histones are then reassembled by chaperone proteins and placed back on the DNA behind the moving RNA polymerase II (see Figure 13.20). These histones may be deacetylated—had their acetyl groups removed—so they bind more tightly to the DNA.

DNA Methylation Inhibits Gene Transcription

Let's now turn our attention to a mechanism that usually silences gene expression. DNA structure can be modified by the covalent attachment of methyl groups ($—CH_3$) by an enzyme called **DNA methylase**. This modification, termed DNA methylation, is common in some eukaryotic species but not all. For example, yeast and *Drosophila* have little or no detectable methylation of their DNA, whereas DNA methylation in vertebrates and plants is relatively abundant. In mammals, approximately 5% of the DNA is methylated. Eukaryotic DNA methylation occurs on the cytosine base. The sequence that is methylated is shown here:

$$\begin{array}{c} CH_3 \\ | \\ 5'—CG—3' \\ 3'—GC—5' \\ \quad | \\ \quad CH_3 \end{array}$$

DNA methylation usually inhibits the transcription of eukaryotic genes, particularly when it occurs in the vicinity of the promoter. In vertebrates and flowering plants, many genes contain sequences called **CpG islands** near their promoters. CpG refers to the nucleotides of C and G in DNA that are connected by a phosphodiester linkage. A CpG island is a cluster of CpG sites. Unmethylated CpG islands are usually correlated with active genes, whereas repressed genes contain methylated CpG islands. In this way, DNA methylation may play an important role in the silencing of particular genes.

How does DNA methylation inhibit transcription? This can occur in two general ways. First, methylation of CpG islands may prevent an activator from binding to an enhancer element, thus inhibiting the initiation of transcription. A second way that methylation inhibits transcription is by altering chromatin structure. Proteins known as **methyl-CpG-binding proteins** bind methylated sequences. Once bound to the DNA, the methyl-CpG-binding protein recruits other proteins to the region that inhibit transcription. As discussed in Chapter 17, DNA methylation is also associated with an inheritance pattern called **epigenetic inheritance** in which gene expression is altered in a way that is fixed during an individual's lifetime. Such epigenetic changes may affect the phenotype of the individual, but they are not permanent over the course of two or more generations.

13.5 Regulation of RNA Processing and Translation in Eukaryotes

Learning Outcomes:

1. Outline the process of alternative splicing and how it increases protein diversity.
2. Describe how RNA interference is used to regulate the expression of genes.
3. Explain how RNA-binding proteins can regulate the translation of specific mRNAs, using the regulation of iron absorption in mammals as an example.

In the preceding sections of this chapter, we focused on gene regulation at the level of transcription in bacteria and eukaryotes. Eukaryotic gene expression is also commonly regulated at the levels of RNA processing and translation. These added levels of regulation provide important benefits to eukaryotic species. First, by regulating RNA processing, eukaryotes can produce more than one mRNA transcript from a single gene. This allows a gene to encode two or more polypeptides, thereby increasing the complexity of eukaryotic proteomes. A second issue is timing. Regulation of transcription in eukaryotes takes a fair amount of time before its effects are observed at the cellular level. During transcription (1) the chromatin must be converted to an open conformation, (2) the gene must be transcribed, (3) the RNA must be processed and exported from the nucleus, and (4) the protein must be made via translation. All four steps take time, on the order of several minutes. One way to achieve faster regulation is to control steps that occur after an RNA transcript is made. In eukaryotes, regulation of translation provides a faster way to regulate the levels of gene products, namely, proteins. A small RNA molecule or RNA-binding protein can bind to an mRNA and affect the ability of the mRNA to be translated into a polypeptide.

During the past few decades, many critical advances have been made in our knowledge of the regulation of RNA processing and translation. Even so, molecular geneticists are still finding new forms of regulation, making this an exciting area of modern research. In this section, we will survey a few of the known mechanisms of RNA processing and translational regulation.

Alternative Splicing of Pre-mRNAs Increases Protein Diversity

In eukaryotes, a pre-mRNA transcript is processed before it becomes a mature mRNA (refer back to Figure 12.9). When a pre-mRNA has multiple introns and exons, splicing may occur in more than one way, resulting in the production of two or more different polypeptides. Such **alternative splicing** is a form of gene regulation that allows an organism to use the same gene to make different proteins at different stages of development, in different cell types, and/or in response to a change in the environmental conditions. Alternative splicing is an important form of gene regulation in complex eukaryotes such as animals and plants. An advantage of alternative splicing is that two or more different polypeptides can be derived from a single gene, thereby increasing the size of the proteome while minimizing the size of the genome.

Let's consider an example of alternative splicing for a pre-mRNA that encodes a protein known as α-tropomyosin, which functions in the regulation of cell contraction in animals. It is located along the thin filaments found in smooth muscle cells, such as those in the uterus and small intestine, and in striated muscle cells that are found in cardiac and skeletal muscle. α-tropomyosin is also synthesized in many types of nonmuscle cells but in lower amounts. Within a multicellular organism, different types of cells must regulate their contractibility in subtly different ways. One way this may be accomplished is by the production of different forms of α-tropomyosin.

Figure 13.21 shows the intron-exon structure of the rat α-tropomyosin pre-mRNA and two alternative ways that the

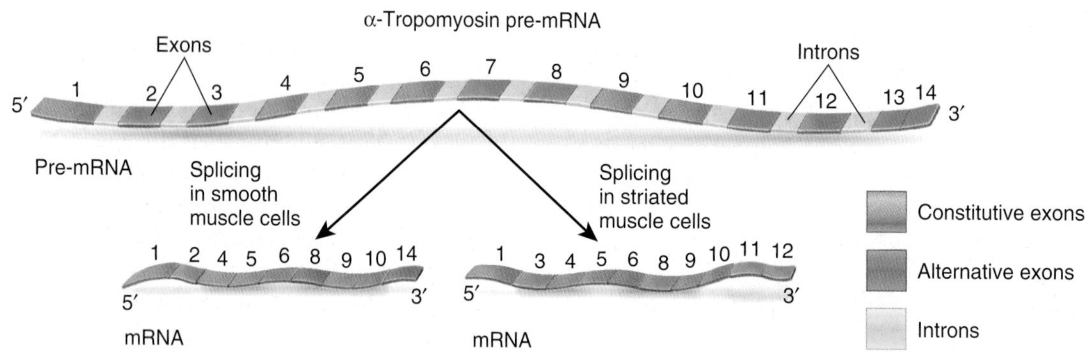

Figure 13.21 Alternative splicing of the rat α-tropomyosin pre-mRNA. The top part of this figure depicts the structure of the rat α-tropomyosin pre-mRNA. Exons are red or green, and introns are yellow. The lower part of the figure describes the final mRNA products in smooth and striated muscle cells after alternative splicing. Note: Exon 8 is found in the final mRNA of smooth and striated muscle cells, but not in the mRNA of certain other cell types. The junction between exons 13 and 14 contains a 3′ splice site that enables exon 13 to be separated from exon 14.

Concept Check: What is the biological advantage of alternative splicing?

pre-mRNA can be spliced. The pre-mRNA contains 14 exons, 6 of which are constitutive exons (shown in red), which are always found in the mature mRNA from all cell types. Presumably, constitutive exons encode polypeptide segments of the α-tropomyosin protein that are necessary for its general structure and function. By comparison, alternative exons (shown in green) are not always found in the mRNA after splicing has occurred. The polypeptide sequences encoded by alternative exons may subtly change the function of α-tropomyosin to meet the needs of the cell type in which it is found. For example, Figure 13.21 shows the predominant splicing products found in smooth muscle cells and striated muscle cells. Exon 2 encodes a segment of the α-tropomyosin protein that alters its function to make it suitable for smooth muscle cells. By comparison, the α-tropomyosin mRNA found in striated muscle cells does not include exon 2. Instead, this mRNA contains exon 3, which is more suitable for that cell type.

Table 13.1 Genome Size and Biological Complexity

Species	Level of complexity	Genome size (million bp)	Approximate number of genes	Percentage of genes alternatively spliced
Escherichia coli	A unicellular bacterium	4.2	4,000	0
Saccharomyces cerevisiae	A unicellular eukaryote	12	6,000	<1
Caenorhabditis elegans	A tiny worm (about 1,000 cells)	97	19,000	2
Drosophila melanogaster	An insect	137	14,000	7
Arabidopsis thaliana	A flowering plant	142	26,000	11
Homo sapiens	A complex mammal	3,000	25,000	70

GENOMES & PROTEOMES CONNECTION

Alternative Splicing Tends to Be More Prevalent in Complex Eukaryotic Species

In the past few decades, many technical advances have improved our ability to analyze the genomes and proteomes of many different species. Researchers have been able to determine the amount of DNA from several species and estimate the total number of genes. In addition, scientists can also estimate the number of polypeptides if information is available about the degree of alternative splicing in a given species.

Table 13.1 compares six species: a bacterium (*Escherichia coli*), a eukaryotic single-celled organism (yeast—*Saccharomyces cerevisiae*), a small nematode worm (*Caenorhabditis elegans*), a fruit fly (*Drosophila melanogaster*), a small flowering plant (*Arabidopsis thaliana*), and a human (*Homo sapiens*). One general trend is that less complex organisms tend to have fewer genes. For example, unicellular organisms have only a few thousand genes, whereas multicellular species have tens of thousands. However, the trend is by no means a linear one. If we compare *C. elegans* and *D. melanogaster*, the fruit fly actually has fewer genes, even though it is morphologically more complex.

A second trend you can see in Table 13.1 concerns alternative splicing. This phenomenon does not occur in bacteria and is rare in *S. cerevisiae*. The frequency of alternative splicing increases from worms to flies to humans. For example, the level of alternative splicing is 10-fold higher in humans than in *Drosophila*. This trend can partially explain the increase in complexity among these species. Even though humans have only about 25,000 different genes, they can make well over 100,000 different proteins because most genes are alternatively spliced in multiple ways.

RNA Interference Blocks the Expression of mRNA

Let's now turn our attention to regulatory mechanisms that affect translation. **MicroRNAs (miRNAs)** and **short-interfering RNAs (siRNAs)** are RNA molecules that are processed to a small size, typically 22 nucleotides in length, and silence the expression of pre-existing mRNAs. The precursors of miRNAs are encoded by genes and usually form a hairpin structure. In most cases, miRNAs are partially complementary to certain cellular mRNAs and inhibit their translation. By comparison, short-interfering RNAs are derived from two RNA molecules that come together to form a double-stranded region. For example, a cellular RNA may bind to an RNA that is transcribed from viral genome. siRNAs are usually a perfect match to specific mRNAs and cause the mRNAs to be degraded.

Insight into the mechanism of miRNA inhibition came from the research of American biologists Andrew Fire and Craig Mello, who discovered the mechanism of action of miRNA (**Figure 13.22**). A pre-miRNA is first synthesized as a single-stranded molecule that folds back on itself to form a hairpin structure. (A pre-siRNA, which is not shown in this figure, would be composed of two RNA molecules that come together to form a double-stranded region.) The double-stranded region is trimmed to a 22-bp sequence by an enzyme called dicer. The 22-bp sequence becomes part of a complex called the **RNA-induced silencing complex (RISC)**, which also includes several proteins. One of the RNA strands is then degraded. The miRNA or siRNA in the complex binds to a target mRNA with a complementary sequence.

Upon binding, two different things may happen. When the miRNA and mRNA are not a perfect match or are only partially complementary, translation is inhibited. Alternatively, when an siRNA and mRNA are a perfect match or highly complementary, the mRNA is cut into two pieces and then degraded. Both miRNA and siRNA have the same effect—the expression of the mRNA is silenced. Fire and Mello called this **RNA interference (RNAi)**, because the miRNA interferes with the proper expression of an mRNA.

Since this study, researchers have discovered that genes encoding miRNAs are widely found in animals and plants. In humans, for example, approximately 200 different genes encode miRNAs. RNAi represents an important mechanism of gene regulation that results in mRNA silencing. In 2006, Fire and Mello were awarded the Nobel Prize in Physiology or Medicine for their studies of RNAi.

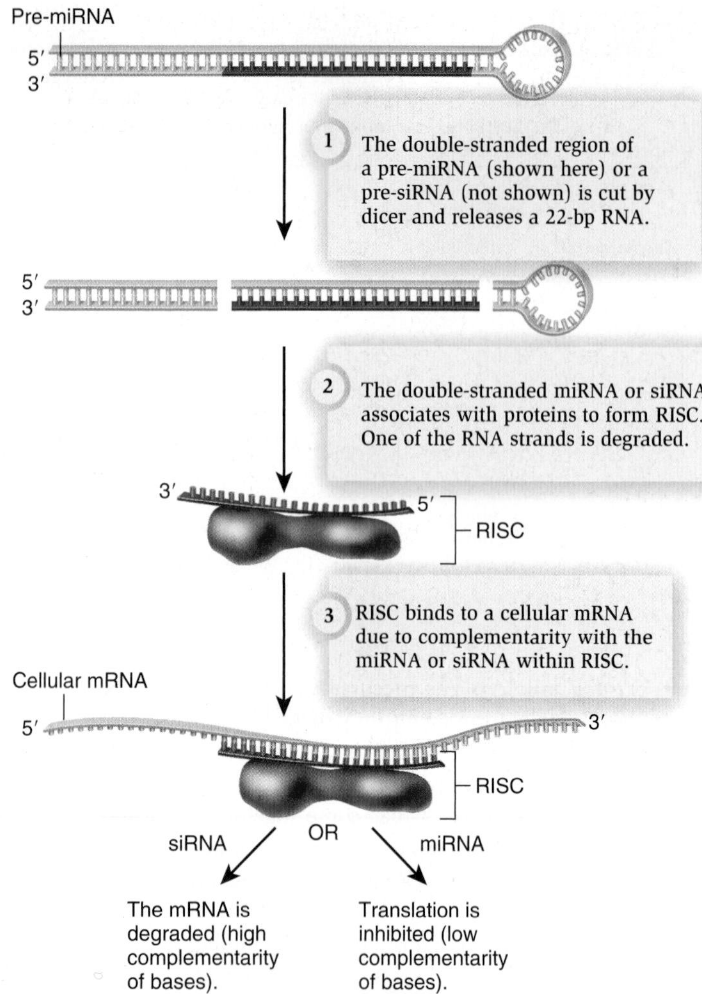

Pre-miRNA
5′
3′
1 The double-stranded region of a pre-miRNA (shown here) or a pre-siRNA (not shown) is cut by dicer and releases a 22-bp RNA.
5′
3′
2 The double-stranded miRNA or siRNA associates with proteins to form RISC. One of the RNA strands is degraded.
3′
5′
RISC
3 RISC binds to a cellular mRNA due to complementarity with the miRNA or siRNA within RISC.
Cellular mRNA
5′
3′
RISC
siRNA
OR
miRNA
The mRNA is degraded (high complementarity of bases).
Translation is inhibited (low complementarity of bases).

Figure 13.22 **Mechanism of action of microRNA (miRNA).** Note: Pre-siRNAs are also acted upon by dicer, but siRNAs are derived from two RNA molecules that form a double-stranded region rather than one RNA molecule that forms a hairpin.

BIOLOGY PRINCIPLE Structure determines function. Because the structure of an miRNA or siRNA is complementary to an mRNA, it is able to inhibit the function of the mRNA.

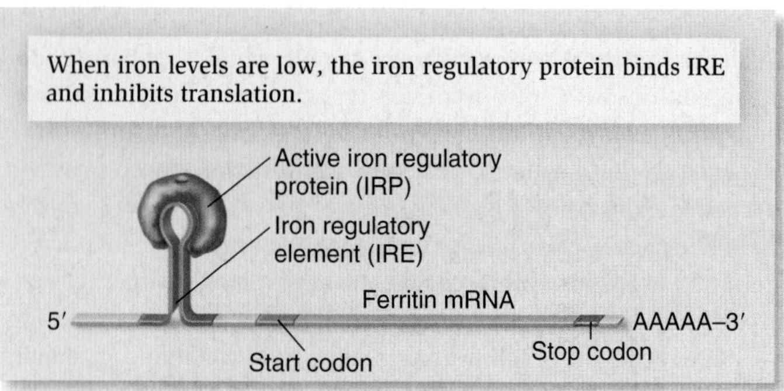

When iron levels are low, the iron regulatory protein binds IRE and inhibits translation.
Active iron regulatory protein (IRP)
Iron regulatory element (IRE)
Ferritin mRNA
5′
AAAAA–3′
Start codon
Stop codon

(a) Low iron levels

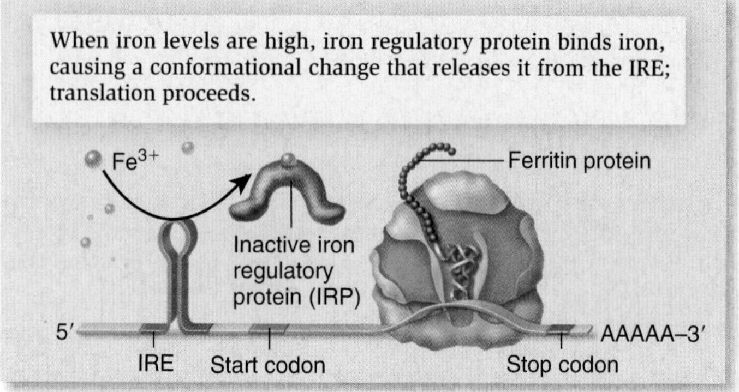

When iron levels are high, iron regulatory protein binds iron, causing a conformational change that releases it from the IRE; translation proceeds.
Fe^{3+}
Ferritin protein
Inactive iron regulatory protein (IRP)
5′
AAAAA–3′
IRE
Start codon
Stop codon

(b) High iron levels

Figure 13.23 **Translational regulation of ferritin mRNA by the iron regulatory protein (IRP).**

Concept Check: Poisoning may occur when a young child finds a bottle of vitamins, such as those that taste like candy, and eats a large number of them. One of the toxic effects involves the ingestion of too much iron. How does the IRP protect people from the toxic effects of too much iron?

The Prevention of Iron Toxicity in Mammals Involves the Regulation of Translation

Another way to regulate translation involves RNA-binding proteins that directly affect the initiation of translation. The regulation of iron absorption provides a well-studied example. Although iron is a vital cofactor for many cellular enzymes, it is toxic at high levels. To prevent toxicity, mammalian cells synthesize a protein called ferritin, which forms a hollow, spherical complex that stores excess iron.

The mRNA that encodes ferritin is controlled by an RNA-binding protein known as the **iron regulatory protein (IRP)**. When iron levels in the cytosol are low and more ferritin is not needed, IRP binds to a regulatory element within the ferritin mRNA known as the **iron regulatory element (IRE)**. The IRE is located between the 5′ cap, where the ribosome binds, and the start codon where translation begins. Due to base pairing, it forms a stem-loop structure. The binding of IRP to the IRE inhibits translation of the ferritin mRNA (Figure 13.23a). However, when iron is abundant in the cytosol, the iron binds directly to IRP, which changes its conformation and prevents it from binding to the IRE. Under these conditions, the ferritin mRNA is translated to make more ferritin protein (Figure 13.23b).

Why is translational regulation of ferritin mRNA an advantage over transcriptional regulation of the ferritin gene? This mechanism of translational control allows cells to rapidly respond to changes in their environment. When cells are confronted with high levels of iron, they can quickly make more ferritin protein to prevent the toxic buildup of iron. This mechanism is faster than transcriptional regulation, which would require the activation of the ferritin gene and the transcription of ferritin mRNA prior to the synthesis of more ferritin protein.

Summary of Key Concepts

13.1 Overview of Gene Regulation

- Most genes are regulated so the level of gene expression can vary under different conditions. By comparison, constitutive genes are expressed at constant levels.
- Gene regulation ensures that gene products are made only when they are needed. An example is the synthesis of the gene products needed for lactose utilization in bacteria (Figure 13.1).
- In eukaryotes, gene regulation leads to the production of different cell types, such as neurons, muscle cells, and skin cells, within an organism (Figure 13.2).
- Eukaryote gene regulation enables gene products to be produced at different developmental stages (Figure 13.3).
- All organisms regulate gene expression at a variety of levels, including transcription, translation, and post-translation. Eukaryotes also regulate RNA processing (Figure 13.4).

13.2 Regulation of Transcription in Bacteria

- Repressors and activators are regulatory transcription factors that bind to the DNA and affect the transcription of genes. Small effector molecules control the ability of regulatory transcription factors to bind to DNA (Figure 13.5).
- An operon is an arrangement of several structural genes controlled by a promoter and operator. The *lac* operon is an example of an inducible operon. The lac repressor exerts negative control by binding to the operator site and preventing RNA polymerase from transcribing the operon. When allolactose binds to the repressor, a conformational change occurs that prevents the repressor from binding to the operator site so transcription can proceed (Figures 13.6, 13.7).
- By constructing a merozygote, Jacob, Monod, and Pardee determined that the *lacI* gene encodes a diffusible protein that represses the *lac* operon (Figures 13.8, 13.9).
- Positive control of the *lac* operon occurs when the catabolite activator protein (CAP) binds to the CAP site in the presence of cAMP. This causes a bend in the DNA, which promotes the binding of RNA polymerase to the promoter (Figure 13.10).
- Glucose inhibits cAMP production, which, in turn, inhibits the expression of the *lac* operon, because CAP cannot bind to the CAP site. This form of regulation provides bacteria with a more efficient utilization of their resources because the bacteria use one sugar at a time (Figure 13.11).
- The *trp* operon is an example of a repressible operon. The presence of tryptophan causes the trp repressor to bind to the *trp* operator and stop transcription. This prevents the excessive buildup of tryptophan in the cell, which would be a waste of energy (Figure 13.12).

13.3 Regulation of Transcription in Eukaryotes: Roles of Transcription Factors and Mediator

- Eukaryotic genes exhibit combinatorial control, meaning that many factors control the expression of a single gene. (See list on p. 269.)
- Eukaryotic promoters consist of a core promoter (containing a TATA box and transcriptional start site) and regulatory elements, such as enhancers or silencers, that regulate the rate of transcription (Figure 13.13).
- General transcription factors (GTFs) are needed for RNA polymerase II to bind to the core promoter, forming a preinitiation complex (Figure 13.14).
- Activators and repressors may regulate RNA polymerase II by interacting with TFIID (a GTF), or via mediator, a protein complex that wraps around RNA polymerase II (Figures 13.15, 13.16).

13.4 Regulation of Transcription in Eukaryotes: Changes in Chromatin Structure and DNA Methylation

- ATP-dependent chromatin-remodeling complexes change the positions and compositions of nucleosomes (Figure 13.17).
- The pattern of covalent modification of the amino terminal tails of histone proteins, also called the histone code, can inhibit or promote transcription (Figure 13.18).
- Eukaryotic genes are usually flanked by nucleosome-free regions (Figure 13.19).
- For eukaryotic genes, a pre-initiation complex forms at a nucleosome-free region. During elongation, nucleosomes are displaced ahead of RNA polymerase and re-form after RNA polymerase has passed (Figure 13.20).
- DNA methylation, which occurs at CpG islands near promoters, usually inhibits transcription by (1) preventing the binding of activator proteins or (2) promoting the binding of proteins that inhibit transcription.

13.5 Regulation of RNA Processing and Translation in Eukaryotes

- In alternative splicing, a single type of pre-mRNA can be spliced in more than one way, producing polypeptides with somewhat different sequences. This is a common way for complex eukaryotes to increase the size of their proteomes (Figure 13.21, Table 13.1).
- MicroRNAs (miRNAs) and short-interfering RNAs (siRNAs) inhibit mRNAs by inhibiting translation or by promoting the degradation of mRNAs, respectively (Figure 13.22).
- RNA-binding proteins can regulate the translation of specific mRNAs. An example is the regulation of iron absorption, in which the iron regulatory protein (IRP) regulates the translation of ferritin mRNA (Figure 13.23).

Assess and Discuss

Test Yourself

1. Genes that are expressed at all times at relatively constant levels are known as ____ genes.
   a. inducible
   b. repressible
   c. positive
   d. constitutive
   e. structural
2. Which of the following is *not* a form of gene regulation in bacteria?
   a. transcriptional
   b. RNA processing
   c. translational
   d. post-translational
   e. All of the above are levels at which bacteria are able to regulate gene expression.
3. Transcription factors that bind to DNA and stimulate transcription are
   a. repressors.
   b. small effector molecules.
   c. activators.
   d. promoters.
   e. operators.
4. In bacteria, the unit of DNA that contains multiple genes under the control of a single promoter is called ___. The mRNA produced from this unit is referred to as ___ mRNA.
   a. an operator, a polycistronic
   b. a template, a structural
   c. an operon, a polycistronic
   d. an operon, a monocistronic
   e. a template, a monocistronic
5. For the *lac* operon, what would be the expected effects of a mutation in the operator site that prevented the binding of the repressor protein?
   a. The operon would always be turned on.
   b. The operon would always be turned off.
   c. The operon would always be turned on, except when glucose is present.
   d. The operon would be turned on only in the presence of lactose.
   e. The operon would be turned on only in the presence of lactose and the absence of glucose.
6. The presence of _____ in the medium prevents CAP from binding to the DNA, resulting in _____ in transcription of the *lac* operon.
   a. lactose, an increase
   b. glucose, an increase
   c. cAMP, a decrease
   d. glucose, a decrease
   e. lactose, a decrease
7. The *trp* operon is considered _____ operon because the structural genes necessary for tryptophan synthesis are not expressed when the levels of tryptophan in the cell are high.
   a. an inducible
   b. a positive
   c. a repressible
   d. a negative
   e. Both c and d are correct.
8. Regulatory elements that function to increase transcription levels in eukaryotes are called
   a. promoters.
   b. silencers.
   c. enhancers.
   d. transcriptional start sites.
   e. activators.
9. DNA methylation in many eukaryotic organisms usually causes
   a. increased translation levels.
   b. decreased translation levels.
   c. increased transcription levels.
   d. decreased transcription levels.
   e. introns to be removed.
10. _____ refers to the phenomenon where a single type of pre-mRNA may give rise to multiple types of mRNAs due to different patterns of intron and exon removal.
    a. Spliceosomes
    b. Variable expression
    c. Alternative splicing
    d. Polycistronic mRNA
    e. Induced silencing

Conceptual Questions

1. What is the difference between inducible and repressible operons? Give an example of each.
2. Transcriptional regulation often involves a regulatory protein that binds to a segment of DNA and a small effector molecule that binds to the regulatory protein. Do the following terms apply to a regulatory protein, a segment of DNA, or a small effector molecule? a. repressor; b. inducer; c. operator site; d. corepressor; e. activator
3. A principle of biology is *the genetic material provides a blueprint for reproduction.* Explain how gene regulation is an important mechanism for reproduction and sustaining life.

Collaborative Questions

1. Discuss the advantages and disadvantages of genetic regulation at the different levels described in Figure 13.4.
2. Discuss the advantages and disadvantages of combinatorial control of eukaryotic genes.

Online Resource

www.brookerbiology.com

Stay a step ahead in your studies with animations that bring concepts to life and practice tests to assess your understanding. Your instructor may also recommend the interactive eBook, individualized learning tools, and more.

Chapter Outline

14.1 Mutation
14.2 DNA Repair
14.3 Cancer
Summary of Key Concepts
Assess and Discuss

Mutation, DNA Repair, and Cancer

14

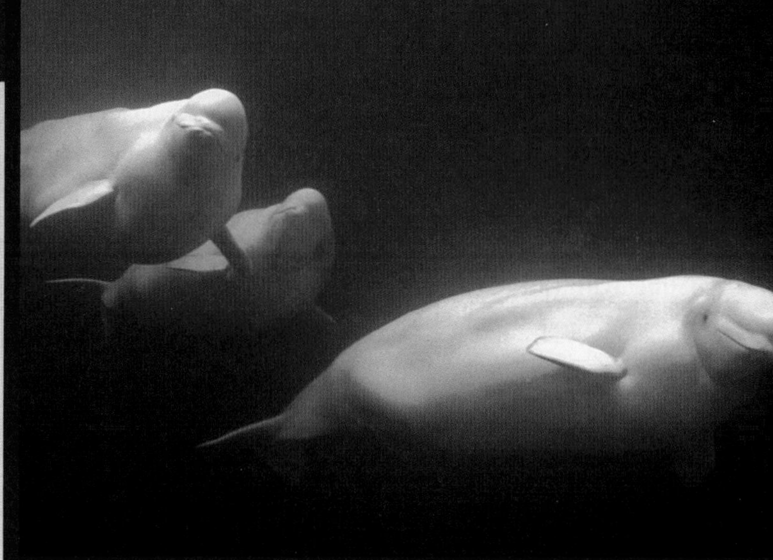

During the past two decades, over 25% of the beluga whales in Canada's St. Lawrence Seaway have died of cancer. Biologists speculate that these deaths are caused by cancer-causing pollutants, such as polycyclic aromatic hydrocarbons (PAHs).

At a summer camp, the children enjoy ice cream, horseback riding, hay rides, swimming, and learning about the habits of owls. Not such an unusual camp, you might be thinking. However, what makes Camp Sundown unique is that the outdoor fun begins at dusk and runs all night. The children at this camp have inherited a disorder called xeroderma pigmentosum (XP), which makes them highly sensitive to sunlight. Their skin will blister or freckle on minimum Sun exposure. Of greater concern, however, is skin cancer. Persons with XP may have a 1,000-fold greater risk of developing skin cancer, though such a risk is greatly decreased if Sun exposure is minimized.

What explains the symptoms of XP? Individuals with this condition are highly susceptible to **mutation**, which is defined as a heritable change in the genetic material. When a mutation occurs, the order of nucleotide bases in a DNA molecule, its base sequence, is changed permanently, an alteration that can be passed from mother to daughter cells during cell division. Mutations that lead to cancer cause particular genes to be expressed in an abnormal way. For example, a mutation could affect the transcription of a gene, or it could alter the functional properties of the polypeptide that is specified by a gene.

Should we be afraid of mutations? Yes and no. On the positive side, mutations are essential to the long-term continuity of life. Mutations provide the foundation for evolutionary change. They supply the variation that enables species to evolve and become better adapted to their environments. On the negative side, however, new mutations are more likely to be harmful than beneficial to the individual. The genes within modern species are the products of billions of years of evolution and have evolved to work properly. Random mutations are more likely to disrupt genes rather than enhance their function. As we will see in this chapter, mutations can cause cancer. In addition, many forms of inherited diseases, such as XP and cystic fibrosis, are caused by gene mutations. For these and many other reasons, understanding the molecular nature of mutations is a compelling area of research.

Because mutations can be harmful, all species have evolved several ways to repair damaged DNA. Such DNA repair systems reverse DNA damage before a permanent mutation can occur. DNA repair systems are vital to the survival of all organisms. If these systems did not exist, mutations would be so prevalent that few species, if any, would survive. Persons with XP have an impaired DNA repair system, which is the underlying cause of their disorder. DNA damage from sunlight is normally corrected by DNA repair systems. In people with XP, damaged DNA often remains unrepaired, which can lead to cancer. In this chapter, we will examine how such DNA repair systems operate. But first, let's explore the molecular basis of mutation.

14.1 Mutation

Learning Outcomes:

1. List the different ways that mutations can alter the amino acid sequence of a polypeptide.
2. Analyze the replica plating experiments of the Lederbergs.
3. Distinguish between mutations in somatic cells and in germ-line cells.
4. Discuss the difference between spontaneous and induced mutations.
5. Analyze the results of an Ames test for determining if a substance is a mutagen.

How do mutations affect traits? To answer this question at the molecular level, we must understand how changes in the DNA sequence of a gene ultimately affect gene function. Most of our understanding of mutation has come from the study of experimental organisms, such as bacteria and *Drosophila*. Researchers can expose these organisms to agents that cause mutations and then study the consequences of the mutations that arise. In addition, because these organisms have a short generation time, researchers can investigate the effects of mutations when they are passed from parent to offspring over many generations.

The structure and amount of genetic material can be altered in a variety of ways. For example, the structure and number of chromosomes can change. We will examine these types of genetic changes in Chapter 15. In this section, we will focus our attention on gene mutations, which are relatively small changes in the sequence of bases in a particular gene. We will also consider how the timing of new mutations during an organism's development has important consequences. Finally, we will explore how environmental agents may bring about mutations and examine a testing method that is used to determine if an agent causes mutations.

Gene Mutations Alter the DNA Sequence of a Gene

Mutations cause two basic types of changes to a gene: (1) the base sequence within a gene can be changed; and (2) one or more base pairs can be added to or removed from a gene. A **point mutation** affects only a single base pair within the DNA. For example, the DNA sequence shown here has been altered by a **base substitution** in which a T (in the top strand) has been replaced by a G and the corresponding A in the bottom strand is replaced with a C:

```
5′-CCCGCTAGATA-3′    ⟶    5′-CCCGCGAGATA-3′
3′-GGGCGATCTAT-5′         3′-GGGCGCTCTAT-5′
```

A point mutation could also involve the addition or deletion of a single base pair to a DNA sequence. For example, in the following sequence, a single base pair (A-T) has been added to the DNA:

```
5′-GGCGCTAGATC–3′    ⟶    5′-GGCAGCTAGATC-3′
3′-CCGCGATCTAG–5′         3′-CCGTCGATCTAG-5′
```

Though point mutations may seem like small changes to a DNA sequence, they can have important consequences when genes are expressed, as we will see next.

Gene Mutations May Affect the Amino Acid Sequence of a Polypeptide

If a mutation occurs within the region of a structural gene that specifies the amino acid sequence, the mutation may alter that sequence in a variety of ways. Table 14.1 considers the potential effects of point mutations. **Silent mutations** do not alter the amino acid sequence of the polypeptide, even though the nucleotide sequence has changed. As discussed in Chapter 12, the genetic code is degenerate, that is, more than one codon can specify the same amino acid. Silent mutations occur in the third base of many codons without changing the type of amino acid it encodes.

Table 14.1 Consequences of Point Mutations Within the Coding Sequence of a Structural Gene

Mutation in the DNA	Effect on polypeptide	Example*
None	None	ATGGCCGGCCCGAAAGAGACC Met-Ala-Gly-Pro-Lys-Glu-Thr
Base substitution	**Silent**—causes no change	ATGGCCGGCCCCAAAGAGACC Met-Ala-Gly-Pro-Lys-Glu-Thr
Base substitution	**Missense**—changes one amino acid	ATGCCCGGCCCGAAAGAGACC Met-Pro-Gly-Pro-Lys-Glu-Thr
Base substitution	**Nonsense**—changes to a stop codon	ATGGCCGGCCCGTAAGAGACC Met-Ala-Gly-Pro-STOP
Addition (or deletion) of single base	**Frameshift**—produces a different amino acid sequence	ATGGCCGGCACCGAAAGAGACC Met-Ala-Gly-Thr-Glu-Arg-Asp

*DNA sequence in the coding strand. This sequence is the same as the mRNA sequence except that RNA contains uracil (U) instead of thymine (T).

A **missense mutation** is a base substitution that changes a single amino acid in a polypeptide sequence. A missense mutation may not alter protein function because it changes only a single amino acid within a polypeptide that is typically hundreds of amino acids in length. A missense mutation that substitutes an amino acid with a chemistry similar to the original amino acid is less likely to alter protein function. For example, a missense mutation that substitutes a glutamic acid for an aspartic acid may not alter protein function because both amino acids are negatively charged and have similar side chain structures (refer back to Figure 3.14).

Alternatively, some missense mutations have a dramatic effect on protein function. A striking example occurs in the human disease known as **sickle cell disease**. This disease involves a missense mutation in the β-globin gene, which encodes one of the polypeptide subunits that make up hemoglobin, the oxygen-carrying protein in red blood cells. In the most common form of this disease, a missense mutation alters the polypeptide sequence such that the sixth amino acid is changed from a glutamic acid to a valine (Figure 14.1). Because glutamic acid is hydrophilic but valine is hydrophobic, this single amino acid substitution alters the structure and function of the hemoglobin protein. The mutant hemoglobin subunits tend to stick to one another when the oxygen concentration is low. The aggregated proteins form fiber-like structures within red blood cells, which causes the cells to lose their normal disc-shaped morphology and become sickle-shaped. It seems amazing that a single amino acid substitution could have such a profound effect on the structure of cells.

Two other types of point mutations cause more dramatic changes to a polypeptide sequence. A **nonsense mutation** involves a change from a normal codon to a stop, or termination, codon. This causes translation to be terminated earlier than expected, producing

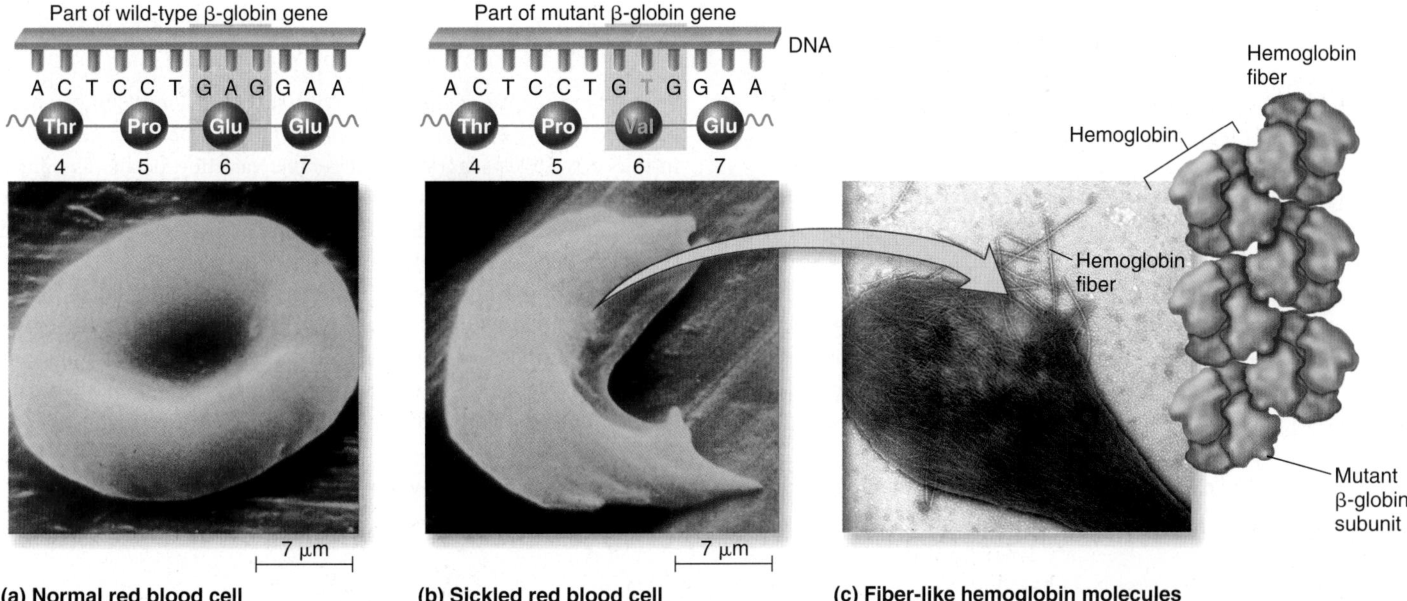

Figure 14.1 **A missense mutation that causes red blood cells to sickle in sickle cell disease.** Scanning electron micrographs of **(a)** normal red blood cell and **(b)** sickled red blood cell. As shown above the micrographs, a missense mutation in the β-globin gene (which codes for a subunit of hemoglobin) changes the sixth amino acid in the β-globin polypeptide from a glutamic acid (Glu) to a valine (Val). **(c)** This micrograph shows how this alteration to the structure of β-globin causes the formation of abnormal fiber-like structures. In normal cells, hemoglobin proteins do not form fibers.

Concept Check: Based on the fiber-like structures seen in part (c), what aspect of hemoglobin structure does a glutamic acid at the sixth position in normal β-globin prevent? Speculate on how the charge of this amino acid may play a role.

a truncated polypeptide (see Table 14.1). Compared with a normal polypeptide, a shorter polypeptide is much less likely to function properly. Finally, a **frameshift mutation** involves the addition or deletion of nucleotides that are not in multiples of three nucleotides. For example, a frameshift mutation could involve the addition or deletion of one, two, four, or five nucleotides. Because the codons are read in multiples of three, these types of insertions or deletions shift the reading frame so a completely different amino acid sequence occurs downstream from the mutation (see Table 14.1). Such a large change in polypeptide structure is likely to inhibit protein function.

Changes in protein function may affect the ability of an organism to survive and to reproduce. Except for silent mutations, new mutations are more likely to produce polypeptides that have reduced rather than enhanced function. However, mutations can occasionally produce a polypeptide that has an enhanced function. Such mutations may change in frequency in a population over the course of many generations due to natural selection. This topic is discussed in Chapter 24.

Gene Mutations That Occur Outside of Coding Sequences Can Influence Gene Expression

Thus far, we have focused our attention on mutations in the coding regions of structural genes. In Chapters 12 and 13, we explored the role of DNA sequences in gene expression. A mutation can occur within a noncoding DNA sequence and affect gene expression (**Table 14.2**). For example, a mutation may alter the sequence within the promoter of a gene, thereby affecting the rate of transcription. A mutation that improves the ability of RNA polymerase to bind to the promoter may enhance transcription, whereas other mutations may inhibit transcription.

Mutations in regulatory elements or operator sites can alter the regulation of gene transcription. For example, in Chapter 13, we considered the roles of regulatory elements such as the *lac* operator site in *E. coli*, which is recognized by the lac repressor protein (refer back to Figure 13.7). Mutations in the *lac* operator site can disrupt the proper regulation of the *lac* operon. An operator mutation may change the DNA sequence so the lac repressor protein does not bind to it. This mutation would cause the operon to be constitutively expressed.

Table 14.2 Effects of Mutations Outside of the Coding Sequence of a Gene

Sequence	Effect of mutation
Promoter	May increase or decrease the rate of transcription
Transcriptional regulatory element/operator site	May alter the regulation of transcription
Splice sites	May alter the ability of pre-mRNA to be properly spliced
Translational regulatory element	May alter the ability of mRNA to be translationally regulated
Intergenic region	Not as likely to have an effect on gene expression

FEATURE INVESTIGATION

The Lederbergs Used Replica Plating to Show That Mutations Are Random Events

As we have seen, mutations affect the expression of genes in a variety of ways. Scientists considered the following question: Do mutations that affect the traits of an individual occur as a result of pre-existing circumstances, or are they random events that may happen in any gene of any individual? In the 19th century, French naturalist Jean Baptiste Lamarck proposed that physiological events (such as use or disuse) determine whether traits are passed along to offspring. For example, his hypothesis suggested that an individual who practiced and became adept at a physical activity, such as the long jump, would pass that quality on to his or her offspring. Alternatively, geneticists in the early 1900s suggested that genetic variation occurs as a matter of chance. According to this view, those individuals whose genes happen to contain beneficial mutations are more likely to survive and pass those genes to their offspring.

These opposing views were tested in bacterial studies in the 1940s and 1950s. One such study, by American microbiologists Joshua and Esther Lederberg, focused on the occurrence of mutations in bacteria (**Figure 14.2**). First, they placed a large number of *E. coli* bacteria onto growth media and incubated them overnight, so each

Figure 14.2 **The experiment performed by the Lederbergs showing that mutations are random events.**

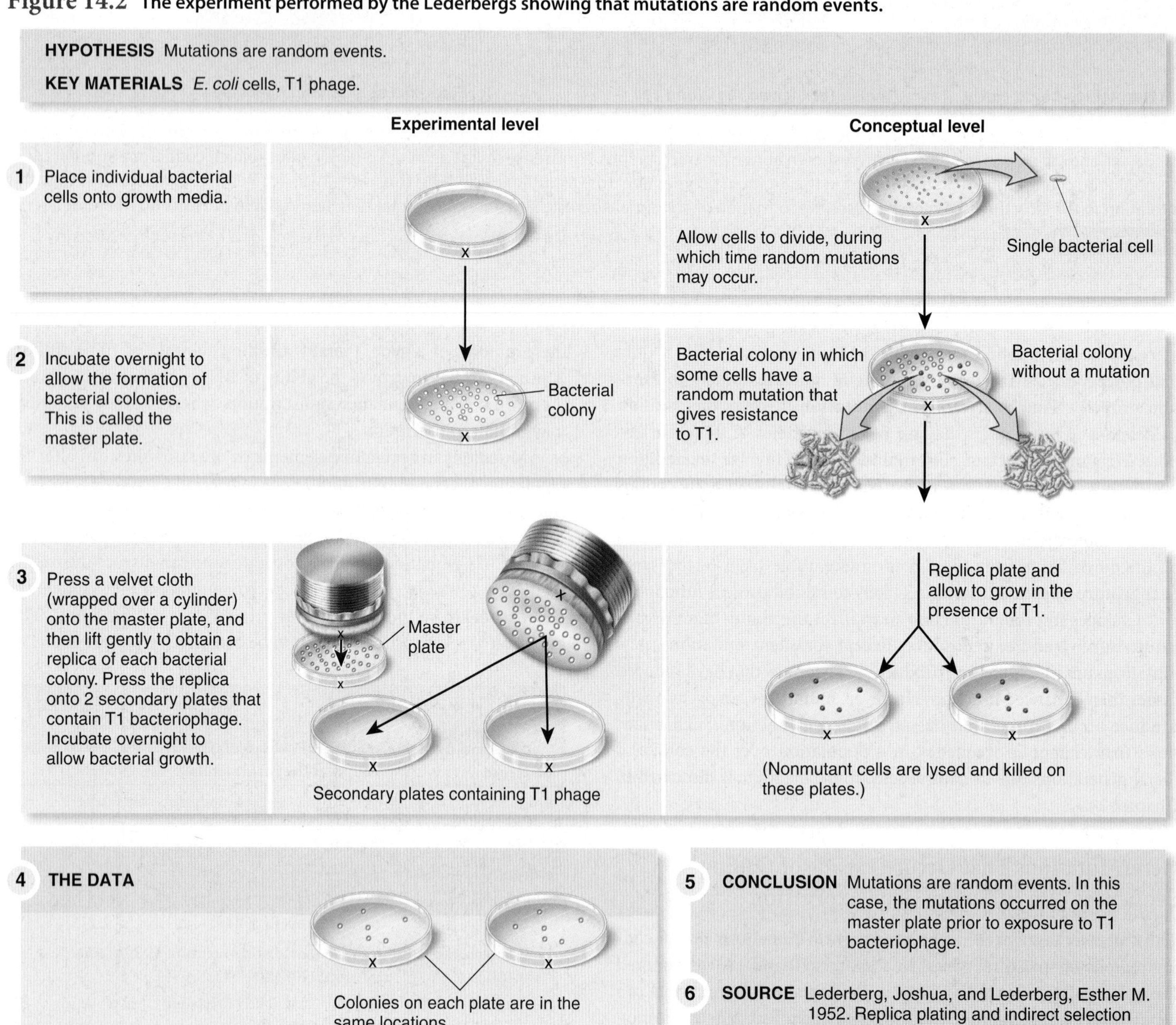

HYPOTHESIS Mutations are random events.
KEY MATERIALS E. coli cells, T1 phage.
Experimental level
Conceptual level
1 Place individual bacterial cells onto growth media.
X
Allow cells to divide, during which time random mutations may occur.
Single bacterial cell
2 Incubate overnight to allow the formation of bacterial colonies. This is called the master plate.
Bacterial colony
Bacterial colony in which some cells have a random mutation that gives resistance to T1.
Bacterial colony without a mutation
3 Press a velvet cloth (wrapped over a cylinder) onto the master plate, and then lift gently to obtain a replica of each bacterial colony. Press the replica onto 2 secondary plates that contain T1 bacteriophage. Incubate overnight to allow bacterial growth.
Master plate
Secondary plates containing T1 phage
Replica plate and allow to grow in the presence of T1.
(Nonmutant cells are lysed and killed on these plates.)
4 THE DATA
Colonies on each plate are in the same locations.
5 CONCLUSION Mutations are random events. In this case, the mutations occurred on the master plate prior to exposure to T1 bacteriophage.
6 SOURCE Lederberg, Joshua, and Lederberg, Esther M. 1952. Replica plating and indirect selection of bacterial mutants. Journal of Bacteriology 63:399–406.

bacterial cell divided many times to form a bacterial colony composed of millions of cells. This is called the master plate. Using a technique known as **replica plating**, a sterile piece of velvet cloth was lightly touched to the master plate to pick up bacterial cells from each colony on the master plate. They then transferred this replica to two secondary plates containing an agent that selected for the growth of bacterial cells with a particular mutation.

In the example shown in Figure 14.2, the secondary plates contained T1 bacteriophages, which are viruses that infect bacteria and cause them to lyse. On these plates, only those rare cells that had acquired a mutation conferring resistance to T1, termed *ton*^{r}, could grow. All other cells were lysed by the proliferation of bacteriophages in the bacteria. Therefore, only a few colonies were observed on the secondary plates. Strikingly, these colonies occupied the same locations on each plate. How did the Lederbergs interpret these results? The data indicated that the *ton*^{r} mutations occurred randomly while the bacterial cells were forming colonies on the nonselective master plate. The presence of T1 bacteriophages in the secondary plates did not cause the mutations to develop. Rather, the T1 bacteriophages simply selected for the growth of *ton*^{r} mutants that were already in the population. These results supported the idea that mutations are random events.

Experimental Questions

1. Explain the opposing views of mutation prior to the Lederbergs' study.
2. What hypothesis was being tested by the Lederbergs? What were the results of the experiment?
3. How did the results of the Lederbergs support or falsify the hypothesis?

Mutations Can Occur in Germ-Line or Somatic Cells

Let's now consider how the timing of a mutation may have an important influence on its potential effects. Multicellular organisms typically begin their lives as a single fertilized egg cell that divides many times to produce all the cells of an adult organism. A mutation can occur in any cell of the body, either very early in life, such as in a gamete (eggs or sperm) or a fertilized egg, or later in life, such as in the embryonic or adult stages. The number and location of cells with a mutation are critical both to the severity of the genetic effect and to whether the mutation can be passed on to offspring.

Geneticists classify the cells of animals into two types: germ-line and somatic cells. The term **germ line** refers to cells that give rise to gametes, such as egg and sperm cells. A germ-line mutation can occur directly in an egg or sperm cell, or it can occur in a precursor cell that produces the gametes. If a mutant human gamete participates in fertilization, all the cells of the resulting offspring will contain the mutation, as indicated by the red color in **Figure 14.3a**. Likewise, when such an individual produces gametes, the mutation may be transmitted to future generations of offspring. Because humans carry two copies of most genes, a new mutation in a single gene has a 50% chance of being transmitted from parent to offspring.

The **somatic cells** constitute all cells of the body except for the germ line. Examples include skin cells and muscle cells. Mutations can also occur within somatic cells at early or late stages of development. What are the consequences of a mutation that happens during the embryonic stage? As shown in **Figure 14.3b**, a mutation occurred within a single embryonic cell. This single somatic cell was the precursor for many cells of the adult. Therefore, in the adult, a patch of tissue contains cells that carry the mutation. The size of any patch depends on the timing of a new mutation. In general, the earlier a mutation occurs during development, the larger the patch. An individual with somatic regions that are genetically different from each other is called a **mosaic**.

Figure 14.4 illustrates a child who had a somatic mutation during an early stage of development. In this case, the child has a streak of white hair while the rest of his hair is black. Presumably, he initially had a single mutation happen in an embryonic cell that ultimately gave rise to the patch that produced the white hair.

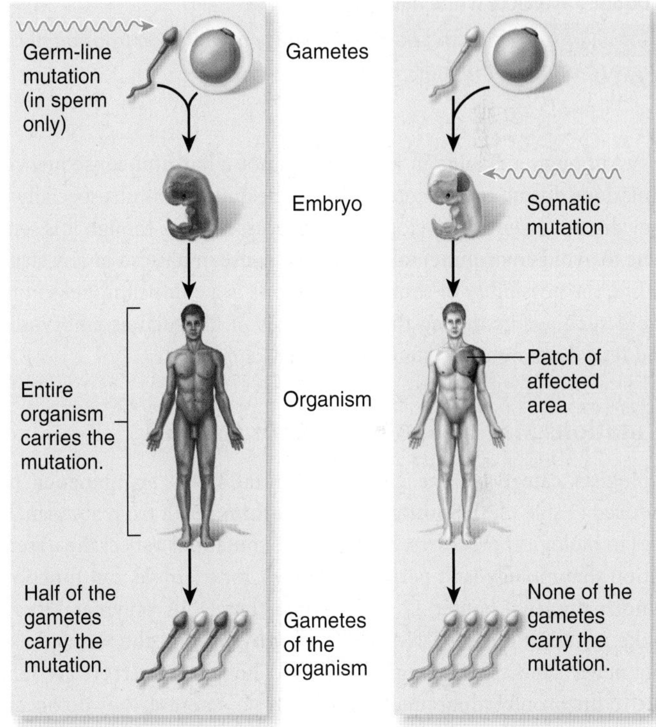

Figure 14.3 The effects of germ-line versus somatic cell mutations. The red color indicates which cells carry the mutation. **(a)** If a mutation is passed via gametes, such germ-line mutations occur in every cell of the body. Because humans have two copies of most genes, a germ-line mutation in one of those two copies is transmitted to only half of the gametes. **(b)** Somatic mutations affect a limited area of the body and are not transmitted to offspring.

BIOLOGY PRINCIPLE Living organisms grow and develop. As a multicellular organism grows and develops, a germ-line mutation is transmitted to all cells of the body, whereas a somatic mutation is found only in a particular region.

Concept Check: *Why are somatic mutations unable to be transmitted to offspring?*

Figure 14.4 **Example of a somatic mutation.** This child has a streak of white hair. This is due to a somatic mutation in a single cell during embryonic development. This cell continued to divide to produce a streak of white hair.

Concept Check: *Can this child with a streak of white hair transmit this trait to his future offspring?*

Although a change in hair color is not a harmful consequence, mutations during early stages of life can be quite harmful, especially if they disrupt essential developmental processes. Even though it is sensible to avoid environmental agents that cause mutations at any stage of life, the possibility of somatic mutations is a compelling reason to avoid such agents during the early stages of life such as embryonic and fetal development, infancy, and early childhood.

Mutations May Be Spontaneous or Induced

Biologists categorize the causes of mutation as spontaneous or induced (Table 14.3). **Spontaneous mutations** result from abnormalities in biological processes. Spontaneous mutations reflect the observation that biology isn't perfect. Enzymes, for example, can function abnormally. In Chapter 11, we learned that DNA polymerase can make mistakes during DNA replication by putting the wrong base in a newly synthesized daughter strand. Though such errors are rare due to the proofreading function of DNA polymerase, they do occur. In addition, normal metabolic processes within the cell may produce toxic chemicals such as free radicals that can react directly with the DNA and alter its structure. Finally, the structure of nucleotides is not absolutely stable. On occasion, the structure of a base may spontaneously change, and such a change may cause a mutation if it occurs immediately prior to DNA replication.

The rates of spontaneous mutations vary from species to species and from gene to gene. Larger genes are usually more likely to incur a mutation than are smaller ones. A common rate of spontaneous mutation among various species is approximately 1 mutation for every 1 million genes per cell division, which equals 1 in 10^6, or simply 10^{-6}. This is the expected rate of spontaneous mutation, which creates the variation that is the raw material of evolution.

Induced mutations are caused by environmental agents that enter the cell and alter the structure of DNA. They cause the mutation

Table 14.3 Some Common Causes of Gene Mutations

Common causes of mutations	Description
Spontaneous:	
Errors in DNA replication	A mistake by DNA polymerase may cause a point mutation.
Toxic metabolic products	The products of normal metabolic processes may be reactive chemicals such as free radicals that can alter the structure of DNA.
Changes in nucleotide structure	On rare occasions, the linkage between purines and deoxyribose can spontaneously break. Changes in base structure (isomerization) may cause mispairing during DNA replication.
Transposons	As discussed in Chapter 21, transposons are small segments of DNA that can insert at various sites in the genome. If they insert into a gene, they may inactivate the gene.
Induced:	
Chemical agents	Chemical substances, such as benzo(*a*)pyrene, a chemical found in cigarette smoke, may cause changes in the structure of DNA.
Physical agents	Physical agents such as UV (ultraviolet) light and X-rays can damage the DNA.

rate to be higher than the spontaneous mutation rate. Agents that cause mutation are called **mutagens**. Mutagenic agents can be categorized as **chemical** or **physical mutagens** (Table 14.4). We will consider their effects next.

Mutagens Alter DNA Structure in Different Ways

Researchers have discovered that an enormous array of agents act as mutagens. We often hear in the news media that we should avoid these agents in our foods and living environments. We even use products such as sunscreens that help us avoid the mutagenic effects of ultraviolet (UV) light from the Sun. The public is often concerned about mutagens for two important reasons. First, mutagenic agents are usually involved in the development of human cancers. Second, because new mutations may be deleterious, people want to avoid mutagens to prevent mutations that may have harmful effects in their future offspring.

How do mutagens affect DNA structure? Some chemical mutagens act by covalently modifying the structure of nucleotides. For example, nitrous acid (HNO_2) deaminates bases by replacing amino groups with keto groups ($—NH_2$ to $=O$). This can change cytosine to uracil. When this altered DNA replicates, the modified base does not pair with the appropriate base in the newly made strand. In this case, uracil pairs with adenine (Figure 14.5).

Similarly, 5-bromouracil and 2-aminopurine, which are called base analogues, have structures that are similar to particular bases in DNA and can substitute for them. When incorporated into DNA, they also cause errors in DNA replication. Other chemical mutagens disrupt the appropriate pairing between nucleotides by alkylating bases within the DNA. During alkylation, methyl or ethyl groups are covalently attached to the bases. Examples of alkylating agents include nitrogen mustards (used as a chemical weapon during World War I) and ethyl methanesulfonate (EMS), which is used as a mutagen in laboratory experiments.

Table 14.4 Examples of Mutagens

Mutagen	Effect(s) on DNA structure
Chemical	
Nitrous acid	Deaminates bases
5-Bromouracil	Acts as a base analogue
2-Aminopurine	Acts as a base analogue
Nitrogen mustard	Alkylates bases
Ethyl methanesulfonate (EMS)	Alkylates bases
Benzo(*a*)pyrene	Inserts next to bases in the DNA double helix and causes additions or deletions
Physical	
X-rays	Causes base deletions, single nicks in DNA strands, crosslinking, and chromosomal breaks
UV light	Promotes pyrimidine dimer formation, which involves covalent bonds between adjacent pyrimidines (C or T)

Some chemical mutagens exert their effects by interfering with DNA replication. For example, benzo(*a*)pyrene, which is found in automobile exhaust, cigarette smoke, and charbroiled food, is metabolized to a compound (benzopyrene diol epoxide) that inserts in between the bases of the double helix, thereby distorting the helical structure. When DNA containing such a mutagen is replicated, single-nucleotide additions and deletions may be incorporated into the newly made strands.

DNA molecules are also sensitive to physical agents such as radiation. In particular, radiation of short wavelength and high energy, known as ionizing radiation, is known to alter DNA structure. Ionizing radiation includes X-rays and gamma rays. This type of radiation can penetrate deeply into biological materials, where it creates free radicals. These molecules can alter the structure of DNA in a variety of ways. Exposure to high doses of ionizing radiation causes base deletions, breaks in one DNA strand, and even a break in both DNA strands.

Nonionizing radiation, such as UV light, contains less energy, and so it penetrates only the surface of biological materials, such as the skin. Nevertheless, UV light is known to cause mutations. For example, UV light can cause the formation of a **thymine dimer**, which is a site where two adjacent thymine bases become covalently crosslinked to each other (**Figure 14.6**).

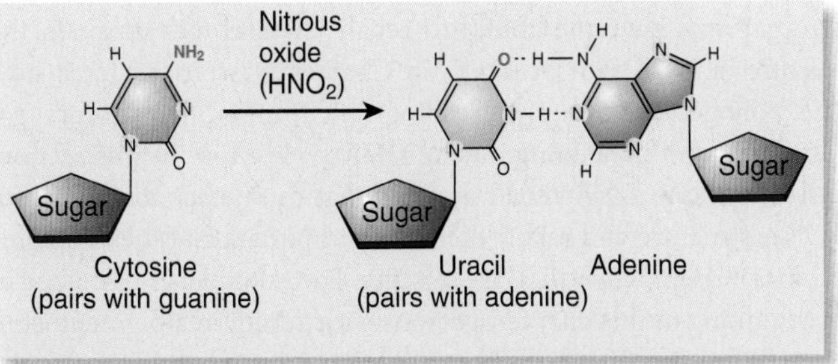

Figure 14.5 Deamination and mispairing of modified bases by a chemical mutagen. Nitrous acid changes cytosine to uracil by replacing NH_2 with an oxygen. During DNA replication, uracil pairs with adenine, thereby creating a mutation in the newly replicated strand.

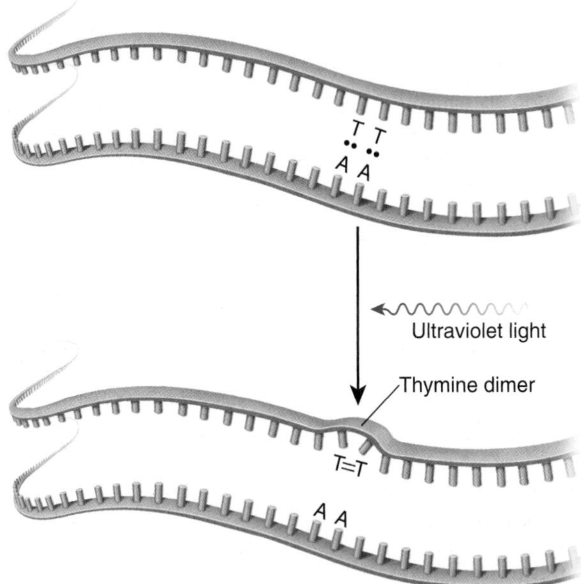

Figure 14.6 Formation and structure of a thymine dimer.

Concept Check: *Why is a thymine dimer harmful?*

Thymine dimers are typically repaired before or during DNA replication. However, if such repair fails to occur, a thymine dimer may cause a mutation when that DNA strand is replicated. When DNA polymerase attempts to replicate over a thymine dimer, proper base pairing does not occur between the template strand and the incoming nucleotides. This mispairing can cause gaps in the newly made strand or the incorporation of incorrect bases. Plants, in particular, must have effective ways to prevent UV damage because they are exposed to sunlight throughout the day.

Testing Methods Determine If an Agent Is a Mutagen

Because mutagens are harmful, researchers have developed testing methods to evaluate the ability of a substance to cause mutation. One commonly used test is the **Ames test**, which was developed by American biochemist Bruce Ames in the 1970s. This test uses a strain of a bacterium, *Salmonella typhimurium*, that cannot synthesize the amino acid histidine. This strain contains a point mutation within a gene that encodes an enzyme required for histidine biosynthesis. The mutation renders the enzyme inactive. The bacteria cannot grow unless histidine has been added to the growth medium. However, a second mutation may correct the first mutation, thereby restoring the ability to synthesize histidine. The Ames test monitors the rate at which this second mutation occurs and thereby indicates whether an agent increases the mutation rate above the spontaneous rate.

Figure 14.7 outlines the steps in the Ames test. The suspected mutagen is mixed with a rat liver extract and the bacterial strain of *S. typhimurium* that cannot synthesize histidine. Because some potential mutagens may require activation by cellular enzymes, the rat liver extract provides a mixture of enzymes that may cause such activation. This step improves the ability to identify agents that cause mutations in mammals. As a control, bacteria that have not been exposed to the mutagen are also tested. After an incubation period in which

Figure 14.7 The Ames test for mutagenicity. In this example, 2 million bacterial cells were placed on plates lacking histidine. Two colonies arose from the control sample, whereas 44 arose from the sample exposed to a suspected mutagen.

BIOLOGY PRINCIPLE Biology affects our society. Biologists have developed many methods, including the Ames test, for determining if a chemical is a mutagen. The results of these tests have prevented the use of many different chemicals in the production of food and also resulted in warning labels on products such as cigarettes.

Concept Check: Based on the results seen in this figure, what is the rate of mutation that is caused by the suspected mutagen?

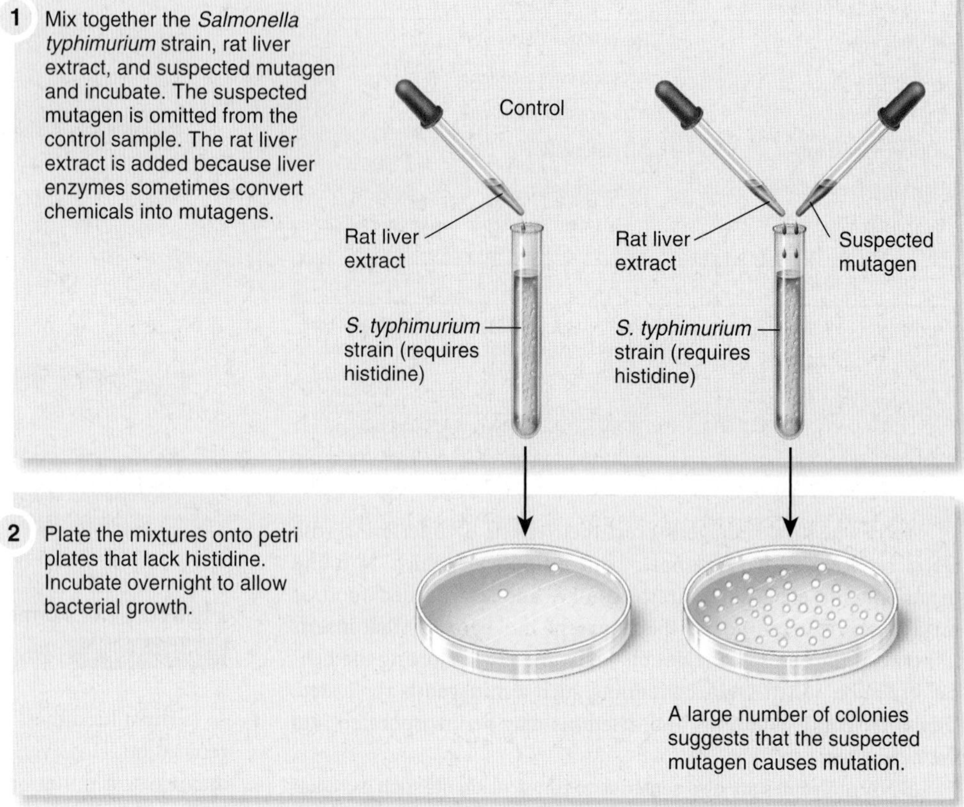

mutations may occur, a large number of bacteria are plated on a growth medium that does not contain histidine. The *S. typhimurium* strain is not expected to grow on these plates. However, if a mutation has occurred that allows a cell to synthesize histidine, the bacterium harboring this second mutation will proliferate during an overnight incubation period to form a visible bacterial colony.

1
Mix together the *Salmonella typhimurium* strain, rat liver extract, and suspected mutagen and incubate. The suspected mutagen is omitted from the control sample. The rat liver extract is added because liver enzymes sometimes convert chemicals into mutagens.
Control
Rat liver extract
S. typhimurium strain (requires histidine)
Rat liver extract
Suspected mutagen
S. typhimurium strain (requires histidine)
2
Plate the mixtures onto petri plates that lack histidine. Incubate overnight to allow bacterial growth.
A large number of colonies suggests that the suspected mutagen causes mutation.

To estimate the mutation rate, the colonies that grow in the absence of histidine are counted and compared with the total number of bacterial cells that were originally placed on the plate for both the suspected-mutagen sample and the control. The control condition is a measure of the spontaneous mutation rate, whereas the other sample measures the rate of mutation in the presence of the suspected mutagen. As an example, let's suppose that 2 million bacteria were plated from both the control and the suspected-mutagen tubes. In the control experiment, 2 bacterial colonies were observed. The spontaneous mutation rate is calculated by dividing 2 (the number of mutants) by 2 million (the number of original cells). This equals 1 in 1 million, or 1×10^{-6}. By comparison, 44 colonies arose from the suspected-mutagen sample (see Figure 14.7). In this case, the mutation rate would be 44 divided by 2 million, which equals 2.2×10^{-5}. The mutation rate in the presence of the mutagen is over 20 times higher than the spontaneous mutation rate.

How do we judge if an agent is a mutagen? Researchers compare the mutation rate in the presence and absence of the suspected mutagen. The experimental approach shown in Figure 14.7 is conducted several times. If statistics reveal that the mutation rate in the suspected-mutagen sample is significantly higher than in the control sample, they may tentatively conclude that the agent is a mutagen. Interestingly, many studies have used the Ames test to compare the urine from cigarette smokers with that from nonsmokers. This research has shown that urine from smokers contains much higher levels of mutagens.

14.2 DNA Repair

Learning Outcomes:

1. List the general features of DNA repair systems.
2. Describe the steps of nucleotide excision repair.
3. Explain the connection between defects in DNA repair systems and the inherited human disease xeroderma pigmentosum.

In the previous section, we considered the causes and consequences of mutation. As we have seen, mutations are random events that often have negative consequences. To minimize mutation, all living organisms have the ability to repair changes that occur in the structure of DNA. For example, in Chapter 11 we considered how DNA polymerase has a proofreading function that helps to prevent mutations from happening during DNA replication. In this section, we will examine DNA repair systems that can detect abnormalities in DNA structure and repair them. The importance of these systems becomes evident when they are missing. For example, as discussed at the beginning of this chapter, persons with xeroderma pigmentosum are highly susceptible to the harmful effects of sunlight because they are missing a single DNA repair system.

How do organisms minimize the occurrence of mutations? Cells contain several DNA repair systems that can fix different types of DNA alterations (**Table 14.5**). Each repair system is composed of one or more

Table 14.5 Common Types of DNA Repair Systems

System	Description
Direct repair	A repair enzyme recognizes an incorrect structure in the DNA and directly converts it back to a correct structure.
Base excision and nucleotide excision repair	An abnormal base or nucleotide is recognized, and a portion of the strand containing the abnormality is removed. The complementary DNA strand is then used as a template to synthesize a normal DNA strand.
Methyl-directed mismatch repair	Similar to excision repair except that the DNA defect is a base pair mismatch in the DNA, not an abnormal nucleotide. The mismatch is recognized, and a strand of DNA in this region is removed. The complementary strand is used to synthesize a normal strand of DNA.

*Other types of repair systems exist; these are common examples.

proteins that play specific roles in the repair mechanism. DNA repair requires two coordinated events. In the first step, one or more proteins in the repair system detect an irregularity in DNA structure. In the second step, the abnormality is repaired. In some cases, the change in DNA structure can be directly repaired. For example, DNA may be modified by the attachment of an alkyl group, such as—CH_2CH_3, to a base. In **direct repair**, an enzyme removes this alkyl group, thereby restoring the structure of the original base. More commonly, however, the altered DNA is removed, and a new segment of DNA is synthesized. In this section, we will examine nucleotide excision repair as an example of how such systems operate. This system, which is found in all species, is an important mechanism of DNA repair.

Nucleotide Excision Repair Removes Segments of Damaged DNA

In **nucleotide excision repair (NER)**, a region encompassing several nucleotides in the damaged strand is removed from the DNA, and the intact undamaged strand is used as a template for the resynthesis of a normal complementary strand. NER can fix many different types of DNA damage, including UV-induced damage, chemically modified bases, missing bases, and various types of crosslinks (such as thymine dimers). The system is found in all species, although its molecular mechanism is better understood in bacteria.

In *E. coli*, the NER system is composed of four key proteins: UvrA, UvrB, UvrC, and UvrD. They are named Uvr because they are involved in ultraviolet light repair of thymine dimers, although these proteins are also important in repairing chemically damaged DNA. In addition, DNA polymerase and DNA ligase are required to complete the repair process.

How does the NER system work? Two UvrA proteins and one UvrB protein form a complex that tracks along the DNA (Figure 14.8). Damaged DNA will have a distorted double helix, which is sensed by the UvrA-UvrB complex. When the complex identifies a damaged site, the two UvrA proteins are released, and UvrC binds to UvrB at the site. The UvrC protein makes incisions in one DNA strand on both sides of the damaged site. After this incision process, UvrC is released. UvrD binds to UvrB. UvrD then begins to separate the DNA strands, and UvrB is released. The action of UvrD unravels the DNA,

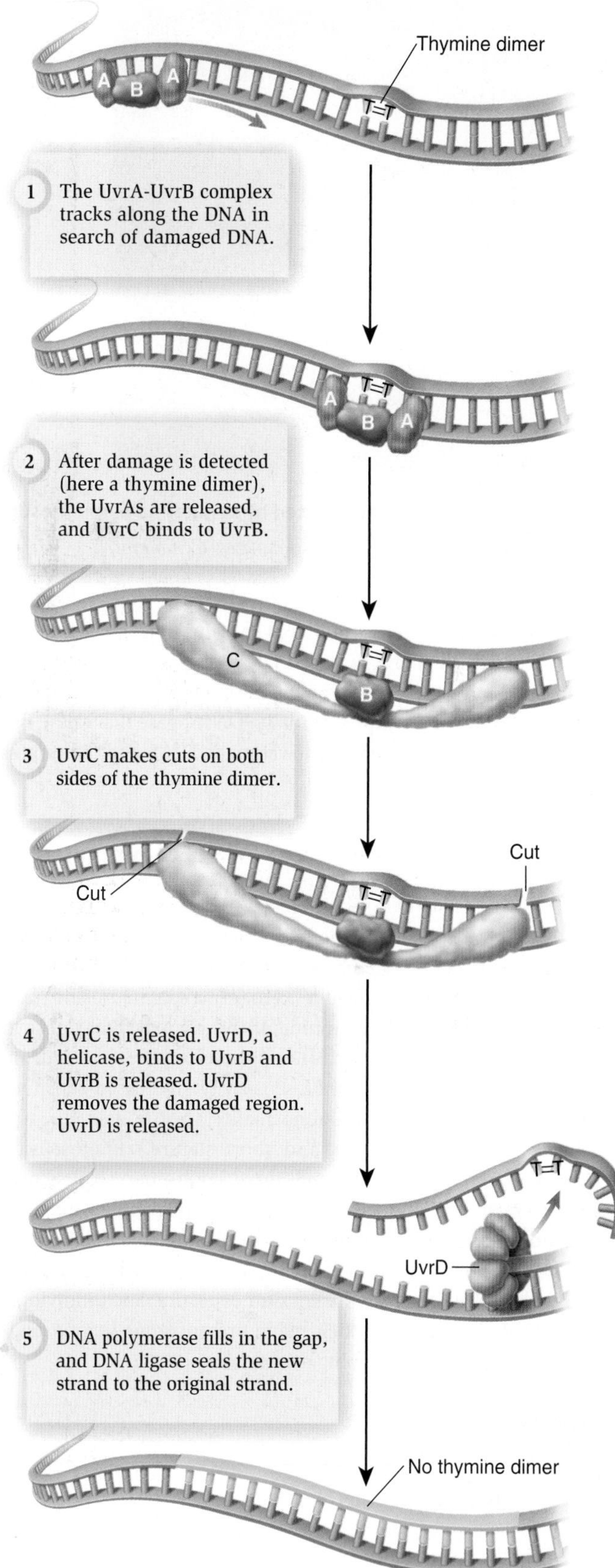

Thymine dimer
A
B
A
T=T
1 The UvrA-UvrB complex tracks along the DNA in search of damaged DNA.
2 After damage is detected (here a thymine dimer), the UvrAs are released, and UvrC binds to UvrB.
C
3 UvrC makes cuts on both sides of the thymine dimer.
Cut
Cut
4 UvrC is released. UvrD, a helicase, binds to UvrB and UvrB is released. UvrD removes the damaged region. UvrD is released.
UvrD
5 DNA polymerase fills in the gap, and DNA ligase seals the new strand to the original strand.
No thymine dimer

Figure 14.8 Nucleotide excision repair in *E. coli*.

Concept Check: *Which components of NER are responsible for removing the damaged DNA?*

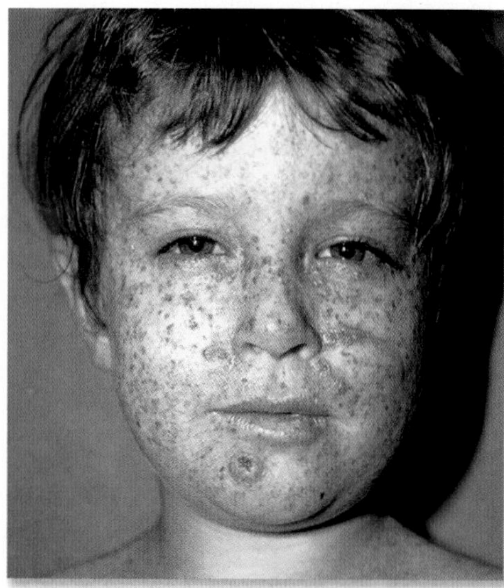

Figure 14.9 An individual affected by xeroderma pigmentosum.

Concept Check: *Why is this person so sensitive to sunlight?*

which removes a short DNA strand that contains the damaged region. UvrD is released. After the damaged DNA strand is removed, a gap is left in the double helix. DNA polymerase fills in the gap using the undamaged strand as a template. Finally, DNA ligase makes the final covalent connection between the newly made DNA and the original DNA strand.

Human Genetic Diseases Occur When a Component of the NER System Is Missing

Thus far, we have considered the NER system in *E. coli*. In humans, NER systems were discovered by the analysis of genetic diseases that affect DNA repair. These include xeroderma pigmentosum (XP), which was discussed at the beginning of this chapter, and Cockayne syndrome (CS) and PIBIDS. (PIBIDS is an acronym for a syndrome with symptoms that include photosensitivity [increased sensitivity to sunlight], ichthyosis [a skin abnormality], brittle hair, impaired intelligence, decreased fertility, and short stature.) Photosensitivity is a common characteristic in all three syndromes because of an inability to repair UV-induced lesions. Therefore, people with any of these syndromes must avoid prolonged exposure to sunlight, as do the children at Camp Sundown. Figure 14.9 shows a photograph of a child with XP who had significant Sun exposure. Such individuals have pigmentation abnormalities, many precancerous lesions, and a high predisposition to developing skin cancer.

14.3 Cancer

Learning Outcomes:

1. Outline the steps in the development of cancer.
2. Describe the general functions of oncogenes.
3. List the four common types of genetic changes that convert proto-oncogenes into oncogenes.
4. Identify the two general functions of the proteins encoded by tumor-suppressor genes.
5. Describe three common ways that tumor-suppressor genes are silenced.

Cancer is a disease of multicellular organisms characterized by uncontrolled cell division. Worldwide, cancer is the second leading cause of death in humans, exceeded only by heart disease. In the United States, approximately 1.5 million people are diagnosed with cancer each year; over 0.5 million will die from the disease. Overall, about one in four Americans will die from cancer.

In about 10% of cancers, a higher predisposition to develop the disease is an inherited trait. Most cancers, though, perhaps 90%, do not involve genetic changes that are passed from parent to offspring. Rather, cancer is usually an acquired condition that typically occurs later in life. At least 80% of all human cancers are related to exposure to **carcinogens**, agents that increase the likelihood of developing cancer. Most carcinogens, such as UV light and certain chemicals in cigarette smoke, are mutagens that promote genetic changes in somatic cells. These genetic changes can alter gene expression in a way that ultimately affects cell division, leading to cancer. In this section, we will explore such genetic abnormalities.

How does cancer occur? In most cases, the development of cancer is a multistep process (Figure 14.10). Cancers originate from a single cell. This single cell and its lineage of daughter cells undergo a series of mutations that cause the cells to grow abnormally. At an early stage, the cells form a **tumor**, which is an overgrowth of cells. For most types of cancer, a tumor begins as a precancerous or **benign** growth. Such tumors do not invade adjacent tissues and do not spread throughout the body. This may be followed by additional mutations that cause some cells in the tumor to lose their normal growth regulation and become **malignant**. At this stage, the individual has cancer. Cancerous tumors invade healthy tissues and may spread through the bloodstream or surrounding body fluids, a process called **metastasis**. If left untreated, malignant cells will cause the death of the organism.

Over the past few decades, researchers have identified many genes that promote cancer when they are mutant. By comparing the function of each mutant gene with the corresponding nonmutant gene found in healthy cells, these genes have been placed into two categories. In some cases, a mutation causes a gene to be overactive—have an abnormally high level of expression. This overactivity contributes to the uncontrolled cell growth that is observed in cancer cells. This type of mutant gene is called an **oncogene**. Alternatively, when a **tumor-suppressor gene** is normal (that is, not mutant), it encodes a protein that helps to prevent cancer. However, when a mutation eliminates its function, cancer may occur. Thus, the two categories of cancer-causing genes are based on the effects of mutations. Oncogenes are the result of mutations that cause overactivity, whereas cancer-causing mutations in tumor-suppressor genes are due to a loss of activity. In this section, we will begin with a discussion of oncogenes and then consider tumor-suppressor genes.

Oncogenes May Result from Mutations That Cause the Overactivity of Proteins Involved with Cell Division

Over the past four decades, researchers have identified many oncogenes. A large number of oncogenes encode proteins that function in cell growth-signaling pathways. Cell division is regulated, in part, by growth factors, which are molecules that regulate cell division. A growth factor binds to a receptor, which results in receptor activation (Figure 14.11). This stimulates an intracellular signal transduction pathway that activates transcription factors. In this way, the

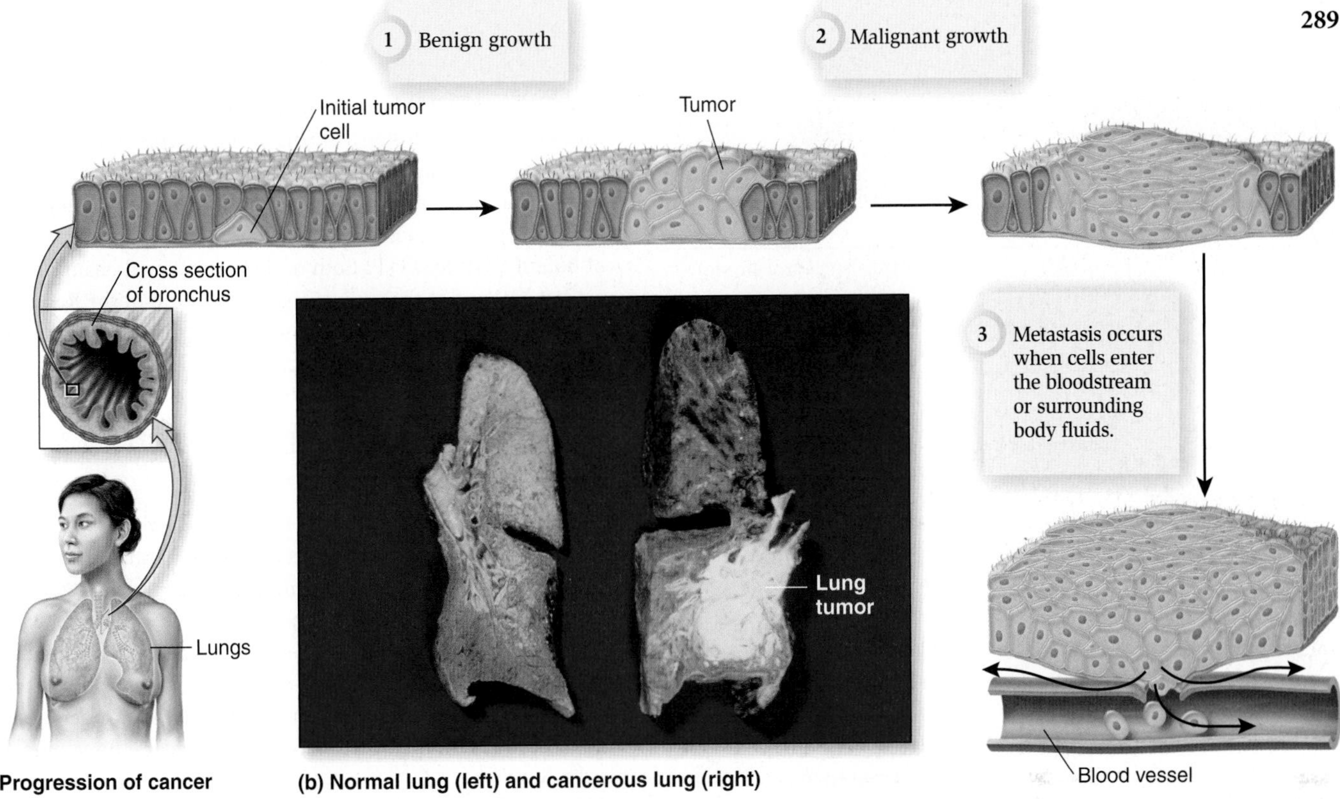
1 Benign growth
2 Malignant growth
Initial tumor cell
Tumor
Cross section of bronchus
3 Metastasis occurs when cells enter the bloodstream or surrounding body fluids.
Lung tumor
Lungs
Blood vessel
(a) Progression of cancer
(b) Normal lung (left) and cancerous lung (right)

Figure 14.10 **Cancer: Its progression and effects.** **(a)** In a healthy individual, an initial mutation converts a normal cell into a tumor cell. This cell divides to produce a benign tumor. Additional genetic changes in the tumor cells may occur, leading to a malignant tumor. At a later stage in malignancy, the tumor cells invade surrounding tissues, and some malignant cells may metastasize by traveling through the bloodstream to other parts of the body. **(b)** On the left of the photo is a human lung that was obtained from a healthy nonsmoker. The lung shown on the right has been ravaged by lung cancer. This lung was taken from a person who was a heavy smoker.

BioConnections: *Look back at Figure 10.17. Explain how a disruption in the arrangement of epithelial cells could promote metastasis.*

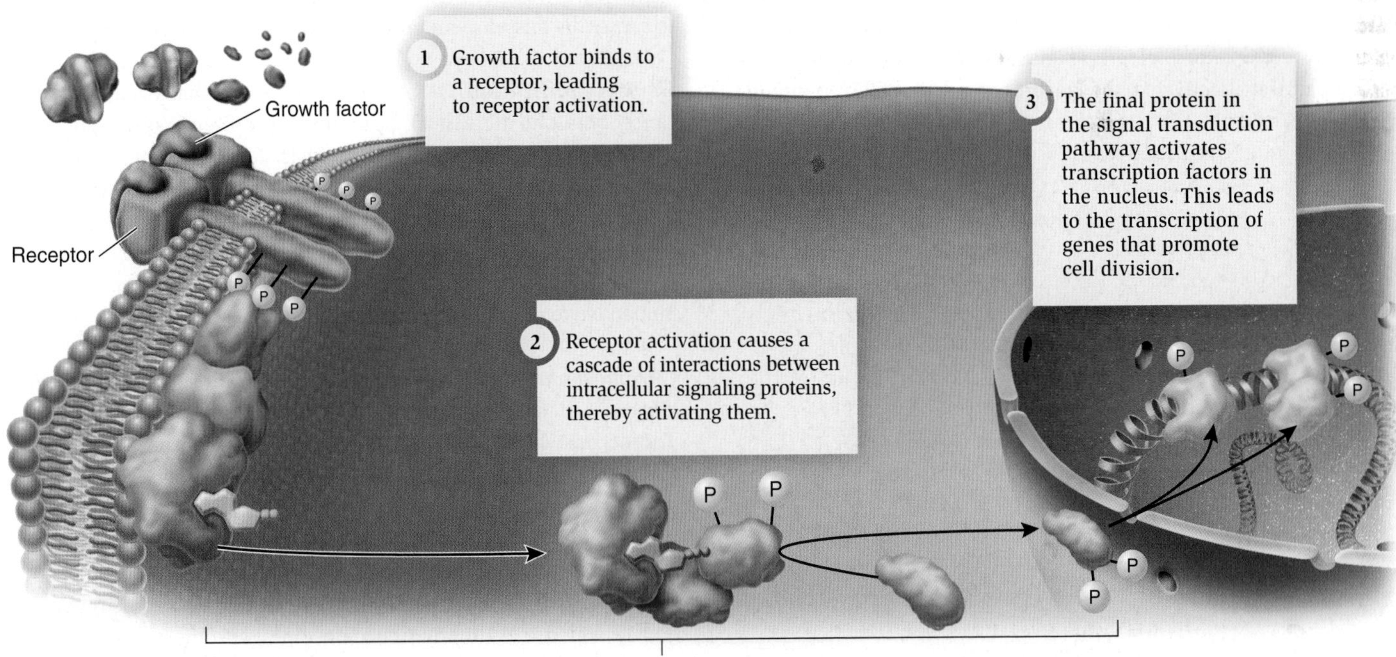
1 Growth factor binds to a receptor, leading to receptor activation.
Growth factor
Receptor
2 Receptor activation causes a cascade of interactions between intracellular signaling proteins, thereby activating them.
3 The final protein in the signal transduction pathway activates transcription factors in the nucleus. This leads to the transcription of genes that promote cell division.
P
Signal transduction pathway

Figure 14.11 **General features of a growth factor-signaling pathway that promotes cell division.** A detailed description of this pathway is found in Chapter 9 (Figure 9.10).

Concept Check: *How does the presence of a growth factor ultimately affect the function of the cell?*

BioConnections: *Look back at Figure 9.10. Could drugs that inhibit protein kinases be used to combat cancer? Explain.*

Table 14.6 Examples of Genes Encoding Proteins of Growth Factor Signaling Pathways That Can Become Oncogenes

Gene*	Cellular function
erbB	Growth factor receptor for EGF (epidermal growth factor)
ras	Intracellular signaling protein
raf	Intracellular signaling protein
src	Intracellular signaling protein
fos	Transcription factor
jun	Transcription factor

*The genes described in this table are found in humans as well as other vertebrate species. Most of the genes have been given three-letter names that are abbreviations for the type of cancer the oncogene causes or the type of virus in which the gene was first identified.

transcription of specific genes is activated in response to a growth factor. After they are made, the gene products promote cell division.

Eukaryotic species produce many different growth factors that play a role in cell division. Likewise, cells have several different types of signal transduction pathways, which are composed of proteins that respond to growth factors and promote cell division. Mutations in the genes that encode these signal transduction proteins can change them into oncogenes (**Table 14.6**).

How does an oncogene promote cancer? In some cases, an oncogene may keep a cell division-signaling pathway in a permanent "on" position. One way oncogenes keep cell division turned on is by producing a functionally overactive protein. As a specific example, let's consider how a mutation alters an intracellular signaling protein called Ras (refer back to Figure 9.10). The Ras protein is a GTPase that hydrolyzes GTP to GDP + P_i (**Figure 14.12**). When a signal transduction pathway is activated, the Ras protein releases GDP and binds GTP. When GTP is bound, the activated Ras protein promotes cell division. The Ras protein returns to its inactive state by hydrolyzing its bound GTP, and cell division is inhibited. Mutations that convert the normal *ras* gene into an oncogenic *ras* either decrease the ability of Ras protein to hydrolyze GTP or increase the rate of exchange of bound GDP for GTP. Both of these functional changes result in a greater amount of the active GTP-bound form of the Ras protein. In this way, these mutations keep the signaling pathway turned on when it should not be, resulting in uncontrolled cell division.

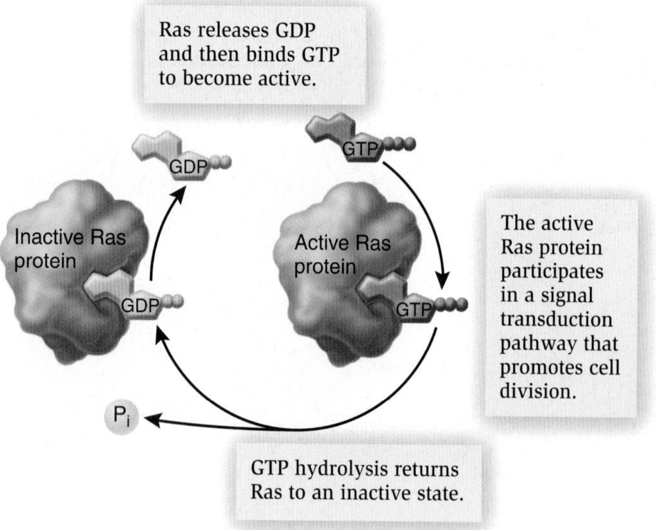

Ras releases GDP and then binds GTP to become active.
GDP
GTP
Inactive Ras protein
GDP
Active Ras protein
GTP
The active Ras protein participates in a signal transduction pathway that promotes cell division.
P_i
GTP hydrolysis returns Ras to an inactive state.

Figure 14.12 The function of Ras, a protein that is part of signal transduction pathways. When GTP is bound, the activated Ras protein promotes cell division. When GTP is hydrolyzed to GDP and P_i, Ras is inactivated, and cell division is inhibited.

Mutations in Proto-Oncogenes Convert Them to Oncogenes

Thus far, we have examined the functions of proteins that cause cancer when they become overactive, resulting in uncontrolled cell division. Let's now consider the common types of genetic changes that create such oncogenes. A **proto-oncogene** is a normal gene that, if mutated, can become an oncogene. Several types of genetic changes may convert a proto-oncogene into an oncogene. **Figure 14.13** describes four common types: missense mutations, gene amplifications, chromosomal translocations, and retroviral insertions.

Missense Mutation A missense mutation (Figure 14.13a), which changes a single amino acid in a protein, alters the function of the encoded protein in a way that promotes cancer. This type of mutation is responsible for the conversion of the *ras* gene into an oncogene. An example is a mutation in the *ras* gene that changes a specific glycine to a valine in the Ras protein. This mutation decreases the ability of the Ras protein to hydrolyze GTP, which promotes cell division (see Figure 14.12). Experimentally, chemical mutagens have been shown to cause this missense mutation, thereby leading to cancer.

Gene Amplification Another genetic event that occurs in some cancer cells is an increase in the number of copies of a proto-oncogene (Figure 14.13b). An abnormal increase in the number of genes results in too much of the encoded protein. Many human cancers are associated with the amplification of particular proto-oncogenes. In 1982, American molecular biologist Mark Groudine discovered that the *myc* gene, which encodes a transcription factor, was amplified in a human leukemia.

Chromosomal Translocation A third type of genetic alteration that can lead to cancer is a chromosomal translocation (Figure 14.13c). This occurs when one segment of a chromosome becomes attached to a different chromosome. In 1960, American pathologist Peter Nowell discovered that a form of leukemia called chronic myelogenous leukemia (CML)—a type of cancer involving blood cells—was correlated with the presence of a shortened version of a human chromosome. This shortened chromosome is the result of a chromosome translocation in which two different chromosomes, chromosomes 9 and 22, exchange pieces. This activates a proto-oncogene, *abl*, in an unusual way (**Figure 14.14**). In healthy individuals, the *bcr* gene and the *abl* gene are located on different chromosomes. In CML, these chromosomes break and rejoin in a way that causes the promoter and the first part of *bcr* to fuse with part of *abl*. This abnormal fusion event produces a **chimeric gene** composed of two gene fragments. This

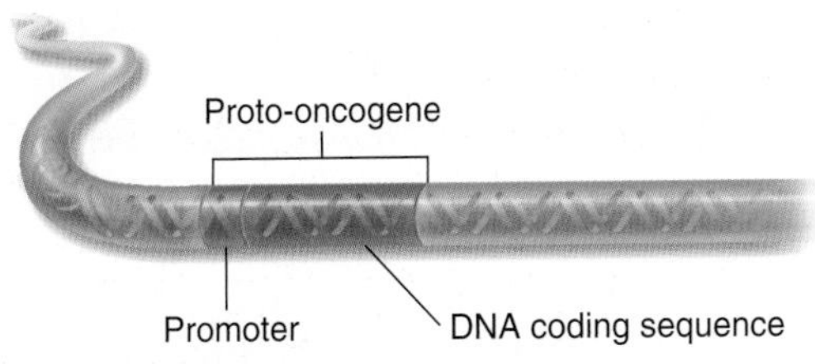

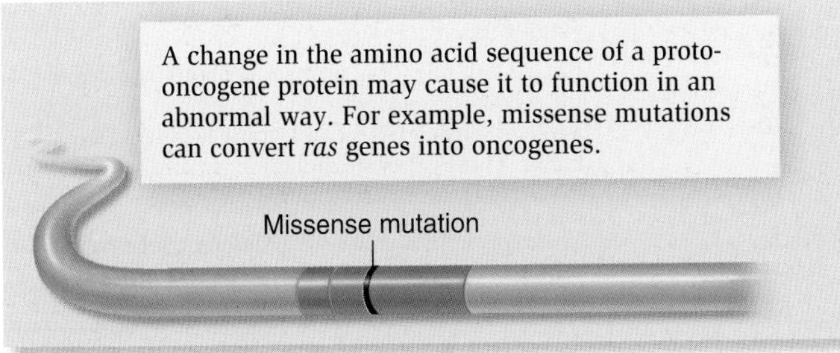

A change in the amino acid sequence of a proto-oncogene protein may cause it to function in an abnormal way. For example, missense mutations can convert *ras* genes into oncogenes.
Missense mutation

(a) Missense mutation

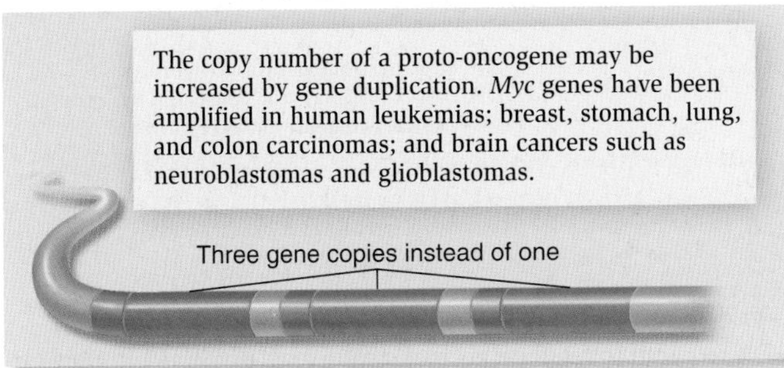

The copy number of a proto-oncogene may be increased by gene duplication. *Myc* genes have been amplified in human leukemias; breast, stomach, lung, and colon carcinomas; and brain cancers such as neuroblastomas and glioblastomas.
Three gene copies instead of one

(b) Gene amplification

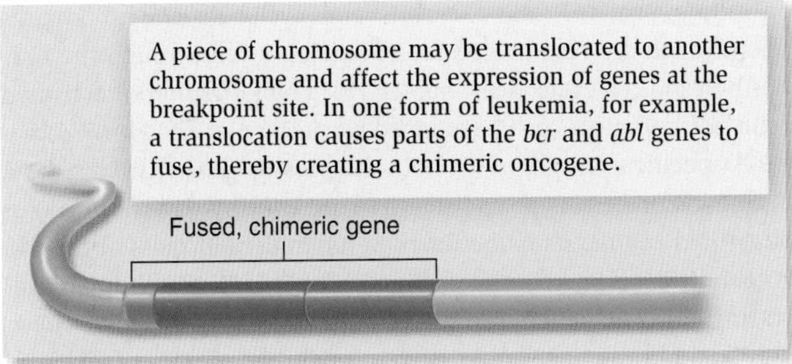

A piece of chromosome may be translocated to another chromosome and affect the expression of genes at the breakpoint site. In one form of leukemia, for example, a translocation causes parts of the *bcr* and *abl* genes to fuse, thereby creating a chimeric oncogene.
Fused, chimeric gene

(c) Chromosomal translocation

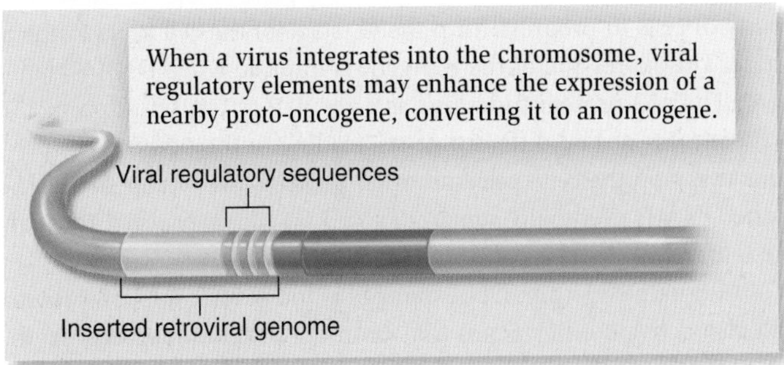

When a virus integrates into the chromosome, viral regulatory elements may enhance the expression of a nearby proto-oncogene, converting it to an oncogene.
Viral regulatory sequences
Inserted retroviral genome

(d) Retroviral insertion

Figure 14.13 **Genetic changes that convert proto-oncogenes to oncogenes.** In addition to these, other types of genetic changes may also convert proto-oncogenes into oncogenes.

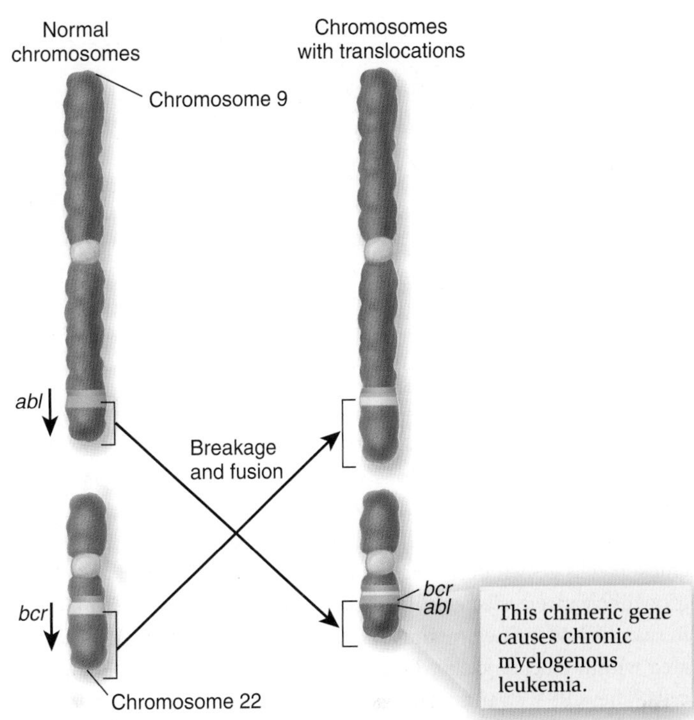

Figure 14.14 **The formation of a chimeric gene found in people with certain forms of leukemia.** The fusion of the *bcr* and *abl* genes creates a chimeric gene that encodes an abnormal fusion protein, leading to leukemia. The blue regions are the promoters for the *bcr* and *abl* genes.

Concept Check: The bcr gene is normally expressed in blood cells. Explain how this observation is related to the type of cancer that the translocation causes.

chimeric gene acts as an oncogene that encodes an abnormal fusion protein whose functional overactivity leads to leukemia.

Retroviral Insertion Certain types of viruses convert proto-oncogenes into oncogenes during the viral replication cycle (see Figure 14.13d). Retroviruses insert their DNA into the chromosomal DNA of the host cell. The viral genome contains promoter and regulatory elements that cause a high level of expression of viral genes. On occasion, the viral DNA may insert into a host chromosome in such a way that a viral promoter and regulatory elements are next to a proto-oncogene. This may result in the overexpression of the proto-oncogene, thereby promoting cancer. This is one way for a virus to cause cancer. Alternatively, a virus may cause cancer because it carries an oncogene in the viral genome. This phenomenon is described next.

Some Types of Cancer Are Caused by Viruses

The great majority of cancers are caused by mutagens that alter the structure and expression of genes that are found in somatic cells. A few viruses, however, are known to cause cancer in plants and animals, including humans (**Table 14.7**).

In 1911, the first cancer-causing virus to be discovered was isolated from chicken sarcomas by American pathologist Peyton Rous. A **sarcoma** is a tumor of connective tissue such as bone or cartilage.

Table 14.7 Examples of Viruses That Cause Cancer

Virus	Description
Rous sarcoma virus	Causes sarcomas in chickens
Simian sarcoma virus	Causes sarcomas in monkeys
Abelson leukemia virus	Causes leukemia in mice
Hardy-Zuckerman 4 feline sarcoma virus	Causes sarcomas in cats
Hepatitis B	Causes liver cancer in several species, including humans

The virus was named the Rous sarcoma virus (RSV). In the 1970s, research involving RSV led to the identification of a viral gene that acts as an oncogene. Researchers investigated RSV by using it to infect chicken cells grown in the laboratory. This causes the chicken cells to grow like cancer cells, continuously and in an uncontrolled manner. Researchers identified mutant RSV strains that infected and proliferated within chicken cells without transforming them into malignant cells. These RSV strains were missing a gene that is found in the form of the virus that does cause cancer. This gene was called the *src* gene because it causes sarcoma.

Americans, biologist Harold Varmus and microbiologist Michael Bishop, in collaboration with molecular biologist Peter Vogt, later discovered that normal (nonviral-infected) chicken cells also contain a copy of the *src* gene in their chromosomes. It is a proto-oncogene. When the *src* gene is incorporated into a viral genome, it is overexpressed because it is transcribed from a very active viral promoter. This ultimately produces too much of the Src protein in infected cells and promotes uncontrolled cell division.

Tumor-Suppressor Genes Prevent Mutation or Cell Proliferation

Thus far, we have examined one category of genes that promote cancer, namely oncogenes. We now turn our attention to the second category of genes, those called tumor-suppressor genes. The functioning of a normal (nonmutant) tumor-suppressor gene prevents cancerous growth. The proteins encoded by tumor-suppressor genes usually have one of two functions: maintenance of genome integrity or negative regulation of cell division (**Table 14.8**).

Maintenance of Genome Integrity Some tumor-suppressor genes encode proteins that maintain the integrity of the genome by monitoring and/or repairing alterations in the genome. The proteins encoded by these genes are vital for the prevention of abnormalities such as gene mutations, DNA breaks, and improperly segregated chromosomes. Therefore, when these proteins are functioning properly, they minimize the chance that a cancer-causing mutation will occur. In some cases, the proteins encoded by tumor-suppressor genes prevent a cell from progressing through the cell cycle if an abnormality is detected. These are termed **checkpoint proteins** because their role is to check the integrity of the genome and prevent a cell from progressing past a certain point in the cell cycle. Checkpoint proteins are not always required to regulate normal, healthy cell division, but they can stop cell division if an abnormality is detected.

Table 14.8 Functions of Selected Tumor-Suppressor Genes

Gene	
	Maintenance of genome integrity
p53	p53 is a transcription factor that acts as a sensor of DNA damage. It can promote DNA repair, prevent the progression through the cell cycle, and promote apoptosis.
BRCA-1 *BRCA-2*	BRCA-1 and BRCA-2 proteins are both involved in the cellular defense against DNA damage. They play a role in sensing DNA damage and facilitate DNA repair. These genes are mutant in persons with certain inherited forms of breast cancer.
XPD	This represents one of several different genes whose products function in DNA repair. These genes are defective in patients with xeroderma pigmentosum.
	Negative regulation of cell division
Rb	The Rb protein is a negative regulator that represses the transcription of genes required for DNA replication and cell division.
NF1	The NF1 protein stimulates Ras to hydrolyze its GTP to GDP. Loss of NF1 function causes the Ras protein to be overactive, which promotes cell division.
p16	The p16 protein is a negative regulator of cyclin-dependent kinases (cdks).

How do checkpoint proteins stop the cell cycle? One way is by controlling proteins called cyclins and cyclin-dependent kinases (cdks), which are responsible for advancing a cell through the four phases of the cell cycle (see Chapter 15). The formation of activated cyclin/cdk complexes can be stopped by checkpoint proteins.

A specific example of a tumor-suppressor gene that encodes a checkpoint protein is *p53*, discovered in 1979 by American biologist Arnold Levine. Its name refers to the molecular mass of the p53 protein, which is 53 kDa (kilodaltons). About 50% of all human cancers, including malignant tumors of the lung, breast, esophagus, liver, bladder, and brain, as well as leukemias and lymphomas (cancer of the lymphatic system), are associated with mutations in this gene.

As shown in **Figure 14.15**, p53 is a protein that controls the ability of cells to progress from the G_1 stage of the cell cycle to the S phase. The expression of the *p53* gene is induced when DNA is damaged. The p53 protein functions as a regulatory transcription factor that activates several different genes, leading to the synthesis of proteins that stop the cell cycle and other proteins that repair the DNA. If the DNA is eventually repaired, a cell may later proceed through the cell cycle.

Alternatively, if the DNA damage is too severe, the p53 protein will also activate other genes that promote programmed cell death. This process, called **apoptosis**, involves cell shrinkage and DNA degradation. As described in Chapter 9, enzymes known as caspases are activated during apoptosis (refer back to Figure 9.20). They function as proteases that are sometimes called the "executioners" of the cell. Caspases digest selected cellular proteins such as microfilaments,

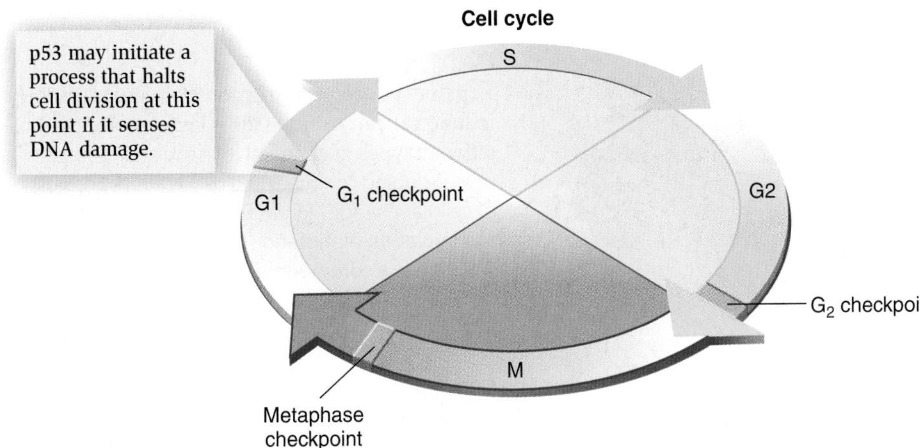
Cell cycle
S
p53 may initiate a process that halts cell division at this point if it senses DNA damage.
G1
G_1 checkpoint
G2
G_2 checkpoint
M
Metaphase checkpoint

Figure 14.15 The cell cycle and checkpoints. As discussed in Chapter 15, eukaryotic cells progress through a cell cycle composed of G_1, S, G_2, and M phases (look ahead to Figure 15.2). The yellow bars indicate common checkpoints that stop the cell cycle if genetic abnormalities are detected. The p53 protein stops a cell at the G_1 checkpoint if it senses DNA damage.

Concept Check: Why is it an advantage for an organism to have checkpoints that can stop the cell cycle?

which are components of the cytoskeleton. This causes the cell to break into small vesicles that are eventually phagocytized by cells of the immune system. It is beneficial for a multicellular organism to kill an occasional cell with cancer-causing potential.

When checkpoint genes such as *p53* are rendered inactive by mutation, the division of normal healthy cells may not be adversely affected. For example, mice that are missing the *p53* gene are born healthy. This indicates that checkpoint proteins such as p53 are not necessary for normal cell growth and division. However, these mice are very sensitive to mutagens such as UV light and easily develop cancer. The loss of p53 function makes it more likely that undesirable genetic changes will occur that could cause cancerous growth.

Negative Regulation of Cell Division A second category of tumor-suppressor genes encodes proteins that are negative regulators or inhibitors of cell division. Their function is necessary to properly halt cell division. If their function is lost, cell division is abnormally accelerated.

An example of such a tumor-suppressor gene is the *Rb* gene. It was the first tumor-suppressor gene to be identified in humans by studying patients with a disease called retinoblastoma, a cancerous tumor that occurs in the retina of the eye. Some people have an inherited predisposition to develop this disease within the first few years of life. By comparison, the noninherited form of retinoblastoma, which is caused by environmental agents, is more likely to occur later in life.

Based on these differences, in 1971, American geneticist Alfred Knudson proposed a two-hit hypothesis for the occurrence of retinoblastoma. According to this hypothesis, retinoblastoma requires two mutations to occur. People have two copies of the *Rb* gene, one from each parent. Individuals with the inherited form of the disease already have received one mutant gene from one of their parents. They need only one additional mutation to develop the disease. Because the retina has more than 1 million cells, it is relatively likely that a mutation may occur in one of these cells at an early age, leading to the disease. However, people with the noninherited form of the disease must have two *Rb* mutations in the same retinal cell to cause the disease. Because two mutations are less likely than a single mutation, the noninherited form of this disease is expected to occur much later in life, and only rarely. Since Knudson's original work, molecular studies have confirmed the two-hit hypothesis for retinoblastoma.

The Rb protein negatively controls a regulatory transcription factor called E2F that activates genes required for cell cycle progression from G_1 to S phase. The binding of the Rb protein to E2F inhibits its activity and prevents cell division (**Figure 14.16**). When a normal cell is supposed to divide, cyclins bind to cyclin-dependent kinases (cdks). This binding activates the kinases, which catalyze the transfer of a phosphate to the Rb protein. The phosphorylated form of the Rb protein is released from E2F, thereby allowing E2F to activate genes

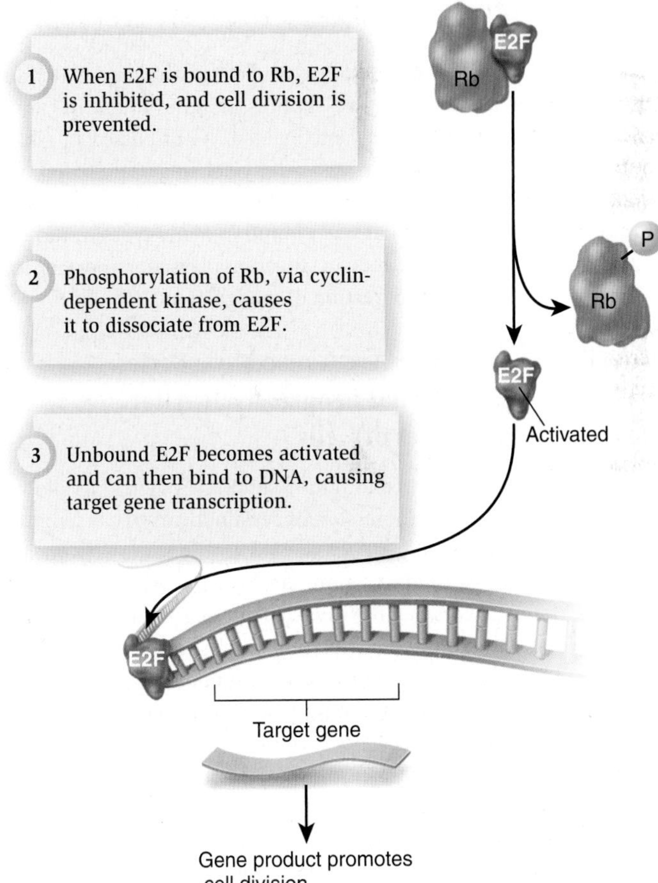
1 When E2F is bound to Rb, E2F is inhibited, and cell division is prevented.
E2F
Rb
2 Phosphorylation of Rb, via cyclin-dependent kinase, causes it to dissociate from E2F.
P
Rb
E2F
Activated
3 Unbound E2F becomes activated and can then bind to DNA, causing target gene transcription.
E2F
Target gene
Gene product promotes cell division.

Figure 14.16 Function of the Rb protein. The Rb protein inhibits the function of E2F, which turns on genes that cause a cell to divide. When cells are supposed to divide, Rb is phosphorylated by cyclin-dependent protein kinase, which allows E2F to function. If Rb protein is inactivated due to a mutation, E2F will always be active, and the cell will be stimulated to divide uncontrollably.

Concept Check: Would cancer occur if both copies of the Rb gene and both copies of the E2F gene were rendered inactive due to mutations?

needed to progress through the cell cycle. When both copies of Rb are defective due to mutations, the E2F protein is always active. This explains why uncontrolled cell division occurs in retinoblastoma.

Gene Mutations, Chromosome Loss, and DNA Methylation Can Inhibit the Expression of Tumor-Suppressor Genes

Cancer biologists want to understand how tumor-suppressor genes are inactivated, because this knowledge may ultimately help them to prevent or combat cancer. How are tumor-suppressor genes silenced? The function of tumor-suppressor genes is lost in three common ways. First, a mutation can occur within a tumor-suppressor gene to inactivate its function. For example, a mutation could abolish the function of the promoter for a tumor-suppressor gene or introduce an early stop codon in its coding sequence. Either of these would prevent the expression of a functional protein.

Chromosome loss is a second way that the function of a tumor-suppressor gene is lost. Chromosome loss may contribute to the progression of cancer if the missing chromosome carries one or more tumor-suppressor genes.

Recently, researchers have discovered a third way that these genes may be inactivated. Tumor-suppressor genes found in cancer cells are sometimes abnormally methylated. As discussed in Chapter 13, transcription is inhibited when CpG islands near a promoter region are methylated. Such DNA methylation near the promoters of tumor-suppressor genes has been found in many types of tumors, suggesting that this form of gene inactivation plays an important role in the formation and/or progression of malignancy.

Most Forms of Cancer Are Caused by a Series of Genetic Changes That Progressively Alter the Growth Properties of Cells

The discovery of oncogenes and tumor-suppressor genes has allowed researchers to study the progression of certain forms of cancer at the molecular level. Multiple genetic changes to the same cell lineage, perhaps in the range of 10 or more, are usually needed for cancer to occur. Many cancers begin with a benign genetic alteration that, over time and with additional mutations, leads to malignancy. Furthermore, a malignancy can continue to accumulate genetic changes that make it even more difficult to treat because the cells divide faster or invade surrounding tissues more readily.

As an example, let consider lung cancer, which is diagnosed in approximately 170,000 men and women each year in the United States. More than 1.2 million cases are diagnosed worldwide. Nearly 90% of these cases are caused by tobacco smoking and are thus preventable. Unlike other cancers for which early diagnosis is possible, lung cancer is usually detected only after it has become advanced and is difficult if not impossible to cure. The 5-year survival rate for lung cancer patients is approximately 15%.

What is the cellular basis for lung cancer? Most cancers of the lung are **carcinomas**—cancers of epithelial cells (Figure 14.17). (Epithelial cells, which form the lining of all internal and external body surfaces, are described in Chapter 10.) The top images in this figure show the normal epithelium found in a healthy lung. The rest of the figure shows the progression of a carcinoma that is due to mutations

Figure 14.17 Progression of changes leading to lung cancer. Lung tissue is largely composed of different types of connective tissue and epithelial cells, including columnar and basal cells. A progression of cellular changes in basal cells, caused by the accumulation of mutations, leads to basal cell carcinoma, a common type of lung cancer.

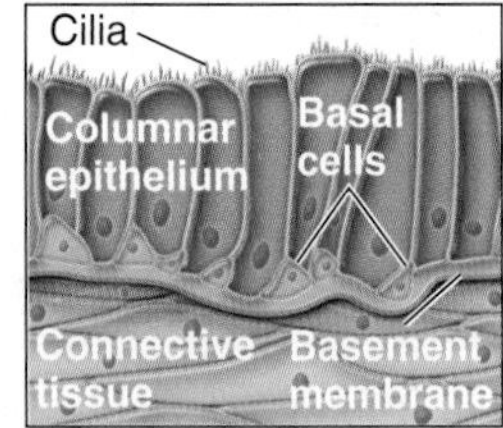

Normal lung epithelium

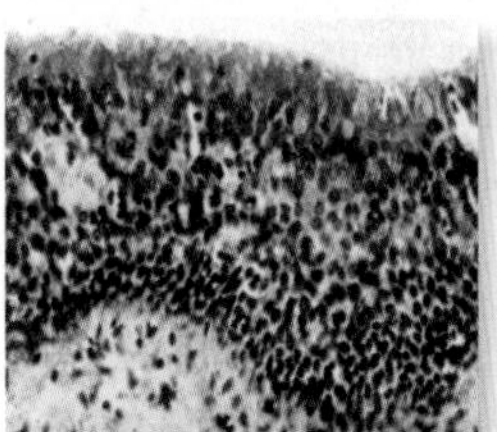
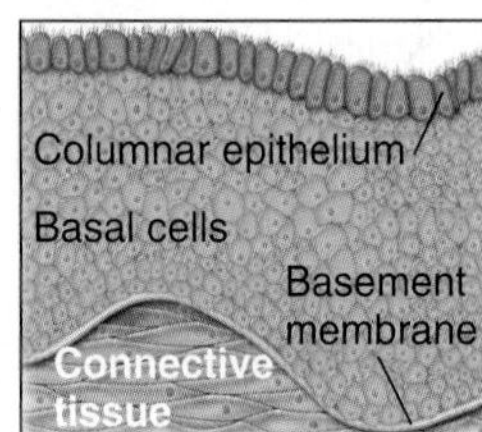

Hyperplasia

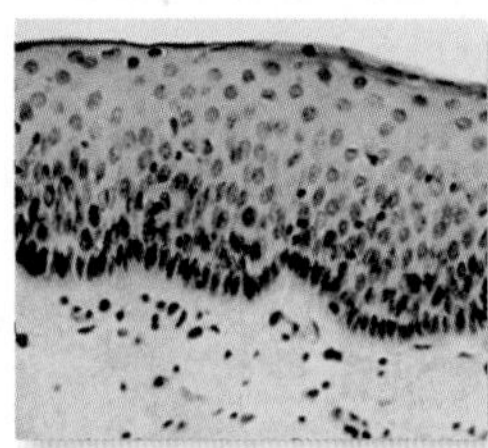
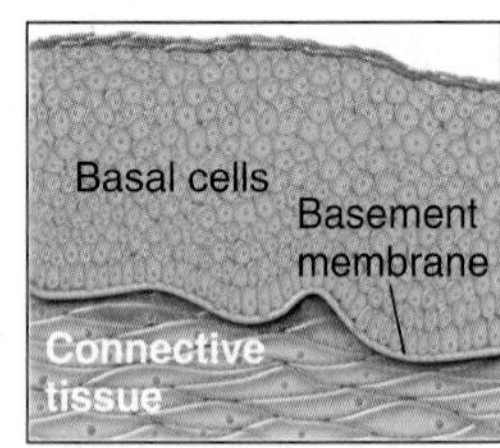

Loss of ciliated cells

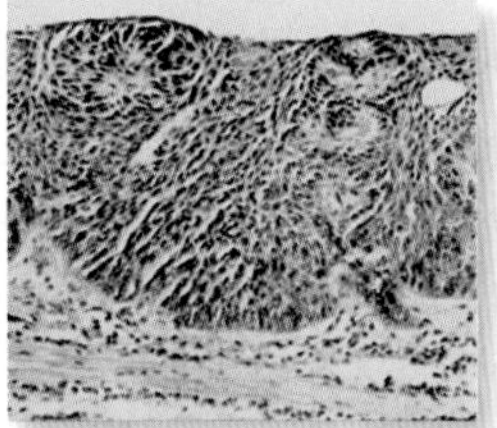
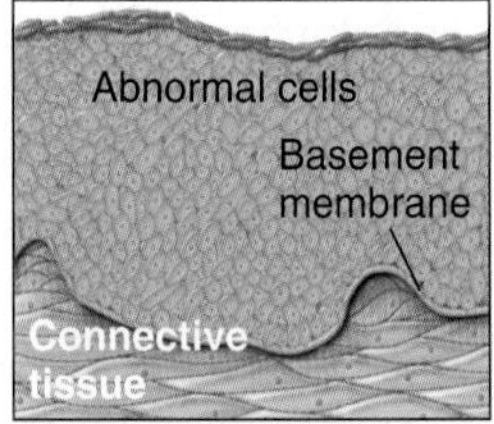

Dysplasia (initially precancerous, then cancerous)

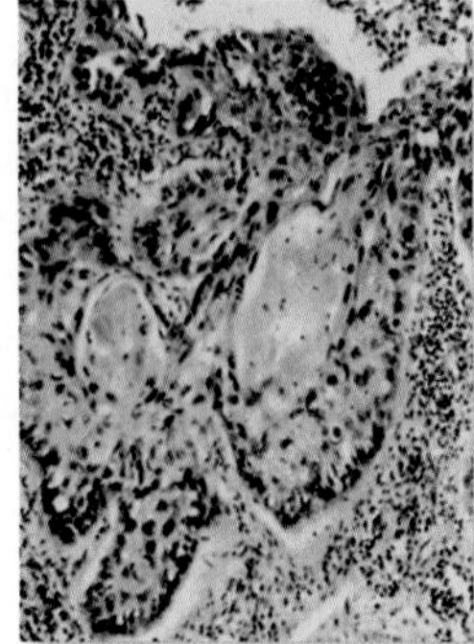
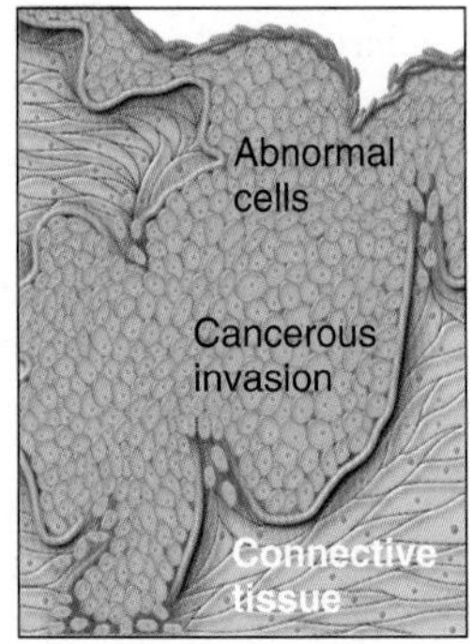

Invasive cancerous cells that can metastasize

in basal cells, a type of epithelial cell. Keep in mind that cancer occurs due to the accumulation of mutations in a cell lineage, beginning with an initial mutant cell that then divides multiple times to produce a population of many daughter cells (refer back to Figure 14.10). As mutations accumulate in a lineage of basal cells, their numbers increase dramatically. This causes a thickening of the epithelium, a condition called hyperplasia. The proliferation of such basal cells causes the loss of the ciliated, columnar epithelial cells that normally line the airways and help remove mucus and its trapped particles from the lungs. As additional mutations accumulate in this cell lineage, the basal cells develop more abnormal morphologies, a condition known as dysplasia. In the early stages of dysplasia, the abnormal basal cells are precancerous. If the source of chronic irritation (usually cigarette smoke) is eliminated, the abnormal cells are likely to disappear. Alternatively, if smoking continues, these abnormal cells may accumulate additional genetic changes and lose the ability to stop dividing. Such cells have become cancerous—the person has basal cell carcinoma.

The basement membrane is a sheetlike layer of extracellular matrix that provides a barrier between the lung cells and the bloodstream. If the cancer cells have not yet penetrated the basement membrane, they will not have metastasized, that is, spread into the blood and to other parts of the body. If the entire tumor is removed at this stage, the patient should be cured. The lower images in Figure 14.17 show a tumor that has broken through the basement membrane. The metastasis of these cells to other parts of the body will likely kill the patient, usually within a year of diagnosis.

The cellular changes that lead to lung cancer are correlated with genetic changes. These include the occurrence of mutations that create oncogenes and inhibit tumor-suppressor genes. The order of mutations is not absolute. It takes time for multiple changes to accumulate, so cancer is usually a disease of older people. Reducing your exposure to mutagens such as cigarette smoke throughout your lifetime helps minimize the risk of mutations to your genes that could promote cancer.

GENOMES & PROTEOMES CONNECTION

Mutations in Approximately 300 Human Genes May Promote Cancer

Researchers have identified a large number of genes that are mutated in cancer cells. Though not all of these mutant genes have been directly shown to affect the growth rate of cells, such mutations are likely to be found in tumors because they provide some type of growth advantage for the cell population from which the cancer developed. For example, certain mutations may affect the functions of proteins that enable cells to metastasize to neighboring locations. These mutations may not affect growth rate, but they provide the growth advantage that cancer cells are not limited to growing in a particular location. They can migrate to new locations.

How many genes can contribute to cancer when they become mutant? Researchers have estimated that about 300 different genes may play a role in the development of human cancer. With an approximate genome size of 20,000 to 25,000 genes, this observation indicates that over 1% of our genes have the potential to promote cancer if their function is altered by a mutation.

Summary of Key Concepts

14.1 Mutation

- A mutation is a heritable change in the genetic material; gene mutations are relatively small changes in the base sequence of a gene.
- Point mutations affect a single base pair and can alter the coding sequence of genes in several ways, including silent, missense, nonsense, and frameshift mutations (Table 14.1).
- Sickle cell disease is caused by a missense mutation that results in a single amino acid substitution in β-globin (Figure 14.1).
- Gene mutations also alter gene function by changing DNA sequences that are not within the coding region (Table 14.2).
- The Lederbergs used replica plating to show that mutations are random events (Figure 14.2).
- Germ-line mutations affect gametes, whereas mutations in somatic cells affect only a part of the body and cannot be passed to offspring (Figures 14.3, 14.4).
- Spontaneous mutations are the result of errors in natural biological processes. Induced mutations are caused by agents in the environment that alter DNA structure (Table 14.3).
- Mutagens are chemical or physical agents that lead to mutations in DNA (Table 14.4, Figures 14.5, 14.6).
- The Ames test is a method of testing whether an agent is a mutagen (Figure 14.7).

14.2 DNA Repair

- DNA repair systems involve proteins that sense DNA damage and repair it before a mutation occurs (Table 14.5).
- In nucleotide excision repair (NER) certain proteins recognize various types of DNA damage, such as thymine dimers. A region in the damaged strand is excised, and a new strand is synthesized, using the intact strand as a template (Figure 14.8).
- Certain inherited diseases in humans, such as xeroderma pigmentosum, are due to defects in the NER system (Figure 14.9).

14.3 Cancer

- Cancer is due to the accumulation of mutations in a lineage of cells that leads to uncontrolled cell growth (Figure 14.10).
- Mutations in proto-oncogenes that result in overactivity produce cancer-causing genes called oncogenes.
- Oncogenes often encode proteins involved in cell-signaling pathways that promote cell division (Figures 14.11, 14.12, Table 14.6).
- Four common types of genetic changes, namely, missense mutations, gene amplifications, chromosomal translocations, and retroviral insertions, can change proto-oncogenes into oncogenes (Figures 14.13, 14.14).
- Some types of cancer are caused by viruses (Table 14.7).
- The normal function of tumor-suppressor genes is to prevent cancer. Loss-of-function mutations in such genes promote cancer.

Tumor-suppressor genes often encode proteins that maintain the integrity of the genome or function as negative regulators of cell division (Table 14.8).

- Checkpoint proteins such as p53 monitor the integrity of the genome and prevent the cell from progressing through the cell cycle if abnormalities are detected (Figure 14.15).
- The Rb protein is an inhibitor of the cell cycle, because it negatively controls E2F, a transcription factor that promotes cell division (Figure 14.16).
- Tumor-suppressor genes can be inactivated by gene mutations, chromosome loss, and DNA methylation.
- Most forms of cancer, such as lung cancer, involve multiple genetic changes that lead to malignancy (Figure 14.17).
- Over 300 human genes, or over 1% of our genes, are known to be associated with cancer when they become mutant.

Assess and Discuss

Test Yourself

1. Point mutations that do not alter the amino acid sequence of the resulting gene product are called ___ mutations.
   a. frameshift
   b. natural
   c. silent
   d. nonsense
   e. missense
2. Some point mutations lead to an mRNA that produces a shorter polypeptide. This type of mutation is known as a ___ mutation.
   a. neutral
   b. silent
   c. missense
   d. nonsense
   e. chromosomal
3. A mutation in which of the following regions is least likely to affect gene function?
   a. promoter
   b. coding region
   c. splice junction
   d. intergenic region
   e. regulatory site
4. Mutagens can cause mutations by
   a. chemically altering DNA nucleotides.
   b. disrupting DNA replication.
   c. altering the genetic code of an organism.
   d. all of the above.
   e. a and b only.
5. The mutagenic effect of UV light is
   a. the alteration of cytosine bases to adenine bases.
   b. the formation of adenine dimers that interfere with genetic expression.
   c. the breaking of the sugar-phosphate backbone of the DNA molecule.
   d. the formation of thymine dimers that disrupt DNA replication.
   e. the deletion of thymine bases along the DNA molecule.
6. The Ames test
   a. provides a way to determine if any type of cell has experienced a mutation.
   b. provides a way to determine if an agent is a mutagen.
   c. allows researchers to experimentally disrupt gene activity by causing a mutation in a specific gene.
   d. provides a way to repair mutations in bacterial cells.
   e. does all of the above.
7. Xeroderma pigmentosum
   a. is a genetic disorder that results in uncontrolled cell growth.
   b. is a genetic disorder in which the NER system is not fully functional.
   c. is a genetic disorder that results in the loss of pigment in certain patches of skin.
   d. results from the lack of DNA polymerase proofreading.
   e. is both b and d.
8. If a mutation eliminated the function of UvrC, which aspect of nucleotide excision repair would not work?
   a. sensing a damaged DNA site
   b. endonuclease cleavage of the damaged strand
   c. removal of the damaged strand
   d. synthesis of a new strand, using the undamaged strand as a template
   e. none of the above
9. Cancer cells are said to be metastatic when they
   a. begin to divide uncontrollably.
   b. invade healthy tissue.
   c. migrate to other parts of the body.
   d. cause mutations in other healthy cells.
   e. do all of the above.
10. Oncogenes can be produced by
    a. missense mutations.
    b. gene amplification.
    c. chromosomal translocation.
    d. retroviral insertion.
    e. all of the above.

Conceptual Questions

1. Is a random mutation more likely to be beneficial or harmful? Explain your answer.
2. Distinguish between spontaneous and induced mutations. Which are more harmful? Which are avoidable?
3. A principle of biology is that the *genetic material provides a blueprint for reproduction*. Explain how mutations may cause alterations to the genetic material that are detrimental for reproduction and sustaining life.

Collaborative Questions

1. Discuss the pros and cons of mutation.
2. A large amount of research is aimed at studying mutation. However, there is not an infinite amount of research dollars. Where would you put your money for mutation research?
   a. testing of potential mutagens
   b. investigating molecular effects of mutagens
   c. investigating DNA repair mechanisms
   d. some other area

Online Resource

www.brookerbiology.com

Stay a step ahead in your studies with animations that bring concepts to life and practice tests to assess your understanding. Your instructor may also recommend the interactive eBook, individualized learning tools, and more.

Chapter Outline

15.1 The Eukaryotic Cell Cycle
15.2 Mitotic Cell Division
15.3 Meiosis and Sexual Reproduction
15.4 Variation in Chromosome Structure and Number
Summary of Key Concepts
Assess and Discuss

The Eukaryotic Cell Cycle, Mitosis, and Meiosis

15

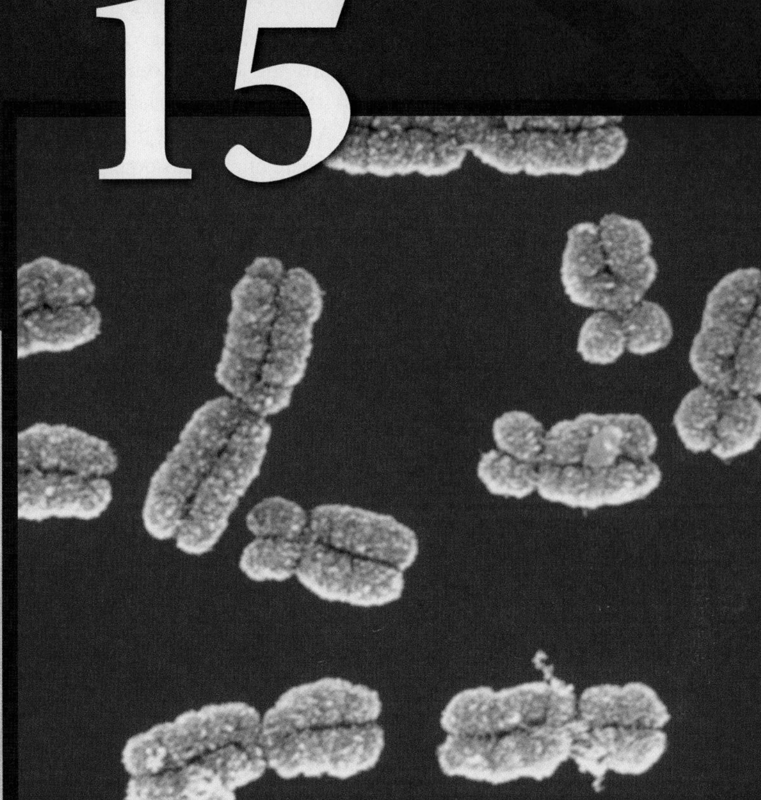

A scanning electron micrograph of human chromosomes. These chromosomes are highly compacted and found in a dividing cell.

Over 10,000,000,000,000! Researchers estimate the adult human body contains somewhere between 10 trillion to 50 trillion cells. It is almost an incomprehensible number. Even more amazing is the accuracy of the process that produces these cells. After a human sperm and egg unite, the fertilized egg goes through a long series of cell divisions to produce an adult with over 10 trillion cells. Let's suppose you randomly removed a cell from your arm and compared it with a cell from your foot. If you examined the chromosomes found in both cells under the microscope, they would look identical. Likewise, the DNA sequences along those chromosomes would also be the same, barring rare mutations. Similar comparisons could be made among the trillions of cells in your body. When you consider how many cell divisions are needed to produce an adult human, the precision of cell division is truly remarkable.

What accounts for this high level of accuracy? As we will examine in this chapter, **cell division**, the reproduction of cells, is a highly regulated process that distributes and monitors the integrity of the genetic material. The eukaryotic cell cycle is a series of phases needed for cell division. The cells of eukaryotic species follow one of two different sorting processes so that new daughter cells receive the correct number and types of chromosomes. The first sorting process we will explore, called mitosis, is needed so two daughter cells receive the same amount of genetic material as the mother cell that produced them. We will then examine another sorting process, called meiosis, which is needed for sexual reproduction. In meiosis, cells that have two sets of chromosomes produce daughter cells with a single set of chromosomes. Lastly, we will explore variation in the structure and number of chromosomes. As you will see, a variety of mechanisms that alter chromosome structure and number have important consequences for the organisms that carry them.

15.1 The Eukaryotic Cell Cycle

Learning Outcomes:

1. Describe the features of chromosomes and how sets of chromosomes are examined microscopically.
2. Outline the phases of the eukaryotic cell cycle.
3. Explain how cyclins and cdks work together to advance a cell through the eukaryotic cell cycle.

Life is a continuum in which new living cells are formed by the division of pre-existing cells. The Latin axiom *omnis cellula e cellula*, meaning "Every cell originates from another cell," was first proposed in 1858 by Rudolf Virchow, a German biologist. From an evolutionary perspective, cell division has a very ancient origin. All living organisms, from unicellular bacteria to multicellular plants and animals, have been produced by a series of repeated rounds of cell growth and division extending back to the beginnings of life nearly 4 billion years ago.

A **cell cycle** is a series of events that leads to cell division. In all species, it is a highly regulated process so cell division occurs at the appropriate time. As discussed in Chapter 18, bacterial cells produce more cells via binary fission. The cell cycle in eukaryotes is more complex, in part, because eukaryotic cells have sets of chromosomes that need to be sorted properly. In this section, we will examine the phases of the eukaryotic cell cycle and see how the cell cycle is controlled by proteins that carefully monitor the division process to ensure its accuracy. But first, we need to consider some general features of chromosomes in eukaryotic species.

Chromosomes Are Inherited in Sets and Occur in Homologous Pairs

To understand the chromosomal composition of cells and the behavior of chromosomes during cell division, scientists observe cells and chromosomes with the use of microscopes. **Cytogenetics** is the field of genetics that involves the microscopic examination of chromosomes. When a cell prepares to divide, the chromosomes become more tightly compacted, a process that decreases their apparent length and increases their diameter. A consequence of this compaction is that distinctive shapes and numbers of chromosomes become visible under a light microscope.

Microscopic Examination of Chromosomes Figure 15.1 shows the general procedure for preparing and viewing chromosomes from a eukaryotic cell. In this example, the cells were obtained from a sample of human blood. Specifically, the chromosomes within leukocytes (white blood cells) were examined. A sample of the blood cells was treated with drugs that stimulated them to divide. The actively dividing cells were centrifuged to concentrate them into a pellet, which was then mixed with a hypotonic solution that caused the cells to swell. The expansion of the cells causes the chromosomes to spread out from each other, making it easier to see each individual chromosome.

Next, the cells were concentrated by a second centrifugation and treated with a fixative, which chemically fixes them in place so the chromosomes can no longer move around. The cells were then exposed to a chemical dye, such as Giemsa, that binds to the chromosomes and stains them. This gives chromosomes a distinctive banding pattern that greatly enhances their contrast and ability to be uniquely identified; in this case, the bands are called G bands. The cells were then placed on a slide and viewed with a light microscope. In a cytogenetics laboratory, the microscopes are equipped with an electronic camera to photograph the chromosomes. On a computer screen, the chromosomes can be organized in a standard way, usually from largest to smallest. A photographic representation of the chromosomes, as in the photo in step 5 of Figure 15.1, is called a **karyotype**. A

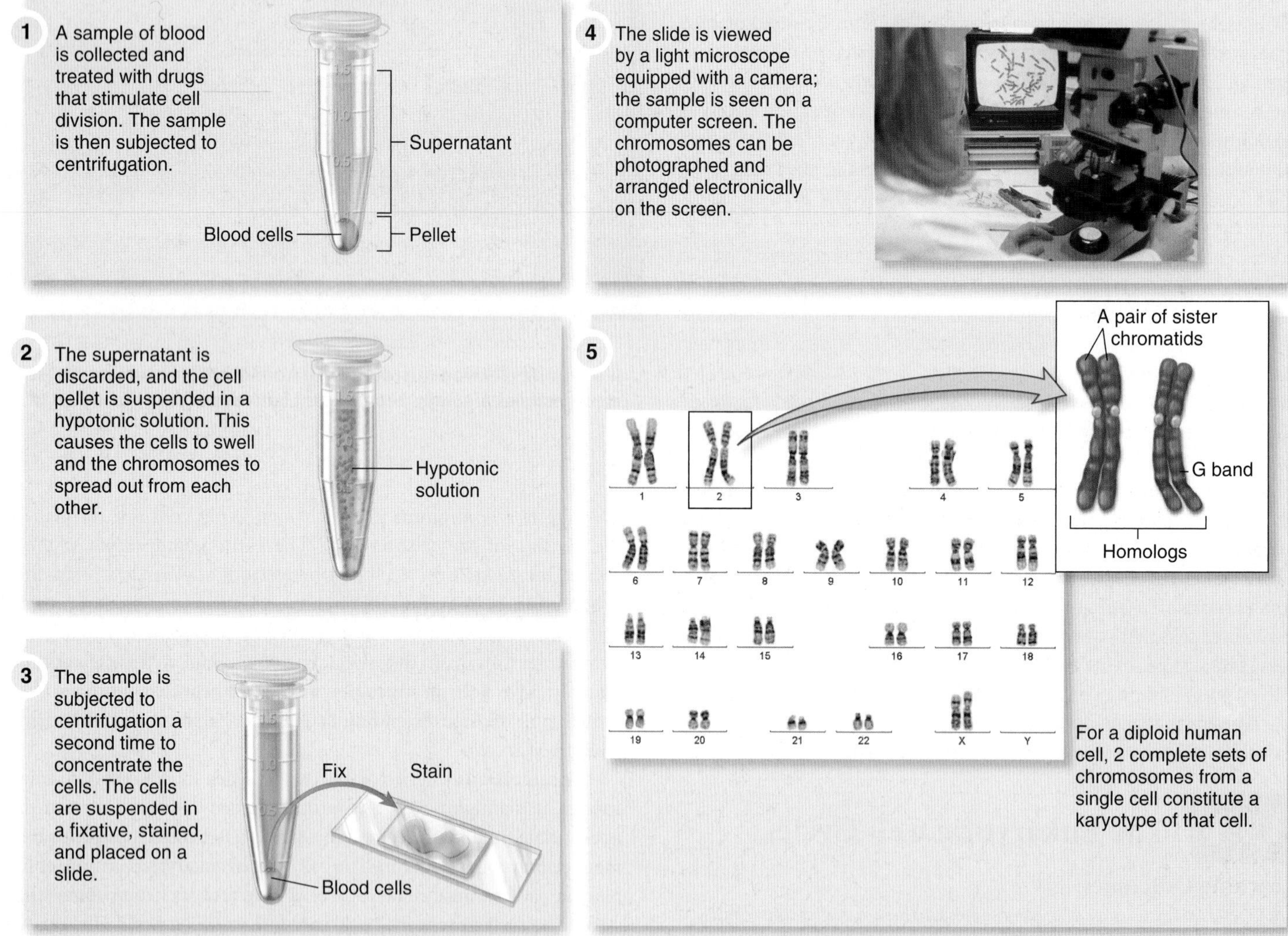
1 A sample of blood is collected and treated with drugs that stimulate cell division. The sample is then subjected to centrifugation.
Supernatant
Blood cells
Pellet
2 The supernatant is discarded, and the cell pellet is suspended in a hypotonic solution. This causes the cells to swell and the chromosomes to spread out from each other.
Hypotonic solution
3 The sample is subjected to centrifugation a second time to concentrate the cells. The cells are suspended in a fixative, stained, and placed on a slide.
Fix
Stain
Blood cells
4 The slide is viewed by a light microscope equipped with a camera; the sample is seen on a computer screen. The chromosomes can be photographed and arranged electronically on the screen.
5
A pair of sister chromatids
G band
Homologs
1
2
3
4
5
6
7
8
9
10
11
12
13
14
15
16
17
18
19
20
21
22
X
Y
For a diploid human cell, 2 complete sets of chromosomes from a single cell constitute a karyotype of that cell.

Figure 15.1 **The procedure for making a karyotype.** In this example, the chromosomes were treated with the Giemsa stain, and the resulting bands are called G bands.

Concept Check: *Researchers usually treat cells with drugs that stimulate them to divide prior to the procedure for making a karyotype. Why is this useful?*

karyotype reveals the number, size, and form of chromosomes found within an actively dividing cell. It should also be noted that the chromosomes viewed in actively dividing cells have already replicated. The two copies are still joined to each other and referred to as a pair of **sister chromatids** (see inset to Figure 15.1).

Sets of Chromosomes What type of information is learned from a karyotype? By studying the karyotypes of many species, scientists have discovered that eukaryotic chromosomes occur in sets. Each set is composed of several different types of chromosomes. For example, one set of human chromosomes contains 23 different types of chromosomes (see Figure 15.1). By convention, the chromosomes are numbered according to size, with the largest chromosomes having the smallest numbers. For example, human chromosomes 1, 2, and 3 are relatively large, whereas 21 and 22 are the two smallest. This numbering system does not apply to the **sex chromosomes**, which determine the sex of the individual. Sex chromosomes in humans are designated with the letters X and Y; females are XX and males are XY. The chromosomes that are not sex chromosomes are called **autosomes**. Humans have 22 different types of autosomes.

A second feature of many eukaryotic species is that most cells contain two sets of chromosomes. The karyotype shown in Figure 15.1 contains two sets of chromosomes, with 23 different chromosomes in each set. Therefore, this human cell contains a total of 46 chromosomes. Each cell has two sets because the individual inherited one set from the father and one set from the mother. When the cells of an organism carry two sets of chromosomes, that organism is said to be **diploid**. Geneticists use the letter n to represent a set of chromosomes. Diploid organisms are referred to as $2n$, because they have two sets of chromosomes. For example, humans are $2n$, where $n = 23$. Most human cells are diploid. An exception involves **gametes**, the sperm and egg cells. Gametes are **haploid**, or $1n$, which means they contain one set of chromosomes.

Homologous Pairs of Chromosomes When an organism is diploid, the members of a pair of chromosomes are called **homologs** (see inset to Figure 15.1). The term **homology** refers to any similarity that is due to common ancestry. Pairs of homologous chromosomes are evolutionarily derived from the same chromosome. However, homologous chromosomes are not usually identical because over many generations they have accumulated some genetic changes that make them distinct.

How similar are homologous chromosomes to each other? Each of the two chromosomes in a homologous pair is nearly identical in size and contains a very similar composition of genetic material. A particular gene found on one copy of a chromosome is also found on the homolog. Because one homolog is received from each parent, the two homologs may vary in the way that a gene affects an organism's traits. As an example, let's consider a gene in humans called *OCA2*, which plays a major role in determining eye color. The *OCA2* gene is found on chromosome 15. One copy of chromosome 15 might carry the form of this eye color gene that confers brown eyes, whereas the gene on the homolog could confer blue eyes. The topic of how genes affect an organism's traits will be considered in Chapter 16.

The DNA sequences on homologous chromosomes are very similar. In most cases, the sequence of bases on one homolog differs by less than 1% from the sequence on the other homolog. For example, the DNA sequence of chromosome 1 that you inherited from your mother is likely to be more than 99% identical to the DNA sequence of chromosome 1 that you inherited from your father. Nevertheless, keep in mind that the sequences are not identical. The slight differences in DNA sequence provide important variation in gene function. Again, if we use the eye color gene *OCA2* as an example, a minor difference in DNA sequence distinguishes two forms of the gene, brown versus blue.

The striking similarity between homologous chromosomes does not apply to the sex chromosomes (for example, X and Y). These chromosomes differ in size and genetic composition. Certain genes found on the X chromosome are not found on the Y chromosome, and vice versa. The X and Y chromosomes are not considered homologous chromosomes, although they do have short regions of homology.

The Cell Cycle Is a Series of Phases That Lead to Cell Division

Eukaryotic cells that are destined to divide progress through the cell cycle, a series of changes that involves growth, replication, and division, and ultimately produces new cells. Figure 15.2 provides an overview of the cell cycle. In this diagram, the mother cell has three pairs of chromosomes, for a total of six individual chromosomes. Such a cell is diploid ($2n$) and contains three chromosomes per set ($n = 3$). The paternal set is shown in blue, and the homologous maternal set is shown in red.

The phases of the cell cycle are $\mathbf{G_1}$ (first gap), **S** (synthesis of DNA, the genetic material), $\mathbf{G_2}$ (second gap), and **M phase** (mitosis and cytokinesis). The G_1 and G_2 phases were originally described as gap phases to indicate the periods between DNA synthesis and mitosis. In actively dividing cells, the G_1, S, and G_2 phases are collectively known as **interphase**. During interphase, the cell grows and copies its chromosomes in preparation for cell division. Alternatively, cells may exit the cell cycle and remain for long periods of time in a phase called $\mathbf{G_0}$ (G zero). The G_0 phase is an alternative to proceeding through G_1. A cell in the G_0 phase has postponed making a decision to divide or, in the case of terminally differentiated cells (such as muscle cells in an adult animal), will never divide again. G_0 is a nondividing phase.

G_1 Phase The G_1 phase is a period in a cell's life when it may become committed to divide. Depending on the environmental conditions and the presence of signaling molecules, a cell in the G_1 phase may accumulate molecular changes that cause it to progress through the rest of the cell cycle. Cell growth typically occurs during the G_1 phase.

S Phase During the S phase, each chromosome is replicated to form a pair of sister chromatids (see Figure 15.1). When S phase is completed, a cell has twice as many chromatids as the number of chromosomes in the G_1 phase. For example, a human cell in the G_1 phase has 46 distinct chromosomes, whereas the same cell in G_2 would have 46 pairs of sister chromatids, for a total of 92 chromatids.

G_2 Phase During the G_2 phase, a cell synthesizes the proteins necessary for chromosome sorting and cell division. Some cell growth may occur.

1 Prior to cell division, a mother cell has 6 chromosomes, 2 sets of 3 each.
2 Chromosome replication produces 6 pairs of sister chromatids.
3 Replication is completed. Cell prepares to divide.
4 Nucleus breaks apart, and replicated chromosomes condense in preparation for mitosis.
5 Sister chromatids separate during mitosis, and 2 cells are formed during cytokinesis.
Two daughter cells form, each containing 6 chromosomes.
S
Interphase
G_1
G_2
M
Mitosis
Cytokinesis
Telophase
Anaphase
Metaphase
Prometaphase
Prophase

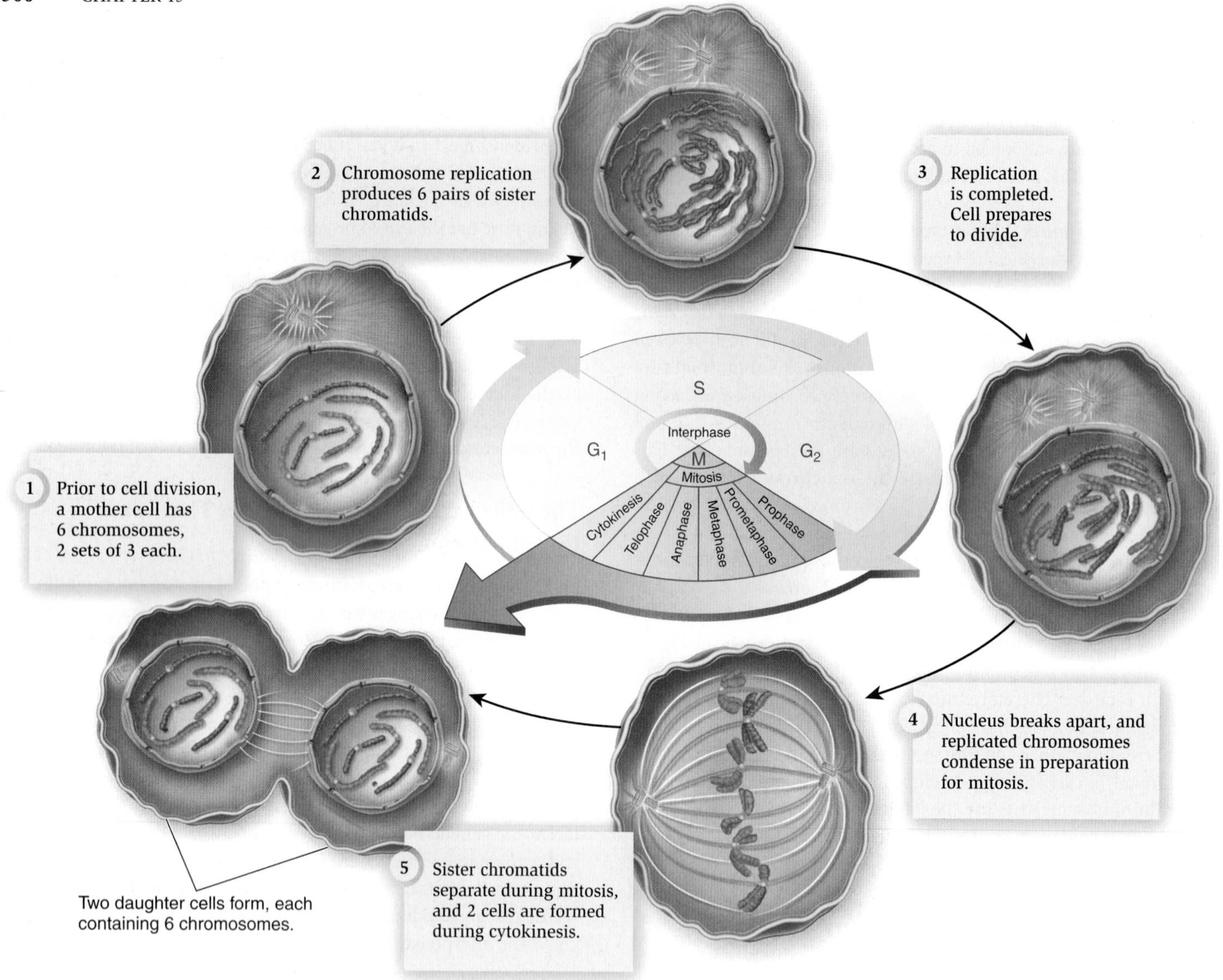

Figure 15.2 **The eukaryotic cell cycle.** Dividing cells progress through a series of phases denoted G_1, S, G_2, and M. This diagram shows the progression of a cell through the cell cycle to produce two daughter cells. The original diploid cell had three pairs of chromosomes, for a total of six individual chromosomes. During S phase, these have replicated to yield 12 chromatids. After mitosis is complete, two daughter cells each contain six individual chromosomes. The width of the phases shown in this figure is not meant to reflect their actual length. G_1 is typically the longest phase of the cell cycle, whereas M phase is relatively short.

BIOLOGY PRINCIPLE Cells are the simplest units of life. The cells of eukaryotic species are made via cell division during the eukaryotic cell cycle.

Concept Check: *Which phases make up interphase?*

M Phase The first part of M phase is **mitosis**. The purpose of mitosis is to divide one cell nucleus into two nuclei, distributing the duplicated chromosomes so each daughter cell receives the same complement of chromosomes. As noted previously, a human cell in the G_2 phase has 92 chromatids, which are found in 46 pairs. During mitosis, these pairs of chromatids are separated and sorted so each daughter cell receives 46 chromosomes. In most cases, mitosis is followed by **cytokinesis**, which is the division of the cytoplasm to produce two distinct daughter cells.

The length of the cell cycle varies considerably among different cell types, ranging from several minutes in quickly growing embryos to several months in slow-growing adult cells. For fast-dividing mammalian cells in adults, such as skin cells, the length of the cycle is often in the range of 10 to 24 hours. The various phases within the cell cycle also vary in length. G_1 is often the longest and also the most variable phase, and M phase is the shortest. For a cell that divides in 24 hours, the following lengths of time for each phase are typical:

- G_1 phase: 11 hours
- S phase: 8 hours
- G_2 phase: 4 hours
- M phase: 1 hour

What factors determine whether or not a cell will divide? First, cell division is controlled by external factors, such as environmental conditions and signaling molecules. The effects of growth factors on cell division are discussed in Chapter 9 (refer back to Figure 9.10). Second, internal factors affect cell division. These include cell cycle control molecules and checkpoints, as we will discuss next.

The Cell Cycle Is Controlled by Checkpoint Proteins

The progression through the cell cycle is a process that is highly regulated to ensure that the nuclear genome is intact and that the conditions are appropriate for a cell to divide. As discussed in Chapter 14, this is necessary to minimize the occurrence of mutations, which could have harmful effects and potentially lead to cancer. Proteins called **cyclins** and **cyclin-dependent kinases (cdks)** are responsible for advancing a cell through the phases of the cell cycle. Cyclins are so named because their amount varies throughout the cell cycle. To be active, the kinases controlling the cell cycle must bind to (are dependent on) a cyclin. The number of different types of cyclins and cdks varies from species to species.

Figure 15.3 gives a simplified description of how cyclins and cdks work together to advance a cell through G_1 and mitosis. During G_1, the amount of a particular cyclin termed G_1 cyclin increases in response to sufficient nutrients and growth factors. The G_1 cyclin binds to a cdk to form an activated G_1 cyclin/cdk complex. Once activated, cdk functions as a protein kinase that phosphorylates other proteins needed to advance the cell to the next phase in the cell cycle. For example, certain proteins involved with DNA synthesis are phosphorylated and activated, thereby allowing the cell to replicate its DNA in S phase. After the cell passes into the S phase, G_1 cyclin is degraded. Similar events advance the cell through other phases of the cell cycle. A different cyclin, called mitotic cyclin, accumulates late in G_2. It binds to a cdk to form an activated mitotic cyclin/cdk complex. This complex phosphorylates proteins that are needed to advance the cell into M phase.

Three critical regulatory points called **checkpoints** are found in the cell cycle of eukaryotic cells (see Figure 15.3). At these checkpoints, a variety of proteins, referred to as checkpoint proteins, act as sensors to determine if a cell is in the proper condition to divide. The G_1 checkpoint, also called the **restriction point**, determines if conditions are favorable for cell division. In addition, G_1-checkpoint proteins can sense if the DNA has incurred damage. What happens if DNA damage is detected? The checkpoint proteins prevent the formation of active cyclin/cdk complexes, thereby stopping the progression of the cell cycle.

A second checkpoint exists in G_2. This checkpoint also checks the DNA for damage and ensures that all of the DNA has been replicated. In addition, the G_2 checkpoint monitors the levels of proteins that are needed to progress through M phase. A third checkpoint, called the metaphase checkpoint, senses the integrity of the spindle apparatus. As we will see later, the spindle apparatus is involved in chromosome sorting. Metaphase is a step in mitosis during which all of the chromosomes should be attached to the spindle apparatus. If a chromosome is not correctly attached, the metaphase checkpoint will stop the cell cycle. This checkpoint prevents cells from incorrectly sorting their chromosomes during division.

Checkpoint proteins delay the cell cycle until problems are fixed or prevent cell division when problems cannot be fixed. A primary aim of checkpoint proteins is to prevent the division of a cell that may have incurred DNA damage or harbors abnormalities in chromosome number. As discussed in Chapter 14, when the functions of checkpoint genes are lost due to mutation, the likelihood increases that undesirable genetic changes will occur that can cause additional mutations and cancerous growth.

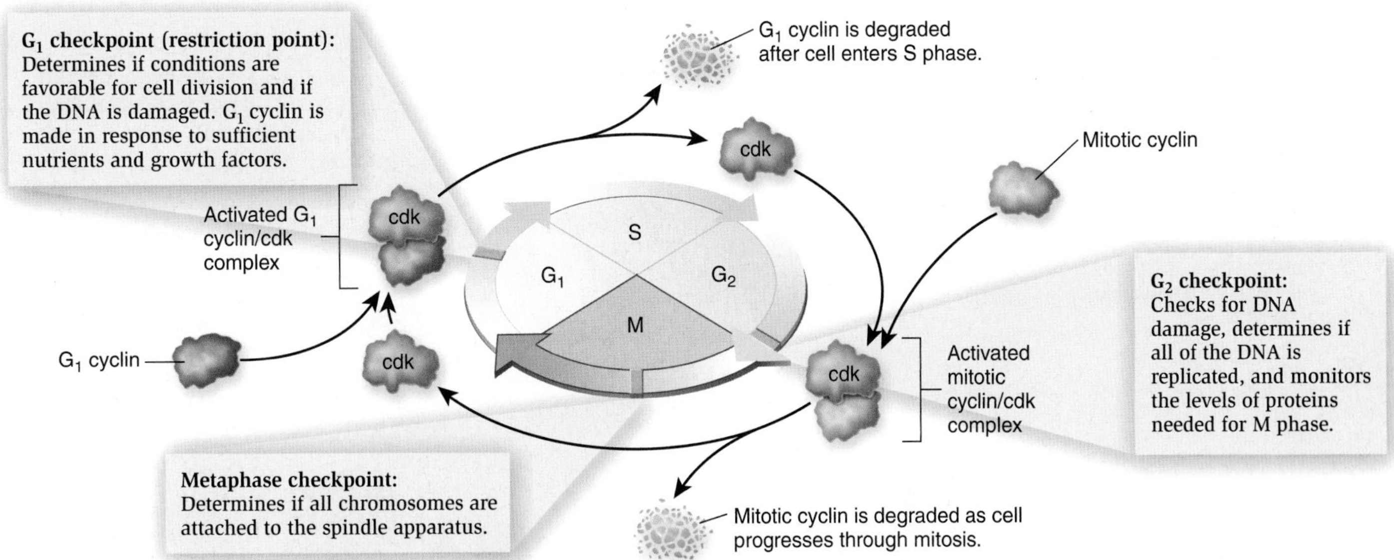
G1 checkpoint (restriction point): Determines if conditions are favorable for cell division and if the DNA is damaged. G1 cyclin is made in response to sufficient nutrients and growth factors.
G1 cyclin is degraded after cell enters S phase.
cdk
Mitotic cyclin
Activated G1 cyclin/cdk complex
S
G1
G2
M
G1 cyclin
Activated mitotic cyclin/cdk complex
G2 checkpoint: Checks for DNA damage, determines if all of the DNA is replicated, and monitors the levels of proteins needed for M phase.
Metaphase checkpoint: Determines if all chromosomes are attached to the spindle apparatus.
Mitotic cyclin is degraded as cell progresses through mitosis.

Figure 15.3 Checkpoints in the cell cycle. This is a general diagram of the eukaryotic cell cycle. Progression through the cell cycle requires the formation of activated cyclin/cdk complexes. Cells make different types of cyclin proteins, which are typically degraded after the cell has progressed to the next phase. The formation of activated cyclin/cdk complexes is regulated by checkpoint proteins.

BioConnections: *Look back at Figure 14.15. How do checkpoint proteins prevent cancer?*

FEATURE INVESTIGATION

Masui and Markert's Study of Oocyte Maturation Led to the Identification of Cyclins and Cyclin-Dependent Kinases

During the 1960s, researchers were intensely searching for the factors that promote cell division. In 1971, Japanese zoologist Yoshio Masui and American biologist Clement Markert developed a way to test whether a substance causes a cell to progress from one phase of the cell cycle to the next. They chose to study frog oocytes—cells that mature into egg cells. At the time of their work, researchers had already determined that frog oocytes naturally become dormant in the G_2 phase of the cell cycle for up to eight months (**Figure 15.4**). During mating season, female frogs produce a hormone called progesterone. After progesterone enters an oocyte and binds to intracellular receptors, the oocyte progresses from G_2 to the beginning of M phase, where the chromosomes condense and become visible under the microscope. This phenomenon is called maturation. When a sperm fertilizes the egg, M phase is completed, and the zygote continues to undergo cellular divisions.

Because progesterone is a signaling molecule, Masui and Markert speculated that this hormone affects the functions and/or amounts of proteins that trigger the oocyte to progress through the cell cycle. To test this hypothesis, they developed the procedure described in **Figure 15.5**, using the oocytes of the leopard frog (*Rana pipiens*). They began by exposing oocytes to progesterone in vitro and then incubated these oocytes for 2 hours or 12 hours. As a control, they also used oocytes that had not been exposed to progesterone. These three types of cells were called the donor oocytes.

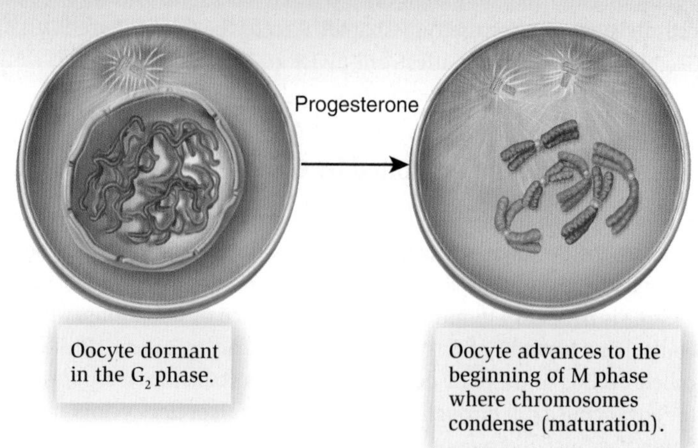

Figure 15.4 **Oocyte maturation in certain species of frogs.**

Next, they used a micropipette to transfer a small amount of cytosol from the three types of donor oocytes to recipient oocytes that had not been exposed to progesterone. As seen in the data, the recipient oocytes that had been injected with cytosol from the control donor oocytes or from oocytes that had been incubated with progesterone for only 2 hours did not progress to M phase. However, cytosol from donor oocytes that had been incubated with progesterone for

Figure 15.5 **The experimental approach of Masui and Markert to identify cyclin and cyclin-dependent kinase (cdk).**

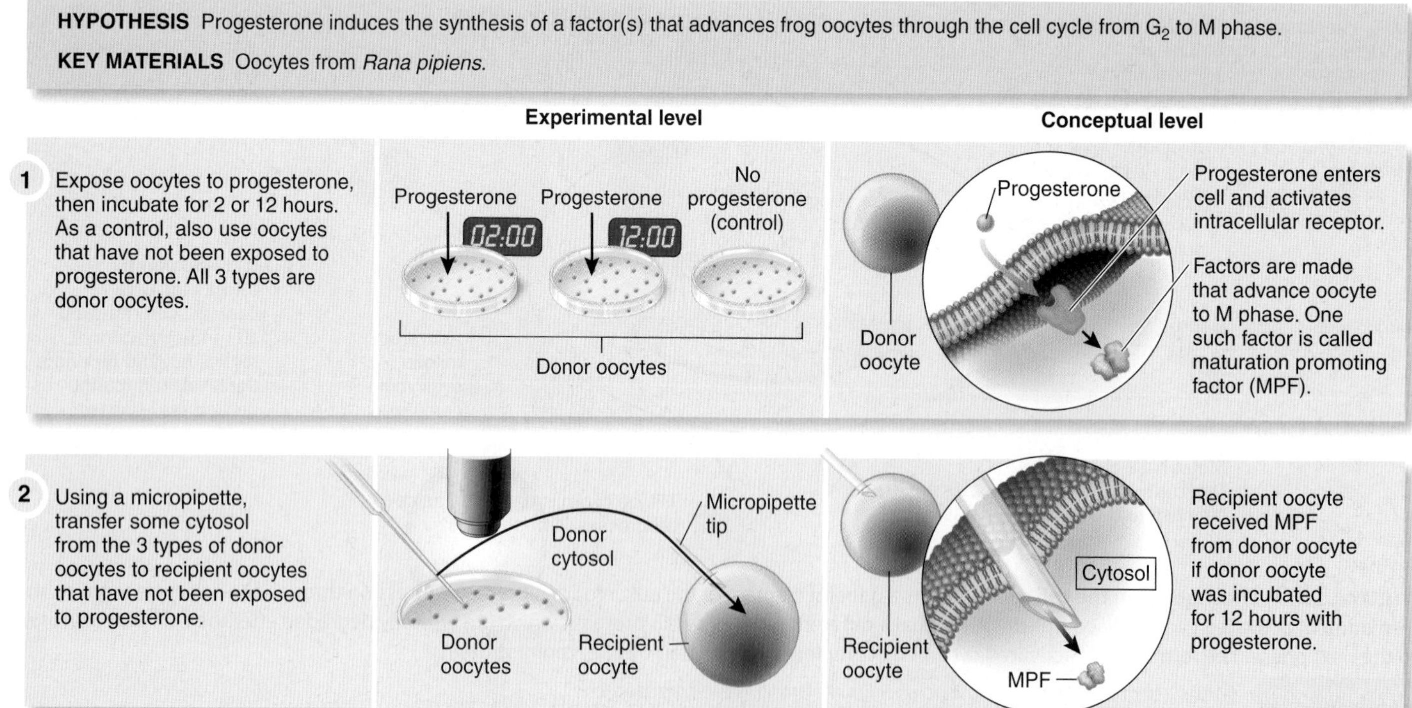

3 Incubate for several hours, and observe the recipient oocytes under the microscope to determine if the recipient oocytes advance to M phase. Advancement to M phase can be determined by the condensation of the chromosomes.

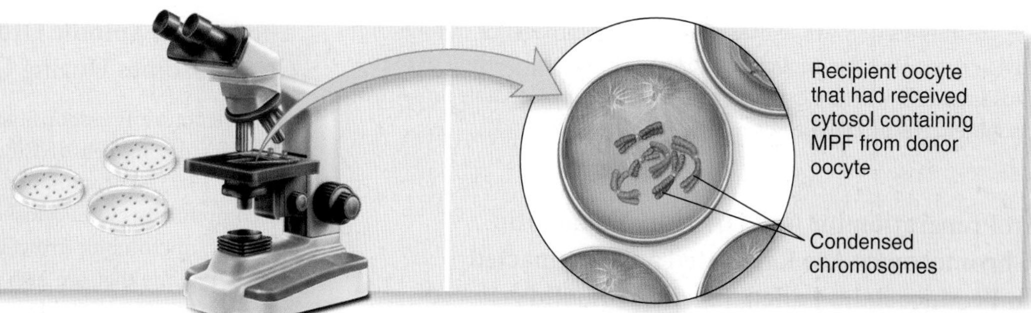

4 THE DATA

Donor oocytes	Recipient oocytes proceeded to M phase?
Control, no progesterone exposure	No
Progesterone exposure, incubation for 2 hours	No
Progesterone exposure, incubation for 12 hours	Yes

5 **CONCLUSION** Exposure of oocytes to progesterone for 12 hours results in the synthesis of a factor(s) that advances frog oocytes through the cell cycle from G_2 to M phase.

6 **SOURCE** Masui, Y., and Markert, C.L. 1971. Cytoplasmic control of nuclear behavior during meiotic maturation of frog oocytes. *Journal of Experimental Zoology* 177:129–145.

12 hours caused the recipient oocytes to advance to M phase. Masui and Markert concluded that a cytosolic factor, which required more than 2 hours to be synthesized after progesterone treatment, had been transferred to the recipient oocytes and induced maturation. The factor that caused the oocytes to progress (or mature) from G_2 to M phase was originally called the maturation-promoting factor (MPF).

After MPF was discovered in frogs, it was found in all eukaryotic species that researchers studied. MPF is important in the division of all types of cells, not just oocytes. It took another 17 years before Manfred Lohka, Marianne Hayes, and James Maller were able to purify the components that make up MPF. This was a difficult undertaking because these components are found in very small amounts in the cytosol and are easily degraded during purification procedures. We now know that MPF is a complex made of a mitotic cyclin and a cyclin-dependent kinase (cdk), as described in Figure 15.3.

Experimental Questions

1. At the time of Masui and Markert's study shown in Figure 15.5, what was known about the effects of progesterone on oocytes?
2. What hypothesis did Masui and Markert propose to explain the function of progesterone? Explain the procedure used to test the hypothesis.
3. How did the researchers explain the difference between the results using 2-hour-exposed donor oocytes versus 12-hour-exposed donor oocytes?

15.2 Mitotic Cell Division

Learning Outcomes:

1. Describe how the replication of eukaryotic chromosomes produces sister chromatids.
2. Explain the structure and function of the mitotic spindle.
3. Outline the key events that occur during the phases of mitosis.

We now turn our attention to a mechanism of cell division and its relationship to chromosome replication and sorting. During the process of **mitotic cell division**, a cell divides to produce two new cells (the daughter cells) that are genetically identical to the original cell (the mother cell). Mitotic cell division involves mitosis—the division of one nucleus into two nuclei—and then cytokinesis in which the mother cell divides into two daughter cells.

Why is mitotic cell division important? One purpose is **asexual reproduction**, a process in which genetically identical offspring are produced from a single parent. Certain unicellular eukaryotic organisms, such as baker's yeast (*Saccharomyces cerevisiae*) and the amoeba, increase their numbers in this manner. A second important reason for mitotic cell division is the production and maintenance of multicellularity. Organisms such as plants, animals, and most fungi are derived from a single cell that subsequently undergoes repeated cellular divisions to become a multicellular organism.

In this section, we will explore how the process of mitotic cell division requires the replication, organization, and sorting of chromosomes. We will also examine how a single cell is separated into two daughter cells by cytokinesis.

In Preparation for Cell Division, Eukaryotic Chromosomes Are Replicated and Compacted to Produce Pairs Called Sister Chromatids

We now turn our attention to how chromosomes are replicated and sorted during cell division. In Chapter 11, we examined the molecular process of DNA replication. Figure 15.6 describes the process at the chromosomal level. Prior to DNA replication, the DNA of each eukaryotic chromosome consists of a linear DNA double helix that is found in the nucleus and is not highly compacted. When the DNA is replicated, two identical copies of the original double helix are produced. As discussed earlier, these copies, along with associated proteins, lie side-by-side and are termed sister chromatids. When a cell prepares to divide, the sister chromatids become highly compacted and readily visible under the microscope. As shown in Figure 15.6b, the two sister chromatids are tightly associated at a region called the **centromere**. A protein called cohesin is necessary to hold the sister chromatids together. In addition, the centromere serves as an attachment site for a group of proteins that form the **kinetochore**, a structure necessary for sorting each chromosome.

The Mitotic Spindle Organizes and Sorts Chromosomes During Cell Division

What structure is responsible for organizing and sorting the chromosomes during cell division? The answer is the **mitotic spindle** (Figure 15.7). It is composed of microtubules—protein fibers that are components of the cytoskeleton (refer back to Table 4.1). In animal cells, microtubule growth and organization starts at two **centrosomes**, regions that are also referred to as microtubule organizing centers (MTOCs). A single centrosome duplicates during interphase. After they separate from each other during mitosis, each centrosome defines a **pole** of the spindle apparatus, one within each of the future daughter cells. The centrosome in animal cells has a pair of **centrioles**. Each one is composed of nine sets of triplet microtubules. However, centrioles are not found in many other eukaryotic species, such as plants, and are not required for spindle formation.

Each centrosome organizes the construction of the microtubules by rapidly polymerizing tubulin proteins. The three types of spindle microtubules are termed astral, polar, and kinetochore microtubules (see Figure 15.7). The astral microtubules, which extend away from the chromosomes, are important for positioning the spindle apparatus within the cell. The polar microtubules project into the region between the two poles. Polar microtubules that overlap with each other play a role in the separation of the two poles. Finally, the kinetochore microtubules are attached to kinetochores, which are bound to the centromere of each chromosome.

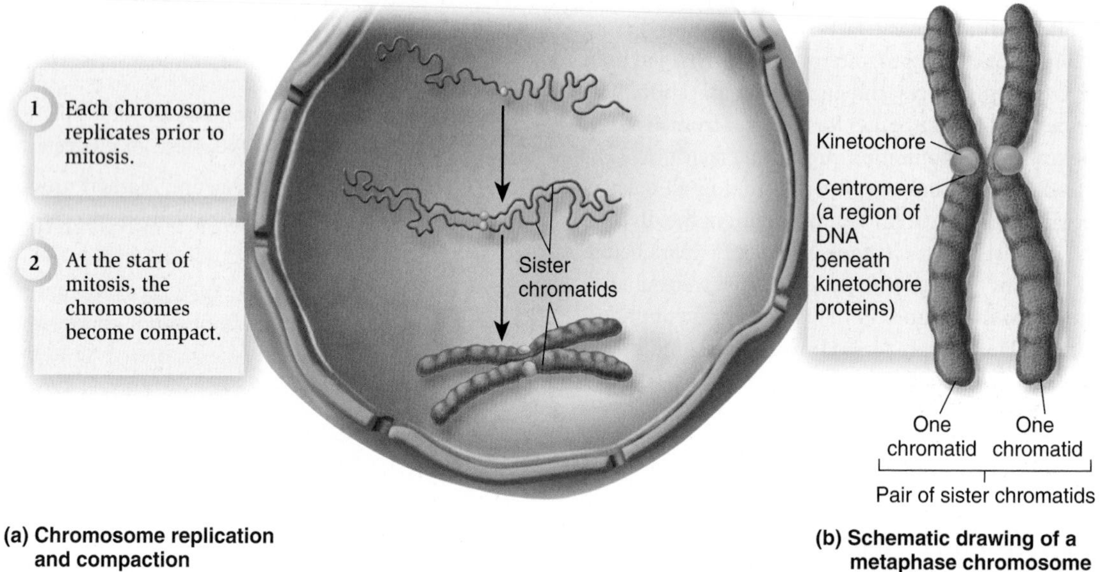

Figure 15.6 Replication and compaction of chromosomes into pairs of sister chromatids. **(a)** Chromosomal replication produces a pair of sister chromatids. While the chromosomes are elongated, they are replicated to produce two copies that are connected and lie parallel to each other. This is a pair of sister chromatids. Later, when the cell is preparing to divide, the sister chromatids condense into more compact structures that are easily seen with a light microscope. **(b)** A schematic drawing of a metaphase chromosome. This structure has two chromatids that lie side-by-side. The two chromatids are held together by cohesin proteins (not shown in this drawing). The kinetochore is a group of proteins that are attached to the centromere and play a role during chromosome sorting.

BIOLOGY PRINCIPLE The genetic material provides a blueprint for reproduction. The process of mitosis ensures that each daughter cell receives a complete copy of the genetic material.

Concept Check: *Look back at the karyotype in Figure 15.1. In this micrograph, is each of the 46 objects a pair of sister chromatids?*

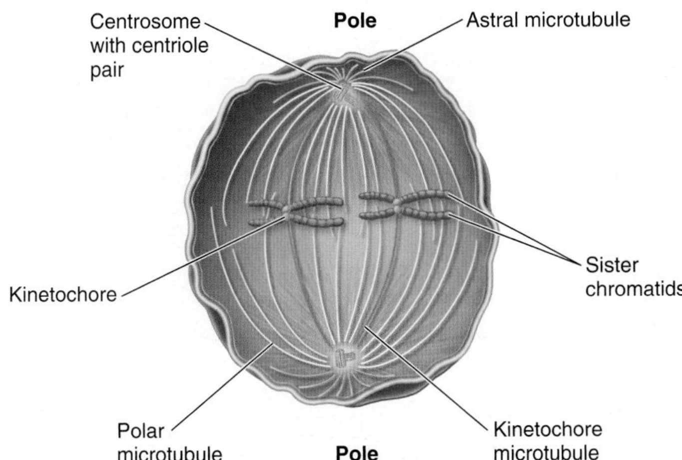

Figure 15.7 **The structure of the mitotic spindle.** The mitotic spindle in animal cells is formed by the centrosomes, which produce three types of microtubules. The astral microtubules emanate away from the region between the poles. The polar microtubules project into the region between the two poles. The kinetochore microtubules are attached to the kinetochores of sister chromatids. Note: For simplicity, this diagram shows only one pair of homologous chromosomes. Eukaryotic species typically have multiple chromosomes per set.

Concept Check: *What are the functions of the three types of microtubules?*

The Transmission of Chromosomes Requires a Sorting Process Known as Mitosis

Mitosis is the sorting process for dividing one cell nucleus into two nuclei. The duplicated chromosomes are distributed so each daughter cell receives the same complement of chromosomes. Mitosis was first observed microscopically in the 1870s by a German biologist, Walther Flemming, who coined the term (from the Greek *mitos*, meaning thread). He studied the large, transparent skin cells of salamander larvae as they were dividing and noticed that chromosomes are constructed of "threads" that are doubled in appearance along their length. These double threads divided and moved apart, one going to each of the two daughter nuclei. By this mechanism, Flemming pointed out, the two daughter cells receive an identical group of threads, the same as the number of threads in the mother cell.

Figure 15.8 depicts the process of mitosis in an animal cell, though the process is quite similar in a plant cell. Mitosis occurs as a continuum of phases known as prophase, prometaphase, metaphase, anaphase, and telophase. In the simplified diagrams shown along the bottom of Figure 15.8, the original cell contains six chromosomes. One set of chromosomes is depicted in red, whereas the homologous set is blue. These different colors represent maternal and paternal chromosomes.

Interphase Prior to mitosis, the cells are in interphase, which consists of the G_1, S, and G_2 phases of the cell cycle. The chromosomes have replicated in S phase and are decondensed and found in the nucleus (Figure 15.8a). The nucleolus, which is the site where the components of ribosomes assemble into ribosomal subunits, is visible during interphase.

Prophase At the start of mitosis, in **prophase**, the chromosomes have already replicated to produce 12 chromatids, joined as six pairs of sister chromatids that have condensed into highly compacted structures readily visible by light microscopy (Figure 15.8b). As prophase proceeds, the nuclear envelope begins to dissociate into small vesicles. The nucleolus is no longer visible.

Prometaphase During **prometaphase**, the nuclear envelope completely fragments into small vesicles, and the mitotic spindle is fully formed (Figure 15.8c). As prometaphase progresses, the centrosomes move apart and demarcate the two poles. Once the nuclear envelope has dissociated, the spindle fibers can interact with the sister chromatids. How do the sister chromatids become attached to the spindle apparatus? Initially, microtubules are rapidly formed and can be seen under a microscope growing out from the two poles. As it grows, if a microtubule happens to make contact with a kinetochore, it is said to be "captured" and remains firmly attached to the kinetochore. Alternatively, if a microtubule does not collide with a kinetochore, the microtubule eventually depolymerizes and retracts to the centrosome. This random process is how sister chromatids become attached to kinetochore microtubules. As the end of prometaphase nears, the two kinetochores on each pair of sister chromatids are attached to kinetochore microtubules from opposite poles. As these events are occurring, the sister chromatids are seen under the microscope to undergo jerky movements as they are tugged, back and forth, between the two poles by the kinetochore microtubules.

Metaphase Eventually, the pairs of sister chromatids are aligned in a single row along the **metaphase plate**, a plane halfway between the poles. When this alignment is complete, the cell is in **metaphase** of mitosis (Figure 15.8d). The chromatids can then be equally distributed into two daughter cells.

Anaphase During **anaphase**, the connections between the pairs of sister chromatids are broken (Figure 15.8e). Each chromatid, now an individual chromosome, is linked to only one of the two poles by one or more kinetochore microtubules. As anaphase proceeds, the kinetochore microtubules shorten, pulling the chromosomes toward the pole to which they are attached. In addition, the two poles move farther away from each other. This occurs because the overlapping polar microtubules lengthen and push against each other, thereby pushing the poles farther apart.

Telophase During **telophase**, the chromosomes have reached their respective poles and decondense. The nuclear envelope now re-forms to produce two separate nuclei. In Figure 15.8f, two nuclei are being produced that contain six chromosomes each.

Cytokinesis In most cases, mitosis is quickly followed by cytokinesis, in which the two nuclei are segregated into separate daughter cells. Whereas the phases of mitosis are similar between plant and animal cells, the process of cytokinesis is quite different. In animal cells, cytokinesis involves the formation of a **cleavage furrow**, which constricts like a drawstring to separate the cells (Figure 15.9a). In plants, vesicles from the Golgi apparatus move along microtubules to the center of the cell and coalesce to form a **cell plate** (Figure 15.9b), which then forms a cell wall between the two daughter cells.

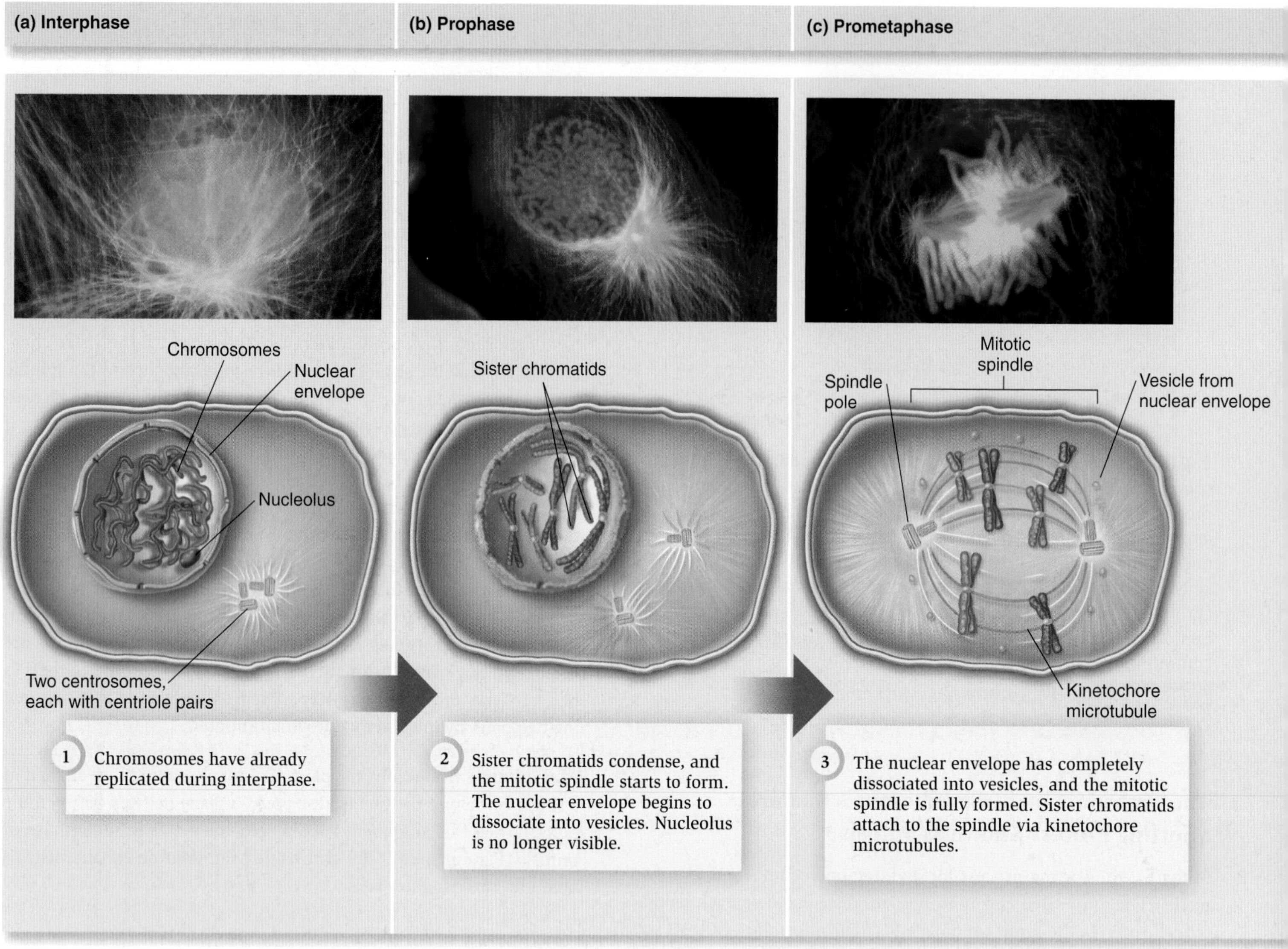

(a) Interphase

Chromosomes

Nuclear envelope

Nucleolus

Two centrosomes, each with centriole pairs

1 Chromosomes have already replicated during interphase.

(b) Prophase

Sister chromatids

2 Sister chromatids condense, and the mitotic spindle starts to form. The nuclear envelope begins to dissociate into vesicles. Nucleolus is no longer visible.

(c) Prometaphase

Mitotic spindle

Spindle pole

Vesicle from nuclear envelope

Kinetochore microtubule

3 The nuclear envelope has completely dissociated into vesicles, and the mitotic spindle is fully formed. Sister chromatids attach to the spindle via kinetochore microtubules.

Figure 15.8 The process of mitosis in an animal cell. The top panels illustrate the cells of a newt progressing through mitosis. The bottom panels are schematic drawings that emphasize the sorting and separation of the chromosomes in which the diploid mother cell had six chromosomes (three in each set). At the start of mitosis, these have already replicated into 12 chromatids. The final result is two daughter cells, each containing six chromosomes.

Concept Check: *With regard to chromosome composition, how does the mother cell compare with the two daughter cells?*

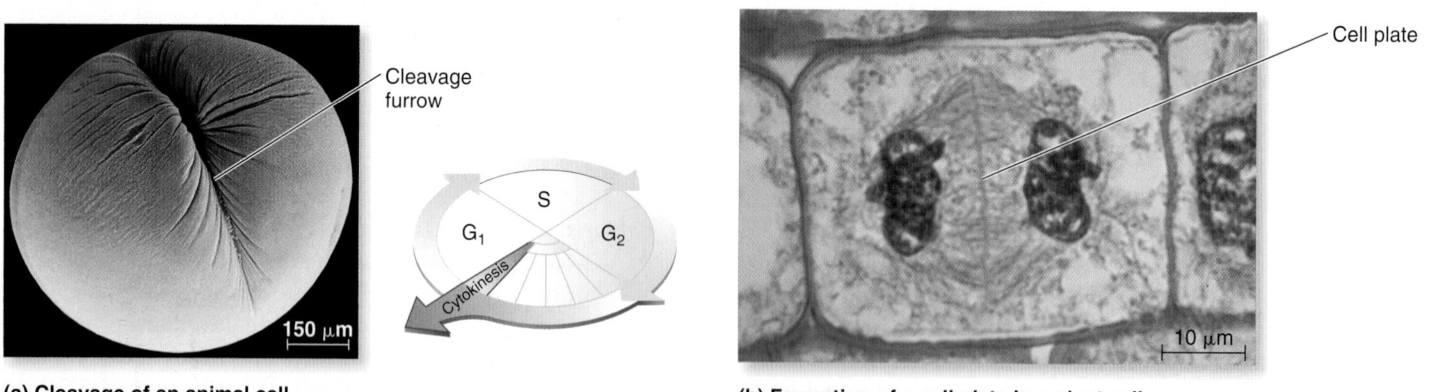

(a) Cleavage of an animal cell

(b) Formation of a cell plate in a plant cell

Figure 15.9 Micrographs showing cytokinesis in animal and plant cells.

Concept Check: *What are the similarities and differences between cytokinesis in animal and plant cells?*

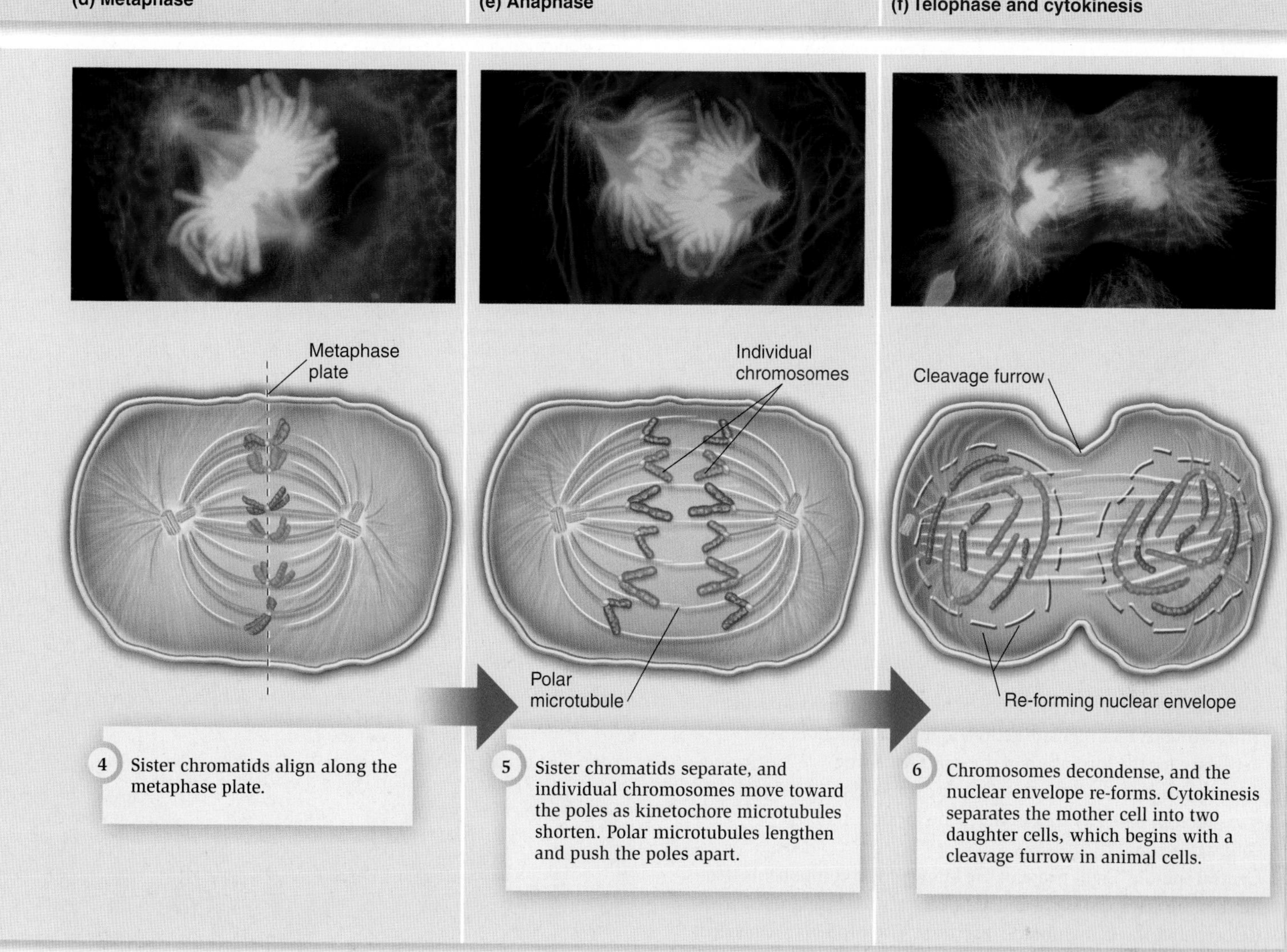

(d) Metaphase
(e) Anaphase
(f) Telophase and cytokinesis
Metaphase plate
Individual chromosomes
Cleavage furrow
Polar microtubule
Re-forming nuclear envelope
4 Sister chromatids align along the metaphase plate.
5 Sister chromatids separate, and individual chromosomes move toward the poles as kinetochore microtubules shorten. Polar microtubules lengthen and push the poles apart.
6 Chromosomes decondense, and the nuclear envelope re-forms. Cytokinesis separates the mother cell into two daughter cells, which begins with a cleavage furrow in animal cells.

What are the results of mitosis and cytokinesis? These processes ultimately produce two daughter cells with the same number of chromosomes as the mother cell. Barring rare mutations, the two daughter cells are genetically identical to each other and to the mother cell from which they were derived. The critical consequence of this sorting process is to ensure genetic consistency from one cell to the next. The development of multicellularity relies on the repeated process of mitosis and cytokinesis.

Genomes & Proteomes Connection

The Genomes of Diverse Animal Species Encode Approximately 20 Proteins Involved in Cytokinesis

To understand how a process works at the molecular and cellular level, researchers often try to identify the genes within a given species that encode proteins necessary for the process. Cytokinesis has been analyzed in this way. By comparing the results from vertebrates, insects, and worms, researchers have identified approximately 20 proteins that are involved with cytokinesis in nearly all animal cells. In any given species, cytokinesis may also involve additional proteins beyond these 20, but these other proteins are not needed among all animal species. Evolutionary biologists would say the 20 proteins that are common to all animals are highly conserved, meaning their structure and function has been retained during the evolution of animals. The 20 conserved proteins are likely to play the most fundamental roles in the process of cytokinesis.

What are the functions of these 20 proteins? To appreciate their functions, we need to take a closer look at cytokinesis in animal cells (**Figure 15.10**). Animal cells produce a contractile ring that is attached to the plasma membrane to create the cleavage furrow. The contractile ring, which encircles a region of the mitotic spindle called the central spindle, is a network of actin (a cytoskeletal protein) and myosin (a motor protein). The motor activity of myosin moves actin filaments in a way that causes the contractile ring to constrict. Once the contractile ring becomes very small, membrane vesicles are inserted into the constricted site to achieve division of the plasma membranes in the two resulting cells.

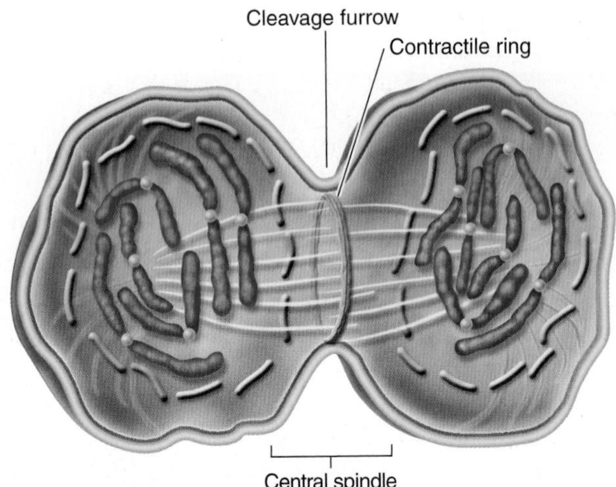

Figure 15.10 **A closer look at cytokinesis in animal cells.**

BioConnections: *Look back at Table 4.1 and Figure 4.11, which describe the structure and function of actin and myosin. How do you think the contractile ring contracts?*

The 20 conserved proteins perform one of four possible functions.

1. Contractile ring: Seven proteins, including actin, myosin, and other proteins that regulate actin and myosin function, are necessary for the formation of the contractile ring.
2. Signal transduction: Five proteins are components of a signal transduction pathway that initiates the formation of the contractile ring.
3. Central spindle: Eight proteins are known to be components that bind to the central spindle and are necessary for cytokinesis.
4. Cell separation via membrane insertion: Two proteins are needed for the final separation of the two daughter cells.

The 20 conserved proteins should be considered a minimum estimate. As we gain a deeper understanding of cytokinesis at the molecular level, it is likely that additional proteins may be discovered.

15.3 Meiosis and Sexual Reproduction

Learning Outcomes:

1. Describe the processes of synapsis and crossing over.
2. Outline the key events that occur during the phases of meiosis.
3. Compare and contrast mitosis and meiosis, focusing on key steps that account for the different outcomes of these two processes
4. Distinguish between the life cycles of diploid-dominant species, haploid-dominant species, and species that exhibit an alternation of generations.

We now turn our attention to **sexual reproduction**, a process in which two haploid gametes unite to form a diploid cell called a **zygote**. For multicellular species such as animals and plants, the zygote then grows and divides by mitotic cell divisions into a multicellular organism with many diploid cells.

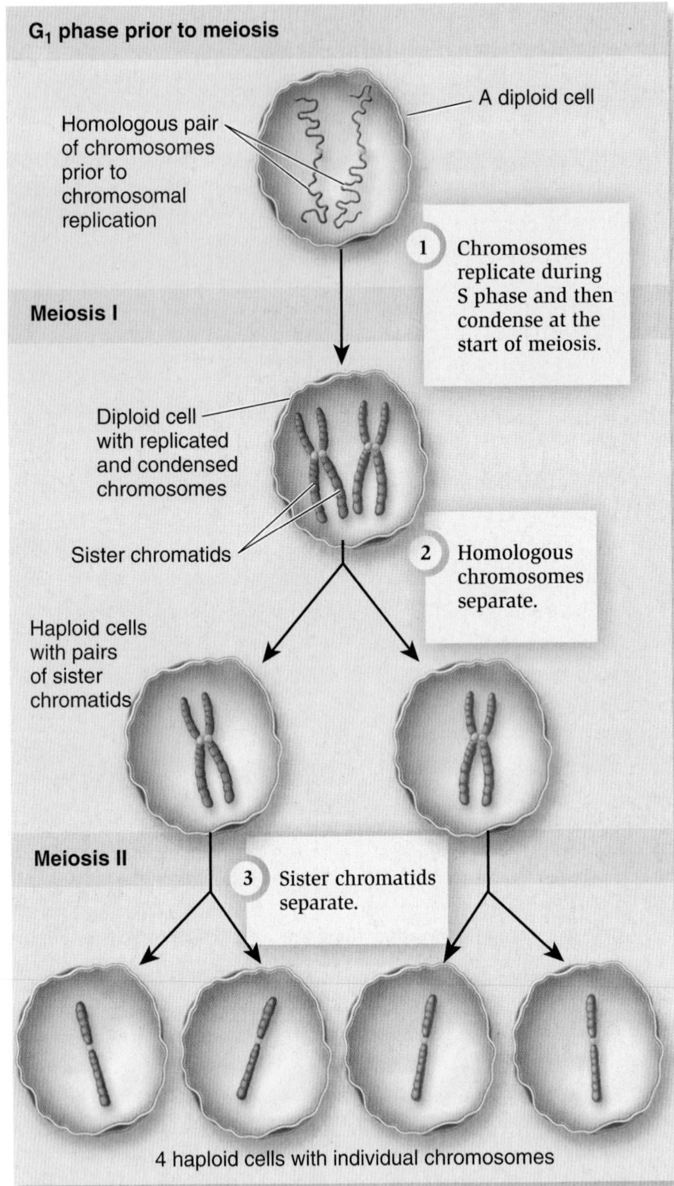

Figure 15.11 **How the process of meiosis reduces chromosome number.** This simplified diagram emphasizes the reduction in chromosome number as a diploid cell divides by meiosis to produce four haploid cells.

As discussed earlier, a diploid cell contains two homologous sets of chromosomes, whereas a haploid cell contains a single set. For example, a diploid human cell contains 46 chromosomes, but a human gamete—sperm or egg cell—is a haploid cell that contains only 23 chromosomes. **Meiosis** is the process by which haploid cells are produced from a cell that was originally diploid. The term meiosis, which means "to make smaller," refers to the fewer chromosomes found in cells following this process. For this to occur, the chromosomes must be correctly sorted and distributed in a way that reduces the chromosome number to half its original diploid value. In the case of human gametes, for example, each gamete must receive one chromosome from each of the 23 pairs. For this to happen, two rounds of divisions are necessary, termed meiosis I and meiosis II (**Figure 15.11**). When a cell begins meiosis, it contains chromosomes that are found in homologous pairs. When meiosis is completed, a single diploid cell with homologous pairs of chromosomes has produced four haploid cells.

In this section, we will examine the cellular events of meiosis that reduce the chromosome number from diploid to haploid. In addition, we will briefly consider how this process plays a role in the life cycles of animals, plants, fungi, and protists.

Bivalent Formation and Crossing Over Occur at the Beginning of Meiosis

Like mitosis, meiosis begins after a cell has progressed through the G_1, S, and G_2 phases of the cell cycle. However, two key events occur at the beginning of meiosis that do not occur in mitosis. First, homologous pairs of sister chromatids associate with each other, lying side by side to form a **bivalent**, also called a tetrad (**Figure 15.12**). The process of forming a bivalent is termed **synapsis**. In most eukaryotic species, a protein structure called the synaptonemal complex connects homologous chromosomes during a portion of meiosis. However, the synaptonemal complex is not required for the pairing of homologous chromosomes because some species of fungi completely lack such a complex, yet their chromosomes associate with each other correctly. At present, the precise role of the synaptonemal complex is not clearly understood.

The second event that occurs at the beginning of meiosis, but not usually during mitosis, is **crossing over**, which involves a physical exchange between chromosome segments of the bivalent (Figure 15.12). As discussed in Chapter 17, crossing over increases the genetic variation of sexually reproducing species. After crossing over occurs, the arms of the chromosomes tend to separate but remain adhered at a crossover site. This connection is called a chiasma (plural, chiasmata), because it physically resembles the Greek letter chi, χ. The number of crossovers is carefully controlled by cells and depends on the size of the chromosome and the species. The range of crossovers for eukaryotic chromosomes is typically one or two to a couple dozen. During the formation of sperm in humans, for example, an average chromosome undergoes slightly more than two crossovers, whereas chromosomes in certain plant species may undergo 20 or more.

Meiosis I Separates Homologous Chromosomes

Now that we have an understanding of bivalent formation and crossing over, we are ready to consider the phases of meiosis (**Figure 15.13**). These simplified diagrams depict a diploid cell ($2n$) that contains a total of six chromosomes (as in our look at mitosis in Figure 15.8). Prior to meiosis, the chromosomes are replicated in S phase to produce pairs of sister chromatids. This single replication event is then followed by sequential divisions called meiosis I and II. Like mitosis, each of these is a continuous series of stages called prophase, prometaphase, metaphase, anaphase, and telophase. The sorting that occurs during **meiosis I** separates homologous chromosomes from each other (Figure 15.13a–e).

Prophase I During prophase I, the replicated chromosomes condense, the homologous chromosomes form bivalents, and crossing over occurs. The nuclear envelope then starts to fragment into small vesicles.

Prometaphase I In prometaphase I, the nuclear envelope is completely broken down into vesicles, and the spindle apparatus is entirely formed. The sister chromatids become attached to kinetochore microtubules. However, a key difference occurs between mitosis and meiosis I. In mitosis, a pair of sister chromatids is attached to both poles

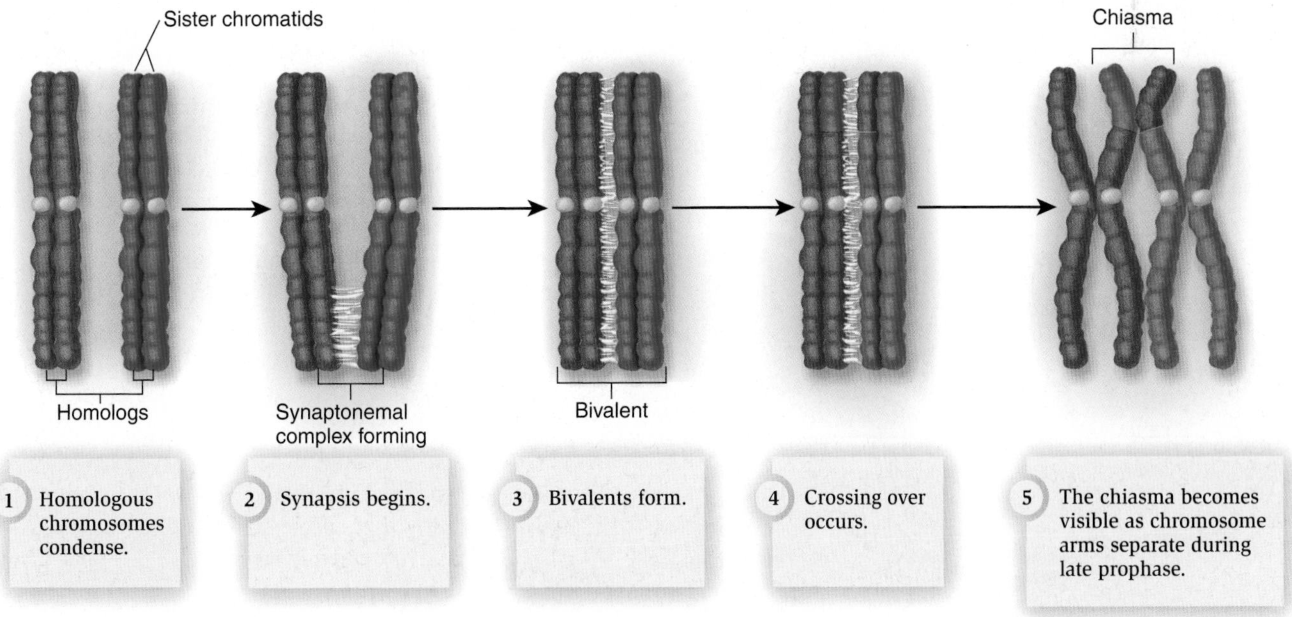

Sister chromatids
Chiasma
Homologs
Synaptonemal complex forming
Bivalent
1 Homologous chromosomes condense.
2 Synapsis begins.
3 Bivalents form.
4 Crossing over occurs.
5 The chiasma becomes visible as chromosome arms separate during late prophase.

Figure 15.12 Formation of a bivalent and crossing over during meiosis I. At the beginning of meiosis, homologous chromosomes pair with each other to form a bivalent, usually with a synaptonemal complex between them. Crossing over then occurs between homologous chromatids within the bivalent. During this process, homologs exchange segments of chromosomes.

Meiosis I

(a) Prophase I

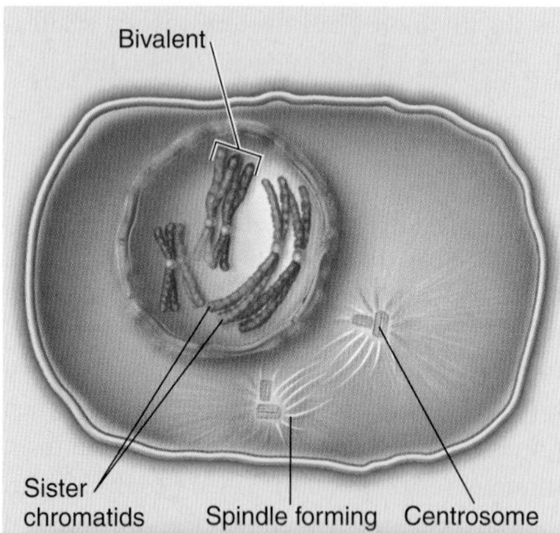

1 Homologous chromosomes synapse to form bivalents, and crossing over occurs. Chromosomes condense, and the nuclear envelope begins to dissociate into vesicles.

(b) Prometaphase I

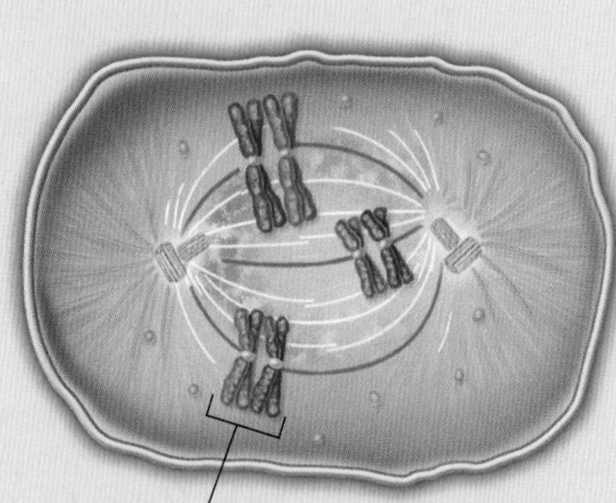

2 The nuclear envelope completely dissociates into vesicles, and bivalents become attached to kinetochore microtubules.

(c) Metaphase I

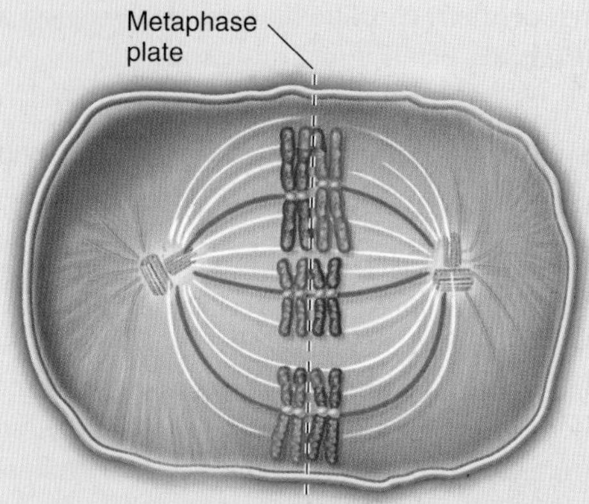

3 Bivalents randomly align along the metaphase plate. Each pair of sister chromatids is attached to one pole.

Meiosis II

(f) Prophase II

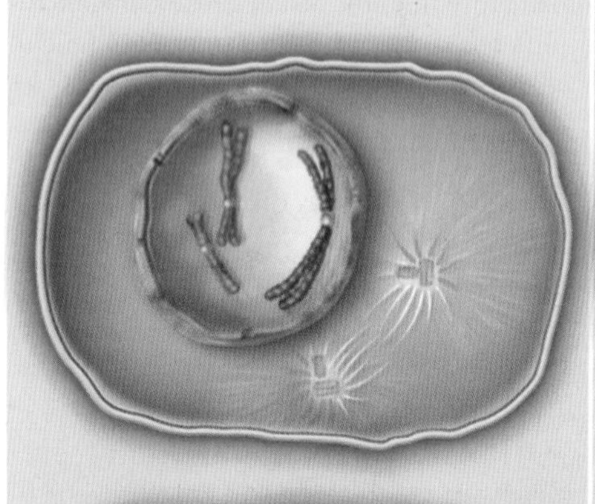

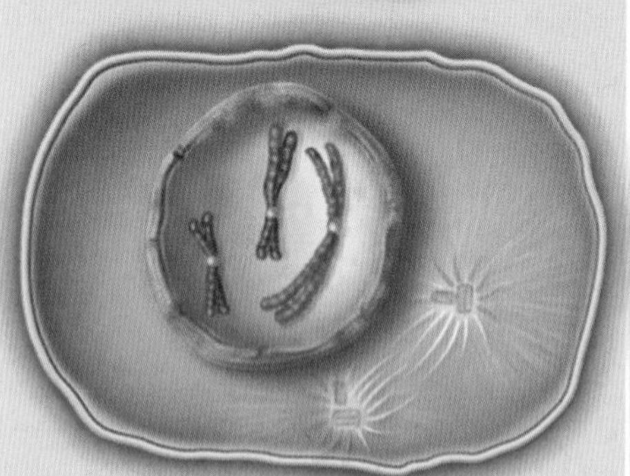

6 Sister chromatids condense, and the spindle starts to form. The nuclear envelope begins to dissociate into vesicles.

(g) Prometaphase II

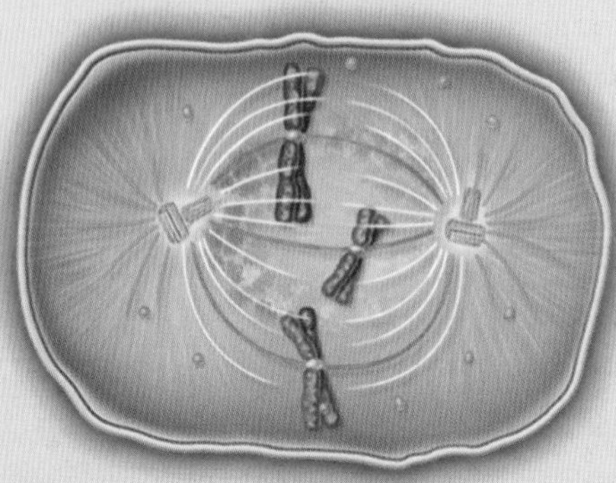

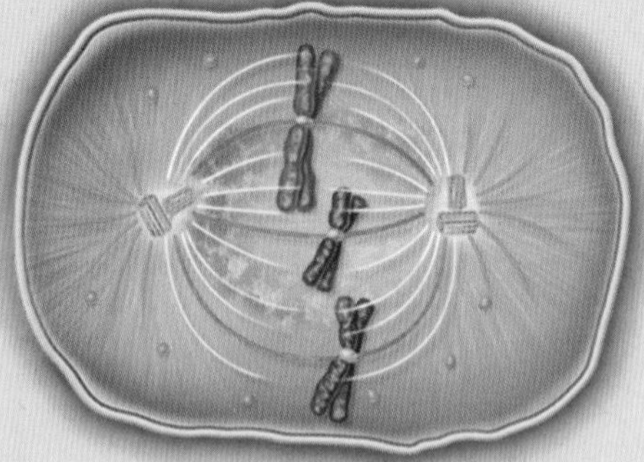

7 The nuclear envelope completely dissociates into vesicles. Sister chromatids attach to the spindle via kinetochore microtubules.

(h) Metaphase II

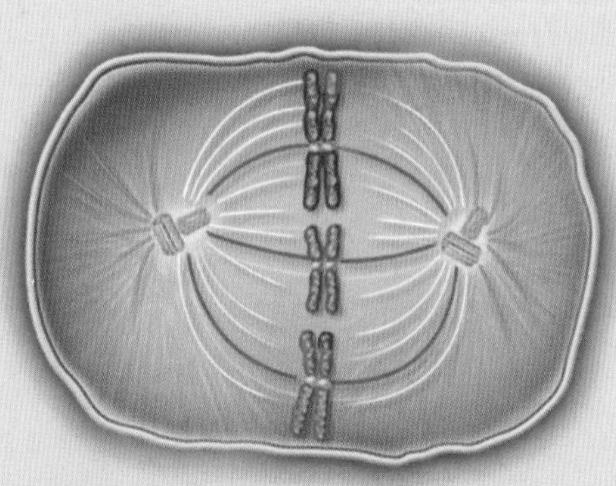

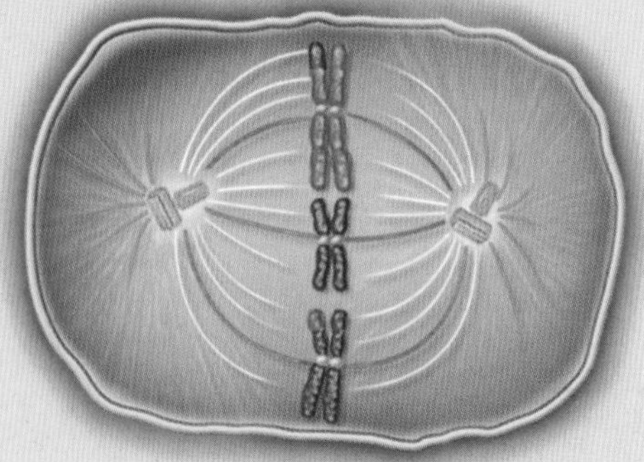

8 Sister chromatids align along the metaphase plate. Each pair of sister chromatids is attached to both poles.

(d) Anaphase I

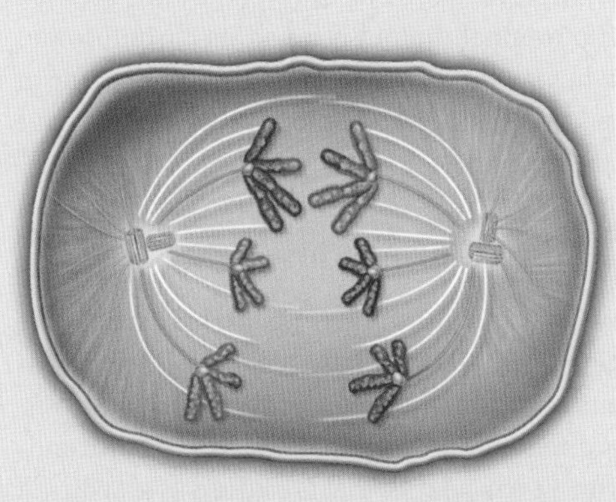

4 Homologous chromosomes separate and move toward opposite poles.

(e) Telophase I and cytokinesis

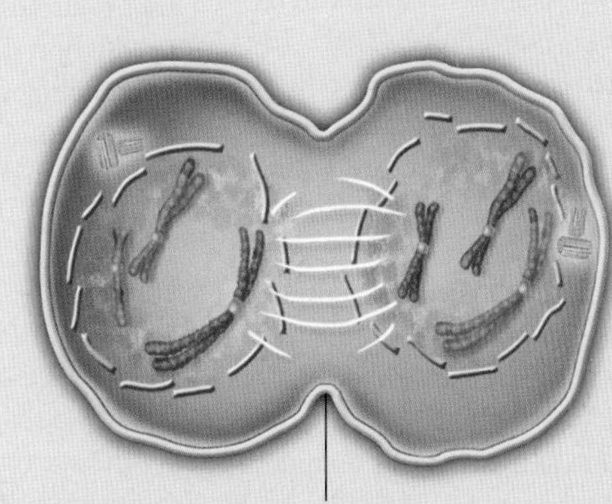

5 The chromosomes decondense, and the nuclear envelope re-forms. The 2 daughter cells are separated by a cleavage furrow.

(i) Anaphase II

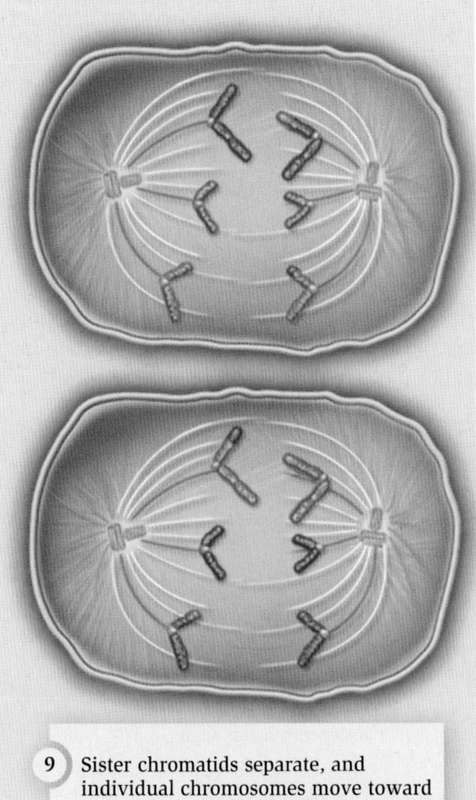

9 Sister chromatids separate, and individual chromosomes move toward the poles as kinetochore microtubules shorten. Polar microtubules lengthen and push the poles apart.

(j) Telophase II and cytokinesis

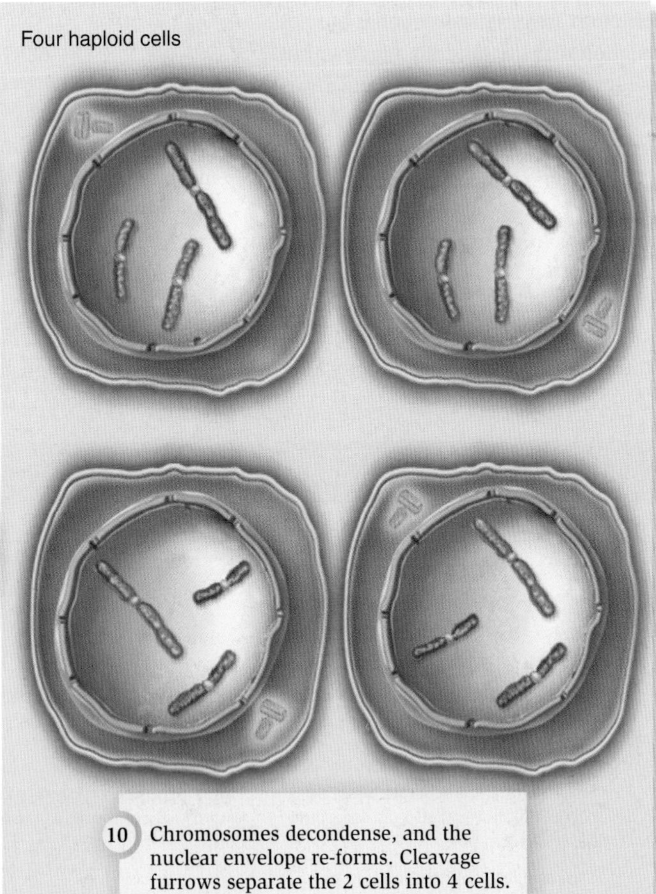

10 Chromosomes decondense, and the nuclear envelope re-forms. Cleavage furrows separate the 2 cells into 4 cells.

Figure 15.13 **The phases of meiosis in an animal cell.**

Concept Check: *Relative to the original cell, what is the end result of meiosis?*

(see Figure 15.8c). In meiosis I, a pair of sister chromatids is attached to just one pole via kinetochore microtubules (Figure 15.13b).

Metaphase I At metaphase I, the bivalents are organized along the metaphase plate. Notice how this pattern of alignment is strikingly different from that observed during mitosis (see Figure 15.8d). In particular, the sister chromatids are aligned in a double row rather than a single row (as in mitosis). Furthermore, the arrangement of sister chromatids within this double row is random with regard to the (red and blue) homologs. (Remember that these different colors represent maternal and paternal chromosomes.) In Figure 15.13c, one of the red homologs is to the left of the metaphase plate, and the other two are to the right, whereas two of the blue homologs are to the left of the metaphase plate and the other one is to the right. In other cells, homologs could be arranged differently along the metaphase plate (for example, three blues to the left and none to the right, or none to the left and three to the right).

Because eukaryotic species typically have many chromosomes per set, maternal and paternal homologs can be randomly aligned along the metaphase plate in a variety of ways. For example, consider that humans have 23 chromosomes per set. The possible number of different, random alignments equals 2^n, where n equals the number of chromosomes per set. The reason why the random alignments equals 2^n is because each chromosome is found in a homologous pair and each member of the pair can align on either side of the metaphase plate. It is a matter of chance which daughter cell of meiosis I will get the maternal chromosome of a homologous pair, and which will get the paternal chromosome. In humans, 2^n equals 2^{23}, or over 8 million possibilities. Because the homologs are genetically similar but not identical, we see from this calculation that the random alignment of homologous chromosomes provides a mechanism to promote a vast amount of genetic diversity among the resulting haploid cells. When meiosis is complete, it is very unlikely that any two human gametes will have the same combination of homologous chromosomes.

Anaphase I The segregation of homologs occurs during anaphase I (Figure 15.13d). The connections between bivalents break, but not the connections that hold sister chromatids together. Each joined pair of chromatids migrates to one pole, and the homologous pair of chromatids moves to the opposite pole, both pulled by kinetochore microtubules.

Telophase I At telophase I, the sister chromatids have reached their respective poles and then decondense. The nuclear envelope now re-forms to produce two separate nuclei.

If we consider the end result of meiosis I, we see that two nuclei are produced, each with three pairs of sister chromatids; this is called a reduction division. The original diploid cell had its chromosomes in homologous pairs, whereas the two cells produced as a result of meiosis I and cytokinesis are considered haploid—they do not have pairs of homologous chromosomes.

Meiosis II Separates Sister Chromatids

Meiosis I is followed by cytokinesis and then **meiosis II** (see Figure 15.13f–j). DNA replication does not occur between meiosis I and meiosis II. The sorting events of meiosis II are similar to those of mitosis, but the starting point is different. For a diploid cell with six chromosomes, mitosis begins with 12 chromatids that are joined as six pairs of sister chromatids (see Figure 15.8). By comparison, the two cells that begin meiosis II each have six chromatids that are joined as three pairs of sister chromatids. Otherwise, the steps that occur during prophase, prometaphase, metaphase, anaphase, and telophase of meiosis II are analogous to a mitotic division. Sister chromatids are separated during anaphase II, unlike anaphase I in which bivalents are separated.

Mitosis and Meiosis Differ in a Few Key Steps

How are the outcomes of mitosis and meiosis different from each other? Mitosis produces two diploid daughter cells that are genetically identical. In our example shown in Figure 15.8, the starting cell had six chromosomes (three homologous pairs of chromosomes), and both daughter cells had copies of the same six chromosomes. By comparison, meiosis reduces the number of sets of chromosomes. In the example shown in Figure 15.13, the starting cell also had six chromosomes, whereas the resulting four daughter cells had only three chromosomes. However, the daughter cells did not contain a random mix of three chromosomes. Each haploid daughter cell contained one complete set of chromosomes, whereas the original diploid mother cell had two complete sets.

How do we explain the different outcomes of mitosis and meiosis? **Table 15.1** emphasizes the differences between certain key steps in mitosis and meiosis that account for the different outcomes of these

Table 15.1 A Comparison of Mitosis, Meiosis I, and Meiosis II

Event	Mitosis	Meiosis I	Meiosis II
DNA replication:	Occurs prior to mitosis	Occurs prior to meiosis I	Does not occur between meiosis I and II
Synapsis during prophase:	No	Yes, bivalents are formed.	No
Crossing over during prophase:	Rarely	Commonly	Rarely
Attachment to poles at prometaphase:	A pair of sister chromatids is attached to kinetochore microtubules from both poles.	A pair of sister chromatids is attached to kinetochore microtubules from just one pole.	A pair of sister chromatids is attached to kinetochore microtubules from both poles.
Alignment along the metaphase plate:	Sister chromatids align.	Bivalents align.	Sister chromatids align.
Type of separation at anaphase:	Sister chromatids separate. A single chromatid, now called a chromosome, moves to each pole.	Homologous chromosomes separate. A pair of sister chromatids moves to each pole.	Sister chromatids separate. A single chromatid, now called a chromosome, moves to each pole.
End result when the mother cell is diploid:	Two daughter cells that are diploid	—	Four daughter cells that are haploid

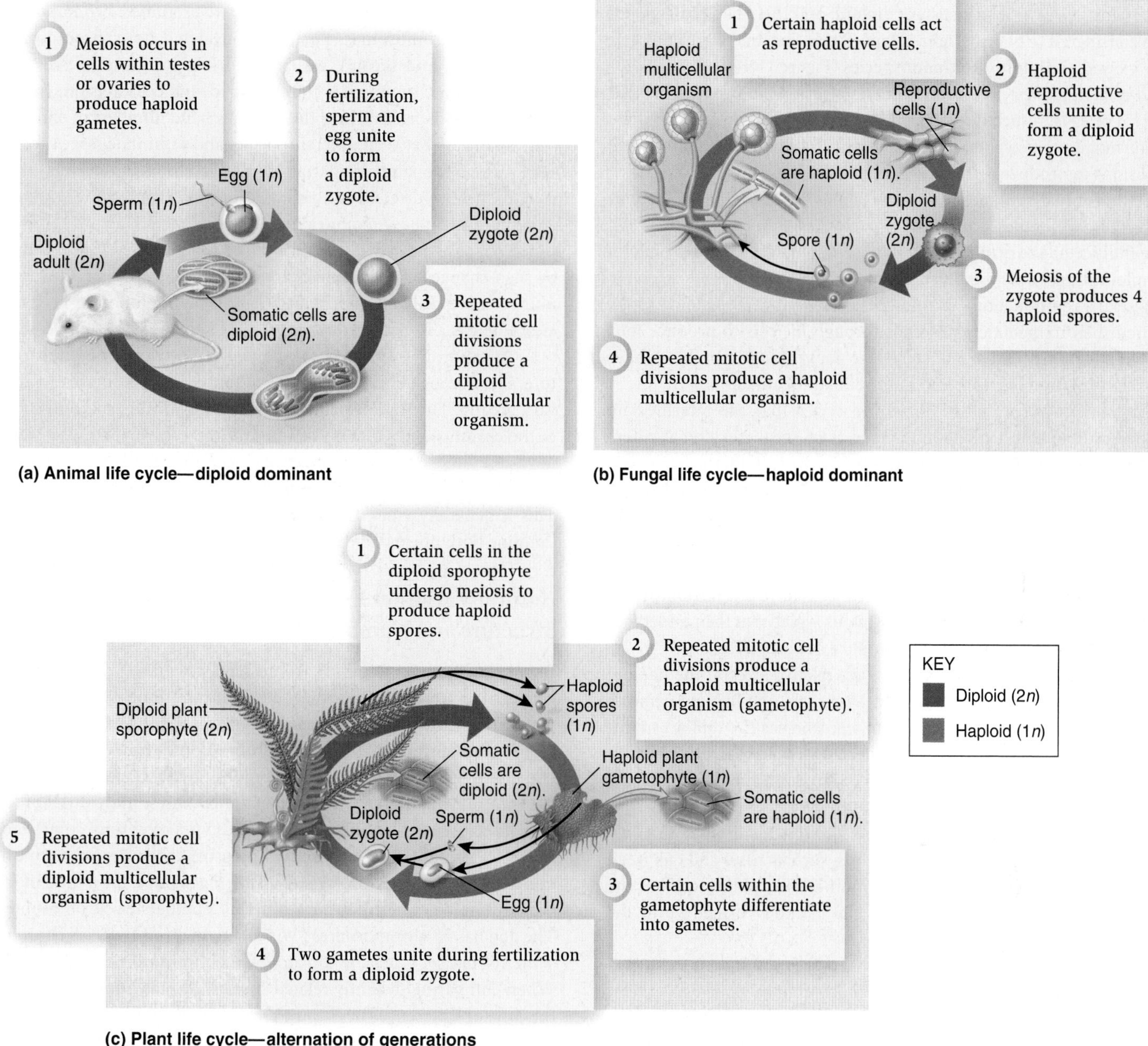
1 Meiosis occurs in cells within testes or ovaries to produce haploid gametes.
2 During fertilization, sperm and egg unite to form a diploid zygote.
Egg (1n)
Sperm (1n)
Diploid adult (2n)
Diploid zygote (2n)
Somatic cells are diploid (2n).
3 Repeated mitotic cell divisions produce a diploid multicellular organism.
(a) Animal life cycle—diploid dominant
1 Certain haploid cells act as reproductive cells.
Haploid multicellular organism
Reproductive cells (1n)
2 Haploid reproductive cells unite to form a diploid zygote.
Somatic cells are haploid (1n).
Diploid zygote (2n)
Spore (1n)
3 Meiosis of the zygote produces 4 haploid spores.
4 Repeated mitotic cell divisions produce a haploid multicellular organism.
(b) Fungal life cycle—haploid dominant
1 Certain cells in the diploid sporophyte undergo meiosis to produce haploid spores.
2 Repeated mitotic cell divisions produce a haploid multicellular organism (gametophyte).
KEY
Diploid (2n)
Haploid (1n)
Diploid plant sporophyte (2n)
Haploid spores (1n)
Somatic cells are diploid (2n).
Haploid plant gametophyte (1n)
Somatic cells are haploid (1n).
Diploid zygote (2n)
Sperm (1n)
Egg (1n)
5 Repeated mitotic cell divisions produce a diploid multicellular organism (sporophyte).
3 Certain cells within the gametophyte differentiate into gametes.
4 Two gametes unite during fertilization to form a diploid zygote.
(c) Plant life cycle—alternation of generations

Figure 15.14 A comparison of three types of sexual life cycles.

Concept Check: *What is the main reason for meiosis in animals? What is the main reason for mitosis in animals?*

two processes. DNA replication occurs prior to mitosis and meiosis I, but not between meiosis I and II. During prophase of meiosis I, the homologs synapse to form bivalents. This explains why crossing over occurs commonly during meiosis, but rarely during mitosis. During prometaphase of mitosis and meiosis II, pairs of sister chromatids are attached to both poles. In contrast, during meiosis I, each pair of sister chromatids (within a bivalent) is attached to a single pole. Bivalents align along the metaphase plate during metaphase of meiosis I, whereas sister chromatids align along the metaphase plate during metaphase of mitosis and meiosis II. At anaphase of meiosis I, the homologous chromosomes separate, but the sister chromatids remain together. In contrast, sister chromatid separation occurs during anaphase of mitosis and meiosis II. Taken together, the steps of mitosis produce two diploid cells that are genetically identical, whereas the steps of meiosis involve two sequential cell divisions that produce four haploid cells that may not be genetically identical.

Sexually Reproducing Species Produce Haploid and Diploid Cells at Different Times in Their Life Cycles

Let's now turn our attention to the relationship between mitosis, meiosis, and sexual reproduction in animals, plants, fungi, and protists. For any given species, the sequence of events that produces another generation of organisms is known as a **life cycle**. For sexually reproducing organisms, this usually involves an alternation between haploid cells or organisms and diploid cells or organisms (Figure 15.14).

Most species of animals are diploid, and their haploid gametes are considered to be a specialized type of cell. For this reason, animals are viewed as **diploid-dominant species** (Figure 15.14a). Certain diploid cells in the testes or ovaries undergo meiosis to produce haploid sperm or eggs, respectively. During fertilization, sperm and egg unite to form a diploid zygote, which then undergoes repeated mitotic cell divisions to produce a diploid multicellular organism.

By comparison, most fungi and some protists are just the opposite; they are **haploid-dominant species** (Figure 15.14b). In fungi, the multicellular organism is haploid ($1n$); only the zygote is diploid. Haploid fungal cells are most commonly produced by mitosis. During sexual reproduction, haploid cells unite to form a diploid zygote, which then immediately proceeds through meiosis to produce four haploid cells called spores. Each spore goes through mitotic cellular divisions to produce a haploid multicellular organism.

Plants and some algae have life cycles that are intermediate between diploid or haploid dominance. Such species exhibit an **alternation of generations** (Figure 15.14c). The species alternate between diploid multicellular organisms called **sporophytes**, and haploid multicellular organisms called **gametophytes**. Meiosis in certain cells within the sporophyte produces haploid spores, which divide by mitosis to produce the gametophyte. Particular cells within the gametophyte differentiate into haploid gametes. Fertilization occurs between two gametes, producing a diploid zygote that then undergoes repeated mitotic cell divisions to produce a sporophyte.

Among different plant species, the relative sizes of the haploid and diploid organisms vary greatly. In mosses, the haploid gametophyte is a visible multicellular organism, whereas the diploid sporophyte is smaller and remains attached to the haploid organism. In other plants, such as ferns (Figure 15.14c), both the diploid sporophyte and haploid gametophyte can grow independently. The sporophyte is considerably larger and is the organism we commonly think of as a fern. In seed-bearing plants, such as roses and oak trees, the diploid sporophyte is the large multicellular plant, whereas the gametophyte is composed of only a few cells and is formed within the sporophyte.

When comparing animals, plants, and fungi, it's interesting to consider how gametes are made. Animals produce gametes by meiosis. In contrast, plants and fungi produce reproductive cells by mitosis. The gametophyte of plants is a haploid multicellular organism that is created by mitotic cellular divisions of a haploid spore. Within the multicellular gametophyte, certain cells become specialized as gametes.

15.4 Variation in Chromosome Structure and Number

Learning Outcomes:

1. Describe how chromosomes can vary in size, centromere location, and number.
2. Identify the four ways that the structure of a chromosome can be changed via mutation.
3. Compare and contrast changes in the number of sets of chromosomes and changes in the number of individual chromosomes.
4. Give examples of how changes in chromosome number affect the characteristics of animals and plants.

In the previous sections of this chapter, we examined two important features of chromosomes. First, we considered how chromosomes occur in sets, and second, we explored two sorting processes that determine the chromosome number following cell division. In this section, we will examine how the structures and numbers of chromosomes can vary between different species and within the same species.

Why is the study of chromosomal variation important? First, geneticists have discovered that variations in chromosome structure and number can have major effects on the characteristics of an organism. We now know that several human genetic diseases are caused by such changes. In addition, changes in chromosome structure and number have been an important factor in the evolution of new species, which is a topic we will consider in Chapter 25.

Chromosome variation can be viewed in two ways. The structure and number of chromosomes among different species tend to vary greatly. There is also considerable variety in the size and shape of the chromosomes of a given species. On relatively rare occasions, however, the structure or number of chromosomes changes so that an individual is different from most other members of the same species. This is generally viewed as an abnormality. In this section, we will examine both normal and abnormal types of chromosome variation.

Natural Variation Exists in Chromosome Structure and Number

Before we begin to examine chromosome variation, we need to have a reference point for a normal set of chromosomes. To determine what the normal chromosomes of a species look like, a cytogeneticist microscopically examines the chromosomes from several members of the species. Chromosome composition within a given species tends to remain relatively constant. In most cases, individuals of the same species have the same number and types of chromosomes. For example, as mentioned previously, the usual chromosome composition of human cells is two sets of 23 chromosomes, for a total of 46. Other diploid species may have different numbers of chromosomes. The dog has 78 chromosomes (39 per set), the fruit fly has 8 chromosomes (4 per set), and the tomato has 24 chromosomes (12 per set). When comparing distantly related species, such as humans and fruit flies, major differences in chromosomal composition are observed.

The chromosomes of a given species also vary considerably in size and shape. Cytogeneticists have various ways to classify and identify chromosomes in their metaphase form. The three most commonly used features are size, location of the centromere, and banding patterns, which are revealed when the chromosomes are treated with stains. Based on centromere location, each chromosome is classified as **metacentric** (near the middle), **submetacentric** (off center), **acrocentric** (near one end), or **telocentric** (at the end) (Figure 15.15). Because the centromere is not exactly in the center of a chromosome, each chromosome has a short arm and a long arm. The short arm is designated with the letter "p" (for the French *petite*), and the long arm is designated with the letter "q." In the case of telocentric chromosomes, the short arm may be nearly nonexistent. When preparing a karyotype (see Figure 15.1), the chromosomes are aligned with the short arms on top and the long arms on the bottom.

Because different chromosomes often have similar sizes and centromeric locations, cytogeneticists must use additional methods to

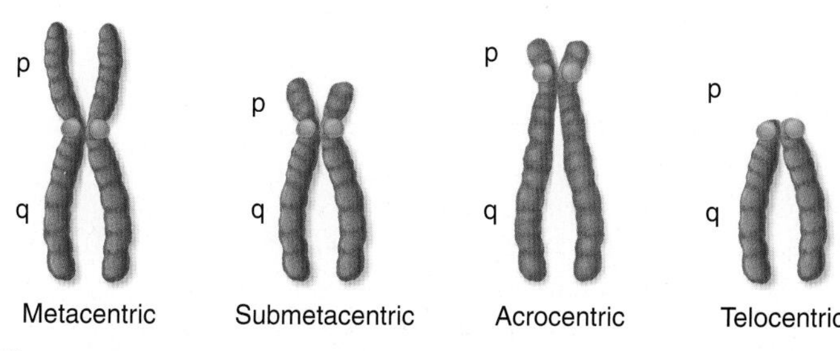

Figure 15.15 A comparison of centromeric locations among metaphase chromosomes.

accurately identify each type of chromosome within a karyotype. For detailed identification, chromosomes are treated with stains to produce characteristic banding patterns. Cytogeneticists use several different staining procedures to identify specific chromosomes. An example is Giemsa stain, which produces G bands (see Figure 15.1). The alternating pattern of G bands is unique for each type of chromosome.

The banding pattern of eukaryotic chromosomes is useful in two ways. First, individual chromosomes can be distinguished from each other, even if they have similar sizes and centromeric locations. Also, banding patterns are used to detect changes in chromosome structure that occur as a result of mutation.

Mutations Can Alter Chromosome Structure

Let's now consider how the structures of chromosomes can be modified by a mutation, a heritable change in the genetic material. Chromosomal mutations, which involve the breaking and rejoining of chromosomes, are categorized as deletions, duplications, inversions, and translocations (Figure 15.16).

Deletions and duplications are changes in the total amount of genetic material in a single chromosome. When a **deletion** occurs, a segment of chromosomal material is removed. The affected chromosome becomes deficient in a significant amount of genetic material. In a **duplication**, a section of a chromosome occurs two or more times in a row.

What are the consequences of a deletion or duplication? The possible effects depend on their size and whether they include genes or portions of genes that are vital to the development of the organism. When deletions or duplications have an effect, they are usually detrimental. Larger changes in the amount of genetic material tend to be more harmful because more genes are missing or duplicated.

Inversions and translocations are chromosomal rearrangements. An **inversion** is a change in the direction of the genetic material along a single chromosome. When a segment of one chromosome has been inverted, the order of G bands is opposite to that of a normal chromosome (Figure 15.16c). A **translocation** occurs when one segment of a chromosome becomes attached to a different chromosome. In a **simple translocation**, a single piece of chromosome is attached to another chromosome (Figure 15.16d). In a **reciprocal translocation**, two different types of chromosomes exchange pieces, thereby producing two abnormal chromosomes carrying translocations (Figure 15.16e).

Variation Occurs in the Number of Chromosome Sets and the Number of Individual Chromosomes

Variations in chromosome number can be categorized in two ways: variation in the number of sets of chromosomes and variation in the number of particular chromosomes within a set. The suffix -ploid or

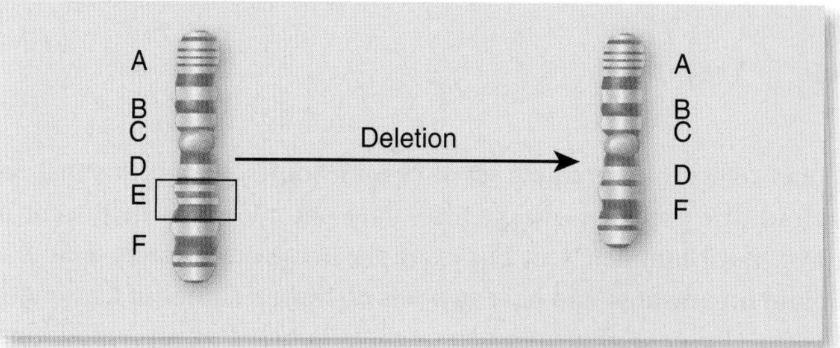

(a) Deletion: Removes a segment of chromosome.

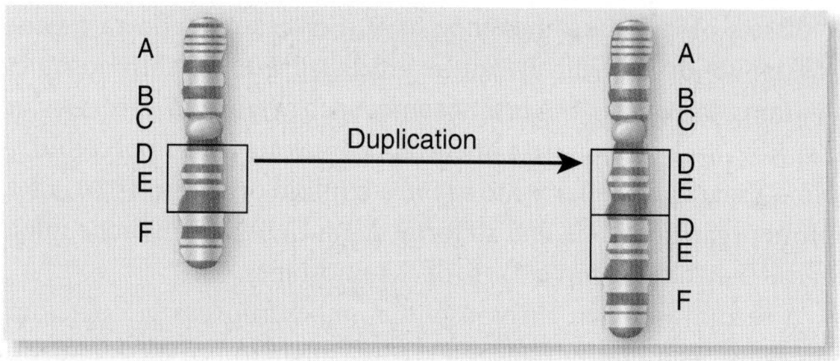

(b) Duplication: Doubles a particular region.

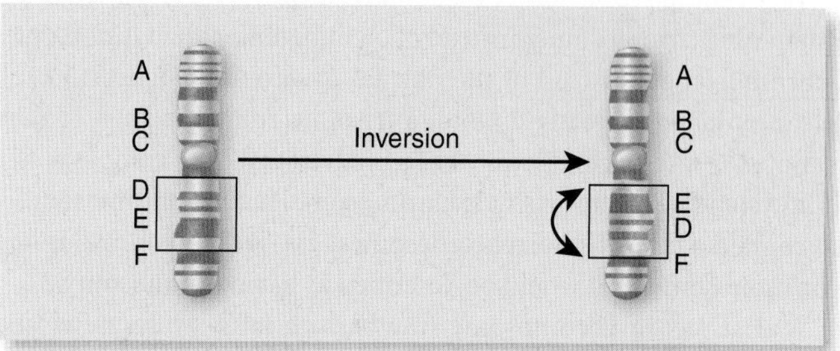

(c) Inversion: Flips a region to the opposite orientation.

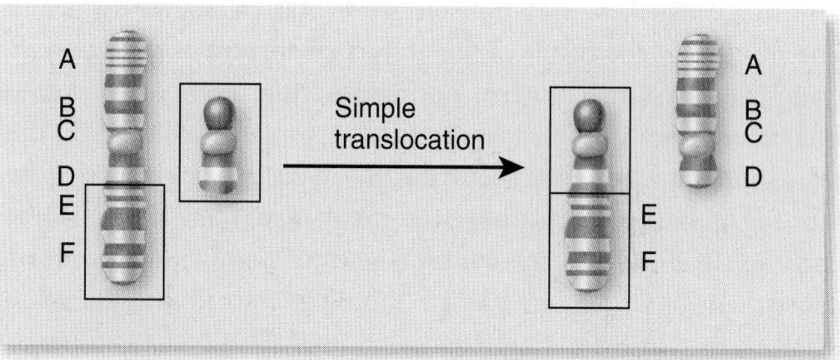

(d) Simple translocation: Moves a segment of 1 chromosome to another chromosome.

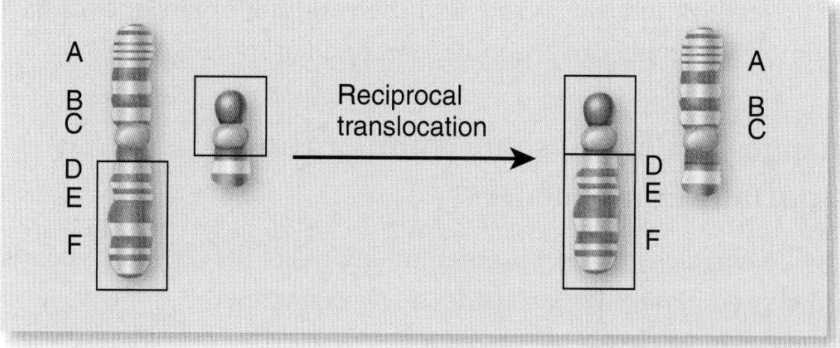

(e) Reciprocal translocation: Exchanges pieces between 2 different chromosomes.

Figure 15.16 Types of changes in chromosome structure. The letters alongside the chromosomes are placed there as frames of reference.

Concept Check: *Which types of changes shown here do not affect the total amount of genetic material?*

-ploidy refers to a complete set of chromosomes. Organisms that are **euploid** (the prefix eu- means true) have chromosomes that occur in one or more complete sets. For example, in a species that is diploid, a euploid organism would have two sets of chromosomes in its somatic cells. In *Drosophila melanogaster*, for example, a normal individual has eight chromosomes. The species is diploid, having two sets of four chromosomes each (Figure 15.17a). Organisms can vary in the number of sets of chromosomes they have. For example, on rare occasions, an abnormal fruit fly can be produced with 12 chromosomes, containing three sets of 4 chromosomes each (Figure 15.17b). Organisms with three or more sets of chromosomes are called **polyploid**. A diploid organism is referred to as $2n$, a **triploid** organism as $3n$, a **tetraploid** organism as $4n$, and so forth. All such organisms are euploid because they have complete sets of chromosomes.

A second way that chromosome number can vary is a phenomenon called **aneuploidy**. This refers to an alteration in the number of particular chromosomes, so the total number of chromosomes is not an exact multiple of a set. For example, an abnormal fruit fly could contain nine chromosomes instead of eight because it had three copies of chromosome 2 instead of the normal two copies (Figure 15.17c). Instead of being perfectly diploid, a trisomic animal is $2n + 1$. Such an animal is said to be trisomic and have **trisomy** 2. By comparison, a fruit fly could be lacking a single chromosome, such as chromosome 3, and contain a total of seven chromosomes ($2n - 1$). This animal is said to be monosomic and would be described as having **monosomy** 3.

Variations in chromosome number are fairly widespread and have a significant effect on the characteristics of plants and animals. For these reasons, researchers have wanted to understand the mechanisms that cause these variations. In some cases, a change in chromosome number is the result of the abnormal sorting of chromosomes during cell division. The term **nondisjunction** refers to an event in which the chromosomes do not separate properly during cell division. Nondisjunction can occur during meiosis I or meiosis II and produces haploid cells that have too many or too few chromosomes. Figure 15.18 illustrates the consequences of nondisjunction during meiosis I. In this case, one pair of homologs moved into the cell on the left instead of separating from each other. This results in the production of aneuploid cells, with either too many or too few chromosomes. If such a cell becomes a gamete that fuses with another gamete during fertilization, the zygote and the resulting organism will have an abnormal number of chromosomes in all of its cells.

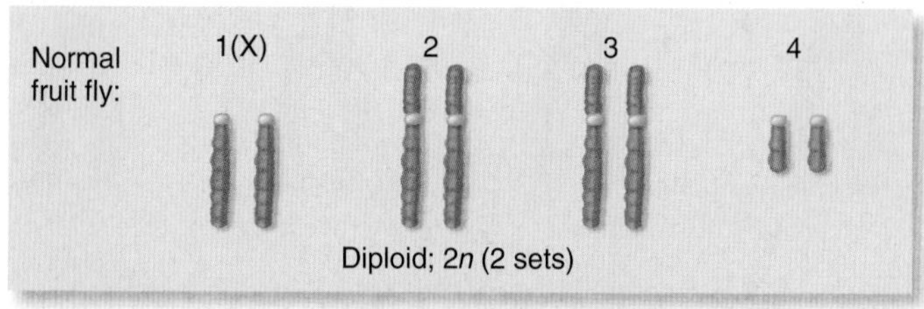

(a) Normal fruit fly chromosome composition

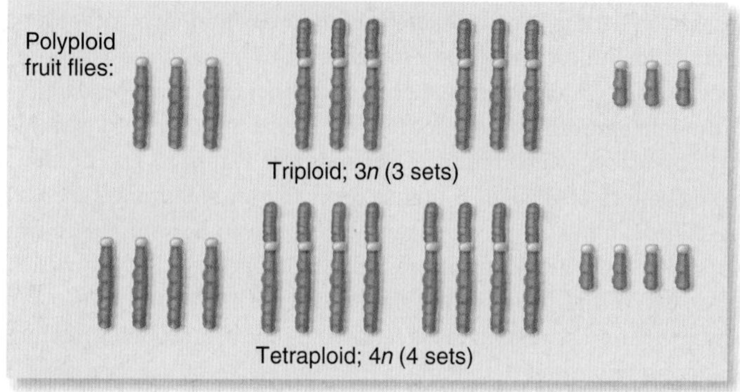

(b) Polyploidy

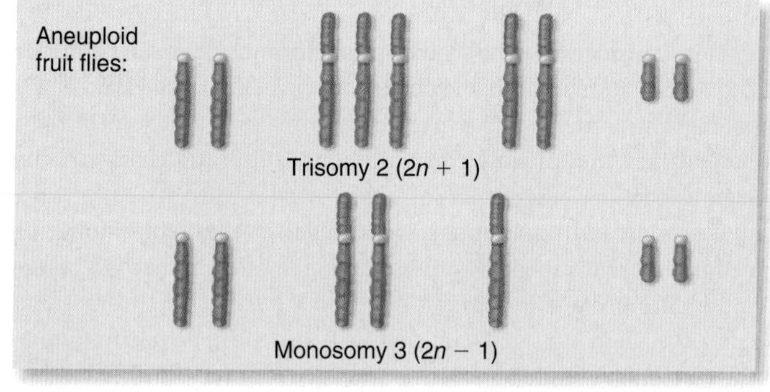

(c) Aneuploidy

Figure 15.17 Types of variation in chromosome number. **(a)** The normal diploid number of chromosomes in *Drosophila*. The X chromosome is also called chromosome 1. Examples of chromosomes of **(b)** polyploid flies and **(c)** aneuploid flies.

Changes in Chromosome Number Have Important Consequences

How do changes in chromosome number affect the characteristics of animals and plants? Let's consider a few examples.

Changes in Chromosome Number in Animals In many cases, animals do not tolerate deviations from diploidy well. For example, polyploidy in mammals is generally a lethal condition. However, a few cases of naturally occurring variations from diploidy do occur in animals. Male bees, which are produced from unfertilized eggs, contain a single set of chromosomes and are therefore haploid organisms. By comparison, fertilized eggs become female bees, which are diploid. A few examples of vertebrate polyploid animals have been discovered. Interestingly, on rare occasions, animals that are morphologically very similar to each other can be found as a diploid species as well as a separate polyploid species. This situation occurs among certain amphibians and reptiles. Figure 15.19 shows photographs of a diploid and a tetraploid frog. As you can see, they look very similar.

One important reason that geneticists are so interested in aneuploidy is its relationship to certain inherited disorders in humans. Even though most people are born with 46 chromosomes, alterations in chromosome number occur at a surprising frequency during gamete formation. About 5% to 10% of all fertilized human eggs result in an embryo with an abnormality in chromosome number. In most cases, these abnormal embryos do not develop properly and result in a spontaneous abortion very early in pregnancy. Approximately 50% of all spontaneous abortions are due to alterations in chromosome number.

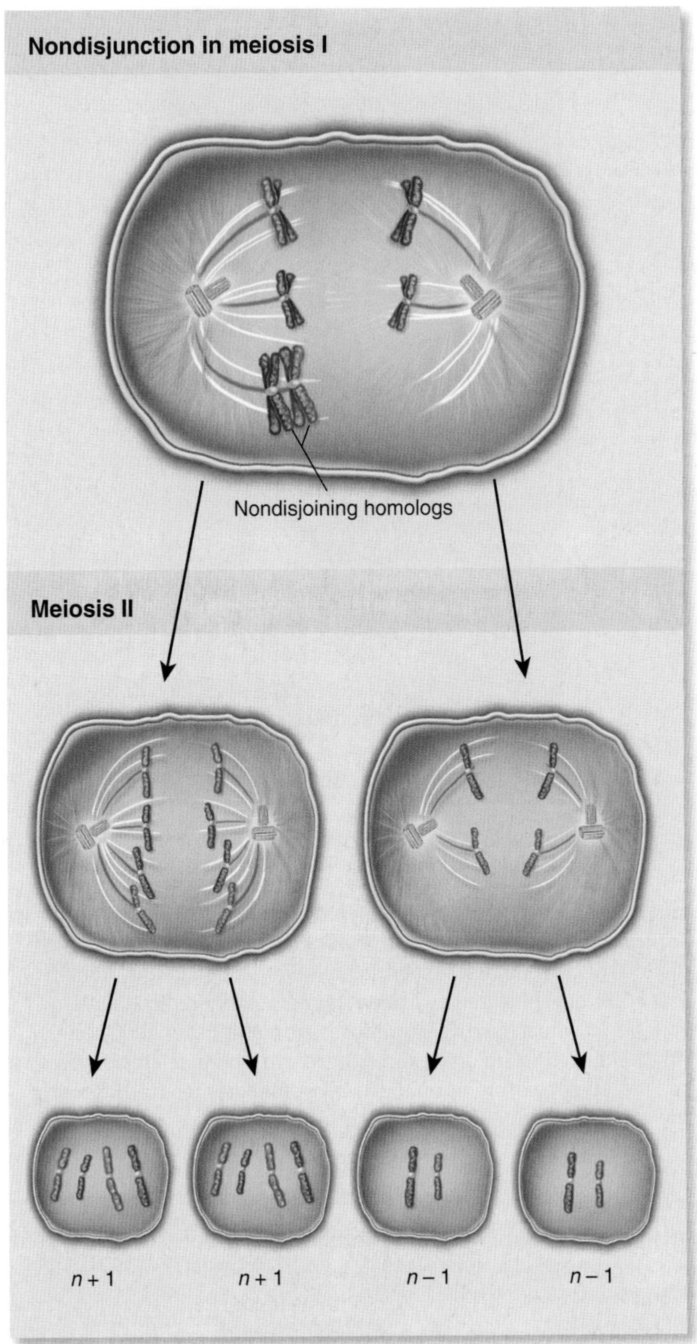

Figure 15.18 **Nondisjunction during meiosis I.** For simplicity, this cell shows only three pairs of homologous chromosomes. One of the three pairs does not disjoin properly, and both homologs have moved into the cell on the left. The resulting haploid cells shown at the bottom are all aneuploid, resulting in gametes with four chromosomes and two chromosomes, instead of three.

In some cases, an abnormality in chromosome number produces an offspring that can survive. Several human disorders are the result of abnormalities in chromosome number. The most common are trisomies of chromosomes 21, 18, or 13, or abnormalities in the number of the sex chromosomes (**Table 15.2**). These syndromes are most likely due to nondisjunction. For example, Turner syndrome (XO) may occur when a gamete that is lacking a sex chromosome due to nondisjunction has fused with a gamete carrying an X chromosome. By comparison, triple X syndrome (XXX) occurs when a gamete carrying two X chromosomes fuses with a gamete carrying a single X chromosome.

Most of the known trisomies involve chromosomes that are relatively small, so they carry fewer genes. Trisomies of the other human chromosomes and most monosomies are presumed to be lethal and have been found in spontaneously aborted embryos and fetuses.

Human abnormalities in chromosome number are influenced by the age of the parents. Older parents are more likely to produce children with abnormalities in chromosome number, possibly because meiotic nondisjunction is more likely to occur in older cells. **Down syndrome**, which was first described by the English physician John Langdon Down in 1866, provides an example. This disorder is caused by the inheritance of three copies of chromosome 21 (see Table 15.2). The incidence of Down syndrome rises with the age of either parent. In males, however, the rise occurs relatively late in life, usually past the age when most men have children. By comparison, the likelihood of having a child with Down syndrome rises dramatically during the later reproductive ages of women.

(a) *Hyla chrysoscelis* (diploid)

(b) *Hyla versicolor* (tetraploid)

Figure 15.19 **Differences in chromosome number in two closely related frog species.** The frog in **(a)** is diploid, whereas the frog in **(b)** is tetraploid. These frogs are in the act of performing their mating calls, which is why the skin under their mouths is protruding as a large bubble.

Table 15.2 Aneuploid Conditions in Humans

Condition	Frequency (# of live births)	Syndrome	Characteristics
Autosomal			
Trisomy 21	1/800	Down	Mental impairment, abnormal pattern of palm creases, slanted eyes, flattened face, short stature
Trisomy 18	1/6,000	Edward	Mental and physical impairment, facial abnormalities, extreme muscle tone, early death
Trisomy 13	1/15,000	Patau	Mental and physical impairment, wide variety of defects in organs, large triangular nose, early death
Sex chromosomal			
XXY	1/1,000 (males)	Klinefelter	Sexual immaturity (no sperm), breast swelling (males)
XYY	1/1,000 (males)	Jacobs	Tall
XXX	1/1,500 (females)	Triple X	Tall and thin, menstrual irregularity
XO	1/5,000 (females)	Turner	Short stature, webbed neck, sexually undeveloped

(a) Wheat, *Triticum aestivum* (hexaploid)

(b) Diploid daylily (left) and tetraploid daylily (right)

Figure 15.20 Examples of polyploid plants. **(a)** Cultivated wheat, *Triticum aestivum*, is a hexaploid. It was derived from three different diploid species of grasses that originally were found in the Middle East and were cultivated by ancient farmers in that region. Modern varieties of wheat have been produced from this hexaploid species. **(b)** Differences in euploidy exist in these two closely related daylily species. The flower stems on the left are diploid, whereas those with the larger flowers on the right are tetraploid.

Changes in Chromosome Number in Plants In contrast to animals, plants commonly exhibit polyploidy. Polyploidy is also important in agriculture. In many instances, polyploid strains of plants display characteristics that are helpful to humans. They are often larger in size and more robust. These traits are clearly advantageous in the production of food. For example, the species of wheat that we use to make bread, *Triticum aestivum*, is a hexaploid (containing six sets of chromosomes) that arose from the union of diploid genomes from three closely related species (**Figure 15.20a**). During the course of its cultivation, two diploid species must have interbred to produce a tetraploid, and then a third species interbred with the tetraploid to produce a hexaploid. Plant polyploids tend to exhibit a greater adaptability, which allows them to withstand harsher environmental conditions. Polyploid ornamental plants commonly produce larger flowers than their diploid counterparts (**Figure 15.20b**).

Although polyploidy is often beneficial in plants, aneuploidy in all eukaryotic species usually has detrimental consequences on the characteristics of an organism. Why is aneuploidy usually detrimental? To answer this question, we need to consider the relationship between gene expression and chromosome number. For many, but not all genes, the level of gene expression is correlated with the number of genes per cell. Compared with a diploid cell, if a gene is carried on a chromosome that is present in three copies instead of two, approximately 150% of the normal amount of gene product is usually made. Alternatively, if only one copy of that gene is present due to a missing chromosome, only 50% of the gene product is typically made. For some genes, producing too much or too little of the gene product may not have adverse effects. However, for other genes, the over- or underexpression may interfere with the proper functioning of cells.

Summary of Key Concepts

15.1 The Eukaryotic Cell Cycle

- Cytogeneticists examine cells microscopically to determine their chromosome composition. A micrograph that shows the alignment of chromosomes from a given cell is called a karyotype. Eukaryotic chromosomes are inherited in sets. A diploid cell has two sets of chromosomes. The members of each pair are called homologs (Figure 15.1).
- The eukaryotic cell cycle consists of four phases called G_1 (first gap), S (synthesis of DNA), G_2 (second gap), and M phase (mitosis and cytokinesis). The G_1, S, and G_2 phases are collectively known as interphase (Figure 15.2).
- An interaction between cyclin and cyclin-dependent kinase is necessary for cells to progress through the cell cycle. Checkpoint proteins sense the environmental conditions and the integrity of the genome and control whether or not the cell progresses through the cell cycle (Figure 15.3).
- Masui and Markert studied the maturation of frog oocytes to identify a substance necessary for oocytes to progress through the cell cycle. This substance was initially called maturation promoting factor (MPF) and was later identified as a complex of mitotic cyclin and cyclin-dependent kinase (cdk) (Figures 15.4, 15.5).

15.2 Mitotic Cell Division

- In the process of mitotic cell division, a cell divides to produce two new cells (the daughter cells) that are genetically identical to the original cell.
- During S phase, eukaryotic chromosomes are replicated to produce a pair of identical sister chromatids that remain attached to each other (Figure 15.6).
- The mitotic spindle is a network of microtubules that plays a central role in chromosome sorting during cell division (Figure 15.7).
- Mitosis occurs in five phases called prophase, prometaphase, metaphase, anaphase, and telophase. During prophase, the chromosomes condense, and the nuclear envelope begins to dissociate. The spindle apparatus is completely formed by the end of prometaphase. At metaphase, the chromosomes are aligned in a single row along the metaphase plate of the spindle. During anaphase, the sister chromatids separate from each other and move to opposite poles; the poles themselves also move farther apart. During telophase, the chromosomes decondense, and the nuclear envelope re-forms (Figure 15.8).
- Cytokinesis, which occurs after mitosis, is the division of the cytoplasm to produce two distinct daughter cells. In animal cells, cytokinesis involves the formation of a cleavage furrow. In plant cells, two separate cells are produced by the formation of a cell plate. Among all animals, 20 different proteins are required for cytokinesis to occur (Figures 15.9, 15.10).

15.3 Meiosis and Sexual Reproduction

- The process of meiosis begins with a diploid cell and produces four haploid cells with one set of chromosomes each (Figure 15.11).
- During prophase of meiosis, homologous pairs of sister chromosomes synapse, and crossing over occurs. After crossing over, chiasmata—the site where crossing over occurs—become visible (Figure 15.12).
- Meiosis consists of two divisions—meiosis I and II—each composed of prophase, prometaphase, metaphase, anaphase, and telophase. During meiosis I, the homologs are separated into two different cells, and during meiosis II, the sister chromatids are separated into four different cells (Figure 15.13, Table 15.1).
- Animals are diploid-dominant species, whereas most fungi and some protists are haploid-dominant. Plants alternate between diploid and haploid forms (Figure 15.14).

15.4 Variation in Chromosome Structure and Number

- Chromosomes are classified as metacentric, submetacentric, acrocentric, and telocentric, based on their centromere location. Each type of chromosome can be uniquely identified by its banding pattern after staining (Figure 15.15).
- Deletions, duplications, inversions, and translocations are different ways in which mutations alter chromosome structure (Figure 15.16).
- A euploid organism has chromosomes that occur in complete sets. A polyploid organism has three or more sets of chromosomes. An organism that has one too many (trisomy) or one too few (monosomy) chromosomes is termed aneuploid. Aneuploidy can be caused by nondisjunction, an event in which the chromosomes do not separate properly during cell division (Figures 15.17, 15.18).
- Polyploid animals are relatively rare, but polyploid plants are common and tend to be larger and more robust than their diploid counterparts (Figures 15.19, 15.20).
- Aneuploidy in humans is responsible for several types of human genetic diseases, including Down syndrome (Table 15.2).

Assess and Discuss

Test Yourself

1. In which phase of the cell cycle are chromosomes replicated?
   a. G_1 phase
   b. S phase
   c. M phase
   d. G_2 phase
   e. none of the above
2. If two chromosomes are homologous, they
   a. look similar under the microscope.
   b. have very similar DNA sequences.
   c. carry the same types of genes.
   d. may carry different versions of the same gene.
   e. are all of the above.
3. Checkpoints during the cell cycle are important because they
   a. allow the organelle activity to catch up to cellular demands.
   b. ensure the integrity of the cell's DNA.
   c. allow the cell to generate sufficient ATP for cellular division.
   d. are the only time DNA replication can occur.
   e. do all of the above.

4. Which of the following is a reason for mitotic cell division?
   a. asexual reproduction
   b. gamete formation in animals
   c. multicellularity
   d. all of the above
   e. both a and c
5. A replicated chromosome is composed of
   a. two homologous chromosomes held together at the centromere.
   b. four sister chromatids held together at the centromere.
   c. two sister chromatids held together at the centromere.
   d. four homologous chromosomes held together at the centromere.
   e. one chromosome with a centromere.
6. Which of the following is *not* an event of anaphase of mitosis?
   a. The nuclear envelope breaks down.
   b. Sister chromatids separate.
   c. Kinetochore microtubules shorten, pulling the chromosomes to the pole.
   d. Polar microtubules push against each other, moving the poles farther apart.
   e. All of the above occur during anaphase.
7. A student is looking at cells under the microscope. The cells are from an organism that has a diploid number of 14. In one particular case, the cell has seven replicated chromosomes (sister chromatids) aligned at the metaphase plate of the cell. Which of the following statements accurately describes this particular cell?
   a. The cell is in metaphase of mitosis.
   b. The cell is in metaphase of meiosis I.
   c. The cell is in metaphase of meiosis II.
   d. All of the above are correct.
   e. Both b and c are correct.
8. Which of the following statements accurately describes a difference between mitosis and meiosis?
   a. Mitosis may produce diploid cells, whereas meiosis produces haploid cells.
   b. Homologous chromosomes synapse during meiosis but do not synapse during mitosis.
   c. Crossing over commonly occurs during meiosis, but it does not commonly occur during mitosis.
   d. All of the above are correct.
   e. Both a and c are correct.
9. During crossing over in meiosis I,
   a. homologous chromosomes are not altered.
   b. homologous chromosomes exchange genetic material.
   c. chromosomal damage occurs.
   d. genetic information is lost.
   e. cytokinesis occurs.
10. Aneuploidy may be the result of
   a. duplication of a region of a chromosome.
   b. inversion of a region of a chromosome.
   c. nondisjunction during meiosis.
   d. interspecies breeding.
   e. all of the above.

Conceptual Questions

1. Distinguish between homologous chromosomes and sister chromatids.
2. The *Oca2* gene, which influences eye color in humans, is found on chromosome 15. How many copies of this gene are found in the karyotype of Figure 15.1? Is it one, two, or four?
3.  A principle of biology is that *cells are the simplest unit of life.* Explain how mitosis is a key process in the formation of new cells.

Collaborative Questions

1. Why is it necessary for chromosomes to condense during mitosis and meiosis? What do you think might happen if chromosomes did not condense?
2. A diploid eukaryotic cell has 10 chromosomes (five per set). As a group, take turns having one student draw the cell as it would look during a phase of mitosis, meiosis I, or meiosis II; then have the other students guess which phase it is.

Online Resource

www.brookerbiology.com

Stay a step ahead in your studies with animations that bring concepts to life and practice tests to assess your understanding. Your instructor may also recommend the interactive eBook, individualized learning tools, and more.

Chapter Outline

16.1 Mendel's Laws of Inheritance
16.2 The Chromosome Theory of Inheritance
16.3 Pedigree Analysis of Human Traits
16.4 Sex Chromosomes and X-Linked Inheritance Patterns
16.5 Variations in Inheritance Patterns and Their Molecular Basis
16.6 Genetics and Probability
Summary of Key Concepts
Assess and Discuss

Simple Patterns of Inheritance

16

An African girl with albinism. This condition results in very light skin and hair color.

Ntombi knew she looked different as long as she can remember. Born in Nigeria in 1991, she has accepted her appearance, though she still finds the occasional stare from strangers to be disturbing. Ntombi has albinism, a condition characterized by a total or a partial lack of pigmentation of the skin, hair, and eyes. As a result, she has very fair skin, blond hair, and blue eyes.* In contrast, her parents and three brothers have dark skin, black hair, and brown eyes, as do most of her relatives and most of the people in the city where she lives. Ntombi is very close to her aunt, who also has albinism.

Cases like Ntombi's have intrigued people for many centuries. How do we explain the traits that are found in people, plants, and other organisms? Can we predict what types of offspring two parents will produce? To answer such questions, researchers have studied the characteristics among related individuals and tried to make some sense of the data. Their goal is to understand **inheritance**—the acquisition of traits by their transmission from parent to offspring.

The first systematic attempt to understand inheritance was carried out by German plant breeder Joseph Kolreuter between 1761 and 1766. In crosses between two strains of tobacco plants, Kolreuter found that the offspring were usually intermediate in appearance between the two parents. He concluded that parents make equal genetic contributions to their offspring and that their genetic material blends together as it is passed to the next generation. This interpretation was consistent with the concept known as **blending inheritance**, which was widely accepted at that time. In the late 1700s, Jean Baptiste Lamarck, a French naturalist, hypothesized that physiological events (such as use or disuse) could modify traits and such modified traits would be inherited by offspring. For example, an individual who became adept at archery would pass that skill to his or her offspring. Overall, the prevailing view prior to the 1800s was that hereditary traits were rather malleable and could change and blend over the course of one or two generations.

In contrast, microscopic observations of chromosome transmission during mitosis and meiosis in the second half of the 19th century provided compelling evidence for **particulate inheritance**—the idea that the determinants of hereditary traits are transmitted in discrete units or particles from one generation to the next. Remarkably, this idea was first put forward in the 1860s by a researcher who knew nothing about chromosomes. Gregor Mendel, remembered today as the "father of genetics," used statistical analysis of carefully designed plant breeding experiments to arrive at the concept of a gene, which is broadly defined as a unit of heredity. Forty years later, through the convergence of Mendel's work and that of cell biologists, this concept became the foundation of the modern science of genetics.

In this chapter, we will consider inheritance patterns and how the transmission of genes is related to the transmission of chromosomes. We will first consider the fundamental genetic patterns known as Mendelian inheritance and the relationship of these patterns to the behavior of chromosomes during meiosis. We will then examine the distinctive inheritance patterns of genes located on the X chromosome, paying special attention to the work of Thomas Hunt Morgan, whose investigation of these patterns confirmed that genes are on chromosomes. Finally, we will discuss the molecular basis of Mendelian inheritance and its variations and consider how probability calculations can be used to predict the outcome of crosses.

*In contrast to popular belief, most people with albinism have blue eyes, not pink eyes. This is particularly the case among Africans with albinism.

16.1 Mendel's Laws of Inheritance

Learning Outcomes:

1. List the advantages of using the garden pea to study inheritance.
2. Describe the difference between dominant and recessive traits.
3. Distinguish between genotype and phenotype.
4. Predict the outcome of genetic crosses using a Punnett square.
5. Define Mendel's law of segregation and law of independent assortment.

Gregor Johann Mendel (**Figure 16.1**) grew up on a small farm in northern Moravia, then a part of the Austrian Empire and now in the Czech Republic. At the age of 21, he entered the Augustinian monastery of St. Thomas in Brno, and was ordained a priest in 1847. Mendel then worked for a short time as a substitute teacher, but to continue teaching he needed a license. Surprisingly, he failed the licensing exam due to poor answers in physics and natural history, so he enrolled at the University of Vienna to expand his knowledge in these two areas. Mendel's training in physics and mathematics taught him to perceive the world as an orderly place, governed by natural laws that could be stated as simple mathematical relationships.

In 1856, Mendel began his historic studies on pea plants. For 8 years, he analyzed thousands of pea plants that he grew on a small plot in his monastery garden. In 1866, he published the results of his work in a paper entitled "Experiments on Plant Hybrids." This paper was largely ignored by scientists at that time, partly because of its title and because it was published in a somewhat obscure journal (*The Proceedings of the Brünn Society of Natural History*). Also, Mendel was clearly ahead of his time. During this period, biology had not yet become a quantitative, experimental science. In addition, the behavior of chromosomes during mitosis and meiosis, which provides a framework for understanding inheritance patterns, had yet to be studied. Prior to his death in 1884, Mendel reflected, "My scientific work has brought me a great deal of satisfaction and I am convinced it will be appreciated before long by the whole world." Sixteen years later, in 1900, Mendel's work was independently rediscovered by three biologists with an interest in plant genetics: Hugo de Vries of Holland, Carl Correns of Germany, and Erich von Tschermak of Austria. Within a few years, the influence of Mendel's landmark studies was felt around the world.

In this section, we will examine Mendel's experiments and how they led to the formulation of the basic genetic principles known as Mendel's laws. We will see that these principles apply not only to the pea plants Mendel studied, but also to a wide variety of sexually reproducing organisms, including humans.

Mendel Chose the Garden Pea to Study Inheritance

When two individuals of the same species with different characteristics are bred or crossed to each other, this is called **hybridization**, and the offspring are referred to as hybrids. For example, a hybridization experiment could involve breeding a purple-flowered plant to a white-flowered plant. Mendel was particularly intrigued by the consistency with which offspring of such crosses showed characteristics of one or the other parent in successive generations. His intellectual foundation in physics and the natural sciences led him to consider that this regularity might be rooted in natural laws that could be expressed mathematically. To uncover these laws, he carried out quantitative experiments in which he carefully analyzed the numbers of offspring carrying specific traits.

Figure 16.1 **Gregor Johann Mendel, the father of genetics.**

Mendel chose the garden pea, *Pisum sativum*, to investigate the natural laws that govern inheritance. Why did he choose this species? Several properties of the garden pea were particularly advantageous for studying inheritance. First, it was available in many varieties that differed in characteristics, such as the appearance of seeds, pods, flowers, and stems. Such general features of an organism are called **characters**. Figure 16.2 illustrates the seven characters that Mendel eventually chose to follow in his breeding experiments. Each of these characters was found in two discrete variants. For example, one character he followed was height, which had the variants known as tall and dwarf. Another was seed color, which had the variants yellow and green. A **trait** is an identifiable characteristic of an organism. The term trait usually refers to a variant for a character.* For example, seed color is a character, and green and yellow seed colors are traits.

A second important feature of garden peas is they are normally self-fertilizing. In **self-fertilization**, a female gamete is fertilized by a male gamete from the same plant. Like many flowering plants, peas have male and female sex organs in the same flower (**Figure 16.3**). Male gametes (sperm cells) are produced within pollen grains, which are formed in structures called stamens. Female gametes (egg cells) are produced in structures called ovules, which form within an organ called an ovary. For fertilization to occur, a pollen grain must land on the receptacle called a stigma, enabling a sperm to migrate to an ovule and fuse with an egg cell. In peas, the stamens and the ovaries are enclosed by a modified petal, an arrangement that greatly favors self-fertilization. Self-fertilization makes it easy to produce plants that breed true for a given trait, meaning the trait does not vary from generation to generation. For example, if a pea plant with yellow seeds breeds true for seed color, all the plants that grow from these seeds will also produce yellow seeds. A variety that continues to exhibit the same trait after several generations of self-fertilization is called a **true-breeding line**. Prior to conducting the studies described in this chapter, Mendel had already established that the seven characters he chose to study were true-breeding in the strains of pea plants he had obtained.

A third reason for using garden peas in hybridization experiments is the ease of making crosses: The flowers are quite large and easy to manipulate. In some cases, Mendel wanted his pea plants to self-fertilize, but in others, he wanted to cross plants that differed with respect to some character, a process called hybridization, or **cross-fertilization**. In garden peas, cross-fertilization requires

*Geneticists may also use the term trait to refer to a character.

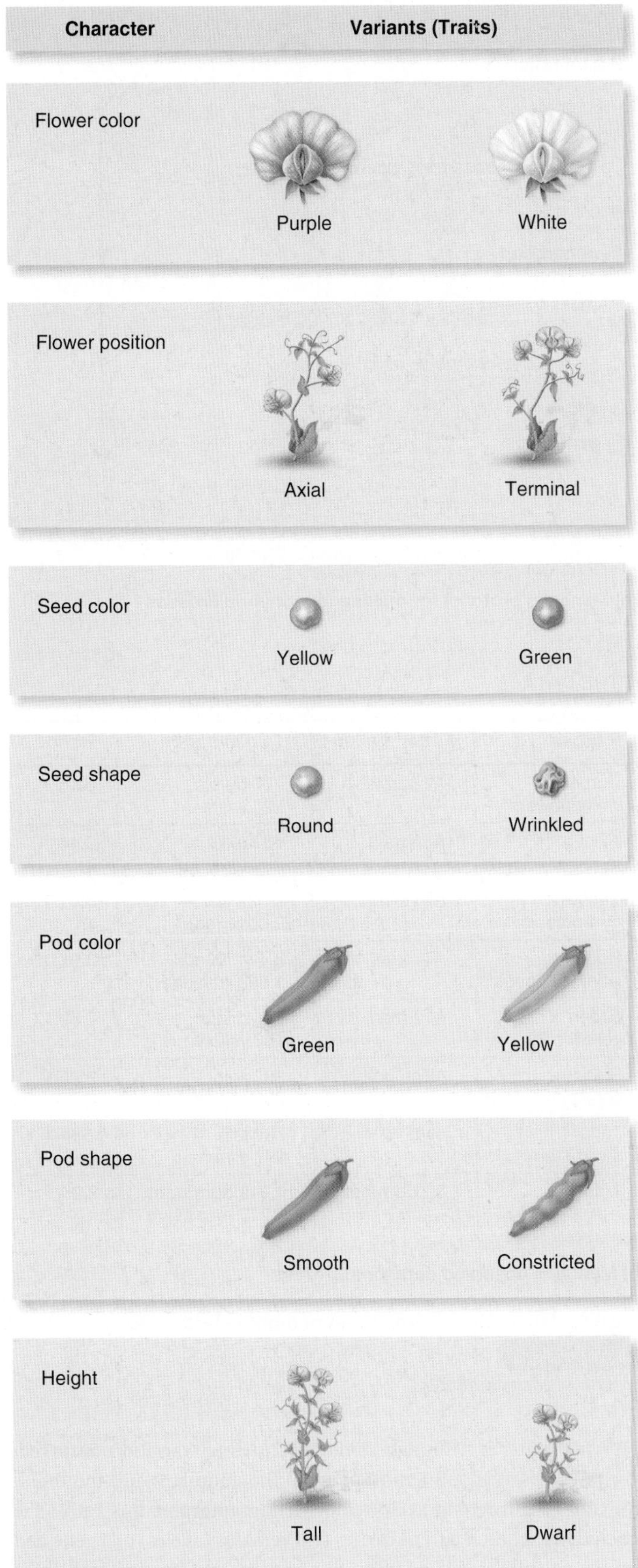

Character
Variants (Traits)
Flower color
Purple
White
Flower position
Axial
Terminal
Seed color
Yellow
Green
Seed shape
Round
Wrinkled
Pod color
Green
Yellow
Pod shape
Smooth
Constricted
Height
Tall
Dwarf

Figure 16.2 The seven characters that Mendel studied.

BIOLOGY PRINCIPLE The genetic material provides a blueprint for reproduction. The traits that Mendel studied in pea plants are governed by the genetic material of this species.

Concept Check: Is having blue eyes a character, a variant, or both?

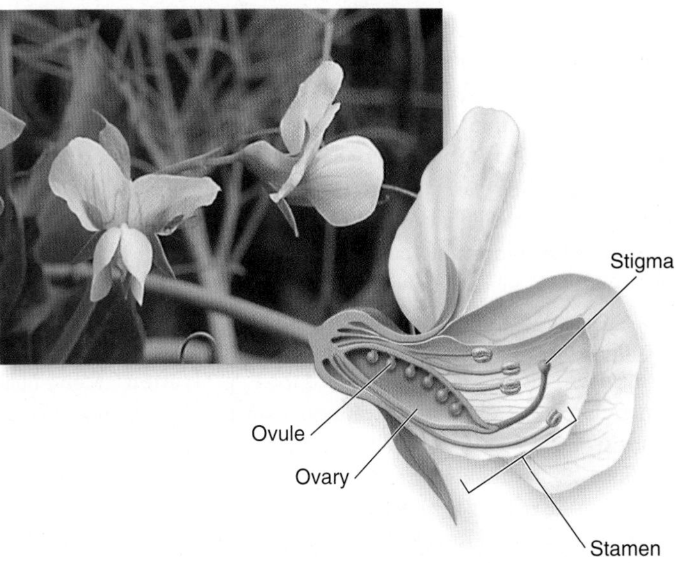

Figure 16.3 Flower structure in pea plants. The pea flower produces both male and female gametes. Sperm form in the pollen produced within the stamens; egg cells form in ovules within the ovary. A modified petal encloses the stamens and stigma, encouraging self-fertilization.

placing pollen from one plant onto the stigma of a flower on a different plant (**Figure 16.4**). Mendel would pry open an immature flower and remove the stamens before they produced pollen, so the flower could not self-fertilize. He then used a paintbrush to transfer pollen from another plant to the stigma of the flower that had its stamens removed. In this way, Mendel was able to cross-fertilize any two of his true-breeding pea plants and obtain any type of hybrid he wanted.

By Following the Inheritance Pattern of Single Traits, Mendel's Work Revealed the Law of Segregation

Mendel began his investigations by studying the inheritance patterns of pea plants that differed in a single character. A cross in which an experimenter follows the variants of only one character is called a **monohybrid cross**. As an example, we will consider a monohybrid

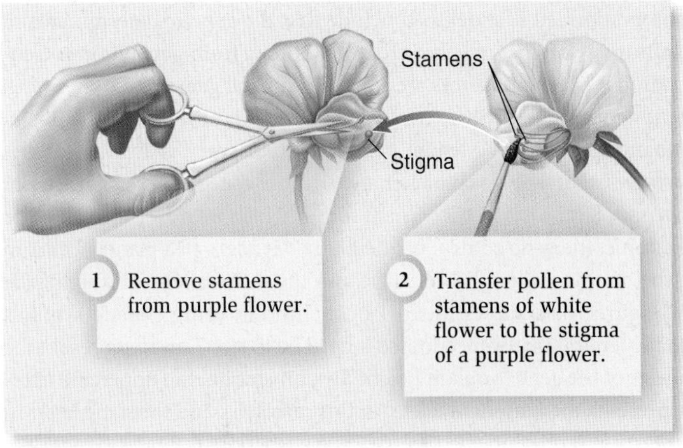

1 Remove stamens from purple flower.

Figure 16.4 A procedure for cross-fertilizing pea plants.

Concept Check: Why are the stamens removed from the purple flower?

cross in which Mendel followed the tall and dwarf variants for height (**Figure 16.5**). The left side of Figure 16.5a shows his experimental approach. The true-breeding parents are termed the **P generation** (parental generation), and their offspring constitute the **F_1 generation** (first filial generation, from the Latin *filius*, meaning son). When the true-breeding parents differ in a single character, their F_1 offspring are called single-trait hybrids, or **monohybrids**. When Mendel crossed true-breeding tall and dwarf plants, he observed that all plants of the F_1 generation were tall.

Next, Mendel followed the transmission of this character for a second generation. To do so, he allowed the F_1 monohybrids to self-fertilize, producing a generation called the **F_2 generation** (second filial generation). The dwarf trait reappeared in the F_2 offspring: Three-fourths of the plants were tall and one-fourth were dwarf. Mendel obtained similar results for each of the seven characters he studied, as shown in the data of Figure 16.5b. A quantitative analysis of his data allowed Mendel to postulate three important ideas about the properties and transmission of these traits from parents to offspring: (1) traits may exist in two forms: dominant and recessive; (2) an individual carries two genes for a given character, and genes have variant forms, which are now called **alleles**; and (3) the two alleles of a gene separate during gamete formation so that each sperm and egg receives only one allele.

Dominant and Recessive Traits Perhaps the most surprising outcome of Mendel's work was that the data argued strongly against the prevailing notion of blending inheritance. In each of the seven cases, the F_1 generation displayed a trait distinctly like one of the two parents rather than an intermediate trait. Using genetic terms that Mendel originated, we describe the alternative traits as dominant and recessive. The term **dominant** describes the displayed trait, whereas the term **recessive** describes a trait that is masked by the presence of a dominant trait. Tall stems and purple flowers are examples of dominant traits; dwarf stems and white flowers are examples of recessive traits. In this case, we say that tall is dominant over dwarf, and purple is dominant over white.

Genes and Alleles Mendel's results were consistent with particulate inheritance, in which the determinants of traits are inherited as unchanging, discrete units. In all seven cases, the recessive trait reappeared in the F_2 generation: Most F_2 plants displayed the dominant trait, whereas a smaller proportion showed the recessive trait. This observation led Mendel to conclude that the genetic determinants of traits are "unit factors" that are passed intact from generation to generation. These unit factors are what we now call **genes** (from the Greek *genos*, meaning birth), a term coined by the Danish botanist Wilhelm Johannsen in 1909. Mendel postulated that every individual carries two genes for a given character and that the gene for each character in his pea plant exists in two variant forms, which we now call alleles. For example, the gene controlling height in Mendel's pea plants occurs in two variants, called the tall allele and the dwarf allele. The right side of Figure 16.5a shows Mendel's conclusions, using genetic symbols (italic letters) that were adopted later. The letters T and t represent the alleles of the gene for plant height. By convention, the uppercase letter represents the dominant allele (in this case, tall), and the same letter in lowercase represents the recessive allele (dwarf).

Segregation of Alleles When Mendel compared the numbers of F_2 offspring exhibiting dominant and recessive traits, he noticed a recurring pattern. Although some experimental variation occurred, he always observed a 3:1 ratio between the dominant and the recessive trait (Figure 16.5b). How did Mendel interpret this ratio? He concluded that each parent carries two versions (alleles) of a gene and that the two alleles carried by an F_1 plant separate, or segregate, from each other during gamete formation, so each sperm or egg carries only one allele. The diagram in **Figure 16.6** shows that the segregation of the F_1 alleles should result in equal numbers of gametes carrying the dominant allele (T) and the recessive allele (t). If these gametes combine with one another randomly at fertilization, as shown in the figure, this would account for the 3:1 ratio of the F_2 generation. Note

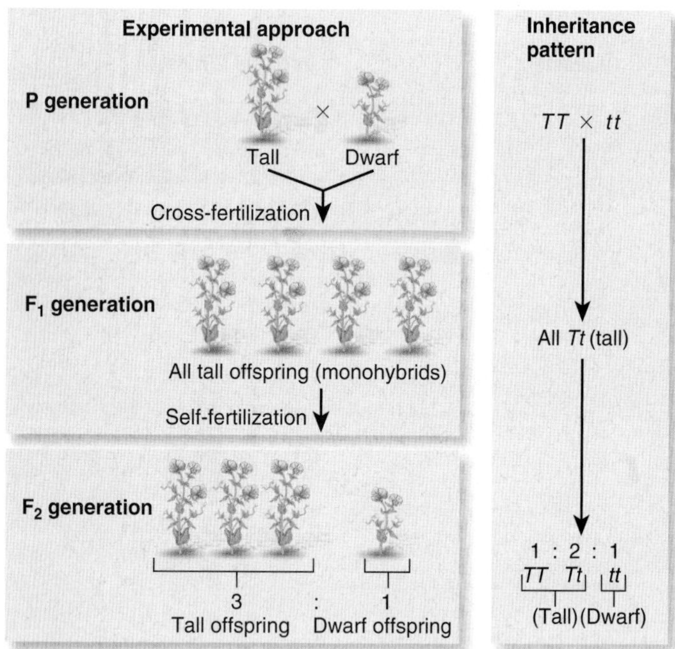

(a) Mendel's protocol for making monohybrid crosses

THE DATA

P cross	F_1 generation	F_2 generation	Ratio
Purple × white flowers	All purple	705 purple, 224 white	3.15:1
Axial × terminal flowers	All axial	651 axial, 207 terminal	3.14:1
Yellow × green seeds	All yellow	6,022 yellow, 2,001 green	3.01:1
Round × wrinkled seeds	All round	5,474 round, 1,850 wrinkled	2.96:1
Green × yellow pods	All green	428 green, 152 yellow	2.82:1
Smooth × constricted pods	All smooth	882 smooth, 299 constricted	2.95:1
Tall × dwarf stem	All tall	787 tall, 277 dwarf	2.84:1
Total	**All dominant**	**14,949 dominant, 5,010 recessive**	**2.98:1**

(b) Mendel's observed data for all 7 traits

Figure 16.5 Mendel's analyses of monohybrid crosses.

Concept Check: Why do offspring of the F_1 generation exhibit only one variant of each character?

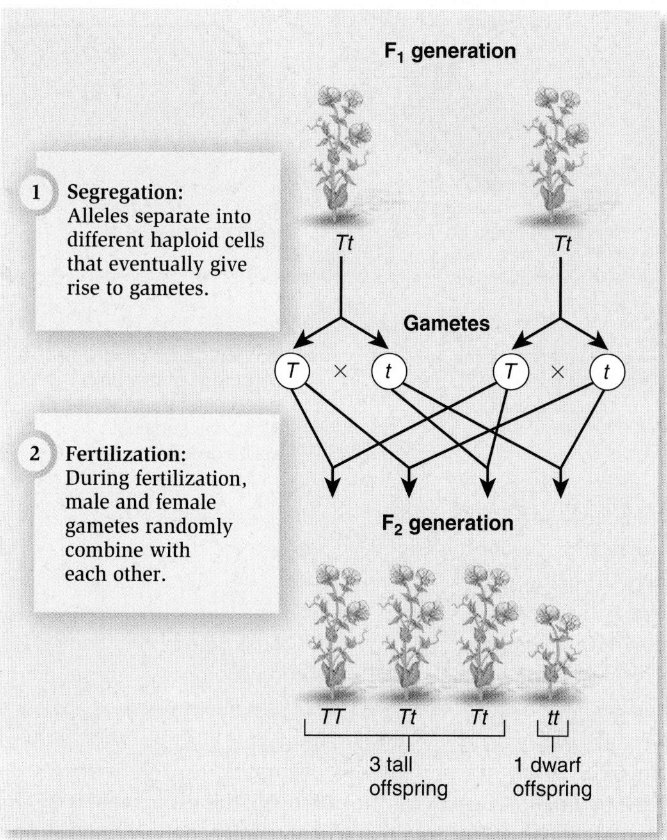

Figure 16.6 How the law of segregation explains Mendel's observed ratios. The segregation of alleles in the F_1 generation gives rise to gametes that carry just one of the two alleles. These gametes combine randomly during fertilization, producing the allele combinations *TT*, *Tt*, and *tt* in the F_2 offspring. The combination *Tt* occurs twice as often as either of the other two combinations because it can be produced in two different ways. The *TT* and *Tt* offspring are tall, whereas the *tt* offspring are dwarf.

Concept Check: What is ratio of the T allele to the t allele in the F_2 generation? Does this ratio differ from the 3:1 phenotype ratio? If so, explain why.

that a *Tt* individual can be produced by two different combinations of alleles—the *T* allele can come from the male gamete and the *t* allele from the female gamete, or vice versa. This accounts for the observation that *Tt* offspring are produced twice as often as either *TT* or *tt*. The idea that *the two alleles of a gene separate (segregate) during the formation of gametes so that every gamete receives only one allele* is known today as Mendel's **law of segregation**.

Genotype Describes an Organism's Genetic Makeup, Whereas Phenotype Describes Its Characteristics

To continue our discussion of Mendel's results, we need to introduce a few more genetic terms. The term **genotype** refers to the genetic composition of an individual. In the example shown in Figure 16.5a, *TT* and *tt* are the genotypes of the P generation, and *Tt* is the genotype of the F_1 generation. In the P generation, both parents are true-breeding plants, which means that each has identical copies of the allele of the gene for height. An individual with two identical alleles of a gene is said to be **homozygous** with respect to that gene. In the specific cross we are considering, the tall plant (*TT*) is homozygous for *T*, and the dwarf plant (*tt*) is homozygous for *t*. In contrast, a **heterozygous** individual carries two different alleles of a gene. Plants of the F_1 generation are heterozygous, with the genotype *Tt*, because every individual carries one copy of the tall allele (*T*) and one copy of the dwarf allele (*t*). The F_2 generation includes both homozygous individuals (homozygotes) and heterozygous individuals (heterozygotes).

The term **phenotype** refers to the characteristics of an organism that are the result of the expression of its genes. In the example in Figure 16.5a, one of the parent plants is phenotypically tall, and the other is phenotypically dwarf. Although the F_1 offspring are heterozygous (*Tt*), they are phenotypically tall because each of them has a copy of the dominant tall allele. In contrast, the F_2 plants display both phenotypes in a ratio of 3:1. Later in the chapter, we will examine the underlying molecular mechanisms that produce phenotypes, but in our discussion of Mendel's results, the term simply refers to a visible characteristic such as flower color or height.

A Punnett Square Is Used to Predict the Outcome of Crosses

A common way to predict the outcome of simple genetic crosses is to make a **Punnett square**, a method originally proposed by the British geneticist Reginald Punnett. To construct a Punnett square, you must know the genotypes of the parents. What follows is a step-by-step description of the Punnett-square approach, using a cross of heterozygous tall plants.

Step 1. *Write down the genotypes of both parents.* In this example, a heterozygous tall plant is crossed to another heterozygous tall plant. The plant providing the pollen is considered the male parent and the plant providing the eggs, the female parent. (In self-pollination, a single individual produces both types of gametes.)

Male parent: *Tt*
Female parent: *Tt*

Step 2. *Write down the possible gametes that each parent can make.* Remember the law of segregation tells us that a gamete contains only one copy of each allele.

Male gametes: *T* or *t*
Female gametes: *T* or *t*

Step 3. *Create an empty Punnett square.* The number of columns equals the number of male gametes, and the number of rows equals the number of female gametes. Our example has two rows and two columns. Place the male gametes across the top of the Punnett square and the female gametes along the side.

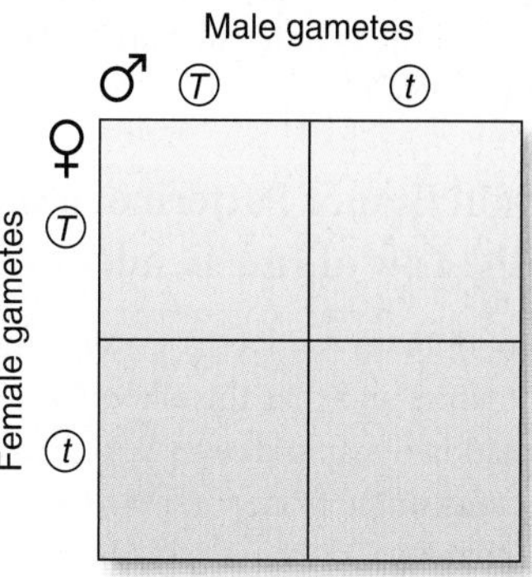

Step 4. *Fill in the possible genotypes of the offspring by combining the alleles of the gametes in the empty boxes.*

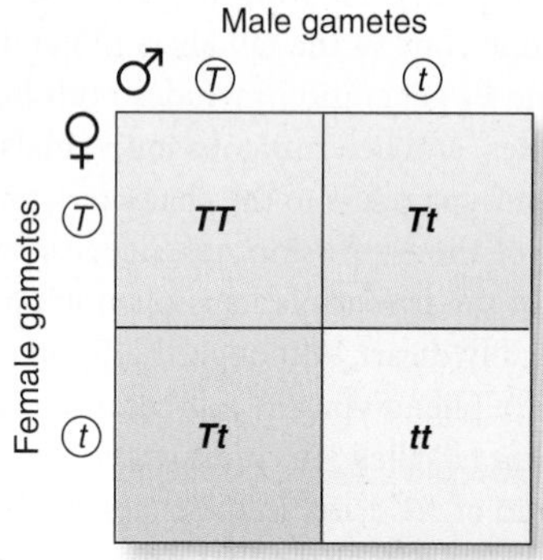

Step 5. *Determine the relative proportions of genotypes and phenotypes of the offspring.* The genotypes are obtained directly from the Punnett square. In this example, the genotypic ratios are 1*TT* : 2*Tt* : 1*tt*. To determine the phenotypes, you must know which allele is dominant. For plant height, *T* (tall) is dominant to *t* (dwarf). The genotypes *TT* and *Tt* are tall, whereas the genotype *tt* is dwarf. Therefore, our Punnett square shows us that the phenotypic ratio is expected to be 3 tall: 1 dwarf. Keep in mind, however, these are predicted ratios for large numbers of offspring. If only a few offspring are produced, the observed ratios could deviate significantly from the predicted ratios. We will examine the topics of sample size and genetic prediction later in this chapter.

A Testcross Is Used to Determine an Individual's Genotype

When a character has two variants, one of which is dominant over the other, we know that an individual with a recessive phenotype is homozygous for the recessive allele. A dwarf pea plant, for example, must have the genotype *tt*. But an individual with a dominant phenotype may be either homozygous or heterozygous—a tall pea plant may have the genotype *TT* or *Tt*. How can we distinguish between these two possibilities? Mendel devised a method called a **testcross** to address this question. In a testcross, the researcher crosses the individual of interest to a homozygous recessive individual and observes the phenotypes of the offspring.

Figure 16.7 shows how this procedure can be used to determine the genotype of a tall pea plant. If the testcross produces some dwarf offspring, as shown in the Punnett square on the right side, these offspring must have two copies of the recessive allele, one inherited from each parent. Therefore, the tall parent must be a heterozygote, with the genotype *Tt*. Alternatively, if all of the offspring are tall, as shown in the Punnett square on the left, the tall parent is likely to be a homozygote, with the genotype *TT*.

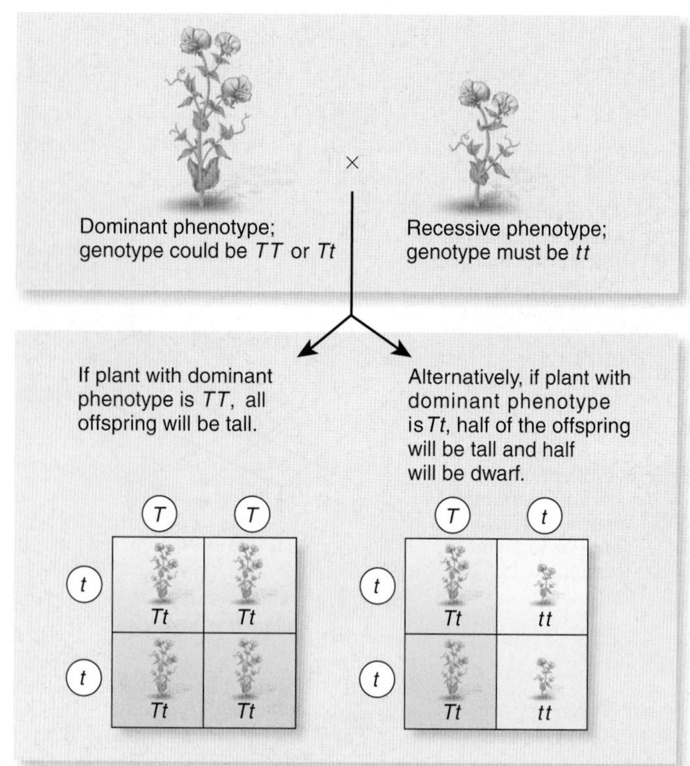

×
Dominant phenotype; genotype could be TT or Tt
Recessive phenotype; genotype must be tt
If plant with dominant phenotype is TT, all offspring will be tall.
Alternatively, if plant with dominant phenotype is Tt, half of the offspring will be tall and half will be dwarf.
T T
t
t
Tt Tt
Tt Tt
T t
t
t
Tt tt
Tt tt

Figure 16.7 **A testcross.** The purpose of this experiment is to determine if the organism with the dominant phenotype, in this case a tall pea plant, is a homozygote (*TT*) or a heterozygote (*Tt*).

Concept Check: *Let's suppose you had a plant with purple flowers and unknown genotype and conducted a testcross to determine its genotype. You obtained 41 plants: 20 with white flowers and 21 with purple flowers. What was the genotype of the original purple-flowered plant?*

Analyzing the Inheritance Pattern of Two Characters Demonstrated the Law of Independent Assortment

Mendel's analysis of monohybrid crosses suggested that traits are inherited as discrete units and that the alleles for a given gene segregate during the formation of haploid cells. To obtain additional insights into how genes are transmitted from parents to offspring, Mendel conducted crosses in which he simultaneously followed the inheritance of two different characters. A cross of this type is called a **dihybrid cross.** We will examine a dihybrid cross in which Mendel simultaneously followed the inheritance of seed color and seed shape (**Figure 16.8**). He began by crossing strains of pea plants that bred true for both characters. The plants of one strain had yellow, round seeds, and plants of the other strain had green, wrinkled seeds. He then allowed the F_1 offspring to self-fertilize and observed the phenotypes of the F_2 generation.

What are the possible patterns of inheritance for two characters? One possibility is that the two genes are linked in some way, so variants that occur together in the parents are always inherited as a unit. In our example, the allele for yellow seeds (*Y*) would always be inherited with the allele for round seeds (*R*), and the alleles for green seeds (*y*) would always be inherited with the allele for wrinkled seeds (*r*), as shown in Figure 16.8a. A second possibility is that the two genes are independent of one another, so their alleles are randomly distributed into gametes (Figure 16.8b). By following the transmission pattern of two characters simultaneously, Mendel could determine whether the genes that determine seed shape and seed color assort (are distributed) together as a unit or independently of each other.

What experimental results could Mendel predict for each of these two models? The two homozygous plants of the P generation can produce only two kinds of gametes, *YR* and *yr*, so in either case the F_1 offspring would be heterozygous for both genes; that is, they would have the genotypes *YyRr*. Because Mendel knew from his earlier experiments that yellow was dominant over green and round over wrinkled, he could predict that all the F_1 plants would have yellow, round seeds. In contrast, as shown in Figure 16.8, the ratios he

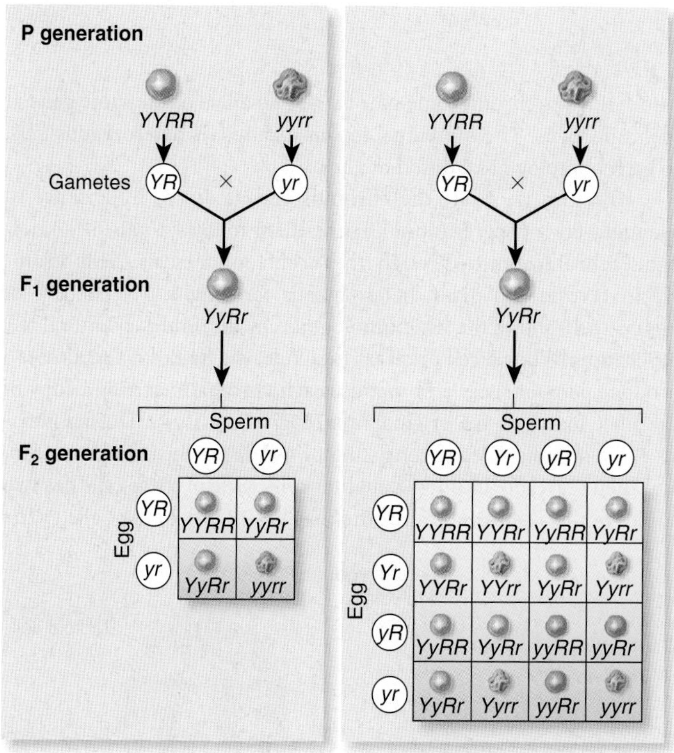

P cross	F_1 generation	F_2 generation
Yellow, round seeds × Green, wrinkled seeds	Yellow, round seeds	315 yellow, round seeds 101 yellow, wrinkled seeds 108 green, round seeds 32 green, wrinkled seeds

(c) The data observed by Mendel

Figure 16.8 Two hypotheses for the assortment of two different genes. In a cross between two true-breeding pea plants, one with yellow, round seeds and one with green, wrinkled seeds, all of the F_1 offspring have yellow, round seeds. When the F_1 offspring self-fertilize, the two hypotheses predict different phenotypes in the F_2 generation: **(a)** linked assortment, in which parental alleles stay associated with each other, or **(b)** independent assortment, in which each allele assorts independently. **(c)** Mendel's observations supported the independent assortment hypothesis.

Concept Check: What ratio of offspring phenotypes would have occurred if the linked hypothesis had been correct?

obtained in the F_2 generation would depend on whether the alleles of both genes assort together or independently.

If the parental genes are linked, as in Figure 16.8a, the F_1 plants could produce gametes that are only *YR* or *yr*. These gametes would combine to produce offspring with the genotypes *YYRR* (yellow, round), *YyRr* (yellow, round), or *yyrr* (green, wrinkled). The ratio of phenotypes would be 3 yellow, round to 1 green, wrinkled. Every F_2 plant would be phenotypically like one P-generation parent or the other. None would display a new combination of the parental traits. However, if the alleles assort independently, the F_2 generation would show a wider range of genotypes and phenotypes, as shown by the large Punnett square in Figure 16.8b. In this case, each F_1 parent produces four kinds of gametes—*YR*, *Yr*, *yR*, and *yr*—instead of two, so the square is constructed with four rows on each side and shows 16 possible genotypes. The F_2 generation includes plants with yellow, round seeds; yellow, wrinkled seeds; green, round seeds; and green, wrinkled seeds, in a ratio of 9:3:3:1.

The actual results of this dihybrid cross are shown in Figure 16.8c. Crossing the true-breeding parents produced **dihybrid** offspring—offspring that are hybrids with respect to both traits. These F_1 dihybrids all had yellow, round seeds, confirming that yellow and round are dominant traits. This result was consistent with either hypothesis. However, the data for the F_2 generation were consistent only with the independent assortment hypothesis. Mendel observed four phenotypically different types of F_2 offspring, in a ratio that was reasonably close to 9:3:3:1.

In his original studies, Mendel reported that he had obtained similar results for every pair of characters he analyzed. His work supported the idea, now called the **law of independent assortment**, that *the alleles of different genes assort independently of each other during gamete formation.* Independent assortment means that a specific allele for one gene may be found in a gamete regardless of which allele for a different gene is found in the same gamete. In our example, the yellow and green alleles assort independently of the round and wrinkled alleles. The union of gametes from F_1 plants carrying these alleles produces the F_2 genotype and phenotype ratios shown in Figure 16.8b.

As we will see in Chapter 17, not all dihybrid crosses exhibit independent assortment. In some cases, the alleles of two genes that are physically located near each other on the same chromosome do not assort independently.

16.2 The Chromosome Theory of Inheritance

Learning Outcomes:

1. Outline the principles of the chromosome theory of inheritance.
2. Relate the behavior of chromosomes during meiosis to Mendel's laws of inheritance.

Mendel's studies with pea plants eventually led to the concept of a gene, which is the foundation for our understanding of inheritance. However, at the time of Mendel's work, the physical nature and location of genes were a complete mystery. The idea that inheritance has a physical basis was not even addressed until 1883, when German biologist August Weismann and Swiss botanist Karl Nägeli championed the idea that a substance in living cells is responsible for the transmission of hereditary traits. This idea challenged other researchers to identify the genetic material. Several scientists, including the German biologists Eduard Strasburger and Walther Flemming, observed dividing cells under the microscope and suggested that the chromosomes are the carriers of the genetic material. As we now know, the genetic material is the DNA within chromosomes.

In the early 1900s, the idea that chromosomes carry the genetic material dramatically unfolded as researchers continued to study the processes of mitosis, meiosis, and fertilization. It became increasingly clear that the characteristics of organisms are rooted in the continuity of cells during the life of an organism and from one generation to the next. Several scientists noted striking parallels between the segregation and assortment of traits noted by Mendel and the behavior of chromosomes during meiosis. Among these scientists were the German biologist Theodor Boveri and the American biologist Walter Sutton, who independently proposed the chromosome theory of inheritance. According to this theory, the inheritance patterns of traits can be explained by the transmission of chromosomes during meiosis and fertilization.

A modern view of the **chromosome theory of inheritance** consists of a few fundamental principles:

1. Chromosomes contain DNA, which is the genetic material. Genes are found within the chromosomes.
2. Chromosomes are replicated and passed from parent to offspring. They are also passed from cell to cell during the development of a multicellular organism.
3. The nucleus of a diploid cell contains two sets of chromosomes, which are found in homologous pairs. The maternal and paternal sets of homologous chromosomes are functionally equivalent; each set carries a full complement of genes.
4. At meiosis, one member of each chromosome pair segregates into one daughter nucleus, and its homolog segregates into the other daughter nucleus. During the formation of haploid cells, the members of different chromosome pairs segregate independently of each other.
5. Gametes are haploid cells that combine to form a diploid cell during fertilization, with each gamete transmitting one set of chromosomes to the offspring.

In this section, we will relate the chromosome theory of inheritance to Mendel's laws of inheritance.

Mendel's Law of Segregation Is Explained by the Segregation of Homologous Chromosomes During Meiosis

Now that you have an understanding of the basic tenets of the chromosome theory of inheritance, let's relate these ideas to Mendel's laws of inheritance. To do so, it will be helpful to introduce another genetic term. The physical location of a gene on a chromosome is called the gene's **locus** (plural, loci). As shown in Figure 16.9, each member of a homologous chromosome pair carries an allele of the same gene at the same locus. The individual in this example is heterozygous (*Tt*), so each homolog has a different allele.

How can we relate the chromosome theory of inheritance to Mendel's law of segregation? Figure 16.10 follows a pair of homologous chromosomes through the events of meiosis. This example involves a pea plant, heterozygous for height, *Tt*. The top of Figure 16.10 shows the two homologous chromosomes prior to DNA replication. When a cell prepares to divide, the homologs replicate to produce pairs of sister chromatids. Each chromatid carries a copy of the allele found on the original homolog, either *T* or *t*. During meiosis I, the homologs, each consisting of two sister chromatids, pair up and then segregate into two daughter cells. One of these cells has two copies of the *T* allele, and the other has two copies of the *t* allele. The

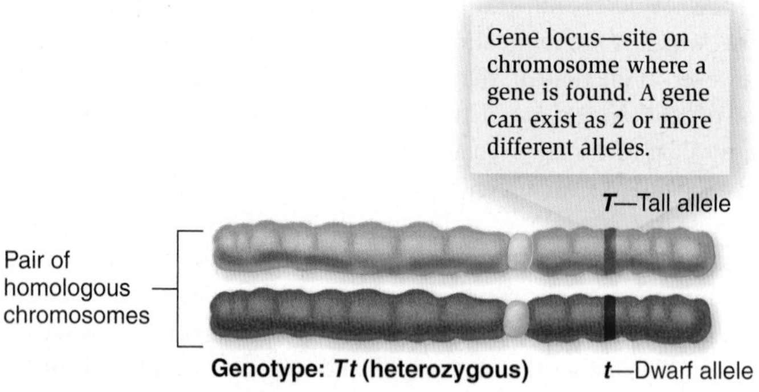
Gene locus—site on chromosome where a gene is found. A gene can exist as 2 or more different alleles.
T—Tall allele
Pair of homologous chromosomes
Genotype: Tt (heterozygous)
t—Dwarf allele

Figure 16.9 A gene locus. The locus (location) of a gene is the same for each member of a homologous pair, whether the individual is homozygous or heterozygous for that gene. This individual is heterozygous (*Tt*) for a gene for plant height.

BioConnections: Look back at Section 15.3 in Chapter 15. Explain the relationship between sexual reproduction and homologous chromosomes.

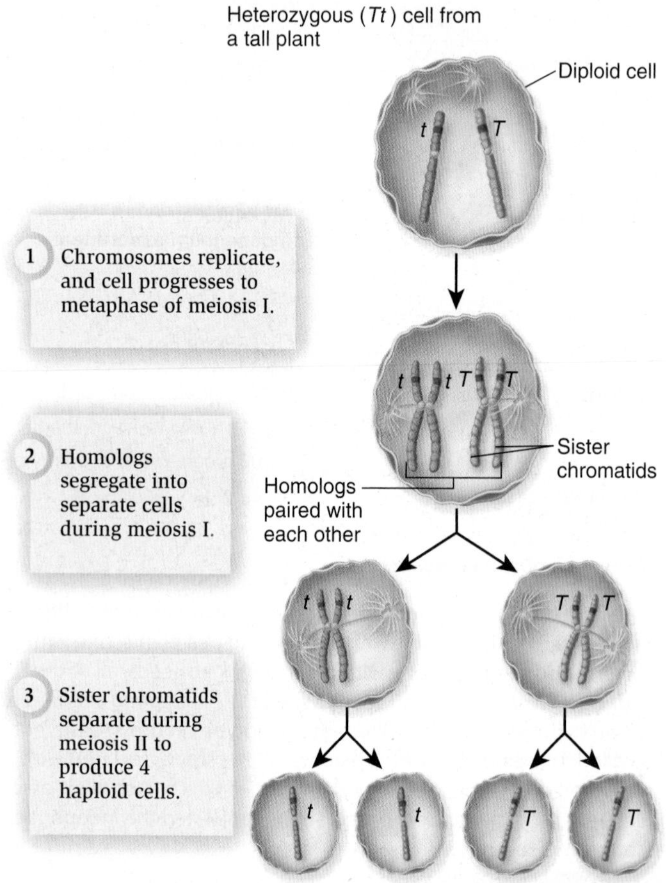
Heterozygous (Tt) cell from a tall plant
Diploid cell
t
T
1 Chromosomes replicate, and cell progresses to metaphase of meiosis I.
t t T T
2 Homologs segregate into separate cells during meiosis I.
Homologs paired with each other
Sister chromatids
t t
T T
3 Sister chromatids separate during meiosis II to produce 4 haploid cells.
t
t
T
T
Four haploid cells

Figure 16.10 The chromosomal basis of allele segregation. This example shows a pair of homologous chromosomes in a cell of a pea plant. The blue chromosome was inherited from the male parent, and the red chromosome was inherited from the female parent. This individual is heterozygous (*Tt*) for a height gene. The two homologs segregate from each other during meiosis, leading to segregation of the tall allele (*T*) and the dwarf allele (*t*) into different haploid cells. Note: For simplicity, this diagram shows a single pair of homologous chromosomes, though eukaryotic cells typically have several different pairs of homologous chromosomes.

BioConnections: When we say that alleles segregate, what does the word segregate mean? How is this related to meiosis, shown in Figure 15.11?

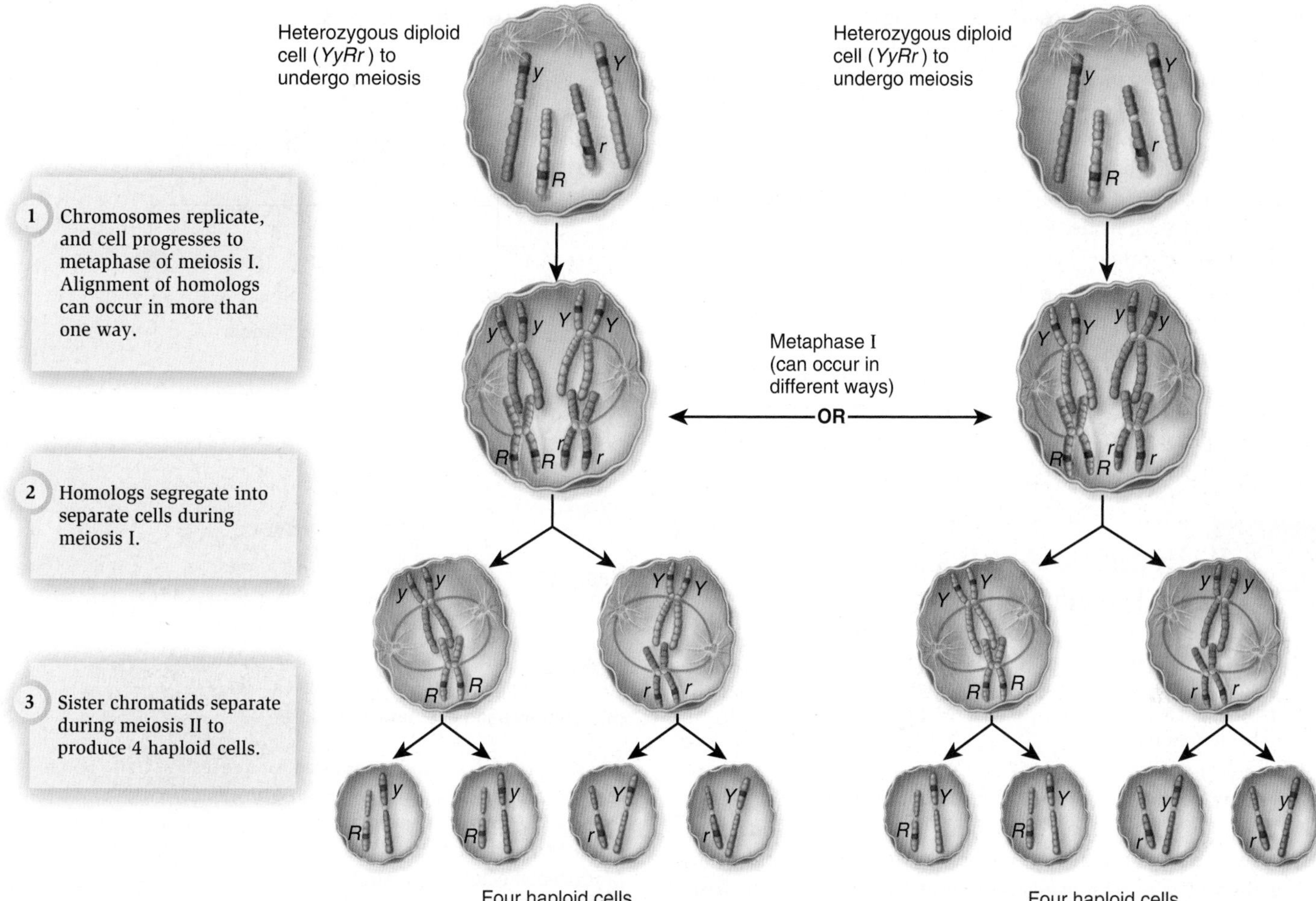
Heterozygous diploid cell (*YyRr*) to undergo meiosis
Heterozygous diploid cell (*YyRr*) to undergo meiosis
1 Chromosomes replicate, and cell progresses to metaphase of meiosis I. Alignment of homologs can occur in more than one way.
Metaphase I (can occur in different ways)
OR
2 Homologs segregate into separate cells during meiosis I.
3 Sister chromatids separate during meiosis II to produce 4 haploid cells.
Four haploid cells
Four haploid cells

Figure 16.11 **The chromosomal basis of independent assortment.** The alleles for seed color (*Y* or *y*) and seed shape (*R* or *r*) in peas are on different chromosomes. During metaphase of meiosis I, different arrangements of the two chromosome pairs lead to different combinations of the alleles in the resulting haploid cells. On the left, the chromosome carrying the dominant *R* allele has segregated with the chromosome carrying the recessive *y* allele. On the right, the two chromosomes carrying the dominant alleles (*R* and *Y*) have segregated together. Note: For simplicity, this diagram shows only two pairs of homologous chromosomes, though eukaryotic cells typically have several different pairs of homologous chromosomes.

Concept Check: Let's suppose that a cell is heterozygous for three different genes (Aa Bb Cc) and that each gene is on a different chromosome. How many different ways can these three pairs of homologous chromosomes align themselves during metaphase I, and how many different types of gametes can be produced?

sister chromatids separate during meiosis II, which produces four haploid cells. The end result of meiosis is that each haploid cell has a copy of just one of the two original homologs. Two of the cells have a chromosome carrying the *T* allele, and the other two have a chromosome carrying the *t* allele at the same locus. If the haploid cells shown at the bottom of Figure 16.10 give rise to gametes that combine randomly during fertilization, they produce diploid offspring with the genotypic and phenotypic ratios shown earlier in Figure 16.6.

Mendel's Law of Independent Assortment Is Explained by the Independent Alignment of Different Chromosomes During Meiosis

How can we relate the chromosome theory of inheritance to Mendel's law of independent assortment? **Figure 16.11** shows the alignment and segregation of two pairs of chromosomes in a pea plant. One pair carries the gene for seed color: The yellow allele (*Y*) is on one chromosome, and the green allele (*y*) is on its homolog. The other pair of chromosomes carries the gene for seed shape: One member of the pair has the round allele (*R*), whereas its homolog carries the wrinkled allele (*r*). Therefore, this individual is heterozygous for both genes, with the genotype *YyRr*.

When meiosis begins, the DNA in each chromosome has already replicated, producing two sister chromatids. At metaphase I of meiosis, the two pairs of chromosomes randomly align themselves along the metaphase plate. This alignment can occur in two equally probable ways, shown on the two sides of the figure. On the left, the chromosome carrying the *y* allele is aligned on the same side of the metaphase plate as the chromosome carrying the *R* allele; *Y* is aligned with *r*. On the right, the opposite has occurred: *Y* is aligned with *R*, and *y* is with *r*. In each case, the chromosomes that aligned on the same side of the metaphase plate segregate into the same daughter cell. In this way, the random alignment of chromosome pairs during meiosis I leads to the independent assortment of alleles found on different chromosomes. For two genes found on different chromosomes, each with two variant alleles, meiosis produces four allele combinations in equal numbers (*Ry*, *rY*, *RY*, and *ry*), as seen at the bottom of the figure.

If a *YyRr* (dihybrid) plant undergoes self-fertilization, any two gametes can combine randomly during fertilization. Because four kinds of gametes are made, 4^2 or 16 possible allele combinations are possible in the offspring. These genotypes, in turn, produce four phenotypes in a 9:3:3:1 ratio, as seen earlier in Figure 16.8. This ratio is the expected outcome when a heterozygote for two genes on different chromosomes undergoes self-fertilization.

But what if two different genes are located on the same chromosome? In this case, the transmission pattern may not conform to the law of independent assortment. We will discuss this phenomenon, known as linkage, in Chapter 17.

16.3 Pedigree Analysis of Human Traits

Learning Outcomes:

1. Apply pedigree analysis to deduce inheritance patterns in humans.
2. Distinguish between recessively inherited disorders and dominantly inherited disorders.

As we have seen, Mendel conducted experiments by making selective crosses of pea plants and analyzing large numbers of offspring. Later geneticists also relied on crosses of experimental organisms, especially fruit flies (*Drosophila melanogaster*). However, geneticists studying human traits cannot use this approach, for ethical and practical reasons. Instead, human geneticists must rely on information from family trees, or pedigrees. In this approach, called **pedigree analysis**, an inherited trait is analyzed over the course of a few generations in one family. The results of this method may be less definitive than the results of breeding experiments because the small size of human families may lead to large sampling errors. Nevertheless, a pedigree analysis often provides important clues concerning human inheritance.

Pedigree analysis has been used to understand the inheritance of human genetic diseases that follow simple Mendelian patterns. Many genes that play a role in disease exist in two forms: the normal allele and an abnormal allele that has arisen by mutation. The disease symptoms are associated with the mutant allele. Pedigree analysis allows us to determine whether the mutant allele is dominant or recessive and to predict the likelihood of an individual being affected.

Let's consider a recessive condition to illustrate pedigree analysis. The pedigree in **Figure 16.12** concerns a human genetic disease known as cystic fibrosis (CF), which involves a mutation in a gene that encodes the cystic fibrosis transmembrane regulator (the *CFTR* gene, also see Figure 1.16). Approximately 3% of Americans of European descent are heterozygous carriers of the recessive *CFTR* allele. Carriers are usually phenotypically normal. Individuals who are homozygous for the *CFTR* allele exhibit the disease symptoms, which include abnormalities of the lungs, pancreas, intestine, and sweat glands. A human pedigree, like the one in Figure 16.12, shows the

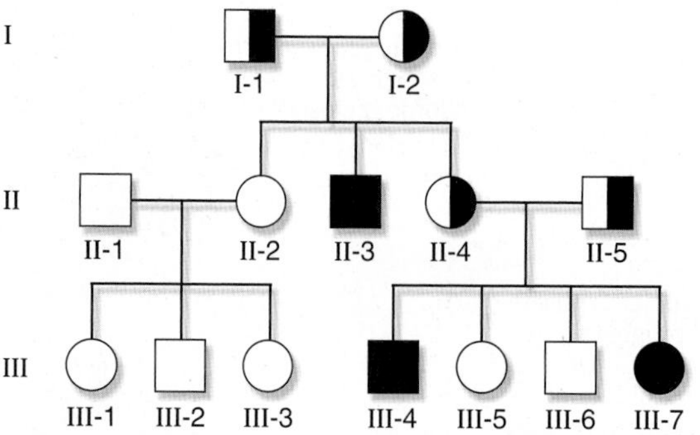

(a) Human pedigree showing cystic fibrosis

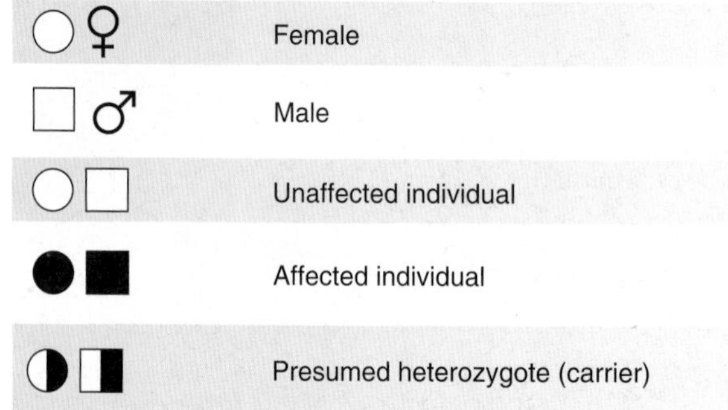

(b) Symbols used in a human pedigree

Figure 16.12 A family pedigree for a recessive trait. Some members of the family in this pedigree are affected with cystic fibrosis. Phenotypically normal individuals I-1, I-2, II-4, and II-5 are presumed to be heterozygotes (carriers) because they have produced affected offspring.

Concept Check: Let's suppose a genetic disease is caused by a mutant allele. If two affected parents produce an unaffected offspring, can the mutant allele be recessive?

oldest generation (designated by the Roman numeral I) at the top, with later generations (II and III) below it. A male (represented by a square) and a female (represented by a circle) who produce offspring are connected by a horizontal line; a vertical line connects parents with their offspring. Siblings (brothers and sisters) are denoted by downward projections from a single horizontal line, from left to right in the order of their birth. For example, individuals I-1 and I-2 are the parents of individuals II-2, II-3, and II-4, who are all siblings. Individuals affected by the disease, such as individual II-3, are depicted by filled symbols.

Why does this pedigree indicate a recessive pattern of inheritance for CF? The answer is that two unaffected individuals can produce an affected offspring. Such individuals are presumed to be heterozygotes (designated by a half-filled symbol). However, the same unaffected parents can also produce unaffected offspring (depicted by an unfilled symbol), because an individual must inherit two copies of the mutant allele to exhibit the disease. A recessive mode of inheritance is also characterized by the observation that all of the offspring of two affected individuals are affected themselves. However, for genetic diseases like CF that limit survival or fertility, there are rarely if ever cases where two affected individuals produce offspring.

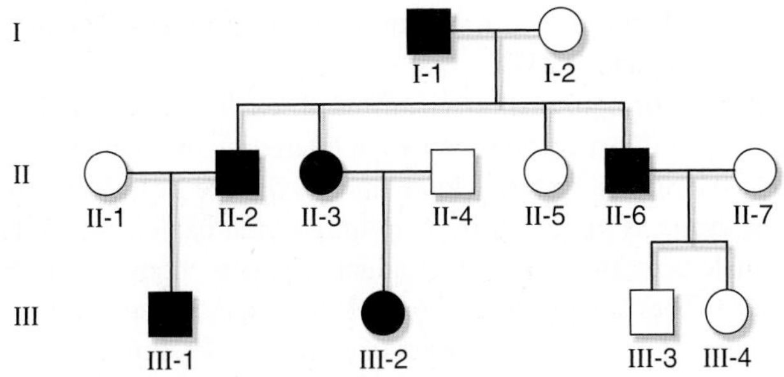

Figure 16.13 **A family pedigree for a dominant trait.** Huntington disease is caused by a dominant allele.

Concept Check: *What observation in a pedigree suggests a dominant pattern of inheritance?*

Although many of the alleles causing human genetic diseases are recessive, some are known to be dominant. Let's consider Huntington disease, a condition that causes the degeneration of brain cells involved in emotions, intellect, and movement. The symptoms of Huntington disease, which usually begin to appear when people are 30 to 50 years old, include uncontrollable jerking movements of the limbs, trunk, and face; progressive loss of mental abilities; and the development of psychiatric problems. If you examine the pedigree shown in Figure 16.13, you will see that every affected individual has one affected parent. This pattern is characteristic of most dominant disorders. However, affected parents do not always produce affected offspring. For example, II-6 is a heterozygote that has passed the normal allele to his offspring, thereby producing unaffected offspring (III-3 and III-4).

Most human genes are found on the paired chromosomes known as **autosomes**, which are the same in both sexes. Mendelian inheritance patterns involving these autosomal genes are described as autosomal inheritance patterns. Huntington disease is an example of a trait with an autosomal dominant inheritance pattern, whereas cystic fibrosis displays an autosomal recessive pattern. However, some human genes are located on sex chromosomes, which are different in males and females. These genes have their own characteristic inheritance patterns, which we will consider next.

16.4 Sex Chromosomes and X-Linked Inheritance Patterns

Learning Outcomes:

1. Describe different systems of sex determination in animals and plants.
2. Predict the outcome of crosses when genes are located on sex chromosomes.
3. Explain why X-linked recessive traits are more likely to occur in males.

In the first part of this chapter, we discussed Mendel's experiments that established the basis for understanding how traits are transmitted from parents to offspring. We also examined the chromosome theory of inheritance, which provided a framework for explaining Mendel's observations. Mendelian patterns of gene transmission are observed for most genes located on autosomes in a wide variety of eukaryotic species.

We will now turn our attention to genes located on **sex chromosomes**. As you learned in Chapter 15, this term refers to a distinctive pair of chromosomes that are different in males and females and that determine the sex of the individuals. Sex chromosomes are found in many but not all species with two sexes. The study of sex chromosomes proved pivotal in confirming the chromosome theory of inheritance. The distinctive transmission patterns of genes on sex chromosomes helped early geneticists show that particular genes are located on particular chromosomes. Later, other researchers became interested in these genes because some of them were found to cause inherited diseases in humans.

In this section, we will consider several mechanisms by which sex chromosomes in various species determine an individual's sex. We will then explore the inheritance patterns of genes on sex chromosomes and see that recessive alleles are expressed more frequently in males than in females. Last, we will examine some of the early research involving sex chromosomes that provided convincing evidence for the chromosome theory of inheritance.

In Many Species, Sex Differences Are Due to the Presence of Sex Chromosomes

Some early evidence supporting the chromosome theory of inheritance involved a consideration of sex determination. In 1901, American biologist C. E. McClung suggested that the inheritance of particular chromosomes is responsible for determining sex in fruit flies. Following McClung's initial observations, several mechanisms of sex determination were found in different species of animals. Some examples are described in Figure 16.14. All of these mechanisms involve chromosomal differences between the sexes, and most involve a difference in a single pair of sex chromosomes.

In the X-Y system of sex determination, which operates in mammals, the somatic cells of males have one X and one Y chromosome, whereas female somatic cells contain two X chromosomes (Figure 16.14a). For example, the 46 chromosomes carried by human cells consist of 22 pairs of autosomes and one pair of sex chromosomes (either XY or XX). Which chromosome, the X or Y, determines sex? In mammals, the presence of the Y chromosome causes maleness. This is known from the analysis of rare individuals who carry chromosomal abnormalities. For example, mistakes that occasionally occur during meiosis may produce an individual who carries two X chromosomes and one Y chromosome. Such an individual develops into a male. A gene called the *SRY* gene located on the Y chromosome of mammals plays a key role in the developmental pathway that leads to maleness.

The X-O system operates in many insects (Figure 16.14b). Unlike the X-Y system in mammals, the presence of the Y chromosome in the X-O system does not determine maleness. Females in this system have a pair of sex chromosomes and are designated XX. In some insect species that follow the X-O system, the male has only one sex chromosome, the X. In other X-O insect species, such as *Drosophila melanogaster*, the male has both an X chromosome and a Y chromosome. In all cases, an insect's sex is determined by the ratio between its X chromosomes and its sets of autosomes. If a fly has one X chromosome and is diploid for the autosomes ($2n$), this ratio is 1/2, or 0.5. This fly will become a male whether

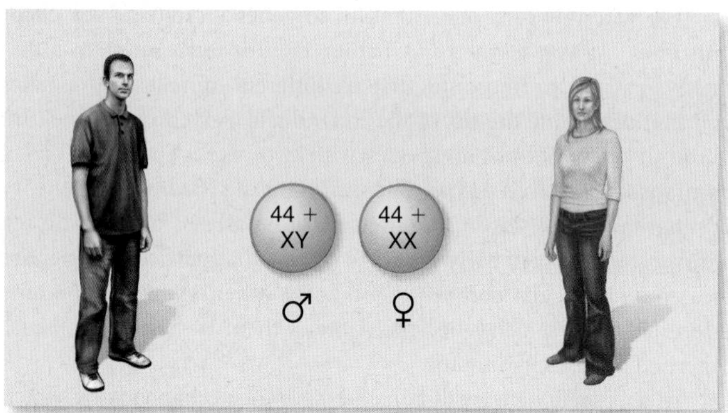

(a) The X-Y system in mammals

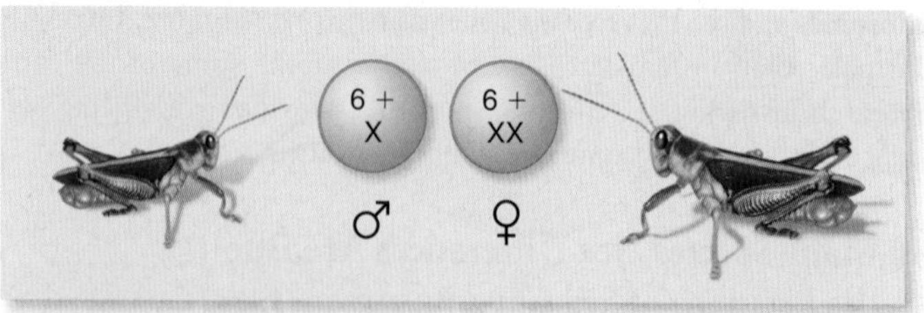

(b) The X-O system in certain insects

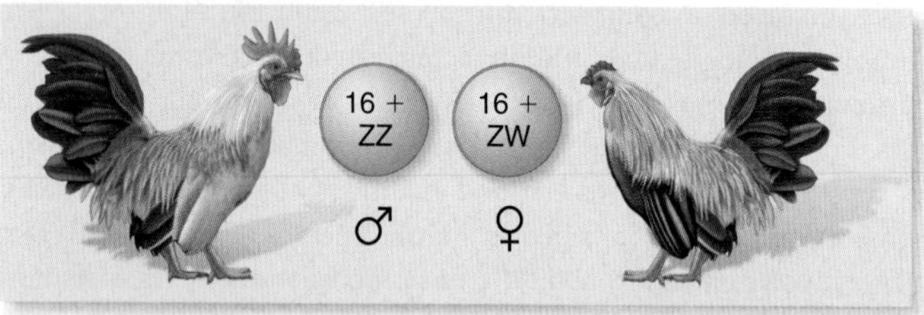

(c) The Z-W system in birds

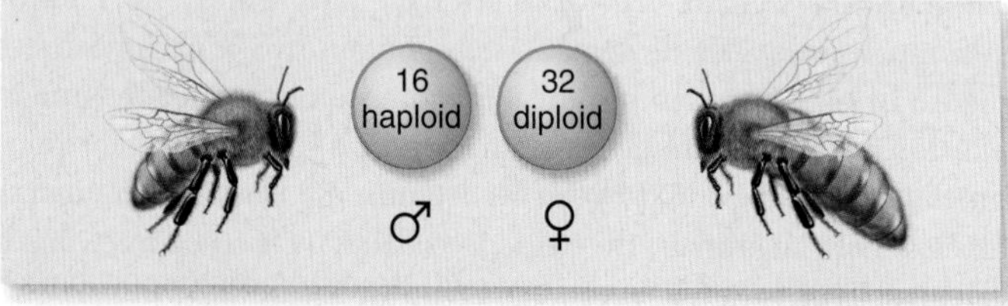

(d) The haplodiploid system in bees

Figure 16.14 Different mechanisms of sex determination in animals. The numbers shown in the circles indicate the numbers of autosomes.

Concept Check: If a person is born with only one X chromosome and no Y chromosome, would you expect that person to be a male or a female? Explain your answer.

or not it receives a Y chromosome. On the other hand, if a diploid fly receives two X chromosomes, the ratio is 2/2, or 1.0, and the fly becomes a female.

Thus far, we have considered examples where females have two similar copies of a sex chromosome, the X. However, in some animal species, such as birds and some fish, the male carries two similar chromosomes (Figure 16.14c). This is called the Z-W system to distinguish it from the X-Y system found in mammals. The male is ZZ, and the female is ZW.

Not all chromosomal mechanisms of sex determination involve a special pair of sex chromosomes. An interesting mechanism known as the haplodiploid system is found in bees (Figure 16.14d). The male bee, or drone, is produced from an unfertilized haploid egg. Therefore, male bees are haploid individuals. Females, both worker bees and queen bees, are produced from fertilized eggs and are diploid.

Although sex in many species of animals is determined by chromosomes, other mechanisms are also known. In certain reptiles and fish, sex is controlled by environmental factors such as temperature. For example, in the American alligator (*Alligator mississippiensis*), temperature controls sex development. When eggs of this alligator are incubated at 33°C, nearly all of them produce male individuals. When the eggs are incubated at a temperature significantly below 33°C, they produce nearly all females, whereas increasing percentages of females are produced above 34°C.

Most species of flowering plants, including pea plants, have a single type of diploid plant, or sporophyte, that makes both male and female gametophytes. However, the sporophytes of some species have two sexually distinct types of individuals, one with flowers that produce male gametophytes, and the other with flowers that produce female gametophytes. Examples include hollies, willows, poplars, and date palms. Sex chromosomes, designated X and Y, are responsible for sex determination in many such species. The male plant is XY, whereas the female plant is XX. However, in some plant species with separate sexes, microscopic examination of the chromosomes does not reveal distinct types of sex chromosomes.

In Humans, Recessive X-Linked Traits Are More Likely to Occur in Males

In humans, the X chromosome is rather large and carries over 1,000 genes, whereas the Y chromosome is quite small and has less than 100 genes. Therefore, many genes are found on the X chromosome but not on the Y; these are known as **X-linked genes**. By comparison, fewer genes are known to be Y linked, meaning they are found on the Y chromosome but not on the X. **Sex-linked genes** are found on one sex chromosome but not on the other. Because fewer genes are found on the Y chromosome, the term usually refers to X-linked genes. In mammals, a male cannot be described as being homozygous or heterozygous for an X-linked gene, because these terms apply to genes that are present in two copies. Instead, the term **hemizygous** is used to describe an individual with only one copy of a particular gene. A male mammal is said to be hemizygous for an X-linked gene.

Many recessive X-linked alleles cause diseases in humans, and these diseases occur more frequently in males than in females. As an example, let's consider the X-linked recessive disorder called classical hemophilia (hemophilia A). In individuals with hemophilia, blood does not clot normally, and a minor cut may bleed for a long time. Common accidental injuries that are minor in most people pose a threat of severe internal or external bleeding for hemophiliacs. Hemophilia A is caused by a recessive X-linked allele that encodes a defective form of a clotting protein. If a mother is a heterozygous carrier of hemophilia A, each of her children has a 50% chance of inheriting the recessive allele. The following Punnett square shows a cross between an unaffected father and a heterozygous mother. X^H designates an X chromosome carrying the normal allele, and $X^{h\text{-}A}$ is the X chromosome that carries the recessive allele for hemophilia A.

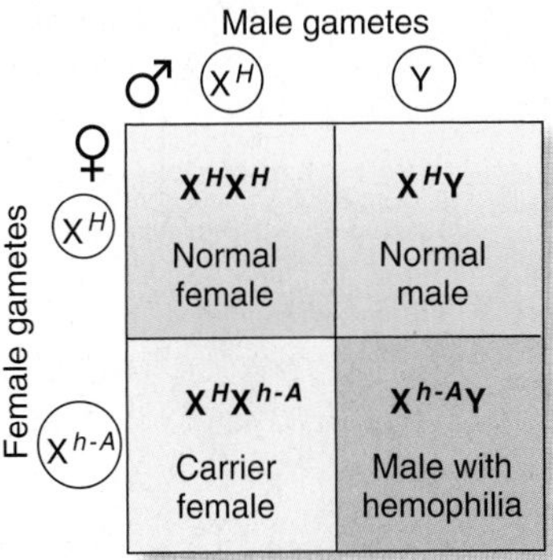

Although each child has a 50% chance of inheriting the hemophilia allele from their mother, only 1/2 of the sons will exhibit the disorder. Because a son inherits only one X chromosome, a son who inherits the abnormal allele from his mother will have hemophilia. However, a daughter inherits an X chromosome from both her mother and her father. In this example, a daughter who inherits the hemophilia allele from her mother also inherits a normal allele from her father. This daughter will have a normal phenotype, but if she passes the abnormal allele to her sons, they will have hemophilia.

Feature Investigation

Morgan's Experiments Showed a Correlation Between a Genetic Trait and the Inheritance of a Sex Chromosome in *Drosophila*

The distinctive inheritance pattern of X-linked alleles provides a way of demonstrating that a specific gene is on an X chromosome. An X-linked gene was the first gene to be located on a specific chromosome. In 1910, the American geneticist Thomas Hunt Morgan began work on a project in which he reared large populations of fruit flies, *Drosophila melanogaster*, in the dark to determine if their eyes would atrophy from disuse and disappear in future generations. Even after many consecutive generations, the flies showed no noticeable changes. After 2 years of looking at many flies, Morgan happened to discover a male fly with white eyes rather than the normal red. The white-eye trait must have arisen from a new mutation that converted a red-eye allele into a white-eye allele.

To study the inheritance of the white-eye trait, Morgan followed an approach similar to Mendel's in which he made crosses and quantitatively analyzed their outcome. In the experiment described in **Figure 16.15**, Morgan crossed his white-eyed male to a red-eyed female. All of the F_1 offspring had red eyes, indicating that red is dominant to white. The F_1 offspring were then mated to each other to obtain an F_2 generation. As seen in the data table, this cross produced 1,011 red-eyed males, 782 white-eyed males, and 2,459

Figure 16.15 **Morgan's crosses of red-eyed and white-eyed *Drosophila*.**

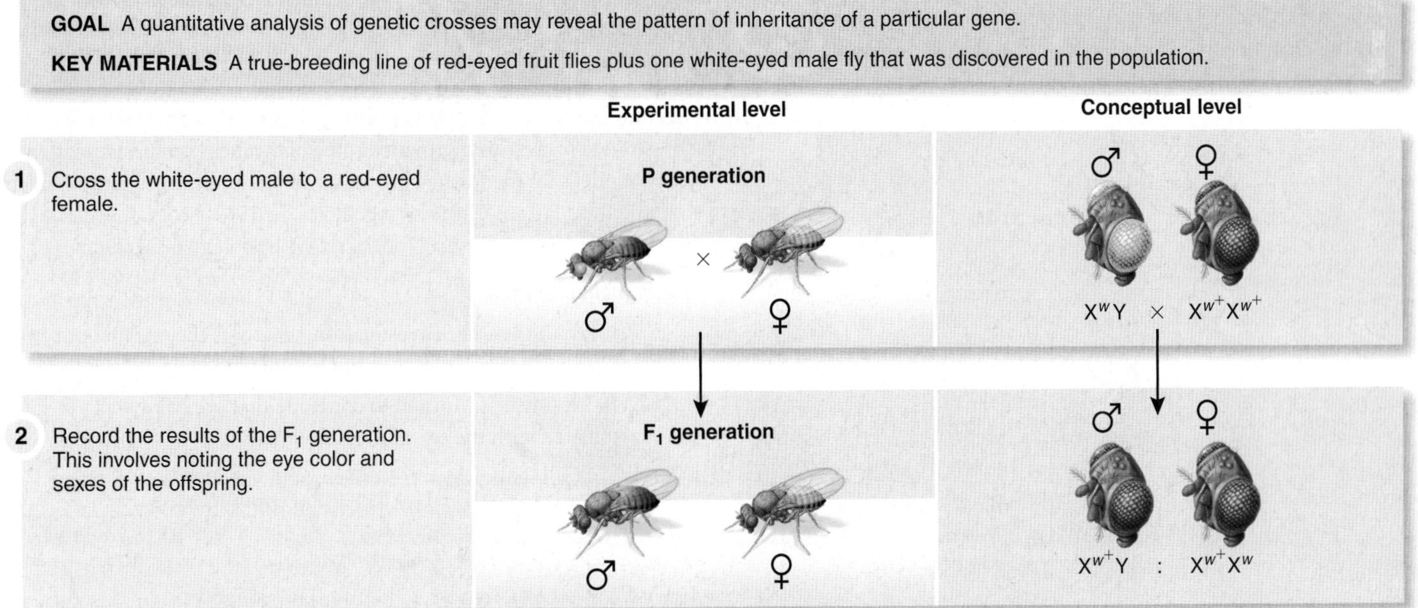

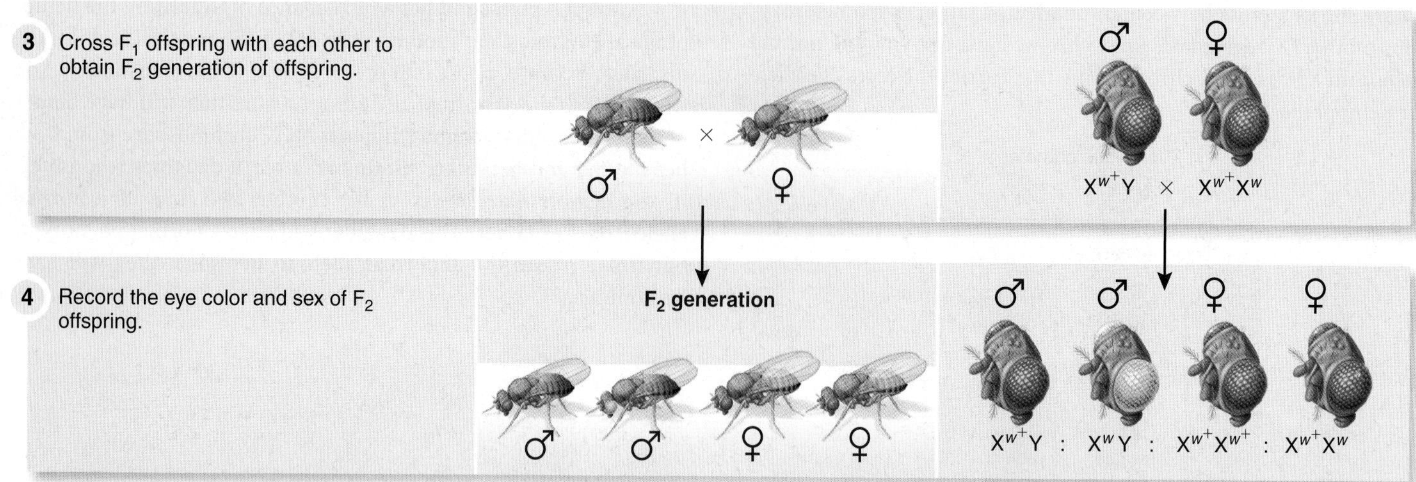

3 Cross F_1 offspring with each other to obtain F_2 generation of offspring.
♂ ♀
♂ × ♀
$X^{w^+}Y$ × $X^{w^+}X^{w}$
4 Record the eye color and sex of F_2 offspring.
F_2 generation
♂ ♂ ♀ ♀
♂ ♂ ♀ ♀
$X^{w^+}Y$: $X^{w}Y$: $X^{w^+}X^{w^+}$: $X^{w^+}X^{w}$

5 THE DATA

Cross	Results	
Original white-eyed male to a red-eyed female	F_1 generation	All red-eyed flies
F_1 males to F_1 females	F_2 generation	1,011 red-eyed males 782 white-eyed males 2,459 red-eyed females 0 white-eyed females

6 CONCLUSION The data are consistent with an inheritance pattern in which an eye-color gene is located on the X chromosome.

7 SOURCE Morgan, Thomas H. 1910. Sex limited inheritance in *Drosophila. Science* 32:120–122.

red-eyed females. Surprisingly, no white-eyed females were observed in the F_2 generation.

How did Morgan interpret these results? The results suggested a connection between the alleles for eye color and the sex of the offspring. As shown in the conceptual column of Figure 16.15 and in the following Punnett square, his data were consistent with the idea that the eye-color alleles in *Drosophila* are located on the X chromosome. X^{w^+} is the chromosome carrying the normal allele for red eyes, and X^{w} is the chromosome with the mutant allele for white eyes.

F_1 male is $X^{w^+}Y$
F_1 female is $X^{w^+}X^{w}$

Female gametes ♀ \ Male gametes ♂	X^{w^+}	Y
X^{w^+}	$X^{w^+}X^{w^+}$ Red, female	$X^{w^+}Y$ Red, male
X^{w}	$X^{w^+}X^{w}$ Red, female	$X^{w}Y$ White, male

The Punnett square predicts that the F_2 generation will not have any white-eyed females. This prediction was confirmed by Morgan's experimental data. However, it should also be pointed out that the experimental ratio of red eyes to white eyes in the F_2 generation is (2,459 + 1,011):782, which equals 4.4:1. This ratio deviates significantly from the ratio of 3:1 predicted in the Punnett square. The lower than expected number of white-eyed flies is explained by a decreased survival of white-eyed flies.

Following this initial discovery, Morgan carried out many experimental crosses that located specific genes on the *Drosophila* X chromosome. This research provided some of the most persuasive evidence for Mendel's laws and the chromosome theory of inheritance, which are the foundations of modern genetics. In 1933, Morgan became the first geneticist to receive a Nobel Prize (Physiology or Medicine).

Experimental Questions

1. Prior to the Feature Investigation, what was the original purpose of Morgan's experiments with *Drosophila*?
2. What results led Morgan to conclude that eye color was associated with the sex of the individual?
3. What crosses between fruit flies could yield female offspring with white eyes?

16.5 Variations in Inheritance Patterns and Their Molecular Basis

Learning Outcomes:

1. Relate dominant and recessive traits to protein function.
2. Describe pleiotropy and explain why it occurs.
3. Predict the outcome of crosses that exhibit incomplete dominance and codominance.
4. Discuss how the environment plays a critical role in determining the outcome of traits.

The term **Mendelian inheritance** describes the inheritance patterns of genes that segregate and assort independently. In the first section of this chapter, we considered the inheritance pattern of traits affected by a single gene that is found in two variants, one of which is dominant over the other. This pattern is called **simple Mendelian inheritance**, because the phenotypic ratios in the offspring clearly demonstrate Mendel's laws. In this section, we will discuss the molecular basis of dominant and recessive traits and see how the molecular expression of a gene can have widespread effects on an organism's phenotype. In addition, we will examine the inheritance patterns of genes that segregate and assort independently but do not display a simple dominant/recessive relationship. The transmission of these genes from parents to offspring does not usually produce the ratios of phenotypes we would expect on the basis of Mendel's observations. This does not mean that Mendel was wrong. Rather, the inheritance patterns of many traits are different from the simple patterns he chose to study. As described in **Table 16.1**, our understanding of gene function at the molecular level explains both simple Mendelian inheritance and other, more complex inheritance patterns that conform to Mendel's laws. This modern knowledge also sheds light on the role of the environment in producing an organism's phenotype, which we will discuss at the end of the section.

Protein Function Explains the Phenomenon of Dominance

As we discussed at the beginning of this chapter, Mendel studied seven characters that were found in two variants each (see Figure 16.2). The dominant variants are caused by the common alleles for these traits in pea plants. For any given gene, geneticists refer to a prevalent allele in a population as a **wild-type allele**. In most cases, a wild-type allele encodes a protein that is made in the proper amount and functions normally. By comparison, alleles that have been altered by mutation are called **mutant alleles**; these tend to be rare in natural populations. In the case of Mendel's seven characters in pea plants, the recessive alleles are due to rare mutations.

How do we explain why one allele is dominant and another allele is recessive? By studying genes and their gene products at the molecular level, researchers have discovered that a recessive allele is often defective in its ability to express a functional protein. In other words, mutations that produce recessive alleles are likely to decrease or eliminate the synthesis or functional activity of a protein. These are called loss-of-function alleles. To understand why many loss-of-function alleles are recessive, we need to take a quantitative look at protein function.

In a simple dominant/recessive relationship, the recessive allele does not affect the phenotype of the heterozygote. In this type of relationship, a single copy of the dominant (wild-type) allele is sufficient to mask the effects of the recessive allele. How do we explain the dominant phenotype of the heterozygote? **Figure 16.16** considers the example of flower color in a pea plant. The gene encodes an enzyme (protein P) that is needed to convert a colorless molecule into a purple pigment. The *P* allele is dominant because one *P* allele encodes enough of the functional protein—50% of the amount found in a normal homozygote—to provide a normal phenotype. Therefore, the *PP* homozygote and the *Pp* heterozygote both make enough of the purple pigment to yield purple flowers. The *pp* homozygote cannot make any of the functional protein required for pigment synthesis, so its flowers are white.

Table 16.1 Different Types of Mendelian Inheritance Patterns and Their Molecular Basis

Type	Description
Simple Mendelian inheritance	**Inheritance pattern:** Pattern of traits determined by a pair of alleles that display a dominant/recessive relationship and are located on an autosome. The presence of the dominant allele masks the presence of the recessive allele. **Molecular basis:** In many cases, the recessive allele is nonfunctional. Though a heterozygote may produce 50% of the functional protein compared with a dominant homozygote, this is sufficient to produce the dominant trait.
X-linked inheritance	**Inheritance pattern:** Pattern of traits determined by genes that display a dominant/recessive relationship and are located on the X chromosome. In mammals and fruit flies, males are hemizygous for X-linked genes. In these species, X-linked recessive traits occur more frequently in males than in females. **Molecular basis:** In a female with one recessive X-linked allele (a heterozygote), the protein encoded by the dominant allele is sufficient to produce the dominant trait. A male with a recessive X-linked allele (a hemizygote) does not have a dominant allele and does not make any of the functional protein.
Incomplete dominance	**Inheritance pattern:** Pattern that occurs when the heterozygote has a phenotype intermediate to the phenotypes of the homozygotes, as when a cross between red-flowered and white-flowered plants produces pink-flowered offspring. **Molecular basis:** Fifty percent of the protein encoded by the functional (wild-type) allele is not sufficient to produce the normal trait.
Codominance	**Inheritance pattern:** Pattern that occurs when the heterozygote expresses both alleles simultaneously. For example, a human carrying the A and B alleles for the ABO antigens of red blood cells produces both the A and the B antigens (has an AB blood type). **Molecular basis:** The codominant alleles encode proteins that function slightly differently from each other. In a heterozygote, the function of each protein affects the phenotype uniquely.

Genotype	*PP*	*Pp*	*pp*
Amount of functional protein P produced	100%	50%	0%
Phenotype	**Purple**	**Purple**	**White**
The relationship of the normal (dominant) and mutant (recessive) alleles displays simple Mendelian inheritance.			

Colorless precursor molecule → Protein P → Purple pigment

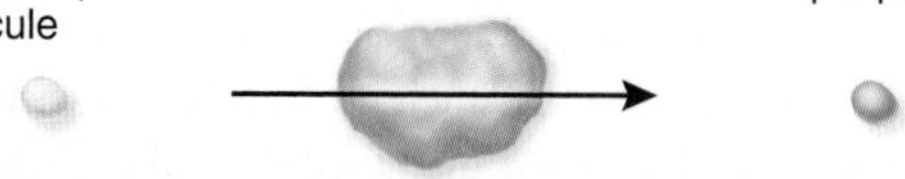

Figure 16.16 How genes give rise to traits during simple Mendelian inheritance. In many cases, the amount of protein encoded by a single dominant allele is sufficient to produce the normal phenotype. In this example, the gene encodes an enzyme that is needed to produce a purple pigment. A plant with one or two copies of the normal allele produces enough pigment to produce purple flowers. In a *pp* homozygote, the complete lack of the normal protein (enzyme) results in white flowers.

This explanation—that 50% of the normal protein is enough—is true for many dominant alleles. In such cases, the normal homozygote is making much more of the protein than necessary, so if the amount is reduced to 50%, as it is in the heterozygote, the individual still has plenty of this protein to accomplish whatever cellular function it performs. In other cases, however, an allele may be dominant because the heterozygote actually produces more than 50% of the normal amount of functional protein. This increased production is due to the phenomenon of gene regulation. The normal gene is "up-regulated" in the heterozygote to compensate for the lack of function of the defective allele.

GENOMES & PROTEOMES CONNECTION

Recessive Alleles That Cause Diseases May Have Multiple Effects on Phenotype

The idea that recessive alleles usually cause a substantial decrease in the expression of a functional protein is supported by analyses of many human genetic diseases. Keep in mind that many genetic diseases are caused by rare mutant alleles. Table 16.2 lists several examples of human genetic diseases in which a recessive allele fails to produce a specific cellular protein in its active form. Over 7,000 human disorders are caused by mutations in single genes; most of these are recessive alleles. With a human genome size of 20,000 to 25,000 genes, this means that roughly one-third of our genes are known to cause some kind of abnormality when mutations alter the expression or functionality of their gene product.

Single-gene disorders often illustrate the phenomenon of **pleiotropy**, which means that a mutation in a single gene can have multiple effects on an individual's phenotype. Pleiotropy occurs for several reasons, including the following:

1. The expression of a single gene can affect cell function in more than one way. For example, a defect in a microtubule protein may affect cell division and cell movement.
2. A gene may be expressed in different cell types in a multicellular organism.
3. A gene may be expressed at different stages of development.

In this genetics unit, we tend to discuss genes as they affect a single trait. This educational approach allows us to appreciate how genes function and how they are transmitted from parents to offspring. However, this focus may also obscure how amazing genes really are. In all or nearly all cases, the expression of a gene is pleiotropic with regard to the characteristics of an organism. The expression of any given gene influences the expression of many other genes in the genome, and vice versa. Pleiotropy is revealed when researchers study the effects of gene mutations.

As an example of a pleiotropic effect, let's consider cystic fibrosis (CF), which we discussed earlier as an example of a recessive human disorder (see Figure 16.12). In the late 1980s, the gene for CF was identified. The normal allele encodes a protein called the cystic fibrosis transmembrane conductance regulator (CFTR), which regulates ionic balance by allowing the transport of chloride ions (Cl^-) across epithelial cell membranes. The mutation that causes CF diminishes the function of this Cl^- transporter, affecting several parts of the body in different ways. Because the movement of Cl^- affects water trans-

Table 16.2 Examples of Recessive Human Genetic Diseases

Disease	Protein produced by the normal gene*	Description
Phenylketonuria	Phenylalanine hydroxylase	Inability to metabolize phenylalanine. Can lead to severe mental impairment and physical degeneration. The disease can be prevented by following a phenylalanine-free diet beginning early in life.
Cystic fibrosis	A chloride-ion transporter	Inability to regulate ion balance in epithelial cells. Leads to a variety of abnormalities, including production of thick lung mucus and chronic lung infections.
Tay-Sachs disease	Hexosaminidase A	Inability to metabolize certain lipids. Leads to paralysis, blindness, and early death.
α_1-Antitrypsin deficiency	α_1-Antitrypsin	Inability to prevent the activity of protease enzymes. Leads to a buildup of certain proteins that cause liver damage and emphysema.
Hemophilia A	Coagulation factor VIII	A defect in blood clotting due to a missing clotting factor. An accident may cause excessive bleeding or internal hemorrhaging.

*Individuals who exhibit the disease are homozygous (or hemizygous) for a recessive allele that results in a defect in the amount or function of the normal protein.

port across membranes, the most severe symptom of CF is the production of thick mucus in the lungs, which occurs because of a water imbalance. In sweat glands, the normal Cl^- transporter has the function of recycling salt out of the glands and back into the skin before it can be lost to the outside world. Persons with CF have excessively salty sweat due to their inability to recycle salt back into their skin cells. A common test for CF is the measurement of salt on the skin. Another effect is seen in the reproductive systems of males who are homozygous for the mutant allele. Some males with CF are infertile because the vas deferens, the tubules that transport sperm from the testes, are absent or undeveloped. Presumably, a normally functioning Cl^- transporter is needed for the proper development of the vas deferens in the embryo. Taken together, we can see that a defect in CFTR has multiple effects throughout the body.

Incomplete Dominance Results in an Intermediate Phenotype

We will now turn our attention to examples in which the alleles for a given gene do not show a simple dominant/recessive relationship. In some cases, a heterozygote that carries two different alleles exhibits a phenotype that is intermediate between the corresponding homozygous individuals. This phenomenon is known as **incomplete dominance**.

In 1905, Carl Correns discovered this pattern of inheritance for alleles affecting flower color in the four-o'clock plant (*Mirabilis jalapa*). Figure 16.17 shows a cross between two four-o'clock plants: a red-flowered homozygote and a white-flowered homozygote. The allele for red flower color is designated C^R, and the white allele is C^W. These alleles are designated with superscripts rather than upper- and lowercase letters because neither allele is dominant. The offspring of this cross have pink flowers—they are C^RC^W heterozygotes with an intermediate phenotype. If these F_1 offspring are allowed to self-fertilize, the F_2 generation has 1/4 red-flowered plants, 1/2 pink-flowered plants, and 1/4 white-flowered plants. This is a 1:2:1 phenotypic ratio rather than the 3:1 ratio observed for simple Mendelian inheritance. What is the molecular explanation for this ratio? In this case, the red allele encodes a functional protein needed to produce a red pigment, whereas the white allele is a mutant allele that is nonfunctional. In the C^RC^W heterozygote, 50% of the protein encoded by the C^R allele is not sufficient to produce the red-flower phenotype, but it does provide enough pigment to give pink flowers.

The degree to which we judge an allele to exhibit incomplete dominance may depend on how closely we examine an individual's phenotype. An example is an inherited human disease called phenylketonuria (PKU). This disorder is caused by a rare mutation in a gene that encodes an enzyme called phenylalanine hydroxylase. This enzyme is needed to metabolize the amino acid phenylalanine, which is found in milk, eggs, and other protein-rich foods in our diet. If left untreated, phenylalanine builds up, affecting various systems in the body. Homozygotes carrying the mutant allele suffer severe symptoms, including mental impairment, seizures, microcephaly (small head), poor development of tooth enamel, and decreased body growth. By comparison, heterozygotes appear phenotypically normal. For this reason, geneticists consider PKU to be a recessive disorder. However, biochemical analysis of the blood of heterozygotes shows they typically have a phenylalanine blood level twice as high as that

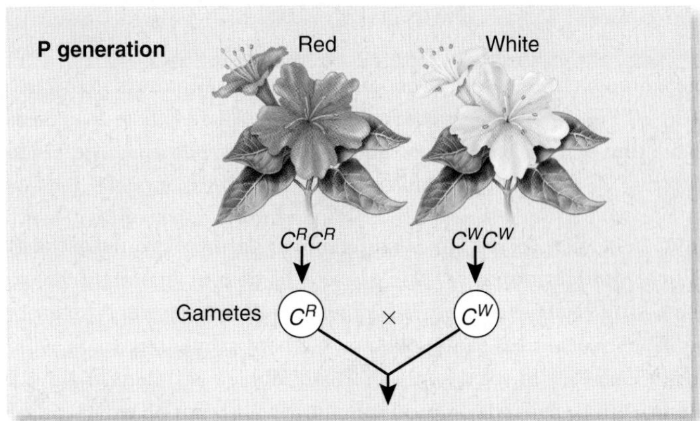

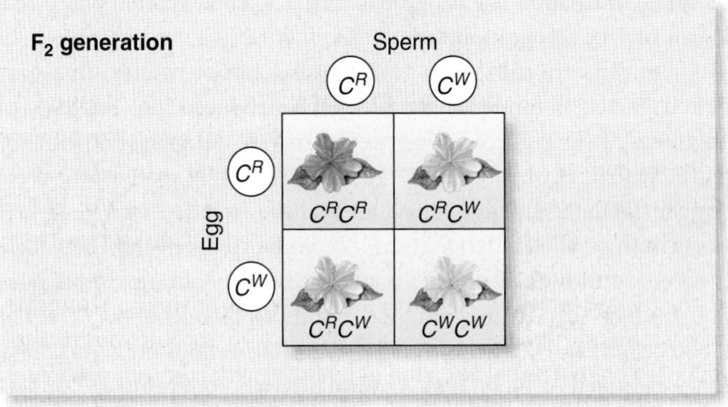

Figure 16.17 **Incomplete dominance in the four-o'clock plant.** When red-flowered and white-flowered homozygotes (C^RC^R and C^WC^W) are crossed, the resulting heterozygote (C^RC^W) has an intermediate phenotype of pink flowers.

of an individual carrying two normal copies of the gene. Individuals with PKU (homozygous recessive) typically have phenylalanine blood levels 30 times higher than normal. Therefore, at this closer level of examination, heterozygotes exhibit an intermediate phenotype relative to the homozygous dominant and recessive individuals. At this closer level of inspection, the relationship between the normal and mutant alleles is defined as incomplete dominance.

ABO Blood Type Is an Example of Multiple Alleles and Codominance

Although diploid individuals have only two copies of most genes, the majority of genes have three or more variants in natural populations. We describe such genes as occurring in **multiple alleles**. Particular

Table 16.3 The ABO Blood Group

Blood type:	O	A	B	AB
Genotype:	*ii*	I^AI^A or I^Ai	I^BI^B or I^Bi	I^AI^B
Surface antigen:	Neither A nor B	A	B	A and B
Antibodies:	Against A and B	Against B	Against A	Neither

phenotypes depend on which two alleles each individual inherits. ABO blood types in humans are an example of phenotypes produced by multiple alleles.

As shown in **Table 16.3**, human red blood cells have structures on their plasma membrane known as surface antigens, which are constructed from several sugar molecules that are connected to form a carbohydrate tree. The carbohydrate tree is attached to lipids or membrane proteins to form glycolipids or glycoproteins, respectively. As noted in Chapter 5, glycolipids and glycoproteins often play a role in cell surface recognition.

Antigens are substances (in this case, carbohydrates) that may be recognized as foreign material when introduced into the body of an animal. Let's consider two types of surface antigens, known as A and B, which may be found on red blood cells. The synthesis of these antigens is determined by enzymes that are encoded by a gene that exists in three alleles, designated I^A, I^B, and *i*, respectively. The *i* allele is recessive to both I^A and I^B. A person who is *ii* homozygous does not produce surface antigen A or B and has blood type O. The red blood cells of an I^AI^A homozygous or I^Ai heterozygous individual have surface antigen A (blood type A). Similarly, a homozygous I^BI^B or heterozygous I^Bi individual produces surface antigen B (blood type B). A person who is I^AI^B heterozygous makes both antigens, A and B, on every red blood cell (blood type AB). The phenomenon in which a single individual expresses two alleles is called **codominance**.

What is the molecular explanation for codominance? Biochemists have analyzed the carbohydrate tree produced in people of differing blood types. The differences are shown schematically in Table 16.3. In type O, the carbohydrate tree is smaller than in type A or type B because a sugar has not been attached to a specific site on the tree. People with blood type O have a loss-of-function mutation in the gene that encodes the enzyme that attaches a sugar at this site. This enzyme, called a glycosyl transferase, is inactive in type O individuals. In contrast, the type A and type B antigens have sugars attached to this site, but each of them has a different sugar. This difference occurs because the enzymes encoded by the I^A allele and the I^B allele have slightly different active sites. As a result, the enzyme encoded by the I^A allele attaches a sugar called *N*-acetylgalactosamine to the carbohydrate tree, whereas the enzyme encoded by the I^B allele attaches galactose. *N*-Acetylgalactosamine is represented by an orange hexagon in Table 16.3, and galactose by a green triangle. The importance of surface antigens and blood type for safe blood transfusions is discussed in Chapter 53.

The Environment Plays a Vital Role in the Making of a Phenotype

In this chapter, we have been mainly concerned with the effects of genes on phenotypes. In addition, phenotypes are shaped by an organism's environment. An organism cannot exist without its genes or without an environment in which to live. Both are indispensable for life. An organism's genotype provides the plan to create a phenotype, and the environment provides nutrients and energy so that plan can be executed.

The **norm of reaction** is the phenotypic range that individuals with a particular genotype exhibit under differing environmental conditions. To evaluate the norm of reaction, researchers study members of true-breeding strains that have the same genotypes and subject them to different environmental conditions. For example, **Figure 16.18** shows the norm of reaction for genetically identical plants raised at different temperatures. As shown in the figure, these plants attain a maximal height when raised at 75°F. At 50°F and 85°F, the plants are substantially shorter. Growth cannot occur below 40°F or above 95°F.

The norm of reaction can be quite dramatic when we consider environmental influences on certain inherited diseases. A striking example is the human genetic disease PKU. As we discussed earlier in the chapter, this disorder is caused by a rare mutation in the gene that encodes the enzyme phenylalanine hydroxylase, which is needed to metabolize the amino acid phenylalanine. People with one or two functional copies of the gene can eat foods containing the amino acid phenylalanine and metabolize it correctly. However, individuals with two copies of the mutant gene cannot metabolize phenylalanine. When these individuals eat a standard diet containing phenylalanine, this amino acid accumulates within their bodies and becomes highly toxic. Under these conditions, PKU homozygotes manifest a variety of detrimental symptoms, including mental impairment, underdeveloped teeth, and foul-smelling urine. In contrast, when these individuals are identified at birth and given a restricted diet that is free of phenylalanine, they develop normally (**Figure 16.19**). This is a dramatic example of how genes and the environment interact to

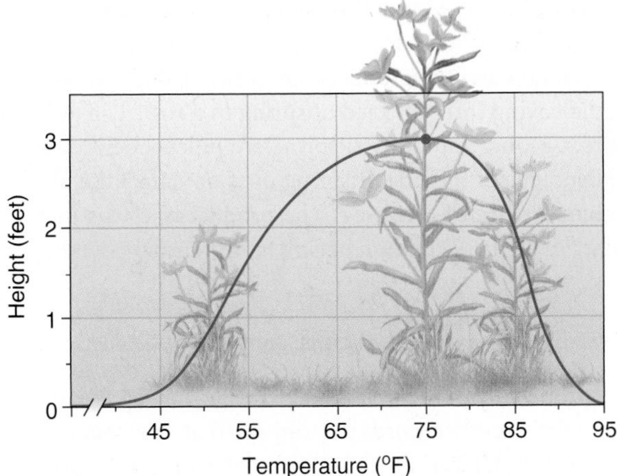

Figure 16.18 The norm of reaction. The norm of reaction is the range of phenotypes that a population of organisms with a particular genotype exhibit under different environmental conditions. In this example, genetically identical plants were grown at different temperatures in a greenhouse and then measured for height.

BIOLOGY PRINCIPLE Living organisms interact with their environment. As seen in this figure, the environment plays a key role in the outcome of traits.

Concept Check: *Could you study the norm of reaction in a wild population of squirrels?*

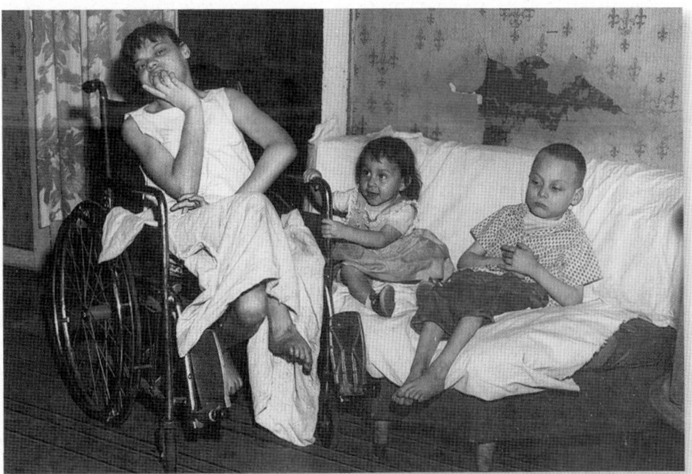

Figure 16.19 Environmental influences on the expression of PKU within a single family. All three children in this photo have inherited the alleles that cause PKU. The child in the middle was raised on a phenylalanine-free diet and developed normally. The other two children, born before the benefits of such a diet were known, were raised on diets containing phenylalanine. These two children have symptoms of PKU, including mental impairment.

BIOLOGY PRINCIPLE Biology affects our society. The study of genetic diseases has sometimes led to treatment options that prevent the harmful consequences of certain mutations, such as the one associated with PKU.

determine an individual's phenotype. In the United States, most newborns are tested for PKU, which occurs in about 1 in 10,000 babies. A newborn who is found to have this disorder can be raised on a phenylalanine-free diet and develop normally.

16.6 Genetics and Probability

Learning Outcomes:

1. Explain the concept of probability.
2. Apply the product rule and sum rule to problems involving genetic crosses.

As we have seen throughout this chapter, Mendel's laws of inheritance can be used to predict the outcome of genetic crosses. How is this useful? In agriculture, plant and animal breeders use predictions about the types and relative numbers of offspring their crosses will produce in order to develop commercially important crops and livestock. Also, people are often interested in the potential characteristics of their future children. This has particular importance to individuals who may carry alleles that cause inherited diseases. Of course, no one can see into the future and definitively predict what will happen. Nevertheless, genetic counselors can help couples predict the likelihood of having an affected child. This probability is one factor that may influence a couple's decision about whether to have children.

Earlier in this chapter, we considered how a Punnett square can be used to predict the outcome of simple genetic crosses. In addition to Punnett squares, we can apply the tools of mathematics and probability to solve more complex genetic problems. In this section, we will examine a couple of ways to calculate the outcomes of genetic crosses using these tools.

Genetic Predictions Are Based on the Rules of Probability

The chance that an event will have a particular outcome is called the **probability** of that outcome. The probability of a given outcome depends on the number of possible outcomes. For example, if you draw a card at random from a 52-card deck, the probability that you will get the jack of diamonds is 1 in 52, because there are 52 possible outcomes for the draw. In contrast, only two outcomes are possible when you flip a coin, so the probability is one in two (1/2, or 0.5, or 50%) that the heads side will be showing when the coin lands. The general formula for the probability (*P*) that a random event will have a specific outcome is

$$P = \frac{\text{Number of times an event occurs}}{\text{Total number of possible outcomes}}$$

Thus, for a single coin toss, the chance of getting heads is

$$P_{\text{heads}} = \frac{1 \text{ heads}}{(1 \text{ heads} + 1 \text{ tails})} = \frac{1}{2}$$

Earlier in this chapter, we used Punnett squares to predict the fractions of offspring with a given genotype or phenotype. In a cross between two pea plants that were heterozygous for the height gene (*Tt*), our Punnett square predicted that one-fourth of the offspring would be dwarf. We can make the same prediction by using a probability calculation.

$$P_{\text{dwarf}} = \frac{1\ tt}{(1\ TT + 2\ Tt + 1\ tt)} = \frac{1}{4} = 25\%$$

A probability calculation allows us to predict the likelihood that a future event will have a specific outcome. However, the accuracy of this prediction depends to a great extent on the number of events we observe—in other words, on the size of our sample. For example, if we toss a coin six times, the calculation we just presented for P_{heads} suggests we should get heads three times and tails three times. However, each coin toss is an independent event, meaning that every time we toss the coin there is a random chance that it will come up heads or tails, regardless of the outcome of the previous toss. With only six tosses, we would not be too surprised if we got four heads and two tails instead of the expected three heads and three tails. The deviation between the observed and expected outcomes is called the **random sampling error**. With a small sample, the random sampling error may cause the observed data to be quite different from the expected outcome. By comparison, if we flipped a coin 1,000 times, the percentage of heads would be fairly close to the predicted 50%. With a larger sample, we expect the sampling error to be smaller.

The Product Rule Is Used to Predict the Outcome of Independent Events

Punnett squares allow us to predict the likelihood that a genetic cross will produce an offspring with a particular genotype or phenotype. To predict the likelihood of producing multiple offspring with particular genotypes or phenotypes, we can use the **product rule**, which states that *the probability that two or more independent events will occur is equal to the product of their individual probabilities*. As we have already discussed, events are independent if the outcome of one event does not affect the outcome of another. In our previous coin-toss example, each toss is an independent event—if one toss comes up heads, another toss still has an equal chance of coming up either heads or tails. If we toss a coin twice, what is the probability that we will get heads both times? The product rule says that it is equal to the probability of getting heads on the first toss (1/2) times the probability of getting heads on the second toss (1/2), or one in four (1/2 × 1/2 = 1/4).

To see how the product rule can be applied to a genetics problem, let's consider a rare recessive human trait known as congenital analgesia. (Congenital refers to a condition present at birth; analgesia means insensitivity to pain.) People with this trait can distinguish between sensations such as sharp and dull, or hot and cold, but they do not perceive extremes of sensation as painful. The first known case of congenital analgesia, described in 1932, was a man who made his living entertaining the public as a "human pincushion." For a phenotypically normal couple, each heterozygous for the recessive allele causing congenital analgesia, we can ask, "What is the probability that their first three offspring will have the disorder?" To answer this question, we must first determine the probability of a single offspring having the abnormal phenotype. By using a Punnett square, we would find that the probability of an individual offspring being homozygous recessive is 1/4. Thus, each of this couple's children has a one in four chance of having the disorder.

We can now use the product rule to calculate the probability of this couple having three affected offspring in a row. The phenotypes of the first, second, and third offspring are independent events; that is, the phenotype of the first offspring does not affect the phenotype of the second or third offspring. The product rule tells us that the probability of all three children having the abnormal phenotype is

$$\frac{1}{4} \times \frac{1}{4} \times \frac{1}{4} = \frac{1}{64} = 0.016$$

The probability of the first three offspring having the disorder is 0.016, or 1.6%. In other words, we can say that this couple's chance of having three children in a row with congenital analgesia is very small—only 1.6 out of 100.

The product rule can also be used to predict the outcome of a cross involving two or more genes. Let's suppose a pea plant with the genotype *TtYy* was crossed to a plant with the genotype *Ttyy*. We could ask the question, "What is the probability that an offspring will have the genotype *ttYy*?" If the two genes independently assort, the probability of inheriting alleles for one gene is independent of the probability for other gene. Therefore, we can separately calculate the probability of the desired outcome for each gene. By constructing two small Punnett squares, we can determine the probability of genotypes for each gene individually, as shown here.

Cross: *TtYy* × *Ttyy*

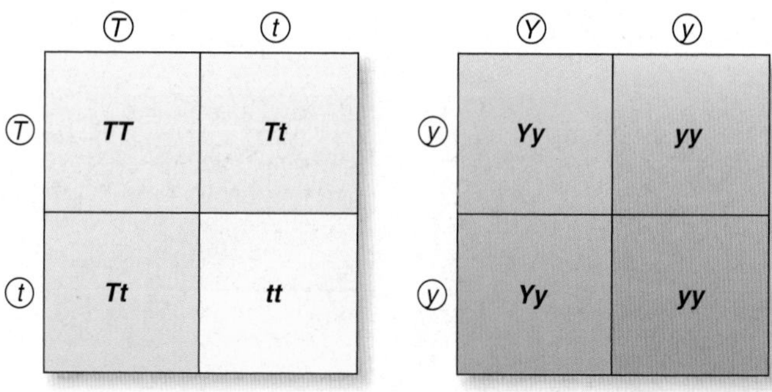

Probability that an offspring will be *tt* is 1/4, or 0.25.

Probability that an offspring will be *Yy* is 1/2, or 0.5.

We can now use the product rule to determine the probability that an offspring will be *ttYy*;

$$P = (0.25)(0.5) = 0.125, \text{ or } 12.5\%$$

The Sum Rule Is Used to Predict the Outcome of Mutually Exclusive Events

Let's now consider a second way to predict the outcome of particular crosses. In a cross between two heterozygous (*Tt*) pea plants, we may want to know the probability of a particular offspring being a homozygote. In this case we are asking, "What is the chance that this individual will be either homozygous *TT* or homozygous *tt*?" To answer an "either/or" question, we use the sum rule, which applies to events with mutually exclusive outcomes. When we say that outcomes are mutually exclusive, we mean they cannot occur at the same time. A pea plant can be tall or dwarf, but not both at the same time. The tall and dwarf phenotypes are mutually exclusive. Similarly, a plant with the genotype *TT* cannot be *Tt* or *tt*. Each of these genotypes is mutually exclusive of the other two. According to the **sum rule**, *the probability that one of two or more mutually exclusive outcomes will occur is the sum of the probabilities of the individual outcomes.*

To find the probability that an offspring will be either homozygous *TT* or homozygous *tt*, we add together the probability that it will be *TT* and the probability that it will be *tt*. If we constructed a Punnett square, we would find that the probability for each of these genotypes is one in four. We can now use the sum rule to determine the probability of an individual having one of these genotypes.

$$\frac{1}{4} + \frac{1}{4} = \frac{1}{2}$$

(probability of *TT*) (probability of *tt*) (probability of either *TT* or *tt*)

This calculation predicts that in crosses of two *Tt* parents, half of the offspring will be homozygotes—either *TT* or *tt*.

Summary of Key Concepts

16.1 Mendel's Laws of Inheritance

- Mendel studied seven characters found in garden peas that existed in two variants each (Figures 16.1, 16.2).
- Mendel allowed his peas to self-fertilize, or he carried out cross-fertilization, also known as hybridization (Figures 16.3, 16.4).
- By following the inheritance pattern of a single character (a monohybrid cross) for two generations, Mendel determined the law of segregation, which states that two alleles of a gene segregate during the formation of eggs and sperm so that every gamete receives only one allele (Figures 16.5, 16.6).
- The genotype is the genetic makeup of an organism. Alleles are alternative versions of the same gene. Phenotype is a description of the traits that an organism displays.
- A Punnett square is constructed to predict the outcome of crosses.
- A testcross is conducted to determine if an individual displaying a dominant trait is a homozygote or a heterozygote (Figure 16.7).
- By conducting a dihybrid cross, Mendel determined the law of independent assortment, which states that the alleles of different genes assort independently of each other during gamete formation. In a dihybrid cross, this yields a 9:3:3:1 ratio in the F_2 generation (Figure 16.8).

16.2 The Chromosome Theory of Inheritance

- The chromosome theory of inheritance explains how the behavior of chromosomes during meiosis accounts for Mendel's laws of inheritance. Each gene is located at a particular site, or locus, on a chromosome (Figures 16.9, 16.10, 16.11).

16.3 Pedigree Analysis of Human Traits

- The inheritance patterns in humans are determined from a pedigree analysis (Figures 16.12, 16.13).

16.4 Sex Chromosomes and X-Linked Inheritance Patterns

- Many species of animals and some species of plants have separate male and female sexes. In many cases, sex is determined by differences in sex chromosomes (Figure 16.14).
- X-linked genes are found on the X chromosome but not the Y chromosome. Recessive X-linked alleles in humans can cause diseases, such as hemophilia, which are more likely to occur in males.
- Morgan's experiments showed that an eye-color gene in *Drosophila* is located on the X chromosome (Figure 16.15).

16.5 Variations in Inheritance Patterns and Their Molecular Basis

- Several inheritance patterns have been discovered that obey Mendel's laws but yield differing ratios of offspring compared with Mendel's crosses (Table 16.1).
- Recessive inheritance is often due to a loss-of-function mutation. In many simple dominant/recessive relationships, the heterozygote has a dominant phenotype because 50% of the normal protein is sufficient to produce that phenotype (Figure 16.16).
- Mutant genes are responsible for many inherited diseases in humans. In many cases, the effects of a mutant gene are pleiotropic, meaning the gene affects several different aspects of bodily structure and function (Table 16.2).
- Incomplete dominance occurs when a heterozygote has a phenotype that is intermediate between either homozygote. This occurs because 50% of the functional protein is not enough to produce the same phenotype as a homozygote (Figure 16.17).
- ABO blood type is produced by the expression of a gene that exists in multiple alleles in a population. The I^A and I^B alleles show codominance, a phenomenon in which both alleles are expressed in the heterozygous individual (Table 16.3).
- Genes and the environment interact to determine an individual's phenotype. The norm of reaction is a description of how a phenotype may change under different environmental conditions (Figures 16.18, 16.19).

16.6 Genetics and Probability

- Probability is the likelihood that an event will occur in the future. Random sampling error is the deviation between observed and expected values.
- The product rule states that the probability of two or more independent events occurring is equal to the product of their individual probabilities. The sum rule states that the probability of two or more mutually exclusive events occurring is equal to the sum of their individual probabilities.

Assess and Discuss

Test Yourself

1. Based on Mendel's experimental crosses, what is the expected F_2 phenotypic ratio of a monohybrid cross?
   a. 1:2:1
   b. 2:1
   c. 3:1
   d. 9:3:3:1
   e. 4:1
2. During which phase of cellular division does Mendel's law of segregation physically occur?
   a. mitosis
   b. meiosis I
   c. meiosis II
   d. all of the above
   e. b and c only
3. An individual that has two different alleles of a particular gene is said to be
   a. dihybrid.
   b. recessive.
   c. homozygous.
   d. heterozygous.
   e. hemizygous.
4. Which of Mendel's laws cannot be observed in a monohybrid cross?
   a. segregation
   b. dominance/recessiveness
   c. independent assortment
   d. codominance
   e. All of the above can be observed in a monohybrid cross.
5. During a ____ cross, an individual with the dominant phenotype and unknown genotype is crossed with a ___ individual to determine the unknown genotype.
   a. monohybrid, homozygous recessive
   b. dihybrid, heterozygous
   c. test, homozygous dominant
   d. monohybrid, homozygous dominant
   e. test, homozygous recessive
6. A woman is heterozygous for an X-linked trait, hemophilia A. If she has a child with a man without hemophilia A, what is the probability that the child will be a male with hemophilia A? (Note: The child could be a male or female.)
   a. 100%
   b. 75%
   c. 50%
   d. 25%
   e. 0%
7. A gene that affects more than one phenotypic trait is said to be
   a. dominant.
   b. wild type.
   c. dihybrid.
   d. pleiotropic.
   e. heterozygous.
8. A hypothetical flowering plant species produces blue, light blue, and white flowers. To determine the inheritance pattern, the following crosses were conducted with the results indicated:

   blue × blue → all blue

   white × white → all white

   blue × white → all light blue

   What type of inheritance pattern does this represent?
   a. simple Mendelian
   b. X-linked
   c. codominance
   d. incomplete dominance
   e. pleiotropy
9. Genes located on a sex chromosome are said to be
   a. X-linked.
   b. dominant.
   c. hemizygous.
   d. sex linked.
   e. autosomal.
10. A man and woman are both heterozygous for the recessive allele that causes cystic fibrosis. What is the probability that their first 2 offspring will have the disorder?
    a. 1
    b. 1/4
    c. 1/16
    d. 1/32
    e. 0

Conceptual Questions

1. Describe one observation in a human pedigree that rules out a recessive pattern of inheritance. Describe an observation that rules out a dominant pattern.
2. A cross is made between individuals of the following genotypes: *AaBbCCDd* and *AabbCcdd*. What is the probability that an offspring will be *AAbbCCDd*? Hint: Don't waste your time making a really large Punnett square. Make four small Punnett squares instead and use the product rule.
3. A principle of biology is that *living organisms interact with their environment.* Discuss how the environment plays a key role in determining the outcome of an individual's traits.

Collaborative Questions

1. Discuss the principles of the chromosome theory of inheritance. Which principles do you think were deduced via light microscopy, and which were deduced from crosses? What modern techniques could be used to support the chromosome theory of inheritance?
2. When examining a human pedigree, what observations do you look for to distinguish between X-linked recessive inheritance versus autosomal recessive inheritance? How would you distinguish X-linked dominant inheritance from autosomal dominant inheritance from an analysis of a human pedigree?

Online Resource

www.brookerbiology.com

Stay a step ahead in your studies with animations that bring concepts to life and practice tests to assess your understanding. Your instructor may also recommend the interactive eBook, individualized learning tools, and more.

Chapter Outline

17.1 Gene Interaction

17.2 Genes on the Same Chromosome: Linkage, Recombination, and Mapping

17.3 Extranuclear Inheritance: Organelle Genomes

17.4 Epigenetic Inheritance

Summary of Key Concepts

Assess and Discuss

Complex Patterns of Inheritance

17

A calico cat. The inheritance of variegated coat colors in the calico cat can't be predicted by Mendel's laws.

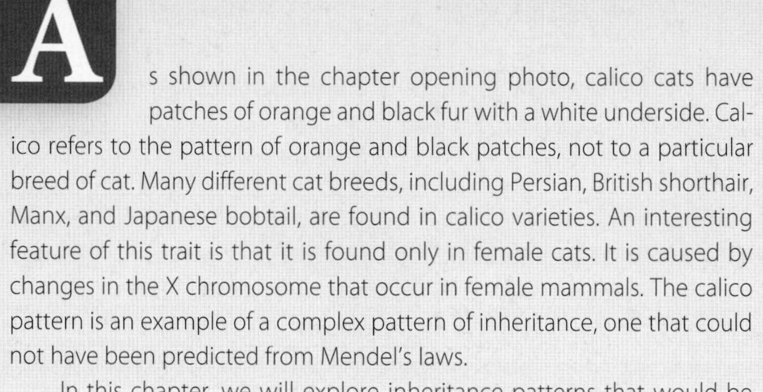

As shown in the chapter opening photo, calico cats have patches of orange and black fur with a white underside. Calico refers to the pattern of orange and black patches, not to a particular breed of cat. Many different cat breeds, including Persian, British shorthair, Manx, and Japanese bobtail, are found in calico varieties. An interesting feature of this trait is that it is found only in female cats. It is caused by changes in the X chromosome that occur in female mammals. The calico pattern is an example of a complex pattern of inheritance, one that could not have been predicted from Mendel's laws.

In this chapter, we will explore inheritance patterns that would be difficult if not impossible to predict based solely on Mendel's laws of inheritance, which we discussed in Chapter 16. Some of them even violate the law of segregation or the law of independent assortment. Studies of complex inheritance patterns have helped us appreciate more fully how genes influence phenotypes. Such research has revealed an astounding variety of ways that inheritance occurs. The picture that emerges is of a wonderful web of diverse mechanisms by which genes give rise to phenotypes. **Table 17.1** provides a summary of Mendelian inheritance and the types of inheritance patterns we will consider in this chapter.

Table 17.1 Different Types of Inheritance Patterns

Type	Description
Mendelian	Inheritance patterns in which a single gene affects a single trait and the alleles obey the law of segregation. These patterns include simple Mendelian inheritance, X-linked inheritance, incomplete dominance, and codominance (refer back to Table 16.1).
Epistasis	A type of gene interaction in which the alleles of one gene mask the effects of a dominant allele of another gene.
Continuous variation	Inheritance pattern in which the offspring display a continuous range of phenotypes. This pattern is produced by the additive interactions of several genes, along with environmental influences.
Linkage	Inheritance pattern involving two or more genes that are close together on the same chromosome. Linked genes do not assort independently.
Extranuclear inheritance	Inheritance pattern of genes found in the genomes of mitochondria or chloroplasts. Usually these genes are inherited from the mother.
X inactivation	Phenomenon of female mammals in which one X chromosome is inactivated in every somatic cell, producing a mosaic phenotype.
Genomic imprinting	Inheritance pattern in which an allele from one parent is silenced in the somatic cells of the offspring, but the allele from the other parent is expressed.

17.1 Gene Interaction

Learning Outcomes:

1. Describe how the alleles of one gene can mask or be epistatic to the alleles of a different gene.
2. Explain why polygenic traits usually show a continuum of phenotype variation.

The study of single genes was pivotal in establishing the science of genetics. This focus allowed Mendel to formulate the basic laws of inheritance for traits with a simple dominant/recessive inheritance pattern. Likewise, this approach helped later researchers understand inheritance patterns involving incomplete dominance and codominance, as well as traits that are influenced by an individual's sex. However, all or nearly all traits are influenced by many genes. For example, in both plants and animals, height is affected by genes that encode proteins involved in the production of growth hormones, cell division, the uptake of nutrients, metabolism, and many other functions. Variation in any of the genes involved in these processes is likely to influence an individual's height.

If height is controlled by many genes, how was Mendel able to study the effects of a single gene that produced tall or dwarf pea plants? The answer lies in the genotypes of his strains. Although many genes affect the height of pea plants, Mendel chose true-breeding strains that differed with regard to only one of those genes. As a hypothetical example, let's suppose that pea plants have 10 genes affecting height, which we will call *K, L, M, N, O, P, Q, R, S,* and *T*. The genotypes of two hypothetical strains of pea plants may be:

Tall strain: *KK LL MM NN OO PP QQ RR SS TT*

Dwarf strain: *KK LL MM NN OO PP QQ RR SS tt*

In this example, the tall and dwarf strains differ at only a single gene. One strain is *TT* and the other is *tt*, and this accounts for the difference in their height. If we make crosses of tall and dwarf plants, the genotypes of the F_2 offspring will differ with regard to only one gene; the other nine genes will be identical in all of them. This approach allows a researcher to study the effects of a single gene even though many genes may affect a single character.

In this section, we will examine situations in which a single character is controlled by two or more different genes, each of which has two or more alleles. This phenomenon is called **gene interaction**. As you will see, allelic variation at two or more genes may affect the outcome of traits in different ways. First we will look at a gene interaction in which an allele of one gene prevents the phenotypic expression of an allele of a different gene. Then we will discuss an interaction in which multiple genes have additive effects on a single character. These additive effects, together with environmental influences, account for the continuous phenotypic variation that we see in most traits.

In an Epistatic Gene Interaction, the Alleles of One Gene Masks the Phenotypic Effects of a Different Gene

In some gene interactions, the alleles of one gene mask the expression of the alleles of another gene. This phenomenon is called **epistasis** (from the Greek *ephistanai*, meaning stopping). An example is the unexpected gene interaction discovered by English geneticists William Bateson and Reginald Punnett in the early 1900s, when they were studying crosses involving the sweet pea, *Lathyrus odoratus*. A cross between a true-breeding purple-flowered plant and a true-breeding white-flowered plant produced an F_1 generation with all purple-flowered plants and an F_2 generation with a 3:1 ratio of purple- to white-flowered plants. Mendel's laws predicted this result. The surprise came when the researchers crossed two different true-breeding varieties of white-flowered sweet peas (**Figure 17.1**). All of the F_1 generation plants had purple flowers! When these plants were

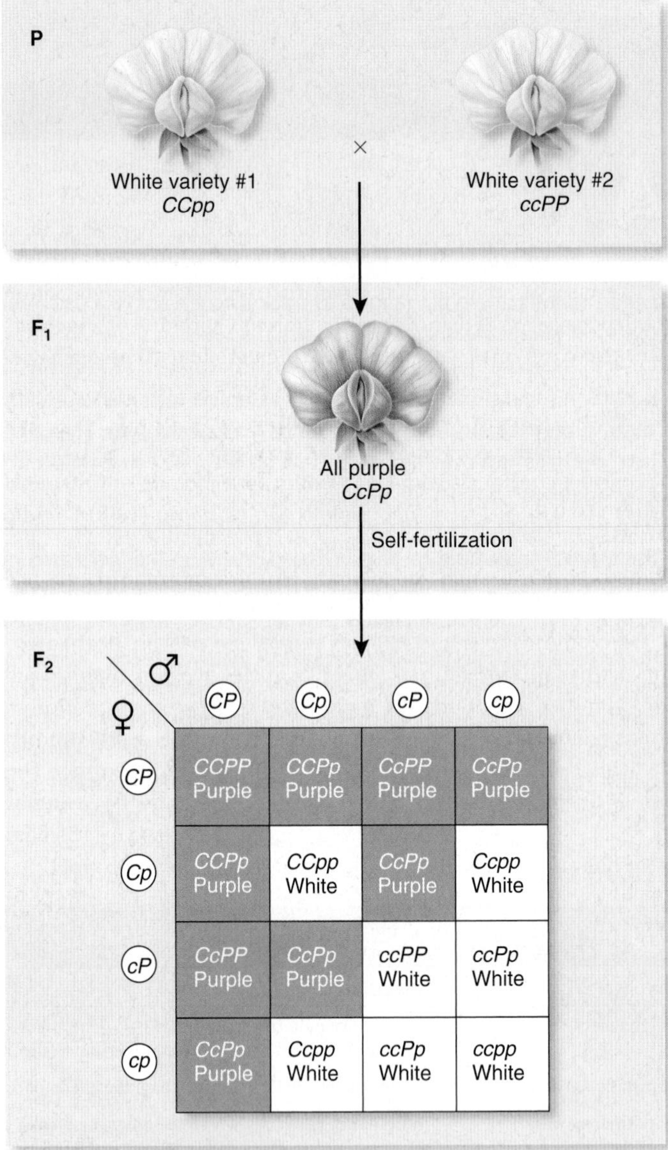

Figure 17.1 **Epistasis in the sweet pea.** The color of the sweet pea flower is controlled by two genes, each with a dominant and a recessive allele. Each of the dominant alleles (*C* and *P*) encodes an enzyme required for the synthesis of purple pigment. A plant that is homozygous recessive for either gene (*cc* or *pp*) cannot synthesize the pigment and will have white flowers.

Concept Check: In a Ccpp individual, which functional enzyme is missing? Is it the enzyme encoded by the C or P gene?

allowed to self-fertilize, the F_2 generation had purple-flowered and white-flowered plants in a 9:7 ratio. From these results, Bateson and Punnett deduced that two different genes were involved. To have purple flowers, a plant must have one or two dominant alleles for each of these genes. The relationships among the alleles are as follows:

C (one allele for purple) is dominant to *c* (white)

P (an allele of a different gene for purple) is dominant to *p* (white)

cc masks *P*, or *pp* masks *C*, in either case producing white flowers

A plant that was homozygous for either *c* or *p* would have white flowers even if it had a dominant purple-producing allele for the other gene.

How do we explain these results at the molecular and cellular level? Epistatic interactions often arise because two or more different proteins are involved in a single cellular function. For example, two or more proteins may be part of a metabolic pathway leading to the formation of a single product. This is the case for the formation of a purple pigment in the sweet pea strains we have been discussing:

Colorless precursor $\xrightarrow{\text{Enzyme C}}$ Colorless intermediate $\xrightarrow{\text{Enzyme P}}$ Purple pigment

In this example, a colorless precursor molecule must be acted on by two different enzymes to produce the purple pigment. Gene *C* encodes a functional protein called enzyme C, which converts the colorless precursor into a colorless intermediate. The recessive *c* allele results in a lack of production of enzyme C in the homozygote. Gene *P* encodes the functional enzyme P, which converts the colorless intermediate into the purple pigment. Like the *c* allele, the *p* allele results in an inability to produce a functional enzyme. A plant homozygous for either of the recessive alleles does not make any functional enzyme C or enzyme P. When either of these enzymes is missing, the plant cannot make the purple pigment and has white flowers. Note that the results observed in Figure 17.1 do not conflict with Mendel's laws of segregation or independent assortment. Mendel investigated the effects of only a single gene on a given character. The 9:7 ratio is due to a gene interaction in which two genes affect a single character.

Polygenic Inheritance and Environmental Influences Produce Continuous Phenotypic Variation

Until now, we have discussed the inheritance of characters with clearly defined phenotypic variants, such as red or white eyes in fruit flies or round or wrinkled seed shape in garden peas. These are known as **discrete traits**, because the phenotypes do not overlap. For most traits, however, the phenotypes cannot be sorted into discrete categories. Traits that show continuous variation over a range of phenotypes are called **quantitative traits**. In humans, quantitative traits include height, weight, skin color, metabolic rate, and heart size. In the case of domestic animals and plant crops, many of the traits that people consider desirable are quantitative in nature, such as the number of eggs a chicken lays, the amount of milk a cow produces, and the number of apples on an apple tree. Consequently, much of our modern understanding of quantitative traits comes from agricultural research.

Quantitative traits are usually **polygenic**, which means that multiple genes contribute to the outcome of the trait. For many polygenic traits, genes contribute to the phenotype in an additive way. As a hypothetical example, let's suppose that three different genes (*W1*, *W2*, and *W3*) affect weight in turkeys; each gene can occur in heavy (*W*) and light (*w*) alleles. A heavy allele contributes an extra pound to an individual's weight compared with the effect of a light allele. A turkey homozygous for all the heavy alleles (*W1W1 W2W2 W3W3*) would weigh 6 pounds more than an individual homozygous for all the light alleles (*w1w1 w2w2 w3w3*). A turkey heterozygous for all three genes (*W1w1 W2w2 W3w3*) would have an intermediate weight that would be 3 pounds lighter than the homozygous turkey carrying all of the heavy alleles, because the heterozygote carries three light alleles.

An individual's environment is another important factor affecting the expression of quantitative traits. As we learned in Chapter 16, the environment plays a vital role in the phenotypic expression of genes. Environmental factors often have a major effect on quantitative traits. For example, an animal's diet affects its weight, and the amount of rain and sunlight that fall on an apple tree affect how many apples it produces.

Because quantitative traits are polygenic and greatly influenced by environmental conditions, the phenotypes among different individuals may vary substantially in any given population. As an example, let's consider skin pigmentation in humans. This character is influenced by several genes that tend to interact in an additive way. As a simplified example, let's consider a population in which skin pigmentation in people is controlled by three genes, which we will designate *A*, *B*, and *C*. Each gene may exist as a dark allele, designated A^D, B^D, or C^D, or a light allele, designated A^L, B^L, or C^L, respectively. All of the alleles encode enzymes that cause the synthesis of skin pigment, but the enzymes encoded by dark alleles cause more pigment synthesis than the enzymes encoded by light alleles.

Figure 17.2 considers a hypothetical case in which people who were heterozygous for all three genes produced a large population of offspring. The bar graph shows the genotypes of the offspring, grouped according to the total number of light and dark alleles. As shown by the shading of the figure, skin pigmentation increases as the number of dark alleles increases. Offspring who have all light alleles or who have all dark alleles—that is, those who are homozygous for all three genes—are fewer in number than those with some combination of light and dark alleles. As seen in the bell-shaped curve above the bar graph, the phenotypes of the offspring fall along a continuum. This continuous phenotypic variation, which is typical of quantitative traits, is produced by genotypic differences together with environmental effects. A second bell-shaped curve (the dashed line) depicts the expected phenotypic range if the same population of offspring had been raised in a sunnier environment, which increases pigment production. This curve illustrates how the environment can also have a significant influence on the range of phenotypes.

In our discussion of genetics, we tend to focus on discrete traits because this makes it easier to relate a specific genotype with a phenotype. This is usually not possible for continuous traits. For example, as depicted in the middle bar of Figure 17.2, seven different genotypes can produce individuals with a medium amount of pigmentation. It is important to emphasize that the majority of traits in all organisms are continuous, not discrete. Most traits are influenced by multiple genes, and the environment has an important influence on the phenotypic outcome.

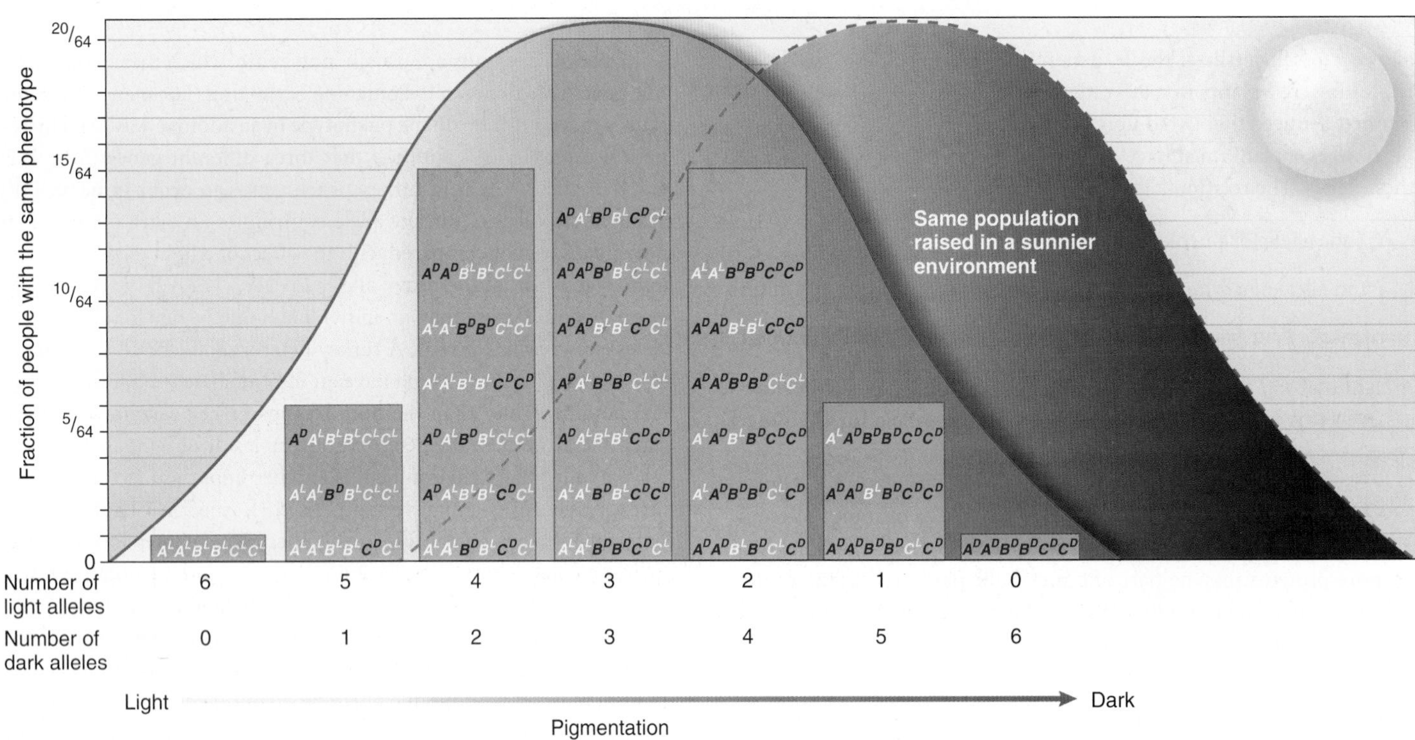

Figure 17.2 **Continuous variation in a polygenic trait.** Skin color is a polygenic character that displays a continuum of phenotypes. The bell curve on the left (solid line) shows the range of skin pigmentation in a hypothetical human population. The bar graphs below the curve show the additive effects of three genes that affect pigment production in this population; each bar shows the fraction of people with a particular number of dark alleles (A^D, B^D, and C^D) and light alleles (A^L, B^L, and C^L). The bell curve on the right (dashed line) represents the expected range of phenotypes if the same population was raised in a sunnier environment.

BIOLOGY PRINCIPLE The genetic material provides the blueprint for reproduction. The genetic blueprint usually involves the expression of several different genes that affect the same trait. In many cases, this results in continuous variation.

17.2 Genes on the Same Chromosome: Linkage, Recombination, and Mapping

Learning Outcomes:

1. Describe how linkage violates the law of independent assortment.
2. Explain how experimental crosses can demonstrate linkage.
3. Calculate the distance between genes that are linked on the same chromosome.

In Chapter 16, we learned that the independent assortment of alleles is due to the random alignment of homologous chromosomes during meiosis (refer back to Figure 16.11). But what happens when the alleles of different genes are on the same chromosome and do not independently assort? A typical chromosome contains many hundreds or even a few thousand different genes. When two genes are close together on the same chromosome, they tend to be transmitted as a unit, a phenomenon known as **linkage**. A group of genes that usually stay together during meiosis is called a **linkage group**, and the genes in the group are said to be linked. In a dihybrid cross, linked genes that are close together on the same chromosome do not follow the law of independent assortment.

In this section, we begin by examining the first experimental cross that demonstrated linkage. This pattern was subsequently explained by Thomas Hunt Morgan, who proposed that different genes located close to each other on the same chromosome tend to be inherited together. We will also see how crossing over between such genes provided the first method of mapping genes along chromosomes.

FEATURE INVESTIGATION

Bateson and Punnett's Crosses of Sweet Peas Showed That Genes Do Not Always Assort Independently

The first study showing linkage between two different genes was a cross of sweet peas carried out by William Bateson and Reginald Punnett in 1911. A surprising result occurred when they conducted a two-factor cross involving flower color and pollen shape (Figure 17.3). One of the parent plants had purple flowers (*PP*) and long pollen (*LL*); the other had red flowers (*pp*) and round pollen (*ll*). As Bateson and Punnett expected, the F_1 plants all had purple flowers and long pollen (*PpLl*). The unexpected result came in the F_2 generation.

Figure 17.3 A cross of sweet peas showing that independent assortment does not always occur.

HYPOTHESIS The alleles of different genes assort independently of each other.

KEY MATERIALS True-breeding sweet pea strains that differ with regard to flower color and pollen shape.

	Experimental level	Conceptual level
1 Cross a plant with purple flowers and long pollen to a plant with red flowers and round pollen.	Purple flowers, long pollen × Red flowers, round pollen	*PPLL* × *ppll*
2 Observe the phenotypes of the F_1 offspring.	Purple flowers, long pollen	*PpLl*
3 Allow the F_1 offspring to self-fertilize.	Purple flowers, long pollen × Purple flowers, long pollen	Meiosis *PL* and *pl* gametes — more frequent *Pl* and *pL* gametes — less frequent
4 Observe the phenotypes of the F_2 offspring.	Purple flowers, long pollen 15.6 : Purple flowers, round pollen 1.0 : Red flowers, long pollen 1.4 : Red flowers, round pollen 4.5	Fertilization F_2 offspring having phenotypes of purple flowers with long pollen or red flowers with round pollen occurred more frequently than expected from Mendel's law of independent assortment.

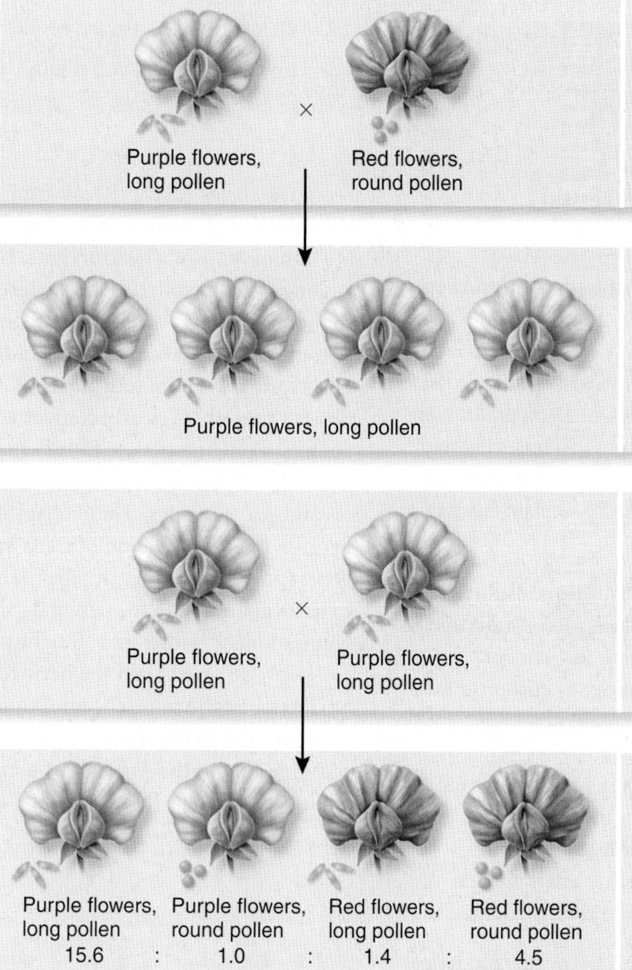

5 **THE DATA**

Phenotypes of F_2 offspring	Observed number	Observed ratio	Expected number	Expected ratio
Purple flowers, long pollen	296	15.6	240	9
Purple flowers, round pollen	19	1.0	80	3
Red flowers, long pollen	27	1.4	80	3
Red flowers, round pollen	85	4.5	27	1

6 **CONCLUSION** The data are not consistent with the law of independent assortment.

7 **SOURCE** Bateson, William, and Punnett, Reginald C. 1911. On the inter-relations of genetic factors. *Proceedings of the Royal Society of London, Series B*, 84:3–8.

Although the F_2 offspring displayed the four phenotypes predicted by Mendel's laws, the observed numbers of offspring did not conform to the predicted 9:3:3:1 ratio (refer back to Figure 16.8). Rather, as seen in the data in Figure 17.3, the F_2 generation had a much higher proportion of the two phenotypes found in the parental generation: purple flowers with long pollen, and red flowers with round pollen. How did Bateson and Punnett explain these results? They suggested that the transmission of flower color and pollen shape was somehow coupled, so these traits did not always assort independently. Although the law of independent assortment applies to many other genes, in this example, the hypothesis of independent assortment was rejected.

Experimental Questions

1. What hypothesis was Bateson and Punnett testing when conducting the crosses in the sweet pea?
2. What were the expected results of Bateson and Punnett's cross?
3. How did the observed results differ from the expected results?

Linkage and Crossing Over Produce Parental and Recombinant Types

Bateson and Punnett realized their results did not conform to Mendel's law of independent assortment. However, they did not know why the genes were not assorting independently. A few years later, Thomas Hunt Morgan obtained similar ratios in crosses of fruit flies while studying the transmission pattern of genes in *Drosophila*. Like Bateson and Punnett, Morgan observed many more F_2 offspring with the combination of traits found in the parental generation than predicted on the basis of independent assortment. To explain his data, Morgan proposed three ideas:

1. When different genes are located on the same chromosome, the traits determined by those genes are more likely to be inherited together. This violates the law of independent assortment.
2. Due to crossing over during meiosis, homologous chromosomes can exchange pieces of chromosomes and create new combinations of alleles.
3. The likelihood of a crossover occurring in the region between two genes depends on the distance between the two genes. Crossovers between homologous chromosomes are much more likely to occur between two genes farther apart along a chromosome compared to two genes that are closer together.

To illustrate the first two ideas, **Figure 17.4** considers a series of crosses involving two genes linked on the same chromosome in *Drosophila*. The two genes are located on an autosome, not on a sex chromosome. The P generation cross is between flies that are homozygous for alleles that affect body color and wing shape. The female is homozygous for the dominant wild-type alleles that produce gray body color (b^+b^+) and straight (normal) wings (c^+c^+); the male is homozygous for recessive mutant alleles that produce black body color (*bb*) and curved wings (*cc*). The symbols for the genes are based on the name of the mutant allele; the dominant wild-type allele is indicated by a superscript plus sign ($^+$). The chromosomes next to the flies in Figure 17.4 show the arrangement of these alleles. If the two genes are on the same chromosome, we know the arrangement of alleles in the P generation flies because these flies are homozygous for both genes ($b^+b^+c^+c^+$ for one parent and *bbcc* for the other parent). In the P generation female on the left, b^+ and c^+ are linked, whereas *b* and *c* are linked in the male on the right.

Let's now look at the outcome of the crosses in Figure 17.4. As expected, the F_1 offspring (b^+bc^+c) all had gray bodies and straight wings, confirming that these are the dominant traits. In the next cross, F_1 females were mated to males that were homozygous for both recessive alleles (*bbcc*). Recall from Chapter 16 that a testcross was conducted to determine if an individual with a dominant phenotype is a homozygote or a heterozygote. However, in the crosses we are discussing here, the purpose of the testcross is to determine if the genes for body color and wing shape are linked. If the genes are on different chromosomes and assort independently, this testcross will produce F_2 offspring with the four possible phenotypes in a 1:1:1:1 ratio. The observed numbers clearly conflict with this prediction. The two most abundant phenotypes are those with the combinations of characteristics in the P generation: gray bodies and straight wings or black bodies and curved wings. These offspring are termed **nonrecombinants**, because this combination of traits has not changed from the parental generation. The smaller number of offspring that have a combination of traits not found in the parental generation—gray bodies and curved wings or black bodies and straight wings—are called **recombinants**.

How do we explain the occurrence of recombinants when genes are linked on the same chromosome? As shown beside the flies of the F_2 generation in Figure 17.4, each recombinant individual has a chromosome that is the product of a crossover. The crossover occurred while the F_1 female fly was making egg cells.

As shown below, four different egg cells are possible:

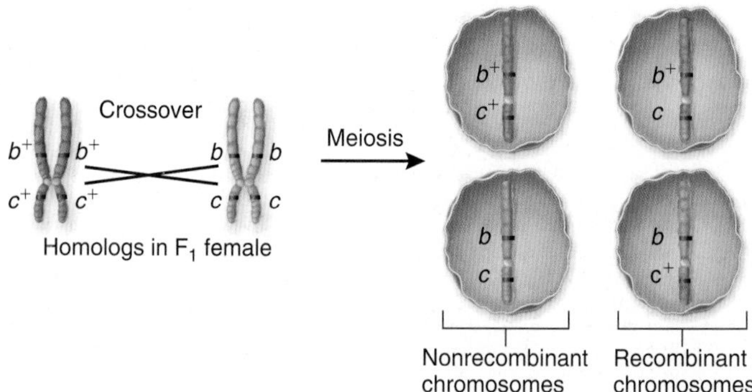

Due to crossing over, two of the four egg cells produced by meiosis have recombinant chromosomes. What happens when eggs containing such chromosomes are fertilized in the testcross? Each of the male fly's sperm cells carries a chromosome with the two recessive alleles. If the egg contains the recombinant chromosome carrying the b^+ and *c* alleles, the testcross will produce an F_2 offspring with a gray body and curved wings. If the egg contains the recombinant chromosome carrying the *b* and c^+ alleles, F_2 offspring will have a black body and straight wings. Therefore, crossing over in the F_1 female can explain the occurrence of both types of F_2 recombinant offspring.

Morgan's third idea regarding linkage was that the frequency of crossing over between linked genes depends on the distance between

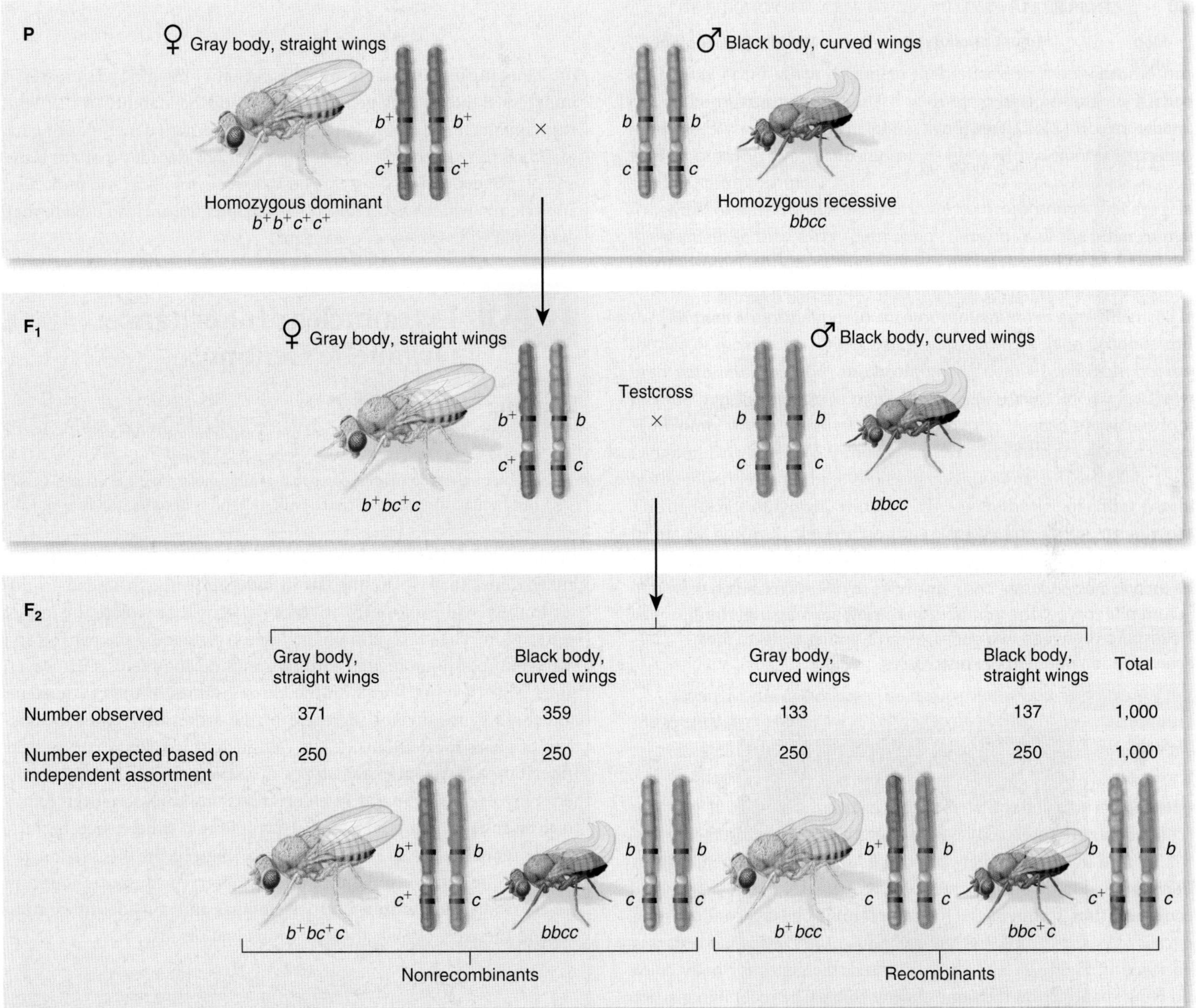

Figure 17.4 **Linkage and recombination of alleles.** An experimenter crossed $b^+b^+c^+c^+$ and *bbcc* flies to produce F_1 heterozygotes. F_1 females were then testcrossed to *bbcc* males. The large number of nonrecombinant phenotypes in the F_2 generation suggests that the two genes are linked on the same chromosome. F_2 recombinant phenotypes occur because the alleles can be rearranged by crossing over. Note: The b^+ and c^+ alleles are dominant, and the *b* and *c* alleles are recessive.

Concept Check: *In which fly or flies did crossing over occur to produce the recombinant offspring of the F_2 generation?*

BioConnections: *Look back at Figure 15.13. When does crossing over occur?*

them. This suggested a method for determining the relative positions of genes on a chromosome, as we will discuss next.

Recombination Frequencies Provide a Method for Mapping Genes Along Chromosomes

The study of the arrangement of genes in a species' genome is called **genetic mapping**. As depicted in Figure 17.5, the linear order of genes along a chromosome is shown in a chart known as a **genetic map**. Each gene has its own physical location, or locus, on a chromosome. For example, the gene for black body color (*b*) discussed earlier is located near the middle of the chromosome, whereas the gene for curved wings (*c*) is closer to one end. The first genetic map, showing five genes on the *Drosophila* X chromosome, was constructed in 1911 by American geneticist Alfred Sturtevant, an undergraduate student who studied in Morgan's laboratory.

Genetic mapping allows us to estimate the relative distances between linked genes based on the likelihood that a crossover will occur between them. This likelihood is proportional to the distance between the genes, as Morgan first proposed. If the genes are very close together, a crossover is unlikely to occur in the region between them. However, if the genes are very far apart, a crossover is more likely to be initiated in the region between the genes and thereby recombine the alleles. Therefore, in a testcross involving two genes on the same chromosome, the percentage of recombinant offspring is correlated with the distance between the genes. If a two-factor testcross produces many recombinants, the

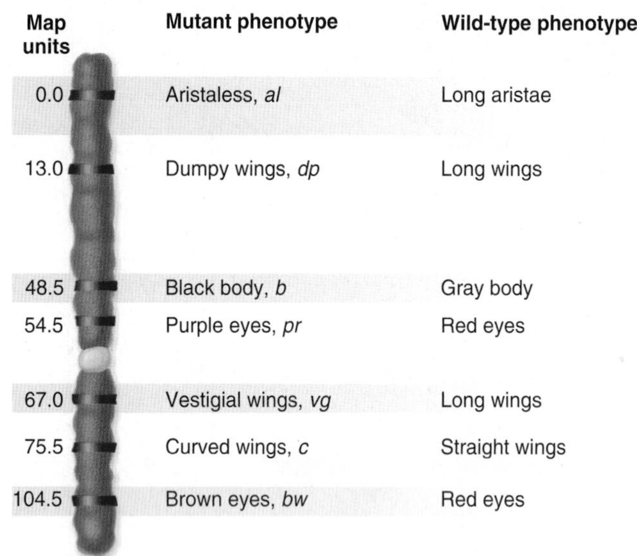

Map units
Mutant phenotype
Wild-type phenotype
0.0
Aristaless, al
Long aristae
13.0
Dumpy wings, dp
Long wings
48.5
Black body, b
Gray body
54.5
Purple eyes, pr
Red eyes
67.0
Vestigial wings, vg
Long wings
75.5
Curved wings, c
Straight wings
104.5
Brown eyes, bw
Red eyes

Figure 17.5 A simplified genetic map. This map shows the relative locations of a few genes along chromosome number 2 in *Drosophila melanogaster*. The name of each gene is based on the mutant phenotype. The numbers on the left are map units (mu). The distance between two genes, in mu's, corresponds to their recombination frequency in testcrosses.

Concept Check: How would you set up a testcross to determine the distance between the al and dp genes? What would be the genotypes of the P, F_1, and F_2 generations?

experimenter concludes that the two genes are far apart. If very few recombinants are observed, the two genes must be close together.

To find the distance between two genes, the experimenter must determine the frequency of crossing over between them, called their **recombination frequency**. This is accomplished by conducting a testcross. As an example, let's refer back to the *Drosophila* testcross described in Figure 17.4. As we discussed, the genes for body color and wing shape are on the same chromosome. The recombinants are the result of crossing over during egg formation in the F_1 female. We can use the data from the testcross shown in Figure 17.4 to estimate the distance between these two genes. The **map distance** between two genes is defined as the number of recombinants divided by the total number of offspring times 100.

$$\text{Map distance} = \frac{\text{Number of recombinants}}{\text{Total number of offspring}} \times 100$$

$$= \frac{133+137}{371+359+133+137} \times 100$$

$$= 27.0 \text{ map units}$$

The units of distance are called **map units (mu)**, or sometimes **centiMorgans (cM)** in honor of Thomas Hunt Morgan. One map unit is equivalent to a 1% recombination frequency. In this example, 270 out of 1,000 offspring are recombinants, so the recombination frequency is 27%, and the two genes are 27.0 mu apart.

Genetic mapping has been useful for analyzing the genes of organisms that are easily crossed and produce many offspring in a short time. It has been used to map the genes of several plant species and of certain species of animals, such as *Drosophila*. However, for most organisms, including humans, genetic mapping via crosses is impractical due to long generation times or the inability to carry out experimental crosses. Fortunately, many alternative methods of gene mapping have been developed in the past few decades that are faster and do not depend on crosses. These newer cytological and molecular approaches, which we will discuss in Chapter 20, are also used to map genes in a wide variety of organisms.

17.3 Extranuclear Inheritance: Organelle Genomes

Learning Outcomes:

1. Describe the general features of mitochondrial and chloroplast genomes.
2. Predict the outcome of crosses that exhibit maternal inheritance.
3. List human diseases associated with mutations in mitochondrial genes.

In the previous section, we examined the inheritance patterns of linked genes that violate the law of independent assortment. In this section, we will explore inheritance patterns that violate the law of segregation. The segregation of genes is explained by the pairing and segregation of homologous chromosomes during meiosis. However, some genes are not found on the chromosomes in the cell nucleus, and these genes do not segregate in the same way. The transmission of genes located outside the cell nucleus is called **extranuclear inheritance**. Two important types of extranuclear inheritance patterns involve genes found in chloroplasts and mitochondria. Extranuclear inheritance is also called cytoplasmic inheritance because these organelles are in the cytoplasm of the cell. In this section, we will examine the transmission patterns observed for genes found in the chloroplast and mitochondrial genomes and consider how mutations in these genes may affect an individual's traits.

Genomes & Proteomes Connection

Chloroplast and Mitochondrial Genomes Are Relatively Small, but Contain Genes That Encode Important Proteins

As we discussed in Chapter 4, mitochondria and chloroplasts are found in eukaryotic cells because of an ancient endosymbiotic relationship. They contain their own genetic material, called the mitochondrial genome and chloroplast genome, respectively (**Figure 17.6**). Mitochondrial and chloroplast genomes are composed of a single, circular DNA molecule. The mitochondrial genome of many mammalian species has been analyzed and usually contains a total of 37 genes. Twenty-four genes encode tRNAs and rRNAs, which are needed for translation inside the mitochondrion, and 13 genes encode proteins that are involved in oxidative phosphorylation. As discussed in Chapter 7, the primary function of the mitochondrion is the synthesis of ATP via oxidative phosphorylation. Among different species of plants, chloroplast genomes typically contain about 110 to 120 genes. Many of these genes encode proteins that are vital to the process of photosynthesis, which we discussed in Chapter 8.

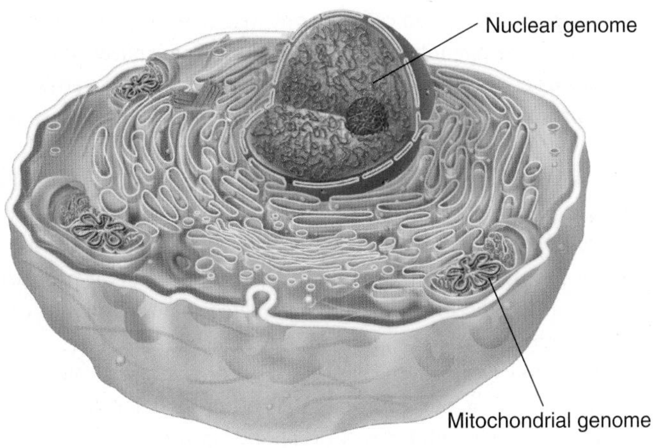

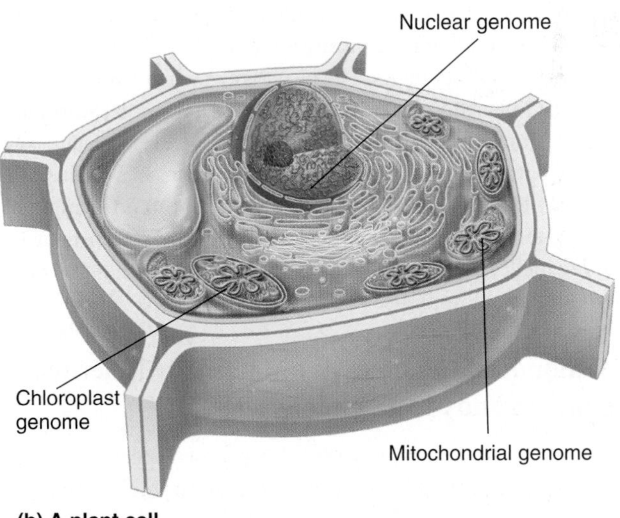

Figure 17.6 The locations of genetic material in animal and plant cells. The chromosomes in the cell nucleus are collectively known as the nuclear genome. Mitochondria and chloroplasts have small circular chromosomes called the mitochondrial and chloroplast genomes, respectively.

BioConnections: *Look back at Figure 4.27. What is the evolutionary origin of mitochondria and chloroplasts in eukaryotic cells?*

Chloroplast Genomes Are Often Maternally Inherited

One of the first experiments showing an extranuclear inheritance pattern was carried out by German botanist Carl Correns in 1909. Correns discovered that leaf pigmentation in the four-o'clock plant (*Mirabilis jalapa*) follows a pattern of inheritance that does not obey Mendel's law of segregation. Four-o'clock leaves may be green, white, or variegated. Correns observed that the pigmentation of the offspring depended solely on the pigmentation of the female parent, a phenomenon called **maternal inheritance** (Figure 17.7). If the female parent had white leaves, all of the offspring had white leaves. Similarly, if the female was green, so were all of the offspring. The offspring of a variegated female parent could be green, white, or variegated.

What accounts for maternal inheritance? At the time, Correns did not understand that chloroplasts contain genetic material. Subsequent

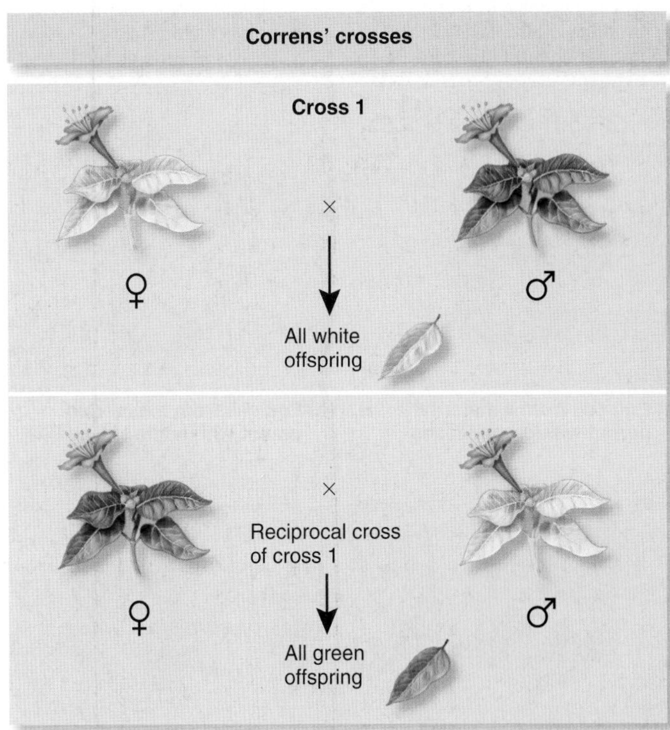

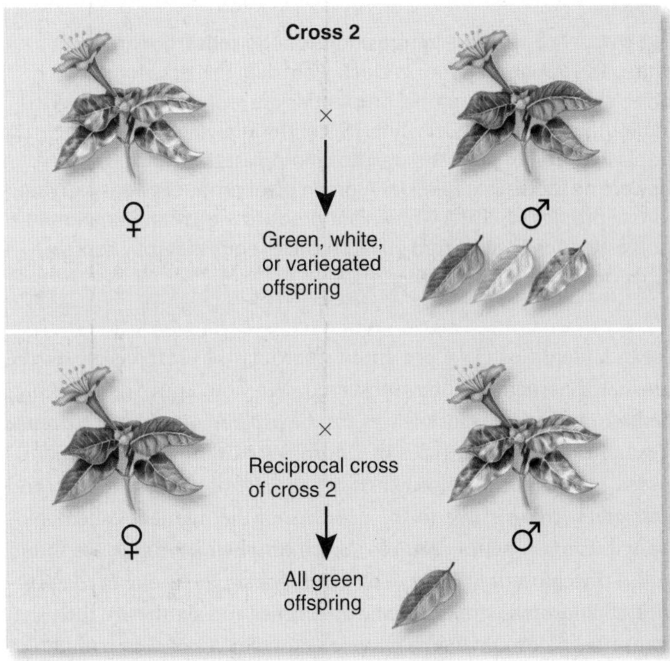

Figure 17.7 Maternal inheritance in the four-o'clock plant. In four-o'clocks, the egg contains all of the proplastids, which develop into chloroplasts, that are inherited by the offspring. The phenotype of the offspring is determined by the maternal parent. The green phenotype is due to the presence of normal chloroplasts. The white phenotype is due to chloroplasts with a mutant allele that greatly reduces green pigment production. The variegated phenotype is due to a mixture of normal and mutant chloroplasts.

Concept Check: *In this example, where is the gene located that causes the green color of four-o'clock leaves? How is this gene transmitted from parent to offspring?*

Normal proplastid will produce chloroplasts with a normal amount of green pigment.

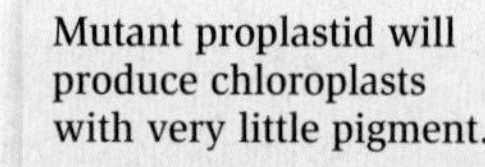
Mutant proplastid will produce chloroplasts with very little pigment.

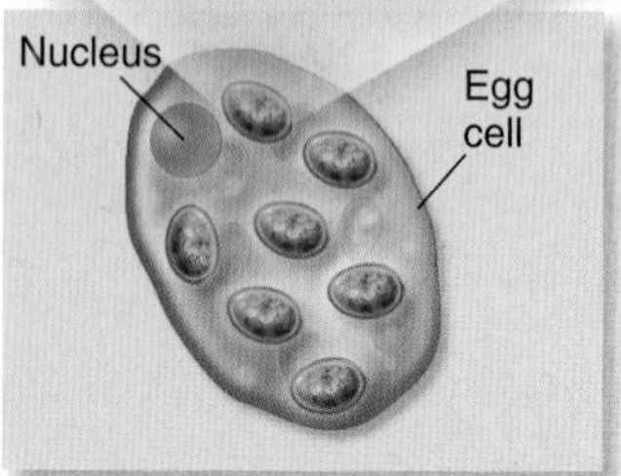
Nucleus
Egg cell

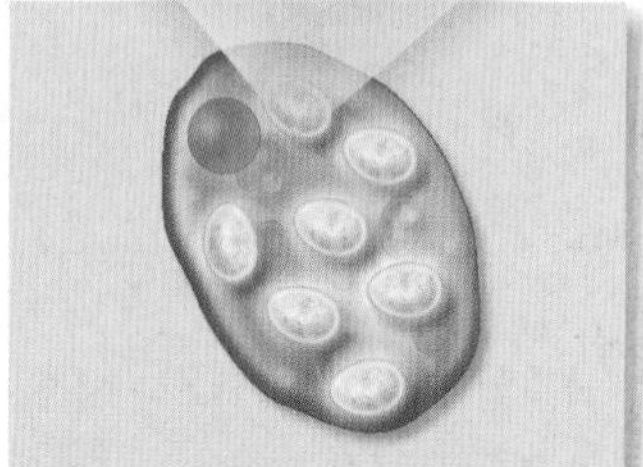

(a) Egg cell from a maternal parent with green leaves

(b) Egg cell from a maternal parent with white leaves

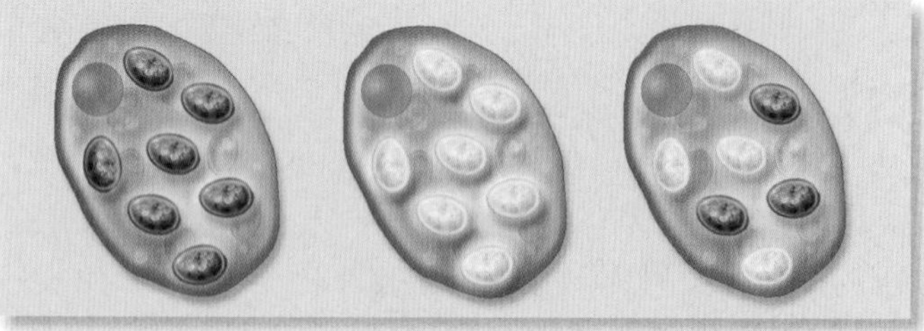

(c) Possible egg cells from a maternal parent with variegated leaves

Figure 17.8 Plastid compositions of egg cells from green, white, and variegated four-o'clock plants. In this drawing of four-o'clock egg cells, normal proplastids are represented as green and mutant proplastids as white. (Note: This drawing is schematic. Proplastids do not differentiate into chloroplasts in egg cells, and they are not actually green.) **(a)** A green plant produces eggs carrying normal proplastids. **(b)** A white plant produces eggs carrying mutant proplastids. **(c)** A variegated plant produces eggs that may contain either or both types of proplastids.

research identified DNA present in chloroplasts as responsible for the unusual inheritance pattern observed. We now know that the pigmentation of four-o'clock leaves can be explained by the occurrence of genetically different types of chloroplasts in the leaf cells. As discussed in Chapter 8, chloroplasts are the site of photosynthesis, and their green color is due to the presence of the pigment called chlorophyll. Certain genes required for chlorophyll synthesis are found within the chloroplast DNA. The green phenotype is due to the presence of chloroplasts that have normal genes and synthesize the usual quantity of chlorophyll. The white phenotype is caused by a mutation in a gene within the chloroplast DNA that prevents the synthesis of most of the chlorophyll. (Enough chlorophyll is made for the plant to survive.) The variegated phenotype occurs in leaves that have a mixture of the two types of chloroplasts.

Leaf pigmentation follows a maternal inheritance pattern because the chloroplasts in four-o'clocks are transmitted only through the cytoplasm of the egg (Figure 17.8). In most species of plants, the egg cell provides most of the zygote's cytoplasm, whereas the much smaller male gamete often provides little more than a nucleus. Therefore, chloroplasts are most often inherited via the egg. Recall from Chapter 4 that chloroplasts are derived from proplastids. In four-o'clocks, the egg cell contains several proplastids that are inherited by the offspring. The sperm cell does not contribute any proplastids. For this reason, the phenotype of a four-o'clock plant reflects the types of proplastids it inherits from the maternal parent. If the maternal parent transmits only normal proplastids, all offspring will have green leaves (Figure 17.8a). Alternatively, if the maternal parent transmits only mutant proplastids, all offspring will have white leaves (Figure 17.8b). Because an egg cell contains several proplastids, an offspring from a variegated maternal parent may inherit only normal proplastids, only mutant proplastids, or a mixture of normal and mutant proplastids. Consequently, the offspring of a variegated maternal parent can be green, white, or variegated individuals (Figure 17.8c).

How do we explain the variegated phenotype at the cellular level? This phenotype is due to events that occur after fertilization. As a zygote containing both types of proplastids grows via cellular division to produce a multicellular plant, some cells may receive mostly those that develop into normal chloroplasts. Further division of these cells gives rise to a patch of green tissue. Alternatively, as a matter of chance, other cells may receive all or mostly mutant chloroplasts that are defective in chlorophyll synthesis. The result is a patch of white tissue.

In seed-bearing plants, maternal inheritance of chloroplasts is the most common transmission pattern. However, certain species exhibit a pattern called **biparental inheritance**, in which both the pollen and the egg contribute chloroplasts to the offspring. Others exhibit **paternal inheritance**, in which only the pollen contributes these organelles. For example, most types of pine trees show paternal inheritance of chloroplasts.

Mitochondrial Genomes Are Maternally Inherited in Humans and Most Other Species

Mitochondria are found in nearly all eukaryotic species. As with the transmission of chloroplasts in plants, maternal inheritance is the most common pattern of mitochondrial transmission in eukaryotes, although some species do exhibit biparental or paternal inheritance.

In humans, mitochondria are maternally inherited. Researchers have discovered that mutations in human mitochondrial genes cause a variety of rare diseases (**Table 17.2**). These are usually chronic

Table 17.2 Examples of Human Mitochondrial Disease

Disease	Causes and symptoms
Leber's hereditary optic neuropathy (LHON)	A mutation in one of several mitochondrial genes that encode electron transport proteins. The main symptom is loss of vision.
Neurogenic muscle weakness	A mutation in a mitochondrial gene that encodes a subunit of mitochondrial ATP synthase, which is required for ATP synthesis. Symptoms involve abnormalities in the nervous system that affect the muscles and eyes.
Maternal myopathy and cardiomyopathy	A mutation in a mitochondrial gene that encodes a tRNA for leucine. The primary symptoms involve muscle abnormalities, most notably in the heart.
Myoclonic epilepsy and ragged-red muscle fibers	A mutation in a mitochondrial gene that encodes a tRNA for lysine. Symptoms include epilepsy, dementia, blindness, deafness, and heart and kidney malfunctions.

degenerative disorders that affect organs and cells, such as the brain, eyes, heart, muscle, kidney, and endocrine glands, that require high levels of ATP. For example, Leber's hereditary optic neuropathy (LHON) affects the optic nerve and leads to the progressive loss of vision in one or both eyes. LHON is caused by point mutations in several different mitochondrial genes.

17.4 Epigenetic Inheritance

Learning Outcomes:

1. Describe the process of X inactivation and how it affects the phenotype of heterozygous females.
2. Explain the molecular basis of genomic imprinting and predict how it affects inheritance patterns.

We will end our discussion of complex inheritance patterns by considering examples in which the timing and control of gene expression result in inheritance patterns that are determined by the sex of the individual or by the sex of the parents. We will consider two patterns, called X inactivation and genomic imprinting, which are types of **epigenetic inheritance**. In epigenetic inheritance, modification of a gene or chromosome occurs during egg formation, sperm formation, or early stages of embryo growth. This modification does not alter the DNA sequence, but it affects gene expression in a way that is fixed during an individual's lifetime. Such epigenetic changes affect the phenotype of the individual, but they are not permanent over the course of two or more generations. For example, a gene may undergo an epigenetic change that inactivates the gene for an individual's entire life, so it is never expressed in that individual. However, when the same individual produces gametes, the gene may become activated and remain active during the lifetime of an offspring that inherits the gene.

In Female Mammals, One X Chromosome Is Inactivated in Each Somatic Cell

In 1961, the British geneticist Mary Lyon proposed the epigenetic phenomenon of **X inactivation**, in which one of the two copies of the X chromosome in the somatic cells of female mammals is inactivated, meaning that its genes are not expressed. X inactivation is based on two lines of evidence. The first came from microscopic studies of mammalian cells. In 1949, Canadian physicians Murray Barr and Ewart Bertram identified a highly condensed structure in the cells of female cats that was not found in the cells of male cats. This structure was named a **Barr body** after one of its discoverers (Figure 17.9). In 1960, Asian American geneticist Susumu Ohno correctly proposed that a Barr body is a highly condensed X chromosome. Lyon's second line of evidence was the inheritance pattern of variegated coat colors in certain female mammals. A classic case is the calico cat, which has randomly distributed patches of black and orange fur (see chapter opening photo).

How do we explain this patchwork phenotype? According to Lyon's hypothesis, the calico pattern is due to the permanent inactivation of one X chromosome in each cell that forms a patch of the cat's skin, as shown in Figure 17.10. The gene involved is an X-linked gene that occurs as an orange allele, X^O, and a black allele, X^B. A female cat heterozygous for this gene will be calico. (The cat's white underside is due to a dominant allele of a different autosomal gene.) At an early stage of embryonic development, one of the two X chromosomes is randomly inactivated in each of the cat's somatic cells, including those that will give rise to the hair-producing skin cells. As the embryo grows and matures, the pattern of X inactivation is maintained during subsequent cell divisions. Skin cells derived from a single embryonic cell in which the X^B-carrying chromosome has been inactivated produce a patch of orange fur, because they express only the X^O allele that is carried on the active chromosome. Alternatively, a group of skin cells in which the chromosome carrying X^O has been inactivated express only the X^B allele, producing a patch of black fur. If female mammals are heterozygous for X-linked genes, approximately half of their somatic cells express one allele, whereas the rest of their somatic cells express the other allele. The result is an animal with randomly distributed patches of black and orange fur. These heterozygotes are called **mosaics** because they are composed of two types of cells.

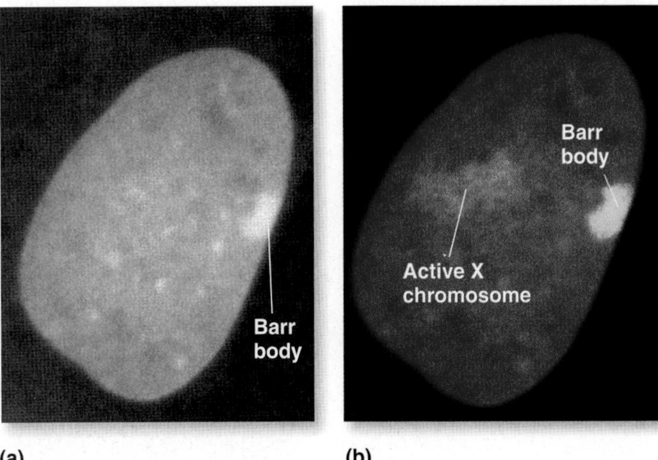

Figure 17.9 X-chromosome inactivation in female mammals. **(a)** A Barr body is seen on the periphery of a human nucleus (during interphase) after staining with a DNA-specific dye. Because it is compact, the Barr body is the most brightly stained. **(b)** The same nucleus was labeled using a yellow fluorescent probe that recognizes the X chromosome. The Barr body is more compact than the active X chromosome, which is to the left of the Barr body.

Concept Check: *How is the Barr body different from the other X chromosome in this cell?*

For many X-linked traits in humans, females who are heterozygous for recessive X-linked alleles usually show the dominant trait because the expression of the dominant allele in 50% of their cells is sufficient to produce the dominant phenotype. For example, let's consider the recessive X-linked form of hemophilia that we discussed in Chapter 16 (hemophilia A). This type of hemophilia is caused by a defect in a gene that encodes a blood-clotting protein, called factor VIII, which is made by cells in the liver and secreted into the bloodstream. In a heterozygous female, approximately half of her liver cells make and secrete this clotting factor, which is sufficient to prevent

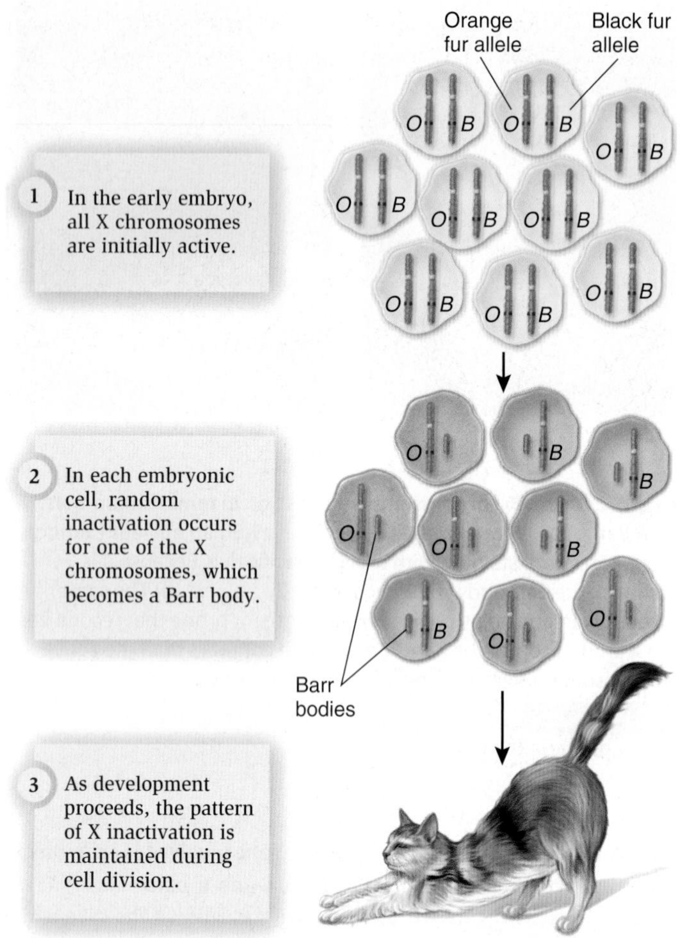

Orange fur allele
Black fur allele
1 In the early embryo, all X chromosomes are initially active.
2 In each embryonic cell, random inactivation occurs for one of the X chromosomes, which becomes a Barr body.
Barr bodies
3 As development proceeds, the pattern of X inactivation is maintained during cell division.

Figure 17.10 **X-chromosome inactivation in a calico cat.** The calico pattern is due to random X-chromosome inactivation in a female that is heterozygous for an X-linked gene with black and orange alleles. The cells at the top of this figure represent a small mass of cells making up the very early embryo. In these cells, both X chromosomes are active. At an early stage of embryonic development, one X chromosome is randomly inactivated in each cell. The initial inactivation pattern is maintained in the descendents of each cell as the embryo matures into an adult. The pattern of orange and black fur in the adult cat reflects the pattern of X inactivation in the embryo.

BIOLOGY PRINCIPLE The genetic material provides a blueprint for reproduction. In this example, the process of X inactivation produces a variegated pattern of fur color.

hemophilia. Therefore, she exhibits the dominant trait of normal blood clotting.

On rare occasions, a female who is heterozygous may show mild or even severe disease symptoms. How is this possible? X inactivation in humans occurs when an embryo is 10 days old. At this stage, the liver contains only about a dozen cells. In most females who are heterozygous for the normal and hemophilia alleles, roughly half of their liver cells express the normal allele. However, on rare occasions, all or most of the dozen embryonic liver cells may inactivate the X chromosome carrying the dominant normal allele. Following growth and development, such a female will have a very low level of factor VIII and as a result shows symptoms of hemophilia.

Why does X inactivation occur? Researchers have proposed that X inactivation achieves **dosage compensation**, a process that equalizes the expression of X-linked genes in male and female mammals. The inactivation of one X chromosome in the female reduces the number of expressed copies (doses) of X-linked genes from two to one. As a result, the expression of X-linked genes in females and males is roughly equal.

The X Chromosome Has an X Inactivation Center That Controls Compaction into a Barr Body

After Lyon's hypothesis was confirmed, researchers became interested in the genetic control of X inactivation. The cells of humans and other mammals have the ability to count their X chromosomes and allow only one of them to remain active. Additional X chromosomes are converted to Barr bodies. In normal females, two X chromosomes are counted and one is inactivated. In normal males, one X chromosome is counted and none inactivated.

On rare occasions, people are born with abnormalities in the number of their sex chromosomes (refer back to Table 15.2). In the disorders known as Turner syndrome, triple X syndrome, and Klinefelter syndrome, the cells inactivate the number of X chromosomes necessary to leave a single active chromosome (**Table 17.3**). For example, in triple X syndrome, in which an extra X chromosome is found in each cell, two X chromosomes are converted to Barr bodies. In spite of X inactivation, people with these three syndromes do exhibit some phenotypic abnormalities. The symptoms associated with these disorders may be due to effects that occur prior to X inactivation or because not all of the genes on the Barr body are completely silenced.

Although the genetic control of inactivation is not entirely understood at the molecular level, a short region on the X chromosome called the **X inactivation center (Xic)** is known to play a critical role. Finnish-born American geneticist Eeva Therman and German-born American geneticist Klaus Patau determined that X inactivation is accomplished by counting the number of Xics and inactivating all X chromosomes except for one. In cells with two X chromosomes, if one of them is missing its Xic due to a chromosome mutation, neither X chromosome will be inactivated, because only one Xic is counted. Having two active X chromosomes is a lethal condition for a human female embryo.

Table 17.3 Relationship Between X Inactivation and the Number of X Chromosomes

Phenotype	Chromosome composition	Number of Barr bodies	Number of active X chromosomes
Normal female	XX	1	1
Normal male	XY	0	1
Turner syndrome (female)	XO	0	1
Triple X syndrome (female)	XXX	2	1
Klinefelter syndrome (male)	XXY	1	1

The expression of a specific gene within the Xic is required for compaction of the X chromosome into a Barr body. This gene, discovered in 1991, is named *Xist* (for X inactive specific transcript). The *Xist* gene product is a long RNA molecule that does not encode a protein. Instead, the role of *Xist* RNA is to coat one of the two X chromosomes during the process of X inactivation. After coating, proteins associate with the *Xist* RNA and promote compaction of the chromosome into a Barr body. The *Xist* gene on the Barr body continues to be expressed after other genes on this chromosome have been silenced. The expression of the *Xist* gene also maintains a chromosome as a Barr body during cell division. Whenever a somatic cell divides in a female mammal, the Barr body is replicated to produce two Barr bodies.

For Imprinted Genes, the Gene from Only One Parent Is Expressed

As we have seen, X inactivation is a type of epigenetic inheritance in which a chromosome is modified in the early embryo, permanently altering gene expression in that individual. Other types of epigenetic inheritance occur when genes or chromosomes are modified in the gametes of a parent, permanently altering gene expression in the offspring. **Genomic imprinting**, which was discovered in the early 1980s, refers to an inheritance pattern in which a segment of DNA is imprinted or marked so that gene expression occurs only from the genetic material inherited from one parent. It occurs in numerous species, including insects, plants, and mammals.

Genomic imprinting may involve a single gene, a part of a chromosome, an entire chromosome, or even all of the chromosomes inherited from one parent. It is permanent in the somatic cells of a given individual, but the marking of the DNA is altered from generation to generation. Imprinted genes do not follow a Mendelian pattern of inheritance because imprinting causes the offspring to distinguish between maternally and paternally inherited alleles. Depending on how a particular gene is marked by each parent, the offspring expresses either the maternal or the paternal allele, but not both.

One of the first imprinted genes to be identified is a gene called *Igf2* that is found in mice and other mammals. This gene encodes a growth hormone called insulin-like growth factor 2 (Igf2) that is needed for proper growth. If a normal copy of this gene is not expressed, a mouse will be dwarf. The *Igf2* gene is known to be located on an autosome, not on a sex chromosome. Because mice are diploid, they have two copies of this gene, one from each parent.

Researchers have discovered mutations in the *Igf2* gene that block the function of the Igf2 hormone. When mice carrying normal or mutant alleles are crossed to each other, a bizarre result is obtained (**Figure 17.11**). If the male parent is homozygous for the normal allele and the female is homozygous for the mutant allele (left), all the offspring grow to a normal size. In contrast, if the male is homozygous for the mutant allele and the female is homozygous for the normal allele (right), all the offspring are dwarf. The reason this result is so surprising is that the normal and dwarf offspring have the same genotype (*Igf2 Igf2⁻*) but different phenotypes! In mice, the *Igf2* gene is imprinted in such a way that only the paternal allele is expressed,

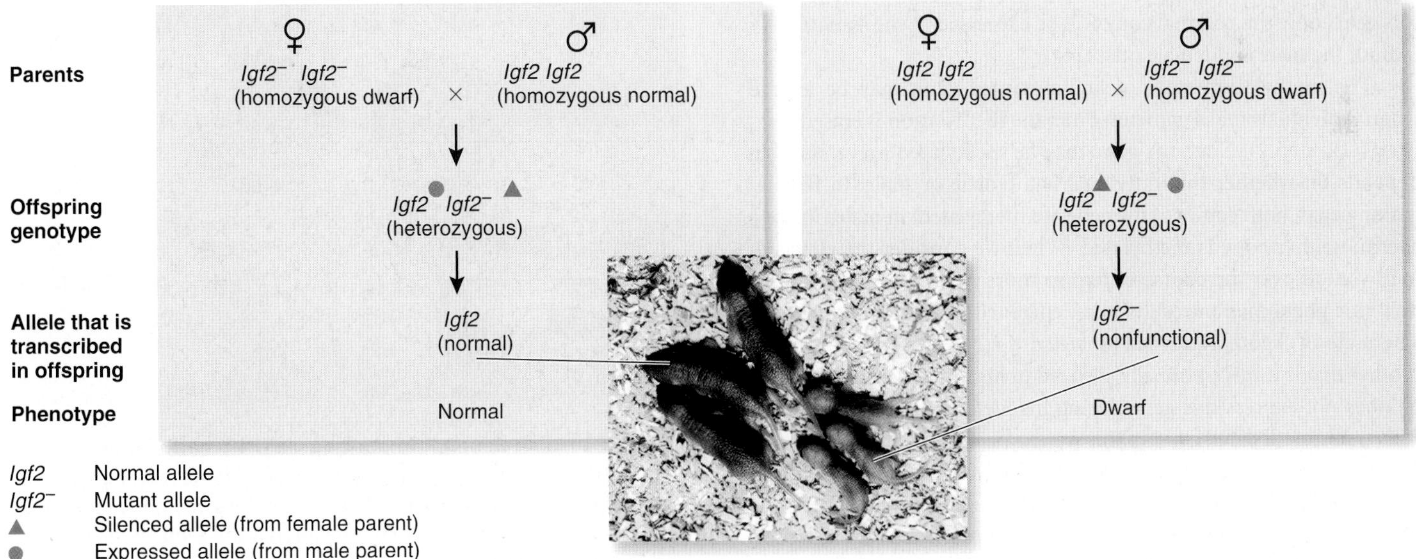

Figure 17.11 An example of genomic imprinting in the mouse. In the cross on the left, a homozygous male with the normal *Igf2* allele was crossed to a homozygous female carrying a defective allele, *Igf2⁻*. Offspring are phenotypically normal because the paternal allele is expressed. In the cross on the right, a homozygous male carrying the defective allele was crossed to a homozygous normal female. In this case, offspring are dwarf because the paternal allele is defective and the maternal allele is not expressed. The photograph shows normal-size (left) and dwarf littermates (right) derived from a cross between a wild-type female (*Igf2 Igf2*) and a heterozygous male carrying a loss-of-function allele (*Igf2 Igf2⁻*) (courtesy of A. Efstratiadis). The loss-of-function allele was made using methods described in Chapter 20.

Concept Check: *If you cross an Igf2 Igf2⁻ male mouse to an Igf2 Igf2 female mouse, what would be the expected results?*

which means it is transcribed into mRNA. The maternal allele is not transcribed. The baby mice shown on the left side of the photograph of Figure 17.11 are normal because they express a functional paternal allele. In contrast, the baby mice on the right are dwarf because the paternal gene is a mutant allele that results in a nonfunctional hormone.

The Transcription of an Imprinted Gene Depends on Methylation

Why is the maternal gene encoding Igf2 not transcribed into mRNA? To answer this question we need to consider the regulation of gene transcription in eukaryotes. As discussed in Chapter 13, DNA methylation, which is the attachment of methyl (—CH_3) groups to bases of DNA, can alter gene transcription. Researchers have discovered that DNA methylation is the marking process that occurs during the imprinting of certain genes, including the *Igf2* gene. For most genes, DNA methylation silences gene expression by inhibiting the initiation of transcription or by causing the chromatin in a region to become more compact. In contrast, for a few imprinted genes, DNA methylation may enhance gene expression by attracting activator proteins to the promoter or by preventing the binding of repressor proteins.

Figure 17.12 shows the imprinting process in which a maternal gene is methylated. The left side of the figure follows the marking process during the life of a female individual; the right side follows the same process in a male. Both individuals received a methylated gene from their mother and a nonmethylated copy of the same gene from their father. Via cell division, the zygote develops into a multicellular organism. Each time a somatic cell divides, enzymes in the cell maintain the methylation of the maternal gene, but the paternal gene remains unmethylated. If methylation inhibits transcription of this gene, only the paternal copy will be expressed in the somatic cells of both the male and female offspring.

The methylation state of an imprinted gene may be altered when individuals make gametes. First, the methylation is erased (Figure 17.12, step 2). Next, the gene may be methylated again, but that depends on whether the individual is a female or male. In females making eggs, both copies of the gene are methylated; in males making sperm, neither copy is methylated. When we consider the effects of methylation over the course of two or more generations, we can see how this phenomenon results in an epigenetic transmission pattern. The male in Figure 17.12 has inherited a methylated gene from his mother that is transcriptionally silenced in his somatic cells. Although he does not express this gene during his lifetime, he can pass on an active, nonmethylated copy of this exact same gene to his offspring.

Genomic imprinting is a recently discovered phenomenon that has been shown to occur for a few genes in mammals. For some genes, such as *Igf2*, the maternal allele is silenced, but for other genes, the paternal allele is silenced. Although several hypotheses have been advanced, biologists are still trying to understand the reason for this curious marking process.

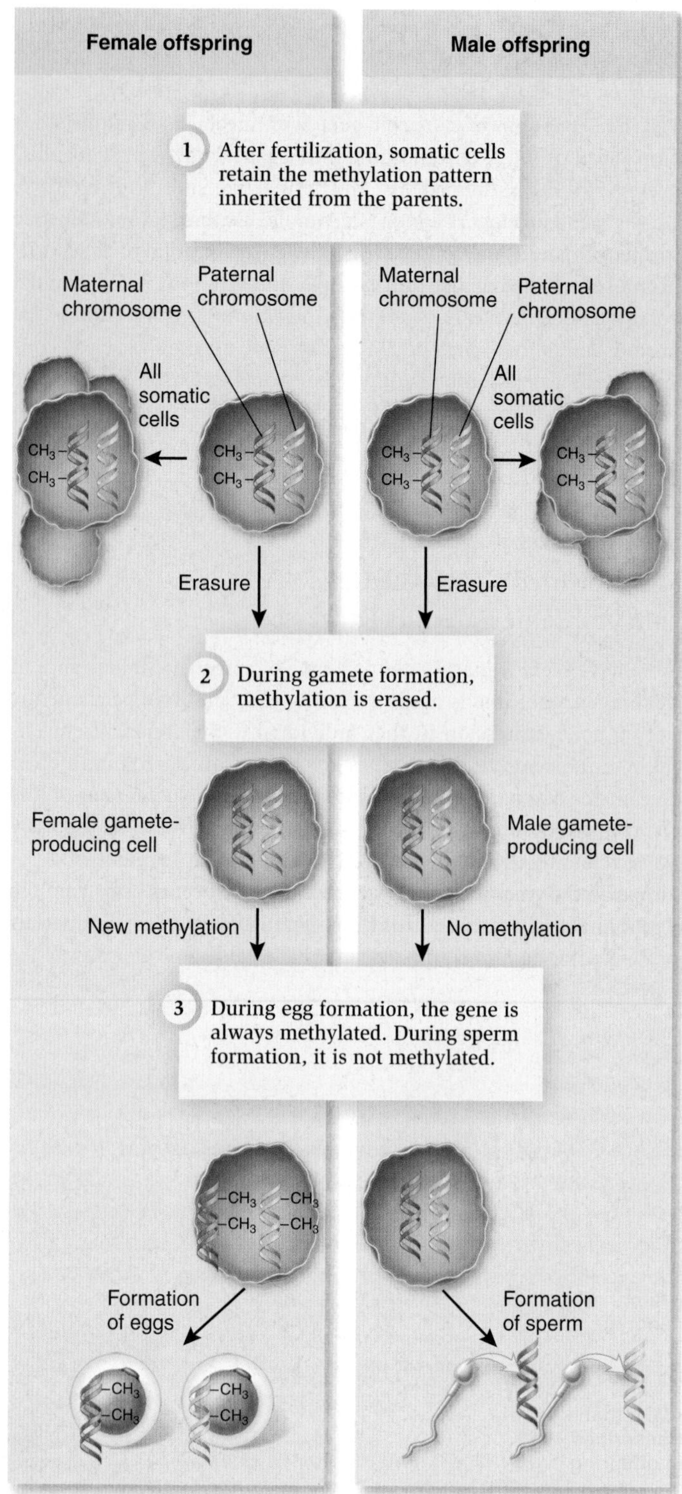

Figure 17.12 **Genomic imprinting via DNA methylation.** The cells at the top of this figure have a methylated gene inherited from the mother and a nonmethylated version of the same gene inherited from the father. This pattern of methylation is the same in male and female offspring and is maintained in their somatic cells. The methylation is erased during gamete formation, but in females, the gene is methylated again at a later stage in the formation of eggs. Therefore, females always transmit a methylated, transcriptionally silent copy of this gene, whereas males transmit a nonmethylated, transcriptionally active copy.

Summary of Key Concepts

- A variety of inheritance patterns are now known and some of these do not obey one or both of Mendel's laws of inheritance (Table 17.1).

17.1 Gene Interaction

- Epistasis is a gene interaction that occurs when the alleles of one gene mask the effects of the alleles of a different gene (Figure 17.1).
- Quantitative traits such as height and weight are polygenic, which means that multiple genes govern the trait. Often, the alleles of such genes contribute in an additive way to the phenotype and are greatly affected by the environment. This produces continuous variation in the trait, which is graphed as a bell-shaped curve (Figure 17.2).

17.2 Genes on the Same Chromosome: Linkage, Recombination, and Mapping

- When two different genes are on the same chromosome, they are said to be linked. Linked genes tend to be inherited as a unit, unless crossing over separates them (Figures 17.3, 17.4).
- Genetic mapping allows us to determine the order of genes along a chromosome and the relative distances between them, based on the frequency of crossing over observed in testcrosses (Figure 17.5).

17.3 Extranuclear Inheritance: Organelle Genomes

- Mitochondria and chloroplasts carry a small number of genes. The inheritance of such genes is called extranuclear inheritance (Figure 17.6).
- Chloroplasts in the four-o'clock plant are transmitted via the egg, a pattern called maternal inheritance (Figures 17.7, 17.8).
- Several human diseases are known to be caused by mutations in mitochondrial genes, which follow a maternal inheritance pattern (Table 17.2).

17.4 Epigenetic Inheritance

- Epigenetic inheritance refers to a modification of a gene or chromosome that affects the expression of one or more genes but does not alter the DNA sequence. Such changes may remain through the life of an organism, but are not permanent over the course of two or more generations.
- X inactivation in female mammals occurs when one X chromosome in every somatic cell is randomly inactivated. If the female is heterozygous for an X-linked gene, this can lead to a mosaic phenotype, with some of the somatic cells expressing one allele and some expressing the other (Figures 17.9, 17.10, Table 17.3).
- In genomic imprinting, offspring express either a maternal or paternal allele, depending on how a particular gene is marked, or imprinted. During gamete formation, DNA methylation of an allele from one parent is a mechanism to achieve imprinting (Figures 17.11, 17.12).

Assess and Discuss

Test Yourself

1. When two genes are located on the same chromosome they are said to be
   a. homologous.
   b. allelic.
   c. epistatic.
   d. linked.
   e. polygenic.
2. Based on the ideas proposed by Morgan, which of the following statements concerning linkage is *not* true?
   a. Traits determined by genes located on the same chromosome are likely to be inherited together.
   b. Crossing over between homologous chromosomes can create new allele combinations.
   c. A crossover is more likely to occur in a region between two genes that are close together than in a region between two genes that are farther apart.
   d. The probability of crossing over depends on the distance between the genes.
   e. Genes that tend to be transmitted together are physically located on the same chromosome.
3. In linkage mapping, 1 map unit is equivalent to
   a. 100 base pairs.
   b. 1 base pair.
   c. 10% recombination frequency.
   d. 1% recombination frequency.
   e. 1% the length of the chromosome.
4. Extranuclear inheritance occurs because
   a. certain genes are found on the X chromosome.
   b. chromosomes in the nucleus may be transferred to the cytoplasm.
   c. some organelles contain DNA.
   d. the nuclear membrane breaks down during cell division.
   e. both a and c.
5. In many organisms, organelles such as the mitochondria are transmitted only by the egg. This phenomenon is known as
   a. biparental inheritance.
   b. paternal inheritance.
   c. X-linked inheritance.
   d. maternal inheritance.
   e. both c and d.
6. Modification of a gene during gamete formation or early development that alters the way the gene is expressed during the individual's lifetime but is not permanent over two or more generations is called
   a. maternal inheritance.
   b. epigenetic inheritance.
   c. epistasis.
   d. multiple allelism.
   e. alternative splicing.
7. A male mouse that is homozygous for the normal allele of the *Igf2* gene is mated to a female that is heterozygous, carrying one normal copy and one defective copy of the gene. What would be the expected outcome of this cross?
   a. all normal offspring
   b. 1/2 normal and 1/2 dwarf
   c. all dwarf
   d. 3/4 normal and 1/4 dwarf
   e. none of the above

8. When a gene is inactivated during gamete formation and that gene is maintained in an inactivated state in the somatic cells of offspring, such an inheritance pattern is called
   a. linkage.
   b. X inactivation.
   c. epistasis.
   d. genomic imprinting.
   e. polygenic inheritance.
9. Calico coat pattern in cats is the result of
   a. X inactivation.
   b. epistasis.
   c. organelle heredity.
   d. genomic imprinting.
   e. maternal inheritance.
10. Genomic imprinting can be explained by
   a. DNA methylation of genes during gamete formation.
   b. epistasis.
   c. the spreading of X inactivation from the Xic locus.
   d. the inheritance of alleles that contribute additively to a trait.
   e. none of the above.

Conceptual Questions

1. Two genes (called gene *A* and gene *B*) are located on the same chromosome and are 12 mu apart. An *AABB* individual was crossed to an *aabb* individual. The F_1 (*AaBb*) offspring were crossed to *aabb* individuals. What percentage of the F_2 offspring would you expect to be *Aabb*?
2. Certain forms of human color blindness are inherited as X-linked recessive traits. Heterozygous females are not usually color blind, but on rare occasions, a female may exhibit partial color blindness or may be color blind in just one eye. Explain how this could happen.
3. A principle of biology is that *Biology is as an experimental science.* Describe how the phenomenon of linkage enables biologists to conduct experiments that are aimed at the mapping of genes along a chromosome.

Collaborative Questions

1. As discussed in Chapter 16, Mendel studied seven traits in pea plants, and the garden pea happens to have seven different chromosomes. It has been pointed out that Mendel was very lucky not to have conducted crosses involving two traits that are closely linked on the same chromosome because the results would have confounded his theory of independent assortment. It has even been suggested that Mendel may not have published data involving traits that were linked. An article by Blixt ("Why Didn't Gregor Mendel Find Linkage?" *Nature* 256:206[1975]) considers this issue. Look up this article and discuss why Mendel did not find linkage.
2. Discuss the similarities and differences between X inactivation and genomic imprinting.

Online Resource

www.brookerbiology.com

Stay a step ahead in your studies with animations that bring concepts to life and practice tests to assess your understanding. Your instructor may also recommend the interactive eBook, individualized learning tools, and more.

Chapter Outline

18.1 Genetic Properties of Viruses

18.2 Viroids and Prions

18.3 Genetic Properties of Bacteria

18.4 Gene Transfer Between Bacteria

Summary of Key Concepts

Assess and Discuss

Genetics of Viruses and Bacteria

18

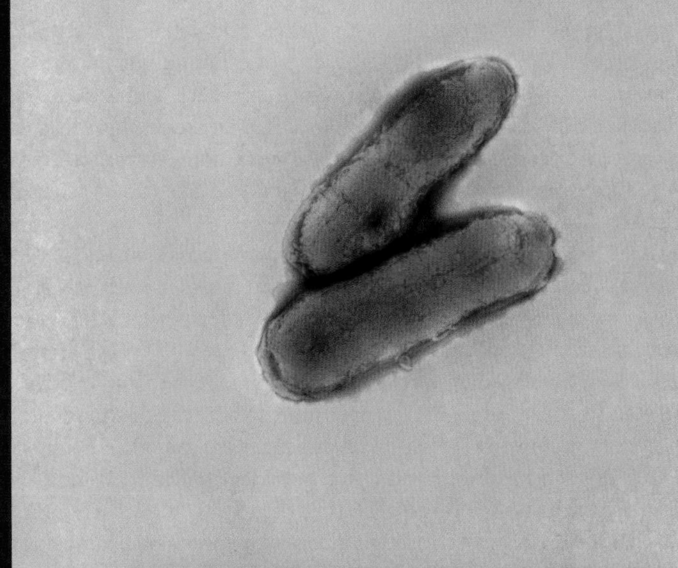

A colorized micrograph of *Haemophilus influenzae*, type b. This bacterium is a common cause of meningitis—a serious infection of the fluid in the spinal cord and the fluid that surrounds the brain.

While studying for his calculus test, Jason was having trouble concentrating due to a severe headache and fever. He thought he must be coming down with a cold. Though he had taken some aspirin, it didn't seem to be working. As he was eating some potato chips, one dropped in his lap. When he tried to look down to see where the chip had fallen, he realized that his neck was extremely stiff; he could barely move his head to look downward. Also, the brightness of his desk light seemed freakishly painful to his eyes. Over the course of that evening, Jason became confused and lethargic, and his roommate urged him to see a doctor. Fortunately, Jason took his advice and went to the college clinic. The diagnosis was bacterial meningitis—an inflammation of the protective membranes that cover the brain and spinal cord, collectively called the meninges. Although a relatively rare disease, bacterial meningitis is up to six times more common among people living in close quarters such as college dormitories. Because Jason sought help early enough, his disease could be treated with antibiotics. Had he not gotten help, the disease could have progressed to the point of causing severe brain damage and even death.

Jason's story highlights a primary reason why biologists are so interested in viruses and bacteria. Infectious diseases caused by viruses and bacteria are a leading cause of human suffering and death, accounting for one-quarter to one-third of deaths worldwide. The spread of infectious diseases results from human behavior, and in recent times, it has been accelerated by changes in land-use patterns, increased trade and travel, and the inappropriate use of antibiotic drugs. Although the incidence of fatal infectious diseases in the U.S. is low compared to the worldwide average, an alarming increase in more deadly strains of viruses and bacteria has occurred over the past few decades. Since 1980, the number of deaths in the U.S. due to infectious diseases has approximately doubled.

In this chapter, we turn our attention to the genetic analyses of viruses and bacteria. We will begin by examining viruses and other nonliving particles that infect living cells. All organisms are susceptible to infection by one or more types of viruses, which use the host's cellular machinery to replicate. Once a cell is infected, the genetic material of a virus orchestrates a series of events that ultimately leads to the production of new virus particles. We will consider the biological complexity of viruses and explore viral reproductive cycles. We will also examine some of the simplest and smallest infectious agents, called viroids and prions.

In the remaining sections of this chapter, we will examine the bacterial genome and the methods used in its investigation. Like their eukaryotic counterparts, bacteria have genetic differences that affect their cellular traits, and techniques of modern microbiology make many of these differences, such as sensitivity to antibiotics and differences in nutritional requirements, easy to study. Although bacteria reproduce asexually by cell division, their genetic variety is enhanced by the phenomenon called gene transfer, in which genes are passed from one bacterial cell to another. Like sexual reproduction in eukaryotes, gene transfer enhances the genetic diversity observed among bacterial species. In this chapter, we will explore three interesting ways that bacteria transfer genetic material.

18.1 Genetic Properties of Viruses

Learning Outcomes:

1. Compare and contrast how viruses differ with regard to their host range, structure, and genomes.
2. List the six steps in viral reproductive cycles and distinguish between the lysogenic and lytic cycles.
3. Describe how emerging viruses such as HIV arise and spread through populations.
4. Explain how an understanding of virus structure and reproduction can aid in the development of drugs to combat viruses.
5. Outline two hypotheses regarding the origin of viruses.

Viruses are nonliving particles with nucleic acid genomes. Why are viruses considered nonliving? They do not exhibit all of the properties associated with living organisms (refer back to Figure 1.4). Viruses are not composed of cells, and by themselves, they do not use energy or carry out metabolism, maintain homeostasis, or even reproduce. A virus or its genetic material must be taken up by a living cell to replicate.

The first virus to be discovered was tobacco mosaic virus (TMV). This virus infects several species of plants and causes mosaic-like patterns in which normal-colored patches are interspersed with light green or yellowish patches on the leaves (**Figure 18.1**). TMV damages leaves, flowers, and fruit but almost never kills the plant. In 1883, German chemist Adolf Mayer determined that this disease could be spread by spraying the sap from one plant onto another. By subjecting this sap to filtration, the Russian scientist Dmitri Ivanovski demonstrated that the disease-causing agent was not a bacterium. Sap that had been passed through filters with pores small enough to prevent the passage of bacterial cells was still able to spread the disease. At first, some researchers suggested the agent was a chemical toxin. However, the Dutch botanist Martinus Beijerinck ruled out this possibility by showing that sap could continue to transmit the disease after many plant generations. A toxin would have been diluted after many generations, but Beijerinck's results indicated the disease agent was multiplying in the plant. Around the same time, animal viruses were discovered in connection with a disease of cattle called foot-and-mouth disease. In 1900, the first human virus, the virus that causes yellow fever, was identified.

Figure 18.1 **A plant infected with tobacco mosaic virus.**

Since these early studies, microbiologists, geneticists, and molecular biologists have taken a great interest in the structure, genetic composition, and replication of viruses. In this section, we will discuss the structure of viruses and examine viral reproductive cycles in detail, paying particular attention to human immunodeficiency virus (HIV), the virus that causes acquired immunodeficiency syndrome (AIDS) in humans.

Viruses Are Remarkably Varied, Despite Their Simple Structure

A **virus** is a small infectious particle that consists of nucleic acid enclosed in a protein coat. Researchers have identified and studied over 4,000 different types of viruses. Although all viruses share some similarities, such as small size and the reliance on a living cell for replication, they vary greatly in their characteristics, including their host range, structure, and genome composition. Some of the major differences are described next, and characteristics of selected viruses are shown in **Table 18.1**.

Differences in Host Range A cell that is infected by a virus is called a **host cell**, and a species that can be infected by a specific virus is called a host species for that virus. Viruses differ greatly in their **host range**—the number of species and cell types they can infect. Table 18.1 lists a few examples of viruses with widely different ranges of host species. Tobacco mosaic virus, which we discussed earlier, has a broad host range. TMV is known to infect over 150 different species of plants. By comparison, other viruses have a narrow host range, with some infecting only a single species. Furthermore, a virus may infect only a specific cell type in a host species. **Figure 18.2** shows some viruses that infect particular human cells and cause disease.

Structural Differences Viruses cannot be resolved by even the best light microscope. Although the existence of viruses was postulated in the 1890s, viruses were not observed until the 1930s when the electron microscope was invented. Viruses range in size from about 20 to 400 nm in diameter (1 nm [nanometer] = 10^{-9} meters). For comparison, a typical bacterium is 1,000 nm in diameter, and the diameter of most eukaryotic cells is 10 to 1,000 times that of a bacterium. Adenoviruses, which cause infections of the respiratory and gastrointestinal tracts, have an average diameter of 75 nm. Over 50 million adenoviruses could fit into an average-sized human cell.

What are the common structural features of all viruses? As shown in **Figure 18.3**, all viruses have a protein coat called a **capsid**, which encloses a genome consisting of one or more molecules of nucleic acid (DNA or RNA). Capsids are composed of one or several different protein subunits called capsomers. Capsids have a variety of shapes, including helical and polyhedral. Figure 18.3a shows the structure of TMV, which has a helical capsid made of identical capsomers. Figure 18.3b shows an adenovirus, which has a polyhedral capsid. Protein fibers with a terminal knob are located at the corners of the polyhedral capsid. Many viruses that infect animal cells, such as the influenza virus shown in Figure 18.3c, have a **viral envelope** enclosing the capsid. The envelope consists of a lipid bilayer that is

Table 18.1 Hosts and Characteristics of Selected Viruses

Virus or group of viruses	Host	Effect on host	Nucleic acid*	Genome size (kb)†	Number of genes†
Phage fd	*E. coli*	Slows growth	ssDNA	6.4	10
Phage λ	*E. coli*	Can exist harmlessly in the host cell or cause lysis	dsDNA	48.5	71
Phage T4	*E. coli*	Causes lysis	dsDNA	169	288
Phage Qβ	*E. coli*	Slows growth	ssRNA	4.2	4
Tobacco mosaic virus (TMV)	Many plants	Causes mottling and necrosis of leaves and other plant parts	ssRNA	6.4	6
Baculoviruses	Insects	Most baculoviruses are species specific; they usually kill the insect	dsDNA	133.9	154
Parvovirus	Mammals	Causes respiratory, flulike symptoms	ssDNA	5.0	5
Influenza virus	Mammals	Causes classical "flu," with fever, cough, sore throat, and headache	ssRNA	13.5	11
Epstein-Barr virus	Humans	Causes mononucleosis, with fever, sore throat, and fatigue	dsDNA	172	80
Adenovirus	Humans	Causes respiratory symptoms and diarrhea	dsDNA	34	35
Herpes simplex type II	Humans	Causes blistering sores around the genital region	dsDNA	158.4	77
HIV (type I)	Humans	Causes AIDS, an immunodeficiency syndrome eventually leading to death	ssRNA	9.7	9

*The abbreviations ss and ds refer to single stranded and double stranded, respectively.

†Several of the viruses listed in this table are found in different strains that vary with regard to genome size and number of genes. The numbers reported in this table are typical values. The abbreviation kb refers to kilobase, which equals 1,000 bases.

derived from a cellular membrane of the host cell and is embedded with virally encoded spike glycoproteins.

In addition to encasing and protecting the genetic material, the capsid and envelope enable viruses to infect their hosts. In many viruses, protein fibers with a knob (Figure 18.3b) or spike glycoproteins (Figure 18.3c) help them bind to the surface of a host cell. Viruses that infect bacteria, called **bacteriophages**, or simply **phages**, may have more complex protein coats, with accessory structures used for anchoring the virus to a host cell and injecting the viral nucleic acid (Figure 18.3d). As discussed later, the tail fibers of such bacteriophages are needed to attach the virus to the bacterial cell wall.

Genome Differences The genetic material in a virus is called a **viral genome**. The composition of viral genomes varies markedly among different types of viruses, as suggested by the examples in Table 18.1. The nucleic acid of some viruses is DNA, whereas in others it is RNA. These are referred to as DNA viruses and RNA viruses, respectively. It is striking that some viruses use RNA for their genome, whereas all living organisms use DNA. In some viruses, the nucleic acid is single stranded, whereas in others, it is double stranded. The genome can be linear or circular, depending on the type of virus. Some kinds of viruses have more than one copy of the genome.

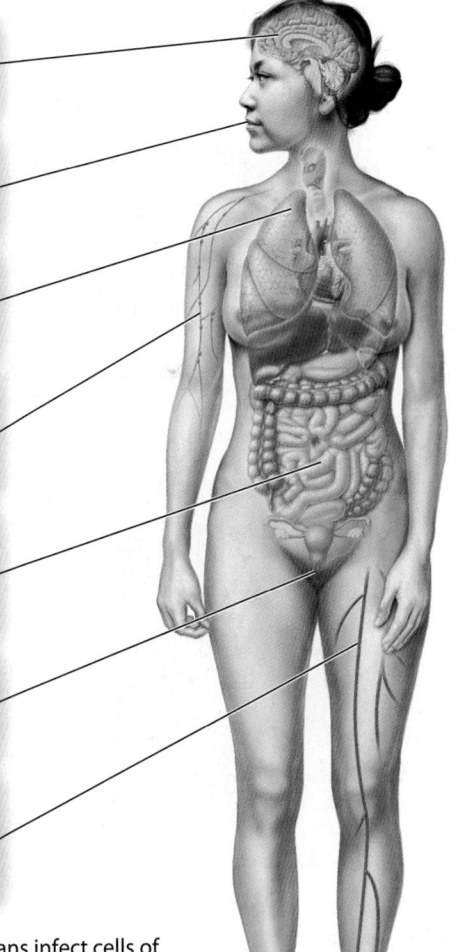

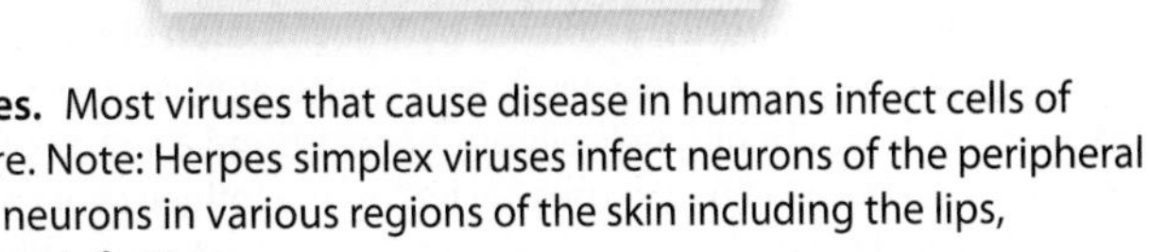

Figure 18.2 Some viruses that cause human diseases. Most viruses that cause disease in humans infect cells of specific tissues, as illustrated by the examples in this figure. Note: Herpes simplex viruses infect neurons of the peripheral nervous system. Herpes simplex type I commonly infects neurons in various regions of the skin including the lips, whereas type II is the most common cause of genital herpes infections.

BIOLOGY PRINCIPLE Biology affects our society. By studying viruses, biologists have developed vaccines and drugs to help prevent their spread.

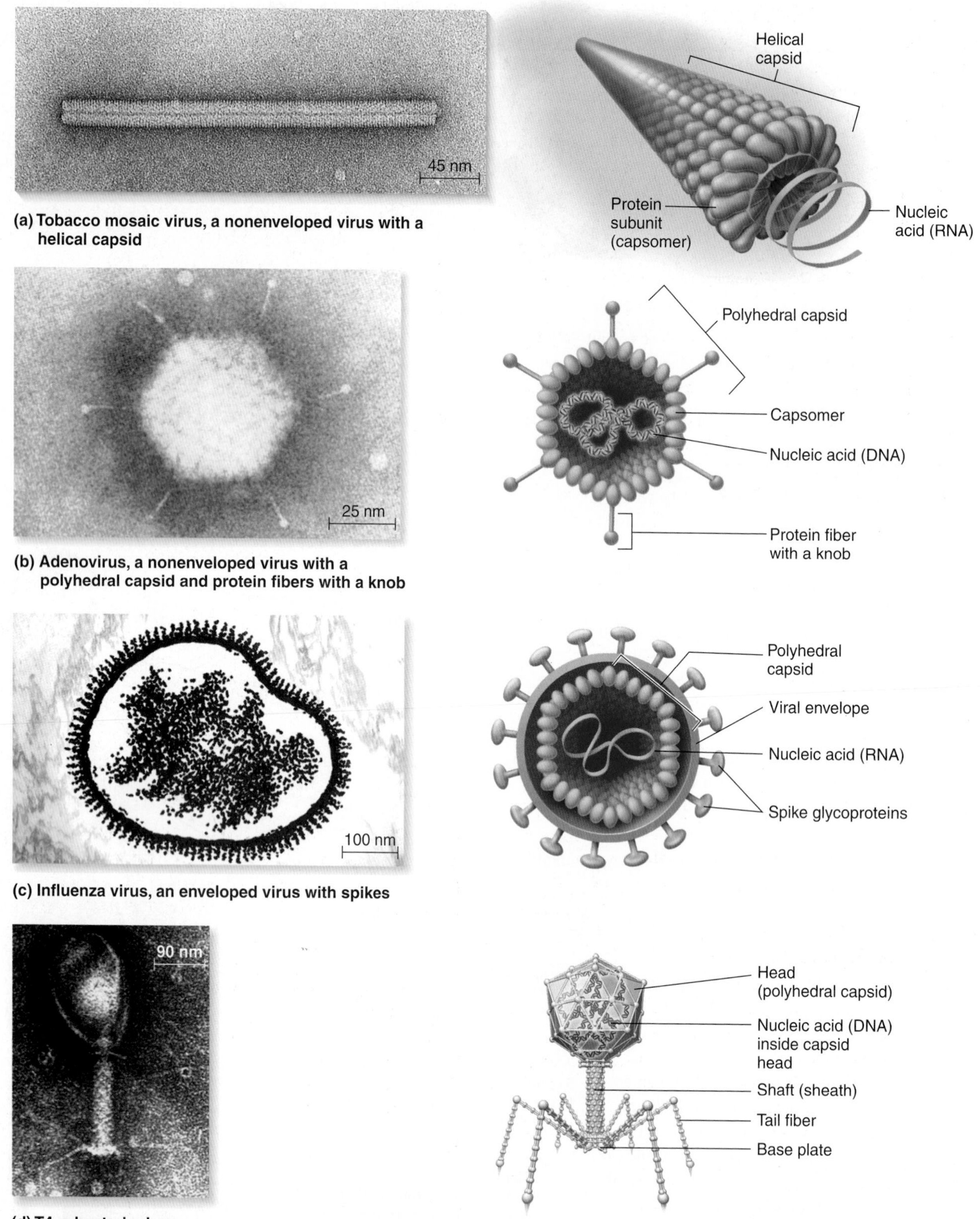

Figure 18.3 Variations in the structure of viruses, as shown by transmission electron microscopy. All viruses contain nucleic acid (DNA or RNA) surrounded by a protein capsid. They may or may not have an outer envelope surrounding the capsid. **(a)** Tobacco mosaic virus (TMV) has a capsid made of 2,130 identical protein subunits, helically arranged around a strand of RNA. **(b)** Adenoviruses have polyhedral capsids containing protein fibers with a knob. **(c)** Many animal viruses, including the influenza virus, have an envelope composed of a lipid bilayer and spike glycoproteins. The lipid bilayer is obtained from the host cell when the virus buds from the plasma membrane. **(d)** Some bacteriophages, such as T4, have protein coats with accessory structures such as tail fibers that facilitate invasion of a bacterial cell.

Concept Check: What features vary among different types of viruses?

Viral genomes also vary considerably in size, ranging from a few thousand to more than a hundred thousand nucleotides in length (see Table 18.1). For example, the genomes of some simple viruses, such as phage Qβ, are only a few thousand nucleotides in length and contain only a few genes. Other viruses, particularly those with a complex structure, such as phage T4, contain many more genes. These extra genes encode many different proteins that are involved in the formation of the elaborate structure shown in Figure 18.3d.

Viral Reproductive Cycles Consist of a Few Basic Steps

When a virus infects a host cell, the expression of viral genes leads to a series of steps, called a **viral reproductive cycle**, which results in the production of new viruses. The details of the steps differ among various types of viruses, and even the same virus may have the capacity to follow alternative cycles. Even so, by studying the reproductive cycles of hundreds of different viruses, researchers have determined that the viral reproductive cycle consists of five or six basic steps.

To illustrate the general features of viral reproductive cycles, Figure 18.4 considers these steps for two types of viruses. Figure 18.4a shows the cycle of phage λ (lambda), a bacteriophage with double-stranded DNA as its genome, and Figure 18.4b depicts the cycle of HIV, an enveloped animal virus containing single-stranded RNA. The descriptions that follow compare the reproductive cycles of these two very different viruses.

Step 1: Attachment In the first step of a viral reproductive cycle, the virus must attach to the surface of a host cell. This attachment is usually specific for one or just a few types of cells because proteins in the virus recognize and bind to specific molecules on the cell surface. In the case of phage λ, the phage tail fibers bind to proteins in the outer bacterial cell membrane of *E. coli* cells. In the case of HIV, spike glycoproteins in the viral envelope bind to protein receptors in the plasma membrane of human blood cells called helper T cells.

Step 2: Entry After attachment, the viral genome enters the host cell. Attachment of phage λ stimulates a conformational change in the phage coat proteins, so the shaft (also called the sheath) contracts, and the phage injects its DNA into the bacterial cytoplasm (refer back to Figure 11.3). In contrast, the envelope of HIV fuses with the plasma membrane of the host cell, so both the capsid and its contents are released into the cytosol. Some of the HIV capsid proteins are then removed by host cell enzymes, a process called uncoating. This releases two copies of the viral RNA and two molecules of an enzyme called reverse transcriptase into the cytosol. As discussed shortly, reverse transcriptase is needed for step 3.

Once a viral genome has entered the cell, one or several viral genes are expressed immediately due to the action of host cell enzymes and ribosomes. Expression of these key genes leads quickly to either step 3 or step 4 of the reproductive cycle, depending on the specific virus. The genome of some viruses, including both phage λ and HIV, can integrate into a chromosome of the host cell. For such viruses, the cycle may proceed from step 2 to step 3 as described next, delaying the production of new viruses. Alternatively, the cycle may proceed directly from step 2 to step 4 and quickly lead to the production of new viruses.

Step 3: Integration Viruses capable of integration carry a gene that encodes an enzyme called **integrase**. For integration to occur, this gene is expressed soon after entry so the integrase protein is made. Integrase cuts the host's chromosomal DNA and inserts the viral genome into the chromosome. In the case of phage λ, the double-stranded DNA that entered the cell can be directly integrated into the double-stranded DNA of the chromosome. Once integrated, the phage DNA in a bacterium is called a **prophage**. When a bacterial cell divides, the prophage DNA is copied and transmitted to daughter cells along with the bacterial chromosomal DNA. While it exists as a prophage, this type of viral reproductive cycle is called the **lysogenic cycle**. As discussed later, new phages are not made during the lysogenic cycle, and the host cell is not destroyed. On occasion, a prophage can be excised from the bacterial chromosome and proceed to step 4.

How can an RNA virus integrate its genome into the host cell's DNA? For this to occur, the viral genome must be copied into DNA. HIV accomplishes this by means of a viral enzyme called **reverse transcriptase**, which is carried within the capsid and released into the host cell along with the viral RNA. Reverse transcriptase uses the viral RNA strand to make a complementary copy of DNA. The complementary DNA is then used as a template to make double-stranded viral DNA. This process is called reverse transcription because it is the reverse of the usual transcription process, in which a DNA strand is used to make a complementary strand of RNA. The viral double-stranded DNA enters the host cell nucleus and is inserted into a host chromosome via integrase. Once integrated, the viral DNA in a eukaryotic cell is called a **provirus**. Viruses that follow this mechanism are called **retroviruses**.

Step 4: Synthesis of Viral Components The production of new viruses by a host cell involves the replication of the viral genome and the synthesis of viral proteins that make up the protein coat. In the case of a bacteriophage that has been integrated into the host chromosome, the prophage must be excised as described in step 3 before the synthesis of new viral components can occur. An enzyme called excisionase is required for this process. Following excision, host cell enzymes make many copies of the phage DNA and transcribe the genes within these copies into mRNA. Host cell ribosomes translate this viral mRNA into viral proteins. The expression of phage genes also leads to the degradation of the host chromosomal DNA.

In the case of HIV, the provirus DNA is not excised from the host chromosome. Instead, it is transcribed in the nucleus to produce many copies of viral RNA. These viral RNA molecules enter the cytosol, where they are used to make viral proteins and serve as the genome for new viral particles.

Step 5: Viral Assembly After all of the necessary components have been synthesized, they must be assembled into new viruses. Some viruses with a simple structure self-assemble—viral components spontaneously bind to each other to form a complete virus particle. An example of a self-assembling virus is TMV, which we examined earlier (see Figure 18.3a). TMV capsid proteins assemble around a TMV RNA molecule, which becomes trapped inside the hollow capsid.

Other viruses, including the two shown in Figure 18.4, do not self-assemble. The correct assembly of phage λ requires the help of noncapsid proteins not found in the completed phage particle. Some

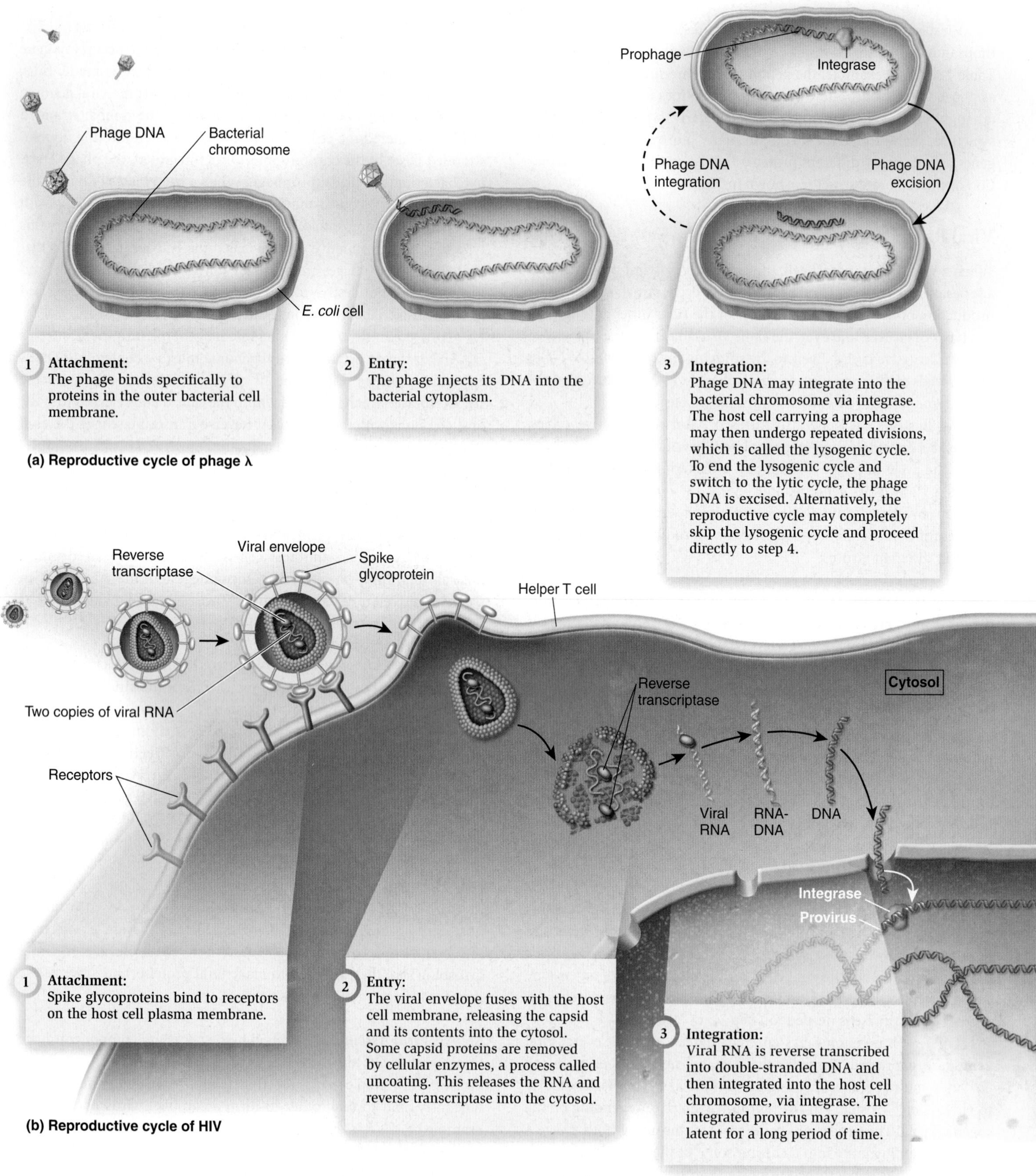

Phage DNA
Bacterial chromosome
E. coli cell
Prophage
Integrase
Phage DNA integration
Phage DNA excision
1 Attachment:
The phage binds specifically to proteins in the outer bacterial cell membrane.
2 Entry:
The phage injects its DNA into the bacterial cytoplasm.
3 Integration:
Phage DNA may integrate into the bacterial chromosome via integrase. The host cell carrying a prophage may then undergo repeated divisions, which is called the lysogenic cycle. To end the lysogenic cycle and switch to the lytic cycle, the phage DNA is excised. Alternatively, the reproductive cycle may completely skip the lysogenic cycle and proceed directly to step 4.
(a) Reproductive cycle of phage λ
Reverse transcriptase
Viral envelope
Spike glycoprotein
Helper T cell
Two copies of viral RNA
Receptors
Reverse transcriptase
Cytosol
Viral RNA
RNA-DNA
DNA
Integrase
Provirus
1 Attachment:
Spike glycoproteins bind to receptors on the host cell plasma membrane.
2 Entry:
The viral envelope fuses with the host cell membrane, releasing the capsid and its contents into the cytosol. Some capsid proteins are removed by cellular enzymes, a process called uncoating. This releases the RNA and reverse transcriptase into the cytosol.
3 Integration:
Viral RNA is reverse transcribed into double-stranded DNA and then integrated into the host cell chromosome, via integrase. The integrated provirus may remain latent for a long period of time.
(b) Reproductive cycle of HIV

Figure 18.4 **Comparison of the steps of two viral reproductive cycles.** **(a)** The reproductive cycle of phage λ, a bacteriophage with a double-stranded DNA genome. **(b)** The reproductive cycle of HIV, an enveloped animal virus with a single-stranded RNA genome.

BioConnections: *Look back at Figure 5.23. How does the release of HIV resemble exocytosis?*

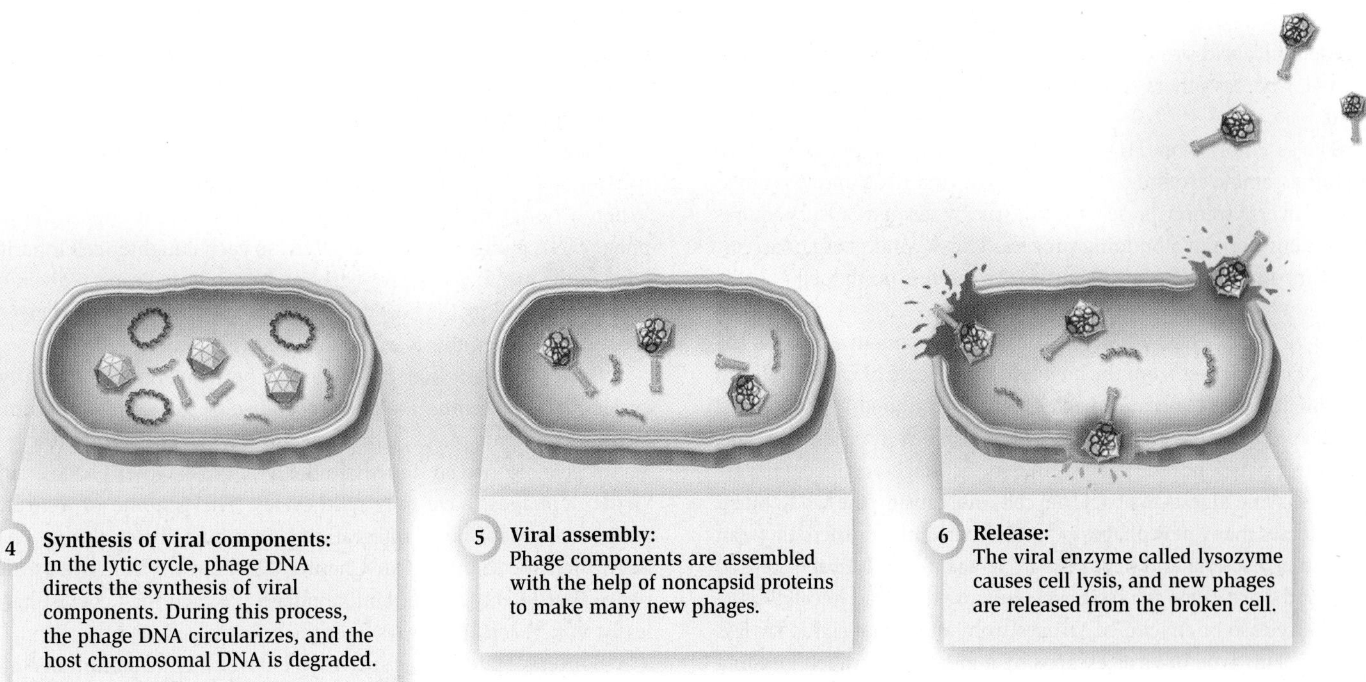

4 Synthesis of viral components:
In the lytic cycle, phage DNA directs the synthesis of viral components. During this process, the phage DNA circularizes, and the host chromosomal DNA is degraded.
5 Viral assembly:
Phage components are assembled with the help of noncapsid proteins to make many new phages.
6 Release:
The viral enzyme called lysozyme causes cell lysis, and new phages are released from the broken cell.

Capsid proteins

Spike glycoproteins

Reverse transcriptase

Viral RNA

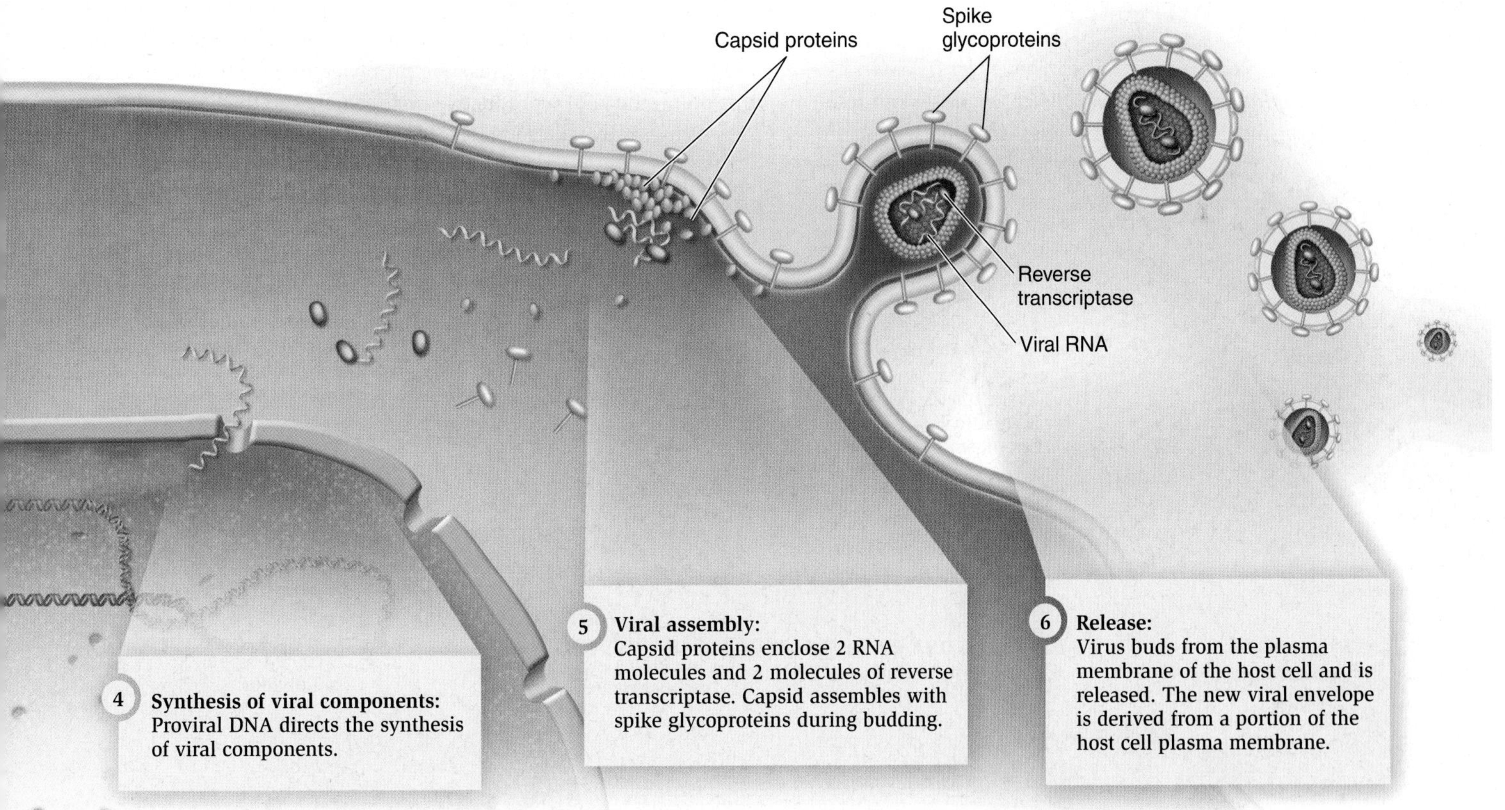

4 **Synthesis of viral components:**
Proviral DNA directs the synthesis of viral components.

5 **Viral assembly:**
Capsid proteins enclose 2 RNA molecules and 2 molecules of reverse transcriptase. Capsid assembles with spike glycoproteins during budding.

6 **Release:**
Virus buds from the plasma membrane of the host cell and is released. The new viral envelope is derived from a portion of the host cell plasma membrane.

of these noncapsid proteins function as enzymes that modify capsid proteins, whereas others serve as scaffolding for the assembly of the capsid.

The assembly of an HIV virus occurs in two stages. First, capsid proteins assemble around two molecules of viral RNA and two molecules of reverse transcriptase. Next, the newly formed capsid acquires its outer envelope in a budding process. This second phase of assembly occurs during step 6, as the virus is released from the cell.

Step 6: Release The last step of a viral reproductive cycle is the release of new viruses from the host cell. The release of bacteriophages is a dramatic event. Because bacteria are surrounded by a rigid cell wall, the phages must burst, or lyse, their host cell in order to escape. After the phages have been assembled, a phage-encoded enzyme called lysozyme digests the bacterial cell wall, causing the cell to burst. Lysis releases many new phages into the environment, where they can infect other bacteria and begin the cycle again. Collectively, steps 1, 2, 4, 5, and 6 are called the **lytic cycle** because they lead to cell lysis.

The release of enveloped viruses from an animal cell is far less dramatic. This type of virus escapes by a mechanism called budding that does not lyse the cell. In the case of HIV, a newly assembled virus particle associates with a portion of the plasma membrane containing HIV spike glycoproteins. The membrane enfolds the viral capsid and eventually buds from the surface of the cell. This is how the virus acquires its envelope, which is a piece of host cell membrane studded with viral glycoproteins.

Latency in Bacteriophages As we saw in step 3, viruses can integrate their genomes into a host chromosome. In some cases, the prophage or provirus may remain inactive, or **latent**, for a long time. Most of the viral genes are silent during latency, and the viral reproductive cycle does not progress to step 4.

Latency in bacteriophages is also called lysogeny. When this occurs, both the prophage and its host cell are said to be lysogenic. When a lysogenic bacterium prepares to divide, it copies the prophage DNA along with its own DNA, so each daughter cell inherits a copy of the prophage. A prophage can be replicated repeatedly in this way without killing the host cell or producing new phage particles. As mentioned earlier, this is called the lysogenic cycle.

Bacteriophages that can alternate between lysogenic and lytic cycles are called **temperate phages** (Figure 18.5). Phage λ is an example of a temperate phage. Upon infection, it can either enter the lysogenic cycle or proceed directly to the lytic cycle. Other phages, called **virulent phages**, have only lytic cycles. The genome of a virulent phage is not capable of integration into a host chromosome. Phage T2, which we examined in Chapter 11 (Figure 11.3), is a virulent phage that infects *E. coli*. Unlike phage λ, which may coexist harmlessly with *E. coli*, T2 always lyses the infected cell.

For temperate phages such as λ, environmental conditions influence whether or not viral DNA is integrated into a host chromosome and how long the virus remains in the lysogenic cycle. If nutrients are readily available, phage λ usually proceeds directly to the lytic cycle after its DNA enters the cell. Alternatively, if nutrients are in short supply, the lysogenic cycle is often favored because sufficient material may not be available to make new viruses. If more nutrients become available later, this may cause the prophage to become activated. At this point, the viral reproductive cycle switches to the lytic cycle, and new viruses are made and released.

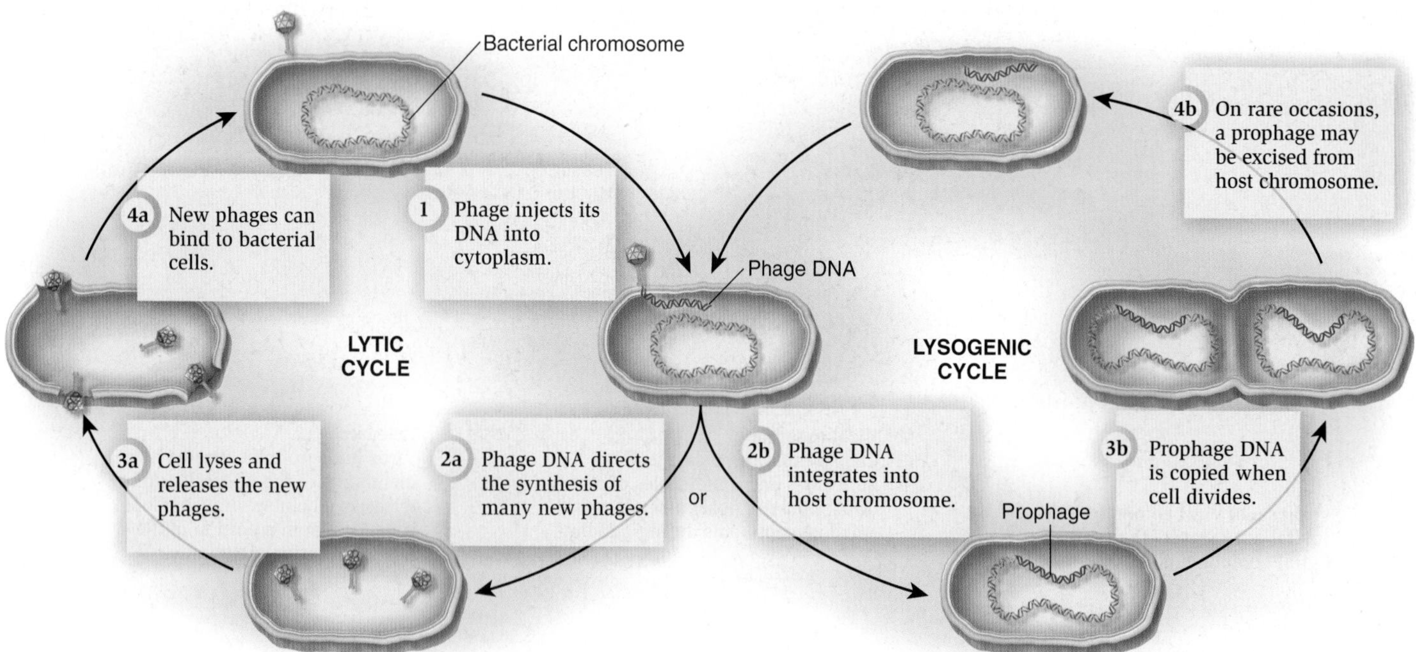

Figure 18.5 **Lytic and lysogenic cycles of bacteriophages.** Some phages, such as phage λ, may follow either a lytic or a lysogenic reproductive cycle. During the lytic cycle, new phages are made, and the bacterial cell lyses. During the lysogenic cycle, the integrated phage DNA, or prophage, is replicated along with the DNA of the host cell. Environmental conditions influence how long the phage remains in the lysogenic cycle.

Concept Check: *From the perspective of the virus, what are the primary advantages of the lytic and lysogenic cycles?*

Latency in Human Viruses Latency among human viruses can occur in two different ways. For HIV, latency occurs because the virus has integrated into the host genome and may remain dormant for a long time. In addition, the genomes of other viruses can exist as an **episome**—a genetic element that replicates independently of the chromosomal DNA but also can occasionally integrate into chromosomal DNA. Examples of viral genomes that exist as episomes include different types of herpesviruses that cause cold sores (usually herpes simplex type I), genital herpes (usually herpes simplex type II), and chickenpox (varicella-zoster). A person infected with a given type of herpesvirus may have periodic outbreaks of disease symptoms when the virus switches from the latent, episomal form to the active form that produces new virus particles.

As an example, let's consider the herpesvirus called varicella-zoster. The initial infection by this virus causes chickenpox, after which the virus may remain latent for many years as an episome. The disease called shingles occurs when varicella-zoster switches from the latent state and starts making new virus particles. Shingles begins as a painful rash that eventually erupts into blisters. The blisters follow the path of the neurons that carry the latent varicella-zoster virus. The blisters often form a ring around the back of the patient's body, which is why the disease is called shingles—the blisters line up like the shingles on a house.

Emerging Viruses, Such as HIV, Have Arisen Recently and May Rapidly Spread Through a Population

A key reason researchers have been interested in viral reproductive cycles is the ability of many viruses to cause diseases in humans and other hosts. **Emerging viruses** are ones that have arisen recently or are likely to have a greater probability of causing infection. Such viruses, which may lead to significant loss of human life, often cause public alarm. New strains of influenza virus arise fairly regularly due to mutations. An example is the strain H1N1, also called swine flu. In the U.S., despite attempts to minimize deaths by vaccination, over 30,000 people die annually from influenza.

During the past few decades, the most devastating example of an emerging virus has been **human immunodeficiency virus (HIV)**, the causative agent of acquired immune deficiency syndrome (AIDS). HIV is primarily spread by sexual contact between infected and uninfected individuals, but it can also be spread by the transfusion of HIV-infected blood, by the sharing of needles among drug users, and from infected mother to unborn child. Since AIDS was first recognized in 1981, the total number of AIDS deaths is over 30 million, making it one of the most deadly diseases in human history. More than 0.5 million of these deaths have occurred in the U.S. In 2009, about 33 million people were living with HIV; approximately 2.6 million of them were infected that year. In that same year, 1.8 million died from AIDS. Worldwide, nearly 1 in every 100 adults between ages 15 and 49 is infected. In the U.S., about 56,000 new HIV infections occur each year, 70% of which are in men and 30% in women.

The devastating effects of AIDS result from viral destruction of helper T cells, a type of white blood cell that plays an essential role in the immune system of mammals. Figure 18.6 shows HIV virus particles invading a helper T cell. As described in Chapter 53, helper T cells interact with other cells of the immune system to facilitate the production of antibodies and other molecules that target and kill foreign invaders of the body. When large numbers of helper T cells are destroyed by HIV, the function of the immune system is seriously compromised, and the individual becomes susceptible to infectious diseases called opportunistic infections that would not normally occur in a healthy person. For example, *Pneumocystis jiroveci*, a fungus that causes a type of pneumonia, is easily destroyed by a healthy immune system. However, in people with AIDS, *Pneumocystis jiroveci* pneumonia can be fatal.

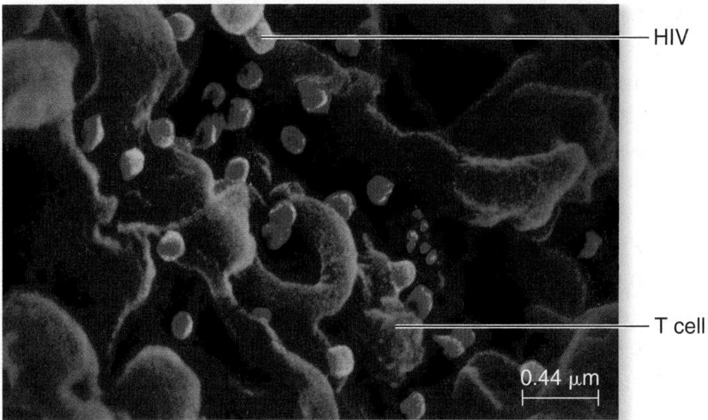

Figure 18.6 Micrograph of HIV invading a human helper T cell. This is a colorized scanning electron micrograph. The surface of the T cell is purple, and HIV particles are red.

An insidious feature of HIV replication, described earlier in Figure 18.4b, is that reverse transcriptase, the enzyme that copies the RNA genome into DNA, lacks a proofreading function. In Chapter 11, we learned that DNA polymerase can identify and remove mismatched nucleotides in newly synthesized DNA. Because reverse transcriptase lacks this function, it makes more errors and thereby tends to create mutant strains of HIV. This undermines the ability of the body to combat HIV because mutant strains may be resistant to the body's defenses. In addition, mutant strains of HIV may be resistant to antiviral drugs, as described next.

Drugs Have Been Developed to Combat the Proliferation of HIV

A compelling reason to understand the reproductive cycle of HIV and other disease-causing viruses is that such knowledge may be used to develop drugs that stop viral proliferation. For example, in the U.S., the highest rate of AIDS-related deaths was approximately 17 per year per 100,000 people in 1994 and 1995. The current rate is about 4 to 5 deaths per year per 100,000 people, owing in part to the use of new antiviral drugs. These drugs inhibit viral proliferation, though they cannot eliminate the virus from the body.

One approach to the design of antiviral treatments has been to develop drugs that specifically bind to proteins encoded by the viral genome. For example, azidothymidine (AZT) mimics the structure of a normal nucleotide and binds to the enzyme reverse transcriptase. In this way, AZT inhibits reverse transcription, thereby inhibiting viral replication. Another way to combat HIV involves the use of antiviral drugs that inhibit proteases, enzymes that are needed during the assembly of the HIV capsid. Certain HIV proteases cut capsid proteins, which makes them smaller and able to assemble into a capsid structure. If the proteases do not function, the capsid will not assemble, and new HIV particles will not be made. Several drugs known as protease inhibitors have been developed that bind to HIV proteases and inhibit their function.

A major challenge in AIDS research is to discover drugs that inhibit viral proteins without also binding to host cell proteins and inhibiting normal cellular functions. A second challenge is to develop drugs to which mutant strains will not become resistant. As mentioned, HIV readily accumulates mutations during viral replication. A current strategy is to treat HIV patients with a "cocktail" of three or more HIV drugs, making it less likely that any mutant strain will overcome all of the inhibitory effects.

Another approach to fighting AIDS and other infectious diseases is **vaccination**—inoculation with a substance or group of substances that causes the immune system to respond to and destroy infectious agents such as bacteria or viruses. Vaccinations have been successful in the prevention of other viral diseases, such as influenza and chickenpox. In the case of HIV, the ideal vaccine should be both inexpensive and easy to store and administer, and it must confer long-lasting protection against HIV infection by sexual contact or by exposure to infected blood. Importantly, due to the high mutation rate of HIV, the vaccine must protect against exposure to many different strains of the virus.

Several Hypotheses Have Been Proposed to Explain the Origin of Viruses

Because viruses are such small particles, there is no fossil record of their evolution. Researchers must rely on analyses of modern viruses to develop hypotheses about their origin. Viral genomes follow the same rules of gene expression as the genomes of their host cells. Viral genes have promoter sequences similar to those of their host cells, and the translation of viral proteins relies on the genetic code. Viruses depend entirely on host cells for their proliferation. No known virus makes its own ribosomes or generates the energy it requires to make new viruses. Therefore, many biologists have argued that cells must have evolved before viruses.

How did viruses come into existence? A common hypothesis for the origin of viruses is that they evolved from macromolecules inside living cells. The precursors of the first viruses may have been plasmids—small, circular DNA molecules that exist independently of chromosomal DNA. (Plasmids are described later in this chapter.) Biologists have hypothesized that such DNA molecules may have acquired genes that code for proteins that facilitate their own replication. Though many biologists favor the idea that viruses originated from primitive plasmids or other chromosomal elements, some have suggested they are an example of regressive evolution—the reduction of a trait or traits over time. This hypothesis proposes that viruses are degenerate cells that have retained the minimal genetic information essential for reproduction.

A new and interesting hypothesis is that viruses did not evolve from living cells but instead evolved in parallel with cellular organisms. As discussed in Chapter 22, the precursors of cellular DNA genomes may have been RNA molecules that could replicate independently of cells. This stage of evolution, termed the RNA world, could have involved the parallel evolution of both viruses and cellular organisms.

18.2 Viroids and Prions

Learning Outcomes:

1. Distinguish between a viroid and a virus.
2. Describe the structure of prions and explain how they cause disease.

Some nonliving infectious agents are even simpler than viruses. Viroids are composed solely of RNA, and prions are composed solely of protein. In this section, we will begin by examining viroids, infectious agents that cause diseases in plants. Next, we will discuss infectious proteins known as prions, which cause devastating neurological diseases in humans and other mammals. Unlike other agents of infection, prions have no genes and cannot be copied by the replication machinery of a cell. Instead, they increase their numbers by inducing changes in other protein molecules within living cells.

Viroids Are RNA Molecules That Infect Plant Cells

In 1971, Swiss-born American plant pathologist Theodor Diener discovered that the agent of potato spindle tuber disease is a small RNA molecule devoid of any protein. He coined the term **viroid** for this newly discovered infectious particle. Viroids are composed solely of a single-stranded circular RNA molecule that is a few hundred nucleotides in length.

Viroids infect plant cells, where they depend entirely on host enzymes for their replication. Some viroids are replicated in the host cell nucleus, others in the chloroplast. In contrast to viral genomes, the RNA genomes of viroids do not code for any proteins. How do viroids affect plant cells? The RNA of some viroids is known to possess ribozyme activity, and some researchers think this activity may damage plants by interfering with the function of host cell molecules. However, the mechanism by which viroids induce disease is not well understood.

Since Diener's initial discovery, many more viroids have been characterized as the agents of diseases affecting many economically important plants, including potato, tomato, cucumber, orange, coconut, grape, avocado, peach, apple, pear, and plum. Some viroids have devastating effects, as illustrated by the case of the coconut cadang-cadang viroid, which has killed more than 20 million coconut trees in Southeast Asia and New Guinea (Figure 18.7). Other viroids produce less severe damage, causing necrosis on leaves, shortening of stems, bark cracking, and delays in foliation, flowering, and fruit ripening. A few viroids induce mild symptoms or no symptoms at all.

Prions Are Infectious Proteins That Cause Neurodegenerative Diseases

Before we end our discussion of nonliving, infectious particles, let's consider an unusual infectious agent that causes a group of rare, fatal brain diseases affecting humans and other mammals. Until the 1980s, biologists thought that any infectious agent, whether living or nonliving, must have genetic material. It seemed logical that genetic material is needed to store the information to produce new infectious particles.

In the 1960s, British researchers Tikvah Alper and John Stanley Griffith discovered that preparations from animals with certain

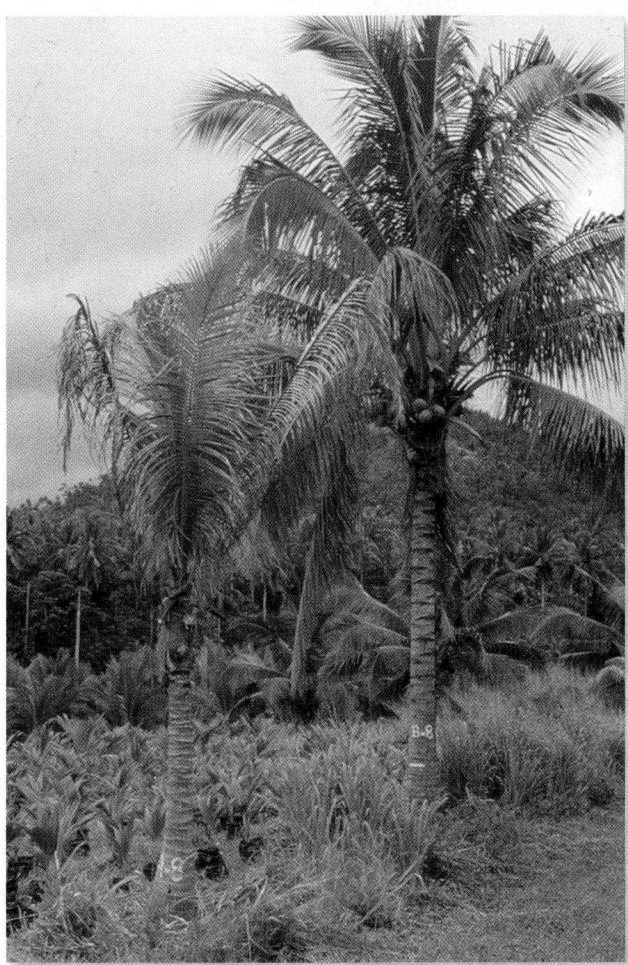

Figure 18.7 Effects of a viroid. The palm tree (left foreground) in Papua, New Guinea, has been infected with the coconut cadang-cadang viroid and shows symptoms of stunting and yellowing.

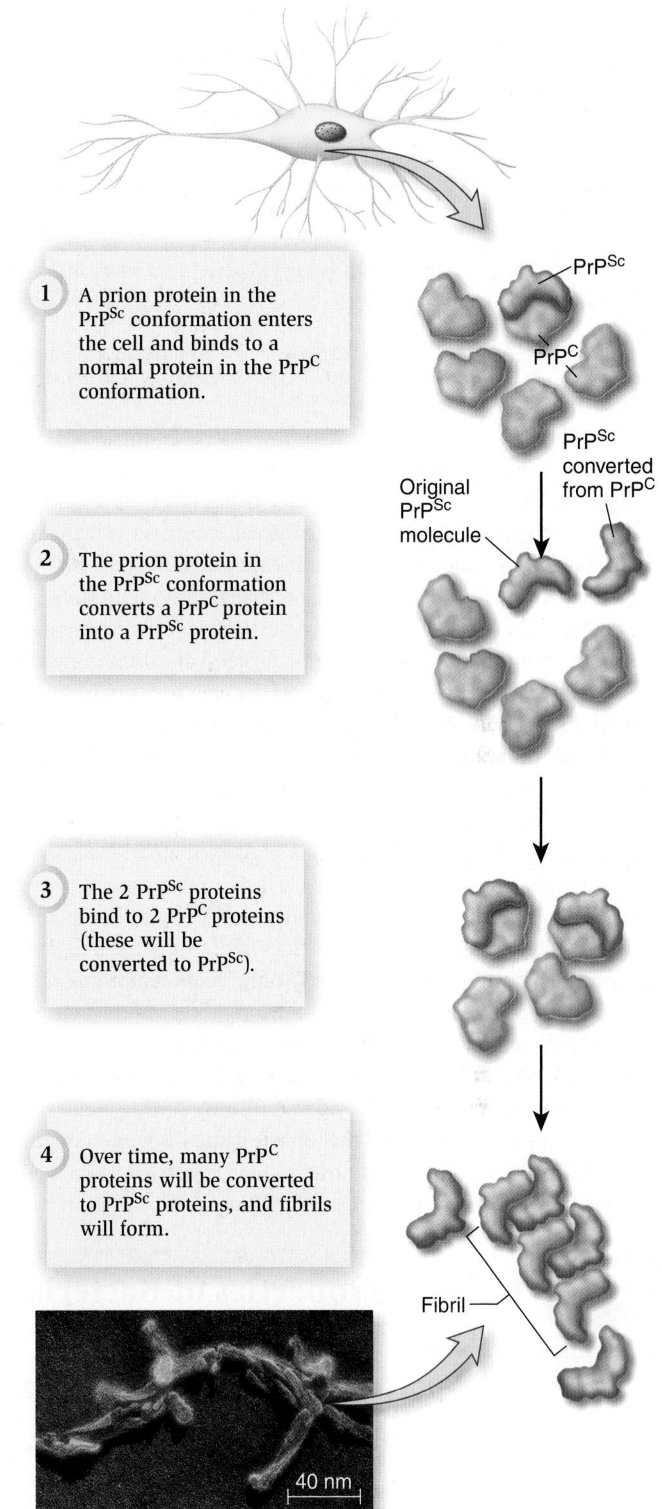

1 A prion protein in the PrP^{Sc} conformation enters the cell and binds to a normal protein in the PrP^{C} conformation.
2 The prion protein in the PrP^{Sc} conformation converts a PrP^{C} protein into a PrP^{Sc} protein.
3 The 2 PrP^{Sc} proteins bind to 2 PrP^{C} proteins (these will be converted to PrP^{Sc}).
4 Over time, many PrP^{C} proteins will be converted to PrP^{Sc} proteins, and fibrils will form.
PrP^{Sc}
PrP^{C}
Original PrP^{Sc} molecule
PrP^{Sc} converted from PrP^{C}
Fibril
40 nm

Figure 18.8 A proposed molecular mechanism of prion diseases. A healthy neuron normally contains only the PrP^{C} conformation of the prion protein. The abnormal PrP^{Sc} conformation catalyzes the conversion of PrP^{C} proteins into PrP^{Sc} proteins, thereby causing the symptoms of prion diseases.

BIOLOGY PRINCIPLE Structure determines function. The two conformations of prion proteins have different structures. The abnormal conformation has the functional ability to convert normal prion proteins to the abnormal conformation.

neurodegenerative diseases remained infectious even after exposure to radiation that would destroy any DNA or RNA. They suggested that the infectious agent was a protein. In the early 1970s, American neurologist Stanley Prusiner, moved by the death of a patient from such a neurodegenerative disease, began to search for the causative agent. In 1982, Prusiner isolated a disease-causing particle composed entirely of protein, which he called a **prion**. The term was based on his characterization of the particle as a proteinaceous infectious agent. In 1997, Prusiner was awarded the Nobel Prize in Physiology or Medicine for his work on prions.

Prion diseases arise from the ability of the prion protein to induce abnormal folding in normal protein molecules (**Figure 18.8**). The prion protein exists in a disease-causing conformation designated PrP^{Sc}. The superscript Sc refers to scrapie, an example of a prion disease. A normal conformation of this same protein, which does not cause disease, is termed PrP^{C}. The superscript C stands for cellular. The normal protein is encoded by an individual's genome, and the protein is expressed at low levels in certain types of neurons.

How does someone contract a prion disease? A healthy person may become "infected" with the abnormal protein by eating meat of an animal with the disease. Unlike most other proteins in the diet, the

Table 18.2	Examples of Neurodegenerative Diseases Caused by Infectious Prions
Disease	**Description**
Scrapie	A disease of sheep and pigs characterized by intense itching. The animals tend to scrape themselves against trees or other objects, followed by neurodegeneration.
Mad cow disease	Begins with changes in posture and temperament, followed by loss of coordination and neurodegeneration.
Chronic wasting disease	A disease of deer (genus *Odocoileus*) and Rocky Mountain elk (*Cervus elaphus*). A consistent symptom is weight loss over time. The disease is progressive and fatal.

prion protein escapes digestion in the stomach and small intestine and is absorbed into the bloodstream. After being taken up by neurons, the prion protein gradually converts the cell's normal proteins to the abnormal conformation. As a prion disease progresses, the PrP^{Sc} proteins are deposited as dense aggregates that form tough fibrils in the cells of the brain and peripheral nervous tissues, causing the disease symptoms. Some of the abnormal prion proteins are also secreted from infected cells, where they travel through the bloodstream. In this way, a prion disease can spread through the body like many viral diseases.

Prions cause several types of fatal neurodegenerative diseases affecting humans, livestock, and wildlife (**Table 18.2**). Prion diseases are termed transmissible spongiform encephalopathies (TSE). The postmortem examination of the brains of affected individuals reveals a substantial destruction of brain tissue. The brain has a spongy appearance. Most prion diseases progress fairly slowly. Over the course of a few years, symptoms proceed from a loss of motor control to dementia, paralysis, wasting, and eventually death. These symptoms are correlated with an increase in the level of prion protein in the neurons of infected individuals. No current treatment can halt the progression of any of the TSEs. For this reason, great public alarm occurs when an outbreak of a TSE is reported. For example, in 2003, a single report of a cow with the form of TSE that affects cattle—bovine spongiform encephalitis (BSE), commonly known as mad cow disease—in the U.S. prompted several countries to restrict the import of American beef.

18.3 Genetic Properties of Bacteria

Learning Outcomes:

1. Outline the key features of a bacterial chromosome.
2. Explain the two processes that serve to compact the bacterial chromosome.
3. Outline the structure and functions of plasmids.
4. Diagram the process of cell division in bacteria.

Bacteria are usually unicellular organisms. Individual cells may exist as single units or remain associated with each other after cell division, forming pairs, chains, or clumps. Bacteria are widespread on Earth, and numerous species are known to cause various types of infectious diseases. We begin this section by exploring the structure and replication of the bacterial genome and the organization of DNA sequences along a bacterial chromosome. We then examine how the chromosome is compacted to fit inside a bacterium and how it is transmitted during asexual reproduction.

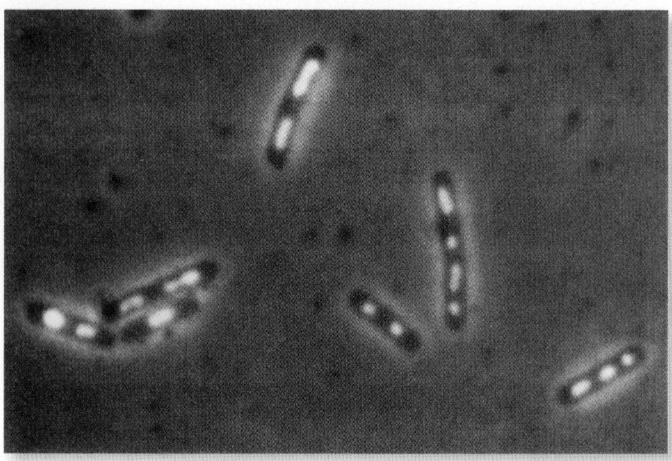

Figure 18.9 **Nucleoid regions within the bacterium *Bacillus subtilis*.** In the light micrograph shown here, the nucleoid regions are fluorescently labeled and seen as bright, oval-shaped areas within the bacterial cytoplasm. Two or more nucleoid regions are usually found within each cell.

Concept Check: *How many nucleoid regions are in the bacterial cell to the far right?*

BioConnections: *Look back at Figures 4.4 and 4.5. How is a nucleoid different from a nucleus found in a eukaryotic cell?*

Bacteria Typically Have Circular Chromosomes That Carry a Few Thousand Genes

The genes of bacteria are found within structures known as bacterial chromosomes. Although a bacterial cell usually has a single type of chromosome, it may have more than one copy of that chromosome. The number of copies depends on the bacterial species and on growth conditions, but a bacterium typically has one to four identical chromosomes. Each bacterial chromosome is tightly packed within a distinct **nucleoid region** of the cell (**Figure 18.9**). Unlike the eukaryotic nucleus, the bacterial nucleoid region is not a separate cellular compartment bounded by a membrane. The DNA in the nucleoid region is in direct contact with the cytoplasm of the cell.

Like eukaryotic chromosomes, bacterial chromosomes contain molecules of double-stranded DNA along with many different proteins. Unlike eukaryotic chromosomes, however, bacterial chromosomes are usually circular and tend to be much shorter, typically only a few million base pairs (bp) long (**Figure 18.10**). For example, the chromosome of *Escherichia coli* has approximately 4.6 million bp, and the *Haemophilus influenzae* chromosome has roughly 1.8 million bp. By comparison, an average human chromosome is over 100 million bp in length. The genomes of bacteria, archaea, and eukaryotes are compared in Chapter 21.

A typical bacterial chromosome contains a few thousand unique genes that are found throughout the chromosome. Gene sequences, primarily those that encode proteins, account for the largest part of bacterial DNA. Other nucleotide sequences in the chromosome play a role in DNA replication, gene expression, and chromosome structure. One of these sequences is the origin of replication, which is a few hundred base pairs long. Bacterial chromosomes have a single origin of replication that functions as an initiation site for the assembly of several proteins that are required for DNA replication (refer back to Figure 11.15b).

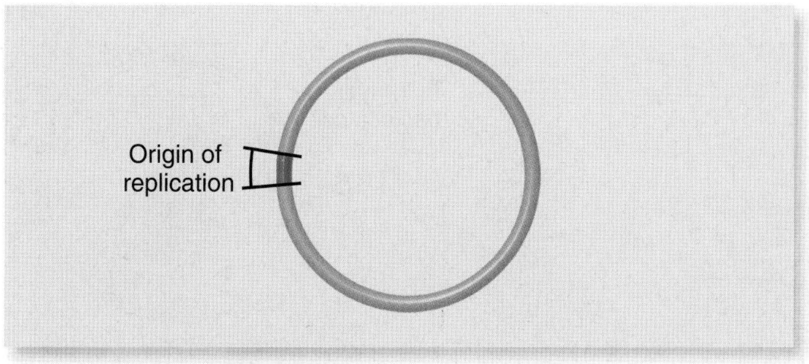

Key features

- Most, but not all, bacterial species contain circular chromosomal DNA.
- A typical chromosome is a few million base pairs in length.
- Most bacterial species contain a single type of chromosome, but it may be present in multiple copies.
- Several thousand different genes are interspersed throughout the chromosome.
- One origin of replication is required to initiate DNA replication.

Figure 18.10 **The organization of nucleotide sequences in bacterial chromosomal DNA.**

The Formation of Chromosomal Loops and DNA Supercoiling Make the Bacterial Chromosome Compact

Bacterial cells are much smaller than most eukaryotic cells (refer back to Figure 4.1). *E. coli* cells, for example, are approximately 1 μm wide and 2 μm long. To fit within a bacterial cell, the DNA of a typical bacterial chromosome must be compacted about 1,000-fold. How does this occur? The compaction of a bacterial chromosome, shown in **Figure 18.11**, occurs by two processes: the formation of loops and DNA supercoiling.

Unlike eukaryotic DNA, bacterial DNA is not wound around histone proteins to form nucleosomes. However, the binding of proteins to bacterial DNA is important in the formation of **loop domains**—chromosomal segments that are folded into loops. As seen in Figure 18.11, DNA-binding proteins anchor the bases of the loops in place. The number of loops varies according to the size of a bacterial chromosome and the species. The *E. coli* chromosome has 50 to 100 loop domains, each with about 40,000 to 80,000 bp. This looping compacts the circular chromosome about 10-fold. A similar process of loop-domain formation occurs in eukaryotic chromatin compaction, which is described in Chapter 11 (Figure 11.25).

DNA supercoiling is a second important way to compact the bacterial chromosome. Because DNA is a long, thin molecule, twisting can dramatically change its conformation. This compaction is similar to what happens to a rubber band if you twist it in one direction. Because the two strands of DNA already coil around each other, the formation of additional coils due to twisting is referred to as supercoiling. Bacterial enzymes called topoisomerases twist the DNA and control the degree of DNA supercoiling.

Plasmids Are Small, Circular Pieces of Extrachromosomal DNA

In addition to chromosomal DNA, bacterial cells commonly contain **plasmids**, small, circular pieces of DNA that exist separately from the bacterial chromosome (**Figure 18.12**). Plasmids occur naturally in many strains of bacteria and in a few types of eukaryotic cells, such as yeast. The smallest plasmids consist of just a few thousand base pairs and carry only a gene or two. The largest are in the range of 100,000 to 500,000 bp and carry several dozen or even hundreds of genes. A plasmid has its own origin of replication that allows it to be replicated independently of the bacterial chromosome. The DNA sequence of the origin of replication influences how many copies of the plasmid are found within a cell. Some origins are said to be very strong because they result in many copies of the plasmid, perhaps as many as 100 per cell. Other origins of replication have sequences that are much weaker, so the number of copies is relatively low, such as one or two per cell.

Why do bacteria have plasmids? Certain genes within a plasmid usually provide some type of growth advantage to the cell or may aid

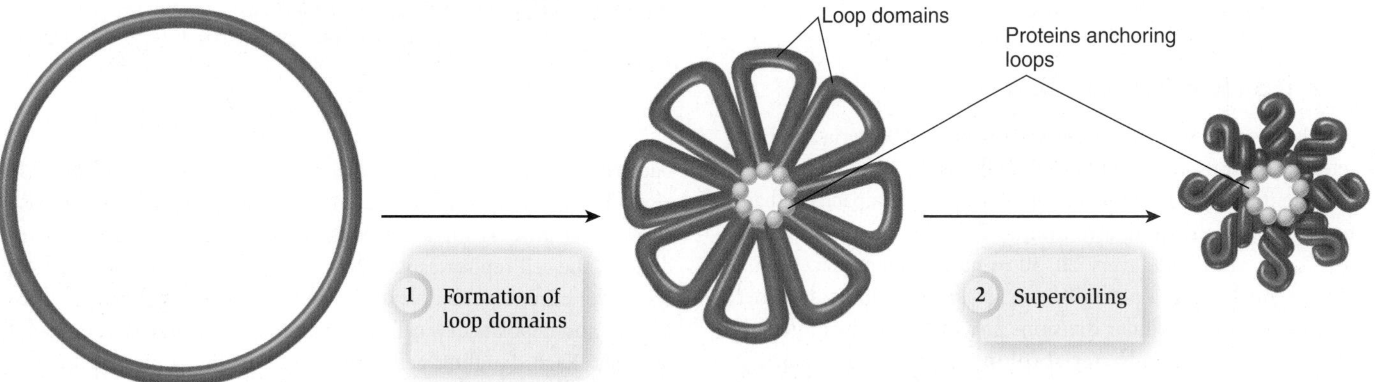

Figure 18.11 **The compaction of a bacterial chromosome.** As a way to compact the large, circular chromosome, segments are organized into smaller loop domains by binding to proteins at the bases of the loops. These loops are made more compact by DNA supercoiling. Note: This is a simplified drawing; bacterial chromosomes typically have 50 to 100 loop domains.

Concept Check: *Describe how the loop domains are held in place.*

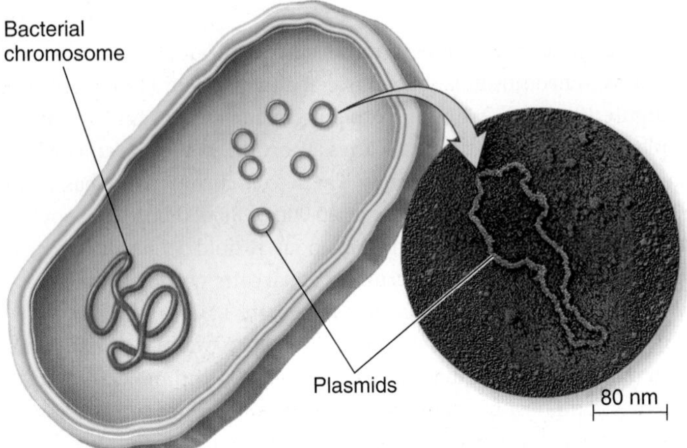

Figure 18.12 Plasmids in a bacterial cell. Plasmids are small, circular DNA molecules that exist independently of the bacterial chromosome.

Concept Check: *Describe the similarities and differences between a bacterial chromosome and a plasmid.*

in survival under certain conditions. By studying plasmids in many different species, researchers have discovered that most plasmids fall into five different categories:

1. Resistance plasmids, also known as R factors, contain genes that confer resistance against antibiotics and other types of toxins.
2. Degradative plasmids carry genes that enable the bacterium to digest and utilize an unusual substance. For example, a degradative plasmid may carry genes that allow a bacterium to digest an organic solvent such as toluene.
3. Col-plasmids contain genes that encode colicins, which are proteins that kill other bacteria.
4. Virulence plasmids carry genes that turn a bacterium into a pathogenic strain.
5. Fertility plasmids, also known as F factors, allow bacteria to transfer genes to each other, a topic described later in this chapter.

On occasion, a plasmid may integrate into the bacterial chromosome. Such plasmids, which can integrate or remain independent of the chromosome, are also termed episomes.

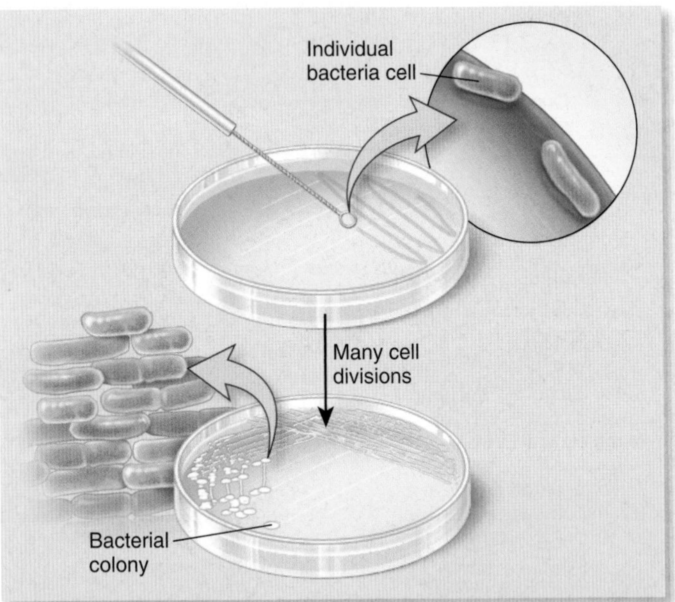

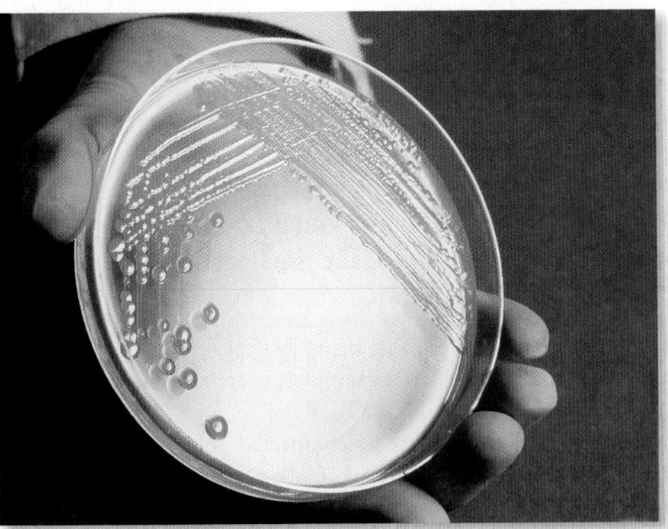

Figure 18.13 Growth of a bacterial colony. Through successive cell divisions, a single bacterial cell of *E. coli* forms a genetically identical group of cells called a bacterial colony.

Concept Check: *Let's suppose a bacterial strain divides every 30 minutes. If a single cell is placed on a plate, how many cells will be in the colony after 16 hours?*

Most Bacteria Reproduce by Binary Fission

Thus far, we have considered the genetic material of bacteria and how the bacterial chromosome is compacted to fit inside the cell. Let's now turn our attention to the process of cell division. The capacity of bacteria to divide is really quite astounding. The cells of some species, such as *E. coli*, can divide every 20–30 minutes. When placed on a solid growth medium in a petri dish, an *E. coli* cell and its daughter cells undergo repeated cellular divisions and form a clone of genetically identical cells called a **bacterial colony** (Figure 18.13). Starting with a single cell that is invisible to the naked eye, a visible bacterial colony containing 10 to 100 million cells forms in less than a day!

As described in Chapter 15, the division of eukaryotic cells requires a sorting process called mitosis, because eukaryotic chromosomes occur in sets and each daughter cell must receive the correct number and types of chromosomes. By comparison, a bacterial cell usually has only a single type of chromosome. Cell division of most bacterial species occurs by a much simpler process called **binary fission**, during which a cell divides into two daughter cells. Figure 18.14 shows this process for a cell with a single chromosome. Before it divides, the cell replicates its DNA. This produces two identical copies of the chromosome. Next, the cell's plasma membrane is drawn inward and deposits new cell-wall material, separating the two daughter cells. Each daughter cell receives one of the copies of the original chromosome. Therefore, except when a mutation occurs, each daughter cell contains an identical copy of the mother cell's genetic material. Like all other types of asexual reproduction, binary fission does not involve genetic contributions from two different parents.

Plasmids replicate independently of the bacterial chromosome. During binary fission, the plasmids are distributed to daughter cells so each daughter cell usually receives one or more copies of the plasmid.

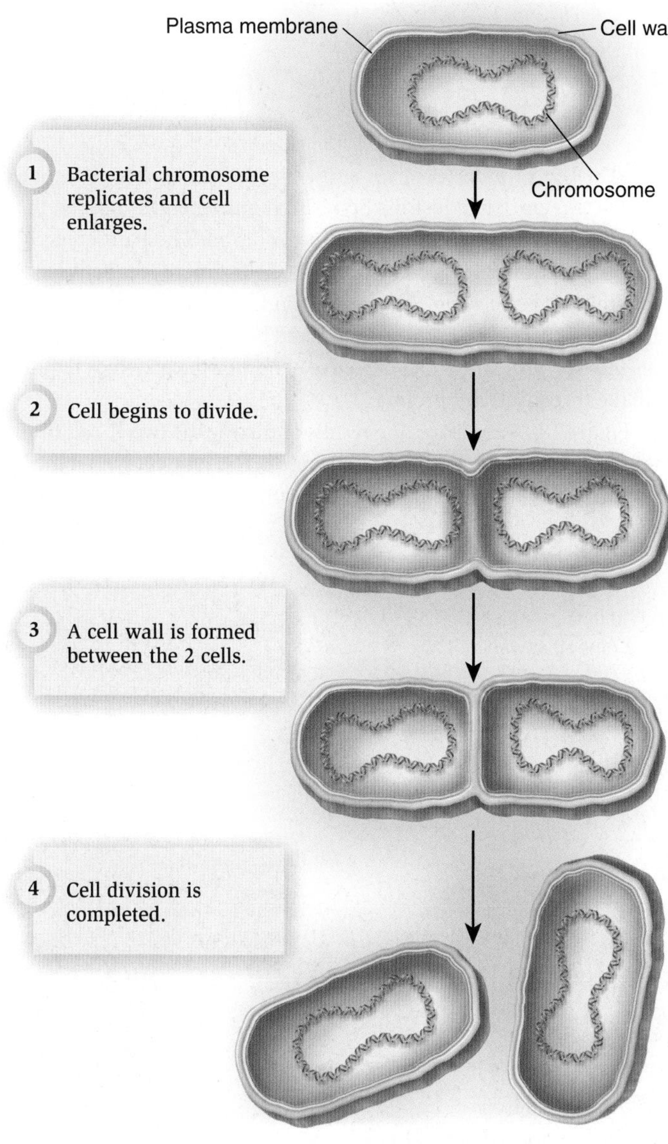

Mother cell
Plasma membrane
Cell wall
Chromosome
1 Bacterial chromosome replicates and cell enlarges.
2 Cell begins to divide.
3 A cell wall is formed between the 2 cells.
4 Cell division is completed.
Two daughter cells

Figure 18.14 **Bacterial cell division.** Bacteria reproduce by binary fission. Before a bacterium divides, the bacterial chromosome is replicated to produce two identical copies. These two copies segregate from each other, with one copy going to each daughter cell.

BIOLOGY PRINCIPLE Living organisms grow and develop. After cell division, a bacterial cell increases in size.

18.4 Gene Transfer Between Bacteria

Learning Outcomes:

1. Compare and contrast the three forms of gene transfer—conjugation, transformation, and transduction—in bacteria.
2. Analyze the results of Lederberg and Tatum and explain how they led to the discovery of conjugation.
3. Define the process of horizontal gene transfer.

Even though bacteria reproduce asexually, they exhibit a great deal of genetic diversity. Within a given bacterial species, the term **strain** refers to a lineage that has genetic differences compared to another strain. For example, one strain of *E. coli* may be resistant to an antibiotic, whereas another strain may be sensitive to the same antibiotic.

How does genetic diversity arise in an asexual species? It comes primarily from two sources. First, mutations can occur that alter the bacterial genome and affect the traits of bacterial cells. Second, diversity can arise by **gene transfer**, in which genetic material is transferred from one bacterial cell to another. Through gene transfer, genetic variation that arises in one bacterium can be spread to other strains and even to other species. For example, an antibiotic-resistance gene may be transferred from a resistant strain to a sensitive strain.

Gene transfer occurs in three different ways, termed conjugation, transformation, and transduction (Table 18.3). The process known as **conjugation** involves a direct physical interaction between two bacterial cells. During conjugation, one bacterium acts as a donor and transfers DNA to a recipient cell. In the process of **transformation**, DNA is released into the environment and taken up by another bacterial cell. **Transduction** occurs when a virus infects a bacterial cell and then transfers some of that cell's DNA to another bacterium. These three types of gene transfer have been extensively investigated in research laboratories, and their molecular pathways continue to be studied with great interest. In this section, we will examine these mechanisms in greater detail and consider the experiments that led to their discovery.

Table 18.3 Mechanisms of Gene Transfer Between Bacterial Cells

Mechanism	Description
Conjugation: Donor cell, Recipient cell	Requires direct contact between a donor cell and a recipient cell. The donor cell transfers a strand of DNA to the recipient. In the example shown here, DNA from a plasmid is transferred to the recipient cell. The end result is that both donor and recipient cells have a plasmid.
Transformation: Donor cell (dead), Recipient cell 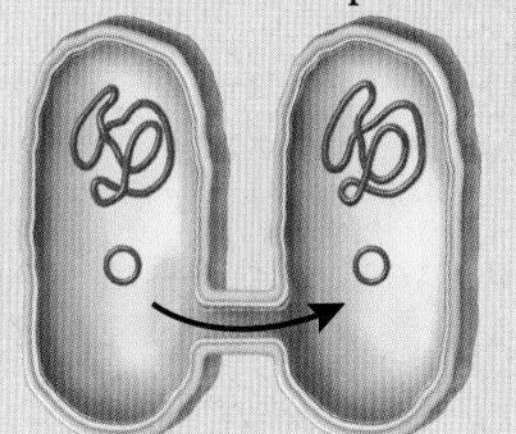	A fragment of DNA from a donor cell is released into the environment. This may happen when a bacterial cell dies. This DNA fragment is taken up by a recipient cell, which incorporates the DNA into its chromosome.
Transduction: Donor cell (infected by a virus), Recipient cell 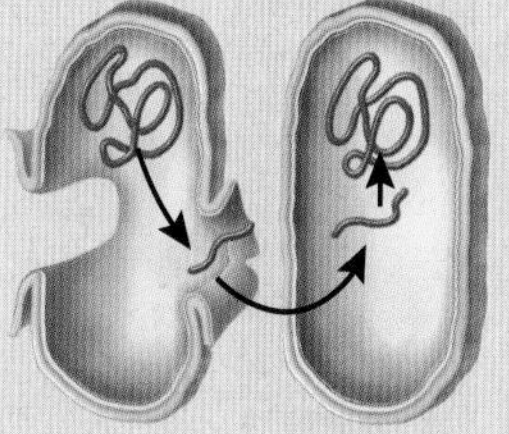	When a virus (in this example a phage) infects a donor cell, it causes the bacterial chromosome of the donor cell to break up into fragments. A fragment of bacterial chromosomal DNA is incorporated into a newly made phage. The phage then transfers this fragment of DNA to a recipient cell.

Feature Investigation

Lederberg and Tatum's Work with *E. coli* Demonstrated Gene Transfer Between Bacteria and Led to the Discovery of Conjugation

In 1946 and 1947, Joshua Lederberg and Edward Tatum carried out the first experiments that showed gene transfer from one bacterial strain to another (**Figure 18.15**). The researchers studied strains of *E. coli* that had different nutritional requirements for growth. They designated one strain $met^-bio^-thr^+pro^+$ because its growth required the amino acid methionine (met) and the vitamin biotin (bio) in the growth medium. This strain did not require the amino acids threonine (thr) or proline (pro) for growth. Another strain, designated $met^+bio^+thr^-pro^-$, had just the opposite requirement. It needed threonine and proline, but not methionine or biotin. These differences in nutritional requirements correspond to allelic differences between the two strains. The $met^-bio^-thr^+pro^+$ strain had defective genes encoding enzymes necessary for methionine and biotin synthesis, whereas the $met^+bio^+thr^-pro^-$ strain had defective genes for the enzymes required to make threonine and proline.

Figure 18.15 compares the results of mixing the two *E. coli* strains with the results when they were not mixed. The tube shown on the left contained only $met^-bio^-thr^+pro^+$ cells, and the tube on the right had

Figure 18.15 **Experiment of Lederberg and Tatum demonstrating gene transfer in *E. coli*.**

HYPOTHESIS Genetic material can be transferred from one bacterial strain to another.

KEY MATERIALS Two bacterial strains, one that was $met^-bio^-thr^+pro^+$ and the other that was $met^+bio^+thr^-pro^-$.

Experimental level | **Conceptual level**

1 In 3 separate tubes, add the $met^-bio^-thr^+pro^+$ strain, the $met^+bio^+thr^-pro^-$ strain, or a mixture of both strains.

Incubate several hours.

2 Remove 10^8 cells from each tube and spread onto plates that lack methionine, biotin, threonine, and proline.

3 Incubate overnight to allow growth of bacterial colonies.

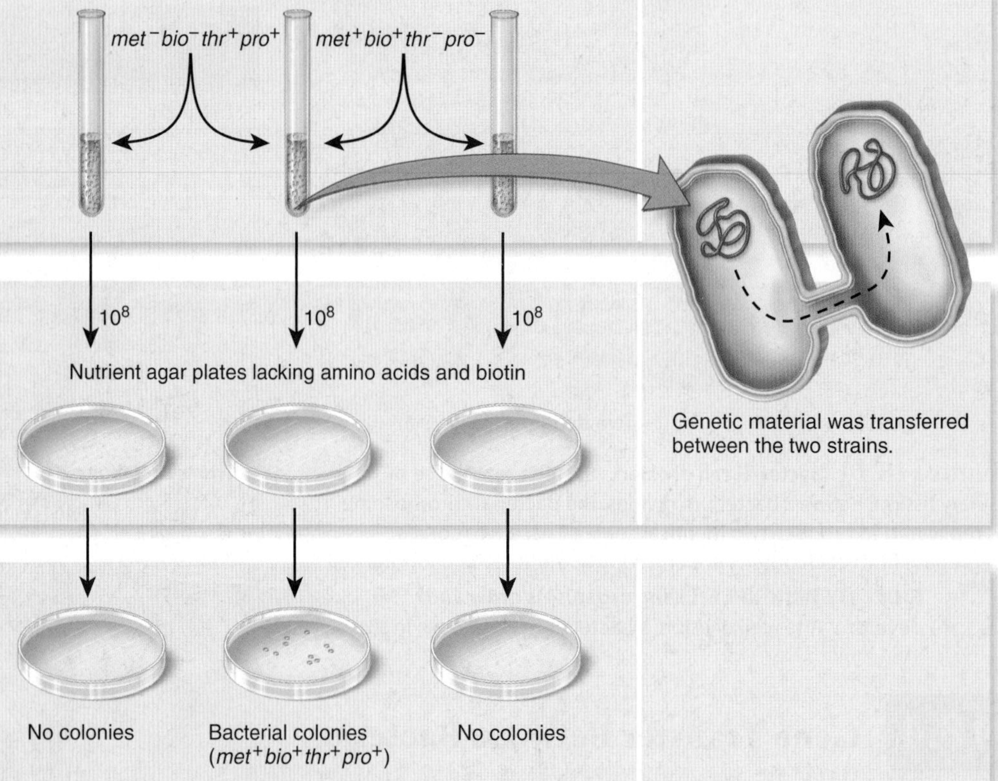

4 **THE DATA**

Strain	Number of colonies after overnight growth
$met^-bio^-thr^+pro^+$	0
$met^+bio^+thr^-pro^-$	0
Both strains together	~10

5 **CONCLUSION** Gene transfer has occurred from one bacterial strain to another.

6 **SOURCES** Lederberg, Joshua, and Tatum, Edward L. 1946. Novel genotypes in mixed cultures of biochemical mutants of bacteria. *Cold Spring Harbor Symposia on Quantitative Biology* 11:113–114.

Tatum, Edward L., and Lederberg, Joshua. 1947. Genetic recombination in the bacterium *Escherichia coli*. *Journal of Bacteriology* 53:673–684.

only $met^+ bio^+ thr^- pro^-$ cells. The middle tube contained a mixture of the two kinds of cells. In each case, the researchers applied about 100 million (10^8) cells to plates containing a growth medium lacking amino acids and the vitamin biotin. When the unmixed strains were applied to these plates, no colonies were observed to grow. This result was expected because the plates did not contain the methionine and biotin that the $met^- bio^- thr^+ pro^+$ cells needed for growth or the threonine and proline that the $met^+ bio^+ thr^- pro^-$ cells required. The striking result occurred when the researchers plated 10^8 cells from the tube containing the mixture of the two strains. In this case, approximately 10 cells multiplied and formed visible bacterial colonies on the plates. Because these cells multiplied without supplemental amino acids or vitamins, their genotype must have been $met^+ bio^+ thr^+ pro^+$. Mutations cannot account for the occurrence of this new genotype because colonies were not observed on the other two plates, which had the same number of cells and also could have incurred mutations.

To explain the results of their experiment, Lederberg and Tatum hypothesized that some genetic material had been transferred between the two strains when they were mixed. This transfer could have occurred in two ways. One possibility is that the genes providing the ability to synthesize threonine and proline ($thr^+ pro^+$) were transferred to the $met^+ bio^+ thr^- pro^-$ strain. Alternatively, the genes providing the ability to synthesize methionine and biotin ($met^+ bio^+$) may have been transferred to the $met^- bio^- thr^+ pro^+$ cells. The experimental results cannot distinguish between these two possibilities, but they provide compelling evidence that at least one of them occurred.

How did the bacteria in Lederberg and Tatum's experiment transfer genes between strains? Two mechanisms seemed plausible. Either genetic material was released from one strain and taken up by the other, or cells of the two different strains made contact with each other and directly transferred genetic material. To distinguish these two scenarios, American microbiologist Bernard Davis conducted experiments using the same two strains of *E. coli*. The apparatus he used, known as a U-tube, is shown in **Figure 18.16**. The tube had a filter with pores big enough for pieces of DNA to pass through, but too small to permit the passage of bacteria. After filling the tube with a liquid medium, Davis added $met^- bio^- thr^+ pro^+$ bacteria on one side of the filter and $met^+ bio^+ thr^- pro^-$ bacteria on the other. The application of pressure or suction promoted the movement of liquid through the pores. Although the two kinds of bacteria could not mix, any genetic material released by one of them would be available to the other.

After allowing the bacteria to incubate in the U-tube, Davis placed cells from each side of the tube on growth plates lacking methionine, biotin, threonine, and proline. No bacterial colonies grew on these plates. How did Davis interpret these results? He proposed that without physical contact, the two *E. coli* strains could not transfer genetic material from one cell to the other. The conceptual level of Figure 18.15, step 1, shows the physical connection that explains Lederberg and Tatum's results. Conjugation is the process of gene transfer that requires direct cell-to-cell contact. It has been subsequently observed in other species of bacteria. Many, but not all, species of bacteria can conjugate.

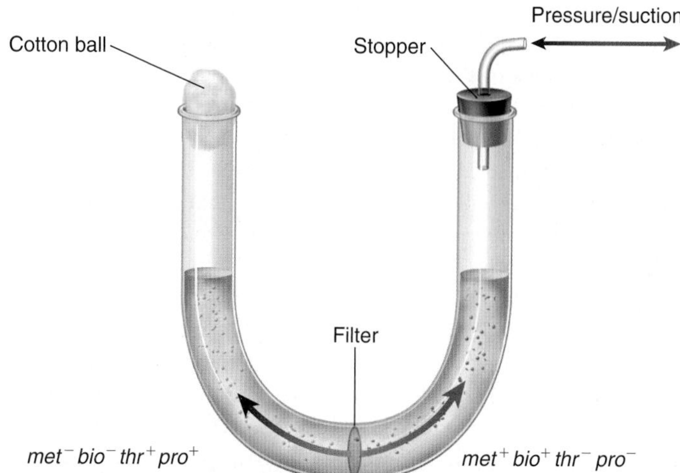

Figure 18.16 A U-tube apparatus like the one used by Bernard Davis. Bacteria of two different strains were suspended in the liquid in the tube and separated by a filter. The liquid was forced through the filter by alternating suction and pressure. The pores in the filter were too small for the passage of bacteria, but they allowed the passage of DNA.

BIOLOGY PRINCIPLE Biology is an experimental science. The U-tube apparatus allowed Davis to determine if direct cell-to-cell contact was needed for gene transfer.

Concept Check: Would the results have been different if the pore size was larger and allowed the passage of bacterial cells?

Experimental Questions

1. What hypothesis did Lederberg and Tatum test?
2. During the Lederberg and Tatum experiment, the researchers compared the growth of mutant strains under two scenarios: mixed strains or unmixed strains. When the unmixed strains were plated on the experimental growth medium, why were no colonies observed to grow? When the mixed strains were plated on the experimental growth medium, 10 colonies were observed. What was the significance of these colonies?
3. The gene transfer seen in the Lederberg and Tatum experiment could have occurred in one of two ways: taking up DNA released into the environment or contact between two bacterial cells allowing for direct transfer. Bernard Davis conducted an experiment to determine the correct process. Explain how his results indicated the correct gene transfer process.

During Conjugation, DNA Is Transferred from a Donor Cell to a Recipient Cell

In the early 1950s, Americans Joshua and Esther Lederberg, Irish physician William Hayes, and Italian geneticist Luca Cavalli-Sforza independently discovered that only certain bacterial strains can donate genetic material during conjugation. For example, only about 5% of *E. coli* strains found in nature act as donor strains. Further research showed that a strain that is incapable of acting as a donor can acquire this ability after being mixed with a donor strain. Hayes correctly proposed that donor strains contain a type of plasmid called a fertility factor, or **F factor**, that can be transferred to recipient strains. Also, other donor *E. coli* strains were later identified that can transfer portions of the bacterial chromosome at high frequencies. After a segment of the chromosome is transferred, it then inserts, or recombines, into the chromosome of the recipient cell. Such donor strains were named *Hfr* (for High frequency of recombination). In our discussion, we will focus on donor strains that carry F factors.

The micrograph in Figure 18.17a shows two conjugating *E. coli* cells. The cell on the left is designated F^+, meaning that it has an F factor. This donor cell is transferring genetic material to the recipient cell on the right, which lacks an F factor and is designated F^-. F factors carry several genes that are required for conjugation and also may carry genes that confer a growth advantage for the bacterium.

Figure 18.17b describes the events that occur during conjugation in *E. coli*. The process is similar in other bacteria that are capable of conjugating, although the details vary somewhat from one species to another. Recall from Chapter 4 that many bacteria have appendages called pili that allow them to attach to surfaces and to each other (refer back to Figure 4.4). **Sex pili** are made by F^+ cells that bind specifically to F^- cells. They are so named because conjugation has sometimes been called bacterial mating. However, this term is a bit misleading because the process does not involve equal genetic contributions from two gametes and it does not produce offspring. Instead, bacterial mating is a form of gene transfer that alters the genetic composition of the recipient cell. Donor strains have genes responsible for the formation of sex pili. In F^+ strains, the genes are located on the F factor. In *E. coli* and some other species, F^+ cells make very long pili that attempt to make contact with nearby F^- cells. Once contact is made, the pili shorten, drawing the donor and recipient cells closer together.

After the pili have shortened, contact between donor and recipient cell stimulates the donor cell to begin the transfer process. First, a conjugation bridge is formed that provides a direct passageway for DNA transfer. One strand of F factor DNA is cut at the origin of transfer and then travels through the conjugation bridge into the recipient cell. The other strand remains in the donor cell, and the complementary strand is synthesized, thereby restoring the F factor DNA to its original double-stranded condition. In the recipient cell, the two ends of the newly acquired F factor DNA strand are joined to form a circular molecule, and its complementary strand is synthesized to produce a double-stranded F factor. If conjugation is successful, the end result is that the recipient cell has acquired an F factor, converting it from an F^- to an F^+ cell. The genetic composition of the donor strain has not been changed.

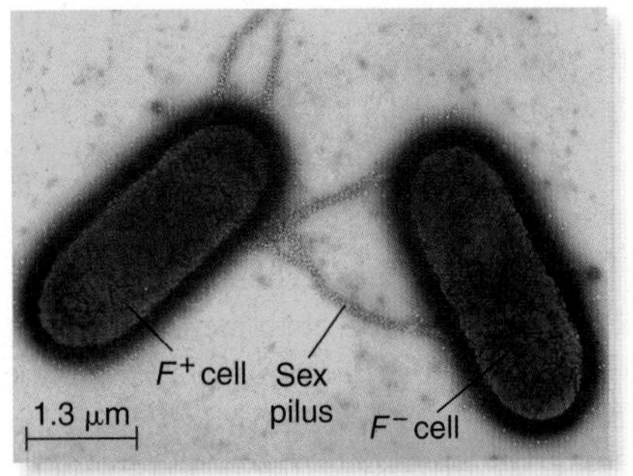

(a) Micrograph of conjugating cells

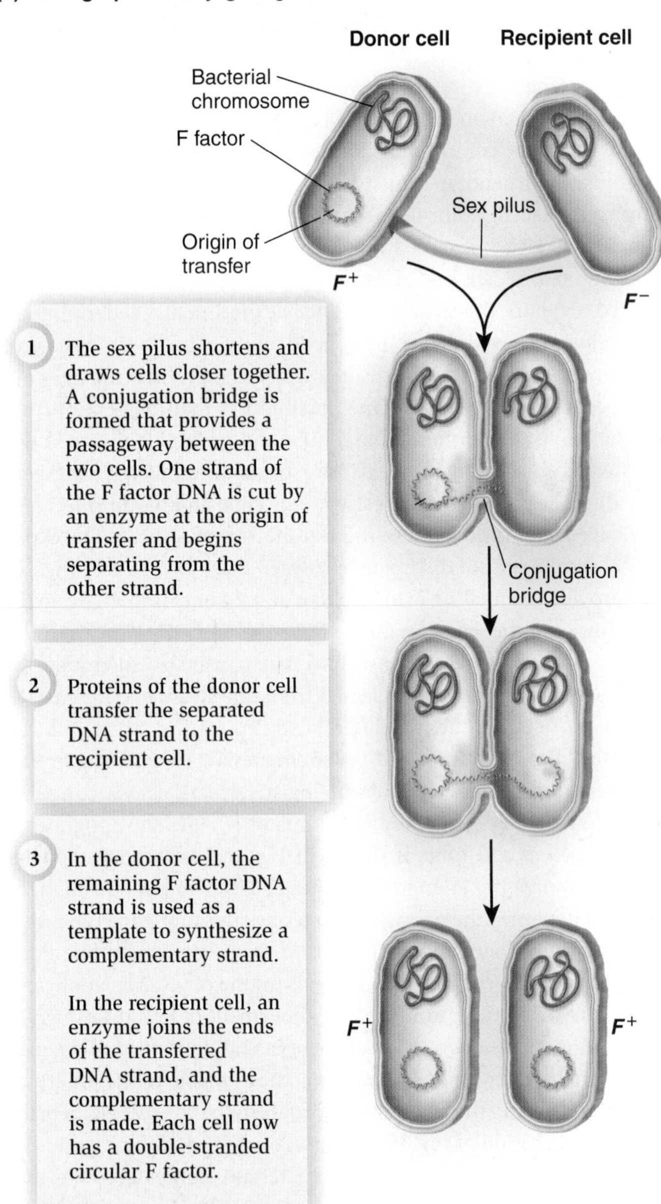

Donor cell
Recipient cell
Bacterial chromosome
F factor
Origin of transfer
Sex pilus
F^+
F^-
1 The sex pilus shortens and draws cells closer together. A conjugation bridge is formed that provides a passageway between the two cells. One strand of the F factor DNA is cut by an enzyme at the origin of transfer and begins separating from the other strand.
Conjugation bridge
2 Proteins of the donor cell transfer the separated DNA strand to the recipient cell.
3 In the donor cell, the remaining F factor DNA strand is used as a template to synthesize a complementary strand.
In the recipient cell, an enzyme joins the ends of the transferred DNA strand, and the complementary strand is made. Each cell now has a double-stranded circular F factor.
F^+
F^+

(b) Transfer of an F factor

Figure 18.17 **Bacterial conjugation.** **(a)** A micrograph of two *E. coli* cells that are conjugating. The cell on the left, designated F^+, is the donor; the cell on the right, designated F^-, is the recipient. The two cells make contact via sex pili made by the F^+ cell. **(b)** The transfer of an F factor during conjugation. At the end of conjugation, both the donor cell and the recipient cell are F^+.

Concept Check: *If a donor cell has only one F factor, explain how the donor and recipient cell both contain one F factor following the transfer of an F factor during conjugation.*

In Transformation, Bacteria Take Up DNA from the Environment

In contrast to conjugation, the process of gene transfer known as bacterial transformation does not require direct contact between bacterial cells. Frederick Griffith first discovered this process in 1928 while working with strains of *Streptococcus pneumoniae*. We discussed early experiments involving transformation in Chapter 11 (refer back to Figures 11.1 and 11.2).

How does a bacterial cell become transformed? First, it imports a strand of DNA from the environment. This DNA strand, which is typically derived from a dead bacterial cell, may then insert or recombine

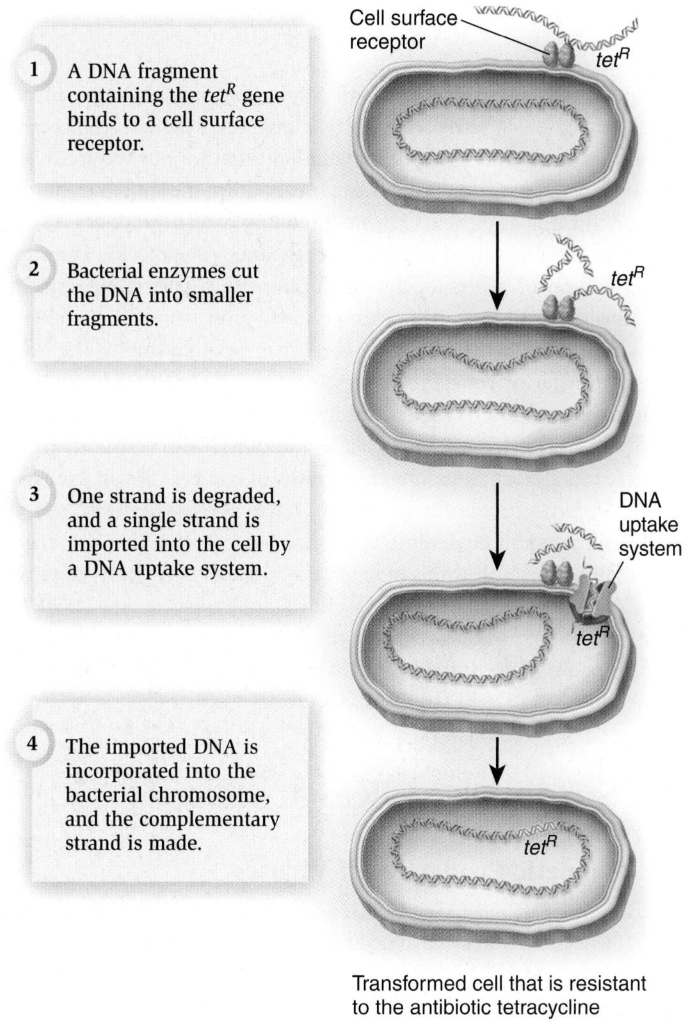

1 A DNA fragment containing the tet^R gene binds to a cell surface receptor.

2 Bacterial enzymes cut the DNA into smaller fragments.

3 One strand is degraded, and a single strand is imported into the cell by a DNA uptake system.

4 The imported DNA is incorporated into the bacterial chromosome, and the complementary strand is made.

Figure 18.18 Bacterial transformation. This process has transformed a bacterium that was sensitive to the antibiotic tetracycline into one that is resistant to this antibiotic.

BioConnections: Look back at Figures 11.1 and 11.2. How did the phenomenon of transformation allow researchers to demonstrate that DNA is the genetic material?

into the bacterial chromosome. The live bacterium then carries genes from the dead bacterium—the live bacterium has been transformed. Not all bacterial strains have the ability to take up DNA. Those that do have this ability are described as naturally **competent**, and they have genes that encode proteins called competence factors. Competence factors facilitate the binding of DNA fragments to the bacterial cell surface, the uptake of DNA into the cytoplasm, and the incorporation of the imported DNA into the bacterial chromosome. Temperature, ionic conditions, and the availability of nutrients affect whether or not a bacterium will be competent to take up genetic material.

In recent years, biologists have unraveled some of the steps that occur when competent bacterial cells are transformed by taking up genetic material from the environment. In the example shown in Figure 18.18, the DNA released from a dead bacterium carries a gene, tet^R, that confers resistance to the antibiotic tetracycline. First, a large fragment of the DNA binds to a cell surface receptor on the outside of a bacterial cell that is sensitive to tetracycline. Enzymes secreted by the bacterium cut this large fragment into fragments small enough to enter the cell. In our example, one of the two strands of a fragment of DNA containing the tet^R gene is degraded. The other strand enters the bacterial cytoplasm via a DNA uptake system that transports the DNA across the plasma membrane. Finally, the imported DNA strand is incorporated into the bacterial chromosome, and the complementary strand is synthesized. Following transformation, the recipient cell has been transformed from a tetracycline-sensitive cell to a tetracycline-resistant cell.

In Transduction, Viruses Transfer Genetic Material from One Bacterium to Another

A third mechanism of gene transfer is transduction, in which bacteriophages transfer bacterial genes from one bacterium to another. As discussed earlier in this chapter, a bacteriophage (or simply phage) is a virus that uses the cellular machinery of a bacterium for its own replication. The new viral particles made in this way usually contain only viral genes. On rare occasions, however, a phage may pick up a piece of DNA from the bacterial chromosome. When a phage carrying a segment of bacterial DNA infects another bacterium, it transfers this segment into the chromosome of its new bacterial host.

Transduction is actually an error in a phage lytic cycle, as shown in Figure 18.19. In this example, a phage called P1 infects an *E. coli* cell that has a gene (his^+) for histidine synthesis. Phage P1 causes the host cell chromosome to be degraded into small pieces. New phage DNA and proteins are synthesized. When new phages are assembled, coat proteins may accidentally enclose a piece of host DNA that carries this gene. This produces a phage carrying bacterial chromosomal DNA. In the example shown in Figure 18.19, this transducing phage is released and binds to an *E. coli* cell that lacks the his^+ gene. It inserts the bacterial DNA fragment into the recipient cell, which then integrates this fragment into its own chromosome. In this case, gene transfer by transduction converts a his^- strain of *E. coli* to a his^+ strain.

Genomes & Proteomes Connection

Horizontal Gene Transfer Is the Transfer of Genes Between the Same or Different Species

The term **horizontal gene transfer** refers to a process in which an organism incorporates genetic material from another organism without being the offspring of that organism. Conjugation, transformation, and transduction are examples of horizontal gene transfer. In contrast, vertical gene transfer occurs when genes are passed from one generation to the next—from parents to offspring and from mother cells to daughter cells.

Conjugation, transformation, and transduction occasionally occur between cells of different bacterial species. In recent years, analyses of bacterial genomes have shown that a sizeable fraction of bacterial genes are derived from horizontal gene transfer. For example, roughly 17% of the genes of *E. coli* and of *Salmonella typhimurium* have been acquired from other species by horizontal gene transfer during the past 100 million years. Many of these acquired genes affect traits that give cells a selective advantage, including genes that confer antibiotic resistance, the ability to degrade toxic compounds, and the ability to withstand extreme environments. Some horizontally

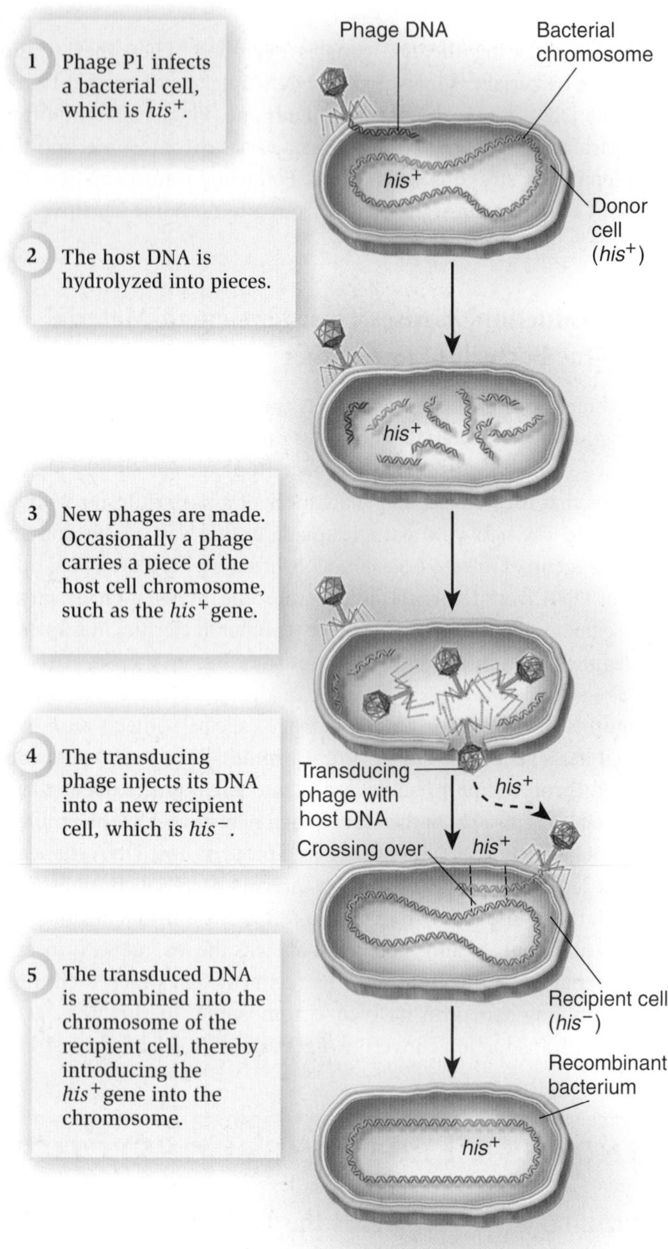
1 Phage P1 infects a bacterial cell, which is *his*+.
Phage DNA
Bacterial chromosome
his+
Donor cell (his+)
2 The host DNA is hydrolyzed into pieces.
his+
3 New phages are made. Occasionally a phage carries a piece of the host cell chromosome, such as the *his*+ gene.
4 The transducing phage injects its DNA into a new recipient cell, which is *his*−.
Transducing phage with host DNA
his+
Crossing over
his+
5 The transduced DNA is recombined into the chromosome of the recipient cell, thereby introducing the *his*+ gene into the chromosome.
Recipient cell (his−)
Recombinant bacterium
his+
The recombinant bacterium has a genotype (his+) that is different from the original recipient bacterial cell (his−).

Figure 18.19 Bacterial transduction by P1 phage.

Concept Check: Is transduction a normal part of the phage life cycle? Explain.

transferred genes confer pathogenicity, turning a harmless bacterial strain into one that can cause disease. Geneticists have suggested that horizontal gene transfer has played a major role in the evolution of different bacterial species. In many cases, the acquisition of new genes allows a bacterium to survive in a new type of environment and can eventually lead to the formation of a new species.

A second reason why horizontal gene transfer is important is its medical relevance. Let's consider the topic of antibiotic resistance. Antibiotics are widely prescribed to treat bacterial infections in humans. They are also used in agriculture to control bacterial diseases. Unfortunately, the widespread use of antibiotics has greatly increased the prevalence of antibiotic-resistant strains of bacteria, strains that have a selective advantage over those that are susceptible to antibiotics. Resistant strains carry genes that counteract the action of antibiotics in various ways. A resistance gene may encode a protein that breaks down the antibiotic, pumps it out of the cell, or prevents it from inhibiting cellular processes.

The term **acquired antibiotic resistance** refers to the common phenomenon of a previously susceptible strain becoming resistant to a specific antibiotic. This change may result from genetic alterations in the bacteria's own genome, but it is often due to the horizontal transfer of resistance genes from a resistant strain. As often mentioned in the news media, antibiotic resistance has increased dramatically worldwide over the past few decades, with resistant strains reported in almost all pathogenic strains of bacteria. As an example, some *Staphylococcus aureus* strains have developed resistance to methicillin and all penicillins. Evidence suggests that these so-called methicillin-resistant strains of *Staphlococcus aureus* (MRSA) acquired the methicillin-resistance gene by horizontal gene transfer, possibly from a strain of *Enterococcus faecalis*. MRSA strains cause skin infections that are more difficult to treat than ordinary "staph" infections caused by nonresistant strains of *S. aureus*.

Summary of Key Concepts

18.1 Genetic Properties of Viruses

- Viruses are nonliving particles that do not exhibit all of the properties associated with living organisms. A virus or its genetic material must be taken up by a living cell to replicate. Tobacco mosaic virus (TMV) was the first virus to be discovered (Figure 18.1).
- Viruses vary with regard to their host range, structure, and genome composition (Table 18.1, Figures 18.2, 18.3).
- The viral reproductive cycle consists of a series of basic steps, including attachment, entry, integration, synthesis, assembly, and release (Figure 18.4).
- Some bacteriophages can follow two different reproductive cycles: the lytic cycle and lysogenic cycle (Figure 18.5).
- Emerging viruses, such as human immunodeficiency virus (HIV), can cause significant loss of human life. HIV, the causative agent of the disease AIDS, is a retrovirus whose reproductive cycle involves the integration of the viral genome into a chromosome in the host cell. HIV accomplishes this by means of a viral enzyme called reverse transcriptase (Figure 18.6).
- Drugs to combat viral proliferation are often developed to specifically inhibit viral proteins.

18.2 Viroids and Prions

- Viroids are RNA molecules that infect plant cells (Figure 18.7).
- Prions are proteins that induce abnormal folding in normal proteins. They cause several fatal neurodegenerative diseases in humans (Figure 18.8, Table 18.2).

18.3 Genetic Properties of Bacteria

- Bacteria typically have a single type of circular chromosome found in the nucleoid region of the cell. The chromosome contains many genes and one origin of replication (Figures 18.9, 18.10).
- The bacterial chromosome is made more compact by the formation of loop domains and by DNA supercoiling (Figure 18.11).

- Plasmids are small, circular DNA molecules that exist independently of the bacterial chromosome (Figure 18.12).
- When placed on solid growth media, a single bacterial cell will divide many times to produce a colony composed of many cells (Figure 18.13).
- Bacterial cells reproduce by asexual reproduction in a process called binary fission, during which a cell divides to form two daughter cells (Figure 18.14).

18.4 Gene Transfer Between Bacteria

- Three common modes of gene transfer among bacteria are conjugation, transformation, and transduction (Table 18.3).
- Lederberg and Tatum's work demonstrated the transfer of bacterial genes between different strains of *E. coli* by conjugation. Davis showed that direct contact was needed for this type of gene transfer (Figures 18.15, 18.16).
- During conjugation, a strand of DNA from an F factor is transferred from a donor to a recipient cell by physical contact (Figure 18.17).
- Transformation is the process in which a segment of DNA from the environment is taken up by a competent cell and incorporated into the bacterial chromosome (Figure 18.18).
- Bacterial transduction is a form of gene transfer in which a bacteriophage transfers a segment of bacterial chromosomal DNA from one bacterial cell to another cell (Figure 18.19).
- Horizontal gene transfer is a process in which an organism incorporates genetic material from another organism without being the offspring of that organism. It has played a large role in the evolution of bacterial species.

Assess and Discuss

Test Yourself

1. The ___________ is the protein coat of a virus that surrounds the genetic material.
   a. host c. capsid e. capsule
   b. prion d. viroid
2. Among the viruses identified, the characteristics of their genomes show many variations. Which of the following does not describe a typical characteristic of viral genomes?
   a. The genetic material may be DNA or RNA.
   b. The nucleic acid may be single stranded or double stranded.
   c. The genome may carry just a few genes or several dozen.
   d. The number of copies of the genome may vary.
   e. All of the above describe typical variation in viral genomes.
3. During viral infection, attachment is usually specific to a particular cell type because
   a. the virus is attracted to the appropriate host cells by proteins secreted into the extracellular fluid.
   b. the virus recognizes and binds to specific molecules in the cytoplasm of the host cell.
   c. the virus recognizes and binds to specific molecules on the surface of the host cell.
   d. the host cell produces channel proteins that provide passageways for viruses to enter the cytoplasm.
   e. the virus releases specific proteins that make holes in the membrane large enough for the virus to enter.
4. HIV, a retrovirus, has a high mutation rate because
   a. the DNA of the viral genome is less stable than other viral genomes.
   b. the viral enzyme reverse transcriptase has a high likelihood of making replication errors.
   c. the viral genome is altered every time it is incorporated into the host genome.
   d. antibodies produced by the host cell mutate the viral genome when infection occurs.
   e. all of the above.
5. A ___ is an infectious agent composed solely of RNA, whereas a ___ is an infectious agent composed solely of protein.
   a. retrovirus, bacteriophage d. retrovirus, prion
   b. viroid, virus e. viroid, prion
   c. prion, virus
6. Genetic diversity is maintained in bacterial populations by all of the following except
   a. binary fission. d. transduction.
   b. mutation. e. conjugation.
   c. transformation.
7. Bacterial cells divide by a process known as
   a. mitosis. c. meiosis. e. glycolysis.
   b. cytokinesis. d. binary fission.
8. Gene transfer in which a bacterial cell takes up bacterial DNA from the environment is also called
   a. conjugation. d. transformation.
   b. binary fission. e. transduction.
   c. asexual reproduction.
9. A bacterial cell can donate DNA during conjugation when it
   a. produces competence factors.
   b. contains an F factor.
   c. is virulent.
   d. has been infected by a bacteriophage.
   e. all of the above.
10. A bacterial species that becomes resistant to certain antibiotics may have acquired the resistance genes from another bacterial species. The phenomenon of acquiring genes from another organism without being the offspring of that organism is known as
   a. hybridization. d. vertical gene transfer.
   b. integration. e. competence.
   c. horizontal gene transfer.

Conceptual Questions

1. How are viruses similar to living cells, and how are they different?
2. Describe the three mechanisms of gene transfer in bacteria.
3. A principle of biology is *the genetic material provides the blueprint for reproduction.* Discuss how horizontal gene transfer alters the blueprint.

Collaborative Questions

1. Discuss the possible origin of viruses. Which idea(s) do you think is (are) the most likely?
2. Conjugation is sometimes called "bacterial mating." Discuss how conjugation is similar to sexual reproduction in eukaryotes and how it is different.

Online Resource

www.brookerbiology.com

Stay a step ahead in your studies with animations that bring concepts to life and practice tests to assess your understanding. Your instructor may also recommend the interactive eBook, individualized learning tools, and more.

Developmental Genetics

19

Chapter Outline

19.1 General Themes in Development
19.2 Development in Animals
19.3 Development in Plants
Summary of Key Concepts
Assess and Discuss

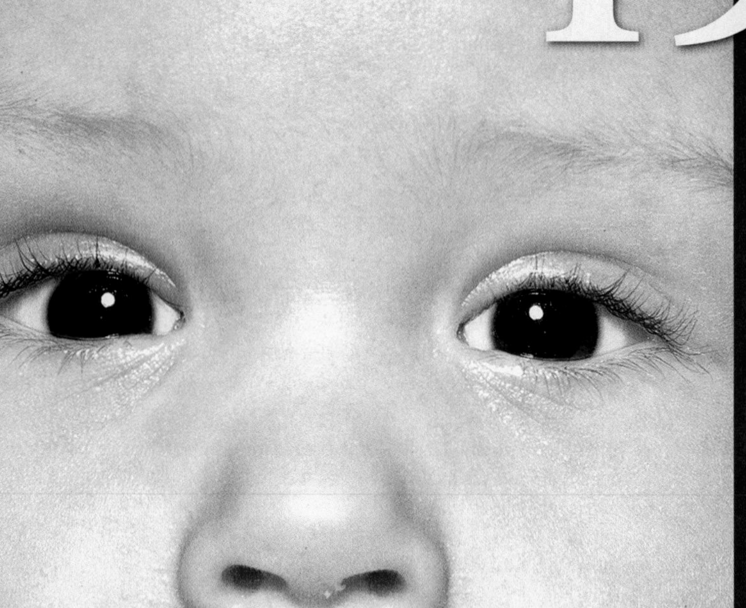

A child with aniridia. The *Pax6* gene plays an important role in eye development. A mutation in the *Pax6* causes aniridia, which results in an absence of the iris.

Take a close look at the child in the chapter-opening photo. Do you notice anything unusual? Though it may not be immediately apparent, this child has a disorder called aniridia, in which the iris in each eye does not develop properly. The iris is the part of the eye, usually blue, green, or brown, that regulates the amount of light entering the eye. In aniridia, the place where each iris should be located appears black, giving the appearance of very large pupils. People with aniridia cannot adjust the amount of light entering their eyes, which results in a decreased quality of vision and leads to eye diseases such as glaucoma and cataracts. In addition, other structures within the eye, such as the retina and optic nerve, may not develop correctly. What is the underlying cause of aniridia? It is due to a mutation in a gene called *Pax6*, which is responsible for the development of the eye. People with aniridia are heterozygotes, having one normal copy of the *Pax6* gene and one mutant copy; it is a dominant disorder. The proper development of the eye requires two normal copies of the *Pax6* gene. This disorder illustrates how genes play a key role in the development of our bodies.

In biology, the term **development** refers to a series of changes in the state of a cell, tissue, organ, or organism. Development is the underlying process that gives rise to the structures and functions of living organisms. The structure or form of an organism is called its **morphology**. As we have learned throughout this textbook, an important principle of biology is that structure (morphology) determines function.

How do developmental changes occur? Since the 1940s, the genetic makeup of an organism has emerged as the fundamental factor behind development. The science of **developmental genetics** is concerned with understanding how gene expression controls the process of development.

In this chapter, we will examine how the sequential actions of genes provide a program for the development of an organism from a fertilized egg to an adult. Utilizing a few experimental organisms such as the fruit fly, a nematode worm, the mouse, and the plant *Arabidopsis*, scientists are working to identify and characterize the genes required to run developmental programs and are exploring how proteins encoded by these genes control the course of development. In this chapter, we will begin with an overview that emphasizes the general principles of development. We will then examine specific examples of development in animals and plants, focusing on the role of genes in embryonic development. Chapters 39 and 52 also consider plant and animal development, respectively, with an emphasis on structure and function.

19.1 General Themes in Development

Learning Outcomes:

1. List several invertebrate, vertebrate, and plant model organisms and describe the reasons why biologists study them.
2. Outline the process of pattern formation in plants and animals.
3. Describe the four ways that cells respond to positional information.
4. Explain how morphogens and cell-to-cell contacts convey positional information.
5. Outline how development in animals occurs in four phases that are controlled by transcription factors.

Animals and plants begin to develop when a sperm and an egg unite to produce a **zygote**, a diploid cell that divides and develops into a multicellular **embryo**, and eventually into an adult organism. During the early stages of development, cells divide and begin to arrange themselves into ordered units. As this occurs, each cell also becomes **determined**, which means it is committed to become a particular cell type, such as a muscle or intestinal cell. The commitment to become a specific cell type occurs long before a cell differentiates. During the process of **cell differentiation**, a cell's morphology and function have changed, usually permanently, into a highly specialized cell type. In an adult, each cell type plays its own particular role. In animals, for example, muscle cells enable an organism to move, and intestinal cells facilitate the absorption of nutrients. This division of labor among various cells of an organism works collectively to promote its survival.

The genomes of living organisms contain a set of genes that constitute a program of development. In unicellular species, the program controls the structure and function of the cell. In multicellular species such as animals and plants, the program not only controls cellular features but also determines the arrangement of cells. In this section, we will examine some of the general concepts associated with the development of multicellular species.

Developmental Biologists Use Model Organisms to Study Development

The development of even a simple multicellular organism involves many types of changes in form and function. For this reason, the research community has focused its efforts on only a few **model organisms**—organisms studied by many different researchers so they can compare their results and determine scientific principles that apply more broadly to other species.

With regard to animal development, the two organisms that have been the most extensively investigated are two invertebrate species: the fruit fly (*Drosophila melanogaster*) and the nematode worm (*Caenorhabditis elegans*) (**Figure 19.1a,b**). Why have these two organisms been chosen as models to investigate development? *Drosophila* has been studied for a variety of reasons. First, researchers have exposed this organism to mutagens and identified many mutant organisms with altered developmental pathways. Second, in all of its life stages, *Drosophila* has distinct morphological features and is large enough to easily identify the effects of mutations. *C. elegans* is used by developmental geneticists because of its simplicity. The adult organism is a small transparent worm about 1 mm in length and composed of only about 1,000 somatic cells. Starting with a fertilized egg, the pattern of cell division and the fate of each cell within the embryo are completely known. This pattern is essentially identical from one worm to another, which allows researchers to predict the fate of cells in this organism.

Embryologists have also studied the morphological features of development in vertebrate species. Historically, amphibians and birds have been studied extensively, because their eggs are rather large and easy to manipulate. From a morphological point of view, the developmental stages of the chicken (*Gallus gallus*) and the African clawed frog (*Xenopus laevis*) have been described in great detail. More recently, a few vertebrate species have been the subject of genetic studies of development. These include the house mouse (*Mus musculus*) and the small aquarium zebrafish (*Danio rerio*) (**Figure 19.1c,d**).

In the study of plant development, the model organism for genetic analysis is a small flowering plant known as thale cress (*Arabidopsis thaliana*), which is typically called *Arabidopsis* by researchers (**Figure 19.1e**). *Arabidopsis* is an annual (a plant that lives out its entire life cycle during a single growing season) belonging to the wild mustard family. It occurs naturally throughout temperate regions of the world. *Arabidopsis* has a short generation time of about 6 weeks and a small genome size of 12×10^7 bp, which is similar in size to that of *Drosophila*

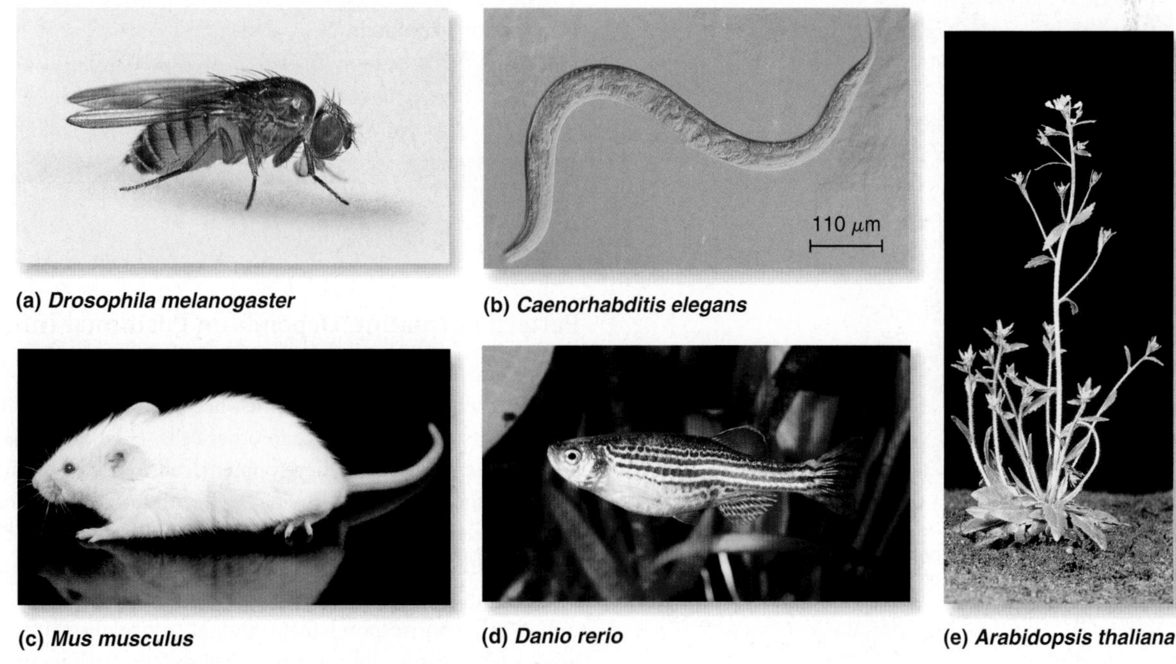

(a) ***Drosophila melanogaster*** **(b)** ***Caenorhabditis elegans*** **(c)** ***Mus musculus*** **(d)** ***Danio rerio*** **(e)** ***Arabidopsis thaliana***

Figure 19.1 **Model organisms used to study developmental genetics.**

BIOLOGY PRINCIPLE Biology is an experimental science. Biologists often focus their attention on model organisms with the expectation that the results will apply more broadly to many species.

and *C. elegans*. A flowering *Arabidopsis* plant is small enough to be grown in the laboratory and produces a large number of seeds.

Both Animals and Plants Develop by Pattern Formation

Development in animals and plants produces a body plan or pattern. At the cellular level, the body pattern is due to the arrangement of cells and their differentiation. The process, called **pattern formation**, gives rise to the formation of a body with a particular morphology. Pattern formation in animals is usually organized along three axes: the **dorsoventral axis**, the **anteroposterior axis**, and the **left-right axis** (Figure 19.2a). In addition, many animal bodies are then segmented into separate sections containing specific body parts such as wings or legs.

Pattern formation in plants is quite different, being organized along a **root-shoot axis** and in a **radial pattern**, in which the cells found in roots and shoots form concentric rings of tissues (Figure 19.2b). The root-shoot axis is determined at the first division of the fertilized egg. As we'll see later, the identification of mutant alleles that disrupt development has permitted great insight into the genes controlling pattern formation.

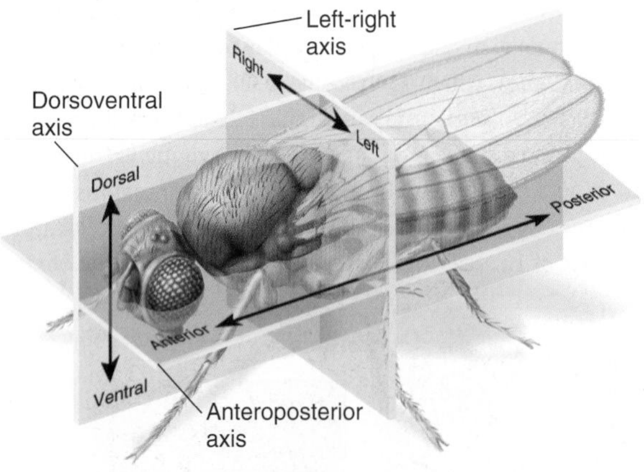

(a) Body plan found in many animals

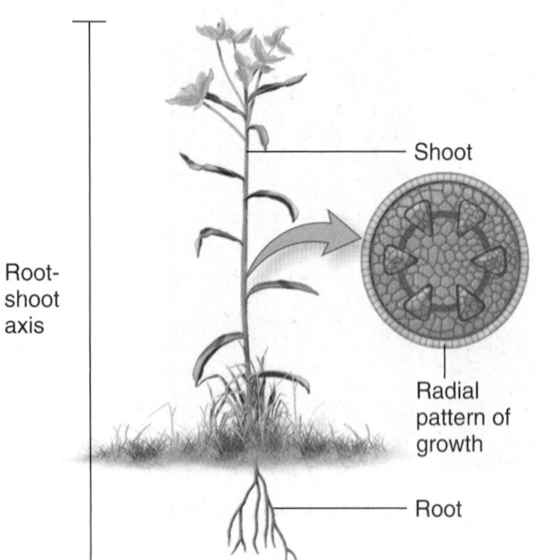

(b) Body plan found in many seed-bearing plants

Figure 19.2 Body plan axes in animals and plants.

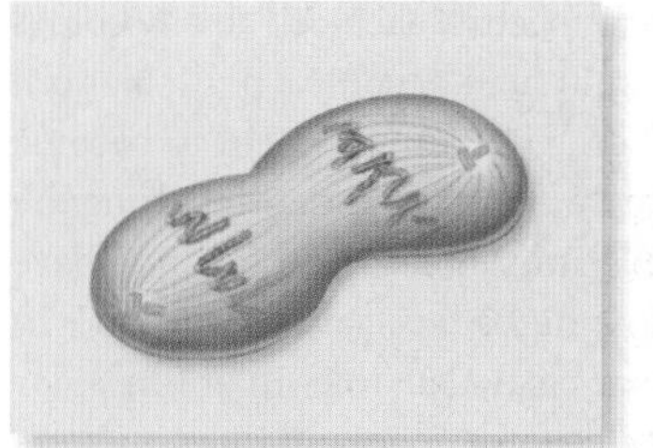

(a) Cell division

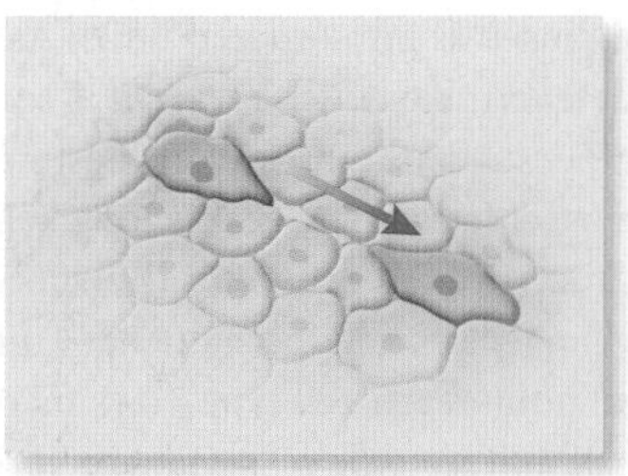

(b) Cell migration

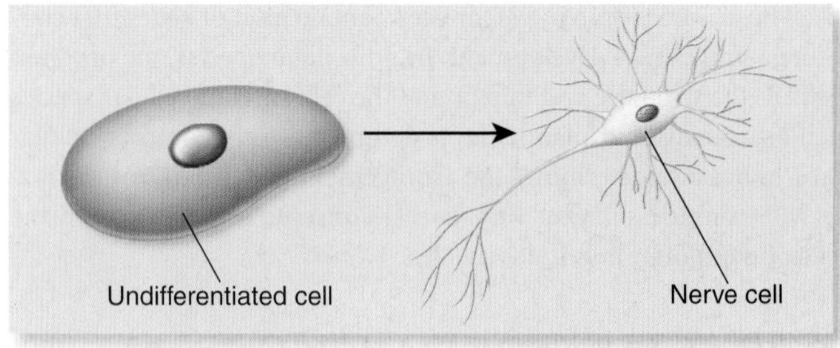

(c) Cell differentiation

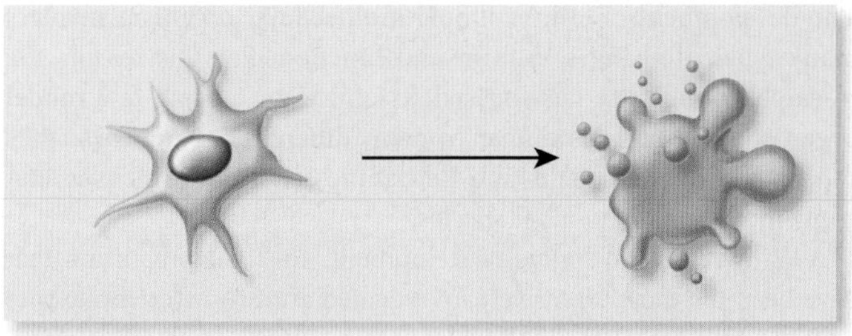

(d) Cell death (apoptosis)

Figure 19.3 Four types of cellular responses to positional information in animals.

Concept Check: *Which of these four responses do you expect to be more prevalent in the early stages of development, and which would become more prevalent in later stages?*

BioConnections: *Look back at Figure 9.18. What are the stages of apoptosis?*

Pattern Formation Depends on Positional Information

For an organism to develop the correct morphological features or pattern, each cell of the body must become the appropriate cell type based on its position relative to other cells. How does this occur? At appropriate times during development, cells receive **positional information**—molecules that provide a cell with information regarding its location relative to other cells of the body. Later in this chapter, we will examine how the expression of genes at the correct times provides this information.

A cell may respond to positional information in one of four ways: cell division, cell migration, cell differentiation, and cell death (Figure 19.3). First, positional information may stimulate a cell to divide. Second, positional information in animals may cause the migration of a cell or group of cells in a particular direction from one region of the embryo to another. (Cell migration does not occur during development in plants.) Third, it may cause a cell to differentiate into a specific cell type such as a neuron. Finally, positional information may promote **apoptosis**, or programmed cell death, which is described in Chapter 9. Apoptosis plays a key role in sculpting the bodies of animals. In vascular plants, certain cells undergo programmed cell death to form tracheids, specialized cells that function in water transport.

As an example of how the coordination of these four processes is required for pattern formation, Figure 19.4 shows the growth and development of a human arm during the embryonic stage. Cell division with accompanying cell growth increases the size of the limb. Cell migration is also important in this process. For example, embryonic cells that eventually form muscles in the arm and hand must migrate from outside the limb to reach their correct location within the limb. As development proceeds, cell differentiation produces the various tissues that will eventually be found in the fully developed limb. Some cells become neurons, others muscle cells, and still others become epidermal cells, forming the outer layer of skin. Finally, apoptosis is important in the formation of fingers. If apoptosis did not occur, a human hand would have webbed fingers.

Morphogens and Cell-to-Cell Contacts Convey Positional Information

How does positional information lead to the development of a body plan? Though the details of pattern formation vary widely among different species, two main molecular mechanisms are commonly used to communicate positional information. One of these mechanisms involves molecules called morphogens. **Morphogens** impart positional information and promote developmental changes at the cellular level. Many morphogens are proteins, but they can also be small signaling molecules. A morphogen influences the fate of a cell by promoting cell division, cell migration, cell differentiation, or apoptosis. A key feature of morphogens is they act in a concentration-dependent

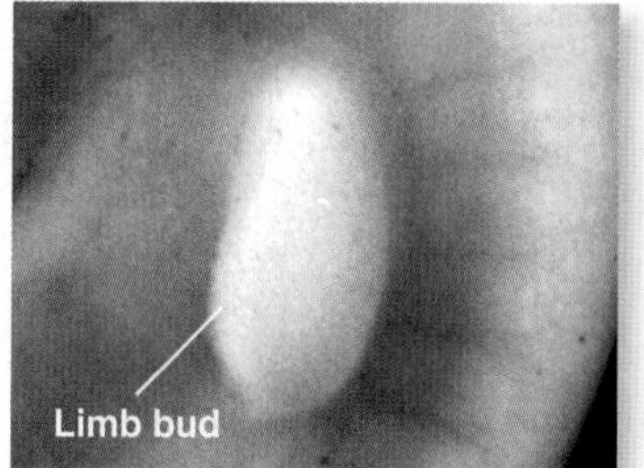

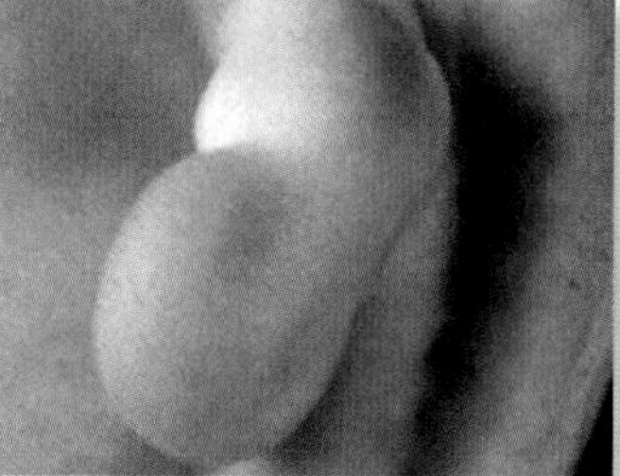
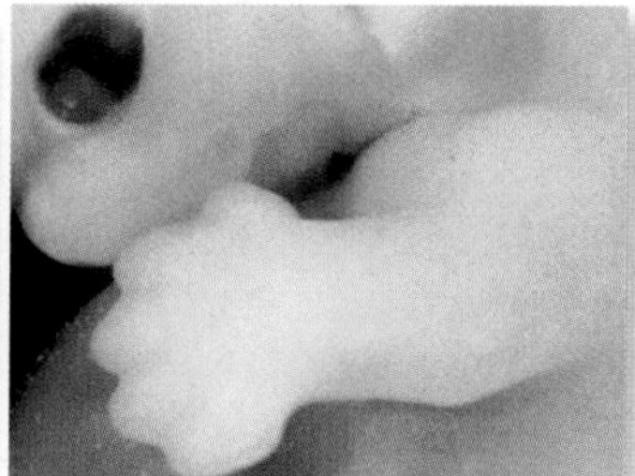
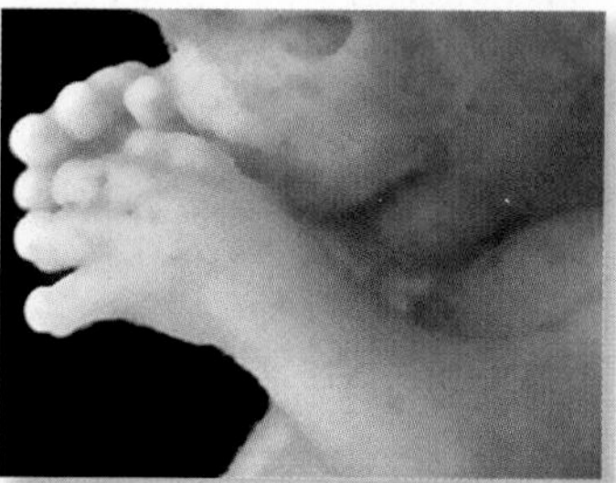

(a) Limb development in a human embryo

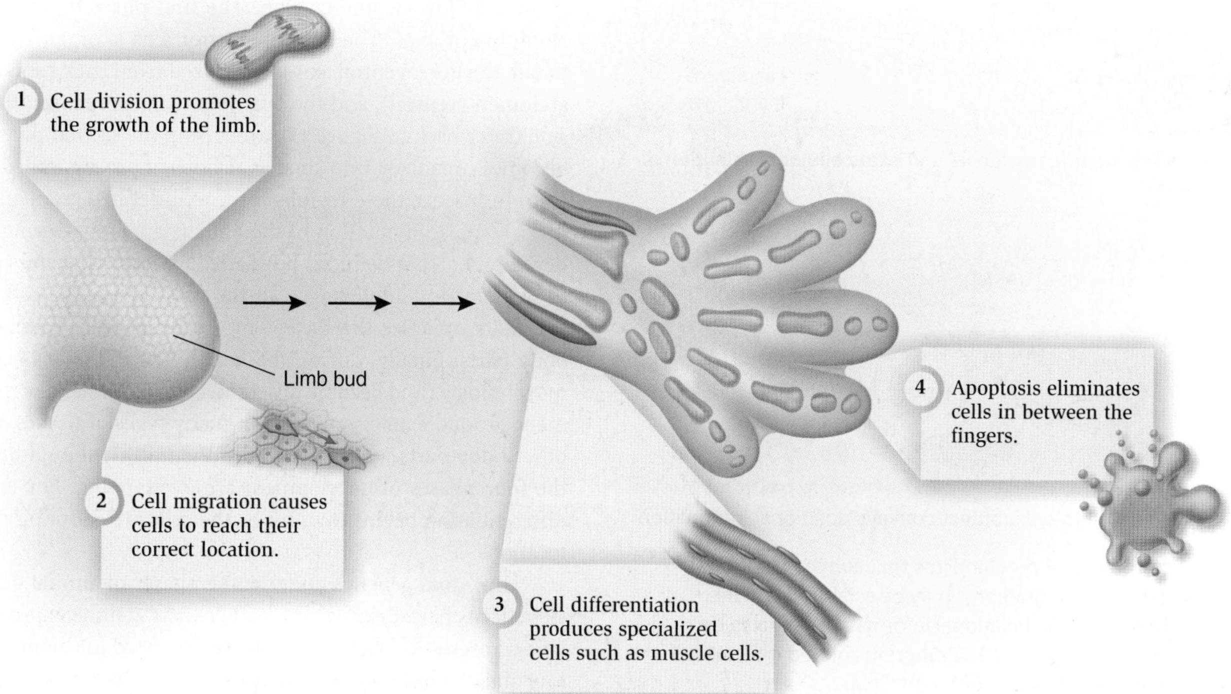

(b) Four cellular processes that promote limb formation

Figure 19.4 Limb development in humans. **(a)** Photographs of limb development in human embryos. The limb begins as a protrusion called a limb bud that eventually forms an arm and hand. **(b)** The development of a human hand from an embryonic limb bud.

Concept Check: *How would finger formation be affected if apoptosis did not occur?*

manner. At a high concentration, a morphogen restricts a cell into a particular developmental pathway, whereas at a lower concentration, it does not. There is often a critical **threshold concentration** above which the morphogen exerts its effects.

Morphogens typically are distributed asymmetrically along a concentration gradient. Morphogen gradients may be established in the **oocyte**, a cell that matures into an egg cell (Figure 19.5a). In addition, a morphogen gradient can be established in the embryo by secretion and diffusion (Figure 19.5b). A certain cell or group of cells may synthesize and secrete a morphogen at a specific stage of development. After secretion, the morphogen diffuses to neighboring cells, as in Figure 19.5b, or it may be transported to cells that are distant from the cells that secrete the morphogen. The morphogen may then influence the fate of cells exposed to it. The process by which a cell or group of cells governs the fate of other cells is known as **induction**.

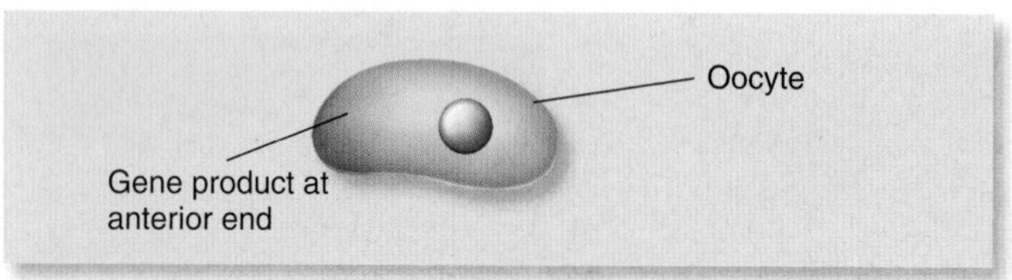

(a) Asymmetric distribution of morphogens in the oocyte

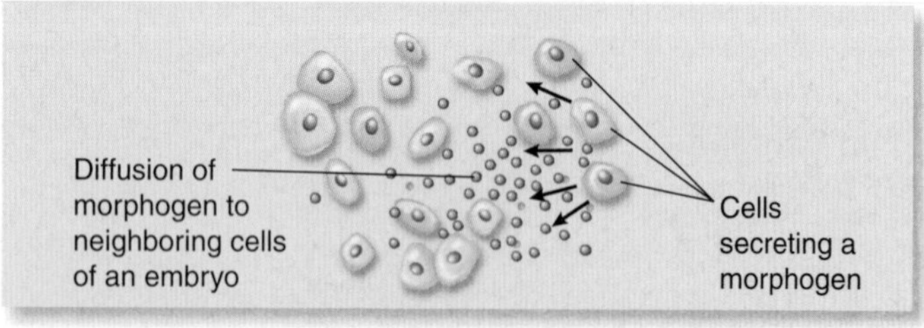

(b) Induction: Asymmetric synthesis and extracellular distribution of a morphogen

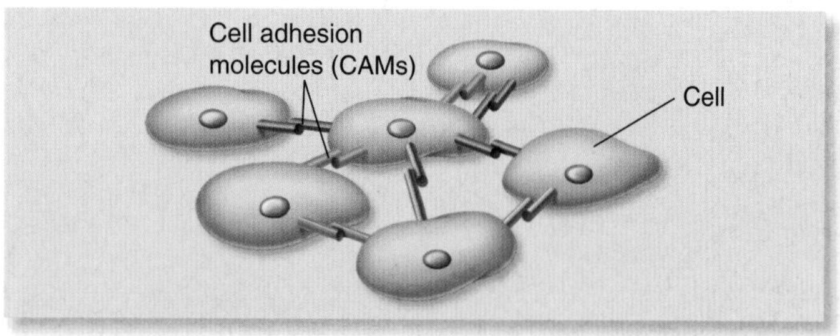

(c) Cell adhesion: Cell-to-cell contact conveys positional information

Figure 19.5 Molecular mechanisms that convey positional information. Morphogen gradients may be established in the **(a)** oocyte or **(b)** embryo. **(c)** Positional information may also be conveyed by cell-to-cell contact. The different colored molecules in the plasma membrane represent different CAMs.

Concept Check: Why is positional information important during development?

BioConnections: Look back at Chapter 9. Discuss the role of cell receptors in responding to positional information.

Another mechanism to convey positional information is **cell adhesion** (Figure 19.5c). Each animal cell makes its own collection of surface receptors that enable it to adhere to other cells and to the extracellular matrix (ECM). Such receptors, known as **cell adhesion molecules (CAMs)**, are described in Chapter 10 (refer back to Figure 10.8). The positioning of a cell within a multicellular organism is strongly influenced by the combination of contacts it makes with other cells and with the ECM.

The phenomenon of cell adhesion and its role in multicellular development was first recognized by American biologist Henry V. Wilson in 1907. He took multicellular sponges and passed them through a sieve, dissociating them into individual cells. Remarkably, the cells actively migrated until they adhered to one another to form a new sponge, complete with the chambers and canals that characterize a sponge's internal structure! When sponge cells from different species were mixed, they sorted themselves properly, adhering only to cells of the same species. Overall, these results indicate that cells possess specific CAMs, which are critical in cell-to-cell recognition. Cell adhesion plays an important role in governing the position that an animal cell will adopt during development.

Pattern Formation Occurs in Phases That Are Controlled by Transcription Factors

The formation of a body, in both animals and plants, occurs in a series of overlapping organizational phases. As an overview of this process, let's consider four general phases of pattern formation in an animal (Figure 19.6). This example involves human development, but research suggests that pattern formation in all complex animals follows a similar plan. The first phase organizes the body along major axes. The anteroposterior axis is organized from head to tail, the dorsoventral axis is organized from back (dorsal) to front/abdomen (ventral), and the left-right axis is organized from side to side (refer back to Figure 19.2a). During the second phase, the body becomes organized into smaller regions, a process called **segmentation**. In insects, these regions form well-defined segments. In mammals, some segmentation of the body is apparent during embryonic development, but defined boundaries are lost as the embryo proceeds to the fetal and adult stages. In the third phase, the cells within the segments organize themselves in ways that will produce particular body parts. Finally, during the fourth phase, the cells change their morphologies and become differentiated. This final phase of development produces an organism with many types of tissues, organs, and other body parts with specialized functions. It should be noted that the four phases of development are overlapping. For example, cell differentiation begins to occur as the cells are adopting their correct locations.

How does genetics underlie the phases of animal development? Geneticists have discovered a parallel between the expression of specific transcription factors, which are described in Chapter 13, and the four major phases of animal development. As noted in Figure 19.6, a hierarchy of transcription factors controls whether or not certain genes are expressed at a specific phase of development in a particular cell type, a phenomenon called **differential gene regulation**. Many morphogens, particularly those that act during early phases of development, function as transcription factors.

Hierarchy of transcription factors

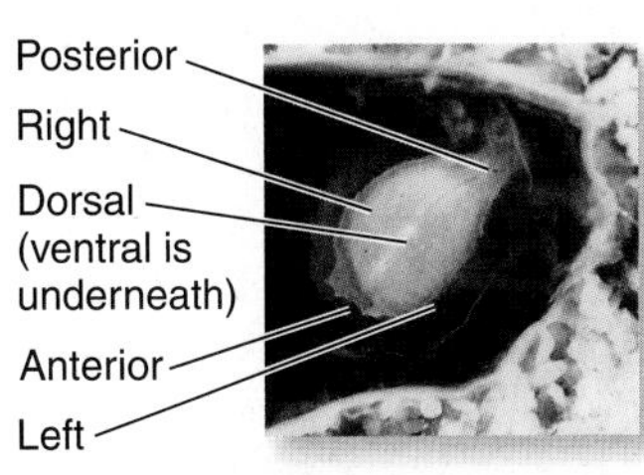

1 **Phase 1:** Transcription factors determine the formation of the body axes and control the expression of transcription factors of phase 2.

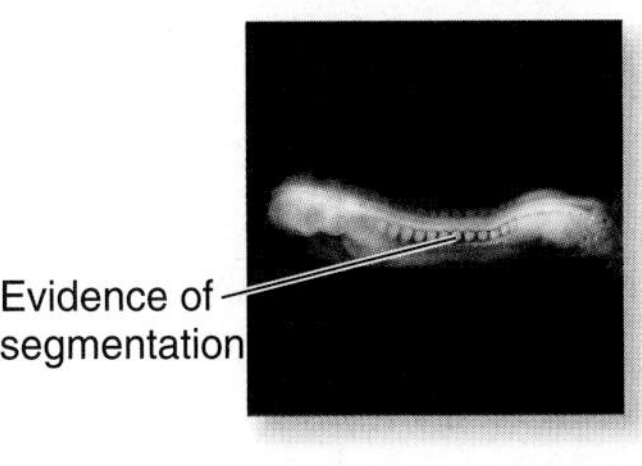

2 **Phase 2:** Transcription factors cause the embryo to become subdivided into regions that have properties of individual segments. They also control transcription factors of phase 3.

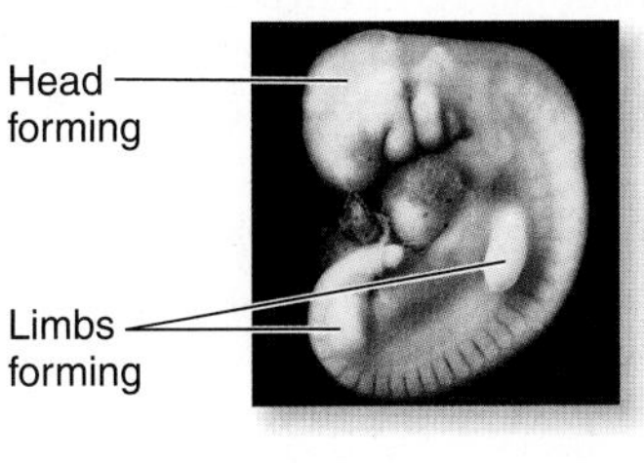

3 **Phase 3:** Transcription factors cause each segment and groups of segments to develop specific characteristics. They also control transcription factors of phase 4.

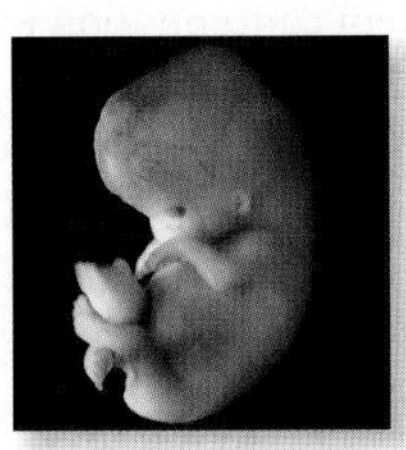

4 **Phase 4:** Transcription factors cause cells to differentiate into specific cell types such as skin, nerve, and muscle cells.

Figure 19.6 Pattern formation in a human embryo. As shown here, pattern formation in animals occurs in four phases controlled by a hierarchy of transcription factors. The ideas in this scenario are based largely on analogies between pattern formation in *Drosophila* and mammals. Many of the transcription factors that are likely to control the early phases of pattern formation in mammals have yet to be identified.

Concept Check: *During which of the four phases of development would you expect cell division and cell migration to be the most prevalent?*

19.2 Development in Animals

Learning Outcomes:

1. Describe how animal development occurs in four overlapping phases.
2. Explain how the analysis of mutants has been an important tool in our understanding of development in animals.
3. Distinguish the functions of maternal effect genes, segmentation genes, and homeotic genes in animal development.
4. Define the general properties of stem cells.
5. Describe how transcription factors control cell differentiation.

In this section, we will begin by examining the general stages of *Drosophila* development and then focus our attention on its embryonic stage. During this stage, the overall body plan is determined. We will see how the differential expression of particular genes within the embryo controls pattern formation. Although the roles of genes in the organization of mammalian embryos are not as well understood as they are in *Drosophila*, the analysis of the genomes of mammals and many other species has revealed many interesting parallels in the developmental program of all animals.

This section will end with an examination of cell differentiation. This process is better understood in mammals than in *Drosophila* because researchers have been studying mammalian cells in the laboratory for many decades. To explore cell differentiation, we will consider mammals as our primary example.

Embryonic Development Determines the Pattern of Structures in the Adult

As a way to appreciate the phases of pattern formation in animals, we will largely focus on development in *Drosophila*. However, as described in Chapter 52, animal development is quite varied among different species. Figure 19.7 illustrates a simplified sequence of events in *Drosophila* development. Let's examine these steps before we consider the differential gene regulation that causes them to happen.

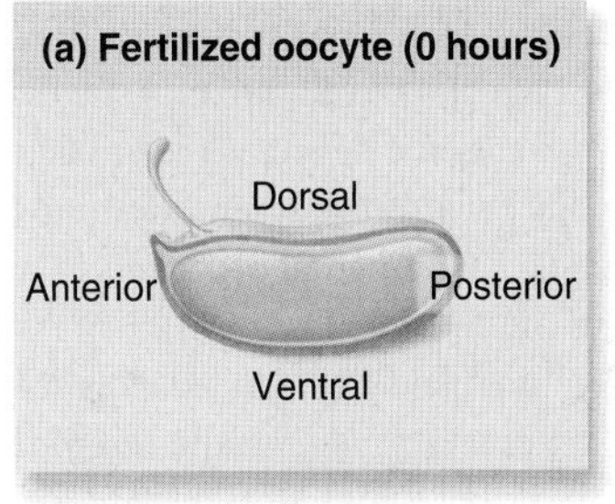

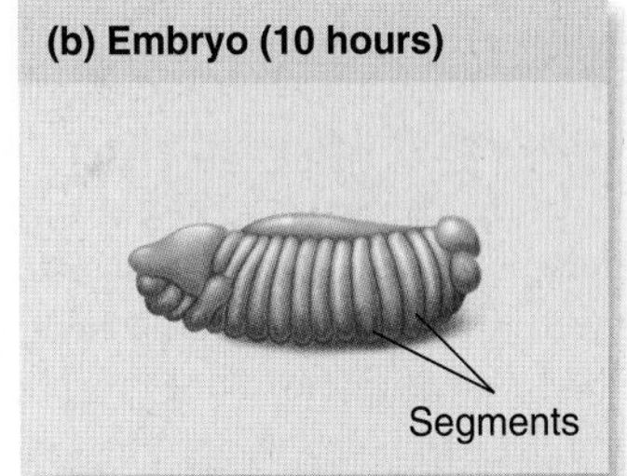

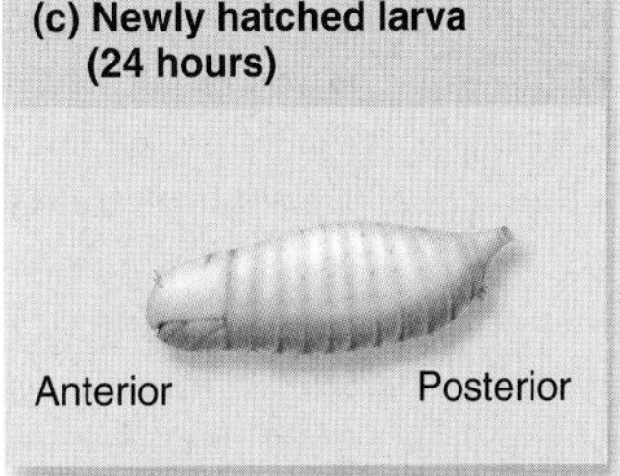

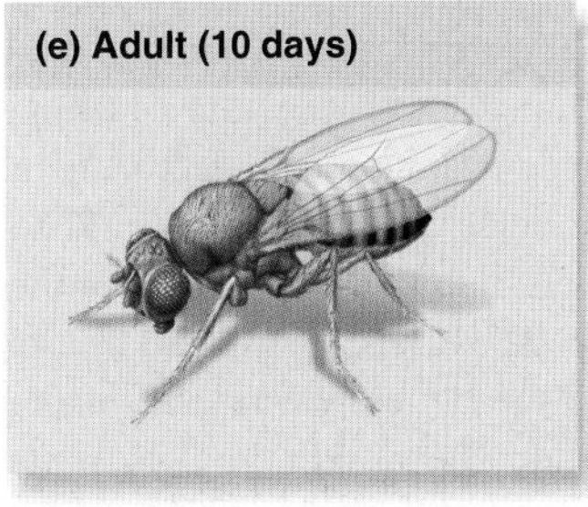

Figure 19.7 Developmental stages of the fruit fly *Drosophila*.

The oocyte is critical to establishing the pattern of development that will ultimately produce an adult organism. It is an elongated cell that contains positional information. As shown in Figure 19.7a, the fertilized egg already has anterior and posterior ends that correspond to those found in the adult (compare Figure 19.7a and e).

A key process in the embryonic development of *Drosophila* is the formation of a segmented body pattern. The embryo is subdivided into visible segments grouped into three general areas: the head, the thorax, and the abdomen. Figure 19.7b shows the segmented pattern of a *Drosophila* embryo about 10 hours after fertilization. A *Drosophila* embryo then develops into a **larva** (Figure 19.7c), a free-living organism that is morphologically very different from the adult. Many animal species do not have larval stages. *Drosophila* undergoes three successive larval stages. After the third larval stage, the organism becomes a **pupa** (Figure 19.7d), a transitional stage between the larva and the adult. Through a process known as **metamorphosis**, the larva transforms into a pupa and then a pupa becomes a mature adult that emerges from the pupal case (Figure 19.7e). Each segment in the adult has its own characteristic structures. For example, the wings are on a thoracic segment. From beginning to end, this process takes about 10 days.

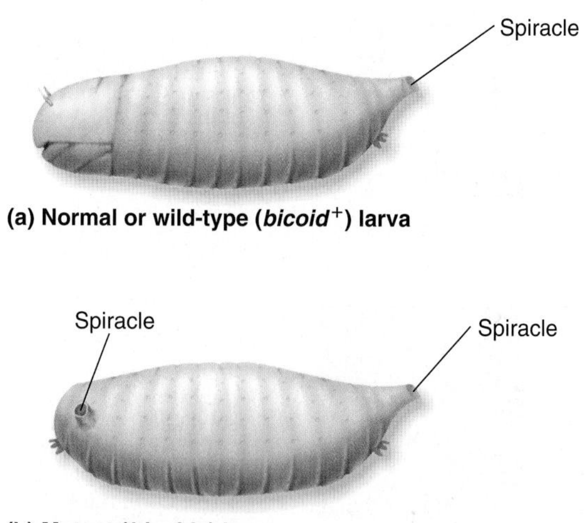

Figure 19.8 The bicoid mutation in *Drosophila*. **(a)** A normal larva from a *bicoid*$^+$ mother. **(b)** An abnormal larva from a *bicoid*$^-$ mother, in which both ends of the larva develop posterior structures. For example, both ends develop a spiracle, a small pore that normally is found only at the posterior end.

Concept Check: What would you expect to be the phenotype of a larva in which the bicoid gene product accumulated in both the anterior region and the posterior region of the oocyte?

Phase 1 Pattern Formation: Maternal Effect Genes Promote the Formation of the Body Axes

The first phase in *Drosophila* pattern formation is the establishment of the body axes, which occurs before the embryo becomes segmented. The morphogens necessary to establish these axes are distributed prior to fertilization. In most invertebrates and some vertebrates, certain morphogens, which are important in early developmental stages, are deposited asymmetrically within the egg as it develops (refer back to Figure 19.5a). Later, after the egg has been fertilized and development begins, these morphogens initiate developmental programs that govern the formation of the body axes of the embryo.

As an example of one morphogen that plays a role in axis formation, let's consider the product of a gene in *Drosophila* called *bicoid*. Its name is derived from the observation that a mutation that inactivates the gene results in a larva with two posterior ends (**Figure 19.8**). During normal oocyte development, the *bicoid* gene product accumulates in the anterior region of the oocyte. This gene product later acts as a morphogen to cause the development of the anterior end of the embryo.

How does the *bicoid* gene product accumulate in the anterior region of the oocyte? The answer involves specialized nurse cells that are found next to the oocyte, which matures in a follicle within the ovary of a female fly. Nurse cells supply oocytes with the products from **maternal effect genes**. They are so named because only the mother's gene product affects the phenotype of the resulting offspring. For example, the *bicoid* gene is a maternal effect gene in *Drosophila* that is transcribed in the nurse cells. The *bicoid* mRNA is then transported from the nurse cells into the anterior end of the oocyte and trapped there (**Figure 19.9a**). Prior to fertilization, the *bicoid* mRNA is highly concentrated near the anterior end of the oocyte (**Figure 19.9b**). After fertilization, the *bicoid* mRNA is translated, and a gradient of Bicoid protein is established across the zygote (**Figure 19.9c**). This gradient starts a progression of developmental events that will provide the positional information that causes the end of the zygote with a high Bicoid protein concentration to become the anterior region of the embryo.

The Bicoid protein is a morphogen that functions as a transcription factor to activate particular genes at specific times. The ability of Bicoid to activate a given gene depends on its concentration. Due to its asymmetric distribution, the Bicoid protein activates genes only in certain regions of the embryo. For example, a high concentration of Bicoid stimulates the expression of a gene called *hunchback* (that also encodes a transcription factor) in the anterior half of the embryo, but its concentration is too low in the posterior half to activate the *hunchback* gene. The ability of Bicoid to activate genes in certain regions but not others plays a role in the second phase of pattern formation—segmentation.

The Study of *Drosophila* Mutants Has Identified Genes That Control the Development of Segments

As described in Figure 19.7b, the second phase of pattern formation is the development of segments. The normal *Drosophila* embryo is subdivided into 15 segments: three head segments, three thoracic segments, and nine abdominal segments (**Figure 19.10**). Each segment of the embryo gives rise to unique morphological features in the adult. For example, the second thoracic segment (T2) produces a pair of legs and a pair of wings.

In the 1970s, German biologist Christiane Nüsslein-Volhard and American developmental biologist Eric Wieschaus undertook a systematic search for *Drosophila* mutants with disrupted development. They focused their search on **segmentation genes**, genes that alter the segmentation pattern of the *Drosophila* embryo and larva. Based on

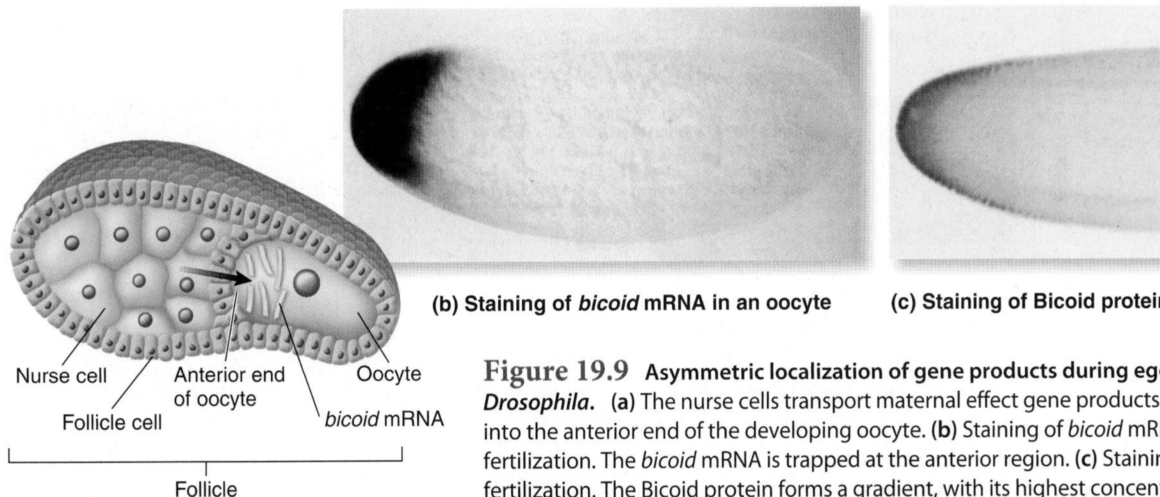

Figure 19.9 Asymmetric localization of gene products during egg development in *Drosophila*. **(a)** The nurse cells transport maternal effect gene products such as *bicoid* mRNA into the anterior end of the developing oocyte. **(b)** Staining of *bicoid* mRNA in an oocyte prior to fertilization. The *bicoid* mRNA is trapped at the anterior region. **(c)** Staining of Bicoid protein after fertilization. The Bicoid protein forms a gradient, with its highest concentration near the anterior end.

Concept Check: *What is the function of the Bicoid protein? After fertilization, in which part of the resulting zygote would its function be highest?*

the characteristics of abnormal larva, they identified three classes of segmentation genes: gap genes, pair-rule genes, and segment-polarity genes. When a mutation inactivates a **gap gene**, several adjacent segments are missing in the larva—a gap occurs. A defect in a **pair-rule gene** causes alternating segments or parts of segments to be absent. Finally, **segment-polarity gene** mutations cause portions of segments to be missing and cause adjacent regions to become mirror images of each other. The role of these segmentation genes during normal *Drosophila* development is described next.

Phase 2 Pattern Formation: Segmentation Genes Act Sequentially to Divide the *Drosophila* Embryo into Segments

The study of segmentation genes has revealed how segments are formed. To make a segment, particular genes act sequentially to govern the fate of a given region of the body. A simplified scheme of gene expression that leads to a segmented pattern in the *Drosophila* embryo is shown in Figure 19.11. Many more genes are actually involved in this process.

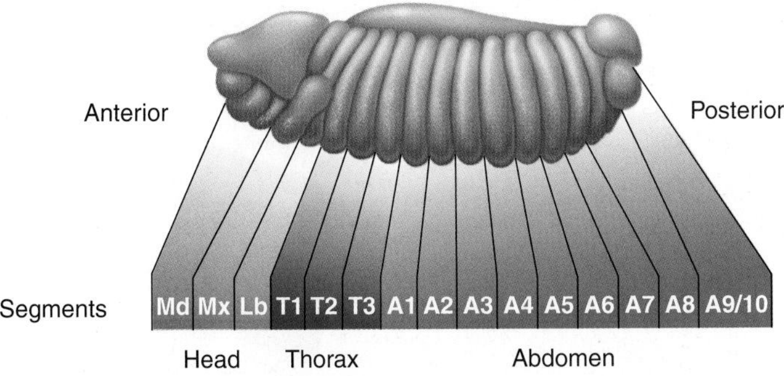

Figure 19.10 The organization of segments in the *Drosophila* embryo. The normal *Drosophila* embryo is subdivided into 15 segments, each of which gives rise to unique structures in the adult.

In general, the products of maternal effect genes such as *bicoid*, which promote the formation of body axes, activate gap genes. This activation is seen as broad bands of gap proteins in the embryo (Figure 19.11, step 2). Next, products from the gap genes and maternal effect genes function as transcription factors to activate the pair-rule genes in alternating stripes in the embryo (Figure 19.11, step 3). Once the pair-rule genes are activated, their gene products then regulate the segment-polarity genes. As you follow the progression from maternal effect genes to segment-polarity genes, notice that a body pattern is emerging in the embryo that matches the segmentation pattern found in the larva and adult animal. As you can see in step 4 of Figure 19.11, the 15 locations where a segment-polarity gene is expressed correspond to portions of segments in the adult fly. To appreciate this phenomenon, notice that the embryo at this stage is curled up and folded back on itself. If you imagine that the embryo was stretched out linearly, the 15 stripes seen in this embryo correspond to portions of the 15 segments of an adult fly.

Phase 3 Pattern Formation: Homeotic Genes Control the Development of Segment Characteristics

Thus far, we have considered how the *Drosophila* embryo becomes organized along axes and then into a segmented body pattern. During the third phase of pattern formation, each segment begins to develop its own unique characteristics (see Figure 19.6, phase 3). Geneticists use the term **fate** to describe the ultimate morphological features that a cell or group of cells adopts. For example, the fate of cells in segment T2 in *Drosophila* is to develop into a thoracic segment containing two legs and two wings. In *Drosophila*, the cells in each segment of the body have their fate determined at a very early stage of embryonic development, long before the morphological features become apparent.

Our understanding of developmental fate has been greatly aided by the identification of mutant genes that alter cell fates. In animals, the first mutant of this type was described by the German entomologist

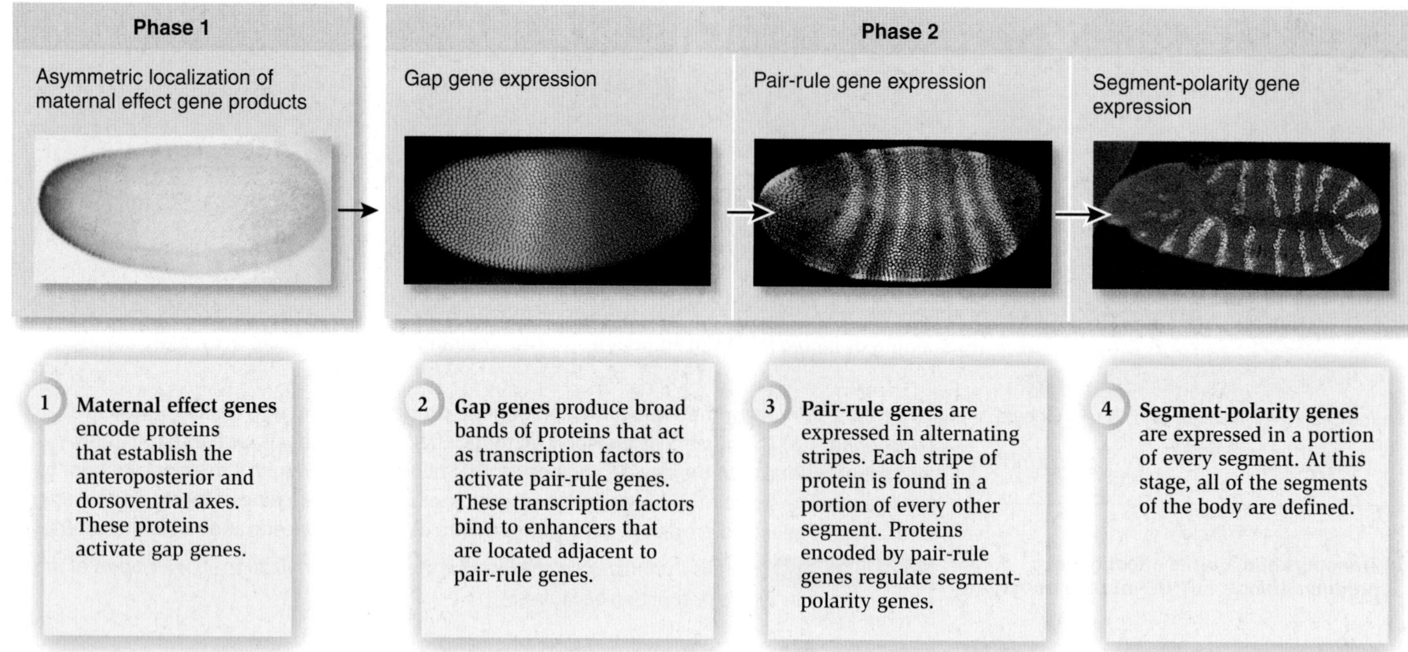

Phase 1
Asymmetric localization of maternal effect gene products
Phase 2
Gap gene expression
Pair-rule gene expression
Segment-polarity gene expression
1 Maternal effect genes encode proteins that establish the anteroposterior and dorsoventral axes. These proteins activate gap genes.
2 Gap genes produce broad bands of proteins that act as transcription factors to activate pair-rule genes. These transcription factors bind to enhancers that are located adjacent to pair-rule genes.
3 Pair-rule genes are expressed in alternating stripes. Each stripe of protein is found in a portion of every other segment. Proteins encoded by pair-rule genes regulate segment-polarity genes.
4 Segment-polarity genes are expressed in a portion of every segment. At this stage, all of the segments of the body are defined.

Figure 19.11 **Overview of segmentation in *Drosophila*.** The micrographs depict the progression of *Drosophila* development during the first few hours following fertilization. The micrographs also show the expression of protein products of a maternal effect gene (step 1) and segmentation genes (steps 2–4). In step 1, the protein is stained brown and is found in the left side of the early embryo, which is the anterior end. In step 2, one protein encoded by a gap gene is stained in green and another is stained in red. The yellow region is where the two different gap proteins overlap. In step 3, a protein encoded by a pair-rule gene is stained in light blue. In step 4, a protein encoded by a segment-polarity gene is stained pink. When comparing steps 3 and 4, note that the embryo has undergone a 180° turn, folding back on itself.

BIOLOGY PRINCIPLE New properties of life emerge from complex interactions. The feature of segmentation is an emergent property that has arisen due to complex interactions among segmentation gene products.

Concept Check: *How many pink stripes can you count in the embryo in step 4? How does this number compare with the number of segments in the embryo in Figure 19.10?*

Ernst G. Kraatz in 1876. He observed a sawfly (*Cimbex axillaris*) in which part of an antenna was replaced with a leg. During the late 19th century, the English zoologist William Bateson collected many of these types of observations and published them in a book. He coined the term homeotic to describe changes in which one body part is replaced by another. These abnormalities are caused by mutant alleles of **homeotic genes**—genes that specify the fate of a particular segment or region of the body.

As an example, Figure 19.12 shows a normal fly and one with mutations in a complex of homeotic genes called the *bithorax* complex. In a normal fly, two wings are found on the second thoracic segment, and two halteres, which together function as a balancing organ that resembles a pair of miniature wings, are found on the third thoracic segment. In this mutant fly, the third thoracic segment has the characteristics of the second, so the fly has no halteres and four wings. The term *bithorax* refers to the duplicated characteristics of the second thoracic segment. Edward Lewis, an American pioneer in the genetic study of development, became interested in the bithorax phenotype and began investigating it in 1946. He discovered that the mutant chromosomal region contains a complex of genes that play a role in the third phase of development.

(a) Normal fly with two wings

(b) Mutant fly with four wings

Figure 19.12 **The bithorax mutation in *Drosophila*.** **(a)** A normal fly has two wings on the second thoracic segment, and two halteres on the third thoracic segment. **(b)** This fly carries mutations in a complex of genes called the *bithorax* complex. In this fly, the third thoracic segment has the same characteristics as the second thoracic segment, thereby producing a fly with four wings instead of two.

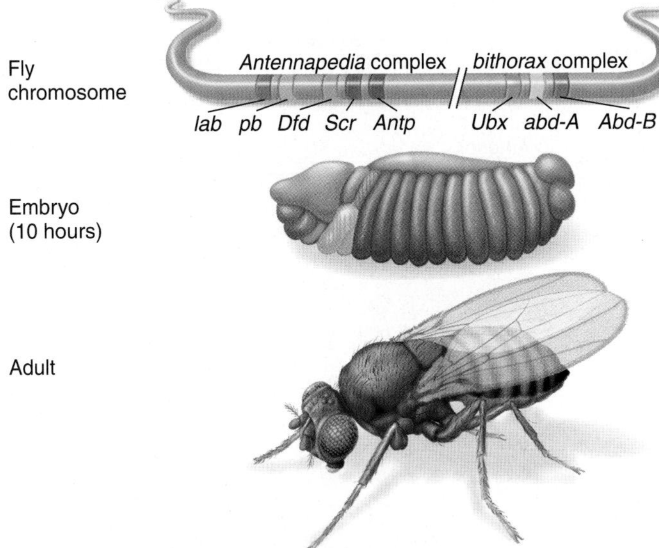

Figure 19.13 **Expression pattern of homeotic genes in *Drosophila*.** The order of homeotic genes, *labial* (*lab*), *proboscipedia* (*pb*), *deformed* (*Dfd*), *sex combs reduced* (*Scr*), *antennapedia* (*Antp*), *ultrabithorax* (*Ubx*), *abdominal A* (*abd-A*), and *abdominal B* (*Abd-B*), correlates with their spatial order of expression in the embryo. (Note: These genes were discovered and named by different researchers and the capitalization of the names is not consistent.)

Drosophila has eight homeotic genes that are found in two clusters called the *Antennapedia* complex and the *bithorax* complex (Figure 19.13). Both of these complexes are located on the same chromosome, but a long stretch of DNA separates them. As you can see in Figure 19.13, the order of homeotic genes along the chromosome correlates with their expression along the anteroposterior axis of the body. This phenomenon is called the **colinearity rule**. For example, *lab* (for labial) is expressed in the anterior segment and governs the formation of mouth structures. The *Antp* (for antennapedia) gene is expressed in the thoracic region during embryonic development and controls the formation of thoracic structures such as legs.

As we have seen in Figure 19.12, the role of homeotic genes in determining the identity of particular segments has been revealed by mutations that alter their function. As a second example, a mutation in the *Antp* gene has been identified in which the gene is incorrectly expressed in an anterior segment (Figure 19.14). A fly with this mutation has the bizarre trait in which it develops legs where antennae are normally found!

How do homeotic genes work at the molecular level? Homeotic genes encode homeotic proteins that function as transcription factors. The coding sequence of homeotic genes contains a 180-bp sequence known as a **homeobox** (Figure 19.15a). This sequence was first discovered in the *Antp* and *Ubx* genes, and it has since been found in many *Drosophila* homeotic genes. The homeobox is also found in other genes affecting pattern formation. The homeobox encodes a region of the protein called a **homeodomain**, which can bind to DNA (Figure 19.15b). The arrangement of α helices in the homeodomain promotes the binding of the protein to the DNA.

The primary function of homeotic proteins is to activate the transcription of specific genes that promote developmental changes in the animal. The homeodomain protein binds to DNA sequences called enhancers, which are described in Chapter 13. These enhancers are found in the vicinity of specific genes that control development.

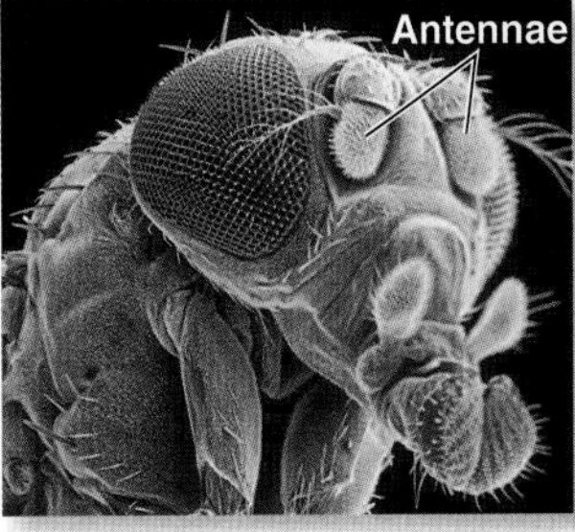

(a) Normal fly

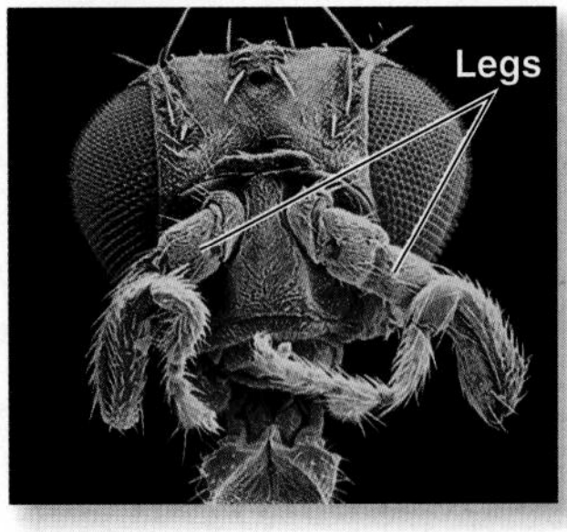

(b) Mutant fly

Figure 19.14 **The Antennapedia mutation in *Drosophila*.** **(a)** A normal fly with antennae. **(b)** This fly has a mutation in which the *Antp* gene is expressed in the embryonic segment that normally gives rise to antennae. The abnormal expression of *Antp* causes this region to have legs rather than antennae.

Concept Check: What phenotype would you expect if the Antp gene was expressed where abd-A is supposed to be expressed?

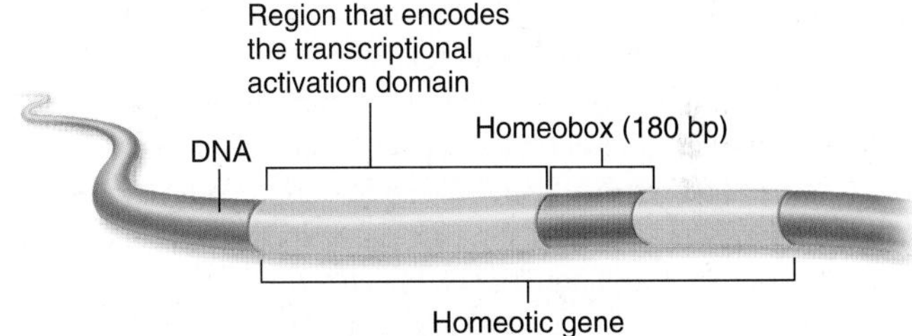

(a) Homeotic gene containing homeobox

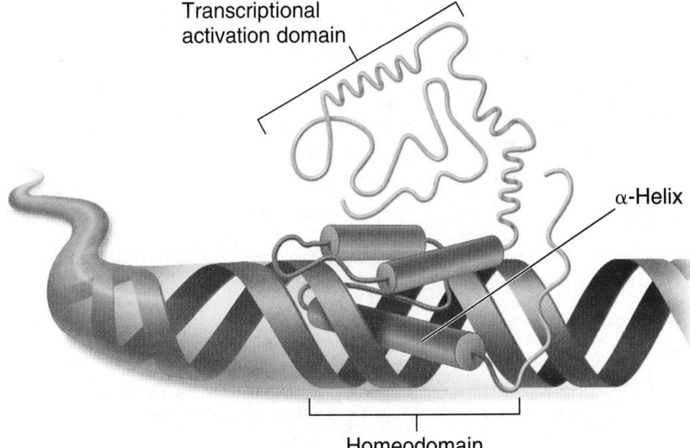

(b) Homeodomain binding to DNA

Figure 19.15 **Molecular features of homeotic genes and proteins.** **(a)** A homeotic gene (shown mostly in green) contains a 180-bp sequence called the homeobox (shown in blue). **(b)** Homeotic genes encode proteins that function as transcription factors. The homeobox encodes a region of the protein called a homeodomain, which binds to the DNA at a regulatory site such as an enhancer. The region of the protein called the transcriptional activation domain activates RNA polymerase to begin transcription.

BIOLOGY PRINCIPLE **Structure determines function.** The structure of a homeotic protein contains one domain (the homeodomain) that allows it to bind to an enhancer and another domain (the transcriptional activation domain) that activates RNA polymerase.

Most homeotic proteins also contain a transcriptional activation domain (see Figure 19.15b). After the homeodomain binds to an enhancer, the transcriptional activation domain of the homeotic protein activates RNA polymerase to begin transcription. Some homeotic proteins also function as repressors of certain genes.

GENOMES & PROTEOMES CONNECTION

A Homologous Group of Homeotic Genes Is Found in Nearly All Animals

Homologous genes are evolutionarily derived from the same ancestral gene and have similar DNA sequences. Researchers have found that homeotic genes in vertebrate species are homologous to genes that control development in simpler invertebrate species such as *Drosophila*. For example, in the mouse and other mammals, including humans, homeotic genes are organized into four clusters, designated *HoxA*, *HoxB*, *HoxC*, and *HoxD*, located on four different chromosomes. These homeotic genes are called ***Hox* genes**, an abbreviation for homeobox-containing genes. Thirty-eight genes are found in the four clusters, which represent 13 different gene types. As shown in **Figure 19.16**, several *Hox* genes in fruit flies and the mouse and other mammals are evolutionarily related. Among the first six types of *Hox* genes in the mouse, five of them are homologous to genes found in the *Antennapedia* complex of *Drosophila*. Among the last seven (genes numbered 7–13), three are homologous to the genes of the *bithorax* complex.

Like the *Antennapedia* and *bithorax* complexes in *Drosophila*, the arrangement of *Hox* genes along the mouse chromosome follows the colinearity rule, reflecting their pattern of expression from the anterior to the posterior end (**Figure 19.17**). Research has shown that the *Hox* genes play a role in determining the fates of regions along the anteroposterior axis. Nevertheless, additional research is necessary to understand the individual role that each of the *Hox* genes plays during embryonic development.

How widespread are *Hox* genes? Sponges, which are the simplest animals with no true tissues, do not have *Hox* genes, though they have copies of an evolutionarily related gene called an *NK-like* gene. Otherwise, all other types of animals have *Hox* genes, but in different numbers. The role of *Hox* genes in the evolution of animal body plans is discussed in Chapter 25 (look ahead to Figure 25.15).

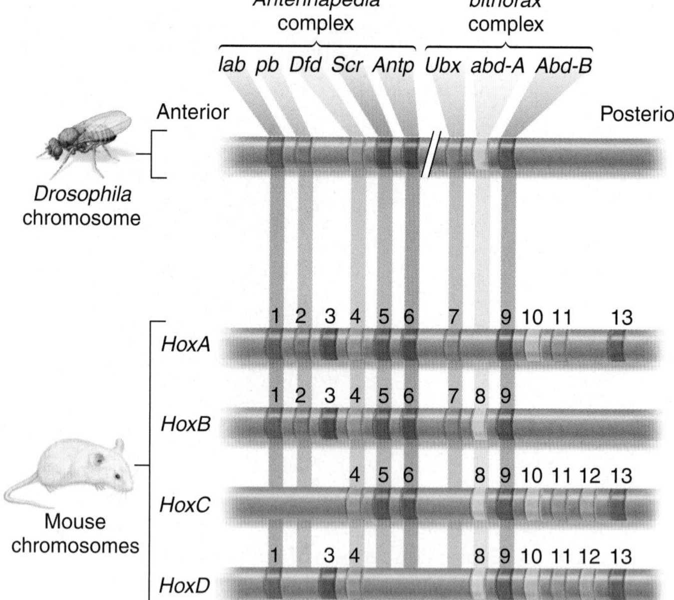

Figure 19.16 **A comparison of homeotic genes in *Drosophila* and the mouse.** The mouse and other mammals have four gene complexes, *HoxA–D*, that are homologous to certain homeotic genes found in *Drosophila*. Thirteen different types of homeotic genes are found in the mouse, although each *Hox* complex does not contain all 13 genes. In this drawing, homologous genes are aligned in columns. For example, *lab* is the homolog to *HoxA-1*, *HoxB-1*, and *HoxD-1*.

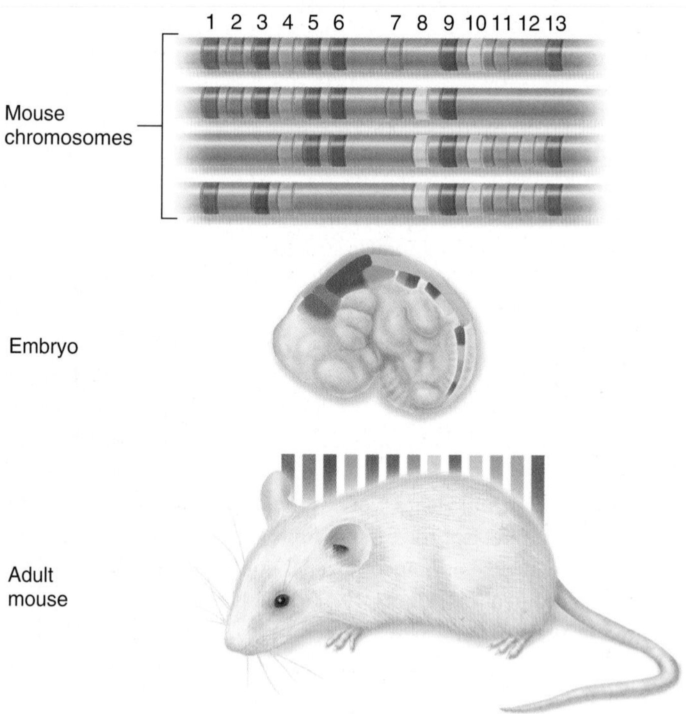

Figure 19.17 **Expression pattern of *Hox* genes in the mouse.** This schematic illustration shows *Hox* gene expression in the embryo and in the corresponding regions in the adult. The order of *Hox* gene expression, from anterior to posterior, parallels the order of genes along four different chromosomes.

BioConnections: *Look ahead to Figure 25.15. How is the number of Hox genes related to the complexity of animal bodies?*

Phase 4 Pattern Formation: Stem Cells Can Divide and Differentiate into Specialized Cell Types

Thus far we have focused our attention on patterns of gene expression that occur during the early stages of development. These genes control the basic body plan of the organism. During the fourth phase of pattern formation, the emphasis shifts to cell differentiation (see Figure 19.6, phase 4).

Although invertebrates have been instrumental in our understanding of pattern formation in animals, cell differentiation has been studied more extensively in mammals. One reason is because researchers have been able to grow mammalian cells in the laboratory for many decades. The availability of laboratory-grown cells makes it much easier to analyze the process of cell differentiation.

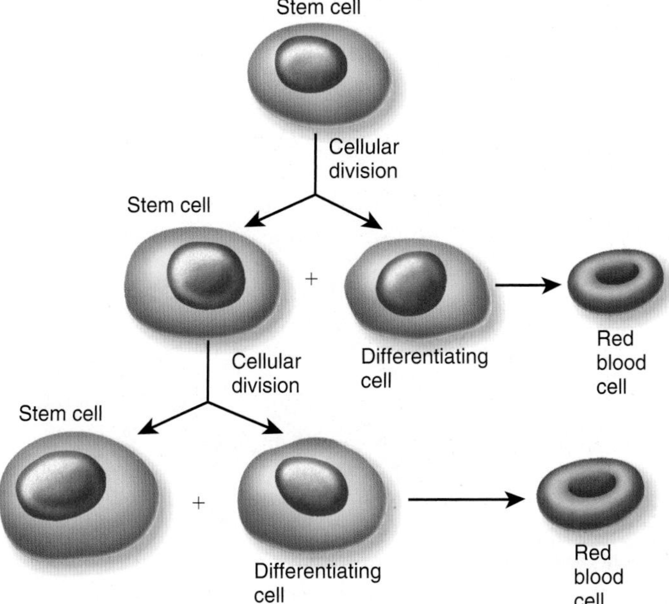

Figure 19.18 **Growth pattern of stem cells.** When a stem cell divides, one of the two daughter cells may remain a stem cell, while the other cell can differentiate into a specialized cell type, such as the red blood cells shown here.

Concept Check: What are the two common characteristics of stem cells?

By studying mammalian cells in the laboratory, geneticists have determined that the morphological differences between two different types of differentiated cells, such as muscle cells and neurons, arise through gene regulation. Though muscle cells and neurons contain the same set of genes, they regulate the expression of their genes in very different ways. Certain genes that are transcriptionally active in muscle cells are inactive in neurons, and vice versa. Therefore, muscle cells and neurons express different proteins that affect the characteristics of the respective cells in distinct ways. In this manner, differential gene regulation underlies cell differentiation.

General Properties of Stem Cells To understand the process of cell differentiation in a multicellular organism, we need to consider the special properties of **stem cells**, undifferentiated cells that divide and supply the cells that constitute the bodies of all animals and plants. Stem cells have two common characteristics. First, they have the capacity to divide, and second, their daughter cells can differentiate into one or more specialized cell types. The two daughter cells that are produced from the division of a stem cell can have different fates (Figure 19.18). One of the cells may remain an undifferentiated stem cell, and the other daughter cell can differentiate into a specialized cell type. With this asymmetric pattern of division and differentiation, stem cells continue dividing throughout life and generate a population of specialized cells. For example, in mammals, this mechanism is needed to replenish cells that have a finite life span, such as skin cells and red blood cells.

Stem Cells During Development In mammals, stem cells are commonly categorized according to their developmental stage and their ability to differentiate (Figure 19.19). The ultimate stem cell is the fertilized egg, which, via multiple cellular divisions, gives rise to an entire organism. A fertilized egg is said to be **totipotent** because it produces

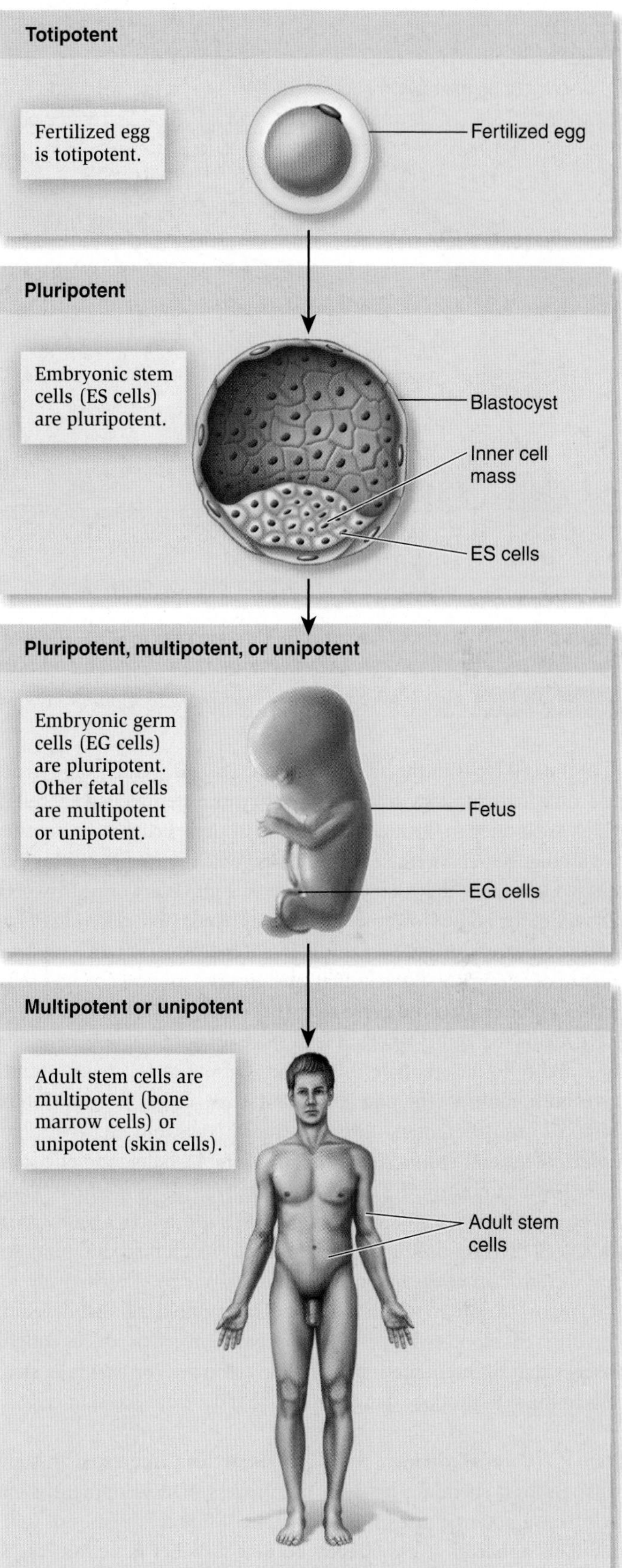

Figure 19.19 **Occurrence of stem cells at different stages of mammalian development.**

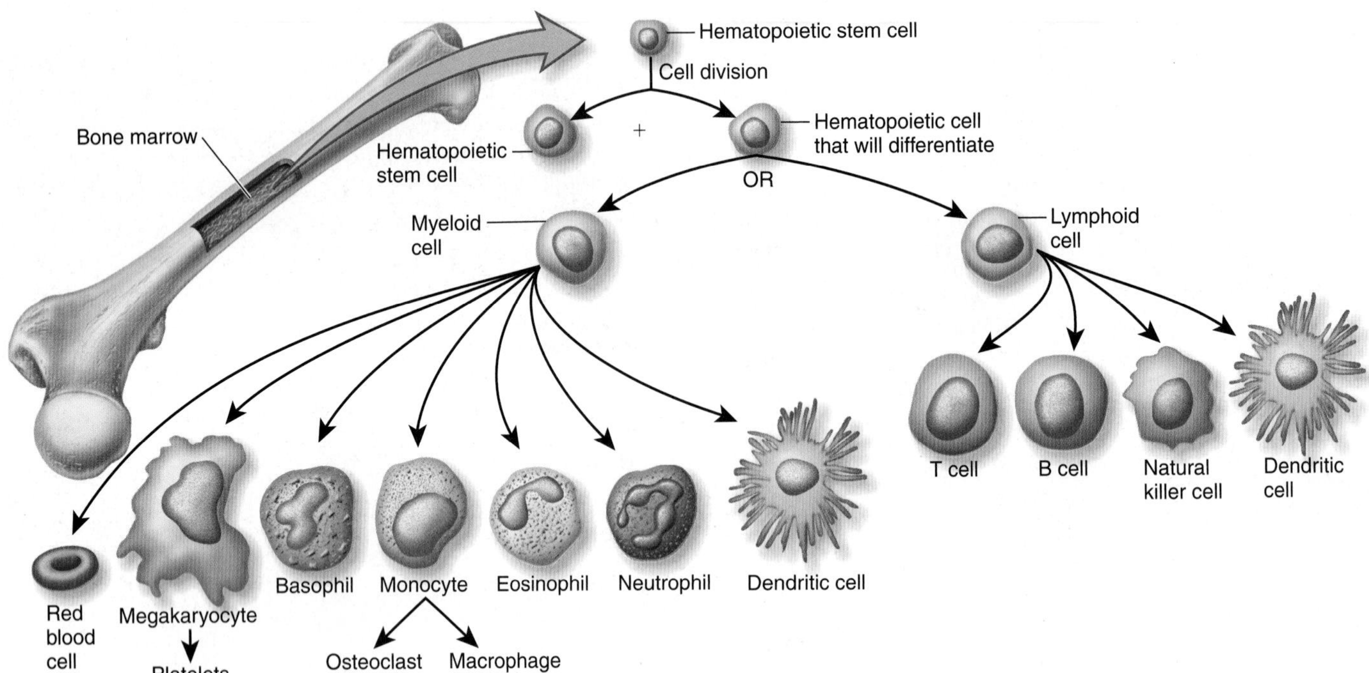

Figure 19.20 **Fates of hematopoietic stem cells (HSCs).** HSCs can follow two pathways in which cell division produces a myeloid cell and one that produces a lymphoid cell. Each develops into different blood cell types.

Concept Check: *Are hematopoietic stem cells totipotent, pluripotent, multipotent, or unipotent?*

all of the cell types in the adult organism. The early embryonic structure called the blastocyst contains **embryonic stem cells (ES cells)**, which are located in the inner cell mass. The inner cell mass is a cluster of cells that give rise to the embryo. Embryonic stem cells are **pluripotent**, which means they can also differentiate into every or nearly every cell type of the body. However, a single embryonic stem cell by itself has lost the ability to produce an entire, intact individual. At an early fetal stage of development, the cells that later give rise to sperm or eggs cells, known as the **embryonic germ cells (EG cells)**, also are pluripotent.

During the embryonic and fetal stages of development, cells lose their ability to differentiate into a wide variety of cell types. Adults have both multipotent and unipotent stem cells. A **multipotent** stem cell can differentiate into several cell types, but far fewer than a pluripotent embryonic stem cell. For example, hematopoietic stem cells (HSCs) found in the bone marrow give rise to multiple blood cell types (**Figure 19.20**). Multipotent HSCs can follow a pathway in which cell division produces a myeloid cell, which then differentiates into various cells of the blood and immune systems. Alternatively, an HSC follows a path in which it becomes a lymphoid cell that develops into different blood cell types. A **unipotent** stem cell produces daughter cells that differentiate into only one cell type. For example, stem cells in the skin produce daughter cells that develop into skin cells.

Stem Cells in Medicine Why are researchers interested in stem cells? Beyond shedding light on the process of development, stem cells have a potential use in the treatment of human diseases or injuries. This application has already become a reality in certain cases. For example, bone marrow transplants are used to treat patients with certain forms of cancer, such as leukemia. When bone marrow from a healthy person is injected into the body of a patient who has had her/his immune system wiped out via radiation, the stem cells within the transplanted marrow have the ability to proliferate and differentiate into various types of blood cells within the body of the patient.

Renewed interest in the use of stem cells in the potential treatment of many other diseases has been fostered by studies in 1998 in which researchers obtained ES cells from blastocysts and EG cells from aborted fetuses and successfully propagated them in the laboratory. Because ES and EG cells are pluripotent, they could potentially be used to treat a wide variety of diseases associated with cell and tissue damage (**Table 19.1**). Much progress has been made in testing the use of stem cells in animal models. However, more research is needed before the use of stem cells to treat such diseases in humans is realized.

From an ethical perspective, the primary issue that raises debate is the source of stem cells for research and potential treatments. Most ES cells have been derived from human embryos that were produced from in vitro fertilization, a method of assisted conception in which fertilization occurs outside of the mother's body and a limited number

Table 19.1 Some Potential Uses of Stem Cells to Treat Diseases

Cell/tissue type	Disease treatment
Nerve	Implantation of cells into the brain to treat Parkinson disease; treatment of spinal cord injuries
Skin	Treatment of burns and skin disorders
Cardiac	Repair of heart damage associated with heart attacks
Cartilage	Repair of joints damaged by injury or arthritis
Bone	Repair or replacement of damaged bone
Liver	Repair or replacement of liver tissue damaged by injury or disease
Skeletal muscle	Repair or replacement of damaged muscle

of the resulting embryos are transferred to the uterus. Most EG cells are obtained from aborted fetuses, either those that spontaneously aborted or those in which the decision to abort was not related to donating the fetal tissue to research. Some feel that it is morally wrong to use such tissue in research and/or the treatment of disease. Furthermore, some people fear this technology could lead to intentional abortions for the sole purpose of obtaining fetal tissues for transplantation. Others feel the embryos and fetuses that have been the sources of ES and EG cells were not going to become living individuals, and therefore it is beneficial to study these cells and to use them in a positive way to treat human diseases and injury. It is not clear whether these two opposing viewpoints can reach a common ground.

If stem cells could be obtained from adult cells and induced to become as pluripotent cells in the laboratory, an ethical dilemma may be avoided, because most people do not have serious moral objections to current procedures that use adult cells such as bone marrow transplantation. In 2006, work by Japanese physician Shinya Yamanaka and colleagues showed that adult mouse fibroblasts (a type of connective tissue cell) could become pluripotent by the introduction of four different genes that encode transcription factors. In 2007, Yamanaka's laboratory and two other research groups were able to show that such induced pluripotent stem cells can differentiate into all cell types when injected into mouse blastocysts and grown into baby mice. These results indicate that adult cells can be reprogrammed to become embryonic stem cells.

FEATURE INVESTIGATION

Davis, Weintraub, and Lassar Identified Genes That Promote Muscle Cell Differentiation

A key question regarding the study of stem cells is, "What causes a stem cell to differentiate into a particular cell type?" Researchers have discovered that certain proteins function as master transcription factors that cause cells to differentiate into specific types of cells. The investigation described here was one of the first studies to reveal this phenomenon.

In 1987, Robert Davis, Harold Weintraub, and Andrew Lassar conducted a study to identify genes that promote skeletal muscle cell differentiation. The initial strategy for their experiments was to identify genes that are expressed only in differentiating skeletal muscle cells, not in nonmuscle cells. Though methods of gene cloning are described in Chapter 20, we will briefly consider these scientists' cloning methods in order to understand their approach. The researchers began with two different laboratory cell lines that could differentiate into muscle cells. From these two cell lines, they cloned and identified about 10,000 different genes that were transcribed into mRNA. Next, they compared the expressed genes in these two muscle cell lines with genes that were expressed in a nonmuscle cell line. Their comparison revealed 26 genes that were expressed only in the two muscle cell lines but not in the nonmuscle cell line. To narrow their search further, they compared these 26 genes with other nonmuscle cell lines they had available. Among the 26, only 3 of them, which they termed *MyoA*, *MyoD*, and *MyoH*, were expressed exclusively in the two muscle cell lines.

In the experiment shown in **Figure 19.21**, the scientists' goal was to determine if any of these three genes could cause nonmuscle cells to differentiate into muscle cells. Using techniques described in Chapter 20, the coding sequence of each gene was placed next to an active promoter that caused a high level of transcription, and then the genes were introduced into fibroblasts, which are a type of cell that normally differentiates into osteoblasts (bone cells), chondrocytes (cartilage cells), adipocytes (fat cells), and smooth muscle cells, but never differentiates into skeletal muscle cells in vivo. The cells were plated on growth media and allowed to grow for 3 to 5 days. When the cloned *MyoD* gene was introduced into fibroblast cells in a laboratory, the fibroblasts differentiated into skeletal muscle cells. These cells contained large amounts of myosin, which is a protein expressed in muscle cells. The other two cloned genes (*MyoA* and *MyoH*) did not cause muscle cell differentiation or promote myosin production.

Figure 19.21 Davis, Weintraub, and Lassar and the promotion of skeletal muscle cell differentiation in fibroblasts by the expression of *MyoD*.

HYPOTHESIS Muscle differentiation is induced by particular genes.

KEY MATERIALS Three cloned genes had been identified that were expressed only in differentiating muscle cell lines. The researchers also had fibroblast cell lines, which do not normally differentiate into muscle cells.

	Experimental level	Conceptual level
1 In 3 separate tubes, add each of the 3 cloned genes, designated *MyoA*, *MyoD*, and *MyoH*.	*MyoA* *MyoD* *MyoH*	*MyoA* *MyoD* *MyoH* DNA

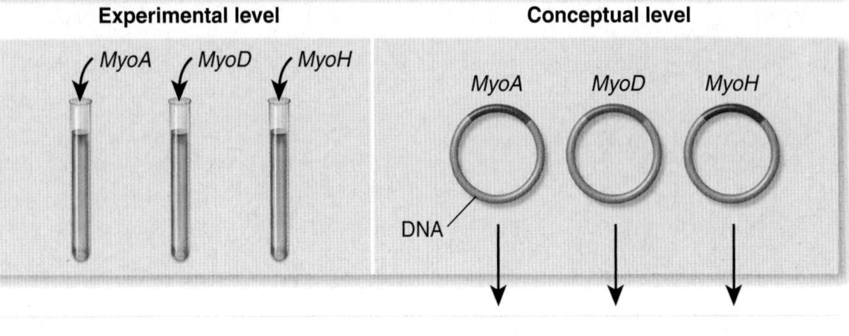

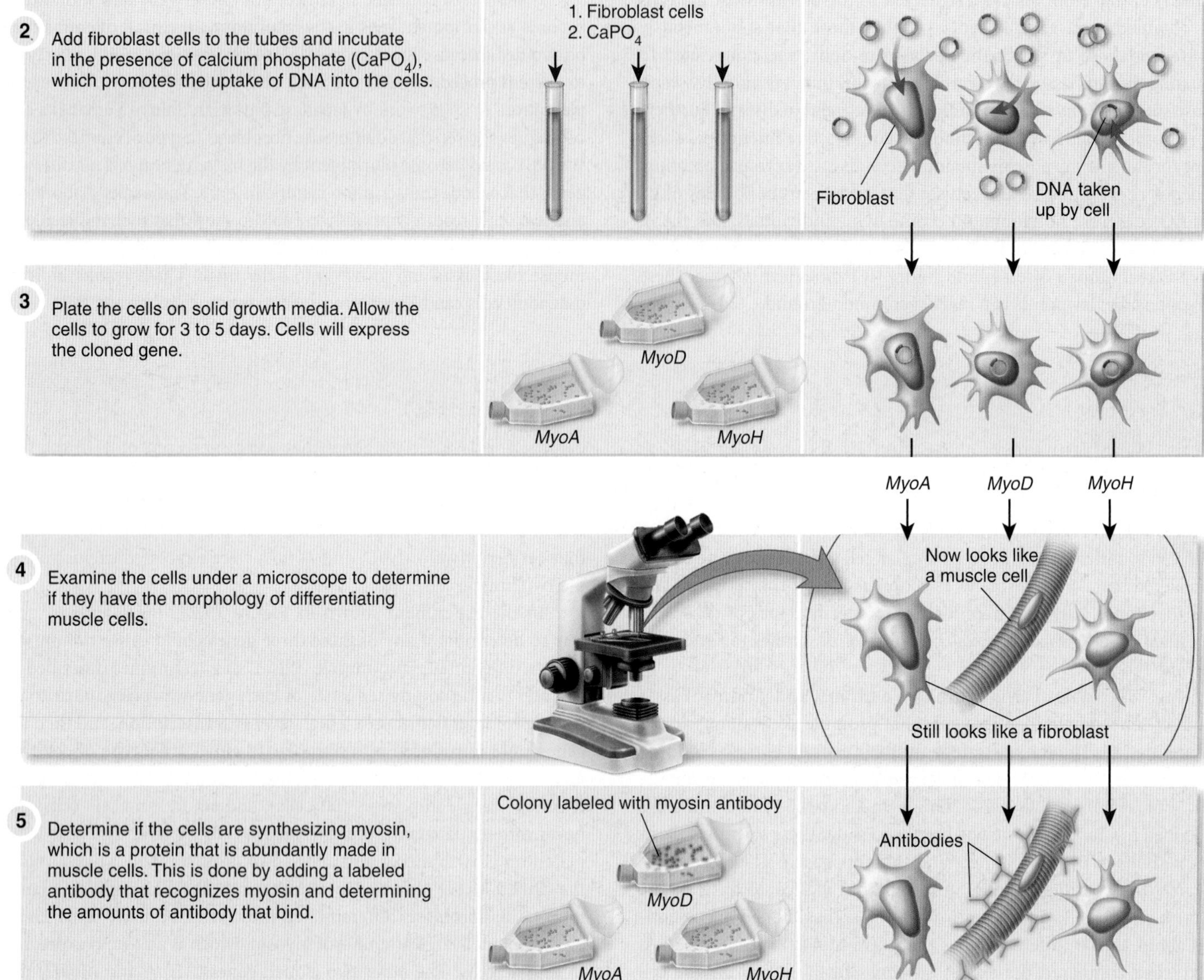

2 Add fibroblast cells to the tubes and incubate in the presence of calcium phosphate ($CaPO_4$), which promotes the uptake of DNA into the cells.
1. Fibroblast cells
2. $CaPO_4$
Fibroblast
DNA taken up by cell
3 Plate the cells on solid growth media. Allow the cells to grow for 3 to 5 days. Cells will express the cloned gene.
MyoD
MyoA
MyoH
MyoA
MyoD
MyoH
4 Examine the cells under a microscope to determine if they have the morphology of differentiating muscle cells.
Now looks like a muscle cell
Still looks like a fibroblast
5 Determine if the cells are synthesizing myosin, which is a protein that is abundantly made in muscle cells. This is done by adding a labeled antibody that recognizes myosin and determining the amounts of antibody that bind.
Colony labeled with myosin antibody
MyoD
MyoA
MyoH
Antibodies

6 THE DATA

Results from step 4:

DNA added	Microscopic morphology of cells
MyoA	Fibroblasts
MyoD	Muscle cells
MyoH	Fibroblasts

Results from step 5:

DNA added	Colonies labeled with antibody that binds to myosin?
MyoA	No
MyoD	Yes
MyoH	No

7 **CONCLUSION** The *MyoD* gene encodes a protein that causes cells to differentiate into skeletal muscle cells.

8 **SOURCE** Davis, Robert L., Weintraub, Harold, and Lassar, Andrew B. 1987. Expression of a single transfected cDNA converts fibroblasts to myoblasts. *Cell* 51:987–1000.

Since this initial discovery, researchers have found that *MyoD* belongs to a small group of genes that initiate muscle cell development. These myogenic genes encode transcription factors. They are found in all vertebrates and have been identified in several invertebrates, such as *Drosophila* and *C. elegans*. In all cases, myogenic genes are activated during skeletal muscle cell development.

Experimental Questions

1. What was the goal of the research conducted by Davis, Weintraub, and Lassar?
2. How did Davis, Weintraub, and Lassar's research identify the candidate genes for muscle differentiation?
3. Once the researchers identified the candidate genes for muscle differentiation, how did they test the effect of each gene on cell differentiation? What were the results of the study?

19.3 Development in Plants

Learning Outcomes:

1. Outline the stages of pattern formation in plants.
2. Explain the ABC model for flower development.

Because all eukaryotic organisms share an evolutionary history, animals and plants have many common features, including the types of events that occur during development. However, the general morphology of plants is quite different from animals. Plant morphology exhibits two key features (see Figure 19.2b). The first is the root-shoot axis. Most plant growth occurs via cell division near the tips of the shoots and the bottoms of the roots. Second, this growth occurs in a well-defined radial pattern, which means that growth in the stems and roots occurs in concentric rings of tissues (**Figure 19.22**).

At the cellular level too, plant development shows some differences from animal development. For example, cell migration does not occur during plant development. In addition, the development of a plant does not rely on morphogens that are deposited in the oocyte, as in many animals. In plants, an entirely new individual can be regenerated from many types of somatic cells—cells that do not give rise to gametes. Such somatic cells of plants are totipotent.

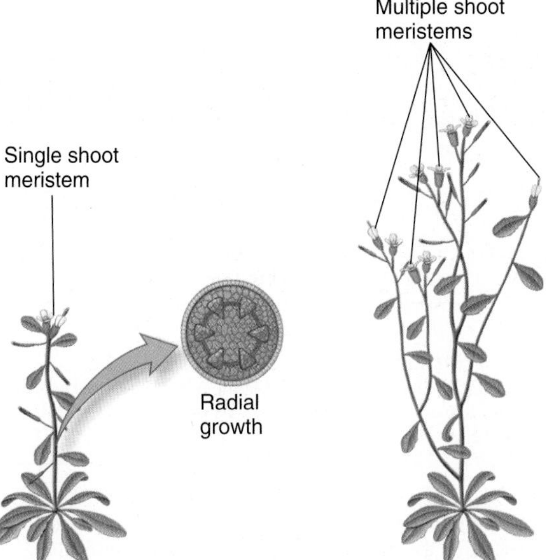

Figure 19.22 Pattern of shoot growth in plants. Early in development, as shown here in *Arabidopsis*, a single shoot promotes the formation of early leaves on the plant. Later, buds will form from this main shoot that will later form branches.

In spite of these apparent differences, the underlying molecular mechanisms of pattern formation in plants still share striking similarities with those in animals. Like animals, plants use the mechanism of differential gene regulation to coordinate the development of a body plan. Like their animal counterparts, a plant's developmental program relies on the use of transcription factors, determining when and how much gene products are made. In this section, we will consider pattern formation in plants and examine how transcription factors play a key role in plant development.

Plant Development Occurs from Meristems That Are Formed in the Embryo

How does pattern formation occur in plants? **Figure 19.23** illustrates a common order of events that takes place in the embryonic development of flowering plants such as *Arabidopsis*. After fertilization, the first cellular division is asymmetrical and produces a smaller apical cell and a larger basal cell (Figure 19.23a). In 2009, Danish geneticist Martin Bayer and colleagues conducted experiments indicating that the sperm carries mRNA molecules that are critical for this asymmetric cell division. The apical cell gives rise to most of the embryo and later develops into the shoot of the plant. In *Arabidopsis*, the basal cell gives rise to the root, along with a structure called the suspensor, which channels nutrients from the parent plant to the young embryo (Figure 19.23b).

At the heart stage, which is composed of only about 100 cells, the basic organization of the plant has been established (Figure 19.23c). Plants have organized groups of actively dividing stem cells called **meristems**. As discussed earlier, stem cells retain the ability both to divide and to differentiate into multiple cell types. The meristem produces offshoots of proliferating and differentiating cells. The **root apical meristem** gives rise only to the root, whereas the **shoot apical meristem** produces all aerial parts of the plant, which include the stem as well as lateral structures such as leaves and flowers.

The heart stage then progresses to the formation of a seedling that has two cotyledons, which are embryonic leaves that store nutrients for the developing embryo and seedling. In the seedling shown in Figure 19.23d, you can see three main regions. The **apical region** produces the leaves and flowers of the plant. The **central region** creates the stem. Finally, the **basal region** produces the roots. Each of

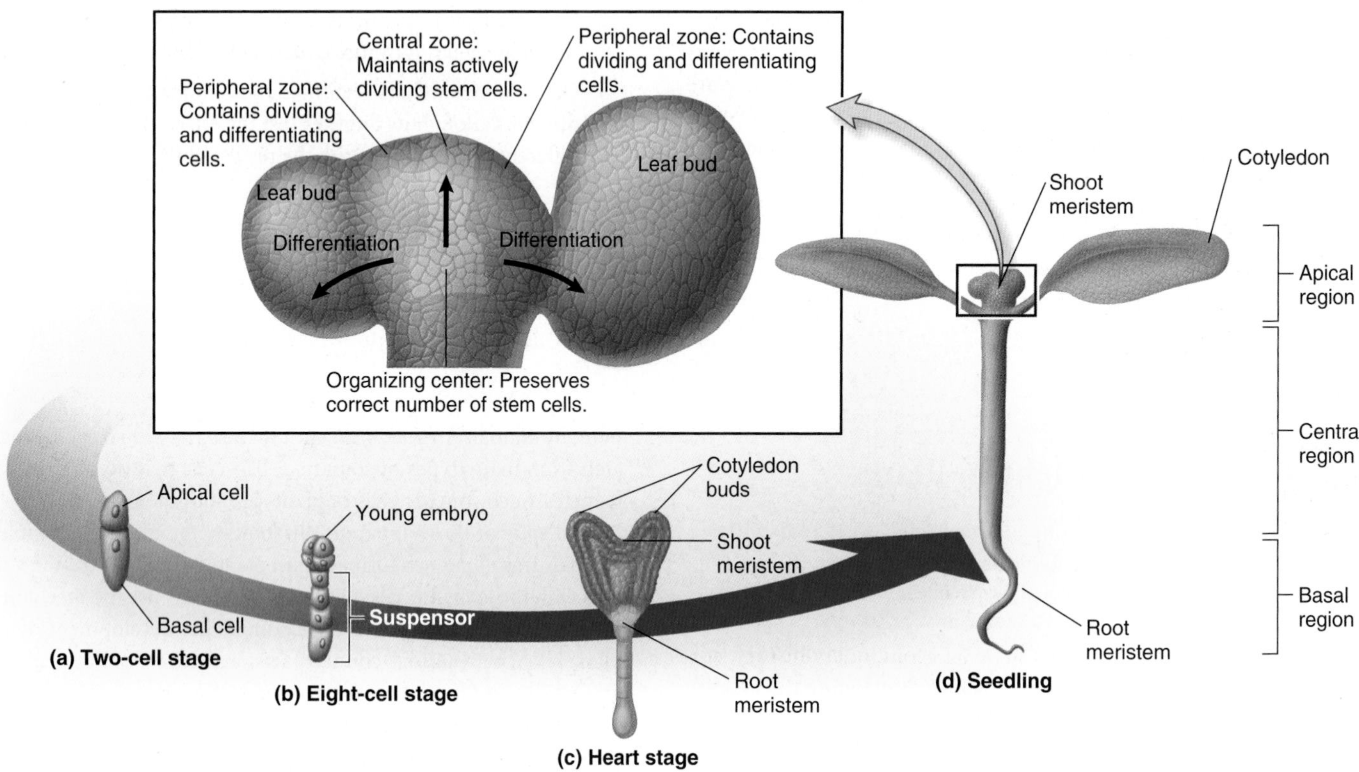

Figure 19.23 **Developmental steps in the formation of a plant embryo.** **(a)** The two-cell stage consists of the apical cell and basal cell. **(b)** The eight-cell stage consists of a young embryo and a suspensor. The suspensor channels nutrients to the young embryo from the parent plant. **(c)** At the heart stage, all of the plant tissues have begun to form. The shoot meristem is located between the future cotyledons, and the root meristem is on the opposite side. **(d)** A seedling showing apical, central, and basal regions. The inset shows the organization of the shoot meristem. Note: The steps shown in parts (a), (b), and (c) occur during seed formation, and the embryo would be enclosed within a seed.

BIOLOGY PRINCIPLE Living organisms grow and develop. This diagram depicts how a plant begins as a fertilized egg that grows and develops into a seedling.

Concept Check: *Where are stem cells found in a growing plant?*

these three regions develops differently, as indicated by their unique cell division patterns and distinct morphologies.

As seen in the inset to Figure 19.23d, the shoot meristem is organized into three areas: the organizing center, the central zone, and the peripheral zone. The **organizing center** ensures the proper organization of the meristem and preserves the correct number of actively dividing stem cells. The **central zone** is an area where undifferentiated stem cells are always maintained. The **peripheral zone** contains dividing cells that eventually differentiate into plant structures. For example, the peripheral zone may form a bud that will produce a leaf or flower.

By analyzing mutations that disrupt the developmental process, researchers have discovered that the apical, central, and basal regions of a growing plant express different sets of genes. A category of genes called **apical-basal-patterning genes** are important in early stages of plant development. A few examples are described in Table 19.2. Mutations in apical-basal-patterning genes cause dramatic effects in one of these three regions. For example, the *Aintegumenta* gene is necessary for apical development. When it is defective, the growth of lateral buds is defective.

Table 19.2 Examples of *Arabidopsis* Apical-Basal-Patterning Genes

Region: *Gene*	**Description**
Apical:	
Aintegumenta	Encodes a transcription factor that is expressed in the peripheral zone. Its expression maintains the growth of lateral buds.
Central:	
Scarecrow	Encodes a transcription factor that plays a role in the asymmetric division that produces the radial pattern of growth in the stem. The Scarecrow protein also affects cell division patterns in roots and plays a role in sensing gravity.
Basal:	
Monopterous	Encodes a transcription factor. When the *Monopterous* gene is defective, the plant embryo cannot initiate the formation of root structures, although root structures can be formed postembryonically. This gene seems to be required for organizing root formation in the embryo.

Plant Homeotic Genes Control Flower Development

Although William Bateson coined the term homeotic to describe mutations in animals in which one body part is replaced by another, the first known homeotic mutations were described in plants. Naturalists in ancient Greece and Rome, for example, recorded their observations of double flowers in which stamens were replaced by petals. In current research, geneticists are studying these types of mutations to better understand developmental pathways in plants. Many homeotic mutations affecting flower development have been identified in *Arabidopsis* and also in the snapdragon (*Antirrhinum majus*).

A normal *Arabidopsis* flower is composed of four concentric whorls of structures (**Figure 19.24a**). The first, outer whorl contains four **sepals**, which protect the flower bud before it opens. The second whorl is composed of four **petals**, and the third whorl contains six **stamens**, structures that make the male gametophyte, pollen. Finally, the fourth, innermost whorl contains two carpels that are fused together. The **carpels** produce, enclose, and nurture the female gametophytes.

By analyzing the effects of many different homeotic mutations in *Arabidopsis*, British plant biologist Enrico Coen and his American colleague, plant geneticist, Elliot Meyerowitz, proposed the ABC model for flower development in 1991. In this model, three classes of genes, called *A*, *B*, and *C*, govern the formation of sepals, petals, stamens, and carpels. More recently, a fourth category, called the *E* genes, was found to be required for this process. Figure 19.24a illustrates how these genes affect normal flower development in *Arabidopsis*. In whorl 1, gene *A* product is made. This promotes sepal formation. In whorl 2, *A*, *B*, and *E* gene products are made, which promotes petal formation. In whorl 3, the expression of genes *B*, *C*, and *E* causes stamens to be made. Finally, in whorl 4, the products of *C* and *E* genes promote carpel formation.

What happens in certain homeotic mutants that undergo transformations of particular whorls? According to the original ABC model, genes *A* and *C* repress each other's expression, and gene *B* functions independently. In a mutant defective in gene *A* expression, gene *C* is also expressed in whorls 1 and 2. This produces a carpel-stamen-stamen-carpel arrangement in which the sepals have been transformed into carpels and the petals into stamens (**Figure 19.24b**). When gene *B* is defective, a flower cannot make petals or stamens. Therefore, a gene *B* defect yields a flower with a sepal-sepal-carpel-carpel arrangement. When gene *C* is defective, gene *A* is expressed in all four whorls. This results in a sepal-petal-petal-sepal pattern. If the expression of *E* genes is defective, the flower consists entirely of sepals.

Overall, the genes described in Figure 19.24 promote a pattern of development that leads to sepal, petal, stamen, or carpel structures. But what happens if genes *A*, *B*, and *C* are all defective? This produces a flower composed entirely of leaves (**Figure 19.24c**). These results indicate that the leaf structure is the default pathway and that the *A*, *B*, and *C* genes cause development to deviate from a leaf structure in order to make something else. In this regard, the sepals, petals, stamens, and carpels of plants can be viewed as modified leaves.

Like the *Drosophila* homeotic genes, plant homeotic genes are part of a hierarchy of gene regulation. Most plant homeotic genes belong to a family of genes called MADS box genes, which encode transcription factor proteins that contain a DNA-binding domain called a MADS domain. The *Arabidopsis* homeotic genes do not contain a sequence similar to the homeobox found in animal homeotic genes.

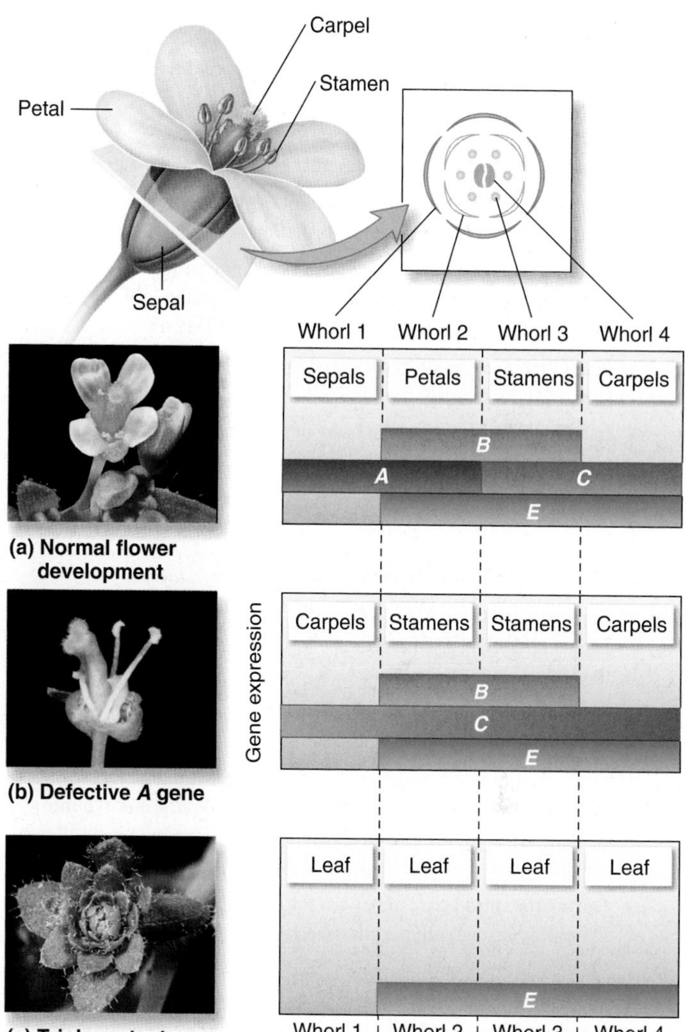

Figure 19.24 Normal and mutant homeotic gene action in *Arabidopsis*. **(a)** A normal flower composed of four concentric whorls of structures: sepals, petals, stamens, and carpels. To the right is the ABC model of homeotic gene action that has been revised to include *E* genes. **(b)** A homeotic mutant defective in gene *A* in which the sepals have been transformed into carpels and the petals have been transformed into stamens. **(c)** A triple mutant defective in the *A*, *B*, and *C* genes, producing a flower with all leaves.

Concept Check: *What pattern would you expect if the B gene was expressed in whorls 2, 3, and 4?*

Summary of Key Concepts

- Development refers to a series of changes in the state of a cell, tissue, organ, or organism. Developmental genetics seeks to understand how gene expression controls the development process.

19.1 General Themes in Development

- A cell's fate becomes determined before the cell differentiates into a specialized cell type.
- *Drosophila, C. elegans*, mice, zebrafish, and *Arabidopsis* are model organisms studied by developmental geneticists (Figure 19.1).
- The process that gives rise to an animal or plant with a particular body structure is called pattern formation (Figure 19.2).
- Pattern formation depends on positional information. Four responses to positional information are cell division, cell migration, cell differentiation, and apoptosis (Figures 19.3, 19.4).
- Positional information is conveyed by morphogens and cell adhesion (Figure 19.5).
- In animals and plants, a hierarchy of transcription factors controls pattern formation. In animals, pattern formation occurs in four general phases (Figure 19.6).

19.2 Development in Animals

- Embryonic development in *Drosophila* occurs in a sequence of stages, from a fertilized egg to an adult. The basic body plan is established in the embryo (Figure 19.7).
- Maternal effect genes control the formation of body axes, the first phase in *Drosophila* pattern formation (Figures 19.8, 19.9).
- In the second phase of pattern formation, the sequential expression of three categories of segmentation genes divides the embryo into segments. Mutations that alter *Drosophila* development have allowed scientists to understand the normal process (Figures 19.10, 19.11).
- During the third phase of pattern formation, each segment begins to develop its own unique characteristics. Homeotic genes control the development of a particular segment or group of segments (Figures 19.12, 19.13, 19.14, 19.15).
- Invertebrates and vertebrates have a homologous set of homeotic genes. In vertebrates, these are called *Hox* genes (Figures 19.16, 19.17).
- In the fourth phase of pattern formation, stem cells divide and differentiate into specialized cell types (Figure 19.18).
- Stem cells are categorized according to their developmental stage and their ability to differentiate. In mammals, a fertilized egg is totipotent; certain embryonic and fetal cells are pluripotent; and stem cells in the adult are multipotent or unipotent (Figures 19.19, 19.20).
- Stem cells have the potential to be used to treat a variety of human disorders (Table 19.1).
- The differentiation of cell types within certain tissues or organs are controlled by master transcription factors. An example is *MyoD*, a gene that initiates skeletal muscle cell development (Figure 19.21).

19.3 Development in Plants

- Plants grow along a root-shoot axis and in a well-defined radial pattern. Cell migration does not occur in plants (Figure 19.22).
- Plant stem cells called meristems promote the development of plant structures such as roots, stems, leaves, and flowers (Figure 19.23).
- The apical, central, and basal regions of the growing plant express different sets of pattern formation genes (Table 19.2).
- Four classes of homeotic genes in plants, *A*, *B*, *C*, and *E*, control flower development (Figure 19.24).

Assess and Discuss

Test Yourself

1. The process whereby a cell's morphology and function have changed is called
   a. determination.
   b. cell fate.
   c. differentiation.
   d. genetic engineering.
   e. both a and c.
2. Pattern formation in plants is along the _______ axis.
   a. dorsoventral
   b. anteroposterior
   c. left-right
   d. root-shoot
   e. all of the above are correct.
3. Positional information is important in determining the fate of a cell in a multicellular organism. Animal cells respond to positional information by
   a. dividing.
   b. migrating.
   c. differentiating.
   d. undergoing apoptosis.
   e. all of the above.
4. Morphogens are
   a. molecules that disrupt normal development.
   b. molecules that convey positional information and promote changes in development.
   c. mutagenic agents that cause apoptosis.
   d. receptors that allow cells to adhere to the extracellular matrix.
   e. both a and c.
5. What group of proteins plays a key role in controlling the program of developmental changes?
   a. motor proteins
   b. transporters
   c. transcription factors
   d. restriction endonucleases
   e. cyclins
6. Using the following list, describe the proper sequence for the events of animal development:
   1. Tissues, organs, and other body structures in each segment are formed.
   2. Axes of the entire animal are determined.
   3. Cells become differentiated.
   4. The entire animal is divided into segments.

   a. 2, 3, 4, 1
   b. 1, 2, 4, 3
   c. 2, 4, 3, 1
   d. 3, 2, 4, 1
   e. 2, 4, 1, 3

7. Homeotic genes in *Drosophila*
   a. determine the structural and functional characteristics of different segments of the developing fly.
   b. encode motor proteins that transport morphogens throughout the embryo.
   c. are dispersed randomly throughout the genome.
   d. are expressed in similar levels in all parts of the developing embryo.
   e. Both a and c are correct.
8. Which of the following genes do *not* play a role in the process whereby segments are formed in the fruit fly embryo?
   a. homeotic genes
   b. gap genes
   c. pair-rule genes
   d. segment-polarity genes
   e. All of the above play a role in segmentation.
9. An embryonic stem cell that can give rise to any type of cell of an adult organism but cannot produce an entire, intact individual is called
   a. totipotent. c. multipotent. e. antipotent.
   b. pluripotent. d. unipotent.
10. During plant development, the leaves and the flowers of the plant are derived from
    a. the central region. d. the apical region.
    b. the basal region. e. both a and d.
    c. the suspensor.

Conceptual Questions

1. If you observed fruit flies with the following developmental abnormalities, would you guess that a mutation had occurred in a segmentation gene or a homeotic gene? Explain your guess.
   a. Three abdominal segments were missing.
   b. One abdominal segment had legs.
2. The *myoD* gene in mammals plays a role in muscle cell differentiation. The *Hox* genes are homeotic genes that play a role in the differentiation of particular regions of the body. Compare and contrast the functions of these genes.
3. A principle of biology is that *living organisms grow and develop*. Discuss how maternal effect genes, segmentation genes, and homeotic genes control the process of development in animals.

Collaborative Questions

1. It seems that developmental genetics boils down to a complex network of gene regulation. Starting with maternal effect genes and ending with master transcription factors, try to draw/describe how this network is structured for *Drosophila*. How many genes do you think are necessary to describe a complete developmental network for the fruit fly? How many genes do you think are needed for a network to specify one segment?
2. Is it possible for a phenotypically normal female fly to be homozygous for a loss-of-function allele in the *bicoid* gene? What would be the phenotype of the offspring that such a fly would produce if it were mated to a male that was homozygous for the normal *bicoid* allele?

Online Resource

www.brookerbiology.com

Stay a step ahead in your studies with animations that bring concepts to life and practice tests to assess your understanding. Your instructor may also recommend the interactive eBook, individualized learning tools, and more.

Genetic Technology

Chapter Outline

20.1 Gene Cloning
20.2 Genomics: Techniques for Studying Genomes
20.3 Biotechnology
Summary of Key Concepts
Assess and Discuss

20

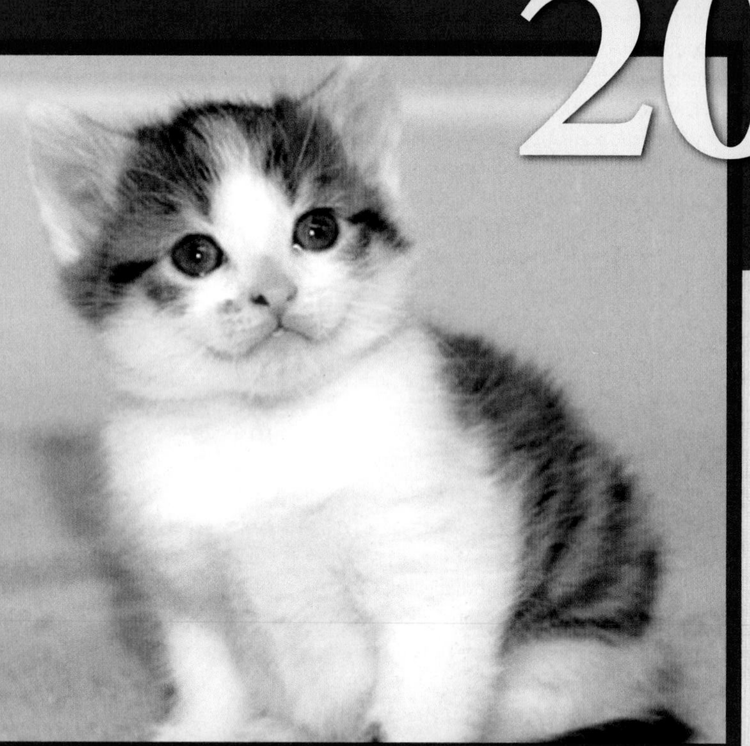

The first cloned pet. In 2002, the cat shown here called CC (for carbon copy) or copy cat was produced by reproductive cloning, a procedure described later in this chapter.

Jacob was diagnosed with type I diabetes, also called insulin-dependent diabetes mellitus, when he was 13 years old. Since then, he's been taking regular injections of insulin, a protein hormone that is normally made by cells in the pancreas. Jacob's pancreas fails to make enough of this hormone. His insulin prescription says Humulin on the bottle, which stands for human insulin. However, you might be surprised to discover that his insulin is not made by human cells. It's actually made by laboratory strains of the bacterium *Escherichia coli* that have been genetically modified to synthesize a hormone that is identical in structure to human insulin. This is just one example of how researchers have been able to apply **recombinant DNA technology**—the use of laboratory techniques to bring together fragments of DNA from multiple sources—to benefit humans. Such technology produces recombinant DNA, which refers to any DNA molecule that has been manipulated so it contains DNA from two or more sources.

In the early 1970s, the first successes in making recombinant DNA molecules were accomplished independently by two groups at Stanford University: David Jackson, Robert Symons, and Paul Berg; and Peter Lobban and A. Dale Kaiser. Both groups were able to isolate and purify pieces of DNA in a test tube and then covalently link two or more DNA fragments. Shortly thereafter, techniques were developed to introduce these recombinant DNA molecules into living cells. Once inside a host cell, the recombinant molecules were replicated to produce many identical copies. The process of making multiple copies of a particular gene is known as **gene cloning**. In the first part of the chapter, we will explore recombinant DNA technology and gene cloning, techniques that have enabled geneticists to probe relationships between gene sequences and phenotypic consequences. Such studies have been fundamental to our understanding of gene structure and function.

In the second part of this chapter, we will consider the topic of **genomics**—the molecular analysis of the entire genome of a species. In recent years, molecular techniques have progressed to the point where researchers can study the structure and function of many genes as large integrated networks. For example, the expression of all genes in a genome can be compared between normal and cancerous cells. This information helps us to understand how changes in gene expression cause uncontrolled cell growth.

In the last section of this chapter, we will explore the topic of **biotechnology**—the use of living organisms or the products of living organisms for human benefit. **Genetic engineering**, the direct manipulation of genes for practical applications, is playing an ever-increasing role in the production of strains of microorganisms, plants, and animals with characteristics that are useful to people. These include bacteria that make hormones such as human insulin, crops that produce their own insecticides, and farm animals that make human medicines.

20.1 Gene Cloning

Learning Outcomes:

1. Outline the steps of gene cloning using vectors.
2. Distinguish between a genomic library and a cDNA library.
3. Explain how gel electrophoresis is used to separate DNA fragments.
4. Describe the steps of gene cloning using polymerase chain reaction (PCR).

As mentioned, the term gene cloning refers to the process of making many copies of a particular gene. Why is gene cloning useful? Figure 20.1 provides an overview of the steps and goals of gene

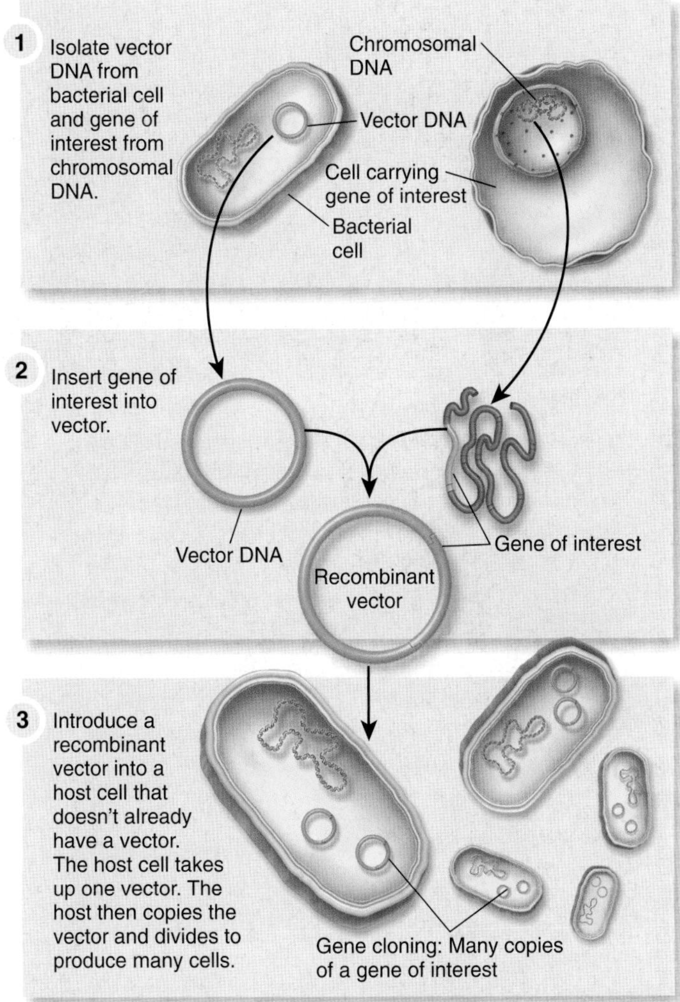
1 Isolate vector DNA from bacterial cell and gene of interest from chromosomal DNA.
Chromosomal DNA
Vector DNA
Cell carrying gene of interest
Bacterial cell
2 Insert gene of interest into vector.
Vector DNA
Gene of interest
Recombinant vector
3 Introduce a recombinant vector into a host cell that doesn't already have a vector. The host cell takes up one vector. The host then copies the vector and divides to produce many cells.
Gene cloning: Many copies of a gene of interest

Gene cloning is done to achieve one of two main goals:

Producing large amounts of DNA of a specific gene	Expressing the cloned gene to produce the encoded protein
Examples	*Examples*
• Cloned genes provide enough DNA for DNA sequencing. The sequence of a gene can help us understand how a gene works and identify mutations that cause diseases.	• Large amounts of the protein can be purified to study its structure and function.
• Cloned DNA can be used as a probe to identify the same gene or similar genes in other organisms.	• Cloned genes can be introduced into bacteria or livestock to make pharmaceutical products such as insulin.
	• Cloned genes can be introduced into plants and animals to alter their traits.
	• Cloned genes can be used to treat diseases—a clinical approach called gene therapy.

Figure 20.1 Gene cloning. The process of gene cloning is used to produce large amounts of a gene or its protein product.

BIOLOGY PRINCIPLE Biology is an experimental science. The technique of gene cloning allows researchers to study genes and gene products in greater detail.

cloning. The process is usually done with one of two goals in mind. One goal is that a researcher or clinician wants many copies of the gene, perhaps to study the DNA directly or to use the DNA as a tool. For example, geneticists may want to determine the sequence of a gene from a person with a disease to see if the gene carries a mutation. Alternatively, the goal may be to obtain a large amount of the gene product—mRNA or protein. Along these lines, biochemists use gene cloning to obtain large amounts of proteins to study their structure and function. In modern molecular biology, gene cloning has provided the foundation for critical technical advances in a variety of disciplines, including molecular biology, genetics, cell biology, biochemistry, and medicine. In this section, we will examine the procedures that are used in gene cloning.

Step 1: Vector DNA and Chromosomal DNA Are the Starting Materials of Gene Cloning

In the first step of gene cloning, a key material is a type of DNA known as a **vector** (from the Latin, for "one who carries") (see Figure 20.1). Vector DNA acts as a carrier of the DNA segment that is to be cloned. In cloning experiments, a vector may carry a small segment of chromosomal DNA, perhaps only a single gene. By comparison, a chromosome carries up to a few thousand genes. When a vector is introduced into a living cell, it can replicate, and so the DNA that it carries is also replicated. This produces many identical copies of the inserted gene.

The vectors commonly used in gene cloning experiments were originally derived from two natural sources: plasmids or viruses. As discussed in Chapter 18, **plasmids** are small, circular pieces of DNA that are found naturally in many strains of bacteria and exist independently of the bacterial chromosome. Commercially available plasmids have been genetically engineered for effective use in cloning experiments. They contain unique sites into which geneticists can easily insert pieces of DNA. An alternative type of vector used in cloning experiments is a **viral vector**. Viruses can infect living cells and propagate themselves by taking control of the host cell's metabolic machinery. When a chromosomal gene is inserted into a viral vector, the gene is replicated whenever the viral DNA is replicated. Therefore, viruses can be used as vectors to carry other pieces of DNA.

Another material necessary for cloning a gene is the gene itself, which we will call the gene of interest. The source of the gene is the chromosomal DNA that carries the gene. The preparation of chromosomal DNA involves breaking open cells and extracting and purifying the DNA using biochemical separation techniques such as chromatography and centrifugation.

Step 2: Cutting Chromosomal and Vector DNA into Pieces and Linking Them Together Produces Recombinant Vectors

The second step in a gene cloning experiment is the insertion of the gene of interest into the vector (see Figure 20.1). How is this accomplished? DNA molecules are cut and pasted to produce recombinant vectors. To cut DNA, researchers use enzymes known as **restriction enzymes** or restriction endonucleases. These enzymes, which were discovered by Swiss geneticist Werner Arber and his American colleagues microbiologists Hamilton Smith and Daniel Nathans in the 1960s and 1970s, are made naturally by many different species of bacteria. Restriction enzymes protect bacterial cells from invasion by viruses by degrading the viral DNA into small fragments. Currently, several hundred different restriction enzymes from various bacterial species have been identified and are available commercially to molecular biologists.

The restriction enzymes used in cloning experiments bind to a specific base sequence and then cleave the DNA backbone at two defined locations, one in each strand. The sequences recognized by restriction enzymes are called **restriction sites**. Most restriction enzymes recognize sequences that are palindromic, which means the sequence is identical when read in the opposite direction in the complementary strand (**Table 20.1**). For example, the sequence recognized by the restriction enzyme *Eco*RI is 5′–GAATTC–3′ in the top strand. Read in the opposite direction in the bottom strand, this sequence is also 5′–GAATTC–3′. Such restriction enzymes are useful in cloning because they digest DNA into fragments with **sticky ends**. These DNA fragments have single-stranded ends that hydrogen-bond to other DNA fragments that are cut with the same enzyme because the sticky ends have complementary sequences.

Figure 20.2 shows the action of a restriction enzyme and the insertion of a gene into a vector. This vector carries the amp^R and *lacZ* genes, whose useful functions will be discussed later. The restriction enzyme binds to specific sequences in both the vector and chromosomal DNA. It then cleaves the DNA backbones, producing DNA fragments with sticky ends (Figure 20.2, step 1). The sticky ends of a piece of chromosomal DNA and the vector DNA can hydrogen-bond with each other (Figure 20.2, step 2). However, this interaction is not stable, because it involves only a few hydrogen bonds between complementary bases. To establish a permanent connection between two DNA fragments, the sugar-phosphate backbones of the DNA strands must be covalently linked. This linkage is catalyzed by DNA ligase (Figure 20.2, step 3). Recall from Chapter 11 that DNA ligase is an enzyme that catalyzes the formation of a covalent bond between adjacent DNA fragments.

In some cases, the two ends of the vector simply ligate back together, restoring it to its original circular structure; this forms a recircularized vector. In other cases, a fragment of chromosomal

Table 20.1 Examples of Restriction Enzymes Used in Gene Cloning

Restriction enzyme*	Bacterial source	Sequence recognized†
*Eco*RI	*Escherichia coli* (strain RY13)	↓ 5′-GAATTC-3′ 3′-CTTAAG-5′ ↑
*Sac*I	*Streptomyces achromogenes*	↓ 5′-GAGCTC-3′ 3′-CTCGAG-5′ ↑

*Restriction enzymes are named according to the species in which they are found. The first three letters are italicized because they indicate the genus and species names. Because a species may produce more than one restriction enzyme, the enzymes are designated I, II, III, etc., to indicate the order in which they were discovered in a given species.

†The arrows show the locations in the upper and lower DNA strands where the restriction enzymes cleave the DNA backbone.

1 Cut vector and chromosomal DNA with *Eco*RI, a restriction enzyme that recognizes the sequence **GAATTC** and cuts at the arrows. **CTTAAG**

lacZ
*Eco*RI site
GAATTC
CTTAAG
amp^R
Origin of replication
Restriction sites
Gene of interest

Vector DNA has one *Eco*RI site.

Chromosomal DNA has many *Eco*RI sites.

The restriction enzyme opens up the vector and cuts the chromosomal DNA into many fragments with short single-stranded regions called sticky ends.

Sticky ends
G
CTTAA
AATTC
G
Sticky end

2 Allow sticky ends to hydrogen-bond with each other due to complementary sequences.

Gene of interest
Gap
Gap
Gap
Gap
Gap
GAATTC
CTTAAG
Gap

In this example, a fragment of DNA carrying the gene of interest has hydrogen-bonded to the vector. Four gaps are found where covalent bonds in the DNA backbone are missing.

3 Add DNA ligase to close the gaps by catalyzing the formation of covalent bonds in the DNA backbone.

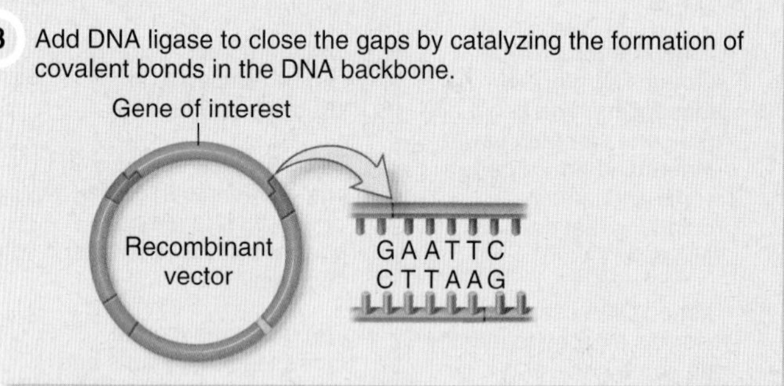

Figure 20.2 Step 2 of gene cloning: The actions of a restriction enzyme and DNA ligase to produce a recombinant vector.

Concept Check: In the procedure shown in this figure, has the gene of interest been cloned?

BioConnections: Look back at Figure 18.12. What are the general properties of plasmids?

1 Mix plasmid DNA with many *E. coli* cells that have been treated with agents that make them permeable to DNA.

Gene of interest

Part of *lacZ*

Part of *lacZ*

amp^R

Origin of replication

In this example, the gene of interest was inserted into a plasmid. This disrupts the *lacZ* gene and renders it nonfunctional. It is also possible for any other chromosomal DNA fragment to be inserted into the plasmid, or the plasmid may recircularize without an insert.

This shows a bacterial cell that has taken up the plasmid carrying the gene of interest. Many bacterial cells fail to take up a plasmid.

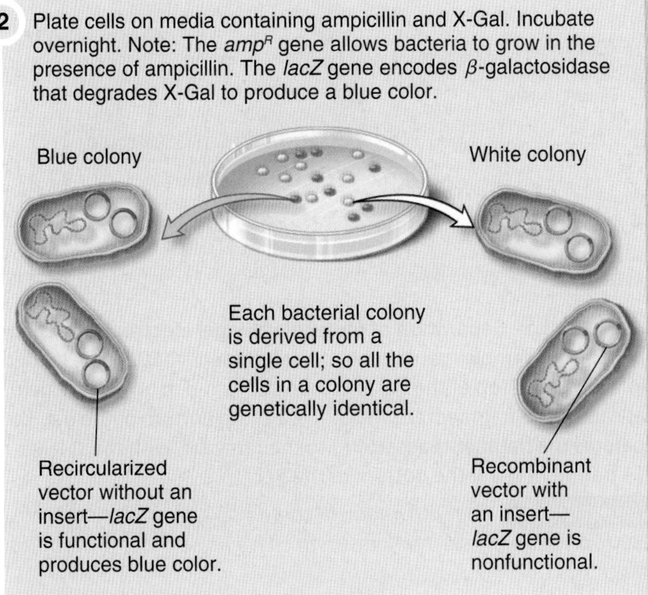

2 Plate cells on media containing ampicillin and X-Gal. Incubate overnight. Note: The ampR gene allows bacteria to grow in the presence of ampicillin. The lacZ gene encodes β-galactosidase that degrades X-Gal to produce a blue color.
Blue colony
White colony
Each bacterial colony is derived from a single cell; so all the cells in a colony are genetically identical.
Recircularized vector without an insert—lacZ gene is functional and produces blue color.
Recombinant vector with an insert—lacZ gene is nonfunctional.

Figure 20.3 **Step 3 of gene cloning: Introduction of a recombinant vector into a host cell.** For cloning to occur, a recombinant vector is introduced into a host cell, which copies the vector and divides to produce many cells. This produces many copies of the gene of interest.

Concept Check: *In this cloning experiment, what is the purpose of having the lacZ gene in the vector?*

DNA may become ligated to both ends of the vector. When this happens, a segment of chromosomal DNA has been inserted into the vector. The result is a vector containing a piece of chromosomal DNA, which is called a **recombinant vector** or a hybrid vector. We then have a recombinant vector that is ready to be cloned. A recombinant vector may contain the gene of interest or it may contain a different piece of chromosomal DNA.

Step 3: Putting Recombinant Vectors into Host Cells and Allowing Those Cells to Propagate Achieves Gene Cloning

The third step in gene cloning is the actual cloning of the gene of interest. In this step, the goal is for the recombinant vector carrying the desired gene to be taken up by bacterial cells that have been treated with agents that render them permeable to DNA molecules (**Figure 20.3**). Bacterial cells with the ability to take up DNA are described as **competent**. Some bacterial cells take up a single plasmid, whereas most cells fail to take up a plasmid. The bacteria are then streaked on petri plates containing a bacterial growth medium and ampicillin.

In the experiment shown here, the bacterial cells were originally sensitive to ampicillin. The plasmid already carries an antibiotic-resistance gene, called the *amp*^{R} gene. What is the purpose of this gene in a cloning experiment? Such a gene is called a **selectable marker** because the presence of the antibiotic selects for the growth of cells expressing the *amp*^{R} gene. The *amp*^{R} gene encodes an enzyme known as β-lactamase that degrades the antibiotic ampicillin, which normally kills bacteria. Bacteria that have not taken up a plasmid are killed by the antibiotic. In contrast, any bacterium that has taken up a plasmid carrying the *amp*^{R} gene grows and divides many times to form a bacterial colony containing tens of millions of cells. Because each cell in a single colony is derived from the same original cell that took up a single plasmid, all cells within a colony contain the same type of plasmid DNA.

The experimenter also needs a way to distinguish bacterial colonies that contain cells with a recombinant vector from those containing cells with a recircularized vector. In a recombinant vector, a piece of chromosomal DNA has been inserted into a region of the vector that contains the *lacZ* gene, which encodes the enzyme β-galactosidase. The insertion of chromosomal DNA into the vector disrupts the *lacZ* gene, thereby preventing the synthesis of β-galactosidase. By comparison, a recircularized vector has a functional *lacZ* gene. The functionality of *lacZ* can be determined by providing the growth medium with a colorless compound, X-Gal, which is cleaved by β-galactosidase into a blue dye. Bacteria grown in the presence of X-Gal form blue colonies if they produce a functional β-galactosidase enzyme and white colonies if they do not. In this experiment, therefore, bacterial colonies containing recircularized vectors form blue colonies, whereas colonies containing recombinant vectors carrying a segment of chromosomal DNA are white.

After a bacterial cell has taken up a recombinant vector, two subsequent events lead to the production of many copies of that vector. First, when the vector has a highly active origin of replication, the bacterial host cell produces many copies of the recombinant vector per cell. Second, the bacterial cells divide approximately every 20 minutes. Following overnight growth, a population of many millions of bacteria are obtained from a single cell. Each of these bacterial cells contains many copies of the cloned gene. For example, a bacterial colony may comprise 10 million cells, with each cell containing 50 copies of the recombinant vector. Therefore, this bacterial colony has 500 million copies of the cloned gene!

A DNA Library Is a Collection of Many Different DNA Fragments

In a typical cloning experiment, such as the one described in Figures 20.2 and 20.3, the treatment of chromosomal DNA with restriction enzymes actually yields tens of thousands of different DNA fragments. Therefore, after the DNA fragments are ligated individually to vectors, a researcher has a collection of many recombinant vectors, with each vector containing a particular fragment of chromosomal DNA. A collection of recombinant vectors containing DNA fragments from a given organism is known as a **DNA library** (Figure 20.4). Researchers make DNA libraries using the methods shown in Figures 20.2 and 20.3 and then use those libraries to obtain clones of the genes of interest to them.

Two types of DNA libraries are commonly made. The library is called a **genomic library** when the inserts are derived from chromosomal DNA. Alternatively, researchers can isolate mRNA and use the enzyme reverse transcriptase, which is described in Chapter 18, to make DNA molecules using mRNA as a starting material. Such DNA is called **complementary DNA**, or **cDNA**. A **cDNA library** is a collection of recombinant vectors that have cDNA inserts. From a research perspective, an important advantage of cDNA is that it lacks introns—intervening sequences that are not translated into proteins. Because introns can be quite large, it is much simpler for researchers to insert cDNAs into vectors rather than work with chromosomal DNA if they

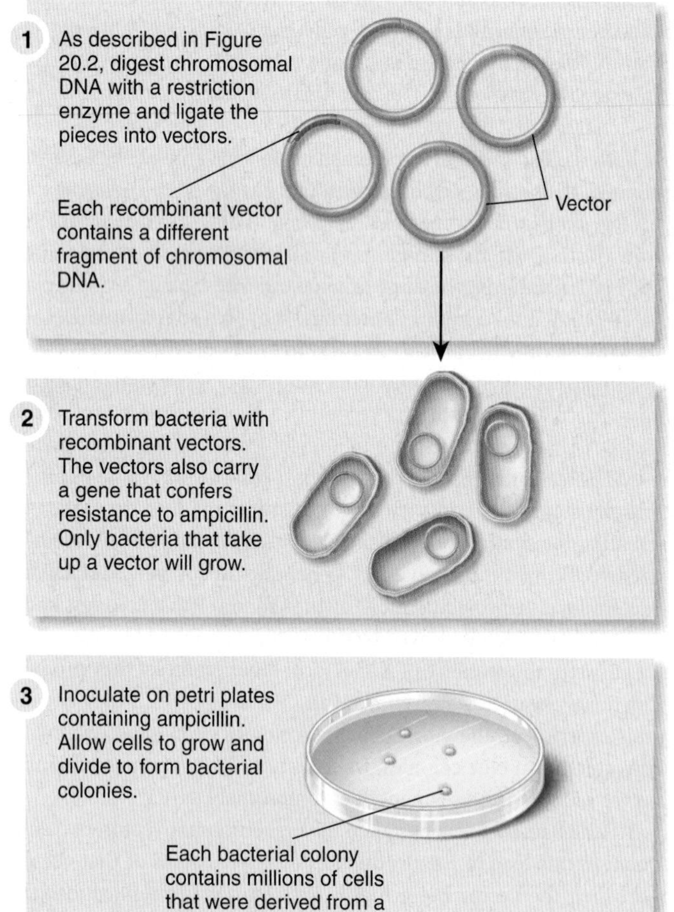

1 As described in Figure 20.2, digest chromosomal DNA with a restriction enzyme and ligate the pieces into vectors.
Each recombinant vector contains a different fragment of chromosomal DNA.
Vector
2 Transform bacteria with recombinant vectors. The vectors also carry a gene that confers resistance to ampicillin. Only bacteria that take up a vector will grow.
3 Inoculate on petri plates containing ampicillin. Allow cells to grow and divide to form bacterial colonies.
Each bacterial colony contains millions of cells that were derived from a single transformed cell.

Figure 20.4 A DNA library. Each colony in a DNA library contains a vector with a different piece of chromosomal DNA.

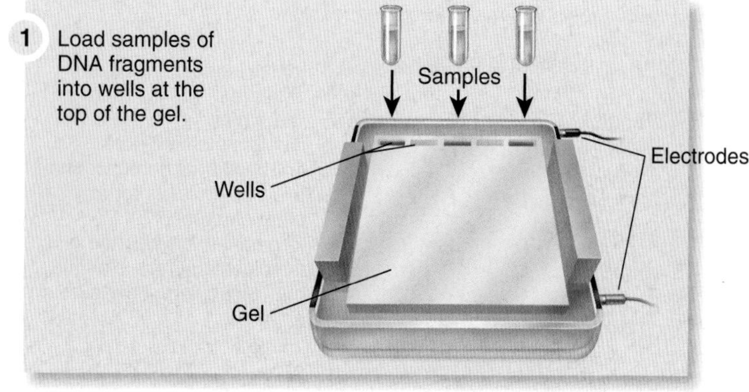

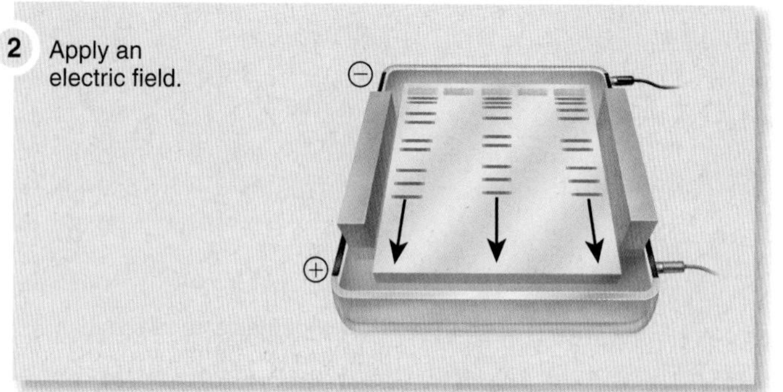

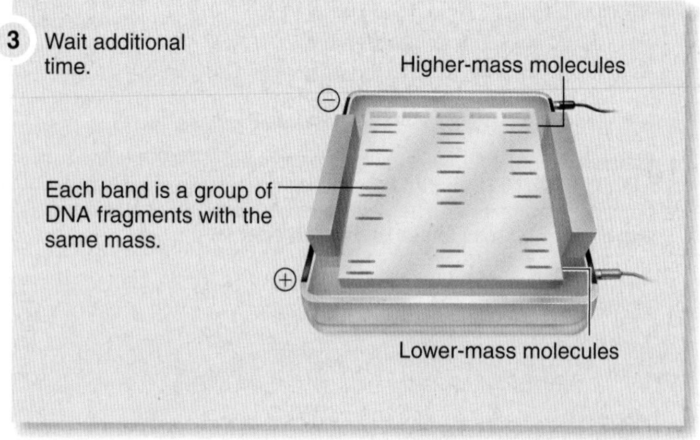

Figure 20.5 Separation of molecules by gel electrophoresis. In this example, samples containing many fragments of DNA are loaded into wells at the top of the gel and then subjected to an electric field that causes the fragments to move toward the bottom of the gel. This separates the fragments according to their masses, with the smaller DNA fragments near the bottom of the gel.

Concept Check: *One DNA fragment contains 600 bp and another has 1,300 bp. Following electrophoresis, which would be closer to the bottom of a gel?*

want to focus their attention on the coding sequence of a gene. In addition, because bacteria do not splice out introns, cDNAs are an advantage if researchers want to express their gene of interest in bacteria.

Gel Electrophoresis Separates Macromolecules, Such as DNA

Gel electrophoresis is a technique for separating macromolecules, such as DNA and proteins, as they migrate through a gel. This method is often used to evaluate the results of a cloning experiment. For example, gel electrophoresis is used to determine the sizes of DNA fragments that have been inserted into recombinant vectors.

Gel electrophoresis can separate molecules based on their charge, size/length, and mass. In the example shown in Figure 20.5, gel electrophoresis is used to separate different fragments of chromosomal DNA based on their masses. The gel is a flat semisolid gel with depressions at the top called wells where samples are added. An electric field is applied to the gel, which causes charged molecules, either proteins or nucleic acids, to migrate from the top of the gel toward the bottom—a process called electrophoresis. DNA is negatively charged and moves toward the positive end of the gel, which is at the bottom in this figure. Smaller DNA fragments move more quickly through the gel than larger fragments, and therefore are located closer to the bottom of the gel than the larger ones. As the gel runs, the DNA fragments are separated into distinct bands within the gel. The fragments in each band can then be stained with a dye for identification.

Polymerase Chain Reaction (PCR) Is Also Used to Make Many Copies of DNA

As we have seen, one method of cloning involves an approach in which the gene of interest is inserted into a vector, introduced into a host cell, and then propagated. Another technique to copy DNA without the aid of vectors and host cells is a process called **polymerase chain reaction (PCR)**, which was developed by American biochemist Kary Mullis in 1985 (Figure 20.6). The goal of PCR is to make many copies of DNA in a defined region, perhaps encompassing a gene or part of a gene. Several reagents are required for the synthesis of DNA. First, two different primers are needed that are complementary to sequences at each end of the DNA region to be amplified. These primers are usually

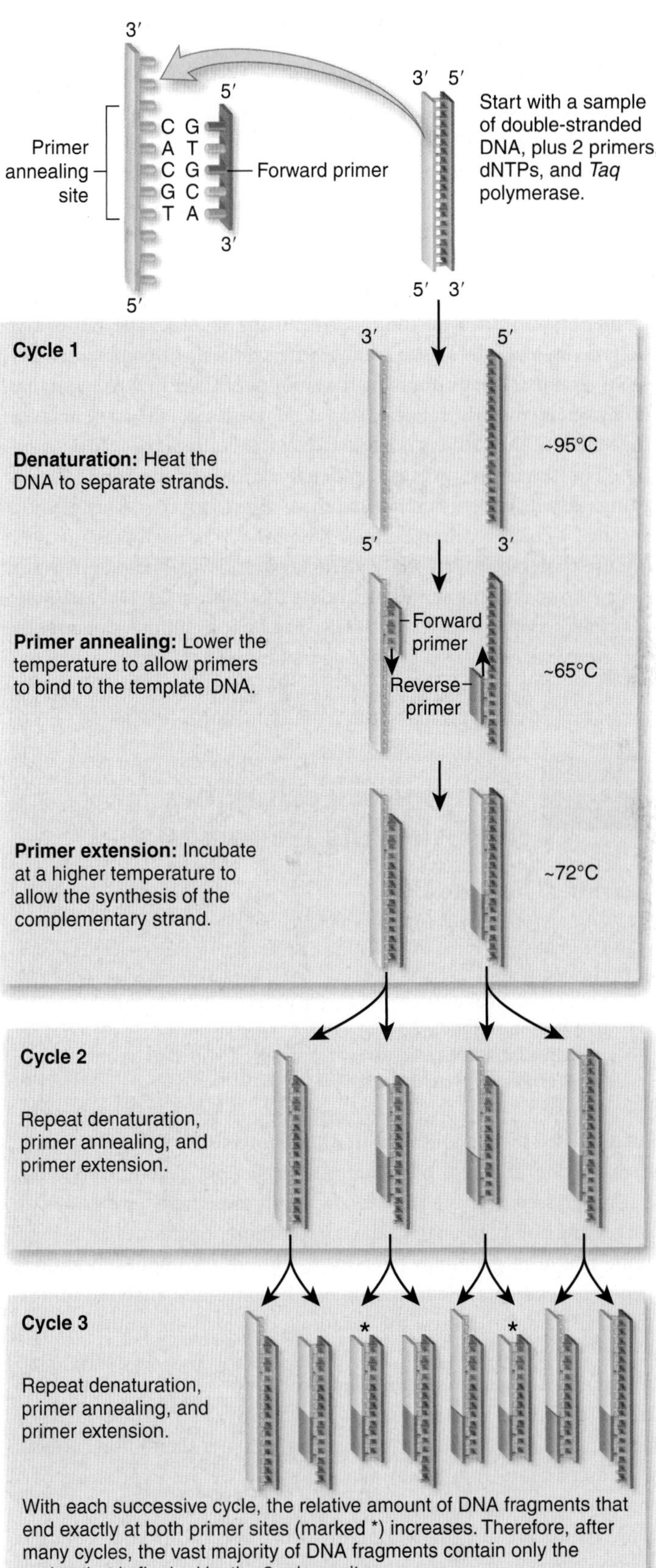

3′
5′
Primer annealing site
C G
A T
C G
G C
T A
3′
5′
Forward primer
3′ 5′
Start with a sample of double-stranded DNA, plus 2 primers, dNTPs, and *Taq* polymerase.
5′ 3′
Cycle 1
3′
5′
Denaturation: Heat the DNA to separate strands.
~95°C
5′
3′
Forward primer
Primer annealing: Lower the temperature to allow primers to bind to the template DNA.
Reverse primer
~65°C
Primer extension: Incubate at a higher temperature to allow the synthesis of the complementary strand.
~72°C
Cycle 2
Repeat denaturation, primer annealing, and primer extension.
Cycle 3
Repeat denaturation, primer annealing, and primer extension.
*
*
With each successive cycle, the relative amount of DNA fragments that end exactly at both primer sites (marked *) increases. Therefore, after many cycles, the vast majority of DNA fragments contain only the region that is flanked by the 2 primer sites.

Figure 20.6 Polymerase chain reaction (PCR). During each PCR cycle, the steps of denaturation, primer annealing, and primer extension take place. The net result of PCR is the synthesis of many copies of DNA in the region that is flanked by the two primers. To conduct this type of PCR experiment, researchers must have prior knowledge about the base sequence of the template DNA so they can make primers with base sequences that are complementary to the ends of the template DNA. Note: The temperatures shown in this figure are approximate and may vary depending on the primer sequence and length.

Concept Check: *Why do the PCR primers bind specifically to the primer-annealing sites?*

BioConnections: *Look back at Figure 11.18. Why are primers needed in a PCR experiment?*

about 20 nucleotides long. One primer is called the forward primer, and the other is the reverse primer. PCR also requires all four deoxynucleoside triphosphates (dNTPs) and a heat-stable form of DNA polymerase called *Taq* polymerase. *Taq* polymerase is isolated from the bacterium *Thermus aquaticus*, which lives in hot springs and can tolerate temperatures up to 95°C. A heat-stable form of DNA polymerase is necessary because PCR involves the use of high temperatures that would inactivate most other natural forms of DNA polymerase.

To make copies, a sample of chromosomal DNA, called the template DNA, is heated to separate (denature) the DNA into single-stranded molecules. Then the primers bind to the DNA as the temperature is lowered (see Figure 20.6). The binding of the primers to the specific sites in the template DNA is called annealing. After the primers have annealed, the temperature is slightly raised and *Taq* polymerase uses dNTPs to catalyze the synthesis of complementary DNA strands, thereby doubling the amount of DNA in the region that is flanked by the primers. This step is called primer extension because the length of the primers is extended by the synthesis of DNA.

The sequential process of denaturation—primer annealing—primer extension is then repeated many times in a row. This method is called a chain reaction because the products of each previous step are used as reactants in subsequent steps. A device that controls the temperature and automates the timing of each step, known as a thermocycler, is used to carry out PCR. The PCR technique can amplify the sample of DNA by a staggering amount. After 30 cycles of denaturation, primer annealing, and primer extension, a DNA sample increases 2^{30}-fold, approximately a billionfold, in a few hours!

20.2 Genomics: Techniques for Studying Genomes

Learning Outcomes:

1. Distinguish between genomics and functional genomics.
2. Describe the features of BACs and YACs and how they are used to map chromosomes.
3. Outline the steps of DNA sequencing using the dideoxy chain-termination method.
4. Explain what a DNA microarray is and how it is used to identify the genes expressed by an organism.

As discussed throughout Unit III, the genome is the complete genetic composition of a cell, organism, or species. As genetic technology has progressed over the past few decades, researchers have gained an increasing ability to analyze the composition of genomes as a whole unit. The term genomics refers to the use of techniques to study a genome. Segments of chromosomes are cloned and analyzed in progressively smaller pieces, the locations of which are known on the intact chromosomes. This is the mapping phase of genomics. The mapping of a genome ultimately progresses to the determination of the complete DNA sequence, which provides the most detailed description available of an organism's genome at the molecular level. By comparison, **functional genomics** studies the expression of a genome. For example, functional genomics can be used to analyze which genes are turned on or off in normal versus cancer cells. In this section, we will consider a few of the methods that are used in genomics and functional genomics.

BAC Cloning Vectors Are Used to Make Contigs of Chromosomes to Map a Genome

In general, most plasmid and viral vectors can accommodate inserts only a few thousand to perhaps tens of thousands of nucleotides in length. For large eukaryotic genomes, which may contain over 1 billion bp (base pairs), cloning an entire genome is much easier when a cloning vector can accept very large chromosomal DNA inserts. If a plasmid or viral vector has a DNA insert that is too large, it will have difficulty with DNA replication and is likely to suffer deletions in the insert. By comparison, a type of cloning vector known as a **bacterial artificial chromosome (BAC)** can reliably contain much larger inserted DNA fragments. BACs are derived from large plasmids called F factors (see Chapter 18). They can typically contain inserts up to 500,000 bp. BACs are used in genomic research in the same way as other types of vectors. Similarly, yeast artificial chromosomes (YACs) are used as vectors in yeast. An insert in a YAC can be several hundred thousand to perhaps 2 million bp in length.

The term **mapping** refers to the process of determining the relative locations of genes or other DNA segments along a chromosome. After many large fragments of chromosomal DNA have been inserted into BACs or YACs, the first step of mapping is to determine the relative locations of the cloned pieces as they would occur in an intact chromosome. This is called **physical mapping**. Each vector contains a clone with regions that overlap those of its neighbors (**Figure 20.7**). For example, clones 7 and 8 both carry gene *K*. These overlapping regions allow researchers to identify the order of the clones along the chromosome. To obtain a complete physical map of a chromosome, researchers assemble the clones into a series of contiguous overlapping DNA fragments known as a **contig**.

The Dideoxy Chain-Termination Method Is Used to Determine the Base Sequence of DNA

Another phase of genomic research is **DNA sequencing**, which is a method of determining the base sequence of DNA. Scientists can learn a great deal of information about the function of a gene if its nucleotide sequence is known. For example, the investigation of genetic sequences has been vital in our understanding of the genetic basis of human diseases.

A commonly used method for DNA sequencing, developed in 1977 by English biochemist Frederick Sanger and colleagues, is known as the **dideoxy chain-termination method**, or more simply, **dideoxy sequencing**. Dideoxy sequencing is based on knowledge of DNA replication. As described in Chapter 11, DNA polymerase connects adjacent deoxynucleoside triphosphates (dNTPs) by catalyzing a covalent linkage between the 5′-phosphate on one nucleotide and the 3′—OH group on the previous nucleotide. Chemists, however, can synthesize nucleotides, called dideoxynucleoside triphosphates (ddNTPs), that are missing the —OH group at the 3′ position (**Figure 20.8**). What happens if a ddNTP is incorporated during DNA replication? If a ddNTP is added to a growing DNA strand, the strand can no longer grow because the 3′—OH group, the site of attachment for the next nucleotide, is missing. This ending of DNA synthesis is called chain termination.

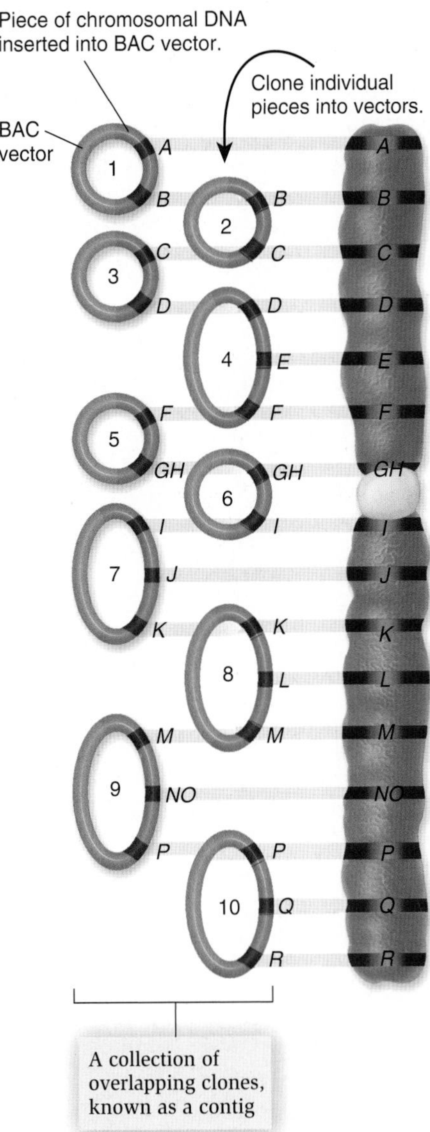
Piece of chromosomal DNA inserted into BAC vector.
Clone individual pieces into vectors.
BAC vector
A collection of overlapping clones, known as a contig

Figure 20.7 A contig. As shown here, a contig is a collection of clones that have overlapping pieces of DNA from a particular chromosome. The numbers denote the order of the members of the contig. The chromosome is labeled with letters that denote the locations of particular genes. The members of the contig have overlapping regions that have the same genes, which allows them to be placed in their proper order.

Concept Check: What does it mean when we say that two members of a contig have overlapping regions?

Before describing the steps of this DNA sequencing protocol, let's first consider the DNA segment that is analyzed in a sequencing experiment. The segment of DNA to be sequenced, the target DNA, must be obtained in large amounts by using the gene cloning techniques that were described earlier in this chapter. In **Figure 20.9a**, the target DNA was inserted into a vector next to a primer-annealing site, the place where a primer will bind. The target DNA is initially double stranded, but Figure 20.9a shows the DNA after it has been denatured into a single strand by heat treatment.

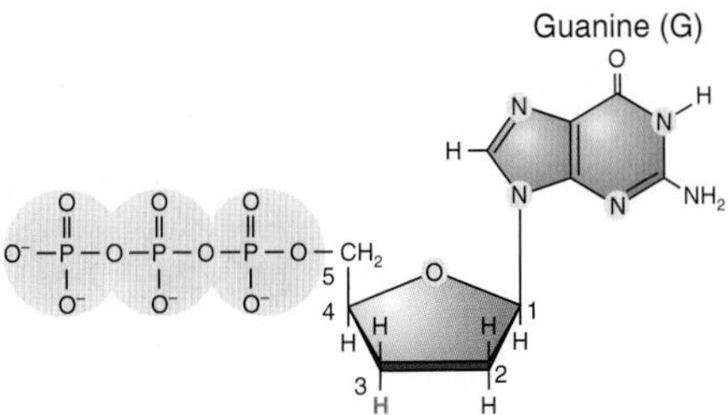
Guanine (G)

Figure 20.8 Structure of a dideoxynucleotide. This figure shows the structure of dideoxyguanosine triphosphate (ddGTP). It has a hydrogen, shown in red, instead of a hydroxyl group at the 3′ position. The prefix, dideoxy-, means it has two (di) missing (de) oxygens (oxy) compared with ribose, which has—OH groups at both the 2′ and 3′ positions.

Let's now examine the steps involved in DNA sequencing. Many copies of this single-stranded template DNA are placed into a tube and mixed with primers that bind to the primer-annealing site. DNA polymerase and all four types of regular dNTPs are also added. In addition, a low concentration of the four possible dideoxynucleoside triphosphates—ddGTP, ddATP, ddTTP, or ddCTP—is added. Each type of ddNTP is tagged with a different colored fluorescent molecule; typically ddA is green, ddT is red, ddG is yellow, and ddC is blue. The tube is then incubated to allow DNA polymerase to synthesize strands that are complementary to the target DNA sequence. However, addition of a ddNTP causes DNA synthesis to terminate early, preventing further elongation of the strand. For example, let's consider ddTTP. Synthesis of new DNA strands stops at the sixth or thirteenth position after the annealing site if a ddTTP, instead of a dTTP, is incorporated into the growing DNA strand (Figure 20.9a). This means the target DNA has a complementary A at the sixth and thirteenth positions. Eventually, a set of fluorescently tagged strands results, with the color of the ddNTP representing the last nucleotide added to the strand.

After the samples have been incubated for several minutes, the newly made DNA strands are separated according to their lengths by subjecting them to gel electrophoresis. This can be done using a slab gel, as shown in Figure 20.9a, or more commonly by running them through a gel-filled capillary tube. The shorter strands move to the bottom of the gel more quickly than the longer ones. Electrophoresis is continued until each band emerges from the bottom of the gel, where a laser excites the fluorescent dye. A fluorescence detector records the amount of fluorescence emission at four wavelengths, corresponding to the four dyes.

An example of a printout from a fluorescence detector is shown in **Figure 20.9b**. The peaks of fluorescence correspond to the DNA sequence that is complementary to the target DNA. The height of the fluorescent peaks is not always the same, because ddNTPs get incorporated at some sites more readily than at others. Note that ddG is usually labeled with a yellow dye, but it is converted to black ink on the printout for ease of reading. Though improvements in automated sequencing continue to be made, a typical sequencing run can provide a DNA sequence that is approximately 700 to 900 bases long, and perhaps even longer.

Researchers are also developing alternative methods to the dideoxy chain termination in order to sequence DNA. For example,

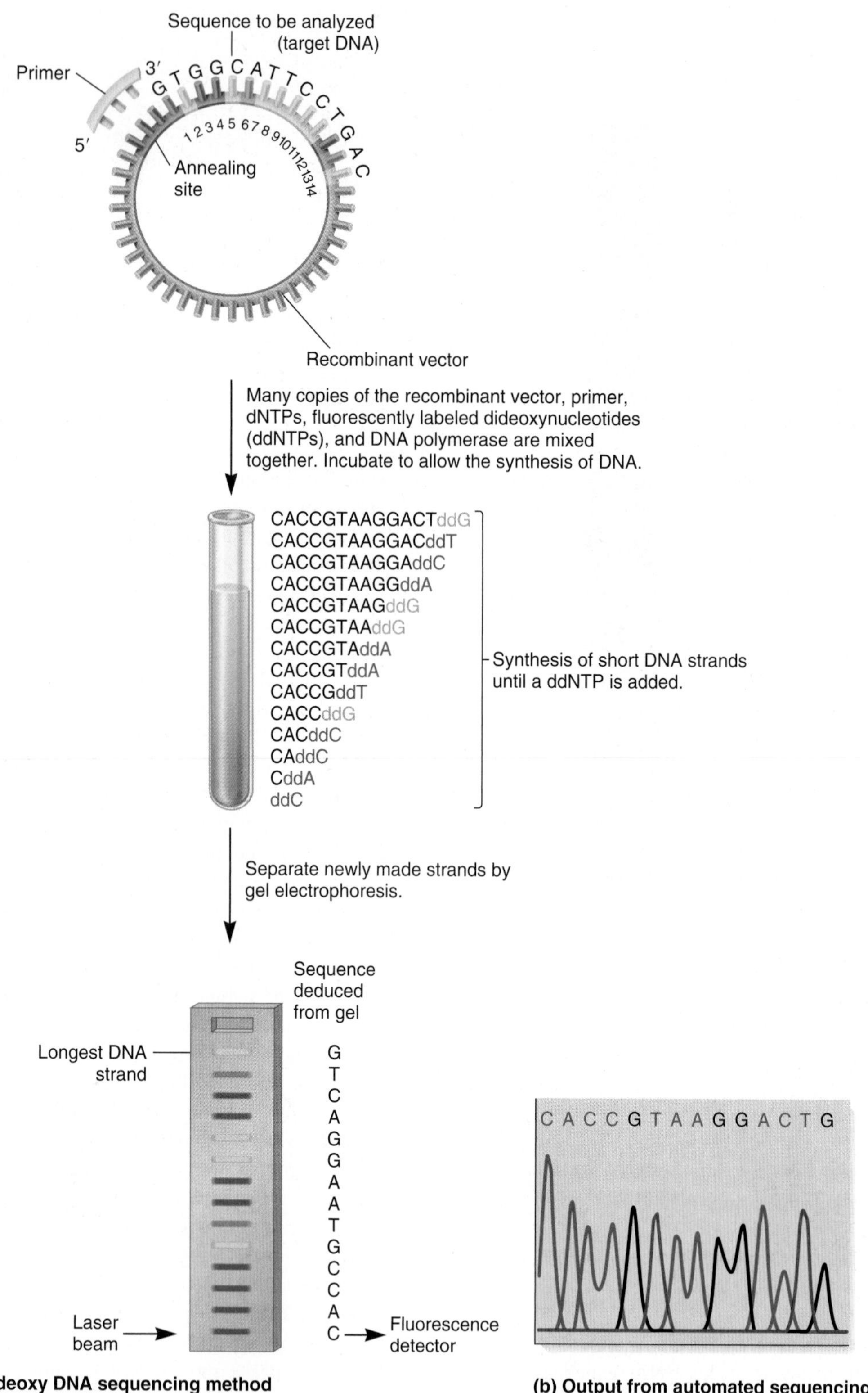

Figure 20.9 **DNA sequencing by the dideoxy method.** **(a)** The procedure of dideoxy sequencing using fluorescently labeled ddNTPs. **(b)** This method uses a fluorescence detector that measures the four kinds of ddNTPs as they emerge from the gel.

Concept Check: *What happens when a ddNTP is incorporated into a growing DNA strand?*

pyrosequencing is a newer method of DNA sequencing that is based on the detection of released pyrophosphate (PP_i) during DNA synthesis.

GENOMES & PROTEOMES CONNECTION

A Microarray Can Identify Which Genes Are Transcribed by a Cell

Let's now turn our attention to functional genomics. Researchers have developed an exciting new technology, called a **DNA microarray**, that is used to monitor the expression of thousands of genes simultaneously. A DNA microarray is a small silica, glass, or plastic slide that is dotted with many different sequences of single-stranded DNA, each corresponding to a short sequence within a known gene. Each spot contains multiple copies of a known DNA sequence. For example, one spot in a microarray may correspond to a sequence within the β-globin gene; another might correspond to a different gene, such as a gene that encodes a glucose transporter. A single slide contains tens of thousands of different spots in an area the size of a postage stamp. These microarrays are typically produced using a technology that "prints" spots of DNA sequences onto a slide, similar to the way that an inkjet printer deposits ink on paper.

What is the purpose of using a DNA microarray? In the experiment shown in Figure 20.10, the goal is to determine which genes are transcribed into mRNA from a particular sample of cells. In other words, which genes in the genome are expressed? To conduct this experiment, the mRNA was isolated from the cells and then used to make fluorescently labeled cDNA. The labeled cDNAs were then incubated with a DNA microarray. The single-stranded DNA in the microarray corresponds to the coding strand—the strand that has a sequence that is similar to mRNA. Those cDNAs that are complementary to the DNAs in the microarray hybridize, thereby remaining bound to the microarray. The array is washed and then analyzed using a microscope equipped with a computer that scans each spot and generates an image of the spots' relative fluorescence.

If the fluorescence intensity in a spot is high, a large amount of cDNA was in the sample that hybridized to the DNA at this location. For example, if the β-globin gene was expressed in the cells being tested, a large amount of cDNA for this gene would be made, and the fluorescence intensity for that spot would be high. Because the DNA sequence of each spot is already known, a fluorescent spot identifies cDNAs that are complementary to those DNA sequences. Furthermore, because the cDNA was generated from mRNA, this technique identifies genes that have been transcribed in a particular cell type under a given set of conditions. However, the amount of protein encoded by an mRNA may not always correlate with the amount of mRNA, due to variation in the rates of mRNA translation and protein degradation.

Thus far, the most common use of DNA microarrays is to study gene expression patterns. In addition, the technology of DNA microarrays has found several other important uses as described in Table 20.2.

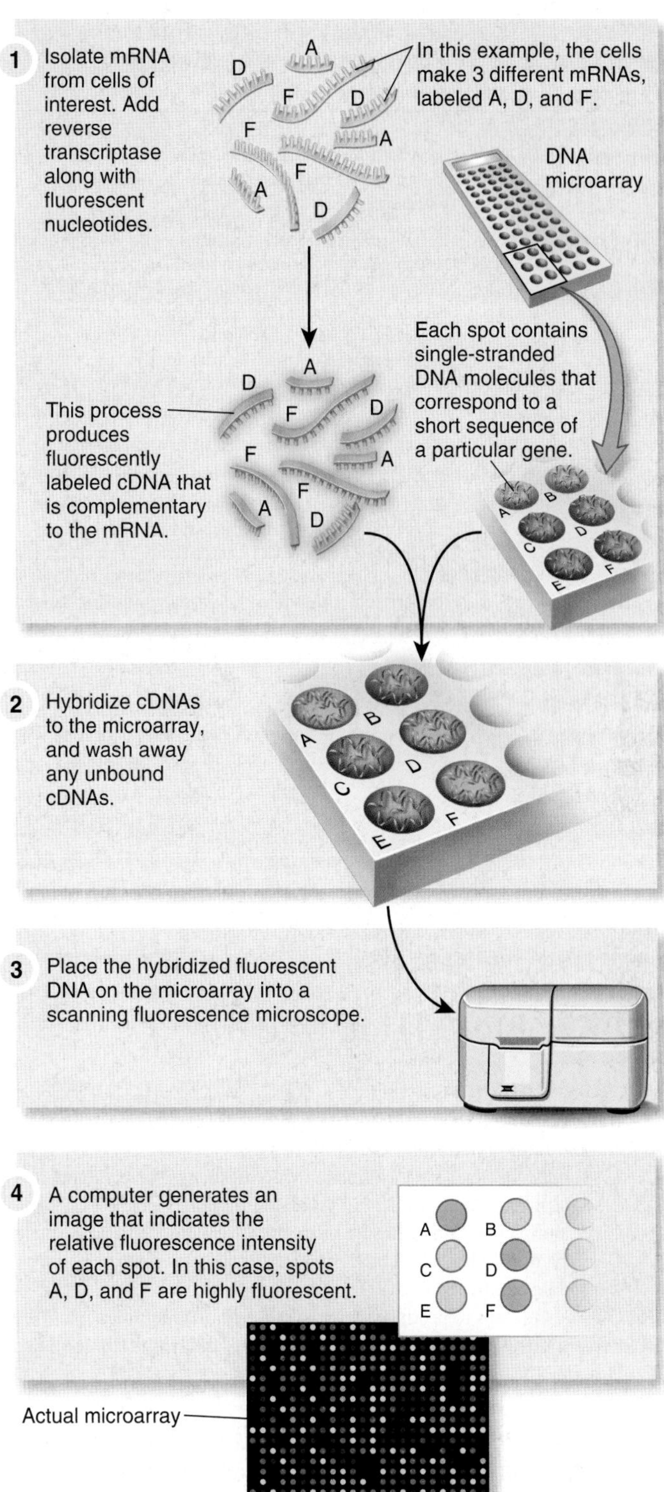

Figure 20.10 Identifying transcribed genes within a DNA microarray. In this simplified example, only three cDNAs specifically hybridize to spots on the microarray. Those genes were expressed in the cells from which the mRNA was isolated. In an actual experiment, the array typically contains hundreds or thousands of different cDNAs and tens of thousands of different spots.

Concept Check: If a fluorescent spot appears on a microarray, what information does this provide regarding gene expression?

Table 20.2 Applications of DNA Microarrays

Application	Description
Cell-specific gene expression	A comparison of microarray data using cDNAs derived from mRNA of different cell types can identify genes that are expressed in a cell-specific manner.
Gene regulation	Because environmental conditions play an important role in gene regulation, a comparison of microarray data using cDNA derived from mRNA from cells exposed to two different environmental conditions may reveal genes that are induced under one set of conditions and repressed under another set.
Elucidation of metabolic pathways	Genes that encode proteins that participate in a common metabolic pathway are often expressed together and can be revealed from a microarray analysis.
Tumor profiling	Different types of cancer cells exhibit striking differences in their gene expression profiles, which can be revealed by a DNA microarray analysis. This approach is gaining use as a tool to classify tumors that are sometimes morphologically indistinguishable.
Genetic variation	A mutant allele may not hybridize to a spot on a microarray as well as a wild-type allele. Therefore, microarrays are gaining use as a tool for detecting genetic variation. This application has been used to identify disease-causing alleles in humans and to identify mutations that contribute to quantitative traits in plants and other species.
Microbial strain identification	Microarrays can distinguish between closely related bacterial species and subspecies.

20.3 Biotechnology

Learning Outcomes:

1. Describe how recombinant microorganisms are used to make medicines and reduce pollutants in the environment.
2. Distinguish between a gene knockout versus a gene replacement, and describe how both are used in the study of human diseases.
3. Give examples of using transgenic livestock to make medical products.
4. Describe the method of using *Agrobacterium tumefaciens* to produce transgenic plants.
5. Outline the steps to produce Dolly, the first cloned mammal.
6. Explain the technique of DNA fingerprinting, and list its common uses.
7. Describe the technique of gene therapy.

Biotechnology is defined as the use of living organisms, or products from living organisms, as a way to benefit humans. Although the term has become associated with molecular genetics, biotechnology is not a new topic. Its use started about 12,000 years ago, when humans began to domesticate animals and plants for the production of food. Since that time, many species of microorganisms, plants, and animals have become routinely used in agriculture and medicine. Beginning in the 1970s, genetic engineering has provided new ways to make use of living organisms to benefit humans.

In this section, we will consider the applications of biotechnology that involve the genetic engineering of microorganisms, plants, and animals. We will also examine several topics that you often hear about in the news, such as the cloning of mammals, DNA fingerprinting, and gene therapy.

Important Medicines Are Produced by Recombinant Microorganisms

Several important medicines are now produced by recombinant organisms. These include tissue plasminogen activator (TPA), used to dissolve blood clots in heart attack patients; factor VIII, used to treat people with the disease hemophilia; and insulin, used to treat people with diabetes. In 1982, the U.S. Food and Drug Administration approved the sale of human insulin made by recombinant bacteria. In healthy individuals, insulin is produced by the beta cells of the pancreas. Insulin functions to regulate several physiological processes, including the uptake of glucose into muscle cells. Persons with insulin-dependent diabetes cannot synthesize an adequate amount of insulin due to a defect in their beta cells. Today, people like Jacob, who was introduced at the beginning of the chapter, are usually treated with human insulin that is made by genetically engineered bacteria. Prior to 1982, insulin was isolated from pancreases removed from cattle and pigs. Unfortunately, in some cases, diabetic individuals developed an allergic response to cow and pig insulin. These patients had to use expensive combinations of insulin from human cadavers and other animals. Now, they can use human insulin made by recombinant bacteria.

Insulin is a hormone composed of two polypeptides: the A and B chains, that are held together with disulfide bonds. To produce this hormone, the coding sequence of the A or B chains is inserted into a plasmid vector next to the coding sequence of the *E. coli* protein β-galactosidase (**Figure 20.11**). After such vectors are introduced into bacteria, the cells produce many copies of a fusion protein (β-gal-insulin) comprising β-galactosidase and the A or B chain. Why is this step done? The reason is because the A and B chains are rapidly degraded when expressed in bacterial cells by themselves. The fusion proteins, however, are not. These fusion proteins are then extracted from the bacterial cells and treated with a chemical, cyanogen bromide (CNBr), which cleaves a peptide bond at a methionine (Met) that is found at the end of the β-galactosidase sequence, thereby separating β-galactosidase from the A or B chain. The A and B chains are purified and mixed together under conditions in which they can fold and associate with each other via disulfide bonds to form a functional insulin hormone molecule.

Microorganisms Can Reduce Pollutants

Bioremediation is the use of living organisms, typically microorganisms or plants, to detoxify pollutants in the environment. As its name suggests, it is a biological remedy for pollution. During microbial bioremediation, enzymes produced by a microorganism modify a toxic pollutant by altering or transforming its structure. In many cases, the toxic pollutant is degraded, yielding less complex, nontoxic metabolites.

Since the early 1900s, microorganisms have been used in sewage treatment plants to degrade sewage. More recently, the field of bioremediation has expanded into the treatment of hazardous and refractory chemical wastes—chemicals that are difficult to degrade and are

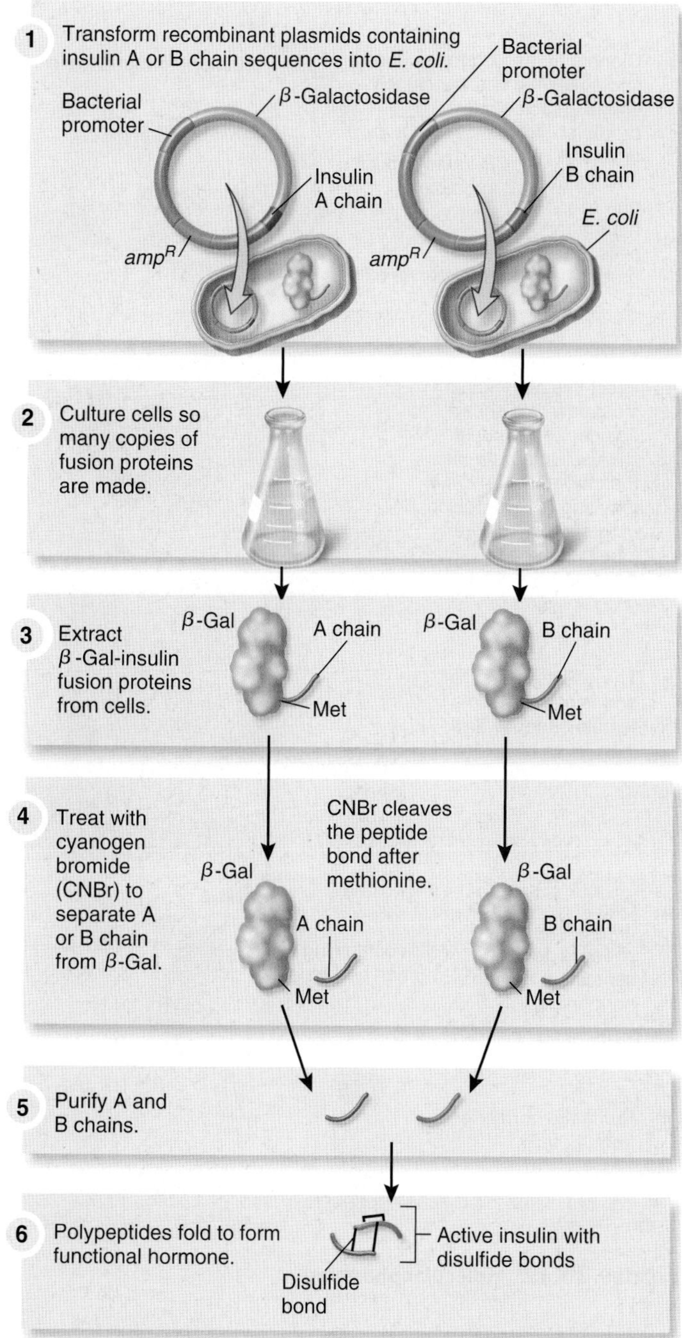
1 Transform recombinant plasmids containing insulin A or B chain sequences into E. coli.
Bacterial promoter
β-Galactosidase
Insulin A chain
ampR
Bacterial promoter
β-Galactosidase
Insulin B chain
E. coli
ampR
2 Culture cells so many copies of fusion proteins are made.
3 Extract β-Gal-insulin fusion proteins from cells.
β-Gal
A chain
Met
β-Gal
B chain
Met
4 Treat with cyanogen bromide (CNBr) to separate A or B chain from β-Gal.
CNBr cleaves the peptide bond after methionine.
β-Gal
A chain
Met
β-Gal
B chain
Met
5 Purify A and B chains.
6 Polypeptides fold to form functional hormone.
Disulfide bond
Active insulin with disulfide bonds

Figure 20.11 The use of bacteria to make human insulin.

BIOLOGY PRINCIPLE Biology affects our society. This method of making insulin, which was developed by researchers, is now used to make insulin for most people suffering from insulin-dependent diabetes.

Concept Check: *Why are the A and B chains made as fusion proteins with β-galactosidase?*

usually associated with industrial activity. These pollutants include heavy metals, petroleum hydrocarbons, as well as pesticides, herbicides, and organic solvents. Many new applications are being tested that use microorganisms to degrade these pollutants. The field of bioremediation has been fostered by better knowledge of how pollutants are degraded by microorganisms, the identification of new and useful strains of microbes, and the ability to enhance bioremediation through genetic engineering.

In 1980, in a landmark case (*Diamond v. Chakrabarty*), the U.S. Supreme Court ruled that a live, recombinant microorganism is patentable as a "manufacture or composition of matter." The first recombinant microorganism to be patented was an "oil-eating" bacterium that contained a laboratory-constructed plasmid. This strain could oxidize the hydrocarbons commonly found in petroleum. It grew faster on crude oil than did any of the natural strains that were tested. Was it a commercial success? The answer is No, because this recombinant strain metabolized only a limited number of toxic compounds. Unfortunately, the strain did not degrade many higher-molecular-weight compounds that tend to persist in the environment.

Bioremediation is a developing industry. Recombinant microorganisms can provide an effective way to decrease the levels of toxic chemicals in our environment. This approach requires careful studies to demonstrate that recombinant organisms are effective at reducing pollutants and safe when released into the environment.

Gene Replacements and Knockouts Are Used to Understand Gene Function and Human Disease

Let's now turn our attention to the genetic engineering of animals. Researchers can introduce a cloned gene into an oocyte, a fertilized egg, or embryonic cells to produce animals that carry the cloned gene. The term **transgenic** is used to describe an organism that carries genes from another organism that were introduced using molecular techniques such as gene cloning. Transgenic organisms are also called **genetically modified organisms (GMOs)**.

Such GMOs are typically made in two ways. In some cases, the cloned gene inserts randomly into the genome and results in **gene addition**—the insertion of cloned gene into the genome. Alternatively, a cloned gene may recombine with the normal gene on a chromosome, a phenomenon called **gene replacement**. For eukaryotic species that are diploid, only one of the two copies is initially replaced. In other words, the initial gene replacement produces a heterozygote carrying one normal copy of the gene and one copy that has been replaced with a cloned gene. Heterozygotes can be crossed to each other to obtain homozygotes, which carry both copies of the cloned gene. If the cloned gene carries a mutation that inactivates the normal gene's function, such a homozygote is said to have undergone a **gene knockout**. The inactive cloned gene has replaced both copies of the normal gene, and the normal gene's function is said to be inoperative, or knocked out. Gene replacements and gene knockouts have become powerful tools for understanding gene function.

A particularly exciting avenue of gene replacement research is its application in the study of human disease. As an example, let's consider the disease cystic fibrosis (CF), which is one of the most common and severe inherited human disorders. In humans, the defective gene that causes CF has been identified. Likewise, the homologous gene in mice was later identified. Using the technique of gene replacement, researchers have produced mice that are homozygous for the same type of mutation that is found in humans with CF. Such mice exhibit disease symptoms resembling those found in humans, namely, respiratory infections and digestive abnormalities. Why is this approach useful? These mice can be used as model organisms to study this human disease. Furthermore, these mice models have been used to test the effects of various therapies in the treatment of the disease.

Biotechnology Holds Promise in Producing Transgenic Livestock

The technology of producing transgenic mice has been extended to other animals, and much research is under way to develop transgenic species of livestock, including fish, sheep, pigs, goats, and cattle. A novel avenue of research involves the production of medically important proteins in the mammary glands of livestock. This approach is sometimes called **molecular pharming**. (The word pharming refers to the use of genetically engineered farm animals or crops to make pharmaceuticals.) Several human proteins have been successfully produced in the milk of domestic livestock such as sheep, goats, and cattle. These include factor IX to treat a certain type of hemophilia, tissue plasminogen activator (TPA) to dissolve blood clots, and α_1-antitrypsin to treat emphysema.

How are human proteins, such as hormones, made so they are produced in the milk of livestock? As we learned in Chapter 13, gene regulation may promote the expression of genes in certain cell types. Researchers have identified specific genes that are transcribed only in lactating mammary cells. One example is the gene that encodes β-lactoglobulin, a protein that is found in the milk of sheep and cows. One strategy for producing a human hormone in the milk of livestock is to clone the human hormone gene into a plasmid vector next to the β-lactoglobulin promoter (Figure 20.12). The plasmid is then injected into an oocyte, such as a sheep oocyte, where it integrates into the genome. The egg is then fertilized by exposure to sperm and implanted into the uterus of a female sheep. The resulting offspring carries the cloned gene. If the offspring is a female, the protein encoded by the human gene will be expressed within the mammary gland and secreted into the milk. The milk is then obtained from the animal, and the human protein isolated and purified.

Compared with the production of proteins in bacteria, one advantage of molecular pharming is that certain proteins are more likely to function properly when expressed in mammals. This may be due to post-translational modifications of proteins that occur in mammals but not in bacteria. In addition, certain proteins may be degraded rapidly or folded improperly when expressed in bacteria. Furthermore, the yield of recombinant proteins in milk can be quite large. Each dairy cow, for example, produces about 10,000 liters of milk per year. In some cases, a transgenic cow can produce approximately 1 g/L of the transgenic protein in its milk.

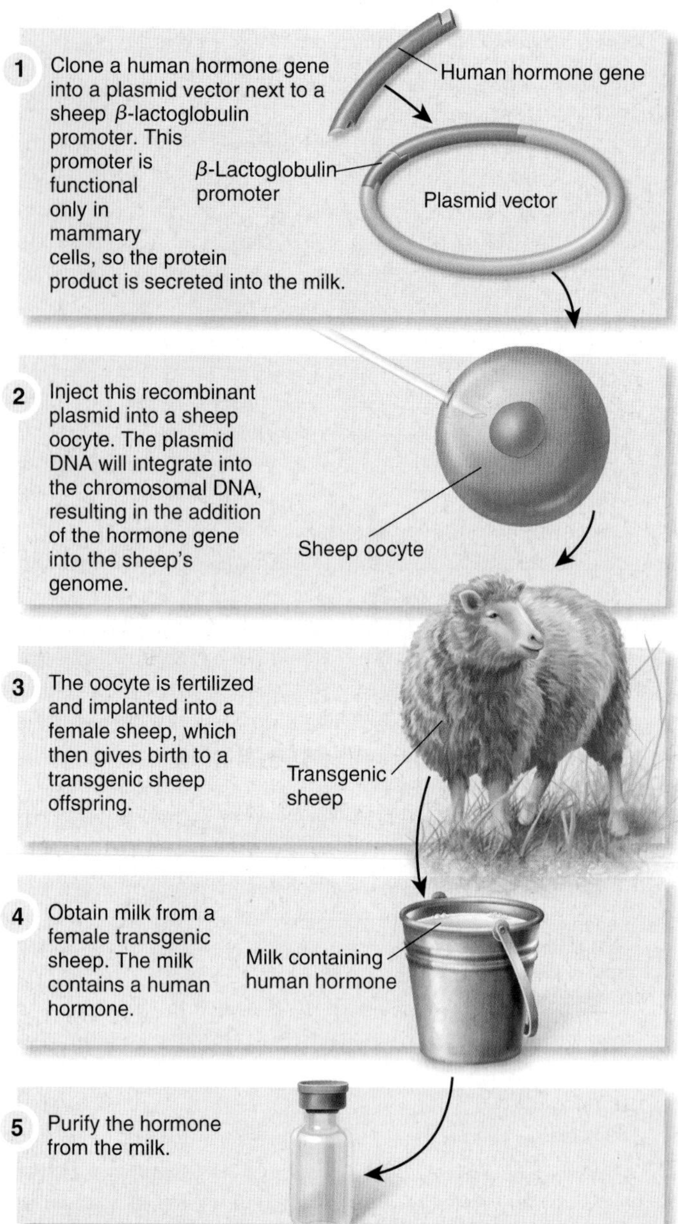

1 Clone a human hormone gene into a plasmid vector next to a sheep β-lactoglobulin promoter. This promoter is functional only in mammary cells, so the protein product is secreted into the milk.
Human hormone gene
β-Lactoglobulin promoter
Plasmid vector
2 Inject this recombinant plasmid into a sheep oocyte. The plasmid DNA will integrate into the chromosomal DNA, resulting in the addition of the hormone gene into the sheep's genome.
Sheep oocyte
3 The oocyte is fertilized and implanted into a female sheep, which then gives birth to a transgenic sheep offspring.
Transgenic sheep
4 Obtain milk from a female transgenic sheep. The milk contains a human hormone.
Milk containing human hormone
5 Purify the hormone from the milk.

Figure 20.12 Molecular pharming.

Agrobacterium tumefaciens Is Used to Make Transgenic Plants

The production of transgenic plants is somewhat easier than producing transgenic animals because many somatic tissues from plants are totipotent, which means that an entire organism can be regenerated from such tissues. Therefore, a transgenic plant can be made by the introduction of cloned genes into somatic tissue, such as the tissue of a leaf. After the cells of a leaf have become transgenic, an entire plant can be regenerated by treating the leaf with plant growth hormones that cause it to form roots and shoots. The result is an entire plant that is transgenic.

Molecular biologists often use the bacterium *Agrobacterium tumefaciens* to produce transgenic plants. This bacterium contains a plasmid known as the **Ti plasmid**, for tumor-inducing plasmid. The plasmid has a region called the T DNA (for transferred DNA) that is transferred from the bacterium to the plant cell and becomes integrated into the plant cell's genome. After this occurs, genes within the T DNA are expressed that cause uncontrolled plant cell growth. This produces a crown gall tumor, a bulbous growth on the plant. Natural strains of *A. tumefaciens* infect many different species of plants, and researchers have developed newer strains that have an even greater capacity to infect different plant species.

Researchers have modified the Ti plasmid to use it as a vector to introduce a gene of interest into plant cells. In the modified Ti plasmid, the genes that cause a crown gall tumor have been removed or inactivated. In addition, the T DNA carries a selectable marker gene to allow selection of plant cells that have taken up the T DNA. *Kan*^{R}, a gene that provides resistance to the antibiotic kanamycin, is commonly used as a selectable marker. Last, the Ti plasmid used in cloning experiments is modified to contain unique restriction sites for the convenient insertion of any gene of interest.

Figure 20.13 shows the general strategy for producing transgenic plants via T DNA–mediated gene transfer. A gene of interest is inserted into the T DNA of a genetically engineered Ti plasmid and then introduced into *A. tumefaciens*. Plant cells are then exposed to *A. tumefaciens* carrying the Ti plasmid. After allowing time for T DNA transfer, the plant cells are grown on a solid medium that contains the antibiotics kanamycin and carbenicillin. Kanamycin kills any plant cells that have not taken up the T DNA, and carbenicillin kills *A. tumefaciens*. Therefore, the only surviving cells are those plant cells that have integrated the T DNA into their genome. Because the T DNA also contains the gene of interest, the selected plant cells are expected to have received this gene as well. The cells are then transferred to a medium that contains the plant growth hormones necessary for the regeneration of entire plants.

Many transgenic plants have been approved for human consumption. Their production has become routine practice for several agriculturally important plant species, including alfalfa, corn, cotton, soybean, tobacco, and tomato. Transgenic plants have been given characteristics that are agriculturally useful, such as those that improve plant quality and insect, disease, or herbicide resistance. In terms of plant quality, gene additions have been made to improve the nutritional value of some plants, such as making the canola grain produce more oil.

Frequently, transgenic research has sought to produce plant strains that are resistant to insects, disease, and herbicides. A successful example of the use of transgenic plants has involved the introduction of genes from *Bacillus thuringiensis* (Bt). This bacterium produces toxins that kill certain types of caterpillars and beetles and has been widely used as an insecticide for several decades. These toxins are proteins that are encoded in the genome of *B. thuringiensis*. Researchers have succeeded in cloning toxin genes from *B. thuringiensis* and transferring those genes into plants. Such Bt varieties of plants produce the toxins themselves and therefore are resistant to many types of caterpillars and beetles. This allows farmers to reduce the use of conventional insecticides and improve yield. Examples of commercialized crops include Bt corn (Figure 20.14a) and Bt cotton. Since their introduction in 1996, the commercial use of these two Bt crops has steadily increased (Figure 20.14b).

The use of transgenic agricultural plants has been strongly opposed by some people. What are the perceived risks? One potential risk is that transgenes in commercial crops could endanger native species. For example, Bt crops may kill pollinators of native species. Another concern is that the planting of transgenic crops could lead to the evolution of resistant insects that would make Bt crops ineffective. To prevent this from happening, researchers are producing transgenic strains that carry more than one toxin gene, which makes it more difficult for insect resistance to arise. A third concern is the potential for transgenic plants to elicit allergic reactions. Despite these and other concerns, many farmers are embracing transgenic crops, and their use continues to rise.

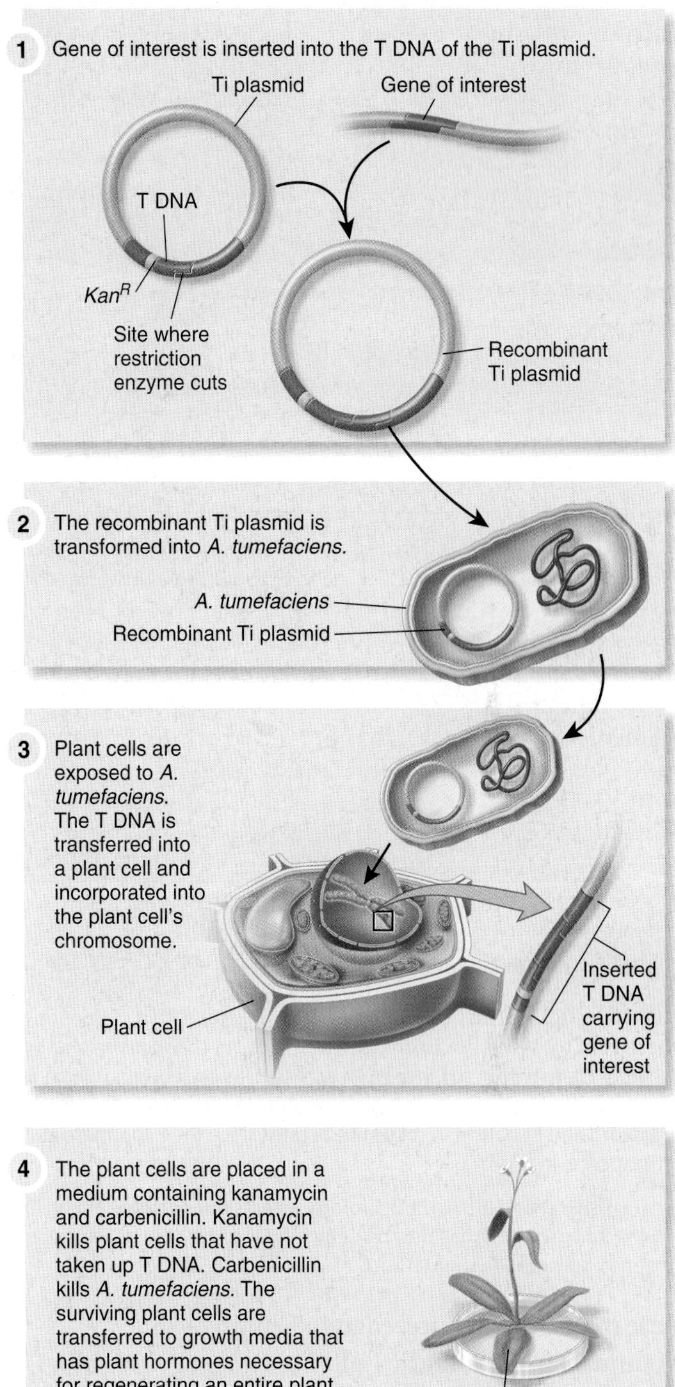
1 Gene of interest is inserted into the T DNA of the Ti plasmid.
Ti plasmid
Gene of interest
T DNA
KanR
Site where restriction enzyme cuts
Recombinant Ti plasmid
2 The recombinant Ti plasmid is transformed into A. tumefaciens.
A. tumefaciens
Recombinant Ti plasmid
3 Plant cells are exposed to A. tumefaciens. The T DNA is transferred into a plant cell and incorporated into the plant cell's chromosome.
Plant cell
Inserted T DNA carrying gene of interest
4 The plant cells are placed in a medium containing kanamycin and carbenicillin. Kanamycin kills plant cells that have not taken up T DNA. Carbenicillin kills A. tumefaciens. The surviving plant cells are transferred to growth media that has plant hormones necessary for regenerating an entire plant.
Plant with cloned gene

Figure 20.13 **Using the Ti plasmid and *Agrobacterium tumefaciens* to produce transgenic plants.**

Concept Check: Which region of the Ti plasmid is transferred to the plant cell?

(a) A field of Bt corn

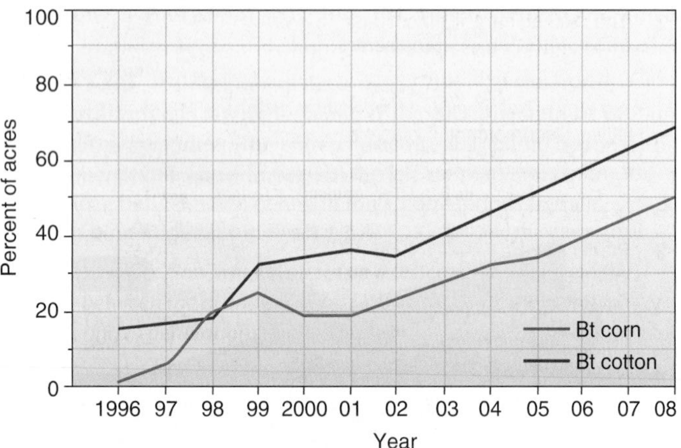

(b) Bt corn and Bt cotton usage since 1996

Figure 20.14 **The production of Bt crops.** **(a)** A field of Bt corn. These corn plants carry a toxin gene from *Bacillus thuringiensis* that provides them with resistance to insects such as corn borers, which are a major pest of corn plants. **(b)** A graph showing the increase in usage of Bt corn and Bt cotton in the United States since their commercial introduction in 1996.

BIOLOGY PRINCIPLE **Biology affects our society.** The field of agriculture has been greatly influenced by the development of genetically modified organisms (GMOs) such as Bt corn.

Researchers Have Succeeded in Cloning Mammals from Somatic Cells

We now turn our attention to cloning as a way of genetically manipulating plants and animals. The term cloning has several different meanings. At the beginning of this chapter we discussed gene cloning, which involves methods that produce many copies of a gene. The cloning of a multicellular organism, called **reproductive cloning**, is a different matter. By accident, this happens in nature. Identical twins are genetic clones that began from the same fertilized egg. Similarly, researchers can take mammalian embryos at an early stage of development (for example, the two- to eight-cell stage), separate the cells, implant them into the uterus of a female, and obtain multiple births of genetically identical individuals.

As previously noted, the reproductive cloning of new individuals is relatively easy in the case of plants, which can be cloned from somatic cells. Until recently, this approach had not been possible with

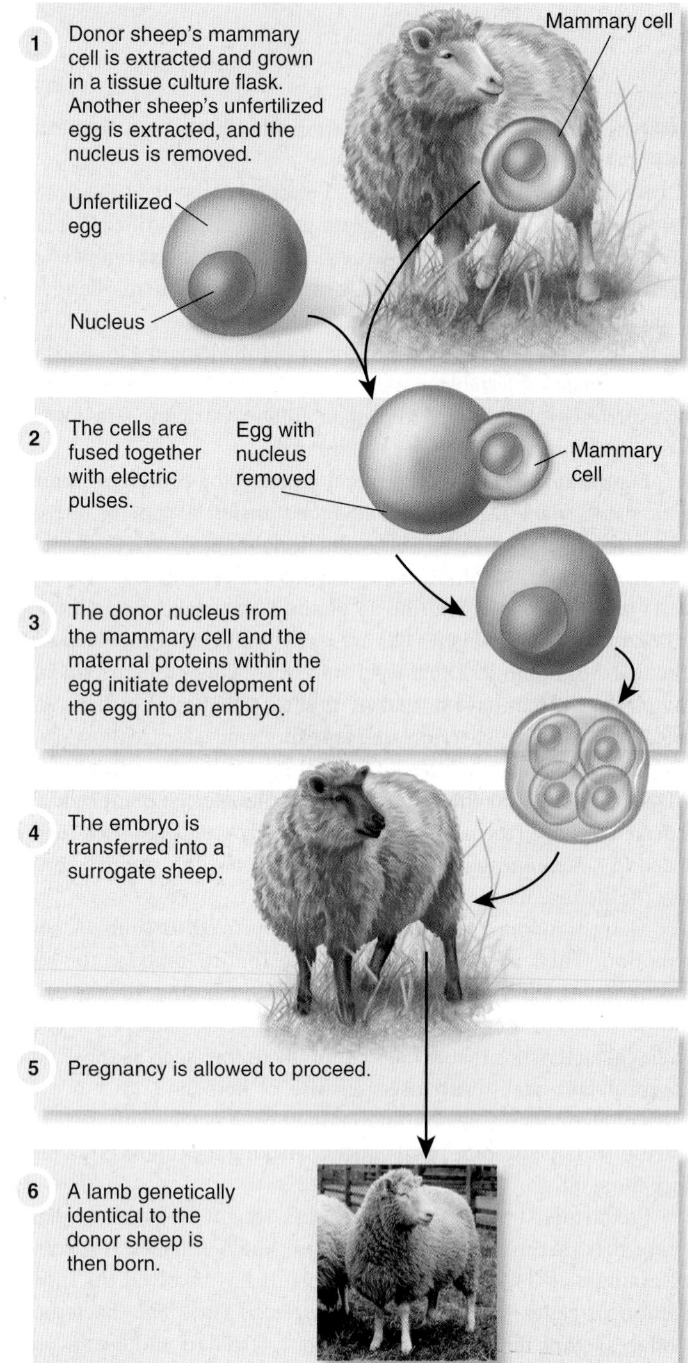

Figure 20.15 **Protocol for the successful cloning of sheep.** In this procedure, the genetic material from a somatic cell is used to make a cloned mammal, in this case, the sheep Dolly.

Concept Check: Did all of Dolly's DNA come from a mammary cell? Explain.

mammals. Scientists believed that chromosomes within the somatic cells of mammals had incurred irreversible genetic changes that rendered them unsuitable for reproductive cloning. However, this hypothesis has proven to be incorrect. In 1996, English embryologist Ian Wilmut and his colleagues at the University of Edinburgh produced clones of sheep using the genetic material from somatic cells. As you may have heard, they named the first cloned lamb Dolly.

How was Dolly produced? As shown in **Figure 20.15**, the researchers removed mammary cells from an adult female sheep and grew them in the laboratory. They then extracted the nucleus from a sheep oocyte and fused the diploid mammary cell with the enucleated oocyte cell. Fusion was promoted by electric pulses. After fusion, the zygote was implanted into the uterus of an adult sheep. One hundred and forty-eight days later, Dolly was born.

Dolly was (almost) genetically identical to the sheep that donated the mammary cell. Dolly and the donor sheep were (almost) genetically identical in the same way that identical twins are. They carry the same set of genes and look remarkably similar. The reason they may not be completely identical is because Dolly and her somatic cell donor may have some minor genetic differences due to possible differences in their mitochondrial DNA.

Mammalian reproductive cloning is still at an early stage of development. Nevertheless, producing Dolly showed it is technically possible. In recent years, cloning using somatic cells has been achieved in several mammalian species, including sheep, cows, mice, goats, pigs, and cats. In 2002, the first pet was cloned, which was named CC for carbon copy (see chapter opening photo). The cloning of mammals provides the potential for many practical applications. With regard to livestock, cloning would enable farmers to use the somatic cells from their best animals to produce genetically homogeneous herds. This could be advantageous in terms of agricultural yield, although such a genetically homogeneous herd may be more susceptible to certain diseases.

Although reproductive cloning may have practical uses in the field of agriculture, our society has become greatly concerned with the possibility of human cloning. This prospect has raised serious ethical questions. Some people feel it is morally wrong and threatens the basic fabric of parenthood and family. Others feel it is a modern technology that offers a new avenue for reproduction, one that could be offered to infertile couples, for example. Human cloning is a complex subject with many more viewpoints than these two. In the public sector, the sentiment toward human cloning has been generally negative. Many countries have issued a complete ban on human cloning, whereas others permit limited research in this area. In the future, our society will have to wrestle with the legal and ethical aspects of cloning as it applies not only to animals but also to people.

DNA Fingerprinting Is Used for Identification and Relationship Testing

DNA fingerprinting is a technology for identifying and distinguishing among individuals based on variations in their DNA. Like the human fingerprint, the DNA of each individual is a distinctive characteristic that provides a means of identification. When subjected to traditional DNA fingerprinting, selected fragments of chromosomal DNA produce a series of bands on a gel (**Figure 20.16a**). The unique pattern of these bands is usually a distinguishing feature of each individual.

DNA fingerprinting is now done using PCR, which amplifies **short tandem repeat sequences (STRs)**—short DNA sequences that are repeated many times in a row. Such tandem repeat sequences, which are noncoding regions of chromosomal DNA, are found at specific locations in the genomes of all species. The number of repeats at each location tends to vary from one individual to the next. This variation provides a way to distinguish one individual from another. Using primers that are complementary to DNA sequences that flank each STR, the STRs from a sample of DNA are amplified by PCR and then separated by gel electrophoresis according to their molecular masses. As in DNA sequencing, the amplified STR fragments are fluorescently labeled. A laser excites the fluorescent molecule within an STR, and a detector records the amount of fluorescence emission for each STR. The DNA fingerprint yields a series of peaks, each peak having a characteristic molecular mass (**Figure 20.16b**). In this automated approach, the pattern of peaks is an individual's DNA fingerprint.

What are the uses of DNA fingerprinting? First, DNA fingerprinting has gained acceptance as a precise method of identification. In medicine, it is used to identify different species of bacteria and

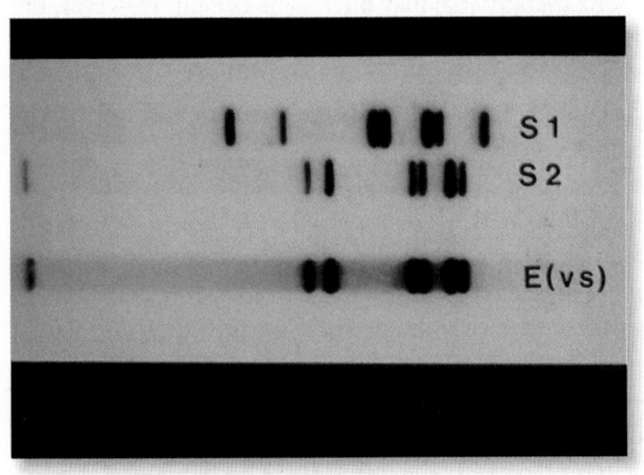

(a) Traditional DNA fingerprinting

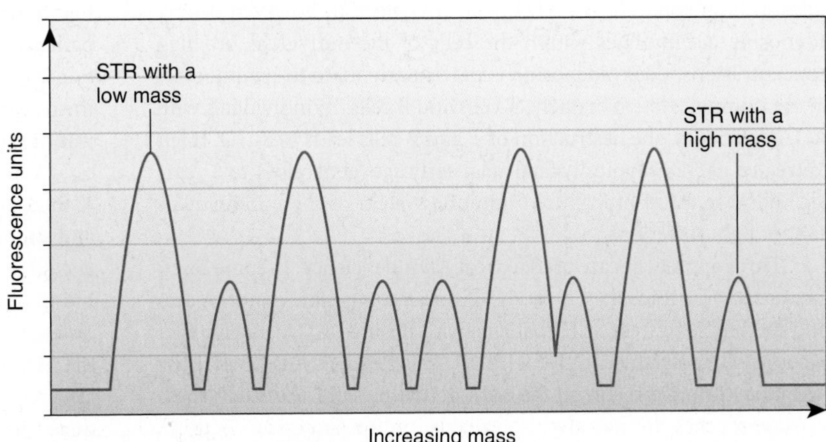

(b) Automated DNA fingerprinting

Figure 20.16 DNA fingerprinting. **(a)** Chromosomal DNA found at a crime scene (E) was compared with DNA from two different individuals, suspect 1 (S 1) and suspect 2 (S 2), using traditional DNA fingerprinting. Their DNA appears as a series of bands on a gel. **(b)** Automated DNA fingerprinting measures the masses of fluorescently labeled DNA segments called short tandem repeat sequences (STRs) from a selected individual. A printout from the fluorescence detector is shown here. Two individuals are almost always different in the pattern of peaks they exhibit.

Concept Check: *Which suspect's DNA fingerprint (S 1 or S 2) matches the DNA evidence collected at a crime scene?*

fungi, and it can even distinguish among closely related strains of the same species. This is useful so clinicians can treat patients with the appropriate antibiotic or fungicide.

A second common use is forensics—the use of scientific techniques in the investigation of crimes. DNA fingerprinting can be used as evidence that an individual was at a crime scene. Forensic DNA was first used in the U.S. court system in 1986. When a DNA sample taken from a crime scene matches the DNA fingerprint of an individual, the probability that a match could occur simply by chance can be calculated. Each STR size is given a probability score based on its observed frequency within a reference human population (Caucasian, Asian, and so on). An automated DNA fingerprint contains many peaks, and the probability scores for each peak are multiplied together to arrive at the likelihood that a particular pattern of peaks would be observed. For example, if a DNA fingerprint contains 20 fluorescent peaks, and the probability of an individual having each peak is 1/4, then the likelihood of having that pattern would be $(1/4)^{20}$, or roughly 1 in 1 trillion. Therefore, a match between two samples is rarely a matter of random chance.

Another important use of DNA fingerprinting is to establish paternity and other family relationships. Persons who are related genetically have some peaks in common. The number they share depends on the closeness of their genetic relationship. For example, offspring are expected to receive half of their peaks from one parent and half from the other. Therefore, DNA fingerprinting can be used as evidence in paternity cases.

Feature Investigation

Blaese and Colleagues Performed the First Gene Therapy to Treat ADA Deficiency

Gene therapy is the introduction of cloned genes into living cells in an attempt to cure disease. Many current research efforts in gene therapy are aimed at alleviating human diseases caused by defective genes. Examples include cystic fibrosis, sickle cell disease, phenylketonuria, hemophilia, and severe combined immunodeficiency (SCID). Human gene therapy is still at an early stage of development. Relatively few patients have been successfully treated with gene therapy in spite of a large research effort. In addition, experimental gene therapy in humans has been associated with adverse reactions that have raised questions about its safety. In 1999, a patient in a clinical trial died from a reaction to gene therapy treatment.

Adenosine deaminase (ADA) deficiency is an autosomal recessive disorder due to a lack of the enzyme adenosine deaminase. When present, this enzyme deaminates the nucleoside deoxyadenosine, which is an important step in the proper metabolism of nucleosides. If both copies of the *ADA* gene are defective, however, deoxyadenosine accumulates within the cells of the individual. At high concentrations, deoxyadenosine is particularly toxic to lymphocytes in the immune system, namely, T cells and B cells. In individuals with ADA deficiency, the destruction of T and B cells leads to SCID. If left untreated, SCID is typically fatal at an early age (generally 1 to 2 years old) because the compromised immune system of these individuals cannot fight infections.

Three approaches are used to treat ADA deficiency. In some cases, a bone marrow transplant is received from a compatible donor. A second method is to treat SCID patients with purified ADA enzyme that is coupled to polyethylene glycol (PEG). This PEG-ADA is taken up by lymphocytes and can correct the ADA deficiency. Unfortunately, these two approaches are not always available and/or successful. A third, more recent approach is to treat ADA patients with gene therapy.

On September 14, 1990, the first human gene therapy was approved for a young girl suffering from ADA deficiency. This work was carried out by a large team of researchers, including American physicians R. Michael Blaese, Kenneth Culver, W. French Anderson, and several collaborators at the National Institutes of Health (NIH). The general aim of this therapy was to remove lymphocytes from the blood of the girl with SCID, introduce the normal *ADA* gene into the cells via a retrovirus, and then return them to her bloodstream.

Figure 20.17 outlines the protocol for the experimental treatment. The researchers removed lymphocytes from the girl and cultured them in a laboratory. The lymphocytes were then infected with a recombinant retrovirus that contained the normal *ADA* gene. During the reproductive cycle of the retrovirus, the retroviral genetic material was inserted into the host cell's DNA. Therefore, because this retrovirus contained the normal *ADA* gene, this gene also was inserted into the chromosomal DNA of the girl's lymphocytes, thereby correcting the defect in ADA. After this occurred in the laboratory, the lymphocytes were reintroduced back into the patient.

In this clinical trial, two U.S. patients were enrolled, and a third patient was later treated in Japan. The data in Figure 20.17 show the results of this trial for one of the three patients. Over the course of 2 years, this patient was given 11 infusions of lymphocytes that had been corrected with the normal *ADA* gene. Even after 4 years, this patient's lymphocytes were still making ADA. These results suggest that this first gene therapy trial may offer some benefit. However, the patients in this study were also treated with PEG-ADA as an additional therapy to prevent the adverse symptoms of SCID. Therefore, the researchers could not determine whether gene transfer into T cells was of significant clinical benefit.

Another form of SCID, termed SCID-X1, is inherited as an X-linked trait. SCID-X1 is characterized by a block in T-cell growth and differentiation. This block is caused by mutations in the gene encoding the γ_c cytokine receptor, which plays a key role in the recognition of signals that are needed to promote the growth, survival, and differentiation of T cells. A gene therapy trial for SCID-X1 similar to the trial shown in Figure 20.17 was initiated in 2000. A normal γ_c cytokine receptor gene was cloned into a retroviral vector and then introduced into SCID-X1 patients' lymphocytes. The lymphocytes were then reintroduced back into their bodies. At a 10-month follow-up, T cells expressing the normal γ_c cytokine receptor were detected in two patients. Most importantly, the T-cell counts in these two patients had risen to levels that were comparable to those in normal individuals.

This clinical trial was the first clear demonstration that gene therapy can offer clinical benefits, providing in these cases what seemed to be a complete correction of the disease phenotype. However, in a

Figure 20.17 **The first human gene therapy for adenosine deaminase (ADA) deficiency by Blaese and colleagues.**

BioConnections: Look back at Figure 18.4b. How does the genetic material of a retrovirus get incorporated into the host cell's genome?

HYPOTHESIS Infecting lymphocytes with a retrovirus containing the normal *ADA* gene will correct the inherited deficiency of the mutant *ADA* gene in patients with ADA deficiency.

KEY MATERIALS A retrovirus with the normal *ADA* gene.

Experimental level | **Conceptual level**

1 Remove *ADA*-deficient lymphocytes from the patient with severe combined immunodeficiency disease (SCID).

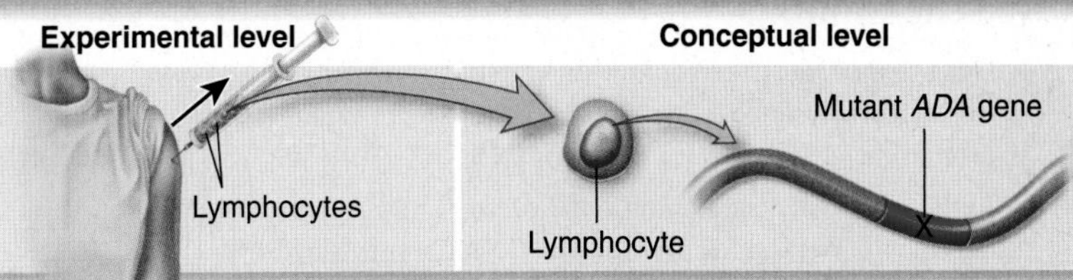

2 Culture the cells in a laboratory.

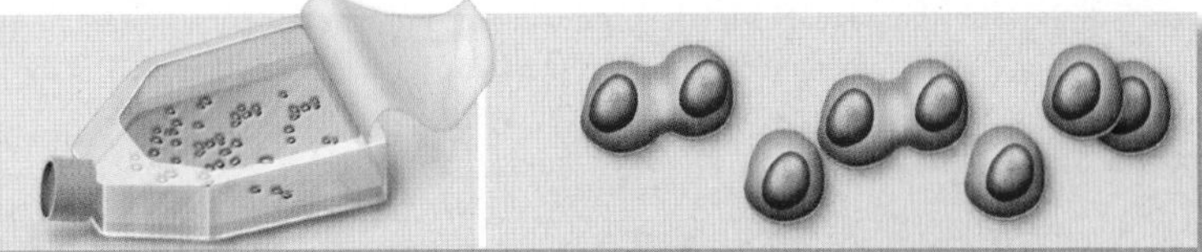

3 Infect the lymphocytes with a retrovirus that contains the normal *ADA* gene. As described in Chapter 18, retroviruses insert their DNA into the host cell chromosome as part of their reproductive cycle.

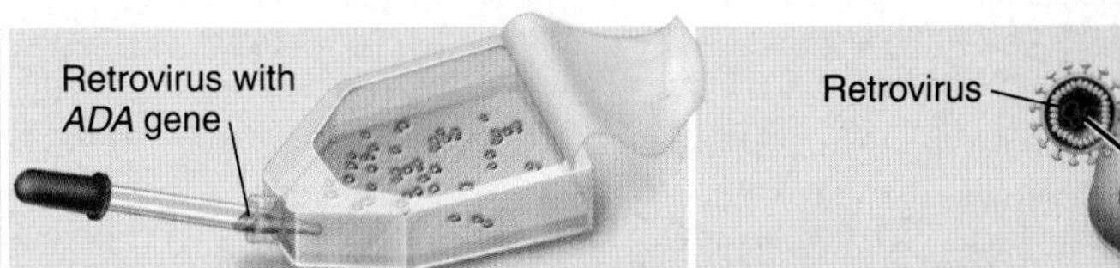

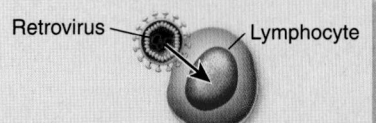

4 Infuse the *ADA*-gene-corrected lymphocytes back into the SCID patient.

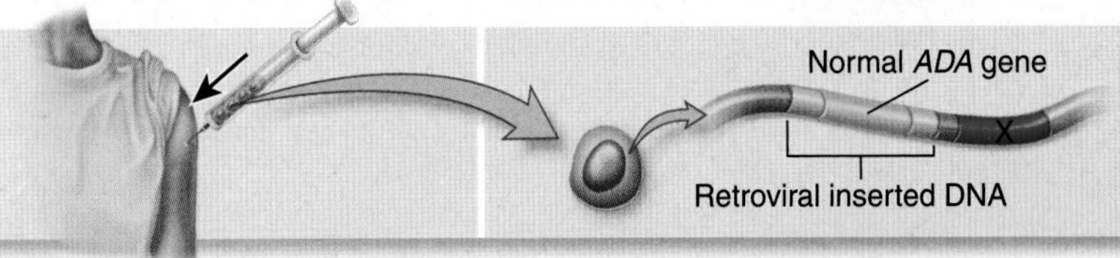

5 **THE DATA**

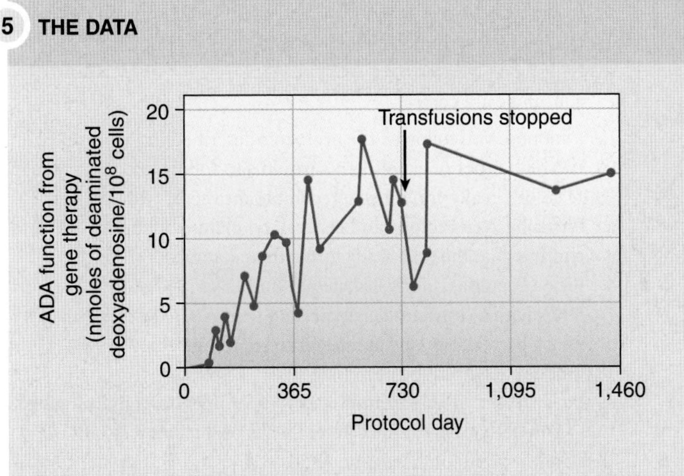

6 **CONCLUSION** The introduction of a cloned *ADA* gene into lymphocytes via gene therapy resulted in higher ADA function, even after 4 years.

7 **SOURCE** Blaese, R. Michael et al. 1995. T lymphocyte-directed gene therapy for ADA-SCID: Initial trial results after 4 years. *Science* 270:475–480.

French study involving 10 SCID-X1 patients, an unexpected and serious side effect occurred. Within 3 years of gene therapy treatment, 3 out of the 10 treated children developed leukemia—a form of cancer involving the proliferation of white blood cells. In these cases, the disease was caused by the integration of the retroviral vector next to an oncogene, a gene that promotes cancer, in the patients' genome. The development of leukemia in these patients has halted many clinical trials involving gene therapy.

Experimental Questions

1. What is gene therapy? What is ADA deficiency?
2. In the investigation of Figure 20.17, how did the researchers treat ADA deficiency?
3. How successful was the gene therapy for ADA deficiency?

Summary of Key Concepts

- Recombinant DNA technology is the use of laboratory techniques to bring together DNA fragments from two or more sources.

20.1 Gene Cloning

- Gene cloning, the process of making multiple copies of a gene, is used to obtain many copies of a particular gene or large amounts of the protein encoded by the gene (Figure 20.1).
- In the process of gene cloning, both a vector and chromosomal DNA are cut with restriction enzymes. The DNA fragments hydrogen-bond to each other at their sticky ends, and the pieces are covalently linked together via DNA ligase, producing recombinant vectors. When a recombinant vector is introduced into a bacterial cell, the cell replicates the vector and divides to produce many cells (Table 20.1, Figures 20.2, 20.3).
- A collection of recombinant vectors, each with a particular piece of chromosomal DNA, is introduced into bacterial cells to produce a DNA library. If the DNA molecules are derived from mRNA, this creates a cDNA (complementary DNA) library (Figure 20.4).
- Gel electrophoresis is used to separate macromolecules by using an electric field that causes them to pass through a gel matrix. Gel electrophoresis typically separates molecules according to their charges, sizes, and masses (Figure 20.5).
- Polymerase chain reaction (PCR) is another common technique for making many copies of a gene without the use of vectors and host cells. Primers are used that flank the region of DNA to be amplified (Figure 20.6).

20.2 Genomics: Techniques for Studying Genomes

- Genomics is the study of genomes as whole units; functional genomics studies the expression of a genome.
- For genomics, large fragments of chromosomal DNA are cloned into vectors such as BACs. Researchers may construct a contig, which is a collection of clones that cover a contiguous region of a chromosome. This is a type of mapping—determining the relative locations of genes or other DNA segments along a chromosome (Figure 20.7).
- The dideoxy method of DNA sequencing uses ddNTPs to determine the base sequence of a segment of DNA (Figures 20.8, 20.9).
- A DNA microarray is a small silica, glass, or plastic slide dotted with different sequences of single-stranded DNA, each corresponding to a short sequence within a known gene. It is used to study gene expression patterns (Figure 20.10, Table 20.2).

20.3 Biotechnology

- Biotechnology is the use of living organisms, or the products of living organisms, for human benefit (Figure 20.11).
- Microorganisms are used to produce human medicines and to reduce pollutants in the environment, a phenomenon called bioremediation.
- Transgenic organisms, also called genetically modified organisms (GMOs), are made via gene addition or gene replacement. Transgenic livestock have been genetically engineered to produce human hormones in their milk (Figure 20.12).
- The Ti plasmid in *Agrobacterium tumefaciens* has been used to produce transgenic plants such as Bt corn, which produces its own insecticide (Figures 20.13, 20.14).
- The reproductive cloning of mammals has been achieved by fusing a somatic cell with an egg that has had its nucleus removed (Figure 20.15).
- DNA fingerprinting is a technology for identifying individuals based on the occurrence of segments of DNA called STRs, which are variable in length among different individuals (Figure 20.16).
- Gene therapy is a method to treat human diseases by the introduction of cloned genes into cells. The first gene therapy involved a disease called severe combined immunodeficiency syndrome (SCID) (Figure 20.17).

Assess and Discuss

Test Yourself

1. Vectors used to clone genes were derived originally from
   a. proteins.
   b. plasmids.
   c. viruses.
   d. all of the above.
   e. b and c only.
2. Restriction enzymes used in most cloning experiments
   a. are used to cut DNA into pieces for gene cloning.
   b. are naturally produced by bacteria cells to prevent viral infection.
   c. produce sticky ends on DNA fragments.
   d. All of the above are correct.
   e. Only a and c are correct.
3. DNA ligase is needed in a cloning experiment
   a. to promote hydrogen bonding between sticky ends.
   b. to catalyze the covalent attachment of the backbone of DNA strands.
   c. to digest the chromosomal DNA into small pieces.
   d. Only a and b are correct.
   e. a, b, and c are correct.
4. Let's suppose you followed the protocols described in Figures 20.2 and 20.3. Which order of experiments would you follow to confirm that a white colony really contained a recombinant vector with an insert?
   a. Pick a bacterial colony and restreak on plates containing X-Gal to confirm that the cells really form white colonies.
   b. Pick a bacterial colony, isolate plasmid DNA, digest the plasmid DNA with a restriction enzyme, and then run the DNA on a gel.
   c. Pick a bacterial colony and test it to see if β-galactosidase is functional within the bacterial cells.
   d. Pick a bacterial colony and retest it on ampicillin-containing plates to double-check that the cells are really ampicillin resistant.
   e. c and d should both be conducted.
5. Why is *Taq* polymerase used in PCR rather than other DNA polymerases?
   a. *Taq* polymerase is a synthetic enzyme that produces DNA strands at a faster rate than natural polymerases.
   b. *Taq* polymerase is a heat-stable form of DNA polymerase that can function after exposure to the high temperatures necessary for PCR.
   c. *Taq* polymerase is easier to isolate than other DNA polymerases.
   d. *Taq* polymerase is the DNA polymerase commonly produced by most eukaryotic cells.
   e. All of the above are correct.

6. Let's suppose you want to clone a gene that has never been analyzed before by DNA sequencing. Which of the following statements do you agree with the most?
   a. Do PCR to clone the gene because it is much faster.
   b. Do PCR to clone the gene because it is very specific and gives a high yield.
   c. You can't do PCR because you can't make forward and reverse primers.
   d. Do cloning by insertion into a vector because it will give you a higher yield.
   e. Do cloning by insertion into a vector because it is easier than PCR.
7. The method of determining the base sequence of DNA is termed
   a. PCR.
   b. gene cloning.
   c. DNA fingerprinting.
   d. DNA sequencing.
   e. gene mapping.
8. During bioremediation, microorganisms are used to
   a. clone genes from eukaryotic organisms.
   b. introduce normal (nonmutant) genes into individuals with genetic diseases.
   c. decrease pollutants in the environment.
   d. produce useful products such as insulin.
   e. do all of the above.
9. Organisms that carry genes that were introduced using molecular techniques are called
   a. transgenics.
   b. recombinant DNA.
   c. mutants.
   d. genetically modified organisms.
   e. both a and d.
10. DNA fingerprinting is used
    a. to precisely identify an organism, such as the identification of specific strains of bacteria.
    b. as a forensics tool to provide evidence in a criminal case.
    c. to determine genetic relationships between individuals.
    d. to determine the identity of an individual.
    e. for all of the above.

Conceptual Questions

1. Explain how using one restriction enzyme to cut both a plasmid and a gene of interest will allow the gene to be inserted into the plasmid.
2. Explain and draw the structural feature of a dideoxyribonucleotide that causes chain termination.
3. A principle of biology is that *biology affects our society.* Describe 2 examples of how genetic engineering has played an important role in agriculture.

Collaborative Questions

1. Discuss three important advances that have resulted from gene cloning.
2. Discuss the ethical issues associated with genetic engineering and reproductive cloning.

Online Resource

www.brookerbiology.com

Stay a step ahead in your studies with animations that bring concepts to life and practice tests to assess your understanding. Your instructor may also recommend the interactive eBook, individualized learning tools, and more.

Genomes, Proteomes, and Bioinformatics

21

Chapter Outline

21.1 Bacterial and Archaeal Genomes
21.2 Eukaryotic Genomes
21.3 Proteomes
21.4 Bioinformatics
Summary of Key Concepts
Assess and Discuss

Genomes and computer technology. The amount of data derived from the analyses of genomes is so staggering in size and complexity that researchers have turned to computers to unravel the amazing information that genomes contain.

Imagine a book that is incredibly long and has taken billions of years to write. Such an analogy applies to the human genome, which is written in only four letters—A, T, G, and C—and is over 3 billion base pairs long. If the entire human genome were typed in a textbook like this, with about 3,000 letters per page, it would be over 1 million pages long! Our genome contains many unsolved mysteries that researchers are trying to unravel, such as: "What are the functions of every gene in the genome? How do gene mutations cause disease? How are humans evolutionarily related to other species?" As you will learn, genomes are full of surprises. For example, did you know that most of your DNA has no known function?

The unifying theme of biology is evolution. The genome of every living species is the product of approximately 4 billion years of evolution. We can understand the unity of modern organisms by realizing that all species evolved from an interrelated group of ancestors. Throughout this textbook, a recurring theme is a series of "Genomes & Proteomes Connections"—topics that highlight the evolutionary connections among all forms of life and underscore how the genetic material produces the form and function of living organisms. By now, you may feel familiar with the concept of a **genome**, the complete genetic composition of a cell, organism, or species. The genome of each species is critical to its existence in several ways:

- The genome stores information in the form of genes, which provide a genetic blueprint to produce the characteristics of organisms.
- The genome is copied and transmitted from generation to generation.
- The accumulation of genetic changes (mutations) over the course of many generations produces the evolutionary changes that alter species and produce new species.

An extension of genome analysis is the study of proteomes. The term **proteome** refers to the entire complement of proteins that a cell or organism makes. The function of most genes is to encode proteins, which are the key determinants of cell structure and function. Analyzing the proteomes of organisms allows researchers to understand many aspects of biology, including the structure and function of cells, the complexity of multicellular organisms, and the interactions between organisms and their environment.

In the first two sections of this chapter, we will consider genome characteristics of bacteria, archaea, and eukaryotes. We then turn our attention to proteomes and examine the roles of the proteins that a species can make. From a molecular perspective, genomes and proteomes contain extensive and complex information, which is studied using computer technology. In the last section of this chapter, we will consider how the field of bioinformatics, which employs computers and statistical techniques to analyze biological information, is critical to the study of genomes and proteomes.

21.1 Bacterial and Archaeal Genomes

Learning Outcomes:

1. List the key characteristics of bacterial and archaeal genomes.
2. Describe the method of shotgun DNA sequencing.

The past decade has seen remarkable advances in our overall understanding of the entire genome of many species. As genetic technology has progressed, researchers have gained an increasing ability to analyze the composition of genomes as a whole unit. For many species, we now know their complete DNA sequence, which provides the most detailed description available of an organism's genome at the molecular level. In this section, we will survey the sizes and composition of genomes in selected species of bacteria and archaea.

The Genomes of Bacteria and Archaea Typically Consist of a Circular Chromosome with a Few Thousand Genes

Geneticists have made great progress in the study of bacterial and archaeal genomes. Some of the key features of bacterial chromosomes are described in Chapter 18 (refer back to Figure 18.10). Why are researchers interested in the genomes of bacteria and archaea? First, bacteria cause many different diseases that affect humans as well as other animals and plants. Studying the genomes of bacteria reveals important clues about the process of infection, which may also help us find ways to combat bacterial infection. A second reason for studying bacterial and archaeal genomes is that the information we learn about these microscopic organisms often applies to larger and more complex organisms. For example, basic genetic mechanisms, such as DNA replication and gene regulation, were first understood in the bacterium *Escherichia coli*. That knowledge provided a critical foundation to understand how these processes work in humans and other eukaryotic species. A third reason is evolution. The origin of the first eukaryotic cell probably involved a union between an archaeal and a bacterial cell, as we will explore in Chapter 22. The study of bacterial and archaeal genomes helps us understand how all living species evolved. Finally, another reason to study the genomes of bacteria is because we use them as tools in research and biotechnology, which was discussed in Chapter 20.

As of 2011, the genomes of over 700 bacterial and archaeal species have been completely sequenced and analyzed. The chromosomes of bacteria and archaea are usually a few million base pairs in length. Genomic researchers refer to 1 million base pairs as 1 megabase pair, abbreviated Mb. Most bacteria and archaea contain a single type of chromosome, though multiple copies may be present in a single cell. However, some bacteria are known to have different chromosomes. For example, *Vibrio cholerae*, the bacterium that causes the diarrheal disease known as cholera, has two different chromosomes in each cell, one 2.9 Mb and the other 1.1 Mb.

Bacterial and archaeal chromosomes are usually circular. For example, the two chromosomes in *V. cholerae* are circular, as is the single type of chromosome found in *E. coli*. However, linear chromosomes are found in some species, such as *Borrelia burgdorferi*, the bacterium that causes Lyme disease, the most common tick-borne disease in the U.S. Certain bacterial species may even contain both linear and circular chromosomes. *Agrobacterium tumefaciens*, which infects plants and causes a disease called crown gall, has one linear chromosome (2.1 Mb) and one circular chromosome (3.0 Mb).

Table 21.1 compares the sequenced genomes from several bacterial and archaeal species. They range in size from 1.7 to 5.2 Mb.

Table 21.1 Examples of Bacterial and Archaeal Genomes That Have Been Sequenced*

Species	Genome size (Mb)†	Number of genes‡	Description
Methanobacterium thermoautotrophicum	1.7	1,869	An archaeon that produces methane
Haemophilus influenzae	1.8	1,743	One of several different bacterial species that causes respiratory illness and meningitis
Sulfolobus solfataricus	3.0	3,032	An archaeon that metabolizes sulfur-containing compounds
Lactobacillus plantarum	3.3	3,052	A type of lactic acid bacterium used in the production of cheese and yogurt
Mycobacterium tuberculosis	4.4	4,294	The bacterium that causes the respiratory disease tuberculosis
Escherichia coli	4.6	4,289	A naturally occurring intestinal bacterium; certain strains can cause human illness
Bacillus anthracis	5.2	5,439	The bacterium that causes the disease anthrax

*Bacterial and archaeal species often exist in different strains that may differ slightly in their genome size and number of genes. The data are from common strains of the indicated species. The species shown in this table have only one type of chromosome.

†Mb equals 1 million base pairs, or a megabase pair.

‡The number of genes is an estimate based on the analysis of genome sequences.

The total number of genes is correlated with the total genome size. Roughly 1,000 genes are found for every megabase pair of DNA. Compared with eukaryotic genomes, bacterial and archaeal genomes are less complex. Their chromosomes lack centromeres and telomeres and have a single origin of replication. Also, chromosomes of bacteria and archaea have relatively little repetitive DNA, whereas repetitive sequences are usually abundant in eukaryotic genomes.

In addition to one or more chromosomes, bacteria often have plasmids, circular pieces of DNA that exist independently of the bacterial chromosome. Plasmids are typically small, in the range of a few thousand to tens of thousands of base pairs in length, though some can be quite large, even hundreds of thousands of base pairs. The various functions of plasmids are described in Chapter 18, and their use as vectors in gene cloning is discussed in Chapter 20.

FEATURE INVESTIGATION

Venter, Smith, and Colleagues Sequenced the First Genome in 1995

The first genome to be entirely sequenced was that of the bacterium *Haemophilus influenzae*. This bacterium causes a variety of diseases in humans, including respiratory illnesses and bacterial meningitis. *H. influenzae* has a relatively small genome consisting of approximately 1.8 Mb of DNA in a single circular chromosome.

A common strategy for sequencing an entire genome is called **shotgun DNA sequencing**. In this approach, researchers use a DNA sequencing method, such as the dideoxy chain-termination method (see Figure 20.9), to randomly sequence many DNA fragments from the genome. As a matter of chance, some of the fragments are overlapping—the end of one fragment contains the same DNA region as the beginning of another fragment. Computers are used to align the overlapping regions and assemble the DNA fragments into a contiguous sequence identical to that found in the intact chromosome. The advantage of shotgun DNA sequencing is that it does not require extensive mapping, a process that can be time-consuming. A disadvantage is that researchers may waste time sequencing the same region of DNA more times than necessary in order to assemble the sequence correctly.

To obtain a complete sequence of a genome with the shotgun approach, how do researchers decide how many fragments to sequence? We can calculate the probability that a base will not be sequenced (P) using this equation:

$$P = e^{-m}$$

where e is the base of the natural logarithm ($e = 2.72$), and m is the number of sequenced bases divided by the total genome size. For example, in the case of *H. influenzae*, with a genome size of 1.8 Mb, if researchers sequenced 9.0 Mb, $m = 5$ (that is, 9.0 Mb divided by 1.8 Mb):

$$P = e^{-m} = e^{-5} = 0.0067, \text{ or } 0.67\%.$$

This means that if we randomly sequence 9.0 Mb, which is five times the length of a single genome, we are likely to miss only 0.67% of the genome. With a genome size of 1.8 Mb, we would miss about 12,000 nucleotides out of approximately 1.8 million. Such missed sequences are typically on small DNA fragments that, as a matter of random chance, did not happen to be sequenced. The missing links in the genome are identified using mapping methods, which are described in Chapter 20.

In their discovery based investigation, American biologists Craig Venter and Hamilton Smith and their colleagues used a shotgun DNA sequencing approach (**Figure 21.1**). The researchers isolated chromosomal DNA from *H. influenzae* and used sound waves to break the DNA into small fragments of approximately 2,000 bp in length. These fragments were randomly inserted into vectors, allowing the DNA to be propagated in *E. coli*. Each *E. coli* clone carried a vector with a different piece of DNA from *H. influenzae*. The complete set of vectors, each containing a different fragment of DNA, is called a **DNA library** (refer back to Figure 20.4). The researchers then subjected many of these clones to the procedure of DNA sequencing. They sequenced a total of approximately 10.8 Mb of DNA.

The outcome of this genome-sequencing project was a very long DNA sequence. In 1995, Venter, Smith, and colleagues published the entire DNA sequence of *H. influenzae*. The researchers then analyzed the genome sequence using a computer to obtain information about the properties of the genome. Questions they asked included, "How many genes does the genome contain, and what are the likely functions of those genes?" Later in this chapter, we will learn how scientists can answer such questions with the use of computers. The data in Figure 21.1 summarize these researchers' results. The *H. influenzae* genome is composed of 1,830,137 bp of DNA. The computer analysis predicted 1,743 genes. Based on their similarities to sequences of genes identified in other species, the researchers also predicted the functions of proteins encoded by nearly two-thirds of those genes.

Figure 21.1 **Determination of the complete genome sequence of *Haemophilus influenzae* by Venter, Smith, and colleagues.**

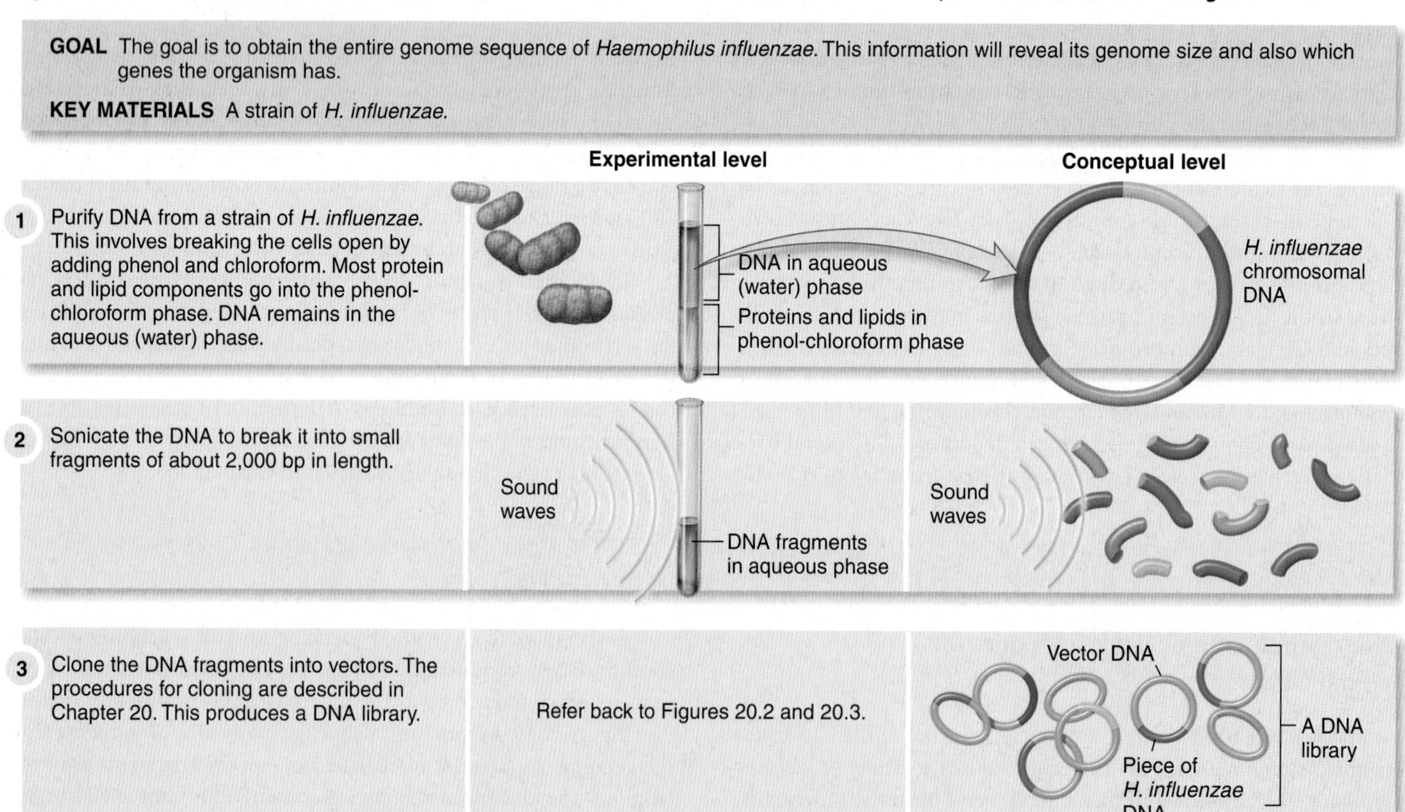

GOAL The goal is to obtain the entire genome sequence of Haemophilus influenzae. This information will reveal its genome size and also which genes the organism has.
KEY MATERIALS A strain of H. influenzae.
Experimental level
Conceptual level
1 Purify DNA from a strain of H. influenzae. This involves breaking the cells open by adding phenol and chloroform. Most protein and lipid components go into the phenol-chloroform phase. DNA remains in the aqueous (water) phase.
DNA in aqueous (water) phase
Proteins and lipids in phenol-chloroform phase
H. influenzae chromosomal DNA
2 Sonicate the DNA to break it into small fragments of about 2,000 bp in length.
Sound waves
DNA fragments in aqueous phase
Sound waves
3 Clone the DNA fragments into vectors. The procedures for cloning are described in Chapter 20. This produces a DNA library.
Refer back to Figures 20.2 and 20.3.
Vector DNA
Piece of H. influenzae DNA
A DNA library

4 Subject many clones to the procedure of dideoxy sequencing, also described in Chapter 20. A total of 10.8 Mb was sequenced. | Refer back to Figure 20.9. | CCCAGTCCCATGCCATGGCCCAGTCCC Produces a large number of sequences with overlapping regions.

5 Use tools of bioinformatics, described in the last section of this chapter, to identify various types of genes in the genome. | CCATGCCATGGCCCCCATGC | Explores the genome sequence and identifies and characterizes genes.

6 THE DATA

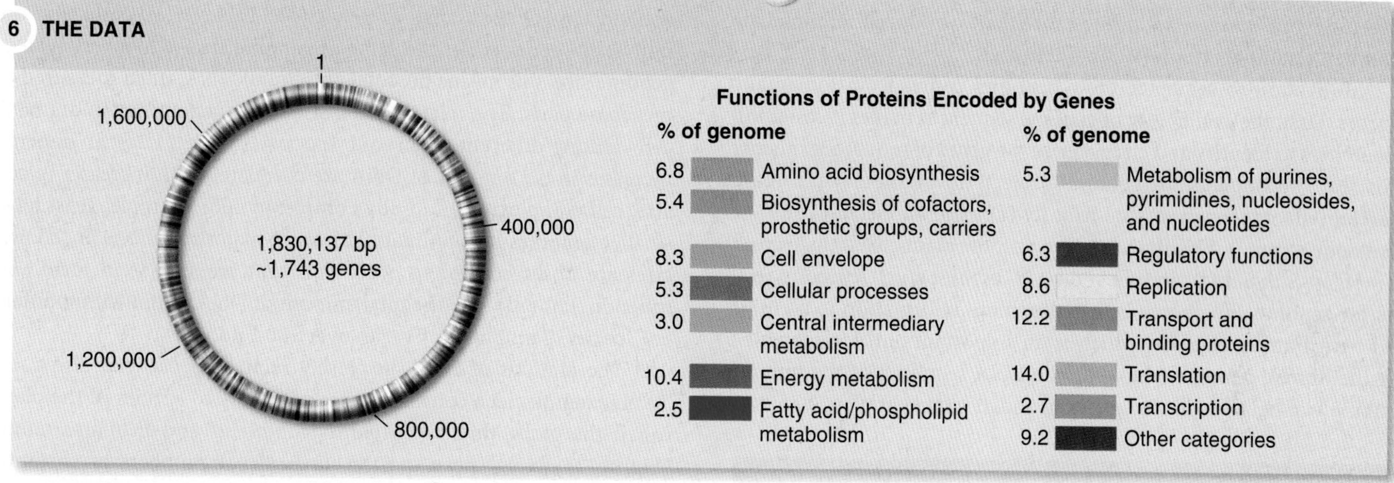

7 **CONCLUSION** *H. influenzae* has a genome size of 1.83 Mb with approximately 1,743 genes. The functions of many of those genes could be inferred by comparing them to genes in other species.

8 **SOURCE** Fleischmann et al. 1995. Whole-genome random sequencing and assembly of *Haemophilus influenzae* Rd. *Science* 269:496–512.

The diagram shown in the data of Figure 21.1 places proteins in various categories based on their predicted function. These results gave the first comprehensive "genome picture" of a living organism!

Experimental Questions

1. What was the goal of the experiment conducted by Venter, Smith, and their colleagues?
2. How does shotgun DNA sequencing differ from procedures that involve mapping? Name an advantage and a disadvantage of the shotgun DNA sequencing approach.
3. What were the results of the study described in Figure 21.1?

21.2 Eukaryotic Genomes

Learning Outcomes:

1. Describe the key features of eukaryotic genomes.
2. Name the two major types of transposable elements and describe how they differ.
3. Explain how gene duplications can occur and lead to the formation of gene families.
4. Outline the goals and results of the Human Genome Project.

Thus far, we have examined bacterial and archaeal genomes. In this section, we turn to eukaryotes, which include protists, fungi, animals, and plants. As you will learn, their genomes are larger and more complex than their bacterial and archaeal counterparts. In addition to genes, eukaryotic genomes often have abundant amounts of noncoding sequences, whose function is largely unknown. For example, eukaryotic genomes typically have a substantial amount of short repeated sequences called repetitive DNA. We will explore how certain types of repetitive DNA sequences are formed by a process called transposition. We will also examine how the duplication of genes can lead to families of related genes.

The Nuclear Genomes of Eukaryotes Are Sets of Linear Chromosomes That Vary Greatly in Size and Composition Among Different Species

As discussed in Chapter 15, the genome located in the nucleus of eukaryotic species is usually found in sets of linear chromosomes. In humans, for example, one set contains 23 linear chromosomes—22 autosomes and one sex chromosome, X or Y. In addition, certain organelles in eukaryotic cells contain a small amount of DNA. These include the mitochondrion, which plays a role in ATP synthesis, and the chloroplast (found in plants and algae), which carries out photosynthesis. The genetic material in these organelles is referred to as the mitochondrial or the chloroplast genome to distinguish it from the nuclear genome, which is located in the cell nucleus. In this chapter, we will focus on the nuclear genome of eukaryotes.

Sizes of Nuclear Genomes In the past decade or so, the DNA sequence of entire nuclear genomes has been determined for over 100 eukaryotic species, including more than two dozen mammalian genomes. Examples are shown in **Table 21.2**.

Motivation to sequence these genomes comes from four main sources. First, the availability of genome sequences makes it easier for researchers to identify and characterize the genes of model organisms. This was the impetus for genome projects involving baker's yeast (*Saccharomyces cerevisiae*), the fruit fly (*Drosophila melanogaster*), a nematode worm (*Caenorhabditis elegans*), the plant called thale cress (*Arabidopsis thaliana*), and the mouse (*Mus musculus*). A second reason for genome sequencing is to gather more information to identify and treat human diseases, which is an important aim for sequencing the human genome. Knowing the DNA sequence of the human genome will help to identify genes in which mutation plays a role in disease. Third, by sequencing the genomes of agriculturally important species, new strains of livestock and plant species with improved traits can be developed. Fourth, biologists are increasingly relying on genome sequences as a way to establish evolutionary relationships.

Table 21.2 Examples of Eukaryotic Nuclear Genomes That Have Been Sequenced

Species	Nuclear genome size (Mb)	Number of genes	Description
Saccharomyces cerevisiae (baker's yeast)	12.1	6,294	One of the simplest eukaryotic species; it has been extensively studied by researchers to understand eukaryotic molecular biology.
Caenorhabditis elegans (nematode worm)	100	~19,000	A model organism used to study animal development.
Drosophila melanogaster (fruit fly)	180	~14,000	A model organism used to study many genetic phenomena, including development.
Arabidopsis thaliana (thale cress)	120	~26,000	A model organism studied by plant biologists.
Oryza sativa (rice)	440	~40,000	A cereal grain with a relatively small genome; it is very important worldwide as a food crop.
Mus musculus (mouse)	2,500	~20,000–25,000	A model mammalian organism used to study genetics, cell biology, and development.
Homo sapiens (humans)	3,200	~20,000–25,000	Our own genome, the sequencing of which will help to elucidate our understanding of inherited traits and to aid in the identification and treatment of diseases.

Note: The genome size refers to the number of megabase (Mb) pairs in one set of chromosomes. For species with sex chromosomes, it would include both sex chromosomes.

Relationship Between Genome Sizes and Repetitive Sequences Eukaryotic genomes are generally larger than bacterial and archaeal genomes, both in terms of the number of genes and genome size. The genomes of simpler eukaryotes, such as yeast, carry several thousand different genes, whereas the genomes of more complex eukaryotes contain tens of thousands of genes (see Table 21.2). Note that the number of genes is not the same as genome size. When we speak of genome size, this means the total amount of DNA, often measured in megabase pairs. The relative sizes of nuclear genomes vary dramatically among different eukaryotic species (**Figure 21.2a**). In general, increases in the amount of DNA are correlated with increases in cell size, cell complexity, and body complexity. For example, yeast have smaller genomes than animals. However, major variations in genome sizes are observed among organisms that are similar in form and function. For example, the total amount of DNA found within different species of amphibians varies over 100-fold.

The amount of DNA in closely related species can also vary. As an example, let's consider two closely related species of the plant called the globe thistle, *Echinops bannaticus* and *Echinops nanus* (**Figure 21.2b,c**). These species have similar numbers of chromosomes, but *E. bannaticus* has nearly double the amount of DNA compared to *E. nanus*. What is the explanation for the larger genome of

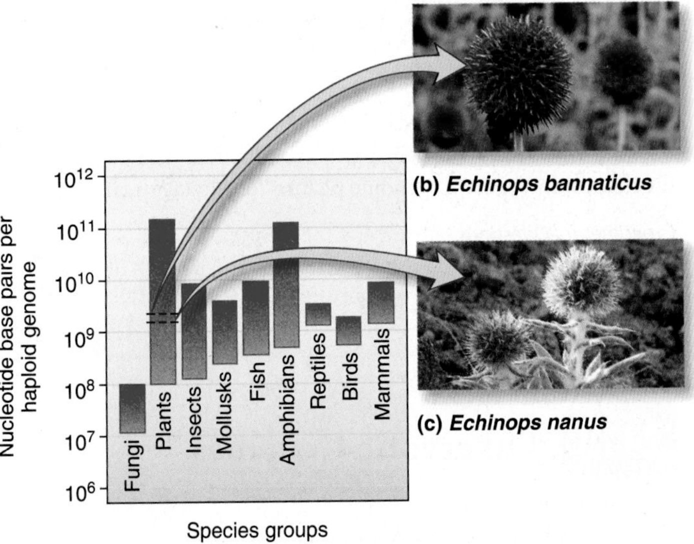

Figure 21.2 Genome sizes among selected groups of eukaryotes. **(a)** Genome sizes among various groups of eukaryotes are shown on a log scale. As an example for comparison, two closely related species of globe thistle are pictured. These species have similar characteristics, but *Echinops bannaticus* **(b)** has nearly double the amount of DNA as does *E. nanus* **(c)** due to the accumulation of repetitive DNA sequences.

Concept Check: *What are two reasons why the groups of species shown in (a) vary in their total amount of DNA?*

E. bannaticus? The genome of *E. bannaticus* is not likely to contain twice as many genes. Rather, its genome composition includes many **repetitive sequences**, which are short DNA sequences that are present in many copies throughout the genome. Repetitive sequences are often abundant in eukaryotic species.

Types of Repetitive Sequences Repetitive sequences fall into two broad categories, moderately and highly repetitive. Sequences that are repeated a few hundred to several thousand times are called **moderately repetitive sequences.** In some cases, these sequences are multiple copies of the same gene. For example, the genes that encode ribosomal RNA (rRNA) are found in many copies. The cell needs a large amount of rRNA for its cellular ribosomes. This is accomplished by having and expressing multiple copies of the genes that encode rRNA. In addition, other types of functionally important sequences can be moderately repetitive. For example, multiple copies of origins of replication are found in eukaryotic chromosomes. Other moderately repetitive sequences may play a role in the regulation of gene transcription and translation.

Highly repetitive sequences are those that are repeated tens of thousands to millions of times throughout the genome. Each copy of a highly repetitive sequence is relatively short, ranging from a few nucleotides to several hundred nucleotides in length. Most of these sequences have no known function, and whether or not they benefit the organism is a matter of debate. A widely studied example is the *Alu* family of sequences found in humans and other primates. The *Alu* sequence is approximately 300 bp long. This sequence derives its name from the observation that it contains a site for cleavage by a restriction enzyme known as *Alu*I. It represents about 10% of the total human DNA and occurs (on average) approximately every 5,000–6,000 bases. Evolutionary studies suggest that the *Alu* sequence arose 65 million years ago (mya) from a section of a single ancestral gene known as the 7SL RNA gene. Remarkably, over the course of 65 million years, the *Alu* sequence has been copied and inserted into the human genome so often that it now occurs more than 1 million times! The mechanism for the proliferation of *Alu* sequences will be described later.

Some highly repetitive sequences, like the *Alu* family, are interspersed throughout the genome. However, other highly repetitive sequences are clustered together in a tandem array, in which a very short nucleotide sequence is repeated many times in a row. In *Drosophila*, for example, 19% of the chromosomal DNA is highly repetitive DNA found in tandem arrays. An example is shown here:

AATATAATATATAATATAATATAATATAT
TTATATTATATATTATATTATATTATATA

In this particular tandem array, two related sequences, AATAT and AATATAT (in the top strand), are repeated multiple times. Highly repetitive sequences, which contain tandem arrays of short sequences, can be quite long, sometimes more than 1 million bp in length!

Figure 21.3 shows the composition of the relative classes of DNA sequences found in the nuclear genome of humans. Surprisingly, exons, the coding regions of structural genes, and the genes that give rise to rRNA and tRNA make up only about 2% of our genome! The other 98% is composed of noncoding sequences. Though we often think of genomes as being the repository of sequences that code for proteins, most eukaryotic genomes are largely composed of other types of sequences. Intron DNA comprises about 24% of the human genome, and unique noncoding DNA, whose function is largely unknown, constitutes 15%. Repetitive DNA makes up 59% of the DNA in the genome. Much of the repetitive DNA is derived from transposable elements, stretches of DNA that can move from one location to another, which are described next.

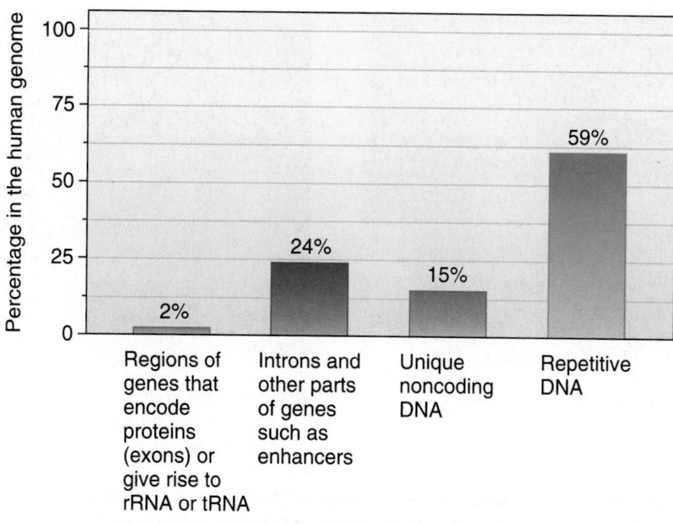

Figure 21.3 The composition of DNA sequences that are found in the nuclear genome of humans. Only about 2% of our genome codes for proteins. Most of our genome is made up of repetitive sequences.

BIOLOGY PRINCIPLE The genetic material provides the blueprint for reproduction. Only a small percentage of the human genome is involved with encoding the proteins that are largely responsible for human traits.

Transposable Elements Move from One Chromosomal Location to Another

During a process called **transposition**, a short segment of DNA moves from its original site to a new site in the genome. Such segments are known as **transposable elements (TEs)**. They range from a few hundred to several thousand base pairs in length. TEs have sometimes been referred to as "jumping genes," because they are inherently mobile. American cytogeneticist Barbara McClintock first identified transposable elements in the late 1940s from her studies with corn plants (Figure 21.4). She identified a segment of DNA that could move into and out of a gene that affected the color of corn kernels, producing a speckled appearance. Since that time, biologists have discovered many different types of TEs in nearly all species examined.

Though McClintock identified TEs in corn in the late 1940s, her work was met with great skepticism because many researchers had trouble believing that DNA segments could be mobile. The advent of molecular technology in the 1960s and 1970s allowed scientists to understand more about the characteristics of TEs that enable their movement. Most notably, research involving bacterial TEs eventually

(a) Barbara McClintock

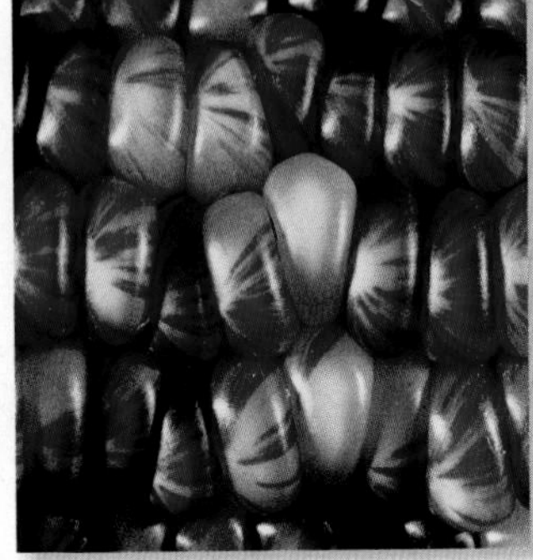

(b) Speckled corn kernels caused by transposable elements

Figure 21.4 Barbara McClintock, who discovered transposable elements. As shown in part **(b)**, when a transposable element is found within a pigment gene in corn, its frequent movement disrupts the gene, causing the kernel color to be speckled.

progressed to a molecular understanding of the transposition process. In 1983, more than 30 years after her initial discovery, McClintock was awarded the Nobel Prize in Physiology or Medicine.

Researchers have studied TEs from many species, including bacteria, archaea, and eukaryotes. They have discovered that TEs fall into two groups, based on different mechanisms of movement.

DNA Transposons Transposable elements that move via a DNA molecule are called **DNA transposons**. Both ends of DNA transposons usually have inverted repeats (IRs)—DNA sequences that are identical (or very similar) but run in opposite directions (**Figure 21.5a**), such as the following:

```
5′-CTGACTCTT-3′  and  5′-AAGAGTCAG-3′
3′-GACTGAGAA-5′       3′-TTCTCAGTC-5′
```

Depending on the particular TE, inverted repeats range from 9 to 40 bp in length. In addition, DNA transposons may contain a central region that encodes **transposase**, an enzyme that facilitates transposition.

As shown in **Figure 21.5b**, transposition of DNA transposons occurs by a cut-and-paste mechanism. Transposase first recognizes the inverted repeats (IR) in the transposon and then removes the DNA transposon from its original site. Next, the transposase/transposon complex moves to a new location, where transposase cleaves the target DNA and inserts the transposon into the site. Transposition may occur when a cell is in the process of DNA replication. If a TE is removed from a site that has already replicated and is inserted into a chromosomal site that has not yet replicated, the transposon will increase in number after DNA replication is complete. This is one way for transposons to become more prevalent in a genome.

Retroelements Another category of TE moves via an RNA intermediate. This form of transposition is very common but is found only in eukaryotic species. These types of elements are known as **retroelements** or **retrotransposons**. The *Alu* sequence in the human genome is an example of a retroelement. Some retroelements contain

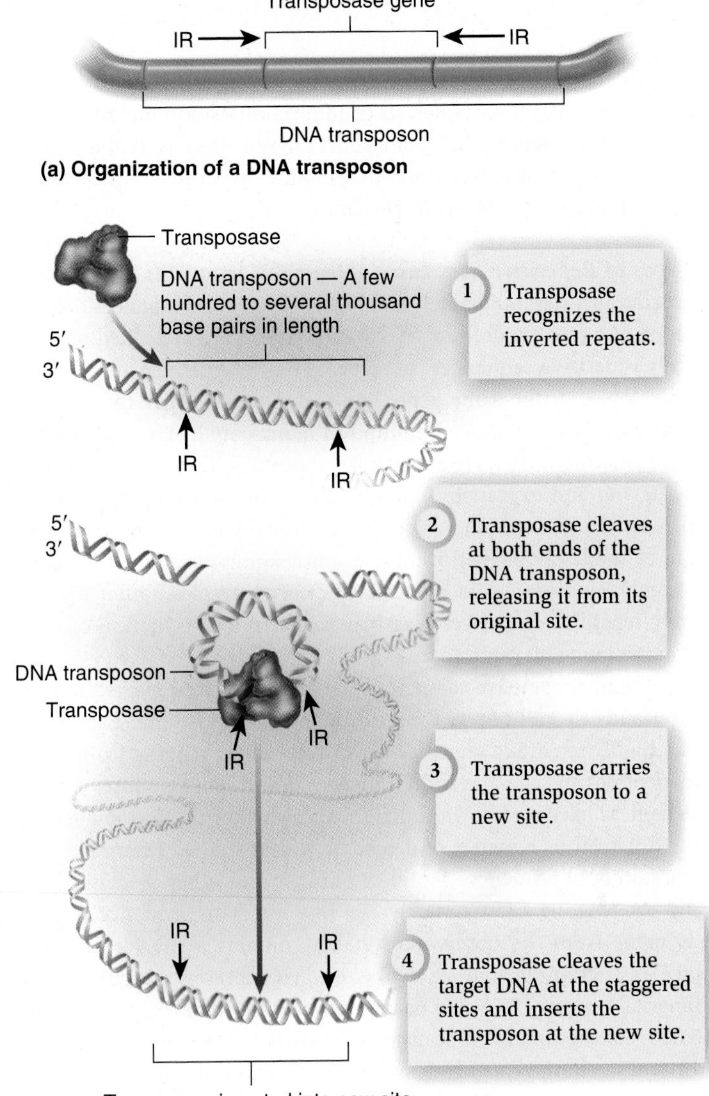

Transposase gene
IR
IR
DNA transposon
Transposase
DNA transposon — A few hundred to several thousand base pairs in length
5′
3′
IR
IR
1 Transposase recognizes the inverted repeats.
5′
3′
2 Transposase cleaves at both ends of the DNA transposon, releasing it from its original site.
DNA transposon
Transposase
IR
IR
3 Transposase carries the transposon to a new site.
IR
IR
4 Transposase cleaves the target DNA at the staggered sites and inserts the transposon at the new site.
Transposon inserted into new site

(a) Organization of a DNA transposon

(b) Cut-and-paste mechanism of transposition

Figure 21.5 DNA transposons and their mechanism of transposition. **(a)** DNA transposons contain inverted repeat (IR) sequences at each end and may contain a gene that encodes transposase in the middle. **(b)** Transposition occurs by a cut-and-paste mechanism.

Concept Check: *What is the role of the inverted repeats in the mechanism of transposition?*

genes that encode the enzymes reverse transcriptase and integrase, which are needed in the transposition process (**Figure 21.6a**). Recall from Chapter 18 that reverse transcriptase uses RNA as a template to synthesize a complementary copy of DNA. Retroelements may also contain repeated sequences called terminal repeats at each end that facilitate their recognition.

The mechanism of retroelement movement is shown in **Figure 21.6b**. First, the enzyme RNA polymerase transcribes the retroelement into RNA. Reverse transcriptase uses this RNA as a template to synthesize a double-stranded DNA molecule. The ends of the double-stranded DNA are then recognized by integrase, which catalyzes the insertion of the DNA into the host chromosomal DNA. The integration of retroelements can occur at many locations within the

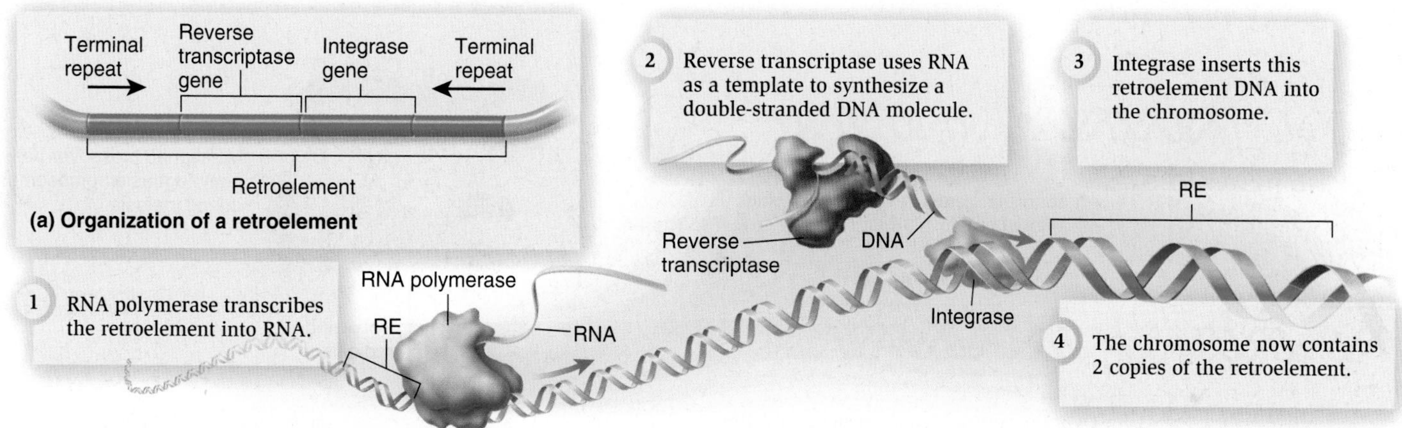

Figure 21.6 **Retroelements and their mechanism of transposition.** Retroelements are found only in eukaryotic species. **(a)** Some retroelements contain terminal repeats and genes that encode the enzymes reverse transcriptase and integrase, which are needed in the transposition process. **(b)** The process that adds a copy of a retroelement into a host chromosome. Note: In addition to using RNA as a template, some forms of reverse transcriptase can also use a DNA template to make a complementary DNA strand. As depicted here, these forms can make double-stranded DNA using a strand of RNA as a starting material. However, some forms of reverse transcriptase can make DNA only using an RNA template. In those cases, reverse transcriptase makes a complementary DNA strand from the RNA template and then the opposite DNA strand is made via a host-cell DNA polymerase.

Concept Check: Based on their mechanism of movement, which type of TEs do you think would proliferate more rapidly in a genome, DNA transposons (see Figure 21.5b) or retroelements?

genome. Furthermore, because a single retroelement can be copied into many RNA transcripts, retroelements may accumulate rapidly within a genome. This explains how the *Alu* element in the human genome was able to proliferate and constitute 10% of our genome.

Role of Transposable Elements What is the biological significance of TEs? The question is not resolved. According to the **selfish DNA hypothesis**, TEs exist solely because they have characteristics that allow them to insert themselves into the host cell DNA. In other words, they resemble parasites in that they inhabit the host without offering any advantage. They can proliferate within the host as long as they do not harm the host to the extent that they significantly disrupt its survival. However, TEs can do harm. For example, if they jump into the middle of an important gene and thereby disrupt its function, this may have a negative effect on the phenotype of an organism.

Other biologists have argued that TEs may provide benefits to a given species. For example, bacterial TEs often carry an antibiotic-resistance gene that provides the organism with a survival advantage. In addition, TEs may cause greater genetic diversity by promoting chromosomal rearrangements. As discussed next, such rearrangements can cause a misaligned crossover during meiosis and promote the formation of a gene family.

Gene Duplications Provide Additional Material for Genome Evolution, Sometimes Leading to the Formation of Gene Families

Let's now turn our attention to a way that the number of genes in a genome can increase. These gene duplications are important because they provide raw material for the addition of more genes into a species' genome. Such duplications can produce **homologous genes**, two or more genes that are derived from the same ancestral gene (**Figure 21.7a**). Over the course of many generations, each version of the gene accumulates different mutations, resulting in genes with similar but not identical DNA sequences.

How do gene duplications occur? One mechanism that produces gene duplications is a misaligned crossover (**Figure 21.7b**). In this example, two homologous chromosomes have paired with each other during meiosis, but the homologs are misaligned. If a crossover occurs, this produces one chromosome with a gene duplication, one with a gene deletion, and two normal chromosomes. Each of these chromosomes are segregated into different haploid cells. If a haploid cell carrying the chromosome with the gene duplication participates in fertilization with another gamete, an offspring with a gene duplication is produced. In this way, gene duplications can form and be transmitted to future generations. The presence of multiple copies of the same transposable element in a genome can foster this process because the chromosomes may misalign while attempting to align TEs that are at different locations in the same chromosome.

During evolution, gene duplications can occur several times. Two or more homologous genes within a single species are also called paralogous genes, or **paralogs**. Multiple gene duplications followed by the accumulation of mutations in each paralog can result in a **gene family**—a group of paralogs that carry out related functions. A well-studied example is the globin gene family found in animals. The globin genes encode polypeptides that are subunits of proteins that function in oxygen binding. Hemoglobin, which is made in red blood cells, carries oxygen throughout the body. In humans, the globin gene family is composed of 14 paralogs that were originally derived from a single ancestral globin gene (**Figure 21.8**). According to an evolutionary analysis, the ancestral globin gene duplicated between 500 and 600 mya. Since that time, additional duplication events and chromosomal rearrangements have occurred to produce the current number of 14 genes on three different human chromosomes. Four of these

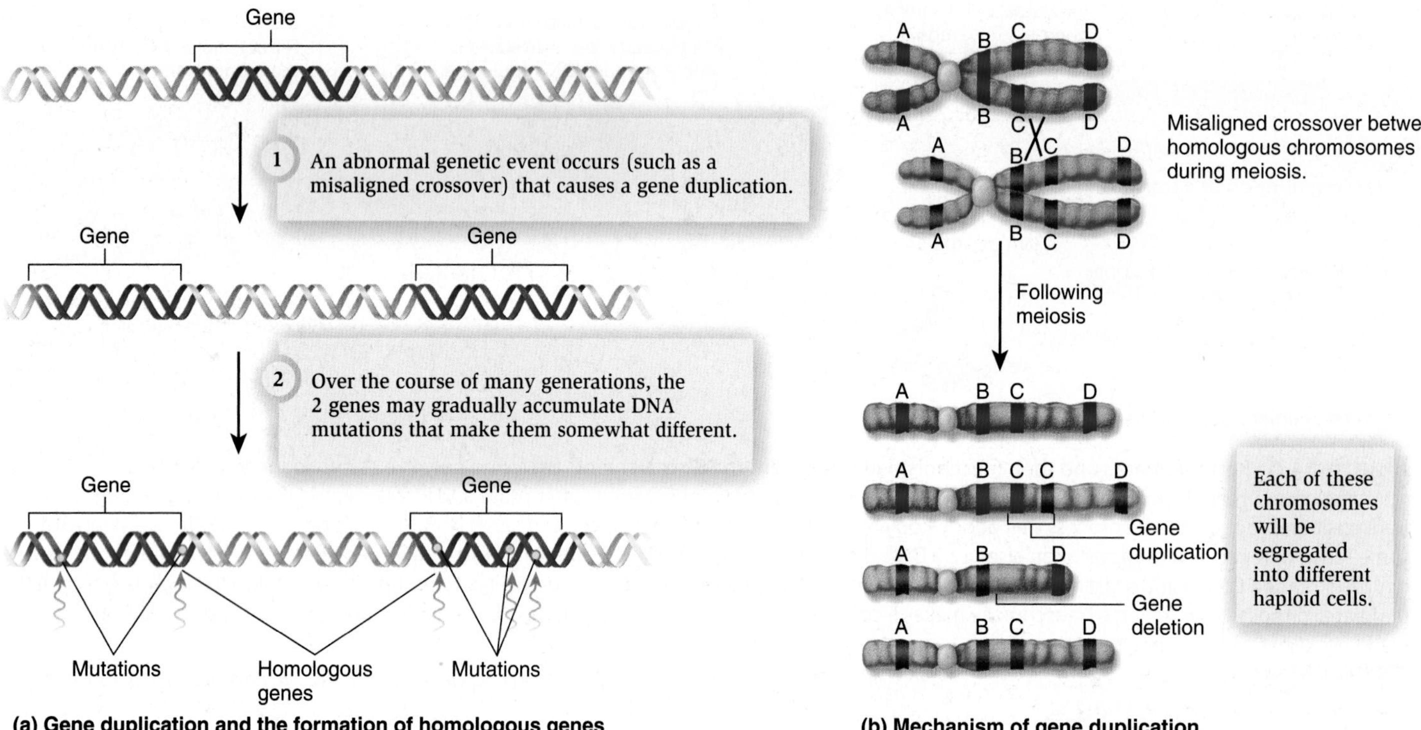

Figure 21.7 Gene duplication and the evolution of homologous genes. **(a)** A gene duplication produces two copies of the same gene. Over time, these copies accumulate different random mutations, which results in homologous genes with similar but not identical DNA sequences. **(b)** Mechanism of gene duplication. If two homologous chromosomes misalign during meiosis, a crossover will produce a chromosome with a gene duplication.

BIOLOGY PRINCIPLE Populations of organisms evolve from one generation to the next. In this example, evolution involves two different types of genetic changes. First, a gene duplication occurs and, second, each copy of the gene accumulates different mutations.

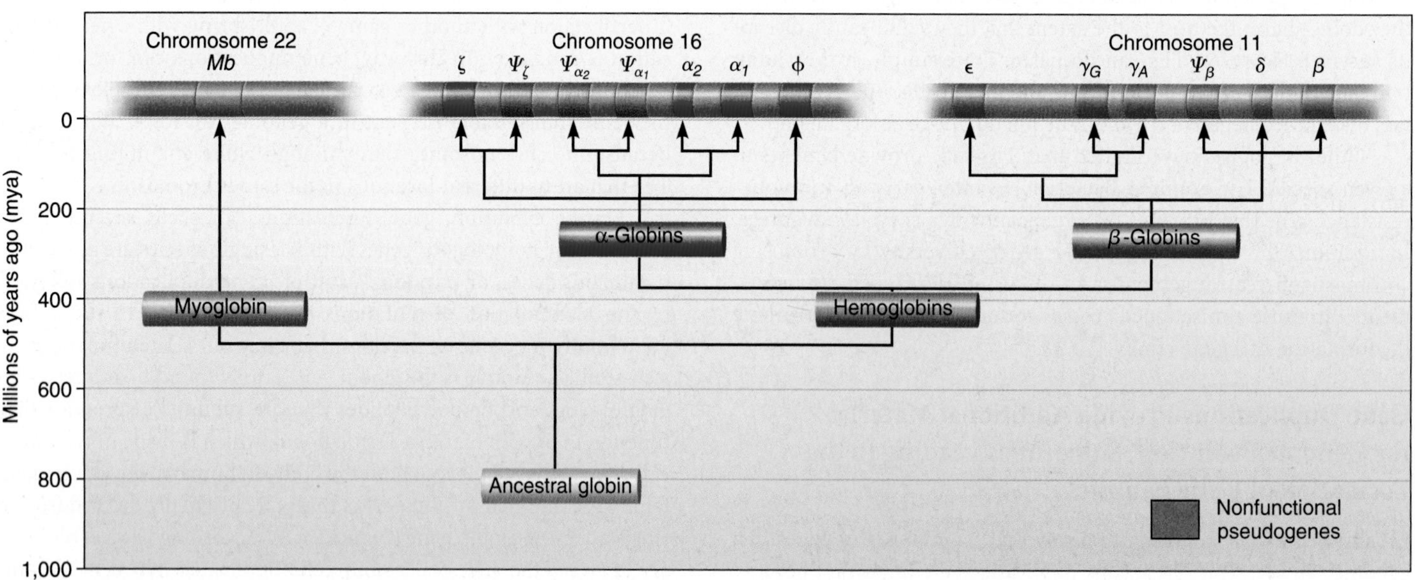

Figure 21.8 The evolution of the globin gene family in humans. The globin gene family evolved from a single ancestral globin gene.

BioConnections: Look back to Figure 13.3. How do different gene family members vary in their affinity for oxygen?

are pseudogenes—genes that have been produced by gene duplication but have accumulated mutations that make them nonfunctional, so they are not transcribed into RNA.

The accumulation of different mutations in the various family members has produced globins that are specialized in their function. For example, myoglobin binds and stores oxygen in muscle cells, whereas the hemoglobins bind and transport oxygen via red blood cells. Also, different globin genes are expressed during different stages of development. The zeta (ζ)-globin and epsilon (ε)-globin genes are expressed very early in embryonic life. During the second trimester of gestation, the alpha (α)-globin and gamma (γ)-globin genes are turned on. Following birth, the γ-globin genes are turned off, and the β-globin gene is turned on. These differences in the expression of the globin genes reflect the differences in the oxygen transport needs of humans during the embryonic, fetal, and postpartum stages of life (refer back to Figure 13.3).

The Human Genome Project Has Stimulated Genomic Research

Before ending our discussion of genomes, let's consider the **Human Genome Project**, a research effort to identify and map all human genes. Scientists had been discussing how to undertake this project since the mid-1980s. In 1988, the National Institutes of Health (NIH) in Bethesda, Maryland, established an Office of Human Genome Research with James Watson as its first director. The Human Genome Project officially began on October 1, 1990, and was largely finished by the end of 2003. It was an international consortium that included research institutions in the U.S., U.K., France, Germany, Japan, and China. From its outset, the Human Genome Project had the following goals:

1. *To identify all human genes.* This involved mapping the locations of genes throughout the entire genome. The data from the Human Genome Project suggest that humans have about 20,000 to 25,000 different genes.
2. *To obtain the DNA sequence of the entire human genome.* The first draft of a nearly completed DNA sequence was published in February 2001, and a second draft was published in 2003. The entire genome is approximately 3.2 billion base pairs in length.
3. *To develop technology for the generation and management of human genome information.* Some of the efforts of the Human Genome Project have involved improvements in molecular genetic technology, such as gene cloning, DNA sequencing, and so forth. The Human Genome Project has also developed computer tools to allow scientists to easily access up-to-date information from the project and analytical tools to interpret genomic information.
4. *To analyze the genomes of model organisms.* These include *E. coli*, *S. cerevisiae*, *D. melanogaster*, *C. elegans*, *A. thaliana*, and *M. musculus*.
5. *To develop programs focused on understanding and addressing the ethical, legal, and social implications of the results obtained from the Human Genome Project.* The Human Genome Project raised many ethical issues regarding genetic information and genetic engineering. Who should have access to genetic information? Should employers, insurance companies, law enforcement agencies, and schools have access to our genetic makeup? Another controversial topic is gene patenting. Should the blueprints of life be considered patentable, like any other invention, or should there be limits to the patenting of genes? In the U.S., genes can be patented for a variety of reasons. For example, the patenting of genes has been associated with the commercial development of diagnostic tests for genetic diseases. Some argue that patenting fosters greater investment in research and development; others say it can impede basic research and scientific innovation. The answers to such questions are complex and will require discussion among many groups.

Some current and potential applications of the Human Genome Project include the improved diagnosis and treatment of genetic diseases such as cystic fibrosis, Huntington disease, and Duchenne muscular dystrophy. The project may also enable researchers to identify the genetic basis of common disorders such as cancer, diabetes, and heart disease, which involve alterations in several genes.

21.3 Proteomes

Learning Outcomes:

1. Distinguish between a proteome and a genome.
2. List the major categories of proteins in a proteome.
3. Explain how the number of proteins in an organism's proteome can be greater than the number of genes in its genome.

Thus far in this chapter, we have considered the genome characteristics of many different species, including humans. Because most genes encode proteins, a logical next step is to examine the functional roles of the proteins that a species can make. As mentioned, the entire collection of proteins that a cell or organism produces is called a proteome. As we move through the 21st century, a key challenge facing molecular biologists is the study of proteomes. Much like the study of genomes, this will require the collective contributions of many scientists, as well as improvements in technologies to investigate the complexities of the proteome. In this section, we will begin by considering the functional categories of proteins and then examine their relative abundance in the proteome. We also will explore the molecular mechanisms that cause an organism's proteome to be much larger than its genome.

The Proteome Is a Diverse Array of Proteins with Many Kinds of Functions

As we have seen, the genomes of simple, unicellular organisms such as bacteria and yeast contain thousands of structural genes, whereas the genomes of complex, multicellular organisms contain tens of thousands. Such genome sizes can produce proteomes with tens of thousands to hundreds of thousands of different proteins. To bring some order to this large amount of complex information, researchers often organize proteins into different categories based on their functions. **Table 21.3** describes some general categories of protein function and provides examples of each type. Many approaches are used to categorize proteins. Table 21.3 shows just one of the more general ways to categorize protein function. Note that the data of Figure 21.1

Table 21.3 Categories of Proteins Found in the Proteome

Function	Examples
Metabolic enzymes—accelerate chemical reactions in the cell.	Hexokinase: Phosphorylates glucose during the first step in glycolysis. Glycogen synthetase: Uses glucose to synthesize a large carbohydrate known as glycogen.
Structural proteins—provide shape and protection to cells.	Tubulin: Forms cytoskeletal structures known as microtubules. Collagen: Forms large fibers in the extracellular matrix (ECM) of animals.
Motor proteins—facilitate intracellular movements and the movements of whole cells.	Myosin: Involved in muscle cell contraction. Kinesin: Involved in the movement of chromosomes during cell division.
Cell-signaling proteins—allow cells to respond to environmental signals and send signals to each other.	Insulin: Influences target cell metabolism and growth. Insulin receptor: Recognizes insulin and initiates a cellular response.
Transport proteins—involved in the transport of ions and molecules across membranes and throughout the body.	Lactose permease: Transports lactose across the bacterial cell membrane. Hemoglobin: Found in red blood cells and transports oxygen throughout the body.
Gene expression and regulatory proteins—involved in transcription, mRNA modification, translation, and gene regulation.	Transcription factors: Regulate the expression of genes. Ribosomal proteins: Components of ribosomes, which are needed for the synthesis of new proteins.
Defense and protective proteins—help cells and organisms to fight disease and survive environmental stress.	Antibodies: Fight viral and bacterial infections in vertebrate species. Chaperones such as heat shock proteins: Play a role in protein folding, thereby helping cells cope with stresses such as abrupt increases in temperature.

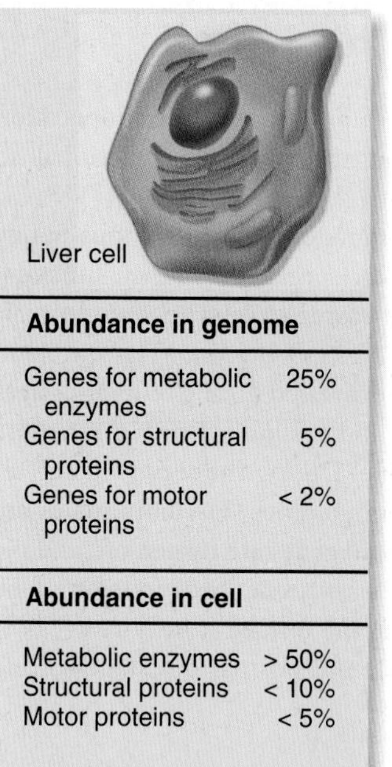

Liver cell
Abundance in genome
Genes for metabolic enzymes 25%
Genes for structural proteins 5%
Genes for motor proteins < 2%
Abundance in cell
Metabolic enzymes > 50%
Structural proteins < 10%
Motor proteins < 5%
Skeletal muscle cell
Abundance in genome
Genes for metabolic enzymes 25%
Genes for structural proteins 5%
Genes for motor proteins < 2%
Abundance in cell
Metabolic enzymes < 10%
Structural proteins 20–30%
Motor proteins 25–40%

Figure 21.9 A comparison of the proteomes in human liver and skeletal muscle cells. Because all cells of the human body carry the same genome, the percentages of proteins encoded in the genome are the same in each cell type. However, the relative amounts of proteins made in different cell types can be vastly different, as seen here.

Concept Check: What genetic process explains the differences in protein abundance in liver cells versus muscle cells?

displays the functions of proteins encoded by genes in a different, more detailed way.

The relative abundance of proteins can be viewed at two levels. First, we can consider abundance in the genome—the numbers of genes in the genome that encode a particular type or category of protein. For example, if an entire genome encodes 10,000 different types of proteins and 1,500 of these are different types of transporters, we would say that 15% of the genome is composed of transporters. However, such an analysis ignores the phenomenon that genes are expressed at different levels. In other words, various proteins are made in different amounts. Therefore, a second way to view protein abundance is to consider abundance in the cell—the amount of a given protein or protein category actually made by a living cell. For example, less than 1% of human genes encode proteins, such as collagen, that are found in the extracellular matrix (ECM). Even so, these genes are highly expressed in certain cells, so a large amount of this type of protein is made compared with other types.

Figure 21.9 is a general comparison of protein abundance in two cell types in humans: liver and muscle cells. Liver cells play a key role in metabolism, whereas muscle cells are involved in bodily movements. Both liver and muscle cells have the same genes. Therefore, at the level of the genome, the percentages of the different protein categories are identical. However, at the cellular level, the relative abundance of certain protein categories is quite different. Liver cells make a large number of different enzymes that play a role in the metabolism of fats, proteins, and carbohydrates. By comparison, their level of structural and motor proteins is relatively small. In contrast, muscle cells have fairly low levels of enzymes but make a high percentage of structural and motor proteins. These differences in protein composition between liver and muscle cells are largely due to differential gene regulation.

The Number of Different Proteins in a Species' Proteome Is Larger Than the Number of Genes in Its Genome

From the sequencing and analysis of genomes, researchers can identify all or nearly all of the genes of a given species. For example, the human genome is believed to contain between 20,000 and 25,000

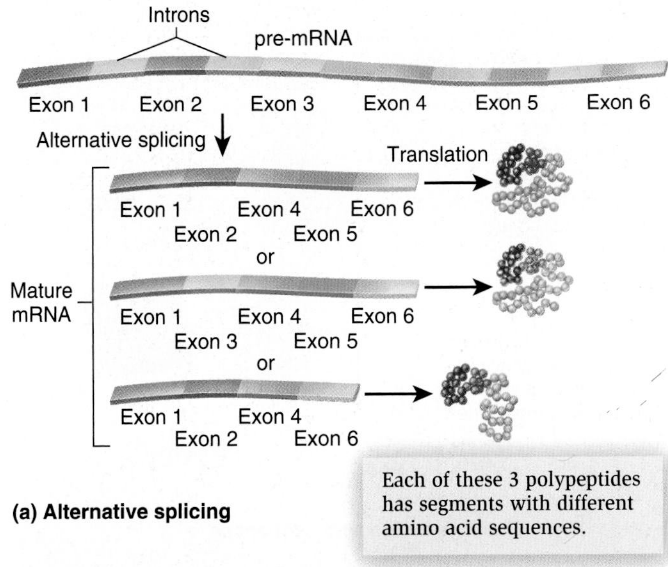

(a) Alternative splicing

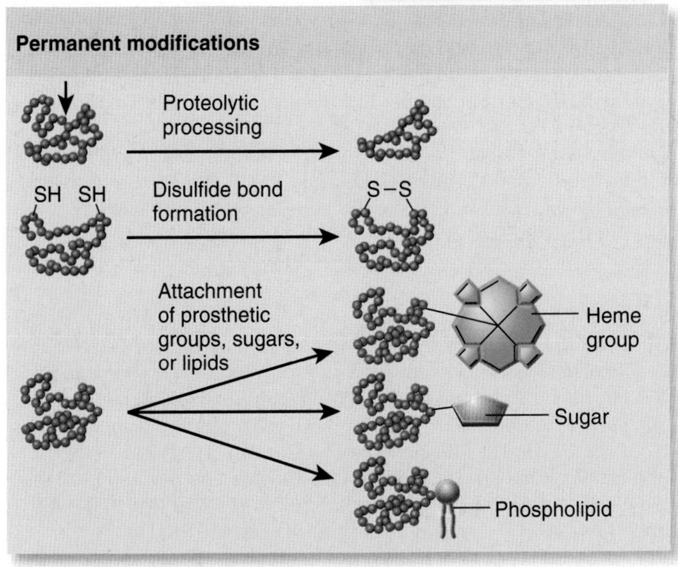

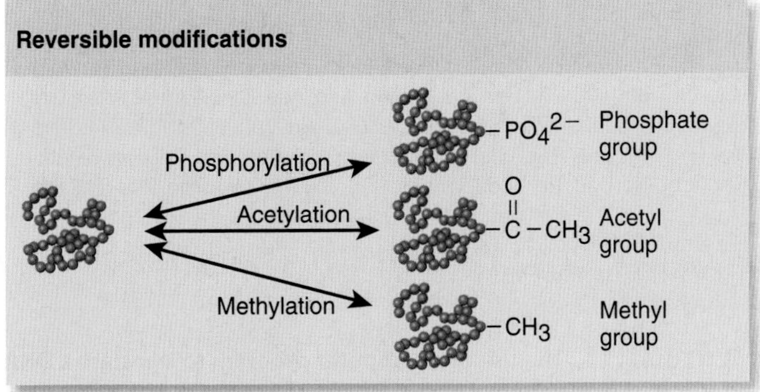

(b) Post-translational covalent modification

Figure 21.10 Cellular mechanisms that increase protein diversity. **(a)** Following alternative splicing, the pattern of exons in the resulting mature mRNA can be different, producing multiple types of polypeptides from the same gene. **(b)** In post-translational covalent modification, after a protein is made, it can be modified in a variety of ways, some of which are permanent and some reversible.

BioConnections: Look back at Figure 9.13. How is reversible phosphorylation used in cell signaling?

different genes that encode proteins. Even so, humans can make many more than 25,000 different types of proteins. How is this possible? The larger size of the proteome relative to the genome is primarily due to two types of cellular processes: alternative splicing and post-translational covalent modification, as described next.

Alternative Splicing Due to **alternative splicing**, different mRNA molecules can be produced from the same pre-mRNA transcript (refer back to Figure 13.21). For many genes, a single pre-mRNA can be spliced in more than one way, resulting in the production of two or more polypeptides with different amino acid sequences (**Figure 21.10a**). The splicing is often cell specific or may be related to environmental conditions. Alternative splicing is widespread, particularly among more complex eukaryotes (refer back to Table 13.1). It can lead to the production of several or perhaps dozens of different polypeptide sequences from the same pre-mRNA. This greatly increases the number of proteins in a species' proteome, while minimizing the size of the genome.

Post-translational Covalent Modification A second process that greatly diversifies the composition of a proteome is the phenomenon of **post-translational covalent modification**, the modification of the structure of a protein after its translation (**Figure 21.10b**). Such modifications can be permanent or reversible. Permanent modifications are often involved with the assembly and construction of functional proteins. These alterations include proteolytic processing (the cleavage of a polypeptide to a smaller unit), disulfide bond formation, and the attachment of prosthetic groups (such as heme, a component of hemoglobin), sugars, or lipids. In Chapter 6, we also considered the attachment of ubiquitin to a protein, which targets it for degradation. By comparison, reversible modifications, such as phosphorylation, acetylation, and methylation, often transiently affect the function of a protein (see Figure 21.10b). Molecules are covalently attached and later removed by cellular enzymes. Because a given type of protein may be subjected to several different types of modifications, this can greatly increase the forms of a particular protein that are found in a cell at any given time.

21.4 Bioinformatics

Learning Outcomes:

1. Describe how scientists use bioinformatics.
2. Define database.
3. Explain how the BLAST program works and why it is useful.

In the previous sections, we learned that the number of genes in a genome and the number of proteins made by a given cell type are extremely large. In the 1960s and 1970s, when the tools of molecular biology first became available, researchers tended to focus on the study of just one or a few genes and proteins at a time. Although this is a useful approach, scientists came to realize that certain properties of life emerge from complex interactions involving the expression of many genes and the functioning of many different proteins. The study of such complex interactions is called **systems biology**. This new perspective, coupled with the wealth of data produced by the genomics projects described previously in the chapter, challenged researchers to invent new tools to study many genes and proteins simultaneously.

As a very general definition, **bioinformatics** is the use of computers, mathematical tools, and statistical techniques to record, store, and analyze biological information. Why do we need bioinformatics? Simply put, the main issues are size and speed. Earlier in this chapter, we learned that the human genome has been sequenced and that it is approximately 3.2 billion base pairs long. A single person, or even a group of talented mathematicians, could not, in a reasonable length of time, analyze such an enormous amount of data. Instead, the data are put into computers, and scientists devise computational procedures to study and evaluate it.

In this section, we will consider the branch of bioinformatics that focuses on using molecular information to study biology. This area, also called **computational molecular biology**, uses computers to characterize the molecular components of living things. Molecular genetic data, which comes in the form of DNA, RNA, or protein sequences, are particularly amenable to computer analysis. In this section, we will first survey the fundamental concepts that underlie the analysis of genetic sequences. We will then consider how these methods are used to provide knowledge about how biology works at the molecular level.

Sequence Files Are Stored and Analyzed by Computers

The first steps in bioinformatics are to collect and store data in a computer. As an example, let's consider a gene sequence as a type of data. The gene sequence must first be determined experimentally using the technique of DNA sequencing. After the sequence is obtained, the next step is to put those data into a file on a computer. Typically, genetic sequence data are entered into a computer file by laboratory instruments that can read experimental data—such as data from a DNA-sequencing experiment—and enter the sequence directly into a computer.

Genetic sequence data in a computer data file can then be investigated in many different ways, corresponding to the many questions a researcher might ask about the sequence and its functional significance, including the following:

1. Does a sequence contain a gene?
2. Does a gene sequence contain a mutation that might cause a disease?
3. Where are functional sequences, such as promoters, regulatory sites, and splice sites, located within a particular gene?
4. From the sequence of a structural gene, what is the amino acid sequence of the polypeptide encoded by that gene?
5. Is there an evolutionary relationship between two or more genetic sequences?

To answer these and many other questions, computer programs have been written to analyze genetic sequences in particular ways. As an example, let's consider a computer program aimed at translating a DNA sequence into an amino acid sequence. Figure 21.11 shows a short computer data file of a DNA sequence that is presumed to be part of the coding sequence of a structural gene. In this figure, only the coding strand of DNA is shown. A computer program can analyze this sequence and print out the possible amino acid sequences that this DNA sequence would encode. The program relies on the genetic code (refer back to Table 12.1). In the example shown in Figure 21.11, the computer program shows the results for all three possible reading frames, beginning at nucleotide 1, 2, or 3, respectively. In a newly

Computer DNA sequence file

5′ GTGTCCACGC GGTCCTGGAA AACCCAGGCT TGGGCAGGAA
ACTCTCTGAC TTTGGACAGG AAACAAGCTA TATTGAAGAC
AACTGCAATC AAAATGGTGC CATATCACTG ATCTTCTCAC
TCAAAGAAGA AGTTGGTGCA TTGGCCAAAG TATTGCGCTT
ATTTGAGGAG AATGATGTAA ACCTGACCCA CATTGAATCT
AGACCTTCTC GTTTAAAGAA AGATGAGTAT GAATTTTTCA
CCCATTTGGA TAAACGTAGC CTGCCTGCTC TGACAAACAT
CATCAAGATC TTGAGGCATG ACATTGGTGC CACTGTCCAT
GAGCTTTCAC GAGATAAGAA GAAAGACACA GTGCCCTGGT
TTCCCAAG 3′

Run a computer program that translates this DNA sequence into an amino acid sequence in all 3 reading frames.

Possible amino acid sequences

5′ → 3′ Frame 1
Val Ser Thr Arg Ser Trp Lys Thr Gln Ala Trp Ala Gly Asn Ser Leu Thr Leu Asp Arg Lys Gln Ala Ile Leu Lys Thr Thr Ala Ile Lys Met Val Pro Tyr His **STOP** Ser Ser His Ser Lys Lys Lys Leu Val His Trp Pro Lys Tyr Cys Ala Tyr Leu Arg Arg Met Met **STOP** Thr **STOP** Pro Thr Leu Asn Leu Asp Leu Leu Val **STOP** Arg Lys Met Ser Met Asn Phe Ser Pro Ile Trp Ile Asn Val Ala Cys Leu Leu **STOP** Gln Thr Ser Ser Arg Ser **STOP** Gly Met Thr Leu Val Pro Leu Ser Met Ser Phe His Glu Ile Arg Arg Lys Thr Gln Cys Pro Gly Ser Gln

5′ → 3′ Frame 2
Cys Pro Arg Gly Pro Gly Lys Pro Arg Leu Gly Gln Glu Thr Leu **STOP** Leu Trp Thr Gly Asn Lys Leu Tyr **STOP** Arg Gln Leu Gln Ser Lys Trp Cys His Ile Thr Asp Leu Leu Thr Gln Arg Arg Ser Trp Cys Ile Gly Gln Ser Ile Ala Leu Ile **STOP** Gly Glu **STOP** Cys Lys Pro Asp Pro His **STOP** Ile **STOP** Thr Phe Ser Phe Lys Glu Arg **STOP** Val **STOP** Ile Phe His Pro Phe Gly **STOP** Thr **STOP** Pro Ala Cys Ser Asp Lys His His Gln Asp Leu Glu Ala **STOP** His Trp Cys His Cys Pro **STOP** Ala Phe Thr Arg **STOP** Glu Glu Arg His Ser Ala Leu Val Pro Lys

5′ → 3′ Frame 3
Val His Ala Val Leu Glu Asn Pro Gly Leu Gly Arg Lys Leu Ser Asp Phe Gly Gln Glu Thr Ser Tyr Ile Glu Asp Asn Cys Asn Gln Asn Gly Ala Ile Ser Leu Ile Phe Ser Leu Lys Glu Glu Val Gly Ala Leu Ala Lys Val Leu Arg Leu Phe Glu Glu Asn Asp Val Asn Leu Thr His Ile Glu Ser Arg Pro Ser Arg Leu Lys Lys Asp Glu Tyr Glu Phe Phe Thr His Leu Asp Lys Arg Ser Leu Pro Ala Leu Thr Asn Ile Ile Lys Ile Leu Arg His Asp Ile Gly Ala Thr Val His Glu Leu Ser Arg Asp Lys Lys Lys Asp Thr Val Pro Trp Phe Pro

Figure 21.11 **The use of a computer program to translate a DNA sequence into an amino acid sequence.** The top part of this figure shows the sequence of a segment of the coding strand of a structural gene (divided into groups of 10 bases for ease of reading). A computer program translates the DNA sequence into an amino acid sequence based on the genetic code. The program produces three different amino acid sequences, as shown at the bottom of the figure. In this example, reading frame 3 is likely to be the correct one because it does not contain any stop codons.

Concept Check: *Why is it helpful to use a computer program to translate a genetic sequence rather than doing it by hand?*

Table 21.4 Examples of Major Genetic Databases

Nucleotide sequences	DNA sequence data are collected into three internationally collaborating databases: GenBank (a U.S. database), EMBL (European Molecular Biology Laboratory Nucleotide Sequence Database), and DDBJ (DNA Data Bank of Japan). These databases receive sequence and sequence annotation data from genome projects, sequencing centers, individual scientists, and patent offices. New and updated entries are exchanged daily.
Amino acid sequences	Protein sequence data are collected into a few international databases, including Swiss-Prot (Swiss protein database), PIR (Protein Information Resource), TrEMBL (translated sequences from the EMBL database), and Genpept (translated peptide sequences from the GenBank database).

obtained DNA sequence, a researcher would not know the proper reading frame—the series of codons read in groups of three, which begins with the start codon. Therefore, the computer program provides all three. If you look at the results, reading frames 1 and 2 include several stop codons, whereas reading frame 3 does not. From these results, reading frame 3 is likely to be the correct one. Also, for a new DNA sequence, a researcher may not know which DNA strand is the coding strand. Therefore, the sequence of the other DNA strand, which is not shown in this figure, would also be analyzed by this computer program.

The Scientific Community Has Collected Computer Data Files and Stored Them in Large Databases

Over the past several decades, the amount of genetic information generated by researchers and clinicians has become enormous. With these advances, scientists have realized that another critical use of computers is to store the staggering amount of data produced from genetic research.

A large amount of data that is collected, stored in a single location, and organized for rapid search and retrieval is called a **database**. The files within databases are often annotated, which means they contain a concise description of each gene sequence, the name of the organism from which the sequence was obtained, and the function of the encoded protein, if it is known. The file may also provide a published reference that contains the sequence.

The research community has collected genetic information from thousands of research laboratories and created several large databases. Table 21.4 describes some of the major genetic databases in use worldwide, all of which can be accessed online. These databases enable researchers to access and compare genetic sequences that are obtained by many laboratories. Later in this chapter, we will learn how researchers use databases to analyze genetic sequences.

Many programs are freely available over the Internet to utilize the information within databases. For example, the National Center for Biotechnology Information (NCBI), which is a part of the U.S. National Institutes of Health, manages a website called "Tools for Data Mining," where anyone can run various types of programs that are used to analyze genetic sequences (www.ncbi.nlm.nih.gov/Tools). Like conventional mining, in which a precious mineral is extracted from a large area of land, **data mining** is the extraction of useful information and often previously unknown relationships from sequence files and large databases.

Computer Programs Can Identify Homologous Sequences

Let's now turn our attention to genes that are evolutionarily related. Organisms that are closely related evolutionarily tend to have genes with similar DNA sequences. As an example, let's consider the gene that encodes β-globin. As described earlier in Figure 21.8, β-globin is a polypeptide found in hemoglobin, which carries oxygen in red blood cells. The β-globin gene is found in humans and other vertebrates.

Figure 21.12a compares a short region of the β-globin gene from the laboratory mouse (*Mus musculus*) with that from a laboratory rat (*Rattus norvegicus*). As you can see, the gene sequences are similar

```
Mouse
GGGCAGGTTGGTATCCAGGTTACAAGGCAGCTCACAAGTAGAAGCTGGGTGCTTGGAGAC
||||||||||||||||||||||||||| |||||  ||||||||| | |||||||||||||
GGGCAGGTTGGTATCCAGGTTACAAGGTAGCTCCTAAGTAGAAGTTTGGTGCTTGGAGAC
Rat
```

(a) A comparison of one DNA strand of the mouse and rat β-globin genes

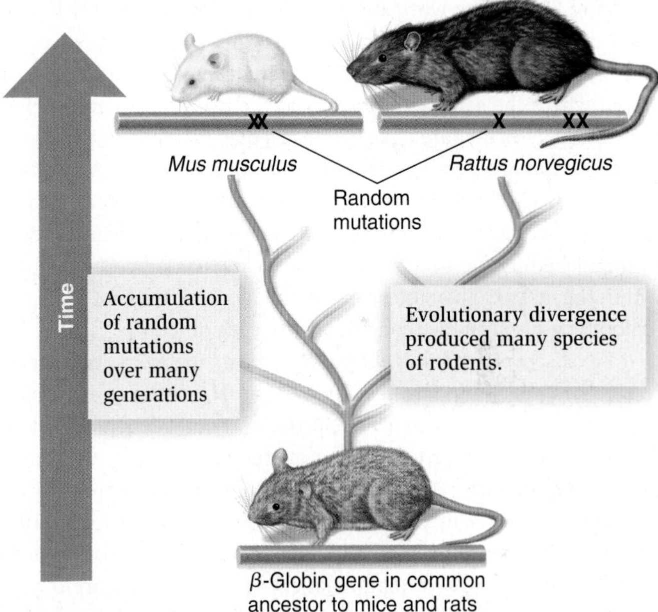

(b) The formation of homologous β-globin genes during evolution of mice and rats

Figure 21.12 Structure and formation of the homologous β-globin genes in mice and rats. (a) A comparison of a short region of the gene that encodes β-globin in laboratory mice (*Mus musculus*) and rats (*Rattus norvegicus*). Only one DNA strand is shown. Bases that are identical between the two sequences are connected by a vertical line. **(b)** Schematic view of the formation of these homologous β-globin genes during evolution. An ancestral β-globin gene was found in an ancestor common to both mice and rats. This ancestral species later diverged into different species, which gave rise to modern rodent species. During this process, the β-globin genes accumulated different random mutations, resulting in DNA sequences that are slightly different from each other.

Concept Check: Is it possible for orthologs from two different species to have exactly the same DNA sequence? Explain.

but not identical. In this 60-nucleotide sequence, five differences are observed. The reason for the sequence similarity is that the genes are derived from the same ancestral gene. This idea is shown schematically in **Figure 21.12b**. An ancestral gene was found in a rodent species that was a common ancestor to both mice and rats. During evolution, this ancestral species diverged into different species, which eventually gave rise to several modern rodent species, including mice and rats. Following divergence, the β-globin genes accumulated distinct mutations that produced somewhat different base sequences for this gene. Therefore, in mice and rats, the β-globin genes have homologous sequences—sequences that are similar because they are derived from the same ancestral gene, but not identical because each species has accumulated a few different random mutations. Homologous genes in different species are also called orthologous genes or **orthologs**. Analyzing orthologs helps biologists understand the evolutionary relationships among modern species, a topic that we will consider in more detail in Units IV and V.

How do researchers, with the aid of computers, determine if two genes are homologous to each other? To evaluate the similarity between two sequences, a matrix can be constructed. **Figure 21.13** illustrates the use of a simplified dot matrix to evaluate two sequences. In Figure 21.13a, the word BIOLOGY is compared with itself. Each point in the grid corresponds to one position of each sequence. The matrix allows all such pairs to be compared simultaneously. Dots are placed where the same letter occurs at the two corresponding positions. Sequences that are alike produce a diagonal line on the matrix. Figure 21.13b compares two similar but different sequences: BIOLOGY and ECOLOGY. This comparison produces only a partial diagonal line. Overall, the key observation is that regions of similarity are distinguished by the occurrence of many dots along a diagonal line within the matrix. This same concept holds true when base sequences of homologous genes are compared with each other.

To discover whether homology exists between two genes, researchers must compare relatively long DNA sequences. For such long sequences, a dot matrix approach is not adequate. Instead, dynamic computer programming methods are used to identify optimal alignments between genetic sequences. This approach was first proposed by Saul Needleman and Christian Wunsch in 1970. Dynamic programming methods are theoretically similar to a dot matrix, but they involve mathematical operations that are beyond the scope of this textbook. In their original work, Needleman and Wunsch demonstrated that whale myoglobin and human β-globin genes have similar sequences.

A Database Can Be Searched to Identify Similar Sequences and Thereby Infer Homology and Gene Function

Why is it useful to identify homology between different genes? Because they are derived from the same ancestral gene, homologous genes usually carry out similar or identical functions. In many cases, the first way to identify the function of a newly determined gene sequence is to find a homologous gene whose function is already known. An example is the gene that is altered in cystic fibrosis patients. After this gene was identified in humans, bioinformatic methods revealed that it is homologous to several genes found in other species. A few of the homologous genes were already known to encode proteins that function in the transport of ions and small molecules across the plasma membrane. This observation provided an important clue that cystic fibrosis involves a defect in membrane transport.

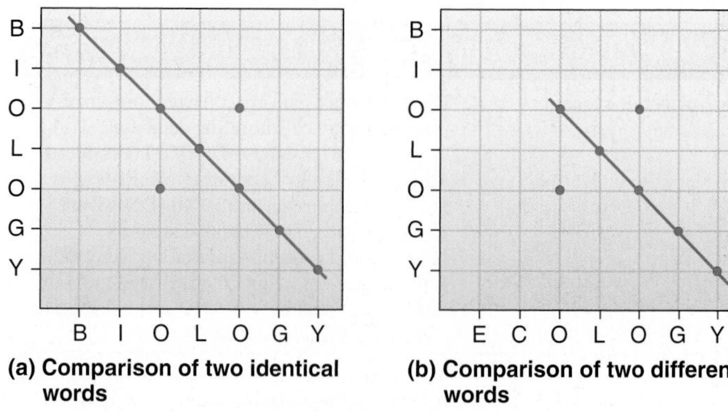

(a) Comparison of two identical words **(b) Comparison of two different words**

Figure 21.13 **The use of a simple dot matrix.** In these comparisons, a diagonal line indicates sequence similarity. **(a)** The word BIOLOGY is compared with itself. Dots are placed where the same letter occurs at the two corresponding positions. Notice the diagonal line that is formed. **(b)** Two similar but different sequences, BIOLOGY and ECOLOGY, are compared with each other. Notice that only a partial line is formed by this comparison.

The ability of computer programs to identify homology between genetic sequences provides a powerful tool for predicting the function of genetic sequences. In 1990, American mathematician Stephen Altschul, his colleague biologist David Lipman, and their coworkers developed a program called **BLAST** (for basic local alignment search tool). The BLAST program has been described by many biologists as the single most important tool in computational molecular biology. This computer program can start with a particular genetic sequence—either a nucleotide or an amino acid sequence—and then locate homologous sequences within a large database.

As an example of how the BLAST program works, let's consider the human enzyme phenylalanine hydroxylase, which functions in the metabolism of phenylalanine, an amino acid. Recessive mutations in the gene that encodes this enzyme are responsible for the disease called phenylketonuria (PKU). The computational experiment shown in **Table 21.5** started with the amino acid sequence of this protein and used the BLAST program to search the Swiss-Prot database, which contains hundreds of thousands of different protein sequences. The BLAST program can determine which sequences in the Swiss-Prot database are the closest matches to the amino acid sequence of human phenylalanine hydroxylase. Table 21.5 shows some of the results, which includes 10 of the matches to human phenylalanine hydroxylase that were identified by the program. Because this enzyme is found in nearly all eukaryotic species, the program identified phenylalanine hydroxylase from many different species. The column to the right of the match number shows the percentage of amino acids that are identical between the species indicated and the human sequence. Because the human phenylalanine hydroxylase sequence is already in the Swiss-Prot database, the closest match of human phenylalanine hydroxylase is to itself. The next nine sequences are in order of similarity. The next most similar sequence is from the orangutan, a primate that is a close relative of humans. This is followed by two other mammals, the mouse and rat, and then five vertebrates that are not mammals. The 10th best match is from *Drosophila*, an invertebrate.

Table 21.5 Results from a BLAST Program Comparing Human Phenylalanine Hydroxylase with Database Sequences

Match	Percentage of identical amino acids*	Species	Function of sequence†
1	100	Human (*Homo sapiens*)	Phenylalanine hydroxylase
2	99	Orangutan (*Pongo pygmaeus*)	Phenylalanine hydroxylase
3	95	Mouse (*Mus musculus*)	Phenylalanine hydroxylase
4	95	Rat (*Rattus norvegicus*)	Phenylalanine hydroxylase
5	89	Chicken (*Gallus gallus*)	Phenylalanine hydroxylase
6	82	Pipid frog (*Xenopus tropicalis*)	Phenylalanine hydroxylase
7	82	Green pufferfish (*Tetraodon nigroviridis*)	Phenylalanine hydroxylase
8	82	Zebrafish (*Danio rerio*)	Phenylalanine hydroxylase
9	80	Japanese pufferfish (*Takifugu rubripes*)	Phenylalanine hydroxylase
10	75	Fruit fly (*Drosophila melanogaster*)	Phenylalanine hydroxylase

*The number indicates the percentage of amino acids that are identical with the amino acid sequence of human phenylalanine hydroxylase. Note: These matches were randomly selected from a long list of matches.

†In some cases, the function of the sequence was determined by biochemical assay. In other cases, the function was inferred due to the high degree of sequence similarity with other species.

As you may have noticed, the order of the matches follows the evolutionary relatedness of the various species to humans. The similarity between any two sequences is related to the time that has passed since they diverged from a common ancestor. Among the species listed in this table, humans are most closely related to the orangutan, other mammals, other vertebrates, and finally invertebrates.

Overall, Table 21.5 is an example of the remarkable computational abilities of current computer technology. In less than a minute, the amino acid sequence of human phenylalanine hydroxylase can be compared with hundreds of thousands of different sequences to yield the data shown in this table! The main power of the BLAST program is its use with newly identified sequences, in which a researcher does not know the function of a gene or an encoded protein. When the BLAST program identifies a match to a sequence whose function is already known, it is likely that the newly identified sequence has an identical or similar function.

Summary of Key Concepts

21.1 Bacterial and Archaeal Genomes

- The genome is the complete genetic makeup of a cell, organism, or species.
- Bacterial and archaeal genomes are typically a single circular chromosome with a few million base pairs of DNA. Such genomes usually have a few thousand different genes. Bacteria often have plasmids (Table 21.1).
- Venter, Smith, and colleagues used a shotgun DNA sequencing strategy to sequence the first genome, that of the bacterium *Haemophilus influenzae* (Figure 21.1).

21.2 Eukaryotic Genomes

- The nuclear genomes of eukaryotic species are composed of sets of linear chromosomes with a total length of several million to billions of base pairs. They typically contain several thousand to tens of thousands of genes (Table 21.2).
- Genome sizes vary among eukaryotic species. In many cases, variation is due to the accumulation of noncoding regions of DNA, particularly repetitive DNA sequences (Figures 21.2, 21.3).
- Much of the repetitive DNA is derived from transposable elements, segments of DNA that can move from one site to another through a process called transposition (Figure 21.4).
- Transposable elements fall into two groups that move by different molecular mechanisms. DNA transposons move by a cut-and-paste mechanism facilitated by the enzyme transposase. Retroelements move to new sites in the genome via RNA intermediates (Figures 21.5, 21.6).
- Gene duplication may occur by a misaligned crossover during meiosis. This is one mechanism that can produce a gene family, two or more homologous genes in a species that have related functions (Figures 21.7, 21.8).
- The Human Genome Project, an international effort to map and sequence the entire human genome, was completed by an international consortium in 2003.

21.3 Proteomes

- A proteome is the collection of proteins that a given cell or species makes.
- Proteins are often placed into broad categories based on their functions. These include metabolic enzymes, structural proteins, motor proteins, cell-signaling proteins, transport proteins, proteins involved with gene expression and regulation, and those involved with protection and defense (Table 21.3).
- Protein abundance refers to the number of genes encoding a particular type of protein or the amount of a protein made by a cell (Figure 21.9).
- Protein diversity increases via mechanisms such as alternative splicing and post-translational covalent modifications (Figure 21.10).

21.4 Bioinformatics

- Bioinformatics involves the use of computers, mathematical tools, and statistical techniques to record, store, and analyze biological information, particularly genetic data such as DNA and protein sequences.
- Genetic information is stored in large genetics databases that are analyzed using computer programs (Figure 21.11, Table 21.4).
- Homologous genes are derived from the same ancestral gene and have accumulated random mutations that make their sequences slightly different (Figure 21.12).
- Scientists use computer programs to determine if gene sequences are homologous (Figure 21.13).
- Computer programs, such as BLAST, identify homologous genes that are found in a database (Table 21.5).

Assess and Discuss

Test Yourself

1. The entire collection of proteins produced by a cell or organism is
   a. a genome.
   b. bioinformatics.
   c. a proteome.
   d. a gene family.
   e. a protein family.
2. Important reasons for studying the genomes of bacteria and archaea include all of the following except
   a. it may provide information that helps us understand how bacteria infect other organisms.
   b. it may provide a basic understanding of cellular processes that allow us to determine eukaryotic cellular function.
   c. it may provide the means of understanding evolutionary processes.
   d. it will reveal the approximate number of genes that an organism has in its genome.
   e. all of the above are important reasons.
3. The enzyme that allows short segments of DNA to move within a cell from one location in the genome to another is
   a. transposase.
   b. DNA polymerase.
   c. protease.
   d. restriction endonuclease.
   e. DNA ligase.
4. A gene family includes
   a. one specific gene found in several different species.
   b. all of the genes on the same chromosome.
   c. two or more homologous genes found within a single species.
   d. genes that code for structural proteins.
   e. both a and c.
5. Which of the following was *not* a goal of the Human Genome Project?
   a. identify all human genes
   b. sequence the entire human genome
   c. address the legal and ethical implications resulting from the project
   d. develop programs to manage the information gathered from the project
   e. be able to clone a human
6. Bioinformatics is
   a. the analysis of DNA by molecular techniques.
   b. the use of computers to analyze and store biological information.
   c. a collection of gene sequences from a single individual.
   d. cloning.
   e. all of the above.
7. Using bioinformatics, evolutionary relationships among species can be characterized by identifying and analyzing
   a. phenotypes of selected organisms.
   b. homologous DNA sequences from different organisms.
   c. fossils of ancestral species.
   d. all of the above.
   e. a and b only.
8. Repetitive sequences
   a. are short DNA sequences that are found many times throughout the genome.
   b. may be multiple copies of the same gene found in the genome.
   c. are more common in eukaryotes.
   d. all of the above
   e. a and c only
9. The BLAST program is a tool for
   a. inserting many DNA fragments into a cell at the same time.
   b. translating a DNA sequence into an amino acid sequence.
   c. identifying homology between a selected sequence and genetic sequences in databases.
   d. all of the above.
   e. both b and c.
10. Let's suppose you used the BLAST program beginning with a DNA sequence from a *Drosophila* hexokinase gene. (Hexokinase is an enzyme involved with glucose metabolism.) Which of the following choices would you expect to be the closest match?
    a. a *Drosophila* globin gene
    b. a human hexokinase gene
    c. a house fly hexokinase gene
    d. an *Arabidopsis* hexokinase gene
    e. an amoeba hexokinase gene

Conceptual Questions

1. Briefly describe whether or not each of the following could be appropriately described as a genome.
   a. the *E. coli* chromosome
   b. human chromosome 11
   c. a complete set of 10 chromosomes in corn
   d. a copy of the single-stranded RNA packaged into human immunodeficiency virus (HIV)
2. Describe two main reasons why the proteomes of eukaryotes species are usually much larger than their genomes.
3. A principle of biology is that *the genetic material provides a blueprint for reproduction.* Explain the roles of the genome and proteome with regard to this principle.

Collaborative Questions

1. Compare and contrast the genomes of bacteria, archaea, and eukaryotes.
2. The following is a DNA sequence from one strand of a gene. (The bases are shown in groups of 10 each.) Go to the NCBI website (www.ncbi .nlm.nih.gov/Tools) and select the Basic Local Alignment Search Tool (BLAST) program to determine which gene it is and in which species it is found.

```
GTGAAGGCTC ATGGCAAGAA AGTGCTCGGT GCCTTTAGTG
ATGGCCTGGC TCACCTGGAC AACCTCAAGG GCACCTTTGC
CACACTGAGT GAGCTGCACT GTGACAAGCT GCACGTGGAT
CCTGAGAACT TCAGGGTGAG TCTATGGGAC GCTTGATGTT
TTCTTTCCCC TTCTTTTCTA TGGTTAAGTT CATGTCATAG
GAAGGGGATA AGTAACAGGG TACAGTTTAG AATGGGAAAC
AGACGAATGA TTGCATCAGT GTGGAAGTCT CAGGATCGTT
TTAGTTTCTT TTATTTGCTG TTCATAACAA TTGTTTTCTT
TTGTTTAATT CTTGCTTTCT TTTTTTTTCT TCTCCGCAAT
```

Online Resource

www.brookerbiology.com

Stay a step ahead in your studies with animations that bring concepts to life and practice tests to assess your understanding. Your instructor may also recommend the interactive eBook, individualized learning tools, and more.

UNIT IV

EVOLUTION

Evolution is a heritable change in one or more characteristics of a population from one generation to the next. This process not only alters the characteristics of populations, it also leads to the formation of new species.

We will begin Unit IV with a discussion of the hypotheses that have been proposed to explain the origin of life on Earth, and then examine a timeline for the evolution of species from 4 billion years ago to the present. In Chapter 23, you will be introduced to the fundamental concepts of evolution, with an emphasis on natural selection. We will examine observations of evolutionary change, which includes (1) the fossil record, (2) a comparison of the characteristics of modern species, and (3) an analysis of molecular data. Chapter 24 continues our discussion of evolution at the molecular level and focuses on how changes in allele and genotype frequencies from one generation to the next are driven by a variety of different factors. By comparison, Chapter 25 shifts the emphasis of evolution to the level of species. We will examine how species are identified and discuss the mechanisms by which new species arise via evolution. Finally, in Chapter 26, we will examine how biologists determine the evolutionary relationships among different species and produce "trees" that describe those relationships.

22

23

24

25

26

The following biology principles will be emphasized in this unit:

- ***Populations of organisms evolve from one generation to the next.*** *This concept will be emphasized throughout the entire unit.*
- ***Living organisms interact with their environment.*** *As discussed in Chapters 23 and 24, natural selection is a process in which certain individuals have greater reproductive success. This success is often due to their ability to survive in a given environment.*
- ***Structure determines function.*** *Chapters 23 and 24 will also consider how structural features change during the evolution of new species. Such changes are related to changes in function.*
- ***All species (past and present) are related by an evolutionary history.*** *Chapter 26 is devoted to examining how biologists determine evolutionary relationships among different species.*
- ***Biology is an experimental science.*** *Every chapter has a Feature Investigation that describes a pivotal experiment that provided insights into our understanding of evolution.*

The Origin and History of Life on Earth

22

Chapter Outline

22.1 Origin of Life on Earth
22.2 The Fossil Record
22.3 History of Life on Earth
Summary of Key Concepts
Assess and Discuss

A fossil fish. This 50-million-year-old fossil of a unicorn fish (*Naso rectifrons*) is an example of the many different kinds of organisms that have existed during the history of life on Earth.

The amazing origin of the universe is difficult to comprehend. Astronomers think the universe began with a cosmic explosion called the Big Bang about 13.7 billion years ago (bya), when the first clouds of the elements hydrogen and helium were formed. Over a long time period, gravitational forces collapsed these clouds to create stars that converted hydrogen and helium into heavier elements, including carbon, nitrogen, and oxygen, which are the atomic building blocks of life on Earth. These elements were returned to interstellar space by exploding stars called supernovas, which created clouds in which simple molecules such as water, carbon monoxide, and hydrocarbons formed. The clouds then collapsed to make a new generation of stars and solar systems.

Our solar system began about 4.6 bya after one or more local supernova explosions. According to one widely accepted scenario, hundreds of planetesimals consisting of bodies such as asteroids and comets, occupied the region where Venus, Earth, and Mars are now found. The Earth, which is estimated to be 4.55 billion years old, grew from the aggregation of such planetesimals over a period of 100 to 200 million years. For the first half billion years or so after its formation, the Earth was too hot to allow liquid water to accumulate on its surface. By 4 bya, the Earth had cooled enough for the outer layers of the planet to solidify and for oceans to begin to form.

The period between 4.0 and 3.5 bya marked the emergence of life on our planet. The first forms of life that we know about produced well-preserved microscopic fossils, such as those found in western Australia. These fossils, estimated to be about 3.5 billion years old, resemble modern cyanobacteria, which are photosynthetic bacteria (**Figure 22.1**). Researchers cannot travel back through time and observe how the first life-forms came into being. However, plausible hypotheses regarding how life first arose have emerged from our understanding of modern life.

This chapter, the first in the Evolution unit, emphasizes when particular forms of life arose. The first section surveys a variety of hypotheses regarding (1) the origin of organic molecules on Earth, (2) the formation of complex molecules such as DNA, RNA, and proteins, (3) the formation of primitive cell-like structures, and (4) the process that gave rise to the first living cells. We will then consider fossils, the preserved remains of organisms that existed in the past. Starting 3.5 bya, the formation of fossils, such as the one shown in the chapter opening photo, has provided biologists with evidence of the history of life on Earth from its earliest beginnings to the present day. The last section provides a broad overview of the geologic time scale and the major events in the history of life on Earth.

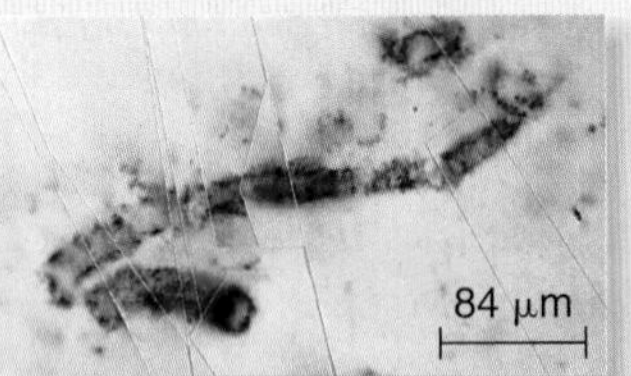

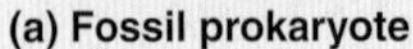

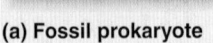

(a) Fossil prokaryote

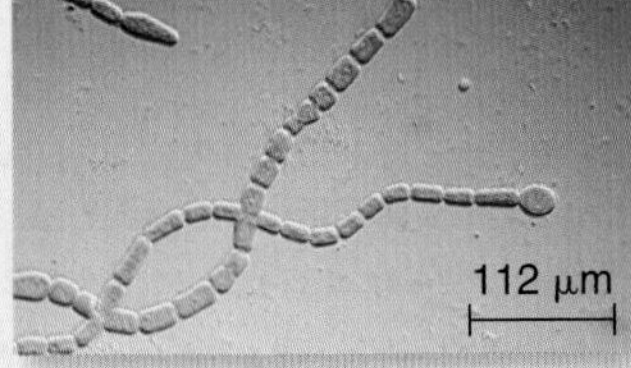

(b) Modern cyanobacteria

Figure 22.1 **Earliest fossils and living cyanobacteria.** **(a)** A fossilized prokaryote about 3.5 billion years old that is thought to be an early cyanobacterium. **(b)** A modern cyanobacterium, which has a similar morphology. Cyanobacterial cells are connected to each other to form chains, as shown here.

22.1 Origin of Life on Earth

Learning Outcomes:

1. Outline the four overlapping stages that are hypothesized to have led to the origin of life.
2. List various hypotheses about how complex organic molecules formed.
3. Analyze the results of Bartel and Szostak that indicated that chemical evolution is possible.
4. Explain the concept of an RNA world and how it could have evolved into a DNA/RNA/protein world.

As we have seen, living cells are complex collections of molecules and macromolecules. DNA stores genetic information, RNA acts as an intermediary in the process of protein synthesis and plays other important roles, and proteins form the foundation for the structure and activities of living cells. Life as we know it requires this interplay between DNA, RNA, and proteins for its existence and perpetuation. On modern Earth, every living cell is made from a pre-existing cell.

But how did life get started? As described in Chapter 1, living organisms have several characteristics that distinguish them from nonliving materials. Because DNA, RNA, and proteins are the central players in the enterprise of life, scientists who are interested in the origin of life have focused much of their attention on the formation of these macromolecules and their building blocks, namely, nucleotides and amino acids. To understand the origin of life, we can view the process as occurring in four overlapping stages:

Stage 1: Nucleotides and amino acids were produced prior to the existence of cells.

Stage 2: Nucleotides became polymerized to form RNA and/or DNA, and amino acids become polymerized to form proteins.

Stage 3: Polymers became enclosed in membranes.

Stage 4: Polymers enclosed in membranes acquired cellular properties.

Researchers have followed a variety of experimental approaches to determine how life may have begun, including the synthesis of organic molecules in the laboratory without the presence of living cells or cellular material. This work has led researchers to propose a variety of hypotheses regarding the origin of life. In this section, we will examine the origin of life at each of these stages and consider a few scientific viewpoints that wrestle with the question, "How did life on Earth begin?"

Stage 1: Organic Molecules Formed Prior to the Existence of Cells

Let's begin our inquiry into the first stage of the origin of life by considering how nucleotides and amino acids may have been made prior to the existence of living cells. In the 1920s, the Russian biochemist Alexander Oparin and the Scottish biologist J.B.S. Haldane independently proposed that organic molecules, such as nucleotides and amino acids, arose spontaneously under the conditions that occurred on early Earth. According to this hypothesis, the spontaneous appearance of organic molecules produced what they called a "primordial soup," which eventually gave rise to living cells.

The conditions on early Earth, which were much different from today, may have been more conducive to the spontaneous formation of organic molecules. Current hypotheses suggest that organic molecules, and eventually macromolecules, formed spontaneously. This is termed prebiotic (before life) or abiotic (without life) synthesis. These slowly forming organic molecules accumulated because there was little free oxygen gas, so they were not spontaneously oxidized, and there were as yet no living organisms, so they were also not metabolized. The slow accumulation of these molecules in the early oceans over a long period of time formed what is now called the **prebiotic soup**. The formation of this medium was a key event that preceded the origin of life.

Though most scientists agree that life originated from the assemblage of nonliving matter on early Earth, the mechanism of how and where these molecules originated is widely debated. Many intriguing hypotheses have been proposed, which are not mutually exclusive. A few of the more widely debated ideas are the reducing atmosphere hypothesis, the extraterrestrial hypothesis, and the deep-sea vent hypothesis.

Reducing Atmosphere Hypothesis Based largely on geological data, many scientists in the 1950s proposed that the atmosphere on early Earth was rich in water vapor (H_2O), hydrogen gas (H_2), methane (CH_4), and ammonia (NH_3). These components, along with a lack of atmospheric oxygen (O_2), produce a reducing atmosphere because methane and ammonia readily give up electrons to other molecules, thereby reducing them. Such oxidation-reduction reactions, or redox reactions, are required for the formation of complex organic molecules from simple inorganic molecules.

In 1953, American chemist Stanley Miller, a student in the laboratory of the physical chemist Harold Urey, was the first scientist to use experimentation to test whether the prebiotic synthesis of organic molecules is possible. His experimental apparatus was intended to simulate the conditions on early Earth that were postulated in the 1950s (**Figure 22.2**). Water vapor from a flask of boiling water rose into another chamber containing hydrogen gas (H_2), methane (CH_4), and ammonia (NH_3). Miller inserted two electrodes that sent electrical discharges into the chamber to simulate lightning bolts. A condenser jacket cooled some of the gases from the chamber, causing droplets to form that fell into a trap. He then took samples from this trap for chemical analysis. In his first experiments, he observed the formation of hydrogen cyanide (HCN) and formaldehyde (CH_2O). Such molecules are precursors of more complex organic molecules. These precursors also combined to make larger molecules such as the amino acid glycine. At the end of 1 week of operation, 10–15% of the carbon had been incorporated into organic compounds. Later experiments by Miller and others demonstrated the formation of sugars, a few types of amino acids, lipids, and nitrogenous bases found in nucleic acids (for example, adenine).

In a study published in 2011, researchers analyzed samples that Miller had preserved from a 1958 experiment in which he used a mixture of CH_4, NH_3, hydrogen sulfide (H_2S), and carbon dioxide (CO_2). For unknown reasons, Miller had not analyzed what products were made in this experiment. When these preserved samples were analyzed using modern technology, they were found to contain 23 different amino acids and 4 amines (another type of organic molecule), more organic compounds than seen in Miller's classic experiments.

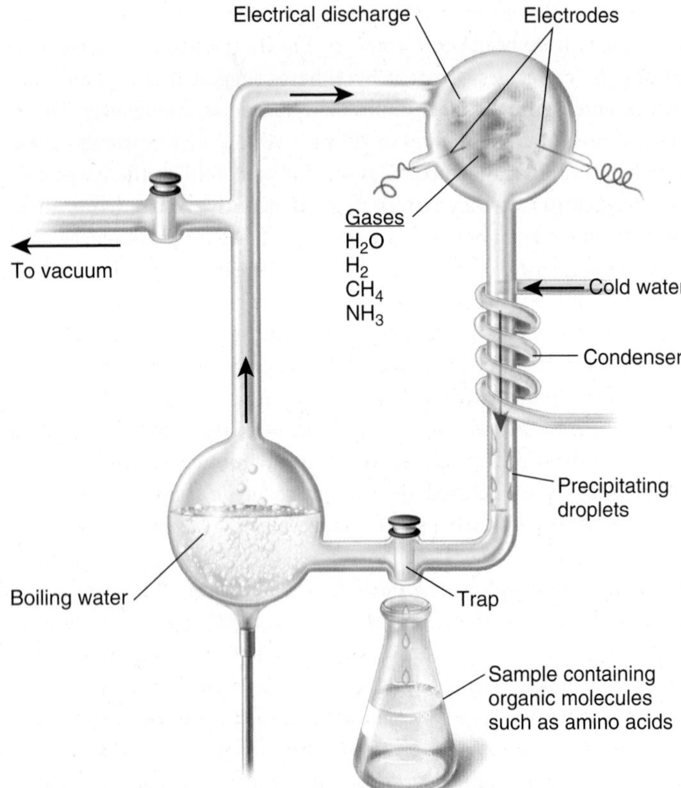

Figure 22.2 Testing the reducing atmosphere hypothesis for the origin of life—the Miller and Urey experiment.

BIOLOGY PRINCIPLE Biology is an experimental science. By conducting experiments, researchers were able to demonstrate the feasibility of the synthesis of organic molecules prior to the emergence of living cells.

Concept Check: With regard to the origin of life, why are biologists interested in the prebiotic synthesis of organic molecules?

Why were these studies important? The work of Miller and Urey was the first attempt to apply scientific experimentation to our quest to understand the origin of life. Their pioneering strategy showed that the prebiotic synthesis of organic molecules is possible, although it could not prove that it really happened that way. In spite of the importance of these studies, critics of the so-called reducing atmosphere hypothesis have argued that Miller and Urey were wrong about the composition of early Earth's environment. More recently, many scientists have suggested that the atmosphere on early Earth was not reducing, but instead was a neutral environment composed mostly of carbon monoxide (CO), carbon dioxide (CO_2), nitrogen gas (N_2), and H_2O. These newer ideas are derived from studies of volcanic gas, which has much more CO_2 and N_2 than CH_4 and NH_3, and from the observation that ultraviolet (UV) radiation destroys CH_4 and NH_3, so these molecules would have been short-lived on early Earth, which had high levels of UV radiation. Nevertheless, since the experiments of Miller and Urey, many newer investigations have shown that organic molecules can be made under a variety of conditions. For example, organic molecules can be made prebiotically from a neutral environment composed primarily of CO, CO_2, N_2, and H_2O.

Extraterrestrial Hypothesis Many scientists have argued that sufficient organic molecules may have been present in the materials from asteroids and comets that reached the surface of early Earth in the form of meteorites. A significant proportion of meteorites belong to a class known as carbonaceous chondrites. Such meteorites may contain a substantial amount of organic carbon, including amino acids and nucleic acid bases. Based on this observation, some scientists have postulated that such meteorites could have transported a significant amount of organic molecules to early Earth.

Opponents of this hypothesis argue that most of this material would have been destroyed by the intense heating that accompanies the passage of large bodies through the atmosphere and their subsequent collision with the surface of the Earth. Though some organic molecules are known to reach the Earth via such meteorites, the degree to which heat would have destroyed many of the organic molecules remains a matter of controversy.

Deep-Sea Vent Hypothesis In 1988, the German lawyer and organic chemist Günter Wächtershäuser proposed that key organic molecules may have originated in deep-sea vents, which are cracks in the Earth's surface where superheated water rich in metal ions and hydrogen sulfide (H_2S) mixes abruptly with cold seawater. These vents release hot gaseous substances from the interior of the Earth at temperatures in excess of 300°C (572°F). Supporters of this hypothesis propose that biologically important molecules may have been formed in the temperature gradient between the extremely hot vent water and the cold water that surrounds the vent (Figure 22.3a).

Experimentally, the temperatures within this gradient are known to be suitable for the synthesis of molecules that form components of biological molecules. For example, the reaction between iron and H_2S yields pyrites and H_2 and has been shown to provide the energy necessary for the reduction of N_2 to NH_3. Nitrogen is an essential component of both nucleic acids and amino acids—the molecular building blocks of life. But N_2, which is found abundantly on Earth, is chemically inert, so it is unlikely to have given rise to life. The conversion of N_2 to NH_3 at deep-sea vents may have led to the production of amino acids and nucleic acids.

Interestingly, complex biological communities are found in the vicinity of modern deep-sea vents. Various types of fish, worms, clams, crabs, shrimp, and bacteria are found in significant abundance in those areas (Figure 22.3b). Unlike most other forms of life on our planet, these organisms receive their energy from chemicals in the vent and not from the Sun. In 2007, American scientist Timothy Kusky and colleagues discovered 1.43 billion-year-old fossils of deep-sea microbes near ancient deep-sea vents. This study provided more evidence that life may have originated on the bottom of the ocean. However, debate continues as to the primary way that organic molecules were made prior to the existence of life on Earth.

Stage 2: Organic Polymers May Have Formed on the Surface of Clay

The preceding three hypotheses provide reasonable mechanisms whereby small organic molecules could have accumulated on early Earth. Scientists hypothesize that the second stage in the origin of life was a period in which simple organic molecules polymerized to

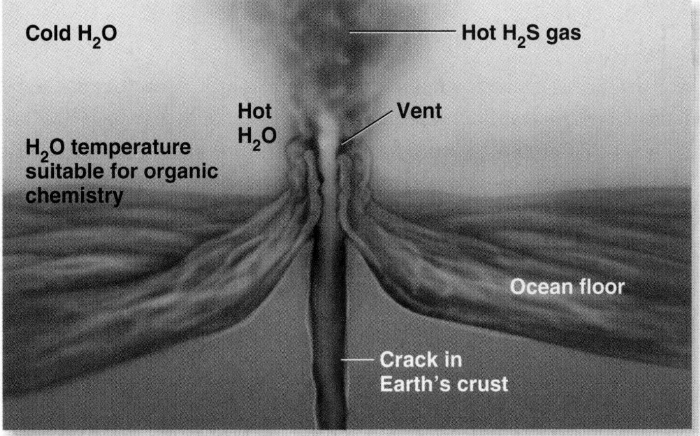

(a) Deep-sea vent hypothesis

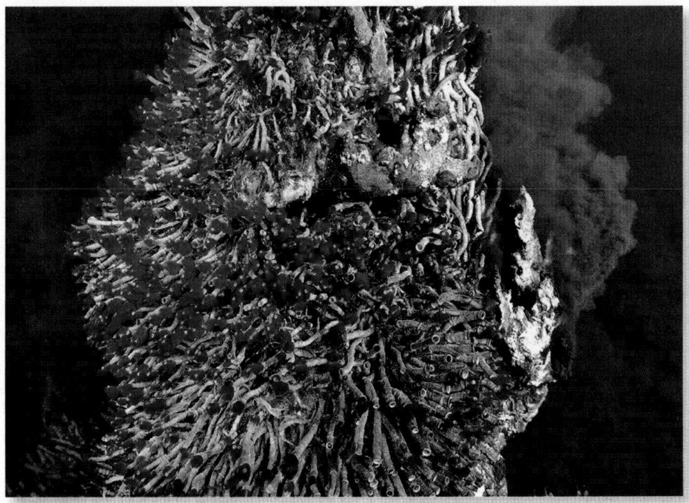

(b) A deep-sea vent community

Figure 22.3 The deep-sea vent hypothesis for the origin of life. **(a)** Deep-sea vents are cracks in the Earth's surface that release hot gases such as hydrogen sulfide (H_2S). This heats the water near the vent and results in a gradient between the very hot water adjacent to the vent and the cold water farther from the vent. The synthesis of organic molecules occurs in this gradient. **(b)** Photograph of a biological community near a deep-sea vent, which includes giant tube worms and crabs.

Concept Check: What properties of deep-sea vents made them suitable for the prebiotic synthesis of molecules?

form more complex organic polymers such as DNA, RNA, or proteins. Most ideas regarding the origin of life assume that polymers with lengths of at least 30–60 monomers are needed to store enough information to make a viable genetic system. Because hydrolysis competes with polymerization, many scientists have speculated that the synthesis of polymers did not occur in a watery prebiotic soup, but instead took place on a solid surface or in evaporating tidal pools.

In 1951, Irish X-ray crystallographer John Bernal first suggested that the prebiotic synthesis of polymers took place on clay. In his book *The Physical Basis of Life*, he wrote that "clays, muds and inorganic crystals are powerful means to concentrate and polymerize organic molecules." Many clay minerals are known to bind organic molecules such as nucleotides and amino acids. Experimentally, many research groups have demonstrated the formation of nucleic acid polymers and polypeptides on the surface of clay, given the presence of monomer building blocks. During the prebiotic synthesis of RNA, the purine bases of the nucleotides interact with the silicate surfaces of the clay. Cations, such as Mg^{2+}, bind the nucleotides to the negative surfaces of the clay, thereby positioning the nucleotides in a way that promotes bond formation between the phosphate of one nucleotide and the ribose sugar of an adjacent nucleotide. In this way, polymers such as RNA may have been formed.

Though the formation of polymers on clay remains a reasonable hypothesis, studies by American chemist Luke Leman and his colleagues English chemist Leslie Orgel and Iranian-American chemist M. Reza Ghadiri indicate that polymers can also form in aqueous solutions, which is contrary to popular belief. Their work in 2004 showed that carbonyl sulfide, a simple gas present in volcanic gases and deep-sea vent emissions, can bring about the formation of peptides from amino acids under mild conditions in water. These results indicate that the synthesis of polymers could have taken place in the prebiotic soup.

Stage 3: Cell-Like Structures May Have Originated When Polymers Were Enclosed by a Boundary

The third stage in the origin of living cells is hypothesized to be the formation of a boundary that separated the internal polymers such as RNA from the environment. The term **protobiont** is used to describe an aggregate of prebiotically produced molecules and macromolecules that acquired a boundary, such as a lipid bilayer, that allowed it to maintain an internal chemical environment distinct from that of its surroundings. What characteristics make protobionts possible precursors of living cells? Scientists envision the existence of four key features:

1. A boundary, such as a membrane, separated the internal contents of the protobiont from the external environment.
2. Polymers inside the protobiont contained information.
3. Polymers inside the protobiont had catalytic functions.
4. The protobionts eventually developed the capability of self-replication.

Protobionts were not capable of precise self-reproduction like living cells, but could divide to increase in number. Such protobionts are thought to have exhibited basic metabolic pathways in which the structures of organic molecules were changed. In particular, the polymers inside protobionts must have gained the catalytic ability to link organic building blocks to produce new polymers. This would have been a critical step in the process that eventually provided protobionts with the ability to self-replicate. According to this scenario, metabolic pathways became more complex, and the ability of protobionts to self-replicate became more refined over time. Eventually, these structures exhibited the characteristics that we attribute to living cells. As described next, researchers have hypothesized that protobionts may have exhibited different types of structures, such as coacervates and liposomes.

In 1924 Alexander Oparin hypothesized that living cells evolved from **coacervates**, droplets that form spontaneously from the association of charged polymers such as proteins, carbohydrates, or nucleic acids surrounded by water. Their name derives from the Latin *coacervare*, meaning to assemble together or cluster. Coacervates measure

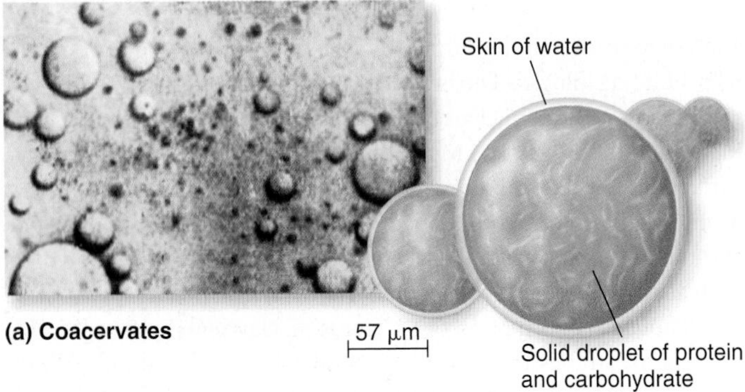

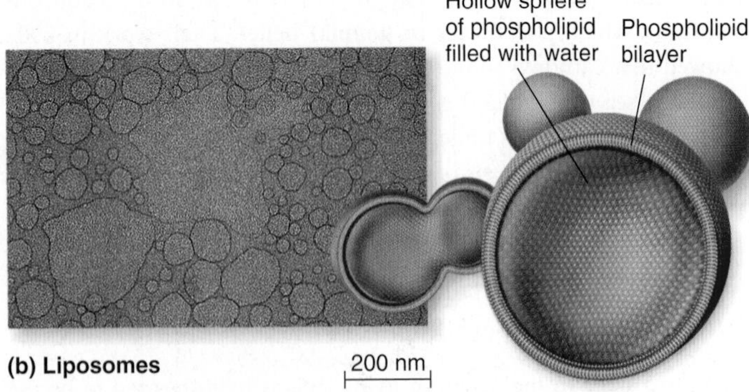

Figure 22.4 Protobionts and their lifelike functions. Primitive cell-like structures such as coacervates and liposomes could have given rise to living cells. **(a)** A micrograph and illustration of coacervates, which are droplets of protein and carbohydrate surrounded by a skin of water molecules. **(b)** An electron micrograph and illustration of liposomes. Each liposome is made of a phospholipid bilayer surrounding an aqueous compartment.

Concept Check: *Which protobiont seems most similar to real cells? Explain.*

BioConnections: *Look back at Figure 3.12. What is the physical/chemical reason why phospholipids tend to form a bilayer?*

1–100 μm (micrometers) across, are surrounded by a tight skin of water molecules, and possess osmotic properties (**Figure 22.4a**). This boundary allows the selective absorption of simple molecules from the surrounding medium.

Enzymes trapped within coacervates can perform primitive metabolic functions. For example, researchers have made coacervates containing the enzyme glycogen phosphorylase. When glucose-1-phosphate was made available to the coacervates, it was taken up into them, and starch was produced. The starch merged with the wall of the coacervates, which increased in size and eventually divided into two. When the enzyme amylase was included, the starch was broken down to maltose, which was released from the coacervates.

As a second possibility, protobionts may have resembled **liposomes**—vesicles surrounded by a lipid bilayer (**Figure 22.4b**). When certain types of lipids are dissolved in water, they spontaneously form liposomes. As discussed in Chapter 5, lipid bilayers are selectively permeable (refer back to Figure 5.11), and some liposomes can even store energy in the form of an electrical gradient. Such liposomes can discharge this energy in a neuron-like fashion, showing rudimentary signs of excitability, which is characteristic of living cells.

In 2003, Danish chemist Martin Hanczyc, American chemist Shelly Fujikawa, and Canadian American biologist Jack Szostak showed that clay can catalyze the formation of liposomes that grow and divide, a primitive form of self-replication. Furthermore, if RNA was on the surface of the clay, the researchers discovered that liposomes that enclosed RNA were formed. These experiments are compelling because they showed that the formation of membrane vesicles containing RNA molecules is a plausible explanation for the emergence of cell-like structures based on simple physical and chemical properties.

Stage 4: Cellular Characteristics May Have Evolved via Chemical Selection, Beginning with an RNA World

The majority of scientists favor RNA as the first macromolecule that was found in protobionts. Unlike other polymers, RNA exhibits three key functions:

1. RNA has the ability to store information in its nucleotide base sequence.
2. Due to base pairing, its nucleotide sequence has the capacity for self-replication.
3. RNA can perform a variety of catalytic functions. The results of many experiments have shown that some RNA molecules can function as **ribozymes**—RNA molecules that catalyze chemical reactions.

By comparison, DNA and proteins are not as versatile as RNA. DNA has very limited catalytic activity, and proteins are not known to undergo self-replication. RNA can perform functions that are characteristic of proteins and, at the same time, can serve as genetic material with replicative and informational functions.

How did the RNA molecules that were first made prebiotically evolve into more complex molecules that produced cell-like characteristics? Researchers propose that a process called chemical selection was responsible. **Chemical selection** occurs when a chemical within a mixture has special properties or advantages that cause it to increase in number relative to other chemicals in the mixture. (As we will discuss in Chapter 23, natural selection is a similar process except that it describes the changing of a population of living organisms over time due to survival and reproductive advantages.) Chemical selection results in **chemical evolution**—a population of molecules changes over time to become a new population with a different chemical composition.

Scientists speculate that initially the special properties that enabled certain RNA molecules to undergo chemical selection were its ability to self-replicate and to perform other catalytic functions. As a way to understand the concept of chemical selection, let's consider a hypothetical scenario showing two steps of chemical selection. Step 1 of **Figure 22.5** shows a group of protobionts that contain RNA molecules that were made prebiotically. RNA molecules inside these protobionts can be used as templates for the prebiotic synthesis of complementary RNA molecules. Such a process of self-replication, however, would be very slow because it would not be catalyzed by enzymes in the protobiont. In a first step of chemical selection, the sequence of one of the RNA molecules has undergone a mutation that gives it the catalytic ability to attach nucleotides together, using RNA molecules as a template. This protobiont would have an advantage

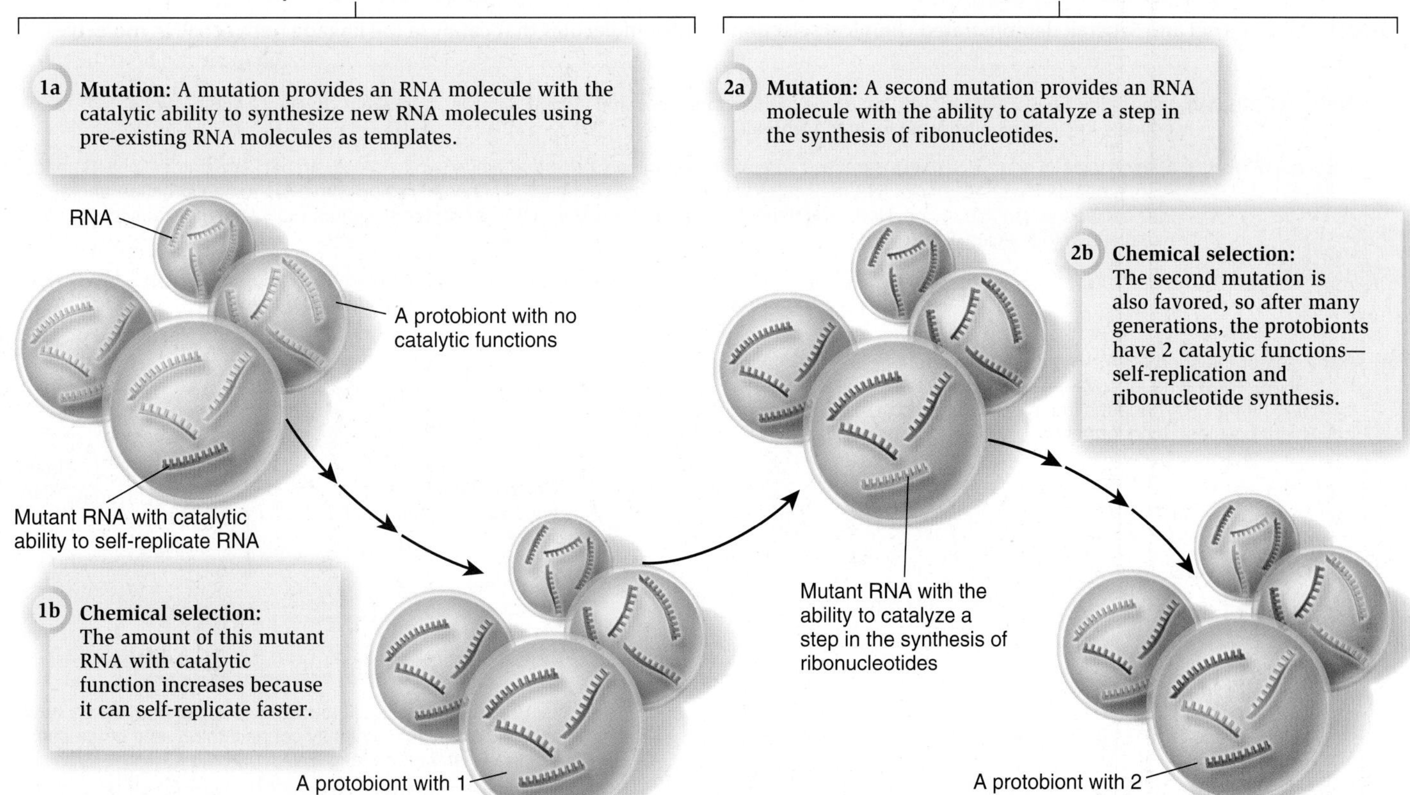
First step of chemical selection
Second step of chemical selection
1a Mutation: A mutation provides an RNA molecule with the catalytic ability to synthesize new RNA molecules using pre-existing RNA molecules as templates.
RNA
A protobiont with no catalytic functions
Mutant RNA with catalytic ability to self-replicate RNA
1b Chemical selection: The amount of this mutant RNA with catalytic function increases because it can self-replicate faster.
A protobiont with 1 catalytic function
2a Mutation: A second mutation provides an RNA molecule with the ability to catalyze a step in the synthesis of ribonucleotides.
2b Chemical selection: The second mutation is also favored, so after many generations, the protobionts have 2 catalytic functions—self-replication and ribonucleotide synthesis.
Mutant RNA with the ability to catalyze a step in the synthesis of ribonucleotides
A protobiont with 2 catalytic functions

Figure 22.5 A hypothetical scenario illustrating the process of chemical selection. This figure shows a two-step scenario. In the first step, RNAs that can self-replicate are selected, and in the second step, RNAs with the ability to catalyze a step in ribonucleotide synthesis are selected.

Concept Check: What is meant by the term chemical selection?

over the others because it would be capable of faster self-replication of its RNA molecules. Over time, due to its enhanced rate of replication, protobionts carrying such RNA molecules would increase in number compared with the others. Eventually, the group of protobionts shown in the figure contains only this type of catalytic RNA.

In the second step of chemical selection (Figure 22.5, right side), a second mutation in an RNA molecule could produce the catalytic function that would help to promote the synthesis of ribonucleotides, the building blocks of RNA. For example, a hypothetical ribozyme may catalyze the attachment of a base to a ribose, thereby catalyzing one of the steps necessary for making a ribonucleotide. This protobiont would not solely rely on the prebiotic synthesis of ribonucleotides, which also is a very slow process. Therefore, the protobiont having the ability to both self-replicate and synthesize ribonucleotides would have an advantage over a protobiont that could only self-replicate. Over time, the faster rate of self-replication and ribonucleotide synthesis would cause an increase in the numbers of the protobionts with both functions.

The **RNA world** is a hypothetical period on early Earth when both the information needed for life and the catalytic activity of living cells were contained solely in RNA molecules. In this scenario, lipid membranes enclosing RNA exhibited the properties of life due to RNA genomes that were copied and maintained through the catalytic function of RNA molecules. Over time, scientists envision that mutations occurred in these RNA molecules, occasionally introducing new functional possibilities. Chemical selection would have eventually produced an increase in complexity in these cells, with RNA molecules accruing activities such as the ability to link amino acids together into proteins and other catalytic functions.

But is an RNA world a plausible scenario? As described next in the Feature Investigation, chemical selection of RNA molecules in the laboratory can result in chemical evolution.

FEATURE INVESTIGATION

Bartel and Szostak Demonstrated Chemical Evolution in the Laboratory

Remarkably, scientists have been able to perform experiments in the laboratory that can select for RNA molecules with a particular function. American biologist David Bartel and Jack Szostak conducted the first study of this type in 1993 (**Figure 22.6**). Using molecular techniques, they synthesized a mixture of 10^{15} RNA molecules that we will call the long RNA molecules. Each long RNA in this mixture contained two regions. The first region at the 5′ end was a constant region that formed a stem-loop. Its sequence was identical among all 10^{15} molecules. The constant region was next to a second region that was

Figure 22.6 **Bartel and Szostak demonstrated chemical selection for RNA molecules that catalyze the linkage between RNA molecules.**

HYPOTHESIS Among a large pool of RNA molecules, some of them may contain the catalytic ability to make a covalent bond between nucleotides; these can be selected for in the laboratory.

KEY MATERIALS Many copies of short RNA were synthesized that had a tag sequence that binds tightly to column packing material called beads. Also, a population of 10^{15} long RNA molecules was made that contained a constant region with a stem-loop structure and a 220-nucleotide variable region. Note: The variable regions of the long RNAs were made using a PCR step that caused mutations in this region.

Experimental level | **Conceptual level**

1 Mix together the short RNAs with the 10^{15} different long RNAs. Allow time for covalent connections to form if the long RNA happens to have the catalytic activity for covalent bond formation.

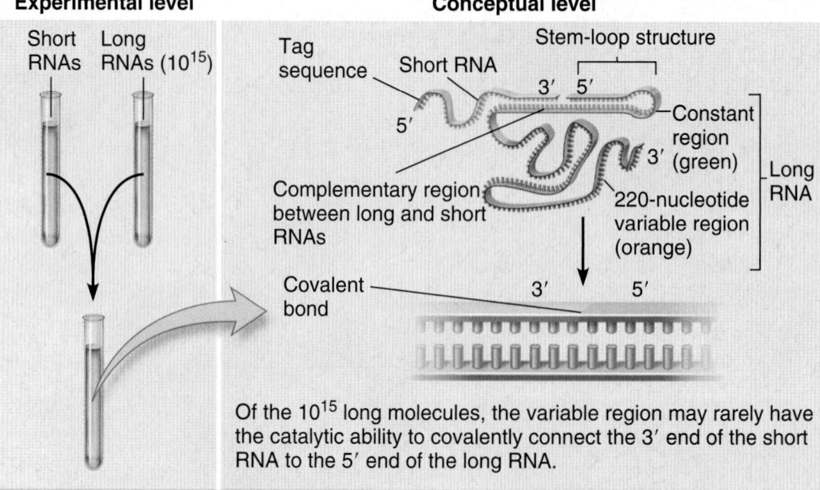
Short RNAs
Long RNAs (10^{15})
Tag sequence
Short RNA
Stem-loop structure
3′ 5′
5′
Constant region (green)
3′
Long RNA
Complementary region between long and short RNAs
220-nucleotide variable region (orange)
Covalent bond
3′
5′
Of the 10^{15} long molecules, the variable region may rarely have the catalytic ability to covalently connect the 3′ end of the short RNA to the 5′ end of the long RNA.

2 Pass the mixture through a column of beads that binds the tag sequence found on the short RNA. Add additional liquid to flush out long RNAs that are not covalently attached to short RNAs.

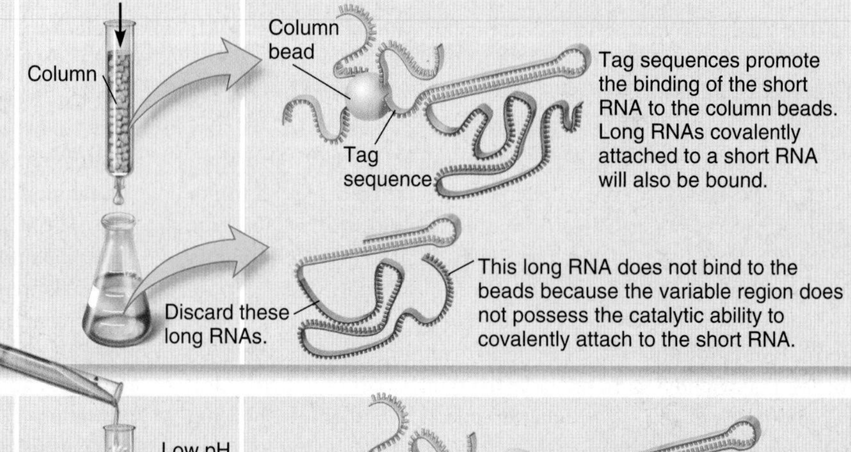
Column
Column bead
Tag sequence
Tag sequences promote the binding of the short RNA to the column beads. Long RNAs covalently attached to a short RNA will also be bound.
Discard these long RNAs.
This long RNA does not bind to the beads because the variable region does not possess the catalytic ability to covalently attach to the short RNA.

3 Add a low pH solution to prevent the tag sequence from binding to the beads. This causes the tightly bound RNAs to be flushed out of the column.

Low pH wash

4 The flushed-out RNAs are termed pool #1. Use pool #1 to make a second batch of long RNA molecules. This involved a PCR step using reverse transcriptase to make cDNA. The PCR primers recognized the beginning and end of the long RNA sequence and copied only this region. The cDNA was then used as a template to make long RNA via RNA polymerase.

Pool #1

This involved using PCR.
See Figure 20.6 for a description of PCR.

5 Repeat procedure to generate 10 consecutive pools of RNA molecules.

Pool #1 Pool #2 Pool #3 Pool #4 Pool #5 Pool #6 Pool #7 Pool #8 Pool #9 Pool #10

6 Test a sample of the original population and each of the 10 pools for the catalytic ability to make a covalent bond between adjacent nucleotides.

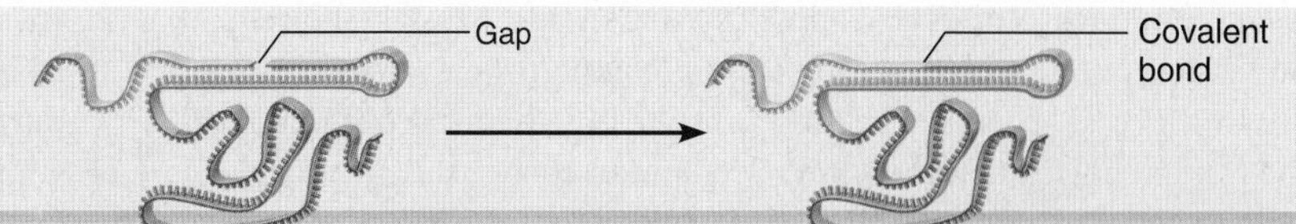

7 THE DATA

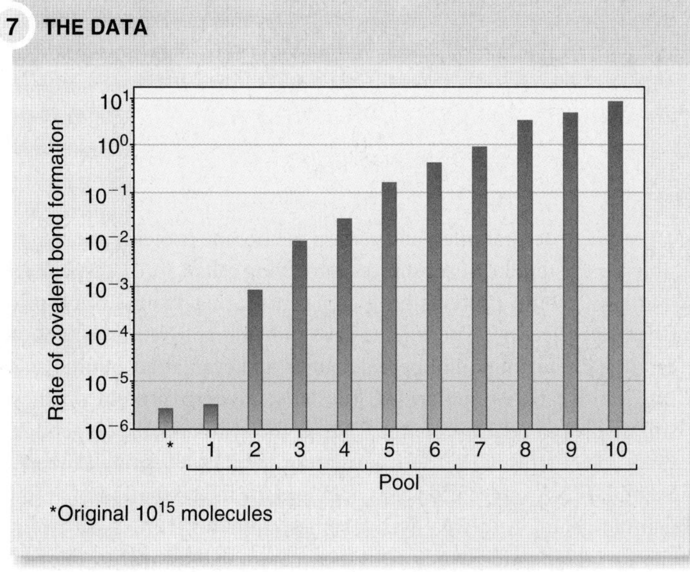

*Original 10^{15} molecules

8 CONCLUSION The increase in covalent bond formation from pool 1 to pool 10 indicates that chemical selection can occur.

9 SOURCE Bartel, David P., and Szostak, Jack W. 1993. Isolation of new ribozymes from a large pool of random sequences. *Science* 261:1411–1418.

220 nucleotides in length. A key feature of the second region is that its sequence varied among the long RNA molecules. The researchers hypothesized that this variation could occasionally result in a long RNA molecule with the ability to catalyze a covalent bond between two adjacent nucleotides.

They also made another type of RNA molecule, which we will call the short RNA, with two important properties. First, the short RNA had a region that was complementary to a site in the constant region of the long RNA molecules. Second, the short RNA had a tag sequence that caused it to bind tightly to column material referred to as beads. The short RNAs did not have a variable region; they were all the same.

To begin this experiment, the researchers incubated a large number of the long and short RNA molecules together. During this incubation period, long and short RNA molecules hydrogen-bonded to each other due to their complementary regions. Although hydrogen bonding is not permanent, this step allowed the long and short RNAs to recognize each other for a short time. The researchers reasoned that a long RNA with the catalytic ability to form a covalent bond between nucleotides would make this interaction more permanent by catalyzing a bond between the long and short RNA molecules. Following this incubation, the mixture of RNAs was passed through a column with beads that specifically bound the short RNA. The aim of this approach was to select for longer RNA molecules that had covalently bonded to the short RNA molecule (see the Conceptual level of step 2).

The vast majority of long RNAs would not have the catalytic ability to catalyze a permanent covalent bond between nucleotides. These would pass out of the column at step 2, because hydrogen bonding between the long and short RNAs is not sufficient to hold them together for very long. Such unbound long RNAs would be discarded. Long RNAs with the ability to catalyze a covalent bond to the short RNA would remain bound to the column beads at step 2. These catalytic RNAs were then flushed out at step 3 to generate a mixture of RNAs termed pool #1. The researchers expected this pool to contain several different long RNA molecules with varying abilities to catalyze a covalent bond between nucleotides.

To further the chemical selection process, the scientists used the first pool of long RNA molecules flushed out at step 3 to make more long RNA molecules. This was accomplished via polymerase chain reaction (PCR). This next batch also had the constant and variable regions but did not have the short RNA covalently attached. Because the variable regions of these new RNA molecules were derived from the variable regions of pool #1 RNA molecules, they were expected to have catalytic activity. The researchers reasoned that additional variation might occasionally produce an RNA molecule with improved catalytic activity. This second batch of long RNA molecules (pool #2) was subjected to the same steps as was the first batch of 10^{15} molecules. In this case, the group of long molecules flushed out at step 3 was termed pool #2. This protocol was followed eight more times to generate 10 consecutive pools of RNA molecules. During this work, the researchers analyzed the original random collection of 10^{15} RNA molecules and each of the 10 pools for the catalytic ability to covalently link RNA molecules. As seen in the data, each successive pool became enriched for molecules with higher catalytic activity. Pool #10 showed catalytic activity that was approximately 3 million times higher than the original random pool of molecules!

Like the work of Miller and Urey, Bartel and Szostak showed the feasibility of another phase of the prebiotic process that led to life. In this case, chemical selection resulted in chemical evolution. The results showed that chemical selection can change the functional characteristics of a group of RNA molecules over time by increasing the proportion of those molecules with enhanced function.

Experimental Questions

1. What is chemical selection? What hypothesis did Bartel and Szostak test?
2. In conducting the selection experiment among pools of long RNA molecules with various catalytic abilities, what was the purpose of using the short RNA molecules?
3. What were the results of the experiment conducted by Bartel and Szostak? How did this study influence our understanding of the evolution of life on Earth?

The RNA World Was Superseded by the Modern DNA/RNA/Protein World

Assuming that an RNA world was the origin of life, researchers have asked the question, "Why and how did the RNA world evolve into the DNA/RNA/protein world we see today?" The RNA world may have been superseded by a DNA/RNA world or an RNA/protein world before the emergence of the modern DNA/RNA/protein world. Let's now consider the advantages of a DNA/RNA/protein world as opposed to the simpler RNA world and explore how this modern biological world might have come into being.

Information Storage RNA can store information in its base sequence. If so, why did DNA take over that function, as is the case in modern cells? During the RNA world, RNA had to perform two roles: the storage of information and the catalysis of chemical reactions. Scientists have speculated that the incorporation of DNA into cells would have relieved RNA of its informational role, thereby allowing RNA to perform a greater variety of other functions. For example, if DNA stored the information for the synthesis of RNA molecules, such RNA molecules could bind cofactors, have modified bases, or bind peptides that might enhance their catalytic function. Cells with both DNA and RNA would have had an advantage over those with just RNA, and so they would have been selected. Another advantage of DNA is its stability. Compared with RNA, DNA strands are less likely to spontaneously break.

A second issue is how DNA came into being. Scientists have proposed that an ancestral RNA molecule had the ability to make DNA using RNA as a template. This function, known as reverse transcription, is described in Chapter 18 in the discussion of retroviruses. Interestingly, modern eukaryotic cells can use RNA as a template to make DNA. For example, an RNA sequence in the enzyme telomerase copies the ends of chromosomes, thus preventing progressive shortening of the chromosomes (refer back to Figure 11.22).

Metabolism and Other Cellular Functions Now let's consider the origin of proteins. The emergence of proteins as catalysts may have been a great benefit to early cells. Due to the different chemical properties of the 20 amino acids, proteins have vastly greater catalytic ability than do RNA molecules, again providing a major advantage to cells that had both RNA and proteins. In modern cells, proteins have taken over most, but not all, catalytic functions. In addition, proteins can perform other important tasks. For example, cytoskeletal proteins carry out structural roles, and certain membrane proteins are responsible for the uptake of substances into living cells.

How would proteins have come into being in an RNA world? Chemical selection experiments have shown that RNA molecules can catalyze the formation of peptide bonds and even attach amino acids to primitive tRNA molecules. Similarly, modern protein synthesis still involves a central role for RNA in the synthesis of polypeptides. First, mRNA provides the information for a polypeptide sequence. Second, tRNA molecules act as adaptors for the formation of a polypeptide chain. And finally, ribosomes containing rRNA provide a site for polypeptide synthesis. Furthermore, rRNA within the ribosome acts as a ribozyme to catalyze peptide bond formation. Taken together, the analysis of translation in modern cells is consistent with an evolutionary history in which RNA molecules were instrumental in the emergence and formation of proteins.

22.2 The Fossil Record

Learning Outcomes:

1. Describe how fossils are formed.
2. Explain how radiometric dating is used to estimate the age of a fossil.
3. List several factors that affect the completeness of the fossil record.

We will now turn our attention to a process that has given us a window into the history of life over the past 3.5 billion years. **Fossils** are the preserved remains of past life on Earth. They can take many forms, including bones, shells, and leaves, and the impression of cells or other evidence, such as footprints or burrows. Scientists who study fossils are called **paleontologists** (from the Greek *palaios*, meaning ancient). Because our understanding of the history of life is derived primarily from the fossil record, it is important to appreciate how fossils are formed and dated and to understand why the fossil record cannot be viewed as complete.

Fossils Are Formed Within Sedimentary Rock

How are fossils usually formed? Many of the rocks observed by paleontologists are sedimentary rocks that were formed from particles of older rocks broken apart by water or wind. These particles, such as gravel, sand, and mud, settle and bury living and dead organisms at the bottoms of rivers, lakes, and oceans. Over time, more particles pile up, and sediments at the bottom of the pile eventually become rock. Gravel particles form rock called conglomerate, sand becomes sandstone, and mud becomes shale. Most fossils are formed when organisms are buried quickly, and then during the process of sedimentary rock formation, their hard parts are gradually replaced over millions of years by minerals, producing a recognizable representation of the original organism (see, for example, the chapter opening photo).

The relative ages of fossils can sometimes be revealed by their locations in sedimentary rock formations. Because sedimentary rocks are formed particle by particle and bed by bed, the layers are piled one on top of the other. In a sequence of layered rocks, the lower rock layers are usually older than the upper layers. Paleontologists often study changes in life-forms over time by studying the fossils in layers from bottom to top (**Figure 22.7**). The more ancient life-forms are found in the lower layers, and newer species are found in the upper layers. However, such an assumption can occasionally be misleading when geological processes such as folding have flipped the layers.

The Analysis of Radioisotopes Is Used to Date Fossils

A common way to estimate the age of a fossil is by analyzing the decay of radioisotopes within the accompanying rock, a process called **radiometric dating**. As discussed in Chapter 2, elements may be found in multiple forms, called isotopes, that differ in the number of neutrons they contain. A radioisotope is an unstable isotope of an element that decays spontaneously, releasing radiation at a constant rate. The **half-life** is the length of time required for a radioisotope to decay to exactly one-half of its initial quantity. Each radioisotope has its own unique half-life (**Figure 22.8a**). Within a sample of rock, scientists can measure the amount of a given radioisotope as well as the amount of the decay product—the isotope that is produced when the original isotope decays. For dating geological materials, several types of isotope decay patterns are particularly useful: carbon to nitrogen, potassium to argon, rubidium to strontium, and uranium to lead (**Figure 22.8b**).

To determine the age of a rock using radiometric dating, paleontologists need to have a way to set the clock—extrapolate back to a starting point in which a rock did not have any amount of the decay product. Except for fossils less than 50,000 years old, in which carbon-14 (^{14}C) dating can be employed, fossil dating is not usually conducted on the fossil itself or on the sedimentary rock in which the fossil is found. Most commonly, igneous rock—rock formed through the cooling and solidification of lava—in the vicinity of the sedimentary rock is dated. Why is igneous rock chosen? One reason is that igneous rock derived from an ancient lava flow initially contains uranium-235 (^{235}U) but no lead-207 (^{207}Pb). The decay product of ^{235}U is ^{207}Pb. By comparing the relative proportions of ^{235}U and ^{207}Pb in a sample, the age of igneous rock can be accurately determined.

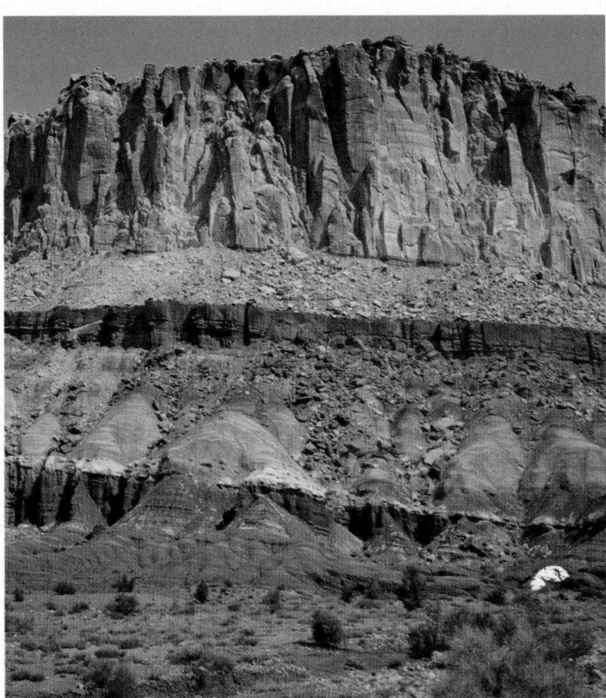

Figure 22.7 An example of layers of sedimentary rock that contain fossils.

Concept Check: Which rock layer in this photo is most likely to be the oldest?

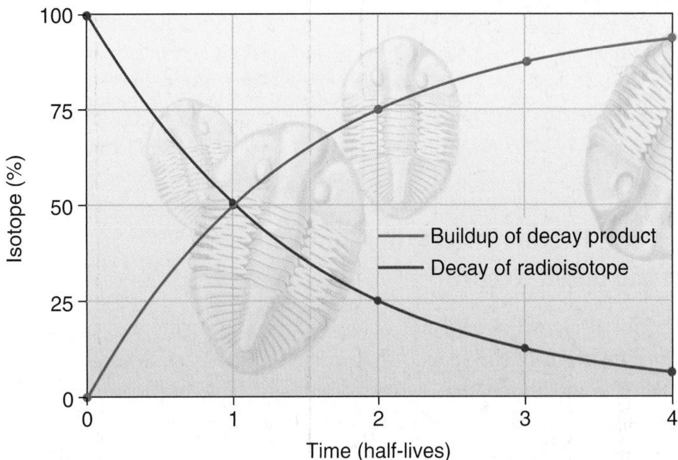

(a) Decay of a radioisotope

Radioisotope	Decay product	Half-life (years)	Useful dating range (years)
Carbon-14	Nitrogen-14	5,730	100–50,000
Potassium-40	Argon-40	1.3 billion	100,000–4.5 billion
Rubidium-87	Strontium-87	47 billion	10 million–4.5 billion
Uranium-235	Lead-207	710 million	10 million–4.5 billion
Uranium-238	Lead-206	4.5 billion	10 million–4.5 billion

(b) Radioisotopes that are useful for geological dating

Figure 22.8 Radiometric dating of fossils. **(a)** A rock can be dated by measuring the relative amounts of a radioisotope and its decay product within the rock. **(b)** These five isotopes are particularly useful for the dating of fossils.

Concept Check: If you suspected a fossil is 50 million years old, which pair of radioisotopes would you choose to analyze?

Table 22.1 Factors That Affect the Fossil Record

Factor	Description
Anatomy	Organisms with hard body parts, such as animals with a skeleton or thick shell, are more likely to be preserved than are organisms composed of soft tissues.
Size	The fossil remains of larger organisms are more likely to be found than those of smaller organisms.
Number	Species that existed in greater numbers or over a larger area are more likely to be preserved within the fossil record than those that existed in smaller numbers or in a smaller area.
Environment	Inland species are less likely to become fossilized than are those that lived in a marine environment or near the edge of water because sedimentary rock is more likely to be formed in or near water.
Time	Organisms that lived relatively recently or existed for a long time are more likely to be found as fossils than organisms that lived very long ago or for a relatively short time.
Geological processes	Due to the chemistry of fossilization, certain organisms are more likely to be preserved than are other organisms.
Paleontology	Certain types of fossils may be more interesting to paleontologists. In addition, a significant bias exists with regard to the locations where paleontologists search for fossils. For example, they tend to search in regions where other fossils have already been found.

Several Factors Affect the Completeness of the Fossil Record

The fossil record should not be viewed as a complete and balanced representation of the species that existed in the past. Several factors affect the likelihood that extinct organisms have been preserved as fossils and will be identified by paleontologists (**Table 22.1**). First, certain organisms are more likely than others to become fossilized. Organisms with hard shells or bones tend to be over-represented. Factors such as anatomy, size, number, and the environment and time in which they lived also play important roles in determining the likelihood that organisms will be preserved in the fossil record. In addition, geological processes may favor the fossilization of certain types of organisms. Finally, unintentional biases arise that are related to the efforts of paleontologists. For example, scientific interests may favor searching for and analyzing certain species over others. For example, researchers have been greatly interested in finding the remains of dinosaurs.

Although the fossil record is incomplete, it has provided a wealth of information regarding the history of the types of life that existed on Earth. The rest of this chapter will survey the emergence of life-forms from 3.5 bya to the present.

22.3 History of Life on Earth

Learning Outcomes:

1. List the types of environmental changes that have affected the history of life on Earth.
2. Describe the cell structure and energy utilization of the first living organisms that arose during the Archaean eon.
3. Explain how the origin of eukaryotic cells involved a union between bacterial and archaeal cells.
4. Describe the key features of multicellular organisms, which arose during the Proterozoic eon.
5. Outline the major events and changes in species diversity during the Paleozoic, Mesozoic, and Cenzoic eras.

Thus far, we have considered hypotheses of how the first cells came into existence, and we have also examined the characteristics of fossils. The first known fossils of single-celled organisms were preserved approximately 3.5 bya. In this section, we will begin with a brief description of the geological changes on Earth that have affected the emergence of new forms of life and then examine some of the major changes in life that have occurred since it began.

Many Environmental Changes Have Occurred Since the Origin of the Earth

The **geological timescale** is a time line of the Earth's history and major events from its origin approximately 4.55 bya to the present (**Figure 22.9**). This time line is subdivided into four eons—the Hadean, Archaean, Proterozoic, and Phanerozoic—and then further subdivided into eras. The first three eons are collectively known as the Precambrian because they preceded the Cambrian era, a geological era that saw a rapid increase in the diversity of life. The names of several eons and eras end in *-zoic* (meaning animal life), because we often recognize these time intervals on the basis of animal life. We will examine these time periods later in this chapter.

The changes that occurred in living organisms over the past 4 billion years are the result of two interactive processes. First, as discussed in the next several chapters, genetic changes in organisms can affect their characteristics. Such changes can influence organisms' abilities to survive and reproduce in their native environment. Second, the environment on Earth has undergone dramatic changes that have profoundly influenced the types of organisms that have existed during different periods of time. In some cases, an environmental change has allowed new types of organisms to flourish. Alternatively, environmental changes have resulted in **extinction**—the complete loss of a species or group of species. Major types of environmental changes are described next.

Temperature During the first 2.5 billion years of its existence, the surface of the Earth gradually cooled. However, during the last 2 billion years, the Earth has undergone major fluctuations in temperature, producing Ice Ages that alternate with warmer periods. Furthermore, the temperature on Earth is not uniform, which produces a range of environments where the temperatures are quite different, such as tropical rain forests and the arctic tundra.

Atmosphere The chemical composition of the gases surrounding the Earth has changed substantially over the past 4 billion years. One notable change involves the amount of oxygen. Prior to 2.4 bya, relatively little oxygen gas was in the atmosphere, but at that time, levels of oxygen in the form of O_2 began to rise significantly. The emergence of organisms that are capable of photosynthesis added oxygen to the atmosphere. Our current atmosphere contains about 21% O_2.

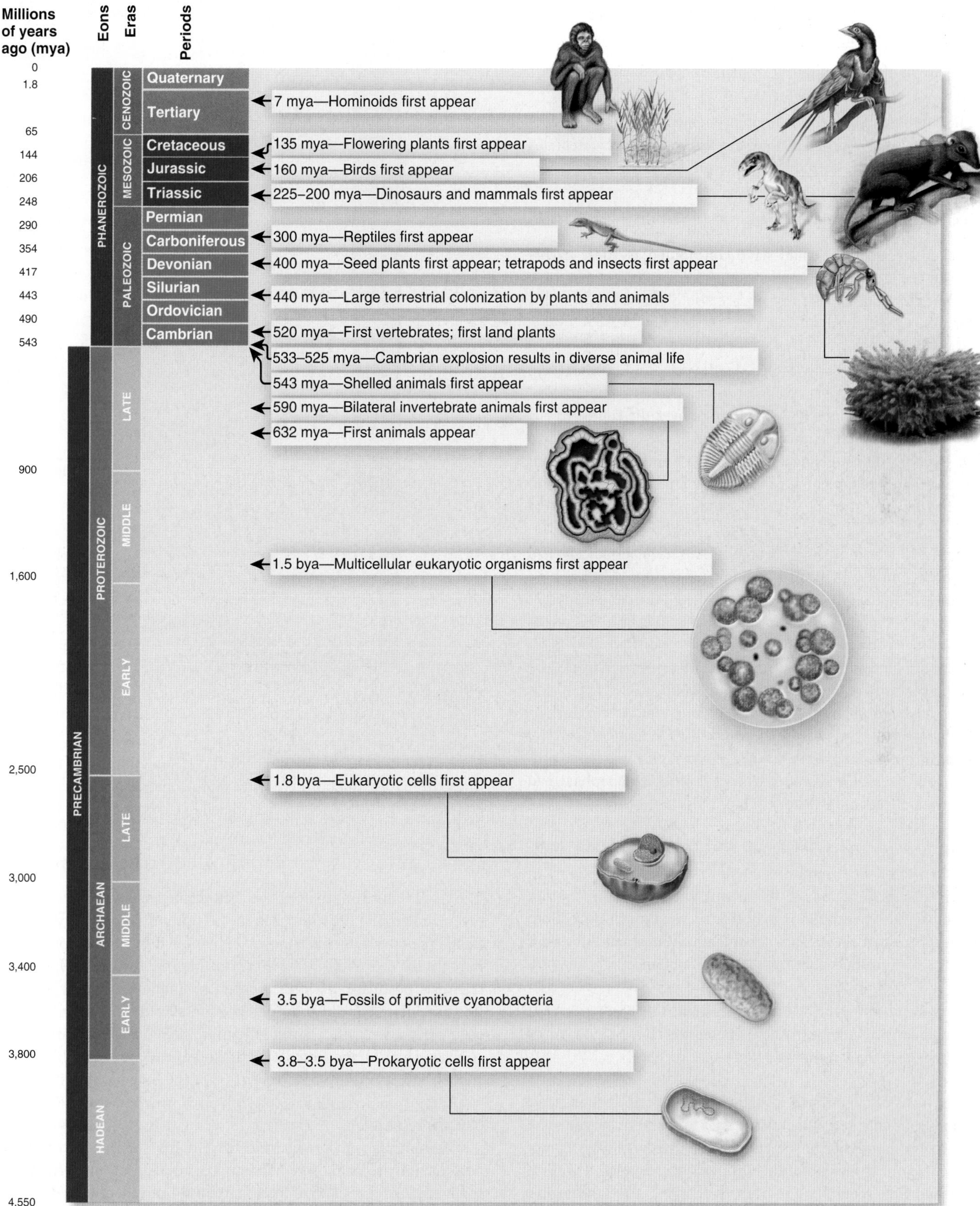

Figure 22.9 The geological timescale and an overview of the history of life on Earth.

Increased levels of oxygen are thought to have a played a key role in various aspects of the history of life, including the following:

- The origin of many animal body plans coincided with a rise in atmospheric O_2.
- The conquest of land by arthropods (about 410 million years ago [mya]) and a second conquest by arthropods and vertebrates (about 350 mya) occurred during periods in which O_2 levels were high or increasing.
- Increases in animal body sizes are associated with higher O_2 levels.

Higher levels of O_2 could have contributed to these events because higher O_2 levels may enhance the ability of animals to carry out aerobic respiration. These events are also discussed later in this chapter and in more detail in Unit VI.

Landmasses As the Earth cooled, landmasses formed that were surrounded by bodies of water. This produced two different environments: terrestrial and aquatic. Furthermore, over the course of billions of years, the major landmasses, known as the continents, have shifted their positions, changed their shapes, and separated from each other. This phenomenon, called **continental drift**, is shown in Figure 22.10.

Floods and Glaciations Catastrophic floods have periodically had major effects on the organisms in the flooded regions. Glaciers have periodically moved across continents and altered the composition of species on those landmasses. As an extreme example, in 1992, American geobiologist Joseph Kirschvink proposed the "Snowball Earth hypothesis," which suggests that the Earth was entirely covered by ice during parts of the period from 790 to 630 mya. This hypothesis was developed to explain various types of geological evidence including sedimentary deposits of glacial origin that are found at tropical latitudes. Although the prior existence of a completely frozen Earth remains controversial, massive glaciations over our planet have had an important effect on the history of life.

Volcanic Eruptions The eruptions of volcanoes harm organisms in the vicinity of the eruption, sometimes causing extinctions. In addition, volcanic eruptions in the oceans lead to the formation of new islands. Massive eruptions may also spew so much debris into the atmosphere that they affect global temperatures and limit solar radiation, which restricts photosynthetic production.

Meteorite Impacts During its long history, the Earth has been struck by many meteorites. Large meteorites have significantly affected the Earth's environment.

The effects of one or more of the changes described above have sometimes caused large numbers of species to go extinct at the same time. Such events are called **mass extinctions**. Five large mass extinctions occurred near the end of the Ordovician, Devonian, Permian, Triassic, and Cretaceous periods. The boundaries between geological time periods are often based on the occurrences of mass extinctions. A recurring pattern seen in the history of life is the extinction of some species and the emergence of new ones. The rapid extinction of many modern species due to human activities is sometimes referred to as the sixth mass extinction. We will examine mass extinctions and the current biodiversity crisis in more detail in Chapter 60.

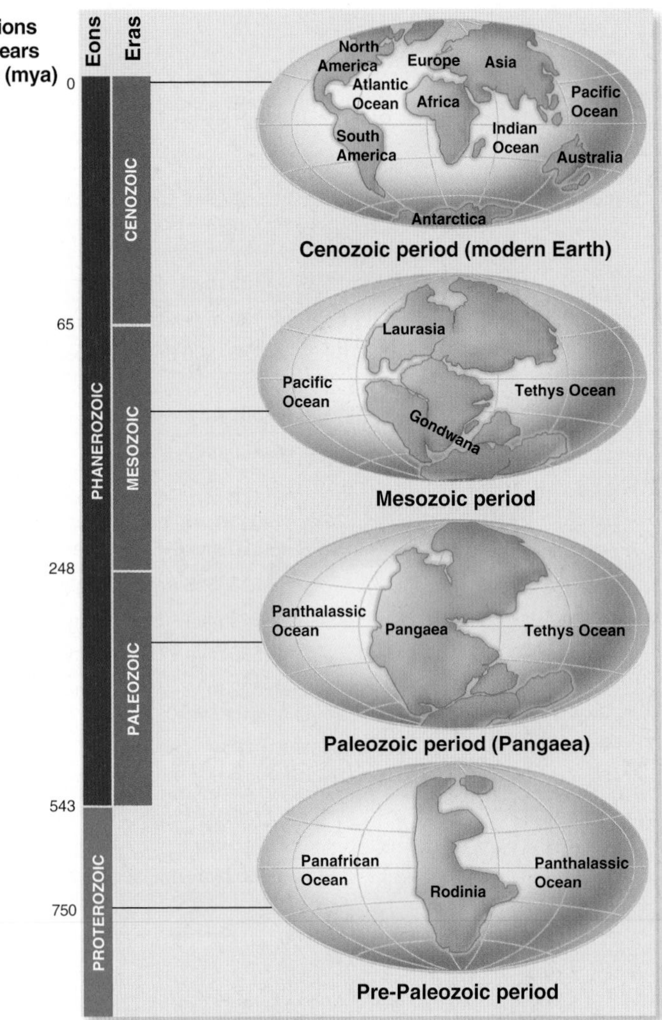

Figure 22.10 **Continental drift.** The relative locations of the continents on Earth have changed dramatically over time.

Prokaryotic Cells Arose During the Archaean Eon

The Archaean (from the Greek, meaning ancient) was an eon when diverse microbial life flourished in the primordial oceans. As mentioned previously, the first known fossils of living cells were preserved in rocks that are about 3.5 billion years old (see Figure 22.1), though scientists postulate that cells arose many millions of years prior to this time. Based on the morphology of fossilized remains, these first cells were prokaryotic. During the more than 1 billion years of the Archaean eon, all life-forms were prokaryotic. Because Earth's atmosphere had very little free oxygen (O_2), the single-celled microorganisms of this eon almost certainly used only anaerobic (without oxygen) respiration.

Organisms with prokaryotic cells are divided into two groups: bacteria and archaea. Bacteria are more prevalent on modern Earth, though many species of archaea have also been identified. Archaea are found in many different environments, with some occupying extreme environments such as hot springs. Both bacteria and archaea share fundamental similarities, indicating that they are derived from a common ancestor. Even so, certain differences suggest that these two types of prokaryotes diverged from each other quite early in the history of life. In particular, bacteria and archaea show some interesting differences in metabolism, lipid composition, and genetic pathways (look ahead to Chapter 26, Table 26.1).

Biologists Are Undecided About Whether Heterotrophs or Autotrophs Came First

An important factor that greatly influenced the emergence of new species is the availability of energy. As we learned in Unit II, all organisms require energy to survive and reproduce. Organisms may follow two different strategies to obtain energy. Some are **heterotrophs**, which means their energy is derived from the chemical bonds within organic molecules they consume. Because the most common sources of organic molecules today are other organisms, heterotrophs typically consume other organisms or materials from other organisms. Alternatively, many organisms are **autotrophs**, which directly harness energy from either inorganic molecules or light. Among modern species, plants are an important example of autotrophs. Plants can directly absorb light energy and use it (via photosynthesis) to synthesize organic molecules such as glucose. On modern Earth, heterotrophs ultimately rely on autotrophs for the production of food.

Were the first forms of life heterotrophs or autotrophs? The answer is not resolved. Some biologists have speculated that autotrophs, such as those living near deep-sea vents, may have arisen first. These organisms would have used chemicals that were made near the vents as an energy source to make organic molecules. Alternatively, many scientists have hypothesized that the first living cells were heterotrophs. They reason that it would have been simpler for the first primitive cells to use the organic molecules in the prebiotic soup as a source of energy.

If heterotrophs came first, why were cyanobacteria preserved in the earliest fossils, rather than heterotrophs? One possible reason is related to their manner of growth. Certain cyanobacteria promote the formation of a layered structure called a **stromatolite** (Figure 22.11). The aquatic environment where these cyanobacteria survive is rich in minerals such as calcium. The cyanobacteria grow in large mats that form layers. As they grow, they deplete the carbon dioxide (CO_2) in the surrounding water. This causes calcium carbonate in the water to gradually precipitate over the bacterial cells, calcifying the older cells in the lower layers and also trapping grains of sediment. Newer cells produce a layer on top. Over time, many layers of calcified cells and sediment are formed, thereby producing a stromatolite. This process still occurs today in places such as Shark Bay in western Australia, which is renowned for the stromatolites along its beaches (Figure 22.11).

The emergence and proliferation of ancient cyanobacteria had two critical consequences. First, the autotrophic nature of these bacteria enabled them to produce organic molecules from CO_2. This prevented the depletion of organic foodstuffs that would have been exhausted if only heterotrophs existed. Second, cyanobacteria produce oxygen (O_2) as a waste product of photosynthesis. During the Archaean and Proterozoic eons, the activity of cyanobacteria led to the gradual rise in O_2 discussed earlier. The increase in O_2 spelled doom for many anaerobic species, which became restricted to a few anoxic (without oxygen) environments, such as deep within the soil. However, O_2 enabled the formation of new bacterial and archaeal species that used aerobic (with oxygen) respiration (see Chapter 7). In addition, aerobic respiration is likely to have played a key role in the emergence and eventual explosion of eukaryotic life-forms, which typically have high energy demands. These eukaryotic life-forms are described next.

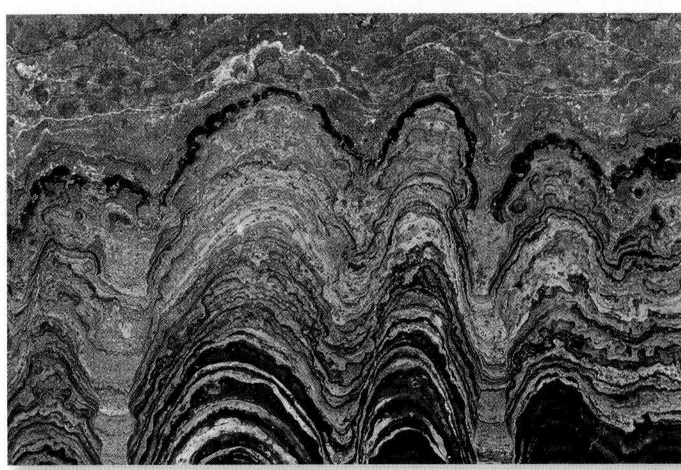

(a) Fossil stromatolite

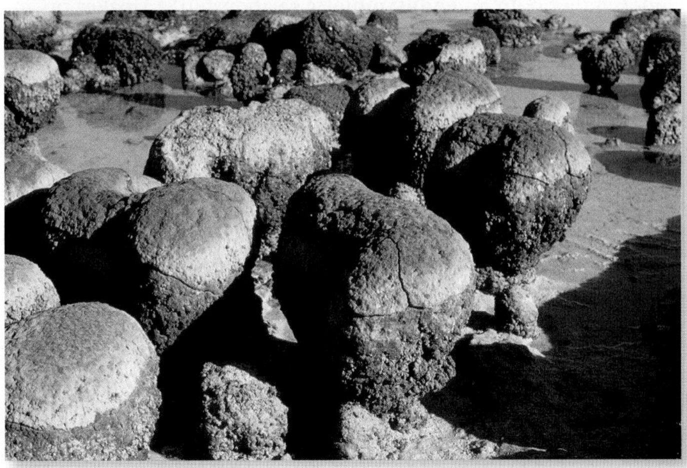

(b) Modern stromatolites

Figure 22.11 Fossil and modern stromatolites: Evidence of autotrophic cyanobacteria. Each stromatolite is a rocklike structure, typically 1 meter in diameter. **(a)** Section of a fossilized stromatolite. These layers are mats of mineralized cyanobacteria, one layer on top of the other. The existence of fossil stromatolites provides evidence of early autotrophic organisms. **(b)** Modern stromatolites that have formed in western Australia.

Genomes & Proteomes Connection

The Origin of Eukaryotic Cells Involved a Union Between Bacterial and Archaeal Cells

Eukaryotic cells arose during the Proterozoic eon, which began 2.5 bya and ended 543 mya (see Figure 22.9). The manner in which the first eukaryotic cell originated is not entirely understood. In modern eukaryotic cells, genetic material is found in three distinct organelles. All eukaryotic cells contain DNA in the nucleus and mitochondria, and plant and algal cells also have DNA in their chloroplasts. To address the issue of the origin of eukaryotic species, scientists have examined the DNA sequences found in these three organelles. From such studies, the nuclear, mitochondrial, and chloroplast genomes appear to be derived from once-separate cells that came together.

Nuclear Genome From a genome perspective, both bacteria and archaea have contributed substantially to the nuclear genome of eukaryotic cells. Eukaryotic nuclear genes encoding proteins involved in metabolic pathways and lipid biosynthesis appear to be derived from ancient bacteria, whereas genes involved with transcription and translation appear to be derived from an archaeal ancestor. To explain the origin of the nuclear genome, several hypotheses have been proposed. The most widely accepted involves an association between ancient bacteria and archaea, which is hypothesized to be endosymbiotic. In an **endosymbiotic** relationship, a smaller organism (the endosymbiont) lives inside a larger organism (the host).

Researchers have suggested that an archaeal species evolved the ability to invaginate its plasma membrane, which could have two results (**Figure 22.12**). First, it could eventually lead to the formation of an extensive internal membrane system and enclose the genetic material in a nuclear envelope. Second, the ability to invaginate the plasma membrane would provide a mechanism to take up materials from the environment via endocytosis, which is described in Chapter 5. In the scenario described in Figure 22.12, an ancient archaeon engulfed a bacterium via endocytosis, maintaining the bacterium in its cytoplasm as an endosymbiont. Over time, some genes from the bacterium were transferred to the archaeal host cell, and the resulting genetic material eventually became the nuclear genome.

Mitochondrial and Chloroplast Genomes As discussed in Chapter 4, the analyses of genes from mitochondria, chloroplasts, and bacteria are consistent with the endosymbiosis theory, which proposes that mitochondria and chloroplasts originated from bacteria that took up residence within a primordial eukaryotic cell (refer back to Figure 4.28). Mitochondria found in eukaryotic cells are likely derived from a bacterial species that resembled modern α-proteobacteria, a diverse group of bacteria that carry out oxidative phosphorylation to make ATP. One possibility is that an endosymbiotic event involving an ancestor of this bacterial species produced the first eukaryotic cell and that the mitochondrion is a remnant of that event. Alternatively, endosymbiosis may have produced the first eukaryotic cell, and then a subsequent endosymbiosis resulted in mitochondria (see Figure 22.12). DNA-sequencing data indicate that chloroplasts were derived from a separate endosymbiotic relationship between a primitive eukaryotic cell and a cyanobacterium. As discussed in Chapter 28, plastids, such as chloroplasts, have arisen on several independent occasions via primary, secondary, and tertiary endosymbiosis (see Figure 28.13).

Interestingly, an endosymbiotic relationship involving two different proteobacteria was reported in 2001. In mealybugs, bacteria survive within the cytoplasm of large host cells of a specialized organ called a bacteriome. Recent analysis has shown that different species of bacteria inside the host cells share their own endosymbiotic relationship. In particular, γ-proteobacteria live endosymbiotically inside β-proteobacteria. Such an observation demonstrates that an endosymbiotic relationship can occur between two bacterial species.

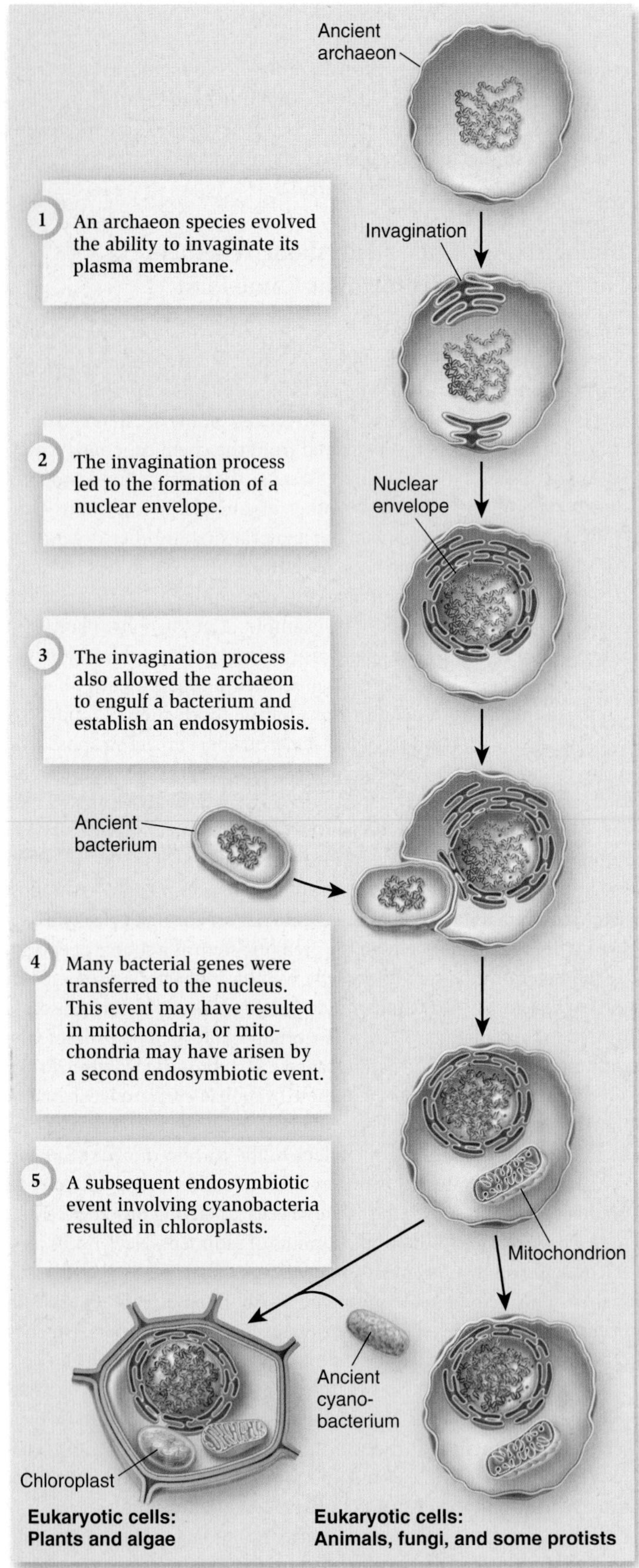
Ancient archaeon
1 An archaeon species evolved the ability to invaginate its plasma membrane.
Invagination
2 The invagination process led to the formation of a nuclear envelope.
Nuclear envelope
3 The invagination process also allowed the archaeon to engulf a bacterium and establish an endosymbiosis.
Ancient bacterium
4 Many bacterial genes were transferred to the nucleus. This event may have resulted in mitochondria, or mitochondria may have arisen by a second endosymbiotic event.
5 A subsequent endosymbiotic event involving cyanobacteria resulted in chloroplasts.
Mitochondrion
Ancient cyano-bacterium
Chloroplast
Eukaryotic cells: Plants and algae
Eukaryotic cells: Animals, fungi, and some protists

Figure 22.12 **Possible endosymbiotic relationships that gave rise to the first eukaryotic cells.**

BioConnections: Look back at Figure 5.24. Explain how endocytosis played a role in endosymbiosis.

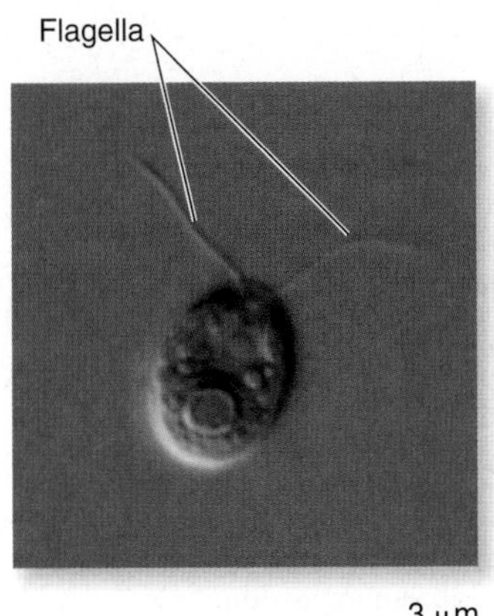

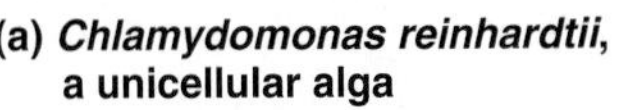

(a) *Chlamydomonas reinhardtii*, a unicellular alga

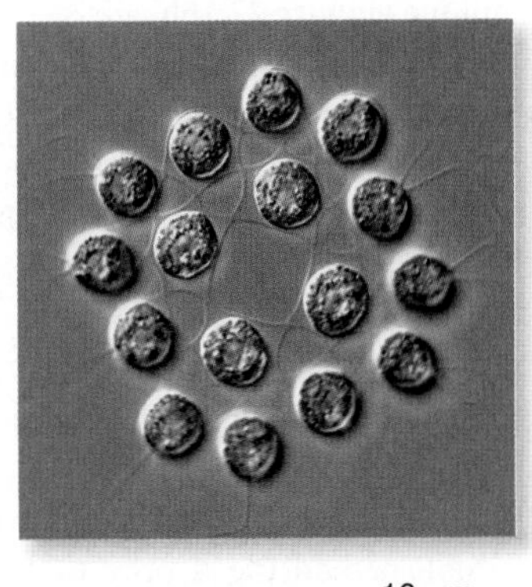

(b) *Gonium pectorale*, composed of 16 identical cells

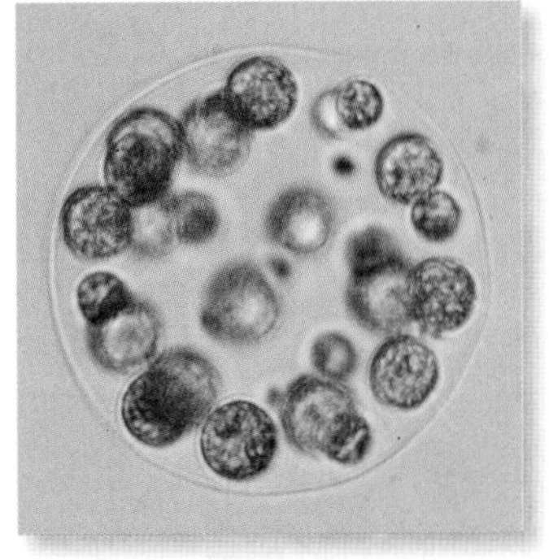

(c) *Pleodorina californica*, composed of 64 to 128 cells, has 2 cell types, somatic and reproductive

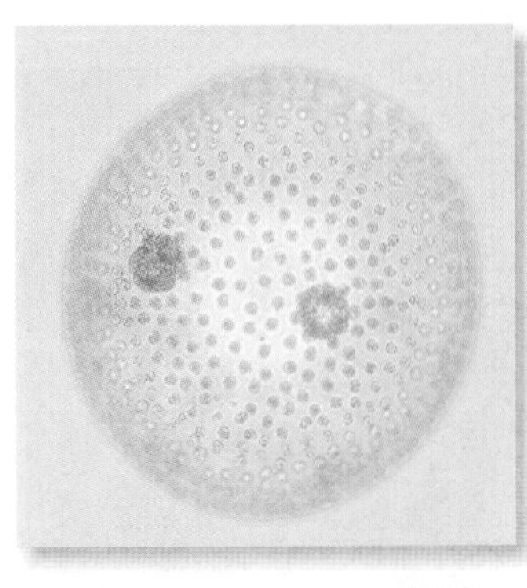

(d) *Volvox aureus*, composed of about 1,000 to 2,000 cells, has 2 cell types, somatic and reproductive

Figure 22.13 **Variation in the level of multicellularity among volvocine algae.**

BIOLOGY PRINCIPLE **New properties of life emerge from complex interactions.** The formation of different cell types is an emergent property of multicellularity.

Multicellular Eukaryotes and the Earliest Animals Arose During the Proterozoic Eon

The first multicellular eukaryotes are thought to have emerged about 1.5 bya, in the middle of the Proterozoic eon. The oldest fossil evidence for multicellular eukaryotes was an organism that resembled modern red algae; this fossil was dated at approximately 1.2 billion years old.

Simple multicellular organisms are believed to have originated in one of two different ways. One possibility is that several individual cells found each other and aggregated to form a colony. Cellular slime molds, discussed in Chapter 28, are examples of modern organisms in which groups of single-celled organisms can come together to form a small multicellular organism. According to the fossil record, such organisms have remained very simple for hundreds of millions of years.

Alternatively, another way that multicellularity can occur is when a single cell divides and the resulting cells stick together. This pattern occurs in many simple multicellular organisms, such as algae and fungi, as well as in species with more complex body plans, such as plants and animals. Biologists cannot be certain whether the first multicellular organisms arose by an aggregation process or by cell division and adhesion. However, the development of complex, multicellular organisms now occurs by cell division and adhesion.

An interesting example showing changes in the level of complexity from unicellular organisms to more complex multicellular organisms is found among evolutionarily related species of volvocine green algae. These algae exist as unicellular species, as small clumps of cells of the same cell type, or as larger groups of cells with two distinct cell types. **Figure 22.13** compares four species of volvocine algae. *Chlamydomonas reinhardtii* is a unicellular alga (Figure 22.13a). It is called a biflagellate because each cell has two flagella. *Gonium pectorale* is a multicellular organism composed of 16 cells (Figure 22.13b). This simple multicellular organism is formed from a single cell by cell division and adhesion. All of the cells in this species are biflagellate. Other volvocine algae have evolved into larger and more complex organisms. *Pleodorina californica* has 64–128 cells (Figure 22.13c), and *Volvox aureus* has about 1,000–2,000 cells (Figure 22.13d). A feature of these more complex organisms is they have two cell types: somatic and reproductive cells. The somatic cells are biflagellate cells, but the reproductive cells are not. When comparing *P. californica* and *V. aureus*, *V. aureus* has a higher percentage of somatic cells than *P. californica*.

Overall, an analysis of these four species of algae illustrates three important principles found among complex multicellular species:

1. Multicellular organisms arise from a single cell that divides to produce daughter cells that adhere to one another.
2. The daughter cells can follow different fates, thereby producing multicellular organisms with different cell types.
3. As organisms get larger, a greater percentage of the cells tend to be somatic cells. The somatic cells carry out the activities required for the survival of the multicellular organism, whereas the reproductive cells are specialized for the sole purpose of producing offspring.

Toward the end of the Proterozoic eon, multicellular animals emerged. The first animals were invertebrates—animals without a backbone. Most animals, except for organisms such as sponges and jellyfish, exhibit bilateral symmetry—a two-sided body plan with a right and left side that are mirror images. Because each side of the body has appendages such as legs, one advantage of bilateral symmetry is that it facilitates locomotion. Bilateral animals also have anterior and posterior ends, with the mouth at the anterior end, as described in Chapter 19. In southern China in 2004, Chinese paleontologist Jun-Yuan Chen, American paleobiologist David Bottjer, and their colleagues discovered a fossil of the earliest known ancestor of animals with bilateral symmetry. This minute creature, with a shape like a flattened helmet, is barely visible to the naked eye (**Figure 22.14**). The fossil is approximately 580–600 million years old.

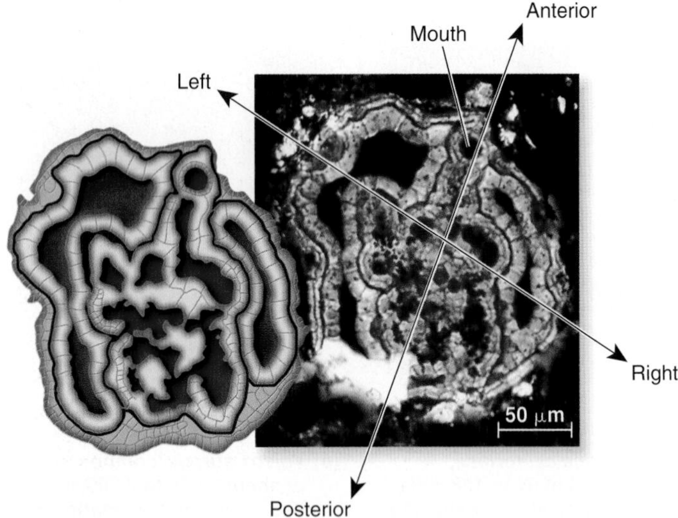

Figure 22.14 Fossil of an early invertebrate animal showing bilateral symmetry. This fossil of an early animal, *Vernanimalcula guizhouena*, dates from 580 to 600 mya.

Concept Check: *Name three other species that exhibit bilateral symmetry.*

Phanerozoic Eon: The Paleozoic Era Saw the Diversification of Invertebrates and the Colonization of Land by Plants and Animals

The proliferation of multicellular eukaryotic life has been extensive during the Phanerozoic eon, which started 543 mya and extends to the present day. Phanerozoic means "well-displayed life," referring to the abundance of fossils of plants and animals that have been identified from this eon. As described in Figure 22.9, the Phanerozoic eon is subdivided into three eras: the Paleozoic, Mesozoic, and Cenozoic. Because they are relatively recent and we have many fossils from these eras, each of them is further subdivided into periods. We will consider each era with its associated conditions and prevalent forms of life separately.

The term Paleozoic means ancient animal life. The Paleozoic era covers approximately 300 million years, from 543 to 248 mya, and is subdivided into six periods: the Cambrian, Ordovician, Silurian, Devonian, Carboniferous, and Permian. Periods are usually named after regions where rocks and fossils of that age were first discovered.

Cambrian Period (543–490 mya) The climate in the Cambrian period was generally warm and wet, with no evidence of ice at the poles. During this time, the diversity of animal species increased rapidly, an event called the **Cambrian explosion**. However, recent evidence suggests that many types of animal groups present during the Cambrian period actually arose prior to this period.

Many fossils from the Cambrian period were found in the Canadian Rockies in a rock bed called the Burgess Shale, which was discovered by American paleontologist Charles Walcott in 1909. At this site, both soft- and hard-bodied (shelled) invertebrates were buried in an underwater mudslide and preserved in water that was so deep and oxygen-free that decomposition was minimal (**Figure 22.15a**). The excellent preservation of the softer tissues is what makes this deposit unique (**Figure 22.15b**).

By the middle of the Cambrian period, all of the existing major types of marine invertebrates were present, plus many others that no longer exist. These include over 100 major animal groups with significantly different body plans. Examples that still exist include echinoderms (sea urchins and starfish), arthropods (insects, spiders, and crustaceans), mollusks (clams and snails), chordates (organisms with a dorsal nerve chord), and vertebrates (animals with backbones). Interestingly, although many new species of animals have arisen since this time, these later species have not shown a major reorganization of body plan, but instead exhibit variations on themes that were established during or prior to the Cambrian explosion.

The cause of the Cambrian explosion is not understood. Because it occurred shortly after marine animals evolved shells, some scientists have speculated that the changes observed in animal species may have allowed them to exploit new environments. Alternatively, others have suggested that the increase in diversity may be related to atmospheric oxygen levels. During this period, oxygen levels were increasing, and perhaps more complex body plans became possible only after the atmospheric oxygen surpassed a certain threshold. In addition, as atmospheric oxygen reached its present levels, an ozone (O_3) layer was produced that screens out harmful ultraviolet radiation, thereby allowing complex life to live in shallow water and eventually on land. Another possible contributor to the Cambrian explosion was an "evolutionary arms race" between interacting species. The ability of predators to capture prey and the ability of prey to avoid predators may have been a major factor that resulted in a diversification of animals into many different species.

Ordovician Period (490–443 mya) As in the Cambrian period, the climate of the early and middle parts of the Ordovician period was warm, and the atmosphere was moist. During this period, a diverse group of hard-shelled marine invertebrates, including trilobites and brachiopods, appeared in the fossil record (**Figure 22.16**). Marine communities consisted of invertebrates, algae, early jawless fishes (a type of early vertebrate), mollusks, and corals. Fossil evidence also suggests that early land plants and arthropods may have first invaded the land during this period.

Toward the end of the Ordovician period, the climate changed rather dramatically. Large glaciers formed, which drained the relatively shallow oceans, causing the water levels to drop. This resulted in a mass extinction in which as much as 60% of the existing marine invertebrates became extinct.

Silurian Period (443–417 mya) In contrast to the dramatic climate changes observed during the Ordovician period, the climate during the Silurian was relatively stable. The glaciers largely melted, which caused the ocean levels to rise. No new major types of invertebrate animals appeared during this period, but significant changes were observed among existing vertebrate and plant species. Many new types of fishes appeared in the fossil record. In addition, coral reefs made their first appearance during this period.

The Silurian marked a major colonization of land by terrestrial plants and animals. For this to occur, certain species evolved adaptations that prevented them from drying out, such as an external cuticle. Ancestral relatives of spiders and centipedes became prevalent.

(a) The Burgess Shale

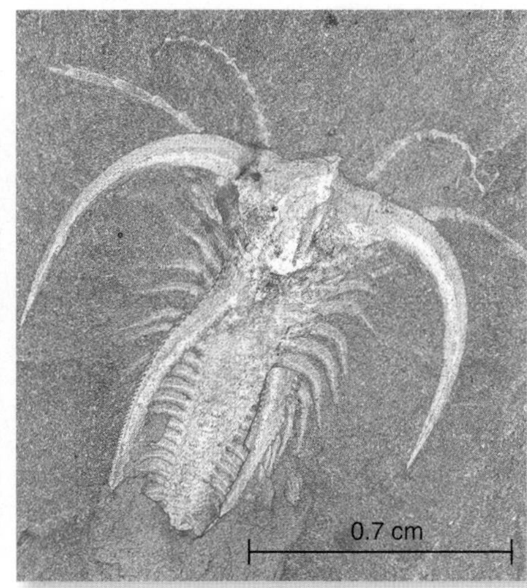

(b) A fossilized arthropod, *Marrella*

Figure 22.15 **The Cambrian explosion and the Burgess Shale.** **(a)** This photograph shows the original site in the Canadian Rockies discovered by Charles Walcott. Since its discovery, this site has been made into a quarry for the collection of fossils. **(b)** A fossil of an extinct arthropod, Marrella, which was found at this site.

(a) Trilobite

(b) Brachiopod

Figure 22.16 **Shelled, invertebrate fossils of the Ordovician period.** Trilobites existed for millions of years before becoming extinct about 250 mya. Many species of brachiopods exist today.

The earliest fossils of vascular plants, which have tissues that are specialized for the transport of water, sugar, and salts throughout the plant body, were observed in this period.

Devonian Period (417–354 mya) In the Devonian period, generally dry conditions occurred across much of the northern landmasses. However, the southern landmasses were mostly covered by cool, temperate oceans.

The Devonian saw a major increase in the number of terrestrial species. At first, the vegetation consisted primarily of small plants, only a meter tall or less. Later, ferns, horsetails, and seed plants, such as gymnosperms, also emerged. By the end of the Devonian, the first trees and forests were formed. A major expansion of terrestrial animals also occurred. Insects first appeared in the fossil record, and other invertebrates became plentiful. In addition, the first tetrapods—vertebrates with four legs—are believed to have arisen in the Devonian. Early tetrapods included amphibians, which lived on land but required water in which to lay their eggs.

In the oceans, many types of invertebrates flourished, including brachiopods, echinoderms, and corals. This period is sometimes called the Age of Fishes, as many new types of fishes emerged. During a period of approximately 20 million years near the end of the Devonian period, a prolonged series of extinctions eliminated many marine species. The cause of this mass extinction is not well understood.

Carboniferous Period (354–290 mya) The term Carboniferous refers to the rich deposits of coal, a sedimentary rock primarily composed of carbon, that were formed during this period. The Carboniferous had the ideal conditions for the subsequent formation of coal. It was a cooler period, and much of the land was covered by forest swamps. Coal was formed over many millions of years from compressed layers of rotting vegetation.

Plants and animals further diversified during the Carboniferous period. Very large plants and trees became prevalent. For example, tree ferns such as *Psaronius* grew to a height of 15 meters or more (**Figure 22.17**). The first flying insects emerged. Giant dragonflies with

Figure 22.17 A giant tree fern, *Psaronius*, from the Carboniferous period. This genus became extinct during the Permian. The illustration is a re-creation based on fossil evidence. The inset shows a fossilized section of the trunk, also known as petrified wood.

Figure 22.18 *Megazostrodon*, the first known mammal of the Triassic period. The illustration is a re-creation based on fossilized skeletons. The *Megazostrodon* was 10 to 12 cm long.

BioConnections: Look ahead to Table 34.1. What are the common characteristics of mammals?

a wingspan of over 2 feet inhabited the forest swamps. Terrestrial vertebrates also became more diverse. Amphibians were very prevalent. One innovation that seemed particularly beneficial was the amniotic egg. In reptiles, the amniotic egg was covered with a leathery or hard shell, which prevented the desiccation of the embryo inside. This innovation was critical for the emergence of reptiles during this period.

Permian Period (290–248 mya) At the beginning of the Permian, continental drift had brought much of the total land together into a supercontinent known as Pangaea (see Figure 22.10). The interior regions of Pangaea were dry, with great seasonal fluctuations. The forests of fernlike plants were replaced with gymnosperms. Species resembling modern conifers first appeared in the fossil record. Amphibians were prevalent, but reptiles became the dominant vertebrate species.

At the end of the Permian period, the largest known mass extinction in the history of life on Earth occurred; 90–95% of marine species and a large proportion of terrestrial species were eliminated. The cause of the Permian extinction is the subject of much research and controversy. One possibility is that glaciation destroyed the habitats of terrestrial species and lowered ocean levels, which would have caused greater competition among marine species. Another hypothesis is that enormous volcanic eruptions in Siberia produced large ash clouds that abruptly changed the climate on Earth.

Phanerozoic Eon: The Mesozoic Era Saw the Rise and Fall of the Dinosaurs

The Permian extinction marks the division between the Paleozoic and Mesozoic eras. Mesozoic means "middle animals." It was a time period that saw great changes in animal and plant species. This era is sometimes called the Age of Dinosaurs, which flourished during this time. The climate during the Mesozoic era was consistently hot, and terrestrial environments were relatively dry. Little if any ice was found at either pole. The Mesozoic is divided into three periods: the Triassic, Jurassic, and Cretaceous.

Triassic Period (248–206 mya) Reptiles were plentiful in this period, including new groups such as crocodiles and turtles. The first dinosaurs emerged during the middle of the Triassic, as did the first mammals, such as the small *Megazostrodon* (**Figure 22.18**). Gymnosperms were the dominant land plant. Volcanic eruptions near the end of the Triassic are thought to have caused global warming, resulting in mass extinctions that eliminated many marine and terrestrial species.

Jurassic Period (206–144 mya) Gymnosperms, such as conifers, continued to be the dominant vegetation. Mammals were not prevalent. Reptiles continued to be the dominant land vertebrate. These included dinosaurs, which were predominantly terrestrial reptiles that shared certain anatomical features, such as an erect posture. Some dinosaurs attained enormous sizes, including the massive *Brachiosaurus*, which reached a length of 25 m (80 ft) and weighed up to 100 tons! Modern birds are descendents of a dinosaur lineage called theropod (meaning "beast-footed") dinosaurs. *Tyrannosaurus rex* is one of the best known theropod dinosaurs. An early birdlike animal, *Archaeopteryx* (**Figure 22.19**), emerged in the Jurassic period. However, paleontologists are debating whether or not *Archaeopteryx* is a true ancestor of modern birds.

Cretaceous Period (144–65 mya) On land, dinosaurs continued to be the dominant animals. The earliest flowering plants, called angiosperms, which form seeds within a protective chamber, emerged and began to diversify.

The end of the Cretaceous witnessed another mass extinction, which brought an end to many previously successful groups of organisms. Except for the lineage that gave rise to birds, dinosaurs abruptly died out, as did many other species. As with the Permian extinction, the cause or causes of this mass extinction are still debated. One plausible hypothesis suggests that a large meteorite hit the region that is now the Yucatan Peninsula of Mexico, lifting massive amounts of debris into the air and thereby blocking the sunlight from reaching the Earth's surface. Such a dense haze could have cooled the Earth's

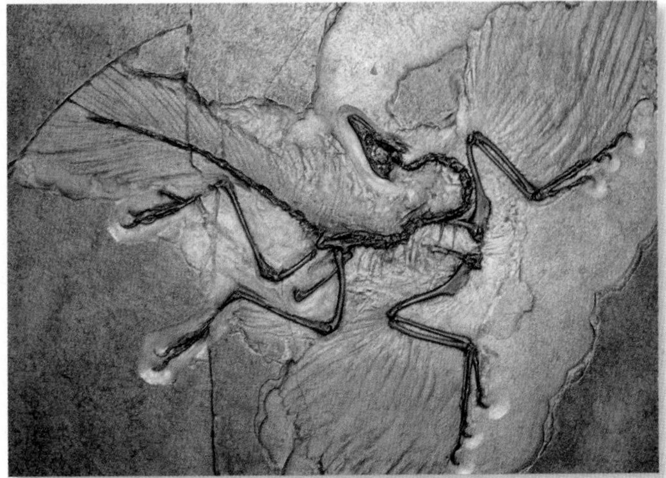

Figure 22.19 **A fossil of an early birdlike animal, *Archaeopteryx*, which emerged in the Jurassic period.**

surface by 11–15°C (20–30°F). Evidence also points to strong volcanic eruptions as a contributing factor for this mass extinction.

Phanerozoic Eon: Mammals and Flowering Plants Diversified During the Cenozoic Era

The Cenozoic era spans the most recent 65 million years. It is divided into two periods: the Tertiary and Quaternary. In many parts of the world, tropical conditions were replaced by a colder, drier climate. During this time, mammals became the largest terrestrial animals, which is why the Cenozoic is sometimes called the Age of Mammals. However, the Cenozoic era also saw an amazing diversification of many types of organisms, including birds, fishes, insects, and flowering plants.

Tertiary Period (65–1.8 mya) On land, the mammals that survived from the Cretaceous began to diversify rapidly during the early part of the Tertiary period. Angiosperms became the dominant land plant, and insects became important for their pollination. Fishes also diversified, and sharks became abundant.

Toward the end of the Tertiary period, about 7 mya, hominoids came into existence. **Hominoids** include humans, chimpanzees, gorillas, orangutans, and gibbons, plus all of their recent ancestors. The subset of hominoids called hominins includes modern humans, extinct human species (for example, of the *Homo* genus), and our immediate ancestors. In 2002, a fossil of the earliest known hominin, *Sahelanthropus tchadensis*, was discovered in Central Africa. This fossil was dated at between 6 and 7 million years old. Another early hominin genus, called *Australopithecus*, first emerged in Africa about 4 mya. Australopithecines walked upright and had a protruding jaw, prominent eyebrow ridges, and a small braincase.

Quaternary Period (1.8 mya–present) Periodic Ice Ages have been prevalent during the last 1.8 million years, covering much of Europe and North America. This period has witnessed the widespread extinction of many species of mammals, particularly larger ones. Certain species of hominins became increasingly more like living humans. Near the beginning of the Quaternary period, fossils were discovered of *Homo habilis*, or handy man, so called because stone tools were found with the fossil remains. Fossils that are classified as *Homo sapiens*—modern humans—first appeared about 170,000 years ago. The evolution of hominins is discussed in more detail in Chapter 34.

Summary of Key Concepts

- Life began on Earth from nonliving material between 3.5 and 4.0 bya (Figure 22.1).

22.1 Origin of Life on Earth

- Life on Earth is hypothesized to have occurred in four overlapping stages. The first stage involved the synthesis of organic molecules to form a prebiotic soup. Possible scenarios of how this occurred are the reducing atmosphere, extraterrestrial, and deep-sea vent hypotheses (Figures 22.2, 22.3).
- The second stage was the formation of polymers from simple organic molecules. This may have occurred on the surface of clay.
- The third stage occurred when polymers became enclosed in structures called protobionts that separated them from the external environment (Figure 22.4).
- In the fourth stage, polymers enclosed in membranes acquired properties of cells, such as self-replication and other catalytic functions (Figure 22.5).
- In the hypothesized period called the RNA world, the first living cells used RNA for both information storage and catalytic functions.
- Bartel and Szostak demonstrated that chemical selection for RNA molecules, which can catalyze covalent bond formation, is possible experimentally (Figure 22.6).
- The RNA world was eventually superseded by the modern DNA/RNA/protein world.

22.2 The Fossil Record

- Fossils, which are preserved remnants of past life-forms, are formed in sedimentary rock (Figure 22.7).
- Radiometric dating is one way of estimating the age of a fossil. Fossils provide an extensive record of the history of life, though the record is incomplete (Figure 22.8, Table 22.1).

22.3 History of Life on Earth

- The geological time scale, which is divided into four eons and many eras and periods, charts the major events that occurred during the history of life on Earth (Figure 22.9).
- The formation of species, as well as mass extinctions, are correlated with changes in temperature, amount of O_2 in the atmosphere, landmass locations, floods and glaciation, volcanic eruptions, and meteorite impacts (Figure 22.10).
- During the Archaean eon, bacteria and archaea arose. The proliferation of cyanobacteria led to a gradual rise in O_2 levels (Figure 22.11).
- Eukaryotic cells arose during the Proterozoic eon. This origin involved a union between bacterial and archaeal cells that is hypothesized to have been endosymbiotic. The origin of mitochondria and chloroplasts was an endosymbiotic relationship (Figure 22.12).
- Multicellular eukaryotes arose about 1.5 bya during the Proterozoic eon. Multicellularity now occurs via cell division and the adherence of the resulting cells to each other. A multicellular organism can produce multiple cell types (Figure 22.13).

- The first bilateral animal emerged toward the end of the Proterozoic eon (Figure 22.14).
- The Phanerozoic eon is subdivided into the Paleozoic, Mesozoic, and Cenozoic eras. During the Paleozoic era, invertebrates greatly diversified, particularly during the Cambrian explosion, and the land became colonized by plants and animals. Terrestrial vertebrates, including tetrapods, became more diverse (Figures 22.15, 22.16, 22.17).
- Dinosaurs were prevalent during the Mesozoic era, particularly during the Jurassic period. Mammals and birds also emerged (Figures 22.18, 22.19).
- During the Cenozoic era, mammals diversified, and flowering plants became the dominant plant species. The first hominoids emerged approximately 7 mya. Fossils classified as *Homo sapiens*, our species, appeared about 170,000 years ago.

Assess and Discuss

Test Yourself

1. The prebiotic soup was
   a. the assemblage of unicellular prokaryotes that existed in the oceans of early Earth.
   b. the accumulation of organic molecules in the oceans of early Earth.
   c. the mixture of organic molecules found in the cytoplasm of the earliest cells on Earth.
   d. a pool of nucleic acids that contained the genetic information for the earliest organisms.
   e. none of the above.
2. Which of the following is *not* a characteristic of protobionts necessary for the evolution of living cells?
   a. a membrane-like boundary separating the external environment from an internal environment
   b. polymers capable of functioning in information storage
   c. polymers capable of catalytic activity
   d. self-replication
   e. compartmentalization of metabolic activity
3. RNA is believed to be the first functional macromolecule in protobionts because it
   a. is easier to synthesize compared with other macromolecules.
   b. has the ability to store information, self-replicate, and perform catalytic activity.
   c. is the simplest of the macromolecules commonly found in living cells.
   d. All of the above are correct.
   e. Only a and c are correct.
4. The movement of landmasses that have changed their positions, shapes, and association with other landmasses is called
   a. glaciation.
   b. Pangaea.
   c. continental drift.
   d. biogeography.
   e. geological scale.
5. Paleontologists estimate the dates of fossils by
   a. the layer of rock in which the fossils are found.
   b. analysis of radioisotopes found in nearby igneous rock.
   c. the complexity of the body plan of the organism.
   d. all of the above.
   e. a and b only.
6. The fossil record does not give us a complete picture of the history of life because
   a. not all past organisms have become fossilized.
   b. only organisms with hard skeletons can become fossilized.
   c. fossils of very small organisms have not been found.
   d. fossils of early organisms are located too deep in the crust of the Earth to be found.
   e. all of the above.
7. The endosymbiosis hypothesis explaining the evolution of eukaryotic cells is supported by
   a. DNA-sequencing analysis comparing bacterial genomes, mitochondrial genomes, and eukaryotic nuclear genomes.
   b. naturally occurring examples of endosymbiotic relationships between bacterial cells and eukaryotic cells.
   c. the presence of DNA in mitochondria and chloroplasts.
   d. all of the above.
   e. a and b only.
8. Which of the following explanations of multicellularity in eukaryotes is seen in the development of complex, multicellular organisms today?
   a. endosymbiosis
   b. aggregation of cells to form a colony
   c. division of cells with the resulting cells adhering together
   d. multiple cell types aggregating to form a complex organism
   e. none of the above
9. The earliest fossils of vascular plants were formed during the ___________ period.
   a. Ordovician
   b. Silurian
   c. Devonian
   d. Triassic
   e. Jurassic
10. The appearance of the first hominoids dates to the ___________ period.
   a. Triassic
   b. Jurassic
   c. Cretaceous
   d. Tertiary
   e. Quaternary

Conceptual Questions

1. What are the four stages that led to the origin of living cells?
2. How are the ages of fossils determined? In your answer, you should discuss which types of rocks are analyzed and explain the concepts of radiometric dating and half-life.
3. Two principles of biology are (1) *living organisms interact with their environment* and (2) *populations of organisms evolve from one generation to the next.* Describe two examples in which changes in the global climate affected the evolution of species.

Collaborative Questions

1. Discuss possible hypotheses of how organic molecules were first formed.
2. Discuss the key features of a protobiont. What distinguishes a protobiont from a living cell?

Online Resource

www.brookerbiology.com

Stay a step ahead in your studies with animations that bring concepts to life and practice tests to assess your understanding. Your instructor may also recommend the interactive eBook, individualized learning tools, and more.

Chapter Outline

23.1 The Theory of Evolution

23.2 Evidence of Evolutionary Change

23.3 The Molecular Processes That Underlie Evolution

Summary of Key Concepts

Assess and Discuss

23 An Introduction to Evolution

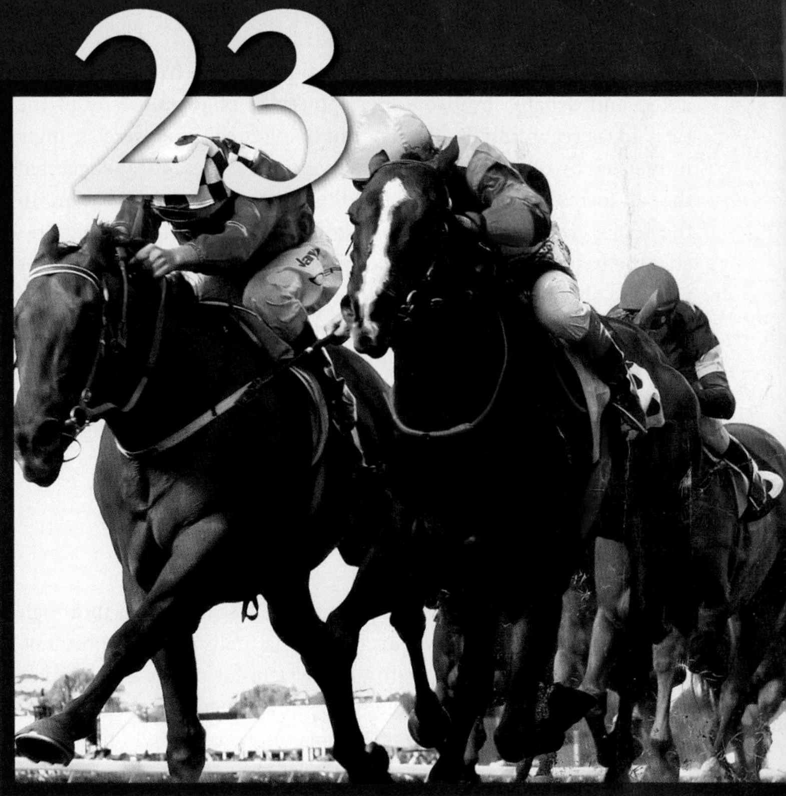

Selective breeding. The horses in this race have been bred for a particular trait, in this case, speed. Such a practice, called selective breeding, can dramatically change the traits of organisms over several generations.

Organic life beneath the shoreless waves
Was born and nurs'd in Ocean's pearly caves
First forms minute, unseen by spheric glass,
Move on the mud, or pierce the watery mass;
These, as successive generations bloom,
New powers acquire, and larger limbs assume;
Whence countless groups of vegetation spring,
And breathing realms of fin, and feet, and wing.

From *The Temple of Nature* by Erasmus Darwin, grandfather of Charles Darwin. Published posthumously in 1803.

The term **evolution** is used to describe a heritable change in one or more characteristics of a population from one generation to the next. Evolution can be viewed on a small scale **(microevolution)** as it relates to changes in a single gene or allele frequencies in a population over time, or it can be viewed on a larger scale **(macroevolution)** as it relates to the formation of new species or groups of related species.

It is helpful to begin our discussion of evolution with a working definition of a species. Biologists often define a **species** as a group of related organisms that share a distinctive form. Among species that reproduce sexually, such as plants and animals, members of the same species are capable of interbreeding in nature to produce viable and fertile offspring. The term **population** refers to all members of a species that live in the same area at the same time and have the opportunity to interbreed. As we will see in Chapter 25, some of the emphasis in the study of evolution is on understanding how populations change over the course of many generations to produce new species.

In the first part of this chapter, we will examine the development of evolutionary thought and some of the basic tenets of evolution, particularly those proposed by the British naturalist Charles Darwin in the mid-1800s. The theory of evolution has been refined over the past 150 years or so, but the fundamental principle of evolution remains unchanged and has provided a cornerstone for our understanding of biology. Ukrainian-born American geneticist Theodosius Dobzhansky, an influential evolutionary scientist of the 1900s, once said, "Nothing in biology makes sense except in the light of evolution." The extraordinarily diverse and often seemingly bizarre array of species on our planet can be explained within the context of evolution. As is the case with all scientific theories, evolution is called a theory because it is supported by a substantial body of evidence and because it explains a wide range of observations. The theory of evolution provides answers to many questions related to the diversity of life. In biology, theories such as this are viewed as scientific knowledge.

In the second part of this chapter, we will survey the extensive data that illustrate the processes by which evolution occurs. These data not only support the theory of evolution but also allow us to understand the interrelatedness of different species, whose similarities are often due to descent from a common ancestor. Much of the early evidence supporting evolution came from direct observations and comparisons of living and extinct species. More recently, advances in molecular genetics, particularly those related to DNA sequencing and genomics, have revolutionized the study of evolution. Scientists now have information that allows us to understand how evolution involves changes in the DNA sequences of a given species. These changes affect both a species' genes and the proteins they encode. **Molecular evolution** refers to the process of evolution at the level of genes and proteins. Comparisons of gene or protein sequences in different organisms can reveal evolutionary relationships that cannot be seen in morphology. A major focus of this textbook, namely genomes and proteomes, is rooted in an understanding of these changes. In the last section of this chapter, we consider some of the exciting new ways of exploring evolutionary change at the molecular level. In the following chapters of this unit, we will examine how such changes are acted upon by evolutionary factors in ways that alter the traits of a given species and may eventually lead to the formation of new species.

23.1 The Theory of Evolution

Learning Outcomes:

1. Define the theory of evolution.
2. Describe the factors that led Darwin to the theory of evolution.
3. Explain the process of natural selection.

Undoubtedly, the question, "Where did we come from?" has been asked and debated by people for thousands of years. Many of the early ideas regarding the existence of living organisms were strongly influenced by religion and philosophy. Some of these ideas suggested that all forms of life have remained the same since their creation. In the 1600s, however, scholars in Europe began a revolution that created the basis of empirical and scientific thought. **Empirical thought** relies on observation to form an idea or hypothesis rather than trying to understand life from a nonphysical or spiritual point of view. As described in this section, the shift toward empirical thought encouraged scholars to look for the basic rationale behind a given process or phenomenon. This perspective played a key role in developing the theory of evolution.

The Work of Several Scientists Set the Stage for Darwin's Ideas

In the mid- to late-1600s, the first scientist to carry out a thorough study of the living world was an English naturalist named John Ray, who developed an early classification system for plants and animals based on anatomy and physiology. He established the modern concept of a species, noting that organisms of one species do not interbreed with members of another and used it as the basic unit of his classification system. Ray's ideas on classification were later expanded by the Swedish naturalist Carolus Linnaeus. How did their work contribute to the development of evolutionary theory? Neither Ray nor Linnaeus proposed that evolutionary change promotes the formation of new species. However, their systematic classification of plants and animals helped scholars of this period perceive the similarities and differences among living organisms.

Late in the 1700s, a small number of European scientists began to quietly suggest that life-forms are not fixed and unchanging. A French zoologist, George Buffon, actually proposed that living things change through time. However, Buffon was careful to hide his views in a 44-volume series of books on natural history. Around the same time, a French naturalist named Jean-Baptiste Lamarck suggested an intimate relationship between variation and evolution. By examining fossils, he realized that some species had remained the same over the millennia and others had changed. Lamarck hypothesized that species change over the course of many generations by adapting to new environments. He believed that living things evolved in a continuously upward direction, from dead matter, through simple to more complex forms, toward "human perfection." According to Lamarck, organisms altered their behavior in response to environmental change. He thought that behavioral changes could modify traits and hypothesized that these modified traits were inherited by offspring. He called this idea the **inheritance of acquired characteristics**. For example, according to Lamarck's hypothesis, giraffes developed their elongated necks and front legs by feeding on the leaves at the top of trees. The exercise of stretching up to the leaves altered the neck and legs, and Lamarck presumed that these acquired characteristics were transmitted to offspring. However, further research has rejected Lamarck's idea that acquired traits can be inherited. Even so, Lamarck's work was important in promoting the idea of evolutionary change.

Interestingly, Erasmus Darwin, the grandfather of Charles Darwin, was a contemporary of Buffon and Lamarck and an early advocate of evolutionary change. He was a physician, a plant biologist, and also a poet (see poem at the beginning of the chapter). He was aware that modern species were different from similar types of fossilized organisms and also saw how plant and animal breeders used breeding practices to change the traits of domesticated species (see chapter opening photo). He knew that offspring inherited features from their parents and went so far as to say that life on Earth could have descended from a common ancestor.

Darwin Suggested That Existing Species Are Derived from Pre-existing Species

Charles Darwin played a central role in developing the theory that existing species have evolved from pre-existing ones. Darwin's unique perspective and his ability to formulate evolutionary theory were shaped by several different fields of study, including ideas of his time about geological and biological processes.

Two main hypotheses about geological processes predominated in the early 19th century. Catastrophism was first proposed by French zoologist and paleontologist Georges Cuvier to explain the age of the Earth. Cuvier suggested that the Earth was just 6,000 years old and that only catastrophic events had changed its geological structure. This idea fit well with certain religious teachings. Alternatively, uniformitarianism, proposed by Scottish geologist James Hutton and popularized by fellow Scotsman geologist Charles Lyell, suggested that changes in the Earth are directly caused by recurring events. For example, they suggested that geological processes such as erosion existed in the past and happened at the same gradual rate as they do now. For such slow geological processes to eventually lead to substantial changes in the Earth's characteristics, a great deal of time was required. Hutton and Lyell were the first to propose that the age of the Earth is well beyond 6,000 years. The ideas of Hutton and Lyell helped to shape Darwin's view of the world.

Darwin's thinking was also influenced by a paper published in 1798 called *Essay on the Principle of Population* by Thomas Malthus, an English economist. Malthus asserted that the population size of humans can, at best, increase linearly due to increased land usage and improvements in agriculture, whereas our reproductive potential is exponential (for example, doubling with each generation). He argued that famine, war, and disease, especially among the poor, keep population growth within existing resources. The relevant message from Malthus's work was that not all members of any population will survive and reproduce.

Darwin's ideas, however, were most influenced by his own experiences and observations. His work as a young man aboard the HMS *Beagle*, a survey ship, lasted from 1831 to 1836 and involved a careful examination of many different species (**Figure 23.1**). The main mission

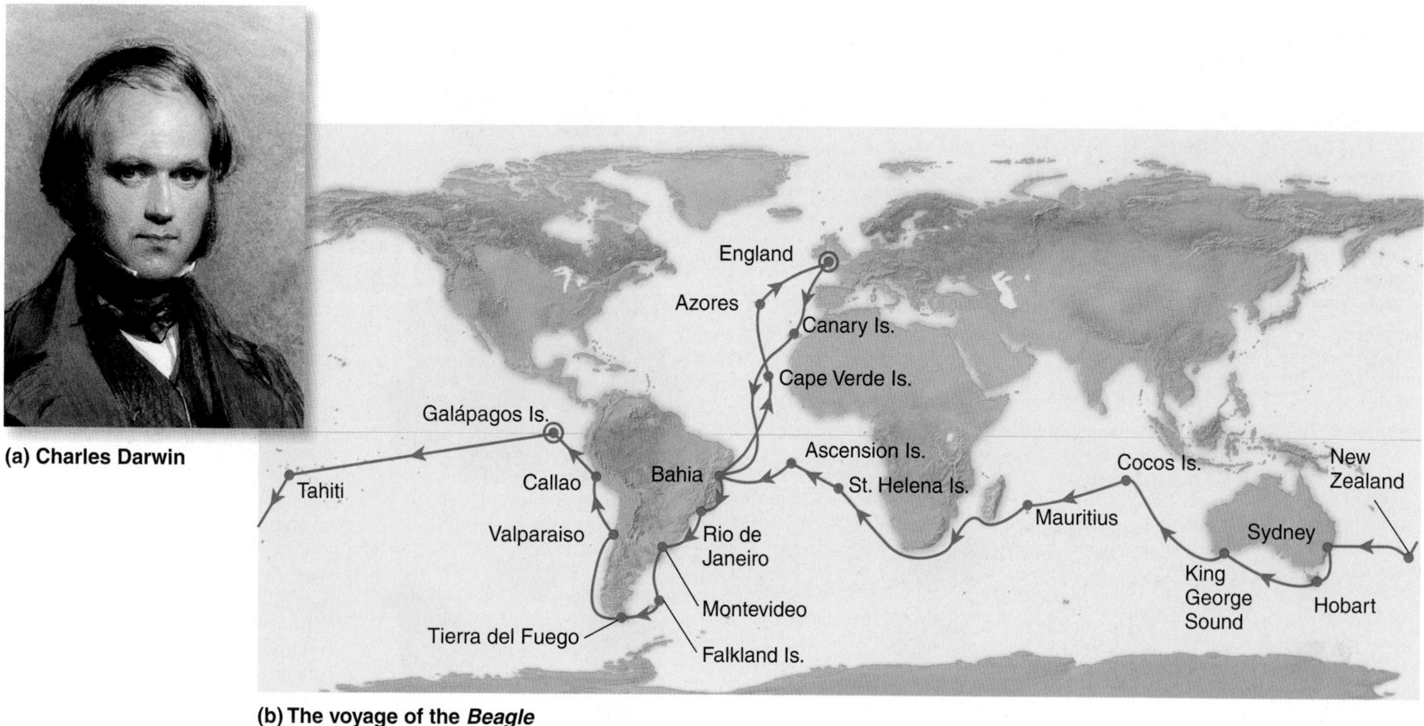

Figure 23.1 **Charles Darwin and the voyage of the *Beagle,* 1831–1836.** **(a)** A portrait of Charles Darwin (1809–1882) at age 31. **(b)** Darwin's voyage on the *Beagle*, which took almost 5 years to circumnavigate the world.

of the *Beagle* was to map the coastline of southern South America and take oceanographic measurements. As the ship's naturalist, Darwin's job was to record information about the weather, geological features, plants, animals, fossils, rocks, minerals, and indigenous people.

Though Darwin made many interesting observations on his journey, he was particularly struck by the distinctive traits of island species. For example, Darwin observed several species of finches found on the Galápagos Islands, a group of volcanic islands 600 miles from the coast of Ecuador. Though it is often assumed that Darwin's personal observations of these finches directly inspired his theory of evolution, this is not the case. Initially, Darwin thought the birds were various species of blackbirds, grosbeaks, and finches. Later, however, the bird specimens from the islands were given to the British ornithologist John Gould, who identified them as several new finch species. Gould's observations helped Darwin in the later formulation of his theory.

As seen in **Table 23.1**, the finches differed widely in the size and shape of their beaks and in their feeding habits. For example, the ground and vegetarian finches have sturdy, crushing beaks they use to crush various sizes of seeds or buds. The tree finches have grasping beaks they use to pick up insects from trees. The mangrove, woodpecker, warbler, and cactus finches have pointed, probing beaks. They use their beaks to search for insects in crevices. The cactus finches use their probing beaks to open cactus fruits and eat the seeds. One species, the woodpecker finch, even uses twigs or cactus spines to extract insect larvae from holes in dead tree branches. Darwin clearly saw the similarities among these species, yet he noted the differences that provided them with specialized feeding strategies. It is now known these finches all evolved from a single species similar to the dull-colored grassquit finch (*Tiaris obscura*), commonly found along the Pacific Coast of South America. Once they arrived on the Galápagos Islands, the finches' ability to survive and reproduce in their new habitat depended, in part, on changes in the size and shape of their beaks over many generations. These specializations enabled succeeding generations to better obtain particular types of food.

With an understanding of geology and population growth, and his observations from his voyage on the *Beagle*, Darwin had formulated his theory of evolution by the mid-1840s. He had also catalogued and described all of the species he had collected on his *Beagle* voyage except for one type of barnacle. Some have speculated that Darwin may have felt that he should establish himself as an expert on one species before making generalizations about all of them. Therefore, he spent several additional years studying barnacles. During this time, the geologist Charles Lyell, who had greatly influenced Darwin's thinking, strongly encouraged Darwin to publish his theory of evolution. In 1856, Darwin began to write a long book to explain his ideas. In 1858, however, Alfred Wallace, a British naturalist working in the East Indies, sent Darwin an unpublished manuscript to read prior to its publication. In it, Wallace proposed the same ideas concerning evolution. In response to this, Darwin decided to use some of his own writings on this subject, and two papers, one by Darwin and one by Wallace, were published in the *Proceedings of the Linnaean Society of London*. These papers were not widely recognized. A year later, however, Darwin finished his book *On the Origin of Species* (1859), which described his ideas in greater detail and included observational support. This book, which received high praise from many scientists and scorn from others, started a great debate concerning evolution.

Table 23.1 A Comparison of Beak Type and Diet Among the Galápagos Finches That Darwin Studied

Type of finch/diet	Species	Type of beak
Ground finches		
Ground finches have beaks shaped to crush various sizes of seeds; large beaks can crush large seeds, whereas smaller beaks are better for crushing small seeds.	Large ground finch (*Geospiza magnirostris*)	Crushing
	Medium ground finch (*G. fortis*)	
	Small ground finch (*G. fuliginosa*)	
	Sharp-billed ground finch (*G. difficilis*)	
Vegetarian finch		
Vegetarian finches have crushing beaks to pull buds from branches.	Vegetarian finch (*Platyspiza crassirostris*)	Crushing
Tree finches		
Tree finches have grasping beaks to pick insects from trees. Those with heavier beaks can also break apart wood in search of insects.	Large tree finch (*Camarhynchus psittacula*)	Grasping
	Medium tree finch (*Camarhynchus pauper*)	
	Small tree finch (*Camarhynchus parvulus*)	
Tree and warbler finches		
These finches have probing beaks to search for insects in crevices and then to pick them up. The woodpecker finch can also use a cactus spine for probing.	Mangrove finch (*Cactospiza heliobates*)	Probing
	Woodpecker finch (*Camarhynchus pallidus*)	
	Warbler finch (*Certhidea olivacea*)	
Cactus finches		
Cactus finches have probing beaks to open cactus fruits and take out seeds.	Large cactus finch (*G. conirostris*)	Probing
	Cactus finch (*G. scandens*)	

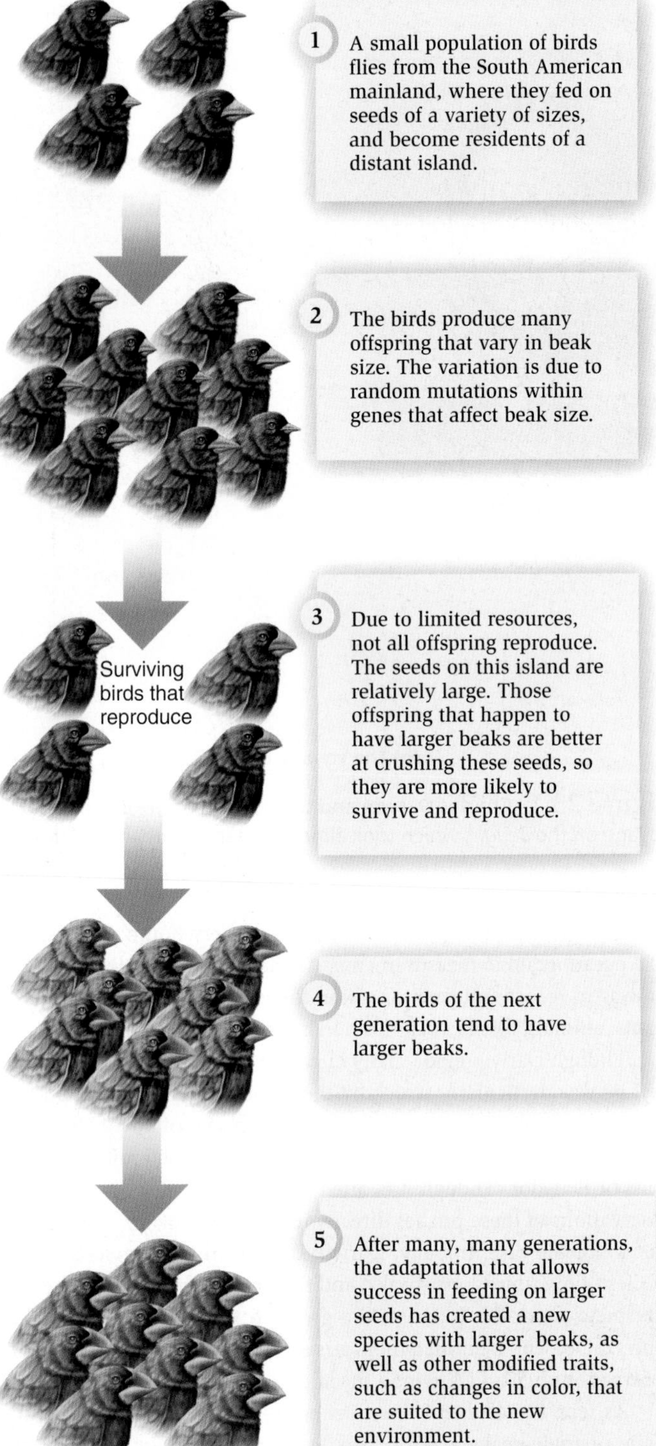

1 A small population of birds flies from the South American mainland, where they fed on seeds of a variety of sizes, and become residents of a distant island.
2 The birds produce many offspring that vary in beak size. The variation is due to random mutations within genes that affect beak size.
Surviving birds that reproduce
3 Due to limited resources, not all offspring reproduce. The seeds on this island are relatively large. Those offspring that happen to have larger beaks are better at crushing these seeds, so they are more likely to survive and reproduce.
4 The birds of the next generation tend to have larger beaks.
5 After many, many generations, the adaptation that allows success in feeding on larger seeds has created a new species with larger beaks, as well as other modified traits, such as changes in color, that are suited to the new environment.

Figure 23.2 Evolutionary adaptation to a new environment via natural selection. The example shown here involves a species of finch adapting to a new environment on a distant island. According to Darwin's theory of evolution, the process of adaptation can lead to the formation of a new species with traits that are better suited to the new environment.

Concept Check: The phrase "an organism evolves" is incorrect. Explain why.

BioConnections: Look back at Figure 22.5. How is natural selection similar to chemical selection? How are they different?

Although some of his ideas were incomplete because the genetic basis of traits was not understood at that time, Darwin's work remains a foundation of our understanding of biology.

Natural Selection Changes Populations from Generation to Generation

Darwin hypothesized that existing life-forms on our planet result from the modification of pre-existing life-forms. He expressed this concept of evolution as "the theory of descent with modification through variation and natural selection." The term evolution refers to change. What factors bring about evolutionary change? According to Darwin's ideas, evolution occurs from generation to generation due to two interacting factors, genetic variation and natural selection:

1. Variation in traits may occur among individuals of a given species. The heritable traits are then passed from parents to offspring. The genetic basis for variation within a species was not understood at the time Darwin proposed his theory of evolution. We now know that such variation is due to different types of genetic changes such as random mutations in genes. Even though Darwin did not fully appreciate the genetic basis of variation, he and many other people before him observed that offspring resemble their parents more than they do unrelated individuals. Therefore, he assumed that some traits are passed from parent to offspring.
2. In each generation, many more offspring are usually produced than will survive and reproduce. Often times, resources in the environment are limiting for an organism's survival. During the process of **natural selection**, individuals with heritable traits that make them better suited to their native environment tend to flourish and reproduce, whereas other individuals are less likely to survive and reproduce. As a result of natural selection, certain traits that favor reproductive success become more prevalent in a population over time.

As an example, we can consider a population of finches that migrates from the South American mainland to a distant island (**Figure 23.2**). Variation exists in the beak sizes among the migrating birds. Let's suppose the seeds produced on the distant island are larger than those produced on the mainland. Those birds with larger beaks would be better able to feed on these larger seeds and therefore would be more likely to survive and pass that trait to their offspring. What are the consequences of this selection process? In succeeding generations, the population tends to have a greater proportion of finches with larger beaks. Alternatively, if a trait happens to be detrimental to an individual's ability to survive and reproduce, natural selection is likely to eliminate this type of variation. For example, if a finch in the same environment had a small beak, this bird would be less likely to acquire food, which would decrease its ability to survive and pass this trait to its offspring. Natural selection may ultimately result in a new species with a combination of multiple traits that are quite different from those of the original species, such as finches with larger beaks and changes in coloration. In other words, the newer species has evolved from a pre-existing one. Let's look at a scientific study involving one such change in a population over time.

FEATURE INVESTIGATION

The Grants Observed Natural Selection in Galápagos Finches

Since 1973, British evolutionary biologists Peter Grant, Rosemary Grant, and their colleagues have studied natural selection in finches found on the Galápagos Islands. For over 30 years, the Grants have focused much of their work on one of the Galápagos Islands known as Daphne Major (**Figure 23.3a**). This small island (0.34 km^2) has a moderate degree of isolation (it is 8 km from the nearest island), an undisturbed habitat, and a resident population of *Geospiza fortis*, the medium ground finch (**Figure 23.3b**).

To study natural selection, the Grants observed various traits in finches over the course of many years. One trait they observed is beak size. The medium ground finch has a relatively small crushing beak, allowing it to more easily feed on small, tender seeds (see Table 23.1). The Grants quantified beak size among the medium ground finches of Daphne Major by carefully measuring beak depth—a measurement of the beak from top to bottom (**Figure 23.4**). The small size of the island made it possible for them to measure a large percentage of birds and their offspring. During the course of their studies, they compared the beak depths of parents and offspring by examining many broods over several years and found that the depth of the beak was transmitted from parents to offspring, regardless of environmental conditions, indicating that differences in beak depths are due to genetic differences in the population. In other words, they found that beak depth was a heritable trait.

(a) Daphne Major

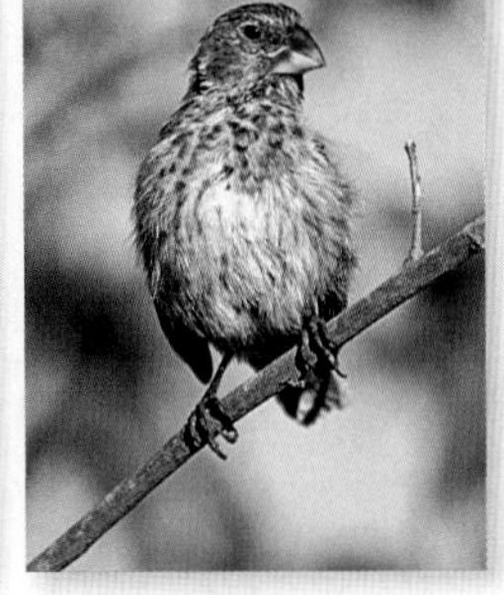

(b) Medium ground finch

Figure 23.3 The Grants' investigation of natural selection in finches. **(a)** Daphne Major, one of the Galápagos Islands. **(b)** One of the medium ground finches (*Geospiza fortis*) that populate this island.

BIOLOGY PRINCIPLE Populations of organisms evolve from one generation to the next. This study was aimed at analyzing how beak size may change from one generation to the next.

By measuring many birds every year, the Grants were able to assemble a detailed portrait of natural selection in action. In the study shown in Figure 23.4, they measured beak depth from 1976 to 1978. In the wet year of 1976, the plants of Daphne Major produced an abundance of the small, tender seeds that these finches could easily eat. However, a severe drought occurred in 1977. During this year,

Figure 23.4 The Grants and natural selection of beak size among the medium ground finch.

HYPOTHESIS Dry conditions produce larger seeds and may result in larger beaks in succeeding generations of *Geospiza fortis* due to natural selection.

KEY MATERIALS A population of *G. fortis* on the Galápagos Island called Daphne Major.

	Experimental level	Conceptual level
1 In 1976, measure beak depth in parents and offspring of the species *G. fortis*.	Capture birds and measure beak depth. 	This is a way to measure a trait that may be subject to natural selection.
2 Repeat the procedure on offspring that were born in 1978 and had reached mature size. A drought had occurred in 1977 that caused plants on the island to produce mostly large dry seeds and relatively few small seeds.	Capture birds and measure beak depth. 	This is a way to measure a trait that may be subject to natural selection.

3 **THE DATA**

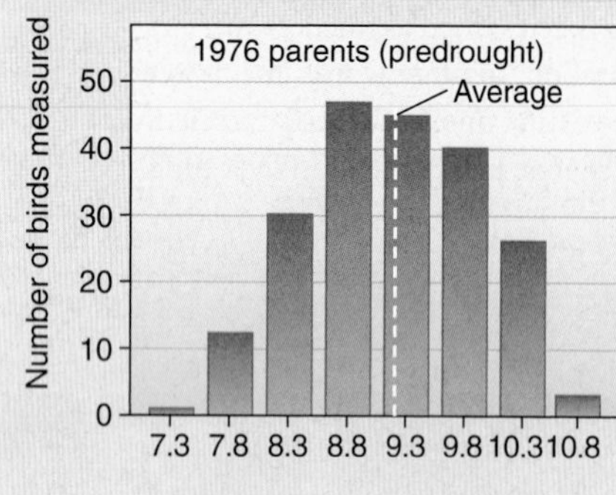

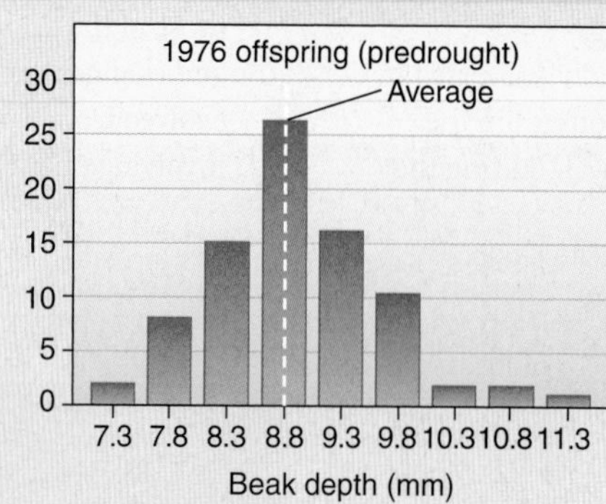

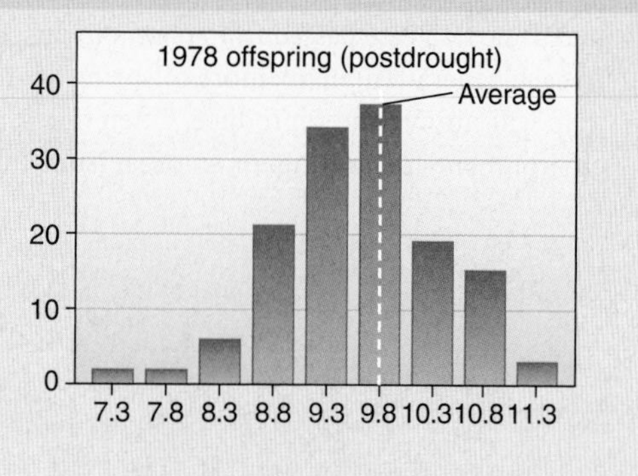

4 **CONCLUSION** Because a drought produced larger seeds, birds with larger beaks were more likely to survive and reproduce. The process of natural selection produced postdrought offspring that had larger beaks compared to predrought offspring.

5 **SOURCE** Grant, B. Rosemary, and Grant, Peter R. 2003. What Darwin's Finches Can Teach Us about the Evolutionary Origin and Regulation of Biodiversity. *Bioscience* 53:965–975.

the plants on Daphne Major tended to produce few of the smaller seeds, which the finches rapidly consumed. Therefore, the finches resorted to eating larger, drier seeds, which are harder to crush. As a result, birds with larger beaks were more likely to survive and reproduce because they were better at breaking open the large seeds. As shown in the data, the average beak depth of birds in the population increased substantially, from 8.8 mm in predrought offspring to 9.8 mm in postdrought offspring. How do we explain these results? According to evolutionary theory, birds with larger beaks were more likely to survive and pass this trait to their offspring. Overall, these results illustrate the power of natural selection to alter the features of a trait—in this case, beak depth—in a given population over time.

Experimental Questions

1. What features of Daphne Major made it a suitable field site for studying the effects of natural selection?
2. Why is beak depth in finches a good trait for a study of natural selection? What environmental conditions were important to allowing the Grants to collect information concerning natural selection?
3. What were the results of the Grants' study following the drought in 1977? What effect did these results have on the theory of evolution?

23.2 Evidence of Evolutionary Change

Learning Outcomes:

1. Summarize the different types of evidence for evolutionary change, including the fossil record, biogeography, convergent traits, selective breeding, and homologies.
2. Provide examples of three types of homologies.

Evidence that supports the theory of evolution has been gleaned from many sources (**Table 23.2**). As we have already seen, the Grants were able to observe changes in a finch population as a result of a drought. Historically, the first evidence of biological evolution came from studies of the fossil record, the distribution of related species on our planet, selective breeding experiments, and the comparison of similar anatomical features in different species. More recently, additional evidence that illustrates the process of evolution has been found at the molecular level. By comparing DNA sequences from many different species, evolutionary biologists have gained great insight into the relationship between the evolution of species and the associated changes in the genetic material. In this section, we will survey the various types of evidence that show the process of evolutionary change.

Table 23.2 Evidence of Biological Evolution

Type of evidence	Description
Studies of natural selection	By following the characteristics of populations over time, researchers have observed how natural selection alters such populations in response to environmental changes (see Figure 23.4).
Fossil record	When fossils are compared according to their age, from oldest to youngest, successive evolutionary change becomes apparent.
Biogeography	Unique species found on islands and other remote areas have arisen because the species in these locations have evolved in isolation from the rest of the world.
Convergent evolution	Two different species from different lineages sometimes become anatomically similar because they occupy similar environments. This indicates that natural selection results in adaptation to a given environment.
Selective breeding	The traits in domesticated species have been profoundly modified by selective breeding (also called artificial selection) in which breeders choose the parents that have desirable traits.
Homologies	
Anatomical	Homologous structures are structures that are anatomically similar to each other because they evolved from a structure in a common ancestor. In some cases, such structures have lost their original function and become vestigial.
Developmental	An analysis of embryonic development often reveals similar features that point to past evolutionary relationships.
Molecular	At the molecular level, certain characteristics are found in all living cells, suggesting that all living species are derived from an interrelated group of common ancestors. In addition, species that are closely related evolutionarily have DNA sequences that are more similar to each other than they are to distantly related organisms.

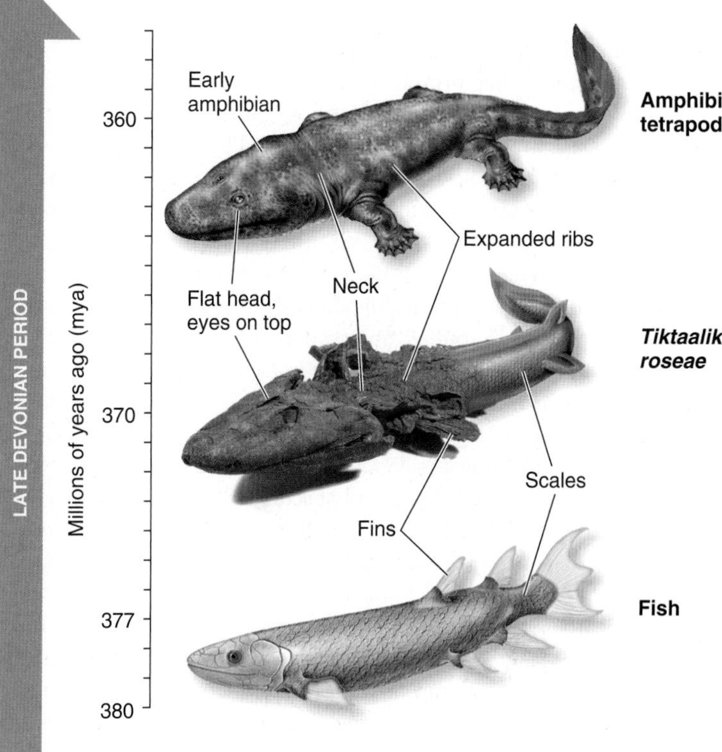

Figure 23.5 A transitional form in the tetrapod lineage. This figure shows two early tetrapod ancestors, a Devonian fish and the transitional form *Tiktaalik roseae*, as well as one of their descendants, an early amphibian. An analysis of the fossils shows that *T. roseae*, also known as a fishapod, had both fish and amphibian characteristics, so it was likely able to survive brief periods out of the water.

Fossils Show Successive Evolutionary Change

As discussed in Chapter 22, the fossil record has provided biologists with evidence of the history of life on Earth. Today, scientists have access to a far more extensive fossil record than was available to Darwin and other scientists of his time. Even though the fossil record is still incomplete, the many fossils that have been discovered provide detailed information regarding evolutionary change in a series of related organisms. When fossils are compared according to their age, from oldest to youngest, successive evolutionary change becomes apparent.

Let's consider a couple of examples in which paleontologists have observed evolutionary change. In 2005, fossils of *Tiktaalik roseae*, nicknamed fishapod, were discovered by paleontologists Ted Daeschler, Neil Shubin, and Farish Jenkins. The discovery of fishapod illuminates one of several steps that led to the evolution of tetrapods, which are animals with four legs. *T. roseae* is called a **transitional form** because it displays an intermediate state between an ancestral form and the form of its descendants (**Figure 23.5**). In this case, the fishapod is a transitional form between fishes, which have fins for locomotion, and tetrapods, which are four-limbed animals. Unlike a true fish, *T. roseae* had a broad skull, a flexible neck, and eyes mounted on the top of its head like a crocodile. Its interlocking rib cage suggests it had primitive lungs. Perhaps the most surprising discovery was that its pectoral fins (those on the side of the body) revealed the beginnings of a primitive wrist and five finger-like bones. These appendages would have allowed *T. roseae* to support its body on shallow river bottoms and lift its head above the water to search for prey and perhaps even move out of the water for short periods. During the Devonian period (417–354 mya), this could have been an important advantage in the marshy floodplains of large rivers.

One of the best-studied observations of evolutionary change through the fossil record is that of the horse family, modern members of which include horses, zebras, and donkeys. These species, which are large, long-legged animals adapted to living in open grasslands, are the remaining descendants of a long lineage that produced many species that have subsequently become extinct since its origin approximately 55 mya. Examination of the horse lineage through fossils provides a particularly interesting case of how evolution involves adaptation to changing environments.

The earliest known fossils of the horse family revealed that the animals were small with short legs and broad feet (**Figure 23.6**). Early horses, such as *Hyracotherium*, lived in wooded habitats and are thought to have browsed on leaves. The fossil record has revealed changes in size, foot anatomy, and tooth morphology among this group of related species over time. Early horses were the size of dogs, whereas modern horses typically weigh more than a half ton. *Hyracotherium*, an early horse, had four toes on its front feet and three on its hind feet. The toes were encased in fleshy pads. By comparison, the feet of modern horses have a single toe, enclosed in a tough, bony hoof. The fossil record shows an increase in the length of the central toe, the development of a bony hoof, and the loss of the other toes. Finally, the teeth of *Hyracotherium* were relatively small compared with those of modern horses. Over the course of millions of years, horse molars have increased in size and developed a complex pattern of ridges.

How do evolutionary biologists explain these changes in horse characteristics? The changes can be attributed to natural selection, which acted on existing variation and resulted in adaptations to changes in global climates. Over North America, where much of horse evolution occurred, changes in climate caused large areas of dense forests to be replaced with grasslands. The increase in size and changes in foot structure enabled horses to escape predators more easily and travel greater distances in search of food. The changes seen in horses' teeth are consistent with a shift from eating the tender leaves of bushes and trees to eating grasses and other vegetation that are abrasive and require more chewing.

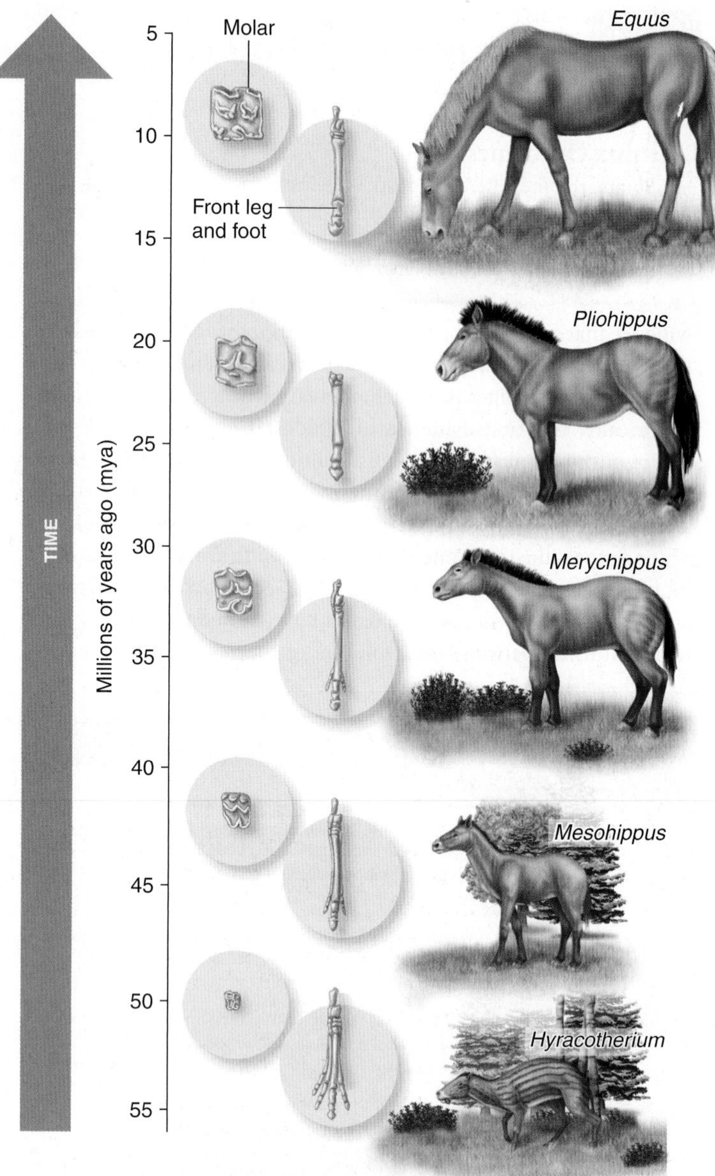

Figure 23.6 Evolutionary changes in horse morphology. Some major changes observed in the fossil record relate to body size, foot anatomy, and tooth morphology. These anatomical changes are hypothesized to be due to adaptations to a changing environment over the last 55 million years. Note: This figure is meant to emphasize general anatomical changes in horse morphology. The evolutionary pathway that produced modern horses involves several branches and is described in Chapter 26.

BIOLOGY PRINCIPLE All species (past and present) are related by an evolutionary history. This diagram shows the morphological changes that occurred in the evolution of the horse.

Biogeography Indicates That Species in a Given Area Have Evolved from Pre-existing Species

Biogeography is the study of the geographic distribution of extinct and living species. Patterns of past evolution are often found in the natural geographic distribution of related species. From such studies, scientists have discovered that isolated continents and island groups have evolved their own distinct plant and animal communities. As mentioned earlier, Darwin observed several species of finches found on the Galápagos Islands that had unique characteristics, such as beak shapes, when compared with similar finches found on the mainland. Scientists now hypothesize these island species evolved from mainland birds that had migrated to the islands and then became adapted to a variety of new feeding habits (see Figure 23.2).

Islands, which are isolated from other large landmasses, provide numerous examples in which geography has played a key role in the evolution of new species. Islands often have many species of plants and animals that are **endemic**, which means they are naturally found only in a particular location. Most endemic island species have closely related relatives on nearby islands or the mainland. For example, consider the island fox (*Urocyon littoralis*), which lives on the Channel Islands located off the coast of Santa Barbara in southern California (**Figure 23.7**). This type of fox is found nowhere else in the world. It weighs about 3–6 pounds and feeds largely on insects, mice, and fruits. The island fox evolved from the mainland gray fox (*Urocyon cinereoargenteus*), which is much larger, usually 7–11 pounds. During the last Ice Age, about 16,000–18,000 years ago, the Santa Barbara channel was frozen and narrow enough for ancestors of the mainland gray fox to cross over to the Channel Islands. When the Ice Age ended, the ice melted and sea levels rose, causing the foxes to be cut off from the mainland. Over the last 16,000–18,000 years, the population of foxes on the Channel Islands evolved into the smaller island fox, which is now considered a different species from the larger gray fox. The gray fox is still found on the mainland. The smaller size of the island fox is an example of island dwarfing, a phenomenon in which the size of large animals on an isolated island shrinks dramatically over many generations. It is the result of natural selection in which a smaller size provides a survival and reproductive advantage, probably because of limited food and other resources.

The evolution of major animal groups is also correlated with known changes in the distribution of landmasses on the Earth. The first mammals arose approximately 200 mya, when the area that is now Australia was still connected to the other continents. However, the first placental mammals, which have a long internal gestation and give birth to well-developed offspring, evolved much later, after continental drift had separated Australia from the other continents (refer back to Figure 22.10). Except for a few species of bats and rodents that have migrated to Australia more recently, Australia lacks any of the larger, terrestrial placental mammals. How do biologists explain this observation? It is consistent with the idea that placental mammals first arose somewhere other than Australia, and that the barrier of a large ocean prevented most terrestrial placental mammals from migrating there. On the other hand, Australia has more than 100 species of kangaroos, koalas, and other marsupials, most of which are not found on any other continent. Marsupials are a group of mammal species in which young are born in a very immature condition and then develop further in the mother's abdominal pouch, which covers the mammary glands. Evolutionary theory is consistent with the idea that the existence of these unique Australian species is due to their having evolved in isolation from the rest of the world for millions of years.

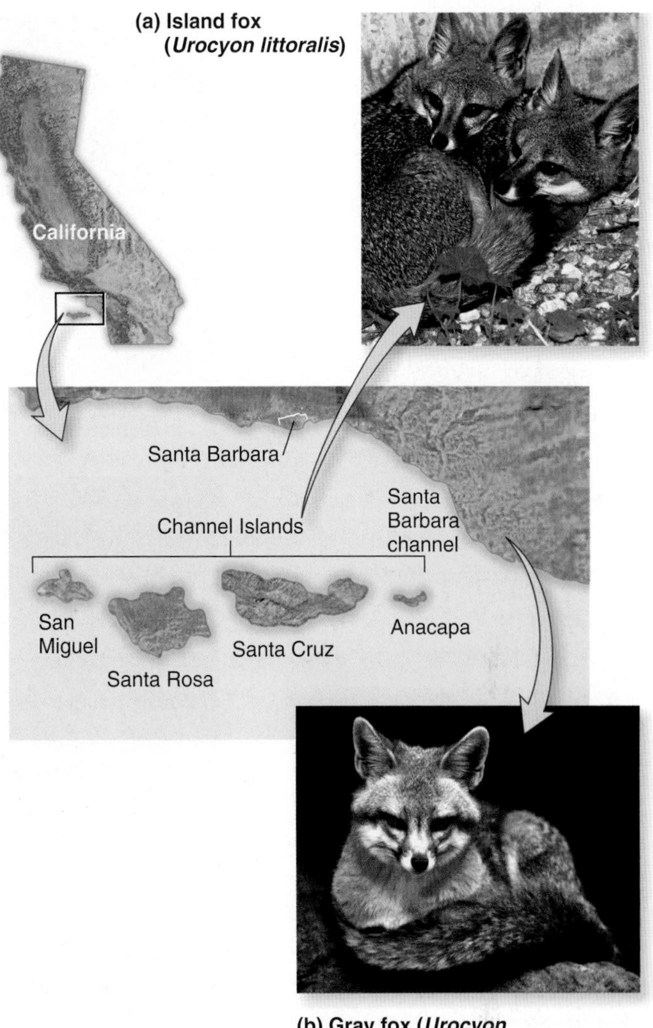

Figure 23.7 The evolution of an endemic island species from a mainland species. **(a)** The smaller island fox found on the Channel Islands evolved from **(b)** the gray fox found on the California mainland.

Concept Check: Explain how geography played a key role in the evolution of the island fox.

Convergent Evolution Suggests Adaptation to the Environment

The process of natural selection is also evident in the study of plants and animals that have similar characteristics, even though they are not closely related evolutionarily. This similarity is the result of **convergent evolution**, in which two species from different lineages have independently evolved similar characteristics because they occupy similar environments. For example, both the giant anteater (*Myrmecophaga tridactyla*), found in South America, and the echidna (*Tachyglossus aculeatus*), found in Australia, have a long snout and tongue. Both species independently evolved these adaptations that enable them to feed on ants (**Figure 23.8a**). The giant anteater is a placental mammal, whereas the echidna is an egg-laying mammal known as a monotreme, so they are not closely related evolutionarily.

Another example of convergent evolution involves aerial rootlets found in vines such as English ivy (*Hedera helix*) and wintercreeper (*Euonymus fortunei*) (**Figure 23.8b**). Based on differences in their structures, these aerial rootlets appear to have developed independently as an effective means of clinging to the support on which a vine attaches itself.

A third example of convergent evolution is revealed by the molecular analysis of fishes that live in very cold water. Antifreeze

(a) The long snouts and tongues of the giant anteater (left) and the echidna (right) allow them to feed on ants.

(b) The aerial rootlets of English ivy (left) and wintercreeper (right) enable them to climb up supports.

(c) The sea raven (left) and the longhorn sculpin (right) have antifreeze proteins that enable them to survive in frigid waters.

Figure 23.8 Examples of convergent evolution. All three pairs of species shown in this figure are not closely related evolutionarily but occupy similar environments, suggesting that natural selection results in similar adaptations to a particular environment.

Concept Check: Can you think of another example in which two species that are not closely related have a similar adaptation?

proteins enable certain species of fishes to survive the subfreezing temperatures of Arctic and Antarctic waters by inhibiting the formation of ice crystals in body fluids. Researchers have determined that these fishes are an interesting case of convergent evolution (Figure 23.8c). Among different species of fishes, one of five different genes has independently evolved to produce antifreeze proteins. For example, in the sea raven (*Hemitripterus americanus*), the antifreeze protein is rich in the amino acid cysteine, and the secondary structure of the protein is in a β sheet conformation. In contrast, the antifreeze protein in the longhorn sculpin (*Trematomus nicolai*) is encoded by an entirely different gene. The antifreeze protein in this species is rich in the amino acid glutamine, and the secondary structure of the protein is largely composed of α helices.

The similar characteristics in the examples shown in Figure 23.8—for example, the snouts of the anteater and the echidna—are called **analogous structures** or convergent traits. They represent cases in

(a) Bulldog

(b) Greyhound

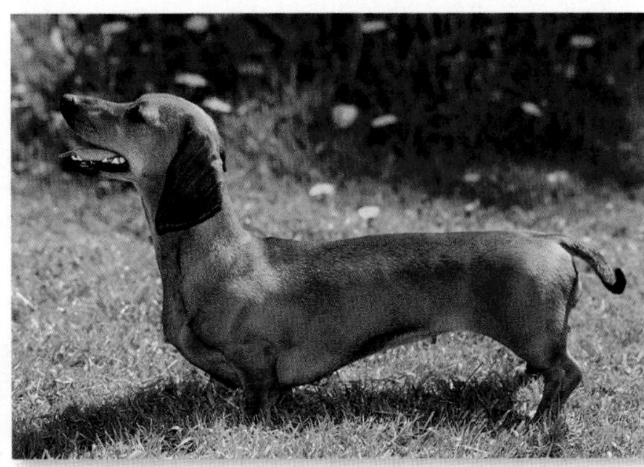

(c) Dachshund

Figure 23.9 Common breeds of dogs that have been obtained by selective breeding. By selecting individuals carrying the alleles that influence traits desirable to humans, dog breeders have produced breeds with distinctive features. All the dogs in this figure carry the same kinds of genes (for example, genes that affect their size, shape, and fur color). However, the alleles for many of these genes are different among these dogs, thereby allowing dog breeders to select for or against them and produce breeds with strikingly different phenotypes.

which characteristics have arisen independently, two or more times, because different species have occupied similar types of environments on the Earth.

Selective Breeding Is a Human-Driven Form of Selection

The term **selective breeding** refers to programs and procedures designed to modify traits in domesticated species. This practice, also called **artificial selection**, is related to natural selection. In forming his theory of evolution, Charles Darwin was influenced by his observations of selective breeding by pigeon breeders. The primary difference between natural and artificial selection is how the parents are chosen. Natural selection occurs because of genetic variation in reproductive success. Organisms that are able to survive and reproduce are more likely to pass their genes to future generations. Environmental factors often determine which individuals will be successful parents. In artificial selection, the breeder chooses as parents those individuals with traits that are desirable from a human perspective.

The underlying phenomenon that makes selective breeding possible is genetic variation. Within a population, variation may exist in a trait of interest. For selective breeding to be successful, the underlying cause of the phenotypic variation is usually related to differences in **alleles**, variant forms of a particular gene, that determine the trait. The breeder chooses parents with desirable phenotypic characteristics. For centuries, humans have employed selective breeding to obtain domesticated species with interesting or agriculturally useful characteristics. For example, many common breeds of dog are the result of selective breeding strategies (Figure 23.9). All dogs are members of the same species, *Canis lupus*, subspecies *familiaris*, so they can interbreed to produce offspring. Selective breeding can dramatically modify the traits in a species. When you compare certain breeds of dogs (for example, a greyhound and a dachshund), they hardly look like members of the same species! Recent work in 2007 by American geneticist Nathan Sutter and colleagues indicates that the size of dogs may be determined by alleles in the *Igf1* gene that encodes a growth hormone called insulin-like growth factor 1. A particular allele of this gene was found to be common to all small breeds of dogs and nearly absent from very large breeds, suggesting that this allele is a major contributor to body size in small breeds of dogs.

Likewise, most of the food we eat—including products such as grains, fruits, vegetables, meat, milk, and juices—is obtained from species that have been profoundly modified by selective breeding strategies. For example, certain characteristics in the wild mustard plant (*Brassica oleracea*) have been modified by selective breeding to produce several varieties of domesticated crops, including broccoli, Brussels sprouts, cabbage, and cauliflower (Figure 23.10). The wild mustard plant is native to Europe and Asia, and plant breeders began to modify its traits approximately 4,000 years ago. As seen here, certain traits in the domestic strains differ dramatically from those of the original wild species. These varieties are all members of the same species. They can interbreed to produce viable offspring. For example, in the grocery store you may have seen broccoflower, a vegetable produced from a cross between broccoli and cauliflower.

As another example, Figure 23.11 shows the results of a selective breeding experiment on corn begun at the University of Illinois Agricultural Experiment Station in 1896, several years before the rediscovery of Mendel's laws. This study began with 163 ears of corn with an oil content ranging from 4 to 6%. In each of 80 succeeding generations, corn plants were divided into two separate groups. In one group, members with the highest oil content in the kernels were chosen as parents of the next generation. In the other group, members with the lowest oil content were chosen. After many generations, the oil content in the first group rose to over 18%. In the other group, it dropped to less than 1%. These results show that selective breeding can modify a trait in a very directed manner.

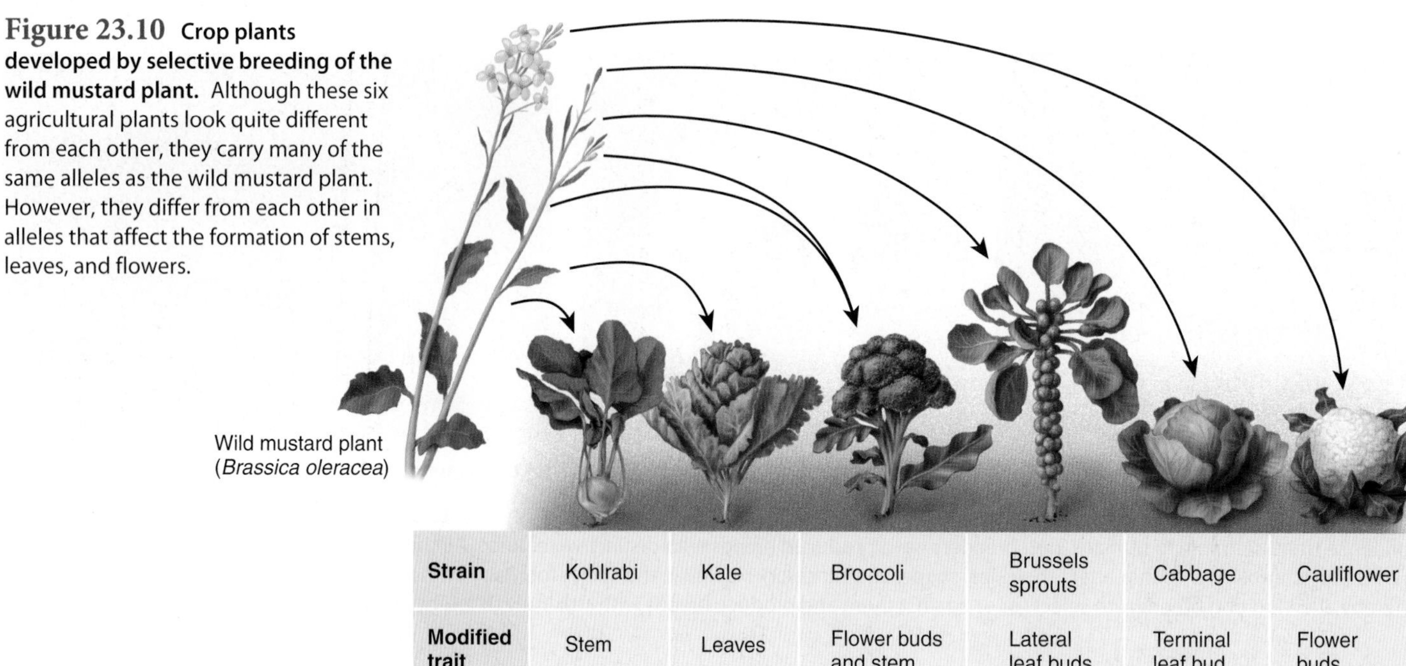

Strain	Kohlrabi	Kale	Broccoli	Brussels sprouts	Cabbage	Cauliflower
Modified trait	Stem	Leaves	Flower buds and stem	Lateral leaf buds	Terminal leaf bud	Flower buds

Figure 23.10 Crop plants developed by selective breeding of the wild mustard plant. Although these six agricultural plants look quite different from each other, they carry many of the same alleles as the wild mustard plant. However, they differ from each other in alleles that affect the formation of stems, leaves, and flowers.

A Comparison of Homologies Shows Evolution of Related Species from a Common Ancestor

Let's now consider other widespread observations of the process of evolution among living organisms. In biology, the term **homology** refers to a similarity that occurs due to descent from a common ancestor. Two species may have a similar trait because the trait was originally found in a common ancestor. As described next, such homologies may involve anatomical, developmental, or molecular features.

Anatomical Homologies As noted by Theodosius Dobzhansky, many observations regarding the features of living organisms simply cannot be understood in any meaningful scientific way except as a result of evolution. A comparison of vertebrate anatomy is a case in point. An examination of the limbs of modern vertebrate species reveals similarities that indicate the same set of bones has undergone evolutionary changes, becoming modified to perform different functions in different species. As seen in **Figure 23.12**, the forelimbs of vertebrates have a strikingly similar pattern of bone arrangements. These are termed **homologous structures**—structures that are similar to each other because they are derived from a common ancestor. The forearm has developed different functions among various vertebrates, including grasping, walking, flying, swimming, and climbing. The theory of evolution explains how these animals have descended from a common ancestor and how natural selection has resulted in modifications to the structure of the original set of bones in ways that ultimately allowed them to be used for several different functions.

Another result of evolution is the phenomenon of **vestigial structures**, anatomical features that have no current function but resemble structures of their presumed ancestors (**Table 23.3**). An interesting case is found in humans. People have a complete set of muscles for moving their ears, even though most people are unable to do so. By comparison, many modern mammals can move their ears, and presumably this was an important trait in a distant human ancestor. Why would organisms have structures that are no longer useful? Within the context of evolutionary theory, vestigial structures are evolutionary relics. Organisms having vestigial structures share a common ancestry with organisms in which the structure is functional. Natural selection maintains functional structures in a population of individuals. However, if a species changes its lifestyle so the structure loses its purpose, the selection that would normally keep the structure in a functional condition is no longer present. When this occurs, the structure may degenerate over the course of many generations due to the accumulation of mutations that limit its size and shape. Natural selection may eventually eliminate such traits due to the inefficiency and cost of producing unused structures.

Developmental Homologies Another example of homology is the way that animals undergo embryonic development. Species that differ substantially at the adult stage often bear striking similarities during early stages of embryonic development. These temporary similarities are called developmental homologies. In addition, evolutionary history is revealed during development in certain organisms, such as vertebrates. For example, if we consider human development, several features are seen in the embryo that are not present at birth. Human embryos have rudimentary gill ridges like a fish embryo, even though human embryos receive oxygen via the umbilical cord. The presence of gill ridges indicates that humans evolved from an aquatic species that had gill slits. A second observation is that every human embryo has a bony tail. It is difficult to see the advantage of such a structure in utero, but easier to understand its presence assuming that an ancestor of the human lineage possessed a tail. These observations, and many others, illustrate that closely related species share similar developmental pathways.

Molecular Homologies Our last examples of homology due to evolution involve molecular studies. Similarities between organisms at the molecular level due to descent from a common ancestor are

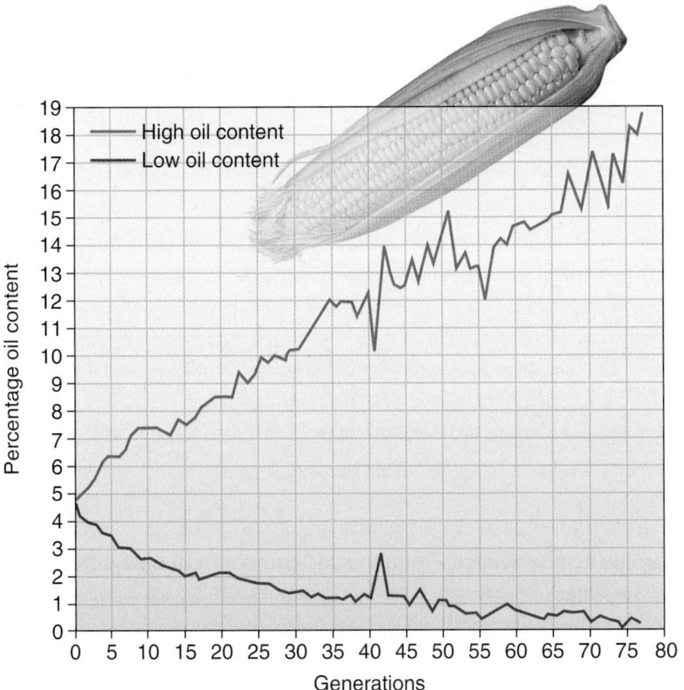

Figure 23.11 Results of selective breeding for oil content in corn plants. In this example, corn plants were selected for breeding based on high or low oil content of the kernels. Over the course of many generations, this had a major influence on the amount of corn oil (an agriculturally important product) made by the two groups of plants.

BIOLOGY PRINCIPLE Populations of organisms evolve from one generation to the next. This study illustrates how a trait in corn, namely oil content, may change over time due to human intervention.

Concept Check: *When comparing Figures 23.9, 23.10, and 23.11, what general effects of artificial selection do you observe?*

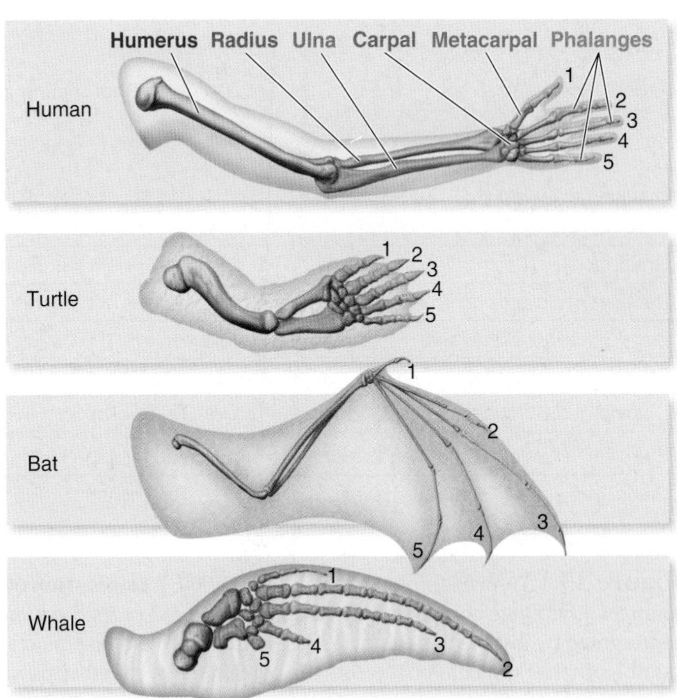

Figure 23.12 An example of anatomical homology: Homologous structures found in vertebrates. The same set of bones is found in the human arm, turtle arm, bat wing, and whale flipper, although their relative sizes and shapes differ significantly. This homology suggests that all of these animals evolved from a common ancestor.

BIOLOGY PRINCIPLE Structure determines function. These homologous sets of bones have evolved into somewhat different structures due to differences in their functions in humans, turtles, bats, and whales.

called **molecular homologies**. For example, all living species use DNA to store information and rely on the genetic code to translate mRNA into proteins. Furthermore, certain biochemical pathways are found in all or nearly all species, although minor changes in the structure and function of proteins involved in these pathways have occurred. For example, all species that use oxygen, which constitutes the great majority of species on our planet, have similar proteins that together make up an electron transport chain and an ATP synthase (refer back to Figure 7.8). In addition, nearly all living organisms can break down glucose via a metabolic pathway that is described in Chapter 7. How do we explain these types of observations? Taken together, they indicate that such molecular phenomena arose very early in the origin of life and have been passed to all or nearly all modern forms.

A very compelling observation at the molecular level indicating that modern life-forms are derived from an interrelated group of common ancestors is revealed by analyzing genetic sequences. The same type of gene is often found in diverse organisms. Furthermore, the degree of similarity between genetic sequences from different species reflects the evolutionary relatedness of those species.

Table 23.3 Examples of Vestigial Structures in Animals

Organism	Vestigial structure(s)
Humans	Tail bone and muscles to wiggle ears in adult
Boa constrictors	Skeletal remnants of hip and hind leg bones
Whales	Skeletal remnants of a pelvis
Manatees	Fingernails on the flippers
Hornbills and cuckoos	Fibrous cords that were derived from the common carotid arteries. In certain families of birds, both of the common carotid arteries are nonfunctional, fibrous cords. Their vascular function has been assumed by other vessels.

As an example, let's consider the *p53* gene, which encodes the p53 protein—a checkpoint protein of the cell cycle (see Chapter 14, Figure 14.15). **Figure 23.13** shows a short amino acid sequence that makes up part of the p53 protein from a variety of species, including five mammals, one bird, and three fish. The top sequence is the human p53 sequence, and the right column describes the percentages of amino acids within the entire sequence that are identical

	Short amino acid sequence within the p53 protein	Percentages of amino acids in the whole p53 protein that are identical to human p53
Human (*Homo sapiens*)	Val Pro Ser Gln Lys Thr Tyr Gln Gly Ser Tyr Gly Phe Arg Leu Gly Phe Leu His Ser Gly Thr	100
Rhesus monkey (*Macaca mulatta*)	Val Pro Ser Gln Lys Thr Tyr His Gly Ser Tyr Gly Phe Arg Leu Gly Phe Leu His Ser Gly Thr	95
Green monkey (*Cercopithecus aethiops*)	Val Pro Ser Gln Lys Thr Tyr His Gly Ser Tyr Gly Phe Arg Leu Gly Phe Leu His Ser Gly Thr	95
Rabbit (*Oryctolagus cuniculus*)	Val Pro Ser Gln Lys Thr Tyr His Gly Asn Tyr Gly Phe Arg Leu Gly Phe Leu His Ser Gly Thr	86
Dog (*Canis lupus familiaris*)	Val Pro Ser Pro Lys Thr Tyr Pro Gly Thr Tyr Gly Phe Arg Leu Gly Phe Leu His Ser Gly Thr	80
Chicken (*Gallus gallus*)	Val Pro Ser Thr Glu Asp Tyr Gly Gly Asp Phe Asp Phe Arg Val Gly Phe Val Glu Ala Gly Thr	53
Channel catfish (*Ictalurus punctatus*)	Val Pro Val Thr Ser Asp Tyr Pro Gly Leu Leu Asn Phe Thr Leu His Phe Gln Glu Ser Ser Gly	48
European flounder (*Platichthys flesus*)	Val Pro Val Val Thr Asp Tyr Pro Gly Glu Tyr Gly Phe Gln Leu Arg Phe Gln Lys Ser Gly Thr	46
Congo puffer fish (*Tetraodon miurus*)	Val Pro Val Thr Thr Asp Tyr Pro Gly Glu Tyr Gly Phe Lys Leu Arg Phe Gln Lys Ser Gly Thr	41

Figure 23.13 An example of genetic homology: A comparison of a short amino acid sequence within the p53 protein from nine different animals. This figure compares a short region of the p53 protein, a tumor suppressor that plays a role in preventing cancer. Amino acids are represented by three-letter abbreviations. The orange-colored amino acids in the sequences are identical to those in the human sequence. The numbers in the right column indicate the percentage of amino acids within the whole p53 protein that is identical with the human p53 protein, which is 393 amino acids in length. For example, 95% of the amino acids, or 373 of 393, are identical between the p53 sequence found in humans and in Rhesus monkeys.

Concept Check: *In the sequence shown in this figure, how many amino acid differences occur between the following pairs: Rhesus and green monkeys, Congo puffer fish and European flounder, and Rhesus monkey and Congo puffer fish? What do these differences tell you about the evolutionary relationships among these four species?*

BioConnections: *Look back at Table 21.5. How are genetic sequences that are retrieved from a database using the BLAST program correlated with the evolutionary relatedness of the species?*

to those in the entire human sequence. Amino acids in the other species that are identical to those in humans are highlighted in orange. The sequences from the two monkeys are the most similar to those in humans, followed by the other two mammalian species (rabbit and dog). The three fish sequences are the least similar to the human sequence, but the fish sequences tend to be similar to each other.

Taken together, the data shown in Figure 23.13 illustrate two critical points about gene evolution. First, specific genes are found in a diverse array of species such as mammals, birds, and fishes. Second, the sequences of closely related species tend to be more similar to each other than they are to distantly related species. The mechanisms for this second observation are discussed in the next section.

23.3 The Molecular Processes That Underlie Evolution

Learning Outcomes:

1. Explain how paralogs and orthologs are produced.
2. Describe how new types of genes arise via exon shuffling.
3. Distinguish between vertical evolution and horizontal gene transfer.

Historically, the study of evolution was based on a comparison of the anatomies of extinct and modern species to identify similarities between related species. However, the advent of molecular approaches for analyzing DNA sequences has revolutionized the field of evolutionary biology. Now we can analyze how changes in the genetic material are associated with changes in phenotype. In this section, we will examine some of the molecular changes in the genetic material that reveal evolutionary change.

Homologous Genes Are Derived from a Common Ancestral Gene

Two or more genes derived from the same ancestral gene are called **homologous genes**. The analysis of homologous genes reveals evidence of evolutionary change at the molecular level. How do homologous genes arise? As an example, let's consider a gene in two different species of bacteria that encodes a transport protein involved in the uptake of metal ions into bacterial cells. Homologous genes that are in different species are termed **orthologs**. Millions of years ago, these two species had a common ancestor (**Figure 23.14**). Over time, the common ancestor diverged into additional species, eventually evolving into *Escherichia coli*, *Clostridium acetylbutylicum*, and many other species. Since this divergence, the metal transporter gene has accumulated mutations that altered its sequence, though

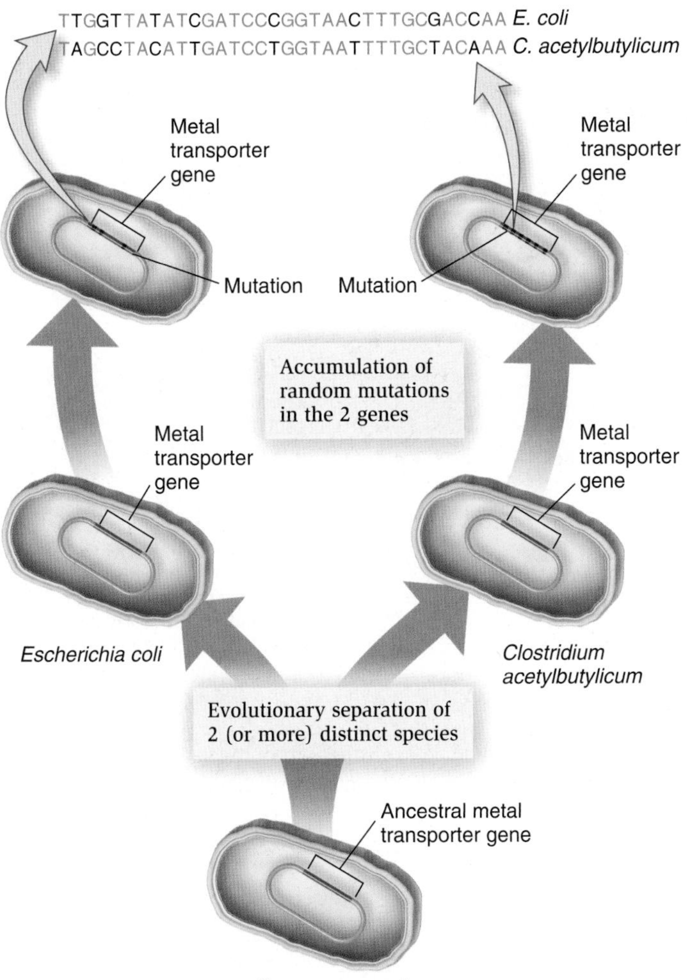

Figure 23.14 The evolution of orthologs, homologous genes from different species. After the two species diverged from each other, the genes accumulated random mutations that resulted in similar, but not identical, gene sequences called orthologs. These orthologs in *E. coli* and *C. acetylbutylicum* encode metal transporters. Only one of the two DNA strands is shown from each of the genes. Bases that are identical between the two genes are shown in orange.

Concept Check: *Why do these orthologs have similar gene sequences? Why aren't they identical?*

the similarity between the *E. coli* and the *C. acetylbutylicum* genes remains striking. In this case, the two sequences are similar because they were derived from the same ancestral gene, but they are not identical due to the independent accumulation of different random mutations.

Gene Duplications Produce Gene Families

Evidence of evolutionary change is also found within a single species. Two or more homologous genes within a single species are termed **paralogs** of each other. Rare gene duplication events produce multiple copies of a gene and ultimately lead to the formation of a **gene family**—a set of paralogs within the genome of a single species. A well-studied example of a gene family is the globin gene family in humans, which is composed of 14 genes that are hypothesized to be derived from a single ancestral globin gene (refer back to Figure 21.8). According to an evolutionary analysis, the ancestral globin gene first duplicated between 500 and 600 mya. Since that time, additional duplication events and chromosomal rearrangements have produced the current number of 14 genes on three different human chromosomes.

What is the advantage of a gene family? Even though all of the globin polypeptides are subunits of proteins that play a role in oxygen binding, the accumulation of changes in the various family members has produced globins that differ in the timing of their expression and in their functional properties. The various globin genes are expressed at different stages of development in humans. The functional differences of the globin proteins correlate with the oxygen transport needs of humans during the embryonic, fetal, and postpartum stages of life (refer back to Figure 13.3).

What is the evolutionary significance of the globin gene family regarding adaptation? On land, egg cells and small embryos are very susceptible to drying out if they are not protected in some way. Species such as birds and reptiles lay eggs with a protective shell around them. Most mammals, however, have become adapted to a terrestrial environment by evolving internal gestation. The ability to develop young internally has been an important factor in the survival and proliferation of humans and other mammals. The embryonic and fetal forms of hemoglobin allow the embryo and fetus to capture oxygen from the bloodstream of the mother.

Genomes & Proteomes Connection

New Genes in Eukaryotes Have Evolved via Exon Shuffling

Thus far, we have considered how evolutionary change results in the formation of homologous genes, either orthologs or paralogs. Evolutionary mechanisms are also revealed when exons, the parts of genes that encode protein domains, are compared within a single species. Many proteins, particularly those found in eukaryotic species, have a modular structure composed of two or more domains with different functions. By comparing the modular structure of eukaryotic proteins with the genes that encode them, geneticists have discovered that each domain tends to be encoded by one exon or by a series of two or more adjacent exons.

During the evolution of eukaryotic species, many new genes have been produced by a type of mutation known as **exon shuffling**. During this process, an exon and parts of the flanking introns from one gene are inserted into another gene, thereby producing a new gene that encodes a protein with an additional domain (Figure 23.15). This process may also involve the duplication and rearrangement of exons. Exon shuffling can result in novel genes that express proteins with new combinations of functional domains. Such proteins may alter traits in the organism and therefore be subjected to natural selection.

Exon shuffling may occur by more than one mechanism. One possibility is that a double crossover could promote the insertion of an exon into another gene (as seen in Figure 23.15). Alternatively, transposable elements, described in Chapter 21, may promote the movement of exons into other genes.

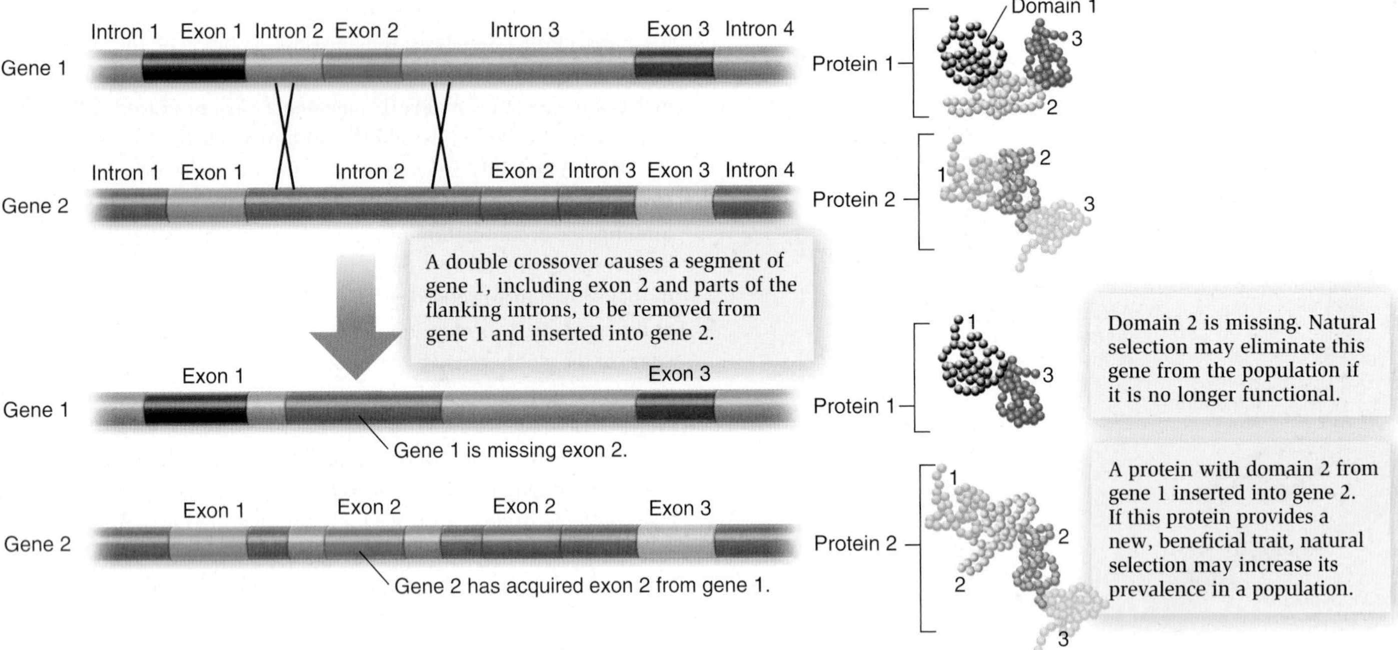

Figure 23.15 The process of exon shuffling. In this example, a segment of one gene containing an exon and part of the flanking introns has been inserted into another gene. A rare, abnormal double crossover event may cause this to happen. Exon shuffling results in proteins that have new combinations of domains and new combinations of functions.

Concept Check: *What is the evolutionary advantage of exon shuffling?*

Horizontal Gene Transfer Contributes to the Evolution of Species

At the molecular level, the type of evolutionary change depicted in Figures 23.13 through 23.15 is called **vertical evolution**. In these cases, new species arise from pre-existing species by the accumulation of genetic changes, such as gene mutations, gene duplications, and exon shuffling. Vertical evolution involves genetic changes in a series of ancestors that form a lineage. In addition to vertical evolution, species accumulate genetic changes by **horizontal gene transfer**—a process in which an organism incorporates genetic material from another organism without being the offspring of that organism. Horizontal gene transfer can involve the exchange of genetic material between members of the same species or different species.

How does horizontal gene transfer occur? Figure 23.16 illustrates one possible mechanism for horizontal gene transfer. In this example, a paramecium, which is a eukaryotic organism, has engulfed a bacterial cell. During the degradation of the bacterium in a phagocytic vesicle, a bacterial gene escapes to the nucleus of the cell, where it is inserted into one of the chromosomes. In this way, a gene has been transferred from a bacterial species to a eukaryotic species. By analyzing gene sequences among many different species, researchers have discovered that horizontal gene transfer is a common phenomenon. This process can occur from bacteria and archaea to eukaryotes, from eukaryotes to bacteria and archaea, between different species of bacteria and archaea, and between different species of eukaryotes. Therefore, when we view evolution, it is not simply a matter of one species evolving into one or more new species via the accumulation of random mutations. It also involves the horizontal transfer of genes

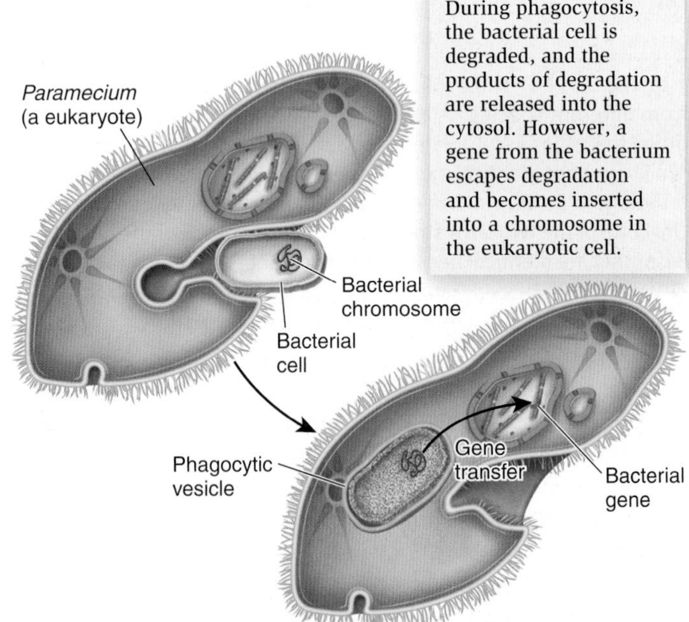

Figure 23.16 Horizontal gene transfer from a bacterium to a eukaryote. In this example, a bacterium is engulfed by a paramecium (a ciliated protist), and a bacterial gene is transferred to one of the paramecium's chromosomes.

BioConnections: *Look back at Table 18.3. What are three mechanisms of gene transfer that could result in horizontal gene transfer between two different bacterial species?*

among different species, enabling those species to acquire new traits that foster the evolutionary process.

Horizontal gene transfer among bacterial species is relatively widespread. As discussed in Chapter 18, bacterial species may carry out three natural mechanisms of gene transfer known as conjugation, transformation, and transduction. By analyzing the genomes of bacterial species, scientists have determined that many genes within a given bacterial genome are derived from horizontal gene transfer. Genome studies have suggested that as much as 20–30% of the variation in the genetic composition of modern bacterial species can be attributed to this process. The roles of the genes acquired by horizontal gene transfer are quite varied, though they commonly involve functions that are beneficial for survival and reproduction. These include genes that confer antibiotic resistance, the ability to degrade toxic compounds, and pathogenicity (the ability to cause disease).

Evolution at the Genomic Level Involves Changes in Chromosome Structure and Number

Thus far, we have considered several ways a species might acquire new genetic variation. These include mutations within pre-existing genes, gene duplications that produce gene families, exon shuffling, and horizontal gene transfer. Evolution also involves changes in chromosome structure and number. When comparing the chromosomes of closely related species, changes in chromosome structure and/or number are common.

As an example, Figure 23.17 compares the banding patterns of the three largest chromosomes in humans and the corresponding chromosomes in chimpanzees, gorillas, and orangutans. (Refer back to Chapter 15, Figure 15.1 for an example of chromosome banding.) The banding patterns in the chromosomes are strikingly similar because these species are closely related evolutionarily. Chromosome 1 looks very similar in all species. Even so, you can see some interesting differences. Humans have one large chromosome 2, but this chromosome is divided into two separate chromosomes in the other three species. This explains why human cells have 23 pairs of chromosomes, whereas cells of chimpanzees, gorillas, and orangutans have 24. The fusion of the two smaller chromosomes during the development of the human lineage may have caused this difference in chromosome number. Another interesting change in chromosome structure is seen in chromosome 3. The banding patterns among humans, chimpanzees, and gorillas are very similar, but the orangutan has a large inversion that flips the order of the bands in the centromeric region. As discussed in Chapter 25, changes in chromosome structure and number may affect the ability of two organisms to breed with one another. In this way, such changes have been important in the establishment of new species.

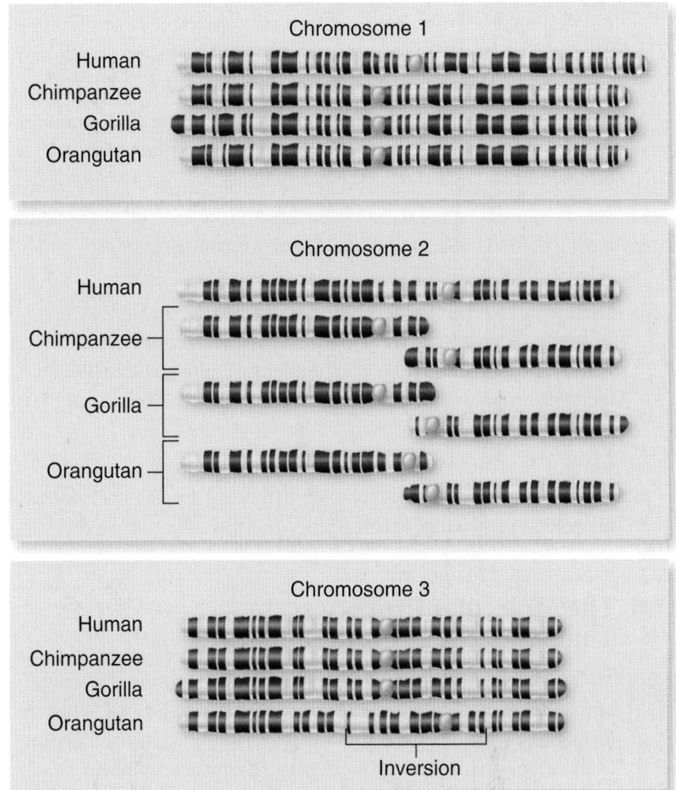

Figure 23.17 Evolutionary changes in chromosome structure and number found in primates. This figure is a comparison of the three largest human chromosomes and the corresponding chromosomes in the chimpanzee, gorilla, and orangutan. It is a schematic drawing of Giemsa-stained chromosomes. The differences between these chromosomes illustrate the changes that have occurred during the evolution of these related primate species.

Concept Check: Describe two changes in chromosome structure that have occurred among these chromosomes.

Summary of Key Concepts

23.1 The Theory of Evolution

- Evolution is a heritable change in one or more characteristics of a population from one generation to the next.
- Charles Darwin proposed the theory of evolution based on his understanding of geology and population growth and his observations of species in their natural settings. His voyage on the *Beagle*, during which he studied many species, including finches on the Galápagos Islands, was particularly influential in the development of his ideas (Figure 23.1, Table 23.1).
- Darwin expressed his theory of evolution as descent with modification through variation and natural selection. As a result of natural selection, genetic variation changes from generation to generation to produce populations of organisms with traits (adaptations) that favor greater reproductive success (Figure 23.2).
- The Grants' research on finches showed how differences in beak size (a heritable trait) were driven by natural selection (Figures 23.3, 23.4).

23.2 Evidence of Evolutionary Change

- Evidence of evolutionary change is found in studies of natural selection, the fossil record, biogeography, convergent evolution, selective breeding, and homologies (Table 23.2).
- Fossils provide evidence of evolutionary change in a series of related organisms. The fossil record often reveals transitional forms that link past ancestors to modern species (Figures 23.5, 23.6).
- Biogeography provides information on the geographic distribution of related species. When populations become isolated on islands or continents, they often evolve into new species (Figure 23.7).
- In convergent evolution, independent adaptations result in similar characteristics, or analogous structures, because different species occupy similar environments (Figure 23.8).
- Selective breeding, the selecting and breeding of individual organisms having desired traits, is a human-driven form of selection (Figures 23.9, 23.10, 23.11).
- Homologies are similarities that occur due to descent from a common ancestor. The set of bones in the forearms of vertebrates is an example of an anatomical homology. Homologies can also be seen during embryonic development and at the molecular level (Figures 23.12, 23.13).
- Vestigial structures, structures that were functional in an ancestor but no longer have a useful function in modern species, are evidence of evolutionary change (Table 23.3).

23.3 The Molecular Processes That Underlie Evolution

- Molecular evolution is the process of evolution at the level of genes and proteins. Molecular processes that underlie evolution include the formation of orthologs and paralogs, exon shuffling, horizontal gene transfer, and changes in chromosome structure and number.
- Orthologs are homologous genes in different species that have accumulated random mutations over time (Figure 23.14).
- Paralogs are homologous genes in the same species that are produced by gene duplication events. Gene duplication can result in the formation of a gene family such as the globin gene family, which supported the evolutionary adaptation of internal gestation.
- Exon shuffling is a process in which exons from one gene are inserted into another gene, producing proteins that have additional functions (Figure 23.15).
- Another mechanism that produces genetic variation is horizontal gene transfer, in which genetic material is transferred from one organism to another organism that is not its offspring. Such genetic changes are subject to natural selection (Figure 23.16).
- Molecular evolution can also involve changes in chromosome structure and number (Figure 23.17).

Assess and Discuss

Test Yourself

1. A change in one or more characteristics of a population that is heritable and occurs from one generation to the next is called
   a. natural selection.
   b. sexual selection.
   c. population genetics.
   d. evolution.
   e. inheritance of acquired characteristics.
2. Lamarck's vision of evolution differed from Darwin's in that Lamarck believed
   a. living things evolve in an upward direction.
   b. behavioral changes modify heritable traits.
   c. genetic differences among individuals in the population allow for evolution.
   d. a and b only.
   e. none of the above.
3. Which of the following scientists influenced Darwin's views on the nature of population growth?
   a. Cuvier c. Lyell e. Wallace
   b. Malthus d. Hutton
4. An evolutionary change in which a population of organisms changes its characteristics over many generations in ways that make it better suited to its environment is
   a. natural selection. d. evolution.
   b. an adaptation. e. both a and c.
   c. an acquired characteristic.
5. Vestigial structures are anatomical structures
   a. that have more than one function.
   b. that were functional in an ancestor but no longer have a useful function.
   c. that look similar in different species but have different functions.
   d. that have the same function in different species but have very different appearances.
   e. of the body wall.
6. Which of the following is an example of a developmental homology seen in human embryonic development and other vertebrate species that are not mammals?
   a. gill ridges c. tail e. all of the above
   b. umbilical cord d. both a and c
7. Two or more homologous genes found within a particular species are called
   a. homozygous. c. paralogs. e. duplicates.
   b. orthologs. d. alleles.

8. The phenomenon of exon shuffling
   a. produces new gene products by changing the pattern of intron removal in pre-mRNA.
   b. produces new genes by inserting exons and flanking introns into a different gene sequence, thereby introducing a new domain in a protein.
   c. deletes one or more bases in a single gene.
   d. rearranges the introns in a particular gene, producing new proteins.
   e. both a and d.
9. Horizontal gene transfer
   a. is a process in which an organism incorporates genetic material from another organism without being the offspring of that organism.
   b. can involve the exchange of genetic material among individuals of the same species.
   c. can involve the exchange of genetic material among individuals of different species.
   d. can be all of the above.
   e. can be a and b only.
10. Genetic variation can occur as a result of
   a. random mutations in genes.
   b. exon shuffling.
   c. gene duplication.
   d. horizontal gene transfer.
   e. all of the above.

Conceptual Questions

1. Evolution that results in adaptation is rooted in two phenomena: genetic variation and natural selection. In a very concise way (three sentences or less), describe how genetic variation and natural selection can bring about evolution.
2. What is convergent evolution? How does it support the theory of evolution?
3. A principle of biology is that *populations of organisms evolve from one generation to the next.* Explain how the homologous forelimbs of vertebrates indicate that populations evolve from one generation to the next.

Collaborative Questions

1. The term natural selection is sometimes confused with the term evolution. Discuss the meanings of these two terms. Explain how the terms are different and how they are related to each other.
2. Make a list of the observations made by biologists that support the theory of evolution. Which of the observations on your list do you find the most convincing and the least convincing?

Online Resource

www.brookerbiology.com

Stay a step ahead in your studies with animations that bring concepts to life and practice tests to assess your understanding. Your instructor may also recommend the interactive eBook, individualized learning tools, and more.

Population Genetics

24

Chapter Outline

24.1 Genes in Populations
24.2 Natural Selection
24.3 Sexual Selection
24.4 Genetic Drift
24.5 Migration and Nonrandom Mating
Summary of Key Concepts
Assess and Discuss

Colorful African cichlids. Color is a factor that influences the choice of mates in populations of cichlids.

Kimbareta, age 19, lives in the Democratic Republic of Congo (formerly Zaire) with his parents, one brother, and two sisters. Kimbareta has sickle cell disease, which causes his red blood cells to occasionally form a crescent or sickled shape. The sickled cells may block the flow of blood through his vessels. This results in tissue and organ damage along with painful episodes (also called crises) involving his arms, legs, chest, and abdomen. In some people with this disease, stroke may even occur. Sickle cell disease follows a recessive pattern of inheritance. It is caused by a mutation in a gene that encodes β-globin, a subunit of hemoglobin that carries oxygen in the red blood cells.

Many different recessive diseases have been identified by geneticists. Most of them are very rare. However, in the village where Kimbareta lives, sickle cell disease is surprisingly common. Nearly 2% of the inhabitants have the disease—an incidence that is similar to other places in the country. How can we explain such a high occurrence of a serious inherited disease? If natural selection tends to eliminate detrimental genetic variation, as we saw in Chapter 23, why does the sickle cell allele persist in this population? As we will see later, biologists have discovered that the effect of the allele in heterozygotes is the underlying factor. Heterozygotes, who carry one copy of the sickle cell allele and one copy of the more common (non-disease-causing) allele, have an increased resistance to malaria.

Population genetics is the study of genes and genotypes in a population. A **population** is a group of individuals of the same species that occupy the same environment at the same time. For sexually reproducing species, members of a given population can interbreed with one another. The central issue in population genetics is genetic variation—its extent within populations, why it exists, how it is maintained, and how it changes over the course of many generations. Population genetics helps us understand how underlying genetic variation is related to phenotypic variation.

Population genetics emerged as a branch of genetics in the 1920s and 1930s. Its mathematical foundations were developed by theoreticians who extended the principles of Darwin and Mendel by deriving equations to explain the occurrence of genotypes within populations. These foundations can be largely attributed to British evolutionary biologists J. B. S. Haldane and Ronald Fisher, and American geneticist Sewall Wright. As we will see, several researchers who analyzed the genetic composition of natural and experimental populations provided support for their mathematical theories. More recently, population geneticists have used techniques to probe genetic variation at the molecular level. In addition, the staggering improvements in computer technology have aided population geneticists in the analysis of data and the testing of genetic hypotheses.

We will begin this chapter by exploring the extent of genetic variation that occurs in populations and how such variation is subject to change. In many cases, genetic changes are associated with evolutionary adaptations, which are characteristics of a species that have evolved over a long period of time by the process of natural selection. In the second half of the chapter, we will examine the various evolutionary mechanisms that promote genetic change in a population, including natural selection, genetic drift, migration, and nonrandom mating.

24.1 Genes in Populations

Learning Outcomes:

1. Define a gene pool.
2. Distinguish between allele and genotype frequency.
3. Use the Hardy-Weinberg equation to calculate allele and genotype frequencies of a given population.
4. List the conditions that must be met for a population to be in Hardy-Weinberg equilibrium.
5. Describe the factors that cause microevolution to happen.

Population genetics is an extension of our understanding of Darwin's theory of natural selection, Mendel's laws of inheritance, and newer studies in molecular genetics. All of the alleles for every gene in a given population make up the **gene pool**. Each member of the population receives its genes from its parents, which, in turn, are members of the gene pool. Individuals that reproduce contribute to the gene pool of the next generation. Population geneticists study the genetic variation within the gene pool and how such variation changes from one generation to the next. The emphasis is often on understanding the variation in alleles among members of a population. In this section, we will examine some of the general features of populations and gene pools.

Populations Are Dynamic Units

Recall that a population is a group of individuals of the same species that occupy the same environment at the same time and can interbreed with one another. Certain species occupy a wide geographic range and are divided into discrete populations due to geographic isolation. For example, distinct populations of a given species may be located on different sides of a physical barrier, such as a mountain.

Populations change from one generation to the next. How might populations become different? Populations may change in size and geographic location. As the size and locations of a population change, their genetic composition generally changes as well. Some of the genetic changes involve adaptation, in which a population becomes better suited to its environment, making it more likely to survive and reproduce. For example, a population of mammals may move from a warmer to a colder geographic location. Over the course of many generations, natural selection may change the population such that the fur of most animals is thicker and provides better insulation against the colder temperatures.

GENOMES & PROTEOMES CONNECTION

Genes Are Usually Polymorphic

The term **polymorphism** (from the Greek, meaning many forms) refers to the presence of two or more variants or traits for a given character within a population. Figure 24.1 illustrates a striking example of polymorphism in the elder-flowered orchid (*Dactylorhiza sambucina*). Throughout the range of this species in Europe, both yellow- and red-flowered individuals are prevalent.

Polymorphism in a character is usually due to two or more alleles of a gene that influences the character. Geneticists also use the term polymorphism to describe the variation in the DNA sequence of genes. A gene that commonly exists as two or more alleles in a population is a **polymorphic gene**. To be considered polymorphic, a gene must exist in at least two alleles, and each allele must occur at a frequency that is greater than 1%. By comparison, a **monomorphic gene** exists predominantly as a single allele in a population. When 99% or more of the alleles of a given gene are identical in a population, the gene is considered to be monomorphic.

What types of molecular changes cause genes to be polymorphic? A polymorphism may involve various types of changes, such as a deletion of a significant region of the gene, a duplication of a

Figure 24.1 An example of polymorphism: The two color variations found in the orchid *Dactylorhiza sambucina*.

region, or a change in a single nucleotide. This last type of variation is called a **single-nucleotide polymorphism (SNP)**. SNPs ("snips") are the smallest type of genetic variation that can occur within a given gene and also the most common. For example, the sickle cell allele discussed at the beginning of the chapter involves a single-nucleotide change in the β-globin gene, which encodes a subunit of the oxygen-carrying protein called hemoglobin. The non-disease-causing allele and sickle cell allele represent a SNP of the β-globin gene:

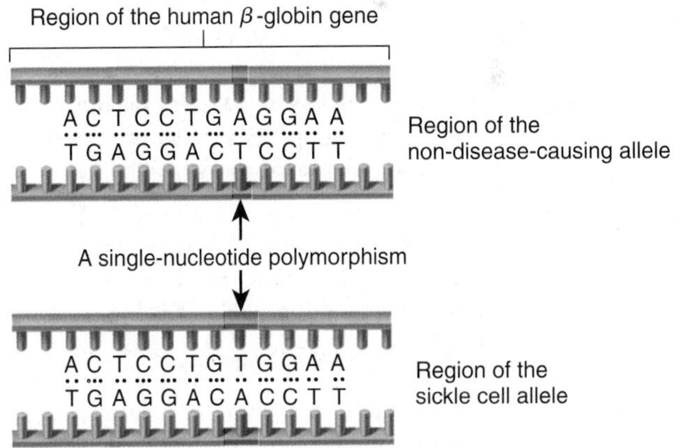

Relative to the non-disease-causing allele, this is a single-nucleotide substitution of an A (in the top strand) to a T in the sickle cell allele.

SNPs represent 90% of all variation in human DNA sequences that occurs among different people. In human populations, a gene that is 2,000–3,000 bp in length, on average, contains 10 different SNPs. Likewise, SNPs with a frequency of 1% or more are found very frequently among genes of nearly all species. Polymorphism is the norm for relatively large, healthy populations of nearly all species, as evidenced by the occurrence of SNPs within most genes.

Why do we care about SNPs? One reason is their importance in human health. By analyzing SNPs in human genes, researchers have determined that these small variations in DNA sequences can affect the function of the proteins encoded by the genes. These effects on the proteome, in turn, may influence how humans develop diseases, such as heart disease, diabetes, and sickle cell disease. Variations in SNPs in the human population are also associated with how people respond to viruses, drugs, and vaccines. The analysis of SNPs may be instrumental in the current and future development of **personalized medicine**—a medical practice in which information about a patient's genotype is used to tailor her or his medical care. For example, an analysis of a person's SNPs may be used to select between different types of medication or customize the dosage. In addition, SNP analysis may reveal that a person has a high predisposition to develop a particular disease, such as heart disease. Such information may be used to initiate preventative measures to minimize the chances of developing the disease.

Population Genetics Is Concerned with Allele and Genotype Frequencies

One approach to analyzing genetic variation in populations is to consider the frequency of specific alleles and genotypes in a quantitative way. Two fundamental calculations are central to population genetics: **allele frequency** and **genotype frequency**. Allele and genotype frequency are defined as follows:

$$\text{Allele frequency} = \frac{\text{Number of copies of a specific allele in a population}}{\text{Total number of all alleles for that gene in a population}}$$

$$\text{Genotype frequency} = \frac{\text{Number of individuals with a particular genotype in a population}}{\text{Total number of individuals in a population}}$$

Although allele and genotype frequencies are related, make sure you clearly distinguish between them. As an example, let's consider a population of 100 four-o'clock plants (*Mirabilis jalapa*) with the following genotypes:

49 red-flowered plants with the genotype C^RC^R

42 pink-flowered plants with the genotype C^RC^W

9 white-flowered plants with the genotype C^WC^W

When calculating an allele frequency for a diploid species, remember that homozygous individuals have two copies of a given allele, whereas heterozygotes have only one. For example, in tallying the C^W allele, each of the 42 heterozygotes has one copy of the C^W allele, and each white-flowered plant has two copies. Therefore, the allele frequency for C^W (the white color allele) equals

$$\text{Frequency of } C^W = \frac{(C^RC^W) + 2(C^WC^W)}{2(C^RC^R) + 2(C^RC^W) + 2(C^WC^W)}$$

$$\text{Frequency of } C^W = \frac{42 + (2)(9)}{(2)(49) + (2)(42) + (2)(9)}$$

$$= \frac{60}{200} = 0.3, \text{ or } 30\%$$

This result tells us that the allele frequency of C^W is 0.3. In other words, 30% of the alleles for this gene in the population are the white color (C^W) allele.

Let's now calculate the genotype frequency of C^WC^W homozygotes (white-flowered plants).

$$\text{Frequency of } C^WC^W = \frac{9}{49 + 42 + 9}$$

$$= \frac{9}{100} = 0.09, \text{ or } 9\%$$

We see that 9% of the individuals in this population have the white-flower genotype.

The Hardy-Weinberg Equation Relates Allele and Genotype Frequencies in a Population

In 1908, Godfrey Harold Hardy, an English mathematician, and Wilhelm Weinberg, a German physician, independently derived a simple mathematical expression, now called the Hardy-Weinberg equation, that describes the relationship between allele and genotype frequencies when a population is not evolving. Let's examine the Hardy-Weinberg equation using the population of four-o'clock plants that we just considered. If the allele frequency of C^R is denoted by the symbol p and the allele frequency of C^W by q, then

$$p + q = 1$$

For example, if $p = 0.7$, then q must be 0.3. In other words, if the allele frequency of C^R equals 70%, the remaining 30% of alleles must be C^W, because together they equal 100%.

For a gene that exists in two alleles, the **Hardy-Weinberg equation** states that

$$p^2 + 2pq + q^2 = 1$$

If we apply this equation to our flower color gene, then

p^2 = the genotype frequency of C^RC^R homozygotes

$2pq$ = the genotype frequency of C^RC^W heterozygotes

q^2 = the genotype frequency of C^WC^W homozygotes

If $p = 0.7$ and $q = 0.3$, then

$$\text{Frequency of } C^RC^R = p^2 = (0.7)^2 = 0.49$$

$$\text{Frequency of } C^RC^W = 2pq = 2(0.7)(0.3) = 0.42$$

$$\text{Frequency of } C^WC^W = q^2 = (0.3)^2 = 0.09$$

In other words, if the allele frequency of C^R is 70% and the allele frequency of C^W is 30%, the expected genotype frequency of C^RC^R is 49%, C^RC^W is 42%, and C^WC^W is 9%.

Figure 24.2 uses a Punnett square to illustrate the relationship between allele frequencies and the way that gametes combine to produce genotypes. To be valid, the Hardy-Weinberg equation carries the assumption that two gametes combine randomly with each

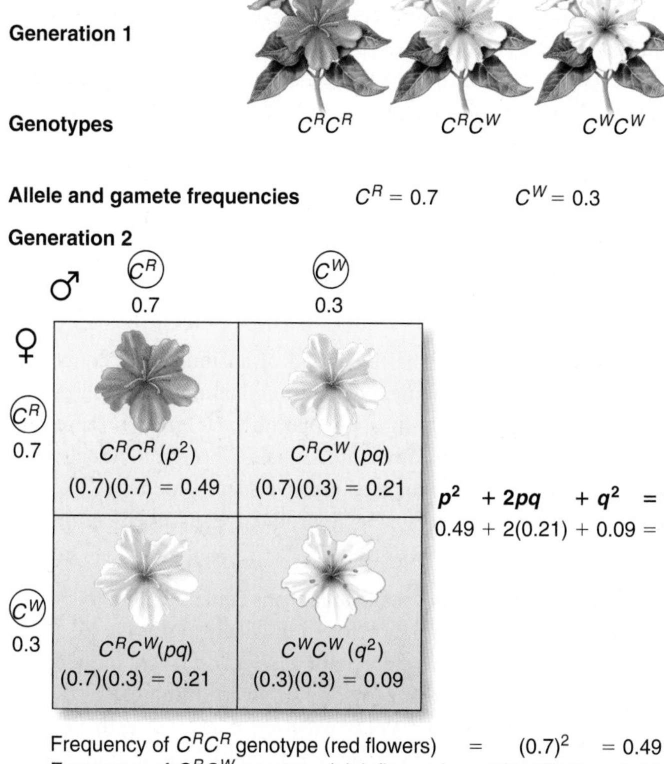

Figure 24.2 Calculating allele and genotype frequencies with the Hardy-Weinberg equation. A population of four-o'clock plants has allele and gamete frequencies of 0.7 for the C^R allele and 0.3 for the C^W allele. Knowing the allele frequencies allows us to calculate the genotype frequencies in the population.

Concept Check: *What would be the frequency of pink flowers in a population in which the allele frequency of C^R is 0.4 and the population is in Hardy-Weinberg equilibrium? Assume that C^R and C^W are the only two alleles.*

other to produce offspring. In a population, the frequency of a gamete carrying a particular allele is equal to the allele frequency in that population. For example, if the allele frequency of C^R equals 0.7, the frequency of a gamete carrying the C^R allele also equals 0.7. The probability of producing a C^RC^R homozygote with red flowers is $0.7 \times 0.7 = 0.49$, or 49%. The probability of inheriting both C^W alleles, which produces white flowers, is $0.3 \times 0.3 = 0.09$, or 9%. Two different gamete combinations produce heterozygotes with pink flowers. An offspring could inherit the C^R allele from the pollen and C^W from the egg, or C^R from the egg and C^W from the pollen. Therefore, the frequency of heterozygotes is $pq + pq$, which equals $2pq$. In our example, this is $2(0.7)(0.3) = 0.42$, or 42%. Note that the frequencies for all three genotypes total 100%.

The Hardy-Weinberg equation predicts that allele and genotype frequencies will remain the same, generation after generation, provided that a population is in equilibrium. To be in equilibrium, evolutionary mechanisms that can change allele and genotype frequencies are not acting on a population. For this to occur, the following conditions must be met:

- No new mutations occur to alter allele frequencies.
- No natural selection occurs; that is, no survival or reproductive advantage exists for any of the genotypes.
- The population is so large that allele frequencies do not change due to random chance.
- No migration occurs between different populations, altering the allele frequencies.
- Random mating occurs; that is, the members of the population mate with each other without regard to their genotypes.

Why is the Hardy-Weinberg equilibrium a useful concept? An equilibrium is a null hypothesis, which suggests that evolutionary change is not occurring. In reality, however, populations rarely achieve an equilibrium, though in large natural populations with little migration and negligible natural selection, the Hardy-Weinberg equilibrium may be nearly approximated for certain genes. Sometimes, when researchers experimentally examine allele and genotype frequencies for one or more genes in a given species, they discover that the frequencies are not in Hardy-Weinberg equilibrium. In such cases, they assume that one or more of the conditions are being violated—in other words, mechanisms of evolutionary change are affecting the population. Conservation biologists and wildlife managers may wish to determine why such disequilibrium has occurred because it may affect the future survival of the species. Next, we will take a look at the mechanisms that cause evolutionary change.

Microevolution Involves Changes in Allele Frequencies from One Generation to the Next

The term **microevolution** is used to describe changes in a population's gene pool, such as changes in allele frequencies, from generation to generation. What causes microevolution to happen? Such change is rooted in two related phenomena (**Table 24.1**). First, the introduction of new genetic variation into a population is one essential aspect of microevolution. New alleles of preexisting genes arise by random mutation and, as discussed in Chapter 23, new genes can be introduced into a population by gene duplication, exon shuffling, and horizontal gene transfer. Such mutations, albeit rare, provide a continuous source of new variation to populations. In 1926, the Russian geneticist Sergei Chetverikov was the first to suggest that random mutations are the raw material for evolution. However, due to their low rate of occurrence, mutations by themselves do not play a major role in changing allele frequencies in a population over time. They do not significantly disrupt a Hardy-Weinberg equilibrium.

The second phenomenon that is required for evolution to occur is one or more mechanisms that alter the prevalence of a given allele or genotype in a population. These mechanisms are natural selection, genetic drift, migration, and nonrandom mating (see Table 24.1). Over the course of many generations, these mechanisms may promote widespread genetic changes in a population. In the remainder of this chapter, we will examine how natural selection, genetic drift, migration, and nonrandom mating affect the type of genetic variation that occurs when a gene exists as two alleles in a population.

Table 24.1 Factors That Govern Microevolution

Sources of new genetic variation*	
New mutations within genes that produce new alleles	Random mutations within pre-existing genes introduce new alleles into populations, but at a very low rate. New mutations may be neutral, deleterious, or beneficial. Because mutations are rare, the change from one generation to the next is generally very small. For alleles to rise to a significant percentage in a population, evolutionary mechanisms, such as natural selection, genetic drift, and migration, must operate on them.
Gene duplication†	Abnormal crossover events and transposable elements may increase the number of copies of a gene. Over time, the additional copies accumulate random mutations and constitute a gene family.
Exon shuffling‡	Abnormal crossover events and transposable elements may promote gene rearrangements in which one or more exons from one gene are inserted into another gene. The protein encoded by such a gene may display a novel function that is acted on by evolutionary mechanisms.
Horizontal gene‡ transfer	Genes from one species may be introduced into another species. The transferred gene may be acted on by evolutionary mechanisms.
Evolutionary mechanisms that alter the frequencies of existing genetic variation	
Natural selection	The process in which individuals that possess certain traits are more likely to survive and reproduce than individuals without those traits. Over the course of many generations, beneficial traits that are heritable become more common and detrimental traits become less common.
Genetic drift	A change in genetic variation from generation to generation due to random chance. Allele frequencies may change as a matter of chance from one generation to the next. Genetic drift has a greater influence in a small population.
Migration	Migration can occur between two populations that have different allele frequencies. The introduction of migrants into a recipient population may change the allele frequencies of that population.
Nonrandom mating	The phenomenon in which individuals select mates based on their phenotypes or genetic lineage. This alters the relative proportion of homozygotes and heterozygotes that is predicted by the Hardy-Weinberg equation, but it does not change allele frequencies.

* These are examples that affect single genes. Other events, such as crossing over, independent assortment, and changes in chromosome structure and number, may alter the genetic variation among many genes.
† Described in Chapter 21. See Figures 21.7 and 21.8.
‡ Described in Chapter 23. See Figures 23.15 and 23.16.

24.2 Natural Selection

Learning Outcomes:

1. Explain how natural selection can result in a population that is better adapted to its environment and more successful at reproduction.
2. Calculate the fitness values of given genotypes.
3. List and distinguish between four different types of natural selection.

Recall from Chapter 23 that **natural selection** is the process in which individuals with certain heritable traits tend to survive and reproduce at higher rates than those without those traits. As a result, favorable heritable traits become more common, while detrimental heritable traits become less common. Keep in mind that natural selection itself is not evolution. Rather it is a key mechanism that causes evolution to happen. Over time, natural selection results in **adaptations**—changes in populations of living organisms that increase their ability to survive and reproduce in a particular environment. In this section, we will examine various ways that natural selection produces such adaptations.

Natural Selection Favors Individuals with Greater Reproductive Success

Reproductive success is the likelihood of an individual contributing fertile offspring to the next generation. Natural selection occurs because some individuals in a population have greater reproductive success compared to other individuals. Those individuals having heritable traits that favor reproductive success are more likely to pass those traits to their offspring. Reproductive success is commonly attributed to two categories of traits:

1. Certain characteristics make organisms better adapted to their environment and therefore more likely to survive to reproductive age. Therefore, natural selection favors individuals with characteristics that provide a survival advantage.
2. Reproductive success may involve traits that are directly associated with reproduction, such as the abilities to find a mate and produce viable gametes and offspring. Traits that enhance the ability of individuals to reproduce, such as brightly colored plumage in male birds, are often subject to natural selection.

As discussed in Chapter 23, Charles Darwin and Alfred Wallace independently proposed the theory of evolution by natural selection. A modern description of the principles of natural selection can relate our knowledge of molecular genetics to the process of evolution:

1. Within a population, allelic variation arises from random mutations that cause differences in DNA sequences. A mutation that creates a new allele may alter the amino acid sequence of the encoded protein. This, in turn, may alter the function of the protein.
2. Some alleles encode proteins that enhance an individual's survival or reproductive capability over that of other members of the population. For example, an allele may produce a protein that is more efficient at a higher temperature, conferring on the individual a greater probability of survival in a hot climate.
3. Individuals with beneficial alleles are more likely to survive and contribute their alleles to the gene pool of the next generation.
4. Over the course of many generations, allele frequencies of many different genes may change through natural selection, thereby significantly altering the characteristics of a population. The net result of natural selection is a population that is better adapted to its environment and more successful at reproduction.

Fitness Is a Quantitative Measure of Reproductive Success

As mentioned earlier, Haldane, Fisher, and Wright developed mathematical relationships to explain the phenomenon of natural selection. To begin our quantitative discussion of natural selection, we need to consider the concept of **fitness**, which is the relative likelihood that one genotype will contribute to the gene pool of the next generation compared with other genotypes. Although this property often correlates with physical fitness, the two ideas should not be confused. Fitness is a measure of reproductive success. An extremely fertile individual may have a higher fitness than a less fertile individual that appears more physically fit.

To examine fitness, let's consider an example of a hypothetical gene existing in *A* and *a* alleles. We can assign fitness values to each of the three possible genotypes according to their relative reproductive success. For example, let's suppose the average reproductive successes of the three genotypes are

AA produces 5 offspring
Aa produces 4 offspring
aa produces 1 offspring

By convention, the genotype with the highest reproductive success is given a fitness value of 1.0. Fitness values are denoted by the variable *w*. The fitness values of the other genotypes are assigned values relative to this 1.0 value.

Fitness of *AA*: $w^{AA} = 1.0$

Fitness of *Aa*: $w^{Aa} = 4/5 = 0.8$

Fitness of *aa*: $w^{aa} = 1/5 = 0.2$

Variation in fitness occurs because certain genotypes result in individuals that have a greater reproductive success than other genotypes.

Likewise, the effects of natural selection can be viewed at the level of a population. The average reproductive success of members of a population is called the **mean fitness of the population**. Over many generations, as individuals with higher fitness values become more prevalent, natural selection also increases the mean fitness of the population. In this way, the process of natural selection results in a population of organisms that is well adapted to its native environment and more likely to be successful at reproduction.

Natural Selection Follows Different Patterns

By studying species in their native environments, population geneticists have discovered that natural selection can occur in several ways. In most of the examples described next, natural selection leads to adaptations in which certain members of a species are more likely to survive to reproductive age.

Directional Selection During **directional selection**, individuals at one extreme of a phenotypic range have greater reproductive success in a particular environment. Different phenomena may initiate the process of directional selection. A common reason for directional selection is that a population may be exposed to a prolonged change in its living environment. Under the new environmental conditions, the relative fitness values may change to favor one genotype, which will promote the elimination of other genotypes. As an example, let's suppose a population of finches on a mainland already has genetic variation that affects beak size (refer back to Figure 23.2). A small number of birds migrate to an island where the seeds are generally larger than on the mainland. In this new environment, birds with larger beaks have a higher fitness because they are better able to crack open the larger seeds and thereby survive to reproduce. Over the course of many generations, directional selection would produce a population of birds carrying alleles that promote larger beak size.

Another way that directional selection may arise is that a new allele may be introduced into a population by mutation, and the new allele may confer a higher fitness in individuals that carry it (**Figure 24.3**). What are the long-term effects of such directional selection? If the homozygote carrying the favored allele has the highest fitness value, directional selection may cause this favored allele to eventually predominate in the population, perhaps even leading to a monomorphic gene.

Stabilizing Selection A type of natural selection called **stabilizing selection** favors the survival of individuals with intermediate phenotypes and selects against those with extreme phenotypes. Stabilizing selection tends to decrease genetic diversity. An example of stabilizing selection involves clutch size (number of eggs laid) in birds, which was first studied by British biologist David Lack in 1947. Under stabilizing selection, birds that lay too many or too few eggs per nest have lower fitness values than do those that lay an intermediate number. When a bird lays too many eggs, many offspring die due to inadequate parental care and food. In addition, the strain on the parents themselves may decrease their likelihood of survival and consequently their ability to produce more offspring. Having too few offspring, however, does not contribute many individuals to the next generation. Therefore, the most successful parents are those that produce an intermediate clutch size. In the 1980s, Swedish evolutionary biologist Lars Gustafsson and his colleagues examined the phenomenon of stabilizing selection in the collared flycatcher (*Ficedula albicollis*) on the Swedish island of Gotland. They discovered that Lack's hypothesis concerning an optimal clutch size appears to be true for this species (**Figure 24.4**).

Diversifying Selection **Diversifying selection** (also known as disruptive selection) favors the survival of two or more different genotypes that produce different phenotypes. In diversifying selection, the fitness values of a particular genotype are higher in one environment and lower in a different one, whereas the fitness values of the second genotype vary in an opposite manner. Diversifying selection is likely to occur in populations that occupy heterogeneous environments, so some members of the species are more likely to survive in each type of environmental condition.

An example of diversifying selection involves colonial bentgrass (*Agrostis capillaris*) (**Figure 24.5**). In certain locations where this grass is found, such as South Wales, isolated places occur where the soil is contaminated with high levels of heavy metals due to mining. The relatively recent metal contamination has selected for the proliferation of mutant strains of *A. capillaris* that are tolerant of the heavy metals (Figure 24.5a). Such genetic changes enable these mutant strains to grow on contaminated soil but tend to inhibit their growth on normal, noncontaminated soil. These metal-resistant plants often grow on contaminated sites that are close to plants that grow on uncontaminated land and do not show metal tolerance.

Balancing Selection Contrary to a popular misconception, natural selection does not always cause the elimination of "weaker" or less-fit

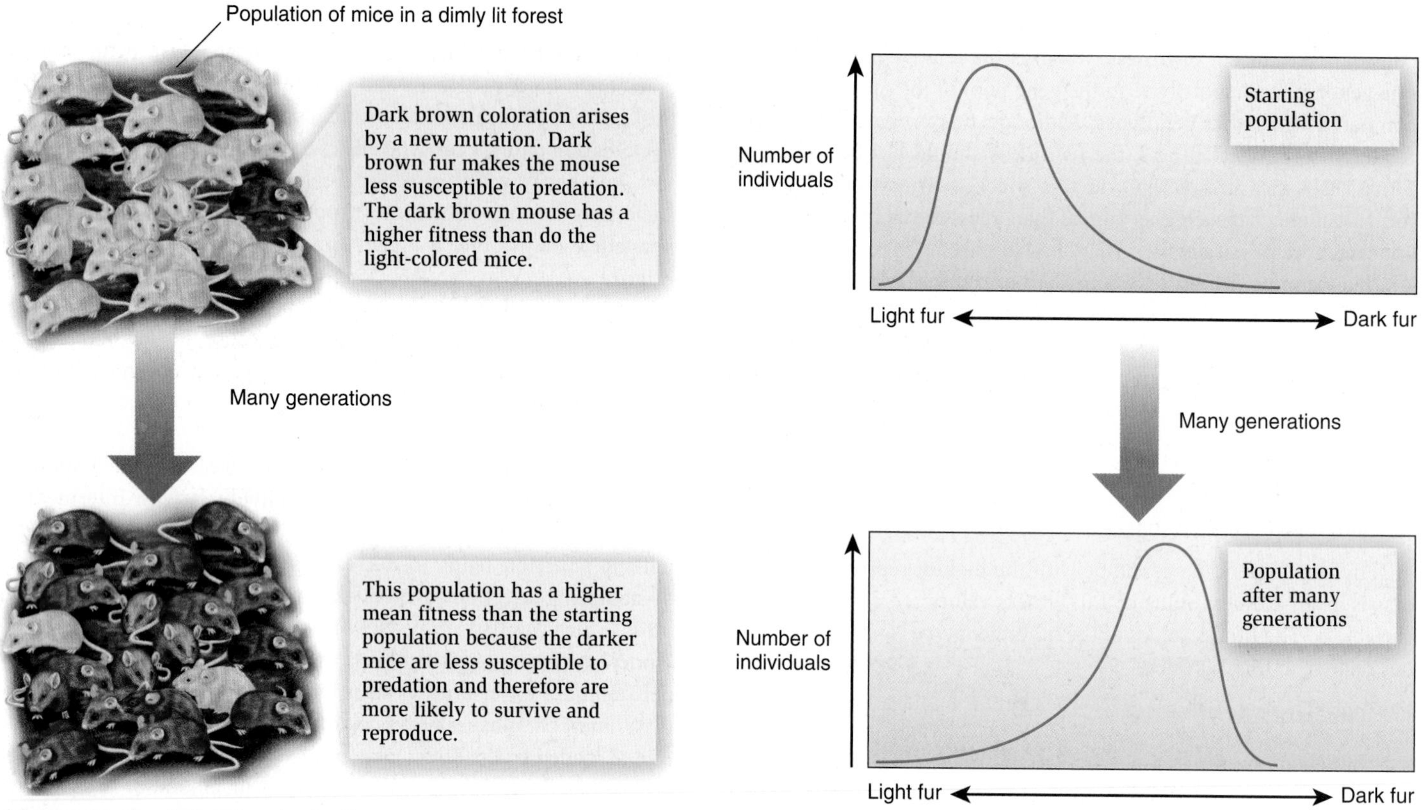

Population of mice in a dimly lit forest
Dark brown coloration arises by a new mutation. Dark brown fur makes the mouse less susceptible to predation. The dark brown mouse has a higher fitness than do the light-colored mice.
Many generations
This population has a higher mean fitness than the starting population because the darker mice are less susceptible to predation and therefore are more likely to survive and reproduce.
Number of individuals
Starting population
Light fur
Dark fur
Many generations
Number of individuals
Population after many generations
Light fur
Dark fur

(a) An example of directional selection

(b) Graphical representation of directional selection

Figure 24.3 **Directional selection.** This pattern of natural selection selects for one extreme of a phenotype that confers the highest fitness in the population's environment. **(a)** In this example, a mutation causing darker fur arises in a population of mice. This new genotype confers higher fitness, because mice with darker fur can evade predators and are more likely to survive and reproduce. Over many generations, directional selection favors the prevalence of individuals with darker fur. **(b)** These graphs show the change in fur color phenotypes before and after directional selection.

Concept Check: Let's suppose the climate on an island abruptly changed such that the average temperature was 10°C higher. The climate change is permanent. How would directional selection affect the genetic diversity in a population of mice on the island (1) over the short run and (2) over the long run?

alleles. **Balancing selection** is a type of natural selection that maintains genetic diversity in a population. Over many generations, balancing selection results in a **balanced polymorphism**, in which two or more alleles are kept in balance and therefore are maintained in a population over many generations.

How does balancing selection maintain a polymorphism? Population geneticists have identified two common ways that balancing selection occurs. First, for genetic variation involving a single gene, balancing selection can favor the heterozygote over either corresponding homozygote. This situation is called **heterozygote advantage**. Heterozygote advantage sometimes explains the persistence of alleles that are deleterious in a homozygous condition.

Number of nests
Starting population
Few
Number of eggs
Many
Many generations
Number of nests
Population after many generations
Few
Number of eggs
Many

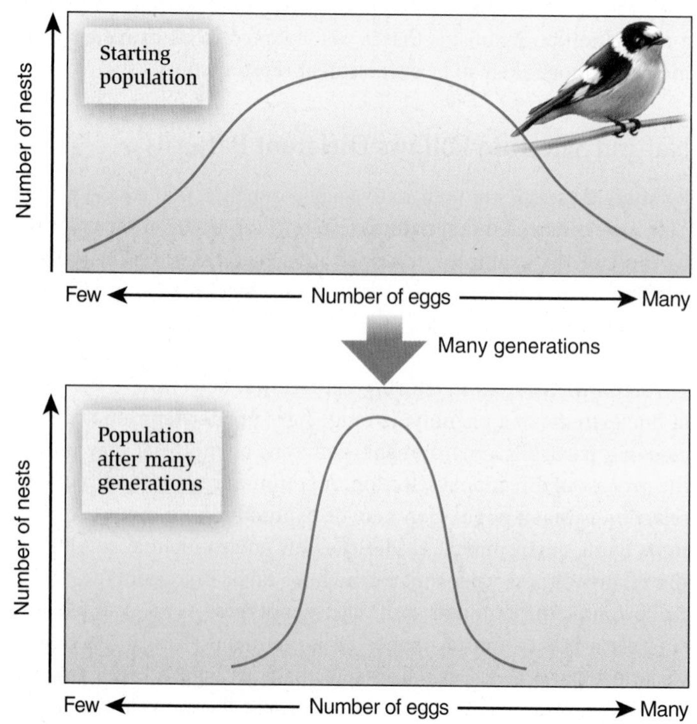

Figure 24.4 **Stabilizing selection.** In this pattern of natural selection, the extremes of a phenotypic distribution are selected against. Those individuals with intermediate traits have the highest fitness. These graphs show the results of stabilizing selection on clutch size in a population of collared flycatchers (*Ficedula albicollis*). This process results in a population with less diversity and more uniform traits.

Concept Check: Why does stabilizing selection decrease genetic diversity?

(a) Growth of *Agrostis capillaris* on contaminated soil

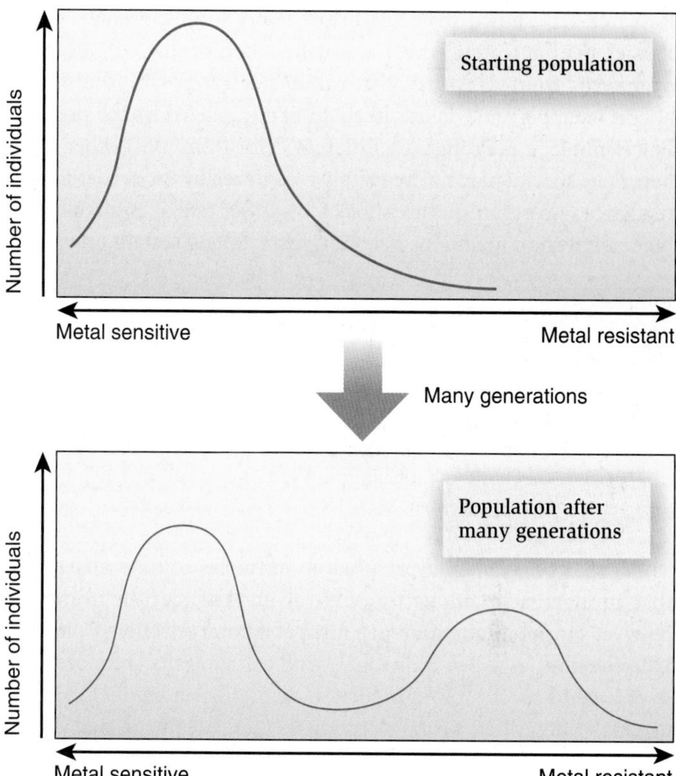

(b) Graphical representation of disruptive selection

Figure 24.5 Diversifying selection. This pattern of natural selection selects for two different phenotypes, each of which is most fit in a particular environment. **(a)** In this example, random mutations have resulted in metal-resistant alleles in colonial bentgrass (*Agrostis capillaris*) that allow it to grow on contaminated soil. In uncontaminated soils, the grass does not show metal tolerance. The existence of both metal-resistant and metal-sensitive alleles in the population is an example of diversifying selection due to heterogeneous environments. **(b)** Graphs showing the change in phenotypes in this bentgrass population before and after diversifying selection.

BIOLOGY PRINCIPLE Populations of organisms evolve from one generation to the next. In this example, the frequencies of metal-resistant alleles become more prevalent when populations of *A. capillaris* are exposed to toxic metals in the soil.

A classic example of heterozygote advantage involves the H^S allele of the human β-globin gene. A homozygous H^SH^S individual, such as Kimbareta, discussed at the beginning of the chapter, has sickle cell disease. This disease causes the red blood cells to form a sickle shape. Sickle-shaped cells deliver less oxygen to the body's tissues and can block the flow of blood through the vessels. The H^SH^S homozygote has a lower fitness than a homozygote with two copies of the more common β-globin allele, H^AH^A. Heterozygotes, H^AH^S, do not typically show symptoms of the disease, but they have an increased resistance to malaria. Compared with H^AH^A homozygotes, heterozygotes have the highest fitness because they have a 10–15% better chance of surviving if infected by the malarial parasite *Plasmodium falciparum*. Therefore, the H^S allele is maintained in populations where malaria is prevalent, such as the Democratic Republic of Congo, even though the allele is detrimental in the homozygous state (**Figure 24.6**). This balanced polymorphism results in a higher mean fitness of the population. In areas where malaria is endemic, a population composed of all H^AH^A individuals would have a lower mean fitness.

Negative frequency-dependent selection is a second way that natural selection produces a balanced polymorphism. In this pattern of natural selection, the fitness of a genotype decreases when its frequency becomes higher. In other words, common individuals have a lower fitness, and rare individuals have a higher fitness. Therefore, common individuals are less likely to reproduce, whereas rare individuals are more likely to reproduce, thereby producing a balanced polymorphism in which no genotype becomes too rare or too common.

Negative frequency-dependent selection is thought to maintain polymorphisms among species that are preyed upon by predators.

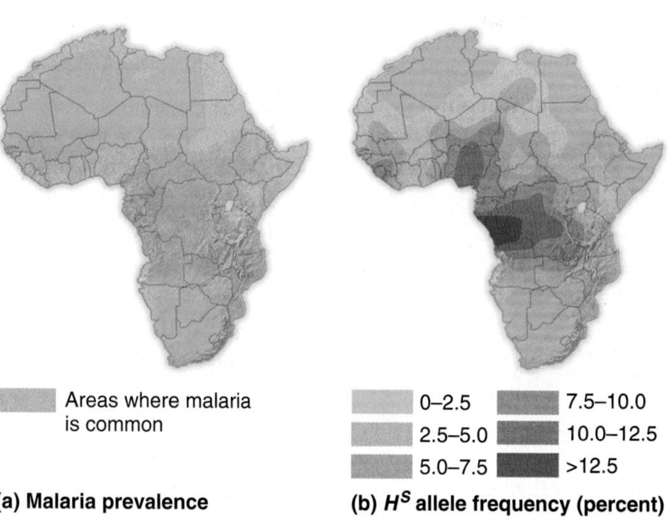

(a) Malaria prevalence **(b) H^S allele frequency (percent)**

Figure 24.6 Balancing selection and heterozygote advantage. **(a)** The geographic prevalence of malaria in Africa. **(b)** The frequency of the H^S allele of the β-globin gene in the same area. In the homozygous condition, the H^S allele causes sickle cell disease. This allele is maintained in human populations in areas where malaria is prevalent, because the heterozygote (H^AH^S) has a higher fitness than either of the corresponding homozygotes (H^AH^A or H^SH^S).

Concept Check: *If malaria was eradicated, what would you expect to happen to the frequencies of the H^A and H^S alleles over the long run?*

Research has shown that certain predators form a mental "search image" for their prey, which is usually based on the common type of prey in an area. A prey that exhibits a rare polymorphism that affects its appearance is less likely to be recognized by the predator. For example, a prey that is a different color from most other members of its species may not be readily recognized by the predator. Such relatively rare organisms are subject to a lower rate of predation. This type of selection maintains polymorphism among certain prey.

24.3 Sexual Selection

Learning Outcomes:

1. Define the term sexual selection.
2. Distinguish between intrasexual and intersexual selection.
3. Analyze the results of Seehausen and van Alphen and explain how they relate to sexual selection.

Thus far, we have largely focused on examples of natural selection that produce adaptations for survival in particular environments. Now let's turn our attention to a form of natural selection, called **sexual selection**, in which individuals with certain traits are more likely to engage in successful reproduction than other individuals. Darwin originally described sexual selection as "the advantage that certain individuals have over others of the same sex and species solely with respect to reproduction." In this section, we will explore how sexual selection alters traits that play a key role in reproduction.

Sexual Selection Is a Type of Natural Selection Pertaining to Traits That Are Directly Involved with Reproduction

In many species of animals, sexual selection affects the characteristics of males more intensely than those of females. Unlike females, which tend to be fairly uniform in their reproductive success, male success tends to be more variable, with some males mating with many females and others not mating at all. Sexual selection results in the evolution of traits, called secondary sex characteristics, that favor reproductive success. The process can result in **sexual dimorphism**—a significant difference between the appearances of the two sexes within a species.

Sexual selection operates in one of two ways. In **intrasexual selection**, members of one sex, usually males, directly compete for the opportunity to mate with individuals of the opposite sex. Examples of traits that result from intrasexual selection in animals include horns in male sheep, antlers in male moose, and the enlarged claw of male fiddler crabs (**Figure 24.7a**). In fiddler crabs (*Uca paradussumieri*), males enter the burrows of females that are ready to mate. If another male attempts to enter the burrow, the male already inside stands in the burrow shaft and blocks the entrance with his enlarged claw. Males with the largest claws are more likely to be successful at driving off their rivals and being able to mate and therefore more likely to pass on their genes to future generations.

In **intersexual selection**, also called mate choice, members of one sex, usually females, choose their mates from individuals of the other sex on the basis of certain desirable characteristics. This type of sexual selection often results in showy characteristics in males. **Figure 24.7b** shows a classic example that involves the Indian peafowl (*Pavo cristatus*), the national bird of India. Male peacocks have long and brightly colored tail feathers, which they fan out as a mating behavior. Female peahens select among males based on feather color and pattern as well as the physical prowess of the display.

A less obvious type of intersexual selection is cryptic female choice, in which the female reproductive system influences the relative success of sperm. As an example of cryptic female choice, the female genital tract of certain animals selects for sperm that tend to be genetically unrelated to the female. Sperm from males closely related to the female, such as brothers or cousins, are less successful than are sperm from genetically unrelated males. The selection for sperm may occur over the journey through the reproductive tract. The egg itself may even have mechanisms to prevent fertilization by genetically related sperm. Cryptic female choice occurs in species in which females may mate with more than one male, such as many species of reptiles and ducks. A similar mechanism is found in many plant species in which pollen from genetically related plants, perhaps from the same flower, is unsuccessful at fertilization, whereas pollen from unrelated plants is successful. One possible advantage of cryptic female choice is that it inhibits inbreeding (described later in this chapter). At the population level, cryptic female choice may promote genetic diversity by favoring interbreeding among genetically unrelated individuals.

Sexual selection is sometimes a combination of both intrasexual and intersexual selection. During breeding season, male elk (*Cervus elaphus*) become aggressive and bugle loudly to challenge other male elk. Males spar with their antlers, which usually turns into a pushing match to determine which elk is stronger. Female elk then choose the strongest bulls as their mates.

Sexual selection can explain the existence of traits that could decrease an individual's chances of survival but increase their chances of reproducing. For example, the male guppy (*Poecilia reticulata*) is brightly colored compared with the female (**Figure 24.7c**). In nature, females prefer brightly colored males. However, brightly colored males are more likely to be seen and eaten by predators. In places with few predators, the males tend to be brightly colored. In contrast, where predators are abundant, brightly colored males are less plentiful because they are subject to predation. In this case, the relative abundance of brightly and dully colored males depends on the balance between sexual selection, which favors bright coloring, and escape from predation, which favors dull coloring.

Many animals have secondary sexual characteristics, and evolutionary biologists generally agree that sexual selection is responsible for such traits. But why should males compete, and why should females be choosy? Researchers have proposed various hypotheses to explain the underlying mechanisms. One possible reason is related to the different roles that males and females play in the nurturing of offspring. In some animal species, the female is the primary caregiver, whereas the male plays a minor role. In such species, mating behavior may influence the fitness of both males and females. Males increase their fitness by mating with multiple females. This increases their likelihood of passing their genes on to the next generation. By comparison, females may produce relatively fewer offspring, and their reproductive success may not be limited by the number of available males. In these circumstances, females will have higher fitness if they choose males that are good defenders of their territory and

(a) Intrasexual selection

(b) Intersexual selection

(c) Sexual selection balanced by predation

Figure 24.7 Examples of the results of sexual selection, a type of natural selection. **(a)** An example of intrasexual selection. The enlarged claw of the male fiddler crab is used in direct male-to-male competition. In this photograph, a male inside a burrow is extending its claw out of the burrow to prevent another male from entering and mating with the female. **(b)** An example of intersexual selection. Female peahens choose male peacocks based on the males' colorful and long tail feathers and the robustness of their display. **(c)** Male guppies (on the right) are brightly colored to attract a female (on the left), but brightly colored males are less common where predation is high. Note: These photos also illustrate the concept of sexual dimorphism.

Concept Check: Male birds of many species have loud and elaborate courtship songs. Is this likely to be the result of intersexual or intrasexual selection? Explain.

have alleles that confer a survival advantage to their offspring. One measure of alleles that confer higher fitness is age. Males that live to an older age are more likely to carry beneficial alleles. Many research studies involving female choice have shown that females tend to select traits that are more likely to be well developed in older males than in immature ones. For example, in certain species of birds, females tend to choose males with a larger repertoire of songs, which is more likely to occur in older males. Sexual selection is governed by the same processes involved in the evolution of traits that are not directly related to sex. Sexual selection can occur by directional, stabilizing, diversifying, or balancing selection. For example, the evolution of the large and brightly colored tail of the male peacock reflects directional selection.

Feature Investigation

Seehausen and van Alphen Found That Male Coloration in African Cichlids Is Subject to Female Choice

In 1998, population geneticists Ole Seehausen and Jacques van Alphen investigated the possible role of sexual selection as it pertains to male coloration of two species of cichlid—a tropical freshwater fish popular among aquarium enthusiasts. The Cichlidae family is composed of more than 3,000 species that vary in body shape, coloration, behavior, and feeding habits, making it one of the largest and most diverse vertebrate families. By far, the greatest diversity of these fish is found in Lake Victoria, Lake Malawi, and Lake Tanganyika in East Africa, where more than 1,800 species are found.

Cichlids have complex mating behavior, and females play an important role in choosing males with particular characteristics, such as color (see chapter opening photo). In some locations, *Pundamilia pundamilia* and *P. nyererei* do not readily interbreed and behave like two distinct biological species, whereas in other places, they behave like a single interbreeding species with two color morphs. Males of both species have blackish underparts and blackish vertical bars on their sides (**Figure 24.8a**). *P. pundamilia* males are grayish white on top and on the sides, and they have a metallic blue and red dorsal

(a) Males of two species in normal light

(b) Males of two species in artificial light

Figure 24.8 Male coloration in African cichlids. **(a)** Two males (*Pundamilia pundamilia*, top, and *Pundamilia nyererei*, bottom) under normal illumination. **(b)** The same species under orange monochromatic light, which obscures their color differences.

Figure 24.9 **A study by Seehausen and van Alphen evaluating the effects of male coloration on female choice in African cichlids.**

HYPOTHESIS Female African cichlids choose mates based on the males' coloration.

KEY MATERIALS Two species of cichlid, *Pundamilia pundamilia* and *P. nyererei*, were chosen. The males differ with regard to their coloration. A total of 8 males and 8 females (4 males and 4 females from each species) were tested.

Experimental level | **Conceptual level**

1 Place 1 female and 2 males in an aquarium. Each male is within a separate glass enclosure. The enclosures contain 1 male from each species.

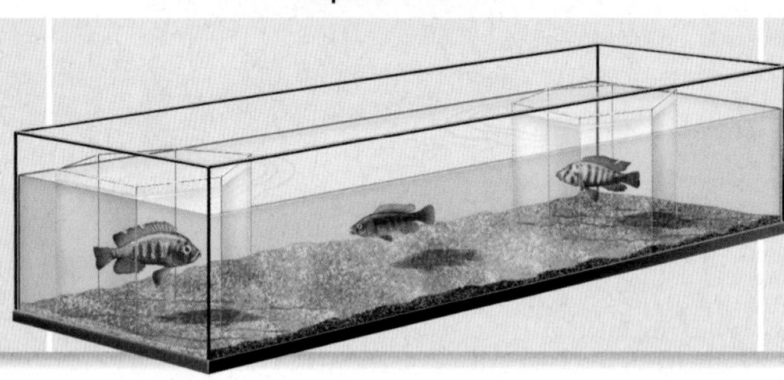

This is a method to evaluate sexual selection via female choice in 2 species of cichlid.

2 Observe potential courtship behavior for 1 hour. If a male exhibited lateral display (a courtship invitation) and then the female approached the enclosure that contained the male, this was scored as a positive encounter. This protocol was performed under normal light and under orange monochromatic light.

3 **THE DATA**

Female	Male	Light condition	Percentage of positive encounters*
P. pundamilia	*P. pundamilia*	Normal	16
P. pundamilia	*P. nyererei*	Normal	2
P. nyererei	*P. nyererei*	Normal	16
P. nyererei	*P. pundamilia*	Normal	5
P. pundamilia	*P. pundamilia*	Monochromatic	20
P. pundamilia	*P. nyererei*	Monochromatic	18
P. nyererei	*P. nyererei*	Monochromatic	13
P. nyererei	*P. pundamilia*	Monochromatic	18

*A positive encounter occurred when a male's lateral display was followed by the female approaching the male.

4 **CONCLUSION** Under normal light, where colors can be distinguished, *P. pundamilia* females prefer *P. pundamilia* males, and *P. nyererei* females prefer *P. nyererei* males.

5 **SOURCE** Seehausen, O., and van Alphen, J.J.M. 1998. The effect of male coloration on female mate choice in closely related Lake Victoria cichlids (*Haplochromis nyererei* complex). *Behav. Ecol. Sociobiol.* 42:1–8.

fin—the uppermost fin. *P. nyererei* males are red-orange on top and yellow on their sides.

Seehausen and van Alphen hypothesized that females choose males for mates based, in part, on the males' coloration. The researchers took advantage of the observation that colors are obscured under orange monochromatic light. As seen in **Figure 24.8b**, males of both species look similar under these conditions. In their study, a female of one species was placed in an aquarium that contained one male of each species within an enclosure (**Figure 24.9**). The males were within glass enclosures to avoid direct competition with each other, which would have likely affected female choice. The goal of the experiment was to determine which of the two males a female would prefer. Courtship between a male and female begins when a male swims toward a female and exhibits a lateral display (that is, shows the side of his body to the female). If the female is interested, she will approach the male, and then the male will display a quivering motion. Such courtship behavior was examined under normal light and under orange monochromatic light.

As seen in the data, Seehausen and van Alphen found that the females' preference for males was dramatically different depending on the illumination conditions. Under normal light, *P. pundamilia* females preferred *P. pundamilia* males, and *P. nyererei* females preferred *P. nyererei* males. However, such mating preference was lost when colors were masked by artificial light. If the light conditions in their native habitats are similar to the normal light used in this experiment, female choice would be expected to separate cichlids into two populations, with *P. pundamilia* females mating with *P. pundamilia* males and *P. nyererei* females mating with *P. nyererei* males. In this case, sexual selection appears to have followed a diversifying mechanism in which certain females prefer males with one color pattern, whereas other females prefer males with a different color pattern. A possible outcome of such sexual selection is that it can separate one large population into smaller populations that selectively breed with each other and eventually become distinct species. We will discuss the topic of species formation in more depth in Chapter 25.

Experimental Questions

1. What hypothesis is tested in the Seehausen and van Alphen experiment?
2. Describe the experimental design for this study, illustrated in Figure 24.9. What was the purpose of conducting the experiment under the two different light conditions?
3. What were the results of the experiment in Figure 24.9?

24.4 Genetic Drift

Learning Outcomes:

1. Define genetic drift and explain its effects on allele frequencies over time.
2. Compare and contrast the bottleneck and founder effects.
3. Explain how neutral mutations can spread through a population.

Thus far, we have focused on natural selection as a mechanism that can promote widespread genetic changes in a population. Let's now turn our attention to a second important way the gene pool of a population can change. In the 1930s, Sewall Wright played a large role in developing the concept of **genetic drift** (also called random genetic drift), which refers to changes in allele frequencies due to random chance. The term genetic drift is derived from the observation that allele frequencies may "drift" randomly from generation to generation as a matter of chance.

Changes in allele frequencies due to genetic drift happen regardless of the fitness of individuals that carry those alleles. For example, an individual with a high fitness value may, by chance, not encounter a member of the opposite sex. Likewise, random chance can influence which alleles happen to be found in the gametes that fuse with each other in a successful fertilization. In this section, we will examine how genetic drift alters allele frequencies in populations.

Genetic Drift Has a Greater Effect in Small Populations

What are the effects of genetic drift? Over the long run, genetic drift favors either the elimination or the fixation of an allele, that is, when its frequency reaches 0% or 100% in a population, respectively. However, the number of generations it takes for an allele to be lost or fixed greatly depends on the population size. **Figure 24.10** illustrates the potential consequences of genetic drift in one large (N = 1,000) and two small (N = 10) populations. This simulation

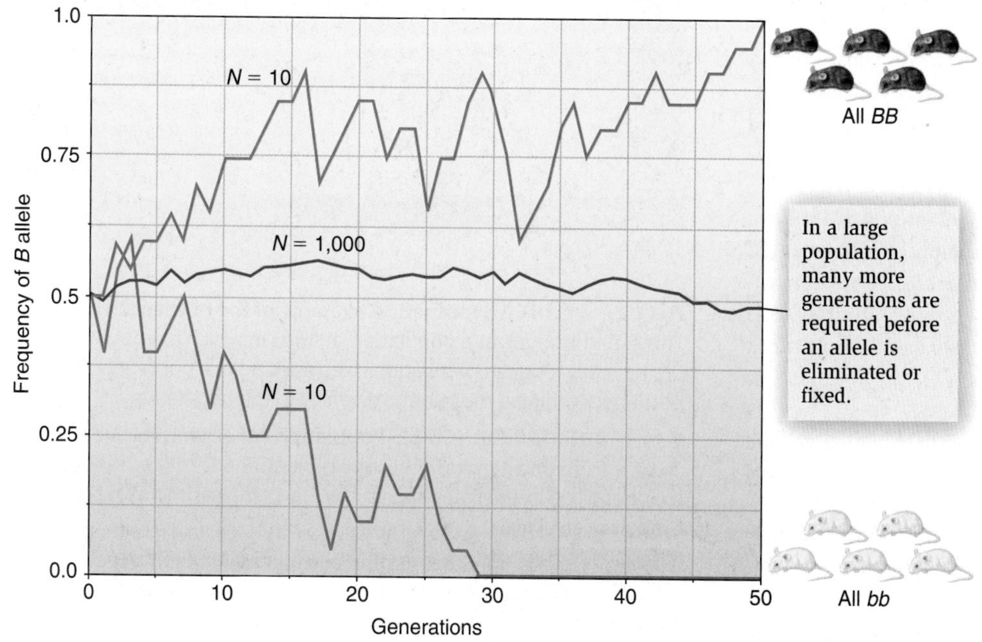

Figure 24.10 Genetic drift and population size. This graph shows three hypothetical simulations of genetic drift and their effects on small and large populations of black (*B* allele) and white (*b* allele) mice. In all cases, the starting allele frequencies are B = 0.5 and b = 0.5. The red lines illustrate two populations of mice in which N = 10; the blue line shows a population in which N = 1,000.

involves the frequency of hypothetical B and b alleles of a gene for fur color in a population of mice—B is the black allele, and b is the white allele.

At the beginning of this hypothetical simulation, which runs for 50 generations, all three populations had identical allele frequencies: $B = 0.5$ and $b = 0.5$. In the small populations, the allele frequencies fluctuated substantially from generation to generation. Eventually, in one population, the b allele was eliminated; in another, it was fixed at 100%. These small populations would then consist of only black mice or white mice, respectively. At this point, the gene has become monomorphic and cannot change any further. By comparison, the frequencies of B and b in the large population fluctuated much less. As discussed in Chapter 16, the relative effect of random chance, also termed random sampling error, is much smaller when the sample size is large. Nevertheless, genetic drift can eventually lead to allele loss or fixation even in large populations, but this will take many more generations to occur than it does in small populations.

In nature, genetic drift may rapidly alter allele frequencies when the size of a population dramatically decreases. Two examples of this phenomenon are the bottleneck effect and the founder effect, which are described next.

Bottleneck Effect A population can be dramatically reduced in size by events such as earthquakes, floods, drought, and human destruction of habitat. These occurrences may eliminate most members of the population without regard to their genetic composition. The population is said to have passed through a bottleneck. The change in allele frequencies of the resulting population due to genetic drift is called the **bottleneck effect**. Some alleles may be overrepresented whereas others may even be eliminated. Such changes may happen for two reasons. First, the surviving population often has allele frequencies that differ from those of the original population that was much larger. Second, as we saw in Figure 24.10, genetic drift acts more quickly to reduce genetic variation when the population size is small. Eventually, a population that has gone through a bottleneck may regain its original size. However, the new population is likely to have less genetic variation than the original one.

A hypothetical example of the bottleneck effect is shown with a population of frogs in **Figure 24.11**. In this example, a starting population of frogs is found in three phenotypes: yellow, dark green, and striped. Due to a bottleneck caused by a drought, the dark green variety is lost from the population.

As a real-life example, the Northern elephant seal (*Mirounga angustirostris*) has lost much of its genetic variation. This was caused by a bottleneck effect in which the population decreased to approximately 20 to 30 surviving members in the 1890s due to hunting. The species has rebounded in numbers to over 100,000, but the bottleneck reduced its genetic variation to very low levels.

Founder Effect Another common phenomenon in which genetic drift may rapidly alter allele frequencies is the **founder effect**. This occurs when a small group of individuals separates from a larger population and establishes a colony in a new location. For example, a few individuals may migrate from a large population on a continent and become the founders of an island population. The founder effect differs from a bottleneck in that it occurs in a new location,

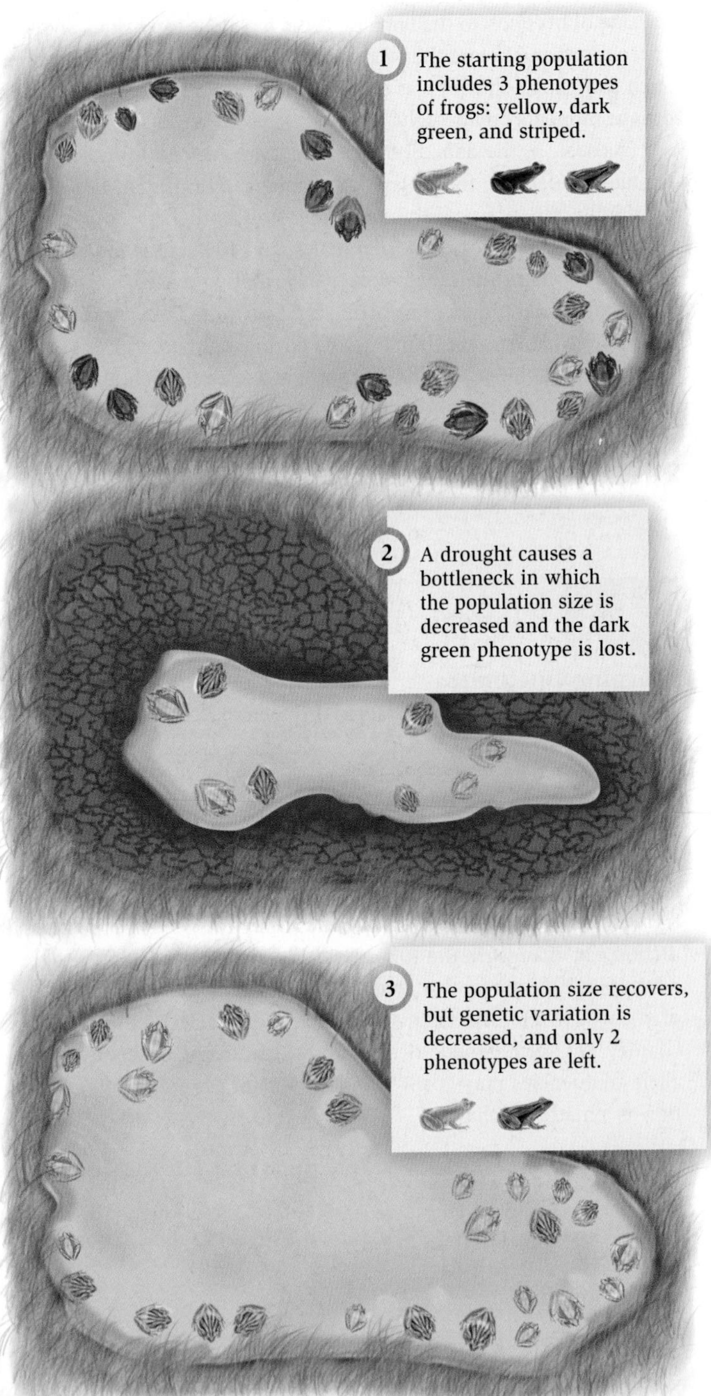

1 The starting population includes 3 phenotypes of frogs: yellow, dark green, and striped.
2 A drought causes a bottleneck in which the population size is decreased and the dark green phenotype is lost.
3 The population size recovers, but genetic variation is decreased, and only 2 phenotypes are left.

Figure 24.11 A hypothetical example of the bottleneck effect. This example involves a population of frogs in which a drought dramatically reduced population size, resulting in a bottleneck. The bottleneck reduced the genetic diversity in the population.

BIOLOGY PRINCIPLE Populations of organisms evolve from one generation to the next. Genetic drift randomly changes allele frequencies and (in the long run) leads to a loss or fixation of an allele.

Concept Check: How does the bottleneck effect undermine the efforts of conservation biologists who are trying to save species nearing extinction?

although both effects are related to a reduction in population size. The founder effect has two important consequences. First, the founding population, which is relatively small, is expected to have less genetic variation than the larger original population from which it was derived. Second, as a matter of chance, the allele frequencies in the founding population may differ markedly from those of the original population.

Population geneticists have studied many examples in which isolated populations were founded via colonization by members of another population. For example, in the 1960s, American geneticist Victor McKusick studied allele frequencies in the Amish of Lancaster County, Pennsylvania. At that time, this group included about 8,000 people, descended from just three couples that immigrated to the U.S. in 1770. Among this population of 8,000, a genetic disease known as Ellis–van Creveld syndrome (a recessive form of dwarfism) was found at a frequency of 0.07, or 7%. By comparison, this disorder is extremely rare in other human populations, even the population from which the founding members had originated. Evidence suggests that the high frequency in the Lancaster County population can be traced to one couple, one of whom carried the mutated gene that causes the syndrome.

Genetic Drift Plays an Important Role in Promoting Genetic Change

In 1968, Japanese evolutionary biologist Motoo Kimura proposed that much of the DNA sequence variation seen in genes in natural populations is the result of genetic drift rather than natural selection. Genetic drift is a random process that does not preferentially select for any particular allele—it can alter the frequencies of both beneficial and deleterious alleles. Much of the time, genetic drift promotes **neutral variation**—changes in genes and proteins that do not have an effect on reproductive success.

According to Kimura, most variation in DNA sequences is due to the accumulation of neutral mutations that have attained high frequencies in a population via genetic drift. For example, a new mutation within a gene that changes a glycine codon from GGG to GGC would not affect the amino acid sequence of the encoded protein. Both genotypes are equal in fitness. However, such new mutations can spread throughout a population due to genetic drift (Figure 24.12). This phenomenon has been called **non-Darwinian evolution** and also "survival of the luckiest." Kimura agreed with Darwin that natural selection is responsible for adaptive changes in a species during evolution. The long neck of the giraffe is the result of natural selection. His main idea is that much of the variation in DNA sequences is explained by neutral variation rather than adaptive variation.

The sequencing of genomes from many species is consistent with Kimura's proposal. When we examine changes of the coding sequence within structural genes, we find that nucleotide substitutions are more prevalent in the third base of a codon than in the first or second base. Mutations in the third base are often neutral; that is, they do not change the amino acid sequence of the protein (refer back to Table 12.1). In contrast, random mutations at the first or second base are more likely to be harmful than beneficial and tend to be eliminated from a population.

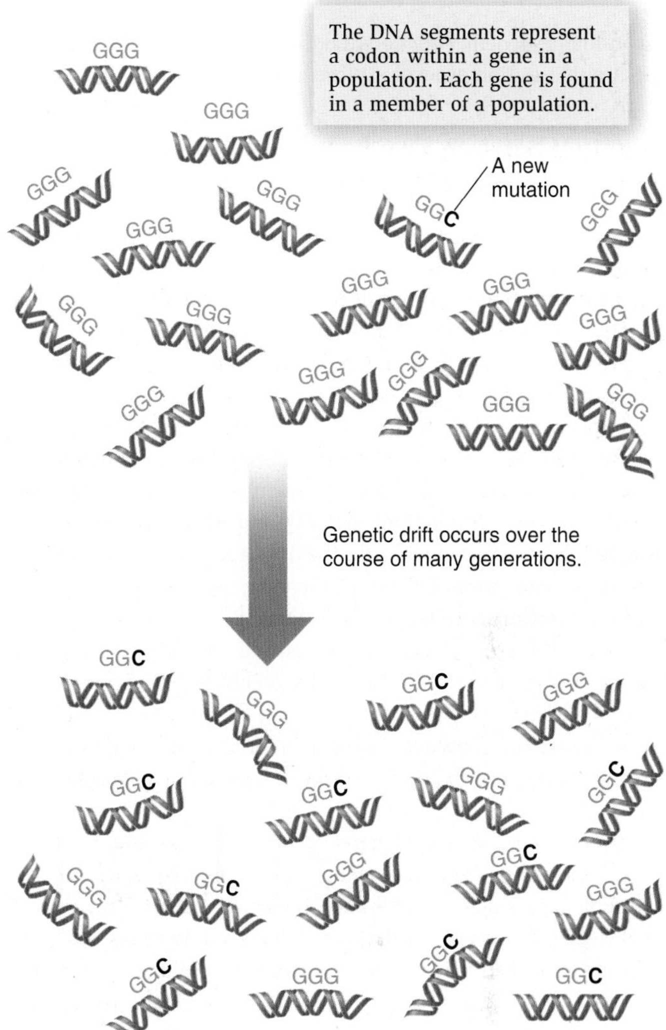

The DNA segments represent a codon within a gene in a population. Each gene is found in a member of a population.
A new mutation
GGG
GGC
Genetic drift occurs over the course of many generations.

Figure 24.12 **Neutral evolution in a population.** In this example, a mutation within a gene changes a glycine codon from GGG to GGC, which does not affect the amino acid sequence of the encoded protein. Each gene shown represents a copy of the gene in a member of a population. Over the course of many generations, genetic drift may cause this neutral allele to become prevalent in the population, perhaps even monomorphic.

BioConnections: *Look back at the genetic code described in Table 12.1. Describe three different genetic changes that you would expect to be neutral.*

24.5 Migration and Nonrandom Mating

Learning Outcomes:

1. Describe how gene flow affects genetic variation in neighboring populations.
2. Define inbreeding and explain how it may have detrimental consequences.

Thus far, we have considered how natural selection and genetic drift are key mechanisms that cause evolution to happen. In addition, migration between neighboring populations and nonrandom mating may influence genetic variation and the relative proportions of genotypes. In this section, we will explore how these mechanisms work.

Migration Between Two Populations Tends to Increase Genetic Variation

Earlier in this chapter, we considered how migration to a new location by a relatively small group can result in a founding population with an altered genetic composition due to genetic drift. In addition, migration between two different established populations can alter allele frequencies. As an example, let's consider two populations of a particular species of deer that are separated by a mountain range running north and south (**Figure 24.13**). On rare occasions, a few deer from the western population may travel through a narrow pass between the mountains and become members of the eastern population. If the two populations are different with regard to genetic variation, this migration will alter the frequencies of certain alleles in the eastern population. Of course, this migration could occur in the opposite direction as well and would then affect the western population. This transfer of alleles into or out of a population, called **gene flow**, occurs whenever individuals move between populations having different allele frequencies.

What are the consequences of migration? First, migration tends to reduce differences in allele frequencies between neighboring populations. Population geneticists can evaluate the extent of migration between two populations by analyzing the similarities and differences between their allele frequencies. Populations that frequently mix their gene pools via migration tend to have similar allele frequencies, whereas the allele frequencies of isolated populations are more disparate, due to the effects of natural selection and genetic drift. Second, migration tends to increase genetic diversity within populations. As discussed earlier in this chapter, new mutations are relatively rare events. Therefore, a new mutation may arise in only one population, and migration may then introduce this new allele into a neighboring population.

Nonrandom Mating Affects the Relative Proportion of Homozygotes and Heterozygotes in a Population

As mentioned earlier, one of the conditions required to establish Hardy-Weinberg equilibrium is random mating, which means that members of a population choose their mates irrespective of their genotypes or phenotypes. In many species, including human populations, this condition is violated. Such **nonrandom mating** takes different forms. Assortative mating occurs when individuals with similar phenotypes are more likely to mate. If the similar phenotypes are due to similar genotypes, assortative mating tends to increase the proportion of homozygotes and decrease the proportion of heterozygotes in the population. The opposite situation, where dissimilar phenotypes mate preferentially, causes heterozygosity to increase.

Another form of nonrandom mating involves the choice of mates based on their genetic history rather than their phenotypes. Individuals may choose a mate that is part of the same genetic lineage. The mating of two genetically related individuals, such as cousins, is called **inbreeding**. This sometimes occurs in human societies and is more likely to take place in nature when population size becomes very small.

In the absence of other evolutionary factors, nonrandom mating does not affect allele frequencies in a population. However, it will alter the balance of genotypes predicted by the Hardy-Weinberg equilibrium. As an example, let's consider a human pedigree involving a mating between cousins (**Figure 24.14**). Individuals III-2 and III-3 are cousins and have produced the daughter labeled IV-1. She is said to be inbred, because her parents are genetically related. The parents of an inbred individual have one or more common ancestors. In the pedigree of Figure 24.14, I-2 is the grandfather of both III-2 and III-3.

Inbreeding increases the relative proportions of homozygotes and decreases the likelihood of heterozygotes in a population. Why does this happen? An inbred individual has a higher chance of being homozygous for any given gene because the same allele for that gene could be inherited twice from a common ancestor. For example, individual I-2 is a heterozygote, *Cc*. The *c* allele could pass from I-2 to II-2 to III-2 and finally to IV-1 (see red lines in Figure 24.14). Likewise, the *c* allele could pass from I-2 to II-3 to III-3 and then to IV-1. Therefore, IV-1 has a chance of being homozygous because she inherited both copies of the *c* allele from a common ancestor to both of her parents. Inbreeding does not favor any particular allele—it does not favor *c* over *C*—but it does increase the likelihood that an individual will be homozygous for any given gene.

Although inbreeding by itself does not affect allele frequencies, it may have negative consequences with regard to recessive alleles. Rare recessive alleles that are harmful in the homozygous condition are found in all populations. Such alleles do not usually pose a problem because heterozygotes carrying a rare recessive allele are also rare, making it very unlikely that two such heterozygotes will mate with each other. However, related individuals share some of their genes, including recessive alleles. Therefore, if inbreeding occurs, homozygous offspring are more likely to be produced. For example, rare recessive diseases in humans are more frequent when inbreeding occurs.

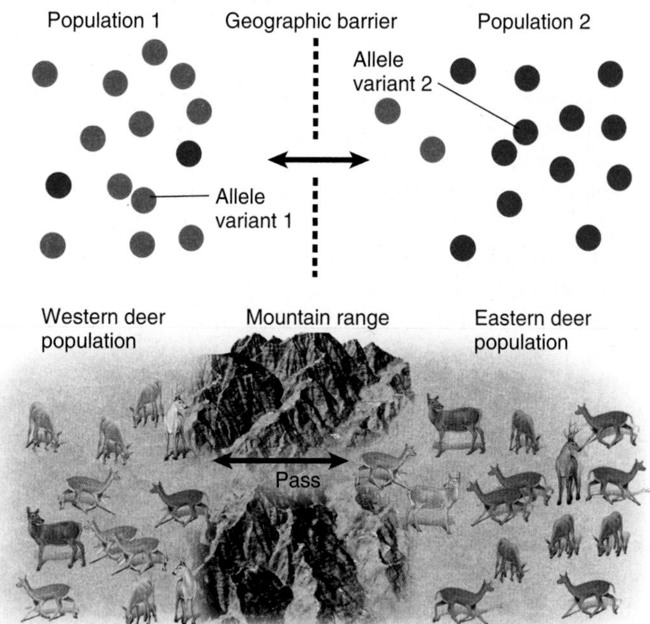

Figure 24.13 Migration and gene flow. In this example, two populations of a deer species are separated by a mountain range. On rare occasions, a few deer from one population travel through a narrow pass and become members of the other population. If the two populations differ in regard to genetic variation, this migration will alter the frequencies of alleles in the populations.

Concept Check: *How does migration affect the genetic compositions of populations?*

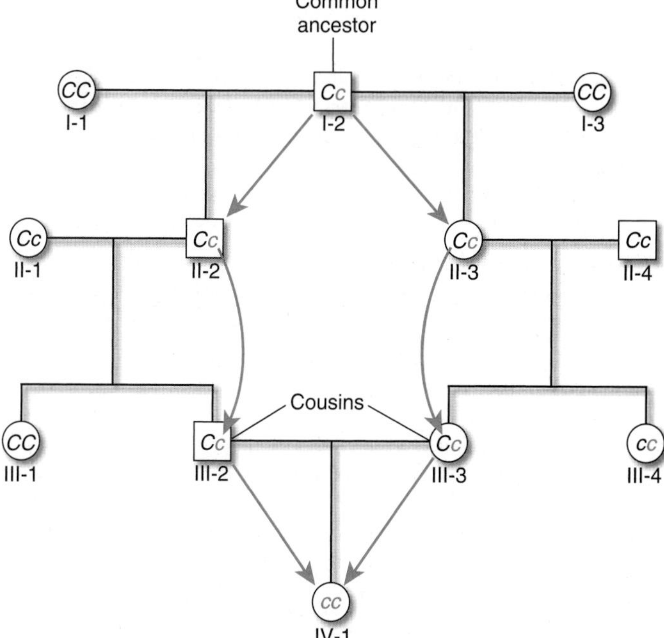

Figure 24.14 A human pedigree containing inbreeding. The parents of individual IV-1 are genetically related (cousins), and, therefore, individual IV-1 is inbred. Inbreeding increases the likelihood that an individual will be homozygous for any given gene. The red arrows show how IV-1 could become homozygous by inheriting the same allele (*c*) from the common ancestor (I-2) to both of her parents.

BioConnections: Many inherited human diseases show a recessive pattern of inheritance (see Table 16.2). Explain whether inbreeding would increase or decrease the likelihood of such diseases.

In natural populations, inbreeding lowers the mean fitness of the population if homozygous offspring have lower fitness values. This can be a serious problem as natural populations become smaller due to human destruction of habitat. As the population shrinks, inbreeding becomes more likely because individuals have fewer potential mates from which to choose. The inbreeding, in turn, produces homozygotes that are less fit, thereby decreasing the reproductive success of the population. This phenomenon is called **inbreeding depression**. Conservation biologists sometimes try to circumvent this problem by introducing individuals from one population into another. For example, the endangered Florida panther (*Puma concolor coryi*) suffers from inbreeding-related defects, which include poor sperm quality and quantity, and morphological abnormalities. To alleviate these effects, panthers from Texas have been introduced into the Florida population of panthers.

Summary of Key Concepts

24.1 Genes in Populations

- Population genetics is the study of genes and genotypes in a population. A population is a group of individuals of the same species that occupy the same environment and can interbreed. All of the alleles for every gene in a population constitute a gene pool.
- Polymorphism, which is very common in nearly all populations, refers to two or more variants of a character in a population. A monomorphic gene exists as a single allele (>99%) in a population (Figure 24.1).
- Allele frequency is the number of copies of a specific allele divided by the total number of all alleles in a population. Genotype frequency is the number of individuals with a given genotype divided by the total number of individuals in a population.
- The Hardy-Weinberg equation ($p^2 + 2pq + q^2 = 1$) predicts that allele and genotype frequencies will remain in equilibrium if no new mutations are formed, no natural selection occurs, the population size is very large, migration does not occur, and mating is random (Figure 24.2).
- Sources of new genetic variation include random gene mutations, gene duplications, exon shuffling, and horizontal gene transfer. Natural selection, genetic drift, migration, and nonrandom mating may alter allele and genotype frequencies and cause a population to evolve (Table 24.1).

24.2 Natural Selection

- Natural selection is the process in which individuals with certain heritable traits that favor survival and reproduction tend to become more prevalent in a population. Fitness, the relative likelihood that a genotype will contribute to the gene pool of the next generation, is a measure of reproductive success.
- Directional selection is the process in which one extreme of a phenotypic distribution is favored (Figure 24.3).
- Stabilizing selection is the process in which an intermediate phenotype is favored (Figure 24.4).
- Diversifying selection is the process in which two or more phenotypes are favored. An example is a population that occupies a diverse environment (Figure 24.5).
- Balancing selection maintains genetic polymorphism in a population. Examples include heterozygote advantage and negative frequency-dependent selection (Figure 24.6).

24.3 Sexual Selection

- Sexual selection is a form of natural selection in which individuals with certain traits are more likely than others to engage in successful mating. In intrasexual selection, members of one sex compete for the opportunity to mate with individuals of the opposite sex. In intersexual selection, members of one sex choose their mates on the basis of certain desirable characteristics (Figure 24.7).
- Seehausen and van Alphen discovered that female cichlids' choice of mates is influenced by male coloration. This is an example of sexual selection (Figures 24.8, 24.9).

24.4 Genetic Drift

- Genetic drift involves changes in allele frequencies over time due to chance events. It occurs more rapidly in small populations and leads to either the elimination or the fixation of alleles (Figure 24.10).
- In the bottleneck effect, an environmental event dramatically reduces a population size and the allele frequencies of the resulting population change due to genetic drift (Figure 24.11).
- The founder effect occurs when a small population moves to a new geographic location and genetic drift alters the genetic composition of that population.
- Kimura proposed that genetic drift promotes the accumulation of neutral genetic changes that do not affect reproductive success. Much of the genetic variation in DNA sequences in populations appears to be the result of genetic drift rather than natural selection (Figure 24.12).

24.5 Migration and Nonrandom Mating

- Gene flow occurs when individuals migrate between populations with different allele frequencies. It reduces differences in allele frequencies between populations and enhances genetic diversity (Figure 24.13).
- Inbreeding, a form of nonrandom mating in which genetically related individuals have offspring with each other, tends to increase the proportion of homozygotes relative to heterozygotes. When the resulting homozygotes have lower fitness, this phenomenon is called inbreeding depression (Figure 24.14).

Assess and Discuss

Test Yourself

1. Population geneticists are interested in the genetic variation in populations. The most common type of genetic change that causes polymorphism in a population is
   a. a deletion of a gene sequence.
   b. a duplication of a region of a gene.
   c. a rearrangement of a gene sequence.
   d. a single-nucleotide substitution.
   e. an inversion of a segment of a chromosome.
2. The Hardy-Weinberg equation characterizes the allele and genotype frequencies
   a. of a population that is experiencing selection for mating success.
   b. of a population that is extremely small.
   c. of a population that is very large and not evolving.
   d. of a community of species that is not evolving.
   e. of a community of species that is experiencing selection.
3. In the Hardy-Weinberg equation, what portion of the equation would be used to calculate the frequency of individuals that do not exhibit a recessive disease but are carriers of a recessive allele?
   a. q c. $2pq$ e. both b and d
   b. p^2 d. q^2
4. By itself, which of the following is not likely to have a major influence on allele frequencies?
   a. natural selection d. inbreeding
   b. genetic drift e. both c and d
   c. mutation
5. Which of the following statements is correct regarding mutations?
   a. Mutations are not important in evolution.
   b. Mutations provide the source for genetic variation, but other evolutionary factors are more important in determining allele frequencies in a population.
   c. Mutations occur at such a high rate that they promote major changes in the gene pool from one generation to the next.
   d. Mutations are of greater importance in smaller populations than in larger ones.
   e. Mutations are of greater importance in larger populations than in smaller ones.
6. In a population of fish, body coloration varies from a light shade, almost white, to a very dark shade of green. If changes in the environment resulted in decreased predation of individuals with the lightest coloration, this would be an example of ____________ selection.
   a. diversifying c. directional e. artificial
   b. stabilizing d. sexual
7. Considering the same population of fish described in question 6, if the stream environment included several areas of sandy, light-colored bottom areas and a lot of dark-colored vegetation, both the light- and dark-colored fish would have selective advantage and increased survival in certain places. This type of scenario could explain the occurrence of
   a. genetic drift. d. stabilizing selection.
   b. diversifying selection. e. sexual selection.
   c. mutation.
8. The microevolutionary factor most sensitive to population size is
   a. mutation. c. selection. e. all of the above.
   b. migration. d. genetic drift.
9. Kimura's proposal regarding neutral mutations differs from Darwinian evolution in that
   a. natural selection does not exist.
   b. most of the genetic variation in a population is due to neutral mutations, which do not affect reproductive success.
   c. neutral variation alters survival and reproductive success.
   d. neutral mutations are not affected by population size.
   e. both b and c.
10. Populations that experience inbreeding may also experience
   a. a decrease in fitness due to an increased frequency of recessive genetic diseases.
   b. an increase in fitness due to increases in heterozygosity.
   c. very little genetic drift.
   d. no apparent change.
   e. increased mutation rates.

Conceptual Questions

1. The percentage of individuals exhibiting a recessive disease in a population is 0.04, which is 4%. Based on a Hardy-Weinberg equilibrium, what percentage of individuals would be expected to be heterozygous carriers?
2. Compare and contrast the four patterns of natural selection that lead to environmental adaptation. You should also discuss sexual selection.
3. A principle of biology is that *populations of organisms evolve from one generation to the next.* Explain how genetic drift results in evolution.

Collaborative Questions

1. Antibiotics are commonly used to combat bacterial and fungal infections. During the past several decades, however, antibiotic-resistant strains of microorganisms have become alarmingly prevalent. This has undermined the ability of physicians to treat many types of infectious disease. Discuss how the following processes that alter allele frequencies may have contributed to the emergence of antibiotic-resistant strains:
   a. random mutation
   b. genetic drift
   c. natural selection
2. Discuss the similarities and differences among directional, disruptive, balancing, and stabilizing selection.

Online Resource

www.brookerbiology.com

Stay a step ahead in your studies with animations that bring concepts to life and practice tests to assess your understanding. Your instructor may also recommend the interactive eBook, individualized learning tools, and more.

Chapter Outline

25.1 Identification of Species
25.2 Mechanisms of Speciation
25.3 The Pace of Speciation
25.4 Evo-Devo: Evolutionary Developmental Biology
Summary of Key Concepts
Assess and Discuss

Origin of Species and Macroevolution

25

Two different species of zebras. Grevy's zebra (*Equus grevyi*) is shown on the left, and Grant's zebra (*Equus quagga boehmi*), which has fewer and thicker stripes, is shown on the right.

The origin of living organisms has been described by philosophers as the great "mystery of mysteries." Perhaps that is why so many different views have been put forth to explain the existence of living species. At the time of Aristotle (4th century B.C.E.), most people believed that some living organisms came into being by spontaneous generation—the idea that nonliving materials can give rise to living organisms. For example, it was commonly believed that worms and frogs could arise from mud, and mice could come from grain. By comparison, many religious teachings contend that species were divinely made and have remained the same since their creation. In contrast to these ideas, the work of Charles Darwin provided the scientific theory of evolution by descent with modification. Darwin's work, and that of subsequent biologists, helps us to understand the diversity of life, and in particular, it presents a logical explanation for how new species can evolve from pre-existing species.

This chapter provides an exciting way to build on the information that we have considered in previous chapters. In Chapter 22, we examined how the first primitive cells in an RNA world could have evolved into prokaryotic cells and eventually eukaryotes. Chapter 23 surveyed the tenets on which the theory of evolution is built, and in Chapter 24, we viewed microevolution—evolution on a small scale as it relates to allele frequencies in a population. In this chapter, we will consider evolution on a larger scale. **Macroevolution** refers to evolutionary changes that produce new species and groups of species.

To biologists, the concept of a **species** has come to mean a group of related organisms that share a distinctive set of attributes in nature. Members of the same species share an evolutionary history, which makes them more genetically similar to each other than they are to members of a different species. You may already have an intuitive sense of this concept. It is obvious that zebras and mice are different species. However, as we will learn in the first section of this chapter, the distinction between different, closely related species is often blurred in natural environments. Two closely related species may look very similar, as the chapter opening photo illustrates. Species identification has several practical uses. For example, it allows biologists to plan for the preservation and conservation of endangered species. In addition, it is often important for a physician to correctly identify the bacterial species that is causing a disease in a patient so the proper medication can be prescribed.

In this chapter, we will also focus on the mechanisms that promote the formation of new species, a phenomenon called **speciation**. Such macroevolution typically occurs by the accumulation of microevolutionary changes, those that occur in single genes (see Chapter 24). We will also consider how macroevolution can happen at a fast or slow pace and explore how variations in the genes that control development play a role in the evolution of new species.

25.1 Identification of Species

Learning Outcomes:

1. Outline the characteristics that biologists use to distinguish different species.
2. Describe different species concepts.
3. Compare and contrast prezygotic and postzygotic isolating mechanisms.

How many different species are on Earth? The number is astounding. A study done by American biologist E. O. Wilson and colleagues in 1990 estimated the known number of species at approximately 1.4 million. Currently, about 1.3 million species have been identified and catalogued. However, a vast number of species have yet to be classified. This is particularly true among bacteria and archaea, which are difficult to categorize into distinct species. Also, new invertebrate and even vertebrate species are still being found in the far reaches of pristine habitats. Common estimates of the total number of species range from 5 to 50 million!

When studying natural populations, evolutionary biologists are often confronted with situations in which some differences between two populations are apparent, but it is difficult to decide whether the two populations truly represent separate species. When two or more geographically restricted groups of the same species display one or more traits that are somewhat different but not enough to warrant their placement into different species, biologists sometimes classify such groups as **subspecies**. Similarly, many bacterial species are subdivided into **ecotypes**. Each ecotype is a genetically distinct population adapted to its local environment. In this section, we will consider the characteristics that biologists examine when deciding if two groups of organisms constitute different species.

Each Species Is Established Using Characteristics and Histories That Distinguish It from Other Species

As mentioned, a species is a group of organisms that share a distinctive set of attributes in nature. In the case of sexually reproducing species, members of one species usually cannot successfully interbreed with members of other species. Members of the same species share an evolutionary history that is distinct from other species. Although this may seem like a reasonable way to characterize a given species, biologists would agree that distinguishing between species is a more difficult undertaking. What criteria do we use to distinguish species? How many differences must exist between two populations to classify them as different species? Such questions are often difficult to answer.

The characteristics that a biologist uses to identify a species depend, in large part, on the species in question. For example, the traits used to distinguish insect species are quite different from those used to identify different bacterial species. The relatively high level of horizontal gene transfer among bacteria presents special challenges in the grouping of bacterial species. Among bacteria, it is sometimes very difficult and perhaps arbitrary to divide closely related organisms into separate species.

The most commonly used characteristics for identifying species are morphological traits, the ability to interbreed, molecular features, ecological factors, and evolutionary relationships. A comparison of these concepts will help you appreciate the various approaches that biologists use to identify the bewildering array of species on our planet.

Morphological Traits One way to establish that a population constitutes a unique species is based on their physical characteristics. Organisms are classified as the same species if their anatomical traits appear to be very similar. Likewise, microorganisms can be classified according to morphological traits at the cellular level. By comparing many different morphological traits, biologists may be able to decide that certain populations constitute a unique species.

Although an analysis of morphological traits is a common way for biologists to establish that a particular group constitutes a species, this approach has drawbacks. First, researchers may have difficulty deciding how many traits to consider. In addition, quantitative traits, such as size and weight, that vary in a continuous way among members of the same species are not easy to analyze. Another drawback is that the degree of dissimilarity that distinguishes different species may not show a simple relationship. The members of the same

(a) Frogs of the same species

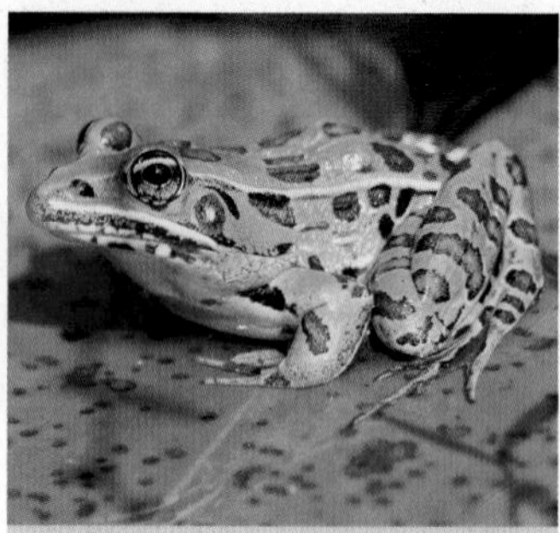

(b) Frogs of different species

Figure 25.1 Difficulties of using morphological traits to identify species. In some cases, members of the same species appear quite different. In other cases, members of different species look very similar. **(a)** Two frogs of the same species, the dyeing poison frog (*Dendrobates tinctorius*). **(b)** Two different species of frog, the Northern leopard frog (*Rana pipiens*, left) and the Southern leopard frog (*Rana utricularia*, right).

Concept Check: Can you think of another example of two different species that look very similar?

species sometimes look very different, and conversely, members of different species sometimes look remarkably similar to each other. For example, **Figure 25.1a** shows two different frogs of the species *Dendrobates tinctorius*, commonly called the dyeing poison frog. This species exists in many different-colored morphs, which are individuals of the same species that have noticeably dissimilar appearances. In contrast, **Figure 25.1b** shows two different species of frogs, the Northern leopard frog (*Rana pipiens*) and the Southern leopard frog (*Rana utricularia*), which look fairly similar.

Reproductive Isolation Why would biologists describe two species, such as the Northern leopard frog and Southern leopard frog, as being different if they are morphologically similar? One reason is that biologists have discovered that they are unable to breed with each other in nature. Therefore, a second way of identifying a species is by its ability to interbreed. In the late 1920s, geneticist Theodosius Dobzhansky proposed that each species is reproductively isolated from other species. Such **reproductive isolation** prevents one species from successfully interbreeding with other species. In 1942, German evolutionary biologist Ernst Mayr expanded on the ideas of Dobzhansky to provide a reproductive definition of a species. According to Mayr, a key feature of sexually reproducing species is that, in nature, the members of one species have the potential to interbreed with one another to produce viable, fertile offspring but cannot successfully interbreed with members of other species. As discussed later in this section, reproductive isolation among species of plants and animals occurs by an amazing variety of different mechanisms.

Reproductive isolation has been used to distinguish many plant and animal species, especially those that look alike but do not interbreed. Even so, this criterion suffers from four main problems. First, in nature, it may be difficult to determine if two populations are reproductively isolated, particularly if the populations have nonoverlapping geographic ranges. Second, biologists have noted many cases in which two different species can interbreed in nature yet consistently maintain themselves as separate species. For example, different species of yucca plants, such as *Yucca pallida* and *Yucca constricta*, do interbreed in nature yet typically maintain populations with distinct characteristics. For this reason, they are viewed as distinct species. A third drawback of reproductive isolation is that it does not apply to asexual species such as bacteria. Likewise, some species of plants and fungi reproduce only asexually. Finally, a fourth drawback is that it cannot be applied to extinct species. For these reasons, reproductive isolation has been primarily used to distinguish closely related species of modern animals and plants that reproduce sexually.

Molecular Features Molecular features are now commonly used to determine if two different populations are different species. Evolutionary biologists often compare DNA sequences within genes, gene order along chromosomes, chromosome structure, and chromosome number in order to identify similarities and differences among different populations. For example, researchers may compare the DNA sequence of the *16S rRNA* gene between different bacterial populations as a way of determining if the two populations represent different species. When the sequences are very similar, such populations would probably be judged as the same species. However, it may be difficult to draw the line when separating groups into different species. How much difference must be present for species to be considered separate? Is a 2% difference in their genome sequences sufficient to warrant placement into two different species, or do we need a 5% difference?

Ecological Factors A variety of factors related to an organism's habitat are used to distinguish one species from another. For example, certain species of warblers are distinguished by the habitat in which they forage for food. Some species search the ground for food, others forage in bushes or small trees, and some species primarily forage in tall trees. Such habitat differences are used to distinguish different species that look morphologically similar.

Many bacterial species have been categorized as distinct based on ecological factors. Bacterial cells of the same species are likely to use the same types of resources (such as sugars and vitamins) and grow under the same types of conditions (such as temperature and pH). However, a drawback of this approach is that different groups of bacteria sometimes display very similar growth characteristics, and even the same species may show great variation in the growth conditions it will tolerate.

Evolutionary Relationships In Chapter 26, we will examine the methods that are used to produce evolutionary trees that describe the relationships between ancestral species and modern species. In some cases, such relationships are based on an analysis of the fossil record. For example, in Chapter 26, we will consider how the fossil record was used to construct a tree that shows the ancestors that led to modern horse species. Alternatively, another way of establishing evolutionary relationships is by the analysis of DNA sequences. Researchers obtain samples of cells from different individuals and compare the genes within those cells to see how similar or different they are.

Biologists Have Proposed Different Species Concepts

A **species concept** is a way of defining the concept of a species and/or of providing an approach to distinguish one species from another. However, even Darwin realized the difficulty in defining a species. In 1859, he said, "No one definition [of species] has as yet satisfied all naturalists; yet every naturalist knows vaguely what he means when he speaks of a species." Since 1942, over 20 different species concepts have been proposed by a variety of evolutionary biologists. Ernst Mayr proposed one of the first species concepts, called the **biological species concept**. According to Mayr's concept, a species is a group of individuals whose members have the potential to interbreed with one another in nature to produce viable, fertile offspring but cannot successfully interbreed with members of other species. The biological species concept emphasizes reproductive isolation as the most important criterion for delimiting species.

Another example is the **evolutionary lineage concept** proposed by American paleontologist George Gaylord Simpson in 1961. A **lineage** is a series of species that forms a line of descent, with each new species the direct result of speciation from an immediate ancestral species. According to Gaylord, species should be defined based on the separate evolution of lineages. A third example is the **ecological species concept**, described by American evolutionary biologist Leigh Van Valen in 1976. According to this viewpoint, each species occupies an ecological niche, which is the unique set of habitat resources that a species requires, as well as its influence on the environment and other species.

Most evolutionary biologists would agree that different methods are needed to distinguish the vast array of species on Earth. Even so, some evolutionary biologists have questioned whether it is valid to have many different species concepts. In 1998, American zoologist Kevin de Queiroz suggested that there is only a single general species concept, which concurs with Simpson's evolutionary lineage concept and includes all previous concepts. According to de Queiroz's **general lineage concept**, each species is a population of an independently evolving lineage. Each species has evolved from a specific series of ancestors and, as a consequence, forms a group of organisms with a particular set of characteristics. Multiple criteria are used to determine if a population is part of an independent evolutionary lineage, and thus a species, which is distinct from others. Typically, researchers use analyses of morphology, reproductive isolation, DNA sequences, and ecology to determine if a population or group of populations is distinct from others. Because of its generality, the general lineage concept has received significant support.

Reproductive Isolating Mechanisms Help to Maintain the Distinctiveness of Each Species

Thus far we have considered various ways of differentiating species. In our discussion, you may have realized that the identification of a species is not always a simple matter. The phenomenon of reproductive isolation has played a major role in the way biologists study plant and animal species, partly because it identifies a possible mechanism for the process of forming new species. For this reason, much research has been done to try to understand **reproductive isolating mechanisms**, the mechanisms that prevent interbreeding between different species. Why do reproductive isolating mechanisms occur? Populations do not intentionally erect these reproductive barriers. Rather, reproductive isolation is a consequence of genetic changes that occur usually because a species becomes adapted to its own particular environment. The view of evolutionary biologists is that reproductive isolation typically evolves as a by-product of genetic divergence. Over time, as a species evolves its own unique characteristics, some of those traits are likely to prevent breeding with other species.

Reproductive isolating mechanisms fall into two categories: **prezygotic isolating mechanisms**, which prevent the formation of a zygote, and **postzygotic isolating mechanisms**, which block the development of a viable and fertile individual after fertilization has taken place. **Figure 25.2** summarizes some of the more common ways that reproductive isolating mechanisms prevent reproduction between different species. When two species do produce offspring, such an offspring is called an **interspecies hybrid**.

Prezygotic Isolating Mechanisms We will consider five types of prezygotic isolating mechanisms.

Habitat Isolation: One obvious way to prevent interbreeding is for members of different species to never come in contact with each other. This phenomenon, called habitat isolation, may involve a geographic barrier to interbreeding. For example, a large body of water may separate two different plant species that live on nearby islands.

Temporal Isolation: In temporal isolation, species happen to reproduce at different times of the day or year. In the northeastern U.S., for example, the two most abundant field crickets, *Gryllus veletis* and *Gryllus pennsylvanicus* (spring and fall field crickets, respectively), do not differ in song or habitat and are morphologically very similar (**Figure 25.3**). How do the two species maintain reproductive isolation? *G. veletis* matures in the spring, whereas *G. pennsylvanicus* matures in the fall. This minimizes interbreeding between the two species.

Behavioral Isolation: In the case of animals, mating behavior and anatomy often play key roles in promoting reproductive isolation. An example of the third type of prezygotic isolation, behavioral isolation, is found between the western meadowlark (*Sturnella neglecta*) and eastern meadowlark (*Sturnella magna*). Both species are nearly identical in shape, coloration, and habitat, and their ranges overlap in the central U.S. (**Figure 25.4**). For many years, they were thought to be the same species. When biologists discovered that the western meadowlark is a separate species, it was given the species name *S. neglecta* to reflect the long delay in its recognition. In the zone of overlap, very little interspecies mating takes place between western and eastern meadowlarks, largely due to differences in their songs. The song of the western meadowlark is a long series of flutelike gurgling notes that go down the scale. By comparison, the eastern meadowlark's song is a simple series of whistles, typically about four or five notes. These differences in songs enable meadowlarks to recognize potential mates as members of their own species.

Mechanical Isolation: A fourth type of prezygotic isolation, called mechanical isolation, occurs when morphological features such as size or incompatible genitalia prevent two species from interbreed-

Species 1

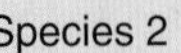

Species 2

Prezygotic isolating mechanisms
Habitat isolation: Species occupy different habitats, so they never come in contact with each other.
Temporal isolation: Species have different mating or flowering seasons or times of day or become sexually mature at different times of the year.
Behavioral isolation: Sexual attraction between males and females of different animal species is limited due to differences in behavior or physiology.
Attempted mating
Mechanical isolation: Morphological features such as size and incompatible genitalia prevent 2 members of different species from interbreeding.
Gametic isolation: Gametic transfer takes place, but the gametes fail to unite with each other. This can occur because the male and female gametes fail to attract, because they are unable to fuse, or because the male gametes are inviable in the female reproductive tract of another species. In plants, the pollen of one species usually cannot generate a pollen tube to fertilize the egg cells of another species.
Fertilization
Postzygotic isolating mechanisms
Hybrid inviability: The egg of one species is fertilized by the sperm from another species, but the fertilized egg fails to develop past the early embryonic stages.
Hybrid sterility: An interspecies hybrid survives, but it is sterile. For example, the mule, which is sterile, is produced from a cross between a male donkey (Equus asinus) and a female horse (Equus caballus).
Hybrid breakdown: The F1 interspecies hybrid is viable and fertile, but succeeding generations (F2, and so on) become increasingly inviable. This is usually due to the formation of less-fit genotypes by genetic recombination.
Interspecies hybrid

Figure 25.2 Reproductive isolating mechanisms. These mechanisms prevent successful breeding between different species. They can occur prior to fertilization (prezygotic) or after fertilization (postzygotic).

BioConnections: Look back at Figure 24.9. Is female choice an example of a prezygotic or postzygotic isolating mechanism?

(a) Spring field cricket (*Gryllus veletis*)

(b) Fall field cricket (*Gryllus pennsylvanicus*)

Figure 25.3 Temporal isolation. Interbreeding between these two species of crickets does not usually occur because *Gryllus veletis* matures in the spring, whereas *Gryllus pennsylvanicus* matures in the fall.

Concept Check: Is this an example of a prezygotic or a postzygotic isolating mechanism?

North America

(b) Eastern meadowlark (*Sturnella magna*)

Western meadowlark
Eastern meadowlark
Zone of overlap

(a) Western meadowlark (*Sturnella neglecta*)

Figure 25.4 Behavioral isolation. (a) The western meadowlark (*Sturnella neglecta*) and **(b)** eastern meadowlark (*Sturnella magna*) are very similar in appearance. The red region in this map shows where the two species' ranges overlap. However, very little interspecies mating takes place due to differences in their songs.

BIOLOGY PRINCIPLE Populations of organisms evolve from one generation to the next. For these two species of meadowlarks, one evolutionary change that took place is that their mating songs became different.

ing. For example, male dragonflies use a pair of special appendages to grasp females during copulation. When a male tries to mate with a female of a different species, his grasping appendages do not fit her body shape.

Gametic Isolation: A fifth type of prezygotic isolating mechanism occurs when two species attempt to interbreed, but the gametes fail to unite in a successful fertilization event. This phenomenon, called gametic isolation, is widespread among plant and animal species. In aquatic animals that release sperm and egg cells into the water, gametic isolation is important in preventing interspecies hybrids. For example, closely related species of sea urchins may release sperm and eggs into the water at the same time. Researchers have discovered that sea urchin sperm have a protein on their surface called bindin, which mediates sperm-egg attachment and membrane fusion. The structure of bindin is significantly different among different sea urchin species, thereby ensuring that fertilization occurs only between sperm and egg cells of the same species.

In flowering plants, gametic isolation is commonly associated with pollination. As discussed in Chapter 39, plant fertilization is initiated when a pollen grain lands on the stigma of a flower and sprouts a pollen tube that ultimately reaches an egg cell (look ahead to Figure 39.4). When pollen is released from a plant, it could be transferred to the stigma of many different plant species. In most cases, when a pollen grain lands on the stigma of a different species, it either fails to generate a pollen tube or the tube does not grow properly and reach the egg cell.

Postzygotic Isolating Mechanisms Let's now turn to postzygotic mechanisms of reproductive isolation, of which there are three common types.

Hybrid Inviability: The mechanism of hybrid inviability occurs when an egg of one species is fertilized by a sperm from another species, but the fertilized egg cannot develop past the early embryonic stages.

Hybrid Sterility: A second postzygotic isolating mechanism is hybrid sterility, in which an interspecies hybrid may be viable but sterile. A classic example of hybrid sterility is the mule, which is produced by a mating between a male donkey (*Equus asinus*) and a female horse (*Equus ferus caballus*) (**Figure 25.5**). All male mules and most female mules are sterile. Why are mules usually sterile? Two reasons explain the sterility. Because the horse has 32 chromosomes per set and a donkey has 31, a mule inherits 63 chromosomes (32 + 31). Due to the uneven number, all of the chromosomes cannot pair evenly. Also, the chromosomes of the horse and donkey have structural differences, which either prevent them from pairing correctly or lead to chromosomal abnormalities if crossing over occurs during meiosis. For these reasons, mules usually produce inviable gametes. Note that the mule has no species name because it is not considered a species due to this sterility.

Hybrid Breakdown: Finally, interspecies hybrids may be viable and fertile, but the subsequent generation(s) may harbor genetic abnormalities that are detrimental. This third mechanism, called hybrid breakdown, can be caused by changes in chromosome structure. The chromosomes of closely related species may have structural differences from each other, such as inversions. In hybrids, a crossover

Male donkey (*Equus asinus*)

×

Female horse (*Equus ferus caballus*)

Mule

Figure 25.5 Hybrid sterility. When a male donkey (*Equus asinus*) mates with a female horse (*Equus ferus caballus*), their offspring is a mule, which is usually sterile.

Concept Check: Is this an example of a prezygotic or a postzygotic isolating mechanism?

may occur in the region that is inverted in one species but not the other. This will produce gametes with too little or too much genetic material. Such hybrids often have offspring with developmental abnormalities.

Postzygotic isolating mechanisms tend to be uncommon in nature compared with prezygotic mechanisms. Why are postzygotic mechanisms rare? One explanation is that they are more costly in terms of energy and resources used. For example, a female mammal would use a large amount of energy to produce an offspring that is sterile. Evolutionary biologists hypothesize that natural selection has favored prezygotic isolating mechanisms because they do not waste a lot of energy.

25.2 Mechanisms of Speciation

Learning Outcomes:

1. Describe how allopatric speciation can occur and how it can lead to adaptive radiation.
2. Outline three different mechanisms of sympatric speciation.

Speciation, the formation of a new species, is caused by genetic changes in a particular group that make it different from the species from which it was derived. As discussed in Chapter 24, mutations in genes can be acted on by natural selection and other evolutionary mechanisms to alter the genetic composition of a population. New species commonly evolve in this manner. In addition, interspecies matings, changes in chromosome number, and horizontal gene transfer may also cause new species to arise. In all of these cases, the underlying cause of speciation is the accumulation of genetic changes that ultimately promote enough differences so we judge a population to constitute a unique species.

Even though genetic changes account for the phenotypic differences observed among living organisms, such changes do not fully explain the existence of many distinct species on our planet. Why does life often diversify into the more or less discrete populations that we recognize as species? Two main explanations have been proposed:

1. In some cases, speciation may occur due to abrupt events, such as changes in chromosome number, that cause reproductive isolation.
2. More commonly, species arise as a consequence of adaptation to different ecological niches. For sexually reproducing organisms, reproductive isolation is typically a by-product of that adaptation.

Depending on the species involved, one or both factors may play a dominant role in the formation of new species. In this section, we will consider how reproductive isolating mechanisms and adaptation to particular environments are critical aspects of the speciation process.

Geographic and Habitat Isolation Can Promote Allopatric Speciation

Cladogenesis is the splitting or diverging of a population into two or more species. In the case of sexually reproducing organisms, the process of cladogenesis requires that gene flow becomes interrupted between two or more populations, limiting or eliminating reproduction between members of different populations. **Allopatric speciation** (from the Greek *allos*, meaning other, and the Latin *patria*, meaning homeland) is the most prevalent way for cladogenesis to occur. This form of speciation occurs when a population becomes isolated from other populations and evolves into one or more species. Typically, this isolation may involve a geographic barrier such as a large area of land or body of water.

In some cases, geographic separation may be caused by slow geological events that eventually produce quite large geographic barriers. For example, a mountain range may emerge and split one species that occupies the lowland regions, or a creeping glacier may divide a population. **Figure 25.6** shows an interesting example in which geological separation promoted speciation. A fish called the Panamic porkfish (*Anisotremus taeniatus*) is found in the Pacific Ocean, whereas the porkfish (*Anisotremus virginicus*) is found in the Caribbean Sea. These two species were derived from an ancestral species whose population was split by the formation of the Isthmus of Panama about 3.5 mya. Before that event, the waters of the Pacific Ocean and Caribbean Sea mixed freely. Since the formation of the isthmus, the two populations have been geographically isolated and have evolved into distinct species.

Allopatric speciation can also occur when a small population moves to a new location that is geographically isolated from the main population. For example, a storm may force a small group of birds from a mainland to a distant island. In this case, migration between the island and the mainland population is an infrequent event. In a relatively short period of time, the small founding population on the island may evolve into a new species. How does speciation occur rapidly? Because the environment on the island may differ significantly from the mainland environment, natural selection may rapidly alter

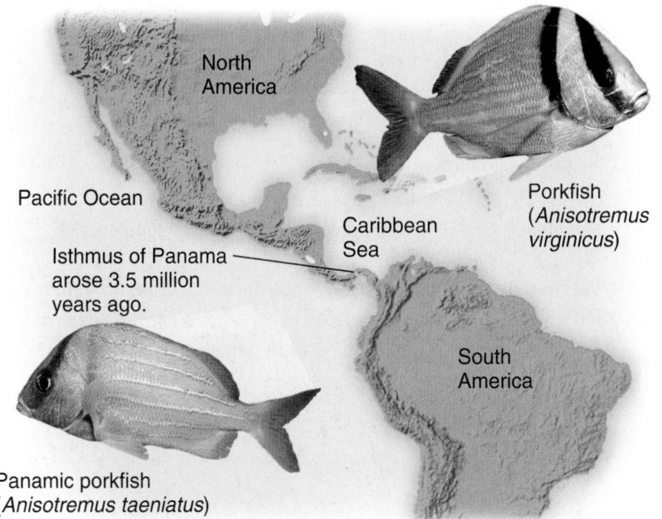

Figure 25.6 Allopatric speciation. An ancestral fish population was split into two by the formation of the Isthmus of Panama about 3.5 mya. Since that time, different genetic changes occurred in the two populations. These changes eventually led to the formation of different species: the Panamic porkfish (*Anisotremus taeniatus*) is found in the Pacific Ocean, and the porkfish (*Anisotremus virginicus*) is found in the Caribbean Sea.

BIOLOGY PRINCIPLE All species (past and present) are related by an evolutionary history. These two species of fish look similar because they share a common ancestor that existed in the fairly recent past.

the genetic composition of the population, leading to adaptation to the new environment. In addition, as discussed in Chapter 24, a form of genetic drift known as the founder effect can have a larger influence in small founding populations.

The Hawaiian Islands are a showcase of allopatric speciation. The islands' extreme isolation coupled with their phenomenal array of ecological niches has enabled a small number of founding species to evolve into a vast assortment of different species. Biologists have investigated several examples of **adaptive radiation**, in which a single ancestral species has evolved into a wide array of descendant species that differ in their habitat, form, or behavior. For example, approximately 1,000 species of *Drosophila* are found dispersed throughout the Hawaiian Islands. Evolutionary studies suggest that these evolved from a single colonization by one species of fruit fly! Natural selection resulted in changes in body form and function that produced the amazing diversity of *Drosophila* species that are now found on the islands.

As shown in Figure 25.7, an example of adaptive radiation is seen with a family of birds called honeycreepers (*Drepanidinae*). Researchers estimate that the honeycreepers' ancestor arrived in

(a) Migration of ancestor to the Hawaiian Islands

(b) Examples of Hawaiian honeycreepers

Figure 25.7 Adaptive radiation. (a) The honeycreepers' ancestor is believed to be related to a Eurasian rosefinch that arrived on the Hawaiian Islands approximately 3–7 mya. Since that time, at least 54 different species of honeycreepers (*Drepanidinae*) have evolved on the islands. **(b)** Adaptations to feeding have produced honeycreeper species with notable differences in beak morphology.

BioConnections: Look back at Figure 24.5b. Discuss how diversifying selection played a role in the diversity of honeycreepers on the Hawaiian Islands.

Hawaii 3–7 mya. This ancestor was a single species of finch, possibly a Eurasian rosefinch (genus *Carpodacus*) or, less likely, the North American house finch (*Carpodacus mexicanus*). At least 54 different species of honeycreepers, many of which are now extinct, evolved from this founding event to fill available niches in the islands' habitats. Natural selection resulted in the formation of many species with different feeding strategies. Seed eaters have stouter, stronger bills capable of cracking tough husks. Insect-eating honeycreepers have thin, warbler-like bills adapted for picking insects from foliage or strong, hooked bills to root out wood-boring insects. The curved bills of nectar-feeding honeycreepers enable them to extract nectar from the flowers of Hawaii's endemic plants.

Before ending our discussion of allopatric speciation, let's consider a common situation in which geographic separation is not complete. The zones where two populations can interbreed are known as **hybrid zones.** Figure 25.8 shows a hybrid zone along a mountain pass that connects two deer populations. For speciation to occur, the amount of gene flow within hybrid zones must become very limited. How does this happen? As the two populations accumulate different genetic changes, the ability of individuals from different populations to mate with each other in the hybrid zone may decrease. For example, natural selection in the western deer population may favor an increase in body size that is not favored in the eastern population. Over time, as this size difference between members of the two populations becomes greater, breeding in the hybrid zone may decrease. Larger individuals may not interbreed easily with smaller ones due to mechanical isolation. In addition, larger individuals may prefer larger individuals as mates, and smaller individuals may also prefer each other. Once gene flow through the hybrid zone is greatly diminished, the two populations are reproductively isolated. Over the course of many generations, such populations may evolve into distinct species.

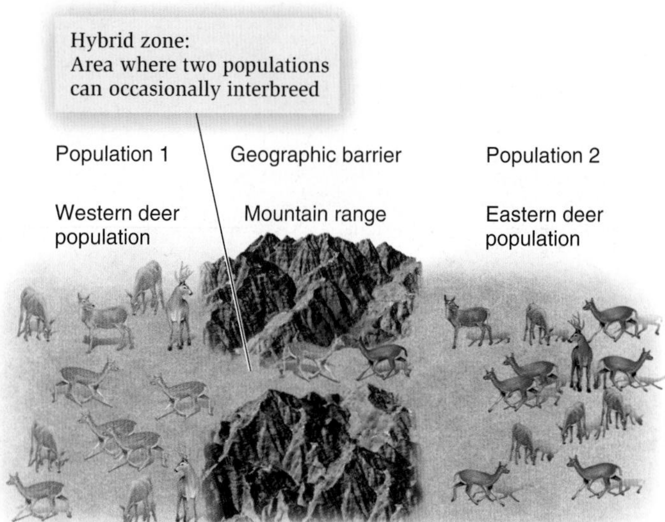

Figure 25.8 Hybrid zones. Two populations of deer are separated by a mountain range. A hybrid zone exists in a mountain pass, where occasional interbreeding may occur.

Feature Investigation

Podos Found That an Adaptation to Feeding May Have Promoted Reproductive Isolation in Finches

In 2001, American evolutionary biologist Jeffrey Podos analyzed the songs of Darwin's finches on the Galápagos Islands to determine how environmental adaptation may contribute to reproductive isolation. As in honeycreepers, the differences in beak sizes and shapes among the various species of finches are adaptations to different feeding strategies. Podos hypothesized that changes in beak morphology could also affect the songs that the birds produce, thereby having the potential to affect mate choice. The components of the vocal tract of birds, including the trachea, larynx, and beak, work collectively to produce a bird's song. Birds actively modify the shape of their vocal tracts during singing, and beak movements are normally very rapid and precise.

Podos focused on two aspects of a bird's song. The first feature is the frequency range, which is a measure of the minimum and maximum frequencies in a bird's song, measured in kilohertz (kHz). The second feature is the trill rate. A trill is a series of notes or group of notes repeated in succession. Figure 25.9 shows a graphical depiction of the songs of Darwin's finches. As you can see, the song patterns of these finches are quite different from each other.

To quantitatively study the relationship between beak size and song, Podos first captured male finches on Santa Cruz, one of the Galápagos Islands, and measured their beak sizes (Figure 25.10). The birds were banded and then released into the wild. The banding provided a way of identifying the birds whose beaks had already been measured. After release, the songs of the banded birds were recorded on a tape recorder, and their range of frequencies and trill rate were analyzed. Podos then compared the data for the Galápagos finches to a large body of data that had been collected on many other bird species. This comparison was used to evaluate whether beak size, in this case, beak depth—the measurement of the beak from top to bottom, at its base—constrained either the frequency range and/or the trill rate of the finches.

The results of this comparison are shown in the data of Figure 25.10. As seen here, the relative constraint on vocal performance became higher as the beak depth became larger. This means that birds with larger beaks had a narrower frequency range and/or a slower trill rate. Podos proposed that as jaws and beaks became adapted for strength to crack open larger, harder seeds, they became less able to perform the rapid movements associated with certain types of songs. In contrast, the finches with smaller beaks adapted to probe for insects or eat smaller seeds had less constraint on their vocal performance. From the perspective of evolution, the changes observed in song patterns for the Galápagos finches could have played an important role in promoting reproductive isolation, because song pattern is an important factor in mate selection in birds. Therefore, a by-product of beak adaptation for feeding is that it also appears to have affected song pattern, possibly promoting reproductive isolation and eventually the formation of distinct species.

Figure 25.9 Differences in the songs of Galápagos finches. These spectrograms depict the frequency of each bird's song over time, measured in kilohertz (kHz). The songs are produced in a series of trills that have a particular pattern and occur at regular intervals. Notice the differences in frequency and trill rate between different species of birds.

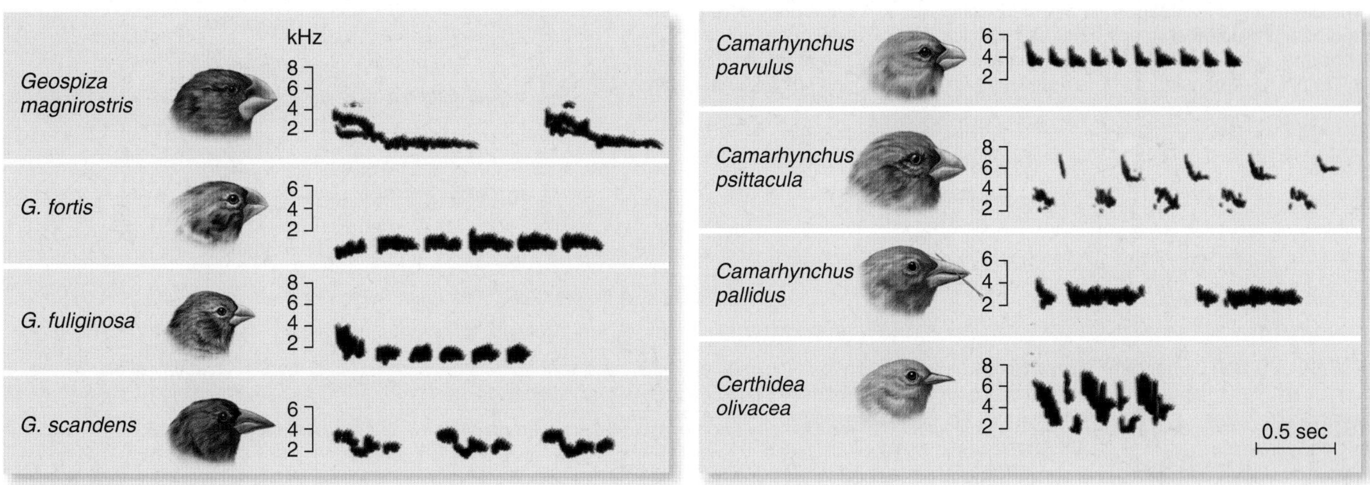

Figure 25.10 Study by Podos investigating the effects of beak depth on song among different species of Galápagos finches.

HYPOTHESIS Changes in beak morphology that are an adaptation to feeding may also affect the songs of Galápagos finches and thereby lead to reproductive isolation between species.

KEY MATERIALS This study was conducted on finch populations of the Galápagos Island of Santa Cruz.

	Experimental level	Conceptual level
1 Capture male finches and measure their beak depth. Beak depth is measured at the base of beak, from top to bottom.	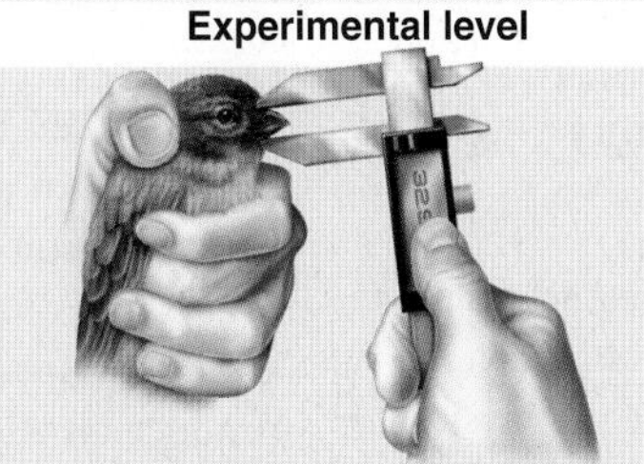	This is a measurement of phenotypic variation in beak size.
2 Band the birds and release them back into the wild.	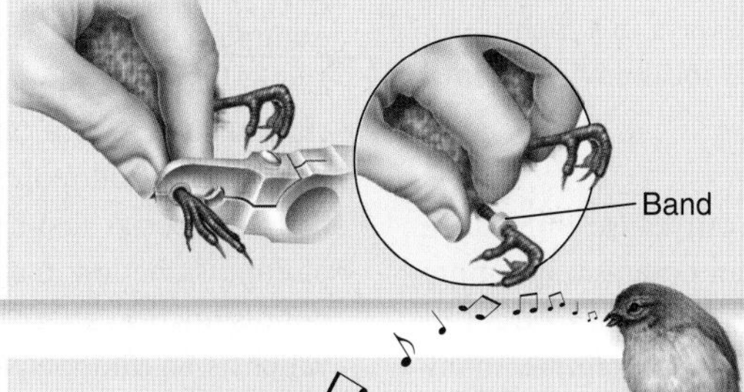 	Banding allows identification of birds with known beak depths.
3 Record the bird's songs on a tape recorder.		This is a measurement of phenotypic variation in song.
4 Analyze the songs with regard to frequency range and trill rate.	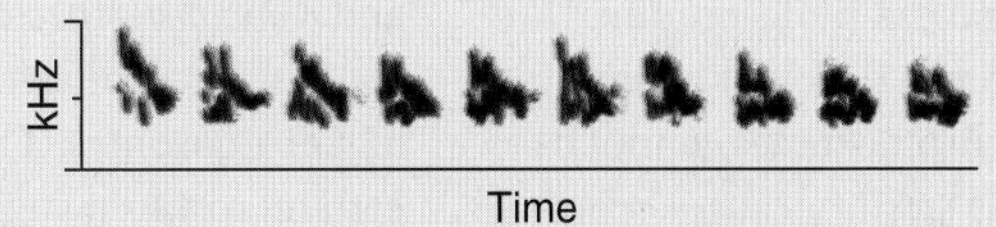 	The frequency range is the value between high and low frequencies. The trill rate is the number of repeats per unit time.

5 THE DATA

The data for the Galápagos finches were compared to a large body of data that had been collected on many other bird species. The relative constraint on vocal performance is higher if a bird has a narrower frequency range and/or a slower trill rate. These constraints were analyzed with regard to each bird's beak depth.

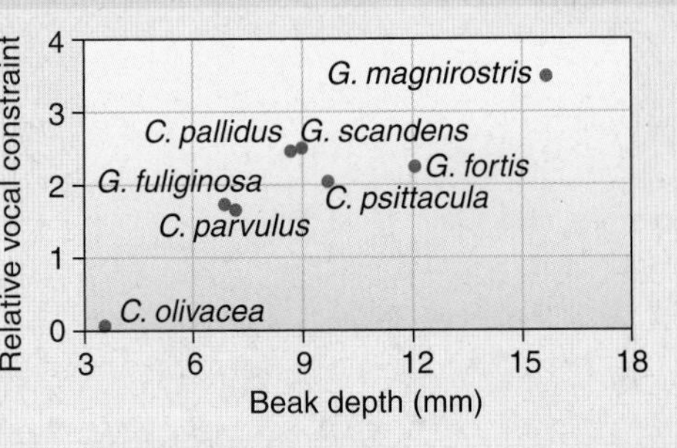

6 CONCLUSION Larger beak size, which is an adaptation to cracking open large, hard seeds, constrains vocal performance. This may affect mating song patterns and thereby promote reproductive isolation and, in turn, speciation.

7 SOURCE Podos, Jeffrey. 2001. Correlated evolution of morphology and vocal signal structure in Darwin's finches. *Nature* 409:185–188.

Experimental Questions

1. What did Podos hypothesize regarding the effects of beak size on a bird's song? How could changes in beak size and shape lead to reproductive isolation among the finches?
2. How did Podos test the hypothesis that beak morphology caused changes in the birds' songs?
3. Did the results of Podos's study support his original hypothesis? Explain. What is meant by the phrase "by-product of adaptation," and how does it apply to this particular study?

Sympatric Speciation Occurs When Populations Are in Direct Contact

Sympatric speciation (from the Greek *sym*, meaning together) occurs when members of a species that are within the same range diverge into two or more different species even though there are no physical barriers to interbreeding. Although sympatric speciation is believed to be less common than allopatric speciation, particularly in animals, evolutionary biologists have discovered several ways in which it can occur. These include polyploidy, adaptation to local environments, and sexual selection.

Polyploidy A type of genetic change that can cause immediate reproductive isolation is **polyploidy**, in which an organism has more than two sets of chromosomes. Plants tend to be more tolerant of changes in chromosome number than animals. For example, many crops and decorative species of plants are polyploid. How does polyploidy occur? One mechanism is complete nondisjunction of chromosomes, which increases the number of chromosome sets in a given species (autopolyploidy). Such changes can result in an abrupt sympatric speciation. For example, nondisjunction could produce a tetraploid plant with four sets of chromosomes from a species that was diploid with two sets. A cross between a tetraploid and a diploid produces a triploid offspring with three sets of chromosomes. Triploid offspring are usually sterile because an odd number of chromosomes cannot be evenly segregated during meiosis. This hybrid sterility causes reproductive isolation between the tetraploid and diploid species.

Another mechanism that leads to polyploidy is interspecies breeding. An **alloploid** organism contains at least one set of chromosomes from two or more different species. This term refers to the occurrence of chromosome sets (ploidy) from the genomes of different (allo-) species. Interbreeding between two different species may produce an allodiploid, an organism that has only one set of chromosomes from each species. Species that are close evolutionary relatives are most likely to breed and produce allodiploid offspring. For example, closely related species of grasses may interbreed to produce allodiploids. An organism containing two or more complete sets of chromosomes from two or more different species is called an allopolyploid. An allopolyploid can be the result of interspecies breeding between species that are already polyploid, or it can occur as a result of nondisjunction in an allodiploid organism. For example, complete nondisjunction in an allodiploid could produce an allotetraploid, which is an allopolyploid with two complete sets of chromosomes from two species for a total of four sets.

The formation of an allopolyploid can also abruptly lead to reproductive isolation, thereby promoting speciation. As an example, let's consider the origin of a natural species of a plant called the common hemp nettle, *Galeopsis tetrahit*. This species is thought to be an allotetraploid derived from two diploid species, *Galeopsis pubescens* and *Galeopsis speciosa* (**Figure 25.11a**). These two diploid species contain 16 chromosomes each ($2n = 16$), whereas *G. tetrahit* contains 32 chromosomes. Though the origin of *G. tetrahit* is not completely certain, research suggests it may have originated from an interspecies cross between *G. pubescens* and *G. speciosa*, which initially produced

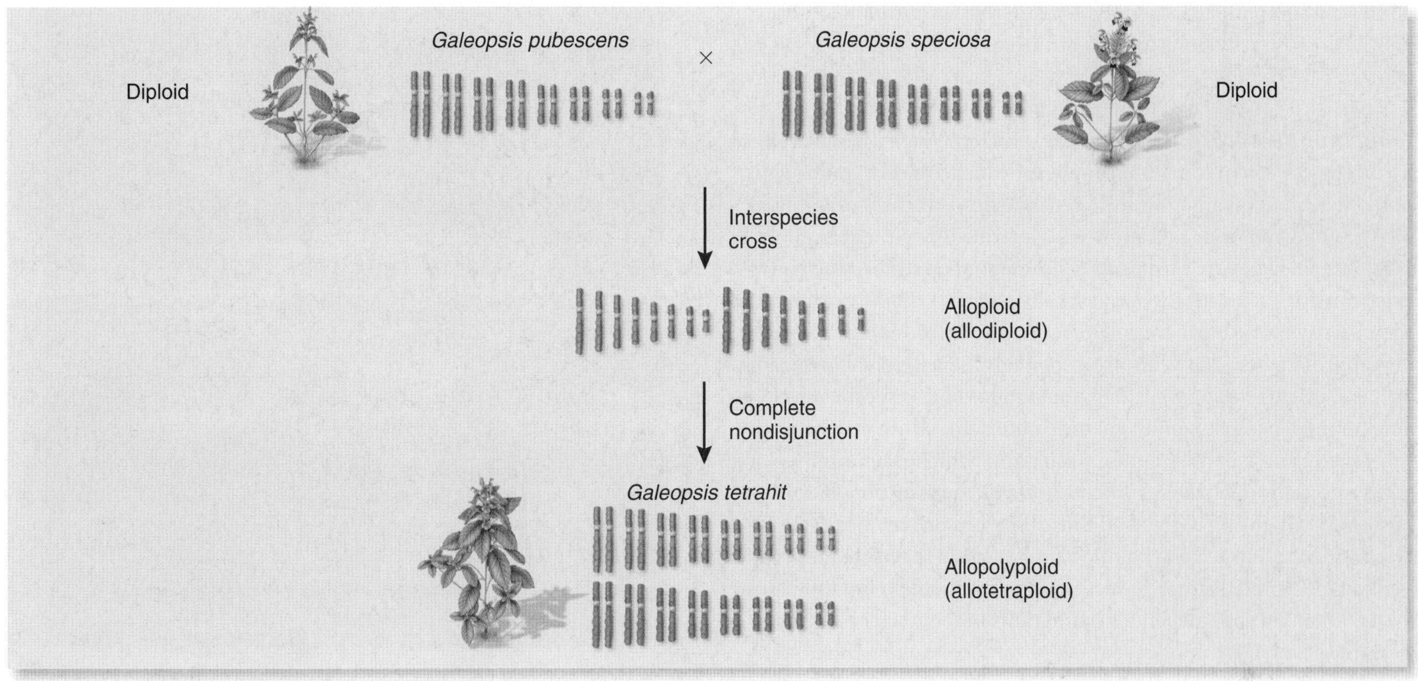

(a) Possible formation of *G. tetrahit*

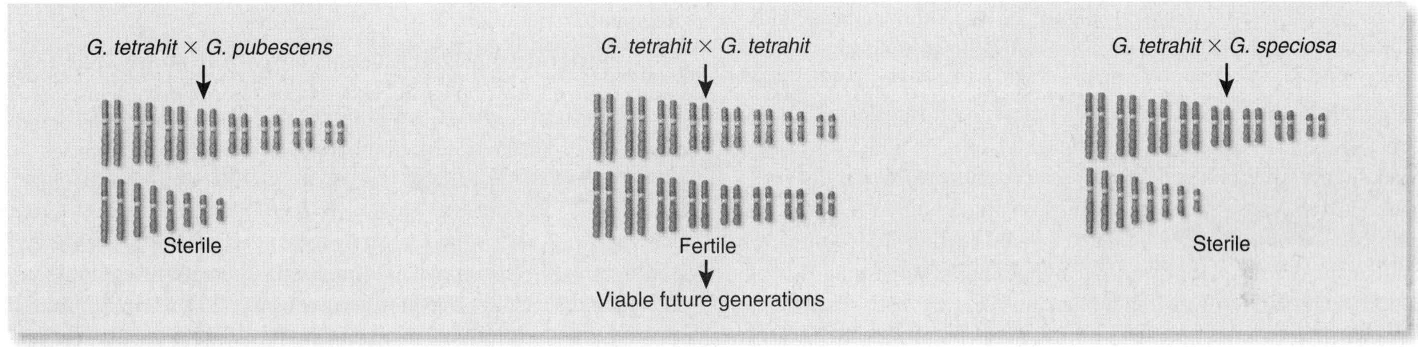

(b) Outcome of breeding among *G. tetrahit*, *G. pubescens*, and *G. speciosa*

Figure 25.11 **Polyploidy and sympatric speciation.** **(a)** *Galeopsis tetrahit* may have arisen by an interspecies cross between *Galeopsis pubescens* and *Galeopsis speciosa*, which was followed by a subsequent nondisjunction event. **(b)** Polyploidy may have caused reproductive isolation between these three natural species of hemp nettle. If *G. tetrahit* is mated with either of the other two species, the resulting offspring would be monoploid for one chromosome set and diploid for the other set, making them sterile. Therefore, *G. tetrahit* is reproductively isolated from the diploid species, making it a new species.

Concept Check: Suppose that G. tetrahit was crossed to G. pubescens to produce an interspecies hybrid as shown at the left side of part (b). If this interspecies hybrid was crossed to G. tetrahit, how many chromosomes do you think an offspring would have? The answer you give should be a range, not a single number.

an allodiploid with 16 chromosomes (one set from each species). The allodiploid then underwent complete nondisjunction to become an allotetraploid carrying four sets of chromosomes—two from each species.

How do these genetic changes cause reproductive isolation? The allotetraploid, *G. tetrahit*, is fertile, because all of its chromosomes occur in homologous pairs that can segregate evenly during meiosis. However, a cross between *G. tetrahit* and a diploid, *G. pubescens* or *G. speciosa*, produces an offspring that is monoploid for one chromosome set and diploid for the other set (**Figure 25.11b**). The chromosomes of the monoploid set cannot be evenly segregated during meiosis. These offspring are expected to be sterile, because they will produce gametes that have incomplete sets of chromosomes. This hybrid sterility causes the allotetraploid to be reproductively isolated from both diploid species. Therefore, this process could have led to the formation of a new species, *G. tetrahit*, by sympatric speciation.

Polyploidy is so frequent in plants that it is a major mechanism of their speciation. In ferns and flowering plants, about 40–70% of the species are polyploid. By comparison, polyploidy can occur in animals, but it is much less common. For example, less than 1% of reptiles and amphibians are polyploids derived from diploid ancestors. The reason why polyploidy is not usually tolerated in animals is not understood.

Adaptation to Local Environments In some cases, populations that occupy different local environments, which are continuous with each other, may diverge into different species. An early example of this type of sympatric speciation was described by American biologists Jeffrey Feder, Guy Bush, and colleagues. They studied the North American apple maggot fly (*Rhagoletis pomenella*). This fly originally fed on native hawthorn trees. However, the introduction of apple trees approximately 200 years ago provided a new local environment for this species. The apple-feeding populations of this species develop more rapidly because apples mature more quickly than hawthorn fruit. The result is partial temporal isolation, which is an example of prezygotic reproductive isolation. Although the two populations—those that feed on apple trees and those that feed on hawthorn trees—are considered subspecies, evolutionary biologists speculate they may eventually become distinct species due to reproductive isolation and the accumulation of independent mutations in the two populations.

American entomologist Sara Via and colleagues have studied the beginnings of sympatric speciation in pea aphids (*Acyrthosiphon pisum*), a small, plant-eating insect. Pea aphids in the same geographic area can be found on both alfalfa (*Medicago sativa*) and red clover (*Trifolium pratenae*) (**Figure 25.12**). Although pea aphids on these two host plants look identical, they show significant genetic differences and are highly ecologically specialized. Pea aphids that are found on alfalfa exhibit a lower fitness when transferred to red clover, whereas pea aphids found on red clover exhibit a lower fitness when transferred to alfalfa. The same traits involved in this host specialization cause these two groups of pea aphids to be substantially reproductively isolated. Taken together, the observations of the North American apple maggot fly, pea aphids, and other insect species suggest that diversifying selection (described in Chapter 24) occurs because some members within the same range evolve to feed on a different host. This may be an important mechanism of sympatric speciation among insects.

Sexual Selection Another mechanism that may promote sympatric speciation is sexual selection. As discussed in Chapter 24, one type of sexual selection is mate choice (refer back to Figures 24.8 and 24.9). Ole Seehausen and Jacques van Alphen found that male coloration in African cichlids is subject to female choice. In this case, sexual selection appears to have followed a diversifying mechanism in which certain females prefer males with one color pattern, and other females prefer males with a different color pattern. A possible outcome of such sexual selection is that it can separate one large sympatric population into smaller populations that eventually become distinct species because they selectively breed among themselves.

Figure 25.12 Pea aphids, a possible example of sympatric speciation in progress. Some pea aphids prefer alfalfa, whereas others prefer red clover. These two populations may be in the process of sympatric speciation.

BIOLOGY PRINCIPLE Populations of organisms evolve from one generation to the next. Populations of pea aphids are evolving based on preference for different food sources—alfalfa or red clover. The populations may eventually evolve into separate species.

Concept Check: *How may host preference eventually lead to speciation?*

25.3 The Pace of Speciation

Learning Outcome:

1. Compare and contrast the concepts of gradualism and punctuated equilibria.

Throughout the history of life on Earth, the rate of evolutionary change and speciation has not been constant. Even Darwin himself suggested that evolution can be fast or slow. **Figure 25.13** illustrates two contrasting views concerning the rate of evolutionary change. These ideas are not mutually exclusive but represent two different ways to consider the tempo of evolution. The concept of **gradualism** suggests that each new species evolves continuously over long spans of time (Figure 25.13a). The principal idea is that large phenotypic differences that produce new species are due to the gradual accumulation of many small genetic changes. By comparison, the concept of **punctuated equilibrium**, advocated in the 1970s by American paleontologist and evolutionary biologist Niles Eldredge and Stephen Jay Gould, suggests that the tempo of evolution is more sporadic (Figure 25.13b). According to this hypothesis, species exist relatively unchanged for many generations. During this equilibrium period, genetic changes are likely to accumulate, particularly neutral changes. However, genetic changes that significantly alter phenotype do not substantially change the overall composition of a population. These long periods of equilibria are punctuated by relatively short periods (that is, on a geological timescale) during which the frequencies of certain phenotypes in a population change substantially at a far more rapid rate.

A rapid rate of evolution could commonly occur via allopatric speciation in which a small group migrates away from a larger

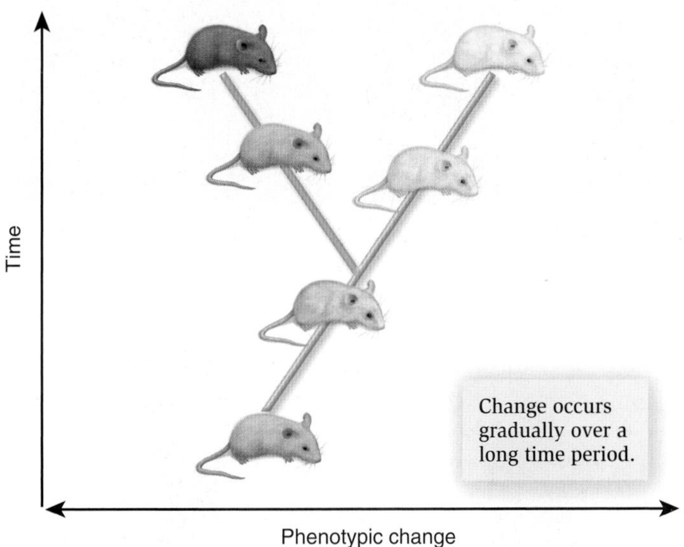
Time
Change occurs gradually over a long time period.
Phenotypic change

(a) Gradualism

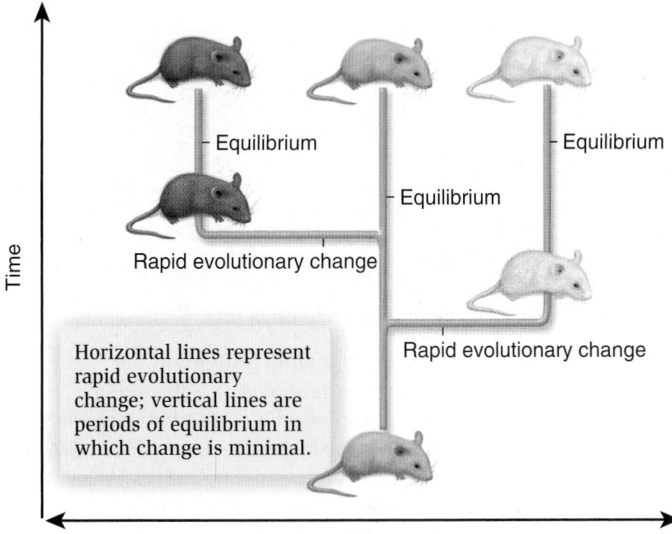
Time
Equilibrium
Equilibrium
Equilibrium
Rapid evolutionary change
Rapid evolutionary change
Horizontal lines represent rapid evolutionary change; vertical lines are periods of equilibrium in which change is minimal.
Phenotypic change

(b) Punctuated equilibrium

Figure 25.13 A comparison of gradualism and punctuated equilibrium. (a) During gradualism, the phenotypic characteristics of a species gradually change due to the accumulation of small genetic changes. **(b)** During punctuated equilibrium, long periods of equilibrium in which species exist essentially unchanged are punctuated by relatively short periods of evolutionary change during which phenotypic characteristics may change rapidly.

BIOLOGY PRINCIPLE Populations of organisms evolve from one generation to the next. Gradualism and punctuated equilibrium are two different views regarding the pace of evolution.

population to a new environment in which different alleles provide better adaptation to the surroundings. By natural selection, the small population may rapidly evolve into a new species. In addition, events such as polyploidy may abruptly produce individuals with new phenotypic traits. On an evolutionary timescale, these types of events can be rather rapid, because a few genetic changes can have a major influence on phenotype.

In conjunction with genetic changes, species may also be subjected to sudden environmental shifts that quickly drive the gene pool in a particular direction via natural selection. For example, the climate may change or a new predator may infiltrate the geographic range of the species. Natural selection may lead to a rapid evolution of the gene pool by favoring those alleles that allow members of the population to survive the climatic change or to have phenotypic characteristics that allow them to avoid the predator.

Which viewpoint is correct, punctuated equilibrium or gradualism? Both have merit. The occurrence of punctuated equilibrium is often supported by the fossil record. New species seem to arise rather suddenly in a layer of rocks, persist relatively unchanged for a very long period of time, and then become extinct. In such cases, scientists hypothesize that the period during which a previous species evolved into a new species was so short that few, if any, of the transitional forms of the species were preserved as fossils. Even so, these rapid periods of change were probably followed by long periods that likely involved the additional accumulation of many small genetic changes, consistent with gradualism.

Finally, another issue associated with the speed of speciation is generation time. Species of large animals with long generation times tend to evolve much more slowly than do microbial species with short generations. Many new species of bacteria will come into existence during our lifetime, whereas new species of large animals tend to arise on a much longer timescale. This is an important consideration because bacteria have great environmental effects. They are decomposers of organic materials and pollutants in the environment, and they play a role in many diseases of plants and animals, including humans.

25.4 Evo-Devo: Evolutionary Developmental Biology

Learning Outcomes:

1. Describe how the spatial expression of genes, such as *BMP4* and *Gremlin*, affects pattern formation.
2. Explain the relationship between the number of *Hox* genes and the body plan of an animal species.
3. Outline how differences in the growth rates of body parts can change the characteristics of species.
4. Describe how the study of the *Pax6* gene suggests that the eyes of different animal species evolved from a common ancestor.

As we have learned, the origin of new species involves genetic changes that lead to adaptations to environmental niches and/or to reproductive isolating mechanisms that prevent closely related species from interbreeding. These genetic changes result in morphological and physiological differences that distinguish one species from another. In recent years, many evolutionary biologists have begun to investigate how genetic variation produces species and groups of species with novel shapes and forms. The underlying reasons for such changes are often rooted in the developmental pathways that control an organism's morphology.

Evolutionary developmental biology (referred to as **evo-devo**) is an exciting and relatively new field of biology that compares the

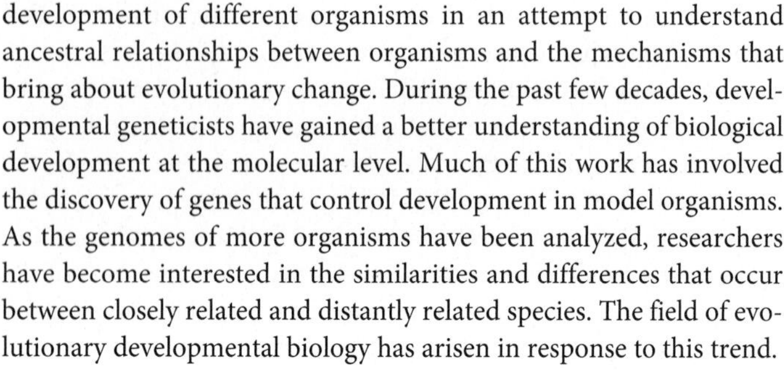

development of different organisms in an attempt to understand ancestral relationships between organisms and the mechanisms that bring about evolutionary change. During the past few decades, developmental geneticists have gained a better understanding of biological development at the molecular level. Much of this work has involved the discovery of genes that control development in model organisms. As the genomes of more organisms have been analyzed, researchers have become interested in the similarities and differences that occur between closely related and distantly related species. The field of evolutionary developmental biology has arisen in response to this trend.

How do new morphological forms come into being? For example, how does a nonwebbed foot evolve into a webbed foot? How does a new organ, such as an eye, come into existence? As we will learn, such novelty arises through genetic changes, also called genetic innovations. Certain types of genetic innovations have been so advantageous they have resulted in groups of new species. For example, the innovation of wings resulted in the evolution of many different species of birds. In this section, we will see that proteins that control developmental changes, such as cell-signaling proteins and transcription factors, often play a key role in promoting the morphological changes that occur during evolution.

The Spatial Expression of Genes That Affect Development Has a Dramatic Effect on Phenotype

In Chapter 19, we considered the role of genetics in the development of plants and animals. As we learned, genes that play a role in development influence cell division, cell migration, cell differentiation, and cell death. The interplay among these four processes produces an organism with a specific body pattern, a process called **pattern formation**. As you might imagine, developmental genes are very important to the phenotypes of individuals. They affect traits such as the shape of a bird's beak, the length of a giraffe's neck, and the size of a plant's flower. In recent years, the study of development has indicated that developmental genes are key players in the evolution of many types of traits. Changes in such genes affect traits that can be acted on by natural selection. Furthermore, variation in the expression of these genes may be commonly involved in the acquisition of new traits that promote speciation.

As an example, let's compare the formation of a chicken's foot with that of a duck. Developmental biologists have discovered that the morphological differences between a nonwebbed and a webbed foot are due to the differential expression of two different cell-signaling proteins called bone morphogenetic protein 4 (BMP4) and gremlin. The *BMP4* gene is expressed throughout the developing limb of both the chicken and duck; this is shown in **Figure 25.14a**, in which the BMP4 protein is stained purple. The BMP4 protein causes cells to undergo apoptosis and die. The gremlin protein, which is stained brown in **Figure 25.14b**, inhibits the function of BMP4, thereby allowing cells to survive. In the developing chicken limb, the *Gremlin* gene is expressed throughout the limb, except in the regions between each digit. Therefore, in these regions, the cells die, and a chicken develops a nonwebbed foot (**Figure 25.14c**). By comparison, in the duck, *Gremlin* is expressed throughout the entire limb, including the interdigit regions, and the duck develops a webbed foot. Interestingly, researchers have been able to introduce gremlin protein into the interdigit regions of developing chicken limbs. This produces a chicken with webbed feet!

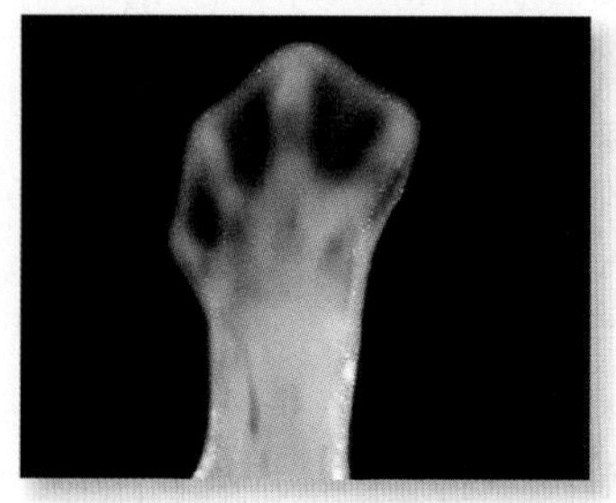

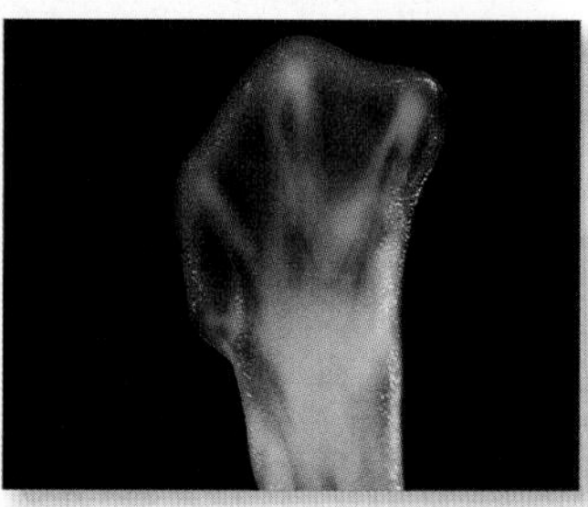

(a) BMP4 protein levels - similar expression in chicken and duck

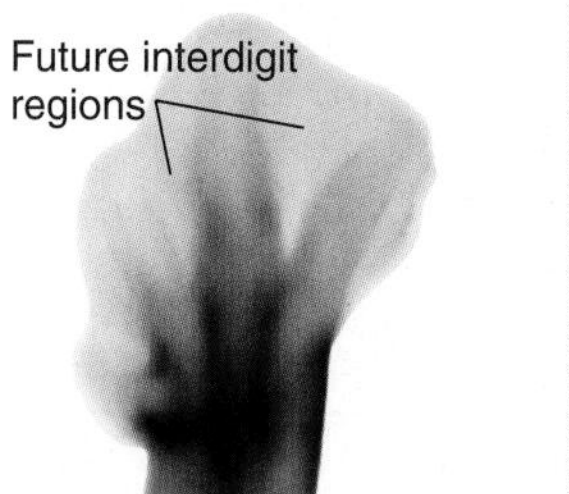

(b) Gremlin protein levels - not expressed in interdigit region in chicken

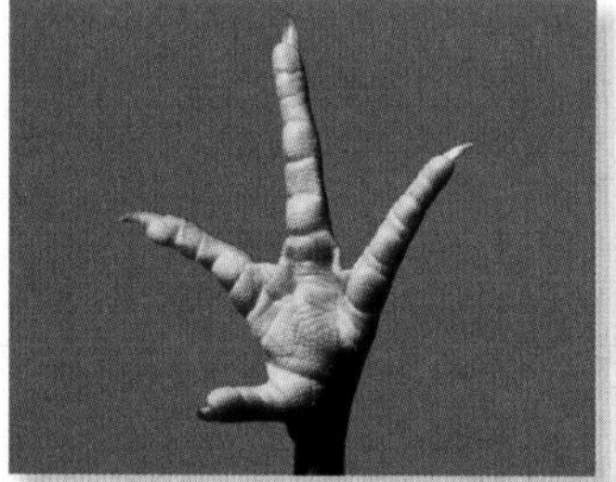

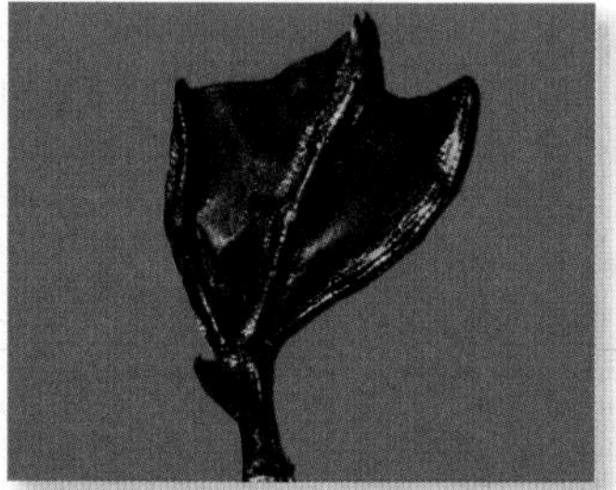

(c) Comparison of a chicken foot and a duck foot

Figure 25.14 **The role of cell-signaling proteins in the morphology of birds' feet.** This figure shows how changes in developmental gene expression can affect webbing between the toes. **(a)** Expression of the *BMP4* gene in the developing limbs. BMP4 protein is stained purple here and is expressed throughout the limb. **(b)** Expression of the *Gremlin* gene in the developing limbs. Gremlin protein is stained brown here. Note that *Gremlin* is not expressed in the interdigit regions of the chicken but is expressed in these regions of the duck. Gremlin inhibits BMP4, which causes programmed cell death. **(c)** Because BMP4 is not inhibited in the interdigit regions in the chicken, the cells in this region die, and the foot is not webbed. By comparison, inhibition of BMP4 in the interdigit regions in the duck results in a webbed foot.

Concept Check: *What would you expect to happen to the morphology of the feet of ducks if the Gremlin gene was under expressed?*

How are these observations related to evolution? During the evolution of birds, genetic variation arose such that some individuals expressed the *Gremlin* gene in the regions between each digit, but others did not. This variation determined whether or not a bird's feet were webbed. In terrestrial settings, having nonwebbed feet is an advantage because it enables the individual to hold onto perches, run along the ground, and snatch prey. Therefore, natural selection would favor nonwebbed feet in terrestrial environments. This process explains the occurrence of nonwebbed feet in chickens, hawks, crows, and many other terrestrial birds. In aquatic environments, however, webbed feet are an advantage because they act as paddles for swimming, so genetic variation that produced webbed feet in aquatic birds would have been acted on by natural selection. Over time, this gave rise to the webbed feet now found in a wide variety of aquatic birds, including ducks, geese, and penguins.

The *Hox* Genes Have Been Important in the Evolution of a Variety of Body Plans

The study of developmental genes has revealed interesting trends among large groups of species. *Hox* genes, which are discussed in Chapter 19, are found in nearly all animals, indicating they have originated very early in animal evolution. *Hox* genes are homeotic genes, which specify the fate of a particular segment or region of the body. Developmental biologists have hypothesized that variation in the *Hox* genes has spawned the formation of many new body plans. As shown in **Figure 25.15**, the number and arrangement of *Hox* genes varies considerably among different types of animals. Sponges, the simplest of animals, have at least one gene that is homologous to *Hox* genes. Insects typically have nine or more *Hox* genes. In most cases, multiple *Hox* genes occur in a cluster in which the genes are close to each other along a chromosome. In mammals, *Hox* gene clusters have been duplicated twice during the course of evolution to form four clusters, all slightly different, containing a total of 38 genes.

Researchers propose that increases in the number of *Hox* genes have been instrumental in the evolution of many animal species with greater complexity in body structure. To understand how, let's first consider *Hox* gene function. All *Hox* genes encode transcription factors that act as master control proteins for directing the formation of particular regions of the body. Each *Hox* gene controls a hierarchy of many regulatory genes that regulate the expression of genes encoding

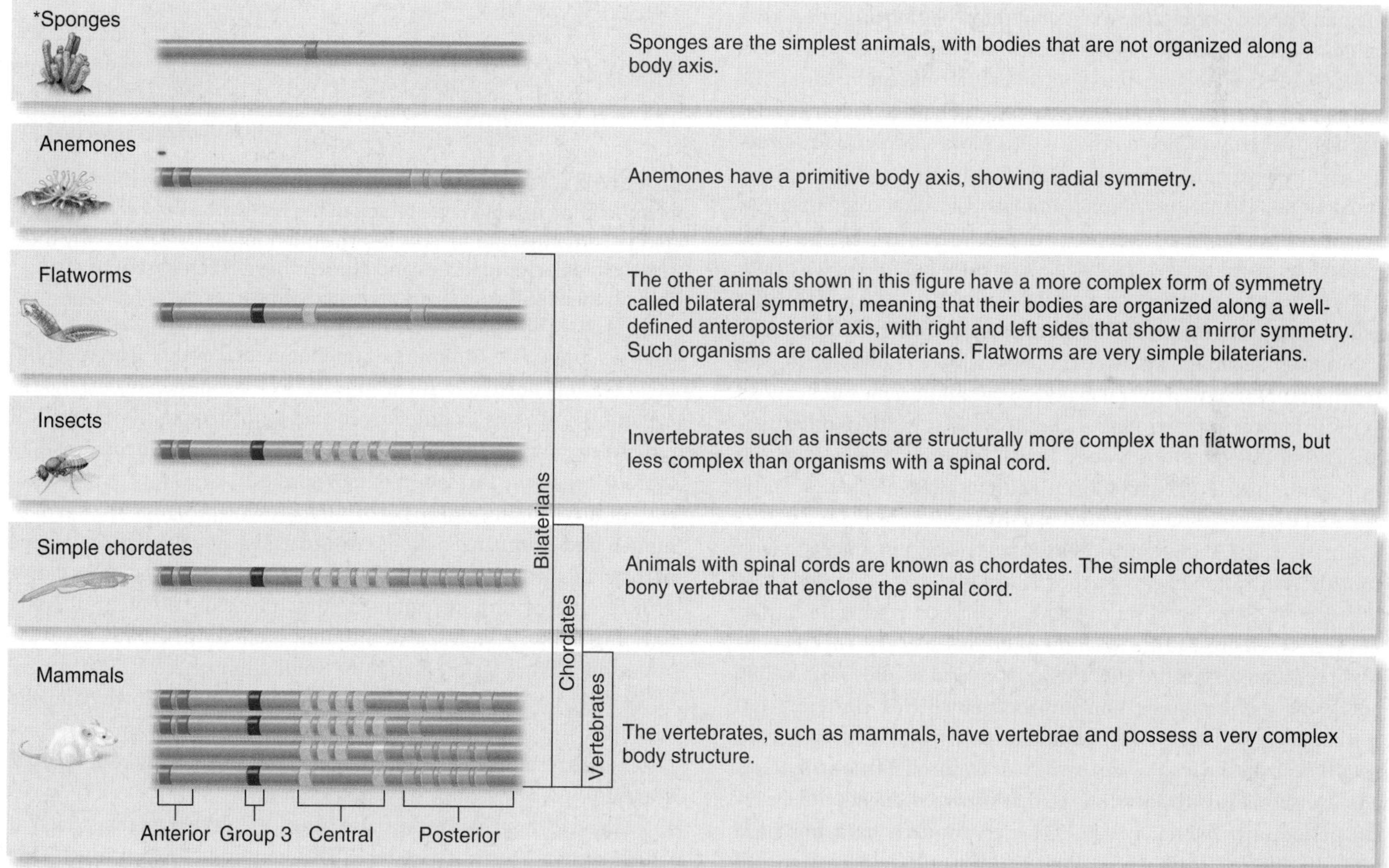

Figure 25.15 ***Hox* gene number and body complexity in different types of animals.** Researchers speculate that the duplication of *Hox* genes and *Hox* gene clusters played a key role in the evolution of more complex body plans in animals. A correlation is observed between increasing numbers of *Hox* genes and increasing complexity of body structure. The *Hox* genes are divided into four groups, called anterior, group 3, central, and posterior, based on their relative similarities. Each group is represented by a different color in this figure. *Note: Sponges, which are the simplest animals with no true tissues, do not have true *Hox* genes, though they have an evolutionarily related gene called an *NK-like* gene. Some species of sponges have more than one copy of this gene.

Concept Check: *What is the relationship between the total number of Hox genes in an animal species and its morphological complexity?*

BioConnections: *Look back at Figures 19.16 and 19.17. How is the expression of Hox genes related to segmentation and the anteroposterior axis?*

proteins that ultimately affect the morphology of the organism. The evolution of complex body plans is associated with an increase not only in the number of regulatory genes—as evidenced by the increase in *Hox* gene complexity during evolution—but also in genes that encode proteins that directly affect an organism's form and function.

How would an increase in *Hox* genes enable more complex body forms to evolve? Part of the answer lies in the spatial expression of the *Hox* genes. Different *Hox* genes are expressed in different regions of the body along the anteroposterior axis (refer back to Figure 19.16). Therefore, an increase in the number of *Hox* genes allows each of these master control genes to become more specialized in the region that it controls. In fruit flies, one segment in the middle of the body can be controlled by a particular *Hox* gene and form wings and legs, whereas a segment in the head region can be controlled by a different *Hox* gene and develops antennae. Therefore, research suggests that one way for new, more complex body forms to evolve is by increasing the number of *Hox* genes, thereby making it possible to form many specialized parts of the body that are organized along a body axis.

Three lines of evidence support the idea that increases in *Hox* gene number have been instrumental in the evolution and speciation of animals with different body patterns. First, as discussed in Chapter 19, *Hox* genes are known to control the fate of regions along the anteroposterior axis. Second, as described in Figure 25.15, a general trend is observed in which animals with a more complex body structure tend to have more *Hox* genes and *Hox* clusters in their genomes than do the genomes of simpler animals. Third, a comparison of *Hox* gene evolution and animal evolution bears striking parallels. Researchers have analyzed *Hox* gene sequences among modern species and made estimates regarding the timing of past events. Using this type of approach, geneticists have estimated when the first *Hox* gene arose by gene innovation. Though the date is difficult to precisely pinpoint, it is well over 600 mya. In addition, gene duplications of this primordial gene produced clusters of *Hox* genes in other species. Clusters such as those found in modern insects were likely to be present approximately 600 mya. A duplication of that cluster is estimated to have occurred around 520 mya.

Interestingly, these estimates of *Hox* gene origins correlate with major diversification events in the history of animals. As described in Chapter 22, the Cambrian period, which occurred from 543 to 490 mya, saw a great diversification of animal species. This diversification occurred after the *Hox* cluster was formed and was possibly undergoing its first duplication to produce two *Hox* clusters. Also, approximately 420 mya, a second duplication produced species with four *Hox* clusters. This event preceded the proliferation of tetrapods—vertebrates with four limbs—that occurred during the Devonian period, approximately 417–354 mya. Modern tetrapods have four *Hox* clusters. This second duplication may have been a critical event that led to the evolution of complex terrestrial vertebrates with four limbs, such as amphibians, reptiles, and mammals.

Variation in Growth Rates Can Have a Dramatic Effect on Phenotype

Another way that genetic variation can influence morphology is by controlling the relative growth rates of different parts of the body during development. The term **heterochrony** refers to evolutionary changes in the rate or timing of developmental events. The speeding up or slowing down of growth appears to be a common occurrence in evolution and leads to different species with striking morphological differences. With regard to the pace of evolution, such changes may rapidly lead to the formation of new species.

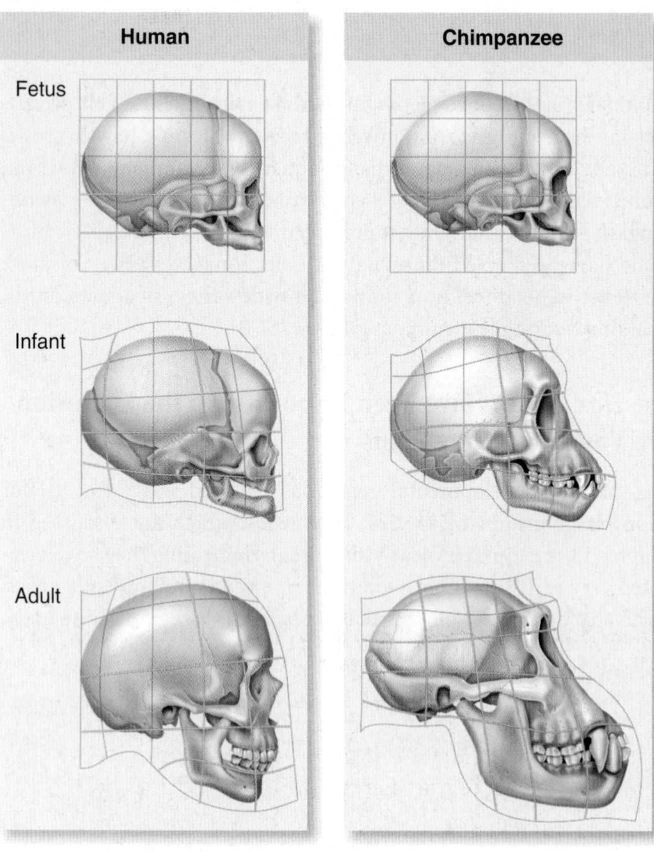

Figure 25.16 Heterochrony. Heterochrony refers to the phenomenon in which one region of the body grows faster than another among different species. The phenomenon explains why the skulls of adult chimpanzees and humans have different shapes even though their fetal skull shapes are quite similar.

As an example, Figure 25.16 compares the progressive growth of human and chimpanzee skulls. At the fetal stage, the size and shape of the skulls look fairly similar. However, after this stage, the relative growth rates of certain regions become markedly different, thereby affecting the shape and size of the adult skull. In the chimpanzee, the jaw region grows faster, giving the adult chimpanzee a much larger and longer jaw. In the human, the jaw grows more slowly, and the region of the skull that surrounds the brain—the cranium—grows faster. The result is that adult humans have smaller jaws but a larger cranium.

Changes in growth rates also affect the developmental stage at which certain species reproduce. This can occur in two ways. One possibility is that the parts of the body associated with reproduction develop faster than the rest of the body. Alternatively, reproduction may occur at the same absolute age, but the development of nonreproductive body parts is slowed down. In either case, the morphological result is the same—reproduction is observed at an earlier stage in one species than it is in another. In such cases, the sexually mature organism may retain traits typical of the juvenile stage of the organism's ancestor, a condition called **paedomorphosis** (from the Greek *paedo*, meaning young or juvenile, and *morph*, meaning the form of an

Figure 25.17 **Paedomorphosis.** Paedomorphosis occurs when an adult species retains characteristics that are juvenile traits in another related species. Cope's giant salamander reproduces at the tadpole stage.

organism). It is particularly common among salamanders. Typically when salamanders mature, they lose their gills and tail fins, features associated with aquatic life. Paedomorphic species retain certain juvenile features as adults, but have the ability to reproduce successfully. For example, Cope's giant salamander (*Dicamptodon copei*) becomes mature and reproduces in the aquatic form, without changing into a terrestrial adult as do other salamander species (**Figure 25.17**). The adult form of Cope's giant salamander has gills and a large paddle-shaped tail, features that resemble those of the larval (tadpole) stage of other salamander species. Such a change in morphology was likely a contributing factor to the formation of this species or an ancestral species to Cope's giant salamander.

Genomes & Proteomes Connection

The Study of the *Pax6* Gene Indicates That Different Types of Eyes Evolved from a Simpler Form

Thus far in this section, we have focused on the roles of particular genes as they influence the development of species with novel shapes and forms. Explaining how a complex organ comes into existence is another major challenge for evolutionary biologists. Although it is relatively easy to comprehend how a limb could undergo evolutionary modifications to become a wing, flipper, or arm, it is more difficult to understand how a body structure, such as a limb, comes into being in the first place. In his book *The Origin of Species*, Charles Darwin addressed this question and admitted that the evolution and development of a complex organ such as the eye was difficult to understand. As noted by Darwin, the eyes of vertebrate species are exceedingly complex, being able to adjust focus, let in different amounts of light, and detect a spectrum of colors. Darwin speculated that such complex eyes must have evolved from a simpler structure through the process of descent with modification. With amazing insight, he suggested that a very simple eye would be composed of two cell types, a photoreceptor cell and an adjacent pigment cell. The photoreceptor cell, which is a type of nerve cell, is able to absorb light and respond to it. The function of the pigment cell is to stop the light from reaching one side of the photoreceptor cell. This primitive, two-cell arrangement would allow an organism to sense both light and the direction from which the light comes.

A primitive eye would provide an additional way for an organism to sense its environment, possibly allowing it to avoid predators or locate food. Vision is nearly universal among animals, which indicates a strong selective advantage for eyesight. Over time, eyes could become more complex by enhancing the ability to absorb different amounts and wavelengths of light and also by refinements in structures such as the addition of lenses that focus the incoming light.

Since the time of Darwin, many evolutionary biologists have wrestled with the question of eye evolution. From an anatomical point of view, researchers have discovered many different types of eyes. For example, the eyes of fruit flies, squid, and humans are quite different from each other. This observation led evolutionary biologists such as Austrian zoologist Luitfried von Salvini-Plawen and German evolutionary biologist Ernst Mayr to propose that eyes may have independently arisen multiple times during evolution. Based solely on morphology, such a hypothesis seemed reasonable and for many years was accepted by the scientific community.

The situation took a dramatic turn when geneticists began to study eye development. Researchers identified a master control gene, *Pax6*[1]. The protein encoded by the *Pax6* gene is a transcription factor that controls the expression of many other genes, including those involved in the development of the eye in both rodents and humans. In mice and rats, a mutation in the *Pax6* gene results in small eyes. A mutation in the human *Pax6* gene causes an eye disorder called aniridia, in which the iris and other structures of the eye do not develop properly. Similarly, *Drosophila* has a gene named *eyeless* that also causes a defect in eye development when mutant. *Eyeless* and *Pax6* are homologous genes; they are derived from the same ancestral gene.

In 1995, Swiss geneticist Walter Gehring and his colleagues were able to show experimentally that the expression of the *eyeless* gene in parts of *Drosophila* where it is normally inactive could promote the formation of additional eyes. For example, using genetic engineering techniques, they were able to express the *eyeless* gene in the region where antennae should form. As seen in **Figure 25.18a**, this resulted in the formation of an eye where antennae are normally found! Remarkably, the expression of the mouse *Pax6* gene in *Drosophila* can also cause the formation of eyes in unusual places. For example, **Figure 25.18b** shows the formation of an eye on the leg of *Drosophila*.

Note that when the mouse *Pax6* master control gene switches on eye formation in *Drosophila*, the eye produced is a *Drosophila* eye, not a mouse eye. Why does this occur? It happens because the *Pax6* master control gene activates genes from the *Drosophila* genome. In *Drosophila*, the *Pax6* homolog called *eyeless* switches on a cascade involving several hundred genes required for eye morphogenesis. In organisms with simpler eyes, the *Pax6* gene would be expected to control a cascade of fewer genes.

Since the discovery of the *Pax6* and *eyeless* genes, homologs of this gene have been discovered in many different species. In all cases where it has been tested, this gene is involved with eye development. Gehring and colleagues have hypothesized that the eyes of many different species have evolved from a common ancestral form consisting

[1]Pax is an abbreviation for paired box. The protein encoded by this gene contains a domain called a paired box.

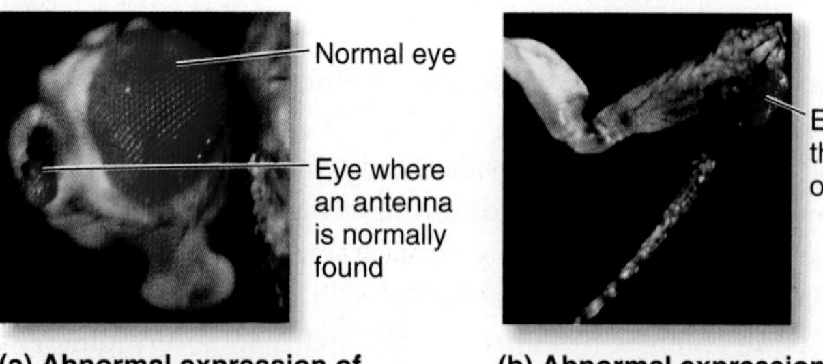

(a) Abnormal expression of the *Drosophila eyeless* gene in the antenna region

(b) Abnormal expression of the mouse *Pax6* gene in a fruit fly leg

Figure 25.18 Formation of additional eyes in *Drosophila* due to the abnormal expression of a master control gene for eye morphogenesis. **(a)** When the *Drosophila eyeless* gene is expressed in the antenna region, eyes are formed where antennae should be located. **(b)** When the mouse *Pax6* gene is expressed in the leg region of *Drosophila*, a small eye is formed there.

Concept Check: *What do you think would happen if the Drosophila eyeless gene was expressed at the tip of a mouse's tail?*

of, as proposed by Darwin, one photoreceptor cell and one pigment cell (**Figure 25.19**). As mentioned, such a very simple eye can accomplish a rudimentary form of vision by detecting light and its direction. Eyes such as these are still found in modern species, such as the larvae of certain types of mollusks. Over time, simple eyes evolved into more complex types of eyes by modifications that resulted in the addition of more types of cells, such as lens cells and nerve cells. Alternatively, other researchers propose that *Pax6* may control only certain features of eye development and that different types of eyes may have evolved independently. Future research will be needed to resolve this controversy.

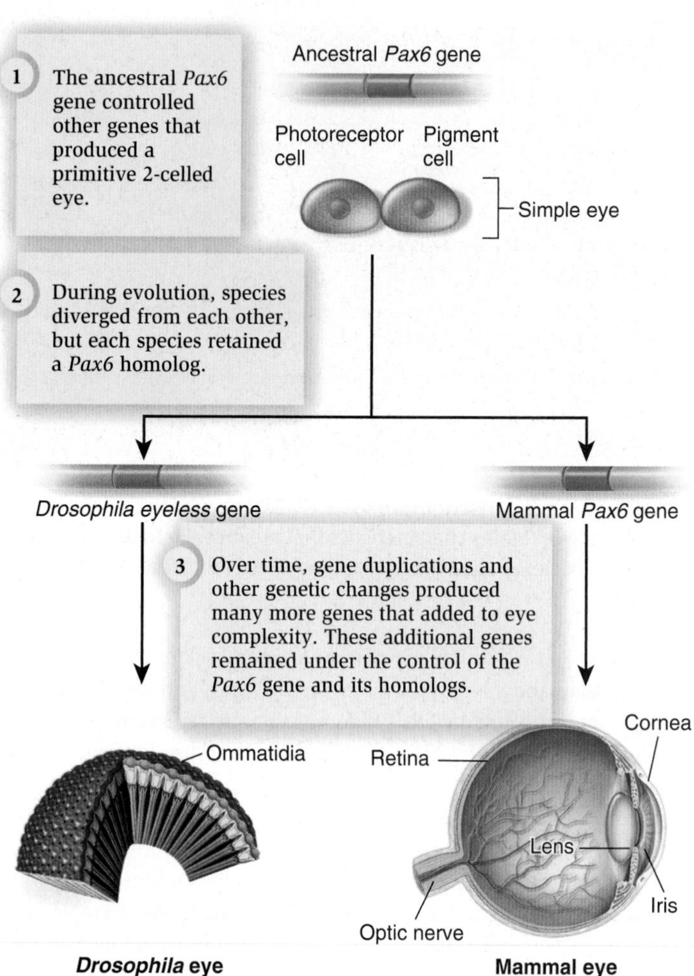

Figure 25.19 Genetic control of eye evolution. In this diagram, genetic changes, under the control of the ancestral *Pax6* gene, led to the evolution of different types of eyes.

Summary of Key Concepts

25.1 Identification of Species

- A species is a group of related organisms that shares a distinctive set of attributes in nature. Speciation is the process by which new species are formed. Macroevolution refers to evolutionary changes that produce new species and groups of species.
- Different characteristics, including morphological traits, reproductive isolation, molecular features, ecological factors, and evolutionary relationships, are used to identify species (Figure 25.1).
- Reproductive isolating mechanisms prevent two different species from breeding with each other (Figure 25.2).
- Prezygotic isolating mechanisms include habitat isolation, temporal isolation, behavioral isolation, mechanical isolation, and gametic isolation (Figures 25.3, 25.4).
- Postzygotic isolating mechanisms include hybrid inviability, hybrid sterility, and hybrid breakdown (Figure 25.5).

25.2 Mechanisms of Speciation

- Allopatric speciation occurs when a population becomes isolated from other populations and evolves into one or more new species. When speciation from a single ancestral species occurs multiple times, the process is called adaptive radiation. If two populations are incompletely separated, interbreeding may occur in hybrid zones (Figures 25.6, 25.7, 27.8).
- Podos hypothesized that changes in beak depth, associated with adaptation to feeding, promoted reproductive isolation by altering the song pattern of finches (Figures 25.9, 25.10).
- Sympatric speciation involves the formation of different species in populations that are not geographically isolated from one another. Polyploidy, adaptation to local environments, and sexual selection are mechanisms that promote sympatric speciation (Figures 25.11, 25.12).

25.3 The Pace of Speciation

- The pace of evolution may seem relatively constant or it may vary. Gradualism involves steady evolution due to many small genetic changes, whereas punctuated equilibrium is a pattern of evolution in which new species arise more rapidly and then remain unchanged for long periods of time (Figure 25.13).

25.4 Evo-Devo: Evolutionary Developmental Biology

- Evolutionary developmental biology compares the development of different species in order to understand ancestral relationships and the mechanisms that bring about evolutionary change. These changes often involve variation in the expression of cell-signaling proteins and transcription factors.

- The spatial expression of genes that affect development can affect phenotypes dramatically, as shown by the expression of the *BMP4* and *Gremlin* genes in birds with nonwebbed or webbed feet (Figure 25.14).
- An increase in the number of *Hox* genes played an important role in the evolution of more complex body forms in animals (Figure 25.15).
- A difference in the relative growth rates of body parts among different species is called heterochrony. Paedomorphosis occurs when an adult organism retains characteristics that are typical of the juvenile stage in another related species (Figures 25.16, 25.17).
- The *Pax6* gene and its homolog in other species are master control genes that control eye development in animals (Figures 25.18, 25.19).

Assess and Discuss

Test Yourself

1. Macroevolution refers to evolutionary changes that
   a. occur in multicellular organisms.
   b. produce new species and groups of species.
   c. occur over long periods of time.
   d. cause changes in allele frequencies.
   e. occur in large mammals.
2. The biological species concept classifies a species based on
   a. morphological characteristics.
   b. reproductive isolation.
   c. the niche the organism occupies in the environment.
   d. genetic relationships between an organism and its ancestors.
   e. both a and b.
3. Which of the following is considered an example of a postzygotic isolating mechanism?
   a. incompatible genitalia
   b. different mating seasons
   c. incompatible gametes
   d. mountain range separating two populations
   e. fertilized egg fails to develop normally
4. Hybrid breakdown occurs when species hybrids
   a. do not develop past the early embryonic stages.
   b. have a reduced life span.
   c. are infertile.
   d. are fertile but produce offspring with reduced viability and fertility.
   e. produce offspring that express the traits of only one of the original species.
5. The evolution of one species into two or more species is called
   a. gradualism.
   b. punctuated equilibrium.
   c. cladogenesis.
   d. horizontal gene transfer.
   e. microevolution.
6. A large number of honeycreeper species on the Hawaiian Islands is an example of
   a. adaptive radiation.
   b. genetic drift.
   c. stabilizing selection.
   d. horizontal gene transfer.
   e. microevolution.
7. A major mechanism of speciation in plants but not in animals is
   a. adaptation to new environments.
   b. polyploidy.
   c. hybrid breakdown.
   d. genetic changes that alter the organism's niche.
   e. both a and d.
8. The concept of punctuated equilibrium suggests that
   a. the rate of evolution is constant, with short time periods of no evolutionary change.
   b. evolution occurs gradually over time.
   c. small genetic changes accumulate over time to allow for phenotypic change and speciation.
   d. long periods of little evolutionary change are interrupted by short periods of major evolutionary change.
   e. both b and c.
9. Researchers suggest that an increase in the number of *Hox* genes
   a. leads to reproductive isolation in all cases.
   b. could explain the evolution of color vision.
   c. allows for the evolution of more complex body forms in animals.
   d. results in the decrease in the number of body segments in insects.
   e. does all of the above.
10. The observation that the mammalian *Pax6* gene and the *Drosophila eyeless* gene are homologous genes that promote the formation of different types of eyes suggests that
    a. *Drosophila* eyes are more complex.
    b. mammalian eyes are more complex.
    c. eyes arose once during evolution.
    d. eyes arose at least twice during evolution.
    e. eye development is a simple process.

Conceptual Questions

1. What is the key difference between prezygotic and postzygotic isolating mechanisms? Give an example of each type. Which type is more costly from the perspective of energy?
2. What are the key differences between gradualism and punctuated equilibrium? How are genetic changes related to these two models?
3. A principle of biology is that *populations of organisms evolve from one generation to the next.* Describe one example in which genes that control development played an important role in the evolution of different species.

Collaborative Questions

1. What is a species? Discuss how geographic isolation can lead to speciation, and explain how reproductive isolation plays a role.
2. Discuss the type of speciation (allopatric or sympatric) that is most likely to occur under each of the following conditions:
   a. A pregnant female rat is transported by an ocean liner to a new continent.
   b. A meadow containing several species of grasses is exposed to a pesticide that promotes nondisjunction.
   c. In a very large lake containing several species of fishes, the water level gradually falls over the course of several years. Eventually, the large lake becomes subdivided into smaller lakes, some of which are connected by narrow streams.

Online Resource

www.brookerbiology.com

Stay a step ahead in your studies with animations that bring concepts to life and practice tests to assess your understanding. Your instructor may also recommend the interactive eBook, individualized learning tools, and more.

Taxonomy and Systematics

26

Chapter Outline

26.1 Taxonomy
26.2 Phylogenetic Trees
26.3 Cladistics
26.4 Molecular Clocks
26.5 Horizontal Gene Transfer
Summary of Key Concepts
Assess and Discuss

The African forest elephant, *Loxodonta cyclotis*. In 2001, biologists decided that this is a unique species of elephant.

Until recently, biologists classified elephants into only two species—the African savanna elephant (*Loxodonta africana*) and the Asian elephant (*Elephas maximus*). However, by analyzing the DNA of African elephants, researchers have revised this classification, and proposed a third species, now called the African forest elephant (*Loxodonta cyclotis*) (see chapter-opening photo). How was this new species identified? This surprising finding was made somewhat by accident in 2001. Elephants in Africa are being killed for their tusks at a high rate, despite the 1989 international ban on ivory sales. Scientists set up a DNA identification system to trace tusks to the region in Africa where the elephants were likely killed, which could give law enforcement officials the leverage they need to target poachers in those areas. By studying the DNA from captured tusks, researchers decided that Africa has two distinctly different *Loxodonta* elephant species. The African forest elephant is found in the forests of central and western Africa. The African savanna elephant, which is larger and has longer tusks, lives on large, dry grasslands. One consequence of this discovery is its effect on conservation efforts, which had previously been based on a single species of African elephants.

The rules for the classification of newly described species, such as the African forest elephant, are governed by the discipline of taxonomy (from the Greek *taxis*, meaning order, and *nomos*, meaning law). **Taxonomy** is the science of describing, naming, and classifying **extant** species, those that still exist today, as well as **extinct** species, those that have died out. Taxonomy results in the ordered division of species into groups based on similarities and dissimilarities in their characteristics. This task has been ongoing for over 300 years. As discussed in Chapter 23, the naturalist John Ray made the first attempt to broadly classify all known forms of life. Ray's ideas were later extended by naturalist Carolus Linnaeus in the mid-1700s, which is considered by some as the official birth of taxonomy.

Systematics is the study of biological diversity and the evolutionary relationships among species, both extant and extinct. In the 1950s, German entomologist Willi Hennig began classifying species in a new way. Hennig proposed that evolutionary relationships should be inferred from features shared by descendants of a common ancestor. Since that time, biologists have applied systematics to the field of taxonomy. Researchers now try to place new species into taxonomic groups based on evolutionary relationships with other species. In addition, previously established taxonomic groups are revised as new data shed light on evolutionary relationships. As in any scientific discipline, taxonomy should be viewed as a work in progress.

In this chapter, we will begin with a discussion of taxonomy and the concept of taxonomic groups. We will then examine how biologists use systematics to determine evolutionary relationships among species, looking in particular at how these relationships are portrayed in diagrams called phylogenetic trees. We will then explore how analyses of morphological data and molecular genetic data are used to understand the evolutionary history of life on Earth.

26.1 Taxonomy

Learning Outcomes:

1. Identify the three domains of life.
2. Explain the hierarchy of groupings in taxonomy.
3. Describe how species are named using binomial nomenclature.

A hierarchy is a system of organization that involves successive levels. In biological taxonomy, every species is placed into several different nested groups within a hierarchy. For example, a leopard and a fruit fly are both classified as animals, though they differ in many characteristics. By comparison, leopards and lions are placed together into a group with a smaller number of species called felines (more formally named Felidae), which are predatory cats. The felines are a subset of the animal group, which has species that share many similar features. The species that are placed together into small taxonomic groups are likely to share many of the same characteristics. In this section, we will consider how biologists use a hierarchy to group similar species.

Species Are Subdivided into Three Domains of Life

Modern taxonomy places species into progressively smaller hierarchical groups. Each group at any level is called a **taxon** (plural, taxa). The taxon called the **kingdom** was originally the highest and most inclusive. Linnaeus had classified all life into two kingdoms, plants and animals. In 1969, American ecologist Robert Whittaker proposed a five-kingdom system in which all life was classified into the kingdoms Monera, Protista, Fungi, Plantae, and Animalia. However, as biologists began to learn more about the evolutionary relationships among these groups, they found that this classification did not correctly reflect the relationships among them.

In the late 1970s, based on information in the sequences of genes, American biologist Carl Woese proposed the idea of creating a category called a **domain**. In the taxonomy hierarchy, a domain is above a kingdom. Under this system, all forms of life are grouped within three domains: **Bacteria**, **Archaea**, and **Eukarya** (Figure 26.1). The terms Bacteria and Archaea are capitalized when referring to the domains, but are not capitalized when referring to individual species. A single bacterial cell is called a bacterium, and a single archaeal cell is an archaeon.

The domain Eukarya formerly consisted of four kingdoms called **Protista**, **Fungi**, **Plantae**, and **Animalia**. However, researchers later discovered that Protista is not a separate kingdom but instead is a very broad collection of species. Taxonomists now place eukaryotes into seven groups called supergroups. In the taxonomy of eukaryotes, a **supergroup** lies between a domain and a kingdom (see Figure 26.1). As discussed in Chapter 28, all seven supergroups contain a distinctive group of protists. In addition, kingdoms Fungi and Animalia are within the supergroup Opisthokonta, because they are closely related to the protists in this supergroup. Kingdom Plantae is within the supergroup called Land plants and relatives. Plants are closely related to green algae, which are protists in this supergroup. **Table 26.1** compares a variety of molecular and cellular characteristics among the domains Bacteria, Archaea, and Eukarya.

Table 26.1 Distinguishing Cellular and Molecular Features of Domains Bacteria, Archaea, and Eukarya*

Characteristic	Bacteria	Archaea	Eukarya
Chromosomes	Usually circular	Circular	Usually linear
Nucleosome structure	No	No	Yes
Chromosome segregation/cell division	Binary fission	Binary fission	Mitosis/meiosis
Introns in genes	Rarely	Rarely	Commonly
Ribosomes	70S	70S	80S
Initiator tRNA	Formylmethionine	Methionine	Methionine
Operons	Yes	Yes	No
Capping of mRNA	No	No	Yes
RNA polymerases	One	Several	Three
Promoters of structural genes	−35 and −10 sequences	TATA box	TATA box
Cell compartmentalization	No	No	Yes
Membrane lipids	Ester-linked	Ether-linked	Ester-linked

*The descriptions in this table are meant to represent the general features of most species in each domain. Some exceptions are observed. For example, certain bacterial species have linear chromosomes, and operons occasionally are found in eukaryotes, such as the nematode worm *Caenorhabditis elegans*.

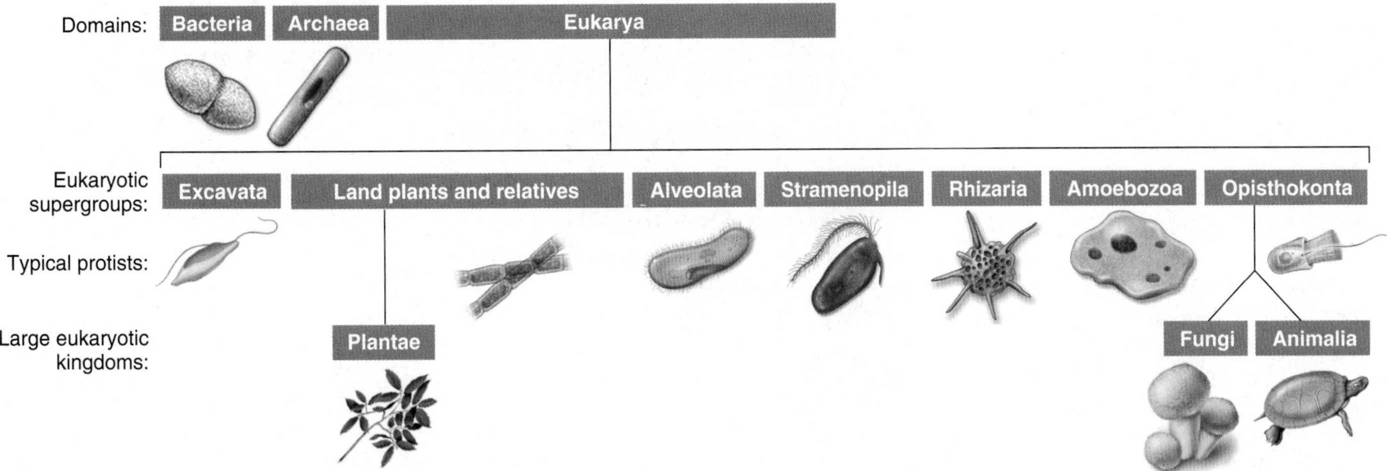

Figure 26.1 A classification system for living and extinct organisms. All organisms are grouped into three domains: Bacteria, Archaea, or Eukarya. Eukaryotes are divided into seven supergroups. The division of eukaryotes into supergroups is a subject of current investigation and debate and should be viewed as a work in progress.

BioConnections: *Look back at Figure 4.4. Which of the three domains contains organisms with prokaryotic cells?*

Every Species Is Placed into a Taxonomic Hierarchy

Why is it useful to categorize species into groups? The three domains of life contain millions of different species. Subdividing them into progressively smaller taxonomic groups makes it easier for biologists to appreciate the relationships among such a large number of species.

Below the domain and the supergroup is the kingdom, which is divided into **phyla** (singular, phylum). Each phylum is divided into **classes**, then **orders**, **families**, **genera** (singular, genus), and **species**. As noted in Chapter 25, species may be divided into subspecies, often based on geographical distribution. Each of these taxa contains progressively fewer species that are more similar to each other than they are to the members of the taxa above them in the hierarchy. For example, the taxon Animalia, which is at the kingdom level, has a larger number of fairly diverse species than does the class Mammalia, which contains fewer species that are relatively similar to each other.

To further understand taxonomy, let's consider the classification of the gray wolf (*Canis lupus*) (**Figure 26.2**). The gray wolf is placed in the domain Eukarya, the supergroup Opisthokonta, and then within the kingdom Animalia, which includes over 1 million species of all animals. Next, the gray wolf is classified in the phylum Chordata. The 50,000 species of animals in this group all have four common features at some stage of their development. These are a notochord (a cartilaginous rod that runs along the back of all chordates at some point in their life cycle), a tubular nerve or spinal cord located above the notochord, gill slits or arches, and a postanal tail. Examples of animals in the phylum Chordata include fishes, reptiles, and mammals.

The gray wolf is in the class Mammalia, which includes about 5,000 species of mammals. Two distinguishing features of animals in this group are hair, which helps the body maintain a warm, constant body temperature, and mammary glands, which produce milk to nourish the young. There are 26 orders of mammals; the order that includes the gray wolf is called Carnivora and has about 270 species that are meat-eating animals with prominent canine teeth. The gray wolf is placed in the family Canidae, which is a relatively small family of 34 species, including different species of wolves, jackals, foxes, wild dogs, and the coyote and domestic dog. All species in the family Canidae are doglike animals. The smallest grouping that contains the gray wolf is the genus *Canis*, which has four species of jackals, the coyote, and two types of wolves. The species *Canis lupus* encompasses several subspecies, including the domestic dog (*Canis lupus familiaris*).

Binomial Nomenclature Is Used to Name Species

As originally advocated by Linnaeus, **binomial nomenclature** is the standard method for naming species. The scientific name of every species has two names, its genus name and its unique specific epithet. For the gray wolf, the genus is *Canis* and the species epithet is *lupus*.

Taxonomic group	Gray wolf found in	Number of species
Domain	Eukarya	~4–10 million
Supergroup	Opisthokonta	>1 million
Kingdom	Animalia	>1 million
Phylum	Chordata	~50,000
Class	Mammalia	~5,000
Order	Carnivora	~270
Family	Canidae	34
Genus	*Canis*	7
Species	*lupus*	1

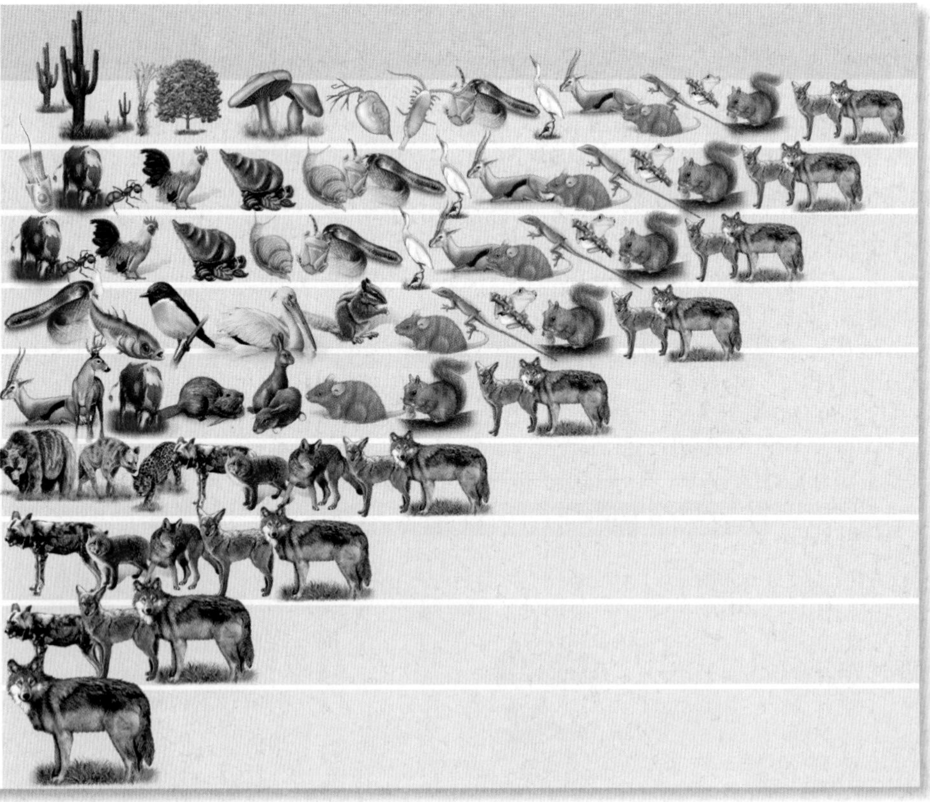

Figure 26.2 **A taxonomic classification of the gray wolf (*Canis lupus*).**

BIOLOGY PRINCIPLE All species (past and present) are connected by an evolutionary history. A goal of taxonomy is to relate the diversity of species according to their evolutionary relationships.

Concept Check: *Which group is broader, a phylum or a family?*

The genus name is always capitalized, but the specific epithet is not. Both names are italicized. After the first mention, the genus name is abbreviated to a single letter. For example, we would write that *Canis lupus* is the gray wolf, and in subsequent sentences, the species would be referred to as *C. lupus*.

When naming a new species, genus names are always nouns or treated as nouns, whereas species epithets may be either nouns or adjectives. The names often have a Latin or Greek origin and refer to characteristics of the species or to features of its habitat. For example, the genus name of the newly discovered African forest elephant, *Loxodonta*, is from the Greek *loxo*, meaning slanting, and *odonta*, meaning tooth. The species epithet *cyclotis* refers to the observation that the ears of this species are rounder than those of *L. africana*.

The rules for naming animal species, such as *Canis lupus* and *Loxodonta africana*, were established by the International Commission on Zoological Nomenclature (ICZN). The ICZN provides and regulates a uniform system of nomenclature to ensure that every animal has a unique and universally accepted scientific name. Who is allowed to identify and name a new species? As long as ICZN rules are followed, new animal species can be named by anyone, not only scientists. The rules for naming plants are described in the International Code of Botanical Nomenclature (ICBN), and the naming of bacteria and archaea is overseen by the International Committee on Systematics of Prokaryotes (ICSP).

26.2 Phylogenetic Trees

Learning Outcomes:

1. Define phylogeny and explain its basis for the construction of phylogenetic trees.
2. Compare and contrast cladogenesis and anagenesis as patterns of speciation.
3. Describe how homology is used to construct phylogenetic trees.

As mentioned, systematics is the study of biological diversity and evolutionary relationships. By studying the similarities and differences among species, biologists can construct a **phylogeny**, which is the evolutionary history of a species or group of species. To propose a phylogeny, biologists use the tools of systematics. For example, the classification of the gray wolf described in Figure 26.2 is based on systematics. Therefore, one use of systematics is to place species into taxa and to understand the evolutionary relationships among different taxa.

In this section, we will consider the features of diagrams or trees that describe the evolutionary relationships among various species, both extant and extinct. As you will learn, such trees are usually based on morphological or genetic data.

A Phylogenetic Tree Depicts Evolutionary Relationships Among Species

A **phylogenetic tree** is a diagram that describes the evolutionary relationships among various species, based on the information available to and gathered by systematists. Phylogenetic trees should be viewed as hypotheses that are proposed, tested, and later refined as additional data become available. Let's look at what information a phylogenetic tree contains and the form in which it is presented. Figure 26.3 shows a hypothetical phylogenetic tree of the relationships among various flowering plant species, in which the species are labeled A through K. The vertical axis represents time, with the oldest species at the bottom.

New species can be formed by **anagenesis**, in which a single species evolves into a different species, or more commonly by **cladogenesis**, in which a species diverges into two or more species. The branch points in a phylogenetic tree, also called **nodes**, illustrate times when cladogenesis has occurred. For example, approximately 12 mya, species A diverged into species A and species B. Figure 26.3 also shows anagenesis in which species C evolved into species G. The tips of branches may represent species that became extinct in the past, such as species B and E, or living species, such as F, I, G, J, H, and K, which are at the top of the tree. Species A and D are also extinct but gave rise to species that are still in existence.

By studying the branch points of a phylogenetic tree, researchers can group species according to common ancestry. A **clade** consists of a common ancestral species and all of its descendant species. For example, the group highlighted in light green in Figure 26.3 is a clade derived from the common ancestral species labeled D. Likewise, the entire tree forms a clade, with species A as a common ancestor. Therefore, smaller and more recent clades are nested within larger clades that have older common ancestors.

A Central Goal of Systematics Is to Construct Taxonomic Groups Based on Evolutionary Relationships

A key goal of modern systematics is to create taxonomic groups that reflect evolutionary relationships. Systematics attempts to organize species into clades, which means that each group includes an ancestral species and all of its descendants. A **monophyletic group** is a taxon that is a clade. Ideally, every taxon, whether it is a domain, supergroup, kingdom, phylum, class, order, family, or genus, should be a monophyletic group.

What is the relationship between a phylogenetic tree and taxonomy? The relationship depends on how far back we go to identify a common ancestor. For broader taxa, such as a kingdom, the common ancestor existed a very long time ago, on the order of hundreds of millions or even billions of years ago. For smaller taxa, such as a family or genus, the common ancestor occurred much more recently, on the order of millions or tens of millions of years ago. This concept is shown in a very schematic way in Figure 26.4. This small, hypothetical kingdom is a clade that contains 64 living species. (Actual kingdoms are obviously larger and exceedingly more complex.) The diagram emphasizes the taxa that contain the species designated number 43. The common ancestor that gave rise to this kingdom of organisms existed approximately 1 billion years ago. Over time, more recent species arose that subsequently became the common ancestors to the phylum, class, order, family, and genus that contain species number 43.

How does research in systematics affect taxonomy? As researchers gather new information, they sometimes discover that some of the current taxonomic groups are not monophyletic. Figure 26.5 compares a monophyletic group with taxonomic groups that are not.

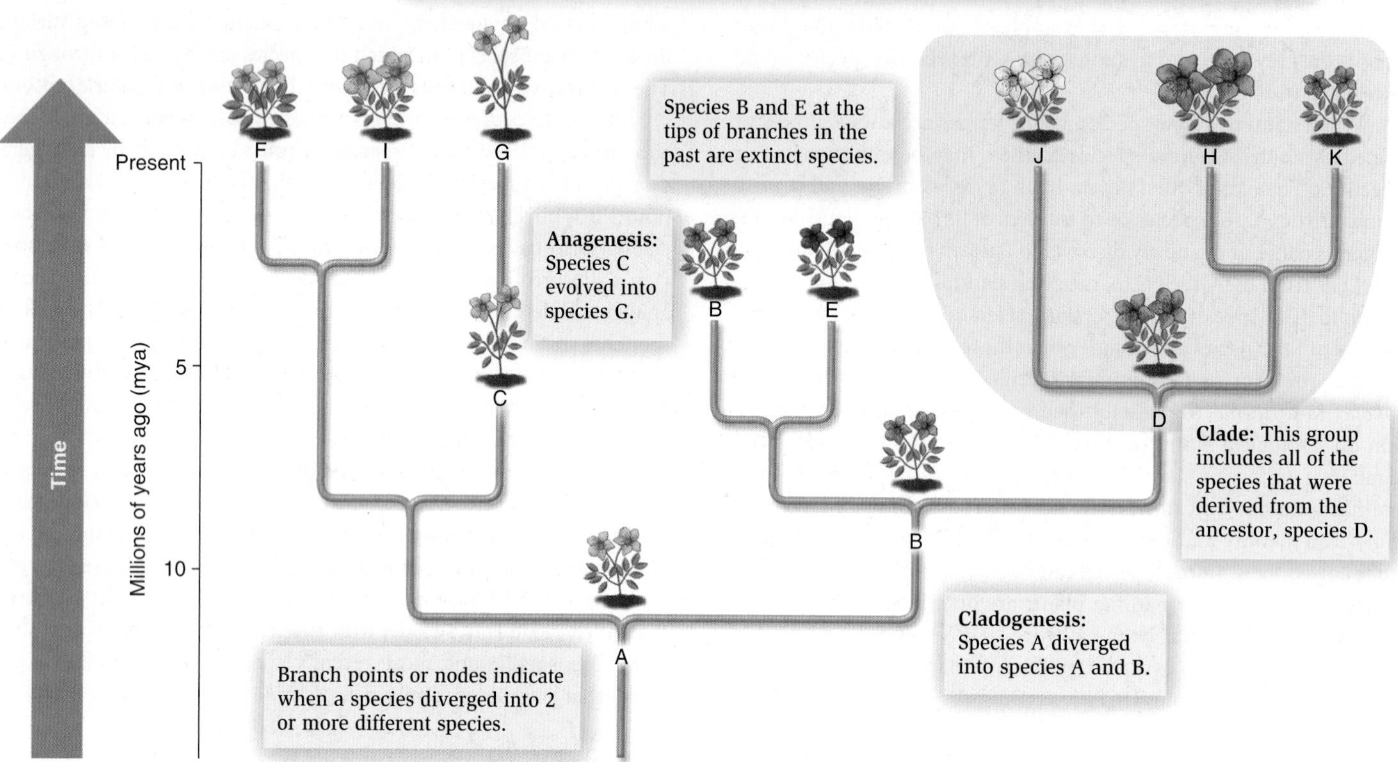

Figure 26.3 How to read a phylogenetic tree. This hypothetical tree shows the proposed relationships between various flowering plant species. Species are placed into clades, groups of organisms containing an ancestral organism and all of its descendants.

Concept Check: *Can two different species have more than one common ancestor?*

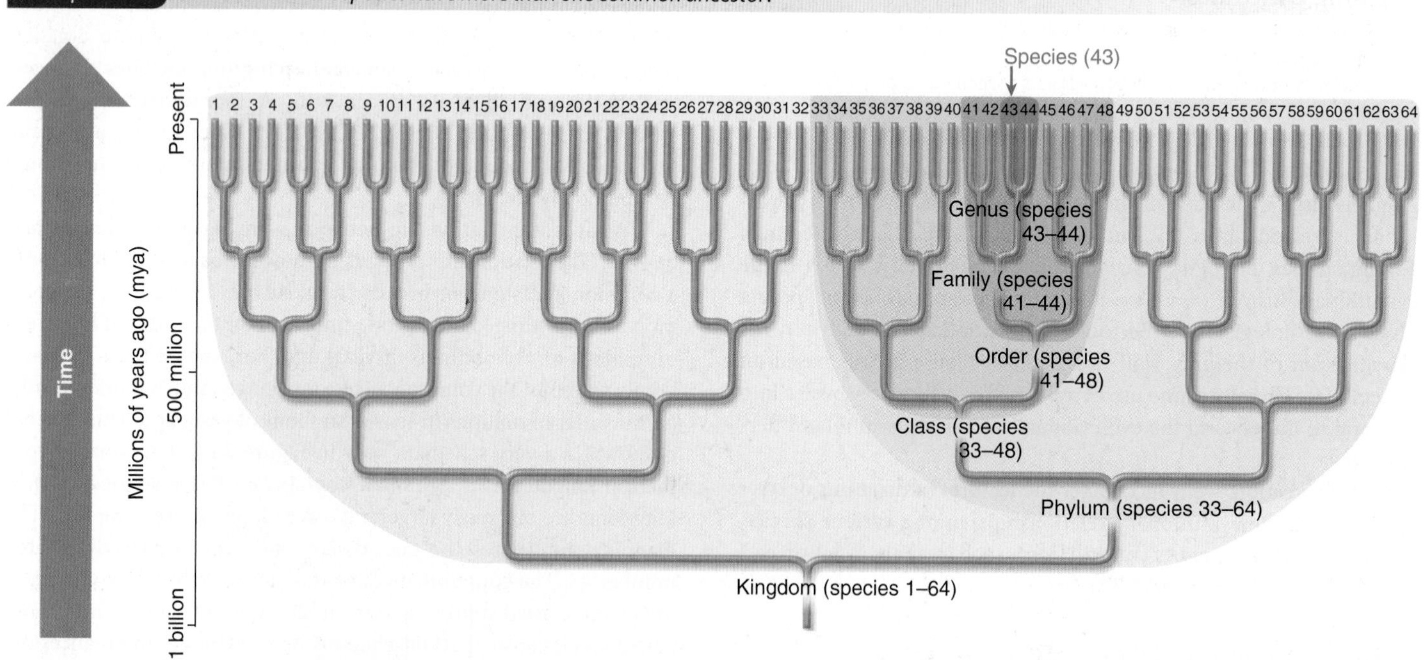

Figure 26.4 Schematic relationship between a phylogenetic tree and taxonomy, when taxonomy is correctly based on evolutionary relationships. The shaded areas highlight the kingdom, phylum, class, order, family, and genus for species number 43. All of the taxa are clades. Broader taxa, such as phyla and classes, are derived from more ancient common ancestors. Smaller taxa, such as families and genera, are derived from more recent common ancestors. These smaller taxa are subsets of the broader taxa.

Concept Check: *Which taxon would have a more recent common ancestor, a phylum or an order?*

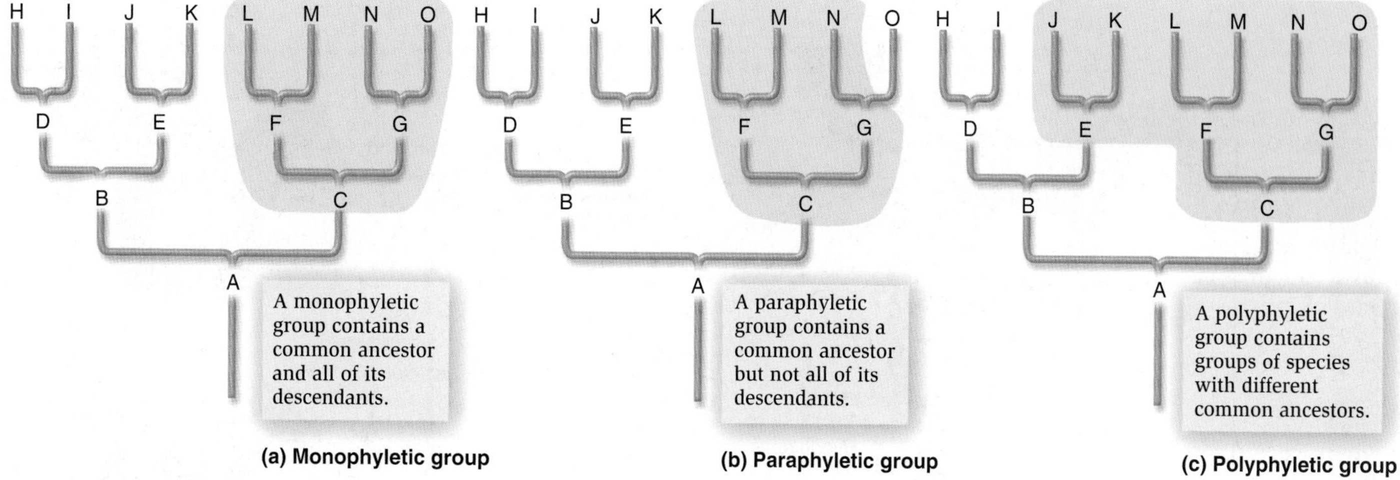

Figure 26.5 **A comparison of monophyletic, paraphyletic, and polyphyletic taxonomic groups.**

A **paraphyletic group** contains a common ancestor and some, but not all, of its descendants (Figure 26.5b). In contrast, a **polyphyletic group** consists of members of several evolutionary lines and does not include the most recent common ancestor of the included lineages (Figure 26.5c).

Over time, as we learn more about evolutionary relationships, taxonomic groups are being reorganized in an attempt to recognize only monophyletic groups in taxonomy. For example, traditional classification schemes once separated birds and reptiles into separate classes (Figure 26.6a). In this scheme, the reptile class (officially named Reptilia) contained orders that included turtles, lizards and snakes, and crocodiles, with birds constituting a different class. Research has indicated that the reptile taxon was paraphyletic, because birds were excluded from the group. This group can be made monophyletic by including birds as a class within the reptile clade and elevating the other groups to a class status (Figure 26.6b).

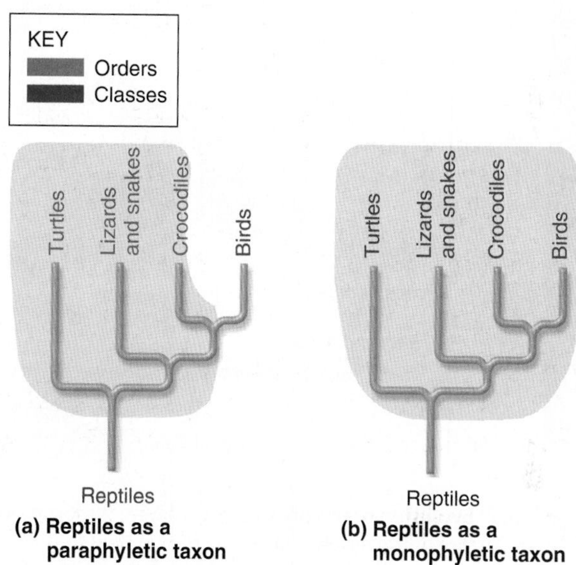

Figure 26.6 **An example of a taxon that is not monophyletic.** **(a)** The class of reptiles as a paraphyletic taxon. **(b)** The group can be made monophyletic if birds and the other orders were classified as classes within the reptile clade.

The Study of Systematics Is Usually Based on Morphological or Genetic Homology

As discussed in Chapter 23, the term **homology** refers to a similarity that occurs due to descent from a common ancestor. Such features are said to be homologous. For example, the arm of a human, the wing of a bat, and the flipper of a whale are homologous structures (refer back to Figure 23.12). Similarly, genes found in different species are homologous if they have been derived from the same ancestral gene (refer back to Figure 23.13).

In systematics, researchers identify homologous features that are shared by some species but not by others, which allows them to group species based on their shared similarities. Researchers usually study homology by examining morphological features or genetic data. In addition, the data they gather are viewed in light of geographic data. Many organisms do not migrate extremely long distances. Species that are closely related evolutionarily are relatively likely to inhabit neighboring or overlapping geographic regions, though many exceptions are known to occur.

Morphological Analysis The first studies in systematics focused on morphological features of extinct and living species. Morphological traits continue to be widely used in systematic studies, particularly in those studies pertaining to extinct species and those involving groups that have not been extensively studied at the molecular level. To establish evolutionary relationships based on morphological homology, many traits have to be analyzed to identify similarities and differences.

By studying morphological features of extinct species in the fossil record, paleontologists can propose phylogenetic trees that chart the evolutionary lineages of species, including those that still exist. In this approach, the trees are based on morphological features that change over the course of many generations. As an example, Figure 26.7 depicts a current hypothesis of the evolutionary changes that led to the development of the modern horse. This figure shows representative species from various genera. Many morphological features were used to propose this tree. Because hard parts of the body are more commonly preserved in the fossil record, this tree is largely based on

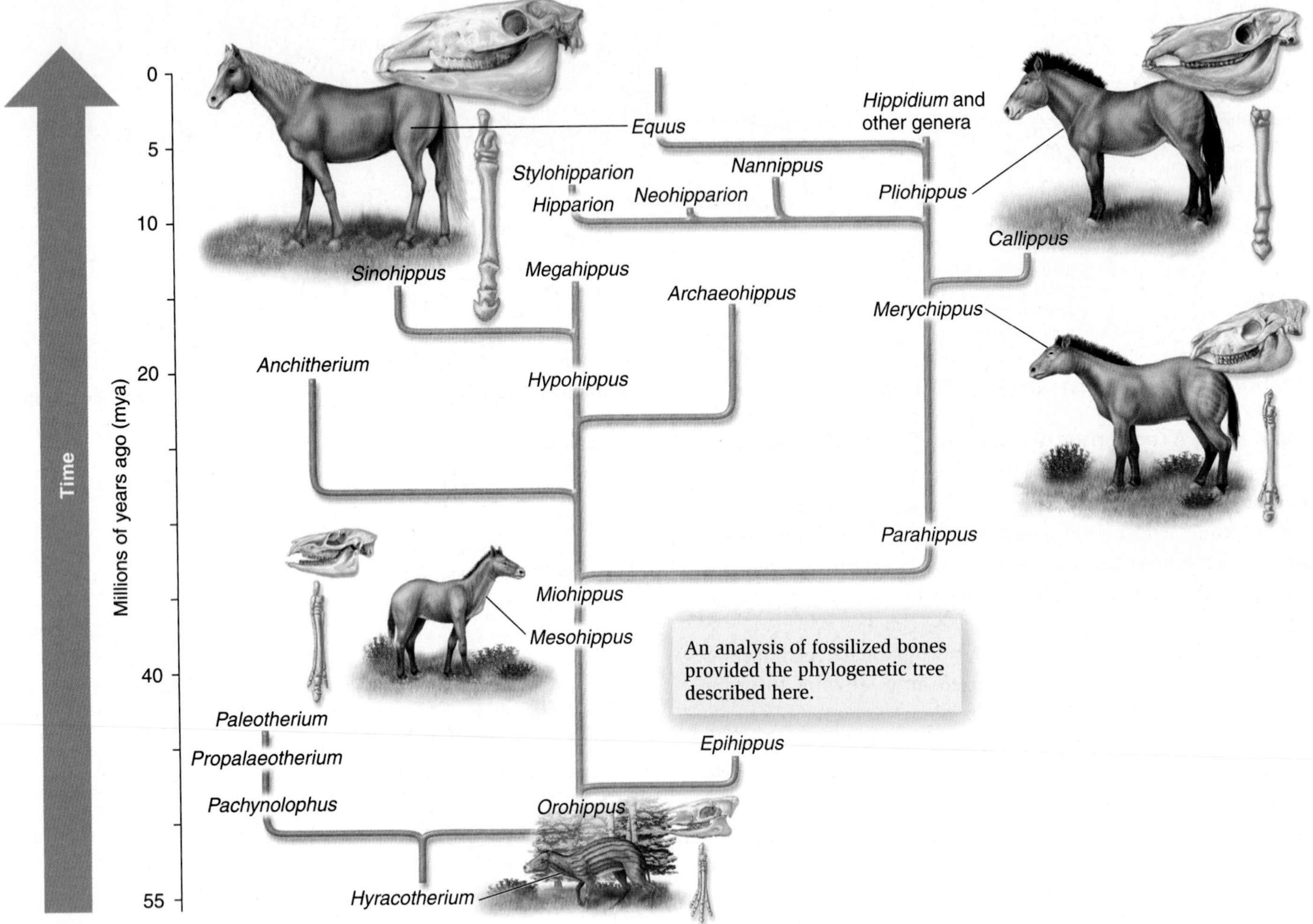

Figure 26.7 Evolution of the horses. An analysis of morphological traits was used to produce this phylogenetic tree, which shows the evolutionary history of the modern horse. As shown next to the horses, three important morphological changes in these genera were larger size, fewer toes, and a shift toward teeth suited for grazing.

BIOLOGY PRINCIPLE Structure determines function. The changes in structural features during horse evolution are related to changes in their functional needs. During this time, horse populations shifted from feeding on leaves in forested regions to feeding on abrasive grasses in more wide-open spaces.

the analysis of skeletal changes in foot structure, lengths and shapes of various leg bones, skull shape and size, and jaw and tooth morphology. Over an evolutionary time scale, the accumulation of many genetic changes has had a dramatic effect on species' characteristics. In the genera depicted in this figure, a variety of morphological changes occurred, such as an increase in size, a reduction in the number of toes, and modifications in the jaw and teeth consistent with a dietary shift from tender leaves to fibrous grasses (see Figure 23.6).

Similar morphological features due to convergent evolution may occasionally confound an evolutionary analysis. As described in Chapter 23, convergent evolution results in analogous structures, characteristics that arise independently in different lineages because the species have evolved in similar environments. For example, the giant anteater and the echidna have long snouts and tongues that enable these animals to feed on ants (refer back to Figure 23.8a). These traits are not derived from a common ancestor. Rather, they arose independently during evolution due to adaptation to similar environments. The use of analogous structures in systematics can cause errors if a researcher assumes that a particular trait arose only once and that all species having the trait are derived from a common ancestor.

Molecular Systematics The field of **molecular systematics** involves the analysis of genetic data, such as DNA sequences or amino acid sequences, to identify and study genetic homologies and propose phylogenetic trees. In 1963, Austrian biologist Emile Zuckerkandl and American chemist Linus Pauling were the first to suggest that molecular data could be used to establish evolutionary relationships. How can a comparison of genetic sequences help to establish evolutionary relationships? As discussed later in this chapter, DNA sequences change over the course of many generations due to the accumulation of mutations. Therefore, when comparing homologous sequences in different species, DNA sequences from closely related species are more similar to each other than they are to sequences from distantly related species.

26.3 Cladistics

Learning Outcomes:

1. Distinguish between shared primitive characters and shared derived characters.
2. Outline the steps of using a cladistics approach to construct a phylogenetic tree, and explain how the principle of parsimony is used to choose among phylogenetic trees.
3. Describe how maximum likelihood is also used to discriminate among phylogenetic trees.
4. Explain how DNA can be analyzed to explore relationships among extant and extinct species.

Cladistics is the classification of species based on evolutionary relationships. A cladistic approach produces phylogenetic trees by considering the possible pathways of evolutionary changes that involve characteristics that are shared or not shared among various species. Such trees are known as **cladograms**. In this section, we will consider how biologists produce phylogenetic trees.

Species Differ with Regard to Primitive and Derived Characters

A cladistic approach compares homologous features, also called **characters**, which may exist in two or more **character states**. For example, among different species, a front limb, which is a character, may exist in different character states such as a wing, an arm, or a flipper. The various character states are either shared or not shared by different species.

To understand the cladistic approach, let's take a look at a simplified phylogeny (Figure 26.8). We can place the living species that currently exist into two groups: D and E, and F and G. The most recent common ancestor to D and E is B, whereas species C is the most recent common ancestor to F and G. With these ideas in mind, let's focus on the front limbs (flippers versus legs) and eyes.

A character that is shared by two or more different taxa and inherited from ancestors older than their last common ancestor is called a **shared primitive character**, or **symplesiomorphy**. Such characters are viewed as being older—ones that occurred earlier in evolution. With regard to species D, E, F, and G, having two eyes is a shared primitive character. It originated prior to species B and C.

By comparison, a **shared derived character**, or **synapomorphy**, is a character that is shared by two or more species or taxa and has originated in their most recent common ancestor. With regard to species D and E, having two front flippers is a shared derived character that originated in species B, their most recent common ancestor (see Figure 26.8). Compared with shared primitive characters, shared derived characters are more recent traits on an evolutionary timescale. For example, among mammals, only some species, such as whales and dolphins, have flippers. In this case, flippers were derived from the two front limbs of an ancestral species. The word "derived" indicates that evolution involves the modification of traits in pre-existing species. In other words, populations of organisms with new traits are derived from changes in pre-existing populations. The basis of the cladistic approach is to analyze many shared derived characters among groups of species to deduce the pathway that gave rise to those species.

Note that the terms "primitive" and "derived" do not indicate the complexity of a character. For example, the flippers of a dolphin do not appear more complex than the front limbs of ancestral species A (see Figure 26.8), which were limbs with individual toes. Derived characters can be similar in complexity, less complex, or more complex than primitive characters.

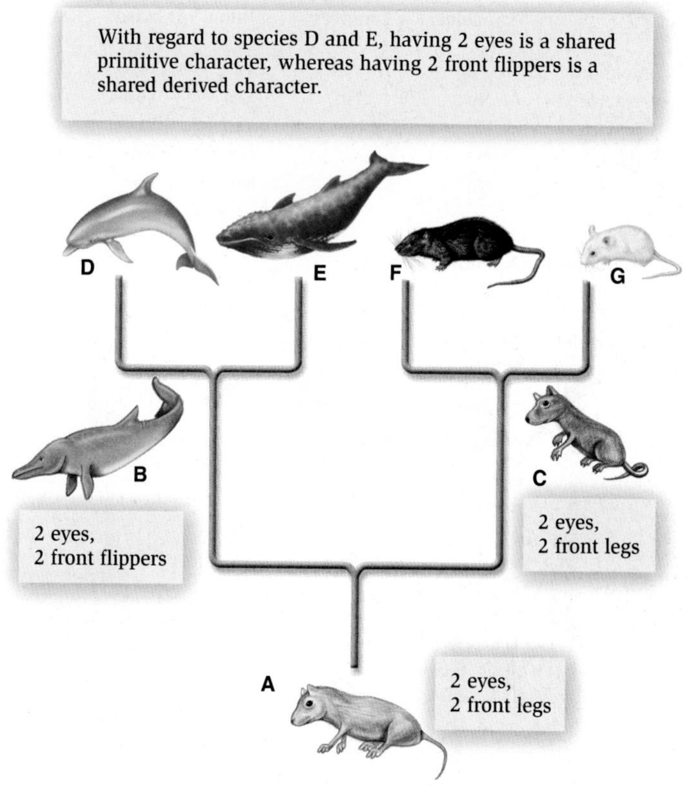

With regard to species D and E, having 2 eyes is a shared primitive character, whereas having 2 front flippers is a shared derived character.

Figure 26.8 **A comparison of shared primitive characters and shared derived characters.**

A Cladistic Approach Produces a Cladogram Based on Shared Derived Characters

To understand how shared derived characters are used to propose a phylogenetic tree, **Figure 26.9a** compares several traits among five species of animals. The proposed cladogram shown in **Figure 26.9b** is consistent with the distribution of shared derived characters among these species. A branch point is where two species differ in a character. The oldest common ancestor, which would now be extinct, had a notochord and was an ancestor to all five species. Vertebrae are a shared derived character of the lamprey, salmon, lizard, and rabbit, but not the lancelet, which is an invertebrate. By comparison, a hinged jaw is a shared derived character of the salmon, lizard, and rabbit, but not of the lamprey or lancelet.

In a cladogram, an **ingroup** is the group whose evolutionary relationships we wish to understand. By comparison, an **outgroup** is a species or group of species that is assumed to have diverged before the species in the ingroup. An outgroup lacks one or more shared derived characters that are found in the ingroup. A designated outgroup can be closely related or more distantly related to the ingroup. In the tree shown in Figure 26.9, if the salmon, lizard, and rabbit are an ingroup, the lamprey is an outgroup. The lamprey has a notochord and vertebrae but lacks a character shared by the ingroup, namely, a hinged jaw. Thus, for the ingroup, the notochord and vertebrae

	Lancelet	Lamprey	Salmon	Lizard	Rabbit
Notochord	Yes	Yes	Yes	Yes	Yes
Vertebrae	No	Yes	Yes	Yes	Yes
Hinged jaw	No	No	Yes	Yes	Yes
Tetrapod	No	No	No	Yes	Yes
Mammary glands	No	No	No	No	Yes

(a) Characteristics among species

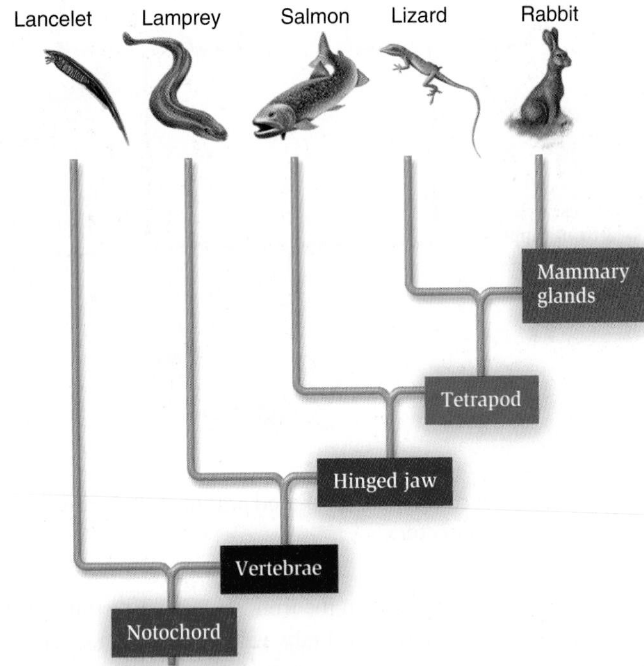

(b) Cladogram based on morphological traits

Figure 26.9 **Using shared primitive characters and shared derived characters to propose a phylogenetic tree. (a)** A comparison of characteristics among these species. **(b)** This phylogenetic tree illustrates both shared primitive and shared derived characters in a cladogram of five animal species.

Concept Check: What shared derived character is common to the salmon, lizard, and rabbit, but not the lamprey?

are shared primitive characters, whereas the hinged jaw is a shared derived character not found in the outgroup.

Likewise, the concept of shared derived characters can apply to molecular data, such as a gene sequence. Let's consider an example to illustrate this idea. Our example involves molecular data obtained from seven different hypothetical plant species called A–G. In these species, a homologous region of DNA was sequenced as shown here:

```
   12345678910
A: GATAGTACCC
B: GATAGTTCCC
C: GATAGTTCCG
D: GGTATTACCC
E: GGTATAACCC
F: GGTAGTACCA
G: GGTAGTACCC
```

The cladogram of Figure 26.10 is a hypothesis of how these DNA sequences arose. A mutation that changes the sequence of nucleotides is comparable to a modification of a character. For example, let's designate species D as an outgroup and species A, B, C, F, and G as the ingroup. In this case, a G (guanine) at the fifth position is a shared derived character. The genetic sequence carrying this G is derived from an older primitive sequence.

Now that we have an understanding of shared primitive and derived characters, let's consider the steps a researcher would follow to propose a cladogram using a cladistics approach.

1. **Choose the species in whose evolutionary relationships you are interested.** In a simple cladogram, such as those described in this chapter, individual species are compared with each other. In more complex cladograms, species may be grouped into larger taxa (for example, families) and compared with each other. If such grouping is done, the results are not reliable if the groups are not clades.
2. **Choose characters for comparing the species selected in step 1.** As mentioned, a character is a general feature of an organism and may come in different versions called character states. For example, a front limb is a character in mammals, which could exist in different character states such as a wing, an arm, or a flipper.
3. **Determine the polarity of character states.** In other words, determine if a character state came first and is primitive or came later and is a derived character. This information may be available by examining the fossil record, for example, but is usually done by comparing the ingroup with the outgroup. For a character with two character states, an assumption is made that a character state shared by the outgroup and ingroup is primitive. A character state shared only by members of the ingroup is derived.
4. **Analyze cladograms based on the following principles:**
   - All species (or higher taxa) are placed on tips in a phylogenetic tree, not at branch points.
   - Each cladogram branch point should have a list of one or more shared derived characters that are common to all species above the branch point unless the character is later modified.
   - All shared derived characters appear together only once in a cladogram unless they arose independently during evolution more than once.
5. **Among many possible options, choose the cladogram that provides the simplest explanation for the data.** A common approach is to use a computer program that generates many possible cladograms. Analyzing the data and choosing among the possibilities are key aspects of this process. As described later, different theoretical approaches, such as the principle of parsimony, can be used to choose among possible phylogenies.
6. **Provide a root to the phylogenetic tree by choosing a noncontroversial outgroup.** In this textbook, most phylogenetic trees are rooted, which means that a single node at the bottom of the tree corresponds to a common ancestor for all of the species or groups of species in the tree. A method for rooting trees is the use of a noncontroversial outgroup.

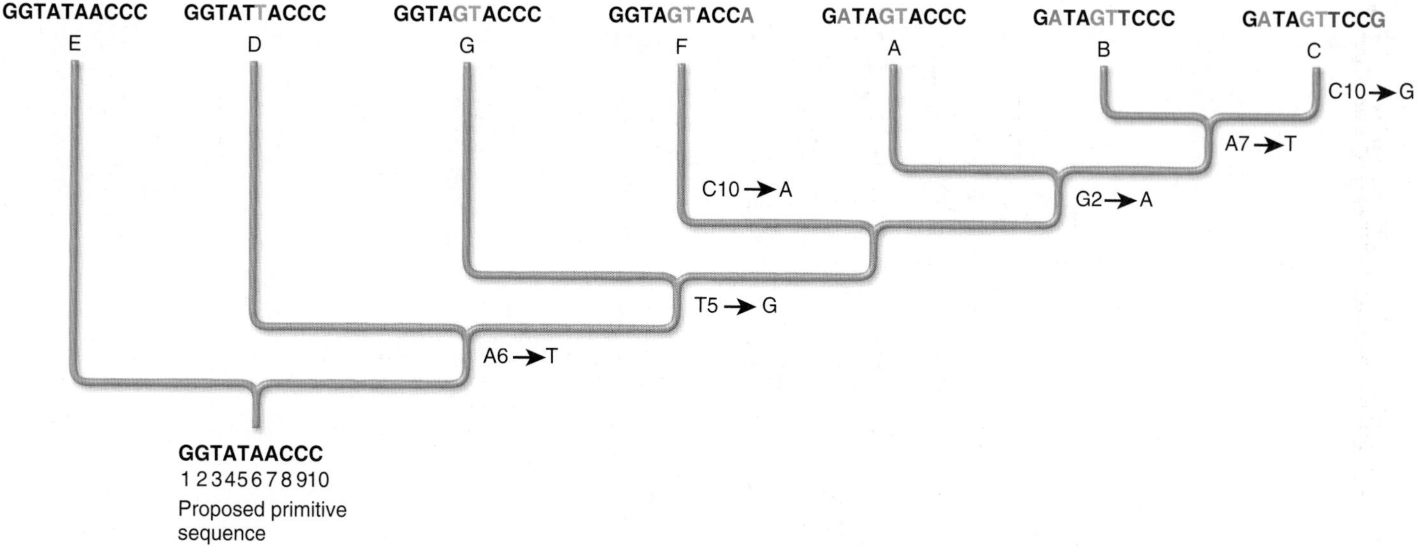

Figure 26.10 The use of shared derived characters applied to molecular data. This phylogenetic tree illustrates a cladogram involving homologous gene sequences found in seven hypothetical plant species. Mutations that alter a primitive DNA sequence are shared among certain species but not others. Note: A, T, G, and C refer to nucleotide bases, and the numbers refer to the position of the base in the nucleotide sequences. For example, A6 refers to an adenine at the sixth position.

Concept Check: *What nucleotide change is a shared derived character for species A, B, and C, but not for species G?*

Such an outgroup typically shares morphological traits and/or DNA sequence similarities with the members of the ingroup to allow a comparison between the ingroup and outgroup. Even so, the outgroup must be noncontroversial in that it shows enough distinctive differences with the ingroup to be considered a clear outgroup. For example, if the ingroup was a group of mammalian species, an outgroup could be a reptile.

The Principle of Parsimony Is Used to Choose from Among Possible Cladograms

One approach for choosing among possible cladograms is to assume that the best hypothesis is the one that requires the fewest number of evolutionary changes. This concept, called the **principle of parsimony**, states that the preferred hypothesis is the one that is the simplest for all the characters and their states. For example, if two species possess a tail, we would initially assume that a tail arose once during evolution and that both species have descended from a common ancestor with a tail. Such a hypothesis is simpler, and more likely to be correct, than assuming that tails arose twice during evolution and that the tails in the two species are not due to descent from a common ancestor.

The principle of parsimony can also be applied to gene sequence data, in which case the most likely hypothesis is the one requiring the fewest base changes. Let's consider a hypothetical example involving molecular data from four taxa (A–D), where A is presumed to be the outgroup.

```
   12345

A: GTACA (outgroup)
B: GACAG
C: GTCAA
D: GACCG
```

Given that B, C, and D are the ingroup, three hypotheses for phylogenetic trees are shown in Figure 26.11, although more are possible. Tree 1 requires seven mutations, and tree 2 requires six, whereas tree 3 requires only five. Therefore, tree 3 requires the smallest number of mutations and is considered the most parsimonious. Based on the principle of parsimony, the tree with the fewest number of base changes is the hypothesis that is the most likely to accurately reflect the evolutionary history of the taxon (or ingroup) in question. In practice, when researchers have multiple sequences that are longer than the ones shown here, computer programs are used to find the most parsimonious tree.

Maximum Likelihood Is Also Used to Discriminate Among Possible Phylogenetic Trees

In addition to the principle of parsimony, evolutionary biologists also apply other approaches to the evaluation of phylogenetic trees. These methods involve the use of an evolutionary model—a set of assumptions about how evolution is likely to happen. For example, mutations affecting the third base in a codon are often neutral because they don't affect the amino acid sequence of the encoded protein and therefore don't affect the fitness of an organism. As discussed in Chapter 24, such neutral mutations are more likely to become prevalent in a population than are mutations in the first or second base. Therefore, one possible assumption of an evolutionary model is that neutral mutations are more likely than nonneutral mutations.

According to an approach called **maximum likelihood**, researchers may ask the question: What is the probability that an evolutionary model and a proposed phylogenetic tree would give rise to observed molecular data? To answer this question, they must devise rules about how DNA sequences change over time. For example, one rule may

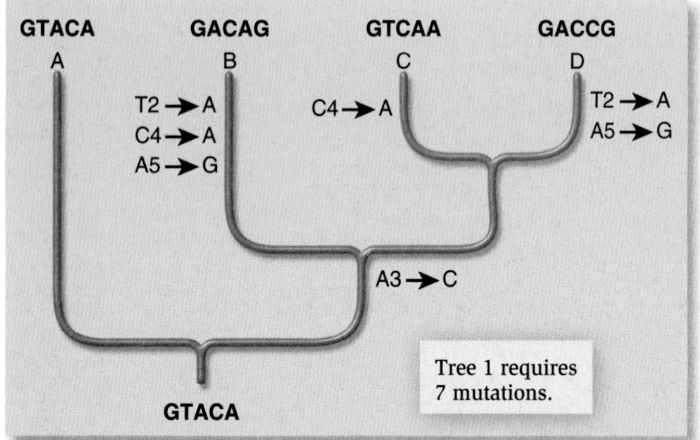

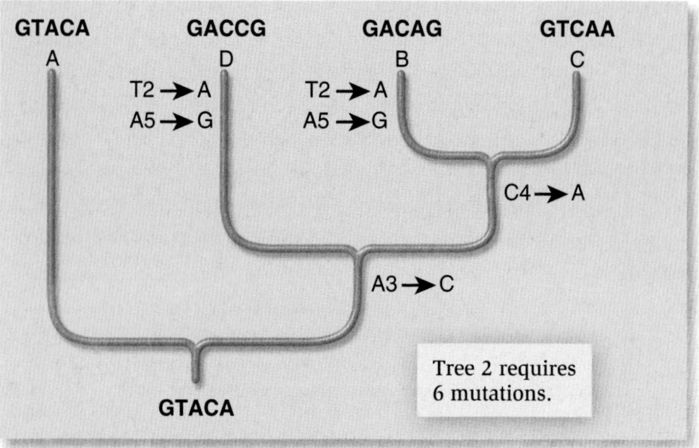

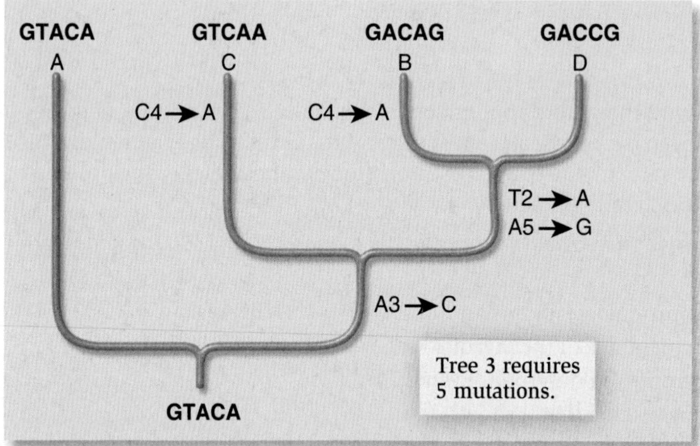

Figure 26.11 **Using the principle of parsimony and molecular genetic data to choose a phylogenetic tree.** Shown are three possible phylogenetic trees for the evolution of a short DNA sequence (although many more are possible). Changes in nucleotide sequence are indicated for each tree. For example, T2 → A means that the second base, a T, was changed to an A. According to the principle of parsimony, tree number 3 is the more likely choice because it requires only five mutations.

be that neutral mutations are more likely to occur than nonneutral mutations. A second rule might be that the rate of change of DNA sequences is relatively constant from one generation to the next in a particular lineage. With a set of probability rules, researchers can analyze different possible trees and predict the relative probabilities for each of them. The phylogenetic tree that gives the highest probability of producing the observed data is preferred to any trees that give a lower probability.

Feature Investigation

Cooper and Colleagues Compared DNA from Extinct Flightless Birds and Existing Species to Propose a New Phylogenetic Tree

Genetic sequence information is primarily used for studying relationships among existing species. However, in some cases, DNA can be obtained from extinct organisms. Starting with small tissue samples from extinct species, scientists have discovered that it is occasionally possible to obtain DNA sequence information. This is called ancient DNA analysis or molecular paleontology. Since the mid-1980s, some researchers have become excited about the information derived from sequencing DNA of extinct specimens. Debate has centered on how long DNA can remain intact after an organism has died. Over time, the structure of DNA is degraded by hydrolysis and the loss of purines. Nevertheless, under certain conditions (cold temperature, low oxygen, and so on), DNA samples may be stable for as long as 50,000–100,000 years. In most studies involving extinct specimens, the ancient DNA is extracted from bone, dried muscle, or preserved skin. In recent years, this approach has been used to study evolutionary relationships between living and extinct species.

As shown in **Figure 26.12**, Alan Cooper, Cécile Mourer-Chauviré, Geoffrey Chambers, Arndt von Haeseler, Allan Wilson, and Svante Pääbo investigated the evolutionary relationships among some extant and extinct species of flightless birds. In this example of discovery-based science, the researchers gathered data with the goal of proposing a hypothesis about the evolutionary relationships among several bird species. The kiwis and moas are two groups of flightless birds that existed in New Zealand during the Pleistocene. Species of kiwis still exist, but the moas are now extinct. Eleven known species of moas formerly existed. In this study, the researchers investigated the phylogenetic relationships between four extinct species of moas, which were available as museum samples; three species of New Zealand kiwis; and living species of other flightless birds, including the emu and the cassowary (both found in Australia and/or New Guinea), the ostrich (found in Africa and formerly Asia), and two rheas (found in South America).

Samples from the various species were subjected to polymerase chain reaction (PCR) to amplify a region of the gene that encodes an RNA found in the mitochondrial small ribosomal subunit (SSU rRNA). This provided enough DNA for sequencing. The data in Figure 26.12 illustrate a comparison of the sequences of a continuous region of the SSU rRNA gene from these species. The first line shows the DNA sequence for one of the four extinct moa species. Below it are the sequences of several of the other species they analyzed. When the other sequences are identical to the first sequence, a dot is placed in the corresponding position. When the sequences are different, the changed nucleotide base (A, T, G, or C) is placed there. In a few regions, the genes are different lengths. In these cases, a dash is placed to indicate missing nucleotides.

As you can see from the large number of dots, the gene sequences among these flightless birds are very similar, though some differences occur. If you look carefully at the data, you will notice that the sequence from the kiwi (a New Zealand species) is actually more similar to the sequence from the ostrich (an African species) than it is to that of the moa, which was once found in New Zealand. Likewise, the kiwi is more similar to the emu and cassowary (found in Australia and New Guinea) than to the moa. How were these results interpreted? The researchers concluded that the kiwis are more closely related to African and Australian flightless birds than they are to the moas. From these results, they concluded that New Zealand was colonized twice by ancestors of flightless birds. The researchers used a maximum likelihood analysis to propose a new phylogenetic tree that illustrates the revised relationships among these living and extinct species (**Figure 26.13**).

Figure 26.12 **DNA analysis of phylogenetic relationships among living and extinct flightless birds by Cooper and colleagues.**

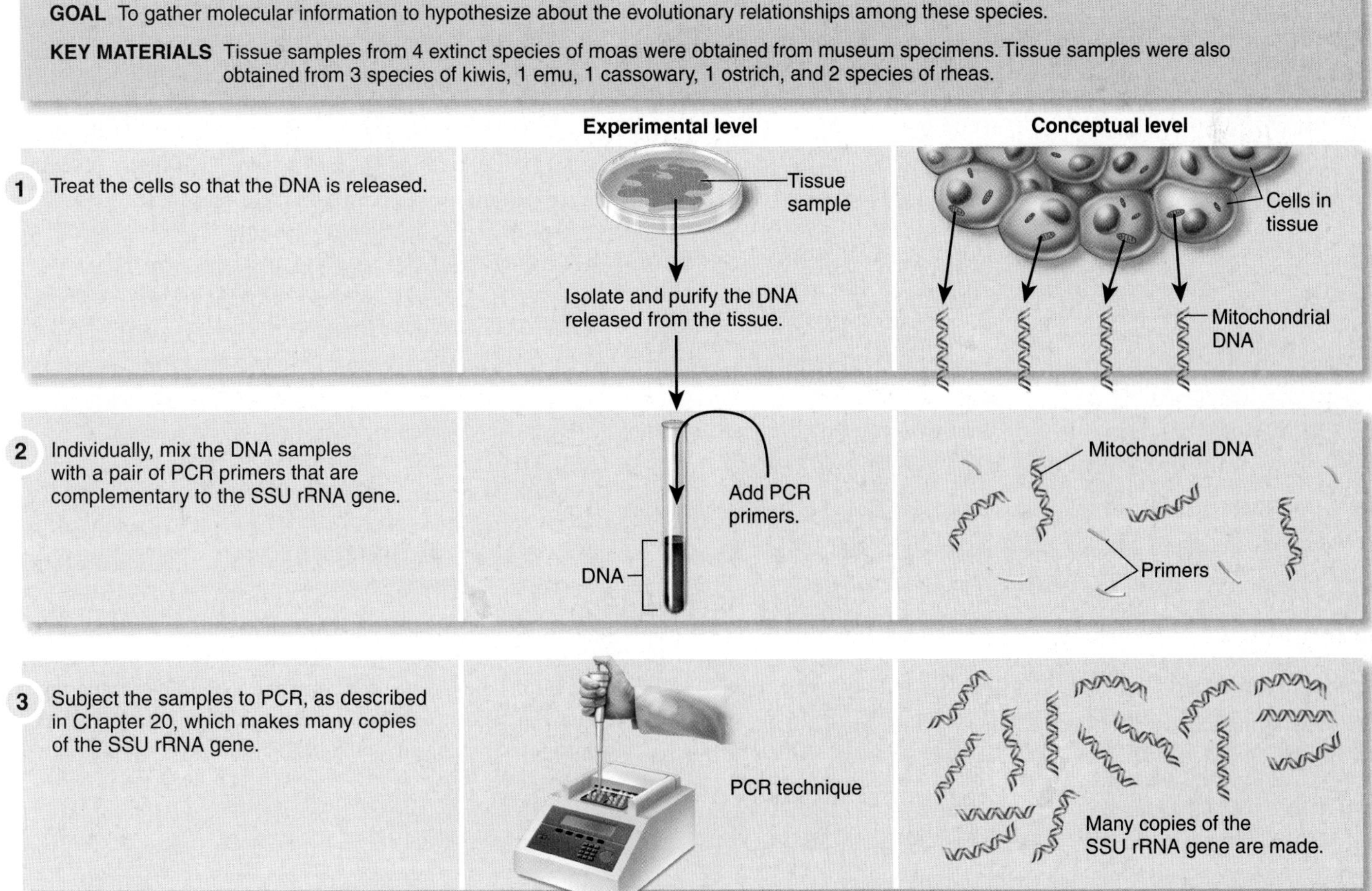
GOAL To gather molecular information to hypothesize about the evolutionary relationships among these species.
KEY MATERIALS Tissue samples from 4 extinct species of moas were obtained from museum specimens. Tissue samples were also obtained from 3 species of kiwis, 1 emu, 1 cassowary, 1 ostrich, and 2 species of rheas.
Experimental level
Conceptual level
1 Treat the cells so that the DNA is released.
Tissue sample
Isolate and purify the DNA released from the tissue.
Cells in tissue
Mitochondrial DNA
2 Individually, mix the DNA samples with a pair of PCR primers that are complementary to the SSU rRNA gene.
Add PCR primers.
DNA
Mitochondrial DNA
Primers
3 Subject the samples to PCR, as described in Chapter 20, which makes many copies of the SSU rRNA gene.
PCR technique
Many copies of the SSU rRNA gene are made.

4 Subject the amplified DNA fragments to DNA sequencing, as described in Chapter 20.	Sequence the amplified DNA.	The amplification of the SSU rRNA gene allows it to be subjected to DNA sequencing.
5 Align the DNA sequences to each other, using computer techniques described in Chapter 21.	 Align sequences, using computer programs.	Align sequences to compare the degree of similarity.

6 THE DATA

```
Moa 1      GCTTAGCCCTAAATCCAGATACTTACCCTACACAAGTATCCGCCCGAGAACTACGAGCACAAACGCTTAAAACTCTAAGGACTTGGCGGTGCCCCAAACCCA
Kiwi 1     ················T·G·······GT···CT·····C····························································T·······
Emu        ··············TT······C···T···CAG··C·······T·······················································T·······
Cassoway   ··············TT······CG·TA···CTG··································································T·······
Ostrich    ·······T······AT···········C··CT···································································T·······
Rhea 1     ···············T···········C··CT···································································T·······

Moa 1      CCTAGAGGAGCCTGTTCTATAATCGATAATCCACGATACACCCGACCATCCCTCGCCCGT-GCAGCCTACATACCGCCGTCCCCAGCCCGCCT--AATGAAA
Kiwi 1     ···························C·············A······T··T···AAC–A·······T···········G·····T·····AA·····G·
Emu        ···························C·············A······T··T···AA·–A··················G···········––·······
Cassowary  ···························C·············AG·····T··T···AA·TA··················G···········––·G···G·
Ostrich    ·································T·····A····C····T···A––T··················G···········C––····G·
Rhea 1     ···························C··················T··T····A·–································TA·G·····

Moa 1      G–AACAATAGCGAGCACAACAGCCCTCCCCCGCTAACAAGACAGGTCAAGGTATAGCATATGAGATGGAAGAAATGGGCTACATTTTCTAACATAGAACACC
Kiwi 1     ·–·····C·····A·········TA··–··A·······················C·································A······T··T
Emu        ·–················T····AC––TT·························G·········································T·T·
Cassowary  ·–·············T·······AC––T··························G·········································T···
Ostrich    ·–················T·····A––································GAG·································T··A
Rhea 1     ·–·····C···AG···T··T···TA–––·······························G·····························TC·······A·

Moa 1      C––––––––––––ACGAAAGAGAAGGTGAAACCCTCCTCAAAAGGCGGATTTAGCAGTAAAATAGAACAAGAATGCCTATTTTAAGCCCGGCCCTGGGGC
Kiwi 1     –···················A··GGT······T·–C···T·G·····················C····T····GA·T···········–·T······A····
Emu        –······················AG·T······T·AC·T···G····················C····T····GA·T·········A–·T···T··A····
Cassowary  –··················A··G·T······T·A····T·G·····················C·········GA·T·········A–·········A····
Ostrich    –·····················G·TA·····T·A······G··························T····GA·T··········–T····T··A····
Rhea 1     –·················G·····GGCA·······–AC····CG··················G···G·TC····A···C·C······–·········A····
```

7 CONCLUSION This discovery-based investigation led to a hypothesis regarding the evolutionary relationships among these bird species, which is described in Figure 26.13.

8 SOURCE Cooper, Alan et al. 1992. Independent origins of New Zealand moas and kiwis. *Proceedings of the National Academy of Sciences* 89:8741–8744.

Experimental Questions

1. What is molecular paleontology? What was the purpose of the study conducted by Cooper and colleagues?
2. What birds were examined in the Cooper study, and what are their geographic distributions? Why were the different species selected for this study?
3. What results did Cooper and colleagues obtain by comparing these DNA sequences? How did the results of this study affect the proposed phylogeny of flightless birds?

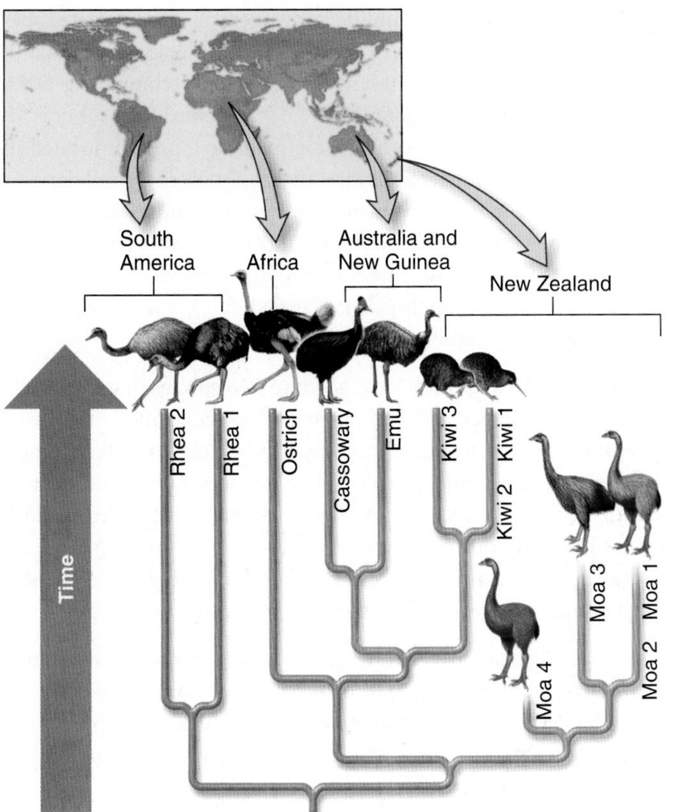

Figure 26.13 A revised phylogenetic tree of flightless birds. This tree is based on a comparison of DNA sequences from extinct and living flightless birds, as described in Figure 26.12.

Concept Check: With regard to geography, why are the results of Cooper and his colleagues surprising?

26.4 Molecular Clocks

Learning Outcomes:

1. Explain how molecular clocks are used in the dating of evolutionary events.
2. Compare and contrast the use of different genes to produce phylogenetic trees.

As we have seen, researchers employ various methods to choose a phylogeny that describes the evolutionary relationships among various species. Researchers are interested not only in the most likely pathway of evolution (the branches of the trees), but also the timing of evolutionary change (the lengths of the branches). How can researchers determine when different species diverged from each other in the past? As shown earlier in Figure 26.7, the fossil record can sometimes help researchers apply a timescale to a phylogeny.

Another way to infer the timing of past events is by analyzing genetic sequences. The **neutral theory of evolution** proposes that most genetic variation that exists in populations is due to the accumulation of neutral mutations—changes in genes and proteins that are not acted on by natural selection. The reasoning behind this concept is that favorable mutations are likely to be very rare, and detrimental mutations are likely to be eliminated from a population by natural selection. A large body of evidence supports the idea that much of the genetic variation observed in living species is due to the accumulation of neutral mutations. From an evolutionary point of view, if neutral mutations occur at a relatively constant rate, they can act as a **molecular clock** on which to measure evolutionary time. In this section, we will consider the concept of a molecular clock and its application in phylogenetic trees.

The Timing of Evolutionary Change May Be Inferred from Molecular Clock Data

Figure 26.14 illustrates the concept of a molecular clock. The graph's y-axis is a measure of the number of nucleotide differences in a homologous gene between different pairs of species. The x-axis plots the amount of time that has elapsed since each pair of species shared a common ancestor. As discussed in the figure, the number of nucleotide differences is lower when two species shared a common ancestor in the more recent past than it is in pairs that shared a more distant common ancestor. The explanation for this phenomenon is that the gene sequences of various species accumulate independent mutations after they have diverged from each other. A longer period of time since their divergence allows for a greater accumulation of mutations, which makes their sequences more different.

Figure 26.14 suggests a linear relationship between the number of nucleotide changes and the time of divergence. For example, a linear

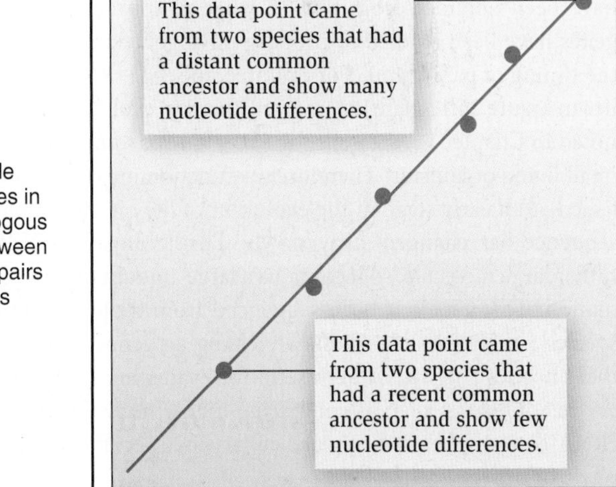

Figure 26.14 A molecular clock. According to the concept of a molecular clock, neutral mutations accumulate at a relatively constant rate over evolutionary time. When comparing the same homologous gene between pairs of different species, those species that diverged more recently tend to have fewer differences than do those whose common ancestor occurred in the distant past.

BioConnections: Look back at Table 12.1, which shows the genetic code. Propose three changes to a codon sequence that you think would be a neutral mutation.

relationship predicts that a pair of species with, say, 20 nucleotide differences in a given gene sequence would have a common ancestor that is roughly twice as old as that of a pair showing 10 nucleotide differences. Although actual data sometimes show a relatively linear relationship over a defined time period, evolutionary biologists do not think that molecular clocks are perfectly linear over very long periods of time. Several factors can contribute to nonlinearity of molecular clocks. These include differences in the generation times of the species being analyzed and variation in the mutation rates of genes between different species.

To obtain reliable data, researchers must calibrate their molecular clocks. How much time does it take to accumulate a certain percentage of nucleotide changes? To perform such a calibration, researchers must have information regarding the date when two species diverged from a common ancestor. Such information could come from the fossil record, for instance. The genetic differences between those species are then divided by the amount of time since their last common ancestor to calculate a rate of evolutionary change. For example, fossil evidence suggests that humans and chimpanzees diverged from a common ancestor approximately 6 mya. The percentage of nucleotide differences between mitochondrial DNA of humans and chimpanzees is 12%. From these data, the molecular clock for changes in mitochondrial DNA sequences of primates is calibrated at roughly 2% nucleotide changes per million years.

Different Genes Are Analyzed to Study Phylogeny and Evaluate the Timing of Evolutionary Change

For evolutionary comparisons, the DNA sequences of many genes have been obtained from a wide range of sources. Many different genes have been studied to propose phylogenetic trees and evaluate the timing of past events. For example, the SSU rRNA described earlier in Figure 26.12 gene is commonly used in evolutionary studies. As noted in Chapter 12, the gene for SSU rRNA is found in the genomes of all living organisms. Therefore, its function must have been established at an early stage in the evolution of life on this planet, and its sequence has changed fairly slowly. Furthermore, SSU rRNA is a rather large molecule, so it contains a large amount of sequence information. This gene has been sequenced from thousands of different species (see Figure 12.16). Slowly changing genes such as the gene that encodes SSU rRNA are useful for evaluating distant evolutionary relationships, such as comparing higher taxa. For example, SSU rRNA data can be used to place eukaryotic species into their proper phyla or orders.

Other genes have changed more rapidly during evolution because of a greater tolerance of neutral mutations. For example, the mitochondrial genome and DNA sequences within introns can more easily incur neutral mutations (compared to the coding sequences of genes), and so their sequences change frequently during evolution. More rapidly changing DNA sequences have been used to study recent evolutionary relationships, particularly among eukaryotic species such as species of large animals that have long generation times and therefore tend to evolve more slowly. In these cases, slowly evolving genes may not be very useful for establishing evolutionary relationships because two closely related species may have identical or nearly identical DNA sequences for such genes.

Figure 26.15 shows a simplified phylogeny of closely related species of primates. A molecular clock was used to give a timescale to this phylogenetic tree. The tree was proposed by comparing DNA sequence changes in the gene for cytochrome oxidase subunit II, one of several subunits of cytochrome oxidase, a protein located in the mitochondrial inner membrane that is involved in cellular respiration. This gene tends to change fairly rapidly on an evolutionary timescale. The vertical scale on Figure 26.15 represents time, and the branch points labeled with letters represent common ancestors. Let's take a look at three branch points (labeled A, D, and E) and relate them to the accumulation of neutral mutations.

Ancestor A: This ancestor diverged into two species that ultimately gave rise to siamangs and the other five species. Since this divergence, there has been a long time (approximately 23 million years) for the siamang genome to accumulate a relatively high number of random neutral changes that would be different from the random changes that have occurred in the genomes of the other five species (see the yellow bar in Figure 26.15). Therefore, the gene in the siamangs is fairly different from the genes in the other five species.

Ancestor D: This ancestor diverged into two species that eventually gave rise to humans and chimpanzees. This divergence occurred a moderate time ago, approximately 6 mya, as illustrated by the red bar. The differences in gene sequences between humans and chimpanzees are relatively moderate.

Ancestor E: This ancestor diverged into two species of chimpanzees. Since the divergence of species E into two species, approximately 3 mya, the time for the molecular clock to "tick" (that is, accumulate random mutations) is relatively short, as depicted by the green bar in Figure 26.15. Therefore, the two existing species of chimpanzees have fewer differences in their gene sequences compared to other primates.

26.5 Horizontal Gene Transfer

Learning Outcome:

1. Explain how horizontal gene transfer affects evolution and how it affects the relationships among different taxa.

Thus far, we have considered various ways to propose phylogenetic trees, which describe the relationships between ancestors and their descendents. The type of evolution depicted in previous figures, which involves changes in groups of species due to descent from a common ancestor, is called vertical evolution. Since the time of Darwin, vertical evolution has been the traditional way that biologists view the evolutionary process. However, over the past couple of decades researchers have come to realize that evolution is not so simple. In addition to vertical evolution, horizontal gene transfer has also played a significant role in the phylogeny of living species.

As described in Chapters 1 and 23, **horizontal gene transfer** is used to describe any process in which an organism incorporates genetic material from another organism without being the offspring of that organism. As discussed next, this phenomenon has reshaped the way that biologists view the evolution of species.

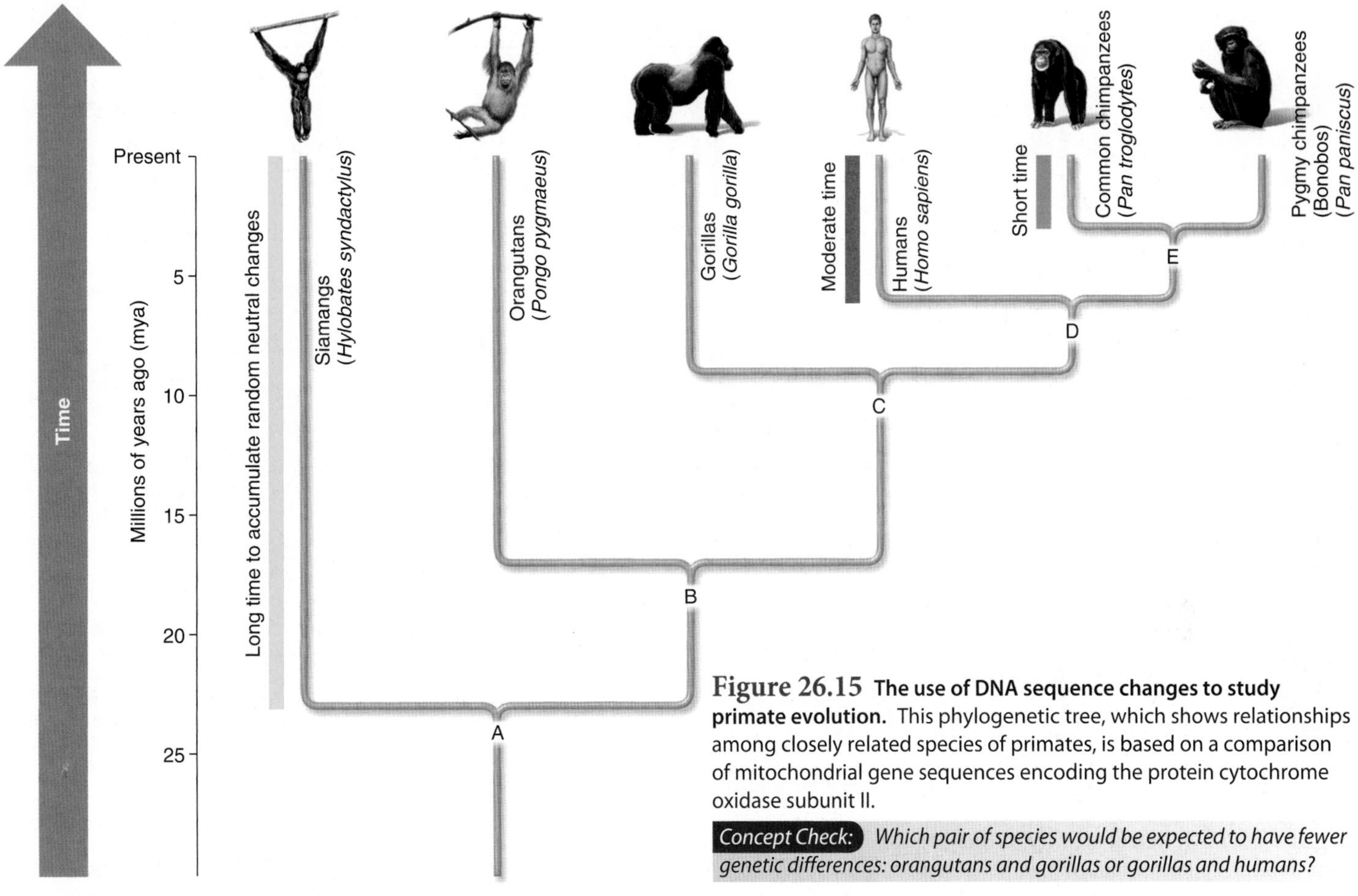

Figure 26.15 The use of DNA sequence changes to study primate evolution. This phylogenetic tree, which shows relationships among closely related species of primates, is based on a comparison of mitochondrial gene sequences encoding the protein cytochrome oxidase subunit II.

Concept Check: Which pair of species would be expected to have fewer genetic differences: orangutans and gorillas or gorillas and humans?

GENOMES & PROTEOMES CONNECTION

Due to Horizontal Gene Transfer, the "Tree of Life" Is Really a "Web of Life"

Horizontal gene transfer has played a major role in the evolution of many species. As discussed in Chapter 18, bacteria can transfer genes via conjugation, transformation, and transduction. Bacterial gene transfer can occur between strains of the same species or, occasionally, between cells of different bacterial species. The transferred genes may encode proteins that provide a survival advantage, such as resistance to antibiotics or the ability to metabolize an organic molecule in the environment. Horizontal gene transfer is also fairly common among certain unicellular eukaryotes. However, its relative frequency and importance in the evolution of multicellular eukaryotes remains difficult to evaluate.

Scientists have debated the role of horizontal gene transfer in the earliest stages of evolution, prior to the divergence of the bacterial and archaeal domains. The traditional viewpoint was that the three domains of life—Bacteria, Archaea, and Eukarya—arose from a single type of prokaryotic (or pre-prokaryotic) cell called the universal ancestor. However, genomic research has suggested that horizontal gene transfer may have been particularly common during the early stages of evolution on Earth, when all species were unicellular. Horizontal gene transfer may have been so prevalent that the universal ancestor may have actually been an ancestral community of cell lineages that evolved as a whole. If that were the case, the tree of life cannot be traced back to a single ancestor.

Figure 26.16 illustrates a schematic scenario for the evolution of life that includes the roles of both vertical evolution and horizontal gene transfer. This has been described as a "web of life" rather than a "tree of life." In this scenario, instead of a universal ancestor, a community of primitive cells transferred genetic material primarily in a horizontal fashion. Horizontal gene transfer was also prevalent during the early evolution of bacteria and archaea, and when eukaryotes first emerged as unicellular species. In living bacteria and archaea, it remains a prominent way to foster evolutionary change. By comparison, the region of the diagram that contains most eukaryotic species has a more treelike structure. Researchers have speculated that multicellularity and sexual reproduction have presented barriers to horizontal gene transfer in most eukaryotes. For a gene to be transmitted to eukaryotic offspring, it would have to be transferred into a eukaryotic cell that is a gamete or a cell that gives rise to gametes. Horizontal gene transfer has become less common in eukaryotes, particularly among multicellular species, though it does occur occasionally.

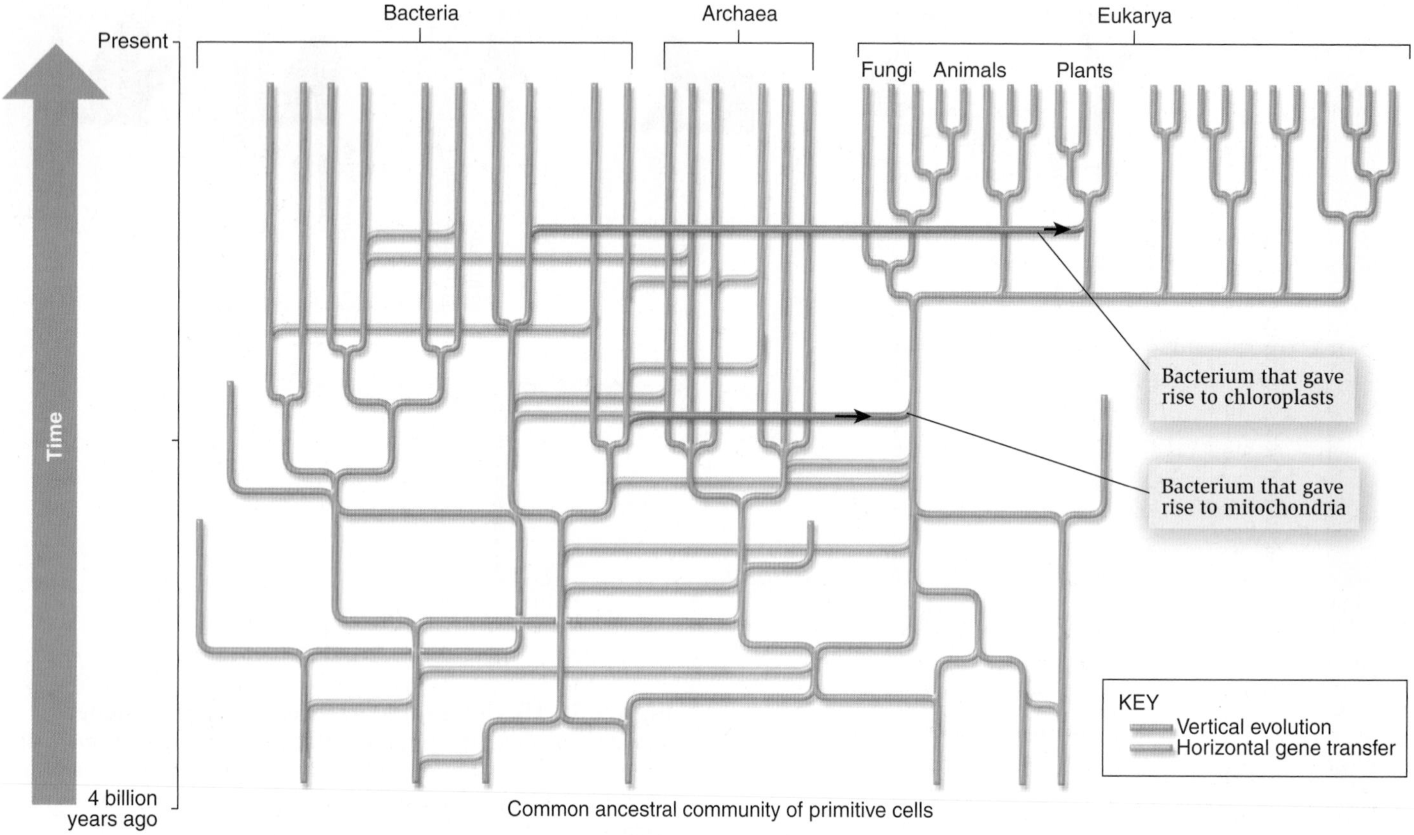

Figure 26.16 A web of life. This phylogenetic tree shows not only the vertical evolution of life but also the contribution of horizontal gene transfer. In this scenario, horizontal gene transfer was prevalent during the early stages of evolution, when all organisms were unicellular, and continues to be a prominent factor in the speciation of Bacteria and Archaea. Note: This tree is meant to be schematic. Also, although the introduction of chloroplasts into the eukaryotic domain is shown as a single event, such events have occurred multiple times and by different mechanisms.

Concept Check: *How does the phenomenon of horizontal gene transfer muddle the concept of monophyletic groups?*

Summary of Key Concepts

26.1 Taxonomy

- Taxonomy is the field of biology concerned with the describing, naming, and classifying living and extinct organisms. Systematics is the study and classification of evolutionary relationships among organisms through time.
- Taxonomy places all living organisms into progressively smaller hierarchical groups called taxa (singular, taxon). The broadest groups are the three domains, called Bacteria, Archaea, and Eukarya, followed by supergroups, kingdoms, phyla, classes, orders, families, genera, and species (Figures 26.1, 26.2, Table 26.1).
- Binomial nomenclature is a naming convention that provides each species with two names: its genus name and species epithet.

26.2 Phylogenetic Trees

- The evolutionary history of a species is its phylogeny. A phylogenetic tree is a diagram that describes the phylogeny of particular species and should be viewed as a hypothesis. (Figure 26.3).
- A central goal of systematics is to construct taxa and phylogenetic trees based on evolutionary relationships. Smaller taxa, such as families and genera, are derived from more recent common ancestors than are broader taxa such as kingdoms and phyla (Figure 26.4).
- Ideally, all taxa should be monophyletic, consisting of the most recent common ancestor and all of its descendants, though previously established taxa sometimes turn out to be paraphyletic or polyphyletic (Figures 26.5, 26.6).
- Both morphological and genetic data are used to propose phylogenetic trees. Molecular systematics, which involves the analysis of genetic sequences, has led to major revisions in taxonomy (Figure 26.7).

26.3 Cladistics

- In the cladistic approach to creating a phylogenetic tree, also called a cladogram, species are grouped together according to shared derived characters (Figure 26.8).
- An ingroup is the group of interest, whereas an outgroup is a species or group of species that lacks one or more shared derived characters (synapomorphies). A comparison of the ingroup and outgroup is

used to determine which character states are derived and which are primitive (Figures 26.9, 26.10).

- The cladistic approach produces many possible cladograms. The most likely phylogenetic tree is chosen by a variety of methods, including the analysis of fossils, the principle of parsimony, and maximum likelihood (Figure 26.11).
- Cooper and colleagues analyzed DNA sequences from extinct and living flightless birds and proposed a new phylogenetic tree showing that New Zealand was colonized twice by ancestors of flightless birds (Figures 26.12, 26.13).

26.4 Molecular Clocks

- The neutral theory of evolution proposes that most genetic variation is due to neutral mutations. Assuming that neutral mutations occur at a relatively constant rate, genetic data can act as a molecular clock with which to measure the timing of evolutionary changes (Figure 26.14).
- Slowly changing genes are useful for analyzing distant evolutionary relationships, whereas rapidly changing genes are used to analyze more recent evolutionary relationships, particularly among eukaryotes that have long generation times and evolve more slowly (Figure 26.15).

26.5 Horizontal Gene Transfer

- Horizontal gene transfer is the phenomenon in which an organism incorporates genetic material from another organism without being the offspring of that organism. Due to horizontal gene transfer, the tree of life may more accurately be described as a web of life (Figure 26.16).

Assess and Discuss

Test Yourself

1. The study of biological diversity based on evolutionary relationships is
   a. paleontology. c. systematics. e. both a and b.
   b. evolution. d. phylogeny.
2. Which of the following is the correct order of the taxa used to classify organisms?
   a. kingdom, domain, phylum, class, order, family, genus, species
   b. domain, kingdom, class, phylum, order, family, genus, species
   c. domain, kingdom, phylum, class, family, order, genus, species
   d. domain, kingdom, phylum, class, order, family, genus, species
   e. kingdom, domain, phylum, order, class, family, species, genus
3. When considering organisms within the same taxon, which level includes organisms with the greatest similarity?
   a. kingdom c. order e. genus
   b. class d. family
4. Which of the following characteristics is not shared by bacteria, archaea, and eukaryotes?
   a. DNA is the genetic material.
   b. Messenger RNA encodes the information to produce proteins.
   c. All cells are surrounded by a plasma membrane.
   d. The cytoplasm is compartmentalized into organelles.
   e. both a and d
5. The branch points or nodes in a phylogenetic tree depict which of the following?
   a. anagenesis d. a and b only
   b. cladogenesis e. b and c only
   c. horizontal gene transfer
6. The evolutionary history of a species is its
   a. systematics. c. evolution. e. embryology.
   b. taxonomy. d. phylogeny.
7. A taxon composed of all species derived from a common ancestor is referred to as
   a. a phylum.
   b. a monophyletic group or clade.
   c. a genus.
   d. an outgroup.
   e. all of the above.
8. A goal of modern taxonomy is to
   a. classify all organisms based on morphological similarities.
   b. classify all organisms in monophyletic groups.
   c. classify all organisms based solely on genetic similarities.
   d. determine the evolutionary relationships only between similar species.
   e. none of the above
9. The concept that the preferred hypothesis is the one that is the simplest is
   a. phenetics. d. maximum likelihood.
   b. cladistics. e. both b and d.
   c. the principle of parsimony.
10. Research indicates that horizontal gene transfer is less prevalent in eukaryotes because of
    a. the presence of organelles. d. all of the above.
    b. multicellularity. e. b and c only.
    c. sexual reproduction.

Conceptual Questions

1. Explain how species' names follow a binomial nomenclature. Give an example.
2. What is a molecular clock? How is it used in depicting phylogenetic trees?
3. A principle of biology is that *populations of organisms evolve from one generation to the next.* What are some advantages and potential pitfalls of using changes in morphology to construct phylogenetic trees?

Collaborative Questions

1. Discuss how taxonomy is useful. Make a list of some practical applications that are derived from taxonomy.
2. Discuss systematics and how it is used to propose a phylogenetic tree. Discuss the rationale behind using the principle of parsimony.

Online Resource

www.brookerbiology.com

Stay a step ahead in your studies with animations that bring concepts to life and practice tests to assess your understanding. Your instructor may also recommend the interactive eBook, individualized learning tools, and more.

UNIT V

DIVERSITY

Biological diversity encompasses the variety of living things that exist now, as well as all the life forms that lived in the past. Knowing about species that lived in the past and how they are related to modern microorganisms, plants, and animals aids our comprehension of evolutionary process, which is the source of organismal variation. Knowing about the many different kinds of modern organisms also helps us to understand how life forms are structured in ways that allow them to function differently in nature (described in Units VI and VII) and how species interact with each other and with their environments (described in Unit VIII—Ecology).

Unit V begins with Bacteria and Archaea, the oldest, simplest, and most numerous of Earth's life-forms, whose prominent ecological roles are described in Chapter 27. In Chapter 28, we survey the surprisingly diverse Protists, which affect humans and other organisms in many important ways. Chapter 29 explores the evolutionary origin of the first plants, a process that explains the features and functions of the seed plants that are vital sources of human food, fiber, and medicine, as described in Chapter 30. The mysteries of the fungi, essential to the brewing and baking industries as well as ecological stability, are revealed in Chapter 31. An overview of the diversity and evolutionary history of the animals, provided in Chapter 32, provides the basis for exploring the simplest animals, the invertebrates in Chapter 33. More complex animals, including humans and their closest relatives, are the focus of Chapter 34, which reveals how our species arose.

The following biology principles will be emphasized in this unit:

- ***Cells are the simplest units of life:*** *Many of Earth's present and past species have bodies consisting of only one or a few cells, whereas other species display more complex bodies composed of many cells.*
- ***Living organisms interact with their environment:*** *Chapters 27 and 29, for example, explain how ancient microorganisms and plants dramatically changed the composition of Earth's atmosphere, and how their modern descendants continue to influence today's atmosphere and climate.*
- ***All species (past and present) are related by an evolutionary history:*** *This concept is emphasized throughout the unit.*

Chapter Outline

27.1 Diversity and Evolution

27.2 Structure and Movement

27.3 Reproduction

27.4 Nutrition and Metabolism

27.5 Ecological Roles and Biotechnology Applications

Summary of Key Concepts

Assess and Discuss

Archaea and Bacteria

27

Cyanobacterial bloom. A visible cyanobacterial bloom gives a blue-green coloration to this lake water.

BIOLOGY PRINCIPLE Biology affects our society. Cyanobacterial blooms affect society by poisoning humans and other organisms that people value.

One late-summer afternoon, the veterinarian was about to close the clinic at the end of a busy day when an emergency case arrived. A very sick dog had collapsed after taking a lakeside walk with its owner. Responding to the vet's questions, the owner reported that the thirsty dog had consumed lake water thick with blue-green material. The vet deduced that the blue-green substance represented a large population of toxin-producing bacteria and that drinking the lake water had poisoned the dog—a conclusion that aided treatment.

Bacteria and archaea are examples of microorganisms, organisms so small they can usually be seen only with the use of a microscope. However, in phosphorus-rich bodies of water, some species of photosynthetic bacteria known as cyanobacteria grow rapidly into large, visible populations (blooms) that color the water blue-green or cyan (see chapter opening photo). The individual cells release small amounts of toxins that help to keep small aquatic animals from eating them; when large populations, known as blooms, occur, toxins can rise to levels that poison humans, pets, livestock, and wildlife. Consequently, public health authorities often warn that people should not swim in waters with visible blue-green blooms, and should not allow pets and livestock to drink such water. People can prevent the formation of harmful cyanobacterial blooms by reducing the input of phosphorus-rich fertilizers, manure, and sewage into bodies of water.

Despite the harmful effects of some species, cyanobacteria provide important benefits to humans and other organisms, such as producing atmospheric oxygen. Many cyanobacteria also have the ability to convert abundant but inert atmospheric nitrogen gas into ammonia, which algae and plants can use to synthesize amino acids and proteins. This process enriches nutrient-poor soils, particularly wet paddy fields where rice is grown in many regions of the world, thereby helping to provide food for billions of people. Some cyanobacteria have been genetically engineered to produce renewable, clean-burning biofuels—fuels derived from renewable, biological sources—that aid the world's energy sustainability. Others produce new types of antibiotics that may help to control disease in humans and other animals. In this chapter, we will survey the diversity, structure, reproduction, metabolism, and ecology of archaea and bacteria. Our survey will reveal additional surprising ways in which these microorganisms affect the lives of humans and the world we inhabit.

27.1 Diversity and Evolution

Learning Outcomes:

1. Make a drawing that shows the evolutionary relationship among Domains Archaea, Bacteria, and Eukarya.
2. Describe how many species of archaea are able to grow in extreme habitats.
3. Understand the medical, environmental, and evolutionary importance of cyanobacteria and proteobacteria.
4. List common mechanisms for horizontal gene transfer.

Life on Earth is classified into three domains. Members of the Domain Eukarya—animals, plants, fungi, and protists—have cells with a eukaryotic structure. In contrast, the **Archaea** (informally, archaea) and **Bacteria** (informally, bacteria) are domains of microorganisms whose cells have a prokaryotic structure. Archaeal and bacterial cells

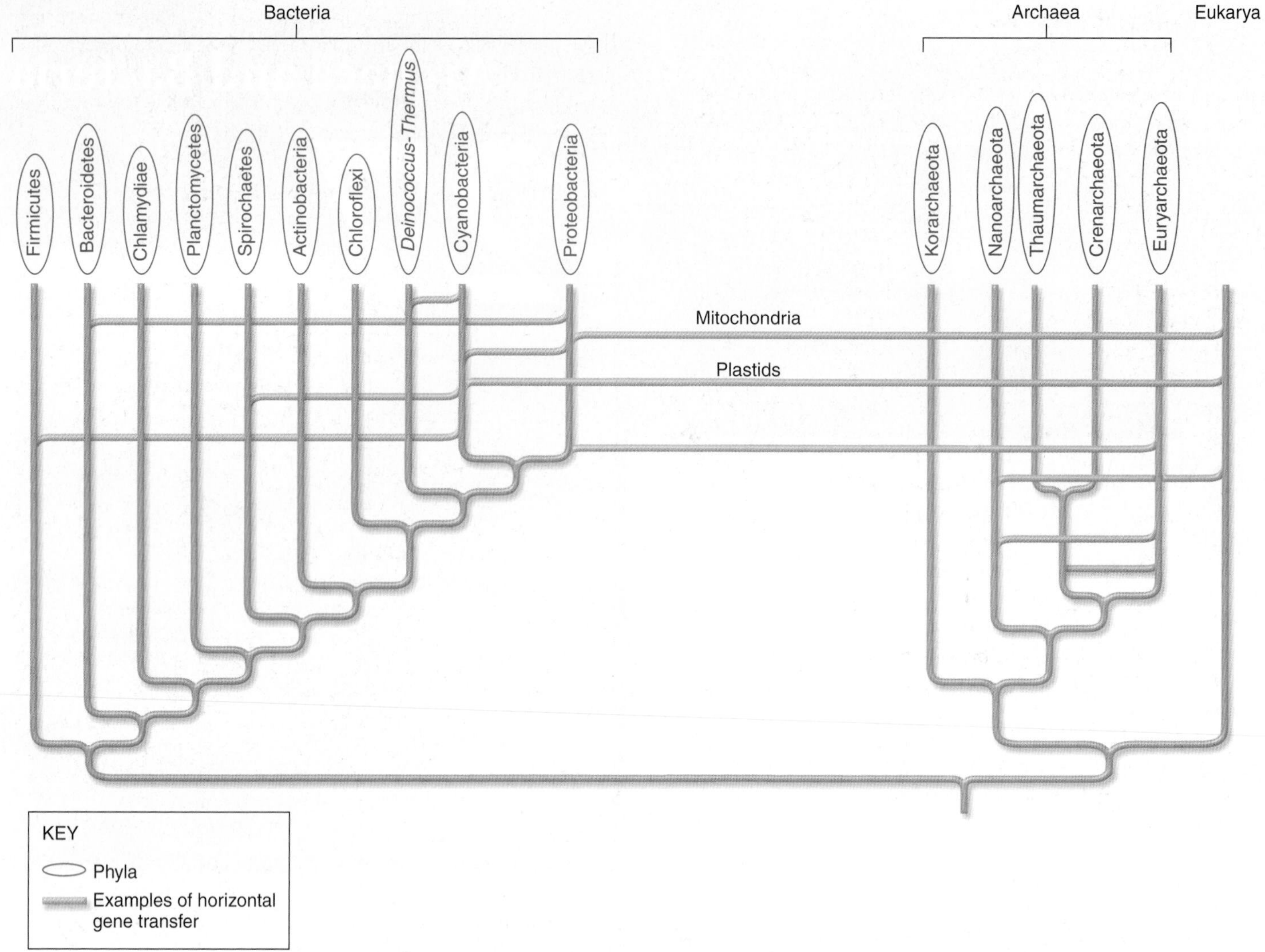

Figure 27.1 **Relationships and diversification of Bacteria and Archaea.** Bacteria and Archaea are the two domains featuring prokaryotic cells. Eukarya is the domain of organisms composed of eukaryotic cells. Each domain has diversified into multiple phyla. Many cases of horizontal gene transfer among phyla and domains are known. These cases include the acquisition of mitochondria and chloroplasts by eukaryotes. Note: The domain Bacteria includes about 50 phyla, but for simplicity, only 10 are shown in this figure.

lack nuclei with porous envelopes and other cellular features typical of eukaryotes (see Chapter 4).

Although archaea and bacteria are sometimes collectively termed "prokaryotes," such an aggregation is not a monophyletic group, but rather is a paraphyletic group—one that does not include all of the descendants of a single common ancestor. That's because Domain Archaea is more closely related to Domain Eukarya than either is to Domain Bacteria (Figure 27.1). Even so, archaea and bacteria display some common features in addition to a prokaryotic cell structure.

Archaea and bacteria include the smallest known cells and are the most abundant organisms on Earth. About half of Earth's total biomass consists of an estimated 10^{30} individual bacteria or archaea. Just a pinch of garden soil can contain 2 billion prokaryotic cells, and about a million occur in 1 mL of seawater. Archaea and bacteria live in nearly every conceivable habitat, including extremely hot or salty waters that support no other life, and they are also Earth's most ancient organisms, having originated more than 3 billion years ago (bya). Their great age and varied habitats have resulted in extraordinarily high metabolic diversity.

Today, many millions of species of archaea and bacteria collectively display more diverse metabolic processes than occur in any other group of organisms. Many of these metabolic processes are important on a global scale, influencing Earth's climate, atmosphere, soils, water quality, and human health and technology. In the past, microbiologists studied diversity by isolating these organisms from nature and growing cultures in the laboratory to observe variation in cell structure and metabolism. Today, biologists also use molecular techniques to assess archaeal and bacterial diversity and infer metabolic functions. Such molecular studies reveal that archaea and bacteria are vastly more diverse than previously realized.

In this section, we will first survey the major kingdoms and phyla of the domains Archaea and Bacteria and then explore how horizontal gene transfer—the transfer of genes between different species—has influenced their evolution.

Domain Archaea Was Ancestral to Domain Eukarya

Organisms classified in the domain Archaea, informally known as archaea, share a number of features with those classified in Eukarya, suggesting common ancestry. For example, histone proteins are typically associated with the DNA of both archaea and eukaryotes, but they are absent from most bacteria. Archaea and eukaryotes share more than 30 ribosomal proteins that are not present in bacteria, and archaeal RNA polymerases are closely related to their eukaryotic counterparts. Even so, archaea possess distinctive membrane lipids, which are formed with ether bonds; in contrast, ester bonds characterize the membrane lipids of bacteria and eukaryotes (Figure 27.2). Ether-bonded membranes are resistant to damage by heat and other extreme conditions, which helps explain why many archaea are able to grow in extremely harsh environments. Also note that archaea use isoprene chains instead of fatty acid chains in making membranes.

Though many archaea occur in soils and surface ocean waters of moderate conditions, diverse archaea occupy habitats with very high salt content, acidity, methane levels, or temperatures that would kill most bacteria and eukaryotes. Organisms that occur primarily in extreme habitats are known as **extremophiles**. One example is the methane producer *Methanopyrus*, which grows best at deep-sea thermal vent sites where the temperature is 98°C. At this temperature, the proteins of most organisms would denature, but those of *Methanopyrus* are resistant to such damage. *Methanopyrus* is so closely adapted to its extremely hot environment that it cannot grow when the temperature is less than 84°C. Such archaea are known as **hyperthermophiles**. Some archaea prefer habitats having both high temperatures and extremely low pH. For example, the archaeal genus *Sulfolobus* was discovered in samples taken from sulfur hot springs having a pH of 3 or lower. Archaea help biologists to better understand the origin of life, the origin of eukaryotes, how life on Earth has evolved in extreme environments, and what kinds of extraterrestrial life might exist.

The domain Archaea includes five phyla: Korarchaeota, Nanoarchaeota, Thaumarchaeota, Crenarchaeota, and Euryarchaeota (see Figure 27.1). Early diverging Korarchaeota are primarily known from DNA sequences found in samples from hot springs. Nanoarchaeota includes the hyperthermophile *Nanoarchaeum equitans*, which appears to be a parasite of the thermal vent crenarchaeote *Ignicoccus*. Thaumarchaeota species that oxidize ammonia are important in global nitrogen cycling. Crenarchaeota includes organisms that live in extremely hot or cold habitats and also some that are widespread in aquatic and terrestrial habitats. The Euryarchaeota includes some hyperthermophiles, diverse methane producers, and extreme halophiles—species able to grow in higher than usual salt concentrations.

Domain Bacteria Includes Cyanobacteria, Proteobacteria and Many Other Phyla

The Domain Bacteria is considerably more diverse than Archaea. Molecular studies suggest the existence of 50 or more bacterial phyla, though many are poorly known. Though some members of domain Bacteria live in extreme environments, most favor moderate conditions. Many bacteria form symbiotic associations with eukaryotes and are thus of concern in medicine and agriculture. The characteristics of 10 prominent bacterial phyla are briefly summarized in Table 27.1.

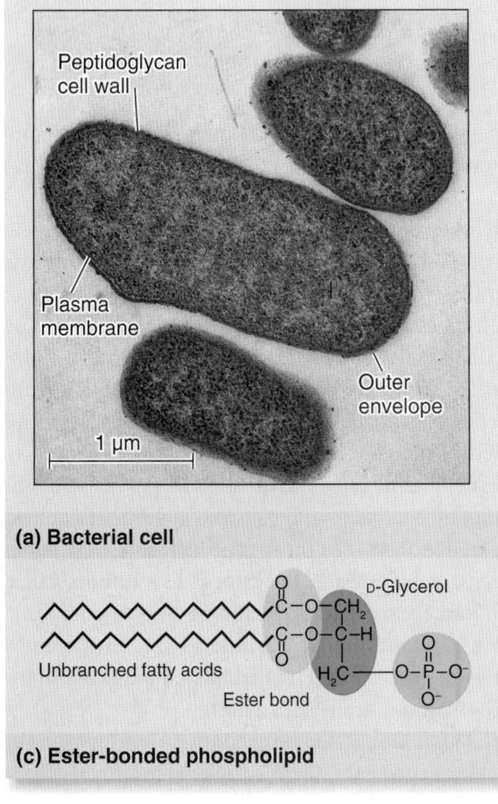

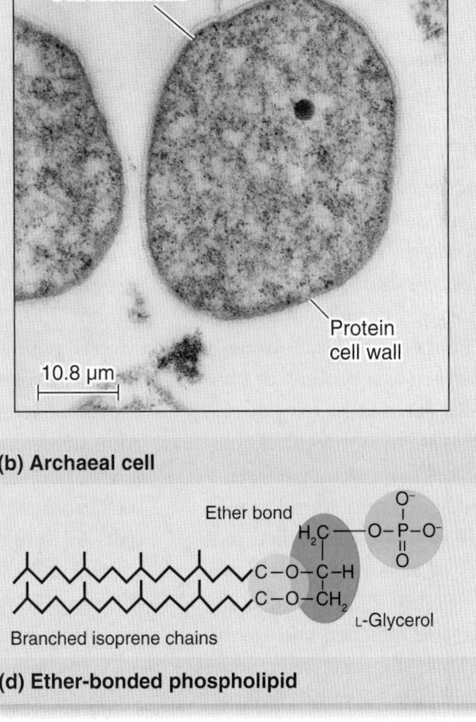

Figure 27.2 Bacteria and Archaea. **(a)** Bacteria and **(b)** archaea both have prokaryotic cell structure, but **(c)** bacterial membrane lipids are formed with ester linkages, whereas **(d)** archaeal membrane lipids feature ether linkages, which are thought to be more stable under extreme environmental conditions. As shown in transmission electron microscopic (TEM) images (a) and (b), most bacteria feature cell walls made of a material known as peptidoglycan that is often enclosed by an outer envelope, whereas archaea lack these features. Most archaea have outer coverings made of protein.

Table 27.1 Representative Bacterial Phyla

Phyla	Characteristics
Firmicutes	Diverse Gram-positive bacteria, some of which produce endospores. The disease-causing *Clostridium difficile* is an example.
Bacteroidetes	Includes representatives of diverse metabolism types; some are common in the human intestinal tract, and others are primarily aquatic.
Chlamydiae	Notably tiny, obligate intracellular parasites. Some cause eye disease in newborns or sexually transmitted diseases.
Planctomycetes	Reproduce by budding rather than binary fission; cell walls lack peptidoglycan; cytoplasm may contain nucleus-like bodies, endocytosis may occur.
Spirochaetes	Motile bacteria having distinctive corkscrew shapes, with flagella held close to the body. They include the pathogens *Treponema pallidum*, the agent of syphilis, and *Borrelia burgdorferi*, which causes Lyme disease.
Actinobacteria	Gram-positive bacteria producing branched filaments; many form spores. *Mycobacterium tuberculosis*, the agent of tuberculosis in humans, is an example. Actinobacteria are notable antibiotic producers; over 500 different antibiotics are known from this group. Some fix nitrogen in association with plants.
Chloroflexi	Known as the green nonsulfur bacteria; conduct photosynthesis without releasing oxygen (anoxygenic photosynthesis).
Deinococcus-Thermus	Extremophiles. The genus *Deinococcus* is known for high resistance to ionizing radiation, and the genus *Thermus* inhabits hot springs. *Thermus aquaticus* has been used in commercial production of Taq polymerase enzyme used in the polymerase chain reaction (PCR), an important procedure in molecular biology laboratories.
Cyanobacteria	The oxygen-producing photosynthetic bacteria (some are also capable of anoxygenic photosynthesis). Photosynthetic pigments include chlorophyll *a* and phycobilins, which often give cells a blue-green pigmentation. Occur as unicells, colonies, unbranched filaments, and branched filaments. Many of the filamentous species produce specialized cells: dormant akinetes and heterocytes in which nitrogen fixation occurs. In waters having excess nutrients, cyanobacteria produce blooms and may release toxins harmful to the health of humans and wild and domesticated animals.
Proteobacteria	A very large group of Gram-negative bacteria, collectively having high metabolic diversity. Includes many species important in medicine, agriculture, and industry such as *Agrobacterium tumifaciens*, *Escherichia coli*, and *Haemophilus influenzae*. *Myxococcus xanthus* is a Gram-negative bacterium that is able to glide across surfaces, forming swarms of thousands of cells. This behavior aids feeding by concentrating digestive enzymes secreted by the bacteria. When food is scarce, the swarms form tiny tree-shaped structures from which tough spores disperse. By this means, cells move to new, food-rich places.

Among these, the Cyanobacteria and the Proteobacteria are particularly diverse and relevant to eukaryotic cell evolution, global ecology, and human affairs.

Cyanobacteria The phylum Cyanobacteria contains photosynthetic bacteria that are abundant in fresh waters, oceans, and wetlands and on the surfaces of arid soils. Cyanobacteria are named for the typical blue-green (cyan) coloration of their cells. Blue-green pigmentation results from the presence of photosynthetic pigments called phycobilins that help chlorophyll absorb light energy. Cyanobacteria are the only bacteria known to generate oxygen as a product of photosynthesis. Ancient cyanobacteria produced Earth's first oxygen-rich atmosphere, which allowed the eventual rise of eukaryotes. The chloroplasts of eukaryotic algae and plants derived from cyanobacteria.

Cyanobacteria display the greatest body diversity found among bacterial phyla (**Figure 27.3**). Some occur as single cells called unicells (Figure 27.3a); others form colonies of cells held together by a thick gluey substance called mucilage (Figure 27.3b), and many cyanobacteria form filaments of cells that are attached end-to-end (Figure 27.3c) including filaments that branch (Figure 27.3d). Some of the filamentous cyanobacteria display hallmarks of multicellularity: cellular attachment, specialized cells, intercellular chemical communication, and programmed cell death.

Proteobacteria Though Proteobacteria share molecular and cell-wall features, this phylum displays amazing diversity of form and metabolism. Genera of this phylum are classified into five major subgroups: alpha (α), beta (β), gamma (γ), delta (δ), and epsilon (ε). As we saw in Chapter 22 (Figure 22.12), the ancestry of mitochondria can be traced to the α-proteobacteria, which also include several genera noted for mutually beneficial relationships with animals and plants. For example, *Rhizobium* and related genera of α-proteobacteria form nutritionally beneficial associations with the roots of legume plants such as beans and peas and are thus agriculturally important (see Figure 38.17). Another α-proteobacterium, *Agrobacterium tumifaciens*, causes destructive cancer-like tumors called galls to develop on susceptible plants, including grapes and ornamental crops (**Figure 27.4**). *A. tumifaciens* induces gall formation by injecting DNA into plant cells, a property that has led to the use of the bacterium in the production of transgenic plants.

The genus *Nitrosomonas*, a soil inhabitant important in the global nitrogen cycle, represents the β-proteobacteria. *Neisseria*

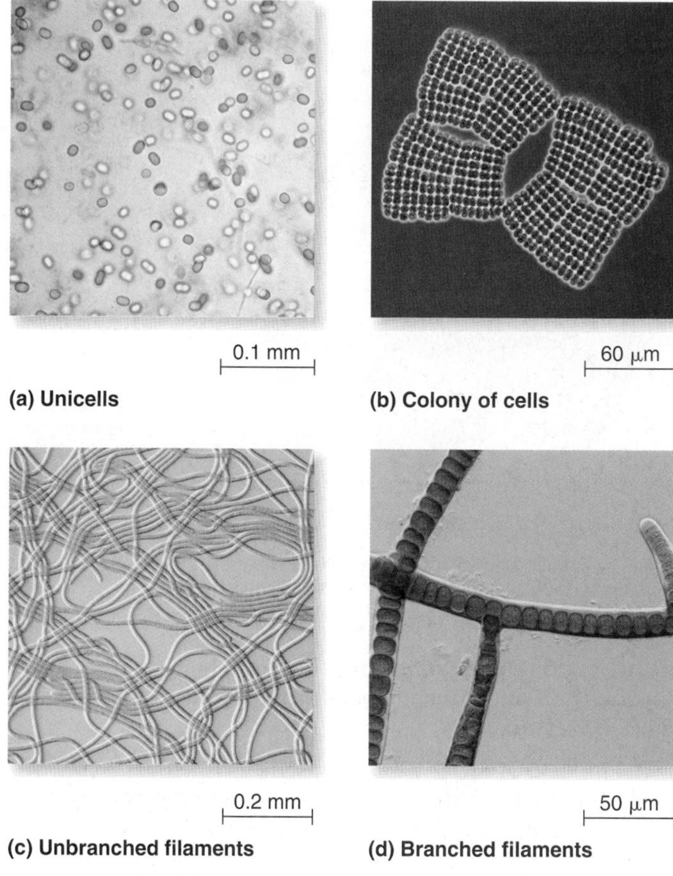

Figure 27.3 **Major body types found in the phylum Cyanobacteria.** **(a)** The genus *Chroococcus* occurs as unicells. **(b)** The genus *Merismopedia* is a flat colony of cells held together by mucilage. **(c)** The genus *Oscillatoria* is an unbranched filament. **(d)** The genus *Stigonema* is a branched filament having a mucilage sheath; sunscreen compounds that protect the cells from damage by ultraviolet (UV) radiation cause the brown color of the sheath.

BioConnections: Look ahead to Figure 28.3d and Figure 31.23. What eukaryotic microbes also display branched filamentous bodies?

gonorrhoeae, the agent of the sexually transmitted disease gonorrhea, is a member of the γ-proteobacteria. *Vibrio cholerae*, another γ-proteobacterium, causes cholera epidemics when drinking water becomes contaminated with animal waste during floods and other natural disasters. The γ-proteobacteria *Salmonella enterica* and *Escherichia coli* strain O157:H7 also cause human disease, so food and water are widely tested for their presence. The δ-proteobacteria includes the colony-forming myxobacteria and predatory bdellovibrios, which drill through the cell walls of other bacteria in order to consume them. *Helicobacter pylori*, which causes stomach ulcers, belongs to the ε-proteobacteria.

Additional bacterial phyla mentioned in this chapter because of their medical, ecological, or evolutionary significance include Firmicutes, Bacteriodetes, Chlamydiae, Planctomycetes, Spirochaetes, Actinobacteria, Chloroflexi, and *Deinococcus-Thermus* (see Table 21.1). Now that we have learned something about the diversity of Archaea and Bacteria, let's consider the evolutionary effects of gene exchanges within and between the domains of life.

Figure 27.4 ***Agrobacterium tumifaciens*** **infection.** This proteobacterium causes cancer-like tumors to grow on plants.

Horizontal Gene Transfer Influences the Evolution of Archaea and Bacteria

Horizontal gene transfer is the process in which an organism receives genetic material from another organism without being the offspring of that organism. This process contrasts with vertical gene transfer, which occurs from parent to progeny, and is increasingly recognized as an important evolutionary mechanism. Horizontal gene transfer occurs most frequently between species that are closely related or that live in close proximity, and it affects the evolutionary process by increasing genetic diversity.

Horizontal gene transfer is common among archaea and bacteria, and it can result in large genetic changes that confer new metabolic capacities. For example, at least 17% of the genes present in the common human gut inhabitant *E. coli* came from other bacteria. Study of nearly 200 genomes has revealed that about 80% of prokaryotic genes have been involved in horizontal transfer at some point in their history. Genes also move among the bacterial, archaeal, and eukaryotic domains. For example, about a third of the genes present in the archaeon *Methanosarcina mazei* originally came from bacteria, and the bacterial phylum Chlamydiae contributed at least 55 genes to plants.

Horizontal gene transfer can occur between different bacterial species via transduction, transformation, and conjugation, as discussed in Chapter 18. In addition, horizontal gene transfer also occurs by means of endosymbiosis, the process in which one species—the endosymbiont—lives in the body or cells of another species—the host. For example, certain γ-proteobacteria occupy the cells of β-proteobacterial hosts, which themselves live within insect cells. Such close proximity increases the odds for gene exchange between distantly related species. Endosymbiosis theory proposes that the mitochondria and chloroplasts of eukaryotic cells originated from α-proteobacteria and cyanobacteria, respectively, by endosymbiosis (see Chapter 22). In these cases, endosymbiosis resulted in the horizontal transfer of many genes from bacterial genomes to eukaryotic nuclei. In this process, so many genes were lost from ancestral bacterial genomes that mitochondria and chloroplasts cannot reproduce outside host eukaryotic cells.

As we have seen, archaea and bacteria share features of prokaryotic cell organization, small size, high environmental populations, and high metabolic diversity, and they exchange genes. Even so, bacteria more frequently interact with other organisms. For example, bacteria play more important roles in animal and plant disease than do archaea. In the next sections we focus on the structure, movement, reproduction, ecology, and biotechnological applications of bacteria and archaea.

27.2 Structure and Movement

Learning Outcomes:

1. Discuss cellular structural adaptations that have increased the complexity of prokaryotic cells.
2. Explain how mucilage influences the behavior of bacterial cells.
3. Describe the structural differences between Gram-positive and Gram-negative bacterial cells.
4. List the different means by which prokaryotic cells can move.

Bulbnose unicorn fish (*Naso tonganus*) living in Australian coastal ocean waters contain cigar-shaped bacterial symbionts (*Epulopiscium*) whose cells are more than 600 μm long, larger than most eukaryotic cells (between 10 and 100 μm in largest dimension). Spherical cells of the bacterial species *Thiomargarita namibiensis*, which lives in African coastal regions, likewise reach record-setting sizes, some being 800 μm in diameter and large enough to be seen without a microscope. However, most bacteria (and archaea) are much smaller: 1–5 μm in diameter. Small cell size limits the amount of materials that can be stored within cells but allows much faster cell division. When nutrients are sufficient, many bacteria can divide many times within a single day. This explains how bacteria can spoil food rapidly and why bacterial infections can spread quickly within the human body. Despite their generally small size, bacteria can display a high level of variation in cell structure and shape, surface and cell-wall features, and movement.

Prokaryotic Cells Display a Surprising Degree of Complexity

Bacteria, like archaea, have a much simpler cellular organization than do eukaryotes. Even so, many prokaryotic cells display cellular structural adaptations that increase their complexity. Features of prokaryotic cellular complexity illustrate the biological principle that structure determines function, partly explain why prokaryotic organisms have such high metabolic diversity, and help us understand how the first eukaryotes arose.

Cyanobacteria and other photosynthetic bacteria, for example, are able to use light energy to produce organic compounds because their

Thylakoids provide a greater surface area for chlorophyll and other molecules involved in photosynthesis.

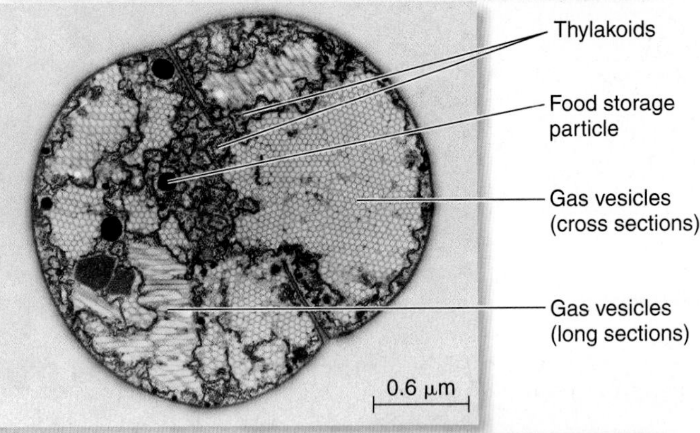

The gas vesicles buoy this photosynthetic organism to the lighted water surface, where it often forms conspicuous scums.

Figure 27.5 Photosynthetic thylakoid membranes and numerous gas vesicles found in a cell of the aquatic cyanobacterial genus *Microcystis*.

BIOLOGY PRINCIPLE Structure determines function. Thylakoids, which contain chlorophyll and other components of the photosynthetic apparatus, are the locations of the light harvesting reactions.

Concept Check: Do you think this cell would tend to float or sink in a water body?

BioConnections: Look back to Figure 4.26 to see thylakoids within plant chloroplasts, and look ahead to Figure 28.12 to see thylakoids of algal plastids. What similarities to Figure 27.5 help to demonstrate that plastids evolved from endosymbiotic cyanobacteria?

cells contain large numbers of thylakoids, flattened tubular membranes that grow inward from the plasma membrane (**Figure 27.5**). The extensive membrane surface of the thylakoids bears large amounts of chlorophyll and other components of the photosynthetic apparatus. This explains why thylakoids are also abundant in plant chloroplasts, which descended from cyanobacterial ancestors. Thylakoids enable photosynthetic bacteria and chloroplasts to take maximum advantage of light energy in their environments. Photosynthetic bacteria of aquatic systems also commonly contain many gas vesicles. These protein-walled structures increase cell buoyancy and thus help the organisms float within well-illuminated surface waters (see Figure 27.5).

In other bacteria, plasma membrane ingrowth has generated additional intriguing adaptations—magnetosomes and nucleus-like bodies—that are sometimes described as bacterial organelles. Magnetosomes are tiny crystals of an iron mineral known as magnetite, each surrounded by a membrane. These structures occur in the bacterium *Magnetospirillum* and related genera (**Figure 27.6**). In each cell, about 15 to 20 magnetosomes occur in a row, together acting as a compass needle that responds to the Earth's magnetic field. Magnetosomes help the bacteria to orient themselves in space and thereby locate the submerged, low-oxygen habitats they prefer. Magnetosome

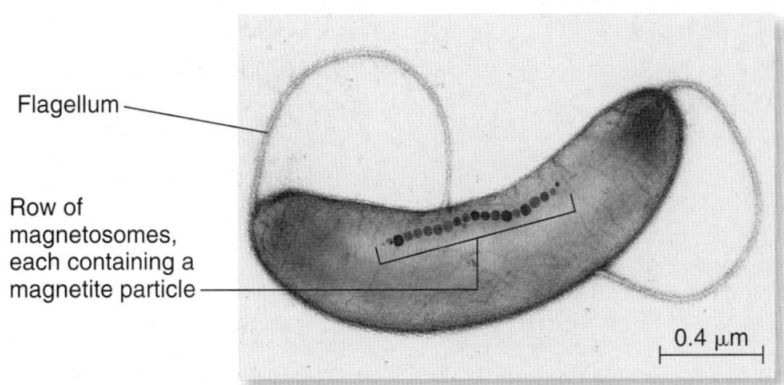

Figure 27.6 Magnetosomes found in the spirillum *Magnetospirillum magnetotacticum*. An internal row of iron-rich magnetite crystals, each enclosed by a membrane derived from the plasma membrane. The row of magnetosomes functions like a compass needle, allowing this bacterium to detect the Earth's magnetic field. This feature allows *M. magnetotacticum* to orient itself in space and thereby locate its preferred habitat, low-oxygen subsurface waters. These and other bacterial cells use flagella to move from less-favorable to more-attractive locations.

BioConnections: Look ahead to Section 43.4, which describes electromagnetic sensing by animals. What animals are like M. magnetotacticum in being able to sense and respond to magnetic fields?

development begins with ingrowth of the plasma membrane to form a row of spherical vesicles. If *Magnetospirillum* cells are grown in media having low iron levels, the vesicles remain empty. But if iron is available, a magnetite crystal forms within each vesicle. Fibrils of an actin-like protein keep the magnetosomes aligned in a row. (Recall from Chapter 4 that actin is a major cytoskeletal protein of eukaryotes.) Mutant bacteria lacking a functional form of this protein produce magnetosomes, but they do not remain aligned in a row. Instead, magnetosomes scatter around mutant cells, disrupting their ability to detect a magnetic field.

Plasma membrane invaginations produce nucleus-like bodies in *Gemmata obscuriglobus* and other members of the bacterial phylum Planctomycetes. In *G. obscuriglobus*, an envelope composed of a double membrane encloses all cellular DNA and some ribosomes. Although this bacterial envelope lacks the nuclear pores characteristic of the eukaryotic nuclear envelope, it likely plays a similar adaptive role in isolating DNA from other cellular influences. This and related bacterial species are also known to accomplish endocytosis by means of membrane coat proteins similar to those present in eukaryotic cells. The cellular diversity and surprising complexity of bacterial cell structure helps us to understand not only how bacteria function in nature, but also how important features of eukaryotic cells first evolved.

Prokaryotic Cells Vary in Shape

Although prokaryotic cells occur in multiple forms, they have five common shapes (Figure 27.7): spheres (**cocci**), elongate rods (**bacilli**), comma-shaped cells (**vibrios**), and spiral-shaped cells that are either flexible (**spirochaetes**) or rigid (**spirilli**; see Figure 27.6). Cytoskeletal proteins similar to those present in eukaryotic cells control these cell shapes. For example, helical strands of an actin-like protein are responsible for the rod shape of bacilli; if this protein is not produced, bacilli become spherical in shape. Cellular shape is an important component of bacterial function in nature. Cocci may have greater surface area-to-volume ratio, which facilitates exchange of materials with the environment, but bacilli can often store more nutrients.

Slimy Mucilage Often Coats Cellular Surfaces

Many bacteria exude a coat of slimy mucilage, sometimes called a glycocalyx, capsule, or extracellular polymeric substances (EPS). Mucilage, which varies in consistency and thickness, is composed of hydrated polysaccharides and protein, as well as lipid and nucleic acids. A capsule helps some disease bacteria evade the defense system of their host. You may recall that Frederick Griffith discovered the transfer of genetic material while experimenting with capsule-producing pathogenic strains and capsule-less nonpathogenic strains of the bacterium *Streptococcus pneumoniae* (refer back to Figure 11.1).

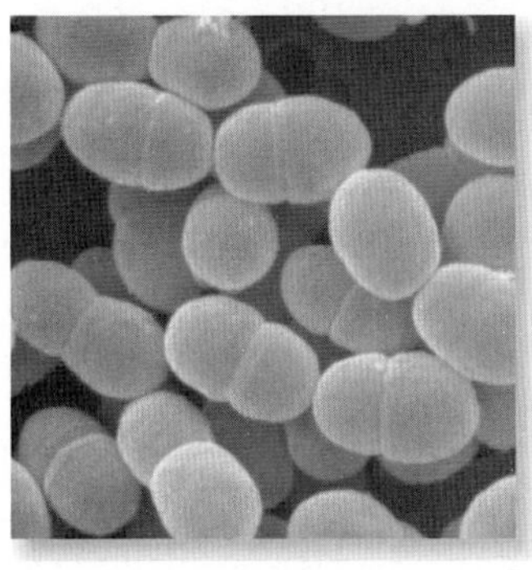

(a) Sphere-shaped cocci (*Lactococcus lactis*)

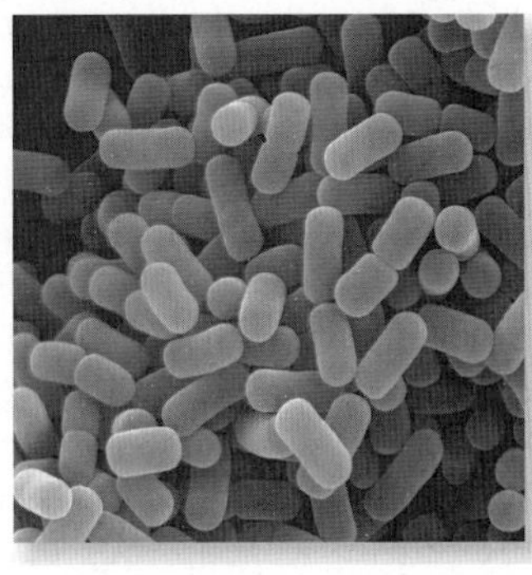

(b) Rod-shaped bacilli (*Lactobacillus plantarum*)

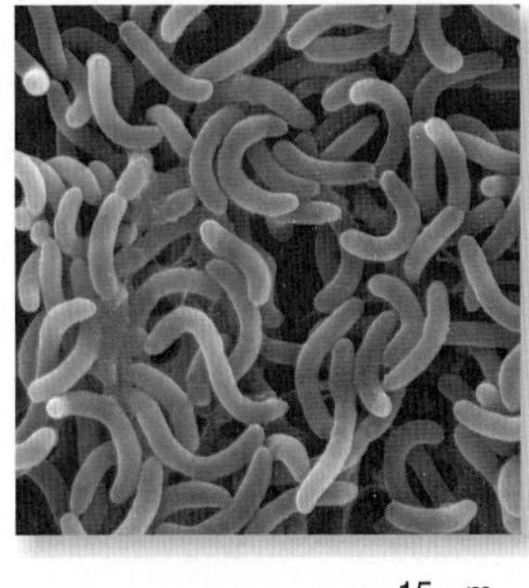

(c) Comma-shaped vibrios (*Vibrio cholerae*)

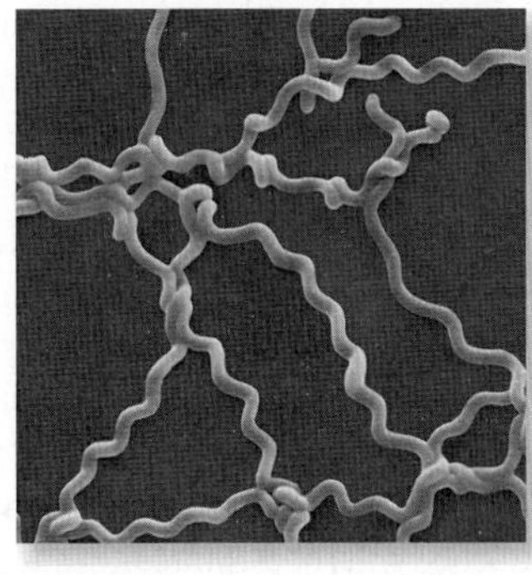

(d) Spiral-shaped spirochaetes (*Leptospira* sp.)

Figure 27.7 Major types of prokaryotic cell shapes. Scanning electron microscopic views.

BIOLOGY PRINCIPLE Cells are the simplest units of life. In the case of unicellular bacteria and archaea, a single cell represents an entire organism.

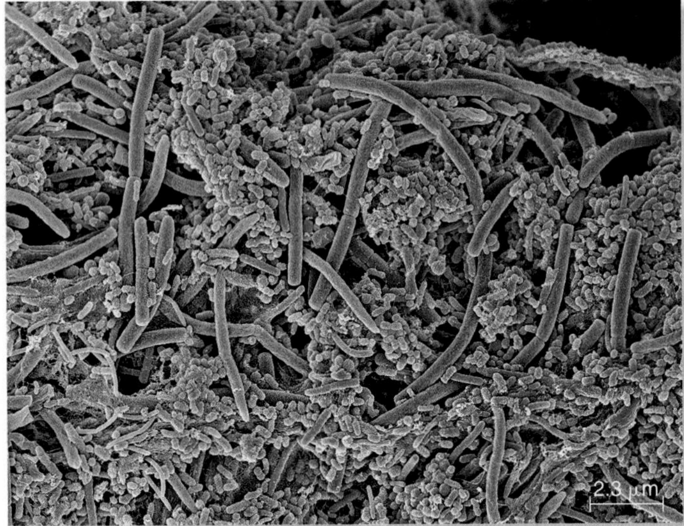

Figure 27.8 A biofilm composed of a community of microorganisms glued by mucilage to a surface. This SEM shows dental plaque, consisting of several types of bacteria—falsely colored purple, green, and blue—attached to a tooth surface by mucilage.

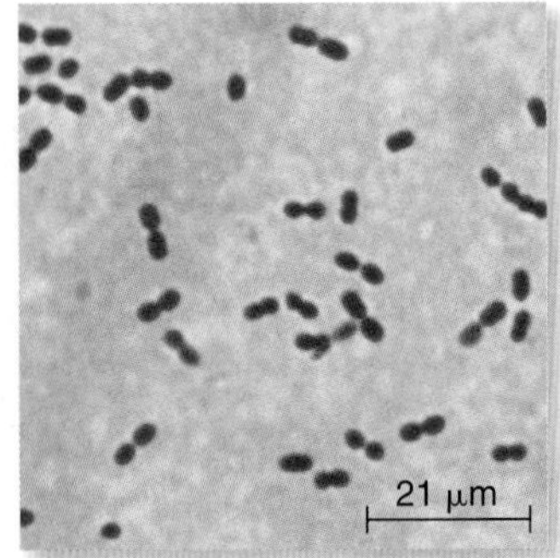

(a) Gram-positive bacteria

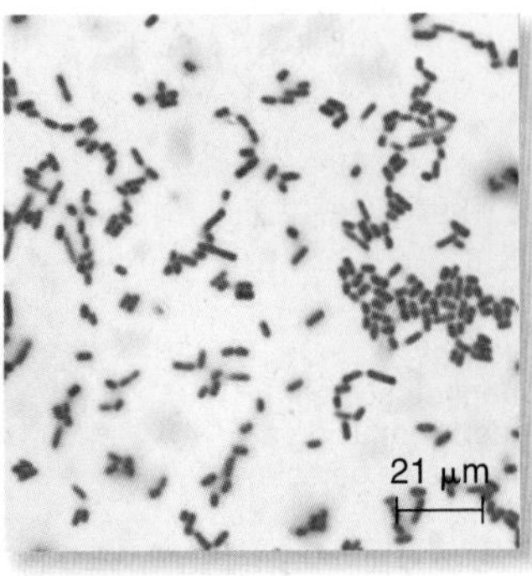

(b) Gram-negative bacteria

Figure 27.9 Gram-positive and Gram-negative bacteria. **(a)** *Streptococcus pneumoniae*, a member of the phylum Firmicutes, stains positive (purple) with the Gram stain. **(b)** *Escherichia coli*, a member of the Proteobacteria, stains negative (pink) when the Gram stain procedure is applied.

The immune system cells of mice are able to destroy this bacterium only if it lacks a capsule.

Mucilage plays many additional roles: holding cells together closely enough for chemical communication and DNA exchange to occur, helping aquatic species to float in water, binding mineral nutrients, and repelling attack. Pigmented slime sheaths (see Figure 27.3d) coat some bacterial filaments, where they help to prevent UV damage.

Biofilms are aggregations of microorganisms that secrete adhesive mucilage, thereby gluing themselves to surfaces. They help microbes to remain in favorable locations for growth; otherwise body or environmental fluids would wash them away. A process known as **quorum sensing** fosters biofilm formation. During quorum sensing, individual microbes secrete small molecules having the potential to influence the behavior of nearby microbes. If enough individuals are present (a quorum), the concentration of signaling molecules builds to a level that causes collective behavior. In the case of biofilms, populations of microbes respond to chemical signals by moving to a common location and producing mucilage.

Biofilms are environmentally and medically important. From a human standpoint, biofilms have both beneficial and harmful consequences. In aquatic and terrestrial environments, they help to stabilize and enrich sand and soil surfaces. Microbial biofilms can also be important in the formation of mineral deposits. Biofilms that form on the surfaces of animal tissues, however, can be harmful. Dental plaque is an example of a harmful biofilm (**Figure 27.8**); if allowed to remain, the bacterial community secretes acids that can damage tooth enamel. Biofilms may also develop in industrial pipelines, where the attached microbes can contribute to corrosion by secreting enzymes that chemically degrade metal surfaces.

Prokaryotic Cells Vary in Cell-Wall Structure

Whether coated with mucilage or not, most prokaryotic cells possess a rigid cell wall outside the plasma membrane. Cell walls maintain cell shape and help protect against attack by viruses or predatory bacteria. Cell walls also help microbes avoid lysing in hypotonic conditions, when the solute concentration is higher inside the cell than outside. The structure and composition of bacterial cell walls are medically important.

Although some archaea lack cell walls, most possess a wall composed of protein. In contrast, the polymer known as **peptidoglycan**, lacking from archaea, is an important component of most bacterial cell walls. Peptidoglycan is composed of carbohydrates that are cross-linked by peptides. Bacterial cell walls occur in two major forms that differ in peptidoglycan thickness, staining properties, and response to antibiotics. Bacteria having these chemically different walls are called Gram-positive or Gram-negative bacteria, after the staining process used to distinguish them (**Figure 27.9**). The stain is named for its inventor, the Danish scientist Hans Christian Gram.

Gram-positive bacteria classified in the phyla Firmicutes and Actinobacteria have walls with a relatively thick peptidoglycan layer (**Figure 27.10a**). By contrast, the Gram-negative cell walls of Cyanobacteria, Proteobacteria, and other species have a thinner peptidoglycan layer and are enclosed by a thin, outer envelope whose outer leaflet is rich in **lipopolysaccharides** (**Figure 27.10b**; see Figure 25.2a). This outer layer envelope of Gram-negative bacteria is a lipid bilayer, but is distinct from the plasma membrane. Peptidoglycan and lipopolysaccharides can affect disease symptoms, the composition of vaccines, and bacterial responses to antibiotics. For example, part of the peptidoglycan covering of the Gram-negative bacterial species *Bordetella pertussis* is responsible for the extensive tissue damage to the respiratory tract associated with whooping cough, and whooping cough vaccines can be improved by including antibodies that reduce the ability of the lipopolysaccharide layer to attach to host cells.

The lipopolysaccharide-rich outer membrane of Gram-negative bacteria helps them to resist the entry of some antibiotics. However, this outer envelope also impedes the secretion of proteins from bacterial cells into the environment, a process that normally allows cells to communicate with each other, as in quorum sensing. Gram-negative bacteria have adapted to the presence of an outer membrane by evolving five or more types of protein systems that function in secretion, known as types I–V secretion systems. We will later see how some of these secretion systems have been modified in ways that allow disease-causing bacteria to attack eukaryotic cells.

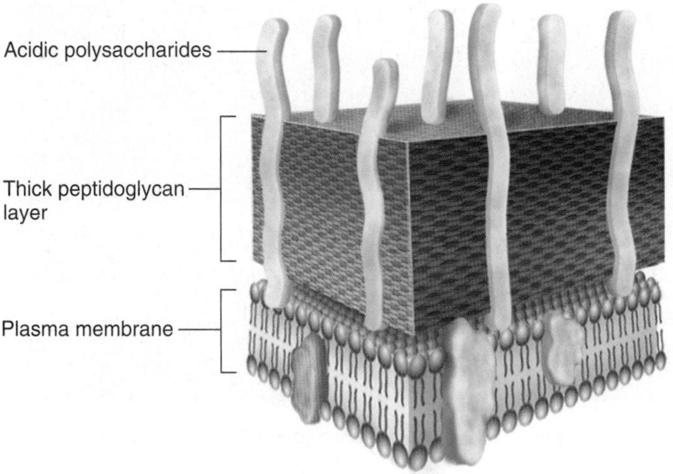

(a) Gram-positive: thick peptidoglycan layer, no outer envelope

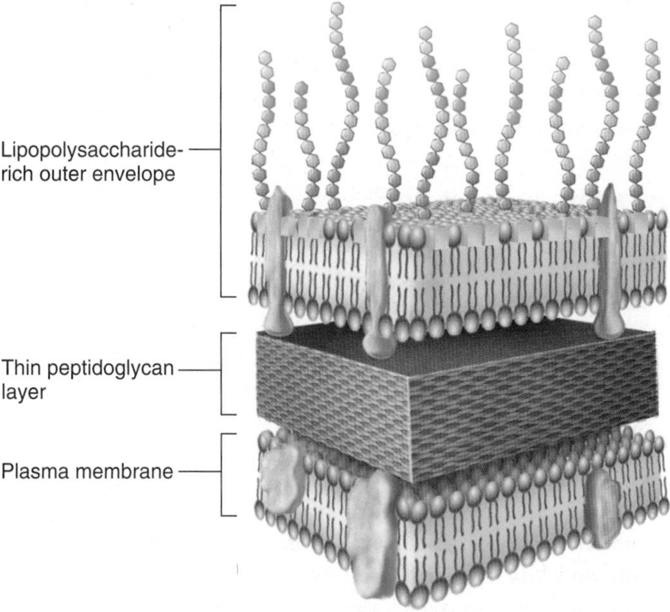

(b) Gram-negative: thinner peptidoglycan layer, with outer envelope

Figure 27.10 **Cell-wall structures of Gram-positive and Gram-negative bacteria.** **(a)** The structure of the cell wall of Gram-positive bacteria. **(b)** The structure of the cell wall and lipopolysaccharide envelope typical of Gram-negative bacteria.

Distinguishing Gram-positive from Gram-negative bacteria is an important factor in choosing the best antibiotics for treating infectious diseases. For example, Gram-positive bacteria are typically more susceptible than Gram-negative bacteria to penicillin and related antibiotics because these antibiotics interfere with synthesis of peptidoglycan, which Gram-positive bacteria require in larger amounts. For this reason, penicillin or related antibiotics such as methicillin are widely used to treat infections caused by Gram-positive bacteria. However, it is of societal concern that some strains of Gram-positive bacteria have become resistant to some antibiotics, an example being methicillin-resistant *Staphylococcus aureus*, or MRSA.

Bacterial and Archaea Display Diverse Types of Movements

Many bacteria and archaea have structures at the cell surface or within cells that enable them to change position in their environment, a process known as motility. Diverse motility adaptations allow microbes to respond to chemical signals emitted from other cells during quorum sensing and mating, and to move to favorable conditions within gradients of light, gases, or nutrients. For example, we have already learned that gas buoyancy vesicles help cyanobacteria to float into well-illuminated waters that allow photosynthesis to occur (see Figure 27.5). In addition, prokaryotic cells may move by twitching, gliding, or swimming by means of flagella.

Bacterial **flagella** (singular, flagellum) differ from eukaryotic flagella in several ways. Although bacterial flagella are largely built of about 30 types of proteins, these lack a plasma membrane covering, an internal cytoskeleton of microtubules made of the protein tubulin, and the motor protein dynein—all features that characterize eukaryotic flagella (see Chapter 4). Unlike eukaryotic flagella, prokaryotic flagella do not repeatedly bend and straighten. Instead, prokaryotic flagella spin, propelled by molecular machines composed of a filament, hook, and motor that work together somewhat like a boat's outboard motor and propeller (**Figure 27.11**). Lying outside the cell, the long, stiff, curved filament acts as a propeller. The hook links the filament with the motor that contains a set of protein rings at the cell

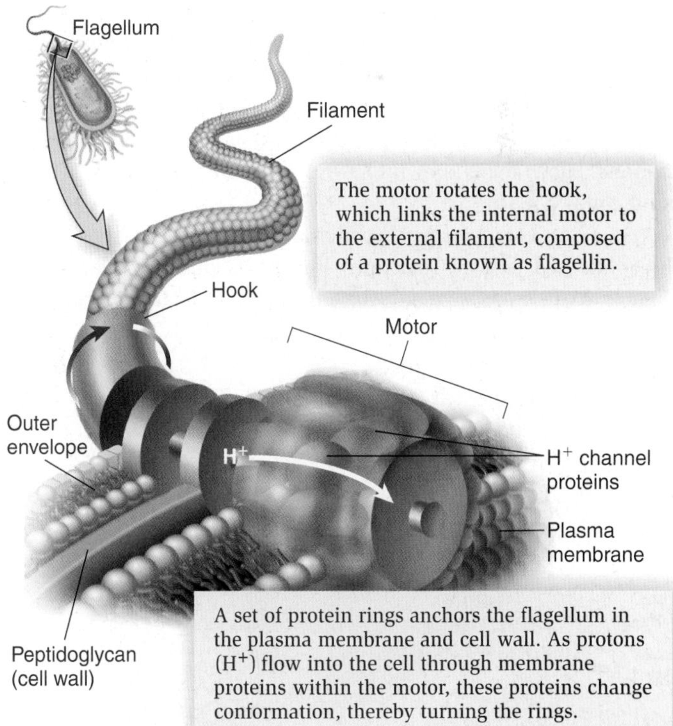

Flagellum
Filament
The motor rotates the hook, which links the internal motor to the external filament, composed of a protein known as flagellin.
Hook
Motor
Outer envelope
H^+
H^+ channel proteins
Plasma membrane
Peptidoglycan (cell wall)
A set of protein rings anchors the flagellum in the plasma membrane and cell wall. As protons (H^+) flow into the cell through membrane proteins within the motor, these proteins change conformation, thereby turning the rings.

Figure 27.11 **Diagram of a prokaryotic flagellum, showing a filament, hook, and motor.**

Concept Check: *Does the filament move more like the arms of a human swimmer or the shaft of a boat propeller?*

Figure 27.12 **Differences in the number and location of flagella.** Depending on the species, microbial cells can produce one or more flagella at the poles or numerous flagella around the periphery. **(a)** *Vibrio parahaemoliticus*, a bacterium that causes seafood poisoning, has a single short flagellum. **(b)** *Salmonella enterica*, another bacterium that causes food poisoning, has many flagella distributed around the cell periphery.

BioConnections: *Look ahead to Figure 28.7b. What heterotrophic eukaryote (like V. parahaemoliticus and S. enterica) moves to its food source in the human gut by means of flagella?*

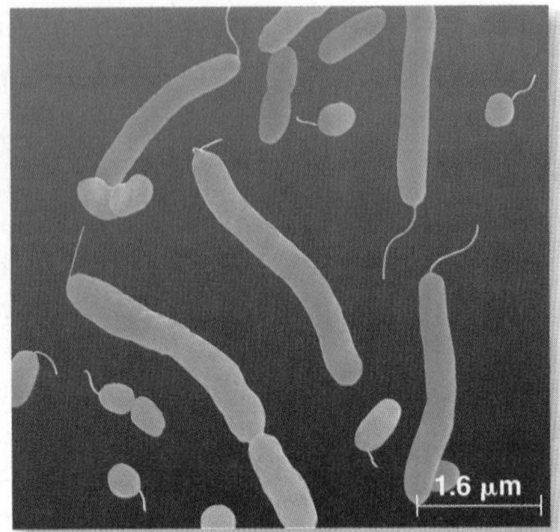

(a) Bacteria with a single short flagellum

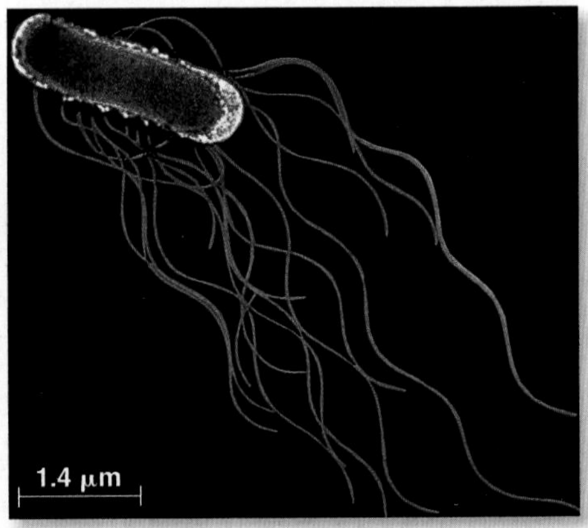

(b) Bacterium with multiple long flagella

surface. Hydrogen ions (protons), which have been pumped out of the cytoplasm, usually via the electron transport system, diffuse back into the cell through channel proteins within the motor. This proton flow powers the turning of the hook and filament at rates of hundreds of revolutions per second. Archaeal flagella also rotate but are much thinner than bacterial flagella, composed of different proteins, and powered differently. Archaeal flagella are powered by the hydrolysis of ATP.

Prokaryotic species differ in the number and location of flagella, which may occur singly or in tufts at one pole or may emerge from around the cell (**Figure 27.12**). Differences in flagellar number and location cause microorganisms to exhibit different modes or rates of swimming. For example, spirochaete flagella are located outside the peptidoglycan cell wall but within the confines of an outer membrane that holds them close to the cell. Rotation of these flagella causes spirochaetes to display characteristic bending, flexing, and twirling motions. Bacteria are known to swim at rates of more than 150 μm per second.

Some prokaryotic species twitch or glide across surfaces, using threadlike cell surface structures known as **pili** (singular, pilus) (**Figure 27.13**). *Myxococcus xanthus* cells, for example, move by alternately extending and retracting pili from one pole or the other. This process allows directional movement toward food materials. If nutrients are low, cells of these bacteria glide together to form tiny treelike colonies, which are part of a reproductive process. These and other motility adaptations help to explain how bacteria and archaea behave in their environments.

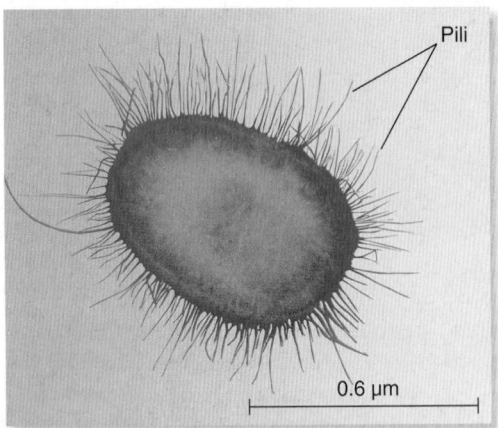

Figure 27.13 **Pili extending from the surface of *Proteus mirabilis*.**

Concept Check: *What type of motion does this cell likely use?*

27.3 Reproduction

Learning Outcomes:

1. Explain how populations of prokaryotic organisms increase in number.
2. Give examples of how some bacteria survive under stressful conditions.
3. Describe how bacteria can be counted in medical and environmental samples.

Bacteria and archaea lack eukaryote-type sexual reproduction involving specialized gametes, gamete fusion (syngamy), and meiosis, though they can exchange some genes by conjugation, transformation, and transduction (described in Chapter 18). Bacteria and archaea commonly reproduce asexually, generally by means of a type of cell division known as binary fission that enlarges populations. In addition, some bacteria produce tough cells that can withstand deleterious conditions for long periods in a dormant condition.

Prokaryotic Cells Generally Divide by Binary Fission

The cells of most prokaryotic cells divide by splitting in two, a process known as **binary fission** (**Figure 27.14a**; refer back to Figure 18.14). Bacterial binary fission generally requires a protein known as FtsZ, which is related to the tubulin protein that makes up eukaryotic microtubules. FtsZ squeezes dividing cells into two progeny cells; if this protein is not functional, most bacterial cells cannot complete binary fission, and the cells become very long. Some archaea likewise use the FtsZ process for dividing, but others utilize different mechanisms. When sufficient nutrients are available, an entire population of identical cells can be produced from a single parental cell by repeated binary fission. This growth process allows microbes to become very numerous in water, food, or animal tissues, potentially causing harm.

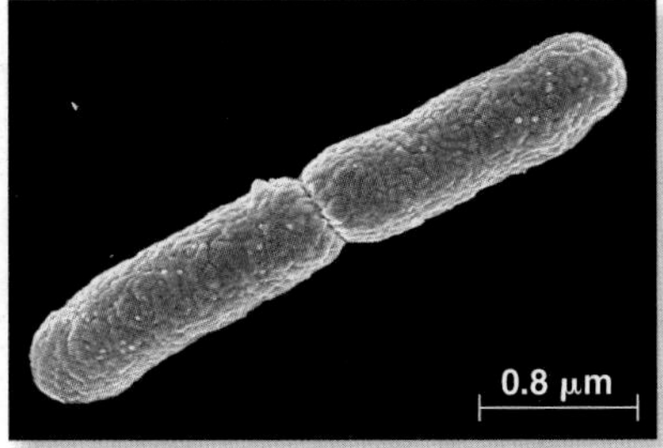

(a) Bacterium undergoing binary fission

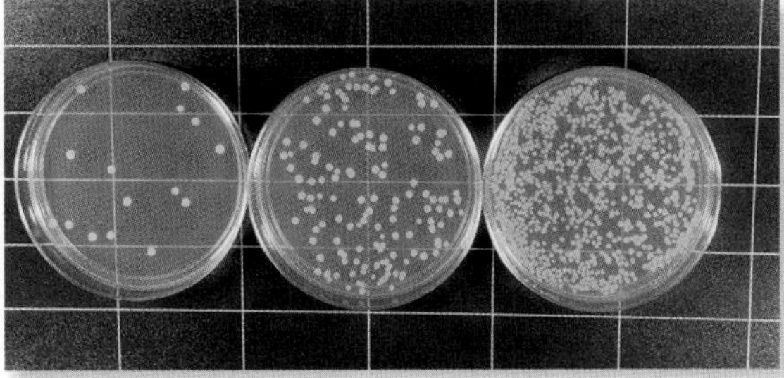

(b) Colonies developed from single cells

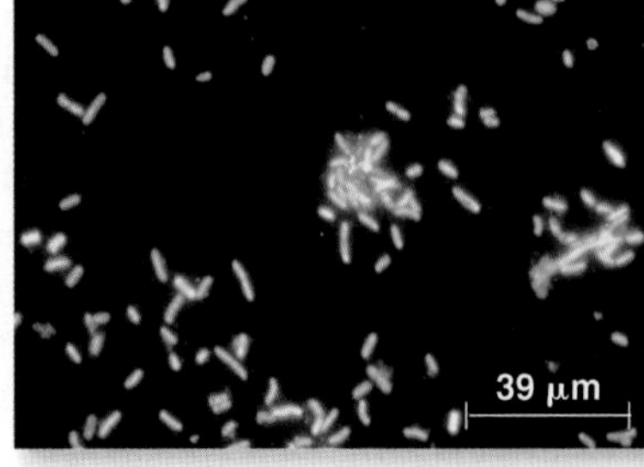

(c) Bacteria stained with fluorescent DNA-binding dye

Figure 27.14 Binary fission and counting microbes. **(a)** Division of a bacterial cell as viewed by scanning electron microscopy. **(b)** When samples are spread onto the surfaces of laboratory dishes containing nutrients, single cells of bacteria or archaea may divide repeatedly to form visible colonies, which can be easily counted. The number of colonies is an estimate of the number of culturable cells in the original sample. **(c)** If a fluorescence microscope is available, cells can be counted directly by applying a fluorescent stain that binds to cell DNA. Each cell glows brightly when illuminated with ultraviolet light.

Concept Check: *Which procedure would you choose to count bacteria in a sample that is known to include many species that have not as yet been cultured?*

BioConnections: *Look back to Figure 18.14. How many cells result from the binary fission of a single mother cell?*

Binary fission is the basis of a widely used method for detecting and counting bacteria in food, water samples, or patient fluids. Microbiologists who study the spread of disease need to quantify bacterial cells in samples taken from the environment. Medical technicians often need to count bacteria in body fluid samples to assess the likelihood of infection. However, because bacterial cells are small and often unpigmented, they are difficult to count directly. One way that microbiologists count bacteria is to place a measured volume of sample into plastic dishes filled with a semisolid nutrient medium. Bacteria in the sample undergo repeated binary fission to form colonies of cells visible to the unaided eye (**Figure 27.14b**). Because each colony represents a single cell that was present in the original sample, the number of colonies in the dish reflects the number of living bacteria in the original sample.

Another way to detect and count prokaryotic cells is to treat samples with a stain that binds bacterial DNA, causing cells to glow brightly when illuminated with ultraviolet light. The glowing cells can be viewed and counted by the use of a fluorescence microscope (**Figure 27.14c**). The fluorescence method must be used when the microbes of interest cannot be cultivated in the laboratory. For many bacteria and archaea, the conditions needed to foster population growth in the laboratory are not known.

Some Bacteria Survive Harsh Conditions as Akinetes or Endospores

Some bacteria produce thick-walled cells that are able to survive unfavorable conditions in a dormant state. These specialized cells develop when bacteria have experienced stress, such as low nutrients or unfavorable temperatures, and are able to germinate into metabolically active cells when conditions improve again. For example, aquatic filamentous cyanobacteria often produce **akinetes**, large, food-filled cells, when winter approaches (**Figure 27.15a**). Akinetes are able to survive winter at the bottoms of lakes, and they produce new filaments in

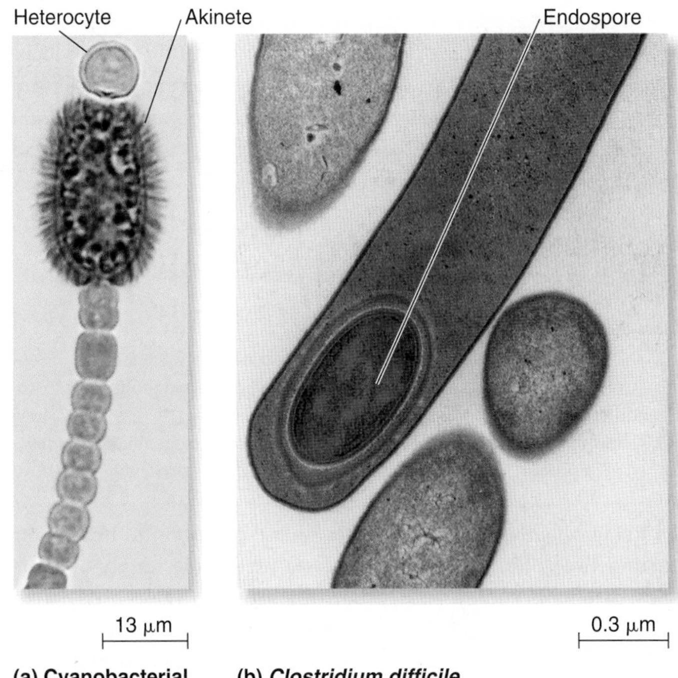

(a) Cyanobacterial akinete (b) *Clostridium difficile*

Figure 27.15 Specialized cells capable of dormancy. **(a)** Akinetes are thick-walled, food-filled cells produced by some cyanobacteria. They are able to resist stressful conditions and generate new populations when conditions improve. As discussed later, the heterocyte is a specialized cell in which nitrogen fixation occurs. **(b)** An endospore with a resistant wall develops within the cytoplasm of the pathogen *Clostridium difficile*.

Concept Check: *How do endospores influence the ability of some bacteria to cause disease?*

spring when they are carried by water currents to the brightly lit surface. Persistence of such akinetes explains how harmful cyanobacterial blooms can develop year after year in overly fertile lakes.

Endospores (Figure 27.15b) are produced inside bacterial cells by the enclosure of DNA and other materials within a tough coat, and then are released when the enclosing cell dies and breaks down. Endospores can remain alive, though in a dormant state, for long periods, then reactivate when conditions are suitable.

The ability to produce endospores allows some Gram-positive Firmicutes bacteria to cause serious diseases. For example, *Bacillus anthracis* causes the disease anthrax, a potential agent in bioterrorism and germ warfare. Most cases of human anthrax result when endospores of *B. anthracis* enter breaks in the skin, causing skin infections that are relatively easily cured by antibiotic treatment. But sometimes the endospores are inhaled or consumed in undercooked, contaminated meat, potentially causing more serious illness or death. *Clostridium botulinum* can contaminate improperly canned food that has not been heated to temperatures high enough to destroy its tough endospores. When the endospores germinate and bacterial cells grow in the food, they produce a deadly toxin, as well as NH_3 and CO_2 gas, which causes can lids to bulge. If humans consume the food, the toxin causes botulism, a severe type of food poisoning that can lead to respiratory and muscular paralysis. Interestingly, the botulism toxin is marketed commercially as Botox, which is injected into the skin, where it paralyzes facial muscles, thereby reducing the appearance of wrinkles. *Clostridium tetani* produces a nerve toxin that causes lockjaw, also known as tetanus, when bacterial cells or endospores from soil enter wounds. The ability of the genera *Bacillus* and *Clostridium* to produce resistant endospores helps to explain their widespread presence in nature and their effect on humans.

27.4 Nutrition and Metabolism

Learning Outcomes:

1. List the major modes of nutrition used by prokaryotic species.
2. Compare and contrast the effects of oxygen on the metabolism of different types of prokaryotic species.
3. Describe in basic terms the process of biological nitrogen fixation, why it is important, and how oxygen interferes with this process.

All living cells require energy and a source of carbon to build their organic molecules. Bacteria and archaea use a wide variety of strategies to obtain energy and carbon for growth (**Table 27.2**). Such microbes can be classified according to their energy source, carbon source, response to oxygen, and presence of specialized metabolic processes.

Types of Nutrition and Responses to Oxygen

Cyanobacteria and some other prokaryotic species are **autotrophs** (from the Greek, meaning self-feeders), organisms that are able to produce all or most of their own organic compounds from inorganic sources. Autotrophs fall into two categories: photoautotrophs and chemoautotrophs. **Photoautotrophs**, including cyanobacteria, use light as a source of energy for the synthesis of organic compounds from CO_2 and H_2O or from H_2S. **Chemoautotrophs** such as the archaeon *Sulfolobus* use energy obtained by chemical modifications of inorganic compounds to synthesize organic compounds. Such chemical modifications include nitrification (the conversion of ammonia to nitrate) and the oxidation of sulfur, iron, or hydrogen.

Table 27.2 Major Types of Archaea and Bacteria Based on Energy and Carbon Source

Type	Energy source	Carbon source	Example
Autotroph			
Photoautotroph	Light	CO_2	Cyanobacteria
Chemoautotroph	Inorganic compounds	CO_2	*Sulfolobus* (Archaea)
Heterotroph			
Photoheterotroph	Light	Organic compounds	Chloroflexi (Bacteria)
Chemoheterotroph	Organic compounds	Organic compounds	Many

Heterotrophs (from the Greek, meaning other feeders) are organisms that require at least one organic compound, and often more, from their environment. Some microorganisms, including the chloroflexi bacteria, are **photoheterotrophs**, meaning that they are able to use light energy to generate ATP, but they must take in organic compounds from their environment as a source of carbon. **Chemoheterotrophs** must obtain organic molecules for both energy and as a carbon source. Among the many types of bacterial chemoheterotrophs is the Gram-positive species *Propionibacterium acnes*, which causes acne, affecting up to 80% of adolescents in the U.S. The genome sequence of *P. acnes* has revealed numerous genes that allow it to break down skin cells and consume the products.

Prokaryotic species differ in their need for and responses to oxygen. Like most eukaryotes (including humans), many prokaryotic organisms are **obligate aerobes**, meaning that they require O_2 in order to survive. In contrast to obligate aerobes, obligate anaerobes, such as the Firmicutes genus *Clostridium*, are poisoned by O_2. People suffering from gas gangrene (caused by *Clostridium perfringens* and related species) are usually treated by placement in a chamber having a high oxygen content (called a hyperbaric chamber), which kills the organisms and deactivates the toxins. **Aerotolerant anaerobes** do not use O_2, but they are not poisoned by it either. These organisms obtain their energy by fermentation or anaerobic respiration, which uses electron acceptors other than oxygen in electron transport processes. Anaerobic metabolic processes include denitrification (the conversion of nitrate into N_2 gas) and the reduction of manganese, iron, and sulfate, which are all important in the Earth's cycling of minerals.

Facultative anaerobes can use O_2 via aerobic respiration, obtain energy via anaerobic fermentation, or use inorganic chemical reactions to obtain energy—shifting between modes depending on environmental conditions. One fascinating example of a facultative anaerobe is the earlier-mentioned species *Thiomargarita namibiensis*, a large proteobacterium. This chemoheterotroph obtains its energy in two ways: by oxidizing sulfide with oxygen when this is available or, when oxygen is low or unavailable, by oxidizing sulfide with nitrate. In either case, the cells convert sulfide to elemental sulfur, which is stored within the cells as large globules.

Some Prokaryotic Species Play Important Roles as Nitrogen Fixers

Many cyanobacteria and some other prokaryotic organisms conduct a specialized metabolic process called biological **nitrogen fixation**. The removal of nitrogen from the gaseous phase is called fixation, and microbes that perform this process are known as nitrogen fixers. During nitrogen fixation, the enzyme nitrogenase converts inert atmospheric gas (N_2) into ammonia (NH_3). As noted in the chapter opener, plants and algae can use ammonia (though not N_2) to produce proteins and other essential nitrogen-containing molecules. As a result, many plants have developed close relationships with nitrogen fixers, which provide ammonia fertilizer to the plant partner. In addition to the aquatic photosynthetic cyanobacteria discussed at the chapter opener, many types of heterotrophic soil bacteria also fix nitrogen. Examples include the proteobacterium *Rhizobium* and its relatives, which live within the roots of protein-rich legume plants (see Chapter 37).

Oxygen can poison nitrogenase, so most nitrogen fixers conduct nitrogen fixation only in low-oxygen conditions. Many cyanobacteria generate low-oxygen conditions in specialized cells known as **heterocytes**, allowing nitrogen fixation to occur in these cells (see Figure 27.15a). Heterocytes display adaptations that reduce nitrogenase exposure to oxygen including thick walls that reduce inward O_2 diffusion, increase cellular reactions that consume oxygen, and downregulate the oxygen-producing components of photosynthesis. The latter adaptation, involving reduction in chlorophyll synthesis, explains why heterocytes are paler in color than neighboring photosynthetic cells.

27.5 Ecological Roles and Biotechnology Applications

Learning Outcomes:

1. Discuss the role of bacteria and archaea in the carbon cycle.
2. List examples of bacterial-eukaryote symbiosis and of pathogenic microbes.
3. Define the concept of microbiomes and explain where they occur and why they are important.
4. List ways in which bacteria contribute to industrial and biotechnology applications.

Bacteria and archaea play many key ecological roles. These include the production and breakdown of organic carbon, beneficial symbionts in plants and animals, and disease agents. In this section, we focus on these diverse ecological roles and also provide examples of ways that humans use the metabolic capabilities of bacteria in biotechnology.

Bacteria and Archaea Play Important Roles in Earth's Carbon Cycle

The Earth's carbon cycle is the sum of all the chemical changes that occur among compounds that contain carbon. (See Chapter 59 for a detailed discussion of the carbon cycle.) One way that bacteria and archaea influence Earth's carbon cycle is by producing and consuming methane. Methane (CH_4)—the major component of natural gas—is a greenhouse gas more powerful than CO_2; CH_4 increases global warming over 20 times more per molecule than does CO_2. In consequence, atmospheric CH_4 has the potential to alter the Earth's climate, and in recent years the level of CH_4 has been increasing in Earth's atmosphere as the result of human activities. Several groups of anaerobic archaea known as **methanogens** convert CO_2, methyl groups, or acetate to CH_4 and release CH_4 from their cells into the atmosphere. Methanogens live in swampy wetlands, in deep-sea habitats, or in the digestive systems of animals including cattle and humans. Marsh gas produced in wetlands is largely composed of CH_4, and large quantities of CH_4 produced long ago are trapped in deep-sea and subsurface Arctic deposits. Certain bacteria known as **methanotrophs** consume CH_4, thereby reducing its concentration in the atmosphere. In the absence of methanotrophs, Earth's atmosphere would be much richer in the greenhouse gas CH_4, which would substantially increase global temperatures.

Bacteria and archaea are also important in producing and degrading complex organic compounds. For example, cyanobacteria and other autotrophic bacteria are important **producers**. Such bacteria, together with algae and plants, use photosynthesis to synthesize the organic compounds used by other organisms for food. **Decomposers**, also known as saprobes, include heterotrophic microorganisms (as well as fungi and animals). These organisms break down dead organisms and organic matter, releasing minerals for uptake by living things. Astonishingly, many bacteria are able to break down antibiotics for use as a source of organic carbon, as discussed next.

Feature Investigation

Dantas and Colleagues Found That Many Bacteria Can Break Down and Consume Antibiotics as a Sole Carbon Source

Many microorganisms naturally secrete antibiotics, chemicals that inhibit the growth of other microorganisms. Antibiotic compounds are evolutionary adaptations that allow bacteria and other microbes to avoid attack or reduce competition for resources. People have taken advantage of high antibiotic production by certain bacteria, particularly species of the phylum Actinomycetes, to make commercial antibiotics in industrial processes.

In nature, many chemoheterotrophic bacteria have taken advantage of widespread natural antibiotic production by utilizing these organic compounds as a source of carbon. In 2008, Gautam Dantas, George Church, and their colleagues reported this conclusion after experimentally testing their hypothesis that soil bacteria might be able to metabolize antibiotics (Figure 27.16). The investigators first cultivated bacteria from 11 different soils in the laboratory, finding diverse phylogenetic types. Almost 90% of the cultured bacteria were Gram-negative Proteobacteria, some closely related to human pathogens, while 7% of the cultures were Gram-positive Actinomycetes. These researchers then tested the ability of the bacteria cultured

Figure 27.16 Diverse bacteria isolated from different soils are able to grow on many types of antibiotics.

HYPOTHESIS The soil bacterial community contains species that can take up and metabolize antibiotics.

KEY MATERIALS Eleven diverse soil samples; 18 types of antibiotics.

	Experimental level	Conceptual level
1 Inoculate soil samples onto growth media in culture dishes.	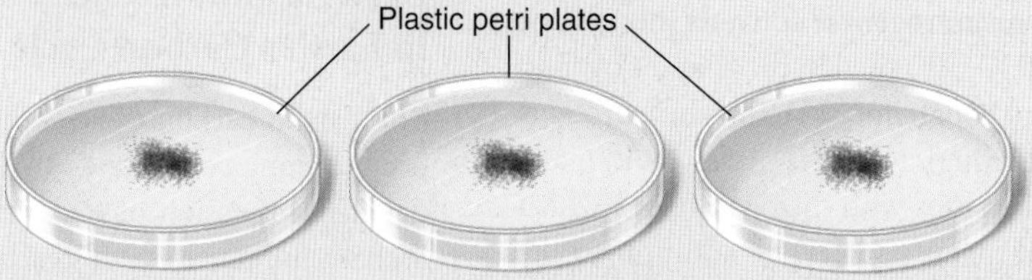 Plastic petri plates	
2 Isolate bacterial species that grow from single cells to visible colonies by repeated binary fission.	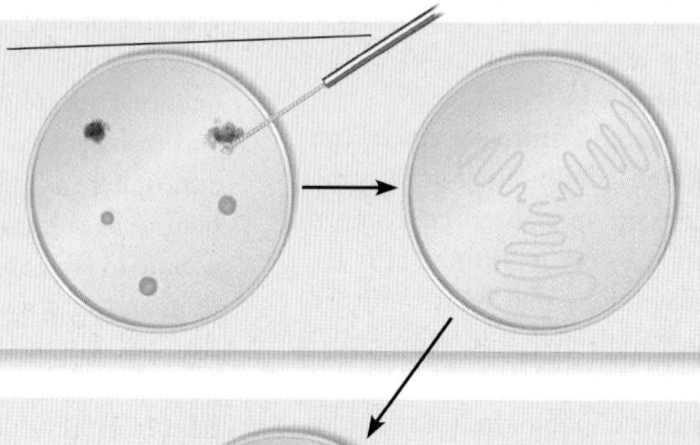 Transfer loop –a device used to move microbial cells	Different species have distinctive colony characteristics (color, shape, size).
3 Grow isolates into large populations for testing on antibiotics.		Bacterial cells undergo repeated binary fission to quickly form colonies large enough to see with the unaided eye.
4 Inoculate each bacterial isolate onto replicate dishes containing a different antibiotic as the only food source.	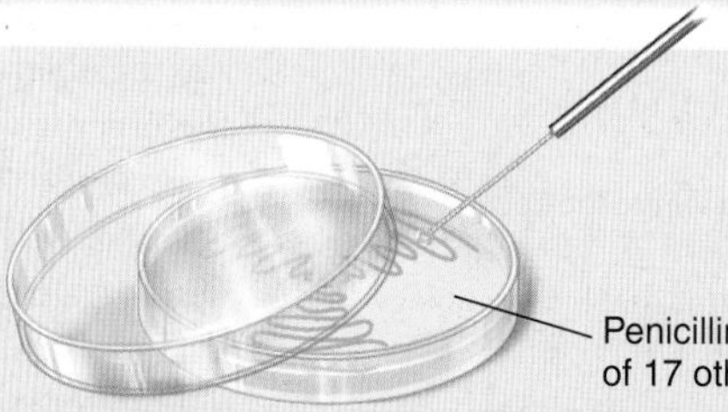 Penicillin G (or one of 17 other antibiotics)	Test the ability of each isolate to grow on a range of antibiotics.
5 Allow time for bacterial population growth; compare growth among dishes.	Strong growth of forest soil #1 bacterial isolate on penicillin G food. Poor growth of urban soil #3 bacterial isolate on dicloxacillin food.	Compare isolate ability to grow on different antibiotics.

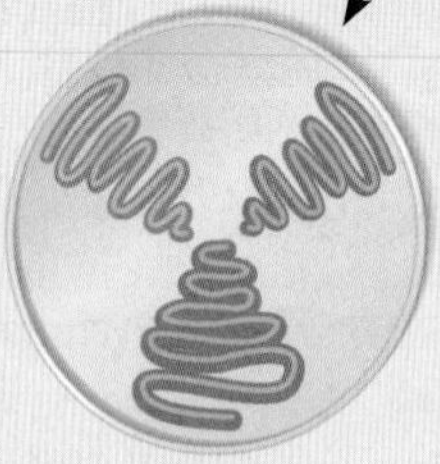

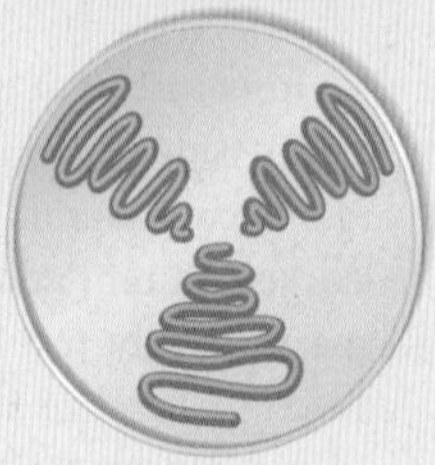

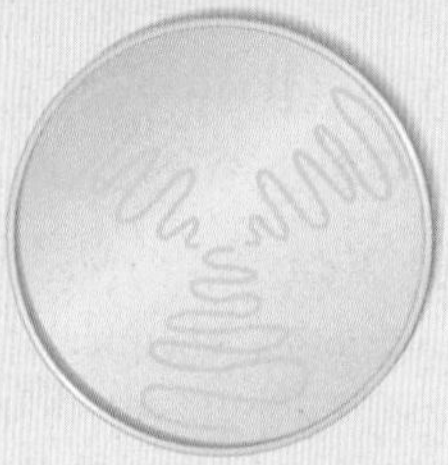

6 THE DATA

Most soils tested contained bacterial species that were able to use antibiotics of many types for food and thus were resistant to those antibiotics.

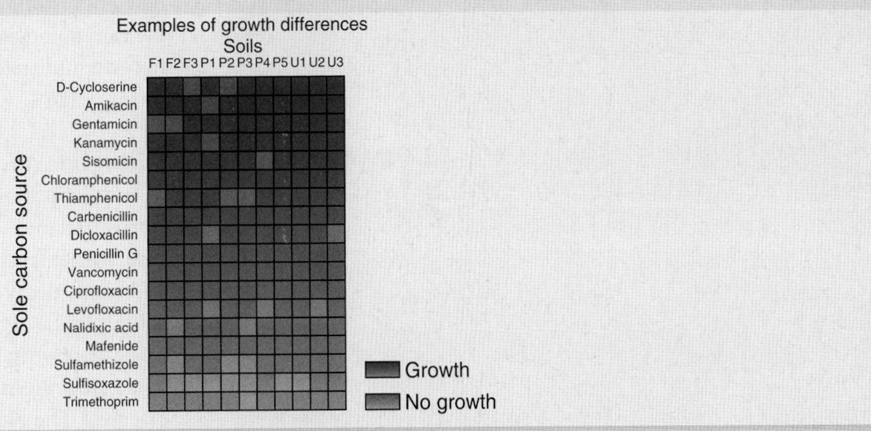

7 CONCLUSION Natural soils contain bacteria that are able to utilize antibiotics produced naturally by other species as food. Soil bacteria are a previously unrecognized source of antibiotic resistance genes that can be transferred to other species.

8 SOURCE Dantas, G., Sommer, M. O. A., Oluwasegun, R. D., and Church, G. M. 2008. Bacteria subsisting on antibiotics. *Science* 320:100–103.

from different soils (isolates) to use various antibiotics as a sole carbon source. The 18 antibiotics tested included penicillin and related compounds, as well as widely prescribed ciprofloxacin (Cipro). Every antibiotic tested supported the growth of bacteria from soil. Importantly, each antibiotic-eating isolate was resistant to several antibiotics at concentrations used in medical treatment of infections.

In today's society, the widespread use of antibiotics in medicine and agriculture is of concern because it is thought to foster increases in antibiotic resistance (see Section 18.4). The experiment by Dantas and associates revealed that natural evolutionary processes—the widespread development by diverse soil bacteria of metabolic processes to utilize many types of antibiotics as food—represent a previously unrecognized source of antibiotic resistance. The study also indicated that natural bacteria are a potential source of antibiotic-resistance genes that could be transferred to disease-causing bacteria.

Experimental Questions

1. What features of soil bacteria attracted the attention of researchers?
2. What processes did researchers use to test their hypothesis that soil bacteria might use antibiotics as a food source?
3. Why was it important to researchers to test the ability of soil bacteria to resist antibiotics in the same concentrations that physicians use to treat infections?

Many Bacteria Live in Symbiotic Associations

An organism that lives in close association with one or more other organisms is said to occur in **symbiosis** (from the Greek, meaning life together with). If symbiotic association is beneficial to both partners, the interaction is known as a **mutualism**. If one partner benefits at the expense of the other, the association is known as a **parasitism**. Many mutualistic bacteria live in associations of two or a few other bacterial species that supply each other with essential nutrients. For example, certain deep-sea archaea are able to metabolize the plentiful methane present in such anaerobic conditions only by partnering with bacteria that reduce sulfate. The marine worm *Olavius algarvensis* has no mouth, gut, or anus and depends on several types of bacteria that live within it to provide food and recycle its wastes. Numerous cases exist of eukaryote associations with mutualistic and parasitic bacteria, but such archaeal symbioses seem rare.

Mutualistic Partnerships Between Bacteria and Eukaryotes Bacteria are involved in many mutually beneficial symbioses in which they provide aquatic or terrestrial eukaryotes with minerals or vitamins or other valuable services. The common green protist seaweed *Ulva* does not display its typical lettuce-leaf shape unless bacterial partners belonging to the phylum Bacteroidetes are present because the bacteria produce a compound that induces normal seaweed development. Bioluminescent bacteria, bacteria that have the ability to produce and emit light (**Figure 27.17**), often form symbiotic relationships with squid and other marine animals. In deep-sea thermal vent communities, bacteria live within the tissues of tubeworms and

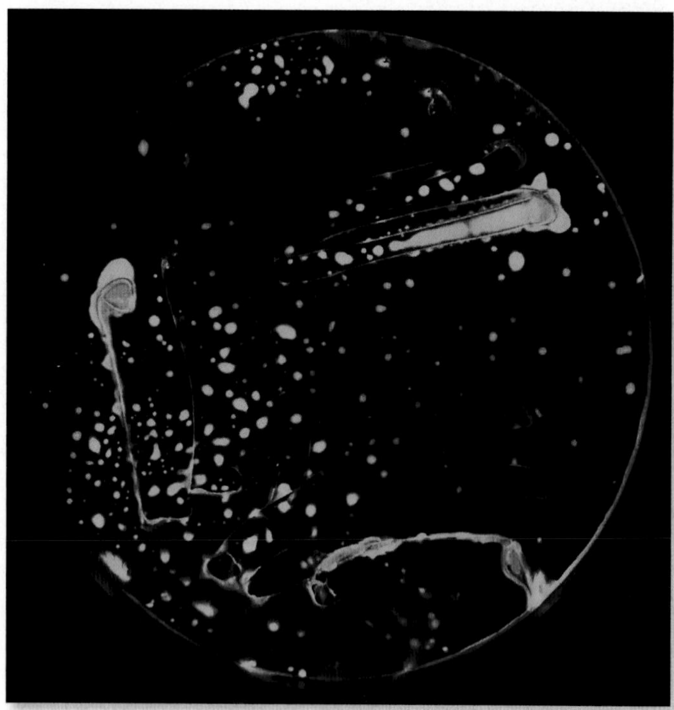

Figure 27.17 Bioluminescent bacteria. These colonies of *Vibrio fischeri* bacteria are growing on nutritive media in a culture plate. The colonies produce so much light that additional light was not needed to make this photo.

mussels, supplying these animals with carbon compounds used as food. One terrestrial example of mutualism is a complex association involving four partners: ants, fungi that the ants cultivate for food, parasitic fungi that attack the food fungi, and Actinobacteria that produce antibiotics. The antibiotics control the growth of the parasitic fungi, preventing them from destroying the ants' fungal food supply. The ants rear the useful bacteria in cavities on their body surfaces; glands near these cavities supply the bacteria with nutrients.

The Human Microbiome Humans likewise harbor symbiotic microbes. On human skin and in our digestive and reproductive systems many types of microbes exist that are known collectively as the human **microbiome**. An estimated 10–100 trillion microbes live in the typical human colon! These microbes provide services using traits that humans do not possess, and the diverse types of metabolism present in the microbiome have coevolved with human metabolism. For this reason, humans and other multicellular organisms, together with their microbiomes, are considered to function and evolve as superorganisms. Recent research has revealed that human gut microbiome communities contain hundreds of prokaryotic species, dominated by the bacterial phyla Firmicutes and Bacteroidetes, and that extensive horizontal gene transfer has occurred among gut microbial species. Studies also reveal that gut microbiomes differ among healthy people, and between healthy people and those having different types of medical conditions. The Human Microbiome Project seeks to understand the relationship between human metabolism and microbiome communities at several sites in the human body, with the goal of opening new opportunities to improve human health. Other projects explore the microbiomes of domesticated animals, plants, and other habitats to better understand bacterial communities and how they affect other forms of life.

Pathogenic Microbes Microorganisms that cause disease in one or more types of host organism are known as **pathogens**. Cholera, leprosy, tetanus, pneumonia, whooping cough, diphtheria, Lyme disease, scarlet fever, rheumatic fever, typhoid fever, bacterial dysentery, and tooth decay are among the many examples of human diseases caused by bacterial pathogens. Bacteria also cause many plant diseases of importance in agriculture, including blights, soft rots, and wilts. How do microbiologists determine which bacteria cause these diseases? The pioneering research of the Nobel Prize–winning German physician Robert Koch provided the answer.

In the mid- to late 1800s, Koch established a series of four steps to determine whether a particular organism causes a specific disease. First, the presence of the suspected pathogen must correlate with occurrence of symptoms. Second, the pathogen must be isolated from an infected host and grown in pure culture if possible. Third, cells from the pure culture should cause disease when inoculated into a healthy host. Fourth and finally, one should be able to isolate the same pathogen from the second infected host. Using these steps, known as **Koch's postulates**, Koch discovered the bacterial causes of anthrax, cholera, and tuberculosis. Subsequent investigators have used Koch's postulates to establish the identities of additional bacteria that cause other infectious diseases. Recent studies have also indicated how pathogenic bacteria attack host cells.

How Pathogenic Bacteria Attack Cells Understanding how disease-causing bacteria attack host cells aids in developing strategies for disease prevention and treatment. Many pathogenic bacteria attack cells by binding to the target cell surfaces and injecting substances that help them utilize cell components. During their evolution, some Gram-negative pathogenic bacteria developed needle-like systems, made of components also found in flagella, that inject proteins into animal or plant cells as part of the infection process. Such structures are known as type III secretion systems, also called injectisomes (**Figure 27.18a**). Examples of bacteria whose injectisomes allow them to attack human cells are *Yersinia pestis* (the agent of bubonic plague), *Salmonella enterica* (which causes the food poisoning called salmonellosis), and *Burkholderia pseudomallei* (the cause of melioidosis, a deadly disease of emerging concern in some parts of the world). These bacteria also induce the host cell to form a plasma membrane pocket that encloses the bacterial cell, bringing it into the host cell. Once within a host cell, pathogenic bacteria use the cell's resources to reproduce and spread to nearby tissues. More than 20 million people are infected by means of injectisomes every year.

Some other Gram-negative bacterial pathogens use a type IV secretion system to deliver toxins or to transform DNA into cells (**Figure 27.18b**). Examples of such bacteria that cause human disease

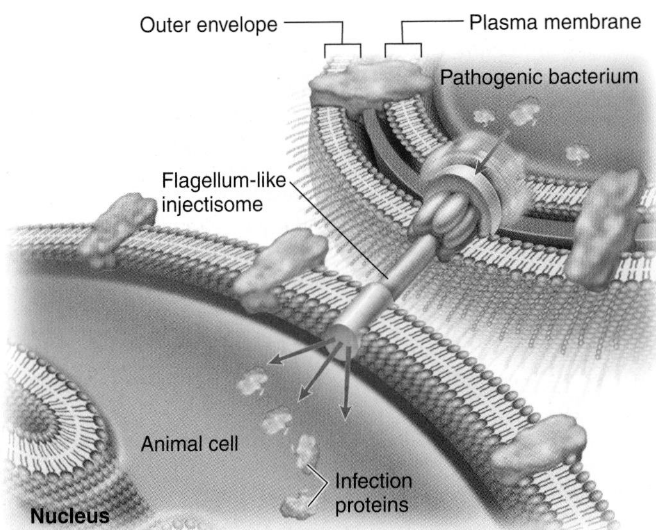

(a) Type III secretion system

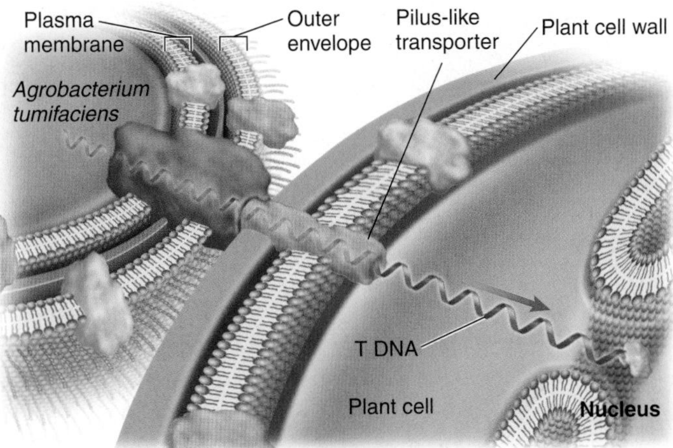

(b) Type IV secretion system

Figure 27.18 Attack systems of pathogenic bacteria. **(a)** The type III secretion system functions like a syringe to inject proteins into host cells, in this case an animal cell, thereby initiating the disease process. **(b)** The type IV secretion system forms a channel through which DNA can be transmitted from a pathogen to a host cell, in this case from the bacterium *Agrobacterium tumifaciens* into a plant cell.

Concept Check: *How does the type IV secretion system illustrate the evolutionary concept of descent with modification?*

include *Helicobacter pylori*, *Legionella pneumophila*, and *Bordetella pertussis*. The plant pathogen *Agrobacterium tumifaciens* uses a type IV secretion system to transfer DNA (T DNA) into plant cells. The bacterial T DNA encodes an enzyme that affects normal plant growth, with the result that cancer-like tumors called galls develop (see Figure 27.4). The type IV system evolved from pili and other components of bacterial mating. As such, it is an example of descent with modification, the evolutionary process by which organisms acquire new features.

Genomes & Proteomes Connection

The Evolution of Bacterial Pathogens

Genomic and proteomic studies have illuminated the evolution of bacterial pathogens, providing insight useful in devising new ways to control infectious disease. Such studies reveal that some pathogens have evolved small, compact genomes encoding specialized metabolic functions, whereas others have acquired large genomes that provide diverse metabolic capabilities. Horizontal gene transfer plays a major role in increasing disease severity.

Mycoplasma pneumoniae, which causes pneumonia in humans, has one of the smallest genomes known to occur among self-replicating organisms. The tiny cells, only 0.3 μm in diameter, possess fewer than 700 protein-coding genes and make only 178 types of protein complexes. The bacterium is a model organism for investigating the minimal cellular machinery required for life. As one way of gaining insight into a minimal proteome, the locations of five types of protein complexes have been mapped in cells by means of specialized microscopic techniques (**Figure 27.19**).

By contrast, *Pseudomonas aeruginosa*—which causes respiratory disease in humans and other animals, and also infects plants—has a larger genome encoding about 5,000 genes. Diverse strains share a common genome core, but also have strain-specific genes acquired by horizontal transfer that confer a wide variety of metabolic abilities. This genomic flexibility allows *P. aeruginosa* to survive in a wide range of environments. As an example of wide metabolic capability, an isolate of *P. aeruginosa* obtained from an infected human also possessed the genes and proteins necessary to degrade tough defensive resins produced by trees and use them as a food source!

Escherichia coli strain O157:H7, which causes deadly outbreaks of food-borne illness and is the leading cause of acute kidney failure in children, evolved from harmless strains by the step-wise horizontal acquisition of toxin genes, a large plasmid that encodes proteins that foster disease (known as virulence factors), and other genetic elements. These genomic features enable this strain to use its flagellar tips to attach to host intestinal epithelium and then attack cells with a type III injectisome. Bacterial toxin produced in the intestine enters the host circulation system and reaches the kidneys, where the toxin inhibits protein synthesis, resulting in severe tissue damage.

Some Bacteria Are Useful in Industrial and Other Applications

Several industries have harnessed the metabolic capabilities of microbes obtained from nature. The food industry uses bacteria to produce chemical changes in food that improve consistency or flavor—to make dairy products, including cheese and yogurt. Cheese makers add pure cultures of certain bacteria to milk. The bacteria consume milk sugar (lactose) and produce lactic acid, which aids in curdling the milk.

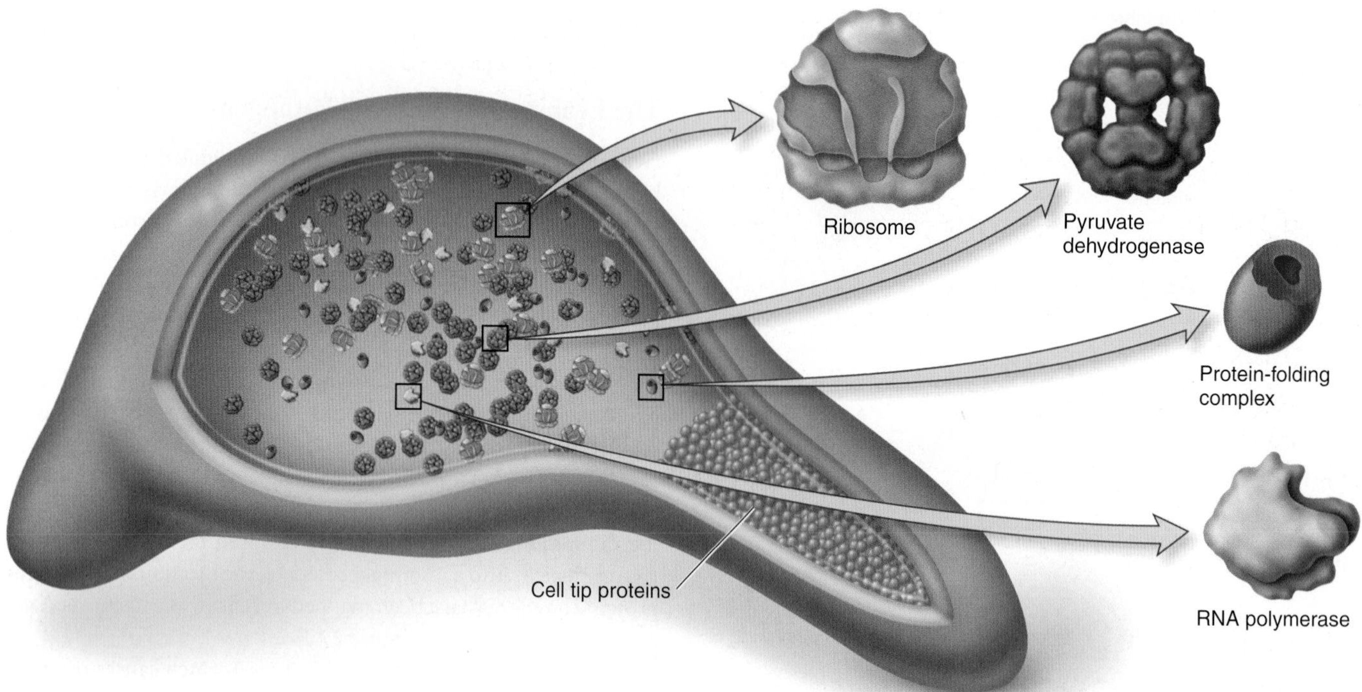

Figure 27.19 **Map of the location of five protein complexes in the tiny pathogen *Mycoplasma pneumoniae*.** The cell tip is rich in proteins (colored green) that help these bacteria attach to host epithelial cells. Other mapped protein complexes are pyruvate dehydrogenase (involved in energy metabolism), ribosomes, RNA polymerase, and a protein-folding complex (colored red).

The chemical industry produces materials such as enzymes, vinegar, amino acids, vitamins, insulin, vaccines, antibiotics, and other useful pharmaceuticals by growing particular bacteria in giant vats. For example, the hot springs bacterial species *Thermus aquaticus* is a source of a form of DNA polymerase widely used in biology laboratories to amplify DNA in polymerase chain reaction (PCR). Industrially grown bacteria produce the antibiotics streptomycin, tetracycline, kanamycin, gentamycin, bacitracin, polymyxin-B, and neomycin.

The new field of synthetic biology utilizes bacteria as chemical factories, by genetically modifying bacterial genomes so that bacteria produce particular useful compounds, such as pharmaceuticals and renewable biofuels. The ability of some microorganisms to break down organic compounds or precipitate metals makes them very useful in treating wastewater, industrial discharges, and harmful substances such as explosives, pesticides, and oil spills. This process, known as **bioremediation**, is used to reduce levels of harmful materials in the environment.

Agriculture employs several species of *Bacillus*, particularly *B. thuringiensis* (Bt), which produce toxins that kill the insects that ingest them, but are harmless to many noninsect species. Tent caterpillars, potato beetles, gypsy moths, mosquitoes, and black flies are among the pests that can be controlled by the Bt toxin. For this reason, toxin genes from *B. thuringiensis* have been cloned and introduced into some crop plants, such as corn and cotton, to reduce conventional pesticide use and increase crop yields (refer back to Figure 20.14).

Summary of Key Concepts

27.1 Diversity and Evolution

- Domains of life include Bacteria, Archaea, and Eukarya (known informally as bacteria, archaea, and eukaryotes). Domain Archaea is more closely related to Domain Eukarya than either is to Domain Bacteria (Figure 27.1).
- Many representatives of the Domain Archaea occur in extremely hot, salty, or acidic habitats. Ether-linked membrane lipids are among the features of archaea that enable their survival in extreme habitats (Figure 27.2).
- The Domain Bacteria includes 50 or more phyla, including Cyanobacteria and Proteobacteria, which are particularly diverse and of great evolutionary and ecological importance (Table 27.1, Figures 27.3, 27.4).
- Widespread horizontal DNA transfer has occurred among bacteria and archaea. Horizontal DNA transfer by means of viral transduction, transformation, or conjugation allows microorganisms to evolve rapidly.

27.2 Structure and Movement

- Bacteria and archaea are composed of prokaryotic cells that are smaller and simpler than those of eukaryotes. Structures such as thylakoids, magnetosomes, and nucleus-like bodies are examples of prokaryotic cell structure complexity (Figures 27.5, 27.6).

- Major prokaryotic cell shape types are spherical cocci, rod-shaped bacilli, comma-shaped vibrios, and coiled spirochaetes and spirilli. Some cyanobacteria display features of multicellular organisms (Figure 27.7).
- Many microbes secrete a coating of slimy mucilage, which plays a role in diseases and in the development of biofilms. Biofilm development is influenced by quorum sensing, a process in which group activity is coordinated by chemical communication (Figure 27.8).
- Most bacterial cell walls contain peptidoglycan, which is composed of carbohydrates cross-linked by peptides. Gram-positive bacterial cells have thick peptidoglycan walls, whereas Gram-negative cells have less peptidoglycan in their walls and are enclosed by an outer lipopolysaccharide envelope (Figures 27.9, 27.10).
- Motility enables microbes to change positions within their environment, which aids in locating favorable conditions for growth. Some bacteria have gas vesicles, which enable them to float, whereas others swim by means of flagella or twitch or glide by the action of pili (Figures 27.11, 27.12, 27.13).

27.3 Reproduction

- Populations of most bacteria and archaea enlarge by binary fission, a simple type of cell division that provides a means by which culturable microbes can be counted (Figure 27.14).
- Some bacteria are able to survive harsh conditions as dormant akinetes or endospores (Figure 27.15).

27.4 Nutrition and Metabolism

- Bacteria and archaea can be grouped according to nutritional type, response to oxygen, or presence of distinctive metabolic features. Major nutritional types are photoautotrophs, chemoautotrophs, photoheterotrophs, and chemoheterotrophs (Table 27.2).
- Obligate aerobes require oxygen, whereas obligate anaerobes are poisoned by oxygen. Aerotolerant anaerobes do not use oxygen but are not poisoned by it. Both obligate and aerotolerant anaerobes obtain their energy by anaerobic respiration. Facultative aerobes are able to live with or without oxygen by using different processes for obtaining energy.
- Nitrogen fixation is an example of a distinctive metabolism displayed only by certain microorganisms.

27.5 Ecological Roles and Biotechnology Applications

- Bacteria and archaea play key roles in Earth's carbon cycle as producers, decomposers, beneficial symbionts, or pathogens. Some bacteria are able to consume antibiotics, a process linked to the evolution and spread of antibiotic resistance (Figure 27.16).
- Methane-producing methanogens and methane-consuming methanotrophs are important in the carbon cycle, and they influence the Earth's climate.
- Pathogenic bacteria obtain organic compounds from living host cells.
- Bacteria attack eukaryotic cells by means of flagella-like type III secretion systems or type IV secretion systems, which evolved from pili (Figure 27.17).
- During their evolution, some pathogenic bacteria have reduced their genomes and proteomes while others have acquired large genomes conferring diverse metabolic capacities; horizontal gene transfer is a major process by which disease severity increases (Figures 27.18, 27.19).
- Many bacteria and archaea are useful in industrial and other applications; others are used to make food products or antibiotics or to clean up polluted environments.

Assess and Discuss

Test Yourself

1. Which of the following features is common to prokaryotic cells?
   a. a nucleus, featuring a nuclear envelope with pores
   b. mitochondria
   c. plasma membranes
   d. mitotic spindle
   e. none of the above
2. The bacterial phylum that typically produces oxygen gas as the result of photosynthesis is
   a. the proteobacteria.
   b. the cyanobacteria.
   c. the Gram-positive bacteria.
   d. all of the listed choices.
   e. none of the listed choices.
3. The Gram stain is a procedure that microbiologists use to
   a. determine if a bacterial strain is a pathogen.
   b. determine if a bacterial sample can break down oil.
   c. infer the structure of a bacterial cell wall and bacterial response to antibiotics.
   d. count bacteria in medical or environmental samples.
   e. do all of the above.
4. Place the following steps in the correct order, according to Koch's postulates:
   I. Determine if pure cultures of bacteria cause disease symptoms when introduced to a healthy host.
   II. Determine if disease symptoms correlate with presence of a suspected pathogen.
   III. Isolate the suspected pathogen and grow it in pure culture, free of other possible pathogens.
   IV. Attempt to isolate pathogen from second-infected hosts.
   a. II, III, IV, I
   b. II, IV, III, I
   c. III, II, I, IV
   d. II, III, I, IV
   e. I, II, III, IV
5. Cyanobacteria play what ecological role?
   a. producers
   b. consumers
   c. decomposers
   d. parasites
   e. none of the listed choices

6. Bacterial structures that are produced by pathogenic bacteria for use in attacking host cells include
   a. type III and IV secretion systems.
   b. magnetosomes.
   c. gas vesicles.
   d. thylakoids.
   e. none of the above.
7. The structures that enable some Gram-positive bacteria to remain dormant for extremely long periods of time are known as
   a. akinetes.
   b. endospores.
   c. biofilms.
   d. lipopolysaccharide envelopes.
   e. pili.
8. By means of what process do populations of bacteria or archaea increase their size?
   a. mitosis
   b. meiosis
   c. conjugation
   d. transduction
   e. none of the above
9. By what means do bacterial cells acquire new DNA?
   a. by conjugation, the mating of two cells of the same bacterial species
   b. by transduction, the injection of viral DNA into bacterial cells
   c. by transformation, the uptake of DNA from the environment
   d. all of the above
   e. none of the above
10. How do various types of bacteria move?
    a. by the use of flagella, composed of a filament, hook, and motor
    b. by means of pili, which help cells twitch or glide along a surface
    c. by using gas vesicles to regulate buoyancy in water bodies
    d. all of the above
    e. none of the above

Conceptual Questions

1. Explain why many microbial populations grow more rapidly than do eukaryotes and how bacterial population growth influences the rate of food spoilage or infection.
2. What processes contribute to antibiotic resistance?
3. A principle of biology is that *living organisms interact with their environment*. What organisms are responsible for the blue-green blooms that often occur in warm weather on lake surfaces? Think carefully; the answer is not just "cyanobacteria," as you might first guess.

Collaborative Questions

1. How would you go about cataloging the phyla of bacteria and archaea that occur in a particular place?
2. How would you go about developing a bacterial product that could be sold for remediation of a site contaminated with materials that are harmful to humans?

Online Resource

www.brookerbiology.com

Stay a step ahead in your studies with animations that bring concepts to life and practice tests to assess your understanding. Your instructor may also recommend the interactive eBook, individualized learning tools, and more.

Chapter Outline

28.1 An Introduction to Protists
28.2 Evolution and Relationships
28.3 Nutritional and Defensive Adaptations
28.4 Reproductive Adaptations
Summary of Key Concepts
Assess and Discuss

Protists

28

Protists such as these green algal cells and their plant descendants produce much of the Earth's oxygen. Each of the cells in this population is surrounded by a halo of protective mucilage.

Protists are eukaryotes that live in moist habitats and are mostly microscopic in size. Despite their small size, protists have a greater influence on global ecology and human affairs than most people realize. For example, the photosynthetic protists known as algae generate at least half of the oxygen in the Earth's atmosphere and produce organic compounds that feed marine and freshwater animals. The oil that fuels our cars and industry is derived from pressure-cooked algae that accumulated on the ocean floor over millions of years. Today, algae are being engineered into systems for cleaning pollutants from water or air and for producing biofuels.

Protists also include some parasites that cause serious human illnesses. For example, in 1993, the waterborne protist *Cryptosporidium parvum* sickened 400,000 people in Milwaukee, Wisconsin, costing $96 million in medical expenses and lost work time. Species of the related protist *Plasmodium*, which is carried by mosquitoes in many warm regions of the world, cause the disease malaria. Every year, nearly 500 million people become ill with malaria, and more than 2 million die of this disease. As we will see, sequencing the genomes of these and other protist species has suggested new ways of battling such deadly pathogens.

In this chapter, we will survey protist diversity, including structural, nutritional, and ecological variations. We begin by exploring ways of informally naming protists by their ecological roles, habitat, and motility. We then will focus on the defining features, classification, and evolutionary importance of the major protist phyla. Next, the nutritional modes and defensive adaptations of protists are discussed, and we conclude by looking at the reproductive adaptations that allow protists to exploit and thrive in a variety of environments.

28.1 An Introduction to Protists

Learning Outcomes:

1. List three features that define protists.
2. Label protists informally by ecological role, habitat, and type of motility.

The term protist comes from the Greek word *protos*, meaning first, reflecting the observation that protists were Earth's first eukaryotes. Protists are eukaryotes that are not classified in the plant, animal, or fungal kingdoms. Protists display two additional common characteristics: They are most abundant in moist habitats, and most of them are microscopic in size. Protists are often informally labeled according to their ecological roles, habitats, or type of motility.

Protists Can Be Informally Labeled According to Their Diverse Ecological Roles

Protists are often labeled according to their ecological roles, which occur in three major types: algae, protozoa, and fungus-like protists. The term **algae** (singular, alga. From the Latin, meaning seaweeds) applies to protists that are generally photoautotrophic, meaning that most can produce organic compounds from inorganic sources by means of photosynthesis. In addition to organic compounds that can be used as food by heterotrophs—organisms that obtain their food from other organisms—photosynthetic algae produce oxygen, which is also needed by most heterotrophs. Thanks to their photosynthetic abilities, algae are increasingly important sources of renewable biofuels. Despite the general feature of photosynthesis, algae do not form a monophyletic group descended from a single common ancestor.

The term **protozoa** (from the Greek, meaning first life) is commonly used to describe diverse heterotrophic protists. Protozoa feed

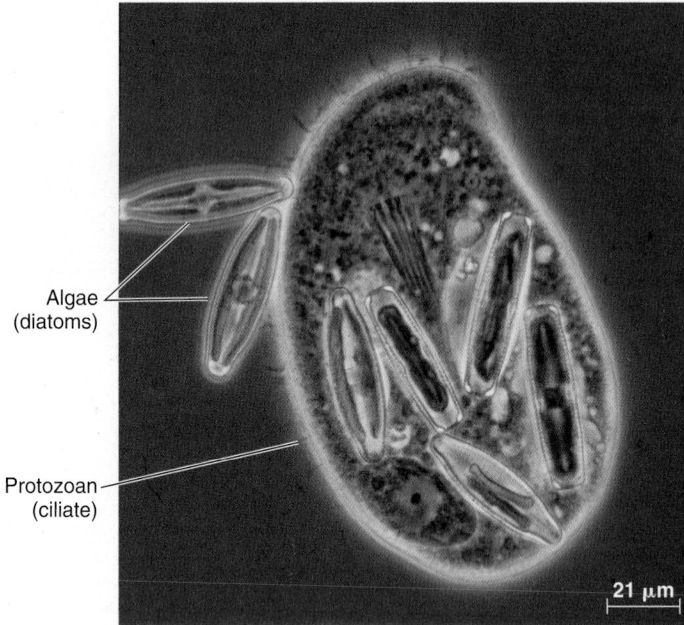

Figure 28.1 A heterotrophic protozoan feeding on photosynthetic algae. The ciliate shown here has consumed several oil-rich, golden-pigmented, silica-walled algal cells known as diatoms. Diatom cells that have avoided capture glide nearby.

BIOLOGY PRINCIPLE Living organisms use energy. The diatoms were ingested by the process of phagocytosis, and their organic components are digested as food.

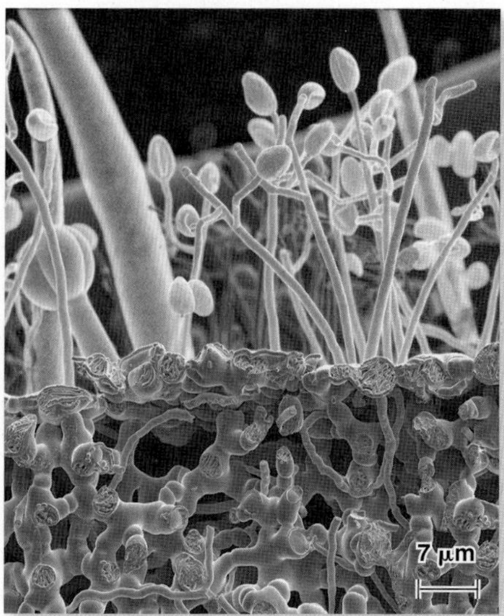

Figure 28.2 A fungus-like protist, *Phytophthora infestans*. This organism causes the disease of potato known as late-blight, or potato-blight. SEM of protist growing on host leaf.

by absorbing small organic molecules or by ingesting prey. For example, the protozoa known as ciliates consume smaller cells such as the single-celled photosynthetic algae known as diatoms (**Figure 28.1**). Like the algae, the protozoa do not form a monophyletic group.

Several types of heterotrophic **fungus-like protists** have bodies, nutrition, or reproduction mechanisms similar to those of the true fungi. For example, fungus-like protists often have threadlike, filamentous bodies and absorb nutrients from their environment, as do the true fungi (see Chapter 31). However, fungus-like protists are not actually related to fungi; their similar features represent cases of convergent evolution, in which species from different lineages have independently evolved similar characteristics (see Chapter 23). Water molds, some of which cause diseases of fish, and *Phytophthora infestans*, which causes diseases of many crops and wild plants, are examples of fungus-like protists (**Figure 28.2**). Various types of slime molds, some of which can be observed on decaying wood in forests, are also fungus-like though not closely related to water molds and *Phytophthora*. These examples illustrate that the terms algae, protozoa, and fungus-like protists, although very useful in describing ecological roles, lack taxonomic or evolutionary meaning.

Protists Can Be Informally Labeled According to Their Diverse Habitats

Although protists occupy nearly every type of moist habitat, they are particularly common and diverse in oceans, lakes, wetlands, and rivers. Even extreme aquatic environments such as Antarctic ice and acidic hot springs serve as habitats for some protists. In such places, protists may swim or float in open water or live attached to surfaces such as rocks or beach sand. These different habitats influence protist structure and size.

Protists that swim or float in fresh or salt water are members of an informal aggregate of organisms known as plankton, which also includes bacteria, viruses, and small animals. The photosynthetic protists in plankton are called **phytoplankton** (plantlike plankton). Planktonic protists are necessarily quite small in size; otherwise they would readily sink to the bottom. Staying afloat is a particularly important characteristic of phytoplankton, which need light for photosynthesis. For this reason, planktonic protists occur primarily as single cells, colonies of cells held together with mucilage, or short filaments of cells linked end to end (**Figure 28.3a–c**).

Many protists live within **periphyton**—communities of microorganisms attached by mucilage to underwater surfaces such as rocks, sand, and plants. Because sinking is not a problem for attached protists, these often produce multicellular bodies, such as branched filaments (**Figure 28.3d**). Photosynthetic protists large enough to see with the unaided eye are known as **macroalgae**, or seaweeds. Although the bodies of some macroalgae are very large single cells (**Figure 28.3e**), most macroalgae are multicellular, often producing large and complex bodies. Macroalgae usually grow attached to underwater surfaces such as rocks, sand, docks, ship hulls, or offshore oil platforms. Seaweeds require sunlight and carbon dioxide for photosynthesis and growth, so most of them grow along coastal shorelines, fairly near the water's surface. Macroalgae serve as refuges for aquatic animals, generate large amounts of organic carbon that enters aquatic food chains, and play additional important ecological roles. Humans harvest some macroalgae for use as food, fertilizers for crops, or as sources of industrial chemicals to make diverse commercial products.

Protists Can Be Informally Labeled According to Their Type of Motility

Microscopic protists have evolved diverse ways to propel themselves in moist environments. Swimming by means of flagella, cilia, amoeboid movement, and gliding are major types of protist movements.

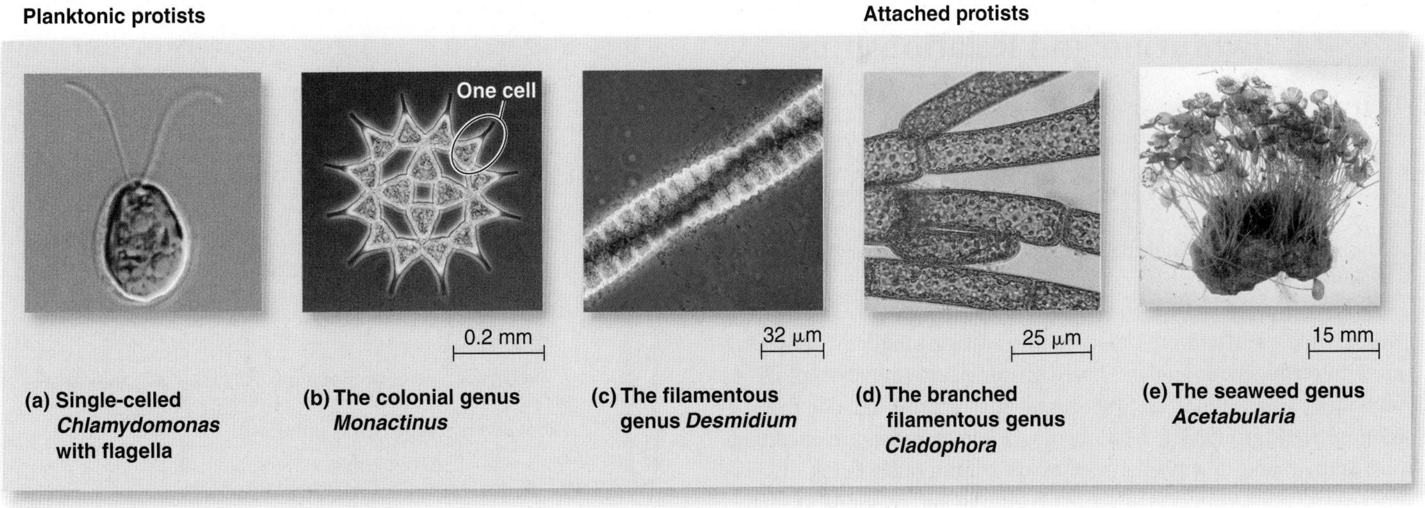

Planktonic protists

Attached protists

0.2 mm

32 μm

25 μm

15 mm

One cell

(a) Single-celled *Chlamydomonas* with flagella

(b) The colonial genus *Monactinus*

(c) The filamentous genus *Desmidium*

(d) The branched filamentous genus *Cladophora*

(e) The seaweed genus *Acetabularia*

Figure 28.3 The diversity of algal body types reflects their habitats. (a) The single-celled flagellate genus *Chlamydomonas* occurs in the phytoplankton of lakes. **(b)** The colonial genus *Monactinus* is composed of several cells arranged in a lacy star shape that helps to keep this alga afloat in water and avoid being consumed by aquatic animals. **(c)** The filamentous genus *Desmidium* occurs as a twisted row of cells. **(d)** The branched filamentous genus *Cladophora* that grows attached to nearshore surfaces is large enough to see with the unaided eye. **(e)** The relatively large seaweed genus *Acetabularia* lives on rocks and coral rubble in shallow tropical oceans. The body of *Acetabularia* is a single very large cell.

Many types of photosynthetic and heterotrophic protists are able to swim because they produce one or a few eukaryotic flagella—cellular extensions whose movement is based on interactions between microtubules and the motor protein dynein (refer back to Figure 4.14). Eukaryotic flagella rapidly bend and straighten, thereby pulling or pushing cells through the water. Protists that use flagella to move in water are commonly known as **flagellates** (Figure 28.3a). Flagellates are typically composed of one or only a few cells and are small—usually from 2 to 20 μm long—because flagellar motion is not powerful enough to keep larger bodies from sinking. Some flagellate protists are sedentary, living attached to underwater surfaces. These protists use flagella to collect bacteria and other small particles for food. Macroalgae and other immobile protists often produce small, flagellate reproductive cells that allow these protists to mate and disperse to new habitats.

An alternative type of protist motility relies on cilia, tiny hairlike extensions on the outsides of cells, mentioned earlier as occurring on the surfaces of some protozoa. Cilia are structurally similar to eukaryotic flagella but are shorter and more abundant on cells (**Figure 28.4**). Protists that move by means of cilia are **ciliates**. Having many cilia allows ciliates to achieve larger sizes than flagellates yet still remain buoyant in water.

A third type of motility is amoeboid movement. This kind of motion involves extending protist cytoplasm into lobes, known as pseudopodia (from the Greek, meaning false feet). Once these pseudopodia move toward a food source or other stimulus, the rest of the cytoplasm flows after them, thereby changing the shape of the entire organism as it creeps along. Protist cells that move by pseudopodia are described as **amoebae** (**Figure 28.5**).

Finally, many diatoms, the malarial parasite genus *Plasmodium*, and some other protists glide along surfaces in a snail-like fashion by secreting protein or carbohydrate slime. With the exception of ciliates, motility classification does not correspond with the phylogenetic classification of protists, our next topic.

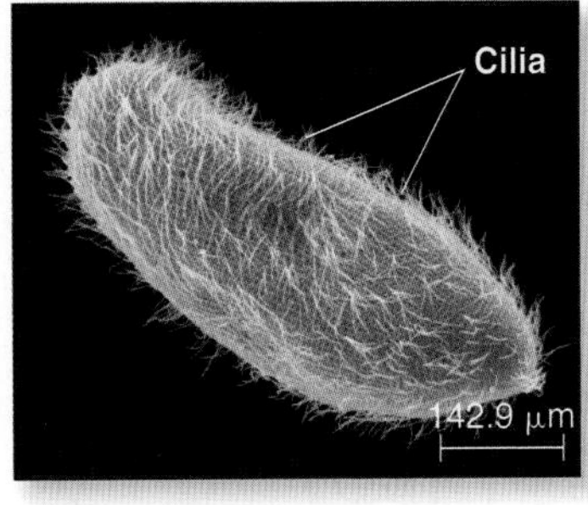

Figure 28.4 A member of the ciliate genus *Paramecium*, showing numerous cilia on the cell surface. SEM view.

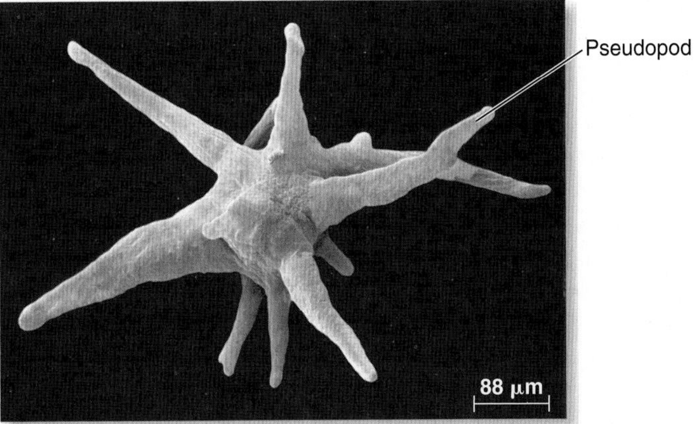

Figure 28.5 A member of the amoebozoan genus *Pelomyxa*, showing pseudopodia. SEM view.

BioConnections: *Look ahead to Figure 33.2. What kind of mobile, amoeba-shaped cells carry materials within the bodies of the early-diverging opisthokont animals known as sponges?*

28.2 Evolution and Relationships

Learning Outcomes:

1. Describe a distinctive structural characteristic for each of seven eukaryotic supergroups.
2. List at least one species of each eukaryotic supergroup that is important to human life.
3. Draw a diagram showing how the process of endosymbiosis has affected eukaryotic diversity.

At one time, protists were classified into a single kingdom. However, modern phylogenetic analyses based on comparative analysis of DNA sequences and cellular features reveal that protists do not form a monophyletic group. The relationships of some protists are uncertain or disputed, and new protist species are continuously being discovered. As a result, concepts of protist evolution and relationships have been changing as new information becomes available.

Even so, molecular and cellular data reveal that many protist phyla can be classified within several eukaryotic **supergroups** that each display distinctive features (Figure 28.6). All of the eukaryotic supergroups include phyla of protists; some, in fact, contain only protist phyla. The supergroup Opisthokonta includes the multicellular animal and fungal kingdoms and related protists, whereas another supergroup includes the multicellular plant kingdom and the protists most closely related to it. The study of such protists helps to reveal how multicellularity originated in animals, fungi, and plants.

In this section, we survey the eukaryotic supergroups, focusing on the defining features and evolutionary importance of the major protist phyla. We will also examine ways in which protists are important ecologically or in human affairs.

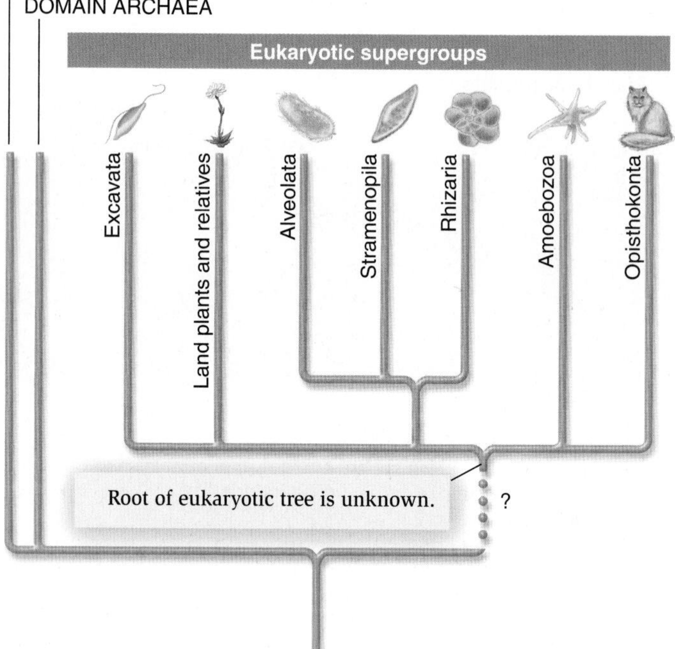

Figure 28.6 A phylogenetic tree showing the major eukaryotic supergroups. Each of the eukaryotic supergroups shown here includes some protist phyla, and most supergroups consist only of protists.

BIOLOGY PRINCIPLE All species (past and present) are related by an evolutionary history. Many more eukaryotic branches exist than are shown in this streamlined diagram.

A Feeding Groove Characterizes Many Protists Classified in the Excavata

The protist supergroup known as the Excavata originated very early among eukaryotes, so this supergroup is important in understanding the early evolution of eukaryotes. The Excavata is named for a feeding groove "excavated" into the cells of many representatives, such as the genus *Jakoba* (phylum Metamonada) (**Figure 28.7**). The feeding groove is an important adaptation that allows these single-celled organisms (informally called excavates) to ingest small particles of food in their aquatic habitats. Once food particles are collected within the feeding groove, they are then taken into cells by a type of endocytosis known as **phagocytosis** (from the Greek, meaning cellular eating). During phagocytosis, a vesicle of plasma membrane surrounds each food particle and pinches off within the cytoplasm. Enzymes within these food vesicles break the food particles down into small molecules that, upon their release into the cytoplasm, can be used for energy.

Phagocytosis is also the basis for an important evolutionary process known as **endosymbiosis**, a symbiotic association in which a smaller species known as the endosymbiont lives within the body of a larger species known as the host. Phagocytosis provides a way for protist cells that function as hosts to take in prokaryotic or eukaryotic cells that function as endosymbionts. Such endosymbiotic cells confer valuable traits and are not digested. Endosymbiosis has played a particularly important role in protist evolution. For example, early in protist history, endosymbiotic bacterial cells gave rise to mitochondria, the organelles that are the major site of ATP synthesis in most eukaryotic cells (see Figure 4.28). Consequently, most protists possess mitochondria, though these may be highly modified in some species.

Some excavate protists have become parasitic within animals, including human hosts. In addition to feeding by phagocytosis, parasitic species attack host cells and absorb food molecules released from them. For example, *Trichomonas vaginalis* causes a sexually transmitted infection of the human genitourinary tract. In this location, *T. vaginalis* consumes bacteria and host epithelial and red blood cells by phagocytosis, as well as carbohydrates and proteins released from damaged host cells. More than 170 million cases of this infection are estimated to occur each year around the globe, and infections can predispose humans to other diseases. *T. vaginalis* has an undulating membrane and flagella that allow it to move over mucus-coated skin (**Figure 28.8a**).

Giardia intestinalis (previously known as *G. lamblia*), another type of excavate protist, contains two active nuclei and produces eight flagella (**Figure 28.8b**). *G. intestinalis* causes giardiasis, an intestinal infection that can result from drinking untreated water or from unsanitary conditions in day-care centers. Nearly 300 million human infections occur every year, and the disease also harms young farm animals, dogs, cats, and wild animals. In the animal body, flagellate cells cause disease and also produce infectious stages known as cysts that are transmitted in feces and can survive several weeks outside a host. When an animal ingests as few as 10 of these cysts, within 15 minutes stomach acids induce the flagellate stage to develop and adhere to cells of the small intestine. *T. vaginalis* and *G. intestinalis* were once thought to lack mitochondria, but they are now known to possess simpler structures that are highly modified mitochondria.

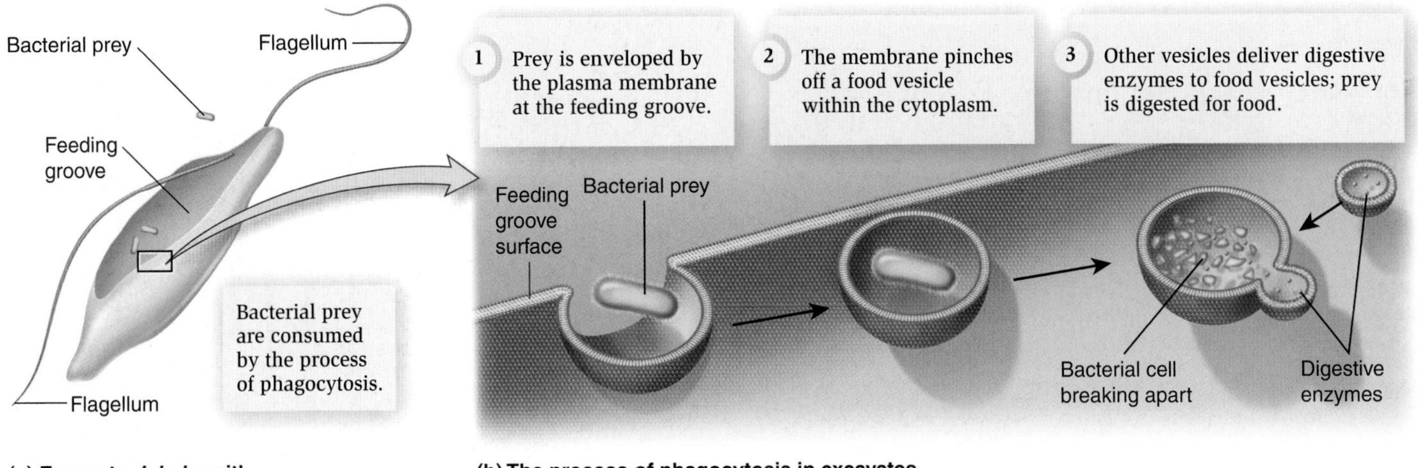

Bacterial prey
Flagellum
Feeding groove
Bacterial prey are consumed by the process of phagocytosis.
Flagellum
1 Prey is enveloped by the plasma membrane at the feeding groove.
2 The membrane pinches off a food vesicle within the cytoplasm.
3 Other vesicles deliver digestive enzymes to food vesicles; prey is digested for food.
Feeding groove surface
Bacterial prey
Bacterial cell breaking apart
Digestive enzymes

(a) Excavate *Jakoba* with feeding groove **(b) The process of phagocytosis in excavates**

Figure 28.7 Feeding groove and phagocytosis displayed by many species of supergroup Excavata. (a) Diagram of *Jakoba libera*, phylum Metamonada, showing flagella emerging from the feeding groove. **(b)** Diagram of phagocytosis, the process by which food particles are consumed at a feeding groove.

Concept Check: What happens to ingested particles after they enter feeding cells?

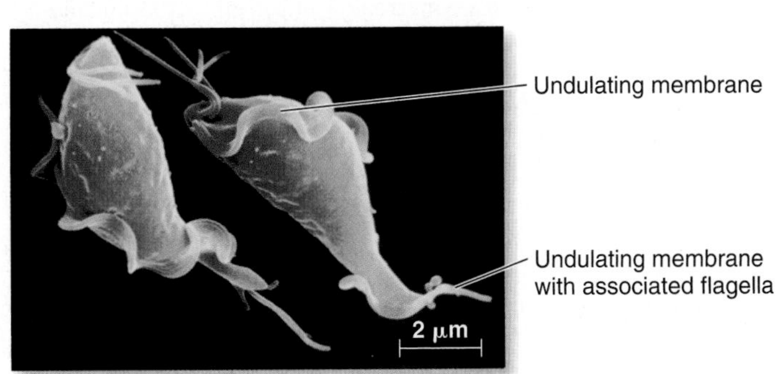

(a) *Trichomonas vaginalis*

(b) *Giardia intestinalis*

Figure 28.8 Parasitic members of the supergroup Excavata. (a) *Trichomonas vaginalis*. **(b)** *Giardia intestinalis*. These specialized heterotrophic flagellates use flagella to disperse across the surfaces of moist host tissues; the flagellates then absorb nutrients from living cells. These images were made with a scanning electron microscope (SEM) that employs electrons rather than visible light, with the result that cellular structures do not appear in color.

Concept Check: How do these two parasitic protists differ in the process of transmission from one human host to another?

GENOMES & PROTEOMES CONNECTION

Genome Sequences Reveal the Different Evolutionary Pathways of *Trichomonas vaginalis* and *Giardia intestinalis*

In 2007, genome sequences were reported for two excavates, *T. vaginalis* and *G. intestinalis*. A comparison of their genomic features reveals similarities and differences in the evolution of parasitic lifestyles. One common feature is that horizontal gene transfer from bacterial or archaeal donors has powerfully affected both genomes. About 100 *G. intestinalis* genes are likely to have originated via horizontal gene transfer. In *T. vaginalis*, more than 150 cases of likely horizontal gene transfer were identified, with most transferred genes encoding metabolic enzymes such as those involved in carbohydrate or protein metabolism. Another similarity between *T. vaginalis* and *G. intestinalis* revealed by comparative genomics is an absence of the cytoskeletal protein myosin, which is present in most eukaryotic cells.

Despite these similarities, the genome sequences of *T. vaginalis* and *G. lamblia* reveal some dramatic differences. The *G. intestinalis* genome is quite compact, only 11.7 megabases (Mb) in size, with relatively simple metabolic pathways and machinery for DNA replication, transcription, and RNA processing. In contrast, the *T. vaginalis* genome is a surprisingly large 160 Mb in size. *T. vaginalis* has a core set of about 60,000 protein-coding genes, one of the greatest coding capacities known among eukaryotes. The additional genes provide an expanded capacity for biochemical degradation. Because most trichomonads inhabit animal intestine, the genomic data suggest that the large genome size of *T. vaginalis* is related to its ecological transition to a new habitat, the urogenital tract.

Euglenoids The excavate protists known as euglenoids possess unique, interlocking ribbon-like protein strips just beneath their plasma membranes (Figure 28.9a). These strips make the surfaces of some euglenoids so flexible that they can crawl through mud. Many

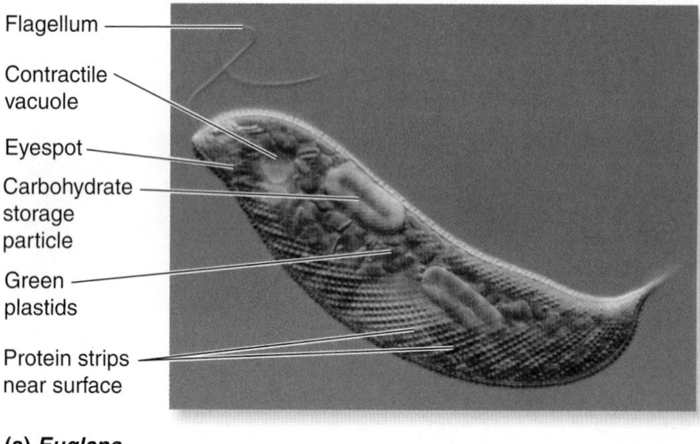

(a) *Euglena*

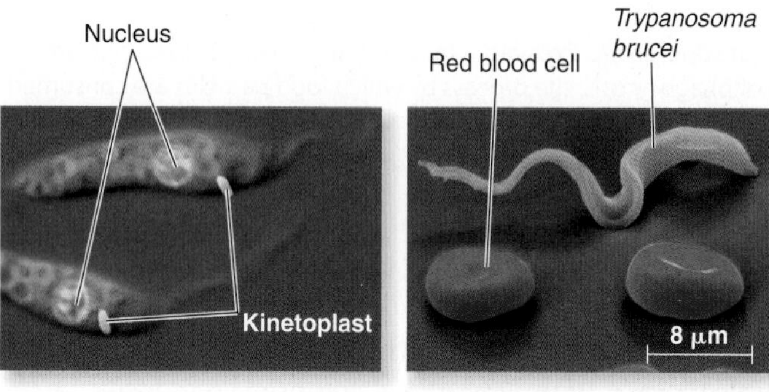

(b) *Leishmania* (c) *Trypanosoma*

Figure 28.9 Representative euglenoids and kinetoplastids. **(a)** *Euglena* has helical protein ribbons near its surface, internal green plastids, white storage carbohydrate granules, and a red eyespot. **(b)** Fluorescence light micrograph of *Leishmania* showing the kinetoplast DNA mass typical of kinetoplastid mitochondria. **(c)** In this artificially colorized SEM, an undulating kinetoplastid (*Trypanosoma*) appears near disc-shaped red blood cells.

euglenoids are colorless and heterotrophic, but *Euglena* and some other genera possess green plastids and are photosynthetic. Plastids are organelles found in plant and algal cells that are distinguished by their synthetic abilities and that were acquired via endosymbiosis. Many euglenoids possess a light-sensing system that includes a conspicuous red structure known as an eyespot, or stigma, and light-detecting molecules located in a swollen region at the base of a flagellum. These structures enable green euglenoids to detect light environments that are optimal for photosynthesis. Most euglenoids produce conspicuous carbohydrate-storage particles that occur in the cytoplasm. Euglenoids are particularly abundant and ecologically significant in wetlands, which are rich in the organic materials that many euglenoids require.

Kinetoplastids The heterotrophic excavate protists known as kinetoplastids are named for a large mass of DNA known as a kinetoplast that occurs in their single large mitochondrion (**Figure 28.9b**). These protists lack plastids, but they do possess an unusual modified peroxisome in which glycolysis takes place; in most eukaryotes, glycolysis occurs in the cytosol. Some kinetoplastids, including *Leishmania* (see **Figure 28.9b**), which causes an ulcerative skin disease and can result in organ damage, and *Trypanosoma brucei*, the causative agent of sleeping sickness, are serious pathogens of humans and other animals (**Figure 28.9c**).

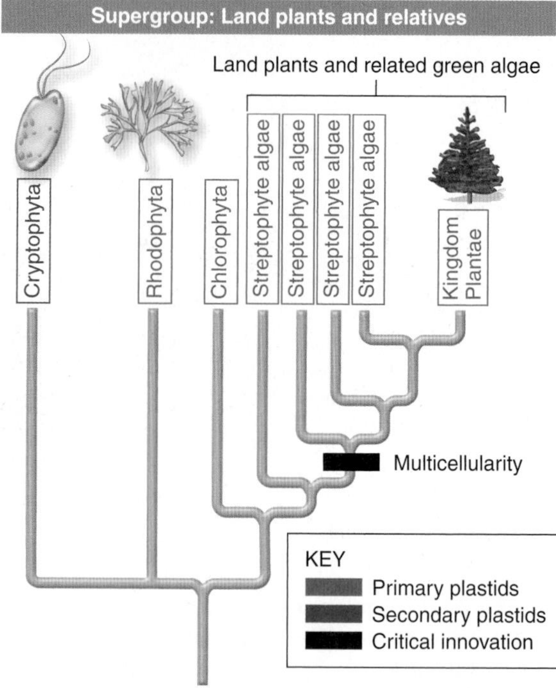

Figure 28.10 A phylogenetic tree of the supergroup that includes land plants and their close protist relatives. Note that plant multicellularity first arose in closely related streptophyte algae. Many chlorophyte and red algae are macroalgae in which multicellularity arose independently.

Land Plants and Related Algae Share Similar Genetic Features

The supergroup that includes land plants also encompasses several protist phyla (**Figure 28.10**). The land plants, also known as the kingdom Plantae (described more fully in Chapters 29 and 30), evolved from green algal ancestors. Together, plants and some closely related green algae form the clade Streptophyta, informally known as streptophytes, whereas most green algae are classified in the phylum Chlorophyta. The red algae, classified in the phylum Rhodophyta, are also regarded as close relatives of green algae and land plants. Recent molecular sequence data and some similarities in cell structure link the plants and green and red algae to additional protist phyla, such as the Cryptophyta.

Green Algae Diverse structural types of green algae (see Figure 28.2) occur in fresh water, the ocean, and on land. Most of the green algae are photosynthetic, and their cells contain the same types of plastids and photosynthetic pigments that are present in land plants. Some green algae are responsible for harmful algal growths, but others are useful as food for aquatic animals, model organisms, and sources of renewable oil supplies. Many green algae possess flagella or the ability to produce them during the development of reproductive cells.

Red Algae Most species of the protists known as red algae are multicellular marine macroalgae (**Figure 28.11**). The red appearance of these algae is caused by the presence of distinctive photosynthetic pigments that are absent from green algae or land plants. Red algae characteristically lack flagella—a feature that has strongly influenced the evolution of this group, resulting in unusually complex life cycles (illustrated in Section 28.4). These life cycles are important to humans because we cultivate red algae in ocean waters for production of billions of dollars worth of food or industrial and scientific materials yearly. For example, the sushi wrappers called nori are composed of the sheetlike red algal genus *Porphyra*, which is grown in ocean farms.

(a) ***Calliarthron*** (b) ***Chondrus crispus***

Figure 28.11 **Representative red algae (Rhodophyta). (a)** The genus *Calliarthron* has cell walls that are impregnated with calcium carbonate. This stony, white material makes the red alga appear pink. **(b)** *Chondrus crispus* is an edible red seaweed.

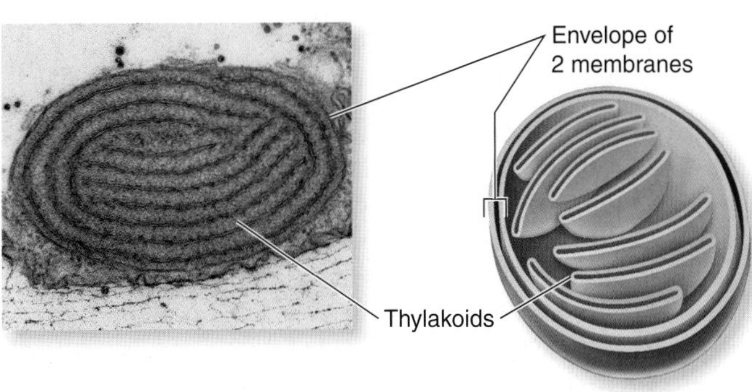

Figure 28.12 **A primary plastid, showing an envelope composed of two membranes.** The plastid shown here is red, but primary plastids can also be green or blue-green in color.

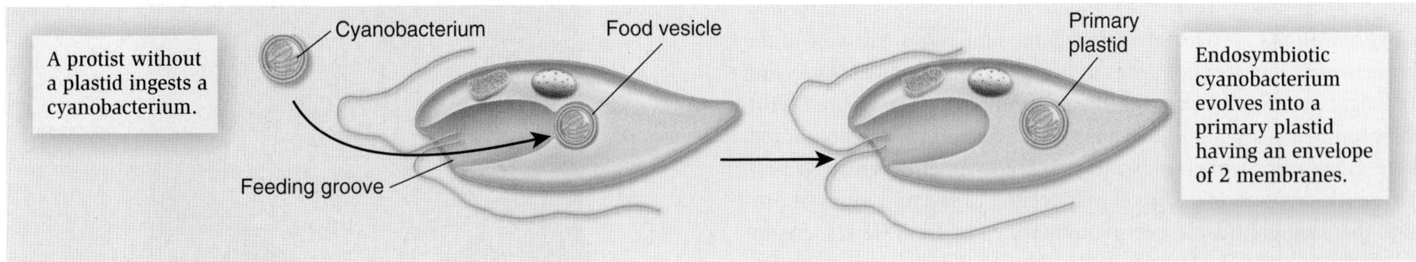

A protist without a plastid ingests a cyanobacterium.
Cyanobacterium
Food vesicle
Feeding groove
Primary plastid
Endosymbiotic cyanobacterium evolves into a primary plastid having an envelope of 2 membranes.

(a) Primary endosymbiosis

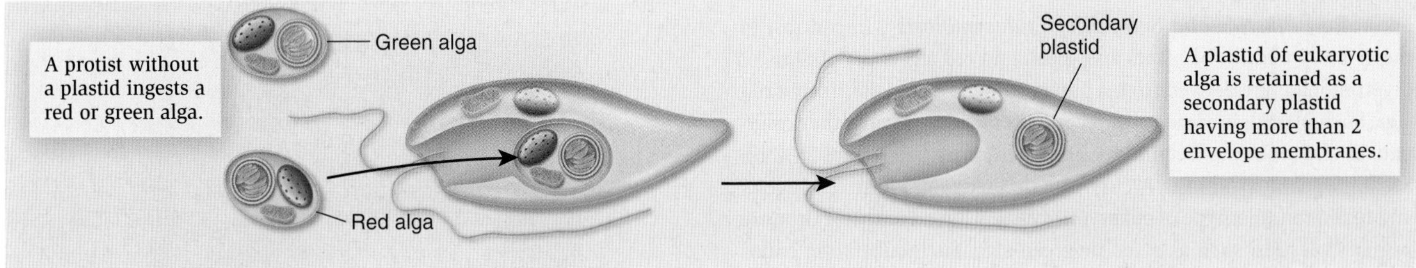

A protist without a plastid ingests a red or green alga.
Green alga
Red alga
Secondary plastid
A plastid of eukaryotic alga is retained as a secondary plastid having more than 2 envelope membranes.

(b) Secondary endosymbiosis

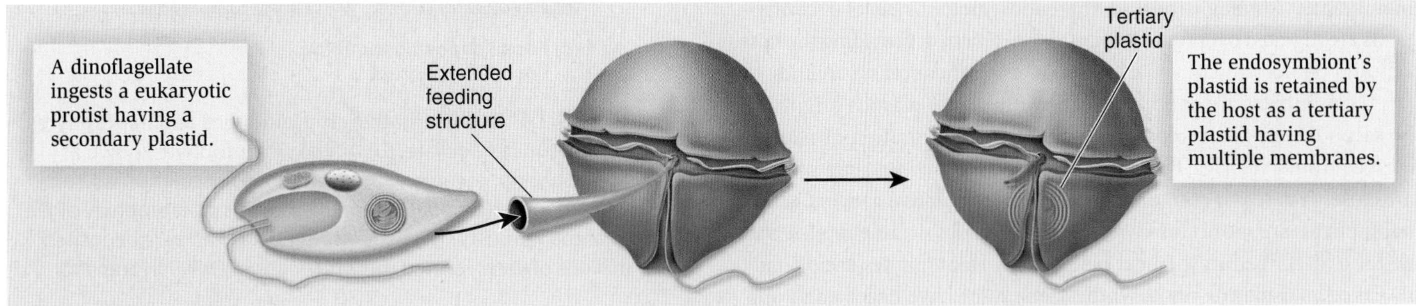

A dinoflagellate ingests a eukaryotic protist having a secondary plastid.
Extended feeding structure
Tertiary plastid
The endosymbiont's plastid is retained by the host as a tertiary plastid having multiple membranes.

(c) Tertiary endosymbiosis

Figure 28.13 **Primary, secondary, and tertiary endosymbiosis. (a)** Primary endosymbiosis involves the acquisition of a cyanobacterial endosymbiont by a host cell without a plastid. During the evolution of a primary plastid, the bacterial cell wall is lost, and most endosymbiont genes are transferred to the host nucleus. **(b)** Secondary endosymbiosis involves the acquisition by a host cell of a eukaryotic endosymbiont that contains one or more primary plastids. During the evolution of a secondary plastid, most components of the endosymbiont cell are lost, but a plastid is often retained within an envelope of endoplasmic reticulum. **(c)** Tertiary endosymbiosis involves the acquisition by a host cell of a eukaryotic endosymbiont that possesses secondary plastids.

Carrageenan, agar, and agarose are complex polysaccharides extracted from red algae that are essential to the food industry and in biology laboratories for cultivating microorganisms and working with DNA.

Primary Plastids and Primary Endosymbiosis The plastids of red algae resemble those of green algae and land plants (and differ from most other algae) in having an enclosing envelope composed of two membranes (Figure 28.12). Such plastids, known as **primary plastids**, are thought to have originated via a process known as **primary endosymbiosis** (Figure 28.13). During primary endosymbiosis, heterotrophic host cells captured cyanobacterial cells via phagocytosis but did not digest them. These endosymbiotic cyanobacteria provided host cells with photosynthetic capability and other useful biochemical pathways and eventually evolved into primary plastids (Figure 28.13a). Endosymbiotic acquisitions of plastids and mitochondria resulted in massive horizontal gene transfer from the endosymbiont to the host nucleus. As a result of such gene transfer, many of the proteins needed by plastids and mitochondria are synthesized in the host cytoplasm and then targeted to these organelles. All cells of plants, green algae, and red algae contain one or more plastids, and most of these organisms are photosynthetic. However, some species (or some of the cells within the multicellular bodies of photosynthetic species) are heterotrophic because photosynthetic pigments are not produced in the plastids. In these cases, plastids play other essential metabolic roles, such as producing amino acids and fatty acids.

Cryptomonads Together with cellular similarities, analyses of the sequences of hundreds of genes indicate that plants, green algae, and red algae are closely related to the cryptomonads (see Figure 28.10). Cryptomonads are unicellular flagellates, most of which contain red, blue-green, or brown plastids and are photosynthetic (Figure 28.14a). However, cryptomonads are closely related to several groups of protists that lack plastids. Occurring in marine and fresh waters, cryptomonads are excellent sources of the fatty-acid-rich food essential to aquatic animals.

Secondary Plastids and Secondary Endosymbiosis In contrast to the primary plastids of plants and green and red algae, the plastids of cryptomonads are derived from a photosynthetic eukaryote, likely a red alga. Such plastids are known as **secondary plastids** because they originate by the process of **secondary endosymbiosis** (see Figure 28.13b). Secondary endosymbiosis occurs when a eukaryotic host cell ingests and retains another type of eukaryotic cell that already has one or more primary plastids, such as a red or green alga. Such eukaryotic endosymbionts are often enclosed by endoplasmic reticulum (ER), explaining why secondary plastids typically have envelopes of more than two membranes. Although most of the endosymbiont's cellular components are digested over time, its plastids are retained, providing the host cell with photosynthetic capacity and other biochemical capabilities.

Haptophytes are another algal phylum whose plastids originated by secondary endosymbiosis involving the incorporation of plastids derived from a red alga. Some experts have proposed that haptophytes are closely related to cryptomonads, but an alternative concept is that haptophytes are more closely related to alveolates, stramenopiles, and rhizaria, described in the next paragraphs. Haptophytes are primarily unicellular marine photosynthesizers; some have flagella and others do not. Some haptophytes are known as the coccolithophorids because they produce a covering of intricate white calcium carbonate discs known as coccoliths (Figure 28.14b). Coccolithophorids often form massive ocean growths that are visible from space and play important roles in Earth's climate by reflecting sunlight and producing compounds that foster cloud formation. In some places, coccoliths produced by huge populations of ancient coccolithophorids accumulated on the ocean floor, together with the calcium carbonate remains of other protists, for millions of years. These deposits were later raised above sea level, forming massive limestone formations or chalk cliffs such as those visible at Dover, on the southern coast of England (Figure 28.14c).

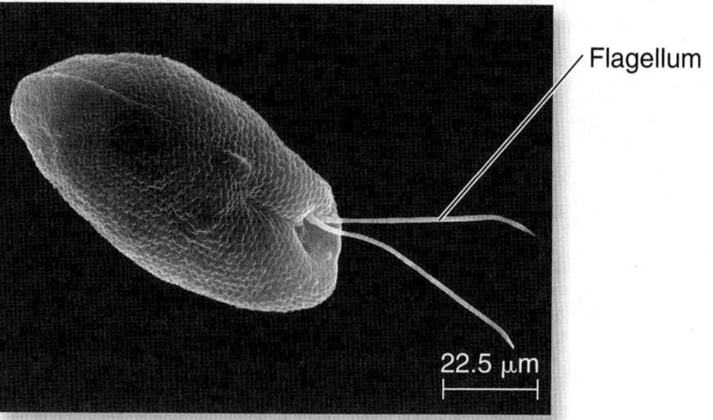

(a) A cryptomonad

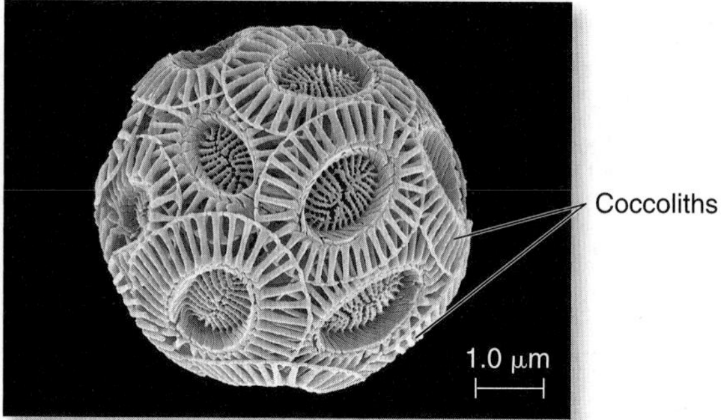

(b) A haptophyte coccolithophorid

(c) Fossil deposit containing coccolithophorids

Figure 28.14 Representative cryptomonads and haptophytes. **(a)** A cryptomonad flagellate. **(b)** A type of haptophyte known as a coccolithophorid, covered with disc-shaped coccoliths made of calcium carbonate. **(c)** Fossil carbonate remains of haptophyte algae and protozoan protists known as foraminifera that were deposited over millions of years formed the white cliffs of Dover in England.

Membrane Sacs Lie at the Cell Periphery of Alveolata

The three supergroups Alveolata, Stramenopila, and Rhizaria seem to form a cluster in recent phylogenetic studies (Figure 28.15). Turning first to Alveolata, we see that it includes three important phyla: (1) the Ciliophora, or ciliates; (2) the Apicomplexa, a medically important group of parasites; and (3) the Dinozoa, informally known as dinoflagellates. Apicomplexans include the malarial agent *Plasmodium* (see Section 28.4), the related protist *Cryptosporidium parvum*, whose effects were noted in the chapter opening, and other serious pathogens of humans and other animals. Dinoflagellates are recognized

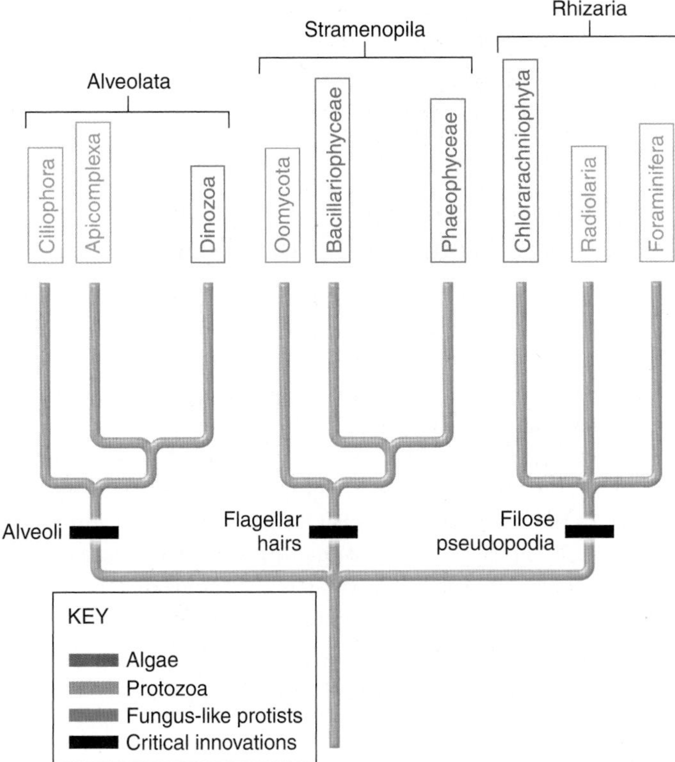

Figure 28.15 A phylogenetic tree illustrating close relationship among the supergroups Alveolata, Stramenopila, and Rhizaria. Some stramenopiles, such as giant kelps, are multicellular.

both for their mutualistic relationship with reef-building corals (look forward to Figure 54.25b) and for the harmful blooms (red tides) that some species produce (see Section 28.3). The Alveolata is named for saclike membranous vesicles known as alveoli that are present at the cell periphery in all of these phyla (**Figure 28.16a**).

The alveoli of some dinoflagellates seem empty, so the cell surface appears smooth. By contrast, the alveoli of many dinoflagellates contain plates of cellulose, which form an armor-like enclosure (**Figure 28.16b**). These plates are often modified in ways that provide an adaptive advantage, such as protection from predators or increased ability to float.

About half of dinoflagellate species are heterotrophic, and half possess photosynthetic plastids of diverse types that originated by secondary or even tertiary endosymbiosis; therefore, these are known as secondary or tertiary plastids. **Tertiary plastids** were obtained by **tertiary endosymbiosis**—the acquisition by hosts of plastids from cells that possessed secondary plastids (see Figure 28.13c). Species having tertiary plastids have received genes by horizontal transfer from diverse genomes.

Flagellar Hairs Distinguish Stramenopila

The supergroup Stramenopila (informally known as the stramenopiles) encompasses a wide range of algae, protozoa, and fungus-like protists that usually produce flagellate cells at some point in their lives (see Figure 28.15). The Stramenopila (from the Greek *stramen*, meaning straw, and *pila*, meaning hair) is named for distinctive strawlike hairs that occur on the surfaces of flagella (**Figure 28.17**). These flagellar hairs function something like oars to greatly increase swimming efficiency. Stramenopiles are also informally known as heterokonts (from the Greek, meaning different flagella), because the two flagella often present on swimming cells have slightly different structures.

Heterotrophic stramenopiles include the fungus-like protist *Phytophthora infestans*, which causes the serious potato disease known as late blight. *P. infestans* is responsible for an estimated $7 billion in crop losses every year. Photosynthetic stramenopiles include diatoms (Bacillariophyceae), whose glasslike silicate cell walls are elaborately ornamented with pores, lines, and other intricate features (**Figure 28.18a**). Vast accumulations of the translucent walls of ancient diatoms, known as diatomite or diatomaceous earth, are mined for use in reflective paint and other industrial products. Recent genome sequencing projects have focused on the processes by which diatoms produce their detailed silicate structures, which may prove useful in industrial microfabrication applications.

Diverse, photosynthetic brown algae (Phaeophyceae) are sources of industrial products such as polysaccharide emulsifiers known as

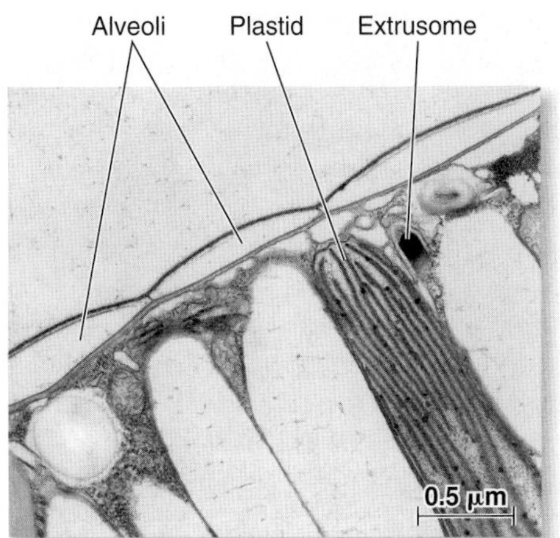

(a) Cross section through characteristic alveoli

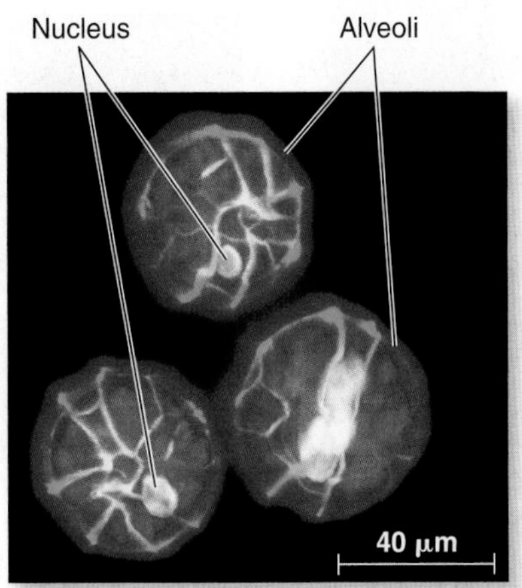

(b) A dinoflagellate with alveoli containing cellulose that here appears blue-white

Figure 28.16 Dinoflagellates of the supergroup Alveolata and their characteristic alveoli. (a) Sac-shaped membranous vesicles known as alveoli lie beneath the plasma membrane of a dinoflagellate, along with defensive projectiles, called extrusomes, that are ready for discharge. **(b)** Fluorescence microscopy reveals that alveoli of the dinoflagellate *Alexandrium catenella* contain cellulose plates, which glow blue when treated with a cellulose-binding dye. The nucleus appears green because DNA has bound a fluorescent dye, and chlorophyll self-fluoresces red.

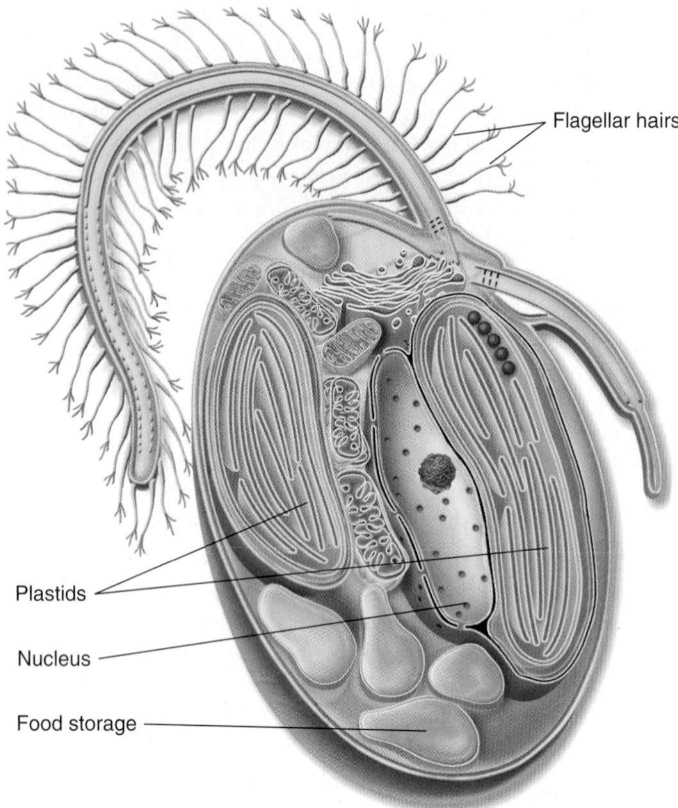

Figure 28.17 A flagellate stramenopile cell, showing characteristic flagellar hairs.

Concept Check: *How do the flagellar hairs aid cell motion?*

(a) Diatom **(b) Kelp forest**

Figure 28.18 Stramenopiles include diatoms and giant kelps. **(a)** SEM of the silicate cell wall of the common diatom *Cyclotella meneghiniana*, showing elaborate ornamentation of the silicate structure. The many pores in the colorless silicate wall lighten the cell, helping to keep it afloat in the water. **(b)** Forests of giant kelps occur along many ocean shores, providing habitat for diverse organisms.

Concept Check: *In what ways are kelp forests economically important?*

alginates. The brown algae known as giant kelps are ecologically important because they form extensive forests in cold and temperate coastal oceans (**Figure 28.18b**). Kelp forests are essential nurseries for fish and shellfish. The reproductive processes of diatoms and kelps are described in Section 28.4.

Spiky Cytoplasmic Extensions Are Present on the Cells of Many Protists Classified in Rhizaria

Several groups of flagellates and amoebae that have thin, hairlike extensions of their cytoplasm—known as filose pseudopodia—are classified into the supergroup Rhizaria (from the Greek *rhiza*, meaning root) (see Figure 28.15). Rhizaria includes the phylum Chlorarachniophyta, whose spider-shaped cells possess secondary plastids obtained from endosymbiotic green algae. Other examples of Rhizaria are the Radiolaria (**Figure 28.19a**) and Foraminifera (**Figure 28.19b**)—two phyla of ocean plankton that produce exquisite mineral shells. Fossil shells of foraminiferans are widely used to infer past climatic conditions. Stable oxygen isotope ratios contained in the shells can be used to reconstruct past water temperatures.

Amoebozoa Includes Many Types of Amoebae with Pseudopodia

The supergroup Amoebozoa includes many types of amoebae that move by extension of pseudopodia (see Figure 28.5). Several types of protists known as slime molds are classified in this supergroup. One example, *Dictyostelium discoideum*, is widely used as a model organism for understanding movement, communication among cells, and development. During reproduction, in response to starvation, single *Dictyostelium* amoebae aggregate into a multicellular slug that produces a cellulose-stalked structure containing many single-celled, asexual spores. In favorable conditions these spores produce new amoebae, which feed on bacteria. A recent study revealed that some *Dictyostelium* clones carry favored bacterial food through these reproductive stages, showing a simple "farming" behavior.

(a) Radiolarian **(b) Foraminiferan**

Figure 28.19 Representatives of supergroup Rhizaria. **(a)** A radiolarian, *Acanthoplegma* spp., showing long filose pseudopodia. **(b)** A foraminiferan, showing calcium carbonate shell with long filose pseudopodia extending from pores in the shell.

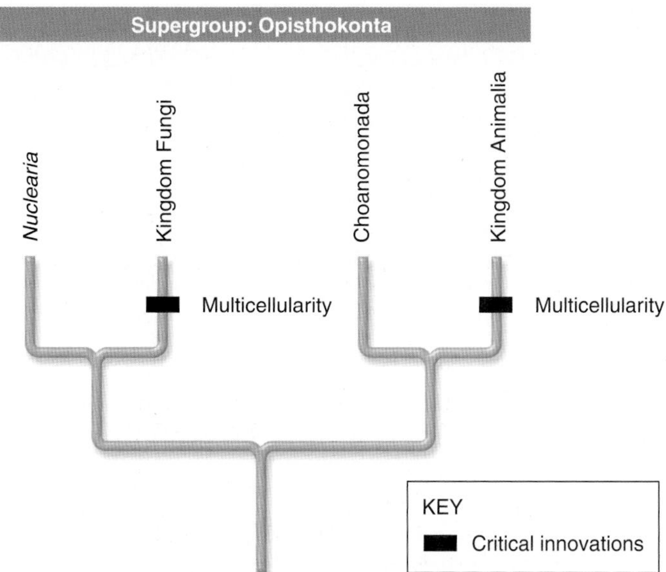

Figure 28.20 **A phylogenetic tree of the supergroup Opisthokonta.** This supergroup includes protist phyla as well as the kingdoms Fungi and Animalia. Multicellularity arose independently in these kingdoms.

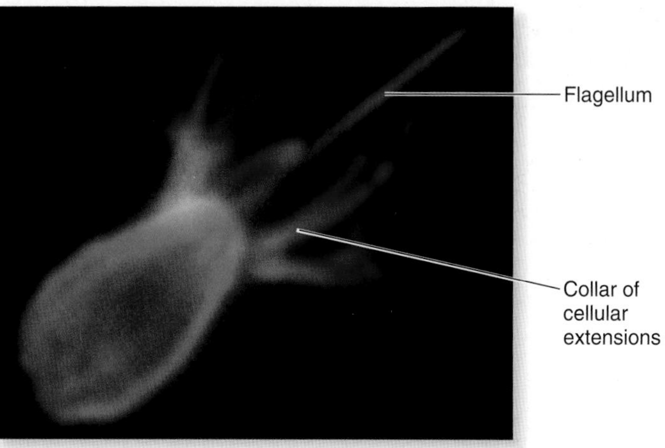

Figure 28.21 **A choanoflagellate of the supergroup Opisthokonta.** This cell has been stained with fluorescent dyes specific for DNA (blue) and the proteins actin (red) and tubulin (green). The single flagellum is stained green because it is rich in tubulin. A collar of cellular extensions that are rich in actin surrounds the flagellum.

Concept Check: What features of the ancient choanoflagellate ancestors of animals were important in the evolution of multicellularity, and what function do such features serve in modern choanoflagellates?

A Single Flagellum Occurs on Swimming Cells of Opisthokonta

The supergroup Opisthokonta includes the animal and fungal kingdoms and related protists (Figure 28.20). This supergroup is named for the presence of a single posterior flagellum on swimming cells. *Nuclearia* is a protist genus that seems particularly closely related to the Kingdom Fungi. The more than 125 species of protists known as choanoflagellates (formally, the Choanomonada) are single-celled or colonial protists featuring a distinctive collar surrounding the single flagellum (Figure 28.21). The collar is made of cytoplasmic extensions that filter bacterial food from water currents generated by flagellar motion.

Choanoflagellates are believed to represent the closest living relatives of animals (look ahead to Figure 32.1). Evolutionary biologists interested in the origin of animals study choanoflagellates for molecular clues to this important event in our evolutionary history. In 2008, American evolutionary biologist Nicole King and associates reported a genome sequence for the choanoflagellate *Monosiga brevicollis* and identified several genes that are present only in choanoflagellates and animals. Some of the shared genes encode cell adhesion and extracellular matrix proteins that help choanoflagellates attach to surfaces and were also essential to the evolution of multicellularity in animals. The choanoflagellate genome also encodes the p53 protein, a regulatory transcription factor that plays essential roles in the animal cell cycle, cancer, and reproduction (refer back to Figure 14.15).

The preceding survey of protist diversity, summarized in Table 28.1, illustrates the enormous evolutionary and ecological importance of protists. Next, we consider the diverse ways in which protists have become adapted to their environments.

28.3 Nutritional and Defensive Adaptations

Learning Outcomes:

1. List four types of protist nutrition.
2. Give examples of major types of protist defensive adaptations.

Wherever you look in moist places, you will find protists playing diverse and important ecological roles. In this section, we will survey nutritional and defensive adaptations that occur widely among protists, that is, in more than one supergroup. Such adaptations help to explain protists' ecological roles.

Protists Display Four Basic Types of Nutrition

Protist nutrition occurs by four basic mechanisms: phagotrophy, osmotrophy, photoautotrophy, and mixotrophy. Heterotrophic protists that feed by ingesting particles, or phagocytosis, are known as **phagotrophs** (see Figure 28.7). Protists that rely on osmotrophy—the uptake of small organic molecules across the cell membrane followed by their digestion—are **osmotrophs**. Protists that feed on nonliving organic material function as decomposers, essential in breaking down wastes and releasing minerals for use by other organisms. Protists that feed on the living cells of other organisms are parasites that may cause disease in other organisms. *Trichomonas vaginalis*, *Giardia intestinalis*, and *Phytophthora infestans* are examples of pathogenic protists. Humans view such protists as pests when they harm us or our agricultural animals and crops, but pathogenic protists also play important roles in nature by controlling the population growth of other organisms.

Table 28.1 Eukaryotic Supergroups and Examples of Constituent Kingdoms, Phyla, Classes, or Species

Supergroup	KINGDOMS, Phyla, classes, or species	Distinguishing features
Excavata	**Metamonada** *Giardia intestinalis* *Trichomonas vaginalis*	Unicellular flagellates, often with feeding groove
	Kinetoplastea (kinetoplastids) *Trypanosoma brucei*	
	Euglenida (euglenoids)	Secondary plastids (when present) derived from endosymbiotic green algae
Land and Plant and Algal Relatives	**Rhodophyta** (red algae) **Chlorophyta** (green algae) KINGDOM PLANTAE and close green algal relatives	Land plants, green algae, and red algae have primary plastids derived from cyanobacteria; such plastids have two envelope membranes.
	Cryptophyta (cryptomonads)	Most cryptomonads (like haptophytes) possess secondary plastids derived from red algae; such plastids have more than two envelope membranes. Cryptomonads, which are closely related to several groups of plastidless protists, are now often classified with plants and green and red algae.
Alveolata	**Ciliophora** (ciliates) **Apicomplexa** (apicomplexans) *Plasmodium falciparum* *Cryptosporidium parvum* **Dinozoa** (dinoflagellates) *Pfiesteria shumwayae*	Peripheral membrane sacs (alveoli); some ciliates harbor endosymbiotic algal cells or organelles; Apicomplexa often have nonphotosynthetic secondary plastids; some Dinozoa have secondary plastids derived from red algae, some have secondary plastids derived from green algae, and some have tertiary plastids derived from diatoms, haptophytes, or cryptomonads.
Stramenopila	Bacillariophyceae (diatoms) Phaeophyceae (brown algae) *Phytophthora infestans* (fungus-like)	Strawlike flagellar hairs; secondary plastids (when present) derived from red algae; fucoxanthin accessory pigment common in autotrophic forms
Rhizaria	**Chlorarachniophyta** **Radiolaria** **Foraminifera**	Thin, cytoplasmic projections; secondary plastids (when present) derived from endosymbiotic green algae
Amoebozoa	*Entamoeba histolytica* **Dictyostelia** (a slime mold phylum) *Dictyostelium discoideum*	Amoeboid movement by pseudopodia
Opisthokonta	*Nuclearia* spp. KINGDOM FUNGI **Choanomonada** (choanoflagellates) KINGDOM ANIMALIA	Swimming cells possess a single posterior flagellum

Photosynthetic protists (algae) are **photoautotrophs**, organisms that can make their own organic nutrients from inorganic sources by harvesting light energy. Because water absorbs much of the red component of sunlight, algae have evolved photosynthetic systems that compensate by capturing more of the blue-green light available underwater. For example, red algae produce the red pigment phycoerythrin, which absorbs blue-green light and transfers energy to chlorophyll *a* (see Figure 28.11). Likewise, blue-green light-absorbing fucoxanthin generates the golden and brown colors of other algae (see Figures 28.1, 28.18b). Carotene (the source of vitamin A) and lutein play similar light-harvesting roles in green algae and were inherited by their land plant descendants, today playing important roles in animal nutrition. Sunlight energy is captured in the bonds of polysaccharide and lipid molecules that function in food storage (see Figure 28.9a), explaining why algae of diverse types are good sources of food for aquatic animals and renewable energy materials.

Mixotrophs are able to use photoautotrophy and phagotrophy or osmotrophy to obtain organic nutrients. The genus *Dinobryon* (Figure 28.22), a photosynthetic stramenopile that lives in the phytoplankton of freshwater lakes, is an example of a mixotroph. These protists may switch back and forth between photoautotrophy and heterotrophy, depending on conditions in their environment. If sufficient light, carbon dioxide, and other minerals are available, *Dinobryon* cells produce their own organic food. If any of these resources limits photosynthesis, or organic food is especially abundant, *Dinobryon* cells can function as heterotrophs, consuming enormous numbers of bacteria. Mixotrophs thus have remarkable nutritional flexibility, explaining why diverse lineages of photosynthetic eukaryotes seem to have mixotrophic capability.

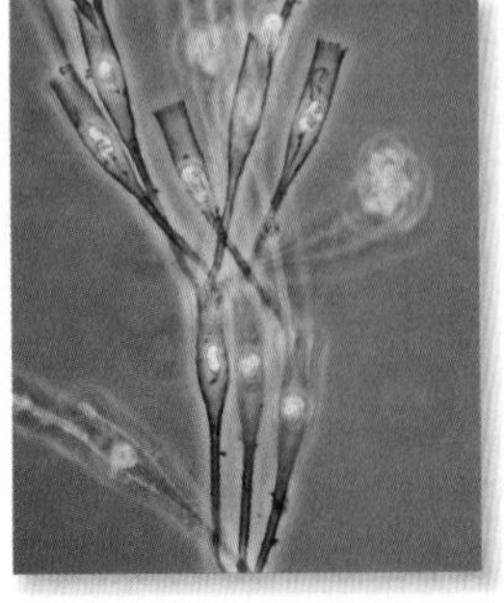

Figure 28.22 A mixotrophic protist. The genus *Dinobryon* is a colonial flagellate that occurs in the phytoplankton of freshwater lakes. The photosynthetic cells have golden photosynthetic plastids and also capture and consume bacterial cells.

Protists Defend Themselves in Diverse Ways

Protists use a wide variety of defensive adaptations to ward off attack. Major types of defenses are cell coverings; sharp projectiles that can be explosively shot from cells; light flashes; and toxic compounds.

Slimy mucilage (see chapter-opening photo) or spiny cell walls (see Figure 28.3b) provide protection from attack by herbivores or pathogens. Cell coverings made of polysaccharide polymers such as cellulose or minerals such as silica also help to prevent osmotic damage or enhance flotation in water.

Evolutionarily diverse protist cells contain structures known as extrusomes (extruded bodies) that are ejected when cells are disturbed, forming spear-like defenses (see Figure 28.16a). Some species of ocean dinoflagellates emit flashes of blue light when disturbed, explaining why ocean waters teeming with these protists display bioluminescence. The light flashes may deter herbivores by startling them, but when ingested, the dinoflagellates make the herbivores also glow, revealing them to hungry fishes. Light flashes benefit dinoflagellates by helping to reduce populations of herbivores that consume the algae.

Various protist species produce **toxins**, compounds that inhibit animal physiology and may function to deter small herbivores. Dinoflagellates are probably the most important protist toxin producers; they synthesize several types of toxins that affect humans and other animals. Why does this happen? Under natural conditions, small populations of dinoflagellates produce low amounts of toxin that do not harm large organisms. Dinoflagellate toxins become dangerous to humans when people contaminate natural waters with excess mineral nutrients such as nitrogen and phosphorus from untreated sewage, industrial discharges, or fertilizer that washes off of agricultural fields. The excess nutrients fuel the development of harmful algal blooms, which then produce sufficient toxin to affect birds, aquatic mammals, fishes, and humans. Toxins can concentrate in organisms. Humans who ingest shellfish that have accumulated dinoflagellate toxins can suffer poisoning.

In the early 1990s, American ecologist JoAnn Burkholder and colleagues reported that the toxic dinoflagellate *Pfiesteria* had caused major fish kills in the nutrient-rich waters of the Chesapeake Bay. These investigators observed that the dinoflagellates consume other algal cells for food, but also produce a toxin that damages fish skin, allowing the dinoflagellates to consume fish flesh. These biologists also discovered that *Pfiesteria*-associated toxin caused amnesia and other nervous system conditions in fishers and scientists who were exposed to it, though the cellular basis of the effect on humans was unclear. The discovery of *Pfiesteria* excited the media, which dubbed it a "killer alga" and which Burkholder herself had referred to as "the cell from hell," and focused attention on nutrient pollution of the Chesapeake Bay, the fundamental cause of *Pfiesteria*'s excessive growth and fish kills. Because of this organism's importance to the fishing industry and human health, teams of aquatic ecologists have continued to study the genus *Pfiesteria* and its toxin, as described next.

FEATURE INVESTIGATION

Burkholder and Colleagues Demonstrated That Strains of the Dinoflagellate Genus *Pfiesteria* Are Toxic to Mammalian Cells

A team of investigators led by JoAnn Burkholder performed an experiment to determine whether or not two strains of *Pfiesteria shumwayae* were toxic to mammalian cells (**Figure 28.23**). Although one of these strains (CCMP 1024C) was thought to be toxic to fish and people, other work suggested that a different strain (CCMP 2089) did not produce toxin, a difference that might have resulted from variation in growth conditions. Neither strain had been tested for its effect on mammalian cells. For the safety of the investigators, the experiment was conducted in a biohazard containment facility.

Figure 28.23 **Burkholder and colleagues demonstrated that some strains of *Pfiesteria shumwayae* are toxic to fish and mammalian cells.**

GOAL To determine the toxicity of two *Pfiesteria shumwayae* strains, grown on different food types, to mammalian cells.

KEY MATERIALS
1) *P. shumwayae* strain CCMP 2089
2) *P. shumwayae* strain CAAE 1024C
3) *Cryptomonas* spp.—algal food for dinoflagellates
4) Juvenile tilapia (*Oreochromis* spp.)—fish food for dinoflagellates
5) Mammalian pituitary cell line

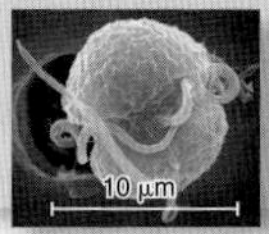

	Experimental level	Conceptual level
1 Grow strains CCMP 2089 and CAAE 1024C with algal food in culture flasks, or with fish as food in tanks. Use a biohazard containment facility to prevent toxins from harming scientists.	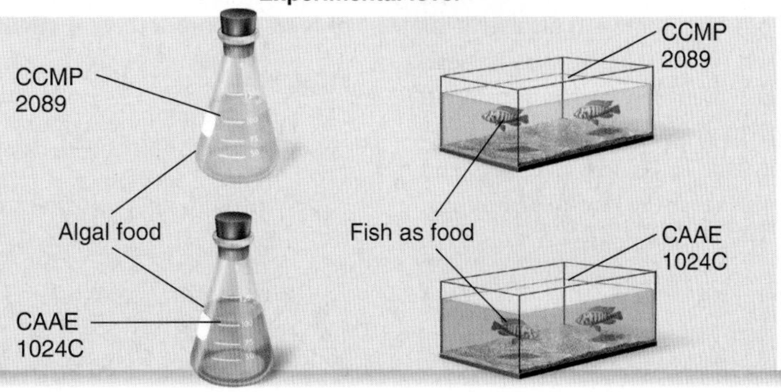	Food source, algae or juvenile fish, might affect the amount of toxin produced by dinoflagellates. The 2 dinoflagellate strains might differ in toxin production response.

2 Transfer dinoflagellates grown under step 1 conditions to new tanks containing juvenile fish.

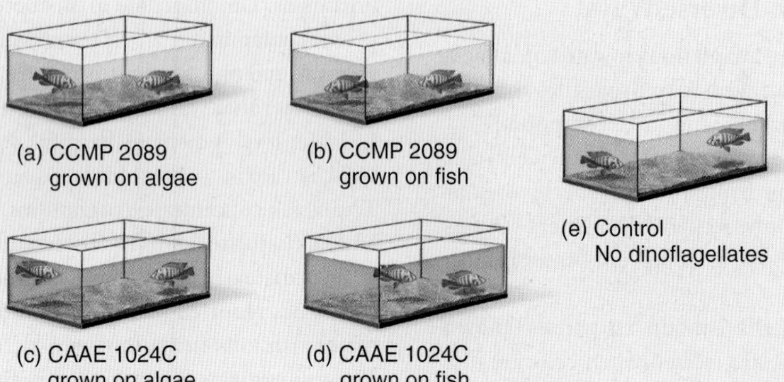

Grow with fish to elicit maximal toxin production.

3 Expose mammalian pituitary cells to dinoflagellates from treatments a–d and control water from tank e in step 2.

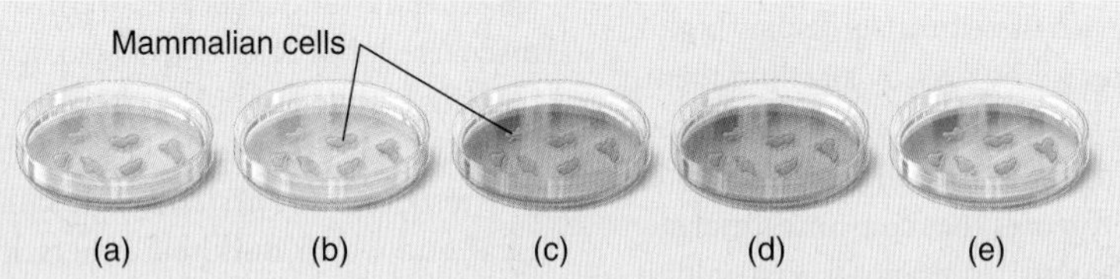

Determine toxicity of dinoflagellates to mammalian cells.

4 THE DATA

Results from step 3:

Numbers of mammalian cells killed per dinoflagellate cell

0, 50, 100, 150, 200, 250, 300

CCMP 2089

CAAE 1024C

Control

(a) Algae-grown (b) Fish-grown (c) Algae-grown (d) Fish-grown (e) No dinoflagellates

5 CONCLUSION Both strains of *P. shumwayae* were toxic to mammalian cells, and both types of food supported the growth of toxic dinoflagellates.

6 SOURCE Burkholder, J.M., et al. 2005. Demonstration of toxicity to fish and to mammalian cells by *Pfiesteria* species: Comparison of assay methods and strains. *Proceedings of the National Academy of Sciences of the United States of America* 102:3471–3476.

In the first step of the experiment, the team provided both *Pfiesteria* strains with two forms of food—cryptomonad algal cells or juvenile fish—because the effect of food type on toxicity was unclear. In a second step, the dinoflagellates grown in step 1 were transferred to tanks with fish to elicit maximal toxin production. Dinoflagellates were not added to a control tank of fish. Toxin was detected, and fish deaths occurred in all of the tanks except the control. In a third step, dinoflagellate samples from the step 2 treatments were added to mammalian cell cultures, and investigators determined the relative levels of toxicity to the mammalian cells. They found that both strains of *P. shumwayae* were toxic to mammalian cells and that both types of food supported the growth of toxin-producing dinoflagellates.

Experimental Questions

1. Why did the investigators test two different strains of *P. shumwayae*?
2. Why did the investigators grow *P. shumwayae* with algae or fish as food?
3. Why did the investigators use a biohazard containment facility?

28.4 Reproductive Adaptations

Learning Outcomes:

1. Briefly describe asexual reproduction and zygotic, sporic, and gametic sexual life cycles in protists.
2. Give examples of how protist life cycles are important to humans.

Diverse reproductive adaptations allow protists to thrive in an amazing variety of environments, including the bodies of hosts in the cases of parasitic protists. These adaptations include specialized asexual reproductive cells, tough-walled dormant cells that allow protists to survive periods of environmental stress, and several types of sexual life cycles.

Protist Populations Increase by Means of Asexual Reproduction

All protists are able to reproduce asexually by mitotic cell divisions of parental cells to produce progeny. When resources are plentiful, repeated mitotic divisions of single-celled protists generate large protist populations. Multicellular protists often generate specialized asexual cells that help disperse the organisms in their environment.

Many protists produce unicellular **cysts** as the result of asexual (and in some cases, sexual) reproduction (**Figure 28.24**). Cysts often have thick, protective walls and can remain dormant through periods of unfavorable climate or low food availability. Dinoflagellates commonly produce cysts that can be transported in the water of a ship's ballast from one port to another, a problem that has caused harmful dinoflagellate blooms to appear in harbors around the world. Ship captains can help to prevent such ecological disasters by heating ballast water before it is discharged from ships.

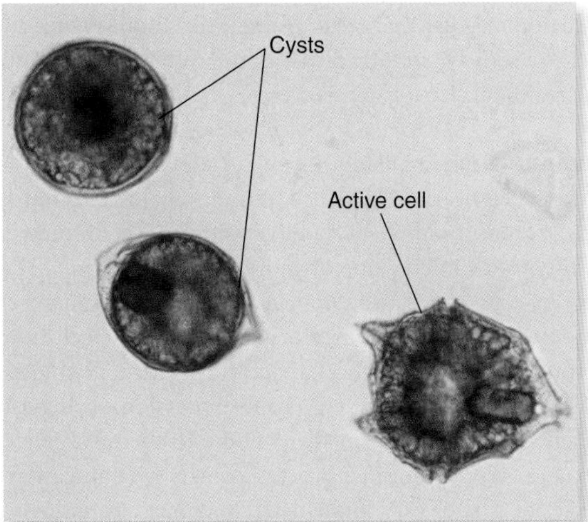

Figure 28.24 Protistan cysts. The round cells are dormant, tough-walled cysts of the dinoflagellate *Peridinium limbatum*. The pointed cell is an actively growing cell of the same species. As cysts develop, the outer cellulose plates present on actively growing cells are cast off.

Concept Check: How can cysts be involved in the spread of harmful algae and disease-causing parasitic protists?

Many disease-causing protists spread from one host to another via cysts. As noted in the chapter opener, the alveolate pathogen *Cryptosporidium parvum* infects humans via waterborne cysts. The amoebozoan *Entamoeba histolytica* infects people who consume food or water that is contaminated with its cysts. Once inside the human digestive system, *E. histolytica* attacks intestinal cells, causing amoebic dysentery.

Sexual Reproduction Provides Multiple Benefits to Protists

Eukaryotic sexual reproduction, featuring gametes, zygotes, and meiosis, first arose among protists. Sexual reproduction has not been observed in some protist phyla but is common in others. Sexual reproduction is generally adaptive because it produces diverse genotypes, thereby increasing the potential for faster evolutionary response to environmental change. Many protists reap additional ecological benefits from sexual reproduction, illustrated by several types of sexual life cycles.

Zygotic Life Cycles Most unicellular protists that reproduce sexually display what is known as a **zygotic life cycle** (**Figure 28.25**). In this type of life cycle, haploid cells develop into gametes. Some protists produce nonmotile eggs and smaller flagellate sperm. However, many other protists have gametes that look similar to each other structurally but have distinctive biochemical features and hence are known as + and – mating types, as shown in Figure 28.25. Gametes fuse (mate) to produce thick-walled diploid zygotes, which give this type of life cycle its name. Such zygotes often have tough cell walls and can survive stressful conditions, much like cysts. When conditions permit, the zygote divides by meiosis to produce haploid cells that increase in number via mitotic cell divisions.

Sporic Life Cycles Many multicellular green and brown seaweeds display a **sporic life cycle**, which is also known as alternation of generations (**Figure 28.26**). Giant kelps and some other protists having sporic life cycles produce two types of multicellular organisms: a haploid gametophyte generation that produces gametes (sperm or eggs) and a diploid sporophyte generation that produces spores by the process of meiosis (Figure 28.26a). This type of life cycle takes its name from the characteristic production of spores as the result of meiosis. Each of the two types of multicellular organisms can adapt to distinct habitats or seasonal conditions, thus allowing protists to occupy more types of environments for longer periods.

Many red seaweeds display a variation of the sporic life cycle that involves alternation of three distinct multicellular generations (Figure 28.26b). This unique type of sexual life cycle has evolved as compensation for the lack of flagella on red algal sperm. Because these sperm are unable to swim to eggs, fertilization occurs only when sperm carried by ocean currents happen to drift close to eggs. As a consequence, fertilization can be rare. Many red algae therefore make millions of spores that are produced by two distinct sporophyte generations. A small sporophyte produces diploid spores, and a larger sporophyte produces haploid spores. Diverse economically valuable red algae possess this type of life cycle, an understanding of which is critical to growing seaweed crops.

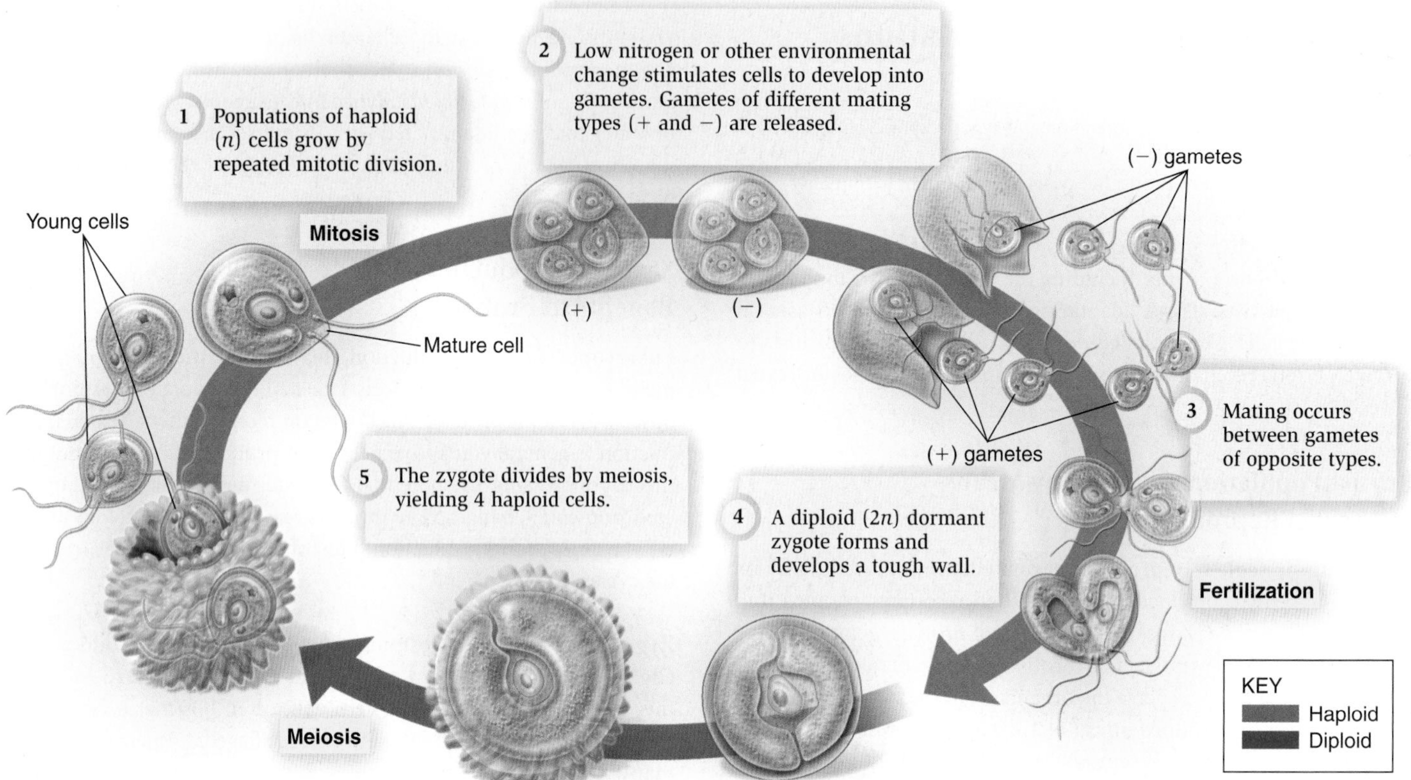

1 Populations of haploid (n) cells grow by repeated mitotic division.
2 Low nitrogen or other environmental change stimulates cells to develop into gametes. Gametes of different mating types (+ and −) are released.
(−) gametes
Young cells
Mitosis
(+)
(−)
Mature cell
3 Mating occurs between gametes of opposite types.
(+) gametes
5 The zygote divides by meiosis, yielding 4 haploid cells.
4 A diploid (2n) dormant zygote forms and develops a tough wall.
Fertilization
KEY
Haploid
Diploid
Meiosis

Figure 28.25 **Zygotic life cycle, illustrated by the unicellular flagellate genus *Chlamydomonas*.** In *Chlamydomonas*, most cells are haploid; only the zygote is diploid.

BioConnections: Look back to Figure 19.1. This figure shows organisms that serve as model systems for the study of plant and animal development. Chlamydomonas is likewise a model system for molecular analysis of flagellar and chloroplast development. What aspects of its life cycle foster such genetic studies?

Gametic Life Cycle Diatoms, which provide food for aquatic animals, are one of the relatively few types of protists known to display a **gametic life cycle** (Figure 28.27), as do animals. In gametic life cycles, all cells except the gametes are diploid, and gametes are produced by meiosis. Sexual reproduction in diatoms not only increases their genetic variability, but it also has another major benefit related to cell size.

In many diatoms, one daughter cell arising from asexual reproduction, which involves mitosis, is smaller than the other, and it is also smaller than the parent cell (Figure 28.27a). This happens because diatom cell walls are composed of two overlapping halves, much like two-part round laboratory dishes having lids that overlap the bottoms. After each mitotic division, each daughter cell receives one-half of the parent cell wall. The daughter cell that inherits a larger, overlapping parental "lid" then produces a new "bottom" that fits inside. This daughter cell will be the same size as its parent. However, the daughter cell that inherits the parental "bottom" uses this wall half as its lid and produces a new, even smaller "bottom." This cell will be smaller than its sibling or parent. Consequently, after many such mitotic divisions, the average cell size of diatom populations often declines over time. If diatom cells become too small, they cannot survive.

Sexual reproduction allows diatom species to attain maximal cell size. Diatom cells mate within a blanket of mucilage, each partner undergoing meiotic divisions to produce gametes. The large, spherical diatom zygotes that result from fertilization (Figure 28.27b) later undergo a series of mitotic divisions to produce new diatom cells having the maximal size for the species.

Ciliate Reproduction Ciliates reproduce asexually by mitosis and forming cysts (**Figure 28.28a**). In addition, ciliates can reproduce sexually by a process known as conjugation. Ciliates are unusual in having two types of nuclei: one or more smaller micronuclei and a single large macronucleus. Macronuclei, which contain many copies of the genome, serve as the source of information for cell function. Both macronuclei and micronuclei divide during asexual mitosis. The diploid micronuclei do not undergo gene expression during growth; instead, their role is to transmit the genome to the next generation during sexual reproduction in a process known as conjugation. Different species of ciliates vary in the detail of conjugation; the process for *Paramecium caudatum* is shown in **Figure 28.28b**.

Parasitic Protists May Use Alternate Hosts for Different Life Stages

Parasitic protists are notable for often using more than one host organism, in which different life stages occur. The malarial parasite genus *Plasmodium* is a prominent example. About 40% of humans

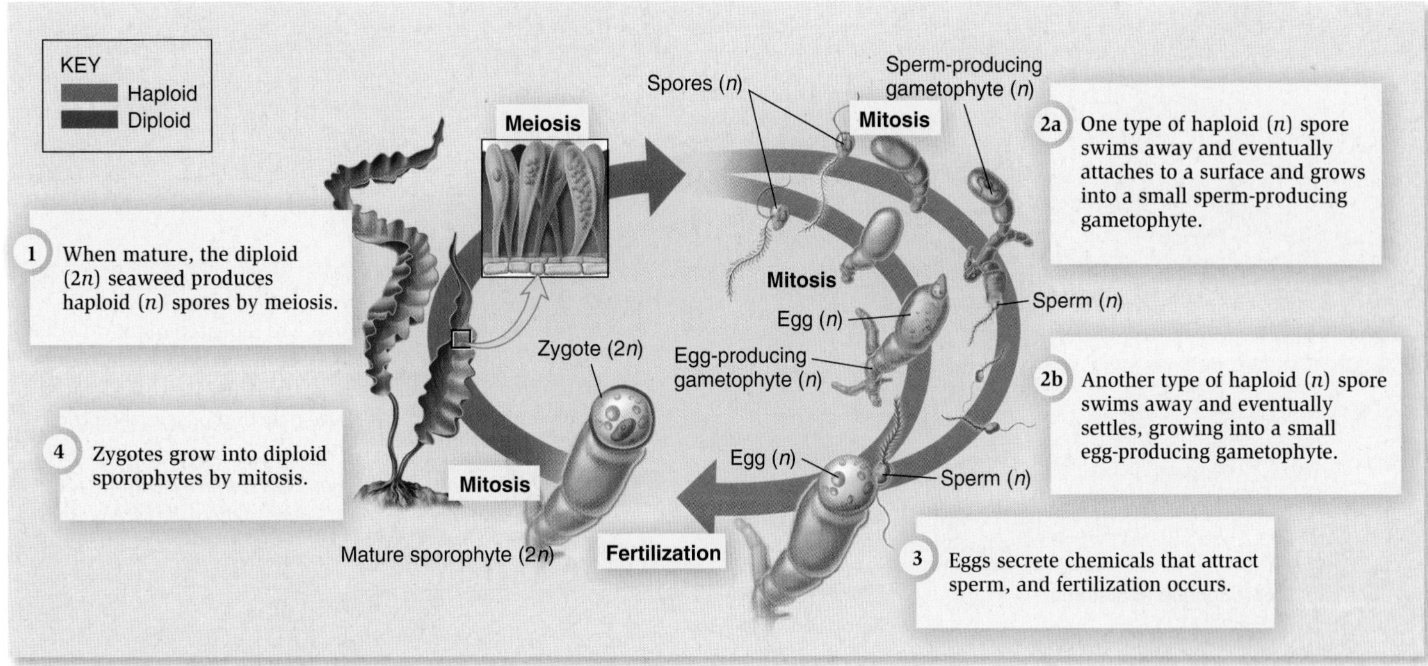

(a) *Laminaria* life cycle—alternation of 2 generations

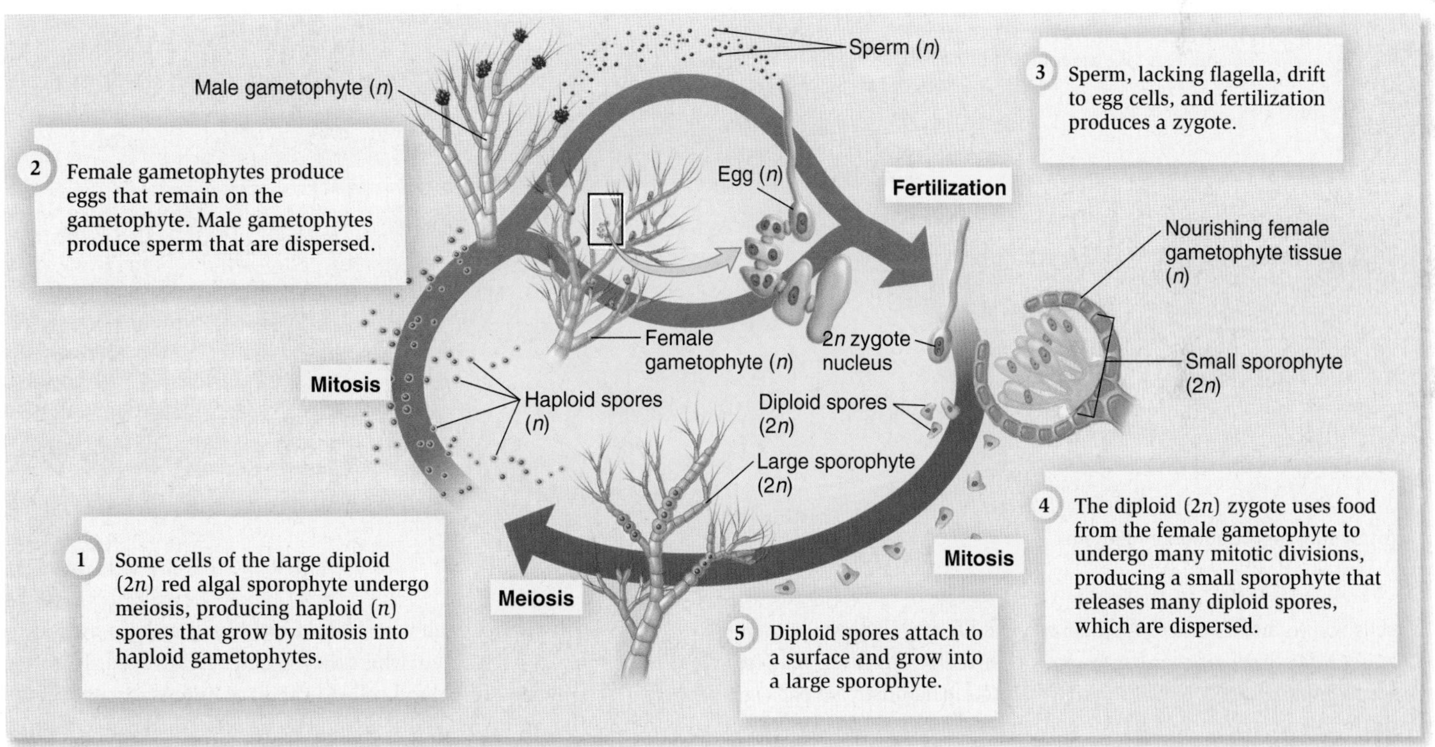

Sperm (n)
2b Another type of haploid (n) spore swims away and eventually settles, growing into a small egg-producing gametophyte.
Male gametophyte (n)
3 Sperm, lacking flagella, drift to egg cells, and fertilization produces a zygote.
2 Female gametophytes produce eggs that remain on the gametophyte. Male gametophytes produce sperm that are dispersed.
Egg (n)
Fertilization
Nourishing female gametophyte tissue (n)
Female gametophyte (n)
2n zygote nucleus
Small sporophyte (2n)
Mitosis
Haploid spores (n)
Diploid spores (2n)
Large sporophyte (2n)
4 The diploid (2n) zygote uses food from the female gametophyte to undergo many mitotic divisions, producing a small sporophyte that releases many diploid spores, which are dispersed.
1 Some cells of the large diploid (2n) red algal sporophyte undergo meiosis, producing haploid (n) spores that grow by mitosis into haploid gametophytes.
Meiosis
Mitosis
5 Diploid spores attach to a surface and grow into a large sporophyte.

(b) *Polysiphonia* life cycle—alternation of 3 generations

Figure 28.26 **Sporic life cycles.** **(a)** Sporic life cycle with two alternating generations, illustrated by the brown seaweed *Laminaria.* **(b)** Sporic life cycle involving three alternating generations, illustrated by the common red seaweed *Polysiphonia.*

live in tropical regions of the world where malaria occurs, and as noted earlier, millions of infections and human deaths result each year. Malaria is particularly deadly for young children. In addition to humans, the malarial parasite's alternate host is the mosquito classified in the genus *Anopheles*, which can also transmit malaria to great apes. Though insecticides can be used to control mosquito populations and though antimalarial drugs exist, malarial parasites can develop drug resistance. Experts are concerned that cases may double in the next 20 years.

When a mosquito bites a human or a great ape, *Plasmodium* enters the bloodstream as an asexual life stage known as a sporozoite (Figure 28.29). Upon reaching a victim's liver, sporozoites enter liver

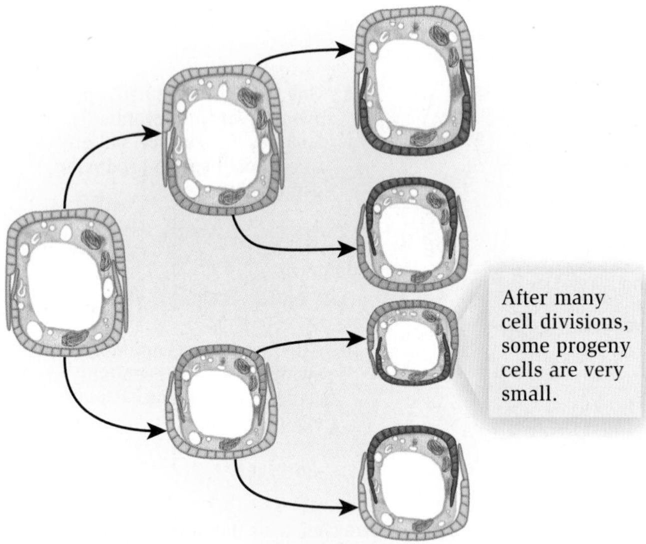

Figure 28.27 Gametic life cycle, as illustrated by diatoms. **(a)** Diatom asexual reproduction involves repeated mitotic division. Because a new bottom cell-wall piece is always synthesized, asexual reproduction may eventually cause the mean cell size to decline in a diatom population. **(b)** Small cell size may trigger sexual reproduction, which regenerates maximal cell size.

BioConnections: Look back to Figure 15.15, which illustrates the life cycles of animals, fungi, and plants. Which of these life cycles is most similar to that of diatoms?

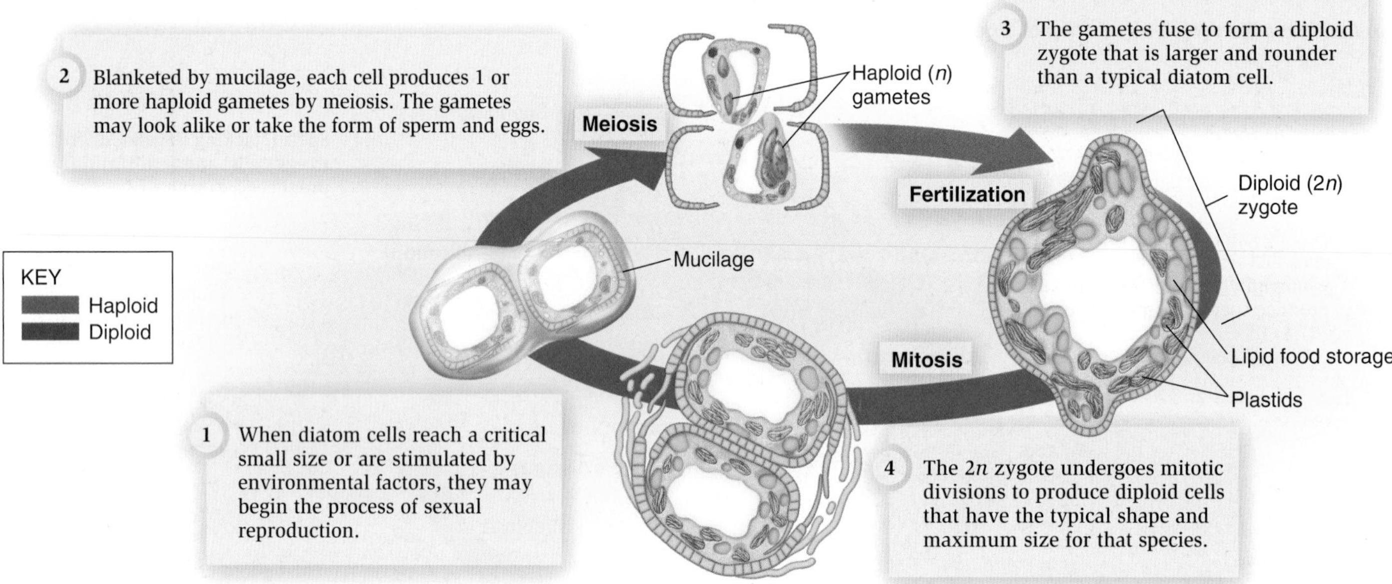

2 Blanketed by mucilage, each cell produces 1 or more haploid gametes by meiosis. The gametes may look alike or take the form of sperm and eggs.
Meiosis
Haploid (n) gametes
3 The gametes fuse to form a diploid zygote that is larger and rounder than a typical diatom cell.
Fertilization
Diploid (2n) zygote
KEY
Haploid
Diploid
Mucilage
Mitosis
Lipid food storage
Plastids
1 When diatom cells reach a critical small size or are stimulated by environmental factors, they may begin the process of sexual reproduction.
4 The 2n zygote undergoes mitotic divisions to produce diploid cells that have the typical shape and maximum size for that species.
(b) Sexual reproduction in diatoms

cells where they divide to form an asexual life stage known as merozoites. Hundreds of merozoites are produced within liver cells (see inset Figure 28.29), which then release into the bloodstream packages of merozoites enclosed by a host-derived cell membrane (see inset Figure 28.29). This membrane protects merozoites from destruction by host immune cells, which would otherwise engulf merozoites by phagocytosis and then destroy the invaders. In the bloodstream, the protective host membranes disintegrate, releasing merozoites. The merozoites have protein complexes at their front ends, or apices, that allow them to invade human red blood cells. (The presence of these apical complexes gives rise to the phylum name Apicomplexa.) Within red blood cells, merozoites release more than 200 proteins, which enable the parasites to commandeer these cells, causing many changes. For example, infected red blood cells form surface knobs that function like molecular Velcro, attaching cells to capillary linings. This process allows infected red blood cells to avoid being transported to the spleen, where they would be destroyed. The attachment of infected red blood cells to capillary linings disrupts circulation in the brain and kidney, a process that can cause death of the animal host.

While living within red blood cells, merozoites form rings, which can be visualized by staining and the use of a microscope, allowing diagnosis. The merozoites consume the hemoglobin in red blood cells, providing resources needed to reproduce asexually. Large numbers of new merozoites synchronously break out of red blood cells at intervals of 48 or 72 hours. These merozoite reproduction cycles correspond to cycles of chills and fever that an infected person experiences. Some merozoites produce sexual structures—gametocytes—which, along with blood, are transmitted to a female mosquito as she bites an infected person.

Macronucleus
Micronucleus
Cyst
Swimming ciliate
Mitotic divison

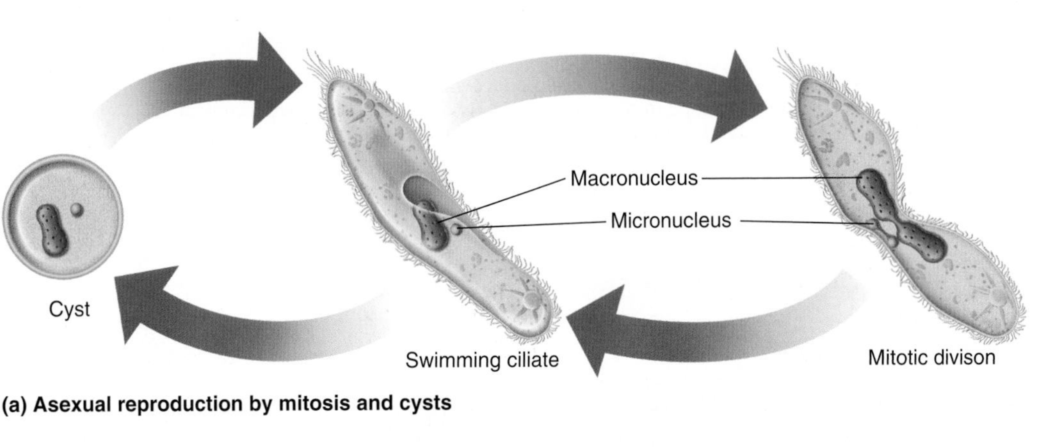

(a) Asexual reproduction by mitosis and cysts

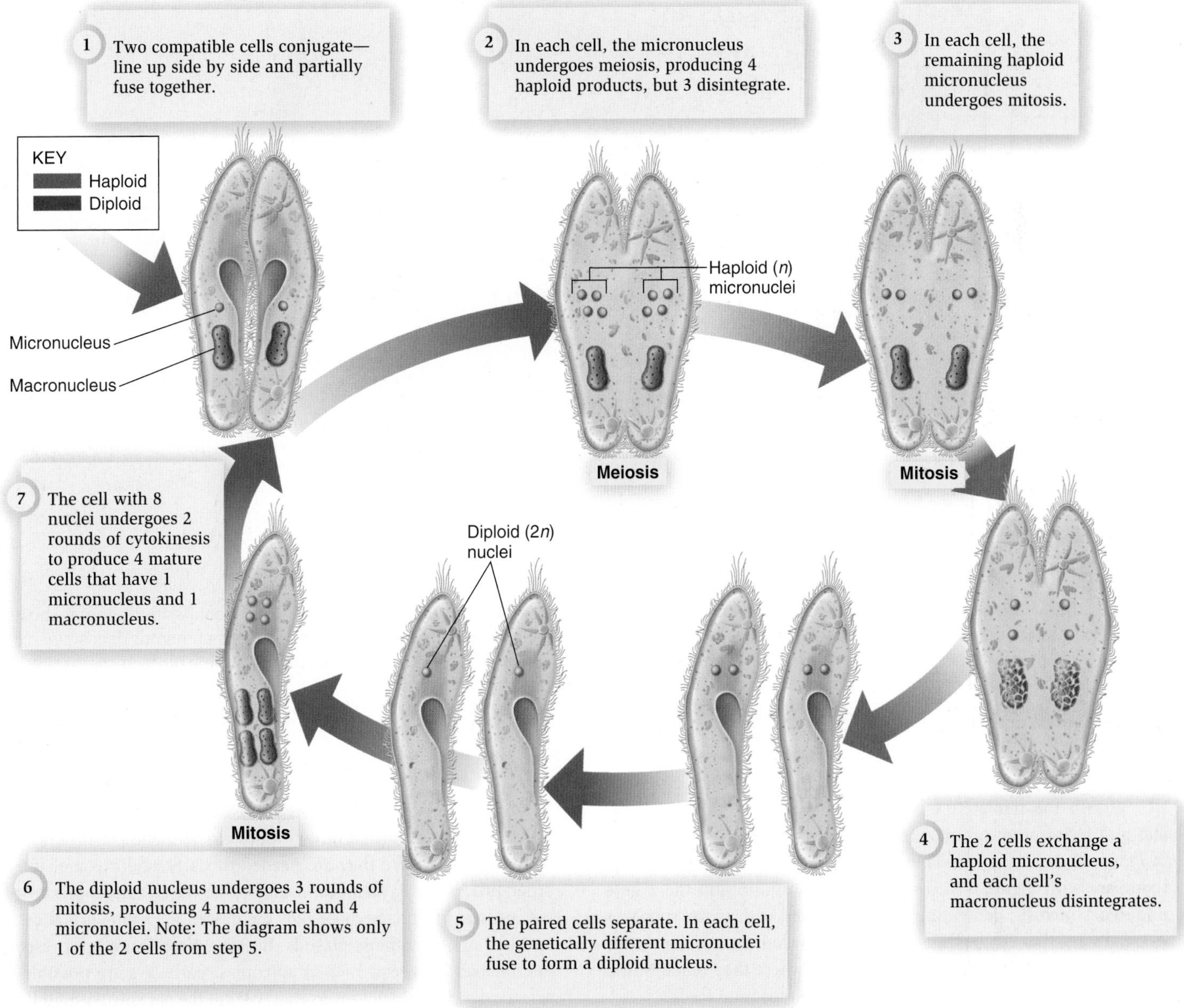

1 Two compatible cells conjugate—line up side by side and partially fuse together.
2 In each cell, the micronucleus undergoes meiosis, producing 4 haploid products, but 3 disintegrate.
3 In each cell, the remaining haploid micronucleus undergoes mitosis.
KEY
Haploid
Diploid
Haploid (n) micronuclei
Micronucleus
Macronucleus
Meiosis
Mitosis
7 The cell with 8 nuclei undergoes 2 rounds of cytokinesis to produce 4 mature cells that have 1 micronucleus and 1 macronucleus.
Diploid (2n) nuclei
Mitosis
4 The 2 cells exchange a haploid micronucleus, and each cell's macronucleus disintegrates.
6 The diploid nucleus undergoes 3 rounds of mitosis, producing 4 macronuclei and 4 micronuclei. Note: The diagram shows only 1 of the 2 cells from step 5.
5 The paired cells separate. In each cell, the genetically different micronuclei fuse to form a diploid nucleus.

(b) Sexual reproduction by conjugation

Figure 28.28 Ciliate reproduction. **(a)** The asexual reproductive process in ciliates. **(b)** The sexual reproductive process, known as conjugation, of the ciliate *Paramecium caudatum*.

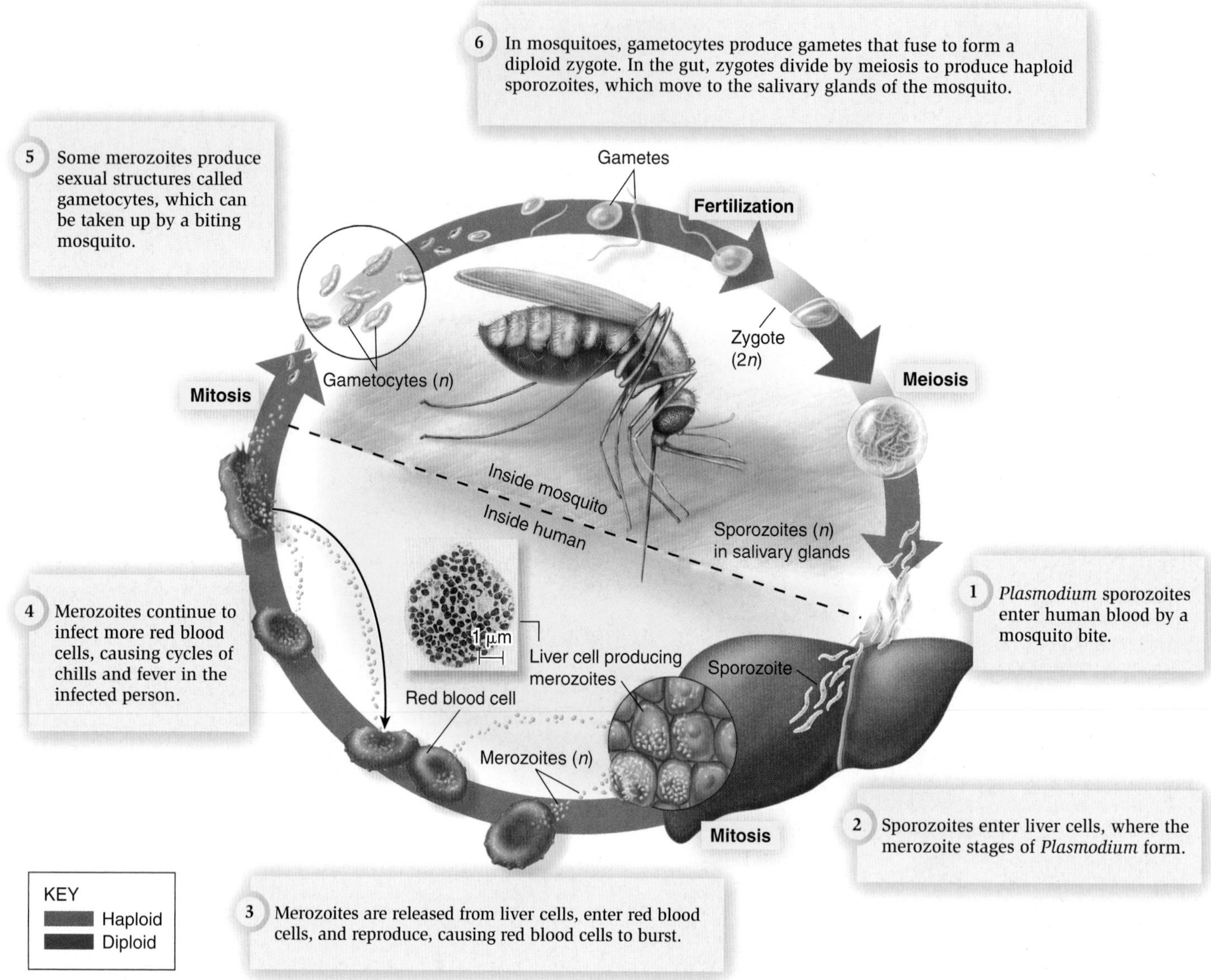

6 In mosquitoes, gametocytes produce gametes that fuse to form a diploid zygote. In the gut, zygotes divide by meiosis to produce haploid sporozoites, which move to the salivary glands of the mosquito.
5 Some merozoites produce sexual structures called gametocytes, which can be taken up by a biting mosquito.
Gametes
Fertilization
Zygote (2n)
Meiosis
Gametocytes (n)
Mitosis
Inside mosquito
Inside human
Sporozoites (n) in salivary glands
1 Plasmodium sporozoites enter human blood by a mosquito bite.
4 Merozoites continue to infect more red blood cells, causing cycles of chills and fever in the infected person.
1 µm
Liver cell producing merozoites
Sporozoite
Red blood cell
Merozoites (n)
Mitosis
2 Sporozoites enter liver cells, where the merozoite stages of Plasmodium form.
KEY
Haploid
Diploid
3 Merozoites are released from liver cells, enter red blood cells, and reproduce, causing red blood cells to burst.

Figure 28.29 **Diagram of the life cycle of *Plasmodium falciparum*, a species that causes malaria.** This life cycle requires two alternate hosts, humans (or great apes) and *Anopheles* mosquitoes. The inset is a TEM view of an infected human liver cell that contains numerous dark-stained merozoites. Such infected cells bud off groups of merozoites enclosed within a host-produced membrane. This membrane protects merozoites from being engulfed and destroyed by phagocytotic cells of the body's immune system.

Concept Check: *In which of the hosts does sexual mating of P. falciparum gametes occur?*

Within the mosquito's body, the gametocytes produce gametes and fertilization occurs, yielding a zygote, the only diploid cell in *Plasmodium*'s life cycle. Thus, *Plasmodium* has a zygotic life cycle (compare with Figure 28.25). Within the mosquito gut, the zygote undergoes meiosis, generating structures filled with many sporozoites, the stage that can be transmitted to a new human host. Sporozoites move to the mosquito's salivary glands, where they remain until they are injected into a human host when the mosquito feeds.

In recent years, genomic information that has added to our knowledge of these life stages is helping medical scientists to develop new ways to prevent or treat malaria. In the case of *P. falciparum*, genomic data have already highlighted potential new pharmaceutical approaches. About 550 (some 10%) of the nuclear-encoded proteins are likely imported into a nonphotosynthetic plastid known as an apicoplast, where they are needed for fatty-acid metabolism and other processes. *P. falciparum* and some other apicomplexan protists possess plastids because they are descended from algal ancestors that had photosynthetic plastids. Because plastids are not present in mammalian cells, enzymes in apicoplast pathways are possible targets for development of drugs that will kill the parasite without harming the host. Mammals also lack calcium-dependent protein kinases (CDPKs), enzymes that are essential to merozoite release from red blood cells and the parasite's sexual development, offering another potential drug target.

Summary of Key Concepts

28.1 An Introduction to Protists

- Protists are eukaryotes that are not classified in the plant, animal, or fungal kingdoms; are abundant in moist habitats; and are mostly microscopic in size.
- Protists are often informally labeled according to their ecological roles: Algae are mostly photosynthetic protists; protozoa are heterotrophic protists that are often mobile; and fungus-like protists resemble true fungi in some ways (Figures 28.1, 28.2).
- Protists are particularly diverse in aquatic habitats, occurring as small floating or swimming phytoplankton, attached members of the periphyton, and more complex macroalgae (seaweeds) (Figure 28.3).
- Microscopic protists propel themselves by means of flagella (flagellates), cilia (ciliates), pseudopodia (amoebae), or by gliding across surfaces (Figures 28.4 28.5).

28.2 Evolution and Relationships

- Modern phylogenetic analysis has revealed that protists do not form a monophyletic group; instead, many can be classified into one of seven major eukaryotic supergroups (Figure 28.6).
- The supergroup Excavata includes flagellate protists characterized by a feeding groove, including the kinetoplastids and euglenoids, some of which are photosynthetic (Figures 28.7, 28.8, 28.9).
- Land plants are related to green algae and red algae (having primary plastids) and probably cryptomonads (featuring secondary plastids). Haptophytes also display secondary plastids (Figures 28.10, 28.11, 28.12, 28.13, 28.14).
- The supergroup Alveolata includes the ciliates, apicomplexans, and dinoflagellates, whose cells feature saclike membrane vesicles called alveoli. Many dinoflagellates display secondary plastids, and some feature tertiary plastids (Figures 28.15, 28.16).
- The supergroup Stramenopila includes protists whose flagella have strawlike hairs that aid in swimming. Stramenopiles include diatoms, giant kelps, and other groups of algae, as well as some fungus-like protists (Figures 28.17, 28.18).
- The supergroup Rhizaria consists of flagellates and amoebae with thin hairlike extensions of cytoplasm called filose pseudopodia. Three prominent phyla are Chlorarachniophyta, with secondary green plastids; mineral-shelled Radiolaria; and Foraminifera, with calcium carbonate shells (Figure 28.19).
- The supergroup Amoebozoa is composed of many types of amoebae and includes slime molds such as *Dictyostelium discoideum*.
- The supergroup Opisthokonta includes organisms that produce swimming cells having a single posterior flagellum. It includes the fungal and animal kingdoms and choanoflagellate protists, which are related to the ancestor of animals (Figures 28.20, 28.21, Table 28.1).

28.3 Nutritional and Defensive Adaptations

- Protists display four basic types of nutrition: phagotrophs feed by ingesting particles; osmotrophs absorb small organic molecules; photoautotrophs make their own organic food by using light energy; and mixotrophs use both photoautotrophy and heterotrophy to obtain nutrients (Figure 28.22).
- Protists possess defensive adaptations such as protective cell coverings, sharp projectiles, light flashes, and toxic compounds. Dinoflagellates are particularly important toxin producers, and aquatic ecologists found that populations of the unicellular dinoflagellate genus *Pfiesteria* kill fishes by using a toxin that can also harm human cells (Figure 28.23).

28.4 Reproductive Adaptations

- Protist populations grow by means of asexual reproduction involving mitosis, and many persist through unfavorable conditions by producing tough-walled cysts (Figure 28.24).
- Sexual reproduction arose among protists. In the zygotic life cycle, haploid cells develop into gametes, which fuse to produce diploid zygotes. These zygotes often have tough cell walls that enable them to survive unfavorable conditions (Figure 28.25).
- In protists displaying a sporic life cycle (also called alternation of generations), a haploid generation produces gametes and a diploid generation produces spores. Each type can adapt to different environments or conditions, allowing protists to occupy multiple habitats (Figure 28.26).
- In the gametic life cycle, all cells but gametes are diploid. Sexual reproduction in diatoms, which have a gametic life cycle, increases genetic variability and allows species to attain maximal cell size (Figure 28.27).
- Ciliate protists display asexual reproduction and sexual reproduction by conjugation (Figure 28.28).
- Parasitic protists may have life cycles involving alternate hosts. One example is *Plasmodium*, the agent of malaria, whose alternate hosts are humans (or great apes) and mosquitoes (Figure 28.29).

Assess and Discuss

Test Yourself

1. If you were studying the evolution of animal-specific cell-to-cell signaling systems, from which of the following would you choose representative species to observe?
   a. Rhodophyta
   b. Excavata
   c. Choanomonada
   d. Radiolaria
   e. Chlorophyta
2. If you were studying the origin of land plant traits, which of the following groups would you study?
   a. green algae
   b. radiolarians
   c. choanoflagellates
   d. diatoms
   e. ciliates
3. Which informal ecological group of protists includes photoautotrophs?
   a. protozoa
   b. algae
   c. fungus-like protists
   d. ciliates
   e. all of the above
4. How would you recognize a primary plastid? It would:
   a. have one envelope membrane.
   b. have two envelope membranes.
   c. have more than two envelope membranes.
   d. lack pigments.
   e. be golden brown in color.

5. What organisms have tertiary plastids?
   a. certain stramenopiles
   b. certain euglenoids
   c. certain cryptomonads
   d. certain opisthokonts
   e. certain dinoflagellates
6. What is surprising about mixotrophs?
   a. They have no plastids, but they occur mixed in communities with autotrophs.
   b. They have mixed heterotrophic and autotrophic nutrition.
   c. Their cells contain a mixture of red and green plastids.
   d. Their cells contain a mixture of haploid and diploid nuclei.
   e. They consume a mixed diet of algae.
7. What advantages do diatoms obtain from sexual reproduction?
   a. increased genetic variability
   b. increased ability of populations to respond to environmental change
   c. evolutionary potential
   d. regeneration of maximal cell size for the species
   e. all of the above
8. What are extrusomes?
   a. hairs on flagella
   b. membrane sacs beneath the cell surface
   c. tough-walled asexual cells
   d. spearlike defensive structures shot from cells under attack
   e. special types of survival cysts
9. How do pigments such as phycoerythrin in red algae and fucoxanthin in brown algae benefit these autotrophic protists?
   a. The pigments provide camouflage, so herbivores cannot see algae.
   b. The pigments absorb blue-green underwater light and transfer the energy to chlorophyll *a* for use in photosynthesis.
   c. The pigments attract aquatic animals that carry gametes between seaweeds.
   d. The pigments absorb ultraviolet (UV) light that would harm the photosynthetic apparatus.
   e. All of the above are correct.
10. What are the alternate hosts of the malarial parasite *Plasmodium falciparum*?
    a. humans (or great apes) and ticks
    b. ticks and mosquitoes
    c. humans (or great apes) and *Anopheles* mosquitoes
    d. humans (or great apes) and all types of mosquitoes
    e. sporophytes and gametophytes

Conceptual Questions

1. Explain why protists are classified into multiple supergroups, rather than a single kingdom or phylum.
2. Why have molecular biologists sequenced the genomes of several parasitic protists?
3. A principle of biology is that *biology affects our society*. Why are the cysts of protists important to epidemiologists, the biologists who study the spread of disease?

Collaborative Questions

1. Imagine you are studying an insect species and you discover that the insects are dying of a disease that results in the production of cysts of the type that protists often generate. Thinking that the cysts might have been produced by a parasitic protist that could be used as an insect control agent, how would you go about identifying the disease agent?
2. Imagine you are part of a marine biology team seeking to catalogue the organisms inhabiting a threatened coral reef. The team has found two new types of macroalgae (seaweeds), each of which occurs during a particular time of the year when the water temperature differs. You suspect that the two macroalgae might be different generations of the same species that have differing optimal temperature conditions. How would you go about testing your hypothesis?

Online Resource

www.brookerbiology.com

Stay a step ahead in your studies with animations that bring concepts to life and practice tests to assess your understanding. Your instructor may also recommend the interactive eBook, individualized learning tools, and more.

Chapter Outline

29.1 Ancestry and Diversity of Modern Plants
29.2 An Evolutionary History of Land Plants
29.3 The Origin and Evolutionary Importance of the Plant Embryo
29.4 The Origin and Evolutionary Importance of Leaves and Seeds
Summary of Key Concepts
Assess and Discuss

Plants and the Conquest of Land

29

A temperate rain forest containing diverse plant phyla in Olympic National Park in Washington State.

When thinking about plants, people envision lush green lawns, shady street trees, garden flowers, or leafy fields of valuable crops. On a broader scale, they might imagine lush rain forests (see chapter-opening photo), vast grassy plains, or tough desert vegetation. Shopping in the produce section of the local grocery store may remind us that plant photosynthesis is the basic source of our food. Just breathing crisp fresh air might bring to mind the role of plants as oxygen producers—the ultimate air fresheners. Do you start your day with a "wake-up" cup of coffee, tea, or hot chocolate? Then you may appreciate the plants that produce these and many other materials we use in daily life: medicines, cotton, linen, wood, bamboo, cork, and paper.

In addition to their importance to humans and modern ecosystems, plants have played dramatic roles in the Earth's past. Throughout their evolutionary history, diverse plants have influenced Earth's atmospheric chemistry, climate, and soils. Plants have also affected the evolution of many other groups of organisms, including humans. In this chapter, we will survey the diversity of modern plant phyla and their distinctive features. This chapter also explains how early plants adapted to land and how plants have continued to adapt to changing terrestrial environments. During this process, we will gain insight into descent with modification and the biological principle that all life is related by an evolutionary history.

29.1 Ancestry and Diversity of Modern Plants

Learning Outcomes:

1. List key derived features that land plants share with their closest algal relatives.
2. Name several characteristics unique to land plants.
3. Compare and contrast the features of vascular and nonvascular plants.
4. List adaptations that enable vascular plants to maintain stable water content.

Several hundred thousand modern species are formally classified into the kingdom Plantae, informally known as the plants or land plants (Figure 29.1). **Plants** are multicellular eukaryotic organisms composed of cells having plastids, and plants primarily live on land. Molecular and other evidence indicates that the plant kingdom evolved from green algal ancestors that primarily lived in aquatic habitats such as lakes or ponds. Together, plants and modern green algal relatives are known as **streptophytes**. Plants are distinguished from algal relatives by the presence of traits that foster survival in terrestrial conditions, which are drier, sunnier, hotter, colder, and less physically supportive than aquatic habitats. In this section, we will examine the modern algae that are most closely related to plants and survey the diverse phyla of living land plants. This process reveals how plants gradually acquired diverse structural, biochemical, and reproductive adaptations that fostered survival on land.

Modern Green Algae Are Closely Related to the Ancestors of Land Plants

Molecular, biochemical, and structural data indicate that the kingdom Plantae originated from a photosynthetic protist ancestor that, if present today, would be classified among the **streptophyte algae** (Figure 29.2). All streptophyte algae have features in common with land plants, but the later-diverging streptophyte algae display several

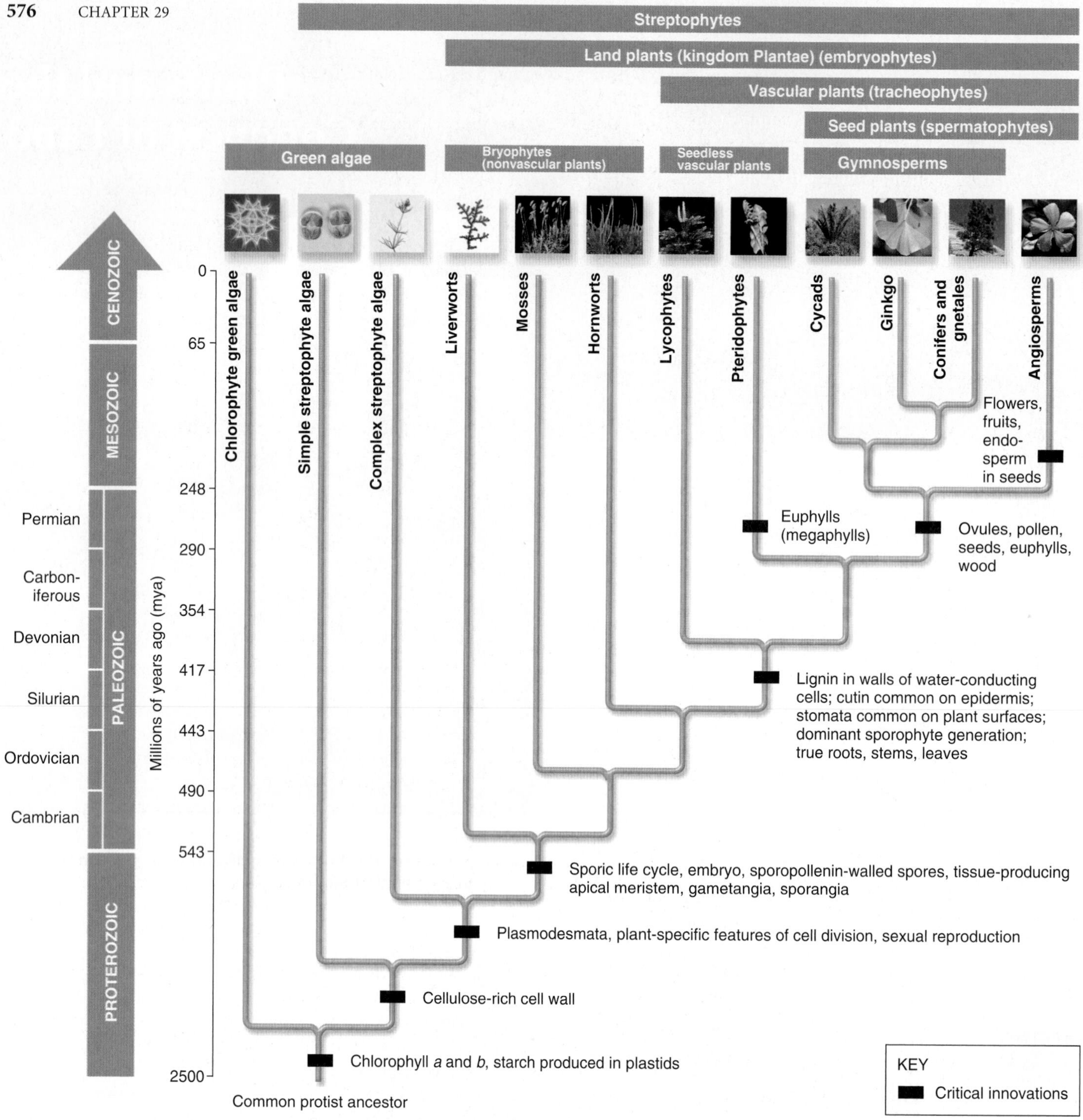

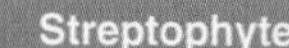

Figure 29.1 **Evolutionary relationships of the modern plant phyla.** Land plants gradually acquired diverse structural, biochemical, and reproductive adaptations, allowing them to better survive in terrestrial habitats. For simplicity's sake, fewer branches of streptophyte algae are shown here than actually exist.

BIOLOGY PRINCIPLE **All species (past and present) are related by an evolutionary history.** This diagram shows maximal evolutionary divergence times indicated by molecular clock and some fossil evidence, suggesting when clades may first have arisen. Other fossil evidence provides minimal divergence times that indicate when clades had become well established; in this case, more numerous and more easily identified fossils were formed.

critical innovations—derived features shared with land plants that fostered plant success on land. Examples of these shared features are a distinctive type of cytokinesis, intercellular connections known as plasmodesmata (see Chapter 10), and sexual reproduction (see Figure 29.1). For this reason, streptophyte algae are good sources of information about the ancestors of land plants.

Streptophyte algae do not form a monophyletic group, but together with land plants, they form a clade—the streptophytes. All

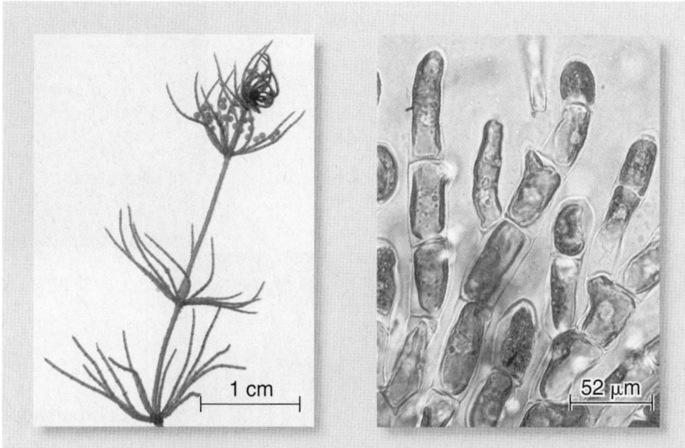

(a) Complex streptophyte algae: *Chara zeylanica* (left) and *Coleochaete pulvinata* (right)

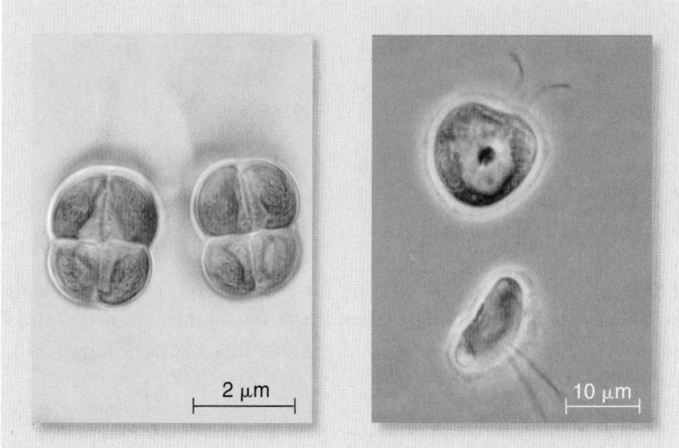

(b) Simple streptophyte algae: *Chlorokybus atmophyticus* (left) and *Mesostigma viridae* (right)

Figure 29.2 **Streptophyte green algal relatives of the land plants.** Streptophyte algae inhabit freshwater lakes and ponds, and they display structural, reproductive, biochemical, and molecular features in common with land plants.

other green algae—the chlorophytes—are related to streptophytes (see Figure 29.1). Streptophytes and chlorophytes share distinctive characters such as green, starch-bearing chloroplasts that contain the accessory pigments chlorophyll *b* and *β*-carotene. Even so, the land plants display several common features that distinguish them from green algae, even their closest algal relatives.

Distinctive Features of the Land Plants

The features that distinguish land plants represent early adaptations to the land habitat. For example, the bodies of all land plants are primarily composed of three-dimensional tissues, defined as close associations of cells of the same type. Tissues provide land plants with an increased ability to avoid water loss at their surfaces. That's because bodies composed of tissues have lower surface area-to-volume ratios than do branched filaments. Land plant tissues arise from one or more actively dividing cells that occur at growing tips. Such localized regions of cell division are known as **apical meristems**. The tissue-producing apical meristems of land plants produce relatively thick, robust bodies able to withstand drought and mechanical stress and produce tissues and organs with specialized functions.

The land plants also have distinctive reproductive features. These include a sporic life cycle involving alternation between two types of multicellular bodies; embryos that depend on maternal tissues during early development; tough-walled reproductive cells known as **spores** that allow dispersal through dry air; and specialized structures that generate, protect, and disperse gametes and spores. The following survey of modern plant phyla and their distinctive features illustrates how these and other fundamental plant features evolved.

Modern Land Plants Can Be Classified into Nine Phyla

Plant systematists use molecular and structural information from living and fossil plants to classify plants into phyla. In this textbook, nine phyla of living land plants are described: (1) the plants informally known as **liverworts** (formally called Hepatophyta), (2) **mosses** (Bryophyta), (3) **hornworts** (Anthocerophyta), (4) **lycophytes** (Lycopodiophyta), (5) **pteridophytes** (Pteridophyta), (6) **cycads** (Cycadophyta), (7) **ginkgos** (Ginkgophyta), (8) **conifers** (Coniferophyta), and (9) the **flowering plants**, also known as **angiosperms** (Anthophyta) (see Figure 29.1). Fossils reveal that additional plant phyla once lived but are now extinct.

Phylogenetic information suggests that the modern plant phyla arose in a particular sequence (see Figure 29.1). Liverworts diverged first, mosses diverged next, and hornworts seem to be closely related to the vascular plants. The vascular plants are distinguished by internal water and nutrient-conducting tissues that also provide structural support. Among modern vascular plants lycophytes diverged earliest, pteridophytes arose next, and then seed plants. Our survey of these modern plant phyla reveals how plants acquired more adaptations to life on land.

Liverworts, Mosses, and Hornworts Are the Simplest Land Plants

Liverworts, mosses, and hornworts are Earth's simplest land plants (Figures 29.3, 29.4, and 29.5), and each forms a distinct, monophyletic group. There are about 6,500 species of modern liverworts, 12,000 or more species of mosses, and about 100 species of hornworts. Collectively, liverworts, mosses, and hornworts are known informally as the **bryophytes** (from the Greek *bryon*, meaning moss, and *phyton*, meaning plant). The bryophytes do not form a clade, but the term bryophyte is useful for expressing common structural, reproductive, and ecological features of liverworts, mosses, and hornworts. For example, the bryophytes are all relatively small in stature and are most common and diverse in moist habitats because they lack traits allowing them to grow tall or reproduce in dry places.

Because bryophytes diverged early in the evolutionary history of land plants (see Figure 29.1), they serve as models of the earliest terrestrial plants. Bryophytes display apical meristems that produce specialized tissues and other features that evolved early in the history of land plants, such as the sporic life cycle. A comparison of the life

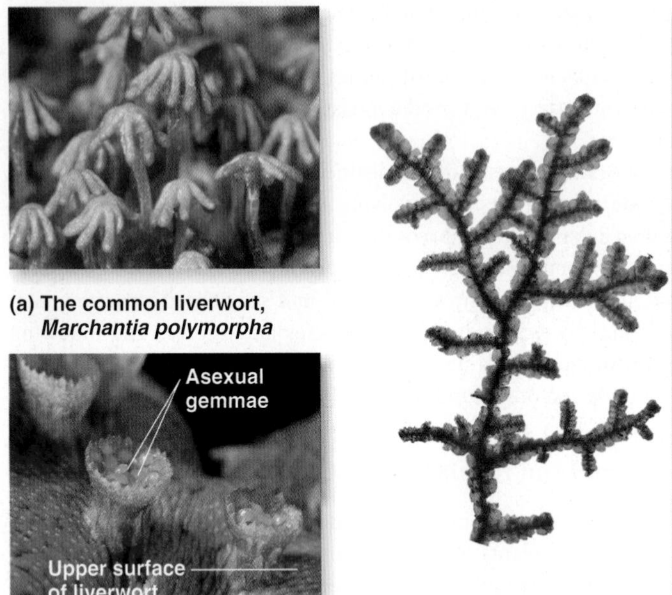

(a) The common liverwort, *Marchantia polymorpha*

(b) Close-up of liverwort structures

(c) A species of leafy liverwort

Figure 29.3 Liverworts. **(a)** The common liverwort, *Marchantia polymorpha*, with raised, umbrella-shaped structures that bear sexually produced sporophytes on the undersides. Mature sporophytes generate spores, then release them. **(b)** A close-up of *M. polymorpha* showing surface cups that contain multicellular, frisbee-shaped asexual structures known as gemmae that are dispersed by wind and grow into new liverworts. **(c)** A species of liverwort with leaflike structures and so known as a leafy liverwort.

Concept Check: Why do you think liverworts produce their reproductive spores on raised structures?

cycle of aquatic algae with that of bryophytes reveals the bryophyte life cycle's adaptive value on land.

The streptophyte algae display a **zygotic life cycle** in which the diploid generation consists of only one cell, the zygote (**Figure 29.6a**). Zygotic life cycles take their name from the observation that the zygote is the only cell that undergoes meiosis (see Figure 28.25). By contrast, sexual reproduction in bryophytes and all other plants follows what is called a **sporic life cycle**, in which generations alternate (see Figures 15.15c and 28.26a). In the sporic life cycle, meiosis results in the formation of spores, reproductive cells that allow organisms to disperse in the environment (**Figure 29.6b**). **Alternation of generations** means that land plants produce two types of multicellular bodies that alternate in time. These two types of bodies are known as the diploid ($2n$), spore-producing **sporophyte** generation and the haploid (n), gamete-producing **gametophyte** generation. Biologists think the plant life cycle originated by a delay in zygotic meiosis, with the result that the diploid generation became multicellular before undergoing meiosis. For the earliest land plants, having a multicellular diploid generation provided several advantages in coping with terrestrial conditions. What are these advantages? A closer look at the process of sexual reproduction in bryophytes reveals the answer and also highlights ways in which bryophytes differ from other plants.

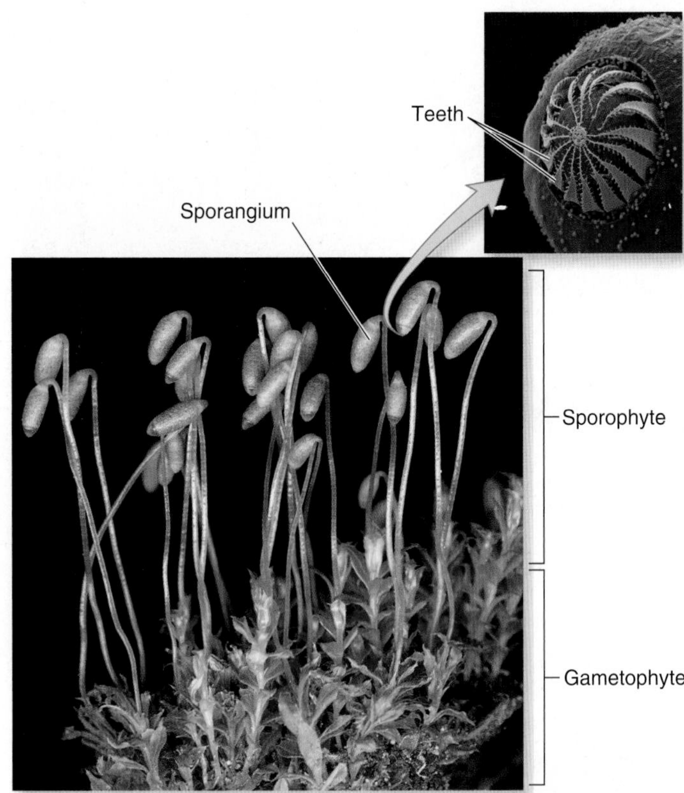

Figure 29.4 Mosses. The common moss genus *Mnium* has a leafy green gametophyte (multicellular body that generates gametes) and an unbranched, dependent sporophyte that bears a spore-producing sporangium at its tip. Inset: This SEM shows that the tips of moss sporangia often have teeth separated by spaces, so that spores are sprinkled into the wind and dispersed over time, rather than being released all at once.

Concept Check: Why might it be advantageous for a moss sporophyte to release spores gradually?

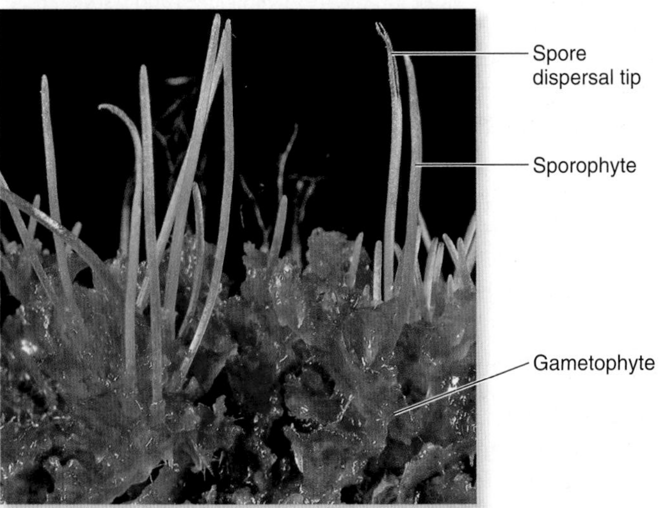

Figure 29.5 Hornworts. Sporophytes of hornworts generally grow up into the air, whereas the gametophytes grow close to the ground. Hornwort sporophytes become mature and open at the top, dispersing spores.

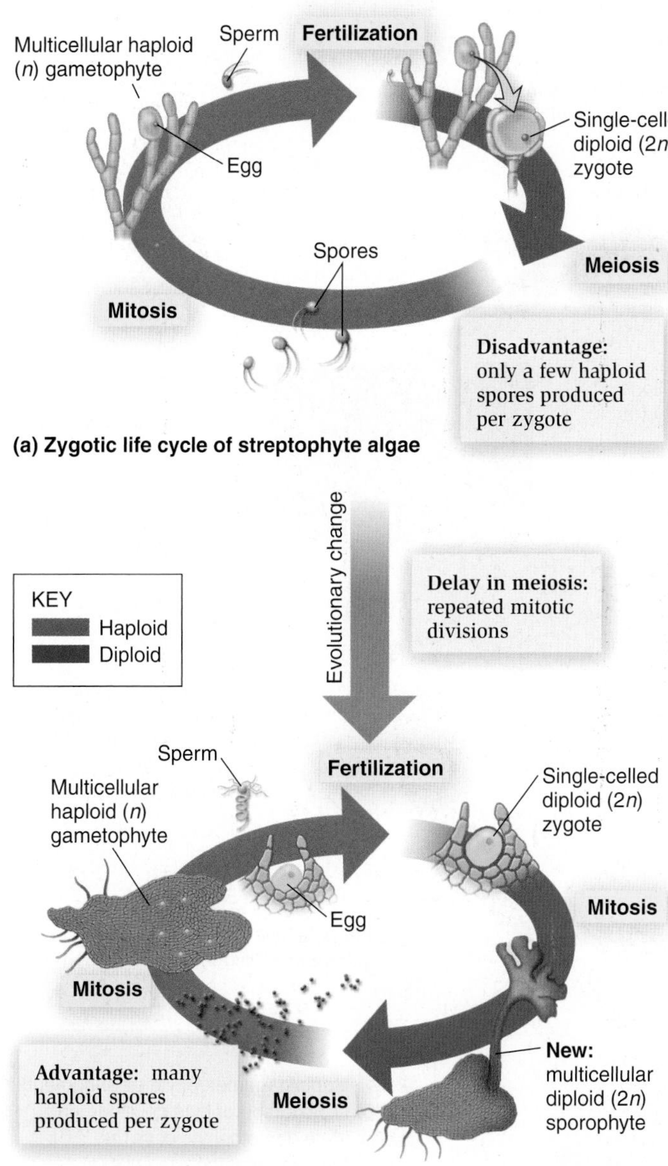

Figure 29.6 **A comparison of the life cycle of primarily aquatic streptophyte algae (part a) with the derived life cycle of primarily terrestrial bryophytes (part b).**

Bryophyte Reproduction Illustrates Early Plant Adaptations to Life on Land

As noted, bryophytes and other land plants display an alternation of generations (**Figure 29.7**). On land, a multicellular diploid sporophyte generation is advantageous because it allows a single plant to disperse widely by using meiosis to produce numerous, genetically variable haploid spores. Each spore has the potential to grow into a gametophyte. In bryophytes, the gametophyte generation is generally green and photosynthetic and thus has a major function of organic food production. The more spores that sporophytes produce, the greater the numbers of gametophytes, helping bryophytes to spread in their environments, thereby increasing in fitness. We next discuss how gametophytes and sporophytes function together during the life cycle of a bryophyte, starting with gamete production.

Gametophytes From an evolutionary and reproductive viewpoint, the role of plant gametophytes is to produce haploid gametes. Because the gametophyte cells are already haploid, meiosis is not involved in producing plant gametes. Instead, plant gametes are produced by mitosis; therefore, all gametes produced from a single gametophyte are genetically identical.

The gametophytes of bryophytes and many other land plants produce gametes in specialized structures known as **gametangia** (from the Greek, meaning gamete containers). Certain cells of gametangia develop into gametes, and other cells form an outer protective jacket of tissue. The gametangial jacket protects delicate gametes from drying out and from microbial attack while they develop. Flask-shaped gametangia that each enclose a single egg cell are known as **archegonia** (singular, archegonium); spherical or elongate gametangia that each produce many sperm are known as **antheridia** (singular, antheridium) (see Figure 29.7).

When the plant sperm are mature, if moist conditions exist, the sperm are released from antheridia into films of water. Under the influence of sex-attractant molecules secreted from archegonia, the sperm swim toward the eggs, twisting their way down the tubular neck of the archegonium. The sperm then fuse with egg cells in the process of fertilization to form diploid zygotes, which grow into embryos. New sporophytes develop from embryos. Fertilization cannot occur in bryophytes unless water is present because the sperm are flagellate and need water to reach eggs. Conditions of uncertain moisture, common in the land habitat, can thus limit plant reproductive success. As we explain next, plant embryos and sporophytes are adaptive responses to this environmental challenge.

Sporophytes One reproductive advantage of the plant life cycle is that zygotes remain enclosed within gametophyte tissues, where they are sheltered and fed (a process described in more detail in Section 29.3). This critical innovation, known as **matrotrophy** (from the Latin, meaning mother, and the Greek, meaning food) gives zygotes a good start while they grow into embryos. Because all groups of land plants possess matrotrophic embryos, they are known as **embryophytes** (see Figure 29.1). Sheltering and feeding embryos is particularly important when embryo production is limited by water availability, as is often the case for bryophytes.

Another reproductive advantage to plants of the sporic life cycle is that, when mature, specialized cells within multicellular sporophytes undergo meiosis to produce many genetically diverse spores. Meiosis occurs within enclosures known as **sporangia** (from the Greek, meaning spore containers), whose tough cell walls protect developing spores from harmful UV radiation and microbial attack. Bryophyte sporangia open in specialized ways that foster dispersal of mature spores into the air, allowing spore transport by wind (see Figures 29.4 and 29.7). Dispersed plant spores have cell walls containing a tough material, known as **sporopollenin** that helps to prevent cellular damage during transport in air. If spores reach habitats favorable for growth, their walls crack open, and new gametophytes develop by mitotic divisions, completing the life cycle.

Spore production is a measure of plant fitness, because plants can better disperse progeny throughout the environment when they produce more spores. The larger the diploid generation, the more

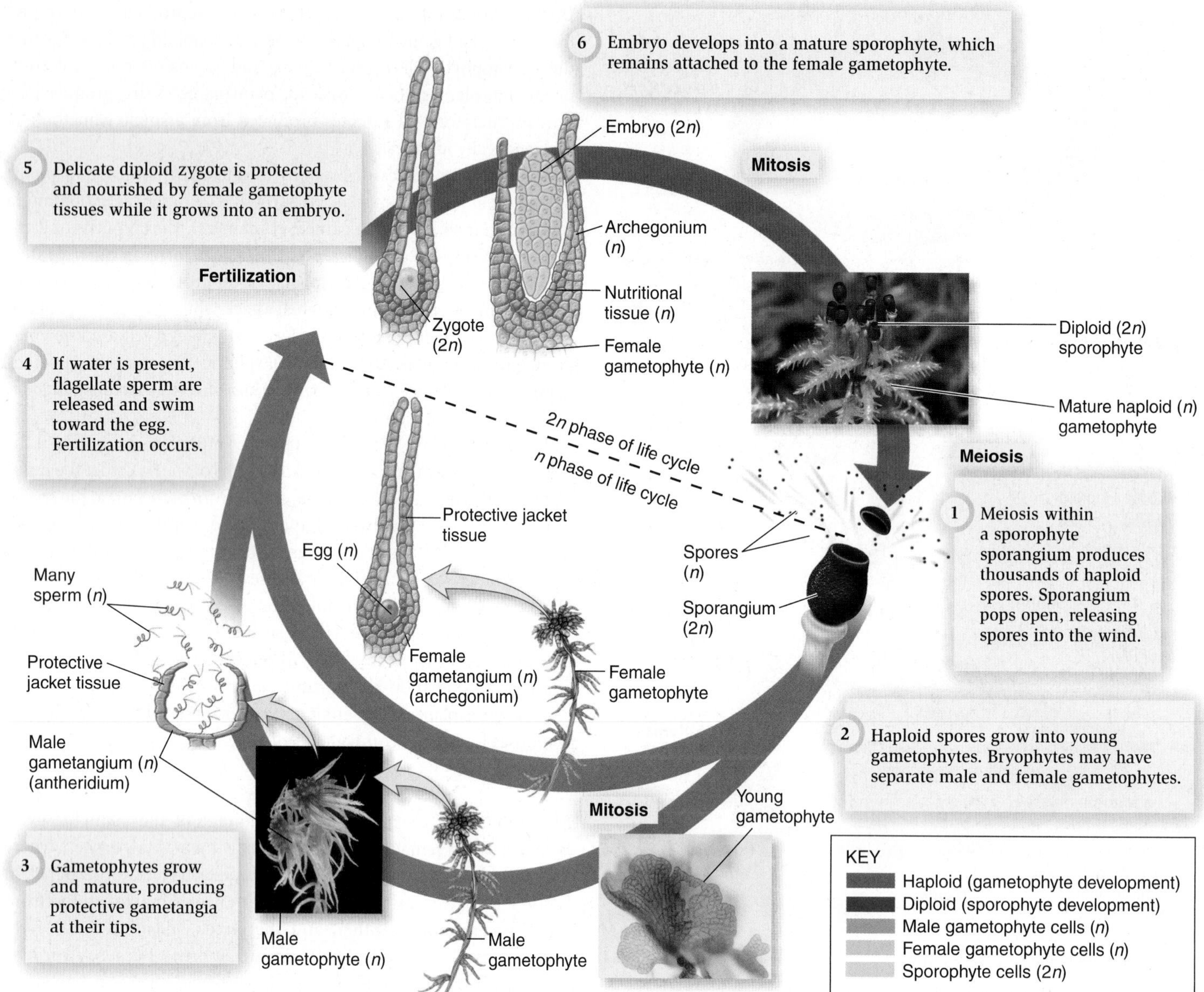

Figure 29.7 **The life cycle of the early diverging moss genus *Sphagnum*.** The life cycle of this organism illustrates reproductive adaptations that likely helped early plants to reproduce on land. Among modern bryophytes, *Sphagnum* is the single most abundant and ecologically important genus.

spores a plant can produce. As a result, during plant evolution, the sporophyte generation has become larger and more complex (**Figure 29.8**).

Bryophytes Display Several Distinguishing Features

As we have seen, bryophytes share several fundamental adaptive traits: alternation of generations, tissue-producing apical meristems, protective gametangia and sporangia, and sporopollenin-walled spores. These same traits are shared with other land plants. Even so, several features distinguish bryophytes from other land plants. First, bryophyte gametophytes are more common in nature, larger, and longer-lived than bryophyte sporophytes. Green patches of moss that you might see in the woods are primarily gametophytes. In order to observe bryophyte sporophytes, you would have to look very closely, because this life stage is quite small and remains attached to gametophytes throughout its short lifetime (see Figures 29.4, 29.5, and 29.7). Plant biologists consider bryophyte gametophytes to be the dominant generation in their life cycle (Figure 29.8a). By contrast, in other plants the sporophyte generation is dominant (Figure 29.8b, c).

Bryophyte sporophytes are small; they remain attached to parental gametophytes, are unable to branch, and have short lifetimes. At their tips, bryophyte sporophytes produce only a single sporangium containing a limited number of spores. By contrast, the sporophytes of other land plants become independent, and because they can branch, they can continue to grow and produce sporangia on lateral branches, often for many years (Figure 29.8c). In all land plants except bryophytes, the sporophyte generation is the dominant

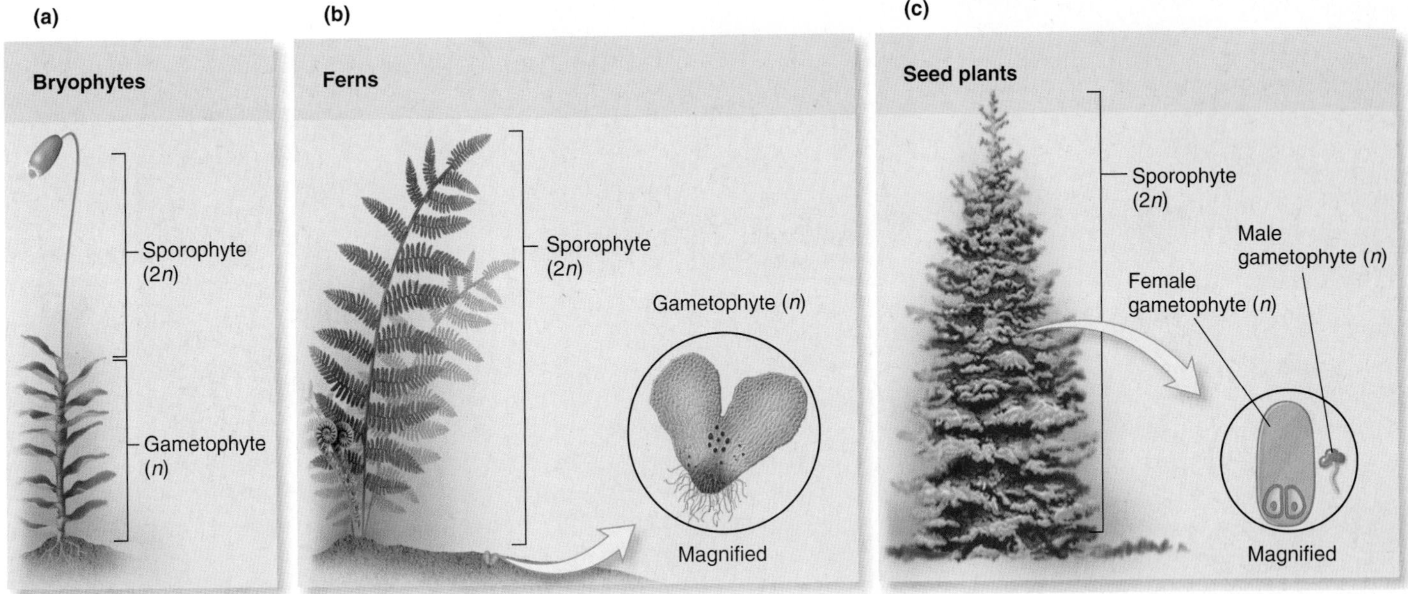

Figure 29.8 **Relative sizes of the sporophyte and gametophyte generations of bryophytes, ferns, and seed plants.**

Concept Check: *What is the advantage to ferns and seed plants of having larger sporophytes than those of bryophytes?*

generation, meaning that it is larger, more complex, and longer-lived than the gametophyte.

Yet another distinguishing feature of bryophytes is that they lack tissues that both provide structural support and serve in conduction of water and nutrients, known as **vascular tissues**. Although the gametophytes of some bryophytes display simple conducting tissues, these do not provide much structural support. Thus, bryophytes are informally known as **nonvascular plants**. As mentioned, other modern plant phyla are collectively and informally known as **vascular plants** because they produce vascular tissues that function in both conduction and support.

Lycophytes and Pteridophytes Are Vascular Plants That Do Not Produce Seeds

If you take a look outside, most of the plants in view are probably vascular plants. Their dominant sporophyte, presence of vascular tissue, and the ability to branch allow most of these plants to grow much taller than bryophytes and to produce more spores. As a result, vascular plants are more prominent than bryophytes in most modern plant communities.

Vascular plants have been important to Earth's ecology for several hundreds of millions of years. Fossils tell us that the first vascular plants appeared later than the earliest bryophytes (see Section 29.2) and that several early vascular plant lineages once existed but became extinct. Molecular data indicate that the lycophytes are the oldest phylum of living vascular plants and that pteridophytes are the next oldest living plant phylum (see Figure 29.1). In the past, lycophytes were very diverse and included tall trees that contributed importantly to coal deposits, but the tree lycophytes became extinct, and now only about 1,000 relatively small species exist (Figure 29.9). Pteridophytes have diversified more recently, and there are about 12,000 species of modern pteridophytes, including horsetails, whisk ferns, and other ferns (Figure 29.10).

Because the lycophytes and pteridophytes diverged prior to the origin of seeds, they are informally known as seedless vascular plants. Together, lycophytes, pteridophytes, and seed-producing plants are known as the **tracheophytes**. The latter term takes its name from **tracheids**, a type of specialized vascular cell that conducts water and minerals and provides structural support. Vascular tissues occur in the major plant organs: stems, roots, and leaves.

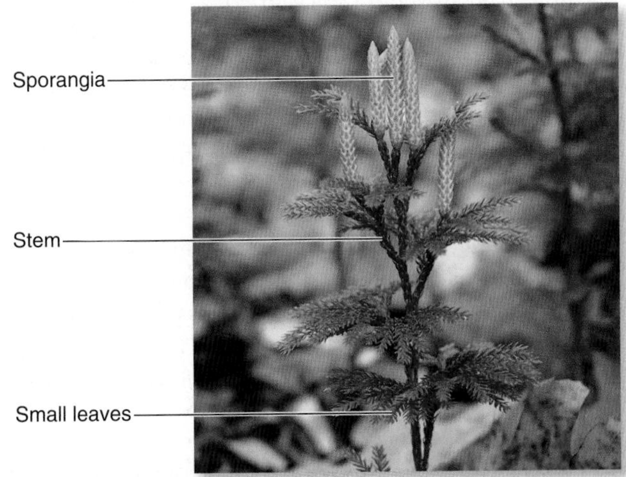

Figure 29.9 **An example of a lycophyte (*Lycopodium obscurum*).** The sporophyte stems bear many tiny leaves, and sporangia generally occur in club-shaped clusters. For this reason, lycophytes are informally known as club mosses or spike mosses, though they are not true mosses. The gametophytes of lycophytes are small structures that often occur underground, where they are better protected from drying.

(a) **A whisk fern**

(b) **The giant horsetail**

(c) **An early-diverging fern**

(d) **A later-diverging fern**

Figure 29.10 **Pteridophyte diversity.** **(a)** The leafless, rootless green stems of the whisk fern (*Psilotum nudum*) branch by forking and bear many clusters of yellow sporangia that disperse spores via wind. The gametophyte of this plant is a tiny pale structure that lives underground in a symbiotic partnership with fungi. **(b)** The giant horsetail (*Equisetum telmateia*) displays branches in whorls around the green stems. The leaves of this plant are tiny, light brown structures that encircle branches at intervals. This plant produces spores in cone-shaped structures, and the wind-dispersed spores grow into small green gametophytes. **(c)** The early-diverging fern *Botrychium lunaria*, showing a green photosynthetic leaf with leaflets and a modified leaf that bears many round sporangia. **(d)** The later-diverging fern *Blechnum capense*, viewed from above, showing a whorl of young leaves that are in the process of unrolling from the bases to the tips. The leaves have many leaflets. The stem of this fern grows parallel to the ground and thus is not shown. Most ferns produce spores in sporangia on the undersides of leaves.

Stems, Roots, and Leaves **Stems** are branching structures that contain vascular tissue and produce leaves and sporangia. Stems contain the specialized conducting tissues known as **phloem** and **xylem**, the latter of which contains tracheids. Together, such conducting tissues enable vascular plants to conduct water, minerals, and organic compounds throughout the plant body. The xylem also provides structural support, allowing vascular plants to grow taller than nonvascular plants. This support function arises from the presence of a compression and decay-resistant waterproofing material known as **lignin**, which occurs in the cell walls of tracheids and some other types of plant cells. Most vascular plants also produce **roots**—organs specialized for uptake of water and minerals from the soil—and **leaves**, flattened plant organs that emerge from stems and generally have a photosynthetic function.

Lycophyte roots and leaves differ from those of pteridophytes. For example, lycophyte roots fork at their tips, whereas roots of pteridophytes branch from the inside like the roots of seed plants (see Figure 35.3). Lycophyte leaves are relatively small and possess only one unbranched vein, whereas pteridophyte leaves are larger and have branched veins, as do those of seed plants (compare Figures 29.9 and 29.10d). The evolutionary origins of leaves are discussed in Section 29.4.

Adaptations That Foster Stable Internal Water Content In relatively dry habitats, lycophytes, pteridophytes, and other vascular plants are able to grow to larger sizes and remain metabolically active for longer periods than can bryophytes. Vascular plants have this advantage because they are better able to maintain stable internal

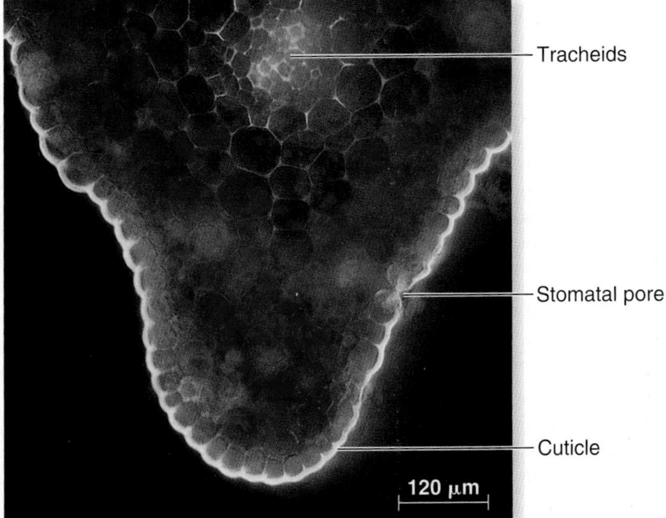

(a) Stem showing tracheophyte adaptations

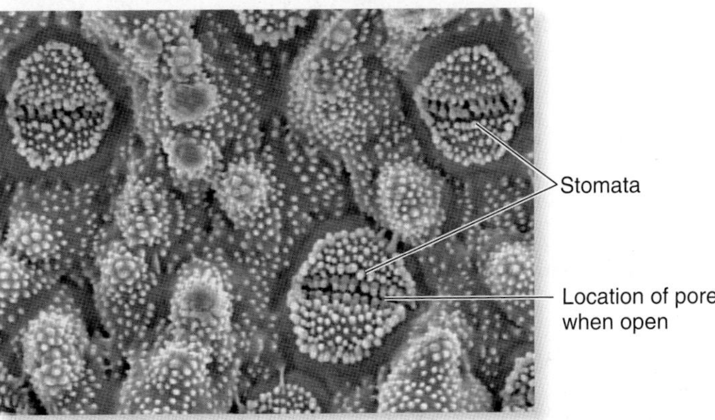

(b) Close-up of stomata

Figure 29.11 **A pteridophyte stem with tracheophyte adaptations for transporting and conserving water.** **(a)** This is a cross section through a stem of the pteridophyte *Psilotum nudum*. When viewed with fluorescence microscopy and illuminated with violet light, an internal core of xylem tracheids glows yellow, as does the surface cuticle. **(b)** Surface pores associated with specialized cells—the complexes known as stomata—allow for gas exchange between plant and atmosphere. This photo naturally lacks color because it was made with a scanning electron microscope (SEM), which uses electrons rather than visible light to form magnified images.

BIOLOGY PRINCIPLE Living organisms maintain homeostasis. The structures illustrated in this figure explain how vascular land plants maintain homeostasis in water content.

water content by means of several adaptations, including conducting tissues (discussed earlier), a waxy cuticle, and stomata. A protective waxy **cuticle** is present on most surfaces of vascular plant sporophytes (Figure 29.11a). The plant cuticle contains a polyester polymer known as cutin, which helps to prevent attack by pathogens, and wax, which helps to prevent desiccation (drying out). The surface tissue of vascular plant stems and leaves contains many **stomata** (singular, stoma or stomate)—pores that are able to open and close (Figure 29.11b). Stomata allow plants to take in the carbon dioxide needed for photosynthesis and release oxygen to the air, while conserving water. When the environment is moist, the pores open, allowing photosynthetic gas exchange to occur. When the environment is very dry, the pores close, thereby reducing water loss (more information about stomata can be found in Chapters 35 and 38). Though a cuticle and stomata occur in bryophytes, they are not as common as in vascular plants, and bryophytes easily become dry. In vascular plants, conducting tissues, a waxy cuticle, and stomata function together to maintain moisture homeostasis, allowing tracheophytes to exploit a wider spectrum of land habitats.

Life Cycle Lycophyte and pteridophyte gametophytes are small, delicate, and easily harmed by exposure to heat and drought. This explains why the gametophytes of lycophytes and pteridophytes are restricted to moist places, including underground, and why they have short lifetimes. Lycophyte and pteridophyte sperm are released from antheridia into water films within which they swim to eggs in archegonia (Figure 29.12, step 6). For this reason, lycophyte and pteridophyte reproduction is limited by dry conditions, as is the case for bryophytes. However, if fertilization occurs, lycophytes and pteridophytes can produce many more spores, because the spore-producing sporophyte generation grows to a much larger size than do bryophyte sporophytes. This fundamental difference has two explanations. First, vascular plant sporophytes depend on maternal gametophytes for only a short time during early embryo development; they eventually become independent by developing a first leaf and roots able to harvest resources needed for photosynthesis (Figure 29.12, step 8). Second, the stems of vascular plant sporophytes are able to produce branches, forming relatively large adult plants having many leaves. Roots obtain large amounts of soil water and minerals, supporting the ability of leaves to generate abundant organic compounds by photosynthesis. Lycophytes and pteridophytes use such resources to produce large numbers of sporangia that eject multitudes of spores. You might have seen clusters of sporangia as dark brown dots or lines on the undersides of fern leaves. Dispersed by the wind, spores may land in a suitable place and grow into new gametophytes, completing the life cycle.

GENOMES & PROTEOMES CONNECTION

Comparison of Plant Genomes Reveals Genetic Changes That Occurred During Plant Evolution

The first complete genome sequence for seedless vascular plants was reported for the lycophyte *Selaginella moellendorffii* by Jody Banks at Purdue University and coworkers in 2011. This advance has allowed plant evolutionary biologists to compare the new sequence with previously sequenced plant genomes, with the goal of identifying genes associated with major evolutionary transitions in plants.

Such genome comparisons revealed that 6,820 gene families, representing a basic set of embryophyte genes, are present in all land plants. The majority of gene families that are involved in flowering plant development were also observed in the lycophyte *Selaginella* and the bryophyte *Physcomitrella*, a moss. A comparison of the genomes of *Selaginella* and *Physcomitrella* indicated that 516 genes were gained and 89 genes were lost during the transition from

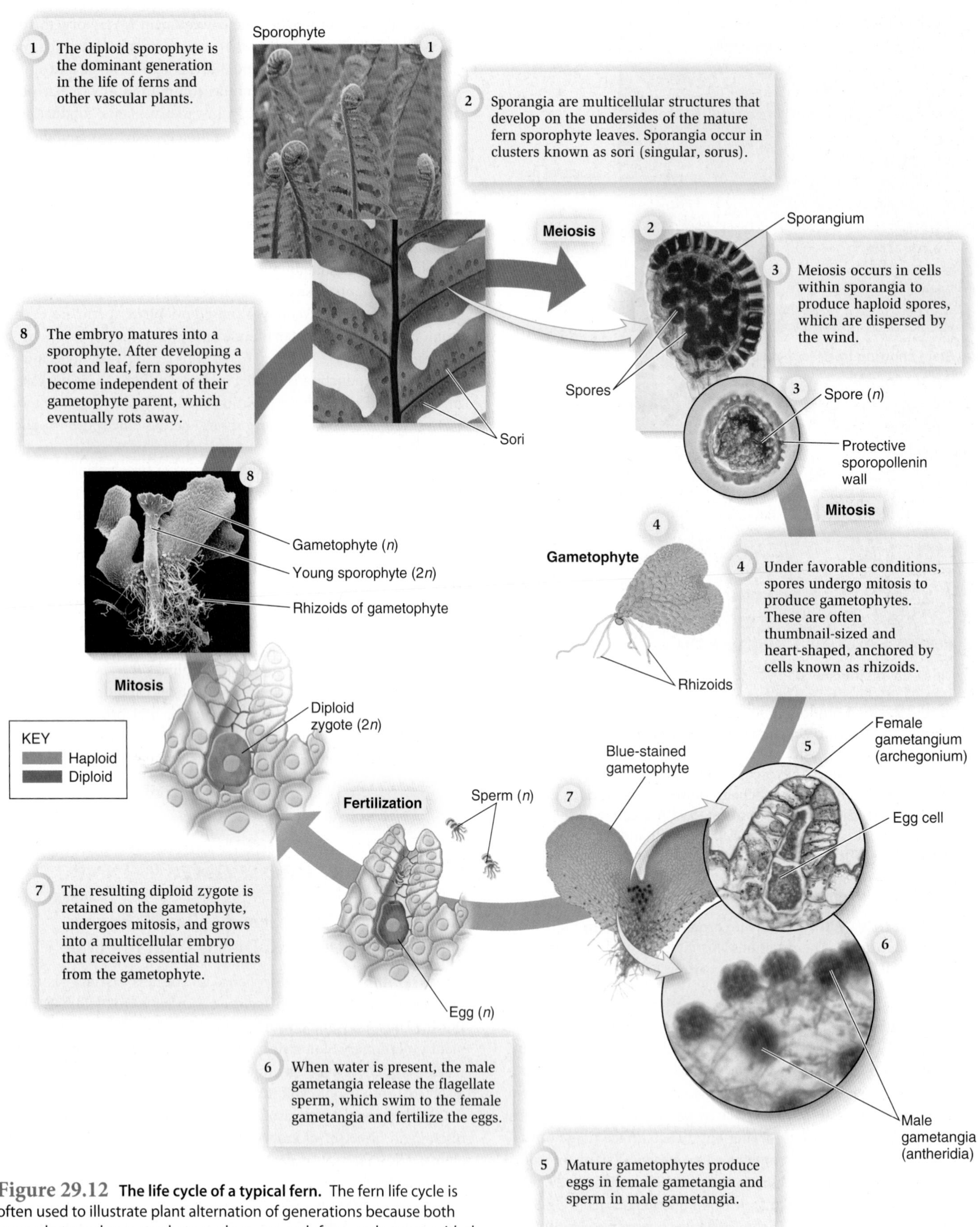

1 The diploid sporophyte is the dominant generation in the life of ferns and other vascular plants.
Sporophyte
1
2 Sporangia are multicellular structures that develop on the undersides of the mature fern sporophyte leaves. Sporangia occur in clusters known as sori (singular, sorus).
Meiosis
2
Sporangium
3 Meiosis occurs in cells within sporangia to produce haploid spores, which are dispersed by the wind.
Spores
Sori
3
Spore (n)
Protective sporopollenin wall
Mitosis
8 The embryo matures into a sporophyte. After developing a root and leaf, fern sporophytes become independent of their gametophyte parent, which eventually rots away.
8
Gametophyte (n)
Young sporophyte (2n)
Rhizoids of gametophyte
4
Gametophyte
4 Under favorable conditions, spores undergo mitosis to produce gametophytes. These are often thumbnail-sized and heart-shaped, anchored by cells known as rhizoids.
Rhizoids
Mitosis
Diploid zygote (2n)
KEY
Haploid
Diploid
Female gametangium (archegonium)
5
Blue-stained gametophyte
7
Sperm (n)
Fertilization
Egg cell
6
7 The resulting diploid zygote is retained on the gametophyte, undergoes mitosis, and grows into a multicellular embryo that receives essential nutrients from the gametophyte.
Egg (n)
6 When water is present, the male gametangia release the flagellate sperm, which swim to the female gametangia and fertilize the eggs.
Male gametangia (antheridia)
5 Mature gametophytes produce eggs in female gametangia and sperm in male gametangia.

Figure 29.12 The life cycle of a typical fern. The fern life cycle is often used to illustrate plant alternation of generations because both sporophyte and gametophyte are large enough for people to see with the unaided eye. Inset 8: © Dr. Richard Kessel & Dr. Gene Shih/Visuals Unlimited

nonvascular, gametophyte-dominant plants (bryophytes) to vascular, sporophyte-dominant plants (lycophytes). By contrast, 1,350 more genes were gained in the evolution of traits specific to angiosperms. These data indicate that the transition from bryophyte to lycophyte was less genetically complex than was the transition from lycophyte to angiosperm.

Comparison of plant genomes has also revealed the great importance of secondary metabolites to all land plants. Secondary metabolites are organic compounds that are not essential for cell structure and growth, but are advantageous to the organism that synthesizes them. An example of a secondary metabolite is caffeine, the stimulatory component of coffee and tea. Caffeine plays a role in plant defense, a common function of secondary metabolites (discussed in more detail in Chapters 30, 39, and 57). The data indicate that a small number of secondary metabolite genes present in a common ancestral vascular plant radiated extensively and independently in the lycophytes and angiosperms. In terms of the complexity of their secondary metabolite genetic repertoire, lycophytes appear to rival the angiosperms, with each group displaying unique compounds. Because many angiosperm secondary metabolites are important to humans and some have medicinal applications (see Chapter 30, Section 30.2), plant biologists suspect that lycophyte secondary metabolites may be potential sources of new pharmaceuticals.

Gymnosperms and Angiosperms Are the Modern Seed Plants

Among the vascular plants, the seed plant phyla dominate most modern landscapes. The modern seed plant phyla commonly known as cycads, ginkgos, conifers, and gnetophytes are collectively known as gymnosperms (Figure 29.13 shows an example). **Gymnosperms** reproduce using both spores and seeds, as do the flowering plants, the angiosperms (Figure 29.14). For this reason, gymnosperms and angiosperms are known informally as the **seed plants**. **Seeds** are complex structures having specialized tissues that protectively enclose embryos and contain stores of carbohydrate, lipid, and protein. Embryos use such food stores to grow and develop. As discussed in Section 29.4, the ability to produce seeds helps to free seed plants from the reproductive limitations experienced by the seedless plants, revealing why seed plants are the dominant plants on Earth today.

In addition to the modern seed plant phyla, several additional phyla once existed and left fossils but are now extinct. Collectively, all of the living and fossil seed plant phyla are formally known as **spermatophytes** (the prefix sperm in this case, from the Greek, meaning seed) (see Figure 29.1). Many spermatophytes have the capacity to produce wood. Wood is composed of xylem, a tissue whose cellulose-rich cell walls also contain lignin. Functioning something like superglue, lignin cements the fibrils of cellulose together, making wood exceptionally strong. Wood production enables plants to increase in girth and become tall. Though not all modern seed plants produce wood, many are trees or shrubs that produce considerable amounts of wood (described more completely in Chapters 30, 35, and 38).

The angiosperms are distinguished by the presence of flowers, fruits, and a specialized seed tissue known as endosperm. A **flower** is a short stem bearing reproductive organs that are specialized in ways that enhance seed production (see Figure 29.14). **Fruits** are structures that develop from flowers, enclose seeds, and foster seed dispersal in the environment. The term angiosperm comes from the Greek, meaning enclosed seeds, reflecting the observation that the flowering plants produce seeds within fruits. **Endosperm** is a nutritive seed tissue that increases the efficiency with which food is stored in the seeds of flowering plants (explained further in Section 29.4). Flowers, fruits, and endosperm are defining features of the angiosperms, and they are integral components of animal nutrition.

Though gymnosperms produce seeds and many are woody plants, they lack flowers, fruits, and endosperm. The term gymnosperm comes from the Greek, meaning naked seeds, reflecting the observation that

Figure 29.13 An example of a gymnosperm, the pine (*Pinus*).

Concept Check: What key characteristics of spermatophytes are displayed in this image?

Figure 29.14 An example of an angiosperm, the bleeding heart plant (genus *Dicentra*).

gymnosperm seeds are not enclosed within fruits. Despite their lack of flowers, fruits, and seed endosperm, the modern gymnosperms are diverse and abundant in many places (see Chapter 30).

29.2 An Evolutionary History of Land Plants

Learning Outcome:

1. Describe how two major events in plant history affected other life on Earth.

A billion years ago, Earth's terrestrial surface was comparatively devoid of life. Green or brown crusts of cyanobacteria most likely grew in moist places, but there would have been very little soil, no plants, and no animal life. The origin of the first land plants was essential to development of the first substantial soils, the evolution of modern plant communities, and the ability of animals to colonize land.

How can we know about events such as the origin and diversification of land plants? One line of information comes from comparing molecular and other features of modern plants. For example, the genome sequence of the moss *Physcomitrella patens*, first reported in 2007, reveals the presence of genes that aid heat and drought tolerance, which are especially useful in the terrestrial habitat. Plant fossils, the preserved remains of plants that lived in earlier times, provide another line of information (**Figure 29.15**). The distinctive plant materials sporopollenin, cutin, and lignin do not readily decay and therefore foster the fossilization of plant parts that contain these tough materials.

The study of fossils and the molecular, structural, and functional features of modern plants have revealed an amazing story—how plants gradually acquired adaptations, allowing them to conquer the land. Next we will explore how seedless plants transformed Earth's ecology, and how an ancient cataclysm led to the diversification of modern angiosperm lineages.

Figure 29.15 **Fossil of *Pseudosalix handleyi*, an angiosperm.**

Concept Check: *What biochemical components of plants favor the formation of fossils?*

Seedless Plants Transformed Earth's Ecology

Several types of decay-resistant tissues evolved in early seedless plants, likely in response to attack by soil bacteria and fungi. When the plants died, some of their organic constituents were not completely degraded to carbon dioxide (CO_2), but instead were buried in sediments that were eventually transformed into rock. Such fossil organic carbon can accumulate and remain buried for very long time periods, with the consequence that the amount of CO_2 in the atmosphere declined. CO_2 is a greenhouse gas, meaning that an increase in its concentration warms the atmosphere, thereby influencing climate. Throughout their evolutionary history, plants have influenced Earth's past climate by reducing the concentration of atmospheric CO_2, and plants continue to do so today.

Ecological Effects of Ancient and Modern Bryophytes Modern bryophytes play important roles by storing CO_2 as decay-resistant organic compounds, suggesting that ancient relatives likewise played this role. Plants of the locally abundant modern moss genus *Sphagnum* contain so much decay-resistant body mass that in many places, dead moss has accumulated over thousands of years, forming deep peat deposits. By storing very large amounts of organic carbon for long periods, *Sphagnum* helps to keep Earth's climate steady. Under cooler than normal conditions, *Sphagnum* grows more slowly and thus absorbs less CO_2, allowing atmospheric CO_2 to rise a bit, warming the climate a little. As the climate warms, *Sphagnum* grows faster and sponges up more CO_2, storing it in peat deposits. Such a reduction in atmospheric CO_2 returns the climate to slightly cooler conditions. In this way, ancient and modern peat mosses have helped to keep the world's climate from changing dramatically. Today, experts are concerned that large regions currently dominated by peat mosses are being affected by changes in land use, harvesting of peat, and other environmental alterations that could reduce peat moss' ability to moderate Earth's climate.

Ecological Effects of Ancient Vascular Plants Fossils tell us that extensive forests dominated by tree-sized lycophytes, pteridophytes, and early seed plants occurred in widespread swampy regions during the warm, moist Carboniferous period (354–290 mya) (**Figure 29.16**). For example, in 2007, Smithsonian paleobotanist William DiMichele and colleagues reported that they had found fossils of an entire coastal forest that was 297 million years old and covered an area of more than 1,000 hectares in what is now Illinois. Large trees related to modern lycophytes dominated this forest, but pteridophytes and representatives of extinct plant phyla were also present. The forest was well preserved because an earthquake had quickly dropped it below sea level into water low in oxygen (O_2). Under such conditions, decay microbes could not completely decompose the forest, which was then buried in sediments that later formed coal. Much of today's coal is similarly derived from the abundant remains of ancient plants, explaining why the Carboniferous is commonly known as the Coal Age. Carboniferous plants converted huge amounts of the atmospheric CO_2 into decay-resistant organic materials such as lignin. Long-term burial of these materials, compressed into coal, together with chemical interactions between soil and the roots of vascular plants, dramatically changed Earth's atmosphere and climate. The removal of large amounts of the greenhouse gas CO_2 from the atmosphere by plants

Giant dragonfly
Giant horsetail (pteridophyte)
Giant lycophyte

Figure 29.16 Reconstruction of a Carboniferous (Coal Age) forest. This ancient forest was dominated by tree-sized lycophytes and pteridophytes, which later contributed to the formation of large coal deposits.

Concept Check: Why did giant dragonflies occur during the Carboniferous, but not now?

had a cooling effect on the climate, which also became drier because cold air holds less moisture than warm air.

Mathematical models of ancient atmospheric chemistry, supported by measurements of natural carbon isotopes, led American paleoclimatologist Robert A. Berner to propose that the Carboniferous proliferation of vascular plants was correlated with a dramatic decrease in atmospheric CO_2, which reached the lowest known levels about 290 mya (Figure 29.17). During this period of very low CO_2, atmospheric O_2 levels rose to the highest known levels, because less O_2 was being used to break down organic carbon into CO_2. High atmospheric O_2 content has been invoked to explain the occurrence of giant Carboniferous dragonflies and other huge insects. The great Carboniferous decline in CO_2 level ultimately caused cool, dry conditions to prevail in the late Carboniferous and early Permian periods. As a result of this relatively abrupt global climate change, many of the tall seedless lycophytes and pteridophytes that had dominated earlier Carboniferous forests became extinct, as did organisms such as the giant dragonflies. Cooler, drier conditions favored extensive diversification of the first seed plants, the gymnosperms. Compared with seedless plants, seed plants were better at reproducing in cooler, drier habitats (as we will see in Section 29.4). As a result, seed plants came to dominate Earth's terrestrial communities, as they continue to do.

An Ancient Cataclysm Marked the Rise of Angiosperms

Diverse phyla of gymnosperms dominated Earth's vegetation through the Mesozoic era (248–65 mya), which is sometimes called the Age of Dinosaurs. In addition, fossils provide evidence that early mammals

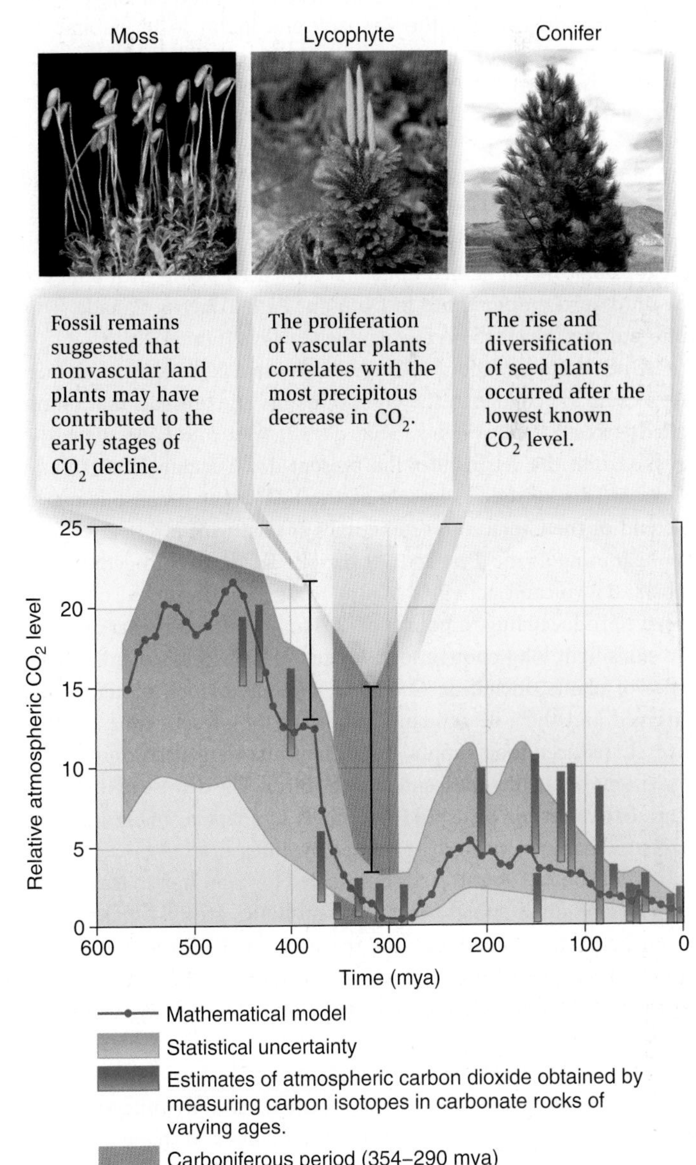
Moss
Lycophyte
Conifer
Fossil remains suggested that nonvascular land plants may have contributed to the early stages of CO_2 decline.
The proliferation of vascular plants correlates with the most precipitous decrease in CO_2.
The rise and diversification of seed plants occurred after the lowest known CO_2 level.
Relative atmospheric CO_2 level
25
20
15
10
5
0
600
500
400
300
200
100
0
Time (mya)
Mathematical model
Statistical uncertainty
Estimates of atmospheric carbon dioxide obtained by measuring carbon isotopes in carbonate rocks of varying ages.
Carboniferous period (354–290 mya)

Figure 29.17 Changes in Earth's atmospheric carbon dioxide levels over geological time. Geological evidence indicates that carbon dioxide levels in Earth's atmosphere were once higher than they are now, but that the rise of land plants caused CO_2 to reach the lowest known level about 300 mya.

Figure 29.18 **Early angiosperms, sources of food for large herbivorous dinosaurs of the Mesozoic era.** In this artist's habitat reconstruction from fossils, the extinct angiosperm *Cobbania corrugata* is shown growing in wetlands that were also inhabited by large dinosaurs such as *Ornithomimus*, whose head is illustrated here.

and flowering plants existed in the Mesozoic. Gymnosperms and early angiosperms were probably sources of food for early mammals as well as for herbivorous dinosaurs. For examples, fossils of an aquatic angiosperm named *Cobbania* have been found with bones of the dinosaur *Ornithomimus* in Dinosaur Park, Alberta, Canada. This dinosaur may have fed on the plant when alive (Figure 29.18).

One fateful day about 65 mya, disaster struck from the sky, causing a dramatic change in the types of plants and animals that dominated terrestrial ecosystems. That day, at least one large meteorite crashed into the Earth near the present-day Yucatán Peninsula in Mexico. This episode is known as the **K/T event** because it marks the end of the Cretaceous (sometimes spelled with a K) period and the beginning of the Tertiary (T) period. The impact, together with substantial volcanic activity that also occurred at this time, is thought to have produced huge amounts of ash, smoke, and haze that dimmed the sun's light long enough to kill many of the world's plants. Many types of plants, including *Cobbania*, became extinct, though some survived and their descendants persist to the present time. With a severely reduced food supply, most dinosaurs were also doomed, the exceptions being their descendants, the birds. The demise of the dinosaurs left room for birds and mammals to adapt to many kinds of terrestrial habitats formerly inhabited by dinosaurs.

After the K/T event, ferns dominated long enough to leave huge numbers of fossil spores, and then surviving groups of flowering plants began to diversify into the space left by the extinction of previous plants. The rise of angiosperms fostered the diversification of beetles (see Chapter 33) and other types of insects that associate with modern plants.

Our brief survey of plant evolutionary history reveals some important concepts. Although environment certainly influenced the diversification of plants, plant diversification has also changed Earth's environment in ways that affected the evolution of other organisms. In addition, plant evolutionary history serves as essential background for a closer focus on the evolution of critical innovations, new features that foster the diversification of phyla. Among the critical innovations that appeared during the evolutionary history of plants, embryos, leaves, and seeds were particularly important.

29.3 The Origin and Evolutionary Importance of the Plant Embryo

Learning Outcomes:

1. Outline how plant embryos might have evolved and why.
2. Analyze the results of Browning and Gunning and explain the role of placental transfer tissues in the movement of nutrients from mother plant to embryo.

The embryo was one of the first critical innovations acquired by land plants (see Figure 29.1). Recall that plant embryos are young sporophytes that develop from zygotes and are enclosed by maternal tissues that provide sustenance. The presence of an embryo is critical to plant reproduction in terrestrial environments. Drought, heat, ultraviolet light, and microbial attack could kill delicate plant egg cells, zygotes, and embryos if these were not protected and nourished by enclosing maternal tissues. The first embryo-producing plants diversified into hundreds of thousands of diverse modern species, as well as many species that have become extinct. A closer look at embryos reveals why their origin and evolution are so important to all land plants.

Plant Embryos Grow Protected Within the Maternal Plant Body

A plant embryo has several characteristic features, some of which were previously described. First, plant embryos are multicellular and diploid (see Section 29.1). Plant embryos develop by repeated mitosis from a single-celled zygote resulting from fertilization (see Figure 29.6b). In addition, we have also learned that plant eggs are fertilized while still enclosed by the maternal plant body and embryos begin their development within the protective confines of maternal tissues (see Figure 29.7). Plant biologists say that plants retain their zygotes and embryos. Third, plant embryo development depends on organic and mineral materials supplied by the mother plant. Nutritive tissues composed of specialized **placental transfer tissues** aid in the transfer of nutrients from mother to embryo. Taking a closer look at placental transfer tissues reveals their valuable role.

Placental transfer tissues function similarly to the placenta present in most mammals, which fosters nutrient movement from the mother's bloodstream to the developing fetus. Plant placental transfer tissues often occur in haploid gametophyte tissues that lie closest to embryos and in the diploid tissues of young embryos themselves. Such transfer tissues contain cells that are specialized in ways that promote the movement of solutes from gametophyte to embryo. For example, the cells of placental transport tissues display complex arrays of finger-like cell-wall ingrowths (Figure 29.19). Because the plant plasma membrane lines this elaborate plant cell wall, the ingrowths vastly increase the surface area of plasma membrane. This increase provides the space needed for abundant membrane transport proteins, which move solutes into and out of cells. With more transport proteins present, materials can move at a faster rate from one cell to another. (Similar finger-like structures in animal intestine and placenta likewise foster nutrient flow by increasing cellular surface area.) Classic experiments have revealed that dissolved sugars, amino acids, and minerals first move from maternal plant cells into the intercellular space between maternal tissues and the embryo. Then, transporter proteins in the membranes of nearby embryo cells efficiently import materials into the embryo.

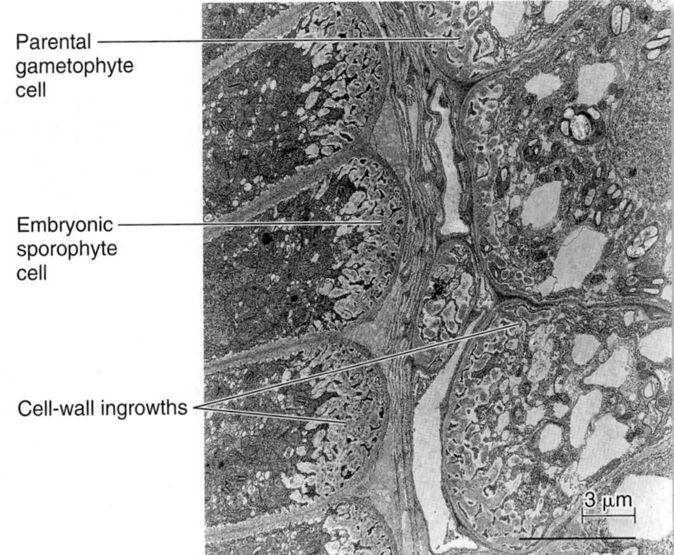

Figure 29.19 **Placental transfer tissue from a plant in the liverwort genus *Monoclea*.**

BIOLOGY PRINCIPLE Structure determines function. This TEM shows that placental transfer tissues contain specialized cells having extensive finger-shaped cell-wall ingrowths. Similar cells occur in all plants and help nutrients to move rapidly from parental gametophytes to embryonic sporophytes, a process that fosters plant reproductive success.

BioConnections: Look ahead to Figures 45.9 and 51.11b. How do similar finger-like projections of animal intestine and placenta likewise foster rapid movement of nutrients?

FEATURE INVESTIGATION

Browning and Gunning Demonstrated That Placental Transfer Tissues Facilitate the Movement of Organic Molecules from Gametophytes to Sporophytes

In the 1970s, plant cell biologists Adrian Browning and Brian Gunning explored placental transfer tissue function. Using a simple moss experimental system, they investigated the rate at which radioactively labeled carbon moves through placental transfer tissues from green gametophytes into young sporophytes. Recall that embryos are very young, few-celled sporophytes and that in mosses and other bryophytes, all stages of sporophyte development are nutritionally dependent on gametophyte tissues. Browning and Gunning investigated nutrient flow into young sporophytes because these slightly older and larger developmental stages were easier to manipulate in the laboratory than were tiny embryos.

In a first step, the investigators grew many gametophytes of the moss *Funaria hygrometrica* in a greenhouse until young sporophytes developed as the result of sexual reproduction (**Figure 29.20**).

In a second step, they placed black glass tubing over young sporophytes as a shade to prevent photosynthesis, enclosed the whole plants—gametophytes and their attached sporophytes—within transparent jars, and supplied the plants with radioactively labeled carbon dioxide ($^{14}CO_2$) for measured time periods known as pulses. Because the moss gametophytes were not shaded, their photosynthetic cells were able to convert the $^{14}CO_2$ into labeled organic compounds, such as sugars and amino acids. Shading prevented the young sporophytes, which possess some photosynthetic tissue, from using the $^{14}CO_2$ to produce organic compounds.

In a third step, the researchers added an excess amount of nonradioactive CO_2 to prevent the further uptake of the $^{14}CO_2$ from their experimental system, a process known as a chase. This process stopped the radiolabeling of photosynthetic products because the vast majority of CO_2 taken up was now unlabeled. (Experiments such as these are known as pulse-chase experiments.) In a final step, Browning and Gunning plucked young sporophytes of different sizes (ages) from the gametophytes and measured the amount of $^{14}CO_2$ present in the separated gametophyte and sporophyte tissues at various times following the chase.

From these data, they were able to calculate the relative amount of organic carbon that had moved from the photosynthetic moss gametophytes to their sporophytes. Browning and Gunning discovered that about 22% of the organic carbon produced by gametophyte photosynthesis was transferred to the young sporophytes during an 8-hour chase period (see Data I in Figure 29.20). They also calculated

Figure 29.20 **Browning and Gunning demonstrated that placental transfer tissues increase plant reproductive success.**

HYPOTHESES 1. Placental transfer tissues allow organic nutrients to flow from plant gametophytes to sporophytes faster than such nutrients move through plant tissues lacking transfer cells.
2. The rate of organic nutrient transfer into larger sporophytes is faster than into smaller sporophytes.

KEY MATERIALS Moss *Funaria hygrometrica*, $^{14}CO_2$ (radiolabeled carbon dioxide)

	Experimental level	Conceptual level
1 Grow moss gametophytes until young sporophytes develop from embryos, and measure sporophyte size.	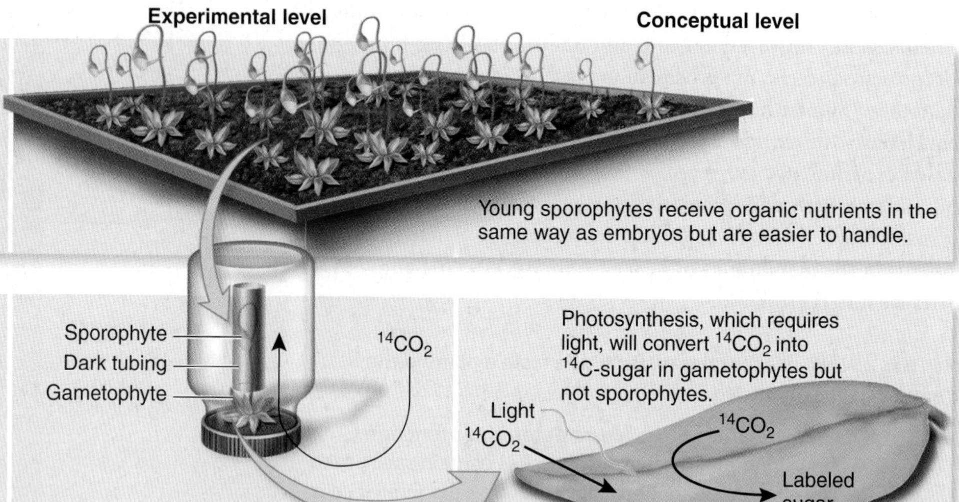	Young sporophytes receive organic nutrients in the same way as embryos but are easier to handle.
2 Shade young sporophytes from light with blackened glass tubing, and enclose whole plant in clear glass jar. Expose plants to $^{14}CO_2$ for 15 minutes. This is called a pulse.	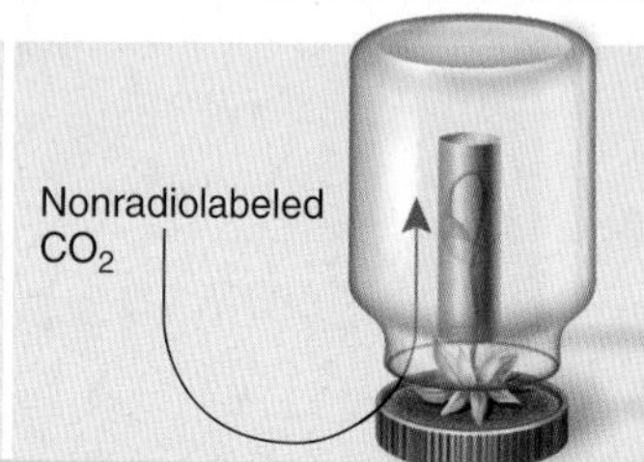	Photosynthesis, which requires light, will convert $^{14}CO_2$ into ^{14}C-sugar in gametophytes but not sporophytes. 
3 Expose plants to a large amount of nonradioactive CO_2. This is called a chase. Incubate up to 8 hours.		Addition of excess nonlabeled CO_2 is known as a chase because it chases away the ability of the cells to make any more radioactive sugars.
4 Pluck young sporophytes of differing sizes from gametophytes. Assay ^{14}C in both sporophytes and gametophytes using a scintillation counter. This was done immediately following the chase, or 2 or 8 hours after the chase.		Determine how much organic carbon flowed into sporophytes during each chase time.

5 THE DATA I

Organic carbon transfer from gametophyte to sporophyte:

Mean ^{14}C content of 5 gametophytes at 0 chase time	Mean ^{14}C lost from gametophytes after 8-hour chase	Mean ^{14}C gained by sporophytes after 8-hour chase
228 units	145 units	51 units

6 THE DATA II

Sporophyte size effect:

Sporophyte size	Mean ^{14}C content of 8 sporophytes after 2-hour chase
5–7 mm	8.47 ± 4.29 units
11–13 mm	9.93 ± 3.94 units
23–25 mm	24.97 ± 5.30 units

7 CONCLUSION Organic carbon moves from photosynthetic gametophytes into developing sporophytes, facilitated by transfer cell-wall ingrowths. Larger sporophytes absorb more organic carbon than smaller ones.

8 SOURCES

Browning, A.J., and Gunning, B.E.S. 1979. Structure and function of transfer cells in the sporophyte haustorium of *Funaria hygrometrica.* Hedw. II. Kinetics of uptake of labelled sugars and localization of absorbed products by freeze-substitution. *Journal of Experimental Botany* 30:1247–1264.

Browning, A.J., and Gunning, B.E.S. 1979. Structure and function of transfer cells in the sporophyte haustorium of *Funaria hygrometrica.* III. Translocation of assimilate into the attached sporophyte and along the seta of attached and excised sporophytes. *Journal of Experimental Botany* 30:1265–1273.

the rate of nutrient transfer from gametophyte to sporophyte and compared this rate with the rate (determined in other studies) at which organic carbon moves within several other plant tissues that lack specialized transfer cells. Browning and Gunning discovered that organic carbon moved from moss gametophytes to young sporophytes nine times faster than organic carbon moves within these other plant tissues (see Data II in Figure 29.20). These investigators inferred that the increased rate of nutrient movement could be attributed to placental transfer cell structure, namely, the fact that cell-wall ingrowths enhanced plasma membrane surface area. By comparing the amount of radioactive carbon accumulated by sporophytes of differing sizes, they also learned that larger sporophytes absorbed $^{14}CO_2$ about three times faster than smaller ones.

These data are consistent with the hypothesis that placental transfer tissues increase plant reproductive success by providing embryos and growing sporophytes with more nutrients than they would otherwise receive. Supplied with these greater amounts of nutrients, sporophytes are able to grow larger than they otherwise would, and eventually they produce more progeny spores.

Experimental Questions

1. What were the goals of the Browning and Gunning investigation?
2. How did Browning and Gunning prevent photosynthesis from occurring in moss sporophytes during the experiment (shown in Figure 29.20), and why did they do this?
3. How did the measurements Browning and Gunning made after adding an excess amount of unlabeled CO_2 lead them to their conclusions?

29.4 The Origin and Evolutionary Importance of Leaves and Seeds

Learning Outcomes:

1. Describe how the leaves of ferns and seed plants likely evolved from branched-stem systems.
2. Discuss how seeds develop from fertilized ovules.
3. Name several advantages that seeds provide.
4. List hypothetical steps in the evolution of seeds.

Like plant embryos, leaves and seeds are critical innovations that increased plant fitness and fostered diversification. Unlike the plant embryo, which likely originated just once at the birth of the plant kingdom, leaves and seeds probably evolved several times during plant evolutionary history. Comparative studies of diverse types of leaves and seeds in fossil and living plants suggest how these critical innovations might have originated.

The Large Leaves of Ferns Evolved from Branched-Stem Systems

Leaves are the solar panels of the plant world. Their flat structure provides a high surface area that helps effectively capture sunlight for use in photosynthesis. Among the vascular plants, lycophytes produce the simplest and most ancient type of leaves. Modern lycophytes have tiny leaves, known as **lycophylls** (also known as microphylls), which typically have only a single unbranched vein (**Figure 29.21a**). Some experts think that these small leaves may have evolved from sporangia.

In contrast, the leaves of ferns and seed plants have extensively branched veins. Such leaves are known as **euphylls** (from the Greek, meaning true leaves) (**Figure 29.21b**). The branched veins of euphylls are able to supply relatively large areas of photosynthetic tissue with water and minerals. Thus, euphylls are typically much larger than lycophylls, explaining why euphylls are also known as megaphylls. Euphylls provide considerable photosynthetic advantage to ferns and seed plants, because they provide more surface for solar energy capture than do small leaves. Hence, the evolution of relatively large leaves allowed plants to more effectively accomplish photosynthesis, enabling them to grow larger and produce more progeny.

Study of fern fossils indicates that euphylls likely arose from leafless, cylindrical, branched-stem systems by a series of steps (**Figure 29.21c**). First, one branch assumed the role of the main axis, while the other was reduced in size, became flattened in one plane, and finally, the spaces between the branches of this flattened system became filled with photosynthetic tissue. Such a process would explain why euphylls have branched vascular systems; individual veins apparently originated from the separate branches of an ancestral branched stem. Plant evolutionary biologists suspect that euphylls arose several times by means of similar, parallel processes, and that leaves of ferns and seed plants are not homologous structures.

Seeds Develop from the Interaction of Ovules and Pollen

The seed plants dominate modern ecosystems, suggesting that seeds offer reproductive advantages. Seed plants are also the plants with the greatest importance to humans, as described in Chapter 30. For these reasons, plant biologists are interested in understanding why seeds are so advantageous and how they evolved. To consider these questions, we must first take a closer look at seed structure and development.

We noted earlier that all plants produce spores by meiosis within sporangia, and seed plants are no exception. However, seed plants produce two distinct types of spores in two types of sporangia, a trait known as **heterospory**, meaning different spores. Microsporangia produce small **microspores** that give rise to male gametophytes, which develop into pollen grains. Megasporangia produce larger **megaspores** that give rise to female gametophytes, which develop and produce eggs while enclosed by protective megaspore walls. The female gametophytes are not photosynthetic, so they need help in feeding the embryos that develop from fertilized eggs. Female gametophytes get this help by remaining attached to the previous sporophyte generation, which provides gametophytes with the nutrients needed for embryo development.

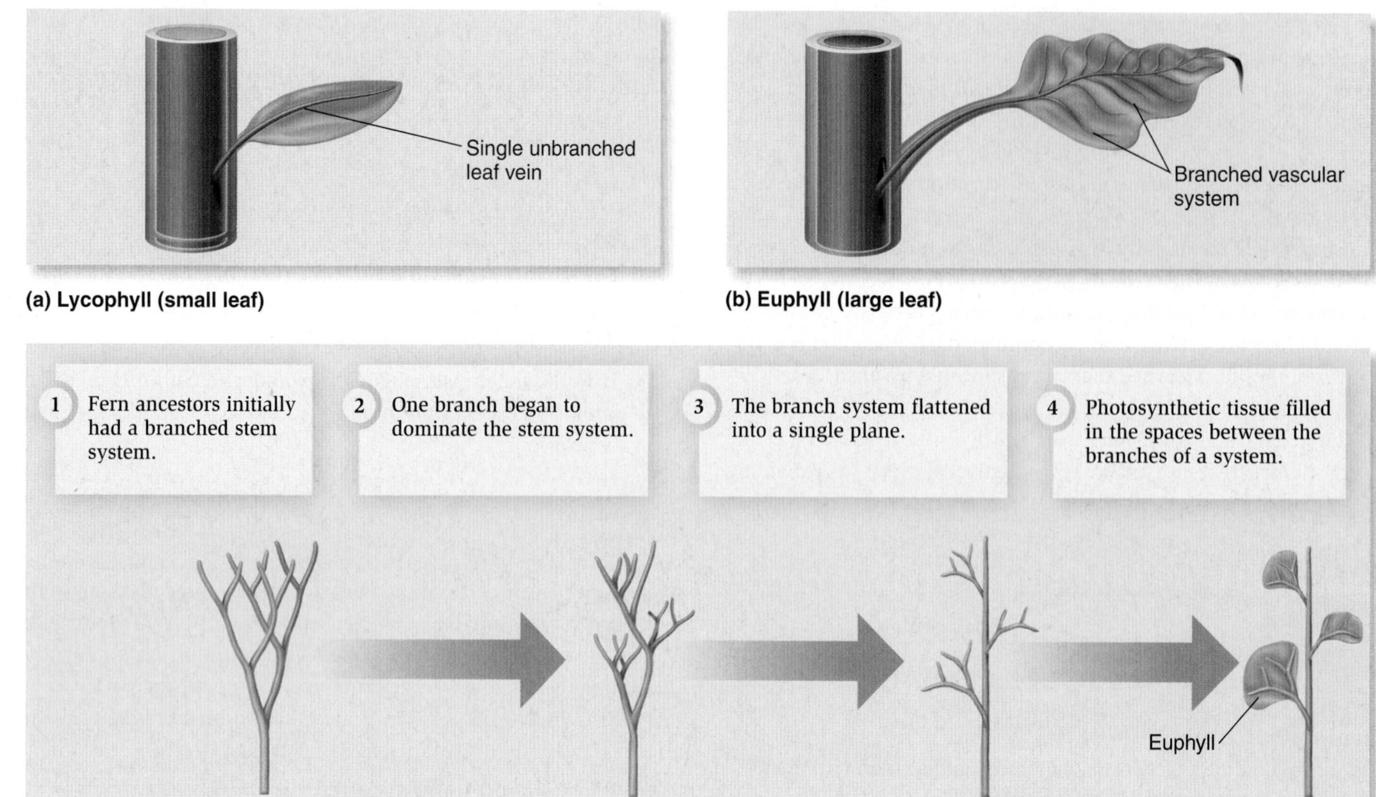
Single unbranched leaf vein
(a) Lycophyll (small leaf)
Branched vascular system
(b) Euphyll (large leaf)
1 Fern ancestors initially had a branched stem system.
2 One branch began to dominate the stem system.
3 The branch system flattened into a single plane.
4 Photosynthetic tissue filled in the spaces between the branches of a system.
Euphyll
(c) Euphyll evolution process in pteridophytes

Figure 29.21 Lycophylls and euphylls. **(a)** Most lycophylls possess only a single unbranched leaf vein with limited conduction capacity, explaining why lycophylls are generally quite small. **(b)** Euphylls possess branched vascular systems with greater conduction capacity, explaining why many euphylls are relatively large. **(c)** Fossil evidence suggests how pteridophyte euphylls might have evolved from branched-stem systems.

Concept Check: Imagine that the leaves of some other plant group evolved from stem systems that were more highly branched (that is, there were more branches per unit length of stem) than those of ferns. How do you think the leaves of such plants might differ from those of ferns?

Plants produce seeds by reproductive structures known as ovules and pollen that are unique to seed plants. An **ovule** is a sporangium that typically contains only a single spore that develops into a very small egg-producing gametophyte, the whole megasporangium enclosed by leaflike structures known as **integuments** (Figure 29.22a). You can think of an ovule as being like a nesting doll with four increasingly smaller dolls inside. The smallest doll corresponds to an egg cell; intermediate-sized dolls represent the gametophyte, spore wall, and megasporangium; and the largest doll represents the integuments. Fertilization converts such layered ovules into seeds. In seed plants, the sperm needed for fertilization are supplied by **pollen**, tiny male gametophytes enclosed by protective sporopollenin walls.

Embryos develop as the result of fertilization, which cannot occur until after **pollination**, the process by which pollen comes close to ovules. Pollination typically occurs by means of wind or animal transport (see Chapter 30). Fertilization occurs in seed plants when a male gametophyte extends a slender pollen tube that carries two sperm toward an egg. The pollen tube enters through an opening in the integument called the micropyle and releases the sperm. The fertilized egg becomes an embryo, and the ovule's integument develops into a protective, often hard and tough **seed coat** (Figure 29.22b,c).

Gymnosperm seeds contain female gametophyte tissue that has accumulated large amounts of protein, lipids, and carbohydrates prior to fertilization. These nutrients are used during both embryo development and seed germination to help nurture the growth of the seedling. Angiosperm seeds also contain this useful food supply, but most angiosperm ovules do not store food materials before fertilization. Instead, angiosperm seeds generally store food only after fertilization occurs, ensuring that the food is not wasted if an embryo does not form. How is this accomplished? The answer is a process known as **double fertilization**. This process produces both a zygote and a food storage tissue known as endosperm, a tissue unique to angiosperms. One of the two sperm delivered by each pollen tube fuses with the egg, producing a diploid zygote, as you might expect. The other sperm nucleus fuses with different gametophyte nuclei to form an unusual cell that has more than the diploid number of chromosomes; this cell generates the endosperm food tissue. Endosperm will be discussed in more detail in Chapter 39.

Seeds allow embryos access to food supplied by the previous sporophyte generation, an option not available to seedless plants. The layered structure of ovules explains why seeds are also layered, with a protective seed coat enclosing the embryo and stored food. These

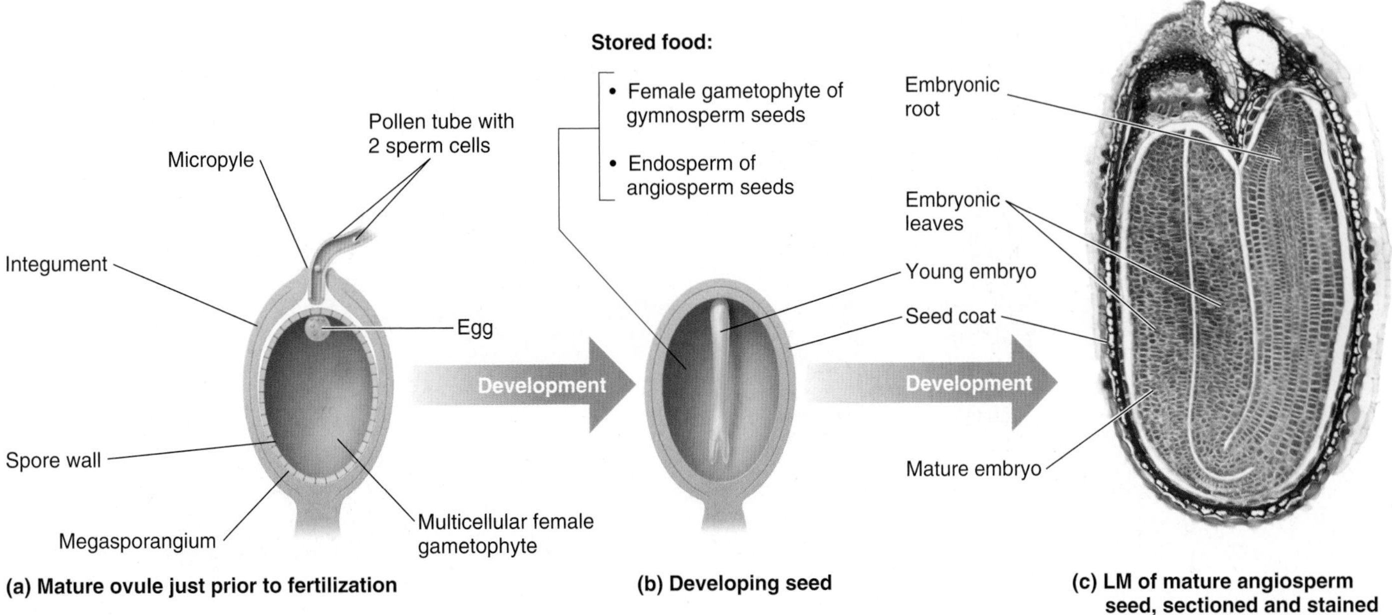

Figure 29.22 **Structure of an ovule developing into a seed.**

Concept Check: *Can you hypothesize why this angiosperm seed with its mature embryo does not show obvious endosperm tissue?*

seed features improve the chances of embryo and seedling survival, thereby increasing seed plant fitness.

Seeds Confer Important Ecological Advantages

Seeds provide plants with numerous ecological advantages. First, many seeds are able to remain dormant in the soil for long periods, until conditions become favorable for germination and seedling growth. Furthermore, seed coats are often adapted in ways that improve dispersal in diverse habitats. For example, many plants produce winged seeds that are effectively dispersed by wind. Other plants produce seeds with fleshy coverings that attract animals, which consume the seeds, digest their fleshy covering, and eliminate them at some distance from the originating plants.

Another advantage of seeds is that they can store considerable amounts of food, which supports embryo growth and helps plant seedlings grow large enough to compete for light, water, and minerals. This is especially important for seeds that must germinate in shady forests. Finally, the sperm of most seed plants can reach eggs without having to swim through water, because pollen tubes deliver sperm directly to ovules. Consequently, seed plant fertilization is not typically limited by lack of water, in contrast to that of seedless plants. Therefore, seed plants are better able to reproduce in arid and seasonally dry habitats. For these reasons, seeds are considered to be a key adaptation to reproduction in a land habitat.

Ovule and Seed Evolution Illustrate Descent with Modification

As we have seen, seed plants reproduce using both spores and seeds, but note that seed plants have not replaced spores with seeds. Rather, seed plants continue to produce spores, and ovules and seeds have evolved from spore-producing structures by descent with modification. Recall that this evolutionary principle involves changes in pre-existing structures and processes. Fossils provide some clues about ovule and seed evolution, and other information can be obtained by comparing reproduction in living lycophytes and pteridophytes.

Most modern lycophytes and pteridophytes release one type of spore that develops into one type of gametophyte. Such plants are considered to be homosporous, and their gametophytes live independently and produce both male and female gametangia (see Figure 29.12). However, some lycophytes and pteridophytes produce and release two distinct kinds of spores: relatively small microspores and larger megaspores, which, respectively, grow into male and female gametophytes. The larger size of megaspores enables them to store food that later supports developing sporophytes. Recall that this reproductive process, known as heterospory, also characterizes all seed plants. The gametophytes produced by heterosporous plants also grow within the confines of microspore and megaspore walls and therefore are known as **endosporic gametophytes.**

An advantage of heterospory is that it mandates cross-fertilization. The eggs and sperm that fuse are derived from different gametophytes and hence from different spores and different meiotic events. This makes it likely that the gametes are of distinct genotypes. Cross-fertilization increases the potential for genetic variation, which aids evolutionary flexibility. Endosporic gametophytes receive protection from environmental damage by surrounding spore walls. From these observations, we can infer that heterospory and endosporic gametophytes were probably also features of seed plant ancestors and constitute early steps toward seed evolution (**Figure 29.23**). Fossils and modern plants also illustrate subsequent stages in seed evolution, such as retaining megaspores within sporangia, rather than releasing them.

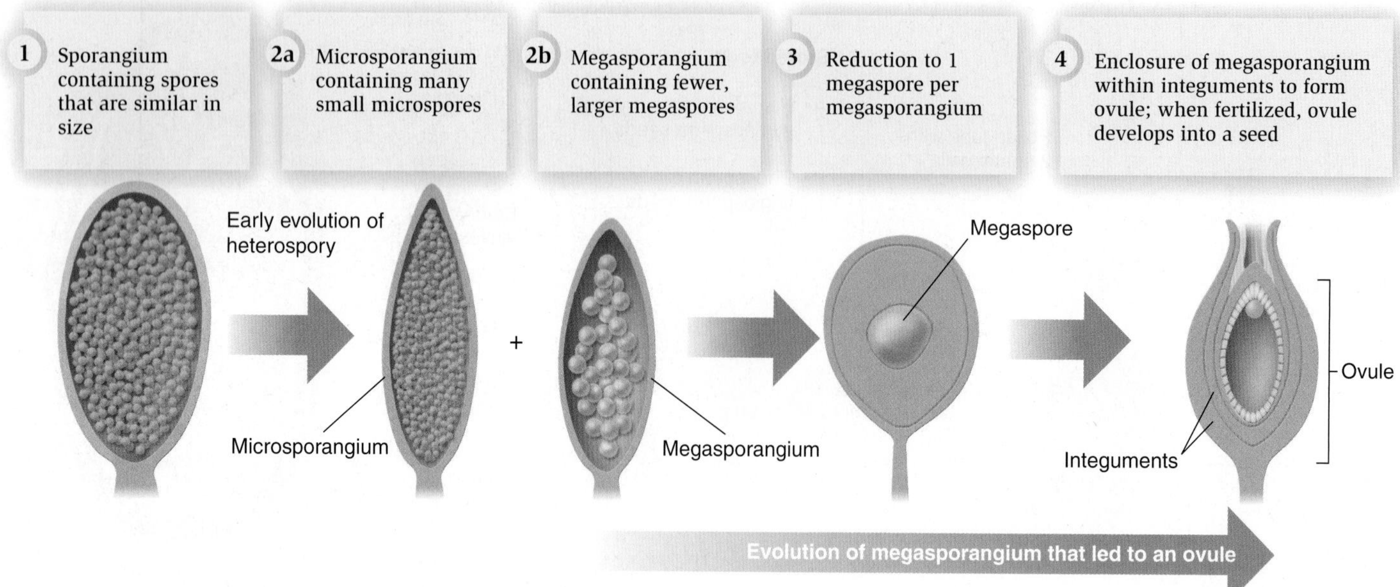
1 Sporangium containing spores that are similar in size
2a Microsporangium containing many small microspores
2b Megasporangium containing fewer, larger megaspores
3 Reduction to 1 megaspore per megasporangium
4 Enclosure of megasporangium within integuments to form ovule; when fertilized, ovule develops into a seed
Early evolution of heterospory
Microsporangium
+
Megasporangium
Megaspore
Integuments
Ovule
Evolution of megasporangium that led to an ovule

Figure 29.23 **Hypothetical stages in the evolution of seeds.** The parallel evolution of heterospory and endosporic gametophytes in some lycophytes and pteridophytes as well as the seed plants suggests that these features were acquired early in the evolution of seeds. Later-occurring events in the origin of seeds included reduction of the number of megaspores to one per megasporangium and enclosure of the megasporangium by protective integuments.

BioConnections: Look ahead to Figure 34.15, which illustrates the amniotic egg produced by many terrestrial animals. How is the plant seed like the amniotic egg?

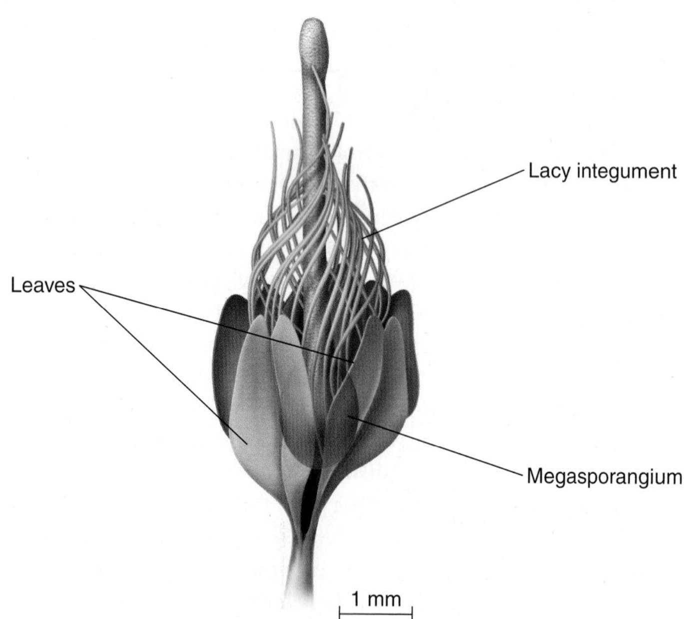
Lacy integument
Leaves
Megasporangium
1 mm

Figure 29.24 **The fossil *Runcaria heinzelinii*, a plant with a probable precursor to an ovule or seed.**

Concept Check: Based on your knowledge of integument function in modern seed plants, can you hypothesize a function for the lacy integument of Runcaria?

A further step in seed evolution may have been the production of only one megaspore per sporangium rather than multiple spores per sporangium, which is common in seedless plants. Reduction of megaspore numbers (from those present in seedless plants) would have allowed plants to channel more nutrients into each megaspore. A final step might have been the retention of megasporangia on parental sporophytes by the development of integuments (**Figure 29.24**). As we have noted, this adaptation would allow food materials to flow from mature photosynthetic sporophytes to their dependent gametophytes and young embryos. Integuments also help ovules to receive pollen.

Fossils provide information about when and how the process of ovule and seed evolution first occurred. Fossil reproductive structures of an extinct Devonian plant named *Runcaria heinzelinii* may represent a precursor to an ovule or seed (see Figure 29.24). These fossil structures had a lacy integument that did not completely enclose the megasporangium. Very early fossil seeds such as *Elkinsia polymorpha* and *Archaeosperma arnoldii* were present by 365 mya. The evolutionary journey illustrated by the transition from aquatic streptophyte algae to bryophytes, to seedless plants, and finally to seed plants reveals how adaptation is related to environmental change, as well as ways in which plants themselves shaped Earth's ecosystems. As a summary of what we have learned in this chapter, **Table 29.1** provides a list of the distinguishing features of land plants and their algal relatives.

Table 29.1 Distinguishing Features of Modern Streptophyte Algae and Land Plants

Streptophyte Algae
Primarily aquatic habitat; zygotic life cycle; sporangia absent; sporophytes absent
LAND PLANTS (EMBRYOPHYTES)
Primarily terrestrial habitat; sporic life cycle consisting of alternation of two multicellular generations—diploid sporophyte and haploid gametophyte; multicellular embryos are nutritionally dependent on maternal gametophyte for at least some time during development; spore-producing sporangia; gamete-producing gametangia; sporopollenin-walled spores
Nonvascular plants (Bryophytes) **(liverworts, mosses, hornworts)**
Dominant gametophyte generation; supportive, lignin-containing vascular tissue absent; true roots, stems, leaves absent; sporophytes unbranched and unable to grow independently of gametophytes
VASCULAR PLANTS (TRACHEOPHYTES) (lycophytes, pteridophytes, spermatophytes)
Dominant sporophyte generation; lignified water-conducting tissue—xylem; specialized organic food-conducting tissue—phloem; sporophytes branched; sporophytes eventually become independent of gametophytes
Lycophytes Leaves generally small with a single, unbranched vein (lycophylls); sporangia borne on sides of stems
PTERIDOPHYTES + SEED PLANTS
Pteridophytes Leaves relatively large with extensively branched vein system (euphylls or megaphylls); sporangia borne on leaves; seeds absent
SEED PLANTS (SPERMATOPHYTES)
Seeds present; leaves are euphylls that evolved independently from those of pteridophytes
GYMNOSPERMS **(cycads, ginkgos, conifers)**
Flowers and fruits absent; seed food stored before fertilization in female gametophyte, endosperm absent
Angiosperms (flowering plants)
Flowers and fruit present; seed food stored after fertilization in endosperm formed by double fertilization

*Key: **Phyla**; LARGER MONOPHYLETIC CLADES (synonyms). All other classification terms are not clades.

Summary of Key Concepts

29.1 Ancestry and Diversity of Modern Plants

- Plants are multicellular eukaryotic organisms composed of cells having plastids; they display many adaptations to life on land. The modern plant kingdom consists of several hundred thousand species classified into nine phyla, informally known as the liverworts, mosses, hornworts, lycophytes, pteridophytes, cycads, ginkgos, conifers, and angiosperms (Figure 29.1).
- The land plants evolved from ancestors that were probably similar to modern complex streptophyte algae (Figure 29.2).
- The monophyletic liverwort, moss, and hornwort phyla are together known informally as the bryophytes. Bryophytes illustrate early-evolved features of land plants, such as a sporic life cycle involving embryos that develop within protective, nourishing gametophyte tissues (Figures 29.3, 29.4, 29.5, 29.6).
- Bryophytes differ from other plants in having a dominant gametophyte generation and a dependent, nonbranching, short-lived sporophyte generation. Bryophytes also lack supportive vascular tissues, in contrast to other modern plant phyla, which are known as the vascular plants or tracheophytes (Figures 29.7, 29.8).
- Lycophytes, pteridophytes, and other vascular plants generally possess stems, roots, and leaves having vascular tissues composed of phloem and xylem, cuticle, and stomata (Figures 29.9, 29.10, 29.11).
- The fern life cycle illustrates the dominant sporophyte characteristic of vascular plants (Figure 29.12).
- Cycads, ginkgos, conifers, and gnetophytes are collectively known as gymnosperms. Gymnosperms produce seeds and inherited an ancestral capacity to produce wood. Angiosperms, the flowering plants, produce seeds, and many also produce wood. Flowers, fruits, and seed endosperm are distinctive features of angiosperms (Figures 29.13, 29.14).

29.2 An Evolutionary History of Land Plants

- Paleobiologists and plant evolutionary biologists infer the history of land plants by analyzing the molecular features of modern plants and by comparing the structural features of fossil and modern plants (Figure 29.15).
- Ancient vascular plants transformed Earth's ecology by altering atmospheric chemistry and climate (Figures 29.16, 29.17, 29.18).
- The K/T meteorite impact event, a probable meteorite collision with Earth that occurred 65 mya, helped cause the extinction of previously dominant dinosaurs and many types of gymnosperms, leaving space into which angiosperms, insects, birds, and mammals diversified.

29.3 The Origin and Evolutionary Importance of the Plant Embryo

- The origin of the plant embryo was a critical innovation that fostered diversification of the land plants. Plant embryos are supported by nutrients supplied by female gametophytes with the aid of specialized placental transfer tissues (Figure 29.19).
- In a classic experiment, Browning and Gunning inferred that placental transfer tissues were responsible for an enhanced flow rate of nutrients from parental gametophytes to embryos (Figure 29.20).

29.4 The Origin and Evolutionary Importance of Leaves and Seeds

- Leaves are specialized photosynthetic organs that evolved more than once during plant evolutionary history. The lycophylls of lycophytes are relatively small leaves having a single unbranched vein. The leaves of ferns and seed plants are known as euphylls and are larger, with an extensively branched vascular system. Fossils indicate that fern euphylls evolved from branched-stem systems (Figure 29.21).
- Seeds develop from ovules, integument-enclosed sporangia that typically contain only a single spore that develops into an egg-producing gametophyte. Pollen produces thin cellular tubes that deliver sperm to eggs produced by female gametophytes. Following pollination and fertilization, ovules develop into seeds. Mature seeds contain stored food and an embryonic sporophyte that develops from the zygote (Figure 29.22).

- Seeds confer many reproductive advantages, including dormancy through unfavorable conditions, greater protection for embryos from mechanical and pathogen damage, seed coat modifications that enhance seed dispersal, and reduction of plant dependence on water for fertilization (Figures 29.23, 29.24).
- The distinctive traits of streptophyte algae and the different phyla of land plants reveal the occurrence of descent with modification (Table 29.1).

Assess and Discuss

Test Yourself

1. The simplest and most ancient phylum of modern land plants is probably
   a. the pteridophytes.
   b. the cycads.
   c. the liverworts.
   d. the angiosperms.
   e. none of the listed choices.
2. An important feature of land plants that originated during the diversification of streptophyte algae is
   a. the sporophyte.
   b. spores, which are dispersed in air and coated with sporopollenin.
   c. tracheids.
   d. plasmodesmata.
   e. fruits.
3. A phylum whose members are also known as bryophytes is commonly known as
   a. liverworts.
   b. hornworts.
   c. mosses.
   d. all of the above.
   e. none of the above.
4. Plants possess a life cycle that involves alternation of two multicellular generations: the gametophyte and
   a. the lycophyte.
   b. the bryophyte.
   c. the pteridophyte.
   d. the lignophyte.
   e. the sporophyte.
5. The seed plants are also known as
   a. bryophytes.
   b. spermatophytes.
   c. pteridophytes.
   d. lycophytes.
   e. none of the above.
6. A waxy cuticle is an adaptation that
   a. helps to prevent water loss from tracheophytes.
   b. helps to prevent water loss from streptophyte algae.
   c. helps to prevent water loss from bryophytes.
   d. aids in water transport within the bodies of vascular plants.
   e. does all of the above.
7. Plant photosynthesis transformed a very large amount of carbon dioxide into decay-resistant organic compounds, thereby causing a dramatic decrease in atmospheric carbon dioxide levels during the geological period known as the
   a. Cambrian.
   b. Ordovician.
   c. Carboniferous.
   d. Permian.
   e. Pleistocene.
8. Which phylum among the plants listed is likely to have the largest leaves?
   a. liverworts
   b. hornworts
   c. mosses
   d. lycophytes
   e. pteridophytes
9. Fern euphylls, also known as megaphylls, probably evolved from
   a. the leaves of mosses.
   b. lycophylls.
   c. branched-stem systems.
   d. modified roots.
   e. none of the listed choices.
10. A seed develops from
    a. a spore.
    b. a fertilized ovule.
    c. a microsporangium covered by integuments.
    d. endosperm.
    e. none of the above.

Conceptual Questions

1. List several common traits that lead evolutionary biologists to infer that land plants evolved from ancestors related to modern streptophyte algae.
2. Why have bryophytes such as mosses been able to diversify into so many species even though they have relatively small, dependent sporophytes?
3. A principle of biology is that *living organisms maintain homeostasis.* Explain how several structural features help vascular plants to maintain stable internal water content.

Collaborative Questions

1. Discuss at least one difference in environmental conditions experienced by early land plants and ancestral complex streptophyte algae.
2. Discuss as many plant adaptations to land as you can.

Online Resource

www.brookerbiology.com

Stay a step ahead in your studies with animations that bring concepts to life and practice tests to assess your understanding. Your instructor may also recommend the interactive eBook, individualized learning tools, and more.

Chapter Outline

30.1 The Evolution and Diversity of Modern Gymnosperms

30.2 The Evolution and Diversity of Modern Angiosperms

30.3 The Role of Coevolution in Angiosperm Diversification

30.4 Human Influences on Angiosperm Diversification

Summary of Key Concepts

Assess and Discuss

The Evolution and Diversity of Modern Gymnosperms and Angiosperms

30

The Madagascar periwinkle (*Catharanthus roseus*), one of the many seed plants on which humans depend.

The seed plants—gymnosperms and angiosperms—are particularly important in our everyday lives because they are the sources of many products, including wood, paper, beverages, food, cosmetics, and medicines. Leukemia, for example, is effectively treated with vincristine, a drug extracted from the beautiful flowering plant known as the Madagascar periwinkle (*Catharanthus roseus*), pictured in the chapter opening photo. Vinblastine—another extract from *C. roseus*—is used to treat lymphatic cancers. Taxol, a compound used in the treatment of breast and ovarian cancers, was first discovered in extracts of the bark of the Pacific yew tree, a gymnosperm known as *Taxus brevifolia*. Vincristine, vinblastine, taxol, and many other plant-derived medicines are examples of plant secondary metabolites, which are distinct from the products of primary metabolism (carbohydrates, lipids, proteins, and nucleic acids). Secondary metabolites play essential roles in protecting plants from disease-causing organisms and plant-eating animals, and they also aid plant growth and reproduction. Though all plants produce secondary metabolites, these natural products are exceptionally diverse in gymnosperms and angiosperms.

In this chapter, we will learn how the hundreds of thousands of modern seed plants play many additional important roles in the lives of humans and modern ecosystems. This chapter builds on the introduction to seeds and seed plants provided in Chapter 29. It begins by focusing on the diversity of modern lineages of gymnosperms and angiosperms. Coevolutionary interactions among angiosperms and animals are presented as major forces influencing the diversification of these groups. This chapter concludes by considering human influences on seed plant evolution and the importance of seed plants in modern agriculture.

30.1 The Evolution and Diversity of Modern Gymnosperms

Learning Outcomes:

1. Describe how gymnosperms evolved from woody but seedless ancestors.
2. Identify three gymnosperm phyla and describe their importance to humans.

Figure 30.1 shows our current understanding of the relationships among modern seed plants, the flowering plants (angiosperms) and three phyla of gymnosperms. Following the early diversification of gymnosperms, an unknown, extinct gymnosperm lineage (not specified in Figure 30.1) gave rise to the angiosperms—the flowering plants. Table 30.1 provides a summary of the critical innovations of all modern seed plants.

Gymnosperms are plants that produce seeds that are exposed rather than enclosed in fruits, as is the case for angiosperms. The word gymnosperm comes from the Greek *gymnos*, meaning naked (referring to the unclothed state of ancient athletes), and *sperma*, meaning seed. Most modern gymnosperms are woody plants that occur as shrubs or trees. Seeds and wood are critical innovations that allow gymnosperms to cope with global climate changes and to live in relatively cold and dry habitats. In this section, we will first consider some fossil plants that help to explain important gymnosperm traits. Then we will survey the structure, reproduction, ecological roles, and human uses of modern gymnosperm phyla.

Modern Gymnosperms Arose from Woody Ancestors

Modern gymnosperms include the famous giant sequoias (*Sequoiadendron giganteum*) native to the Sierra Nevada mountains of the western U.S. Giant sequoias are among Earth's largest organisms,

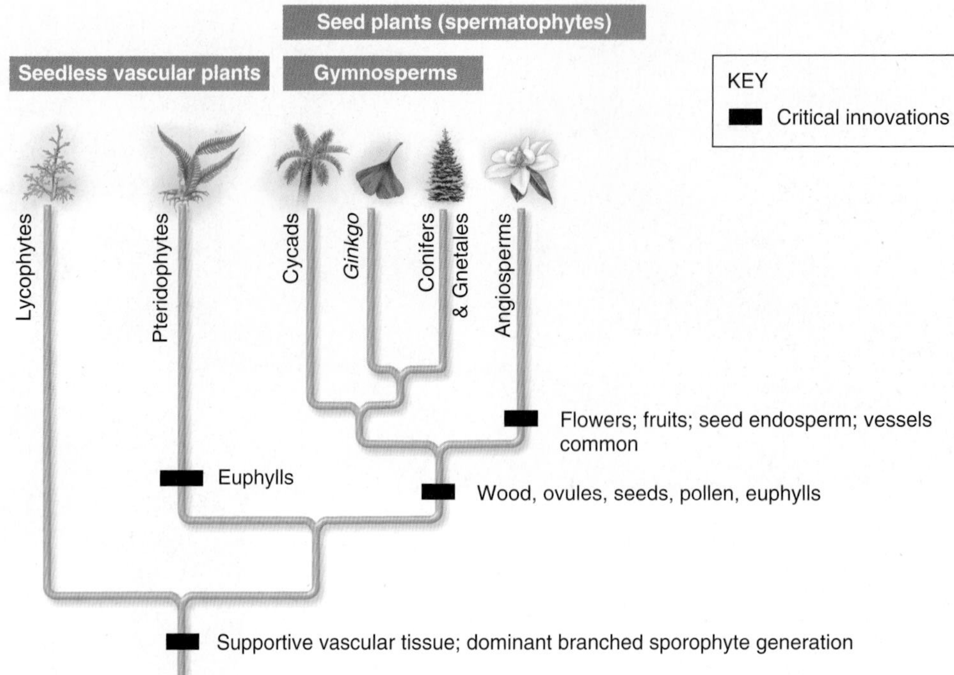

Figure 30.1 **A phylogeny of modern vascular seedless and seed plants.**

Table 30.1 Critical Innovations of Seed Plant Groups

Plant group	Innovation	Advantages
All seed plants	Vascular cambium that makes wood and inner bark	Seed plants have the potential to grow tall and produce many branches and reproductive structures.
	Pollen, ovules, seeds	Pollen allows seed plants to disperse male gametophytes. Ovules provide protection and nutrition to female gametophytes and developing embryos. Seeds allow seed plants to reproduce in arid or shady habitats.
Conifers	Tracheid torus	Fosters water flow in arid or cold conditions
	Scales or needle-shaped leaves	Retard water loss from leaf surface
	Conical shape	Sheds snow, preventing damage
	Resin	Protects against pathogens and herbivores
Angiosperms	Flowers	Foster pollen dispersal, ovule protection, pollination, and seed production
	Fruits	Foster seed dispersal
	Endosperm	Efficiently provides food to embryo of developing seed
	Vessels	Relatively wide diameter fosters water flow
	Many secondary compounds	Provide flower colors and fragrances and protect against herbivores

weighing as much as 6,000 tons and reaching an amazing 100 m in height. The large size of sequoias and other trees is based on the presence of **wood**, a tissue composed of numerous pipelike arrays of empty, water-conducting cells whose walls are strengthened by an exceptionally tough polymer known as lignin. These properties enable woody tissues to transport water upward for great distances and also to provide the structural support needed for trees to grow tall and produce many branches and leaves. In modern seed plants, a special tissue known as the **vascular cambium** produces both thick layers of wood and thinner layers of inner bark. The **inner bark** transports watery solutions of organic compounds. (The structure and function of the vascular cambium, wood, and inner bark are described in more detail in Chapter 35.) Vascular cambium, wood, and inner bark help gymnosperms and woody angiosperms to compete effectively for light and other resources needed for photosynthesis.

Wood first appeared in a group of ancient plants known as the **progymnosperms** (from the Greek, meaning before gymnosperms). Woody progymnosperms, such as the fossil plant *Archaeopteris*, which lived 370 million years ago (mya), were the first trees that had leafy twigs (**Figure 30.2**). The vascular tissue of progymnosperms was arranged in a ring around a central pith of nonvascular tissue. (The vascular tissue of earlier tracheophytes was arranged differently.) This ring of vascular tissue, known as a **eustele**, contained cells that were able to develop into the vascular cambium as seedlings grew into saplings. The vascular cambium then produced wood, allowing saplings to grow into tall trees. Modern seed plants inherited the eustele, explaining why many gymnosperms and angiosperms are also able to produce vascular cambia and wood. Although progymnosperms were woody plants, fossil evidence indicates that they did not produce seeds. This observation reveals that wood originated before the evolution of seeds, which first appeared more than 300 mya. The evolutionary origin of seeds, discussed in Chapter 29, stimulated the rapid diversification of gymnosperms.

Figure 30.2 An early forest in which the only trees were the progymnosperm *Archaeopteris*. This illustration was reconstructed from fossil data.

BioConnections: *Look ahead to Figure 54.26a–e. In what way did ancient Archaeopteris forests differ from most forests of the present time?*

The greatest diversity of gymnosperms occurred during the Mesozoic era, when gymnosperms were the major vegetation present. This period was also known as the Age of Dinosaurs, and gymnosperms are thought to have been the major food for plant-eating dinosaurs during most of their history. Some groups of gymnosperms became extinct before or as a result of the K/T event at the end of the Cretaceous period 65 mya (see Chapter 29). Only a few gymnosperm phyla have survived to modern times: cycads (the Cycadophyta); *Ginkgo biloba*, the only surviving member of a once-large phylum termed Ginkgophyta; and conifers (the Coniferophyta) with about 800 species. These phyla display distinctive reproductive features and play important roles in ecology and human affairs.

Cycads Are Endangered in the Wild But Are Widely Used as Ornamentals

Cycads are regarded as the earliest diverging modern gymnosperm phylum, originating more than 300 mya. Nearly 300 cycad species occur today, primarily in tropical and subtropical regions. However, many species of cycads are rare, and their tropical forest homes are increasingly threatened by human activities. Many cycads are listed as endangered, and commercial trade in cycads is regulated by CITES (Convention on International Trade in Endangered Species of Wild Fauna and Flora), a voluntary international agreement between governments to protect such species.

(a) Emergent cycad stem

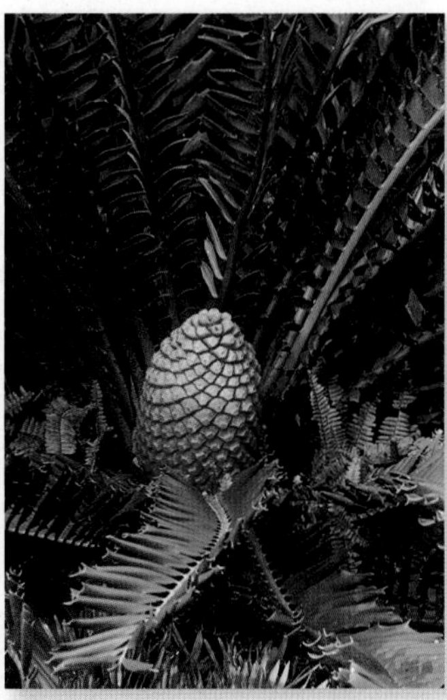

(b) Submergent cycad stem

Figure 30.3 Cycads. Palmlike foliage and conspicuous seed-producing cones are features of most cycads. **(a)** The stems of some cycads emerge from the ground. **(b)** The stems of other cycads are submerged in the ground, so the leaves emerge at ground level.

The structure of cycads is so interesting and attractive that many species are cultivated for use in outdoor plantings or as houseplants. The nonwoody stems of some cycads emerge from the ground much like tree trunks, some reaching 15 m in height, whereas the stems of other cycads are not conspicuous because they are subterranean (**Figure 30.3**). Cycads display spreading, palmlike leaves (*cycad* comes from a Greek word, meaning palm). Mature leaves of the African cycad *Encephalartos laurentianus* can reach an astounding 8.8 m in length!

In addition to underground roots, which provide anchorage and take up water and minerals, many cycads produce coralloid roots. Such roots extend aboveground and have branching shapes resembling corals (**Figure 30.4a**). Coralloid roots harbor light-dependent, photosynthetic cyanobacteria within their tissues. The cyanobacteria, which form a bright blue-green ring beneath root surfaces (**Figure 30.4b**), convert atmospheric nitrogen (N_2) into ammonia (NH_3), providing their plant hosts with nitrogen minerals crucial to their growth (see Chapter 37).

Cycad reproduction is distinctive in several ways. Individual cycad plants produce conspicuous conelike structures that bear either ovules and seeds or pollen (see Figure 30.3). When mature, both types of reproductive structures emit odors that attract beetles. These insects carry pollen to ovules, where the pollen produces tubes that deliver sperm to eggs.

Ginkgo biloba Is the Last Survivor of a Once-Diverse Group

The beautiful tree *Ginkgo biloba* (**Figure 30.5a**) is the single remaining species of a phylum that was much more diverse during the Age of Dinosaurs. *G. biloba* takes its species name from the two-lobed

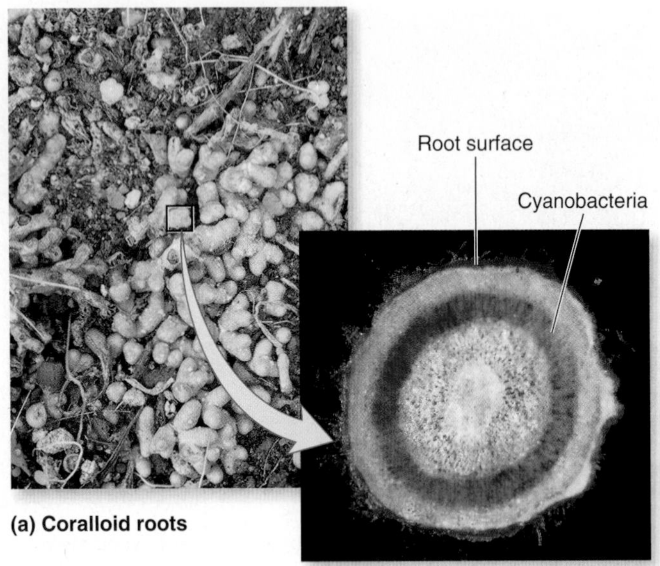

(a) Coralloid roots

(b) Coralloid root cross section

Figure 30.4 Coralloid roots of cycads. (a) Many cycads produce aboveground branching roots that resemble branched corals. (b) This magnified cross section of a coralloid root shows a ring of symbiotic blue-green cyanobacteria, which provide the plant with a form of nitrogen that can be used to make essential cellular compounds.

Concept Check: Why do the coralloid roots grow aboveground?

shape of its leaves, which have unusual forked veins (Figure 30.5b). Today, *G. biloba* may be nearly extinct in the wild; widely cultivated modern *Ginkgo* trees are descended from seeds produced by a tree found in a remote Japanese temple garden and brought to Europe by 17th-century explorers.

G. biloba trees are widely planted along city streets because they are ornamental and also tolerate cold, heat, and pollution better than many other trees. In addition, these trees are long-lived—individuals can live for more than a thousand years and grow to 30 m in height. Individual trees produce either ovules and seeds or pollen, based on a sex chromosome system much like that of humans. Ovule-producing trees have two X chromosomes; pollen-producing trees have one X and one Y chromosome. Wind disperses pollen to ovules, where pollen grains germinate to produce pollen tubes. These tubes grow through ovule tissues for several months, absorbing nutrients that are used for sperm development. Eventually the pollen tubes burst, delivering flagellate sperm to egg cells. After fertilization, zygotes develop into embryos, and the ovule integument develops into a fleshy, bad-smelling outer seed coat and a hard, inner seed coat (Figure 30.5c). For street-side or garden plantings, people usually select the pollen-producing trees to avoid the stinky seeds.

Conifers Are the Most Diverse Modern Gymnosperm Lineage

The conifers (Figure 30.6) are a lineage of trees named for their seed cones, of which pinecones are familiar examples. Modern conifer families include more than 50 genera. Conifers are particularly common in mountain and high-latitude forests and are important sources of wood and paper pulp.

Conifers produce simple pollen cones and more complex ovule-bearing cones (Figure 30.7). The pollen cones of conifers bear many leaflike structures, each bearing a microsporangium in which meiosis occurs and pollen grains develop. The ovule cones, also called seed cones, are composed of many short branch systems that bear ovules. Ovules contain female gametophytes, within which eggs develop.

When conifer pollen is mature, it is released into the wind, which transports pollen to ovules. When released from pollen tubes, sperm fuse with eggs, generating zygotes that grow into the embryos within seeds. Altogether, it takes nearly 2 years for pine (the genus *Pinus*) to complete the processes of male and female gamete development, fertilization, and seed development (see Figure 30.7). The seeds of pine and some other conifers develop wings that aid in wind dispersal (Figure 30.8a). Other conifers, such as yew and juniper, produce seeds or cones with bright-colored, fleshy coatings that are attractive to birds, which help to disperse the seeds (Figure 30.8b,c).

Conifer wood contains many specialized vascular cells known as tracheids that are adapted for efficient water and mineral conduction even in dry conditions. Like the tracheids of other vascular plants,

(a) *Ginkgo biloba* tree

(b) *Ginkgo biloba* leaf

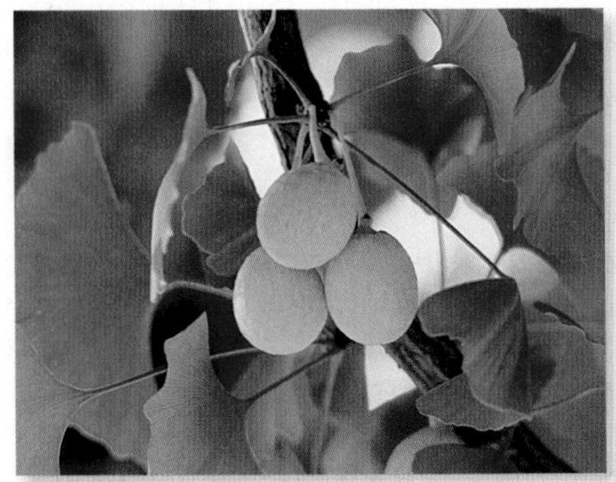

(c) *Ginkgo biloba* seeds

Figure 30.5 *Ginkgo biloba*. (a) A *Ginkgo biloba* tree; (b) fan-shaped leaves with forked veins; and (c) seeds with fleshy, foul-smelling seed coats.

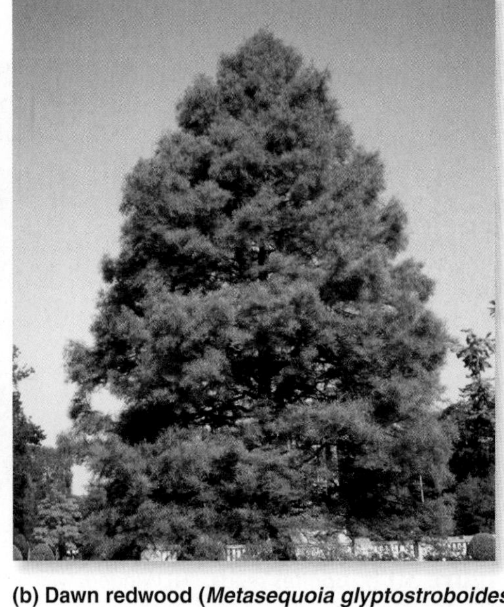

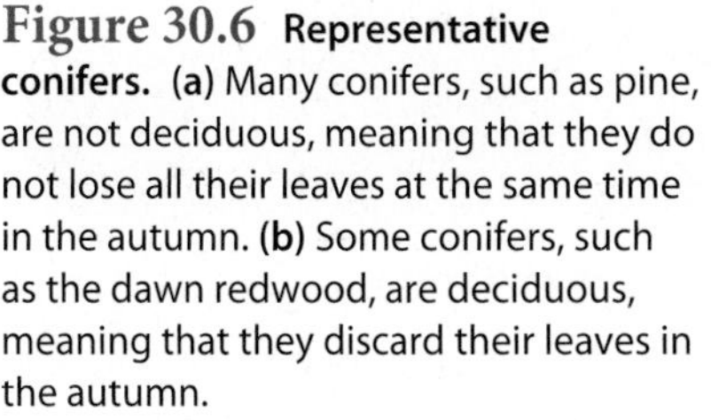

Figure 30.6 Representative conifers. **(a)** Many conifers, such as pine, are not deciduous, meaning that they do not lose all their leaves at the same time in the autumn. **(b)** Some conifers, such as the dawn redwood, are deciduous, meaning that they discard their leaves in the autumn.

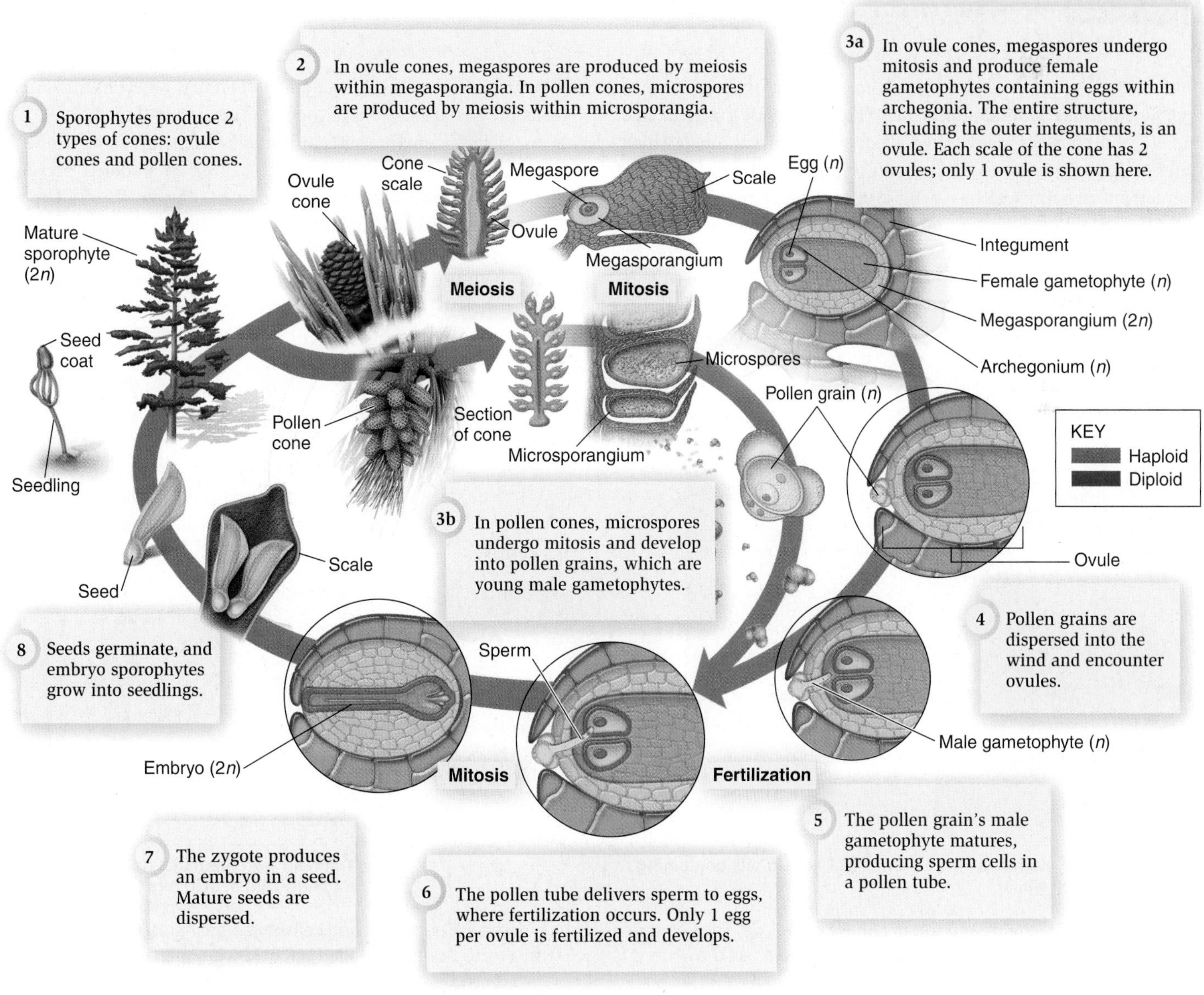

1 Sporophytes produce 2 types of cones: ovule cones and pollen cones.

2 In ovule cones, megaspores are produced by meiosis within megasporangia. In pollen cones, microspores are produced by meiosis within microsporangia.

3a In ovule cones, megaspores undergo mitosis and produce female gametophytes containing eggs within archegonia. The entire structure, including the outer integuments, is an ovule. Each scale of the cone has 2 ovules; only 1 ovule is shown here.

3b In pollen cones, microspores undergo mitosis and develop into pollen grains, which are young male gametophytes.

4 Pollen grains are dispersed into the wind and encounter ovules.

5 The pollen grain's male gametophyte matures, producing sperm cells in a pollen tube.

6 The pollen tube delivers sperm to eggs, where fertilization occurs. Only 1 egg per ovule is fertilized and develops.

7 The zygote produces an embryo in a seed. Mature seeds are dispersed.

8 Seeds germinate, and embryo sporophytes grow into seedlings.

Mature sporophyte (2*n*)
Ovule cone
Cone scale
Megaspore
Ovule
Scale
Megasporangium
Egg (*n*)
Integument
Female gametophyte (*n*)
Megasporangium (2*n*)
Archegonium (*n*)
Meiosis
Mitosis
Microspores
Section of cone
Microsporangium
Pollen cone
Pollen grain (*n*)
KEY
Haploid
Diploid
Ovule
Seed coat
Seedling
Seed
Scale
Sperm
Male gametophyte (*n*)
Embryo (2*n*)
Mitosis
Fertilization

Figure 30.7 **The life cycle of the genus *Pinus*.**

BIOLOGY PRINCIPLE A principle of biology is that living organisms grow and develop. This diagram illustrates the entire seed-to-seed growth and development cycle of conifers.

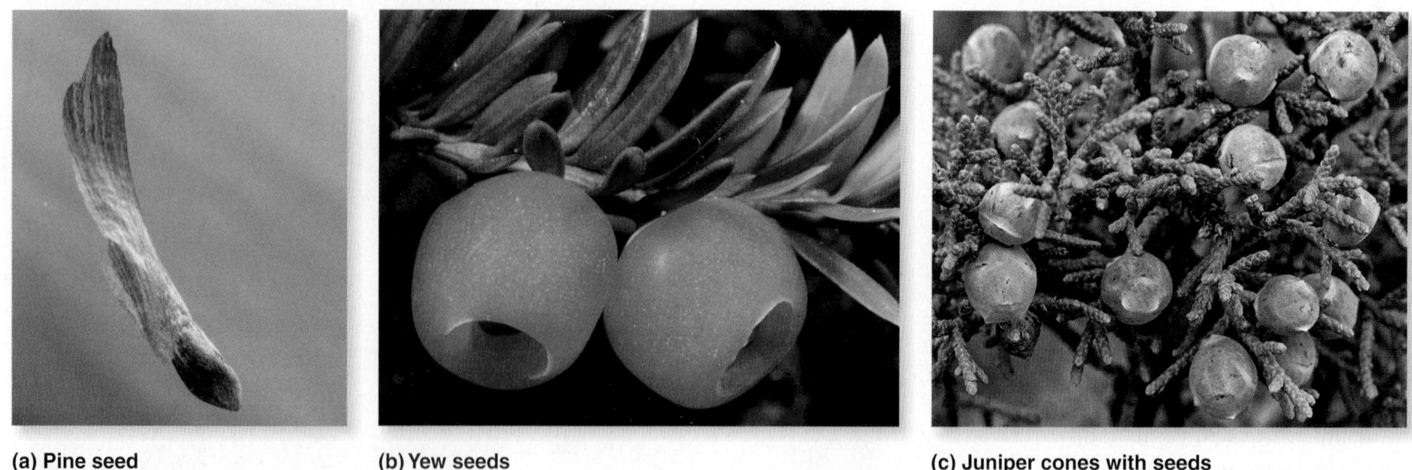

Figure 30.8 Conifer seeds. **(a)** Winged, wind-dispersed seed of the genus *Pinus*. **(b)** Fleshy-coated, bird-dispersed seeds of yew (*Taxus baccata*). **(c)** Fleshy cones of juniper (*Juniperus scopularum*) contain one or more seeds and are dispersed by birds. Juniper seeds are used to flavor gin.

BioConnections: Look forward to Figure 30.21. How are wind-dispersed pine seeds similar to wind-dispersed fruits of the angiosperm maple?

those of conifers are devoid of cytoplasm and occur in long columns that function like plumbing pipelines (**Figure 30.9a**). Tracheid side and end walls possess many thin-walled, circular **pits** through which water moves both vertically and laterally from one tracheid to another. Conifer pits are unusual in having a porous outer region that lets water flow through and a nonporous, flexible central region called the **torus** (plural, tori) that functions like a valve (**Figure 30.9b**). If conifer tracheids become dry, a common event in arid or cold habitats, they fill with air and are no longer able to conduct water. In this case, the torus presses against the pit opening, sealing it (**Figure 30.9c**). The torus valve thereby prevents air bubbles from spreading to the next tracheid. This conifer adaptation localizes air bubbles, preventing them from stopping water conduction in other tracheids. The presence of tori in their tracheids helps to explain why conifers have been so successful for hundreds of millions of years. Conifer wood (and leaves) may also display conspicuous resin ducts, passageways for the

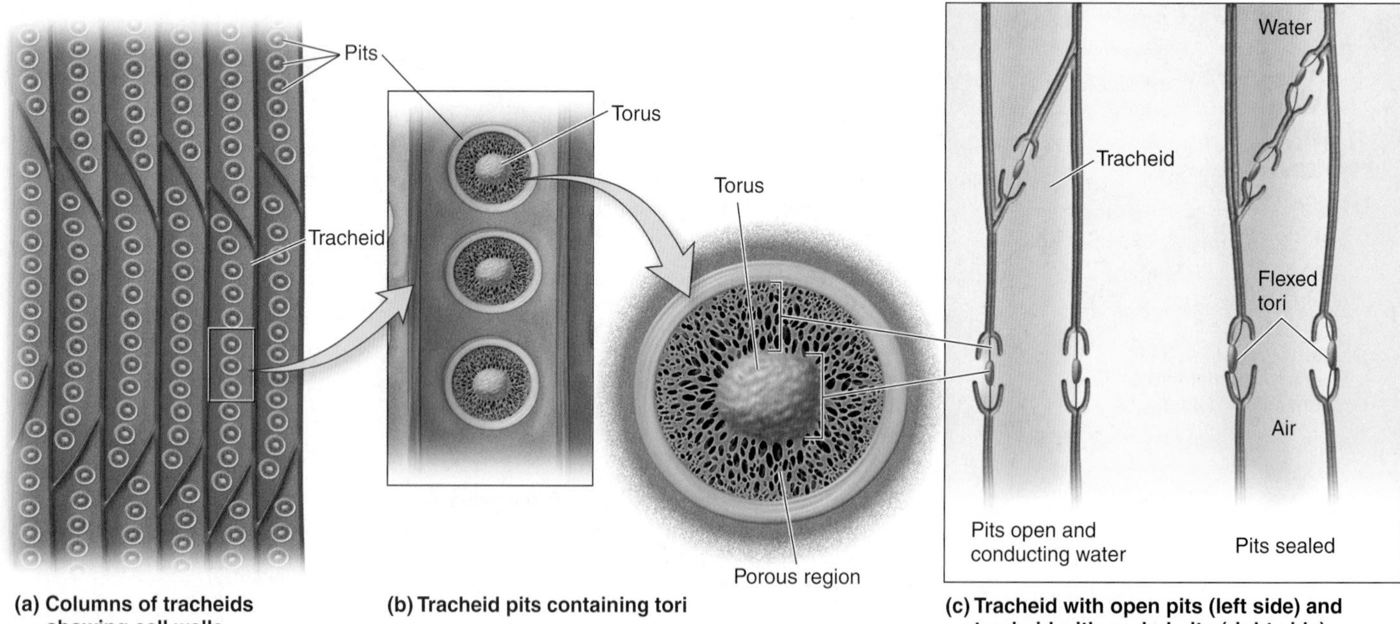

Figure 30.9 Tracheids and tori in conifer wood. **(a)** The lignin-rich cell walls of the water-conducting cells called tracheids. **(b)** Detailed view of a portion of a tracheid that shows the thin-walled areas known as pits, each with a torus. **(c)** A water-filled tracheid with open pits and an air-filled tracheid with pits sealed by the flexed tori.

BIOLOGY PRINCIPLE A principle of biology is that structure determines function. This illustration shows how tori in water-conducting cells of conifers aid survival in arid or cold habitats.

flow of syrup-like resin that helps to prevent attack by pathogens and herbivores. Resin that exudes from tree surfaces may trap insects and other organisms, then harden in the air and fossilize, preserving the inclusions in amber.

Many conifers occur in cold climates and thus display numerous adaptations to such environments. Their conical shapes and flexible branches help conifer trees shed snow, preventing heavy snow accumulations from breaking branches. People who use conifers in landscape plantings also value these traits. Conifer leaf shape and structure are adapted to resist damage from drought that occurs in both summer and winter, when liquid water is scarce. Conifer leaves are often scalelike (**Figure 30.10a**) or needle-shaped (**Figure 30.10b**); these shapes reduce the area of leaf surface from which water can evaporate. In addition, a thick, waxy cuticle coats conifer leaf surfaces (**Figure 30.10c**), retarding water loss and attack by disease organisms.

Many conifers are evergreen; that is, their leaves live for more than 1 year before being shed and are not all shed during the same season. Retaining leaves through winter helps conifers start up photosynthesis earlier than deciduous trees, which in spring must replace leaves lost during the previous autumn. Evergreen leaves thus provide an advantage in the short growth season of alpine or high-latitude environments. However, some conifers do lose all their leaves in the autumn. The bald cypress (*Taxodium distichum*) of southern U.S. floodplains, tamarack (*Larix laricina*) of northern bogs, and dawn redwood (*Metasequoia glyptostroboides*; see Figure 30.6b) are examples of deciduous conifers. Fossils indicate that dawn redwoods once grew abundantly across wide areas of the Northern Hemisphere until a few million years ago and then disappeared. However, in the 1940s, a forester found a living dawn redwood growing in a remote Chinese village, and subsequent expeditions located forests of the conifers. Since then, dawn redwood trees have been widely planted as ornamentals, prized for their attractive foliage and cones. As recently as 1994, botanists found a previously unknown conifer species, *Wollemia nobilis*, in an Australian national park. Like the dawn redwood, *Wollemia* is an attractive tree that is likely to become more widely distributed as the result of human cultivation.

The conifer phylum also includes the Gnetales, an order of three genera, *Gnetum*, *Ephedra*, and *Welwitschia* that feature distinctive adaptations. *Gnetum* is unusual among modern gymnosperms in having broad leaves similar to those of many tropical plants (**Figure 30.11a**). Such leaves foster light capture in the dim forest habitat. More than 30 species of the genus *Gnetum* occur as vines, shrubs, or trees in tropical Africa or Asia. *Ephedra*, native to arid regions of the southwestern U.S., has tiny brown scalelike leaves and green, photosynthetic stems (**Figure 30.11b**). These adaptations help to conserve water by preventing water loss that would otherwise occur from the surfaces of larger leaves. *Ephedra* produces secondary metabolites that aid in plant protection but also affect human physiology. Early settlers of the western U.S. used *Ephedra* to treat colds and other medical conditions. In fact, the modern decongestant drug pseudoephedrine is based on the chemical structure of ephedrine, which was named for and originally obtained from *Ephedra*. Pseudoephedrine sales are now restricted in many places because this compound can be used as a starting point for the synthesis of illegal drugs. Ephedrine has also been used to enhance sports performance, a practice that has elicited medical concern.

(a) Scale-shaped leaves of Eastern red cedar

(b) Needle-shaped leaves of pine

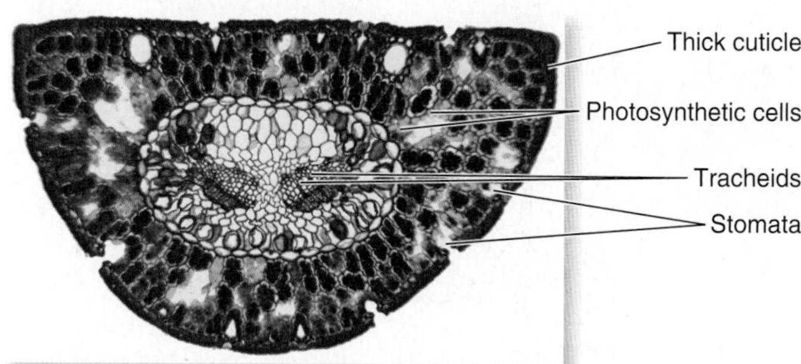

(c) Stained cross section of pine needle, showing the thick cuticle

Figure 30.10 **Conifer leaves.** The leaves of conifers are typically shaped as small scales or long needles, with similar internal structure.

Concept Check: *In what ways are conifer leaves adapted to resist water loss from their surfaces?*

Welwitschia has only one living representative species. *Welwitschia mirabilis* is a strange-looking plant that grows in the coastal Namib Desert of southwestern Africa, one of the driest places on Earth (**Figure 30.11c**). A long taproot anchors a stubby stem that barely emerges from the ground. Two very long leaves grow from the

(a) Genus *Gnetum*

(b) *Ephedra californica*

(c) *Welwitschia mirabilis*

Figure 30.11 Gnetales. **(a)** A tropical plant of the genus *Gnetum*, displaying broad leaves and reproductive structures. **(b)** *Ephedra californica* growing in deserts of North America, showing minuscule brown leaves on green, photosynthetic stems and reproductive structures. **(c)** *Welwitschia mirabilis* growing in the Namib Desert of southwestern Africa, showing long, wind-shredded leaves and reproductive structures.

stem but are rapidly shredded by the wind into many strips. The plant is thought to obtain most of its water from coastal fog that accumulates on the leaves, explaining how *W. mirabilis* can grow and reproduce in such a dry place.

30.2 The Evolution and Diversity of Modern Angiosperms

Learning Outcomes:

1. List four flower organs and their functions, and explain how each flower part may have first evolved.
2. Describe how diversification of flowers and fruits enhances seed production and dispersal.
3. Name three major types of angiosperm secondary metabolites and explain how these affect animals.

More than 124 mya, one extinct gymnosperm group, although it's unclear which one, gave rise to the angiosperms—the flowering plants. Charles Darwin famously referred to the origin of the flowering plants as "an abominable mystery," one that has not been fully solved even today. Angiosperms retained many structural and reproductive features from ancestral seed plants, but also evolved several traits not found or seldom found among other land plants.

Flowers and fruits are two of the defining features of angiosperms (Figure 30.12), because these features do not occur in other modern plants. The term **angiosperm** is from Greek words, meaning enclosed seed, which reflects the presence of seeds within fruits. The seed nutritive material known as endosperm is another defining feature of the flowering plants (see Chapters 29 and 39). Flowers, fruits, and seed endosperm are critical innovations that foster reproduction. Flowers foster seed production, fruits favor seed dispersal, and endosperm food helps embryos within seeds grow into seedlings. In addition, most angiosperms possess distinctive water-conducting cells, known as **vessels**, which are wider than tracheids and therefore increase the efficiency of water flow through plants. Although similar conducting cells occur in some seedless plants and certain gymnosperms, the vessels of angiosperms are thought to have evolved independently.

Although humans obtain wood, medicines, and other valuable products from gymnosperms, we depend even more on the

Figure 30.12 Angiosperm flowers and fruits. Citrus plants display the critical innovations of flowering plants: the flowers and fruits shown here and seed endosperm (not shown).

Concept Check: *What other trait occurs widely among angiosperms but rarely among other plants?*

angiosperms. Our food, beverages, and spices—flavored by an amazing variety of secondary metabolites—primarily come from flowering plants. People surround themselves with ornamental flowering plants and decorative items displaying flowers or fruit. We also commonly use flowers and fruit in ceremonies. In this section, we focus on how flowers, fruits, and secondary metabolites played key roles in angiosperm diversification. We will also learn that features of flowers, fruits, and secondary metabolites are used to classify and identify angiosperm species.

Flower Organs Evolved from Leaflike Structures

Flowers are complex reproductive structures that are specialized for the efficient production of pollen and seeds. The sexual reproduction process of angiosperms depends on flowers. As the flowering plants diversified, flowers of varied types evolved as reproductive adaptations to differing environmental conditions. To understand this process, we can start by considering the basic flower parts and their roles in reproduction.

Flower Parts and Their Reproductive Roles Flowers are produced at stem tips, and may contain four types of organs: sepals, petals, pollen-producing stamens, and ovule-producing carpels (**Figure 30.13**). These flower organs are supported by tissue known as a **receptacle**, located at the tip of a flower stalk—a **pedicel**. The functioning of several genes that control flower organ development explains why carpels are the most central flower organs, why stamens surround carpels, and why petals and sepals are the outermost flower organs (refer back to Figure 19.24).

Many flowers produce attractive **petals** that play a role in **pollination**, the transfer of pollen among flowers. **Sepals** of many flowers are green and form the outer layer of flower buds. By contrast, the sepals of other flowers look similar to petals, in which case both

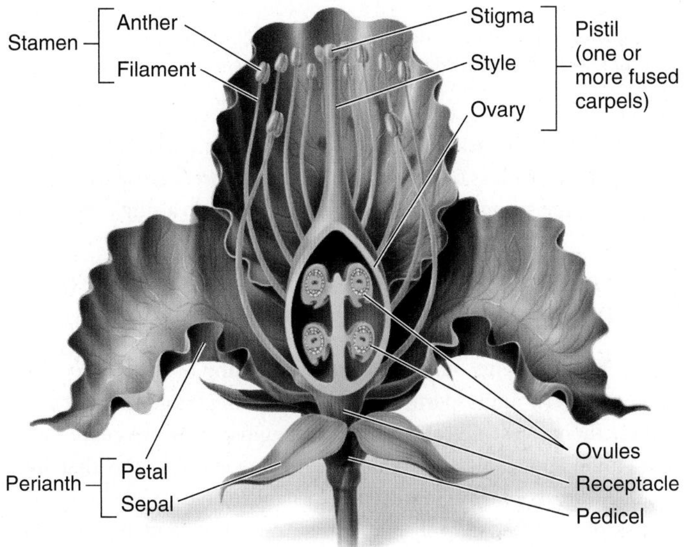

Figure 30.13 **Generalized flower structure.** Although flowers are diverse in size, shape, and color, they commonly have the parts illustrated here.

sepals and petals are known as **tepals**. All of a flower's sepals and petals are collectively known as the **perianth**. Most flowers produce one or more **stamens**, the structures that produce and disperse pollen. Most flowers also contain a single or multiple **carpels**, structures that produce ovules.

Some flowers lack perianths, stamens, or carpels. Flowers that possess all four types of flower organs are known as **complete flowers**, and flowers lacking one or more organ types are known as **incomplete flowers**. Flowers that contain both stamens and carpels are described as **perfect flowers**, and flowers lacking either stamens or carpels are **imperfect flowers**.

Flowers also differ in the numbers of organs they produce. Some flowers produce only a single carpel, others display several separate carpels, and many possess several carpels that are fused together into a compound structure. Both a single carpel and compound carpels are referred to as a **pistil** (from the Latin *pistillum*, meaning pestle) because it resembles the device people use to grind materials to powder in a mortar (see Figure 30.13). Only one pistil is present in flowers that have only one carpel and in flowers with fused carpels. By contrast, flowers possessing several separate carpels display multiple pistils.

Pistil structure can be divided into three regions having distinct functions. A topmost portion of the pistil, known as the **stigma**, receives and recognizes pollen of the appropriate species or genotype. The elongate middle portion of the pistil is called the **style**. The lowermost portion of the pistil is the **ovary**, which encloses and protects ovules.

During the flowering plant life cycle (**Figure 30.14**), the stigma allows pollen of appropriate genetic type to germinate, producing a long pollen tube that grows through the style. The pollen tube thereby delivers two sperm cells to ovules. In the distinctive angiosperm process known as **double fertilization**, one sperm fuses with the egg to form a zygote, and the other sperm fuses with other haploid cells of the female gametophyte. The latter is the first step in the development of a characteristic angiosperm nutritive tissue known as endosperm. Fed by the endosperm, the zygote develops into an embryo, and the ovule develops into a seed. Ovaries (and sometimes additional flower parts) develop into fruits.

Early Flowers Fossils of whole plants with recognizable flowers and fruits are known from geological deposits about 124 million years old, though molecular clock data and fossil pollen grains suggest that angiosperms may have originated earlier. Flowers were a critical innovation that led to extensive angiosperm diversification. Comparative studies of the structures of modern and fossil flowers suggest how modern stamens and carpels might have arisen.

Structural comparison and molecular data indicate that stamens are homologous to gymnosperm microsporophylls, leaflike structures that produce microspores (young pollen). Early fossil flowers and some modern flowers have broad stamens that are leaf-shaped, with elongated, pollen-producing microsporangia on the stamen surface (**Figure 30.15a**). In contrast, the stamens of most modern plants have narrowed to form **filaments**, or stalks, that elevate **anthers**, clusters of microsporangia that produce pollen and then open to release it (see Figure 30.13). Filaments and anthers are adaptations that foster pollen dispersal.

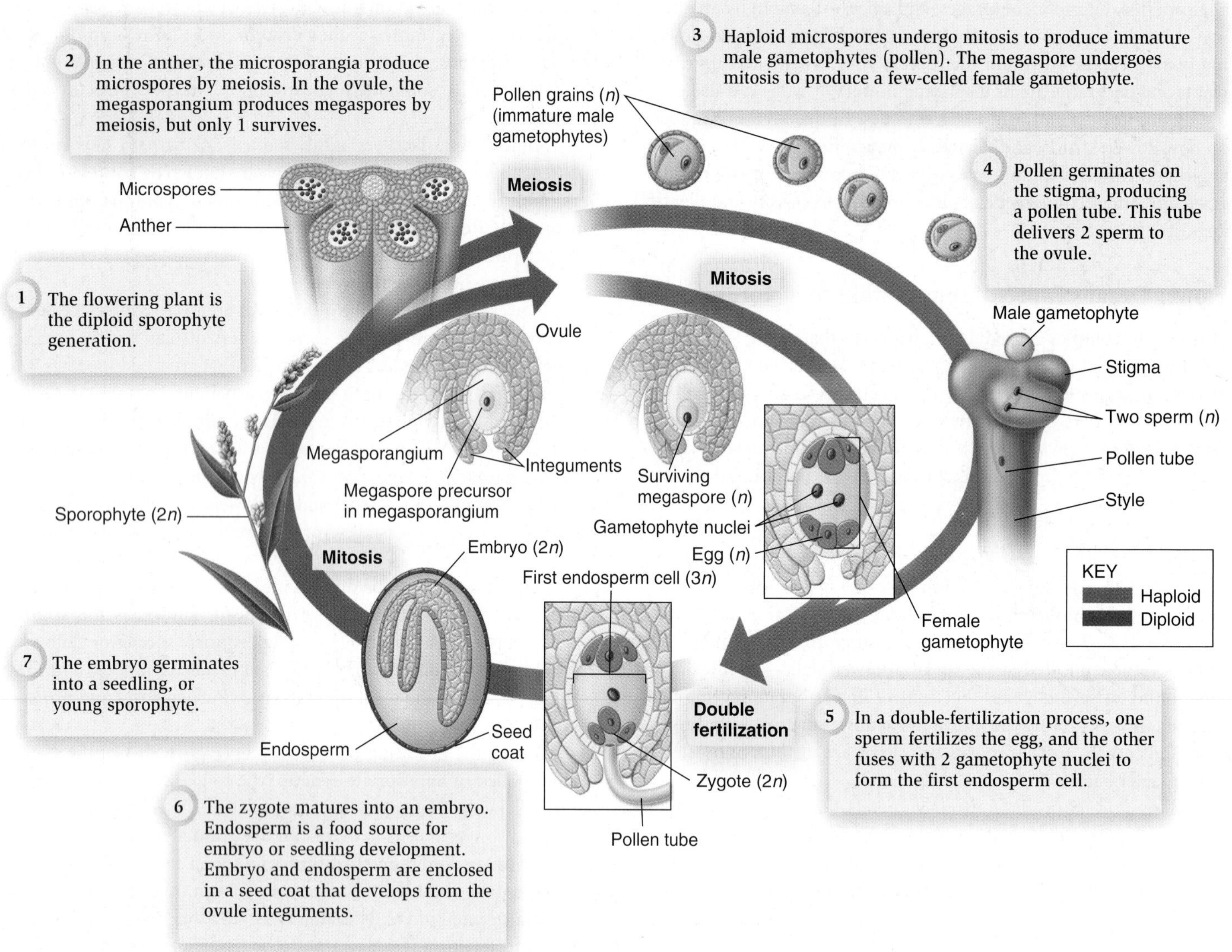

2 In the anther, the microsporangia produce microspores by meiosis. In the ovule, the megasporangium produces megaspores by meiosis, but only 1 survives.
Pollen grains (n) (immature male gametophytes)
3 Haploid microspores undergo mitosis to produce immature male gametophytes (pollen). The megaspore undergoes mitosis to produce a few-celled female gametophyte.
Microspores
Anther
Meiosis
4 Pollen germinates on the stigma, producing a pollen tube. This tube delivers 2 sperm to the ovule.
Mitosis
1 The flowering plant is the diploid sporophyte generation.
Male gametophyte
Ovule
Stigma
Two sperm (n)
Megasporangium
Integuments
Surviving megaspore (n)
Pollen tube
Megaspore precursor in megasporangium
Style
Sporophyte (2n)
Gametophyte nuclei
Mitosis
Embryo (2n)
Egg (n)
First endosperm cell (3n)
KEY
Haploid
Diploid
Female gametophyte
7 The embryo germinates into a seedling, or young sporophyte.
Double fertilization
5 In a double-fertilization process, one sperm fertilizes the egg, and the other fuses with 2 gametophyte nuclei to form the first endosperm cell.
Endosperm
Seed coat
Zygote (2n)
6 The zygote matures into an embryo. Endosperm is a food source for embryo or seedling development. Embryo and endosperm are enclosed in a seed coat that develops from the ovule integuments.
Pollen tube

Figure 30.14 The life cycle of a flowering plant, illustrated by the genus *Polygonum*. Flowering plant life cycles differ in length of the cycle and the number of cells and nuclei occurring in the female gametophyte, the seven-celled, eight-nuclei of *Polygonum* being common.

Plant biologists likewise hypothesize that carpels are homologous to gymnosperm megasporophylls, leaflike structures that bear ovules on their surfaces. In early angiosperms, such leaves folded over ovules, protecting them. In support of this hypothesis is the observation that the carpels of some early-diverging modern plants are leaflike structures that fold over ovules, with the carpel edges stuck together by secretions (**Figure 30.15b**). In contrast, most modern flowers produce carpels whose edges have fused together into a tube whose lower portion (ovary) encloses ovules. Plant biologists hypothesize that such evolutionary change increased ovule protection, which would improve plant fitness.

In contrast, flower sepals and petals have no recognizable homologs in modern gymnosperms. These perianth structures are unique to angiosperms, so plant biologists have long wondered how sepals and petals arose. Recent analyses reveal that the gene expression patterns of the pollen cones of gymnosperms (see Figure 30.7) share features with flower stamens, as expected, but also with the flower perianth. These new data suggest that perianth parts originated from stamen-like structures, by loss of sporangia. The first flowers arose when early stamens, carpels, and perianth parts aggregated into a single structure.

Flowering Plants Diversified into Several Lineages, Including Monocots and Eudicots

Figure 30.16 presents our current understanding of the relationships among modern angiosperm groups. According to gene-sequencing studies, the earliest-diverging modern angiosperms are represented by a single species called *Amborella trichopoda*, a shrub that lives in cloud forests on the South Pacific island of New Caledonia. The flowers of *A. trichopoda* display hypothesized ancient features. For example, the fairly small flowers have stamens with broad filaments and several separate carpels (**Figure 30.17**). *A. trichopoda* also lacks vessels in the water-conducting tissues. In contrast, typical angiosperm vessels are present in later-diverging groups of angiosperms,

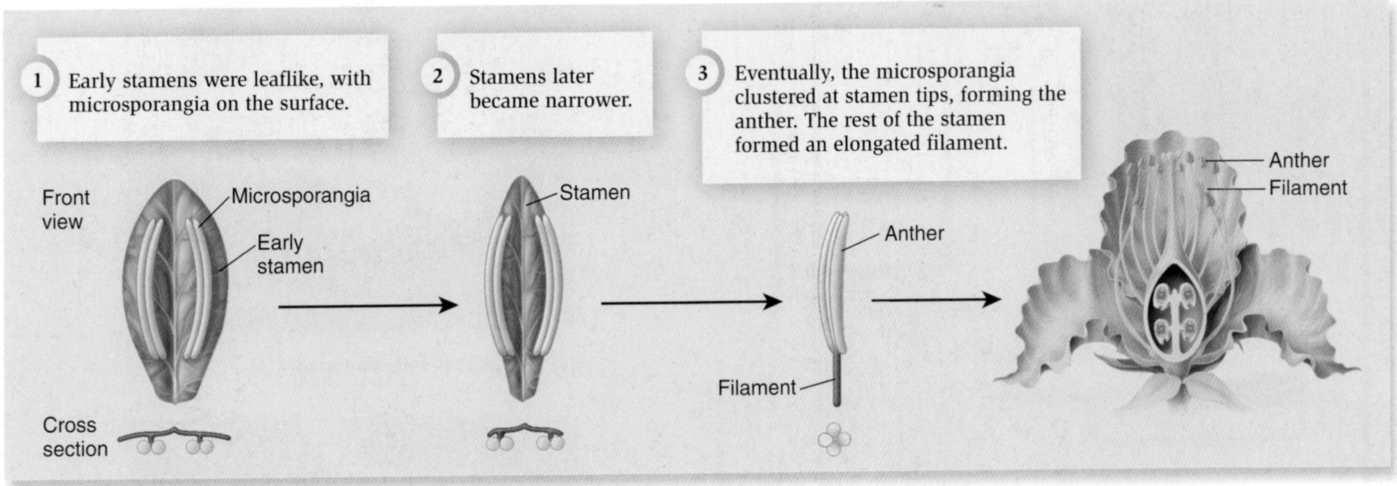

(a) Stamen evolution

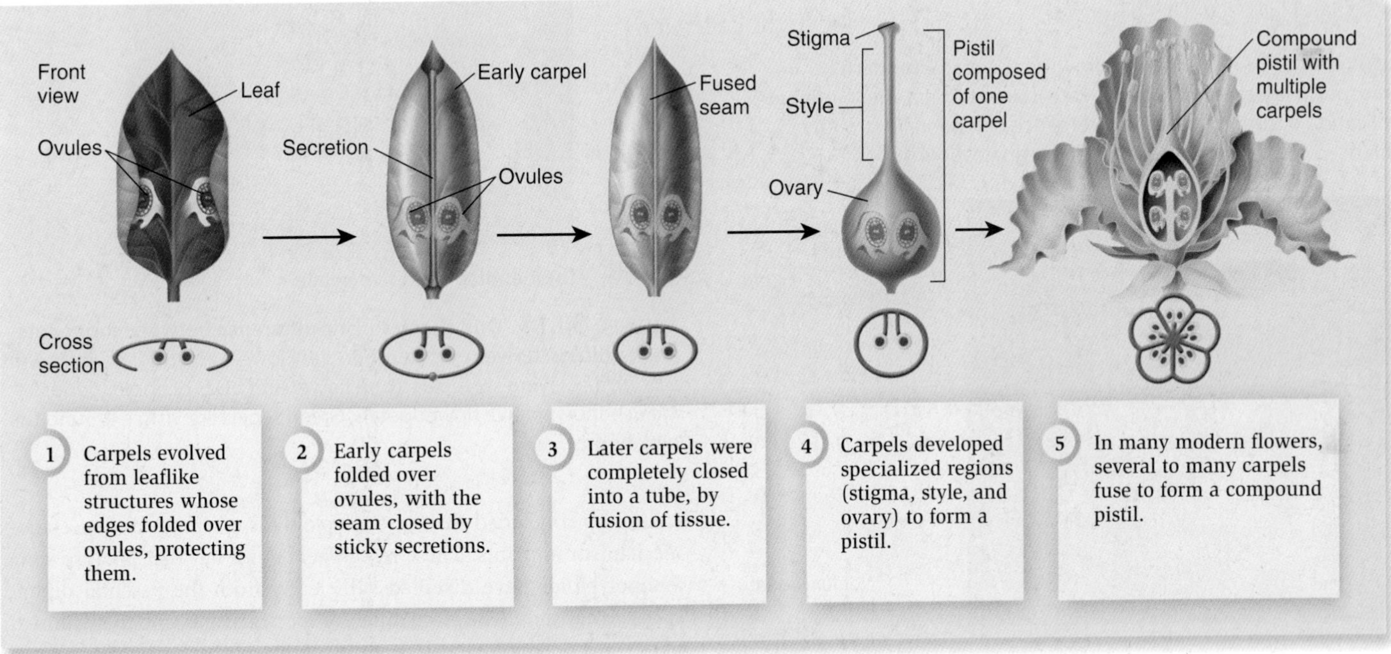

2 Stamens later became narrower.
3 Eventually, the microsporangia clustered at stamen tips, forming the anther. The rest of the stamen formed an elongated filament.
Front view
Leaf
Ovules
Secretion
Early carpel
Ovules
Fused seam
Stigma
Style
Ovary
Pistil composed of one carpel
Compound pistil with multiple carpels
Cross section
1 Carpels evolved from leaflike structures whose edges folded over ovules, protecting them.
2 Early carpels folded over ovules, with the seam closed by sticky secretions.
3 Later carpels were completely closed into a tube, by fusion of tissue.
4 Carpels developed specialized regions (stigma, style, and ovary) to form a pistil.
5 In many modern flowers, several to many carpels fuse to form a compound pistil.

(b) Carpel evolution

Figure 30.15 **Hypothetical evolutionary origin of stamens, carpels, and pistils.** Plant biologists test these models by searching for new fossils or generating additional molecular data.

including water lilies, the star anise plant, and other close relatives (see Figure 30.16). Magnoliids, represented by the genus *Magnolia*, are the next-diverging group. Magnoliids are closely related to two very large and diverse angiosperm lineages: the **monocots** and the **eudicots**.

Monocots and eudicots are named for differences in the number of embryonic leaves called cotyledons. Monocot embryos possess one cotyledon, whereas eudicots possess two cotyledons. Monocots differ from eudicots in several additional ways (look ahead to Table 35.1). For example, monocots typically have flowers with parts numbering three or some multiple of three (**Figure 30.18a**). In contrast, eudicot flower parts often occur in fours, fives, or a multiple of four or five (**Figure 30.18b**).

GENOMES & PROTEOMES CONNECTION

Whole-Genome Duplications Influenced the Evolution of Flowering Plants

Genome doubling, also known as polyploidy, occurs in a wide variety of eukaryotes and has happened frequently during the evolutionary history of plants. Recent molecular analyses indicate that the entire plant genome was duplicated after the divergence of *Amborella* and before the divergence of water lilies. Different patterns of whole-genome duplication occurred in monocots and eudicots after their divergence,

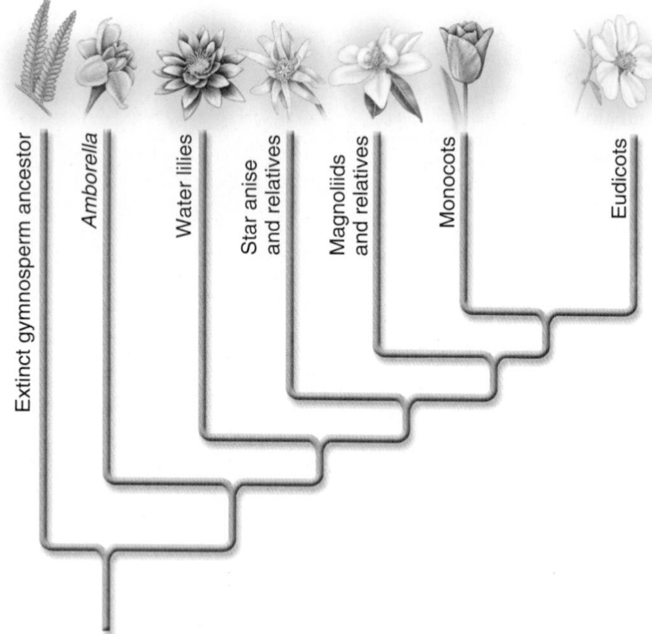

Figure 30.16 A phylogeny showing the major modern angiosperm lineages. Molecular data indicate that a whole-genome duplication occurring after the divergence of *Amborella* strongly influenced the diversification of modern angiosperms.

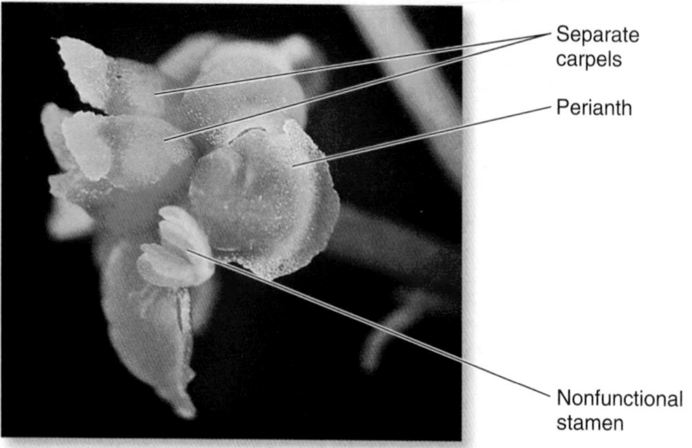

Figure 30.17 *Amborella trichopoda* flower, similar to a hypothesized early flower. This small flower is only about 3–4 mm in diameter. It displays several central, greenish carpels; nonfunctional stamens; and a pink perianth of tepals. This plant species also produces flowers that lack carpels but have many functional stamens.

(a) A monocot with six tepals

(b) A eudicot with five petals

Figure 30.18 One characteristic difference between monocots and eudicots: flower part number. **(a)** Flowers and buds of lily (genus *Lilium*), displaying six tepals. **(b)** A flower and buds of apple (genus *Malus*), showing five flower petals. Green sepals are visible around the pink buds.

and the plant genetic model *Arabidopsis* has undergone two or three whole-genome duplications. Whole-genome duplication has the potential to affect species' evolutionary pathways because it offers the opportunity for many genes to diverge, forming gene families.

Two major types of polyploidy occur in plants. **Autopolyploidy** occurs when homologous chromosome pairs do not separate during meiosis, a process known as nondisjunction. As a result, plants produce diploid spores, gametophytes, and gametes, when these life stages would otherwise be haploid. The mating of two diploid gametes produces a tetraploid plant—one having four sets of chromosomes. As the result of a large study of the occurrence of autoploidy in plants, American plant evolutionary biologists Douglas and Pamela Soltis and their colleagues concluded that autopolyploidy is an important speciation mechanism in plants. These researchers point out that because some autopolyploids have diverged sufficiently from the parental diploid species, they should be given different species names, and that failure to name such polyploids causes natural biodiversity to be underestimated.

A second major type of whole-genome duplication in plants involves an initial hybridization between two species (look back to Figure 25.11). Such a hybrid may not be able to produce viable spores because chromosomes may not pair properly at meiosis. However, if such a hybrid plant then undergoes a whole-genome duplication process, it produces an **allopolyploid**—an organism with two or more complete sets of chromosomes from two or more species. In allopolyploid species homologous chromosomes are available for pairing during meiosis, and viable spores, gametophytes, and gametes can form. Many common crops, including wheat and cotton, are allopolyploids, as are wild plants such as the desert sunflowers *Helianthus anomalus*, *H. deserticola*, and *H. paradoxus*. These sunflower hybrids are better adapted to survive drought conditions than are their parent species, *H. annuis* and *H. petiolaris*.

In addition to allopolyploidy, plants are known to obtain mitochondrial genes from other plant species by means of horizontal gene transfer. Genes have moved from one angiosperm species to another and between angiosperms and nonflowering plants, probably by mitochondrial fusion. For example, of the 31 genes present in the mitochondrial genome of the early-diverging flowering plant *Amborella trichopoda*, at least 20 were obtained from other angiosperms or mosses that grow on *A. trichopoda*'s surfaces. Because plant nuclear genomes commonly take up genes from organelles in the same cell, such foreign mitochondrial genes could end up in the nucleus.

Flower Diversification Has Fostered Efficient Seed Production

During the diversification of flowering plants, flower evolution has involved several types of changes that foster the transfer of pollen from one plant to another. Effective pollination is essential to efficient seed production because it minimizes the amount of energy plants must expend to accomplish sexual reproduction. Fusion of flower organs, clustering of flowers into groups, and reducing the perianth are some examples of changes leading to effective pollination.

Many flowers have fused petals that form floral tubes. Such tubes tend to accumulate sugar-rich nectar that provides a reward for **pollinators**, animals that transfer pollen among plants. The diameters of floral tubes vary among flowers and are evolutionarily tuned to the feeding structures of diverse animals, which range from the narrow tongues of butterflies to the wider bills of nectar-feeding birds (Figure 30.19). Nectar-feeding bats stick their heads into even larger tubular flowers to lap up nectar with their tongues. Orchids provide another example of ways in which flower parts have become fused; stamens and carpels are fused together into a single reproductive column that is surrounded by attractive tepals (Figure 30.20a). This arrangement of flower organs fosters orchid pollination by particular insects and is a distinctive feature of the orchid family.

Many plants produce **inflorescences**, groups of flowers tightly clustered together, which occur in several types. The sunflower family features a type of inflorescence in which many small flowers are clustered into a head (Figure 30.20b). The flowers at the center of a sunflower head function in reproduction and lack showy petals, but flowers at the rim have showy petals that attract pollinators. Flower heads allow pollinators to transfer pollen among a large number of flowers at the same time.

The grass family features flowers with few or no perianths, which explains why grass flowers are not showy (Figure 30.20c). This adaptation fosters pollination by wind, since petals would only get in the way of such pollen transfer.

Diverse Types of Fruits Function in Seed Dispersal

Fruits are structures that develop from ovary walls in diverse ways that aid the dispersal of enclosed seeds. Seed dispersal helps to prevent seedlings from competing with their larger parents for scarce resources such as water and light. Dispersal of seeds also allows plants to colonize new habitats. Diverse fruit types illustrate the many ways in which plants have become adapted for effective seed dispersal. Like flower types, fruit types are useful in classifying and identifying angiosperms.

Many mature angiosperm fruits, such as cherries, grapes, and lemons, are attractively colored, soft, juicy, and tasty (Figure 30.21a–c). Such fruits are adapted to attract animals that consume the fruits, digest the outer portion as food, and eliminate the seeds, thereby dispersing them. Hard seed coats prevent such seeds from being destroyed by the animal's digestive system. Strawberries are aggregate fruits, many fruits that all develop from a single flower having multiple pistils (Figure 30.21d). The ovaries of these

(a) Zinnia flower and butterfly

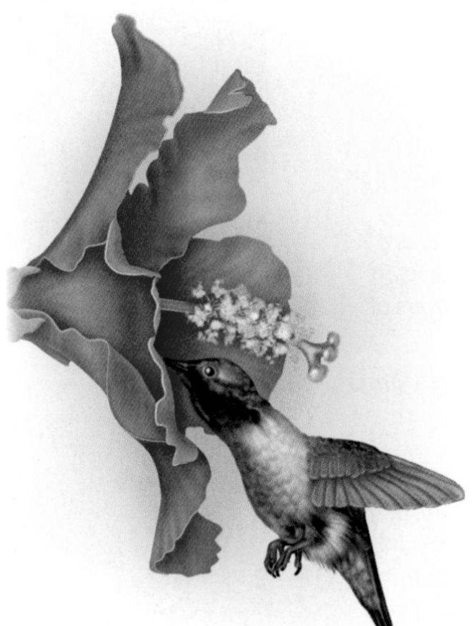

(b) Hibiscus flower and hummingbird

(c) Saguaro cactus flower and bat

Figure 30.19 **Flowers whose perianths form nectar-containing floral tubes of different widths that accommodate different pollinators.** **(a)** This zinnia is composed of an outer rim of showy flowers and a central disc of narrow tubular flowers that produce nectar. Butterflies, but not other pollinators, are able to reach the nectar by means of narrow tongues. **(b)** The hibiscus flower produces nectar in a floral tube whose diameter corresponds to the dimensions of a hummingbird bill. **(c)** The saguaro cactus (*Carnegiea gigantea*) flower forms a floral tube that is wide enough for nectar-feeding bats to get their heads inside. The cactus flower has been drawn here as if it were transparent, to illustrate bat pollination.

(a) An orchid flower with fused pistil and stamens

(b) A sunflower plant showing inflorescence

(c) Grass flowers lacking showy perianth

Figure 30.20 **Evolutionary changes in flower structure.** **(a)** An orchid of the genus *Cattleya* has fused stamens and pistil, and six tepals, one of which is specialized to form a lower lip. **(b)** An inflorescence (head) of sunflower (genus *Helianthus*). This inflorescence includes a rim of flowers with conspicuous petals that attract pollinators and an inner disc of flowers that lack attractive perianths. **(c)** Grass flowers of the genus *Triticum* lack a showy perianth.

Concept Check: *What advantage does the nonshowy perianth of grass flowers provide?*

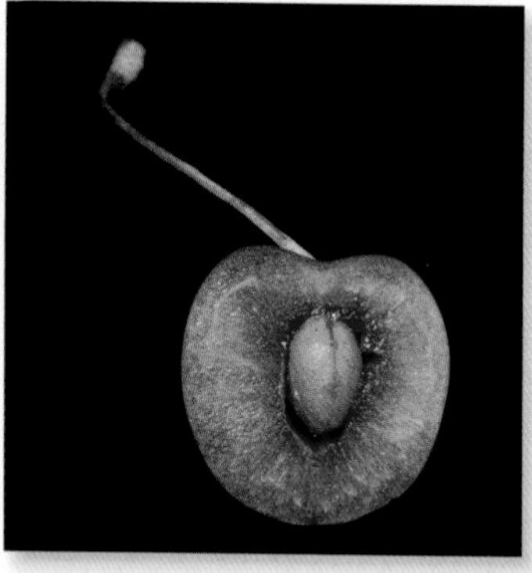

(a) A fleshy fruit (cherry)

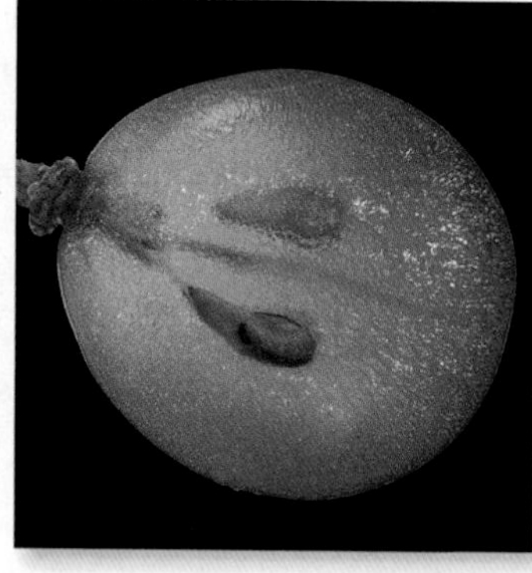

(b) A fleshy berry fruit (grape)

(c) A fleshy fruit (lemon)

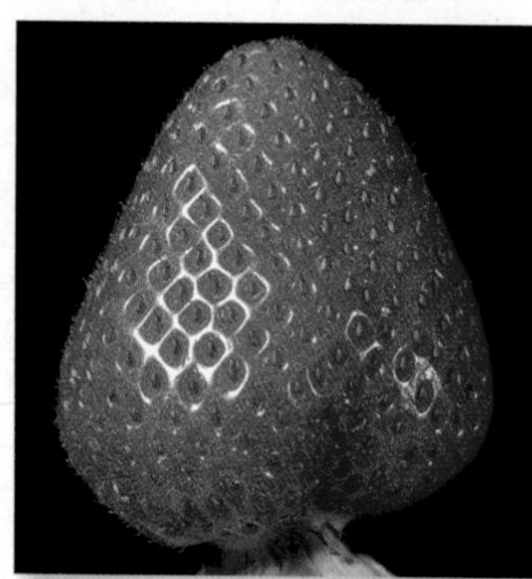

(d) An aggregate fruit (strawberry)

(e) A multiple fruit (pineapple)

(f) Legumes with dry pods (peas)

(g) Fruit with husk (coconut)

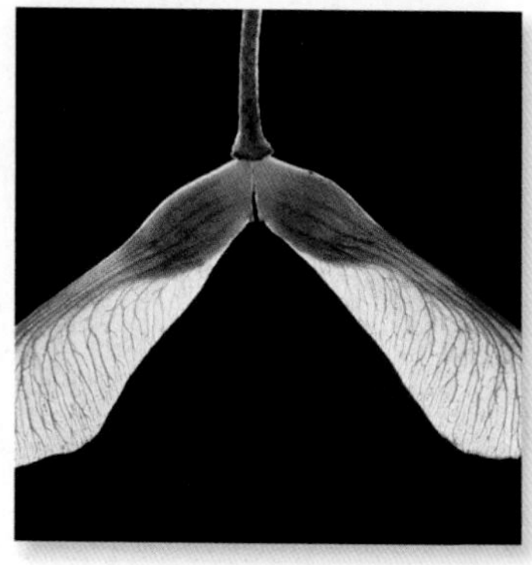

(h) A dry, winged fruit (maple)

Figure 30.21 **Representative fruit types.** **(a–c)** The cherry, grape, and lemon are fleshy fruits adapted to attract animals that consume the fruits and excrete the seeds. **(d)** Strawberry is an aggregate fruit, consisting of many tiny, single-seeded fruits produced by a single flower. The fruits are embedded in the surface of a fleshy receptacle that is adapted to attract animal seed-dispersal agents. **(e)** Pineapple is a large multiple fruit formed by the aggregation of smaller fruits, each produced by one of the flowers in an inflorescence. **(f)** Peas produce legumes, fruits that open on two sides to release seeds. **(g)** Coconut fruits possess a fibrous husk that aids dispersal in water. **(h)** Maple trees produce dry fruits with wings adapted for wind dispersal.

pistils develop into tiny, single-seeded yellow fruits on a strawberry surface; the fleshy, red, sweet portion of a strawberry develops from a flower receptacle. Aggregate fruits allow a single animal consumer, such as a bird, to disperse many seeds at the same time. Pineapples (**Figure 30.21e**) are juicy multiple fruits that develop when many ovaries of an inflorescence fuse together. Such multiple fruits are larger and attract relatively large animals that have the ability to disperse seeds for long distances.

The plant family informally known as **legumes** is named for its distinctive fruits, dry pods that open down both sides when seeds are mature, thereby releasing them (**Figure 30.21f**). Nuts and grains are additional examples of dry fruits. **Grains** are the characteristic single-seeded fruits of cereal grasses such as rice, corn (maize), barley, and wheat. Coconut fruits are adapted for dispersal in ocean currents and can float for months before being cast ashore (**Figure 30.21g**). Maple trees produce dry and thus lightweight fruits having wings, features that foster effective wind dispersal (**Figure 30.21h**). Other plants produce dry fruits with surface burrs that attach to animal fur. These are just a few examples of the diverse mechanisms that flowering plants use to disperse their seeds.

Angiosperms Produce Diverse Secondary Metabolites That Play Important Roles in Structure, Reproduction, and Protection

Secondary metabolism involves the synthesis of organic compounds that are not essential for cell structure and growth but aid organism survival and reproduction. These molecules, called **secondary metabolites**, are produced by various prokaryotes, protists, fungi, some animals, and all plants, but are most diverse in the angiosperms. About 100,000 different types of secondary metabolites are known, most of which are produced by flowering plants. Because secondary metabolites play essential roles in plant structure, reproduction, and protection, diversification of these compounds has influenced flowering plant evolution. Three major classes of plant secondary metabolites occur: (1) terpenes and terpenoids; (2) phenolics, which include flavonoids and related compounds; and (3) alkaloids (**Figure 30.22**).

About 25,000 types of plant terpenes and terpenoids are constructed from different arrangements of the simple hydrocarbon gas isoprene. Taxol, previously mentioned for its use in the treatment of cancer, is a terpene, as are citronella and a variety of other compounds that repel insects. Rubber, turpentine, rosin, and amber are complex terpenoids that likewise serve important roles in plant biology as well as having useful human applications.

Phenolic compounds are responsible for some flower and fruit colors as well as the distinctive flavors of cinnamon, nutmeg, ginger, cloves, chilies, and vanilla. Phenolics absorb ultraviolet radiation, thereby preventing damage to cellular DNA. They also help to defend plants against insects and disease microbes. Some phenolic compounds found in tea, red wine, grape juice, and blueberries are antioxidants that detoxify free radicals, thereby preventing cellular damage.

Alkaloids are nitrogen-containing secondary metabolites that often have potent effects on the animal nervous system. Plants produce at least 12,000 types of alkaloids, and certain species produce many alkaloids. Caffeine, nicotine, morphine, ephedrine, cocaine, and codeine are examples of alkaloids that influence the physiology and behavior of humans and are thus of societal concern. Like flower and fruit structure, secondary metabolites are useful in distinguishing among Earth's hundreds of thousands of flowering plant species.

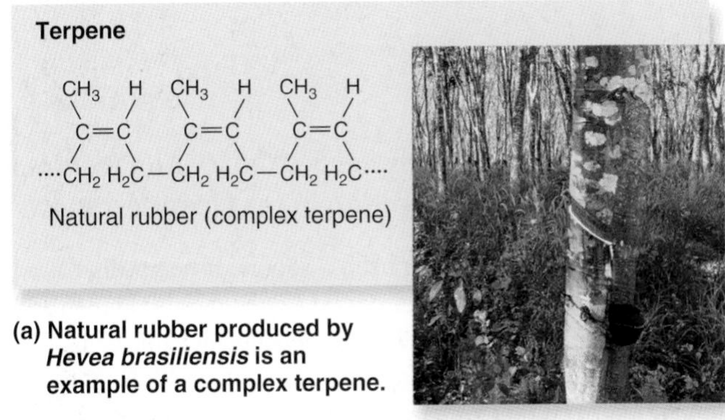

(a) Natural rubber produced by *Hevea brasiliensis* is an example of a complex terpene.

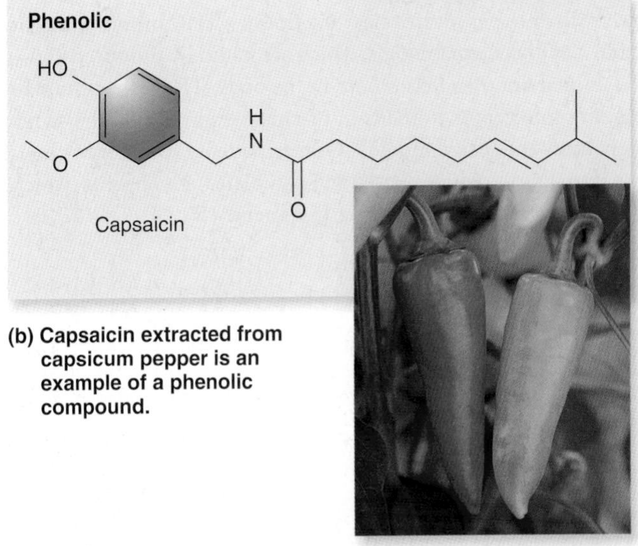

(b) Capsaicin extracted from capsicum pepper is an example of a phenolic compound.

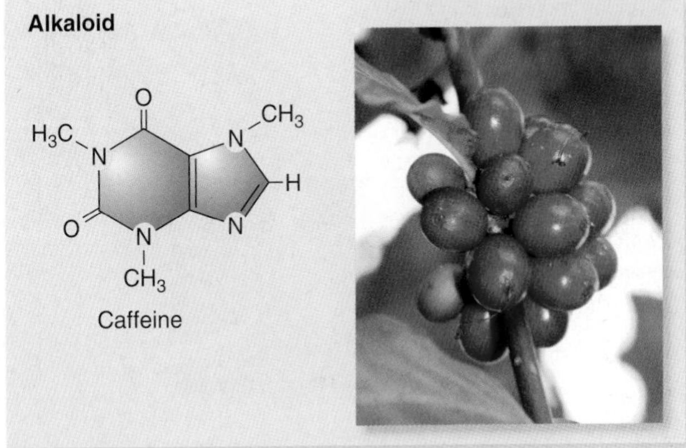

(c) Caffeine produced by *Coffea arabica* is an example of an alkaloid.

Figure 30.22 Major types of plant secondary metabolites. Note that the chemistry of plant secondary metabolites differs from that of the primary compounds produced by all cells. The production by plants of terpenes, phenolics, and alkaloids helps to explain how plants survive and reproduce, and why plants are useful to humans in so many ways.

FEATURE INVESTIGATION

Hillig and Mahlberg Analyzed Secondary Metabolites to Explore Species Diversification in the Genus *Cannabis*

The genus *Cannabis* has long been a source of hemp fiber used for ropes and fabric. People have also used *Cannabis* (also known as marijuana) in traditional medicine and as a hallucinogenic drug. *Cannabis* produces THC (tetrahydrocannabinol), a type of alkaloid called a cannabinoid. THC and other cannabinoids are produced in glandular hairs that cover most of the *Cannabis* plant's surface but are particularly rich in leaves located near the flowers. THC mimics compounds known as endocannabinoids, which are naturally produced and act in the animal brain and elsewhere in the body. THC affects humans by binding to receptor proteins in plasma membranes in the same way as natural endocannabinoids. Cancer patients sometimes choose to use cannabis to reduce nausea and stimulate their appetite, which can decline as a side effect of cancer treatment.

Because humans have subjected cultivated *Cannabis* plants to artificial selection for so long, plant biologists have been uncertain how cultivated *Cannabis* species are related to those in the wild. In the past, plants cultivated for drug production were often identified as *Cannabis indica*, whereas those grown for hemp were typically known as *Cannabis sativa*. However, these species are difficult to distinguish on the basis of structural features, and the relevance of these names to wild cannabis was unknown. At the same time, species identification has become important for biodiversity studies, agriculture, and law enforcement. For these reasons, plant biologists Karl Hillig and Paul Mahlberg hypothesized that ratios of THC to another cannabinoid known as CBD (cannabidiol) might aid in defining *Cannabis* species and identifying plant samples at the species level, as shown in Figure 30.23.

To test their hypothesis, the investigators began by collecting *Cannabis* fruits from nearly a hundred diverse locations around the

Figure 30.23 **Hillig and Mahlberg's analysis of secondary metabolites in the genus *Cannabis*.**

GOAL To determine if cannabinoids aid in distinguishing *Cannabis* species.

KEY MATERIALS *Cannabis* fruits obtained from nearly 100 different worldwide sources.

	Experimental level	Conceptual level
1 Grow multiple *Cannabis* plants from seeds under standard conditions in a greenhouse.		Eliminates differential environmental effects on cannabinoid content.
2 Extract cannabinoids from leaves surrounding flowers.		Extracts were made from tissues richest in cannabinoids; this reduces the chance that cannabinoids present in lower levels would be missed.
3 Analyze cannabinoids by gas chromatography. Determine ratios of THC (tetrahydrocannabinol) to CBD (cannabidiol) in about 200 *Cannabis* plants.	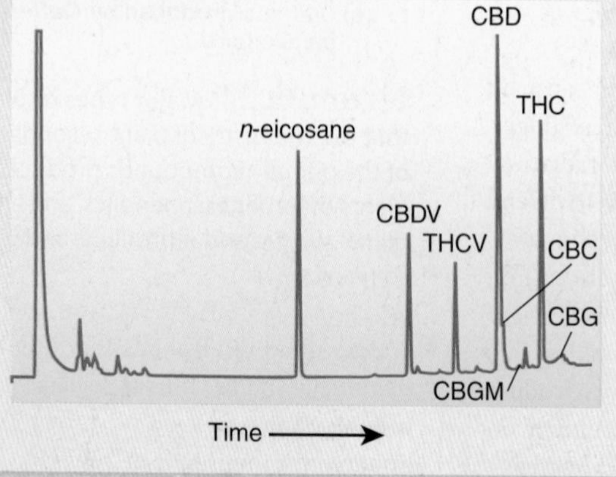	Previous data suggested that ratios of THC to CBD might be different in separate species. OH OH R Tetrahydrocannabinol (THC) ($R = C_5H_{11}\ \Delta^9$) OH R OH Cannabidiol (CBD) ($R = C_5H_{11}$)

4 THE DATA

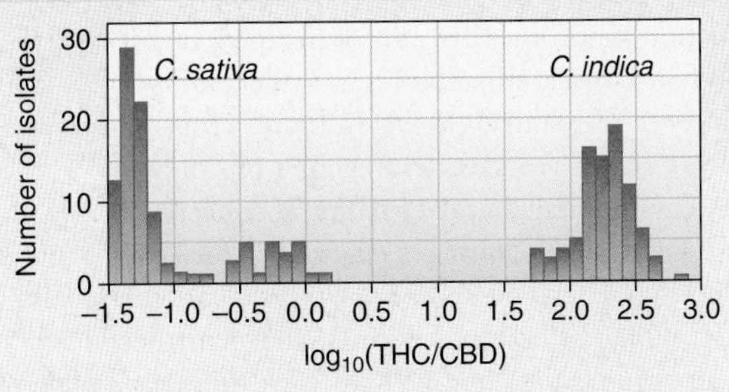

Cannabis plants isolated from diverse sources worldwide formed 2 groups—those having relatively high THC to CBD ratios and those having lower THC to CBD ratios.

Plants having low THC to CBD ratios, often used as hemp fiber sources, corresponded to the species *C. sativa.*

Plants having high THC to CBD ratios, often used as drug sources, corresponded to the species *C. indica.*

5 **CONCLUSION** Differing cannabinoid ratios support a concept of 2 *Cannabis* species.

6 **SOURCE** Hillig, K.W., and Mahlberg, P.G. 2004. A chemotaxonomic analysis of cannabinoid variation in *Cannabis* (Cannabaceae). *American Journal of Botany* 91:966–975.

world and then growing these plants from seed under uniform conditions in a greenhouse. The investigators next extracted cannabinoids, analyzed them by means of gas chromatography (a laboratory technique used to identify components of a mixture), and determined the ratios of THC to CBD. The results, published in 2004, suggested that the wild and cultivated *Cannabis* samples evaluated in this study could be classified into two species: *C. sativa*, displaying relatively low THC levels, and *C. indica*, having relatively high THC levels. As a result of this work, ecologists, agricultural scientists, and forensic scientists can reliably use ratios of THC to CBD to classify samples. Similar studies of plant secondary metabolites offer the benefit of uncovering potential new medicinal compounds or other applications of significance to humans.

Experimental Questions

1. In Figure 30.23, note that investigators obtained nearly a hundred *Cannabis* fruit samples from around the world. Why were so many samples needed?
2. Why did Hillig and Mahlberg grow plants in a greenhouse before conducting the cannabinoid analysis?
3. Why did Hillig and Mahlberg collect samples from the leaves growing nearest the flowers?

30.3 The Role of Coevolution in Angiosperm Diversification

Learning Outcomes:

1. Explain the concept of coevolution.
2. List examples of coevolution between plants and animal pollinators.
3. List examples of coevolution between plants and animal seed dispersal agents.

In the previous section, we learned that flowering plants are commonly associated with animals in ways that strongly influence plant evolution. Likewise, plants have influenced animal evolution in a diversity-generating process known as **coevolution**, which is the process by which two or more species of organisms influence each other's evolutionary pathway. During the diversification of flowering plants, coevolution with animals has been a major evolutionary force. Coevolution is reflected in the diverse forms of most flowers and many fruits and the many ways that plants accomplish effective pollen and seed dispersal. Human attraction to flowers and fruit also is an example of coevolution. This is because human sensory systems are similar to those of various animals that have coevolved with angiosperms.

Pollination Coevolution Influences the Diversification of Flowers and Animals

Animal pollinators transfer pollen from the anthers of one flower to the stigmas of other flowers of the same species. Pollinators thereby foster genetic variability and plant potential for evolutionary change. Insects, birds, bats, and other pollinators learn the characteristics of particular flowers, visiting them preferentially. This animal behavior, known as constancy or fidelity, increases the odds that a flower stigma will receive pollen of the appropriate species. Animal pollinators offer precision of pollen transfer, which reduces the amount of pollen that plants must produce to achieve pollination. By contrast, wind-pollinated plants must produce much larger amounts of pollen because windblown pollen reaches appropriate flowers by chance.

Flowers attract the most appropriate pollinators by means of attractive colors, odors, shapes, and sizes. Secondary metabolites influence the colors and odors of many flowers. Flavonoids, for example, color many blue, purple, or pink flowers. More than 700 types of chemical compounds contribute to floral odors.

Most flowers reward pollinators with food: sugar-rich nectar, lipid- and protein-rich pollen, or both. In this way, flowering plants provide an important biological service, providing food for many types of pollinator animals. However, some flowers "trick" pollinators into visiting or trap pollinators temporarily, thereby achieving pollination without actually rewarding the pollinator. Examples include flowers that look and smell like dead meat, thereby attracting flies, which are fooled but accomplish pollination anyway.

Although many flowers are pollinated by a variety of animals, others have flowers that have become specialized for particular pollinators, and vice versa. These specializations, which have resulted from coevolution, are known as **pollination syndromes** (Table 30.2). For example, odorless red flowers, such as those of hibiscus (see Figure 30.19b), are attractive to birds, which can see the color red but lack a sense of smell. By contrast, bees are not typically attracted to red flowers because bee vision does not extend to the red end of the visible light spectrum. Rather, bees are attracted to blue, purple, yellow, and white flowers having sweet odors. If you are allergic to bee stings or just want to reduce the possibility of being stung, do not dress in bee-attracting flower colors or wear a flowery fragrance when in locales frequented by bees.

Table 30.2 Pollination Syndromes

Animal features	Coevolved flower features
Bees	
Color vision includes ultraviolet (UV), not red	Often blue, purple, yellow, white (not red) colors
Good sense of smell	Fragrant
Require nectar and pollen	Nectar and abundant pollen
Butterflies	
Good color vision	Blue, purple, deep pink, orange, red colors
Sense odors with feet	Light floral scent
Need landing place	Landing place
Feed with long, tubular tongue	Nectar in deep, narrow floral tubes
Moths	
Active at night	Open at night; white or bright colors
Good sense of smell	Heavy, musky odors
Feed with long, thin tongue	Nectar in deep, narrow floral tubes
Birds	
Color vision, includes red	Often colored red
Often require perch	Strong, damage-resistant structure
Poor sense of smell	No fragrance
Feed in daytime	Open in daytime
High nectar requirement	Copious nectar in floral tubes
Hover (hummingbirds)	Pendulous (dangling) flowers
Bats	
Color blind	Light, reflective colors
Good sense of smell	Strong odors
Active at night	Open at night
High food requirements	Copious nectar and pollen provided
Navigate by echolocation	Pendulous or borne on tree trunks

Pollination syndromes are also of practical importance in agriculture and in conservation biology. Fruit growers often import colonies of bees to pollinate flowers of fruit crops and so increase crop yields. In recent years, widespread die-offs of bee colonies have become an environmental and agricultural concern. When bee pollinators are not available, growers cannot produce some fruit crops. Some plants have become so specialized to particular pollinators that if the pollinator becomes extinct, the plant becomes endangered. An example is the Hawaiian cliff-dwelling *Brighamia insignis* (**Figure 30.24**), whose presumed moth pollinator has become extinct. Humans that hand-pollinate *B. insignis* are all that stand between this plant and extinction.

Seed-Dispersal Coevolution Influences the Characteristics of Fruits and Animals

As in the case of pollination, coevolution between plants and their animal seed-dispersal agents has influenced both plant fruit characteristics and those of seed-dispersing animals. In addition, flowering plant fruits provide food for animals, an important biological service. For example, many of the plants of temperate forests produce fruits that are attractive to resident birds. Such juicy, sweet fruits have small seeds that readily pass through bird guts. Many plants signal fruit ripeness by undergoing color changes from unripe green fruits to red, orange, yellow, blue, or black (**Figure 30.25**). Because birds have good color vision, they are able to detect the presence of ripe fruits and consume them before the fruits drop from plants and rot. Apples, strawberries, cherries, blueberries, and blackberries are examples of fruits

Figure 30.24 ***Brighamia insignis*, a plant endangered by the loss of its pollinator.** The pollinator that coevolved with *B. insignis* has become extinct, with the result that the plant is unable to produce seed unless artificially pollinated by humans.

Concept Check: What kind of animal likely pollinated B. insignis?

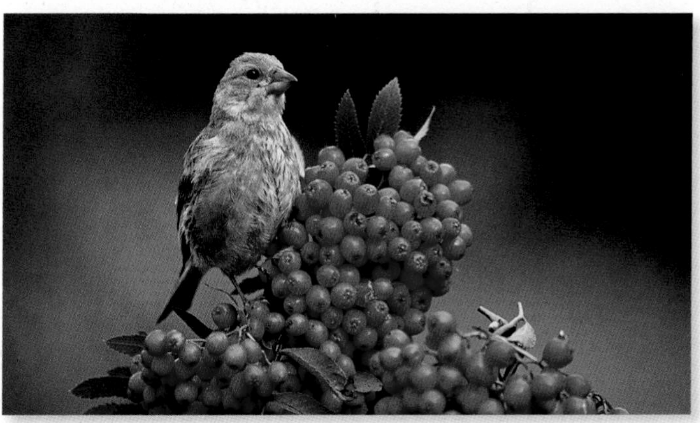

Figure 30.25 Fruits attractive to animal seed-dispersal agents. Color and odor signals alert coevolved animal species that fruits are ripe, thus favoring the dispersal of mature seeds.

whose seed dispersal adaptations have made them attractive food for humans as well. By contrast, the lipid-rich fruits of Virginia creeper (*Parthenocissus quinquefolia*) and some other autumn-fruiting plants energize migratory birds but are not tasty to humans. The Virginia creeper's leaves often turn fall colors earlier than surrounding plants, thereby signaling the availability of nutritious, ripe fruit to high-flying birds. Such lipid-rich fruits must be consumed promptly because they rot easily, in which case seed dispersal cannot occur.

30.4 Human Influences on Angiosperm Diversification

Learning Outcome:

1. Describe how humans created the domesticated wheat, corn, and rice grain crops widely planted around the world today.

By means of the process known as **domestication**, which involves artificial selection for traits desirable to humans, ancient humans transformed wild plant species into new crop species. Cultivated bread wheat (*Triticum aestivum*) was probably among the earliest food crops, having originated more than 8,000 years ago, in what is now southeastern Turkey and northern Syria. Bread wheat originated by a series of steps that included hybridization and whole-genome duplication from wild ancestors (*Triticum boeoticum* and *Triticum dicoccoides*). Among the earliest changes that occurred during wheat domestication was the loss of **shattering**, the process by which ears of wild grain crops break apart and disperse their grains. A mutation probably caused the ears of some wheat plants to remain intact, a trait that is disadvantageous in nature but beneficial to humans. Nonshattering ears would have been easier for humans to harvest than normal ears. Early farmers probably selected seed stock from plants having nonshattering ears and other favorable traits such as larger grains. These ancient artificial selection processes, together with modern breeding efforts, explain why cultivated wheat differs from its wild relatives in shattering and other properties. The accumulation of these trait differences explains why cultivated and wild wheat plants are classified as different species.

About 9,000 years ago, people living in what is now Mexico domesticated a native grass known as teosinte (of the genus *Zea*),

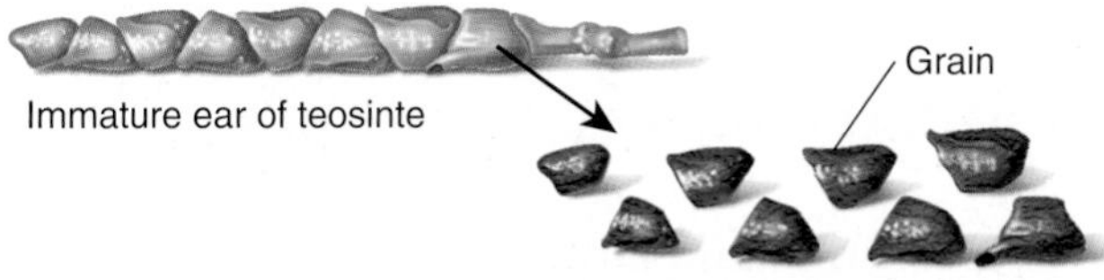

Figure 30.26 Ears and grains of modern corn and its ancestor, teosinte. This illustration shows that domesticated corn ears are much larger than those of the ancestral grass teosinte. In addition, corn fruits are softer and more edible than the grains of teosinte, which are enclosed in a hard casing.

BIOLOGY PRINCIPLE Biology affects our society. The domestication of corn from a wild grass to one of the world's largest production crops is an amazing feat of artificial selection.

Concept Check: In what other way do corn ears differ from those of teosinte?

producing a new species, *Zea mays*, known as corn or maize. The evidence for this pivotal event includes ancient ears that were larger than wild ones and distinctive fossil pollen. Modern ears of corn are much larger than those of teosinte, with many more rows and larger and softer corn grains, and modern corn ears do not shatter, as do those of ancestral teosinte (**Figure 30.26**). These and other trait changes reflect artificial selection accomplished by humans. An analysis of the corn genome, reported in 2005 by Canadian biologist Stephen Wright, U.S. evolutionary biologist Brandon Gaut, and coworkers, suggests that 1,200 genes have been affected by artificial selection.

Molecular analyses indicate that domesticated rice (*Oryza sativa*) originated from ancestral wild species of grasses (*Oryza nivara* and/or *Oryza rifipogon*). As in the cases of wheat and corn, domestication of rice involved loss of ear shattering, in this case resulting from a key amino acid substitution. Ancient humans might have unconsciously selected for this mutation while gathering rice from wild populations, because the mutants would not so easily have shed grains during the harvesting process. Eventually, the nonshattering mutant became a widely planted crop throughout Asia, and today it is the food staple for millions of people.

Although humans generated these and other new plant species, in modern times humans have caused the extinction of plants as the result of habitat destruction and other threats to species. Protecting biodiversity will continue to challenge humans as populations and demands on the Earth's resources increase. Plant biologists are working to identify one or more molecular sequence tools for use in barcoding plants, a process that is also widely used to identify and catalog animals (see the Genomes & Proteomes Connection in Chapter 33). The ability to barcode plants, which would enable researchers to quickly analyze the DNA of a species and identify it based on existing barcodes, is important to organizations like CITES and others that monitor international trade in endangered plant species.

Summary of Key Concepts

30.1 The Evolution and Diversity of Modern Gymnosperms

- The seed plants (also called spermatophytes) consist of the gymnosperms and the angiosperms (Figure 30.1, Table 30.1).
- Gymnosperms are plants that produce exposed seeds rather than seeds enclosed in fruits. Many gymnosperms produce wood by means of a special tissue called vascular cambium. Several phyla of gymnosperms once existed but have become extinct and are known only from fossils. The greatest diversity of gymnosperms occurred during the Mesozoic era, when gymnosperms were the major vegetation present (Figure 30.2).
- The diversity of modern gymnosperms includes three modern phyla: cycads, *Ginkgo biloba*, and the conifers. Nearly 300 species of cycads primarily live in tropical and subtropical regions. Features of cycads include palmlike leaves, nonwoody stems, coralloid roots with cyanobacterial endosymbionts, toxins, and large conelike seed-producing structures (Figures 30.3, 30.4).
- The tree *Ginkgo biloba* is the last surviving species of a phylum that was diverse during the Age of Dinosaurs. Individual trees produce ovules and seeds or pollen, with a sex chromosome system much like that of humans (Figure 30.5).
- Conifers have been widespread and diverse members of plant communities for the past 300 million years and are important sources of wood and paper pulp to humans. Reproduction involves simple pollen cones and complex ovule-producing cones. Many conifers display adaptations that help them to survive in cold climates. Three distinctive genera known as gnetales display distinctive adaptations (Figures 30.6, 30.7, 30.8, 30.9, 30.10, 30.11).

30.2 The Evolution and Diversity of Modern Angiosperms

- Angiosperms inherited seeds and other features from gymnosperm ancestors but display distinctive features not found in other land plants, such as flowers and fruits (Figure 30.12).
- Flowers foster seed production and are adapted in various ways that aid pollination. The major flower organs are sepals and petals (or tepals), stamens, and carpels, which may occur singly or in fused groups. Both single carpels and a group of fused carpels take a distinctive shape known as a pistil, which displays regions of specialized function. The stigma is a receptive surface for pollen, pollen tubes grow through the style, and ovules develop within the ovary. Pollination is the transfer of pollen from a stamen to a pistil, a process distinct from fertilization. Double fertilization, the production of both a zygote and a nutritive tissue known as endosperm, is a key innovation of angiosperms. If pollen germinates on the stigma and pollen tubes successfully deposit sperm near eggs in ovules, double fertilization may occur. This process allows ovules to develop into seeds containing embryos and endosperm, and ovaries to develop into fruits. Stamens and carpels may have evolved from leaflike structures bearing sporangia (Figures 30.13, 30.14, 30.15).
- The two largest and most diverse lineages of flowering plants are the monocots and eudicots (Figures 30.16, 30.17, 30.18).
- Whole-genome duplications arising from autopolyploidy and allopolyploidy and horizontal gene transfer by mitochondrial fusion have influenced plant evolution.
- Flower diversification involved evolutionary changes such as fusion of petals, clustering of flowers into inflorescences, and reduced perianth. These changes improve pollination effectiveness, which enhances seed production (Figures 30.19, 30.20).
- Fruits are structures that enclose seeds and aid in their dispersal. Fruits occur in many types that foster seed dispersal (Figure 30.21).
- Angiosperms produce three main groups of secondary metabolites: (1) terpenes and terpenoids; (2) phenolics, flavonoids, and related compounds; and (3) alkaloids, which play essential roles in plant structure, reproduction, and defense, respectively (Figure 30.22).
- Hillig and Mahlberg demonstrated the use of particular secondary metabolites in distinguishing species of the genus *Cannabis* (Figure 30.23).

30.3 The Role of Coevolution in Angiosperm Diversification

- Coevolutionary interactions between flowering plants and animals that serve as pollen- and seed-dispersal agents played a powerful role in the diversification of both angiosperms and animals (Table 30.2, Figures 30.24, 30.25).
- Human appreciation of flowers and fruits is based on sensory systems similar to those present in the animals with which angiosperms coevolved.

30.4 Human Influences on Angiosperm Diversification

- Humans have produced new crop species by domesticating wild plants. The process of domestication involved artificial selection for traits such as nonshattering ears of wheat, corn, and rice (Figure 30.26).

Assess and Discuss

Test Yourself

1. What feature must be present for a plant to produce wood?
   a. a type of conducting system in which vascular bundles occur in a ring around pith
   b. a eustele
   c. a vascular cambium
   d. all of the above
   e. none of the above
2. Which sequence of critical adaptations reflects the order of their appearance in time?
   a. embryos, vascular tissue, wood, seeds, flowers
   b. vascular tissue, embryos, wood, flowers, seeds
   c. vascular tissue, wood, seeds, embryos, flowers
   d. wood, seeds, embryos, flowers, vascular tissue
   e. seeds, vascular tissue, wood, embryos, flowers

3. How long have ancient and modern groups of gymnosperms been important members of plant communities?
   a. 10,000 years, since the dawn of agriculture
   b. 100,000 years
   c. 300,000 years
   d. 65 million years, since the K/T event
   e. 300 million years, since the Coal Age
4. What similar features do gymnosperms and angiosperms possess that differ from other modern vascular plants?
   a. Gymnosperms and angiosperms both produce flagellate sperm.
   b. Gymnosperms and angiosperms both produce flowers.
   c. Gymnosperms and angiosperms both produce tracheids, but not vessels, in their vascular tissues.
   d. Gymnosperms and angiosperms both produce fruits.
   e. none of the above
5. Which part of a flower receives pollen from the wind or a pollinating animal?
   a. perianth c. filament e. ovary
   b. stigma d. pedicel
6. The primary function of a fruit is to
   a. provide food for the developing seed.
   b. provide food for the developing seedling.
   c. foster pollen dispersal.
   d. foster seed dispersal.
   e. none of the above.
7. What are some ways in which flowers have diversified?
   a. color
   b. number of flower parts
   c. fusion of organs
   d. aggregation into inflorescences
   e. all of the above
8. Flowers of the genus *Fuchsia* produce deep pink to red flowers that dangle from plants, produce nectar in floral tubes, and have no scent. Based on these features, which animal is most likely to be a coevolved pollinator?
   a. bee c. hummingbird e. moth
   b. bat d. butterfly
9. Which type of plant secondary metabolite is best known for the antioxidant properties of human foods such as blueberries, tea, and grape juice?
   a. alkaloids c. carotenoids e. terpenoids
   b. cannabinoids d. phenolics
10. What features of domesticated grain crops might differ from those of wild ancestors?
   a. the degree to which ears shatter, allowing for seed dispersal
   b. grain size
   c. number of grains per ear
   d. softness and edibility of grains
   e. all of the above

Conceptual Questions

1. Make a diagram that shows how plant biologists think flowers arose.
2. Explain why fruits such as apples, strawberries, and cherries are attractive and harmless foods for humans.
3. A principle of biology is that *structure determines function.* Compare the structures of an apple flower and a sunflower, explaining how they relate to differences in pollination and seed dispersal.

Collaborative Questions

1. Where in the world would you have to travel to find wild plants representing all of the gymnosperm phyla, including the three types of gnetophytes?
2. How would you go about trying to solve what Darwin called an "abominable mystery," that is, the identity of the seed plant group that was ancestral to the flowering plants?

Online Resource

www.brookerbiology.com

Stay a step ahead in your studies with animations that bring concepts to life and practice tests to assess your understanding. Your instructor may also recommend the interactive eBook, individualized learning tools, and more.

Fungi

Chapter Outline

31.1 Evolution and Distinctive Features

31.2 Fungal Asexual and Sexual Reproduction

31.3 Diversity of Fungi

31.4 Fungal Ecology and Biotechnology

Summary of Key Concepts

Assess and Discuss

31

The aboveground reproductive parts of the fungus *Armillaria ostoyae*. Because of the large extent of its underground components, this fungus may be the largest organism in the world.

You might think that the largest organism in the world is a whale or perhaps a giant redwood tree. Amazingly, giant fungi would also be good candidates. For example, an individual of the fungus *Armillaria ostoyae* weighs hundreds of tons, is more than 2,000 years old, and spreads over 2,200 acres of Oregon forest soil! Scientists discovered the extent of this enormous fungus when they found identical DNA sequences in soil samples taken over this wide area. Other examples of such huge fungi have been found, and mycologists—scientists who study fungi—suspect that they may be fairly common, existing underfoot yet largely unseen.

Regardless of their size, fungi typically occur within soil or other materials, becoming conspicuous only when the reproductive portions such as mushrooms extend above the surface. Even though fungi can be inconspicuous, they play essential roles in the Earth's environment; are associated in diverse ways with other organisms, including humans; and have many applications in biotechnology. In this chapter, we will explore the distinctive features of fungal structure, growth, nutrition, reproduction, and diversity. In the process, you will learn how fungi are connected to decomposition, forest growth, food production and food toxins, sick building syndrome, and other topics of great importance to humans.

31.1 Evolution and Distinctive Features

Learning Outcomes:

1. Describe the evolutionary relationships of fungi, and identify six phyla described in this chapter.
2. Outline the distinctive features of fungi, including how they obtain food.
3. Discuss how fungal feeding is related to fungal growth.

The eukaryotes known as fungi are so distinct from other organisms that they are placed in their own kingdom, the kingdom Fungi (**Figure 31.1**). Together with certain closely related protists, the kingdom Fungi and the kingdom Animalia (also known as Metazoa) form a eukaryotic supergroup known as Opisthokonta (refer back to Figure 28.6). The kingdom Fungi, also known as the true fungi, diverged from Animalia more than a billion years ago, during the Middle Proterozoic Era. Several types of slime molds, disease-causing oomycetes, and other fungus-like protists—though often studied with fungi—are classified with nonopisthokont protists rather than true fungi (see Chapter 28).

The true fungi form a monophyletic group of more than 100,000 species occurring in more than 15 lineages. Recent environmental sampling and phylogenetic analyses have revealed the existence of major new groups of fungi and have determined that several previously defined fungal phyla are not monophyletic. In consequence, many fungal phyla will need to be formally named, a process that is not yet complete. For that reason, in this chapter we discuss seven lineages of true fungi, using their informal names: cryptomycota, chytrids, microsporidia, zygomycetes, AM fungi, ascomycetes, and basidiomycetes.

The earliest fungi diverged from opisthokont protists and are closely related to the genus *Nuclearia*—an amoeba that feeds by ingesting algal and bacterial cells, a process known as phagotrophy

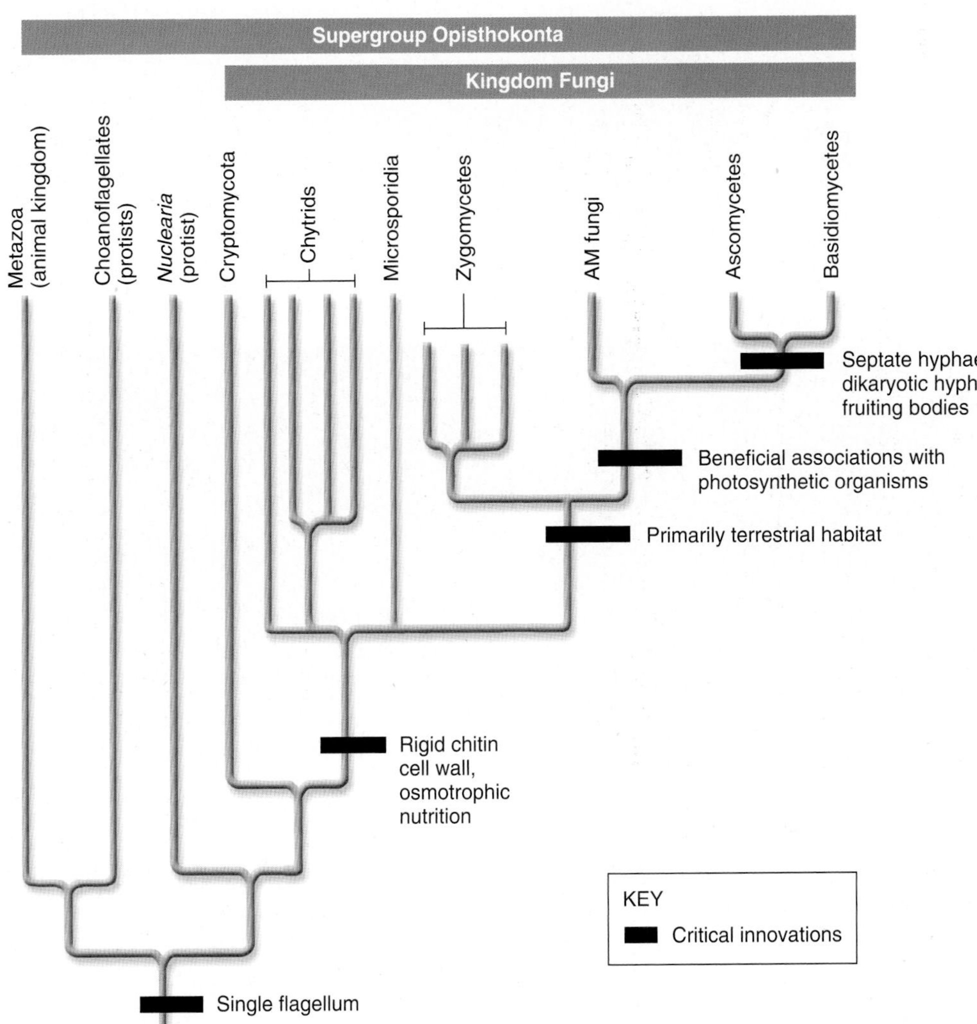

Figure 31.1 Evolutionary relationships of the fungi. The kingdom Fungi arose from a protist ancestor similar to the modern genus *Nuclearia*. More than 15 fungal phyla occur; because these are still being defined and named, seven informal fungal groups are described in this chapter: cryptomycota, chytrids, microsporidia, zygomycetes, AM fungi, ascomycetes, and basidiomycetes.

(see Figure 31.1). The earliest-diverging modern fungi are classified as **cryptomycota**, which occur in diverse genetic types in soil and water. Though little is known about how the cryptomycota live, they have the genetic capacity to produce flagella and they lack a cell wall containing **chitin**, a tough polysaccharide polymer that contains nitrogen. By contrast, rigid chitin-rich cell walls are a key feature of all other fungi. The evolution of a chitin cell wall enables most fungi to resist high osmotic pressure that results when they feed by absorbing small organic molecules, a process known as **osmotrophy**. Fungal cells that possess rigid chitin walls cannot feed by ingesting food particles (phagocytosis; see Figure 28.7). The evolution of a chitin wall signals a key evolutionary transition in fungal nutrition from feeding by phagocytosis to osmotrophy (see Figure 31.1).

Several aquatic lineages of microscopic species, informally known as **chytrids**, produce flagellate reproductive cells. Flagella are useful in moving through aquatic environments, but were lost during the diversification of other fungi, which primarily live in terrestrial habitats. The **microsporidia** are single-celled fungi that parasitize animal cells. Familiar black bread molds represent one of at least three lineages of terrestrial fungi known as **zygomycetes**. The zygomycetes are named for their distinctive large zygotes known as zygospores, and the suffix *mycetes* derives from a Greek word meaning fungus. The arbuscular mycorrhizal fungi, abbreviated **AM fungi**, are well known for their widespread symbiotic associations with plant roots. The **ascomycetes** (also informally known as the sac fungi) and the **basidiomycetes** (club fungi) are later-diverging fungal phyla that display many adaptations to life on land.

Because fungi are closely related to the animal kingdom, fungi and animals display some common features. For example, both are **heterotrophic**, meaning that they cannot produce their own food but must obtain it from the environment. Fungi use an amazing array of organic compounds as food, which is termed their **substrate**. The substrate could be the soil, a rotting log, a piece of bread, a living tissue, or a wide array of other materials. Fungi are also like animals in having **absorptive nutrition**. Both fungi and the cells of animal digestive systems secrete enzymes that break down complex organic materials and absorb the resulting small organic food molecules. In addition, both fungi and animals store surplus food as the carbohydrate glycogen in their cells. Despite these nutritional commonalities, fungal body structure, growth, and reproduction are distinct from that in animals and differ among fungal lineages. Because structure, growth, and reproductive differences are key to understanding fungal

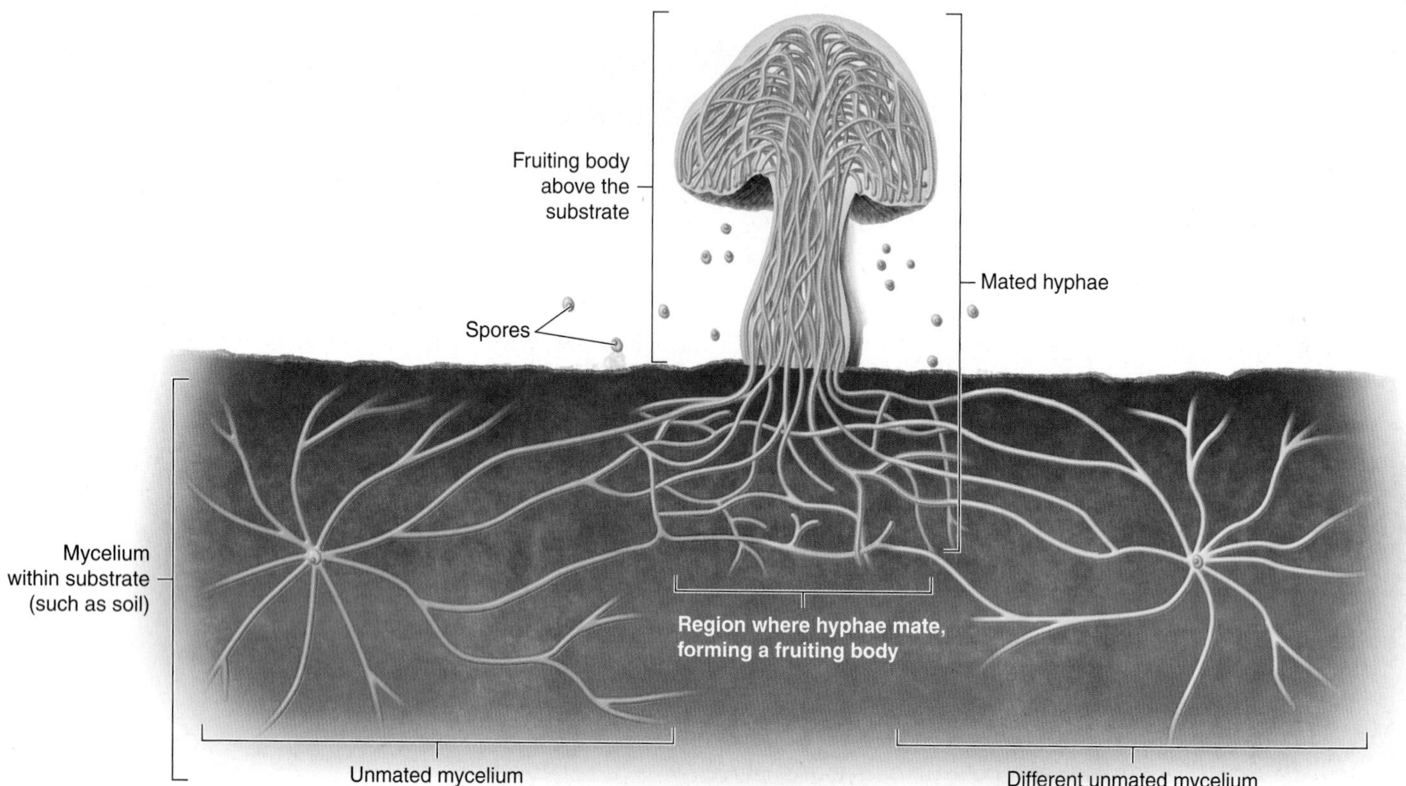

Figure 31.2 Fungal morphology. The greater part of a fungus consists of food-gathering hyphae that grow and branch from a central point to form a diffuse mycelium within a food substrate, such as soil.

BIOLOGY PRINCIPLE Living organisms grow and develop. After a mating process occurs, mated hyphae may aggregate and grow out of the substrate, forming fruiting bodies that produce and disperse spores. In suitable sites, spores may germinate, producing new mycelia.

diversity, we will focus on these features before exploring fungal diversity in more detail (Section 31.3).

Fungi Have a Unique Body Form

Most fungi have a distinctive body known as a **mycelium** (plural, mycelia), which is composed of individual microscopic, branched filaments known as **hyphae** (singular, hypha) (Figure 31.2). Hyphae and mycelia evolved even before fungi made the transition from aquatic to terrestrial habitats. The hyphae of early-diverging fungi are not partitioned into smaller cells. Rather, these hyphae are **aseptate** and multinucleate (Figure 31.3a), a condition that results when nuclei repeatedly divide without intervening cytokinesis. Such aseptate hyphae are described as being coenocytic. By contrast, the hyphae of later-diverging fungi are subdivided into many small cells by cross walls known as **septa** (singular septum) (Figure 31.3b). In such fungi, known as septate fungi, each round of nuclear division is followed by the formation of a septum that is perforated by a small pore. Septate hyphae appeared after the divergence of the AM fungi, but prior to the divergence of ascomycetes from basidiomycetes (see Figure 31.1).

As mentioned, a fungal mycelium may be very extensive, as in the case of *Armillaria ostoyae* (see chapter-opening photo), but is often inconspicuous because the component hyphae are so tiny and spread out in the substrate. The diffuse form of the fungal mycelium makes sense because most hyphae function to absorb organic food from the substrate. By spreading out, hyphae can absorb food from a large volume of substrate. The absorbed food is used for mycelial growth and for reproduction by means of fruiting bodies, which are more conspicuous parts of the fungal body (see Figure 31.2).

Mushrooms are types of fungal reproductive structures called **fruiting bodies** (see Figure 31.2). Fruiting bodies are composed of densely packed hyphae that have undergone a sexual mating process

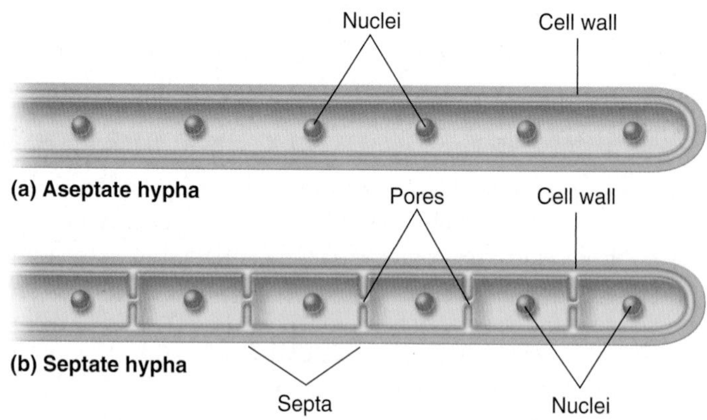

Figure 31.3 Types of fungal hyphae.

BIOLOGY PRINCIPLE Cells are the simplest units of life. This figure compares **(a)** the multinucleate hypha of a septate fungus with **(b)** a hypha of a septate fungus whose cells have a single nucleus.

during which unmated hyphae of different, but compatible, mycelia are attracted to each other and fuse. The resulting mated hyphae differ genetically and biochemically from unmated hyphae. Researchers suspect that mated hyphae secrete signaling substances that cause many such hyphae to cluster together and grow out of the substrate and into the air, where reproductive cells can be more easily dispersed. Amazingly diverse in form, color, and odor, mature fruiting bodies are specialized to produce and disperse reproductive cells known as **spores**. Produced by the process of meiosis and protected by tough walls, spores reflect a major adaptation to the terrestrial habitat. When fungal spores settle in places where conditions are favorable for growth, they produce new mycelia. When the new mycelia undergo sexual reproduction, they produce new fruiting bodies.

Fungi Have Distinctive Growth Processes

If you have ever watched bread or fruit become increasingly moldy over the course of several days, you have observed fungal growth. When a food source is plentiful, fungal mycelia can grow rapidly, adding as much as a kilometer of new hyphae per day. The mycelia grow at their edges as the fungal hyphae extend their tips through the undigested substrate. The narrow dimensions and extensive branching of hyphae provide a very high surface area for absorption of organic molecules, water, and minerals.

Hyphal Tip Growth Cytoplasmic streaming and osmosis are important cellular processes in hyphal growth. Osmosis (see Chapter 5) is the diffusion of water through a membrane, from a solution with a lower solute concentration into a solution with a higher solute concentration. Water enters fungal hyphae by means of osmosis because their cytoplasm is rich in sugars, ions, and other solutes. Water entry swells the hyphal tip, producing the force necessary for tip extension. Masses of tiny vesicles carrying enzymes and cell-wall materials made in the Golgi apparatus collect in the hyphal tip (Figure 31.4). The vesicles then fuse with the plasma membrane. Some vesicles release enzymes that digest materials in the environment, releasing small organic molecules that are absorbed as food. Other vesicles deliver cell-wall materials to the hyphal tip, allowing it to extend.

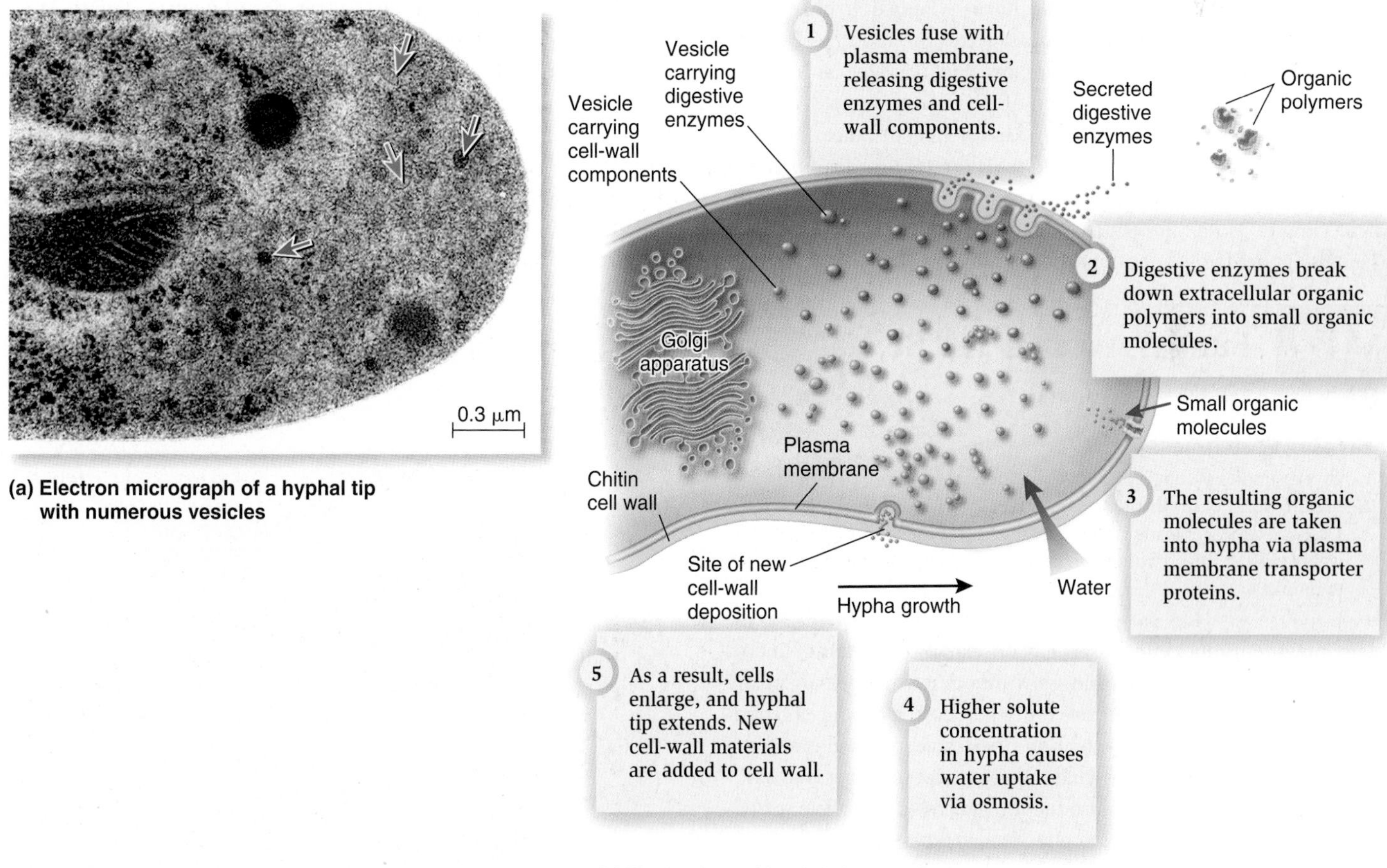

Figure 31.4 Hyphal tip growth and absorptive nutrition. **(a)** TEM showing the hyphal tip of *Aspergillus nidulans*, a fungus commonly used as a genetic model organism. The tip is filled with membrane-bound vesicles that fuse with the plasma membrane. Purple arrowheads show dark-stained vesicles carrying digestive enzymes; green arrowheads point out light-stained vesicles carrying cell-wall materials. **(b)** Diagram of a hyphal tip, with vesicles of the same two types, showing the steps of hyphal tip growth.

Concept Check: What do you think would happen to fungal hyphae that begin to grow into a substrate with a higher solute concentration? How might your answer be related to food preservation techniques such as drying or salting?

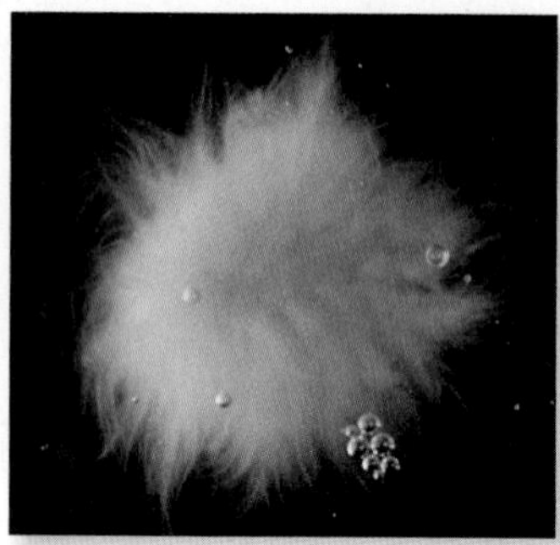

(a) Mycelium growing in liquid medium

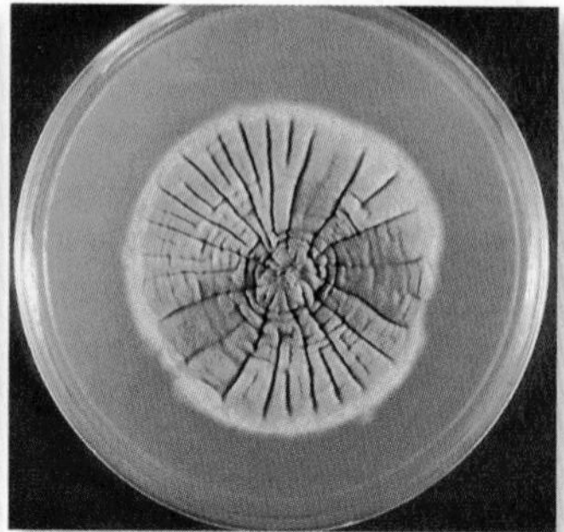

(b) Mycelium growing on flat, solid medium

Figure 31.5 Fungal shape shifting. (a) When a mycelium, such as that of this *Rhizoctonia solani*, is surrounded by food substrate in a liquid medium, it will grow into a spherical form. (b) When the food supply is limited to a two-dimensional supply, as shown by *Neotestudina rosatii* in a laboratory dish, the mycelium will form a disc. Likewise, distribution of the food substrate determines the mycelium shape in nature.

Variations in Mycelium Growth Form Fungal hyphae grow rapidly through a substrate from areas where the food has become depleted to food-rich areas. In nature, mycelia may take an irregular shape, depending on the distribution of the food substrate. A fungal mycelium may extend into food-rich areas for great distances, as noted at the beginning of the chapter. In liquid laboratory media, fungi will grow as a spherical mycelium that resembles a cotton ball floating in water (**Figure 31.5a**). Grown in flat laboratory dishes, the mycelium assumes a more two-dimensional growth form (**Figure 31.5b**).

31.2 Fungal Asexual and Sexual Reproduction

Learning Outcomes:

1. Give examples of fungal asexual reproduction.
2. Identify some of the distinctive sexual reproductive processes in fungi.
3. Describe why people may safely consume some fungal fruiting bodies, whereas other fungal fruiting bodies produce substances that are toxic to humans.

Many fungi reproduce either asexually or sexually by means of microscopic spores, each of which can grow into a new mature organism. Asexual reproduction is a natural cloning process; it produces genetically identical organisms. Production of asexual spores allows fungi that are well adapted to a particular environment to disperse to similar, favorable places. Sexual reproduction generates new allele combinations that may allow fungi to colonize new types of habitats.

Fungi Reproduce Asexually by Dispersing Specialized Cells

Asexual reproduction is particularly important to fungi, allowing them to spread rapidly. To reproduce asexually, fungi do not need to find compatible mates or expend resources on fruiting-body formation and meiosis. More than 17,000 fungal species reproduce primarily or exclusively by asexual means. DNA-sequencing studies have revealed that many types of modern fungi that reproduce only asexually have evolved from ancestors that had both sexual and asexual reproduction.

Figure 31.6 Asexual reproductive cells of fungi. SEM of the asexual spores (conidia) of *Aspergillus versicolor*, which causes skin infections in burn victims and lung infections in AIDS patients. Each of these small cells is able to detach and grow into an individual that is genetically identical to the parent fungus and so is able to grow in similar conditions.

Concept Check: How might you try to protect a burn patient from infection by a conidial fungus?

Many fungi produce asexual spores known as **conidia** (from the Greek *konis*, meaning dust) at the tips of hyphae (**Figure 31.6**). When they land on a favorable substrate, conidia germinate into a new mycelium that produces many more conidia. The green molds that form on citrus fruits are familiar examples of conidial fungi. A single fungus can produce as many as 40 million conidia per hour over a period of 2 days.

Because they can spread so rapidly, asexual fungi are responsible for costly fungal food spoilage, allergies, and diseases. Medically important fungi that reproduce primarily by asexual means include the athlete's foot fungus (*Epidermophyton floccosum*) and the infectious yeast (*Candida albicans*). **Yeasts** are unicellular fungi of various lineages. Asexual reproduction in some yeasts occurs by budding (**Figure 31.7**).

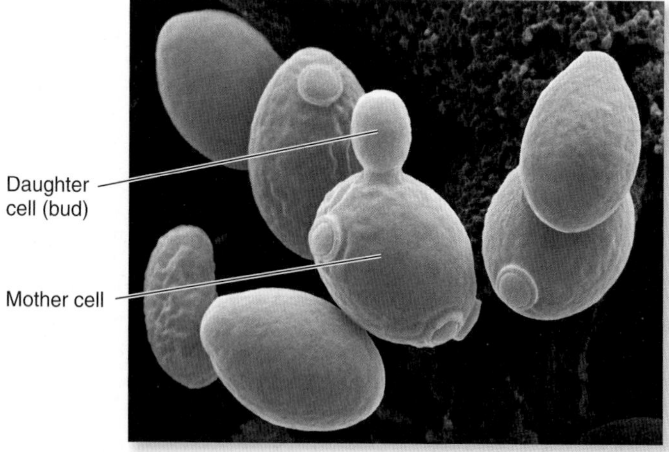

Figure 31.7 The budding yeast *Saccharomyces cerevisiae*. In budding, a small daughter cell is formed on the surface of a larger mother cell, eventually pinching off and forming a new cell.

BioConnections: Look back at Table 13.1, which shows the genome characteristics of some model organisms. How does the genome of S. cerevisiae compare with genomes of other model organisms?

Fungi Have Distinctive Sexual Reproductive Processes

As is typical for eukaryotes, the fungal sexual reproductive cycle involves the union of gametes, the formation of zygotes, and the process of meiosis. In contrast to plants, whose life cycle is an alternation of haploid and diploid generations, and diploid-dominant animals, the fungal life cycle is haploid-dominant (look back to Figure 15.14). Some other aspects of fungal sexual reproduction are unique, including the function of hyphal branches as gametes and the development of fruiting bodies.

Fungal Gametes and Mating Early-diverging fungi that live in the water produce flagellate sperm that swim to nonmotile eggs, as do animals and many protists and plants. By contrast, the gametes of terrestrial fungi are cells of hyphal branches rather than distinguishable male and female gametes. Fungal mycelia occur in multiple mating types that differ biochemically. The compatibility of these mating types is controlled by particular genes. During fungal sexual reproduction, hyphal branches of different, but compatible mycelia are attracted to each other by secreted peptides, and when hyphae have grown sufficiently close, they fuse. This distinctive mating process represents adaptation to terrestrial life.

Fruiting Bodies Under appropriate environmental conditions, such as seasonal change, a mated mycelium may produce a fleshy fruiting body, such as a mushroom. Fungal fruiting bodies typically emerge from the substrate and produce haploid spores (see Figure 31.2). Each spore acquires a tough chitin wall that protects it from drying out and other stresses. Wind, rain, or animals disperse the mature spores, which grow into haploid mycelia. If a haploid mycelium encounters hyphae of an appropriate mating type, hyphal branches will fuse and start the sexual cycle over again.

Mycelium growth requires organic molecules, minerals, and water provided by the substrate, but in most cases, spores are more easily dispersed if released outside of the substrate. The structures of fruiting bodies vary in ways that reflect different adaptations that foster spore dispersal by wind, rain, or animals. For example, mature puffballs have delicate surfaces upon which just a slight pressure causes the spores to puff out into wind currents (**Figure 31.8a**). Birds' nest fungi form characteristic egg-shaped spore clusters. Raindrops splash on these clusters and disperse the spores. The fruiting bodies of stinkhorn fungi smell and look like rotting meat (**Figure 31.8b**), which attracts carrion flies. The flies land on the fungi to investigate the potential meal and then fly away, in the process dispersing spores that stick to their bodies. The fruiting bodies of fungal truffles are unusual in being produced underground. Truffles have evolved a spore dispersal process that depends on animals that eat fungi. Mature truffles emit an odor that attracts wild pigs and dogs, which break up the fruiting structures while digging for them, thereby dispersing the spores (look ahead to Figure 31.19). Collectors use trained leashed pigs or dogs to locate valuable truffles from forests for the market.

Many fungal fruiting bodies such as truffles and morels are edible, and several species of edible fungi are cultivated for human consumption (**Figure 31.9**). However, the bodies of many other fungi produce toxic substances that may deter animals from consuming them (**Figure 31.10**). For example, several fungi that attack stored grains, fruits, and spices produce **aflatoxins** that cause liver cancer

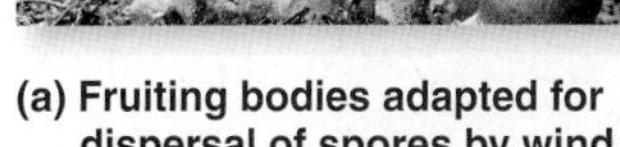

(a) Fruiting bodies adapted for dispersal of spores by wind

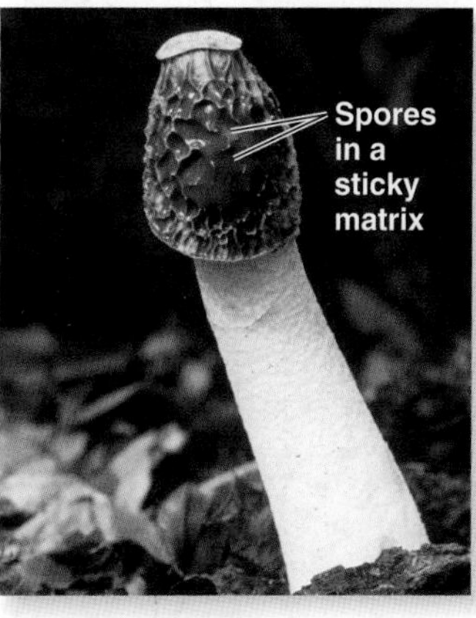

(b) Fruiting body adapted for dispersal of spores by insects

Figure 31.8 Fruiting body adaptations that foster spore dispersal. **(a)** When disturbed by wind gusts or animal movements, spores puff from fruiting bodies of the puffball fungus (*Lycoperdon perlatum*). **(b)** The fruiting bodies of stinkhorn fungi, such as this *Phallus impudicus*, smell and look like dung or rotting meat. This attracts flies, which come into contact with the sticky fungal spores, thereby dispersing them.

Figure 31.9 Several types of edible fungi available in the market.

Figure 31.10 Toxic fruiting body of *Amanita muscaria*. Common in conifer forests, *A. muscaria* is both toxic and hallucinogenic. Ancient people used this fungus to induce spiritual visions and to reduce fear during raids. This fungus produces a toxin, amanitin, which specifically inhibits RNA polymerase II of eukaryotes.

BioConnections: *Look back at Figure 13.14, which illustrates the cellular role of RNA polymerase II in eukaryotes. What effect would the amanitin toxin have on human cells?*

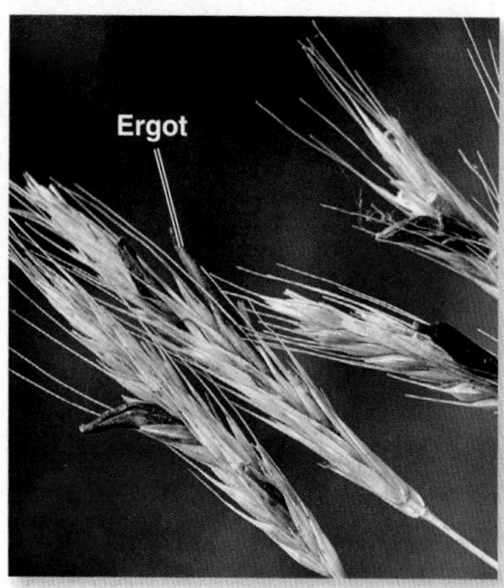

Figure 31.11 **Ergot of rye.** The fungus *Claviceps purpurea* infects rye and other grasses, producing hard masses of mycelia known as ergots in place of some of the grains (fruits).

BIOLOGY PRINCIPLE Biology affects our society. Ergots such as the one illustrated produce alkaloids related to LSD and thus cause psychotic delusions in humans and animals that consume products made with infected rye.

and are a major health concern worldwide. When people consume the forest mushroom *Amanita virosa*, known as the "destroying angel," they ingest a powerful toxin that may cause liver failure so severe that death may ensue unless a liver transplant is performed. Each year, many people in North America are poisoned when they consume similarly toxic mushrooms gathered in the wild. There is no reliable way for nonexperts to distinguish poisonous from nontoxic fungi; it is essential to receive instruction from an expert before foraging for mushrooms in the woods. Therefore, many authorities recommend that it is better to search for mushrooms in the grocery store than in the wild.

Several types of fungal fruiting structures produce hallucinogenic or psychoactive substances. As in the case of fungal toxins, fungal hallucinogens may have evolved as herbivore deterrents, but humans have inadvertently experienced their effects. For example, *Claviceps purpurea*, which causes a disease of rye crops and other grasses known as ergot, produces a psychogenic compound related to LSD (lysergic acid diethylamide) (Figure 31.11). Some experts speculate that cases of hysteria, convulsions, infertility, and a burning sensation of the skin that occurred in Europe during the Middle Ages and that were attributed to witchcraft resulted from ergot-contaminated rye used in foods. Another example of a hallucinogenic fungus is the "magic mushroom" (*Psilocybe*), which is used in traditional rituals in some cultures. Like ergot, the magic mushroom produces a compound similar to LSD. Consuming hallucinogenic fungi is risky because the amount used to achieve psychoactive effects is dangerously close to a poisonous dose.

31.3 Diversity of Fungi

Learning Outcome:

1. Outline the distinguishing features of seven fungal phyla: cryptomycota, chytrids, microsporidia, zygomycetes, AM fungi, ascomycetes, and basidiomycetes.

As earlier noted, the kingdom Fungi is a monophyletic group that arose from a protist ancestor, diversifying first in aquatic habitats, then later in terrestrial environments (see Figure 31.1). Here we describe in more detail the seven fungal lineages listed informally in Table 31.1: cryptomycota, chytrids, microsporidia, zygomycetes, AM fungi, ascomycetes, and basidiomycetes. In this section, we will survey the habitats and characteristics of these groups of fungi, focusing on distinctive ecological, structural, growth, and reproductive features.

Table 31.1 Distinguishing Features of Fungal Phyla

Informal name	Habitat	Ecological role	Reproduction	Examples cited in this chapter
Cryptomycota	Water and soil	Unknown	Flagellate cells	*Rozella allomycis*
Chytrids	Water and soil	Mostly decomposers; some parasites	Flagellate spores or gametes	*Batrachochytrium dendrobatidis*
Microsporidia	Animal cells	Parasites, pathogens	Nonflagellate spores	*Nosema ceranae*
Zygomycetes	Mostly terrestrial	Decomposers and pathogens	Nonflagellate asexual spores produced in sporangia; resistant sexual zygospores	*Rhizopus stolonifer*
AM Fungi	Terrestrial	Form mutually beneficial mycorrhizal associations with plants	Distinctively large, nonflagellate, multinucleate asexual spores	The genus *Glomus*
Ascomycetes	Mostly terrestrial	Decomposers; pathogens; many form lichens; some are mycorrhizal	Asexual conidia; nonflagellate sexual spores (ascospores) in sacs (asci) on fruiting bodies (ascocarps)	*Aleuria aurantia, Venturia inaequalis, Saccharomyces cerevisiae, Tuber melanosporum*
Basidiomycetes	Terrestrial	Decomposers; many are mycorrhizal; less commonly form lichens	Several types of asexual spores; nonflagellate sexual spores (basidiospores) on club-shaped basidia on fruiting bodies (basidiocarps)	*Coprinus disseminatus, Rhizoctonia solani, Armillaria mellea, Puccinia graminis, Ustilago maydis, Phanerochaete chrysosporium, Laccaria bicolor, Amanita muscaria, Phallus impudicus, Lycoperdon perlatum*

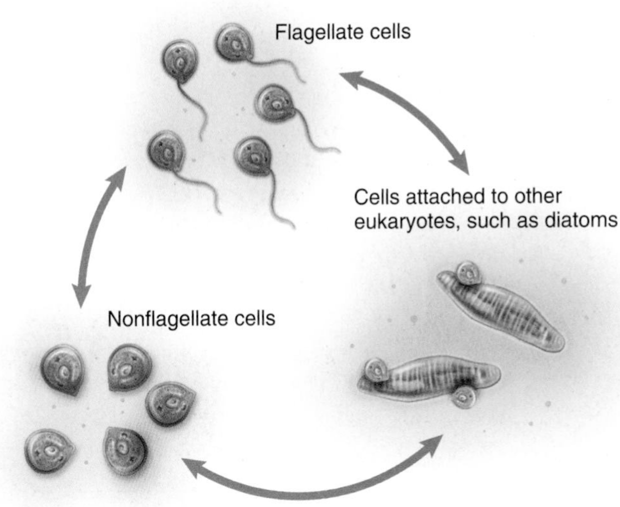

Figure 31.12 Life phases of the recently discovered cryptomycota. These early-diverging fungi lack chitin walls, which characterize all other fungi.

Cryptomycota Occur in Water and Soil and Lack Chitin Walls

Cryptomycota are a recently defined group of genetically diverse, single-celled organisms found in water and soil. An example is *Rozella allomycis*. Cells of cryptomycota lack a rigid chitin cell wall, a feature present in all other fungal groups. The life phases of cryptomycota include cells having a single flagellum and nonflagellate cells that attach to (and possibly feed upon) other organisms, such as diatoms (Figure 31.12). The presence of a single flagellum and DNA evidence link the cryptomycota and chytrids with the ancestry of other opisthokonts—the choanoflagellate protists and animals (see Chapter 28).

Chytrids Primarily Live in Water or Soil But Possess Chitin Walls

Polyphyletic chytrids live in aquatic habitats or in moist soil and produce flagellate reproductive cells. A rigid chitin wall is present. Some chytrids occur as single, spherical cells that may produce hyphae (Figure 31.13). Most chytrids are decomposers, but some are parasites of protists and pathogens of plants or animals. For example, the chytrid *Batrachochytrium dendrobatidis* has been associated with declining frog populations in Australia and the Americas (look ahead to Figure 54.1).

Microsporidia are Unicellular Animal Parasites

Microsporidia are named for their very small size (1–4 μm) and occurrence as single-celled, chitin-walled spores. The chitin wall helps microsporidia to survive in the environment until they enter the bodies of animals; microsporidia are pathogens that can only reproduce inside the cells of an animal host. Microsporidia characteristically contain a coiled threadlike structure that when released helps them to invade diverse types of animal cells, including those

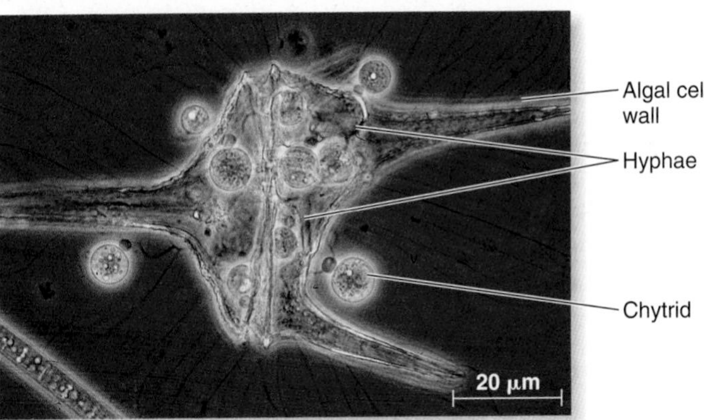

Figure 31.13 Chytrids growing on a freshwater protist. The colorless chytrids produce hyphae that penetrate the cellulose cell walls of the protist *Ceratium hirundinella*, absorbing organic materials. Chytrids use these materials to produce spherical flagellate spores that swim away to attack other algal cells.

of humans. Microsporidia sometimes cause disease, particularly in people whose immune systems are impaired. More than 1,000 species have been described. *Nosema ceranae* (Figure 31.14) has been linked to honeybee decline, in conjunction with an RNA bee virus. Honeybee colony collapse disorder, in which bees suddenly disappear from hives, is a serious agricultural concern worldwide, because the bees are necessary to the pollination of many crops, as well as to the production of honey and beeswax.

Zygomycetes Produce Distinctive Zygospores

The zygomycetes feature a mycelium that is mostly composed of aseptate hyphae (those lacking cross walls) and distinctive asexual and sexual reproductive structures. For example, during asexual reproduction the black bread mold *Rhizopus stolonifer* produces spores in dark-pigmented enclosures known as **sporangia** (singular, sporangium) (Figure 31.15a). A sporangium is a structure that

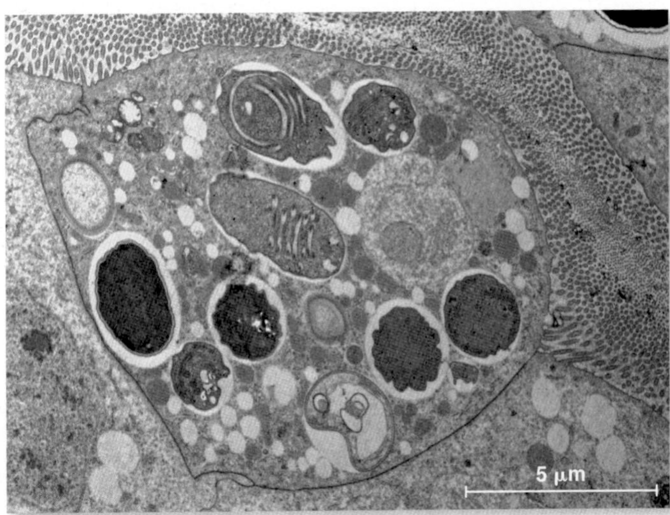

Figure 31.14 The microsporidian fungus *Nosema ceranae*. This fungus is linked with honeybee decline.

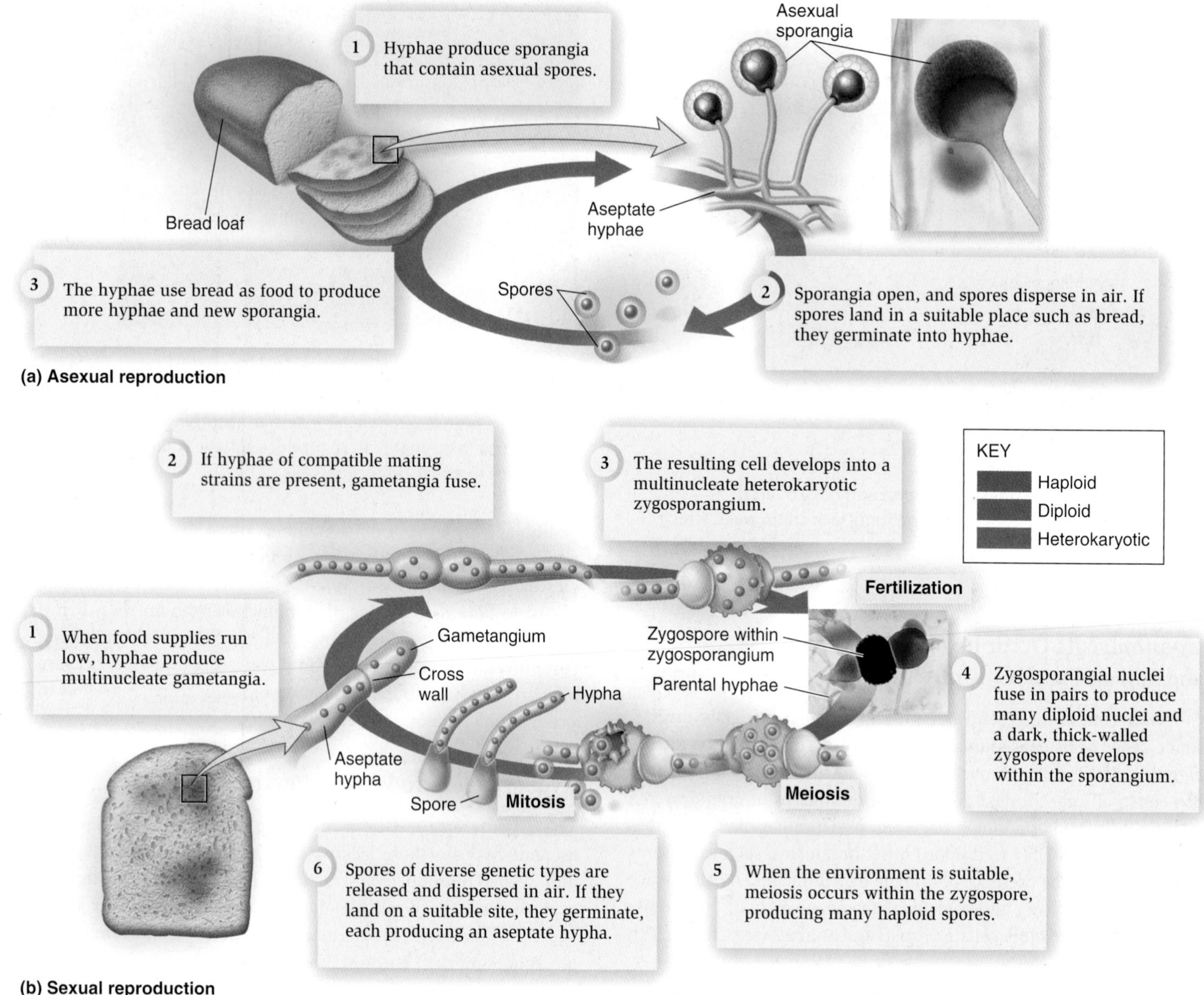

1 Hyphae produce sporangia that contain asexual spores.
Asexual sporangia
Bread loaf
Aseptate hyphae
3 The hyphae use bread as food to produce more hyphae and new sporangia.
Spores
2 Sporangia open, and spores disperse in air. If spores land in a suitable place such as bread, they germinate into hyphae.
(a) Asexual reproduction
2 If hyphae of compatible mating strains are present, gametangia fuse.
3 The resulting cell develops into a multinucleate heterokaryotic zygosporangium.
KEY
Haploid
Diploid
Heterokaryotic
Fertilization
1 When food supplies run low, hyphae produce multinucleate gametangia.
Gametangium
Cross wall
Hypha
Zygospore within zygosporangium
Parental hyphae
4 Zygosporangial nuclei fuse in pairs to produce many diploid nuclei and a dark, thick-walled zygospore develops within the sporangium.
Aseptate hypha
Spore
Mitosis
Meiosis
6 Spores of diverse genetic types are released and dispersed in air. If they land on a suitable site, they germinate, each producing an aseptate hypha.
5 When the environment is suitable, meiosis occurs within the zygospore, producing many haploid spores.
(b) Sexual reproduction

Figure 31.15 **The asexual and sexual life cycles of a zygomycete, the black bread mold *Rhizopus stolonifer*.**

produces spores. Bread mold sporangia form at hyphal tips in such large numbers that they make moldy bread appear black. Zygomycete asexual sporangia may each release up to 100,000 spores into the air! The great abundance of such spores means that bread molds easily unless the baker adds retardant chemicals.

Zygomycetes are named for the zygospore, a distinctive feature of their sexual reproduction (**Figure 31.15b**). In the black bread mold, zygospore production begins with the development of **gametangia** (from the Greek, meaning gamete-bearers). In the zygomycete fungi, gametangia are hyphal branches whose cytoplasm is isolated from the rest of the mycelium by cross walls. These gametangia enclose gametes that are basically a mass of cytoplasm containing several haploid nuclei. When food supplies run low and if compatible mating strains are present, the gametangia of compatible mating types fuse, as do the gamete cytoplasms. The resulting cell becomes a sporangium that contains many haploid nuclei. Eventually these haploid nuclei fuse in pairs, producing many diploid nuclei (zygote nuclei). For this reason, a zygomycete sporangium produced by sexual reproduction is called a zygosporangium. A single dark-pigmented, thick-walled, multinucleate **zygospore** matures within each zygosporangium. The zygospore is capable of surviving stressful conditions, but when the environment is suitable, the diploid nuclei within the zygospore may undergo meiosis and germinate, dispersing many haploid spores. If the haploid spores land in a suitable place, they germinate to form aseptate hyphae that contain many haploid nuclei produced by mitosis. Most zygomycetes live on decaying materials in soil, but some are parasites of plants or animals.

AM Fungi Live with Plant Partners

The microscopic arbuscular mycorrhizal fungi—commonly known as the AM fungi—have aseptate hyphae and reproduce only

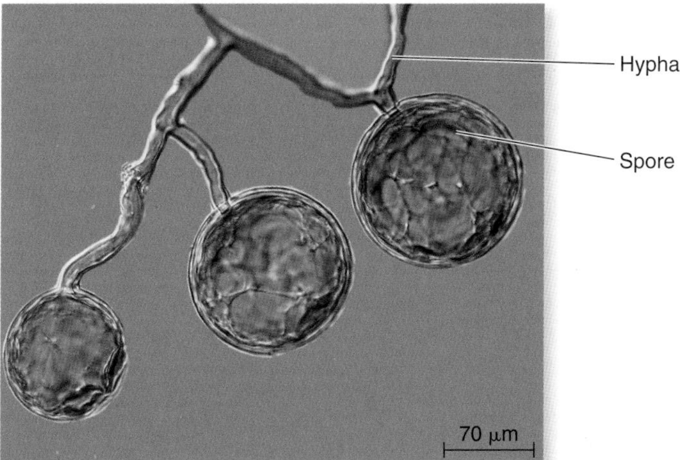

Figure 31.16 **The genus *Glomus*, an example of an AM fungus.** The hyphae of these endomycorrhizal (arbuscular mycorrhizal) fungi are found in roots of many types of plants, aiding them in acquiring water and nutrients. AM fungi produce large, multinucleate spores by asexual processes.

asexually by means of unusually large spores containing many nuclei (**Figure 31.16**). Many vascular plants depend on AM fungi, and these fungi are not known to grow separately from plants or cyanobacterial partners. The ecological importance of partnerships between the AM fungi and their plant partners are described more completely in Section 31.4.

Molecular evidence suggests that AM fungi originated more than 750 mya. Fossils having aseptate hyphae and large spores similar to those of modern AM fungi are known from the time when land plants first became common and widespread, about 460 mya (see Chapter 30). This and other fossil evidence suggests that the ability of early plants to thrive on land may have depended on help from fungal associates, as is common today.

Ascomycetes Produce Sexual Spores in Saclike Asci

Both the ascomycetes and basidiomycetes (discussed later in this section) are composed of hyphae subdivided into cells by septa. In ascomycetes, these septa display simpler pores than those of basidiomycete septa (**Figure 31.17**). Such pores allow cytoplasmic structures and materials to pass through the hyphae.

The sexual reproductive processes of ascomycetes and basidiomycetes are remarkable in producing a **dikaryotic mycelium**, one whose cells contain two nuclei of differing genetic types (**Figure 31.18**). In most sexual organisms, gametes undergo fusion of their cytoplasms—a process known as **plasmogamy**—and then the nuclei fuse in a process known as **karyogamy**. However, in ascomycete and basidiomycete fungi, after plasmogamy the haploid gamete nuclei generally remain separate for a time, rather than immediately undergoing karyogamy. During this time period, the gamete nuclei both divide at each cell division, producing a mycelium whose cells each possess both parental nuclei. Although the nuclei of dikaryotic mycelia remain haploid, alternative forms of many alleles occur in the separate nuclei. Thus, dikaryotic mycelia are functionally diploid. Eventually, dikaryotic mycelia produce fruiting bodies, the next stage of reproduction.

The name ascomycetes derives from unique sporangia known as **asci** (singular, ascus) from the Greek *asco*, meaning bags or sacs). During sexual reproduction asci produce spores known as **ascospores** (see Figure 31.18b). The asci are produced on fruiting bodies known as **ascocarps**. Although many ascomycetes have lost the ability to reproduce sexually, the presence of hyphal septa with simple pores (see Figure 31.17a) and DNA data can be used to identify them as members of this phylum.

Ascomycetes occur in terrestrial and aquatic environments, and they include many decomposers as well as pathogens. Important ascomycete plant pathogens include powdery mildews, chestnut blight (*Cryphonectria parasitica*), Dutch elm disease (the genus *Ophiostoma*), and apple scab (*Venturia inaequalis*). Cup fungi (see the ascocarp photo in Figure 31.18) are common examples of ascomycetes.

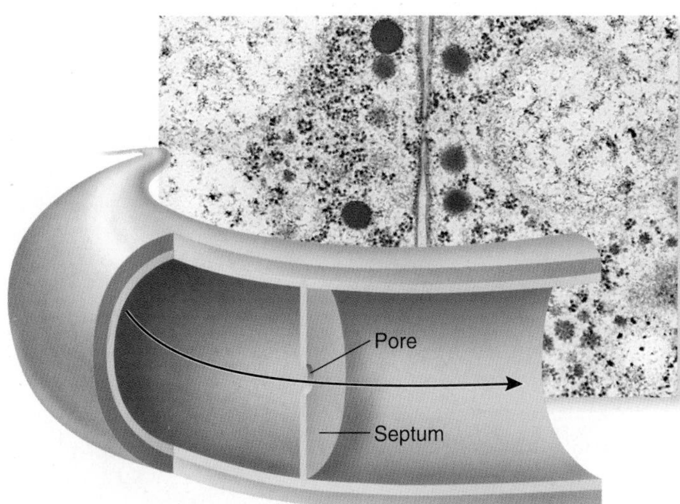

(a) Simple pore—ascomycetes

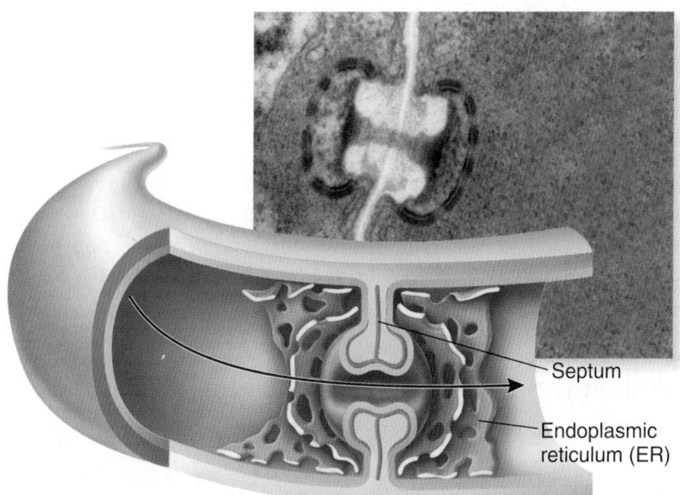

(b) Complex pore—basidiomycetes

Figure 31.17 **Septal pores of ascomycetes and basidiomycetes.** **(a)** The septa of ascomycetes have simple pores at the centers. **(b)** More complex pores distinguish the septa of most types of basidiomycetes.

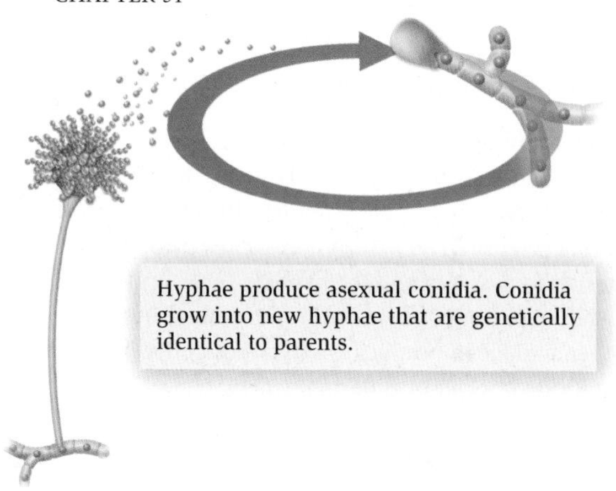

Hyphae produce asexual conidia. Conidia grow into new hyphae that are genetically identical to parents.

(a) Asexual reproduction

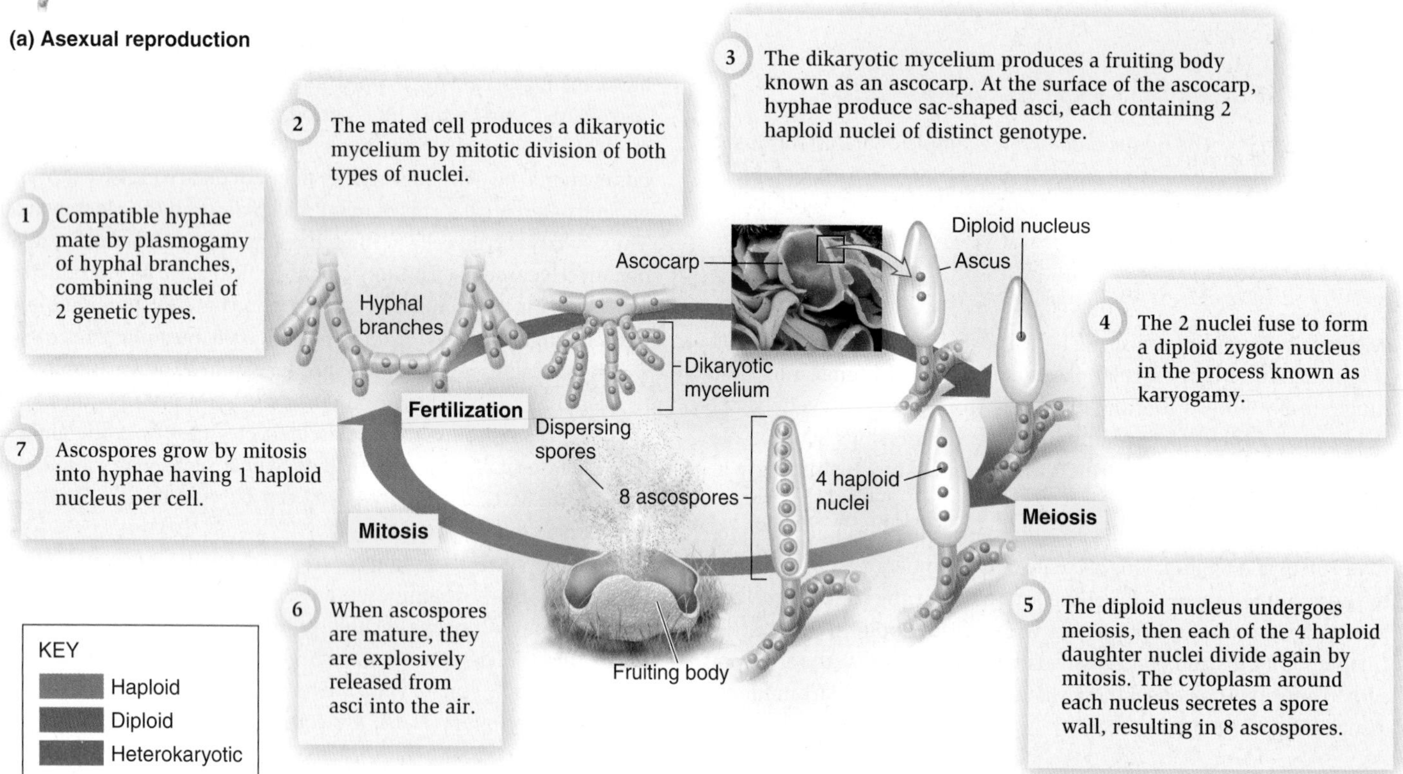

1 Compatible hyphae mate by plasmogamy of hyphal branches, combining nuclei of 2 genetic types.
2 The mated cell produces a dikaryotic mycelium by mitotic division of both types of nuclei.
3 The dikaryotic mycelium produces a fruiting body known as an ascocarp. At the surface of the ascocarp, hyphae produce sac-shaped asci, each containing 2 haploid nuclei of distinct genotype.
4 The 2 nuclei fuse to form a diploid zygote nucleus in the process known as karyogamy.
5 The diploid nucleus undergoes meiosis, then each of the 4 haploid daughter nuclei divide again by mitosis. The cytoplasm around each nucleus secretes a spore wall, resulting in 8 ascospores.
6 When ascospores are mature, they are explosively released from asci into the air.
7 Ascospores grow by mitosis into hyphae having 1 haploid nucleus per cell.
Hyphal branches
Ascocarp
Diploid nucleus
Ascus
Dikaryotic mycelium
Fertilization
Dispersing spores
8 ascospores
4 haploid nuclei
Meiosis
Mitosis
Fruiting body
KEY
Haploid
Diploid
Heterokaryotic

(b) Sexual reproduction of the ascomycete *Aleuria aurantia*

Figure 31.18 The asexual and sexual life cycles of ascomycete fungi. Mating generates dikaryotic hyphae that may form a fruiting body. Nuclei in the dikaryotic surface cells of the fruiting body fuse to form zygotes that undergo meiosis to produce haploid spores.

Many yeasts are also ascomycetes. Edible truffles (Figure 31.19) and morels are the fruiting bodies of particular ascomycetes whose mycelia form partnerships with plant roots, described in Section 31.4. Ascomycetes are the most common fungal components of lichens (see Section 31.4).

Basidiomycetes Produce Diverse Fruiting Bodies

DNA-sequencing comparisons indicate that **basidiomycetes**, together with ascomycetes, are the most recently diverged groups of fungi. The mated dikaryotic mycelia of basidiomycetes can live for hundreds of years and produce many fruiting bodies. The name given to the basidiomycetes derives from **basidia**, the club-shaped cells of fruiting bodies that produce sexual spores known as **basidiospores**

Figure 31.19 The black truffle *Tuber melanosporum*, an ascomycete fungus.

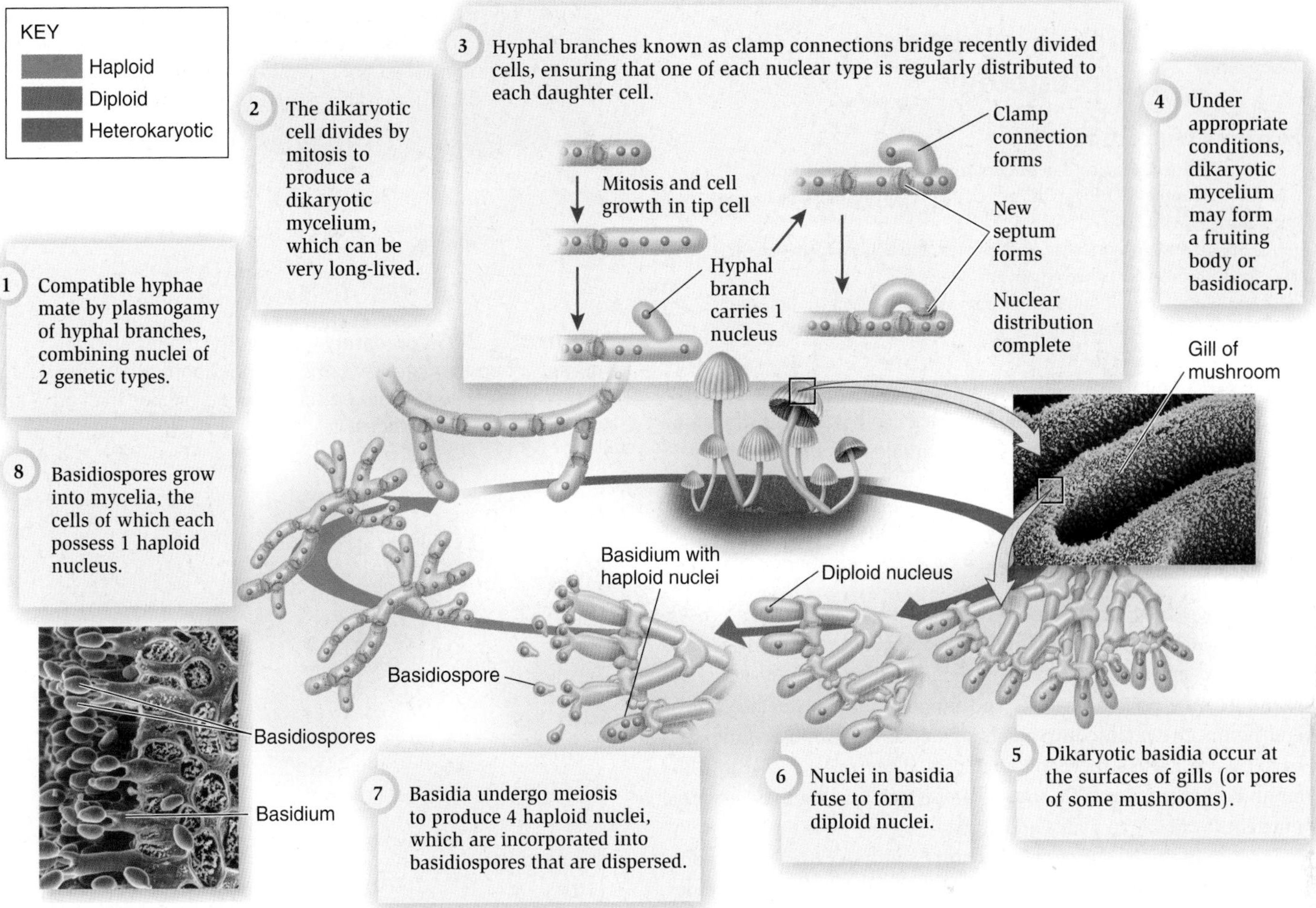

KEY
Haploid
Diploid
Heterokaryotic

1 Compatible hyphae mate by plasmogamy of hyphal branches, combining nuclei of 2 genetic types.

2 The dikaryotic cell divides by mitosis to produce a dikaryotic mycelium, which can be very long-lived.

3 Hyphal branches known as clamp connections bridge recently divided cells, ensuring that one of each nuclear type is regularly distributed to each daughter cell.

Mitosis and cell growth in tip cell
Hyphal branch carries 1 nucleus
Clamp connection forms
New septum forms
Nuclear distribution complete

4 Under appropriate conditions, dikaryotic mycelium may form a fruiting body or basidiocarp.

Gill of mushroom

5 Dikaryotic basidia occur at the surfaces of gills (or pores of some mushrooms).

Diploid nucleus

6 Nuclei in basidia fuse to form diploid nuclei.

Basidium with haploid nuclei
Basidiospore

7 Basidia undergo meiosis to produce 4 haploid nuclei, which are incorporated into basidiospores that are dispersed.

Basidiospores
Basidium

8 Basidiospores grow into mycelia, the cells of which each possess 1 haploid nucleus.

Figure 31.20 **The sexual life cycle of the basidiomycete fungus *Coprinus disseminatus*.**

(Figure 31.20). Basidia are typically located on the undersides of fruiting bodies, which are generally known as **basidiocarps**. Though some basidiomycetes have lost the property of sexual reproduction, they can be identified as members of this phylum by unique hyphal structures known as clamp connections that help distribute nuclei during cell division (see Figure 31.20). Basidiomycetes can also be identified by distinctive septa having complex pores (see Figure 31.17b) and by DNA methods. Basidiomycetes reproduce asexually by various types of spores.

An estimated 30,000 modern basidiomycete species are known. Basidiomycetes are very important as decomposers and in symbiotic associations with plants, producing diverse basidiocarps commonly known as mushrooms, puffballs, stinkhorns, shelf fungi, rusts, and smuts (Figure 31.21). Basidiocarps are also shown in Figures 31.8, 31.9, and 31.10. The fairy rings of mushrooms that sometimes occur in open, grassy areas are ring- or arc-shaped arrays of basidiomycete fruiting bodies.

(a) Corn smut **(b) Shelf fungi**

Figure 31.21 **Fruiting bodies of basidiomycetes.** **(a)** Corn smut (*Ustilago maydis*) produces dikaryotic mycelial masses within the kernels (fruits) of infected corn plants. These mycelia produce many dark spores in which karyogamy and meiosis occur. Masses of these dark spores cause the smutty appearance. When the spores germinate, they produce basidiospores that can infect other corn plants. **(b)** Shelf fungi, such as this sulfur shelf fungus (*Laetiporus sulphureus*), are the fruiting bodies of basidiomycete fungi that have infected trees.

31.4 Fungal Ecology and Biotechnology

Learning Outcomes:

1. Identify the ecological roles of decomposer and disease fungi.
2. Give examples of fungal diseases of plants and animals, including humans.
3. Explain how mycorrhizae, endophytes, and lichens form beneficial associations with other organisms.
4. List several uses of fungi in biochemistry, biological studies, and industrial processes.

Fungi play important ecological roles as decomposers, predators, pathogens, and beneficial symbionts. Such ecological diversity allows humans to utilize fungi in diverse biotechnological applications.

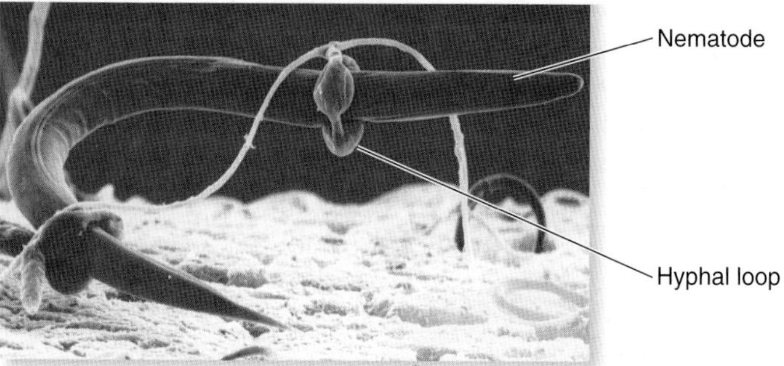

Figure 31.22 A predatory fungus. The fungus *Arthrobotrys anchonia* traps nematode worms in hyphal loops that suddenly swell in response to the animal's presence. Fungal hyphae then grow into the worm's body and digest it.

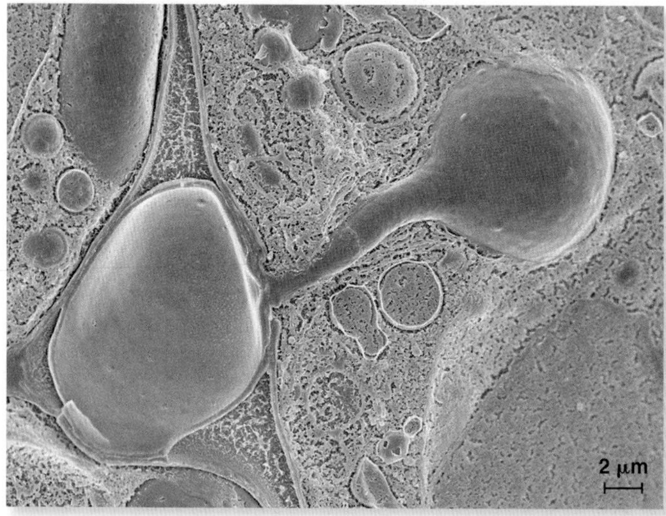

Figure 31.23 Fungal haustoria. Fungi that parasitize plants often produce specialized cells called haustoria that absorb organic food from plant cells.

Decomposer and Predatory Fungi Play Important Ecological Roles

Decomposer fungi are essential components of the Earth's ecosystems. Together with bacteria, they decompose dead organisms and wastes, preventing the buildup of organic debris in ecosystems. For example, only certain bacteria and fungi can break down cellulose and lignin, the main components of wood. Decomposer fungi and bacteria are Earth's recycling engineers. They release CO_2 into the air and other minerals into the soil and water, making these essential nutrients available to plants and algae.

More than 200 species of predatory soil fungi use special adhesive or nooselike hyphae to trap tiny soil animals, such as nematodes, and absorb nutrients from their bodies (Figure 31.22). Such fungi help to control populations of nematodes, some of which attack plant roots. Other fungi obtain nutrients by attacking insects, and certain of these species have been used as biological control agents to kill black field crickets, red-legged earth mites, and other pests.

Pathogenic Fungi Cause Plant and Animal Diseases

One of the most important ways in which fungi affect humans is by causing diseases of crop plants and animals. Five thousand fungal species are known as plant pathogens because they cause serious crop diseases. Plant pathogenic fungi typically display specialized hyphae known as haustoria, whose increased cell membrane surface area aids the absorption of organic food from plant cells (Figure 31.23). Pathogenic fungi use the absorbed organic compounds to grow, attack more plant cells, and produce reproductive spores capable of infecting more plants.

Wheat rust is an example of a common crop disease caused by fungi (Figure 31.24). Rusts are named for the reddish spores that emerge from the surfaces of infected plants. Many types of plants can be attacked by rust fungi, but rusts are of particular concern when new strains attack crops. For example, in late 2004, agricultural scientists discovered that a devastating rust named *Phakopsora pachyrhizi* had begun to spread in the U.S. soybean crop. This rust kills soybean plants by attacking the leaves, causing complete leaf drop in less than 2 weeks. The disease had apparently spread to U.S. farms by means of spores blown on hurricane winds from South America. To control the spread of fungal diseases, agricultural experts work to identify effective fungicidal chemicals and develop resistant crop varieties. Agricultural customs inspectors closely monitor the entry of plants, soil, foods, and other materials that might harbor pathogenic fungi.

Fungi cause several types of disease in animals. For example, *Geomyces destructans* is associated with white nose syndrome of bats, which has killed more than 1 million hibernating bats in the U.S. Athlete's foot and ringworm are common human skin diseases caused by several types of fungi that are known as dermatophytes because they colonize the human epidermis. *Pneumocystis jiroveci* and *Cryptococcus neoformans* are fungal pathogens that infect individuals with weakened immune systems, such those with AIDS, sometimes causing death. **Dimorphic fungi** (from the Greek, meaning two forms) live as spore-producing hyphae in the soil but transform into pathogenic yeasts when mammals inhale their wind-dispersed spores (Figure 31.25). Dimorphic fungi include *Blastomyces dermatitidis*, which causes the disease blastomycosis; *Coccidioides immitis*, the cause of coccidiomycosis; and *Histoplasma capsulatum*, the agent of

Figure 31.24 Wheat rust. The plant pathogenic fungus *Puccinia graminis* grows within the tissues of wheat plants, using plant nutrients to produce rusty streaks of red spores that erupt at the stem and leaf surface where spores can be dispersed. Red spore production is but one stage of a complex life cycle involving several types of spores. Rusts infect many other crops in addition to wheat, causing immense economic damage.

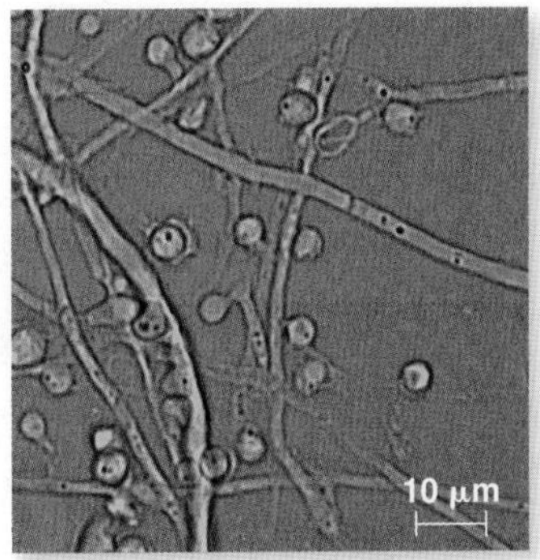

(a) Soil-dwelling hyphal phase

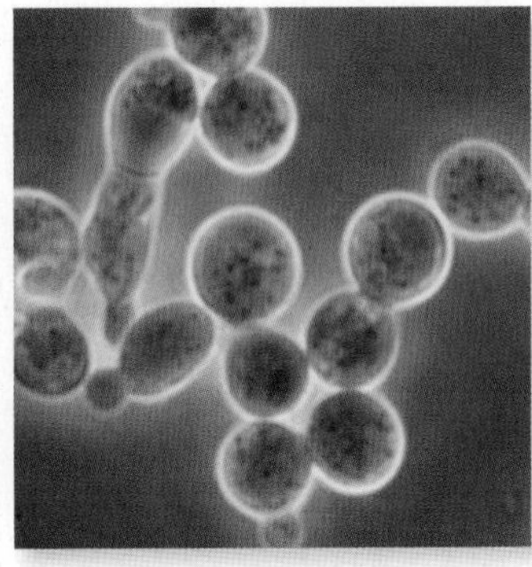

(b) Budding yeast phase in host

Figure 31.25 Dimorphic fungi. **(a)** The soil-dwelling hyphal stage reproduces by airborne spores. **(b)** When a mammal inhales the spores, body heat causes the budding yeast phase to develop and attack host tissues.

histoplasmosis. These fungal diseases affect the lungs and may spread to other parts of the body, causing severe illness. Host body temperature triggers the change from hyphal to yeast form. Instead of producing spores, in the mammalian body, these pathogenic yeasts reproduce by forming buds that more effectively stick to lung cells, spread within lung tissue, and move to other organs. Though fungal diseases that attack humans are of medical concern, in nature, fungal pathogens often help to control populations of other organisms, which is an important ecological role.

Fungi Form Beneficial Associations with Other Species

Symbioses are close associations of one or more other species, and mutualistic symbioses occur when all partners in a close association benefit. Fungi form several types of mutualistic symbiosis with animals, plants, algae, bacteria, and even viruses. For example, leaf-cutting ants, certain termites and beetles, and the salt marsh snail (*Littoraria irrorata*) cultivate particular fungi for food—much as human mushroom growers do. Other fungi obtain organic food molecules from photosynthetic organisms—plants, green algae, or cyanobacteria—that, in turn, receive benefits from the fungi. We focus next on three types of fungi—mycorrhizal fungi, endophytes, and lichen fungi—that are beneficially associated with photosynthetic organisms.

Mycorrhizae Mutualistic symbioses between the hyphae of certain fungi and the roots of most seed plants are known as **mycorrhizae** (from the Greek, meaning fungus roots). Such fungus-root associations are very important in nature and agriculture; more than 80% of terrestrial plants form mycorrhizae. Plants that have mycorrhizal partners receive an increased supply of water and mineral nutrients, primarily phosphate, copper, and zinc. They do so because an extensive fungal mycelium is able to absorb minerals from a much larger volume of soil than roots alone are able to do (**Figure 31.26**). Added together, all the branches of a fungal mycelium in 1 m^3 of soil can reach 20,000 km in total length. Experiments have shown that mycorrhizae greatly enhance plant growth in comparison to plants lacking fungal partners. In return, plants provide fungi with organic food molecules, sometimes contributing as much as 20% of their photosynthetic products.

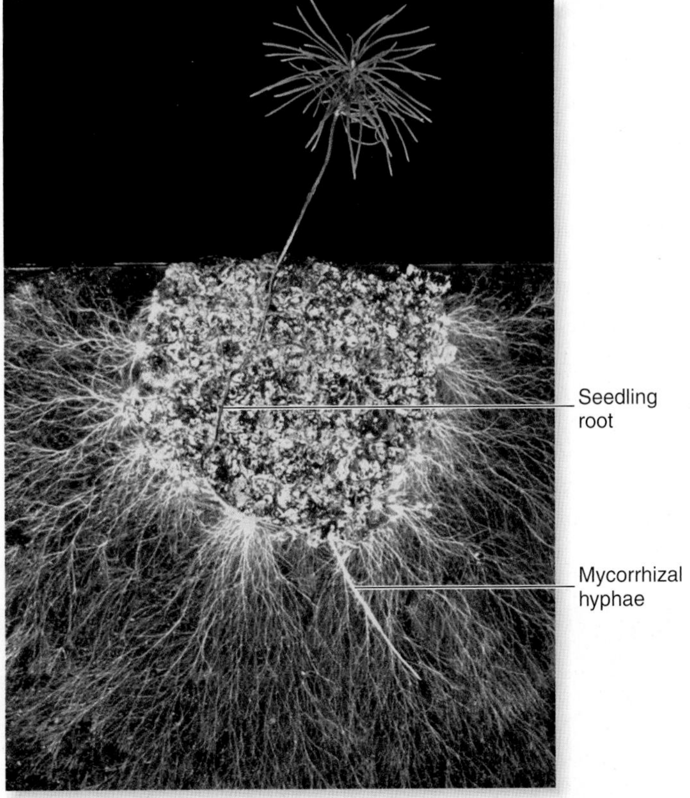

Figure 31.26 Tree seedling with mycorrhizal fungi. Hyphae of a mycorrhizal fungus extend farther into the soil than do plant roots, helping plants to obtain mineral nutrients.

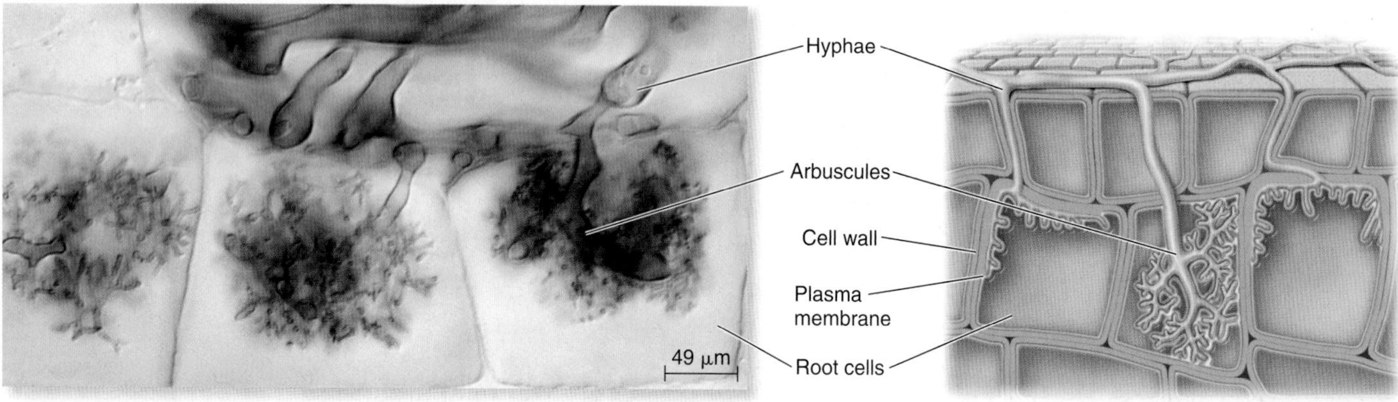

(a) Micrograph of arbuscular mycorrhizae

(b) Hyphae growing between cell walls and plasma membranes

Figure 31.27 Endomycorrhizae. (a) Light micrograph showing black-stained AM fungi within the roots of the forest herb *Asarum canadensis*. Endomycorrhizal fungal hyphae penetrate plant root cell walls, and then branch into the space between root cell walls and plasma membranes. (b) Diagram showing the position of highly branched arbuscules. Hyphal branches or arbuscules are found on the surface of the plasma membrane, which becomes highly invaginated. The result is that both hyphae and plant membranes have very high surface areas.

Concept Check: *What fungal phylum consists entirely of endomycorrhizal fungi that are completely dependent upon plant hosts?*

The two most common types of mycorrhizae are endomycorrhizae, which occur within root tissues, and ectomycorrhizae, which coat roots. **Endomycorrhizae** (from the Greek *endo*, meaning inside) are partnerships between plants and fungi in which the fungal hyphae penetrate the spaces between root cell walls and plasma membranes and grow along the outer surface of the plasma membrane. In such spaces, endomycorrhizal fungi often form highly branched, bushy arbuscules (from the word "arbor," referring to tree shape). As the arbuscules develop, the root plasma membrane also expands. Consequently, the arbuscules and the root plasma membranes surrounding them have a very high surface area that facilitates rapid and efficient exchange of materials: Minerals flow from fungal hyphae to root cells, while organic food molecules move from root cells to hyphae. These fungus-root associations are known as **arbuscular mycorrhizae**, abbreviated **AM** (Figure 31.27). AM fungi are associated with apple and peach trees, coffee shrubs, and many herbaceous plants, including legumes, grasses, tomatoes, and strawberries.

Ectomycorrhizae (from the Greek *ecto*, meaning outside) are mutualistic symbioses between temperate forest trees and soil fungi. The fungi that engage in such associations are known as ectomycorrhizal fungi (Figure 31.28a). The hyphae of ectomycorrhizal fungi coat tree-root surfaces (Figure 31.28b) and grow into the spaces between root cells but do not penetrate the cell membrane (Figure 31.28c). Some species of oak, beech, pine, and spruce trees will not grow unless their ectomycorrhizal partners are also present. Mycorrhizae are thus essential to the success of commercial nursery tree production and reforestation projects. New genetic information has illuminated how mycorrhizal fungi evolved.

(a) Ectomycorrhizal fruiting body

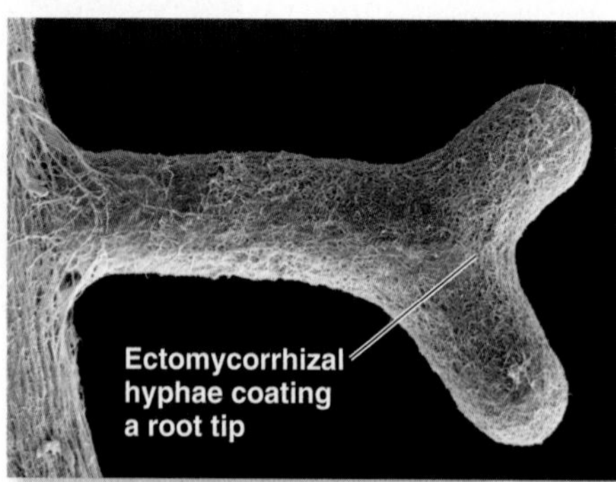

(b) SEM of ectomycorrhizal hyphae

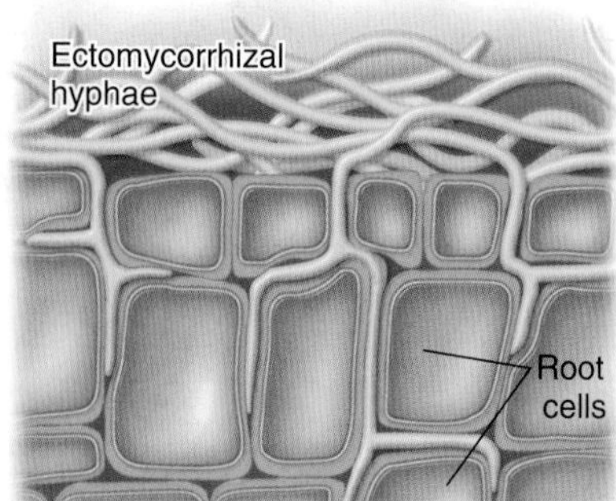

(c) Hyphae invading intercellular spaces

Figure 31.28 Ectomycorrhizae. (a) The fruiting body of the common forest fungus *Laccaria bicolor*. This is an ectomycorrhizal fungus that is associated with tree roots. (b) Ectomycorrhizal fungal hyphae of *L. bicolor* cover the surfaces of young *Pinus resinosa* root tips. (c) Diagram showing that the hyphae of ectomycorrhizal fungi do not penetrate root cell walls but grow within intercellular spaces. In this location, fungal hyphae are able to obtain organic food molecules produced by plant photosynthesis.

Concept Check: *What benefits do plants obtain from the association with fungi?*

GENOMES & PROTEOMES CONNECTION

Genomic Comparison of Forest Fungi Reveals Coevolution with Plants

Genomic sequences have been obtained for many of the basidiomycete fungi that occur in forests, allowing comparisons that reveal how forest fungi have coevolved with plants. Coevolution, a process that has also strongly influenced the diversification of plants and animals (see Chapter 30), is the process by which two or more species of organisms influence each other's evolutionary pathway. The genome of the forest fungus *Serpula lacrymans*, reported in 2011 by Daniel Eastwood, Sarah Watkinson, and associates, has been key to understanding how fungi have evolved many types of symbiotic associations with plants. Certain forest fungi are parasites, whereas others form mutualistic ectomycorrhizal associations. Other forest fungi decompose cellulose and lignin, major components of wood. Cellulose and lignin are the two most abundant biological polymers on Earth and are relatively resistant to microbial decomposition.

About 6% of forest fungi, including *S. lacrymans*, are known as brown rot fungi because they obtain energy by breaking down the cellulose contained in wood, leaving brown-colored lignin. The lignin that brown rot fungi leave behind contributes a substantial amount of organic carbon to forest soils, thereby influencing soil fertility. Other forest fungi, known as white rot fungi, decompose both the cellulose and lignin present in wood, leaving white-colored remains. White rot fungi, such as *Phanerochaete chrysosporium*, feature particularly complex enzymatic pathways, which are needed to break down the many types of chemical bonds present in lignin. Such complex enzymatic pathways are energetically expensive to produce, but allow the fungi ready access to cellulose embedded within a lignin matrix.

Genomic comparisons have revealed that brown rot fungi such as *S. lacrymans* evolved from white rot ancestors by loss of protein families involved in lignin degradation. For example, brown rot fungi lack class II peroxidase enzymes that white rot fungi use to help decompose lignin. Consequently the brown rot cannot break down lignin, but they save the energy they would need to expend in this process. Genomic comparisons also suggest that ectomycorrizal fungi evolved from brown rot ancestors. Patterns of forest fungal divergence match those of forest trees, indicating that forest fungi and plants have coevolved over time.

Fungal Endophytes Other fungi are known as **endophytes** because they live within the leaf and stem tissues of almost all plants, without causing disease. In tropical regions, hundreds of endophytic fungal species can occur within a single tree. The endophytes obtain organic food molecules from plants and, in turn, contribute toxins or antibiotics that deter foraging animals, insect pests, and microbial pathogens. In general, plants that contain nonpathogenic endophytic fungi grow better than plants lacking such fungal partners. Endophytic fungi also help some plants tolerate higher temperatures.

FEATURE INVESTIGATION

Márquez and Associates Discovered That a Three-Partner Symbiosis Allows Plants to Cope with Heat Stress

The endophytic fungus *Curvularia protuberata* commonly lives within aboveground tissues of the grass *Dichanthelium lanuginosum*, which is unusual in its ability to grow on very hot soils in thermal areas of Yellowstone National Park. When the soil reaches 38°C, *D. lanuginosum* plants and *C. protuberata* fungi both die—unless they live together in a symbiosis. In the symbiotic association, the partners can survive temperatures near 65°C!

In 2007, a team of investigators led by Luis Márquez discovered that a virus is also involved, revealing the occurrence of a three-partner symbiosis (Figure 31.29). These biologists were able to isolate the virus from the fungus and named it *Curvularia* thermal tolerance virus (CthTV) to indicate its host and phenotype. The investigators also noticed that some of their fungal cultures contained very little

Figure 31.29 **Márquez and associates discovered that a three-partner symbiosis allows plants to cope with heat stress.**

GOAL To determine if a virus is essential to the protective role of endophytic fungi to host plants under heat stress.

KEY MATERIALS *Curvularia* thermal tolerance virus (CthTV), cultures of the endophytic fungus *Curvularia protuberata* infected with CthTV, *C. protuberata* cultures free of CthTV, and *Dichanthelium lanuginosum* plants.

Experimental level

Conceptual level

1 Plant 25 replicate containers with *D. lanuginosum* lacking fungal symbionts (a) or with *C. protuberata* endophytes that either did (b) or did not (c) have virus.

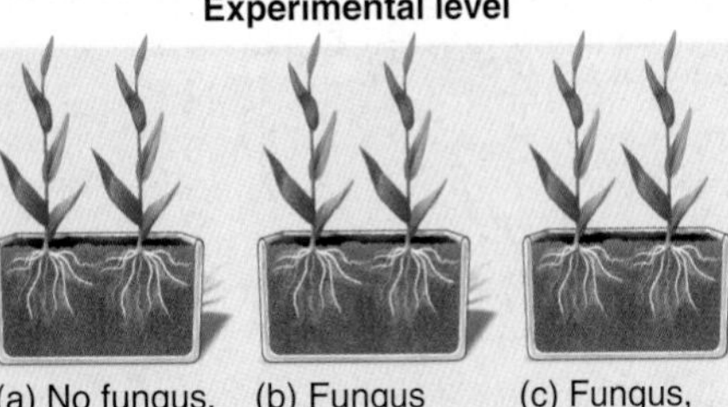

Compare the effects of virus on the ability of the fungus to confer heat stress protection.

2 Expose plants to heat stress treatment (up to 65°C) for 2 weeks in a greenhouse.

Keep environmental conditions constant to reduce experimental error.

3 Count the number of plants that were green (alive), yellow (dying), or brown (dead).

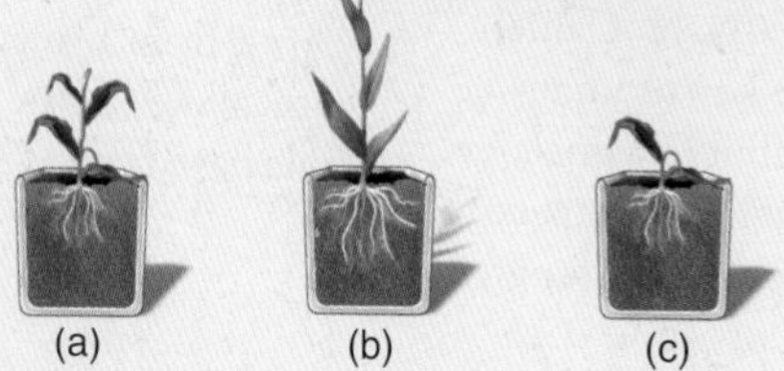

Assess plant survival in the presence or absence of fungus and/or virus.

4 THE DATA

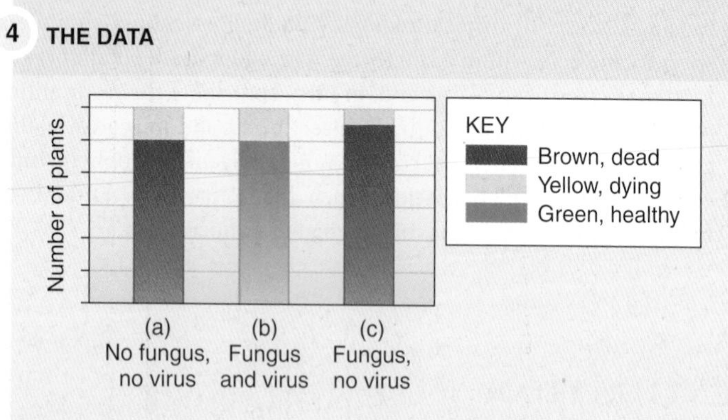

5 CONCLUSION A virus enhances the protective role of endophytic fungi in this grass species. The next step will be to try to determine just how the virus changes the fungus so that the fungus is able to protect the plant from heat stress.

6 SOURCE Márquez, Luis M. et al. 2007. A virus in a fungus in a plant: Three-way symbiosis required for thermal tolerance. *Science* 315:513–515.

virus, so they were able to use drying and freeze-thaw cycles to cure such cultures of the virus. This procedure allowed them to experimentally determine the relative abilities of virus-infected and virus-free *C. protuberata* fungus to tolerate high temperatures and confer this property to plant partners. They found that plants having viral-infected fungal endophytes tolerated high temperatures much better than plants that lacked fungal endophytes or possessed only virus-free fungal endophytes. Márquez and associates also reintroduced virus to their virus-free fungi and found that such fungi acquired the ability to confer heat tolerance to host plants. Finally, the researchers determined that the virus-infected fungus (but not virus-free fungi) could also protect a distantly related crop plant (tomato) from heat stress. These results add to accumulating evidence that multipartner symbioses are more common than previously realized and demonstrated that endophytic fungi may have useful agricultural applications.

Experimental Questions

1. Would you expect plants that grow on unusually hot soils to have endophytic fungi or not?
2. How did Márquez and associates demonstrate that a virus was important in the heat tolerance of the *Dichanthelium lanuginosum*/*Curvularia protuberata* symbiosis?
3. How might the results of the work by Márquez and associates be usefully applied in agriculture?

Lichens Multipartner associations are also represented by **lichens**, which are composed of particular fungi, certain photosynthetic green algae and/or cyanobacteria, and nonphotosynthetic bacteria such as actinomycetes. There are at least 25,000 lichen species, but these did not all descend from a common ancestor. DNA-sequencing studies suggest that lichens evolved independently in at least five separate fungal lineages. Molecular studies also show that some fungi have lost their ancestral ability to form lichen associations.

Lichen bodies take one of three major forms: (1) crustose—flat bodies that are tightly adherent to an underlying surface (**Figure 31.30a**); (2) foliose—flat, leaflike bodies (**Figure 31.30b**); or (3) fruticose—bodies that grow upright (**Figure 31.30c**) or hang down from tree branches. The photosynthetic green algae or cyanobacteria typically occur in a distinct layer close to the lichen's surface (**Figure 31.30d**). Lichen structure differs dramatically from that of the fungal component grown separately, demonstrating that the photosynthetic components influence lichen form.

The photosynthetic partner provides lichen fungi with organic food molecules and oxygen, and, in turn, it receives carbon dioxide, water, and minerals from the fungal partner. Lichen fungi also protect their photosynthetic partners from environmental stress. For example, lichens that occupy exposed habitats of high light intensity often produce bright yellow, orange, or red-colored compounds that absorb excess light, thereby helping to prevent damage to the photosynthetic apparatus (see Figure 31.30a). Lichen fungi also produce distinctive organic acids and other compounds that deter animal and microbial attacks.

Many lichens reproduce by both sexual and asexual means, and about one-third of lichen species reproduce only asexually. Asexual reproductive structures include soredia (singular, soredium), small clumps of hyphae surrounding a few algal cells that can disperse in wind currents. Soredia are lichen clones. By forming soredia, lichen fungi can disperse along with their photosynthetic partners.

Figure 31.30 **Lichen structure.** **(a)** An orange-colored crustose lichen grows tightly pressed to the substrate. **(b)** The flattened, leaf-shaped genus *Umbilicaria* is a common foliose lichen. **(c)** The highly branched genus *Cladonia* is a common fruticose lichen. **(d)** A handmade thin slice of *Umbilicaria* viewed with a light microscope reveals that the photosynthetic algae occur in a thin upper layer. Fungal hyphae make up the rest of the lichen.

The fungal partners of many lichens can undergo sexual reproduction, producing fruiting bodies and sexual spores much like those of related fungi that do not form lichens. DNA studies have shown that some lichen fungi can self-fertilize, which is advantageous in harsh environments where potential mates may be lacking. To produce new lichens, hyphae that grow from sexual spores must acquire new photosynthetic partners, but only particular green algae or cyanobacteria are suitable. However, lichens do not always contain the same algal partners as their parents because lichen fungi may switch algal partners, "trading up" for better algae. Partner switching allows lichens to adjust to changes in their environments.

Lichens often grow on rocks, buildings, tombstones, tree bark, soil, or other surfaces that easily become dry. When water is not available, the lichens are dormant until moisture returns. Thus, lichens may spend much of their time in an inactive state, and for this reason, they often grow very slowly. However, because they can persist for long periods, lichens can be very old; some are estimated to be more than 4,500 years old. Lichens occur in diverse types of habitats, and a number grow in some of the most extreme, forbidding terrestrial sites on Earth—deserts, mountaintops, and the Arctic and Antarctic—places where most plants cannot survive. In these locations, lichens serve as a food source for reindeer and other hardy organisms. Though unpalatable, lichens are not toxic to humans and have also served as survival foods for indigenous peoples in times of shortages.

Soil-building is another important lichen function. Lichen acids help to break up the surfaces of rocks, beginning the process of soil development. Lichens having cyanobacterial partners can also increase soil fertility by adding fixed nitrogen. One study showed that such lichens released 20% of the nitrogen they fixed into the environment, where it is available for uptake by plants.

Lichens are useful as air-quality monitors because they are particularly sensitive to air pollutants such as sulfur dioxide. Air pollutants severely injure the photosynthetic components, causing death of the lichens. The disappearance of lichens serves as an early warning sign of air-pollution levels that are also likely to affect humans. Lichens can also be used to monitor atmospheric radiation levels because they accumulate radioactive substances from the air.

Fungi Have Many Applications in Biotechnology

The ability of fungi to grow on many types of substrates and produce many types of organic compounds reflects their diverse ecological adaptations. Humans have harnessed fungal biochemistry in many types of biotechnology applications. A variety of industrial processes use fungi to convert inexpensive organic compounds into valuable materials such as citric acid used in the soft-drink industry, glycerol, antibiotics such as penicillin, and cyclosporine, a drug widely used to prevent rejection of organ transplants. Enzymes extracted from fungi are used to break down plant materials for renewable bioenergy production. In the food industry, fungi are used to produce the distinctive flavors of blue cheese and other cheeses. Other fungi secrete enzymes that are used in the manufacture of protein-rich tempeh and other food products from soybeans. The brewing and winemaking industries find yeasts essential, and the baking industry depends on the yeast *Saccharomyces cerevisiae* (see Figure 31.7) for bread production.

S. cerevisiae is also widely used as a model organism for fundamental biological studies. Yeasts are useful in the laboratory because they have short life cycles, are easy and safe for lab workers to maintain, and their genomes are similar to those of animals. Some 31% of yeast proteins have human homologs, and nearly 50% of human genes that have been implicated in heritable diseases have homologs in yeast.

Summary of Key Concepts

31.1 Evolution and Distinctive Features

- Fungi form a monophyletic kingdom of heterotrophs that, together with the animal kingdom and certain protists, form the supergroup Opisthokonta (Figure 31.1).
- Fungal cells possess cell walls composed of the polysaccharide chitin. Fungal bodies, known as mycelia, are composed of microscopic branched filaments known as hyphae. Early-diverging fungi have aseptate hyphae that are not subdivided into cells. The hyphae of later-diverging fungi are subdivided into cells by cross walls known as septa (Figures 31.2, 31.3).
- Fungal hyphae feed and grow at their tips (Figure 31.4).
- Mycelial shape depends on the distribution of nutrients in the environment, which determines the direction in which cell division and hyphal growth will occur (Figure 31.5).

31.2 Fungal Asexual and Sexual Reproduction

- Fungi spread rapidly by means of spores produced by asexual or sexual reproduction.
- Asexual reproduction does not involve mating or meiosis, and it occurs by means of asexual spores such as conidia or by budding (Figures 31.6, 31.7).
- Fungi display a haploid-dominant sexual life cycle. During sexual reproduction of terrestrial fungi, hyphal branches (gametes) fuse with those of a different mycelium of compatible mating type. Mated hyphae form fruiting bodies in which haploid spores are produced by meiosis. Dispersed spores germinate to produce haploid fungal mycelia.
- Fungi produce diverse types of fruiting bodies that foster spore dispersal by wind, water, or animals. Although many fungal fruiting bodies are edible, many others produce defensive toxins or hallucinogens (Figures 31.8, 31.9, 31.10, 31.11).

31.3 Diversity of Fungi

- Seven informal fungal groups are cryptomycota, chytrids, microsporidia, zygomycetes, AM fungi, ascomycetes, and basidiomycetes (Table 31.1).
- Cryptomycota and chytrids are among the simplest and earliest-divergent fungi. They commonly occur in aquatic habitats and moist soil, where they produce flagellate reproductive cells (Figures 31.12, 31.13).
- Microsporidia are single-celled fungi that parasitize animal cells (Figure 31.14).
- Zygomycetes are named for their distinctive, large zygospores, which are the result of sexual reproduction. Common black bread mold and related fungi reproduce asexually by means of many small spores (Figure 31.15).
- The AM fungi produce distinctive large, multinucleate spores and form beneficial arbuscular mycorrhizal relationships with many types of plants (Figure 31.16).
- Ascomycetes produce sexual ascospores in saclike asci located at the surfaces of fruiting bodies known as ascocarps. Many are lichen symbionts. The septa of hyphae have simple pores (Figures 31.17, 31.18, 31.19).
- Basidiomycetes produce sexual basidiospores on club-shaped basidia located on the surfaces of fruiting bodies known as basidiocarps. Such fruiting bodies take a wide variety of forms, including mushrooms, puffballs, stinkhorns, shelf fungi, rusts, and smuts. Hyphae display complex septal pores and clamp connections. Mating commonly generates a long-lived dikaryotic mycelium that can produce many fruiting bodies (Figures 31.20, 31.21).

31.4 Fungal Ecology and Biotechnology

- Fungi play important roles in nature as decomposers, predators, and pathogens, and by forming beneficial associations with other organisms. Pathogenic fungi cause plant and animal diseases (Figures 31.22, 31.23, 31.24, 31.25).
- Mycorrhizae are symbiotic associations between fungi and plant roots. Endomycorrhizae commonly form highly branched arbuscules in the spaces between root cell walls and plasma membranes. Ectomycorrhizae coat root surfaces, extending into root intercellular spaces (Figures 31.26, 31.27, 31.28).
- Comparative genomic studies reveal the evolution of forest fungal nutritional variations.
- Endophytic fungi live symbiotically within the tissues of plants (Figure 31.29).
- Lichens are multispecies partnerships between fungi, photosynthetic green algae and/or cyanobacteria, and other bacteria. Lichens can reproduce asexually or sexually. They occur in diverse habitats, including bare rock surfaces, where they help to build soil (Figure 31.30).
- Fungi are useful in the chemical, food processing, waste-treatment, and renewable biofuel industries. The yeast *Saccharomyces cerevisiae* is a model organism and also important to the brewing and baking industries.

Assess and Discuss

Test Yourself

1. Fungal cells differ from animal cells in that fungal cells
   a. lack ribosomes, though these are present in animal cells.
   b. lack mitochondria, though these occur in animal cells.
   c. have chitin cell walls, whereas animal cells lack rigid walls.
   d. lack cell walls, whereas animal cells possess walls.
   e. none of the above

2. Conidia are
   a. cells produced by some fungi as the result of sexual reproduction.
   b. fungal asexual reproductive cells produced by the process of mitosis.
   c. structures that occur in septal pores.
   d. the unspecialized gametes of fungi.
   e. none of the above.
3. What are mycorrhizae?
   a. the bodies of fungi, composed of hyphae
   b. fungi that attack plant roots, causing disease
   c. fungal hyphae that are massed together into stringlike structures
   d. fungi that have symbiotic partnerships with algae or cyanobacteria
   e. mutually beneficial associations of particular fungi and plant roots
4. Where could you find diploid nuclei in an ascomycete or basidiomycete fungus?
   a. in spores
   b. in cells at the surfaces of fruiting bodies
   c. in conidia
   d. in soredia
   e. all of the above
5. Which fungi are examples of hallucinogen producers?
   a. *Claviceps* and *Psilocybe*
   b. *Epidermophyton* and *Candida*
   c. *Pneumocystis jiroveci* and *Histoplasma capsulatum*
   d. *Saccharomyces cerevisiae* and *Phanerochaete chrysosporium*
   e. *Cryphoenectria parasitica* and *Ventura inaequalis*
6. What role do fungal endophytes play in nature?
   a. They are decomposers.
   b. They are human pathogens that cause skin diseases.
   c. They are plant pathogens that cause serious crop diseases.
   d. They live within the tissues of plants, helping to protect them from herbivores, pathogens, and heat stress.
   e. All of the above are correct.
7. What forms do lichens take?
   a. crusts, flat bodies
   b. foliose, leaf-shaped bodies
   c. fruticose, erect or dangling bodies
   d. single cells
   e. a, b, and c
8. Lichens consist of a partnership between fungi and what other organisms?
   a. red algae and brown algae
   b. green algae, cyanobacteria, and heterotrophic bacteria
   c. the roots of vascular plants
   d. choanoflagellates and *Nuclearia*
   e. none of the above
9. How can ascomycetes be distinguished from basidiomycetes?
   a. Ascomycete hyphae have simple pores in their septa and lack clamp connections, whereas basidiomycete hyphae display complex septal pores and clamp connections.
   b. Ascomycetes produce sexual spores in sacs, whereas basidiomycetes produce sexual spores on the surfaces of club-shaped structures.
   c. Ascomycetes are commonly found in lichens, whereas basidiomycetes are less commonly partners in lichen associations.
   d. Ascomycetes are not commonly mycorrhizal partners, but basidiomycetes are commonly present in mycorrhizal associations.
   e. All of the above are correct.
10. Which group of organisms listed is most closely related to the kingdom Fungi?
   a. the animal kingdom
   b. the green algae
   c. the land plants
   d. the bacteria
   e. the archaea

Conceptual Questions

1. Explain three ways that fungi are like animals and two ways in which fungi resemble plants.
2. Explain why some fungi produce toxic or hallucinogenic compounds.
3. A principle of biology is that *living organisms interact with their environment.* Explain three ways in which fungi function as beneficial partners with autotrophs and what benefit the fungi receive from the partnerships.

Collaborative Questions

1. Thinking about the natural habitats closest to you, where can you find fungi, and what roles do these fungi play?
2. Imagine that you are helping to restore the natural vegetation on a piece of land that had long been used to grow crops. You are placed in charge of planting pine seedlings (*Pinus resinosa*) and fostering their growth. In what way could you consider using fungi?

Online Resource

www.brookerbiology.com

Stay a step ahead in your studies with animations that bring concepts to life and practice tests to assess your understanding. Your instructor may also recommend the interactive eBook, individualized learning tools, and more.

An Introduction to Animal Diversity

32

Chapter Outline

32.1 Characteristics of Animals

32.2 Animal Classification

32.3 Molecular Views of Animal Diversity

Summary of Key Concepts

Assess and Discuss

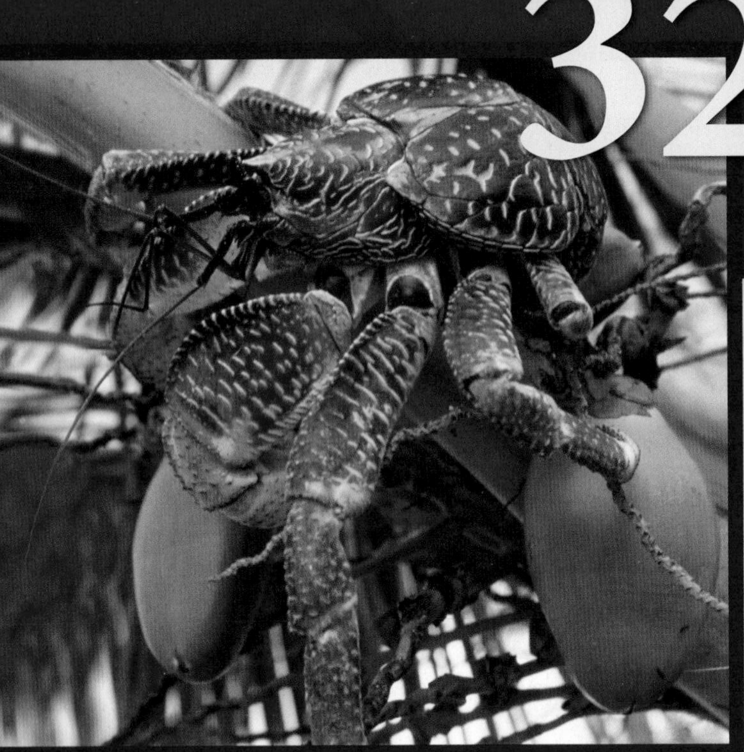

The variety of life forms on Earth is staggering. The robber crab (*Birgus latro*) is the largest terrestrial invertebrate on Earth.

The animal illustrated in the chapter-opening photo is a coconut crab, *Birgus latro*, the largest land-living arthropod in the world. Found on islands throughout the Indian and Pacific Oceans, these crabs climb trees to cut down coconuts and use their powerful claws to open them. They are also rumored to steal shiny items, including pots and pans, from houses or tents, hence their alternative name of robber crab. Although most crabs are marine species, some have adaptations to life on land. The coconut crab has branchiostegal lungs—respiratory organs they use instead of gills—and they have evolved sensory organs that can detect smells in the air. The coconut crab, like all members of its phylum, molts (sheds) its entire outer skeleton, or exoskeleton, many times as it grows.

Animals constitute the most species-rich kingdom. About 1.3 million species have been found and described, and an estimated 2 to 5 million more species await discovery and classification. Beyond being members of this kingdom ourselves, humans depend on animals. Many different kinds of animals and their products are part of our diet. We use a diverse array of animal products for clothing and have traditionally employed animals such as horses and oxen as a source of labor and transportation. Humans enjoy many animal species as companions and depend on other species to test lifesaving drugs. We share parts of our genome with other organisms such as fruit flies, nematodes, and zebra fish—all of which are used as model organisms for understanding aspects of human molecular and developmental biology.

However, we are also in competition with animals such as insects that threaten our food supply and other animals that transmit deadly diseases. Malaria is transmitted by mosquitoes; sleeping sickness, by tsetse flies; and rabies, by a number of animals, including dogs, raccoons, and bats. With such a huge number and diversity of existing animals and with animals featuring so prominently in our lives, understanding animal diversity is of paramount importance. Therefore, researchers have spent a great deal of effort in determining the unique characteristics of different taxonomic groups and identifying their evolutionary relationships.

Since the time of Carolus Linnaeus in the 1700s, scientists have classified animals based on their morphology, that is, on their physical structure. Then, as now, a lively debate has surrounded the question of what constitutes the "correct" animal phylogeny. In the 1990s, animal classifications based on similarities in DNA and rRNA sequences became more common. Quite often, classifications based on morphology and those based on molecular data were similar, but some important differences arose. In this chapter, we will begin by defining the key characteristics of animals and then take a look at the major features of animal body plans that form the basis of classification. We will explore how new molecular data have enabled scientists to revise and refine the animal phylogenetic tree. As more molecular-based evidence becomes available, systematists will likely continue to redraw the tree of animal life. Thus, as you read this chapter, keep in mind that the classification of animals is now, and will continue to be, a work in progress.

32.1 Characteristics of Animals

Learning Outcomes:

1. List the key characteristics of animals that distinguish them from other organisms.
2. Provide a brief overview of the history of animal life on earth.

The Earth contains a dazzling diversity of animal species, living in environments from the deep sea to the desert and exhibiting an amazing array of characteristics. Most animals move and eat multicellular prey, and therefore, they are loosely differentiated from species in other kingdoms. However, coming up with a firm definition of an animal can be tricky because animals are so diverse that biologists can find exceptions to nearly any given characteristic. Even so, a number of key features can help us broadly characterize the group we call animals (Table 32.1).

Animals Are Multicellular Heterotrophs

Animals have several characteristics relating to cell structure, mode of nutrition, movement, and reproduction and development that distinguish them from other organisms.

Cell Structure Like other eukaryotes, plants and fungi, animals are multicellular. However, animal cells lack cell walls and are flexible. This flexibility facilitates movement. Animal cells gain structural support from an extensive extra cellular matrix (ECM) that forms strong fibers outside the cell (refer back to Figure 10.1). Additionally, a group of unique cell junctions—anchoring, tight, and gap junctions—play an important role in holding animal cells in place and allowing communication between cells (refer back to Table 10.3).

Mode of Nutrition Unlike plants, animals are heterotrophs; that is, they cannot synthesize all their organic molecules from inorganic substances and thus must ingest other organisms or their products to sustain life. Many different modes of feeding exist among animals, including suspension feeding (filtering food out of the surrounding water); bulk feeding (such as carnivores and herbivores)—eating large food pieces; and fluid feeding—sucking plant sap or animal body fluids (Figure 32.1). Although fungi and animals both rely on absorptive nutrition—that is, they secrete enzymes that break down complex materials and absorb the resulting small organic molecules—fungi use external digestion to obtain their nutrients. Animals ingest their food into an internal gut and then break it down using enzymes.

Movement Most animals have muscle cells and nerve cells organized into tissues. Muscle tissue is unique to animals, and most animals are capable of some type of locomotion, the ability to move from place to place, in order to acquire food or escape predators. This ability has led to the development of muscular-skeletal systems, systems of sensory structures, and a nervous system that coordinates movement and prey capture. Sessile species such as barnacles, which stay in one place, use bristled appendages to obtain nearby food. However, in many sessile species, although adults are immobile, the larvae can swim.

Table 32.1 Common Characteristics of Animals

Characteristic	Example
Multicellularity	Even relatively simple types of animals such as sponges are multicellular, in contrast to the mostly single-celled eukaryotic microorganisms called protists (see Chapter 28).
Heterotrophs	Animals obtain their food by eating other organisms or their products. This contrasts with plants and algae, most of which are autotrophs and essentially make their own food.
No cell walls	Plant, fungal, and bacterial cells possess a rigid cell wall, but animal cells lack a cell wall and are quite flexible.
Nervous tissue	The presence of a nervous system in most animals enables them to respond rapidly to environmental stimuli.
Movement	Most animals have a muscle system, which, combined with a nervous system, allows them to move in their environment.
Sexual reproduction	Most animals reproduce sexually, with small, mobile sperm uniting with a much larger egg to form a fertilized egg, or zygote.
Extracellular matrix	Proteins such as collagen bind animal cells together to give them added support and strength (see Figure 10.1).
Characteristic cell junctions	Animals have characteristic cell junctions, called anchoring, tight, and gap junctions (see Figures 10.7, 10.9, 10.11).
Special clusters of *Hox* genes	All animals possess *Hox* genes, which function in patterning the body axis (see Figures 19.16, 25.15).
Similar rRNA	Animals have very similar genes that encode for RNA of the small ribosomal subunit (SSU) rRNA (see Figure 12.16).

Reproduction and Development Nearly all members of the animal kingdom reproduce sexually, although certain insects, fish, and lizard species can reproduce asexually. During sexual reproduction, a small, mobile sperm generally unites with a much larger egg to form a fertilized egg, or zygote. Fertilization can occur internally, which is

(a)

(b)

(c)

Figure 32.1 Modes of animal nutrition. **(a)** Suspension feeders, such as this tube worm, filter food particles from the water column. **(b)** Grizzly bears and other bulk feeders tear off large pieces of their food and chew it or swallow it whole. **(c)** Fluid feeders, such as this aphid, suck fluid from their food source.

common in terrestrial species, or externally, which is more common in aquatic species. Similarly, embryos develop inside the mother or outside in the mother's environment. A particularly unusual developmental phenomenon is the occurrence of **metamorphosis**, by which an organism changes from a juvenile to an adult form. Metamorphosis is common in arthropods and is thought to reduce competition for food between juveniles and adults and to disperse the species over long distances For example, caterpillars—the larval form of moths and butterflies—generally feed in one spot, whereas adult moths and butterflies can fly long distances in search of food.

Animal Life Began More Than a Half Billion Years Ago

The history of animal life spans over 590 million years, starting at the end of the Proterozoic eon, when multicellular animals emerged (refer back to Figure 22.9). The first animals to evolve were invertebrates, animals without a vertebral column, or backbone. A profusion of animal phyla appeared during the Cambrian explosion, 533–525 million years ago (mya), including sponges, jellyfish, corals, flatworms, mollusks, annelid worms, the first arthropods, and echinoderms, plus many phyla that no longer exist today (Figure 32.2).

The causes of this sudden increase in animal life at that time are not fully understood, but three reasons have been proposed. First, species proliferation may have been related to a favorable environment, which was warm and wet with no evidence of ice at the poles. At the same time, atmospheric oxygen levels were increasing, permitting increased metabolic rates, and an ozone layer had developed, blocking out harmful ultraviolet radiation and allowing complex life to thrive in shallow water and eventually on land. Second, the evolution of the *Hox* gene complex may have permitted much variation in morphology. Third, as new types of predators evolved, prey developed adaptations that enabled them to avoid their predators, leading to counteradaptations by predators, and so on. This evolutionary "arms race" may have resulted in a proliferation of predator and prey types. These hypotheses are not mutually exclusive and may well have operated at the same time.

Figure 32.2 The profusion of animal life in the Cambrian period, about 520 mya. This artist's reconstruction of marine life shows many different phyla, some of which are now extinct.

Around 520 mya, the first vertebrates, fishes, appeared at roughly the same time as the first plants invaded land. The appearance of land plants represented a viable food source for any organisms that could utilize them. However, the realm of land and air presented organisms with many challenges. For colonization of land to occur, certain species evolved adaptations that prevented them from drying out and enabled them to breathe, move, and reproduce in the new environment, in much the same way as the plant embryo, leaves, seeds, and other adaptations permitted plants to colonize terrestrial habitats (see Chapter 29). For animal species, such features included lungs, a bony skeleton, and internal fertilization. The development of the amniotic egg, which features a tough, protective shell to prevent drying out, enabled animals to be terrestrial for their entire life cycle. The amniotic egg appeared during the Carboniferous period, about 300 mya, and was responsible for the success of the reptiles, which appeared during this period. Reptiles were to dominate the Earth for many millions of years during the rise and fall of the dinosaurs. Mammals also appeared at the same time as dinosaurs, although they were not prevalent. The number and diversity of mammals exploded only after the dinosaurs abruptly died out at the end of the Cretaceous era, about 65 mya.

32.2 Animal Classification

Learning Outcomes:

1. Discuss why choanoflagellates are believed to be the closest living relatives of animals.
2. Describe each of the major morphological and developmental features of animal body plans that form the basis of the classification of animals.

Although animals constitute an extremely diverse kingdom, most biologists agree that the kingdom is monophyletic, meaning that all taxa have evolved from a single common ancestor. Today, scientists recognize about 35 animal phyla. At first glance, many of these phyla seem so distantly related to one another (for example, chordates and jellyfish) that making sense of this diversity with a classification scheme seems very challenging. However, over the course of centuries, scientists have come to some basic conclusions about the evolutionary relationships among animals. In this section, we explore the major features of animal body plans that form the basis of animal phylogeny (Figure 32.3).

Animals Evolved from a Choanoflagellate-like Ancestor

With the monophyletic nature of the animal kingdom in mind, scientists have attempted to characterize the organism from which animals most likely evolved. According to research, the closest living relative of animals is believed to be a flagellated protist known as a choanoflagellate. Choanoflagellates are tiny, single-celled organisms, each with a single flagellum surrounded by a collar composed of cytoplasmic tentacles (Figure 32.4a; look back to Figure 28.21). A number

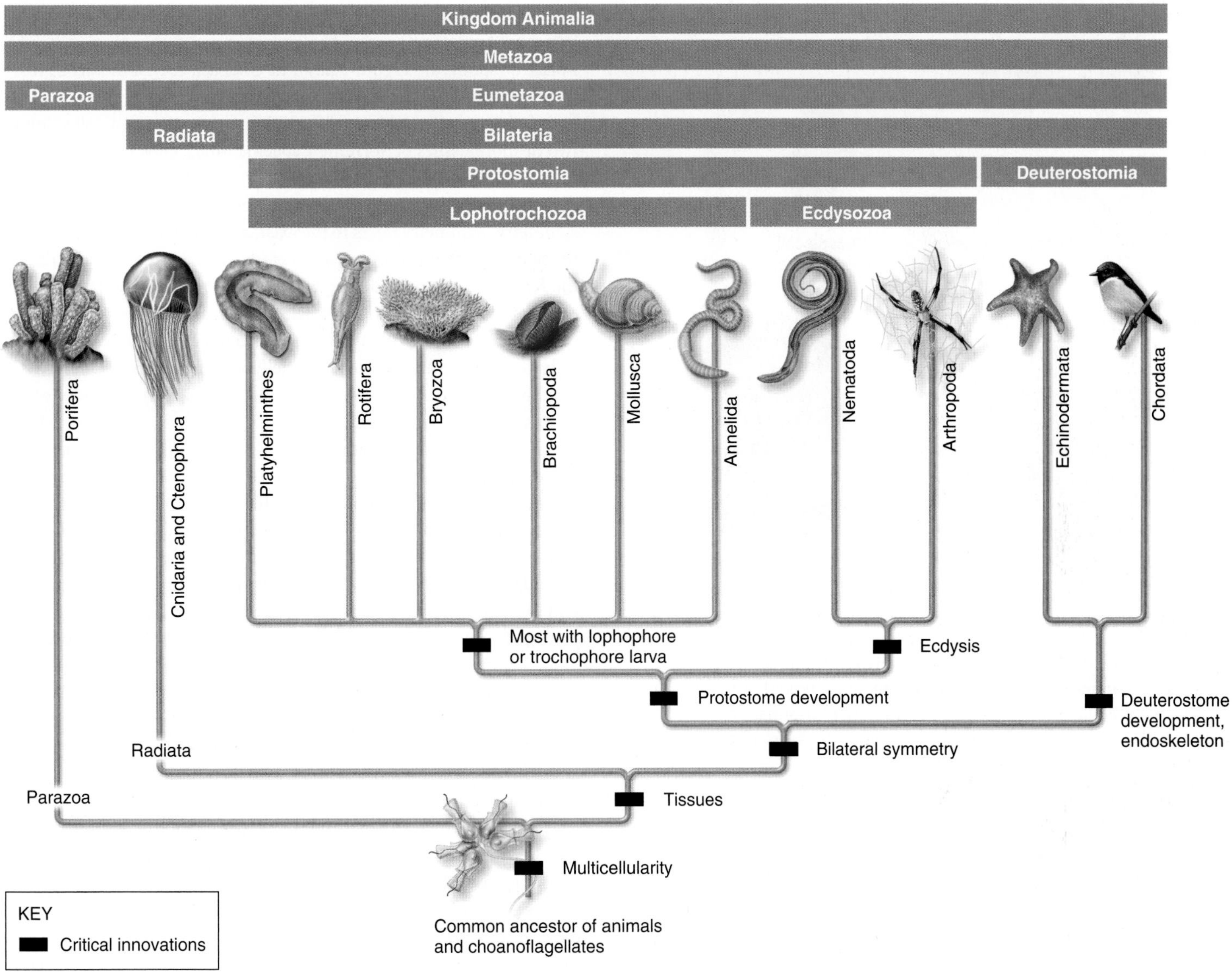

Kingdom Animalia
Metazoa
Parazoa
Eumetazoa
Radiata
Bilateria
Protostomia
Deuterostomia
Lophotrochozoa
Ecdysozoa

Porifera
Cnidaria and Ctenophora
Platyhelminthes
Rotifera
Bryozoa
Brachiopoda
Mollusca
Annelida
Nematoda
Arthropoda
Echinodermata
Chordata

Most with lophophore or trochophore larva
Ecdysis
Protostome development
Deuterostome development, endoskeleton
Bilateral symmetry
Radiata
Parazoa
Tissues
Multicellularity
Common ancestor of animals and choanoflagellates

KEY
Critical innovations

Figure 32.3 **An animal phylogeny based on body plans and molecular data.** Though there are about 35 different animal phyla, we will focus our discussions here and in the next two chapters on the 12 phyla with the greatest numbers of species.

BIOLOGY PRINCIPLE All species (past and present) are related by an evolutionary history. All animals are believed to be derived from a choanoflagellate-like ancestor.

BioConnections: What shared derived character is common to the Eumetazoa? Look back at Figure 26.9.

of species form colonies consisting of many individual organisms forming a cluster of cells on a single stalk. Scientists think that some of these cells may have gradually taken on specialized functions—for example, movement or nutrition—while still maintaining coordination with other cells and cell types.

As shown in Figure 32.4b, colonial choanoflagellate cells bear a striking similarity to cell types called choanocytes, which are found in sponges, the simplest animals. As discussed later, evolutionary changes to this simple body plan resulted in critical innovations that led to more complex body plans found in other animals. Molecular data also point to choanoflagellates as the closest living relatives of animals.

Animal Classification Is Based Mainly on Body Plans

Biologists traditionally classified animal diversity in terms of these three main morphological and developmental features of animal body plans:

1. Presence or absence of different tissue types
2. Type of body symmetry
3. Specific features of embryonic development

We will discuss each of these major features of animal body plans next.

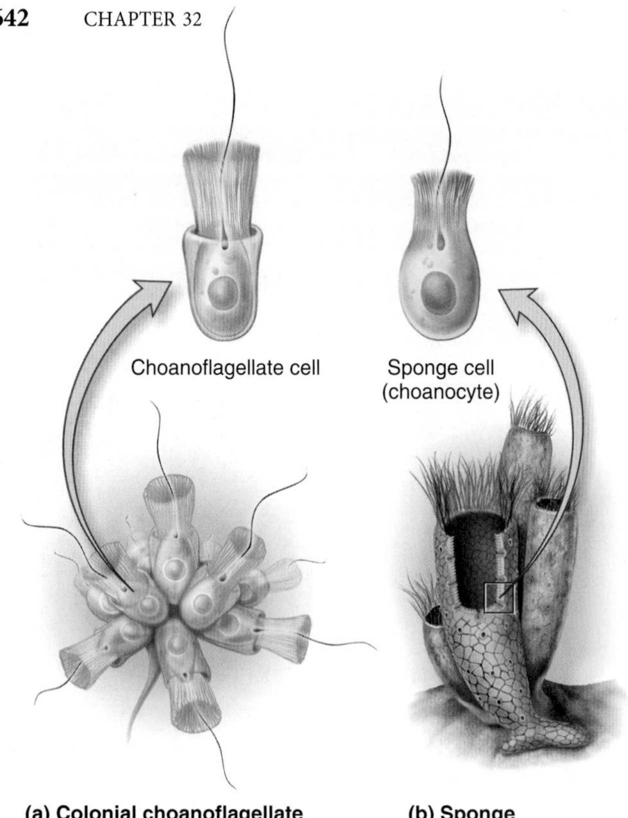

Figure 32.4 Early animal characteristics: A comparison of a colonial choanoflagellate and a sponge. Both types of organisms have very similar types of cells. The structure of sponges and in particular the sponge cell called the choanocyte is described in Chapter 33 (look ahead to Figure 33.2c).

Concept Check: *Why are sponges considered animals but simple choanoflagellates are not?*

Tissues Collectively, animals are known as **Metazoa**. Animals can be divided into two subgroups based on whether or not they have specialized types of tissues, that is, groups of cells that have a similar structure and function. The **Parazoa** (from the Greek, meaning alongside animals) are not generally thought to possess specialized tissue types or organs, although they may have several distinct types of cells. Those cells can change their shape and location, making any associations between them temporary. The Parazoa consist of a single phylum, Porifera (sponges) (Figure 32.5a). In contrast, the **Eumetazoa** (from the Greek, meaning true animals) have one or more types of tissue and, for the most part, have different types of organs—a collection of two or more tissues performing a specific function or set of functions.

Symmetry The Eumetazoa are divided according to their type of symmetry. Symmetry refers to the existence of balanced proportions of the body on either side of a median plane. Radially symmetric animals, the **Radiata**, can be divided equally by any longitudinal plane passing through the central axis (Figure 32.5b). Such animals are often circular or tubular in shape, with a mouth at one end, and include the animals called cnidarians and ctenophores (jellyfish and related species).

Bilaterally symmetric animals, the **Bilateria**, can be divided along a vertical plane at the midline to create two halves (Figure 32.5c). Thus, a bilateral animal has a left side and a right side, which are mirror images, as well as a **dorsal** (upper) and a **ventral** (lower) side, which are not identical, and an **anterior** (head) and a **posterior** (tail) end (refer back to Figure 19.2). Bilateral symmetry is strongly correlated with both the ability to move through the environment and **cephalization**—the localization of sensory structures at the anterior end of the body. Such abilities allow animals to encounter their environment initially with their head, which is best equipped to detect and consume prey and to detect and respond to predators and other dangers. Most animals are bilaterally symmetrical.

Another key difference between the Radiata and Bilateria is that radial animals have two embryonic cell layers, called **germ layers**, whereas bilateral animals have three germ layers. In all animals except the sponges, the growing embryo develops different layers of cells during a process known as **gastrulation** (Figure 32.6).

Fertilization of an egg by a sperm creates a diploid zygote. The zygote then undergoes **cleavage**—a succession of rapid cell divisions with no significant growth that produces a hollow sphere of cells called a **blastula**. In gastrulation, an area in the blastula folds inward, or invaginates, creating in the process a structure called a **gastrula**. The inner layer of cells becomes the **endoderm**, which lines the **archenteron**, or primitive digestive tract. The outer layer, or **ectoderm**,

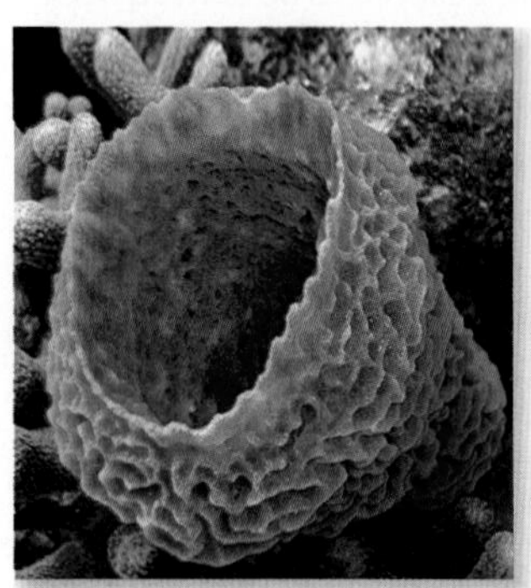

(a) Parazoa: no tissue types

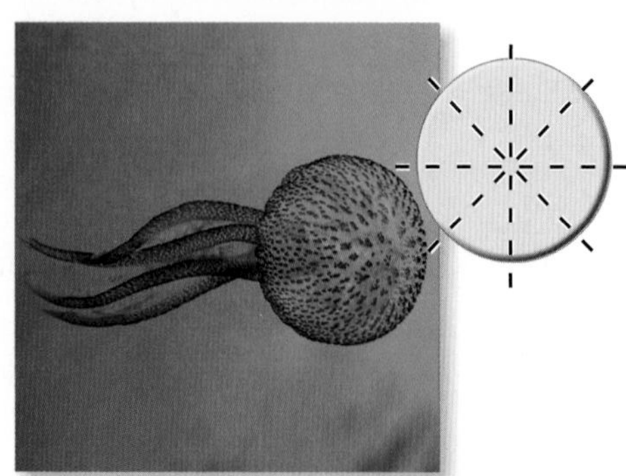

(b) Eumetazoa: two tissue types
Radiata: radial symmetry

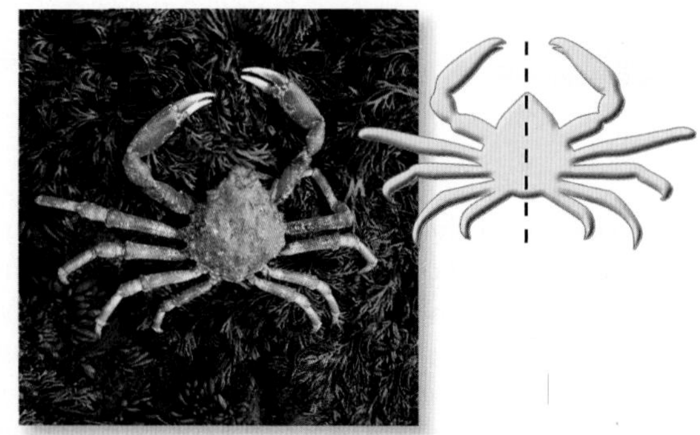

(c) Eumetazoa: three tissue types
Bilateria: bilateral symmetry

Figure 32.5 Early divisions in the animal phylogeny. Animals can be categorized based on **(a)** the absence of different tissue types (Parazoa; the sponges) or **(b, c)** the presence of tissues (Eumetazoa; all other animals). Further categorization is based on the presence of **(b)** radial symmetry (Radiata; the cnidarians and ctenophores) or **(c)** bilateral symmetry (Bilateria; all other animals).

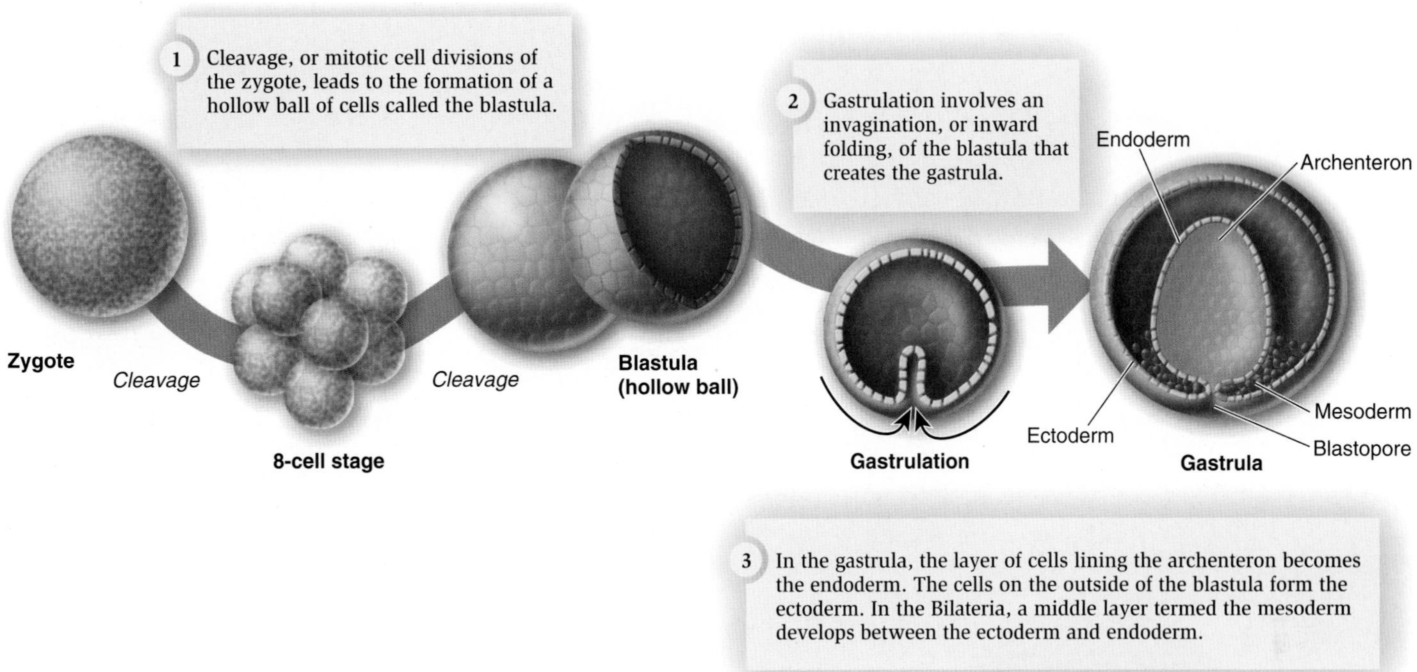

1 Cleavage, or mitotic cell divisions of the zygote, leads to the formation of a hollow ball of cells called the blastula.
2 Gastrulation involves an invagination, or inward folding, of the blastula that creates the gastrula.
Endoderm
Archenteron
Zygote
Cleavage
Cleavage
Blastula (hollow ball)
8-cell stage
Gastrulation
Ectoderm
Gastrula
Mesoderm
Blastopore
3 In the gastrula, the layer of cells lining the archenteron becomes the endoderm. The cells on the outside of the blastula form the ectoderm. In the Bilateria, a middle layer termed the mesoderm develops between the ectoderm and endoderm.

Figure 32.6 **Formation of germ layers.** Note: Radially symmetric animals (Radiata) do not form mesoderm.

BioConnections: Look back to Figure 26.8. Is the existence of three layers in triploblastic animals a shared primitive character or a shared derived character?

covers the surface of the embryo and differentiates into the epidermis and nervous system.

The Bilateria develop a third layer of cells, termed the **mesoderm**, between the ectoderm and endoderm. Mesoderm forms the muscles and most other organs between the digestive tract and the ectoderm (look ahead to Figure 52.7). Because the Bilateria have these three distinct germ layers, they are often referred to as **triploblastic**, whereas the Radiata, which have only ectoderm and endoderm, are termed **diploblastic**.

Specific Features of Embryonic Development The most fundamental feature of embryonic development concerns the development of a mouth and anus (**Figure 32.7a**). In gastrulation, the endoderm forms an indentation, the **blastopore**, which is the opening of the archenteron to the outside. In **protostomes** (from the Greek *protos*, meaning first, and *stoma*, meaning mouth), the blastopore becomes the mouth. If an anus is formed in a protostome, it develops from a secondary opening. In contrast, in **deuterostomes** (from the Greek *deuteros*, meaning second), the blastopore becomes the anus, and the mouth is formed from the secondary opening.

In addition, protostomes and deuterostomes differ in some other embryonic features. In the early stages of embryonic development, repeated cell divisions occur without cell growth, a process known as cleavage. Protostome development is generally characterized by so-called **determinate cleavage**, in which the fate of each embryonic cell is determined very early (**Figure 32.7b**). If one of the cells is removed from a four-cell protostome embryo, neither the single cell nor the remaining three-cell mass can form viable embryos, and development is halted. In contrast, most deuterostome development is characterized by **indeterminate cleavage**, in which each cell produced by early cleavage retains the ability to develop into a complete embryo. For example, when one cell is excised from a four-cell sea urchin embryo, both the single cell and the remaining three can go on to form viable embryos. Other embryonic cells compensate for the missing cells. In human embryos, if individual embryonic cells separate from one another early in development, identical twins can result.

In the developing zygote, cleavage may occur by two mechanisms (**Figure 32.7c**). In **spiral cleavage**, the planes of cell cleavage are oblique to the vertical axis of the embryo, resulting in an arrangement in which newly formed upper cells lie centered between the underlying cells. Many protostomes, including mollusks and annelid worms, exhibit spiral cleavage. The coiled shells of some mollusks result from spiral cleavage. Organisms with spiral cleavage are also known as spiralians. In **radial cleavage**, the cleavage planes are either parallel or perpendicular to the vertical axis of the egg. This results in tiers of cells, one directly above the other. All deuterostomes exhibit radial cleavage, as do insects and nematodes, suggesting it may have been an ancestral condition.

Other Morphological Criteria Have Been Used to Classify Animals

In older phylogenetic trees of animal life, classification was also based on morphological features, such as the possession of a fluid-filled body cavity called a **coelom** or the presence of body segmentation. More recent molecular data suggest that although these features are helpful in describing differences in animal structure, they are not as useful in shedding light on the evolutionary history of animals as previously believed.

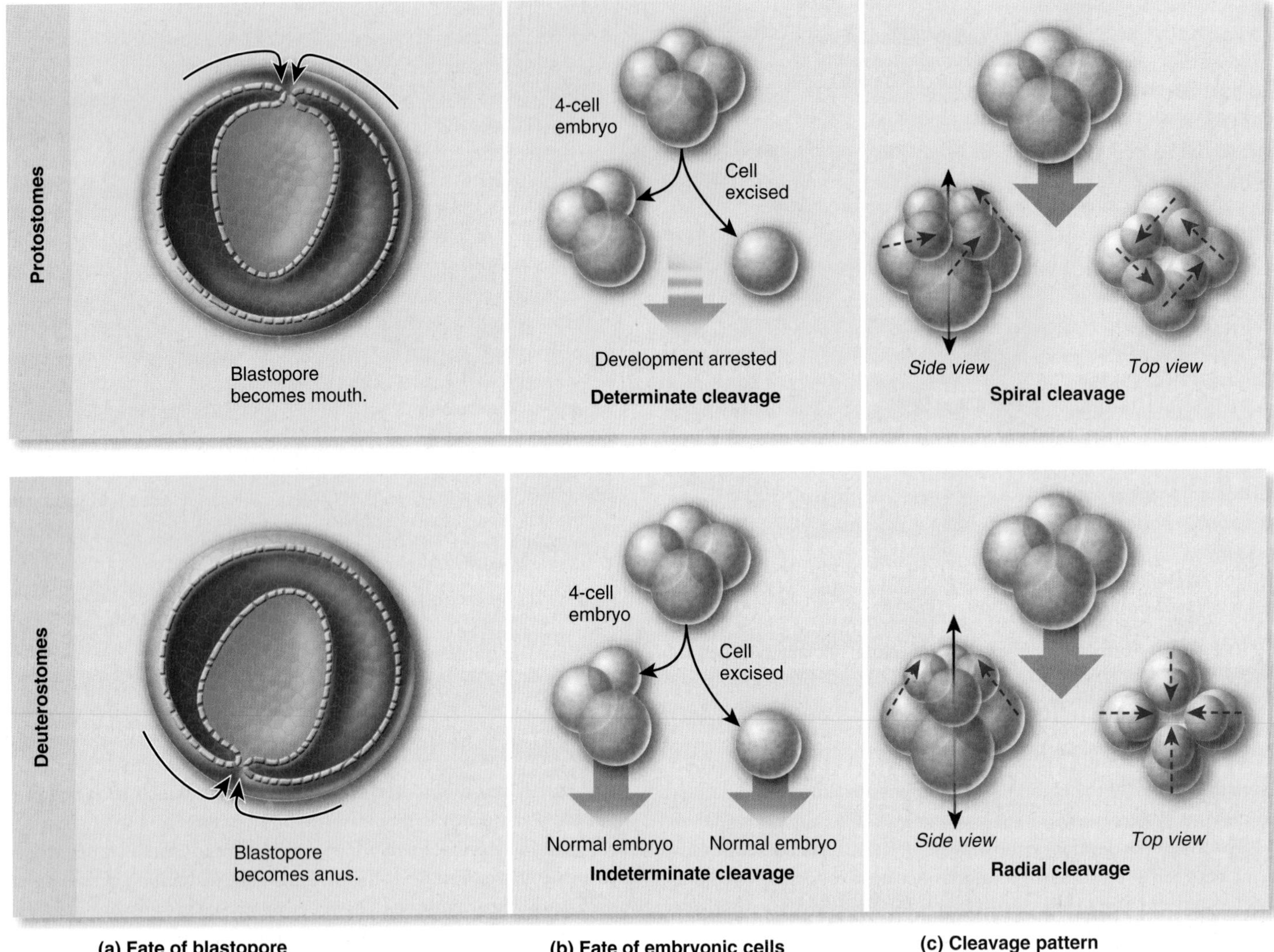

Figure 32.7 **Differences in embryonic development between protostomes and deuterostomes.** **(a)** In protostomes, the blastopore becomes the mouth. In deuterostomes, the blastopore becomes the anus. **(b)** Protostomes have determinate cleavage, whereas deuterostomes have indeterminate cleavage. **(c)** Many protostomes have spiral cleavage, and most deuterostomes have radial cleavage. The dashed arrows indicate the direction of cleavage.

Body Cavity In many animals, the body cavity is completely lined with mesoderm and is called a true coelom. Animals with a true coelom are termed **coelomates** (Figure 32.8a). If the fluid-filled cavity is not completely lined by tissue derived from mesoderm, it is known as a pseudocoelom (Figure 32.8b). Animals with a pseudocoelom, including rotifers and roundworms, are termed **pseudocoelomates**. Some animals, such as flatworms, lack a fluid-filled body cavity and are termed **acoelomates** (Figure 32.8c). Instead of fluid, this region contains mesenchyme, a tissue derived from mesoderm.

A coelom has many important functions, perhaps the most important being that its fluid is relatively incompressible and therefore cushions internal organs such as the heart and intestinal tract, helping to prevent injury from external forces. A coelom also enables internal organs to move and grow independently of the outer body wall. Furthermore, in some soft-bodied invertebrates, such as earthworms, the coelom functions as a **hydrostatic skeleton**—a fluid-filled body cavity surrounded by muscles that gives support and shape to the body of organisms. Muscle contractions at one part of the body push this fluid toward another part of the body. This type of movement can best be observed in an earthworm (look ahead to Figure 44.1a). Finally, in some organisms, the fluid in the body cavity also acts as a simple circulatory system. At one stage, the presence or absence of a coelom or pseudocoelom was a distinction traditionally used in the construction of animal phylogeny, but scientists now believe this feature may not be useful in classification because animals that once possessed coeloms may have lost them over long periods of evolutionary time, as is true for the ancestors of flatworms. In addition, it is also believed that the coelom may have arisen more than once in animal evolution, once in protostomes and once in deuterostomes.

Segmentation Another well-known feature of the animal body plan is the presence or absence of segmentation. In segmentation, the body is divided into regions called segments. It is most obvious in the annelids, or segmented worms, but it is also evident in arthropods

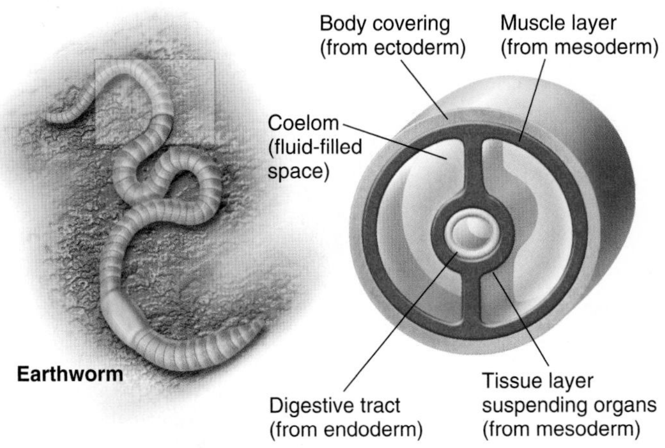

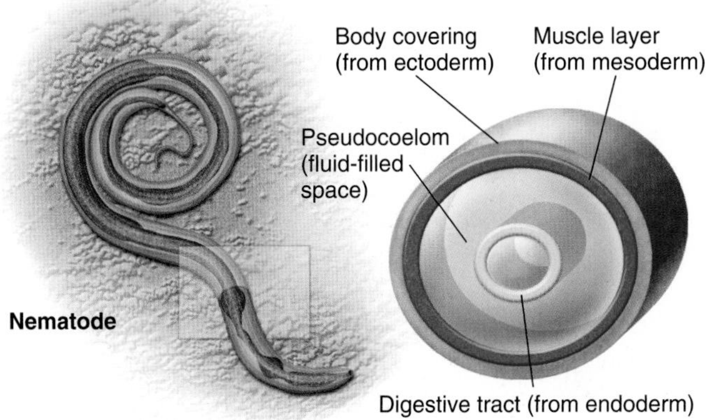

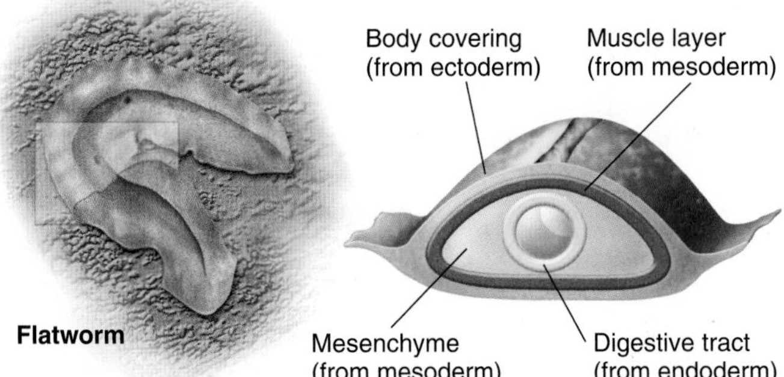

Figure 32.8 The three basic body cavities of bilaterally symmetric animals. Cross sections of each animal are shown on the right.

Concept Check: What advantages does a coelom confer for movement?

and chordates (Figure 32.9). In annelids, each segment contains the same set of blood vessels, nerves, and muscles. Some segments may differ, such as those containing the brain or the sex organs, but many segments are very similar. In chordates, we can see segmentation in the backbone and muscles. The advantage of segmentation is that it allows specialization of body regions. For example, as we will see in Chapter 33, arthropods exhibit a vast degree of specialization of their segments. Many insects have wings and only three pairs of legs, whereas centipedes have no wings and many legs. Crabs, lobsters, and shrimp have highly specialized thoracic appendages that aid in feeding. The presence or absence of segmentation was used in the construction of animal phylogeny, with annelid worms and arthropods thought to be fairly closely related because of their body segmentation. Scientists are beginning to understand the genetic basis for some of these morphological traits. Recent studies have shown that changes in specialization among arthropod body segments can be traced to relatively simple changes in *Hox* genes.

Genomes & Proteomes Connection

Changes in *Hox* Gene Expression Control Body Segment Specialization

***Hox* genes**, which are present in all animals, are involved in determining the spatial patterning of the body and appendages. As described in Chapter 19, animals have several *Hox* genes that are expressed in particular regions of the body. The *Hox* genes are organized into four clusters of 13 genes, each designated with numbers 1 through 13. Some are expressed in anterior segments; others are expressed in posterior segments (refer back to Figure 19.17). In the 1990s, Greek molecular biologist Michalis Averof and coworkers showed how relatively simple shifts in the expression patterns of *Hox* genes along the anteroposterior axis can account for the large variation in arthropod appendage types. More recent work by Averof and colleagues (2010) has showed how specific changes in *Hox* expression could be linked to changes in crustacean maxillipeds—appendages near the mouth that are used for feeding. Maxillipeds arise in the anterior thoracic segments and display a mixture of locomotory and feeding functions. By knocking out *Hox* genes or expressing *Hox* genes in an abnormal position, the researchers could change maxillipeds into leglike appendages or transform leglike appendages into maxillipeds.

Shifts in the patterns of expression of *Hox* genes in the embryo along the anteroposterior axis are similarly prominent in vertebrate evolution. In vertebrates, the transition from one type of vertebra to another, for example, from cervical (neck) to thoracic (chest) vertebrae, is controlled by particular *Hox* genes. The site of the cervicothoracic boundary appears to be influenced by the *HoxC-6* gene (Figure 32.10). Differences in its relative position of expression, which occurs prior to vertebrae development, control neck length in vertebrates. In mice, which have a relatively short neck, the expression of *HoxC-6* begins between vertebrae 7 and 8. In chickens and geese, which have longer necks, the expression begins farther back, between vertebrae 14 and 15, or 17 and 18, respectively. The forelimbs also arise at this boundary in all vertebrates. Interestingly, snakes, which essentially have no neck or forelimbs, do not exhibit this boundary, and *HoxC-6* expression occurs immediately behind their heads. This, in effect, means that snakes got longer by losing their neck and lengthening their chest.

American molecular biologist Sean Carroll has remarked that it is very satisfying to find that the evolution of body forms and novel structures in two of the most successful and diverse animal groups, arthropods and vertebrates, is shaped by the shifting of *Hox* genes. It also reminds us of one of the basic principles of biology—that the genetic material provides a blueprint for reproduction. Much of the diversity in animal phyla can be seen as variations on a common theme.

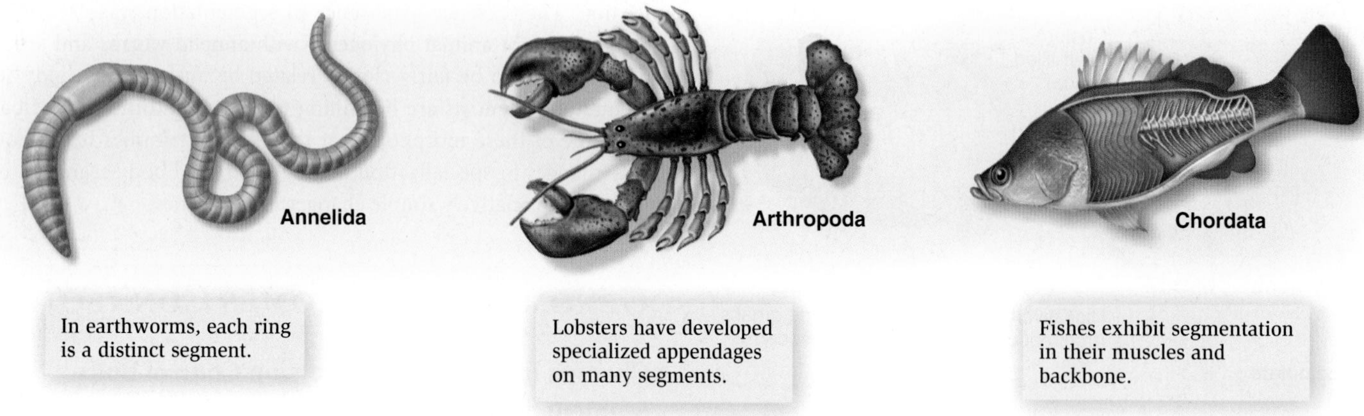

Annelida
Arthropoda
Chordata
In earthworms, each ring is a distinct segment.
Lobsters have developed specialized appendages on many segments.
Fishes exhibit segmentation in their muscles and backbone.

Figure 32.9 Segmentation. Annelids, arthropods, and chordates all exhibit segmentation.

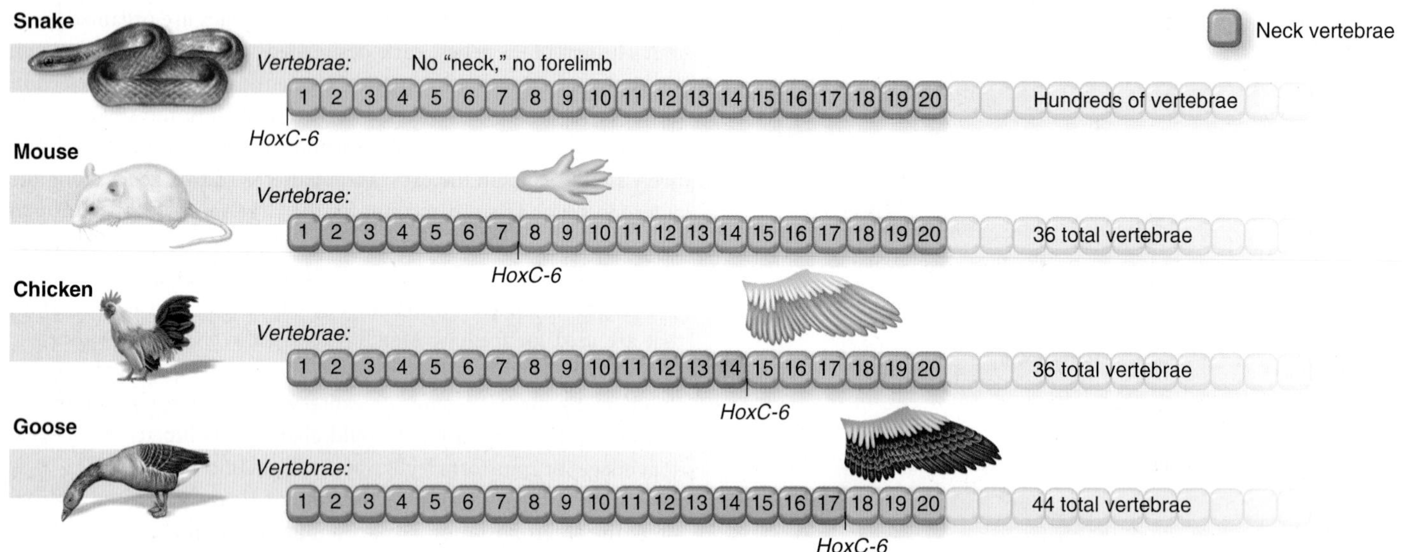

Figure 32.10 Relationship between *HoxC-6* gene expression and neck length. In vertebrates, the transition between neck and trunk vertebrae is controlled by the position of the *HoxC-6* gene. In snakes, the expression of this gene is shifted so far forward that a neck does not develop.

32.3 Molecular Views of Animal Diversity

Learning Outcomes:

1. Discuss the similarities and differences between animal phylogenies based on molecular data and those based on morphological data.
2. List the morphological features of the Ecdysozoa and the Lophotrochozoa.

As discussed in Chapter 26, molecular systematics involves the comparison of genetic data, such as DNA, RNA, and amino acid sequences, from different organisms to estimate their evolutionary relationships based on the degree of similarities between the sequences. More closely related organisms exhibit fewer sequence differences than distantly related organisms. An advantage of the molecular approach over approaches based on morphological data is that genetic sequences among different species are easier to quantify and compare. For example, the DNA sequence contains four easily identified and mutually exclusive characters: the nucleotides A, T, G, and C (RNA has A, U, G, and C). Contrast this with morphological and embryological data, for which characters are scored more subjectively, often based on the qualitative assessment of many traits. Morphological approaches sometimes left us with questions about the evolutionary relationships among animals; biologists now use molecular data to clarify questions posed by earlier data and arrive at an interpretation that agrees most closely with the available evidence.

To perform molecular analyses, scientists have often focused on comparing sequences of nucleotides in the gene that encodes RNA of the small ribosomal subunit (SSU rRNA) (see Chapter 13). SSU rRNA is universal in all organisms, and its base pair sequence has changed very slowly over long periods of time. Researchers have also studied *Hox* genes, which are found in all animals, to study the evolution of body plans (refer back to Figure 25.15). They believe that duplication of

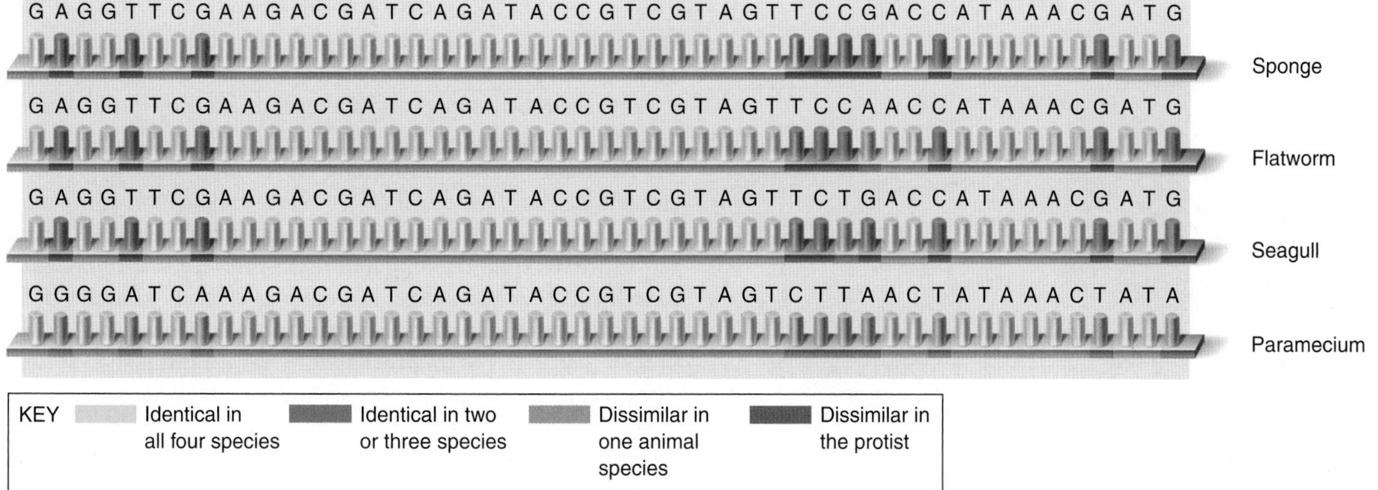

Figure 32.11 Comparison of small subunit (SSU) rRNA gene sequences from three animals and a protist. Note the similarities between the animals, even though they are very different species, and the differences with the protist. This and other comparative studies of gene sequences underscore the likelihood that animals share a common ancestor.

BioConnections: Look back at Figure 12.16. Which color represents sequences of bases that are the most evolutionarily conserved?

Hox genes and gene clusters has led to the evolution of more complex animal body forms. Examination of the genes that regulate early developmental differences has provided insight into the evolution of animal development and how, when, and why animal body plans diversified.

Phylogenies based on SSU rRNA and *Hox* genes are similar and, in many cases, agree with the structure of the morphologically based phylogenetic tree. For example, recent analyses of genes have strengthened the view that animals form a single clade—a group of species derived from a single common ancestor. We can appreciate this by comparing a portion of the sequence of the SSU rRNA genes of a sponge, flatworm, seagull, and paramecium (Figure 32.11) in much the same way as we did in Chapter 12 (refer back to Figure 12.16). The three animal sequences are very similar compared with that of the paramecium (a protist), indicating that the animals share a common ancestor.

However, phylogenies based on molecular data contain some important differences from those based on assessment of body plans. One of the most influential of the molecular studies was a paper by Aguinaldo and colleagues, which established evidence for a new clade of molting animals, the **Ecdysozoa**, consisting of the nematodes and arthropods. This study (discussed in more detail in the Feature Investigation) represented a scientific breakthrough because it underscored the value of molecular phylogenies. According to molecular evidence, the other major protostome clade is the **Lophotrochozoa**, which encompasses the mollusks, annelids, and several other phyla (see Figure 32.14). When some morphologists reviewed their data given this new information, they found there was also morphological support for these new groupings. Let's look at what morphological features make each of these groups unique.

The Ecdysozoa is so named because all of its members secrete a nonliving cuticle, an external skeleton (exoskeleton); think of the hard shell of a beetle or that of a crab. As these animals grow, the exoskeleton becomes too small, and the animal molts, or breaks out of its old exoskeleton, and secretes a newer, larger one (Figure 32.12).

Figure 32.12 Ecdysis. The dragonfly, shown here emerging from a discarded exoskeleton, is a member of the Ecdysozoa—a clade of animals exhibiting ecdysis, the periodic shedding (molting) and re-formation of the exoskeleton.

BIOLOGY PRINCIPLE Living organisms grow and develop. For animals with exoskeletons, growth and development necessitate molting.

Concept Check: What are the main members of the Ecdysozoa?

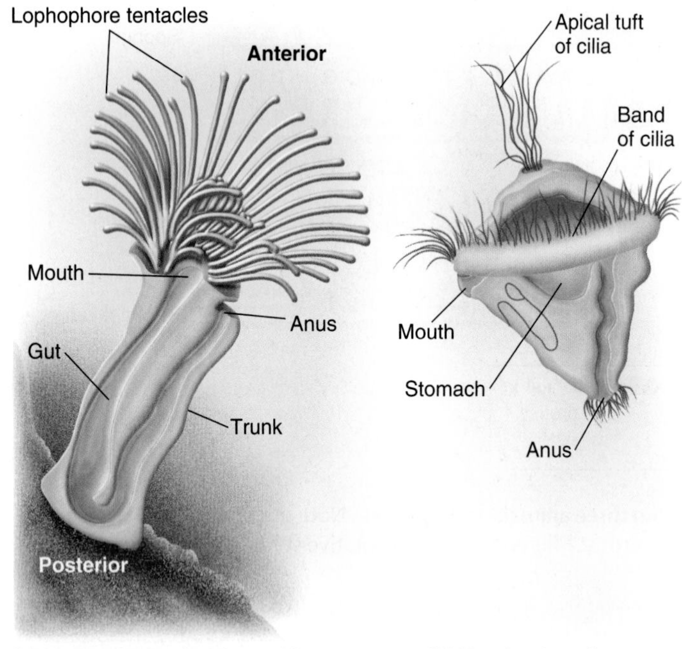

(a) Lophophore of a phoronid worm **(b) Trochophore larva**

Figure 32.13 Characteristics of the Lophotrochozoa. **(a)** A lophophore, a crown of ciliated tentacles, generates a current to bring food particles into the mouth. **(b)** The trochophore larval form is found in several animal lineages.

This molting process is called ecdysis; hence the name Ecdysozoa. Although this group was named for this morphological characteristic, it was first strongly supported as a separate clade by molecular evidence such as similarities in DNA.

Although the Lophotrochozoa clade was organized primarily through analysis of molecular data, its name also stems from two morphological features seen in many organisms of this clade. The "lopho" part is derived from the **lophophore**, a horseshoe-shaped crown of tentacles used for feeding that is present on some phyla in this clade, such as the rotifers, bryozoans, and brachiopods (Figure 32.13a). The "trocho" part refers to the **trochophore larva**, a distinct larval stage characterized by a band of cilia around its middle that is used for swimming (Figure 32.13b). Trochophore larvae are found in several Lophotrochozoa phyla, such as annelid worms and mollusks, indicating their similar ancestry. Other members of the clade, such as the platyhelminthes, have neither of these morphological features and are classified as lophotrochozoans based strictly on molecular data. As a reference, **Table 32.2** summarizes the basic characteristics of the major animal phyla. In Chapter 33, we will discuss the diversity of all the phyla in the Parazoa, Radiata, Lophotrochozoa, Ecdysozoa, and Deuterostomia, the latter including the phylum Echinodermata and invertebrate members of the phylum Chordata; these are all the animals without a backbone. In Chapter 34, we will turn our attention to the members of the phylum Chordata, including fishes, amphibians, reptiles, birds, and mammals, which possess a backbone.

Table 32.2 Summary of the Basic Characteristics of the Major Animal Phyla

Feature	Porifera (sponges)	Cnidaria and Ctenophora (hydra, anemones, jellyfish)	Platyhelminthes (flatworms)	Rotifera (rotifers)	Bryozoa and Brachiopoda (bryozoans and brachiopods)	Mollusca (snails, clams, squids)	Annelida (segmented worms)	Nematoda (roundworms)	Arthropoda (insects, arachnids, crustaceans)	Echinodermata (sea stars, sea urchins)	Chordata (vertebrates and others)
Estimated number of species	8,000	11,000	20,000	2,000	4,500+	110,000	15,000	20,000	1,000,000+	7,000+	52,000+
Level of organization	Cellular; lack tissues and organs	Tissue; lack organs	Organs	Organs	Organs	Organs	Organs	Organs	Organs	Organs	Organs
Symmetry	Absent	Radial	Bilateral	Bilateral	Bilateral	Bilateral	Bilateral	Bilateral	Bilateral	Bilateral larvae, radial adults	Bilateral
Cephalization	Absent	Absent	Present	Present	Reduced	Present	Present	Present	Present	Absent	Present
Germ layers	Absent	Two	Three	Three	Three	Three	Three	Three	Three	Three	Three
Body cavity or Coelom	Absent	Absent	Absent	Pseudocoelom	Coelom	Reduced coelom	Coelom	Pseudocoelom	Reduced coelom	Coelom	Coelom
Segmentation	Absent	Absent	Absent	Absent	Absent	Absent	Present	Absent	Present	Absent	Present

FEATURE INVESTIGATION

Aguinaldo and Colleagues Used SSU rRNA to Analyze the Taxonomic Relationships of Arthropods to Other Taxa

In 1997, American molecular biologists Anna Marie Aguinaldo, James Lake, and colleagues analyzed the relationships of arthropods to other taxa by sequencing the complete gene that encodes SSU rRNA from a variety of representative taxa (**Figure 32.14**). Total genomic DNA was isolated using standard techniques and amplified by the polymerase chain reaction (PCR; refer back to Figure 20.6). PCR fragments were then subjected to DNA sequencing, a technique also described in Chapter 20, and the evolutionary relationships among 50 species were examined. The resulting data indicated the existence of a monophyletic clade—the Ecdysozoa—containing the nematodes and arthropods.

The hypothesis that nematodes are more closely related to arthropods than previously thought has important ramifications. First, it implies that two well-researched model organisms, *Caenorhabditis elegans* (a nematode) and the fruit fly *Drosophila melanogaster* (an arthropod), are more closely related than had been believed. Second,

Figure 32.14 **A molecular animal phylogeny based on sequencing of SSU rRNA.**

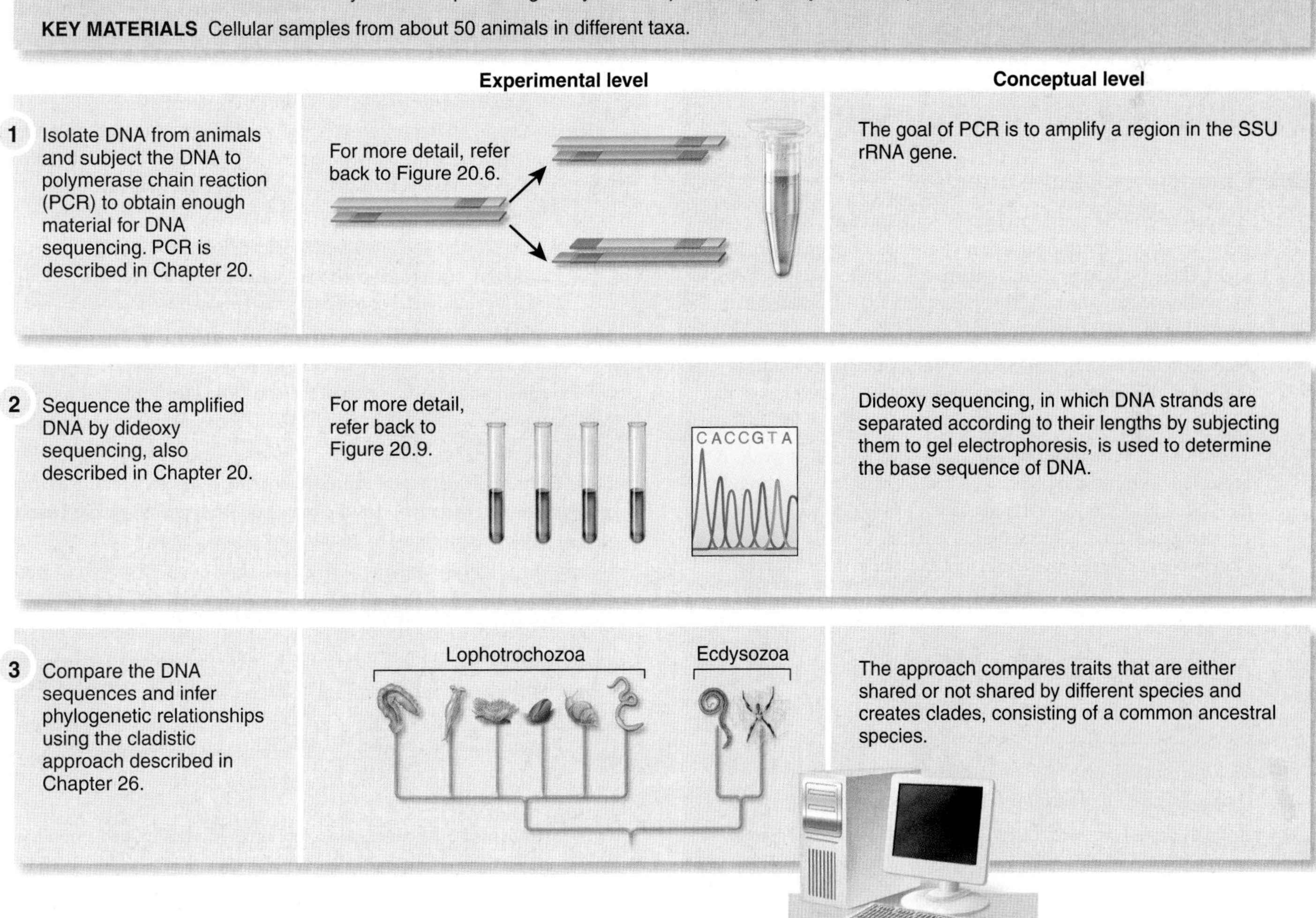

GOAL To determine the evolutionary relationships among many animal species, especially the arthropods.
KEY MATERIALS Cellular samples from about 50 animals in different taxa.
Experimental level
Conceptual level
1 Isolate DNA from animals and subject the DNA to polymerase chain reaction (PCR) to obtain enough material for DNA sequencing. PCR is described in Chapter 20.
For more detail, refer back to Figure 20.6.
The goal of PCR is to amplify a region in the SSU rRNA gene.
2 Sequence the amplified DNA by dideoxy sequencing, also described in Chapter 20.
For more detail, refer back to Figure 20.9.
CACCGTA
Dideoxy sequencing, in which DNA strands are separated according to their lengths by subjecting them to gel electrophoresis, is used to determine the base sequence of DNA.
3 Compare the DNA sequences and infer phylogenetic relationships using the cladistic approach described in Chapter 26.
Lophotrochozoa
Ecdysozoa
The approach compares traits that are either shared or not shared by different species and creates clades, consisting of a common ancestral species.
4 THE DATA
This process resulted in a large group of DNA sequences that were then analyzed with the use of computer programs.

5 **CONCLUSION** The arthropods are most closely related to the nematodes, and both phyla are placed in the clade Ecdysozoa. All other protostomes belong to a new clade called the Lophotrochozoa.

6 **SOURCE** Aguinaldo, A.M. et al. 1997. Evidence for a clade of nematodes, arthropods, and other moulting animals. *Nature* 387 (6632):489–493.

morphological classification had assumed that arthropods and annelids were closely related to each other based on the presence of segmentation. Molecular data provides no evidence that annelids and arthropods form a clade of segmented animals.

Experimental Questions

1. What was the purpose of the study conducted by Aguinaldo and colleagues?
2. What was the major finding of this particular study?
3. What impact does the new view of nematode and arthropod phylogeny have on other areas of research?

Summary of Key Concepts

32.1 Characteristics of Animals

- Animals constitute a very species-rich kingdom, with a number of characteristics that distinguish them from other organisms, including multicellularity, an extracellular matrix, and unique cell junctions, in addition to heterotrophic feeding and internal digestion and the possession of nervous and muscle tissues (Table 32.1).
- Many different feeding modes exist among animals, including predation, herbivory, parasitism, filter-feeding, and decomposition (Figure 32.1).
- The history of animal life on earth spans over 540 million years. A profusion of animal phyla appeared in the Cambrian explosion (530–525 mya). Animals evolved adaptations to deal with the colonization of land, about 520 mya, and the number and diversity of mammals exploded after dinosaurs died out at the end of the Cretaceous period, 65 mya (Figure 32.2).

32.2 Animal Classification

- The animal kingdom is monophyletic, meaning that all taxa have evolved from a single common ancestor (Figure 32.3).
- Biologists hypothesize that animals evolved from a colonial choanoflagellate-like ancestor (Figure 32.4).
- Animals can be categorized according to the absence of different types of tissues (the Parazoa or sponges) and the presence of tissues (Eumetazoa or all other animals). The Eumetazoa can also be divided according to their type of symmetry, whether radial (Radiata, the cnidarians and ctenophores) or bilateral (Bilateria, all other animals) (Figure 32.5).
- The Radiata have two embryonic cell layers (germ layers): the endoderm and the ectoderm. The Bilateria have a third germ layer termed the mesoderm, which develops between the endoderm and the ectoderm (Figure 32.6).
- Animals are also classified according to patterns of embryonic development. In protostomes, the blastopore becomes the mouth; in deuterostomes, the blastopore becomes the anus. Most protostomes have spiral cleavage and all deuterostomes have radial cleavage (Figure 32.7).
- Animals with a coelom, a body cavity that is completely lined with mesoderm, are termed coelomates. Animals that possess a coelom that is not completely lined by tissue derived from mesoderm are called pseudocoelomates. Those animals lacking a fluid-filled body cavity are termed acoelomates (Figure 32.8).
- Segmentation, the division of the body into identical subunits called segments, is a key feature of the animal body plan in several phyla (Figure 32.9).
- Shifts in the pattern of expression of *Hox* genes are prominent in evolution. In vertebrates, the transition from one type of vertebra to another is controlled by certain *Hox* genes (Figure 32.10).

32.3 Molecular Views of Animal Diversity

- Molecular techniques that compare similarities in DNA, RNA, and amino acid sequences support the view that all animals share a common ancestor (Figure 32.11).
- Recent molecular studies propose a division of the protostomes into two major clades: the Ecdysozoa and the Lophotrochozoa (Figure 32.14).
- Members of the Ecdysozoa secrete and periodically shed a nonliving cuticle, typically an exoskeleton, or external skeleton (Figure 32.12).
- The Lophotrochozoa are grouped primarily through analysis of molecular data, but some members are distinguished by two morphological features: the lophophore, a crown of tentacles used for feeding, and the trochophore larva, a distinct larval stage (Figure 32.13). Each animal phylum shows a distinctive set of general characteristics (Table 32.2).

Assess and Discuss

Test Yourself

1. Which of the following is not a distinguishing characteristic of animals?
   a. the capacity to move at some point in their life cycle
   b. possession of cell walls
   c. multicellularity
   d. heterotrophy
   e. All of the above are characteristics of animals.
2. Which is the correct hierarchy of divisions in the animal kingdom, from most inclusive to least inclusive?
   a. Eumetazoa, Metazoa, Protostomia, Ecdysozoa
   b. Parazoa, Radiata, Lophotrochozoa, Deuterostomia
   c. Metazoa, Eumetazoa, Bilateria, Protostomia
   d. Radiata, Eumetazoa, Deuterostomia, Ecdysozoa
   e. none of the above
3. Bilateral symmetry is strongly correlated with
   a. the ability to move through the environment.
   b. cephalization.
   c. the ability to detect prey.
   d. a and b.
   e. a, b, and c.
4. In triploblastic animals, the inner lining of the digestive tract is derived from
   a. the ectoderm.
   b. the mesoderm.
   c. the endoderm.
   d. the pseudocoelom.
   e. the coelom.
5. Pseudocoelomates
   a. lack a fluid-filled cavity.
   b. have a fluid-filled cavity that is completely lined with mesoderm.
   c. have a fluid-filled cavity that is partially lined with mesoderm.
   d. have a fluid-filled cavity that is not lined with mesoderm.
   e. have an air-filled cavity that is partially lined with mesoderm.
6. Protostomes and deuterostomes can be classified based on
   a. cleavage pattern.
   b. destiny of the blastopore.
   c. whether the fate of the embryonic cells is fixed early during development.
   d. all of the above.
7. Indeterminate cleavage is found in
   a. annelids.
   b. mollusks.
   c. nematodes.
   d. vertebrates.
   e. all of the above.
8. Naturally occurring identical twins are possible only in animals that
   a. have spiral cleavage.
   b. have determinate cleavage.
   c. are protostomes.
   d. have indeterminate cleavage.
   e. a, b, and c
9. Genes involved in the patterning of the body axis, that is, in determining characteristics such as neck length and appendage formation, are called
   a. small subunit (SSU) rRNA genes.
   b. *Hox* genes.
   c. metameric genes.
   d. determinate genes.
   e. none of the above.
10. A major finding of recent molecular studies is that
    a. the presence or absence of the mesoderm is not important in molecular phylogeny.
    b. molecular phylogeny suggests that all animals do not share a single common ancestor.
    c. body symmetry, whether radial or bilateral, is not an important determinant in molecular phylogeny.
    d. molecular phylogeny does not include the echinoderms in the deuterostome clade.
    e. molecular phylogeny suggests that the presence or absence of a coelom is not important for classification.

Conceptual Questions

1. Early morphological phylogenies were based on what three features of animal body plans?
2. Why was the evolution of a coelom important?
3. A principle of biology is that *all species (past and present) are related by an evolutionary history*. Annelids, arthropods, and chordates all exhibit segmentation. Are these monophyletic, paraphyletic, or polyphyletic groups? Explain your answer. Refer back to Figure 26.5 to remind yourself about these terms.

Collaborative Questions

1. Discuss the many ways that animals can affect humans, both positively and negatively.
2. Summarize how molecular evidence has enabled scientists to refine their views on animal phylogeny.

Online Resource

www.brookerbiology.com

Stay a step ahead in your studies with animations that bring concepts to life and practice tests to assess your understanding. Your instructor may also recommend the interactive eBook, individualized learning tools, and more.

The Invertebrates

33

Chapter Outline

33.1 Parazoa: Sponges, the First Multicellular Animals
33.2 Radiata: Jellyfish and Other Radially Symmetric Animals
33.3 Lophotrochozoa: The Flatworms, Rotifers, Bryozoans, Brachiopods, Mollusks, and Annelids
33.4 Ecdysozoa: The Nematodes and Arthropods
33.5 Deuterostomia: The Echinoderms and Chordates
Summary of Key Concepts
Assess and Discuss

What is this organism, and how does it feed?

If you thought the organism shown in the opening photograph was an underwater plant, complete with long leaflike structures and roots, you'd be wrong. This organism is an animal, a type of echinoderm called a feather star, and it is related to sea stars. Its long arms catch food particles floating in the ocean current, and tiny tube feet pass these particles into special food gutters that run along the center of each arm and empty into the mouth. The number of arms varies from species to species and may reach 200. Feather stars can creep along the ocean floor by means of rootlike projections called cirri. There are about 550 species of feather stars in existence today, but some fossil formations are packed with feather star fragments, showing how successful the group was in the past.

The history of animal life on Earth has evolved over hundreds of millions of years. Some scientists suggest that changing environmental conditions, such as a buildup of dissolved oxygen and minerals in the ocean or an increase in atmospheric oxygen, eventually permitted higher metabolic rates and increased the activity of a wide range of animals. Others suggest that with the development of sophisticated locomotor skills, a wide range of predators and prey evolved, leading to an evolutionary "arms race" in which predators evolved powerful weapons and prey evolved more powerful defenses against them. Such adaptations and counteradaptations would have led to a proliferation of different lifestyles and taxa. Finally, the evolution of *Hox* genes may have permitted an increase in diversity.

Over the next two chapters, we will survey the wondrous diversity of animal life on Earth. In this chapter, we examine the **invertebrates**, or animals without a backbone, a category that makes up more than 95% of all animal species. We begin by exploring some of the earliest animal lineages, the Parazoa and Radiata. We then turn to the Lophotrochozoa and Ecdysozoa, the two sister groups of protostomes introduced in Chapter 32. Finally, we will examine the deuterostomes, focusing here on the echinoderms and the invertebrate members of the phylum Chordata. The animal classification outlined in Chapter 32 (refer back to Figure 32.3) will serve as the basis for our discussion of animal lineages. It is summarized in **Figure 33.1**. Keep in mind, however, that animal phylogeny is a work in progress, and further revisions, refinements, and perhaps surprises lay ahead, as the genomes of more and more species are sequenced and compared.

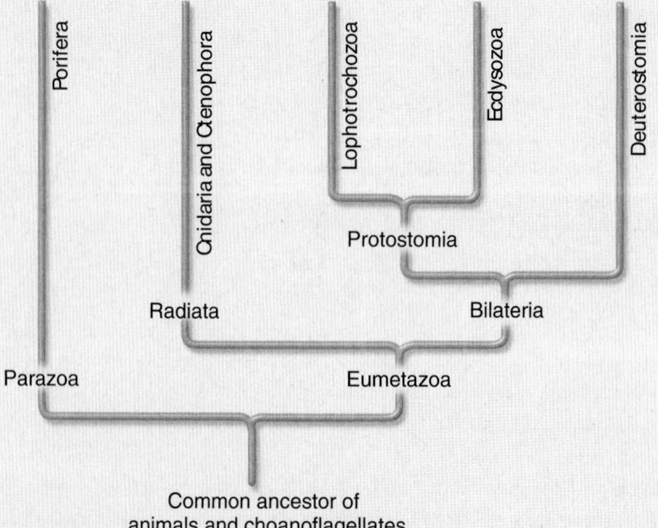

Figure 33.1 An animal phylogeny. This phylogenetic tree summarizes our current understanding about the relationships between animal groups.

33.1 Parazoa: Sponges, the First Multicellular Animals

Learning Outcomes:

1. Outline the body plan and unique characteristics of sponges.
2. Describe how sponges defend themselves against predators.

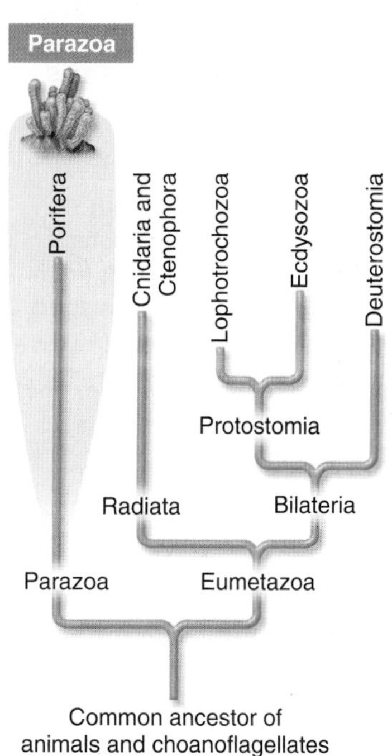

The Parazoa consist of one phylum, Porifera (from the Latin, meaning pore bearers), whose members are commonly referred to as sponges. Sponges lack true tissues—groups of cells that have a similar structure and function. However, sponges are multicellular and possess several types of cells that perform different functions. Biologists have identified approximately 8,000 species of sponges, the vast majority of which are marine. Sponges range in size from only a few millimeters across to more than 2 m in diameter. The smaller sponges may be radially symmetrical, but most have no apparent symmetry. Some sponges have a low encrusting growth form, whereas others grow tall and erect (Figure 33.2a). Although adult sponges are sessile, that is, anchored in place, the larvae are free-swimming.

Choanocytes Help Circulate Water

The body of a sponge looks similar to a vase pierced with small holes or pores (Figure 33.2b). Water is drawn through these pores into a central cavity, the **spongocoel**, and flows out through the large opening at the top, called the osculum. The water enters the pores by the beating action of the flagella of the **choanocytes**, or collar cells, that line the spongocoel (Figure 33.2c). In the process, the choanocytes trap and eat small particulate matter and tiny plankton. As noted in Chapter 32, because of striking morphological and molecular similarities between choanocytes and choanoflagellates, a group of modern protists having a single flagellum, scientists believe that sponges originated from a choanoflagellate-like ancestor.

A layer of flattened epithelial cells similar to those making up the outer layer of other phyla protects the sponge body. In between the choanocytes and the epithelial cells lies a gelatinous, protein-rich matrix called the **mesohyl**. Within this matrix are mobile cells called **amoebocytes** that absorb food from choanocytes, digest it, and carry the nutrients to other cells. Thus, considerable cell-to-cell contact and communication exist in sponges. Sponges are unique among the major animal phyla in using intracellular digestion, the uptake of food particles by cells, as a mode of feeding.

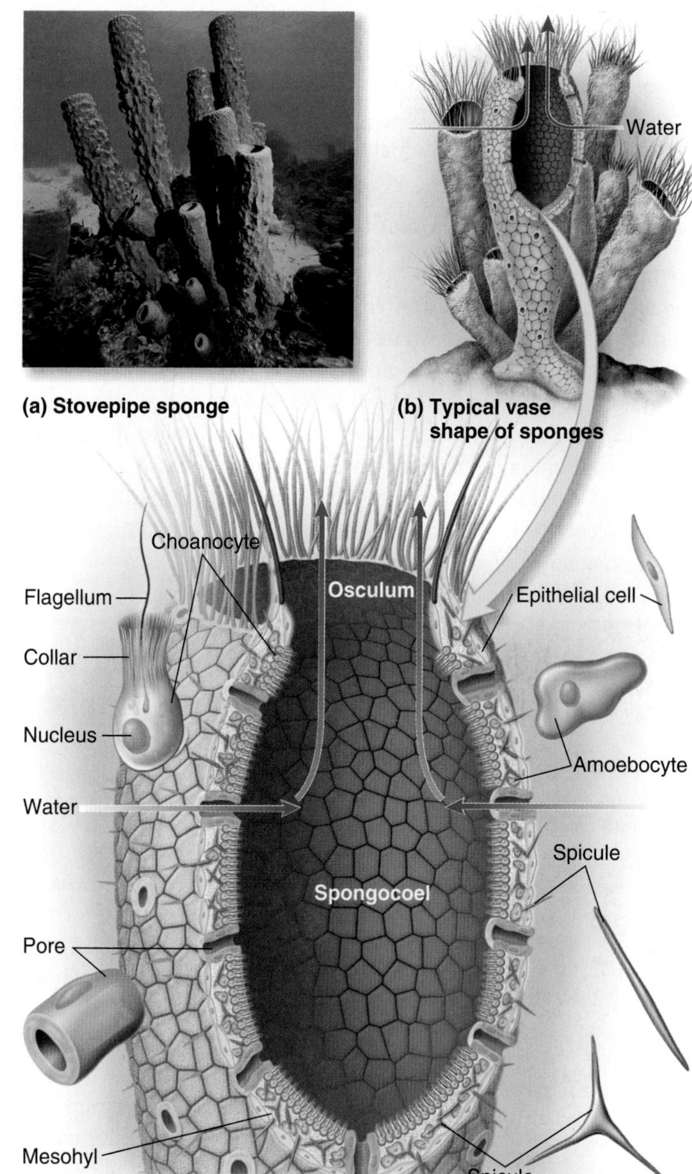

Figure 33.2 Sponge morphology. **(a)** The stovepipe sponge (*Aplysina archeri*) is a common sponge found on Caribbean reefs. **(b)** Many sponges have a vaselike shape. **(c)** A cross section reveals that sponges are truly multicellular animals, having various cell types but no distinct tissues.

Concept Check: *If sponges are soft and sessile, why aren't they eaten by other organisms?*

Sponges Have Mechanical and Chemical Defenses Against Predators

Some amoebocytes can also form tough skeletal fibers that support the body. In many sponges, this skeleton consists of sharp **spicules** formed of protein, calcium carbonate, or silica. For example, some deep-ocean species, called glass sponges, are distinguished by needlelike silica spicules that form elaborate lattice-like skeletons. The presence of such tough spicules may help explain why there is not much predation of sponges. Other sponges have fibers of a tough protein called **spongin** that lend skeletal support. Spongin skeletons are still commercially harvested and sold as bath sponges. Many species produce toxic defensive chemicals, some of which are thought to have possible antibiotic and anti-inflammatory effects in humans.

Sponges Reproduce Sexually and Asexually

Sponges reproduce through both sexual and asexual means. Most sponges are **hermaphrodites** (from the Greek, for the Greek god Hermes and the goddess Aphrodite), individuals that can produce both sperm and eggs. Gametes are derived from amoebocytes or choanocytes. The eggs remain in the mesohyl, and the sperm are released into the water and carried by water currents to fertilize the eggs of neighboring sponges. Zygotes develop into flagellated swimming larvae that eventually settle on a suitable substrate to become sessile adults. In asexual reproduction, a small fragment or bud may detach and form a new sponge.

33.2 Radiata: Jellyfish and Other Radially Symmetrical Animals

Learning Outcomes:

1. Compare and contrast the two body forms of cnidarians.
2. Describe how cnidarians defend themselves.
3. Describe the unique features of ctenophores.

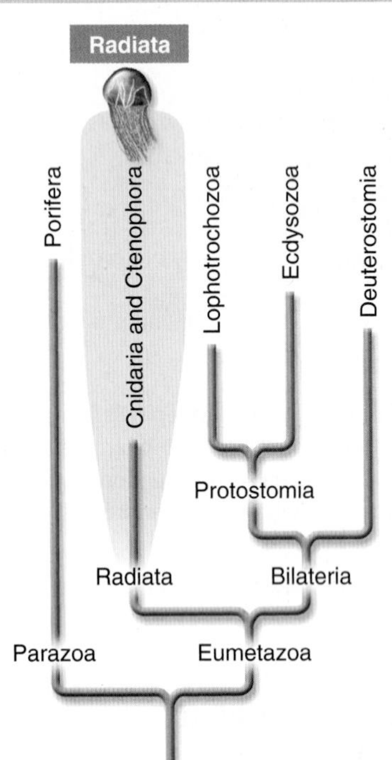

The Radiata consists of two closely related phyla: the Cnidaria (from the Greek *knide*, meaning nettle, and *aria*, meaning related to; pronounced nid-air'-e-ah) and the Ctenophora (from the Greek *ktenos*, meaning comb, and *phora*, meaning bearing; pronounced teen-o-for'-ah). Members of the Radiata phyla, or radiates, are mostly found in marine environments, although a few are freshwater species, such as hydra. The Cnidaria include hydra, jellyfish, box jellies, sea anemones, and corals, and the Ctenophora consist of the comb jellies. The Radiata have only two embryonic germ layers: the ectoderm and the endoderm, which give rise to the epidermis and the gastrodermis, respectively. A gelatinous substance called the **mesoglea** connects the two layers. In jellyfish, the mesoglea is enlarged and forms a transparent jelly, whereas in coral, the mesoglea is very thin. The Radiata is the first clade with true tissues (refer back to Chapter 32).

Both cnidarians and ctenophores possess a **gastrovascular cavity**, a body cavity with a single opening to the external environment, where extracellular digestion takes place. Extracellular digestion, the breakdown of large molecules, takes place outside of the cell, allows the ingestion of larger food particles, and represents a major increase in complexity over the sponges, which use only intracellular digestion. Most radiates have tentacles around the mouth that aid in food detection and capture. Radiates also have true nerve cells arranged as a **nerve net** consisting of interconnected neurons with no central control organ. In nerve nets, nerve impulses pass in either direction along a given neuron.

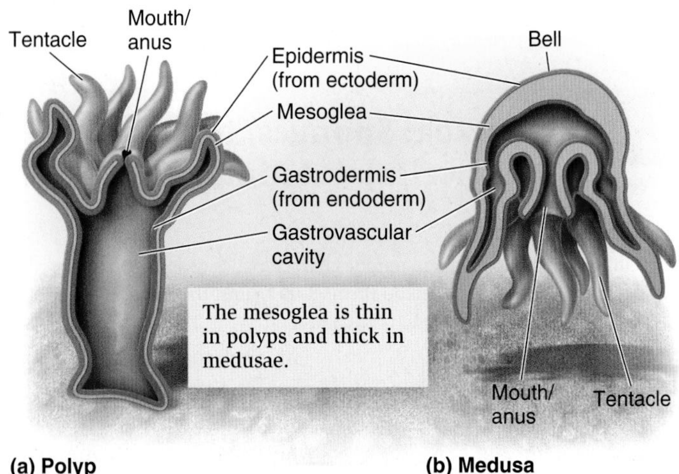

Figure 33.3 **Polyp and medusa forms of cnidarians.** Both **(a)** polyp and **(b)** medusa forms have two layers of cells, an outer epidermis (from ectoderm) and an inner layer of gastrodermis (from endoderm). In between is a layer of mesoglea, which is thin in polyps, such as corals, and thick in medusae, such as most jellyfish.

BIOLOGY PRINCIPLE **Structure determines function.** The inverted, umbrella-shaped medusae are free swimming, whereas the tubular, polyp forms are sedentary.

The Cnidarians Exist in Two Different Body Forms

Most cnidarians exist as two different body forms and associated lifestyles: the sessile **polyp** or the motile **medusa** (**Figure 33.3**). For example, corals exhibit only the polyp form, and jellyfish exist predominantly in the medusa form. Many cnidarians, such as *Obelia*, have a life cycle that prominently features both polyp and medusa stages (**Figure 33.4**).

The polyp form has a tubular body with an opening at the oral (top) end that is surrounded by tentacles and functions as both mouth and anus (see Figure 33.3a). The aboral (bottom) end is attached to the substrate. In the 18th century, the Swiss naturalist Abraham Trembley discovered that when a freshwater hydra was cut in two, each part not only survived but could also regenerate the missing half. Polyps exist colonially, as they do in corals, or alone, as in sea anemones. Corals take dissolved calcium and carbonate ions from seawater and precipitate them as limestone underneath their bodies. In some species, this leads to a buildup of limestone deposits. As each successive generation of polyps dies, the limestone remains in place, and new polyps grow on top. Thus, huge underwater limestone deposits called coral reefs are formed (look ahead to Figure 54.25b). The largest of these is Australia's Great Barrier Reef, which stretches over 2,300 km. Many other extensive coral reefs are known, including the

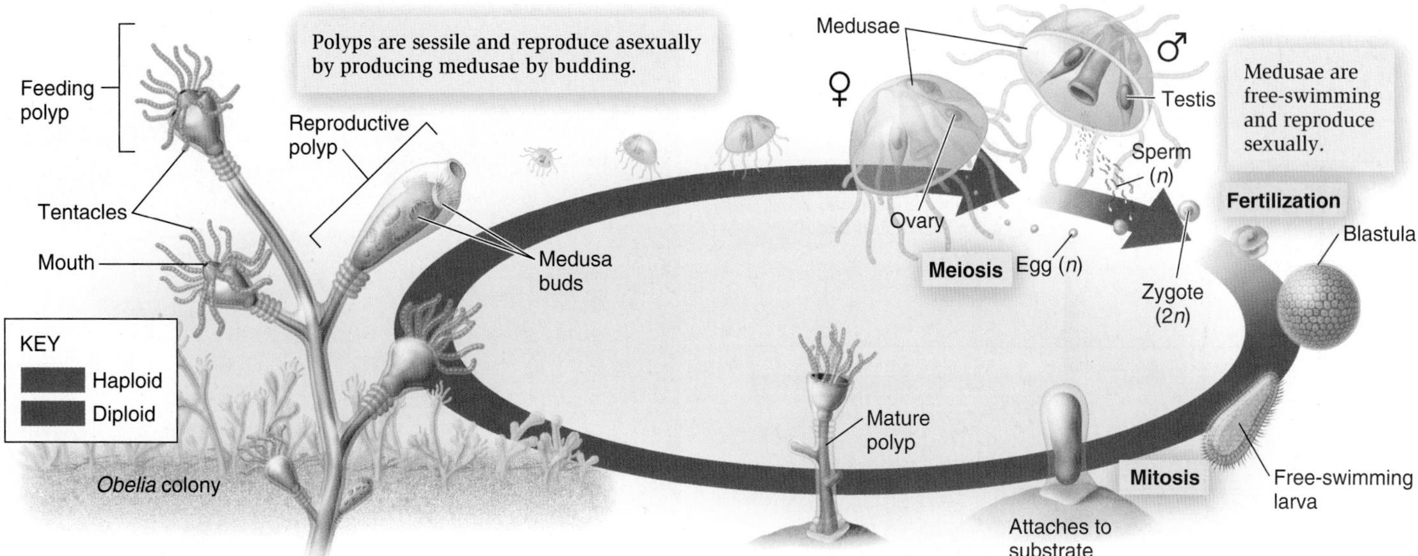

Figure 33.4 Life cycle of *Obelia*, a colonial cnidarian. This species exhibits both polyp and medusa stages.

Concept Check: *What are the dominant life stages of the following types of cnidarians: jellyfish, sea anemone, and Portuguese man-of-war?*

reef system along the Florida Keys, all of which occur in warm water, generally between 20°C and 30°C.

The free-swimming medusa form has an umbrella-shaped body with an opening that serves as both mouth and anus on the concave underside that is surrounded by tentacles (see Figure 33.3b). More mobile medusae possess simple sense organs near the bell margin, including organs of equilibrium called statocysts and photosensitive organs known as **ocelli**. When one side of the bell tips upward, the statocysts on that side are stimulated, and muscle contraction is initiated to right the medusa. The ocelli allow medusae to position themselves in particular light levels.

Cnidarians Have Specialized Stinging Cells

One of the unique and characteristic features of the cnidarians is the existence of stinging cells called **cnidocytes**, which function in defense or the capture of prey (Figure 33.5a). Cnidocytes contain **nematocysts**, powerful capsules with an inverted coiled and barbed thread. Each cnidocyte has a hairlike trigger called a **cnidocil** on its surface. When the cnidocil is touched or a chemical stimulus is detected, the nematocyst is discharged, and its filament penetrates the prey and injects a small amount of toxin. Small prey are immobilized and passed into the mouth by the tentacles. After discharge, the cnidocyte is absorbed, and a new one grows to replace it. The nematocysts of most cnidarians are not harmful to humans, but those on the tentacles of the larger jellyfish and the Portuguese man-of-war (Figure 33.5b) can be extremely painful and even fatal. Tentacles of the largest jellyfish, *Cyanea arctica*, may be over 40 m long.

Contractile fibers and nerves exist in their simplest forms in cnidarians. Contractile fibers are found in both the epidermis and gastrodermis. Although not true muscles, which only arise from the mesoderm and therefore do not appear in diploblastic animals, these fibers can contract to change the shape of the animal. For example, in the presence of a predator, an anemone can expel water very quickly

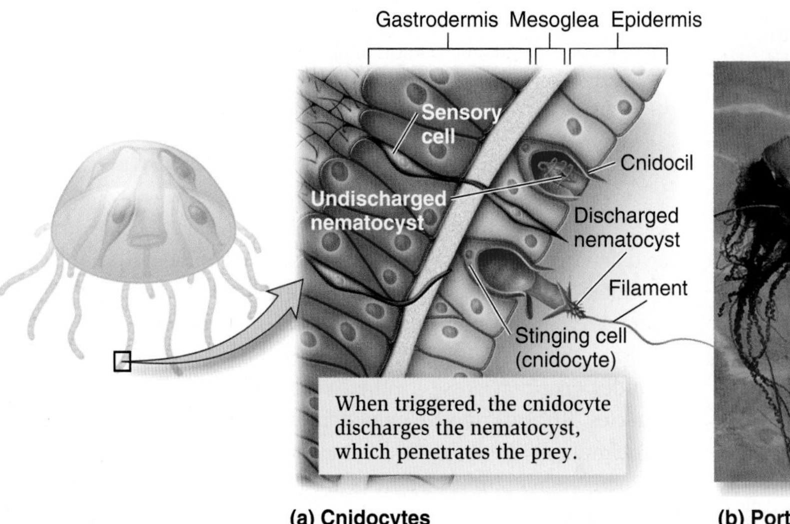

(a) Cnidocytes

(b) Portuguese man-of-war

Figure 33.5 Specialized stinging cells of cnidarians, called cnidocytes. **(a)** Cnidocytes, which contain stinging capsules called nematocysts, are situated in the tentacles. **(b)** The Portuguese man-of-war (*Physalia physalis*) employs cnidocytes that can be lethal to humans.

Concept Check: *How are cnidocytes recycled for re-use once they have been fired?*

Table 33.1 Main Classes and Characteristics of the Cnidaria

Class and examples (est. # of species)	Class characteristics
Hydrozoa: *Obelia*, Portuguese man-of-war, *Hydra*, some corals (2,700)	Mostly marine; polyp stage usually dominant and colonial, reduced medusa stage
Scyphozoa: jellyfish (200)	All marine; medusa stage dominant and large (up to 2 m); reduced polyp stage
Anthozoa: sea anemones, sea fans, most corals (6,000)	All marine; polyp stage dominant; medusa stage absent; many are colonial
Cubozoa: box jellies, sea wasps (20)	All marine; medusa stage dominant; box-shaped

through its open mouth and shrink down to a very small body form. The contractile fibers work against the fluid contained in the body, which thus acts as a hydrostatic skeleton. The nerve net that conducts signals from sensory nerves to muscle cells allows coordination of simple movements and shape changes.

The phylum Cnidaria consists of four classes: Hydrozoa (hydroids including *Obelia* and the Portuguese man-of-war), Scyphozoa (jellyfish), Anthozoa (sea anemones and corals), and Cubozoa (box jellies). The distinguishing characteristics of these classes are shown in **Table 33.1**.

The Ctenophores Have a Complete Gut

Ctenophores, also known as comb jellies, are a small phylum of fewer than 100 species, all of which are marine and look very much like jellyfish (**Figure 33.6**). They have eight rows of cilia on their surface that resemble combs. The coordinated beating of the cilia, rather than muscular contractions, propels the ctenophores. Averaging about 1–10 cm in length, comb jellies are probably the largest animals to use cilia for locomotion. There are even a few ribbon-like species up to 1 m long.

Comb jellies possess two long tentacles but lack stinging cells. Instead, they have colloblasts, cells on the tentacles that secrete a sticky substance onto which small prey adhere. The tentacles are then drawn over the mouth. As with cnidarians, digestion occurs in a gastrovascular cavity, but waste and water are eliminated through two anal pores. Thus, the comb jellies possess the first complete gut. Prey are generally small and may include tiny crustaceans called copepods and small fishes. Comb jellies are often transported around the world in ships' ballast water. *Mnemiopsis leidyi*, a ctenophore species native to the Atlantic coast of North and South America, was accidentally introduced into the Black and Caspian seas in the 1980s. With a plentiful food supply and a lack of predators, *Mnemiopsis* underwent a population explosion and ultimately devastated the local fishing industries.

Figure 33.6 A ctenophore. Ctenophores are called comb jellies because the eight rows of cilia on their surfaces resemble combs.

All ctenophores are hermaphroditic, possessing both ovaries and testes, and gametes are shed into the water to eventually form a free-swimming larva that grows into an adult. There is no polyp stage. Nearly all ctenophores exhibit **bioluminescence**, a phenomenon that results from chemical reactions that give off light rather than heat. Individuals are particularly evident at night, and ctenophores that wash up onshore can make the sand or mud appear luminescent.

33.3 Lophotrochozoa: The Flatworms, Rotifers, Bryozoans, Brachiopods, Mollusks, and Annelids

Learning Outcomes:

1. Describe the unique features of platyhelminthes, rotifers, bryozoans, and brachiopods.
2. Outline the main biological features and list the main classes of the mollusks.
3. List the advantages of segmentation in the annelids.

As we explored in Chapter 32 (refer back to Figure 32.3), molecular data suggest that there are three clades of bilateral animals: the Lophotrochozoa and the Ecdysozoa (collectively known as the protostomes) and the Deuterostomia. In this section, we explore the distinguishing characteristics of the Lophotrochozoa, a diverse group that includes taxa that possess either a lophophore (a crown of ciliated tentacles, Bryozoa and Brachiopoda) or a distinct larval stage called a trochophore (Mollusca and Annelida). Also included in this clade are the Platyhelminthes (some of which have trocophore-like larvae) and the Rotifera (which have a lophophore-like feeding device), both

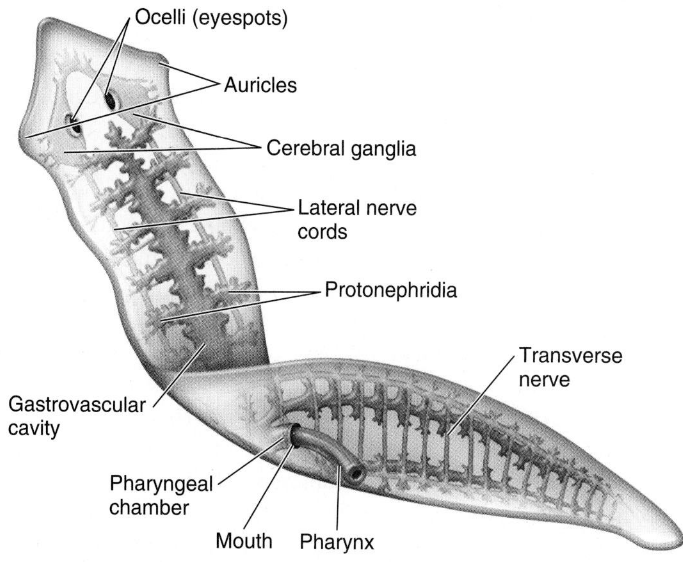

Figure 33.7 Body plan of a flatworm. Flatworm morphology is represented by a planarian, a member of the class Turbellaria.

Concept Check: *How do flatworms breathe?*

Table 33.2 Main Classes and Characteristics of Platyhelminthes

	Class and examples (est. # of species)	Class characteristics
	Turbellaria: planarian (3,000)	Mostly marine; free-living flatworms; predatory or scavengers
	Monogenea: fish flukes (1,000)	Marine and freshwater; usually external parasites of fish; simple life cycle (no intermediate host)
	Trematoda: flukes (11,000)	Internal parasites of vertebrates; complex life cycle with several intermediate hosts
	Cestoda: tapeworms (5,000)	Internal parasites of vertebrates; complex life cycle, usually with one intermediate host; no digestive system; nutrients absorbed across epidermis

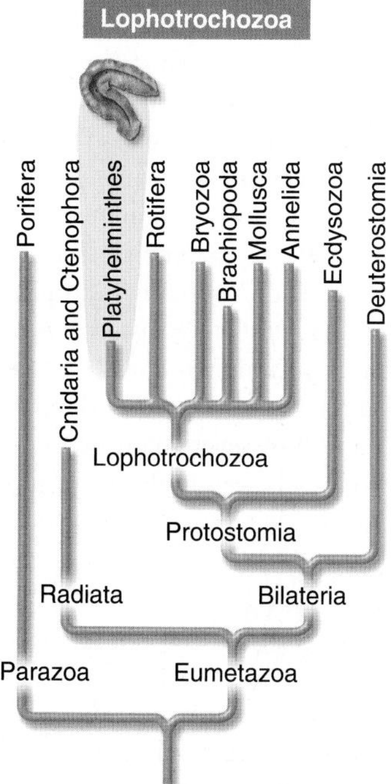

of which share molecular similarities with the other members of the Lophotrochozoa.

The Phylum Platyhelminthes Consists of Flatworms with No Coelom

Platyhelminthes (from the Greek *platy*, meaning flat, and *helminth*, meaning worm), or flatworms, were among the first animals to develop an active predatory lifestyle. Platyhelminthes, and indeed most animals, are bilaterally symmetrical, with a head bearing sensory appendages, a feature called cephalization (**Figure 33.7**).

The Development of Mesoderm The flatworms are also believed to be the first animals to develop three distinctive embryonic germ layers—ectoderm, endoderm, and mesoderm—with mesoderm replacing the simpler gelatinous mesoglea of cnidarians. As such, they are said to be triploblastic. The muscles in flatworms, which are derived from mesoderm, are well developed. The development of mesoderm was therefore a critical evolutionary innovation in animals, leading to the development of more sophisticated organs. Flatworms lack a fluid-filled body cavity in which the gut is suspended, and instead mesoderm fills the body spaces around the gastrovascular cavity; hence, they are described as acoelomates. Flatworms lack a specialized respiratory or circulatory system and must respire by diffusion. Thus, no cell can be too far from the surface, making a flattened shape necessary. The digestive system of flatworms is incomplete, with only one opening, which serves as both mouth and anus, as in cnidarians. Most flatworms possess a muscular pharynx that may be extended through the mouth. The pharynx opens to a gastrovascular cavity, where food is digested. In large flatworms, the gastrovascular cavity is highly branched to distribute nutrients to all parts of the body. The incomplete digestive system of flatworms prevents continuous feeding. Some flatworms are predators, but many species invade other animals as parasites.

Flatworms have a distinct excretory system, consisting of **protonephridia**, two lateral canals with branches capped by **flame cells**. Protonephridia are dead-end tubules lacking internal openings. The flame cells, which are ciliated and waft water through the lateral canals to the outside (look ahead to Figure 49.7), primarily function in maintaining osmotic balance between the flatworm's body and the surrounding fluids. Simple though this system is, its development was key to permitting the movement of animals into freshwater habitats and even moist terrestrial areas.

Cephalization At the anterior end of some free-living flatworms are light-sensitive eyespots, called ocelli, as well as chemoreceptive and sensory cells that are concentrated in organs called auricles. A pair of **cerebral ganglia**, clusters of nerve cell bodies, receives input from photoreceptors in eyespots and sensory cells. From the ganglia, a pair of lateral nerve cords running the length of the body allows rapid movement of information from anterior to posterior. In addition, transverse nerves form a nerve net on the ventral surface similar to that of cnidarians. Thus, flatworms retain the cnidarian-style nervous system, while possessing the beginnings of the more centralized type of nervous system seen throughout much of the rest of the animal kingdom.

In all the Platyhelminthes, reproduction is either sexual or asexual. Most species are hermaphroditic but do not fertilize their own eggs. Flatworms can also reproduce asexually by splitting into two parts, with each half regenerating the missing fragment.

Classes of Flatworms The four classes of flatworms are the Turbellaria, Monogenea, Trematoda (flukes), and Cestoda (tapeworms) (**Table 33.2**). Turbellarians are the only free-living class of flatworms and are widespread in lakes, ponds, and marine environments

(a)

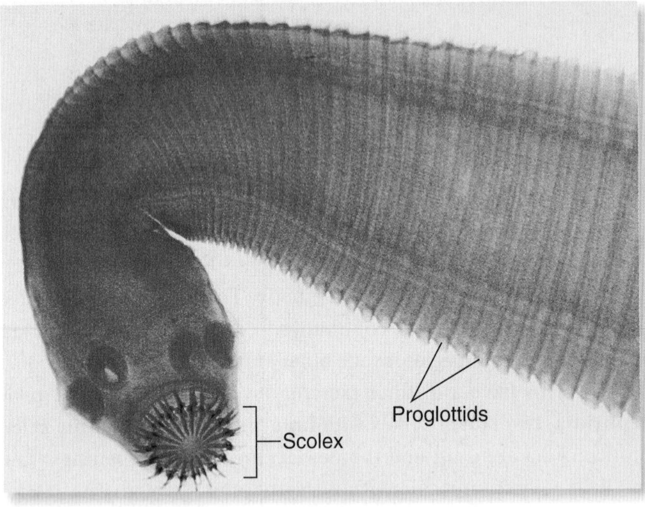

(b)

Figure 33.8 **Flatworms.** **(a)** Many free living marine turbellarians are brightly colored such as this racing stripe flatworm, *Pseudoceros bifurcus*, from Bali, Indonesia. **(b)** A tapeworm, *Taenia pisiformis*, a member of the class Cestoda. Note the tiny hooks and suckers that make up the scolex. Each segment is a proglottid, replete with eggs.

BIOLOGY PRINCIPLE **Biology affects our society.** About 1% of U.S. cattle are infected by beef tapeworms. Consuming beef that is not sufficiently well cooked can lead to infection by these parasites.

(Figure 33.8a). Monogeneans are relatively simple external parasites with just one host species (a fish). Both trematodes and cestodes are internally parasitic in humans and therefore are of great medical and veterinary importance. They possess a variety of organs of attachment, such as hooks and suckers, that enable them to remain embedded within their hosts. For example, cestodes attach to their host by means of an organ at the head end called a scolex (Figure 33.8b). They have no mouth or gastrovascular cavity and absorb nutrients across the body surface. Behind the scolex is a long ribbon of identical segments called proglottids. These are essentially segments of sex organs that develop thousands of eggs. The proglottids are continually shed in the host's feces. Cestodes often require two separate vertebrate host species, such as pigs or cattle, to begin their life cycle and humans to complete their development. Many tapeworms can live inside humans who consume undercooked, infected meat—hence the value of thoroughly cooking meat.

The life history of trematodes is even more complex than that of cestodes, involving multiple hosts. The first host, called the intermediate host, is usually a mollusk, and the final host, or definitive host, is usually a vertebrate, but often a second or even a third intermediate host is involved. In the case of the Chinese liver fluke (*Clonorchis sinensis*) (Figure 33.9), (1) the adult parasite lives and reproduces in the definitive host, a human. (2) The resultant embryos are called miracidia. They are encapsulated to form eggs and pass from the host via the feces. (3) An intermediate host, such as a snail, eats the eggs. The miracidia are released and transform into sporocysts. (4) The sporocysts asexually produce more sporocysts, which are called rediae. (5) The rediae reproduce asexually to produce cercariae. Cercariae bore their way out of the snail and (6) infect their second intermediate host, fishes, by entering via the gills. Here, the cercariae develop into metacercarial cysts (juvenile flukes) and lodge in fish muscle, which the definitive host will eat, allowing the cycle to continue. In the definitive host, the cyst protects the metacercaria from the host's gastric juices. From the small intestine, the metacercariae travel to the liver and grow into adult flukes, and the life cycle begins anew. The life cycle of a trematode thus can involve at least seven stages: adult, miracidium, egg, sporocyst, rediae, cercaria, and metacercaria. The low probability of each larva reaching a suitable host is low, so trematodes produce large numbers of offspring to ensure that some survive.

Blood flukes, genus *Schistosoma*, are the most common parasitic trematodes infecting humans; they cause the disease known as schistosomiasis. Over 200 million people worldwide, primarily in tropical Asia, Africa, and South America, are infected with schistosomiasis. The inch-long adult flukes can live for years in human hosts, and the release of eggs may cause chronic inflammation and blockage in many organs. Untreated schistosomiasis can lead to severe damage to the liver, intestines, and lungs and can eventually lead to death. Sewage treatment and access to clean water can greatly reduce infection rates.

Members of the Phylum Rotifera Have a Pseudocoelom and a Ciliated Crown

Members of the phylum Rotifera (from the Latin *rota*, meaning wheel, and *fera*, meaning to bear) get their name from their ciliated crown, or **corona**, which, when beating, looks similar to a rotating wheel (Figure 33.10). Most rotifers are microscopic animals, usually less than 1 mm long, and some have beautiful colors. There are about 2,000 species of rotifers, most of which inhabit fresh water, with a few marine and terrestrial species. Most often they are bottom-dwelling organisms, living on the pond floor or along lakeside vegetation.

Rotifers have an alimentary canal, a digestive tract with a separate mouth and anus, which means they can feed continuously. The corona creates water currents that propel the animal through the water and that waft small planktonic organisms or decomposing organic material toward the mouth. The mouth opens into a circular, muscular pharynx called a **mastax**, which has jaws for grasping and chewing. The mastax, which in some species can protrude through the mouth to seize small prey, is a structure unique to rotifers. The body of the rotifer bears a jointed foot with one to four toes. **Pedal glands** in the foot secrete a sticky substance that aids in attachment

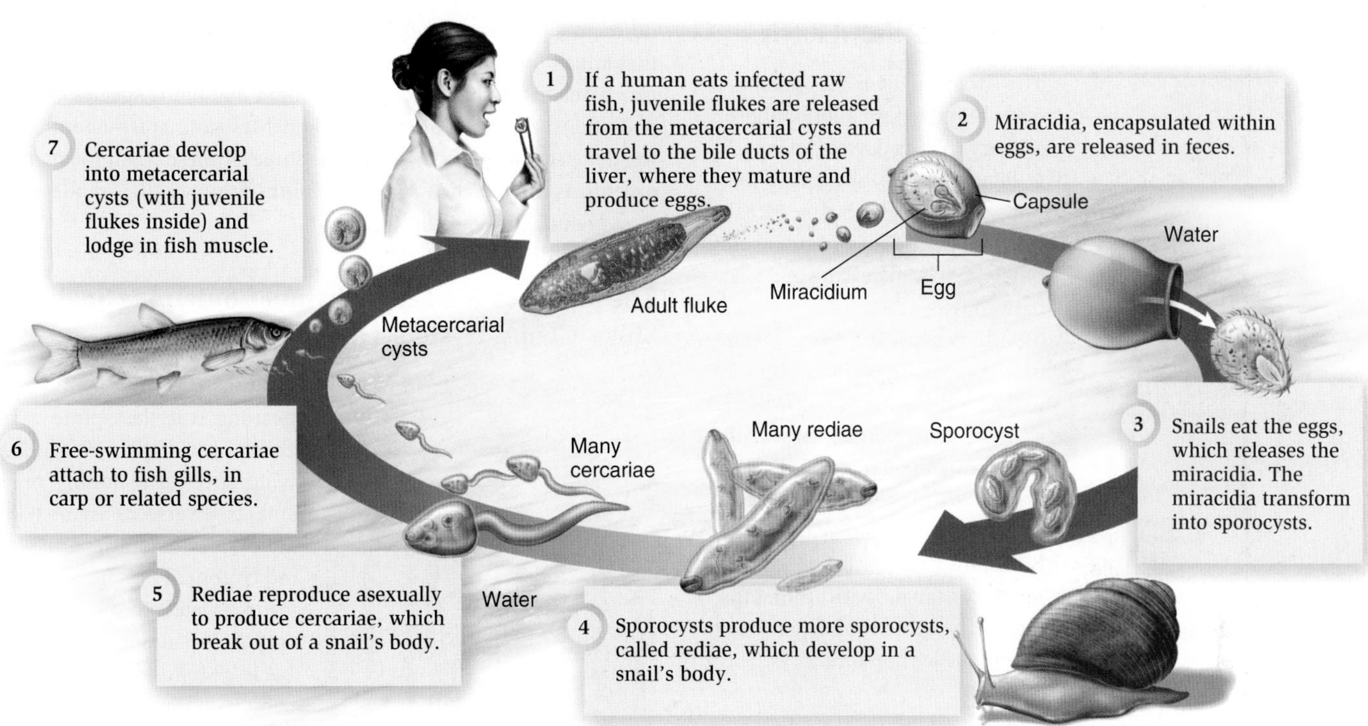

1 If a human eats infected raw fish, juvenile flukes are released from the metacercarial cysts and travel to the bile ducts of the liver, where they mature and produce eggs.
2 Miracidia, encapsulated within eggs, are released in feces.
Capsule
Water
Egg
Miracidium
Adult fluke
3 Snails eat the eggs, which releases the miracidia. The miracidia transform into sporocysts.
Sporocyst
Many rediae
Many cercariae
4 Sporocysts produce more sporocysts, called rediae, which develop in a snail's body.
Water
5 Rediae reproduce asexually to produce cercariae, which break out of a snail's body.
6 Free-swimming cercariae attach to fish gills, in carp or related species.
Metacercarial cysts
7 Cercariae develop into metacercarial cysts (with juvenile flukes inside) and lodge in fish muscle.

Figure 33.9 The complete life cycle of a trematode. This figure shows the life cycle of the Chinese liver fluke (*Clonorchis sinensis*).

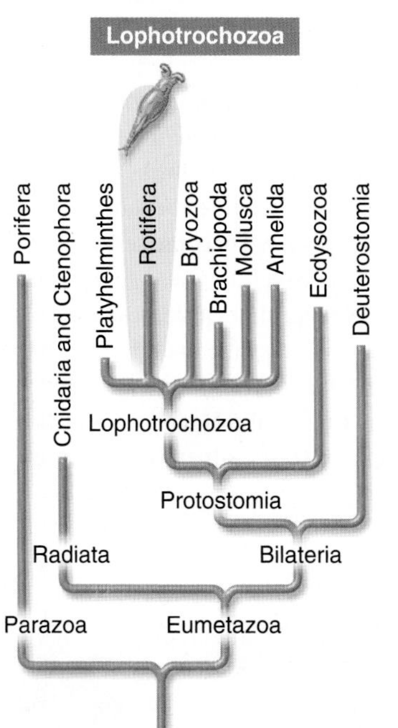

to a substrate. The internal organs lie within a pseudocoelom, a fluid-filled body cavity that is not completely lined with mesoderm. The pseudocoelom serves as a hydrostatic skeleton and as a medium for the internal transport of nutrients and wastes. Rotifers also have a pair of protonephridia with flame bulbs that collect excretory and digestive waste and drain into a bladder, which passes waste to the anus. The nervous system consists of nerves that extend from the sensory organs, especially the eyespots and some bristles on the corona, to the brain.

Reproduction in rotifers is unique. In some species, unfertilized diploid eggs that have not undergone meiotic division develop into females through a process known as **parthenogenesis**. In other species, some unfertilized eggs develop into females, whereas others develop into males that live only long enough to produce and release sperm that fertilize the eggs. The resultant fertilized eggs form zygotes, which have a thick shell and can survive for long periods of harsh conditions, such as if a water supply dries up, before developing into new females. Because the tiny zygotes are easily transported, rotifers show up in the smallest of aquatic environments, such as birdbaths or roof gutters.

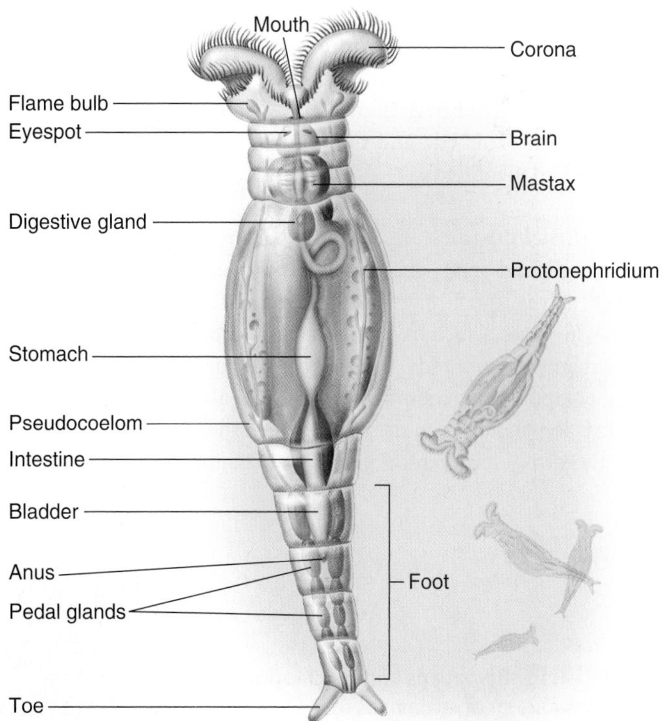

Figure 33.10 Body plan of a common rotifer, *Philodina* genus.

Bryozoa and Brachiopoda Are Closely Related Phyla

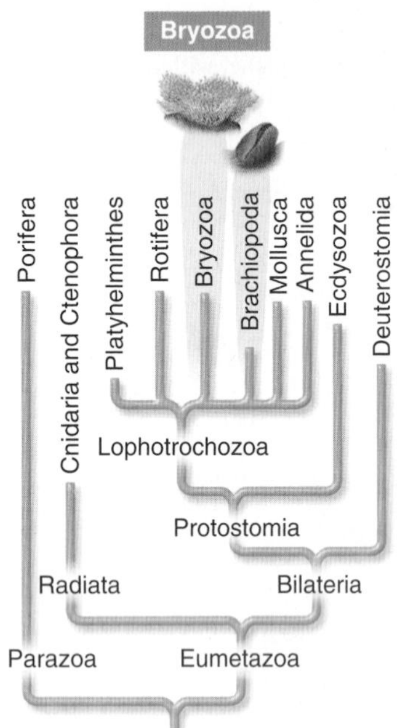

The Bryozoa and the Brachiopoda both possess a lophophore, a ciliary feeding device (refer back to Figure 32.13a), and a true coelom (refer back to Figure 32.8a). The lophophore is a circular fold of the body wall bearing tentacles that draw water toward the mouth. Because a thin extension of the coelom penetrates each tentacle, the tentacles also serve as a respiratory organ. Gases diffuse across the tentacles and into or out of the coelomic fluid and are carried throughout the body. Both phyla have a U-shaped alimentary canal, with the anus located near the mouth but outside of the lophophore.

Phylum Bryozoa The bryozoans (from the Greek *bryon*, meaning moss, and *zoon*, meaning animal) are small colonial animals, most of which are less than 0.5 mm long, that can be found encrusted on rocks in shallow aquatic environments. They look very much like plants. There are about 4,500 species, many of which encrust boat hulls and have to be scraped off periodically. Within the colony, each animal secretes and lives inside a nonliving exoskeleton called a zoecium that is composed of chitin or calcium carbonate (**Figure 33.11a**). For this reason, bryozoans have been important reef-builders. Bryozoans date back to the Paleozoic era, and thousands of fossil forms have been discovered and identified.

Phylum Brachiopoda Brachiopods (from the Greek *brachio*, meaning arm, and *podos*, meaning foot) are marine organisms with two shell halves, much like clams (**Figure 33.11b**). Unlike bivalve mollusks, however, which have a left and right valve (side) of the shell, with the plane of symmetry lying along the line at which the valves join, brachiopods have a dorsal and ventral valve, with the plane of symmetry perpendicular to the line at which the valves join. In other words, the dorsal and ventral valves of brachiopods are of slightly different sizes and shapes. Brachiopods are bottom-dwelling species that attach to the substrate via a muscular pedicle. Although they are now a relatively small group, with about 300 living species, brachiopods flourished in the Paleozoic and Mesozoic eras—about 30,000 fossil species have been identified. Some of these fossil forms tell of organisms that reached 30 cm in length, although their modern relatives are only 0.5–8.0 cm long.

The Mollusca Is a Large Phylum Containing Snails, Slugs, Clams, Oysters, Octopuses, and Squids

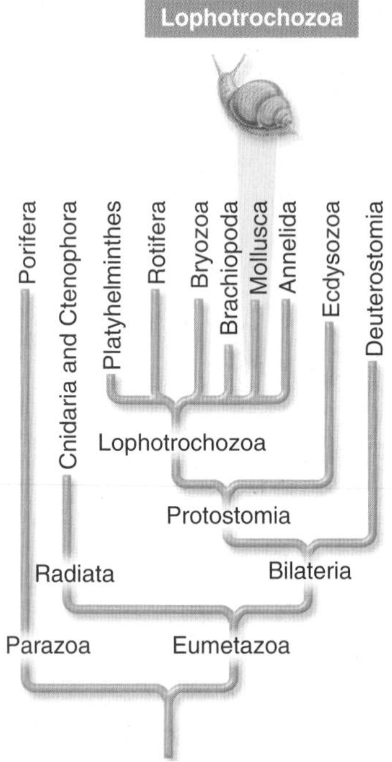

Mollusks (from the Latin *mollis*, meaning soft) constitute a very large phylum, with over 100,000 living species, including organisms as diverse as snails, clams and oysters, octopuses and squid, and chitons. They are an ancient group, as evidenced by the classification of about 35,000 fossil species. Mollusks have a considerable economic, aesthetic, and ecological importance to humans. Many serve as sources of food, including scallops, oysters, clams, and squids. A significant industry involves the farming of oysters to produce cultured pearls, and rare and beautiful mollusk shells are extremely valuable to collectors. Snails and slugs can damage vegetables and ornamental plants, and boring mollusks can penetrate wooden ships and wharfs. Mollusks are intermediate hosts to many parasites, and several exotic species have become serious pests. For example, populations of the zebra mussel (*Dreissena polymorpha*) appear to have been introduced into North America from Asia via ballast water from transoceanic ships. Since their introduction, they have spread rapidly throughout the Great Lakes and an increasing number of inland waterways, adversely impacting native organisms and clogging water intake valves to municipal water-treatment plants around the lakes.

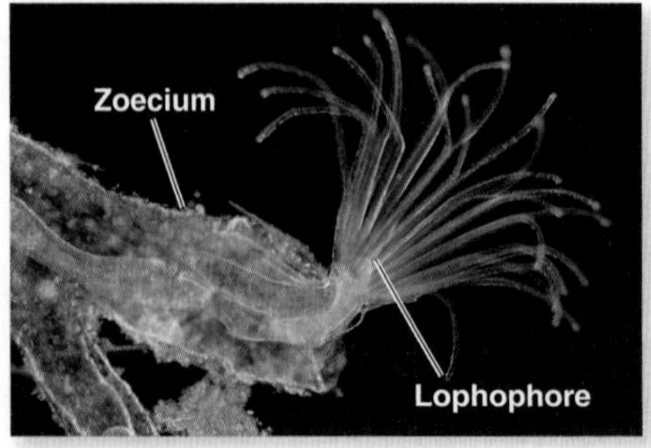

(a) A bryozoan

(b) A brachiopod, the northern lamp shell

Figure 33.11 **Bryozoans and brachiopods.** **(a)** Bryozoans are colonial animals that reside in a nonliving case called a zoecium. **(b)** Brachiopods, such as this northern lamp shell (*Terebratulina septentrionalis*), have dorsal and ventral shells.

Concept Check: *What are the two main functions of the lophophore?*

The Mollusk Body Plan One common feature of the mollusks is their soft body, which exists, in many species, under a protective external shell. Most mollusks are marine, although some have colonized fresh water. Many snails and slugs have moved onto land, but they survive only in humid areas and where the calcium necessary for shell formation is abundant in the soil. The ability to colonize freshwater and terrestrial habitats has led to a diversification of mollusk body plans. Thus, in the amazing diversity of mollusks we see how organismal diversity is related to environmental diversity.

Although great variation in morphology occurs between classes, mollusks have a basic body plan consisting of three parts (Figure 33.12). A muscular **foot** is usually used for movement, and a **visceral mass** containing the internal organs rests atop the foot. The **mantle**, a fold of skin draped over the visceral mass, secretes a shell in those species that form shells. The mantle often extends beyond the visceral mass, creating a chamber called the **mantle cavity**, which houses delicate **gills**, filamentous organs that are specialized for gas exchange. A continuous current of water, often induced by cilia present on the gills or by muscular pumping, flushes out the wastes from the mantle cavity and brings in new oxygen-rich water.

Mollusks are coelomate organisms, but the coelom is confined to a small area around the heart. The mollusks' organs are supplied with oxygen and nutrients via a circulatory system. Mollusks have an **open circulatory system** with a heart that pumps body fluid called hemolymph through vessels and into sinuses. Sinuses are the open, fluid-filled cavities between their internal organs. The organs and tissues are therefore continually bathed in hemolymph. The sinuses coalesce to form an open cavity known as the hemocoel (blood cavity). From these sinuses, the hemolymph drains into vessels that take it to the gills and then back to the heart. Excretory organs called **metanephridia** remove nitrogenous and other wastes. Metanephridia have ciliated funnel-like openings inside the coelom connected to ducts that lead to the exterior mantle cavity. The pores from the metanephridia discharge wastes into this cavity. The anus also opens into the mantle cavity. The metanephridial ducts may also serve to discharge sperm or eggs from the gonads. The nervous system varies from simple ganglia and nerve chords in most species to much larger brains and sophisticated organs of touch, smell, taste, and vision in octopuses.

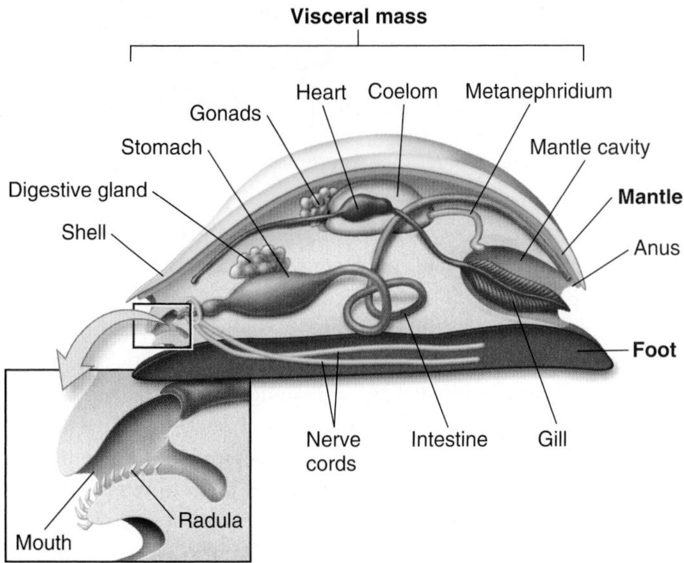

Figure 33.12 The mollusk body plan. The generalized body plan of a mollusk includes the characteristic foot, mantle, and visceral mass.

Concept Check: *Do molluscan hearts pump blood?*

The mollusk's mouth may contain a **radula**, a unique, protrusible, tonguelike organ that has many teeth and is used to eat plants, scrape food particles off rocks, or, if the mollusk is predatory, bore into shells of other species and tear flesh. In the cone shells (genus *Conus*), the radula is reduced to a few poison-injecting teeth on the end of a long proboscis that is cast about in search of prey, such as a worm or even a fish. Some Indo-Pacific cone shell species produce a neuromuscular toxin that can kill humans. Other mollusks, particularly bivalves, have lost their radula and are filter feeders that strain water brought in by ciliary currents.

Most shells are complex three-layered structures secreted by the mantle that continue to grow as the mollusk grows. Shell growth is often seasonal, resulting in distinct growth lines on the shell, much the same as tree rings (Figure 33.13a). Using shell growth patterns, biologists have discovered some bivalves that are over 100 years old. The innermost layer of the shells of oysters, mussels, abalone, and other mollusks is a smooth, iridescent lining called nacre, which is commonly known as mother-of-pearl and is often collected from abalone shells for jewelry. Actual pearl production in mollusks, primarily oysters, occurs when a foreign object, such as a grain of sand, becomes lodged between the shell and the mantle, and layers of nacre are laid down around it to reduce the irritation.

Most mollusks have separate sexes, although some are hermaphroditic. Gametes are usually released into the water, where they mix and fertilization occurs. In some snails, however, fertilization is internal, with the male inserting sperm directly into the female. Internal fertilization was a key evolutionary development, enabling some snails to colonize land, and can be considered a critical innovation that fostered extensive adaptive radiation. In many species, reproduction involves the production of a trochophore larva that develops into a **veliger**, a free-swimming larva that has a rudimentary foot, shell, and mantle.

The Major Molluscan Classes Of the eight molluscan classes, the four most common are the Polyplacophora (chitons), Gastropoda (snails and slugs), Bivalvia (clams and mussels), and Cephalopoda (octopuses, squids, and nautiluses) (Table 33.3). Chitons are marine mollusks with a shell composed of eight separate plates (Figure 33.13b). Chitons are common in the intertidal zone, an area above water at low tide and under water at high tide, and they creep along when covered by the tide. Feeding occurs by scraping algae off rock surfaces. When the tide recedes, the muscular foot holds the chiton tight to the rock surface, preventing desiccation. The class Gastropoda (from the Greek *gaster*, meaning stomach, and *podos*, meaning foot) is the largest group of mollusks and encompasses about 75,000 living species, including snails, periwinkles, limpets, and other shelled members (Figure 33.13c). The class also includes species such as slugs and nudibranchs, whose shells have been greatly reduced or completely lost during their evolution (Figure 33.13d). Most gastropods are marine or freshwater species, but some species,

Figure 33.13 **Mollusks.** **(a)** A bivalve shell, class Bivalvia, with growth rings. Quahog clams (*Mercenaria mercenaria*) can live over 20 years. **(b)** A chiton (*Tonicella lineata*), a polyplacophoran with a shell made up of eight separate plates. **(c)** A gastropod, the tree snail, *Liguus fasciatus*, from the Florida Everglades showing its characteristic coiled shell. **(d)** A nudibranch (*Phyllidia ocellata*). The nudibranchs are a gastropod subclass whose members have lost their shell altogether. **(e)** The highly poisonous blue-ringed octopus (*Hapalochlaena lunulata*), a cephalopod.

(a) A Quahog clam, class Bivalvia

(b) A chiton, class Polyplacophora

(c) A snail, class Gastropoda

(d) A sea slug, class Gastropoda

(e) A blue-ringed octopus, class Cephalopoda

Table 33.3 Main Classes and Characteristics of Mollusks

	Class and examples (est. # of species)	Class characteristics
	Polyplacophora: chitons (860)	Marine; eight-plated shell
	Gastropoda: snails, slugs, nudibranchs (75,000)	Marine, freshwater, or terrestrial; most with coiled shell, but shell absent in slugs and nudibranchs; radula present
	Bivalvia: clams, mussels, oysters (30,000)	Marine or freshwater; shell with two halves or valves; primarily filter feeders with siphons
	Cephalopoda: octopuses, squids, nautiluses (780)	Marine; predatory, with tentacles around mouth, often with suckers; shell often absent or reduced; closed circulatory system; jet propulsion via siphon

including snails and slugs, have also colonized land. Most gastropods are slow-moving animals that are weighed down by their shell. Unlike bivalves, gastropods have a one-piece shell, into which the animal can withdraw to escape predators.

The 780 species of Cephalopoda (from the Greek *kephale*, meaning head, and *podos*, meaning foot) are the most morphologically complex of the mollusks and indeed among the most complex of all invertebrates. Most are fast-swimming marine predators that range from organisms just a few centimeters in size to the colossal squid (*Mesonychoteuthis hamiltoni*), which is known to reach over 13 m in length and 495 kg (1,091 lb) in weight. A cephalopod's mouth is surrounded by many long arms commonly armed with suckers. Octopuses have 8 arms with suckers, and squids and cuttlefish have 10 arms—8 with suckers and 2 long tentacles with suckers limited to their ends. Nautiluses have from 60 to 90 tentacles around the mouth. All cephalopods have a beaklike jaw that allows them to bite their prey, and some, such as the blue-ringed octopus (*Hapalochlaena lunulata*), deliver a deadly poison through their saliva (**Figure 33.13e**). Only one group, the nautiluses, has retained its external shell. In octopuses, the shell is not present, and in squid and cuttlefish, it is greatly reduced and internal. However, the fossil record is full of shelled cephalopods, called ammonites, some of which were as big as truck tires

Figure 33.14 A fossil ammonite. These shelled cephalopods were abundant in the Cretaceous period.

BioConnections: In what period did mollusks arise? Look back to Figure 22.15.

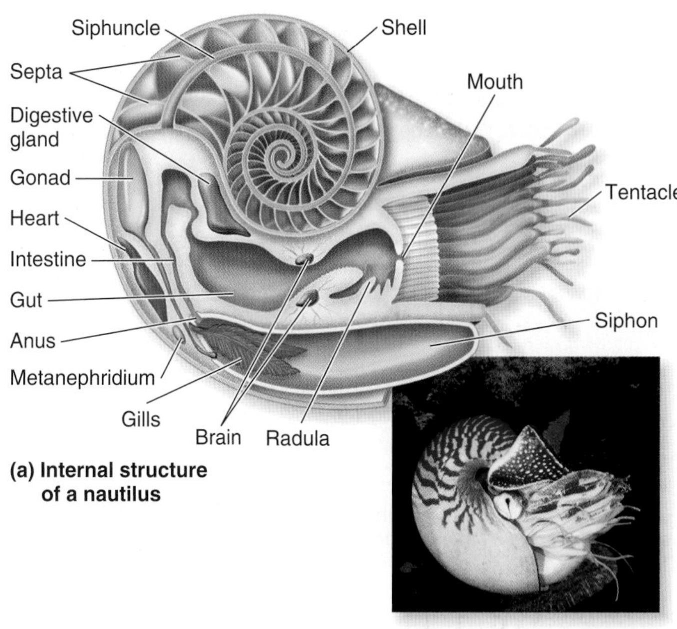

Figure 33.15 The nautilus. (a) A longitudinal section of a nautilus, showing the coiled shell with many chambers. The animal secretes a new chamber each year and lives only in the new one. **(b)** The chambered nautilus (*Nautilus pompilius*).

(Figure 33.14). They became extinct at the end of the Cretaceous period, although the reasons for this are not well understood.

The foot of some cephalopods has become modified into a muscular siphon. Water drawn into the mantle cavity is quickly expelled through the siphon, propelling the organisms forward or backward in a kind of jet propulsion. Such vigorous movement requires powerful muscles and a very efficient circulatory system to deliver oxygen and nutrients to the muscles. Cephalopods are the only mollusks with a **closed circulatory system**, in which blood flows throughout an animal entirely within a series of vessels. One of the advantages of this type of system is that the heart can pump blood through the tissues rapidly, making oxygen more readily available. The blood of cephalopods contains the copper-rich protein hemocyanin for transporting oxygen. Less efficient than the iron-rich hemoglobin of vertebrates, hemocyanin gives the blood a blue color.

The shell of the nautilus is comprised of many individual sealed chambers (Figure 33.15). As it grows, the nautilus secretes a new chamber and seals off the old one with a **septum** (plural, septa). The older chambers are gas-filled and act as buoyancy chambers. A thin strip of living tissue called the siphuncle removes liquid from the old chamber and replaces it with gas. The gas pressure within the chambers is thus only 1 atmosphere, despite the fact that nautiluses may be swimming at 400 m depths at a pressure of about 40 atmospheres. The shell's structure is strong enough to withstand this amount of pressure differential.

Cephalopods have a well-developed nervous system and brain that support their active lifestyle. Their sense organs, especially their eyes, are also very well developed. Many cephalopods (with the exception of nautiluses) have an ink sac that contains the pigment melanin; the sac can be emptied to provide a "smokescreen" to confuse predators. In many species, melanin is also distributed in special pigment cells in the skin, which allows for color changes. Octopuses often change color when disturbed and they can rapidly change color to blend in with their background and escape detection. The central nervous system of the octopus is among the most complex in the invertebrate world. Behavioral biologists have demonstrated that octopuses can behave in sophisticated ways, and scientists are currently debating to what degree they are capable of learning by observation.

Feature Investigation

Fiorito and Scotto's Experiments Showed Invertebrates Can Exhibit Sophisticated Observational Learning Behavior

We tend to think of the ability to learn from others as an exclusively vertebrate phenomenon, especially among species that live in social groups. However, in 1992, Italian researchers Graziano Fiorito and Pietro Scotto demonstrated that octopuses can learn by observing the behavior of other octopuses (Figure 33.16). This was a surprising finding, in part because *Octopus vulgaris*, the species they studied, lives a solitary existence for most of its life.

In their experiments, they used a system of reward (a small piece of fish placed behind the ball that the octopus could not see) and punishment (a small electric shock for choosing the wrong ball) to train octopuses to attack either a red or a white ball. This type of learning is called classical conditioning (see Chapter 55). Because octopuses are

Figure 33.16 **Observational learning in octopuses.**

HYPOTHESIS Octopuses can learn by observing another's behavior.

STUDY LOCATION Laboratory setting with *Octopus vulgaris* collected from the Bay of Naples, Italy.

	Experimental level		Conceptual level
1 Train 2 groups of octopuses, one to attack white balls, one to attack red. These are called the demonstrator octopuses.	Reward choice of correct ball (with fish) and punish choice of incorrect ball (with electric shock). Training is complete when octopus makes no "mistakes" in 5 trials.		Conditions a demonstrator octopus to attack a particular color of ball.
2 In an adjacent tank, allow observer octopus to watch trained demonstrator octopus.			Observer octopus may be learning the correct ball to attack by watching the demonstrator octopus.
3 Drop balls into the tank of the observer octopus. Test the observer octopus to see if it makes the same decisions as the demonstrator octopus.	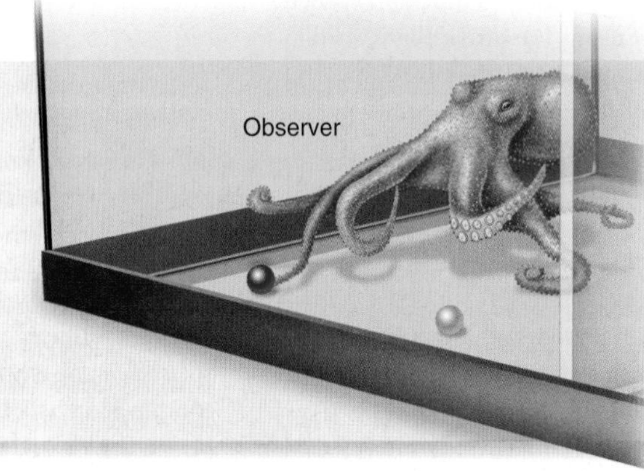		If the observer octopus is learning from the demonstrator octopus, the observer octopus should attack the ball of the same color as the demonstrator octopus was trained to attack.

4 **THE DATA**

Participant	Color of ball chosen in 5 trials*	
	Red	**White**
Observers (watched demonstrator attack red)	4.31	0.31
Observers (watched demonstrator attack white)	0.40	4.10
Untrained (did not watch demonstrations)	2.11	1.94

*Average of 5 trials; data do not always sum to 5, because some trials resulted in no balls being chosen.

5 **CONCLUSION** Invertebrate animals are capable of learning from watching other individuals behave, in much the same way as vertebrate species learn from watching others.

6 **SOURCE** Fiorito, G., and Scotto, P. 1992. Observational learning in *Octopus vulgaris*. *Science* 256:545–547.

color blind, they must distinguish between the relative brightness of the balls. Octopuses were considered to be trained when they made no mistakes in five trials. Observer octopuses in adjacent tanks were then allowed to watch the trained octopuses attacking the balls. In the third part of the experiment, the observer octopuses were themselves tested. In these cases, observers nearly always attacked the same color ball as they had observed the demonstrators attacking. In addition, learning by observation was achieved more quickly than the original training. This remarkable observational learning behavior is considered by some to be the precursor to more complex forms of learning, including problem solving.

Experimental Questions

1. What was the hypothesis tested by Fiorito and Scotto?
2. What were the results of the experiment? Did these results support the hypothesis?
3. What is the significance of performing the experiment on both trained and untrained octopuses?

The Phylum Annelida Consists of the Segmented Worms

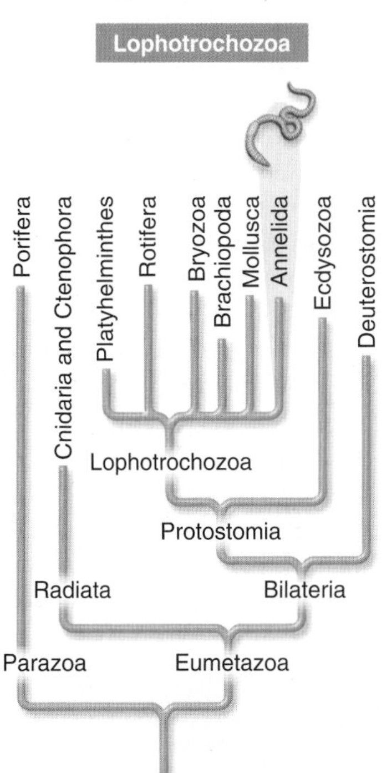

Annelids are a large phylum with about 15,000 described species. Its members include free-ranging marine worms, tube worms, the familiar earthworm, and leeches. They range in size from less than 1 mm to enormous Australian earthworms that can reach a size of 3 m. All annelids except the leeches have chitinous bristles, called **setae**, on each segment. In some, these are situated on fleshy, footlike **parapodia** (from the Greek, meaning almost feet) that are pushed into the substrate to provide traction during movement. In others, the setae are held closer to the body. Many annelid species burrow into soil or into muddy marine sediments and extract nutrients from ingested soil or mud. Some annelids also feed on dead or living vegetation, whereas others are predatory or parasitic.

Benefits of Segmentation If you look at an earthworm, you will see little rings all down its body. Indeed, the phylum name Annelida is derived from the Latin *annulus*, meaning little ring. Each ring is a distinct segment of the annelid's body, with each segment separated from the one in front and the one behind by septa (Figure 33.17). Segmentation, the division of the body into compartments that are often very similar to each other (refer back to Figure 32.9), confers at least three major advantages.

First, many components of the body are repeated in each segment, including blood vessels, nerves, and excretory and reproductive organs. Excretion is accomplished by metanephridia, paired excretory organs in every segment that extract waste from the blood and coelomic fluid, emptying it to the exterior via pores in the skin (look ahead to Figure 49.8). If the excretory organs in one segment fail, the organs of another segment will still function.

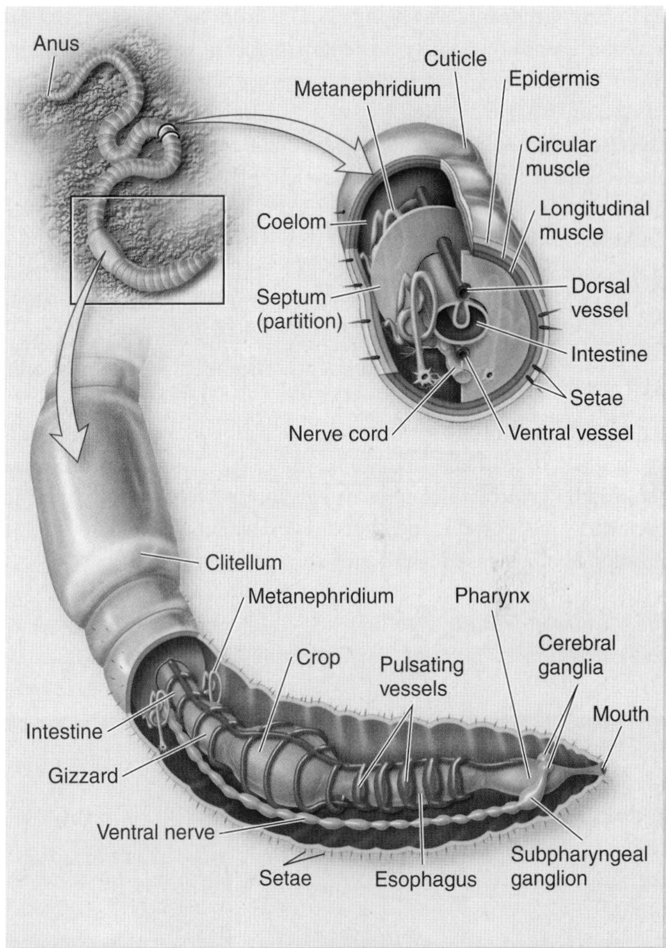

Figure 33.17 The segmented body plan of an annelid, as illustrated by an earthworm. The segmented nature of the worm is apparent internally as well as externally. Individual segments are separated by septa.

Concept Check: What are some of the advantages of segmentation?

Second, annelids possess a fluid-filled coelom that acts as a hydrostatic skeleton. In unsegmented coelomate animals, muscle contractions can distort the entire body during movement. However, such distortion is minimized in segmented animals, which allows for more effective locomotion over solid surfaces. In an earthworm, when the circular muscles around a segment contract against the hydrostatic skeleton, that segment becomes elongated. When the longitudinal muscles contract, the segment becomes compact. Waves of muscular contraction ripple down the segments, which elongate or contract independently (refer to Figure 44.1a).

Third, segmentation also permits specialization of some segments, especially at the annelid's anterior end. Therefore, although segments are often similar, they are not identical. This contrasts with tapeworms, where most segments are identical. As a rule, as we move from more animals with simple body plans to those with more complex body plans, segments become more specialized.

The Annelid Body Plan Annelids have a relatively sophisticated nervous system involving a pair of cerebral ganglia that connect to a subpharyngeal ganglion (Figure 33.17). From there, a large ventral nerve cord runs down the entire length of the body. The ventral nerve cord is unusual because it contains a few very large nerve cells called **giant axons** that facilitate high-speed nerve conduction and rapid responses to stimuli. Annelids have a double transport system. Both the circulatory system and the coelomic fluid carry nutrients, wastes, and respiratory gases, to some degree. Annelids have a closed circulatory system, with dorsal and ventral vessels connected by pairs of pulsating vessels. The dorsal vessel is the main pumping vessel. The blood of most annelid species contains the respiratory pigment hemoglobin. Respiration occurs directly through the permeable skin surface, which restricts annelids to moist environments. The digestive system is complete and unsegmented, with many specialized regions: mouth, pharynx, esophagus, crop, gizzard, intestine, and anus. Sexual reproduction involves two individuals, often of separate sexes, but sometimes hermaphrodites, which exchange sperm via internal fertilization. In some species, asexual reproduction by fission occurs, in which the posterior part of the body breaks off and forms a new individual.

The Major Annelidan Groups Recent evidence published by German evolutionary biologist Torsten Struck and colleagues in 2011 suggests that the phylum Annelida contains two major groups: the Errantia and the Sedentaria.

Members of the Errantia have many long setae bristling out of their body and are supported on footlike parapodia (**Figure 33.18a**). Most of them are free-ranging predators with well-developed eyes and powerful jaws. Many are brightly colored. In turn, most species are important prey for fishes and crustaceans.

In the Sedentaria, setae are in close proximity to the body wall, which facilitates better anchorage in tubes and burrows. Their more sedentary lifestyle is associated with reductions in head appendages. Within the Sedentaria, three types of lifestyles are apparent: tube worms, earthworms, and leeches. Tube worms are marine sedentarians that exhibit beautiful tentacle crowns for filtering food items, such as plankton, from the water column. The bulk of these worms remain hidden in a tube deep in the mud or sand.

Earthworms play a unique and beneficial role in conditioning the soil, primarily due to the effects of their burrows and excretion. Earthworms ingest soil and leaf tissue to extract nutrients and in the process create burrows in the Earth. As plant material and soil pass through the earthworm's digestive system, it is finely ground in the gizzard into smaller fragments. Once excreted, this material—called castings—enriches the soil. Because a worm can eat its own weight in soil every day, worm castings on the soil surface can be extensive. The biologist Charles Darwin was interested in earthworm activity, and his last work, *The Formation of Vegetable Mould, through the Actions of Worms, with Observations on Their Habits*, was the first detailed study of earthworm ecology. In it, he wrote, "All the fertile areas of this planet have at least once passed through the bodies of earthworms."

Leeches are primarily found in freshwater environments, but there are also some marine species as well as terrestrial species that inhabit warm, moist areas such as tropical forests. Leeches have a fixed number of segments, usually 34, though in most species the septa have disappeared. Most leeches feed as blood-sucking parasites

(a)

(b)

(c)

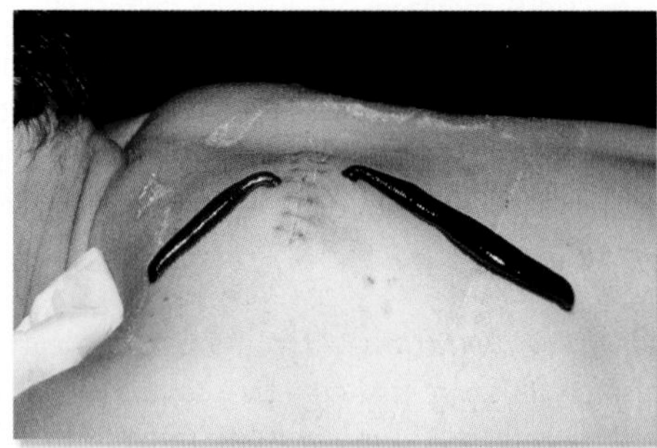

(d)

Figure 33.18 Annelids. **(a)** This free-ranging marine worm from Indonesia is a member of the group Errantia. Members of the group Sedentaria include **(b)** tube worms, **(c)** earthworms, and **(d)** leeches. This species, *Hirudo medicinalis*, is sucking blood from a hematoma, a swelling of blood that can occur after surgery.

of vertebrates. They have powerful suckers at both ends of the body, and the anterior sucker is equipped with razor-sharp jaws that can bore or slice into the host's tissues. The salivary secretion of leeches (hirudin) acts as an anticoagulant to stop the prey's blood from clotting and an anesthetic to numb the pain. Leeches can suck up to several times their own weight in blood. They were once used in the medical field in the practice of bloodletting, the withdrawal of often considerable quantities of blood from a patient in the erroneous belief that this would prevent or cure illness and disease. Even today, leeches may be used after surgeries (**Figure 33.18d**). In these cases, the blood vessels are not fully reconnected and excess blood accumulates, causing a swelling called a hematoma. The accumulated blood blocks the delivery of new blood and stops the formation of new vessels. The leeches remove the accumulated blood, and new capillaries are more likely to form, thereby giving the newly perfused tissues a better chance to heal.

Unlike cestode and trematode flatworms, which are internally parasitic and quite host-specific, leeches are generally external parasites that feed on a broad range of hosts, including fishes, amphibians, and mammals. However, there are always exceptions. *Placobdelloides jaegerskioeldi* is a parasitic leech that lives only in the rectum of hippopotamuses.

33.4 Ecdysozoa: The Nematodes and Arthropods

Learning Outcomes:

1. List the distinguishing characteristics of nematodes.
2. Describe the arthropod body plan and its major features.
3. Give examples of the arthropod subphyla Chelicerata, Myriapoda, Hexapoda, and Crustacea.
4. List the features that help account for the diversity of insect species.

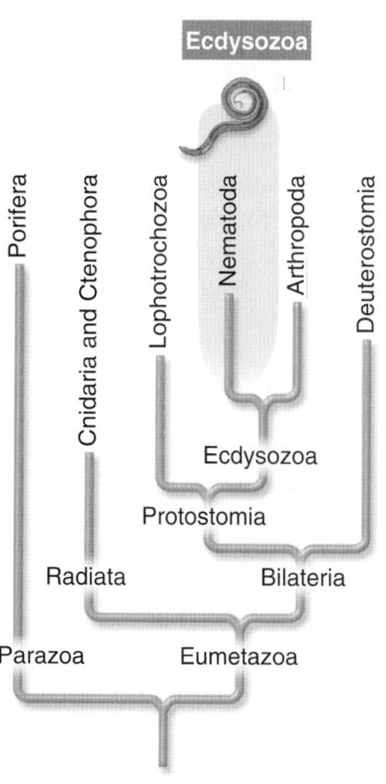

The Ecdysozoa is the sister group to the Lophotrochozoa. Although the separation is supported by molecular evidence, the Ecdysozoa is named for a morphological characteristic, the physical phenomenon of **ecdysis**, or the periodic molting of the exoskeleton (refer back to Figure 32.12). All ecdysozoans possess a **cuticle**, a nonliving cover that serves to both support and protect the animal. Once formed, however, the cuticle typically cannot increase in size, which restricts the growth of the animal inside. The solution for growth is the formation of a new, softer cuticle under the old one. The old one then splits open and is sloughed off, allowing the new, soft cuticle to expand to a bigger size before it hardens. Where the cuticle is thick, as in arthropods, it impedes the diffusion of oxygen across the skin. Such species acquire oxygen by lungs, gills, or a set of branching, air-filled tubes called tracheae. A variety of appendages specialized for locomotion evolved in many species, including legs for walking or swimming and wings for flying. The ability to shed the cuticle opened up developmental options for the ecdysozoans. For example, many species undergo a complete metamorphosis, changing from a wormlike larva into a winged adult. Animals with internal skeletons cannot do this because growth occurs only by adding more minerals to the existing skeleton. Another significant adaptation is the development of internal fertilization, which permitted species to live in dry environments.

Because of these innovations, ecdysozoans are an incredibly successful group. Of the eight ecdysozoan phyla, we will consider the two most common: the nematodes and arthropods. The grouping of nematodes and arthropods is a relatively new concept supported by molecular data and implies that the process of molting arose only once in animal evolution. In support of this, certain hormones that stimulate molting have been discovered to exist only in both nematodes and arthropods. Furthermore, in 2007, American geneticist Julie Dunning Hotopp and colleagues demonstrated the existence of lateral gene transfer (the movement of genes between distantly related organisms) between these groups. Elements of the bacterial genome *Wolbachia pipientis* were found in four insect and four nematode species. This both provides a mechanism for the acquisition of new genes by these eukaryotes and further underscores the idea that nematodes and arthropods are closely related.

The Phylum Nematoda Consists of Small Pseudocoelomate Worms Covered by a Tough Cuticle

The nematodes (from the Greek *nematos*, meaning thread), also called roundworms, are small, thin worms that range from less than 1 mm to about 5 cm (**Figure 33.19**), although some parasitic species measuring 1 m or more have been found in the placenta of sperm whales. Nematodes are ubiquitous organisms that exist in nearly all

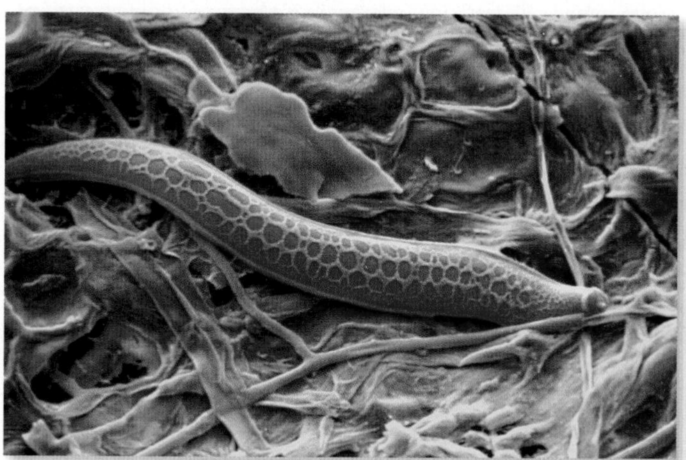

Figure 33.19 **Scanning electron micrograph of a nematode within a plant leaf.**

Concept Check: *Both nematodes and annelids are wormlike in appearance. How are they different?*

habitats, from the poles to the tropics. They are found in the soil, in both freshwater and marine environments, and inside plants and animals as parasites. A shovelful of soil may contain a million nematodes. Over 20,000 species are known, but there are probably at least five times as many undiscovered species.

The Nematode Body Plan Nematodes have several distinguishing characteristics. A tough cuticle covers the body. The cuticle is secreted by the epidermis and is made primarily of **collagen**, a structural protein also present in vertebrates. The cuticle is shed periodically as the nematode grows. Beneath the epidermis are longitudinal muscles but no circular muscles, which means that muscle contraction results in more thrashing of the body than smoother wormlike movement. The pseudocoelom functions as both a fluid-filled skeleton and a circulatory system. Diffusion of gases occurs through the cuticle. Roundworms have a complete digestive tract composed of a mouth, pharynx, intestine, and anus. The mouth often contains sharp, piercing organs called **stylets**, and the muscular pharynx functions to suck in food. Excretion of metabolic waste occurs via two simple tubules that have no cilia or flame cells.

Nematode reproduction is usually sexual, with separate males and females, and fertilization takes place internally. Females are generally larger than males and can produce prodigious numbers of eggs, in some cases, over 100,000 per day. Development in some nematodes is easily observed because the organism is transparent and the generation time is short. For these reasons, the small, free-living nematode *Caenorhabditis elegans* has become a model organism for researchers to study (refer back to Figure 19.1b and Table 21.2). This nematode has 1,090 somatic cells, but 131 die, leaving exactly 959 cells. In 2002, the Nobel Prize in Medicine or Physiology was shared by South African researcher Sydney Brenner, his American colleague Robert Horvitz, and British colleague John Sulston for their studies of the genetic regulation of development and programmed cell death in *C. elegans*. Many diseases in humans, including acquired immunodeficiency syndrome (AIDS), cause extensive cell death, whereas others, such as cancer and autoimmune diseases, reduce cell death so that cells that should die do not. Researchers are studying the process of programmed cell death in *C. elegans* in the hope of finding treatments for these and other human diseases.

Parasitic Nematodes A large number of nematodes are parasitic in humans and other vertebrates. The large roundworm *Ascaris lumbricoides* is a parasite of the small intestine that can reach up to 30 cm in length. Over a billion people worldwide carry this parasite. Although infections are most prevalent in tropical or developing countries, the prevalence of *A. lumbricoides* is relatively high in rural areas of the southeastern U.S. Eggs pass out in feces and can remain viable in the soil for years, although they require ingestion before hatching into an infective stage. Hookworms (*Necator americanus*), so named because their anterior end curves dorsally like a hook, are also parasites of the human intestine. The eggs pass out in feces, and recently hatched hookworms can penetrate the skin of a host's foot to establish a new infection. In areas with modern plumbing, these diseases are uncommon.

Pinworms (*Enterobius vermicularis*), although a nuisance, have relatively benign effects on their hosts. The rate of infection in the U.S., however, is staggering: 30% of children and 16% of adults are believed to be hosts. Adult pinworms live in the large intestine and migrate to the anal region at night to lay their eggs, which causes intense itching. The resultant scratching can spread the eggs from the hand to the mouth. In the tropics, some 250 million people are infected with *Wuchereria bancrofti*, a fairly large (100 mm) worm that lives in the lymphatic system, blocking the flow of lymph, and, in extreme cases, causing elephantiasis, an extreme swelling of the legs and other body parts (**Figure 33.20**). Females release tiny, live young called microfilariae, which are transmitted to new hosts via mosquitoes.

Figure 33.20 Elephantiasis in a human leg. The disease is caused by the nematode parasite *Wuchereria bancrofti*, which lives in the lymphatic system and blocks the flow of lymph.

Concept Check: *What other nematodes are parasitic in humans?*

The Phylum Arthropoda Contains the Spiders, Millipedes and Centipedes, Insects, and Crustaceans, Species with Jointed Appendages

The arthropods (from the Greek *arthron*, meaning joint, and *podos*, meaning foot) constitute perhaps the most successful phylum on Earth. About three-quarters of all described living species present on Earth are arthropods, and scientists have estimated they are also numerically common, with an estimated 10^{18} (a billion billion) individual organisms. The huge success of the arthropods, in terms of their sheer numbers and diversity, is related to features that permit these animals to live in all the major biomes on Earth, from the poles to the tropics, and from marine and freshwater habitats to dry land. Such features include an exoskeleton, segmentation, and jointed appendages.

The Arthropod Body Plan The body of a typical arthropod is covered by a hard cuticle, an **exoskeleton** (external skeleton), made of layers of chitin and protein. The cuticle can be extremely tough in some parts, as in the shells of crabs, lobsters, and even beetles, yet be soft and flexible in other parts, between body segments and segments of appendages, to allow for movement. In the class of arthropods called crustaceans, the exoskeleton is reinforced with calcium carbonate to make it extra hard. The exoskeleton provides protection and also a point of attachment for muscles, all of which are internal. It is also relatively impermeable to water, a feature that may have enabled many arthropods to conserve water and colonize land, in much the same way as a tough seed coat allowed plants to colonize land (see

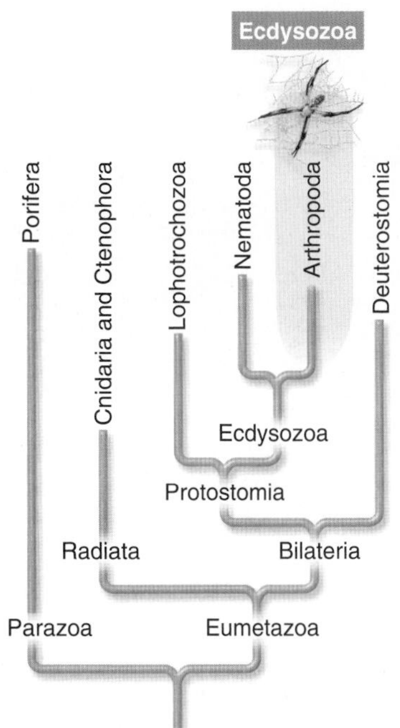

Chapter 29). From this point of view, the development of a hard cuticle was a critical innovation. It also reminds us that the ability to adapt to diverse environmental conditions can itself lead to increased organismal diversity.

Arthropods are segmented, and many of the segments bear jointed appendages. Jointed appendages permit complex movements and functions such as walking, swimming, sensing, breathing, food handling, or reproduction. These appendages are operated by muscles within each segment. In many orders, the body segments have become fused into functional units, or **tagmata**, such as the head, thorax, and abdomen of an insect (Figure 33.21). Cephalization is extensive, and arthropods have well-developed sensory organs, including organs of sight, touch, smell, hearing, and balance. Arthropods have compound eyes composed of many independent visual units called **ommatidia** (singular, ommatidium) (look ahead to Figure 43.14). Together, these lenses render a mosaic-like image of the environment. Some species, particularly some insects, possess additional simple eyes, or ocelli, that are probably only capable of distinguishing light from dark.

The arthropod brain consists of two or three cerebral ganglia connected to several smaller ventral nerve ganglia. Like most mollusks, arthropods have an open circulatory system (look ahead to Figure 47.2), in which hemolymph is pumped from a tubelike heart into the aorta or short arteries and then into the open sinuses that coalesce to form a cavity called the hemocoel. From the hemocoel, gases and nutrients from the hemolymph diffuse into tissues. The hemolymph flows back into the heart via pores, called ostia, that are equipped with valves.

Because the cuticle impedes the diffusion of gases through the body surface, arthropods possess special organs that permit gas exchange. In aquatic arthropods, these consist of feathery gills that have an extensive surface area in contact with the surrounding water. Terrestrial species have a highly developed **tracheal system** (look ahead to Figure 48.7). On the body surface, pores called **spiracles** provide openings to a series of finely branched air tubes within the body called trachea. The tracheal system delivers oxygen directly to tissues and cells, and the circulatory system does not play a role in gas exchange. Some spiders have book lungs, consisting of a series of sheetlike structures, like the pages of a book, extending into a hemolymph-filled chamber on the underside of the abdomen. Gases also diffuse across thin areas of the cuticle.

The digestive system is complex and often includes a mouth, crop, stomach, intestine, and rectum. The stomach has glands called digestive cecae that secrete digestive enzymes. Excretion is accomplished by specialized metanephridia or, in insects and some other taxa, by **Malpighian tubules**, extensive tubes that extend from the digestive tract into body cavity, where they are surrounded by hemolymph (look ahead to Figure 49.9). Nitrogenous wastes are absorbed by the tubules and emptied into the gut, where the intestine and rectum reabsorb water and salts and the waste is excreted through the anus. This excretory system, allowing the retention of water, was another critical innovation that permitted the colonization of land by arthropods.

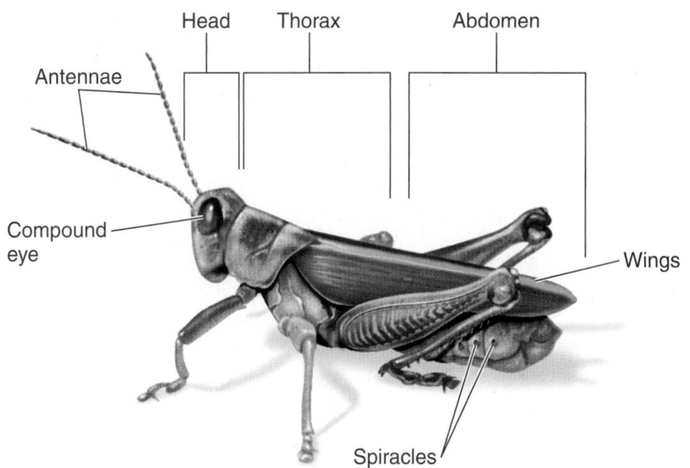

(a) External anatomy

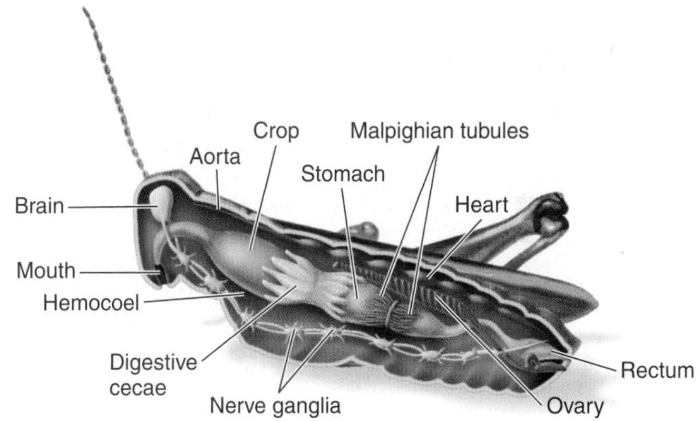

(b) Internal anatomy

Figure 33.21 **Body plan of an arthropod, as represented by a grasshopper.**

BioConnections: Look forward to Figure 49.9. Why did the Malpighian tubule system play a key role in the colonization of land by insects and other arthropods?

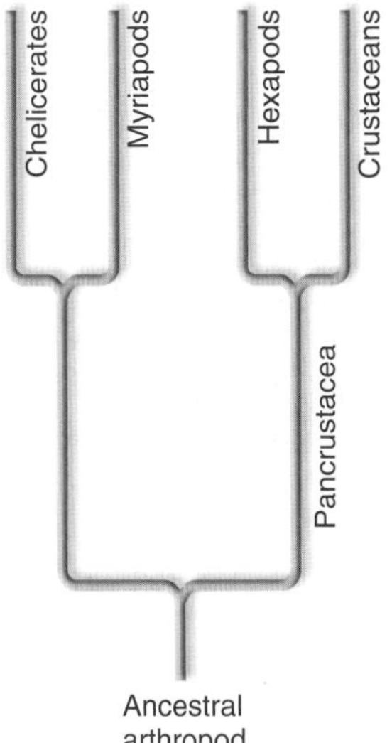

Arthropod Diversity The history of arthropod classification is extensive and active. Although many classifications have been proposed, a 1995 study of the mitochondrial DNA of arthropod species by American geneticist Jeffrey Boore and colleagues suggests a phylogeny with five main subphyla: one now-extinct

Table 33.4 Main Subphyla and Characteristics of Arthropods

Subphyla and examples (est. # of species)	Class characteristics
Chelicerata: spiders, scorpions, mites, ticks, horseshoe crabs, and sea spiders (74,000)	Body usually with cephalothorax and abdomen only; six pairs of appendages, including four pairs of legs, one pair of fangs, and one pair of pedipalps; terrestrial; predatory or parasitic
Myriapoda: millipedes and centipedes (13,000)	Body with head and highly segmented trunk. In millipedes, each segment with two pairs of walking legs; terrestrial; herbivorous. In centipedes, each segment with one pair of walking legs; terrestrial; predatory, poison jaws
Hexapoda: insects such as beetles, butterflies, flies, fleas, grasshoppers, ants, bees, wasps, termites and springtails (>1 million)	Body with head, thorax, and abdomen; mouthparts modified for biting, chewing, sucking, or lapping; usually with two pairs of wings and three pairs of legs; mostly terrestrial, some freshwater; herbivorous, parasitic, or predatory
Crustacea: crabs, lobsters, shrimp (45,000)	Body of two to three parts; three or more pairs of legs; chewing mouthparts; usually marine

subphyla, Trilobita (trilobites), and four living subphyla: Chelicerata (spiders and scorpions), Myriapoda (millipedes and centipedes), Hexapoda (insects and relatives), and Crustacea (crabs and relatives) (**Table 33.4**). Boore's research showed that the Trilobita were among the earliest-diverging arthropods. The lineage then split into two groups. One, often referred to as the Pancrustacea, contains the insects and crustaceans. The other, with no overarching name, contains the myriapods and chelicerates. Molecular evidence thus suggests insects are more closely related to crustaceans than they are to spiders or millipedes and centipedes.

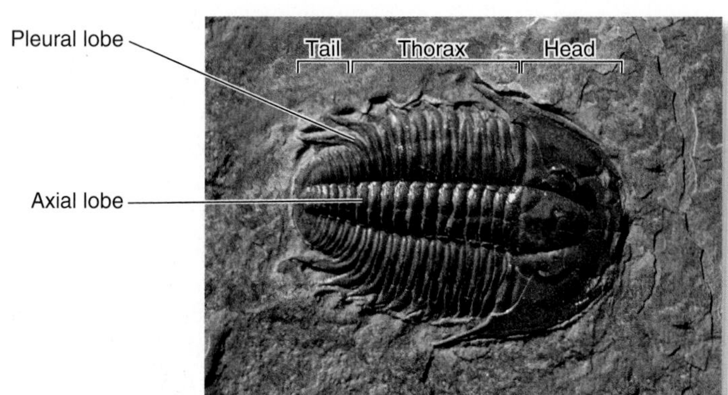

Figure 33.22 A fossil trilobite. About 4,000 fossil species of these early arthropods, including *Huntonia huntonesis* shown here, at about 20 cm long, have been described.

Subphylum Trilobita: Extinct Early Arthropods The trilobites were among the earliest arthropods, flourishing in shallow seas of the Paleozoic era, some 500 mya, and dying out about 250 mya. Most trilobites were bottom feeders and were generally 3–10 cm in size, although some reached almost 1 m in length (**Figure 33.22**). They had three main tagmata: the head, thorax, and tail. Trilobites also had two dorsal grooves that divided the body longitudinally into three lobes—an axial lobe and two pleural lobes—a structural characteristic giving the class its name. Most of the body segments showed little specialization. In contrast, later-diverging arthropods developed specialized appendages on many segments, including appendages for grasping, walking, and swimming.

Subphylum Chelicerata: The Spiders, Scorpions, Mites, and Ticks The Chelicerata consists mainly of the class Arachnida, which contains predatory spiders and scorpions as well as the ticks and mites, some of which are blood-sucking parasites that feed on vertebrates. The two other living classes are the Merostomata, the horseshoe crabs (four species), and the Pycnogonida, the sea spiders (1,000 species), both of which are marine, reflecting the groups' marine ancestry. All species have a body consisting of two tagmata: a fused head and thorax, called a **cephalothorax**, and an abdomen (**Figure 33.23**). All species also possess six pairs of appendages: the

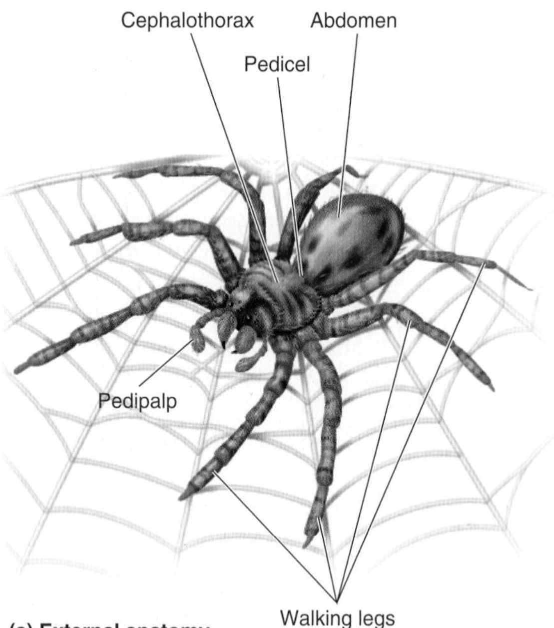

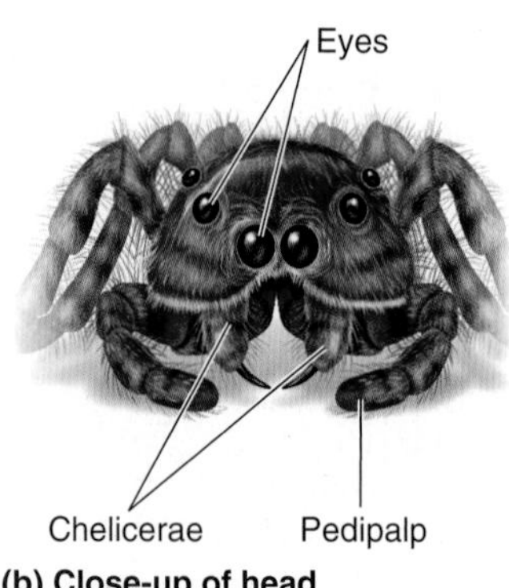

Figure 33.23 Spider morphology. All species possess just two tagmata: a cephalothorax and an abdomen.

(a) Normal web

(b) Web spun by spider fed with prey containing caffeine

(c) Web spun by spider fed with prey containing marijuana

Figure 33.24 **Spider-web construction by normal and drugged spiders.**

BIOLOGY PRINCIPLE **Biology is an experimental science.** Some scientists have suggested using web-spinning spiders to test substances for the presence of drugs or even to indicate environmental contamination.

chelicerae, or fangs; a pair of **pedipalps**, which have various sensory, predatory, or reproductive functions; and four pairs of walking legs.

In spiders (order Araneae), the cephalothorax and abdomen are joined by a **pedicel**, a narrow, waistlike point of attachment (see Figure 33.23). The fangs are supplied with venom from poison glands. Most spider bites are harmless to humans, although they are very effective in immobilizing and/or killing their insect prey. Venom from some species, including the black widow (*Latrodectus mactans*) and the brown recluse (*Loxosceles reclusa*), are potentially, although rarely, fatal to humans. The toxin of the black widow is a neurotoxin, which interferes with the functioning of the nervous system, whereas that of the brown recluse is hemolytic, meaning it destroys red blood cells around the bite. After the spider has subdued its prey, it pumps digestive fluid into the tissues via the fangs and sucks out the partially digested meal.

Spiders have abdominal silk glands, called spinnerets, and many spin webs to catch prey (**Figure 33.24a**). The silk is a protein that stiffens after extrusion from the body because the mechanical shearing causes a change in the organization of the amino acids. Silk is stronger than steel of the same diameter and is more elastic than Kevlar, the material used in bulletproof vests. Each spider family constructs a characteristic size and style of web and can do it perfectly on its first attempt, indicating that web spinning is an innate (instinctual) behavior (see Chapter 55). Spiders also use silk to wrap up prey and to construct egg sacs. Interestingly, spiders that are fed drugged prey spin their webs differently than undrugged spiders (**Figure 33.24b,c**). Some scientists have suggested that web-spinning spiders be used to test substances for the presence of drugs or even to indicate environmental contamination. Not all spiders use silk extensively. Some spiders, including the wolf spider, actively pursue their prey (**Figure 33.25a**).

Scorpions (order Scorpionida) are generally tropical or subtropical animals that feed primarily on insects, though they may eat spiders and other arthropods as well as smaller reptiles and mice. Their pedipalps are modified into large claws, and their abdomen tapers into a

(a) Wolf spider

(b) Scorpion with young

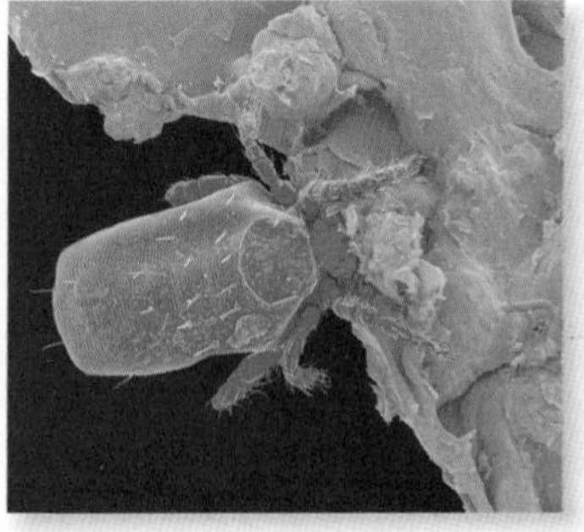

(c) Chigger mite

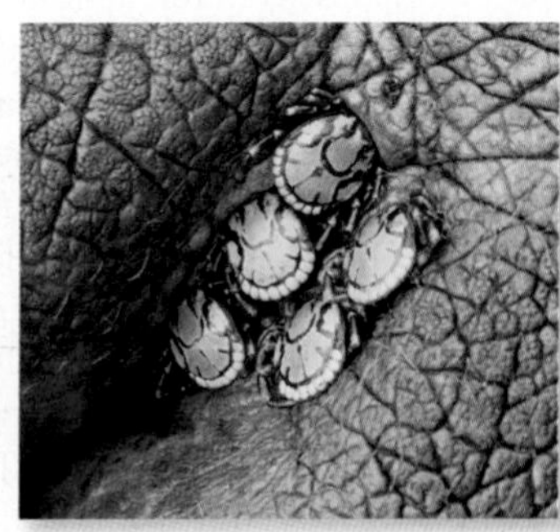

(d) Bont ticks

Figure 33.25 **Common arachnids.** **(a)** This wolf spider (*Lycosa tarantula*) does not spin a web but instead runs after its prey. Note the pedipalps, which look like short legs. **(b)** The Cape thick-tailed scorpion (*Parabuthus capensis*) is highly venomous and carries its white young on its back. **(c)** SEM of a chigger mite (*Trombicula alfreddugesi*) that can cause irritation to human skin and spread disease. **(d)** These South African bont ticks (*Amblyomma hebraeum*) are feeding on a white rhinoceros.

Concept Check: *What is one of the main characteristics distinguishing arachnids from insects?*

stinger, which is used to inject venom. Although the venom of most North American species is generally not fatal to humans, that of the *Centruroides* genus from deserts in the U.S. Southwest and Mexico can be deadly. Fatal species are also found in India, Africa, and other countries. Unlike spiders, which lay eggs, scorpions bear live young that the mother then carries around on her back until they have their first molt (**Figure 33.25b**).

In mites and ticks (order Acari), the two main body segments (cephalothorax and abdomen) are fused and appear as one large segment. Many mite species are free-living scavengers that feed on dead plant or animal material. Other mites are serious pests on crops, and some, like chiggers (*Trombicula alfreddugesi*), are parasites of humans that can spread diseases such as typhus (**Figure 33.25c**). Chiggers are parasites only in their larval stage. Chiggers do not bore into the skin; their bite and salivary secretions cause skin irritation. *Demodex brevis* is a hair-follicle mite that is common in animals and humans. The mite is estimated to be present in over 90% of adult humans. Although the mite causes no irritation in most humans, *Demodex canis* causes the skin disease known as mange in domestic animals, particularly dogs.

Ticks are larger organisms than mites, and all are ectoparasitic, feeding on the body surface, on vertebrates. Their life cycle includes attachment to a host, sucking blood until they are replete, and dropping off the host to molt (**Figure 33.25d**). Ticks can carry a huge variety of viral and bacterial diseases, including Lyme disease, a bacterial disease so named because it was first found in the town of Lyme, Connecticut, in the 1970s.

Subphylum Myriapoda: The Millipedes and Centipedes Myriapods have one pair of antennae on the head and three pairs of appendages that are modified as mouth parts, including mandibles that act like jaws. The millipedes and centipedes, both wormlike arthropods with legs, are among the earliest terrestrial animal phyla known. Millipedes (class Diplopoda) have two pairs of legs per segment, as their class name denotes (from the Latin *diplo*, meaning two, and *podos*, meaning feet), not 1,000 legs, as their common name suggests (**Figure 33.26a**). They are slow-moving herbivorous creatures that eat decaying leaves and other plant material. When threatened, the millipede's response is to roll up into a protective coil. Many millipede species also have glands on their underside that can eject a variety of toxic, repellent secretions. Some millipedes are brightly colored, warning potential predators that they can protect themselves.

Class Chilopoda (from the Latin *chilo*, meaning lip, and *podos*, meaning feet), or centipedes, are fast-moving carnivores that have one pair of walking legs per segment (**Figure 33.26b**). The head has many sensory appendages, including a pair of antennae and three pairs of appendages modified as mouthparts, including powerful claws connected to poison glands. The toxin from venom of some of the larger species, such as *Scolopendra heros*, is powerful enough to cause pain in humans. Most species do not have a waxy waterproofing layer on their cuticle and so are restricted to moist environments under leaf litter or in decaying logs, usually coming out at night to actively hunt their prey.

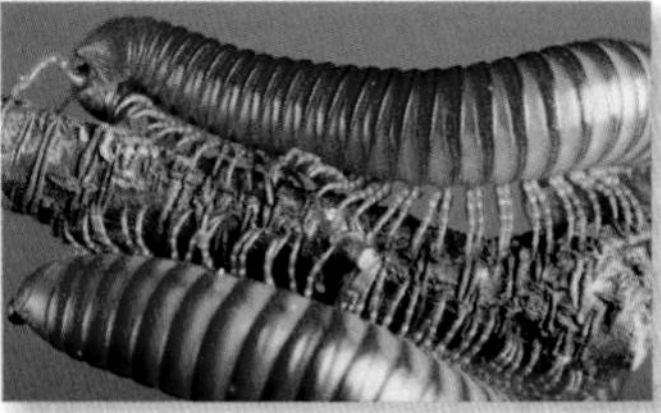
(a) Two millipedes

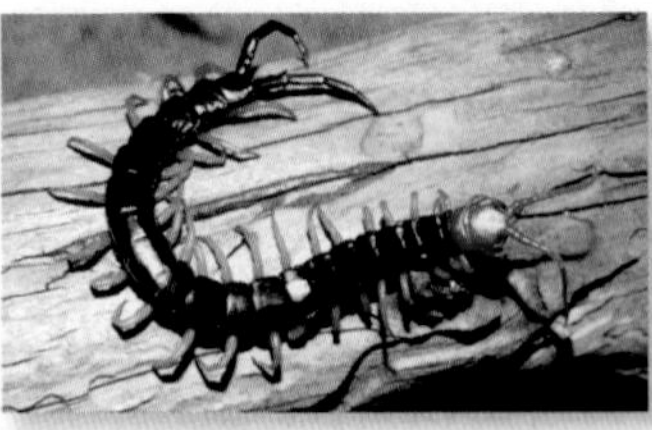
(b) A centipede

Figure 33.26 Millipedes and centipedes. **(a)** Millipedes have two pairs of legs per segment. **(b)** The venom of the giant centipede (*Scolopendra heros*) is known to produce significant swelling and pain in humans.

Subphylum Hexapoda: A Diverse Array of Insects and Close Relatives Hexapods are six-legged arthropods. Most are insects, but there are a few earlier-diverging noninsect hexapods, including soil-dwelling groups such as collembolans, that molecular studies have shown represent a separate but related lineage. Insects are in a class by themselves (Insecta), literally and figuratively. There are more species of insects than all other species of animal life combined. One million species of insects have been described, and, according to best estimates, 2–5 million more species await description. At least 90,000 species of insects have been identified in the U.S. and Canada alone. Genetic barcoding can help resolve many taxonomic dilemmas between closely related species.

Insects are the subject of an entire field of scientific study, **entomology**. They are studied in large part because of their significance as pests of the world's agricultural crops and carriers of some of the world's most deadly diseases. Insects live in all terrestrial habitats, and virtually all species of plants are fed upon by at least one, usually tens, and sometimes, in the case of large trees, hundreds of insect species. Because approximately one-quarter of the world's crops are lost annually to insects, we are constantly trying to find ways to reduce pest densities. Insect pest reduction often involves chemical control (the use of pesticides) or biological control (the use of living organisms). Many species of insects are also important pests or parasites of humans and livestock, both by their own actions and as vectors of diseases such as malaria and sleeping sickness.

In contrast, insects also provide us with many types of essential biological services. We depend on insects such as honeybees, butterflies, and moths to pollinate our crops. Bees also produce honey, and silkworms are the source of silk fiber. Despite the revulsion they provoke in us, fly larvae (maggots) are important in the decomposition process of both dead plants and animals. In addition, we use insects in the biological control of other insects.

Of paramount importance to the success of insects was the evolution of wings, a feature possessed by no other arthropod and indeed no other living animal except birds and bats. Unlike vertebrate wings, however, insect wings are actually outgrowths of the body wall cuticle and are not true segmental appendages. This means that insects still have all their walking legs. Insects are thus like the mythological horse Pegasus, which sprouted wings out of its back while retaining all four legs. In contrast, birds and bats have one pair of appendages (arms) modified for flight, which leaves them considerably less agile on the ground.

The great diversity of insects is illustrated by the fact that there are 35 different orders, some of which have over 100,000 species. The

Table 33.5 Main Orders and Characteristics of Insects

Order and examples (est. # of species)	Order characteristics
Coleoptera: beetles, weevils (500,000)	Two pairs of wings (front pair thick and leathery, acting as wing cases, back pair membranous); armored exoskeleton; biting and chewing mouthparts; complete metamorphosis; largest order of insects
Hymenoptera: ants, bees, wasps (190,000)	Two pairs of membranous wings; chewing or sucking mouthparts; many have posterior stinging organ on females; complete metamorphosis; many species social; important pollinators
Diptera: flies, mosquitoes (190,000)	One pair of wings with hind wings modified into halteres (balancing organs); sucking, piercing, or lapping mouthparts; complete metamorphosis; larvae are grublike maggots in various food sources; some adults are disease vectors
Lepidoptera: butterflies, moths (180,000)	Two pairs of colorful wings covered with tiny scales; long tubelike tongue for sucking; complete metamorphosis; larvae are plant-feeding caterpillars; adults are important pollinators
Hemiptera: true bugs; assassin bug, bedbug, chinch bug, cicada (100,000)	Two pairs of membranous wings; piercing or sucking mouthparts; incomplete metamorphosis; many plant feeders; some predatory or blood feeders; vectors of plant diseases
Orthoptera: crickets, grasshoppers (30,000)	Two pairs of wings (front pair leathery, back pair membranous); chewing mouthparts; mostly herbivorous; incomplete metamorphosis; powerful hind legs for jumping
Odonata: damselflies, dragonflies (6,500)	Two pairs of long, membranous wings; chewing mouthparts; large eyes; predatory on other insects; incomplete metamorphosis; nymphs aquatic; considered early-diverging insects
Siphonaptera: fleas (2,600)	Wingless, laterally flattened; piercing and sucking mouthparts; adults are bloodsuckers on birds and mammals; jumping legs; complete metamorphosis; vectors of plague
Phthiraptera: sucking lice (2,400)	Wingless ectoparasites; sucking mouthparts; flattened body; reduced eyes; legs with clawlike tarsi for clinging to skin; incomplete metamorphosis; very host specific; vectors of typhus
Isoptera: termites (2,000)	Two pairs of membranous wings when present; some stages wingless; chewing mouthparts; social species; incomplete metamorphosis

most common of the orders are discussed in **Table 33.5**. Although all insects have six legs, different orders have slightly different wing structures, and many of the orders are based on wing type (their names often include the root *pter-*, from the Greek *pteron*, meaning wing). In beetles (Coleoptera), only the back pair of wings is functional, as the front wings have been hardened into protective shell-like coverings under which the back pair folds when not in use. Wasps and bees (Hymenoptera) have two pairs of wings hooked together that move as one wing. Flies (Diptera) possess only one pair of wings (the front pair); the back pair has been modified into a small pair of balancing organs, called halteres, that act like miniature gyroscopes. Butterflies (Lepidoptera) have wings that are covered in scales (from the Greek *lepido*, meaning scale); other insects generally have clear, membranous wings. In ant and termite colonies, the queen and the drones (males) retain their wings, whereas female individuals called workers have lost theirs. Other species, such as fleas and lice, are completely wingless.

Insects in different orders have also evolved a variety of mouthparts in which the constituent parts, the mandibles and maxillae, are modified for different functions (**Figure 33.27**). Many of these mouthparts are modified walking appendages and are bilaterally paired. As a result, the jaws of many insects, such as grasshoppers, move in a side-to-side motion, rather than up and down as human jaws do. Grasshoppers, beetles, dragonflies, and many others have mouthparts adapted for chewing. Mosquitoes and many plant pests have mouthparts adapted for piercing and sucking. Butterflies and moths have a coiled tongue (**proboscis**) that can be uncoiled, enabling them to drink nectar from flowers. Some flies have lapping, sponge-like mouthparts that sop up liquid food. Their varied mouthparts are adaptations that allow insects to specialize their feeding on virtually anything: plant matter, decaying organic matter, and other living animals. The biological diversity of insects is therefore related to environmental diversity, in this case, the variety of foods that insects eat. Parasitic insects attach themselves to other species, and there are even some insect parasites (called hyperparasites) that feed on other parasites, as noted by the 18th-century English poet and satirist Jonathan Swift:

Big fleas have little fleas

upon their backs to bite 'em;

and little fleas have lesser fleas

and so, ad infinitum.

All insects have separate sexes, and fertilization is internal. During development, the majority (approximately 85%) of insects undergo a change in body form known as **complete metamorphosis** (from the Greek *meta*, meaning change, and *morph*, meaning form) (**Figure 33.28a**). Animals that undergo complete metamorphosis have four types of stages: egg, larva, pupa, and adult. The dramatic body transformation from larva to adult occurs in the pupa stage. The larval stage is often spent in an entirely different habitat from that of the adult, and larval and adult forms use different food sources. Consequently, they do not compete directly for the same resources. Furthermore, metamorphosis permits existence of a dispersive stage, usually a winged adult, and a feeding stage, often a sac-like caterpillar or larva.

The remaining insects undergo **incomplete metamorphosis**, in which change is more gradual (**Figure 33.28b**). Incomplete metamorphosis has only three types of stages: egg, nymph, and adult. Young insects, called nymphs, look like miniature adults when they hatch from their eggs, but usually don't have wings. As they grow and feed, they shed their exoskeleton and replace it with a larger one several times, each time entering a new instar, or stage of growth. When the insects reach their adult size, they have also grown wings.

Some insects, such as bees, wasps, ants, and termites, have developed complex social behavior and live cooperatively in underground or aboveground nests. Such colonies exhibit a division of labor, in that some individuals forage for food and care for the brood (workers), others protect the nest (soldiers), and some only reproduce (the queen and drones) (**Figure 33.29**).

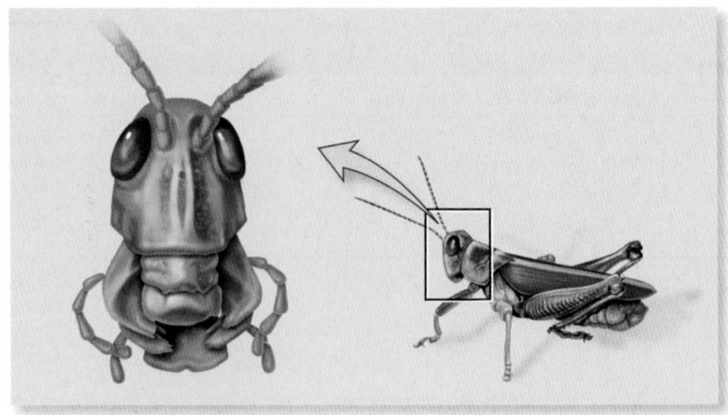

(a) Chewing (grasshopper)

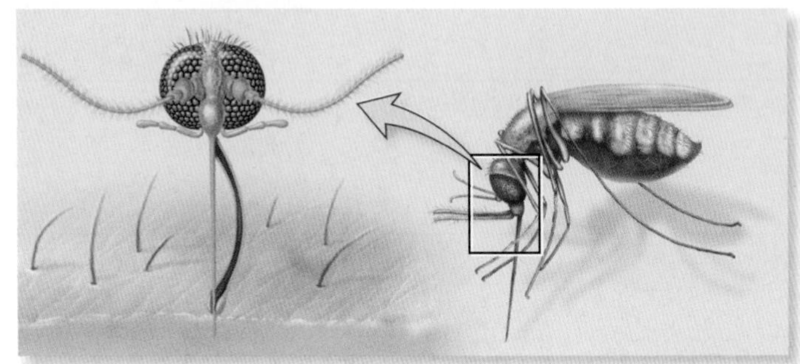

(b) Piercing and blood sucking (mosquito)

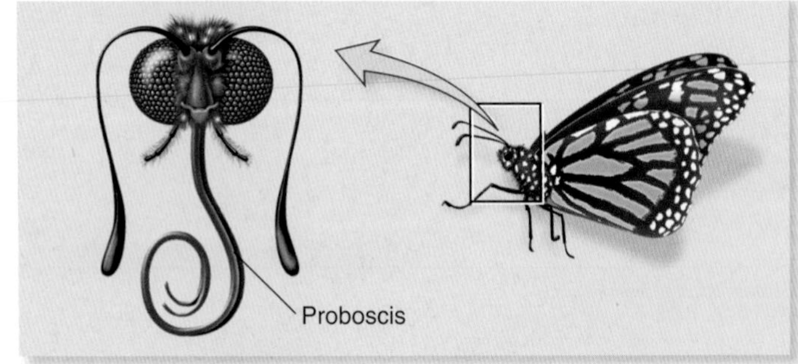

(c) Nectar sucking (butterfly)

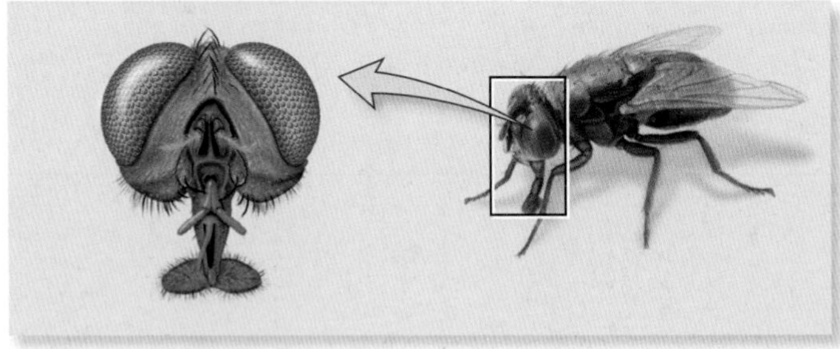

(d) Sponging liquid (housefly)

Figure 33.27 A variety of insect mouthparts. Insect mouthparts have become modified in ways that allow insects to feed by a variety of methods, including **(a)** chewing (Orthoptera, Coleoptera, and others), **(b)** piercing and blood sucking (Diptera), **(c)** nectar sucking (Lepidoptera), and **(d)** sponging up liquid (Diptera).

Concept Check: Insects have a variety of mouthparts. Name two other key insect adaptations.

BioConnections: Look forward to Figure 43.24. Some insects do not taste their food with their mouths, so how do they taste?

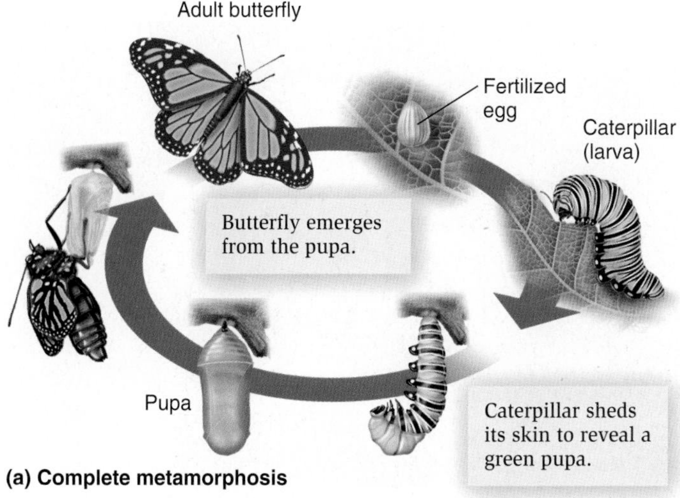

(a) Complete metamorphosis

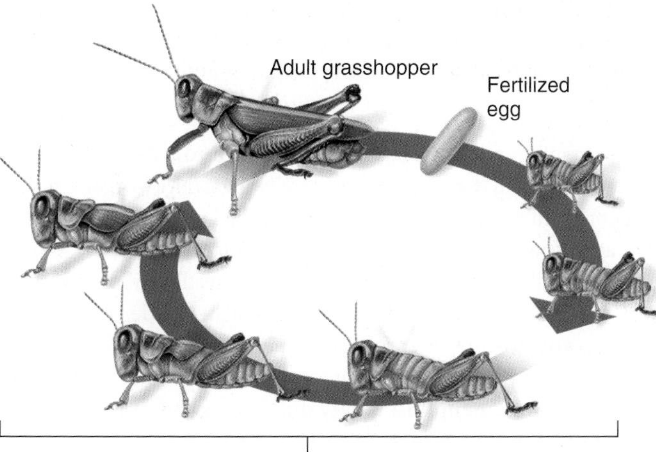

(b) Incomplete metamorphosis

Figure 33.28 Metamorphosis. **(a)** Complete metamorphosis, as illustrated by the life cycle of a monarch butterfly. The adult butterfly has a completely different appearance than the larval caterpillar. **(b)** Incomplete metamorphosis, as illustrated by the life cycle of a grasshopper. The eggs hatch into nymphs, essentially miniature versions of the adult.

(a) Worker and soldier ants **(b) Queen ant**

Figure 33.29 The division of labor in insect societies. Individuals from the same insect colony may appear very different. Among these army ants (*Eciton burchelli*) from Paraguay, there are **(a)** workers that forage for the colony, soldiers (with large mandibles) that protect the colony from predators, and **(b)** the queen, which reproduces and lays eggs.

Genomes & Proteomes Connection

Barcoding: A New Tool for Classification

The International Barcoding of Life project, developed in 2003 by Canadian biologist Dr. Paul Hebert of the University of Guelph, is a broader initiative that seeks to create a digital identification system for all life forms. Hebert made the analogy that the large diversity of products in a grocery store can each be distinguished with a relatively small barcode. Though the diversity of the world's animal species is considerably larger, Dr. Hebert reasoned that all species could be distinguished using their DNA. The complete genome would be too large to analyze rapidly, so Dr. Hebert suggested analyzing a small piece of DNA of all species. The DNA sequence he proposed is the first 684 base pairs of a gene called *CO1*, for cytochrome oxidase, an enzyme in the electron transport of mitochondria (refer back to Figure 7.9). All animals have this gene, and it occurs in the mitochondria. A key element is that although this part of the *CO1* gene varies widely between species, it hardly varies at all between individuals of the same species—only 2%.

Insects are a very species-rich taxon. For example, there are at least 3,500 species of mosquitoes, many of them hard to tell apart. Some mosquitoes transmit deadly diseases such as malaria and yellow fever and are subject to stringent control measures in many countries. Other mosquito species are relatively benign. Distinguishing mosquito species in the field is not easy. The Mosquito Barcoding Initiative aims to catalog each mosquito species by using analysis of its mtDNA and thus build up a DNA bar code database. Field researchers will be able to quickly analyze the DNA of a particular species and identify it based on existing bar codes. Appropriate control measures can then be instigated against the species if it is a disease carrier.

For blood-feeding insects, scientists can also bar code their blood meals, target their feeding preferences, and optimize control measures accordingly. For example, tsetse flies transmit tryptosomiasis, a parasitic disease that causes sleeping sickness in humans, and African animal trypanosomiasis, a disease that leads to serious economic losses in livestock. Tsetse flies are hard to track in nature because of their solitary habits and secretive nature, hiding in bushes and waiting for prey to pass by. Capturing individuals and barcoding their blood meals avoids the necessity of costly and difficult field behavioral studies. If cattle are found to be the source of most blood, spraying them with insecticides is an effective control strategy. If wildlife such as buffalo, giraffe, elephants, and warthogs are the source, then trapping devices are used.

Dr. Hebert foresees the day when all species can be identified by their DNA barcode. A huge advantage is that only a small piece of tissue is necessary. The specimen can come from adult or immature individuals, a great help considering much insect taxonomy is based solely on adults. Many scientists anticipate the day when handheld field barcoding identification devices appear. At the moment, barcoding involves a laboratory analysis taking about an hour and costing $2.00 per sample. As of 2011, over 100,000 species have been barcoded, and the target for 2015 is the barcoding of half a million species.

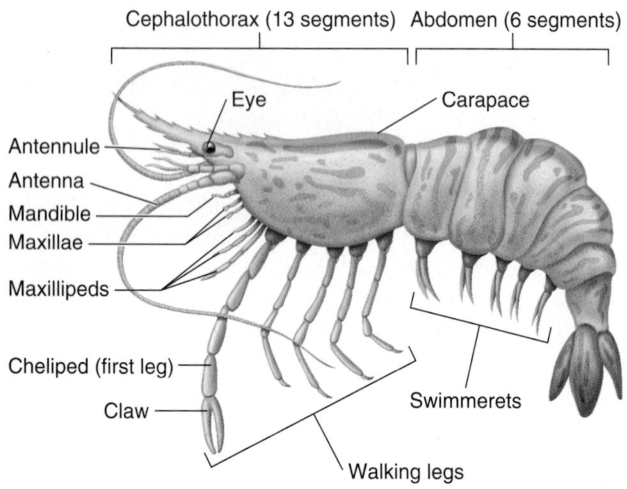

Figure 33.30 Body plan of a crustacean, as represented by a shrimp.

BioConnections: Look forward to Figure 43.10. Where are a crustacean's organs of balance located?

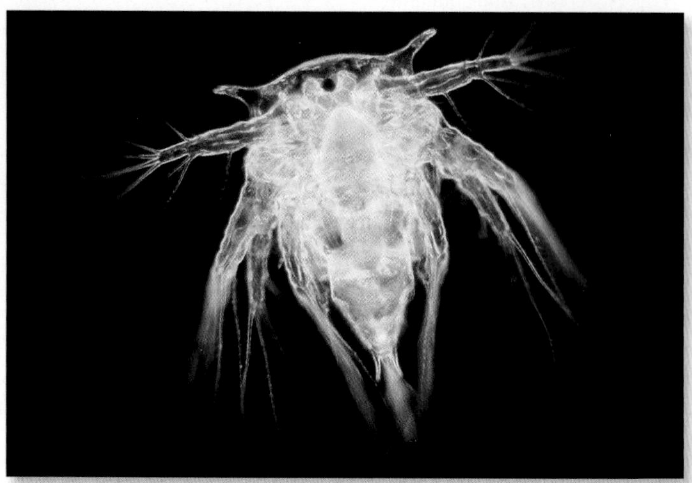

Figure 33.31 Crustacean larva. The nauplius is a distinct larval type possessed by most crustaceans, which molts several times before reaching maturity. Many are less than one hundredth of a millimeter long.

Subphylum Crustacea: Crabs, Lobsters, Barnacles, and Shrimp

The crustaceans are common inhabitants of marine environments, although some species live in fresh water and a few are terrestrial. Many species, including crabs, lobsters, crayfish, and shrimp, are economically important food items for humans; smaller species are important food sources for other predators.

The crustaceans are unique among the arthropods in that they possess two pairs of antennae at the anterior end of the body—the antennule (first pair) and antenna (second pair) (Figure 33.30). In addition, they have three or more sensory and feeding appendages that are modified mouthparts: the mandibles, maxillae, and maxillipeds. These are followed by walking legs and, often, additional abdominal appendages, called swimmerets, and a powerful tail. In some orders, the first pair of walking legs, or chelipeds, is modified to form powerful claws. The head and thorax are often fused together, forming the cephalothorax. In many species, the cuticle covering the head extends over most of the cephalothorax, forming a hard protective fold called the **carapace**. For growth to occur, a crustacean must shed the entire exoskeleton.

Many crustaceans are predators, but others are scavengers, and some, such as barnacles, are filter feeders. Gas exchange typically occurs via gills, and crustaceans, like other arthropods, have an open circulatory system. Crustaceans possess two excretory organs: antennal glands and maxillary glands, both modified metanephridia, which open at the bases of the antennae and maxillae, respectively. Reproduction usually involves separate sexes, and fertilization is internal. Most species carry their eggs in brood pouches under the female's body. Eggs of most species produce larvae that must go through many different molts prior to assuming adult form. The first of these larval stages, called a **nauplius**, is very different in appearance from the adult crustacean (Figure 33.31).

There are many crustacean clades, but most are small and obscure, although many feature prominently in marine food chains, a depiction of feeding relationships between organisms in which each organism of the chain feeds on and derives energy from the member below it. Crustacean clades include the Ostracoda, Copepoda, Cirripedia, and Malacostraca. Ostracods are tiny creatures that superficially resemble clams, and copepods are tiny and abundant planktonic crustaceans, both of which are a food source for filter-feeding organisms and small fish. The clade Cirripedia is composed of the barnacles, crustaceans whose carapace forms calcified plates that cover most of the body (Figure 33.32a). Their legs are modified into feathery filter-feeding structures.

(a) Goose barnacles—order Cirripedia

(b) Pill bug—order Isopoda

(c) Coral crab—order Decapoda

Figure 33.32 Common crustaceans. **(a)** Goose barnacles (*Lepas anatifera*). **(b)** Pill bug, or wood louse (*Armadillium vulgare*). **(c)** Coral crab (*Carpilius maculates*).

Malacostracans are divided into many orders. Euphausiacea are shrimplike krill that grow to about 3 cm and provide a large part of the diet of many whales, seals, penguins, fish, and squid. The order Isopoda contains many small species that are parasitic on marine fishes. There are also terrestrial isopods, better known as pill bugs, or wood lice, that retain a strong connection to water and need to live in moist environments such as leaf litter or decaying logs (Figure 33.32b). When threatened, they curl up into a tight ball, making it difficult for predators to get a grip on them.

The most famous Malacostracan order is the Decapoda, which includes the crabs and lobsters, the largest crustacean species (Figure 33.32c). As their name suggests, these decapods have 10 walking legs (five pairs), although the first pair is invariably modified to support large claws. Most decapods are marine, but there are many freshwater species, such as crayfish, and in hot, moist tropical areas, even some terrestrial species called land crabs (refer back to the opening paragraph in Chapter 32). The larvae of many larger crustaceans are planktonic and grow to about 3 cm. These are abundant in some oceans and are a staple food source for many species.

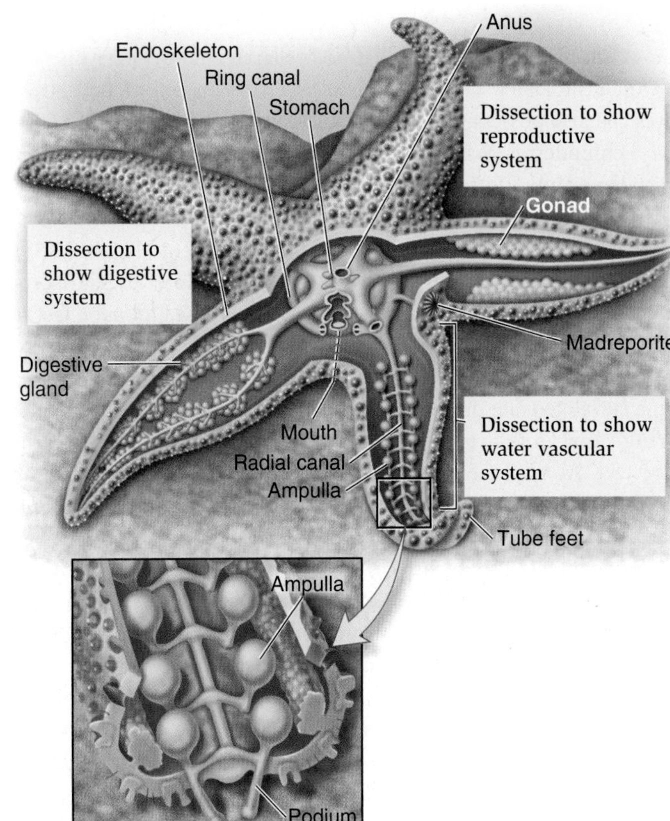

Figure 33.33 **Body plan of an echinoderm, as represented by a sea star.** The arms of this sea star have been dissected to different degrees to show the echinoderm's various organs. The inset shows a close-up view of the tube feet, part of the water vascular system characteristic of echinoderms.

Concept Check: *Echinoderms and chordates are both deuterostomes. What are three defining features of deuterostomes?*

33.5 Deuterostomia: The Echinoderms and Chordates

Learning Outcomes:

1. Identify the distinguishing characteristics of echinoderms.
2. List the four critical innovations in the body plan of chordates.
3. Describe the two invertebrate subphylum of the phylum Chordata and their relationship to the vertebrates.

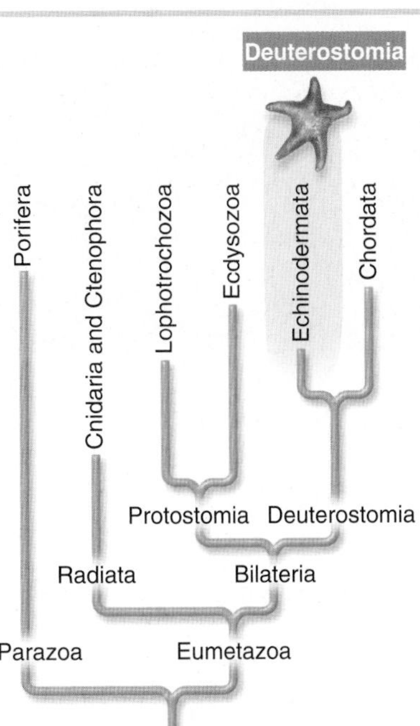

As we explored in Chapter 32, the deuterostomes are grouped together because they share similarities in patterns of development (refer back to Figure 32.7). Molecular evidence also supports a deuterostome clade. All animals in the phylum Chordata (from the Greek *chorde*, meaning string, referring to the spinal cord), which includes the vertebrates, are deuterostomes. Interestingly, so is one invertebrate group, the phylum Echinodermata, which includes the sea stars, sea urchins, and sea cucumbers. Although there are far fewer phyla and species of deuterostomes than protostomes, the deuterostomes are generally much more familiar to us. After all, humans are deuterostomes. We will conclude our discussion of invertebrate biology by turning our attention to the invertebrate deuterostomes. In this section, we will explore the phylum Echinodermata and then introduce the phylum Chordata, looking in particular at its distinguishing characteristics and at its two invertebrate subphyla: the cephalochordates, commonly referred to as the lancelets, and the urochordates, also known as the tunicates. We will discuss the subphylum Vertebrata in Chapter 34.

The Phylum Echinodermata Includes Sea Stars and Sea Urchins, Species with a Water Vascular System

The phylum Echinodermata (from the Greek *echinos*, meaning spiny, and *derma*, meaning skin) consists of a unique grouping of deuterostomes. A striking feature of all echinoderms is their modified radial symmetry. The body of most species can be divided into five parts pointing out from the center. As a consequence, cephalization is absent in most classes. There is no brain and only a simple nervous system consisting of a central nerve ring from which arise radial branches to each limb. The radial symmetry of echinoderms is secondary, present only in adults. The free-swimming larvae have bilateral symmetry and metamorphose into the radially symmetrical adult form.

The Echinoderm Body Plan Most echinoderms have an **endoskeleton**, an internal hard skeleton composed of calcareous plates overlaid by a thin skin (Figure 33.33). The skeleton is covered with spines and jawlike pincers called pedicellariae, the primary purpose of which is to deter settling of animals such as barnacles. These structures can also possess poison glands.

Echinoderms possess a true coelom, and a portion of the coelom has been adapted to serve as a unique **water vascular system**, a network of canals that branch into tiny **tube feet** that function in movement, gas exchange, feeding, and excretion (see inset to Figure 33.33). The water vascular system uses hydraulic power (water pressure generated by the contraction of muscles), which enables the tube feet to extend and contract, allowing echinoderms to move only very slowly.

Water enters the water vascular system through the **madreporite**, a sievelike plate on the animal's surface. From there it flows into a **ring canal** in the central disc, into five radial canals, and into the tube feet. At the base of each tube foot is a muscular sac called an **ampulla**, which stores water. Contractions of the ampullae force water into the tube feet, causing them to straighten and extend. When the foot contacts a solid surface, muscles in the foot contract, forcing water back into the ampulla. Sea stars also use their tube feet in feeding, where they can exert a constant and strong pressure on bivalves, whose adductor muscles open and close the shell. The adductor muscles eventually tire, allowing the shell to open slightly. At this stage, the sea star everts its stomach and inserts it into the opening. It then digests its prey, using juices secreted from extensive digestive glands. Sea stars also feed on sea urchins, brittle stars, and sand dollars, prey that cannot easily escape them.

Echinoderms cannot osmoregulate, so no species have entered freshwater environments. No excretory organs are present. For some species, both respiration and excretion of nitrogenous waste take place by diffusion across their tube feet. Coelomic fluid circulates around the body.

Most echinoderms exhibit **autotomy**, the ability to intentionally detach a body part, such as a limb, that will later regenerate. In some species, a broken limb can even regenerate into a whole animal. Some sea stars regularly reproduce by breaking in two. Most echinoderms reproduce sexually and have separate sexes. Fertilization is usually external, with gametes shed into the water. Fertilized eggs develop into free-swimming larvae, which become sedentary adults.

The Major Echinoderm Classes Although over 20 classes of echinoderms have been described from the fossil record, only 5 main classes of echinoderms exist today: the Asteroidea (sea stars), Ophiuroidea (brittle stars), Echinoidea (sea urchins and sand dollars), Crinoidea (sea lilies and feather stars), and Holothuroidea (sea cucumbers). The key features of the echinoderms and their classes are listed in **Table 33.6**.

The most unusual of the echinoderms are members of the class Holothuroidea, the sea cucumbers. These animals really do look like a cucumber rather than a sea star or sea urchin (**Figure 33.34**). The hard plates of the endoskeleton are less extensive, so the animal appears and feels fleshy. Sea cucumbers possess specialized respiratory structures called respiratory trees that pump water in and out of the anus. They are typically deposit feeders, ingesting sediment and extracting nutrients. When threatened by a predator, a few tropical species of sea cucumber can eject sticky, toxic substances from their anus. If these do not deter the predator, a member of these species can undergo the process of evisceration—ejecting its digestive tract, respiratory structures, and gonads from the anus. If the sea cucumber survives, it can regenerate its organs later.

Table 33.6 Main Classes and Characteristics of Echinoderms

Class and examples (est. # of species)	Class characteristics
Asteroidea: sea stars (1,600)	Five arms; tube feet; predatory on bivalves and other echinoderms; eversible stomach
Ophiuroidea: brittle stars (2,000)	Five long, slender arms; tube feet not used for locomotion; no pedicellariae; browse on sea bottom or filter feed
Echinoidea: sea urchins, sand dollars (1,900)	Spherical (sea urchins) or disc-shaped (sand dollars); no arms; tube feet and moveable spines; pedicellariae present; many feed on seaweeds
Crinoidea: sea lilies and feather stars (700)	Cup-shaped; often attached to substrate via stalk; arms feathery and used in filter feeding; very abundant in fossil record
Holothuroidea: sea cucumbers (1,200)	Cucumber-shaped; no arms; spines absent; endoskeleton reduced; tube feet; browse on sea bottom

Figure 33.34 **The edible sea cucumber, *Holothuria edulis*, on the Great Barrier Reef, Australia.**

Concept Check: What are two unique features of an echinoderm?

The Phylum Chordata Includes All the Vertebrates and Some Invertebrates

The deuterostomes consist of two major phyla: the echinoderms and the chordates. As deuterostomes, both phyla share similar developmental traits. In addition, both have an endoskeleton, consisting in the echinoderms of calcareous plates and in chordates, for the most

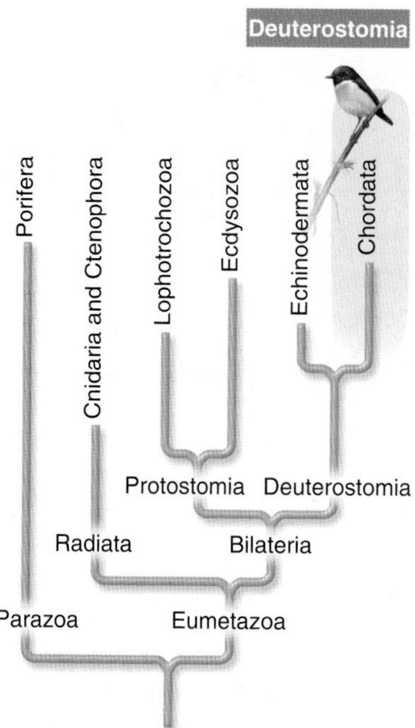

part, of bone. However, the echinoderm endoskeleton functions in much the same way as the arthropod exoskeleton, in that an important function is providing protection. The chordate endoskeleton serves a very different purpose. In early-divergent chordates, the endoskeleton is composed of a single flexible rod situated dorsally, deep inside the body. Muscles move this rod, and their contractions cause the back and tail end to move from side to side, permitting a swimming motion in water. The endoskeleton becomes more complex in different lineages that develop limbs, as we will see in Chapter 34, but it is always internal, with muscles attached. This arrangement permits the possibility of complex movements, including the ability to move on land.

Let's take a look at the four critical innovations in the body plan of chordates that distinguish them from all other animal life (**Figure 33.35**):

1. **Notochord.** Chordates are named for the **notochord**, a single flexible rod that lies between the digestive tract and the nerve cord. Composed of fibrous tissue encasing fluid-filled cells, the notochord is stiff yet flexible and provides skeletal support for all early-diverging chordates. In most chordates, such as vertebrates, a more complex jointed backbone usually replaces the notochord; its remnants exist only as the soft material within the discs between each vertebrae.
2. **Dorsal hollow nerve cord.** Many animals have a long nerve cord, but in nonchordate invertebrates, it is a solid tube that lies ventral to the alimentary canal. In contrast, the nerve cord in chordates is a hollow tube that develops dorsal to the alimentary canal. In vertebrates, the dorsal hollow nerve cord develops into the brain and spinal cord.
3. **Pharyngeal slits.** Chordates, like many animals, have a complete gut, from mouth to anus. However, in chordates, slits develop in the pharyngeal region, close to the mouth, that open to the outside. This permits water to enter through the mouth and exit via the slits, without having to go through the digestive tract. In early-divergent chordates, **pharyngeal slits** function as a filter-feeding device, whereas in later-divergent chordates, they develop into gills for gas exchange. In terrestrial chordates, the slits do not fully form, and they become modified for other purposes.
4. **Postanal tail.** Chordates possess a postanal tail of variable length that extends posterior to the anal opening. In aquatic chordates such as fishes, the tail is used in locomotion. In terrestrial chordates, the tail may be used in a variety of functions. In virtually all other nonchordate phyla, the anus is at the end of the body.

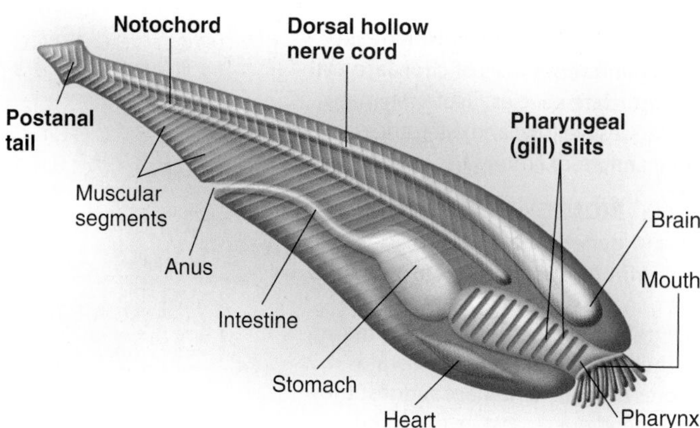

Figure 33.35 Chordate characteristics. The generalized chordate body plan has four main features: notochord, dorsal hollow nerve cord, pharyngeal slits, and postanal tail.

Although few chordates apart from fishes possess all of these characteristics in their adult life, they all exhibit them at some time during development. For example, in adult humans, the notochord becomes the spinal column, and the dorsal hollow nerve cord becomes the central nervous system. However, humans exhibit pharyngeal slits and a postanal tail only during early embryonic development. All the pharyngeal slits, except one, which forms the auditory (Eustachian) tubes in the ear, are eventually lost, and the postanal tail regresses to form the tailbone (the coccyx).

The phylum Chordata consists of the invertebrate chordates—the subphylum Cephalochordata (lancelets) and the subphylum Urochordata (tunicates)—along with the subphylum Vertebrata. Although the Vertebrata is by far the largest of these subphyla, biologists have focused on the Cephalochordata and Urochordata for clues as to how the chordate phylum may have evolved. Comparisons of gene sequences for the small subunit rRNA (SSU rRNA) show that these two subphyla are our closest invertebrate relatives (**Figure 33.36**).

Subphylum Cephalochordata: The Lancelets The cephalochordates (from the Greek *cephalo*, meaning head) look a lot more chordate-like than do tunicates. They are commonly referred to as lancelets, in reference to their bladelike shape and size, about 5–7 cm in length (**Figure 33.37a**). Lancelets are a small subphylum of 26 species, all marine filter feeders, with 4 species occurring in North American waters. Most of them belong to the genus *Branchiostoma.*

The lancelets live mostly buried in sand, with only the anterior end protruding into the water. Lancelets have the four distinguishing chordate characteristics: a clearly discernible notochord (extending well into the head), dorsal hollow nerve cord, pharyngeal slits, and postanal tail (**Figure 33.37b**). They are filter feeders, drawing water through the mouth and into the pharynx, where it is filtered through the pharyngeal slits. A mucous net across the pharyngeal slits traps food particles, and ciliary action takes the food into the intestine, while water exits via the atriopore. Gas exchange generally takes place across the body surface. Although the lancelet is usually sessile, it can leave its sandy burrow and swim to a new spot, using a sequence of serially arranged muscles that appear like chevrons (<<<<) along

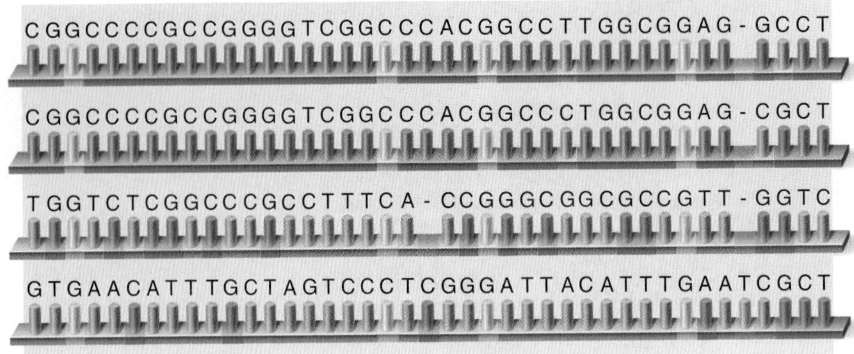

Figure 33.36 Comparison of SSU rRNA gene sequences of chordate and nonchordate species. Note the many similarities (yellow) and differences (green and red) among the sequences.

BIOLOGY PRINCIPLE The genetic material provides a blueprint for sustaining reproduction. The genetic similarities between the invertebrate chordates (represented by the lancelet) and the vertebrates (represented by a human) suggest they are indeed our closest invertebrate relatives.

(a) Lancelet in the sand

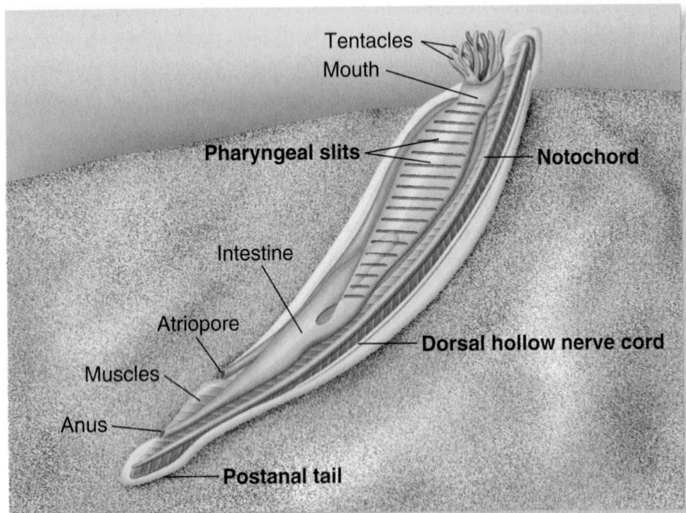

(b) Body plan of the lancelet

Figure 33.37 Lancelets. (a) A bladelike lancelet. **(b)** The body plan of the lancelet clearly displays the four characteristic chordate features.

its sides. These muscles reflect the segmented nature of the lancelet body and permit a fishlike swimming motion.

Subphylum Urochordata: The Tunicates The urochordates (from the Greek *oura*, meaning tail) are a group of 3,000 marine species also known as tunicates. Looking at an adult tunicate, you might never guess that it is a relative of modern vertebrates. The only one of the four distinguishing chordate characteristics that it possesses is pharyngeal slits (**Figure 33.38a**). The larval tunicate, in contrast, looks like a tadpole and exhibits all four chordate hallmarks (**Figure 33.38b**). The larval tadpole swims for only a few days, usually without feeding. Larvae settle on and attach to a rock surface via rootlike extensions called stolons. Here the larvae metamorphose into adult tunicates and in the process lose most of their chordate characteristics. In 1928, the English marine biologist Walter Garstang suggested modern vertebrates arose from a larval tunicate form that had somehow acquired the ability to reproduce. Analysis of molecular data in 2006 led French evolutionary biologist Frédéric Delsuc and colleagues to propose that tunicates are the closest living relatives of vertebrates. These researchers group the cephalochordates more closely with the echinoderms. This means the common ancestor of living deuterostomes was a free-living, bilaterally symmetrical animal with pharyngeal slits, a segmented body, and a dorsal hollow nerve cord. This ancestral line split into two groups, the echinoderm–cephalochordate group and the tunicate–vertebrate group. Echinoderms lost most of their ancestral features, but cephalochordates did not. In this view, tunicates lost their segmentation and most became sedentary, whereas vertebrates did not.

Adult tunicates are marine animals, some colonial and others solitary, that superficially resemble sponges or cnidarians. Tunicates are filter feeders that draw water through the mouth through an **incurrent siphon**, using a ciliated pharynx, and filter it through extensive pharyngeal slits. The food is trapped on a mucous sheet; passes via ciliary action to the stomach, intestine, and anus; and exits through the excurrent siphon. The whole animal is enclosed in a nonliving **tunic** made of a protein and a cellulose-like material called tunicin. Tunicates are also known as sea squirts for their ability to squirt out water from the excurrent siphon when disturbed. They have a rudimentary circulatory system with a heart and a simple nervous system of relatively few nerves connected to sensory tentacles around the incurrent siphon. The animals are mostly hermaphroditic.

As a reference, **Table 33.7** describes the common body characteristics of the various invertebrate animal phyla that we have considered in this chapter.

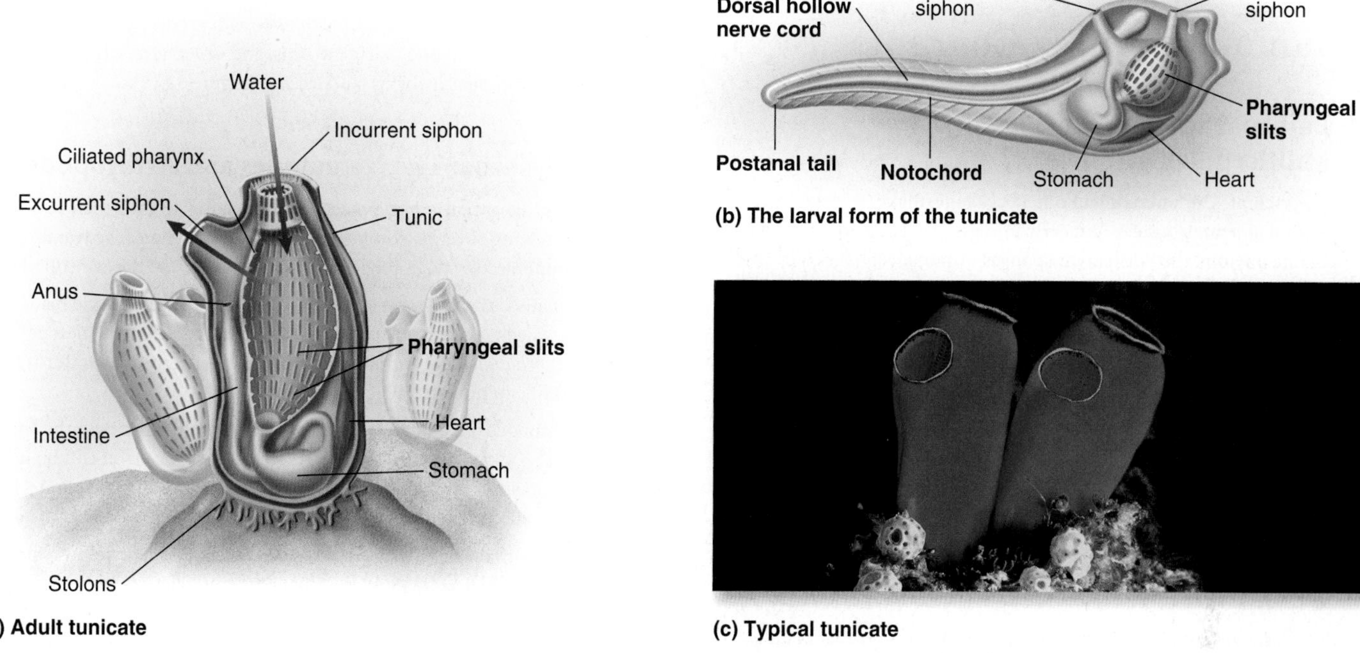

Figure 33.38 **Tunicates.** **(a)** Body plan of the sessile, filter-feeding adult tunicate. **(b)** The larval form, which shows the four characteristic chordate features, has been proposed as a possible ancestor of modern vertebrates. **(c)** The blue tunicate, *Rhopalaea crassa*.

Table 33.7 Summary of the Physical Characteristics of the Major Invertebrate Phyla

Feature	Porifera (sponges)	Cnidaria and Ctenophora (hydra, anemones, jellyfish)	Platyhelminthes (flatworms)	Rotifera (rotifers)	Bryozoa and Brachiopoda	Mollusca (snails, clams, squid)	Annelida (segmented worms)	Nematoda (roundworms)	Arthropoda (insects, arachnids, crustaceans)	Echinodermata (sea stars, sea urchins)	Chordata (vertebrates and others)
Digestive system	Absent	Gastrovascular cavity; ctenophores have complete gut	Gastrovascular cavity	Complete gut (usually)	Complete gut	Complete gut	Complete gut	Complete gut	Complete gut	Usually complete gut	Complete gut
Circulatory system	Absent	Absent	Absent	Absent	Absent; open; or closed	Open; closed in cephalopods	Closed	Absent	Open	Absent	Closed
Respiratory system	Absent	Absent	Absent	Absent	Absent	Gills	Absent	Absent	Trachae; gills; or book lungs	Tube feet; respiratory tree	Gills; lungs
Excretory system	Absent	Absent	Protonephridia with flame cells	Protonephridia	Metanephridia	Metanephridia	Metanephridia	Excretory tubules	Excretory glands resembling metanephridia	Absent	Kidneys
Nervous system	Absent	Nerve net	Brain; cerebral ganglia; lateral nerve chords; nerve net	Brain; nerve cords	No brain; nerve ring	Ganglia; nerve cords	Brain; ventral nerve cord	Brain; nerve cords	Brain; ventral nerve cord	No brain; nerve ring and radial nerves	Well-developed brain; dorsal hollow nerve cord
Reproduction	Sexual; asexual (budding)	Sexual; asexual (budding)	Sexual (most hermaphroditic); asexual (body splits)	Mostly parthenogenetic; males appear only rarely	Sexual (some hermaphroditic); asexual (budding)	Sexual (some hermaphroditic)	Sexual (some hermaphroditic)	Sexual (some hermaphroditic)	Usually sexual (some hermaphroditic)	Sexual (some hermaphroditic); parthenogenetic; asexual by regeneration (rare)	Sexual; rarely parthenogenetic
Support	Endo-skeleton of spicules and collagen	Mesoglea	Parenchyma	Tissue	Exoskeleton	Hydrostatic skeleton and shell	Hydrostatic skeleton	Fluid skeleton	Exoskeleton	Endoskeleton of plates beneath outer skin	Endoskeleton of cartilage or bone

Summary of Key Concepts

33.1 Parazoa: Sponges, the First Multicellular Animals

- Invertebrates, or animals without a backbone, make up more than 95% of all animal species. An early lineage—the Parazoa—consists of one phylum, the Porifera, or sponges. Although sponges lack true tissues, they are multicellular animals possessing several types of cells (Figures 33.1, 33.2).

33.2 Radiata: Jellyfish and Other Radially Symmetrical Animals

- The Radiata consists of two phyla: the Cnidaria (hydra, jellyfish, box jellies, sea anemones, and corals) and the Ctenophora (comb jellies). Radiata have only two embryonic germ layers: the ectoderm and endoderm, with a gelatinous substance (mesoglea) connecting the two layers.
- Cnidarians exist in two forms: polyp or medusa. A characteristic feature of cnidarians is their stinging cells, or cnidocytes, which function in defense and prey capture. Ctenophores possess the first complete gut, and nearly all exhibit bioluminescence (Figures 33.3, 33.4, 33.5, 33.6, Table 33.1).

33.3 Lophotrochozoa: The Flatworms, Rotifers, Bryozoans, Brachiopods, Mollusks, and Annelids

- Most Lophotrochozoa include taxa that possess either a lophophore or trochophore larva. Platyhelminthes, or flatworms, are regarded as the first animals to have the organ-system level of organization (Figure 33.7, Table 33.2).
- The four classes of flatworms are the Turbellaria, Monogenea, Trematoda (flukes), and Cestoda (tapeworms). Flukes and tapeworms are internally parasitic, with complex life cycles (Figures 33.8, 33.9).
- Rotifers are microscopic animals that have a complete digestive tract with separate mouth and anus; the mastax, a muscular pharynx, is a structure unique to the rotifers (Figure 33.10).
- The bryozoa and brachiopods both possess a lophophore, a ciliary feeding structure (Figure 33.11).
- The mollusks, which constitute a large phylum with over 100,000 diverse living species, have a basic body plan with three parts—a foot, a visceral mass, and a mantle—and an open circulatory system (Figures 33.12, 33.13).
- The four most common mollusk classes are the polyplacophora (chitons), gastropoda (snails and slugs), bivalvia (clams and mussels), and cephalopoda (octopuses, squids, and nautiluses) (Table 33.3).
- Cephalopods are among the most complex of all invertebrates. They are the only mollusks with a closed circulatory system; they have a well-developed nervous system and brain and are believed to exhibit learning by observation (Figures 33.14, 33.15, 33.16).
- Segmentation, in which the body is divided into compartments, is a critical evolutionary innovation in the annelids, although specialization of segments is only minimally present at the anterior end (Figure 33.17).
- Annelids are a large phylum with two main groups: Errantia, which includes free ranging marine worms, and Sedentaria, which includes tube worms, earthworms, and leeches (Figure 33.18).

33.4 Ecdysozoa: The Nematodes and Arthropods

- The ecdysozoans are so named for their ability to shed their cuticle, a nonliving cover providing support and protection. The two most common ecdysozoan phyla are the nematodes and the arthropods.
- Nematodes, which exist in nearly all habitats, have a cuticle made of collagen, a structural protein. The small, free-living nematode *Caenorhabditis elegans* is a model organism. Many nematodes are parasitic in humans (Figures 33.19, 33.20).
- Arthropods are perhaps the most successful phylum on Earth. The arthropod body is covered by a cuticle made of layers of chitin and protein, and it is segmented, with segments fused into functional units called tagmata (Figure 33.21).
- The five main subphyla of arthropods are Trilobita (trilobites; now extinct), Chelicerata (spiders, scorpions, and relatives), Myriapoda (millipedes and centipedes), Hexapoda (insects), and Crustacea (crabs and relatives) (Table 33.4, Figures 33.22, 33.23, 33.24, 33.25, 33.26).
- More insect species are known than all other animal species combined. The development of a variety of wing structures and mouthparts was a key to the success of insects (Figure 33.27, Table 33.5).
- Insects undergo a change in body form during development, either complete metamorphosis or incomplete metamorphosis, and have developed complex social behaviors (Figures 33.28, 33.29).
- Most crustacean orders are small and feature prominently in marine food chains. The most well-known order of crustaceans is the Decapoda, which includes the crabs, lobsters, and shrimp (Figures 33.30, 33.31, 33.32).

33.5 Deuterostomia: The Echinoderms and Chordates

- The Deuterostomia includes the phyla Echinodermata and Chordata. A striking feature of the echinoderms is their radial symmetry, which is secondary; the free-swimming larvae are bilaterally symmetrical. Echinoderms possess a unique water vascular system (Figure 33.33).
- Five main classes of echinoderms exist today: the Asteroidea (sea stars), Ophiuroidea (brittle stars), Echinoidea (sea urchins and sand dollars), Crinoidea (sea lilies and feather stars), and Holothuroidea (sea cucumbers) (Table 33.6, Figure 33.34).
- The phylum Chordata is distinguished by four critical innovations: the notochord, dorsal hollow nerve chord, pharyngeal slits, and postanal tail (Figure 33.35).
- The subphylum Cephalochordata (lancelets) and subphylum Urochordata (tunicates) are invertebrate chordates. Genetic studies have shown that tunicates are the closest invertebrate relatives of the vertebrate chordates (subphylum Vertebrata) (Figures 33.36, 33.37, 33.38, Table 33.7).

Assess and Discuss

Test Yourself

1. Choanocytes are
   a. a group of protists that are believed to have given rise to animals.
   b. specialized cells of sponges that function to trap and eat small particles.
   c. cells that make up the gelatinous layer in sponges.
   d. cells of sponges that function to transfer nutrients to other cells.
   e. cells that form spicules in sponges.
2. Why aren't sponges eaten more by predators?
   a. They are protected by silica spicules.
   b. They are protected by toxic defensive chemicals.
   c. They are eaten; it's just that the leftover cells reaggregate into new, smaller sponges.
   d. a and b are correct.
   e. a, b, and c are correct.
3. Which of the following organisms can produce female offspring through parthenogenesis?
   a. cnidarians
   b. flukes
   c. choanocytes
   d. rotifers
   e. annelids
4. What organisms can survive without a mouth, digestive system, or anus?
   a. cnidarians
   b. rotifers
   c. echinoderms
   d. cestodes
   e. nematodes
5. Which phylum does not have at least some members with a closed circulatory system?
   a. Lophophorata
   b. Arthopoda
   c. Annelida
   d. Mollusca
   e. All of the above phyla have some members with a closed circulatory system.
6. A defining feature of the Ecdysozoa is
   a. a segmented body.
   b. a closed circulatory system.
   c. a cuticle.
   d. a complete gut.
   e. a lophophore.
7. In arthropods, the tracheal system is
   a. a unique set of structures that function in ingestion and digestion of food.
   b. a series of branching tubes extending into the body that allow for gas exchange.
   c. a series of tubules that allow waste products in the blood to be released into the digestive tract.
   d. the series of ommatidia that form the compound eye.
   e. none of the above.
8. Characteristics of the class Arachnida include
   a. two tagmata.
   b. six walking legs.
   c. an aquatic lifestyle.
   d. a lobed body.
   e. both b and d.
9. Incomplete metamorphosis
   a. is characterized by distinct larval and adult stages that do not compete for resources.
   b. is typically seen in arachnids.
   c. involves gradual changes in life stages where young resemble the adult stage.
   d. is characteristic of the majority of insects.
   e. always includes a pupal stage.
10. Echinodermata are *not* a member of which clade?
   a. protostomia
   b. bilateria
   c. eumetazoa
   d. metazoa
   e. They are a member of all the above clades.

Conceptual Questions

1. Compare and contrast the five main feeding types discussed in the chapter.
2. Why is external fertilization common in aquatic invertebrates but rare in terrestrial species?
3. A principle of biology is that *living organisms grow and develop.* Explain the difference between complete metamorphosis and incomplete metamorphosis.

Collaborative Questions

1. Revisit the animal phylogeny outlined in Figure 33.1 and discuss the critical innovations that led to the separation of each of the clades shown.
2. Why are there more species of insects than any other taxa?

Online Resource

www.brookerbiology.com

Stay a step ahead in your studies with animations that bring concepts to life and practice tests to assess your understanding. Your instructor may also recommend the interactive eBook, individualized learning tools, and more.

The Vertebrates

34

Chapter Outline

34.1 Vertebrates: Chordates with a Backbone
34.2 Gnathostomes: Jawed Vertebrates
34.3 Tetrapods: Gnathostomes with Four Limbs
34.4 Amniotes: Tetrapods with a Desiccation-Resistant Egg
34.5 Mammals: Milk-Producing Amniotes
Summary of Key Concepts
Assess and Discuss

The star-nosed mole (*Condylura cristata*). This species is a vertebrate, a fascinating group of animals that includes human beings.

The star-nosed mole, *Condylura cristata,* lives in tunnels in wet areas of eastern Canada and the northeastern United States. It is one of the most distinctive mammals anywhere on Earth. The mole lives for the most part in complete darkness and is virtually blind. It feels its way around by means of 22 fleshy appendages, which consist of more than 25,000 minute and highly sensitive sensory receptors called Eimer's organs that allow it to find prey without using sight. The mole has a voracious appetite and needs to eat frequently. In fact, the star-nosed mole has been identified as the world's fastest eating mammal, averaging less than a quarter of a second to identify and consume a food item. Its astoundingly acute sensory abilities thus more than make up for its poor eyesight. The moles can swim under water in search of food and smell their prey by exhaling air bubbles then inhaling them to detect scent.

The star-nosed mole is a **vertebrate** (from the Latin *vertebratus,* meaning joint of the spine), an animal with a backbone. Vertebrates range in size from tiny fishes weighing 0.1 g to huge whales of over 100,000 kg. They occupy nearly all of Earth's habitats, from the deepest depths of the oceans to mountaintops and the sky beyond. Throughout history, humans have depended on many vertebrate species for their welfare, domesticating species such as horses, cattle, pigs, sheep, and chickens; using skin and fur for clothes, and keeping countless species, including cats and dogs, as pets. Many vertebrate species are the subjects of conservation efforts, as we will see in Chapter 60.

In Chapter 33, we discussed two chordate subphyla: the cephalochordates (lancelets) and urochordates (tunicates). The third subphylum of chordates, the Vertebrata, with about 53,000 species, is by far the largest and most dominant group of chordates. In this chapter, we will explore the characteristics of vertebrates and the evolutionary development of the major vertebrate classes, including fishes, amphibians, reptiles, and mammals.

Our current understanding of the relationships between the vertebrate groups is shown in **Figure 34.1**. Nested within the vertebrates are various clades based on morphological characteristics. For example, most vertebrates have jaws and are collectively known as gnathostomes. Many gnathostomes have four limbs for movement and are known as tetrapods. Next, we explain these sequential divisions in the vertebrate lineage.

34.1 Vertebrates: Chordates with a Backbone

Learning Outcomes:

1. List the main distinguishing characteristics of vertebrates.
2. Identify the two classes of existing jawless vertebrates.

The vertebrates retain all chordate characteristics we outlined in Chapter 33, as well as possessing several additional traits, including the following:

1. **Vertebral column.** During development in vertebrates, the notochord is replaced by a bony or cartilaginous column of interlocking **vertebrae** that provides support and also protects the nerve cord, which lies within its tubelike structure.
2. **Cranium.** The anterior end of the nerve cord elaborates to form a more developed brain that is encased in a protective bony or

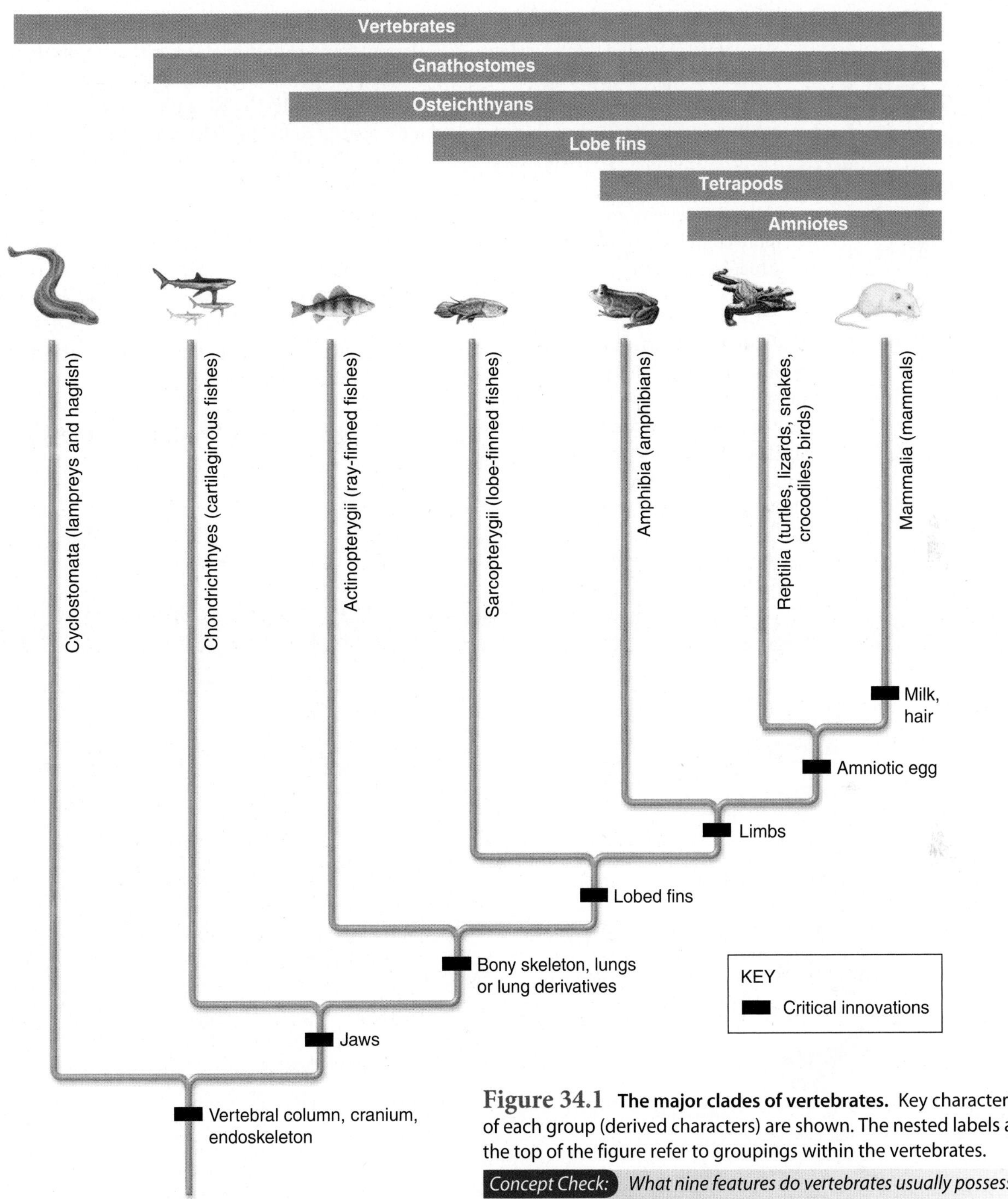

Figure 34.1 The major clades of vertebrates. Key characteristics of each group (derived characters) are shown. The nested labels across the top of the figure refer to groupings within the vertebrates.

Concept Check: What nine features do vertebrates usually possess that invertebrates do not?

cartilaginous housing called the **cranium**. This continues the trend of cephalization—the development of the head end in animals.

3. **Endoskeleton of cartilage or bone.** The cranium and vertebral column are parts of the endoskeleton, the living skeleton of vertebrates that forms within the animal's body. Most vertebrates also have two pairs of appendages, such as fins, legs, or arms. The endoskeleton is composed of either bone or cartilage, materials that are very strong yet more flexible than the chitin found in insects and other arthropods. The endoskeleton also contains living cells that secrete the skeleton, which grows with the animal, unlike the nonliving exoskeleton of arthropods.
4. **Neural crest.** The **neural crest** is a group of embryonic cells found on either side of the neural tube as it develops. The cells disperse throughout the embryo, where they contribute to the development of the skeleton, especially the cranium and other structures, including jaws and teeth.

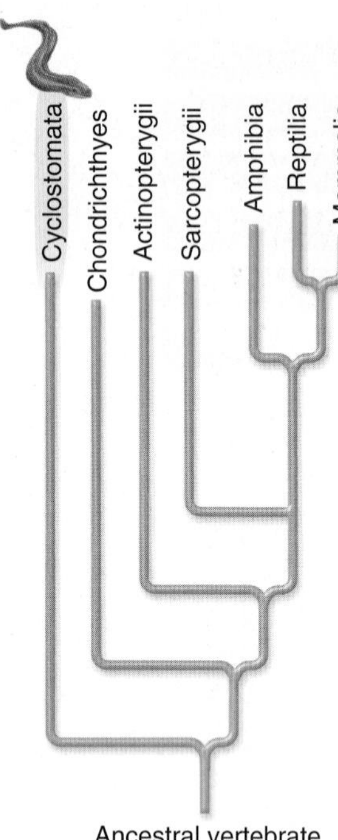

5. **Diversity of internal organs.** Vertebrates possess a great diversity of internal organs, including a liver, kidneys, endocrine glands, and a heart with at least two chambers. The liver is unique to vertebrates, and the vertebrate kidneys, endocrine system, and heart are more complex than are analogous structures in invertebrate taxa.

Although these are the main distinguishing characteristics of vertebrates, there are others. For example, most vertebrates have multiple clusters of *Hox* genes, compared with the single cluster of *Hox* genes in tunicates and lancelets. These additional gene clusters are believed to have permitted increasingly complex morphologies beyond those possessed by invertebrate chordates. Although these features are exhibited in all vertebrate classes, some classes evolved critical innovations that helped them succeed in specific environments such as on land or in the air. For example, birds developed feathers and wings, structures that enable most species to fly. In fact, each of the vertebrate classes is distinctly different from one another, as outlined in **Table 34.1**. One of the earliest innovations was the development of jaws. All vertebrates except some early-diverging fishes possessed jaws. Today, the only jawless vertebrates are hagfish and lampreys, together known as the Cyclostomata. Cyclostomes or "circle mouths" are eel-like animals that do not possess jaws. In addition, hagfish and lampreys share a very distinct type of immune system. Sequencing of RNA libraries in 2010, together with genomic surveys, yields strong support that the Cyclostomata is monophyletic.

The Hagfish, Class Myxini, Are the Simplest Living Cyclostomes

The hagfish are entirely marine cyclostomes that lack eyes, jaws, and fins and even vertebrae (**Figure 34.2**). The hagfish skeleton consists largely of a notochord and a cartilaginous skull that encloses the brain.

Table 34.1 The Main Classes and Characteristics of Living Vertebrates

Class	Examples (approx. # of species)	Main characteristics
Cyclostomata	Lampreys and hagfish (100)	Jawless fishes, no appendages
Chondrichthyes	Sharks, skates, rays (970)	Fishes with cartilaginous skeleton; teeth not fused to jaw; no swim bladder; well-developed fins; internal fertilization; single blood circulation
Actinopterygii	Ray-finned fishes, most bony fish (27,000)	Fishes with ossified skeleton; single gill opening covered by operculum; fins supported by rays, fin muscles within body; swim bladder often present; mucous glands in skin
Sarcopterygii	Lobe-finned fishes, of which coelacanths (2) and lungfishes (6) are the only living members	Fishes with ossified skeleton; bony extensions, together with muscles, project into pectoral and pelvic fins
Amphibia	Frogs, toads, salamanders (6,346)	Adults able to live on land; fresh water needed for reproduction; development usually involving metamorphosis from tadpoles; adults with lungs and double blood circulation; moist skin; shell-less eggs
Testudines	Turtles (310)	Body encased in hard shell; no teeth; head and neck retractable into shell; eggs laid on land
Squamata	Lizards, snakes (7,900)	Lower jaw not attached to skull; skin covered in scales
Crocodilia	Crocodiles, alligators (23)	Four-chambered heart; large aquatic predators; parental care of young
Aves	Birds (10,000)	Feathers; hollow bones; air sacs; reduced internal organs; endothermic; four-chambered heart
Mammalia	Mammals (5,500)	Mammary glands; hair; specialized teeth; enlarged skull; external ears; endothermic; four-chambered heart; highly developed brains; diversity of body forms

Figure 34.2 **The hagfish.**

The lack of a vertebral column leads to extensive flexibility. So how can hagfish be vertebrates without a vertebral column? The strong molecular support for a cyclostome clade suggests hagfish anatomy has degenerated to a remarkable degree and only the cranium, neural crest, and diversity of organs provide a link to the vertebrates. Hagfish live in the cold waters of northern oceans, close to the muddy bottom, feeding on marine worms and other invertebrates. Essentially blind, hagfish have a very keen sense of smell and are attracted to dead and dying fish, which they attach themselves to via toothed plates on the mouth. The powerful tongue then rasps off pieces of tissue. Though the hagfish cannot see approaching predators, they have special glands that produce copious amounts of slime. When provoked, the hagfish's slime production increases dramatically, enough to potentially distract predators or coat their gills and interfere with breathing. Hagfish can sneeze to free their nostrils of their own slime.

The Lampreys, Class Petromyzontida, Are Eel-like Animals That Lack Jaws

Lampreys are similar to hagfish because they lack both a hinged jaw and true appendages. However, lampreys do possess a notochord surrounded by a cartilaginous rod that represents a rudimentary vertebral column. Lampreys can be found in both marine and freshwater environments. Marine lampreys are parasitic as adults. They grasp other fish with their circular mouth (**Figure 34.3a**) and rasp a hole in the fish's side, sucking blood, tissue, and fluids until they are replete (**Figure 34.3b**). Reproduction of all species, whether they live in marine or freshwater environments, is similar. Males and females spawn in freshwater streams, and the resultant larval lampreys bury into the sand or mud, much like lancelets (refer back to Figure 33.37a), emerging to feed on small invertebrates or detritus at night. This stage can last for 3 to 7 years, at which time the larvae metamorphose into adults. In most freshwater species, the adults do not feed at all but quickly mate and die. Marine species migrate from fresh water back to the ocean, until they return to fresh water to spawn and then die. Although many other species of jawless fishes would evolve and thrive for over 300 million years, they all became extinct by the end of the Devonian period. Jawed fishes, which had appeared in the mid-Ordovician period (about 470 mya), radiated in both fresh and salt water.

34.2 Gnathostomes: Jawed Vertebrates

Learning Outcomes:

1. Describe how jaws evolved.
2. Discuss the distinguishing features of sharks.
3. List the three features that distinguish bony fishes from cartilaginous fishes.
4. Outline the differences between the ray-finned fish and the lobe-finned fish.

All vertebrate species that possess jaws are called **gnathostomes** (from the Greek, meaning jaw mouth) (see Figure 34.1). Gnathostomes are a diverse clade of vertebrates that include fishes, amphibians, reptiles, and mammals. The earliest-diverging gnathostomes were fishes. Biologists have identified about 25,000 species of living fishes, more than all other species of vertebrates combined. Most are aquatic, gill-breathing species that usually possess fins and a scaly skin. There are three separate clades of jawed fishes, each of which has distinguishing characteristics (see Table 34.1): the Chondrichthyes (cartilaginous fishes), Actinopterygii (ray-finned fishes), and Sarcopterygii (coelacanths and lungfishes).

(a) Jawless mouth of a sea lamprey

(b) A sea lamprey feeding

Figure 34.3 **The lamprey, a modern jawless fish.** **(a)** The sea lamprey (*Petromyzon marinus*) has a circular, jawless mouth. **(b)** A sea lamprey feeding on a fish.

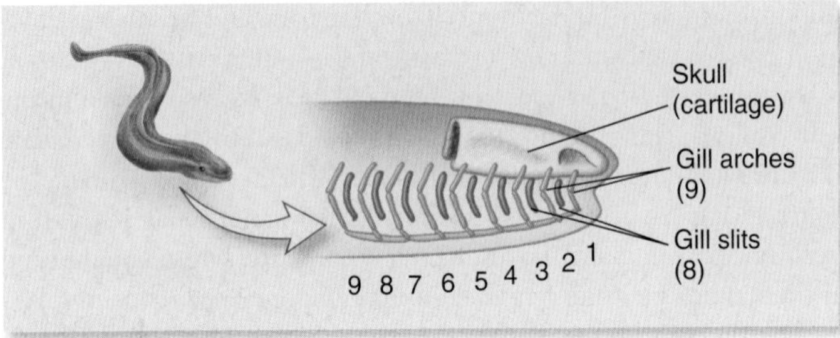

(a) Primitive jawless fishes

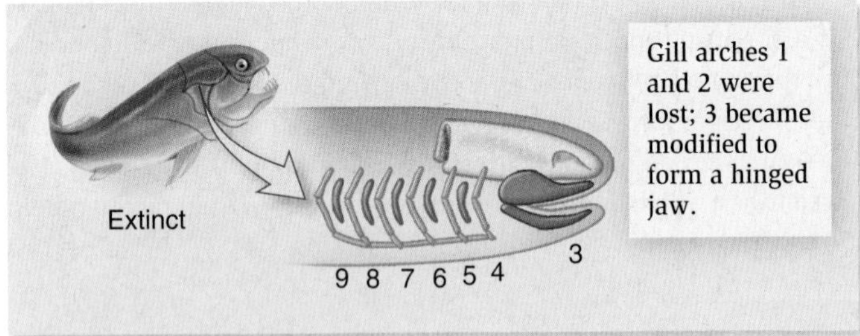

(b) Early jawed fishes (placoderms)

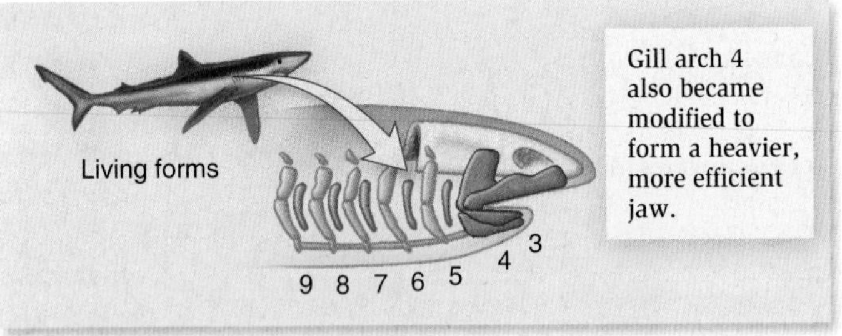

(c) Modern jawed fishes (cartilaginous and bony fishes)

Figure 34.4 The evolution of the vertebrate jaw. **(a)** Primitive fishes and extant jawless fishes such as lampreys have nine cartilaginous gill arches that support eight gill slits. **(b)** In early jawed fishes such as the placoderms, the first two pairs of gill arches were lost, and the third pair became modified to form a hinged jaw. This left six gill arches (4–9) to support the remaining five gill slits, which were still used in breathing. **(c)** In modern jawed fishes, the fourth gill arch also contributes to jaw support, allowing stronger, more powerful bites to be delivered.

BIOLOGY PRINCIPLE Structure determines function. The development of a jaw increased the predatory capabilities of gnathostomes.

The jawed mouth was a significant evolutionary development. It enabled an animal to grip its prey more firmly, which may have increased its rate of capture, and to attack larger prey species, thus increasing its potential food supply. Accompanying the jawed mouth was the development of more sophisticated head and body structures, including two pairs of appendages called fins. Gnathostomes also possess two additional *Hox* gene clusters over the cyclostomes (bringing their total to four), which led to increased morphological complexity.

The hinged jaw developed from the gill arches, cartilaginous or bony rods that help to support gills. Similarities between cells that make up jaws and gill arches support this view. Primitive jawless fishes had nine gill arches surrounding the eight gill slits (Figure 34.4a). During the late Silurian period (about 417 mya), some of these gill arches became modified. The first and second gill arches were lost, and the third and fourth pairs evolved to form the jaws (Figure 34.4b,c). This is how evolution typically works; body features do not appear de novo, but instead, existing features become modified to serve other functions.

By the mid-Devonian period, two classes of jawed fishes, the Acanthodii (spiny fishes) and Placodermi (armored fishes) were common. Some of the placoderms were huge individuals, over 9 m long. Both classes died out by the end of the Devonian as part of one of several mass extinctions that occurred in the Earth's geological and biological history. The reasons for this extinction are not well understood, but other types of jawed fishes present at the same time—the cartilaginous and bony fishes—did not go extinct and in fact flourished in the aftermath of the extinction.

Chondrichthyans Are Fishes with Cartilaginous Skeletons

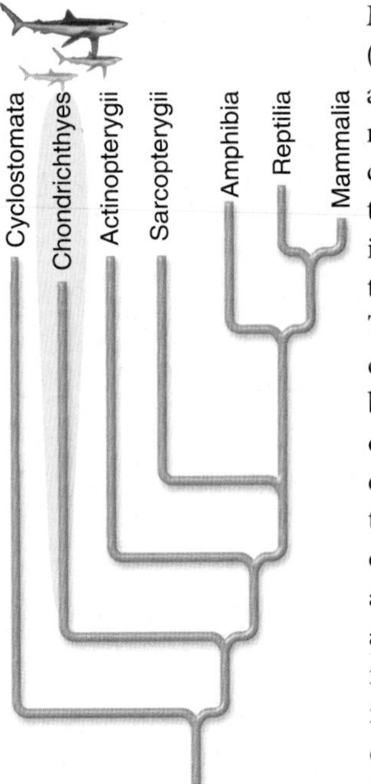

Members of the class Chondrichthyes (the **chondrichthyans**)—sharks, skates, and rays—are also called cartilaginous fishes because their skeleton is composed of flexible cartilage rather than bone. The cartilaginous skeleton is not considered an ancestral character but rather a derived character. This means that the ancestors of the chondrichthyans had bony skeletons, but that members of this class subsequently lost this feature. This hypothesis is reinforced by the observation that during development, the skeleton of most vertebrates is cartilaginous, and then it becomes bony (ossified) as a hard calcium-phosphate matrix replaces the softer cartilage. A change in the developmental sequence of the cartilaginous fishes is believed to prevent the ossification process.

In the Carboniferous period (354–290 mya), sharks were the great predators of the ocean. Aided by fins, sharks became fast, extremely efficient swimmers (Figure 34.5a). Perhaps the most important fin for propulsion is the large and powerful caudal fin, or tail fin, which, when swept from side to side, thrusts the fish forward at great speed. For example, great white sharks (*Carcharodon carcharias*) can swim at over 40 km per hour, and Mako sharks (*Isurus oxyrinchus*) have been clocked at nearly 50 km per hour. The paired pelvic fins (at the back) and pectoral fins (at the front) act like flaps on airplane wings, allowing the shark to dive deeper or rise to the surface. They also aid in steering. In addition, the dorsal fin (on the shark's back) acts as a stabilizer to prevent the shark from rolling in the water as the tail fin pushes it forward.

Sharks were among the earliest fishes to develop teeth. Shark teeth evolved from rough scales on the skin that also contain dentin and enamel. Although shark's teeth are very sharp and hard, they are

(a) Silvertip shark

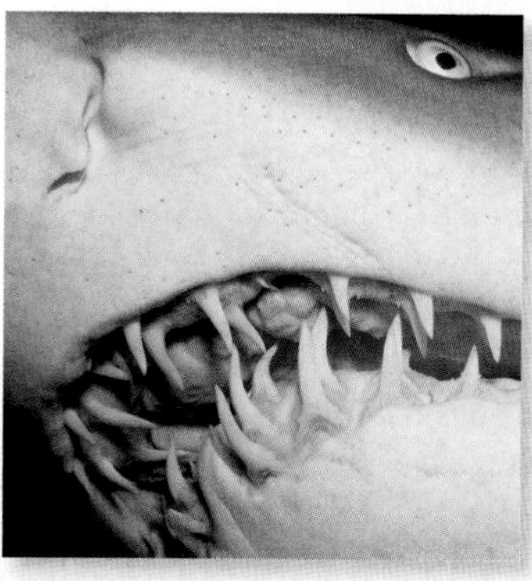

(b) Rows of shark teeth

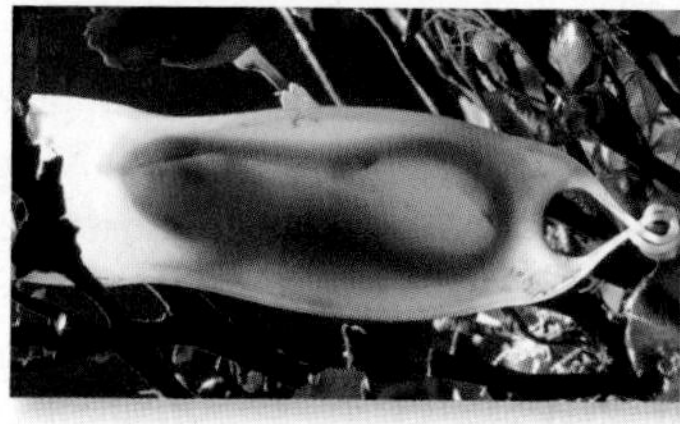

(c) Shark egg pouch

(d) Stingray

Figure 34.5 **Cartilaginous fishes.** **(a)** The silvertip shark (*Carcharhinus albimarginatus*) is one of the ocean's most powerful predators. **(b)** Close-up of the mouth of a sand tiger shark (*Carcharias taurus*), showing rows of teeth. **(c)** This mermaid's purse (egg pouch) of a dogfish shark (*Scyliorhinus canicula*) is entwined in vegetation to keep it stationary. **(d)** Stingrays are essentially flattened sharks with very large pectoral fins.

BioConnections: *Look forward to Figure 40.5. What types of cells are responsible for forming cartilage?*

not set into the jaw, as are human teeth, so they break off easily. Teeth are continually replaced, row by row (Figure 34.5b). Sharks may have 20 rows of teeth, with the front pair in active use and the ones behind ready to grow in as replacements when needed. Tooth replacement time varies from 9 days in the cookie-cutter shark (named for its characteristic of biting round plugs of flesh from its prey) to 242 days in great whites. Some experts estimate that certain sharks can use up to 20,000 teeth in a lifetime.

The Shark Body Plan All chondrichthyans are denser than water, which theoretically means that they would sink if they stopped swimming. Many sharks never stop swimming and maintain buoyancy via the use of their fins and a large oil-filled liver. Another advantage of swimming is that water continually enters the mouth and is forced over the gills, allowing sharks to extract oxygen and breathe. How then do skates and rays breathe when they rest on the ocean floor? These species, and a few sharks such as the nurse shark, use a muscular pharynx and jaw muscles to pump water over the gills. In these and indeed all species of fishes, the heart consists of two chambers, an atrium and a ventricle, that contract in sequence. They employ what is known as a single circulation, in which blood is pumped from the heart to capillaries in the gills to collect oxygen, and then it flows through arteries to the tissues of the body, before returning to the heart (look ahead to Figure 47.4a).

Shark Senses Sharks have a powerful sense of smell, facilitated by sense organs in the nostrils (sharks and other fishes do not use nostrils for breathing). They can see well but cannot distinguish colors. Although sharks have no eardrum, they can detect pressure waves generated by moving objects. Sharks have an extra sense to help them find and track prey. The ampullae of Lorenzini, vesicles and pores found around the shark's head, are sensory organs that detect electromagnetic fields produced by other organisms. All jawed fishes have a row of microscopic organs in the skin, arranged in a line that runs laterally down each side of the body, that can sense movements in the surrounding water. This system of sense organs, known as the **lateral line**, senses pressure waves and sends nervous signals to the inner ear and then on to the brain.

The Shark Life Cycle Fertilization is internal in chondrichthyans, with the male transferring sperm to the female via a pair of **claspers**, modifications of the pelvic fins. Some shark species are **oviparous**, that is, they lay eggs, often inside a protective pouch called a mermaid's purse (Figure 34.5c). In **ovoviparous** species, the eggs are retained within the female's body, but there is no placenta to nourish the young. A few species are **viviparous**; the eggs develop within the uterus, receiving nourishment from the mother via a placenta. Both ovoviparous and viviparous sharks give birth to live young.

The sharks have been a very successful vertebrate group, with many species identified in the fossil record. Although many species died out in the mass extinction at the end of the Permian period (290–248 mya), the survivors underwent a period of further speciation in the Mesozoic era, when most of the 375 modern species appeared.

Skates and rays are essentially flattened sharks that cruise along the ocean floor by using hugely expanded pectoral fins. In addition, their thin and whiplike tails are often equipped with a venomous barb used in defense (Figure 34.5d). Most of the 475 or so species of skates and rays feed on bottom-dwelling crustaceans and mollusks.

Osteichthyans Are Fishes with Bony Skeletons

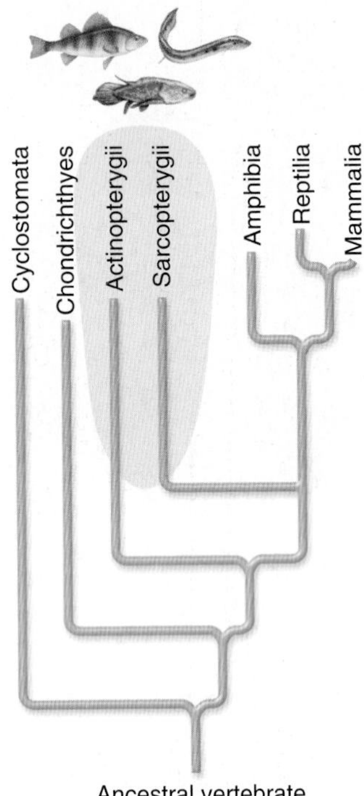

Unlike the cartilaginous fishes, all other gnathostomes have a bony skeleton and belong to the clade known as osteichthyes. This term means "bony fish" and was originally proposed for just that group. With the advent of modern phylogenetic systematics the term **osteichthyans** has expanded to include all vertebrates with a bony skeleton, including tetrapods (refer back to Figure 34.1).

Bony fishes are the most numerous of all types of fishes, with more individuals and more species (about 27,000) than any other. Most authorities now recognize three living classes: the Actinopterygii (ray-finned fishes), the Actinistia (coelacanths), and the Dipnoi (lungfishes). Fishes in all three classes possess a bony skeleton and scale-covered skin. The skin of bony fishes, unlike the rough skin of sharks, is covered by a thin epidermal layer containing glands that produce mucus, an adaptation that reduces drag during swimming. Just as in the cartilaginous fishes, water is drawn over the gills for breathing, but in bony fishes, a protective flap called an **operculum** covers the gills (Figure 34.6). Muscle contractions around the gills and operculum draw water across the gills so that bony fishes do not need to swim continuously to breathe. Some early bony fishes lived in shallow, oxygen-poor waters and developed lungs as an embryological offshoot of the pharynx. These fish could rise to the water surface and gulp air. As we will see, modern lungfishes operate in much the same fashion. Many other fishes, known as bimodal breathers, can breathe through their gills and by gulping air, absorbing oxygen through their digestive tracts or accessory organs. For example, Siamese fighting fish (*Betta splendens,* known as betta), a popular freshwater aquarium fish, is a bimodal breather that is relatively easy to care for, since it can survive without an air pump in its aquarium. In most bony fishes, the lungs evolved into a **swim bladder**, a gas-filled, balloon-like structure that helps the fish remain buoyant in the water even when it is completely stationary. In early diverging fishes, the gut and swim bladder are connected via a duct, and the fishes can fill their swim bladder by gulping air. In later diverging species, the swim bladder is connected to the circulatory system, and gases are transported in and out of the blood, allowing the fishes to change the volume of the swim bladder and so to rise and sink. Therefore, unlike the sharks, many bony fishes can remain motionless and use a "sit-and-wait" ambush style. These three features—bony skeleton, operculum, and swim bladder—distinguish bony fishes from cartilaginous fishes.

Reproductive strategies of bony fishes vary tremendously, but most species reproduce via external fertilization, with the female shedding her eggs and the male depositing sperm on top of them. Although adult bony fishes can maintain their buoyancy, their eggs tend to sink. This is why many species spawn in shallow, more oxygen- and food-rich waters and why coastal areas are important fish nurseries.

Bony fishes have colonized nearly all aquatic habitats. Following the cooling of the newly formed planet Earth, water condensed into rain and over a vast period of time filled what are now the oceans. Later, as water evaporated from the oceans and sodium, potassium, and calcium were added via runoff from the land, the oceans became salty. Therefore, most fishes probably evolved in freshwater habitats and secondarily became adapted to marine environments. This, of course, required the development of physiological adaptations to the different osmotic problems seawater presents compared with fresh water (look ahead to Figure 49.3).

Actinopterygii Are Ray-Finned Fish The most species-rich class of bony fishes is the Actinopterygii, or **ray-finned fishes**, which includes all bony fishes except the coelacanths and lungfishes. In Actinopterygii, the fins are supported by thin, bony, flexible rays and

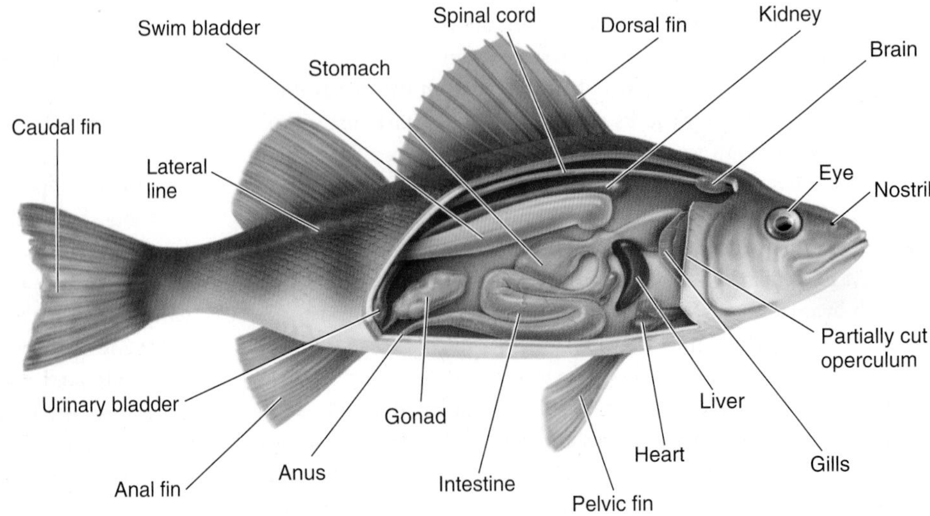

Figure 34.6 Generalized body plan of a bony fish. Bony fish possess a swim bladder and an operculum that covers the gills.

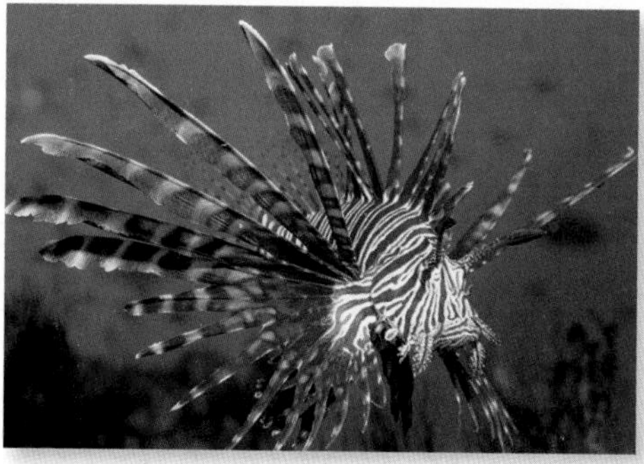

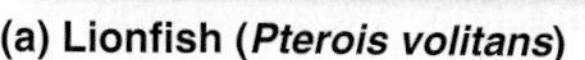

(a) Lionfish (*Pterois volitans*)

(b) Whitemouth moray eel (*Gymnothorax meleagris*)

(c) Leafy sea dragon (*Phycodurus eques*)

Figure 34.7 **The diversity of ray-finned fishes.** Ray-finned fishes exhibit many different sizes and body shapes.

Concept Check: What features distinguish ray-finned fishes from sharks?

are moved by muscles on the interior of the body. The class has a diversity of forms, from lionfish and large predatory moray eels to delicate sea dragons (**Figure 34.7**). Whole fisheries are built around the harvest of species such as cod, anchovies, and salmon.

Sarcopterygii Are Lobe-Finned Fish The Actinistia (coelacanths) and Dipnoi (lungfishes) are both considered Sarcopterygii, or **lobe fins**. The name Sarcopterygii used to refer solely to the lobe-finned fishes, but since it has become clear that terrestrial vertebrates (tetrapods) evolved from such fishes, the definition of the group has been expanded to include both lobe-finned fishes and tetrapods (see Figure 34.1). In the **lobe-finned fishes**, the fins are supported by skeletal extensions of the pectoral and pelvic areas that are moved by muscles within the fins.

The fossil record revealed that the Actinistia, or coelacanths, were a very successful group in the Devonian period, but all fishes of the class were believed to have died off at the end of the Mesozoic era (some 65 mya). You can therefore imagine the scientific excitement when in 1938, a modern coelacanth was discovered as part of the catch of a boat fishing near the Chalumna River in South Africa (**Figure 34.8**). Intensive searches in the area revealed that coelacanths were living in deep waters off the southern African coast and especially off a group of islands near the coast of Madagascar called the Comoros Islands. Another species was found more recently in Indonesian waters.

Early-diverging lobe-finned fishes probably evolved in fresh water and had lungs, but the coelacanth lost lungs and returned to the sea. One distinctive feature of this group is a special joint in the skull that allows the jaws to open extremely wide and gives the coelacanth a powerful bite. As further evidence of the coelacanth's unusual body plan, its swim bladder is filled with oil rather than gas, although it serves a similar purpose—to increase buoyancy.

The Dipnoi, or **lungfishes**, like the coelacanths, are also not currently a very species-rich class, having just three genera and six species (**Figure 34.9**). Lungfishes live in oxygen-poor freshwater swamps and ponds. They have both gills and lungs, the latter of which enable them to come to the surface and gulp air. In fact, lungfish will drown if they are unable to breathe air. When ponds dry out, some species of lungfish can dig a burrow and survive in it until the next rain. Because they also have muscular lobe fins, they are often able to successfully traverse quite long distances over shallow-bottomed lakes that may be drying out.

Figure 34.8 **A lobe-finned fish, the coelacanth (*Latimeria chalumnae*).**

Figure 34.9 **An Australian lungfish (*Neoceratodus forsteri*).**

Concept Check: How are lungfishes similar to coelacanths?

The morphological features of coelacanths, lungfishes, and primitive terrestrial vertebrates, together with the similarity of their nuclear genes, suggest to many scientists that lobe-fin ancestors gave rise to three lineages: the coelacanths, the lungfishes, and the tetrapods. In the next section we will examine the biology of tetrapods in more detail.

34.3 Tetrapods: Gnathostomes with Four Limbs

Learning Outcomes:

1. List adaptations that the transition to life on land required.
2. Describe the different amphibian orders and what differentiates them.

During the Devonian period (from about 417 to 354 mya), a diversity of plants and animals colonized the land. The presence of plants served as both a source of oxygen and a potential food source for animals that ventured out of the aquatic environment. Terrestrial arthropods appeared during the Devonian, as did the first land-living vertebrates.

The transition to life on land involved a large number of adaptations. Paramount among these were adaptations preventing desiccation and making locomotion and reproduction on land possible. We have seen that some fish evolved the ability to breathe air. In this section, we begin by outlining the development of the **tetrapods**, vertebrate animals having four legs or leglike appendages. We will discuss the first terrestrial vertebrates and their immediate descendants, the amphibians. We will then explore the characteristic features and diversity of modern amphibians.

The Origin of Tetrapods Involved the Development of Four Limbs

Over the Devonian period, the fossil records demonstrate the evolution of sturdy lobe-finned fishes to fishes with four limbs. The abundance of light and nutrients in shallow waters encouraged a profusion of plant life and the invertebrates that fed on them. The development of lungs enabled lungfishes to colonize these productive yet often oxygen-poor waters. Here, the ability to move in shallow water clogged with plants and debris was more vital than the ability to swim swiftly through open water and may have favored the progressive development of sturdy limbs. As an animal's weight began to be borne more by the limbs, the vertebral column strengthened, and hip bones and shoulder bones were braced against the backbone for added strength. Such modifications are the result of changes in the expression of genes, especially *Hox* genes (see the Feature Investigation). In particular, *Hox* genes 9–13 work together to specify limb formation from the proximal to the distal direction, meaning from close to the point of attachment to the body to the terminal end of the limb (Figure 34.10; see also Figure 19.17).

One of the transitional forms between fish and tetrapods was *Tiktaalik roseae*, nicknamed fishapod (refer back to Figure 23.5). Fishapods had broad skulls with eyes mounted on the top; lungs; and pectoral fins with five finger-like bones. This is an important species, for it represents a transitional form, displaying an intermediate state between an ancestral form and the form of its descendants.

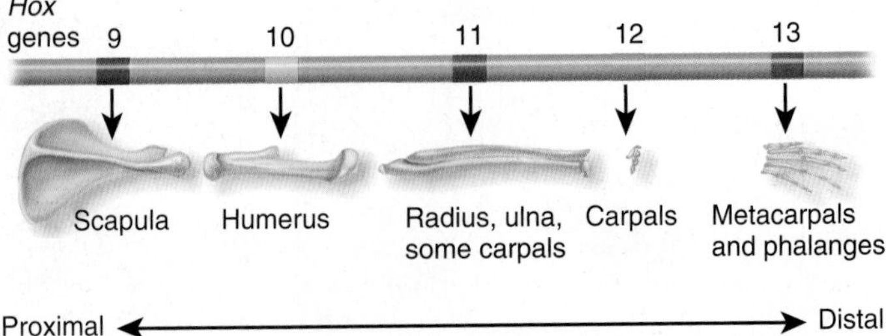

Figure 34.10 The roles of *Hox* genes 9–13 in specifying limb formation from the proximal to distal direction. The axis of limb development in mice is shown, together with the associated genes.

Eventually, species more like modern amphibians evolved, species that were still tied to water for reproduction but increasingly lived on land. In these species, the vertebral column, hip bones, and shoulder bones grew sturdier. Such changes were needed as the animal's weight was no longer supported by water but was borne entirely on the limbs.

By the middle of the Carboniferous period (about 320 mya), species similar to modern amphibians had become common in the terrestrial environment. For example, *Cacops* was a large amphibian, as big as a pony (Figure 34.11). Its skin was heavy and tough, an adaptation that helped prevent water loss; its breathing was accomplished more by lungs than by skin; and it possessed **pentadactyl limbs** (limbs ending in five digits). With a bonanza of terrestrial arthropods to feast on, the amphibians became very numerous and species rich, and the mid-Permian period (some 260 mya) is sometimes known as the Age of Amphibians. However, most of the large amphibians became extinct at the end of the Permian period. This was the largest known mass extinction in Earth's history, with the extinction of 90–95% of marine species and a large proportion of terrestrial species. Most surviving amphibians were smaller organisms resembling modern species.

Figure 34.11 A primitive tetrapod. *Cacops* was a large, early amphibian of the Permian period.

Concept Check: What were the advantages to animals of moving on to land?

FEATURE INVESTIGATION

Davis and Colleagues Provide a Genetic-Developmental Explanation for Limb Length in Tetrapods

The development of limbs in tetrapods was a vital step that allowed animals to colonize land. The diversity of vertebrate limb types is amazing, from fins in fish and marine mammals to different wing types in bats and birds to legs and arms in primates. Early in vertebrate evolution, an ancestral gene complex was duplicated twice to give rise to four groups of genes, called *Hox A*, *B*, *C*, and *D*, which control limb development. In 1995, Allen Davis, Mario Capecchi, and colleagues analyzed the effects of mutations in specific *Hox* genes that are responsible for determining limb formation in mice. The vertebrate forelimb is divided into three zones: humerus (upper arm); radius and ulna (forearm); and carpals, metacarpals, and phalanges (digits). The authors had no specific hypothesis in mind; their goal was to understand the role of *Hox* genes in limb formation. As described in Figure 34.12, they began with strains of mice carrying loss-of-function mutations in *HoxA-11* or *HoxD-11* that, on their own, did not cause dramatic changes in limb formation. They bred the mice and obtained offspring carrying one, two, three, or four

Figure 34.12 **Relatively simple changes in *Hox* genes control limb formation in tetrapods.**

GOAL To determine the role of *Hox* genes in limb development in mice.

KEY MATERIALS Mice with individual mutations in *HoxA-11* and *HoxD-11* genes.

Experimental level | **Conceptual level**

1 Breed mice with individual mutations in *HoxA-11* and *HoxD-11* genes. (The *A* and *D* refer to wild-type alleles; *a* and *d* are mutant alleles.)

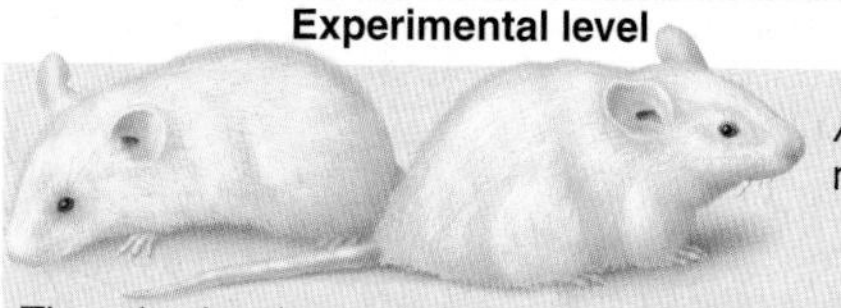

AaDd mice

The mice bred were heterozygous for both genes (*AaDd*).

Based on previous studies, researchers expect mutant mice to produce viable offspring, perhaps with altered limb morphologies.

2 Using molecular techniques described in Chapter 20, obtain DNA from the tail and determine the genotypes of offspring.

The resulting genotypes occur in Mendelian ratios, generating mice with different combinations of wild-type and mutant alleles.

	AD	*Ad*	*aD*	*ad*
AD	*AADD*	*AADd*	*AaDD*	*AaDd*
Ad	*AADd*	*AAdd*	*AaDd*	*Aadd*
aD	*AaDD*	*AaDd*	*aaDD*	*aaDd*
ad	*AaDd*	*Aadd*	*aaDd*	*aadd* ← Double mutant

9:3:3:1 phenotypic ratio expected in a dihybrid cross

3 Stain the skeletons and compare the limb characteristics of the wild-type mice (*AADD*) to those of strains carrying mutant alleles in one or both genes.

Mutant mice may have altered bone morphologies.

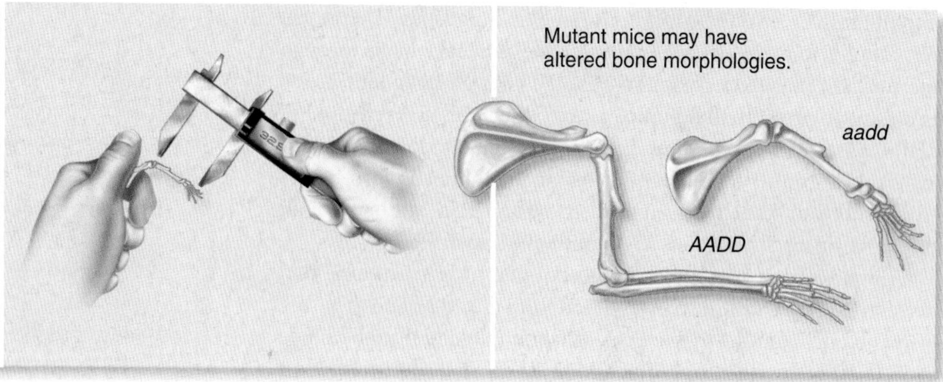

4 THE DATA

Genotype	Carpal bone fusions (% of mice showing the fusion)			
	Normal (none fused)	NL fused to T	T fused to P	NL fused to T and P
AADD	100	0	0	0
AaDD	100	0	0	0
aaDD	33	17	50	0
AADd	100	0	0	0
AAdd	0	17	17	67
AaDd	17	17	33	33

5 CONCLUSION Relatively simple mutations involving two genes can cause large changes in limb development.

6 SOURCE Davis, A.P. et al. 1995. Absence of radius and ulna in mice lacking *Hoxa-11* and *Hoxd-11*. *Nature* 375:791–795.

loss-of-function mutations. Double mutants exhibited dramatically different phenotypes not seen in mice homozygous for the individual mutations. The radius and ulna were almost entirely eliminated.

As seen in the data, the mutations affected the formation of limbs. For example, the wrist contains seven bones: three proximal carpals—called navicular lunate (NL), triangular (T), and pisiform (P)—and four distal carpals (d1–d4). In mice with the genotypes *aaDD* and *AAdd*, the proximal carpal bones are usually fused together. Individuals having one recessive allele (*AADd* and *AaDD*) do not show this defect, but individuals having two recessive alleles (*AaDd*) often do. Therefore, any two mutant alleles (either from both *HoxA-11* and *HoxD-11* or one from each locus) cause carpal fusions. Deformities became even more severe with three mutant alleles (*Aadd* or *aaDd*) or four mutant alleles (*aadd*) (data not shown in the figure). Thus, scientists have shown that relatively simple mutations can control relatively large changes in limb development.

Experimental Questions

1. What was the purpose of the study conducted by Davis and colleagues?
2. How were the researchers able to study the effects of individual genes?
3. Explain the results of the experiment and how this relates to limb development in vertebrates.

Amphibian Lungs and Limbs Are Adaptations to a Semiterrestrial Lifestyle

Amphibians (from the Greek, *amphibios* meaning both ways of life) live in two worlds: They have successfully invaded the land, but most must return to the water to reproduce. One of the first challenges terrestrial animals had to overcome was breathing air when on land. Amphibians can use the same technique as lungfishes: They open their mouths to let in air. Alternatively, they may take in air through their nostrils. They then close and raise the floor of the mouth, creating a positive pressure that pumps air into the lungs. This method of breathing is called **buccal pumping**. In addition, the skin of amphibians is much thinner than that of fishes, and amphibians absorb oxygen from the air directly through their outer moist skin or through the skin lining of the inside of the mouth or pharynx.

Amphibians have a three-chambered heart, with two atria and one ventricle. One atrium receives blood from the body, and the other receives blood from the lungs. Both atria pump blood into the single ventricle, which pumps some blood to the lungs and some to the rest of the body (look ahead to Figure 47.4b). This form of circulation allows the tissues to receive well-oxygenated blood at a higher pressure than is possible via single circulation, because some of the blood that returns to the heart is directly pumped to the tissues without being slowed down by passage through the lung capillaries. Oxygenated and deoxygenated bloods are kept somewhat separate, which enhances the delivery of nutrients and oxygen to the tissues.

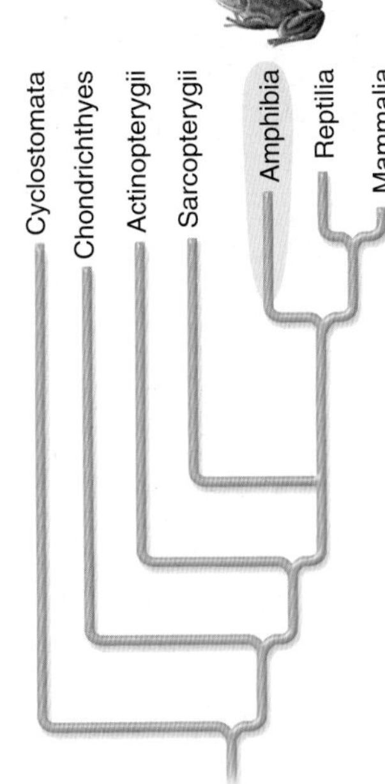

Because the skin of amphibians is so thin, the animals face the problem of desiccation, or drying out. As a consequence, even amphibian adults are more abundant in damp habitats, such as swamps or rain forests, than in dry areas. Also, most amphibians cannot venture too far from water because their larval stages

(a) Gelatinous mass of amphibian eggs

(b) Tadpole

(c) Tadpole undergoing metamorphosis

Figure 34.13 **Amphibian development in the wood frog (*Rana sylvatica*).** **(a)** Amphibian eggs are laid in gelatinous masses in water. **(b)** The eggs develop into tadpoles, aquatic herbivores with a fishlike tail that breathe through gills. **(c)** During metamorphosis, the tadpole loses its gills and tail and develops limbs and lungs.

BioConnections: Look ahead to Figure 47.4. Do frogs breathe through their skin while they are underwater?

are still aquatic. In frogs and toads, fertilization is generally external, with males shedding sperm over the gelatinous egg masses laid by the females in water (**Figure 34.13a**). The fertilized eggs lack a shell and would quickly dry out if exposed to the air. They soon hatch into tadpoles (**Figure 34.13b**), small fishlike animals that lack limbs and breathe through gills. As the tadpole nears the adult stage, the tail and gills are resorbed, and limbs and lungs appear (**Figure 34.13c**). Such a dramatic change in body form, from juvenile to adult, is known as metamorphosis. A few species of amphibians do not require water to reproduce. These species are ovoviparous or viviparous—retaining the eggs in the reproductive tract and giving birth to live young.

Modern Amphibians Include a Variety of Frogs, Toads, Salamanders, and Caecilians

Approximately 6,346 living amphibian species are known, and the vast majority of these, some 5,602 species, are frogs and toads of the order **Anura** (from the Greek, meaning tail-less ones) (**Figure 34.14a**). The other two orders are the **Apoda** (from the Greek, meaning legless ones), the wormlike caecilians; and the **Urodela** (from the Latin, meaning tailed ones), the salamanders. Global warming is currently threatening many anurans with extinction (see chapter opener for Chapter 54).

Adult anurans are carnivores, eating a variety of invertebrates by catching them on a long, sticky tongue. In contrast, the aquatic larvae (tadpoles) are primarily herbivores. Frogs generally have smooth, moist skin and long hind legs, making them excellent jumpers and swimmers. In addition to secreting mucus, which keeps their skin moist, some frogs can also secrete poisonous chemicals that deter would-be predators. Some amphibians advertise the poisonous nature of their skin with warning coloration (look ahead to Figure 57.9b). Others use camouflage as a way of avoiding detection by predators. Toads have a drier, bumpier skin and shorter legs than frogs. They are less impressive leapers than frogs, but toads can better tolerate drier conditions.

Caecilians (order Apoda) are a small order of about 174 species of legless, nearly blind amphibians (**Figure 34.14b**). Most are tropical and burrow in forest soils, but a few live in ponds and streams. They are secondarily legless, which means they evolved from legged ancestors. Caecilians have tiny jaws equipped with teeth and eat worms and other soil invertebrates. In this order, fertilization is internal, and females usually bear live young. The young are nourished inside the mother's body by a thick, creamy secretion known as uterine milk. In most caecilian species, the young grow into adults about 30 cm long, though species up to 1.3 m in length are known.

(a) Tree frog

(b) A caecilian

(c) Mud salamander

Figure 34.14 **Amphibians.** **(a)** Most amphibians are frogs and toads of the order Anura, including this red-eyed tree frog (*Agalychnis callidryas*). **(b)** The order Apoda includes wormlike caecilians such as this species from Colombia, *Caecilia nigricans*. **(c)** The order Urodela includes species such as this mud salamander (*Pseudotriton montanus*).

Concept Check: Do all amphibians produce tadpoles?

The salamanders (order Urodela, about 570 species) possess a tail and have a more elongate body than anurans (**Figure 34.14c**). During locomotion, they seem to sway from side to side, perhaps reminiscent of how the earliest tetrapods may have walked. Like frogs, salamanders often have colorful skin patterns that advertise their distastefulness to predators. Salamanders retain their moist skin by living in damp areas under leaves or logs or beneath lush vegetation. They generally range in size from 10 to 30 cm. Fertilization is usually internal, with females using their cloaca, a common opening for the digestive and urogenital tracts, to pick up sperm packets deposited by males. A very few salamander species do not undergo metamorphosis, and the newly hatched young resemble tiny adults. However, some species, such as Cope's giant salamander (*Dicamptodon copei*), retain the gills and tail fins characteristic of the larval stage into adulthood, and mature sexually in the larval stage, a phenomenon known as paedomorphosis (refer back to Figure 25.16b).

34.4 Amniotes: Tetrapods with a Desiccation-Resistant Egg

Learning Outcomes:

1. Diagram the structure of the amniotic egg.
2. Identify the critical innovations of the amniotes.
3. Describe the distinguishing features of the major amniote classes.
4. List the features that allowed birds to fly.

Although amphibians live successfully in a terrestrial environment, they must lay their eggs in water or in a very moist place, so their shell-less eggs do not dry out on exposure to air. Thus, a critical innovation in animal evolution was the development of a shelled egg that sheltered the embryo from desiccating conditions on land. A shelled egg containing fluids was like a personal enclosed pond for each developing individual. Such an egg evolved in the common ancestor of turtles, lizards, snakes, crocodiles, birds, and mammals—a group of tetrapods collectively known as the **amniotes**. The amniotic egg permitted animals to lay their eggs in a dry place so that reproduction was no longer tied to water. It was truly a critical innovation, untethering animals from water in much the same way as the development of seeds liberated plants from water (see Chapter 29).

In time, the amniotes became very diverse in species and morphology. Mammals are considered amniotes, too, because even though most of them do not lay eggs, they retain other features of amniotic reproduction. In this section, we begin by discussing in detail the morphology of the amniotic egg and other adaptations that permitted animal species to become fully terrestrial. We then discuss the biology of the reptiles, the first group of vertebrates to fully exploit land.

The Amniotic Egg and Other Innovations Permitted Life on Land

The **amniotic egg** (Figure 34.15) contains the developing embryo and the four separate extraembryonic membranes that it produces:

1. The innermost membrane is the **amnion**, which protects the developing embryo in a fluid-filled sac called the amniotic cavity.
2. The **yolk sac** encloses a stockpile of nutrients, in the form of yolk, for the developing embryo.
3. The **allantois** functions as a disposal sac for metabolic wastes.
4. The **chorion**, along with the allantois, provides gas exchange between the embryo and the surrounding air.

Surrounding the chorion is the albumin, or egg white, which also stores nutrients. The **shell** provides a tough, protective covering that is not very permeable to water and prevents the embryo from drying out. However, the shell remains permeable to oxygen and carbon dioxide, so the embryo can breathe. In birds, this shell is hard and

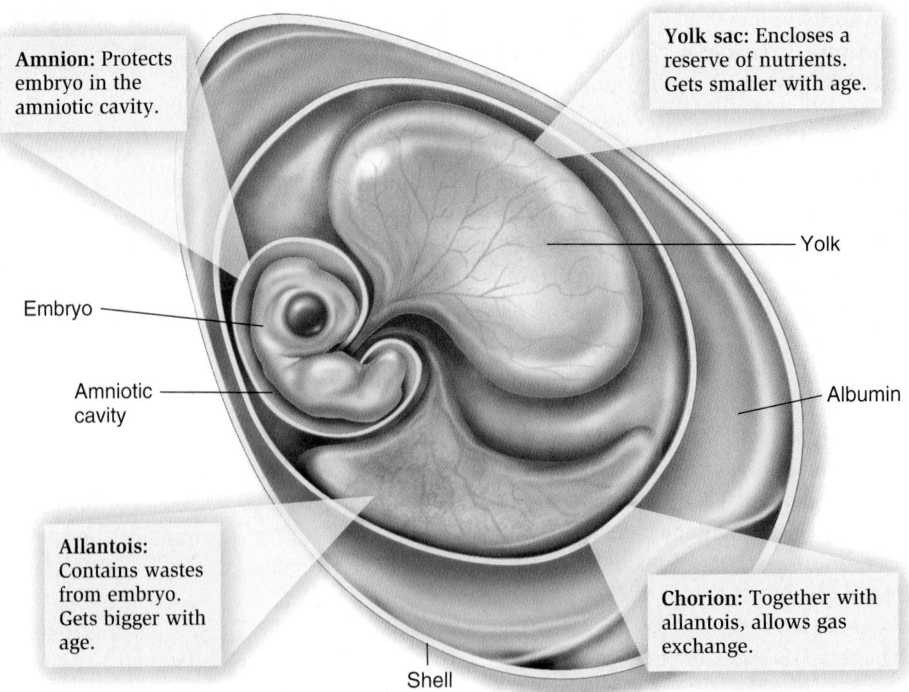
Amnion: Protects embryo in the amniotic cavity.
Yolk sac: Encloses a reserve of nutrients. Gets smaller with age.
Yolk
Embryo
Amniotic cavity
Albumin
Allantois: Contains wastes from embryo. Gets bigger with age.
Chorion: Together with allantois, allows gas exchange.
Shell

Figure 34.15 The amniotic egg.

Concept Check: *What are the other critical innovations of amniotes?*

calcareous, whereas in reptiles and early-diverging mammals such as the platypus and echidna, it is soft and leathery. In most mammals, however, the embryos embed into the wall of the uterus and receive their nutrients directly from the mother.

Along with the amniotic egg, other critical innovations that enabled the conquest of land include the following:

- **Desiccation-resistant skin.** Whereas the skin of amphibians is moist and aids in respiration, the skin of amniotes is thicker and water resistant and contains keratin, a tough protein. As a result, most gas exchange takes place through the lungs.
- **Thoracic breathing.** Amphibians use buccal pumping to breathe, contracting the mouth to force air into the lungs. In contrast, amniotes use thoracic breathing, in which coordinated contractions of muscles expand the rib cage, creating a negative pressure to suck air in and then forcing it out later. This results in a greater volume of air being displaced with each breath than with buccal pumping.
- **Water-conserving kidneys.** The ability to concentrate wastes prior to elimination and thus conserve water is an important role of the amniotic kidneys.
- **Internal fertilization.** Because sperm cannot penetrate a shelled egg, fertilization occurs internally, within the female's body before the shell is secreted. In this process, the male of the species often uses a copulatory organ (penis) to transfer sperm into the female reproductive tract. However, birds usually transfer sperm from cloaca to cloaca.

Reptiles Include Turtles, Lizards, Snakes, Crocodilians, Dinosaurs, and Birds

Early amniote ancestors gave rise to all modern amniotes we know today, from lizards and snakes to birds and mammals. The traditional view of amniotes involved three living classes: the reptiles (turtles, lizards, snakes, and crocodilians), birds, and mammals. As we will see later in the chapter, modern systematists have argued that enough

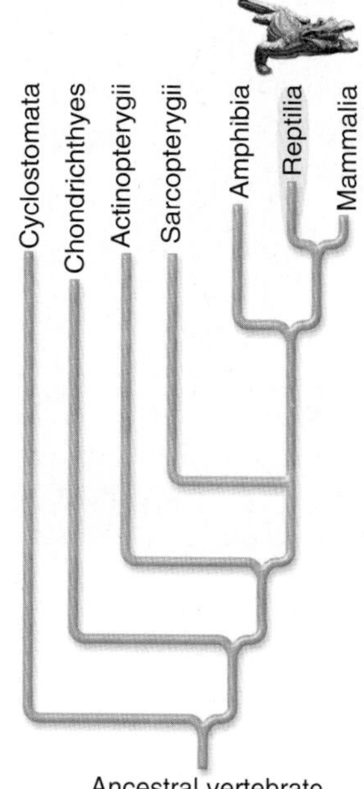

similarities exist between birds and the classic reptiles that birds should be considered part of the reptilian lineage. This is the classification scheme that we will follow in this chapter. The fossil record includes other reptilian classes, all of which are extinct, including two classes of dinosaurs (ornithischian and saurischian dinosaurs), flying reptiles (pterosaurs), and two classes of ancient aquatic reptiles (icthyosaurs and plesiosaurs).

Class Testudines: The Turtles Turtles is an umbrella term for terrestrial species, also called tortoises, and aquatic species, sometimes known as terrapins. The turtle lineage is ancient and has remained virtually unchanged for 200 million years. The major distinguishing characteristic of the turtle is a hard protective shell into which the animal can withdraw its head and limbs. In most species, the vertebrae and ribs are fused to form this shell. All turtles lack teeth but have sharp beaks for biting.

Most turtles are aquatic and have webbed feet. The forelimbs of marine species have evolved to become large flippers. All turtles, even the aquatic species, lay their eggs on land, usually in soft sand. The gender of hatchlings is dependent on temperature, with high temperatures producing more females. Marine species often make long migrations to sandy beaches to lay their eggs (Figure 34.16). Most land turtles are quite slow movers, possibly due to a low metabolic rate and a heavy shell. However, they are very long-lived species, often surviving for 120 years or more. Furthermore, turtles do not appear to show reproductive senescence or aging, reproducing continually throughout their lifetime. Most organs such as the liver, lungs, and kidneys of a centenarian turtle function as effectively as do organs in young individuals, prompting genetic researchers to examine the

(a) Green turtle

(b) Slow worm

(c) Gila monster

Figure 34.16 **A variety of reptiles.** **(a)** A green turtle (*Chelonia mydas*) laying eggs in the sand in Malaysia. **(b)** The slow worm, *Anguis fragilis*, is a type of legless lizard. **(c)** The Gila monster (*Heloderma suspectum*), one of only two venomous lizards, is an inhabitant of the desert Southwest of the U.S. and of Mexico.

Figure 34.17 The kinetic skull. In snakes and lizards, both the top and bottom of the jaw is hinged on the skull, thereby permitting large prey to be swallowed. This Halloween snake (*Pliocercus euryzonus*) is swallowing a Costa Rican rain frog.

Concept Check: If snakes are limbless, how can they be considered tetrapods?

turtle genome for longevity genes. Many turtle species are in danger of extinction, due to egg hunting, harvesting for shells or meat, destruction of habitat and nesting sites, and death in fishing nets.

Class Squamata: Lizards and Snakes The class ***Squamata*** is the largest class within the traditional reptiles, with about 4,900 species of lizards (order Sauria) and 3,000 species of snakes (order Serpentes). Many species have an elongated body form. One of the defining characteristics of the orders is a **kinetic skull**, in which the joints between various parts of the skull are extremely mobile. The lower jaw does not join directly to the skull but rather is connected by a multijointed hinge, and the upper jaw is hinged and movable from the rest of the head. This allows the jaws to open relatively wider than other vertebrate jaws, with the result that lizards, and especially snakes, can swallow large prey (**Figure 34.17**). Nearly all species are carnivores.

A main difference between lizards and snakes is that lizards generally have limbs, whereas snakes do not. Leglessness is a derived condition, meaning snake ancestors possessed legs but later lost them. Also, snakes may be venomous, whereas lizards usually are not. However, there are exceptions to these general rules. Many legless lizard species exist (see Figure 34.16b), and two lizards are venomous: the Gila monster (*Heloderma suspectum*) of the U.S. Southwest (see Figure 34.16c) and the Mexican beaded lizard (*Heloderma horridum*). A more reliable distinguishing morphological characteristic is that lizards have movable eyelids and external ears (at least ear canals), but snakes do not.

Class Crocodilia: The Crocodiles and Alligators The Crocodilia is a small class of large, carnivorous, aquatic animals that have remained essentially unchanged for nearly 200 million years (**Figure 34.18**). Indeed, these animals existed at the same time as the dinosaurs. Most of the 23 recognized species live in tropical or subtropical regions. There are only two extant species of alligators: one living in the southeastern U.S. and one found in China.

Although the class is small, it is evolutionarily very important. Crocodiles have a four-chambered heart, a feature they share with birds and mammals (look ahead to Figure 47.4c). In this regard, crocodiles are more closely related to birds than to any other living reptile class. Their teeth are set in sockets, a feature typical of the dinosaurs and the earliest birds. Similarly, crocodiles care for their young, another trait they have in common with birds. These and other features suggest that crocodiles and birds are more closely related than crocodiles and lizards. As with turtles, the gender of the offspring is dependent on nest temperature.

(a) American alligator

(b) American crocodile

Figure 34.18 Crocodilians. The Crocodilia is an ancient class that has existed unchanged for millions of years. **(a)** Alligators, such as this American alligator (*Alligator mississippiensis*), have a broad snout, and the lower jaw teeth close on the inside of the upper jaw (and thus are almost completely hidden when the mouth is closed). **(b)** Crocodiles, including this American crocodile (*Crocodylus acutus*), have a longer, thinner snout, and the lower jaw teeth close on the outside of the upper jaw (and thus are visible when the mouth is closed).

BioConnections: Look ahead to Figure 47.4c. In what ways are crocodilians similar to birds and mammals?

(a) Ornithischian (*Stegosaurus*)

(b) Saurischian (*Tyrannosaurus*)

Figure 34.19 **Classes of dinosaurs.** **(a)** Herbivorous ornithischians included *Stegosaurus,* and **(b)** carnivorous saurischians included bipedal species such as *Tyrannosaurus rex.*

Classes Ornithischia and Saurischia: The Dinosaurs In 1841, the English paleontologist Richard Owen coined the term **dinosaur** (from the Greek, meaning terrible lizard) to describe some of the wondrous fossil animals discovered in the 19th century. About 215 mya, dinosaurs were the dominant tetrapods on Earth and remained so for 150 million years, far longer than any other vertebrate. The two main classes were the ornithischian, or bird-hipped dinosaurs, which were herbivores such as *Stegosaurus*; and the saurischian, or lizard-hipped dinosaurs, which were fast, bipedal carnivores such as *Tyrannosaurus* (**Figure 34.19**). In contrast to the limbs of lizards, amphibians, and crocodiles, which splay out to the side, the legs of dinosaurs were positioned directly under the body, like pillars, a position that may have helped support their heavy bodies. Because less energy was devoted to lifting the body from the ground, some dinosaurs are believed to have been fast runners. Members of different but closely related classes—the pterosaurs (the first vertebrates to fly) and ichthyosaurs and plesiosaurs (marine reptiles)—were also common at this time.

Dinosaurs were the biggest animals ever to walk on the planet, with some animals weighing up to 50 tonnes (metric tons) or over 100,000 pounds. The variety of the thousands of dinosaur species found in fossil form around the world is staggering. However, perhaps not surprisingly for such long-extinct species, scientists are still hotly debating many details of their lives. For example, an issue still unresolved is whether some dinosaur species were **endothermic**, capable of generating and retaining body heat through their own metabolism, as birds and mammals are, or whether they were **ectothermic**, dependent on external heat as the main source of their body heat, as most reptiles are. Another issue is whether dinosaurs exhibited parental care of their young.

All nonavian dinosaurs, and many other animals, died out abruptly during a mass extinction at the end of the Cretaceous period (about 65 mya). Although widely attributed to climatic change brought about by the impact of a meteorite, scientists continue to debate the cause or causes of this mass extinction. We do not yet know why dinosaurs died out, while many other animals, including birds and small mammals, survived.

Class Aves: The Birds The defining characteristics of birds (class Aves, plural of the Latin *avis,* meaning bird) are that they have feathers and nearly all species can fly. As we will see, the ability to fly has shaped nearly every feature of the bird body. The other vertebrates that have evolved the ability to fly, the bats and the now-extinct pterosaurs, used skin stretched tight over elongated limbs to fly. Such a surface can be irreparably damaged, though some holes may heal remarkably quickly. In contrast, birds use feathers, epidermal outgrowths that can be replaced if damaged. Recent research shows that feathers evolved in dinosaurs before the appearance of birds.

In the rest of this section, we will discuss the likely evolution of birds from dinosaur ancestors, outline the key characteristics of birds, and provide a brief overview of the various bird orders.

Modern Birds Evolved from Small, Feather-Covered Dinosaurs

To trace the evolution of birds, it is necessary to look at transitional forms, the earliest type of animals that had feathers. One of the first known fossils exhibiting the faint impression of feathers was *Archaeopteryx lithographica* (from the Greek, meaning ancient wings and stone picture), found in a limestone quarry in Germany in 1861. The fossil was dated at 150 million years old, which places it during the Jurassic period. Except for the presence of feathers, *Archaeopteryx* appears to have had features similar to those of dinosaurs (**Figure 34.20a**; see also Figure 22.19). First, the fossil had an impression of a long tail with many vertebrae, a dinosaur feature. Some modern birds have long tails, but they are made of feathers, with the actual tailbone being much reduced. Second, the wings had claws halfway down the leading edge, another dinosaur-like character. Among modern birds, only the hoatzin, a South American swamp-inhabiting bird, has claws on its wings, which enable the chicks to climb back into the nest if they fall out. A third dinosaur-like feature is *Archaeopteryx*'s toothed beak. Fourth, the fossils show that *Archaeopteryx* lacked an enlarged breastbone, a feature that modern birds possess to anchor their large flight muscles, so it likely could not fly.

Similarities between the structure of the skull, feet, and hind leg bones have led scientists to conclude that *Archaeopteryx* is closely related to **theropods**, a group of bipedal saurischian dinosaurs. The wings and feathers of *Archaeopteryx* may have enabled it to glide from tree to tree, helped to keep it warm, or cut out the glare when folded over its head when hunting, in much the same way as some herons fold their wings over their heads when they are fishing. Later, the wings and feathers may have taken on functions of flight.

In the mid 1990s, paleontologists unearthed fossils of about the same age as *Archaeopteryx* in China that similarly suggest a close kinship between dinosaurs and modern birds. *Caudipteryx zoui* was a dinosaur-like animal with feathers on its wings and tail and a toothed beak (**Figure 34.20b**). *Confuciusornis sanctus* was a small, flightless but completely feathered dinosaur lacking the long, bony tail and toothed jaw found in other theropod dinosaurs. Its large tail feathers may have functioned in courtship displays (**Figure 34.20c**).

These three species—*Archaeopteryx, Caudipteryx,* and *Confuciusornis*—help trace a lineage from dinosaurs to birds. By the early Cretaceous period, and only a relatively short period after *Archaeopteryx* evolved, the fossil record shows the existence of a huge array of bird types resembling modern species. These were to share the skies with pterosaurs for 70 million years, before eventually having the airways to themselves.

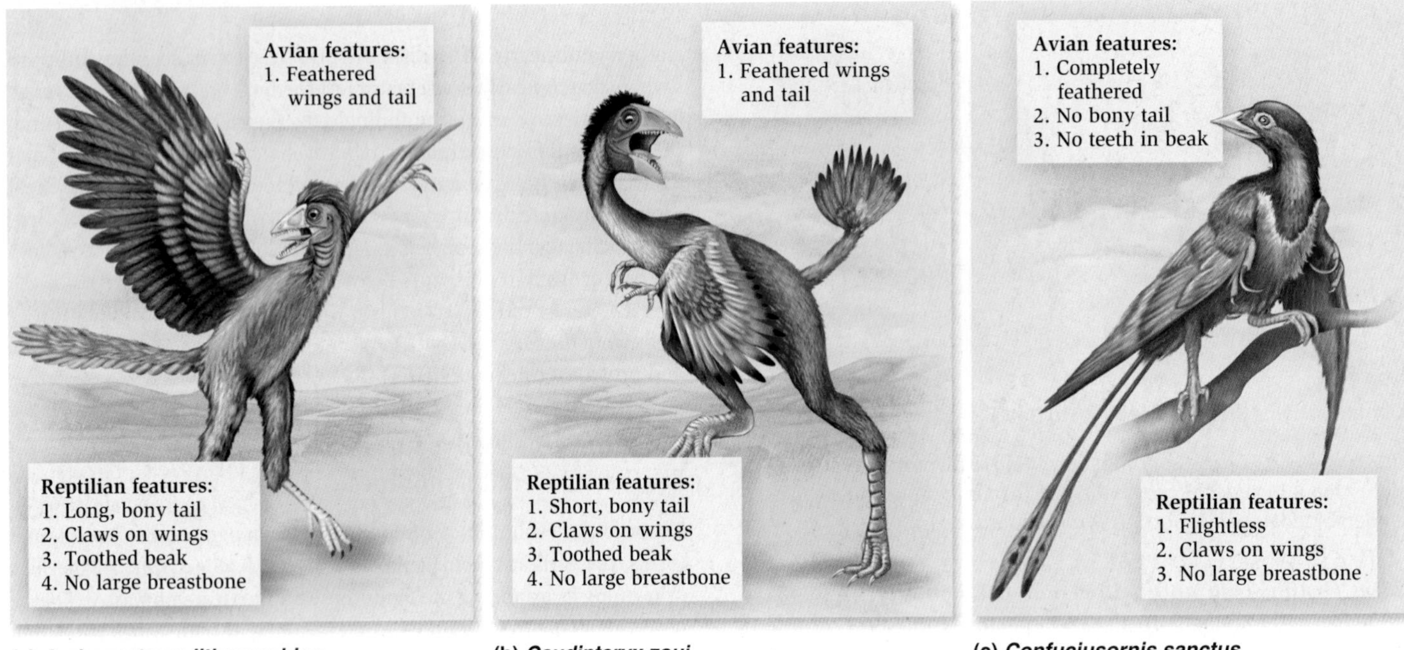

(a) ***Archaeopteryx lithographica*** (b) ***Caudipteryx zoui*** (c) ***Confuciusornis sanctus***

Figure 34.20 Transitional forms between dinosaurs and birds. **(a)** *Archaeopteryx lithographica* was a Jurassic animal with dinosaur-like features as well as wings and feathers. **(b)** *Caudipteryx zoui* was a dinosaur with feathers on its tail and wings. **(c)** *Confuciusornis sanctus* was a birdlike animal with a horny, toothless beak.

Birds Have Feathers, a Lightweight Skeleton, Air Sacs, and Reduced Organs

Modern birds possess many characteristics, including scales on their feet and legs and shelled eggs, that reveal their reptilian ancestry. In addition, however, among living animals birds have four unique features, all of which are associated with flight.

1. **Feathers.** Feathers are modified scales that keep birds warm and enable flight (Figure 34.21a). Soft, downy feathers, which are close to the body, maintain heat, whereas stiffer contour feathers, supported on a modified forelimb, give the wing the airfoil shape it needs to generate lift. Each contour feather develops from a follicle, a tiny pit in the skin. If a feather is lost, a new one can be regrown. The contour feathers consist of many

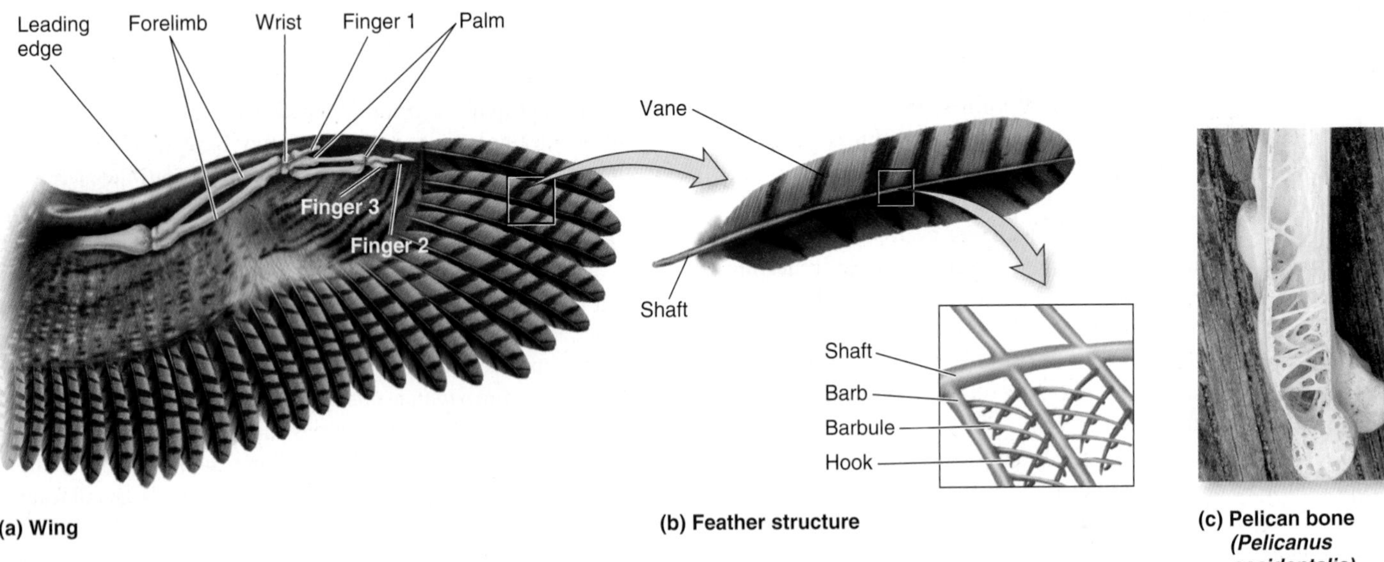

(a) Wing **(b) Feather structure** **(c) Pelican bone *(Pelicanus occidentalis)***

Figure 34.21 Features of the bird wing and feather. **(a)** The wing is supported by an elongated and modified forelimb with three extended fingers. **(b)** Each feather has a hollow shaft that supports many barbs, which, in turn, support barbules that interlock with hooks to give the feather its form. **(c)** The bones of a pelican (*Pelicanus occidentalis*) are hollow but crisscrossed with a honeycomb structure that provides added strength.

Concept Check: *What adaptations in birds help reduce their body weight to enable flight?*

BioConnections: *Look ahead to Figure 48.13. How do birds acquire enough oxygen to fuel their metabolism to support flight?*

paired barbs, each of which supports barbules that contain hooks that interlock with barbules from neighboring barbs to give the feather its shape (Figure 34.21b).

2. **Air sacs.** Flight requires a great deal of energy generated from an active metabolism that requires abundant oxygen. Birds have nine air sacs—large, hollow sacs that may extend into the bones (look ahead to Figure 48.11)—that expand and contract when a bird inhales and exhales, while the lungs remain stationary. Air is therefore being constantly moved across the lungs during inhalation and exhalation. Although making bird breathing very efficient, this process also makes birds especially susceptible to airborne toxins (hence, the utility of the canary in the coal mine; the bird's death signaled the presence of harmful carbon dioxide or methane gas that was otherwise unnoticed by miners).
3. **Reduction of organs.** Some organs are reduced in size or are lacking altogether in birds, which reduces the total mass that the bird carries. For example, birds have only one ovary and can carry relatively few eggs. As a result, they lay fewer eggs than most other reptile species. In fact, the gonads of both males and females are reduced, except during the breeding season, when they increase in size. Most birds also lack a urinary bladder. In addition, the loss of teeth reduces weight at the head end.
4. **Lightweight bones.** Most bird bones are thin and hollow and are crisscrossed internally by tiny pieces of bone to give them a honeycomb structure (Figure 34.21c). An enlarged breastbone, or **sternum**, provides an anchor on which a bird's powerful flight muscles attach. These muscles may contribute up to 30% of the bird's body weight. Birds' skulls are lighter than skulls from mammals of approximately the same weight. The keratin of bird beaks is tough and malleable, and a wide assortment of bird beaks have evolved, with the form dependent on the function of the beak (Figure 34.22).

Birds also have other distinct features, though mammals also possess some of these. For example, birds are endotherms, which ensures rapid metabolism and the quick production of adenosine triphosphate (ATP) that these active organisms need to fuel flight and other activities. In fact, birds' body temperatures are generally 40–42°C, considerably warmer than the human body's average of 37°C. Birds have a double circulation and a four-chambered heart that ensures rapid blood circulation. Rapid flight requires good vision, and bird vision is the best in the vertebrate world. Birds generally have high energy needs, and most birds are carnivores, eating insects or other invertebrates. However, some birds, such as parrots, eat just the more-nutrient-rich fruits and seeds. Bird eggs also need be kept warm for successful development, which entails brooding by an adult bird. Often, the males and females take turns brooding so that one parent can feed and maintain its strength. Picking successful partners is therefore an important task, and birds often engage in complex courtship rituals (look ahead to Figure 55.22).

(a) Cracking beak

(b) Scooping beak

(c) Tearing beak

(d) Probing beak

(e) Nectar-feeding beak

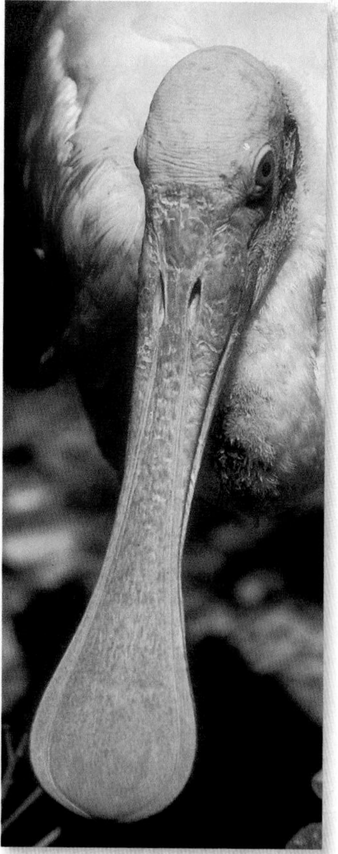
(f) Sieving beak

Figure 34.22 **A variety of bird beaks.** Birds have evolved a variety of beak shapes used in different types of food gathering. **(a)** Hyacinthe macaw (*Anodorhynchus hyacinthinus*)—cracking. **(b)** White pelican (*Pelecanus onocrotalus*)—scooping. **(c)** Verreaux's eagle (*Aquila verreauxii*)—tearing. **(d)** American avocet (*Recurvirostra americana*)—probing. **(e)** Lucifer hummingbird (*Calothorax lucifer*)—nectar feeding. **(f)** Roseate spoonbill (*Ajaia ajaja*)—sieving.

BIOLOGY PRINCIPLE Structure determines function. Each of these beak shapes permits a different method of feeding.

Table 34.2 The Main Orders of Birds, in Order of Species Richness

Order	Examples (approx. # of species)	Main characteristics
Passeriformes	Robins, starlings, sparrows, warblers (5,400)	Perching birds with perching feet; songbirds
Apodiformes	Hummingbirds, swifts (430)	Fast fliers with rapidly beating wings; small bodies
Piciformes	Woodpeckers, toucans (380)	Large with specialized beaks; two toes pointing forward and two backward
Psittaciformes	Parrots, cockatoos (360)	Large, powerful beaks
Chadradriiformes	Seagulls, wading birds (330)	Shorebirds
Columbiformes	Doves, pigeons (330)	Round bodies; short legs
Galliformes	Chickens, pheasants, quail (270)	Often large birds; weak flyers; ground nesters
Accipitriformes	Eagles, hawks, vultures (240)	Large diurnal carnivores; birds of prey; powerful talons; strong beaks
Coraciiformes	Hornbills, kingfishers (200)	Large beaks; cavity nesters
Strigiformes	Owls (170)	Nocturnal carnivores; powerful talons; strong beaks
Anseriformes	Ducks, swans, geese (160)	Able to swim; webbed feet; broad bills
Pelecaniformes	Pelicans, ibises, herons (100)	Large, water inhabiting
Sphenisciformes	Penguins (17)	Flightless; wings modified into flippers for swimming; marine; Southern Hemisphere

There Are Many Orders of Birds, All with the Same Body Plan

Birds are the most species-rich class of terrestrial vertebrates, with 28 orders, 166 families, and about 10,000 species (**Table 34.2**). Despite this diversity, birds lack the variety of body shapes that exist in the other endothermic class of vertebrates, the mammals, some of which can swim, others fly, others walk on four legs, and yet others walk only on two legs. Most birds fly, and therefore, most have the same general body shape. The biggest departures from this body shape are the flightless birds, including the cassowaries, emus, and ostriches. These birds have smaller wing bones, and the keel on the breastbone is greatly reduced or absent. Penguins are also flightless birds whose upper limbs are modified as flippers used in swimming.

34.5 Mammals: Milk-Producing Amniotes

Learning Outcomes:

1. Identify the four features that separate mammals from other vertebrate classes.
2. List the defining characteristics of primates.
3. Discuss the evolution of modern humans, *Homo sapiens*.

Mammals evolved from amniote ancestors earlier than birds. About 225 mya, the first mammals appeared in the mid-Triassic period (refer back to Figure 22.18). They evolved from small mammal-like reptiles that went extinct about 170 mya. Mammals survived,

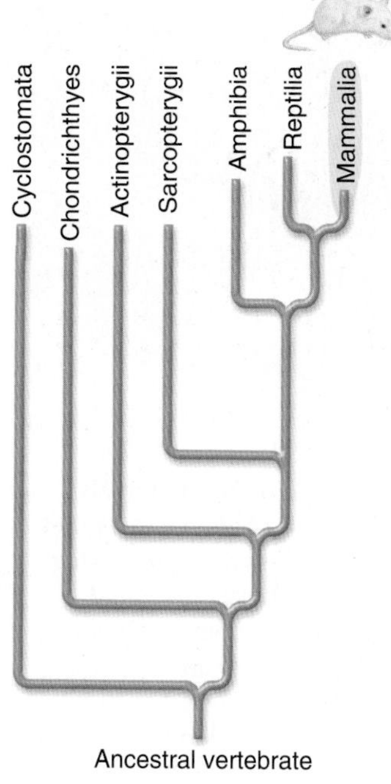

and until recently, most were believed to have been small, insect-eating species that lived in the shadows of dinosaurs. However, in January 2005, two fossils of a 130-million-year-old mammalian genus called *Repenomamus* were discovered that challenge the notion of mammals as small insect eaters. One fossil was of an animal estimated to weigh about 13 kg (30 lbs), about the size of a small dog, which is larger than some dinosaurs living in the same region at the time. The other fossil had the remains of a baby dinosaur in its stomach area.

The extinction of the dinosaurs in the Cretaceous period, some 65 mya, paved the way for mammals to increase in size. Today, biologists have identified about 5,500 species of mammals with a diverse array of lifestyles, from fishlike dolphins to birdlike bats, and from small insectivores such as shrews to large herbivores such as giraffes and elephants. The range of sizes and body forms of mammals is unmatched by any other vertebrate group, and mammals are prime illustrations of the concept that organismal diversity is related to environmental diversity. In this section, we will outline the features that distinguish mammals from other taxa. We will also examine the diversity of mammals that exists on Earth and will end by turning our attention to the evolution of primates and, in particular, humans.

Mammals Have Mammary Glands, Hair, Specialized Teeth, and an Enlarged Skull

Four characteristics distinguish mammals: the possession of mammary glands, hair, specialized teeth, and an enlarged skull.

- **Mammary glands.** Mammals, or the class Mammalia (from the Latin *mamma*, meaning breast), are named after the female's distinctive mammary glands, which secrete milk. Milk is a fluid rich in fat, sugar, protein, and vital minerals, especially calcium. Newborn mammals suckle this fluid, which helps promote rapid growth.
- **Hair.** All mammals have hair, although some have more than others. Whales have hair in utero, but adults are hairless or retain only a few hairs on their snout. Compared with many mammals, humans are relatively hairless. In some animals, the hair is dense and is referred to as fur. In some aquatic species such as beavers, the fur is so dense it cannot be thoroughly wetted, so the hair underneath remains dry. Mammals are endothermic, and their fur is an efficient insulator. Hair can also take on functions other than insulation. Many mammals, including cats, dogs, walruses, and whales, have sensory hairs called vibrissae (**Figure 34.23a**). Hair can be of many colors, to allow the mammals to blend into their background (**Figure 34.23b**). In some cases, as in porcupines and hedgehogs, the hairs become long, stiffened, and sharp (quills) and serve as a defense mechanism (**Figure 34.23c**).
- **Specialized teeth.** Mammals are the only vertebrates with highly differentiated teeth—incisors, canines, premolars, and molars—that are adapted for different types of diets (**Figure 34.24**). Although teeth are generally present in all species, different teeth are larger, smaller, lost, or reduced, depending on diet. Of particular importance to carnivores such as wolves are the

(a) Sensory hairs

(b) Camouflaged coat

(c) Defensive quills

Figure 34.23 **Mammalian hair.** **(a)** The sensory hairs (vibrissae) of the walrus (*Odobenus rosmarus*). **(b)** The camouflaged coat of a bobcat (*Lynx rufus*). **(c)** The defensive quills of the crested porcupine (*Hystrix africaeaustralis*).

(a) Biting teeth

(b) Grinding teeth

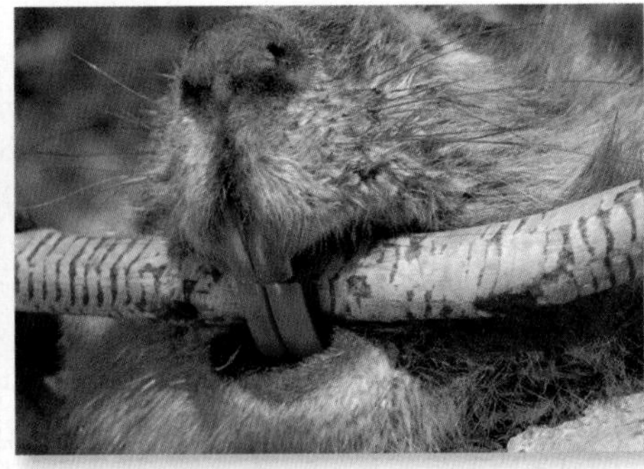

(c) Gnawing teeth

(d) Tusks

(e) Grasping teeth

Figure 34.24 Mammalian teeth. Mammals have different types of teeth, according to their diet. **(a)** The wolf has long canine teeth that bite its prey. **(b)** The deer has a long row of molars that grind plant material. **(c)** The beaver, a rodent, has long, continually growing incisors used to gnaw wood. **(d)** The elephant's incisors are modified into tusks. **(e)** Dolphins and other fishes or plankton feeders have numerous small teeth used to grasp prey.

piercing canine teeth, whereas herbivorous species such as deer depend on their chisel-like incisors to snip off vegetation and their many molars to grind plant material. Only mammals chew their food in this fashion. Rodent incisors grow continuously throughout life, and species such as beavers wear them down by gnawing tough plant material such as wood. Mammals that have different types of teeth are called heterodont; others, such as dolphins, where the teeth are of uniform size and shape, are called homodont.

- **Enlarged skull.** The mammalian skull differs from other amniote skulls in several ways. First, the brain is enlarged and is contained within a relatively large skull. Second, mammals have a single lower jawbone, unlike reptiles, whose lower jaw is composed of multiple bones. Third, mammals have three bones in the middle ear, as opposed to reptiles, which have one bone in the middle ear. Fourth, most mammals, except some seals, have external ears.

In addition to those uniquely mammalian characteristics, some, but not all, mammals possess these additional features:

- **The ability to digest plants.** Apart from tortoises and marine iguanas, certain species of mammals are the only large vertebrates alive today that can exist on a steady diet of grasses or tree leaves; indeed, most large mammals are herbivores. Though mammals cannot digest cellulose, the principal constituent of the cell wall of many plants, some species have a large four-chambered stomach containing cellulose-digesting bacteria. These bacteria can break down the cellulose and make the plant cell contents available to the animal. Others have an extensive cecum or large intestine where digestion occurs.
- **Horns and antlers.** Mammals are the only living class of vertebrates to possess horns or antlers. Many mammals, especially antelopes, cattle, and sheep, have horns, typically consisting of a bony core that is a permanent outgrowth of the skull surrounded by a hairlike keratin sheath, as shown in the large antelope called a kudu (**Figure 34.25a**). Rhinoceros horns are outgrowths of the epidermis, consisting of very tightly matted hair (**Figure 34.25b**). In contrast, deer antlers are made entirely of bone (**Figure 34.25c**). Deer grow a new set of antlers each year and shed them after the mating season. Hooves are also made of keratin and protect an animal's toes from the impact of its feet striking the ground.

Mammals Are the Most Diverse Group of Vertebrates Living on Earth

Modern mammals are incredibly diverse in size and life styles (**Table 34.3**). They vary in size from tiny insect-eating bats, weighing in at only 2 g, to leviathans such as the blue whale, the largest animal ever known, which tips the scales at 100 tonnes (over 200,000 lbs). Mammalian orders are divided into two distinct subclasses (**Figure 34.26**). The subclass Prototheria contains only the order Monotremata, or **monotremes**, which are found in Australia and New Guinea. There are only five species: the duck-billed platypus (**Figure 34.27a**) and four species of echidna, a spiny animal resembling a hedgehog. Monotremes are early-diverging mammals that lay eggs rather than bear live young, lack a placenta, and have mammary glands with poorly developed nipples. The mothers incubate the eggs, and upon hatching, the young simply lap up the milk as it oozes onto the fur.

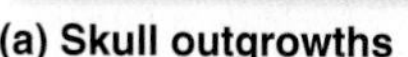

(a) Skull outgrowths

(b) Epidermal outgrowths

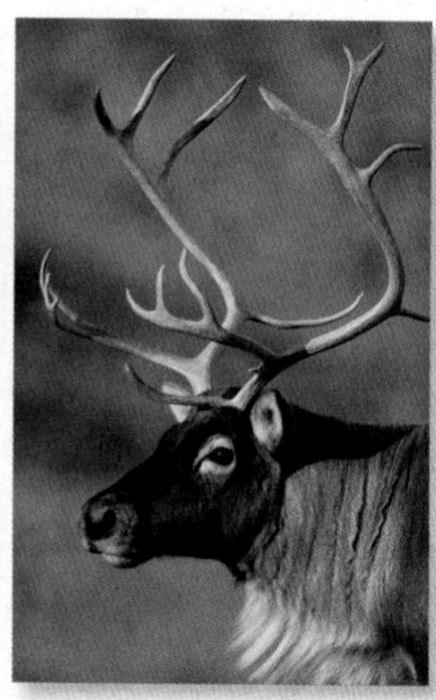

(c) Bony antlers

Figure 34.25 Horns and antlers in mammals. Mammals have a variety of outgrowths that are used for defense or by males as weapons in contests over females. **(a)** The horns of this male kudu (*Tragelaphus strepsiceros*) are bony outgrowths of the skull covered in a keratin sheath. **(b)** The horns of the black rhinoceros (*Diceros bicornis*) are outgrowths of the epidermis, made of tightly matted hair. **(c)** The antlers of the caribou (*Rangifer tarandus*), also known as reindeer, are made entirely of bone and are grown and shed each year.

Table 34.3 The Main Orders of Mammals, in Order of Species Richness

Order	Examples (approx. # of species)	Main characteristics
Rodentia	Mice, rats, squirrels, beavers, porcupines (2,277)	Plant eating; gnawing habit, with two pairs of continually growing incisor teeth
Chiroptera	Bats (1,116)	Insect or fruit eating; small; have ability to fly; navigate by sonar; nocturnal
Eulipotyphla	Shrews, moles, hedgehogs (452)	Insect eaters; primitive placental mammals
Primates	Monkeys, apes, humans (404)	Opposable thumb; binocular vision; large brains
Carnivora	Cats, dogs, weasels, bears, seals, sea lions (286)	Flesh-eating mammals; canine teeth
Artiodactyla	Deer, antelopes, cattle, sheep, goats, camels, pigs (240)	Herbivorous hoofed mammals, usually with two toes, hippopotamus and others with four toes; many with horns or antlers
Diprotodontia	Kangaroos, koalas, opossums, wombats (143)	Pouched mammals mainly found in Australia
Lagomorpha	Rabbits, hares (92)	Powerful hind legs; rodent-like teeth
Cetacea	Whales, dolphins (84)	Marine fishes or plankton feeders; front limbs modified into flippers; no hind limbs; little hair except on snout
Perissodactyla	Horses, zebras, tapirs, rhinoceroses (18)	Hoofed herbivorous mammals with odd number of toes, one (horses) or three (rhinoceroses)
Monotremata	Duck-billed platypuses, echidna (5)	Egg-laying mammals found only in Australia and New Guinea
Proboscidea	Elephants (3)	Long trunk; large, upper incisors modified as tusks

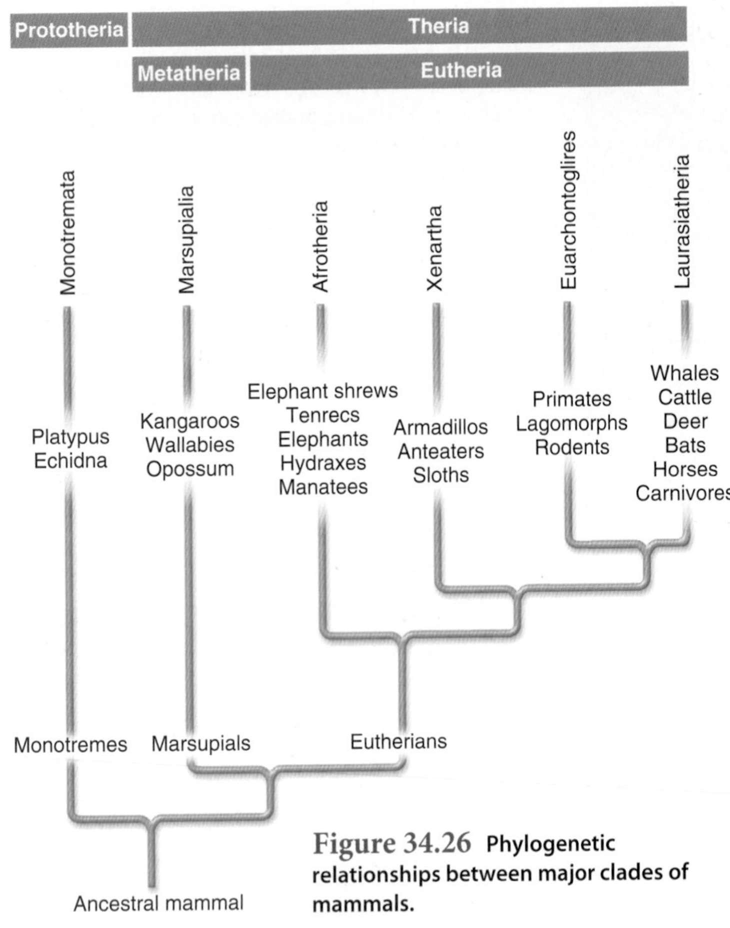

Figure 34.26 Phylogenetic relationships between major clades of mammals.

The subclass Theria contains all remaining live-bearing mammals. The Theria are divided into two clades, the Metatheria and the Eutheria. The clade Metatheria, or the **marsupials**, is a group of seven orders, with about 280 species, including the rock wallaby pictured in Figure 34.27b. Once widespread, members of this order are now largely confined to Australia, although some marsupials exist in South America, and one species—the opossum—is found in North America. Fertilization is internal, and reproduction is viviparous in marsupials. Marsupials have a placenta that nourishes the embryo. Unlike other mammals, however, marsupials are extremely small when they are born (often only 1–2 cm) and make their way to a ventral pouch called a marsupium for further development.

All the other mammalian orders are members of the clade Eutheria and are considered **eutherians**, or placental mammals, such as the orangutans shown in Figure 34.27c. Eutherians have a long-lived and complex placenta, compared with that of marsupials. In eutherians, fertilization is internal, and reproduction is viviparous, but the developmental period, or gestation, of the young is prolonged. Molecular studies suggest four clades of placental mammals that diverged in the Cretaceous. The earliest diverging clade was the Afrotheria, which evolved in the African landmass starting about 110–100 mya. This clade includes the elephant shrews, tenrecs, golden moles, manatees and dugongs, hyraxes, aardvark, and elephants. Shortly thereafter, about 100–95 mya, the Xenartha evolved in South America, where the armadillos, anteaters, and sloths appeared. The other two clades, the Euarchontoglires, containing the primates, lagomorphs, and rodents, and the Laurasiatheria, containing the whales, artiodactyla, bats, horses, and carnivores, both evolved in the northern continent

(a) Prototherian (duck-billed platypus)

(c) Eutherian (orangutan)

(b) Metatherian (rock wallaby)

Figure 34.27 Diversity among mammals. **(a)** Prototherians, such as this duck-billed platypus (*Ornithorhynchus anatinus*), lay eggs, lack a placenta, and possess mammary glands with poorly developed nipples. **(b)** Metatherians, or marsupials, such as this rock wallaby (*Petrogale assimilis*), feed and carry their developing young, or "joeys," in a ventral pouch. **(c)** Gestation lasts longer in eutherians, and their young are more developed at birth, as illustrated by this young orangutan (*Pongo pygmaeus*).

BioConnections: Look ahead to Figure 51.11. The placenta serves as the provisional lungs, intestine, and kidneys of the developing fetus. How much mixing is there of maternal and fetal blood?

of Laurasia and became separate about 95–85 mya. Later, following continental drift, Africa and Arabia collided with Laurasia, and the Isthmus of Panama joined North and South America. These new land bridges facilitated animal movement between once separated continents.

The diversity of mammals is often threatened by human activities such as habitat destruction. In addition, many species are hunted for food. Others, such as wild cats and whales, were hunted for their products (fur and oil, respectively), and still others, such as the oryx, have simply been shot for sport.

Primates Are Mammals with Opposable Thumbs and a Large Brain

The primates, and specifically humans, have had a huge impact on the world. Primates are primarily tree-dwelling species that are believed to have evolved from a group of small, arboreal insect-eating mammals about 85 mya, before dinosaurs went extinct. Primates have several defining characteristics, mostly relating to their tree-dwelling nature:

- **Grasping hands.** All primates have grasping hands, a characteristic that enables them to hold onto branches (see Figure 34.29a,b). Most primate species also possess an opposable thumb, a thumb that can be placed opposite the fingers of the same hand, which gives them a precision grip and enables the manipulation of small objects. All primates except humans also have an opposable big toe.
- **Large brain.** Acute vision and other senses enhancing the ability to move quickly through the trees require the efficient processing of large amounts of information. As a result, the primate brain is large and well developed. In turn, this has facilitated complex social behaviors.
- **At least some digits with flat nails instead of claws.** This feature is believed to aid in the manipulation of objects.
- **Binocular vision.** Primates have forward-facing eyes that are positioned close together on a flattened face, though some other mammals share this characteristic. Jumping from branch to branch requires accurate judgment of distances. This is facilitated by binocular vision in which the field of vision for both eyes overlaps, producing a single image.
- **Complex social behavior and well-developed parental care.** Primates have a tendency toward complex social behavior and increased parental care.

Some of these characteristics are possessed by other animals. For example, binocular vision occurs in owls and some other birds, grasping hands are found in raccoons, and relatively large brains occur in marine mammals. Primates are defined by possessing the whole suite of these characteristics together.

Primates may be classified in several ways. Taxonomists often divide them into two groups: the strepsirrhini and the haplorrhini (**Figure 34.28**). The **strepsirrhini** contain the smaller species such as bush babies, lemurs, and pottos. These are generally nocturnal and smaller-brained primates with eyes positioned a little more toward the side of their heads (**Figure 34.29a**). The strepsirrhini are named for their wet noses with no fur at the tip. The **haplorrhini** have dry noses with a fully furred nose tip and fully forward-facing eyes. This group consists of the larger-brained and diurnal **anthropoidea**: the monkeys (**Figure 34.29b**) and the **hominoidea** (gibbons, orangutans, gorillas, chimpanzees, and humans) (**Figure 34.29c**). The tarsiers also belong in the haplorrhini, despite their small size, based on their forward-facing eyes and DNA similarities to the monkeys and apes.

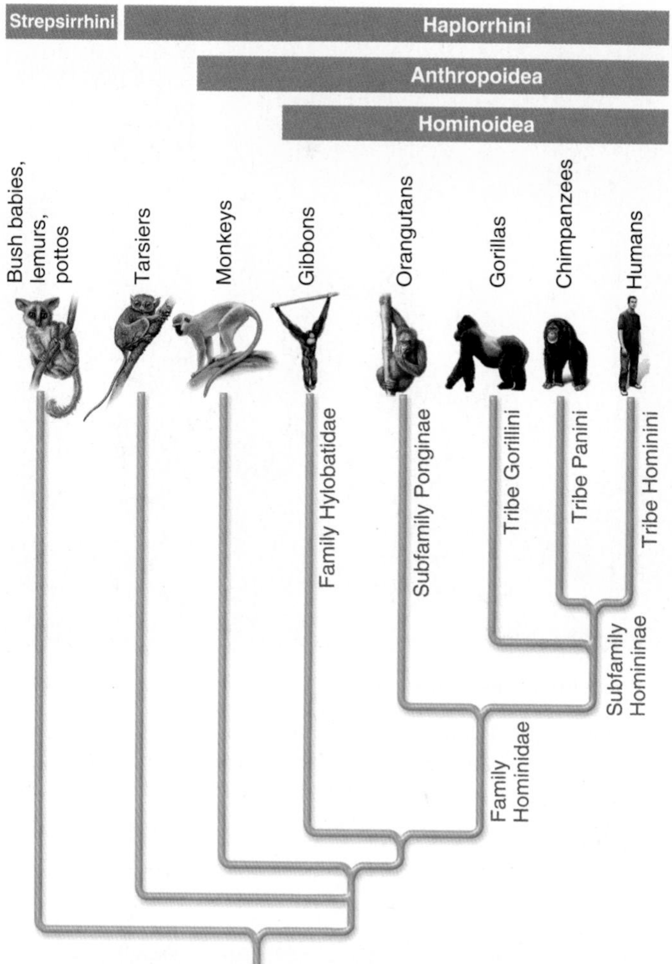

Figure 34.28 Evolutionary tree of the primates.

What differentiates monkeys from hominoids? Most monkeys have tails, but hominoids do not. In addition, apes have more mobile shoulder joints, broader rib cages, and a shorter spine. These features aid in brachiation, a swinging movement in trees. Apes also possess relatively long limbs and short legs and, with the exception of gibbons, are much larger than monkeys. The 20 species of hominoids are split into two groups: the lesser apes (family Hylobatidae), or the gibbons; and the greater apes (family Hominidae), or the orangutans, gorillas, chimpanzees, and humans (**Figure 34.30**). The lesser apes are strictly arboreal, whereas the greater apes often descend to the ground to feed.

Although humans are closely related to chimpanzees and gorillas, they did not evolve directly from them. Rather, all hominoid species shared a common ancestor. Recent molecular studies show that

(a) Strepsirrhini (lesser bush baby)

(b) Anthropoidea (capuchin monkey)

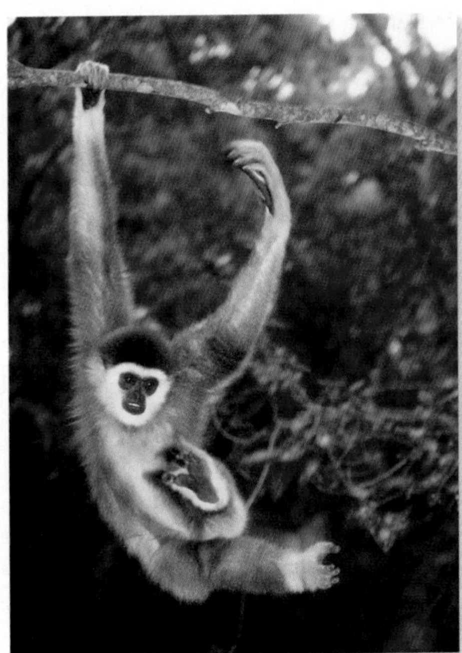
(c) Hominoidea (white-handed gibbon)

Figure 34.29 **Primate classification.** Many authorities divide the primates into two groups: **(a)** the strepsirrhini (smaller, nocturnal species such as this bush baby), and the haplorrhini (larger diurnal species). Haplorrhini comprise **(b)** the monkeys and tarsiers, such as this Capuchin monkey (*Cebus capucinus*), and **(c)** the hominoids, species such as this white-handed gibbon (*Hylobates lar*).

Concept Check: What are the defining features of primates?

(a) Gorilla (*Gorilla gorilla*)

(b) Chimpanzee (*Pan troglodytes*)

(c) Human (*Homo sapiens*)

Figure 34.30 **Members of the family Hominidae.** **(a)** Gorillas, the largest of the living primates, are ground-dwelling herbivores that inhabit the forests of Africa. **(b)** Chimpanzees are smaller, omnivorous primates that also live in Africa. The chimpanzees are close living relatives of modern humans. **(c)** Humans are also members of the family Hominidae. The orangutan is also a member of this group.

gorillas, chimpanzees, and humans are more closely related to one another than they are to gibbons and orangutans, so scientists have split the family Hominidae into groups, including the subfamily Ponginae (orangutans) and the subfamily Homininae (gorillas, chimpanzees, and humans and their ancestors). In turn, the Homininae are split into three tribes: the Gorillini (gorillas), the Panini (chimpanzees), and the Hominini (humans and their ancestors). The sequencing of the chimpanzee genome by the Chimpanzee Sequencing and Analysis Consortium in 2005 allowed detailed comparisons to be made with the human genome.

GENOMES & PROTEOMES CONNECTION

Comparing the Human and Chimpanzee Genetic Codes

A male chimp called Clint who lived at a primate research center in Atlanta provided the DNA used to sequence the chimp genome. In 2005, the Chimpanzee Sequencing and Analysis Consortium published an initial sequence of the chimpanzee genome. The draft sequence followed the 2003 publication of the human genome (see Chapter 21) and allowed scientists to make detailed comparisons between the two species. These comparisons revealed that the sequence of base pairs making up both species' genomes differ by only 1.23%, 10 times less than the difference between the mouse and rat genomes. Comparisons of human and chimpanzee proteomes revealed that 29% of all proteins are identical, with most others differing by one or two amino acid substitutions.

Many of the genetic differences between chimps and humans result from chromosome inversions and duplications. Geneticists have found over 1,500 inversions between the chimp and human genomes. Although many inversions occur in the noncoding regions of the genome, the DNA in these regions may regulate the expression of the genes in the coding regions. Duplications and deletions are also common. For example, one gene that codes for a subunit of a protein found in areas of the brain occurs in multiple copies in a wide range of primates, but humans have the most copies. However, humans appear to have lost a gene called *caspase-12*, which in other primates may protect against Alzheimer disease.

Some interesting genetic differences were apparent between chimps and humans even before their entire genomes were sequenced. In 1998, Indian physician-geneticist Ajit Varki and colleagues investigated a molecule called sialic acid that occurs on cell surfaces and acts as a locking site for pathogens such as malaria and influenza. They found an altered form of the molecule in humans, coded for by a single damaged gene, which may explain why humans are more susceptible to these diseases than are chimpanzees. In 2002, Swedish molecular geneticist Svante Pääbo discovered differences between humans and chimps in a gene called *FOXP2*, which plays a role in speech development. Proteins coded for by this gene differ in just two amino acids of a 715-amino-acid sequence. Researchers propose that the mutations in this gene have been crucial for the development of human speech.

More recently, a team led by American geneticist David Reich in 2006 discovered that the human X chromosome diverged from the chimpanzee X chromosome about 1.2 million years more recently than the other chromosomes. This indicated to the researchers that the human and chimp lineages split apart, then began interbreeding before diverging again. This would explain why many fossils appear to exhibit traits of both humans and chimps, because they may actually have been human-chimpanzee hybrids.

Humans Evolved from Ancestral Primates

About 6 mya in Africa, a lineage that led to humans began to separate from other primate lineages. The evolution of humans should not be viewed as a neat, stepwise progression from one species to another. Rather, human evolution, like the evolution of most species, can be visualized more like a tree, with one or two **hominin** species—members of the Hominini tribe—likely coexisting at the same point in time, with some eventually going extinct and some giving rise to other species (**Figure 34.31**).

The key characteristic differentiating hominins from other apes is that hominins walk on two feet, that is, they are **bipedal**. At about the time when hominins diverged from other ape lineages, the Earth's climate had cooled, and the forests of Africa had given way to grassy savannas. A bipedal method of locomotion and upright stance may

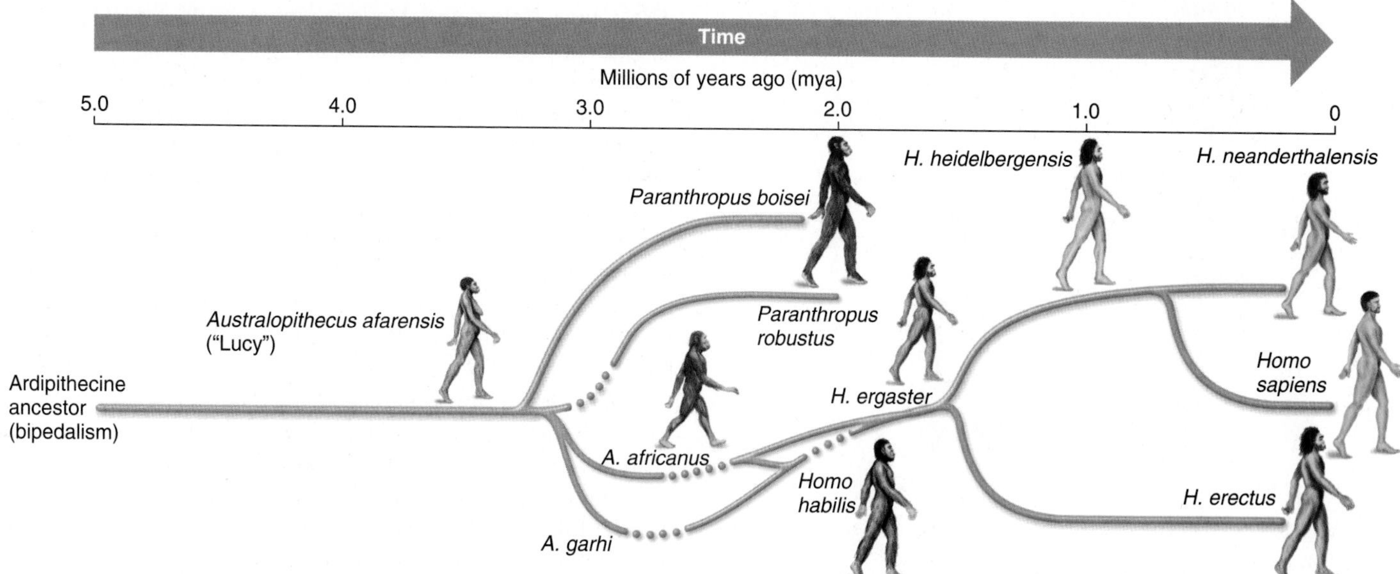

Figure 34.31 A possible scenario for human evolution. In this human family tree (based on the ongoing work of Donald Johanson), several hominin species lived contemporaneously with one another, but only one lineage gave rise to modern humans (*Homo sapiens*).

have been advantageous in allowing hominins to peer over the tall grass of the savanna to see predators or even prey.

Bipedalism is correlated with many anatomical changes in hominins. First, the opening of the skull where the spinal cord enters shifted forward, allowing the spine to be more directly underneath the head. Second, the hominin pelvis became broader to support the additional weight. And third, the lower limbs, used for walking, became relatively larger than those in other apes. These are the types of anatomical changes paleontologists look for in the fossil record to help determine whether fossil remains are hominin. The earliest known hominin, *Sahelanthropus tchadensis*, was discovered in Central Africa in 2002. Another early group of hominins included several species of a smaller-brained genus, *Australopithecus*, which first emerged in Africa about 4 mya. *Australopithecus afarensis* is generally regarded as the common ancestor of most of these species. From there, the evolution of different species becomes somewhat hazy. It is generally agreed that two genera evolved from *Australopithecus*: the robust *Paranthropus* and the more slender *Homo*. The early stages of the evolution of *Homo* species, and their differentiation from at least two possible *Australopithecus*, have not yet been determined with any certainty. However, the later divergence of various *Homo* species is a little better understood.

Australopithecines Since 1924, when the first fossil australopithecine (from the Latin *austral*, meaning southern, and the Greek *pithecus*, meaning ape) was found in South Africa, hundreds of fossils of this group have been unearthed all over southern and eastern Africa, the areas where fossil deposits are best exposed to paleontologists. This was a widespread group, with at least six species. In 1974, American paleontologist Donald Johanson unearthed the skeleton of a female *A. afarensis* in the Afar region of Ethiopia and dubbed her Lucy. (The Beatles' song "Lucy in the Sky with Diamonds" was playing in the camp the night when Johanson was sorting the unearthed bones.) Over 40% of the skeleton had been preserved, enough to provide a good idea of the physical appearance of australopithecines. Compared with modern humans, all were relatively small, about 1–1.5 m in height and around 18 kg in weight (Figure 34.32). Females were much smaller than males, a condition known as sexual dimorphism. Examination of the bones revealed that *A. afarensis* walked on two legs. They possessed a facial structure and brain size (about 500 cubic centimeters [cm^3]) similar to those of a chimp.

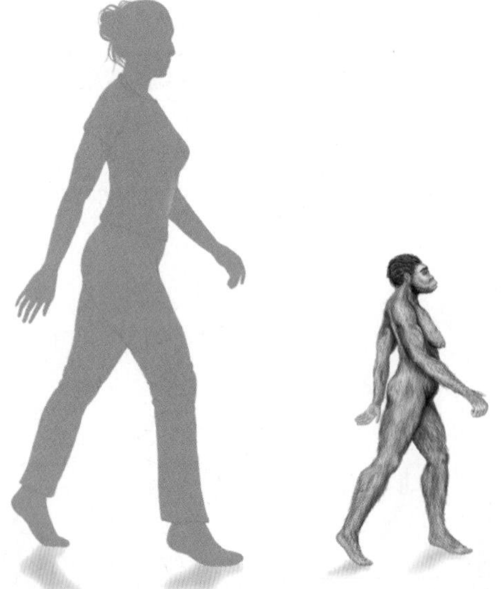

Figure 34.32 A modern woman compared to an australopithecine. Compared with modern humans, Australopithecines, as illustrated by this reconstruction of the famous fossil Lucy, were much smaller and lighter.

In the 1930s, the remains of bigger-boned hominids were found. Two of the larger species now considered to be a separate genus, *Paranthropus*, weighed about 40 kg and lived contemporaneously with australopithecines and members of *Homo* species. *Paranthropus* were vegetarians with enormous jaws used for grinding up tough roots and tubers. Both *Paranthropus* species died out rather suddenly about 1.5–2.0 mya. Although *Australopithecus africanus* was thought to have evolved slightly later than *A. afarensis*, its bones had been found much earlier than those of *A. afarensis*. In the 1920s, Australian anthropologist Raymond Dart described *A. africanus* from infant bones discovered in a cave in Taung, South Africa. The type specimen was called Taung child. The well-preserved skull was small but was well rounded, unlike the skulls of chimpanzees and gorillas. Also, the positioning of the head on the vertebral column suggested bipedalism. These facts suggested to Dart that he had found a transitional form between apes and humans. However, it would take another 20 years and the discovery of more fossils to convince the scientific world to support Dart's view. In 1996, remains of another species, *Australopithecus garhi*, were also found in the Afar region. They were somewhat of a surprise in that the dentition suggested similarities with *Paranthropus boisei*. "Garhi" means surprise in the local Afar language. The position of both *A. garhi* and *A. africanus* as ancestors of modern humans has been the subject of much debate, and they have been viewed as dead-end cousins or the ancestors of the first members of the genus *Homo*.

The Genus Homo and Modern Humans In the 1960s, British paleontologist Louis Leakey found hominin fossils estimated to be about 2 million years old in Olduvai Gorge, Tanzania. Two particularly interesting observations stand out about these fossils. First, reconstruction of the skull showed a brain size of about 680 cm^3, larger than that of *Australopithecus*. Second, the fossils were found with a wealth of stone tools. As a result, Leakey assigned the fossils to a new species, *Homo habilis*, from the Latin, meaning handy man. The discovery of several more *Homo* fossils followed, but there have been no extensive finds, as there were with Lucy. This makes it difficult to determine which *Australopithecus* lineage gave rise to the *Homo* lineage (see Figure 34.31), and scientists remain divided on this point.

Homo habilis lived alongside *Paranthropus* in East Africa but had much smaller jaws and teeth, indicating that it probably ate large quantities of meat. The smaller jaw provided more space in the skull for brain development. *Homo habilis* probably scavenged most of its meat from the kills of large predators. A meatier diet is easier to digest and is rich in nutrients and calories. The human brain uses a lot of energy, 20% of the body's total energy production. The meat-eating habit thus helped propel the evolution of increasing brain size in humans. Cut marks on animal bones of the period reveal that early humans used stone tools to smash open bones and extract the protein-rich bone marrow—a food source that other organisms were unable to obtain.

Although we are not clear exactly how, researchers believe that *H. habilis* probably gave rise to one of the most important species of *Homo*, *Homo ergaster. Homo ergaster* was a hominin that evolved in Africa; it had a human-looking face and skull, with downward-facing nostrils. *Homo ergaster* was also a tool user, and now the tools, such as hand axes, were larger and more sophisticated. *Homo ergaster* evolved in a period of global cooling and drying that reduced tropical forests even more and promoted savanna conditions. Hairlessness and the regulation of body temperature through sweating may also have evolved at this time as adaptations to the sunny environment. A leaner body shape was evident. We know this from so-called Turkana boy, a fossil teenage boy found in Kenya in 1984. Though only 13 years old, scientists predict he would have been about 185 cm (6 ft 1 in.) when adult, much the same height as the Masai tribesman that inhabit the area today. A dark skin probably protected *H. ergaster* from the sun's rays. The pelvis had narrowed, promoting efficiencies in walking upright, and the size of the brain and hence the skull increased, which may have produced more difficulty in childbirth. Mothers had to push increasingly large-brained infants through a narrowed pelvis. Researchers think that as a result, the human gestation period was shortened. Earlier birth leads to prolonged care of human infants compared with that in other apes. Prolonged childcare required well-nourished mothers, who would have benefited from the support of their male partner and other members of a social group. Some anthropologists have suggested this was the beginning of the family.

H. ergaster is thought to have given rise to many species, including *Homo erectus*, *Homo heidelbergensis*, *Homo neanderthalensis*, and *Homo sapiens.* A possible time line and geographic location for these species are given in Figure 34.33. *Homo ergaster* probably was the first type of human to leave Africa, as similar bones have been found in the Eurasian country of Georgia. *H. ergaster* is believed to be a direct ancestor of modern humans, with *Homo heidelbergensis* viewed as an intermediary step. Living contemporaneously with *H. heidelbergensis* was another descendent of *H. ergaster*, *H. erectus.*

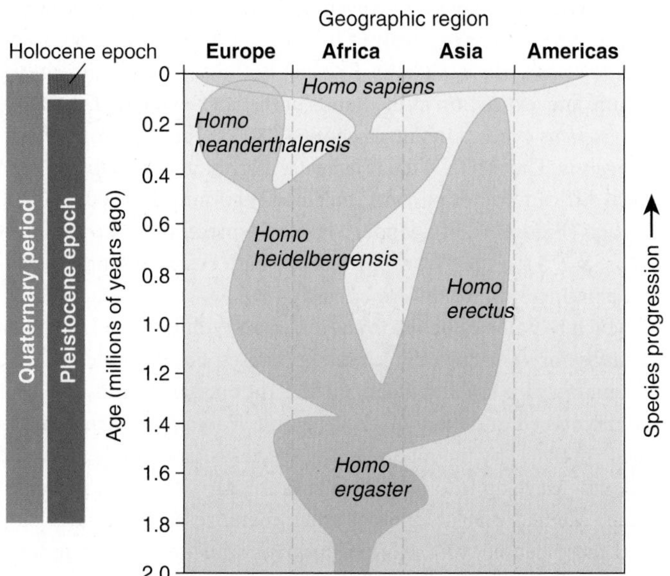

Figure 34.33 One view of the temporal and geographic evolution of hominid populations. The Holocene epoch is a geological time period beginning at the end of the Pleistocene and continuing to the present.

H. erectus was a large hominin, as large as a modern human but with heavier bones and a smaller brain capacity of between 750 and 1,225 cm^3 (modern brain size is about 1,350 cm^3). Fossil evidence shows that *H. erectus* was a social species that used tools, hunted animals, and cooked over fires. The meat-eating habit may have sparked the migration of *H. erectus*, because carnivores had larger ranges than similar-sized herbivores, their prey being scarcer per unit area. *H. erectus* spread out of Africa soon after the species appeared, over a million years ago, and fossils have been found as far away as China and Indonesia. The first fossil was found by Dutch physician Eugene Dubois in 1891 on the Indonesian island of Java. Stone tools are rarely found in these Asian sites, suggesting *H. erectus* based their technology on other materials, such as bamboo, which was abundant at that time. Bamboo is strong yet lightweight and could have been used to make spears. These people may even have used rafts to take to the seas. *H. erectus* went extinct about 100,000 years ago, for reasons that are unclear but may be related to the spread of *H. sapiens* into its range.

Homo heidelbergensis was similar in body form to modern humans. Large caches of their bones were found in Spain, at the bottom of a 14 m (45 ft) shaft known as La Sima de Los Huesos (the pit of bones). Similar remains were also found at Boxgrove in England. Shinbones recovered from Boxgrove suggest males stood around 180 cm (6 ft) and weighed 88 kg (196 pounds). Skulls were large, with brain volumes from 1,100 to 1,400 cm^3, similar to modern humans. Animal bones from these sites showed cut marks from stone blades beneath tooth marks from carnivores. This showed humans were killing large prey before scavengers arrived. Horses, giant deer, and rhinoceroses were common prey and would have required much skill and cooperation to hunt.

Homo heidelbergensis gave rise to two species, *H. neanderthalensis* and *H. sapiens. H. neanderthalensis* was named for the Neander Valley of Germany, where the first fossils of its type were found. In the Pleistocene epoch (see Figure 34.33), glaciers were locked in a cycle of advance and retreat, and the European landscape was often covered with snow. The more slender body form of *H. heidelbergensis* evolved into a shorter, stockier build that was better equipped to conserve heat; we now call this type of human Neanderthal. Neanderthals also possessed a more massive skull and larger brain size than modern humans, about 1,450 cm^3, perhaps associated with their bulk. Males were about 168 cm (5 ft 6 in.) in height and would have been very strong by modern standards. They had a large face with a prominent bridge over the eyebrows, a large nose, and no chin. They lived predominantly in Europe, with a range extending to the Middle East. Their muscular physique was well suited to the rigors of cold climates and hunting prey. Paleontologists have found a high rate of head and neck injuries in Neanderthal bones, similar to that seen in present-day rodeo riders. This suggests that close encounters with large prey often resulted in blows that knocked the hunters off their feet. The hyoid bone, which holds the larynx (voice box) in place, was

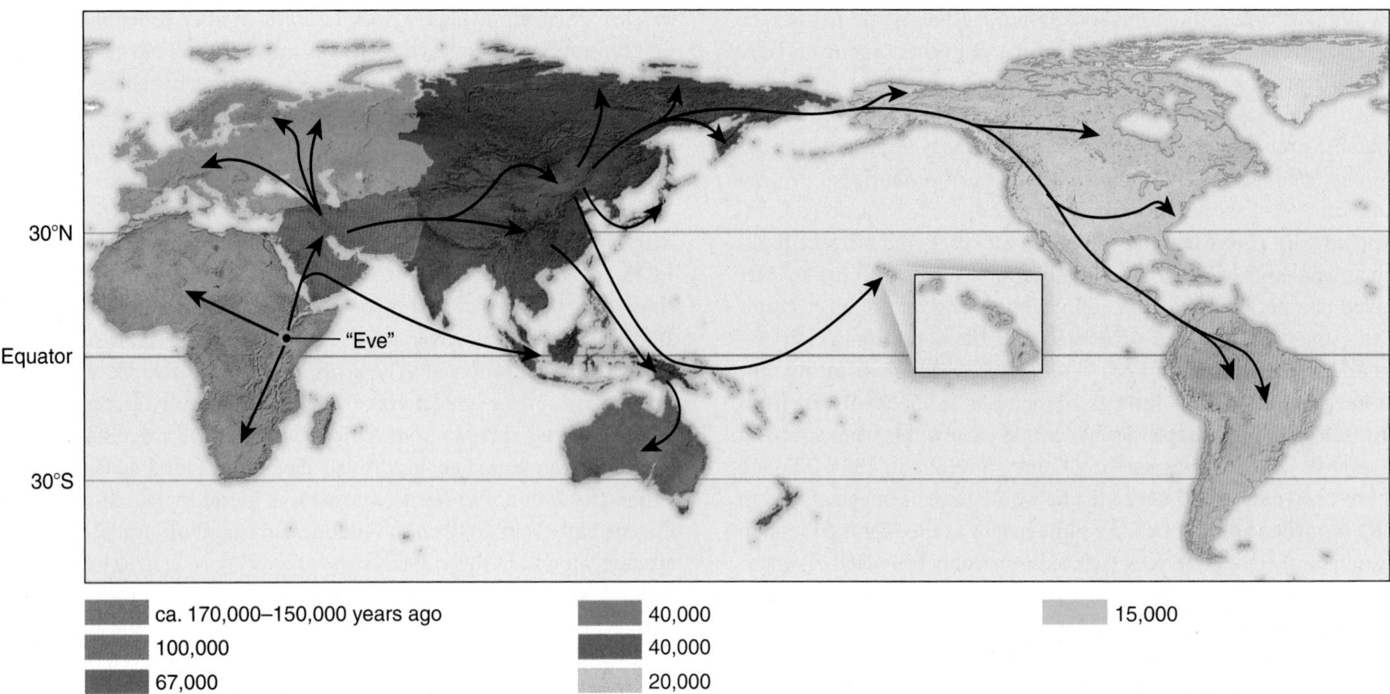

Figure 34.34 The probable origin and spread of *Homo sapiens* throughout the world. This map, based on differences of mtDNA throughout current members of the world's population, suggests *Homo sapiens* originated from "mitochondrial Eve" in east Africa. About 100,000 years ago, the species spread into the Middle East and from there to Europe, Asia, Australia, and the Americas.

well developed, suggesting speech was used. However, about 30,000 years ago, this species disappeared, replaced by another hominin species, *H. sapiens* (from the Latin, meaning wise man), our own species. *H. sapiens* was a taller, lighter-weight species with a slightly smaller brain capacity than that of the Neanderthals. Researchers posit a variety of reasons for why *H. sapiens* thrived while the Neanderthals disappeared, including possessing a more efficient body type with lower energy needs, increased longevity, and differences in social structure and cultural adaptations.

Paleontologists remain divided as to whether *H. sapiens* evolved in Africa and spread from there to other areas of the world, or whether premodern humans such as *H. ergaster* that migrated from Africa evolved to become modern humans in different parts of the world. The first model, the Out of Africa hypothesis, suggests that after the evolution of hominins in Africa, they migrated to other continents three times, once for *H. ergaster*, once for *H. erectus*, and once for *H. sapiens*. Then *H. sapiens* would have gradually replaced species such as *H. erectus* and *H. neanderthalensis* in other parts of the world. Some scientists find this difficult to accept, however, and suggest that human groups have evolved from *H. ergaster* populations in a number of different parts of the world, a model known as the multiregional hypothesis. According to this hypothesis, gene flow between neighboring populations prevented the formation of several different species.

Studies of human mitochondrial DNA (mtDNA), which occurs only in the cellular organelles called mitochondria, and which is passed from mother to offspring, show that all modern people share a common ancestor, dubbed "mitochondrial Eve," dating to about 170,000 years ago. This evidence is consistent with the Out of Africa hypothesis, because the common ancestor would have to be much older than that to support the multiregional hypothesis. Furthermore, 2006 analyses of DNA from Neanderthal bones show it to be distinct from the DNA of *H. sapiens*, even though Neanderthals and humans share 99.5% of their genome. This also suggests that there was little interbreeding between Neanderthals and the *H. sapiens* who migrated into Europe.

Evidence overall appears to support the Out of Africa hypothesis. In this scenario, *H. ergaster* evolved in Africa and spread to Asia and Europe. Later, *H. erectus* evolved in Africa and spread into Asia, and *H. neanderthalensis* evolved in Europe. Both of these species shared the same fate, extinction at the hands of the later-evolving *H. sapiens*. *Homo sapiens* evolved in Africa about 170,000 years ago from *H. heidelbergensis*. The mtDNA data suggest a migration of *H. sapiens* from eastern Africa to other parts of the globe beginning 170,000–150,000 years ago (**Figure 34.34**). Modern humans spread first into the Middle East and Asia, then later into Europe and Australia, finally crossing the Bering Strait to the Americas.

Much remains to be resolved in human evolution, and new data constantly forces us to rethink our hypotheses. For example, in 2004, the remains of a small human on the Indonesian island of Flores were discovered and were given the name *Homo florensiensis*, nicknamed "hobbits" by the media. Many species—for example, deer and elephants—develop into small forms in insular situations, so hobbit humans seemed plausible. Since then, many researchers have suggested these people were modern humans who were suffering from a genetic disorder. Even modern humans on Flores are pygmies. Pathological dwarfism would have made these people even smaller. Only *H. sapiens* tools have been found at the area where the bones occur, suggesting these individuals were indeed dwarf forms of modern humans.

Summary of Key Concepts

34.1 Vertebrates: Chordates with a Backbone

- Vertebrates have several characteristic features, including a vertebral column, cranium, endoskeleton of cartilage or bone, neural crest, and internal organs (Figure 34.1, Table 34.1).
- Early-diverging vertebrates lacked jaws. Today the only jawless vertebrates are the hagfish and lampreys (Figures 34.2, 34.3).

34.2 Gnathostomes: Jawed Vertebrates

- A critical innovation in vertebrate evolution is the hinged jaw, which first developed in fishes. Gnathostomes are vertebrate species that possess a hinged jaw (Figure 34.4).
- The chondrichthyans (sharks, skates, and rays) have a skeleton composed of flexible cartilage and powerful appendages called fins. They are active predators with acute senses and were among the earliest fishes to develop teeth (Figure 34.5).
- Bony fishes consist of the Actinopterygii (ray-finned fishes, the most species-rich class), Actinistia (coelacanths), and the Dipnoi (lungfishes). In Actinopterygii, the fins are supported by thin, flexible rays and moved by muscles inside the body (Figures 34.6, 34.7).
- The lobe fins comprise the lobe-finned fishes (Actinistia and Dipnoi) and the tetrapods. In the lobe-finned fishes, the fins are supported by extensions of the pectoral and pelvic areas and are moved by their own muscles (Figures 34.8, 34.9).

34.3 Tetrapods: Gnathostomes with Four Limbs

- Fossils record the evolution of lobe-finned fishes to fishes with four limbs. Recent research has shown that relatively simple mutations control large changes in limb development (Figures 34.10, 34.11, 34.12).
- Amphibians live on land but return to the water to reproduce. The larval stage undergoes metamorphosis, losing gills and tail for lungs and limbs (Figure 34.13).
- The majority of amphibians belong to the order Anura (frogs and toads). Other orders are the Gymnophiona (caecilians) and Caudata (salamanders) (Figure 34.14).

34.4 Amniotes: Tetrapods with a Desiccation-Resistant Egg

- The amniotic egg permitted animals to become fully terrestrial. Other critical innovations included desiccation-resistant skin, thoracic breathing, water-conserving kidneys, and internal fertilization (Figure 34.15).
- Living reptilian classes include the Testudines (turtles), Lepidosauria (lizards and snakes), Crocodilia (crocodiles), and Aves (birds). The Ornithischia and Saurischia are two extinct classes of dinosaurs (Figures 34.16, 34.17, 34.18, 34.19).
- Three species—*Archaeopteryx*, *Caudipteryx*, and *Confuciusornis*—help trace a lineage from dinosaurs to birds (Figure 34.20).
- The four key characteristics of birds are feathers, a lightweight skeleton, air sacs, and reduced organs. Birds are the most species-rich class of terrestrial vertebrates. The diversity of bird beaks reflects the varied methods they use for feeding (Figures 34.21, 34.22, Table 34.2).

34.5 Mammals: Milk-Producing Amniotes

- The distinguishing characteristics of mammals are mammary glands, hair, specialized teeth, and an enlarged skull. Other unique characteristics of some mammals are the ability to digest plants and horns or antlers. Mammal tooth shape varies according to diet (Figures 34.23, 34.24, 34.25).
- Two subclasses of mammals exist: the Prototheria (monotremes) and the Theria (the live-bearing mammals). The live-bearing mammals are, in turn, divided into the Metatheria (marsupials) and Eutheria (placental mammals). The Eutheria have been divided into four different clades (Table 34.3, Figures 34.26, 34.27).
- Many defining characteristics of primates relate to their tree-dwelling nature and include grasping hands, large brain, nails instead of claws, and binocular vision (Figures 34.28, 34.29, 34.30).
- About 6 mya in Africa, a lineage that led to humans began to separate from other primate lineages. A key characteristic of hominins (extinct and modern humans) is bipedalism. Human evolution can be visualized like a tree, with a few hominin species coexisting at the same point in time, some eventually going extinct, and some giving rise to other species (Figures 34.31, 34.32).
- The Out of Africa hypothesis suggests that the migration of hominins from Africa happened at least three times, with *Homo sapiens* gradually replacing other hominin species in other parts of the world (Figure 34.33). The multiregional hypothesis proposes that human groups evolved in a number of different parts of the world. Most scientists believe the Out of Africa hypothesis is better supported by the data.
- Data from human mitochondrial DNA suggest all humans derive from a "mitochondrial Eve" that originated in east Africa. From there, *H. sapiens* spread to Asia and then to all other parts of the globe (Figure 34.34).

Assess and Discuss

Test Yourself

1. Which of the following is *not* a defining characteristic of vertebrates?
   a. cranium
   b. neural crest
   c. hinged jaw
   d. vertebral column
   e. endoskeleton
2. The presence of a bony skeleton, an operculum, and a swim bladder are all defining characteristics of
   a. Myxini.
   b. lampreys.
   c. Chondrichthyes.
   d. bony fishes.
   e. amphibians.
3. Organisms that lay eggs are said to be
   a. oviparous.
   b. ovoviparous.
   c. viviparous.
   d. placental.
   e. none of the above.
4. Which clade does not include frogs?
   a. vertebrates
   b. gnathostomes
   c. tetrapods
   d. amniotes
   e. lobe fins

5. In some amphibians, the adult retains certain larval characteristics, which is known as
   a. metamorphosis.
   b. parthenogenesis.
   c. cephalization.
   d. paedomorphosis.
   e. hermaphrodism.
6. The membrane of the amniotic egg that serves as a site for waste storage is
   a. the amnion.
   b. the yolk sac.
   c. the allantois.
   d. the chorion.
   e. the albumin.
7. Which characteristic qualifies lizards as gnathostomes?
   a. a cranium
   b. a skeleton of bone or cartilage
   c. a hinged jaw
   d. the possession of limbs
   e. amniotic eggs
8. Which of the following is *not* a distinguishing characteristic of birds?
   a. amniotic egg
   b. feathers
   c. air sacs
   d. lack of certain organs
   e. lightweight skeletons
9. What is *not* a derived trait of primates?
   a. opposable thumb
   b. grasping hands
   c. prehensile tail
   d. flat nails
   e. large brain
10. Despite their small size and nocturnal habits, tarsiers are classed with much larger monkeys and apes as Haplorrhini. This is based on which of the following characteristics?
    a. dry fully furred noses
    b. forward-facing eyes
    c. DNA similarities
    d. a and b
    e. a, b, and c

Conceptual Questions

1. How is vertebrate movement accomplished in a similar way to arthropod movement, and how is it different?
2. Why aren't all reptiles endothermic if both birds and mammals are?
3. A principle of biology is that *all species (past and present) are related by an evolutionary history.* Are birds living dinosaurs?

Collaborative Questions

1. By what means can vertebrates move?
2. Why are amphibians considered good indicator species, which are species whose status provides information on the overall health of an ecosystem?

Online Resource

www.brookerbiology.com

Stay a step ahead in your studies with animations that bring concepts to life and practice tests to assess your understanding. Your instructor may also recommend the interactive eBook, individualized learning tools, and more.

UNIT VI

FLOWERING PLANTS

Flowering plants, also known as the angiosperms, are essential to the lives of humans and most other organisms on Earth. Flowering plants provide most of our food, either directly in the form of vegetables and grains and other fruits, or indirectly as animal fodder. Cotton, linen, and other fibers that we use for clothing; and wood that we use for construction and fuel; as well as powerful cancer drugs and many other medicines come from the flowering plants. Unit VI reveals molecular, biochemical, structural, evolutionary, and ecological features of the hundreds of thousands of flowering plants that support Earth's life.

Unit VI begins with Chapter 35, which provides an overview of flowering plant structure and function, focusing on the seed-to-seed life cycle. By comparing plant bodies with those of animals, you will learn how plants are constructed and how they grow. Building on this background, Chapter 36 explains the genetic and physiological bases of plant behavior—plant responses to external stimuli such as day length that are mediated by internally produced hormones. In this chapter you will learn that like animals, plants have evolved sophisticated sensory systems that monitor environmental conditions and allow plants to respond in predictable ways. Chapter 37 explains the nutritional requirements of plants, a deep understanding of which is critical to human agriculture and our ability to feed increasing populations of humans. In chapter 38, we see how plant water transport influences global climate and how plants import organic food into nonphotosynthetic organs and tissues. After engaging this chapter, you will understand how plant fruits can become so sweet. Chapter 39 focuses on the molecular and cellular bases of plant reproduction, the process that generates seeds and fruits. This final unit chapter ties together key concepts presented throughout Unit VI: the seed-to-seed life cycle of plants and their distinctively structured bodies, plant development and growth in response to environmental and hormonal influences, and plant acquisition and transport of materials that support growth and reproduction.

The following biology principles will be emphasized in this unit:

- ***Living organisms use energy:*** *Though sunlight is the major source of energy for photosynthetic plants, hundreds of flowering plant species as well as diverse tissues occurring in all flowering plants are heterotrophic and thus require a supply of organic food as a source of energy.*
- ***Living organisms interact with their environment:*** *Some of the ways in which plants detect and respond to light and other environmental factors are similar to those operating in microbes or animals, but others are distinctive.*
- ***Living organisms maintain homeostasis:*** *As is the case for animals, it is essential for flowering plants to maintain body water content and energy balance within tolerance limits, or they will die.*
- ***Living organisms grow and develop:*** *Some of the cellular and molecular bases of plant growth and development are also features of microbes and animals, but plants display some unique growth and development modes.*
- ***Structure determines function:*** *Have you ever wondered why and how trees become so tall? The unique features of plant bodies as well as variations occurring among plants explain how flowering plants function in nature and in human agriculture.*
- ***Biology is an experimental science:*** *A modern understanding of flowering plant structure, function, and behavior—such as flower blooming—has been derived from many types of experimental studies, some of which are described in this unit.*

An Introduction to Flowering Plant Form and Function

35

Chapter Outline

35.1 From Seed to Seed—The Life of a Flowering Plant

35.2 How Plants Grow and Develop

35.3 The Shoot System: Stem and Leaf Adaptations

35.4 Root System Adaptations

Summary of Key Concepts

Assess and Discuss

Diverse types of seeds produced by flowering plants. **Seeds, each containing a plant embryo together with food storages and protected by a tough seed coat, are key resources for modern agriculture and ecological restoration projects.**

Anyone who seeks to improve human life by reducing the effects of disease, producing more food, or improving our environment needs to know something about plant form and function. That's because humans depend on flowering plants not only for nutritious food, fibers such as cotton and linen, wood and paper, medicines, and biofuels, but also for plentiful fresh air and clean water. Knowledge of basic plant form and function, including how plants develop from and produce seeds, is essential to understanding these useful plant features.

Seeds, illustrated in the chapter-opening photo, are key to human ability to produce crops to support ever-increasing populations and to restore degraded environments. For this reason, people store seeds of diverse plant species in seed banks, from which plant geneticists and restoration biologists can make withdrawals. Plant geneticists use seeds obtained from seed banks to breed new crop varieties that are resistant to pests and diseases, or that will help farmers cope with changing environmental conditions. Restoration biologists need large amounts of seed of diverse natural species for replanting degraded or destroyed lands. For example, the Kew Botanical Garden in the U.K. stores 1.8 billion seeds of 30,000 plant species in temperature-controlled vaults. The U.S. and other nations also maintain valuable seed banks to help biologists of the future understand how plants evolve in response to environmental change.

This chapter provides an introduction to the seed-producing flowering plants, focusing on fundamental principles of body form and function—anatomy and physiology. These principles are basic to human efforts to use genetic methods to improve crops in ways that enhance human health. Just as medical scientists seek to understand how mutations influence human anatomy and physiology, plant geneticists need to know how genes affect the plant body and its health. The principles of plant form and function are equally important to evolutionary biologists who seek to understand how and why variations in plant body structure arise, as well as to ecologists who want to know how plants, animals, and microbes function together in nature.

We will begin by considering the life of a flowering plant, from seed germination to the production of a new generation, analogous to the life of an animal from birth to reproductive adulthood. The following overview of how plants grow and develop reveals fascinating similarities to animals, but also intriguing differences. Finally, centering our attention on the adult plant body reveals how adaptation to different environments has generated diverse species whose form and function vary dramatically. This background supports subsequent chapters in this unit that provide more information about flowering plant behavior, nutrition, transport, and reproduction and development.

35.1 From Seed to Seed—The Life of a Flowering Plant

Learning Outcomes:

1. List ways in which seed plant reproduction and growth resemble or differ from those of animals.
2. Describe how the two major types of flowering plants—monocots and eudicots—differ in form.
3. Distinguish among annual, biennial, and perennial plants.

Several major events punctuate the lives of flowering plants, also known as the **angiosperms**. When seeds germinate, dormant embryos wake to metabolic activity and begin the process of seedling development. Seedlings grow and develop into mature plants capable of

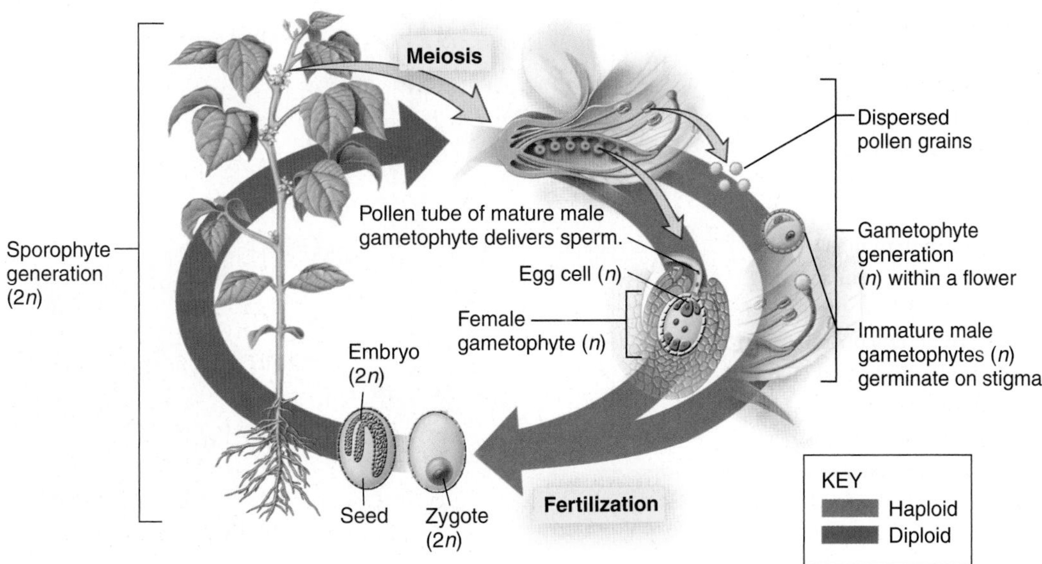

Figure 35.1 The plant sexual cycle. The sexual cycle of flowering plants involves alternation of sporophyte and gametophyte generations. In flowering plants, the sporophyte is the dominant, conspicuous generation, whereas the tiny gametophytes are mostly hidden within flowers. The bean plant shown here produces both male and female gametophytes. Embryos form when eggs produced by female gametophytes are fertilized by sperm produced by male gametophytes. Embryos, which are the young sporophytes of the next generation, are dispersed from plants within seeds.

reproduction. Finally, flowers produce and fruits disperse the next generation of seeds. In this section, we will briefly survey the life of flowering plants, focusing on the basic structural features of each life stage.

Seedlings Develop from Embryos in Seeds

Seeds are reproductive structures produced by flowering plants and other seed plants, usually as the result of sexual reproduction. Seeds contain embryos that develop into young plants—seedlings—when seeds germinate. As in animals, the embryo is an essential stage in the sexual cycle of a plant. The plant sexual cycle explains how embryos typically arise (**Figure 35.1**, see also Figure 30.14).

Unlike animals, sexual reproduction in plants requires two multicellular stages: a gamete-producing **gametophyte** and a spore-producing **sporophyte** (refer back to Figure 30.14). In the life cycle of plants, these two life stages alternate with one another in a process called **alternation of generations**. Flowering plants produce relatively large sporophytes and microscopic gametophytes that grow and develop within flowers. Diploid sporophytes produce haploid spores by the process of meiosis. These spores grow into gametophytes that produce plant gametes—eggs and sperm. Fusion of egg and sperm in the process of fertilization generates a diploid zygote, which undergoes repeated mitotic divisions to form the plant **embryo**.

The plant embryo is a very young sporophyte that lies dormant within seeds, accompanied by a supply of stored food and enclosed by a tough, protective seed coat (**Figure 35.2a**), much like the animal amniotic egg. Like an eggshell, the seed coat protects the delicate plant embryo during the dispersal of seeds from parent plants into the environment. Dispersed seeds may remain dormant in the soil—sometimes for long periods—but germinate when temperature, moisture, and light conditions are favorable. Such conditions activate embryo

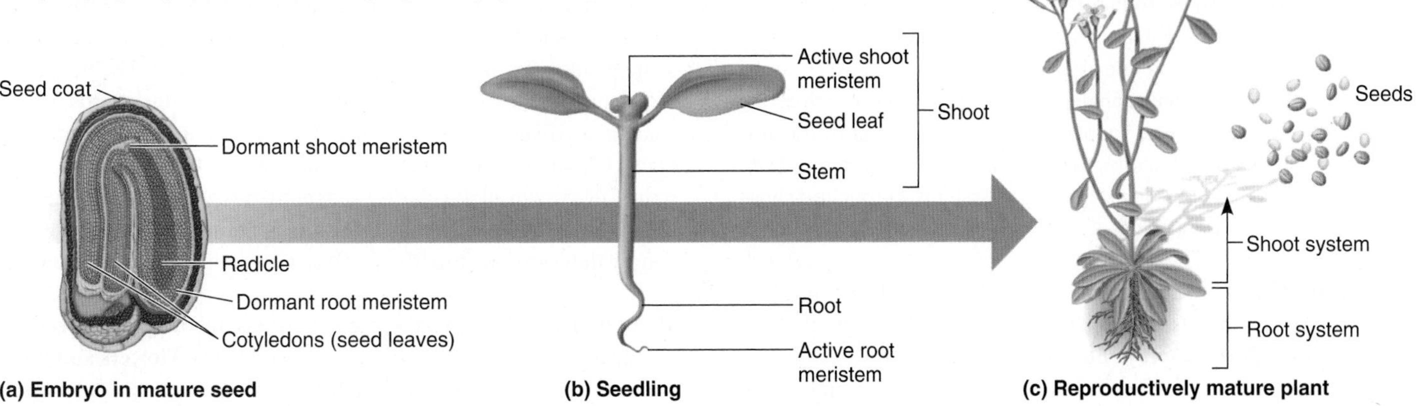

Figure 35.2 The seed-to-seed life of flowering plants. **(a)** Seed embryos possess embryonic leaves, known as cotyledons; a dormant shoot meristem; an embryonic root, known as a radicle; and a dormant root meristem. **(b)** When seeds germinate, the shoot and root meristems become active. Meristem activity allows the radicle to produce the seedling root and the young shoot of the seedling to grow and produce leaves. **(c)** Reproductively mature plants have branched shoot and root systems and bear flowers and fruits that disperse seeds.

BIOLOGY PRINCIPLE Living organisms grow and develop. The plant *Arabidopsis thaliana* is widely used as a model organism for understanding the genetics of plant structure and development.

metabolism, in which stored food is respired for energy needed for cell division and growth. As is the case for animals, **growth** is an increase in weight or size, and **development** is a series of changes in the state of a cell, tissue, organ, or organism. Enlarging plant embryos break the seed coat and grow into seedlings (**Figure 35.2b**). If sufficient resources such as water and minerals are available, seedlings develop into mature plants (**Figure 35.2c**).

The angiosperm plant body is simpler in form than most animal bodies and is composed of only three types of organs: stems, leaves, and roots. **Stems** produce leaves and branches and bear the reproductive structures of mature plants. **Leaves** are flattened structures that emerge from stems and are often specialized in ways that enable photosynthesis. Stems and leaves together make up the plant **shoot** (see Figure 35.2b). Mature plants often possess multiple stems bearing many leaves, which together form the **shoot system**. **Roots** provide anchorage in the soil and also foster efficient uptake of water and minerals. The aggregate of a plant's roots make up the **root system** (see Figure 35.2c).

As in animals, the process of body and organ development involves the differentiation of specialized cells having distinctive structure and function. But unlike animals, plant seedlings and mature plants produce new tissues in specific areas called meristems. A **meristem** (from the Greek *merizein*, meaning to divide) is a region of undifferentiated cells that produces new tissues by cell division. A dormant meristem occurs at the shoot and root tips of seed embryos, and these meristems become active in seedlings (see Figure 35.2a,b). In mature plants, active meristems occur at each stem and root tip. Such meristems are known as shoot and root **apical meristems** because they occur at shoot and root tips, also known as apices.

Mature Sporophytes Develop from Seedlings

As seedlings develop into mature sporophytes, the aboveground shoot typically becomes green and photosynthetic and thus able to produce organic food. Photosynthesis powers the transformation of seedlings into mature plants. The development of mature plants encompasses both **vegetative growth**, a process that increases the size of the shoot and root systems, and reproductive development. Vegetative growth and reproductive development involve **organ systems**, structures that are composed of more than one organ. Branches, buds, flowers, seeds, and fruits are organ systems, analogous to organ systems such as the circulatory, skeletal, and reproductive systems occurring in the animal body. The hierarchy of structure in a mature plant, ranging from specialized cells, tissues, organs, and organ systems to root and shoot systems, is shown in **Figure 35.3**.

Vegetative Growth During their growth, plant shoots produce **buds**—miniature shoots each having a dormant shoot apical meristem. Scaly modified leaves protect the bud contents. Under favorable conditions, the bud scales fall off, and the vegetative buds open. Newly opened buds display young leaves on a short shoot. The shoot apical meristem then becomes active, producing new stem tissue and leaves. In this way, buds generate leafy branches. A bud is an example of an organ system because it contains more than one organ.

Vegetative shoots often display **indeterminate growth**, meaning that apical meristems continuously produce new stem tissues and leaves as long as conditions remain favorable. This process explains how very large plants, such as trees, can develop from seedlings (**Figure 35.4a**). However, plant size is also under genetic control, so some plants remain small even when they are mature. The tiny floating plants of *Lemna* species, commonly known as duckweeds, which sometimes cover the surfaces of ponds in summer, are examples of plants whose small size is genetically determined (**Figure 35.4b**). Indeterminate growth allows plants to adapt their vegetative body structure to environmental conditions. By contrast, animal bodies and flowers display **determinate growth**, which is growth of limited duration.

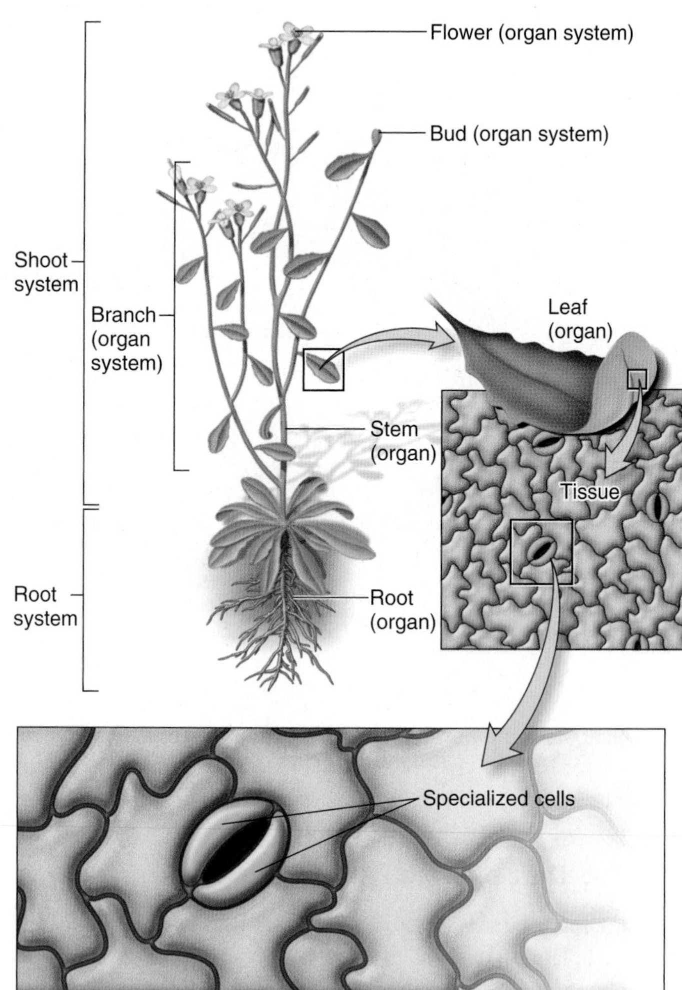

Figure 35.3 Levels of biological organization in a plant. Flowering plant sporophyte bodies consist of a root system and a shoot system. Shoot systems produce organ systems such as buds, flowers, fruits, and seeds, which are composed of organs, tissues, and specialized cells. Root systems are composed of organs, tissues, and specialized cells.

Development Under favorable conditions, mature plants produce reproductive structures: flowers, seeds, and fruits. **Flowers** and floral buds are reproductive shoots that develop when shoot apical meristems produce flower parts instead of new stem tissues and leaves. Flower development occurs under the control of several genes whose roles are well understood (refer back to Figure 19.24). In contrast to shoots, which often show indeterminate growth, flowers are produced by determinate growth. A floral shoot no longer produces new stem growth or leaves. Therefore, vegetative growth and reproductive

(a) *Brachychiton*, a tree native to Australia

(b) Lemna (common duckweed)

Figure 35.4 Plants display indeterminate growth, but vary in size. A large woody angiosperm **(a)** has the same indeterminate growth pattern as a tiny duckweed plant **(b)**.

development are alternative processes. In order to flower, a plant must give up some of its potential to continue vegetative growth.

Flower tissues enclose and protect tiny male and female gametophytes during their growth and development (see Figure 35.1). Female gametophytes produce eggs within structures known as ovules, produced in the ovary of a flower pistil. Male gametophytes begin their development within pollen grains produced in the anthers of a flower stamen. Pollen is dispersed to the flower pistil, whereupon pollen grains may germinate, producing a tube that delivers sperm to eggs. Fertilization generates zygotes, which develop into embryos, and also triggers the process by which ovules develop into seeds and flower parts develop into fruits. **Fruits** thus enclose seeds and function in seed dispersal. Flower buds, flowers, fruits, and seeds are organ systems because they consist of more than one organ. For example, flowers typically contain several leafy organs, including sepals and petals, as well as stamens and pistils, that evolved from leaves (refer back to Figure 30.15).

Flowering Plants Vary in the Structure of Organs and Organ Systems

With some exceptions, flowering plants occur in two groups, informally known as the **eudicots** and the **monocots** (refer back to Figure 30.18). These groups take their names from the number of seed leaves (cotyledons) that are present on seed embryos. For example, bean plants and relatives, which possess two (*di*) seed leaves, are examples of eudicots. Most woody trees, shrubs, and vines are also eudicots. Corn, which has only one (*mono*) seed leaf, is an example of a monocot, as is tiny duckweed (see Figure 35.4b). Eudicots and monocots also vary in the structure of other organs and organ systems. For example, eudicot flowers typically have petals and other parts numbering four, five, or a multiple of those numbers, whereas monocot flower parts usually occur in threes or a multiple of three. Stems, roots, leaves, and pollen of eudicots and monocots also vary in distinctive ways, as shown in **Table 35.1**.

Table 35.1 Distinguishing Features of Eudicots and Monocots, Two Major Groups of Flowering Plants

Feature	Eudicots	Monocots
Number of seed leaves (cotyledons)	Two	One
Number of flower parts	Usually four, five, or multiples of these	Usually three or multiple of three
Stem vascular bundles	Arranged in a ring	Scattered
Root system	Branched taproot	Fibrous; adventitious
Leaf venation	Netted or branched	Often parallel
Pollen	Three pores or slits	One pore or slit

Flowering Plants Vary in Seed-to-Seed Lifetime

The lifetime of a flowering plant can vary from a few weeks to many years. Plants that die after producing seed during their first year of life are known as **annuals**. Corn and the common bean are examples of annual crops whose nutrient-rich seeds are harvested within a few months after planting and must be replanted at the beginning of each new growing season. Plants that do not reproduce during the first year of life but may reproduce within the following year are known as **biennials**. Such plants often store food in fleshy roots during the first year of growth, and this food fuels reproduction during the second or later year of life. Humans use some of these fleshy roots for food, including carrots, parsnips, and sugar beets. Trees are examples of **perennials**, plants that live for more than 2 years, often producing seed each year after they reach reproductive maturity. Many flowering plants use environmental signals to time flowering in ways that enhance seed production. Temperature and day length are examples of environmental factors that determine flowering time. Plant seed-to-seed lifetimes are also influenced by the longevity of their seeds. Seeds of some plants are able to germinate after more than a thousand years of dormancy, whereas other plant seeds are unable to remain alive for long periods.

35.2 How Plants Grow and Develop

Learning Outcomes:

1. Compare and contrast plant body architecture and development with those of animals.
2. Explain how the shoot system differs from the root system.

As plants grow from seedlings, their development depends on four processes that are also essential to animal growth and development: cell division, growth, cell specialization, and programmed cell death. Additional and distinctive aspects of plant growth and development include (1) development and maintenance of a plant-specific architecture throughout life, (2) an increase in length by the activity of apical meristems, (3) maintenance of a population of youthful stem cells in meristems, and (4) expansion of cells in controlled directions, by water uptake.

Plants Display a Distinctive Architecture

In plant biology, the term apical has two distinct meanings. As we have seen, apical refers to the tips or apices of shoots and roots, as in shoot apical meristems or root apical meristems. A second meaning for apical is the part of a plant that typically projects upward, which is the top of the shoot. By contrast, the bottom of a root is termed the basal region. So the shoot apical meristem occurs at the apical pole, and the root apical meristem occurs at the basal pole. This property, known as **apical-basal polarity**, explains why plants produce shoots at their tops and roots at their lower regions. Apical-basal polarity originates during embryo development. As seedlings and maturing plants grow in length by the activity of shoot and root meristems, apical-basal polarity is maintained (**Figure 35.5a**). Animals likewise have anterior and posterior ends, whose development is influenced by *Hox* genes (refer back to Chapter 32). In contrast, plant apical-basal polarity is under the control of genes such as *GNOM*; mutations in such genes result in plant embryos that are cone-shaped or spherical and thus lack normal apical-basal architecture (**Figure 35.5b**).

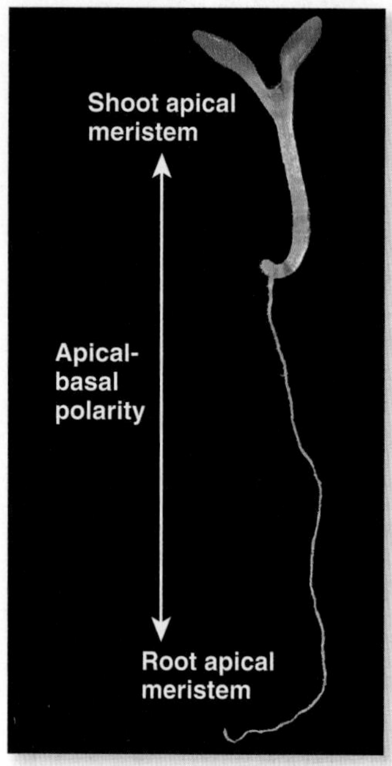

(a) Normal seedling

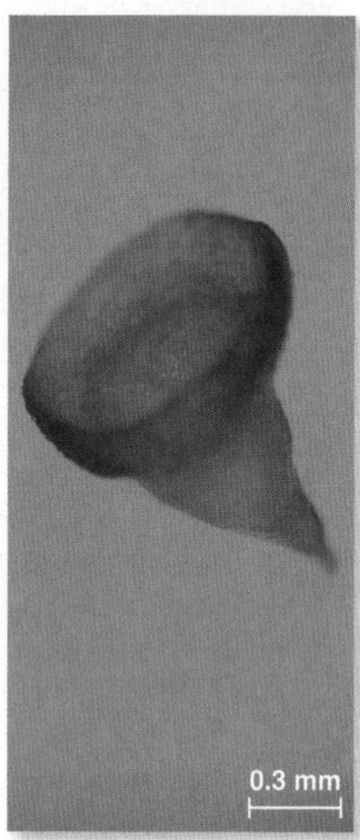

(b) Abnormal *GNOM* mutants

Figure 35.5 Plant apical-basal polarity. (a) Normal plants exhibit apical-basal polarity, as shown by this seedling. Such growth occurs at two meristems, one at the shoot (SAM) and one at the root (RAM). (b) *GNOM* mutants of *Arabidopsis thaliana* lack apical-basal polarity and thus produce abnormal embryos and seedlings.

BioConnections: Look back to Figure 19.17 to see the relationship between Hox gene expression and body organization in the mouse, as a model animal. How is animal body polarity similar to that of plants?

A second architectural feature of the typical plant body is **radial symmetry**. Plant embryos normally display a cylindrical shape, also known as an axis, which is retained in the stems and roots of seedlings and mature plants. A thin slice or cross section of an embryo, stem, or root is typically circular in shape. Most plants produce new leaves or flower parts in circular whorls, or spirals, around shoot tips (**Figure 35.6**). Buds and branches likewise emerge from stems in radial patterns, as do lateral roots from a central root axis. Together, apical-basal polarity and radial symmetry explain why diverse plant species have a fundamentally similar architecture. Although radial symmetry also characterizes early-diverging animals, most animals are bilaterally symmetric (see Figure 32.5).

Primary Meristems Increase Plant Length and Produce Plant Organs

We previously noted that plant embryos grow into seedlings by adding new cells from only two growth points, the **shoot apical meristem (SAM)** and the **root apical meristem (RAM)**. During plant development, the SAM and RAM of the embryo give rise to many apical meristems located in the buds of shoots and at the tips of roots.

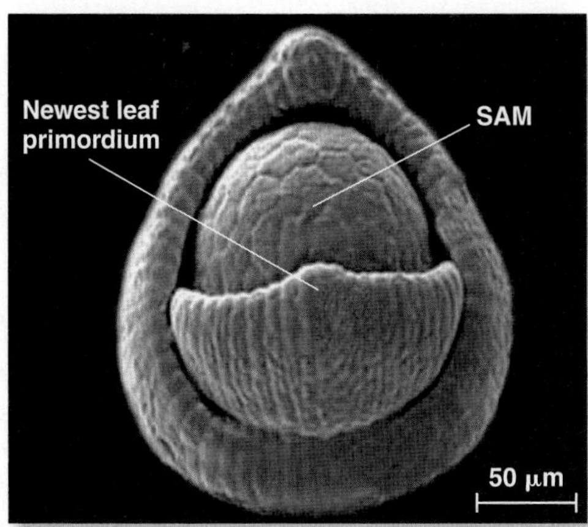

Figure 35.6 Plant radial symmetry. This top-down view of a shoot apical meristem reveals the radial symmetry of the shoot, illustrated by its cylindrical shape. Leaf primordia are produced in circles or spirals around the shoot axis.

Concept Check: *How could you determine that roots have radial symmetry?*

As a SAM grows, it leaves behind meristematic tissues that increase plant length and produce new organs. Such meristems are known as the **primary meristems**, and in a process known as **primary growth**, they ultimately produce primary tissues and organs of diverse types (Table 35.2). Tissues differ in their cellular complexity. Simple primary tissues are those composed of only one or two cell types; complex primary tissues are made of more cell types. As described in Section 35.3, the primary meristems of woody plants also give rise to secondary or lateral meristems. In a process known as **secondary growth**, the secondary meristems increase the girth of woody plant stems and roots by producing secondary tissues (Table 35.2).

Plant biologists have discovered that plant cell specialization and tissue development do not depend on the lineage (the parentage) of a cell or tissue. Chemical influences such as hormones, transcription factor proteins, and microRNAs that move through plants are much more important in determining the type of specialized tissue produced by unspecialized plant cells. Examples of such chemical influences are provided in the rest of this chapter.

Primary Stem Structure and Development New primary stem tissues arise by the cell division activities of shoot apical meristems. A layer of outermost tissue known as the **epidermis** develops at the stem surface. The epidermis produces a waxy surface coating known as the **cuticle**, which helps to reduce water loss from the plant surface and helps to protect plants from damage by ultraviolet (UV) light, animals, and disease microorganisms (refer back to Figure 29.11a).

Beneath the epidermis lies the stem **cortex**, which is largely composed of **parenchyma tissue** (Figure 35.7a). This tissue is composed of only one cell type, thin-walled cells known as **parenchyma cells**. These cells often store starch in plastids and therefore serve as an organic food reserve. Stem parenchyma also has the ability to undergo cell division (meristematic capacity), which aids wound healing when stems are damaged. The cell division capability of stem parenchyma also explains how people are able to grow new plants from stem cuttings. Stems also contain **collenchyma tissue** (Figure 35.7b), composed of flexible **collenchyma cells**, and rigid **sclerenchyma tissue** (Figure 35.7c), composed of two types of tough-walled sclerenchyma cells termed **fibers** and **sclereids**. These tough cells provide strength and protection to the plant stem.

Table 35.2 Examples of Tissues and Specialized Cells Found in Flowering Plants*

Primary Growth	
Simple primary tissues (composed of one or two cell types)	**Plant cell types found in those tissues**
Parenchyma	Parenchyma cells
Collenchyma	Collenchyma cells
Sclerenchyma	Fibers and sclereids
Root endodermis	Endodermal cells
Root pericycle	Pericycle cells
Complex primary tissues (composed of at least two cell types)	**Plant cell types found in those tissues**
Leaf or stem epidermis	Flattened epidermal cells, trichomes, stomatal guard cells
Root epidermis	Flattened epidermal cells, root hairs
Leaf mesophyll	Spongy parenchyma cells, palisade parenchyma cells
Leaf, stem, or root xylem	Tracheids, vessel elements, fibers, parenchyma cells
Leaf, stem, or root phloem	Sieve-tube elements, companion cells, fibers, parenchyma cells
Secondary Growth	
Simple and complex secondary tissues	**Plant cell types found in those tissues**
Secondary xylem (wood)	Tracheids, vessel elements, fibers, parenchyma
Secondary phloem (inner bark)	Sieve-tube elements, companion cells, fibers, parenchyma
Outer bark	Cork cells

*This list does not include all of the tissues and cell types found in flowering plants. Some of these examples will be described later in this chapter.

New water- and food-conducting tissues develop at the core of a young shoot. These conducting tissues are known as **primary vascular tissues** because they develop from new cells produced by the SAM. Vascular tissues occur in two forms—**xylem** and **phloem**—which are composed of several types of specialized tissues and cells (see Section 35.3 and Table 35.2). Newly formed stem xylem and phloem connect with older conducting tissues that extend throughout the stem system. Stem xylem and phloem link to vascular tissues of the root system, forming a continuous route for conduction of water, minerals, and organic compounds through the plant. However, in contrast to the circulatory system of most animals, the plant conduction system is not closed, but instead is open to the environment (plant transport is described more completely in Chapter 38). Primary vascular tissues are typically arranged in elongate clusters known as **vascular bundles** that appear round or oval when cross-cut (Figure 35.7d). In the primary stems of beans and other eudicots, the

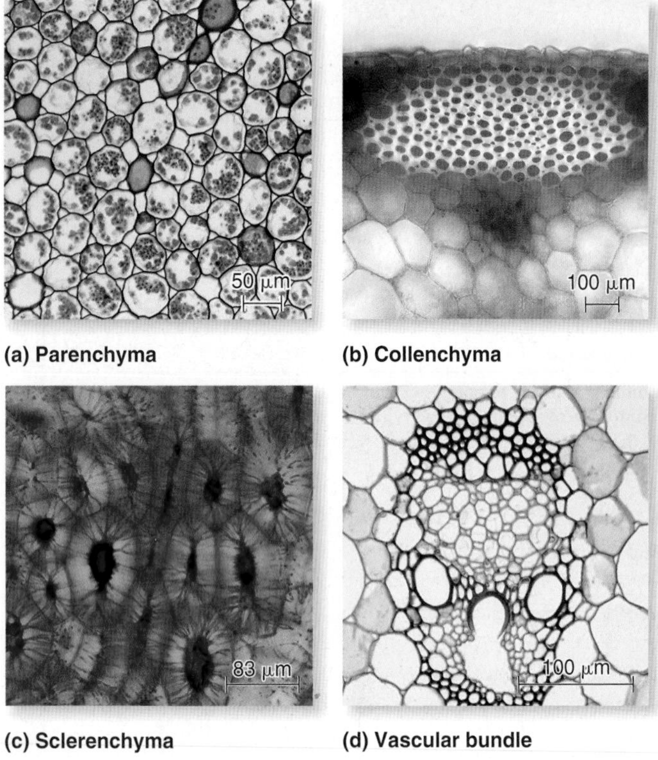

(a) Parenchyma (b) Collenchyma (c) Sclerenchyma (d) Vascular bundle

Figure 35.7 Examples of tissues produced by primary shoot meristems. (a) Parenchyma, (b) collenchyma, (c) sclerenchyma, and (d) a vascular bundle composed of complex xylem and phloem tissues.

vascular bundles are arranged in a ring, which is easily seen in thin slices made across a stem (see Table 35.1). By contrast, in the stems of corn and other monocots, the vascular bundles are scattered.

Leaf Structure and Development Young leaves are produced at the sides of a SAM as small bumps known as **leaf primordia** (see Figure 35.5). As young leaves develop, they acquire vascular tissue that is connected to the stem xylem and phloem (Figure 35.8a) and become flattened, a process that expands the area of leaf surface available for light collection during photosynthesis. In the cases of some leaves, thinness is an adaptation that helps them to shed excess heat. Leaves also become bilaterally symmetrical, meaning that they can be divided into two equal halves in only one direction, from the leaf tip to its base (Figure 35.8b).

Upper and lower leaf tissues develop differently in several ways that foster photosynthesis (Figure 35.8c). For example, the more shaded lower leaf epidermis usually displays larger numbers of pores, known as **stomata** (from the Greek word *stoma*, meaning mouth), than the sunnier upper leaf surface (refer back to Figure 29.11b). When open, stomata allow CO_2 to enter and water vapor and O_2 to escape leaf tissues. Closure of stomata helps to prevent excess water loss from plant surfaces. **Palisade parenchyma** consists of closely packed, elongated cells of the inner leaf that are adapted to absorb sunlight efficiently. **Spongy parenchyma**, located closer to the lower leaf surface, contains rounder cells separated by abundant air spaces. These air spaces foster CO_2 absorption and O_2 release by leaves. Together, the palisade and spongy parenchyma are known as the leaf **mesophyll**.

Leaf veins composed of vascular tissue commonly occur at the junction of palisade and spongy parenchyma, or within the spongy parenchyma (see Figure 35.8c). Leaf parenchyma tissues are typically green and active in photosynthesis, a process that requires

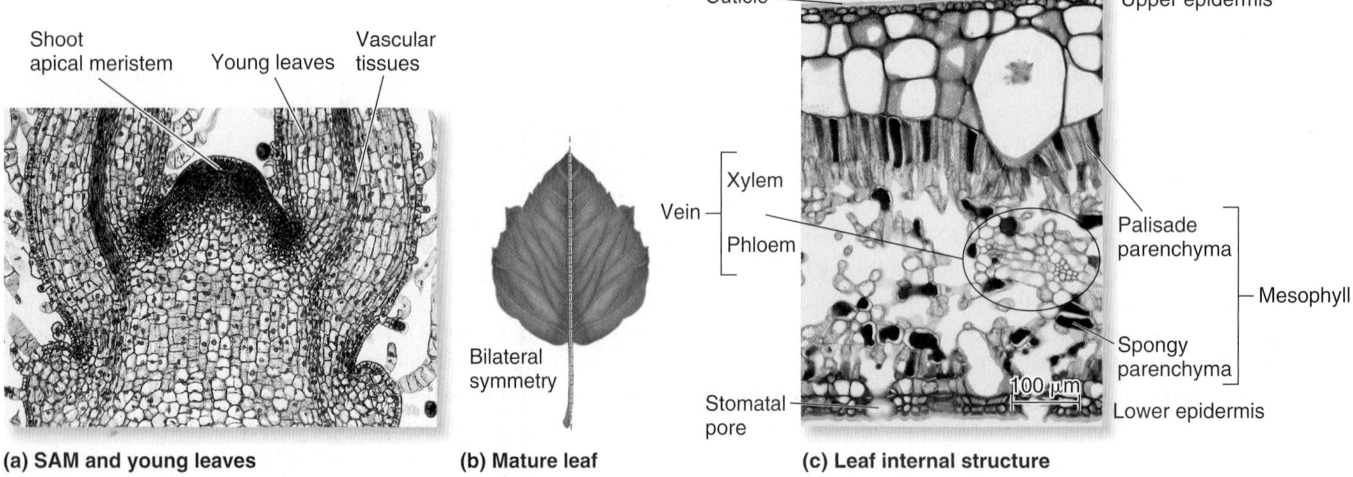

(a) SAM and young leaves (b) Mature leaf (c) Leaf internal structure

Figure 35.8 Leaf development and structure. (a) Young leaves develop at the sides of SAMs, as shown in this thinly sliced, stained shoot tip. Note that the darkly stained vascular tissues of young leaves are connected to those of the stem. (b) Mature leaves are typically thin and flat and show bilateral symmetry. (c) An internal view of a thinly sliced and stained leaf reveals upper to lower surface tissue differentiation. A layer of palisade parenchyma lies just beneath the upper epidermis capped with a waxy cuticle. Veins of conducting tissue (xylem and phloem) are embedded in the photosynthetic mesophyll. Spongy parenchyma lies above the lower epidermis, which displays stomata. These structural features of mature foliage leaves facilitate photosynthesis.

Concept Check: *What advantage do plant leaves obtain by having stomata on the lower epidermal surface?*

water, carbon dioxide, and dissolved minerals. Parenchyma cells can only take up carbon dioxide that has first dissolved into water, a feature inherited from ancient aquatic algal ancestors. For this reason, in order to perform photosynthesis, leaf parenchyma cells must be bathed in water. The xylem tissues of veins conduct water and minerals throughout leaf tissues, fostering photosynthesis. Phloem tissues of leaf veins carry the sugar products of photosynthesis from leaf cells to stem vascular tissues. In this way, sugar produced in leaves can be exported to other parts of the plant. The density of veins in angiosperm leaves is about four times that of other vascular plants. This allows plants to conduct materials more efficiently, which helps to explain why flowering plants are the dominant type of vegetation in most habitats today.

Root System Structure and Development In beans and most other eudicots, a main root develops from the embryonic root and then produces branch roots, also known as lateral roots. Such a root system of eudicots is known as a **taproot system**; this kind of root system has one main root with many branch roots (see Table 35.1). In contrast, the embryonic root of most monocots dies soon after seed germination, and it is replaced by a **fibrous root system** consisting of multiple roots that grow from the stem base (see Table 35.1). Fibrous roots are examples of **adventitious roots**, structures that are produced on the surfaces of stems (and sometimes leaves) of both monocots and eudicots. Roots that develop at the bases of stem cuttings are also adventitious.

As noted earlier, the tips of roots and their branches each possess an apical meristem that adds new cells. Expansion of these new cells allows roots to grow into the soil. As they lengthen, roots produce branches but not from buds, as is the case for stems. Instead, branch roots develop from meristematic tissues located within the root (see Section 35.4). The root system both anchors plants in the soil and plays an essential role in harvesting water and mineral nutrients. Root tissues are usually not green and photosynthetic. They must rely on organic compounds transported from the shoot. The plant root system and shoot system therefore depend on each other.

Plant Meristems Contain Youthful Stem Cells

Plant meristems include undifferentiated cells referred to as **stem cells**. In the late 19th century, the Russian-born American biologist Alexander Maximow coined the term *Stammzelle*, which is derived from the German words *stamm*, meaning stem, such as a plant stem, and *zelle*, meaning cell. Maximow used the term "stem cell" to describe animal cells that remain undifferentiated but are also able to generate specialized tissues. Animal stem cells are currently much in the news because of their potential for use in the treatment of human diseases that cause cell or tissue damage. The term stem cell is now also widely used for cells located within the plant meristem that likewise remain undifferentiated but also can divide and produce the cells that constitute new tissues. In the context of plant development, the term does not mean any cell located in a plant stem, only the undifferentiated cells located within the meristems of the shoot and root.

When plant stem cells divide, they produce two cells: one that remains young and unspecialized plus another cell. This second cell may differentiate into various types of specialized cells, but it often retains the ability to divide (refer back to Figure 19.23). As a result of these properties, stem cell numbers influence the size of a meristem, which, in turn, affects plant growth. Normal function of plant stem cells depends on a plant gene that is closely related to the animal tumor-suppressor gene *Rb* (refer back to Table 14.8). Rb protein functions to regulate the cell cycle in both animals and plants. In animals, mutations in *Rb* cause retinoblastoma, a cancerous tumor of the eye; in plants, mutations cause meristematic cells to fail to produce differentiated cells and organs.

Plant Cells Expand in a Controlled Way by Absorbing Water

As we have observed, meristem production of new cells is an important component of plant growth. In addition, plant growth involves cell expansion, which is a much less important component of animal growth. The diameters of newly formed stem and root cells are usually equal in all dimensions, but many soon begin to extend lengthwise, thereby helping shoots and roots to grow longer. Recall that plant cells typically possess a relatively large vacuole (refer back to Figure 4.20a). Cell extension occurs when water enters the central vacuole by osmosis (**Figure 35.9**). As the central vacuole expands, the cell wall also expands and increases the cell's volume. By taking up water, plant cells can enlarge quickly, allowing rapid plant growth. Bamboo, for instance, can grow taller by 2 m within a week and can grow up to 30 m in less than three months! The importance of water uptake in cell expansion helps to explain why plant growth is so dependent on water supply.

Plant cell walls contain cellulose microfibrils that are held together by crosslinking polysaccharides. When plant cells and their vacuoles absorb water, pressure builds on cell walls. In response to this pressure and under acidic conditions, proteins unique to plants—known as expansins—are produced. **Expansins** unzip crosslinking cell-wall polysaccharides from cellulose microfibrils so that the cell wall can stretch (**Figure 35.10**). As a result, cells enlarge, often by elongating in a particular direction, which is important to plant form. Some plant cells are able to elongate up to 20 times their original length.

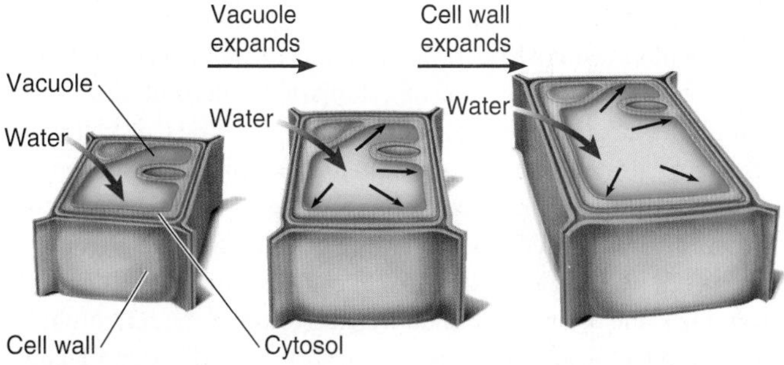

Figure 35.9 **Plant cells expand by taking up water into their vacuoles.**

BioConnections: Look back to Figure 4.20a to see an electron micrograph of a plant vacuole and Figure 4.9, which compares generalized animal and plant cells. How do plant cells differ from those of animal cells in terms of vacuoles?

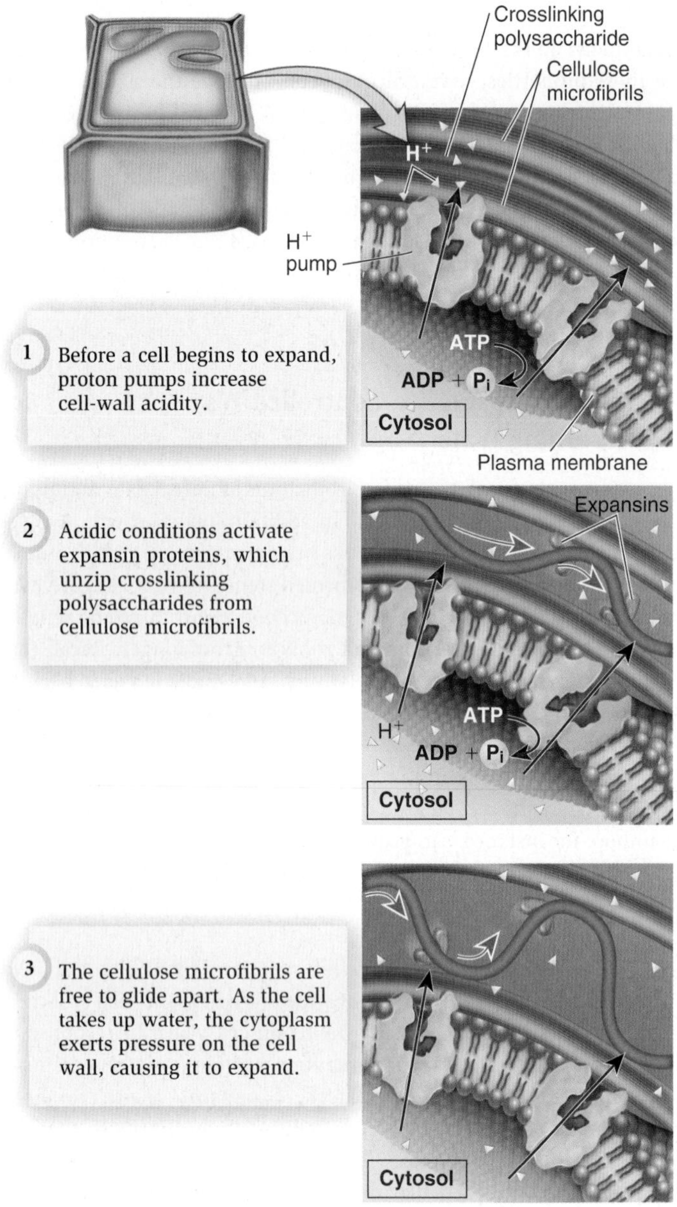

Crosslinking polysaccharide

Cellulose microfibrils

H^+ pump

Plasma membrane

Cytosol

Expansins

1 Before a cell begins to expand, proton pumps increase cell-wall acidity.

2 Acidic conditions activate expansin proteins, which unzip crosslinking polysaccharides from cellulose microfibrils.

3 The cellulose microfibrils are free to glide apart. As the cell takes up water, the cytoplasm exerts pressure on the cell wall, causing it to expand.

Figure 35.10 **A hypothetical model of the process of cell-wall expansion.**

The direction in which a plant cell expands depends on the arrangement of cellulose microfibrils in its cell wall, which is in turn determined by the orientation of cytoplasmic microtubules. These microtubules are thought to influence the positions of cellulose-synthesizing protein complexes located in the plant plasma membrane. The protein complexes connect sugars to form cellulose polymers, spinning cellulose microfibrils onto the cell surface to form the cell wall. As a result, cell-wall cellulose microfibrils encircle cells in the same orientation as underlying cytoplasmic microtubules (Figure 35.11). Because cellulose microfibrils do not extend lengthwise, plant cell walls expand more easily in a direction perpendicular to them. To visualize this process, imagine encircling a spherical balloon with several parallel bands of tape before it has been completely inflated. The tape bands, which do not extend lengthwise, represent encircling cellulose microfibrils. As you add more air, the balloon tends to extend in the direction perpendicular to the bands of tape.

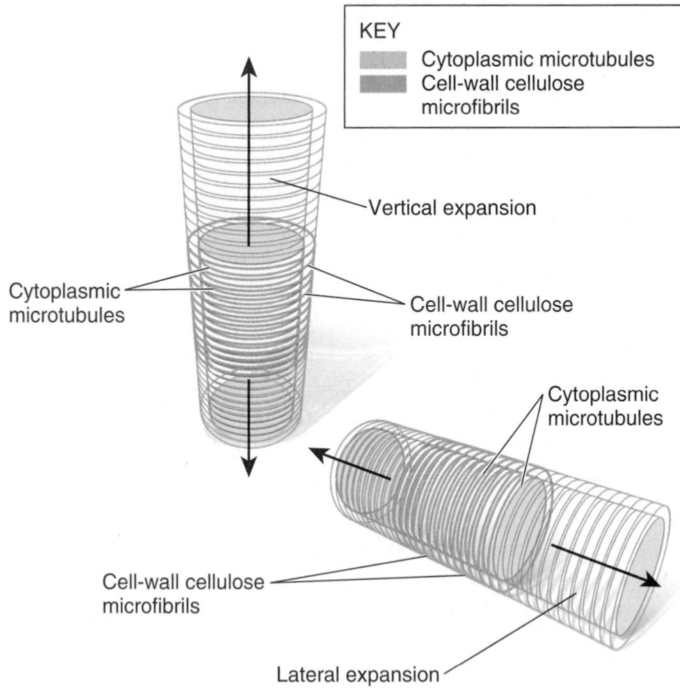

Figure 35.11 **Control of the direction of plant cell expansion by microfibrils and microtubules.** Plant cells enlarge in the direction perpendicular to encircling cell-wall cellulose microfibrils, which run parallel to the orientation of underlying cytoplasmic microtubules.

BioConnections: Look back to Table 4.1 to see images of microtubules. How are plant and animal microtubules different?

Microtubules control not only the direction of cell expansion but also the plane of cell division, which is also critical to plant form. Mutation of the *FASS* gene in the model plant *Arabidopsis* illustrates the importance of microtubule orientation to plant structure. In cells of such mutants microtubules are randomly arranged, causing cells to divide and grow abnormally, and producing plants with stubby organs.

35.3 The Shoot System: Stem and Leaf Adaptations

Learning Outcomes:

1. Discuss why plant shoots are said to have a modular structure.
2. Explain why leaves having different shapes and vein patterns exist in nature.
3. Compare the structure and function of the conducting tissues xylem and phloem.
4. Create a drawing showing how bark and wood originate in woody plants.

As we have seen, the shoot system includes all of a plant's stems, branches, leaves, and buds. It also produces flowers and fruits when the plant has reached reproductive maturity. Thus, the shoot system is essential to plant growth, photosynthesis, and reproduction. In this section, we examine stem and leaf structure and development in more detail. We will observe that features of shoot stems and leaves vary among plants in ways that explain plant ecological function and see how these features are useful to us in distinguishing plant species.

Shoot Systems Have a Modular Structure

More than 200 years ago, the German author, politician, and scientist Johann Wolfgang von Goethe realized that plants are modular organisms, composed of repeated units. Shoots are notably modular (Figure 35.12). Each shoot module consists of four parts: a stem node, an internode, a leaf, and an axillary meristem or bud. A **node** is the stem region from which one or more leaves emerge. An **internode** is the region of stem between adjacent nodes. Differences in numbers and lengths of internodes help to explain why plants differ in height (look back to Figure 35.4). Each time a young leaf is produced at the SAM, a new meristem develops in the upper angle formed where the leaf emerges from the stem. This angle is known as an axil (from the Greek *axilla*, meaning armpit), and the meristem formed there is called an **axillary meristem**. Such axillary meristems generate **axillary buds**, which can produce flowers or branches known as lateral shoots. Such new branches bear a SAM at their tips. SAMs located at the apices of both main and lateral shoots produce new leaves. What causes new leaves to arise? The answer involves chemical messengers known as hormones.

Hormones and MicroRNAs Influence Leaf Development

As we have noted, leaf primordia are surface bumps of tissue that develop at the sides of a SAM (see Figure 35.5). Production of such leaf primordia is under the control of a hormone known as **auxin**. In general, **hormones** are signaling molecules that exert their effects at a site distant from the place where such compounds are produced. Plant hormones are important in coordinating both plant development and plant responses to environmental conditions and are discussed in more detail in Chapter 36.

The outermost epidermal layer of cells at shoot tips produces auxin, which moves from cell to cell by means of specific membrane transport proteins. Auxin accumulates in particular locations because cells of the shoot apex differ in their ability to import and export auxin. When auxin accumulates in a particular apical region, the hormone causes expansin gene expression to increase. When expansin loosens their cell walls (see Figure 35.10), cells expand by taking up water, thereby forming a tissue bulge—a leaf primordium. The development of leaf primordia depletes auxin from nearby tissue, with the result that the next leaf primordium will develop in a different place on the shoot apex, where the auxin level is higher. Such changes in auxin concentration on the surface of the shoot explain why leaf (and flower) primordia develop in spiral or whorled patterns around the shoot tip. The youngest leaf primordia occur closest to the shoot tip, and successively older leaf primordia occur on the sides of the shoot tip (see Figure 35.5).

Hormones also influence the transformation of primordia into leaves. The cells of leaf primordia do not produce a protein known as KNOX, which is produced by other shoot meristematic cells. KNOX is a transcription factor, a protein that regulates gene transcription. The absence of KNOX proteins induces leaf primordia to produce a plant hormone known as **gibberellic acid**. This hormone stimulates both cell division and cell enlargement, causing young leaves to grow.

Other molecules produced by a SAM direct leaf flattening and differentiation of the upper and lower tissues (see Figure 35.8c). These signaling effects were first demonstrated in 1955 by plant developmental biologist Ian Sussex, who made surgical cuts around

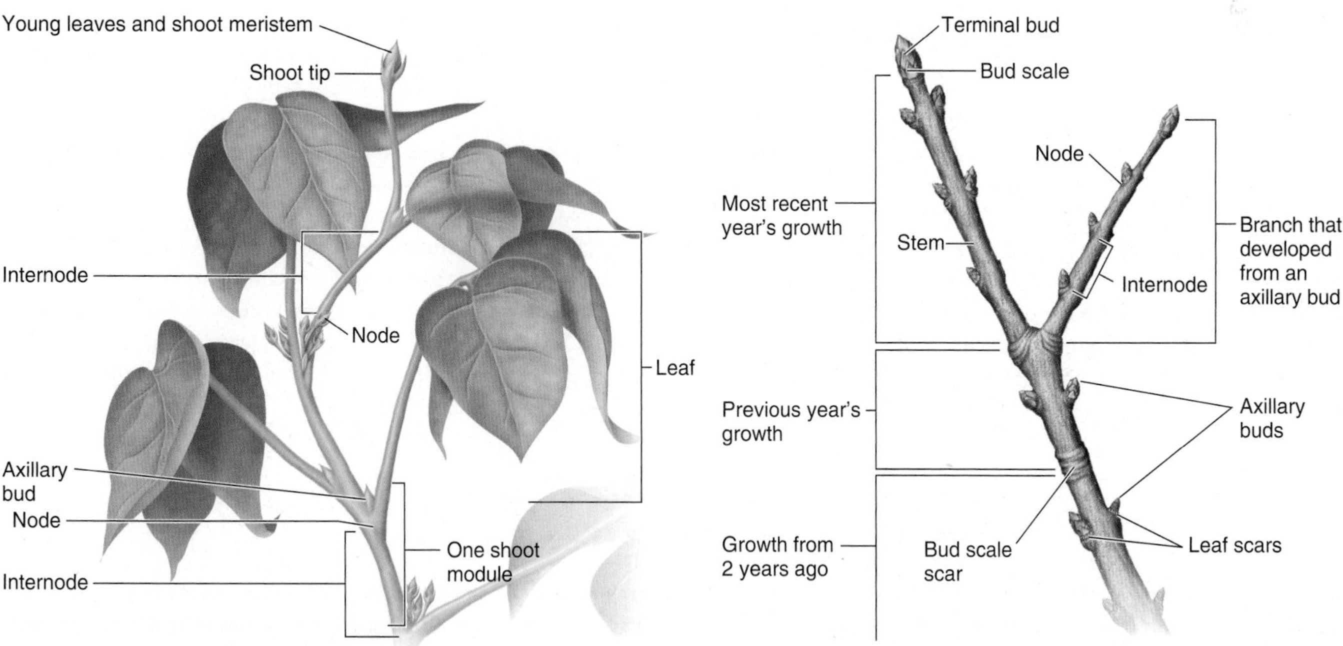

Figure 35.12 The modular organization of plant shoots. **(a)** The top end of an herbaceous stem showing the shoot modules. Each module consists of a node with its associated leaf and axillary meristem or bud and an internode. **(b)** The modular organization shown by a woody stem as it appears during winter. Axillary buds lie above the scars left by leaf fall. Regions between successive sets of bud scale scars mark each year's growth.

Concept Check: *If a twig has five sets of bud scale scars, how old is the twig likely to be?*

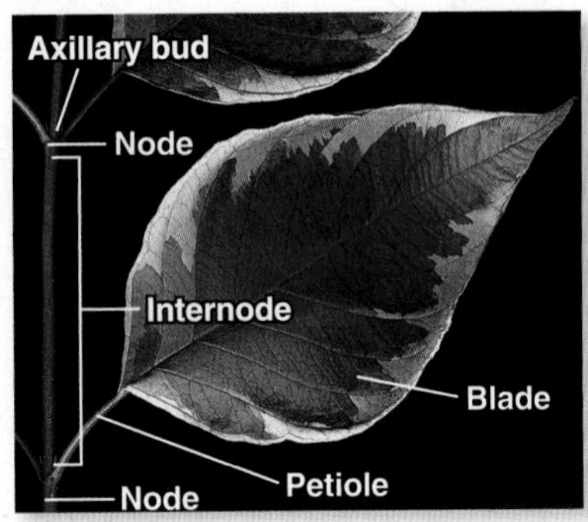

(a) Eudicot stem with simple leaf and pinnate venation

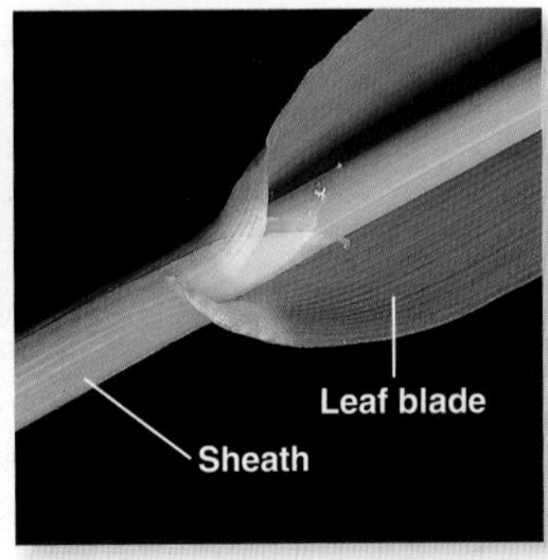

(b) Monocot stem and leaf

(c) A compound leaf

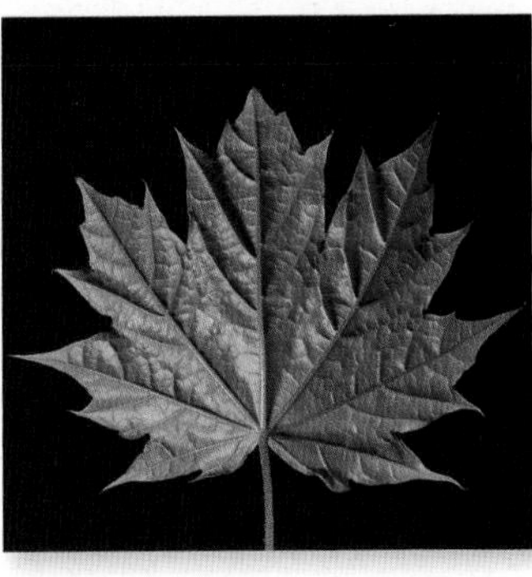

(d) A simple leaf and palmate venation

Figure 35.13 **Examples of variation in leaf form.** **(a)** A simple eudicot leaf, showing blade, petiole, and axillary bud. This leaf has a pinnate venation pattern. **(b)** The leaf of a monocot, showing parallel veins. The base of the leaf encircles the stem. **(c)** A compound leaf divided into leaflets. **(d)** A simple leaf having palmate venation.

leaf primordia, thereby isolating them from chemical substances produced by the meristem. The isolated primordia retained their initial cylindrical, radially symmetric structure rather than becoming flat and bilaterally symmetric, as they should. Later, investigators discovered that a particular set of genes expressed in young primordia causes uppermost leaf tissue layers to develop and that a different set of genes specifies development of lower leaf tissue layers. This differential gene expression is caused by differences in the location of a type of microRNA (miRNA), small RNA molecules that silence the expression of pre-existing mRNAs (refer back to Figure 13.22). Other types of miRNA molecules influence leaf shape, which helps to explain the diversity of leaf shapes produced by different plant species.

Leaf Shape and Surface Features Reflect Adaptation to Environmental Stress

As we have noted, leaf flatness facilitates solar energy collection, and thinness helps leaves to avoid overheating. Leaf shape and surface features also reflect adaptation to stressful environmental conditions.

Leaf Form The flattened portion of a leaf is known as the leaf **blade**. In beans and most other eudicots, blades are attached to the stem by means of a stalk known as a **petiole**, and an axillary bud occurs at the junction of stem and petiole (**Figure 35.13a**). In contrast, corn and other monocots have leaf blades that grow directly from the stem, encircling it to form a leaf sheath (**Figure 35.13b**).

Leaf shape can be simple or compound, each having particular advantages. Simple leaves have only one blade, though the edges may be smooth, toothed, or lobed. Simple leaves are advantageous in shady environments because they provide maximal light absorption surface, but they can overheat in sunny environments. As an evolutionary response to heating stress, the blades of some leaves have become highly dissected into leaflets. Such leaves are known as compound leaves (**Figure 35.13c**). Leaflets can be distinguished from leaves because leaflets lack axillary buds at their bases. Compound leaves are common in hot environments because leaflets foster heat dissipation. During the development of at least some compound leaves, the transcription factor KNOX becomes active shortly after the leaf primordia form, causing these primordia to produce multiple growth points that generate the leaflets. In contrast, during simple leaf development, KNOX is not active, because the expression of other proteins suppresses such *KNOX* gene expression.

Leaf Vein Patterns Leaf vein patterns are known as venation. Eudicot leaves occur in two major venation forms. They may have a single main vein from which smaller lateral veins diverge in a feather-like pattern known as **pinnate** venation (Figure 35.13a). Alternatively, several main veins may spread from a common point on the petiole like the fingers of your hand, a pattern known as **palmate** venation (Figure 35.13d). In eudicot leaves, small veins connect in a netted pattern, but most monocot leaves have a distinctive parallel venation (Figure 35.13b and Table 35.1).

Feature Investigation

Lawren Sack and Colleagues Showed That Palmate Venation Confers Tolerance of Leaf Vein Breakage

In 2008, Lawren Sack and associates studied the adaptive value of leaf venation patterns by comparing water conduction after injury in pinnately and palmately veined leaves (**Figure 35.14**). Leaves with palmate venation have several main veins, while leaves with pinnate venation have just one main vein. The investigators hypothesized that the multiple main veins of palmate leaves could confer greater tolerance of vein breakage of the type that would occur during mechanical injury or insect damage; if one main vein were damaged, water flow could continue through the other main veins. To test the hypothesis, the investigators experimentally cut a main vein in the leaves of several plants belonging to seven different plant species: four having pinnately veined leaves and three having palmately veined leaves. They conducted the experiments in vivo, that is, in the plant's natural forest environment. After the wounds had healed, the investigators measured the extent of water flow within the leaves at two or three places on each leaf. They found that across all species examined, palmately veined leaves tolerated the disruption in water flow better than pinnately veined leaves. Although palmate venation provides redundancy in case a main vein becomes damaged, it is more costly in terms of materials needed to construct the additional main veins. Hence, leaves with pinnate venation are less costly to produce and work well when the potential for vein damage is low.

Figure 35.14 **Sack and colleagues investigated the function of palmate venation.**

HYPOTHESIS Palmate venation provides vascular redundancy, which allows leaves to tolerate vein breakage.

KEY MATERIALS Seven species of trees or shrubs at Harvard Forest, Petersham, MA.

	Experimental level	Conceptual level
1 Identify 7 species, 4 with pinnately veined leaves and 3 with palmately veined leaves. For each species, the researchers analyzed 10 leaves on 3 different plants.	Pinnately veined leaves: Single primary vein connecting directly to petiole; Petiole; *Quercus rubra*, *Betula alleghaniensis*, *Viburnum cassinoides*, *Kalmia latifolia* Palmately veined leaves: More than one primary vein connecting directly to petiole; *Viburnum acerifolium*, *Acer saccharum*, *Acer rubrum*	Locate pinnate and palmate leaves for comparison.
2 With the leaf still attached to the plant, use a scalpel to cut across 1 primary vein in 5 experimental leaves but not 5 control leaves.		This procedure initially cuts off water supply via 1 primary vein.
3 Cover the cuts with medical tape on both top and bottom of leaves. Tape same area of uncut control leaves.	Tape	The tape prevents infection. Controls: Tape is applied to controls for experimental consistency.
4 Fold cardboard over base of cut leaves, forming a splint. Splint the same area of uncut control leaves.	Splint	The cardboard prevents leaf collapse. Control: Cardboard is applied to controls for experimental consistency.
5 Measure water conduction 2–9 weeks after treatment, in 2 regions (A, B) of each pinnate leaf and in 3 regions (A, B, C) of each palmate leaf.	Pinnate leaf (A, B); Palmate leaf (A, B, C)	After 2–9 weeks, the cuts had healed. Measurement of water conduction will determine the effect of the vein cut.

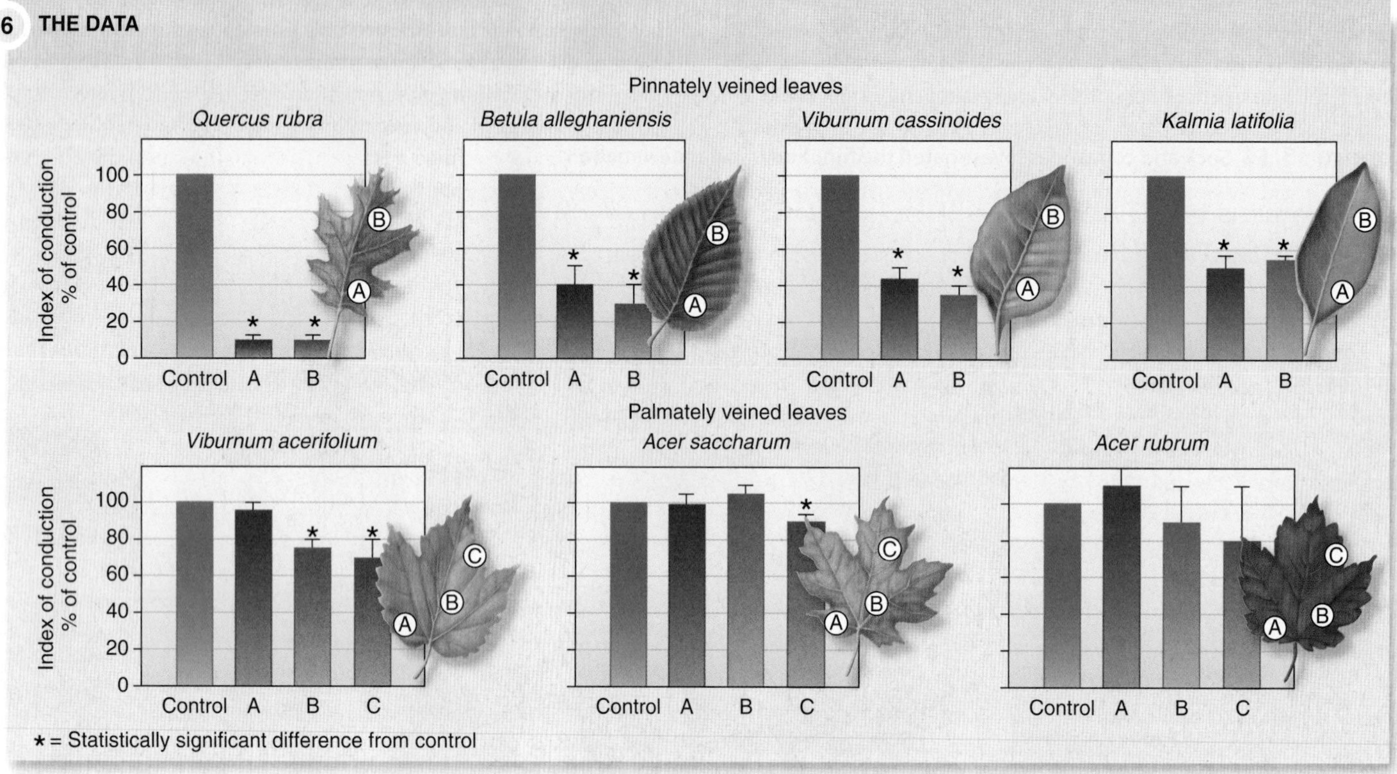

7 CONCLUSION Palmately veined leaves did not suffer as much conduction loss from a primary vein cut as did pinnately veined leaves.

8 SOURCE Sack, L. et al. 2008. Leaf palmate venation and vascular redundancy confer tolerance of hydraulic disruption. *Proceedings of the National Academy of Sciences of the U.S.* 105:1567–1572.

Experimental Questions

1. Why did Sack and associates conduct their studies of palmate venation on plants growing in a forest rather than in a greenhouse?
2. Why did Sack and colleagues splint leaves having cut veins as well as controls?
3. Why did Sack and associates measure leaf water conduction at two or more places on each leaf?

Leaf Surface Features Leaf surfaces also show adaptive features. As we have previously noted, a layer of epidermal tissue occurs at upper and lower leaf surfaces (see Figure 35.8c). These epidermal cells secrete a cuticle composed of protective wax and polyester compounds. The cuticle helps plants to avoid drying in the same way that enclosure in waxed paper keeps food moist. Plants that grow in very arid climates often have thick cuticles, whereas plants native to moist habitats typically have thinner cuticles.

Some leaf epidermal cells may differentiate into spiky or hair-like projections known as **trichomes** (**Figure 35.15**). Blankets of trichomes offer protection from excessive light, UV radiation, extreme air temperature, excess water loss, or attack by herbivores—animals that consume plant tissues. Broken trichomes of the stinging nettle,

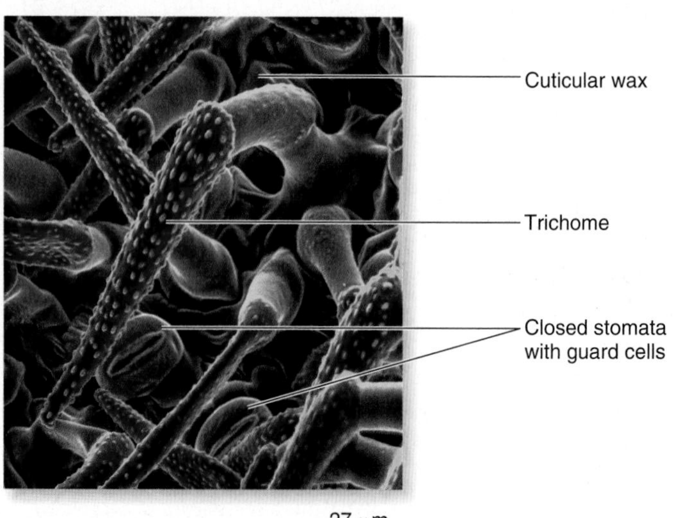

Figure 35.15 Leaf surface features. These features, viewed by SEM, include cuticular wax, trichomes, and stomatal pores with guard cells.

for example, release a caustic substance that irritates animals' skin, causing them to avoid these plants. Leaf epidermal cells include pairs of specialized guard cells located on either side of stomata (see Figure 35.15). These **guard cells** allow stomata to be open during moist conditions and to close when conditions are dry, thereby preventing plants from losing too much water. The genetic basis of guard cell development is becoming increasingly well understood.

Genomes & Proteomes Connection

Genetic Control of Guard-Cell Development

The flowering plant *Arabidopsis thaliana* is a model organism that is widely used to explore the genetic basis for plant structure and development. Several features increase *A. thaliana*'s utility for such studies: It is small in size, it has a fast seed-to-seed life cycle, it produces a relatively large number of seeds, and the genome has been sequenced. Mutants of this plant have been used to identify the genes controlling many aspects of plant structure and development, including the development of specialized stomatal guard cells.

Guard-cell development begins with an unspecialized protodermal cell that divides unequally (**Figure 35.16**). The larger of the two progeny cells eventually becomes a flat, puzzle piece–shaped epidermal cell, and the smaller is called a meristemoid because it functions like a stem cell located in a meristem. Meristemoids undergo one or more unequal cell divisions, producing more puzzle piece–shaped epidermal cells before finally dividing equally to produce a pair of guard cells. Genetic studies of *A. thaliana* have revealed that the meristemoid secretes a protein that inhibits division by adjacent cells but does not affect cells farther away. This process distributes stomata evenly and prevents too many of them from forming, which could increase the loss of water from plant surfaces.

In 2007, two teams, led by Lynn Pillitteri and Cora MacAlister, respectively, independently reported experiments with *A. thaliana* showing how three closely related genes control guard-cell development at three consecutive steps. A gene called *SPEECHLESS* starts the process by establishing the first unequal cell divisions of meristemoids. A protein encoded by the *MUTE* gene then causes meristemoids to stop dividing unequally so that equal divisions can produce the two guard cells (see Figure 35.16). Disabling mutations of these two genes cause the plant epidermis to completely lack stomatal pores (and so lack epidermal "mouths" and be speechless or mute). Finally, the gene *FAMA* directs guard-cell specialization. The proteins encoded by these plant genes are members of a type known as basic helix-loop-helix (bHLH) proteins. Similar bHLH proteins control the development of muscle and nerve cells in animals (refer back to Figure 19.21).

Modified Leaves Perform Diverse Functions

Though most leaves function primarily as photosynthetic organs, some plants produce leaves that are modified in ways that allow them to play other roles. For example, threadlike tendrils that help some plants attach to a supporting structure are modified leaves or leaflets (**Figure 35.17a**). The tough scales that protect buds on plants such as the sycamore from winter damage are modified leaves (**Figure 35.17b**). Poinsettia "petals" are actually modified leaves known as bracts, which are larger and more brightly colored than the flowers they surround and help attract pollinators (**Figure 35.17c**). Cactus spines are actually modified leaves that have taken on a defensive role, leaving photosynthesis to the cactus stem (**Figure 35.17d**).

Stems May Contain Primary and Secondary Vascular Systems

Stems, leaves, roots, buds, flowers, and fruits all contain vascular systems composed of xylem and phloem tissues that conduct water, minerals, and organic compounds. **Herbaceous plants** such as corn and bean produce mostly primary vascular tissues. In contrast, **woody plants** produce both primary and secondary vascular tissues. A comparison of primary and secondary vascular tissues will aid in understanding their roles.

Primary Vascular Tissues **Primary vascular tissues** are composed of primary xylem and phloem. Primary xylem is a complex tissue containing several cell types (see Table 35.2). These include unspecialized parenchyma cells; stiff fibers that provide structural support; and two types of cells facilitate water transport: narrower tracheids and wider

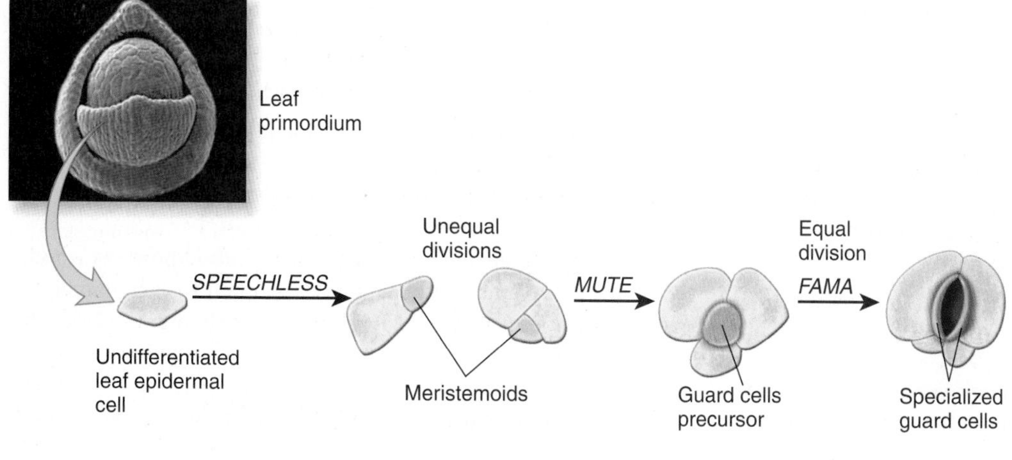

Figure 35.16 **The development of stomatal guard cells, controlled by three genes.**

BIOLOGY PRINCIPLE The genetic material provides a blueprint for reproduction. This sequence shows the gene actions necessary to generate stomata.

Figure 35.17 **Examples of modified leaves.** **(a)** The tendrils of an American vetch plant are modified leaves that help the plant attach to a trellis. **(b)** Bud scales, such as those on this sycamore bud, are modified leaves that protect buds from winter damage. **(c)** The attractive red bracts of poinsettia are modified leaves that function like flower petals to attract pollinator insects to the small flowers. **(d)** Cactus spines, such as these on this giant saguaro, are modified leaves that function in defense.

Concept Check: *Because cactus leaves are so highly modified for defense that they cannot effectively accomplish photosynthesis, how do cacti obtain organic compounds?*

vessel elements (see Figures 38.11 and 38.12). Arranged in pipeline-like arrays, **tracheids** and **vessel elements** conduct water, along with dissolved minerals and certain organic compounds (**Figure 35.18**). Mature tracheids and vessel elements are no longer living cells, and the absence of cytoplasm facilitates water flow. During development, these cells lose their cytoplasm by the process of programmed cell death, a process that resembles apoptosis in animals. However, the cell walls of tracheids and vessel elements don't easily break down or collapse because they are impregnated with a tough polymer known as lignin. The rigid cell walls of tracheids and vessel elements not only foster water conduction but also help support the plant body. Like plastic plumbing pipes, lignin provides the hydrophobic surface needed for water movement, as well as the strength to support trees weighing more than 2,000 metric tons.

In contrast to xylem, living phloem tissue transports organic compounds such as sugars and certain minerals in a watery solution. Phloem tissue includes **sieve-tube elements**, thin-walled living cells that are arranged end to end to form pipelines (**Figure 35.19**). Pores in the end walls of sieve-tube elements allow solutions to move from one cell to another. Phloem tissue also includes companion cells that aid sieve-tube element metabolism, supportive fibers, and parenchyma cells (see Table 35.2). Phloem fibers are tough-walled sclerenchyma cells that are surprisingly long, 20–50 mm, and valued for their high strength. The phloem fibers of hemp (*Cannabis sativa*), flax (*Linum usitatissimum*), jute (*Corchorus capsularis*), kenaf (*Hibiscus cannabinus*), and ramie (*Boehmeria nivea*) are commercially important in the production of rope, textiles, and paper.

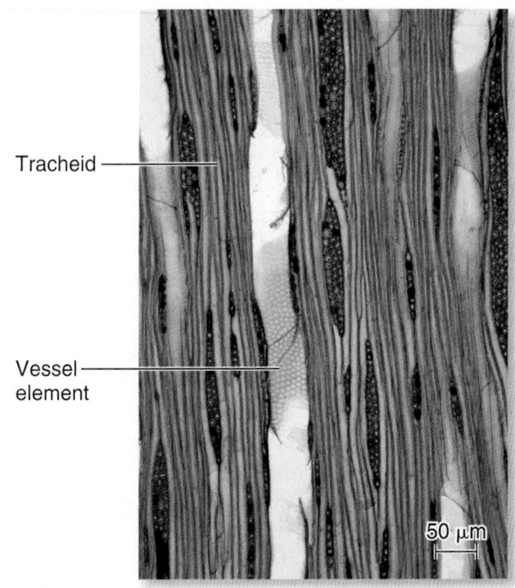

Figure 35.18 **Water-conducting cells of the xylem.** In this thinly sliced portion of a stem, the stained, lignin-impregnated walls of narrow tracheids and wider vessel elements can be distinguished.

BIOLOGY PRINCIPLE **Structure determines function.** Because lignin strengthens these cells, they do not collapse as large volumes of water move through them.

Secondary Vascular Tissues Woody plants begin life as herbaceous seedlings that possess only primary vascular systems. But as these plants mature, they produce secondary vascular tissues and bark. Secondary vascular tissues are composed of secondary xylem and secondary phloem. **Secondary xylem** is also known as **wood**, a component of plants that plays many important roles in human life. Wood is composed of about 25% lignin, 45% cellulose, and 25% other polysaccharides that are together known as hemicelluloses.

Secondary phloem is the **inner bark**. **Outer bark** is protective layers of mostly dead cork cells that cover the outside of woody

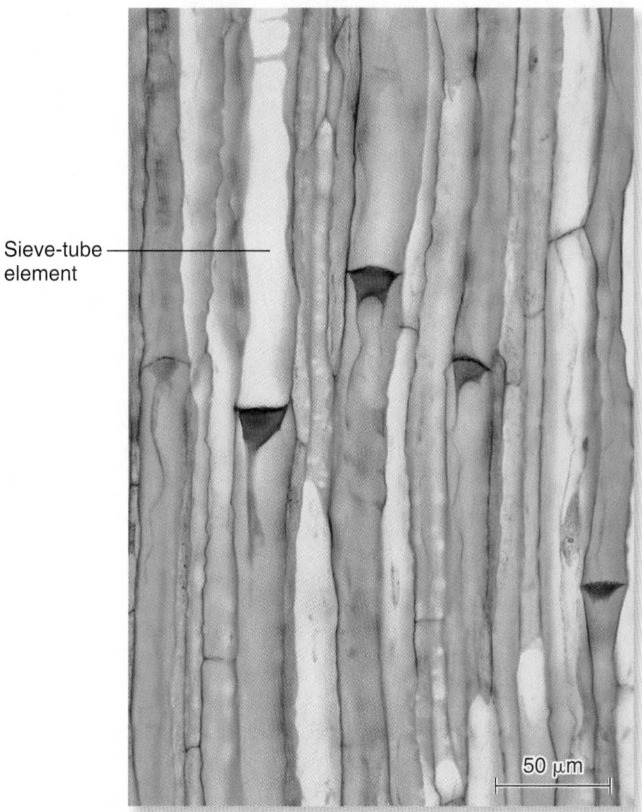

Figure 35.19 Food-conducting cells of the phloem. This thinly sliced portion of a stem shows stained, thin-walled sieve-tube elements that conduct watery solutions of certain minerals and organic compounds such as sugar.

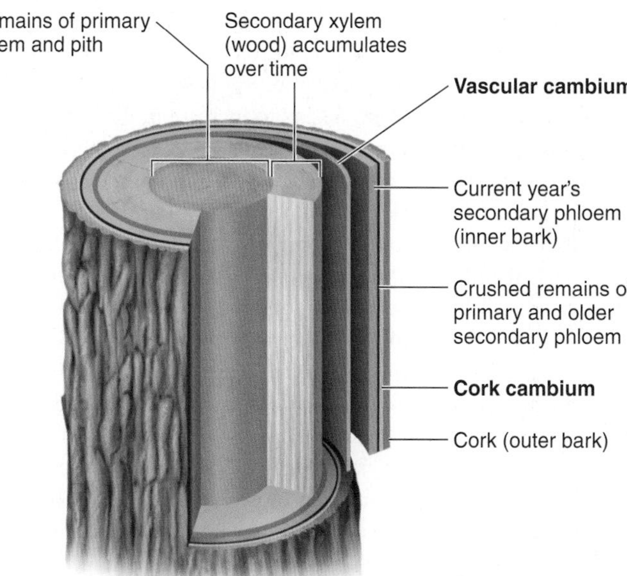

Figure 35.20 Formation of wood and bark by secondary (lateral) meristems. The vascular cambium is a thin cylinder of tissue that produces a thick cylinder of wood (secondary xylem) toward the inside of the stem and a thinner cylinder of inner bark (secondary phloem) toward the outside of the stem. The cork cambium forms an outer coating of protective cork (outer bark).

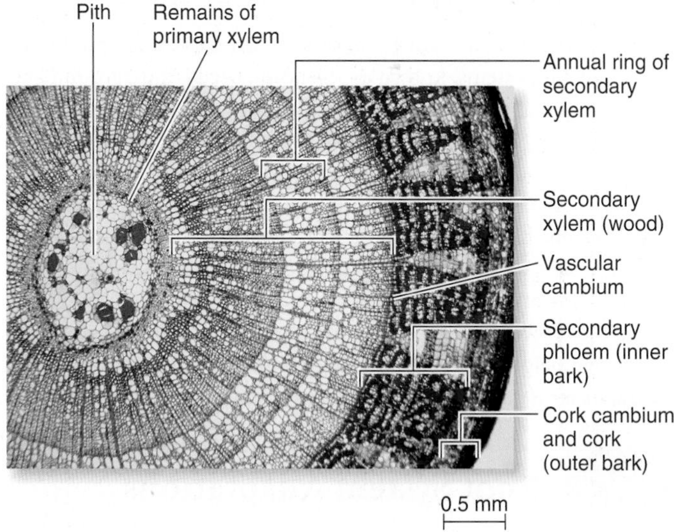

Figure 35.21 The anatomy of a tree trunk. Each year, a new cylinder of wood is produced; this yearly wood production appears as annual rings on the cut surface of a woody stem.

Concept Check: Why do tree trunks have a thicker layer of wood (secondary xylem) than of inner bark (secondary phloem)?

stems and roots. Therefore, bark includes both inner bark (secondary phloem) and outer bark (cork). Woody plants produce secondary vascular tissues by means of **secondary meristems**, also known as **lateral meristems**, which form rings of actively dividing cells that encircle the stem. The two types of secondary meristems are vascular cambium and cork cambium, which are derived from primary meristems.

The secondary meristem known as **vascular cambium** is a ring of dividing cells that produces secondary xylem to its interior and secondary phloem to its exterior (Figure 35.20). Secondary xylem conducts most of a woody plant's water and minerals. Cell divisions that occur in secondary meristems increase the girth of woody stems. During each new growing season, the vascular cambium produces new cylinders of secondary xylem and secondary phloem. In temperate trees, each year's addition of new secondary xylem forms growth rings that can be observed on the cut stem surfaces (Figure 35.21). The growth rings of secondary xylem surround the remains of the primary xylem and a central cylinder of parenchyma cells known as the pith. If environmental conditions favor plant growth, the growth rings formed at that time will be wider than those formed during stressful conditions. Climatologists use growth ring widths in samples of old wood to deduce past climatic conditions, and archeologists use growth ring data to determine the age of wood constructions and artifacts left by ancient cultures.

Secondary xylem may transport water for several years, but usually only the current year's production of secondary phloem is active in food transport. This is because thin-walled sieve elements typically live for only a year. Thus, only a thin layer of phloem, the inner bark, is responsible for most of the sugar transport in a large tree. Deep abrasion of tree bark may damage this thin phloem layer, disrupting a tree's food transport. If a groove is cut all the way around a tree trunk—a process known as girdling—the tree will die because all of its functional phloem transport routes will have been interrupted.

As a young woody stem begins to increase in diameter, its thin epidermis eventually ruptures and is replaced by outer bark, which is composed of protective cork tissues. Cork is produced by a secondary meristem called the **cork cambium**, another ring of actively dividing cells. The cork cambium surrounds the secondary phloem (see Figures 35.20 and 35.21). Together, the cork cambium, layers of cork tissue produced by the cambium, and associated parenchyma cells are known as a **periderm**. The outer bark becomes thicker as woody stems accumulate multiple periderm layers. The outer bark surface is often interrupted by passages known as **lenticels** that allow inner stem tissues to accomplish gas exchange.

Cork cells are dead when mature, and their walls are layered with suberin, a material that helps to prevent both attack by microbial pathogens and water loss from the stem surface. Cork tissues also produce tannins, compounds that protect against pathogens by inactivating their proteins. The cracked surfaces of tree trunks are dead cork tissues of the outer bark. Commercial cork is sustainably harvested from the cork oak tree (*Quercus suber*) for production of flooring material, bottle stoppers, and other items. Additional information about the structure and function of primary and secondary xylem and phloem can be found in Chapter 38.

Modified Stems Display Diverse Forms and Functions

Stems mostly grow upright because light is required for photosynthesis. But some stems, known as rhizomes, occur underground and grow horizontally. For example, potato tubers are the swollen, food-storing tips of rhizomes. Grass stems also grow horizontally, as either rhizomes just beneath the soil surface or stolons, which grow along the soil surface. The leaves and reproductive shoots of grasses grow upward from the point where they are attached to these horizontal stems. Grass blades continue to elongate from their bases even if you cut their tips off, explaining why lawns must be mowed repeatedly during the growing season. The horizontal stems of grasses are adaptations that help to protect vulnerable shoot apical meristems against natural hazards such as fire and grazing animals.

35.4 Root System Adaptations

Learning Outcomes:

1. List ways that root structure has been modified in different plants in response to different habitats.
2. Create a drawing that shows how root systems and branch roots develop.
3. Discuss several examples that demonstrate how mobile proteins or miRNAs influence root development.

Roots play the essential roles of absorbing water and minerals, anchoring plants in soil, and storing nutrients. The external form of roots varies among flowering plants, reflecting adaptation to particular life spans or habitats. In contrast, root internal structure is more uniform. In this section, we first consider variation in root external structure and then focus on root internal structure and development.

(a) Buttress roots

(b) Pneumatophores

Figure 35.22 Modified aboveground roots. **(a)** Buttress roots help to keep tropical trees such as this *Pterocarpus hayesii* from toppling in windstorms. **(b)** Pneumatophores produced by mangroves are roots that extend upward into the air. These roots take up air and then transmit it to underwater roots that grow in oxygen-poor sediments.

Modified Roots Display Diverse External Forms and Functions

As we have observed, the common bean and other eudicots display an underground taproot system, whereas corn and other monocots have a fibrous root system (see Table 35.1). Plants produce several other types of roots that provide adaptive advantages in response to different habitats. For example, corn and many other plants produce supportive prop roots from the lower portions of their stems. Many tropical trees grow in such thin soils that the trees are vulnerable to being blown down in windstorms. Such trees often produce dramatic aboveground buttress roots that help keep trees upright (**Figure 35.22a**). Many mangrove trees that grow along tropical coasts produce pneumatophores (Greek meaning breath bearers), roots that grow upward into the air (**Figure 35.22b**). Functioning like snorkels, pneumatophores absorb oxygen-rich air, which diffuses to submerged roots growing in oxygen-poor sediments. This is necessary because, like animals, all roots require a supply of oxygen in order to produce ATP. Roots use this ATP to power root growth and the uptake of mineral nutrients (see Chapter 37).

Root Internal Growth and Tissue Specialization Occur in Distinct Zones

Studies of gene expression in *Arabidopsis* roots reveal that roots are amazingly complex in their internal structure, having at least 15 distinct cell types. For our purposes, a simpler microscopic examination

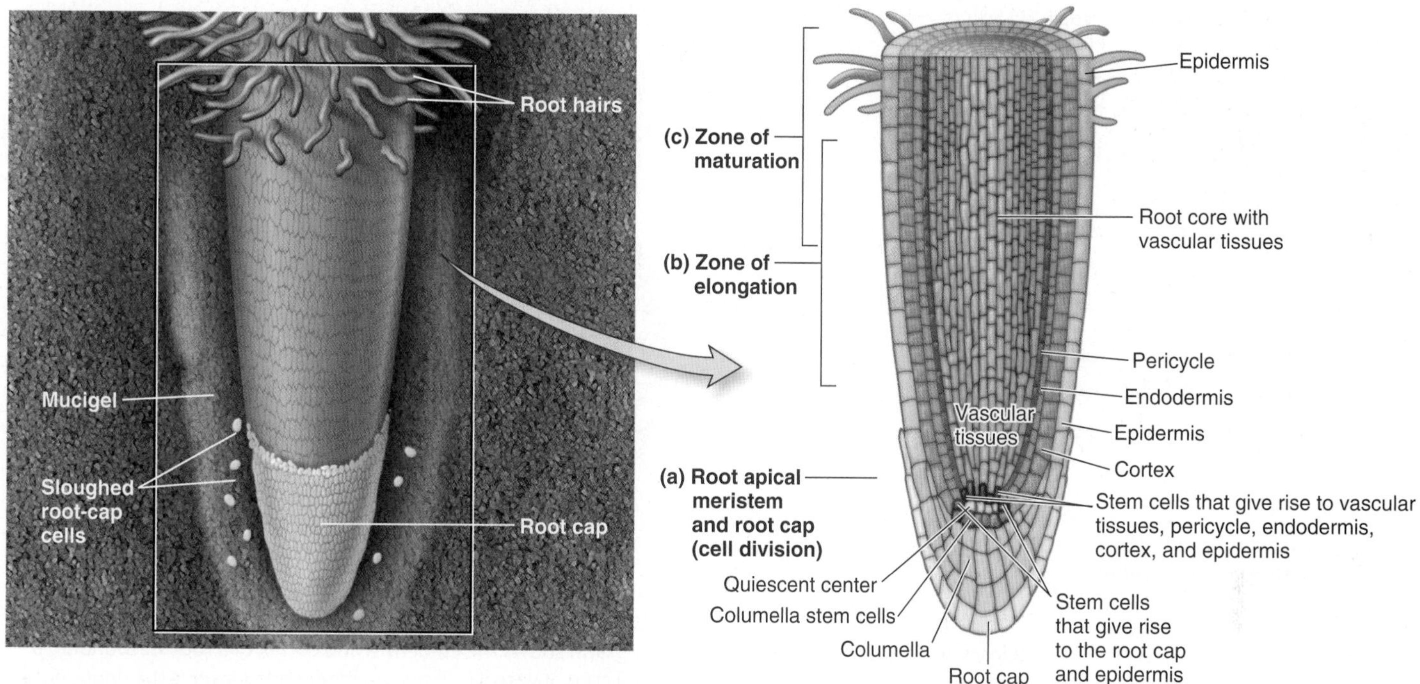

Figure 35.23 Three zones of root growth. A longitudinal view of a typical root reveals three major zones: **(a)** a root apical meristem region that includes stem cells, a quiescent zone, columella, and a root cap; **(b)** a zone of elongation; and **(c)** a zone of maturation, characterized by specialized cells and tissues including epidermal root hairs, cylinders of endodermis and pericycle tissue, and a core of vascular tissue.

of root internal structure reveals three major zones: (1) a root apical meristem (RAM) protected by a root cap, (2) a zone of root elongation, and (3) a zone of maturation in which specialized cells can be observed (Figure 35.23).

Root Apical Meristem and Root Cap As discussed earlier, an apical meristem occurs at the tips of roots and their branches. Like the SAM, the RAM contains stem cells, but these are organized differently in root apices. Root stem cells surround a tiny region of cells that rarely divide, known as the quiescent center. Signals emanating from the quiescent center keep nearby stem cells in an undifferentiated state. Root stem cells farther away from the quiescent center produce new cells in multiple directions. Toward the root tip, stem cells produce columella cells that sense gravity and touch, which helps roots extend downward into the soil and around obstacles such as rocks. At the sides of the quiescent center, stem cells produce a protective root cap and epidermal cells. Root tip epidermal cells secrete a sticky substance called mucigel that lubricates root growth through the soil. Toward the shoot, stem cells generate cells that become internal root tissues.

Zones of Elongation and Maturation Above the RAM lies the **zone of elongation**, in which cells extend by water uptake, thereby dramatically increasing root length (see Figure 35.23). The root elongation zone illustrates the general principle that cell expansion in plants is not necessarily linked directly to cell division. (By contrast, animal cell expansion is more closely linked to cell division.)

Above and overlapping with the root zone of elongation is the **zone of maturation**, where most root cell differentiation and tissue specialization occur. Specialized root tissues include mature vascular tissues at the root core, an enclosing cylinder of cells known as the pericycle, another cell cylinder called the endodermis (meaning inside skin), and epidermal cells at the root surface. Relatively unspecialized parenchyma cells form a cortex that lies between the endodermis and the epidermis. Starting with the epidermis and moving inward, we will take a closer look at these root tissues and factors that control their development.

The zone of maturation can be identified by the presence of numerous microscopic hairs that emerge from the root epidermis. **Root hairs** are specialized epidermal cells that can be as long as 1.3 cm, about the width of your little finger, but are only 10 μm in diameter, less than the width of a finger cell. Their small diameter allows root hairs to obtain water and minerals from soil pores that are too narrow for even the smallest roots to enter. Root hair plasma membranes are rich in transport proteins that use ATP to selectively absorb materials from the soil (shown in Figure 38.3).

The production of hairs from root epidermal cells is controlled by the activity of transcription factors that move between cells and also by cell position (**Figure 35.24**). A root hair will develop from epidermal cells lying over the junction of two cortical cells, whereas no hair will develop if an epidermal cell lies over only one cortical cell. Root hairs are so delicate that they are easily damaged by abrasion as roots grow through the soil, and they live for only 4 or 5 days. As a result, root hairs are absent from older regions above the zone of maturation. To compensate, roots must continually produce new root hairs. The average rate of root hair production has been estimated at more than 100 million per day for some plants. One reason that gardeners use care when transplanting seedlings is to prevent extensive damage to the root hair zone.

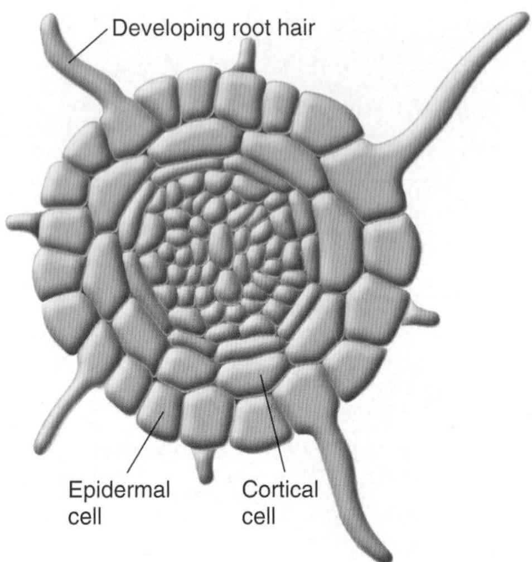

Figure 35.24 **A cross section of a root showing positional influence on root-hair development.**

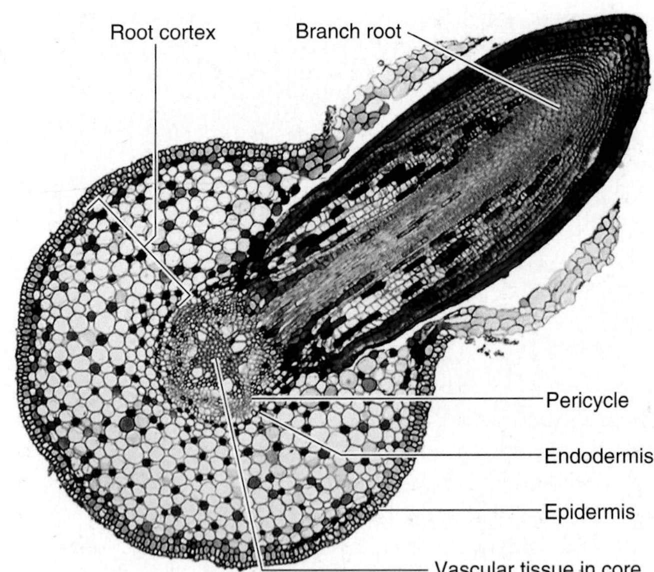

Figure 35.25 **Cross section of a mature root.** This stained light micrograph shows the epidermis and cortex of a root surrounding a central core of vascular tissue. An inner cortex layer is the endodermis, which surrounds a cylinder of meristematic pericycle tissue. The pericycle has produced a young branch (lateral) root that has grown through the cortex and the epidermis.

Concept Check: *Why must lateral roots be produced in this way?*

The epidermis of mature roots encloses a cylinder of parenchyma known as the root cortex (**Figure 35.25**). Much like the stem cortex, root cortex cells are often rich in starch and therefore serve as a food storage site for plants. The root cortex of some plants contains intercellular air spaces that arise from programmed cell death and provide routes for oxygen diffusion within the root. Water and dissolved minerals also diffuse from the environment into roots through spaces between cortex cells, stopping only when they reach a one-cell-thick cylinder of specialized tissue known as **endodermis** ("inside skin"). This endodermal cell layer is an important component of the mechanism by which roots absorb selected minerals (described further in Chapter 37). Endodermal cells specialize in response to a protein transcription factor (SHORT ROOT) that is synthesized in cells of the root core, then transported outward.

A cylinder of tissue having cell division (meristematic) capacity, known as the **pericycle**, encloses the root vascular tissue (see Figure 35.25). The pericycle produces lateral branch roots that force their way through the cortex to the surface. This process differs from the way that stems produce branches by means of buds. In the model plant *Arabidopsis thaliana*, the expression of a gene known as *ARABIDILLO* is known to promote lateral root development. In some roots, the pericycle generates a vascular cambium that produces wood—secondary xylem. Such woody roots also possess a cork cambium that makes a protective covering of suberin-coated cork tissue. Like woody stems, woody roots produce primary vascular tissues in their youth and secondary vascular tissues at maturity. The woody roots of trees are sometimes visible aboveground.

Finally, the primary vascular system in a mature root includes xylem and phloem located at the root core (see Figure 35.25). The xylem exports water and minerals upward to shoots, and the phloem imports a watery solution of organic compounds from the shoots. Root xylem development is regulated by microRNA165/6, which is produced in and secreted from the endodermis. In newly produced, but as-yet-unspecialized root core cells, these microRNAs act as a signal to repress mRNA levels of a particular transcription factor, causing root core cells to differentiate into xylem. MicroRNA 165/6 and the mobile transcription factors that influence endodermal and root hair development are examples of the strong influence that chemical signaling has on the development of specialized plant tissues.

Summary of Key Concepts

35.1 From Seed to Seed—The Life of a Flowering Plant

- Seed embryos, seedlings, and mature plants are components of the sporophyte generation in the plant sexual cycle; tiny gametophytes develop and grow within flowers (Figure 35.1).
- Plant organs are composed of tissues that contain specialized cells. The basic plant organs are roots, stems, and leaves. Shoot systems include stems and stem branches, and stems produce leaves, buds, flowers, and fruits. Root systems include one or more main roots with branches. Buds, flowers, fruits, and seeds are organ systems, composed of more than one organ (Figures 35.2, 35.3).
- The two major groups of flowering plants—eudicots and monocots—differ in the structure of their seed embryos, flowers, stems, roots, leaves, and pollen (Figure 35.4, Table 35.1).

35.2 How Plants Grow and Develop

- The principles of plant growth and development include the presence of a fundamental architecture featuring apical-basal polarity and radial symmetry throughout the life of a plant (Figures 35.5, 35.6).
- Plants grow by producing new cells at meristems and controlled cell enlargement involving water uptake.

- Shoot apical meristems produce primary meristems that increase plant length and produce organs (Table 35.2).
- The simple plant tissues, containing one or two cell types, include parenchyma, collenchyma, and sclerenchyma. Complex plant tissues include the vascular tissues known as xylem and phloem, and the primary vascular tissues occur in vascular bundles (Figure 35.7).
- Leaves develop from primordia at shoot apices. Foliage leaves have internal and external structure that is adapted for photosynthetic functions (Figure 35.8).
- Meristems include youthful stem cells, whose numbers influence plant growth and structure and are thus genetically controlled.
- Plant cells are able to expand under conditions that result in loosening of cell-wall components, and by water uptake into vacuoles. The direction in which plant cells expand is determined by the arrangement of wall cellulose microfibrils, which is influenced by the orientation of microtubules in the nearby cytoplasm (Figures 35.9, 35.10, 35.11).

35.3 The Shoot System: Stem and Leaf Adaptations

- Shoots are modular systems; each module includes a node, internode, leaf, and axillary meristem or bud. An axillary bud develops in leaf axils; such buds may grow into new branches (Figure 35.12).
- Variations in leaf structure reflect adaptations that aid photosynthesis or protect against stress. For example, Sack and colleagues demonstrated that palmate leaves provide conducting system redundancy useful in coping with vein damage (Figures 35.13, 35.14, 35.15).
- Stomatal guard cell differentiation is controlled by several genes (Figure 35.16).
- Leaves not only function in photosynthesis but also play other roles, including attachment, attraction, and protection (Figure 35.17).
- Herbaceous plants are those whose stems produce little or no wood and are mostly composed of primary vascular tissues. The primary vascular tissues are primary xylem and primary phloem (Figures 35.18, 35.19).
- In addition to primary tissues, woody plants—trees, shrubs, and lianas—possess secondary meristems that produce wood and bark. The vascular cambium produces secondary xylem (wood) and secondary phloem (inner bark). The cork cambium produces cork tissues that form outer bark (Figures 35.20, 35.21).
- Stems occur in diverse forms that reflect adaptation to environmental conditions. Examples include grass rhizomes, which grow horizontally underground and are therefore better protected from fire and grazing animals (as well as lawnmowers).

35.4 Root System Adaptations

- Roots occur in multiple forms that reflect adaptation to environmental conditions. Examples of aboveground roots include prop roots, buttress roots, and pneumatophores (Figure 35.22).
- The internal organization of roots is comparatively uniform, and three major zones can be recognized with the use of a microscope: the root apical meristem and root cap, a zone of cell and root elongation, and a zone of tissue maturation. Features of the mature root include epidermal root hairs that aid nutrient uptake, a food-storing cortex, an endodermis that functions in mineral selection, a pericycle that produces lateral (branch) roots (and vascular cambium in the case of woody roots), and an inner core of vascular tissue (Figures 35.23, 35.24, 35.25).

Assess and Discuss

Test Yourself

1. Where would you look to find the gametophyte generation of a flowering plant?
   a. at the shoot apical meristem
   b. at the root apical meristem
   c. in seeds
   d. in flower parts
   e. Flowering plants lack a gametophyte generation.
2. What is a radicle?
   a. an embryonic leaf
   b. an embryonic stem
   c. an embryonic root
   d. a mature root system of a monocot
   e. an organism that has extreme political views
3. Which type of plant is most likely to have food-rich roots that are useful as human food?
   a. an annual
   b. a biennial
   c. a perennial
   d. a centennial
   e. plants that grow along coastal shorelines
4. Which of the following terms best describes the distinctive architecture of plants?
   a. radial symmetry and apical-basal polarity
   b. bilateral symmetry and apical-basal polarity
   c. radial symmetry and absence of apical-basal polarity
   d. bilateral symmetry and absence of apical-basal polarity
   e. absence of symmetry and absence of apical-basal polarity
5. Which is the most accurate description of how plants grow?
   a. by the addition of new cells at meristems that include stem cells
   b. by cell enlargement as the result of water uptake
   c. by both the addition of new cells and cell expansion
   d. by addition of fat cells
   e. all of the above
6. Where would you look for leaf primordia?
   a. at a vegetative shoot tip
   b. at the root apical meristem
   c. at the vascular cambium
   d. at the cork cambium
   e. in a floral bud
7. Which leaf tissues display the greatest amount of air space?
   a. the upper epidermis
   b. the lower epidermis
   c. the palisade parenchyma
   d. the spongy parenchyma
   e. the vascular tissues
8. What are adventitious roots?
   a. roots that develop on plant cuttings that have been placed in water
   b. buttress roots that grow from tree trunks
   c. the only kinds of roots produced by monocots, because their embryonic root dies soon after seed germination
   d. any root that is produced by stem (or sometimes leaf) tissue, rather than developing directly from the embryonic root
   e. all of the above

9. During its development, a tracheid elongates in a direction parallel to the shoot or root axis. Based on this information, what can you say about the orientation of cellulose cell-wall microfibrils and cytoplasmic microtubules in this developing tracheid?
   a. The microfibrils will be oriented perpendicularly (at right angles) to the long axis of the developing tracheid, encircling it, but the cytoplasmic microtubules will be oriented parallel to the direction in which the tracheid is elongating.
   b. Microfibrils and microtubules will both be oriented perpendicularly (at right angles) to the elongating axis of the tracheid.
   c. Microfibrils and microtubules will both be oriented parallel to the direction of tracheid elongation.
   d. Microfibrils will be oriented parallel to the direction of tracheid elongation, but microtubules will be perpendicular (at right angles) to both the microfibrils and the elongating tracheid.
   e. None of the above is correct.
10. What are examples of woody plants?
   a. trees
   b. shrubs
   c. woody vines, known as lianas
   d. all of the above
   e. none of the above

Conceptual Questions

1. What would be the consequences if overall plant architecture were bilaterally symmetric?
2. What would be the consequences if leaves were radially symmetric (shaped like spheres or cylinders)?
3. **A principle of biology is that *structure determines function*. Why are most tall plants woody, rather than herbaceous?**

Collaborative Questions

1. Find a tree stump or a large limb that has recently been cut from a tree (or imagine doing so). Which of the following features could you locate with the unaided eye: the outer bark, the inner bark, the secondary xylem, the vascular cambium, annual rings?
2. Which physical factors would you expect to influence shoot growth most strongly? Which physical factors would you expect to influence underground root growth most strongly?

Online Resource

www.brookerbiology.com

Stay a step ahead in your studies with animations that bring concepts to life and practice tests to assess your understanding. Your instructor may also recommend the interactive eBook, individualized learning tools, and more.

Chapter Outline

36.1 Overview of Plant Behavioral Responses
36.2 Plant Hormones
36.3 Plant Responses to Environmental Stimuli
Summary of Key Concepts
Assess and Discuss

Flowering Plants: Behavior

36

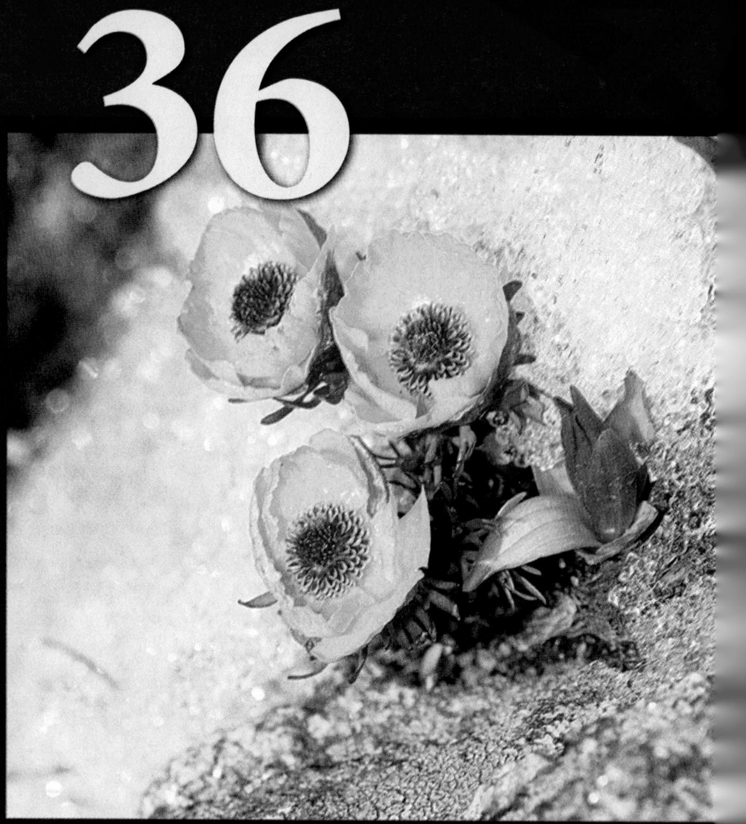

Behavior of the snow buttercup. The snow buttercup (*Ranunculus adoneus*) holds its flowers above the surface of the snow. The flowers move so they always face the Sun during the day, a behavior known as Sun tracking.

The snow buttercup (*Ranunculus adoneus*) grows in deep snowbanks in the high Rocky Mountains, with flower stems protruding above the snow surface toward the Sun, as shown in the chapter-opening photo. Amazingly, snow buttercup flowers change their position so they face the Sun throughout the day, a process known as sun tracking. Experiments have demonstrated that Sun tracking warms snow buttercup flowers, thereby favoring pollen development and germination. These processes foster fertilization, which leads to more effective seed production. In this way, Sun tracking can be understood as an adaptation that increases snow buttercup reproductive fitness. Like Sun-tracking solar panels, the leaves of alfalfa, lupine, soybean, common bean, cotton, and other wild and agricultural plants also track the Sun, a process that aids energy storage by photosynthesis.

Sun tracking is but one example of the many ways in which plants display behavior—that is, responses to stimuli. Plants respond not only to changes in Sun position, but also to day length, which is the period of daily illumination. In addition to light, plants respond to gravity, wind, attack by animals and disease-causing microorganisms, and other environmental stimuli. This chapter provides examples of ways in which understanding plant behavior has been key to increasing agricultural productivity and protecting natural ecosystems for the benefit of humans.

We begin with a survey of the diverse types of stimuli that induce plant behavior and review how cells perceive and respond to stimuli by means of signal transduction pathways. Next, we will focus on plant hormones, which are major types of internal mobile molecules that influence plant behavior. Finally, we consider how responses to environmental stimuli foster plant survival and reproduction.

36.1 Overview of Plant Behavioral Responses

Learning Outcomes:

1. List examples of plant responses to internal and environmental stimuli.
2. Describe the three stages of cell signaling.

Behavior is defined as a response of organisms to an internal or external stimulus. Examples of plant behavior include plant movements, some types of which were described in 1880 by Charles Darwin and his son Francis in their book *The Power of Movement in Plants.* Modern time-lapse photography, which makes plant behavior more obvious to humans, reveals that most plants are constantly in motion, bending, twisting, or rotating in dancelike movements known as nutation (**Figure 36.1a**). Some plants display relatively rapid movements, illustrated by the sensitive plant (*Mimosa pudica*), whose leaves quickly fold when touched then open more slowly (**Figure 36.1b**). Plant shoots typically grow toward light and against the pull of gravity, and most roots grow toward water and in the same direction as the gravitational force. Seeds germinate when they detect the presence of sufficient light and moisture for successful seedling growth. Producing flowers, fruit, and seeds only at the season most favorable for reproductive success is likewise a behavioral response to environmental change. Plants also take protective actions when they sense attack by disease microbes or hungry animals, thereby preventing excessive damage to their own bodies or those of neighboring plants. To gain a more complete understanding of plant behavior, we will begin by surveying the types of stimuli that cause plant responses.

Figure 36.1 **Examples of plant movements.** **(a)** Sixteen superimposed photographs of a shoot of the honeysuckle vine *Lonicera japonica,* taken over a period of 2 hours, reveal the circular movement known as nutation. **(b)** Photographs of the sensitive plant (*Mimosa pudica*) made before and shortly after a touch reveal the rapid process of leaf folding. Even if only one leaflet is touched, electrical signals travel throughout the complex leaf, causing the entire organ to fold. The leaves will eventually unfold.

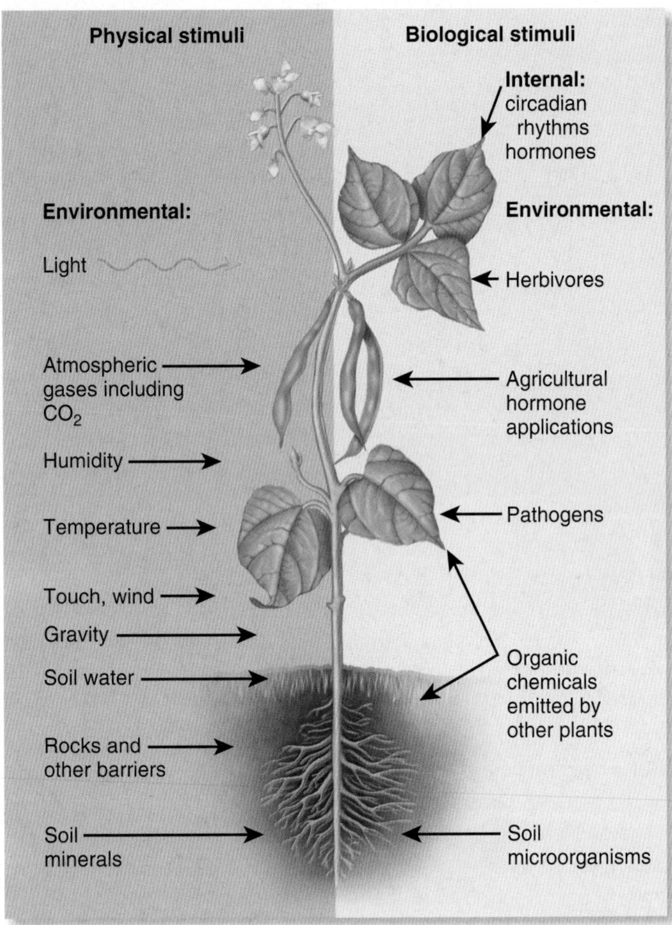

Figure 36.2 **Types of plant stimuli.** Plants respond to both physical and biological stimuli. Stimuli may be internal to the plant or come from the environment.

Plant Behavior Involves Responses to Internal and External Stimuli

Most people are aware that both internal chemical signals and environmental factors influence animal behavior. Bird nesting behavior in spring, for example, involves hormonal changes triggered by seasonal conditions. Plants likewise respond to internal and environmental stimuli (Figure 36.2).

Internal Stimuli Plants respond to two types of internal stimuli: internal biological clocks and mobile chemical signals. Internal biological clocks, known as **circadian rhythms**, occur not only in plants, but also in animals and other organisms. The word circadian comes from the Latin words meaning "about" and "day." Circadian rhythms evolved under the influence of Earth's rotation, which causes the regular alternation of night and day. Leaf movements, flower opening, fragrance emission, and many other behaviors result from the operation of circadian rhythms in plants.

Like animals, plants also respond to internal chemical signals that are produced within the body and move from one location to another, acting at very low concentrations. In plants, these chemical signals may move short distances from cell to cell via plasmodesmata, or long distances within the vascular system. Plant chemical signals include transcription factors and microRNAs (see Chapter 35) as well as chemically different compounds known as **hormones**. Some plant hormones are similar to particular animal hormones. One example is the plant hormone jasmonic acid, which is much like prostaglandins produced by animals. The gaseous hormone nitric oxide (NO) functions in both animals and plants. Other hormones are distinctive to plants. Even so, as in the case of animal hormones, plant hormones often interact with each other and with external stimuli to maintain a stable internal environment (homeostasis) and enable progress through life stages.

Environmental Stimuli Plants sense and respond to many types of physical and biological environmental stimuli (see Figure 36.2). Physical stimuli in natural plant environments include light, atmospheric gases such as CO_2 and water vapor, temperature, touch, wind, gravity, soil water, rocks and other barriers to root growth, and soil minerals. Biological stimuli include herbivores (animals that consume plant parts), airborne pathogens (disease-causing microbes), organic chemicals emitted from neighboring plants, and beneficial or harmful soil microorganisms. Crop plants also respond to applications of agricultural chemicals, which may include hormones. That plants have evolved such a broad array of sensory capacity is not surprising, because all of the listed environmental influences affect plant survival and reproduction.

Plant Responses to Environmental Stimuli Though plants lack the specialized sense organs typical of animals, receptor molecules located in plant cells sense stimuli and cause responses. When many cells of a tissue receive and respond to the same biological or physical

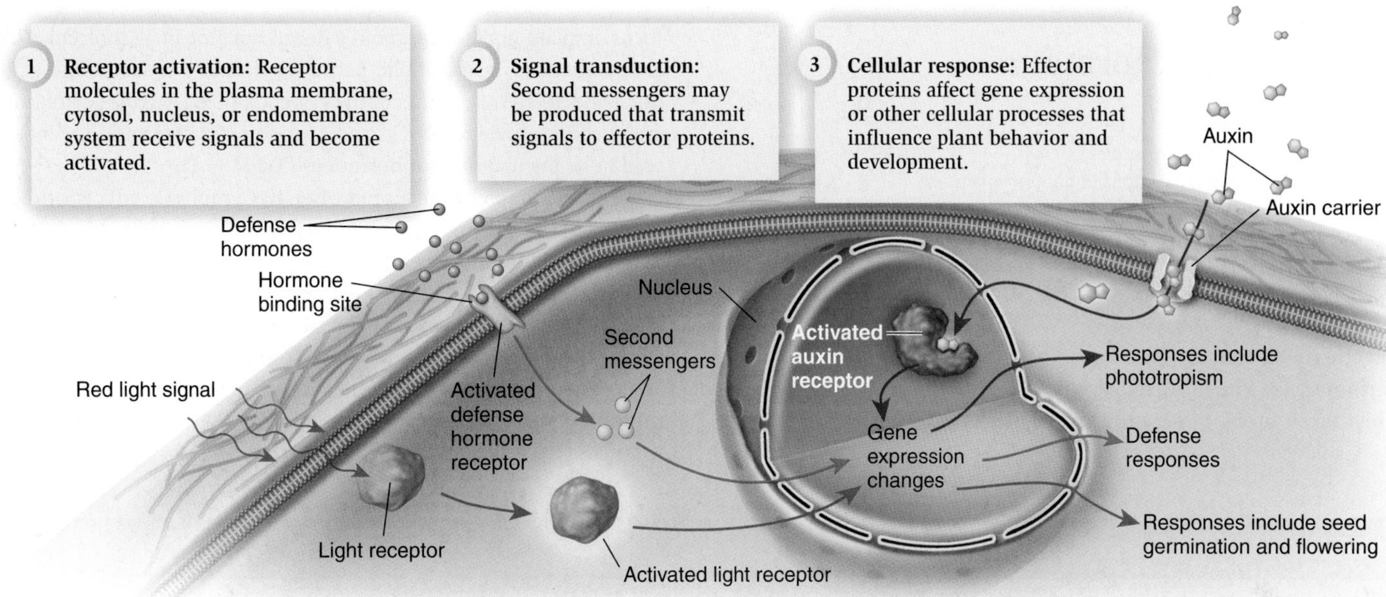

Figure 36.3 An overview of plant cell signaling. Plant cells respond to hormonal signals produced within the plant body, as well as to environmental stimuli. Three different signal transduction processes are shown here: one started by light, a defense hormone having a plasma membrane receptor, and auxin having a receptor located in the nucleus.

BIOLOGY PRINCIPLE Cells are the simplest units of life. Internal and environmental stimuli are received and elicit responses at the cellular level.

BioConnections: Look back to Figure 9.4, which shows an example of a cell-signaling process in an animal cell. Which of the processes shown above is most similar?

stimuli, entire organs or plant bodies display behavior. For example, houseplants tend to grow toward a light source such as a window. This process, known as positive **phototropism**, involves both a cellular perception of light and a growth response of stem tissue to an internal chemical signal. In general, a **tropism** is a growth response that depends on a stimulus that occurs in a particular direction. In the case of phototropism, the plant senses the direction of light and responds by changing the location of a plant hormone known as auxin. The auxin moves into cells where it influences gene expression, causing the stem to bend toward light. We will next review how plant cells more generally receive signals and transmit them intracellularly, a process known as cell signaling, that occurs in all cells (refer back to Chapter 9).

Plant Cell Signaling Involves Receptors, Messengers, and Effectors

Cell signaling is the process in which a cell perceives a physical or chemical signal, thereby switching on an intracellular pathway that leads to a cellular response (Figure 36.3). The process of cell signaling often involves three types of molecules: receptors that may become activated, second messengers that transmit signals, and effectors that cause a cellular response (refer back to Figure 9.4).

Receptors (also known as sensors) are proteins that become activated when they receive a specific type of signal (see Figure 36.3). Receptors occur in diverse cellular locations. Whereas some plant defense receptors are located within the plasma membrane, light receptors may occur in the cytosol and auxin receptors in the nucleus. Some activated receptors directly generate a response, such as an increased flow of ions across a membrane. In contrast, many activated receptors first bind to signaling molecules that initiate an intracellular signaling pathway. For example, the binding of brassinosteroid defense hormones to plasma membrane receptors results in the intracellular production of second messenger molecules that transmit the signal.

Second messengers transmit messages from many types of activated receptors. Cyclic AMP (cAMP), inositol trisphosphate (IP_3), and calcium ions (Ca^{2+}) are major types of second messengers in animal and plant cells. Ca^{2+} is a particularly common messenger in plant cells. Touch and various other stimuli cause Ca^{2+} to flow from storage sites in the endoplasmic reticulum (ER) lumen into the cytosol. Calmodulin or other calcium-binding proteins then bind the Ca^{2+}. Calcium binding alters the structure of such proteins, causing them to interact with other cell proteins or to alter enzymatic function.

Effectors are molecules that directly influence cellular responses. In plants, calcium-dependent protein kinases (CDPKs) are particularly important effector molecules. Cell signaling ends when an effector causes a cellular response, such as opening or closing an ion channel or switching the transcription of particular genes on or off. A single activated receptor can dispatch many second-messenger molecules, which, in turn, can activate scores of effectors, leading to many molecular responses within a single cell.

36.2 Plant Hormones

Learning Outcomes:

1. List the major plant hormones.
2. Explain how auxin functions in phototropism.
3. Explain how cytokinins, gibberellins, and ethylene are relevant to agriculture.
4. Discuss how abscisic acid and brassinosteroids help plants cope with environmental stress.

Together with transcription factors and microRNA molecules, plant hormones are chemical signals transported within the plant body; when taken up by target cells, the signals elicit responses. Here, we focus on plant hormones, about a dozen types of small molecules that are synthesized in metabolic pathways that also make amino acids, nucleotides, sterols, or secondary metabolites. Auxins, cytokinins, gibberellins, ethylene, abscisic acid, and brassinosteroids are considered to be the major plant hormones (**Table 36.1**).

Individual plant hormones often have multiple effects, and different concentrations or combinations of hormones can produce distinct growth or developmental responses. Several plant hormones are known to act by causing the removal of gene repressors, thereby allowing gene expression to occur. This general mechanism allows hormones to cause relatively rapid responses. A closer look at the major types of plant hormones reveals their multifaceted roles.

Table 36.1 The Major Types of Plant Hormones

Type of plant hormone	Chemical structure of an example	Functions (Note: This is a partial, not complete list)
Auxins	Indoleacetic acid (IAA) 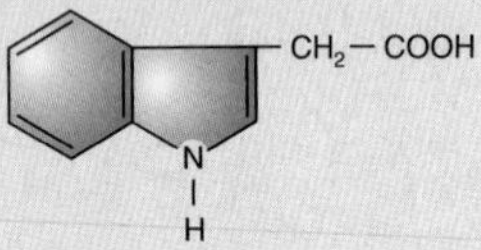 CH_2-COOH; N–H	Establish apical-basal polarity, induce vascular tissue development, mediate phototropism, promote formation of adventitious roots, inhibit leaf and fruit drop, and stimulate fruit development
Cytokinins	Zeatin 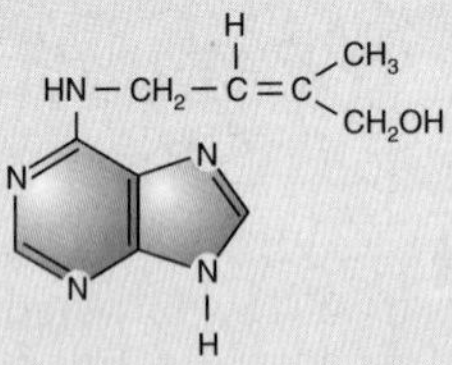 $HN-CH_2-C(H)=C(CH_3)(CH_2OH)$; N, N, N, N–H	Promote cell division, influence cell specialization and plant aging, activate secondary meristem development, promote adventitious root growth, and promote shoot development on callus
Gibberellins	Gibberellic acid 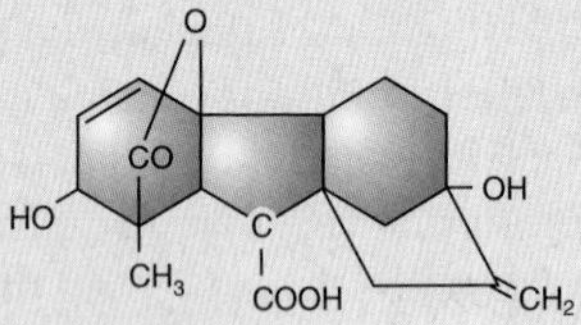 O; CO; HO; OH; CH_3; C; COOH; CH_2	Stimulate cell division and cell elongation, stimulate stem elongation and flowering, and promote seed germination
Ethylene	Ethylene $H_2C=CH_2$	Promotes seedling growth, induces fruit ripening, plays a role in leaf and petal aging and drop, coordinates defenses against osmotic stress and pathogen attack
Abscisic acid	Abscisic acid 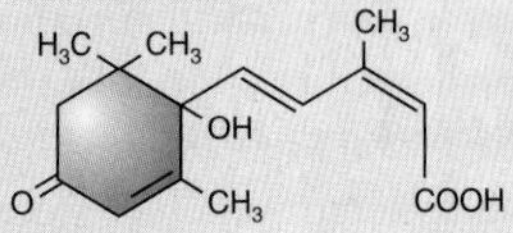 H_3C; CH_3; CH_3; OH; O; CH_3; COOH	Slows or stops metabolism during environmental stress, induces bud and seed dormancy, prevents seed germination in unfavorable conditions, and promotes stomatal closing
Brassinosteroids	Brassinolide 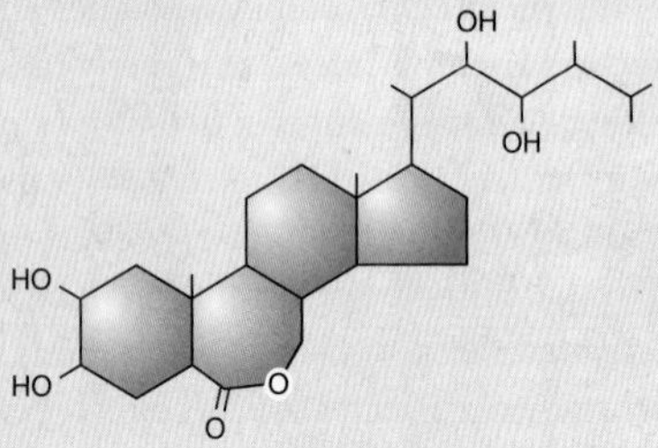 OH; OH; HO; HO; O; O	Promote cell expansion, stimulate shoot elongation, retard leaf drop, stimulate xylem development, and promote stress responses

Auxins Are the Master Plant Hormones

Plants produce several types of **auxins**, which are considered to be the master plant hormones because they influence plant structure, development, and behavior in many ways, often working with other hormones. Indoleacetic acid (IAA) is one plant auxin (see Table 36.1), but other natural and artificial compounds have similar structures and effects. In this section, we will refer to this family of related compounds simply as auxin.

Auxin exerts so many effects because it promotes the expression of diverse genes, together known as **auxin-response genes**. Under low auxin conditions, proteins called Aux/IAA repressors prevent plant cells from expressing these genes. The repressor proteins prevent gene expression by binding to activator proteins at gene promoters. When the auxin concentration is high enough, auxin molecules glue repressors onto a protein complex called TIR1, which causes the breakdown of the repressors. Free of the repressors, the activator proteins enhance the expression of auxin-response genes.

Auxin Transport The way in which auxin is transported into and out of cells is integral to its effects in plants. Auxin is produced in apical shoot tips and young leaves, and it is directionally transported from one living parenchyma cell to another. Auxin in an uncharged form (IAAH) may enter cells from intercellular spaces by means of diffusion; however, the negatively charged form (IAA$^-$) requires the aid of a plasma membrane protein known as the **auxin influx carrier** (AUX1). Several types of proteins, called PIN proteins, transport auxin out of cells. They are named for the pin-shaped shoot apices of plants having mutations in *PIN* genes. Because they transport auxin out of cells, PIN proteins are called **auxin efflux carriers**. They are necessary because in the cytoplasm, auxin occurs as a charged ion that does not readily diffuse out of cells.

In shoots, AUX1 is located at the apical ends of cells, whereas PIN proteins often occur at the basal ends (Figure 36.4a). This polar distribution of auxin carriers explains why auxin primarily flows downward in shoots and into roots, a process called **polar transport** (Figure 36.4b). However, the locations of auxin carriers can also change within cell plasma membranes, allowing lateral or upward transport of auxin. Differences in the presence and positions of auxin carrier proteins explain variations in auxin concentration within plants. By measuring the local auxin concentration, plant cells determine their position within the plant body and respond by dividing, expanding, or specializing.

Auxin Effects In nature, auxin influences plants throughout their lifetimes. Auxin establishes the apical-basal polarity of seed embryos, induces vascular tissue to differentiate, mediates phototropism, promotes formation of adventitious roots, and stimulates fruit

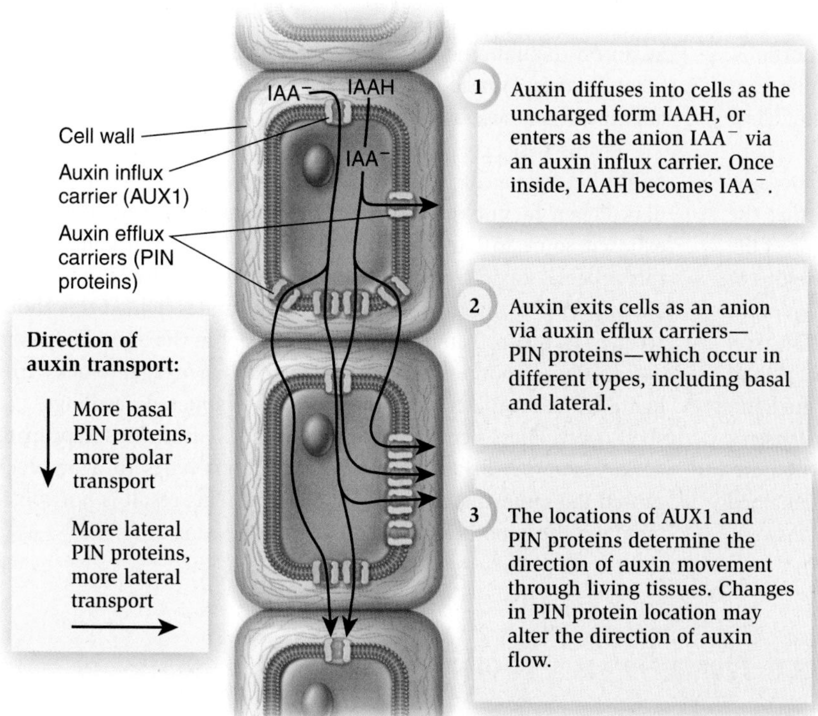

IAA$^-$
IAAH
IAA$^-$
Cell wall
Auxin influx carrier (AUX1)
Auxin efflux carriers (PIN proteins)
Direction of auxin transport:
More basal PIN proteins, more polar transport
More lateral PIN proteins, more lateral transport
1 Auxin diffuses into cells as the uncharged form IAAH, or enters as the anion IAA$^-$ via an auxin influx carrier. Once inside, IAAH becomes IAA$^-$.
2 Auxin exits cells as an anion via auxin efflux carriers—PIN proteins—which occur in different types, including basal and lateral.
3 The locations of AUX1 and PIN proteins determine the direction of auxin movement through living tissues. Changes in PIN protein location may alter the direction of auxin flow.
(a) Cellular mechanism of auxin transport
Lateral auxin movement
Polar auxin movement
Shoot
Root
Auxin circulation at root tip
(b) Auxin transport throughout a plant

Figure 36.4 **Auxin transport.** **(a)** Polar and lateral auxin transport is controlled by the distribution of auxin efflux carriers located in the plasma membrane. When efflux carriers primarily occur at the basal ends of cells, auxin will flow downward. Auxin may flow laterally when auxin efflux carriers occur at the sides of cells. **(b)** In a whole plant, auxin primarily flows downward from shoot tips to root tips, where it then flows upward for a short distance.

Concept Check: *How could auxin carriers be organized to allow auxin to move upward in roots?*

development. Many of auxin's effects are also of practical importance to humans. Auxin is used to produce some types of seedless fruit, retard premature fruit drop in orchards, and stimulate root development on stem cuttings. Auxin produced by intact shoot tips inhibits lateral bud growth, a process known as apical dominance. Gardeners know that if they remove the topmost portion of a plant shoot, nearby lateral buds will begin to grow and produce new branches, allowing plants to become bushier. Such decapitation disrupts the flow of auxin from the shoot apex. Recent research suggests that a class of newly discovered hormones known as strigolactones also regulate shoot branching. Strigolactones are synthesized in roots and shoots and transported to buds, where they are thought to work with auxin. Although there is still much to learn about auxin function, auxin's role in phototropism has been elucidated by a series of experiments, as described next.

The Role of Auxin in Phototropism In the 1880s, Charles Darwin and his son Francis were the first to publish results of experiments on plant phototropism. The Darwins performed their experiments on cereal seedlings, whose tips are protected by a sheath of tissue called a coleoptile. In a simple but elegant experiment, the Darwins covered either the tips or lower portions of coleoptiles with shading materials such as blackened glass tubes, left other seedlings uncovered, and removed the tips of some seedlings. They then compared how those seedlings responded to illumination from the side. The seedlings whose tips were left uncovered grew toward the light, whereas seedlings whose tips were covered or removed did not. The Darwins concluded that seedling tips transmit some "influence" to lower portions, causing them to bend toward the light. You can probably guess what this influence was, but technology available at the time did not allow the Darwins to determine this.

Three decades later, in the 1910s, Danish botanist Peter Boysen-Jensen confirmed the Darwins' results and demonstrated that the influence was a chemical substance that diffused from the tips of the seedlings to other parts. To do this, Boysen-Jensen cut off the tips of oat seedlings and placed either a porous layer of gelatin or a nonporous material such as a sheet of the mineral mica on the cut surface. Then he replaced the tips. Oat seedlings layered with porous gelatin displayed a normal phototropic response, bending toward the light, but those layered with nonporous mica did not. Boysen-Jensen's experiment demonstrated that the phototropic substance was a diffusible chemical, but exactly which one, and how it worked, remained unknown. A series of additional experiments provided some answers.

In the 1920s, the Dutch plant physiologist Frits Went named the substance discovered by Boysen-Jensen auxin (from the Greek word *auxein*, meaning to increase). Although the chemical structure of auxin was not determined until 1934, Went performed experiments that helped explain how auxin works. In a first step, Went cut the tips off of oat seedlings and placed these tips onto agar blocks. Agar, a complex polysaccharide derived from red algae, forms a mesh capable of holding considerable water and dissolved compounds. Agar's permeability to auxin is similar to that of the protein gelatin used by Boysen-Jensen, but agar is much more stable at room temperature and more resistant to microbial breakdown and therefore is easier to use in laboratory experiments. In Went's experiment, the auxin diffused from cut seedling tips into these agar blocks. In the next steps, he treated decapitated seedlings in one of four ways: (1) placed auxin-laden agar blocks off-center on some, (2) placed auxin-laden blocks evenly on others, (3) placed plain agar blocks off-center on some, and (4) left some uncapped. All seedlings were then kept in the dark throughout the experiment. Only seedlings that were capped off-center with an auxin-laden block grew in the direction away from the agar block. This experiment demonstrated that auxin application could substitute for the directional light stimulus and suggested that asymmetric auxin distribution is the mechanism by which light causes plants to bend.

Subsequently, Went and N. O. Cholodny independently proposed that light causes auxin to move to the unlit side of seedling tips, causing cells on that side to elongate more, which results in bending. But other scientists argued that bending could result if light destroys auxin on the illuminated side of a seedling. In the 1950s, American plant biologist Winslow Briggs designed two experiments to test these alternate hypotheses.

In his first experiment, which tested the hypothesis that auxin might be destroyed by light, Briggs first grew corn seedlings in the dark. Then he cut off their tips, put the tips on agar blocks, and exposed some to darkness and others to directional light. During this process, auxin from tips diffused evenly into the agar blocks. If auxin were destroyed by light, agar blocks under lighted tips should receive less auxin than blocks under tips kept in the dark. The auxin-destruction hypothesis also predicts that when agar blocks from lighted tips are placed on one side of decapitated seedlings, they should cause less bending than would blocks from tips kept in the dark. However, Briggs discovered that both types of agar blocks caused the same amount of shoot bending. This result is not consistent with the hypothesis that light destroys auxin.

Feature Investigation

An Experiment Performed by Briggs Revealed the Role of Auxin in Phototropism

In a second experiment, Briggs tested the hypothesis that light causes auxin to move to the shaded side of seedlings (**Figure 36.5**). Briggs set shoot tips onto agar blocks (step 1) and then used a mica sheet (which is impervious to auxin) to completely divide some tips and blocks into halves (step 2A). In other cases, he divided blocks completely but left tips incompletely divided, allowing auxin to diffuse across tips but not the block halves (step 2B). Then Briggs exposed all sets of tips and blocks to directional light. He predicted that auxin would not be able to move across tips having complete mica barricades but that auxin would be able to move across tips that had been only partially divided. Auxin diffused from tips into blocks, but it could not diffuse evenly across blocks divided by mica sheets. When Briggs later placed the agar block halves on decapitated shoots (step 3), those receiving

Figure 36.5 Briggs demonstrated the relationship between light perception and auxin function.

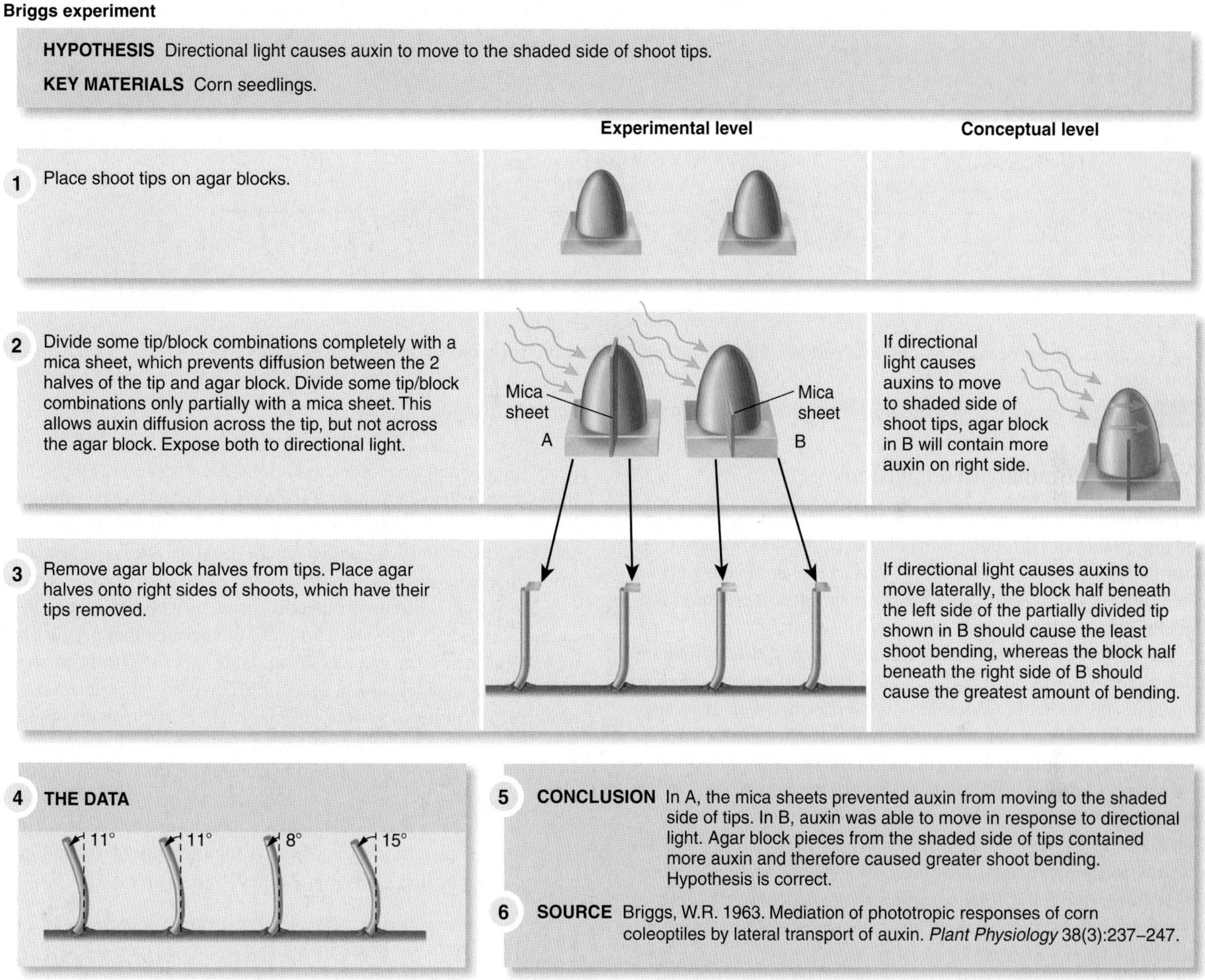

Briggs experiment

HYPOTHESIS Directional light causes auxin to move to the shaded side of shoot tips.

KEY MATERIALS Corn seedlings.

	Experimental level	Conceptual level
1 Place shoot tips on agar blocks.		
2 Divide some tip/block combinations completely with a mica sheet, which prevents diffusion between the 2 halves of the tip and agar block. Divide some tip/block combinations only partially with a mica sheet. This allows auxin diffusion across the tip, but not across the agar block. Expose both to directional light.	Mica sheet A; Mica sheet B	If directional light causes auxins to move to shaded side of shoot tips, agar block in B will contain more auxin on right side.
3 Remove agar block halves from tips. Place agar halves onto right sides of shoots, which have their tips removed.		If directional light causes auxins to move laterally, the block half beneath the left side of the partially divided tip shown in B should cause the least shoot bending, whereas the block half beneath the right side of B should cause the greatest amount of bending.

4 **THE DATA**

5 **CONCLUSION** In A, the mica sheets prevented auxin from moving to the shaded side of tips. In B, auxin was able to move in response to directional light. Agar block pieces from the shaded side of tips contained more auxin and therefore caused greater shoot bending. Hypothesis is correct.

6 **SOURCE** Briggs, W.R. 1963. Mediation of phototropic responses of corn coleoptiles by lateral transport of auxin. *Plant Physiology* 38(3):237–247.

auxin from completely divided tips were bent by the same amount. By contrast, agar block halves from the lit side of partially divided tips induced less bending, whereas halves from the unlit side of partially divided tips caused the most bending (see the Data in Figure 36.5). These experimental results support the hypothesis that unidirectional light causes auxin to accumulate on the shaded side. Modern plant scientists would explain such auxin movement as the result of lateral transport involving PIN proteins.

How might auxin accumulation cause phototropic bending? One widely held hypothesis is that auxin accumulation on the shaded side of a plant shoot causes plasma membrane proton pumps located there to work at a faster rate. In response, the cell wall becomes more acidic, which activates expansins—proteins that break cross-links between cellulose microfibrils and allow cells to elongate (refer back to Figure 35.10). This process might explain how auxin accumulation in cells located on the shaded side of shoot tips causes them to elongate more than do cells on the sunny side, causing the tip to bend toward the light.

Experimental Questions

1. Use the text discussion to draw a diagram illustrating Briggs' experiment in order to determine whether or not light destroys auxin.
2. What is the current hypothesized mechanism by which auxin accumulation causes shoot bending in response to directional light?
3. Figure 36.5 illustrates experiments performed with four seedling tips. Would this number really be enough to allow conclusions to be made about how such seedling tips would generally respond?

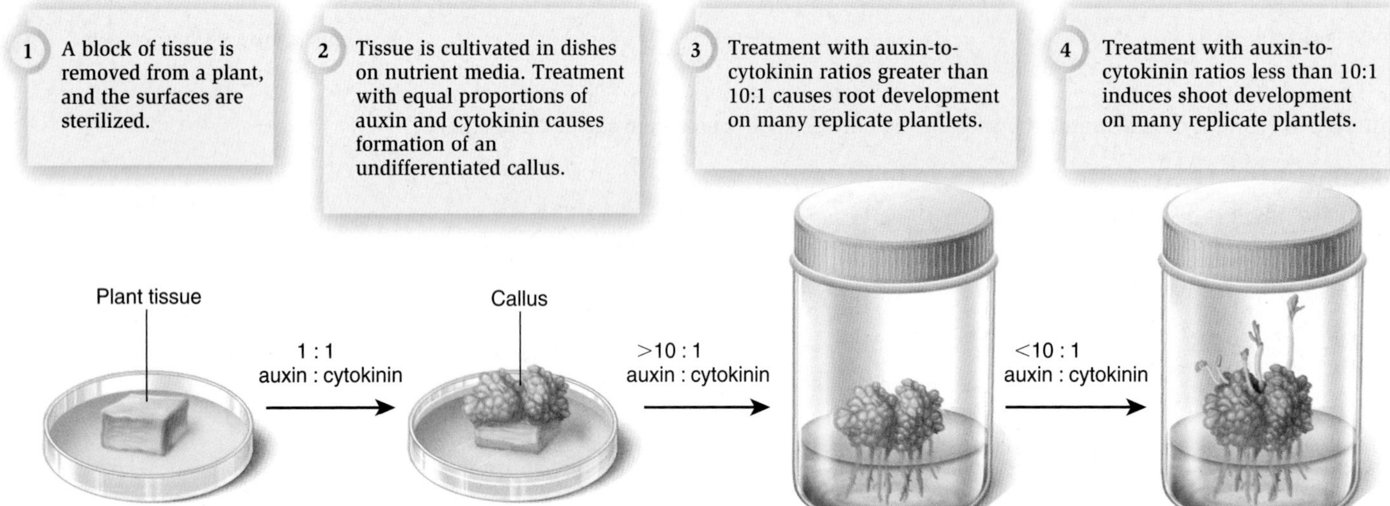

1 A block of tissue is removed from a plant, and the surfaces are sterilized.
2 Tissue is cultivated in dishes on nutrient media. Treatment with equal proportions of auxin and cytokinin causes formation of an undifferentiated callus.
3 Treatment with auxin-to-cytokinin ratios greater than 10:1 causes root development on many replicate plantlets.
4 Treatment with auxin-to-cytokinin ratios less than 10:1 induces shoot development on many replicate plantlets.
Plant tissue
1 : 1
auxin : cytokinin
Callus
>10 : 1
auxin : cytokinin
<10 : 1
auxin : cytokinin

Figure 36.6 The process of plant tissue culture. Plant tissue culture illustrates the effect of different proportions of auxin and cytokinin on plant organ development.

Concept Check: How do commercial growers use this process to produce many identical plants?

Cytokinins Stimulate Cell Division

Like auxins, the plant hormones known as **cytokinins** play varied and important roles throughout the lives of plants. The name of these hormones reflects their major effect—an increase in the rate of plant cytokinesis, or cell division. Root tips are major sites of cytokinin production, but shoots and seeds also make this plant hormone. Transported in the xylem to meristems and other plants parts, cytokinins bind to receptors thought to be located in the plasma membrane. At shoot and root tips, cytokinins influence meristem size, stem cell activity, and vascular tissue development. Cytokinins are also involved in root and shoot growth and branching, the production of flowers and seeds, and leaf aging.

Plant Tissue Culture In the laboratory, cytokinin and auxin are essential to cloning plants. This involves a process, known as **plant tissue culture**, which is used commercially to produce thousands of identical plants having the same desirable characteristics. Plant tissue culture begins with pieces of stem, leaf, or root that have been removed from a plant, and their surfaces are sterilized to prevent growth of microbes (Figure 36.6, step 1). The cleaned plant pieces are then placed into dishes containing nutrients (minerals, vitamins, and sugar) and various proportions of auxin and cytokinin. If the proportions of auxin and cytokinin are about the same (1:1), plant cells undergo division, forming a mass of white tissue known as a callus (step 2). If the callus is then transferred to a new dish containing the same nutrients, with auxin-to-cytokinin proportions greater than 10:1, the callus will form roots (step 3). Auxin-to-cytokinin proportions of less than 10:1 cause the callus to develop green shoots (step 4). Thus, by altering the ratios of auxin and cytokinin, entire plants can be regenerated from a callus. A single callus can be divided into many pieces and each piece treated with these hormones, thereby producing many hundreds of identical new plants.

Gibberellins Stimulate Cell Division and Elongation

The **gibberellins** (also known as gibberellic acids, or GA; see Table 36.1) are another family of plant hormones. Gibberellins are produced in apical buds, roots, young leaves, and seed embryos. In addition to promoting shoot development on laboratory calluses, gibberellins interact with light and other hormones to foster seed germination and enhance stem elongation and flowering in nature. Gibberellins also retard leaf and fruit aging. These multiple effects largely arise from gibberellin's stimulatory effects on cell division and elongation.

More than a hundred different forms of gibberellin have been found. Many kinds of dwarf plants are short because they produce less gibberellin than taller varieties of the same species. The dwarf strain of pea plants Mendel used in some of his breeding experiments is an example. When dwarf varieties of plants are experimentally sprayed with gibberellin, their stems grow to normal heights. However, dwarf wheat and rice crops are valued in agriculture because they can be more productive and less vulnerable to storm damage than taller varieties. Since the discovery of gibberellin, plant scientists have discovered how this plant hormone works at the molecular level and how gibberellin regulation of plant growth evolved, our next topics.

GENOMES & PROTEOMES CONNECTION

Gibberellin Function Arose in a Series of Stages During Plant Evolution

In flowering plants, gibberellin works by helping to liberate repressed transcription factors. In the absence of gibberellin, DELLA proteins bind particular transcription factors needed for the expression of gibberellin-responsive genes (Figure 36.7a). In this way, DELLAs function as brakes that restrain cell division and expansion. When sufficient gibberellin is present, it binds receptor proteins called GID1 (Figure 36.7b). Gibberellin-binding increases the ability of GID1 proteins to interact with DELLA proteins, starting a process that leads to the destruction of DELLAs. In the absence of DELLA proteins, transcription factors are able to bind the promoter regions of gibberellin-responsive genes, allowing their expression. As a result, cell division and expansion occur, leading to growth.

In 2007, Yuki Yasumura, Nicholas Harberd, and their colleagues reported that the gibberellin-DELLA mechanism for regulating the growth of flowering plants arose in a stepwise fashion. They discovered this by comparing the DELLA and GID1 proteins of flowering

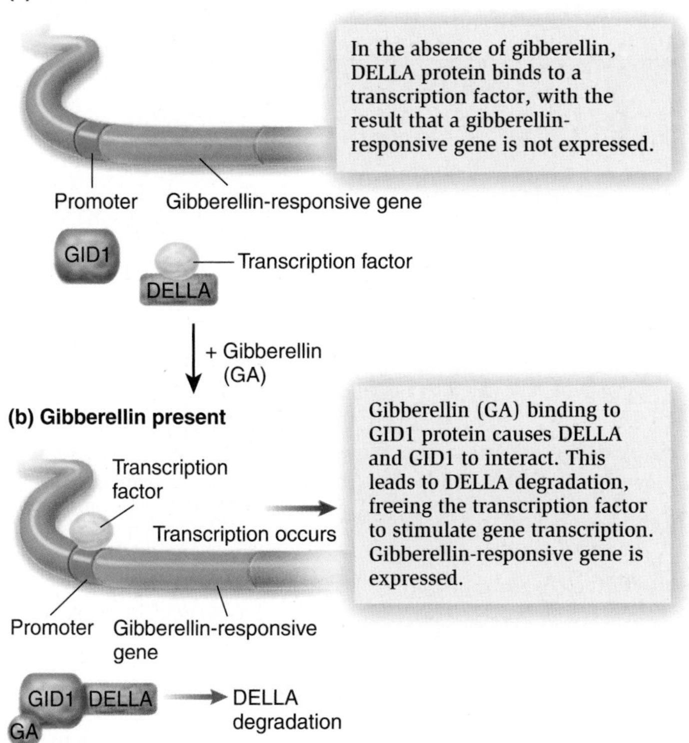

Figure 36.7 Gibberellin works by releasing trapped transcription factors. **(a)** In the absence of gibberellin, DELLA binds transcription factors so that gibberellin-response genes are not expressed. **(b)** When gibberellin binds to the protein GID1, GID1 can bind DELLA proteins, which causes DELLA proteins to be degraded. As a result, transcription factors that had been bound to DELLA proteins are released and can bind to gene promoters, thereby inducing gene expression.

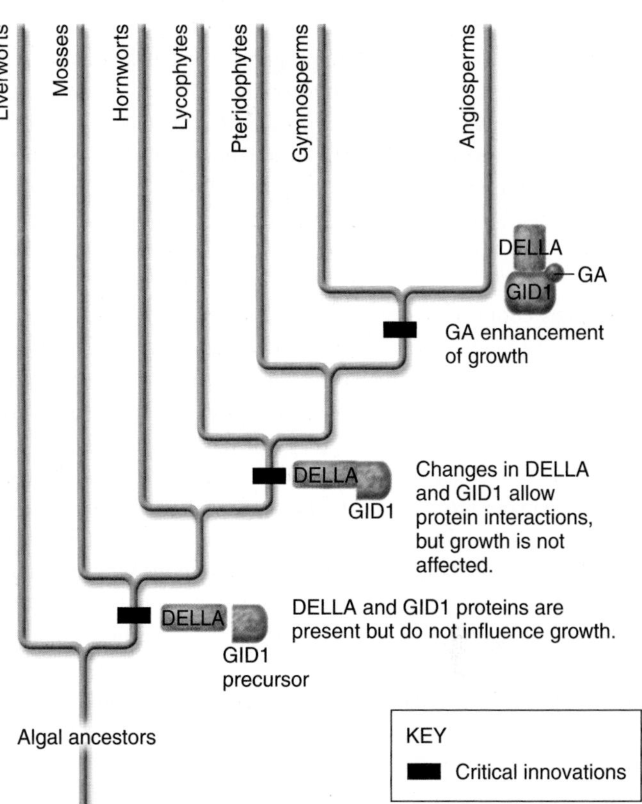

Figure 36.8 Evolution of the gibberellin-DELLA system. Although DELLA and GID1 proteins are present in a bryophyte, they do not interact, nor does gibberellin bind to GID1. In lycophytes, GID1 interacts with DELLA, but the system does not influence growth. The growth responses of gibberellin apparently evolved after the divergence of lycophytes.

BIOLOGY PRINCIPLE All species (past and present) are related by an evolutionary history. The gibberellin-DELLA system that controls growth of flowering plants evolved in a step-by-step fashion.

plants with homologous proteins of a bryophyte and a lycophyte. The seedless lycophytes, the oldest living phylum of vascular plants, first appeared millions of years before the first flowering plants, and the seedless, nonvascular bryophytes arose millions of years before the first vascular plants (refer back to Figure 29.1). The investigators studied the interactions of DELLA proteins with GID1 and evaluated the extent to which gibberellin enhanced DELLA-GID1 binding and growth in these different groups of plants.

The results indicated that the bryophyte possesses DELLA and GID1 proteins, but these proteins don't interact, and bryophyte DELLA does not repress growth. In contrast, differences in the protein structure of lycophyte DELLAs enable them to interact with GID1 and gibberellin, but without detectably influencing growth. Neither the bryophyte nor the lycophyte showed detectable growth responses to gibberellin, though their DELLAs were able to repress growth when expressed in a flowering plant. This pattern led the biologists to propose that DELLA-mediated repression of plant growth evolved after the divergence of lycophytes but prior to the appearance of the first flowering plants (**Figure 36.8**). Though the necessary components (DELLAs and GID1 proteins) were present earlier, they did not assemble into a growth regulation system until later in plant evolutionary history.

Ethylene's Effects Include Cell Expansion

The plant hormone **ethylene** is particularly important in coordinating plant developmental and stress responses. Ethylene is a simple hydrocarbon gas produced during seedling growth, flower development, and fruit ripening (see Table 36.1). In the root tip, ethylene determines how many stem cells remain inactive in the quiescent center and how many cells undergo divisions. This hormone also plays important roles in defense against osmotic stress and pathogen attack, and leaf and petal aging and drop. As a gas, ethylene is able to diffuse through the plasma membrane and cytosol to bind to ethylene receptors localized in the endoplasmic reticulum. When activated by ethylene binding, these receptors inactivate a protein kinase known as CTR1. This action ultimately enables transcription factors to induce the transcription of various genes.

People first noticed the effects of ethylene gas on plants in the 1800s when they observed that street-side trees exposed to leaking street lanterns unexpectedly lost their leaves. A 17-year-old student in St. Petersburg, Russia, Dimitry Neljubov, performed the first experiments to explore the effects of illumination gas on plants. He exposed pea seedlings grown in the laboratory to illumination gas and noticed that the pea seedlings grew sideways rather than upward. Then he tested the individual chemical components of illumination gas for the same effect. After conducting many experiments, in 1901, Neljubov reported that ethylene was the only component of illumination gas that caused the seedlings to grow horizontally and that ethylene was effective in very low concentrations (as low as 0.06 parts per million in air). Later, scientists established that ethylene influences cell expansion, often in association with auxin. Ethylene does this by increasing the disorder of microtubules within cells, thereby causing random orientation of cell-wall microfibrils. As a result, cells exposed to ethylene tend to expand in all directions rather than elongating.

Ethylene's effects on cell expansion explain this hormone's important role in dicot seedling growth. The tender apical meristems of seedlings could be easily damaged during their growth through crusty soil. Ethylene helps seedlings avoid such damage by inducing what is known as the triple response (**Figure 36.9**). First, ethylene prevents the seedling stem and root from elongating. Second, the hormone induces the stem and root to swell radially, thereby increasing in thickness. Together, these responses strengthen the seedling stem and root. Third, the seedling stem bends so that embryonic leaves and the delicate meristem grow horizontally rather than vertically; this is the sideways growth response that Neljubov first observed. The bent portion of the stem, known as a hook, then pushes up through the soil. The hook forms as the result of an imbalance of auxin across the stem axis, which causes cells on one side of the stem to elongate faster than cells on the other side. Ethylene drives this auxin imbalance.

Knowledge of the effects of ethylene on fruit has been very useful commercially. Ripe fruit can be easily damaged during transit, but tomatoes and apples can be picked before they ripen for transport with minimal damage. At their destination, such fruit can be ripened by treatment with ethylene. However, fruit that becomes overripe may exude ethylene, which hastens ripening in nearby, unripe fruit. For this reason, fruit that must be stored for extended periods is kept in ethylene-free environments.

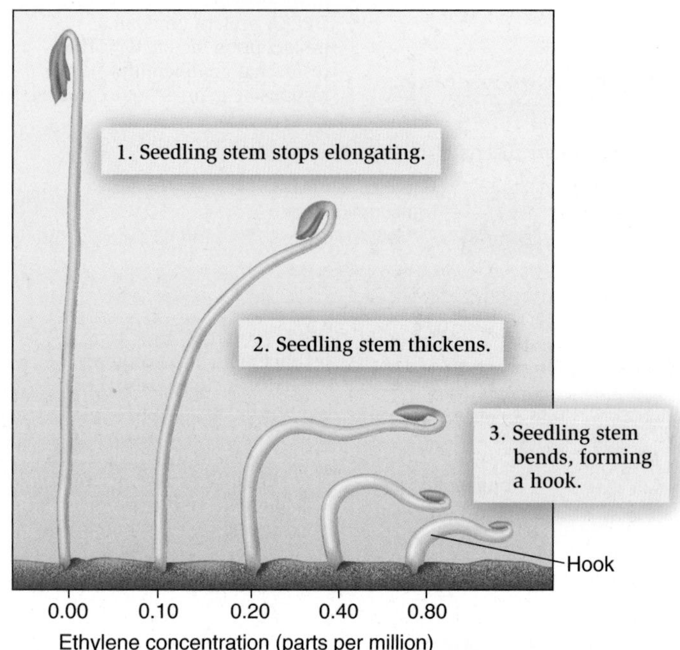

Figure 36.9 **Seedling growth showing the triple response to ethylene.** When applied at levels above a particular concentration (0.80 ppm), ethylene causes seedlings to cease elongation, swell radially, and bend to form a hook that can push upward through the soil. Ethylene produced naturally within seedlings causes the same response.

Concept Check: *What adaptive advantage does this seedling behavior provide?*

Several Hormones Help Plants Cope with Environmental Stresses

Several plant hormones share the property of helping plants respond to environmental stresses such as flooding, drought, high salinity, cold, heat, and attack by disease microorganisms and animal herbivores. These protective hormones include the major plant hormones known as **abscisic acid** and **brassinosteroids** (see Table 36.1). Additional protective hormones are salicylic acid (SA), whose chemical structure is similar to that of aspirin, and a peptide known as systemin. The fragrant compound jasmonic acid and the gas nitric oxide (NO) are also protective plant hormones.

Abscisic Acid Abscisic acid (ABA) was named at a time when plant biologists thought that it played a role in leaf or fruit drop, also known as abscission. Later, they discovered that ethylene actually causes leaf and fruit abscission, whereas abscisic acid slows or stops plant metabolism when growing conditions are poor. For example, ABA may induce bud and seed dormancy. Dormant buds and seeds resume growth only when specific environmental signals reveal the onset of conditions suitable for survival. In preparation for winter, ABA stimulates the formation of tough, protective scales around the buds of perennial plants. Seed coats of apple, cherry, and other plants also accumulate ABA, which prevents seeds from germinating unless temperature and moisture conditions are favorable for seedling growth. Water-stressed roots also produce ABA, which is then transported to shoots, where (together with ABA produced by water-stressed leaf mesophyll) it helps to prevent water loss from leaf surfaces by inducing leaf pores (stomata) to close. ABA receptors are soluble proteins. The binding of ABA to receptors promotes the transcription of ABA-responsive genes.

Brassinosteroids Brassinosteroids are named after the cruciferous plant genus *Brassica* (which includes cabbage and broccoli), from which they were first identified. However, seeds, fruit, shoots, leaves, and flower buds of all types of plants contain brassinosteroids. These plant hormones induce vacuole water uptake and influence enzymes that alter cell-wall carbohydrates, thereby fostering cell expansion. Mutations that affect brassinosteroid synthesis cause plants to exhibit dwarfism. Such plants have small, dark green cells because their tissues are unable to expand. Brassinosteroids also impede leaf drop, help grass leaves to unroll, and stimulate xylem development. They can be applied to crops to help protect plants from heat, cold, high salinity, and herbicide injury.

Brassinosteroids are chemically related to animal steroid hormones, such as human sex hormones. However, unlike animal steroid hormones, which bind to receptors in the nucleus or cytosol, brassinosteroids bind to receptors in the plasma membrane. When they bind brassinosteroids, the membrane receptors initiate a signal transduction pathway that activates transcription factors for brassinosteroid-responsive genes.

36.3 Plant Responses to Environmental Stimuli

Learning Outcomes:

1. Describe how photoreceptors allow plants to respond to light, including day length.
2. Discuss how plant roots respond to gravity and touch.
3. Give examples of how plants respond to flood and drought.
4. Outline how plants protect themselves against pathogenic attack.

Plants encounter many types of environmental challenges and behave accordingly. For example, if buried seeds were to germinate beneath soil layers too deep for light to penetrate, or beneath a cover of established plants, seedlings would not be able to obtain sufficient light for photosynthesis and would die in the ground. A related reproductive challenge for plants is to flower at times of the year that are most beneficial for achieving pollination or seed dispersal. How do plants determine if there is enough light for seeds to germinate and for seedlings to grow? How do plants determine when to flower?

The answer is that plants possess cellular systems for measuring light and determining the seasonal time of year. Using these systems, plant seeds germinate only when there is sufficient light for seedling growth, and flowering occurs during the most advantageous season. Therefore, plants can sense and respond to their light environments. Plants are also able to respond to other physical and biological external stimuli. In this section, we survey plant responses to these stimuli, beginning with light.

Plants Detect Light and Measure Day Length

A plant's ability to measure and respond to light amounts and day length, a process called **photoperiodism**, is based on the presence of light receptors within cells. Such light sensors are known as **photoreceptors**, of which there are several types. Some types of plant photoreceptors also occur in animals, whereas others are particular to plants. Each type of photoreceptor has a light-absorbing component as well as other regions that respond to light absorption by switching on signal transduction pathways. Responses by many cells in a tissue or organ cumulatively result in behaviors such as sun tracking, phototropism, flowering, and seed germination.

Blue-Light Receptors Cryptochromes and phototropins are two types of blue-light receptors—molecules that absorb and respond to blue light. Experiments suggest that receptors called **cryptochromes** help young seedlings determine if their light environment is bright enough to allow photosynthesis. If not, seedlings continue to elongate through the soil, toward the light. Cryptochromes also function as light sensors in animals.

Phototropin is the main blue-light sensor involved in plant phototropism. The light-activated form of this sensor has two components: a protein that has a kinase domain and a flavin pigment that can absorb blue light. In the dark, the flavin is not covalently bound to the protein. However, when the flavin absorbs blue light, it changes conformation and becomes able to covalently bind to the protein. Flavin binding, in turn, changes the conformation of the phototropin protein, allowing it to phosphorylate itself by means of the protein kinase domain. When a plant organ is exposed to directional blue light, phototropin becomes phosphorylated. In this way, a light signal is converted into a chemical signal that influences auxin movement.

Phytochrome, the Red- and Far-Red-Light Receptor Many plant growth and developmental processes are influenced by **phytochrome**, a red- and far-red-light receptor. Phytochrome operates much like a light switch, flipping back and forth between two conformations: P_r that absorbs red light, and P_{fr} that absorbs only far-red light (light having a wavelength longer than that of red light) (**Figure 36.10**). When red

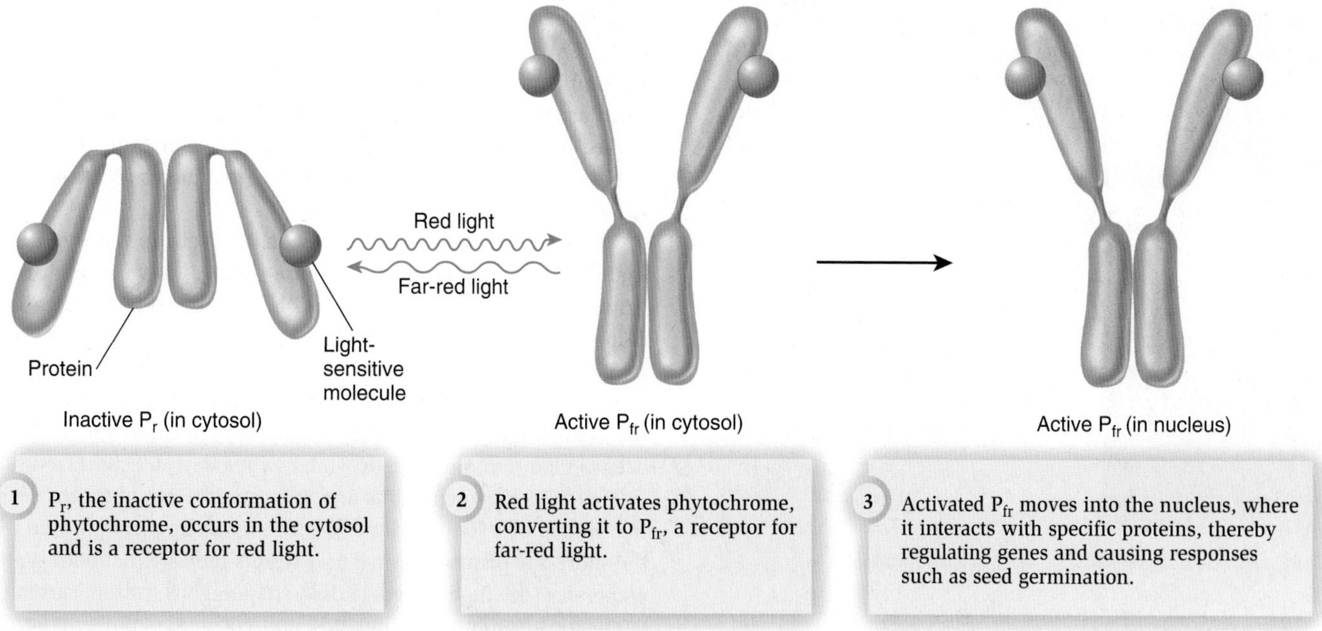
Red light
Far-red light
Protein
Light-sensitive molecule
Inactive P_r (in cytosol)
Active P_{fr} (in cytosol)
Active P_{fr} (in nucleus)
1 P_r, the inactive conformation of phytochrome, occurs in the cytosol and is a receptor for red light.
2 Red light activates phytochrome, converting it to P_{fr}, a receptor for far-red light.
3 Activated P_{fr} moves into the nucleus, where it interacts with specific proteins, thereby regulating genes and causing responses such as seed germination.

Figure 36.10 **How phytochrome acts as a molecular light switch.**

Concept Check: *What kind of light does the active conformation of phytochrome absorb, and what kind of change does such absorption cause?*

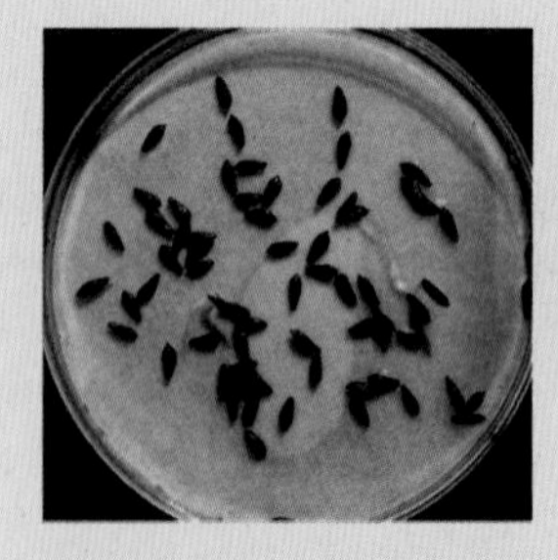

DARKNESS
In darkness, seeds do not germinate because phytochrome remains in the inactive P_r conformation.

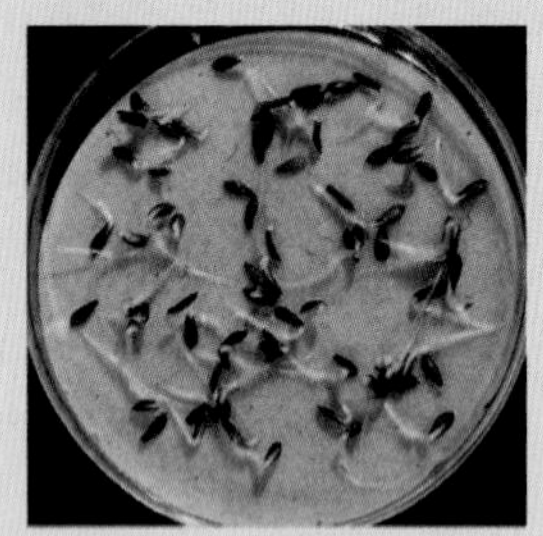

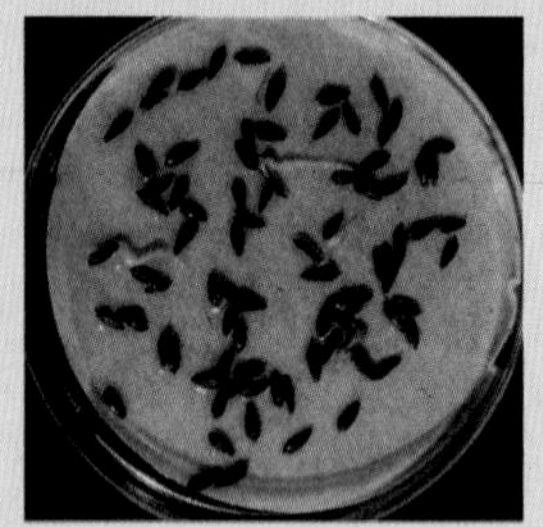

Red
Far red
Exposure to far-red light after red-light exposure converts active P_{fr} to inactive P_r, so seeds do not germinate.

Red
Far red
Red
Exposure to red light after far-red light switches phytochrome back to the active P_{fr} conformation, so seeds germinate.

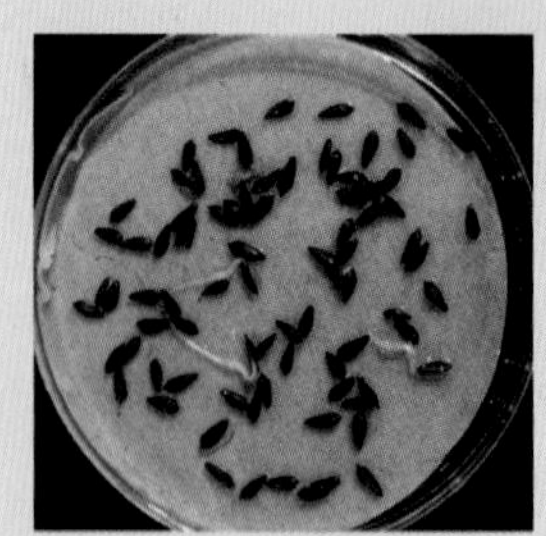

Figure 36.11 **How phytochrome influences seed germination.**

Concept Check: Describe the change in phytochrome that would occur if a deeply buried seed were uncovered enough to receive sunlight.

light is abundant, as in full sunlight, P_r absorbs red light and changes to it to P_{fr}, which activates cellular responses such as seed germination. When left in the dark for a long period, P_{fr} slowly transforms into P_r.

The role of phytochrome as a plant "light switch" has been shown experimentally in studies of lettuce-seed germination. Researchers have found that water-soaked lettuce seeds will not germinate in darkness, but they will germinate if exposed to as little as 1 minute of red light (**Figure 36.11**). This amount of light exposure is sufficient to transform a critical amount of P_r to the active P_{fr} conformation, which stimulates germination. However, if this brief red-light treatment is followed by a few minutes of treatment with far-red light, the lettuce seeds will not germinate. This short period of far-red illumination is enough to convert seed P_{fr} back to the inactive P_r conformation.

The most recent light exposure determines whether the phytochrome occurs in the active or inactive conformation. In nature, if seeds are close enough to the surface that their phytochrome is switched on by red light, the seeds will germinate. But if seeds are buried too deeply for red light to penetrate, they will not germinate. In this way, seeds can sense if they are close enough to the surface to begin the germination process.

Most of our understanding of phytochrome's effect on gene expression comes from the study of the model plant *Arabidopsis thaliana*. This plant has five phytochrome genes (*PHYA–PHYE*). Each of the five types of phytochrome is composed of two proteins, each having a light-sensitive nonprotein pigment called a chromophore (see Figure 36.10). In the dark, phytochrome molecules in the P_r state reside in the cytosol. After exposure to red light, activated phytochrome (P_{fr}) molecules typically move from the cytosol to the nucleus. Within the nucleus, P_{fr} interacts with a transcription factor protein known as PIF3 (phytochrome-interacting factor 3). PIF3 binds to the regulatory elements of several phytochrome-responsive genes, functioning as a positive regulator of some genes and as a negative regulator of others.

Photoperiodism Phytochromes also play a critical role in photoperiodism, the response to relative lengths of darkness that influence the timing of dormancy and flowering. Flowering plants can be classified as long-day, short-day, or day-neutral plants. When scientists named these groups, they thought that plants measured the amount of daylight. Researchers later discovered that plants actually measure night length.

Lettuce, spinach, radish, beet, clover, gladiolus, and iris are examples of **long-day plants** because they flower in spring or early summer, when the night period is shorter (and thus the day length is longer) than a defined period (**Figure 36.12**). In contrast, asters, strawberries, dahlias, poinsettias, potatoes, soybeans, and goldenrods are examples of **short-day plants** because they flower only when the night length is longer than a defined period. Such night lengths occur in late summer, fall, or winter, when days are short. As shown in Figure 36.12, when plants are given an experimental light flash in the middle of a long dark period, the long-day plants flower while the short-day plants do not. These results indicate that both types of plants measure night length. Roses, snapdragons, cotton, carnations, dandelions, sunflowers, tomatoes, and cucumbers flower regardless of the night length, as long as day length meets the minimal requirements for plant growth, and are thus known as **day-neutral plants**.

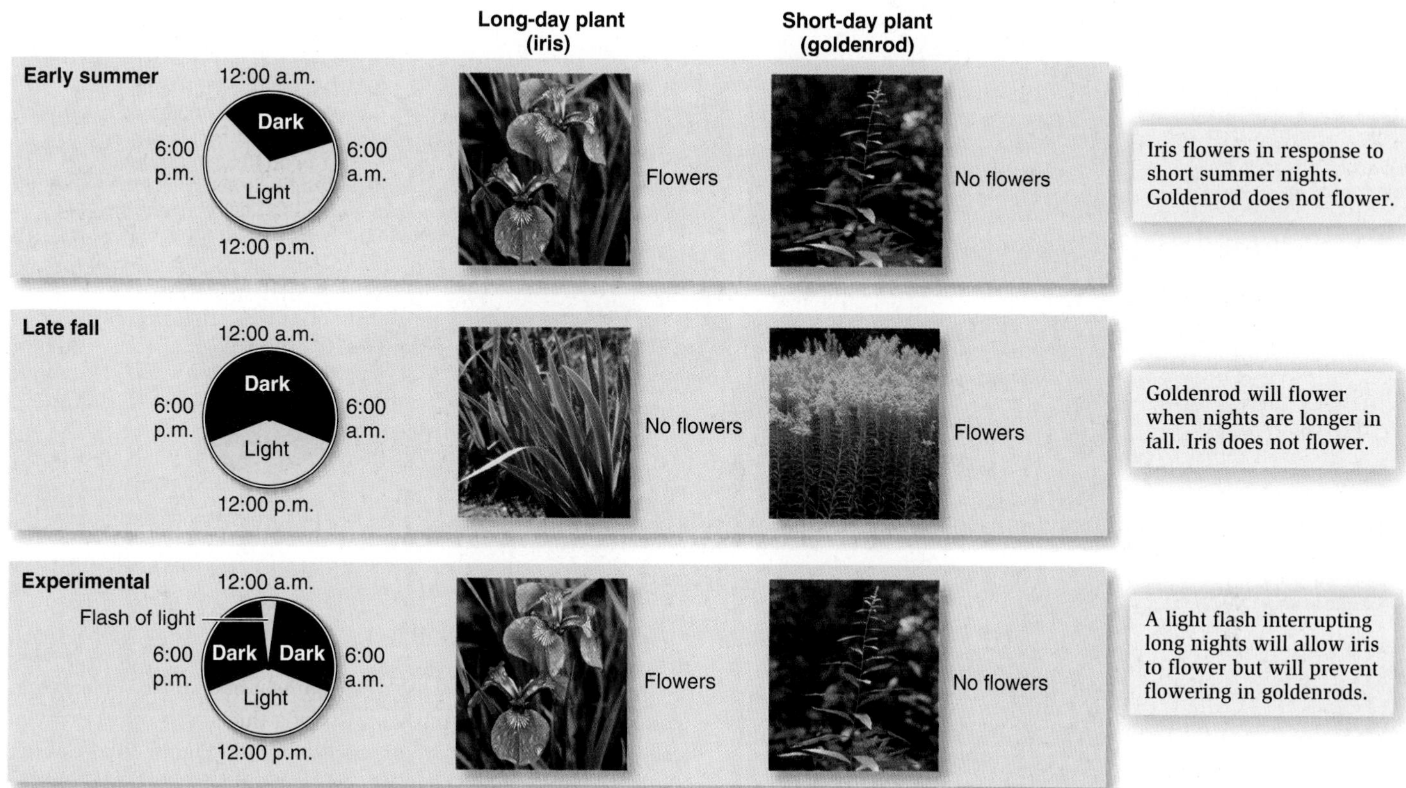

Figure 36.12 Flowering and photoperiodism. Iris is a long-day plant that flowers in response to the short nights of late spring and early summer, whereas goldenrod is a short-day plant that flowers in response to the longer nights of autumn. The length of night is the critical factor, as shown by the effects of light flashes.

Concept Check: *What would happen if you gave these plants a brief exposure to darkness in the middle of the daytime?*

Ornamental plant growers manipulate night length to produce flowers for market during seasons when they are not naturally available. For example, chrysanthemums are short-day plants that usually flower in the fall, but growers use light-blocking shades to increase night length in order to produce flowering plants at any season.

Shading Responses Phytochrome also mediates plant responses to shading. These responses include the extension of leaves from shady portions of a dense tree canopy into the light, and growth that allows plants to avoid being shaded by neighboring plants. These growth responses occur by the elongation of branch internodes. Leaves detect shade as an increased proportion of far-red light to red light. This means that more of the phytochrome in shaded leaves is in the inactive (P_r) state than is the case for leaves in the sun. Activated phytochrome (P_{fr}) inhibits the growth of shoot internodes, but phytochrome in the inactivated state does not, so branches bearing shaded leaves extend toward sunlight.

Figure 36.13 Negative gravitropism in a shoot. This tomato shoot system has resumed upward growth after being placed on its side. Upward growth started about 4 hours after the plant was turned sideways; this photo was taken 20 hours later. Shoots sense gravity by means of starchy statoliths present in stem tissue near the central vascular tissue.

Plants Respond to Gravity and Touch

Have you ever wondered what causes plant stems to generally grow upward and roots downward? The upward growth of shoots and the downward growth of roots are behaviors known as **gravitropism**, growth in response to the force of gravity. Shoots are said to be negatively gravitropic because they often grow in the direction opposite to gravitational force. If a potted plant is turned over on its side, the shoot will eventually bend and begin to grow vertically again (**Figure 36.13**). Most roots are said to be positively gravitropic because they grow in the same direction as the gravitational force.

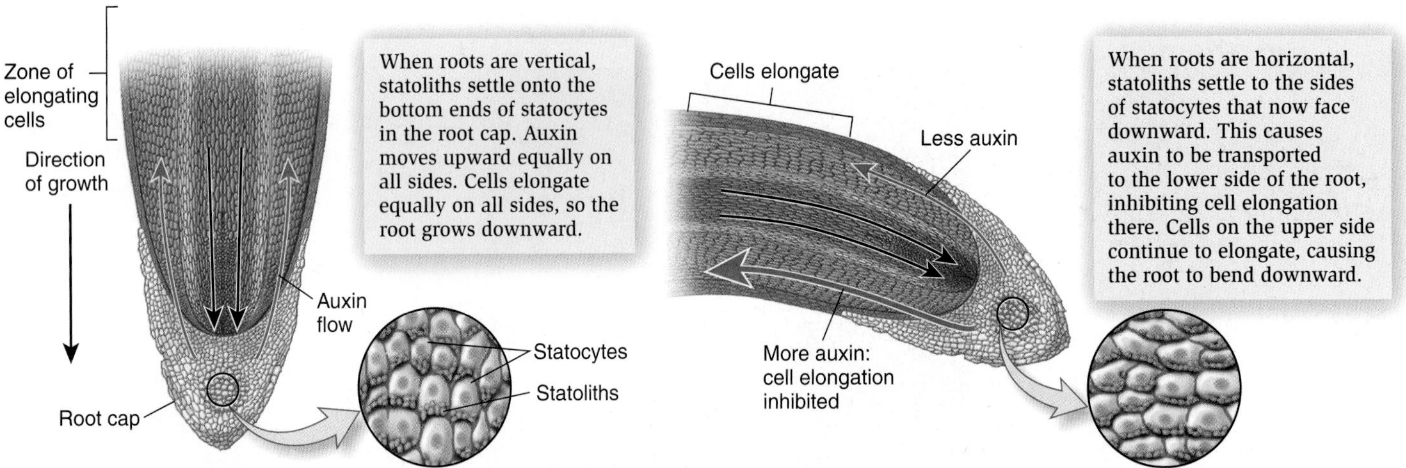

Figure 36.14 Positive gravitropism in a root. Root-tip cells sense gravity by means of starchy statoliths present in cells at the center of the root cap.

Concept Check: *Is there any other environmental signal that roots could use to achieve downward growth?*

Both roots and shoots detect gravity by means of starch-heavy plastids known as **statoliths**, which are located in specialized gravity-sensing cells called statocytes. In shoots, statocytes are located in a tissue known as the endodermis, which forms a sheath around vascular tissues. In roots, gravity-sensing cells primarily occur in the center of the root cap.

Gravity causes the relatively heavy statoliths to sink, which causes changes in calcium ion messengers that affect the direction of auxin transport. This process induces changes in the direction of shoot or root growth. For example, in a root that becomes oriented horizontally, the statoliths are pulled by gravity to the lower sides of statocytes (Figure 36.14). The change in statolith position causes auxin to move to cells on the lower sides of roots. In roots, auxin inhibits cell elongation (in contrast to its action in shoots). Therefore, root growth slows on the lower side, while cell elongation continues normally on the upper side. This process causes the root to bend, so that it eventually grows downward again.

Recent studies suggest that gravity responses are related to touch responses, known as **thigmotropism** (from the Greek *thigma*, meaning touch). For example, when roots encounter rocks or other barriers to their downward growth in the soil, they display a touch response that temporarily supersedes their response to gravity. Such roots grow horizontally until they get around the barrier, whereupon downward growth in response to gravity resumes. Plant shoots also respond to touch; examples include vines with tendrils that wind around or clasp supporting structures. Wind also induces touch responses. In very windy places, trees tend to be shorter than normal, giving them the advantage of being less likely to blow over than are taller trees. In the laboratory, plant scientists have simulated natural touch responses by rubbing plant stems and found that this treatment can result in shorter plants. Touch causes the release of calcium ion messengers that influence gene expression.

More rapid responses to touch, such as leaf folding by the sensitive plant (see Figure 36.1b), are based on changes in the water content of cells within a structure known as a pulvinus (plural, pulvini), a swelling located at the base of attachment of each pair of leaflets in complex leaves. A pulvinus consists of a thick layer of parenchyma cells that surrounds a core of vascular tissue (Figure 36.15). When the leaflet of a sensitive plant is touched, an action potential opens ion channels in parenchyma cells near the lower surfaces of the pulvini. These cells expel potassium and chloride ions, causing water to flow out and the cells to become flattened. This bends the leaflets together, starting the leaflet-folding process. The action potential generated at the touch site also flows through the leaflet, causing many or all of the leaflets to also bend, with the result that the entire leaf folds. Reversal of this process allows the leaf to unfold.

The action of pulvini also explains some plant movements that are unrelated to touch, including sleep movements and sun tracking. Sleep movements are changes in leaf position that occur in response to day-night cycles. Sun tracking, as mentioned at the beginning of this chapter, is the movement of leaves or flowers in response to the Sun's position.

Plants Respond to Physical Stresses Such as Flooding and Drought

Plants display many types of adaptations that help them cope with unfavorable growth conditions, such as flooding and drought. These responses are often mediated by hormones.

The major harmful effect of flooding is that too much water makes roots unable to obtain sufficient oxygen to fuel respiratory processes. Without oxygen from the air, roots cannot produce the ATP needed to absorb minerals from soil. Many plants reduce the effects of flooding by producing **aerenchyma**, a tissue containing large, snorkel-like air channels that allow more oxygen to flow from shoots to the submerged roots (Figure 36.16). In some plants, aerenchyma formation is developmentally programmed, and in others, it is a response to change in environmental conditions. For example, aerenchyma develops in the roots of many plants native to wetland habitats even when the soil is not wet. Aerenchyma can also form

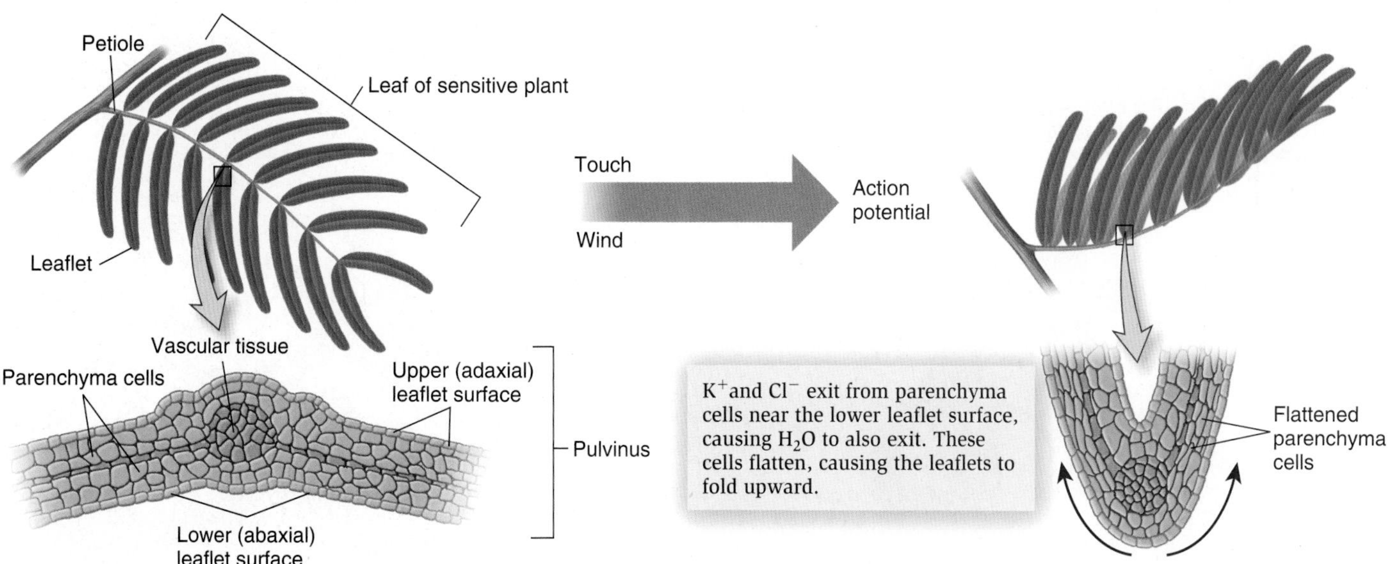

Figure 36.15 **Leaf folding in the sensitive plant: How pulvini change the positions of leaflets.** The electrical signals known as action potentials result from the rapid flow of ions through membrane ion channels. Electrical signals spread from one cell to another through plasmodesmata. Cells near the lower surface of the pulvinus respond to ion flow by losing water, causing them to flatten.

in the roots of plants such as corn as a response to flooding. Aerenchyma formation is regulated by the action of ethylene, which leads to programmed cell death, followed by cell collapse.

Drought is related to several other environmental stresses—high salinity, heat, and cold—all of which reduce the amount of liquid water present in plant cells. Most plants that lose half or more of their water are unable to recover. Thus, plants possess diverse adaptations that reduce water loss, and the hormone abscisic acid often triggers these responses.

Plants Respond to Biological Stresses Such as Herbivore and Pathogen Attack

Plants are vulnerable to attack by animal herbivores and pathogens—disease-causing microorganisms. Structural barriers such as cuticles, epidermal trichomes, and outer bark, described in Chapter 35, help to reduce infection and herbivore attack. These barriers, together with chemical defense compounds, explain why remarkably little natural vegetation is lost to herbivore or pathogen attack. However, agricultural crops can be more vulnerable to attack than their wild counterparts. This is because some protective adaptations have been lost during crop domestication as the result of genetic changes that increase edibility. For this reason, crop scientists are particularly interested in understanding plant defense, with the goal of being able to breed or genetically engineer crop plants that are better protected from pests.

Plant Responses to Herbivores Herbivory defense compounds include diverse secondary metabolites—terpenes and terpenoids; phenolics; and alkaloids (refer back to Figure 30.22)—but also poisonous hydrogen cyanide and other molecules. Some of these substances act directly on herbivores, making plants taste bad so that herbivores learn to avoid them. Other chemical compounds function indirectly. For example, when attacked by insect caterpillars, cruciferous plants release terpenoids that attract the bodyguard wasp (*Cotesia rubecula*), which attacks the caterpillars.

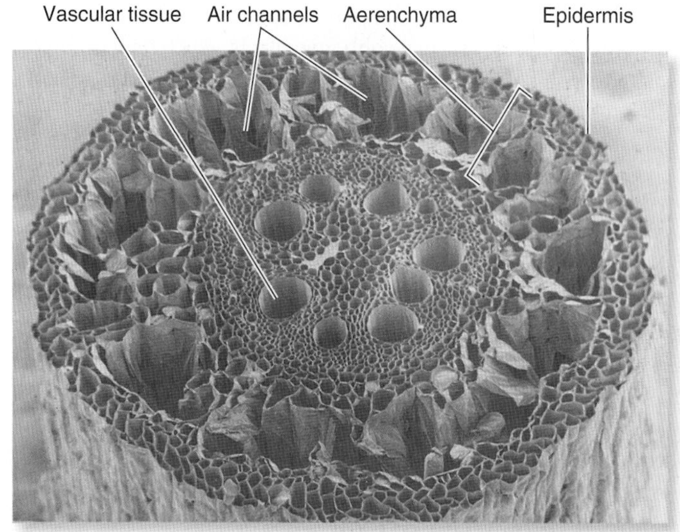

Figure 36.16 **A plant response to flooding.** This illustration of a slice of a root shows air channels within aerenchyma tissue.

BIOLOGY PRINCIPLE **Structure determines function.** Air channels within aerenchyma tissue allow air to flow from shoots to roots even when the plant is partially submerged.

Concept Check: What are two ways in which aerenchyma tissue is formed?

Protective plant hormones are often involved in plant defenses (**Figure 36.17**). For example, when insects wound tomato plants, damaged cells release the peptide hormone systemin. This hormone

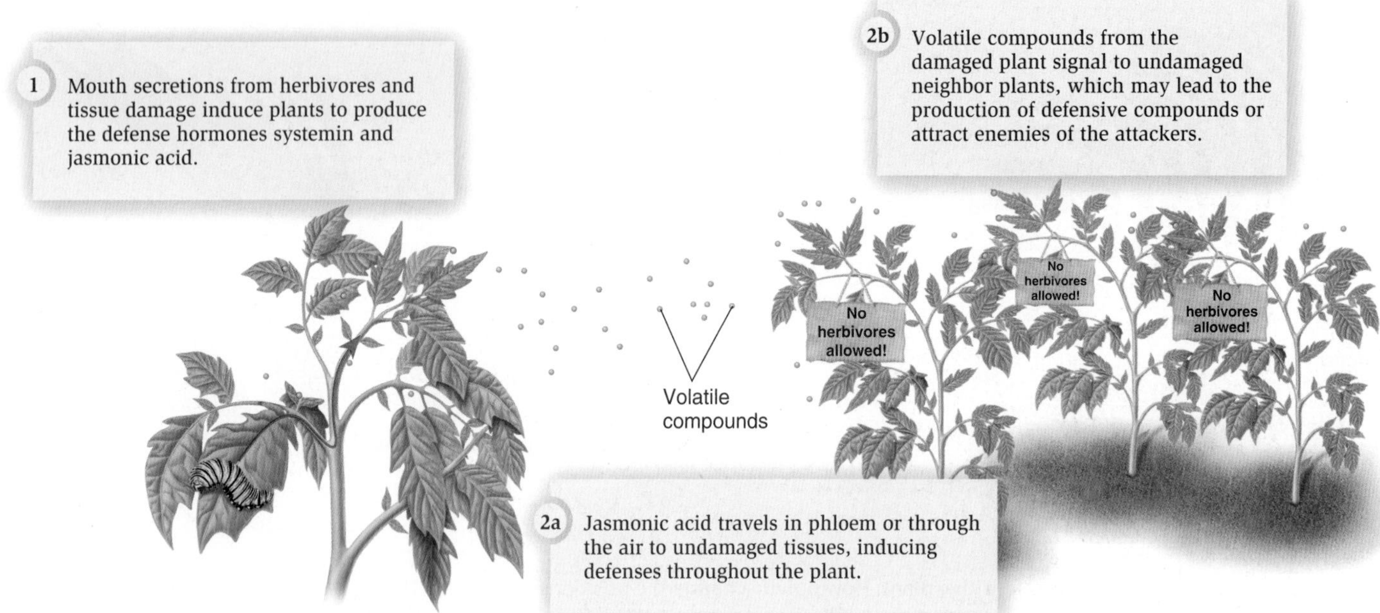

Figure 36.17 Plant responses to herbivore attack. **(1)** Leaf damage induces local defensive responses, including production of compounds that travel through the plant and cause defense responses elsewhere. **(2a)** In addition, attacked plants may release volatile compounds, such as jasmonic acid. These volatile compounds foster the development of defenses in undamaged parts of the same plant. **(2b)** Such volatile compounds may also induce defenses in neighboring plants, which then become less vulnerable to herbivores. Volatile compounds may also attract predators that feed on the herbivore attacker.

Concept Check: *What advantage would a predator obtain by responding to the volatile signals emitted from herbivore-damaged plants? What adaptive advantage could a plant obtain by attracting the enemies of its attackers?*

induces undamaged cells to produce the hormone jasmonic acid (JA), which functions as an alarm system. JA travels in the phloem or through the air to undamaged parts of the plant, inducing widespread, preemptive production of defensive compounds (see Figure 36.17, step 2a). Airborne compounds released from herbivore-damaged plants may also attract the enemies of insect attackers. Nearby plants may detect these airborne signals and respond by similarly arming themselves against attack (see Figure 36.17, step 2b).

Plant Responses to Pathogen Attack Every year, about 15% of crop production is lost to disease caused by certain bacteria, fungi, protists, or viruses. During their evolution, plants have evolved mechanisms for detecting and responding to disease microbes, but microbes can rapidly evolve new disease processes. Agricultural scientists aim to understand how pathogenic microbes attack plants and how plants are able to prevent disease, in order to develop disease-resistant crop varieties.

Plant pathogens produce compounds known as **elicitors**. Such elicitors promote **virulence**, the infection of plant cells and tissues. **Avirulence genes (*Avr* genes)** are microbial genes that encode virulence-enhancing elicitors (**Figure 36.18**, step 1). Some bacteria inject elicitors into plant cells by means of syringe-like systems that are also used to attack animal cells. Fungal pathogens often deliver elicitors into plant cells and absorb nutrients from them by means of penetration structures known as haustoria (from a Latin word, meaning to drink). In response to elicitors, plant cells have evolved first and second lines of defense. The first line of defense consists of plasma membrane receptors that specifically bind microbe-associated molecules, such as bacterial lipopolysaccharides or the flagellar protein flagellin, or the fungal cell wall compound chitin. These plasma membrane receptors, which are typically protein kinases, allow plant cells to sense disease microbes. Researchers have discovered that animal cells use homologous receptor molecules to sense and respond to pathogens.

A second line of defense occurs in the cytosol. Here, microRNAs help to destroy the nucleic acids of invading viruses, in processes similar to those present in animal cells. In addition, receptors that occur in the cytosol allow plant cells to recognize injected elictors. The binding of receptors with elicitors triggers the production of chemical defense signals: hydrogen peroxide (H_2O_2), the gas nitrous oxide (NO), and mobile hormones. Defense responses include deposition of polymers that strengthen plant cell walls at the site of attack, programmed death of infected cells as a way of limiting the spread of infection, and production of compounds that kill pathogens (see Figure 36.18).

Many types of **resistance genes (*R* genes)** encode plant disease receptor proteins. If a plant is genetically unable to produce a receptor that can recognize a pathogen's presence or elicitor, disease may result. By contrast, plants successfully resist disease when the product of a dominant *R* gene (a receptor) recognizes a pathogen's presence by characteristic compounds or its dominant *Avr* gene product (an elicitor). Diverse alleles of *R* genes occur in plant populations, providing plants with a large capacity to cope with different types of pathogens. These receptors, together with plant responses—the

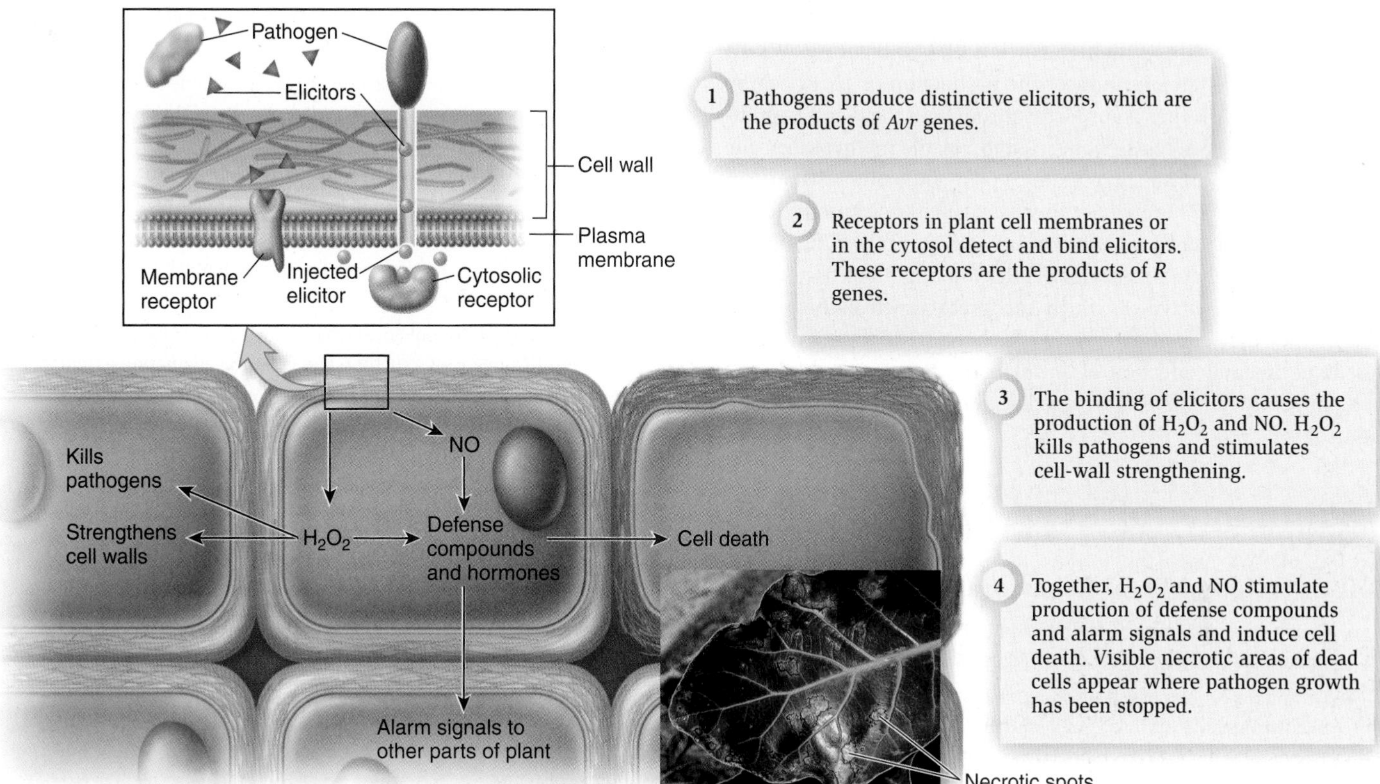
Pathogen
Elicitors
Cell wall
Plasma membrane
Membrane receptor
Injected elicitor
Cytosolic receptor
1 Pathogens produce distinctive elicitors, which are the products of *Avr* genes.
2 Receptors in plant cell membranes or in the cytosol detect and bind elicitors. These receptors are the products of *R* genes.
3 The binding of elicitors causes the production of H_2O_2 and NO. H_2O_2 kills pathogens and stimulates cell-wall strengthening.
4 Together, H_2O_2 and NO stimulate production of defense compounds and alarm signals and induce cell death. Visible necrotic areas of dead cells appear where pathogen growth has been stopped.
Kills pathogens
Strengthens cell walls
H_2O_2
NO
Defense compounds and hormones
Cell death
Alarm signals to other parts of plant
Necrotic spots

Figure 36.18 **Pathogen/plant interactions and the hypersensitive response to pathogen attack.**

BioConnections: *Look back to Figures 27.10b and 27.18b. What component of the Gram-negative bacterial pathogen Agrobacterium tumifaciens might be recognizable by a host plant?*

hypersensitive response and systemic acquired resistance—represent the plant immune system.

The Hypersensitive Response to Pathogen Attack The plant **hypersensitive response (HR)** is a local reaction to pathogen attack that limits the progression of disease (see Figure 36.18, steps 3 and 4). H_2O_2 can kill pathogens and also helps strengthen the cell wall by deposition of cross-binding polymers. The gaseous hormone NO works with H_2O_2 to stimulate the synthesis of hydrolytic enzymes, defensive secondary metabolites, the hormone salicylic acid (SA), and tough lignin in the cell walls of nearby tissues. NO also induces programmed cell death, which deprives pathogens of food, helping to limit the spread of disease. Necrotic spots are brown patches on plant organs that reveal where disease pathogens have attacked plants and where plant tissues have battled back. SA or the related compound methyl salicylate sends alarm signals to other parts of the plant, which respond by preparing defenses.

Systemic Acquired Resistance Whereas the HR is a local reaction, **systemic acquired resistance (SAR)** is an immune response of the whole plant induced by a pathogen attack (**Figure 36.19**). SAR immunizes the plant not only from the inducing pathogen, but also many

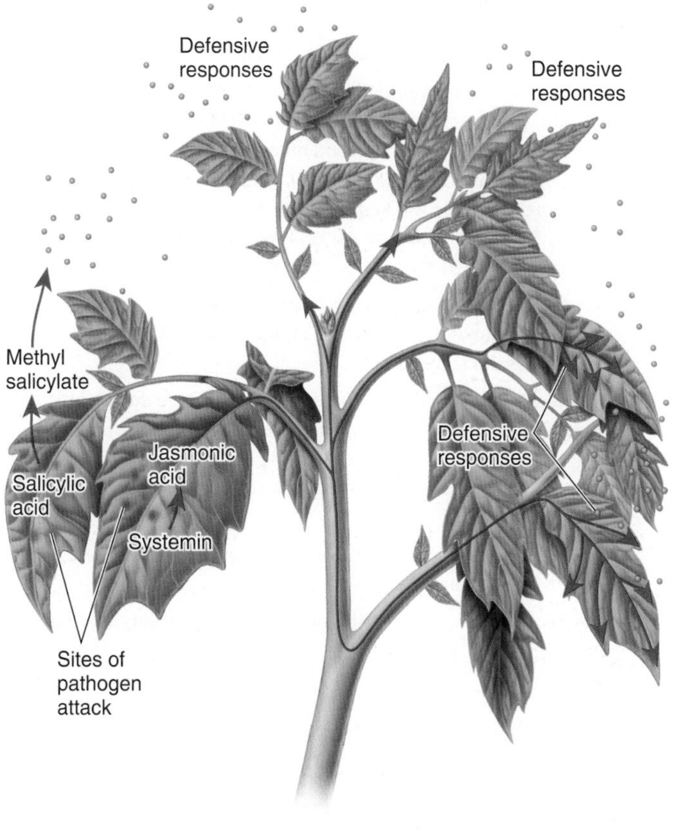

Defensive responses
Defensive responses
Methyl salicylate
Salicylic acid
Jasmonic acid
Systemin
Defensive responses
Sites of pathogen attack

Figure 36.19 **Systemic acquired resistance to pathogen attack.**

Concept Check: *In what way is systemic acquired resistance to pathogens similar to plant responses to herbivore attack?*

others. The SAR system uses the same type of long-distance signaling process that is associated with herbivore attack (see Figure 36.17). At or near a wound, systemin induces the production of JA in nearby vascular tissues. JA is then transported throughout the plant. In addition, the hormone SA can be converted to volatile methyl salicylate, which diffuses into the air surrounding a plant, inducing resistance in noninfected tissues. These signals involve changes in transcription of up to 10% of plant genes, including plant analogs of the *BRCA* genes associated with human breast cancer. Responding plant tissues may produce defensive enzymes, which can break down pathogen cell walls, or they may generate tannins, which are toxic to microorganisms. Experimental infection of *Arabidopsis* leaves with the bacterial pathogen *Pseudomonas syringae* led to induction of SAR within 6 hours.

Summary of Key Concepts

36.1 Overview of Plant Behavioral Responses

- Plants sense and respond to diverse internal stimuli and external signals and thus display behavior (Figures 36.1, 36.2).
- During the process of cell signaling, cellular receptors respond to environmental stimuli as well as to internal hormonal signals. The process involves receptor activation, signal transduction by messengers, and a cellular response (Figure 36.3).
- Cellular responses include changes in gene expression and ion channels that influence plant growth, development, reproduction, chemistry, and movements such as sun tracking and leaf folding.

36.2 Plant Hormones

- Plant hormones interact with environmental stimuli to control plant development, growth, and behavior (Table 36.1).
- Auxin plays an important role in many aspects of plant behavior, including phototropism, as demonstrated by the classic experiments of the Darwins, Went, Briggs, and others. Auxin can be transported downward or upward (polar transport) and sideways (lateral transport) in the plant. The position of auxin influx and efflux carriers determines the direction of auxin transport (Figures 36.4, 36.5).
- The major effect of cytokinins is to increase the rate of cell division. Auxin and cytokinin are used in laboratory and commercial plant tissue culture (Figure 36.6).
- Gibberellin function illustrates the general principle that plant hormones often act to release cellular brakes on gene expression. The components of a system by which gibberellin and cell proteins interact to influence plant growth evolved in a step-by-step fashion (Figures 36.7, 36.8).
- The gaseous hormone ethylene plays an important role in seed germination (Figure 36.9).
- Abscisic acid (ABA) and brassinosteroids help plants to respond to environmental stress.

36.3 Plant Responses to Environmental Stimuli

- Light-sensitive pigments such as cryptochrome, phototropin, and phytochrome allow plants to respond to light stimuli and influence sun tracking, seed germination, and photoperiodic control of flowering (Figures 36.10, 36.11, 36.12).
- Plant shoots and roots respond to gravity (in the process of gravitropism) by means of starch-heavy statoliths located within statocytes. Rapid touch responses (thigmotropism), such as leaf folding in sensitive plants, depend on changes in the water content of cells in structures known as pulvini (Figures 36.13, 36.14, 36.15).
- Many plants cope with physical stresses such as flooding and drought through the production of special tissue (aerenchyma), which is regulated by ethylene (Figure 36.16).
- Plants cope with biological stresses such as herbivore and pathogen attack by means of structural and chemical adaptations. Injured plant parts produce volatile hormones that signal other parts of the same plant and nearby plants to produce defensive responses. Plant pathogens begin their attack by producing elicitor compounds encoded by avirulence (*Avr*) genes. If an attacked plant can produce a compound that interferes with that elicitor, encoded by an *R* (resistance) gene, then the plant can resist infection by that pathogen. The hypersensitive response (HR) is a local defensive response to pathogen attack, whereas systemic acquired resistance (SAR) is a whole-plant immune response (Figures 36.17, 36.18, 36.19).

Assess and Discuss

Test Yourself

1. The major types of plant hormones include
   a. cyclic AMP, IP_3, and calcium ions.
   b. calcium, CDPKs, and DELLA proteins.
   c. auxin, cytokinin, and gibberellin.
   d. cryptochrome, phototropin, and phytochrome.
   e. statoliths, pulvini, and aerenchyma.
2. Phototropism is
   a. the production of flowers in response to a particular day length.
   b. the production of flowers in response to a particular night length.
   c. the growth response of a plant, organ system, or organ to directional light.
   d. the growth response of a plant, organ system, or organ to gravity.
   e. the growth response of a plant, organ system, or organ to touch.
3. What is the most accurate order of events during signal transduction?
   a. first, receptor activation; then, messenger signaling; and last, an effector response
   b. first, an effector response; then, messenger signaling; and last, receptor activation
   c. first, messenger signaling; then, receptor activation; and last, an effector response
   d. first, an effector response; then, receptor activation; and last, messenger signaling
   e. none of the above
4. Which of the plant hormones is known as the "master hormone," and why?
   a. cytokinin, because many plant functions require cell division
   b. gibberellins, because growth is essential to many plant responses
   c. abscisic acid, because it is necessary for leaf and fruit drop
   d. brassinosteroids, because water uptake is so fundamental to plant growth
   e. auxin, because there are many different auxin-response genes

5. Gaseous hormones are able to enter cells without requiring special membrane transporter systems. Which of the major plant hormones is a diffusible gas?
   a. auxin
   b. gibberellin
   c. cytokinin
   d. ethylene
   e. abscisic acid
6. Photoreceptor molecules allow plant cells to detect light of particular wavelengths. Which of these molecules is considered to be a plant photoreceptor?
   a. cryptochrome
   b. phototropin
   c. phytochrome
   d. a, b, and c are all correct
   e. no earlier answer listed is correct
7. Thigmotropism is a plant response to
   a. light.
   b. cold.
   c. touch.
   d. gravity.
   e. drought.
8. Which response is an adaptation to flooding?
   a. geotropism
   b. stomatal closure
   c. photoperiodism
   d. production of aerenchyma
   e. opening aquaporins
9. What are avirulence genes?
   a. plant genes that encode proteins that prevent infection (virulence)
   b. plant genes that cause infection when the proteins they encode bind to pathogen elicitors
   c. pathogen genes that prevent the pathogens from causing plant disease
   d. pathogen genes that encode elicitors that foster disease in plants
   e. none of the above
10. How do plants defend themselves against pathogens?
   a. Plants produce resistance molecules (usually proteins) that bind pathogen elicitors, thereby preventing disease.
   b. Plants display a hypersensitive response that limits the ability of pathogens to survive and spread.
   c. Plants display systemic acquired resistance, whereby an infection induces immunity to diverse pathogens in other parts of a plant.
   d. All of the above are correct.
   e. None of the above

Conceptual Questions

1. Why can plants be said to display behavior?
2. Why do plants produce so many types of resistance (*R*) genes?
3. A principle of biology is that *living organisms interact with their environment.* Because diverse plants exude volatile compounds in response to herbivore or pathogen attack, some experts have written about “talking trees.” Is there any such thing?

Collaborative Questions

1. Why are most wild plants distasteful, and some even poisonous, to people?
2. How could you increase the resistance of a particular crop plant species to particular types of herbivores?

Online Resource

www.brookerbiology.com

Stay a step ahead in your studies with animations that bring concepts to life and practice tests to assess your understanding. Your instructor may also recommend the interactive eBook, individualized learning tools, and more.

Flowering Plants: Nutrition

37

Chapter Outline

37.1 Plant Nutritional Requirements
37.2 The Role of Soil in Plant Nutrition
37.3 Biological Sources of Plant Nutrients
Summary of Key Concepts
Assess and Discuss

The tentacled leaves of the sundews (*Drosera rotundifolia* and *D. intermedia*), shown with a trapped fly, are a plant adaptation for the acquisition of nutrients.

Many types of fascinating carnivorous (meat-eating) plants grow abundantly in wetlands around the world, even though the soils in these places are infertile. How is this possible? Like most plants, carnivorous plants are photosynthetic and thus produce their own organic food from carbon dioxide and water, using sunlight as an energy source. These resources are abundant in wetlands, but wetland soils are low in other nutrients such as nitrogen that are needed for plant growth. Carnivorous plants, such as the sundews pictured in the chapter-opening photo, have adapted by obtaining nutrients from the bodies of trapped insects and other small animals. Carnivorous plants lure animals with enticing fragrances, brightly colored leaves, or glistening sugar-rich drops of nectar. The unsuspecting prey fall into deep, water-filled pitchers; become ensnared by gluelike mucilage; or are trapped within the walls of leafy jails whose doors suddenly snap shut. Decomposition of the animal bodies releases nutrients that plant leaves quickly absorb. Other wild and cultivated plants face similar nutritional challenges and likewise display adaptations that help them acquire sufficient resources for growth and reproduction.

This chapter focuses on plant nutrition, the processes by which plants obtain essential resources. We will begin by describing the resources needed by plants for completion of their seed-to-seed life cycle in good health. Next, we will explore the role of soil as an essential resource for plants. Last, we will examine the biological sources of plant nutrients, focusing on nutritional associations between plants and microorganisms, sources of nutrients for carnivorous plants, and how some plants obtain nutrients from other plants. An understanding of these topics is crucial for those who seek ways to grow more plant-derived food or biofuels for humans without causing environmental harm. Plant nutritional information is also useful to people who tend gardens or houseplants, and those who seek to restore degraded habitats.

37.1 Plant Nutritional Requirements

Learning Outcomes:

1. List the major nutritional resources that most plants need for healthy growth.
2. Describe how some plants have adapted to light limitation in shady habitats.
3. Distinguish between plant macronutrients and micronutrients.

The case of carnivorous plants illustrates the concept that all plants—like all animals—have nutritional requirements. A **nutrient** is a substance that is metabolized by or incorporated into an organism. Photosynthetic plants require carbon dioxide (CO_2), water (H_2O), and more than a dozen elements, such as potassium (as K^+), nitrogen (in the form of NH_4^+ or NO_3^-), and calcium (as Ca^{2+}) to produce organic food by means of photosynthesis. CO_2 is primarily absorbed from air, whereas H_2O and elements are primarily taken up from soil in the form of dissolved ions (Figure 37.1). As is the case for animals, deficiency symptoms develop in plants that receive too little of these substances. The environmental scarcity of nutrients selects for adaptations that help plants to acquire them, illustrated by the existence of carnivorous plants.

Essential elements are chemical elements required by plants that play many roles in plant metabolism, often functioning as

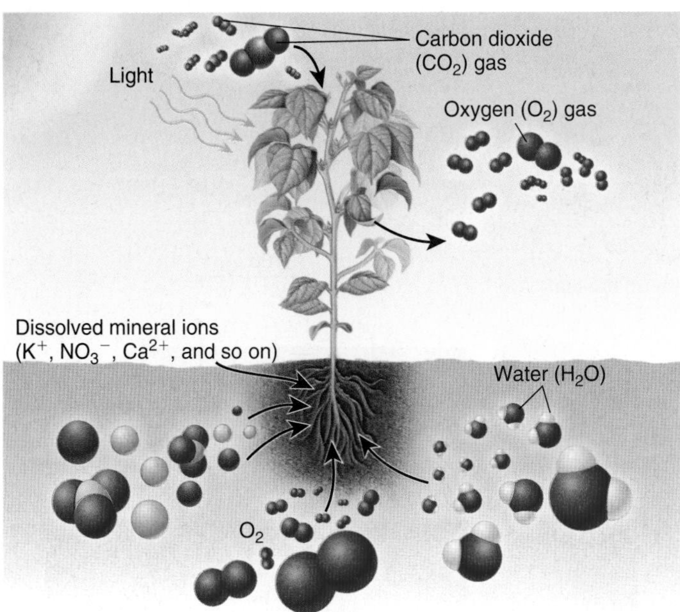

Figure 37.1 **The major types of plant nutrients and their sources.** Elements that are required in relatively large amounts are known as macronutrients, whereas elements required in smaller amounts are known as micronutrients.

enzyme cofactors (**Table 37.1**). Elements that are generally required in amounts of at least 1 g/kg of plant dry mass are known as **macronutrients**. In contrast, elements that are needed in amounts at or less than 0.1 g/kg of plant dry mass are known as **micronutrients**, or trace elements. Because insufficient amounts of light, carbon dioxide, water, and other mineral nutrients can limit the extent of green plant growth, these resources are known as **limiting factors**. If you were in charge of a garden, greenhouse, farm or forest used to generate products such as wood or paper, or if you were overseeing an environmental restoration project, you would want to understand the conditions that foster or limit plant growth. In this section, we focus on plant resource requirements, starting with light.

Light Is an Essential Resource for the Growth of Green Plants

All photosynthetic plants require light energy for the formation of the covalent bonds of organic compounds that make up the plant body. Green plants' use of light energy as an essential resource parallels animals' nutritional requirements for organic food as a source of chemical energy. For the several hundred species of heterotrophic plants that have lost their photosynthetic capacity, such as the Indian pipe (*Monotropa uniflora*), essential nutrients include absorbed organic compounds that replace light as a source of energy (**Figure 37.2**).

In nature, plants must adapt to environments with varying amounts of light and shade. For example, in forests, light availability limits the growth of tree seedlings and other small plants that are shaded by the leafy tree canopy overhead. Conversely, plants growing in deserts or on mountains often experience light so intense that it can damage the photosynthetic components. Plants have evolved

Table 37.1 Plant Essential Nutrients

Element (chemical symbol)	Percent of plant dry mass	Major source	Form taken up by plants	Function(s)
Macronutrients				
Carbon (C)	45	Air	CO_2	Component of all organic molecules
Oxygen (O)	45	Air, soil, water	CO_2, O_2, H_2O	Component of all organic molecules
Hydrogen (H)	6	Water	H_2O	Component of all organic molecules; protons used in chemiosmosis and cotransport
Nitrogen (N)	1.5	Soil	NO_3^-, NH_4^+	Component of proteins, nucleic acids, chlorophyll, coenzymes, and alkaloids
Potassium (K)	1.0	Soil	K^+	Has essential role in cell ionic balance
Calcium (Ca)	0.5	Soil	Ca^{2+}	Component of cell walls; messenger in signal transduction
Magnesium (Mg)	0.2	Soil	Mg^{2+}	Component of chlorophyll; activates some enzymes
Phosphorus (P)	0.2	Soil	HPO_4^{2-}	Component of nucleic acids, ATP, phospholipids, and some coenzymes
Sulfur (S)	0.1	Soil	SO_4^{2-}	Component of proteins, some coenzymes, and defense compounds
Micronutrients				
Chlorine (Cl)	0.01	Soil	Cl^-	Required for water splitting in photosystem; cell ion balance
Iron (Fe)	0.01	Soil	Fe^{3+}, Fe^{2+}	Enzyme cofactor; component of cytochromes
Manganese (Mn)	0.005	Soil	Mn^{2+}	Enzyme cofactor
Boron (B)	0.002	Soil	$B(OH)_3$	Enzyme cofactor; component of cell walls
Zinc (Zn)	0.002	Soil	Zn^{2+}	Enzyme cofactor
Sodium (Na)	0.001	Soil	Na^+	Required to generate PEP in C_4 and CAM plants
Copper (Cu)	0.0006	Soil	Cu^+, Cu^{2+}	Enzyme cofactor
Molybdenum (Mo)	0.00001	Soil	MoO_4^{2-}	Enzyme cofactor
Nickel (Ni)	0.000005	Soil	Ni^{2+}	Enzyme cofactor

CAM, crassulacean acid metabolism; PEP, phosphoenolpyruvate.

Figure 37.2 **The heterotrophic flowering plant, Indian pipe (*Monotropa uniflora*).**

BIOLOGY PRINCIPLE Living organisms use energy. This nongreen heterotrophic plant lacks photosynthetic capacity and therefore must absorb organic compounds for use as an energy source. In contrast, nearby autotrophic green plants use sunlight as a source of energy.

BioConnections: *Look back at Figure 31.5 to see the white body of a fungus, known as a mycelium. How is the nutrition of this (and all fungi) similar to that of the flowering plant Monotropa uniflora?*

adaptations that help them cope with environments that have too little or too much light.

Adaptations to Shade Plants can adapt to shading by producing thin leaves that allow some light to pass through to other leaves, producing more chlorophyll, or producing distinctive Sun and shade leaves (**Figure 37.3**). Sun leaves have a thicker layer of chlorophyll-containing mesophyll and are able to harvest more of the bright sunlight that penetrates deeply into the leaf. In contrast, shade leaves have a thinner mesophyll layer, with more air spaces than do Sun leaves. The arrangement of leaves on plants and stem-branching patterns can also reflect adaptations that reduce shading. For example, many tropical forest trees that must compete for light with closely crowded neighbors are extremely tall, and they produce branches and leaves only at their very tops. Shorter plants native to the shady interiors of tropical rain forests are so well adapted to these moist but dim conditions that they make excellent houseplants (**Figure 37.4**).

Adaptations to Excessive Light Too much light can damage plant chloroplasts by destroying an essential photosynthetic protein, called the D1 protein. Specific carotenoid pigments in the chloroplast absorb some of the excess light energy and dissipate it as harmless heat. Other adaptations prevent ultraviolet (UV) damage. Harmful amounts of UV radiation are absorbed by the plant's surface cuticle, as well as by carotenoid and flavonoid compounds located within leaf cells. These protective compounds are often brightly colored and, when present in large amounts, explain the attractive red or purple colors of some leaves.

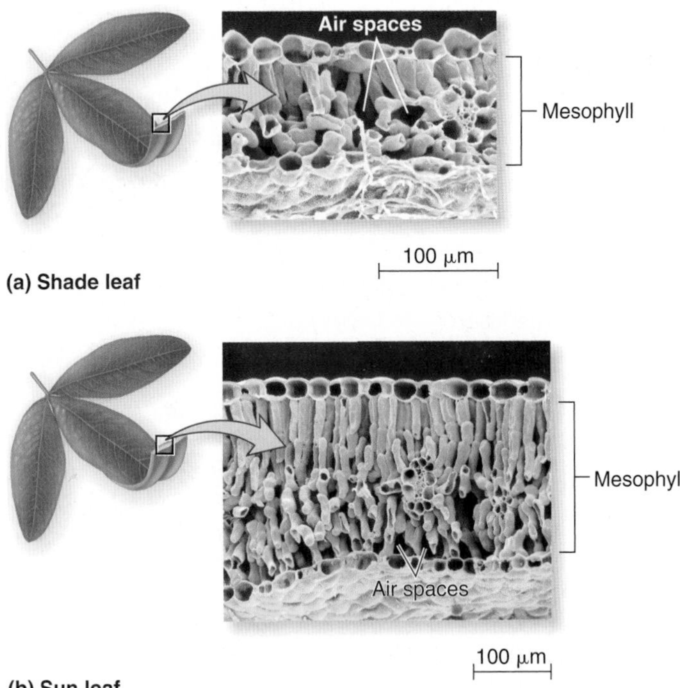

Figure 37.3 **Shade and Sun leaves.**

BIOLOGY PRINCIPLE Structure determines function. These scanning electron micrographs of cut leaves reveal the thinner mesophyll and greater amount of air spaces in **(a)** shade leaves as compared with **(b)** Sun leaves.

Figure 37.4 **The tropical houseplant *Monstera deliciosa*.** *M. deliciosa* makes an attractive houseplant because its large, deep green leaves are adapted to the moist and shady conditions present in the interiors of tropical forests.

Carbon Dioxide Concentration Influences Plant Growth

Although light provides the energy for plant photosynthesis, most plant dry mass originates from carbon dioxide (CO_2). Most plants obtain CO_2 gas from the atmosphere, by absorption through stomata, pores that occur in the plant epidermis. Under experimental conditions, plant photosynthetic rates increase with CO_2 concentration until the Calvin cycle enzyme rubisco has become fully supplied, that is, saturated with CO_2. (To review leaf structure, see Figure 35.8c; to review the role of rubisco and the Calvin cycle in photosynthesis, see Figure 8.15.)

CO_2 Limitation In nature, plants are often unable to obtain enough CO_2 for maximal photosynthesis. As a result, CO_2 limits agricultural crop productivity. This is partly because the modern atmospheric concentration of CO_2 is only 390 μL/L (a small fraction of atmospheric content), whereas considerably more CO_2 would be required to saturate photosynthesis in most plants. Evidence that plant photosynthesis can be limited by CO_2 availability is provided by studies of crops grown in greenhouses where atmospheric gas content can be controlled. When supplied with air enriched with CO_2, tomatoes, cucumbers, leafy vegetables, and some other crops can double their growth rate. In nature, plants that experience hot, dry conditions are particularly vulnerable to low CO_2 levels when their stomata close to conserve water. When stomata are closed, plants cannot absorb CO_2, which limits photosynthesis. Many plants possess structural and biochemical adaptations that help them cope with CO_2 limitation by improving CO_2 absorption.

CO_2-Absorption Adaptations Many plants that live in arid or hot environments display an adaptation known as C_4 photosynthesis, which improves productivity by aiding CO_2 absorption (refer back to Figure 8.19a). About 30,000 plant species utilize C_4 photosynthesis, which is thought to have evolved from ancestral C_3 photosynthesis on more than 40 separate occasions. Today, C_4 plants dominate extensive warm grassland habitats of the world, contributing about 25% of the total terrestrial photosynthesis.

C_4 photosynthesis relies on specializations in the internal structure of leaves. In C_4 plants, the leaf mesophyll harvests light, but does not contain Calvin-cycle enzymes that bind CO_2 to form sugar. Instead, C_4 leaf mesophyll produces the enzyme phosphoenolpyruvate (PEP) carboxylase, which binds CO_2 more readily than does rubisco. PEP carboxylase adds CO_2 to PEP, a three-carbon molecule, to produce the four-carbon compound oxaloacetate. The leaves of C_4 plants also have increased vein density, and specialized bundle-sheath cells surround each vein. In the leaves of most C_4 plants, such as maize (corn), the four-carbon compounds produced in mesophyll move through intracellular connections into the bundle-sheath cells. Within the bundle-sheath cells, enzymes release CO_2 from the C_4 compounds, and the CO_2 is incorporated into organic carbon by Calvin cycle enzymes. Some plants native to arid regions have evolved a type of C_4 photosynthesis that occurs within a single cell. Photosynthetic cells of these plants have two types of plastids, one type containing PEP carboxylase and the other containing rubisco (**Figure 37.5**). C_4 photosynthesis is particularly valuable to plants that occur in hot, dry environments because it allows plants to absorb sufficient CO_2 without losing too much water.

Water Is an Essential Plant Resource

Water is essential to plants for several reasons. As a nutrient, water is the source of most of the hydrogen atoms and much of the oxygen atoms in organic compounds (see Table 37.1). For example, the oxygen in the abundant plant polymer cellulose derives from water taken up from the soil. Hydrogen is incorporated when CO_2 is reduced during the process of photosynthesis, and oxygen is incorporated during hydrolysis reactions. Water is also the solvent for other mineral nutrients and is the main transport medium in plants, allowing movement of minerals and other solutes throughout the plant body via the vascular tissues. Cytoplasmic and vacuolar water also help to support plants by maintaining hydrostatic pressure on the cell wall.

Though plants vary in water content, water typically makes up about 90% of the weight of living plants. Most plants die when their water content falls below half of the amount normal for that particular species. Although many types of desiccation-resistant plants display adaptations that allow them to survive for extended periods in nearly dry conditions, all plants require an adequate supply of water for active metabolism and growth.

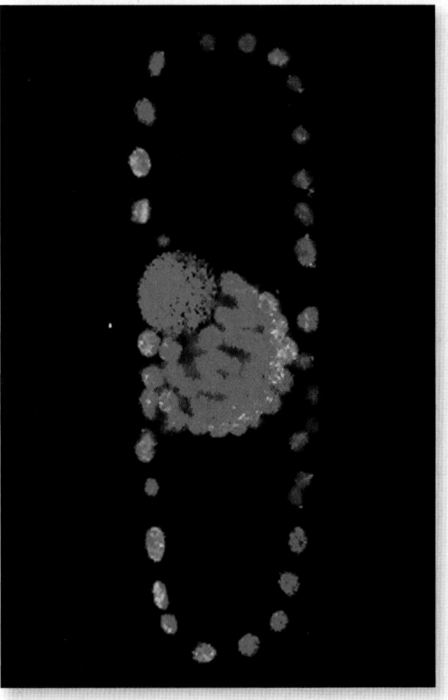

Figure 37.5 A CO_2-acquisition adaptation. In contrast to most C_4 plants, *Bienertia cycloptera* is able to conduct C_4 photosynthesis within the confines of a single cell. In this fluorescence photograph of a *B. cycloptera* cell, the red plastids at the cell periphery function like the mesophyll cells of most C_4 plants. In contrast, the red plastids clustered near the green nucleus function much like the bundle-sheath cells of most C_4 plants. The plastids appear red because chlorophyll emits red light when it fluoresces.

Concept Check: Which plastids should have more rubisco enzyme in them, plastids at the cell periphery or those clustered near the nucleus?

Figure 37.6 **Chlorosis as a symptom of mineral deficiency.** This camellia plant is suffering from an iron deficiency, as revealed by the yellow leaves, a symptom known as chlorosis.

Concept Check: *Does chlorosis always indicate iron deficiency in plants?*

Soil Provides Additional Plant Nutrients

In addition to light, CO_2, and water, plants require additional elements that occur naturally in water and soil or that can be added in the form of fertilizers. Plant biologists have quantitatively analyzed the elemental requirements of plants by growing them hydroponically, that is, by bathing plant roots in a water solution to which elements are added in various combinations and amounts. Hydroponic studies reveal that when plants lack an adequate supply of an essential elemental nutrient, they display characteristic deficiency symptoms. Such symptoms include failure to reproduce, tissue death, and changes in leaf color. Yellowing of leaves, known as **chlorosis**, is a common mineral deficiency symptom, because many elements are needed for chlorophyll production (**Figure 37.6**). Hydroponic studies reveal the relative amounts and roles of essential elements in plant growth more easily than can be accomplished by growing plants in soil, which supplies most plants with most of the elements they need (see Table 37.1). Soil plays several additional roles in plant growth.

37.2 The Role of Soil in Plant Nutrition

Learning Outcomes:

1. List the benefits of soil organic material for plant growth.
2. Explain why plants require a source of fixed nitrogen and how they obtain it.
3. Discuss various plant adaptations that increase the ability to obtain phosphorus.

Soil is an essential resource for most wild and cultivated plants, providing water and other essential nutrients. For this reason, extensive loss of soil by wind and water erosion is of wide concern. Soils vary greatly in fertility, that is, their ability to support plant growth. Thus, plants of many types have had to adapt to the challenges of obtaining nutrients from poor soils. In this section, we explore soil structure and chemistry from the perspective of plant growth and examine how plants take up nutrients from the soil.

The Physical Structure of Soils Affects Their Aeration, Water-Holding Capacity, and Fertility

Natural soils display layers, known as soil horizons (**Figure 37.7a**). The remains of plants that have recently died and other organisms form a layer of litter above the **topsoil**, also known as the A-horizon. Many of the inorganic minerals and organic materials that enrich high-quality topsoil arise from the activities of microorganisms that decompose the litter. In this way, the minerals contained in living plants are eventually recycled to subsequent generations. Beneath the topsoil lie layers called subsoil and soil base, which are largely composed of mineral materials. Bedrock is the bottom layer that supports the soil horizons. Plant roots play an important role in conveying deep-lying minerals to the surface, thereby helping to enrich the topsoil.

Soil horizons vary in composition and thickness, depending on various factors—including climate, vegetation, bedrock type, and human influences. For example, natural grasslands produce deep, rich topsoil that is used for cropland in many regions of the world (**Figure 37.7b**). In dramatic contrast, tropical rain forests often have only thin layers of topsoil; their low-fertility soils are composed mostly of inorganic materials that are not useful to plants (**Figure 37.7c**). Farmers cope with reduced soil fertility by adding organic or inorganic fertilizers. The proportions of organic to inorganic materials and the sizes of inorganic particles are used to classify soils into different types. Soils also display variation in their amount of aeration, water-holding capacity, pH, and mineral content. All of these soil properties affect plant growth. For this reason, we will take a closer look at soil structure and the role of fertilizers.

Soil Organic Matter Soil organic matter, also known as humus, is largely derived from plant detritus, the dead and decaying remains of plants, although animal wastes and decayed animal bodies also contribute to the organic content of soils. Soil organic matter provides many benefits. Organic-rich soils, containing 8% or so organic matter, are less likely to erode—wash or blow away with water or wind. Soil organic matter also binds mineral nutrients, thereby fostering soil fertility, and gives soils a soft consistency that fosters plant root growth and farmers' ability to cultivate the soil. Gardeners often produce their own organic matter-rich compost by layering vegetable waste from the kitchen and yard waste with soil, and turning the pile occasionally to introduce the oxygen necessary for decomposition (**Figure 37.8**). Mature compost is then mixed with garden soil to improve its fertility.

Inorganic Soil Constituents Inorganic materials in soil are derived from the physical and chemical breakdown of rock, a process known as **weathering**. Rock, which is an aggregate of two or more

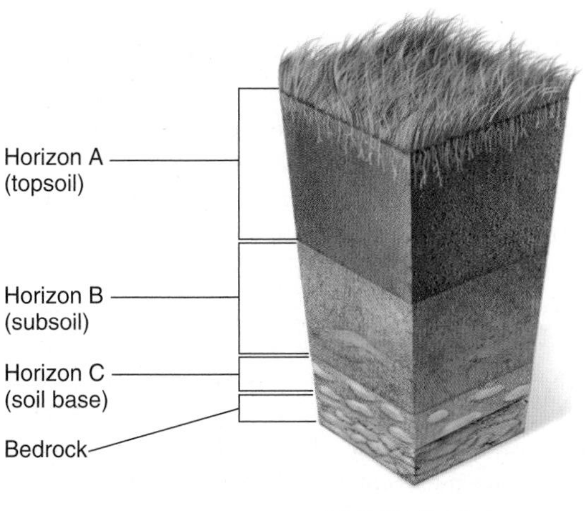

(a) Soil structure

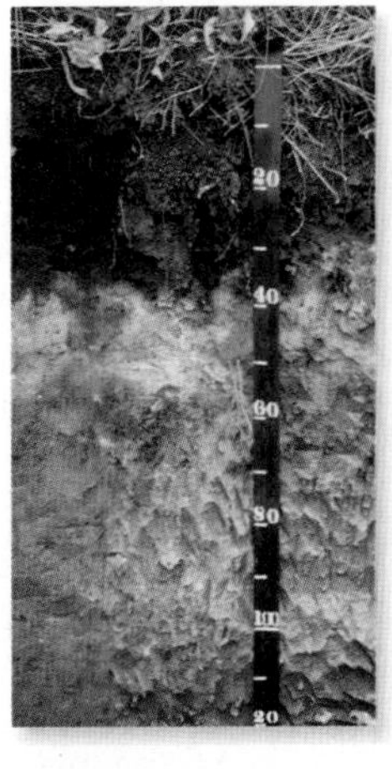

(b) Thick layer of topsoil in cropland

(c) Thin layer of topsoil in rain forest

Figure 37.7 Soil horizons, the structural layers of soil. (a) Diagram showing general soil structure. (b) A vertical view of an agricultural soil, showing a relatively deep layer of dark topsoil. (c) A vertical view of a tropical rain forest soil, showing a thin layer of dark topsoil.

BioConnections: Look ahead to Figure 60.17b, which shows a habitat that was seriously degraded by phosphate-mining, and whose topsoil has had to be replaced. How could damaged land possibly be rehabilitated to support natural vegetation or agriculture, and what are the likely consequences?

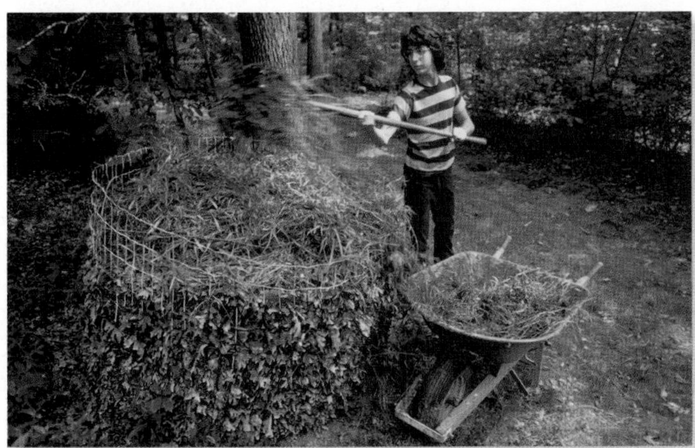

Figure 37.8 Composting. Gardeners produce compost by layering small amounts of soil with vegetable waste from the kitchen and yard waste and periodically turning the pile to introduce the oxygen needed for decomposition. Compost can be used to increase the organic and mineral content of garden soils.

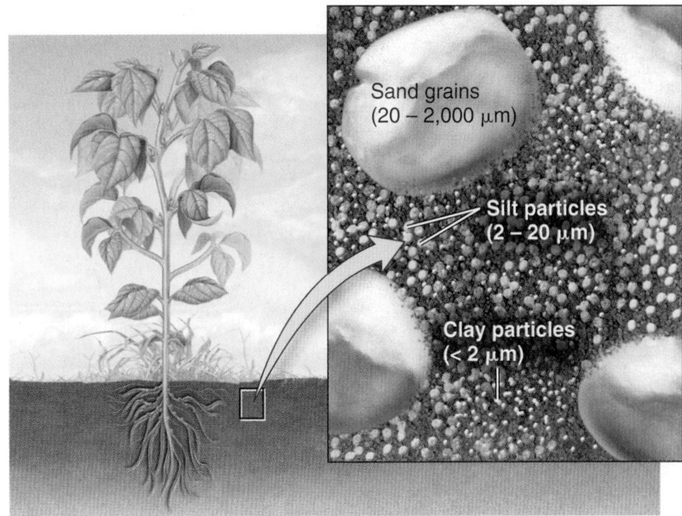

Figure 37.9 The relative sizes of inorganic soil components.

minerals, is physically weathered by changes such as cycles of freezing and thawing. Lichens and plant roots may produce organic acids that contribute to chemical weathering of rocks. During chemical weathering, soluble salts are washed out, and minerals are hydrolyzed or oxidized. **Leaching** is the dissolution and removal of inorganic ions as water percolates through materials. Heavy rainfall can reduce the fertility of soils by leaching large amounts of nutrients from them. The leached elements often end up in natural bodies of water, where they can foster the growth of cyanobacteria, algae, and aquatic plants.

Inorganic soil materials occur as particles that can be categorized according to their size as sand, silt, or clay (Figure 37.9). Sand grains range from 2,000 mm to 20 μm in diameter. Particles of silt range from 20 to 2 μm in diameter, and clay particles are even smaller. Soils can be classified according to their relative content of coarse and fine materials. For example, soils that contain 45% or more sand and 35% or less clay are classified as sandy soils. The other main types of soil are silt, clay, and loam. **Loam** contains a mixture of sand, silt, and clay and is ideal for the cultivation of most plants.

Because of their size differences, sand, silt, and clay particles confer different properties on soils. The relatively large size and irregular shapes of sand particles allow air and water to move rapidly through sandy soils, which are said to be porous (**Figure 37.10**). Sandy soils are well aerated, which is favorable to the growth of plant roots, because they require oxygen for cellular respiration. However, sandy soils hold less water than the same volume of clay, and rapid percolation of water through sandy soils both reduces the amount of water available to the roots and leaches minerals from the soil.

Silt and clay particles fit closely together, so soils containing larger amounts of these materials are less porous than sandy soils.

Water percolates less easily through silty and clay soils, which therefore retain more ionic mineral nutrients than do sandy soils. Clay particles have negative charges on their surfaces that electrostatically bind positively charged ions (cations) such as NH_4^+, Ca^{2+}, and Fe^{2+} (**Figure 37.11a**). To become available to plants, these cations must be detached from the clay particles. Cations having higher valence numbers (such as Fe^{2+}) are bound more tightly than ions having lower valence numbers. In order to be available to plants, cations must be detached from clay particles. Hydrogen ions (H^+, protons) are able to replace mineral cations on the surfaces of clay particles in a process known as **cation exchange** (**Figure 37.11b**). Cation exchange releases cations to soil water, making them available for uptake by plant roots. However, free ions are also more easily washed out of soil. If the H^+ concentration becomes too high, large numbers of mineral ions are released and can be leached from soil by heavy rainfall. Such leached minerals may include heavy metals such as aluminum (Al^{3+}) that would otherwise be bound in soil. Cation exchange is the mechanism by which acid rain, which adds H^+ to soil, causes loss of soil fertility and the pollution of streams with toxic substances, such as aluminum, that can harm human health.

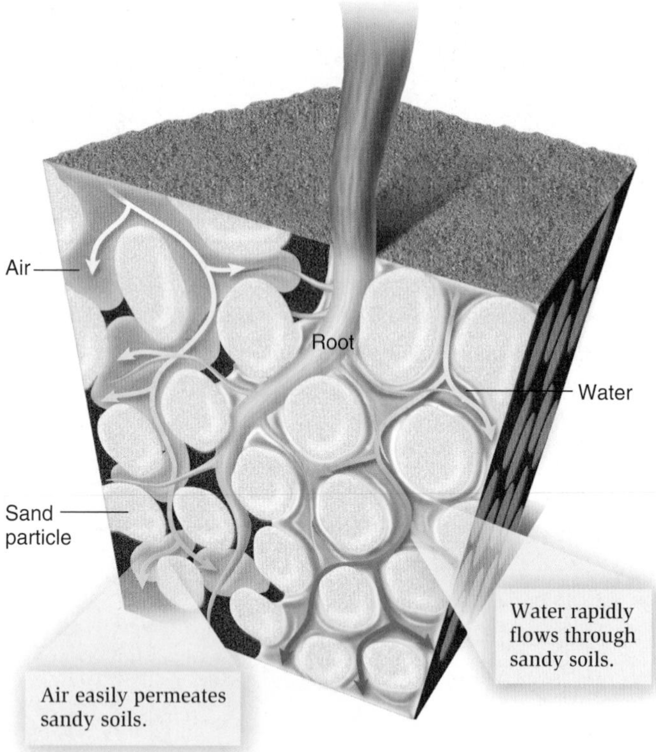

Figure 37.10 **The movement of air and water in sandy soil.**

Concept Check: *In which type of soil is nutrient leaching a more common problem, sandy or clay soils?*

Despite their water- and mineral-retention features, silt- and clay-rich soils may be poorly aerated and therefore unfavorable to root growth. Gardeners often mix organic materials and sand into silt or clay-rich soils to improve their aeration properties. Loam is the preferred soil for agriculture because it combines the aeration provided by sand with the mineral and water retention capacity of silt and clay. Even after achieving a good soil mix, fertilizers may be needed to achieve maximum crop production.

The Role of Fertilizers **Fertilizers** are soil additions that enhance plant growth by providing essential elements. The addition of fertilizer to soils can compensate for deficiencies in soil organic matter or mineral content and thus improve soil fertility. Fertilizers occur in organic and inorganic forms. Organic fertilizers are those in which most of the minerals are bound to organic molecules and are thus released relatively slowly. They play an important role in **organic farming**, the production of crops without the use of inorganic fertilizers, growth substances, and pesticides. Manure and compost are examples of organic fertilizers.

In contrast, inorganic fertilizers consist largely of inorganic minerals, which are immediately useful to plants but can more easily

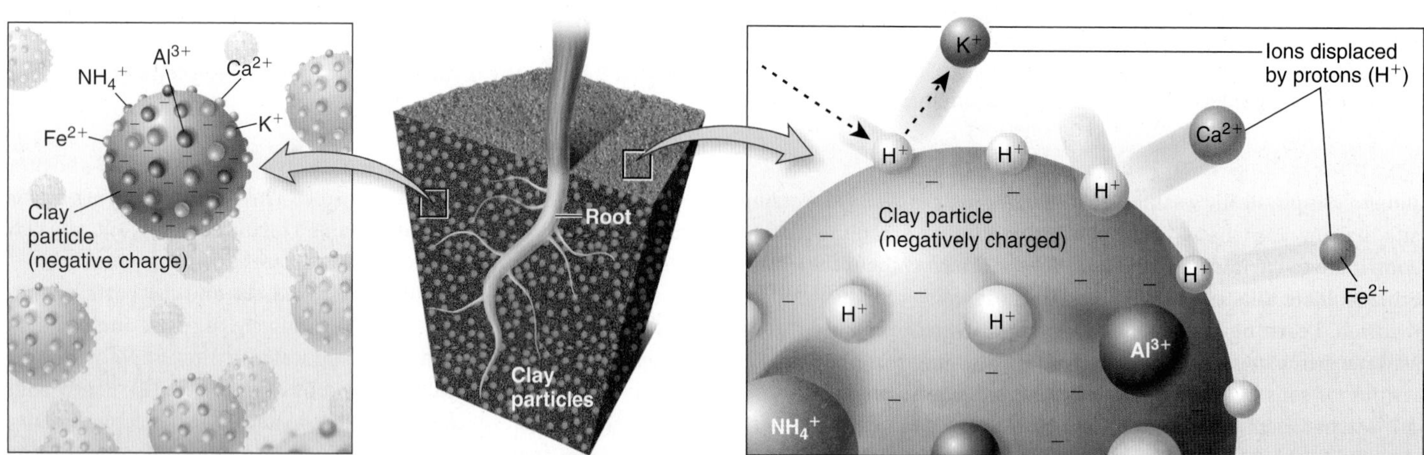

Figure 37.11 **Cation binding and exchange.** **(a)** Clay and organic particles in the soil display negative electrostatic surface charges that bind cations. Bound cations include not only plant mineral nutrients such as ammonium (NH_4^+) but also cations that are not plant nutrients, such as Al^{3+}. **(b)** Cation exchange occurs when protons (H^+) displace other cations, releasing them to soil water. This process makes cations more available for uptake by plant roots, but it also increases the potential for cations to leach away during rains or floods.

leach from soils by heavy rainfall. Nitrogen (N), phosphorus (P), and potassium (K) are the mineral nutrients that most frequently limit crop growth. For this reason, these minerals are the main components of the most common type of commercial inorganic fertilizers. Such fertilizers are available in different ratios of minerals, which are optimal for different types of plants.

Excessive application of fertilizers to fields and lawns is undesirable because minerals not taken up by plant roots are easily washed by rain into waterways, where they can fuel large growths of algae and aquatic plants that can harm other aquatic life-forms. For example, large areas of the Gulf of Mexico and other coastal regions are now known as "dead zones" because microbial decomposition of large algal populations has depleted oxygen from the water, suffocating the animal life. These large populations of algae are fostered by fertilizers that wash from farm soils into rivers, such as the Mississippi River, that drain into coastal oceans. More careful application of fertilizers and planting stream and river edges with vegetation that helps to absorb mineral nutrients can help reduce or prevent dead zones.

Figure 37.12 A soil surface crust that includes nitrogen-fixing, soil-enriching cyanobacteria. Such crusts are widespread in grasslands and other arid regions.

Concept Check: How might soil crusts influence the ecology and economy of regions in which grazing is important?

Plants Require Fixed Nitrogen

Nitrogen is frequently limiting to plant growth in nature and in crop fields, because large amounts of it are required for plant synthesis of amino acids, nucleotides, and alkaloids, among many other cellular constituents. Nitrogen is the largest component of plants by mass after carbon, oxygen, and hydrogen. Although the Earth's atmosphere is 78% nitrogen gas (N_2), plants cannot utilize nitrogen in this form. To be of use to plants, soil nitrogen must occur in a combined form, such as ammonia (NH_3), ammonium ion (NH_4^+), or nitrate ion (NO_3^-) also known as **fixed nitrogen**. Ammonia and its dissolved form—NH_4^+—can be used directly for amino acid production by plants, explaining why ammonia is often applied as a fertilizer to farm fields in springtime. However, in oxygen-rich soils, microorganisms oxidize much of the NH_4^+ to nitrate, so NO_3^- may be the form in which fixed nitrogen enters most plants. Plants use NH_4^+ and NO_3^- to make a wide range of essential organic compounds, including amino acids, nucleic acids, and chlorophyll.

Much of the fixed nitrogen in soils has been recycled from compounds previously utilized by other organisms. Nitrogen flows through the environment in a nitrogen cycle, discussed in Chapter 59 (look ahead to Figure 59.21). New fixed nitrogen can be added to soils by the action of lightning, fire, and air pollution, as well as biological and industrial nitrogen fixation. **Nitrogen fixation** is the process by which atmospheric N_2 is combined with hydrogen to produce NH_3. Most of the fixed nitrogen in soils is produced by **biological nitrogen fixation**, which is performed in nature only by certain prokaryotic organisms. Nitrogen fertilizers applied to crops are produced by **industrial nitrogen fixation**, a human activity.

Biological Nitrogen Fixation by Bacteria Nitrogen-fixing prokaryotic organisms include many types of cyanobacteria, which are photosynthetic organisms that occur in oceans, lakes, and other aquatic systems, as well as in surface soil crusts (Figure 37.12). Various types of nonphotosynthetic bacteria living in water and soil are also able to fix nitrogen. Nitrogen-fixing prokaryotic organisms often excrete a substantial amount of fixed nitrogen, and their death makes still more fixed nitrogen available to plants. Many plants have nitrogen-fixing, prokaryotic symbionts that transfer fixed nitrogen directly to plant cells. Nitrogen-fixation symbioses are so important in nature and in agriculture that they are discussed in more detail in Section 37.3.

All nitrogen-fixing prokaryotic organisms utilize relatively large amounts of ATP and an enzyme known as **nitrogenase** to fix nitrogen (Figure 37.13). This process occurs in three steps. In the first step, a molecule of nitrogen gas (N_2) binds to nitrogenase. In the second step, the bound nitrogen is reduced by the addition of two hydrogen atoms (2 H), a reaction powered by the breakdown of ATP. Such a reduction occurs three times, with the addition of a total of three hydrogen atoms to each nitrogen atom. In a third and final step, two molecules of ammonia (NH_3) are released and dissolve in cell water to form NH_4^+. The nitrogenase enzyme is then free to bind more N_2.

Because the O_2 molecule resembles N_2, oxygen can bind to the active site of nitrogenase. Oxygen-binding disables nitrogenase, thereby stopping nitrogen fixation. Many of the genes involved in prokaryotic nitrogen fixation are known, and crop scientists are working to genetically engineer nitrogen fixation capacity into crop plants such as rice and maize (corn). However, the vulnerability of nitrogenase to oxygen means that this enzyme may need to be altered so that it binds oxygen less readily, or plants must be engineered with some mechanism that protects nitrogenase from oxygen.

Industrial Nitrogen Fixation Worldwide, farmers apply more than 80 million metric tons of nitrogen fertilizer per year. The fixed nitrogen found in fertilizer is produced industrially from N_2 by means of a procedure invented by German chemists Fritz Haber and Carl Bosch in 1909. The reduction of N_2 gas to NH_3 is energetically favorable at room temperature, but the activation energy is very high, so the reaction occurs extremely slowly. Using an iron catalyst, temperatures of 400–650°C (752–1,202°F), and high pressures (150–400 atmospheres), the Haber-Bosch process generates NH_3 rapidly. However,

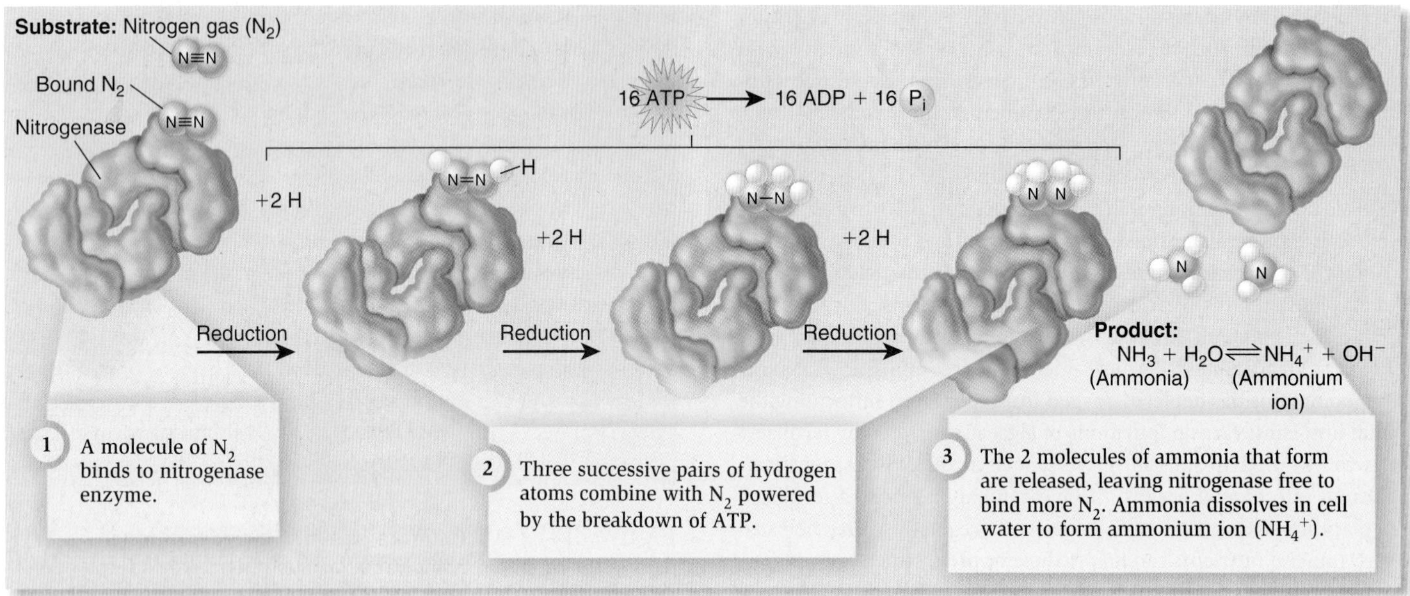

Figure 37.13 **The biological process of nitrogen fixation.**

Concept Check: *What common substance inactivates nitrogenase enzyme by binding to its active site?*

because of its high energy requirements, industrial nitrogen fixation can be costly from the perspective of many of the world's farmers. The high cost of fertilizers helps to explain why agricultural scientists are so interested in the possibility of genetically engineering nitrogen fixation into crop plants.

Plants Display Adaptations for Acquiring Phosphorus

Phosphorus (P) is another soil mineral that often limits plant growth. Plants obtain phosphorus from the ion known as phosphate (PO_4^{3-}), which occurs in the soil in three dissolved forms: H_3PO_4, $H_2PO_4^-$, and HPO_4^{2-}. HPO_4^{2-} (called hydrogen phosphate ion) is the form most commonly absorbed by plants. Phosphate uptake involves ATP-requiring proton cotransport, as is the case for other anions such as nitrate (NO_3^-) and sulfate (SO_4^{2-}). The uptake of these nutrients at the root hair surface is discussed further in Chapter 38.

Although phosphorus can be abundant in soil, it is often unavailable to plants. One reason is that PO_4^{3-} forms tightly bound complexes with clay, iron and aluminum oxides, and calcium carbonate in soils. In addition, soil microbes convert PO_4^{3-} into organic compounds that are not taken up by plants.

Because plants need large supplies of phosphorus for a variety of cell processes, they have evolved various adaptations that increase their ability to obtain PO_4^{3-} from soil. A common adaptation for acquiring PO_4^{3-} is the symbiotic association of plant roots with various types of fungi (see Section 37.3). In addition, plants that grow in soils having low phosphorous content may produce more highly branched roots and more and longer root hairs. Plant roots also secrete protons and organic acids such as citrate and malate into the soil, which help release phosphorus from inorganic complexes. For example, the plant *Lupinus alba* releases as much as 25% of its total photosynthetic carbon into the soil as organic acids—a high price to pay but apparently one that is essential for the plant to obtain sufficient phosphorus. Plants may also secrete phosphatase enzymes from roots. These enzymes release phosphorus from organic compounds in the soil. The general process by which P, N, CO_2, and other minerals are released from organic compounds is called **mineralization**.

Farmers and gardeners apply phosphate-rich fertilizers to crop fields and gardens as a way of preventing phosphorous deficiencies, which reduce yields. Phosphorous fertilizers are obtained from phosphate-rich mineral deposits, but experts have warned that inexpensive sources of PO_4^{3-} will be exhausted within the next 90 years. Consequently, there is much interest in devising ways to maximize the efficiency by which plants are able to take up and use phosphorus. Genetic engineering to produce "smart plants" that can sense the levels of nutrients in the soil may offer some options.

FEATURE INVESTIGATION

Hammond and Colleagues Engineered Smart Plants That Can Communicate Their Phosphate Needs

If farmers could apply fertilizer to crops in the precise amounts needed by plants, not only would farmers save money, but also less fertilizer would run off fields into aquatic habitats, where it can lead to harmful ecological effects. Plant biologists have used genetic engineering to produce smart plants that signal impending nutrient deficiency via a visible marker. Such plants could serve as sentinels, warning farmers of the conditions of an entire field. With this information, farmers could apply just enough mineral nutrients to prevent deficiency, thereby avoiding overapplication of fertilizers.

In 2003, working with the model plant *Arabidopsis*, John Hammond, Philip White, and their associates grew plants hydroponically, which means their roots were in a water solution rather than soil. They identified some of the genes whose expression changes when plants are transferred from nutrient solutions containing sufficient phosphorus to solutions lacking phosphorus. They found that some genes were turned on quickly after phosphorus removal, but other genes took much longer, up to 100 hours or more. This timing is important because genes expressed between 24 and 72 hours after PO_4^{3-} removal are considered useful as phosphorous monitors. During this window of time, plant tissue levels of PO_4^{3-} decreased but had not yet affected plant growth. One gene that met this timing criterion was *SQD1*, which is required for the synthesis of sulfur-containing lipids. Expression of *SQD1* allows plants to respond to low phosphorous levels by replacing plastid phospholipids with sulfur-containing lipids, thereby reducing their phosphorous requirement. This evidence suggested that smart plants could be engineered to communicate impending PO_4^{3-} deficiency when they express *SQD1*.

To make smart plants, the researchers first placed the reporter gene *GUS* under the control of the *SQD1* promoter and transformed this gene into *Arabidopsis* plants (**Figure 37.14**). The researchers then grew these genetically engineered plants for various time periods in hydroponic solutions of differing PO_4^{3-} levels. After different time periods, they removed leaves and chemically treated them with a compound that produces a light blue color when the *GUS* gene is expressed. Some leaves were removed before transfer to the phosphate-deficient solution and served as controls. Because the *GUS* gene was under the control of the *SQD1* promoter, leaves from plants that were developing PO_4^{3-} deficiency turned blue! These smart

Figure 37.14 **The experiment of Hammond and colleagues showed that plants can be engineered to communicate changes in the level of nutrients.**

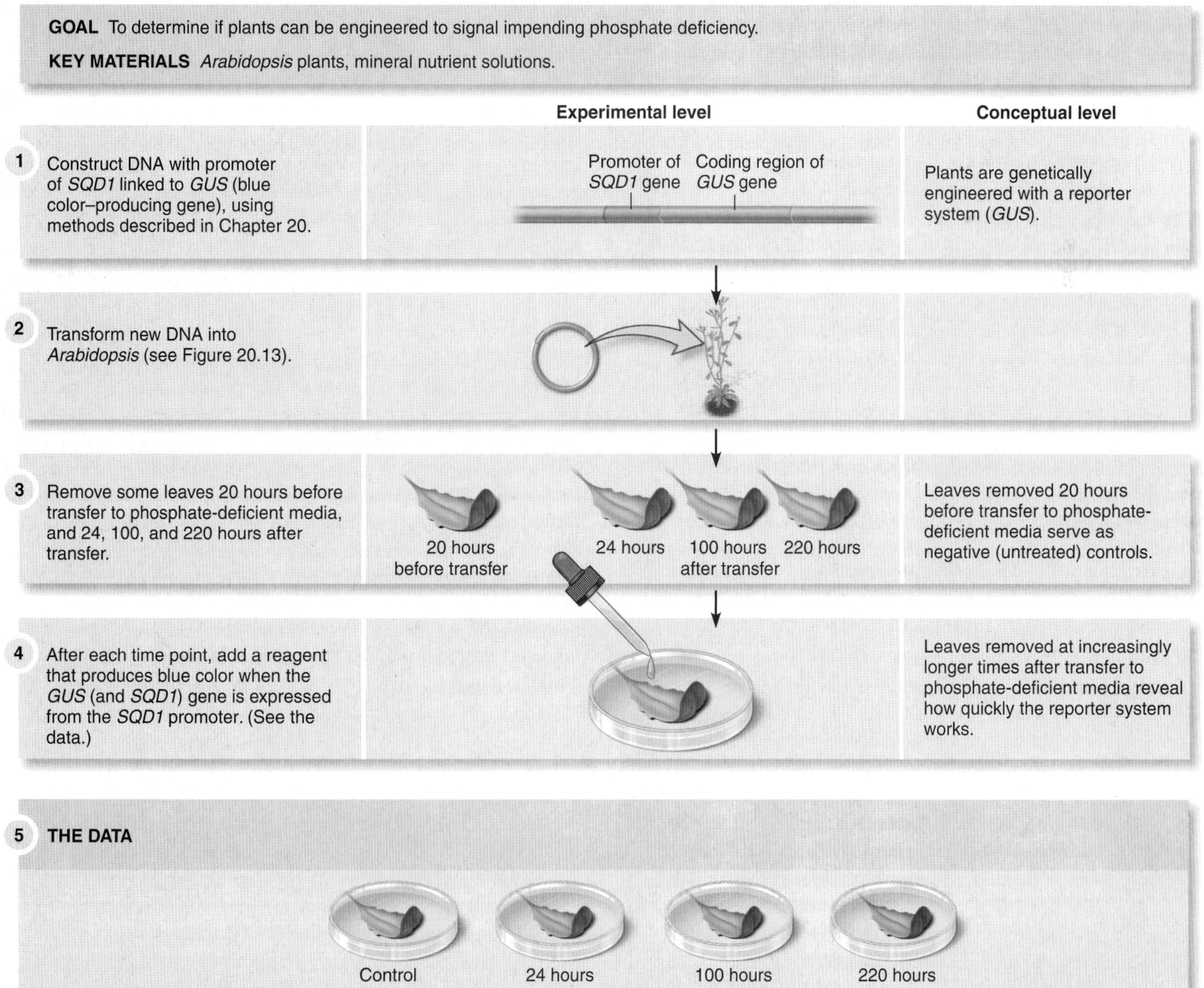

GOAL To determine if plants can be engineered to signal impending phosphate deficiency.
KEY MATERIALS Arabidopsis plants, mineral nutrient solutions.
Experimental level
Conceptual level
1 Construct DNA with promoter of SQD1 linked to GUS (blue color–producing gene), using methods described in Chapter 20.
Promoter of SQD1 gene
Coding region of GUS gene
Plants are genetically engineered with a reporter system (GUS).
2 Transform new DNA into Arabidopsis (see Figure 20.13).
3 Remove some leaves 20 hours before transfer to phosphate-deficient media, and 24, 100, and 220 hours after transfer.
20 hours before transfer
24 hours
100 hours
220 hours
after transfer
Leaves removed 20 hours before transfer to phosphate-deficient media serve as negative (untreated) controls.
4 After each time point, add a reagent that produces blue color when the GUS (and SQD1) gene is expressed from the SQD1 promoter. (See the data.)
Leaves removed at increasingly longer times after transfer to phosphate-deficient media reveal how quickly the reporter system works.
5 THE DATA
Control
24 hours
100 hours
220 hours

6 **CONCLUSION** Plants can be genetically engineered to express color signals in time for farmers to apply fertilizer sufficient to prevent nutrient deficiency.

7 **SOURCE** Hammond, John P. et al. June 2003. Changes in gene expression in *Arabidopsis* shoots during phosphate starvation and the potential for developing smart plants. *Plant Physiology* 132:578–596.

plants were able to communicate impending phosphorous deficiency in time for a farmer to apply fertilizer.

In a later and more extensive study of gene expression in *Arabidopsis*, other investigators discovered that PO_4^{3-} induces 612 genes and represses 254 genes. Some of these genes may encode proteins useful in monitoring plant phosphorous status. If smart plant technology can be developed for crop plants, farmers may be able to monitor and fertilize fields with much greater precision.

Experimental Questions

1. Why did Hammond and colleagues seek to identify genes whose expression changed between 24 and 72 hours after plants experience phosphorous limitation?
2. What advantage do plants obtain when the *SQD1* gene is expressed?
3. How were the investigators able to identify potential sentinel plants that were starting to experience phosphorous deficiency?

37.3 Biological Sources of Plant Nutrients

Learning Outcomes:

1. Explain the importance of mycorrhizal fungi to plants.
2. List the major types of prokaryotic organisms that occur in symbioses with plants, thereby helping them to acquire nitrogen.
3. Distinguish carnivorous plants from parasitic plants, giving examples.

This section focuses on several fascinating ways in which plants use other organisms as sources of nutrients. Biological sources of plant nutrients include symbiotic fungi or bacteria, the animal prey of carnivorous plants, and green plants that serve as hosts for nonphotosynthetic plant parasites.

Mycorrhizal Associations Help Most Plants Obtain Mineral Nutrients

At least 80% of seed plants have symbiotic associations with fungi that live within the tissues of plant roots or that envelop root surfaces (Figures 31.26–31.28). These associations are termed **mycorrhizae**; the prefix *myco* refers to fungi, and *rhiza* means root, so the term literally means "fungus root."

In mycorrhizal associations, soil fungi obtain organic food from the roots of a photosynthetic plant host, while the fungal partner supplies the plant with water and mineral nutrients. Due to the extensive mycelia that fungi produce within the soil, these fungal-root associations provide an exceptionally efficient way for plants to harvest water and minerals, especially phosphate, from a much larger volume of soil than is available to roots by themselves. The presence of lush vegetation on thin, infertile tropical rain forest soils is largely due to the ability of mycorrhizae to rapidly absorb mineral nutrients released by decaying organisms and transmit the nutrients directly to plant roots (**Figure 37.15**). In many tropical rain forests, mineral nutrients occur within the bodies of living organisms, rather than accumulating in the soil where they could easily be leached away by heavy, frequent rains.

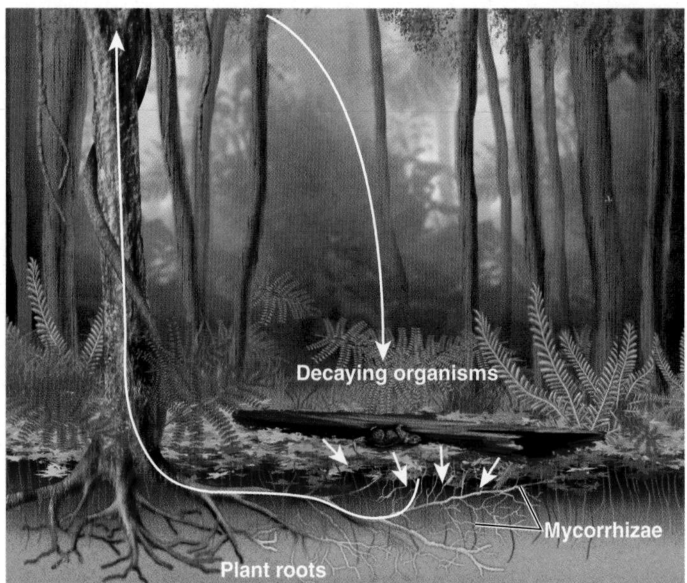

Figure 37.15 **Nutrient acquisition via mycorrhizae.**

BIOLOGY PRINCIPLE New properties emerge from complex interactions. In all forests, but particularly those of tropical regions, mycorrhizal fungi rapidly collect soil minerals released from decaying organisms and transport them directly to plant roots. Such efficient nutrient cycling bypasses the soil, from which mineral ions can be easily leached by heavy rainfall. This process explains how lush forests can grow on thin, infertile soils.

Various species of ghostly pale plants have lost their photosynthetic pigments (see Figure 37.2) and have thus become dependent on organic compounds supplied by fungi that form mycorrhizal associations with a photosynthetic host, such as a nearby tree. In this process, known as mycoheterotrophy, the fungus serves as an underground conduit for the flow of organic nutrients from a green, photosynthetic plant to a heterotrophic plant. Many plant seedlings that grow in the shade of taller plants also use mycoheterotrophy to survive until they are able to obtain enough light for photosynthesis.

Plant-Bacterial Symbioses Provide Some Plants with Fixed Nitrogen

Some kinds of plants have symbiotic relationships with bacteria that provide them with fixed nitrogen. Though many nitrogen-fixing bacteria live freely within the soil, some form nitrogen-fixing partnerships with plants, actually living within plant cells or tissues. Such symbioses are advantageous to both partners. The plants provide organic nutrients to the bacteria, and the bacteria supply the plants with a much higher supply of fixed nitrogen than the plants could obtain from most soils. Representatives of three types of bacteria—cyanobacteria, actinobacteria, and proteobacteria—are symbiotically associated with specific types of plants. (For more information about the characteristics of these bacterial groups, see Chapter 27.)

Plant-Cyanobacteria Symbioses Although cyanobacteria are themselves photosynthetic, organic compounds supplied by plant partners subsidize the high energy costs of nitrogen fixation. This allows the cyanobacteria to fix more nitrogen than they require, secreting the excess to plant partners. Nitrogen-fixing cyanobacteria form symbioses with some bryophytes, ferns, and gymnosperms, as well as the flowering plant *Gunnera*. This plant, commonly known as the giant rhubarb or prickly rhubarb, can produce leaves almost 3 m across (**Figure 37.16**). Nitrogen-fixing symbionts are advantageous

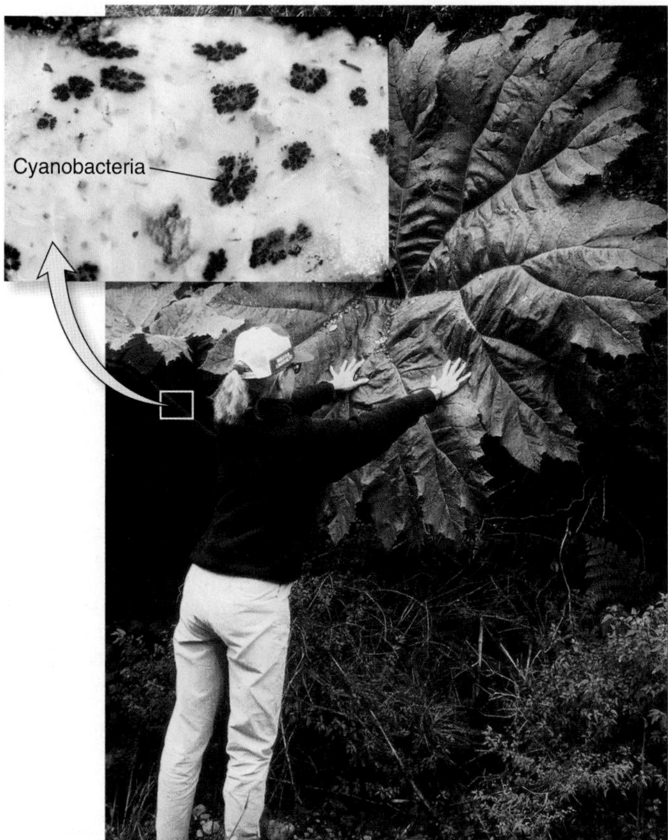

Figure 37.16 ***Gunnera*** **growing on nitrogen-poor soil.** Nitrogen-fixing cyanobacteria that live within cavities in this plant's leaf petioles provide the plant with fixed nitrogen, which explains how such a large plant can grow on infertile soils.

to *Gunnera* because this large plant grows in nitrogen-poor habitats, such as volcanic slopes in Hawaii. *Gunnera* harbors cyanobacteria within stems and leaf petioles. In these locations, the cyanobacteria can use cyclic electron flow to transform light energy into ATP, needed to produce fixed nitrogen. The presence of nitrogen-fixing cyanobacteria helps to explain why *Gunnera* can grow to dramatic size on poor soils.

Woody Plant/Actinobacteria Symbioses In contrast to cyanobacteria, actinobacteria are heterotrophic nitrogen-fixing bacteria. Actinobacteria known as *Frankia* occur in nodules formed on the underground roots of certain shrubs or trees, such as alder (*Alnus*) and myrtle (*Myrica*). These plants receive fixed nitrogen from their bacterial partners, which, in turn, obtain organic nutrients. Woody plants, such as *Ceanothus* shrubs, that have *Frankia* symbionts are able to grow abundantly even in places where soil nitrogen is low. This symbiosis helps to explain why *Ceanothus* covers extensive areas in mountainous regions of the western U.S.

Legume-Rhizobia Symbioses The nitrogen-fixation symbioses most important in nature and to agriculture involve certain proteobacteria that are collectively known as **rhizobia** (from the Greek *rhiza*, meaning root). Rhizobia live within root cells of wild and cultivated legumes, forming legume-rhizobia symbioses. In nature, legume plants are important sources of fixed nitrogen for other plants. When legumes die, they generate soil organic matter that is enriched with fixed nitrogen. Consequently, wild legumes are regarded as particularly valuable members of natural plant communities.

Important legume crops include soybeans, peas, beans, peanuts, clover, and alfalfa. Foods produced from soybeans, peas, beans, and peanuts are valued for their high protein content. Clover and alfalfa are used for animal food and to enrich fields with the fixed nitrogen needed by subsequent food crops. The value of these crops arises from their fixed-nitrogen content. The amount of ammonia produced by legume-rhizobia symbioses nearly equals the world's entire industrial production.

GENOMES & PROTEOMES CONNECTION

Development of Legume-Rhizobia Symbioses

Rhizobia can live independently in the soil, but they fix nitrogen only when they occur within lumpy **nodules** that form on legume roots (**Figure 37.17**). Different species of rhizobia preferentially form symbioses with particular plant species. Because of their agricultural importance, these legume–rhizobia symbioses have been extensively studied, and a great deal is now known about the molecular basis of their development. This information is potentially useful in efforts to genetically engineer nitrogen-fixation capacity into nonlegume crops.

Nodule development involves a series of chemical signals sent back and forth between rhizobia and their host plants (**Figure 37.18**). Legumes start this exchange by secreting particular flavonoid compounds from their roots. Recall that flavonoids are phenolic secondary metabolites that play essential roles in plant structure, reproduction, and protection (see Figure 30.22b). These flavonoids

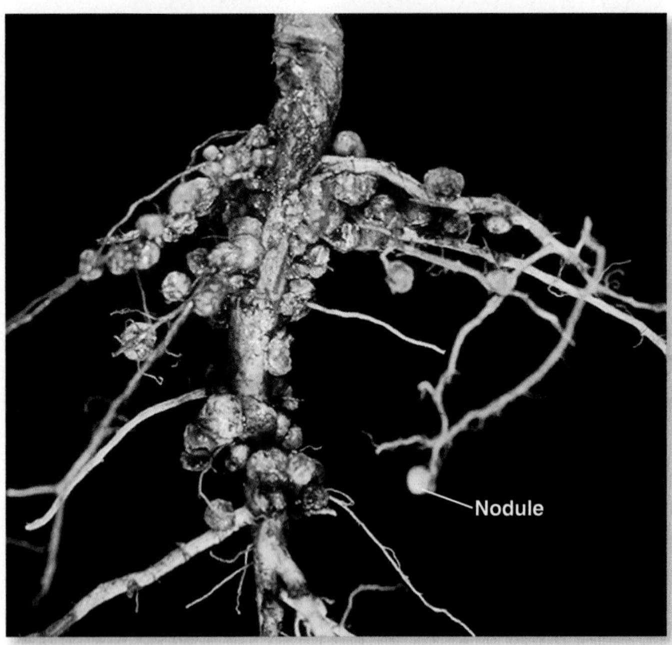

Figure 37.17 **Legume root nodules.** The cells of nodules on the roots of this soybean plant (*Glycine max*) and other legumes contain nitrogen-fixing bacteria known as rhizobia.

bind to receptors in the plasma membranes of compatible soil rhizobia (Figure 37.18, step 1). In response, the rhizobia typically secrete **Nod factors** (nodulation factors). Each rhizobial species produces Nod factors with distinctive structural variations that can be recognized by the preferred host species. These Nod factors function something like keys that unlock doors, allowing bacteria to enter roots via root hairs. These factors bind to receptors in the membranes of root hair cells in the host plant (step 2).

Within minutes after its membrane receptors bind Nod factors, the root hair plasma membrane allows an influx of calcium ions, and a few minutes later, root hair calcium concentrations start oscillating rapidly. Root hairs respond to these calcium changes by swelling at their tips and curling around the rhizobia (step 3). The rhizobia then inject infection proteins into root hairs. In response, the cell wall at the root hair tip changes in a way that allows bacterial enzymes to erode a small hole in the wall, allowing bacterial cells to enter. The

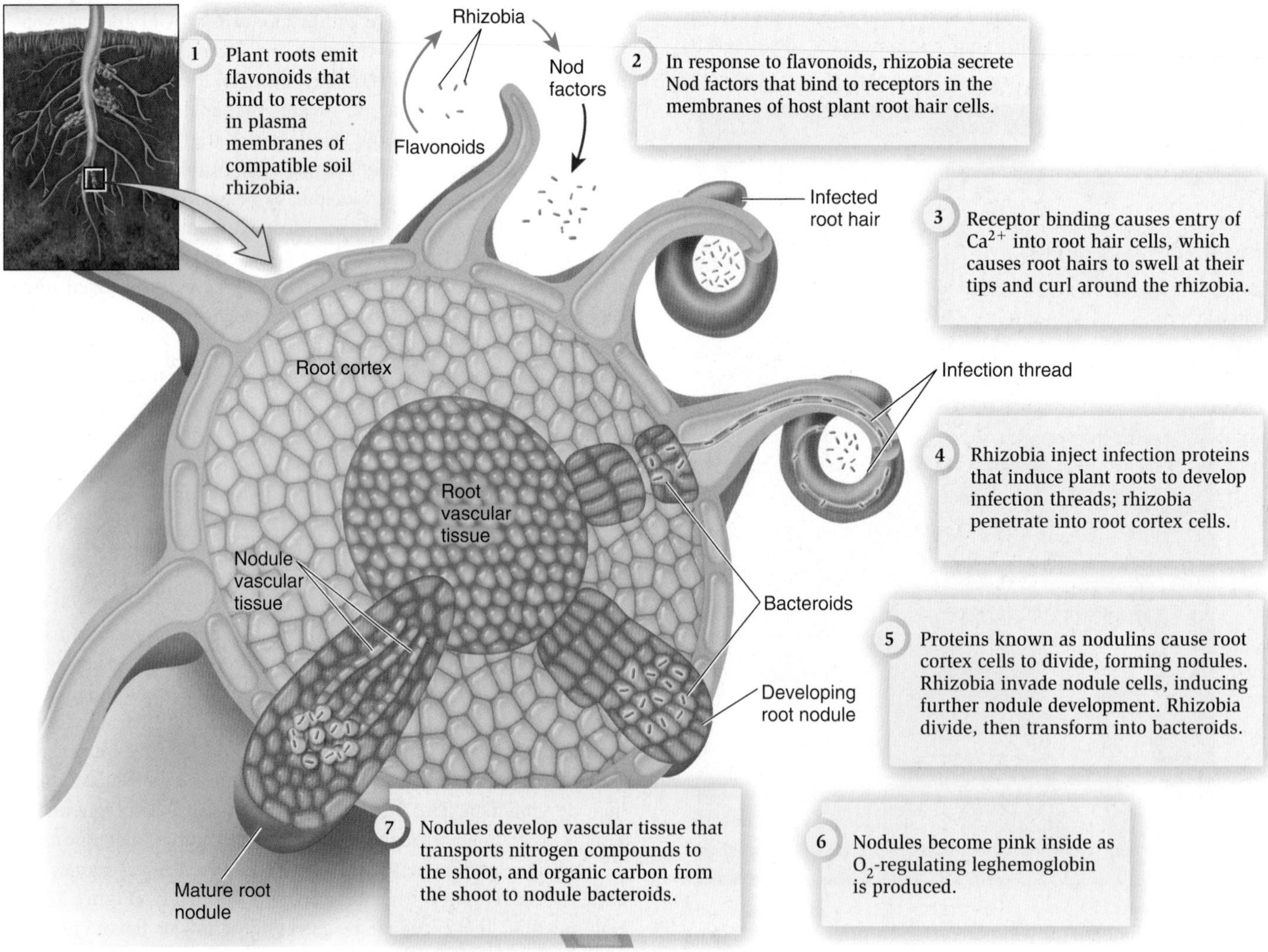

1 Plant roots emit flavonoids that bind to receptors in plasma membranes of compatible soil rhizobia.

Rhizobia

Flavonoids

Nod factors

2 In response to flavonoids, rhizobia secrete Nod factors that bind to receptors in the membranes of host plant root hair cells.

Infected root hair

3 Receptor binding causes entry of Ca^{2+} into root hair cells, which causes root hairs to swell at their tips and curl around the rhizobia.

Root cortex

Infection thread

4 Rhizobia inject infection proteins that induce plant roots to develop infection threads; rhizobia penetrate into root cortex cells.

Root vascular tissue

Nodule vascular tissue

Bacteroids

5 Proteins known as nodulins cause root cortex cells to divide, forming nodules. Rhizobia invade nodule cells, inducing further nodule development. Rhizobia divide, then transform into bacteroids.

Developing root nodule

Mature root nodule

7 Nodules develop vascular tissue that transports nitrogen compounds to the shoot, and organic carbon from the shoot to nodule bacteroids.

6 Nodules become pink inside as O_2-regulating leghemoglobin is produced.

Figure 37.18 **Root nodule development.** The process of root nodule development involves a chemical conversation between the legume plant and nitrogen-fixing bacteria.

plasma membrane forms a tubular infection thread through which rhizobia move into the root cortex. The tip of the infection thread fuses with the plasma membrane of a cortex cell, then the rhizobia are released into the cortex cell cytoplasm, each bacterial cell enclosed by the host membrane (step 4).

Meanwhile, plants produce proteins known as **nodulins** that foster nodule development. Within 18–30 hours after the initial infection, root cortex cells start to divide to form root nodules. Environmental conditions in developing nodules cause rhizobia to undergo changes in their structure and gene expression patterns. These modified rhizobia are known as **bacteroids** (step 5). Bacteroid respiration provides the large amounts of ATP that are necessary for nitrogen fixation.

Legume nodules typically produce **leghemoglobin** (legume hemoglobin), a pink protein that helps to regulate local oxygen concentrations, transporting enough oxygen to bacteroids to support respiration but preventing oxygen from disabling nitrogenase (step 6). Mature nodules also produce vascular tissue that moves nitrogen fixed by bacteroids to the root vascular system for transport throughout the plant. These nodule vascular tissues also supply organic food produced by the legume to their bacteroid partners (step 7). Gene expression studies have revealed that nearly 5,000 gene expression changes are associated with the legume-rhizobia symbiosis.

Carnivorous Plants Are Autotrophs That Obtain Mineral Nutrients from Animals

About 600 species of flowering plants have adapted to low-nitrogen environments by evolving mechanisms for trapping and digesting animals and are therefore known as carnivorous plants. Their leaves are modified in ways that allow them to capture animal prey, primarily insects, though larger animals are sometimes snared as well. (Despite the popular play and movie *Little Shop of Horrors*, there are no wild or cultivated carnivorous plants that, like Audrey II, are large enough to consume humans!) Carnivorous plants are photosynthetic autotrophs that supply their own organic compounds; prey animals are primarily sources of nitrogen. The experimental use of radioactively labeled prey insects has revealed that carnivorous plants obtain as much as 87% of their nitrogen from animals.

The trapping mechanisms used by carnivorous plants are classified as passive or active. Plants with passive trapping mechanisms depend on the prey to fall or wander into the trap. For example, tropical pitcher plants (genus *Nepenthes*) have leaves that are folded and partially fused to form tubes that collect rainwater (**Figure 37.19a**). The interior walls of these pitchers are slippery and have downward-pointing hairs. Insects and other small animals such as lizards and frogs that fall into the pitchers are unable to climb out. Eventually, the trapped animals drown and are digested by microbes living within the pitchers.

Plants with active mechanisms, such as Venus flytraps and sundews (see chapter-opening photo), have traps that are stimulated by touch. Charles Darwin, who was fascinated by carnivorous plants, was one of the first to study the trapping mechanisms of Venus flytraps and sundews. The Venus flytrap (*Dionaea muscipula*) has an active trap formed by two-lobed leaves that are edged with lance-shaped teeth (**Figure 37.19b**). The leaf surface has glands that secrete carbohydrates, which lure prey, as well as glands that secrete digestive enzymes after prey has been trapped. Also present on leaf surfaces are modified hairs, usually three per leaf lobe. If a single hair is touched—perhaps by wind or rain, or debris—and another touch does not occur soon thereafter, nothing happens. But when a fly or similar prey lands on the leaf and brushes against the same hair twice, or touches a second hair within 20–40 seconds, the leaf lobes snap shut around it.

Experimental studies indicate that an action potential—much like that occurring during signal transmission in animal nerves—develops in the stimulated hairs. The plant electrical signal then travels from cell to cell along plasma membranes, via plasmodesmata, at about 10 cm/sec. This signal causes leaf cells to take up ions and water so that the leaf enlarges and changes shape, springing the trap. Digestion of the prey is typically finished within 10 days, whereupon the trap may reopen. Trap leaves can go through three or four digestive cycles during their lifetime.

Sundews (such as *Drosera rotundifolia*) have leaves bearing glandular hairs whose sticky tips glisten in the sunlight. Insects that land on sundew leaves get mired in the sticky mucilage exuded by these hairs, as shown in the chapter-opening photo. As the insects struggle to get away, they become covered with more mucilage and eventually

(a) Pitcher plant (genus *Nepenthes*)

(b) Venus flytrap (*Dionaea muscipula*)

Figure 37.19 Carnivorous plants. **(a)** A pitcher plant passively captures animals that accidentally fall into its water-filled pitcher; **(b)** The Venus flytrap has an active trap that is stimulated by the touch of its prey, in this case, a fly.

smother as their breathing pores become clogged. Darwin discovered that sundew leaves bend after being touched and that glandular hairs not originally in contact with the insects also bend, folding over the prey as you would fold your fingers over an object in your palm. Later, investigators discovered that this bending involves the plant hormone auxin. In response to touch, auxin accumulates in sundew leaf tips and then flows downward, stimulating the cell expansion that causes the leaf bending. The glandular hairs also produce enzymes that digest the prey.

Parasitic Plants Obtain Nutrients from Photosynthetic Plants

More than 4,500 species of plants live as complete or partial **parasites**, organisms that obtain all or much of their water, minerals, and organic compounds from another organism. Dodder and witchweed are prominent examples of plants that are completely parasitic.

Dodder (*Cuscuta pentagona*) lacks roots and does not grow from the soil. Instead, all of the 150 species of this parasite live aboveground (**Figure 37.20**). These parasites twine their yellow or orange stems around green plant hosts, into which they sink peg-shaped, absorptive structures known as haustoria. These haustoria tap into the host plant's vascular system, gaining water, minerals, and sugar, which the parasite uses for growth and reproduction. The long, flexible stems of dodder often loop from one plant to another, such that an individual dodder plant can tap into many different plants at the same time. Dodder reproduces very rapidly by means of broken-off stem fragments and seeds. A single dodder plant can produce more than 16,000 seeds. In consequence, dodder is a widespread agricultural pest that attacks citrus, tomatoes, and many other fruit, vegetable, forage, and flower crops.

Another group of parasitic plants, the witchweeds (genus *Striga*), are serious problems for agriculture worldwide, because these parasites attack major cereal crops: corn, sorghum, rice, and millet. Witchweed seeds lie dormant in soil until secretions from host plant roots stimulate their germination. Genetic engineers are working to find ways to protect crops from the debilitating effects of these crop parasites.

Figure 37.20 A parasitic plant. Dodder (*Cuscuta pentagona*) is an example of a parasitic plant that obtains all of its water, minerals, and organic compounds from one or more green plant hosts.

BioConnections: Look back to Figure 33.8b to see an example of a parasitic animal—a tapeworm—and Table 33.2, which describes characteristics of tapeworms. What is similar about food acquisition in dodder and tapeworms?

Summary of Key Concepts

37.1 Plant Nutritional Requirements

- The nutritional requirements of green plants are light energy, CO_2, H_2O, and several types of elements absorbed from soil or water (Figure 37.1, Table 37.1).
- Like animals and fungi, heterotrophic plants obtain chemical energy by metabolizing organic compounds absorbed from their environment (Figure 37.2).
- Plants display many adaptations that allow them to cope with insufficient or excess amounts of light and inadequate CO_2 (Figures 37.3, 37.4, 37.5).
- Nutrients can limit plant growth and cause nutrient-deficiency symptoms (Figure 37.6).

37.2 The Role of Soil in Plant Nutrition

- Natural soils display layers known as soil horizons. Soils are composed of organic material and inorganic minerals. Soil organic matter is largely derived from plant detritus, animal wastes, and decayed animal bodies (Figures 37.7, 37.8, 37.9).
- Inorganic soil components occur as particles that can be categorized according to their size as sand, silt, or clay. Cation exchange releases cations to soil water, making cations available for uptake by plant roots (Figures 37.9, 37.10, 37.11).
- Biological or industrial processes can convert atmospheric nitrogen gas into fixed nitrogen that plants can utilize. Biological nitrogen fixation can be performed only by certain prokaryotes (Figures 37.12, 37.13).
- Plants display several types of adaptations to cope with phosphate deficiency. Genetically modified smart plants can signal impending phosphate deficiency (Figure 37.14).

37.3 Biological Sources of Plant Nutrients

- Mycorrhizal fungi, which are associated with the roots of most plants, provide plants with water, phosphorus, and other minerals (Figure 37.15).
- Nitrogen-fixing prokaryotes living within the tissues of some plants provide them with fixed nitrogen. Legume-rhizobia associations are

particularly important in nature and in agriculture (Figures 37.16, 37.17, 37.18).

- Carnivorous plants obtain mineral nutrients from the digested bodies of trapped animals. Parasitic plants obtain water, mineral ions, and organic compounds from green plant hosts (Figures 37.19, 37.20).

Assess and Discuss

Test Yourself

1. Which of the following substances can limit plant growth in nature?
   a. sunlight
   b. water
   c. carbon dioxide
   d. fixed nitrogen
   e. all of the above
2. In what form do plants take up most soil minerals?
   a. as ions dissolved in water
   b. as neutral salts
   c. as mineral-clay complexes
   d. linked to particles of organic carbon
   e. none of the above
3. Why do plants need sulfur?
   a. for the construction of cell walls
   b. as an essential component of chlorophyll
   c. to produce proteins and some coenzymes
   d. all of the above
   e. none of the above
4. Soil organic matter provides the benefit of:
   a. allowing water to percolate rapidly through soil.
   b. making soil softer in consistency.
   c. increasing the aluminum content of soil.
   d. causing minerals to be leached more rapidly from soil.
   e. none of the above.
5. Which environments are conducive to heavy leaching of minerals from soils?
   a. those having soils that are composed primarily of sand particles
   b. those having acidic soils
   c. those impacted by acid rain
   d. regions characterized by heavy rainfall
   e. all of the above
6. Which property is *not* characteristic of clay-rich soils?
   a. high mineral nutrient retention
   b. high water retention
   c. high aeration
   d. lower amounts of sand than clay
   e. all of the above
7. Which of the plants listed below is most likely to be heterotrophic (to obtain organic food from the environment)?
   a. a green houseplant
   b. a legume plant such as bean
   c. a carnivorous plant such as the sundew
   d. ghostly white *Monotropa*
   e. none of the above
8. What kinds of organisms occur in nitrogen-fixing symbioses with plants?
   a. cyanobacteria
   b. actinobacteria
   c. rhizobia bacteria
   d. all of the above
   e. none of the above
9. How do legume roots attract rhizobia?
   a. They secrete flavonoids.
   b. They secrete carotenoids.
   c. They secrete alkaloids.
   d. They secrete Nod factors.
   e. none of the above
10. Which plant uses a passive trap to obtain animal prey as a source of mineral nutrients?
    a. the Indian pipe (*Monotropa uniflora*)
    b. the tropical pitcher plant (*Nepenthes* spp.)
    c. the Venus flytrap (*Dionaea muscipula*)
    d. dodder (*Cuscuta* spp.)
    e. all of the above

Conceptual Questions

1. Why are agricultural experts and ecologists alike concerned about overfertilization of crop fields?
2. Draw a diagram showing how rhizobia and legume roots communicate chemically during nodule formation.

Collaborative Questions

1. Imagine that you have bought a farm and want to start growing a crop to sell at a local market. How could you go about determining if the soil needs to be fertilized and with what mineral nutrients?
2. A principle of biology is that *biology affects our society.* Imagine that you own a large farm with a trout stream running through it. How would you protect the water quality of the stream?

Online Resource

www.brookerbiology.com

Stay a step ahead in your studies with animations that bring concepts to life and practice tests to assess your understanding. Your instructor may also recommend the interactive eBook, individualized learning tools, and more.

Flowering Plants: Transport

38

Chapter Outline

38.1 Overview of Plant Transport
38.2 Uptake and Movement of Materials at the Cellular Level
38.3 Tissue-Level Transport
38.4 Long-Distance Transport
Summary of Key Concepts
Assess and Discuss

A shade tree. The evaporation of water from plant leaves cools them and us, and even affects local and global climate.

On hot days, people naturally gravitate to the cool shade beneath trees, as shown in the chapter-opening photo. But most people do not realize that trees are not only Sun umbrellas. Plants actually cool the air around them as water evaporates from their surfaces. That's why grass feels cool when you walk barefoot on it, even on a hot day.

Plants benefit from this evaporation process—known as transpiration—because it cools their surfaces and enables the movement of a continuous stream of water from the soil, through roots and stems, to leaves. This evaporation process not only helps to distribute water throughout the plant body, it also aids the movement of dissolved minerals and organic compounds such as sugars, hormones, and other organic materials over long distances within plants. Transport is therefore crucial for the functions of plant growth, behavior, and nutrition, which were described in the preceding three chapters.

In addition, plant transport plays a critical role in Earth's global climate. On a worldwide basis, plant transpiration annually moves 32×10^3 billion tons of water from the soil into the atmosphere as water vapor. Plant transpiration provides important ecological services. For example, plant-produced atmospheric water vapor is the source of 30% of rain, the rest originating by evaporation from the surfaces of oceans and freshwater bodies. Along with other atmospheric gases (including carbon dioxide and methane), water vapor also works as a greenhouse gas that helps to warm Earth's climate by absorbing the sun's heat. Plant transport processes are also relevant to agriculture, as humans seek to improve crop productivity and the efficiency of water and nutrient use. Conservation biologists understand that plant transport processes are important to the preservation and restoration of natural environments.

To comprehend plant transport more fully, in this chapter, we first survey the materials that move through plants and the general directions of material movements. Next, we will focus more closely on water and solute uptake by plant cells. We will then examine how these materials are moved within plants over short and long distances and explore some of the plant adaptations that allow such transport to be as efficient as possible in a variety of environments. In the process, we will learn why plants are so cool!

38.1 Overview of Plant Transport

Learning Outcomes:

1. Describe how plants transport water and minerals from roots to leaves.
2. Describe how plants transport organic molecules from leaves to nonphotosynthetic parts.

Our previous surveys of angiosperm plant structure, behavior, and nutrition described the interdependence of plant root and shoot systems (Chapters 35–37). We have observed that in most plants, the root system absorbs water and dissolved minerals from the soil and that the shoot system takes up carbon dioxide (CO_2) from the atmosphere via stomata, pores that occur in the surfaces of leaves and other aboveground structures (**Figure 38.1**). Photosynthetic cells use these materials to produce sugar and other organic compounds needed for overall plant growth and reproduction. Nonphotosynthetic plant cells, such as those of roots and flowers, depend upon organic food produced by green tissues. Therefore, plants must transport water and minerals upward from roots to shoots and transport

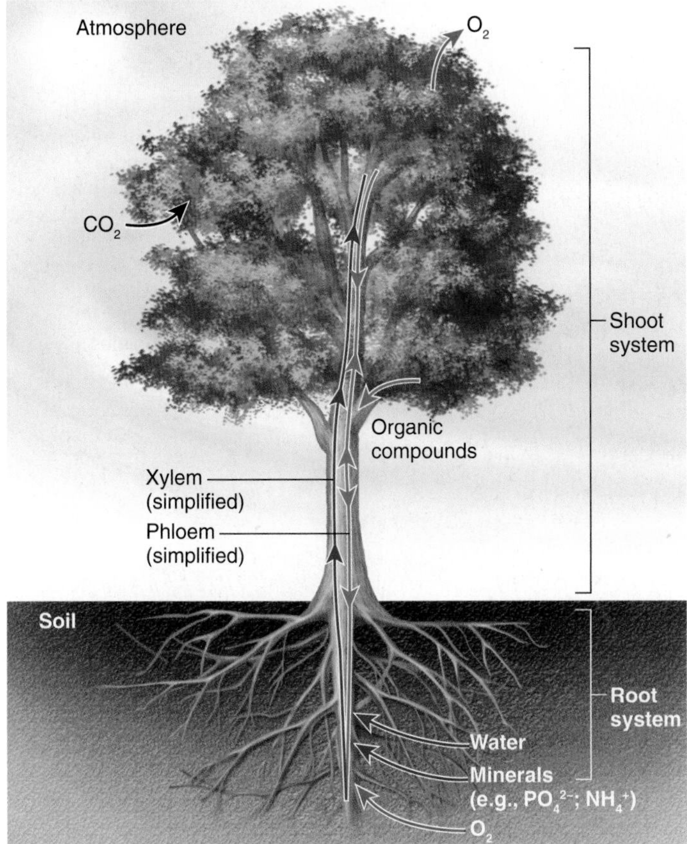

Figure 38.1 **Overview of material uptake and long-distance transport processes in plants.**

organic food from photosynthetic to nonphotosynthetic parts. Since plants can grow to sizeable heights, the tallest trees being over 100 m tall, transport of materials often occurs over long distances.

The long-distance transport of water, dissolved minerals, and sugar throughout the plant body occurs within a continuous system of conducting tissues. Recall that the complex tissues of vascular plants that primarily conduct water and dissolved minerals are known as the xylem, and those that conduct organic substances in a watery sap are termed the phloem. These conducting tissues are key to the ability of vascular plants to thrive in terrestrial habitats, which can sometimes be quite arid. To fully understand how plants accomplish long-distance transport, we will begin by reviewing the processes by which minerals, organic compounds, and water are taken up and move at the cellular level.

38.2 Uptake and Movement of Materials at the Cellular Level

Learning Outcomes:

1. Describe the differences among turgid, flaccid, and plasmolyzed plant cells.
2. Explain how water potential and relative water content are calculated.
3. List ways in which plants cope with cellular osmotic stress.

Chapter 5 described how all cells use both passive and active processes to import or export materials. Here, we briefly review these processes, illustrating how they work in flowering plants.

Passive Transport Does Not Require the Input of Energy

Recall from your earlier studies of plasma membranes that water, gases, and certain small, uncharged compounds can diffuse across membranes in the direction of their concentration gradients. **Passive transport** is the movement of materials into or out of cells down a concentration gradient without the expenditure of energy in the form of ATP. Passive transport across plasma membranes occurs in two ways: by diffusion or facilitated diffusion. **Diffusion** into or out of cells is the movement of molecules through a phospholipid bilayer down a concentration gradient. **Facilitated diffusion** is the transport of molecules across plasma membranes down a concentration gradient with the aid of membrane transport proteins (Figure 38.2a).

The two main types of membrane transport proteins that function in facilitated diffusion are channels and transporters. **Channels** are membrane pores formed by proteins that allow movement of ions and molecules across membranes (see Figure 38.2). **Transporters** are proteins that transport molecules by binding them on one side of the membrane and then changing conformation so that the molecule is released to the other side of the membrane (refer back to Figure 5.19). Transporters increase the rate at which specific mineral ions and organic molecules are able to enter or leave plant cells and vacuoles.

Recall that osmosis is the diffusion of water across a selectively permeable membrane in response to differences in solute concentrations. In the case of plants, water moves from a solution that has a lower solute concentration (soil) to one of higher solute concentration (root cells). Osmotic water uptake into living plant cells is essential to photosynthesis, as well as to cell expansion and structural support. However, the passive diffusion of water does not occur rapidly enough to supply the water needs of rapidly expanding plant cells. In this case, facilitated diffusion of water occurs through protein channels known as **aquaporins**, which occur widely in living things. Thirty-five distinct aquaporin genes have been identified in the genome of the model plant *Arabidopsis*. Aquaporins increase the rate at which water flows into expanding plant cells and their vacuoles. In the same way, many other types of plasma membrane protein channels and transporters facilitate the diffusion of specific mineral ions and organic molecules into and out of plant cells and vacuoles.

ATP Hydrolysis Powers Active Transport

If a substance must be transported across a plasma membrane against its concentration gradient, work must be performed in the process known as active transport. During **active transport**, membrane transporter proteins use energy to move substances against their concentration gradients. An example is the H^+-ATPase proton pump, found in the plasma membranes of plant cells, which uses ATP to pump H^+, which are protons, against a gradient (Figure 38.2b). This proton gradient generates an electrical difference across the membrane, which is known as a **membrane potential**. Energy is released when protons pass back across the plasma membrane, in the direction of their

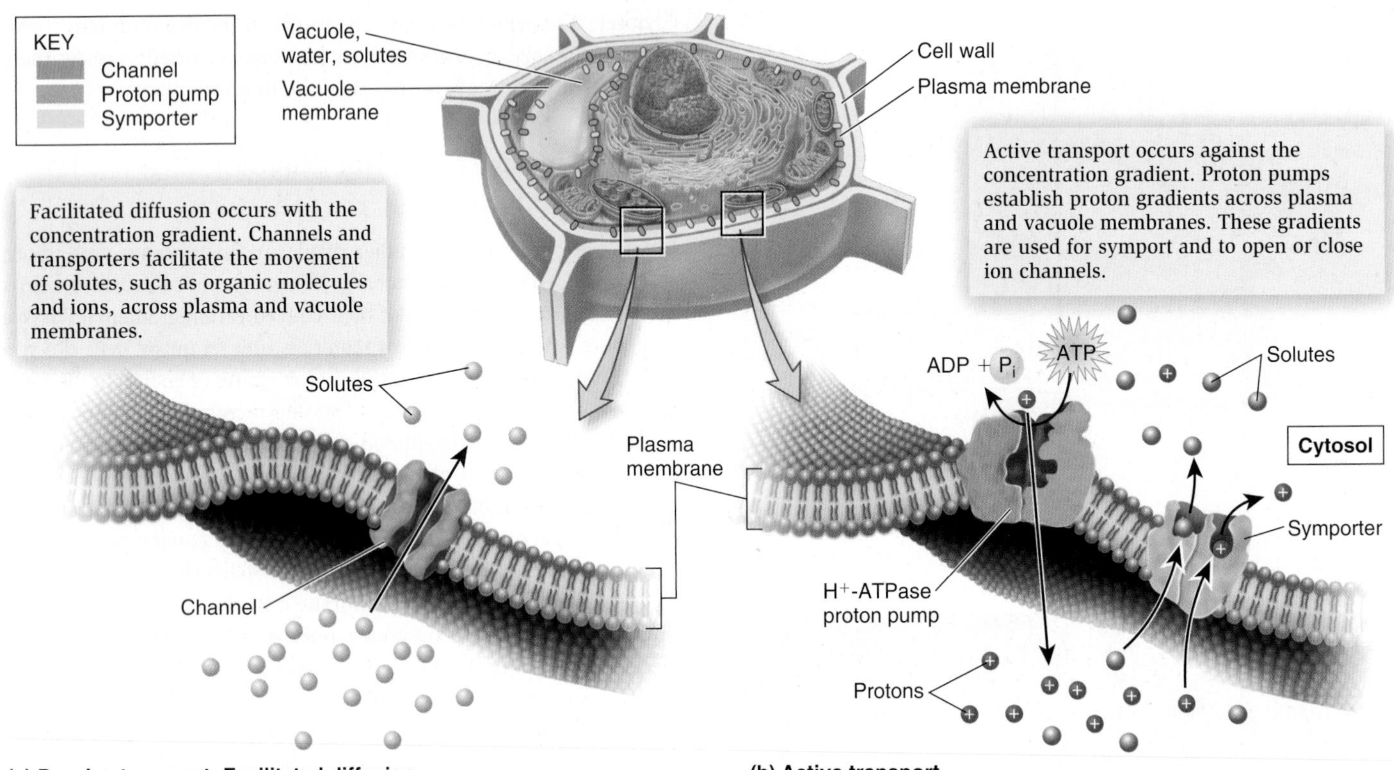

(a) Passive transport: Facilitated diffusion

(b) Active transport

Figure 38.2 Passive and active transport.

electrochemical gradient. This energy can then be used to power other active transport of ions or organic materials. For example, it might be used to open or close ion channels or in the functioning of a proton/solute **symporter**, a protein that transports two substances in the same direction across a membrane. Symporters are needed for the uptake of organic solutes such as sugars, amino acids, and nucleotide bases.

Active transport proteins are particularly abundant in root cell membranes (Figure 38.3), allowing root cells to concentrate dissolved mineral nutrients to more than 75 times their abundance in soil. As a result, soil water flows into root cells by osmosis. We next take a closer look at osmotic water movement into and out of plant cells.

Cellular Water Content Is Influenced by Solute Content and Turgor Pressure

The water content of plant cells depends on osmosis, and osmosis depends on two factors: solute content and turgor pressure. **Turgor pressure** is the hydrostatic pressure required to stop the net flow of water across a plasma membrane due to osmosis. Turgor pressure increases as water enters plant cells, because their cell walls restrict the extent to which the cells can swell.

A plant cell whose cytosol is so full of water that the plasma membrane presses right up against the cell wall is said to be **turgid** (Figure 38.4a). The pressure relationship between the cytosol and cell wall recalls the way that a soccer ball's leather skin presses inward upon the air within, while at the same time the internal air presses on the ball's cover. If you add more air to a limp ball, the ball will stiffen. In the same way, if a nonturgid cell absorbs more water, it

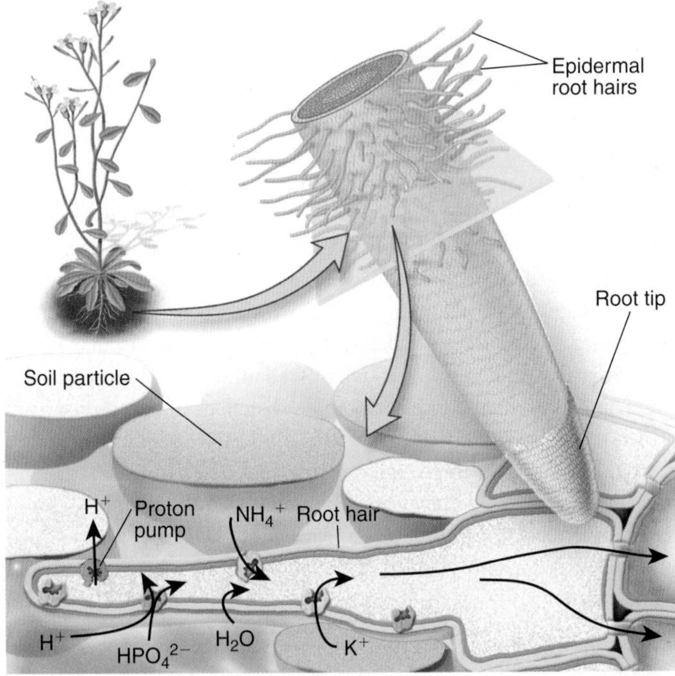

Figure 38.3 **Ion uptake at root-hair membranes.** The H^+-ATPase proton pump establishes an electrochemical gradient that drives the active uptake of solutes. The resulting increase in intracellular solute concentration also drives the osmotic diffusion of water into the cell.

BIOLOGY PRINCIPLE Living organisms use energy. When soil mineral ion concentrations are lower than those within cells, root-hair plasma membranes take up nutrient ions by active transport, which requires energy in the form of ATP.

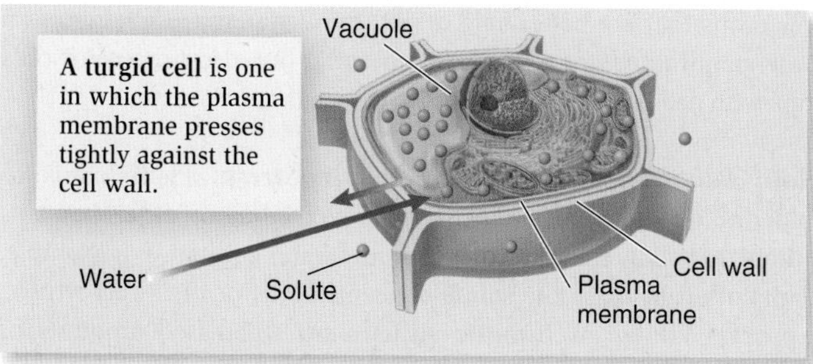

(a) Turgid cell in a hypotonic solution

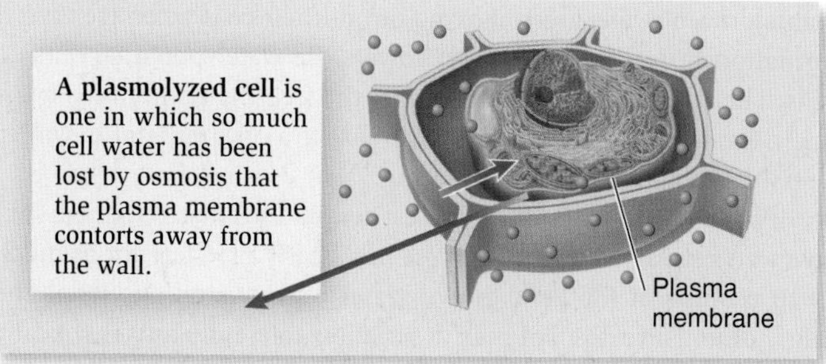

A plasmolyzed cell is one in which so much cell water has been lost by osmosis that the plasma membrane contorts away from the wall.
Plasma membrane

(b) Plasmolyzed cell in a hypertonic solution

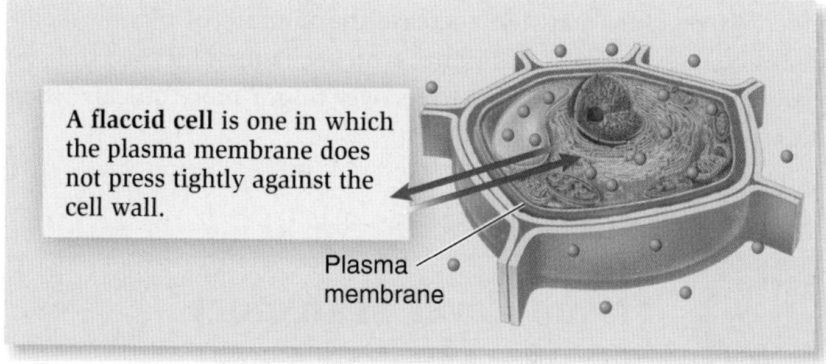

(c) Flaccid cell in an isotonic solution

Figure 38.4 **Turgid, plasmolyzed, and flaccid plant cells.** **(a)** When the concentration of solutes inside a cell is greater than that outside (the cell is surrounded by a hypotonic solution), more water may enter the cell than will leave it. As a result, the plant cell may become swollen, or turgid. **(b)** When the concentration of solutes outside a cell is greater than within it (the cell is surrounded by a hypertonic solution), more water will leave the cell than will enter it. As a result, a plant cell will become plasmolyzed. **(c)** When a cell is bathed in an isotonic solution, it will be flaccid.

will become more rigid as the water exerts pressure on the cell wall. By contrast, a plasmolyzed cell is one that has lost so much water by osmosis that turgor pressure has also been lost. **Plasmolysis** is the condition in which the plasma membrane no longer presses on the cell wall (Figure 38.4b). The concentration of solutes is the same outside and inside a **flaccid cell**, which will have a water content higher than a plasmolyzed cell but lower than a turgid cell (Figure 38.4c).

Together, solute concentration and presence of a cell wall influence an important plant cell property known as **water potential**, the potential energy of water. Water moves from a region of higher water potential to a region of lower water potential. A good analogy is a waterfall, which has high gravitational potential energy at its top. Pressure also influences water potential; a waterfall would flow upward if pressure greater than the force of gravity were applied. Solutes and some other factors also affect water potential. Water potential is measured in pressure units known as megapascals (MPa) (a pascal is equal to 1 newton per square meter). One MPa is equal to 10 times the average air pressure at sea level, about the same pressure that occurs within an inflated bicycle tire. As another reference point, 1 MPa is several times the pressure in typical home plumbing pipes, which you experience when turning on a water faucet.

In the study of plants, the concept of water potential is used to understand the movement of water into and out of cells (cellular water potential) and between entire plants and their environment. The concept of relative water content is used to gauge the water status of whole plants or organs. We begin by considering the water potential of cells.

Cellular Water Potential A water potential equation can be used to predict the direction of cellular water movement, given information about the solute concentrations inside and outside of plant cells, and a measure of pressure at the cell-wall–membrane interface. In this equation, cellular water potential is symbolized by the Greek letter psi (ψ) with the subscript W for water: ψ_W. In its simplest form, total ψ_W is calculated as:

$$\psi_W = \psi_S + \psi_P$$

where ψ_S is solute potential and ψ_P is pressure potential.

Solute potential (ψ_S) is the component of water potential due to the presence of solute molecules. As you might expect, solute potential is proportional to the concentration of solutes in a solution. The solute potential of pure water open to the air, at sea level and room temperature, is defined as zero. When solutes are added, they interact with water molecules, thereby diluting the water and affecting its disorder. As a result, fewer free water molecules are present, which reduces the potential energy of water. Thus, in the absence of a pressure potential, water that contains solutes always has a negative solute potential. The higher the concentration of dissolved solutes, the lower (more negative) the solute potential.

Pressure potential (ψ_P) is the component of water potential due to hydrostatic pressure. In plant cells, the hydrostatic pressure is determined in part by the resistance provided by the cell wall. Because of this resistance, the value for pressure potential can be either positive or negative. For example, a turgid cell has a positive pressure potential, which typically measures about 1 MPa. This high pressure inside turgid plant cells is a testimony to the strength of their cellulose-rich plant cell walls. In contrast to turgid cells, both flaccid and plasmolyzed cells have a pressure potential of zero. Plants or plant organs having many cells with low turgor pressure appear wilted. If your houseplants become wilted, watering will enable the cells to increase their pressure potential, restoring cell turgor and normal plant appearance. Therefore, the water content of an entire organ or plant is influenced by the water potential of its component cells (Figure 38.5).

Relative Water Content The property known as **relative water content (RWC)** is often used to gauge the water content of a plant

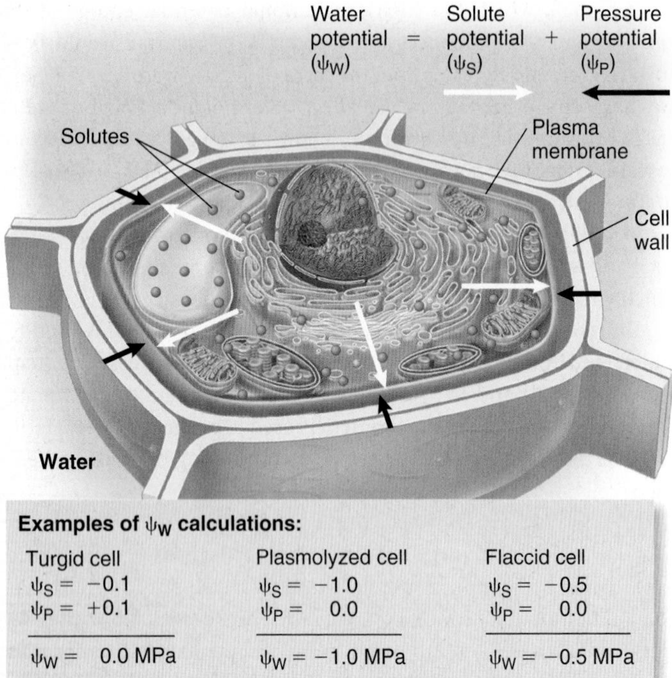

Figure 38.5 **Plant-cell water potential.** The water potential of a plant cell, which predicts the direction of water movement into or out of a cell, can be simplified as the sum of the solute potential and the pressure potential resulting from pressure exerted by the cell-wall–plasma membrane complex. Examples of water potential calculations are shown for a turgid cell, a plasmolyzed cell, and a flaccid cell.

Concept Check: Make a drawing that shows the direction of water movement when a cell of each of these types is placed into a solution of pure water (whose water potential, ψ_W, is defined as 0 MPa).

organ or entire plant and is easy to measure with common equipment. RWC integrates the water potential of all cells within an organ or plant and is thus a measure of relative turgidity. Measurements of RWC can be used to predict a plant's ability to recover from the wilted condition. An RWC of less than 50% spells death for most plants, but some plants can tolerate lower water content for substantial time periods.

A standard method for determining RWC involves three simple weight measurements: fresh weight, turgid weight, and dry weight. Sample tissue taken from a plant under a given set of conditions is first weighed to obtain the fresh weight. Then it is completely hydrated in water within an enclosed, lighted chamber until constant turgid weight is achieved. Finally, the sample is dried to a constant dry weight. Researchers use these measurements to calculate RWC using the equation

$$\text{RWC} = \frac{(\text{fresh weight} - \text{dry weight})}{(\text{turgid weight} - \text{dry weight})} \times 100$$

RWC measurements have been very useful in ecological studies of natural plant adaptation to cold, drought, or salt stress and in agricultural research for developing drought-tolerant crops. Developing new crops that are better able to withstand water stress requires an understanding of not only water potential but also how plant cells cope with cellular osmotic stress, which leads to water stress.

Plant Adaptations to Cellular Osmotic Stress Plants native to cold, dry, or saline environments have evolved many different adaptations that allow them to cope with low water content. For example, plants often increase the solute concentrations of their cell cytosol, a process known as **osmotic adjustment**. Increased amounts of the amino acid proline, sugars, or sugar alcohols such as mannitol decrease the cells' water potential, thus drawing water into cells. By increasing the concentration of solutes inside cells, cold-resistant plants prevent water from moving out of their cells when ice crystal formation in intercellular spaces lowers the water potential outside cells. The additional solutes also lower the freezing point of the cytosol, in the same way that adding antifreeze to a car's radiator in winter keeps the radiator fluid from freezing.

Plants of arid lands often possess adaptations that help them survive water stress. Many can survive in a nearly dry state for as much as 10 months of the year, growing and reproducing only after the rains come. The cytosol of such desiccation-tolerant plants is typically rich in sugars that bind to phospholipids to form a glasslike structure. This helps to stabilize the cellular membranes, preventing them from becoming damaged during plasmolysis (see Figure 38.4b). Plant cells under water stress may also increase the number of plasma membrane aquaporins. These additional protein channels increase the rate of water uptake, allowing cells to recover turgor more quickly when water becomes available.

38.3 Tissue-Level Transport

Learning Outcomes:

1. Define transmembrane transport, symplastic transport, and apoplastic transport.
2. Explain how the root endodermis functions as a diffusion barrier.

Now that we have reviewed how water, dissolved minerals, and organic compounds enter or leave plant cells, we are prepared to consider short-distance transport within and among nearby tissues. Tissue-level transport occurs in three forms: transmembrane transport, symplastic transport, and apoplastic transport (Figure 38.6).

Transmembrane transport involves the export of a material from one cell via membrane proteins, followed by import of the same substance by an adjacent cell (Figure 38.6a). One prominent example of transmembrane transport is the movement of the plant hormone auxin downward in shoots. Auxin travels from one phloem parenchyma cell to another in a linear series with the aid of carrier proteins (refer back to Figure 36.4). This process explains how auxin produced in one part of the plant body can influence more distant tissues.

Symplastic transport is the movement of a substance from the cytosol of one cell to the cytosol of an adjacent cell via membrane-lined channels called plasmodesmata (Figure 38.6b). Plasmodesmata are large enough in diameter to allow transport of proteins and nucleic acids, as well as smaller molecules. Together, all of a plant's

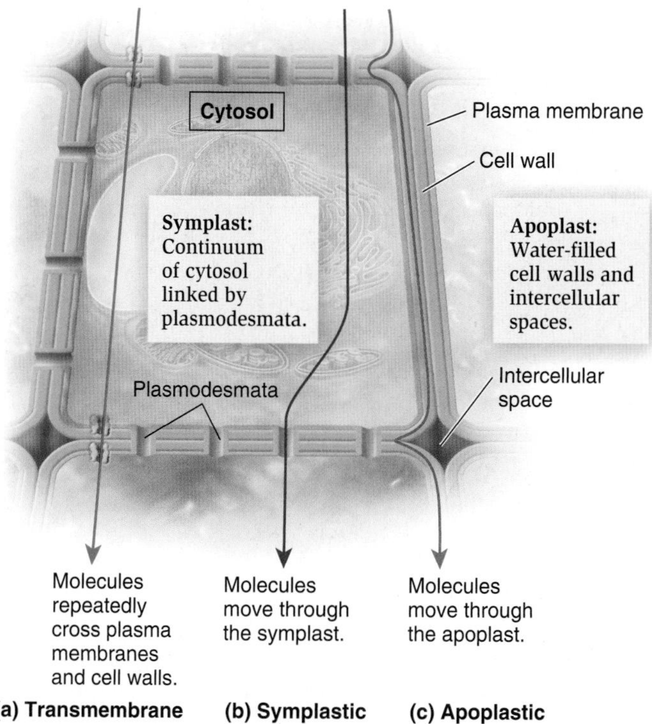

Figure 38.6 **Three routes of tissue-level transport in plants: (a) transmembrane, (b) symplastic, and (c) apoplastic.**

protoplasts (the cell contents without the cell walls) and plasmodesmata form the **symplast**. Symplastic transport has the potential to move molecules widely among the cells and tissues of the plant body.

In contrast to the symplast, the **apoplast** refers to the continuum of water-soaked cell walls and intercellular spaces (Figure 38.6c). **Apoplastic transport** is the movement of solutes along cell walls and the spaces between cells. Water and dissolved minerals often move through plant tissues for short distances by apoplastic transport.

Both symplastic and apoplastic transport play important roles in mineral nutrient transport through the outer tissues of roots (Figure 38.7). As we have noted, the plasma membrane of epidermal root hair cells is rich in channels and transporters that selectively absorb essential mineral ions from soil water. Absorbed ions can move symplastically from the cytosol of root hairs, cortex, and endodermis directly to xylem parenchyma cells. Plasmodesmata make such cell-to-cell, tissue-level transport possible.

Apoplastic transport can also move water and dissolved minerals into root epidermal and cortex tissues. However, apoplastic movement of water and minerals stops at the **endodermis**, a term meaning "inside skin." In roots, the endodermis is a thin cylinder of tissue whose close-fitting cells and specialized cell walls form a barrier to diffusion between the cortex and the central core of vascular tissue. The plant endodermis is functionally analogous to animal epithelial tissues whose cellular tight junctions likewise form diffusion barriers. Materials in the root apoplast cannot penetrate farther into the root unless endodermal cells transport them into their cytosol, a process that requires specific transporter proteins.

Root endodermal cell walls possess ribbon-like strips of waterproof suberin, composed of wax and phenolic polymers. These suberin ribbons, known as **Casparian strips**, prevent apoplastic diffusion of water and solutes through endodermal cell walls and into the root vascular tissues (Figure 38.8). The root endodermis also prevents harmful solutes (such as toxic metal ions) from moving through the apoplast to vascular tissues and being transported to the shoot. For example, aluminum ions (Al^{3+}) are commonly dissolved in soil water, but they are not plant nutrients and are highly toxic to plants. Aluminum ions can penetrate the root epidermis and cortex by moving through the apoplast, but they stop at the endodermis because they are unable to enter the cytosol of endodermal cells.

Endodermal plasma membranes possess specific channels and transporters for essential mineral nutrients (such as K^+), which are thereby able to enter the cytosol of root endodermal cells. By moving through endodermal cytosol, symplastically transported essential

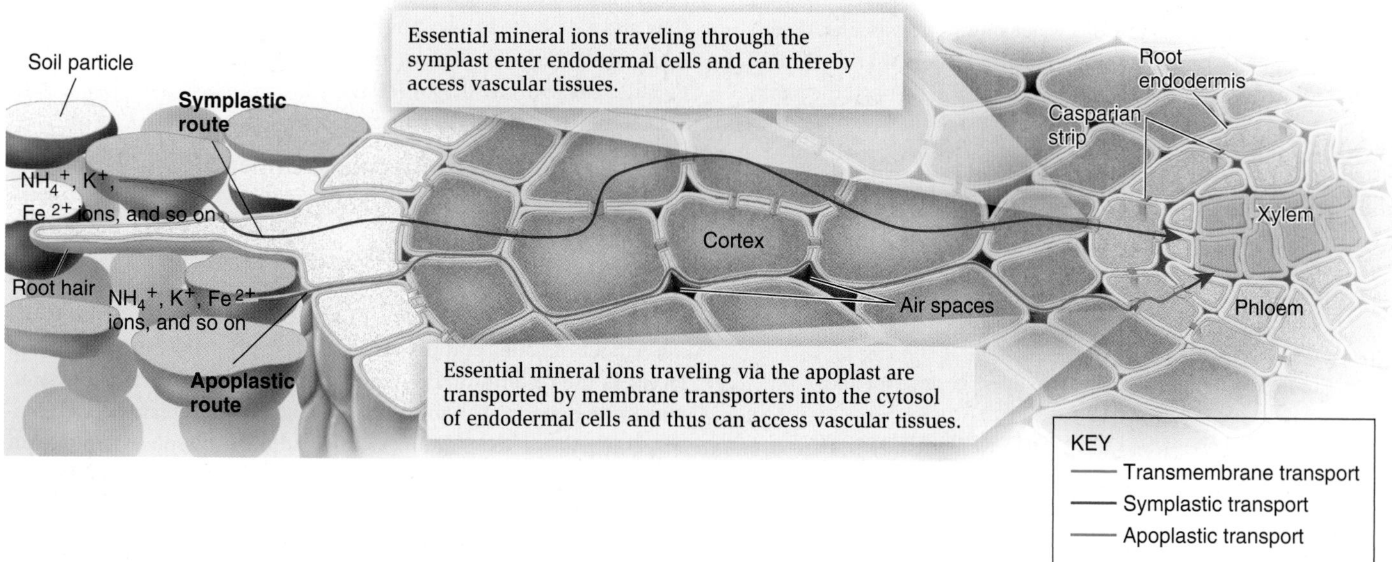

Figure 38.7 **Symplastic and apoplastic transport of mineral ions in roots.**

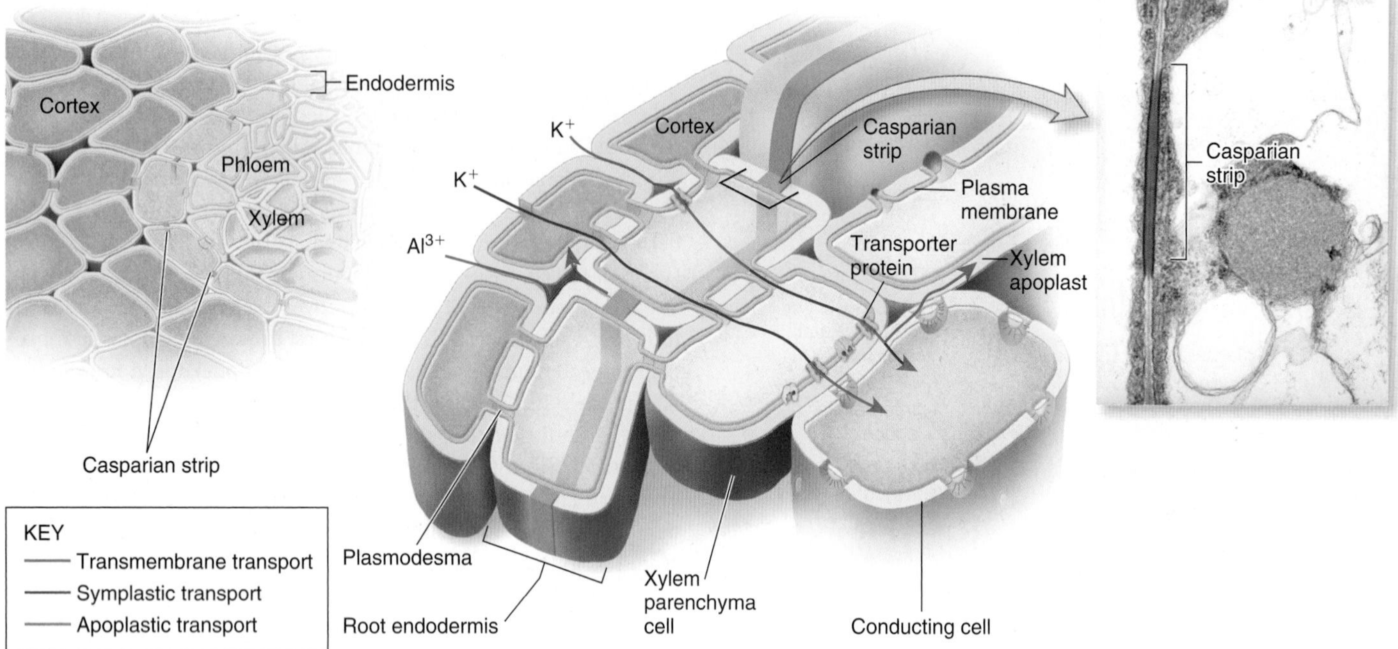

Figure 38.8 **Ion transport pathways across the root endodermis.** Casparian strips in endodermal cell walls prevent apoplastic transport across the root endodermis, limiting entry of harmful soil minerals such as Al^{3+} and exit of useful minerals. Mineral nutrients that are transported into the cytosol of endodermal cells are able to pass through the endodermal barrier to xylem parenchyma cells via plasmodesmata. Once past the endodermis, nutrient ions such as K^+ are moved across plasma membranes to the apoplast of the vascular tissue and are thus able to enter xylem. Inset shows a transmission electron micrograph (TEM) of a Casparian strip in the wall of an endodermal cell.

BioConnections: *Look back to Figure 10.9 to see a diagram of tight junctions between adjacent cells of animal intestinal epithelium. How is the root endodermis similar in structure and function?*

minerals are able to bypass the Casparian strip. Therefore, the root endodermis functions as a molecular filter that allows the passage of beneficial solutes.

Once solutes have moved through endodermal cells, they are transported into the apoplast of the vascular system, which includes conducting cells of the xylem (see Figure 38.8). The endodermis prevents solutes from returning to outer root tissues or the soil, so the solute concentrations of xylem parenchyma cells rise, decreasing their cellular water potential. As a result, water flows into vascular tissues from outer root tissues and the soil. In the next section, we will see how water and solutes are transported for long distances through the plant.

38.4 Long-Distance Transport

Learning Outcomes:

1. Describe how bulk flow occurs in the xylem and phloem of flowering plants.
2. Explain how vessel elements differ from tracheids, and how both function in the xylem of flowering plants.
3. Describe the cohesion-tension theory as an explanation for long-distance water movement in plants.
4. Explain how stomata and leaf abscission help reduce transpirational water loss.
5. Describe how sieve-tube elements and companion cells work together to form a transport system.
6. Create a diagram showing how and why phloem sap moves from source to sink.

Plants rely on long-distance transport to move water and dissolved materials from roots to shoots and among organs. Tall trees are able to transport water and minerals to astounding heights, more than 110 m in some cases. This is possible because plants possess an extensive, branched, long-distance vascular system composed of xylem and phloem tissues. Watery solutions move through these tissues by **bulk flow**, the mass movement of liquid caused by pressure, gravity, or both. Plant conducting tissues are specialized in ways that foster bulk flow and aid plants in adapting to water stress. In this section, we will take a closer look at bulk flow and the major factors involved in long-distance transport by this process.

Bulk Flow Is Water Movement Under the Influence of Pressure and Gravity

Bulk flow (also known as mass flow) occurs when molecules of liquid all move together from one place to another as the result of differences in pressure and/or gravity. One example of bulk flow is leaching, the movement of water and dissolved minerals downward through soil layers as the result of gravity (Chapter 37). Bulk flow is one way in which mineral ions can move through soil toward plant roots. Likewise, once inside the plant, minerals and other dissolved solutes can move through xylem and phloem conducting tissues via bulk flow, which is much faster than diffusion. For example, phloem sap, which contains sugars and other dissolved solutes, moves by bulk flow up to 1 m per hour. The bulk movement of water and solutes within xylem and phloem results from difference in water pressure. However, such

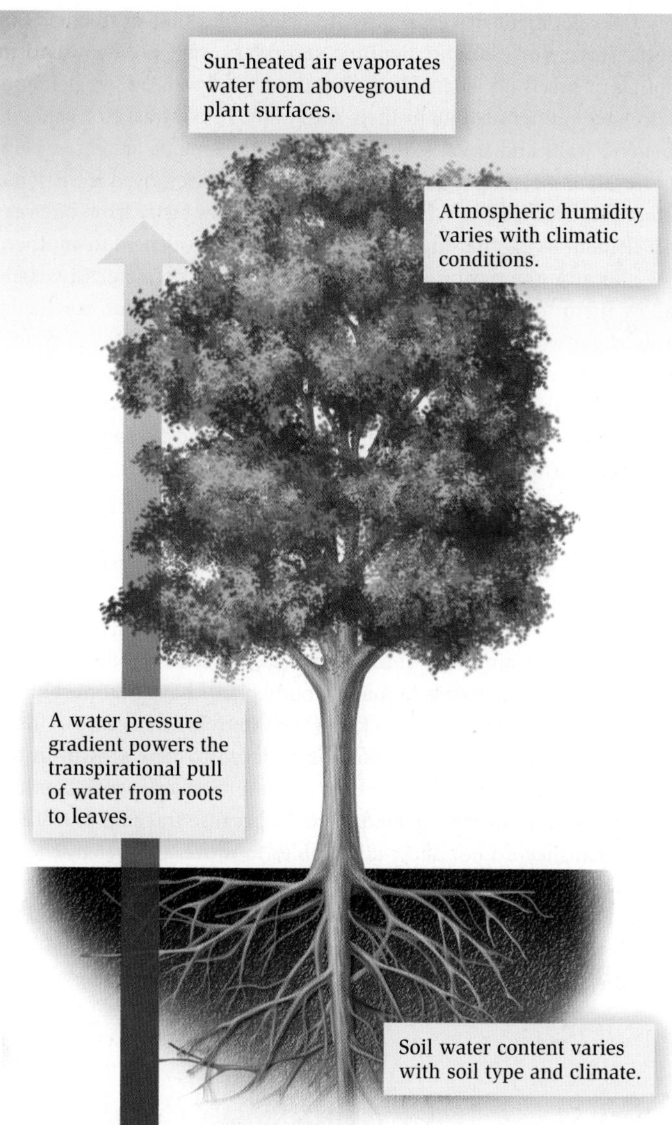

Sun-heated air evaporates water from aboveground plant surfaces.
Atmospheric humidity varies with climatic conditions.
A water pressure gradient powers the transpirational pull of water from roots to leaves.
Soil water content varies with soil type and climate.

Figure 38.9 Upward transport of water in xylem. Water pressure differences between moist soil and drier air drive the upward movement of water in plants.

pressure differences arise from the operation of different processes in xylem as compared with phloem.

Water and dissolved materials move by bulk flow through xylem tissues as the result of differences in root and shoot water pressure. When soil water is abundantly available to roots, water pressure will be higher in root xylem than in shoot xylem. Water pressure is higher in shoot xylem located closer to roots than in xylem located in the upper parts of plant shoots, because water diffuses from the plant through stomata into the atmosphere (Figure 38.9). The extent to which plants lose water to the atmosphere depends on atmospheric humidity; water will diffuse outward more readily when the humidity is lower than when it is higher.

Pressure differences arise in phloem by a different process. Pressure builds up in the phloem sap of leaves because they produce organic solutes such as sugars as the result of photosynthesis.

Figure 38.10 Sugar-maple tapping. In early spring, xylem transports sugar from storage sites to the shoot buds of woody plants such as this sugar maple (*Acer saccharum*).

BIOLOGY PRINCIPLE Biology affects our society. People tap sugar maples by boring holes into the tree trunks and collecting the xylem sap (shown here in buckets). The sap is boiled to evaporate some of the water, leaving the concentrated product—maple syrup.

An increase in sap organic solutes causes water to enter, causing sap pressure to rise. Phloem sap that contains fewer solutes will display lower pressure. Such differences in pressure cause water to flow from regions of higher solute concentration to regions of lower solute concentrations, such as developing fruit. Bulk flow is impeded by obstructions such as cytoplasm. This explains why phloem conducting cells have reduced amounts of cytoplasm and why xylem conducting cells are devoid of cytoplasm altogether.

Although xylem serves as the primary transport system for water and minerals, and phloem for organic compounds dissolved in water, the transport functions of xylem and phloem overlap somewhat. Phloem can aid in the distribution of certain minerals, and xylem sometimes transports organic compounds. For example, in early spring, trees convert starch stored in stem parenchyma cells into sugars that are used during bud expansion and flower development. These sugars are transported in xylem sap. Maple trees produce a copious flow of sugar-rich xylem sap that people have long tapped to make maple syrup (Figure 38.10).

Xylem Is Adapted for Long-Distance Transport of Water and Minerals

Xylem structure plays an essential role in its transport function. The xylem of flowering plants contains several types of specialized cells, some of which remain alive at maturity, and some of which are dead when they are fully functional. Xylem parenchyma cells are alive, but thick-walled supportive fibers may be alive or dead at maturity. Two types of specialized water-conducting cells are always dead and empty of cytosol when mature: tracheids and vessel elements.

Together, tracheids and vessel elements are known as **tracheary elements**. During the development of tracheary elements, a secondary wall is deposited in patterns on the inside of the primary cell wall. This secondary wall is rich in a plastic-like polymer known as lignin. Because lignin is resistant to compression, microbial decay, and water infiltration, it confers strength, durability, and waterproofing. Like the plumbing pipes of a building, tracheary elements do not readily collapse as water moves through them under tension. These characteristics explain why tracheary elements contribute to structural support of the plant body as well as transport.

Tracheids Long and narrow in shape, **tracheids** typically have slanted end walls that fit together to form long tubes (Figure 38.11a). The end walls of tracheids are not lignified, nor are large areas of the side walls of tracheids that occur in plant tissues that are still growing. Such tracheids are extensible because they have rings or spirals of lignin that allow tracheids to continue elongating (Figure 38.11b). In contrast, tracheids that develop in tissues that have already expanded have more lignin, which makes them rigid and unable to elongate any more. Tracheid walls that are extensively lignified display numerous small, lignin-free cell-wall regions known as **pits**. At such pits, the thin primary wall of the tracheid remains readily permeable to water. Water moves from one tracheid to another both vertically and laterally through pits.

Vessels and Vessel Elements Mature **vessel elements** are a second type of water-conducting cell present in xylem tissue. Vessel elements are aligned in pipeline-like files known as **vessels** (Figure 38.12a). Flowering plants are distinguished from other plant groups by the abundance of vessels; nonflowering plants primarily rely on tracheids for water conduction. Vessel elements are larger in diameter than tracheids, conferring greater capacity for bulk flow and therefore represent one of the many ways in which flowering plants are particularly well adapted to life on land.

Development of vessel elements resembles that of tracheids in some ways. For example, lignified secondary walls are deposited in spirals or sheets on the inside of the primary cell wall. Vessel elements also have numerous pits in their side walls. In contrast to tracheids, the end walls and some side walls of vessel elements are extensively perforated, meaning that all cell-wall material is removed from some areas (Figure 38.12b,c). This allows water to flow faster from one vessel element to another than it can flow from one tracheid to another.

Because the perforated end walls and large diameter of vessels allow them to transport more water at a faster rate than tracheids, you might wonder why flowering plants possess two types of water-conducting cells. The answer is that vessels are more vulnerable than tracheids to embolism, meaning blockage by air bubbles. Once an embolism forms in a vessel element, it can extend through the large end-wall perforations into many elements, thereby blocking an entire vessel. Just as air bubbles can cause disruption of blood circulation in people, sometimes leading to death, such bubbles also disrupt water transport in plants, sometimes severely. An embolism can form within a vessel as the result of physical damage, drought, or repeated cycles of freezing and thawing. Air bubbles form frequently during winter, because air does not dissolve in ice. By the end of a cold winter, the functional vessels of many woody plants have become almost completely blocked by air. Blocked vessels in trees often cease to function in water transport and must be replaced by new growth in the spring. Fortunately, even if vessels become blocked, water conduction can still occur via tracheids. This is because tracheid pits are so small that they do not allow air bubbles to move to other tracheids. Thus, an air bubble is confined to the single tracheid in which it first formed, and water continues to flow through nearby tracheids. Tracheids thereby provide a fail-safe conduction route when vessels have become disabled by embolisms.

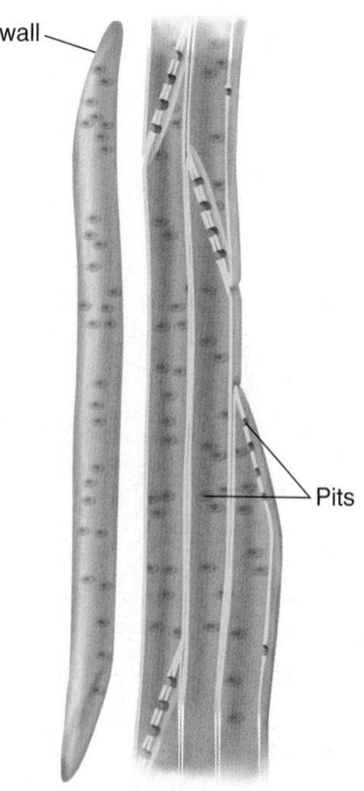

(a) Tracheids

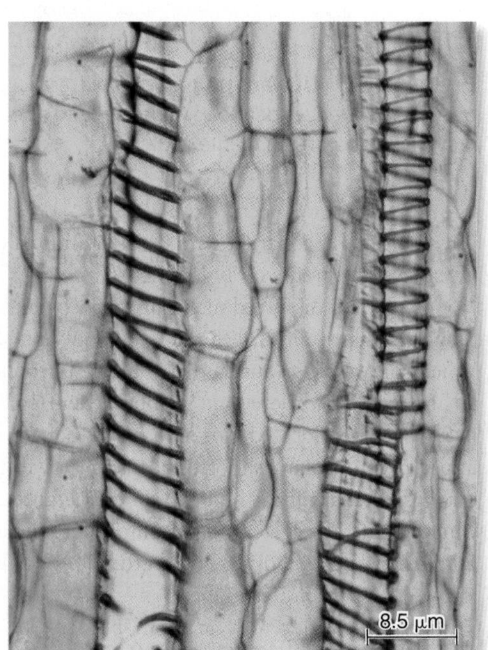

(b) Extensible tracheids

Figure 38.11 Tracheid cells in xylem tissue.
(a) Tracheids are long, tubular cells with slanted end walls. Water and ions move from cell to cell through the pits. **(b)** Light micrograph of extensible tracheids from the xylem of pumpkin.

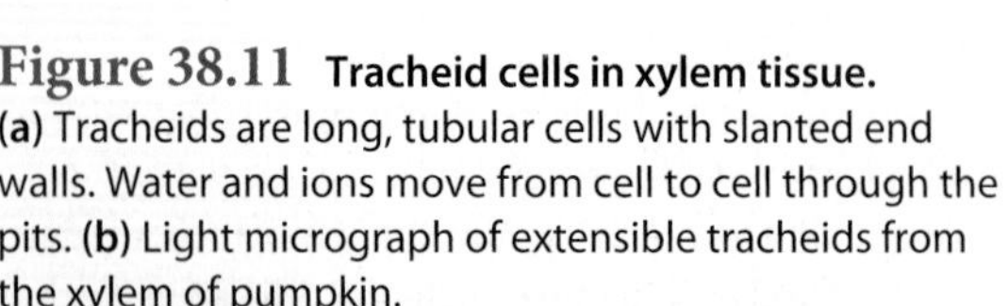

Concept Check: *If you applied a stain specific for lignin to tracheids present in a longitudinal slice of a plant stem that is still growing in length, then observed the cells with a light microscope, what portions of the tracheids would be stained, and what parts would not be stained?*

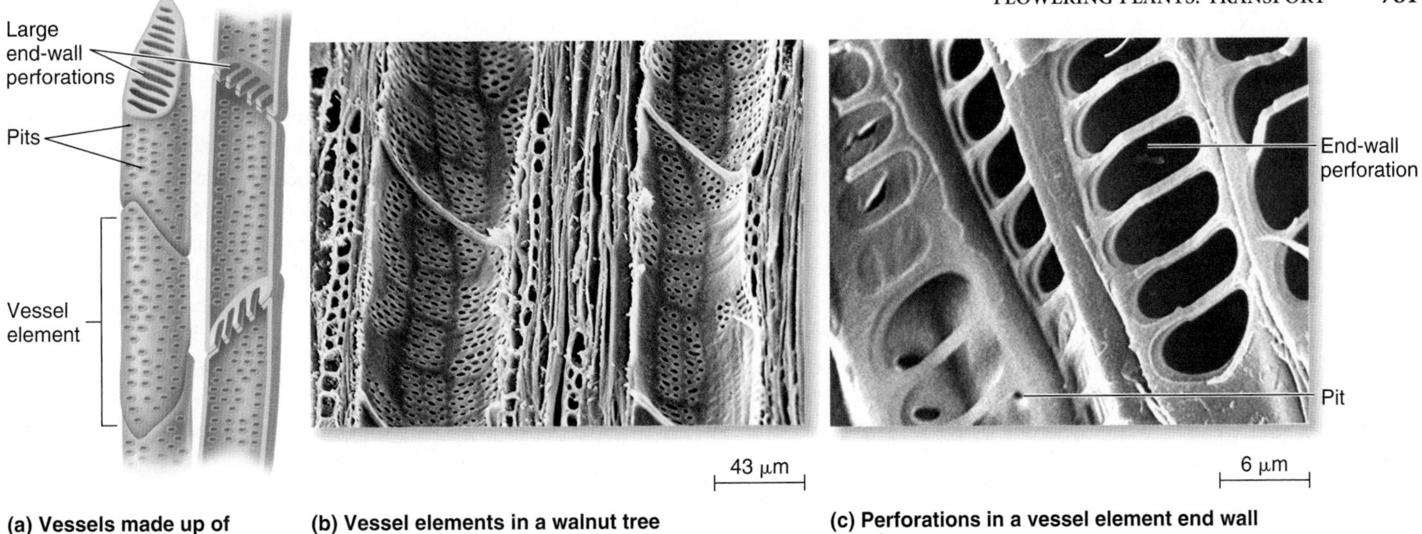

Figure 38.12 **Vessels composed of vessel elements in xylem tissue.** **(a)** This illustration shows the wide diameter of vessel elements with many pits and end-wall perforations. **(b)** SEM of vessels in the wood of the walnut tree (genus *Juglans*). **(c)** SEM of a perforated vessel element end wall from the tulip tree (*Liriodendron tulipifera*).

Concept Check: *Which structural features of vessel elements explain the vulnerability of vessels to embolism, that is, blockage by air bubbles?*

Some plants are able to refill embolized vessels by means of a process known as **root pressure**. At night, the xylem of roots may accumulate high concentrations of ions that are not immediately transported upward to shoots. In this case, the root acts much like a cell rich in solutes, with the result that water gushes in so rapidly that it pushes upward to leaves. Evidence of this process can be observed in the early morning as droplets of water at the edges of leaves, a phenomenon known as **guttation** (Figure 38.13). As the water rushes upward, it can dislodge air bubbles or dissolve them, thereby reversing an embolism. Root pressure refilling has been observed to occur in nonwoody plants such as corn (*Zea mays*) and in some woody plants, including the sugar maple (*Acer saccharum*).

Figure 38.13 **Guttation, the result of root pressure.**

Concept Check: *What functions can root pressure serve in plants?*

Feature Investigation

Holbrook and Associates Revealed the Dynamic Role of Xylem in Transport

Plant biologists once thought that xylem sap moved from one conducting cell to another at a uniform rate, unless blocked by an embolism. After all, botanists hypothesized, the cells are dead and thus lack cytoplasmic components that might influence sap flow rates. However, recent experiments have revealed that xylem sap actually moves more readily from one vessel to another through pits when the solute concentration of xylem sap is relatively high. These studies suggest that this variation results from dynamic changes in pit-wall thickness.

More than 20 years ago, plant biologists noticed that tap water moved more rapidly through cut pieces of stem than did pure water. This observation suggested that ions present in the tap water might have influenced bulk flow rates. However, this hypothesis remained untested until 2001, when N. Michelle Holbrook and colleagues conducted experiments designed to explore this phenomenon, as shown in Figure 38.14. These investigators compared the flow rate of pure water with that of artificial sap, pure water to which they had added various levels of potassium chloride (KCl) or other salts that normally occur in xylem sap. The artificial sap and pure water were introduced into the same experimental tobacco plant through separate flaps cut into the stem. (Using different flaps on the same plant helps to remove experimental variation that might have resulted if investigators had compared results from separate plants.)

These researchers observed that the xylem flow rate was higher when the flap received artificial sap than when it was receiving pure water. Then the investigators supplied pure water to both stem flaps, with the result that the flow rate equalized (see step 2 in Figure 38.14 and also The Data, part a). The investigators then provided artificial sap containing various types of solutes to one flap, along with alternating additions of pure water (step 3). The flow rate increased for artificial sap containing dissolved salts, but not for artificial sap containing the organic molecules sucrose or ethanol (see The Data, part b).

Next, the scientists performed experiments designed to determine if living cells in the xylem tissue were causing these effects. They flushed the stems with boiling water or froze the stems in liquid nitrogen before treating them with artificial sap or water as in step 3. These treatments kill living cells. They found that flow rates through the boiled or frozen stems yielded the same results: The flow rate was higher when the flap received sap containing dissolved salts rather than pure water or organic molecules. These results indicated that living cells did not cause the effect.

The investigators suspected that ions change vessel element structure in some way that allows greater flow through them. In later experiments not shown here, the scientists devised an apparatus for

Figure 38.14 **The Holbrook team's dynamic xylem experiment.**

BIOLOGY PRINCIPLE **Biology is an experimental science.** The illustrated experiment revealed previously unknown xylem responses to the solute content of water.

HYPOTHESIS Dissolved minerals increase water flow through microchannels in xylem vessel pit walls.

KEY MATERIALS Plants with split stems.

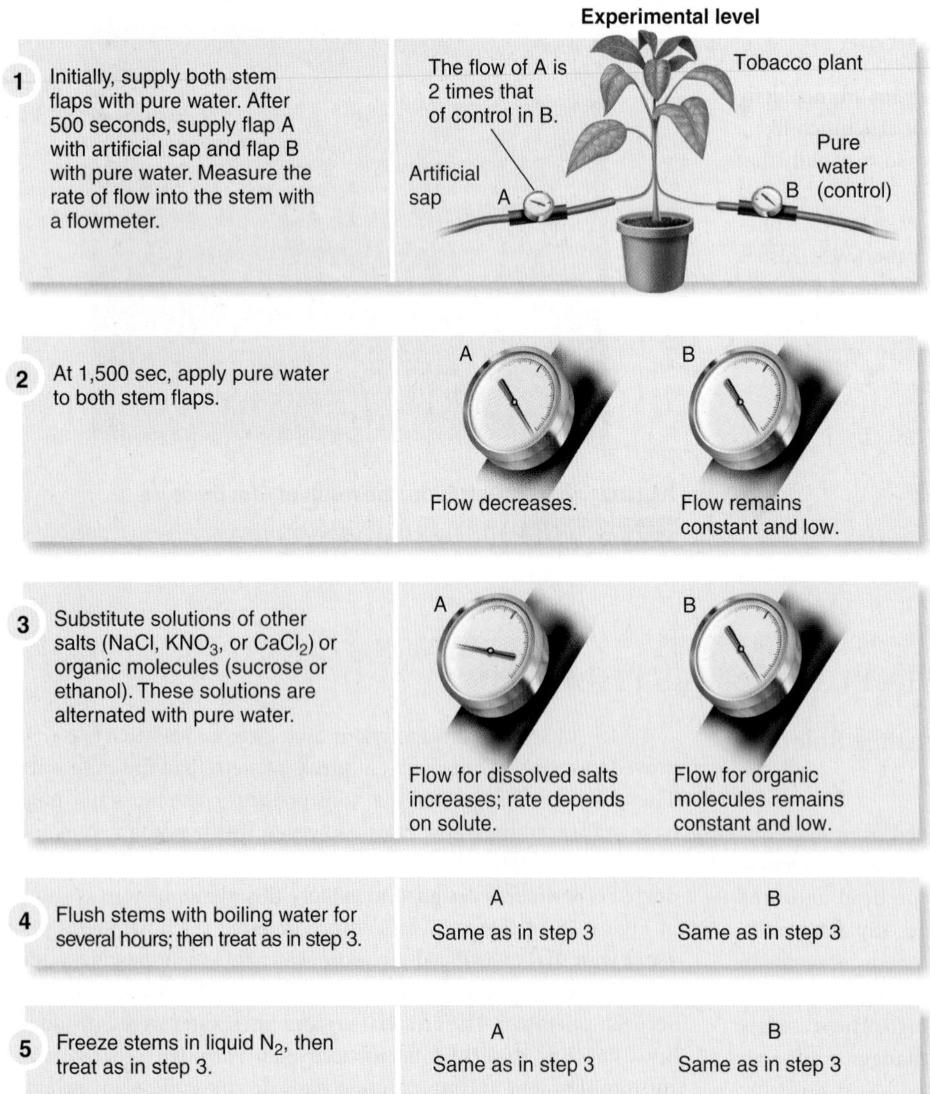

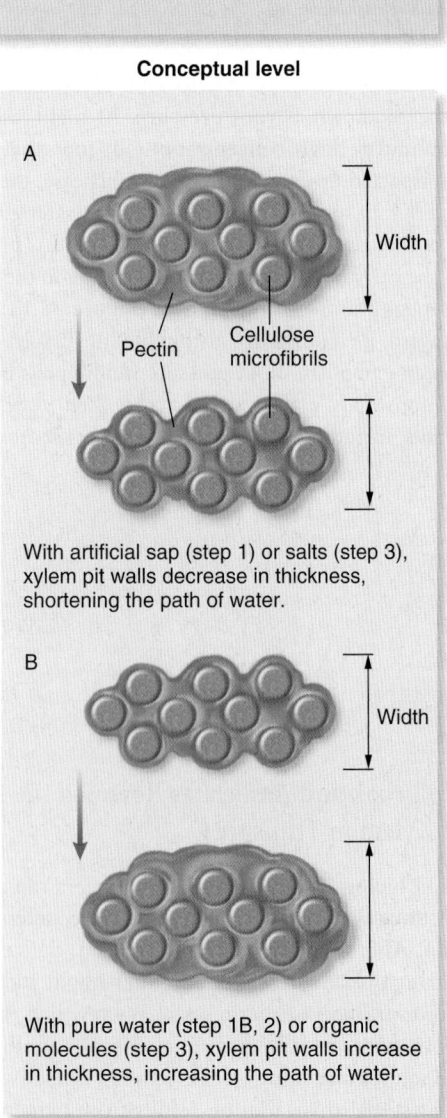

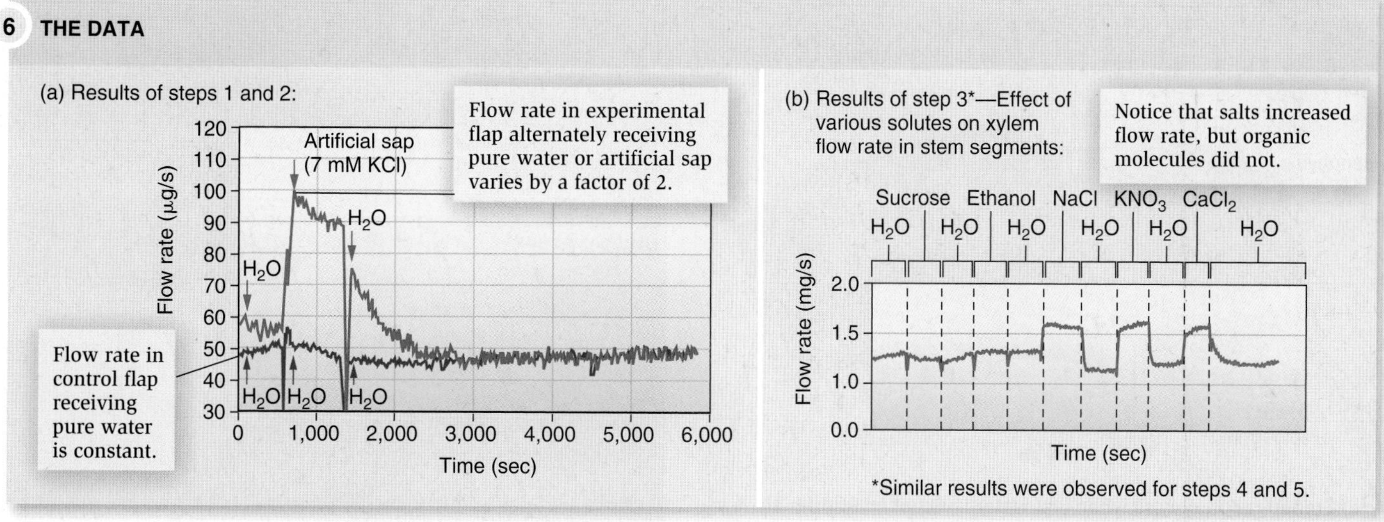

7 **CONCLUSION** Artificial sap and other salts increase the rate of sap flow compared to pure water, likely by shrinking cell-wall pectins, thereby opening micropores in the thin pit walls between adjacent vessels.

8 **SOURCES** Zwieniecki, Maciej A., Melcher, Peter J., and Holbrook, N. Michele. 2001. Hydrogel control of xylem hydraulic resistance in plants. *Science* 291: 1059–1062.
Lee, Jinkee, Holbrook, N. Michele, and Zwieniecki, M.A. 2012. Ion induced changes in the structure of bordered pit membranes. *Frontiers in Plant Science* 3:1–4.

measuring the flow of sap through individual xylem vessels versus sap flow through groups of vessels. They found that flow through an isolated vessel did not respond to changes in sap ion content, whereas flow between adjacent vessels did respond. The researchers interpreted these results as evidence that cell wall pectins shrink in response to sap ions and swell in response to pure water. As a result, walls decrease in thickness in response to sap ions, increasing flow rates, but thicken in response to pure water (see Figure 38.14, step 1, Conceptual level). Researchers speculated that this process hastens the delivery of ion-rich sap to transpiring tissues. Investigators also proposed that this process, which increases sap flow, may help plants compensate for partial loss of xylem sap flow when an air embolism occurs.

Experimental Questions

1. Why did the Holbrook team use the same plant to examine the effect of a pure-water control and an experimental, artificial xylem sap?
2. What caused water to flow from the source of artificial sap (or pure water) into the xylem of the plants used in the experiments conducted by Holbrook and associates?
3. Why were sap flow rates the same in living cells and in dead cells in the experiments conducted by the Holbrook team?

Cohesion-Tension Theory Explains the Role of Transpiration in Long-Distance Water Transport

In warm, dry air, water evaporates from plant surfaces. This evaporation process is known as **transpiration** (from the French *transpirer*, meaning to perspire) (**Figure 38.15a**). Transpiration is capable of pulling water by bulk flow up to the tops of the tallest trees and is the primary way in which water is transported for long distances in plants. Plants expend no energy to transport water and minerals by transpiration. Rather, the Sun's energy indirectly powers this process by generating a water pressure difference between moist soil and drier air (see Figure 38.9).

How does evaporation at plant surfaces influence long-distance water transport in the xylem? To answer this question, we must consider the unique physical properties of water. Liquid water molecules are linked by hydrogen bonds (see Chapter 2). As a result, liquid water is amazingly cohesive, explaining why water tends to form continuous streams (**Figure 38.15b**). Consequently, when water molecules evaporate from plants, water films present in leaves display high surface tension. This tension causes a curved water surface known as a meniscus to form (**Figure 38.15c**) that pulls on neighboring liquid water molecules and eventually on water in the nearest vein, which is connected to the plant's entire water supply. As the result of water's cohesion and the tension exerted on water at the plant's surface, a continuous stream of water can be pulled up through the plant body from the soil, into roots, through stems, and into leaves. This explanation for long-distance water movement in plants is known as the **cohesion-tension theory**. (Recall from Chapter 1 that a scientific theory is a well-established concept, not just a hypothesis.)

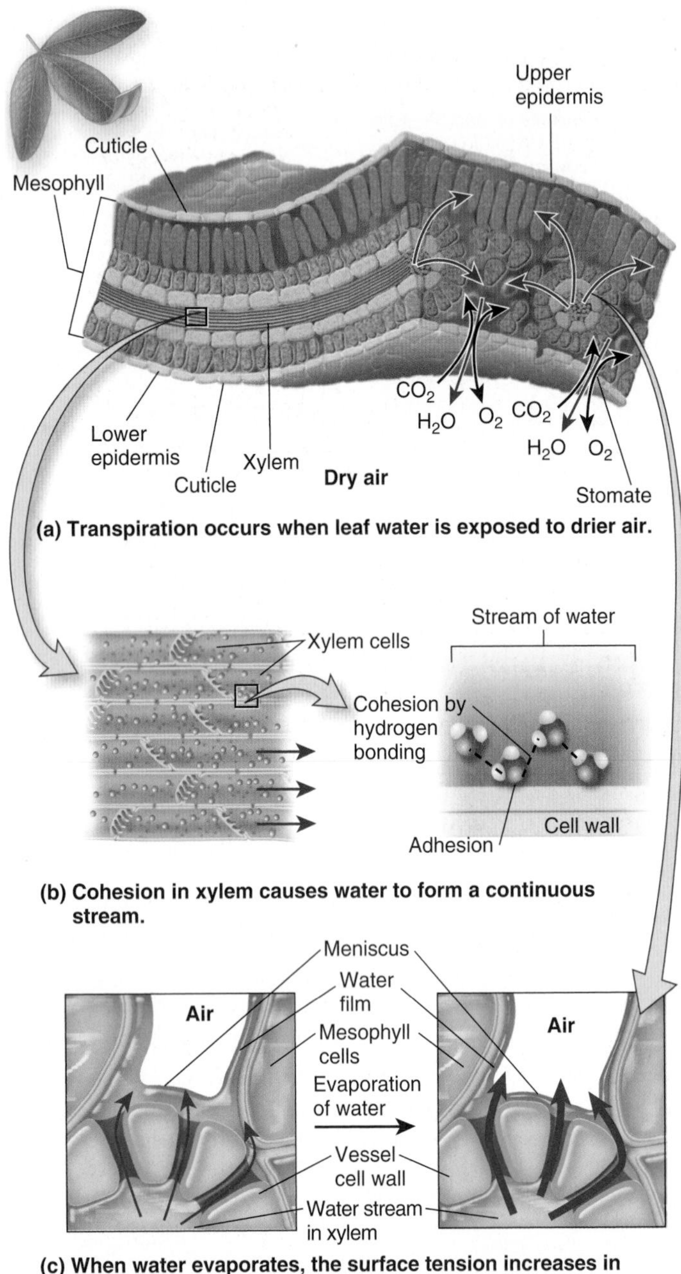

Figure 38.15 **The roles of transpiration, cohesion, adhesion, and tension in long-distance water transport.**

Figure 38.16 **Plant-transpired water vapor mist rising from a tropical rain forest.** This mist visually illustrates the enormous amount of water that is transpired from the surfaces of plants into the atmosphere. Water vapor derived from plant transpiration is an important source of rainfall, and the process of evaporation cools plant surfaces as well as affecting the local and global climate.

Concept Check: *Why does evaporation of water have such a powerful cooling effect?*

Plant transpiration moves huge amounts of water from the soil to the atmosphere. About 99% of the water that enters plants via roots is generally lost as water vapor during transpiration. Each crop season, a single corn plant (*Zea mays*) loses more than 200 L of water, which is more than 100 times the corn plant's mass. A typical tree loses 400 L of water per day! On a regional and global basis, plant transpiration has enormous climate effects. For example, an estimated one-half to three-quarters of rainfall received by the Amazon tropical rain forest actually originates from plant-transpired water vapor, often visible as mist (Figure 38.16). Furthermore, about half of the solar heat received by Amazonian plants is dispersed to the atmosphere during transpiration. This heat dispersal has a cooling effect on regional ground temperature, which would be much higher in the absence of plant transpiration. Such cooling effects result from water's unusually high heat of vaporization, the amount of heat needed to isolate water molecules from the liquid phase and move them to the vapor phase. Most of this energy is needed to break the large numbers of hydrogen bonds that occur in liquid water. The evaporation of large amounts of water from plant surfaces effectively dissipates heat, explaining how plants cool themselves and their environments.

Although evaporation of water from plant surfaces plays an essential role in bulk flow through xylem, if plants lose too much water, they will die. Plant surfaces, including those of leaves, typically produce a cuticle, a wax-containing layer that retards water loss. Only about 5% of water evaporated from plant surfaces emerges through the cuticle, however. More than 90% of the water that evaporates from plants is lost through stomata, surface pores that can be closed to retain water or opened to allow the entry of CO_2 needed for photosynthesis. When the stomata are open, O_2 also exits the plant, as does water vapor when the atmospheric humidity is relatively low. Stomata are often abundantly located on the lower surfaces of leaves. Tobacco leaves, for example, possess an estimated 12,000 stomata per square centimeter of leaf surface! Plants face a constant dilemma: whether to open their stomata for CO_2 intake and suffer the effect of reduced water content or to close stomata to retain water, thereby preventing CO_2 uptake.

Plant Adaptations Help to Reduce Transpirational Water Loss

Under some conditions, almost all plants experience water stress, which is an inadequate amount of water. Water stress is common for plants of the world's arid regions, and their growth is often limited by water availability. Even plants of moist, forested regions of the world experience water stress during drier or colder seasons or under windy conditions. The leaves at the tops of tall trees are generally under considerable water stress because gravity has a substantial effect on their water potential. Earlier, we considered examples of plant cellular adaptations to deal with osmotic stress. Flowering plants have evolved two additional ways to prevent excessive loss of water by transpiration: regulation of stomatal opening and leaf drop.

Stomatal Opening and Closing Plant stomata close to conserve water under conditions of water stress and open when the stress has been relieved, allowing air exchange with the leaf's spongy mesophyll. Stomata are bordered by a pair of **guard cells**, which are sausage-shaped chloroplast-containing cells attached at their ends (Figure 38.17a). The distinctive structural features of guard cells explain how they are able to open and close a pore. As guard cells become fully turgid, their volume expands by 40–100%. This expansion does not occur evenly, however, because the innermost cell walls are thicker and less extensible than are other parts of the guard cell walls. In addition, the cells expand primarily in the lengthwise direction because bands of radially oriented cellulose microfibrils prevent the guard cells from expanding laterally (Figure 38.17b). Thus, when guard cells are turgid, a stomatal pore opens between them, allowing air exchange with the leaf's spongy mesophyll. Conversely, when the guard cells lose their turgor, their volume decreases, and the stomatal pore closes.

What causes the change in turgor? In flowering plants, stomata often open early in the morning, in response to sunlight. This response makes sense, given that light, water, and carbon dioxide are all required for photosynthesis. Blue light stimulates H^+-ATPase proton pumps, leading to guard cell uptake of ions, especially potassium (K^+), and other solutes. As a result of increases in solute concentrations inside guard cells, osmotic water uptake occurs via plasma membrane aquaporins, resulting in cell expansion and stomatal opening (Figure 38.18a).

At night, the reverse process closes flowering plant stomata. Potassium and other solutes are pumped out of guard cells, causing water to exit and deflate guard cells, resulting in pore closure. Flowering plants also close stomata during the daytime under conditions of water stress, a process mediated by the stress hormone abscisic acid (ABA). Water stress causes a 50-fold increase in ABA, which is transported in the xylem sap to guard cells. ABA then binds to a receptor, which elicits a Ca^{2+} second messenger, causing the guard cells to lose solutes and deflate (Figure 38.18b).

Leaf Abscission Angiosperm trees and shrubs of seasonally cold habitats experience water stress every winter, when evaporation from plant surfaces occurs, yet soil water is frozen and therefore unavailable for uptake by roots. Desert plants also experience water stress conditions, but at less predictable times and for much of the year. Both types of plants are adapted to cope with water stress by dropping their leaves, a process known as **leaf abscission**. Dropping leaves lets these plants avoid very low leaf water potentials and the consequent danger of xylem embolism. Leaf abscission also reduces the amount of root mass that plants must produce to obtain water under arid conditions. The ocotillo (*Fouquieria splendens*) of North American deserts can produce leaves after sporadic rains and then drop all of its leaves as a direct response to drought as many as six times a year (Figure 38.19).

The sugar maple (*Acer saccharum*) is an example of the many types of temperate forest trees or shrubs that drop their leaves each autumn and are thus known as deciduous plants. Deciduous plants contrast with evergreen conifers, whose needle- or scale-shaped leaves are adaptations that reduce the area of leaf surface from which water can evaporate, which helps these gymnosperms cope with water stress during the cold season (refer back to Figure 30.10). The broader, thinner leaves produced by many angiosperms are well adapted for

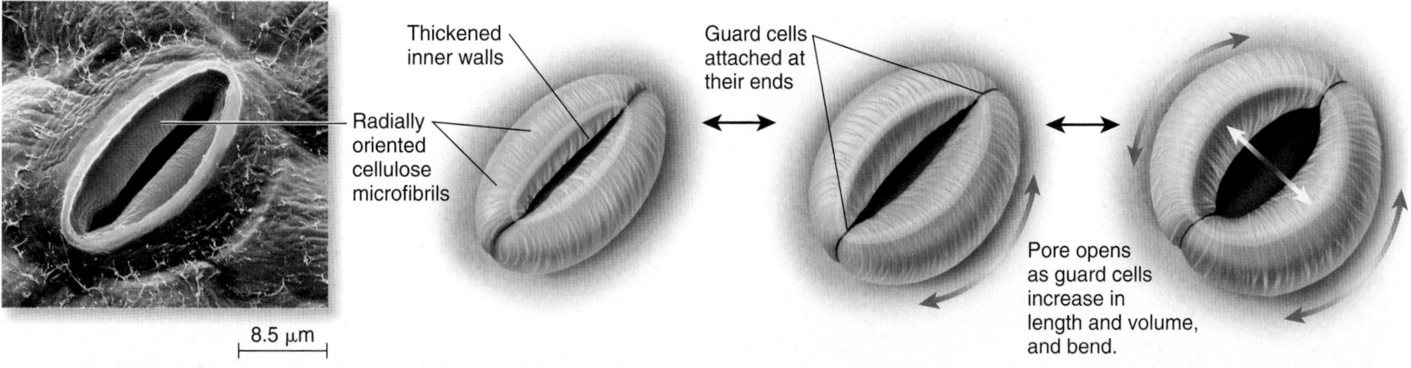

Figure 38.17 **The structure of guard cells.** When flaccid, guard cells close stomata. Turgid guard cells produce a stomatal opening. **(a)** An SEM of a stomate in a rose leaf, showing the two guard cells bordering a partly open pore. **(b)** Thickened inner cell walls and radial orientation of cellulose microfibrils in the guard cell walls explain why they separate when turgid, forming a pore.

Concept Check: *How could you make a physical model that would illustrate how guard cell structure affects its function?*

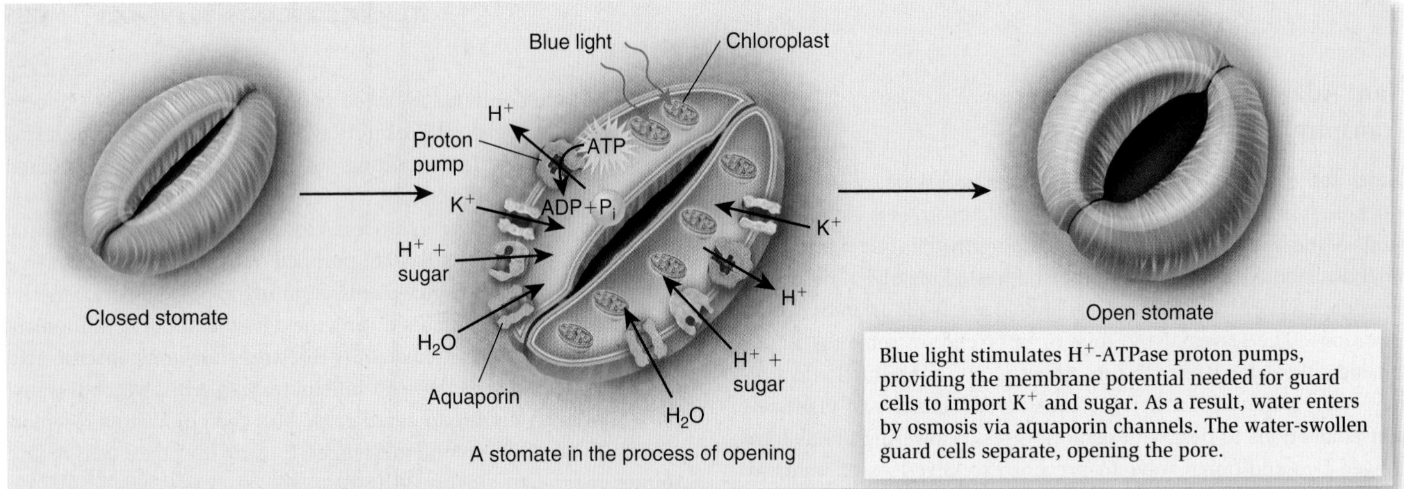

(a) The process of stomate opening

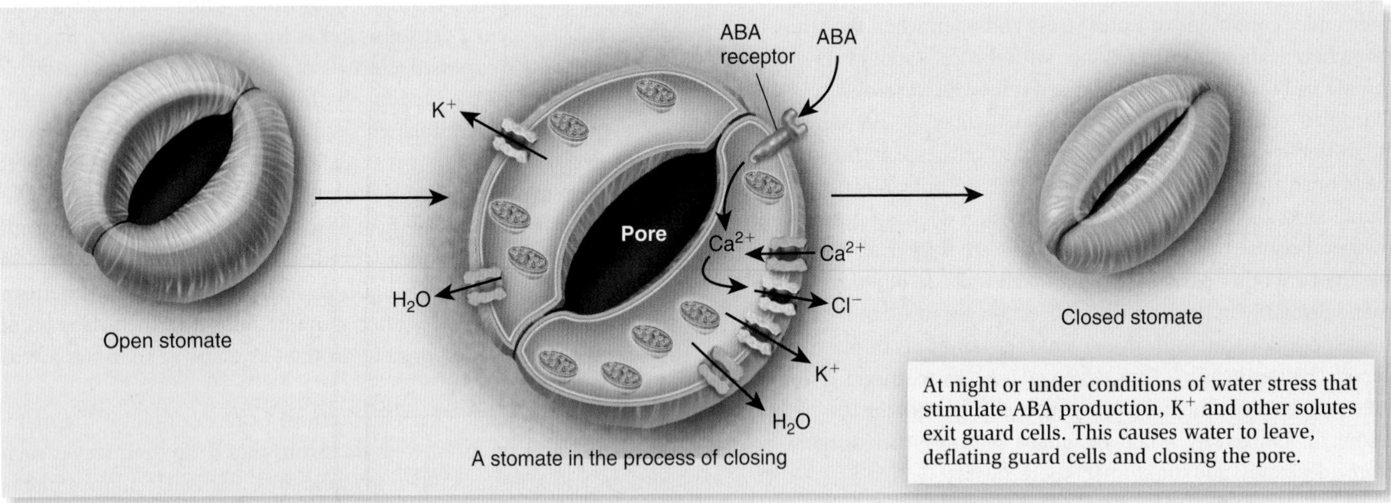

(b) The process of stomate closing

Figure 38.18 How stomata of flowering plants open and close. (a) Angiosperm stomata usually open in response to sunlight. (b) Angiosperm stomata usually close in response to lack of sunlight. They can also close during the day under conditions of water stress, which induces plants to produce more of the hormone abscisic acid (ABA). Guard cell plasma membranes possess ABA receptors, which receive the drought signal.

efficient light-capture, but more vulnerable to the stresses caused by cold. During their evolution, temperate zone angiosperm trees and shrubs have acquired the genetic capacity to predict the onset of cold, dry winter conditions and respond with preemptive leaf abscission. In contrast to the case of the ocotillo, autumn leaf drop in temperate angiosperms is not directly induced by drought.

Leaf abscission is a highly coordinated developmental process. The hormone ethylene stimulates an abscission zone to develop at the bases of leaf petioles (Figure 38.20a). The abscission zone contains

(a) Ocotillo with leaves **(b) Ocotillo without leaves**

Figure 38.19 Leaf abscission as a drought adaptation. The ocotillo (*Fouquieria splendens*), a plant native to North American deserts, is known for its ability to respond to intermittent rain and drought by producing and dropping leaves multiple times within a year.

Concept Check: Why does the ocotillo not drop its leaves at a single predictable time each year, as do temperate angiosperm trees and shrubs?

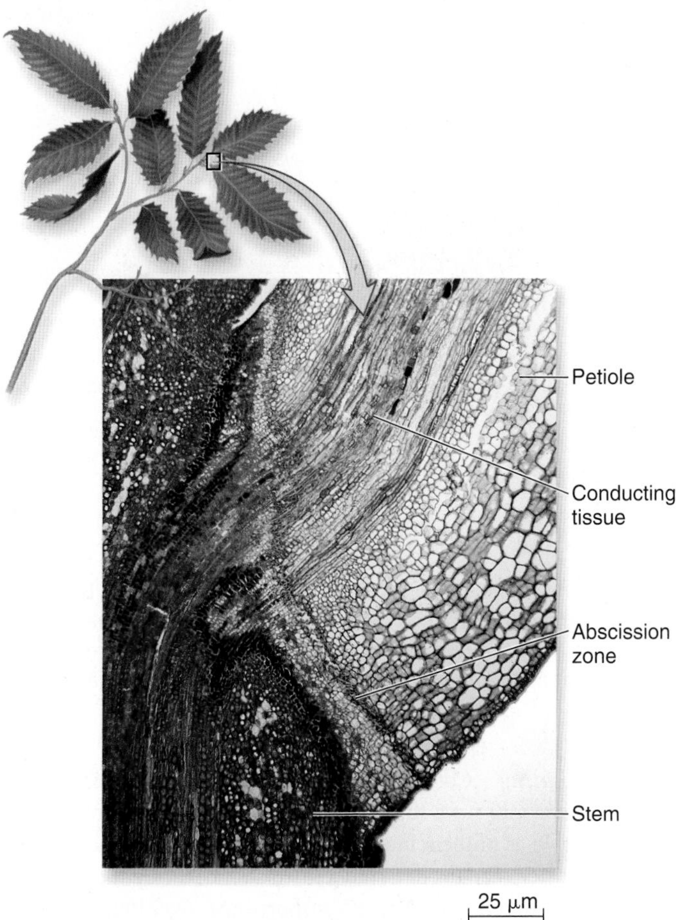

(a) LM of leaf abscission zone at the junction of a petiole and stem, stained with dyes

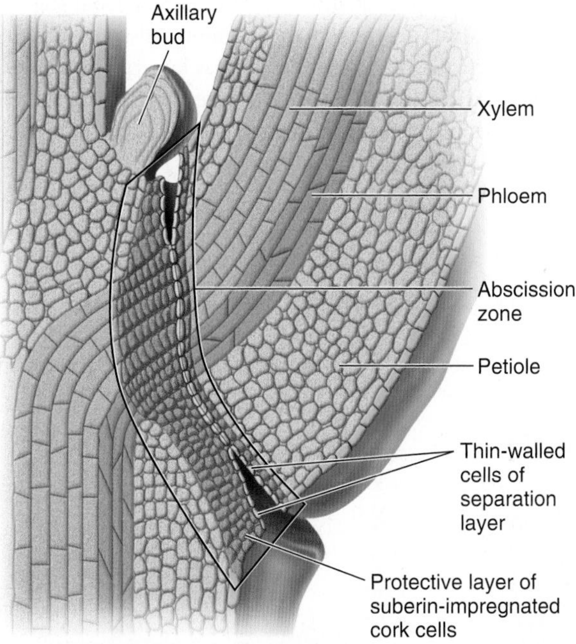

(b) Tissues involved in leaf abscission

Figure 38.20 **Leaf abscission.** Leaf abscission helps maintain plant hydration.

BIOLOGY PRINCIPLE Living organisms maintain homeostasis. This is illustrated by leaf drop in response to, or in anticipation of, arid conditions.

two types of tissues: a separation layer of short, thin-walled cells and an underlying protective layer of suberin-impregnated cork cells (**Figure 38.20b**). Suberin contains both waterproofing wax and phenolic polymers that retard microbial attack. As the abscission layer develops across the vein linking the petiole with the stem, it eventually cuts off the water supply to the leaf. Chlorophyll in the leaves degrades, revealing colorful orange and yellow carotenoid and xanthophyll pigments that were hidden beneath. In addition, some plants synthesize red and reddish blue pigments in response to changing environmental conditions. The presence of these pigments explains colorful autumn vegetation in temperate zones. Enzymes eventually break down the cell-wall components of the separation layer, causing the petiole to break off the stem. The underlying protective layer forms a leaf scar that seals the wound, helping to protect the plant stem from water loss and pathogen attack.

Long-Distance Transport of Organic Molecules Occurs in the Phloem

Phloem plays an essential role in long-distance transport of organic molecules and some minerals in the plant body. Phloem often transports sugars from where they are produced to other sites where they are also used. Recall that primary phloem occurs in the vascular bundles of herbaceous plants and secondary phloem occurs as the inner bark of woody plants (refer back to Figure 35.20). In contrast to xylem, whose transport tissues are dead and empty of cytoplasm at maturity, mature phloem tissues remain alive and retain at least some cytoplasmic components. A closer look at phloem structure and function will help to illuminate these differences.

Phloem Structure Phloem tissues of flowering plants include supporting fibers, parenchyma cells, sieve-tube elements, and adjacent companion cells. **Sieve-tube elements** are arranged end to end to form transport pipes (**Figure 38.21**), analogous to the way that the xylem's vessel elements are aligned to form longitudinal vessels. Together, the sieve-tube elements and companion cells form a system for the transport of soluble organic material made during photosynthesis.

Each sieve-tube element and companion cell pair has a common origin. They are produced by an unequal division of a single precursor cell and are therefore linked by plasmodesmata formed at cytokinesis. The smaller of the two cells develops into a companion cell, whose name reflects its life-support function, and the larger of the two cells develops into a sieve-tube element. The sieve-tube element loses its nucleus and most of its cytoplasm as an adaptation that reduces obstruction to bulk flow. Mature sieve-tube elements retain only a thin film of peripheral cytoplasm that includes some endoplasmic reticulum, plastids, and mitochondria. The end walls of developing sieve-tube elements become perforated by the action of wall-digesting enzymes that enlarge existing plasmodesmata. The perforated end walls of mature sieve-tube elements are known as **sieve plates**, and the numerous perforations are known as **sieve plate pores**. Phloem sap passes through these plates from one sieve-tube element to another.

Mature sieve-tube elements are not dead. However, because they lack a nucleus, they depend on their neighboring companion cell for messenger RNA (mRNA) and proteins, which are supplied via plasmodesmata. For example, when a plant's conducting system

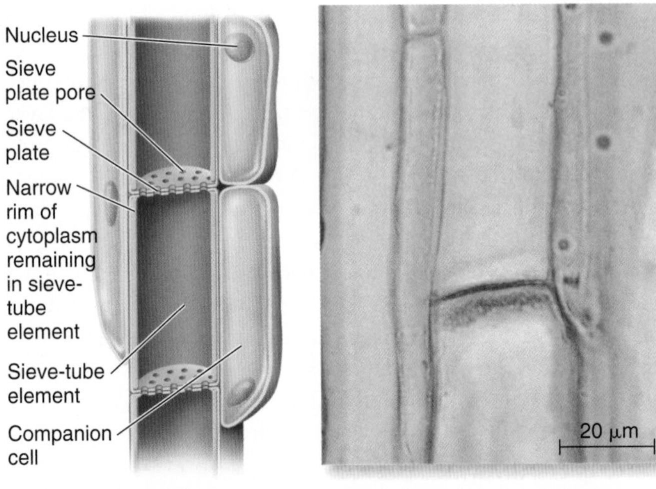

(a) Sieve-tube elements and companion cells

(b) Light micrograph of phloem stained with blue dye, showing sieve-tube elements

Figure 38.21 **Sieve-tube elements and companion cells of phloem.**

BioConnections: Look back to Figure 35.16, which illustrates the development of guard cells. In what way are guard cell development and companion cell development similar?

is damaged, a short-term wound response occurs that involves a protein known as **P protein** (for phloem protein). Masses of this protein accumulate along sieve plates, preventing loss of phloem sap (Figure 38.22). This mass functions much like a clot that helps reduce blood loss from wounded animals. P protein also binds to the cell walls of pathogens, thereby helping to prevent infection at wounds. However, sieve-tube elements cannot produce P protein by themselves. Their companion cells provide either P protein mRNA or the protein itself to sieve-tube elements. In a longer term response, plants deposit the carbohydrate callose as a sealant at the wound site.

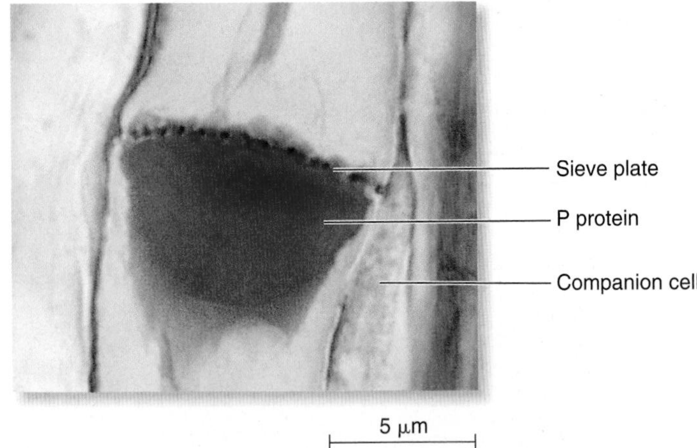

Figure 38.22 **Phloem wound response.** When phloem is damaged, the cytoplasm of a sieve-tube element surges toward the sieve plate, depositing P protein, stained red in this light micrograph. In this location, P protein helps to prevent infection and leakage of solutes.

Phloem Loading Companion cells also play an essential role in moving sugars into sieve-tube elements for long-distance transport, a process known as **phloem loading**. Although glucose and some other monosaccharides can occur in phloem, the disaccharide sucrose is the main form in which most plants transport sugar over long distances. Plant biologists think that sucrose is less vulnerable to metabolic breakdown en route than are monosaccharides.

Two types of phloem loading occur: symplastic and partly apoplastic. Many woody plants transport sucrose from sugar-producing cells of the leaf mesophyll to companion cells and then to sieve-tube elements via plasmodesmata, a process known as symplastic phloem loading (Figure 38.23a). The advantage of symplastic loading is that it does not require ATP; by moving through plasmodesmata, sugar does not have to cross plasma membranes.

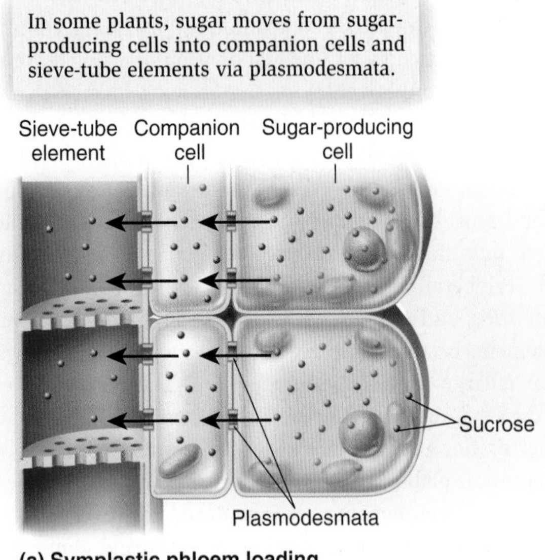

(a) Symplastic phloem loading

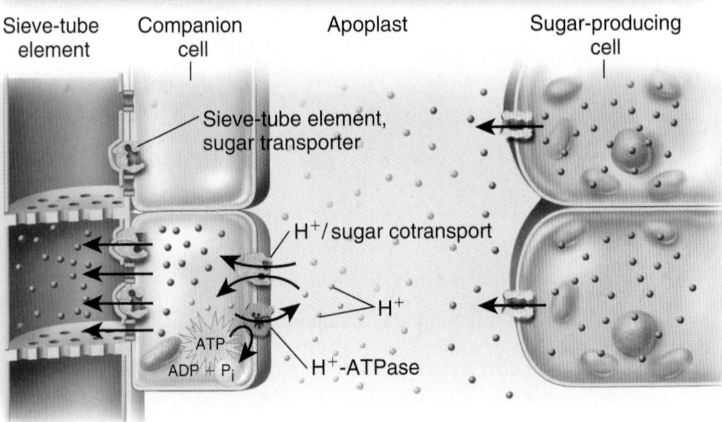

(b) Partly apoplastic phloem loading

Figure 38.23 **Symplastic and partly apoplastic phloem loading.**

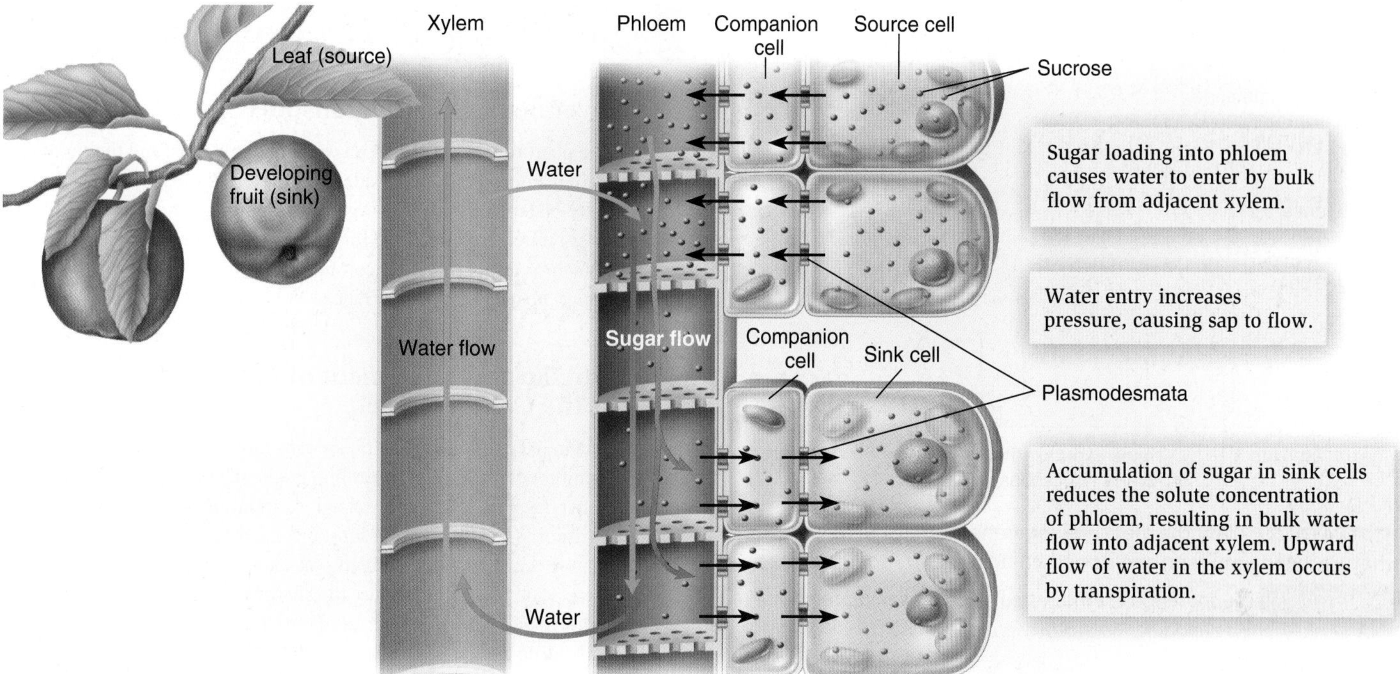

Figure 38.24 **Pressure-flow hypothesis for phloem transport.**

In contrast, most herbaceous plants, including important crop plants and the model plant *Arabidopsis,* load sugar into sieve-tube elements or companion cells from intercellular spaces, often against a concentration gradient. ATP must be used to move the sugar across a plasma membrane into a companion cell or sieve-tube element (Figure 38.23b). Therefore, this second type of phloem loading is partly apoplastic and partly a transmembrane process.

The Pressure-Flow Hypothesis Helps Explain Transport in Phloem Tissues

We have learned that transpiration, driven by the Sun's energy, moves water in plant xylem by the processes of cohesion and tension. Once sugar has been loaded into phloem sieve-tube elements, how does it move within the plant? The most common explanation is that phloem transport is driven by differences in turgor pressure that occur between cells of a **sugar source**, where sugar is produced, and those of a **sugar sink**, where sugar is consumed. Photosynthetic leaf mesophyll is the main sugar source. Roots and developing leaves, seeds, and fruits are examples of sugar sinks. In the process known as **translocation**, phloem transports substances from source to sink. The direction of phloem movement can be horizontal as well as vertical, depending on the relative positions of the sources and sinks.

Because sieve-tube elements near source tissues have higher solute contents than surrounding tissues, water tends to rush into them from nearby xylem, thereby building turgor pressure. The production of turgor pressure requires an intact plasma membrane. This explains why mature phloem sieve-tube elements must be alive in order to function.

In contrast to sieve-tube elements near source tissues, sieve-tube elements near sink tissues have lower solute concentration. The resulting water pressure difference drives the bulk flow of phloem sap from source to sink tissues. This explanation for translocation is known as the **pressure-flow hypothesis** (Figure 38.24). At sink tissues, sugar is typically unloaded through plasmodesmata (see Figure 38.24). Because plasmodesmata are very narrow in comparison with sieve-tube elements, they slow the flow of phloem sap from sieve-tube elements into sink tissues. This reduction in flow rate helps to equalize the distribution of phloem sap, preventing delivery of too much sap to any single sink. When the solute concentration of phloem sap has been sufficiently reduced, water flows from the phloem back into the xylem, where upward transport occurs. The reliance of phloem bulk flow on water supplied by the xylem explains the close proximity of phloem and xylem tissues in vascular bundles and woody stems.

Genomes & Proteomes Connection

Microarray Studies of Gene Transcription Reveal Xylem- and Phloem-Specific Genes

Xylem and phloem tissues have important economic, agricultural, and ecological roles. For example, secondary xylem produced by the vascular cambium, otherwise known as wood, forms the basis of our forest products industry. The ability of plants to respond to environmental factors also relies on vascular tissues to transport hormones. Seeking increased fundamental understanding and enhanced ability to engineer plants, molecular biologists have tried to identify the genes involved in xylem and phloem development. The use of microarrays, also known as gene chips, is providing a genomewide view of gene expression related to xylem and phloem.

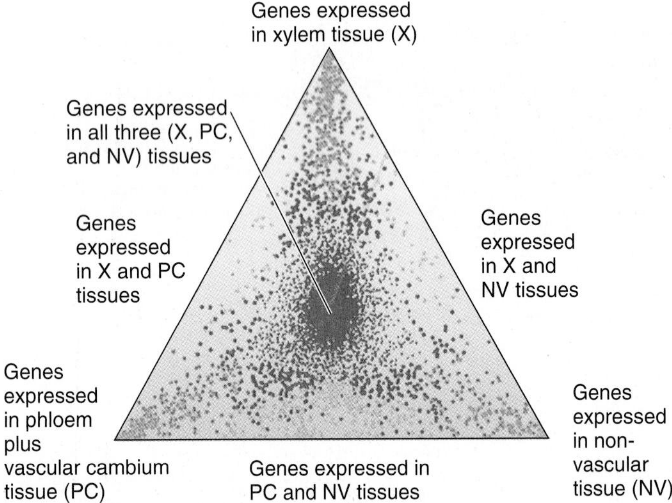

Figure 38.25 **Triangle plot showing gene expression related to immature xylem, phloem, and nonvascular tissue in *Arabidopsis*.**

Concept Check: *Suppose you are interested in identifying proteins that are involved in the development of both xylem and phloem. How might results illustrated here help you?*

Chengsong Zhao, Eric Beers, and their coworkers compared gene expression in immature xylem (X), phloem plus vascular cambium (PC), and nonvascular (NV) tissues dissected from young *Arabidopsis* plants. Recall that the vascular cambium is the meristematic tissue that produces secondary phloem (inner bark) and secondary xylem (wood) (refer back to Figure 35.20). The investigators used a dissecting microscope to view 1-cm-long pieces of tissue and a razor blade to make a longitudinal cut into nonvascular tissues and phloem, isolating the immature xylem. Then they peeled the nonvascular tissue away from the phloem. They ensured that their tissue samples were pure and authentic by checking for the expression of marker genes already known to be tissue-specific. The investigators ascertained that their X and NV tissue samples had very low expression of certain PC genes and that their PC samples had very low expression of certain X and NV genes. They then extracted RNA from the three separate tissues and used it to make labeled nucleic acid for hybridization to gene microarrays containing about 90% of the *Arabidopsis* genome (refer back to Table 21.2).

Binding of labeled nucleic acid revealed which genes were expressed. By comparing the binding patterns on different chips hybridized with nucleic acid from distinct tissues, the researchers were able to determine the genes that were preferentially expressed in X, PC, and/or NV tissues. These differences are revealed by a "triangle plot" that reflects the proportions of genes expressed in the three tested tissues (**Figure 38.25**). Each dot represents a different gene. The dots representing genes expressed primarily in only one tissue type were placed at the triangle corners, and genes expressed in two tissues but not the third are plotted at the triangle edges. Genes expressed in all three tissues lie in the center. The researchers identified 319 X-biased genes, 211 PC-biased genes, and 154 NV-biased genes. Not surprisingly, 17% of the X-expressed genes were related to cell-wall biosynthesis, including lignin production. The investigators then confirmed, via microscopy, that several proteins predicted by microarray data to be located in X, PC, or NV tissues were actually made in these tissues.

Summary of Key Concepts

38.1 Overview of Plant Transport

- The plant root system takes up water and minerals, and the shoot system absorbs carbon dioxide from the air. Photosynthetic cells use these materials to produce organic compounds. The vascular system xylem transports water and minerals from roots to shoots, and the phloem transports organic compounds from photosynthetic to nonphotosynthetic tissues (Figure 38.1).

38.2 Uptake and Movement of Materials at the Cellular Level

- Facilitated diffusion is the transport of molecules across plasma membranes down a concentration gradient with the aid of membrane transport proteins, typically channels and transporters. Active transport—the transport of substances across plasma membranes against concentration gradients—usually requires ATP. Plasma membrane proton pumps use the energy released by ATP hydrolysis to move protons from the cytosol into the intercellular space, generating a membrane potential. Potential energy released by the flow of protons back into the cell can be coupled to the transport of ions and solutes (Figures 38.2, 38.3).
- Turgor pressure increases as water enters plant cells because cell walls restrict the extent to which cells can swell. A cell that is so full of water that the plasma membrane presses closely against the cell wall is turgid. A cell that contains so little water that the plasma membrane pulls away from the cell wall is plasmolyzed. A flaccid cell has a water content between these extremes (Figure 38.4).
- Solute potential and pressure potential arising from the presence of a cell wall are major factors affecting cellular water potential. Water moves from a region of higher water potential to a region of lower water potential (Figure 38.5).
- Plants display a variety of adaptations that help them cope with cellular osmotic stress.

38.3 Tissue-Level Transport

- Tissue-level transport occurs in three forms. Transmembrane transport involves the movement of materials from one cell to another from intercellular spaces, across plasma membranes, and into the cytosol. Symplastic transport allows materials to move from one cell to another through the symplast, the cells' cytosol and plasmodesmata, without crossing plasma membranes. In apoplastic transport, water and solutes move through the apoplast, the water-filled cell walls and intercellular spaces of tissues (Figures 38.6, 38.7).
- In roots, waxy Casparian strips on cell walls of endodermal tissue function as diffusion barriers that reduce the movement of toxic ions and concentrate useful minerals into the plant vascular system (Figure 38.8).

38.4 Long-Distance Transport

- Water and solutes move for long distances by bulk flow within the xylem and phloem. Plant vascular tissues are adapted in ways that reduce resistance to bulk flow. Bulk flow of water upward in xylem is powered by the water pressure difference between moist soil and drier air, the latter resulting from solar heating. Xylem is the main conduit for water and dissolved mineral nutrients, but it may also transport certain organic compounds (Figures 38.9, 38.10).

- The water-conducting cells of xylem, tracheids and vessel elements (together known as tracheary elements), are dead and empty of cytoplasm at maturity. Pits in tracheary element walls allow water entry and exit, and narrow or constrict in response to xylem sap solute content. Vessel elements are wider than tracheids but are more vulnerable to blockage by air bubbles or embolisms. Root pressure, the effects of which include the morning water drops on leaf tips known as guttation, helps some plants to refill embolized vessels (Figures 38.10, 38.11, 38.12, 38.13, 38.14).
- Transpiration is the evaporative loss of water from plant surfaces. The cohesion-tension theory proposes that as the result of water's cohesion and the tension exerted on water at the plant's surface by evaporation, a continuous stream of water can be pulled up through the plant body from the soil, into roots, through stems, and into leaves (Figures 38.15, 38.16).
- Regulation of stomata helps plants prevent excessive water loss by transpiration. Expansion of guard cells causes stomata to open, allowing CO_2 intake. Guard cell deflation causes pores to close, limiting water loss. Plants under existing or predicted water stress often drop their leaves in a process known as abscission, an adaptive response that lets plants avoid very low water potentials and the threat of embolism (Figures 38.17, 38.18, 38.19, 38.20).
- Organic solutes and minerals are transported in phloem sap as the result of osmosis. Phloem sap moves within sieve-tube elements, which are living when mature, but lack a nucleus and are thus dependent on companion cells. Phloem loading, the movement of sugars into sieve-tube elements for long-distance transport, occurs by symplastic or partly apoplastic transport (Figures 38.21, 38.22, 38.23).
- In the process known as translocation, phloem transports substances from source to sink. The pressure-flow hypothesis helps to explain translocation as a process driven by differences in turgor pressure that occur between a sugar source (for example, leaves) and a sugar sink (for example, developing fruit) (Figure 38.24). Genes associated with xylem and phloem development are beginning to be understood (Figure 38.25).

Assess and Discuss

Test Yourself

1. An aquaporin is
   a. a channel protein that allows the influx of K^+ into cells, causing water to also flow in between the phospholipids of a plasma membrane.
   b. a type of blue-colored pore in the epidermal surfaces of plants.
   c. a protein channel in plasma membranes that facilitates the diffusion of water.
   d. a protein transporter in plasma membranes that uses protons to cotransport water.
   e. none of the above.
2. Why is turgor pressure a property of plant cells?
   a. Plant cells possess the necessary chloroplasts.
   b. Plant cells possess a cell wall, necessary for formation of turgor.
   c. Plant cells possess mitochondria, which provide the ATP needed for turgor.
   d. all of the above
   e. none of the above
3. How might plant cells avoid losing too much water in very cold, dry, or saline habitats?
   a. They may balance the osmotic condition of their cytosol with that of the environment.
   b. Their epidermal cells may be coated with waxy cuticle.
   c. They may stabilize their membranes with sugars.
   d. They may produce more aquaporin water channels to take maximum advantage of available moisture.
   e. All of the above are possible.
4. What are ways in which plants accomplish tissue-level transport?
   a. transmembrane transport of solutes from one cell to another
   b. symplastic transport of materials from one cell to another via plasmodesmata
   c. apoplastic transport of water and dissolved solutes through cell walls and intercellular spaces
   d. all of the above
   e. none of the above
5. A root endodermis is
   a. an innermost layer of cortex cells that each display characteristic Casparian strips.
   b. a layer of cells just inside the epidermis of a root.
   c. a layer of cells just outside the epidermis of a root.
   d. a group of specialized cells that occur within the root epidermis.
   e. none of the above.
6. Which of the statements listed best explains how water enters root cells from the soil?
   a. Roots accumulate sugars from shoots, thereby increasing root cell ability to absorb water by osmosis.
   b. Roots actively pump water from the soil using the chemical energy of ATP.
   c. Water enters the spongy spaces between cells and within cell walls.
   d. Membrane-level transport of ions from soil into root hair cells increases the osmotic flow of water into them.
   e. Both c and d.
7. What features of water explain how it can be drawn up a tall tree from roots to leaves?
   a. cohesion, the result of extensive hydrogen bonding
   b. adhesion, water's tendency to stick to surfaces such as the inner walls of tracheid and vessels
   c. high surface tension that develops when water evaporates from intercellular leaf spaces
   d. all of the above
   e. none of the above
8. What feature of vascular plants contributes to their ability to maintain relatively stable internal water content?
   a. a waxy surface cuticle
   b. an extensive root system that mines water from soil
   c. specialized water-conducting tracheary elements composed of dead cells
   d. epidermal pores that open and close
   e. all of the above
9. What structural features of guard cells foster their ability to form an open pore in plant epidermal surfaces?
   a. thickened inner cell walls and radially oriented microfibrils
   b. thickened outer cell walls and radially oriented microfibrils
   c. thickened inner cell walls and longitudinal microfibrils
   d. thickened outer cell walls and longitudinal microfibrils
   e. uniform thickness of cell walls and randomly arranged microfibrils

10. What substances plug wounded sieve-tube elements, thereby preventing the leakage of phloem sap?
    a. X protein and callose
    b. C protein and callose
    c. P protein and callose
    d. P protein and sucrose
    e. none of the previously listed choices

Conceptual Questions

1. Why is it a bad idea to overfertilize your houseplants? If the amount recommended on the package is good, wouldn't more be better?
2. Why is it a bad idea for subsistence farmers (those barely able to grow enough crops to feed themselves) to allow livestock to graze natural vegetation to the point that it disappears?
3. A principle of biology is that *living organisms interact with their environment.* In the desert southwestern U.S., the ocotillo plant is often used as a landscaping plant, thanks to its interesting shape and beautiful floral displays. Imagine that you are responsible for property landscaped with ocotillo but find the plants bare of leaves. Can you assume that the plants are dead and need to be replaced?

Collaborative Questions

1. Imagine that you are part of a team assigned to determine what environmental conditions best suit a new crop so that the crop can be recommended to farmers in appropriate climate regions. What features of the crop plants might you investigate?
2. Take a look outside or imagine a forest or grassland. What can you deduce about the availability of soil water from the types of plants that occur?

Online Resource

www.brookerbiology.com

Stay a step ahead in your studies with animations that bring concepts to life and practice tests to assess your understanding. Your instructor may also recommend the interactive eBook, individualized learning tools, and more.

Chapter Outline

39.1 An Overview of Flowering Plant Reproduction

39.2 Flower Production, Structure, and Development

39.3 Male and Female Gametophytes and Double Fertilization

39.4 Embryo, Seed, Fruit, and Seedling Development

39.5 Asexual Reproduction in Flowering Plants

Summary of Key Concepts

Assess and Discuss

Flowering Plants: Reproduction

39

The reproductive success of dandelions.

Dandelions sometimes seem to be taking over the world, growing abundantly in open, sunny areas. Bright yellow flower heads, as shown in the chapter-opening photo, are one of the secrets of dandelions' success. If you pull a dandelion flower head apart, you can see that it is actually a bouquet of 200 or so small flowers. Each flower produces a tiny, one-seeded fruit equipped with a "parachute" for effective long-distance dispersal by wind. Each dandelion plant can produce up to 5,000 fruits during its lifetime, which explains how dandelions can spread so rapidly across a landscape.

Though most flowering plants produce seeds by means of sexual reproduction, dandelions and some other plants are able to produce seeds by a type of asexual reproduction, a process that does not involve meiosis and fusion of gametes. As a result, the traits of asexually reproducing dandelion parents and their progeny are uniform. Asexual reproduction could be very usefully applied in agriculture, because most crops reproduce only sexually. Each year many U.S. farmers buy and plant hybrid seeds that develop into mature plants having uniform and desirable trait combinations. Such farmers typically do not use seed from one year's crop to plant the next, because sexual reproduction mixes genes into diverse combinations present among the resulting seeds. For this reason, plants that grow from sexually produced seeds do not uniformly express the desirable trait combinations present in their hybrid parents. But if hybrid crop plants could be engineered to produce seed asexually, as dandelions do, farmers might be able to use seed from one crop to plant the next and continue to harvest uniformly desirable crops. Because reproduction is so important to agriculture, plant biologists aim to better understand both sexual and asexual reproduction in flowering plants.

Flowers, fruits, and seeds are essential reproductive features of the diverse types of flowering plants found in nature. This chapter begins with an overview of the reproductive cycle of flowering plants that describes how flowers, fruits, and seeds function. This overview provides essential background for a closer look at flower structure and development and some of the genes that control flower production and appearance. We will also examine the sexual reproductive processes by which plants produce gametes and accomplish fertilization, thereby producing zygotes, embryos, seeds, fruits, and seedlings. Finally, we will take a closer look at the ways in which dandelions and some other plants reproduce without using the sexual process.

39.1 An Overview of Flowering Plant Reproduction

Learning Outcomes:

1. Explain how the sexual life cycle of flowering plants differs from that of earlier-evolved mosses.
2. List the four organs found in many flowers and the functions of each organ.
3. Describe the phenomenon of double fertilization.
4. Describe the parts of an angiosperm seed and explain how each part is produced and functions.

Most flowering plants display **sexual reproduction**, the process by which two gametes fuse to produce offspring that have unique combinations of genes. Flowering plants, also known as angiosperms, inherited their sexual life cycle, known as alternation of generations, from ancestors extending back to the earliest land plants (see Chapters 29 and 30). Though all plants share the same basic life cycle, flowering plants display unique reproductive features. In this section, we

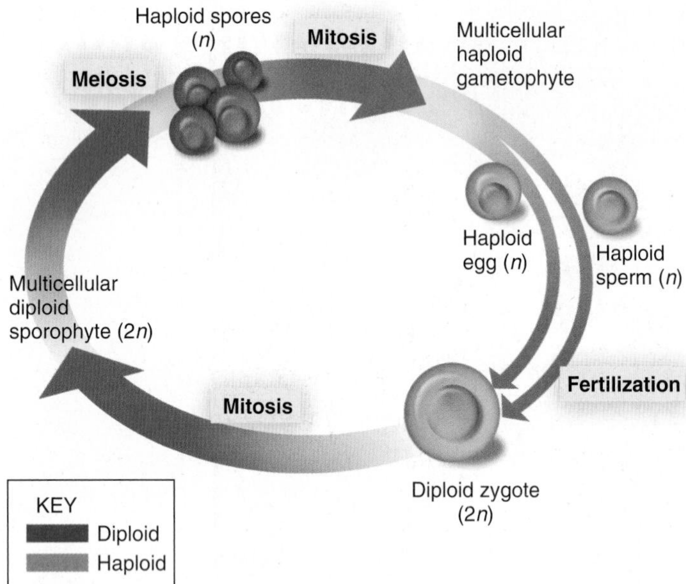

Figure 39.1 Alternation of generations, the plant life cycle.

BioConnections: Look back to Figure 30.14 to see a more detailed illustration of the flowering plant life cycle. How can one recognize and where can one find the gametophyte generation of a flowering plant?

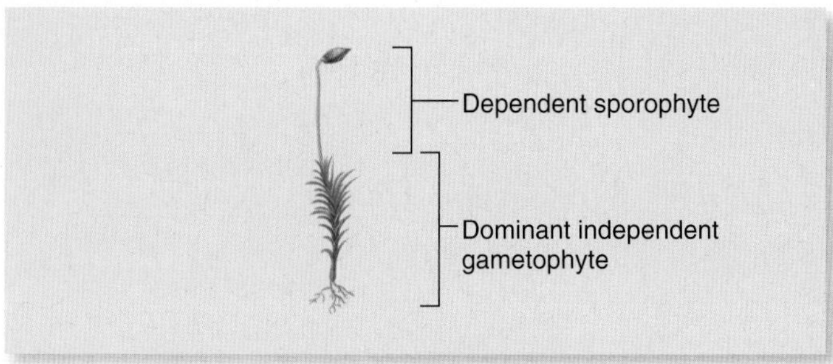

(a) Gametophyte-dominant bryophyte (moss)

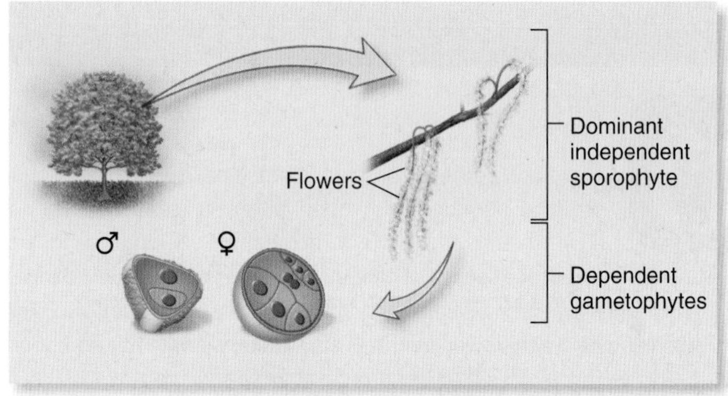

(b) Sporophyte-dominant flowering plant (oak)

Figure 39.2 Evolutionary shift in plant life cycle stage dominance. **(a)** In mosses, the gametophyte is the dominant life cycle stage, and the sporophyte depends on the gametophyte for resources. **(b)** In flowering plants such as oak trees, the sporophyte life cycle stage is dominant. Microscopic flowering plant gametophytes develop and grow within sporophytic flower tissues and depend completely on sporophytes.

Concept Check: What advantages do flowering plants obtain by having such small and dependent gametophytes?

first review the general features of alternation of generations and then consider more specific features of the angiosperm life cycle.

Flowering Plants Display Alternation of Generations

All groups of land plants produce two multicellular life cycle stages, in essence, two distinct plants. These two life cycle stages are the diploid, spore-producing **sporophyte** and the haploid, gamete-producing **gametophyte**. In all groups of plants, haploid spores are typically produced by diploid sporophytes as the result of meiosis. These spores undergo mitotic cell divisions to produce multicellular gametophytes. Certain cells within the gametophytes differentiate into gametes. Thus, meiosis does not directly generate the gametes of plants. The processes of meiosis and fertilization form the transitions between the sporophyte and gametophyte life stages and link them in a cycle (**Figure 39.1**). The land plant life cycle is known as **alternation of generations** because it involves the cycling between distinct sporophyte and gametophyte generations.

During the evolutionary diversification of land plants, the sporophyte generation has become larger and more complex, while the gametophyte generation has become smaller and less complex. To illustrate this, let's compare the life cycle stages of mosses to those of angiosperms. Mosses diverged early in the history of land plants, whereas angiosperms appeared much later. During the intervening time, the relative sizes and dependence of the sporophyte and gametophyte generations changed dramatically. Moss sporophytes are small structures that always grow attached to larger, photosynthetic gametophytes, because moss sporophytes are incapable of independent life (**Figure 39.2a**). Moss sporophytes depend on gametophytes to supply them with essential nutrients.

In contrast, flowering plant sporophytes are notably larger and more complex than gametophytes. A tall oak tree, for example, is a single sporophyte. However, oak gametophytes are few-celled, microscopic structures that develop and grow within flowers (**Figure 39.2b**). In addition, photosynthetic oak seedlings and trees grow independently, but nonphotosynthetic oak gametophytes depend completely on the sporophyte generation for their nutrition. A closer look at flower structure will help us to gain a more complete view of angiosperm gametophyte structure and function.

Flowers Produce and Nurture Male and Female Gametophytes

The literary wit Gertrude Stein famously paraphrased Shakespeare ("a rose by any other name would smell as sweet") when she wrote, "A rose is a rose is a rose." Everyone knows that a rose is a flower, but what, exactly, is a flower? A **flower** is defined as a reproductive shoot, a stem branch that produces reproductive organs instead of leaves. Flowers are organ systems because several different organs typically occur within a flower (see Chapter 35). Flower organs are produced

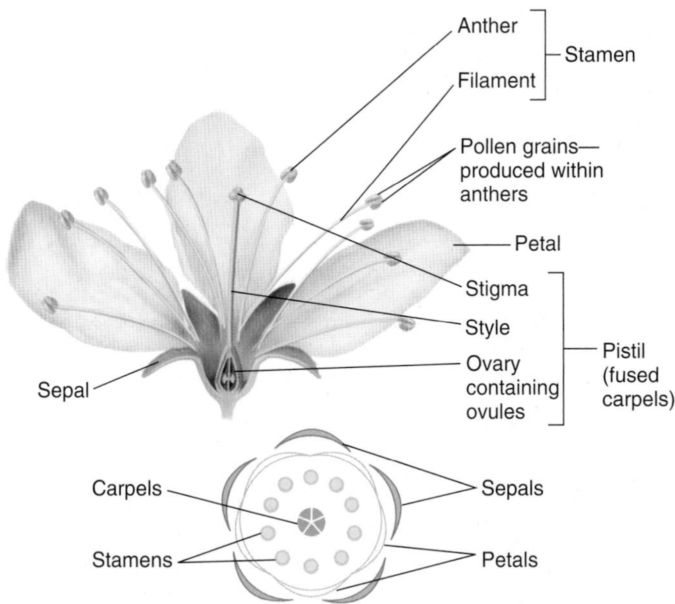

(a) Flower parts

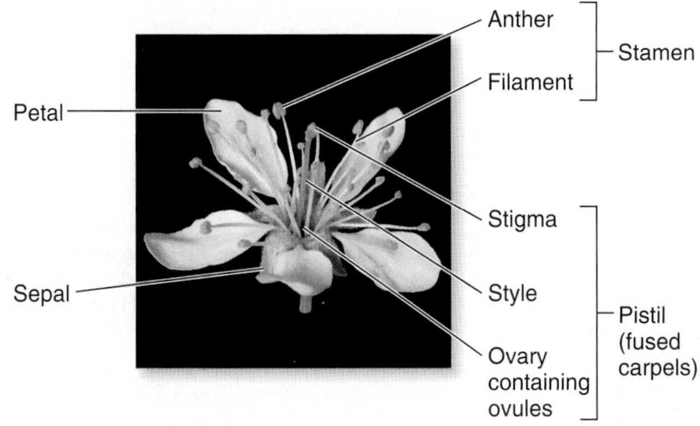

(b) *Prunus americana* (plum)

Figure 39.3 The structure of a typical flower.

Concept Check: Do all flowers have all of the structures illustrated here?

by shoot apical meristems much like those that generate leaves and are thought to have evolved from leaflike structures by descent with modification (refer back to Figure 30.15).

Most flowers contain four types of organs: sepals, petals, stamens, and carpels (Figure 39.3). **Sepals** often function to protect the unopened flower bud. **Petals** usually serve to attract insects or other animals for pollen transport (refer back to Figures 30.19 and 30.20). **Stamens** and **carpels** each produce distinctive types of spores by the process of meiosis. From these spores, tiny multicellular gametophytes develop, and certain gametophytic cells become specialized gametes.

Stamens Stamens produce **male gametophytes** and foster their early development. Most stamens display an elongate stalk, known as a **filament**, which is topped by an anther (see Figure 39.3). Filaments contain vascular tissue that delivers nutrients from the parental sporophyte to the anthers. Each **anther** is a group of four sporangia, structures in which spores are produced. Within the anther's sporangia, many diploid cells undergo meiosis, each producing four tiny, haploid spores. Because they are so small, generally 25–50 μm in diameter, the spores produced within anthers are known as **microspores**. Immature male gametophytes, known as **pollen grains**, develop from microspores. The term pollen comes from a Latin word meaning "fine flour," reflecting the small size of pollen grains. Pollen grains are eventually dispersed through pores or slits in the anthers. At the time of dispersal, the pollen grain is a two- or three-celled immature male gametophyte produced by mitotic division. During a later phase of development, a mature male gametophyte produces **sperm cells**.

Carpels Carpels are vase-shaped structures that produce, enclose, and nurture **female gametophytes**. Carpels contain veins of vascular tissue that deliver nutrients from the parent sporophyte to the developing gametophytes. The term **pistil** (named for its resemblance to the pestle used to grind materials to a powder) refers to a single carpel or several fused carpels (see Figure 39.3). The topmost portion of a pistil, known as a **stigma** (Greek, meaning mark), receives pollen grains. The **style** is the middle portion of the pistil, and an ovary is at the bottom of the pistil. The **ovary** produces and nourishes one or more ovules. An **ovule** consists of a spore-producing structure (a sporangium) and enclosing tissues consisting of modified leaves known as **integument**. Within an ovule, a diploid cell produces four haploid **megaspores** by meiosis, three of which die. The surviving megaspore generates a female gametophyte by mitosis. The female gametophytes of flowering plants typically consist of seven cells, one of which is the female gamete, the **egg cell**. This basic information about male and female gametophytes will help us to understand how they function to produce a young sporophyte within a seed.

Fertilization Triggers the Development of Embryonic Sporophytes, Seeds, and Fruits

In flowering plants, fertilization leads to the production of a young sporophyte that lies within a seed, completing the life cycle (Figure 39.4). Prior to fertilization, pollen grains released from anthers first find their way to the stigma of a compatible flower, a process known as **pollination**. Some plants display **self-pollination**, in which pollen from the anthers of a flower is transferred to the stigma of the same flower or between flowers of the same plant. **Cross-pollination**, which occurs when a stigma receives pollen from a different plant of the same species, is also common. Many flowers are attractive to insects or other animals that transport pollen, whereas oak flowers and those of some other angiosperms are adapted for pollen transport by wind. A few plants move pollen by means of water currents. Flowers are adapted for effective pollination by diverse pollination mechanisms (refer back to Table 30.1).

Pollen Germination When pollen grains land on the stigma, the stigma functions as a gatekeeper, allowing only pollen of appropriate genotype to germinate. During germination, a pollen grain produces a long, thin **pollen tube** that contains two sperm cells. The pollen tube grows through the style toward the ovary. Upon reaching the ovules, the pollen tube grows through the **micropyle**, an

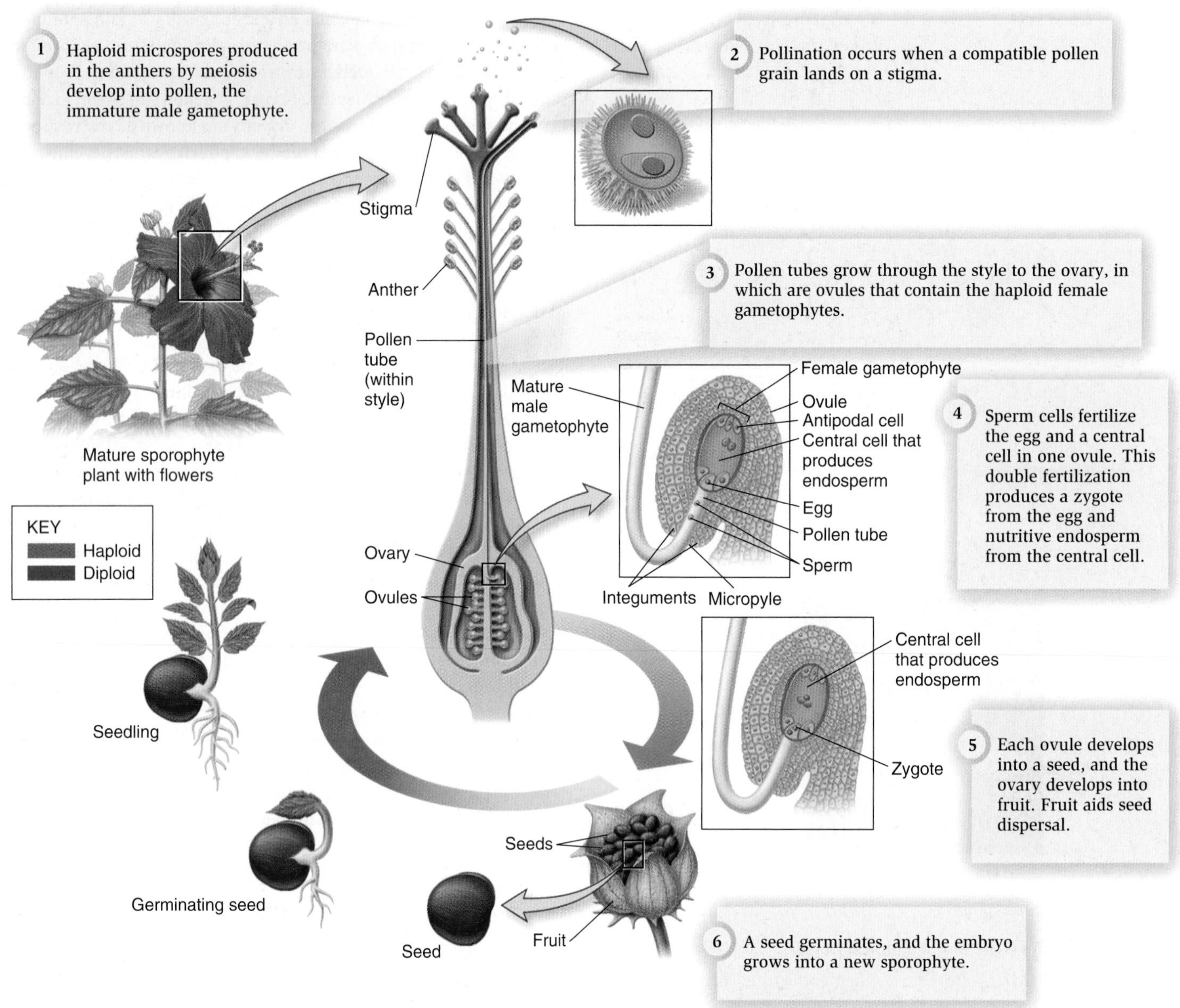

1 Haploid microspores produced in the anthers by meiosis develop into pollen, the immature male gametophyte.
2 Pollination occurs when a compatible pollen grain lands on a stigma.
Stigma
Anther
Pollen tube (within style)
3 Pollen tubes grow through the style to the ovary, in which are ovules that contain the haploid female gametophytes.
Mature sporophyte plant with flowers
KEY
Haploid
Diploid
Mature male gametophyte
Female gametophyte
Ovule
Antipodal cell
Central cell that produces endosperm
Egg
Pollen tube
Sperm
4 Sperm cells fertilize the egg and a central cell in one ovule. This double fertilization produces a zygote from the egg and nutritive endosperm from the central cell.
Ovary
Ovules
Integuments
Micropyle
Central cell that produces endosperm
Zygote
Seedling
5 Each ovule develops into a seed, and the ovary develops into fruit. Fruit aids seed dispersal.
Seeds
Germinating seed
Seed
Fruit
6 A seed germinates, and the embryo grows into a new sporophyte.

Figure 39.4 The life cycle of a flowering plant. The plant reproductive cycle is illustrated here by hibiscus.

Concept Check: *What advantage does the hibiscus flower gain by clustering its stamens around the pistil?*

opening in the ovule, and delivers sperm to the female gametophyte (see Figure 39.4). These sperm unite with haploid cells of the female gametophyte in the process of **fertilization**. Note that pollination and fertilization are distinct processes in flowering plants.

Double Fertilization Angiosperms display a phenomenon known as **double fertilization**. In this process, two different fertilization events occur. One of the two sperm cells delivered by a pollen tube fertilizes the egg cell, thereby forming a diploid **zygote**. This zygote may develop by mitotic division into a young sporophyte, known as an **embryo**. Fertilization thus begins a new cycle of alternation between sporophyte and gametophyte generations. The other sperm delivered by the same pollen tube fuses with two nuclei present in one of the cells of the female gametophyte. The cell formed by this second fertilization undergoes mitosis, eventually producing a nutritive tissue known as the **endosperm**. The embryo and the endosperm are essential parts of maturing seeds. Fertilization not only starts the development of zygotes into embryos but also triggers the transformation of ovules into seeds and ovaries into fruits. Embryo, seed, and fruit development occur at the same time.

Embryos and Seeds An embryo is a young, multicellular, diploid sporophyte that develops from a single-celled zygote by mitosis. Because they are not yet capable of photosynthesis, embryos depend on organic food and other materials supplied by sporophytes. Therefore, embryo development occurs within developing seeds located in a flower ovary. Seeds develop from fertilized ovules. Each developing seed contains an embryo and nutritive endosperm tissue, enclosed and protected by a **seed coat** that develops from the ovule integuments. When embryos and the seed coat have fully matured, they undergo drying, and the seed enters a phase of metabolic slowdown known as **dormancy**. Fully mature, dormant seeds are ready to be dispersed.

Fruit and Seed Dispersal A **fruit** is a structure that encloses and helps to disperse seeds (see Figure 39.4). Seed dispersal benefits plants by reducing competition for resources among seedlings and parental plants, and it allows plants to colonize new sites.

Fruits develop from the flower's ovary and sometimes include other flower parts. Young fruits bearing immature seeds are typically small and green. During the time that embryos and seeds are developing, the fruit also matures. The ovary wall changes into a fruit wall known as a **pericarp** (from the Greek, meaning surrounding the fruit). Mature fruits vary greatly among plant species in size, shape,

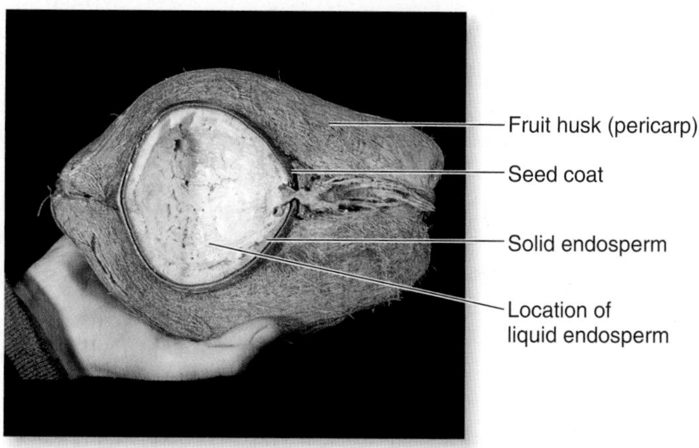

(a) Coconut fruit and seed

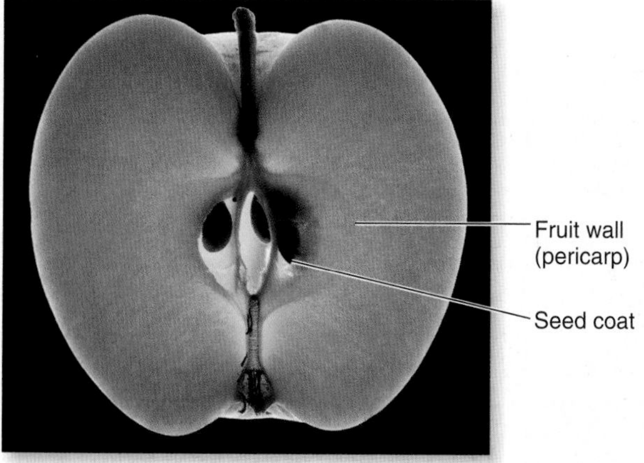

(b) Apple fruit and seed

Figure 39.5 Fruit adaptations for seed dispersal. (a) The coconut fruit's outer wall, called a husk, allows the fruit to float and thus disperse its seed among tropical shores. (b) The apple is a juicy, sweet fruit that attracts animals to consume it, thereby helping to disperse its seed.

color, and water content. These variations represent adaptations for seed dispersal in different ways. For example, single-seeded dandelion fruits are dry and lightweight and bear a fluffy "parachute" derived from the flower's sepals (see the chapter-opening photo). These features foster dispersal by wind. In contrast, coconut fruits feature an airy husk (the pericarp) that keeps them afloat in ocean currents that carry coconuts from one tropical shore to another. Inside the coconut fruit is a single, large seed loaded with liquid and solid endosperm, which people use as coconut milk and coconut meat, respectively (Figure 39.5a). These large amounts of endosperm provide nutrients that sustain coconut seedling growth on infertile, sandy shores. Fruit variation is also extremely important to wild animals and in human agriculture. For example, most fruit crops are juicy and sweet, with relatively small seeds (Figure 39.5b). In nature, these features foster dispersal by birds and other animals that feed on such fruits.

Seed Germination and Seedlings If a dispersed seed encounters favorable conditions, including sufficient sunlight and water, it will undergo seed **germination**. During seed germination, the embryo absorbs water, becomes metabolically active, and grows out of the seed coat, producing a seedling. If the seedling obtains sufficient nutrients from the environment, it grows into a mature sporophyte capable of producing flowers. In the next section, we focus on flower production, structure, and development.

39.2 Flower Production, Structure, and Development

Learning Outcomes:

1. List some examples of particular genes that control flower production or shape.
2. Give an example of how gene expression affects flowering time or flower appearance.
3. Explain recent research findings on how flowers bloom, that is, open up from a closed bud.

Flowers are essential sources of food for many animal pollinators. As the result of coevolutionary relationships with such animals, flowers occur in a spectacular array of colors and forms that attract particular pollinators (refer back to Table 30.1). Flowers also attract humans because we possess sensory systems much like those of animal pollinators. We give bouquets to show love and appreciation; decorate homes, workplaces, and objects with flowers; display flower arrangements on ceremonial occasions; and make perfume from flowers. Consequently, many types of flowers are grown for the florist and perfume industries. Flowers are necessary for the production of grain and other fruit crops. For these and other reasons, biologists investigate how flower development is controlled by environmental signals and changes in gene expression.

Environmental Signals Interact with Genes to Control Flower Production

You've probably noticed that different plants flower at particular times of the year. How do plants know when to flower? Flowering time is controlled by the integration of environmental information such as temperature and day length (photoperiod) with hormonal influences and circadian rhythms (see Chapter 36). These stimuli are perceived and integrated by leaves, which then use a mobile protein known as FT (*f*lowering *t*ime) to signal shoot meristems to produce flowers. FT protein interacts with other proteins in the shoot meristem, initiating the process by which a leaf-producing apical meristem transforms into a reproductive meristem.

Some plants—such as winter wheat—are planted and sprout in fall, are dormant in winter, and flower in the following spring. In these plants a transcription factor known as FLC represses flowering genes, thereby preventing flowers from appearing too soon. Exposure to cold winter conditions causes the production of a small RNA molecule that silences FLC, allowing these plants to flower when the appropriate day length occurs in spring. The process by which cold exposure allows plants to flower in spring is known as vernalization.

Developmental Genes Control Flower Structure

Organ identity genes specify the four basic flower organs—sepals, petals, stamens, and carpels. Other genes determine flower shape, color, odor, or grouping into bunches known as inflorescences.

The Genetic Basis of Flower Organ Identity Sepals, petals, stamens, and carpels occur in four concentric rings known as **whorls**. Sepals (collectively known as the calyx) form the outermost whorl, and petals (together known as the corolla) form an adjacent whorl type. Stamens (together, the androecium) create a third type of whorl, and carpels (the gynoecium) form the innermost whorl (Figure 39.6). The **perianth** consists of the calyx plus the corolla. You may recall that *A*, *B*, *C*, and *E* genes encode transcription factors that control the production and arrangement of these whorls (refer back to Figure 19.24).

Variation in Number of Whorls Flowers that possess all four types of flower whorls—calyx, corolla, androecium (stamens), and gynoecium (one or more carpels)—are known as **complete flowers**.

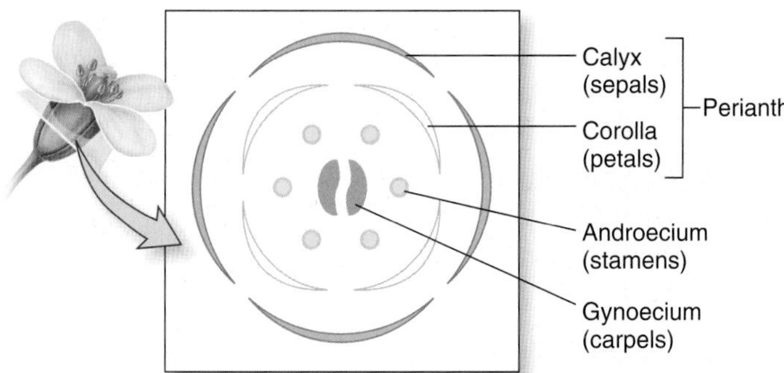

Figure 39.6 **The occurrence of flower parts in concentric whorls.**

In contrast, flowers that lack one or more flower whorls are described as **incomplete flowers**. Flowers having both stamens and carpels are said to be **perfect flowers**, whereas flowers lacking stamens or carpels are described as **imperfect flowers**. An imperfect flower that produces only carpels is known as a carpellate flower (or pistillate flower). Imperfect flowers that produce only stamens are described as staminate flowers.

Corn produces both imperfect staminate and carpellate flowers on an individual plant (Figure 39.7). The flowers of corn start to develop as perfect flowers, but in carpellate flowers, the stamens stop developing. After pollination and fertilization, each carpellate flower produces one of the kernels on a cob of corn. In contrast, staminate flowers of corn, which are found in corn tassels, produce the pollen. Corn is termed **monoecious** (meaning "one house") because it produces staminate and carpellate flowers on the same plant. Holly and willow also produce staminate and carpellate flowers, though on separate plants, and are examples of plants described as **dioecious** (meaning "two houses").

Variation in Flower Organ Number In addition to variation in whorls, flowers vary in number of organs. You may recall that flowering plants occur in two major groups differing in flower structure

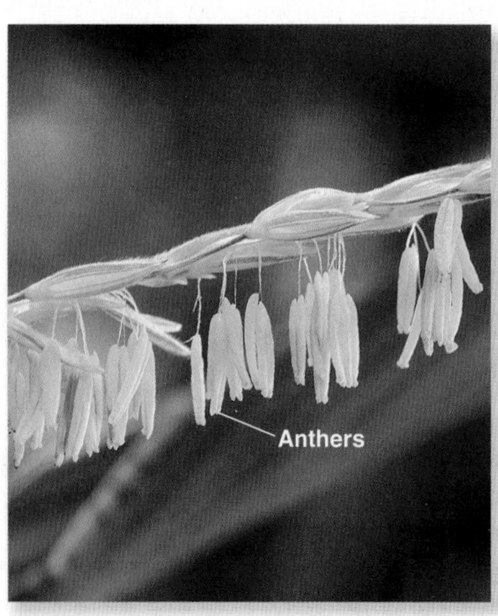

(a) Staminate flowers of *Zea mays* (corn)

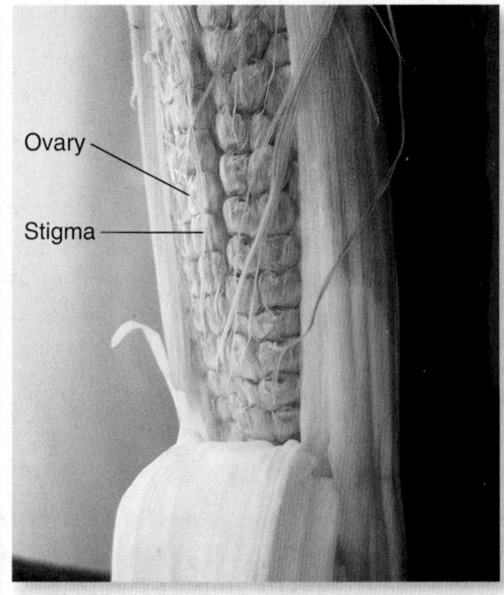

(b) Carpellate flowers of *Zea mays* (corn)

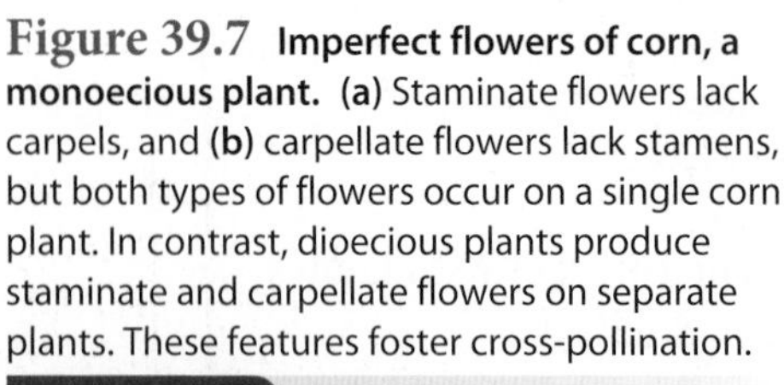

Figure 39.7 **Imperfect flowers of corn, a monoecious plant.** **(a)** Staminate flowers lack carpels, and **(b)** carpellate flowers lack stamens, but both types of flowers occur on a single corn plant. In contrast, dioecious plants produce staminate and carpellate flowers on separate plants. These features foster cross-pollination.

Concept Check: *What inference can you draw from the observation that flowers of corn lack showy petals?*

and other features: the eudicots and the monocots (see Table 35.1). Eudicot flower organs usually occur in fours or fives or a multiple of these numbers. By contrast, monocot flower organs typically occur in threes or a multiple of three. Other flowers possess relatively few organs. For example, the minuscule flowers of the tiny aquatic flowering plant *Lemna gibba* produce no perianths and only two stamens.

Many plants sold for use in gardens have been bred so that the flowers produce multiple organs. Garden roses, for example, typically have many more whorls of petals than do wild roses. This change results from a mutation that causes organs that would have become stamens to instead develop into additional petals.

Variation in Flower Color Different flower parts may vary in color. The calyx and corolla of monocot flowers such as tulip and lily are often similar in appearance and attractive function. In contrast, eudicot flowers tend to have green, leaflike sepals (see Figure 39.3) that are quite distinct from petals, which are often colorful and fragrant. Color variations arise from differences in gene action that influence pigment biosynthesis pathways.

In 1990, American plant molecular biologist Rich Jorgensen and colleagues reported one of the first cases of gene silencing in plants. You may recall that gene silencing is one of the ways in which gene expression is controlled (see Chapter 13). In an attempt to produce flowers of deeper color, the researchers introduced an extra copy of a pigment-producing gene into petunia plants. To their surprise, the extra copy of the gene sometimes produced flowers whose petals had white patches or were completely white (**Figure 39.8**). Adding the extra gene caused the plants to produce short-interfering RNA (siRNA) that not only silenced the expression of the extra gene, but also the natural pigment-producing genes, causing white patches on the petals.

Variation in Flower Fragrance The fragrances of flowers result from secondary metabolites that diffuse into the air from petals and other flower organs. One example is the terpene known as geraniol that gives roses their distinctive fragrance. Diverse fragrances function to attract particular types of animal pollinators. Because humans possess sensory systems similar to those of many pollinators, we are also attracted to many of these fragrances. Genetic variation in the synthesis of different types of secondary metabolites is responsible for the many types of flower scents.

Flower Shape Variation Resulting from Organ Fusion During their development, many flowers undergo genetically controlled fusion between whorls or fusion of the organs within a whorl. For example, pistils are often composed of two or more fused carpels. In addition, stamen filaments often partially fuse with the carpel or form a tube surrounding the pistil, a feature displayed by hibiscus flowers (see Figure 39.4). Each small dandelion flower has five petals that are fused at their sides to form a single strap-shaped structure. Some flowers have petals that are fused together to form a tube that holds nectar consumed by animal pollinators.

Variations in Flower Symmetry and Aggregation Flower shape variation can also result from changes in symmetry. Flowers that possess radial symmetry are described as regular, actinomorphic, or polysymmetric flowers. Flowers having radial symmetry can be divided into two equal parts by more than one plane inserted through the center of the flower. In contrast, flowers that display bilateral symmetry are known as irregular, zygomorphic, or monosymmetric flowers. Flowers having bilateral symmetry can be divided into two equal parts by only a single plane inserted through the center.

Symmetry, like other flower features, is under genetic control. The production of flowers having bilateral symmetry is controlled by transcription factors such those encoded by the *CYCLOIDEA* gene. For example, snapdragon (*Antirrhinum majus*) flowers are normally bilaterally symmetric, but a loss-of-function mutation in the *CYCLOIDEA* gene causes these flowers to display radial symmetry (**Figure 39.9**). Another interesting example is the cut-flower crop plant *Gerbera hybrida*, whose "flowers" are actually **inflorescences**, groups of flowers clustered together. Although many flowers occur singly, many others occur in inflorescences of diverse types that are

(a) Normal purple petunia flowers

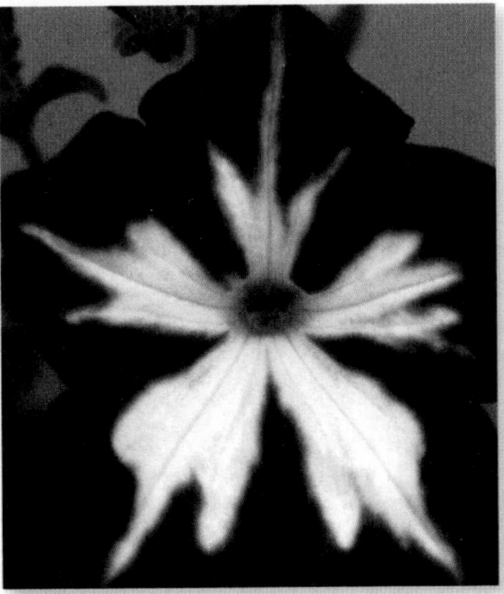

(b) Petunia flower affected by siRNA

Figure 39.8 Gene silencing and flower color in petunias. **(a)** Normal purple petunia (*Petunia hybrida*) flowers produced by the expression of all genes involved in flavonoid synthesis. **(b)** Flower of a genetically engineered plant displays petal tissues in which expression of one of the genes needed for production of purple pigment production has been silenced (suppressed) by siRNA.

(a) Normal snapdragon flower

(b) Snapdragon flower with *CYCLOIDEA* mutation

Figure 39.9 Genetic control of flower symmetry. **(a)** Normal snapdragon flowers, with functioning *CYCLOIDEA* genes, are bilaterally symmetric. **(b)** Snapdragon plants carrying mutations in the *CYCLOIDEA* gene produce flowers that have radial symmetry.

BIOLOGY PRINCIPLE The genetic material provides a blueprint for reproduction. This principle is illustrated by a comparison of normal and mutant snapdragon flowers.

adapted to particular pollination circumstances. As in the case of sunflowers (refer back to Figure 30.20b), *Gerbera* flowers occur in more than one structural type having distinctive functions. During inflorescence development, a *CYCLOIDEA*-like transcription factor is expressed more strongly at the margins so flowers in that position become bilaterally symmetric, whereas the centermost flowers of *Gerbera* are more radially symmetric (**Figure 39.10**). These central flowers produce stamens and therefore function to supply pollen. In contrast, stamens do not develop in the marginal flowers, and their petals fuse to form larger colorful structures that are specialized to attract pollinators.

Figure 39.10 An inflorescence. The clustering of flowers into a type of inflorescence known as a head, displayed here by a Gerbera daisy, is an adaptation that fosters pollination by insects.

Concept Check: List several ways in which the flowers at the rim of the Gerbera inflorescence differ from those at the center.

Feature Investigation

Liang and Mahadevan Used Time-Lapse Video and Mathematical Modeling to Explain How Flowers Bloom

Like the curled limbs of a human embryo within the womb, concave and overlapping petals occupy minimal space within a flower bud. But when flowers bloom, petals rapidly become convex, spreading outward from each other. Various hypotheses had been proposed to explain petal movements during the opening of flowers. One was that growth started at the midrib running along the center of each petal; another proposed that the blooming process was driven by differential growth rates of top and bottom petal surfaces. In 2011, Haiyi Liang and Lakshminarayanan Mahadevan reported a new explanation for blooming based on their studies of the common lily, *Lilium casablanca* (**Figure 39.11**).

The investigators placed cut stems bearing young green buds into water, in an environment of constant humidity, temperature, and light. To evaluate growth during blooming, they first painted small black dots 1 cm apart along the edges and centers of many sepals and petals. By marking many sepals and petals, the researchers repeated, or replicated, the experiment. Replication is key to the experimental process because it helps researchers to avoid making erroneous conclusions that might be reached on the basis of just a few observations. Investigators then used time-lapse video to track changes in positions of the black dots during the opening of young green buds, until the flowers had fully bloomed $4^1/_2$ days later. This process allowed investigators to measure and graph growth strain at the petal edge versus the center. These measurements also enabled them to mathematically model changes in petal shape, starting with equations that describe changes in the shapes of thin sheets (see Figure 39.11).

The biologists observed that by the end of the fourth day, each bud had absorbed one-fifth of a liter of water, increased length by 10% and width by 20%, and turned white. The inner three petals had

Figure 39.11 **Liang and Mahadevan used time-lapse video and mathematical modeling to explain how flowers bloom.**

GOAL To better understand flower blooming.

KEY MATERIALS Asiatic lily, *Lilium casablanca*, cut stems with flower buds.

	Experimental level	Conceptual level
1 Paint small black dots at 1 cm intervals along edges and centers of perianth parts on replicate flower buds.	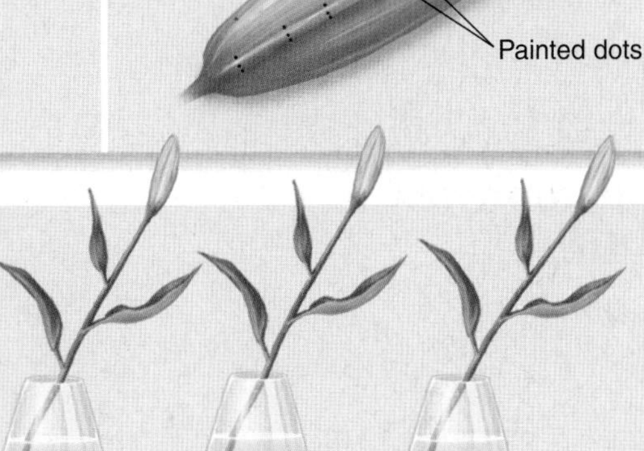	Compare rate of growth at center and edges of petals/sepals.
2 Maintain cut stems at constant temperature, humidity, and light conditions for 4.5 days, until flowers open.	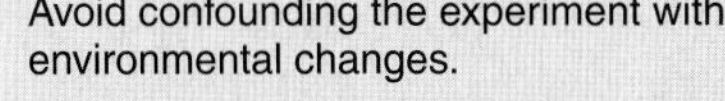	Avoid confounding the experiment with environmental changes.
3 Set up an automatic time-lapse video system that records blooming at intervals of 1 minute.		Generates a record of changes in the positions of dots on petals/sepals.
4 Use equations for growth of thin films to model petal/sepal shape changes.		Compare model to video-based observations.

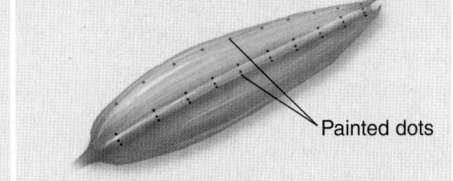

5 THE DATA

(1) The strain of longitudinal growth at petal/sepal edges increases more rapidly than at centers.

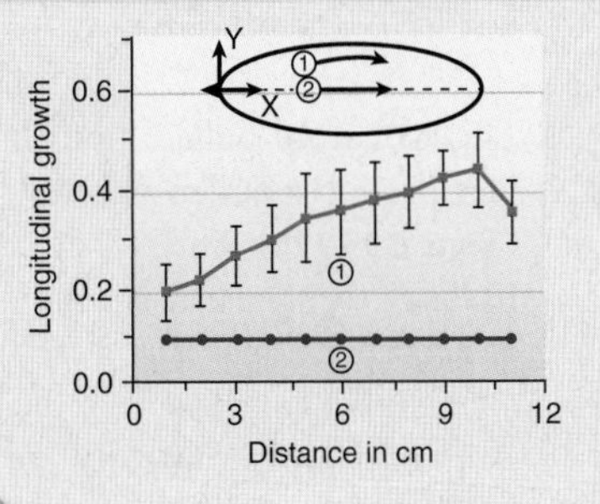

(2) Simulation of the blooming process based on physics and math modeling matches observations made with time-lapse videography.

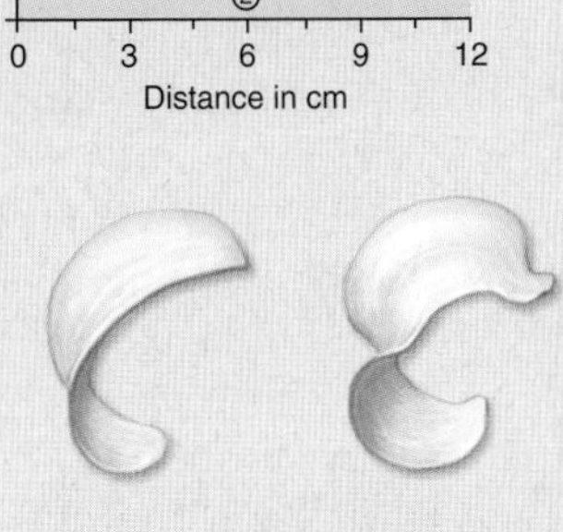

6 CONCLUSION Lily blooming is caused by faster growth at the edges than at centers of petals/sepals.

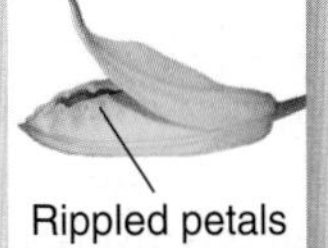

The blooming process suggests new ways to engineer the shapes of plastic sheets.

7 SOURCE Liang, H., and Mahadevan, L. 2011. Growth, geometry, and mechanics of a blooming lily. *Proceedings of the National Academy of Sciences* (USA) 108:5516–5521.

become wrinkled, especially at the edges, a key clue to the mechanism underlying the blooming process. The wrinkling reflected greater growth at the edges than in the petal center. This difference in growth generated stress values large enough to cause the flower to bloom rapidly as individual petals and sepals simultaneously reversed curvature and bent outward, a process predicted by mathematical modeling. The study revealed that flower blooming is based upon a different mechanism than previously thought and illustrates the value of applying physics and mathematical models to biological phenomena.

Experimental Questions:

1. Why do you think Liang and Mahadevan used *Lilium casablanca* for their analysis of flower blooming?
2. How did time-lapse video improve data gathering in this study?
3. What is the value of using mathematical simulations to model the blooming process?

39.3 Male and Female Gametophytes and Double Fertilization

Learning Outcomes:

1. Explain how plant sperm production differs from sperm production in animals.
2. Discuss how the pistil controls pollen germination.
3. Describe the structure of the female gametophyte.
4. Outline the different fates of the two sperm cells transmitted by each pollen tube.

In flowering plants, mature male gametophytes are pollen tubes that deliver sperm cells, the male gametes. Mature female gametophytes are located within ovules and produce egg cells, the female gametes. In this section, we will focus on the structure and function of male and female gametophytes to learn more about how fertilization occurs.

Pollen Grains Are Immature Male Gametophytes

In animals, the cells that will produce gametes—known as the germ line—are set aside from other body cells during early development, but this is not the case for plants. Instead, plants make gamete-producing cells from body cells each time that stamens form, on an "as needed" basis. Plant sperm arise from male gametophytes. Immature male gametophytes, known as pollen grains, are produced within stamens.

Pollen grains develop within sporangia located in the anthers of stamens (see Figure 39.4). Sporangia are structures produced by all plants, within which meiosis generates spores. (Note that in plants, meiosis generates haploid spores, not haploid gametes as in animals.) Inside protective plant sporangia, diploid body cells produce a cluster of four haploid microspores, each having a thin cellulose cell wall. The development of microspores into pollen grains involves two processes that occur at the same time: (1) microspore division to produce a young male gametophyte, and (2) development of a tough pollen wall that protects the gametophyte during pollen transport. Both of these processes are completed before anthers release pollen.

Each microspore nucleus undergoes one or two mitotic divisions to form a young male gametophyte. The first division gives rise to two specialized cells: a tube cell and a generative cell suspended within the tube cell (**Figure 39.12a**). The **generative cell** divides to produce two sperm cells, either before or (more commonly) after pollination. The **tube cell** produces the pollen tube, which delivers sperm to the female gametophyte.

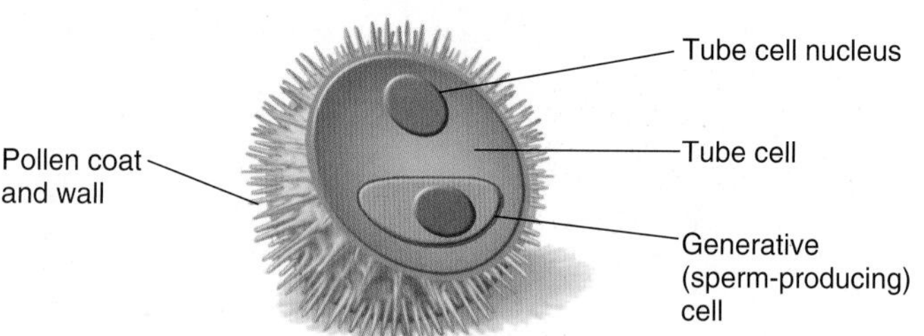

(a) A cut pollen grain showing immature male gametophyte

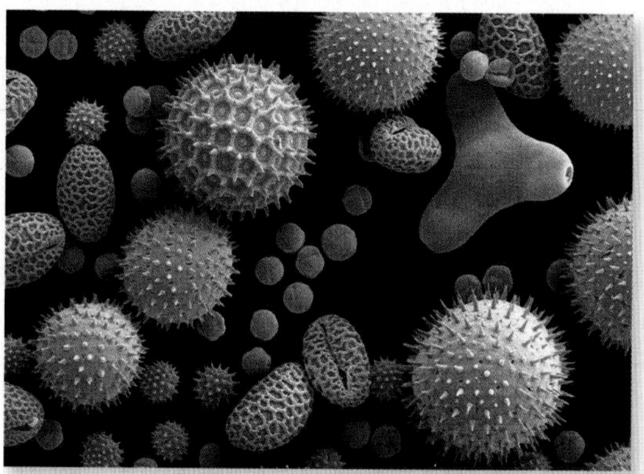

(b) SEM of whole pollen grains showing distinctive pollen wall ornamentation

Figure 39.12 Pollen grains. **(a)** Diagram of cut pollen grain. **(b)** SEM of whole pollen grains of different species.

Concept Check: *What is the maximum number of cells in a mature male gametophyte of a flowering plant?*

A mature pollen grain has a tough wall, and each plant species produces pollen whose wall has a distinctive sculptural shape (**Figure 39.12b**). The pollen wall, which surrounds the plasma membrane of the tube cell, is composed largely of a tough polymer known as sporopollenin. Named for its presence on the surfaces of mature spores and pollen, sporopollenin protects spores and pollen from damage. Development of the pollen wall starts with deposition of a blanket of the carbohydrate callose around each cluster of four microspores after they form by meiosis. The callose blanket seals microspores off from the influences of adjacent sporophyte tissues, thereby aiding pollen differentiation. Callose also provides a surface pattern for sporopollenin deposition and holds microspores together until an anther enzyme degrades the callose, freeing pollen grains from each other. As pollen grains mature, anther cells secrete a pollen coat, a

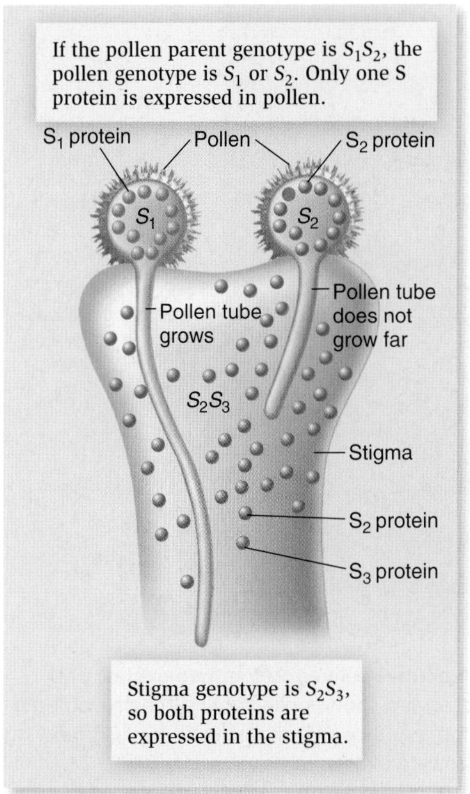

(a) Gametophytic SI: If pollen *S* allele does not match either stigma allele, pollen will germinate.

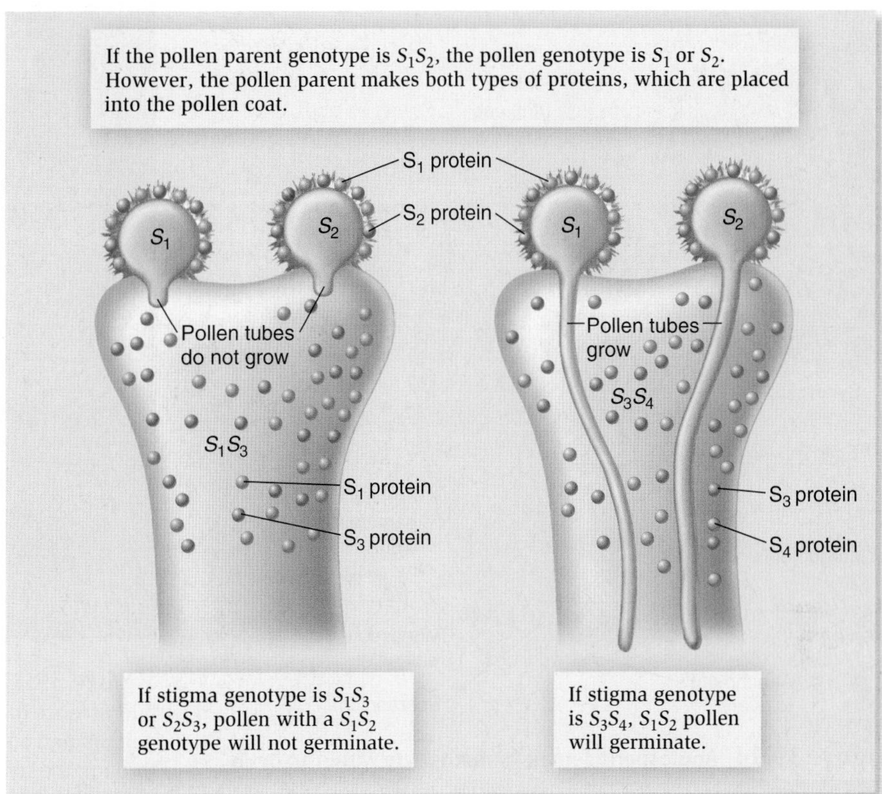

If the pollen parent genotype is S_1S_2, the pollen genotype is S_1 or S_2. However, the pollen parent makes both types of proteins, which are placed into the pollen coat.
S_1 protein
S_2 protein
S_1
S_2
Pollen tubes do not grow
S_1S_3
S_1 protein
S_3 protein
Pollen tubes grow
S_3S_4
S_3 protein
S_4 protein
If stigma genotype is S_1S_3 or S_2S_3, pollen with a S_1S_2 genotype will not germinate.
If stigma genotype is S_3S_4, S_1S_2 pollen will germinate.

(b) Sporophytic SI: If pollen coat S proteins do not match either stigma S protein, pollen tubes will grow.

Figure 39.13 **Self-incompatibility.** Self-incompatibility helps plants to avoid the combination of gametes that are too genetically similar. It is controlled by interactions between chemical constituents of pollen and the pistil. **(a)** In gametophytic self-incompatibility, compatibility between pollen and pistil is determined by the haploid genotype of the pollen. In this case, pollen S protein is located in the cytosol. **(b)** In sporophytic self-incompatibility, compatibility between pollen and pistil is determined by the sporophyte that produced the pollen and contributed S proteins to its coat.

layer of material that covers the sporopollenin-rich pollen wall. Coat materials include additional sporopollenin and pigments that give pollen its typically yellow, orange, or brown coloration, and lipids and proteins that aid in pollen attachment to carpels. Certain of these pollen coat compounds are responsible for allergic reactions in people exposed to particular types of airborne pollen. About 10% of flowering plants are wind-pollinated, and such plants produce copious amounts of pollen. For example, ragweed plants (genus *Ambrosia*), which are commonly associated with allergies, each produce an estimated 1 billion pollen grains during a year.

After Pollination, the Pistil Controls Pollen Germination

Recall that pollination is the process by which pollen is delivered to surfaces of the stigma, the uppermost part of the flower pistil. However, even if a pollen grain reaches the stigma of the right flower species, it may not be able to germinate and produce a sperm-delivery tube. The stigma and the style determine whether or not pollen grains germinate and pollen tubes grow toward ovules. How is this accomplished?

About half of plant species can serve as both mother and father to their progeny, because pollen produced by those plants is able to germinate on pistils of the same plants. Such self-pollinating plants are also termed **self-compatible (SC)**. By contrast, plants that prevent the germination of pollen that is too genetically similar to the pistil are **self-incompatible (SI)**. As in the case of human cultural practices that prevent mating between close relatives, in plants SI helps to decrease the likelihood of recessive disorders in offspring. In many plants, SI involves the *S* gene locus, which encodes S proteins. Each locus contains genetic sequences that determine pollen compatibility traits and pistil compatibility traits. Multiple *S* alleles for both genes occur in plant populations.

Two major types of SI are known: gametophytic SI and sporophytic SI. Gametophytic SI occurs when one or more *S* genes within pollen determine compatibility by encoding S protein located in the pollen cytosol. When the gametophyte controls compatibility, tubes may start to grow from incompatible pollen, but S protein encoded by pistil cells enters the tubes and destroys the pollen tube RNA. This halts tube growth. In contrast, S protein within genetically compatible pollen binds the pistil-produced protein, thereby preventing destruction of pollen tube RNA and allowing tube growth to continue (**Figure 39.13a**).

Sporophytic SI occurs when pollen compatibility is determined by the sporophyte that produces the pollen. This control is exerted when anthers deposit proteins into the pollen coat. When the

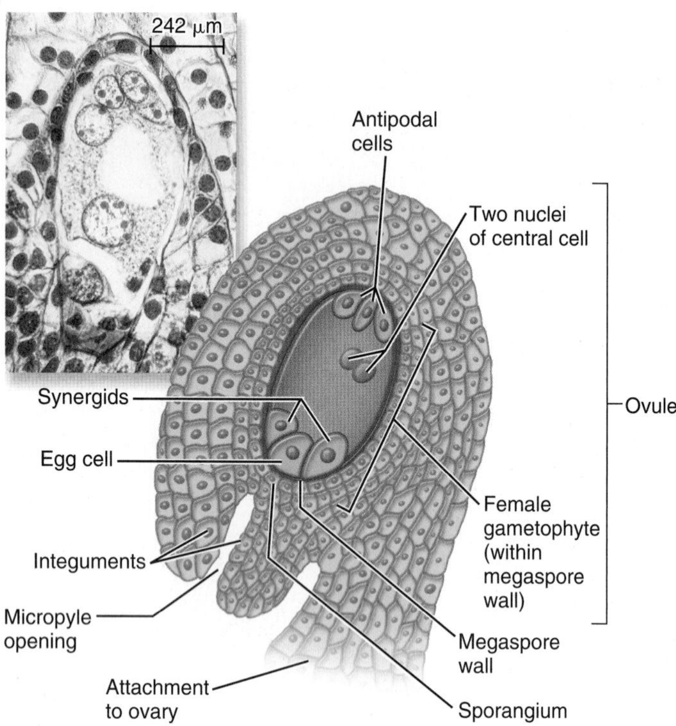

Figure 39.14 Angiosperm female gametophyte within an ovule.

Concept Check: How do female gametophytes obtain nutrients?

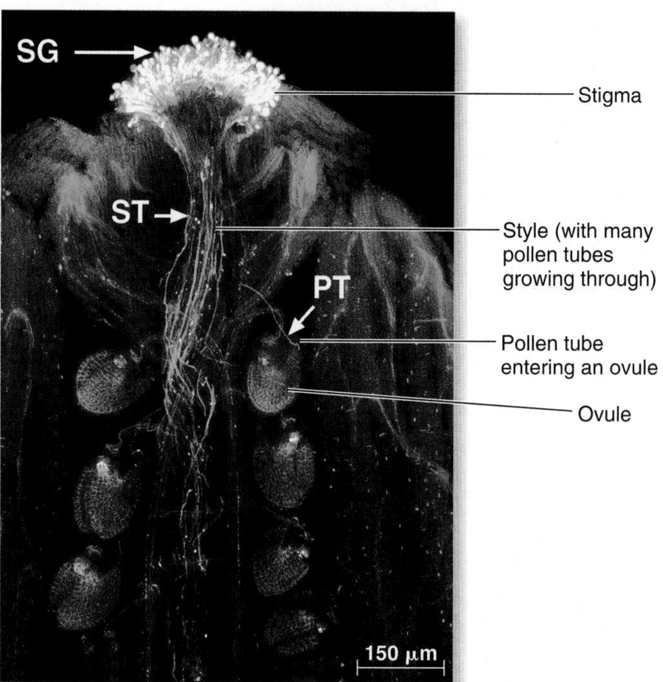

Figure 39.15 Pollen tubes delivering sperm to ovules. This fluorescence microscopic view shows pollen grains (the pale objects) germinating on the stigma surface (SG) and pollen tubes (PT) growing through the style (ST) toward ovules.

BioConnections: Look forward to Figure 51.5, which illustrates the male sexual organ of humans, an adaptation that allows internal fertilization. How is the mature male gametophyte of plants like a human penis?

sporophyte controls compatibility, pollen cannot germinate when S proteins in the plasma membranes of stigma cells recognize (bind) the S proteins of incompatible pollen. In this case, protein binding leads to signal transduction processes that prevent pollen germination. However, pollen can germinate if stigma proteins are unable to bind genetically distinct S proteins in its coat (Figure 39.13b).

A Female Gametophyte Develops Within Each Ovule

Each ovule produces a single female gametophyte that often consists of seven cells and eight nuclei (Figure 39.14). One of these cells is an egg cell, which lies wedged between two cells known as **synergids**. These synergids help move nutrients from the larger sporophyte to the nonphotosynthetic female gametophyte. Synergids also produce small proteins that attract pollen tubes. As a pollen tube approaches, synergids use molecular communication processes to detect the tube and help prepare the egg cell to be fertilized. The other four cells of the angiosperm female gametophyte consist of three antipodal cells, whose functions are not well understood, and a large **central cell** that contains two nuclei. This central cell and the egg cell are involved in the process of double fertilization.

Pollen Tubes Deliver Sperm Cells That Accomplish Double Fertilization

When pollen grains germinate successfully, they take up water, and the tube cell produces a long pollen tube. Within the tube, the generative cell divides by mitosis to produce two sperm cells. To deliver sperm to female gametophytes, the tube must grow at its tip from the stigma, through the style, to reach the ovule (Figure 39.15). Tip growth is controlled by the tube cell nucleus. During tip growth, new cytoplasm and cell-wall material are added to the tip of an elongating cell. Pollen tubes have been observed to grow toward ovules at about 0.5 mm per hour, commonly taking from 1 hour to 2 days to reach their destination. When a pollen tube encounters an ovule, it enters through the micropyle and penetrates a female synergid. The tube then stops growing, and its thin tip wall bursts, releasing the sperm. The bursting process propels the two sperm toward the egg and the central cell. One sperm nucleus fuses with the egg cell to produce a zygote, the first cell of a new sporophyte generation (see step 4 in Figure 39.4). The other sperm fuses with the two nuclei of the central cell to form the first endosperm cell (see step 5 in Figure 39.4). The term "double fertilization" arises from these two processes.

The zygote and fertilized central cell have dramatically different fates. The endosperm absorbs protein, lipid, carbohydrates, vitamins, and minerals from the mother plant and stores these nutrients. The stored nutrients provide material and energy needed by the zygote to develop into an embryo and by the embryo to develop into a seedling. The nutritive role of endosperm explains why a large percentage of human and animal food comes from seed endosperm of grain crops; corn, wheat, rice, and other grain crops generate more than 380 billion pounds of endosperm per year in the U.S. alone. You might wonder why the embryo and endosperm resulting from double fertilization have such different fates despite similar genetic composition. Part of the answer is that during its development, plant endosperm undergoes genomic imprinting by DNA methylation, a process also known to occur in mammals (see Figure 17.12). The imprinting process causes gene expression to occur differently in plant endosperm than it does in the embryo.

39.4 Embryo, Seed, Fruit, and Seedling Development

Learning Outcomes:

1. Discuss the major stages in angiosperm embryo development.
2. List examples of environmental and internal factors that influence seed germination.
3. Explain how seed germination varies among plants.

Seeds and fruits are major components of plant reproduction and are essential to the nutrition of animals, including humans. Seeds contain dormant plant embryos that may develop into seedlings under favorable conditions. As noted earlier, fruits aid seed dispersal, which allows plants to colonize new sites. Embryos, seeds, and fruits mature simultaneously, and their development is coordinated by hormonal signals that were introduced in Chapters 35 and 36. Seedling development is also hormonally regulated.

Embryos Develop from Zygotes

Fueled by endosperm nutrients, angiosperm embryos undergo development in a series of stages known as **embryogenesis** (Figure 39.16). Sometime within a period of days to several weeks following fertilization, a zygote begins to divide. At this point, a zygote is blanketed with a layer of callose, which helps to seal it off from the environment, thereby fostering embryo-specific gene expression. The zygote's first cell division is unequal, producing a smaller cell and a larger cell (Figure 39.16, step 1). This unequal division helps to establish the apical-basal (top-bottom) polarity of the embryo, which persists through the life of the plant. The smaller cell develops into the embryo, whose radial symmetry is established at this point and continues in adult *Arabidopsis* plants. The larger cell develops into a **suspensor**, a short chain of cells anchored near the micropyle at the ovule entrance (Figure 39.16, step 2). The suspensor channels nutrients and hormones into the young embryo, which absorbs them at its surfaces.

Young eudicot embryos are spherical, but they soon become heart-shaped as the seedling leaves, called **cotyledons**, start to develop. At this point, auxin and a mobile protein transcription factor are involved in establishing the young shoot and root at the apical and basal poles, respectively. This process is also influenced by the TOPLESS protein, which helps to repress auxin-response genes that would otherwise promote root development at the apical pole. (Mutants that have lost normal *TOPLESS* function have roots at both apical and basal poles, but no shoots, and are thus topless.) Eudicot embryos such as *Arabidopsis* then become torpedo-shaped, and as the cotyledons grow, they often curl to fit within the developing seed (Figure 39.16, step 4). In contrast, mature monocot embryos are cylindrical, with a single cotyledon and a side notch where the apical meristem forms. Mature embryos then become dormant as seeds mature.

Mature Seeds Contain Dormant Embryos

As seeds mature, they undergo changes leading to dormancy, an adaptation that prevents them from germinating when environmental conditions are not suitable for seedling growth. During this

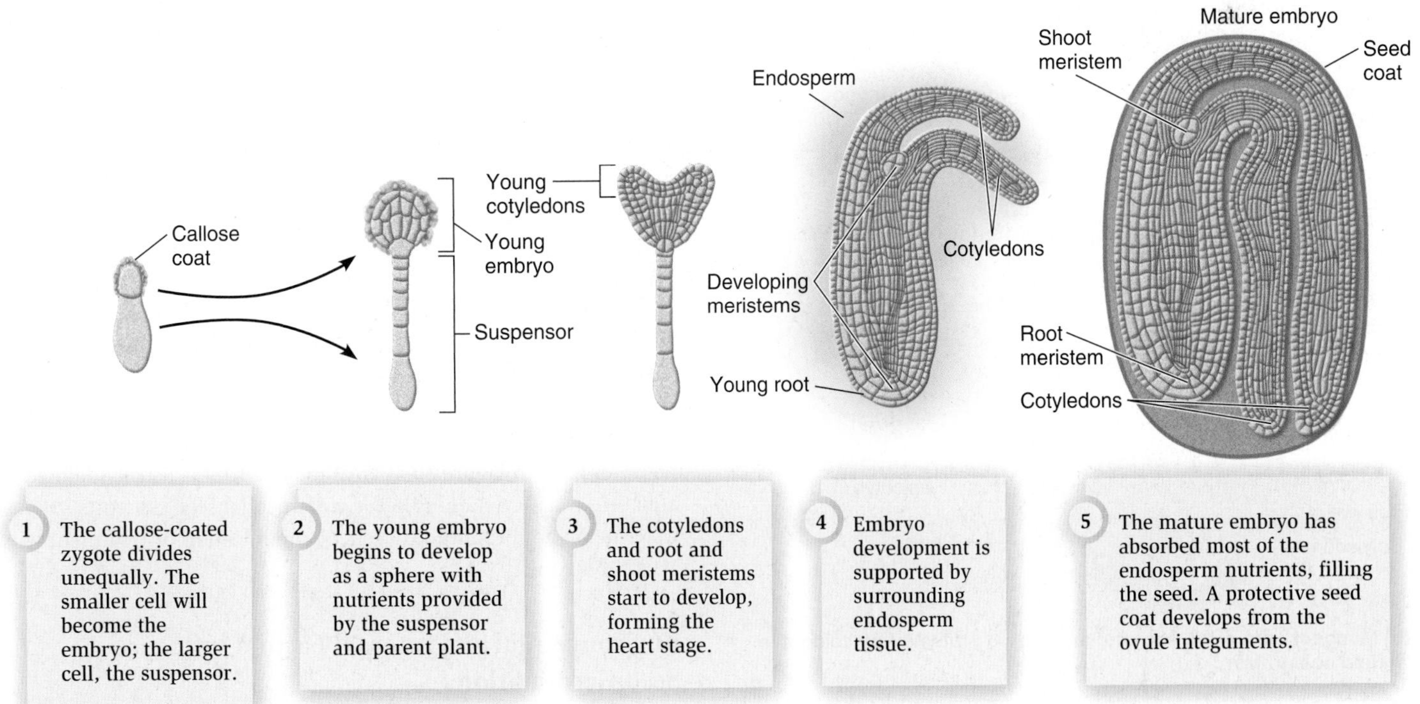

Callose coat
Young cotyledons
Young embryo
Suspensor
Endosperm
Developing meristems
Young root
Cotyledons
Shoot meristem
Mature embryo
Seed coat
Root meristem
Cotyledons
1 The callose-coated zygote divides unequally. The smaller cell will become the embryo; the larger cell, the suspensor.
2 The young embryo begins to develop as a sphere with nutrients provided by the suspensor and parent plant.
3 The cotyledons and root and shoot meristems start to develop, forming the heart stage.
4 Embryo development is supported by surrounding endosperm tissue.
5 The mature embryo has absorbed most of the endosperm nutrients, filling the seed. A protective seed coat develops from the ovule integuments.

Figure 39.16 **Embryogenesis in the eudicot *Arabidopsis*.**

Concept Check: *How would this embryo differ if its TOPLESS genes were nonfunctional?*

process, embryos become dry and thus able to survive in the absence of water. Seed maturation includes transformation of the ovule's integuments into a tough seed coat (see Figure 39.16, step 5). The seed coat restrains seedlings from growing and prevents the entry of water and oxygen, which maintains low seed metabolism. In addition, the coats of some seeds are darkly colored with pigments that may help to prevent damage by UV radiation or microbial attack. Another change leading to seed dormancy is gradual, controlled loss of water from the embryo and other seed tissues. As the result, the water content of dispersed seeds is only 5–15%. Abscisic acid (ABA) is a hormone that induces the activity of genes that help embryo tissues to survive the drying process. Some of these desiccation-tolerance genes encode proteins that form loose coils enclosing cell contents, thereby preventing damage as the cytoplasm becomes almost completely dry.

The structure of mature monocot and eudicot seeds differs. Within eudicot seeds, mature embryos often display an **epicotyl**, the portion of an embryonic stem with two tiny leaves in a first bud that is located above the point of attachment of the cotyledons (**Figure 39.17a**). The **hypocotyl** is the portion of an embryonic stem located below the point of attachment of the cotyledons. An embryonic root, the **radicle**, extends from the hypocotyl. Much of the endosperm has been absorbed into the large cotyledons. In contrast, mature monocot embryos, such as those of corn, feature an epicotyl with a first bud enclosed in a protective sheath known as the **coleoptile**. The young monocot root is enclosed within a protective envelope known as the **coleorhiza** (**Figure 39.17b**). When the seeds of flowering plants are dry and ready for dispersal, they are released from the plant while enclosed in a fruit or released when the fruit breaks open.

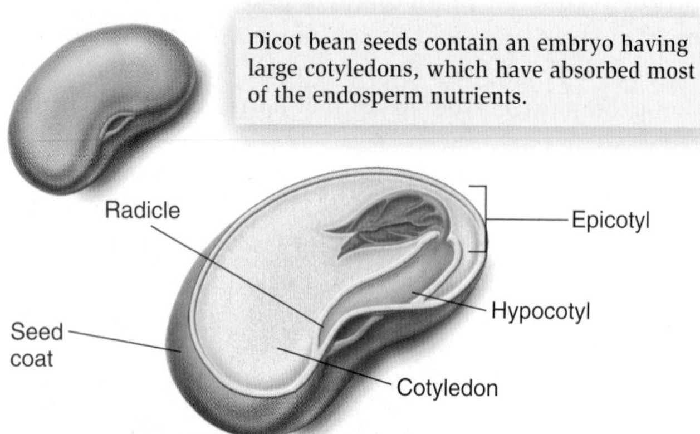

Dicot bean seeds contain an embryo having large cotyledons, which have absorbed most of the endosperm nutrients.
Radicle
Epicotyl
Hypocotyl
Seed coat
Cotyledon

(a) Eudicot bean seed, showing embryo with epicotyl, hypocotyl, and radicle

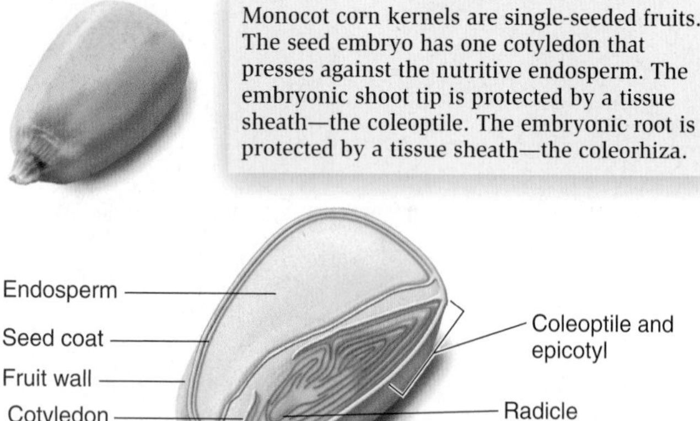

Monocot corn kernels are single-seeded fruits. The seed embryo has one cotyledon that presses against the nutritive endosperm. The embryonic shoot tip is protected by a tissue sheath—the coleoptile. The embryonic root is protected by a tissue sheath—the coleorhiza.
Endosperm
Seed coat
Fruit wall
Cotyledon
Coleoptile and epicotyl
Radicle
Coleorhiza

(b) Monocot corn seed, showing an embryo protected by coleoptile and coleorhiza

Figure 39.17 **Structure of mature seeds and embryos.**

Concept Check: *Why do mature seeds of eudicots lack extensive amounts of endosperm?*

Fruits Develop from Ovaries and Other Flower Parts

All fruits develop from ovaries and sometimes other flower parts. They occur in diverse forms that aid seed dispersal. Some fruits are dry, whereas others are moist and juicy; some open to release seeds, and others do not. Fruits also display a wide variety of sizes, colors, and fragrances. These variations result from differences in the process of fruit development. Plant hormones, including auxin, gibberellic acid, and cytokinins, control this transformation. ABA stimulates cell expansion, and ethylene influences fruit ripening. For instance, ethylene helps to ripen nuts, a type of dry fruit, by inducing plasma membranes to rupture, causing water loss. Under the influence of plant hormones, the pericarp (ripened ovary wall) of peaches, plums, and related fruits swells and softens, and orange or red chromoplasts replace green chloroplasts. As fruits mature, the outer protective cuticle often becomes very thick, contributing to peel toughness, which helps to prevent microbe attack. In addition, many maturing fruits increase their sugar and acid content, which produces the distinctive tastes of ripe fruit. Many fruits also produce fragrant volatile compounds.

Differences in the shape, color, fragrance, and moisture content of wild fruits reflect evolutionary adaptation for effective seed dispersal. Though many fruits and seeds are dispersed by wind or water or by attaching to animal fur, others are consumed by fruit-eating animals that are attracted by fruit color and fragrance. Blackberries provide a good example of fruits adapted for animal dispersal. Blackberry flowers produce many separate pistils, each containing a single ovule (**Figure 39.18a**). Following pollination and fertilization, the ovary of each pistil develops into a sweet, juicy fruitlet containing a single seed. As the individual fruitlets develop, they fuse together at the sides. Consequently, the many fruitlets produced by a single blackberry flower are dispersed together (**Figure 39.18b**). Attracted by the color, birds consume the whole aggregate and excrete the seeds, thereby dispersing many at a time. Many other types of fruits occur and these likewise represent adaptations that foster seed dispersal (refer back to Figure 30.21). Although a fruit is usually defined as a mature ovary containing seeds, commercial seedless fruits such as watermelon are produced by genetic modification or treatment with artificial auxin.

Environmental and Internal Factors Influence Seed Germination

Seeds vary greatly in their ability to germinate after dispersal. Small seeds such as those of dandelions and lettuces germinate quickly if light is available. Other seeds require a period of dormancy before

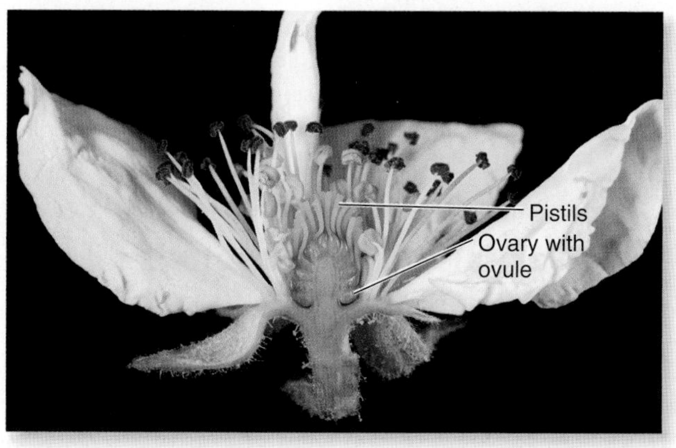

(a) *Rubus allegheniensis* (common blackberry) flower

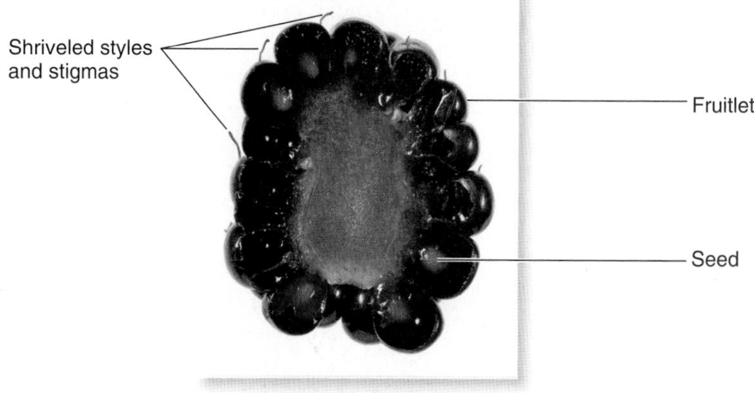

(b) Blackberry fruit

Figure 39.18 Blackberry flower and fruit. **(a)** Each of the many separate pistils in a blackberry flower is able to produce a single one-seed fruit (called a fruitlet) if fertilization occurs. **(b)** Together, the individual fruitlets of the blackberry compose an aggregate fruit that allows many seeds to be efficiently dispersed at the same time by the same animal agent.

BIOLOGY PRINCIPLE Living organisms interact with their environment. This principle is illustrated by the dispersal by birds of many blackberry seeds at the same time.

germination occurs. Some seeds can remain dormant for amazingly long time periods. For example, a lotus (*Nelumbo nucifera*) seed collected from a lake bed in China germinated at the age of 1,300 years, as determined by radiocarbon dating. In 2005, plant scientists germinated a 2,000-year-old date seed found in Israel.

Water is generally required to rehydrate seeds so that embryos can resume their metabolic activity. Water absorption also swells seeds, helping to break the seed coat and allowing embryonic organs to emerge. In some cases, rainfall of sufficient duration to leach germination-inhibiting compounds out of seeds is required. The optimal temperature for germination of most seeds lies between 25°C and 30.25°C (77°F and 86.25°F). This explains why gardeners wait until the soil is warm before planting seeds outdoors in spring. However, some seeds need a period of cold treatment or seed coat abrasion before they will germinate. Such physical stimuli induce the activity of more than 2,000 genes associated with seed germination.

When grass seeds rehydrate, the young shoot secretes the hormone gibberellic acid from the seed cotyledon into the outermost endosperm layer, known as the aleurone. In response, the aleurone secretes digestive enzymes into the central endosperm, releasing sugars from stored starch (**Figure 39.19**). The seedling uses these sugars for growth. This highly coordinated process allows grass seeds to quickly germinate when it rains, an advantage in arid grassland habitats. Humans also use this basic process of germination to make beer. In the process known as malting, beer brewers apply gibberellic acid to barley seeds to induce them to germinate simultaneously. The barley seeds are then baked at a high temperature to stop germination, a process that produces malt. Brewers then treat malt with water and heat, add the dried flowers of the hop plant (the genus *Humulus*), and add yeasts to ferment the plant sugars to alcohol.

Once seeds have germinated, plants vary in the process by which the embryonic shoot emerges. When bean and onion seeds germinate, the hypocotyl forms a hook that first breaches the soil surface and then straightens, thereby pulling the rest of the seedling and cotyledons aboveground (**Figure 39.20a,b**). In contrast, when pea

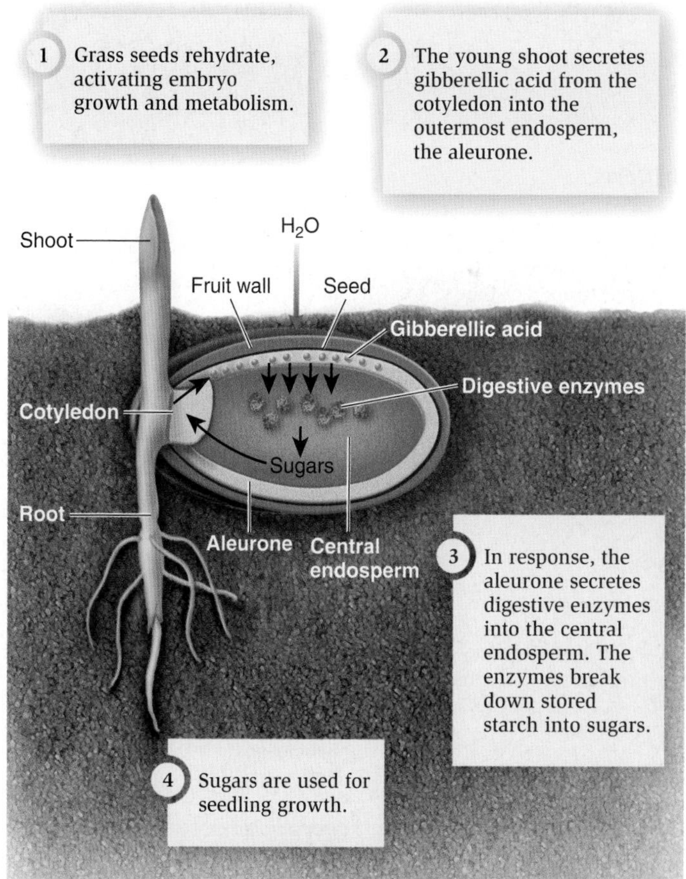

1 Grass seeds rehydrate, activating embryo growth and metabolism.
2 The young shoot secretes gibberellic acid from the cotyledon into the outermost endosperm, the aleurone.
Shoot
H_2O
Fruit wall
Seed
Gibberellic acid
Digestive enzymes
Cotyledon
Sugars
Root
Aleurone
Central endosperm
3 In response, the aleurone secretes digestive enzymes into the central endosperm. The enzymes break down stored starch into sugars.
4 Sugars are used for seedling growth.

Figure 39.19 Germination of grass seeds.

seeds germinate, the epicotyl forms a hook that pulls the shoot tip out of the ground, leaving the cotyledons beneath the soil surface (**Figure 39.20c**). In both cases, the tough hook cells bear the brunt of passage through hard surface soil crusts, thereby protecting the delicate shoot tips. The plant hormone ethylene controls seedling hook

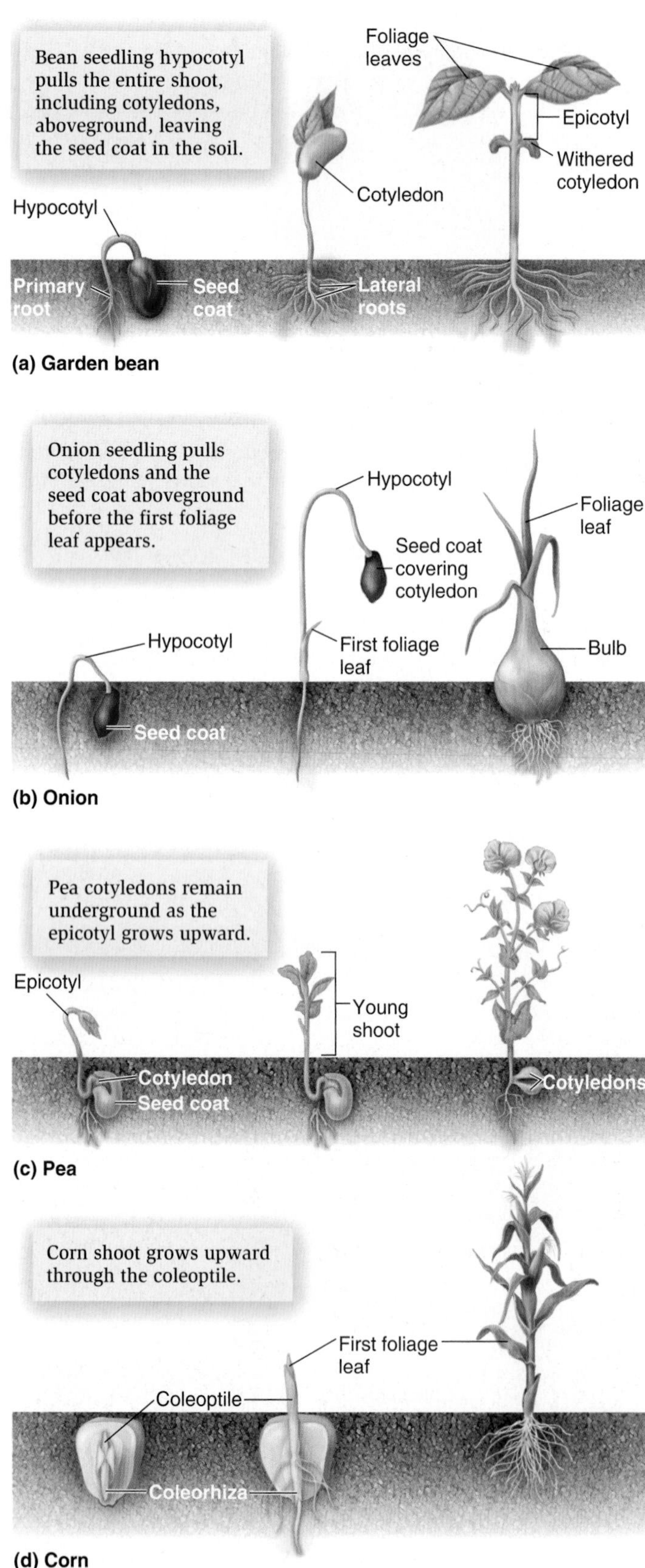

Figure 39.20 Variations in seed germination and seedling growth patterns.

formation, as described in Chapter 36 (refer back to Figure 36.9). However, not all seedlings form hooks. For example, as they grow through the soil, the shoot tips of corn seedlings and those of other grasses are protected by the coleoptile, a protective tube that encloses the first foliage leaves (**Figure 39.20d**).

39.5 Asexual Reproduction in Flowering Plants

Learning Outcomes:

1. Explain the benefits of asexual reproduction in flowering plants.
2. List several examples of ways in which plant asexual reproduction is important to agriculture.

Many plants rely on sexual reproduction. However, a wide variety of angiosperms reproduce primarily by asexual means, and other plants commonly utilize both sexual and asexual reproduction. **Asexual reproduction** is the production of new individuals from a single parent without the occurrence of fertilization. Although sexual reproduction provides beneficial genetic variation, asexual reproduction can be advantageous in other ways. For example, asexual reproduction maintains favorable gene combinations that allow faster population growth in stable environments. Asexual reproduction is also advantageous in stressful habitats where pollinators or mates can be rare, because it allows a single individual to start a new population. Finally, asexual reproduction allows some plants to persist for very long periods of time. Among the oldest known plants are creosote bushes that are asexual clones of a parent that grew from a seed about 12,000 years ago!

The artificial vegetative propagation of plants from cuttings is a form of asexual reproduction that is widely used commercially and by home gardeners. In this section, we will explore the three main mechanisms of plant asexual reproduction: specialized reproductive structures, somatic embryogenesis, and apomixis.

Vegetative Asexual Reproduction Generates Plant Clones from Organs

Roots, stems, and leaves are vegetative plant organs that can function as asexual reproductive structures. For example, root sprouts, such as those produced by aspens, can generate entire groves of genetically identical trees. Sucker shoots, such as those appearing at the bases of banana plants and date palms, and the pieces of tuber-bearing "eyes" (which are buds) that develop into potato plants are examples of vegetative plant organs that have agricultural importance. Attractive horticultural varieties of African violets and other plants can be propagated from leaf cuttings. Such asexual offspring grow into adult plants that have the same valued properties as their parent plant. In contrast, these crops grown from seed would produce diverse progeny, not all of which would have economically prized properties.

Somatic embryogenesis is the production of plant embryos from body (somatic) cells. Embryos can develop from many types of plant cells. Somatic embryos develop normally to the torpedo stage but do not dehydrate and become dormant, as is normal for zygotic

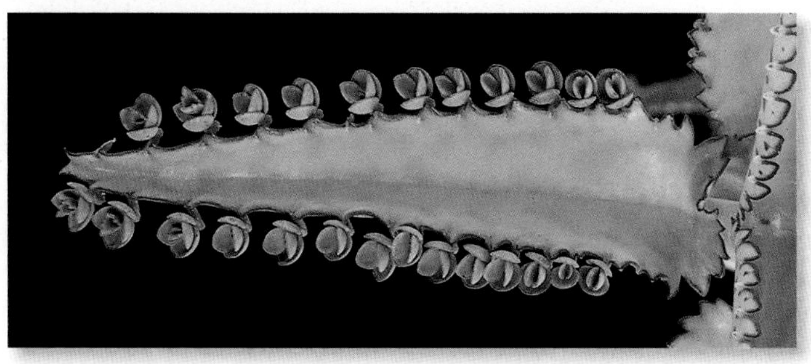

Figure 39.21 **Asexual reproduction via somatic embryogenesis.** The leaves of this *Kalanchoë* plant bear small plantlets around the edges. When mature, these plantlets drop off and, under the right conditions, grow into new plants.

BioConnections: Look back at Figure 33.4, which illustrates the life cycle of the animal Obelia, a colonial cnidarian related to jellyfish and corals. How is the life cycle of this animal similar to that of the Kalanchoë plant?

embryos. Rather, somatic embryos produce root and shoot systems and develop into mature plants. Somatic embryogenesis occurs naturally in citrus, mango, onion, and tobacco plants, but agricultural scientists have made use of it as well.

In the 1950s, British plant biologist F. C. Steward and associates were the first researchers to successfully clone a complex organism, carrot plants, by means of somatic embryogenesis. They used differentiated cells from carrot roots, grew the cells in conditions that caused some cells to lose their specialized properties and develop into embryos, and then cultivated each embryo in conditions that favored development into a mature carrot plant. Many types of plants are now cloned by somatic embryogenesis, which allows commercial growers to produce large numbers of genetically identical individuals.

The common houseplant *Kalanchoë daigremontiana* has leaves bearing many tiny plantlets at their edges. When these detach, they are able to take root and grow into new individuals (**Figure 39.21**). Recent molecular studies have revealed that embryogenesis genes are involved in this process and its evolution.

Genomes & Proteomes Connection

The Evolution of Plantlet Production in *Kalanchoë*

Kalanchoë daigremontiana is informally known as mother of thousands because it produces many plantlets at the edges of leaves. Helena Garces, Neelima Sinha, and their colleagues investigated the evolution of this type of asexual reproduction by studying genes that are involved in the development of organs and embryos in four species of *Kalanchoë*. In addition to *K. daigremontiana*, they investigated *K. marmorata*, which does not produce plantlets; *K. pinnata*, which produces plantlets only under certain stressful conditions; and *K. gastonis-bonnieri*, which produces plantlets both normally and when stressed. In leaves of these species, the biologists looked for expression of the gene *STM*, which encodes a key regulator of leaf production at the shoot meristem. They found that *STM* is expressed in cells at the leaf margins of all *Kalanchoë* species that produce leaf plantlets, but not the species that do not produce plantlets. Then the investigators checked for the expression in leaves of two genes that are involved in embryo development (*LEC1* and *FUS3*). They discovered that these embryo-linked genes were expressed only in the leaf margins of species that normally form plantlets, not the species in which plantlet formation is induced by stress.

In a survey of a larger number of species, these investigators also discovered that *Kalanchoë* species that normally produce plantlets have an altered LEC1 protein. The normal form of LEC1 protein is essential to the process by which embryos become dry and thus tolerant of arid conditions. LEC1 is therefore necessary for the production of viable seeds, those able to germinate. *Kalanchoë* species that produce plantlets only under stressful conditions and those that do not produce plantlets at all were able to produce viable seeds. By contrast, the seeds of plantlet-producing species were not viable.

Together, these data allowed the investigators to infer that the evolution of plantlet formation began when certain leaf cells of some species gained the ability to function like a shoot meristem, thereby producing structures resembling small shoots. In some of the descendents of these species, normal LEC1 function in seeds was lost, as was the ability to produce viable seeds. Some species adapted by expressing the embryo-development process in leaf margin cells, a process that allowed them to produce plantlets. Such plantlets are not affected by loss of LEC1 function because, unlike seeds, they do not undergo a drying process during development.

Apomixis Is Seed Production Without Fertilization

Apomixis (from the Greek, meaning away from mixing, that is, genetic mixing) is a natural asexual reproductive process in which fruits and seeds are produced in the absence of fertilization. More than 300 species of flowering plants, including hawkweeds, dandelions, and some types of citrus, are able to reproduce asexually by apomixis. Dandelions and some other apomictic plants require pollination to stimulate seed development, but others do not. Agricultural scientists are interested in apomixis as a potential method for producing genetically uniform seeds, propagating hybrids, and removing the need for fertilization in crop plants.

Most studies of apomixis have been carried out with dandelions, because they are widespread and populations are composed mainly of individuals that reproduce by apomixis. In these plants, meiosis produces microspores, but most pollen grains have abnormal chromosomes. Such grains produce pollen tubes but not sperm cells, so that fertilization does not follow pollination. However, female gametophytes and the eggs they produce are diploid. This condition arises because during the preceding meiotic divisions that generate megaspores, homologous chromosomes do not pair and meiosis II does not occur. The diploid eggs of apomictic dandelions develop into normal embryos without fertilization and an endosperm develops from the unfertilized central cell. This explains why apomictic dandelions can produce their single-seeded fruits despite the absence of gamete fusion.

Summary of Key Concepts

39.1 An Overview of Flowering Plant Reproduction

- Flowering plants display a sexual life cycle known as alternation of generations. The gamete-producing male and female gametophytes of flowering plants are very small and depend entirely on nurturing sporophytic tissues. Plant gametes arise by the process of mitosis (Figures 39.1, 39.2).
- Flowers are reproductive shoots that develop from a shoot apical meristem. The role of flowers is to promote seed production. A flower shoot generally produces four types of organs: sepals, petals, stamens, and carpels. Stamens produce pollen grains, which are immature male gametophytes. Carpels produce, enclose, and nurture female gametophytes (Figure 39.3).
- Mature male gametophytes, or pollen tubes, each produce two sperm and deliver them to ovules within the ovary. Flowering plants display double fertilization: One of the two sperm released from a pollen tube combines with an egg cell to form a zygote, while the other fuses with two nuclei located in a central cell of the female gametophyte, producing the first cell of endosperm tissue. Endosperm is a nutritive tissue that supports development of an embryonic sporophyte (Figure 39.4).
- Seeds are reproductive structures that contain a dormant embryo enclosed by a protective seed coat that develops from ovule integuments. Fruits are structures that contain seeds and foster seed dispersal. Like flowers and endosperm, fruits are unique features of flowering plants (Figure 39.5).

39.2 Flower Production, Structure, and Development

- Plants flower in response to environmental stimuli, such as temperature and day length, by the conversion of a leaf-producing shoot into a flowering shoot.
- Flowers vary in the type of whorls present, the number of flower organs, color, fragrance, organ fusion, symmetry, and arrangement (whether single or in inflorescences). These variations are related to pollination mechanisms and are genetically controlled (Figures 39.6, 39.7, 39.8, 39.9, 39.10).
- Flower blooming involves dramatic changes in petal shape caused by faster growth at petal edges than at their centers (Figure 39.11).

39.3 Male and Female Gametophytes and Double Fertilization

- Pollen grains are immature male gametophytes protected by a tough sporopollenin wall. Female gametophyte development occurs within an ovule. Mature female gametophytes include an egg and two synergids, a central cell with two nuclei, and three antipodal cells (Figures 39.12, 39.13).
- After pollination, interactions between proteins of pistil cells and those of pollen determine pollen germination (Figure 39.14).
- Germinated pollen delivers two sperm to female gametophytes by means of a long pollen tube. The style plays a role in the guidance, nutrition, and fate of the pollen tube. One sperm nucleus fuses with the egg to produce a zygote, the first cell of a new sporophyte generation; the other sperm nucleus fuses with the two nuclei of the central cell, generating the first cell of the nutritive endosperm tissue (Figure 39.15).

39.4 Embryo, Seed, Fruit, and Seedling Development

- Unequal division of a zygote leads to development of a nutritive suspensor and an embryo. Young eudicot embryos are heart-shaped as two embryonic leaves (cotyledons) develop, and the embryo assumes a torpedo shape as the embryonic root forms. Monocot embryos have only a single cotyledon; the embryonic shoot tip is protected by the coleoptile, and the embryonic root is protected by the coleorhiza (Figures 39.16, 39.17).
- Mature seeds contain embryos that become dry and are protected by desiccation-resistance proteins and a tough seed coat. These adaptations enable seeds to withstand long periods of dormancy, germinating only when conditions are favorable for seedling survival. Mature fruits develop from ovaries and aid in seed dispersal (Figure 39.18).
- Seed germination is influenced by environmental and internal factors. The embryonic root (radicle) is the first organ to emerge, an adaptation that allows rapid water uptake, essential for seedling development (Figures 39.19, 39.20).

39.5 Asexual Reproduction in Flowering Plants

- Asexual reproduction is the production of new individuals from a single parent without the occurrence of fertilization. Vegetative reproduction is the development of whole plants from nonreproductive organs. Somatic embryogenesis is the production of embryos from individual body cells. Apomixis is a mechanism by which some plants produce seeds from flowers without fertilization (Figure 39.21).

Assess and Discuss

Test Yourself

1. Where do the pollen grains of flowering plants develop?
   a. in the anthers of a flower
   b. in the carpels of a flower
   c. while being dispersed by wind, water, or animals
   d. within ovules
   e. within pistils
2. Where do mature male gametophytes of flowering plants primarily develop?
   a. in the anthers of a flower
   b. in the carpels of a flower
   c. while being dispersed in wind, water, or by animals
   d. within ovules
   e. on the surfaces of leaves
3. Where would you find female gametophytes of a flowering plant?
   a. in the anthers of a flower
   b. at the stigma of a pistil
   c. in the style
   d. within ovules in a flower's ovary
   e. in structures that are dispersed by wind, water, or animals

4. How does double fertilization occur in flowering plants?
   a. The two sperm in a pollen tube fertilize the two egg cells present in each female gametophyte.
   b. One of the two sperm in a pollen tube fertilizes the single egg in a female gametophyte, and the other fuses with the two nuclei present in the central cell.
   c. Two sperm, one contributed by each of two different pollen tubes, fertilize the two egg cells in a single female gametophyte.
   d. Two sperm contributed by separate pollen tubes enter a single female gametophyte; one of the sperm fertilizes the egg cell; the other fertilizes the central cell.
   e. None of the above is correct.
5. A seed is
   a. an embryo produced by the fertilization of an egg, which is protected by a seed coat.
   b. a structure that germinates to form a seedling under the right conditions.
   c. an embryo produced by parthenogenesis that is enclosed by a seed coat.
   d. all of the above.
   e. none of the above.
6. What is the likely chemical composition of the chemical stimulus of flowering that is produced by leaves and transported to the shoot apical meristem?
   a. the hormone auxin
   b. the protein STM
   c. the carbohydrate callose
   d. the mineral ion K^+
   e. none of the listed choices
7. How many whorls of organs occur in complete flowers?
   a. two
   b. four
   c. six
   d. eight
   e. ten
8. If an ovary contains eight ovules, how many seeds could potentially result if pollen tubes reach all eight ovules?
   a. one
   b. four
   c. eight
   d. more than 20
   e. none of the listed choices
9. What function(s) does the polysaccharide callose have in the reproduction of flowering plants?
   a. Callose forms a coat that isolates young embryos during their early development.
   b. Callose forms a coat that isolates groups of four microspores during their early development into pollen grains.
   c. Callose helps to pattern the sculptured sporopollenin walls of pollen grains.
   d. all of the above
   e. none of the above
10. From what structure does a fruit pericarp primarily develop?
   a. the style
   b. a stamen filament
   c. the ovary wall
   d. a group of fused sepals
   e. the stigma

Conceptual Questions

1. Why are pollen grain walls composed of sporopollenin?
2. Why are seed coats often tough?
3. A principle of biology is that *all species (past and present) are related by an evolutionary history.* Modern flowering plants likely possessed a single common ancestor. Why, then, do flowers occur in such a diversity of shapes and colors?

Collaborative Questions

1. Observe or view images of orchid flowers. Are these flowers bilaterally symmetric or radially symmetric? Are these flowers more likely to be wind-pollinated or pollinated by animals? What gene might be involved in the production of orchid flower shape?
2. How do plants prevent the production of many offspring expressing deleterious recessive traits?

Online Resource

www.brookerbiology.com

Stay a step ahead in your studies with animations that bring concepts to life and practice tests to assess your understanding. Your instructor may also recommend the interactive eBook, individualized learning tools, and more.

UNIT VII

ANIMALS

Despite the amazing diversity of animal life, fundamental similarities link the millions of animal species. We will explore many of these similarities in this unit. The basic features of animal bodies and the ability of animals to maintain homeostasis will be introduced in Chapter 40. Chapters 41 to 43 will describe major principles of nervous systems in animals. The ability of animals to move through their environments will be covered in Chapter 44. How food is obtained, processed, and utilized in animals is the subject of the following two chapters. Circulatory, respiratory, urinary, and endocrine systems are then covered in Chapters 47 to 50. The unit concludes with Chapters 51 to 53, which cover animal reproduction, development, and immune systems, respectively.

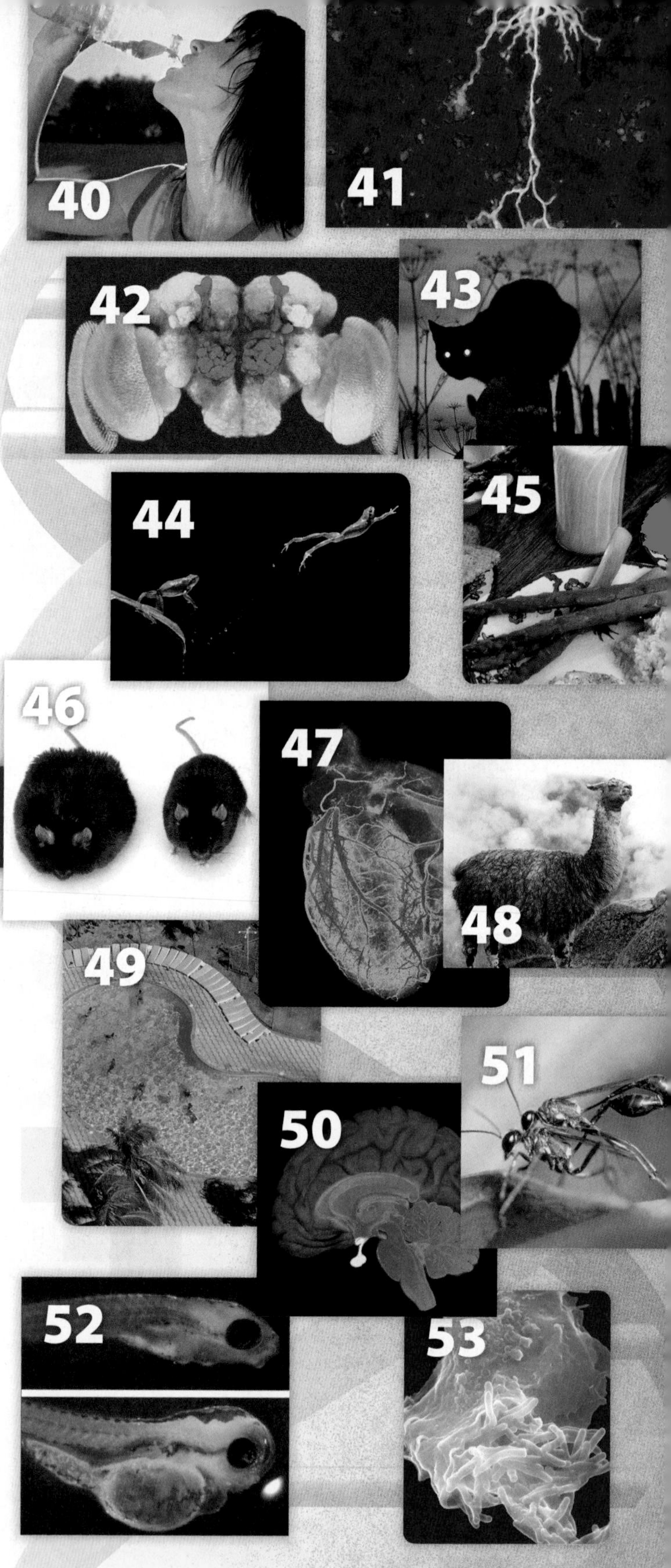

The following biology principles will be emphasized in this unit:

- ***Living organisms maintain homeostasis:*** *This is one of the fundamental principles of biology and a key to understanding virtually all aspects of animal biology; it is introduced in Chapter 40.*
- ***Living organisms use energy:*** *We will see how many of the processes required to achieve and maintain homeostasis require energy; for example, it takes energy to maintain blood flow (Chapter 47) and a stable concentration of sodium in an animal's body fluids (Chapter 49).*
- ***Living organisms grow and develop:*** *In Chapter 52, we will explore the genetic and environmental factors that regulate how an embryo develops into a mature animal.*
- ***Living organisms interact with their environment:*** *How animals sense changes in their external environment will be covered in Chapter 43.*
- ***Structure determines function:*** *The relationship between structure and function in animals will be evident at multiple levels: cells, tissues, organs, and whole bodies.*
- ***All species (past and present) are related by an evolutionary history:*** *This principle will be explored in the Genome and Proteome Connections throughout the unit. As an example, the importance of an ancient family of proteins found in most animals in providing immune defenses will be covered in Chapter 53.*
- ***Biology affects our society:*** *At the end of Chapters 41 to 53 are sections entitled "Impact on Public Health." These sections relate what you've learned in each chapter to several important facets of human health and disease.*
- ***Biology is an experimental science:*** *Every chapter has a Feature Investigation that describes a classic or recent experiment notable not only for its innovation and creativity, but also for moving the field of animal biology forward in a significant way.*

40 41 42 43 44 45 46 47 48 49 50 51 52 53

Chapter Outline

40.1 Organization of Animal Bodies

40.2 The Relationship Between Form and Function

40.3 Homeostasis

Summary of Key Concepts

Assess and Discuss

Animal Bodies and Homeostasis

40

Perspiring and drinking water are both mechanisms that help achieve homeostasis—a stable internal body environment.

Look at the image of the young woman in the chapter-opening photo. What is happening at that moment? Obviously it is a hot day, and the woman is perspiring and thirsty; it appears that she has just been exercising outdoors. To accomplish this seemingly ordinary feat, however, several important things must occur. First, her body must prevent an excessive build-up of heat, because high body temperature can destroy enzymes and damage cellular membranes. One way some animals, including humans, can eliminate heat from the body is to perspire; recall from Chapter 2 that water has a high heat of vaporization, and therefore the evaporation of water from her body surface helps eliminate body heat. This comes with a cost, however, because perspiration depletes some of the body's water, potentially leading to changes in body fluid salt concentrations, blood pressure, and other critical features of a person's physiology. Consequently, structures in her brain and other organs that are sensitive to the body's fluid levels and salt concentrations trigger the sensation of thirst, and the woman drinks water to replenish what was lost through perspiration. The process through which the different aspects of an animal's internal environment (in this case, temperature and body fluids) are maintained within normal limits, even in the face of changing circumstances or external challenges, is known as homeostasis.

Maintenance of homeostasis is one of the fundamental principles of biology and is the key to understanding virtually all aspects of animal biology. Before we can fully appreciate what homeostasis means to animals and how it is achieved, however, we first need to understand some basic features of animal bodies. We begin this chapter with a discussion of how cells are organized into tissues, and tissues into organs, and how fluids exist in different body compartments. In Chapter 10, we looked at the organization of cells into tissues and organs from the perspective of cell biology; here, we look at tissues and organs from the perspective of the whole animal, examining how the properties of life arise from the complex interactions of its components. Next, we will discuss a principle of biology that also helps us understand homeostasis, the concept that structure (form) determines function. We then link these principles together with a detailed look at what homeostasis means for different animals, how it may be challenged, and how it is restored or maintained.

40.1 Organization of Animal Bodies

Learning Outcomes:

1. List the different categories of animal tissue, providing general functions and specific examples of each.
2. Name the various organ systems found in many animals, list the components of each, and describe their general functions.
3. Describe the movement of solutes and water between compartments.

All animal cells share similarities in the ways in which they exchange materials with their surroundings, obtain energy from organic nutrients, synthesize complex molecules, reproduce themselves, and detect and respond to signals in their immediate environment. Animals typically begin life as a single cell—most commonly a fertilized egg—which divides to create two cells, each of which divides in turn, resulting in four cells, and so on. If cell multiplication were the only event occurring, the end result would be a spherical mass of identical cells. As we

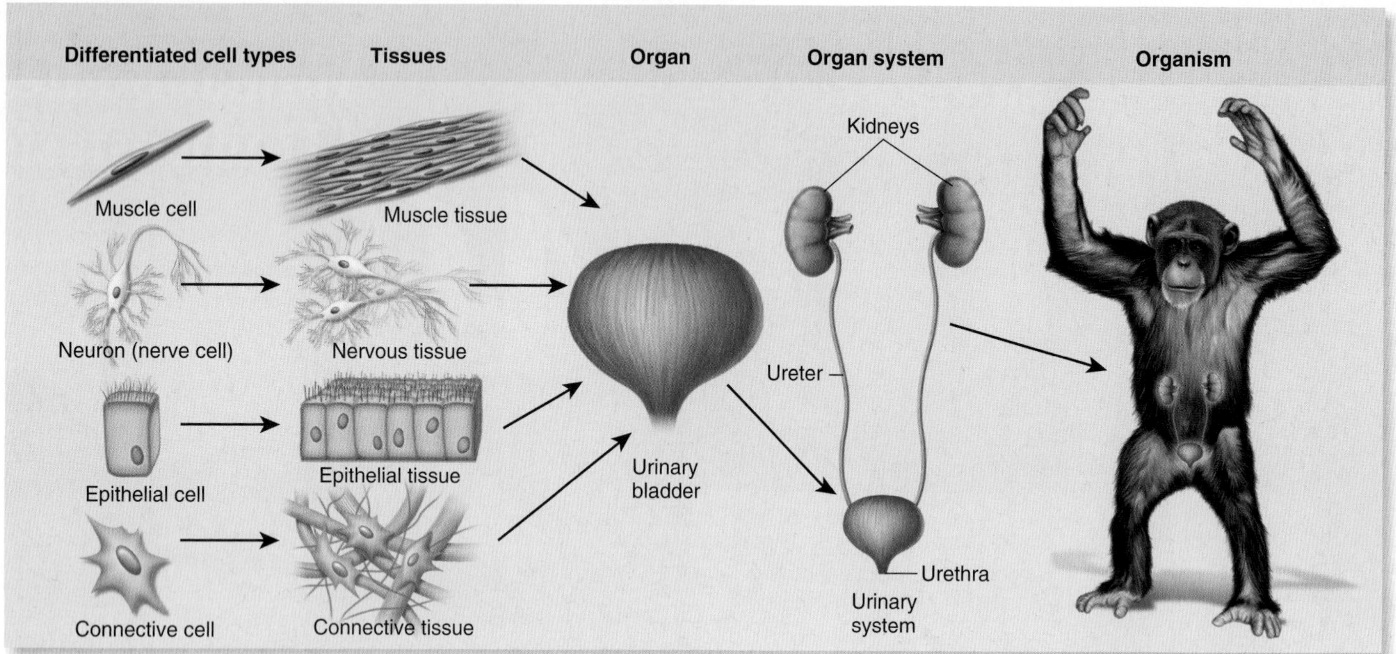

Figure 40.1 **The internal organization of cells, tissues, organs, and organ systems in a mammal. Most animals share the same four tissue types.**

BIOLOGY PRINCIPLE **New properties emerge from complex interactions.** By themselves, none of the four tissues that constitute a bladder or kidney could perform the functions of those organs, but when combined in precise ways, the result is a functional organ system capable of purifying the fluids of an animal's body.

will see in Chapter 52, however, cells become specialized during development to perform a particular function (that is, they differentiate). Examples of differentiated cells are muscle and blood cells. Cells also migrate to new locations within the developing organism and form clusters with other cells. In this way, the cells of an animal's body are arranged in various combinations to form organized, multicellular structures (Figure 40.1). Cells with similar properties group together to form tissues (for example, muscle tissue), which combine with other types of tissues to form organs (a urinary bladder). Organs are anatomically or functionally linked to form organ systems (the urinary system).

Specialized Cells Are Organized into Tissues

A **tissue** is an association of many cells having a similar structure and function. The tissues in a typical animal's body can be classified into four categories, according to their locations and the types of functions they perform: muscle, nervous, epithelial, and connective tissues. Within each of these functional categories, subtypes of tissues perform variations of that function, as illustrated by the three types of muscle tissue.

Muscle Tissues **Muscle tissues** consist of cells specialized to shorten, or contract, generating the mechanical forces that produce body movement, decrease the diameter of a tube, or exert pressure on a fluid-filled cavity. Three types of muscle tissue may be found in animals: skeletal, smooth, and cardiac (Figure 40.2). **Skeletal muscles** are generally linked to bones in vertebrates via bundles of collagen fibers called tendons, and to the exoskeleton of invertebrates. When skeletal muscles are stimulated by signals from the nervous system, they generate force that leads to the contraction of the muscle (see Chapter 44). Contraction of these muscles may be under voluntary control and can produce the types of movements required for locomotion, such as extending limbs or flapping wings. Skeletal muscles may also attach to skin, such as the muscles producing facial expressions. **Smooth muscles** surround hollow tubes and cavities inside the body's organs, such that their contraction can move the contents of those organs. For example, the contraction of smooth muscle in the stomach wall propels partially digested food into the intestines, where it can be digested fully. Smooth muscle also surrounds and forms part of small blood vessels and airway tubes (bronchioles). Contraction in those regions reduces blood flow or movement of air, respectively. Contraction of all smooth muscle is involuntary—that is, it occurs automatically without conscious control. In the third type, **cardiac muscle**, physical and electrical connections between individual cells enable many cells to contract almost simultaneously. Like smooth muscle, cardiac muscle is involuntary. It is found only in the heart, however, where it provides the force that generates pressure sufficient to pump blood through an animal's body.

Nervous Tissues **Nervous tissues** are complex networks of cells specialized to initiate and conduct electrical signals from one part of an animal's body to another part (Figure 40.3). A single nerve cell, called a **neuron**, may connect two or more other neurons and be only a few micrometers long. Such neurons are found widely throughout the vertebrate brain. In contrast, other neurons located in the brain may send extensions along the length of the spinal cord; in a large

Lungs

Heart

Skeletal muscle cell

Bronchiole

Smooth muscle cells

Cardiac muscle cell

Smooth muscle layer

Skeletal muscles are usually attached to bone in vertebrates and provide the force needed for locomotion. All skeletal muscle is under voluntary control.

Smooth muscles often surround hollow tubes, like the bronchioles in a mammal's lungs, where they control the tube's diameter. Smooth muscle contraction is involuntary.

Cardiac muscle is found only in the heart, where muscle cells are interconnected and provide the force needed for a heartbeat. Cardiac muscle contraction is involuntary.

Figure 40.2 **Three types of muscle tissue: skeletal, smooth, and cardiac.** All three types produce force, but they differ in their appearance and in their locations within animals' bodies. (Right inset: © Dr. Richard Kessel/Visuals Unlimited)

Concept Check: *All muscles produce movement, but only skeletal muscle produces locomotion. What is meant by this statement?*

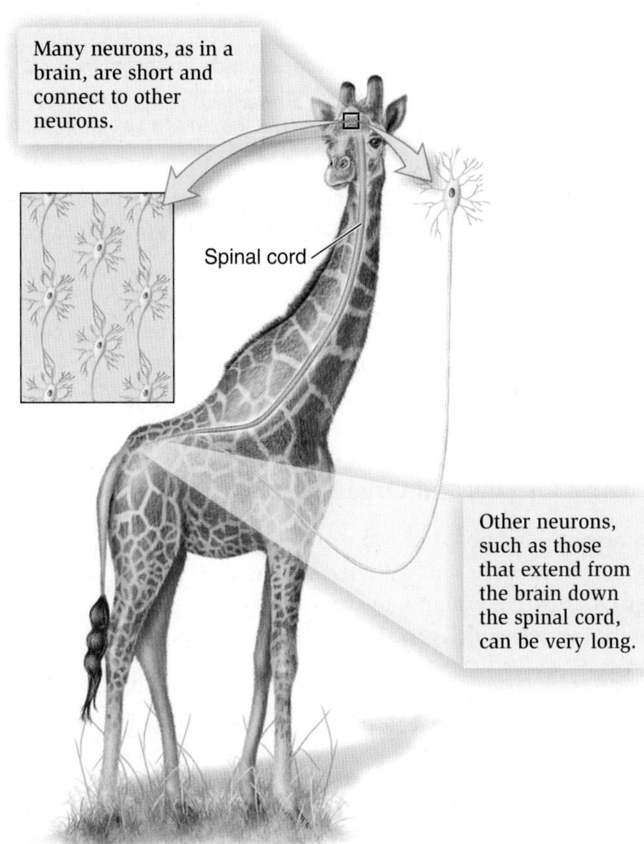

Many neurons, as in a brain, are short and connect to other neurons.
Spinal cord
Other neurons, such as those that extend from the brain down the spinal cord, can be very long.

Figure 40.3 **Variation in the shape and length of neurons.**

BIOLOGY PRINCIPLE Living organisms interact with their environment. Animals interact with their environment in many ways; this largely depends on the use of nervous tissue, which allows animals to sense and respond to changes in the environment.

animal like a giraffe, this distance may extend over 2 m! Depending on where it is generated in an animal's body, an electrical signal produced in one neuron may stimulate or inhibit other neurons to initiate new electrical signals, stimulate muscle tissue to contract, or stimulate glandular cells to release chemicals into the animal's body fluids. Thus, nervous tissue provides a critical means of controlling many diverse activities of the body's cells.

Epithelial Tissues **Epithelial tissues** are sheets of densely packed cells that cover the body or individual organs or line the walls of various cavities inside the body. Epithelial cells are specialized to protect structures and to secrete and absorb ions and organic molecules (see Chapter 10). For example, epithelial tissue can invaginate (fold inward) to form sweat glands that secrete water and ions onto the surface of an animal's skin. Epithelial cells come in a variety of shapes, such as cuboidal (cube-shaped), squamous (flattened), and columnar (elongated). They are typically arranged in epithelial tissues as simple (one layer), stratified (multiple layers), pseudostratified (one layer, but with nuclei located in such a way that it appears stratified), or, in certain cases such as the urinary system, transitional (multiple layers with the ability to expand and contract) epithelia (**Figure 40.4**). Regardless of their shape, organization into tissues, or location, all epithelial cells are asymmetric, or polarized. This means that one side of such a cell is anchored to an extracellular matrix (ECM) called the basal lamina, or basement membrane (see Figure 10.1); this side is called the basal or basolateral membrane. The other side, called the apical membrane, faces the internal or external environment of the animal. Thus, epithelial cells form boundaries between different body compartments, as discussed later in this chapter. In this way, epithelial tissues can function as selective barriers that regulate the exchange of molecules between compartments. For example, epithelial tissues in an animal's skin help to form a barrier that prevents most substances in the external environment from entering the body.

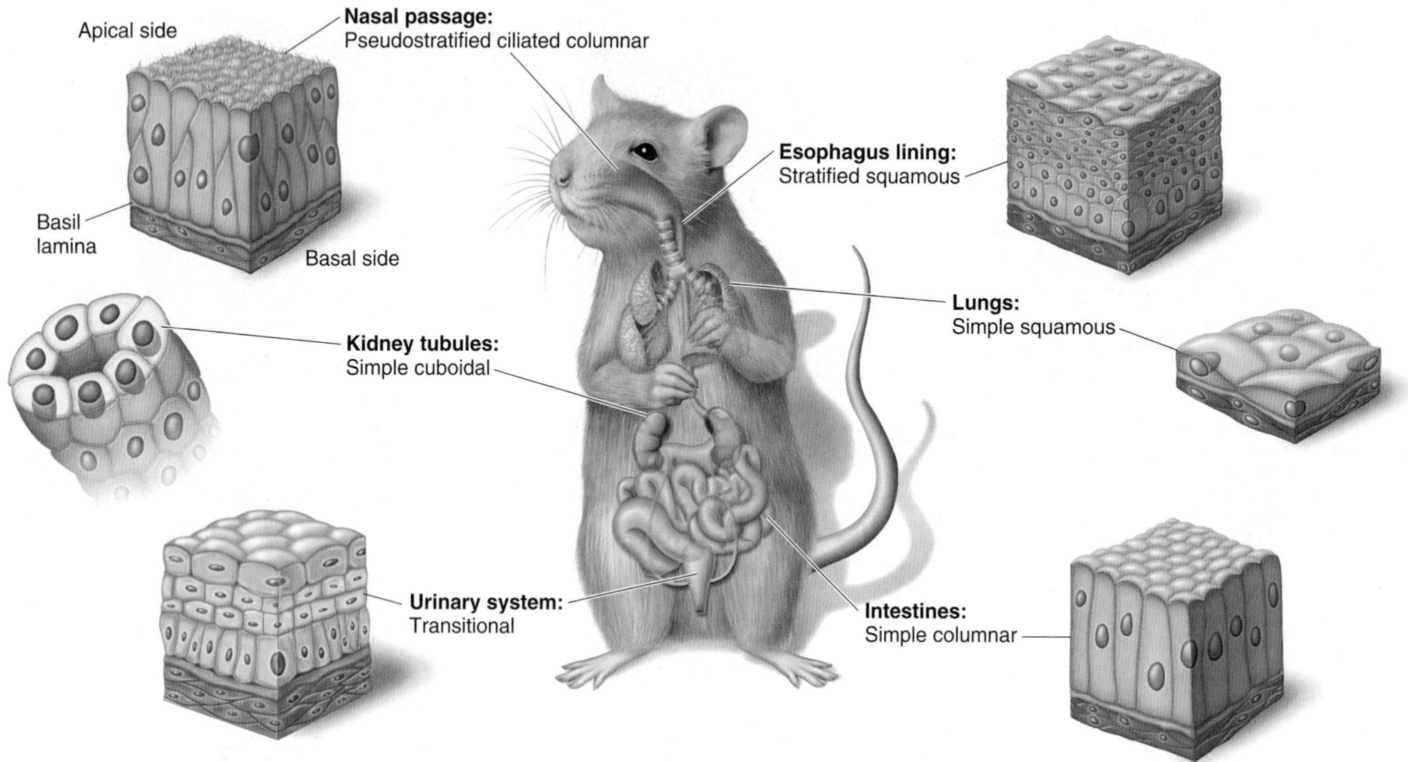

Figure 40.4 Examples of epithelial tissue. There are several types of epithelial tissue, distinguished by their appearance. Epithelial tissue is used to construct body coverings and the protective sheets that line and cover hollow tubes and cavities. The epithelial cells that make up epithelial tissues have an apical and basal (or basolateral) membrane; the apical side typically faces the exterior of the body or the lumen of a structure such as the intestines.

BIOLOGY PRINCIPLE Structure determines function. Note in this illustration how epithelial cells arranged in different ways form tissues with different functions. For example, when arranged as tubules in the kidney, they permit the passage of filtered body fluids. When modified as ciliated columns of cells in the nasal passage, they act as filters of airborne particles and debris.

Connective Tissues As their name implies, **connective tissues** connect, surround, anchor, and support the structures of an animal's body. Connective tissues include blood, adipose (fat-storing) tissue, bone, cartilage, loose connective tissue, and dense connective tissue (Figure 40.5).

An important function of some types of connective tissue cells is to form part of the ECM around cells by secreting a mixture of fibrous proteins and carbohydrates, such as glycosaminoglycans. These carbohydrates may covalently attach to proteins to form proteoglycans (refer back to Figure 10.4). In some cases, the extracellular matrix is rich in minerals. The final characteristics of any type of connective tissue are determined in part by the relative proportions and types of proteins, proteoglycans, and minerals secreted into the ECM. The ECM serves several general functions, which include (1) providing a scaffold to which cells attach and organize themselves into more complex structures, (2) protecting and cushioning parts of the body, (3) providing mechanical strength, and (4) cell signaling—transmitting information to the cells that helps regulate their activity, migration, growth, and differentiation.

The proteins of the ECM of a tissue consist mainly of two types. The first type is insoluble fiber-like proteins such as **collagen** and the rubber bandlike protein **elastin**; these proteins are often referred to as fibers. A second category is adhesive proteins (fibronectin and laminin) that serve to organize the protein and carbohydrate components of the extracellular matrix (refer back to Table 10.1).

Different Tissue Types Combine to Form Organs and Organ Systems

An **organ** is composed of two or more kinds of tissues arranged in various proportions and patterns, such as sheets, tubes, layers, bundles, or strips. For example, the vertebrate stomach (Figure 40.6) consists of the following layers (from outermost to innermost):

- an outer covering of simple squamous epithelial tissue;
- connective tissue layers covering and cementing the organ together;
- layers of smooth muscle tissue whose contraction mechanically breaks up food and propels it through the stomach and into the small intestine;
- nervous tissue that comes in close contact with the smooth muscle tissues and helps regulate their activity;
- an inner lining of simple columnar epithelial tissue that secretes enzymes and acid (important in the digestive process), and protective mucus into the cavity, or lumen, of the stomach.

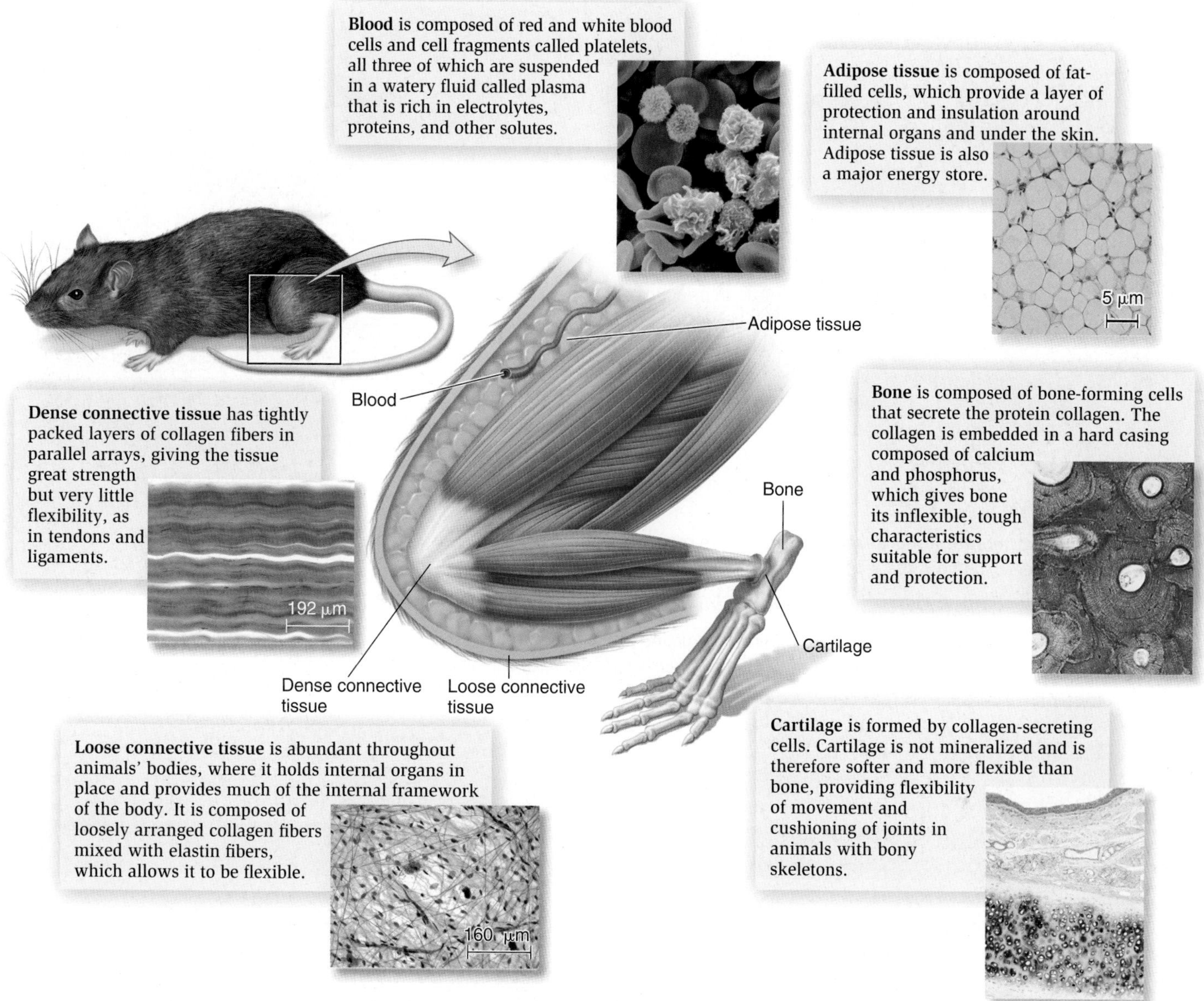
Blood is composed of red and white blood cells and cell fragments called platelets, all three of which are suspended in a watery fluid called plasma that is rich in electrolytes, proteins, and other solutes.
Adipose tissue is composed of fat-filled cells, which provide a layer of protection and insulation around internal organs and under the skin. Adipose tissue is also a major energy store.
5 μm
Adipose tissue
Blood
Dense connective tissue has tightly packed layers of collagen fibers in parallel arrays, giving the tissue great strength but very little flexibility, as in tendons and ligaments.
192 μm
Bone is composed of bone-forming cells that secrete the protein collagen. The collagen is embedded in a hard casing composed of calcium and phosphorus, which gives bone its inflexible, tough characteristics suitable for support and protection.
Bone
Cartilage
Dense connective tissue
Loose connective tissue
Loose connective tissue is abundant throughout animals' bodies, where it holds internal organs in place and provides much of the internal framework of the body. It is composed of loosely arranged collagen fibers mixed with elastin fibers, which allows it to be flexible.
160 μm
Cartilage is formed by collagen-secreting cells. Cartilage is not mineralized and is therefore softer and more flexible than bone, providing flexibility of movement and cushioning of joints in animals with bony skeletons.

Figure 40.5 **Examples of connective tissue in mammals.** Connective tissue connects, surrounds, anchors, and supports other tissues and may exist as isolated cells (blood), clumps of cells (fat), or tough, rigid material (bone and cartilage). The samples have been stained or the micrographs have been computer-colorized to reveal connective tissue.

In an **organ system**, different organs function together to perform an overall function. In the example just described, the stomach is part of the digestive system, along with other structures, such as the mouth, esophagus, small and large intestines, and anus. In another familiar example, the kidneys, the urinary bladder, the tubes leading from the kidneys to the bladder, and the tube leading from the bladder to the exterior of the body constitute the urinary system in mammals (see Figure 40.1). This system helps regulate the composition of body fluids and removes waste products from the blood, which are then excreted in the urine. The organ systems found in animals are listed in Table 40.1.

Organ systems frequently influence each other and depend on each other in many ways. For example, signals from the nervous system and endocrine system strongly influence how much water the mammalian kidney retains as it forms urine, an adaptation that can be lifesaving under certain circumstances. Likewise, nerve signals from the urinary bladder relay information to the cells of the brain when the bladder is full and ready to be emptied. As another example, the circulatory and respiratory systems of many animals operate together to provide oxygen to the cells of the body.

The spatial arrangement of organs into organ systems is part of the overall body plan of animals. Organ systems develop at specific times and locations within the body and along the anteroposterior body axis, as do other structures, such as limbs, tentacles, antennae, and other animal appendages. Scientists have long wondered how the layout of animal bodies is determined during the period when an embryo is developing. Remarkably, organ development in animals appears to be under the control of a highly conserved family of body-plan genes with homologs in all animals, as described next.

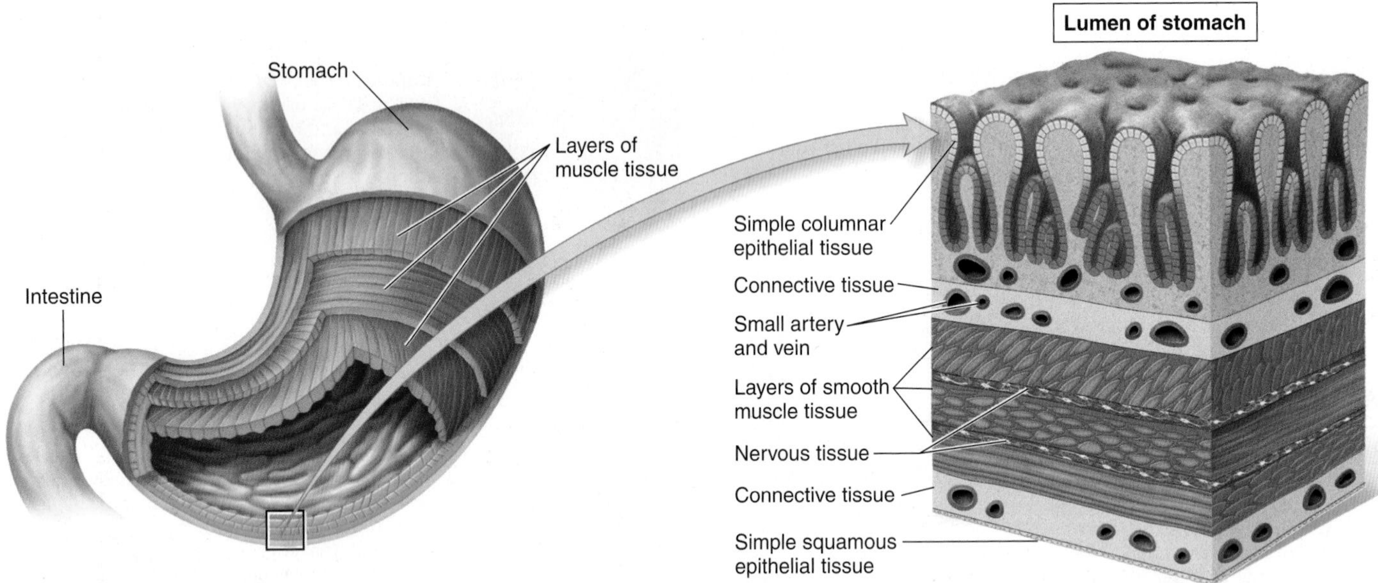

Figure 40.6 **The vertebrate stomach as an example of an organ composed of all four tissue types.** In this illustration, the thickness and appearance of the layers of nervous tissue have been considerably exaggerated for visual clarity.

Concept Check: *Do all animal organs contain all four tissue types? Can you think of an example that does not?*

BioConnections: *Look ahead to Figure 45.9 and compare the structure of the small intestine in that figure with the structure of the stomach shown in this figure. What can you conclude about common functions of the stomach and intestine from this comparison?*

Table 40.1 Organ Systems Found in Animals

Organ system	Major components*	Major functions
Circulatory	Contractile element (heart or vessel); distribution network (blood vessels); blood or hemolymph	Distributes solutes (nutrients, gases, wastes, and so on) to all parts of an animal's body
Digestive	Ingestion structures (mouth, mouthparts); storage structures (crop, stomach); digestive and absorptive structures (stomach, intestines); elimination structures (rectum, anus); accessory structures (pancreas, gallbladder)	Breaks complex foods into absorbable units; absorbs organic nutrients, salts, and water; eliminates solid wastes
Endocrine	All glands, organs, or tissues that secrete hormones	Regulates and coordinates growth, development, metabolism, mineral balance, water balance, blood pressure, behavior and reproduction
Excretory	Filtration system (kidneys or comparable structures); storage sites for soluble wastes (bladder); tubes connecting kidneys and bladder, and bladder to external environment	Eliminates soluble metabolic wastes; regulates body fluid volume and solute concentrations
Immune and lymphatic	Circulating white blood cells; lymph vessels and nodes	Defends against pathogens
Integumentary	Body surfaces (skin)	Protects from dehydration and injury; defends against pathogens; in some animals plays a role in regulation of body temperature
Muscular-skeletal	Force-producing structures (muscles); support structures (bones, cartilage, exoskeleton); connective structures (tendons, ligaments)	Produces locomotion; generates force; propels materials through body organs; supports body
Nervous	Processing (brain); signal delivery (spinal cord, peripheral nerves and ganglia, sense organs)	Regulates and coordinates movement, sensation, organ functions, and learning
Reproductive	Gonads and associated structures	Produces gametes (sperm and egg); in some animals, provides nutritive environment for embryo and fetus
Respiratory	Gas exchange sites (gills, skin, trachea, lungs)	Exchanges oxygen and carbon dioxide with environment; regulates blood pH

*Selected examples only; these do not necessarily pertain to all animals.

Genomes & Proteomes Connection

Organ Development and Function Are Controlled by *Hox* Genes

In previous chapters, you have learned about a family of ancient, highly conserved genes called *Hox* genes that are found in all animals. *Hox* genes determine the timing and spatial patterning of the anteroposterior body axis during development. For example, we saw in Chapter 19 how these genes determine the number and position of legs and wings in *Drosophila. Hox* genes play a similar role in determining the spatial patterning of the vertebrate body and appendages. Recently, scientists have begun exploring the role of *Hox* genes in the development and spatial patterning of the organs that make up animals' organ systems.

By generating mutant mice that fail to express one or more *Hox* genes (the genes are said to have been knocked out; see Chapter 20), researchers have discovered the important role these genes play in determining where within the vertebrate body particular organs form. Recall from Chapter 19 (refer back to Figures 19.16 and 19.17) that mouse *Hox* genes are arranged in four clusters, designated A–D, with multiple genes per cluster. Because these homologous genes (for example, *HoxA-3*, *HoxB-3*, and *HoxD-3*) are found within a single species, they are considered paralogous genes. Such paralogous genes typically act in concert to regulate similar developmental processes. For example, *HoxA-3* is important for development of anterior parts of the body, including the neck. When this gene is knocked out, mouse embryos show defects in neck structure, such as abnormal blood vessels. Also, the organs within the neck—including the thymus, thyroid, and parathyroid glands—do not develop normally. When two or more paralogs of this group are knocked out, certain neck organs fail to form at all. Experimental deletion of *Hox* genes associated with other body segments does not affect the development and function of neck glands and organs. Likewise, investigators have uncovered vital roles of different *Hox* genes in lung development within the thorax and the proper positioning and development of the vertebrate kidneys in the abdomen.

Of particular interest is the discovery that *Hox* genes are important not only for spatial patterning of organs but also for their growth, development, and function. Genes in the *Hox* 1 and 3 groups, for instance, help determine the final branching patterns of the airways of the lungs, the final size of the lungs, and the ability of the lungs to produce secretions that are important for breathing air after birth. Other *Hox* genes have been shown to control cell proliferation, shape changes, apoptosis, cell migration, and cell-cell adhesion within various organs. This is also true in invertebrates, such as the leech, where *Hox* genes are first expressed during organ formation, and in *Drosophila*, where the final shape and size of the heart are partly controlled by *Hox* genes.

Body Fluids Are Distributed into Compartments

All animal bodies are composed primarily of water. Most of the water in an animal's body is contained inside its cells and therefore is called **intracellular fluid** (from the Latin *intra*, meaning inside of). The rest of the water in its body exists outside of the cells and is therefore called

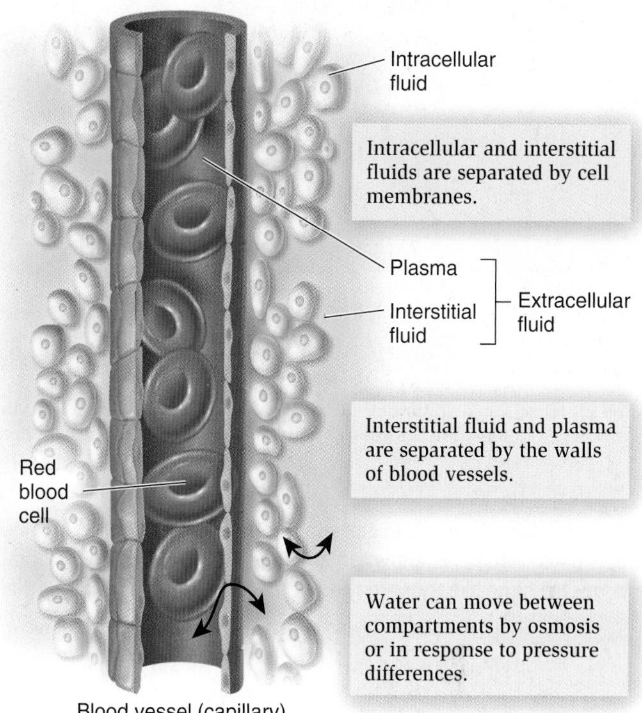

Figure 40.7 **Fluid compartments in a typical vertebrate.** Most of the fluid within an animal's body exists within cells (intracellular fluid). Extracellular fluid is that portion of the body's fluid that lies outside cells (interstitial fluid) and within blood vessels (plasma), such as the capillary shown here. Arrows indicate directions of water movement between adjacent compartments.

Concept Check: *What would happen to the distribution of water in the fluid compartments of an animal's body if a blood vessel were damaged such that it leaked its contents?*

extracellular fluid (from the Latin *extra*, meaning outside of). Extracellular fluid is composed of the fluid part of blood, called **plasma**, and the fluid-filled spaces that surround cells, called **interstitial fluid** (from the Latin *inter*, meaning between) (**Figure 40.7**). In vertebrates and some invertebrates, plasma and interstitial fluid are kept separate within a closed circulatory system. In many invertebrates with open circulatory systems (see Chapter 47), however, plasma and interstitial fluid are intermingled into a single fluid called hemolymph.

The fluids depicted in Figure 40.7 are said to be enclosed in compartments. In a typical vertebrate, the total water volume in the three compartments (intracellular fluid, plasma, and interstitial fluid) accounts for about two-thirds of body weight, with solids comprising the rest. Of the total body water, up to two-thirds is intracellular and one-third extracellular, the majority of which is located in the interstitial compartment.

Plasma membranes separate the intracellular fluid from the extracellular fluid. The two components of extracellular fluid in vertebrates and some invertebrates—the interstitial fluid and the plasma—are separated by the walls of the blood vessels (for example, arteries, capillaries, and veins).

The solute composition of the extracellular fluid is very different from that of the intracellular fluid. Maintaining differences in solute composition across the plasma membrane is an important way in which

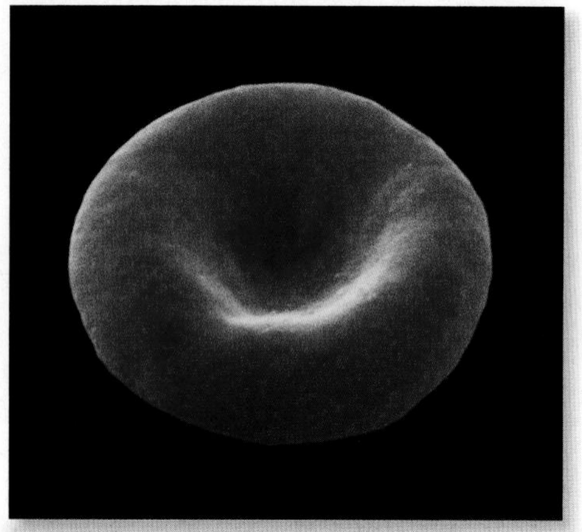

Red blood cell in a solution of normal osmolarity (an isotonic solution)

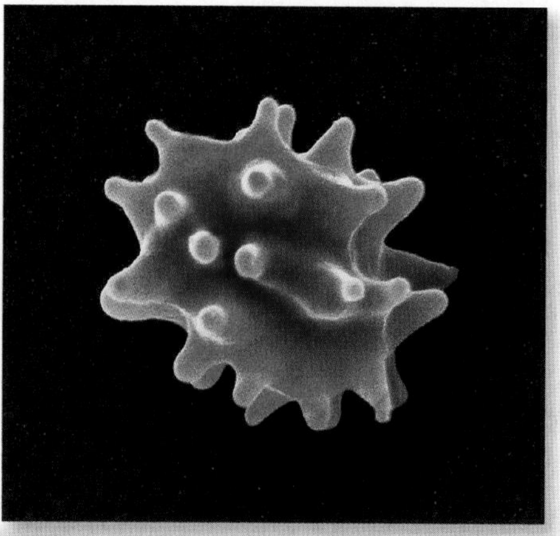

Red blood cell that has lost intracellular fluid when placed in a solution of higher than normal osmolarity (a hypertonic solution)

Red blood cell that has gained intracellular fluid when placed in a solution of lower than normal osmolarity (a hypotonic solution)

Figure 40.8 **Changes in cell shape due to alterations in intracellular fluid volume.** Alterations in intracellular fluid volume can have drastic effects on cell shape, as shown by these scanning electron micrographs of red blood cells. Each cell is approximately 7 μm in diameter. Dramatic changes in shape like those in the middle and right are usually lethal for cells.

Concept Check: *What effect would changes in intracellular fluid volume have on intracellular solute concentration?*

animal cells regulate their own activity. For example, many different proteins that are important in regulating cellular events such as mitosis, cytokinesis, and metabolism are confined to the intracellular fluid.

Movement of Solutes Between Compartments Solutes must move between body fluid compartments in order for cells in an animal's body to maintain concentrations of ions, nutrients, and gases such as oxygen within their normal ranges. Barriers separating adjacent fluid compartments determine which solutes can move between them. Solute movement, in turn, accounts for the differences in composition of the different compartments. We discussed the mechanisms by which solutes move in Chapter 5. Let's summarize those mechanisms, which apply to all animal cells.

Simple diffusion is the net movement of a solute down its concentration gradient without the aid of a transport protein or hydrolysis of ATP. Simple diffusion is one way in which cells gain and lose solutes. Molecules that can cross lipid bilayers are able to passively diffuse into or out of a cell. Examples include nonpolar solutes such as lipids and CO_2, and some very small polar solutes such as ethyl alcohol. Most polar molecules and ions, however, can diffuse through a plasma membrane only if the membrane has channels or transporters of some type that permit the molecule to pass through the bilayer. The rate of diffusion of any solute depends on several factors, notably the concentration gradient of the solute and the area across which it is diffusing. The rate of diffusion of a solute across a membrane of given thickness can be calculated by the Fick diffusion equation, shown here in simplified form:

$$J = KA\,(C_1 - C_2)$$

where J is the rate of diffusion, K is a constant that includes temperature, A is the cross-sectional area of the barrier across which diffusion is occurring, and C_1 and C_2 are the concentrations of the solute at two locations (for example, inside and outside a cell). This equation makes it possible to determine how changes in solute concentrations, temperature, or area can influence the rate at which a substance moves across a plasma membrane. For example, breathing a gas mixture from a tank that is enriched in oxygen will increase the amount of oxygen entering the blood in a mountain climber at high altitude, where oxygen is limited. According to the Fick equation, the difference between C_1 (inhaled oxygen) and C_2 (oxygen in the blood) will be increased by this procedure. Therefore, we can predict that J, the rate of diffusion of oxygen into the blood, will also be increased, an important survival mechanism at very high altitudes.

In contrast to simple diffusion, the movement of most solutes between compartments or across plasma membranes is mediated by transport proteins via facilitated diffusion or active transport (see Chapter 5). Each of these processes is critical to regulation of intracellular and extracellular fluid composition in animal cells. In Chapter 45, we will discuss one example, the mechanism by which cells obtain their most important energy source, glucose. Because glucose is a moderately sized polar molecule, it cannot diffuse through lipid bilayers. Instead, it is transported from the interstitial fluid into cells by membrane-bound proteins.

Movement of Water Between Compartments Water can readily move between adjacent compartments in an animal's body, because barriers such as plasma membranes tend to be highly permeable to water due to the presence of water channels called aquaporins (see Figure 5.18). This movement depends on pressure differences in the fluids of each compartment and on osmosis (see Chapter 5), in which water moves from a region of lower solute concentration to one of higher solute concentration. For cells to function properly, they require a relatively stable internal composition, including ion and protein concentrations, cellular volume, and pH. A decrease in solute concentration outside a cell, for example, would cause water to move by osmosis from outside the cell to inside. In this case, osmosis redistributes fluid from the interstitial to the intracellular compartment. This would cause a cell to become deformed as it swells due to the influx of water. In contrast, an increase in extracellular solute concentration would lead to osmosis of water from inside the cell to outside, causing the cell to shrink. In either case, a swollen or shrunken animal cell generally is more fragile than a normal cell and will die if its membrane ruptures. Figure 40.8 shows examples of mammalian red blood cells in which intracellular fluid levels have been altered. This could occur, for example, if the blood cells were exposed to extracellular fluids that were either more dilute (hypoosmotic or hypotonic) or more concentrated (hyperosmotic or hypertonic) than the fluid inside the blood cell. When red blood cells swell, they may burst, a phenomenon called hemolysis; shrinkage of red blood cells is called crenation and is also potentially destructive to cells (Figure 40.8, middle panel).

40.2 The Relationship Between Form and Function

Learning Outcomes:

1. Provide an example of how the structure of an animal's tissues or organs can help predict their function.
2. Describe the importance of the surface area/volume (SA/V) ratio to animal form and function.

A key principle of biology throughout this unit is that form (structure) and function are closely related. The appearance or structure of an animal's tissues and organs can often help us predict the function of those structures. For example, let's compare the respiratory systems of an insect and a mammal (**Figure 40.9**). The respiratory systems of animals exchange oxygen from the environment with carbon dioxide generated by the body. Although many important differences exist between the respiratory systems of insects and mammals, notably the absence of lungs in insects, certain structural similarities suggest that both systems serve similar functions. In both cases, for example, a series of internal branching tubes composed of epithelial and connective tissues arises from one or more openings that connect with the outside environment (the mouth and nose in the mammal, and the pores called spiracles in the insect). These tubes become smaller and smaller as they continue to branch, eventually terminating in narrow structures that are only one cell thick.

Without knowing anything else about the respiratory systems of these two animals, we can surmise that in both cases these branching tubes serve as conduits for air to flow back and forth between the environment and the internal spaces of the animal. In the insect, the ends of the branching tubes called tracheoles are where oxygen diffuses from the air to the fluid around individual cells (and from there to intracellular fluid) (Figure 40.9a). In the mammal, the ends of the tubes form saclike structures called alveoli across which oxygen diffuses into the bloodstream.

If we examine the mammalian lung in greater detail (Figure 40.9b), we see that the alveoli are composed of extremely thin, squamous epithelial cells. The shape of the cells provides a clue to their function. Their flat, thin structure permits rapid diffusion of gases across the cells. Imagine the resistance to oxygen diffusion if the cells were thick or scaly, like the cells of the body surface of many animals, for example. Therefore, both the gross and microscopic anatomy of the gas-exchange surfaces of respiratory systems facilitates their functions.

An additional structural similarity found in essentially all respiratory surfaces, including gills, is an extensive surface area. In fact, we can expand our discussion to include all cells, tissues, and organs that mediate diffusion or absorption of a solute from one compartment to another, or which require extensive cell-to-cell contacts. Consider, for instance, the finger-like projections of the small intestine of a human, the skin folds of some high-altitude frogs, the cellular extensions on the surface of neurons of a mouse, and the feathery antennae of a moth (**Figure 40.10**). What do these structures have in common? They all have a large surface area, which maximizes their

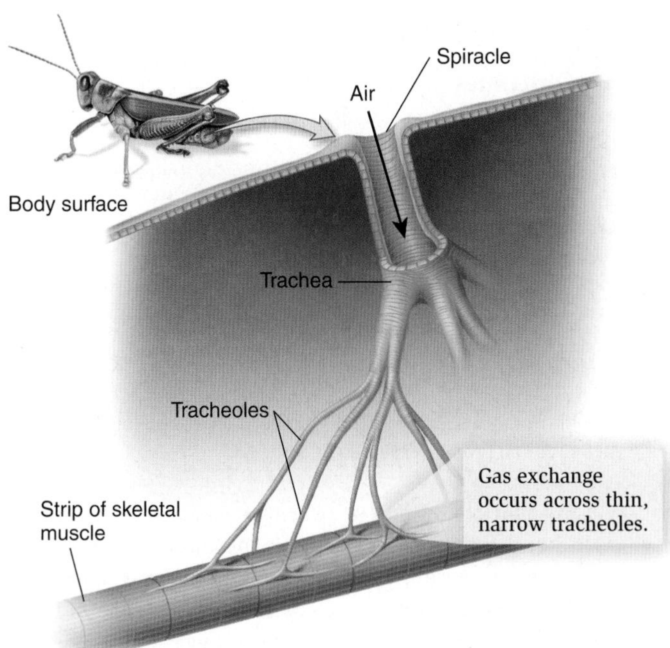

(a) Insect respiratory system

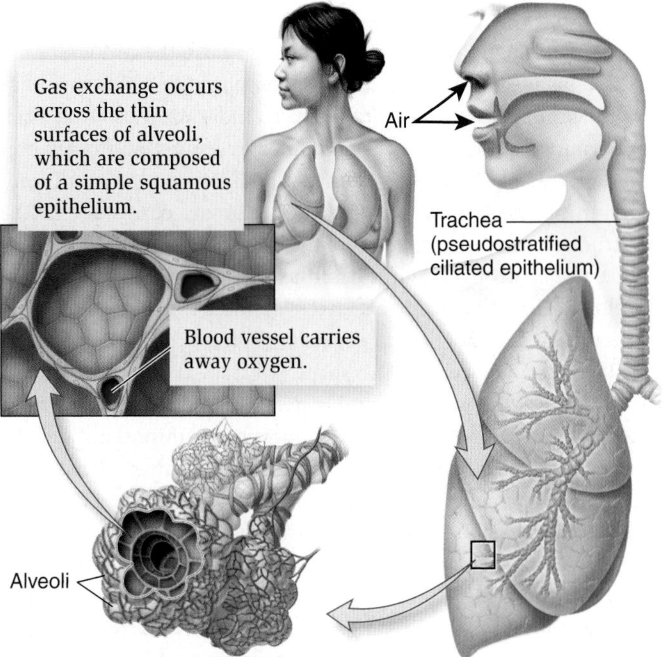

(b) Human respiratory system

Figure 40.9 Comparison of the branching air tubes in (a) an insect and (b) a mammal. Note the similar features of highly branching, internalized hollow tubules that connect to the outside air, suggesting that these systems perform similar functions.

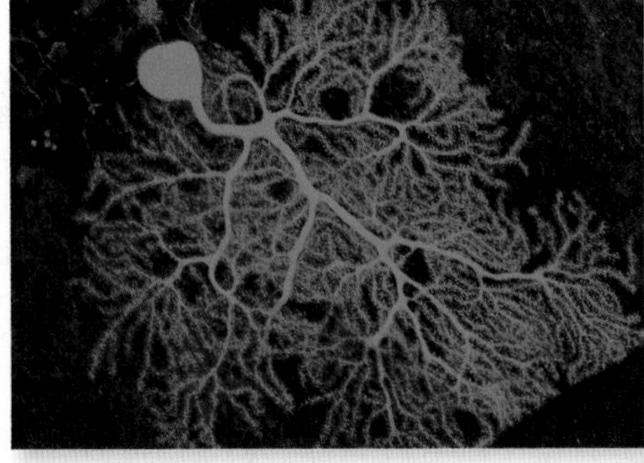

(a) Human intestine

(b) Frog skin

(c) Mouse brain neuron stained with a fluorescent marker

(d) Moth antennae

Figure 40.10 Examples of structures in which extensive surface area is important for function. A large surface area allows **(a)** high rates of transport of solutes across the intestine of a human, **(b)** increased diffusion of oxygen across the folds of skin of a frog living at high altitude, where O_2 is less available, **(c)** extensive communication between neurons in a mouse's brain, and **(d)** detection of airborne chemicals by moth antennae.

Concept Check: *Is surface area important only for animals, or could it also provide advantages to other living organisms?*

ability to absorb solutes (intestine), obtain oxygen by diffusion from the environment (frog skin), communicate with other cells (neurons), or detect airborne molecules (moth antennae). Increasing surface area of a structure, however, comes at the expense of greatly increasing volume if the shape of the structure is not changed (refer back to Figure 4.8). This is because as an object enlarges, its volume grows relatively more than its surface area; surface area increases by the power of two, while volume increases by the power of three as an object enlarges. For example, if an animal's height, length, and width are increased by a factor of 10, its surface area is increased 100 times, but its volume is increased 1,000 times. This relationship is also true when considering an animal's organs and other structures, and could create certain disadvantages. For example, the ability to obtain sufficient oxygen from water requires a great amount of surface area within the fine structure of a fish's gills. Were the gills to increase their volume in the expected proportion just described, they would become far too unwieldy to allow the normal behaviors required for survival of a fish. The challenge of packaging an extensive surface area into a confined space is overcome by changes in shape (see, for example, the way the inner surface of the intestines in Figure 40.10a folds inward to form finger-like extensions). In a fish gill, numerous thin, flat, platelike structures are packed together, one on another, forming a dense array of surfaces available for oxygen diffusion. The ratio between a structure's surface area and the volume in which the structure is contained is called the **surface area-to-volume (SA/V)** ratio (see Figure 4.8). A high SA/V ratio is ideal for exchange of heat, solutes, and water across a surface without contributing greatly to the mass of an animal or body part. This concept will appear throughout this unit as we explore the ways in which animals obtain energy, regulate their metabolism and body temperature, obtain oxygen, and eliminate wastes.

When all of an animal's organ systems operate correctly and body fluid levels and solute composition are maintained within normal limits, the animal is generally considered to be in a healthy condition. In the rest of this chapter, we will examine the process of achieving and maintaining this condition, which is known as homeostasis.

40.3 Homeostasis

Learning Outcomes:

1. Discuss the concept of homeostasis as it applies to the internal environment of animals.
2. Distinguish between conforming and regulating as strategies to maintain homeostasis.
3. List several variables that are regulated within a homeostatic range in vertebrate animals.
4. Name the four components of a homeostatic control system and describe the importance of each to the regulation of an animal's internal environment.
5. Describe how negative feedback, positive feedback, and feedforward regulation contribute to the maintenance of homeostasis in animals.
6. Explain the importance of paracrine and hormonal signaling to homeostasis and provide examples of each.

The environmental conditions in which organisms—including all animals—live are rarely, if ever, constant. Animals are exposed to fluctuations in air and water temperatures, nutrient and water supplies, pH, and, in some cases, oxygen availability. Any one of these environmental changes could be harmful or even fatal if an organism is unable to respond appropriately. However, as you might expect from the incredible diversity of environments in which they exist, animals can adjust in many ways to their surroundings and thrive.

The process of maintaining a relatively stable internal environment despite changes in the external surroundings is known as **homeostasis** (from the Greek *homoios*, meaning similar, and *stasis*, meaning to stand still). The term was coined in the 20th century by the American physician and physiologist Walter Cannon, but the concept itself originated in the 19th century with the French physician and physiologist Claude Bernard, who postulated that a constant *milieu interieur* (internal environment) was a prerequisite for good health.

Some Animals Conform to External Environments; Others Regulate Their Internal Environments

Generally speaking, animals maintain homeostasis in one of two ways: conforming and regulating. Some animals conform to their environments so that some feature of their internal body composition matches their external surroundings. A marine crab, for example, has about the same solute concentration in its body fluids as is found in seawater. Likewise, the body temperatures of many fishes and aquatic invertebrates match the temperatures of the surrounding waters. Energetically speaking, conforming is a cheap strategy for survival. It would take a great deal of energy for a small fish to maintain its body temperature at, say, 37°C when swimming in the waters near Antarctica. However, because they do not actively adjust their internal body composition, conformers are often restricted to living in environments that are relatively unchanging.

Other animals regulate the composition of their fluids and solutes at levels that are different from those of the external environment. Regulating the internal environment requires considerable energy in the form of ATP. For this reason, homeostasis for regulators comes with a high energy price tag. However, the price paid by regulators makes it possible for them to exploit environments that fluctuate significantly, something that is difficult for conformers.

Vertebrates Maintain Most Physiological Variables Within a Narrow Range

In vertebrates, the common physiological variables—concentrations of blood-borne solutes such as minerals, glucose, and oxygen, for example—are usually maintained within a certain range despite fluctuating external environmental conditions (**Table 40.2**). At first glance, homeostasis may appear to be a state of stable balance of physiological variables. However, this simple idea cannot capture the scope of homeostasis. For example, no physiological function is constant for very long, which is why we call them variables. Some variables may fluctuate around an average value during the course of a single day yet still be considered in balance. Homeostasis is a dynamic process, not a static one.

Table 40.2 Selected Examples of Homeostatic Variables in Animals

Variable	Factors that influence homeostasis	Examples of functions
Minerals	Eating food; excreting wastes	
Na^+ and K^+		Establish resting membrane potentials across plasma membranes in all cells and are responsible for transmitting electrical signals in excitable tissues (muscles and nervous tissue).
Ca^{2+}		Important for muscle contraction; neuron function; skeleton and shell formation
Fe^{2+}		Binds and transports oxygen in blood or body fluids (some invertebrates use copper instead of iron)
Energy sources	Eating food; expending energy	
Glucose		Broken down to provide energy for use by all cells, especially brain cells
Fat		Provides an alternate source of energy, particularly for cells not in the nervous system; major component of plasma membranes
ATP		Provides energy to drive most chemical reactions and body functions; modifies function of many proteins by transferring its terminal phosphate group to proteins
Body temperature	Rate of energy expenditure; environmental temperature; behavioral mechanisms (look ahead to Chapter 46)	Determines the rate of chemical reactions in an animal's body
pH of body fluids	Hydrogen ion pumps in cells; buffers in body fluids; rates of energy expenditure; breathing rate	Affects enzymatic activity in all cells
Other variables		
Oxygen and carbon dioxide	Movement of air or water across respiratory surfaces (for example, lungs and gills); rate of energy expenditure	Oxygen circulates in body fluids and enters cells, where it is used during the production of ATP; carbon dioxide is a waste product that is eliminated to the environment, but it is also a key factor that regulates the rate of breathing air or water
Water	Drinking, eating, excretion of wastes, perspiration, osmosis across body surface (skin or gills)	Numerous biological functions including participating in chemical reactions; helping to regulate body temperature; acting as a solvent for biologically important molecules (refer back to Chapter 2)

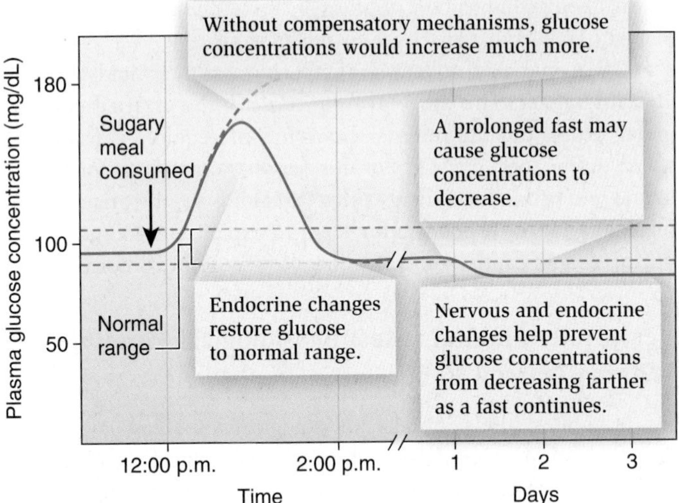

Figure 40.11 An example of a homeostatically controlled variable, glucose levels in human blood. Note that glucose levels in the plasma may rise or fall depending on whether an animal has recently eaten. However, even after a sugary meal or a prolonged fast, homeostatic mechanisms either return glucose levels to normal or enable those levels to remain within the range required for survival. Traditional units for glucose used in the U.S. are given on the *y*-axis; as a reference, a value of 100 mg/dL is equal to 5.5 mM.

Consider an example in your own body. Normally, blood sugar (glucose) remains at fairly steady and predictable concentrations in any healthy individual. After a meal, however, the concentration of glucose in your blood can increase quickly, especially if you have just eaten something sweet. Conversely, if you skip a few meals, your blood glucose concentration may decrease slightly (Figure 40.11). Such fluctuations above and below the normal value might suggest that blood glucose concentrations are not homeostatic, but this is incorrect. Once blood glucose increases or decreases, homeostatic mechanisms restore glucose concentrations back toward normal. In the case of glucose, the endocrine system is primarily responsible for this quick adjustment, but in other examples, a wide variety of control systems may be initiated. In later chapters, we will see how every organ and tissue of an animal's body contributes to homeostasis, sometimes in multiple ways, and usually in concert with each other.

Homeostasis, then, does not imply that a given physiological function or variable is rigidly constant. Instead, homeostasis means that a variable fluctuates within a certain normal range and that once disturbed from that range, compensatory mechanisms restore the variable toward normal.

Homeostatic Control Systems Maintain the Internal Environment

The activities of cells, tissues, and organs must be regulated and coordinated with each other so that any change in the extracellular fluid—the internal environment—initiates a response to correct the change. These compensating regulatory responses are performed by homeostatic control systems. A **homeostatic control system** must have several components, including:

- a **set point**—the normal value for a controlled variable;
- a **sensor**, which monitors the level or activity of a particular variable;
- an **integrator**, which compares signals from the sensor with the set point; and
- an **effector**, which compensates for any deviation between the actual value and the set point.

Figure 40.12 shows an example of a homeostatic control system that regulates body temperature in mammals. This system is somewhat analogous to the heating system of a home. In that case, a sensor and integrator within the thermostat compare the actual room temperature with the set point temperature that was determined by setting the thermostat to a given temperature. If the room temperature becomes cooler than the thermostat setting, the effector (furnace) is activated and adds heat to the room. In a mammal, the sensors are temperature-sensitive neurons in the skin and brain, whereas the integrator is within the brain. Signals from the brain are sent along nerves to the effectors, which include skeletal muscles. If body temperature decreases, the muscles contract vigorously in response to these signals, resulting in shivering—a key way in which mammals' bodies generate heat. We will discuss other heat-conserving and heat-generating mechanisms that contribute to this homeostatic control system in Chapter 46.

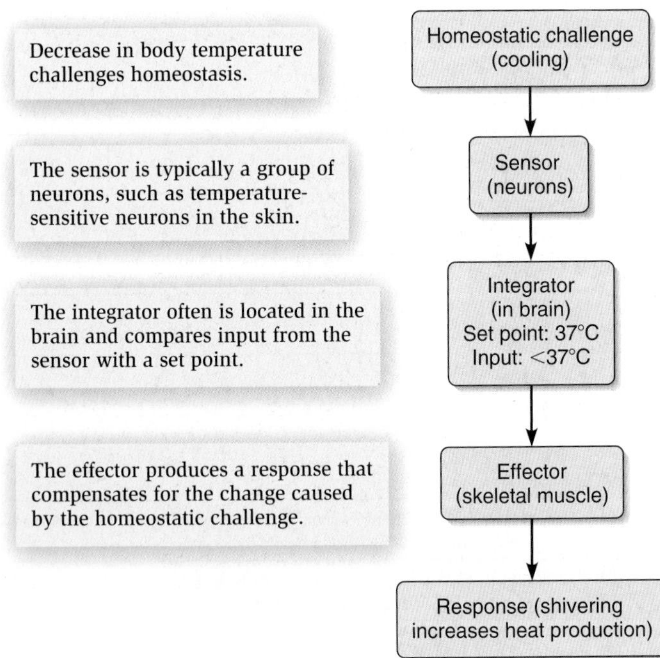

Figure 40.12 An example of a homeostatic control system. The sensor and effector for responding to a decrease in body temperature are shown. Different homeostatic control systems have different sensors and effectors.

Concept Check: Would you expect all animals to have similar set points for a given homeostatic variable?

When a homeostatic control system operates normally, changes in a physiological variable are kept to a minimum. These changes are limited by the process of feedback, as we will now see.

Feedback Is a Key Feature of Homeostasis

Feedback is a fundamental feature of homeostasis and a major way in which disturbances to a physiological variable are minimized. The temperature regulation system just described is an example of a **negative feedback loop**, in which a change in the variable being regulated brings about responses that move the variable in the opposite direction. Thus, a decrease in body temperature leads to responses that increase body temperature—that is, move it back toward its original value.

Negative feedback also prevents homeostatic responses from overcompensating. When the blood pressure of a bleeding animal decreases as more and more blood is lost, for example, pressure sensors in the heart and certain blood vessels detect the change in pressure and send the information to the integrator—the brain (Figure 40.13). In the brain, the signal is compared with the normal set point for blood pressure. The brain responds to this sharp deviation from the blood pressure set point in two ways. First, signals are sent along nerves to the effectors—in this case, the kidneys, heart, and blood vessels. Second, the brain stimulates the release of certain hormones into the blood; these hormones act with the nervous system on the effectors. The result is that the heart beats more rapidly and forcefully, the kidneys produce less urine and thereby retain more body fluid, and the blood vessels preferentially direct blood to the most vital organs such as the brain. These responses raise the animal's blood pressure back toward the set point. Restoring the blood pressure removes the stimulus from the sensor, and this, in turn, shuts off further production of the hormonal and neural responses (negative feedback). If feedback inhibition did not occur, the blood pressure would not only rebound back to the set point but might rise to abnormally high and possibly dangerous levels.

Negative feedback may occur at the organ, cellular, or molecular level. For instance, feedback mechanisms regulate many enzymatic processes. In one example, ATP regulates the rate of its own formation in cells by inhibiting certain intracellular enzymes that catalyze the breakdown of glucose molecules, a key event in the production of ATP.

Not all forms of feedback contribute to homeostasis. In some cases, a **positive feedback loop** may accelerate a process (think of an avalanche that begins with a small snowball rolling down a steep hill). This is contrary to the principle of homeostasis, in which large fluctuations in a variable are minimized and reversed. Not surprisingly, perhaps, positive feedback is far less common than negative feedback in animals. Nonetheless, positive feedback is crucial to some processes in animal biology. One example is the process of birth in mammals (Figure 40.14). Birth is triggered by a positive feedback loop between nerve signals arising from smooth muscle cells of the cervix (the lower narrower part of the uterus) and the mother's brain and pituitary gland (a component of the endocrine system). As the uterus pushes the baby's head against the cervix in late pregnancy, nerve signals from the cervix send information to the brain, which triggers the pituitary gland to release certain hormones. These hormones stimulate the uterus to contract with more force, which in turn sends additional nerve signals to the brain, and so on. Eventually, the uterus contracts strongly enough to deliver the baby—an event that stops the signals from the cervix to the brain and allows the uterus to gradually relax.

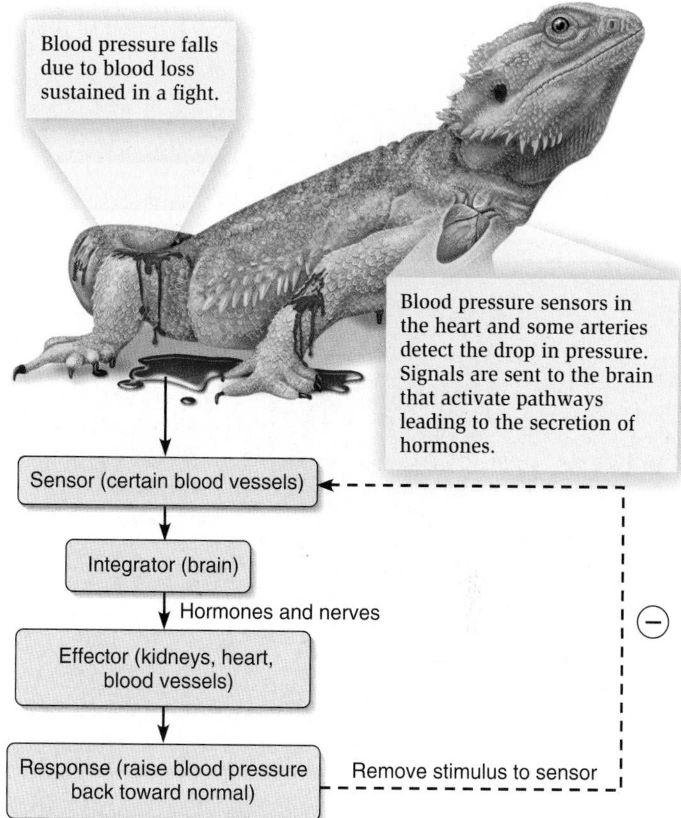

Figure 40.13 Negative feedback as a mechanism by which homeostatic control systems operate. In this example, loss of blood results in a decrease in blood pressure, which could be life-threatening if not corrected. Effectors such as the kidneys, heart, and blood vessels help restore blood pressure toward normal. They do not increase blood pressure above normal, however, because of negative feedback (as denoted by the minus sign).

BioConnections: Look back at Figure 6.13 for an example of negative feedback inhibition that occurs at the molecular level. In both examples, negative feedback contributes to the maintenance of homeostasis.

Feedforward Regulation Prepares for an Upcoming Challenge to Homeostasis

Built into the homeostatic mechanisms of many animals, particularly those with well-developed nervous systems, is another important feature that minimizes large swings in physiological variables. In **feedforward regulation**, an animal's body begins preparing for a change in some variable (for example, blood glucose concentrations) before it even occurs. Consider the anticipatory changes that occur when a hungry dog smells or sees food; indeed, the same phenomenon happens in humans. First, the animal starts to salivate, and its stomach begins to churn. Salivation and movements of the stomach

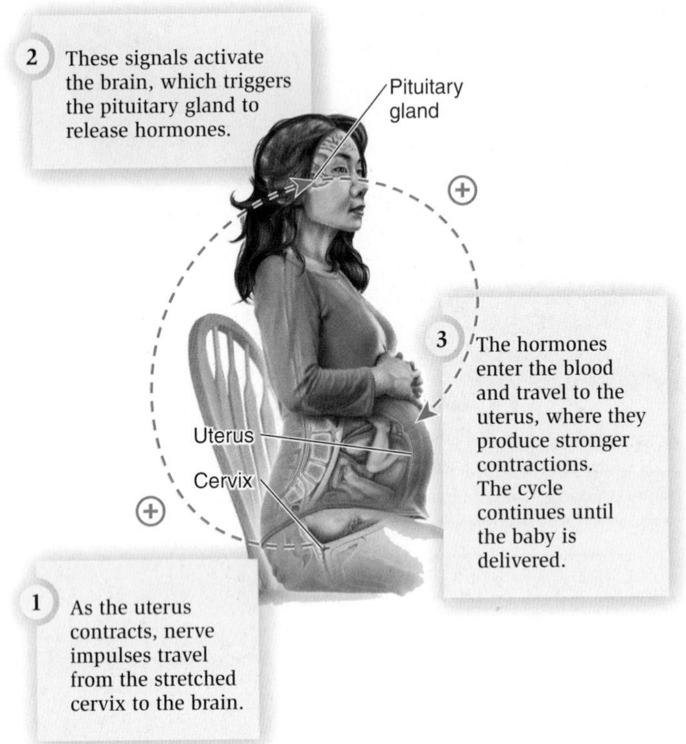

Figure 40.14 **The human birth process as an example of positive feedback.**

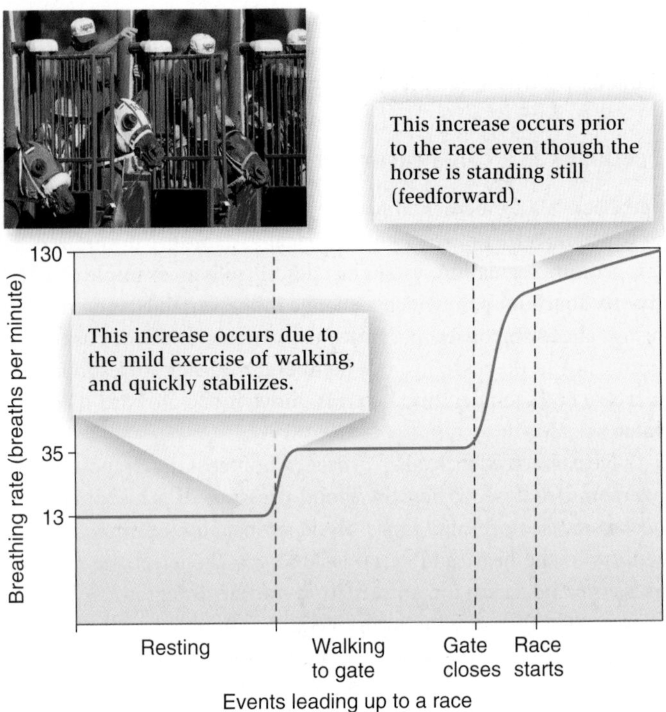

Figure 40.15 **Feedforward control of breathing rate in an animal trained for athletic exercise.** Feedforward processes prepare the body for an ensuing challenge or event, such as the race shown here.

Concept Check: *What kind of similar feedforward response might occur in nature?*

are important components of the digestive process, yet at this stage, the animal has not actually eaten any food. Instead, its digestive system is already preparing for the arrival of food in order to maximize digestive efficiency, speed the flow of nutrients into the blood, and minimize the time required for active cells to replenish energy stores. Therefore, feedforward regulation speeds up the body's homeostatic responses and minimizes fluctuations in the variable being regulated—that is, it reduces the amount of deviation from the set point.

In the preceding example, feedforward control uses sensory detectors that recognize odors and sights. Many examples of feedforward control, however, result from, or are modified by, the phenomenon called learning. The result of this process is that the nervous system learns to anticipate a homeostatic challenge. Familiar examples are the increased heart rate and breathing rate that occur just before an athletic competition—for example, in trained racehorses before the start of a race (**Figure 40.15**). The process of training, in which the horses' bodies learn to prepare for the exertion of the race, allows there to be no delay between the start of exercise and the flow of blood and nutrients to skeletal muscle. The most famous example of feedforward control was first demonstrated in the early 20th century by Russian physiologist Ivan Pavlov, as described next.

Feature Investigation

Pavlov Demonstrated the Relationship Between Learning and Feedforward Processes

The Russian physiologist Ivan Pavlov made numerous contributions to our understanding of digestive processes in mammals, for which he earned a Nobel Prize in Physiology or Medicine in 1904. Today, however, Pavlov is best remembered for work he did later in his career, when he demonstrated that feedforward processes associated with digestion could be conditioned to an irrelevant stimulus, that is, one that normally is not associated with digestive processes.

Pavlov was interested in the factors that increase production of saliva in a hungry animal. Saliva is an important secretion made by glands in the mouth, because among other things it aids in swallowing and contains enzymes that kill pathogens and begin the process of digesting starches in food. Pavlov discovered that dogs accustomed to being fed by the same researcher each day would begin to salivate any time they saw the researcher approaching, even without receiving, seeing, or smelling any food. In other words, the dogs had become conditioned to associate the researcher with food. This did not happen in dogs that were not yet accustomed to the regular feeding schedule of the laboratory. Pavlov hypothesized that any stimulus could elicit the feedforward process of salivation, even one that is not normally associated with feeding, so long as that stimulus was somehow paired with feeding.

To test this hypothesis, Pavlov presented two groups of dogs with either food (the control group) or food plus the ticking of a metronome (the experimental group) (**Figure 40.16**). Each dog was isolated in a room so that it could not see, hear, or smell the researcher or sense any other cues that might interfere with the experiment. Pavlov reasoned that after some period of time, the experimental dogs would become conditioned to the sound of the metronome and would learn to associate it with the arrival of food. If this was correct, the metronome by itself should eventually be sufficient to elicit the feedforward process of salivation.

To measure salivary production rates, Pavlov surgically altered the ducts leading from the salivary glands located under the dogs' tongues such that the ducts opened up outside the dog's chin. A glass funnel was then secured beneath the chin. In Pavlov's initial experiments, the amount of saliva was quantified simply by counting the number of drops coming from a tube leading from the funnel. This was achieved by allowing the saliva to settle onto a mechanical device attached to a rocker arm that caused a deflection on a rotating electrical recorder. Each time a drop was collected, the arm moved and registered a hatch mark on the recorder. The number of hatch marks was equal to the total number of drops of saliva collected in any given period. Pavlov's hypothesis was confirmed when, after a conditioning period of several days, the conditioned stimulus by itself (the metronome) elicited increased salivation in the conditioned dogs in under 10 seconds. The control dogs responded slightly sooner to the appearance of food than the conditioned dogs did to the metronome, but the amount of saliva produced was roughly comparable in both groups of dogs.

In later experiments, Pavlov demonstrated that the response to a conditioned stimulus was not permanent. If a dog that was already conditioned to the sound of a metronome was then repeatedly exposed to that sound without the simultaneous presentation of food, the salivary response to the sound would eventually cease. These experiments demonstrated that certain feedforward processes can be modulated by experience and learning, and they suggested that

Figure 40.16 **Pavlov's experiments showed that feedforward events can be conditioned by learning.**

HYPOTHESIS It is possible to condition a feedforward response to an irrelevant stimulus through learning.

KEY MATERIALS Experimental animals (dogs), metronome, collection ports for saliva, food, soundproof testing rooms.

Experimental level | **Conceptual level**

1 Divide dogs into 2 groups. The control dogs are given food only. The conditioned dogs are exposed to a metronome when food is given.

Food only (control dogs)

Food dish

Separate soundproof testing rooms

Ticking metronome

Start metronome simultaneously with food presentation for several days (conditioned dogs)

Food dish

The dogs exposed to metronome become conditioned to the sound.

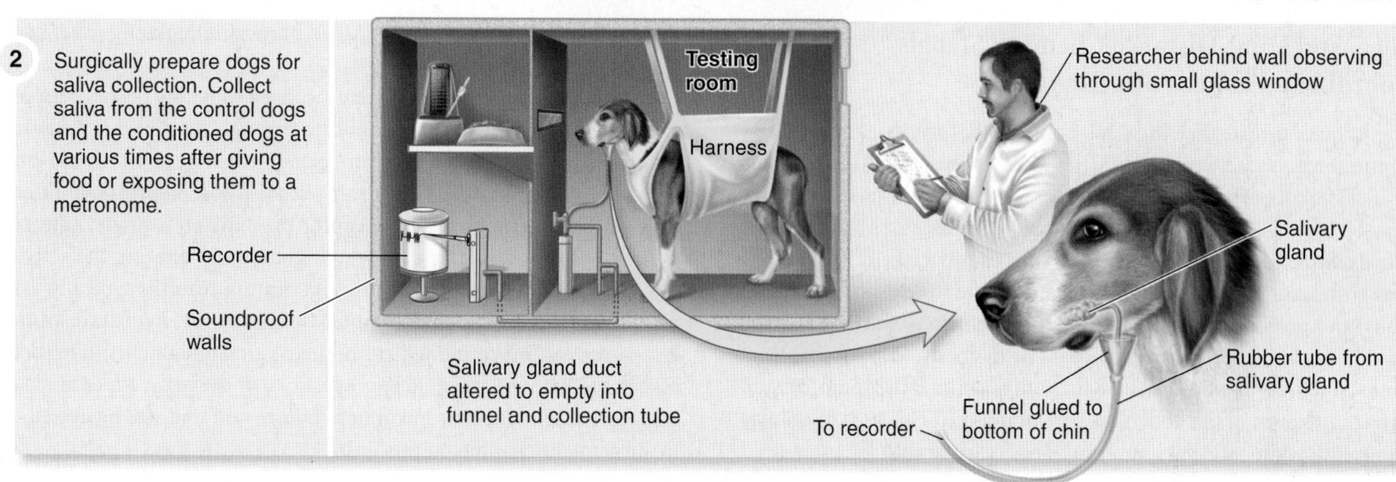

3 THE DATA

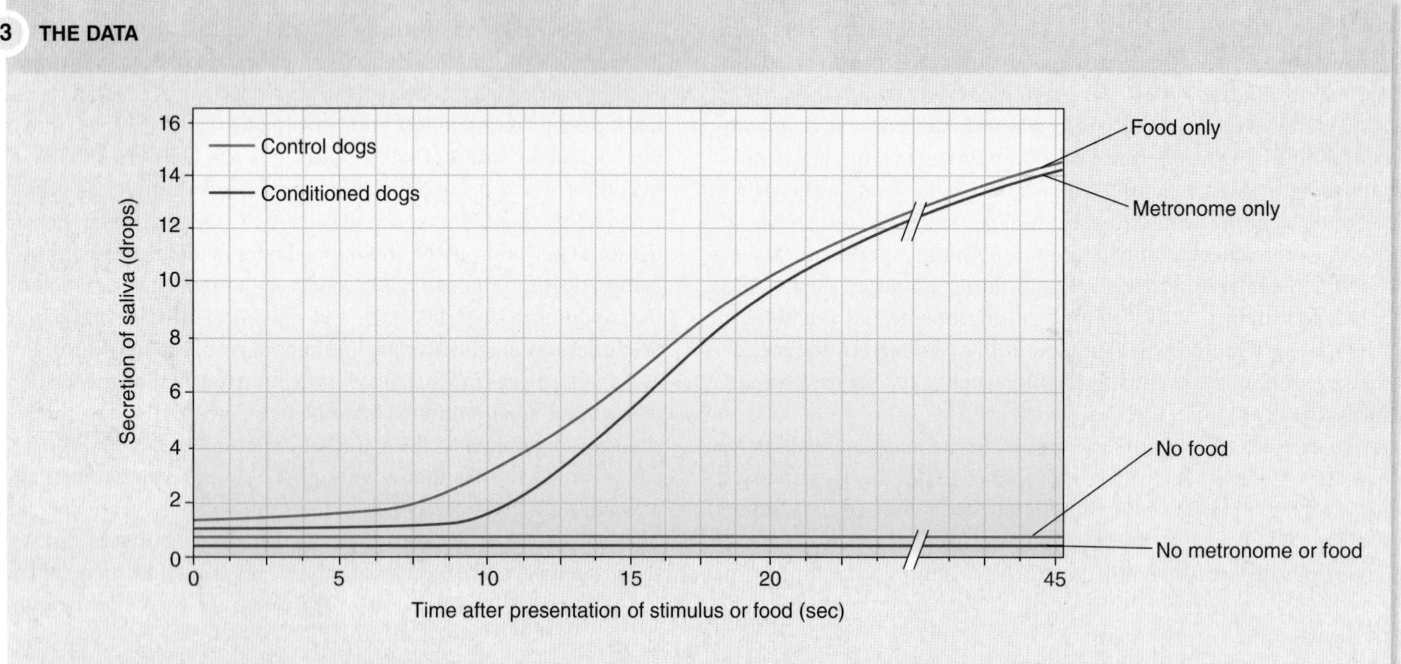

4 **CONCLUSION** The feedforward response of salivation can be modulated by experience and learning, demonstrating the adaptability of animals' physiological responses.

5 **SOURCE** These are idealized data compiled from several of the original published experiments performed by Pavlov and coworkers. The original experiments were reported in Pawlow, J. 1903. Sur la sécrétion psychique des glandes salivares. *Archives Internationales Physiologie* 1:119–135. (Note the alternate spelling of Pavlov's name.)

animals have a much greater ability to recognize and adapt to changes in their environment than previously thought.

Experimental Questions

1. What process did Pavlov study in his experimental animals, and what was his hypothesis?
2. How did Pavlov control for the possibility that other stimuli, such as the smell of the investigator, were somehow causing the feedforward response?
3. What did Pavlov measure, and what was his major finding?

Local and Long-Distance Chemical Signals Coordinate Homeostatic Responses

A common thread that links all homeostatic processes together is communication between cells, whether cells are close to each other or in different parts of an animal's body. Some homeostatic responses may be highly localized, occurring only in the area of a disturbance. For example, damage to an area of skin causes cells in the injured area to release molecules that help contain the injury, prevent infections, and promote tissue repair in the immediate vicinity (see Chapter 53). Local responses provide areas of the body with mechanisms for local self-regulation. It is no benefit to an animal to promote tissue repair in regions of the body that are not injured. This type of cellular communication—in which molecules are released into the interstitial fluid and act on nearby cells—is called **paracrine** signaling (refer back to Figure 9.3d).

Another example of extremely localized signaling occurs between neurons. Most neurons communicate through the release of **neurotransmitters**, small signaling molecules that are synthesized and stored in neurons. When a neuron releases these neurotransmitters, they diffuse to combine with receptor proteins on a single, adjacent neuron (or in some cases a muscle or gland cell), altering the activity of that cell. This type of cell-to-cell communication is typically very rapid, finishing within milliseconds. Consequently, neurotransmitter responses can make immediate homeostatic adjustments, like those associated with reflexes. These are just two of the many types of localized signaling that occur in animals' bodies and that will be described in subsequent chapters in this unit.

In addition to paracrine and neurotransmitter signaling, cells can communicate over long distances by releasing chemical messenger molecules into the blood. This type of signaling is mediated by hormones—chemical messengers produced by the endocrine system of animals. A hormone released in response to a homeostatic disturbance, such as the decrease in blood pressure described earlier, can influence the activities of many different cells, tissues, and organs simultaneously because the hormone is carried throughout the entire blood circulation. Some hormones act quickly—within seconds—whereas others take minutes or even hours for their effects to occur. In subsequent chapters, we will see that hormones are a key part of the regulatory processes that govern the functions of every organ system in a vertebrate's body, and they play key roles in growth, development, and reproduction in invertebrates.

Summary of Key Concepts

40.1 Organization of Animal Bodies

- In an animal's body, differentiated cells with similar properties group to form tissues, which combine with other types of tissues to form organs. Organs are functionally linked to form organ systems (Figure 40.1).
- Muscle tissues consist of cells specialized to contract. There are three categories of muscle tissue: skeletal, smooth, and cardiac. Nervous tissues initiate and conduct electrical signals from one part of an animal's body to another part. Epithelial tissues are specialized to protect structures and to secrete and absorb ions and organic molecules. Connective tissues connect, surround, anchor, and support the structures of an animal's body (Figures 40.2, 40.3, 40.4, 40.5).
- An organ is composed of two or more kinds of tissues. In an organ system, different organs work together to perform an overall function (Figure 40.6, Table 40.1).
- The development, spatial positioning, and function of many body organs are under the control of highly conserved *Hox* genes in vertebrates.
- An animal's body fluids are distributed into three compartments (intracellular fluid, plasma, and interstitial fluid). Alterations in intracellular fluid volume can have drastic effects on cell shape (Figures 40.7, 40.8).

40.2 The Relationship Between Form and Function

- The structure of an animal's tissues and organs are related to the function of those structures (Figure 40.9).
- Extensive surface area maximizes the ability of a tissue or organ to absorb solutes, exchange oxygen and carbon dioxide with the environment, communicate with other cells, and receive sensory information from the environment. The ratio between a structure's surface area and the volume in which the structure is contained is called the surface area/volume (SA/V) ratio (Figure 40.10).

40.3 Homeostasis

- Homeostasis is the process of maintaining a relatively stable internal environment despite changes in the external environment. Some animals conform to their environment, and others regulate internal processes in response to their environment.
- Vertebrates maintain most physiological variables within a certain range despite variations in external environmental conditions (Table 40.2, Figure 40.11).
- Homeostatic control systems regulate the activities of cells, tissues, and organs. Negative feedback loops minimize changes in a variable and prevent homeostatic responses from overcompensating (Figures 40.12, 40.13).
- Positive feedback loops accelerate a process and may not contribute to homeostasis. Feedforward regulation prepares the body for an upcoming challenge to homeostasis (Figures 40.14, 40.15).
- Pavlov's experiments with dogs demonstrated that feedforward responses can be conditioned by experience and learning (Figure 40.16).
- Chemical communication between cells is essential to homeostasis. Local and long-distance chemical signals coordinate homeostatic processes.

Assess and Discuss

Test Yourself

1. Tissue that is specialized to conduct electrical signals from one structure in the body to another structure is __________ tissue.
   a. epithelial
   b. connective
   c. nervous
   d. muscle
2. Structures composed of two or more tissue types arranged in various proportions and patterns are
   a. cells.
   b. tissues.
   c. organs.
   d. organ systems.
   e. organisms.
3. The extracellular matrix of connective tissue
   a. contains different proteins that provide structural support to cells.
   b. provides a scaffolding for the cells of the tissue.
   c. plays a role in cellular communication.
   d. does all of the above.
   e. does a and b only.
4. When examining the structure of many animal organs,
   a. it is apparent that all organ systems are fully functional in all animals.
   b. the function can be predicted based on the structural adaptations.
   c. the function of the structure is difficult to determine because of the lack of variation among the different structures of the body.
   d. all four tissues are equally represented in all organs.
   e. None of the above are correct.
5. Most of the water in an animal's body
   a. lacks any type of ions or other solutes.
   b. is found in the spaces between cells.
   c. is contained inside the cells.
   d. is located in the extracellular fluid.
   e. is unable to move between body compartments.
6. The folds, convolutions, or extensions found in many structures of animals results in
   a. decreased level of activity in that particular structure.
   b. interruption in the normal functioning of the structure.
   c. increased surface area for absorption, communication, or exchange.
   d. higher susceptibility to infection.
   e. none of the above.

7. Adapting to changes in the external environment and maintaining internal variables within physiological ranges is
   a. natural selection.
   b. evolution.
   c. positive feedback.
   d. homeostasis.
   e. both c and d.
8. Which of the following statements regarding negative feedback is not correct?
   a. It helps regulate variables such as body temperature and blood pressure.
   b. It is the mechanism by which birth occurs in mammals.
   c. It is a major feature of homeostatic control systems.
   d. It prevents homeostatic responses from overcompensating.
   e. It may occur at the organ, cellular, or molecular level.
9. The ability of an animal's body to prepare for a change in some variable before the change occurs is called
   a. positive feedback.
   b. negative feedback.
   c. homeostasis.
   d. feedforward regulation.
   e. autoregulation.
10. A hormone differs from a neurotransmitter in that
   a. hormones act extracellularly; neurotransmitters act within the cell that synthesized them.
   b. hormones are released only by neurons; neurotransmitters are released by many different types of cells.
   c. hormones cause only fast responses (seconds or less) to stimuli; neurotransmitters cause slow responses (minutes to hours) to stimuli.
   d. hormones affect only epithelial cells; neurotransmitters affect only muscle cells.
   e. hormones are released into the bloodstream and can activate many cells in many parts of the body; neurotransmitters are released by neurons and affect adjacent cells.

Conceptual Questions

1. Describe the relationship between structure and function in animals, and how surface area and volume are related.
2. Define homeostasis, and give examples of homeostatic variables in animals.
3. Two principles of biology are that *living organisms use energy* and that *living organisms maintain homeostasis.* How are energy use and the maintenance of homeostasis related?

Collaborative Questions

1. Describe the terms negative feedback loop and positive feedback loop, and describe how they differ.
2. Discuss the organization of animal bodies from the cellular to the organ system level.

Online Resource

www.brookerbiology.com

Stay a step ahead in your studies with animations that bring concepts to life and practice tests to assess your understanding. Your instructor may also recommend the interactive eBook, individualized learning tools, and more.

Chapter Outline

41.1 Cellular Components of Nervous Systems

41.2 Electrical Properties of Neurons and the Resting Membrane Potential

41.3 Generation and Transmission of Electrical Signals Along Neurons

41.4 Neurons Communicate Electrically or Chemically at Synapses

41.5 Impact on Public Health

Summary of Key Concepts

Assess and Discuss

Neuroscience I: Cells of the Nervous System

41

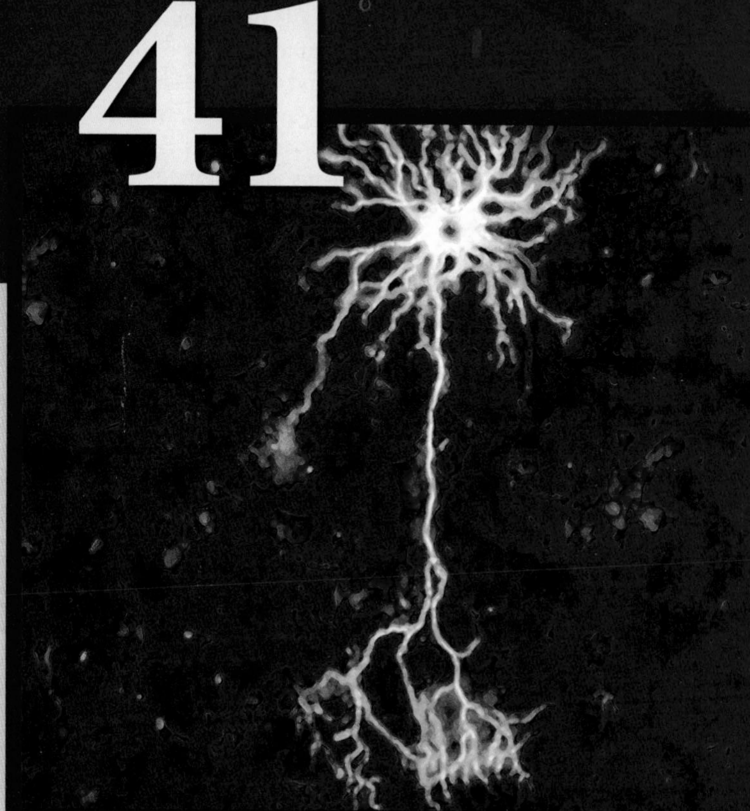

A fluorescently stained human neuron. These types of cells are responsible for sending signals throughout the nervous systems of animals.

As you begin reading, stop and think. Can you describe everything you are doing right now? Your eyes are sensing light reflected off this page, and your brain is interpreting the meanings of the words you are reading. You may be hearing sounds and, in some cases, choosing to ignore them as you concentrate on reading. Your digestive system may be sending signals about hunger, or perhaps if you've just eaten, you feel full and are digesting what you ate. You are breathing, perspiring, feeling the chair on which you're sitting, and your heart is beating. All of these processes and many others are under the control of the **nervous system**, coordinated circuits of cells that sense internal and environmental changes and transmit signals that enable us to respond in an appropriate way. Nervous systems help us to exert control over our bodies. They also allow animals such as ourselves to sense what is going on in the outside world, initiate actions that influence events and respond to demands, and regulate internal processes—all while maintaining homeostasis (see Chapter 40 for a discussion of homeostasis). You are conscious of some of these functions, such as reading this text or feeling hungry. Many others, however, such as maintaining your body temperature and controlling your heart rate, occur without your awareness.

Neuroscience is the scientific study of nervous systems. Neuroscientists are interested in topics such as the structure and function of the brain and the biological basis of consciousness, memory, learning, and behavior. As such, it is an interdisciplinary field that interfaces with other disciplines such as cell and molecular biology, chemistry, physics, psychology, and linguistics. Neuroscience is now experiencing an unprecedented level of new discoveries and rapid growth.

In this chapter, our focus on neuroscience will be at the cellular level. Nervous systems are composed of circuits of **neurons**, highly specialized cells that communicate with each other and with other types of cells by electrical or chemical signals (see chapter-opening photo). In many animals, such as ourselves, neurons become organized into a central processing area of the nervous system called a **brain**. The brain sends commands to various parts of the body through **nerves**—bundles of neuronal cell extensions encased in connective tissue and projecting to and from various tissues and organs.

We begin by investigating the special features of neurons that make them suited for rapid communication between cells. We look at electrical and chemical gradients across the neuronal plasma membrane and how these gradients produce a way for neurons to convey signals. Next, we explore communication between neurons, and how neurons send and receive rapid, brief, and repeated signals. We explore the major features of the main classes of signaling molecules called neurotransmitters, and end by examining the various effects of therapeutic, recreational, and illicit drugs on neurotransmitter action. Chapters 42 and 43 will explore how nervous systems evolved and are organized, how the brain functions, and how animals use their nervous systems to sense the world around them.

41.1 Cellular Components of Nervous Systems

Learning Outcomes:

1. Explain the difference between the central nervous system (CNS) and the peripheral nervous system (PNS).
2. List the cellular components of the nervous system and describe the function of each cell type.
3. Identify the different parts of a typical neuron.
4. Describe the functional relationships among sensory neurons, motor neurons, and interneurons.
5. Explain what a reflex is, and why reflexes are important in animals.

The organization of nervous systems permits extremely rapid responses to changes in an animal's external or internal environment. In complex animals, the **central nervous system (CNS)** consists of a brain and a nerve cord, which in vertebrates extends from the brain through the vertebral column and is called the **spinal cord**. The **peripheral nervous system (PNS)** consists of all neurons and projections of their plasma membranes that are outside of but connect with the CNS, such as projections that end on muscle and gland cells. In certain invertebrates with simple nervous systems, the distinction between the CNS and PNS is less clear or not present.

The evolution of nervous systems has allowed animals to receive information about the environment via their PNS, transmit that information along nerves to a CNS, where that information is interpreted, and, if necessary, initiate a behavioral response via their PNS (**Figure 41.1**). For example, if a hungry hyena receives stimuli such as the smell and taste of food, odor-sensing cells in the nose and taste-sensing cells in the tongue act as receptors for the stimuli and then send signals along nerves to the brain. There, the signals are interpreted and recognized. The brain then sends a signal via nerves that stimulate gland cells in the mouth, which respond by producing saliva in preparation for the arrival and swallowing of food. In this section, we will survey the general properties of the cells of the nervous system.

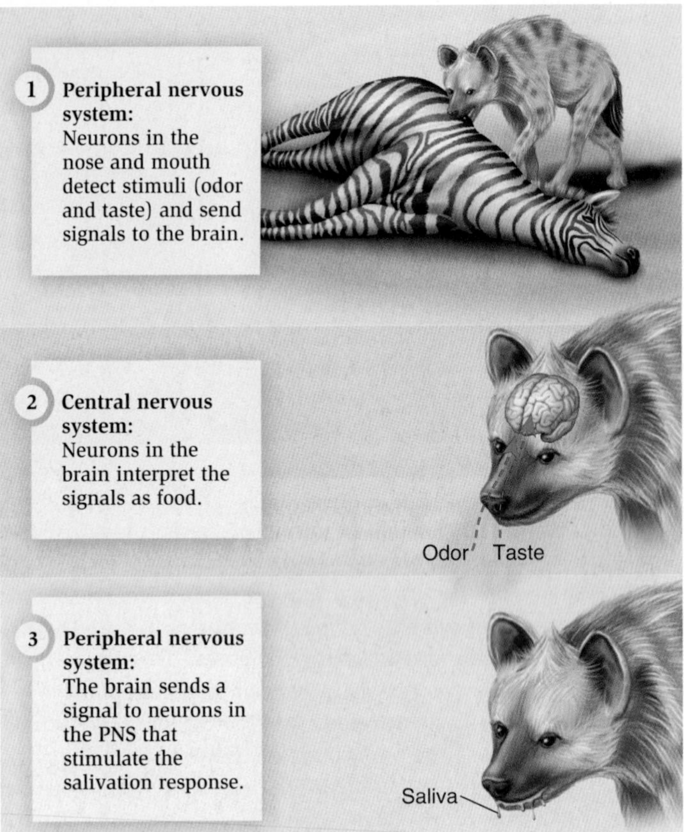

1 Peripheral nervous system:
Neurons in the nose and mouth detect stimuli (odor and taste) and send signals to the brain.
2 Central nervous system:
Neurons in the brain interpret the signals as food.
Odor
Taste
3 Peripheral nervous system:
The brain sends a signal to neurons in the PNS that stimulate the salivation response.
Saliva

Figure 41.1 Roles of the central and peripheral nervous systems. In this example, a hungry hyena senses a smell and taste, which the brain interprets as a potential food source. This initiates a biological response (salivation) that prepares the hyena for eating.

BIOLOGY PRINCIPLE Living organisms interact with their environment. Throughout this and the following two chapters, you will see many examples of how the nervous system of animals is a key way in which this biological principle is manifest. In this example, a hungry animal uses both divisions of its nervous system to interact with its environment.

BioConnections: Look back at Figure 40.16. What type of response is demonstrated by the production of saliva in this figure? Note that saliva is being produced before the animal has consumed any food.

Cells of the Nervous System Are Specialized to Transfer Signals

Nervous systems transfer signals from one part of the body to another and direct the activities of cells, tissues, organs, and glands. Although these are complex tasks, nervous systems have only two unique classes of cells: neurons and glia.

Neurons All animals except sponges have neurons, cells that send and receive electrical and chemical signals to and from each other and other cells throughout the body. The number of neurons in the nervous systems of different species varies widely, partly as a function of the size of an animal's head and brain, but also as a function of the complexity of its behavior. As a comparison, the tiny, short-lived nematode *C. elegans* has 302 neurons in its nervous system, compared with several thousand in a wasp, several hundred thousand in a salamander, 300 million in an octopus, and over 100 billion in a human!

Regardless of the total number, neurons in one animal species look and function much like neurons from any other species. A neuron is composed of a **cell body** (sometimes called the "soma"), which contains the cell nucleus and other organelles (Figure 41.2a,b). There are two types of extensions or projections that arise from the cell body: dendrites and the axon. **Dendrites** (from the Greek word *dendron*, meaning tree) may be single projections of the cell body but more commonly are elaborate treelike structures with numerous branching extensions that provide a large surface area for contacts with other neurons. Electrical and chemical messages from other neurons are received by the dendrites, and electrical signals move toward the cell body (Figure 41.2c). The cell body processes these signals along with some signals that are received directly by the cell body and produces an outgoing signal to a structure called an axon. An **axon** is an extension of the cell body that transmits signals along its length and eventually to neighboring cells. An axon may be only a few micrometers long or as long as 2 m, such as those in very large or long-limbed animals (refer back to Figure 40.3). A typical neuron has a single axon, which may have branches. The part of the axon closest to the cell body is named the **axon hillock**. As we will see later, the axon hillock is important in the generation of the electrical signals that travel along an axon. At the other end of the axon are one or more **axon terminals**, which convey electrical or chemical messages to other cells, such as other neurons or muscle cells.

Within an animal's body, many axons tend to run in parallel bundles to form nerves, which are covered by a protective layer of connective tissue. Nerves enter and leave the CNS and transmit signals between the PNS and the CNS. Along the way, the axon terminals communicate with particular cells of the body.

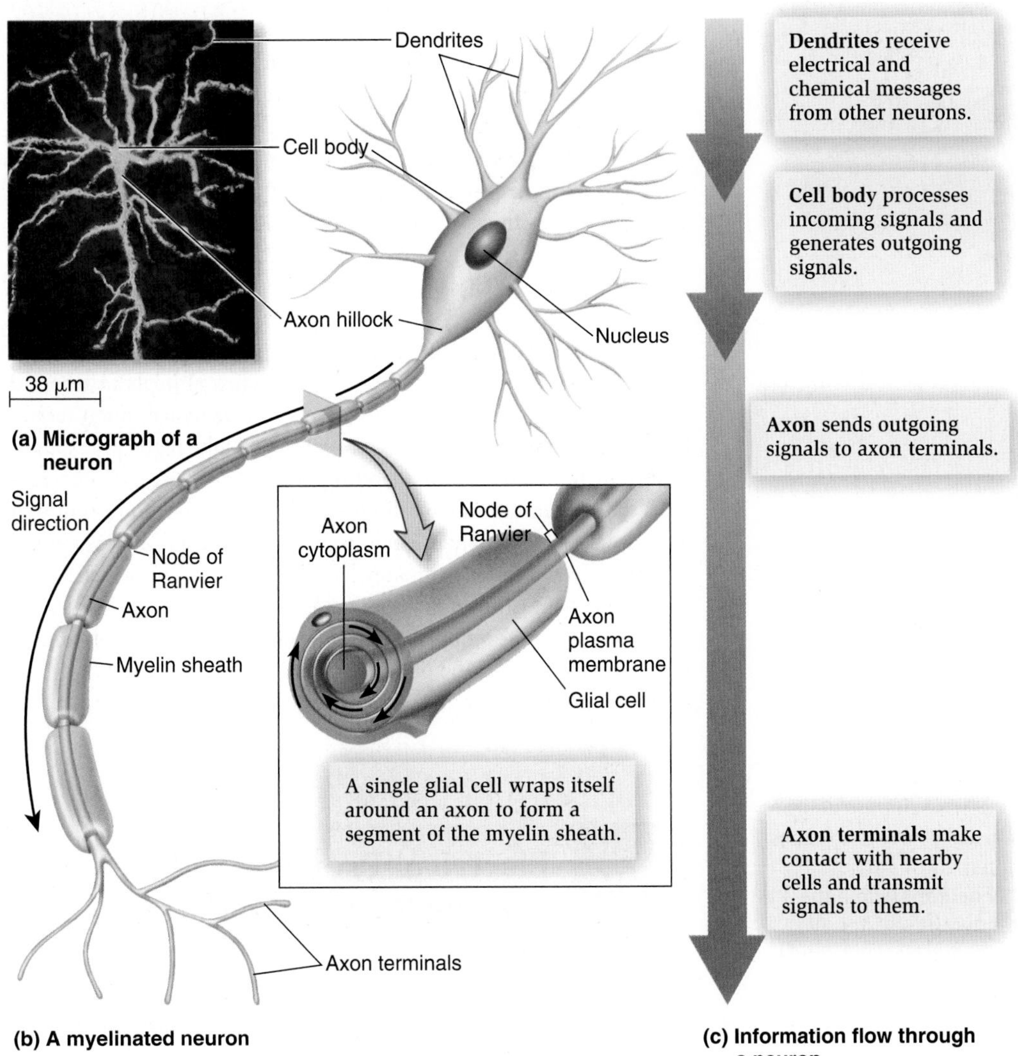

Figure 41.2 Structure and basic function of a typical vertebrate neuron and associated glial cells. **(a)** A stained neuron seen at high magnification (confocal microscopy). **(b)** A diagrammatic representation of a peripheral neuron with glial cells—in this instance, a type of glia called Schwann cells. The Schwann cells wrap their membranes around the axon at regular intervals, creating a myelin sheath that is interrupted by nodes of Ranvier. **(c)** The structures involved in the processing of information by a neuron; information flows in the direction shown.

Glia **Glia** (from the Greek, meaning glue) are cells that surround the neurons and perform numerous roles. One type of glia, called astrocytes, provides metabolic support for neurons and also is involved in forming the blood-brain barrier, which is a physical barrier between blood vessels and most parts of the CNS. This barrier prevents the passage of toxins and other damaging chemicals from the blood into the CNS. Astrocytes also help to maintain a stable concentration of ions in the extracellular fluid. Other glia, called microglia, remove cellular debris produced by damaged or dying cells. In developing embryos, glia form tracks along which neurons migrate to form the nervous system. In addition, some glia function as stem cells to produce more glial cells and neurons.

In vertebrates, specialized glial cells wrap around the axons at regular intervals to form an insulating layer called a **myelin sheath** (see Figure 41.2). The sheath is periodically interrupted by noninsulated gaps called **nodes of Ranvier**. In the human brain and spinal cord, the myelin-producing glial cells are called **oligodendrocytes**. **Schwann cells** are the glial cells that form myelin on axons that travel outside the brain and spinal cord. As we will see later, myelin and the nodes of Ranvier increase the speed with which electrical signals pass down the axon.

Sensory and Motor Neurons as Well as Interneurons Form Pathways in a Nervous System

Neurons can be categorized into three main types: sensory neurons, motor neurons, and interneurons. The structures of each type reflect their specialized functions.

Sensory Neurons As their name suggests, **sensory neurons** detect or sense information from the outside world, such as light, odors, touch, or heat. In addition, sensory neurons detect internal body conditions such as blood pressure or body temperature. Sensory neurons are also called afferent (from the Latin, meaning to bring toward) neurons because they transmit information to the CNS. Many sensory neurons have a long, single axon that branches into a peripheral process and a central process, with the cell body in between (Figure 41.3a). This arrangement allows for the rapid transmission of a sensory signal to the CNS.

Motor Neurons **Motor neurons** transmit signals away from the CNS and elicit some type of response. They are so named because one type of response they cause is movement. In addition, motor neurons

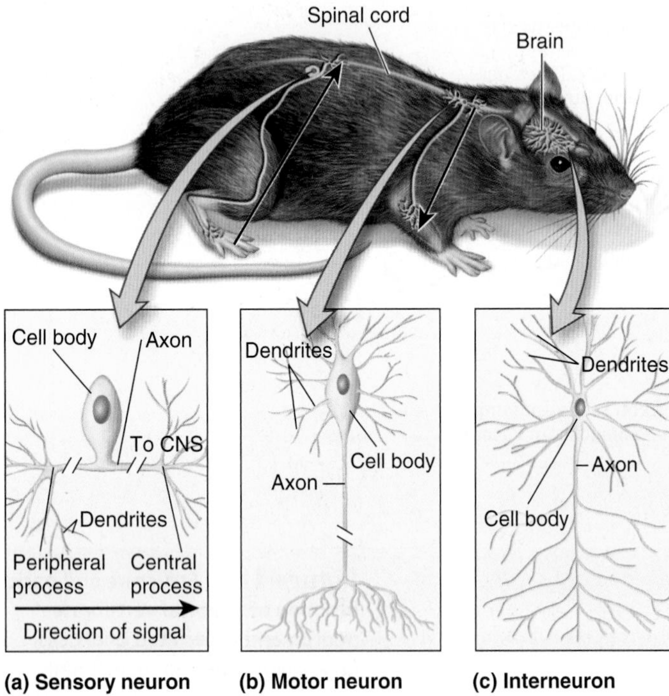

Figure 41.3 Types of neurons. **(a)** The vertebrate sensory neurons are afferent neurons with an axon that bypasses the cell body and projects to the CNS. **(b)** Motor neurons are efferent neurons that transmit signals away from the CNS and usually have long axons that enable them to act on distant cells. **(c)** Interneurons are usually short neurons that connect two or more other neurons within the CNS. Although short, the axons and dendrites may have extensive branches, allowing them to receive many inputs and transmit signals to many neurons.

BIOLOGY PRINCIPLE Cells are the simplest units of life. Cells are the simplest units of an organ system, including the nervous system. However, the variety and complexity of cells within an organ system can be great, as seen here.

may cause other effects such as the secretion of hormones from endocrine glands. Because they send signals away from the CNS, motor neurons are also called efferent (from the Latin, meaning to carry from) neurons. Like sensory neurons, motor neurons tend to have long axons (Figure 41.3b), but do not branch into two main processes.

Interneurons A third type of neuron, called the **interneuron**, forms interconnections between other neurons in the CNS. The signals sent between interneurons are critical in the interpretation of information that the CNS receives, as well as the response that it may elicit. Interneurons tend to have many dendrites, and their axons are typically short and highly branched (Figure 41.3c). This arrangement allows interneurons to form complex connections with many other cells.

Reflex Circuits As a way to understand the interplay between sensory neurons, interneurons, and motor neurons, let's consider a simple example in which these types of neurons form interconnections with each other. Neurons transmit information to each other through a series of connections that form a circuit. An example of a simple circuit is a **reflex arc**, which allows an organism to respond rapidly to inputs from sensory neurons and consists of only a few neurons (Figure 41.4). The stimulus from sensory neurons is sent to the CNS, but there is little or no interpretation of the signal; typically there are very few, if any, interneurons involved, as in the example shown in Figure 41.4. The signal is then transmitted to motor neurons, which elicit a response, such as a knee jerk. Such a response is very quick and automatic.

Reflexes are among the evolutionarily oldest and most important features of nervous systems, because they allow animals to respond quickly to potentially dangerous events. For instance, many vertebrates will immediately cringe, jump, leap, or take flight in response to a loud noise, which could represent sudden danger. Some animals that live in the water will reflexively dive deeper in response to a shadow overhead, which could represent a passing shark or other predator. Many infant primates have strong grasping reflexes that help them hold onto their mothers as they move about. Countless examples of useful reflexes are found in animals, and their importance is evident from the observation that they arose early in evolution and exist in nearly all animals.

41.2 Electrical Properties of Neurons and the Resting Membrane Potential

Learning Outcomes:

1. Explain what a membrane potential is.
2. Describe how the resting potential is established and maintained.
3. Describe how an electrochemical gradient determines the direction in which an ion will move.
4. Define the Nernst equation and how it can be used to make predictions about the ways in which different ions move across a cell membrane.

In the late 18th century, Italian scientists Luigi Galvani and Alessandro Volta experimented with ways to stimulate the contraction of frog leg muscles that had been dissected and placed in saline (NaCl) solutions. The saline solutions approximated the ion concentrations in plasma, kept the muscles alive, and also conducted electricity. They discovered that the stimulation of the nerve that led to the muscle or the stimulation of the muscle itself with any source of electric current caused the muscle to contract. Eventually, Galvani postulated that electric current could somehow be generated by the tissue itself, something he called "animal electricity."

Today, we know that Galvani's animal electricity comes from neurons, which use electrical signals to communicate with other neurons, muscle cells, or gland cells. These signals, often called nerve impulses but properly called action potentials, involve changes in the amount of electric charge across a neuron's plasma membrane. In this section, we first examine the electrical and chemical gradients across the plasma membrane of neurons. Later, we explore how such gradients provide a way for neurons to conduct signals.

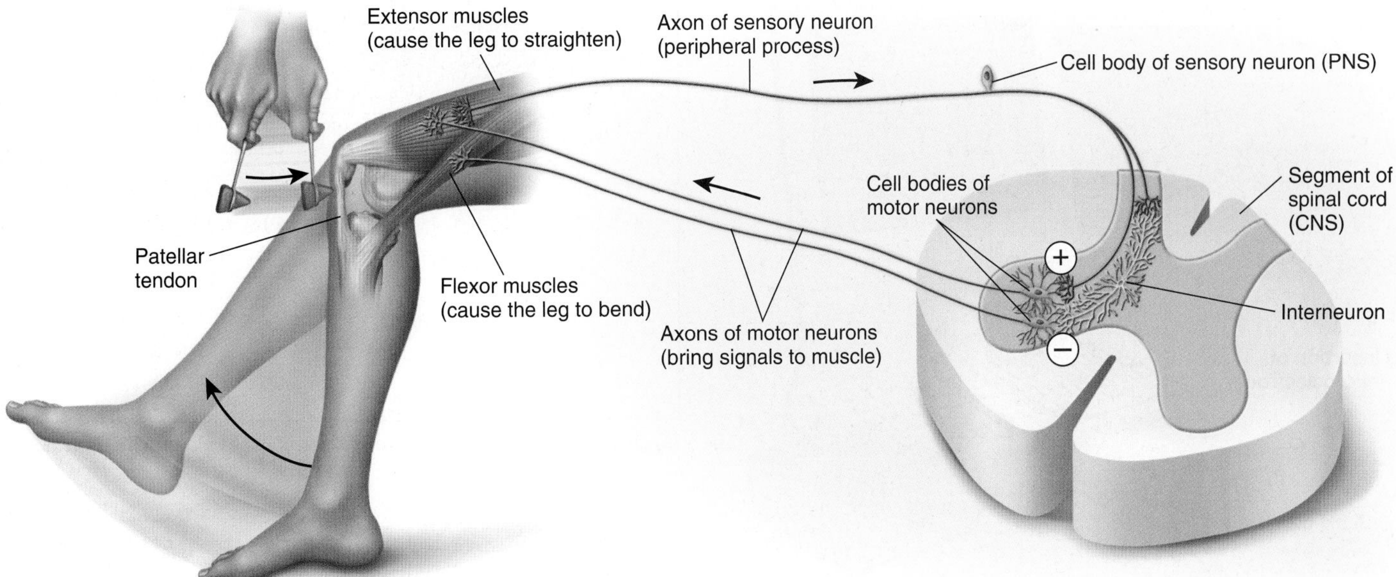

Figure 41.4 **A reflex arc.** The knee-jerk response is an example of a reflex arc. A tap below the kneecap (also known as the patella) stretches the patellar tendon, which acts as a stimulus. This stimulus initiates a reflex arc that activates (+) a motor neuron that causes the extensor muscle on top of the thigh to contract. At the same time, an interneuron inhibits (–) the motor neuron of the flexor muscle, causing it to relax.

Concept Check: *Animals have many types of reflexes. Once initiated, must all reflexes occur to completion, or do you think that in some cases they may be overridden or partially suppressed?*

Neurons Establish Differences in Ion Concentration and Electric Charge Across Their Membranes

Like all cell membranes, the plasma membrane of a neuron acts as a barrier that separates charges. Ion concentrations differ between the interior and exterior of the cell. Such differences in charge act as an electrical force measured in **volts** (V), named after Alessandro Volta. Analogous to a battery, neurons have negative and positive poles, but these are the inside and outside surfaces of the plasma membrane. For this reason, a neuron is said to be electrically **polarized**. The difference between the electric charges along the inside and outside surfaces of a cell membrane is called a potential difference, or **membrane potential**. The **resting potential** refers to the membrane potential of a cell that is not sending action potentials.

How do scientists measure electrical changes in a structure as tiny as a neuron? To do this, the cell is impaled with a microelectrode, which is a recording instrument constructed of a glass pipette with an extremely thin tip of less than 1 μm in diameter. Within the microelectrode is a salt solution that can conduct electric charge (ions). Many invertebrates, such as the squid, have very large neurons, and for that reason were historically used in such studies (**Figure 41.5**). The squid giant axon is part of a large (about 1-mm diameter) neuron that controls muscle movement; it can be dissected from the animal and placed in a solution that has ionic concentrations similar to those of normal extracellular fluid. Two microelectrodes are placed into the solution. One of these is called the recording electrode, and the other is the reference electrode. Although both electrodes are in the same extracellular solution, there is no electrical potential difference between the two; a voltmeter records the potential difference between them as zero. As the recording microelectrode is pushed through the axon membrane into the cell, however, the voltmeter records a voltage difference, which is a measure of the membrane potential.

Let's begin our discussion of electrical signaling by examining how the resting potential is established and maintained. The plasma membrane is not very permeable to cations and anions, so it separates charge by keeping different ions largely inside or outside the cell. When investigators first measured the resting potential of a squid giant axon, they registered a voltage that read about –70 millivolts (mV) inside the cell with respect to the outside bathing medium. This means that the interior of the cell had a more negative charge than the exterior, which turns out to be true of all animal cells in their resting state. A resting potential of –70 mV is tiny compared with the voltages used to provide electric current in a home (approximately 120 V), or even that of a small 1.5-V battery. Nonetheless, this tiny difference in charge across the membrane of a neuron is sufficient to generate an action potential that can travel from one end of a neuron to the other, as we will see later in this chapter.

The resting potential is determined by the ions located along the inner and outer surfaces of the plasma membrane (**Figure 41.6a**). Ions of opposite charges align on either side of the membrane because they are attracted to each other due to electrical forces. Negative ions within the cell are drawn to the positive ions arrayed on the outer surface of the plasma membrane. Although there are more positive charges along the outside surface of a neuron and more negative charges inside, the actual number of ions that contribute to the resting membrane potential is extremely small compared with the total number of ions inside and outside the cell. **Table 41.1** lists the ions that are important in maintaining the resting potential and their intracellular and extracellular concentrations. The ions that are most

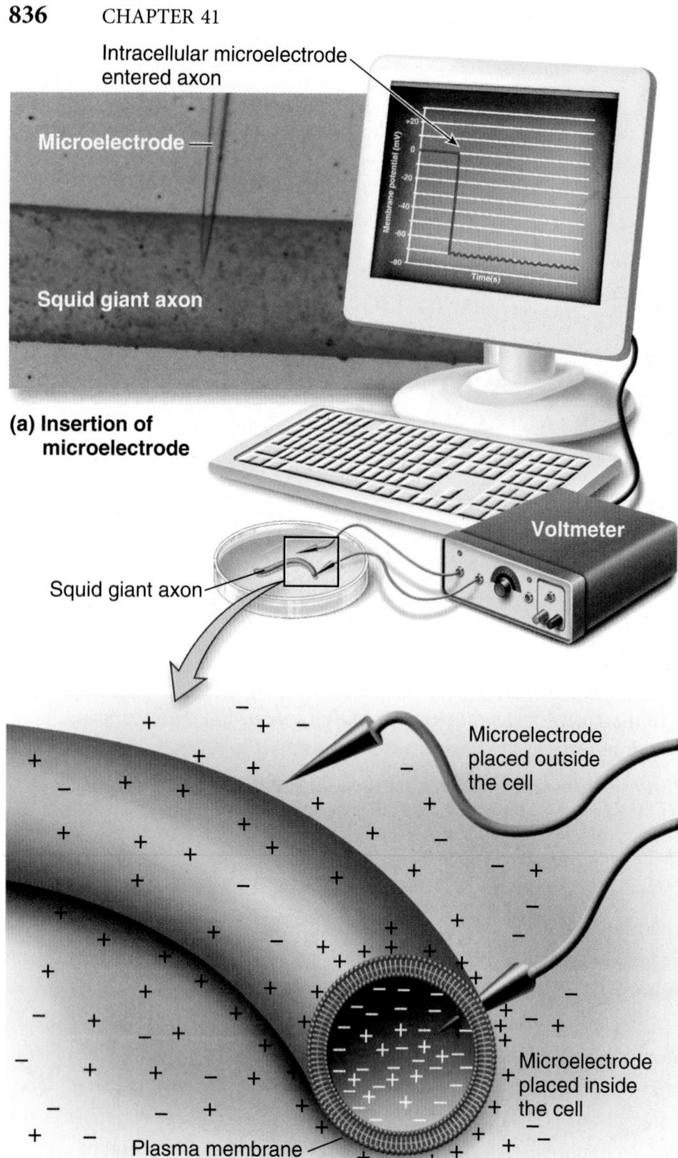

Figure 41.5 Recording the membrane potential of neurons. **(a)** A microelectrode, seen here under a microscope, impales the giant axon from a squid. **(b)** The potential difference across the membrane of the axon is recorded by comparing the electric charge inside and outside the axon. At first, before the intracellular ("recording") microelectrode is lowered into the axon, the potential difference between it and the extracellular ("reference") microelectrode is zero, because both electrodes are in the same solution. Once the intracellular electrode impales the cell, however, a negative potential is registered, indicating that the interior of the axon is negatively charged with respect to the external solution.

Concept Check: What obvious difference (besides its size) can you see between the giant axon of the squid, as illustrated here, and the illustration of the vertebrate axon depicted in Figure 41.2b?

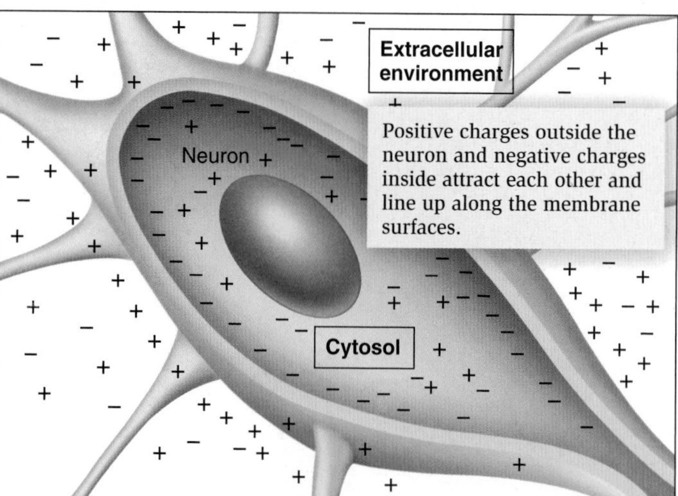

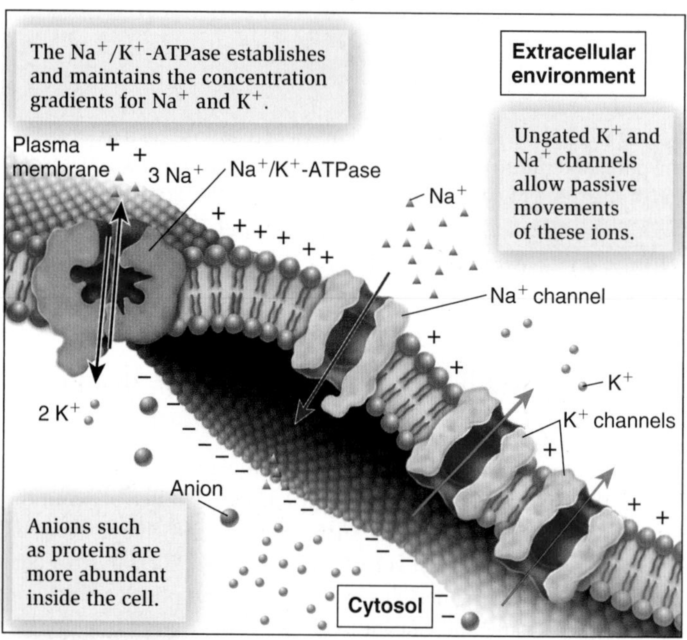

Figure 41.6 The resting membrane potential. The slight excess of negative charges inside the cell is shown in part **(a)**. In part **(b)**, the three major factors that contribute to this charge distribution are shown: the Na^+/K^+-ATPase pump that establishes ion concentrations gradients, ungated ion channels that permit diffusion of Na^+ and especially K^+ across the membrane, and intracellular anions.

BIOLOGY PRINCIPLE Living organisms use energy. The Na^+/K^+-ATPase pump uses the energy stored in ATP to pump ions across the neuronal plasma membrane. Consequently, establishing and maintaining resting membrane potentials accounts for a significant fraction of the energy consumed per day in an animal like ourselves.

critical for establishing the resting potential are Na^+ and K^+ and, to a lesser extent, Cl^-, and intracellular anions such as negatively charged proteins.

Three factors are primarily responsible for the resting membrane potential (**Figure 41.6b**). First, the sodium-potassium (Na^+/K^+-ATPase) pump within the plasma membrane continually moves sodium ions (Na^+) out of the cell and potassium ions (K^+) into the cytosol (refer back to Figure 5.22). The Na^+/K^+-ATPase pump uses the energy of ATP to transport three Na^+ out of the cell for every two K^+ it moves into the cell. Consequently, the pump contributes modestly to a charge difference across the plasma membrane. More importantly, however, it establishes concentration gradients for Na^+ and K^+

Table 41.1 Extracellular and Intracellular Concentrations of Ions for a Typical Mammalian Neuron		
	Concentration (mM)	
Ion	**Extracellular**	**Intracellular**
Na^+	145	15
K^+	5	150
Cl^-	110	7
Anions found in macromolecules	<1	65

by continually transporting them across the membrane in opposite directions. These gradients will determine the directions in which the ions will move by diffusion either into or out of the cell. Second, the plasma membrane contains ion-specific channels that affect the permeability of Na^+ and K^+ across the membrane. Ungated channels that are specific for Na^+ or K^+ influence the resting potential by allowing the passive movement of these ions. An ungated channel, sometimes referred to as a "leak" channel, is one that is open at rest and that does not respond to other stimuli, such as ligand binding. In most neurons, there are anywhere from 10 to 100 times more ungated K^+ channels than ungated Na^+ channels, so at rest, the membrane is more permeable to K^+ than to Na^+. As you will learn shortly, Na^+ tends to diffuse into cells and K^+ diffuses out of cells when the cell is at rest. The greater number of leak channels for K^+, therefore, means that excess positive charges exit the cell. Finally, a third (minor) contribution to resting potential is the presence of negatively charged molecules such as proteins that are more abundant inside the cell. These anions do not readily move through the plasma membrane, so they contribute some negative electric charge to the interior of the cell.

An Electrochemical Gradient Governs the Movement of Ions Across a Membrane

Both the membrane potential and the chemical concentration of ions influence the direction of ion movement across a membrane. The direction that an ion moves depends on the **electrochemical gradient** for that ion, which is the combined effect of both an electrical and chemical gradient. Figure 41.7 considers the concept of an electrochemical gradient for K^+. This hypothetical drawing illustrates two compartments separated by a semipermeable membrane that permits the diffusion of K^+. Figure 41.7a illustrates an electrical gradient. In this case, the chemical concentration of K^+ is equal on both sides of the membrane, but the concentrations of other ions such as Na^+ and Cl^- are unequal on both sides of the membrane and thereby produce an electrical gradient. Because K^+ is positively charged, it is attracted to the side of the membrane with more negative charge. Figure 41.7b shows a chemical gradient in which K^+ concentration is higher on one side than the other. In this scenario, K^+ diffuses from a region of high to low concentration. Finally, Figure 41.7c shows an electrochemical gradient. The electrical gradient would favor the movement of K^+ from left to right, but the chemical gradient would favor movement from right to left. These opposing forces can create an electrochemical equilibrium in which there is no net diffusion of K^+ in either

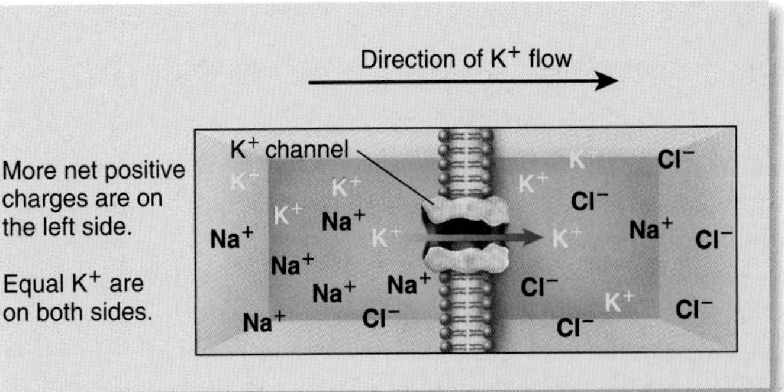

(a) Electrical gradient, no chemical gradient for K^+

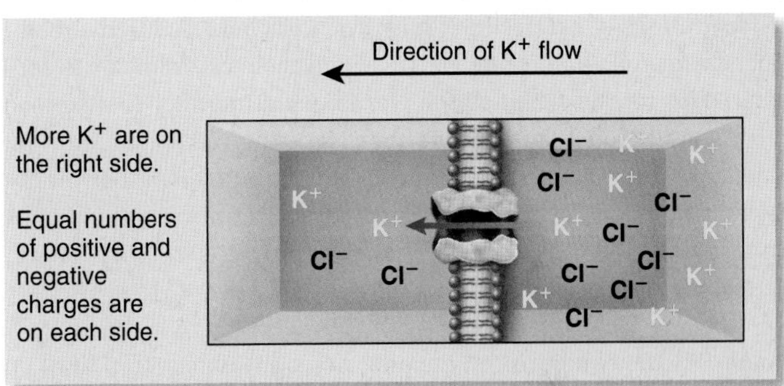

(b) Chemical gradient for K^+, no electrical gradient

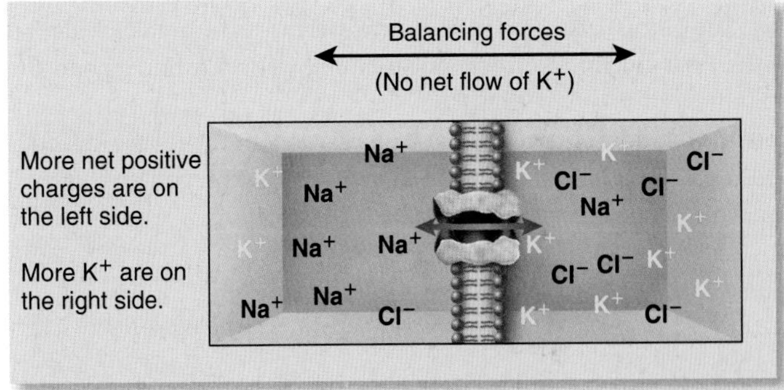

(c) Electrochemical gradient, K^+ equilibrium

Figure 41.7 **Electrical and chemical gradients.** This hypothetical example depicts two chambers separated by a membrane that is permeable to K^+. **(a)** In this example, the compartments initially contain equal concentrations of K^+, but an electrical gradient exists due to an unequal distribution of Na^+ and Cl^-. Potassium ions are attracted to the higher amount of negative charge on the right side of the membrane. **(b)** In this case, there is initially no electrical gradient across the membrane, and the left compartment contains a lower concentration of KCl than in the right compartment. In water, the KCl dissociates to K^+ and Cl^-. Under these conditions, K^+ diffuses down its chemical concentration gradient from right to left. **(c)** This example illustrates opposing electrical and chemical gradients. The right compartment contains a higher chemical concentration of K^+, and the left side has a higher net amount of positive charge. These gradients balance each other, so no net movement of K^+ occurs and a state of electrochemical equilibrium is reached.

Concept Check: *In part (b), does the diffusion of K^+ down its chemical gradient result in an electrical gradient? What will eventually stop the net diffusion of K^+?*

BioConnections: *Look back at Table 5.4 to review the varied biological functions associated with electrochemical ion gradients. Such gradients participate in numerous biological activities besides neuronal membrane potentials.*

direction. The membrane potential at which this occurs for a particular ion at a given concentration gradient is referred to as that ion's **equilibrium potential**.

With two different forces—electrical and chemical gradients—acting on a given ion, is it possible to predict the direction that an ion will move across a membrane at any concentration gradient? In other words, can we compare the relative strengths of the electrical and chemical gradients and predict their net effect? By measuring the membrane potential of isolated neurons in the presence of changing concentrations of extracellular ions, scientists have deduced a mathematical formula that relates electrical and chemical gradients to each other. This formula, named the **Nernst equation** after the German chemist and Nobel laureate Walther Nernst, gives the equilibrium potential for an ion at any given concentration gradient. For monovalent cations such as Na^+ and K^+ at 37°C, the Nernst equation can be expressed as

$$E = 60 \text{ mV} \log_{10} ([X_{\text{extracellular}}]/[X_{\text{intracellular}}])$$

where E is the equilibrium potential; $[X]$ is the concentration of an ion, outside or inside the cell; and 60 mV is a value that depends on temperature, valence of the ion, and other factors. (For anions, the value would be –60 mV.)

The Nernst equation allows neuroscientists to predict when an ion is or is not in electrochemical equilibrium. To understand the usefulness of this equation, consider two examples. First, let's suppose the membrane potential is –88.6 mV and the K^+ concentration is 5 mM outside and 150 mM inside a neuron. If we plug these chemical concentrations into the Nernst equation,

$$E = 60 \text{ mV} \log_{10} (5/150) = 60 \text{ mV} (-1.48)$$

$$= -88.6 \text{ mV}$$

Under these conditions, where the K^+ equilibrium potential equals the membrane potential, K^+ is in electrochemical equilibrium, and no net diffusion of K^+ occurs, even when many K^+ channels are open.

As a second example, let's suppose that the membrane potential is at a typical resting potential of –70 mV and the Na^+ concentration is 100 mM outside and 10 mM inside. If we plug these chemical concentrations into the Nernst equation,

$$E = 60 \text{ mV} \log_{10} (100/10) = 60 \text{ mV} (1)$$

$$= 60 \text{ mV}$$

At a resting potential of –70 mV, the value of +60 mV tells us that Na^+ is not in equilibrium. When the equilibrium potential for a given ion—calculated by the Nernst equation—and the resting membrane potential do not match, there will be a driving force for that ion to diffuse across the membrane. In this example, Na^+ will diffuse into the cell, because both the electrical and chemical gradients favor an inward flow. However, if the membrane potential was +60 mV instead of –70 mV, Na^+ would be in equilibrium at these concentrations, and therefore, there would be no net diffusion of Na^+ in any direction.

By establishing electrochemical gradients and maintaining them with the Na^+/K^+-ATPase pump, neurons have the ability to suddenly allow ions to move across the plasma membrane by opening additional channels that were previously closed. The movement of a charged ion down its concentration gradient results in an electric current. This small current provides the necessary electrical signal that neurons use to communicate with one another, as we will see next.

41.3 Generation and Transmission of Electrical Signals Along Neurons

Learning Outcomes:

1. Distinguish between ligand-gated and voltage-gated ion channels.
2. Compare and contrast graded potentials and action potentials.
3. Describe the events in the conduction of an action potential.
4. Discuss how myelination influences the speed of an action potential.

Communication between neurons begins when one cell receives a stimulus and sends an electrical signal along its plasma membrane via currents generated by ion movements. This signal then influences the next neuron in a circuit. Each signal is brief (only a few milliseconds), but a neuron may receive and transmit millions of signals in its lifetime. In this section, we will survey the amazing ability of neurons to send and receive rapid, brief, and repeated signals.

Signaling by a Neuron Occurs Through Changes in the Membrane Potential

Recall that a cell is polarized because of the separation of charge across its membrane. The resting potential of the membrane is more negative inside than outside. Changes in the membrane potential, therefore, are changes in the degree of polarization. **Depolarization** occurs when the cell membrane becomes less polarized, that is, less negative inside the cell relative to the surrounding fluid. As described later, when a neuron is stimulated by an electrical or chemical stimulus, one or more types of gated membrane channels open, and Na^+ may move into the cell, bringing with them their positive charge. This makes the new membrane potential somewhat less negative than the resting membrane potential. Consequently, the membrane potential is said to be depolarized. In contrast, **hyperpolarization** occurs when the cell membrane becomes more polarized, that is, more negative relative to the extracellular fluid. For example, increased movement of K^+ out of the cell would make the charge along the inside of the cell membrane more negative than it normally is while at the resting potential.

Whereas all cells in an animal's body have a membrane potential, neurons (and muscle cells) are called **excitable cells** because they can generate electrical signals. For a cell to communicate using electrical signals, it must be able to change its membrane potential very rapidly. This is accomplished by gated ion channels, so called because they open and close in a manner analogous to a gate in a fence (Figure 41.8). **Voltage-gated ion channels** open and close in response to changes in voltage across the membrane. **Ligand-gated ion channels**, also known as chemically gated ion channels, open or close when ligands, such as neurotransmitters, bind to them. The opening and closing of ligand-gated and voltage-gated ion channels are responsible for two types of changes in a neuron's membrane potential: graded potentials and action potentials.

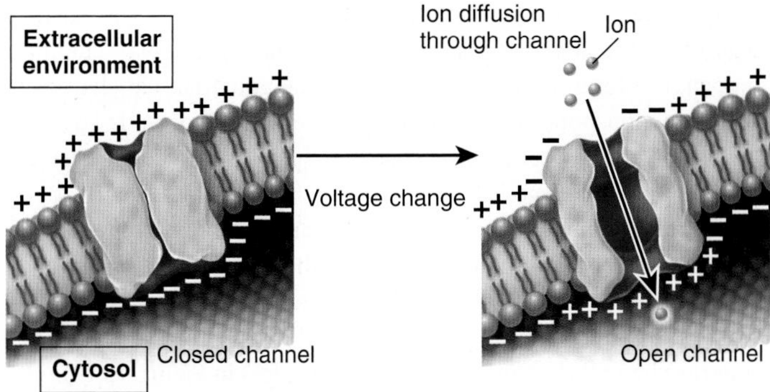

(a) Voltage-gated ion channel

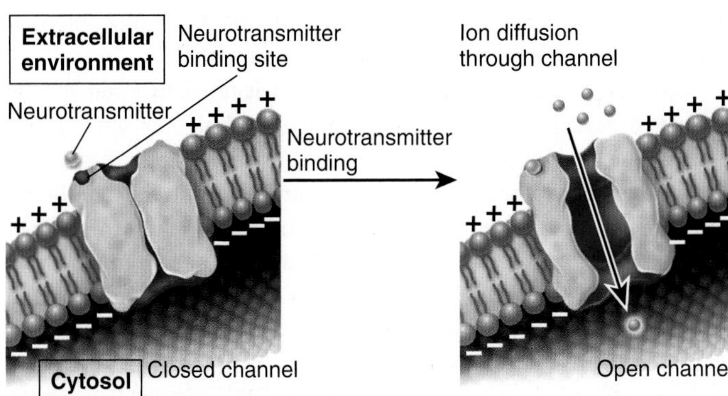

(b) Ligand-gated ion channel

Figure 41.8 Examples of gated ion channels. **(a)** A voltage-gated ion channel allows ions to diffuse into the cell. These channels open or close depending on changes in charge (voltage) across the membrane. **(b)** A ligand-gated ion channel opens or closes in response to ligand binding. In the example here, the binding of a neurotransmitter opens the channel.

BioConnections: Are ions the only substances that can move through channels in plasma membranes? Look back at Figure 5.18 for another important molecule that diffuses through membranes via channels.

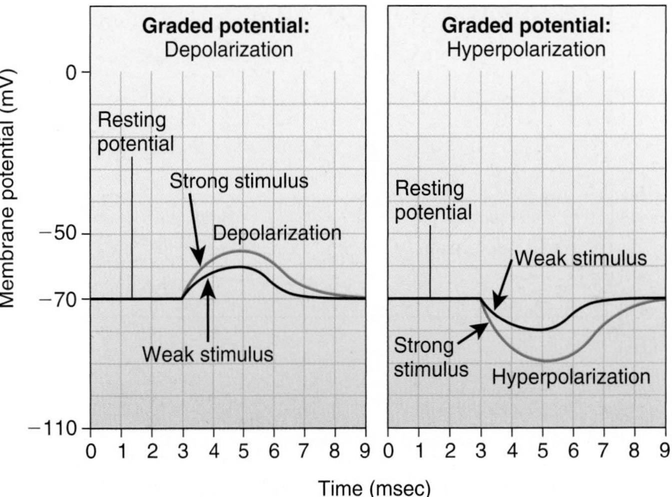

Figure 41.9 Graded membrane potentials. Within a limited range, the change in the membrane potential occurs in proportion to the intensity of the stimulus. If the membrane potential becomes less negative following a stimulus, the resulting change is called a depolarization. If the membrane potential becomes more negative, hyperpolarization results.

Graded Potentials Vary in Size Depending on the Strength of a Stimulus

A **graded potential** is any depolarization or hyperpolarization that varies depending on the strength of the stimulus. A large change in membrane potential occurs when a strong stimulus opens many channels, whereas a weak stimulus causes a small change because only a few channels are opened (Figure 41.9).

Graded potentials occur locally on a particular area of the plasma membrane, such as dendrites or the cell body, where an electrical or chemical stimulus opens ion channels. From this area, a graded potential spreads a small distance across a region of the plasma membrane. In a short time, the membrane potential returns to the resting potential because ion pumps restore the ion concentration gradients, and the ion channels close again.

Graded potentials occur on all neurons and are particularly important for the function of sensory neurons, which must distinguish between strong and weak stimuli coming into the organism from the environment. In addition, though, graded potentials can act as triggers for the second type of electrical signal, the action potential.

Action Potentials Transmit Electrical Signals Down the Length of Axons

Action potentials are the electrical events that carry a signal along an axon. In contrast to a graded potential, an action potential is always a large depolarization; all action potentials in a given neuron have very similar amplitudes, which is the degree to which an action potential changes the membrane potential away from its resting state. Once an action potential has been triggered, it occurs in an all-or-none fashion. In other words, it cannot be graded. Unlike a graded potential, an action potential is actively propagated along the axon, regenerating itself as it travels. Action potentials travel rapidly down the axon to the axon terminals, where they initiate a response at the junction with the next cell.

Figure 41.10 shows the electrical changes that happen in a localized region of an axon when an action potential is occurring. Voltage-gated Na^+ and K^+ channels are both present in very high numbers from the axon hillock to the axon terminals. An action potential begins when a graded potential is large enough to spread to the axon hillock and depolarizes the membrane to a value called the threshold potential (step 2 in Figure 41.10). The **threshold potential** is the membrane potential, typically around −55 to −50 mV, that is sufficient to open voltage-gated Na^+ channels and trigger an action potential.

Voltage-gated Na^+ channels open rapidly when the membrane potential changes from the resting potential (for example, around −70 mV) to the threshold potential. The opening of Na^+ channels involves a change in the conformation of the membrane-spanning region of the channel (see Figure 41.10). When the protein changes shape, the central pore opens, and Na^+ rapidly diffuses into the cell

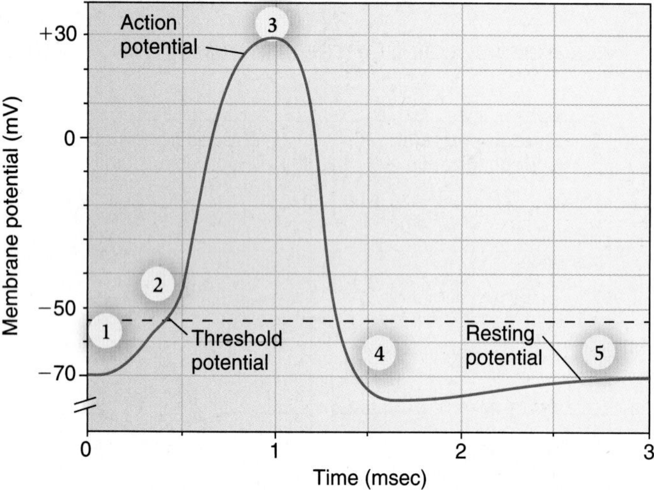

Voltage-gated Na⁺ channels | Voltage-gated K⁺ channels

1 **Resting membrane potential:** The membrane is at the resting potential, occasionally generating graded potentials.

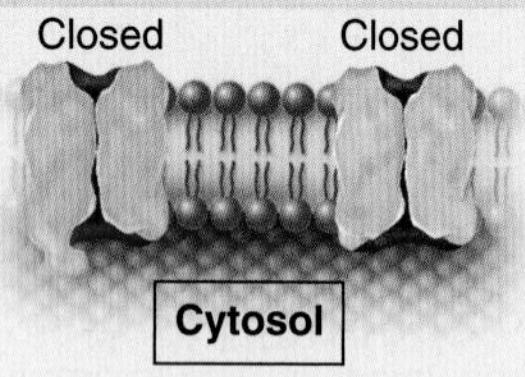

2 **Depolarization to threshold:** An action potential is triggered when the threshold potential of about −55 to −50 mV is reached. Voltage-gated Na^+ channels open. Na^+ diffuses into the cell and depolarizes the membrane.

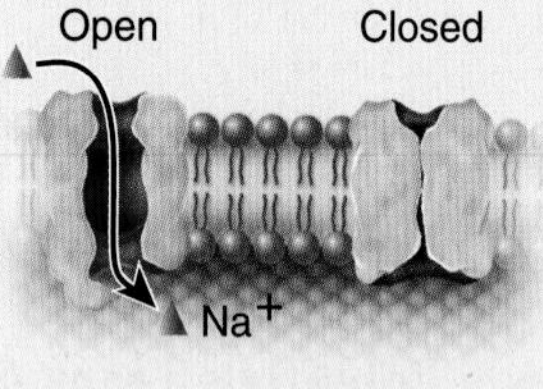

3 **Peak of action potential:** At about +30 mV, voltage-gated Na^+ channels are inactivated, and voltage-gated K^+ channels open. K^+ exits the cell and repolarizes the membrane. At this time, the membrane is in its absolute refractory period.

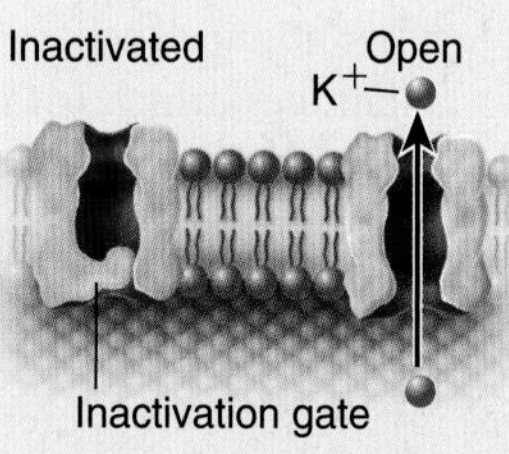

4 **Repolarization:** Voltage-gated Na^+ channels change from inactivated to closed. Voltage-gated K^+ channels remain open, causing a hyperpolarization of the membrane. The membrane is now in its relative refractory period.

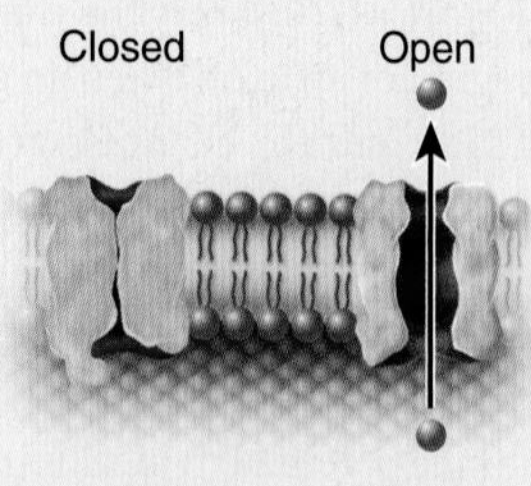

5 **Restoration of resting membrane potential:** Voltage-gated K^+ channels close, and the resting potential of the membrane is restored by the Na^+/K^+-ATPase pump.

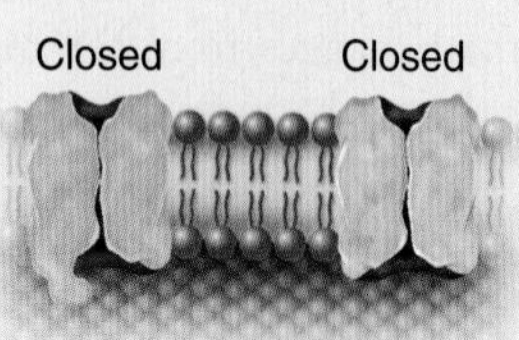

down its electrochemical gradient. This influx of charges further depolarizes the cell, causing even more voltage-gated Na^+ channels to open, resulting in the spike in membrane potential that characterizes the action potential. This positive feedback process is so rapid that the membrane potential reaches its peak positive value in less than 1 millisecond (msec)!

When the membrane potential becomes sufficiently positively polarized, a conformational change in the Na^+ channel blocks the continued flow of Na^+ into the cell. The conformational change involves the **inactivation gate**, a string of amino acids that juts out from the channel protein into the cytosol (step 3 in Figure 41.10). The inactivation gate swings into the channel, thereby preventing any further movement of Na^+ through the channel into the cell. Under these conditions, the Na^+ channel is said to be inactivated. The inactivation gate does not swing out of the channel until the membrane potential returns to a value that is close to the resting potential.

Voltage-gated K^+ channels are also triggered to open by the change in voltage to the threshold potential, but they open about 1 msec later than Na^+ channels. When K^+ channels open, some K^+ leave the cell down their electrochemical gradient, and the membrane potential becomes more negative again. The membrane shows a brief period of hyperpolarization because the K^+ channels remain open and the membrane potential approaches the equilibrium potential for K^+, which is more negative than the resting potential (step 4 in Figure 41.10). When the voltage-gated K^+ channels close, the membrane returns to the resting potential as the Na^+/K^+-ATPase pump restores the original concentration gradients (step 5 in Figure 41.10). At this stage, both the Na^+ and K^+ voltage-gated channels are closed, but they have the ability to reopen if the resting potential increases again to the threshold potential.

The evolution of K^+ channels with a slightly slower opening time than Na^+ channels was a key event that led to the formation of nervous systems. Imagine what would happen if the voltage-gated Na^+ and K^+ channels opened simultaneously: As Na^+ entered the cell down its electrochemical gradient, K^+ would leave the cell down its electrochemical gradient, and they would negate each other's effects on membrane potential.

During the time when the inactivation gate of the Na^+ channel is closed, the membrane is in its **absolute refractory period** (see Figure 41.10), during which time that portion of membrane is unresponsive to another stimulus. A change in voltage cannot open the Na^+ channels while they are inactivated. As the cell repolarizes, the inactivation gate is released and the voltage-gated Na^+ channels revert to the closed state. While the voltage-gated K^+ channels are still open, however, the membrane enters a brief **relative refractory period** (see Figure 41.10). During this time, when the membrane is hyperpolarized, a new action potential may be generated but only

Figure 41.10 Changes that occur during an action potential. The electrical gradient across the plasma membrane of an axon first depolarizes and then repolarizes the cell. The values given for membrane potential are representative and not necessarily the same in all neurons. These changes in the membrane potential are caused by the opening and closing of voltage-gated Na^+ and K^+ channels.

Concept Check: *How would the shape of the action potential be affected if the Na^+ channels were missing their inactivation gate?*

in response to an unusually large stimulus. Therefore, the refractory periods place limits on the frequency with which a neuron can generate and transmit action potentials. As we will see, this also ensures that the action potential does not "retrace its steps" by moving backward toward the cell body.

Many natural toxins work by blocking the actions of Na^+ and K^+ channels. Some animals use these toxins as offensive or defensive mechanisms, and they can be deadly. For instance, tetrodotoxin, a chemical produced by pufferfish, blocks voltage-gated Na^+ channels and, therefore, neuronal activity. An animal that eats a pufferfish dies because the poison paralyzes muscles, including those that depend on nerve stimulation to control breathing and movement. Animals that avoid pufferfish, therefore, have a selection advantage. Eating even a small amount can paralyze or kill a human. Other toxins from the sting of a wasp, bee, or scorpion can block the action of voltage-gated K^+ channels, resulting in abnormal action potentials and loss of normal muscle responses.

Action Potentials Are Conducted Down the Axon with Great Speed

Thus far we have considered the electrical changes that happen when an action potential is initiated. We will now consider how such potentials are conducted down an axon from the axon hillock to the axon terminals (Figure 41.11). Let's begin at the axon hillock. When the neuron receives stimuli from other cells, this causes a graded potential that reaches the axon hillock. If the change in membrane potential is sufficient to reach threshold, an action potential will be triggered. The action potential starts with the abrupt opening of several voltage-gated Na^+ channels just beyond the axon hillock, where many voltage-gated Na^+ channels are found. This action potential, in turn, triggers the opening of nearby Na^+ channels farther along the axon, which allows even more Na^+ to flow into the neuron and depolarize a region closer to the axon terminals, leading to another action potential. In this way, the sequential opening of Na^+ channels along the axon membrane conducts a wave of depolarization from the axon hillock to the axon terminals.

Why doesn't the action potential move backward from the terminals to the hillock? Experimentally, if an axon is stimulated in its middle, action potentials can travel in both directions, toward the cell body and the terminals. The key reason why this doesn't ordinarily happen is the inactivation state of the Na^+ channel, which contributes to the absolute refractory period. Again, let's begin at the hillock. When these voltage-gated Na^+ channels open for 1 msec, they allow Na^+ to rapidly enter the cell, and then they become inactivated. This inactivation prevents the action potential from moving backward (see Figure 41.11). Although the Na^+ channels are inactivated, K^+ channels open, and the resting potential is restored. At the resting potential, the Na^+ channels switch from the inactivated to the closed state. The opening of K^+ channels also travels from the axon hillock to the axon terminals, but it causes a wave of hyperpolarization that helps to reestablish the resting potential.

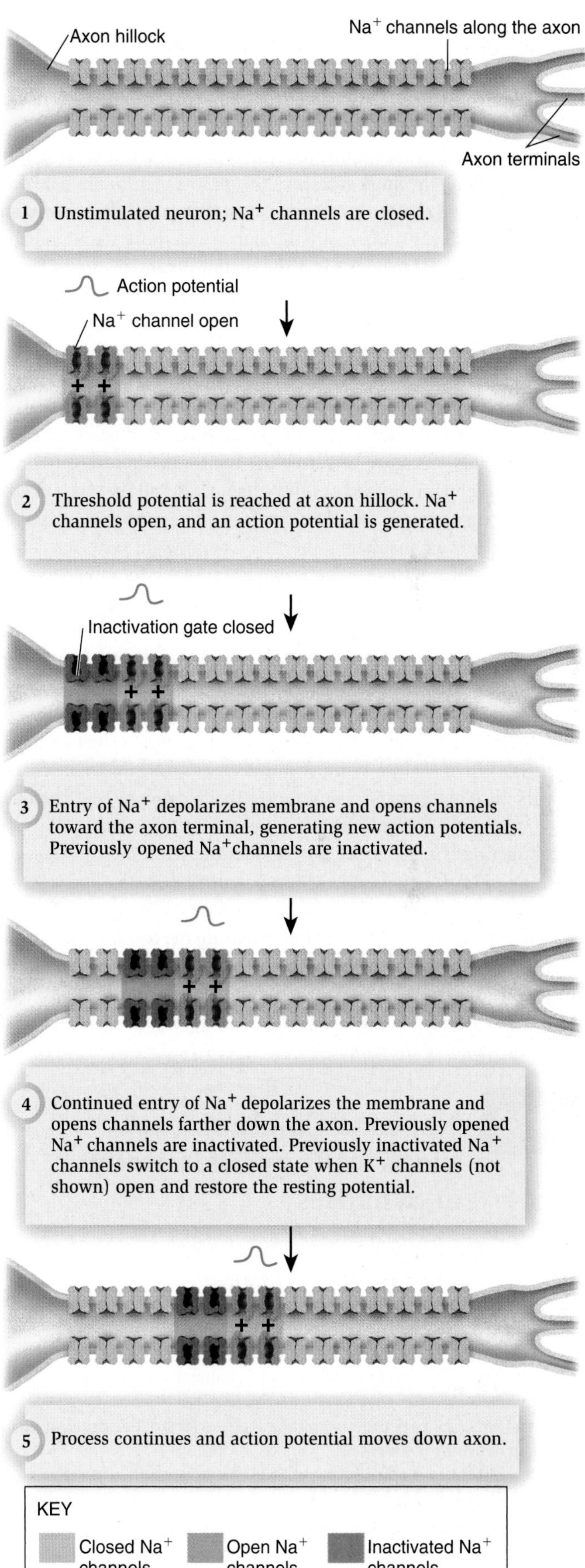

Axon hillock
Na+ channels along the axon
Axon terminals
1 Unstimulated neuron; Na+ channels are closed.
Action potential
Na+ channel open
+ +
2 Threshold potential is reached at axon hillock. Na+ channels open, and an action potential is generated.
Inactivation gate closed
+ +
3 Entry of Na+ depolarizes membrane and opens channels toward the axon terminal, generating new action potentials. Previously opened Na+channels are inactivated.
+ +
4 Continued entry of Na+ depolarizes the membrane and opens channels farther down the axon. Previously opened Na+ channels are inactivated. Previously inactivated Na+ channels switch to a closed state when K+ channels (not shown) open and restore the resting potential.
+ +
5 Process continues and action potential moves down axon.
KEY
Closed Na+ channels
Open Na+ channels
Inactivated Na+ channels

Figure 41.11 **Conduction of the action potential along an axon.** All Na^+ channels shown refer to voltage-gated channels.

An action potential can be conducted down the axon as fast as 100 m/sec or as slow as a centimeter or two per second. The speed is determined by two factors: the axon diameter and the presence or absence of myelin.

The axon diameter influences the rate at which incoming ions can spread along the inner surface of the plasma membrane. The flow of ions meets less resistance in a wide axon than it does in a thin axon, just as water moves more easily through a wide hose than a narrow one. Therefore, in a wider axon the action potential moves faster. The large axons of squids and lobsters, for example, conduct action potentials very rapidly, allowing the animals to move quickly when threatened.

Myelination also influences the speed at which action potentials travel along an axon. Myelinated axons conduct action potentials at a faster rate than do unmyelinated axons. Invertebrate neurons lack the myelinated structure of vertebrate neurons, whereas vertebrate neurons may be myelinated or unmyelinated. Recall that certain glial cells (oligodendrocytes and Schwann cells) wrap around vertebrate axons to form an insulating sheath of membrane. The insulating layer of myelin reduces charge leakage across the membrane of the axon. However, this myelin sheath is not continuous (**Figure 41.12**). As described earlier (see Figure 41.2), the axons of myelinated neurons have exposed areas known as the nodes of Ranvier; these non-myelinated regions are characterized by having many voltage-gated Na^+ channels. The nodes of Ranvier are the only areas of myelinated axons that have enough voltage-gated Na^+ channels to elicit an action potential. When Na^+ ions diffuse into the cell at one node, the charge spreads through the cytosol and causes the opening of Na^+ channels at the next node, where an action potential is regenerated. This type of conduction is called **saltatory conduction** (from the Latin *saltare*, meaning to leap) because the action potential seems to "jump" from one node to the next. In reality, action potentials do not jump from place to place; saltatory conduction speeds up the conduction process because it takes less time for changes in membrane potential to travel from node to node, eliciting action potentials only at the nodes, than it would if each tiny strip of membrane generated action potentials all along the length of the axon.

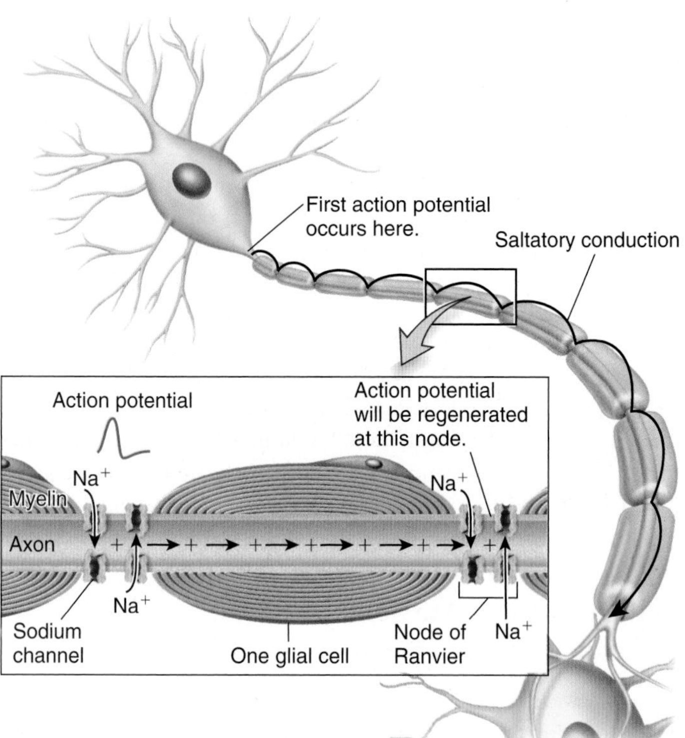

Figure 41.12 **Saltatory conduction along a myelinated axon.** Action potentials are generated only at the nodes of Ranvier, which lack a surrounding sheath of myelin.

Concept Check: *What is the main advantage of saltatory conduction?*

41.4 Neurons Communicate Electrically or Chemically at Synapses

Learning Outcomes:

1. Describe the structural features of a synapse.
2. Distinguish between two types of potentials produced in postsynaptic cells: excitatory postsynaptic potentials (EPSPs) and inhibitory postsynaptic potentials (IPSPs).
3. Describe the difference between spatial summation and temporal summation.
4. List the classes of neurotransmitters and provide brief descriptions of their generalized functions.
5. Describe the two general types of postsynaptic membrane receptors.

Neurons communicate with other cells at a **synapse**, which is a junction where an axon terminal meets another neuron, muscle cell, or gland cell. At a synapse, an electrical or chemical signal passes from the axon terminal to the next cell. A synapse includes the axon terminal of the neuron that is sending the message, the nearby plasma membrane of the receiving cell, and the **synaptic cleft**, or extracellular space between the two cells. The **presynaptic cell** sends the signal, and the **postsynaptic cell** receives it. A given cell may be presynaptic to one cell, and postsynaptic to another (**Figure 41.13**).

By studying neurons from both invertebrates and vertebrates, researchers have identified two types of synapses: electrical and chemical. The first type, the **electrical synapse**, directly passes electric current from the presynaptic to the postsynaptic cell. The electrical signal passes through this type of synapse extremely rapidly, because the plasma membranes of adjacent cells are connected by gap junctions (see Chapter 10) that can move electric charge freely from one cell to the other. Recent research indicates that electrical synapses are more widespread among taxa than originally thought, including vertebrates. The most well-studied examples, however, occur in some aquatic invertebrates such as leeches. In those animals, electrical synapses occur where a group of neurons must fire rapidly and synchronously, such as when an animal must coordinate a number of muscles to swim or escape danger. The second type of synapse is a chemical synapse, in which a chemical called a neurotransmitter is released from the axon terminal and acts as a signal from the presynaptic to the postsynaptic cell. Chemical synapses are more common,

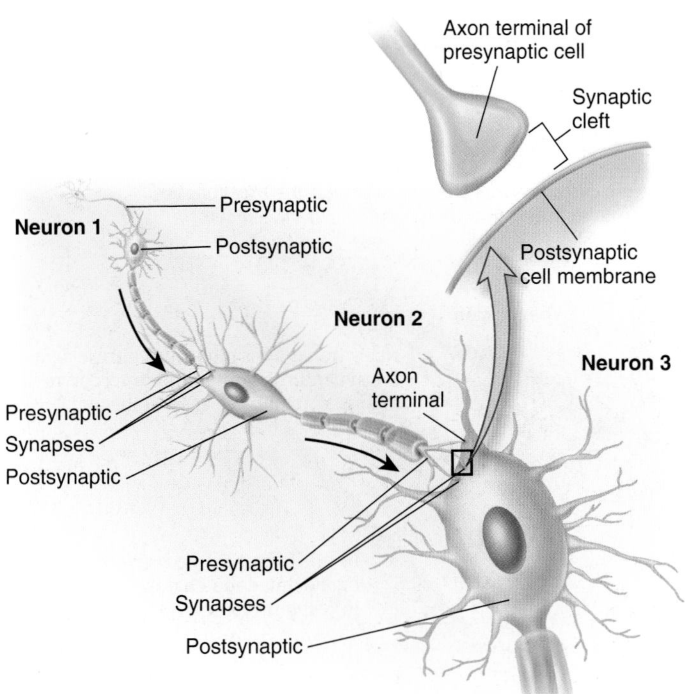

Figure 41.13 **Presynaptic and postsynaptic cells.** The arrows show the direction of signal transmission from one neuron to the next. Note that neuron 2 is postsynaptic with respect to neuron 1 and presynaptic with respect to neuron 3.

particularly in vertebrates. Some neurons release only one type of neurotransmitter, and some neurons can release two or more different ones.

Figure 41.14 describes the steps that occur when two cells communicate via a chemical synapse. The axon terminal of a presynaptic cell contains vesicles—small, membrane-enclosed packets, each containing thousands of molecules of neurotransmitter. The membranes of axon terminals contain a type of ion channel localized primarily to the terminal called voltage-gated Ca^{2+} channels. When an action potential arrives at an axon terminal, these channels open, allowing Ca^{2+} ions to diffuse down their electrochemical gradient into the cell. Calcium binds to a protein associated with the vesicle membrane. This triggers exocytosis, in which the vesicle fuses with the presynaptic membrane, thereby releasing its neurotransmitter molecules into the synaptic cleft. The neurotransmitter molecules diffuse across the 10- to 20-nm-wide synaptic cleft and bind to ion channels or other receptor proteins in the postsynaptic cell membrane.

The binding of neurotransmitter molecules opens or closes ion channels, thereby changing the membrane potential of the postsynaptic cell. In some cases, neurotransmitters directly bind to ion channels and cause the channels to open or close. Others bind to receptor proteins on the neuronal membrane and induce the formation of second messengers in the cell (see Chapter 9), which, in turn, open or close ion channels by any of several mechanisms. Some neurotransmitters are called excitatory, because they depolarize the postsynaptic membrane; the response is an **excitatory postsynaptic potential (EPSP)**. This response is called excitatory because the depolarization of the membrane brings the membrane potential closer to the threshold potential that would trigger an action potential. An EPSP can be caused by the opening of Na^+ channels in a neuronal membrane or by the closing of K^+ channels. In both cases, positive charges would accumulate inside the cell. Conversely, an inhibitory neurotransmitter usually hyperpolarizes the postsynaptic membrane, producing an **inhibitory postsynaptic potential (IPSP)**, which reduces the likelihood of an action potential. An IPSP can be caused by, for example, the opening of Cl^- channels. The equilibrium potential for Cl^- is typically between the resting potential and that of K^+. Therefore, when Cl^- channels are opened, Cl^- moves into cells, bringing them closer to the equilibrium potential for Cl^-.

To end the synaptic signal, neurotransmitter molecules in the synaptic cleft are broken down by enzymes or transported back into the terminal of the presynaptic cell and repackaged into vesicles for reuse. The latter event is called reuptake and is an efficient mechanism for recapturing unused neurotransmitters that were released into the synaptic cleft. As we will see at the end of this chapter, drugs that block the reuptake process can sometimes be used to treat people with disorders of neurotransmission, including depression.

We can appreciate the power of neurotransmitters when we consider the effects of toxins that are known to affect neurotransmitter release. For example, the venom of black widow spiders (genus *Latrodectus*), α-latrotoxin, acts on presynaptic proteins involved in exocytosis to cause massive release of neurotransmitter from motor neurons that stimulate contraction of skeletal muscle. Correspondingly, the symptoms of black widow spider bites in humans are painful muscle cramping and rigidity due to overstimulation of the muscle. Researchers have also demonstrated the function of several proteins involved in vesicle membrane fusion and exocytosis by experimenting with certain toxins that decrease neurotransmitter release. For instance, the toxin secreted by the *Clostridium botulinum* bacterium prevents exocytosis of motor neuron vesicles and release of their neurotransmitter. If ingested in food, the toxin causes botulism, a sometimes fatal form of food poisoning characterized by weak muscles, trouble focusing the eyes, and difficulty swallowing. Researchers have discovered several cosmetic and medical uses for this toxin, however. For example, when injected into a chronically stimulated muscle, it blocks contraction of the muscle by preventing the motor neurons from releasing neurotransmitter, resulting in muscle relaxation. Clinicians may also administer a synthetic form of the toxin (the generic name of which is onabotulinumtoxinA, but is more commonly known as Botox) to patients who wish to reduce facial wrinkles. The effect is to reduce the contractions of the facial muscles that cause frown lines. The procedure is not without risk, however, as the toxin may diffuse from the site of the injection and affect other targets.

Neurons Respond to Multiple Synaptic Inputs

A neuron may receive inputs from many synapses occurring on its dendrites or cell body (Figure 41.15a), and some of these synapses may release their neurotransmitter onto the neuron at the same or nearly the same time. Certain neurotransmitters are excitatory, and others are inhibitory. When do these different inputs lead to an

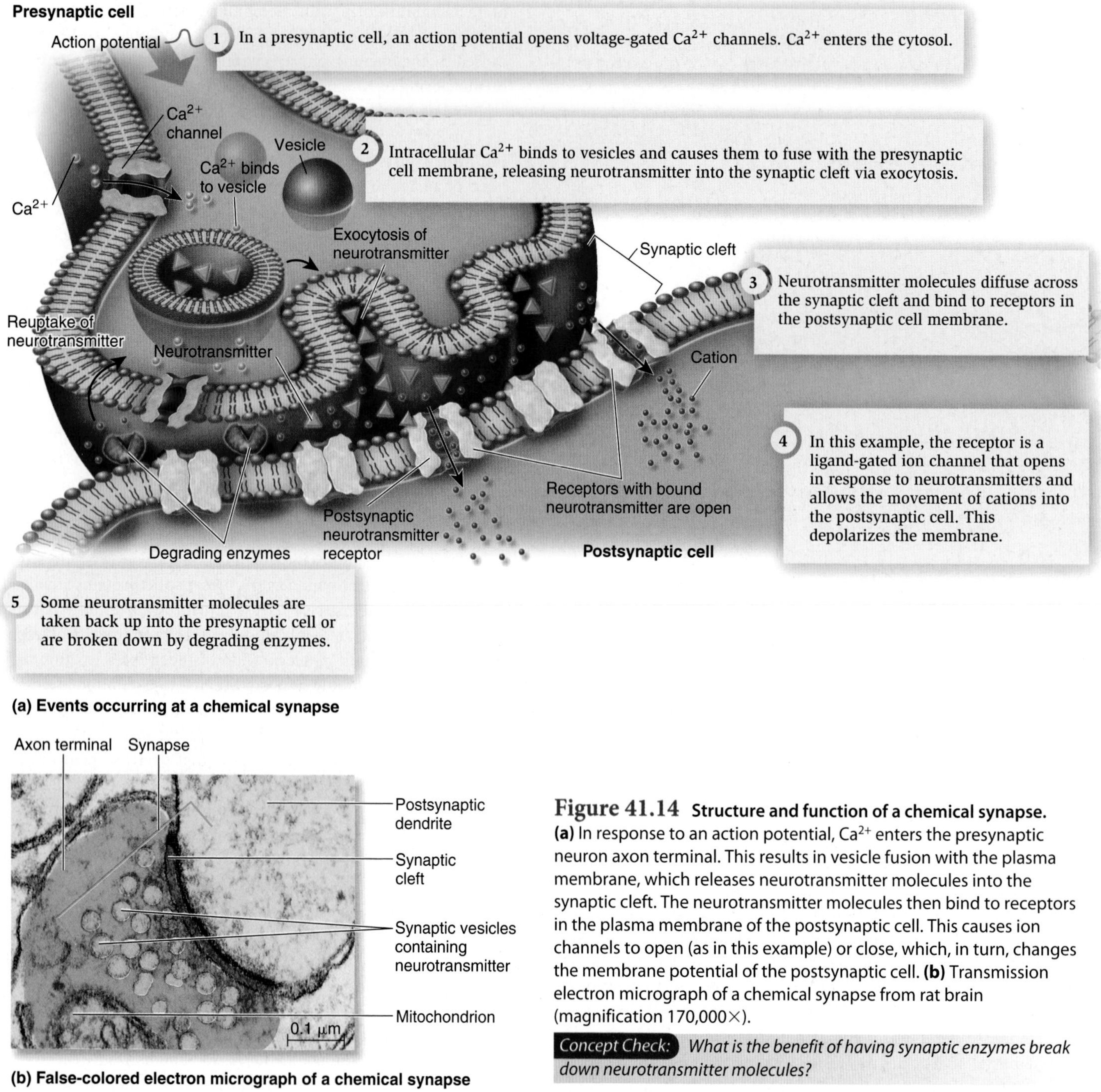

Presynaptic cell

Action potential

1 In a presynaptic cell, an action potential opens voltage-gated Ca^{2+} channels. Ca^{2+} enters the cytosol.

Ca^{2+} channel

Vesicle

Ca^{2+} binds to vesicle

Ca^{2+}

2 Intracellular Ca^{2+} binds to vesicles and causes them to fuse with the presynaptic cell membrane, releasing neurotransmitter into the synaptic cleft via exocytosis.

Exocytosis of neurotransmitter

Synaptic cleft

3 Neurotransmitter molecules diffuse across the synaptic cleft and bind to receptors in the postsynaptic cell membrane.

Reuptake of neurotransmitter

Neurotransmitter

Cation

4 In this example, the receptor is a ligand-gated ion channel that opens in response to neurotransmitters and allows the movement of cations into the postsynaptic cell. This depolarizes the membrane.

Receptors with bound neurotransmitter are open

Postsynaptic neurotransmitter receptor

Degrading enzymes

Postsynaptic cell

5 Some neurotransmitter molecules are taken back up into the presynaptic cell or are broken down by degrading enzymes.

(a) Events occurring at a chemical synapse

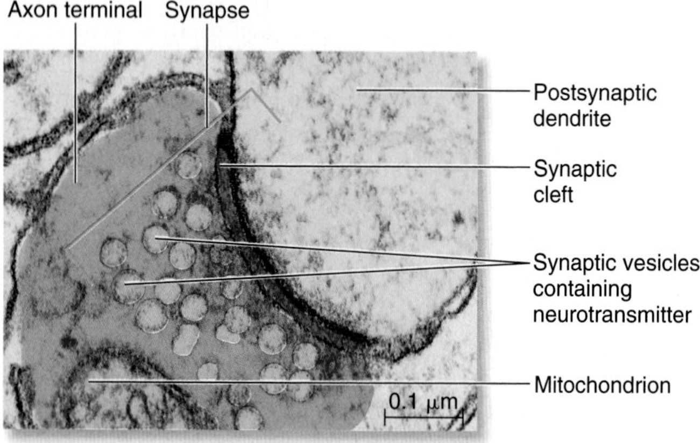

(b) False-colored electron micrograph of a chemical synapse

Figure 41.14 Structure and function of a chemical synapse. **(a)** In response to an action potential, Ca^{2+} enters the presynaptic neuron axon terminal. This results in vesicle fusion with the plasma membrane, which releases neurotransmitter molecules into the synaptic cleft. The neurotransmitter molecules then bind to receptors in the plasma membrane of the postsynaptic cell. This causes ion channels to open (as in this example) or close, which, in turn, changes the membrane potential of the postsynaptic cell. **(b)** Transmission electron micrograph of a chemical synapse from rat brain (magnification 170,000×).

Concept Check: What is the benefit of having synaptic enzymes break down neurotransmitter molecules?

action potential? The effect of a single synapse is usually far too weak to elicit an action potential in the postsynaptic neuron (though a single synapse can cause a muscle cell to contract). However, when two or more EPSPs are generated at one time along different regions of the dendrites and cell body, their depolarizations sum together. The resulting larger depolarization may bring the membrane potential at the axon hillock to the threshold potential, initiating an action potential. This is called **spatial summation** (**Figure 41.15b**). In another example of summation, two or more EPSPs may arrive at the same location in quick succession, such that the first EPSP has not yet decayed away when the next EPSP arrives. In that case, the depolarizations sum and may reach threshold upon arrival at the axon hillock. This is called **temporal summation** (**Figure 41.15c**). Finally, if EPSPs and IPSPs arrive together at a postsynaptic cell, the two types of signals may cancel each other out, and no action potential is elicited (**Figure 41.15d**). From this discussion, you might deduce that when two or more IPSPs arrive together, their hyperpolarizations might sum, too. That is exactly what happens. In that case, the membrane potential of the cell receiving multiple IPSPs moves farther away from threshold (**Figure 41.15e**).

In addition to the number of synapses that stimulate the postsynaptic membrane, the location of the synapses is also important. Synapses initially cause a graded potential in the postsynaptic membrane that spreads only a short distance. If two excitatory synapses are close together and activated at the same time, the resulting depolarization will be larger and will spread farther. This spread increases the chances that the depolarization will reach the axon hillock, where the high concentration of Na^+ channels can trigger an action potential. Synapses that occur far from the axon hillock are less effective than synapses on the cell body nearer the axon hillock.

Several Classes of Neurotransmitters Elicit Responses in Postsynaptic Cells

Neuroscientists have identified more than 100 different neurotransmitters in animals. Generally, neurotransmitters are categorized by size or structure (**Table 41.2**). The changing balance between excitatory and inhibitory neurotransmission controls the state of nervous system circuits at any one time. To understand how neurotransmitters

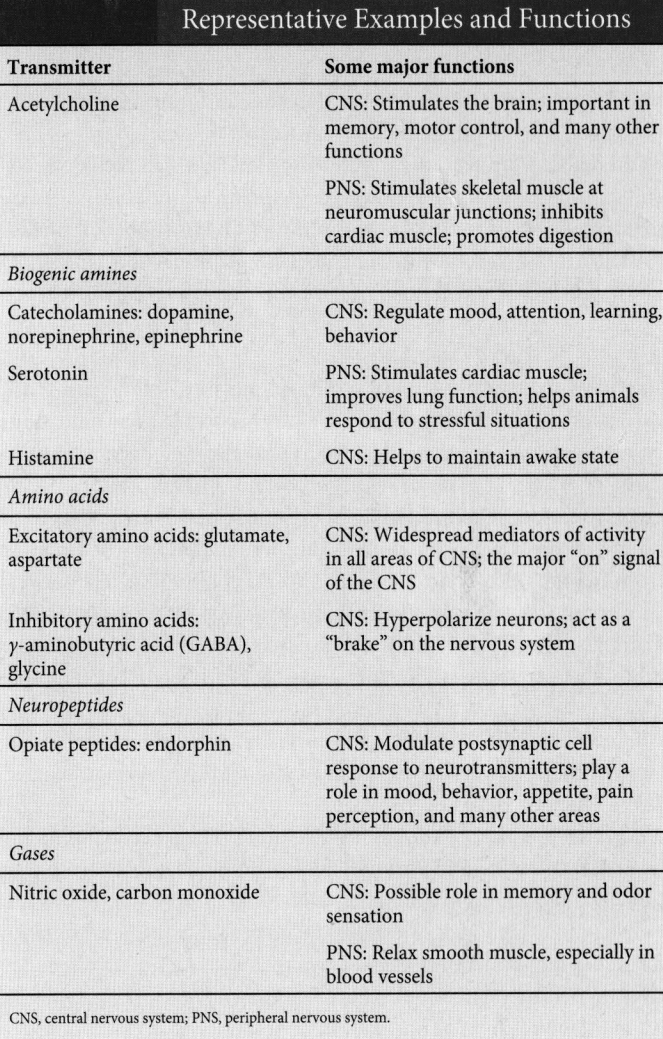

Table 41.2 Classes of Neurotransmitters and Some Representative Examples and Functions

Transmitter	Some major functions
Acetylcholine	CNS: Stimulates the brain; important in memory, motor control, and many other functions PNS: Stimulates skeletal muscle at neuromuscular junctions; inhibits cardiac muscle; promotes digestion
Biogenic amines	
Catecholamines: dopamine, norepinephrine, epinephrine	CNS: Regulate mood, attention, learning, behavior
Serotonin	PNS: Stimulates cardiac muscle; improves lung function; helps animals respond to stressful situations
Histamine	CNS: Helps to maintain awake state
Amino acids	
Excitatory amino acids: glutamate, aspartate	CNS: Widespread mediators of activity in all areas of CNS; the major "on" signal of the CNS
Inhibitory amino acids: γ-aminobutyric acid (GABA), glycine	CNS: Hyperpolarize neurons; act as a "brake" on the nervous system
Neuropeptides	
Opiate peptides: endorphin	CNS: Modulate postsynaptic cell response to neurotransmitters; play a role in mood, behavior, appetite, pain perception, and many other areas
Gases	
Nitric oxide, carbon monoxide	CNS: Possible role in memory and odor sensation PNS: Relax smooth muscle, especially in blood vessels

CNS, central nervous system; PNS, peripheral nervous system.

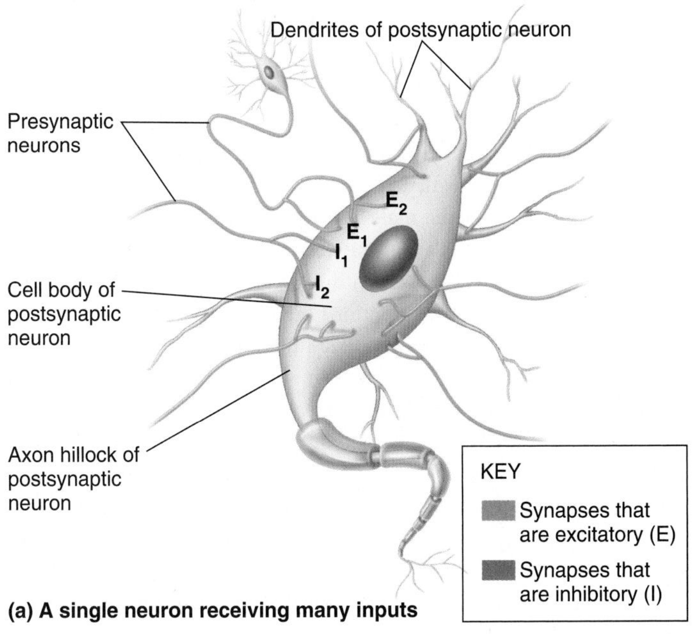

(a) A single neuron receiving many inputs

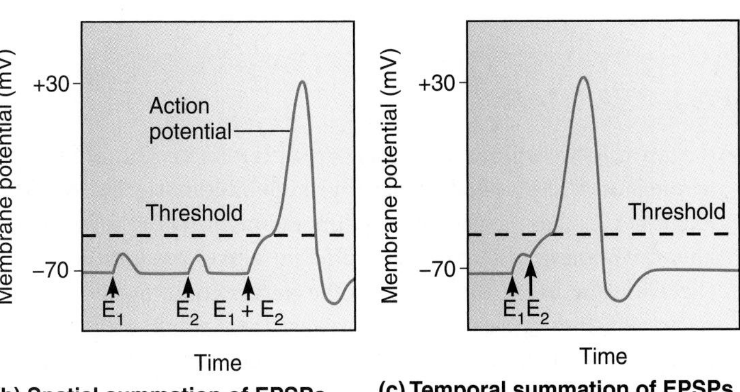

(b) Spatial summation of EPSPs

(c) Temporal summation of EPSPs

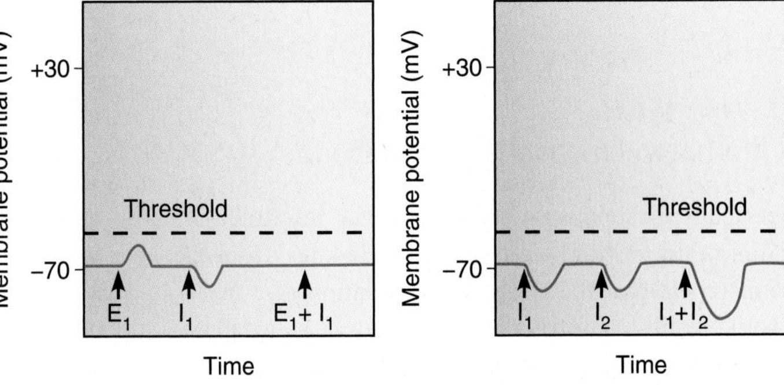

(d) Cancellation of EPSP and IPSP

(e) Spatial summation of IPSPs

Figure 41.15 Integration of synaptic inputs. Membrane potential changes in a postsynaptic membrane are shown under several different circumstances. When multiple excitatory (E) or inhibitory (I) inputs arrive simultaneously or nearly so, the subsequent EPSPs and IPSPs may add (summate) or cancel each other out, depending on the situation.

work, imagine driving a car with one foot on the accelerator and one foot on the brake. To speed up, you could press down on the accelerator, ease up on the brake, or both, whereas to slow down, you could do the opposite. All nervous systems operate in this way, with combined excitatory and inhibitory actions of neurotransmitters.

Next, we highlight major features of the different chemical classes of neurotransmitters found in animals, which include acetylcholine, biogenic amines, amino acids, neuropeptides, and gaseous neurotransmitters. With the exception of acetylcholine, all of these classes contain several different neurotransmitters that are similar in chemical structure but may have different functions in nervous systems.

Acetylcholine Acetylcholine, one of the most widespread neurotransmitters in animals, is released at the synapses of **neuromuscular junctions** in vertebrates, where a neuron contacts skeletal or cardiac muscle. It is also released at synapses within the brain and elsewhere. Acetylcholine acts as an excitatory neurotransmitter in the brain and on skeletal muscle cells, but it is inhibitory when released from neurons that control cardiac muscle contraction. Therefore, it stimulates skeletal muscle but inhibits cardiac muscle. As we will see later, a single type of neurotransmitter can exert both excitatory and inhibitory effects depending on the type of receptor to which it binds.

Biogenic Amines The biogenic amines are compounds containing amine groups that are formed from amino acids. Common biogenic amines include the catecholamines—dopamine, norepinephrine, and epinephrine—and serotonin and histamine. The catecholamines are formed from the amino acid tyrosine, serotonin is formed from tryptophan, and histamine from histidine.

In addition to widespread physiological effects such as control of heart and lung function, catecholamines and serotonin are psychoactive; that is, they affect mood, attention, behavior, and learning. In humans, for example, abnormally high or low levels of catecholamines and serotonin have been associated with a variety of mental illnesses, including schizophrenia, attention-deficit disorder, and depression. Histamine is well known as a component of allergic reactions in people, but this is not related to its neurotransmitter functions. In the brain, neurons that produce histamine are important in modulating sleep; they are most active during waking and are nearly inactive during sleep. This explains why certain antihistamines (drugs that block the ability of histamine to bind to its receptor, thereby inhibiting its action) used to treat colds and allergies also induce drowsiness.

Amino Acids The amino acids glutamate, aspartate, glycine, and γ-aminobutyric acid (GABA) function as neurotransmitters. Glutamate is the most widespread excitatory neurotransmitter found in animal nervous systems, whereas GABA is the most common inhibitory neurotransmitter. GABA hyperpolarizes the postsynaptic membrane by opening chloride channels, allowing negatively charged Cl^- to diffuse into the cell (see Table 41.1); in this way, GABA acts as the major "brake" on the CNS.

Neuropeptides Neuropeptides are short chains of 2 to about 15 amino acids. Like the other neurotransmitters discussed, neuropeptides can be excitatory or inhibitory. Neuropeptides are often called **neuromodulators**, because they can alter or modulate the response of the postsynaptic neuron to other neurotransmitters. For example, a cell exposed to a neuropeptide may increase the number of receptors for another neurotransmitter, which makes the cell more responsive to that neurotransmitter. One group of neuropeptides is called the opiate peptides because opium-like drugs, such as morphine, bind to their receptors. Opiate peptides include the endorphins, a group of peptides that decrease pain and cause natural feelings of euphoria (extreme happiness and feeling of invulnerability).

Gaseous Neurotransmitters Certain gaseous molecules such as nitric oxide (NO) and carbon monoxide (CO) act locally in many tissues and sometimes function as neurotransmitters. Unlike other neurotransmitters, they are not sequestered into vesicles and are produced locally as required. Gaseous neurotransmitters are short-acting and influence other cells by diffusion. In humans, NO is responsible for relaxing the smooth muscle surrounding blood vessels, including those in the penis. When a male becomes sexually aroused, NO levels increase in this tissue, dilating the vessels and increasing blood flow into the penis, producing an erection. Several drugs used to treat male sexual dysfunction enhance erections by increasing or mimicking the action of NO on smooth muscle. The functions of CO are still uncertain, but scientists think it may act as a neurotransmitter in the sense of smell in some animals, such as the terrestrial mollusk *Limax maximus*.

The discovery that the actions of neurons are mediated in large part by neurotransmitters was one of the most significant discoveries in the history of neuroscience. That discovery laid the foundation for our understanding of the nervous system and of many human diseases. Amazingly, the existence of neurotransmitters was demonstrated in a remarkably simple experiment that arose from a dream, as we see next.

FEATURE INVESTIGATION

Otto Loewi Discovered Acetylcholine

German physiologist Otto Loewi was interested in how neurons communicate with skeletal muscle. He knew from the work of other researchers that the electrical stimulation of a nerve in a frog's leg would result in contraction of the muscle associated with that nerve, so it appeared that neurons communicated with the muscle by electrical signals. In 1921, he turned his studies to another type of muscle, the heart. As we will see in Chapter 47, all vertebrate and some invertebrate hearts receive both excitatory and inhibitory signals from different nerves that regulate the rhythm and intensity of the heartbeat. Loewi hypothesized that because different nerves produced opposite effects on the heart, the effects of the nerves could not be a direct electrical action on heart muscle, because there would be no way for the heart muscle to discern between the same type of signal (that is, electricity) from two different nerves. Instead, perhaps the neurons in each nerve released different chemicals of some type, and it was these chemicals that exerted opposite actions on the heart.

As shown in **Figure 41.16**, Loewi removed the hearts from two frogs and placed the hearts in baths containing saline (an isotonic NaCl solution). When a frog's heart is maintained in this solution, it will continue to beat for several hours before eventually dying.

Initially, Loewi began by examining the major inhibitory nerve of the vertebrate heart, called the vagus nerve. When Loewi dissected the hearts from the two frogs, he left the vagus nerve intact in one heart, but removed it from the second heart. Next, Loewi used an electrode to electrically stimulate the vagus nerve attached to the first frog's heart. As expected, this resulted in a decrease in the rate at which the first heart contracted. He then removed some of the saline solution from within and around the first heart and transferred it to the solution that was bathing the second (unstimulated) heart. The rate and force of beating of the second heart quickly decreased, even though it had no vagus nerve and was not exposed to any electrical stimulation. Loewi concluded that a chemical substance was released from the nerve of the first heart into the surrounding fluid and that when this chemical was added to the second heart, it reproduced the effects of electrical stimulation that were observed on the first heart.

Loewi initially named this substance *vagusstoff* (vagus substance), after the vagus nerve he stimulated. It was later renamed acetylcholine when its chemical nature (acetic acid bonded to choline) was determined. Acetylcholine was the first neurotransmitter discovered. Loewi's research opened the door for what we now know about chemical transmission at synapses, and the enormous pharmaceutical industry, which builds on this knowledge to treat neurological disorders (diseases of the nervous system).

Interestingly, as Loewi described later, the idea for his experiment, which would be largely responsible for his earning a share of the 1936 Nobel Prize in Physiology or Medicine, came to him in a dream. He woke in the middle of the night, scribbled down his idea, and returned to sleep. The next morning, he despaired to find that he couldn't read his sleepy scribbling! Incredibly, he had the dream again the following night and, not taking any chances, got up and went directly to his laboratory.

Figure 41.16 **Loewi's experimental discovery of chemical neurotransmission.**

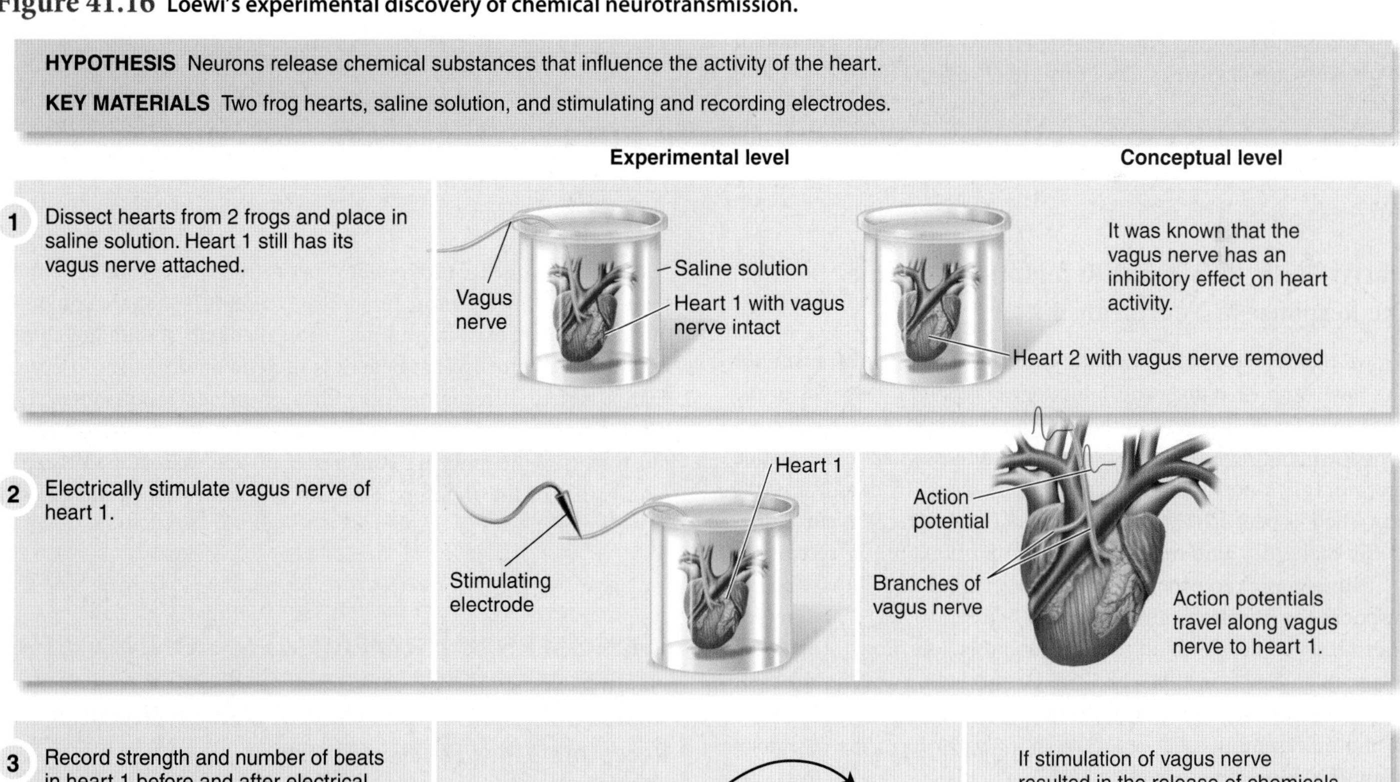
HYPOTHESIS Neurons release chemical substances that influence the activity of the heart.
KEY MATERIALS Two frog hearts, saline solution, and stimulating and recording electrodes.
Experimental level
Conceptual level
1 Dissect hearts from 2 frogs and place in saline solution. Heart 1 still has its vagus nerve attached.
Vagus nerve
Saline solution
Heart 1 with vagus nerve intact
Heart 2 with vagus nerve removed
It was known that the vagus nerve has an inhibitory effect on heart activity.
2 Electrically stimulate vagus nerve of heart 1.
Stimulating electrode
Heart 1
Action potential
Branches of vagus nerve
Action potentials travel along vagus nerve to heart 1.
3 Record strength and number of beats in heart 1 before and after electrical stimulation of vagus nerve. Next, remove a sample of the saline solution in and around heart 1, and transfer to heart 2. Record activity of heart 2. This was done using mercury manometers that were connected to each heart. The manometers measure pressure, which is due to the contractile force of the heart beating.

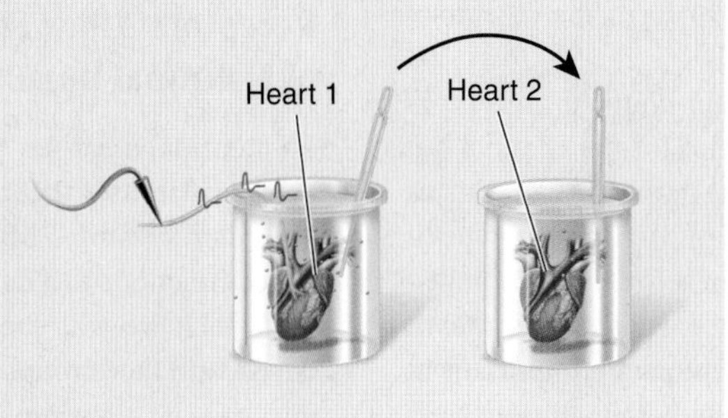
Heart 1
Heart 2

If stimulation of vagus nerve resulted in the release of chemicals onto heart 1, then these same chemicals (some of which may diffuse into the saline solution) should have an identical effect on heart 2.

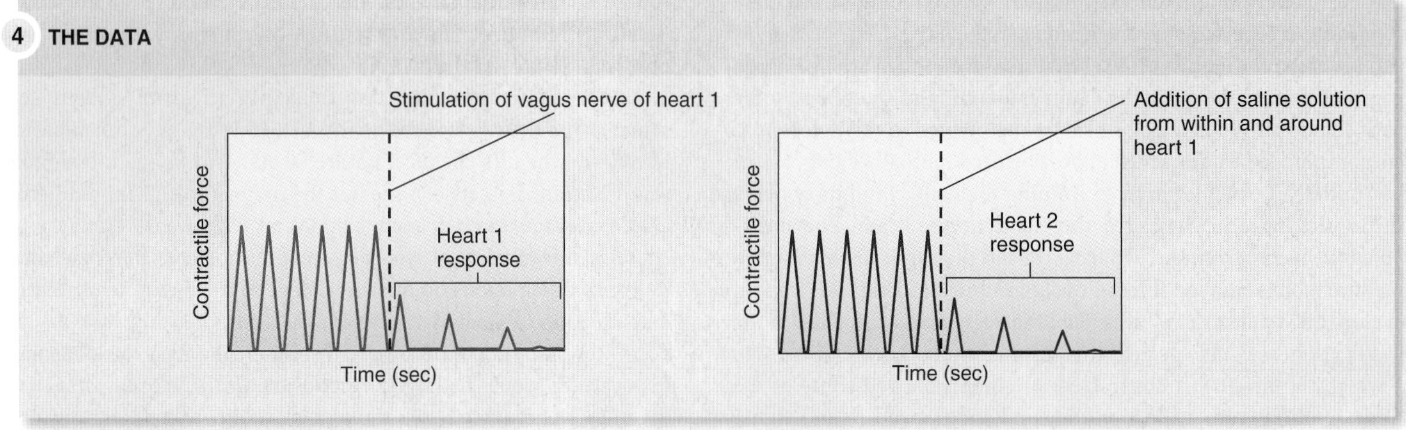

5 CONCLUSION Electrical stimulation causes the vagus nerve to secrete chemicals that decrease heart contractions.

6 SOURCE Loewi, O. 1921. On humoral transmission of the action of heart nerves. Pflügers archives. *European Journal of Physiology* 189:239–242.

Experimental Questions

1. What observations led Loewi to develop his hypothesis of how nerves stimulate or inhibit heart muscle contractions?
2. Describe Loewi's experimental design to test his hypothesis.
3. What were the results of Loewi's experiment? Did the results support his hypothesis?

Postsynaptic Membrane Receptors Determine the Response to Neurotransmitters

As we mentioned earlier, in some cases the same neurotransmitter can have both excitatory and inhibitory effects. The response of the postsynaptic cell depends on the type of receptor present in the postsynaptic membrane. The two major types of postsynaptic membrane receptors are ionotropic and metabotropic, and many neurotransmitters, such as acetylcholine, act on both (**Figure 41.17**). Neurotransmitter molecules bind to the extracellular portion of these receptors.

Ionotropic receptors are ligand-gated ion channels that open in response to neurotransmitter binding (Figure 41.17a). When neurotransmitter molecules bind to these receptors, ions flow through the channels to cause an EPSP or IPSP. Acetylcholine and amino acids bind to ionotropic receptors. Ionotropic receptors are composed of multiple subunits that associate in a ring to form the receptor's channel.

Metabotropic receptors are G-protein-coupled receptors (GPCRs) (refer back to Figure 9.7). They do not form a channel but instead are coupled to an intracellular signaling pathway that initiates changes in the postsynaptic cell (Figure 41.17b). A common type of response is the phosphorylation of ion channels for sodium, potassium, or calcium ions, which are present in the plasma membrane. In Chapter 43, we will see that metabotropic receptors are important in activating sensory cells that respond to visual and other stimuli.

As mentioned earlier, many neurotransmitters bind to more than one type of receptor. In addition, receptors that are composed of subunits may exist in multiple forms made up of different combinations of subunits. What benefit is there to an organism to express such a variety of receptor types for a given neurotransmitter? One helpful way of understanding this diversity and its importance is illustrated by the amazing complexity of one well-studied receptor family, that of the widespread amino acid neurotransmitter GABA, as we see next.

GENOMES & PROTEOMES CONNECTION

Varied Subunit Compositions of Neurotransmitter Receptors Allow Precise Control of Neuronal Regulation

As mentioned earlier, γ-aminobutyric acid (GABA) is an inhibitory neurotransmitter that opens Cl^- channels. Though cells can possess different types of GABA receptors, we will focus here on one type, which functions as a ligand-gated ion channel. This ionotropic receptor, which we will simply refer to as the GABA-A receptor, binds GABA and thereby opens the channel. This event allows Cl^- to diffuse into the cell, causing a hyperpolarization of the plasma membrane

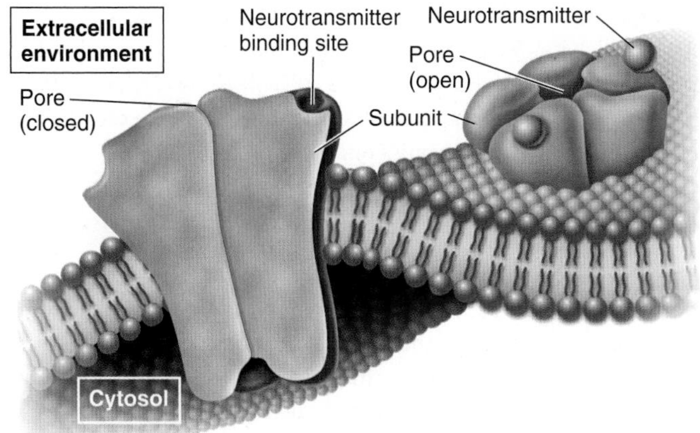

(a) Ionotropic receptor

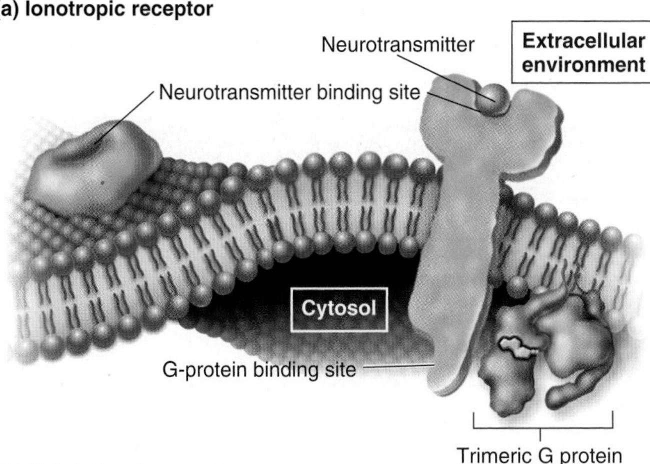

(b) Metabotropic receptor

Figure 41.17 The two major categories of postsynaptic receptors. (a) Ionotropic receptors have several subunits. Neurotransmitters bind to ionotropic receptors and directly open ion channels in the membrane. **(b)** Metabotropic receptors are G-protein-coupled receptors, which are discussed in Chapter 9. Neurotransmitters bind to metabotropic receptors and initiate a signaling pathway that typically opens or closes ion channels.

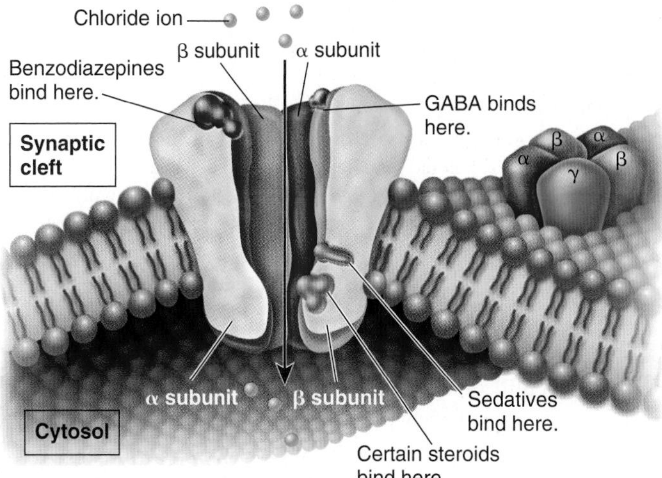

Figure 41.18 The structure and function of the GABA-A receptor. Each type of GABA-A receptor has five subunits that form an anion channel allowing the passage of chloride ions. When GABA binds to the receptor, chloride ions move through the open channel and hyperpolarize the cell. Various other molecules bind to different sites on the receptor. These include molecules that naturally occur in animals' bodies, such as certain steroid hormones, and drugs, such as sedatives and benzodiazepines, used to treat anxiety and other disorders.

BIOLOGY PRINCIPLE Structure determines function. Neurotransmitter receptors are excellent examples of this biological principle at the molecular level. The three-dimensional structure of the receptor determines its ability to bind a particular neurotransmitter and, as shown in this figure, various other ligands that can modulate the receptor's activity.

and shifting the membrane potential toward the equilibrium potential for Cl^- (usually between about –70 to –90 mV) (**Figure 41.18**). In this way, GABA binding to this receptor decreases the likelihood that a neuron will generate an action potential.

The GABA-A receptor is a good example of how receptor subunits can influence a postsynaptic response to a neurotransmitter. GABA-A receptor proteins are usually composed of five subunits (designated α, β, γ, and so on). The genomes of humans and other mammals have a group of homologous genes that encode at least 19 different GABA-A receptor subunits. In addition, subunit variation can be further increased by alternative splicing (refer back to Figure 13.21). This amazing variety in subunits allows cells to express dozens of different kinds of GABA-A receptors.

What are the selection advantages of having evolved such a variety of different GABA-A receptor subunits? Though the answer is not entirely understood, each type of subunit has its own unique properties that can fine-tune the function of the GABA-A receptor so that it works optimally in the neuron in which it is expressed. The various subunits may differ in their affinity for GABA and the rate of Cl^- movement through the channel.

In addition, neuroscientists have been particularly interested in whether the various subunits differ in their ability to recognize molecules other than GABA (see Figure 41.18). This work has shown that the subunits of the GABA receptor bind a variety of other molecules, including naturally occurring ones such as certain steroid hormones. Presumably, the binding of these molecules enhances or reduces the effectiveness of GABA in activating the receptor. This knowledge has proven beneficial in understanding how certain drugs exert their actions. For example, ethanol—found in alcoholic drinks—binds to one of the GABA-A receptor subunits expressed in brain and motor neurons and enhances the actions of GABA. This may explain in part why alcohol depresses the activity of the brain and impairs motor coordination, among other effects. Other subunits of the GABA-A receptor bind drugs including benzodiazepines such as alprazolam (Xanax), which are used to treat chronic or severe anxiety. Apparently, the inhibitory effects of GABA are part of the mechanism for achieving a balance between anxiety and calmness. The ability of the receptor to bind numerous ligands, and the many different combinations of subunits in the receptor, provide an enormous degree of control over precisely how this neurotransmitter system regulates the activity of the brain.

41.5 Impact on Public Health

Learning Outcomes:

1. Describe how disorders of neurotransmission can affect human health.
2. List several "recreational" and illicit drugs and describe the corresponding effects these drugs have on the nervous system.
3. Give examples of disorders of neural conduction and how they arise.

When neurons fail to develop properly or their function is impaired, the consequences can be devastating, affecting mood, behavior, and even the ability to think or move. Over 100 neurological disorders have been identified in humans, and therapeutic drugs to treat them are among the most widely prescribed medicines today (**Table 41.3**). All so-called recreational drugs, including alcohol and tobacco, as well as illicit drugs such as marijuana, cocaine, heroin, LSD, and methamphetamine, exert their effects by altering neurotransmission. The use of these drugs, therefore, can result in symptoms similar to those of neurological disorders. Note in Table 41.3 that certain therapeutic drugs used to treat neurological disorders also exert many of their actions on neurotransmission.

Disorders of Neurotransmission Can Impact Mood

Several neurological disorders result from disrupted neurotransmission between cells. Genetic processes involved in the production of neurotransmitters or malfunction of synaptic events can increase or decrease activity at synapses, which, in turn, affects emotions and behavior. The most common mood disorder is **major depressive disorder** or, simply, depression. This illness results in prolonged periods of sadness, despair, and lack of interest in daily activities without alternating episodes of euphoria. Depression affects 5–12% of men

Table 41.3 Representative Effects of Common Therapeutic, Illicit, or Recreational Drugs on Neurotransmitter Action and Mood

Name of drug	Actions on neurotransmission	Effects on mood	Effects of abuse or overdose
Illicit or Recreational (Note: Some of these drugs have some therapeutic value under certain conditions)			
Alcohol (ethanol)	Enhances inhibitory GABA transmission; increases dopamine transmission; inhibits glutamate transmission	Relaxation; euphoria; sleepiness	Liver damage; brain damage
Amphetamines "uppers," "crystal meth," "pep pills"	Stimulate the release of dopamine and norepinephrine	Euphoria; increased activity	High blood pressure; psychosis
Cocaine	Blocks norepinephrine and dopamine reuptake	Intense euphoria followed by depression	Convulsions; hallucinations; death from overdose
LSD (lysergic acid diethylamide)	Binds to serotonin receptors	Hallucinations; sensory distortions	Unpredictable and irrational behavior
Marijuana (tetrahydrocannabinol)	Binds to receptors for natural cannabinoids	Increased sense of well-being; decreased short-term memory; decreased goal-directed behavior; increased appetite	Delusions; paranoia; confusion
Narcotics: Heroin, morphine, meperidine (Demerol), codeine	Bind to opiate receptors	Pain relief; euphoria; sedation	Slowed breathing; death from overdose
Nicotine	Initially stimulates but then depresses activity in adrenal medulla and neurons in the peripheral nervous system; increases dopamine in brain	Increased attention; decreased irritability	Heart disease and lung disease
PCP (phencyclidine) "angel dust," "ozone"	Blocks channel for excitatory amino acid neurotransmitters; increases dopamine activity	Violent behavior; feelings of power; numbness; disorganized thoughts	Psychosis; convulsions; coma; death
Therapeutic			
Tricyclic antidepressants (for example, Elavil, Anafranil)	Block the reuptake of norepinephrine from synapses	Relieve depression and obsessive-compulsive disorder	Drowsiness; confusion
Selective serotonin reuptake inhibitors (SSRIs) (for example, Prozac, Zoloft, Paxil, Lexapro)	Block the reuptake of serotonin from synapses	Relieve depression and obsessive-compulsive disorder	Insomnia; anxiety; headache
Monoamine oxidase inhibitors (for example, Parnate, Nardil)	Block the breakdown of biogenic amine neurotransmitters	Relieve depression	Liver damage; hyperexcitability
Antianxiety drugs: Benzodiazepines [for example, Xanax, Valium, Librium, Rohypnol ("date rape drug," "roofies")]	Bind to GABA receptors and increase inhibitory neurotransmission	Relieve anxiety; cause sleepiness and in some cases amnesia	Drowsiness; memory loss in some cases
Antipsychotic drugs: Phenothiazines (for example, Thorazine, Mellaril, Stelazine), and atypical antipsychotics (Abilify, Risperdal)	Block dopamine receptors	Ease schizophrenic symptoms	Decreased control of movement

and 10–25% of women at some time during their lives. This condition is thought to result from decreased activity of synapses that release biogenic amines, such as serotonin, which changes neuronal activity within specific areas in the brain involved in processing emotion. Drugs used to treat major depression include the **selective serotonin reuptake inhibitors (SSRIs)**, such as Prozac, Zoloft, and Paxil, which reduce the reuptake of serotonin into the presynaptic terminal after it is released. This allows serotonin to accumulate in the synaptic cleft, counteracting the deficit that causes the alteration in mood.

In many respects, the use of such drugs to treat depression reflects a profound change in the public's attitude toward mental illness. Historical attitudes toward mood disorders held that individuals who were depressed lacked the ability to cope with stressful events in their lives. Only relatively recently has it become accepted that mood disorders are typically caused by changes in the balance of neurotransmitters in the brain. Evidence of a genetic basis for depression is that mood disorders occur more frequently in certain families. Drugs can be very effective in treating these disorders. Patients taking SSRIs often report decreased sadness, increased energy, and a greater interest in daily activities.

Many Illicit Drugs Disrupt Normal Neurotransmission

Many illicit drugs work at the synapse to either enhance or interfere with the normal mechanisms of neurotransmission (see Table 41.3). In the presynaptic terminal, such drugs can decrease neurotransmitter release by reducing Ca^{2+} entry into the cell or by preventing the exocytosis of vesicles containing stored neurotransmitters. In the synaptic cleft, drugs can slow the rate at which the neurotransmitter is broken down into an inactive form or taken back up into the presynaptic neuron, thereby prolonging the action of the neurotransmitter in the synaptic cleft. Some substances act on the postsynaptic membrane by either preventing the neurotransmitter from binding to its receptor or by acting as a substitute for the neurotransmitter by stimulating the receptor.

In effect, these drugs produce changes or imbalances in neurotransmission similar to those observed in some neurological disorders. These substances can induce euphoria, increase activity, alter mood, and produce hallucinations. They can also have potentially life-threatening effects and may be highly addictive.

Some drugs, such as cocaine, block the removal of dopamine and norepinephrine from the synaptic cleft by preventing their reuptake into the presynaptic terminal. Morphine and marijuana mimic the actions of biological substances already in the brain, binding to receptors on the postsynaptic membrane. With these drugs, the resulting effects are much stronger than are the effects of natural neurotransmitters. It is no surprise that many of these drugs are mind-altering. They do, after all, change the ways in which neurons communicate with each other.

Disorders of Conduction May Result in Motor Problems and Abnormal Neuronal Development

Some human diseases are caused by the inability of certain axons to properly conduct an action potential. This occurs most commonly because an axon fails to become myelinated or because a myelinated axon becomes demyelinated.

In **congenital hypothyroidism**, axons fail to become wrapped with myelin during fetal development, which leads to slow conduction speeds and abnormal connections between brain neurons. This results in profound mental defects that cannot be reversed unless treatment begins immediately after birth. Congenital hypothyroidism is caused by insufficient levels of thyroid hormones in the fetus. Among their many actions, thyroid hormones stimulate the formation of myelin during fetal development. However, thyroid hormones cannot be synthesized without the element iodine, which is part of the structure of the hormones. The iodine in the fetus comes from the mother's diet. If a mother's dietary intake of iodine is too low, the fetus will not have enough iodine to make its own thyroid hormones, and therefore, the fetus will not be able to make normal amounts of myelin. Congenital hypothyroidism is rare in the U.S. and many other countries since the advent of iodized table salt; however, it is not uncommon in many parts of the world.

Unlike congenital hypothyroidism, **multiple sclerosis (MS)** usually begins between the ages of 20 and 50 in individuals with apparently healthy nervous systems. With MS, the patient's own immune system, for reasons unknown, attacks and destroys myelin as if it were a foreign substance. Eventually, these repeated attacks leave multiple scarred (sclerotic) areas of tissue in the nervous system and impair the function of myelinated neurons that control movement, speech, memory, and emotion. Multiple sclerosis is a serious and unpredictable disease, characterized by flare-ups followed by periods of remission in which symptoms are reduced or absent. No cure is currently available, but certain drugs may slow its progression and reduce the severity of symptoms. This disease affects roughly 2.5 million people worldwide, about 75% of them women.

Summary of Key Concepts

41.1 Cellular Components of Nervous Systems

- The central nervous system (CNS) is composed of a brain and a nerve cord. The peripheral nervous system (PNS) consists of all neurons and their projections that are outside of and connect with the CNS. Nerves transmit signals between the PNS and CNS (Figure 41.1).
- The two major classes of cells in nervous systems are neurons and glia. In neurons, information flows from dendrites to the cell body and then to the axon and axon terminal. Types of neurons include sensory neurons, motor neurons, and interneurons. A neuron's structure is a reflection of its function (Figures 41.2, 41.3).
- The most basic circuit is a reflex arc, which occurs rapidly in response to inputs from sensory neurons and consists of only one or a few afferent and efferent cells (Figure 41.4).

41.2 Electrical Properties of Neurons and the Resting Membrane Potential

- Neuronal membranes are electrically polarized. The membrane potential is determined by the differential distribution and differential permeability of ions across the plasma membrane. The resting potential is the membrane potential of a cell that is not sending electrical signals. Neurons use electrical signals to communicate with other neurons, muscle cells, or gland cells. These signals involve changes in the amount of electric charge across a cell's plasma membrane (Figures 41.5, 41.6, Table 41.1).

- Diffusion of ions through membrane channels occurs as a result of the concentration gradient of an ion across the membrane and the electric charge across the membrane. Ions move in response to an electrochemical gradient (Figure 41.7).
- The Nernst equation gives the equilibrium potential for an ion at any given concentration gradient.

41.3 Generation and Transmission of Electrical Signals Along Neurons

- Gated ion channels enable a cell to communicate by changing its membrane potential rapidly. The opening and closing of voltage-gated and ligand-gated ion channels cause two types of changes in the neuron's membrane potential—graded potentials and action potentials (Figures 41.8, 41.9).
- Graded potentials can trigger an action potential, an event that carries an electrical signal along an axon, from the axon hillock to the axon terminal. Axon diameter and myelination influence the speed of an action potential (Figures 41.10, 41.11, 41.12).

41.4 Neurons Communicate Electrically or Chemically at Synapses

- In an electrical synapse, an electrical current is conducted from one cell to another via gap junctions. In a chemical synapse, a neurotransmitter carries the signal from the presynaptic to the postsynaptic cell. Many excitatory postsynaptic potentials (EPSPs) generated at one time can sum together and bring the membrane potential to the threshold potential, initiating an action potential (Figures 41.13, 41.14, 41.15).
- Chemical classes of neurotransmitters found in animals include acetylcholine, biogenic amines, amino acids, neuropeptides, and gaseous neurotransmitters. The discovery that the actions of neurons are mediated in large part by neurotransmitters was one of the most significant discoveries in the history of neuroscience (Table 41.2, Figure 41.16).
- The receptors of the postsynaptic neuron determine the types of signals that pass from one neuron to the other. The two major types of postsynaptic receptors are ionotropic and metabotropic (Figures 41.17, 41.18).

41.5 Impact on Public Health

- Most neurological conditions can be classified as disorders of either neurotransmission or conduction. Mood disorders caused by disrupted neurotransmission include major depressive disorder. Drugs used in the treatment of neurological disorders and many recreational and illicit drugs usually alter neurotransmission (Table 41.3).
- Some neurological conditions are caused by the inability of the axon to conduct an action potential. This occurs most commonly because axons fail to become myelinated (congenital hypothyroidism) or because myelinated axons become demyelinated (multiple sclerosis).

Assess and Discuss

Test Yourself

1. In vertebrates, the brain and the spinal cord are parts of
   a. the peripheral nervous system.
   b. the enteric nervous system.
   c. the central nervous system.
   d. the autonomic nervous system.
   e. the endocrine system.
2. The structures of a neuron that function mainly in receiving signals from other neurons are
   a. the myelin sheaths.
   b. the axons.
   c. the axon terminals.
   d. the dendrites.
   e. the K^+ channels.
3. The glial cells that form the myelin sheath in the peripheral nervous system are called
   a. astrocytes.
   b. oligodendrocytes.
   c. microglia.
   d. neurons.
   e. Schwann cells.
4. Neurons that function mainly in connecting other neurons in the central nervous system are
   a. sensory neurons.
   b. efferent neurons.
   c. motor neurons.
   d. afferent neurons.
   e. interneurons.
5. The difference in charges across the plasma membrane of an unstimulated neuron is called
   a. the membrane potential.
   b. the resting membrane potential.
   c. homeostasis.
   d. the graded potential.
   e. the action potential.
6. Which of the following contributes to the resting membrane potential?
   a. negatively charged ions inside and outside of the cell
   b. active transport of ions across the membrane
   c. concentration of Na^+ and K^+ inside and outside of the cell
   d. all of the above
   e. b and c only
7. A neuron has reached a threshold potential when it has depolarized to the point where
   a. most voltage-gated K^+ channels open.
   b. sufficient numbers of voltage-gated Na^+ channels open to initiate a positive feedback cycle, contributing to further depolarization.
   c. voltage-gated K^+ channels close.
   d. voltage-gated Na^+ channels close.
   e. both b and c occur.
8. The speed of transmission of an action potential along an axon is influenced by
   a. the presence of myelin.
   b. an increased concentration of Ca^{2+}.
   c. the diameter of the axon.
   d. all of the above.
   e. a and c only.

9. Gap junctions are characteristic of
   a. electrical synapses.
   b. chemical synapses.
   c. acetylcholine synapses.
   d. GABA synapses.
   e. synapses between motor neurons and muscle cells.
10. The response of the postsynaptic cell is determined by
   a. the type of neurotransmitter released at the synapse.
   b. the type of receptors the postsynaptic cell has.
   c. the number of Na^+ channels in the postsynaptic membrane.
   d. the number of K^+ channels in the postsynaptic membrane.
   e. all of the above.

Conceptual Questions

1. Describe the difference between graded and action potentials.
2. In certain diseases, such as kidney failure, the Na^+ concentration in the body's extracellular fluid can become altered. What effect might a high extracellular Na^+ concentration have on neurons?
3. A principle of biology is that *structure determines function.* How can this be applied to neurons?

Collaborative Questions

1. Discuss how nervous systems are organized into central and peripheral nervous systems in animals.
2. Name the parts of a neuron, and give a brief description of their major characteristics.

Online Resource

www.brookerbiology.com

Stay a step ahead in your studies with animations that bring concepts to life and practice tests to assess your understanding. Your instructor may also recommend the interactive eBook, individualized learning tools, and more.

Neuroscience II: Evolution, Structure, and Function of the Nervous System

42

Chapter Outline

42.1 The Evolution and Development of Nervous Systems
42.2 Structure and Function of the Nervous Systems of Humans and Other Vertebrates
42.3 Cellular Basis of Learning and Memory
42.4 Impact on Public Health
Summary of Key Concepts
Assess and Discuss

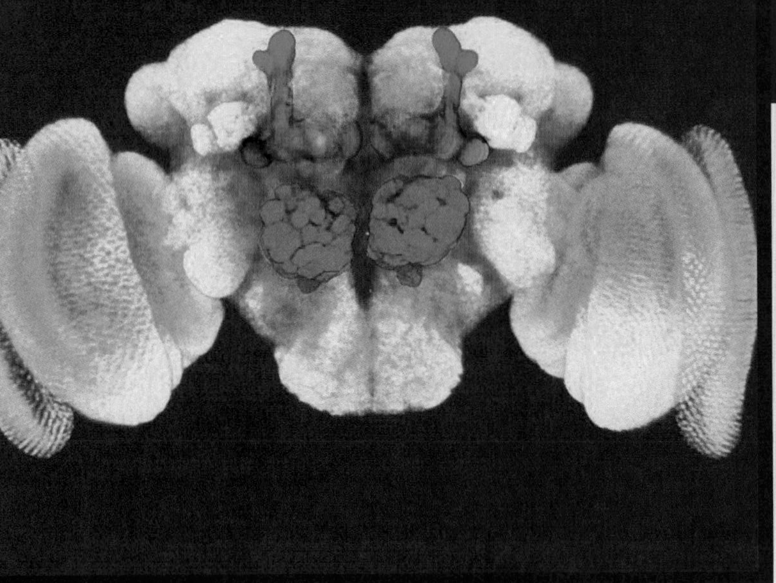

Three-dimensional reconstruction of the brain of a fruit fly. The brains of animals are organized into anatomic structures with specialized functions.

It will take you approximately 2–3 seconds to read this sentence. During that time, many of the 100 billion or so neurons of your brain will have fired off millions of action potentials. Some of those signals will help process the visual information reaching your eyes as you scan the page. Others will activate centers of learning and memory to understand the meanings of the words you've read. Still other signals will help filter out extraneous inputs—such as background noise—that might distract you from your task. The complexity of the seemingly simple task of reading a single sentence emphasizes the enormous level of activity that goes on continually in the brain, even at rest.

The beauty of the brain lies in its incredible complexity. The human brain, for example, has several thousand miles of interconnected neurons and hundreds of trillions of synapses, resulting in a total surface area that if spread out, would cover more than four soccer fields. The brain allows us to move, think, and experience sensation and emotion. Groups of neurons also coordinate homeostatic functions such as breathing, blood circulation, and body temperature. When we examine the way that groups of neurons communicate, we begin to understand the complex mental functions of nervous systems, including learning, memory, and motivation.

Neuroscience—the study of nervous systems—is an area of intense research activity worldwide. The challenge of neuroscience is to transform the astounding complexity of nervous systems into manageable proportions. For this reason, most neuroscience researchers do not study the human nervous system, because of its great complexity and the inherent difficulties in doing research on humans. Many animals, however—including the nematode *Caenorhabditis elegans,* the California sea slug (*Aplysia californica*), the fruit fly *Drosophila melanogaster* (see chapter-opening photo), the zebrafish (*Danio rerio*), and the house mouse (*Mus musculus*)—provide excellent opportunities to study how neurons work and how groups of neurons cooperate to produce animal behavior. The genomes of all of these animals have been sequenced, and neuroscientists are identifying genes that are critical for the structure and function of nervous systems. Many of the relevant genes are homologous to human genes, so the study of the molecular control of the nervous system in these model organisms has the potential to reveal new treatments for many genetically inherited neurological diseases in humans.

In this chapter, we will first survey a variety of nervous systems, which allow animals to sense and respond to environmental changes. We will then examine the nervous system of humans. However, keep in mind that we still have much to learn about the organization, connectivity, and functions of nervous system structures. Our own nervous system is fascinating and mysterious, and the study of how it functions will ultimately tell us much about what makes us human.

42.1 The Evolution and Development of Nervous Systems

Learning Outcomes:

1. List the different types of nervous systems found in animals.
2. Describe the general anatomical organization of the brain in vertebrates.
3. Discuss the changes in brain complexity that accompanied the evolution of mammals.

Animal nervous systems are the products of hundreds of millions of years of evolution. They provide advantages to animals that promote reproductive success. For example, nervous systems allow animals to sense their environment and respond to changes in an appropriate way. In addition, nervous systems form connections with muscles and facilitate movement, which has allowed animals to travel across distances to obtain food. Likewise, nervous systems help animals to avoid predation and other environmental dangers; defend themselves; form social bonds that enhance the chances of survival for both the individual and the group; and even perform the complex tasks of thinking, learning, remembering, and planning.

Studying the evolution and development of nervous systems helps us understand how particular nervous systems are adapted to different functions. At the structural level, the organization of nervous systems ranges from a relatively simple network of a few cells to the complexity of the human brain. The characteristics of an animal's nervous system determine the behaviors that it displays. In this section, we will survey the nervous systems of invertebrates and vertebrates, and also examine the brains of vertebrates in greater detail.

Nervous Systems Evolved to Sense and Respond to Changes in the Environment

Precisely when nervous systems first arose and whether or not the nervous systems of most or all animals can be traced back to a common ancestor are questions of active investigation by neuroscientists. For example, recent genetic studies have uncovered remarkable similarities in the expression and activation of genes coding for proteins that regulate neuronal development across taxa in bilaterally symmetric animals. Those studies suggest that the patterning of nervous system development in bilaterians may be traced to a common ancestor that lived more than 500 mya!

Today, all animals except sponges have a nervous system. Interestingly, though, researchers have recently discovered that sponges express dozens of genes that are similar to genes expressed in human neurons, particularly those that code for proteins that regulate synaptic function. The functions of the sponge genes are uncertain, but the proteins coded for by the genes interact in ways that are reminiscent of human synaptic proteins. Thus, the origin of nervous systems almost certainly can be traced to genes of evolutionarily ancient organisms; as animals evolved, these genes were modified and formed the basis of all future nervous systems.

The simplest nervous system is the **nerve net** of the radially symmetric cnidarians (jellyfish, hydras, and anemones) (**Figure 42.1a**). The neurons are arranged in a network of connections between the inner and outer body layers of the animals. A characteristic feature of nerve nets is that activation of neurons in any one region leads to activation of most or all other neurons, with the excitation spreading in all directions at once. Many of these neurons stimulate contractile cells to contract. This allows the organism to move large areas of its body simultaneously, thereby coordinating simple movements such as swimming. Recent research has identified regions of specialized

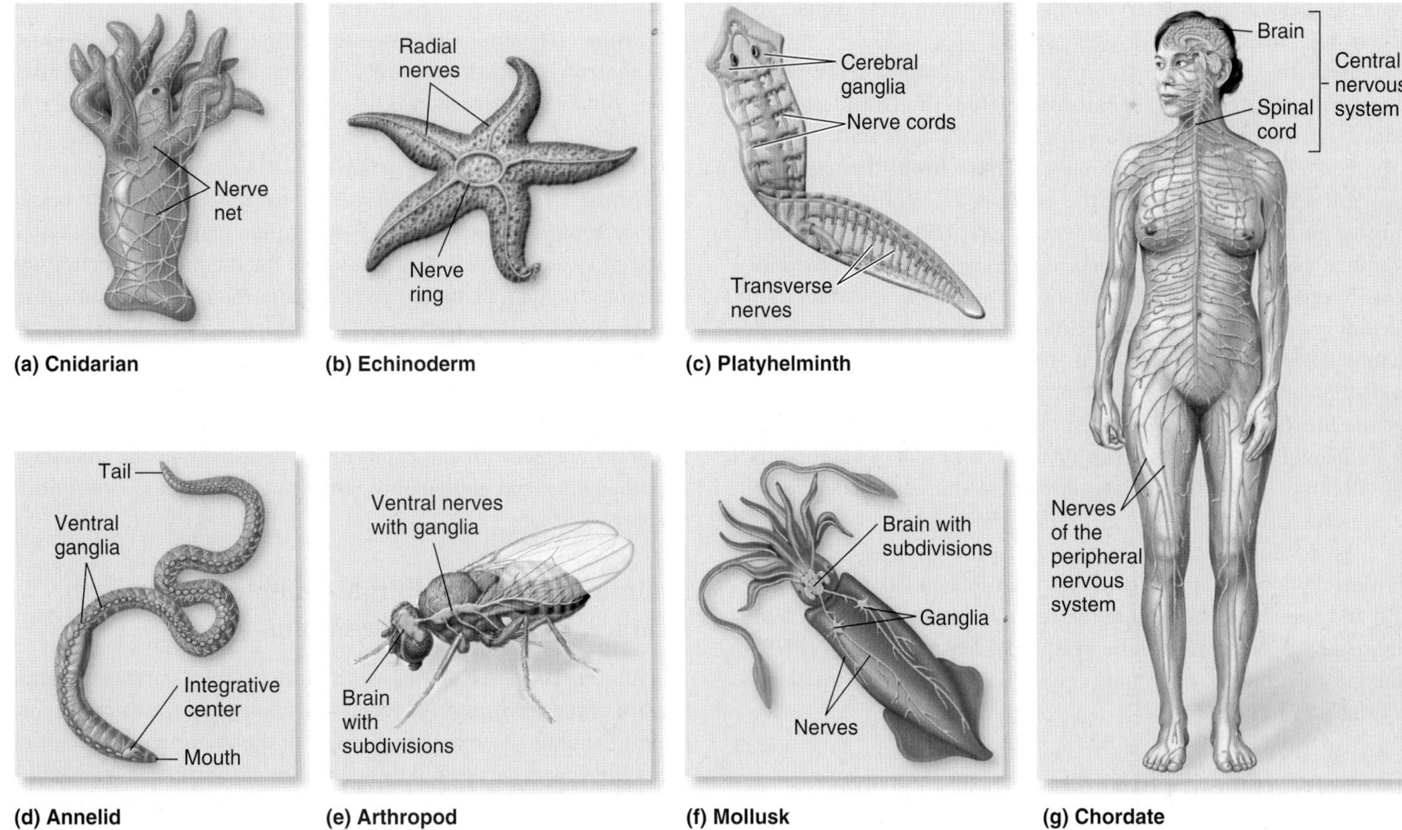

Figure 42.1 **Representative nervous systems throughout the animal kingdom.**

BioConnections: Nervous systems are one of the defining features of animals. In Chapter 32, you learned that several other features define animals and distinguish them from other living organisms. What were some of these features?

function in the nerve nets of some cnidarians, such as local sensory neurons in the outer body wall, thereby tracing the origin of specialized nervous system structures and function further back in evolutionary time than previously thought. Some cnidarians, such as the jellyfish, have two nerve nets: one for moving tentacles and one for swimming.

Sea stars and other echinoderms also have a simple, but slightly more complex nervous system than cnidarians. A nerve ring surrounds the mouth and is connected to larger radial nerves extending into the arms (**Figure 42.1b**). This arrangement allows the mouth and arms to operate independently.

In Platyhelminthes such as *Planaria*, collections of neurons in the animal's head form **cerebral ganglia** (singular, ganglion). These are groups of neuronal cell bodies that perform the basic functions of integrating inputs from sense organs, such as the eyes, and controlling motor outputs such as those involved in swimming (**Figure 42.1c**). Two lateral nerve cords extend along the ventral surface of the animal from the anterior end to the posterior end and are connected to each other by transverse nerves. In annelids (segmented worms), the basic structure is similar, except that more neurons are present and the single ventral nerve cord has ganglia located in each body segment (**Figure 42.1d**). In the head, cerebral ganglia form an integrative center, which functions as a rudimentary brain to control body movements. The other ventral ganglia along the nerve cord receive sensory information from a particular body segment and control local movements. In the simpler types of mollusks, such as the snail, the nervous system is very similar to that of the annelids. The head contains a pair of anterior ganglia; paired nerve cords extend from these ganglia to the eyes, muscular foot, and digestive system.

During the evolution of animals, more complex body types have been associated with **cephalization** (from the Greek *cephalo*, meaning head), the concentration of sense organs at the anterior end of the body, forming an increasingly complex **brain** that controls sensory and motor functions of the entire body. Within the brain, neuronal pathways provide the integrative functions necessary for an animal to make more sophisticated responses to its environment. Brains are found in all vertebrates and most invertebrates, and they are usually composed of more than one anatomical and functional region. For instance, in *Drosophila* (**Figure 42.1e** and the chapter-opening photo), the brain has several subdivisions with separate functions, such as a region devoted to learning and memory. Some mollusks, such as the squid and octopus, have brains with well-developed subdivisions that allow these animals to coordinate the complex visual and motor behaviors necessary for their predatory lifestyle (**Figure 42.1f**). In chordates, the brain is connected to a dorsally located **spinal cord**. As discussed in Chapter 41, the brain and spinal cord constitute the **central nervous system (CNS)** (**Figure 42.1g**). Nerves from the **peripheral nervous system (PNS)** route information into and out of the CNS at separate regions along the spinal cord.

Brains of Vertebrates Have Three Basic Divisions

Development of the vertebrate brain begins with the formation of a central fold in the embryo called the **neural tube**. This hollow tube is the structure from which the entire nervous system develops (you can look ahead to Figure 52.11 for additional details). Increased cell proliferation leads to bending and folding of the neural tube during embryonic development, resulting in bulges that become separate divisions of the nervous system. The anterior end develops into the brain, while the posterior portion becomes the spinal cord.

In vertebrates, the brain has three major anatomic divisions, the **hindbrain**, **midbrain**, and **forebrain** (**Figure 42.2**). Fossils of jawless fishes that lived 400 mya show that their brains were already organized into the three basic divisions that have been retained in all modern vertebrates.

Let's look at the development of the human brain. At 4 weeks, the human embryo exhibits just the hindbrain, midbrain, and forebrain (Figure 42.2a). Just a week later, the hindbrain and forebrain have each formed two separate subdivisions (Figure 42.2b). The hindbrain subdivides into the metencephalon and the myelencephalon. The forebrain subdivides into the telencephalon and the diencephalon. The midbrain, by contrast, does not subdivide and is termed the mesencephalon. By the time the human brain is fully developed, some of these structures have further divided and specialized (Figure 42.2c). The development of brain subdivisions increases the capacity of the brain to perform complex, distinct functions.

Hindbrain The hindbrain includes the medulla oblongata, pons, and cerebellum. It coordinates many basic reflexes and bodily functions, such as breathing, that maintain the normal homeostatic processes of the animal. It is also partly responsible for monitoring and coordinating body movements.

Midbrain The midbrain processes several types of sensory inputs, and controls sophisticated tasks such as coordinating eye movement with visual inspection of the environment; it also plays a role in alertness.

Forebrain The forebrain initiates motor functions and processes sensory inputs. In humans and other mammals, it consists of a group of structures that are responsible for the higher functions of conscious thought, planning, and emotion. Many of these functions are attributed to the **cerebrum**. The surface layer of the cerebrum, which is only a few millimeters thick, is called the **cerebral cortex**. As we will learn later, this thin area is critical to thought, learning, and movement, among other functions. Other structures of the forebrain are beneath the cerebrum. These include the thalamus, hypothalamus, and epithalamus, which we will discuss later in this chapter.

Evolution of Increased Brain Complexity Involved a Larger, Highly Folded Cerebrum

As evolution produced animals with more complex nervous systems, the size of the cerebrum also increased, making up a greater proportion of the brain. As mentioned, many of the important functions of the cerebrum are carried out by neurons along its outer surface in the cerebral cortex. Therefore, increased complexity of the brain is also correlated with an increased surface area of the cerebral cortex. During the evolution of mammals, this increase in surface area occurred more rapidly than an expansion in the size of the skull. How could

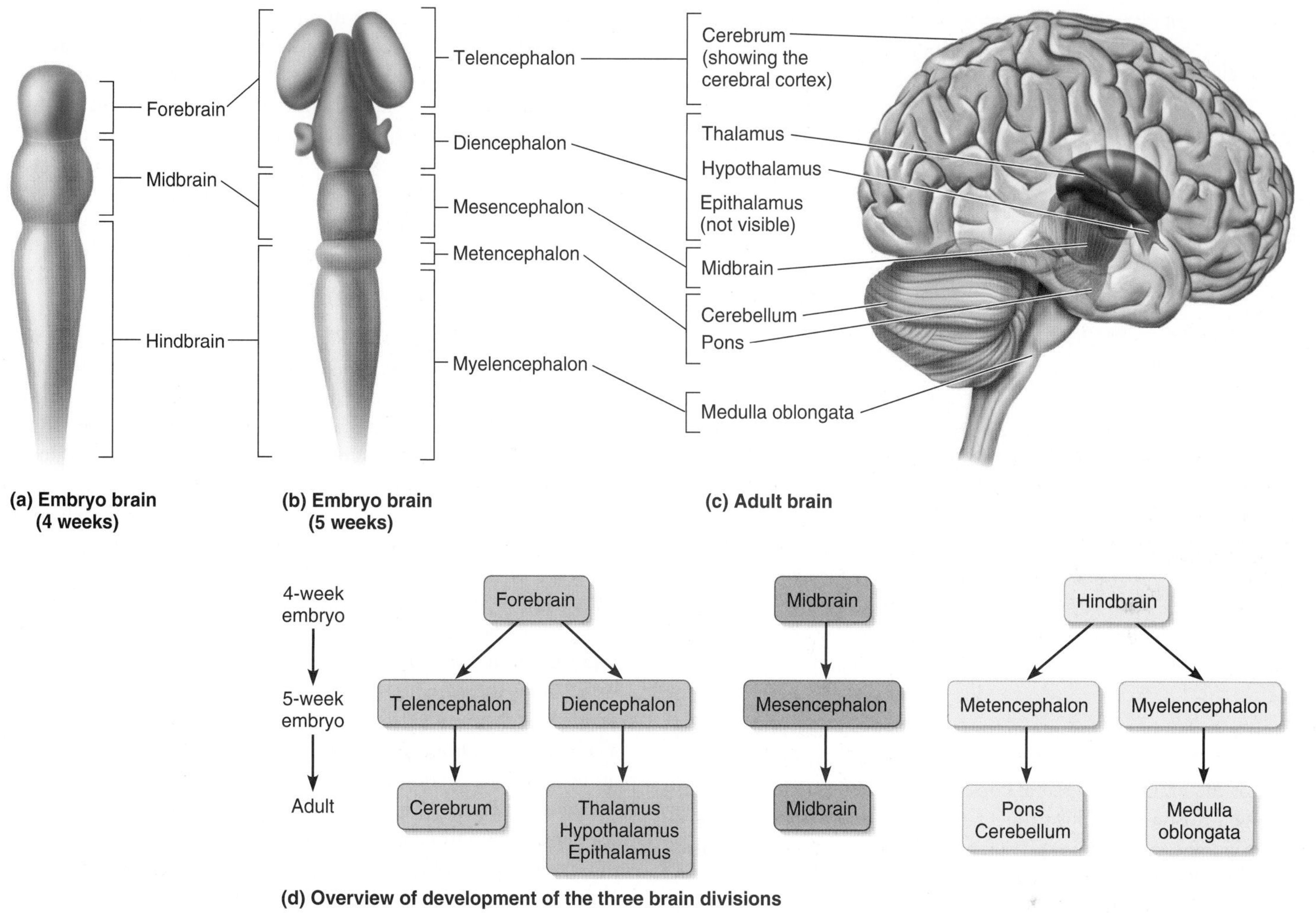

Forebrain
Midbrain
Hindbrain

(a) Embryo brain (4 weeks)

Telencephalon
Diencephalon
Mesencephalon
Metencephalon
Myelencephalon

(b) Embryo brain (5 weeks)

Cerebrum (showing the cerebral cortex)
Thalamus
Hypothalamus
Epithalamus (not visible)
Midbrain
Cerebellum
Pons
Medulla oblongata

(c) Adult brain

4-week embryo: Forebrain → 5-week embryo: Telencephalon, Diencephalon → Adult: Cerebrum; Thalamus, Hypothalamus, Epithalamus

4-week embryo: Midbrain → 5-week embryo: Mesencephalon → Adult: Midbrain

4-week embryo: Hindbrain → 5-week embryo: Metencephalon, Myelencephalon → Adult: Pons, Cerebellum; Medulla oblongata

(d) Overview of development of the three brain divisions

Figure 42.2 Development of the human brain. The structures shown here that occur during embryonic development at **(a)** 4 weeks and **(b)** 5 weeks are compared with **(c)** how they appear in the adult. Most structures beneath the cerebrum are not shown. **(d)** This flowchart gives an overview of the development of the three brain divisions.

this occur? The answer is that the external surface of the cerebrum in animals with increasingly complex brains is highly convoluted, forming many folds called gyri (singular, gyrus), separated by grooves called sulci (singular, sulcus). Compare the relatively smooth-looking surface of the cerebrum of a rat with the highly folded one of a dolphin or human in Figure 42.3.

As body size increases across the animal kingdom, you might expect that brain mass would increase proportionately—that the brain of an elephant would be proportionately larger than that of a bat, for instance. That is generally the case, with a few important exceptions (**Figure 42.4**). In particular, the masses of the human and dolphin brains are greater than would be expected on the basis of body mass.

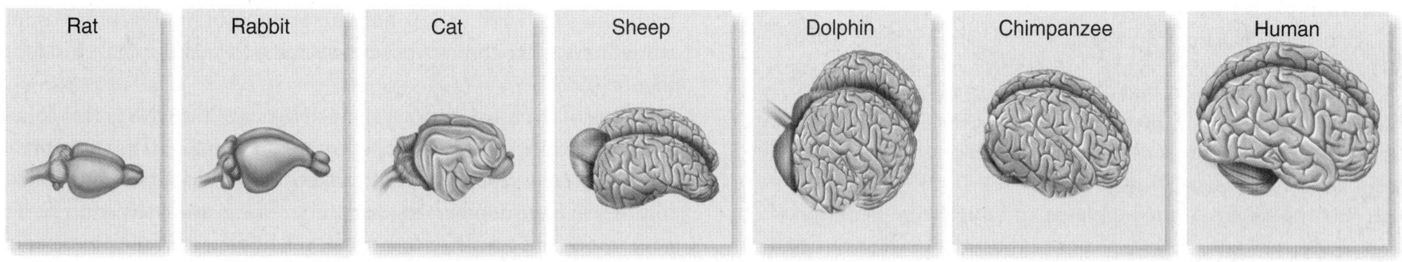

Figure 42.3 The degree of cerebral cortex folding in different mammalian species. The brains are not shown to scale.

BIOLOGY PRINCIPLE Structure determines function. The increased folding of the cerebrums of certain mammals increases the surface area of this part of the brain, allowing for more neuronal connections and complex behaviors including conscious thought.

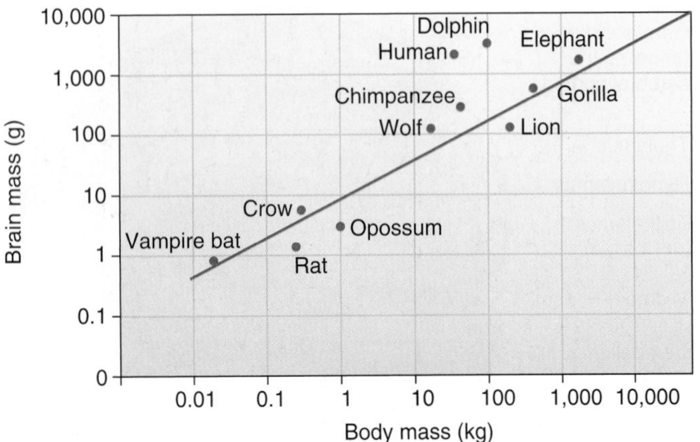

Figure 42.4 **Brain mass as a function of body mass in mammals.** For most mammals, brain weight is in proportion to body mass. However, humans and dolphins have a much higher brain mass relative to their body mass than do other mammals.

Concept Check: *Scientists have determined that the brain mass of Homo neanderthalensis (Neanderthal man) was greater than that of our own. Does this mean that Neanderthals were more intelligent and capable of more complex behaviors than we are?*

Brain mass and the amount of folding are correlated with more complex behaviors. Why is this so? As we've indicated, the outer surface of the brain, the cerebral cortex, plays a key role in conscious thought, reasoning, and learning. Greater size and folding provide a larger number of neurons and synapses and more surface area, which allows greater processing and interpretation of information. Even so, evidence does not suggest that people with small differences in brain size differ in intelligence. Also, it would be wrong to assume that an animal with a small brain is profoundly limited in its behavioral repertoire. A bat with a 0.9-g brain and an elephant with a 2,500-g brain can both perform a great variety of interesting and complex behaviors, such as navigating across great distances and interacting with fellow members of their species.

42.2 Structure and Function of the Nervous Systems of Humans and Other Vertebrates

Learning Outcomes:

1. Outline the anatomical organization of the human nervous system.
2. Describe the organization of the peripheral nervous system.
3. Distinguish between the somatic and autonomic nervous systems.
4. Briefly describe the major structures and functions of the human hindbrain, midbrain, and forebrain.

The nervous system of the human is amazingly complex—the brain alone has over 100 billion neurons and as many as a few hundred billion glia cells. Moreover, complexity is defined by more than just numbers of cells. Within the human brain, for example, are enormous numbers of connections between neurons—a single neuron in the cerebellum may have as many as 100,000 synapses with other cells! In this section, we will examine the human nervous system, with an emphasis on the functions of the major parts of the brain and spinal cord.

The Nervous System Is Composed of the Central and Peripheral Nervous Systems

In vertebrates, the CNS and PNS are connected anatomically and functionally (Figure 42.5). The CNS receives information about the internal or external environment from the PNS. The CNS interprets that information and may initiate a response that is then carried out by the PNS. For example, suppose you accidentally lean against a newly painted fence. Neuronal endings in your skin, which are part of the PNS, would transmit tactile (touch) information through axons that bring information directly into the spinal cord. From there, the information travels to your brain, where the sensation is analyzed and identified as something sticky. Signals are sent from your brain, down your spinal cord, and through the neurons of the PNS to your muscles, causing you to move away.

Within the nervous system, groups of neurons may associate with each other and perform a particular function. In the CNS, the cell bodies of neurons that are involved in a similar function may be grouped into a structure called a **nucleus** (plural, nuclei), which may include thousands of cells. For instance, cell bodies that regulate body temperature and those that recognize visual information are located in separate nuclei in the brain. In the context of the vertebrate nervous system, the term **ganglion** is used to refer to a group of neuronal cell bodies located in the PNS (see Figure 42.5).

Within the vertebrate nervous system, many myelinated axons may run in parallel bundles. (Myelination is described in Chapter 41; see Figure 41.2.) Such a structure is called a **tract** when it is found in the CNS. Tracts convey information from region to region within the brain and between the brain and the spinal cord. Bundles of myelinated axons also form outside the CNS, in the PNS, in which case they are called **nerves**. The cell bodies that give rise to the axons of nerves may be within the PNS or the CNS; in other words, a given nerve may carry information from outside the CNS into the CNS, from the CNS to structures outside the CNS, and commonly both. Connections between the PNS and the CNS occur at the brain or spinal cord. **Cranial nerves** are directly connected to the brain, primarily to sites within the hindbrain and midbrain. By comparison, **spinal nerves** are connections between the PNS and spinal cord (see Figure 42.5).

One of the most obvious characteristics of the CNS is that some parts look white, and others appear gray (Figure 42.6). The **white matter** gets its color from myelin; it consists of myelinated axons bundled together in large numbers to form tracts. The **gray matter** is darker in appearance and consists of neuronal cell bodies, dendrites, and some unmyelinated axons. The cerebral cortex is composed of gray matter

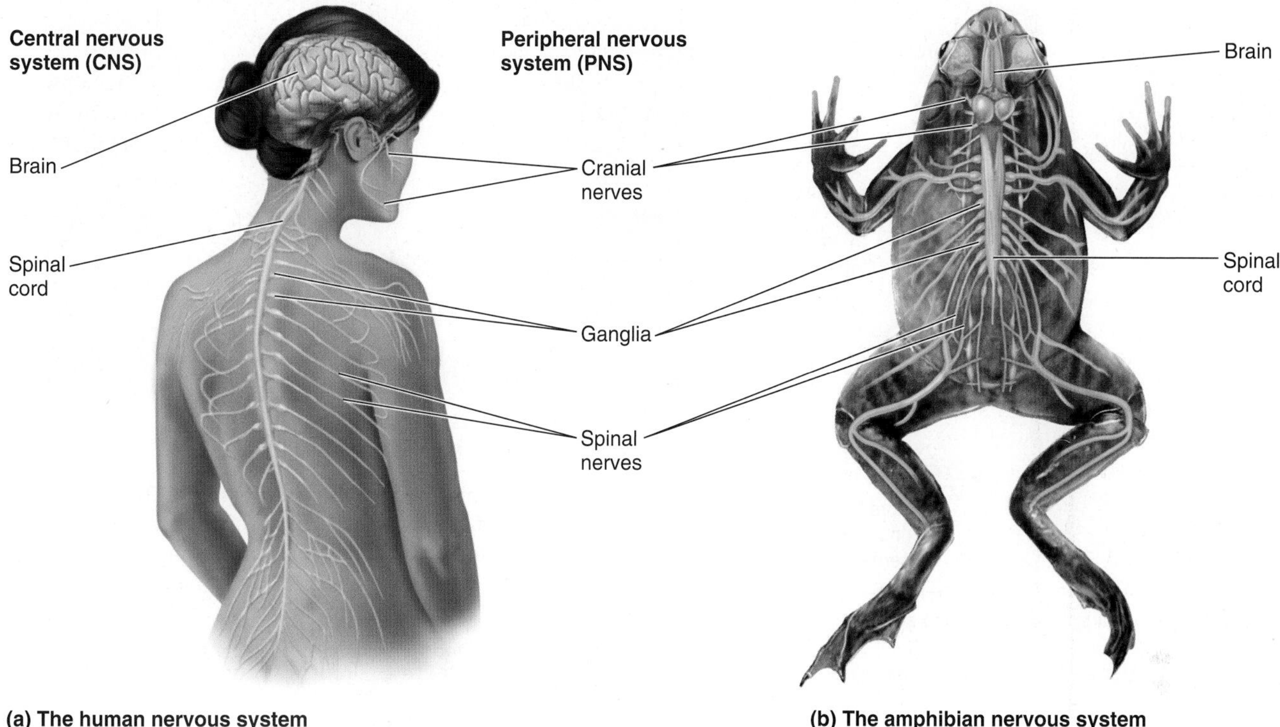

Figure 42.5 **Organization of the vertebrate nervous system.** The CNS consists of the brain and spinal cord, both of which are encased in bone (not shown). The PNS includes cranial nerves, ganglia, and spinal nerves, which carry information to and from the CNS, and many other neurons throughout the body. Note the similarities between two widely divergent vertebrates, **(a)** humans and **(b)** amphibians (a frog).

BIOLOGY PRINCIPLE **All species (past and present) are related by an evolutionary history.** The similarities between the nervous systems of these two very different-looking animals point to a common ancestor.

that sits on top of a large collection of white matter pathways. In the spinal cord, the gray matter is located in the center and forms two dorsal horns and two ventral horns (Figure 42.6a). Each dorsal horn connects to a dorsal root, which is part of a spinal nerve. Dorsal roots receive incoming information from sensory (afferent) nerves of the PNS. The ventral horn connects to the ventral root, which is also part of a spinal nerve that transmits outgoing information to motor (efferent) nerves. A central canal runs through the spinal cord, carrying a nutritive and protective fluid, as described shortly.

Unlike the PNS, the CNS is encased in protective structures including bone (the skull and backbone) and three layers of sheath-like membranes called **meninges** (Figure 42.7). The outermost membrane, the dura mater (from the Latin, meaning hard mother), is a thick protective layer that lies just inside the skull and vertebrae. The middle membrane is called the arachnoid mater (from the Latin, meaning spidery mother) because it has numerous weblike tissue connections to the innermost membrane, the pia mater (from the Latin, meaning thin mother). The pia mater is a very thin membrane that lies on the surface of the brain and spinal cord, folding with the brain's surface.

Between the arachnoid mater and pia mater is the subarachnoid space. This space is filled with **cerebrospinal fluid**, which surrounds the exterior of the brain and spinal cord and absorbs physical shocks to the brain that result from sudden movements or blows to the head. The cerebrospinal fluid contains nutrients, hormones, and other substances that are taken up by cells of the brain. The fluid is also a reservoir for metabolic waste products that are then carried away by the circulatory system. In addition to the subarachnoid space, the cerebrospinal fluid also fills a series of connected cavities called the ventricles that lie deep within the brain and connect to the central canal that extends the length of the spinal cord (see Figure 42.6). These fluid-filled structures provide a cushion of support and protection for the CNS.

The PNS Carries Information to and from the CNS

The PNS of vertebrates is subdivided into two major functional and anatomical components: the somatic nervous system and the autonomic nervous system. Both divisions have sensory (afferent) nerves and motor (efferent) nerves.

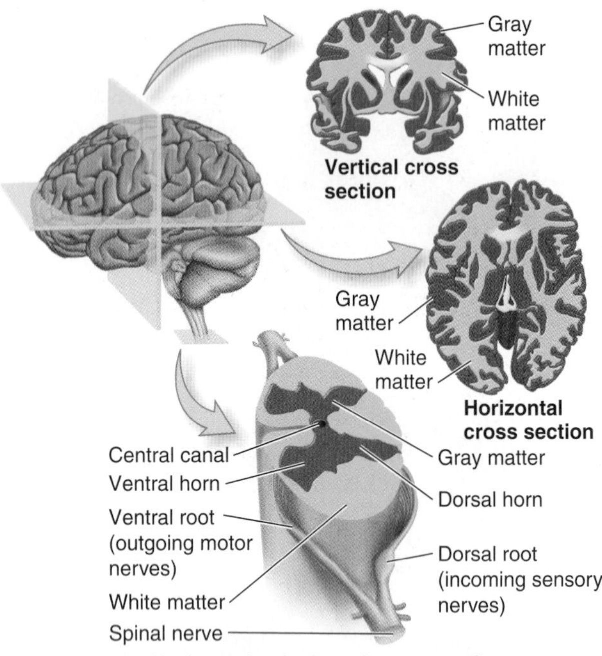

(a) Gray and white matter in the brain and spinal cord

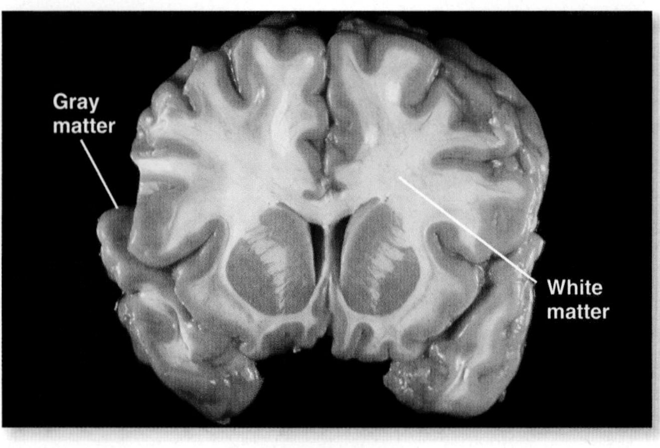

(b) Light microscope image of a vertical cross section of the human brain

Figure 42.6 **Gray matter and white matter in the CNS.** **(a)** The gray matter is composed of groups of cell bodies, dendrites, and unmyelinated axons. The white matter consists of tracks of myelinated axons. **(b)** Photograph of a vertical cross section through an adult human brain. In these images, gray matter is darkened for better visibility.

Concept Check: *Is a spinal nerve composed of axons from afferent or efferent neurons, or both?*

Somatic Nervous System The primary function of the **somatic nervous system** is to sense the external environment and control skeletal muscles. The sensory neurons of the somatic nervous system receive stimuli, such as heat, light, odors, chemicals (in food), sounds, and touch, and transmit signals to the CNS. The motor neurons of the somatic nervous system control skeletal muscles. The cell bodies of these motor neurons are actually located within the CNS. The axons from these cells leave the spinal cord and project directly onto skeletal muscle without any intermediary synapses along the way.

The somatic nervous system is said to be voluntary because many of the responses can be controlled consciously. For example, we use our somatic nervous system to walk and hold a pencil. However, not all responses are voluntary. An example is a reflex arc, such as the knee-jerk response, which is automatic (refer back to Figure 41.4).

Autonomic Nervous System The **autonomic nervous system** regulates homeostasis and organ function. For example, it is involved in regulating heart rate, blood pressure, glucose homeostasis, and the amount of stomach acid secreted in response to a meal. Though the autonomic nervous system is predominantly composed of motor neurons, it also has sensory neurons that detect internal body conditions. For example, baroreceptors are sensory neurons located in the heart and several blood vessels that detect changes in blood pressure. For the most part, the autonomic nervous system is not subject to voluntary control. We usually cannot consciously change our heart rate or blood pressure.

The efferent pathways of the autonomic nervous system involve two motor neurons. The cell body of the first neuron is within the CNS and synapses on a second neuron in ganglia outside the spinal cord; these ganglia, therefore, are part of the PNS. This second neuron sends its axon to an effector cell, where it alters that cell's function. These neurons control smooth muscles, cardiac muscle, and glands.

The efferent nerves of the autonomic system are subdivided into the sympathetic and parasympathetic divisions (Figure 42.8). Both

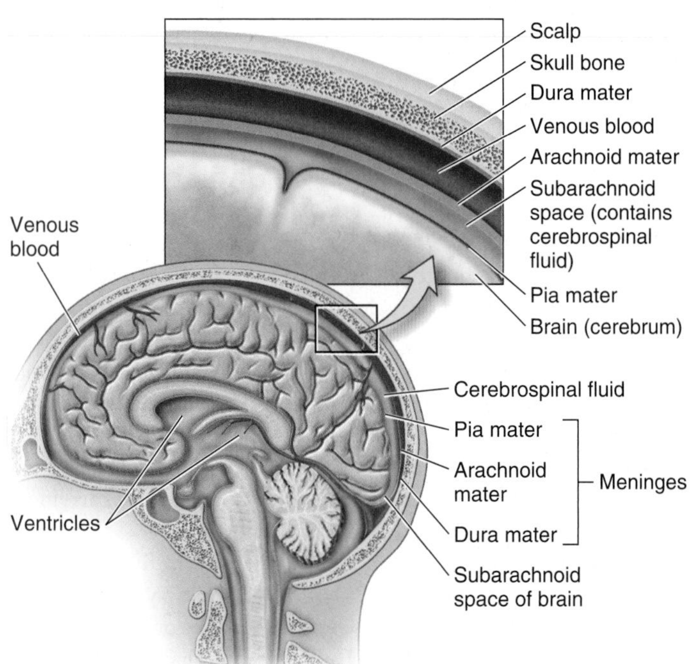

Figure 42.7 **The meninges and ventricles of the CNS.** The thickness of the meninges are exaggerated for illustration purposes. Note that the cerebrospinal fluid encases the entire CNS and also fills the ventricles.

Concept Check: *In a procedure known as a lumbar puncture (commonly referred to as a spinal tap), physicians use a needle to withdraw a small amount of cerebrospinal fluid from the bottom of the spine to help diagnose specific illnesses. What effects might this procedure have on a patient?*

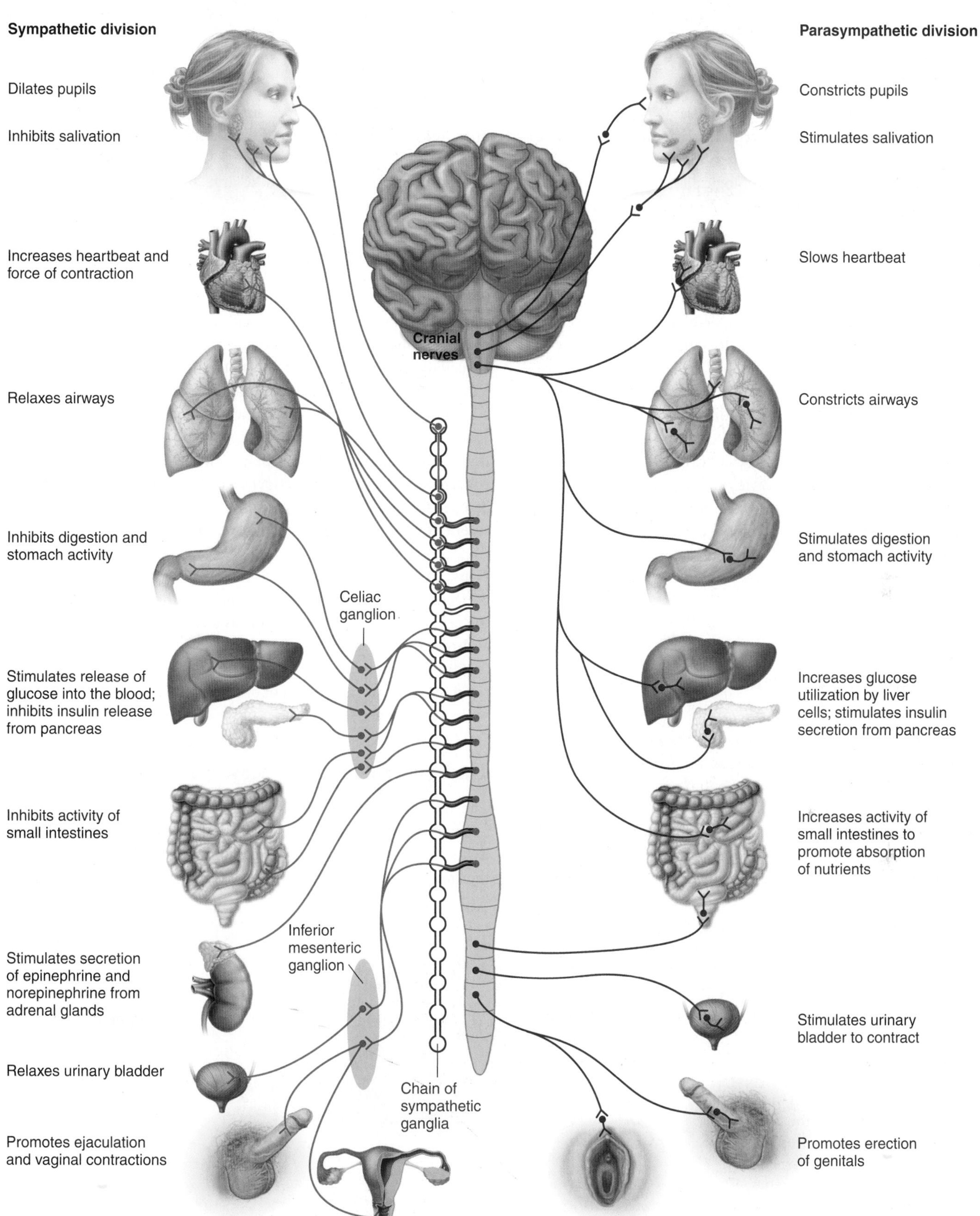

Figure 42.8 **Sympathetic and parasympathetic divisions of the autonomic nervous system.** For simplicity, only some of the major functions of each division are shown in this figure. The sympathetic and parasympathetic systems tend to have opposite effects, and most parts of the body receive inputs from both divisions. Nerves from the sympathetic division make connections with a chain of ganglia, most, but not all, of which are alongside the spinal cord. Nerves from the parasympathetic division make connections in ganglia near or in their targets (for clarity, the ganglia are only shown near the targets).

BIOLOGY PRINCIPLE **Living organisms maintain homeostasis.** The autonomic nervous system is an excellent illustration of one way in which homeostasis is maintained; many organs and other structures in the body are controlled in opposite ways by the two divisions of this branch of the nervous system. Therefore, the function of these structures can be modulated in two directions; for example, heart rate can be accelerated or slowed to match an animal's immediate metabolic requirements.

divisions of the autonomic system act on the same organs and usually have opposing actions. The **sympathetic division** is responsible for rapidly activating systems that prepare the body for danger or stress. Imagine, for example, the physiological responses that would occur if a person was hiking and came upon a grizzly bear. This is the **fight-or-flight response**, which is characterized by increased heart rate, stronger pumping action of the heart, relaxed (opened) airways and faster breathing, inhibition of digestive activity, increased blood flow to skeletal muscles, and increased secretion of energy-supplying substances such as glucose and fats into the blood by the liver and adipose tissue. These features prepare us to confront (fight) or avoid (flight) a perceived or real threat.

The **parasympathetic division** of the autonomic nervous system is involved in maintaining and restoring body functions. It is particularly active during restful periods or after a meal, which is why it is sometimes said to mediate the **rest-or-digest response**. Neurons of the parasympathetic division promote digestion and absorption of food from the gut, slow the heart rate, and decrease the amount of fuel supplied to the blood from the liver and adipose tissue. A summary of these and other major functions of the two divisions of the autonomic nervous system can be found in Figure 42.8.

The Hindbrain Is Important for Homeostasis and Coordination of Bodily Functions

Let's now turn our attention to the structure and function of the human brain (Figure 42.9). We will begin with the evolutionarily oldest structures of the brain, some of which are located in the hindbrain and control the basic processes that sustain life.

Medulla Oblongata The **medulla oblongata** is located between the pons and the anterior part of the spinal cord. It coordinates many processes that maintain homeostasis. It is involved in the control of heart rate, breathing, blood pressure, digestion, swallowing, and vomiting, and gives rise to several of the cranial nerves.

Cerebellum The **cerebellum**, a large structure that sits dorsal to the medulla oblongata, receives sensory inputs from the cerebral cortex and the auditory and visual areas of the brain. It also receives inputs from the spinal cord and inner ears that convey information about the position of the head and limbs and thereby helps maintain balance and coordinate hand-eye movements. In addition, the cerebellum controls the use of multiple muscles at one time and synchronizes fine motor activities such as texting, making a jump shot in basketball, or touching the fingers to the tip of the nose with your eyes closed. When the cerebellum is damaged or injured, such as in an accident, a person may find it difficult to maintain balance and fine-tune motor functions. Although historically scientists have thought that the cerebellum does not play a role in learning, memory, and conscious thought, recent evidence has strongly suggested that the cerebellum may indeed have significant cognitive functions, the full extent of which remains to be discovered.

Pons The **pons** sits anterior to the medulla oblongata and beneath the cerebellum. Major tracts pass through the pons into and out of the cerebellum, so the pons serves as a relay between the cerebellum and other areas of the brain. In addition to this integrative motor function, the pons contains nuclei that play a very important role in regulating the rate and depth of breathing. The pons also gives rise to some of the cranial nerves.

The Midbrain Processes Sensory Inputs

The midbrain lies anterior to the pons. It processes several types of sensory inputs, including vision, olfaction, and audition. It has tracts that pass this information to other parts of the brain for further processing and interpretation. As one example, the midbrain is

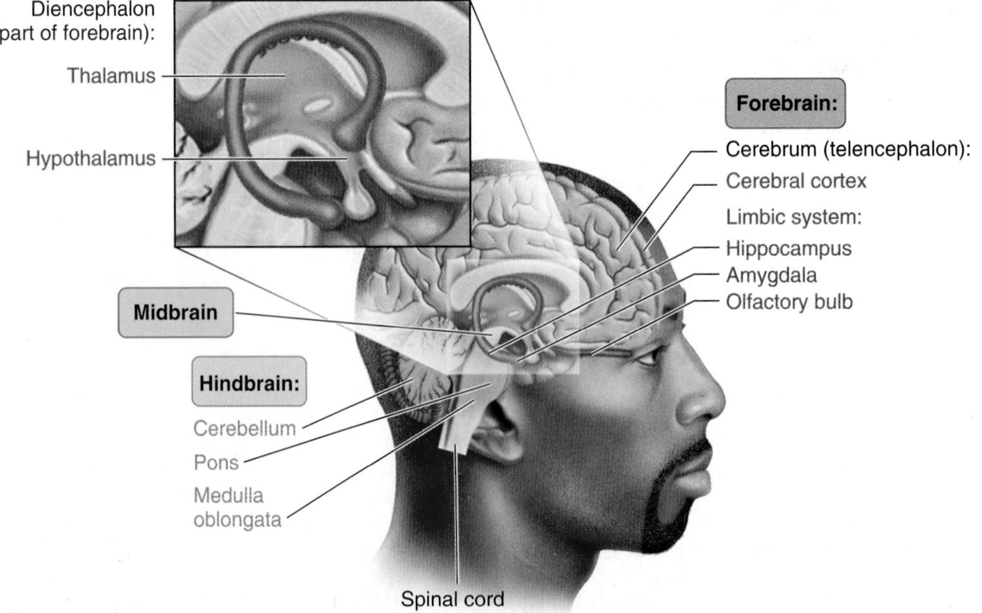

Figure 42.9 Major structures of the human brain. An overview of the brain, showing several internal structures (not all structures are visible in this plane, such as the basal nuclei). The limbic system consists of the olfactory bulbs, amygdala, and hippocampus, which are part of the cerebrum (many neuroscientists consider parts of the thalamus and hypothalamus as part of the limbic system). The midbrain, pons, and medulla oblongata collectively comprise the brainstem.

Concept Check: How would an injury to the cerebellum affect a person's behavior?

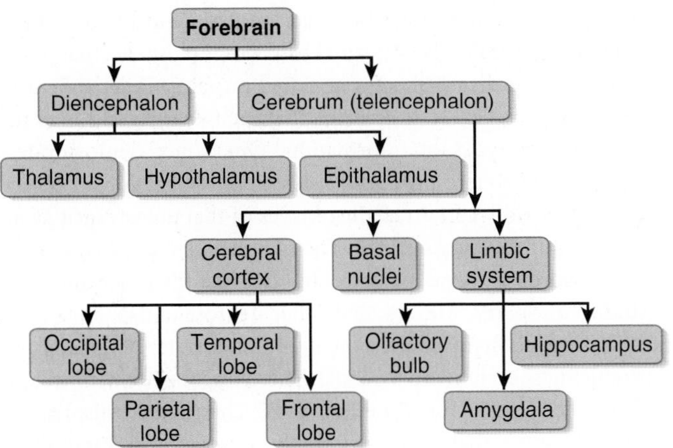

Figure 42.10 Structures that make up the human forebrain.

responsible for activating neural pathways that change the diameter of the pupil of the eye in response to a change in the amount of ambient light. If the midbrain were damaged, this pupillary reflex would not occur normally or at all.

The medulla oblongata, the pons, and the midbrain collectively constitute the **brainstem**. In addition to the functions just described, all three major parts of the brainstem contain additional nuclei that together form the **reticular formation**. This is a network of nuclei and tracts that extends throughout much of the brainstem and sends signals to many other brain regions. The reticular formation maintains and controls consciousness, alertness, and sleep, plus essential functions such as regulation of the respiratory and cardiovascular systems. Because of the importance of the brainstem's functions, damage to it is catastrophic and may result in coma or death.

The Forebrain Is Responsible for Movement, Sensory Function, and Higher Functions of Thought, Learning, and Emotion

The forebrain comprises the diencephalon and cerebrum (Figure 42.10). The diencephalon is made up of the thalamus, hypothalamus, and epithalamus. The cerebrum consists of the basal nuclei, limbic system, and cerebral cortex.

Diencephalon In vertebrates, the **thalamus** plays a major role in relaying sensory information to appropriate parts of the cerebrum and, in turn, sending outputs from the cerebrum to other parts of the brain. It receives input from all sensory systems except olfaction. One type of processing performed by neurons in the thalamus is that of filtering out sensory information in a way that allows us to pay attention to important cues while temporarily ignoring less important ones. This filtering mechanism begins in the reticular formation, which relays information about sensory inputs to the thalamus. Together, these brain regions permit selective attention to certain stimuli. A good example of this occurs when you focus on what someone is saying to you in a crowded room, despite the presence of background sounds, sights, and activities that could otherwise be distracting. Another example is a new parent's ability to sleep through a thunderstorm but awaken immediately to the cry of a baby in the next room. The thalamus also directs feedback about motor activities that it receives from the cerebellum and other structures to the cerebral cortex, which can then adjust its outgoing motor signals if necessary. Last, the thalamus is involved in the perception of pain and the degree of mental arousal in the cerebral cortex.

The **hypothalamus**, located below the thalamus at the floor of the forebrain, controls functions of the digestive and reproductive systems, body temperature (thermoregulation), and many basic behaviors such as eating and drinking. This area has great importance for homeostasis of the body and the control of behavior. Though small in size, it is composed of many nuclei, each with its own vital functions. A major role of the hypothalamus is the production of hormones, which travel to the pituitary gland located just beneath the brain. The pituitary gland, in turn, regulates hormone secretion from other glands in the body, including the thyroid, gonads, and adrenal glands. In addition to producing hormones, the hypothalamus is sensitive to the actions of yet other hormones. For example, certain hormones produced by cells in the stomach, intestine, adipose tissue, gonads, and elsewhere act within the hypothalamus to facilitate the expression of feeding, drinking, sexual, and aggressive behaviors. Finally, a small pair of hypothalamic nuclei called the suprachiasmatic nuclei acts as the "master clock" of the CNS, establishing circadian rhythms, which control the expression of behavioral, physiological, and hormonal rhythms over the 24-hour day.

The **epithalamus** is a collection of structures that have various roles in the control of food and water intake, the integration of olfactory and visceral inputs with emotion and memory centers of the brain, and in some vertebrates rhythmic and seasonal behaviors. One of these structures, the pineal gland, is located in the center of the brain and secretes a hormone called **melatonin** into the blood. Production of melatonin is regulated by the length of the light period in each day. Although the function of melatonin in humans is still debated, it has been suggested to play a role in daily cycles such as our sleep/wake rhythm.

Cerebrum As mentioned, the cerebrum consists of the cerebral cortex, basal nuclei, and limbic system. One of the most recognizable features of the cerebrum, however, is its division into two halves, or **hemispheres**. Each hemisphere is connected to the other by a major tract called the **corpus callosum** (Figure 42.11a). In the 1950s, American neuroscientists Roger Sperry and Ronald Meyers examined the separate functions of the hemispheres in laboratory animals by performing split-brain surgeries in which they severed the corpus callosum. The animals that underwent such surgery maintained their overall health and functioning; therefore, the surgery was considered safe for humans. This became important in 1961, when the procedure was used for the first time to treat patients with severe epilepsy, a disorder characterized by uncontrolled electrical activity that begins in one place in the brain and can spread via the corpus callosum to the other side. Cutting the connection between the hemispheres reduced the severity of epileptic seizures.

As it turned out, split-brain surgery also provided an opportunity for the researchers to make critical observations about the

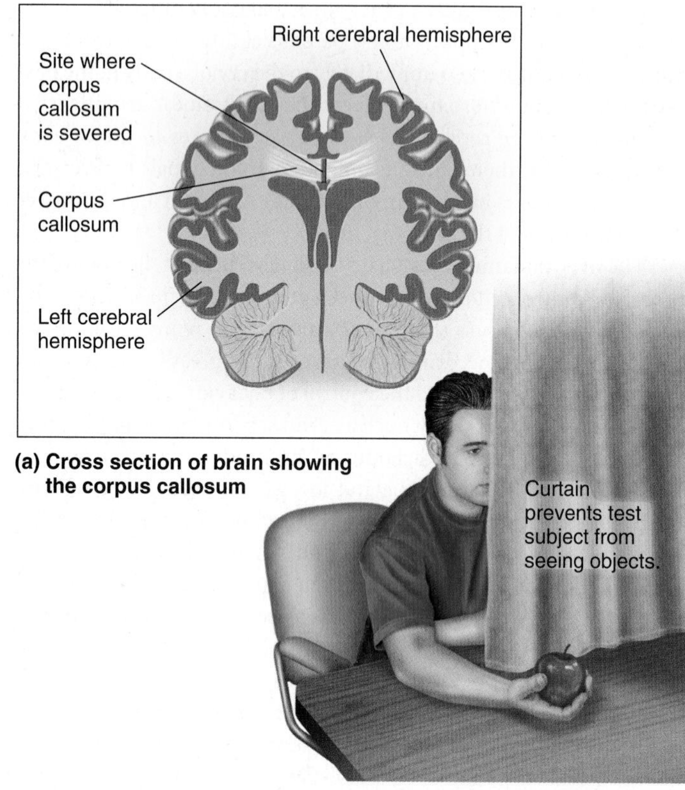

(a) Cross section of brain showing the corpus callosum

(b) Testing of split-brain patient

Figure 42.11 The hemispheres of the human brain. (a) The cerebral hemispheres and their connection by the corpus callosum. (Note: The left hemisphere controls the right side of the body, and the right hemisphere controls the left side.) **(b)** Split-brain patient being tested for hemispheric dominance. By using this apparatus, Roger Sperry and his collaborators showed that the left and right cerebral hemispheres have different capabilities. When a split-brain patient held an object in his right hand but could not see it or touch it with his left hand, he could give it a name (for example, an apple). When he held another object in his left hand, he could describe it (for example, smooth), but could not name it.

Concept Check: With her eyes closed, a split-brain patient was given a rock to hold, and she described it as a rock. Which hand was it in?

importance of communication between the two hemispheres. Split-brain humans generally show normal behavior and intellectual function, because both hemispheres can function fairly independently. However, psychological tests revealed that the two sides of the brain process different types of information. One study demonstrated that the left hemisphere produces a descriptive word for an object but does not identify certain characteristics of that object, such as its shape and texture (Figure 42.11b). The right hemisphere, in contrast, cannot use words to name the object but can identify other qualities. The studies of Sperry, Meyers, and other neuroscientists have concluded that the left hemisphere is involved in understanding language and producing speech in most people. Therefore, the left hemisphere is said to be dominant for those functions. The right hemisphere is dominant for nonverbal memories, recognizing faces, and interpreting emotions. In 1981, Sperry received the Nobel Prize in Physiology or Medicine for his insight regarding specialization in each hemisphere of the brain.

The cerebral cortex is the surface layer of gray matter that covers the cerebrum (see the darkly shaded outer rim of the brain hemispheres in Figure 42.11a). Within the cortex are identifiable regions whose neurons play roles in sensory, motor, or other functions. Although the cerebral cortex is only a few millimeters thick, it contains about 10% of all the neurons in the human brain. The cerebral cortex is divided into four lobes in each hemisphere of the brain: the frontal, parietal, occipital, and temporal lobes (Figure 42.12). Each lobe has a number of functions, many of which are still being actively investigated by researchers. Among other things, the **frontal lobe** is important for voluntary initiation of movement, decision making, controlling impulses, making plans, exhibiting judgment, short-term (working) memory, and for conscious thought and social awareness. The primary motor cortex is located at the posterior part of the frontal lobes running in a band roughly from ear to ear (see the red highlighted area in Figure 42.13). Slightly posterior to the motor cortex and at the beginning of the **parietal lobe** is the region known as the somatosensory cortex (blue highlighted region in Figure 42.13). The somatosensory cortex and parietal lobe receive and interpret sensory input from somatic pathways, including touch from the surface of the body. In addition, the parietal lobe plays an important role in spatial

Figure 42.12 The four lobes of the cerebral cortex as seen on the right hemisphere, and some major functions they control.

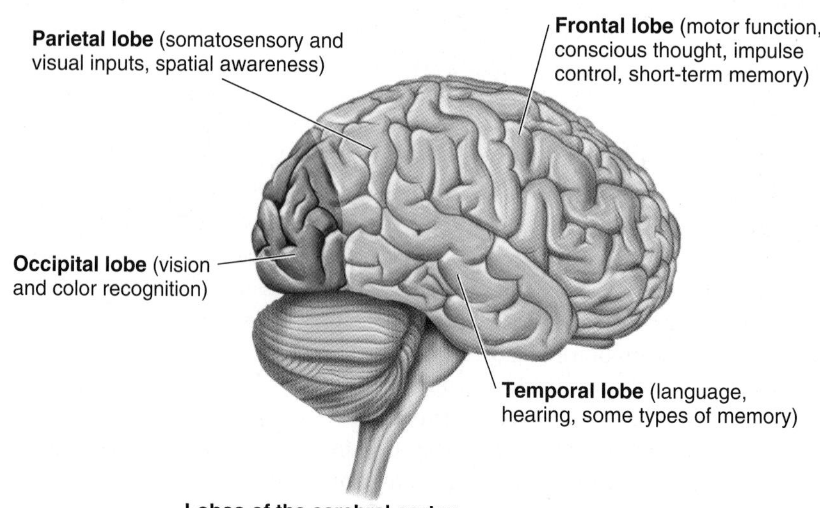

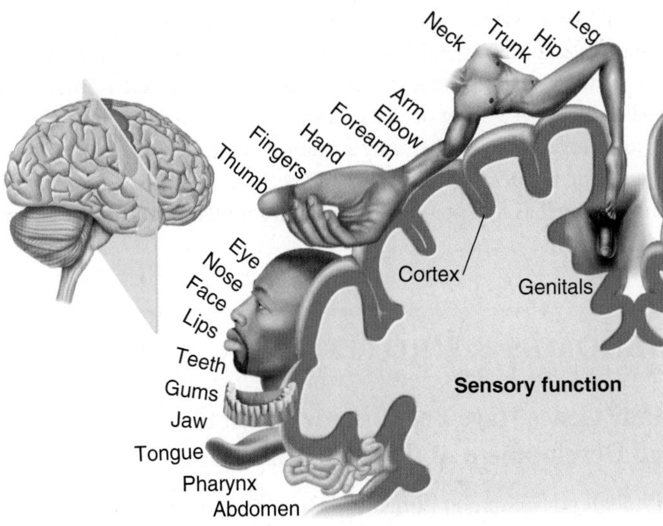

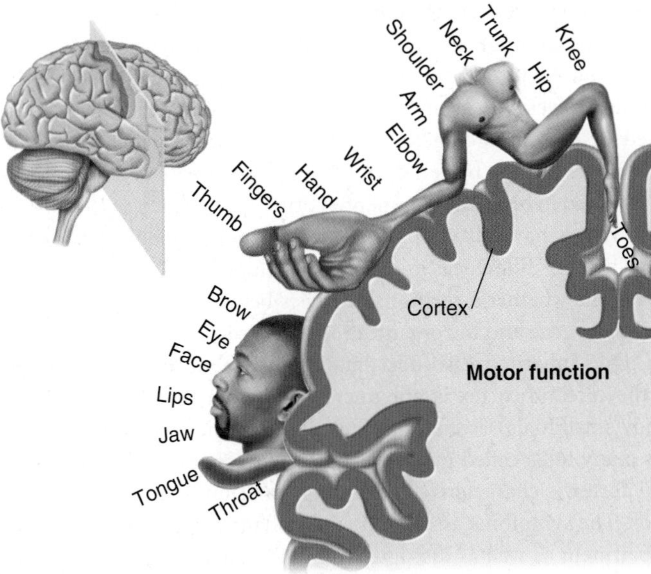

Figure 42.13 Maps of human body parts along the cerebral cortex. These maps represent how the cortex interprets sensory information from these body parts and controls body movements of these parts (motor function). The relative sizes of body parts reflect the relative amount of cortex devoted to them. The blue highlighted region is the somatosensory cortex; the red region is the primary motor cortex.

Table 42.1 Major Functions of Brain Regions in Humans

Region		Major Functions
Hindbrain		
	Medulla oblongata and pons	Coordinate homeostatic functions such as breathing, heart rate, digestion; form part of reticular formation that controls sleep and alertness; give rise to cranial nerves
	Cerebellum	Fine-tuning of complex body movements; maintenance of balance
Midbrain		Processes visual, auditory, and olfactory sensory inputs; forms part of reticular formation
Forebrain		
Diencephalon		
	Thalamus	Routes sensory information (except olfaction) to discrete parts of cerebrum; filters irrelevant sensory information; directs outgoing motor information from cerebral cortex to spinal cord; involved in pain perception and mental arousal
	Hypothalamus	Regulates activities of gastrointestinal and reproductive systems; controls function of pituitary gland; regulates body temperature, appetite, thirst, aggressive behavior, sexual behavior, and body rhythms
	Epithalamus	Produces cerebrospinal fluid; plays a role in food and water intake; contains the pineal gland, which may regulate sleep/wake and body rhythms
Cerebrum		
	Cerebral cortex	Voluntary motor control; perception of sensory inputs; attention; integration of sensory and motor information; generation of speech; decision making; impulse control; judgment; planning; conscious thought; learning; memory; and emotion
	Basal nuclei	Planning, fine-tuning, initiating, inhibiting and learning movements
	Limbic system	Formation and expression of emotions; perception of odors; learning and memory

awareness, that is, our ability to use visual cues to orient ourselves in space. The **occipital lobe** controls many aspects of visual perception and color recognition. The **temporal lobe** is necessary for language, hearing, and some types of memory.

An amazing finding is that sensory inputs enter and motor outputs exit the cerebral cortex in a pattern that forms a map of the body (see Figure 42.13). The regions of the body are represented in proportion to the amount of cortical area devoted to them. For instance, a larger part of the cerebral cortex is devoted to the lips than to other areas of the face. The lips have more nerve endings and are more sensitive to touch than these other areas. Other cortical functions are also mapped in this way; for example, a map that reflects different sound frequencies (the pitch of sound) exists in the temporal lobes. The organization of the cerebral cortex may not be permanent, however, because the map may change depending on the amount of use or disuse of a given part of the body, as discussed in the Feature Investigation later in this chapter. Some of the major functions of the hindbrain, midbrain, and forebrain are summarized in **Table 42.1.**

The **basal nuclei** are a group of nuclei that surround the thalamus and lie beneath the cerebral cortex. Like the cerebellum, the basal nuclei are involved in planning, learning, and fine-tuning movements. They also function via a complex circuitry to initiate or inhibit movements.

Parkinson disease (often called Parkinson's disease) is a relatively common neurological disorder that affects the basal nuclei. People with Parkinson disease have trouble initiating movement, such as beginning to move their legs when they wish to walk. They are capable of walking once movement has begun, but they move slowly with muscle tremors and a shuffling, jerky gait. These symptoms result from the gradual deterioration of dopamine-releasing neurons in an area of the midbrain called the substantia nigra, the neurons of which send axons to the basal nuclei. People in the early stages of Parkinson disease can be treated with L-dopa, a molecule that enters the blood and travels to the basal nuclei. There, axon terminals from remaining healthy cells originating in the substantia nigra take up the L-dopa and convert it into dopamine, which is then released onto cells of the basal nuclei. L-Dopa, therefore, increases the amount of dopamine in the basal nuclei and reduces the Parkinson symptoms.

The **limbic system** refers to a collection of evolutionarily older structures that form an inner layer at the base of the forebrain. These include structures such as the **olfactory bulbs** (which process information about smells), **amygdala**, and **hippocampus**. Many neuroscientists also consider parts of the diencephalon as part of the limbic system, because of the extensive connections between these regions. The limbic system is primarily involved in the formation and expression of emotions, and it plays an important role in learning, memory, and the perception and recognition of smells. The expression of emotions occurs early in childhood before the more advanced functions of the cerebral cortex are evident. Thus, even very young babies can express fear, distress, and anger as well as bond emotionally with their parents.

Deep within the brain, the amygdala is one of the limbic areas critical for understanding and remembering emotional situations. This area also is involved in the ability to recognize emotional expression in others. Emotions are not unique to humans, however, and some are clearly present in other primates and mammals. Being able to express and detect emotions imparts a selective advantage by enabling animals to establish and maintain relationships. Emotions such as fear help an animal defend itself against danger by avoiding conflict. Likewise, anger is associated with aggression, a key behavior by which many animals defend themselves or their territories.

Adjacent to the amygdala and forming a loop within the medial regions within the brain, the hippocampus is composed of several layers of cells that are connected together in a circuit. Its main function appears to be establishing memories for spatial locations, facts, and sequences of events. Damage to certain parts of the hippocampus in humans results in an inability to form new memories, a devastating condition that prevents recognition of other people or even an awareness of daily events. Experiments with laboratory animals have also demonstrated the importance of the hippocampus for memory and learning in other mammals. In a particularly well-studied example, rats are placed into a pool of milky water containing a hidden platform. The animals swim until they find the platform, on which they can safely stand. The time it takes to find the platform in subsequent trials is shorter as they learn and remember its whereabouts. This type of spatial learning depends on activity in the hippocampus. Rats with parts of their hippocampus destroyed fail to improve their times with repeated trials. The hippocampus also receives extensive inputs from the olfactory bulbs, which may explain why smells are such potent triggers of memory in humans and why many animals use their sense of smell as a major way to learn and remember aspects of their environments.

GENOMES & PROTEOMES CONNECTION

Many Genes Have Been Important in the Evolution and Development of the Cerebral Cortex

Although not unique to humans, the elaborate cerebral cortex is one of the defining features of the human brain, because it is responsible for much of what we call our individual personalities. Researchers are now beginning to understand the molecular mechanisms by which this important brain structure evolved and how it develops.

A number of genes are now known to be involved in the evolution and development of the cerebral cortex. Some have been identified by examining genetic mutations in developmentally disabled individuals; others, by comparing human genes with genes known to be involved in brain development in other species such as *Drosophila*. Researchers have also compared these genes in many species that show notable differences in cerebral structure. This last approach can determine whether a relationship exists between the expression of a particular gene and the organization of the cerebral cortex.

One inherited disorder that involves abnormal development of the cerebral cortex is polymicrogyria (from the Greek, meaning many small folds). Recall that the surface of the cerebrum normally has many folds called gyri. In people with polymicrogyria, the cerebral cortex is characterized by multiple and unusually small surface folds. The symptoms associated with polymicrogyria include mental impairment as well as disrupted gait and language production. One type of polymicrogyria is a recessively inherited condition for which eight different mutations of a single gene are known. This gene, called *GPR56*, encodes a G-protein-coupled receptor (described in Chapter 9), which has large extracellular loops. All eight mutations that produce polymicrogyria alter these extracellular loops of the receptor, and scientists think that this alters the ability of the G-protein-coupled receptor to bind its ligand.

Two other genes, called *microcephalin* (*MCPH1*) (from the Greek, meaning small head) and *ASPM* (*abnormal spindle-like microcephaly-associated gene*), have been shown to be determinants of brain size. For example, mutations of these genes in the human population produce individuals with much smaller frontal lobes. Interestingly, the sequences of these genes in several primates, including humans, as well as in other mammals such as dogs and sheep, have shown that the proteins produced by the normal *MCPH1* and *ASPM* genes have undergone greater changes in humans and great apes than in other species. Therefore, these genes may have been under greater selection pressure in animals with larger cerebral cortexes, suggesting that the genes play a key role in the evolution of the cerebral cortex.

42.3 Cellular Basis of Learning and Memory

Learning Outcomes:

1. Define the terms learning and memory and describe their relationship to one another.
2. Discuss how memory is related to changes in the strength of connections between neurons.
3. Describe the evidence that shows the brain is capable of neurogenesis.
4. Discuss the similarities and differences in the technologies of CT, MRI, and fMRI.

In the past few decades, an exciting advance in neuroscience has occurred—researchers have begun to understand complex behaviors, such as learning and memory, at the cellular level. Though it is difficult to separate the two concepts, **learning** can be defined as the process by which new information is acquired. Learning is an evolutionary adaptation that allows past experiences to affect ongoing and future behavior. **Memory** is the ability to retain, retrieve, and use information that was previously learned. Memory connects our experiences throughout life. Our behavior is largely controlled by what we have learned and remember from past experiences. Neuroscientists want to understand how the brain learns and how it captures memories. In this section, we will examine some current ideas about how this may be achieved at the cellular level and consider experimental approaches that neuroscientists follow when investigating such complicated phenomena.

Learning and Memory Occur via Changes Within Neurons and Their Connections with Each Other

Beginning in the 1960s, research along two fronts led to key insights regarding the cellular basis of memory. Norwegian neuroscientist Terje Lømo and British researcher Timothy Bliss focused their efforts on the hippocampus. As described earlier, this is a key region of the brain involved with learning and memory. Lømo and Bliss conducted experiments on anesthetized rabbits in which they monitored signal transmission across particular regions of the hippocampus. Their key discovery involved the effects of multiple stimuli. Experimentally, a series of short, electrical stimulations to a neuron was shown to strengthen, or potentiate, its communication at a synapse with an adjacent cell for minutes or hours. Such multiple stimuli caused neurons to communicate more readily; responses were stronger and more prolonged. This phenomenon was termed **long-term potentiation (LTP)**. LTP is the long-lasting strengthening of the connection between neurons. Later work showed that LTP occurs naturally in the hippocampus and can last from hours to days, and even years.

Austrian-born American neuroscientist Eric Kandel also was interested in learning and the formation of memory. In the 1960s, however, he took a different approach by studying a simpler organism called the California sea slug or sea hare (*Aplysia californica*). He chose this organism for several key reasons. First, it has only about 20,000 neurons, making it easier to identify pathways that are involved in specific types of behavior. Second, some of the neurons in this organism are extremely large, which facilitated the study of action potentials via microelectrodes (as described in Chapter 41). In addition, the large size of the neurons made it technically simpler to inject substances into them and study their effects. Finally, another advantage is that Kandel and colleagues could isolate proteins and mRNA from these large neurons and identify the biochemical and genetic changes that occur when the animal responds to a stimulus.

Much of Kandel's work focused on one type of learning involving the gill-withdrawal reflex, a simple protective reflex that is thought to involve less than 100 neurons in the CNS. The gill and siphon are organs involved in respiration, located in the animal's mantle cavity and protected by muscular appendages called parapodia (**Figure 42.14a**). When the siphon is gently touched with a fine probe, the sea slug closes the siphon and retracts its gills into the mantle cavity for protection (**Figure 42.14b**). Though called a reflex, this behavior is subject to learning. For example, if the touching of the siphon is accompanied with a brief electrical shock to the tail, the sea slug can learn to withdraw its gill in response to a subsequent shock without the siphon being touched. This is similar in some ways to the famous conditioning experiments of Ivan Pavlov, described

(a) Sea slug

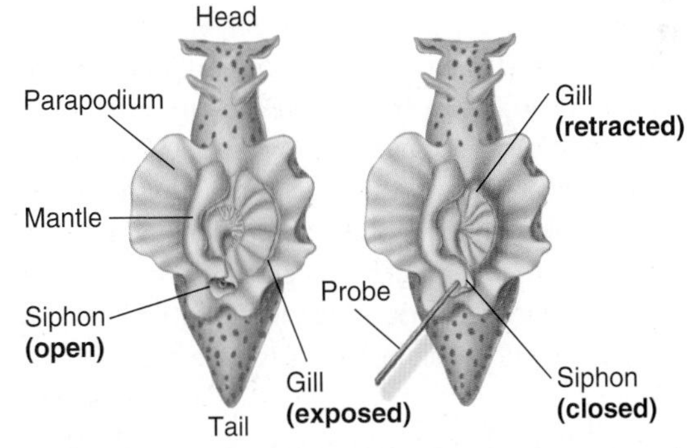

(b) Gill-withdrawal reflex

Figure 42.14 The gill-withdrawal reflex in the sea slug. **(a)** Photo of *Aplysia punctata* (closely related to *A. californica*) in its natural habitat; note the parapodium and tail. **(b)** When touched with a probe, the siphon closes and the gill retracts. In this drawing, the parapodia are moved apart for a better view of the gill.

BIOLOGY PRINCIPLE Biology is an experimental science. The use of animals such as *Aplysia* as model organisms has revealed numerous fundamentally important properties of the nervous system in all animals. Despite the obvious differences in complexity between a sea slug and a human, much of what is learned in these relatively simple animal models is fully applicable to the function of human neurons and the nervous system.

in Chapter 40. Interestingly, a single tail shock paired with a touch of the siphon will result in conditioning that lasts for a few minutes. Amazingly, though, multiple trials over several days result in a lasting memory—a shock given 3 weeks later (without siphon touch) still results in the gill-withdrawal reflex!

Over the course of many years, the work of Kandel and colleagues revealed many clues regarding the cellular basis of learning and memory. As in vertebrates, memory in the sea slug occurs in two forms: short-term memory and long-term memory (**Figure 42.15**). Short-term memory lasts for minutes or hours. This type of memory is typically caused by a single stimulus. Kandel found that short-term memory does not require the synthesis of new proteins. Rather, a single stimulus activates intracellular second-messenger pathways that make it easier for neurons involved in a particular behavior to communicate with each other. For example, as shown in Figure 42.15a, a single stimulus may lead to the activation of protein kinases such as protein kinase A (PKA) in the presynaptic (sensory) cell. PKA, in turn, can phosphorylate proteins such as ion channels, which leads to release of an increased amount of neurotransmitter. These changes enhance the transmission of a signal between the presynaptic and postsynaptic cells.

Kandel and colleagues also discovered that repeated stimuli result in long-term memory, which lasts days or weeks. Such repeated stimuli require the synthesis of new proteins (Figure 42.15b). Long-term memory involves the activation of genes in the presynaptic cell, which leads to the synthesis of mRNA and the translation of the encoded proteins. Once made, such proteins cause the formation of additional synaptic connections. These connections also allow the presynaptic and postsynaptic cells to communicate with each other more readily. Such a change in strength of the connection between two neurons, which occurs as a result of learning, is termed **synaptic plasticity**.

Kandel's work provides a foundation for our ability to understand how learning and memory may occur at the cellular level. Short-term memory may involve changes in pre-existing cellular proteins that make it easier for neurons to communicate. Long-term memory results in protein synthesis that causes physical changes in the synapse itself, also affecting communication. Later studies by Kandel and others showed that such changes also occur in vertebrates such as the mouse. For his work on learning and memory, Kandel shared the Nobel Prize in Physiology or Medicine in 2000.

Neurogenesis May Also Contribute to Learning and Memory

Until fairly recently, neuroscientists had thought that the brain of vertebrates did not produce new neurons in adulthood, that is, that the brain was incapable of **neurogenesis**, the production of new neurons by cell division. However, in 1983, a study showing neurogenesis in an adult vertebrate was carried out by Argentinean biologist Fernando Nottebohm and colleagues. Their research revealed that an increase in the number of neurons in certain brain areas of the canary occurred during the mating season. In the late 1990s, evidence also showed that the primate and human CNS, like other parts of the body, contain stem cells, cells with the potential to differentiate into a variety of cells (refer back to Chapter 19). American researchers

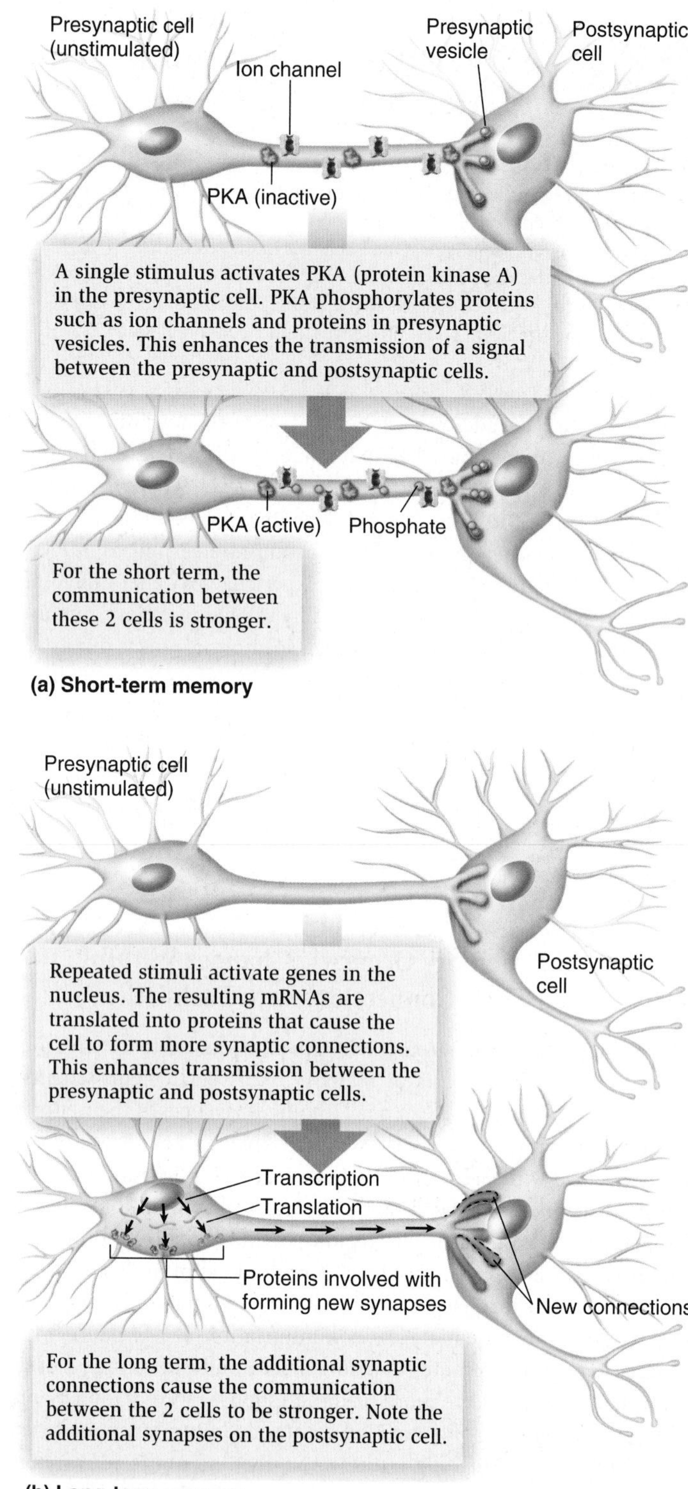
Presynaptic cell (unstimulated)
Ion channel
Presynaptic vesicle
Postsynaptic cell
PKA (inactive)
A single stimulus activates PKA (protein kinase A) in the presynaptic cell. PKA phosphorylates proteins such as ion channels and proteins in presynaptic vesicles. This enhances the transmission of a signal between the presynaptic and postsynaptic cells.
PKA (active)
Phosphate
For the short term, the communication between these 2 cells is stronger.
(a) Short-term memory
Presynaptic cell (unstimulated)
Repeated stimuli activate genes in the nucleus. The resulting mRNAs are translated into proteins that cause the cell to form more synaptic connections. This enhances transmission between the presynaptic and postsynaptic cells.
Postsynaptic cell
Transcription
Translation
Proteins involved with forming new synapses
New connections
For the long term, the additional synaptic connections cause the communication between the 2 cells to be stronger. Note the additional synapses on the postsynaptic cell.
(b) Long-term memory

Figure 42.15 **Cellular changes associated with short-term and long-term memory in the sea slug.**

Elizabeth Gould and Bruce McEwen demonstrated the appearance of new neurons in the hippocampus and olfactory bulbs of adult marmosets and rhesus monkeys. In 1998, American researcher Fred Gage and Swedish physician Peter Eriksson made a key discovery: They found evidence of new hippocampal cells in the brains of

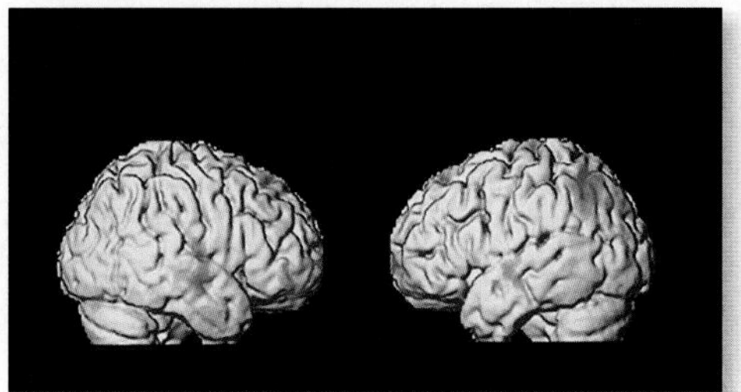

(a) Brain activity of a person thinking about a task that requires finger movements

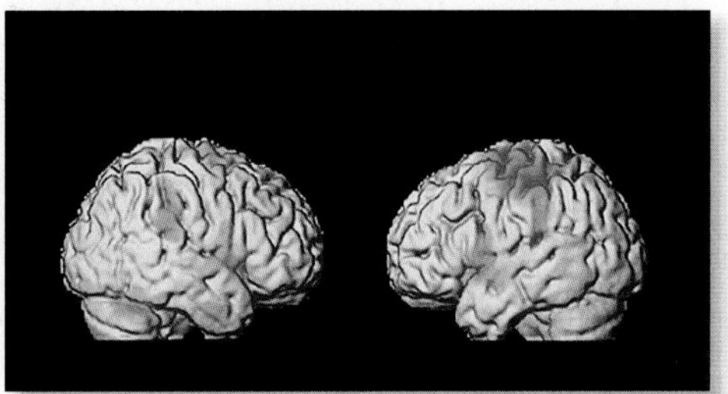

(b) Brain activity when the same person is performing this task

Figure 42.16 Exploring the functional activity of brain regions using fMRI scans. Red color indicates higher O_2 use; both hemispheres are shown.

Concept Check: What can we conclude from part (a) about the metabolic cost of thinking?

recently deceased adult cancer patients. Those patients had previously been treated with bromodeoxyuridine (BrdU) to combat their cancer. BrdU is also taken up by noncancerous cells, but only those that are actively dividing. Its presence in cells can be detected with special stains on sections of brain and other tissue. Gage conducted such staining procedures on the brains of the deceased patients and observed the presence of BrdU in the hippocampus, demonstrating recent neurogenesis.

A key question is whether the neurogenesis observed in adult brains is involved in learning and memory. This question is hotly debated and not resolved. However, some evidence suggests that it could play a role. For example, studies have shown that the hippocampus of adult rhesus monkeys grows new neurons when the animals are placed in enriching environments, and the formation of these neurons slows when animals are chronically stressed. Also, other studies in the rat suggested that new neurons are retained in the hippocampus in response to training in particular tasks that require hippocampus function.

Brain Images Are Used to Assess Brain Structure and Function

Because of the enormous number of neurons and connections between neurons in the vertebrate brain, a key challenge in neuroscience is to understand how such complexity results in sophisticated forms of learning, memory, and responses to environmental conditions. Several imaging techniques allow doctors and researchers to examine the structure and activity level of the brain without anesthesia or surgery. The earliest technique to be developed was computerized tomography (CT). A **CT scan** involves the use of an X-ray beam and a series of detectors that rotate around the head, producing slices of images that are reconstructed into three-dimensional images based on differences in the density of brain tissue. CT scans can easily visualize the ventricles and differences between white and gray matter, but they cannot examine the brain in great detail.

A more sensitive method, called **magnetic resonance imaging (MRI)**, was developed in the 1980s. The patient's head is placed in a device that contains a magnet powerful enough to generate a magnetic field many thousands of times greater than that of the Earth. This stabilizes the spinning, or resonance, of atomic nuclei (usually hydrogen atoms in water molecules) so that most of the nuclei align with the magnetic field. When body tissue is stimulated with a beam of radio waves, its atoms absorb the energy of the waves and the resonance of their nuclei changes, thereby altering their alignment with the magnetic field. When the radio wave pulse stops, the atoms release their energy, which is recorded by a detector. This information is analyzed by a computer, and an image is produced. MRI images allow detection of structures as small as 0.10 mm. For example, they can provide information about abnormal tissue, such as brain tumors, which respond to magnetic and radio frequency pulses differently than normal tissue. MRIs are widely used in medicine to check for injured tissue, cancers, and other abnormalities throughout the body. In 2003, American chemist Paul Lauterbur and British physicist Peter Mansfield received the Nobel Prize in Physiology or Medicine for their work in developing this method.

With certain modifications, MRI can be used to assess the functional activity of areas within the brain. This technique, which is widely used by neuroscientists, is called functional MRI (fMRI). It takes advantage of the observation that blood flow, and therefore oxygen delivery, increases to areas where neurons are more active. This increase in oxygenation is detected via fMRI. In this way, fMRI determines which neurons in particular areas of the brain are active when an individual performs certain intellectual or motor tasks (Figure 42.16). The principle is similar to the standard MRI, except that the higher oxygen content of active tissue alters the resonance of local hydrogen atoms.

The use of fMRI has revealed many fascinating aspects of the activities of different brain regions, notably in people who have suffered brain damage or loss of sensory inputs. For example, individuals who are blind from birth might be predicted to have occipital lobes that are less functional or active than are those in sighted persons (recall that the occipital lobes play the major role in visual processing). However, the work of American researcher Harold Burton and others has revealed with fMRI that the occipital lobes of blind persons are active but have become adapted to other sensory functions such as tactile signals from the fingers, including those arising from Braille reading. Amazingly, this reassignment of occipital function occurs to some extent even in individuals who have lost their vision later in life. Most likely, this does not represent a brand new function of the occipital lobes, but rather an expanded ability of an existing function that remains relatively minor in sighted persons.

The plasticity of the brain revealed by the work of Burton and coworkers is not restricted to clinical situations, as just described. MRI and fMRI are also revealing differences in brain structure and function in individuals due to the types of activities in which they regularly engage, as described next.

Feature Investigation

Gaser and Schlaug Showed That the Sizes of Certain Brain Structures Differ Between Musicians and Nonmusicians

MRI and fMRI have been extremely useful in revealing which brain areas are involved in a particular function. They have also shown that the human brain is surprisingly adaptable. A number of studies have been carried out on musicians, because they practice extensively throughout their lives, enabling researchers to study the effects of repeated use on brain function.

American neuroscientist Christian Gaser and German neuroscientist Gottfried Schlaug used MRI to examine the sizes of brain structures in three groups of people—professional musicians, amateur musicians, and nonmusicians (**Figure 42.17**). Individuals were assigned to each group based on their reported history of musical training: professional musicians with over 2 hours of musical practice time each day, amateur musicians who played a musical instrument regularly but not professionally (practicing about 1 hour/day), and those who never played a musical instrument regularly. The researchers hypothesized that repeated exposure to musical training would increase the size of brain areas associated with visual, motor, and auditory skills, because each of these activities is used to read, make, and interpret music. The results showed that brain areas involved in hearing, moving the fingers, and coordinating movements with vision and hearing were larger in professionals than in amateur musicians, and larger in amateurs than in nonmusicians. The region of the brain that controls finger movements was particularly well developed in the professional musicians, an interesting finding because all of the musicians in this study played keyboard instruments such as the piano.

In another study, American researchers Vincent Schmithorst and Scott Holland used fMRI to determine if musicians' brains were activated differently than nonmusicians' brains when they heard

Figure 42.17 **Gaser and Schlaug's study of the size of visual, motor, and auditory nuclei in the brains of musicians and nonmusicians.**

HYPOTHESIS Musical training is associated with structural differences in the brain.

KEY MATERIALS Volunteer subjects with different degrees of musical training.

	Experimental level	Conceptual level
1 Establish 3 groups of subjects with different musical backgrounds.	Interview subjects for musical history and assign to 1 of 3 groups.	**Controls (nonmusicians)** No musical training. **Amateur musicians** Play an instrument about 1 hr/day and are not employed as musicians. **Professional musicians** Employed as musicians and practice their instrument > 2 hr/day.
2 Control for possible factors that might affect results.	Make group assignments in a way that matches subjects' age and other characteristics.	(see table below)

	Controls	Amateurs	Professionals
Age (average)	23	26	27
Type of instrument	None	Keyboard	Keyboard
Mental IQ	119	123	118
Sex	Male	Male	Male

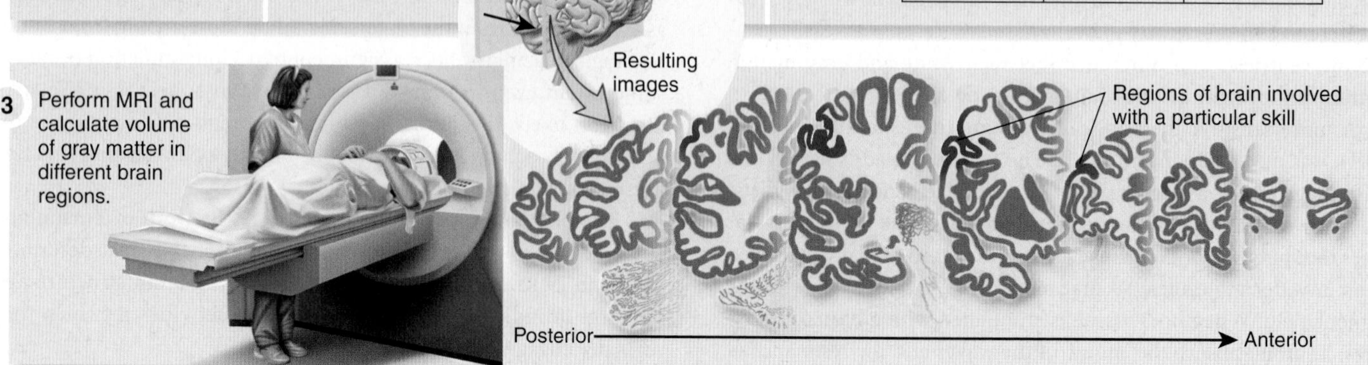

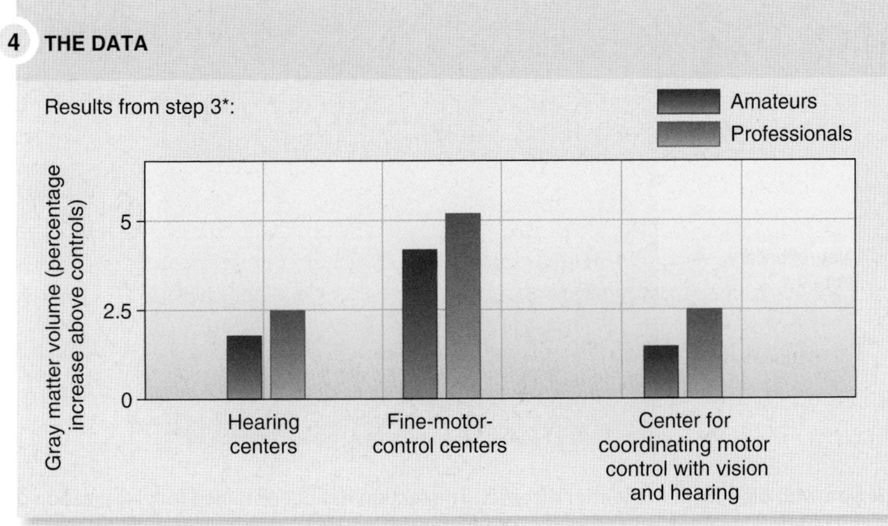

*Controls are not shown separately because the data are expressed relative to controls.

5 CONCLUSION Musical training is associated with increased volumes of brain regions involved in hearing, fine-motor control, and the coordination of motor and sensory information.

6 SOURCE Gaser, C., and Schlaug, G. 2003. Brain structures differ between musicians and non-musicians. *Journal of Neuroscience* 23:9240–9245.

music. They found that one area of the cerebral cortex was selectively activated by melodies only in musicians. This study differed from that of Gaser and Schlaug, because Schmithorst and Holland examined the activity of brain areas as well as their sizes. Their results showed that listening to music activates certain neurons and pathways in the brains of musicians but not in nonmusicians.

The human studies of Gaser, Schlaug, Schmithorst, and Holland have not determined the underlying reason(s) for increased brain size. One possibility is that people with increased brain size in these regions are more likely to become musicians. Alternatively, musical training may actually cause certain regions of the brain to grow larger and alter their neuronal pathways. In other research studies involving experimental animals, groups of animals have been randomly separated into those learning a task versus controls, which do not learn the task. Such experiments have shown increases in the size of brain regions that are associated with learning and memory. The increased size may result from formation of new synapses, growth of blood vessels to the region, and/or production of more glial cells.

Experimental Questions

1. What was the hypothesis proposed by Gaser and Schlaug?
2. How did Gaser and Schlaug test this hypothesis? What were the results of their experiment?
3. How did the research of Schmithorst and Holland influence the findings of Gaser and Schlaug? How did their experiments differ from that of Gaser and Schlaug?

42.4 Impact on Public Health

Learning Outcomes:

1. List the broad groups of diseases affecting the human nervous system.
2. Discuss the impact of meningitis and Alzheimer disease on public health.

Most neurological disorders can be classified into several broad groups (**Table 42.2**). These disorders collectively affect hundreds of millions of people around the world. We will consider two of these disorders—meningitis and Alzheimer disease—that result from very different causes and affect millions of individuals worldwide.

Table 42.2 Categories of Diseases Affecting the Human Central Nervous System

Category	Examples
Infectious diseases	Meningitis, encephalitis
Neurodegenerative disorders	Alzheimer disease
Movement disorders	Parkinson disease
Seizure disorders	Epilepsy
Sleep disorders	Sleep apnea (brain fails to regulate breathing during sleep)
Tumors	Glioma (a tumor arising from glial cells)
Headache disorders	Migraine (severe recurring headache)
Mood disorders	Major depressive disorder; bipolar disorder (see Chapter 41)
Demyelinating disorders	Multiple sclerosis (see Chapter 41)
Injury-related disorders	Brain and spinal cord injuries due to accidents

Meningitis Is an Infectious Disease That Attacks the Meninges

An essential response to infection is inflammation. This response increases the permeability of blood vessels in infected areas, allowing immune cells to be delivered to the site of an infection. When

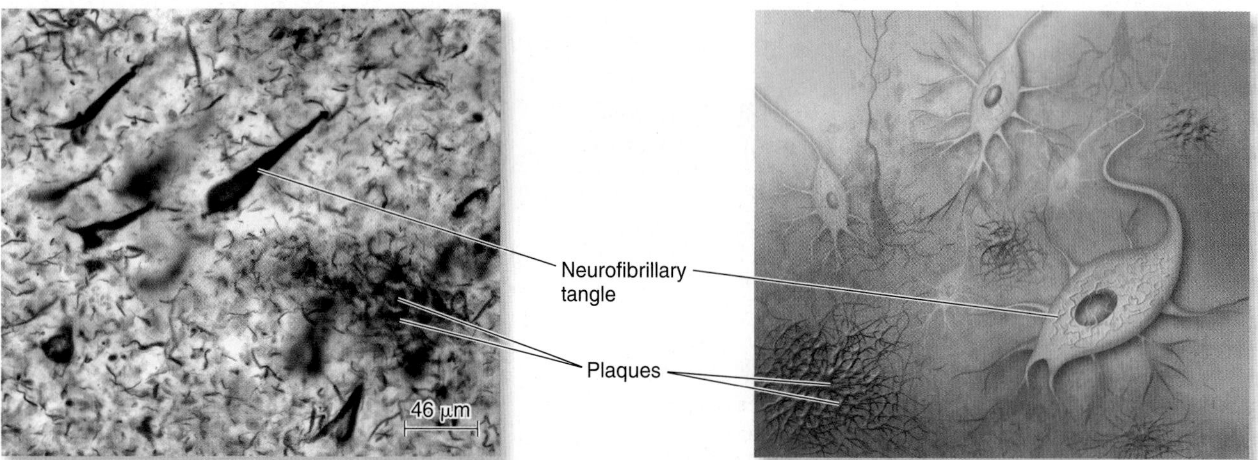

Figure 42.18 **Cellular section from the brain of a person who died from Alzheimer disease.** The section has been stained for visualization of proteins found in plaques and neurofibrillary tangles. An illustration of plaques and tangles is shown for comparison.

that infection occurs within the meninges (causing **meningitis**), fluid accumulates in the subarachnoid space. This accumulation compresses the underlying brain tissue and its blood vessels, interrupting oxygen flow to the neurons of the cerebral cortex. If not treated, the resulting loss of oxygen (and nutrients) causes neuronal death and the loss of function of brain regions associated with those neurons.

The initial symptoms of meningitis include severe headaches, fever, or seizures. Many patients with meningitis also develop a stiff neck because the inflammation proceeds down the spinal cord. If the infection progresses untreated, it may lead to unconsciousness and even death within hours.

Several different viruses or bacterial species can cause meningitis. It usually results from an untreated infection in neighboring regions, such as the sinuses behind the eyes, nose, or ears. Meningitis can be confirmed by using a long needle to sample the cerebrospinal fluid in the spinal cord and analyzing the pressure and contents of the fluid. Large numbers of white blood cells, which are the body's infection-fighting cells, indicate infection in the cerebrospinal fluid and meninges. If the infection is the result of bacterial invasion, meningitis can be treated with bacteria-killing agents such as antibiotics. Antibiotics do not kill viruses, but fortunately the viral form of meningitis is usually less serious than the bacterial form and runs its course after several days or weeks.

Meningitis strikes roughly 25,000 people a year in the U.S. and can affect people of any age. Its incidence in children has greatly declined since the widespread use of a vaccine against the bacterium *Haemophilus influenzae type b* (Hib) began in the U.S. in the early 1990s. Despite the vaccine, meningitis is still a dangerous and prevalent disease worldwide, and it tends to occur in individuals living in close quarters, such as military barracks and college dormitories, where infections may spread rapidly. Occasionally, meningitis can become epidemic. For example, 250,000 people in sub-Saharan Africa were infected and 25,000 died in 1996, and nearly 75,000 people in Southeast Asia died of meningitis in 2004.

Alzheimer Disease Impairs Memory, Thought, and Language

Alzheimer disease (AD) (also called Alzheimer's disease) is the leading cause of dementia worldwide. It is characterized by a loss of memory and cognitive function. AD is a progressive incurable disease that begins with small memory lapses, leading in later stages to problems with language and abstract thinking, and finally to loss of normal motor control and eventual death. The disease usually appears after age 65, though some inherited forms can strike people in their 30s and 40s.

Although psychological testing can help to diagnose AD, historically a definitive diagnosis is possible only after death when the brain is examined microscopically. Brains of Alzheimer patients show two noticeable changes: plaques and neurofibrillary tangles (Figure 42.18). Plaques are extracellular deposits of an abnormal protein, β-amyloid, that forms large, sticky aggregates. These plaques were first noted in 1906 by German physician Alois Alzheimer, after whom the disease was named. Neurofibrillary tangles are intracellular, twisted accumulations of cytoskeletal fibers. Scientists are unsure how these changes influence intellectual function and memory. AD is also associated with the degeneration and death of neurons, particularly in the hippocampus and parietal lobes, which is why it is considered a neurodegenerative disease.

Researchers have identified variation in a few genes whose products are associated with the likelihood of developing AD later in life, but the underlying changes that result in the expression of these and other possible AD-related genes are still the subject of considerable research. Although genetics undoubtedly plays a role in AD, it is not the only possible cause. For example, when one identical twin develops AD, the other appears to be at increased risk but does not always develop the disease, even if he or she survives to very old age. Moreover, evidence suggests that severe head injuries, metabolic diseases such as diabetes, and heart and blood vessel disease may predispose a person to AD in later life.

Currently, AD cannot be prevented or cured. However, three major clinical approaches are currently being tested to prevent or slow down its progression. These approaches are designed to (1) induce a person's immune system to destroy β-amyloid as soon as it is formed, (2) prevent the formation of β-amyloid with drugs that block its synthesis, or (3) prevent the accumulation of β-amyloid into large aggregates using antiaggregation drugs. Each of these approaches holds great promise but is still unproven.

Until a cure for AD is found, its impact on public health remains enormous. Currently, about 4–5 million Americans have AD, and this number is expected to grow to nearly 16 million by 2050. The prevalence of the disease is about 3% for people between the ages of 65 and 74, and 25–50% for people older than 85. Estimated costs associated with providing health care and housing for AD patients (30% of whom live in nursing homes), as well as lost productivity in the workplace, total a staggering $100 billion in the U.S. per year. This number will rise substantially now that the oldest members of the population spike known as the baby boom generation have reached age 65.

Summary of Key Concepts

42.1 The Evolution and Development of Nervous Systems

- All multicellular animals except sponges have a nervous system. Simpler nervous systems include the nerve net of cnidarians and cerebral ganglia, which integrate inputs from sense organs. As animal bodies become more complex, the integrative center in the head becomes a brain that is larger and capable of more functions (Figure 42.1).
- In all vertebrates, the three major divisions of the brain are the hindbrain, midbrain, and forebrain. Human embryos develop these divisions by 4 weeks (Figure 42.2).
- Additional folding of the brain and increased mass allow for expansion of regions associated with conscious thought, reasoning, and learning (Figures 42.3, 42.4).

42.2 Structure and Function of the Nervous Systems of Humans and Other Vertebrates

- In humans and other vertebrates, the brain and spinal cord are the central nervous system (CNS). The neurons and all axons outside the CNS, including the cranial and spinal nerves, constitute the peripheral nervous system (PNS). The CNS relies on the PNS for sensory input, and the PNS relies on commands from the CNS (Figure 42.5).
- The gray matter of the CNS is composed of dendrites, cell bodies, and unmyelinated axons. The white matter consists of tracts of myelinated axons. The meninges are protective coverings of the CNS. Cerebrospinal fluid fills the subarachnoid space and ventricles (Figures 42.6, 42.7).
- The PNS can be subdivided into the somatic and autonomic nervous systems. The somatic nervous system senses external environmental conditions and controls skeletal muscles and skin. The autonomic nervous system senses internal body conditions and controls homeostasis. The efferent part of the autonomic nervous system is divided into two components: sympathetic (fight or flight) and parasympathetic (rest or digest) (Figure 42.8).
- The evolutionarily oldest structures of the brain, some of which are located in the hindbrain, control the basic processes that sustain life. These structures include the medulla oblongata, cerebellum, and pons (Figure 42.9).
- The midbrain processes several types of sensory inputs, including vision, olfaction, and audition. The medulla oblongata, the pons, and the midbrain collectively constitute the brainstem. They also form the reticular formation, a network of neurons that sends signals to many other brain regions (Figure 42.9).
- The forebrain is made of the thalamus, hypothalamus, and epithalamus (diencephalon) and the cerebrum (Figure 42.10). The cerebrum is divided into two hemispheres. Each hemisphere is specialized to perform certain aspects of behavior and can operate independently (Figure 42.11).
- The cerebrum consists of the basal nuclei, limbic system, and cerebral cortex. Each side of the human cerebral cortex is divided into four lobes, each of which has a number of functions (Figures 42.12, 42.13, Table 42.1).

42.3 Cellular Basis of Learning and Memory

- Learning is the process by which new information is acquired; memory involves the retention of that information.
- Repeated stimuli result in long-term potentiation, in which the connections between adjacent neurons become stronger. Studies of the sea slug indicate that short-term memory is caused by a single stimulus that activates second-messenger pathways. Long-term memory is caused by repeated stimuli that activate genes, which results in more synaptic connections, a phenomenon called synaptic plasticity (Figures 42.14, 42.15).
- Imaging techniques such as CT scans, MRI, and fMRI allow us to examine the structure and activity of the brain (Figures 42.16, 42.17).

42.4 Impact on Public Health

- Disorders of the human central nervous system can be placed into several broad categories (Table 42.2).
- Meningitis is a potentially life-threatening infectious disease in which the meninges become inflamed. Alzheimer disease is a progressive disorder characterized by the formation of senile plaques and neurofibrillary tangles in brain tissue. Both are examples of neurological disorders with a large effect on public health (Figure 42.18).

Assess and Discuss

Test Yourself

1. A nerve net consists of
   a. bilateral neurons that extend from the head of the animal to the tail.
   b. a group of neurons that are interconnected and are activated all at once.
   c. a single nerve cord with ganglia in each body segment.
   d. a central nervous system with peripheral nerves associated with different body structures.
   e. none of the above.
2. The division of the vertebrate brain that includes the cerebellum is
   a. the hindbrain.
   b. the telencephalon.
   c. the midbrain.
   d. the forebrain.
   e. the diencephalon.
3. In general, the brains of more complex vertebrates
   a. are larger.
   b. have fewer neurons.
   c. have more folds in the cerebral cortex.
   d. use less oxygen.
   e. both a and c
4. The white matter of the CNS is composed of
   a. dendrites.
   b. unmyelinated axons.
   c. myelinated axons.
   d. cell bodies.
   e. a and b only.
5. The division of the nervous system that controls voluntary muscle movement is
   a. the autonomic nervous system.
   b. the sensory division.
   c. the somatic nervous system.
   d. the parasympathetic division.
   e. the sympathetic division.
6. Which of the following is *not* a response to activation of the sympathetic division of the autonomic nervous system?
   a. increased breathing rate
   b. decreased heart rate
   c. increased blood flow to the skeletal muscles
   d. increased blood glucose levels
   e. All of the above are characteristic responses to activation of the sympathetic division of the autonomic nervous system.
7. The ___ acts as a relay for the cerebrum.
   a. medulla
   b. pons
   c. hypothalamus
   d. midbrain
   e. thalamus
8. The _______ is a portion of the limbic system that is important for memory formation.
   a. amygdala
   b. hippocampus
   c. pons
   d. thalamus
   e. mesencephalon
9. In humans, the ________ hemisphere of the cerebrum is dominant in nonverbal processing.
   a. right
   b. left
   c. both a and b
10. ____________ is a progressive disease that causes a loss of memory and intellectual and emotional function.
   a. Meningitis
   b. Parkinson disease
   c. Amnesia
   d. Alzheimer disease
   e. Stroke

Conceptual Questions

1. One of the most important and fundamental features of all nervous systems is the reflex. Describe why reflexes are adaptive.
2. Explain the differences between white matter and gray matter.
3. A principle of biology is that *new properties emerge from complex interactions.* How is this principle evident by the structure and function of animal nervous systems?

Collaborative Questions

1. Describe the basic features of two different types of nervous systems found in animals.
2. List the three major divisions of the brain of vertebrates and briefly describe the function of each in humans.

Online Resource

www.brookerbiology.com

Stay a step ahead in your studies with animations that bring concepts to life and practice tests to assess your understanding. Your instructor may also recommend the interactive eBook, individualized learning tools, and more.

Chapter Outline

43.1 An Introduction to Sensory Receptors
43.2 Mechanoreception
43.3 Thermoreception and Nociception
43.4 Electromagnetic Reception
43.5 Photoreception
43.6 Chemoreception
43.7 Impact on Public Health
Summary of Key Concepts
Assess and Discuss

Neuroscience III: Sensory Systems

43

Vision in the dark. As with many animals that are active at night, a cat's eye contains a reflective layer of tissue called a tapetum lucidum. This increases their ability to see in the dark.

What does the world look, sound, smell, and feel like to other animals? We can never know exactly how an animal perceives its environments. Biologists can perform experiments, however, to determine the capabilities of an animal's sensory systems. For instance, researchers have examined the structure of a goldfish's eye, conducted behavioral studies to determine if a goldfish can discriminate between different colors, and measured the electrical responses of neurons in the animal's visual system to different visual stimuli. From such studies, we now know that goldfish and probably most animals can discriminate between light of different wavelengths, or colors.

Despite the prevalence of such abilities as light, odor, sound, taste, and touch detection across animal taxa, the sensory experience of different animals may differ radically from our own. We do not see the color patterns of flowers produced by the reflection of ultraviolet light the way honeybees do, or hear the very low frequency sounds (such as those produced by earthquakes) that elephants, whales, and alligators hear. We cannot use echoes of our own sounds to locate flying insects the way some bats do. We cannot detect the presence of chemicals using our entire body surface, the way an earthworm can as it seeks food. We have a sensitive ability to detect touch, but we cannot detect the electric field generated by the muscles and hearts of marine animals, although many sharks and catfish can. We can see relatively well at night, but would have better night vision if we had light-reflective substances in the back of our eyes like some animals, as does the cat in the chapter-opening photo.

Senses allow living organisms to perceive their environments. In neuroscience, a **sense** is broadly defined as a system that consists of specialized cells that detect a specific type of chemical, energy, or physical stimulus (also known as a modality) and send signals to the central nervous system (CNS), where the signals are received and interpreted. The senses allow animals to perceive subtle and complex aspects of their environments. They are the windows through which animals experience the world around them. The nervous systems of most animals also have the ability to sense signals arising from within an animal's body, such as hunger or pain. However, neuroscientists do not always agree on the actual number of different senses. Common examples include sight, smell, taste, touch, hearing, balance, and the ability to sense heat, cold, and pain.

In this chapter, we will examine how nervous systems collect incoming sensory information and how membrane potentials of specialized neurons change in response to sensory inputs. We will learn that other structures of the nervous system may modify or enhance this neural activity before sending it to the brain, where it is interpreted. Finally, we will discuss how problems with sensory systems—in particular vision and hearing deficits—can affect human health.

43.1 An Introduction to Sensory Receptors

Learning Outcomes:

1. Describe the relationship between sensation and perception.
2. Discuss how sensory receptors pass along the intensity of a stimulus.
3. List the classes of sensory receptors and the stimuli to which they respond.

Sensory systems convert chemical or physical stimuli from an animal's body or the external environment into a signal that causes a change in the membrane potential of sensory neurons. **Sensory transduction** is the process by which incoming stimuli are converted into neural signals. Sensory transduction involves cellular changes, such as opening of ion channels, which cause either graded potentials or action potentials in neurons.

Perception is an awareness of the sensations that are experienced. For instance, touching a hot object generates a thermal sensation, which initiates a neuronal response, giving us the perception that this stimulus is hot. Not all sensations are consciously perceived by an organism. Most of the time, for example, we are not aware of the touch of our clothing. The brain also processes sensory information in areas that do not generate conscious thought. For instance, certain neurons constantly monitor blood pressure and the concentrations of oxygen, glucose, and other substances in the blood, but we are not aware that this is occurring.

We begin our study of sensory systems by examining the specialized cells that receive sensory inputs. A **sensory receptor** is a cell that recognizes an internal or external (environmental) stimulus and initiates sensory transduction by creating graded potentials (described in Chapter 41) in itself or an adjacent cell. Sensory receptor is a term that neuroscientists use to describe certain types of cells, which are either neurons or specialized epithelial cells, that respond to internal or environmental stimuli (Figure 43.1). When a response is strong enough, sensory receptors initiate electrical responses to stimuli, such as chemicals, light, heat, and sound, which lead to action potentials that are sent to the CNS.

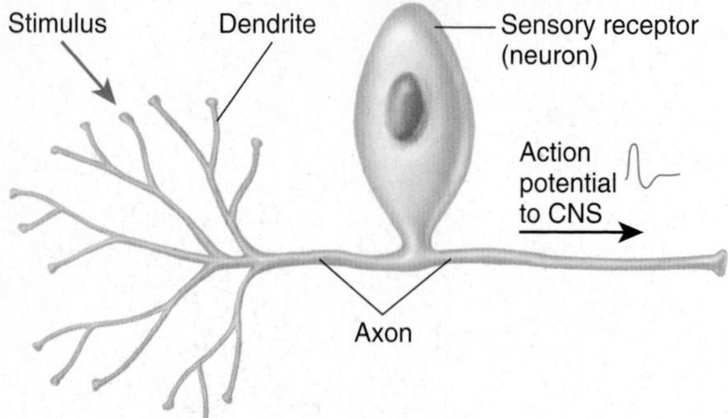

(a) A neuron as a sensory receptor

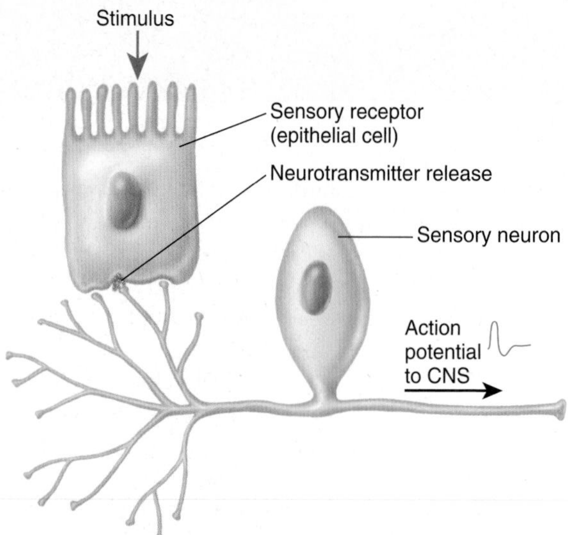

(b) A specialized epithelial cell as a sensory receptor

Figure 43.1 Sensory receptors. **(a)** Many sensory receptors are neurons that directly sense stimuli. **(b)** Other sensory receptors are specialized epithelial cells that sense stimuli and secrete neurotransmitter that stimulates nearby sensory neurons. In both cases, when stimulated, the neurons send action potentials to the CNS, where the signals are interpreted.

BioConnections: What is the difference between the sensory receptor described in this chapter and the membrane receptors described in Chapter 9?

An Intense Stimulus Generates More Frequent Action Potentials

How do sensory receptors pass along the intensity of a stimulus? Let's consider an example involving weak and strong stimuli to the sense of touch (Figure 43.2). Sensory transduction begins when the specialized endings of a sensory receptor respond to a stimulus. Such a stimulus—in this case, the touch of a glass rod—opens ion channels that allow sodium ions (Na^+) to diffuse down their electrochemical gradient into the cell, depolarizing the sensory receptor. The amount of depolarization is directly related to the intensity of the stimulus, because a stronger stimulus opens more ion channels.

The first response of a sensory receptor is usually a graded change in the membrane potential of the cell body that is proportional to the intensity of the stimulus (see Figure 43.2). The membrane potential, known as the **receptor potential** in these cells, becomes less and less negative as the strength of the stimulus increases. When a stimulus is strong enough, it depolarizes the membrane to the threshold potential at the axon hillock and produces an action potential in a sensory neuron (refer back to Figure 41.10).

Recall from Chapter 41 that action potentials proceed in an all-or-none fashion, regardless of the nature or strength of the stimulus that elicits them. How, then, can action potentials provide information about the intensity of a stimulus? The answer is that the strength of the stimulus is indicated by the frequency of action potentials generated. A particularly strong stimulus generates many action potentials in a short period of time. As a result, the frequency of action potentials is higher when the stimulus is strong than when it is weak. The action potentials are transmitted into the CNS and carried to the brain for interpretation. The brain interprets a higher frequency of action potentials as a more intense stimulus.

The CNS Processes Each Sense Within Its Own Pathway

Different stimuli produce different sensations because they activate specific neural pathways that are dedicated to processing only that type of stimulus. We know that we are seeing light because the signals generated by visual sensory receptors in the eye are transmitted along a neural pathway that sends action potentials into areas of the brain that are devoted to processing vision. For this reason, the brain interprets such signals as visual stimuli. The brain can separate and identify each sense because each one uses its own dedicated pathway.

Sensory receptors can be divided into general classes based on the type of stimulus or modality to which they respond. Each type

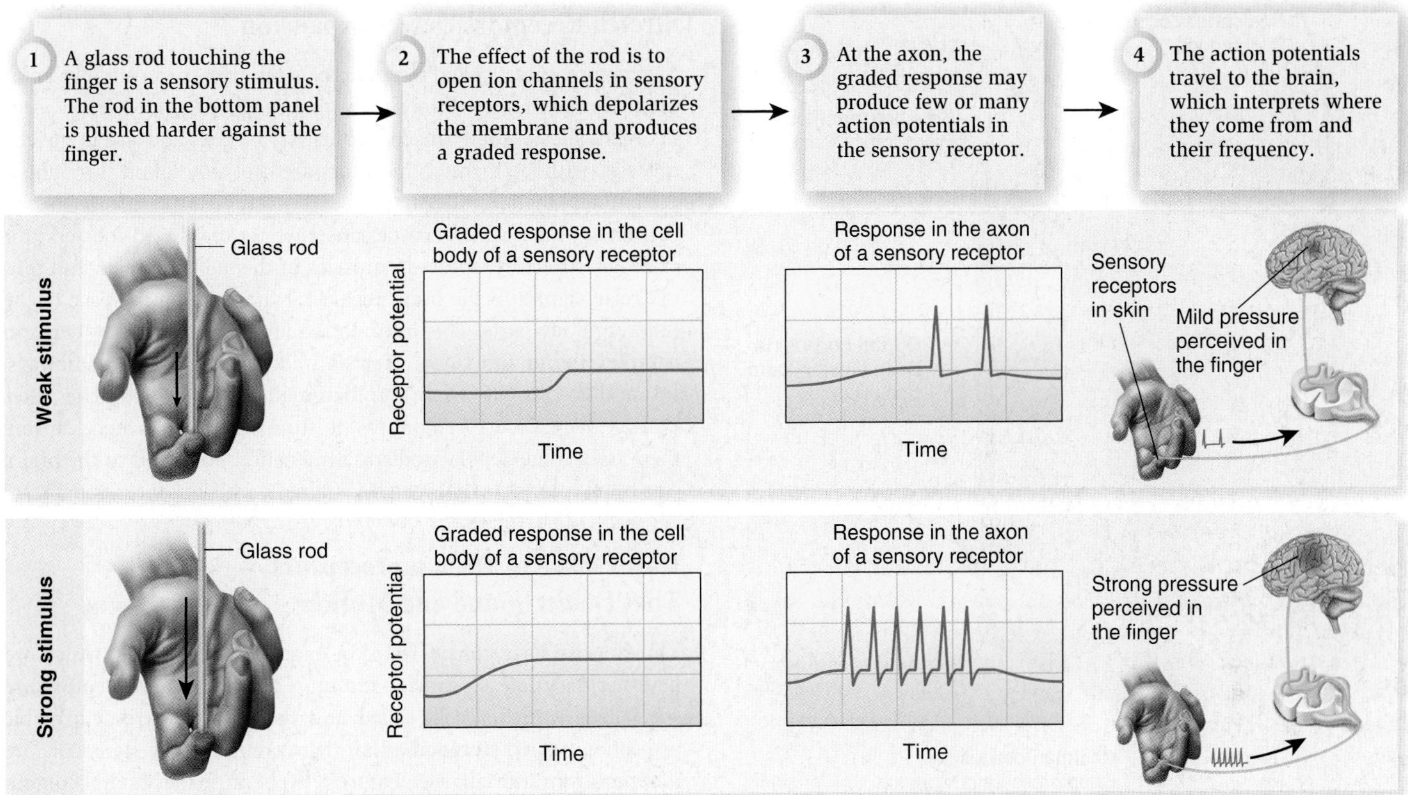
1 A glass rod touching the finger is a sensory stimulus. The rod in the bottom panel is pushed harder against the finger.
2 The effect of the rod is to open ion channels in sensory receptors, which depolarizes the membrane and produces a graded response.
3 At the axon, the graded response may produce few or many action potentials in the sensory receptor.
4 The action potentials travel to the brain, which interprets where they come from and their frequency.
Weak stimulus
Glass rod
Graded response in the cell body of a sensory receptor
Receptor potential
Time
Response in the axon of a sensory receptor
Time
Sensory receptors in skin
Mild pressure perceived in the finger
Strong stimulus
Glass rod
Graded response in the cell body of a sensory receptor
Receptor potential
Time
Response in the axon of a sensory receptor
Time
Strong pressure perceived in the finger

Figure 43.2 **Transduction of a sensory stimulus of two different intensities.** In this example, the sensory receptor is a neuron. Note the faster and larger graded response following the stronger stimulus.

uses a different mechanism to detect stimuli and to transmit the information to different regions of the CNS. **Mechanoreceptors** transduce mechanical energy such as touch, pressure, stretch, movement, and sound. **Thermoreceptors** detect cold and heat. **Nociceptors**, or pain receptors, detect extreme heat, cold, and pressure, as well as certain potentially damaging molecules such as acids. **Electromagnetic receptors** sense radiation within a portion of the range of the electromagnetic spectrum, including visible, ultraviolet, and infrared light, as well as electrical and magnetic fields in some animals. **Photoreceptors** are electromagnetic receptors that detect visible light. **Chemoreceptors** recognize specific chemical compounds in the air, water, body fluids, or food. Most of the remaining sections of this chapter will examine the structures and functions of these types of sensory receptors and the organs in which they are found.

43.2 Mechanoreception

Learning Outcomes:

1. List the types of mechanoreceptors that detect touch, stretch, or movement and how this relates to hearing and balance.
2. Describe the structure of the mammalian ear and how mechanical forces move through it.
3. Give examples of adaptations for hearing in animals that inhabit different environments.
4. Describe how body position and movement are detected by sense organs.

Mechanoreceptors are cells that detect physical stimuli such as touch, pressure, stretch, movement, and sound. Physically touching or deforming a mechanoreceptor cell opens ion channels in its plasma membrane (see Figure 43.2). As discussed in this section, some mechanoreceptors are neurons that send action potentials to the CNS in response to physical stimuli. Other mechanoreceptors are specialized epithelial cells that contain hairlike structures that bend in response to mechanical forces.

Skin Receptors Detect Touch and Pressure

Several types of receptors in the skin detect touch, deep pressure, or the bending of hairs on the skin. Some of these specialized receptors consist of neuronal dendrites covered in dense connective tissue. In mammals, these receptors are located at different depths below the surface of the skin, which makes them suitable for responding to different types of stimuli (Figure 43.3). For example, **Meissner corpuscles** lie just beneath the skin surface and sense touch and light pressure. They are found throughout the skin but are concentrated in areas sensitive to light touch, such as the fingertips, lips, eyelids, and genitals. In contrast, **Pacinian corpuscles** and **Ruffini corpuscles** are located much deeper beneath the surface, particularly in the soles of the feet and the palms of the hands. These corpuscles respond best to deep pressure or vibration. All skin corpuscles contain sensory receptor neurons that generate action potentials when the structure of the corpuscle is deformed. Other skin mechanoreceptors located in the hair follicles respond to movements of hairs and whiskers.

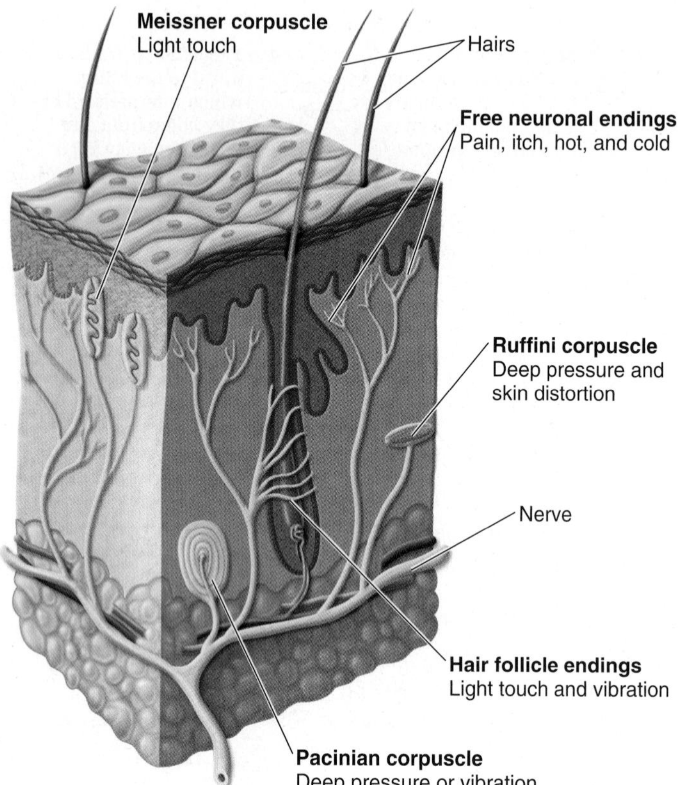

Figure 43.3 **Examples of sensory receptors in the skin of mammals.**

Concept Check: *Several types of touch receptors in the skin respond to different stimuli. What different qualities of touch are you aware of?*

Stretch Receptors Detect Expansion

Mechanoreceptors called **stretch receptors** are neuron endings commonly found in the walls of organs that can be distended. They occur in organs such as the stomach and urinary bladder and also in skeletal muscles. Although stretch receptors are probably found throughout the animal kingdom, they have been best studied in crustaceans and mammals. In decapod crustaceans such as crabs and lobsters, for example, stretch receptors in muscles of the tail, abdomen, and thorax relay signals to the brain regarding the position in space of the different body parts. This information allows the animal to perform complex motor functions, such as walking backward or sideways. In another example, when the mammalian stomach stretches after a meal, the stretch receptors in the stomach are deformed, causing them to become depolarized and send action potentials to the brain. The brain interprets the signals as fullness, which reduces appetite.

Hair Cells Are Mechanoreceptors That Detect Sound and Motion

Thus far, we have considered skin and stretch receptors, which are neurons that detect physical stimuli. Other mechanoreceptors are specialized epithelial cells called **hair cells**, which have deformable projections called **stereocilia** that detect motion. The stereocilia are different from true cilia (see Figure 4.13) because they do not contain motor proteins in their structure, and are bent by movements of fluid or other physical stimuli (**Figure 43.4**).

Hair cells contain ion channels that open or close when the stereocilia bend and thereby change the cell's membrane potential. When the plasma membrane depolarizes, this opens voltage-gated calcium (Ca^{2+}) channels and results in the release of neurotransmitter

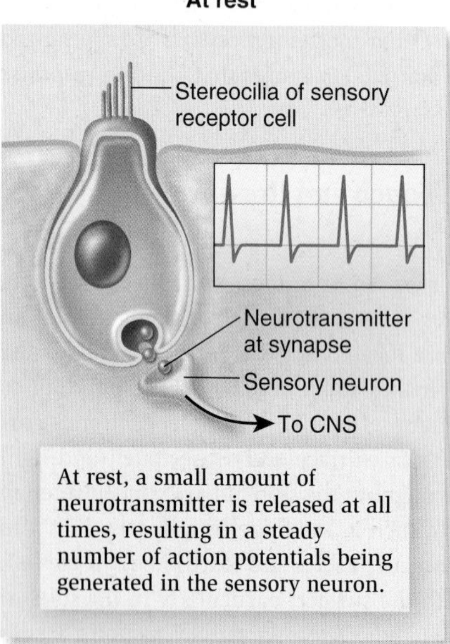

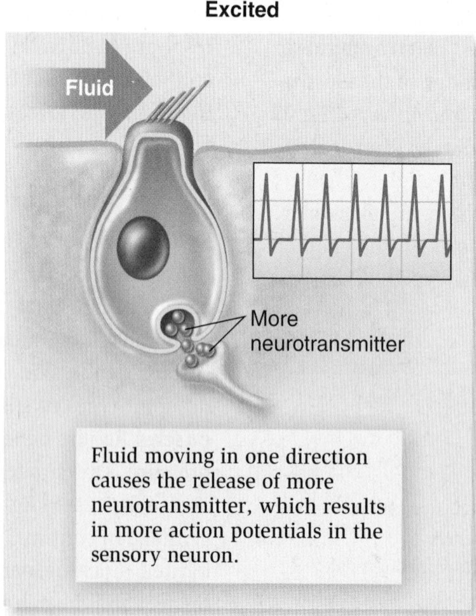

Excited
Fluid
More neurotransmitter
Fluid moving in one direction causes the release of more neurotransmitter, which results in more action potentials in the sensory neuron.

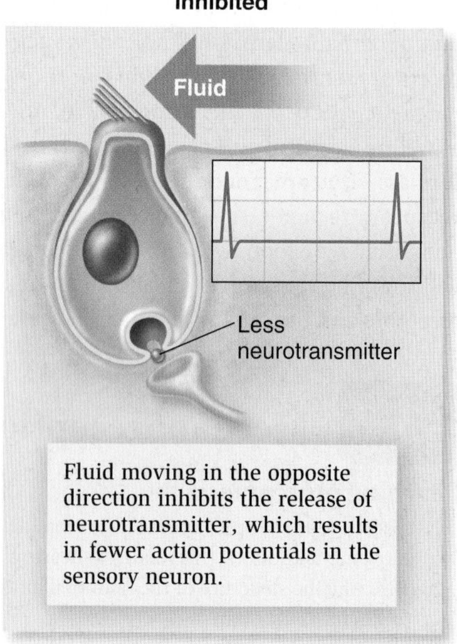

Figure 43.4 **The response of hair cells to mechanical stimulation.** The stereocilia inside these hair cells are hairlike projections of the plasma membrane that contain actin filaments.

BioConnections: *How are stereocilia different from cilia, which are described in Chapter 4?*

molecules from the hair cells. The neurotransmitter then binds to protein receptors in adjacent sensory neurons and can result in action potentials being sent to the CNS. Even when unstimulated, hair cells usually release a small amount of neurotransmitter onto nearby sensory neurons, resulting in a resting level of action potentials in the sensory neurons. In the example shown in Figure 43.4, bending of the stereocilia in one direction in response to fluid movement increases the release of neurotransmitter from the hair cell, exciting the sensory receptors, whereas bending in the other direction decreases the release of the same neurotransmitter, inhibiting the sensory receptors. The result is an increase or decrease, respectively, in the number of action potentials produced in the sensory neurons.

Hair cells are found in the hearing and equilibrium (balance) organs of many invertebrates and vertebrates, where they detect sound or changes in head position. They are also found along the body surface of fishes and some amphibians, where they detect external water currents, as described later.

Audition (Hearing) Involves the Reception of Sound Waves

The sense of hearing, called **audition**, is the ability to detect and interpret sound waves. This sense is critical for the survival and reproduction of many types of animals. For example, a mother seal locates her pup by hearing its calls, and a male bird sings an elaborate song to attract a mate. Hearing is also important for detecting the approach of danger—a predator, a thunderstorm, an automobile—and locating its source.

Sound travels through air or water in waves. The distance from the peak of one wave to the next is a **wavelength**. The number of complete wavelengths that occur in 1 sec is called the **frequency** of the sound, which is measured in number of waves per second, or Hertz (Hz), after the German physicist and pioneer of radio wave research, Heinrich Hertz. The length and frequency of sound waves impart certain characteristics to the stimulus. Short wavelengths have high frequencies that are perceived as a high **pitch** or tone, and long wavelengths have lower frequencies and a lower pitch. The human hearing range is 20–20,000 Hz.

The sense of hearing is present in vertebrates and arthropods, but not in other phyla. Arthropods do not appear to have more than a general sensitivity to sound, although some exceptions exist. For example, some species of moths have sound-sensitive membranes that detect the high frequencies emitted by their chief predators, bats. The sense of hearing, however, is especially well developed in vertebrates (notably birds and mammals); we turn now to a detailed discussion of the mammalian ear and the mechanism by which it detects sound.

Structure of the Mammalian Ear The mammalian ear has three main compartments: the outer, middle, and inner ears (**Figure 43.5**). The **outer ear** consists of the external ear, or pinna (plural, pinnae), and the auditory canal. Mammals have a wide variety of shapes and sizes of the external ear, which partly reflects their different abilities to capture sound waves. The outer ear is separated from the **middle ear** by the tympanic membrane (eardrum). The middle ear contains three small bones called ossicles (named the malleus, incus, and stapes) that link movements of the eardrum with the oval window. The oval window is another membrane similar to the eardrum that separates the middle ear from the **inner ear**. The inner ear is composed of the **cochlea** (from the Latin, meaning snail)—a coiled chamber of bone containing the hair cells and the membrane-like round window—and the vestibular system, which plays a role in balance, as described later. These structures in the inner ear generate the signals that travel via the auditory nerve to the brain.

Both the tympanic membrane and the oval window must be able to vibrate when sound waves meet the eardrum. For this to occur properly, the air pressure in the outer and middle ear compartments must be equal. The pressure in the outer ear is the atmospheric pressure.

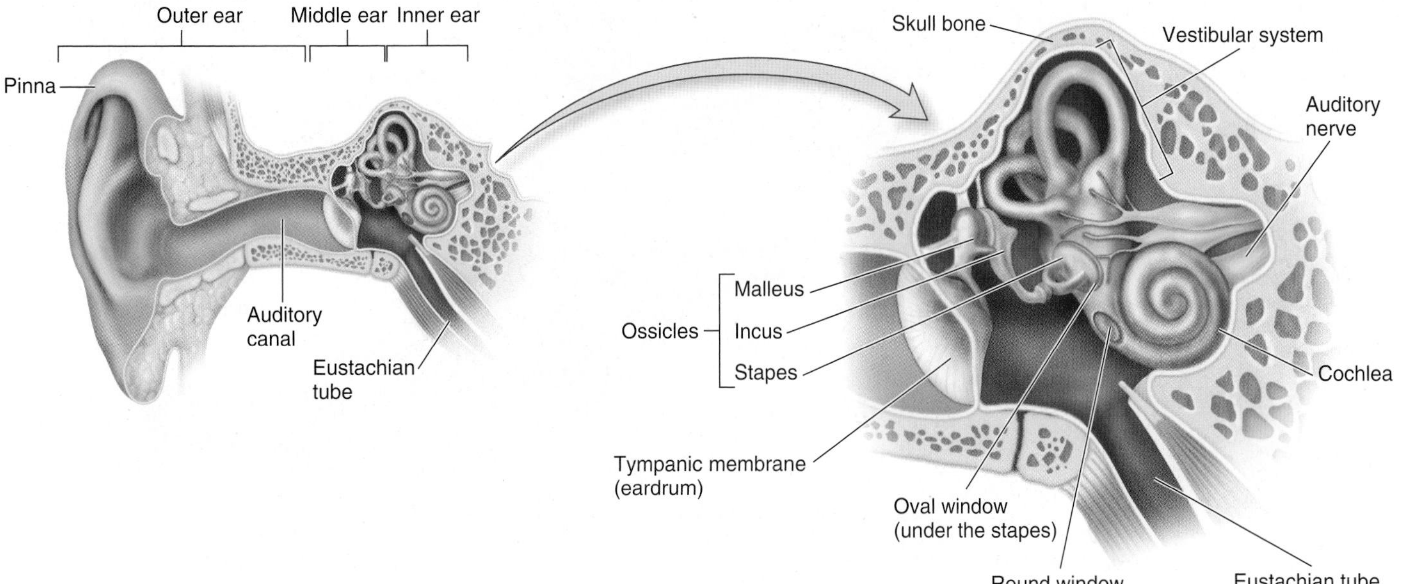

Figure 43.5 The structure of the human ear. The three main compartments are the outer, middle, and inner ear.

The **Eustachian** (or **auditory) tube**, which connects the middle ear to the pharynx, maintains atmospheric pressure in the middle ear. If you change altitude quickly, as during a plane takeoff, the pressure in your outer ear changes because atmospheric pressure decreases with altitude. After a while, the pressures in the outer and middle ear equalize. However, if the pressure in your outer ear becomes greater than the pressure in your middle ear, which occurs when the plane descends, your eardrum will be pushed or bowed inward. The eardrum is less able to vibrate in response to sounds when it is deformed under pressure, and thus your hearing becomes muffled. Swallowing or yawning opens the Eustachian tube and equalizes the pressure in the middle ear with atmospheric pressure. The popping sound you hear when the pressure is equalized is the sudden return of the eardrum to its normal position.

Generation of Electrical Signals in the Mammalian Ear To understand how mammals hear, let's first consider how mechanical forces move through the human ear. Sound waves entering the outer ear cause the tympanic membrane to vibrate back and forth (Figure 43.6). The malleus, incus, and stapes transfer the vibration of the tympanic membrane to the oval window, causing it to vibrate against the cochlea. This sends pressure waves, which also travel in a back-and-forth manner, through a fluid called perilymph. Perilymph is found within two narrow passages in the cochlea called the vestibular and tympanic canals. The waves travel from the vestibular canal to the tympanic canal and eventually strike the round window, where they dissipate. Along the way, the waves cause the vibration of a membrane called the **basilar membrane**, which as you will see shortly is a key part of the mechanism by which different frequency sounds are sensed. Sounds of very low frequency (longer wavelength) create pressure waves that take the complete route through the vestibular and tympanic canals (see green arrows in Figure 43.6). Sounds of higher frequency (shorter wavelength) produce pressure waves that follow a different route, passing from the vestibular canal through a tube between the canals called the cochlear duct (as shown by the blue arrows in Figure 43.6). They then pass through the basilar membrane, before reaching the tympanic canal.

Within the cochlea, mechanical vibrations are changed, or transduced, into electrical signals. This happens in a structure called the **organ of Corti** (named after the Italian anatomist Alfonso Corti), which rests on top of the basilar membrane. To understand how this works, we need to look at a cross section through the cochlea (Figure 43.7). The organ of Corti contains supporting cells and rows of hair cells. The stereocilia of the hair cells are embedded in a gelatinous tectorial membrane. The back-and-forth vibration of the basilar membrane bends the stereocilia in one direction and then the other. When bent in one direction, the hair cells depolarize and release neurotransmitter, which activates adjacent sensory neurons that then send action potentials to the CNS via the auditory nerve. When bent in the other direction, the hair cells hyperpolarize and shut off the release of neurotransmitter. In this way, the frequency of action potentials generated by the sensory neurons is determined by the up-and-down vibration of the basilar membrane.

The basilar membrane is lined with protein fibers that span its width. These fibers function much like the strings of a guitar. The fibers near the oval and round windows at the base of the cochlea

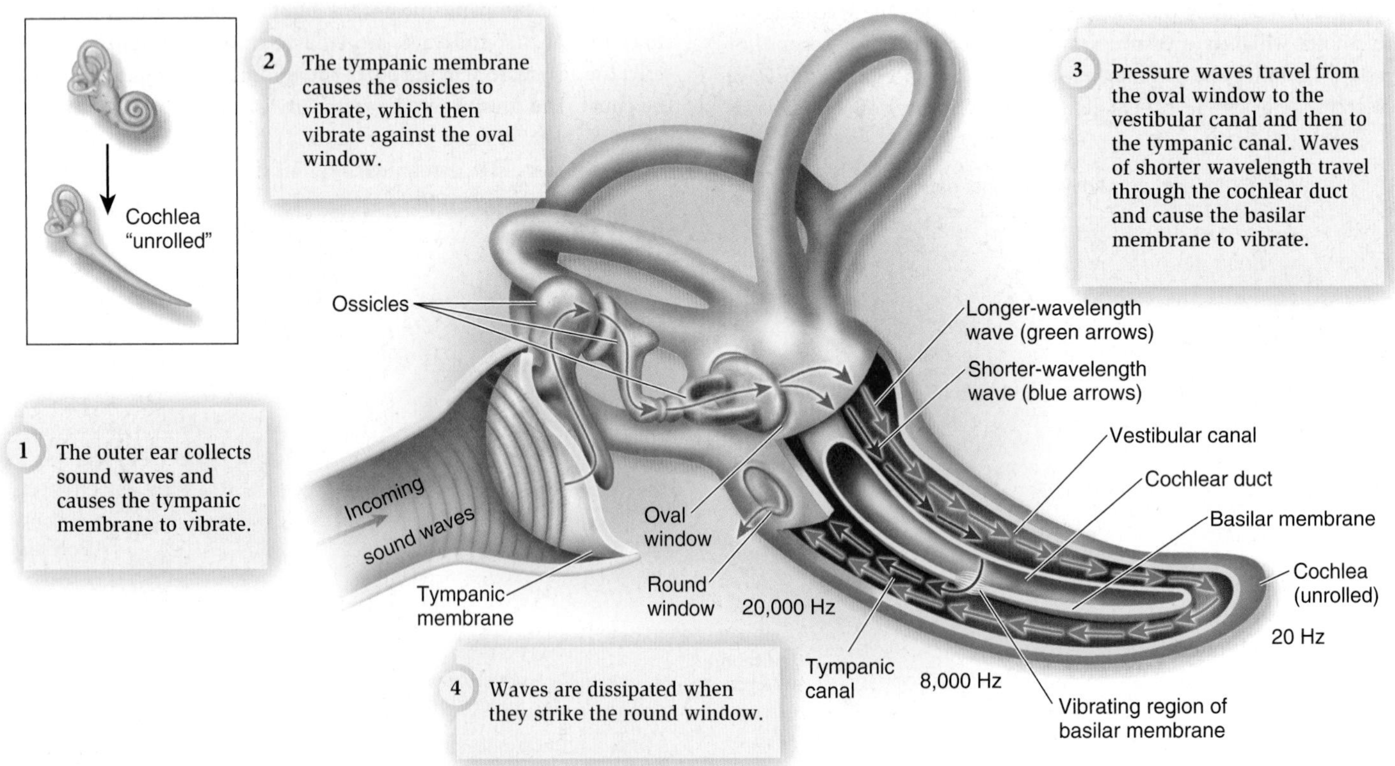

Figure 43.6 **Movement of sound waves through the human ear.**

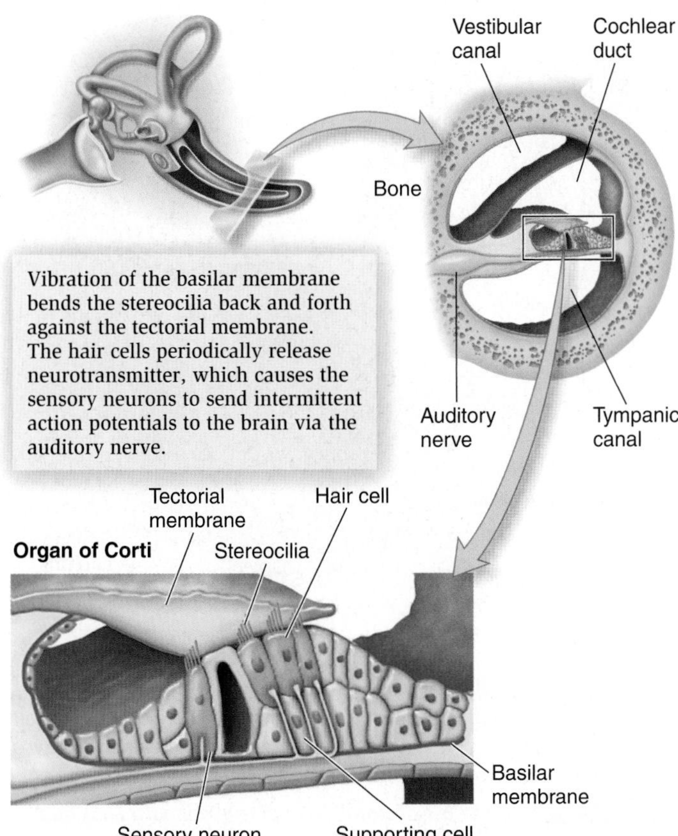
Vestibular canal
Cochlear duct
Bone
Vibration of the basilar membrane bends the stereocilia back and forth against the tectorial membrane. The hair cells periodically release neurotransmitter, which causes the sensory neurons to send intermittent action potentials to the brain via the auditory nerve.
Auditory nerve
Tympanic canal
Tectorial membrane
Hair cell
Organ of Corti
Stereocilia
Basilar membrane
Sensory neuron
Supporting cell

Figure 43.7 Transduction of mechanical vibrations to action potentials in the organ of Corti.

are short and rigid, and they vibrate in response to high-frequency waves. Longer and more resilient fibers are near the other end of the cochlea and vibrate to lower-frequency waves. For this reason, hair cells closer to the oval and round windows respond to high-pitched sounds, whereas those at the opposite end are triggered by lower-pitched sounds. When we hear a great number of sound frequencies at once, such as at a musical concert, the waves traveling through the cochlea activate hair cells all along the basilar membrane in a physical representation of the music! These cells stimulate sensory neurons, which send multiple action potentials to the auditory areas of the brain for processing. The most incredible feature of this process, however, is that the mammalian ear and brain can "tune in" to all of these frequencies simultaneously.

Adaptations for Hearing Allow Animals to Live in a Wide Range of Habitats

Different animals are capable of hearing sounds of different pitches. As noted, humans can hear between 20 and 20,000 Hz (conversation averages 90–300 Hz). Insectivorous bats, toothed whales, and some species of moths may have the highest-frequency range (100,000–240,000 Hz), and baleen whales and elephants may have the lowest-frequency range (to nearly 1 Hz). These adaptations increase the animals' ability to communicate and survive. For instance, high-pitched sounds are useful to bats that need to locate small prey such as flying insects, whereas low-pitched sounds carry great distances through water or air and are especially useful for animals with large territories.

Locating a Sound A vital feature of hearing is the ability to locate the origin of a sound. For example, this ability makes the difference between a successful and an unsuccessful hunter. How does an animal locate a sound? Under most circumstances, sound does not arrive at both ears simultaneously. Sound waves coming from the right, for example, excite the sensory receptors in the right ear first and the left ear some milliseconds later, and therefore, the brain receives action potentials from the auditory nerves of each ear at slightly different times. The brain interprets the time difference to determine the origin of the sound.

Animals such as owls that rely on hearing to pinpoint prey tend to be extremely good at identifying the direction of a sound. An interesting experiment demonstrated this by outfitting owls in a dark room with small headphones. Just as in a human hearing test, sounds could be sent to either headphone or to both. If the investigator sent a high-pitched noise that mimicked the sounds of a mouse first to the left headphone, and then a single millisecond later to the right headphone, the owl turned its head to the left, because the owl's brain perceived the sound to be coming from that direction. If the noises reached both headphones simultaneously, the owl behaved as if the signal was coming from directly in front of its head.

Hearing in Air and Water Amphibians have the special challenge of hearing both on land and underwater, their two natural environments. The ears of amphibians have several interesting specializations that are adapted to these requirements. First, they do not have external ears. The tympanic membrane is located on the outer surface of the head behind the eye (Figure 43.8). This arrangement allows them to swim through the water without being impeded by pinnae and to receive sound waves from any direction. Amphibians also have unusually wide Eustachian tubes. When they are on land, sound waves can pass from one ear through the Eustachian tube to the pharynx and then travel up the Eustachian tube of the other ear, where they hit the backside of the tympanic membrane.

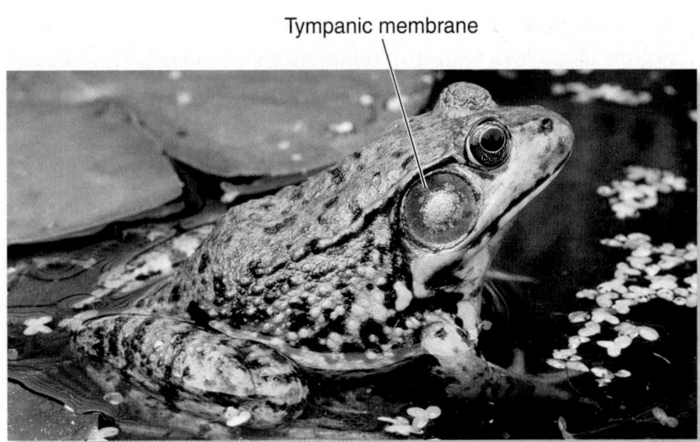
Tympanic membrane

Figure 43.8 The tympanic membrane of a frog.

Echolocation Bats in the air, whales and dolphins in the sea, and shrews in underground tunnels generate high-frequency sound waves to determine the location of an object. In this phenomenon, called **echolocation**, the sound waves bounce off a distant object, like an echo, and return to the animal. The time it takes for the sound to return indicates the distance of the object. Echolocation can be useful in situations where vision is limited, such as in the dark. Although the sound of an echo may be familiar to you, it is hard to imagine how other animals can use echoes to locate objects with such precision. To get a feel for how this works, imagine shouting "hello" across a canyon. After a few seconds, you would hear the echo return to you. Now imagine that you knew the speed of sound in air (approximately 344 m/sec at sea level) and had a stopwatch that measured the length of time it took for the echo to return. From this, you could calculate the distance across the canyon (accounting for the fact that sound first traveled in one direction, then back again, so you would have to divide your result by 2). Bats and other echolocating animals perform this feat instantly, without the aid of mathematics or calculators, and can do it while they and the object they are tracking are both in motion!

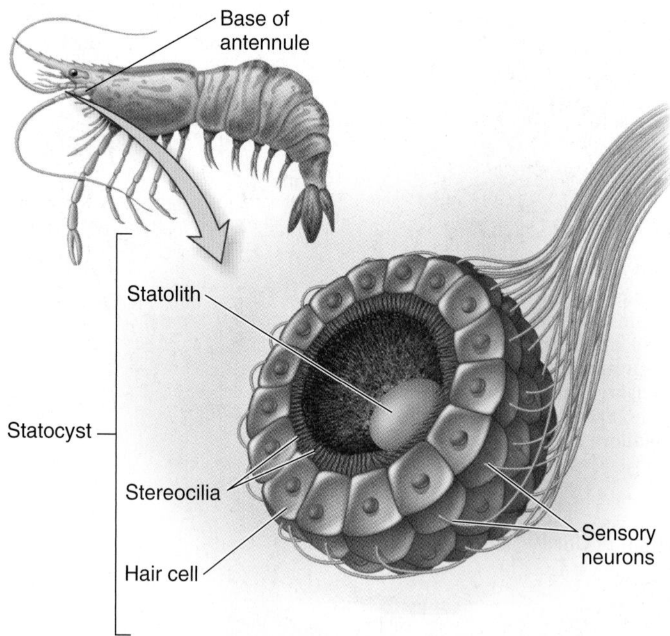

Figure 43.9 **Sensing of balance in aquatic invertebrates.** Statocysts located near the antennae consist of a sphere of sensory hair cells surrounding one or more stony statoliths. When the animal moves, gravity shifts the statolith and stimulates the hair cells beneath it.

BioConnections: Is the use of statoliths unique to animals? Look back to Figure 36.14 for a hint.

The Sense of Balance Is Mediated by Statocysts in Invertebrates and the Vestibular System in Vertebrates

Let's now turn our attention to another form of mechanoreception, the sense of balance, also called equilibrium. Balance is part of a broader sense called proprioception, which is the ability to sense the position, orientation, and movement of the body. Being able to sense body position is vital for the survival of animals. This is how a lobster, for example, rights itself when flipped over by a predator or how a bird maintains its balance while flying.

Many aquatic invertebrates have sensory organs called **statocysts** that send information to the brain about the position of the animal in space (Figure 43.9). Statocysts are small round structures made of an outer sphere of hair cells and one or more **statoliths**, which are tiny granules of sand or other dense objects. When the animal moves, gravity alters the statoliths' position. If the animal turns on its left side, for example, the movement of statoliths stimulates a new set of hair cells to release neurotransmitter, generating action potentials in sensory neurons that inform the brain of the change in body position.

Several experiments have demonstrated the importance of statoliths. In one particularly dramatic example, researchers replaced the statoliths of crayfish with iron filings. Moving a magnet to different positions around the animal displaced the filings, causing the animal to change its position, and even to swim upside down when the magnet was placed directly above its head.

The organ of balance in vertebrates, known as the **vestibular system**, is located in the inner ear next to the cochlea (Figure 43.10). The vestibular system is composed of a series of fluid-filled sacs and tubules, which provide information about either linear or rotational movements. The utricle and saccule, the two sacs nearest the cochlea, detect linear movements of the head (Figure 43.10a), such as those that occur when an animal runs and jumps or changes from lying down to sitting up. The hair cells within these structures are embedded in a gelatinous substance that contains granules of calcium carbonate called **otoliths** (from the Latin, meaning ear stones), which are analogous to statoliths. When the head moves forward, the heavy otoliths are temporarily "left behind" as they are dragged forward more slowly, and the weight of the otoliths bends the stereocilia of the hair cells in the direction opposite that of the linear movement. This changes the membrane potential of the hair cells and alters the electrical responses of nearby sensory neurons. These signals are sent to the brain, which uses them to interpret how the head has moved.

Three **semicircular canals** connect to the utricle at bulbous regions called the **ampullae** (singular, ampulla). The function of the semicircular canals is to detect rotational motions of the head (Figure 43.10b). The hair cells in the semicircular canals are embedded in the ampullae in a gelatinous cone called the **cupula**. When the head moves, the fluid in the canal shifts in the opposite direction. This movement of fluid pushes on the cupula and bends the stereocilia of the hair cells in the direction of the fluid flow, which is opposite that of the motion. The three canals are oriented at right angles to each other, and each canal is maximally sensitive to motion in its own plane. For example, the canal that is oriented horizontally would respond most to rotations such as shaking the head "no," whereas the other canals respond to "yes" motions and to tipping the ear to the shoulder. Overall, by comparing the signals from the three canals, the brain can interpret the motion of the head in three dimensions.

The vestibular system of vertebrates provides conscious information about body position and movement. It also supplies unconscious information for reflexes that maintain normal posture, control head and eye movements, and assist in locomotion. Motion sickness

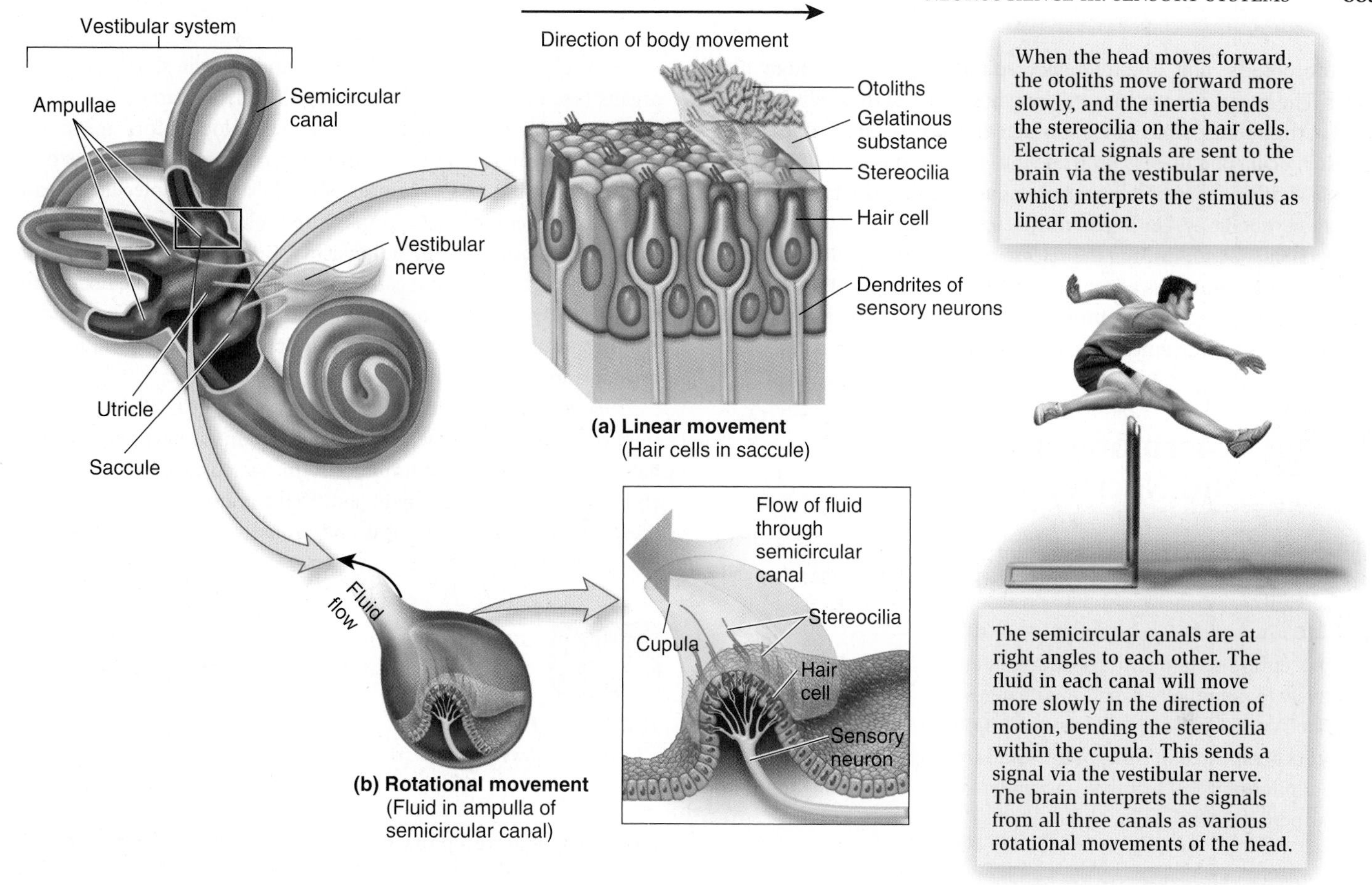

Vestibular system
Ampullae
Semicircular canal
Vestibular nerve
Utricle
Saccule
Direction of body movement
Otoliths
Gelatinous substance
Stereocilia
Hair cell
Dendrites of sensory neurons
(a) Linear movement
(Hair cells in saccule)
When the head moves forward, the otoliths move forward more slowly, and the inertia bends the stereocilia on the hair cells. Electrical signals are sent to the brain via the vestibular nerve, which interprets the stimulus as linear motion.
Fluid flow
Flow of fluid through semicircular canal
Cupula
Stereocilia
Hair cell
Sensory neuron
(b) Rotational movement
(Fluid in ampulla of semicircular canal)
The semicircular canals are at right angles to each other. The fluid in each canal will move more slowly in the direction of motion, bending the stereocilia within the cupula. This sends a signal via the vestibular nerve. The brain interprets the signals from all three canals as various rotational movements of the head.

Figure 43.10 **The vertebrate vestibular system.**

Concept Check: *Note the orientation of the three semicircular canals with respect to each other. Why are the canals oriented in three different planes?*

in humans results when the vestibular system has not adapted to unfamiliar patterns of movement, such as spinning around or sailing aboard a ship. In these cases, the sense from your vestibular system conflicts with information coming from your vision. Vertigo, an illusion of movement or spinning when one is stationary, can result from malfunction of the vestibular system.

Mechanoreceptors in the Lateral Line System Detect Movements in Water

Fishes and some toads detect changes in their environment through a **lateral line** system (**Figure 43.11**). This sensory system has hair cells that detect changes in water currents brought about by waves, nearby

Lateral line
Stimulus
Scale
Pore
Lateral line canal
Nerve
Lateral line organ
Cupula
Stereocilia
Hair cell
Sensory neuron endings
1 A stimulus such as water motion enters the pores and travels through the canal.
2 The stimulus bends the cupula and stereocilia within the cupula, which causes hair cells to release neurotransmitter and thereby stimulate sensory neurons.

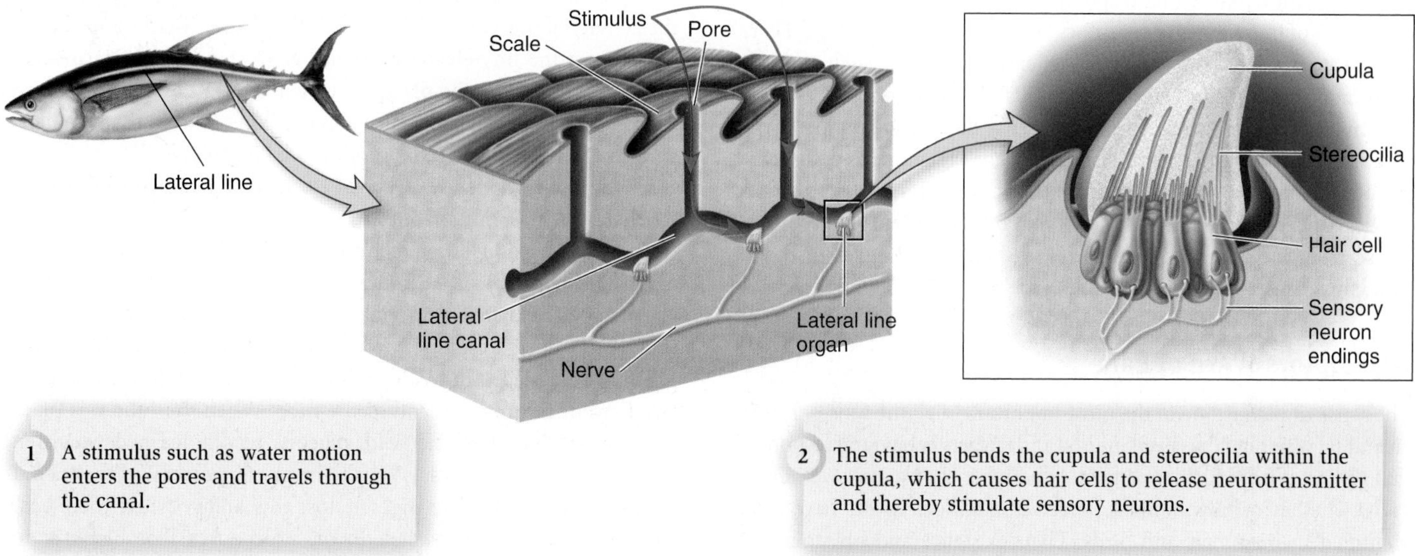

Figure 43.11 **Mechanoreceptors in the lateral line system of fish that detect changes in water movement.**

moving objects, and low-frequency sounds traveling through the water. The lateral line organ runs along both sides of the body and the head of the animal. Small pores let water enter into a lateral line canal. The stereocilia of hair cells protrude into a gelatinous structure called a cupula within the lateral line organ (similar to the cupula of the vertebrate vestibular system). When the cupula is moved by the water, the stereocilia bend, causing the release of neurotransmitter from the hair cell. This stimulates a response in sensory neurons at the base of the hair cells. The response provides information to the brain about changes in water movement, such as the approach of a predator.

43.3 Thermoreception and Nociception

Learning Outcome:

1. Explain how thermoreception and nociception are vital to the safety and survival of an animal.

The perception of temperature and pain enables animals to respond effectively to potentially dangerous changes in their environments. As described in this section, these sensory stimuli are related in that their receptors are located in some of the same areas (skin; see Figure 43.3), share similar physical features, and under certain conditions result in similar perceptions.

Thermoreceptors Detect Temperature

Sensing the outside temperature is important for animals because their body temperature is affected by the external temperature. This is particularly true for ectotherms, animals whose body temperature changes with the environmental temperature. Animals can survive at body temperatures only within certain limits, because cell membranes and the proteins in cells function optimally only within a particular temperature range. Thermoreceptors respond to cold or hot temperatures by activating or inhibiting enzymes within their plasma membranes, which alters membrane channels. There are two types of thermoreceptors: those that respond to hot and those that respond to cold. Both of these types of receptors are free neuronal endings. Thermoreceptors are often linked with reflexive behaviors, such as when an animal steps on a hot surface and pulls its foot away.

In addition to skin receptors that sense the outside temperature, thermoreceptors in the brain also detect changes in core body temperature. Activation of skin or brain thermoreceptors triggers physiological and behavioral adjustments that help maintain homeostasis. These changes, described in Chapter 46, include changes in blood flow, shivering, and behaviors such as seeking shade or sunlight.

Nociceptors Warn of Pain

The sense of pain is one of the most important of all the senses. It tells an animal whether it has been injured and triggers behavioral responses that protect it from further danger. Although in many cases we cannot know whether or how animals perceive pain, nociceptors have been identified in mammals and birds. Their existence and physiological significance in other phyla are likely but are still being investigated.

Nociceptors are free neuronal endings in the skin and internal organs (see Figure 43.3). They respond to tissue damage or to stimuli that are about to cause tissue damage. Nociceptors are unusual because they can respond not only to external stimuli, such as extreme temperatures, but also to internal stimuli, such as molecules released into the extracellular space from injured cells. Damaged cells release a number of substances, including acids and small signaling molecules called prostaglandins, that cause inflammation and make nociceptors more sensitive to painful stimuli. Anti-inflammatory drugs such as aspirin and ibuprofen reduce pain by preventing the production of prostaglandins.

Signals arising from nociceptors travel to the CNS and reach the cerebrum, where the type or cause of the pain is interpreted. The signals are also sent to the limbic system, which holds memories and emotions associated with pain, and to the brainstem reticular formation, which increases alertness and arousal—an important response to a painful stimulus.

43.4 Electromagnetic Reception

Learning Outcome:

1. Identify ways that animals use electromagnetic receptors to sense their environments.

Electromagnetic receptors detect radiation within a wide range of the electromagnetic spectrum, including those wavelengths that correspond to visible light, ultraviolet light, and infrared light, as well as electrical and magnetic stimuli. Photoreceptors are specialized electromagnetic receptors that respond to light and are covered in Section 43.5. Here, we will examine the ability of some animals to sense electrical and magnetic fields and also heat in the form of infrared radiation.

The ability to detect the presence of nearby prey or predators can be especially challenging in certain animals that inhabit low-light environments. The more ways that an animal has to detect other animals, the better it can avoid danger or obtain a meal. Many fishes living in dark waters can detect weak electrical signals given off by the activity of muscles and nerves of other fishes. Sharks and rays can even detect the tiny electrical signals generated by the heartbeats of their prey. Similarly, the platypus, which lives in the murky waters of streams and ponds, has receptors on the skin of its bill that can detect very small electrical currents produced by its prey.

Homing pigeons use a type of electromagnetic sensing to return to their starting points from as far away as 1,500 km. This navigational feat is made possible by small particles of magnetite (iron oxide) in their beaks that indicate direction by acting as a compass. The magnetic particles respond to the Earth's magnetic field and alter the activity of neurons that project to the brain. In one experiment, pigeons were placed individually in large tubes and trained that food was present in only one end of the tube. When the tube was placed in a changeable magnetic field, pigeons readily learned which end contained food based solely on the magnetic polarity of the tube. In another experiment, the pigeons lost this ability when their beaks were anesthetized or cooled down, procedures that block action potentials from being sent to the brain. This demonstrates that their

Figure 43.12 Infrared sensing. Sensory pits enable a white-lipped pit viper (*Cryptelytrops albolabris*) to detect the heat given off by its prey.

magnetic sensing ability is located in the beak and communicated by nerves to the brain.

Magnetic field sensing is not unique to birds. Magnetite has also been found in the heads of migratory fishes such as rainbow trout. However, this probably does not entirely explain the extraordinary ability of migratory animals to navigate great distances, because other cues, such as smell and visual recognition of landmarks, also appear to play roles in this process.

Venomous snakes known as pit vipers (a group that includes copperheads and rattlesnakes) can localize prey in the dark with detectors that sense the heat emitted from animals as infrared radiation. These detectors are located in pits on each side of the head between the eyes and nostrils (Figure 43.12). Within the pit, a thin, nerve-rich, temperature-sensitive membrane becomes activated in response to infrared waves emitted by live animals. When the snake detects the heat of the animal, it localizes its prey by moving its head back and forth until both pits detect the same intensity of radiation. This indicates that the prey is centered in front of the snake.

Electrical, magnetic, and infrared sensing are adaptations for long-distance migration or low-light environments. When light is available, however, photoreception becomes a dominant sensory ability in many animals, as described next.

43.5 Photoreception

Learning Outcomes:

1. Describe the structure of invertebrate visual organs.
2. Describe the structure of the vertebrate (single-lens) eye and how it forms images.
3. Compare and contrast the structure and function of rods and cones.
4. Explain the mechanisms by which photoreceptors respond to light in a single-lens eye.
5. Outline the neural pathway by which visual signals travel to reach the brain.

Although it is a form of electromagnetic reception, photoreception is such an important and widespread sense that we will cover it separately here. Visual systems employ specialized neurons called photoreceptors, which detect photons of light arriving from the sun or other light sources, or reflecting off an object. A photon is the fundamental unit of electromagnetic radiation and has the properties of both a particle and a wave. The properties of light are described in Chapter 8. In this section, we will examine the organs found in animals, usually called **eyes**, that detect light and send signals to the brain. The amazing features of these organs reflect the importance of vision in the animal world.

Eyecups and Compound Eyes Are Found in Certain Invertebrates

The ability to sense light is an ancient adaptation found even in many unicellular organisms, where it may provide a selection advantage for photosynthesis or protection. Typically, such organisms may have a single eyespot that contains a small number of light-sensitive molecules, but lack the ability to discern the direction of a light source or to interpret a visual image.

A slightly more sophisticated visual organ is found in free-living flatworms such as *Planaria*. These animals have two concave structures called **eyecups** (Figure 43.13). Each eyecup contains the endings of photoreceptor cells and a layer of pigment cells that shields the photoreceptors from one side. The left and right eyecups receive light from different directions. This allows the eyecups to detect not only

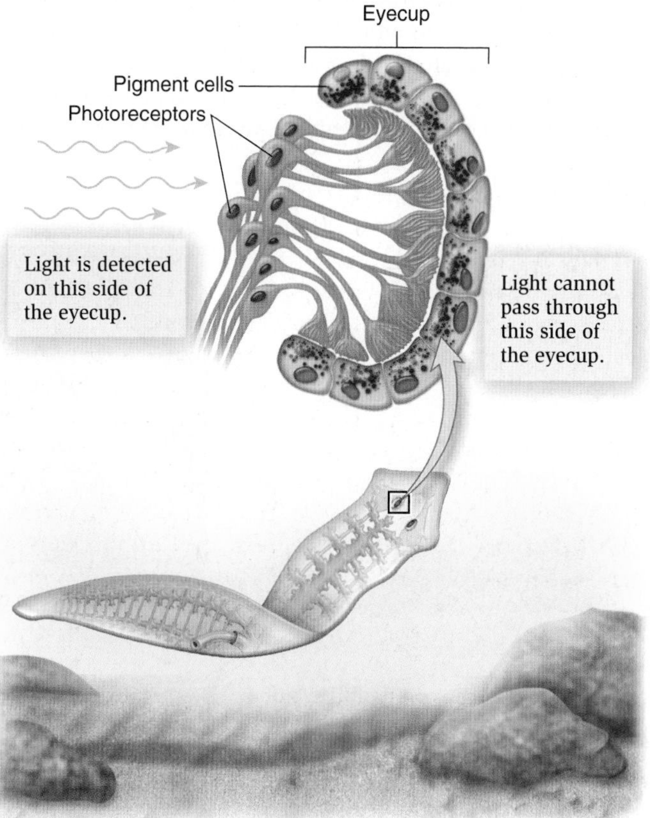

Figure 43.13 The eyecup of a flatworm. The orientation of the eyecup allows light to stimulate photoreceptors from primarily one direction. This type of eye senses only the presence or absence of light and does not form visual images.

the presence or absence of light, but also its direction. The nervous system compares the amount of light detected by each eyecup, and the flatworm moves toward darkness, a behavior that protects it from predators. However, this type of photoreceptor does not form visual images of the environment.

In contrast, arthropods and some annelids have image-forming **compound eyes** (Figure 43.14a), which consist of several hundred to more than 10,000 light detectors called **ommatidia** (singular, ommatidium) (Figure 43.14b). Each ommatidium makes up one facet of the eye. Within the ommatidium, a **lens**—composed in this case of a cornea and a crystalline cone—focuses light onto a long central structure called a rhabdom (Figure 43.14c). The rhabdom is a column of light-sensitive microvilli that project from the cell membranes of the photoreceptor cells of the ommatidium (Figure 43.14d). The light-sensitive molecules required for vision are located in the microvilli; the extensive surface area imparted by the microvilli provides the animal with increased sensitivity to light. Pigmented cells surrounding the photoreceptor cells absorb excess light and thereby isolate each ommatidium from its neighbors.

Each ommatidium senses the intensity and color of light. Combined with the different inputs from neighboring ommatidia, the compound eye forms an image that the brain interprets. Animals such as bees and fruit flies, with large numbers of ommatidia, presumably have sharper vision than those with fewer sensory cells, such as grasshoppers.

As anyone who has tried to swat a fly knows, the compound eye is extremely sensitive to movement and helps flying insects evade birds and other predators. Behavioral studies have shown, however, that the resolving power of even the best compound eye is considerably less than that of the single-lens eye, which we will consider next.

Vertebrates and Some Invertebrates Have a More Complex Single-Lens Eye

Single-lens eyes are found in vertebrates and also in some mollusks, such as squid, octopus, and some snails, and in some annelids. In such eyes, different patterns of light emitted from images in the animal's field of view are transmitted through a small opening, or **pupil**, through the lens, to a sheetlike layer of photoreceptors called the **retina** at the back of the eye (Figure 43.15a). This forms a visual image of the environment on the retina. The activation of these photoreceptors triggers electrical changes in neurons that pass out of the eye through the **optic nerves**, carrying the signals to the brain. The brain then interprets the visual image that was transmitted.

As illustrated in Figure 43.15a, the vertebrate eye has a strong outer sheath called the **sclera** (the white of the eye). Between the sclera and the retina is a layer of blood vessels called the choroid. At the front of the eye, the sclera is continuous with a thin, clear layer known as the **cornea**. Within the eye are two cavities, the anterior and posterior cavities. The anterior cavity is the part of the eye between the lens and the cornea. It is subdivided into the anterior chamber and posterior chamber. The anterior chamber is located between the cornea and the **iris**—the circle of pigmented smooth muscle responsible for eye color. This chamber is filled with a thin liquid called the **aqueous humor**. The posterior chamber is the space between the single lens and the iris. The larger posterior cavity between the lens and the retina contains the thicker **vitreous humor**, which helps maintain the shape of the eye.

The hole in the center of the iris is the pupil. The size of the pupil changes when the muscles of the iris reflexively relax or contract to allow more or less light to enter the eye.

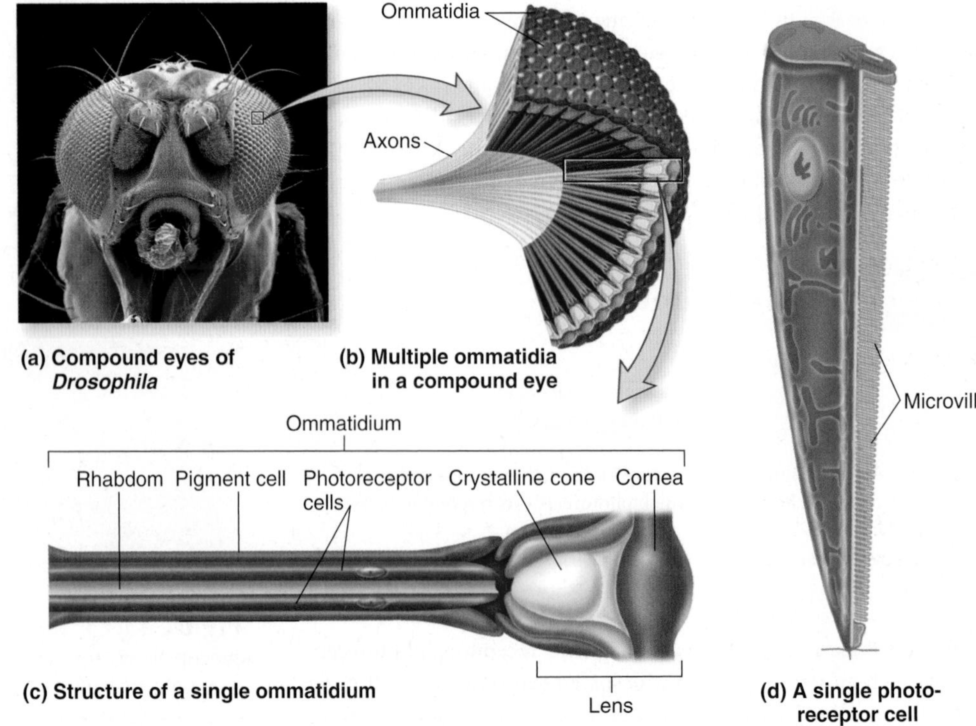

Figure 43.14 The compound eye of insects. **(a)** Close-up of the eyes of a fruit fly (*Drosophila melanogaster*). **(b)** Each eye has approximately 1,000 ommatidia, which form a sheet on the surface of the eye. **(c)** Each ommatidium has a lens that directs light to the photosensitive rhabdom. **(d)** Extending from each photoreceptor cell into the rhabdom are many light-sensitive microvilli.

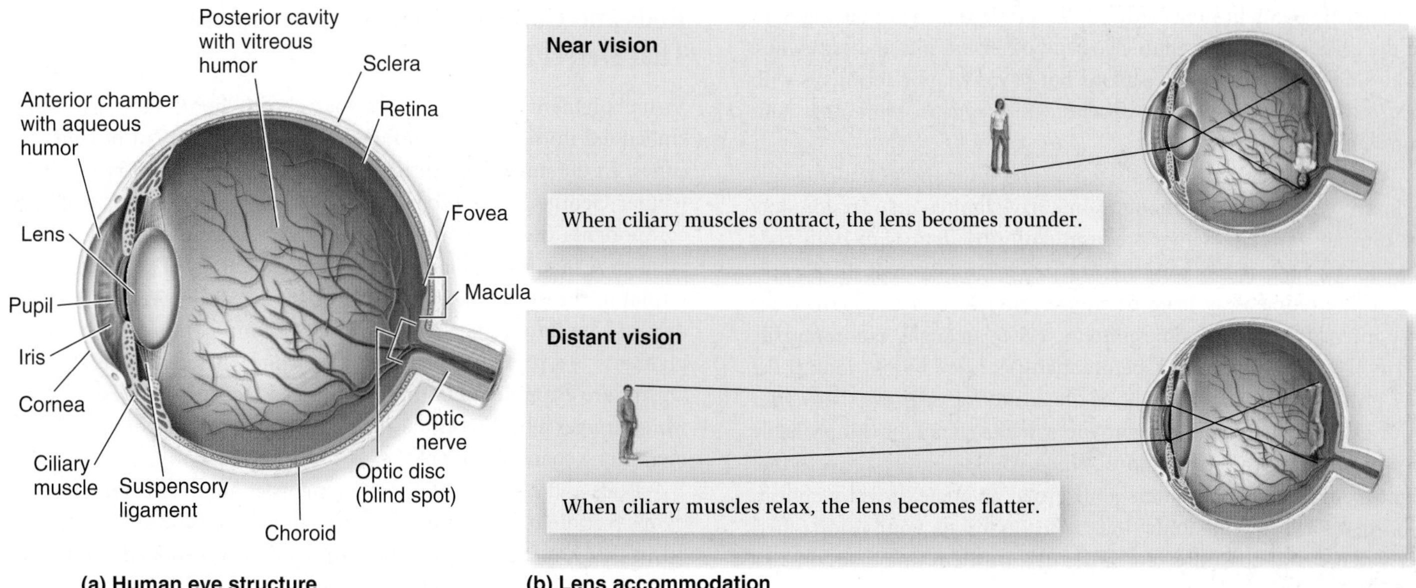

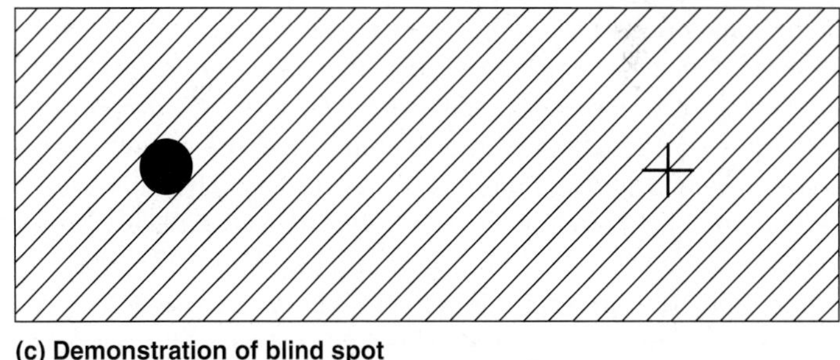

Figure 43.15 **The vertebrate single-lens eye.** **(a)** The structure of the human eye. **(b)** Changes in lens shape during accommodation. When an object is near, the ciliary muscles contract and the lens becomes rounder, causing light to bend more. When the object is far away, the ciliary muscles relax and the lens flattens. **(c)** Demonstration of the blind spot. First, hold this picture up in front of your face, or place the book on a table and stand over it. Next, close your left eye and stare at the black spot with your right eye while you move the picture toward and away from your face. At some point, light reflecting off the plus (+) sign will fall directly on your blind spot, and it will seem to disappear.

Because light radiates in all directions from a light source, light must be bent (refracted) inward toward the photoreceptors at the back of the eye. This is accomplished by the cornea and the lens. Whenever light passes from one medium to another medium of a different density, light waves will bend (try looking at a pencil in a glass partly filled with water). The cornea, which is at the interface between the air and the aqueous humor, initially refracts the light. The light then passes through the thicker lens, where it is refracted again and focused onto the layer of photoreceptors, the retina, at the back of the posterior chamber. The bending of the incoming light results in an upside-down and laterally inverted image on the retina, but the brain adjusts for this, and the image is perceived correctly.

The lens is adjusted to focus light that comes from different distances. In fishes and amphibians, the lens is moved forward or backward. In the avian and mammalian eye, the lens remains stationary but changes shape to become more or less round. When the lens is stretched, it flattens, and light passing through it bends less than when it is round. Contraction and relaxation of the ciliary muscles adjust the lens according to the angle at which light enters the eye, a process called **accommodation** (Figure 43.15b).

How does the retina form such sharp images? The region on the retina directly in line with the pupil and lens is called the macula. Near the center of the macula is the **fovea**, which contains the highest density of photoreceptors for color. The fovea is responsible for the sharpness with which we and many other animals see in daylight. However, the retina also has limitations in forming images. In the eye, as mentioned, the image initiates signals that travel from the retina to the brain through neurons that exit the eye in the optic nerve. In vertebrates, the point on the retina where the optic nerve leaves the eye is called the **optic disc**. The optic disc does not have any photoreceptors, forming a "blind spot" where light does not activate a response (Figure 43.15c). Invertebrates with single-lens eyes do not have a blind spot, because the photoreceptors in their eyes are at the front of the retina; consequently, the optic nerve does not pass through the layer of photoreceptors before leaving the eye.

Rods and Cones Are Photoreceptor Cells

Vertebrates have two types of photoreceptors with names that are derived from their shapes: rods and cones (Figure 43.16). **Rods** are very sensitive to low-intensity light and can respond to as little as one photon, but they do not discriminate different colors. Rods are used mostly at night, and they send signals to the brain that generate a black-and-white visual image. **Cones** are less sensitive to low levels of light but, unlike rods, can detect color. Rods and cones are cells with three functional parts: the outer segment, inner segment, and synaptic terminal. The **outer segment** of the cell contains folds of membranes that form stacks like discs (Figure 43.17). These discs contain the pigment molecules that absorb light. The **inner segment** of the cell contains the cell nucleus and other cytoplasmic organelles. Rods and cones do not have axons but have synaptic terminals with neurotransmitter-containing vesicles, which synapse with other neurons within the retina.

Nocturnal animals (those active predominantly at night) rely primarily on rod vision, though some have limited color vision, too. In diurnal animals (active predominantly by day) with both rods and cones, such as humans, the rods are located around the periphery of the retina away from the fovea. Therefore, it is easiest to see low-intensity light if it comes into the eye at an angle. You can easily verify this. In early evening, before many stars are visible, look at the sky until you notice a star out of the corner of your eye. Now shift your gaze to where you thought you saw the star. You will probably not be able to locate it anymore. When you look away again so that light from the dim star enters your eye at an angle, it will reappear. This demonstrates that under low-light conditions, your vision is better when the light is directed to the part of the retina that contains only rods.

Cones are used in daylight by most diurnal vertebrate species and by some insects such as the honeybee, which can detect the yellow color of pollen. Compared to rods, the human retina has fewer cones, which are clustered in and around the fovea. Cones provide sharp images because of their density at the fovea. Although they are less sensitive to light than rods, this is less critical in daylight because the amount of light reaching the eyes at this time far exceeds what is needed to stimulate any photoreceptor cell. Because the two types of photoreceptor are specialized for either night or day vision, neither rods nor cones function at peak efficiency at twilight. This accounts for our relatively poor vision at this time.

Rods and Cones Contain Visual Pigments That Detect Light

Visual pigments are molecules that absorb light; they are found embedded in the disc membranes of the outer segment of rods and cones. In the mid-20th century, American biologist and Nobel Prize winner George Wald discovered that these pigments consist of two components bonded together. The first is **retinal**, a derivative of vitamin A that is capable of absorbing light energy. The discovery of retinal in the visual pigment explains the need for vitamin A in the diet and its importance in vision. The second component of visual pigments is a protein called **opsin**, of which there are several types. Opsins are examples of G-protein-coupled receptors (see Figure 9.7), which trigger a biochemical cascade that changes the permeability of membrane channels to ions.

Rods and cones have visual pigments containing different types of opsin protein. These pigments are named according to the type of opsin they contain. In rods, the visual pigment is named **rhodopsin** (Figure 43.18). Cones contain several types of visual pigments called **cone pigments**, or **photopsins**. In humans, photopsins are composed of retinal plus one of three possible opsin proteins. Each type of opsin protein determines the wavelength of light that the retinal in a cone can absorb. For this reason, each cone pigment can respond best to red, green, or blue light. Any given cone cell makes only one type of cone pigment. Many different shades of these colors can be perceived, however, because the brain uses information about the proportion of each type of cone that was stimulated to generate all other colors.

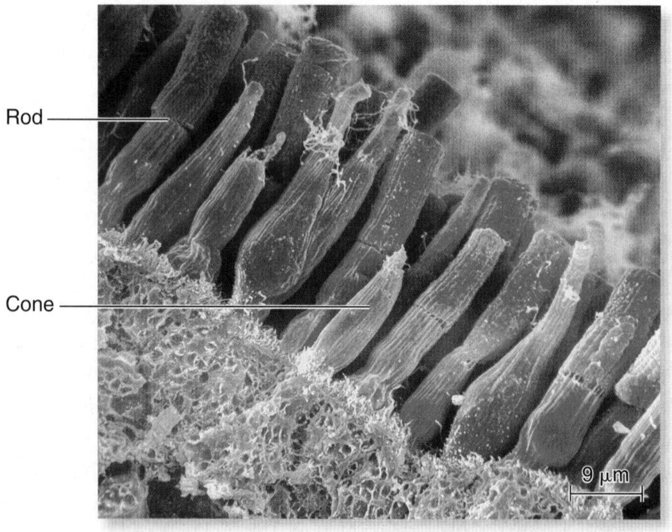

Figure 43.16 Rod and cone photoreceptors. Rods are shown as green and cones as blue in this false-color scanning electron micrograph (SEM).

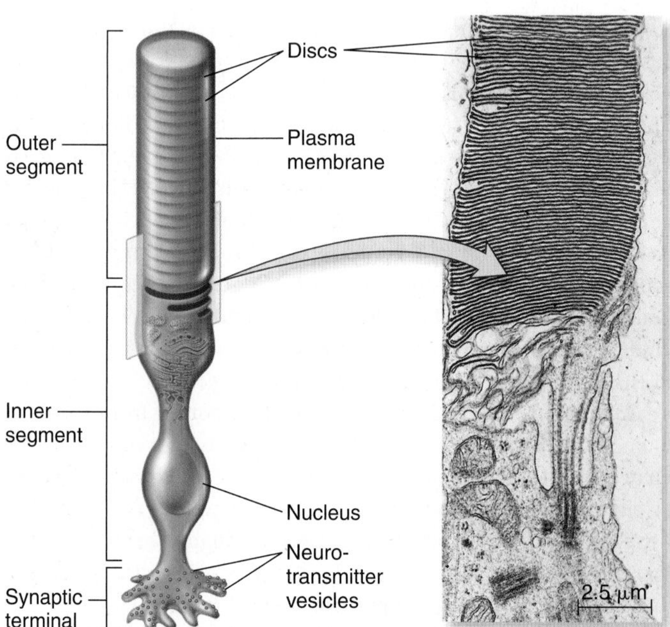

Figure 43.17 Structure of a photoreceptor. The illustration shows the structure of a rod photoreceptor and its appearance in a transmission electron micrograph (TEM). Note the multiple stacks of membranous discs in the outer segment of the cell.

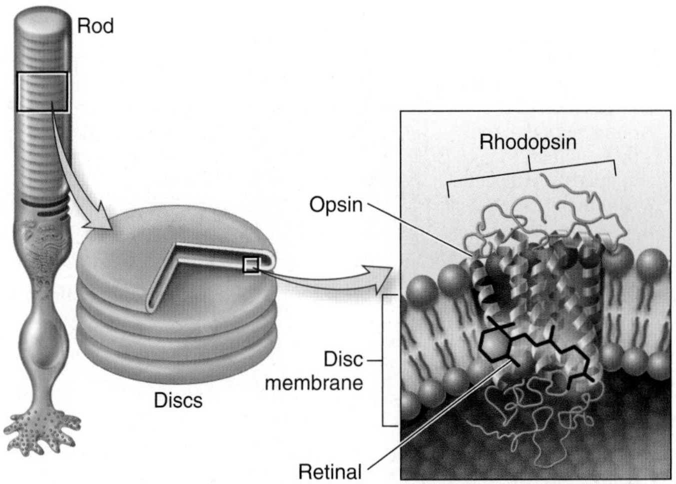

Figure 43.18 A visual pigment. The visual pigment rhodopsin is found within the membrane of the rod photoreceptor discs. It is composed of a transmembrane protein, opsin, that is bonded to a molecule of retinal, a derivative of vitamin A that is capable of absorbing light energy.

BIOLOGY PRINCIPLE Structure determines function. The stacks of discs in the photoreceptor outer segment are an excellent example of a structural adaptation that increases surface area without significantly increasing volume, thereby allowing for greater light-capturing ability and improving the function of the cell as a sensory cell.

Genomes & Proteomes Connection

Color Vision Is an Ancient Adaptation in Animals

Color vision requires the presence of at least two types of opsins with optimal sensitivities to light of different wavelengths. Genome analyses and behavioral and neurophysiological testing of many animals have confirmed that color vision is widespread across invertebrate and vertebrate taxa, and arose early in animal evolution. It appears that the primordial color sensitivity was to short wavelengths in the ultraviolet/blue region of visible light (refer back to Figure 8.4), and from there additional opsins evolved with sensitivities to medium and longer wavelengths. Primates and amphibians have three types of color-sensitive photoreceptors; most other vertebrates that have been studied, however, have four. It is impossible for us to understand how the world must appear to such an animal, which presumably sees shades of colors we cannot. Some insects, including the butterfly *Papilio xuthus*, appear to have up to six types of light receptors! A well-studied model is the honeybee, which has four color-sensitive photoreceptors with sensitivities ranging from 300 nm to 600 nm (refer back to Figure 8.4); they can see ultraviolet wavelengths that humans can't. A rainbow would appear very differently to a honeybee than it does to us; to the bee, the rainbow would extend past the blue edge and stop short of the red edge.

It is debatable why color vision provided a selective advantage to animals. Some investigators believe that it produced more acute vision due to an overall improved contrast and was not originally related to such commonly observed phenomena today, including colorful plumage displays or brightly colored flowers. Coloration in plants may have evolved in response to the color sensitivity in animals, not vice versa, but this is unclear.

An intriguing exception to vertebrates are most mammals and in particular primates. When mammals first evolved, they were most likely entirely nocturnal animals and therefore there would have been no selection pressure to retain sensitive color vision; recall that cones are less sensitive to low levels of light and therefore color vision is limited at night. Most mammals today are dichromatic, that is, they only have two types of opsins and therefore see a more limited color palette than do other vertebrates. Primates, however, are an exception, because a third opsin reappeared in the course of primate evolution, one that was sensitive to middle (green) wavelengths. It has been postulated that this may have imparted an advantage that allowed fruit and leaf-eating primates to better discern orange, red, and yellow fruits against a background of green leaves, but again, this is uncertain.

About 92% of human males and over 99% of females have normal trichromatic color vision. However, problems in color vision may result from defects in the cone pigments arising from mutations in the opsin genes. The most common is red-green color blindness, which occurs predominantly in men (1 in 12 males compared with 1 in 200 females). Individuals with red-green color blindness either lack the red or green cone pigments entirely or, more commonly, have one or both of them in an abnormal form. In one form, for example, an abnormal green pigment responds to red light as well as green, making it difficult to discriminate between the two colors.

Researchers have determined that color blindness results from a recessive mutation in one or more genes encoding the opsins. Genes encoding the red and green opsins are located very close to each other on the X chromosome, but the gene encoding the blue opsin is located on a different chromosome. In males, the presence of only one X chromosome means that a single recessive allele from the mother results in red-green color blindness, even though the mother herself is not color blind (Figure 43.19).

Photons Change Photoreceptor Activity by Altering the Conformation of Visual Pigments

Photoreceptors differ from other sensory receptor cells because at rest in the dark their membrane potential is slightly depolarized, whereas in response to a light stimulus, it is hyperpolarized rather than depolarized (Figure 43.20). In the dark, the cell membranes of the outer segments of resting cells are highly permeable to sodium ions. Sodium ions (Na^+) flow into the cytosol of the cell through open Na^+ channels in the outer segment membrane. The Na^+ channels are gated by intracellular cyclic guanosine monophosphate (cGMP). In the dark, cytosolic concentrations of cGMP are high, keeping Na^+ channels open and depolarizing the cell. This depolarization results in a continuous release of the neurotransmitter glutamate from the synaptic terminal of the photoreceptor. The photoreceptor synapses with a postsynaptic cell that is the next neuron in the visual pathway. This initiates a series of events within the retina that is interpreted by

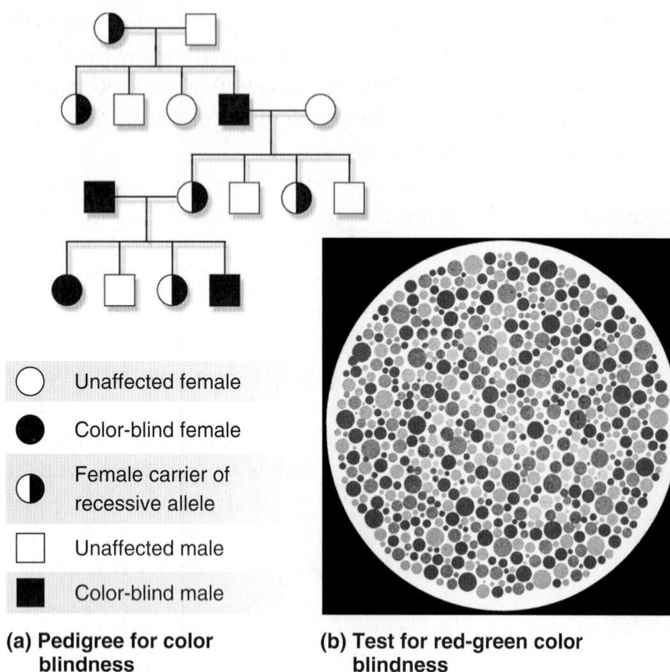

Figure 43.19 **Color blindness.** **(a)** A pedigree for red-green color blindness showing all possible offspring. **(b)** A standard eye test to screen for red-green color blindness. People with red-green color blindness will not see the number 74 hidden in this picture.

Concept Check: *Why is red-green color blindness rare in females?*

the brain as an absence of light. In contrast, when exposed to light, the Na^+ channels in the outer segment membranes of the photoreceptor close. The resulting reduction in sodium ion concentrations leads to a hyperpolarization of the cell. In response, the release of glutamate is stopped. This results in a series of cellular activations within the retina and brain that is interpreted as a visual image.

Let's take a more detailed look at the signal transduction pathway that allows a photoreceptor to respond to light (Figure 43.21). When the photoreceptor is exposed to light, the retinal within the visual pigment absorbs a photon. The energy of the photon alters the retinal from *cis*-retinal to *trans*-retinal, an isomer with a slightly different conformation due to a rotation at one of the molecule's double bonds (see Figure 43.21). This change results in retinal briefly dissociating from the opsin protein, causing the opsin to change its three-dimensional shape and activate a G protein called transducin, located in the disc membrane. The activated transducin, in turn, activates another disc protein, the enzyme phosphodiesterase.

The action of phosphodiesterase results in the closure of Na^+ channels in the outer segment membrane. Remember that in the dark, these channels are kept open by intracellular cGMP. However, when phosphodiesterase is activated, it reduces the amount of cytosolic cGMP by converting it to GMP. This results in the dissociation of cGMP from the channels, so the channels close, and sodium ions stop moving into the cell. The membrane potential of the cell becomes less positive than it was in the dark. Therefore, the response of the cell is a hyperpolarization that is proportional to the intensity of the light. The final result is a decrease in glutamate release from the photoreceptor (see Figure 43.20), ultimately leading to a visual image.

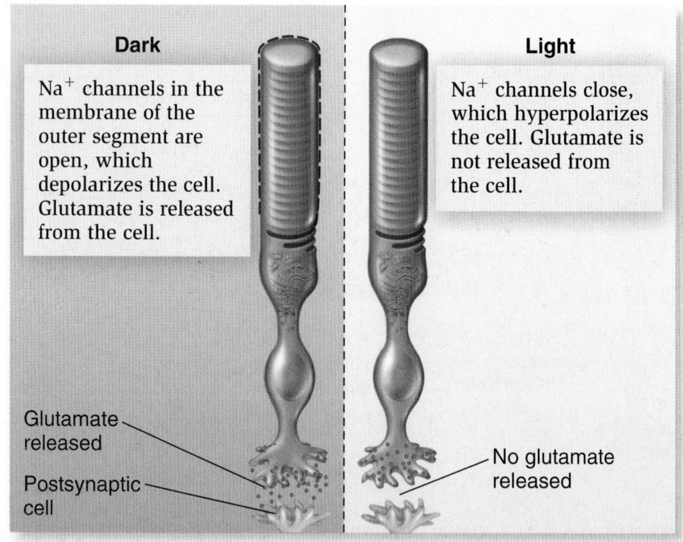

Figure 43.20 **Membrane potential response of photoreceptors to dark and light.**

The sequential activation of enzymes following activation of a single photoreceptor results in an amplification of the original signal (refer back to Figure 9.14). Because of this property, we are able to detect extremely low levels of light.

The Visual Image Is Refined in the Retina

Thus far, we have considered the structure of the vertebrate eye and how photoreceptors transduce light. We will now turn our attention to the neural pathway through which the visual signal travels to reach the brain. To do so, we must consider the cellular organization of the retina. The vertebrate retina has several layers of cells (Figure 43.22). The photoreceptors (rods and cones) form the deepest layer, closest to the inside of the sclera. Immediately behind the photoreceptors is a pigmented epithelium that absorbs light that missed the photoreceptors; this prevents scattering of light within the retina, which could degrade the sharpness of vision. Because the photoreceptors are positioned at the back of the retina, light must pass through two transparent layers of cells before it reaches them. The middle layer contains **bipolar cells**, so named because one end (or "pole") of the cell synapses with the photoreceptors, and the other end relays responses to a top layer of cells, the **ganglion cells**. The ganglion cells send their axons out of the eye into the optic nerve. In addition, two other types of cells, horizontal and amacrine cells, are interspersed across the retina.

The pathway for light reception begins at the rods and cones. These photoreceptor cells release neurotransmitter molecules that affect the membrane potential of bipolar cells. The membrane potential of the bipolar cells determines the amount of neurotransmitter that they release, which, in turn, controls the membrane potential of ganglion cells. When a threshold potential is reached in ganglion cells, action potentials are sent out of the eye via the optic nerve to the brain. These signals travel along pathways that include the thalamus, brainstem, cerebellum, and the cerebral cortex. Visual information is further refined and interpreted within the vision centers of the cortex.

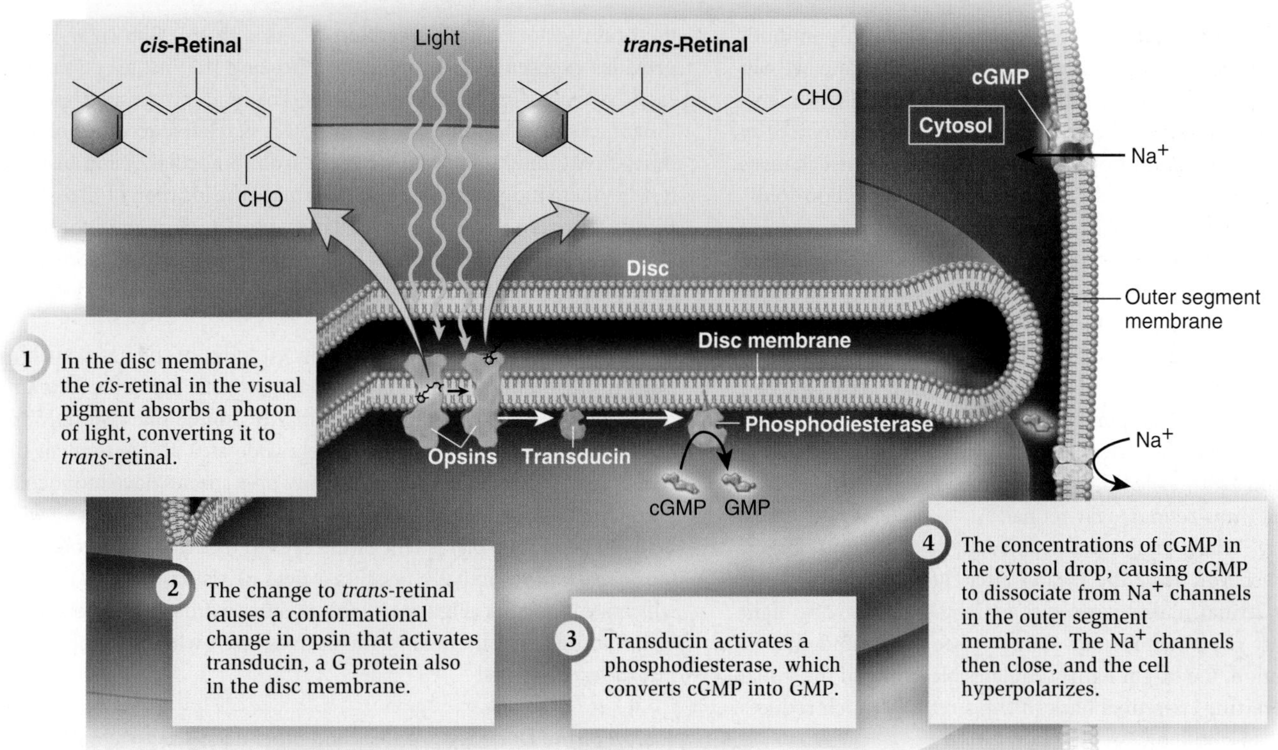
cis-Retinal
CHO
Light
trans-Retinal
CHO
cGMP
Cytosol
Na+
Disc
Disc membrane
Outer segment membrane
1 In the disc membrane, the cis-retinal in the visual pigment absorbs a photon of light, converting it to trans-retinal.
Opsins
Transducin
Phosphodiesterase
Na+
cGMP
GMP
2 The change to trans-retinal causes a conformational change in opsin that activates transducin, a G protein also in the disc membrane.
3 Transducin activates a phosphodiesterase, which converts cGMP into GMP.
4 The concentrations of cGMP in the cytosol drop, causing cGMP to dissociate from Na+ channels in the outer segment membrane. The Na+ channels then close, and the cell hyperpolarizes.

Figure 43.21 **Signal transduction pathway in photoreceptor (rod) cells in response to light.**

The cortex responds to such characteristics of the visual scene as whether something is moving, how far away it is, how one color compares with another, and the nature of the image (for example, a face). The cortex does not form a picture in the brain, but forms a spatial and temporal pattern of electrical activity that is perceived as an image.

Horizontal and amacrine cells modify electrical signals as they pass from the photoreceptors to the ganglion cells. These cells adjust the signal significantly, enhancing an animal's ability to visualize a scene by emphasizing the differences between images. Horizontal cells make connections between photoreceptors and help to define the boundaries of an image. Amacrine cells are important in adjusting the eye to different light intensities and increasing the sensitivity of the eye to moving images. The ability of the retina to refine the image is especially well developed in birds and reptiles. These animals have complex retinas that process the image extensively before it is interpreted in the brain.

Vertebrate Eyes Are Adapted to Environmental Conditions and Life Histories

Many vertebrates show unique modifications of their visual systems that are the result of evolutionary adaptations to environmental conditions. Other adaptations have occurred as a result of behavioral requirements for obtaining food or attracting a mate.

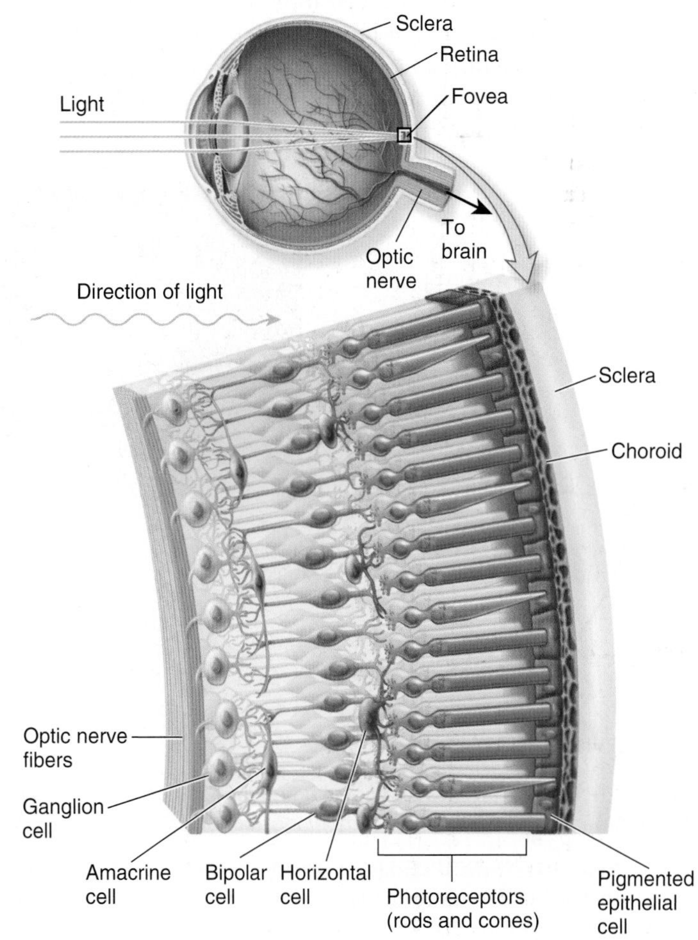
Sclera
Retina
Fovea
Light
To brain
Optic nerve
Direction of light
Sclera
Choroid
Optic nerve fibers
Ganglion cell
Amacrine cell
Bipolar cell
Horizontal cell
Photoreceptors (rods and cones)
Pigmented epithelial cell

Figure 43.22 **The arrangement of cells in the retina.** Light passes through two layers of cells before it reaches the photoreceptors. Amacrine and horizontal cells integrate the responses of the bipolar and ganglion cells. The ganglion cells generate action potentials to carry the information to the brain.

Variation in Light Intensity Over the course of a day, the amount of light that stimulates the eye varies widely. The vertebrate eye can adjust to the differences in illumination in part by adjusting the diameter of the pupil. In addition, the eye can adapt to light differences by changing the relative amounts of *cis*- and *trans*-retinal. Such changes can alter the light sensitivity of the eye by as much as a million-fold!

In a dark environment, a large amount of *cis*-retinal is available for light absorption. When light stimulates *cis*-retinal to change into *trans*-retinal, the visual pigment no longer responds to additional light and is said to be bleached. Bleaching makes the eye less sensitive to low levels of light. When you enter a dark movie theater on a bright day, for example, most of your visual pigments have been bleached by the sunlight and are temporarily unavailable for use. For a few minutes, you are unable to detect low levels of light. This process of adjusting to dim light, called dark adaptation, involves the gradual reconversion of *trans*-retinal to *cis*-retinal.

Conversely, when you move from a dark room into bright sunlight, as when you leave the theater after the matinee, the large amount of *cis*-retinal makes your eyes extremely sensitive to light and temporarily overwhelms your ability to see colors and shapes. In light adaptation, the bright light stimulates bleaching of the visual pigments, converting *cis*-retinal back to *trans*-retinal, which reduces the response to bright light.

Differences in Eye Placement Except for some of the ray-finned fishes, blunt-headed cetaceans, and most amphibians, vertebrate animals have some degree of **binocular** (or stereoscopic) **vision**. Animals with both eyes located at or near the front of the head, such as primates and raptors, have greater binocular vision, because the overlapping images coming into both eyes are processed together in the brain to form one perception (**Figure 43.23**). Binocular vision provides excellent depth perception because the images come into each eye from slightly different angles. The brain processes those tiny differences to determine where an object is relative to other objects in its environment. Predators benefit from binocular vision because it helps them judge distance and determine the location of their prey. Binocular vision is present in predatory birds such as the snowy owl (Figure 43.23a) and mammals (as well as in predatory insects such as mantids), and also in arboreal animals that must judge distances between tree limbs.

In contrast, animals with eyes on the sides of the head, such as most fishes, blunt-headed cetaceans, amphibians, herbivorous mammals, and insects, have strictly or primarily **monocular vision**. Monocular vision allows an animal to see a wide area at one time, at the cost of reduced depth perception. Many prey species have monocular vision, perhaps because it helps them scan for predators across a wide field of vision. The placement of the eyes in the American woodcock (*Scolopax minor*) actually permits a field of vision of 360°, most of which is monocular (Figure 43.23b); in other words, these and similar birds can see directly behind themselves, even when digging in the dirt for earthworms!

Vision in the Deep Sea Fishes and other deep-sea vertebrates have color vision that is limited primarily to the color blue. Light with longer wavelengths that would be seen as red or orange does not usually penetrate more than ~6 m into the water, whereas the higher-energy, shorter-wavelength light, which we see as blue, can penetrate to greater depths. Aquatic animals that live in the deep sea are usually capable of seeing only blue, because they generally have only one opsin, which is responsive to blue light. Deep-dwelling fishes tend to

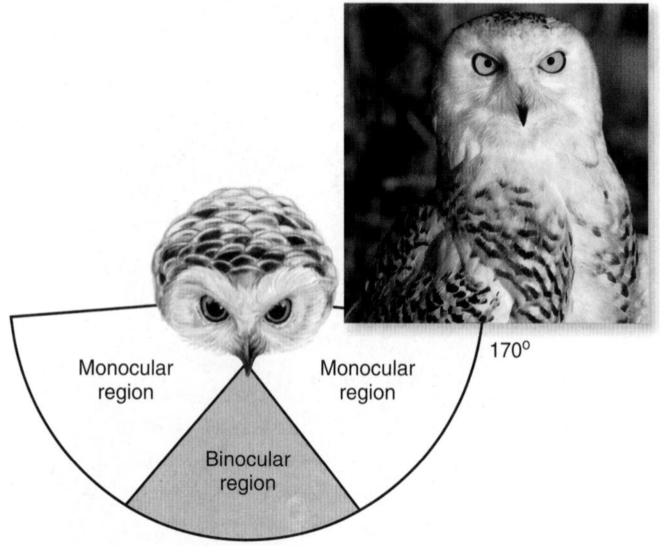

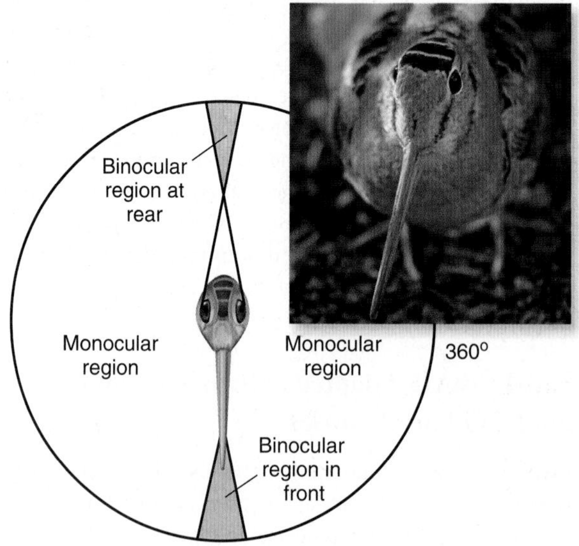

Figure 43.23 Examples of binocular and monocular fields of vision. Visual fields are shown for **(a)** the snowy owl (*Bubo scandiacus*) and **(b)** the American woodcock (*Scolopax minor*). Monocular regions are white; binocular regions are shaded.

BIOLOGY PRINCIPLE Living organisms interact with their environment. Different animals may interact with their environments in ways that are hard for us to imagine. Not only do different animals have vastly different visual fields, but some see better at night than we do, some see wavelengths of light that we cannot see, and some can resolve images traveling at speeds that would appear blurry to us.

be drab-colored, because most wavelengths of light do not penetrate that far and therefore could not reflect off the surface of the fishes; thus, there was no selection pressure for evolving bright coloration. In those deep-dwelling fishes with more than one type of opsin, the additional visual pigments detect the bioluminescence (self-generated light, like that of a firefly) produced by their own or other species.

By contrast, fishes that live near the water's surface sometimes have four or five different opsins, giving them excellent color vision. Not surprisingly, shallow water and surface-dwelling fishes are often very colorful, because light of all wavelengths penetrates shallow water. These fishes have adapted by using coloration for protection (camouflage) or for identification.

Vision with Speed Peregrine falcons, like other raptors, have exceptionally good eyesight. They dive at speeds up to 150 to 200 miles per hour to capture prey, and their visual system is specialized to maintain great sensitivity during the dive. The retina of a falcon has two foveas, one deep and one shallow. The deep fovea has a central pit and provides the highest visual accuracy when objects are far away and to the side. However, when the falcon uses the deep fovea, it does not have good depth perception. The shallow fovea provides the best images of nearby objects and good depth perception. When the falcon dives, it first uses the deep fovea to locate the prey and steer toward it. When the falcon gets closer, it switches to the shallow fovea. Some details of the image are sacrificed, but the improved depth perception allows it to gauge when it will reach its prey.

Vision in the Dark: The Tapetum Lucidum Did you ever wonder why cats' eyes seem to glow in photographs, as in the chapter-opening photo? What you see is light being reflected off the tapetum lucidum, a reflective layer of tissue located beneath the photoreceptors at the back of the eye in some animals. When light enters the retina, some of it passes into the layer containing the photoreceptors, but some light also passes through this layer. Normally, the pigmented epithelium at the back of the eye absorbs this light. However, in animals with a tapetum lucidum, the light reflects off this structure and back to the photoreceptors. This gives the photoreceptors a "second chance" to capture light, making it easier to detect low levels of light. Many nocturnal animals, both vertebrate and invertebrate, have a tapetum lucidum, which increases their ability to see in the dark.

43.6 Chemoreception

Learning Outcomes:

1. Describe olfaction and gustation in insects.
2. Explain how olfactory receptors respond to the binding of odor molecules.
3. Outline how receptor cells within taste buds respond to the binding of food molecules.

Chemoreception includes the senses of smell (**olfaction**) and taste (**gustation**), both of which involve detecting chemicals in air, water, or food. These chemicals bind to chemoreceptors, which, in turn, initiate electrical responses in other neurons that pass into the brain. Amazingly, the binding of a single molecule to a receptor cell can sometimes be perceived as an odor! Airborne molecules that bind to olfactory receptors must be small enough to be carried in the air and into the nose. Taste molecules can be heavier because they are conveyed in food and liquid.

Taste and smell are closely related. In fact, the distinction is largely meaningless for aquatic animals, because for them all chemoreception comes through the water. Even in terrestrial animals, about 80% of the perception of taste is actually due to activation of olfactory receptors. (This is why food loses its flavor when the sense of smell is impaired, such as when you have a cold.) In this section, we explore chemoreception in insects and then in mammals.

Olfaction and Taste in Insects Involve Chemoreceptors in Sensory Hairs

Insects are highly dependent on odor and taste for finding food and mates. In insects, chemoreceptors are neurons that are located on sensory hairs on the proboscis (coiled tongue), legs, feet, and antennae. Each sensory hair on the proboscis and feet has a pore at the tip through which the substance passes. As the example in Figure 43.24a

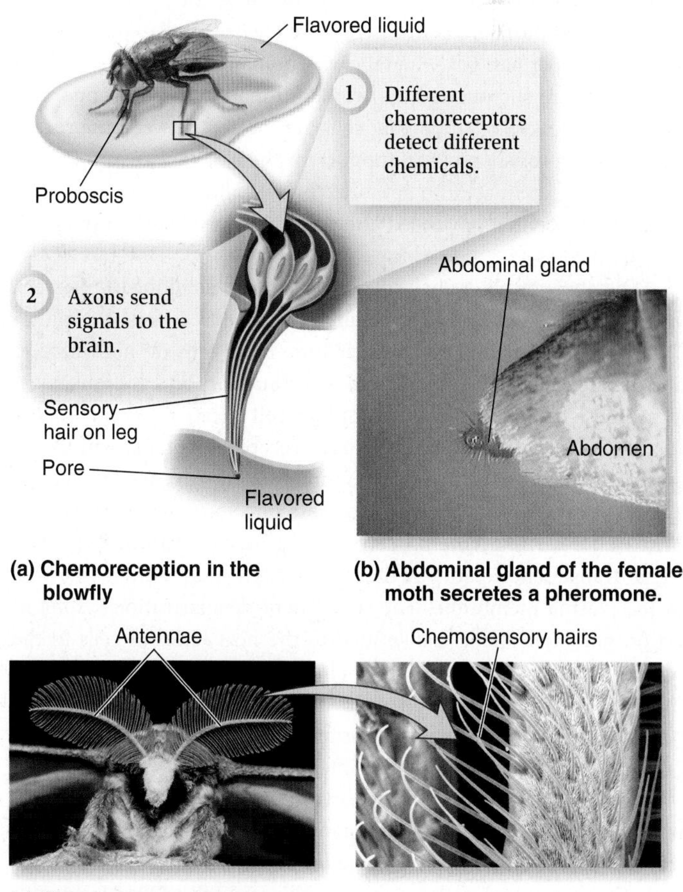
Flavored liquid
1 Different chemoreceptors detect different chemicals.
Proboscis
2 Axons send signals to the brain.
Sensory hair on leg
Pore
Flavored liquid
Abdominal gland
Abdomen
(a) Chemoreception in the blowfly
(b) Abdominal gland of the female moth secretes a pheromone.
Antennae
Chemosensory hairs
(c) Chemosensory hairs on antennae of the male moth detect the pheromone.

Figure 43.24 Chemoreception in insects. **(a)** Chemosensory cells of the blowfly located on the proboscis, legs, and feet sense different chemicals. **(b)** Female moths secrete a pheromone from abdominal glands, which is detected by **(c)** chemosensory hairs in a male's antennae that bind odor molecules.

shows, the blowfly has four separate chemoreceptors within each hair, and each of these neurons responds to different molecules. Dendrites of the chemoreceptor cells inside the pore bind to the molecules and initiate a sensory transduction pathway that opens ion channels in the membrane. This depolarizes the plasma membrane of the chemoreceptor cell and generates action potentials, which are sent to the brain for interpretation.

In certain moths, males have elaborate antennae that can sense pheromones, extremely potent signaling molecules given off by a female. The female secretes a sex-attractant pheromone into the air from an abdominal gland (**Figure 43.24b**), and the chemosensory hairs on the male's antennae (**Figure 43.24c**) can detect extremely low concentrations of it from several kilometers away. This highly sensitive detection system enables the male to locate the female in the dark.

Mammalian Olfactory Receptors Respond to the Binding of Odor Molecules

The olfactory sensitivity of mammals varies widely depending on their supply of olfactory receptor cells, which ranges from 5 or 6 million in humans to 100 million in rabbits and 220 million in dogs. Olfactory receptors are neurons that are located in the epithelial tissue at the upper part of the nasal cavity in mammals (**Figure 43.25a**). These cells are surrounded by two additional cell types: supporting cells and basal cells. Supporting cells are located between the receptor cells and provide physical support for the olfactory receptors. The basal cells differentiate into new olfactory receptors every 30–60 days, replacing those that have died after prolonged exposure of their cell endings.

Olfactory receptors have dendrites from which long, thin extensions called cilia extend into a mucous layer that covers the epithelium. Despite the similarity in structure, these cells do not function like the mechanoreceptor hair cells of the auditory and vestibular systems; these cilia are different from hair cell stereocilia, which bend. Instead, olfactory receptor cells have receptor proteins within the plasma membranes of the cilia (**Figure 43.25b**). Airborne molecules dissolve in the mucus and bind to these olfactory receptor proteins. When an odor molecule binds to its receptor protein, it initiates a signal transduction pathway that ultimately opens sodium channels in the plasma membrane. The subsequent depolarization results in action potentials being transmitted to the next series of cells in the olfactory bulbs of the brain. The olfactory bulbs, which are part of the limbic system, are a collection of neurons that act as an initial processing center of olfactory information and relay it to the cerebral cortex for further processing and interpretation.

The relative size of the olfactory bulbs indicates the importance of olfaction to an animal. In humans, the olfactory bulbs make up only about 5% of the weight of the brain, whereas in nocturnal animals like rats and mice, they can comprise as much as 20%. Even with their relatively limited olfactory sensitivity, however, humans have the capacity to detect up to 10,000 different odors, and other mammals are thought to respond to many more. Exactly how this is possible remained a mystery until 1991, when two scientists uncovered the molecular basis of olfaction.

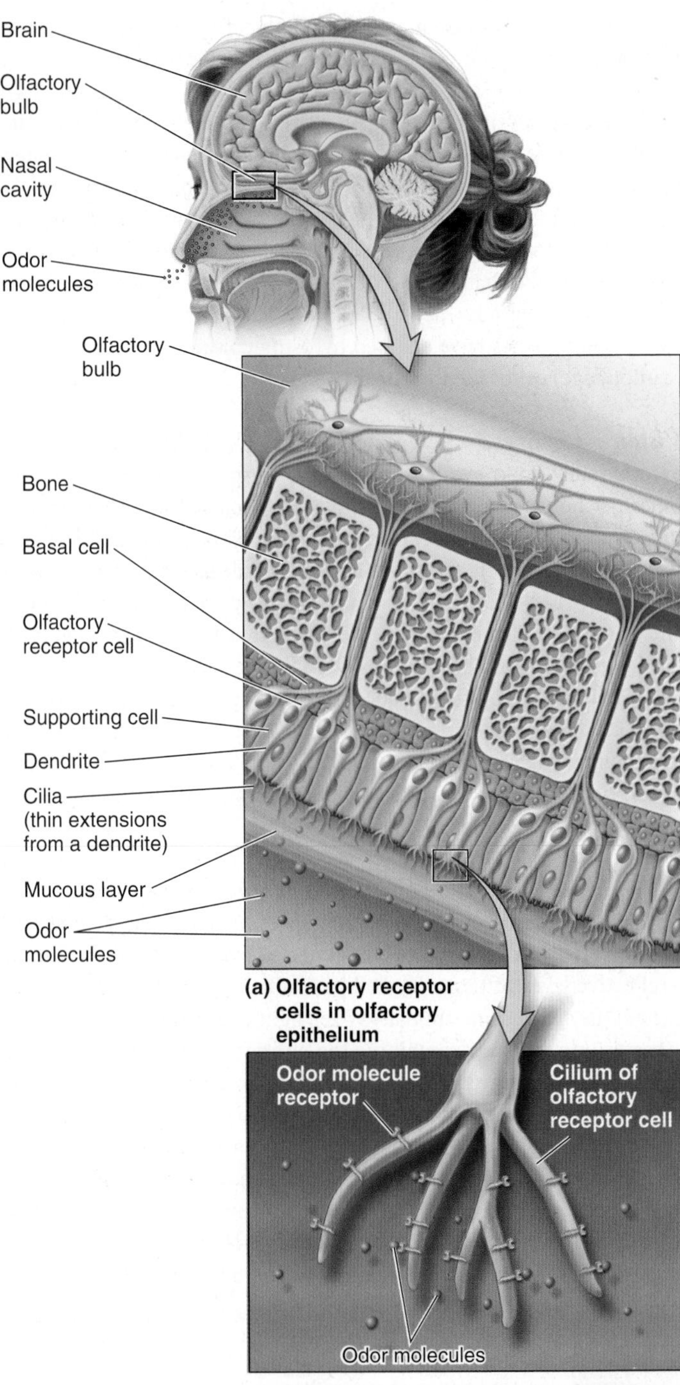

Figure 43.25 **Olfactory structures in the human nose.** Odor molecules dissolve in a layer of mucus that coats the olfactory receptor cells. The molecules bind to protein receptors in the membrane of cilia that extend from the olfactory receptor cells. Action potentials in the olfactory receptor cells are conducted to cells in the olfactory bulb, and from there to the brain for interpretation. Basal cells periodically differentiate into new olfactory receptor cells, replacing dead or damaged cells.

FEATURE INVESTIGATION

Buck and Axel Discovered a Family of Olfactory Receptor Proteins That Bind Specific Odor Molecules

How does the olfactory system discriminate among thousands of different odors? American neuroscientists Linda Buck and Richard Axel set out to study this question. When they began, two hypotheses were proposed to explain this phenomenon. One possibility was that many different types of odor molecules might bind to one or just a few types of receptor proteins, with the brain responding differently depending on the number or distribution of the activated receptors. The second hypothesis was that olfactory receptor cells can make many different types of receptor proteins, each type binding a particular odor molecule or group of odor molecules.

To begin their study, Buck and Axel assumed that olfactory receptor proteins would be highly expressed in the olfactory receptor cells, but not in other parts of the body. Based on previous work, they also postulated that the receptor proteins would be members of the large family of G-protein-coupled receptors (GPCRs). As shown in **Figure 43.26**, they isolated olfactory receptor cells from rats and then used a homogenizer to break open the cells to release the mRNA. The mRNA was purified and then used to make complementary DNA (cDNA) using the enzyme reverse transcriptase. This generated a large pool of cDNAs, representing all of the genes that were expressed in olfactory receptor cells at the time of mRNA collection. To determine if any of these cDNAs encoded GPCRs, they used primers that recognized regions within previously identified GPCR genes that are

Figure 43.26 **Buck and Axel identified olfactory receptor proteins in olfactory receptor cells.**

HYPOTHESES 1. Many different types of odor molecules bind to just a few types of receptor proteins. 2. Odor molecules are detected by many specific olfactory receptor proteins belonging to the family of G-protein-coupled receptors (GPCRs).

KEY MATERIALS Laboratory rats (*Rattus norvegicus*), PCR reagents, DNA-sequencing gels.

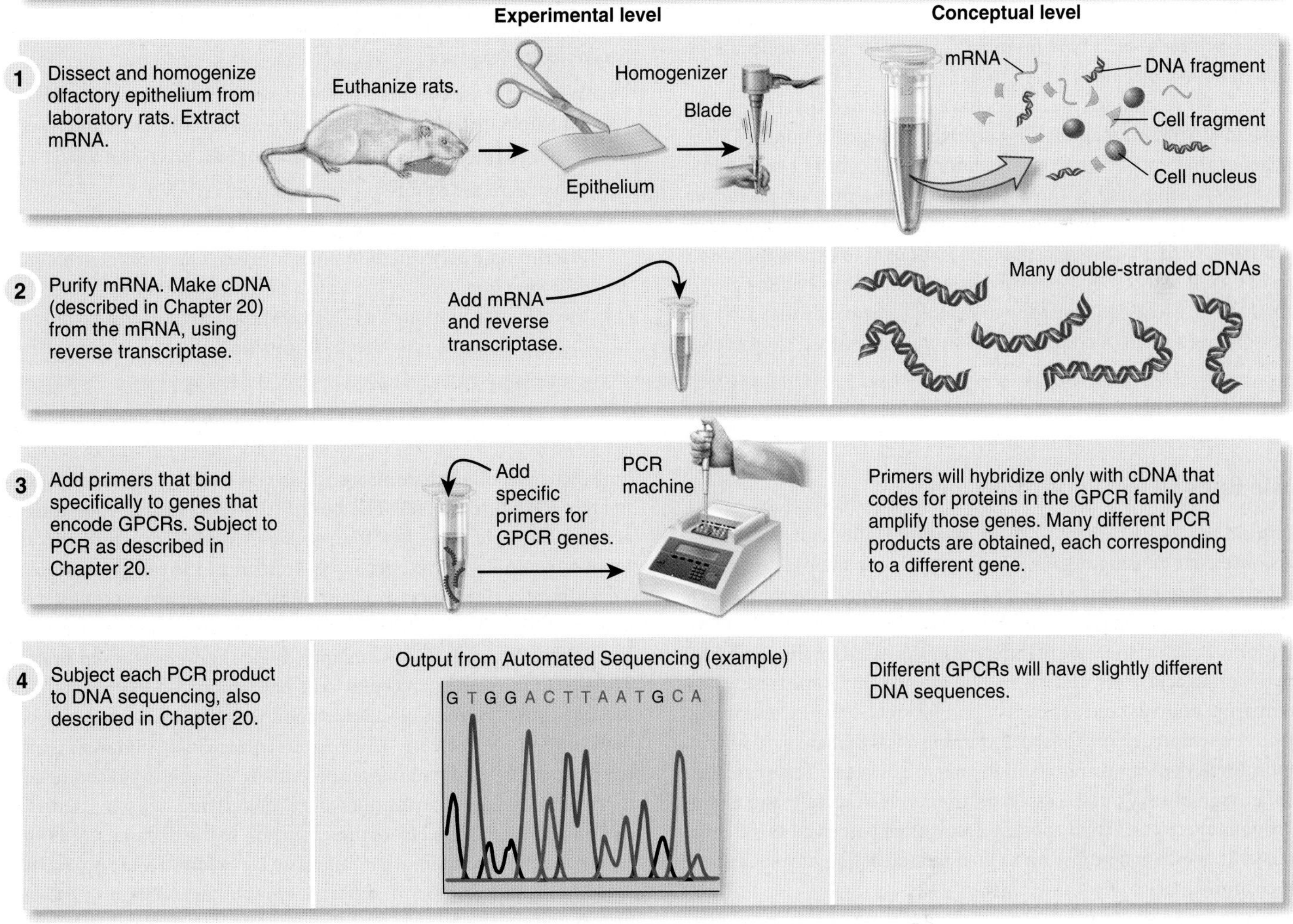

	Experimental level	Conceptual level
1 Dissect and homogenize olfactory epithelium from laboratory rats. Extract mRNA.	Euthanize rats. Homogenizer Blade Epithelium	mRNA DNA fragment Cell fragment Cell nucleus
2 Purify mRNA. Make cDNA (described in Chapter 20) from the mRNA, using reverse transcriptase.	Add mRNA and reverse transcriptase.	Many double-stranded cDNAs
3 Add primers that bind specifically to genes that encode GPCRs. Subject to PCR as described in Chapter 20.	Add specific primers for GPCR genes. PCR machine	Primers will hybridize only with cDNA that codes for proteins in the GPCR family and amplify those genes. Many different PCR products are obtained, each corresponding to a different gene.
4 Subject each PCR product to DNA sequencing, also described in Chapter 20.	Output from Automated Sequencing (example) G T G G A C T T A A T G C A	Different GPCRs will have slightly different DNA sequences.

5 **THE DATA**

The DNA sequencing revealed 18 different GPCRs that were specifically expressed in the receptor cells of the olfactory epithelium.

6 **CONCLUSION** Olfactory receptor cells express many different receptor proteins that account for an animal's ability to detect a wide variety of odors.

7 **SOURCE** Buck, L., and Axel, R. 1991. A novel multigene family may encode odorant receptors: a molecular basis for odor recognition. *Cell* 65:175–187.

conserved. A conserved region is a DNA sequence that rarely changes among different members of a gene family (refer back to Chapter 21). The primers were used in the PCR technique to amplify cDNAs that encoded GPCRs. This produced many PCR products that were then subjected to DNA sequencing. As shown in The Data in Figure 43.26, Buck and Axel initially identified 18 different genes, each encoding a GPCR with a slightly different amino acid sequence. Further research showed that these 18 genes were expressed in nasal epithelia, but not in other parts of the rat's body. These results were consistent with the second hypothesis, namely, that organisms make a large number of receptor proteins, each type binding a particular odor molecule or group of odor molecules.

Since these studies, researchers have determined that this family of olfactory genes in mammals is surprisingly large. In humans, over 600 genes that encode olfactory receptor proteins have been identified, though about half of these are pseudogenes that are no longer functional. This value underscores the importance of olfaction even in humans. Each olfactory receptor cell is thought to express only one type of GPCR that recognizes its own specific odor molecule or group of molecules. Most odors are due to multiple chemicals that activate many different types of odor receptors at the same time. We perceive odors based on the combination of receptors that become activated, and then send signals to the brain. This ability to combine different inputs allows us to perceive several thousand different odors despite having only several hundred different olfactory receptor proteins.

The research of Buck and Axel explained, in part, how animals detect a myriad of odors. In 2004, they received the Nobel Prize in Physiology or Medicine for this pioneering work. Even so, many intriguing questions remain. We do not yet understand the pattern of gene regulation in the receptor cells or how the brain actually interprets the signals it receives from those cells.

Experimental Questions

1. What were the two major hypotheses to explain how animals discriminate between different odors? How did Buck and Axel test the hypothesis of multiple olfactory receptor proteins?
2. What were the results of Buck and Axel's study?
3. Considering the two hypotheses explaining how animals discriminate between different odors (see Question 1), which one was supported by the results of this experiment? With the evidence presented by Buck and Axel, what is the current hypothesis explaining the discrimination of odors in animals?

Taste Buds Detect Food Molecules

Chemical senses are present in all animal phyla, although not all animals have taste-sensing organs. Many animals use their taste (gustation) sense to select appropriate foods. For example, butterflies select nectar based on the sugars found in a particular flower, and carnivorous animals detect the taste of different meats based on the combination of amino acids, fats, and sugars that are present. Some freshwater and marine animals, such as catfish and lobsters, have exceptionally sensitive chemoreceptors for specific amino acids that are particularly important as neurotransmitters in the nervous system. Taste can also help an animal seek out necessary nutrients, such as salt, and avoid poisonous chemicals. Toxic substances are often perceived as bitter or distasteful, which may cause an animal to stop eating any such substance immediately.

Taste buds are structures containing chemosensory cells that detect particular molecules in food. Humans have about 9,000 taste buds, whereas other mammals such as pigs and rabbits have approximately 15,000 to 18,000. The bumps that you see on your tongue are not the taste buds but the papillae, elevated structures on the tongue that collect food molecules in depressions called taste pores. These pores contain many taste buds with sensory receptor cells (Figure 43.27). The sensory receptor cells form a complex with several supporting cells that is organized like the wedges of an orange. The tips of the sensory receptor cells have microvilli that extend into the taste pore. Here, molecules in food that have dissolved in saliva bind to receptor proteins. This triggers intracellular signals that alter ion permeability and membrane potentials. The sensory receptor cells then release neurotransmitters onto underlying sensory neurons. Action potentials travel from these neurons to the thalamus and other regions of the forebrain, where the taste is perceived and recognized.

There are a number of different types of taste cells, and they are distributed on the tongue in different areas, but these areas overlap considerably. Each type of cell has a specific transduction mechanism

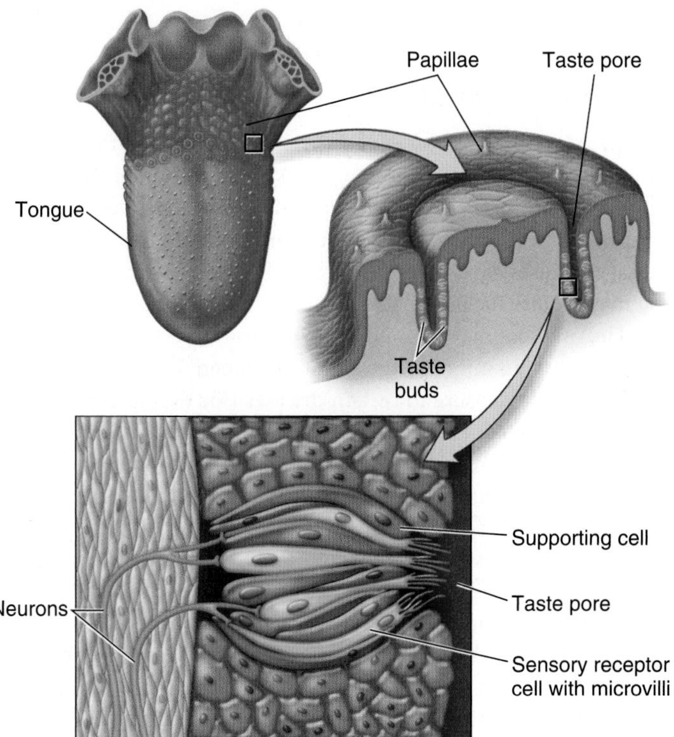

Figure 43.27 Structures involved in the sense of taste. This sense occurs in taste buds, which contain the sensory receptor cells that respond to dissolved food molecules.

Concept Check: Why have many animals, including humans, evolved the ability to sense salty, sweet, sour, bitter, and umami?

that allows it to detect specific chemicals present in the foods and fluids we ingest. Their activation results in the perception of sweet, sour, salty, and bitter tastes. In addition, a newly recognized fifth taste perception called umami (after the Japanese word for "delicious") is associated with the presence in ingested food of glutamate and other similar amino acids, and is usually described as imparting flavorfulness to food. It may account for the widely recognized effects of monosodium glutamate (MSG) in enhancing the flavor of food.

The senses of taste and smell are enhanced when we are hungry, a phenomenon that most likely occurs in other animals as well. Once we have eaten, we are less aware of the smell and taste of food. The importance of this for survival is clear: A hungry animal needs to eat. An improved sense of smell aids in locating food, and a heightened sense of taste encourages an animal to eat. Afterward, these senses become temporarily dampened so as not to distract the animal from its other needs. The mechanism by which these changes occur is uncertain, but it may involve a temporary alteration in the number of smell and taste receptors, or in their ability to bind ligands.

43.7 Impact on Public Health

Learning Outcome:

1. Identify common types of visual and hearing deficits in humans and the causes of each.

Sensory disorders are among the most common neurological problems found in humans and range from mild (needing eyeglasses or a hearing aid) to severe (blindness or deafness). In this section, we present an overview of a few representative sensory disorders that have a major impact on the human health.

Visual Disorders Include Glaucoma, Macular Degeneration, and Cataracts

Visual disorders affect an enormous fraction of the world's population. In the U.S. alone, more than 10 million people have severe eye problems that cannot be corrected by eyeglasses, and more than 1 million people are blind (42 million worldwide). The costs to the U.S. government associated with blindness and severe visual loss amount to nearly $4 billion yearly. Although vision loss has many causes, three disorders account for over half of all cases: glaucoma, macular degeneration, and cataracts.

Normally, the fluid that makes up the aqueous humor in the eye is produced and reabsorbed (drained) in a circulation that keeps the fluid levels constant. In **glaucoma**, drainage of aqueous humor becomes blocked, and the pressure inside the eye increases as the fluid level rises. If untreated, this eventually damages cells in the retina and leads to irreversible loss of vision (Figure 43.28a). The cause of glaucoma is not always known, but in some cases it is due to trauma to the eye, chronic use of certain medicines, or diseases such as diabetes. Some forms of glaucoma may have a genetic component. Glaucoma can be treated with eye drops that reduce fluid production in the eye or with laser surgery to reshape the drainage structures in the eye. The American Foundation for the Blind estimates that 400,000 new cases of glaucoma are diagnosed each year in the U.S., and up to 10 million individuals have elevated eye pressure. Glaucoma accounts for roughly 10% of all cases of blindness in the U.S.

In **macular degeneration**, photoreceptor cells in and around the macula (the region that contains the fovea) are lost. Because this is the region where cones are most densely packed, this condition is associated with loss of sharpness and color vision (Figure 43.28b). It usually does not occur before age 60, but in some cases it is hereditary and can occur at any age. Macular degeneration is the leading cause of blindness in the U.S., accounting for roughly 25% of all cases, but its causes remain obscure.

Cataracts, the accumulation of protein in the lens, cloud the lens and cause blurring, poor night vision, and difficulty focusing on nearby objects (Figure 43.28c). By age 65, as many as 50% of individuals have one or more cataracts in either eye, and this jumps to 70% by age 75. Many cataracts are small enough not to affect vision. In fact, many people do not even realize they have cataracts until they undergo an eye exam. The causes of cataracts are not all known but include trauma, medicinal drugs, diabetes, and heredity. The treatment, when needed, is to wear a powerful contact lens to help do the job formerly done by the person's own lens or to have the affected lens surgically removed and replaced with a plastic lens. Nearly 1.4 million such surgeries are performed each year in the U.S. Without the lens, which helps protect the retina by absorbing some of the high-energy ultraviolet light from the Sun, the retina must be protected during the day by wearing dark sunglasses.

(a) How the world is seen by a person with glaucoma

(b) How the world is seen by a person with macular degeneration

(c) How the world is seen by a person with cataracts

Figure 43.28 Appearance of the visual field in a person with (a) glaucoma, (b) macular degeneration, and (c) cataracts.

Damaged Hair Cells Within the Cochlea Can Cause Deafness

Deafness (hearing loss) is usually caused by damage to the hair cells within the cochlea, although some cases result from functional problems in brain areas that process sound or in the nerves that carry information from the hair cells to the brain. When the hair cells are damaged, noises have to be louder to be detected, and an affected person may require the use of hearing aids to amplify incoming sounds.

Hearing loss may be mild or severe and may result from many causes, including injury to the ear or head, hereditary defects of the inner ear, and exposure to certain diseases (for example, rubella) or toxins during fetal life. By far, however, the most significant cause of hearing loss is repeated, long-term exposure to loud noise. The amplitude of a sound wave (that is, the distance between the peak and the trough of the wave) determines its loudness; as the amplitude of a sound wave gets larger, the loudness of the sound increases. The loudness of sound is measured in decibels. The decibel (dB) scale is logarithmic; every increase of 10 dB is actually 10 times louder than the previous level. Normal conversation is about 60 dB, a chainsaw is 108 dB, and a jet plane taking off can reach 150 dB. Estimates of noise-induced hearing loss, which usually results from job-related activities, range from 7 to 21% of all cases of hearing loss worldwide; this makes it the leading occupational disorder in the U.S. It is, therefore, one of the most significant disabilities in the U.S. in terms of numbers of people affected and costs to society. Scientists estimate that nearly 40 million people are exposed daily to dangerous noise levels in the U.S. as a result of their occupation.

Recent research has led to a better understanding of the mechanism by which noise impairs hearing. Chronic exposure to loud sounds appears to produce a state of metabolic exhaustion in the hair cells of the cochlea. As a result, the cells become fatigued and are unable to maintain normal biochemical processes. One consequence of this is a buildup of free radicals. These compounds oxidize lipids in cellular membranes, damaging the membranes in the process. Mitochondrial membranes appear to be particularly susceptible to these free radicals. Once mitochondria are destroyed, a cell's ability to produce the ATP needed to fulfill its energy demands is compromised, and the cell dies. As hair cells die, the ear becomes less sensitive to sound.

Researchers are investigating drugs that might prevent the formation of free radicals in the cells of the ear, but such drugs are not yet available. If a cochlea is severely damaged, it can be surgically replaced with artificial cochlear implants. These devices generate electrical signals in response to sound waves that can stimulate the auditory nerve, which communicates with the brain. Cochlear implants cannot restore hearing to normal, but they can make it possible to hear conversations.

Summary of Key Concepts

43.1 An Introduction to Sensory Receptors

- Sensory transduction is the process by which incoming stimuli are converted to neural signals. Perception is an awareness of the sensations that are experienced. Sensory receptors are either neurons or specialized epithelial cells that respond to stimuli and begin the process of sensory transduction. A sensory receptor often responds to a stimulus by eliciting a graded response proportional to the intensity of the stimulus (Figures 43.1, 43.2).
- Sensory receptors can be divided into the general classes of mechanoreceptors, thermoreceptors, nociceptors, electromagnetic receptors, photoreceptors, and chemoreceptors.

43.2 Mechanoreception

- Mechanoreceptors respond to physical stimuli such as touch, pressure, stretch, movement, and sound. In mammals, several types of receptors in the skin detect stimuli such as touch and pressure, including Meissner corpuscle, Ruffini corpuscle, and Pacinian corpuscle (Figure 43.3).
- Hair cells have projections called stereocilia that respond to movements of fluid or other stimuli and release neurotransmitter that may result in action potentials in an adjacent sensory neuron (Figure 43.4).
- Audition (hearing) is the ability to sense sound waves and is well developed in vertebrates. The mammalian ear has three main compartments: the outer, middle, and inner ear (Figure 43.5).
- Sound waves move through outer ear, tympanic membrane, malleus, incus, and stapes to the cochlea of the ear, where they may cause the basilar membrane to vibrate. This causes the stereocilia of hair cells to bend back and forth, which results in intermittent action potentials being sent to the brain, leading to the perception of sound (Figures 43.6, 43.7).
- Many adaptations for hearing improve an animal's ability to sense sound (Figure 43.8).
- Statocysts in certain invertebrates allow an animal to sense its body position or equilibrium. The vestibular system in vertebrates allows animals to sense linear and rotational movement (Figures 43.9, 43.10).
- The lateral line system of fishes consists of hair cells that detect water movements (Figure 43.11).

43.3 Thermoreception and Nociception

- Thermoreceptors in the skin and brain allow an animal to sense external and internal temperatures, respectively. Nociceptors in the skin and internal organs sense pain, an important survival adaptation in animals.

43.4 Electromagnetic Reception

- Some animals can detect electrical and magnetic stimuli; in some cases, this ability is used to locate prey (Figure 43.12).

43.5 Photoreception

- The eyecup in flatworms is a simple eye that detects light and its direction but does not form an image. The compound eye found in many invertebrates consists of many ommatidia that focus light (Figures 43.13, 43.14).
- The single-lens eye is found in vertebrates and certain invertebrates. In such eyes, images in the animal's field of view emit patterns of light that are transmitted through the pupil and the lens to a layer of photoreceptors called the retina. This forms a visual image of the environment on the retina. Activation of photoreceptors triggers electrical changes in neurons that pass out of the eye through the optic nerves, carrying the signals to the brain, which then interprets the visual image that was transmitted. Stretching and flattening of the lens aids in focusing on objects at varying distances (Figure 43.15).
- Rods and cones are photoreceptors found in the vertebrate eye. Rods do not discriminate colors; cones can detect color. The visual pigment in rods and cones consists of retinal and a protein called opsin (Figures 43.16, 43.17, 43.18).
- Red-green color blindness is due to a defect in a type of opsin found in cone pigments (Figure 43.19).
- Absorption of photons by retinal triggers a signal transduction pathway that causes hyperpolarization of the cell leading to a visual image (Figures 43.20, 43.21).
- The retina is composed of layers of cells, including photoreceptors, that receive light input and convey it to the visual centers of the brain via the optic nerve (Figure 43.22).
- The vertebrate eye has several adaptations that aid in proper vision for particular animals (Figure 43.23).

43.6 Chemoreception

- Invertebrates have taste receptors on their proboscis, legs, feet, and antennae. Some insects can detect pheromones (Figure 43.24).
- In mammals, olfactory receptors are neurons located in epithelial tissue at the upper part of the nasal cavity. They contain cilia with protein receptors that bind specific odor molecules. The binding initiates a signal transduction pathway that ultimately depolarizes the plasma membrane and transmits action potentials to the olfactory bulb of the brain. Buck and Axel discovered that olfactory receptors have many different types of protein receptors for odor molecules (Figures 43.25, 43.26).
- Taste buds contain chemosensory cells that detect molecules in food. There are different types of taste cells, the activation of which results in the perception of five tastes: sweet, sour, salty, bitter, and umami (Figure 43.27).

43.7 Impact on Public Health

- Common visual disorders include glaucoma, macular degeneration, and cataracts. Prolonged exposure to intense sound can damage hair cells within the cochlea and lead to deafness (Figure 43.28).

Assess and Discuss

Test Yourself

1. The process whereby incoming sensory stimulation is converted to electrical signals is known as
   a. an action potential.
   b. a threshold potential.
   c. perception.
   d. sensory transduction.
   e. reception.
2. Photoreceptors are examples of
   a. nociceptors.
   b. baroreceptors.
   c. mechanoreceptors.
   d. electromagnetic receptors.
   e. thermoreceptors.
3. The sensory receptors for audition (hearing) are located in
   a. the organ of Corti.
   b. the Eustachian tube.
   c. the vestibular system.
   d. the tympanic membrane.
   e. a, b, and c.
4. Statocysts are sensory organs for
   a. hearing found in many invertebrates.
   b. equilibrium found in mammals.
   c. equilibrium found in many invertebrates.
   d. water current changes found in fish.
   e. hearing found in amphibians.
5. The eyecups of planaria can
   a. focus light to form an image.
   b. detect light.
   c. detect the direction of light.
   d. all of the above.
   e. b and c only.
6. The light detectors of a compound eye are called
   a. retinas.
   b. opsins.
   c. cones.
   d. ommatidia.
   e. rods.
7. In the mammalian eye, light from near or far objects is focused on the retina when
   a. the lens moves forward or backward.
   b. the lens changes shape.
   c. the eyeball changes shape.
   d. the cornea changes shape.
   e. none of the above
8. The level of glutamate release from photoreceptors of the vertebrate eye would be highest when
   a. a person is standing in full sunlight.
   b. a person is in a completely dark room.
   c. a person is in a dimly lit room.
   d. Na^+ channels of the photoreceptor are closed.
   e. both a and d
9. Cone pigments detect different wavelengths of light due to
   a. their location in the retina.
   b. the amount of light they absorb.
   c. the type of retinal they have.
   d. the type of opsin protein they have.
   e. interactions with bipolar cells.
10. The stimulation for olfaction involves odorant molecules
    a. bending the cilia of olfactory receptor cells.
    b. binding to protein receptors of olfactory receptor cells.
    c. entering the cytoplasm of olfactory receptor cells.
    d. opening K^+ channels of olfactory receptor cells.
    e. all of the above

Conceptual Questions

1. Distinguish between sensory transduction and perception.
2. Explain how the mammalian ear is adapted to distinguish sounds of different frequencies.
3. A principle of biology is that *living organisms interact with their environment.* This is certainly a fundamental principle of sensory neuroscience. Of the ways in which you use your senses to interact with the environment, which do you think is the least important?

Collaborative Questions

1. Discuss two different types of mechanoreceptors.
2. Discuss different types of eyes found in animals.

Online Resource

www.brookerbiology.com

Stay a step ahead in your studies with animations that bring concepts to life and practice tests to assess your understanding. Your instructor may also recommend the interactive eBook, individualized learning tools, and more.

Chapter Outline

44.1 Types of Animal Skeletons

44.2 The Vertebrate Skeleton

44.3 Skeletal Muscle Structure and the Mechanism of Force Generation

44.4 Skeletal Muscle Function

44.5 Animal Locomotion

44.6 Impact on Public Health

Summary of Key Concepts

Assess and Discuss

Muscular-Skeletal Systems and Locomotion

44

In 1991, Olympic athlete Mike Powell established a record long jump of 8.95 m, roughly 5 times the height of a person. A jump of nearly 9 m sounds impressive, and it is for us, but how do humans compare with other animals? Red kangaroos can hop up to 12 m while running at top speed. Some tree frogs, however, like the one shown in the chapter-opening photo, can leap distances up to 1.4 m without a running start, yet the frog is only 4.5 cm long and weighs only 8 g! This is a sensational leap for an animal that size, roughly 30 times its body length. The jumping abilities of frogs are related to their relatively large leg muscle mass, the elongation of the bones in the legs and feet, and elastic elements in the connective tissue associated with their muscle and bone. Indeed, the jumping ability of frogs makes them excellent model organisms for the study of muscle function in vertebrates.

Muscles are composed of highly specialized cells that have the ability to contract in response to stimuli. The three types of muscle—skeletal, smooth, and cardiac—were introduced in Chapter 40. In this chapter, you will learn about the structure and function of skeletal muscle and how skeletal muscle controls movements such as those required for **locomotion**, the movement of an animal from place to place.

For skeletal muscles to produce locomotion, they must exert a force on an animal's skeleton. We begin the chapter, therefore, with an overview of animal skeletons, looking in particular at the vertebrate skeleton, an endoskeleton. Then we examine the structure and function of skeletal muscle and how the interaction of two muscle proteins, actin and myosin, produce muscle contraction. Next, we consider various modes of locomotion in animals. We conclude with a consideration of important bone and muscle diseases in humans.

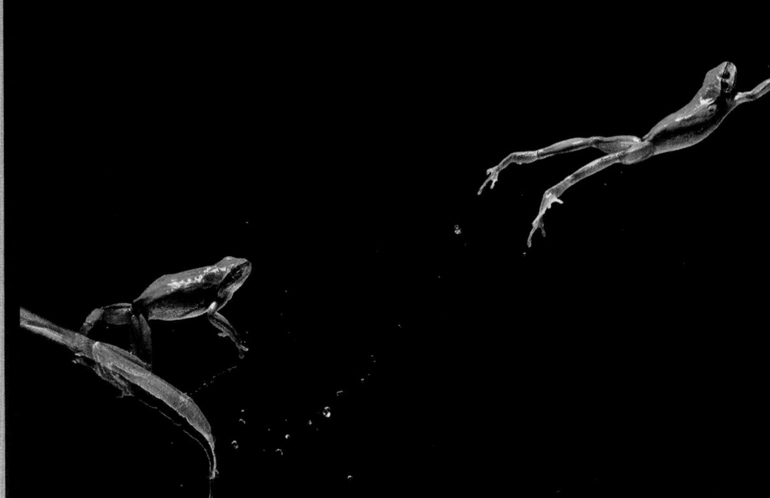

European tree frog (*Hyla arborea*) using its muscular-skeletal system to leap from a leaf.

44.1 Types of Animal Skeletons

Learning Outcomes:

1. Distinguish among hydrostatic skeletons, exoskeletons, and endoskeletons.
2. List the advantages and disadvantages of an exoskeleton.

When we think of the word skeleton, an image of the vertebrate system of bones usually comes to mind. However, invertebrates possess a skeleton as well, although it is not made of bone. Therefore, a broader definition of a **skeleton** is a structure that serves one or more functions related to support, protection, and locomotion. Using this definition, the three types of skeletons found in animals are hydrostatic skeletons, exoskeletons, and endoskeletons. The first two are found in invertebrates, and endoskeletons are found in some sponges and all echinoderms and vertebrates.

Hydrostatic Skeletons Consist of Fluid-Filled Body Compartments

Many soft-bodied invertebrates use hydrostatic pressure to support their bodies and generate movements. Hydrostatic pressure is the pressure of a fluid at rest in a contained space. The combination of muscles and a water-based fluid in the body constitutes a **hydrostatic skeleton**. Water is nearly incompressible, which means that if an animal exerts a force on the fluid that fills its body cavities, it can use the resulting hydrostatic pressure to move the body, much like a balloon partly filled with water can be deformed by squeezing it along its length. In cnidarians such as *Hydra*, for example, contractile cells in

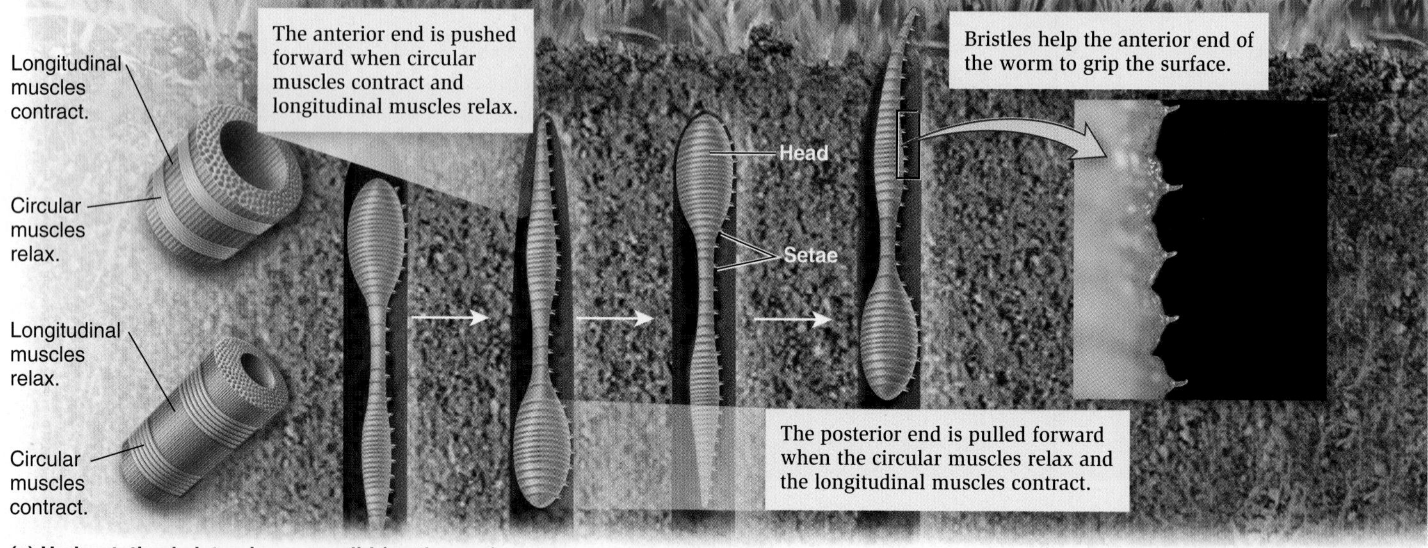

Longitudinal muscles contract.
Circular muscles relax.
Longitudinal muscles relax.
Circular muscles contract.
The anterior end is pushed forward when circular muscles contract and longitudinal muscles relax.
Head
Setae
Bristles help the anterior end of the worm to grip the surface.
The posterior end is pulled forward when the circular muscles relax and the longitudinal muscles contract.

(a) Hydrostatic skeleton in an annelid (earthworm)

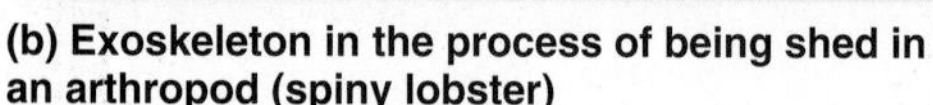

(b) Exoskeleton in the process of being shed in an arthropod (spiny lobster)

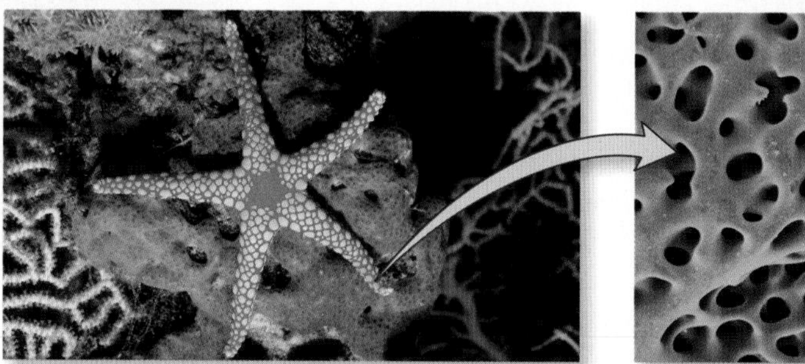

(c) Echinoderm (sea star) and (right) SEM of its endoskeleton

Figure 44.1 **Types of skeletons.** **(a)** Hydrostatic skeleton. By alternately contracting and relaxing circular and longitudinal muscles, earthworms use differences in hydrostatic pressure to achieve locomotion. Clinging setae (bristles) along the body surface (shown in top-right inset) help prevent backsliding. **(b)** Exoskeleton. An arthropod's skeleton covers and protects its body, but it must be periodically shed and replaced to allow growth, as shown in this photo. **(c)** Endoskeleton. Echinoderms such as the sea star (starfish) have endoskeletons of bony plates made of calcium carbonate (shown in bottom-right inset).

BIOLOGY PRINCIPLE Living organisms grow and develop. Part of the growth and development of animals includes their skeletons, as shown in this figure. For animals with exoskeletons, however, this may create a brief period when the animal is particularly vulnerable to predation.

Concept Check: *Endoskeletons do not provide protection for the body surface of animals. How can the lack of an external skeleton also be an advantage for animals?*

the body wall can exert a pressure on the fluid in the gastrovascular cavity. Depending on the direction of the force, the body and tentacles will either be elongated or shortened by the movement of water into and out of different regions of the cavity. These changes contribute to the animal's ability to capture prey, retain food in its cavity, extend or contract its tentacles, and even in a limited way to move about.

Annelids or segmented worms, such as earthworms, move forward by passing a wave of muscular contractions along the length of their body, segment by segment (Figure 44.1a). The muscles are arranged in two orientations: circular and longitudinal (lengthwise). Contraction of circular muscles squeezes and lengthens the body, whereas contraction of the longitudinal muscles shortens and widens it. The worm is "squeezed" forward by differences in hydrostatic pressure created along its body by the muscles. Bristles known as setae along its body surface help the animal grip the ground to prevent backsliding.

Exoskeletons Are on the Outside of an Animal's Body

Arthropods have an **exoskeleton**, an external skeleton that surrounds and protects most of the body surface (Figure 44.1b). Exoskeletons provide support for the body, protection from the environment and predators, and protection for internal organs. The arthropod skeleton is made of a polysaccharide called chitin, and in crustaceans such as lobsters and shrimp, it is sometimes strengthened with calcium and other minerals. Exoskeletons are often tough, durable, and segmented to allow movement. However, to allow growth they must be periodically shed, regrown, and strengthened again, a process called ecdysis, or molting. A disadvantage of exoskeletons is that when an animal is molting, its new exoskeleton is temporarily soft, making the animal more vulnerable to predators and the environment.

Exoskeletons vary enormously in their complexity, thickness, and durability. The differences in exoskeletons are usually adaptations that enhance an animal's survival. Think, for example, of the difference between the wing of a butterfly and the outer skeleton of a lobster. A butterfly's exoskeleton must be light enough for the animal to fly, whereas the thick, tough exoskeleton of the lobster provides a very effective defense against predators in the sea. Exoskeletons may seem somehow primitive compared with the endoskeletons of vertebrates (discussed next), particularly because of the requirement for molting. Even so, arthropods are among the most successful of all animal phyla living today, having survived for hundreds of millions of years and inhabiting nearly every possible ecological niche on the planet. Clearly, exoskeletons have been sufficient for one of the planet's greatest success stories.

Endoskeletons Are Internal Support Structures

Like exoskeletons, **endoskeletons** provide support and protection. Unlike exoskeletons, however, endoskeletons are internal structures and do not protect the body surface. Some endoskeletons do, however, protect internal organs such as those in the thorax of vertebrates.

Endoskeletons are found in some species of sponges and all echinoderms and vertebrates. Minerals such as calcium, magnesium, phosphate, and carbonate supply the hardening material that gives the skeleton its firm structure. The endoskeletons of sponges and echinoderms consist of spiky networks of proteins and minerals, and mineralized platelike structures, respectively. Beneath the body surface of echinoderms, for example, arrays of mineralized plates made largely of calcium carbonate extend into the spines and arms that radiate from the main body (Figure 44.1c).

Vertebrate skeletons, by contrast, are composed of either cartilage or bone or both. Next we consider the structure and function of the vertebrate skeleton in greater detail.

44.2 The Vertebrate Skeleton

Learning Outcomes:

1. List the major functions of the vertebrate skeleton.
2. Describe the composition of vertebrate bone.

Vertebrate skeletons are composed of either cartilage (refer back to Figure 40.5)—as in cartilaginous fishes (sharks, rays, and skates)—or both cartilage and bone, as in bony fishes, amphibians, reptiles, birds, and mammals. First we consider the skeleton as a whole before examining bones in greater detail.

The Vertebrate Skeleton Performs Several Important Functions

In the vertebrate skeleton, bones are connected in ways that allow for support, protection, and movement. The vertebrate endoskeleton is often considered in two parts: the axial and appendicular skeletons (Figure 44.2). The axial skeleton is composed of the bones that form the main longitudinal axis of an animal's body, including the skull, vertebrae, sternum, and ribs. The appendicular skeleton consists of the limb or fin bones and the bones that connect them to the axial skeleton. A **joint** is formed where two or more bones come together. Some joints permit free movement (for example, the shoulders), whereas others allow no movement (fused joints like those interlocking the skull bones) or only limited movement (such as those of the vertebral column). Figure 44.2 illustrates three types of joints that allow different types of movements: pivot, hinge, and ball-and-socket joints.

The skeleton of vertebrates serves several functions in addition to support, protection, and movement. For example, blood cells and platelets, the latter of which help blood to clot (see Chapter 47), are formed within the marrow of the ilia, vertebrae, the ends of the femurs, and other bones. In addition, calcium and phosphate homeostasis is achieved in large part through exchanges of these ions between bone and blood. For example, if dietary intake of calcium is low, calcium is removed from bone and added to the blood, so that all of the vital cellular activities that depend on calcium, such as nerve and muscle function, can continue to operate normally. When dietary calcium is restored to normal, calcium is redeposited in bone. This calcium cycling is under the control of hormones such as parathyroid hormone produced by the parathyroid glands (see Chapter 50). About 99% of all the calcium in a typical vertebrate's body exists in bone. This represents a huge reservoir of calcium ions for the blood.

Bone Consists of a Mixture of Organic and Mineral Components

Bone is a living, dynamic tissue with both organic and mineral components. Organic materials include cells that form bone—called osteoblasts and osteocytes—and cells that break it down—called osteoclasts. The organic part of bone is secreted by osteoblasts and osteocytes, and consists of the protein collagen, which has a unique triple helical structure that gives bone both strength and flexibility (refer back to Figure 10.2). The mineral component is composed of a crystalline mixture of primarily calcium (Ca^{2+}) and phosphate (PO_4^{2-}) and other ions that provide bone its rigidity. These ions must be obtained in an animal's diet, absorbed into the blood, and deposited in bone.

A proper ratio of organic and mineral components is required for normal bone function. Bone lacking sufficient mineral, for example, is easily fractured. Bone is formed at high rates during an animal's growth periods, but even in adulthood bone is continuously formed, broken down, and re-formed. The skeleton is continually changing—the one in your body right now is completely remodeled from the one that was in your body a few years ago. Similarly, the skeleton you will have a few years from now will be different from the one you have today.

Bones cannot move by themselves but instead provide the scaffold upon which skeletal muscles act to cause body movement. We turn now to a discussion of skeletal muscle and the mechanism by which it generates force.

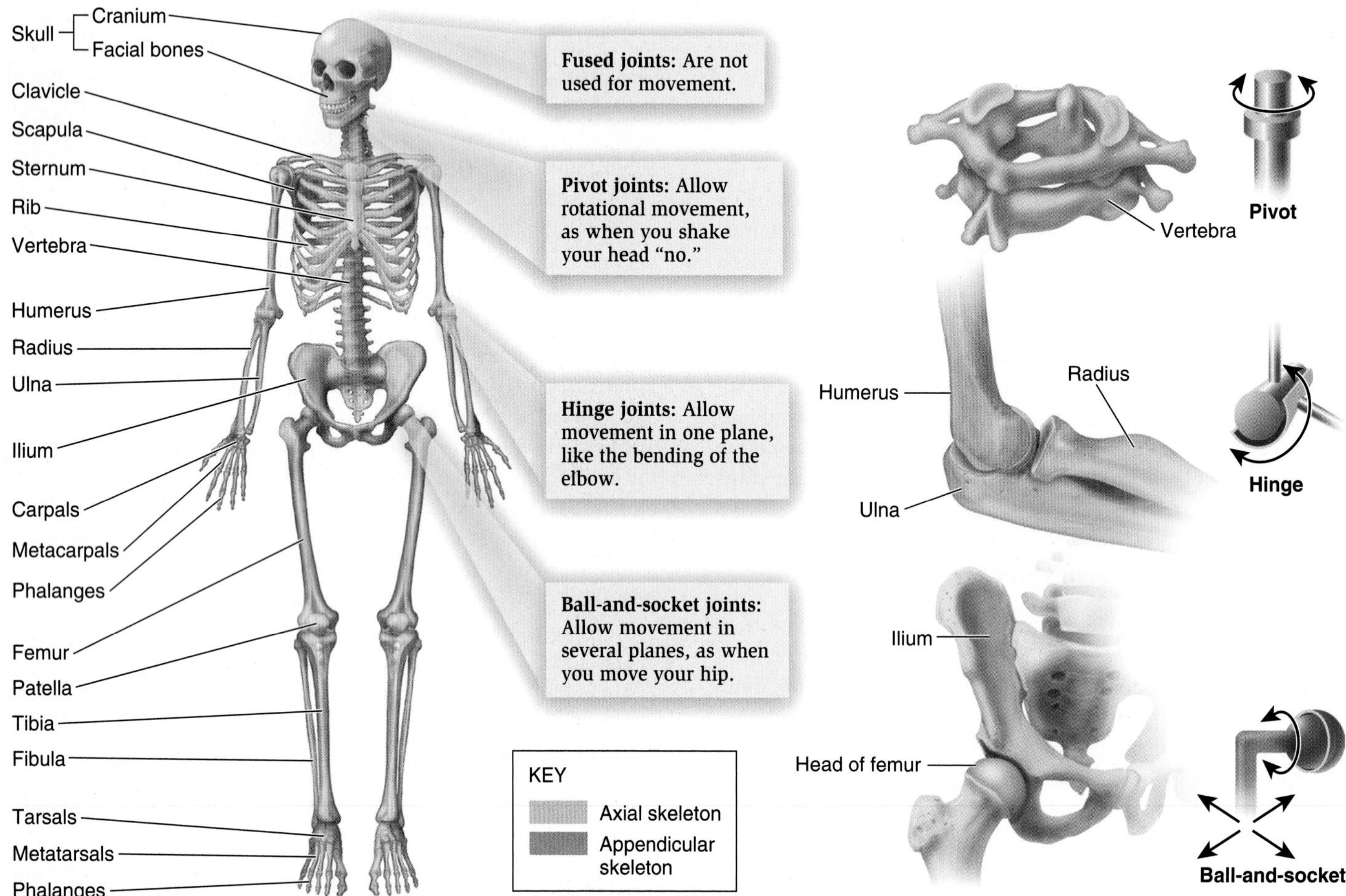

Figure 44.2 **The adult human skeleton.** This diagram shows the axial (beige) and appendicular (green) parts of the skeleton in an adult human, an animal with an endoskeleton. The adult human skeleton consists of 206 separate bones. Three examples of movable joints, pivot, hinge, and ball-and-socket, are shown.

44.3 Skeletal Muscle Structure and the Mechanism of Force Generation

Learning Outcomes:

1. List the three types of muscle found in vertebrates and where they are found in the body.
2. Identify the structural components of a muscle down to the level of the sarcomere.
3. Explain the sliding filament mechanism of muscle contraction.
4. Explain how tropomyosin and troponin help regulate muscle contraction.
5. Describe how electrical excitation and muscle contraction are linked.
6. Describe the structural features of a neuromuscular junction and the role of acetylcholine in mediating skeletal muscle excitation.

Vertebrates have three types of muscle that are classified according to their structure, function, and control mechanisms. They are cardiac muscle, smooth muscle, and skeletal muscle. **Cardiac muscle** is found only in the heart and provides the force required for the heart to pump blood. Cardiac muscle will be discussed in Chapter 47. **Smooth muscle** surrounds and forms part of the lining of hollow organs and tubes, including those of the digestive tract, urinary bladder, uterus, blood vessels, and airways. Contraction of the smooth muscle in hollow organs may propel the contents forward or churn them up, as when the stomach contracts after a meal. In other cases, smooth muscle regulates the flow of substances by changing the tube diameter, as in the widening or narrowing of small blood vessels. Smooth muscle contraction is not under voluntary control. Instead, it is controlled by the autonomic nervous system, hormones, and local chemical signals. Some smooth muscles, such as those found in regions of the digestive tract, have the ability to contract on their own, even in the absence of such signals.

Skeletal muscle is found throughout the body and is directly involved in locomotion. In vertebrates, but not invertebrates, skeletal muscle is electrically excitable—it can generate action potentials in response to a stimulus (invertebrate skeletal muscle cells have graded membrane potentials but do not have action potentials). The action potentials of vertebrate skeletal muscle cells result in increased concentrations of cytosolic calcium ions, which trigger force generation. Before understanding how this is possible, however, let's begin with an overview of skeletal muscle structure and function.

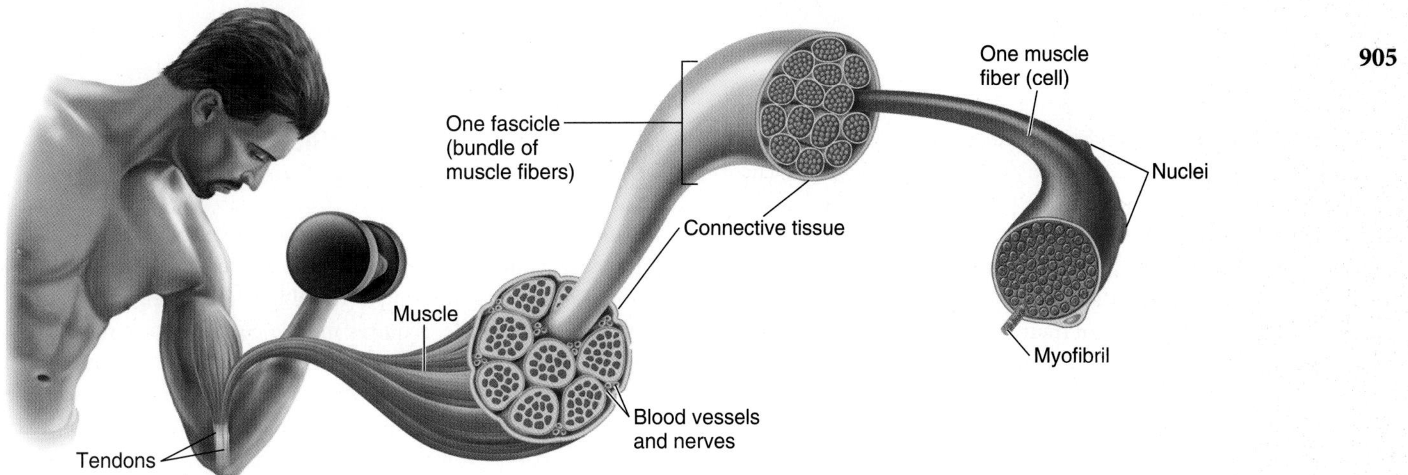

Figure 44.3 **Skeletal muscle structure.** Skeletal muscles attach to bone by tendons, which are bundles of collagen fibers. Each muscle consists of bundles of muscle fibers (muscle cells) bound together by connective tissue.

Concept Check: *What would happen to the ability of a muscle to move a bone if its tendon were torn, for example, due to injury?*

A Skeletal Muscle Is a Contractile Organ That Supports and Moves Bones

A skeletal muscle is a grouping of cells—called **muscle fibers**—bound together in bundles (called fascicles) by a succession of connective tissue layers (Figure 44.3). Skeletal muscles are usually linked to bones by bundles of collagen fibers known as tendons. The transmission of force from contracting muscle to bone can be likened to a number of people pulling on a rope attached to a heavy object. Each person corresponds to a single muscle fiber, the rope corresponds to the tendons, and the bone is the heavy object.

Some tendons are very long, with the site of tendon attachment to bone far removed from the end of the muscle. For example, some of the muscles that move the fingers are in the forearm. You can wiggle your fingers and feel the movement of the muscles in your lower arm. These muscles are connected to the finger bones by long tendons.

Skeletal muscle fibers increase in size during growth from infancy to adulthood, but few new fibers are formed during that time. As described later, enlargement of skeletal muscles in adults, as from weightlifting, is also primarily a function of enlarged fibers, not the formation of new ones.

Muscle Fibers Contain Myofibrils Composed of Arrays of Filaments

Each skeletal muscle fiber is a single cell with multiple nuclei, containing numerous cylindrical bundles known as **myofibrils** (Figure 44.4). Myofibrils extend from one end of the fiber to the other and are

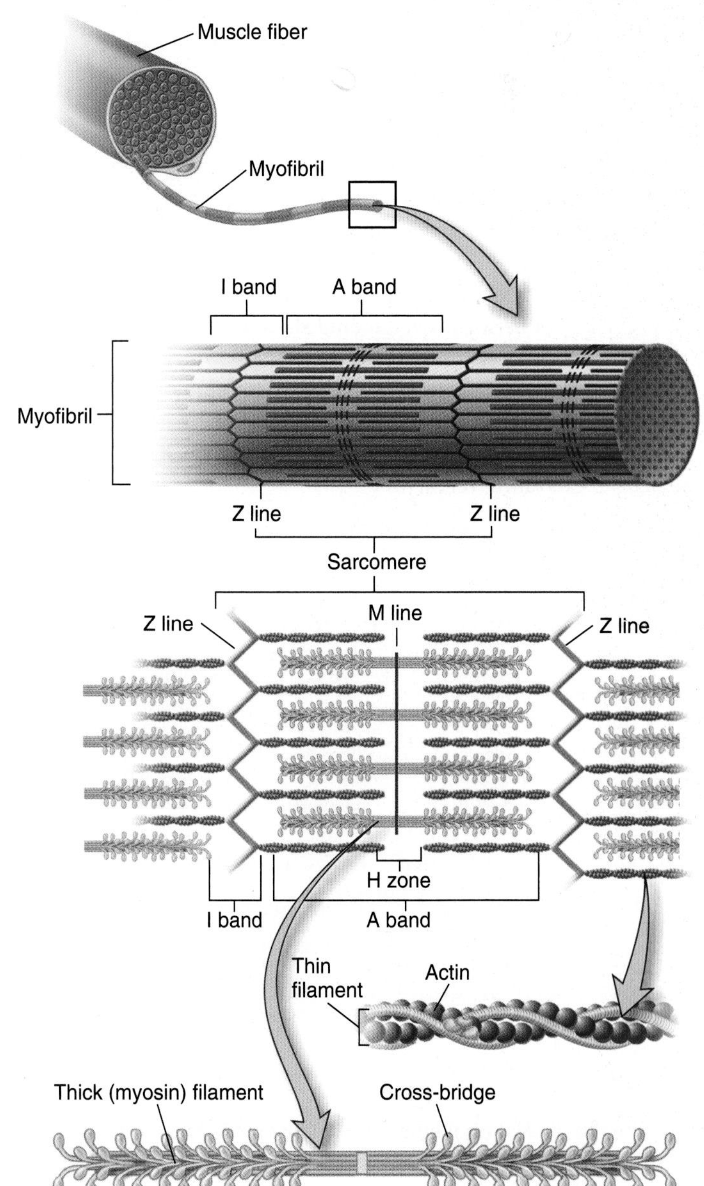

Figure 44.4 **Myofibril structure.** Each muscle fiber consists of numerous myofibrils containing thick and thin filaments. Their arrangement produces a striated banding pattern seen on microscopy. Each repeating unit of the arrayed filaments constitutes a sarcomere.

BIOLOGY PRINCIPLE **Cells are the simplest units of life.** A muscle (an organ) consists of many cells called fibers arranged in a very precise way. As the cells change their shape by shortening, the shape of the entire organ changes too. This is true for all types of muscle, not just skeletal muscle shown here and in Figure 44.3.

linked to the tendons at the ends of the fiber. Each myofibril contains numerous functional structures called thick and thin filaments. These filaments are arranged in a repeating pattern running the length of the myofibril. One complete unit of this repeating pattern is known as a **sarcomere** (from the Greek *sarco*, meaning muscle, and *mer*, meaning part). The **thick filaments** are composed almost entirely of the motor protein **myosin**. The **thin filaments**, which are about half the diameter of the thick filaments, contain the cytoskeletal protein **actin** and associated proteins. Because the arrangement of thick and thin filaments looks like a series of light and dark bands when viewed by microscope, skeletal muscle is also known as **striated muscle**.

The thick filaments are located in the middle of each sarcomere, where their orderly parallel arrangement produces a wide, dark band known as the **A band** (see Figure 44.4). Each sarcomere contains two sets of thin filaments, each of which is anchored to a network of proteins at the **Z line**. The thin filaments overlap a portion of the thick filaments. Two successive Z lines define the boundaries of one sarcomere.

A light band known as the **I band** lies between the A bands of two adjacent sarcomeres. Each I band contains those portions of the thin filaments that do not overlap the thick filaments, and each I band is bisected by a Z line.

Two additional features are present in the A band of each sarcomere. The **H zone** is a narrow, light region in the center of the A band. It corresponds to the space between the two sets of thin filaments in each sarcomere. A narrow, dark band in the center of the H zone, known as the **M line**, corresponds to proteins that link the central regions of adjacent thick filaments.

The spaces between overlapping thick and thin filaments are bridged by projections known as **cross-bridges**, which are regions of myosin molecules that extend from the surface of the thick filaments toward the thin filaments (see Figure 44.4). As we see next, during muscle contraction, the cross-bridges make contact with the thin filaments and exert force on them.

Skeletal Muscle Shortens When Thin Filaments Slide over Thick Filaments

Movement requires shortening of muscles to pull against the attached tendons and bones. However, a muscle can generate force, or contract, without producing movement. Holding a heavy weight at a constant position, for example, requires muscle contraction, but not muscle shortening. Therefore, as used in muscle physiology, the term contraction refers to activation of the cross-bridges within muscle fibers, which initiates the generation of force. When these mechanisms are turned off, contractions end, allowing the muscle fiber to relax.

How a muscle fiber actually shortens is known as the **sliding filament mechanism** of muscle contraction (**Figure 44.5**). In this mechanism, the sarcomeres shorten, but neither the thick nor the thin filaments change in length. Instead, the thick filaments remain stationary while the thin filaments slide, pulling on the Z lines and shortening the sarcomere.

Let's look at how the structure of the thin and thick filaments allows them to produce this sliding movement (**Figure 44.6**). In thin filaments, actin molecules form polymers that are arranged into two intertwined helical chains. These chains are closely associated with two proteins called tropomyosin and troponin that play important roles in regulating contraction.

Myosin proteins have a three-domain structure composed of two intertwined tails, two hinges, and two heads (Figure 44.6a). The hinges are flexible regions that connect the heads to the tails. Each filament is made up of many myosin proteins associated in a parallel array, with the hinges and heads extending out to the sides, forming cross-bridges. Each head contains two binding sites—one for actin and one for ATP. The hydrolysis of ATP provides the energy for the cross-bridge to move via a bending motion at the hinge.

The sliding filament movement is propelled by the myosin cross-bridges (Figure 44.6b). During shortening, each cross-bridge attaches to an actin molecule in a thin filament and moves in a motion much

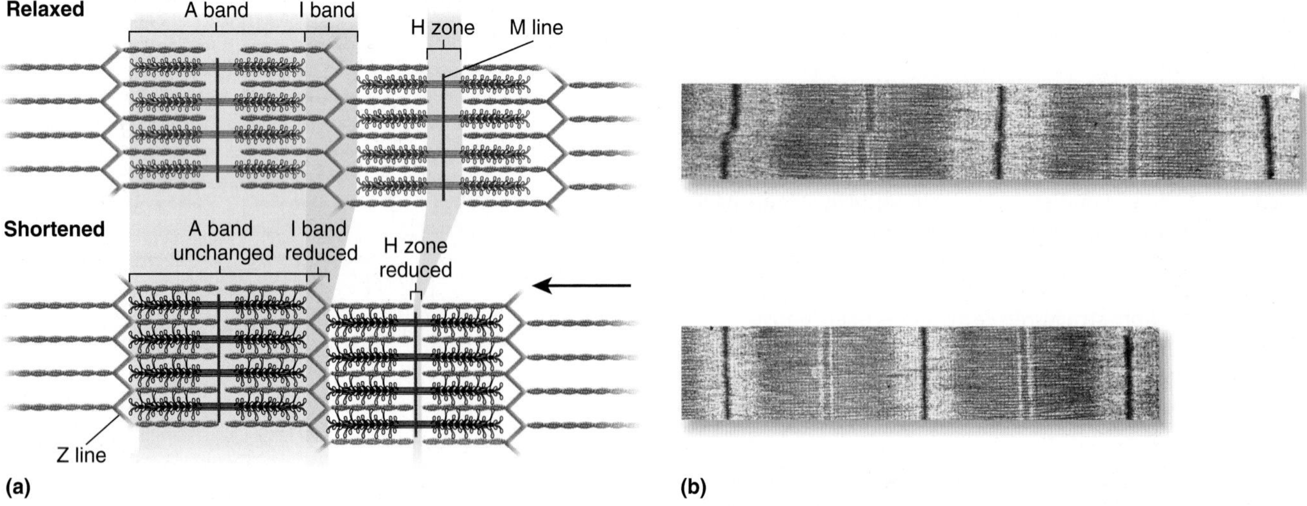

Figure 44.5 **Sliding filament mechanism of muscle contraction.** **(a)** The sliding of thin filaments past the overlapping thick filaments shortens the sarcomere but does not change the lengths of the filaments themselves. **(b)** Transmission electron micrograph of a section of skeletal muscle in the relaxed and shortened states.

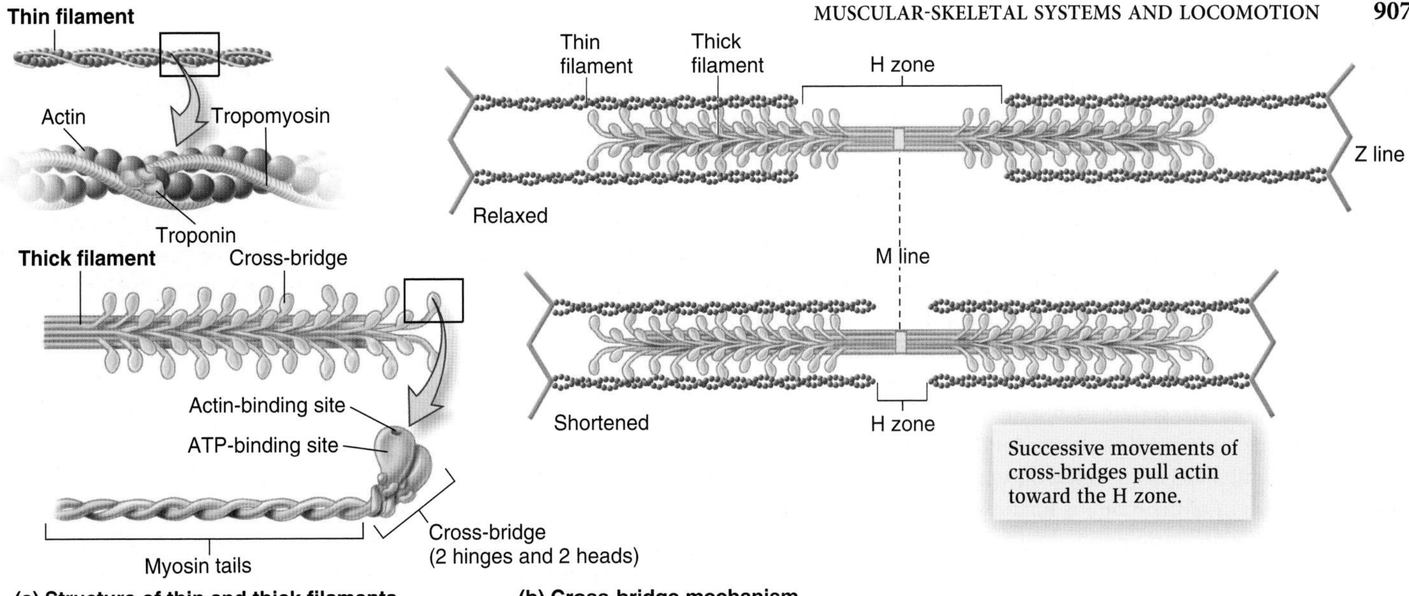

Figure 44.6 **Structure and function of the thin and thick filaments.** **(a)** Thin filaments are composed of two intertwined actin molecules and associated proteins tropomyosin and troponin. Thick filaments are made of the protein myosin, which has two intertwined tails, two hinges, two heads, an actin-binding site, and an ATP-binding site. The end of a myosin molecule that contains the actin-binding sites is bent at an angle to form the cross-bridges. **(b)** When the cross-bridges on myosin molecules bind to actin filaments, the thin filaments are pulled toward each other, shortening the sarcomere.

BIOLOGY PRINCIPLE **Structure determines function.** It is the precise structural arrangement of the proteins that make up thick and thin filaments that gives them their function. Without this structure, actin and myosin would not produce a coordinated force resulting in skeletal muscle cell shortening.

like your fist bending at your wrist. Because of the opposing orientation of the thick filaments, the movement of a cross-bridge forces the thin filaments toward the center of the sarcomere (the H zone), thereby shortening it. One stroke of a cross-bridge produces only a very small movement of a thin filament relative to a thick filament. As long as a muscle fiber is stimulated to contract, however, each cross-bridge repeats its motion many times, resulting in continued sliding of the thin filaments. Thus, the ability of a muscle fiber to generate force and movement depends on the amount of interaction of actin and myosin.

The protein myosin was first discovered in extracts of frog leg muscle in the 1860s by the German physiologist Willi Kühne. We now know that eukaryotic genomes encode a family of related myosin proteins, as described next.

Genomes & Proteomes Connection

Did an Ancient Mutation in Myosin Play a Role in the Development of the Human Brain?

Myosins are among the most ancient of eukaryotic proteins and also among the most highly conserved in amino acid sequences in the animal kingdom. Nonetheless, small differences in the sequences of genes that arose from a single primordial myosin gene have led to tissue-specific expression of various myosin proteins, each with a characteristic ability to bind actin and hydrolyze ATP. One recently discovered member of this gene family, called *MYH16*, codes for a heavy chain of a myosin molecule that is expressed mainly in the muscles of the jaw in primates. In 2004, American researcher Hansell Stedman and colleagues discovered that this gene, although present in humans, did not code for a functional protein because of a mutation that deleted two base pairs from it. This mutation was present in 100% of people tested worldwide but was not found in the eight nonhuman primate species tested, which included chimpanzees.

Based on genetic comparisons and estimates of genetic divergence among species, the researchers estimated that the mutation occurred approximately 2.4 mya. Significantly, the genus *Homo*, with its smaller jaw and larger braincase, is also thought to have first appeared about 2.4 mya, leading Stedman to suggest that the loss of the MYH16 protein led to the smaller, less-muscular jaws characteristic of humans (**Figure 44.7**). Because jaw muscles are attached to skull bones, massive jaw muscles may have placed a mechanical constraint on skull growth in early hominins (as well as in modern nonhuman primates). Smaller jaw-closing muscles would require less bone formation on the outside of the skull (in particular the illustrated anchor points obscuring the sutures or growth plates). This in turn would

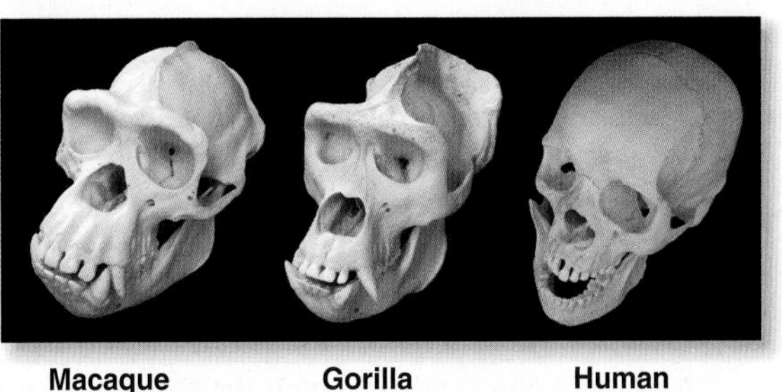

Figure 44.7 **Comparison of jaw size and area of muscle attachment (shown in orange) in modern primates.** Note: Skulls are not to scale. (© Hansell H. Stedman, M.D.)

prevent the early cessation of brain growth seen in the non-human primates. This may have eliminated a major constraint on the evolution of the larger brains seen in modern humans.

Stedman's hypothesis is compelling, given that the myosin mutation affects the jaw-closing muscles and is present only in humans. Another research group, using statistical analyses of additional gene sequences made available through the Chimpanzee genome project, concluded that the mutation may have arisen earlier—perhaps 4.0 to 5.4 mya, before small-jawed species of *Homo* appear in the fossil record. Stedman's hypothesis will require confirmation by future research. Scientists are certain that a human-specific mutation in MYH16 arose at some point in the evolution of hominins. Whether the mutation facilitated expansion of the cranium, which in turn permitted enlargement of the brain, is unknown. The hypothesis, however, does provide an intriguing piece of the puzzle of how modern humans may have evolved.

The Cross-Bridge Cycle Requires ATP and Calcium Ions

Let's now turn our attention to how actin and myosin interact to promote muscle contraction and shortening. The sequence of events that occurs between the time when a cross-bridge binds to a thin filament and when it is set to repeat the process is known as a **cross-bridge cycle**.

Cross-bridge cycling occurs when the Ca^{2+} concentration exceeds a critical threshold in the cytosol. This usually occurs when neural input results in the release of Ca^{2+} from intracellular storage sites (described in detail shortly). In other words, the contraction of skeletal muscle fibers is under nervous control.

Figure 44.8 illustrates the chemical and physical events that occur during the four steps of the cross-bridge cycle. These steps are cross-bridge binding, power stroke (moving), detaching, and resetting. As the cycle begins, the myosin cross-bridges are in an energized state, which is produced by the hydrolysis of their bound ATP to ADP and inorganic phosphate (P_i). The sequence of storage and release of energy by myosin is analogous to the operation of a mousetrap: Energy is first stored in the trap by cocking the spring (ATP hydrolysis) and is then released by the springing of the trap (head binds to actin and moves in power stroke). Let's look at each step of the process more closely.

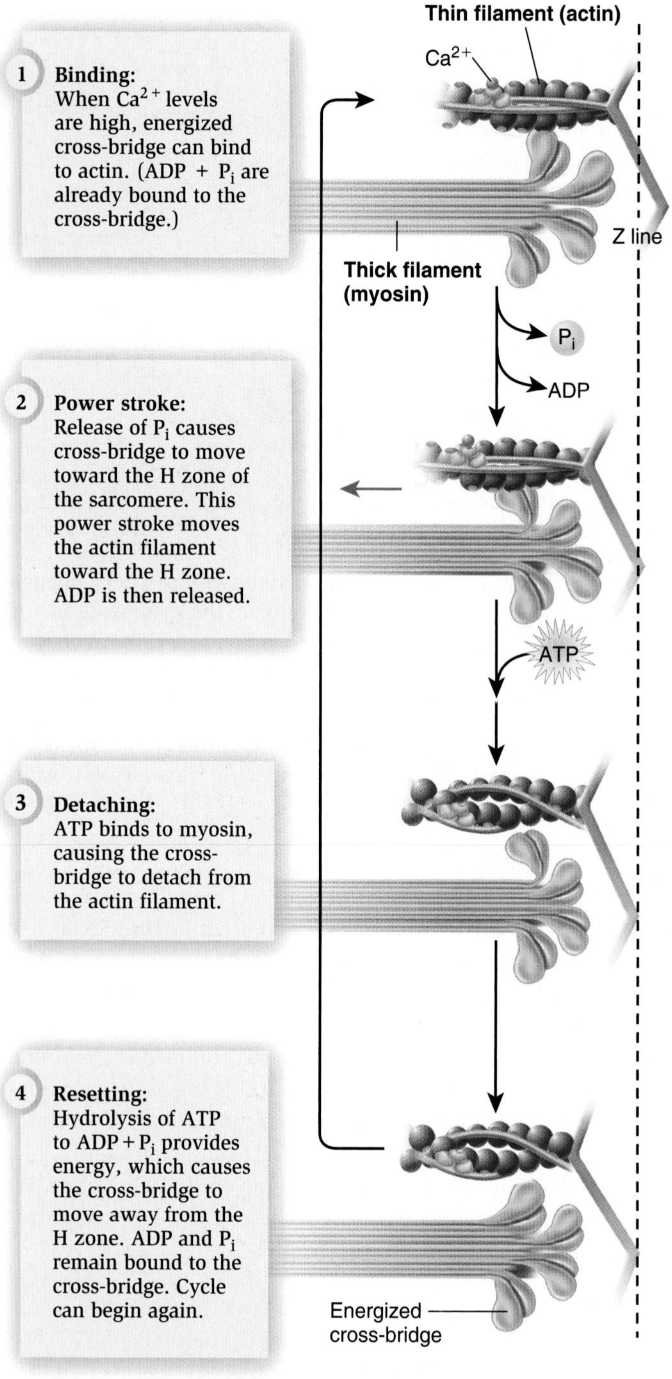

Step 1: Ca^{2+} concentrations increase, triggering the cross-bridge to bind to actin.

When the Ca^{2+} concentration in the cytosol increases, the first step in the cross-bridge cycle is triggered. In this step, an energized myosin cross-bridge, with its associated ADP and P_i, binds to an actin molecule on a thin filament.

$$\text{Actin} + [\text{Energized myosin} \cdot \text{ADP} \cdot P_i] \longrightarrow [\text{Actin} \cdot \text{Myosin} \cdot \text{ADP} \cdot P_i \text{ complex}]$$

Step 2: Release of phosphate (P_i) fuels the power stroke: The cross-bridge and thin filaments move.

The binding of an energized myosin cross-bridge to actin in step 1 triggers the release of P_i. This release causes a conformational change in the hinge of the myosin molecule. This causes the cross-bridge to rotate toward the H zone at the center of the sarcomere. This process is known as the **power stroke**, which moves the actin filament. At the same time, ADP is released.

$$[\text{Actin} \cdot \text{Myosin} \cdot \text{ADP} \cdot P_i \text{ complex}] \longrightarrow [\text{Actin} \cdot \text{Myosin complex}] + \text{ADP} + P_i$$

1 **Binding:** When Ca^{2+} levels are high, energized cross-bridge can bind to actin. (ADP + P_i are already bound to the cross-bridge.)

2 **Power stroke:** Release of P_i causes cross-bridge to move toward the H zone of the sarcomere. This power stroke moves the actin filament toward the H zone. ADP is then released.

3 **Detaching:** ATP binds to myosin, causing the cross-bridge to detach from the actin filament.

4 **Resetting:** Hydrolysis of ATP to ADP + P_i provides energy, which causes the cross-bridge to move away from the H zone. ADP and P_i remain bound to the cross-bridge. Cycle can begin again.

Thin filament (actin)
Ca^{2+}
Thick filament (myosin)
Z line
P_i
ADP
ATP
Energized cross-bridge

Figure 44.8 **The four stages of cross-bridge cycling in skeletal muscle.**

Concept Check: *Rigor mortis is a condition of contracted skeletal muscle that occurs shortly after death. Why would the muscle remain contracted after death, and why does such contraction eventually stop?*

Step 3: ATP binds to myosin, causing the cross-bridge to detach.

During the power stroke, myosin is bound very firmly to actin; this linkage must be broken to allow the cross-bridge to be reenergized and repeat the cycle. The binding of a new molecule of ATP to the myosin cross-bridge breaks the link between actin and myosin.

$$[\text{Actin} \cdot \text{Myosin complex}]\ \text{ATP} \longrightarrow \text{Actin} + [\text{Myosin} \cdot \text{ATP}]$$

The dissociation of actin and myosin by ATP is an example of allosteric regulation of protein activity. When ATP binds at one site on myosin (called an allosteric site; refer back to Chapter 6), myosin's affinity for actin bound at another site decreases. ATP is not hydrolyzed in this step. Instead, ATP acts here only as an allosteric modulator of the myosin head, weakening the binding of myosin to actin and leading to their dissociation.

Step 4: ATP hydrolysis reenergizes and resets the cross-bridge.

After actin and myosin dissociate, the ATP bound to myosin is hydrolyzed. This hydrolysis reenergizes myosin, causing it to reset to the position that allows actin binding.

$$\text{Actin} + [\text{Myosin} \cdot \text{ATP}] \longrightarrow \text{Actin}\ [\text{Energized myosin} \cdot \text{ADP} \cdot P_i]$$

If Ca^{2+} is still present at this time, the cross-bridge can reattach to a new actin molecule in the thin filament, and the cross-bridge cycle repeats, causing the muscle fiber to further shorten.

As we have seen, ATP performs two different roles in the cross-bridge cycle. First, the energy released from ATP hydrolysis provides the energy for cross-bridge movement. Second, ATP binding breaks the link formed between actin and myosin during the cycle, allowing a new cycle to start. Although the precise mechanisms may differ slightly between vertebrates and invertebrates, biologists think that all skeletal muscle functions according to steps similar to those just described. As mentioned, all skeletal muscle function requires the actions of calcium ions, which we describe next.

The Regulation of Muscle Contraction by Calcium Ions Is Mediated by Tropomyosin and Troponin

How does the presence of Ca^{2+} in the muscle fiber cytosol regulate the cycling of cross-bridges? The answer requires a closer look at the two additional thin filament proteins mentioned earlier, tropomyosin and troponin.

Tropomyosin is a rod-shaped molecule composed of two intertwined proteins (Figure 44.9a). Tropomyosin molecules are arranged end to end along the actin thin filament. In the absence of Ca^{2+}, they partially cover the myosin-binding site on each actin molecule, thereby preventing cross-bridges from making contact with actin. Each tropomyosin molecule is held in this blocking position by **troponin**, a smaller, globular-shaped protein with three subunits that is bound to both tropomyosin and actin. In this way, troponin and tropomyosin block access to myosin-binding sites on actin molecules in the relaxed muscle fiber.

One of the three subunits of troponin is capable of binding Ca^{2+}. Such binding produces a change in the shape of troponin, which—through troponin's linkage to tropomyosin—allows tropomyosin to move away from the myosin-binding site on each actin molecule (Figure 44.9b), which permits cross-bridge cycling to occur. Conversely, release of Ca^{2+} from troponin reverses the process, blocking the myosin-binding site and turning off contractile activity.

Thus, the concentration of cytosolic Ca^{2+} determines whether or not Ca^{2+} is bound to troponin molecules, which, in turn, determines

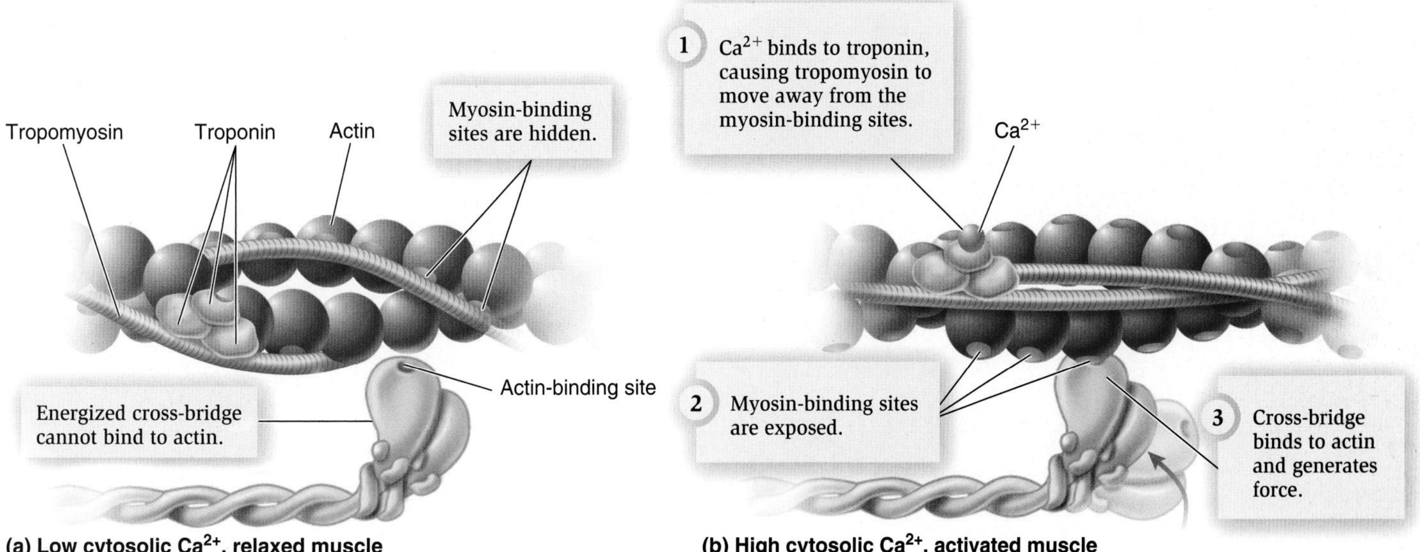

Figure 44.9 **Role of Ca^{2+}, tropomyosin, and troponin in cross-bridge cycling.** **(a)** When cytosolic Ca^{2+} is low, the myosin-binding sites on actin are blocked by tropomyosin. **(b)** When cytosolic Ca^{2+} increases, Ca^{2+} binds to troponin, which, in turn, allows tropomyosin to move away from the myosin-binding sites on actin.

the number of actin sites available for cross-bridge binding. The cytosolic concentration of Ca^{2+}, however, is very low in resting muscle. Let's see how the Ca^{2+} concentration is increased so that contraction can occur.

Contraction of Skeletal Muscle Is Coupled with Electrical Excitation

Like neurons, vertebrate skeletal muscle cells are capable of generating and propagating action potentials in response to an appropriate stimulus. This causes an increase in the concentration of cytosolic Ca^{2+}, which triggers contraction of a muscle fiber. The sequence of events by which an action potential in the plasma membrane of a muscle fiber leads to cross-bridge activity by the mechanisms just described is called **excitation-contraction coupling**. The electrical activity in the plasma membrane does not act directly on the contractile proteins but instead acts as a stimulus to increase cytosolic Ca^{2+} concentration. These increased concentrations continue to activate the contractile apparatus long after electrical activity in the membrane has ceased.

The source of the increased cytosolic Ca^{2+} that occurs following a muscle action potential is the fiber's **sarcoplasmic reticulum**, which acts as a Ca^{2+} reservoir. The sarcoplasmic reticulum, which is a specialized form of the endoplasmic reticulum found in most cells, is composed of interconnected sleevelike compartments and sacs around each myofibril (**Figure 44.10**). Separate tubular structures, the **transverse tubules (T-tubules)**, are invaginations of the plasma membrane that open to the extracellular fluid. The T-tubules course around each myofibril and conduct action potentials from the outer surface of the muscle fiber inside to the myofibrils. An action potential causes the opening of calcium channels in the lateral sacs of the sarcoplasmic reticulum, which allows Ca^{2+} to flow into the cytosol and bind to troponin, initiating muscle contraction.

A contraction continues until Ca^{2+} is removed from troponin and the cytosol. This is achieved by ATP-driven ion pumps in the

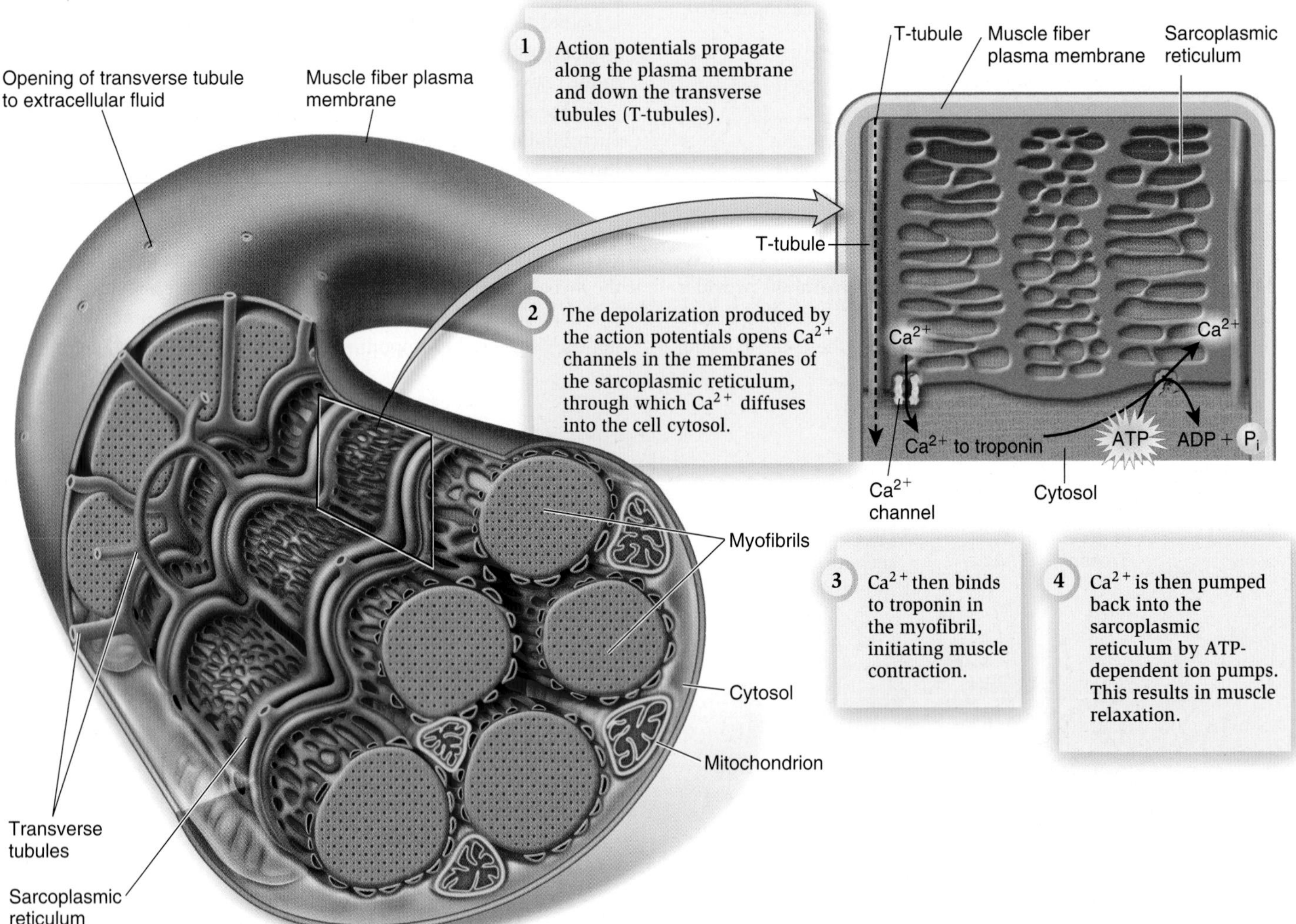

Figure 44.10 **Structure and function of the sarcoplasmic reticulum, transverse tubules (T-tubles), and myofibrils in a skeletal muscle fiber.**

BioConnections: Where else have you learned about voltage-gated Ca^{2+} channels and their role in cell-to-cell communication? Look back to Figure 41.14 for a hint.

sarcoplasmic reticulum that decrease the Ca^{2+} concentration in the cytosol back to its resting concentration.

Electrical Stimulation of Skeletal Muscle Occurs at the Neuromuscular Junction

We have seen that an action potential in the plasma membrane of a skeletal muscle fiber is the signal that triggers contraction. How are these action potentials initiated? The mechanism by which action potentials are initiated in a skeletal muscle involves stimulation by a motor neuron. These are neurons located in the central nervous system (CNS) that transmit signals away from the CNS and directly or indirectly control muscles.

The site where a motor neuron's axon synapses with a muscle fiber is known as a **neuromuscular junction** (Figure 44.11). Near the surface of the muscle fiber, the axon divides into several short processes, or terminals, containing synaptic vesicles filled with the neurotransmitter acetylcholine (ACh) (see Chapter 41). The region of the muscle fiber plasma membrane that lies directly under the axon terminal is called the **motor end plate**; it is folded into what are known as junctional folds, which contain many ACh receptors. These folds therefore increase the total surface area available for the membrane to respond to ACh. The extracellular space between the axon terminal and the motor end plate is called the synaptic cleft.

When an action potential in a motor neuron arrives at the axon terminal, it releases its stored ACh, which crosses the synaptic cleft and binds to receptors in the junctional folds of the muscle fiber. The ACh receptor is a ligand-gated ion channel (see Chapter 5). The binding of ACh opens the channel and causes an influx of Na^+ into the muscle fiber, which causes the muscle fiber to depolarize, resulting in an action potential that spreads along the membrane of the muscle fiber and through the T-tubules. Most neuromuscular junctions are located near the middle of a muscle fiber, and newly generated muscle action potentials propagate from this region toward the ends of the fiber and throughout the T-tubule network. Overstimulation of a muscle fiber is prevented by the action of **acetylcholinesterase**. This enzyme breaks down excess ACh in the synaptic cleft to inactive forms that cannot bind the ACh receptor.

Now that we have learned about the structural characteristics of skeletal muscle and the events that initiate and produce force generation, let's explore how skeletal muscle is adapted to meet the varied functional demands of vertebrates.

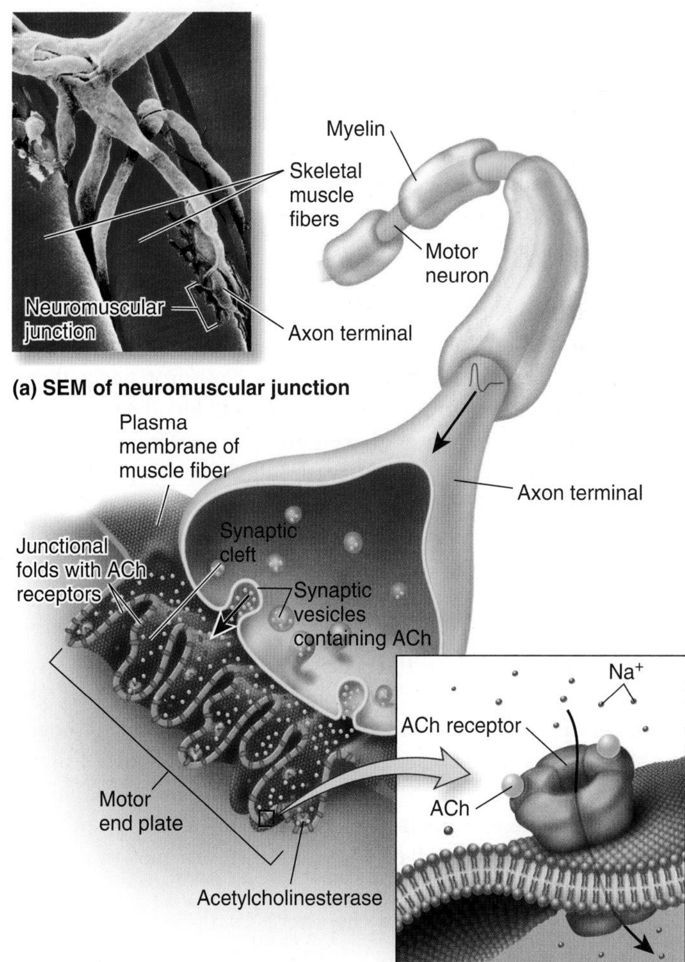

Figure 44.11 **The neuromuscular junction.** The structure of a neuromuscular junction **(a)** as seen in a colorized scanning electron micrograph and **(b)** as depicted schematically. Action potentials in the motor neuron cause exocytosis of ACh-containing synaptic vesicles. ACh binds to receptors in the plasma membrane of the junctional folds of the skeletal muscle cell. This initiates Na^+ entry and, consequently, an action potential in the muscle cell. Excess ACh is removed by the enzyme acetylcholinesterase.

Concept Check: *Why does Na^+ enter the muscle cell after the ACh receptor is activated?*

BioConnections: *What is the function of the myelin indicated in this figure? Refer back to Figures 41.2 and 41.12 for help.*

44.4 Skeletal Muscle Function

Learning Outcomes:

1. Outline the general characteristics of the three types of skeletal muscle fibers.
2. Explain how muscles adapt to exercise.
3. Describe how antagonistic muscles work at a joint.

Animals use skeletal muscle for a wide variety of activities, such as locomotion, stretching, chewing, breathing, and maintaining balance, to name a few. Therefore, it is not surprising that not all skeletal muscle fibers share the same mechanical and metabolic characteristics. In this section, we will consider how different types of fibers can be classified on the basis of their rates of shortening (as either fast or slow) and the way in which they produce the ATP needed for contraction (as oxidative or glycolytic).

Skeletal Muscle Fibers Are Adapted for Different Types of Movement

Different muscle fibers contain forms of myosin that differ in the maximal rates at which they can hydrolyze ATP. This, in turn, determines the maximal rates of cross-bridge cycling and muscle shortening.

Fibers containing myosin with low ATPase activity are called **slow fibers**. Those containing myosin with higher ATPase activity are classified as **fast fibers**. Although the rate of cross-bridge cycling is about four times faster in fast fibers than in slow fibers, the maximal force produced by both types of cross-bridges is approximately the same.

The second means of classifying skeletal muscle fibers is based on the type of metabolic pathways available for synthesizing ATP. Fibers that contain numerous mitochondria and have a high capacity for oxidative phosphorylation are classified as **oxidative fibers**. Most of the ATP production by such fibers depends on blood flow to deliver oxygen and nutrients to the muscle. Not surprisingly, therefore, these fibers are surrounded by many small blood vessels. They also contain large amounts of **myoglobin**, an oxygen-binding protein that increases the availability of oxygen in the fiber by providing an intracellular reservoir of oxygen. The large amounts of myoglobin present in oxidative fibers give these fibers a dark-red color. Oxidative fibers are often referred to as red muscle fibers. The benefit of red muscle fibers is they can maintain sustained action over a long period of time without fatigue.

By contrast, **glycolytic fibers** have few mitochondria but possess both a high concentration of the proteins involved in glycolysis (glycolytic enzymes; refer back to Figure 7.3) and large stores of glycogen, the storage form of glucose. Corresponding to their limited use of oxygen, these fibers are surrounded by relatively few blood vessels and contain little myoglobin. The lack of myoglobin is responsible for the pale color of glycolytic fibers and their designation as white muscle fibers.

On the basis of these two characteristics, three major types of skeletal muscle fibers have been distinguished:

1. **Slow-oxidative fibers** have low rates of myosin ATPase activity but have the ability to make large amounts of ATP. These fibers are used for prolonged, regular types of movement, such as steady, long flight; long-distance swimming; or the maintenance of posture. These muscles, for example, are what give the red color to the dark meat of ducks, which use the muscles for flight. Long-distance runners have a high proportion of these fibers in their leg muscles. These types of activities require muscles that do not fatigue easily.
2. **Fast-oxidative fibers** have high myosin ATPase activity and can make large amounts of ATP. Like slow-oxidative fibers, these fibers do not fatigue quickly and can be used for long-term activities. They are also particularly suited for rapid actions, such as the rapid trilling sounds made by the throat muscles in songbirds or the clicking sounds generated by the shaking of a rattlesnake's tail.
3. **Fast-glycolytic fibers** have high myosin ATPase activity but cannot make as much ATP as oxidative fibers, because their source of ATP is glycolysis. These fibers are best suited for rapid, intense actions, such as a cheetah's short sprint at maximum speed. Sloths, by contrast, have little or no fast-glycolytic fibers in their leg muscles, which is not surprising given a sloth's very sedentary lifestyle. Fast-glycolytic fibers fatigue more rapidly than fast-oxidative fibers. The breast meat of chickens, for example, appears white because, unlike ducks, chickens do not fly except for very short distances and therefore do not require oxidative pectoral muscles. The fast-glycolytic muscles of chickens, however, are ideal for short flights in the air that help them quickly escape predators, and disrupt scent trails on the ground that would otherwise help predators locate the chickens. When they land, chickens use slow-oxidative fibers in their leg muscles to run long distances as they continue to elude a predator.

Different muscle groups within the animal body have different proportions of each fiber type interspersed with one another; many activities require the action of all three types of fibers at once. This is important when you consider the wide range of animal activities related to locomotion alone, including walking, climbing, running, swimming, flying, crawling, crouching, jumping, and maintaining balance and posture. Depending on the needs of an animal at any given moment, the motor nerve inputs can be adjusted to stimulate different ratios of fiber types. When you lift a heavy weight for a brief time, the fast-glycolytic fibers in your arm muscles are activated in large numbers. When a crab uses its pincers to grab prey, fast-glycolytic muscles snap the claws closed quickly, but then slow-oxidative fibers maintain a tight grip for as long as required. The characteristics of the three types of skeletal muscle fibers are summarized in **Table 44.1**.

Muscles Adapt to Exercise

The regularity with which a muscle is used, as well as the duration and intensity of the activity and whether it includes resistance, affects the properties of the muscle. For example, increased amounts of resistance exercise—such as weightlifting—results in hypertrophy (increase in size) of muscle fibers. Because the number of fibers in a muscle remains essentially constant throughout adult life, the increases in muscle size that occur with resistance exercise do not result from increases in the number of muscle fibers, but primarily from increases in the size of each fiber. These fibers undergo an increase in fiber diameter due to the increased synthesis of actin and myosin filaments, which form more myofibrils.

Exercise of relatively low intensity but long duration—popularly called aerobic exercise, including running and swimming—produces increases in the number of mitochondria in the fibers that are needed

Table 44.1 Characteristics of the Three Types of Skeletal Muscle Fibers

	Slow-oxidative	Fast-oxidative	Fast-glycolytic
Primary source of ATP production	Oxidative phosphorylation	Oxidative phosphorylation	Glycolysis
Mitochondria	Many	Many	Few
Blood supply	High	High	Moderate
Myoglobin content	High (red muscle)	High (red muscle)	Low (white muscle)
Rate of fatigue	Slow	Intermediate	Fast
Myosin-ATPase activity	Low	High	High
Rate of contraction	Slow	Fast	Fast

in this type of activity. In addition, the number of blood vessels around these fibers increases to supply the greater energy demands of active muscle. All of these changes increase endurance with a minimum of fatigue.

By contrast, short-duration, high-intensity exercise, such as weightlifting, primarily affects fast-glycolytic fibers, which are used during strong contractions. In addition, glycolytic activity is enhanced by elevated synthesis of glycolytic enzymes. The results of such high-intensity exercise are the increased strength and bulging muscles of a conditioned weight lifter. Such muscles, although very powerful, have little capacity for endurance and therefore fatigue rapidly.

A decline or cessation of muscular activity results in the condition called **atrophy**, a reduction in the size of the muscle. Likewise, if the neurons to a skeletal muscle are destroyed or the neuromuscular junctions become nonfunctional, the denervated muscle fibers will become progressively smaller in diameter. This condition is known as denervation atrophy. Even with an intact nerve supply, a muscle can atrophy if it is not used for a long period of time, as when a broken limb is immobilized in a cast.

The mechanism by which changes occur in skeletal muscle during exercise is an active area of research, but a recent discovery has provided an intriguing clue, as described next.

FEATURE INVESTIGATION

Evans and Colleagues Activated a Gene to Produce "Marathon Mice"

In the course of investigating possible ways to reverse or prevent obesity in humans and other mammals, American biologist Ron Evans and his colleagues at the Salk Institute in California discovered one way in which the ratios of oxidative and glycolytic fibers change in skeletal muscle. Evans was interested in a gene that codes for a transcription factor called PPAR-δ. Activation of this protein results in the expression of genes that enable skeletal muscle or other cells to more efficiently burn fat instead of glucose for energy. Evans hypothesized that mice in which PPAR-δ was chronically activated at high concentrations would lose weight due to increased fat burning, as shown in Figure 44.12.

To test this hypothesis, Evans created transgenic mice (see Chapter 20) which contained the *PPAR-δ* gene in a modified form. The modified gene had a skeletal muscle-specific promoter so that it would be expressed only in skeletal muscle cells. The region of the gene that encoded PPAR-δ was linked to a region of another gene that encoded a viral protein domain called VP16. This domain also facilitates gene activation. The researchers expected the combination of PPAR-δ and VP16 to strongly activate genes that enable cells to more efficiently burn fat instead of glucose for energy.

In the first part of the experiment, Evans monitored the body weights of the transgenic mice after they reached adulthood. The transgenic mice gained significantly less weight than the wild-type mice when fed high-fat diets (which normally cause mice—like humans—to gain weight), confirming Evans' hypothesis. As is sometimes the case in scientific discovery, an unexpected finding arose from this study. When Evans examined several tissues in these mice under the microscope, he observed that the skeletal muscle of the transgenic mice showed a dramatic shift from glycolytic fibers to

Figure 44.12 **Evans and colleagues' activation of a gene to produce "marathon mice."**

HYPOTHESES 1. Increased expression of genes that lead to increased fat oxidation in skeletal muscle cells will prevent obesity in mice.
2. Transgenic mice have a greater capacity for prolonged exercise than do wild-type mice.

KEY MATERIALS Mice, light and electron microscopes, motorized treadmills.

	Experimental level	Conceptual level
1 Prepare a modified gene containing a skeletal muscle-specific promoter and a coding sequence that links *VP16* and *PPAR-δ*. See Chapter 20 for gene cloning methods.	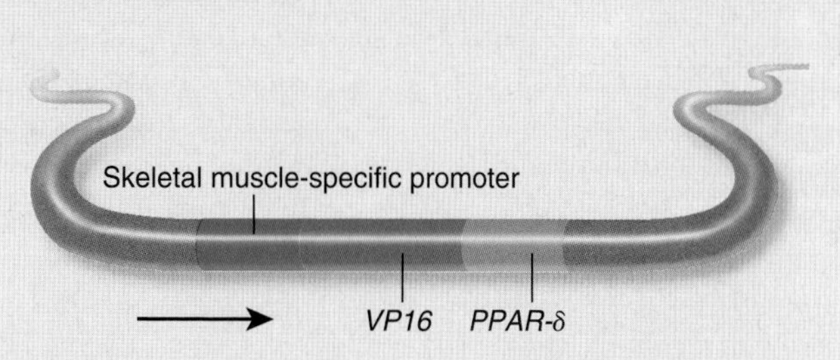	Skeletal muscle-specific promoter ensures gene is turned on only in skeletal muscle. VP16 domain is a domain that always activates transcription. *PPAR-δ* codes for a transcription factor that specifically activates genes that allow cells to efficiently burn fat.
2 Make transgenic mice expressing the *VP16–PPAR-δ* gene.	 See Chapter 20 for a discussion of gene addition.	All of the cells will carry this gene, but only skeletal muscle cells will express the gene.

3 Perform the following tests:

(a) Feed wild-type control mice and transgenic mice a normal-fat diet (4%) and then switch to a high-fat diet (35%). Weigh mice weekly.
(b) Examine the muscle fibers in the mice.
(c) Test their endurance on a treadmill.

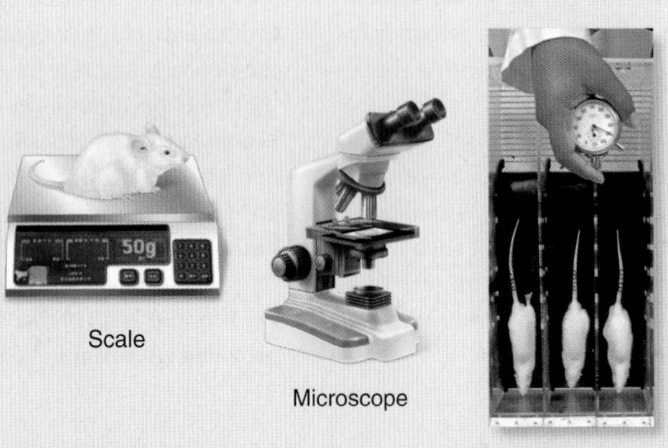

(a) Eating a high-fat diet is known to cause obesity in mice and other mammals.
(b) The appearance of skeletal muscle can be examined by light and electron microscopy.
(c) The treadmills are motorized to keep mice moving until they become exhausted, at which time the treadmills are stopped.

4 THE DATA

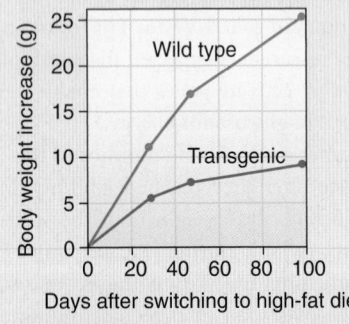

Body weight increase (g): 0, 5, 10, 15, 20, 25
Wild type
Transgenic
Days after switching to high-fat diet: 0, 20, 40, 60, 80, 100

(a) Weight gain

Characteristics of skeletal muscle in transgenic mice:

- redder than wild type
- more myoglobin
- more mitochondria
- more slow-oxidative fibers

(b) Difference in skeletal muscle

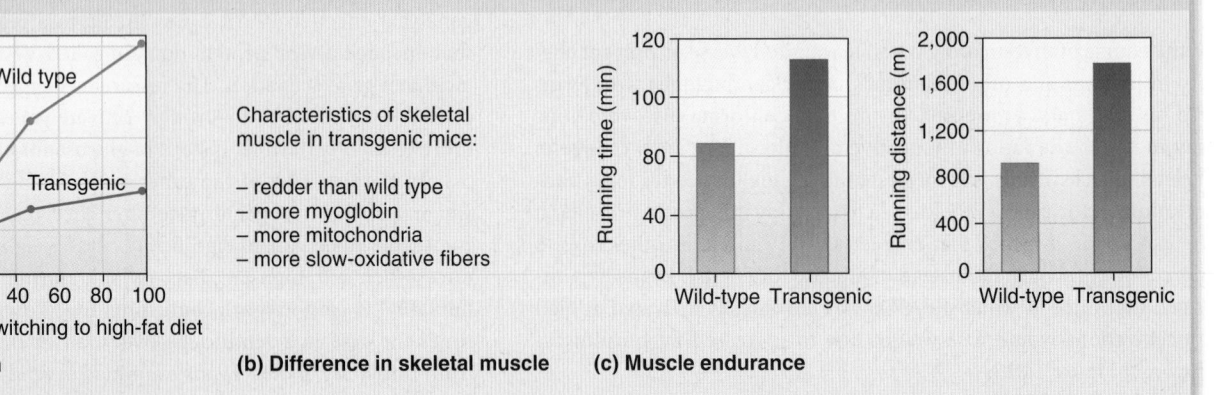

Running time (min): 0, 40, 80, 100, 120
Wild-type Transgenic

Running distance (m): 0, 400, 800, 1,200, 1,600, 2,000
Wild-type Transgenic

(c) Muscle endurance

5 CONCLUSION *PPAR-δ* contributes to both weight loss and endurance in mice. The fiber-type switching associated with exercise does not require exercise, because increasing fat oxidation in skeletal muscle cells resulted in more oxidative fibers even without exercise training.

6 SOURCE Wang, Y.X. et al. 2004. Gene targeting turns mice into long-distance runners. *Public Library of Science Biology* 2:322.

slow-oxidative fibers. The muscle in transgenic mice appeared redder than it did in wild-type mice. It contained more myoglobin and mitochondria, and had higher concentrations of oxidative enzymes capable of providing the cells with sustained levels of ATP. These changes occurred even though the mice had not been subjected to exercise training.

Based on these observations, Evans also tested the hypothesis that the transgenic mice would have a greater capacity for prolonged exercise than wild-type mice. When the transgenic mice were challenged with an endurance exercise test, Evans discovered that they outperformed age- and weight-matched wild-type mice by a factor of nearly twofold! They could sustain a high level of activity on a miniature treadmill for nearly twice as long as wild-type mice (hence the nickname "marathon mice"). This effect occurred in transgenic mice even without prior exercise training. In other words, simply increasing the ratio of oxidative to glycolytic fibers gave the mice greater ability to sustain aerobic activity.

The results of these experiments indicate that increasing the amount of activated PPAR-δ facilitates an oxidative state in skeletal muscle fibers that somehow signals them to convert to types that are best suited for oxidative metabolism. Therefore, the switch in fiber type that occurs in exercise training may not require exercise per se, and it may be mediated in part by proteins that activate or induce *PPAR-δ* expression. These results may have important implications for enhancing physical endurance in humans, as well as for possible treatments for various muscle diseases and for protection against obesity.

Experimental Questions

1. What is the normal function of the PPAR-δ protein in mice?
2. What was the hypothesis proposed by Evans in relation to PPAR-δ and obesity?
3. How did Evans and his colleagues test this hypothesis, and what did they observe?

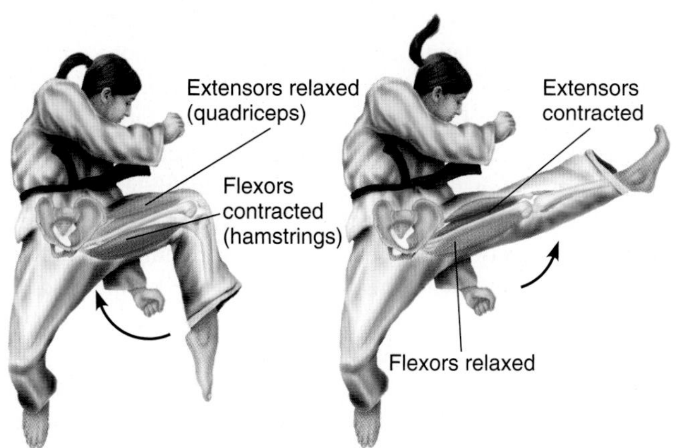

Figure 44.13 **Actions of flexors and extensors.** The figure shows how skeletal muscles cause flexion or extension of a limb. When the flexor muscle contracts, the extensor relaxes, and vice versa.

Skeletal Muscles Can Bend or Straighten a Limb

As mentioned previously, contracting muscle exerts a force on bones through its connecting tendons. When the force is great enough, the bone moves as the muscle shortens. A contracting muscle exerts only a pulling force, so as the muscle shortens, the attached bones are pulled toward or away from each other. Muscles that bend a limb at a joint (that is, reduce the angle between two bones) are called **flexors**, whereas muscles that straighten a limb (increase the angle between two bones) are called **extensors**. Groups of muscles that produce oppositely directed movements at a joint are known as **antagonists**. For example, in Figure 44.13, we can see that contraction of the hamstrings flexes the leg at the knee joint, relaxing the quadriceps, whereas contraction of the quadriceps causes the leg to extend and the hamstrings to relax. Both antagonistic muscles exert only a pulling force when they contract.

44.5 Animal Locomotion

Learning Outcomes:

1. Explain the mechanisms of animal locomotion in water, on land, and in air.
2. Compare the relative energy costs of swimming, running, and flying.
3. Explain how evolution has shaped the structures used for locomotion.

Locomotion, the movement of an animal from place to place, may take many forms, as described earlier. In all cases, animals experience certain constraints to locomotion. For example, all animals must overcome frictional forces (drag) generated by the air, water, or surface of the Earth. In addition, all forms of locomotion require energy to provide thrust, defined as the forward motion of an animal in any environment, and/or lift, which is movement against gravity.

Although the precise mechanism may differ among animals, locomotion with few exceptions (such as the rhythmic beating of cilia in ctenophores) results from muscular contractions that exert force on one of the three types of skeletons discussed at the beginning of this chapter. In this section, we will examine the similarities and differences between locomotion in water, on land, and in air.

Aquatic Animals Must Overcome the Resistance of Water

The greatest challenge to locomotion that aquatic animals face is the density of water, which is much greater than that of air. This is apparent when you compare waving your hand through the air and underwater. Water's resistance to movement increases exponentially as the speed of locomotion increases, which is one reason why many fishes swim at relatively slow speeds. Overcoming this resistance requires considerable muscular effort. Most swimming animals, including fishes, amphibians, reptiles, diving birds, and marine mammals, have evolved streamlined bodies that reduce drag and so make swimming more efficient. Other animals that swim only occasionally or spend time on the surface of the water have evolved adaptations, such as the webbed feet of ducks, which assist their muscles in generating greater thrust through the water.

Although the density of water creates challenges to locomotion, it also provides certain benefits. An energetic advantage to swimming is that fishes and other swimmers do not need to provide as much lift to overcome gravity. Because the density of water is similar to that of an animal's body, water provides buoyancy, which helps support the animal's weight.

The mechanism of swimming is similar among many different vertebrates. Most fishes, for example, contract posterior skeletal muscles to move the tail end of the animal from side to side. This pushes water backward and propels the fish forward. Other muscles and fins provide additional thrust and enable changes in direction. Likewise, amphibians and marine reptiles rely predominantly on their hind legs, their tail, or undulations of the posterior parts of the body for propulsion through the water. Cetaceans (whales and dolphins) use up-and-down thrusts of their tail flukes to provide propulsion. Confining most of the swimming muscles to the rear of an animal's body has certain advantages. With the rear end devoted to movement, the front end is free to explore the environment, fight off aggressors, or find food. By contrast, pinnipeds such as seals and walruses use their flippers to provide thrust and their tails to steer. In these animals, a flexible tail assists with locomotion when on land.

In contrast to swimming vertebrates, many aquatic invertebrates move through water by means other than swimming. Cephalopods such as squids use propulsion provided by the ejection of water to give them brief but rapid bursts of speed. Many other aquatic invertebrates move passively on water currents or crawl along rocks and other underwater surfaces.

Energetically, swimming is the "cheapest" form of locomotion in animals that are adapted to it, due to streamlining, the relatively slow speed of most swimmers, and the buoyancy of water. By contrast, terrestrial animals face considerable energetic costs to locomotion, as we examine next.

Walking and Running on Land Are Energetically Costly

In contrast to swimming and flying, locomotion on land is, on average, the most energetically costly means of locomotion (Figure 44.14). Whereas gravity is not an important factor for locomotion in swimming animals, terrestrial animals must overcome gravity each time

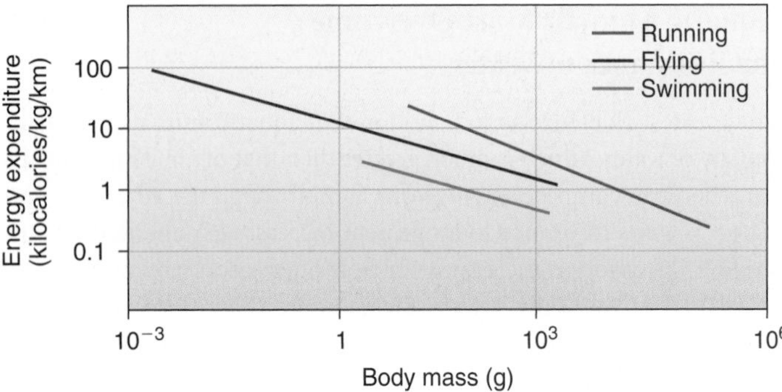

Figure 44.14 Energy costs of locomotion. The energy costs of three different modes of locomotion for animals of different sizes are shown. The *y*-axis gives the energy costs expressed as kilocalories expended per kilogram (kg) per kilometer (km). Energy costs are highest for runners compared with similarly sized fliers and swimmers. Note: Only a portion of the full range of body sizes of swimmers is shown.

Concept Check: Does this graph indicate that smaller animals expend more energy than larger animals when moving a similar distance?

they take a step. Of even greater importance to walking and running animals, though, is the requirement to accelerate and decelerate the limbs with every step. In essence, each step is like starting a movement from scratch, without the luxury of occasionally gliding through water or air as fishes and birds do. This challenge is even greater when an animal moves uphill or over rough terrain.

Apart from gastropods, which move along the surface of the Earth on a layer of secreted mucus, and snakes, which undulate along the ground on a portion of their ventral body surface, most terrestrial animals limit the amount of contact with the ground while moving, thereby minimizing the amount of friction they encounter. Tetrapods usually have only two feet on the ground at any time when walking, and for brief moments, an animal such as a horse galloping at full speed has all four feet off the ground.

Having fewer legs on the ground surface at any one time helps increase speed but can compromise stability. Arthropods, for example, have at least six legs. This apparently provides excellent stability but reduces maximal speed (although cockroaches can attain rapid speeds by running on only two legs). At the other extreme are animals that move by jumping—fleas, certain spiders, click beetles, grasshoppers, frogs, and kangaroos.

Flying Has Evolved in Four Different Lineages

Flying is a highly successful means of locomotion and is hypothesized to have evolved in four different lineages: pterosaurs (extinct reptiles that were the first vertebrates to fly), insects, birds, and bats (mammals). Flying provides numerous advantages: animals can escape land-based predators, scan their surroundings over great distances, and inhabit environments such as high cliffs that may be inaccessible to nonflying animals. The mechanics of flying, however, require animals to overcome gravity and air resistance, which makes flying more energetically costly than swimming but still less costly than running on land (see Figure 44.14). Many migratory birds can travel hundreds of miles daily for a week or longer. A recent study by researchers in New Zealand, using satellite transmitters, tracked the nonstop, week-long migration of four bar-tailed godwits (*Limosa lapponica*) over a distance of roughly 10,000 km, from New Zealand to the coast of South Korea. No terrestrial animal could possibly match such a feat by walking or running. (To grasp how astonishing this feat is, imagine running across the entire continent of North America and back again in 7 days without any rest or food!) As with swimming, resistance to flight is decreased by streamlined bodies. However, earthbound animals have one advantage over their flying cousins—they can grow to much larger sizes than animals that fly. The vast majority of flying animals have a mass between about 1 mg and 1 kg. Only a few large birds have masses exceeding 10 kg. Although this represents a wide range, it falls far short of the sizes achieved by earthbound or aquatic animals.

In flying vertebrates, lift and thrust are provided by pectoral and other muscles that move the wings. The pectoral muscles are so powerful and massive that they constitute as much as 15–20% of a bird's total body mass and up to 30% in hummingbirds, which use their wings not only to fly but also to hover. The requirement for large, strong pectoral muscles is one reason why the body mass of flying vertebrates is limited. The extinct pterosaurs would seem to be an exception because some species were known to have had wingspans of nearly 10 m. However, scientists think that these large animals were unable to generate the force required to lift their massive bodies off the Earth and instead glided off of trees or cliffs to fly.

In birds and bats, the wings are modifications of the forelimbs. In general, bat wings are far more maneuverable than bird wings, because unlike birds, bats have digits at the end of their forelimbs/wings. This allows bats to precisely alter the shape of their wings and provide for rapid, fine-tuned changes in direction, even at high speeds. Bird wings are more similar to those of a fixed-wing aircraft. The largest birds, such as hawks and eagles, are able to glide because of the great surface area of their large wings. By using a bird's momentum to propel it forward, gliding provides a considerable energy savings. Bats and small birds, however, can glide for only very brief moments.

In this chapter, we have seen how skeletal muscles and, in vertebrates, bones work together to provide animals with protection and to enable them to move around in their environments. Unfortunately, muscles and bone are frequently the sites of significant diseases with great effects on human health. We conclude the chapter with a discussion of a few of these diseases, each resulting from a different cause.

44.6 Impact on Public Health

Learning Outcomes:

1. Describe the impact of the bone diseases rickets and osteoporosis on human health.
2. Distinguish between the muscle diseases myasthenia gravis and muscular dystrophy.

A number of diseases can affect bone structure and function in humans. Bone disease may involve defects in either the mineral or organic components of bone. Poor bone formation and structure may result from inadequate nutrition, hormonal imbalances, aging, or skeletal muscle atrophy, to name just a few of the common causes.

In addition, many diseases or disorders directly affect the contraction of skeletal muscle. Some of them are temporary and not serious, such as muscle cramps, whereas others are chronic and severe, such as the disease muscular dystrophy. Also, some diseases result from defects in parts of the nervous system that control contraction of the muscle fibers rather than from defects in the fibers themselves. One example is poliomyelitis (polio), a viral disease in which the destruction of motor neurons leads to skeletal muscle paralysis that may result in death from respiratory failure. Polio, once thought to be nearly eliminated from the human population through vaccination, is still a threat in parts of the world.

Other diseases that affect skeletal muscle function result when normal processes go awry. For example, the immune system that normally protects the body from invaders may turn on itself, or faulty genes may produce an abnormal protein. This section looks at a few of these conditions in more detail.

Rickets and Osteoporosis Affect the Bones of Millions of People

Bone diseases are fairly common, particularly among individuals older than age 50. Two major abnormalities can occur in bone. The first is improper mineral deposition in bone, usually due to inadequate dietary calcium intake or inadequate absorption of calcium from the small intestine. Without adequate minerals, bone becomes soft and easily deformed, as occurs in the weight-bearing bones of the legs of children with **rickets** (or **osteomalacia**, as it is called in adults) (Figure 44.15a). These disorders are best prevented or treated with vitamin D, because this vitamin is the most important factor in promoting absorption of calcium from the small intestine.

The second major abnormality we will discuss is a more common disease called **osteoporosis**, in which both the mineral and organic portions of bone are reduced (Figure 44.15b). This disease, which affects four times as many women as men, occurs when the normal balance between bone formation and bone breakdown is disrupted.

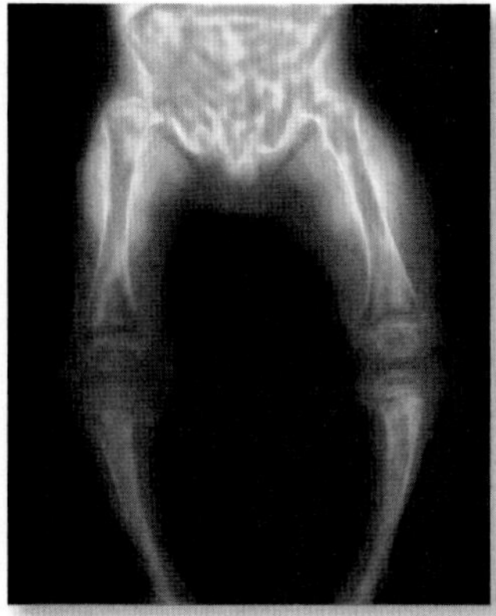

(a) X-ray image of leg bones of a child with rickets

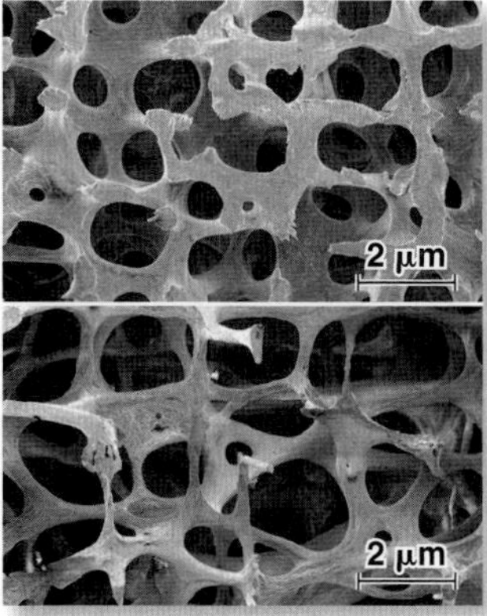

(b) Histologic appearance of normal bone (top) and bone from a person with osteoporosis (bottom)

Figure 44.15 Human bone diseases.

One cause of osteoporosis is prolonged disuse of muscles. In ways that are not completely clear, the force produced by active skeletal muscle contractions helps maintain bone mass. When muscles are not or cannot be used—due to paralysis or long-term immobilizing illnesses or even during prolonged space flight in which the lack of gravity reduces the need for muscles to work hard—bone mass declines.

More commonly, osteoporosis may result from hormonal imbalances. Some hormones—for example, estrogen—stimulate bone formation. When estrogen concentrations decline after menopause (the period of time when a woman's reproductive cycles cease), bone density may decline, increasing the risk of bone fractures. By contrast, some hormones—such as parathyroid hormone (see Chapter 50)—act to demineralize bone, releasing Ca^{2+} as part of the way in which the body normally maintains mineral homeostasis in the blood. If such hormones are present in excess, however, they can cause enough demineralization of bone to result in osteoporosis. This may happen in rare cases when the glands that make these hormones malfunction and overproduce the hormones.

Osteoporosis can be minimized with adequate calcium and vitamin D intake and weight-bearing exercise programs. In some cases, postmenopausal women may be given estrogen to replace what their bodies are no longer producing, but this therapy is controversial due to the potential adverse effects of estrogen on cardiovascular health and possible increased risk of developing breast cancer. Osteoporosis is the most prevalent bone disease in the U.S., affecting up to 15–30 million individuals. It results in annual national expenditures of approximately $15–20 billion in hospital and other medical costs.

Myasthenia Gravis Is an Autoimmune Disease That Affects Skeletal Muscles

Myasthenia gravis is a disease characterized by skeletal muscle fatigue and weakness. It ranges in severity, but at its worst, it can make chewing, swallowing, talking, and even breathing difficult. It affects an estimated 10,000–30,000 Americans, usually not becoming apparent until adolescence.

In myasthenia gravis, skeletal muscle function is reduced because the body's immune system produces antibodies—proteins that attack foreign matter—that bind to and inactivate ACh receptors on skeletal muscle cells. Because the destruction of ACh receptors is brought about by the body's own defense mechanisms gone awry, myasthenia gravis is considered an autoimmune disease (see Chapter 53). Thus, although motor nerve signals to the muscles are normal, the muscles cannot fully respond to the ACh released from the nerves because the receptors are not functional.

A number of approaches are currently used to treat the disease. One is to administer inhibitors of acetylcholinesterase, the enzyme that normally breaks down ACh at the neuromuscular junction. The inhibition of this enzyme prolongs the time that acetylcholine is available at the synapse, which to some degree compensates for reduced ACh receptor abundance. Other therapies are directed at suppressing the immune system to prevent destruction of the ACh receptors by antibodies. This is accomplished either by surgically removing certain glands that constitute part of the immune system or with drug therapy, but it has the disadvantage of making the person more susceptible to infections. Another therapy, called plasmapheresis, involves removing the liquid fraction of blood (plasma), which contains antibodies, and reinfusing the blood after the antibodies have been removed. A combination of these treatments has greatly reduced the mortality rate for myasthenia gravis.

Muscular Dystrophy Is a Rare Genetic Disease That Causes Muscle Degeneration

The group of diseases collectively called muscular dystrophy affects 1 of every 3,500 American males; it is much less common in females. **Muscular dystrophy** is associated with the progressive degeneration of skeletal (and cardiac) muscle fibers, weakening the muscles and leading ultimately to death from lung or heart failure. The signs and symptoms become evident at about 2 to 6 years of age, and most affected individuals do not survive beyond the age of 30.

The most common form of muscular dystrophy, called **Duchenne muscular dystrophy**, is an X-linked recessive disorder resulting from a defective gene on the X chromosome. Females have two X chromosomes, and males only one (plus one Y). Therefore, a heterozygote female with one abnormal and one normal allele will not generally develop Duchenne muscular dystrophy. If she passes the abnormal allele to a son, however, he will have the disease. In Duchenne muscular dystrophy, the affected gene codes for a protein known as dystrophin, which is absent in patients with the disease. Dystrophin is a large protein that links cytoskeletal proteins to the plasma membrane. Scientists think it is involved in maintaining the structural integrity of the plasma membrane in muscle fibers. In the absence of dystrophin, the plasma membrane of muscle fibers is disrupted, causing extracellular fluid to enter the cell. Eventually, the cell ruptures and dies.

Summary of Key Concepts

44.1 Types of Animal Skeletons

- Skeletons are structures that provide support and protection and also function in locomotion.
- Three types of skeletons are found in animals. In hydrostatic skeletons, found in many soft-bodied invertebrates, muscle contractions create differences in hydrostatic pressure that support the body and generate movements. In arthropods, exoskeletons are protective external structures that must be shed to accommodate growth of the animal. Endoskeletons—found in some species of sponges and all echinoderms and vertebrates—are internal structures that grow with the animal but do not protect its body surface (Figure 44.1).

44.2 The Vertebrate Skeleton

- Vertebrate endoskeletons are made up of the axial and appendicular skeletons. A joint is formed where two or more bones of a vertebrate endoskeleton come together (Figure 44.2).
- In addition to the functions of support, protection, and locomotion, the vertebrate skeleton produces blood cells and constitutes a reservoir for ions crucial to homeostasis. A proper ratio of organic and mineral components is required for normal bone function.

44.3 Skeletal Muscle Structure and the Mechanism of Force Generation

- A skeletal muscle is a grouping of cells, called muscle fibers, bound together by connective tissue layers. Within each cell are cylindrical bundles known as myofibrils, each of which contains thick filaments of myosin and thin filaments composed of actin and two other proteins, arranged in repeating units called sarcomeres. Regions of the thick filaments that extend from the surface are called cross-bridges (Figures 44.3, 44.4).
- During muscle contraction, the sarcomeres shorten by a process known as the sliding filament mechanism. In muscle contraction, the thick filaments remain stationary while the thin filaments slide past them propelled by action of the cross-bridges (Figures 44.5, 44.6).
- Mutations in a myosin gene of the jaw muscle may have allowed the human brain to become larger (Figure 44.7).
- In the cross-bridge cycle, the binding of myosin cross-bridges to actin causes a change in the shape of the myosin molecule. As a result, the two filaments slide past each other, shortening the sarcomere and contracting the muscle fiber. Release of the cross-bridge from actin and return of myosin to its original conformation require ATP hydrolysis (Figure 44.8).
- Tropomyosin and troponin, the two proteins associated with actin, play a critical role in the regulation of muscle contraction. The binding of Ca^{2+} to troponin allows tropomyosin to move away from the myosin-binding sites, initiating cross-bridge binding (Figure 44.9).
- The concentration of cytosolic Ca^{2+} determines the number of actin sites available for cross-bridge binding. The source of the cytosolic Ca^{2+} involved in a muscle fiber action potential is the muscle fiber's sarcoplasmic reticulum. Transverse tubules (T-tubules) conduct action potentials from the plasma membrane at the outer surface of the muscle fiber to the myofibrils, allowing Ca^{2+} to be released from the sarcoplasmic reticulum to the cytosol (Figure 44.10).
- Electrical stimulation of skeletal muscle occurs at a neuromuscular junction, in which a motor neuron's axon and a muscle fiber are in close proximity (Figure 44.11).

44.4 Skeletal Muscle Function

- Three major types of skeletal muscle fibers have been distinguished. Slow-oxidative fibers have low rates of myosin ATP hydrolysis; they do not fatigue easily and are used for prolonged, regular activities. Fast-oxidative fibers have high myosin activity, do not fatigue quickly, and are particularly suited for rapid, long-term actions. Fast-glycolytic fibers have high myosin activity but cannot make as much ATP as oxidative fibers; they are best suited for rapid, short-term actions (Table 44.1).
- Increased expression of *PPAR-δ* in mice results in increased slow-oxidative muscle, greater exercise endurance, and weight loss (Figure 44.12).
- Muscles that bend a limb at a joint are called flexors, whereas muscles that straighten a limb are called extensors. Groups of muscles that produce oppositely directed movements at a joint are known as antagonists (Figure 44.13).

44.5 Animal Locomotion

- Locomotion, the movement of an animal from place to place, may take many forms, including swimming, walking, running, crawling, hopping, and flying.

- Due to streamlining, the relatively slow speed of most swimmers, and the buoyancy of water, swimming is energetically the most efficient form of locomotion. Locomotion on land is, on average, the most energetically costly means of locomotion. The energy expenditure required for flight is intermediate between those for swimming and land-based locomotion (Figure 44.14).

44.6 Impact on Public Health

- Several health conditions affect bone or muscle structure and function in humans. Rickets (osteomalacia in adults) is characterized by soft, deformed bones, usually resulting from insufficient dietary intake or absorption of calcium. In osteoporosis, bone density is reduced when bone formation fails to keep pace with normal bone breakdown. Myasthenia gravis is an autoimmune disease characterized by skeletal muscle fatigue and weakness. Muscular dystrophy is an ultimately fatal genetic disease associated with the progressive degeneration of skeletal and cardiac muscle fibers (Figure 44.15).

Assess and Discuss

Test Yourself

1. The hydrostatic skeleton common in soft-bodied invertebrates is composed of
   a. muscle and cartilage.
   b. cartilage and a neural net.
   c. muscle and water-based fluid.
   d. cartilage and water-based fluid.
   e. bone and muscle.
2. Which of the following is *not* a function of the vertebrate skeleton?
   a. structural support
   b. protection of internal organs
   c. calcium reserve
   d. blood cell production
   e. All of the above are functions of the vertebrate skeleton.
3. The protein that provides strength and flexibility to bone is
   a. actin. c. myoglobin. e. elastin.
   b. myosin. d. collagen.
4. In a sarcomere, the ________ contain(s) thin filaments and no thick filaments.
   a. A band c. I band e. both a and d
   b. M line d. H zone
5. The function of ATP during muscle contraction is to
   a. cause an allosteric change in myosin so it detaches from actin.
   b. provide the energy necessary for the movement of the cross-bridge.
   c. expose the myosin-binding sites on the thin filaments.
   d. do all of the above.
   e. do a and b only.
6. The function of calcium ions in skeletal muscle contraction is to
   a. cause an allosteric change in myosin so it detaches from actin.
   b. provide the energy necessary for the movement of the cross-bridge.
   c. expose the myosin-binding sites on the thin filaments.
   d. bind to tropomyosin.
   e. do a and c only.
7. Stimulation of a muscle fiber by a motor neuron occurs at
   a. the neuromuscular junction.
   b. the transverse tubules.
   c. the myofibril.
   d. the sarcoplasmic reticulum.
   e. none of the above.
8. Muscle fibers that have a high number of mitochondria, contain large amounts of myoglobin, and exhibit low rates of ATP hydrolysis are called __________ fibers.
   a. slow-glycolytic d. fast-oxidative
   b. fast-glycolytic e. slow-oxidative
   c. intermediate
9. Which of the following statements about movement and locomotion is *incorrect*?
   a. Terrestrial animals and flying animals expend energy to provide lift.
   b. Swimming animals expend energy to provide thrust but not lift.
   c. Flexors and extensors are examples of muscles called agonists.
   d. Flexors cause bending at a joint.
   e. Extensors cause straightening of a limb.
10. For animals adapted to it, swimming is energetically the cheapest type of locomotion because of
    a. streamlined body forms in aquatic organisms.
    b. slow speed of movement of some swimmers.
    c. buoyancy of water.
    d. a and c only.
    e. a, b, and c.

Conceptual Questions

1. Distinguish between exoskeletons and endoskeletons.
2. List and briefly describe the steps in the cross-bridge cycle.
3. 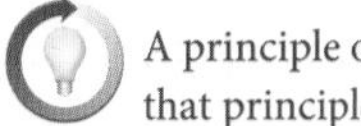 A principle of biology is that *living organisms use energy*. How is that principle illustrated in this chapter?

Collaborative Questions

1. Discuss the structure and function of the three different types of skeletons found in animals.
2. Discuss the three types of muscle tissues found in vertebrates.

Online Resource

www.brookerbiology.com

Stay a step ahead in your studies with animations that bring concepts to life and practice tests to assess your understanding. Your instructor may also recommend the interactive eBook, individualized learning tools, and more.

Nutrition and Animal Digestive Systems

45

Chapter Outline

45.1 Overview of Animal Nutrition and Ingestion

45.2 Principles of Digestion and Absorption of Food

45.3 Vertebrate Digestive Systems

45.4 Mechanisms of Digestion and Absorption in Vertebrates

45.5 Regulation of Digestion

45.6 Impact on Public Health

Summary of Key Concepts

Assess and Discuss

A balanced meal containing many nutrients, including carbohydrates, lipids, proteins, vitamins, minerals, and water.

What would 50,000,000 calories worth of food look like if it were assembled all in one place? (Try to imagine about 170,000 cheeseburgers, or slices of pizza.) That is roughly the amount of calories you will consume in your lifetime (in food labeling, one calorie is actually equivalent to one kilocalorie, or one Calorie with a capital C). All that energy is required for the trillions of body cells to perform varied activities such as synthesizing proteins, making cellular organelles, and maintaining concentration gradients of ions across cellular membranes. These and other activities are necessary for homeostasis and require a lifelong supply of food energy and chemical building blocks. These materials must be consumed by animals in the form of nutrients. A **nutrient** is any organic or inorganic substance that is taken in by an organism and is required for survival, growth, development, tissue repair, or reproduction; the process of consuming and using food and nutrients is called **nutrition**. All organisms require nutrients to survive. Animals receive their nutrients by consuming food, as shown by the nutritionally balanced meal in the chapter-opening photo.

Food processing in animals occurs in four phases: ingestion, digestion, absorption, and egestion (**Figure 45.1**). **Ingestion** is the act of taking food into the body via a structure such as a mouth. From there, the food moves into a digestive cavity or canal. If the nutrients in food are in a form that cannot be directly used by cells, they must be broken down into smaller molecules, a process known as **digestion**. This is followed by the process of **absorption**, in which ions, water, and small molecules diffuse or are transported from the digestive cavity into an animal's circulatory system or body fluids. **Egestion** (or defecation) is the process by which animals pass undigested material out of the body.

In this chapter, we will begin by looking at the types of nutrients that animals require. Next, we will discuss the diverse ways in which animals obtain, digest, and absorb those nutrients, and how these processes are regulated. We end by considering some of the common ways in which the function of the digestive system may go awry in humans.

45.1 Overview of Animal Nutrition and Ingestion

Learning Outcomes:

1. List the major categories of nutrients consumed by animals and some of their general functions.
2. Identify four groups of essential nutrients, listing several examples of each.
3. Describe some examples of the diversity of animal feeding habits.
4. Identify the three broad categories of strategies that animals use to get food and relate this to specialized adaptations such as teeth.

Animals require both organic and inorganic nutrients. Organic nutrients fall into five categories: carbohydrates, proteins, lipids, nucleic acids, and vitamins. These provide energy and the building blocks of new molecules or serve as cofactors in many enzymatic reactions. Inorganic nutrients include water and minerals such as calcium, copper, and iron. The importance of water to life was covered in Chapter 2. Minerals may function as cofactors in enzymatic reactions and other processes.

Because of similarities in the ways their organ systems function, animals share many nutritional requirements. However, the types of foods eaten by animals differ depending on an animal's physiology.

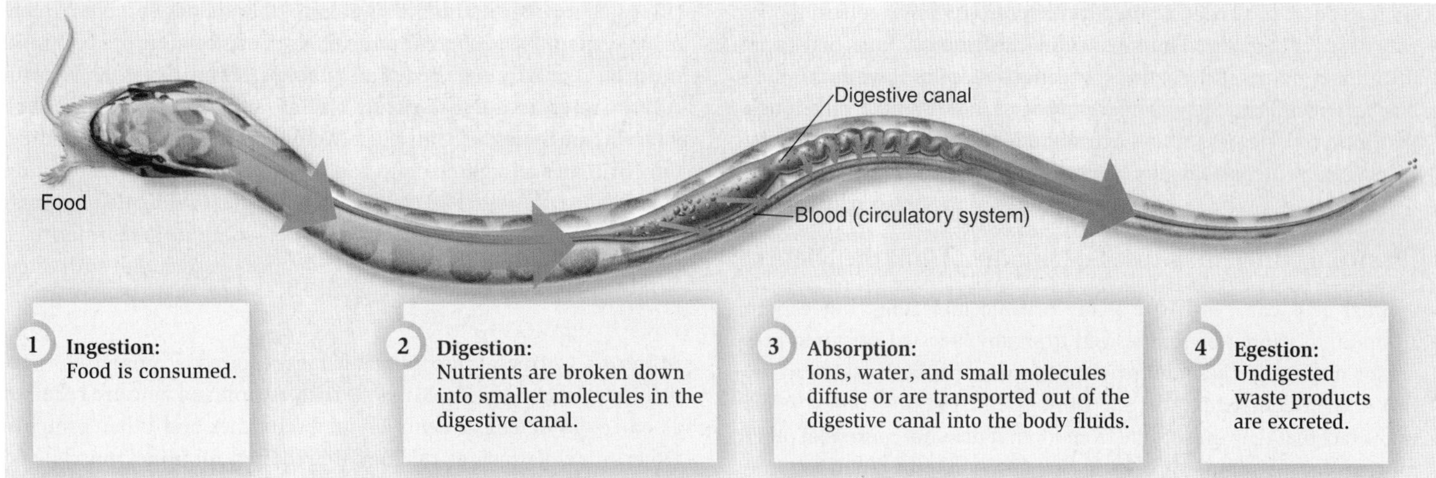

Figure 45.1 An overview of the four phases of food use in animals, as shown in a snake.

The digestive systems of **herbivores**, animals that eat only plants, contain microorganisms that assist in the digestion of cellulose, so these animals are well adapted to subsist on plants. By contrast, **carnivores** are primarily adapted to consume animal flesh or fluids, and **omnivores**, such as humans, eat both plant and animal products.

The amount of each type of nutrient required by an animal may also differ, depending on its activity level, or metabolic rate, which we will discuss in Chapter 46. Generally, highly active and energetic animals, such as most birds and mammals, require a proportionally greater amount of nutrients each day than do relatively inactive or sedentary animals, such as nonmotile invertebrates. However, exceptions to this general rule exist, as when some mammals enter hibernation and reduce their energy use and thus nutrient requirements. In this section, we will examine the various organic and inorganic nutrients consumed by animals and some of the major functions of these molecules. We will then see that despite the enormous number of animal species and the highly varied environments in which they live, the major ways in which animals ingest food can be classified into just a few categories.

Animals Require Nutrients for Energy and the Synthesis of New Molecules

Although different species consume a wide variety of foods, all animals require the same fundamental organic molecules (**Table 45.1**; see also Chapter 3 for a discussion of the chemical nature of organic molecules). Carbohydrates supply energy-yielding glucose and the carbon required for building organic molecules. Proteins supply amino acids that, in addition to building new protein, can be used as an energy source. Lipids supply components for membrane-building and thermal insulation and also provide energy. Nucleic acids supply some of the components required for DNA, RNA, and ATP synthesis. The other category of organic nutrients—vitamins—are required in small quantities and function as coenzymes in various reactions, as described later.

Ingested organic molecules are used for two general purposes: to provide energy and to make new molecules. As described in Chapter 6, the bonds of organic molecules may be broken down to release energy; this energy can be used in the synthesis of ATP, the

Table 45.1 Major Nutrients in Animals, Their Dietary Sources, and Some of Their Functions and Symptoms of Deficiency in Vertebrates

Class of nutrient	Dietary sources	Functions in vertebrates	Symptoms of deficiency
Carbohydrates	All food sources, especially starchy plants	Energy source; component of some proteins; source of carbon	Muscle weakness; weight loss
Proteins	All food sources, especially meat, legumes, cereals, roots	Provide amino acids to make new proteins; build muscle; some amino acids used as energy source	Weight loss; muscle loss; weakness; weakened immune system; increased likelihood of infections
Lipids	All food sources, especially fatty meats, dairy products, plant oils	Major component of cell membranes; energy source; thermal insulator; building blocks of some hormones	Hair loss; dry skin; weight loss; hormonal and reproductive disorders
Nucleic acids	All food sources	Provide sugars, bases and phosphates that can be used to make DNA, RNA, and ATP	None; components of nucleic acids can be synthesized by cells from amino acids and sugars

energy source of all cells. Indirectly, therefore, organic molecules provide the energy required for most of the chemical reactions that occur in animals' bodies. Alternatively, organic molecules serve as building blocks to synthesize new cellular molecules. For example, all animals that have muscles use amino acids obtained from food to make the specialized proteins that allow their muscle fibers to contract.

Essential Nutrients Must Be Obtained from the Diet

Animal cells can synthesize many organic molecules, but certain compounds cannot be synthesized from any ingested or stored precursor molecule. These **essential nutrients** must be obtained in the diet in their complete form. The word essential in this context refers to the fact that they must come from food. It does not mean that other nutrients are less important, because many synthesized nutrients are required for an animal's survival. The essential nutrients can be classified into four groups: essential amino acids, essential fatty acids, vitamins, and minerals.

Essential Amino Acids Nine **essential amino acids** are required in the diet of humans and many but not all other animals—isoleucine, leucine, lysine, methionine, phenylalanine, histidine, threonine, tryptophan, and valine. These amino acids are required for building proteins but cannot be synthesized by the animal's cells, unlike the other 11 amino acids that also make up proteins. Also, animal cells do not store amino acids. Therefore, without a recurring supply of these nine amino acids, protein synthesis in each cell in an animal's body would slow down or stop completely. Carnivores and omnivores readily obtain all of the essential amino acids, because meat (animal muscle) contains all 20 amino acids. Unlike animal meat, most plants do not contain every essential amino acid in sufficient quantities to supply a human's nutritive requirements. Therefore, people who follow a strict vegetarian diet must find ways to balance the protein content of the plant matter they eat. By contrast, some herbivores, such as cows, have evolved the capacity to synthesize the essential amino acids, which allows them to subsist entirely on a plant diet.

Essential Fatty Acids The **essential fatty acids** are certain unsaturated fatty acids, such as linoleic acid, that cannot be synthesized by animal cells. Linoleic acid is vital to an animal's health because it is converted in cells to another fatty acid, called arachidonic acid. This fatty acid is the precursor for production of several compounds important in many aspects of animal physiology; such compounds include the prostaglandins, which play roles in pain, blood clotting, and smooth muscle contraction. Some animals—such as felines—cannot synthesize arachidonic acid from linoleic acid, so arachidonic acid is an essential fatty acid in those species. Unsaturated fatty acids are found primarily in plants, which provide a dietary source of essential fatty acids for both herbivores and omnivores. Strict carnivores such as felines, however, obtain their essential fatty acids from fishes or from the adipose (fat) tissue of birds and mammals.

Vitamins **Vitamins** are important organic nutrients that serve as coenzymes for many metabolic and biosynthetic reactions. The two categories of vitamins are water-soluble and fat-soluble. Water-soluble vitamins, such as vitamin C, are not stored in the body and must be regularly ingested. Fat-soluble vitamins, such as vitamin A, are stored in adipose tissue. Not all animals require the same vitamins in their diet, however. Among vertebrates, for example, only primates and guinea pigs cannot synthesize their own vitamin C and must therefore consume it in the diet. **Table 45.2** summarizes the vitamins, their dietary sources, some of their important functions, and some health consequences associated with their deficiencies.

Minerals **Minerals** are inorganic ions required by animals for normal functioning of cells. Minerals such as iron and zinc are required as cofactors or constituents of some enzymes and other proteins. Other minerals such as calcium are required for bone, muscle, and nervous system function, and still others—notably sodium and potassium—contribute to changes in electrical differences across plasma membranes and therefore are critical for heart, skeletal muscle, and neuronal activity. **Table 45.3** summarizes some of the most important minerals and their functions. Many minerals are required in only trace amounts, far less than 1 mg/day in a relatively large mammal such as humans. Nonetheless, without regular consumption of these small amounts, serious health problems arise.

Some minerals can be stored in an animal's body, reducing the risk of deficiency when the mineral is not available in the diet. Calcium, for example, is stored in huge quantities in bone in vertebrates and in the shells of some invertebrates. If the dietary intake of this mineral falls, calcium ion concentrations in the body fluids may decrease. Because calcium ions are important for proper cellular function, under such conditions, some of it is released from these storage sites and made available to the rest of the body.

Not all minerals are used the same way or at the same rate by all animals. For instance, copper binds oxygen in the fluids of some invertebrates such as horseshoe crabs, whereas in all vertebrates and in most other invertebrates, iron serves this function. Factors that affect mineral usage are an animal's species, age, weight, overall health status, and the types of food it eats.

Animals Have Very Diverse Feeding Habits

Several factors determine how and when an animal obtains food, including its energetic demands, whether or not it is capable of locomotion, its local environment, and the structure of its digestive system. Unlike plants, which are autotrophs that make their own food, all animals are heterotrophic and must consume their food. As stated earlier, most animals can be grouped into one of three broad dietary categories: herbivores, carnivores, or omnivores. Although broadly useful, the three dietary categories are limited when describing the diversity of animal feeding habits. For example, some animals eat protists, algae, and/or fungi as a major source of food. Also, other animals are almost strictly carnivores at one time of year but herbivores at other times. Many nonmigratory birds, for example, feed on insects and worms during the summer but switch to eating whatever vegetation, buds, or seeds they can find during the winter. Similarly,

Table 45.2 Vitamins Required by Animals

Class of nutrient	Dietary sources	Functions in vertebrates	Symptoms of deficiency
Water-soluble vitamins			
Biotin	Liver; legumes; soybeans; eggs; nuts; mushrooms; some green vegetables	Coenzyme for gluconeogenesis and fatty acid and amino acid metabolism	Skin rash; nausea; loss of appetite; mental disorders (depression or hallucinations)
Folic acid	Green vegetables; nuts; legumes; whole grains; organ meats (especially liver; kidney; heart)	Coenzyme required for synthesis of nucleic acids	Anemia (a lower than normal number of red blood cells in the blood); depression; birth defects
Niacin	Legumes; nuts; milk; eggs; meat	Involved in many oxidation/reduction reactions	Skin rashes; diarrhea; mental confusion; memory loss
Pantothenic acid	Nearly all foods	Part of coenzyme A, which is involved in numerous synthetic reactions, including formation of cholesterol	Burning sensation in hands and feet; GI symptoms; depression
Vitamin B_1 (thiamine)	Meats; legumes; whole grains	Coenzyme involved in metabolism of sugars and some amino acids	Beriberi (muscular weakness, anemia, heart problems, loss of weight)
Vitamin B_2 (riboflavin)	Dairy foods; meats; organ meats; cereals; some vegetables	Respiratory coenzyme; required for metabolism of fats, carbohydrates, and proteins	Seborrhea (excessive oil secretion from skin glands resulting in skin lesions)
Vitamin B_6 (pyridoxine)	Meats; liver; fish; nuts; whole grains; legumes	Coenzyme for over 100 enzymes that participate in amino acid metabolism, lipid metabolism, and heme synthesis	Seborrhea; nerve disorders; depression; confusion; muscle spasms
Vitamin B_{12}	Meats; liver; eggs; some shellfish; dairy foods	Required for red blood cell formation	Anemia; nervous system disorders leading to sensory problems; balance and gait problems; loss of bladder and bowel control
Vitamin C (ascorbic acid)	Citrus fruits; green vegetables; tomatoes; potatoes	Antioxidant and free radical scavenger; aids in iron absorption; helps maintain healthy connective tissue and gums	Scurvy (connective tissue disease associated with skin lesions, weakness, poor wound healing, tooth decay); bleeding gums
Fat-soluble vitamins			
Vitamin A (retinol)	Liver; green and yellow vegetables; some fruits in small amounts	Component of visual pigments; regulatory molecule affecting transcription; important for reproduction and immunity	Night blindness due to loss of visual ability; skin lesions; impaired immunity
Vitamin D	Fish oils; fish; egg yolk; liver; synthesized in skin via sunlight	Required for calcium and phosphorus absorption from intestine; bone growth	Rickets (weakened, deformed bones) in children; osteomalacia (weak bones) in adults
Vitamin E	Meats; vegetable oils; grains; nuts; seeds; small amounts in some fruits and vegetables	Antioxidant; inhibits prostaglandin synthesis	Unknown, possibly skeletal muscle atrophy; peripheral nerve disorders
Vitamin K	Legumes; green vegetables; some fruits; some vegetable oils (olive oil, soybean oil); liver; synthesized by hindgut bacteria	Component of blood-clotting mechanism	Reduced blood clotting ability

a coyote prefers to eat only meat but will consume plants and fruits if hungry enough. Animals like these are said to be opportunistic; they have a strong preference for one type of food but can adjust their diet if the need arises.

An animal's life stage may also influence its diet. Mammals begin life as milk-drinkers and later switch to consuming plants, animals, or both. Caterpillars eat leaves, but after metamorphosis, most species of moths or butterflies are strictly fluid-drinkers, typically consuming the nectar in flowers. Other animals that undergo metamorphosis also change their diet as adults. Tadpoles, for example, are mostly herbivores, whereas frogs eat insects and therefore are carnivores. This type of food resource preference serves to reduce competition between the different life stages of a species.

The three major categories also do not indicate what type of plant or animal is consumed. Grasses, cereals, and fruits are all types of plant matter, but each has a different energy and nutrient content. Some carnivores eat only flesh, whereas others drink only the blood of other animals. To more fully understand the nutrition of animals, therefore, we need to investigate the behaviors by which they obtain food.

Table 45.3 Minerals Required by Animals

Class of nutrient	Dietary sources	Functions in vertebrates	Symptoms of deficiency
Calcium (Ca)	Dairy products; cereals; legumes; whole grains; green leafy vegetables; bones (eaten by some animals)	Bone and tooth formation; exocytosis of stored secretions in nerves and other cells; muscle contraction; blood clotting	Muscular disorders; loss of bone; reduced growth in children
Chlorine (Cl)	Meats; dairy foods; blood; natural deposits of salt	Participates in electrical, acid-base, and osmotic balance across cell membranes, notably nerve and heart cells	Muscular and nerve disorders
Chromium (Cr)	Liver; seafood; some nuts; meats; mushrooms; some vegetables	Required for proper glucose metabolism, possibly by aiding the action of the hormone insulin	Disorders of lipid and glucose balance in blood
Copper (Cu)	Fish; shellfish; nuts; legumes; liver; and other organs	Required for hemoglobin production and melanin synthesis; required for connective tissue formation; serves as oxygen-binding component in some invertebrates	Anemia; bone changes
Iodine (I)	Seaweed; seafood; milk; iodized salt	Required for formation of thyroid hormones	Inability to make thyroid hormones, resulting in enlarged thyroid gland
Iron (Fe)	Liver and other organs; some meats; eggs; legumes; leafy green vegetables	Oxygen-binding component of hemoglobin; cofactor in some enzymes	Anemia
Magnesium (Mg)	Hay; grasses; whole grains; green leafy vegetables	Cofactor for many enzymes that use ATP as a substrate	Changes in nervous system function
Manganese (Mn) and molybdenum (Mo)	Nuts; whole grains; legumes; vegetables; liver	Cofactors for many enzymes	Poor growth; abnormal skeletal formation; nervous system disorders (convulsions)
Phosphorus (P)	Dairy foods; grains; legumes; nuts; meats	Bone and tooth formation; component of DNA, RNA, and ATP	Bone loss; muscle weakness
Potassium (K)	Meats; fruits; vegetables; dairy foods; grains	See Chlorine	Muscle weakness; serious heart irregularities; GI symptoms
Selenium (Se)	Seafood; eggs; chicken; soybeans; grains	Antioxidant; cofactor for some enzymes	Keshan disease (damage to and loss of heart muscle)
Sulfur (S)	Proteins from any source	Component of two amino acids (methionine and cysteine)	Inability to synthesize many proteins
Sodium (Na)	Many fruits; vegetables; meats; and natural salt deposits	See Chlorine	Muscle cramps; changes in nerve activities
Zinc (Zn)	Widely found in meats; fish; shellfish (oysters); grains	Many functions related to tissue repair; sperm development; cofactor for many metabolic enzymes; required for certain transcription factors to bind to DNA	Stunted growth; loss of certain sensations like taste; impaired immune function; skin lesions

Animals Have Evolved Multiple Behaviors for Obtaining Food

The ways in which an animal obtains its food are related to its environment. Not surprisingly, for example, a sessile (nonmotile) marine invertebrate and a mobile terrestrial vertebrate face different challenges and opportunities. Let's consider three broad categories of strategies that animals use to get food.

Suspension feeders sieve water, filtering out the organic matter and expelling the rest (Figure 45.2). Bivalve mollusks filter seawater and capture floating bits of organic material on mucus-covered cilia located in their gills, which move the material into the animal's mouth (Figure 45.2a). Sea squirts (tunicates) have a sticky mucous net in their pharynx that traps suspended food particles (refer back to Figure 33.38a). Baleen whales have specialized plates made of keratin suspended like a stiff, frayed comb from the roof of the mouth (Figure 45.2b). The whale engulfs a mouthful of seawater and then uses its tongue to squeeze the water back out through the plates, which act like a sieve, trapping small copepods, protozoa, and other small animals or organic matter.

Carnivores, herbivores, and omnivores can be considered **bulk feeders**, organisms that eat food in large pieces. Carnivores, which inhabit aquatic and terrestrial environments, are generally **predators**, such as piranhas and wolves (that hunt and kill live prey), or **scavengers**, such as vultures (that eat the remains of dead animals). When present, carnivores' teeth show a variety of adaptations for grasping or seizing, biting, slicing, and, in some cases, chewing (Figure 45.3a,b). For example, sharp canine teeth and jagged molars help to slice and tear animal flesh. Carnivores that chew do so mostly to break up food into pieces small enough to swallow. Chewing food also has the advantage of increasing the total surface area of the food that is available to the action of digestive enzymes.

Most carnivores, however, do not extensively chew their food. Some carnivores have mouth parts adapted to tearing off chunks of

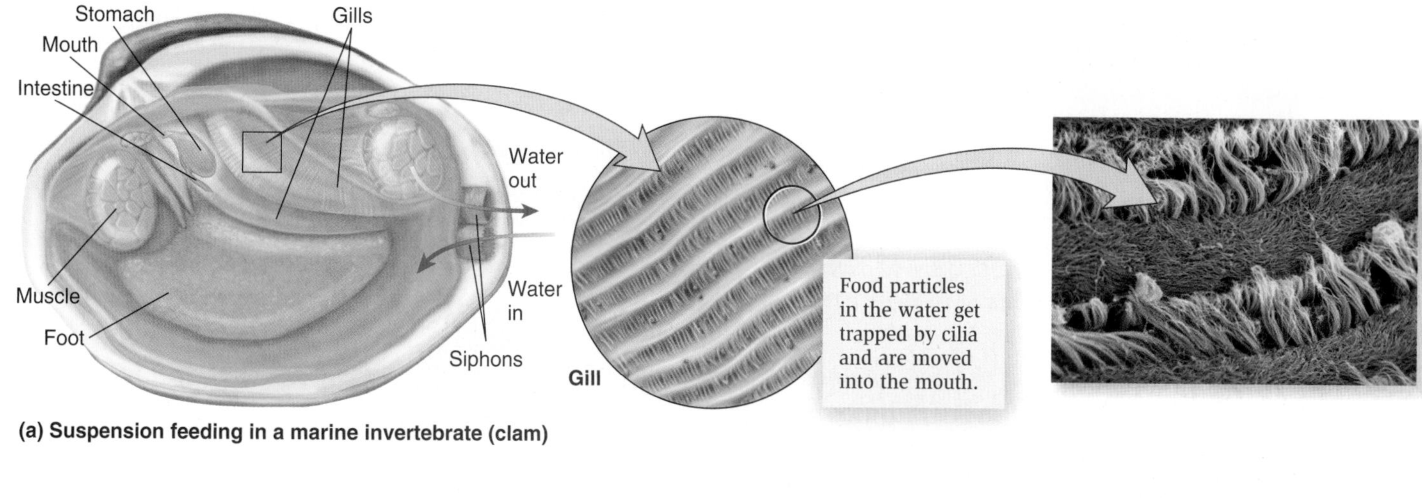

Food particles in the water get trapped by cilia and are moved into the mouth.

(a) Suspension feeding in a marine invertebrate (clam)

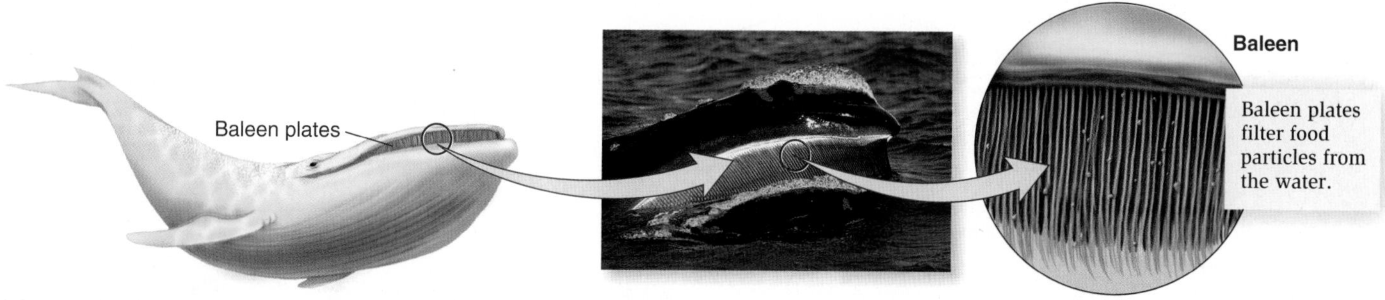

Baleen
Baleen plates
Baleen plates filter food particles from the water.

(b) Suspension feeding in a large vertebrate (baleen whale)

Figure 45.2 **Suspension feeding.** Suspension feeders include species of both invertebrates and vertebrates. **(a)** Bivalves, such as the clam shown here, use a siphon mechanism to move water across their gills. Cilia located along the gills trap and move food particles to the mouth. **(b)** Baleen whales, such as the Southern Right Whale shown here, engulf water with their mouth and then use their tongue to force the water back out across the comblike baleen. The baleen filters tiny organisms or bits of organic matter from the water.

BIOLOGY PRINCIPLE Living organisms use energy. An implication of this principle is that living organisms must obtain energy in order to use it. Animals obtain energy from food. As you read this chapter, think about the energy required by animals to obtain the food they require. That energy may be used to move cilia, as in the clam, or to sustain muscle activity in animals foraging or hunting over great distances.

flesh, which they then swallow. The behavior of consuming entire chunks of food without chewing occurs in species from all the vertebrate classes. Birds of prey, for example, lack teeth and use their sharp beaks to tear away pieces of flesh that can be engulfed all at once. One advantage to swallowing whole chunks of food is that it allows some carnivores—particularly those that hunt and eat in packs—to quickly get another bite when competing with other hungry animals. Some carnivores even swallow their prey whole. The reverse-oriented teeth of snakes, for example, seize prey and prevent it from escaping. Their ability to digest their food is not hindered, because food can still be fully digested if swallowed whole.

Herbivores and algae eaters, however, may have powerful jaw muscles and large, broad molar teeth that are highly adapted for grinding tough, fibrous plants and cell walls (**Figure 45.3c**). Incisors and canine teeth are poorly developed or absent in many herbivores. When present, they are usually for defense (as in hippos and some primates) or for nipping grass (as in horses) or other vegetation. Food is chewed against the molars in a rotary motion, which grinds up the tough plant cell walls and releases the more digestible intracellular contents. In herbivorous fishes, the teeth are often adapted for cropping or scraping plants or algae off the substrate (**Figure 45.3d**). The food is ground up not in the mouth, but in the pharynx or throat, which contains specialized teeth suited for that purpose. In this way, chewing food does not interfere with gill breathing. In some cases, the teeth of herbivorous fishes such as the parrotfish have evolved into a fused beaklike structure that aids in rasping algae off of coral and other hard surfaces, and even chopping off bits of coral itself (Figure 45.3d).

In omnivores, such as humans, the teeth are a mix of the types present in carnivores and herbivores (**Figure 45.3e**). For example, incisors and canines help cut and slice food, whereas relatively flat molars assist in chewing. Chewing occurs in a vertical motion, by which food is crushed. This is particularly useful for consuming hard foods such as nuts.

Fluid-feeders lick or suck fluid from plants or animals and so do not need teeth except, perhaps, to puncture an animal's skin. Fluid-feeding has evolved independently in many types of animals, including worms, insects, fishes, birds, and mammals. Many birds and bats

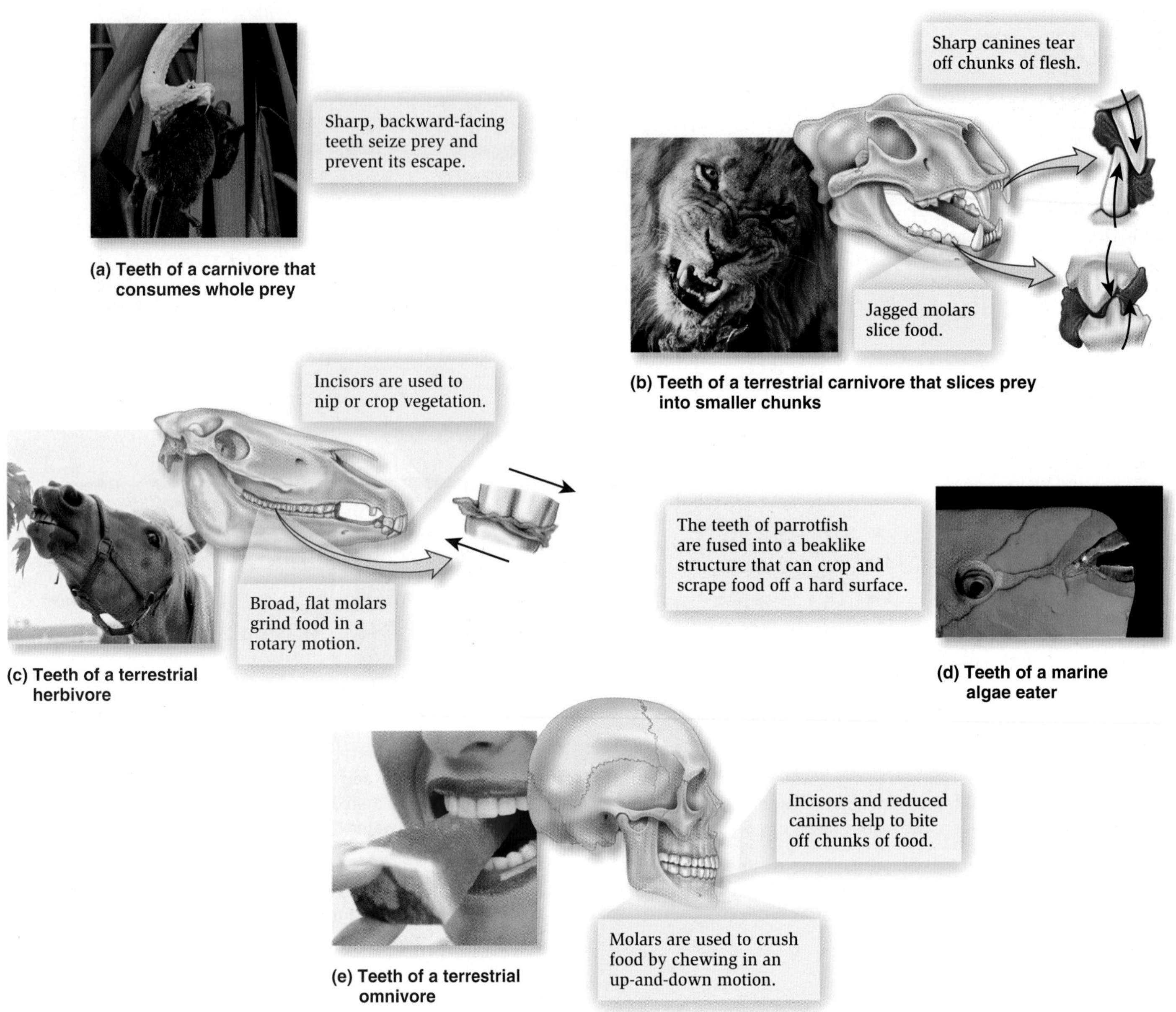

Sharp, backward-facing teeth seize prey and prevent its escape.
(a) Teeth of a carnivore that consumes whole prey
Sharp canines tear off chunks of flesh.
Jagged molars slice food.
(b) Teeth of a terrestrial carnivore that slices prey into smaller chunks
Incisors are used to nip or crop vegetation.
Broad, flat molars grind food in a rotary motion.
(c) Teeth of a terrestrial herbivore
The teeth of parrotfish are fused into a beaklike structure that can crop and scrape food off a hard surface.
(d) Teeth of a marine algae eater
Incisors and reduced canines help to bite off chunks of food.
Molars are used to crush food by chewing in an up-and-down motion.
(e) Teeth of a terrestrial omnivore

Figure 45.3 **Examples of teeth in bulk feeders.** The shape and size of teeth reflect the way in which animals feed and the types of food they consume. In carnivores, teeth may be used for **(a)** seizing prey that will be swallowed whole or **(b)** slicing food into pieces small enough to be swallowed. The teeth of herbivores are used for **(c)** nipping and grinding vegetation, and teeth of algae eaters can **(d)** scrape or crop food off coral or other hard substrates. **(e)** Omnivores typically have a mix of teeth resembling those found in both carnivores and herbivores. The molars may be adapted for crushing hard foods.

have specialized beaks, tongues, or mouths that enable them to reach the nectar of a particular flower (**Figure 45.4a**). When the fluid is obtained from an animal, the fluid-feeder usually has a specialized mouthpart, such as the piercing needle-like extension of a mosquito's mouth (refer back to Figure 33.27b) or the tiny bladelike jaws of a blood-sucking leech (**Figure 45.4b**). Fluid-feeders that consume the blood of other animals have developed a fascinating set of behaviors to ensure a full meal. The European medicinal leech (*Hirudo medicinalis*), for example, secretes a local anesthetic at the site of the bite to dull the prey's pain. An enzyme digests the host's connective tissue to allow the leech to firmly embed its mouth into its host's flesh, and a locally acting chemical is secreted to keep the host's blood vessels open. Finally, to ensure that the blood does not clot before the leech has drunk its fill, the saliva of leeches and other blood-drinking animals such as the vampire bat *Desmodus rotundus* contains an anticoagulant. Scientists are studying these anticoagulants to develop anticlotting drugs for people with certain forms of heart and blood vessel disease.

Regardless of what an animal eats, the useful parts of the eaten material must be digested into molecules that its cells can absorb. Next, we turn to a discussion of how animals digest their food and absorb the nutrients.

(a) Nectar-feeding hummingbird

(b) Blood-sucking leech

Figure 45.4 **Strategies of fluid-feeders.** **(a)** Hummingbirds extend their tongues through long, thin beaks to consume nectar from flowers. **(b)** Blood-drinking animals use teeth, specialized mouthparts, or, like this leech, bladelike jaws to prick the skin of their host and suck or lap up their blood.

Concept Check: *To obtain sufficient nutrients, blood-sucking animals such as mosquitoes, leeches, and some bats must consume large amounts of blood with each meal. For mosquitoes and bats, what effect might this have on their ability to take flight after a meal?*

45.2 Principles of Digestion and Absorption of Food

Learning Outcomes:

1. Compare the processes of intracellular and extracellular digestion, and explain why one is far more common than the other.
2. Describe the general structure of an alimentary canal.
3. Distinguish between passive and active absorption of food.

Once food has been ingested, some of the nutrients from the food must be broken down (digested) so that they can be absorbed by the cells of the digestive tract. In this section, we will examine some of the major principles of digestion and absorption in animals, beginning with where digestion takes place.

Digestion Can Occur Intracellularly or Extracellularly

Food is digested either inside cells (intracellularly) or outside cells (extracellularly). Intracellular digestion occurs only in some very simple invertebrates such as sponges and single-celled organisms and to a limited extent in cnidarians. It involves using phagocytosis to bring food particles directly into a cell, where the food is segregated from the rest of the cytoplasm in food vacuoles. Once inside these vacuoles, hydrolytic enzymes digest the food into monomers (the building blocks of polymers), which then are moved out of the vacuole to be used directly by that cell. Intracellular digestion cannot support the metabolic demands of an active animal for long, because only tiny bits of food can be phagocytosed at one time. It also does not provide a mechanism for storing large quantities of food so that an animal can digest it slowly while going about its other activities.

Most animals digest food via extracellular digestion in a cavity of some sort. Extracellular digestion protects the interior of the cells from the actions of hydrolytic enzymes and allows animals to consume large prey or large pieces of plants. Food enters the digestive cavity, where it is stored, slowly digested, and absorbed gradually over long periods of time, ranging from hours (for example, after a human eats a pizza) to weeks (after a python eats a gazelle).

In the simplest form of extracellular digestion—seen in invertebrates such as flatworms and cnidarians—the digestive cavity has one opening that serves as both an entry and an exit port (Figure 45.5). The digestive cavity of these animals is called a **gastrovascular cavity**, because not only does digestion occur within it, but fluid movements in the cavity also serve as a circulatory—or vascular—system to distribute digested nutrients throughout the animal's body. Food within a gastrovascular cavity is partially digested by enzymes that are secreted into the cavity by the cells lining the cavity. As the food particles become small enough, they are phagocytosed by the lining cells and further digested intracellularly. Undigested material that remains in the gastrovascular cavity is expelled.

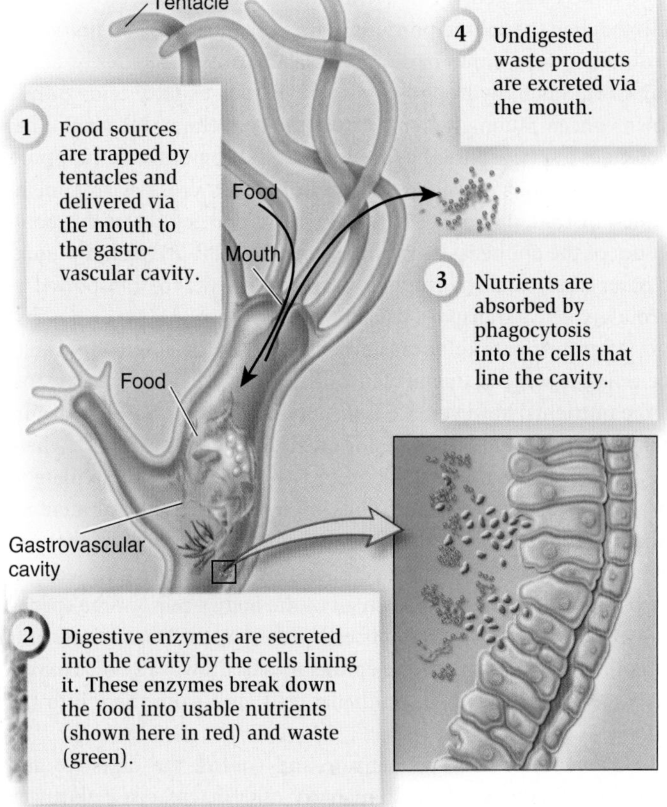

Tentacle
1 Food sources are trapped by tentacles and delivered via the mouth to the gastro-vascular cavity.
Food
Mouth
Food
Gastrovascular cavity
4 Undigested waste products are excreted via the mouth.
3 Nutrients are absorbed by phagocytosis into the cells that line the cavity.
2 Digestive enzymes are secreted into the cavity by the cells lining it. These enzymes break down the food into usable nutrients (shown here in red) and waste (green).

Figure 45.5 **Extracellular digestion in a gastrovascular cavity.** In animals with gastrovascular cavities, such as the cnidarian *Hydra* illustrated here, digestion occurs extracellularly within the cavity.

Most Digestive Cavities Are Tubes with Specialized Regions and Openings at Opposite Ends

In contrast to the gastrovascular cavities of simple invertebrates, all other animals possess digestive systems that consist of a single elongated tube, or **alimentary canal**, with an opening at both ends through which food passes from one end to the other. Along its length, the tube usually contains smooth muscle. When the muscle contracts, it helps churn up the ingested food so that it is mechanically broken into smaller fragments. The canal is lined on its interior surface by a layer of epithelial cells. These cells synthesize and secrete digestive enzymes and other factors into the hollow cavity of the alimentary canal, called the **lumen**, and they secrete certain hormones into the blood that help regulate digestive processes. The cells are also involved with transporting digested material out of the canal.

The alimentary canal has several specialized regions along its length that vary according to species. Because of these specializations, digestive processes requiring acidic conditions can be segregated from those requiring higher pH, and undigested food can be stored in one region while digestion continues in another area. The ability of some animals to store food in the stomach, for example, allows them to eat less frequently, leaving time for other activities.

Absorption of Food May Be Passive or Active

Once food is digested, the nutrients must be absorbed by the epithelial cells that line specialized portions of the alimentary canal. This occurs in different ways, either by simple or facilitated diffusion or by active transport. Small, hydrophobic molecules such as fatty acids diffuse down concentration gradients across the epithelium. Ions and other molecules are transported by facilitated diffusion or active transport. Minerals are ions and therefore do not readily cross plasma membranes. Instead, they are usually actively transported across the membranes of the epithelial cells of the canal by ATP-driven ion pumps. In other cases, small, hydrophilic organic nutrients are transported by secondary active transport, usually with Na^+.

After nutrients enter the epithelial cells of the alimentary canal, the cells use some of the nutrients for their own requirements. Most of the nutrients, however, are transported across the basolateral surface of the epithelial cells (the side facing the interstitial fluid), where they can enter into nearby blood or lymph vessels and circulate to the other cells of the body. Thus, nutrients enter the alimentary canal in food, are digested within the canal into monomers that can be transported into epithelial cells, and from there are released into the blood, where they can reach all of the body's cells. In the special case of water, osmotic gradients established by the transport of ions and other nutrients out of the epithelial cells draw water by osmosis from the canal, across the epithelial cells, and from there into the blood.

The mechanisms that activate and control the digestive and absorptive functions of the alimentary canal have been extensively studied in vertebrates and have great importance for human health. In the rest of this chapter, we will explore the structure and function of the digestive systems of vertebrates.

45.3 Vertebrate Digestive Systems

Learning Outcomes:

1. Describe the general structure of the vertebrate digestive system.
2. Describe how food moves through regions of the alimentary canal, and how each part contributes to the processes of digestion and absorption.
3. Outline how microorganisms can help digest cellulose in ruminants and other herbivores.

The vertebrate **digestive system** consists of the alimentary canal—also known as the gastrointestinal (or GI) tract—plus several accessory glands and organs (Figure 45.6). As illustrated in the example in Figure 45.6, the human GI tract consists of the oral cavity, pharynx, esophagus, stomach, small and large intestines, and anus. The accessory structures, not all of which are found in all vertebrates, are the tongue, teeth, salivary glands, liver, gallbladder, and pancreas. The differences between the digestive systems of various vertebrates reveal much about their respective feeding strategies. In this section, we will look at some of the most important differences as we discuss the form and function of each part of the vertebrate digestive system.

Alimentary Canals Are Divided into Functional Regions

The alimentary canal is one continuous tube that changes in appearance and function along its length, with three general sections. The first section, at the anterior end, functions primarily in the ingestion of food. It contains the oral cavity, salivary glands, pharynx (throat), and esophagus. The middle portion, which functions in the storage and initial digestion of food, contains one or more food storage or digestive organs, including the crop, gizzard, and stomach(s), depending on species. This section also contains the upper part of the small intestine—where most of the digestion and absorption of food takes place—and accessory structures that connect with the intestine, including the pancreas, liver, and gallbladder. The third section, the posterior part of the canal, functions in final digestion and absorption and the elimination of nondigestible wastes. It consists of the remainder of the small intestine, and in most vertebrates other than fishes, a large intestine. Undigested material is defecated through an opening called an anus, or in many amphibians, reptiles, and birds, a cloaca (a common opening for the digestive and urogenital tracts).

From the midesophagus to the anus or cloaca, the GI tract has the same general structure, with a lumen that is lined by a layer of epithelial cells. Included in the epithelial layer are secretory cells that release a protective coating of mucus into the lumen of the tract, and for this reason this layer of cells is also referred to as the mucosa. Other cells in the mucosal layer release hormones into the blood in response to the presence of food. Passing through the mucosal layer are ducts from secretory glands that release acid, enzymes, water, and ions into the lumen. From the stomach onward, the epithelial cells are linked along the edges of their luminal surfaces by tight junctions that prevent digestive enzymes and undigested food from moving between the cells and out of the alimentary canal. In some places,

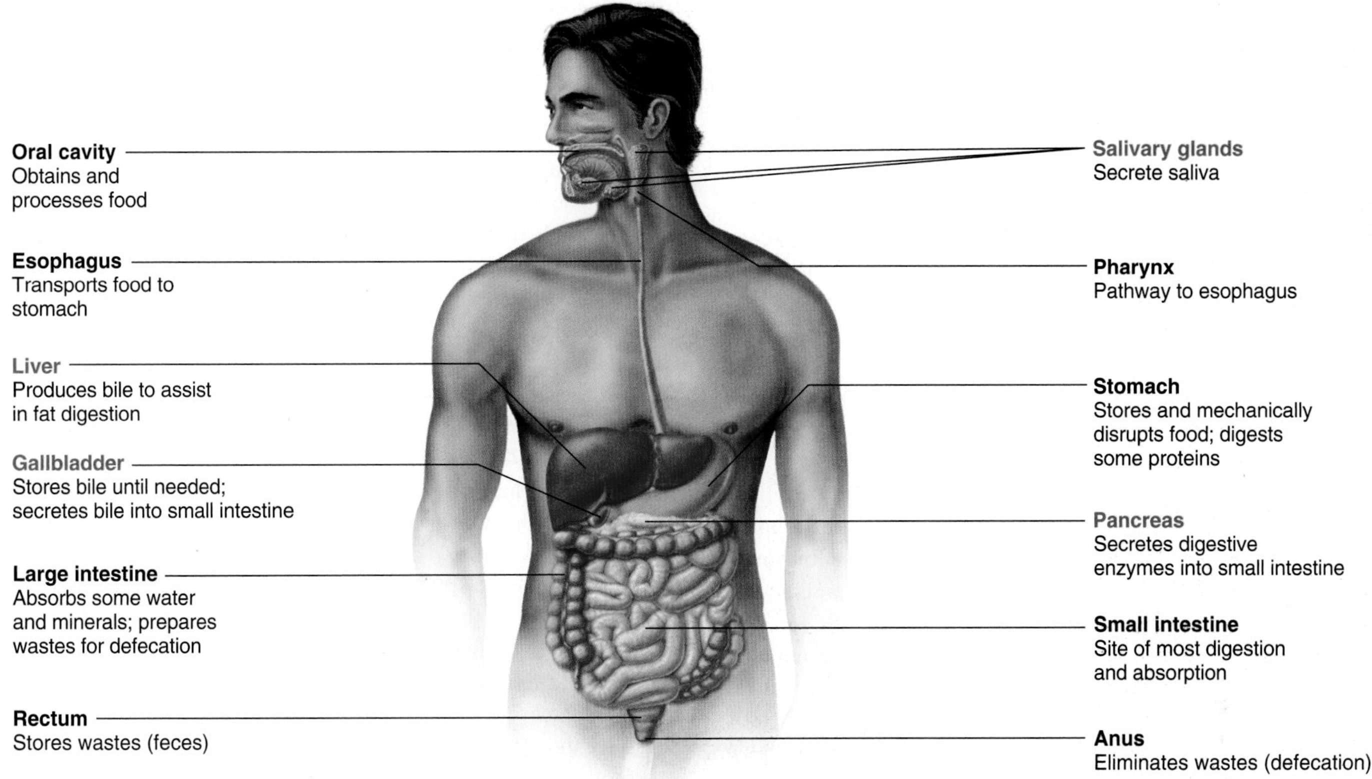

Figure 45.6 **A vertebrate digestive system, as shown in the human.** This figure shows the organs of the gastrointestinal tract (labeled in black) and the major accessory structures (labeled in red). Not all vertebrates share identical features of the digestive system; for example, some fishes lack a stomach, and many birds lack a gallbladder.

such as the small intestine, the luminal surface is highly convoluted, a feature that increases the surface area available for digestion and absorption.

The epithelial cell layer is surrounded by layers of tissue made up of smooth muscles, neurons, connective tissue, and blood vessels. The neurons are activated by nerves coming from the central nervous system that respond to the sight and smell of food. The neurons of the gastrointestinal tract are also activated directly by the presence of food in the tract. Contraction of the muscles is controlled by these neurons and results in mechanical mixing of the contents within the stomach and intestine. This helps speed up digestion and also brings digested foods into contact with the epithelium to facilitate absorption.

Food Processing and Polysaccharide Digestion Begin in the Mouth

Ingestion and the start of digestion begin once food enters the mouth. In terrestrial vertebrates, the presence of food stimulates salivary glands in and around the mouth, cheeks, tongue, and throat to produce a flow of saliva—a watery fluid containing proteins, mucus, and antibacterial agents that keeps the mouth moist and clean; unlike terrestrial vertebrates, fishes, which lack true salivary glands, secrete mucus from specialized cells in their mouth and pharynx. Saliva production in some animals can also be increased simply by the smell or sight of food, a feedforward response discussed in Chapter 40.

Saliva has several functions, not all of which pertain to all vertebrates:

1. To moisten and lubricate food to facilitate swallowing.
2. To dissolve food particles to facilitate the ability of specialized chemical-sensing structures called taste buds to taste food.
3. To kill ingested bacteria with a variety of antibacterial compounds, including antibodies.
4. To initiate digestion of polysaccharides through the action of a secreted enzyme called **amylase**.

Digestion is the least important of these functions in most vertebrates, few of which produce salivary amylase. In humans and other primates, salivary amylase is present and accounts for only a very small percent of total polysaccharide digestion.

The other functions of saliva, however, are very important. For example, imagine trying to swallow unchewed food with a perfectly dry mouth. Also, the antibiotic properties of saliva help cleanse the mouth. In people who have had cancerous salivary glands removed, the teeth and gums often become so diseased that tooth loss may occur.

Peristalsis Moves Swallowed Food Through the Esophagus to a Storage Organ

Once food has been sufficiently processed in the mouth, it is swallowed. The swallowed food moves into the next segments of the alimentary canal, the **pharynx** (throat) and **esophagus**. These structures, although not contributing to digestion or absorption, serve as a pathway to storage organs such as the stomach.

Pharynx and Esophagus The muscles in the walls of the pharynx and esophagus contribute to swallowing. In the pharynx, swallowing begins as a voluntary action but continues in the esophagus by the process of **peristalsis**—rhythmic, spontaneous waves of muscle contraction that begin near the mouth and end at the stomach. When most vertebrates eat, the mouth and stomach are roughly horizontal with respect to each other. In fact, the head of a terrestrial grazing animal is usually lower than its stomach when eating. The wavelike action of peristalsis ensures that food is pushed toward the stomach and does not sit in the esophagus or even move backward into the mouth if the head is lowered.

The Crop In some animals, food moves directly from the esophagus to a storage organ called the **crop**, which is a dilation of the lower esophagus (Figure 45.7). Crops are found in most birds (and are also found in many invertebrates, including insects and some worms). Food is stored and softened by watery secretions in the crop, but little or no digestion occurs there. Because they process large amounts of tough food, birds that eat primarily grains and seeds have larger crops than birds that eat insects and worms. The material that birds regurgitate to their young comes from the crop. In some species, such as pigeons and doves, the cells that line the crop wall secrete a lipid-rich watery solution called crop milk or pigeon milk into the material to be regurgitated.

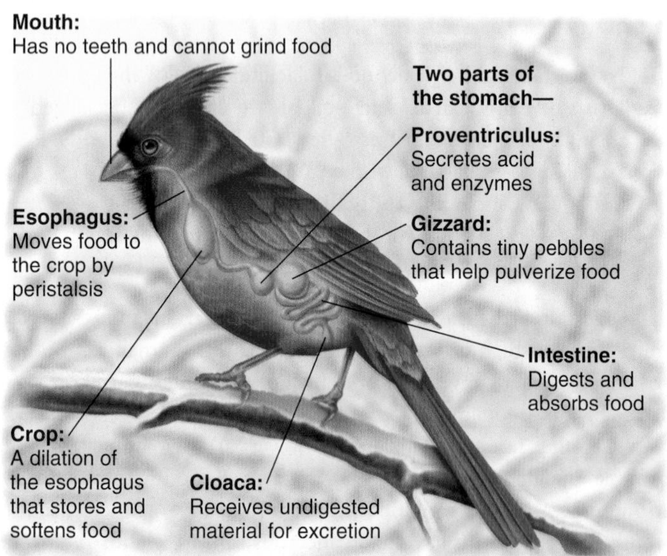

Figure 45.7 The alimentary canal of birds. The avian alimentary canal contains specialized regions for storing and softening food (the crop) and pulverizing food (the gizzard). The gizzard and proventriculus constitute the stomach. Undigested material is excreted through the cloaca.

Concept Check: Smooth, polished stones have been found in the stomach region of fossilized skeletons of ancient sauropod dinosaurs. What does this suggest about the alimentary canal of such animals?

Food Processing Continues and Protein Digestion Begins in the Stomach

Once food passes through the esophagus and crop, if present, it reaches a storage organ such as a stomach. The **stomach** is a saclike organ that most likely evolved as a means of storing food (refer back to Figure 40.6). Stomach-like organs are found in all vertebrate classes, but a true stomach (one that produces hydrochloric acid) is absent in many species of herbivorous fishes. In addition to its storage function, the muscular nature of the stomach helps it to mechanically break up large chunks of food into smaller, more easily digestible fragments. Lastly, the stomach partially digests some of the macromolecules in food and regulates the rate at which the contents empty into the small intestine. Glands within the stomach wall secrete hydrochloric acid (HCl) and an inactive molecule called pepsinogen into the stomach lumen. One function of the acid is to convert pepsinogen into the active enzyme **pepsin**, which is a protease and begins the digestion of protein. Why do stomach cells secrete pepsinogen instead of pepsin? The answer is that if the cells produced active pepsin, they would digest their own cellular proteins.

Within the stomach lumen, HCl kills many of the microorganisms that may have been ingested with food. The acid also helps dissolve the particulate matter in food. In addition, the acid environment in the stomach (or gastric) lumen alters the ionization of polar molecules, especially proteins. This disrupts the structural framework of the tissues in food and makes the proteins more accessible to pepsin. The proteins released by the dissolving action of HCl are partially digested in the stomach by pepsin. By contrast, no significant digestion of carbohydrates or lipids occurs in the stomach.

In birds, the stomach is divided into two parts: the proventriculus and the gizzard (see Figure 45.7). The **proventriculus** is the glandular portion of the stomach, and it secretes acid and pepsinogen. Partially digested and acidified food then moves to the **gizzard**, a muscular structure with a rough inner lining that grinds food into smaller fragments.

The gizzard contains sand or tiny stones swallowed by the bird. The gritty sand and stones take the place of teeth and help mash and grind ingested food. Eventually, the pebbles in the gizzard become smaller as they are worn away, and they are excreted. Thus, birds must occasionally restock the gizzard with new grinding stones. Grain-eating birds, particularly chickens and other fowl, many passerines (perching birds), and pigeons and doves, generally have more muscular gizzards than do insectivorous birds, because of the difficulty in breaking down plant cell walls. In other birds, such as owls, gizzards help compress the nondigestible parts of their meals (bones, teeth, fur, feathers) into a pellet that can be regurgitated. Gizzards, incidentally, are not unique to birds. Certain reptiles that are closely related to birds, such as crocodiles, also contain muscular gizzards. In addition, some species of herbivorous fishes (for example, members of the family Acanthuridae) ingest quantities of inorganic grit with their meals, which help to grind up food in a portion of the stomach that is modified into a strong, muscular grinding organ like a gizzard.

Digestive actions of the stomach reduce food particles to **chyme**, a solution that contains water, salts, molecular fragments of proteins, nucleic acids, polysaccharides, droplets of fat, and various other small molecules. Virtually none of these molecules, except water, can cross the epithelium of the stomach wall; therefore, little or no absorption of organic nutrients occurs in the stomach.

Herbivores Use Microorganisms to Aid Digestion

Cellulose, the main macromolecule of the plant cell wall (refer back to Figures 3.8 and 10.6), is a very important part of the diet of herbivores and many omnivores. Cellulose is also called fiber or roughage, and is useful in eliminating solid wastes. The human digestive system is not equipped with the enzymes to digest cellulose, and it therefore passes through the system intact. Herbivores called **ruminants** (sheep, goats, llamas, and cows) also lack the enzymes, but they are able to digest cellulose with the help of microorganisms living within their digestive tract. The microorganisms break down the cellulose into monosaccharides that can be absorbed along with other by-products of microbial digestion, such as fatty acids and some vitamins. In this way, bacteria and protists predigest the food, and the animal absorbs the broken-down cellulose and uses its sugar as a food source.

Ruminants have a complex stomach consisting of several chambers, beginning with three outpouchings of the lower esophagus composed of the rumen, reticulum, and omasum, in sequence (Figure 45.8). The rumen and reticulum contain the microorganisms that digest cellulose, and the omasum absorbs some of the water and salts released from the chewed and partially digested food. The tough, partially digested food (called the cud) is occasionally regurgitated, rechewed, and swallowed again. Eventually, the partially digested food, the microorganisms, and the by-products of microbial digestion (including useful products such as fatty acids) reach the true stomach, the abomasum, which contains the acid and proteases typical of the vertebrate stomach. From the abomasum, the material passes to the intestines, where digestion and absorption are completed. The microorganisms are killed by the secreted stomach acid, digested, and eliminated. Some microorganisms remain in the rumen and quickly multiply to replenish their populations, ensuring that a well-balanced mutualistic relationship is maintained.

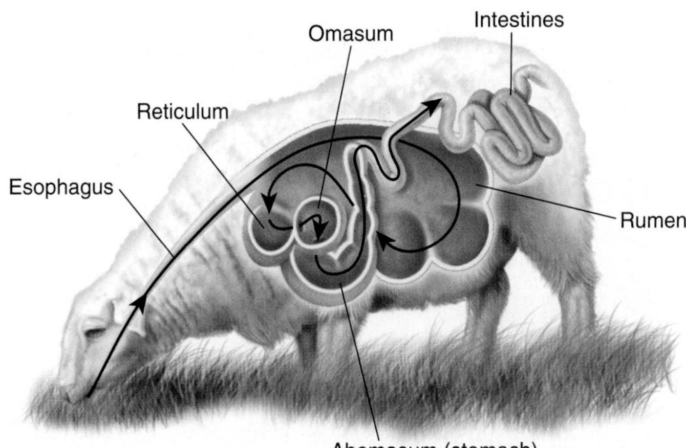

Figure 45.8 Digestive tract of a ruminant. Ruminants have a complex arrangement of three modified pouches arising from the esophagus: the rumen, reticulum, and omasum. The rumen and reticulum act as storage and processing sites (in large ruminants, the rumen may store up to 95 L of undigested food); the omasum absorbs some water and salts. Digestion by acid and pepsin takes place in the abomasum, which then connects with the intestines.

Most Digestion and Absorption Occurs in the Small Intestine

Nearly all digestion of food and absorption of nutrients and water occur in the **small intestine**, the tube that leads from the stomach to the large intestine or, in some animals, directly to the anus or cloaca. Hydrolytic enzymes break down molecules of organic nutrients into monomers. Some of these hydrolytic enzymes are on the apical membrane of the intestinal epithelial cells; others are secreted by the pancreas and enter the intestinal lumen. The products of digestion are absorbed across the epithelial cells and enter the blood. Vitamins and minerals, which do not require enzymatic digestion, are also absorbed in the small intestine. Water is absorbed by osmosis from the small intestine in response to the movement of nutrients across the intestinal epithelium. We will discuss the mechanisms of digestion and absorption in more detail later.

The ability of the small intestine to carry out the bulk of digestion and absorption is aided by mucosal infoldings and specializations along its length. Finger-like projections known as **villi** (singular, villus) extend into the lumen of the vertebrate small intestine (Figure 45.9). The surface of each villus is covered with a layer of epithelial cells whose plasma membranes form small projections called **microvilli**, known collectively as the **brush border**. The combination of folded mucosa, villi, and microvilli increases the small intestine's surface area about 600-fold above that of a flat-surfaced tube having the same length and diameter. The small intestine is small in diameter compared with the large intestine, but it is very long—3 m in an adult human (the small intestine is almost twice as long if removed from the abdomen, because the muscles relax). This brings the total surface area of the human small intestine to about 300 m^2—roughly the size of a tennis court! This enormous surface area means that the likelihood of an ingested food particle encountering a digestive enzyme and being absorbed across the epithelium is very high, so digestion and absorption proceed rapidly.

The center of each intestinal villus is occupied by capillaries, the smallest blood vessels in the body, and by a special type of vessel called a **lacteal**, which is part of the lymphatic system (see Figure 45.9). Most of the fat absorbed in the small intestine exists as bulky protein-bound particles that are too large to enter capillaries. Consequently, absorbed fat enters the larger, wider lacteals. Material absorbed by the lacteals eventually empties into the circulatory system. Other nutrients are absorbed directly into the capillaries and from there into veins.

The length of the small intestine varies among species. Both terrestrial and aquatic herbivores generally have much longer small intestines than do carnivores, which provides an added opportunity for plant material to be digested and absorbed. Even within an individual animal, the length of the small intestine can change. For example, in a bird that switches from eating insects and worms in summer to buds and other nutrient-poor vegetation in winter, the small intestine grows and elongates and increases its total absorptive surface area to meet the digestive challenges associated with an herbivorous diet.

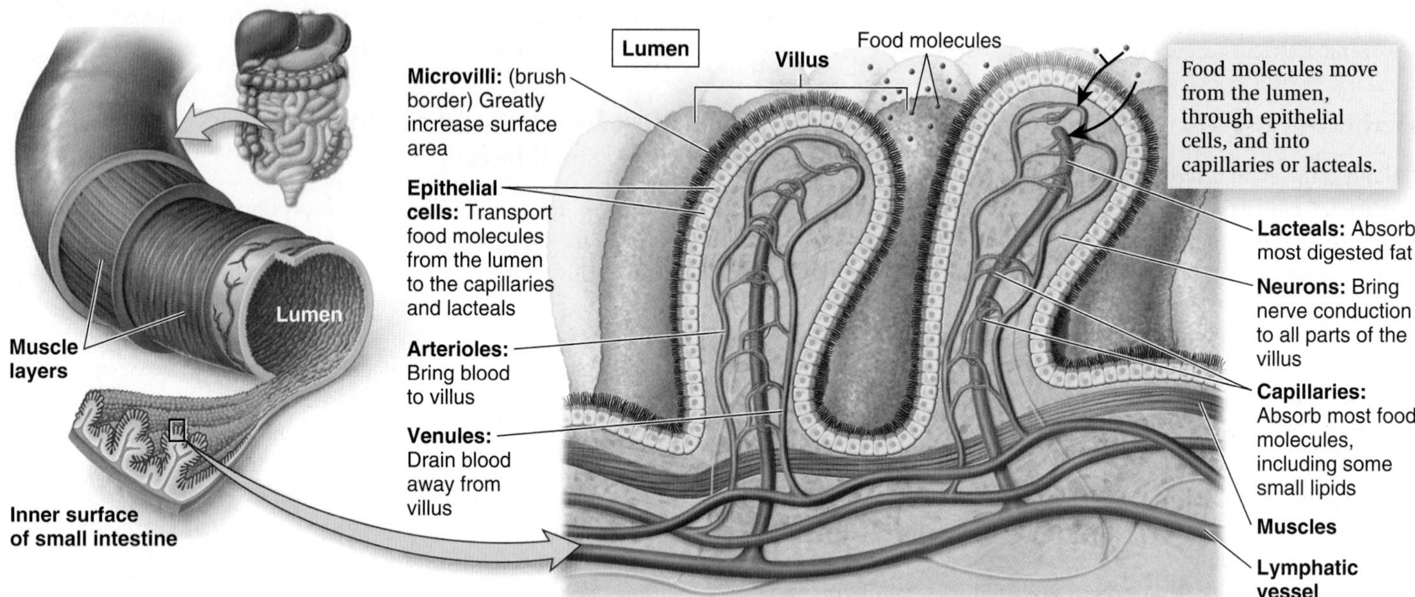

Figure 45.9 **The specialized arrangement of tissues in the small intestine.** The small intestine is folded into numerous villi, which increase the surface area for digestion and absorption. Within each villus are capillaries and lymphatic vessels (lacteals) into which absorbed nutrients are transported. The epithelial cells of the villi have extensions from their surface called microvilli. The microvilli constitute the brush border of the intestine and greatly add to the total surface area.

BIOLOGY PRINCIPLE **Structure determines function.** Note the arrangement of muscle layers in the small intestine. Some are oriented in circles around the lumen, while others are lengthwise (longitudinal). This structure allows the intestine to be squeezed and shortened, not necessarily at the same time, maximizing the mixing and churning of the lumen contents. Also, the structure of the villi and microvilli provide extensive surface area in a compact volume, increasing the efficiency of digestion and absorption.

The Pancreas and Liver Secrete Substances That Aid Digestion

As the chyme moves through the small intestine, two major organs—the pancreas and liver—secrete substances that flow via ducts into the first portion of the intestine, which is called the **duodenum** (**Figure 45.10**). The **pancreas**, a complex organ located behind and below the stomach in humans (see Figure 45.6), has several functions, but in this chapter, we will focus on those that are directly involved in digestion. The gland secretes digestive enzymes and a fluid rich in bicarbonate ions (HCO_3^-). The HCO_3^- neutralizes the acidity of chyme, which would otherwise inactivate the pancreatic enzymes in the small intestine and could also damage the intestinal epithelium.

The **liver** is the site of bile production. **Bile** contains HCO_3^-, cholesterol, phospholipids, a number of organic wastes, and a group of substances collectively termed **bile salts**. The HCO_3^-, like that from the pancreas, helps neutralize acid (H^+) from the stomach, and the bile salts break up dietary fat and increase its accessibility to digestive enzymes.

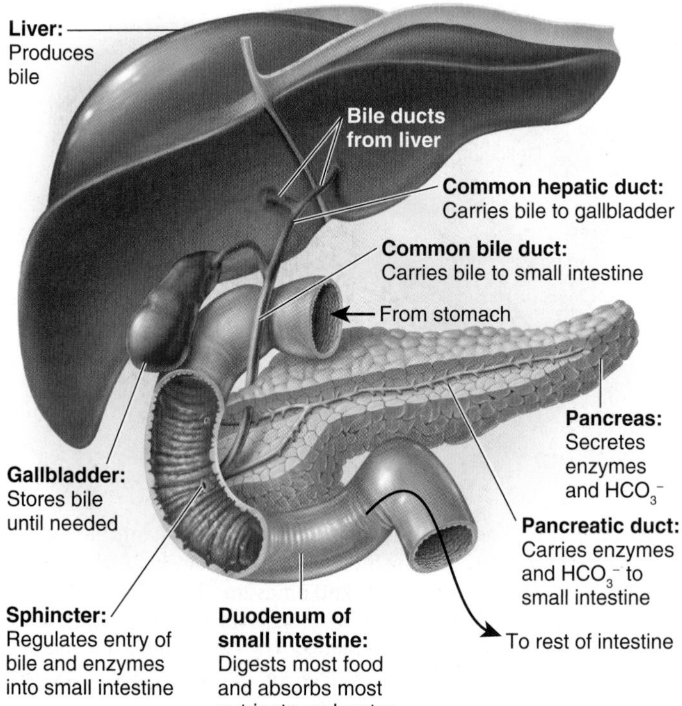

Figure 45.10 **The arrangement and functions of the vertebrate liver, gallbladder, pancreas, and small intestine.** Bile drains from the liver into the gallbladder through the common hepatic duct. During a meal bile is secreted from the gallbladder and enters the duodenum of the small intestine through the common bile duct. Simultaneously, secretions from the pancreas travel through the pancreatic duct, which joins with the common bile duct from the gallbladder and empties into the small intestine. A muscular sphincter controls the entrance to the small intestine.

Concept Check: *What advantage does an animal gain by having a gallbladder?*

The liver secretes bile into small ducts that join to form the common hepatic duct. Between meals, secreted bile is stored in the **gallbladder**, a small sac underneath the liver. During a meal, the smooth muscles in the gallbladder contract, injecting the bile solution into a connecting duct called the common bile duct (see Figure 45.10). The opening of a sphincter allows the bile to flow into the lumen of the small intestine. The gallbladder, therefore, is a storage organ that allows the release of large amounts of bile to be precisely timed to the consumption of fats. However, many animals such as horses and doves that secrete bile do not have a gallbladder. In humans, the gallbladder can be surgically removed without impairing bile secretion by the liver or its flow into the intestinal tract. People without a gallbladder can still digest fat but may need to limit the amount of fat they eat at one time because bile secretion can no longer be timed to a meal.

The digested nutrients, along with water, are absorbed across the plasma membranes of the brush border cells. Peristalsis slowly propels the remaining contents through the posterior two portions of the small intestine, called the jejunum and the ileum, where further absorption occurs. Finally, the remaining material enters the large intestine or the anus or cloaca.

The Large Intestine Concentrates Undigested Material, Which Is Then Egested via the Anus

The size of the large intestine varies greatly among different vertebrates, and the organ is vestigial or even absent in many animals, notably fishes. In humans, the large intestine is a tube about 6 cm in diameter and 1–1.5 m long. Its first portion, the **cecum**, forms a small pouch from which extends the appendix, a finger-like projection having no certain essential function but that may contribute to the body's immune defense mechanisms. The next part of the large intestine in humans and other mammals is called the **colon**. The terminal portion of the colon in humans is S-shaped, forming the sigmoid colon, which empties into the **rectum**, a short segment of the large intestine that ends at the anus. The characteristic appearance of the mammalian colon is not widely found in other vertebrates, most of whom have a simple, straight large intestine.

The primary functions of the large intestine are to store and concentrate fecal material before defecation and to absorb some of the remaining salts and water that were not absorbed in the small intestine. Because most substances are absorbed in the small intestine, only a small volume of water and salts, along with undigested material (feces), is passed on to the large intestine. **Defecation** occurs when contractions of the rectum and relaxation of associated sphincter muscles expel the feces through the final portion of the canal, the **anus** (see Figure 45.6).

Located within the large intestine are large populations of various bacteria, some of which provide benefits to animals. For example, some bacteria release by-products such as certain vitamins into the lumen of the large intestine, which can then be absorbed across the intestinal epithelium. Although this source of vitamins generally provides only a small part of the normal daily requirement, it may make a significant contribution when dietary vitamin intake is low. Sometimes people develop a vitamin deficiency if treated with antibiotics that inhibit these species of bacteria. Other bacterial products include gas (flatus), which is a mixture of nitrogen and carbon dioxide, with small amounts of hydrogen, methane, and hydrogen sulfide. Bacterial processing of undigested polysaccharides produces these gases, except for nitrogen, which is derived from swallowed air. Certain foods (for example, beans) contain large amounts of carbohydrates that cannot be digested by intestinal enzymes but are readily metabolized by bacteria in the large intestine, producing gas.

45.4 Mechanisms of Digestion and Absorption in Vertebrates

Learning Outcomes:

1. Describe the mechanisms of digestion and absorption of carbohydrates, proteins, and fats in vertebrates, and the importance of enzymes in some of these processes.
2. Explain why some nutrients do not require digestion prior to being absorbed.

The preceding sections provided an overview of nutrition and the basic features of digestive systems. We turn now to a more detailed description of how carbohydrates, proteins, and lipids are processed in the vertebrate digestive system and how the end products of digestion are absorbed across intestinal cells.

Carbohydrates Are Digested and Absorbed in the Small Intestine

In omnivores such as humans, most of the ingested carbohydrates are the polysaccharides starch and cellulose from plants and glycogen from animals. The remainder consists of simple carbohydrates, such as the monosaccharides fructose and glucose in fruit, and disaccharides, such as lactose in milk. Humans also add the disaccharide sucrose (table sugar) to their food. Certain other animals consume sucrose from sources such as maple sap and sugarcane.

Although a small amount of starch is digested in the mouth by salivary amylase, most starch digestion takes place in the small intestine by amylase secreted into the intestine by the pancreas. The products of starch digestion via amylase are molecules of the disaccharide maltose (**Figure 45.11**). Maltose, along with any ingested sucrose and lactose, is broken down into monosaccharides—fructose, glucose, and galactose—by enzymes located on the brush border of the small intestine epithelial cells. The monosaccharides are then absorbed into the epithelial cells. Fructose crosses the apical (luminal surface) membrane of the epithelial cells by facilitated diffusion, whereas glucose and galactose undergo secondary active transport coupled to sodium ions. Monosaccharides then leave the epithelial cells by way of facilitated diffusion transporters located in the basolateral membrane of the epithelial cells and enter the blood. The transport of substances from the lumen to the blood is called transepithelial transport because it occurs across a layer of epithelial cells. The bloodstream distributes the monosaccharides and other absorbed nutrients to the cells of the body.

Do you feel ill after drinking milk or eating dairy products? If so, you are among the majority of people who cannot adequately digest lactose, the chief disaccharide in milk. A small percentage of humans,

Figure 45.11 Digestion and absorption of carbohydrates in the small intestine. Digestion and absorption occur in the same cells but are shown separately here for clarity. For simplicity, the microvilli (brush border) are not shown.

Concept Check: The absorption of many nutrients requires secondary active transport. What can we conclude from that about the energetic cost of absorption?

BioConnections: Are the transport processes shown here, including facilitated diffusion and secondary active transport, unique to animal cells? See Figure 38.2 for help if needed.

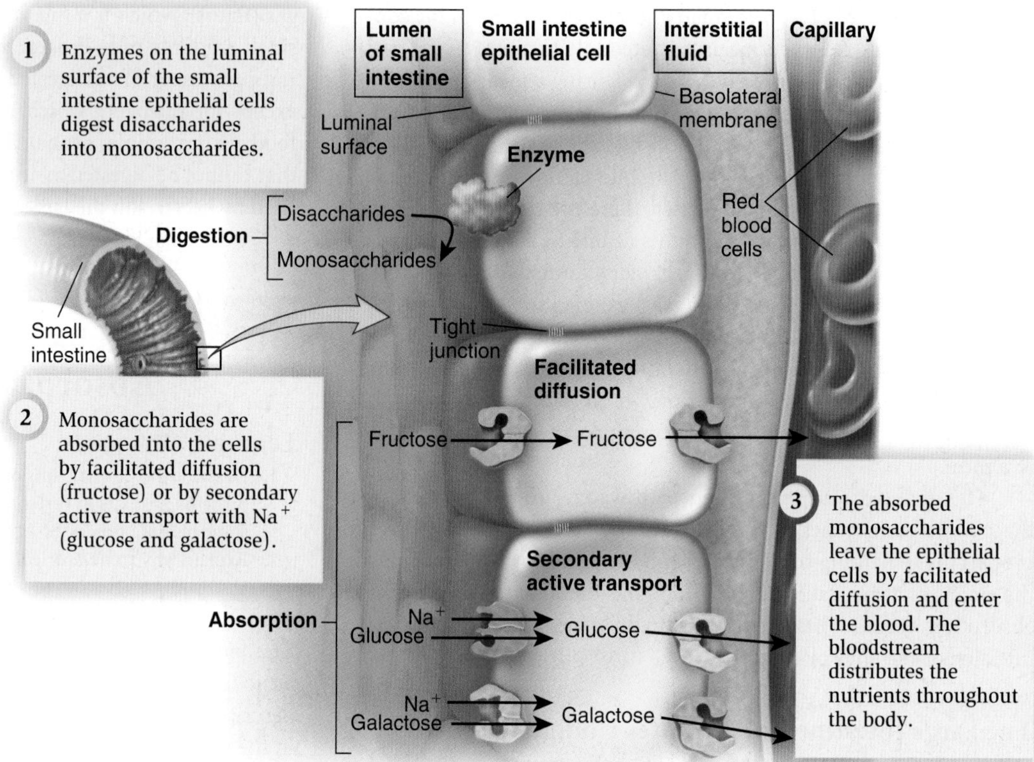
1 Enzymes on the luminal surface of the small intestine epithelial cells digest disaccharides into monosaccharides.
Lumen of small intestine
Small intestine epithelial cell
Interstitial fluid
Capillary
Luminal surface
Basolateral membrane
Enzyme
Red blood cells
Digestion
Disaccharides
Monosaccharides
Small intestine
Tight junction
Facilitated diffusion
2 Monosaccharides are absorbed into the cells by facilitated diffusion (fructose) or by secondary active transport with Na+ (glucose and galactose).
Fructose
Fructose
3 The absorbed monosaccharides leave the epithelial cells by facilitated diffusion and enter the blood. The bloodstream distributes the nutrients throughout the body.
Secondary active transport
Absorption
Na+
Glucose
Glucose
Na+
Galactose
Galactose

however, retain the ability to fully digest lactose throughout life. Next, we'll examine this phenomenon along with its genetic and evolutionary basis.

Genomes & Proteomes Connection

Genetics Explains Lactose Intolerance

One of the defining features of mammals is that they produce and consume milk, a nutrient-rich solution containing all the macro- and micronutrients required by their offspring. With rare exceptions, the milk of all mammals contains lactose. Lactose is digested by the intestinal enzyme lactase, which cleaves lactose into glucose and galactose. Humans with lactose intolerance cannot adequately digest lactose because their lactase is either inactive, absent, or present in small amounts.

Milk is the sole food of most mammals shortly after birth, and the primary food for various lengths of time thereafter until weaning, the transition from consuming mother's milk to eating a diet of solid foods. Once weaned, mammals never again drink milk, except, of course, for humans. The popular notion of adult cats lapping up milk from a bowl is a misconception; most cats are lactose-intolerant. Some visits to the veterinarian are in fact related to GI symptoms caused by well-meaning owners who fed milk to their adult pets.

Because the only dietary source for lactose is milk, it is not surprising that older mammals lose the ability to digest this disaccharide. This occurs because the gene that encodes the enzyme lactase is shut off at the age of weaning or shortly thereafter. The developmental mechanisms that turn off lactase production and activity are not firmly established, but they are known to involve decreased transcription of the lactase gene.

If an adult mammal were to drink milk, the undigested lactose would remain in the gut. As a result, water that normally would be absorbed by osmosis with the digested monosaccharides formed from lactose would also remain in the gut. Farther along the alimentary canal, microorganisms in the large intestine digest some of the lactose for their own use and, in the process, release by-products such as hydrogen and other gases. The combination of water retention and bacterial action results in GI symptoms such as diarrhea, gas, and cramps.

An estimated 90% of the world's human population cannot fully digest lactose after early childhood. In other words, lactose intolerance is not a disorder, it is a normal condition that evolved in all mammals, including humans. Why are some people able to consume dairy products without getting ill? Researchers have discovered that the ability of human adults to digest lactose is clearly linked to genetic background. Lactose intolerance is an example of a human polymorphism, a genetic trait that varies among people. Other than individuals who trace their ancestry to northern Europe and a few isolated regions in western Africa, nearly every population of humans shows a considerable degree of lactose intolerance, reaching nearly 100% in most of south and east Asia. One hypothesis for this phenomenon appears to be a behavioral and cultural change that occurred in Neolithic times, when certain populations domesticated cattle and added cow's milk to their diet. Adults who were able to digest lactose—presumably due to some mutation affecting the expression of the lactase gene—enjoyed a selective advantage and tended to thrive and pass on their genes more frequently than those whose digestive systems could not handle milk. In other words, natural selection may have resulted in certain populations carrying mutations that caused the lactase gene to be expressed after weaning. Eventually, a substantial percentage of people in those regions of the world retained sufficient intestinal lactase to use milk as a staple food.

To understand how this trait was passed on, let's examine the gene that codes for lactase. Surprisingly, the coding regions of the lactase gene and the lactase core promoter are identical whether individuals are lactose tolerant or intolerant. Recently, however, Finnish investigators have uncovered two single-nucleotide changes located in presumed regulatory sites that control the expression of the lactase gene. These changes are associated with prolonged lactase expression, allowing it to occur after weaning. People carrying these mutations are lactose-tolerant and can consume milk products through adulthood. By comparison, adults who lack these mutations—and are therefore lactose-intolerant—can consume dairy products only in small amounts or not at all. However, their ability to do so is greatly improved if the product has been commercially treated with purified lactase to predigest the lactose.

Proteins Are Digested in the Stomach and Small Intestine, and Absorbed in the Small Intestine

Proteins are broken down to peptide fragments in the stomach by pepsin, and in the small intestine by the proteases **trypsin** and **chymotrypsin**. The pancreas secretes the latter two enzymes as inactive precursors, which prevents the active enzymes from digesting the pancreas itself. Once the inactive form of trypsin enters the small intestine, it is cleaved into the active molecule by the enzyme enteropeptidase, whose active site is located on the luminal membranes of the intestinal cells. Trypsin then activates the inactive forms of chymotrypsin.

The peptide fragments produced by trypsin and chymotrypsin are further digested into individual amino acids by the actions of specific proteases located on the membranes of the brush border in the small intestine. These proteases cleave off one amino acid at a time from the N-terminus and C-terminus of peptides, respectively. Individual amino acids then enter the epithelial cells by secondary active transport coupled to sodium ions. Amino acids leave these cells and enter the blood by facilitated diffusion across the basolateral membrane. As with carbohydrates, protein digestion and absorption are largely completed in the duodenum.

Lipid Digestion and Absorption Occur in the Small Intestine

Most ingested lipid is in the form of triglycerides (fats). Fat digestion occurs almost entirely in the small intestine. The major digestive enzyme in this process is **lipase**, secreted by cells of the pancreas into the small intestine. The lipase reaction catalyzes the splitting of bonds in triglycerides, producing two free fatty acids and a monoglyceride:

$$\text{Triglyceride} \rightarrow 2 \text{ Free fatty acids} + 1 \text{ Monoglyceride}$$

Fats are poorly soluble in water and aggregate into large lipid droplets, as you can see if you shake a salad dressing made of oil and vinegar. Because lipase is a water-soluble enzyme, its digestive action in the small intestine can take place only at the surface layer of a lipid droplet. The rate of digestion is substantially increased by the process of **emulsification**, which disrupts the large lipid droplets into many tiny droplets, increasing their total surface area and exposure to lipase. The muscular contractions of the stomach and small intestine provide mechanical disruption of the fat droplets, and bile salts and phospholipids coat the outer surface of the small droplets and prevent them from recombining back into larger ones. The resulting suspension of small lipid droplets is called an emulsion.

Although emulsification speeds up digestion of fats, absorption of the poorly soluble products of the lipase reaction is facilitated by a second action of bile salts, the formation of **micelles** (**Figure 45.12a**). Micelles consist of bile salts, phospholipids, fatty acids, and monoglycerides clustered together. Micelles continuously break down and re-form near the epithelium of the intestine. When a micelle breaks down, the small lipid molecules are released into the solution and diffuse across the intestinal epithelium. As the lipids diffuse into epithelial cells, micelles release more lipids into the aqueous phase. Thus, the micelles keep most of the insoluble fat digestion products in small soluble aggregates while gradually releasing very small quantities of lipids to diffuse into the intestinal epithelium. Note that it is not the micelle that is absorbed but rather the individual lipid molecules that are released from the micelle.

During their passage through the epithelial cells, fatty acids and monoglycerides are resynthesized into triglycerides in the smooth endoplasmic reticulum (SER). This process lowers the concentration of cytosolic free fatty acids and monoglycerides in the epithelial cells, and so maintains a diffusion gradient for these molecules from the lumen into the cell. The resynthesized triglycerides aggregate into **chylomicrons**, large droplets coated with proteins that perform an emulsifying function similar to that of bile salts (**Figure 45.12b**). In addition to triglycerides, chylomicrons contain phospholipids, cholesterol, and fat-soluble vitamins that have been absorbed by the same process that led to fatty acid and monoglyceride movement into the epithelial cells of the small intestine.

Chylomicrons are released by exocytosis from the epithelial cells and enter lacteals (see Figure 45.9). The fluid from the lacteals eventually empties into a large vein and from there into the general blood circulation.

Vitamins, Minerals, and Water Are Not Digested but Must Be Absorbed

As stated earlier, vitamins, minerals, and water do not require digestion, and they are absorbed in their complete form. Most water-soluble vitamins are absorbed by diffusion or active transport in the small intestine. The fat-soluble vitamins—A, D, E, and K—follow the pathway for lipid absorption described in the previous section. Any interference with the secretion of bile or the action of bile salts in the intestine decreases the absorption of fat-soluble vitamins.

Water is the most abundant substance in chyme. Small amounts of ingested water are absorbed in the stomach, but the stomach has a small surface area available for diffusion and lacks the solute-absorbing mechanisms that create the osmotic gradients necessary for water absorption. The great majority of water absorption occurs in the small intestine. The epithelial membranes of the small intestine are very permeable to water, and water diffuses across the epithelium whenever an osmotic gradient is established by the active absorption

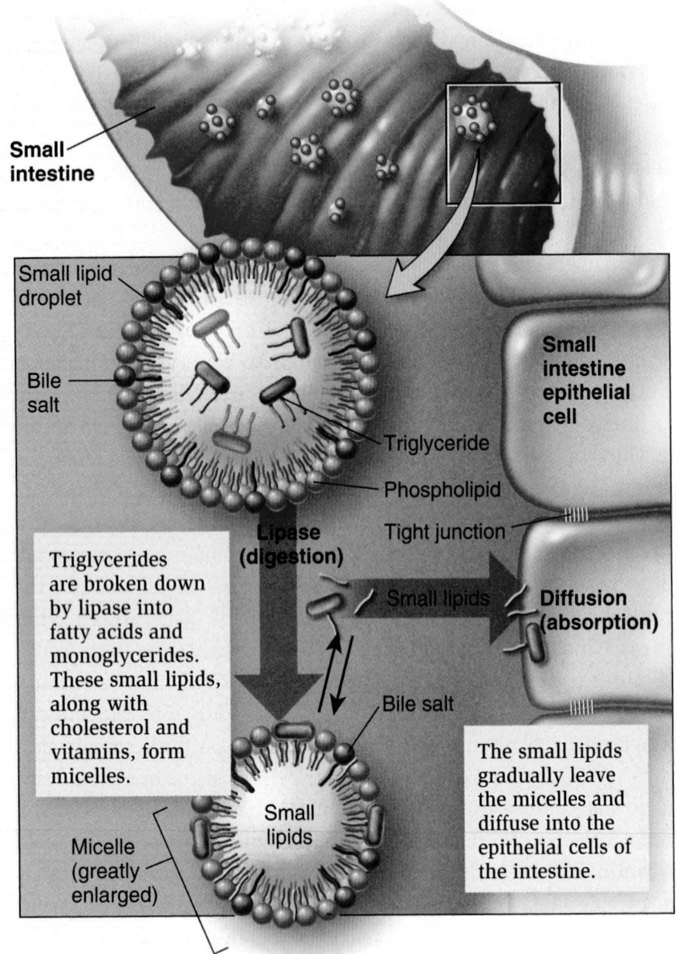

(a) Digestion of emulsified fats into micelles, and absorption into intestinal cells

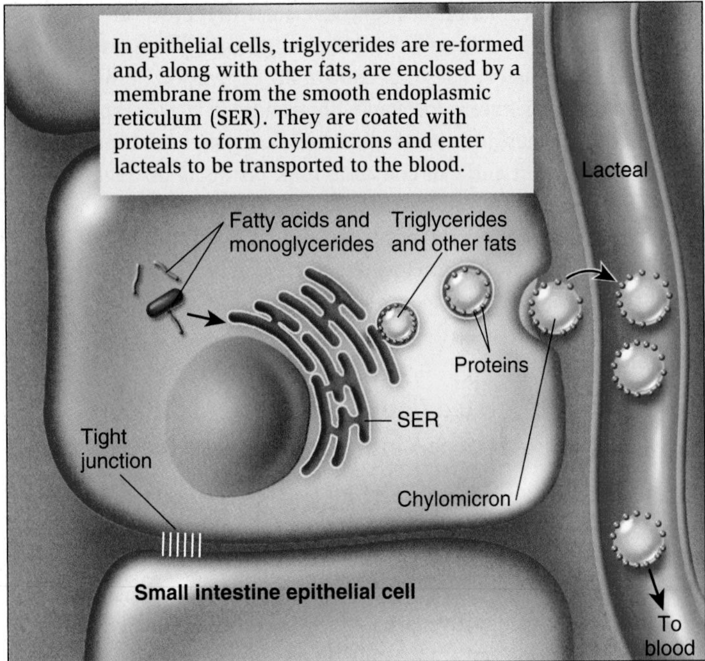

In epithelial cells, triglycerides are re-formed and, along with other fats, are enclosed by a membrane from the smooth endoplasmic reticulum (SER). They are coated with proteins to form chylomicrons and enter lacteals to be transported to the blood.
Lacteal
Fatty acids and monoglycerides
Triglycerides and other fats
Proteins
SER
Tight junction
Chylomicron
Small intestine epithelial cell
To blood

(b) Synthesis of triglycerides and the formation and release of chylomicrons

Figure 45.12 **Digestion and absorption of emulsified fat in the small intestine.**

of solutes, particularly Na^+, Cl^-, and HCO_3^-. Other minerals, such as potassium, magnesium, and calcium, are present in smaller concentrations and are also absorbed by this mechanism, as are trace elements such as zinc and iodide. The mechanisms of absorption of these molecules generally involve transport proteins and/or ion pumps.

45.5 Regulation of Digestion

Learning Outcomes:

1. Explain how the nervous system controls different features of the digestive process.
2. Name three major hormones important for the regulation of digestion in vertebrates and describe the role of each.

The digestive systems of animals are under complex control regulated in part by other organ systems, especially the nervous and endocrine systems. Neurotransmitters from neurons and hormones from endocrine glands control the volume of saliva produced, the amount of acid produced in the stomach, the timing and amount of secretions from the gallbladder and pancreas, and the rate and strength of muscle contractions along the alimentary canal. In this section, we examine the major mechanisms by which the nervous and endocrine systems control the activity of the vertebrate digestive system.

The Nervous System Controls Muscle and Secretory Activity

The nervous system can affect the activities of the digestive system in two major ways: (1) local control of muscle and glandular activity by the neurons within the alimentary canal and (2) long-distance regulation by the brain.

Within the walls along the length of the alimentary canal is a highly branched, interconnected collection of neurons that function in local control of the digestive system. These neurons interact with nearby smooth muscles, glands, and epithelial cells, and stimulation of neuronal activity at one point along the alimentary canal can lead to impulses that are transmitted up and down the canal. When food enters the small intestine, for example, the intestine is stretched. This directly activates the neurons in the intestinal wall. Impulses are sent from these neurons to the muscles of the stomach, where they decrease the contractions of the stomach. This slows the rate at which chyme moves from the stomach into the small intestine, giving the intestine sufficient time for digestion and absorption. In this way, the alimentary canal can regulate its own function independent of the brain.

In long-distance regulation, the brain communicates with neurons in the walls of the stomach and intestines and thereby influences the movement and secretory activity of the alimentary canal. For example, emotional stress, a brain-related event, can affect digestive processes. Likewise, the sight, smell, and taste of food activate digestive functions even before food reaches the stomach. These stimuli when processed by the nervous system act via nerves from the brain as a feedforward mechanism so that saliva production, stomach activity, and digestion can begin as soon as the food is first ingested.

Hormones Regulate the Rate of Digestion

Hormones are chemical messengers secreted by specialized cells into the blood, where they travel to all parts of the body and act on various target cells (see Chapter 50). The hormones that control the digestive system are secreted mainly by cells scattered throughout the epithelium of the stomach and small intestine. One surface of each hormone-producing cell is exposed to the lumen of the GI tract. At this surface, chemical substances in chyme stimulate cells in the stomach epithelium to release a hormone called **gastrin**, which reaches all the parts of the stomach through the bloodstream (**Figure 45.13**). The presence of gastrin stimulates smooth muscle contraction in the stomach, which helps move chyme into the small intestine. Gastrin also stimulates acid production by stomach epithelial cells. In the small intestine, the arrival of chyme stimulates release of the hormones cholecystokinin (CCK) and secretin from intestinal epithelial cells. Cells of the pancreas respond to CCK by secreting digestive enzymes and to secretin by secreting acid-neutralizing bicarbonate ions (HCO_3^-) into the small intestine. CCK also stimulates contraction of the gallbladder and therefore bile release.

Secretin holds a special place in the modern history of biology. It is a polypeptide whose discovery provided the first clear understanding of how the GI tract and its accessory structures communicate, as well as initiating the field of endocrinology—the study of hormones.

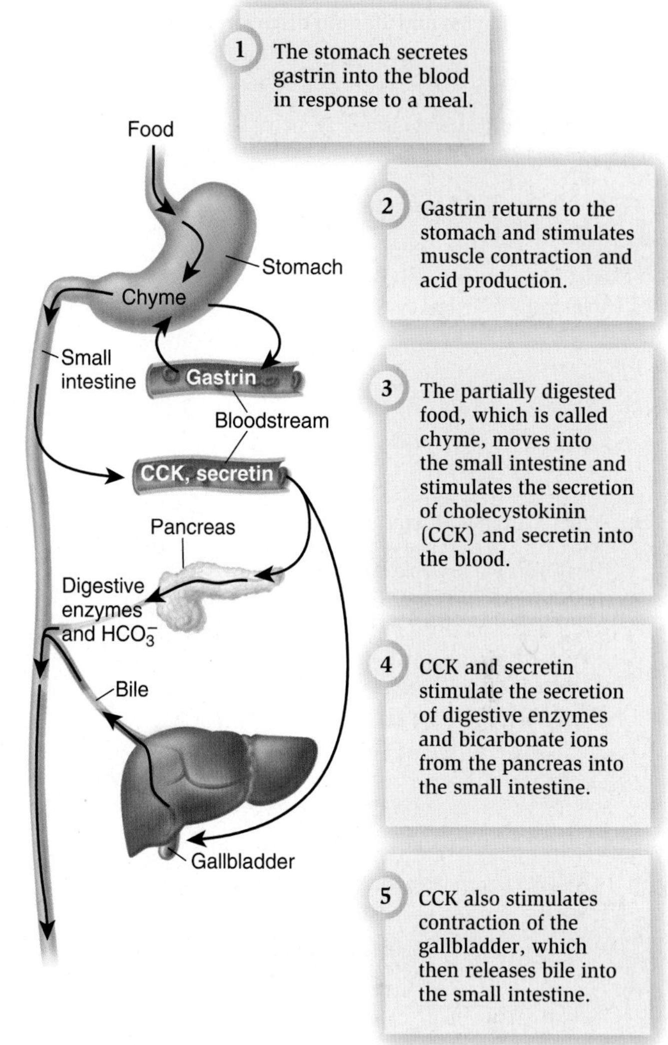

1 The stomach secretes gastrin into the blood in response to a meal.
Food
Stomach
Chyme
Small intestine
Gastrin
Bloodstream
CCK, secretin
Pancreas
Digestive enzymes and HCO_3^-
Bile
Gallbladder
2 Gastrin returns to the stomach and stimulates muscle contraction and acid production.
3 The partially digested food, which is called chyme, moves into the small intestine and stimulates the secretion of cholecystokinin (CCK) and secretin into the blood.
4 CCK and secretin stimulate the secretion of digestive enzymes and bicarbonate ions from the pancreas into the small intestine.
5 CCK also stimulates contraction of the gallbladder, which then releases bile into the small intestine.

Figure 45.13 **Hormonal regulation of digestion in the stomach and small intestine.**

BioConnections: In Chapter 40 you learned about the process of negative feedback, which helps maintain homeostasis in animals. Can you make a prediction about the possible feedback effects, if any, of CCK on the stomach? Do you think CCK might stimulate or inhibit smooth muscle activity and acid production in the stomach?

FEATURE INVESTIGATION

Bayliss and Starling Discovered a Mechanism by Which the Small Intestine Communicates with the Pancreas

The dependence of digestive function on the nervous system and the brain was established by the pioneering work of Ivan Pavlov. Pavlov's work in the early 1900s suggested that nerves conveyed signals between the gastrointestinal tract and other structures, such as the brain and pancreas, and in this way, the tract could communicate with these structures. Such communication is essential for the synchronized release of pancreatic enzymes and HCO_3^- with the arrival of food in the intestine, as we saw in Figure 45.13. Shortly after Pavlov's work, two English scientists, William Bayliss and Ernest Starling, hypothesized that nerves were not the only structures controlling activity of the tract.

To test this hypothesis, Bayliss and Starling carefully dissected away and cut the nerves to the small intestine of an anesthetized dog, as shown in **Figure 45.14**. Next, they directly injected a small quantity of acid (HCl) into the intestinal lumen, because it was known at the time that the acidic contents arriving from the stomach somehow triggered secretion of digestive enzymes and HCO_3^- from the pancreas. They

Figure 45.14 **Bayliss and Starling discovered the mechanism by which the small intestine and pancreas work together in digestion.**

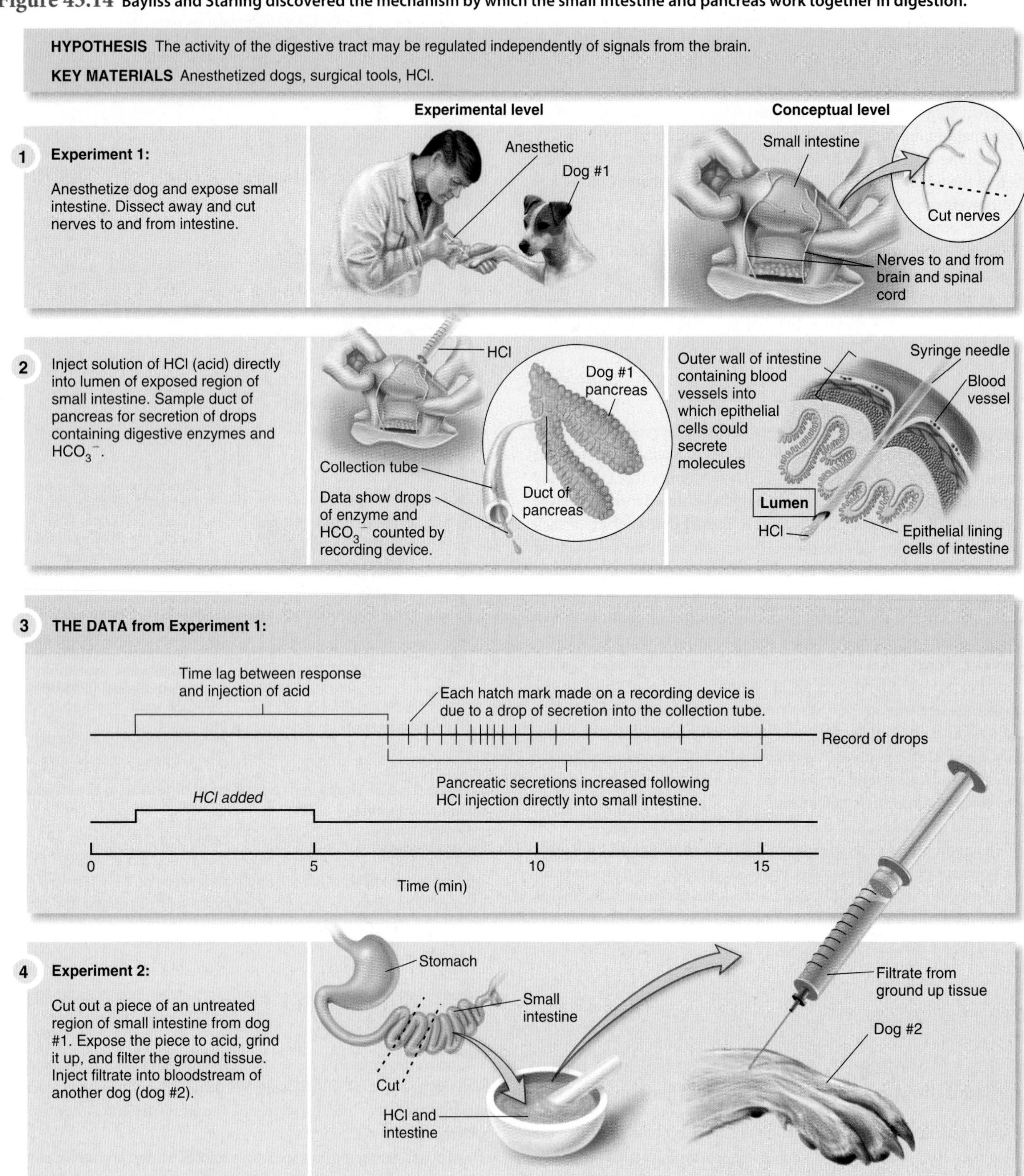

5 Test for pancreatic secretions into small intestine of dog #2.

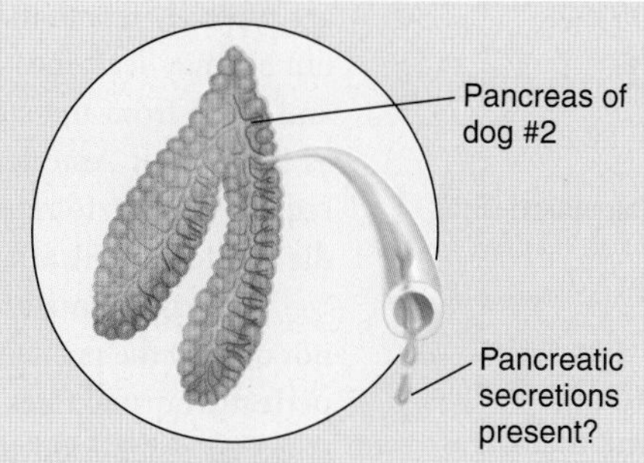

If the filtrate from dog #1 contained chemical signals released upon exposure to acid, those signals should stimulate the pancreas of dog #2 to secrete enzymes and HCO_3^-.

6 **THE DATA from Experiment 2:**

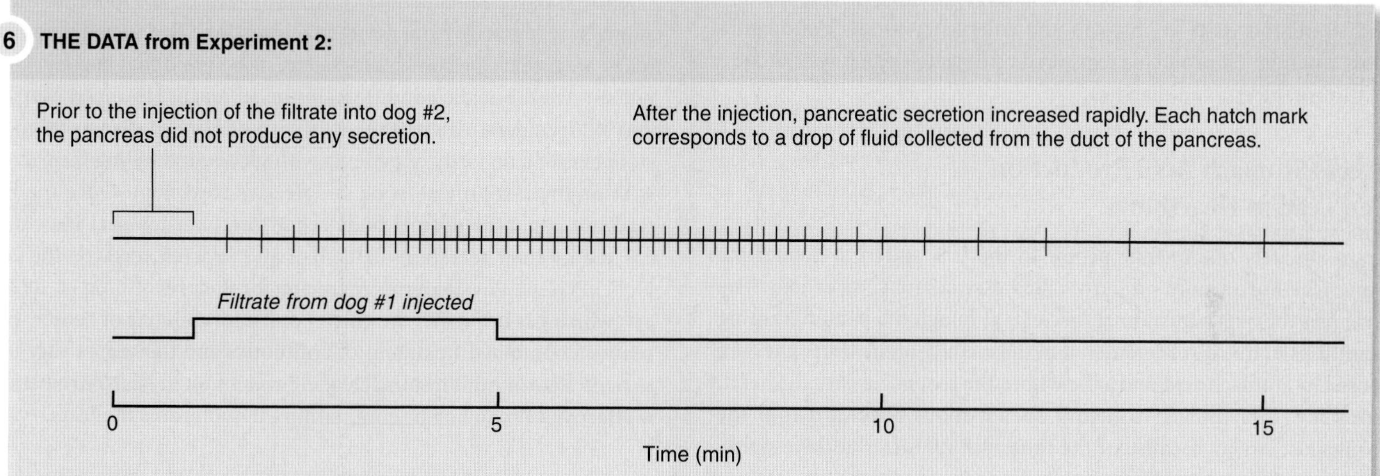

Prior to the injection of the filtrate into dog #2, the pancreas did not produce any secretion.
After the injection, pancreatic secretion increased rapidly. Each hatch mark corresponds to a drop of fluid collected from the duct of the pancreas.
Filtrate from dog #1 injected
0
5
10
15
Time (min)

7 **CONCLUSION** The small intestine contains molecules whose secretion into the blood is stimulated by acid. The molecules travel through the blood to the pancreas, causing secretion of pancreatic enzymes and HCO_3^-.

8 **SOURCE** Bayliss, W.M., and Starling, E.H. 1902. The mechanism of pancreatic secretion. *Journal of Physiology* 28:325–353.

placed a tube into the duct that carries the secretions from the pancreas to the small intestine, so they could collect the pancreatic secretions. Each drop of secretion was recorded as a hatch mark on a chart recorder. They discovered that even in the absence of intact nerves, simply adding acid into the intestinal lumen resulted in increased pancreatic secretions after a short lag period. Bayliss and Starling postulated that the lag was due to the time required for the intestinal cells to produce and secrete an unknown factor that entered the blood and eventually reached the pancreas, where it activated HCO_3^- secretion.

To confirm their results and to counter the suggestion made by Pavlov and others that their first experiment merely failed to sever all the intestinal nerves, Bayliss and Starling performed a follow-up experiment. First, they removed a section of a dog's small intestine, added acid to it, and then ground it up in the acid. The mashed tissue was then filtered to produce an extract of any secretions that may have been produced by the intestine in response to the acid. Next, they injected the filtered extract into the bloodstream of another anesthetized dog whose pancreatic duct was opened as in the first experiment. What they discovered was remarkable: The extract derived from the acid-stimulated intestine of first dog caused the almost immediate secretion of pancreatic enzymes and HCO_3^- from the second dog. This demonstrated that the small intestine secreted some chemical or chemicals in response to the presence of acid and that these secretions could reach the pancreas—presumably through the bloodstream—and stimulate the gland to release its contents. Bayliss and Starling called these chemicals hormones, from a Greek verb meaning to excite or arouse to action, and they named the hormone that activated the pancreas secretin. Thus was born not only the concept of a hormone but the recognition that these chemical messengers play a vital role in regulating the digestive system.

Experimental Questions

1. Explain the first experiment conducted by Bayliss and Starling, which indicated that other factors besides signals from nerves may influence digestive gland secretion.
2. What criticism did Bayliss and Starling need to address to provide more conclusive evidence that the secretion was not due to neural regulation, and how did they address it?
3. What conclusions did the investigators draw from their second experiment?

45.6 Impact on Public Health

Learning Outcomes:

1. Describe the impact of heartburn, ulcers, and diarrhea on public health.
2. Identify the primary cause of ulcers.

As we have seen, the functioning of the vertebrate digestive system is extraordinarily fine-tuned by the brain, nerves, and hormones. When people are well nourished, therefore, you might assume that digestive problems would be rare. However, each year in the U.S. alone, GI complaints account for approximately 40 million visits to doctors and 10 million visits to hospitals and emergency rooms. Among the most common GI problems registered by hospitals and physicians are heartburn, ulcers, and diarrhea.

Excess Stomach Acid Production Can Lead to Heartburn

Approximately one in four people in the U.S. suffers at some time from **heartburn** (more properly called gastroesophageal reflux). The term heartburn is a misnomer, because although the painful burning sensations of this disorder are felt in the vicinity of the heart, they are caused by stomach acid and arise within the esophagus, not the heart. Normally, the stomach contents do not move backward into the esophagus, largely because of the muscular sphincter at the esophageal/stomach juncture. However, under some conditions, the sphincter either does not close entirely or is forced open by the pressure of material in the stomach. When this happens, the acid in the stomach lumen enters the esophagus and irritates nerve endings there.

Many circumstances may contribute to heartburn. Overeating, for example, enlarges the volume of the stomach to the point of forcing its contents through the esophageal sphincter. Lying down after a big meal removes the effect of gravity on food in the stomach and may allow some acid to leak backward. Heartburn is also associated with smoking and consumption of alcohol, citrus fruits (which are acidic), chocolate (which contains caffeine, a known inducer of heartburn), and fatty foods (which take longer to digest than other foods and therefore remain in the stomach longer). One of the most common causes is pregnancy. Toward the middle of pregnancy, the growing fetus pushes up on the abdominal contents, which tends to force material from the stomach into the esophagus.

Common antacids contain calcium carbonate ions, which buffer the acid in the stomach and esophagus. In severe cases, heartburn can damage the walls of the esophagus enough to cause a chronic cough and pain, or even perforate the esophagus. Antacids may not be sufficient to treat these patients, who are instead given drugs that inhibit the stomach's ability to produce acid. These drugs are among the most widely prescribed medications in the U.S.

Erosion of the Walls of the Alimentary Canal Causes an Ulcer

Erosion of any portion of the alimentary canal due to any cause is called an **ulcer**. Most ulcers occur in the stomach (gastric ulcers), and the duodenum, but occasionally also in the lower esophagus, because these sites have the greatest acid concentrations (**Figure 45.15**). Ulcers are typically less than an inch wide; if left untreated, contents of the lumen may leak into the surrounding body cavity, where enzymes and acids from the stomach can do considerable damage. As many as 20 million Americans have an ulcer. Each year, 40,000 patients require surgery to repair tissue damaged by ulcers, and around 6,000 die due to complications from the disease.

Acid is essential for ulcer formation, but we now know that it is not usually the primary cause. In fact, many patients with ulcers have perfectly normal rates of acid production. Contrary to popular belief, stress or spicy food is probably not a major cause of ulcers either. So what is the main cause of ulcers? In the 1980s, two Australian scientists, Barry Marshall and J. Robin Warren, proposed that most stomach ulcers arise from a bacterial infection. This idea did not gain quick acceptance because many scientists believed that bacteria could not survive the acidic conditions in the stomach. However, Marshall and Warren demonstrated the existence of *Helicobacter pylori* in the stomachs of a majority of patients with ulcers, and killing the bacteria with antibiotics proved to be an effective treatment. Once the bacteria are gone, the normal body repair mechanisms heal the wound created by their secretions. Marshall and Warren received the 2005 Nobel Prize in Physiology or Medicine for their revolutionary discovery, which has improved the quality of life for millions of people. Currently, a combined treatment of antibiotics and the same drugs used to neutralize or inhibit stomach acid production in heartburn patients is effective in reducing or eliminating the symptoms of ulcers.

Diarrhea Is the Most Common Gastrointestinal Disorder Worldwide

According to the World Health Organization, worldwide there are over 2 billion cases of **diarrhea**—loose, watery stools occurring at least three times per day—every year. Nearly all episodes of diarrhea resolve in a day or two. Typically they result from infection with a pathogen, such as a virus or bacterium. Other times, however, diarrhea may be caused by food sensitivities (such as lactose intolerance, discussed earlier), reactions to medications, stress-related disorders, or parasites that inhabit the rectum and large intestine.

Most cases of pathogen-related diarrhea in the U.S. result from exposure to one of several related types of bacteria, including those of the genus *Salmonella*, *Shigella*, and *Escherichia*. Often, these infections are handled by the body's immune system within a day or two, but sometimes medical attention is required. One cause of diarrhea stands apart as particularly dangerous, however. Cholera is a disease caused by the bacterium *Vibrio cholerae*, usually ingested by consuming contaminated food or water. Each year at least 2,000 people die of cholera worldwide, with another 100,000 people contracting the disease but surviving. Nearly all cases of cholera occur in Africa, parts of Asia (notably China), and India, although scattered outbreaks have occurred nearly everywhere except the U.S., with the most recent worldwide pandemic lasting from 1961 to 1971. *V. cholerae* releases a toxin that alters the permeability of salts in the large intestine, resulting in a massive flow of these ions, followed by water, into the intestinal lumen. As in all cases of diarrhea, the chief concern with cholera is the loss of nutrients and water and the dehydration that ensues. In addition to killing the bacteria with antibiotics, therefore, the major treatment of cholera is the same as for any cause of diarrhea, which includes drinking solutions of salts and water to replace those that were lost in the feces.

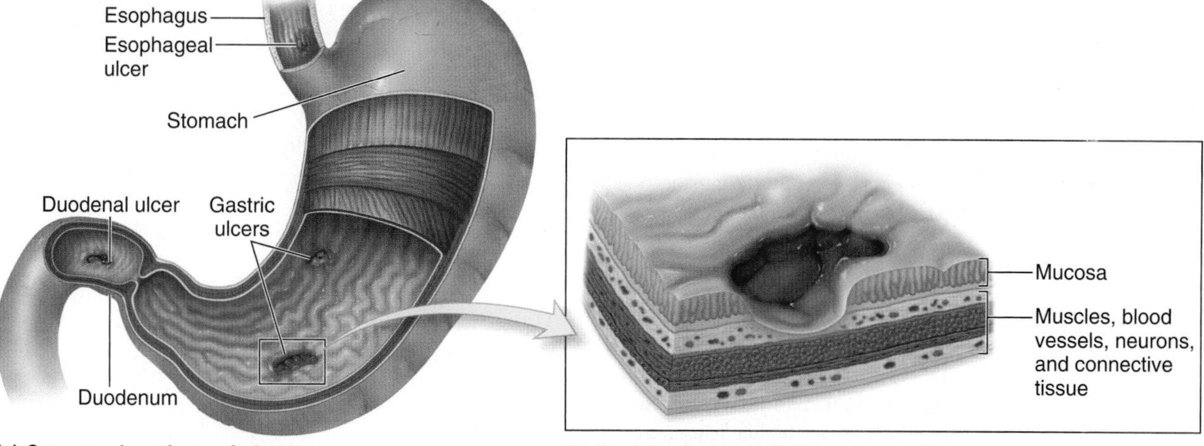

(a) Common locations of ulcers

(b) Gastric ulcer penetrating mucosal layer

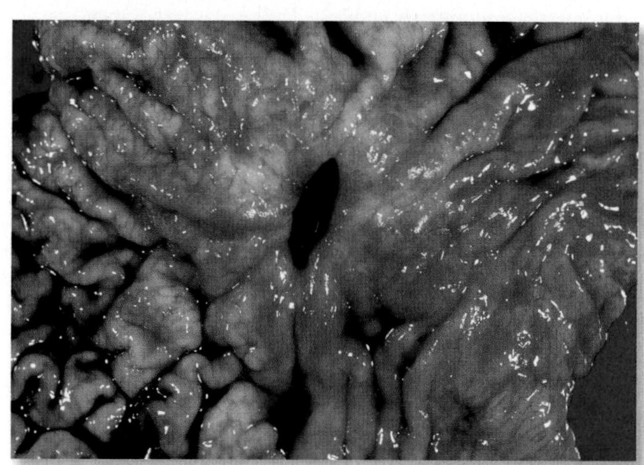

(c) Human gastric ulcer

Figure 45.15 Gastric and duodenal ulcers. (a) Common locations of ulcers in humans. (b) Illustration of a gastric ulcer eroding through the mucosal epithelium. (c) Photograph of an actual gastric ulcer that has penetrated through the stomach wall.

Summary of Key Concepts

45.1 Overview of Animal Nutrition and Ingestion

- The four phases of food use in animals are ingestion, digestion, absorption, and egestion. Animals require organic nutrients—carbohydrates, proteins, lipids, and nucleic acids—and inorganic nutrients in the form of water and minerals. Differences in nutritional demands reflect an animal's physiology and environment (Figure 45.1).
- Essential nutrients include eight essential amino acids that cannot be synthesized by humans and many other animals; vitamins, which are organic nutrients that serve as coenzymes for metabolic and biosynthetic reactions; and minerals, which are inorganic ions that serve many functions (Tables 45.1, 45.2, 45.3).
- Herbivores eat only plants, carnivores consume animal flesh or fluids, and omnivores eat both plant and animal products.
- Animals differ in the way they obtain food. Suspension feeders sift water to filter out organic matter and expel the rest. Bulk feeders eat food in large pieces, and display a variety of adaptations in the size and shape of teeth. Fluid-feeders lick or suck fluid from plants or animals (Figures 45.2, 45.3, 45.4).

45.2 Principles of Digestion and Absorption of Food

- Intracellular digestion occurs in single-celled organisms and simple invertebrates. Most animals digest food via extracellular digestion. Flatworms and cnidarians have a gastrovascular cavity with one opening that serves as both entry and exit port (Figure 45.5).
- Nearly all other animals have an alimentary canal, open at the mouth and anus or cloaca and segregated into specialized regions, through which food passes from one end to the other.
- Once food has been digested, the nutrients must be absorbed via simple diffusion, facilitated diffusion, or primary or secondary active transport.

45.3 Vertebrate Digestive Systems

- The vertebrate digestive system consists of the alimentary canal plus associated glands and organs. The anterior portion of the canal contains the oral cavity, pharynx (throat), and esophagus. The middle portion contains food storage or digestive organs (crop, gizzard, and/or stomach), and the upper part of the small intestine (duodenum). The posterior part of the alimentary canal contains the remainder of the small intestine, the large intestine, the rectum, and the anus or cloaca (Figure 45.6).

- Saliva facilitates swallowing, dissolves food particles, and initiates digestion. Peristalsis moves food through the pharynx and esophagus to the stomach, where digestion begins (Figure 45.7).
- The stomach stores food, partially digests proteins, and regulates the rate at which chyme empties into the small intestine. Ruminants have a complex stomach with several chambers that facilitates cellulose digestion (Figure 45.8).
- Nearly all digestion and absorption occur in the small intestine. The combination of villi and microvilli increases the small intestine's surface area and maximizes the efficiency of digestion and absorption (Figure 45.9).
- The pancreas secretes digestive enzymes and a bicarbonate-rich fluid that neutralizes the acidic chyme. The liver secretes bile acids in the digestion of fat (Figure 45.10).
- The large intestine concentrates undigested material, which is then eliminated from the anus or cloaca.

45.4 Mechanisms of Digestion and Absorption in Vertebrates

- Carbohydrate digestion occurs in the small intestine, where maltose and other disaccharides are digested into monomers by enzymes and transported across epithelial cells into the blood (Figure 45.11).
- In humans, the ability to digest lactose after childhood is linked to a genetic mutation.
- Proteins are broken down to peptide fragments in the stomach by pepsin and further broken down into amino acids in the small intestine by trypsin and chymotrypsin.
- Lipid digestion and absorption occurs entirely in the small intestine by the action of pancreatic lipase (Figure 45.12).
- Most water-soluble vitamins are absorbed by diffusion or active transport. Fat-soluble vitamins follow the pathway for lipid absorption.

45.5 Regulation of Digestion

- The digestive systems of animals are regulated primarily by the nervous and endocrine systems. The nervous system affects the digestive system in two major ways: (1) local control by the nerves in the alimentary canal and (2) long-distance regulation by the brain.
- Several hormones work together to regulate the rate of digestion. Bayliss and Starling demonstrated that secretin, a hormone from the small intestine, triggers secretions from the pancreas into the small intestine (Figures 45.13, 45.14).

45.6 Impact on Public Health

- Heartburn, ulcers, and diarrhea are common disorders involving the GI tract (Figure 45.15).

Assess and Discuss

Test Yourself

1. The process of enzymatically breaking down large molecules into smaller molecules that can be used by cells is
   a. absorption. c. ingestion.
   b. secretion. d. digestion.
2. The term "essential nutrients" refers to
   a. all the carbohydrates, proteins, and lipids ingested by an organism.
   b. nutrients that an animal cannot manufacture.
   c. nutrients that must be obtained from the diet in their complete form.
   d. all of the above.
   e. b and c only.
3. An animal that has a strong preference for a particular food but can adjust its diet if necessary is said to be
   a. herbivorous. c. carnivorous. e. both b and d.
   b. omnivorous. d. opportunistic.
4. Which of the following statements is *not* correct?
   a. Fluid-feeders include animals that consume animal or plant fluids.
   b. Suspension feeding occurs only in invertebrates.
   c. Predators are animals that kill and consume live prey.
   d. Herbivory occurs in both aquatic and terrestrial species.
   e. Blood-sucking animals may contain anticoagulants in their saliva.
5. The pancreas connects to which part of the alimentary canal?
   a. esophagus c. small intestine e. large intestine
   b. stomach d. cecum
6. Which of the following statements regarding the vertebrate stomach is not correct?
   a. Its cells secrete the active protease enzyme pepsin.
   b. It is a saclike organ that evolved to store food.
   c. Its cells secrete hydrochloric acid.
   d. It is the initial site of protein digestion.
   e. Little or no absorption of nutrients occurs there.
7. Absorption in the small intestine is increased by
   a. the many villi present on the inner surface of the small intestine.
   b. the brush border formed by microvilli on the cells of the villi.
   c. the presence of numerous transporter molecules on the epithelial cells.
   d. all of the above.
   e. a and b only.
8. In birds, the secretion of acid and pepsinogen occurs in
   a. the crop. d. the cloaca.
   b. the gizzard. e. the gallbladder.
   c. the proventriculus.
9. Bile is produced by
   a. the liver. d. the small intestine.
   b. the gallbladder. e. the cecum.
   c. the pancreas.

10. Which of the following is a function of the large intestine?
    a. It participates in cellulose digestion by microorganisms that exist in the cecum of herbivores.
    b. It stores and concentrates fecal material.
    c. Its cells absorb salts and water that remain in chyme after it leaves the small intestine.
    d. Its cells absorb certain vitamins produced by bacteria.
    e. All of the above are functions of the large intestine.

Conceptual Questions

1. Distinguish between digestion and absorption.
2. Explain the functions of the crop and gizzard in birds. Why don't humans have a crop or gizzard?
3. A principle of biology is that *structure determines function.* Explain some of the differences between the teeth of herbivores and carnivores and how these differences are adaptive for the animal's diet.

Collaborative Questions

1. Describe several strategies that animals use to obtain food.
2. Define "nutrient." On the basis of that definition, do you consider water a nutrient? Briefly discuss the essential nutrients, vitamins, and minerals that animals must obtain through their diet.

Online Resource

www.brookerbiology.com

Stay a step ahead in your studies with animations that bring concepts to life and practice tests to assess your understanding. Your instructor may also recommend the interactive eBook, individualized learning tools, and more.

Control of Energy Balance, Metabolic Rate, and Body Temperature

46

Chapter Outline

46.1 Nutrient Use and Storage
46.2 Regulation of the Absorptive and Postabsorptive States
46.3 Energy Balance
46.4 Regulation of Body Temperature
46.5 Impact on Public Health
Summary of Key Concepts
Assess and Discuss

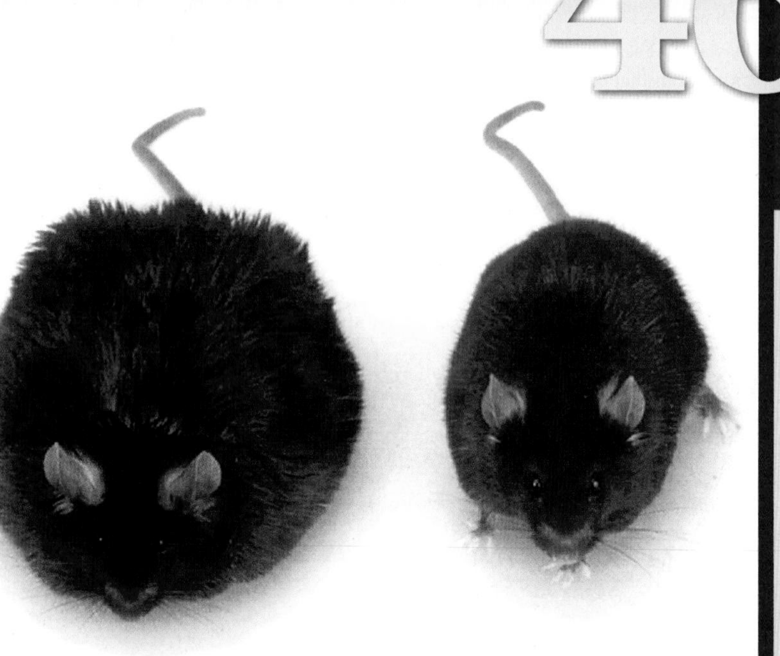

A genetically obese, leptin-deficient mouse and a normal mouse.

In 1997, two young cousins being treated for extreme obesity were brought to the attention of researchers studying a newly discovered hormone called leptin. Leptin had been demonstrated in laboratory rodents to inhibit appetite. The researchers hypothesized that the children—who weighed approximately 30 kg by age 2, and in one case, 86 kg by age 9—were not producing leptin. This hypothesis was confirmed by molecular analyses and traced to a mutation in the children's leptin gene. Subsequently, several other individuals with mutations in leptin were identified, leading in each case to extreme childhood obesity. Treatment of these individuals with leptin has proven beneficial in restoring normal body weight. Although leptin deficiency is rare and is only one of many possible causes of obesity in humans, animal models of leptin deficiency, such as the mice shown in the chapter-opening photo, represent a remarkable breakthrough in our understanding of the genetic bases of the control of appetite and metabolism. How nutrients that supply energy are processed, stored, and used by animals' bodies in times of food abundance or fasting is a subject of this chapter.

In Chapter 45, we saw that all animals require nutrients to assemble the macromolecules that make up body tissues. Some of the ingested nutrients, such as carbohydrates, lipids, and proteins, also represent a form of fuel, or energy, that can be used to generate ATP within cells. Most animals, however, do not have a constant supply of nutrients. For example, insects, cephalopods, and all vertebrates sleep or have periods of greatly reduced activity for part of the day, during which time they do not eat. In addition, environmental changes may reduce the food supply, leading to long fasts. As a consequence of the irregular and sometimes unpredictable flow of nutrients into the body, animals have evolved an array of mechanisms to adequately maintain levels of important fuel molecules even during a fast. In the first two sections of this chapter, we will explore how ingested nutrients are stored in the body for such times of need and the mechanisms by which these stores are tapped.

Metabolism refers to all the bodily activities and chemical reactions in an organism that maintain life. **Metabolic rate** describes the rate at which an organism uses energy to power these reactions. Animals may have widely different metabolic rates, which determine the amount of nutrients they require. The greater an animal's metabolic rate, the more heat it generates as a by-product of breaking down nutrients and using the energy of their chemical bonds to synthesize ATP. Some of this heat escapes to the environment, and some is used to warm an animal's body. Metabolism and body temperature are therefore closely related, and we will examine this relationship in the next two sections of the chapter. We will also discuss energy balance, the balance between energy consumption and expenditure. We conclude with the public health impact of human disorders associated with metabolism, including obesity.

46.1 Nutrient Use and Storage

Learning Outcomes:

1. Describe the processes of absorption of glucose, triglycerides, and amino acids during the absorptive state.
2. Discuss the two major ways that vertebrates can increase their blood glucose concentrations during the postabsorptive period.
3. Explain what glucose sparing is, and why it is important.

Once nutrients have been ingested, they are either used or stored. The handling of nutrients can be divided into two alternating phases. The **absorptive state** occurs when ingested nutrients enter the blood from the gastrointestinal (GI) tract. The **postabsorptive state** occurs when the GI tract is empty of nutrients and the body's own stores must supply energy. During the absorptive period, some ingested nutrients supply the immediate energy requirements of the body. The rest are added to the body's energy stores to be called upon during the next postabsorptive period. An average meal in a human requires about 4 hours for complete absorption. Therefore, our usual three-meal-a-day pattern places us in the postabsorptive state during the late morning and afternoon and part of the night. Total-body energy stores are adequate for the average human to withstand a fast of several weeks. By contrast, some animals can barely survive a single skipped meal—particularly if they have low energy reserves—because their relative metabolic requirements are much greater than our own. In this section, we will focus on nutrient absorption, use, and storage in vertebrates, with a closer look at mammals.

In the Absorptive State, Nutrients Are Absorbed and Used for Energy or Stored

The categories of nutrients that are absorbed during the absorptive state include or are formed from carbohydrates, lipids, proteins, nucleic acids, vitamins, minerals, and water. We will look in depth at the first three of these; **Figure 46.1** gives an overview of what happens to these nutrients. Digested carbohydrates are absorbed as monosaccharides, including glucose (refer back to Figure 45.11). Digested lipids (fats) are absorbed after first being resynthesized into triglycerides in intestinal epithelial cells (refer back to Figure 45.12). Proteins are broken down into amino acids, which are then absorbed.

Absorbed Carbohydrates: Energy or Storage The chief carbohydrate monomer absorbed from the GI tract of vertebrates is glucose, which is one of the body's two major energy sources during the absorptive state (triglycerides being the other). Much of the absorbed glucose enters cells and is enzymatically broken down, resulting in the formation of hydrogen ions, carbon dioxide, and water and, in the process, releasing the energy required to form ATP from ADP and inorganic phosphate (Figure 46.1a). Because skeletal muscle makes up a large fraction of body mass in most vertebrates, it is a major consumer of glucose, particularly when an animal is active. In all vertebrates, skeletal muscle also incorporates some of the glucose into the polymer glycogen, which is stored in the muscles to be used later. If more glucose is absorbed than is required for immediate energy, a portion of the excess is incorporated into glycogen in the liver, and the remainder into triglycerides in fat cells. The structures of glycogen and triglycerides are described in Figures 3.8 and 3.9, respectively.

Absorbed Triglycerides: Storage Triglycerides are too large to diffuse across the plasma membranes of the intestinal epithelial cells. As described in Chapter 45 (refer back to Figure 45.12), they are digested into fatty acids and monoglycerides in the lumen of the small intestine and then resynthesized into triglycerides once they diffuse into the intestinal epithelial cells. The triglycerides and other ingested lipids (for example, cholesterol) are packaged into chylomicrons, which enter lymph and from there the blood circulation. As blood moves through adipose tissue, a blood vessel enzyme called lipoprotein lipase releases the fatty acids from the triglycerides in the chylomicrons. The released fatty acids then enter adipose cells and combine with glycerol to re-form triglycerides (Figure 46.1b). These triglycerides are stored in fat cells until the body requires additional fuel.

As with glucose, some of the ingested fat is not stored but is used by most organs other than the brain during the absorptive state to provide energy. The relative amounts of carbohydrate and fat used for energy during the absorptive period depend largely on the composition of a meal.

Absorbed Amino Acids: Protein Building Amino acids are taken up by all body cells, where they are used to synthesize proteins (Figure 46.1c). All cells require a constant supply of amino acids,

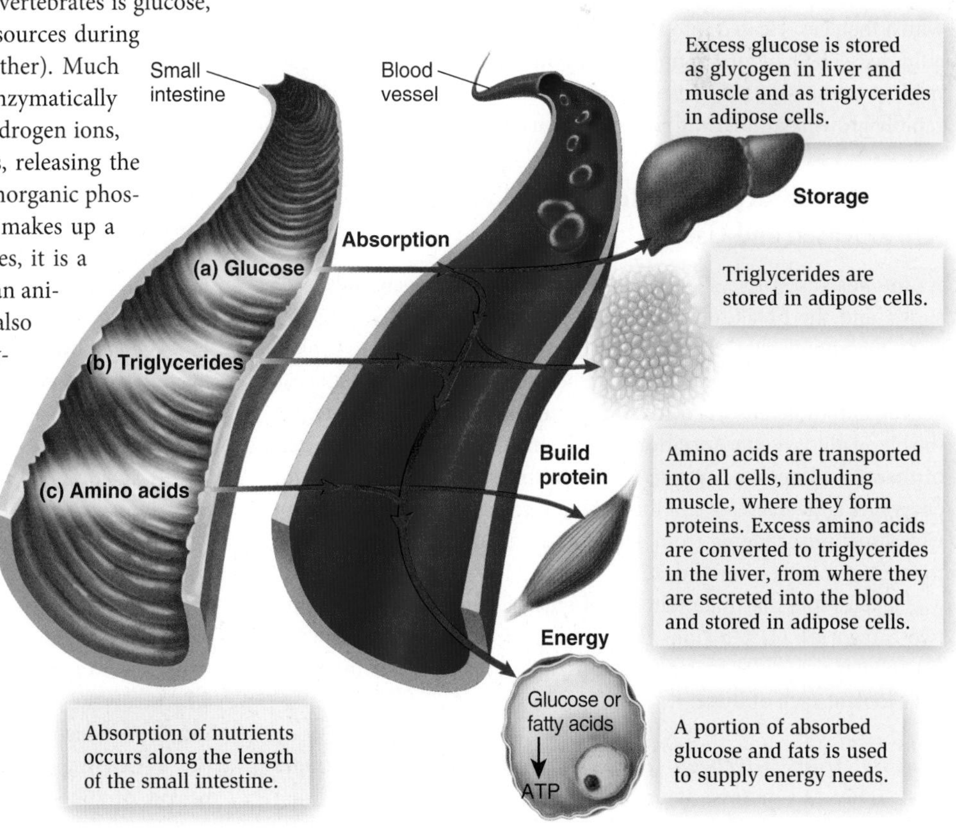
Small intestine
Blood vessel
Excess glucose is stored as glycogen in liver and muscle and as triglycerides in adipose cells.
Storage
Absorption
(a) Glucose
Triglycerides are stored in adipose cells.
(b) Triglycerides
Build protein
Amino acids are transported into all cells, including muscle, where they form proteins. Excess amino acids are converted to triglycerides in the liver, from where they are secreted into the blood and stored in adipose cells.
(c) Amino acids
Energy
Glucose or fatty acids
ATP
Absorption of nutrients occurs along the length of the small intestine.
A portion of absorbed glucose and fats is used to supply energy needs.

Figure 46.1 Events of the absorptive state. The products of digestion are absorbed into the blood along the length of the small intestine. (Note that for simplicity, triglycerides are shown being absorbed directly into blood, not into lacteals; see Figures 45.9 and 45.12 for details.) These nutrients are used for immediate energy demands, or they are deposited in cells as energy stores or as macromolecules important in cell function, such as for building proteins.

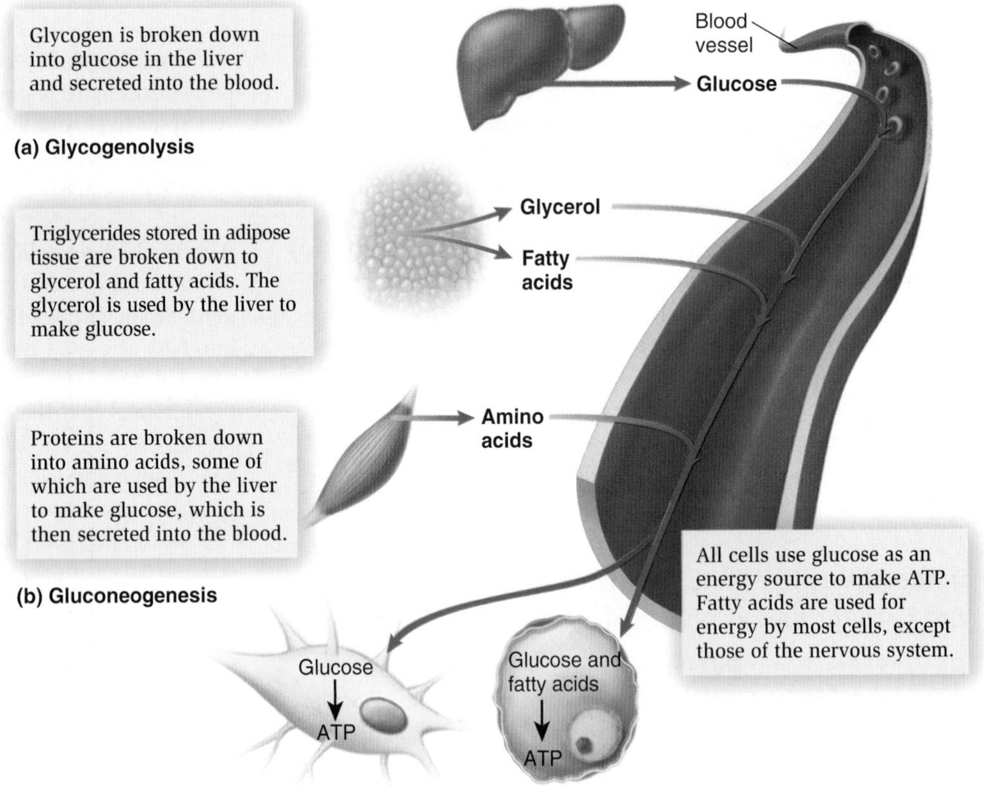

Glycogen is broken down into glucose in the liver and secreted into the blood.

(a) Glycogenolysis

Triglycerides stored in adipose tissue are broken down to glycerol and fatty acids. The glycerol is used by the liver to make glucose.

Proteins are broken down into amino acids, some of which are used by the liver to make glucose, which is then secreted into the blood.

(b) Gluconeogenesis

All cells use glucose as an energy source to make ATP. Fatty acids are used for energy by most cells, except those of the nervous system.

Figure 46.2 **Events of the postabsorptive state.** In this state, macromolecules formed during the absorptive state are broken down to supply smaller molecules that can be used for energy. **(a)** This begins with glycogenolysis—the breakdown of glycogen into glucose. **(b)** In gluconeogenesis, the breakdown products of triglycerides and proteins, namely, glycerol, fatty acids, and amino acids, can be used for energy either directly or indirectly by their conversion in the liver to glucose.

because proteins are constantly being synthesized and degraded. However, unlike excess glucose and fatty acids, which are stored as glycogen and triglycerides, excess amino acids that are ingested are not stored as protein. Instead, excess amino acids are enzymatically converted by liver cells into fatty acids and from there into triglycerides, which then get released into the blood and are taken up and stored in adipose cells. Therefore, eating large amounts of protein does not normally increase stores of body protein. An exception would be a young, rapidly growing animal with its continuous increase in body protein.

In the Postabsorptive State, Stored Nutrients Are Released and Used

As the postabsorptive state begins, synthesis of glycogen and lipids slows, and the breakdown of these substances begins. During this state, macromolecules formed during the absorptive state are broken down to supply monomers that can be used for energy. No glucose is available to be absorbed from the intestines during this time, yet the blood glucose concentration must be maintained because the cells of the central nervous system (CNS) normally use only glucose for energy. A large decrease in blood glucose concentrations can disrupt CNS functions, ranging from subtle impairment of mental function to seizures, coma, or even death.

The events that maintain blood glucose concentration fall into two categories: (1) reactions that provide glucose to the blood and (2) cellular use of fat for energy, thus sparing glucose for the CNS. Let's look at each of these.

Glucose from Glycogen and Noncarbohydrate Precursors Vertebrates can increase their blood glucose concentrations during the postabsorptive period in two major ways: by breaking down glycogen and by making new glucose. First, the glycogen that was formed during the absorptive period can be broken back down into molecules of glucose by hydrolysis, a process known as **glycogenolysis** (Figure 46.2a). This occurs primarily in the liver, after which the glucose is released into the blood, where it can travel to all cells. Muscle glycogen is also broken down into glucose by glycogenolysis, but this glucose is used exclusively by muscle cells and not secreted into the blood.

The amount of liver glycogen available to provide glucose during the postabsorptive period varies among animals, but it is generally sufficient to maintain blood glucose concentrations for only a brief time, such as an overnight fast in a human. Therefore, a second mechanism for maintaining blood glucose concentrations is required when the postabsorptive period continues longer. In the process of **gluconeogenesis** (literally, creation of new glucose), enzymes in the liver convert noncarbohydrates into glucose, which is then secreted into the blood (Figure 46.2b). This process occurs in all vertebrates but appears to be especially important in mammals.

A major precursor for gluconeogenesis is glycerol, which is released from triglycerides in adipose tissue by the breakdown process called **lipolysis**. In lipolysis, enzymes within fat cells hydrolyze triglycerides into fatty acids and glycerol, both of which enter the bloodstream. The fatty acids diffuse into cells, where they are used as an alternate energy source to glucose (except for the CNS, which continues to require glucose). The glycerol is taken up by the liver, where enzymes convert it into glucose, which is then released back into the bloodstream.

When the postabsorptive period continues for an extended period of time—as when an animal fails to find food—protein

Table 46.1 Relative Changes in the Use and Generation of Energy Sources During the Absorptive and Postabsorptive Periods

	Glucose absorption from GI tract	Glucose use by cells	Synthesis of triglycerides	Use of fatty acids for energy by cells	Breakdown of glycogen in muscle and liver	Blood concentrations of glucose
Absorptive period	High	High	High	Low/moderate	Low	Normal
Postabsorptive period	Absent	Moderate (some glucose is spared for CNS use)	Low	High	High	Normal

becomes an important source of blood glucose. Large quantities of protein in muscle and other tissues can be broken down to amino acids without serious tissue damage or loss of function. The amino acids enter the blood and are taken up by the liver. In the liver, the amino group is removed from each amino acid, and the remaining molecule is converted into glucose by a stepwise series of enzyme-catalyzed reactions. This process, however, has limits. Continued protein loss can result in the death of cells throughout the body because they depend on proteins for such vital processes as plasma membrane function, enzymatic activity, and formation of organelles.

Lipid Metabolism by Other Tissues: Glucose Sparing Another way that glucose is made available to the organs and tissues that require it the most—such as the brain—is by having other organs and tissues reduce their dependence on glucose. They do this by increasing their use of fat as an energy supply during this period. This metabolic adjustment, called **glucose sparing**, reserves (or spares) the glucose produced by the liver by glycogenolysis and gluconeogenesis for use by the CNS.

The essential step in glucose sparing is lipolysis, the breakdown of adipose tissue triglycerides, which, as stated earlier, liberates fatty acids and glycerol into the blood. In vertebrates, the circulating fatty acids are taken up and used to provide energy by almost all tissues, excluding the nervous system, whose cells do not express the enzymes required to break down fatty acids for energy.

Of the vertebrate body's tissues and organs, the liver is unique in that most of the fatty acids entering it during the postabsorptive state are not used by that organ for energy. Instead they are processed into three small compounds collectively called **ketones**, or ketone bodies. Ketones are released into the blood during prolonged fasting. They provide an important energy source for the many tissues, including the brain, which are capable of oxidizing ketones through the citric acid cycle.

The use of fatty acids and ketones during fasting provides energy for the body, sparing the available glucose for the brain. Moreover, as just mentioned, the brain can use ketones as energy, and it does so increasingly as ketones build up in the blood during the first few days of a fast. The survival value of this phenomenon is significant. When the brain reduces its glucose requirement by using ketones, much less protein breakdown is required to supply amino acids for gluconeogenesis. Protein stores last longer, enabling the animal to survive a long fast without serious tissue damage.

The combined effects of glycogenolysis, gluconeogenesis, and glucose sparing are so efficient that, after several days of complete fasting, human blood concentrations of glucose decrease by only a small percentage. Table 46.1 summarizes the relative changes in these and other variables during the absorptive and postabsorptive periods.

46.2 Regulation of the Absorptive and Postabsorptive States

Learning Outcomes:

1. Explain how insulin acts to control blood glucose concentration.
2. Describe how the human body regulates the use and storage of glucose.
3. Describe the effect of exercise on energy demands and the mechanisms that are activated to meet those demands.

Tight control mechanisms are required to maintain homeostatic levels of energy-providing molecules in the blood. These controls come in two forms: endocrine and nervous. Cells of the endocrine system produce blood-borne long-distance signaling molecules called hormones. Several hormones work together with signals arising from cells of the nervous system in this coordination. In this section, we will learn that one common function of the endocrine and nervous systems is to regulate the processes of glycogenolysis and gluconeogenesis so that glucose is made available to cells at all times.

Insulin Is a Key Regulator of Metabolism

The blood concentration of **insulin**, a hormone made by the pancreas, increases during the absorptive state and decreases during the postabsorptive state. Insulin regulates metabolism primarily by regulating the blood glucose concentration. It does this by promoting the transport of glucose from extracellular fluid into cells, where it can be used for metabolism. Glucose is a relatively large, polar molecule that cannot cross plasma membranes without the aid of a transporter protein. Insulin increases glucose uptake by binding to a cell-surface receptor and stimulating an intracellular signaling pathway. This pathway increases the availability of transport proteins called glucose transporters (GLUTs) in the plasma membrane (Figure 46.3). These GLUTs are located within preformed vesicles stored in the cytosol of cells. When these vesicles are stimulated, they fuse with the plasma membrane, making more GLUTs available to transport glucose into the cell. Consequently, insulin functions to lower the blood glucose concentration.

Insulin exerts its effects mainly on muscle cells and adipose cells, because these cells have insulin receptors in their plasma membranes. However, there are many types of GLUTs, and only one requires insulin for its activity, as described next.

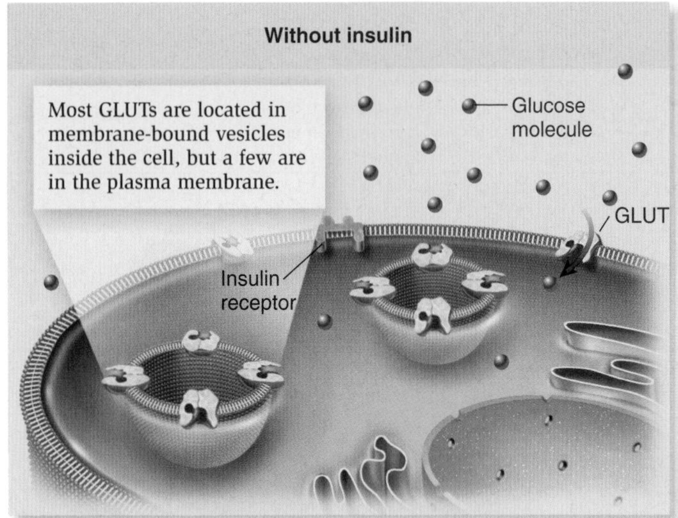

Without insulin
Most GLUTs are located in membrane-bound vesicles inside the cell, but a few are in the plasma membrane.
Glucose molecule
GLUT
Insulin receptor

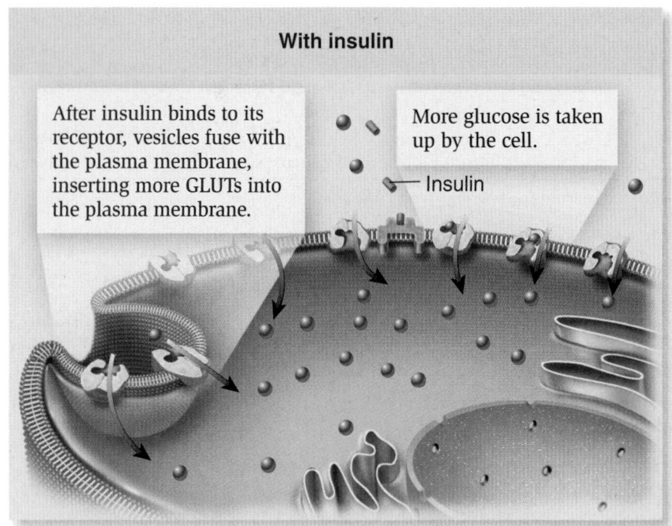

Figure 46.3 **The effect of insulin on glucose transport (GLUT) proteins.** The presence of insulin results in the fusion of intracellular vesicles containing GLUT proteins with the plasma membrane, where they facilitate glucose uptake into the cell.

Concept Check: *What benefit is there to having GLUT proteins premanufactured and stored in intracellular vesicles?*

BioConnections: *In what other contexts have you learned about the fusion of intracellular vesicles with a cell's plasma membrane (e.g., see Figure 5.23)?*

Genomes & Proteomes Connection

GLUT Proteins Transport Glucose in Animal Cells

All animal cells use glucose for energy and thus require transporters to move glucose across their plasma membranes. GLUTs are evolutionarily ancient, and their structure is very similar across phyla. For example, the coding sequence of one *Drosophila* GLUT gene is nearly 70% identical to that of one of the human GLUTs. In more closely related phyla, such as birds and mammals, GLUTs are even more similar, with up to 95% of their amino acid sequences being the same. These and other considerations indicate that the GLUTs arose by accumulated mutations of a common ancestral gene.

GLUTs in mammals make up a family of at least 14 related proteins that share similar structures but are expressed in different tissues. The different GLUTs vary in their ability to bind glucose. For example, some GLUTs have high affinity for glucose, and others have low affinity. High affinity means the protein binds glucose even at very low concentrations of glucose. Let's look at the properties of three mammalian GLUTs, named GLUT1, GLUT3, and GLUT4. Muscle and fat cells express the protein GLUT4, which has a low affinity for glucose, but one that is sufficient for the concentration of glucose normally found in blood. This is also the only GLUT protein whose movement to the plasma membrane requires the cell to be stimulated by insulin. Consequently, as the glucose concentration increases in the blood after a meal, insulin recruits more GLUT4 molecules from vesicles inside the cell to the plasma membrane of muscle and fat cells.

By comparison, GLUT1 and GLUT3 are found predominantly in the brain, where they act in concert to mediate the transport of glucose from blood vessels to the interstitial fluid of the brain, and from there into brain cells. GLUT1 and GLUT3 have much higher affinity for glucose than do other GLUT proteins. This means that neurons of the brain can transport glucose into their cytosol even when the concentration of glucose in the extracellular fluid is very low. In addition, insulin is not required for GLUT1 and GLUT3 to be present in the plasma membrane, unlike the situation for GLUT4. As a result of these properties of GLUT1 and GLUT3, neurons of the brain would still receive adequate energy for survival even if an animal's blood glucose concentration decreased significantly, perhaps due to disease or starvation.

Expressing multiple types of the same functional class of protein means that different parts of an animal's body can meet their own particular metabolic demands. Moreover, these demands may change during development. For example, high-affinity GLUTs such as GLUT1 and GLUT3 are present in high numbers in embryonic cells when there is a greater requirement for glucose but smaller numbers during other stages of life.

Because insulin, through its actions on GLUT4, is the key regulatory molecule that controls blood glucose concentrations, it is important to understand the regulation of insulin production and release, as described next.

The Blood Glucose Concentration Is Maintained Within a Normal Range

Like all homeostatic variables in an animal's body, blood glucose concentrations are controlled by a system of checks and balances. In the absorptive state, after a meal has been eaten, the blood glucose concentration increases; in the postabsorptive state, depending on how long it lasts, the concentration may become too low. What homeostatic mechanisms keep the blood glucose concentration within a normal range?

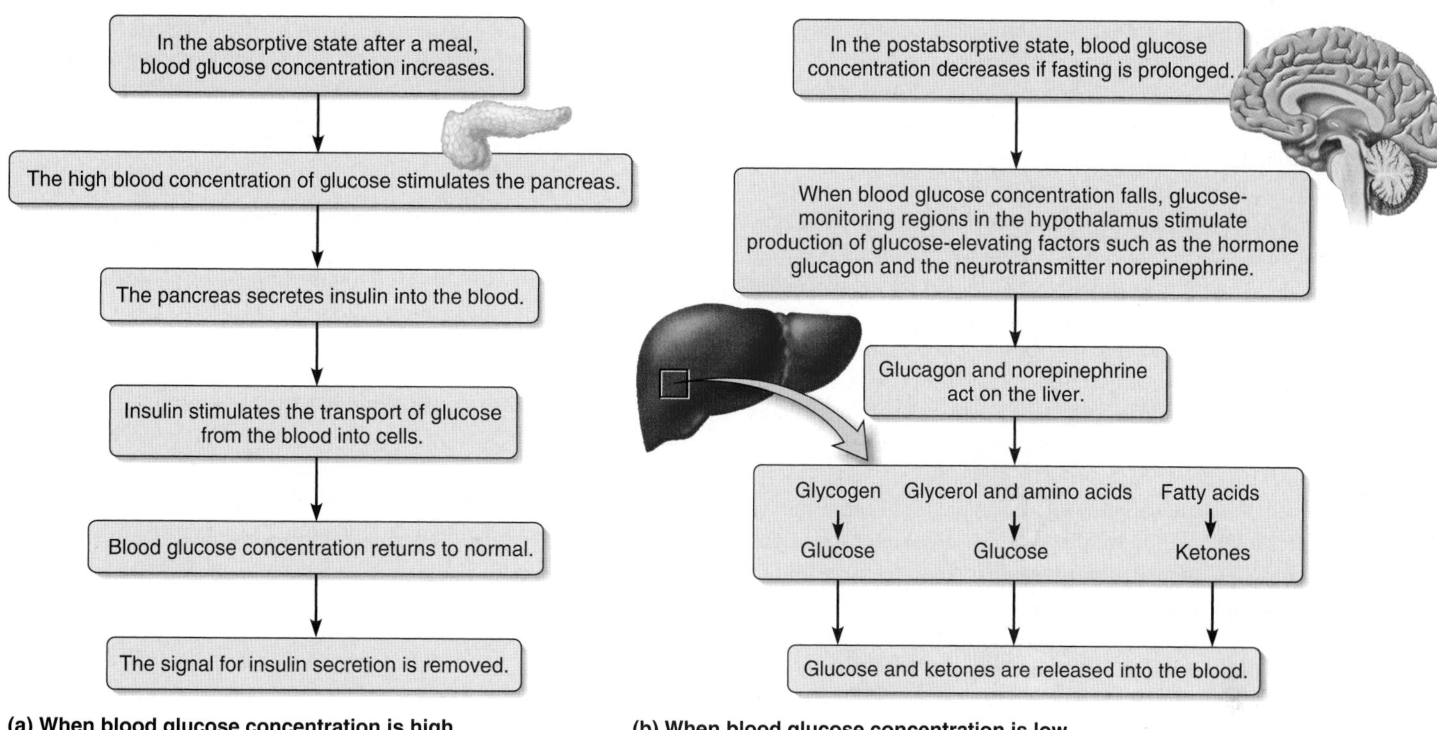

Figure 46.4 **Maintenance of blood glucose concentrations within a normal range.**

BIOLOGY PRINCIPLE Living organisms maintain homeostasis. Most animals cannot feed constantly and thus the ability to tap into stored reserves of energy allows them to maintain the concentration of glucose within a homeostatic range. This, in turn, ensures that homeostasis can be maintained, because cells are supplied with energy even when an animal has not recently ingested nutrients.

Concept Check: What are the sources of the glycerol, fatty acids, and amino acids used to make glucose?

Increased Glucose in the Absorptive State The primary factor controlling the secretion of insulin from the pancreas is the blood glucose concentration (**Figure 46.4a**). An increase in an animal's blood glucose concentration directly stimulates the cells of the pancreas to secrete insulin in proportion to the amount of glucose in the blood. Later, after insulin helps the cells of the body take up glucose, the resulting decrease in the blood glucose concentration removes the signal for insulin secretion. This is an example of a negative feedback loop, as described in Figure 40.13.

In addition to blood glucose concentration, inputs from the nervous system to the pancreas also play a role in the regulation of insulin secretion. During a meal, signals from the parasympathetic division of the autonomic nervous system (see the rest-or-digest response in Chapter 42) stimulate the secretion of insulin into the blood, ultimately reducing circulating glucose concentrations back to normal.

Decreased Glucose in the Postabsorptive State Several factors act in concert to prevent blood glucose from decreasing below the normal homeostatic range, even during a short fast. Otherwise, glucose could decrease so much—a condition called hypoglycemia—that despite the high-affinity GLUT1 and GLUT3 proteins in brain cells, there would not be enough glucose to keep brain cells alive.

If for any reason—such as a prolonged fast—the blood glucose concentration decreases below the normal homeostatic range for an animal, neurons that respond to changes in the extracellular concentration of glucose are activated within the hypothalamus (**Figure 46.4b**). Signals from the hypothalamus then stimulate the production of glucose-elevating factors. These include numerous hormones, notably **glucagon**, a protein hormone secreted by the pancreas that stimulates the processes of glycogenolysis, gluconeogenesis, and the processing of fatty acids to ketones in the liver. In addition, certain other hormones from various endocrine glands, as well as the neurotransmitter norepinephrine released from neurons of the sympathetic division of the autonomic nervous system, stimulate adipose tissue to release fatty acids into the blood. The fatty acids diffuse across plasma membranes and provide another source of energy for the synthesis of ATP. The overall effect is to increase the blood concentrations of glucose, fatty acids, and ketones during the postabsorptive period or during a prolonged fast.

More Energy Is Required During Exercise or Stress

We think of exercise as something humans do for fun or fitness, but in its broadest sense, **exercise** can be defined as any physical activity that increases an animal's metabolic rate. Generally, an animal becomes active to seek something, such as food, shelter, or a mate, or to elude something, such as a predator or a storm. The types of exercise animals engage in, therefore, can be quite varied. When a cheetah sprints after a small antelope, for example, the activity of both predator and prey is brief and intense, perhaps lasting only a few seconds.

By contrast, a tuna may never stop swimming, and a migrating bird may fly a hundred miles a day or more over a span of weeks.

For all types of exercise, including ones which result from an animal's behavioral responses to stress, nutrients must be available to provide the energy required for such things as skeletal muscle contraction, increased heart and lung activity, and increased activity of the nervous system. These energy-providing nutrients include glucose and fatty acids as well as the muscle's own glycogen. The liver supplies the blood with the glucose used during exercise by breaking down its glycogen stores and by gluconeogenesis. This occurs even in the absorptive state, and thus the blood glucose concentration increases above normal when an animal exercises at such times. In addition, an increase in adipose tissue lipolysis releases fatty acids into the blood, which provides an additional source of energy for the exercising muscle.

These events are mediated by the same hormones and nerves responsible for the regulation of the postabsorptive state. For example, inputs from the sympathetic division to the pancreas inhibit insulin secretion during exercise or acutely stressful situations. Consequently, glucose transport into muscle and fat cells is decreased, which tends to increase blood glucose concentrations (this is part of the fight-or-flight response described in Chapter 42). Because the brain does not depend on insulin for glucose transport across neuronal membranes, as just described, more of the body's supply of glucose is available to the brain at such times, while other cells can use fatty acids for energy. Therefore, the body uses all available forms of energy in response to fasting, exercise, and stress.

46.3 Energy Balance

Learning Outcomes:

1. Define the basal metabolic rate (BMR) and how it is measured.
2. List the factors that influence metabolic rate.
3. Describe the role of hormones in regulating metabolic rate, appetite, and body mass.

Animals have a wide range of metabolic demands that depend on numerous factors. Active animals, such as migrating birds, burn fuel at a greater relative rate than inactive animals, such as hibernating mammals. Likewise, juveniles typically burn fuel at a greater relative rate than mature animals. Recall that the amount of energy an organism uses in a given period of time to power its metabolic requirements is called its metabolic rate.

A fundamental characteristic of energy is that it can be neither created nor destroyed, but it can be converted from one form to another (the first law of thermodynamics; see Chapter 6). The breakdown of organic molecules liberates energy in their chemical bonds and transfers it to the bonds in ATP. This is the energy that cells use to perform various biological activities such as muscle contraction, active transport, and molecular synthesis. We refer to these functions as work. Not all of the energy is used to do work, however. Some of it appears as heat, which contributes to an animal's body temperature or is dissipated to the environment. In this section, we examine how metabolic rate is measured in animals; how metabolic rate is influenced by activity, digestion, and size; and how a balance is achieved between energy consumption and expenditure.

Metabolic Rate Can Be Measured by Calorimetry and Compared Among Different Animals

The standard unit of energy is the joule (J), but biologists have historically quantified the energy of metabolism in calories. A **calorie** (equivalent to 4.187 J) is the amount of heat required to raise the temperature of 1 gram of water 1 degree Celsius. Most biological activities, however, require much greater amounts of energy than a calorie, and consequently, the more common unit of measurement is the kilocalorie (1,000 calories, or **kcal**). (In food labeling, a Calorie with a capital C is also the same as kcal.) Biologists often measure and compare the metabolic rates of different animals to learn, for example, how some animals are capable of hibernating, how an animal's body temperature influences its metabolic rate, and how hormones and other factors alter an animal's metabolism.

The most common measure used to compare the metabolic rates of different species is the **basal metabolic rate (BMR)**. The BMR is called the metabolic cost of living, and most of it can be attributed to the routine functions of the heart, liver, kidneys, and brain. In the basal condition, the animal is at rest in the postabsorptive state and at a standard temperature. For endotherms, animals that generate their own internal heat through their metabolism and maintain a relatively narrow range of body temperatures, the standard temperature is within the range that causes the animal to neither generate heat (for example, by shivering) nor lose heat (for example, by perspiration). This temperature range is called an animal's thermoneutral zone. The BMR of ectotherms, animals whose body temperature changes with the environmental temperature, must be measured at a standard temperature for each species—one that approximates the average temperature that a species normally encounters. In this case, the term standard metabolic rate (SMR) is used instead of BMR, because the basal condition in ectotherms is harder to define than for endotherms. Because BMR and SMR apply only to animals in the postabsorptive state and at a standard temperature, any animal that has recently eaten or been active has a higher metabolic rate than its basal metabolic rate.

Two methods used to obtain a good estimate of BMR are direct and indirect calorimetry. **Direct calorimetry** was invented by French chemist and biologist Antoine Lavoisier in 1780 (**Figure 46.5**). In Lavoisier's procedure, an animal was placed in an enclosed, insulated chamber surrounded by ice. As the animal metabolized fuel, it generated heat, which dissipated into the chamber and melted the ice. The amount of water collected from an opening at the bottom of the chamber could be used to estimate the amount of heat generated by the animal. Today, direct calorimetry is measured using more sophisticated instrumentation, but the principle remains the same. It provides an accurate measure of metabolism, because metabolic rate and heat production are directly related. However, the method is not very practical, particularly with large animals.

The second method of measuring BMR, **indirect calorimetry**, is based on the principle that animals require oxygen to metabolize foodstuffs. The more fuel being metabolized—that is, the greater the BMR—the more oxygen must be consumed by the animal. By measuring the rate at which an animal uses oxygen, therefore, we can obtain a good estimate of BMR. Indirect calorimetry can also be used to compare the metabolic rates of an animal during rest and activity,

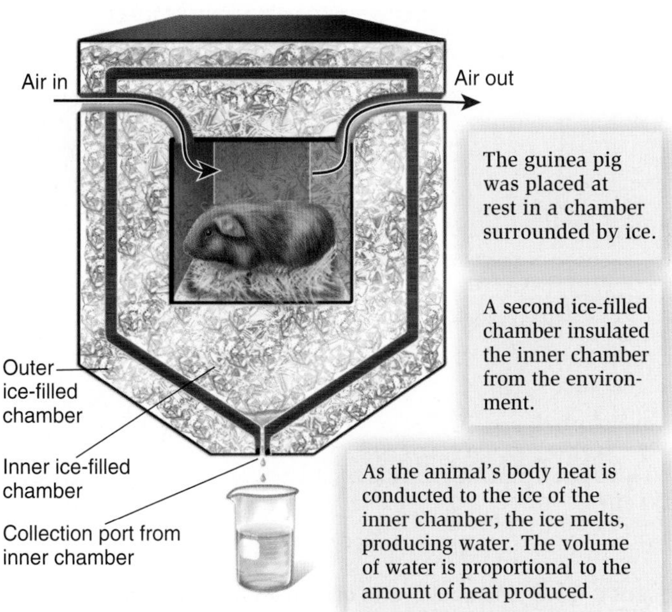

Figure 46.5 **Measuring BMR: Direct calorimetry.** Direct calorimetry determines an animal's basal metabolic rate, such as that of this guinea pig, by determining how much the heat given off by its body raises the temperature of a surrounding ice-filled chamber.

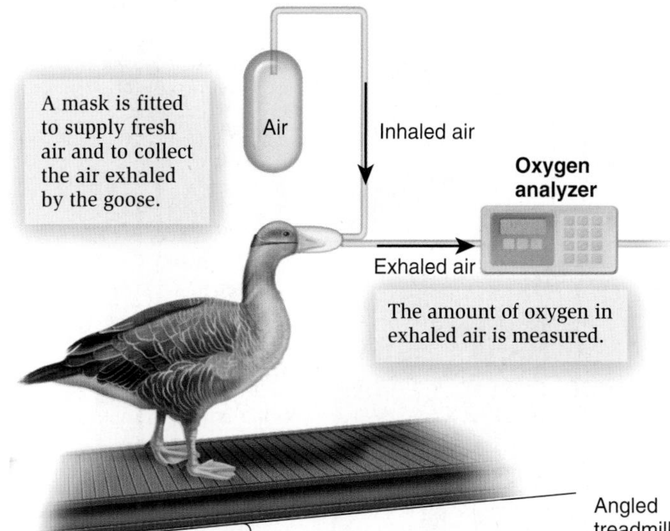

Figure 46.6 **Measuring BMR: Indirect calorimetry.** Many animals, such as this goose, can be trained to walk on a treadmill, which allows scientists to compare metabolism during rest and exercise. Oxygen consumption can be determined by sampling the air exhaled into a tightly fitting mask. One-way valves prevent inhaled and exhaled air from mixing.

BIOLOGICAL PRINCIPLE Biology is an experimental science. This figure illustrates how biologists can learn about general physiological principles using a variety of animal models. For example, many geese and other birds often fly for long periods at very high altitudes where oxygen availability is limited. Using experimental procedures like the one shown here allows scientists to understand how such animals can thrive under conditions that would be very challenging for humans.

Concept Check: If the air had been sampled from the mask just before starting the treadmill while the goose was not yet exercising, would this have been a good estimate of BMR?

when oxygen consumption increases (Figure 46.6). One limitation to this method is that a small percentage of fuel is metabolized anaerobically—that is, without oxygen—and thus indirect calorimetry underestimates actual metabolic rate.

Activity, Digestion, and Size Influence Metabolic Rate

Not all tissues in the body use oxygen and produce heat at the same rate. Some structures, such as skin, consume relatively little oxygen under resting conditions, whereas others, such as the brain, heart, and liver, have high rates of metabolism even when an animal is sleeping. Moreover, the metabolic rates of different tissues can vary depending on their activity. For example, the metabolism of the GI tract increases when food is being digested, and that of skeletal muscle increases during exercise (Figure 46.7).

The primary factor that increases metabolic rate is altered skeletal muscle activity. Even small increases in muscle contraction significantly increase metabolic rate; strenuous activity increases it even more. For example, the total daily expenditure of kilocalories may vary for a healthy adult human from approximately 1,350 kcal for a small person at rest to more than 7,000 kcal for a cyclist competing in the Tour de France. Changes in muscle activity also affect metabolic rate during sleep due to decreased muscle activity, and during exposure to cold temperatures due to increased muscle activity from shivering.

Eating and digesting food also increase the metabolic rate. Particularly in mammals that eat meat, this may increase metabolic rate (and associated heat production) by 10–50% for a few hours after eating. You may have noticed this **food-induced thermogenesis** after consuming a large meal, such as Thanksgiving dinner. Ingested protein—for example, turkey—produces the greatest effect, whereas carbohydrate and fat produce less. The increased heat is believed to result partly from the processing of the absorbed nutrients by the liver and from the energy expended by the GI tract in digestion and absorption. Food-induced thermogenesis is observed in nearly all vertebrates, but it is most notable in certain reptiles that eat large and infrequent meals. Another factor affecting metabolic rate is body size. In general, a large animal uses greater amounts of energy than does a small animal because the large animal has more mass and more cells, all of which consume fuel and generate heat. The metabolic rate and heat generation of an elephant are clearly greater than that of a mouse, for instance. However, when the metabolic rate of an elephant and a mouse are scaled to their respective body masses, we find that the energy expenditure per gram of body mass in a mouse is much higher than the comparable calculation in an elephant. **Mass-specific BMR** is the amount of energy expended per gram of body mass. Mass-specific BMR is a relative term that allows scientists to compare the metabolic rates among animals of different sizes. Research has shown that the relationship between mass-specific BMR and body

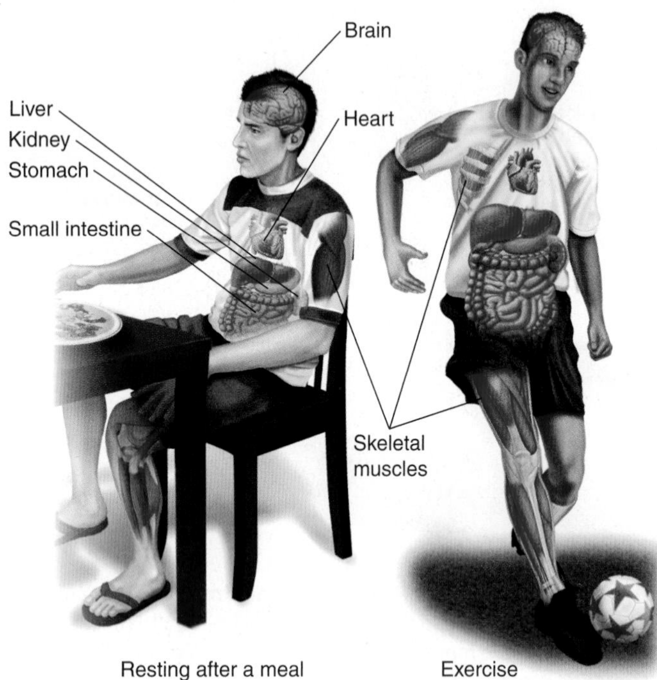

Figure 46.7 Changes in metabolic rate of selected structures in a mammal's body during different activities. Red areas are regions of high metabolism; blue areas are regions of low metabolism. Note that during exercise, skeletal muscle becomes more active, and areas associated with food digestion and absorption become less active. At all times, however, the heart, liver, and brain are highly active.

Concept Check: Why does metabolism decrease in the GI tract during exercise?

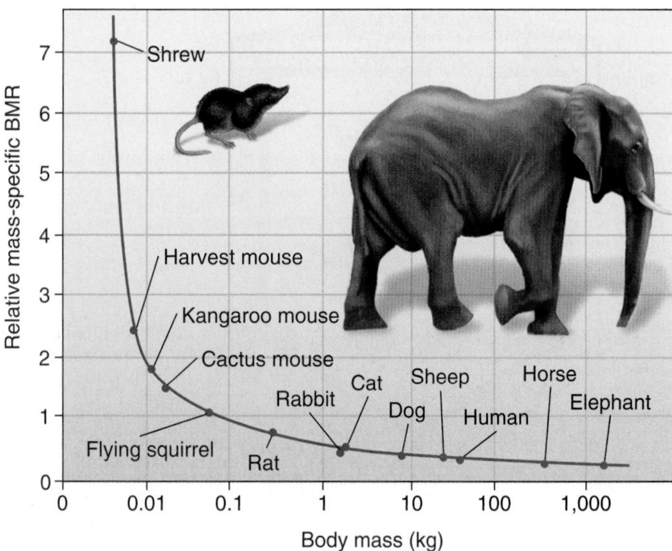

Figure 46.8 Metabolic rates of animals that differ in size. Metabolism can be scaled to body mass by measuring oxygen consumption and normalizing it to the animal's body mass (mass-specific BMR). Note that when expressed in this way, the mass-specific BMR of a shrew is higher than that of an elephant, even though the total oxygen consumption and heat output of the elephant would be much greater. The values on the *y*-axis are relative units of metabolism. Sizes of the animals shown are not to scale.

BIOLOGY PRINCIPLE Living organisms use energy. Many factors affect the relative energy usage by animals, including such things as an animal's life history, age, and as shown here, its relative body mass.

Concept Check: Can the relationship between body size and metabolic rate be used to propose hypotheses about metabolic rates of extinct animals?

mass follows an exponential curve (**Figure 46.8**). One possible explanation for this phenomenon is that because the ratio of an animal's surface area to its body mass is greater in smaller animals than in larger animals, smaller animals lose heat more rapidly than larger ones. According to this hypothesis, smaller animals must generate more heat per gram of body mass than larger animals to replace their heat loss. However, although this hypothesis appears to provide an explanation for the relationship between metabolism and body size in endotherms, it does not explain the observation that the same relationship exists in almost all animals, including ectotherms.

The smallest endotherms—hummingbirds, shrews, mice and other rodents, and some bats—face the special challenge of fueling their very high mass-specific BMRs. This becomes difficult and even impossible during cold months or any time when food is unavailable. Many such animals have evolved a strategy of lowering their internal body temperature to just a few degrees above that of the environment, a process called **torpor**. Torpor may occur on a nightly basis, while the animal continues to be active during the daylight hours, as in hummingbirds, or it can extend for months, in which case it is called **hibernation**. A reduction in body temperature decreases the metabolic demands of all body cells. BMR in a small hibernating rodent, for example, may decrease to less than 1% of what it would be at the normal body temperature of around 38°C. This allows the animal to conserve energy for remarkably long times. Ground squirrels of the genus *Spermophilus*, for example, may hibernate for up to 8 months. Torpor-like conditions are not unique to small vertebrates, however, as even some large animals such as bears enter extended periods of rest or hibernation during months when food is not available.

Hormones and the Nervous System Control Food Intake

When the daily amount of energy an animal consumes is equal to the amount of energy it expends, the animal's body weight remains stable. Tipping the balance in either direction causes weight gain or loss, that is, the total body mass increases or decreases. Normally, energy is stored in the form of fat in adipose tissue.

Body weight in an adult animal is usually regulated around a predetermined set point that differs among species. Body weight is maintained by adjusting caloric intake and energy expenditure in response to changes in body weight. This mechanism usually works very precisely in those animals in which it has been studied. For example, a mammal that eats less one day will eat more the next day to compensate for the previous day's deficit. Similarly, if an animal is overfed one

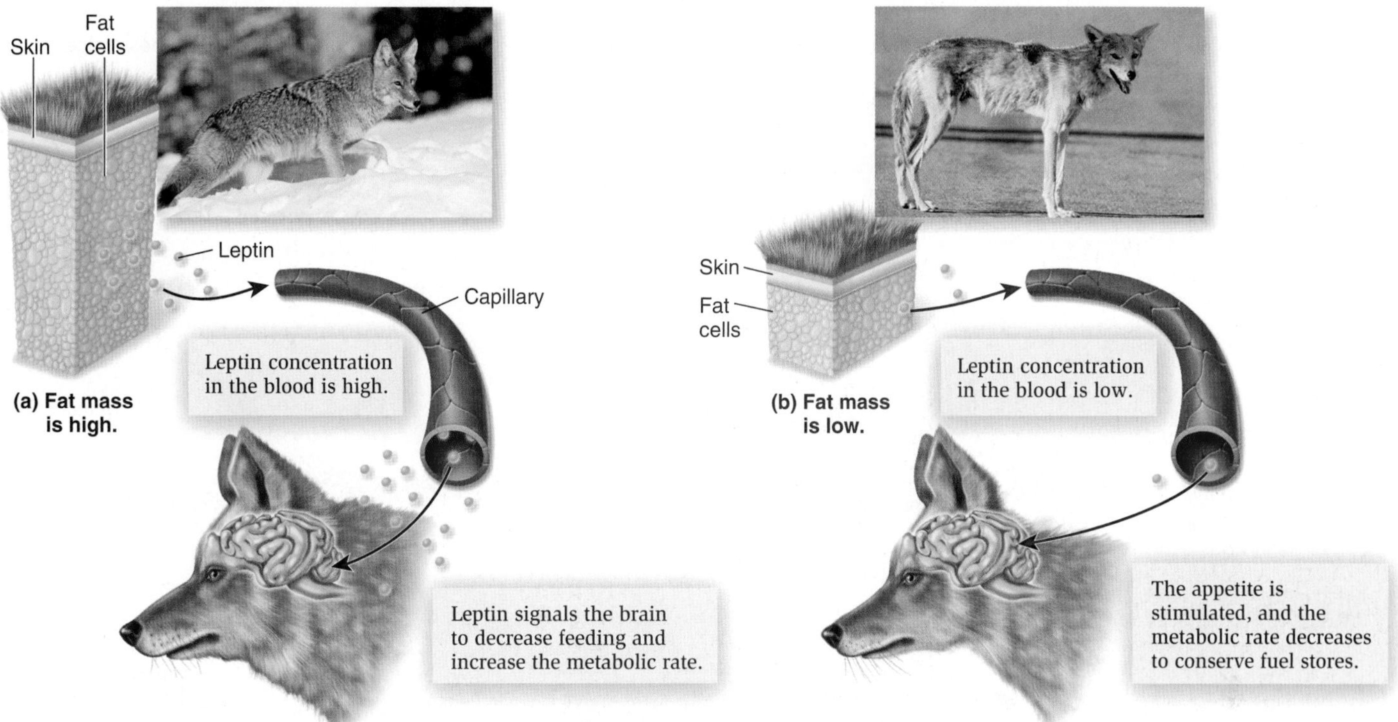

Figure 46.9 The role of leptin in regulating appetite and metabolic rate. In animals such as this coyote, changes in the blood leptin level result directly from changes in fat mass. Animals with more fat make more leptin.

day, it may eat less the next day. Scientists currently think that appetite and metabolism change when food intake is more or less than the amount required to maintain the body's set point. This phenomenon explains why some dieters initially lose weight easily and then become stuck at a plateau, because appetite increases and metabolic rate decreases to compensate for the weight loss. Conversely, it also helps explain why some very thin people have difficulty gaining weight.

Short-term control of feeding generally involves a feeling of **satiety**, that is, fullness. As an animal's stomach and small intestine stretch to accommodate food, nerves send inhibitory signals from these structures to the appetite center in the hypothalamus. At the same time, the stomach and small intestine release hormones into the blood that reach the hypothalamus and suppress appetite. These satiety signals remove the sensation of hunger and set the time period before hunger returns again.

Long-term control of food intake is mediated by many different brain molecules, by hormones, and by emotional state, particularly in humans. One hormone that has received considerable attention in recent years for its ability to control appetite and metabolic rate is **leptin** (from the Greek *leptos*, meaning thin). Leptin has been identified in all classes of vertebrates but has been most extensively studied in mammals. In these animals, leptin is produced by adipose cells in proportion to fat mass: As more fat is stored in adipose cells in the body, more leptin is secreted into the blood. Leptin acts on the hypothalamus to decrease appetite and increase metabolic rate (**Figure 46.9a**). In this way, the brain is made aware of how much fat is stored in the body at all times, and it can adjust appetite and metabolic rate appropriately if fat stores decline or increase. If an animal fasts for a period of time, its adipose cells shrink as they release their stored fat into the blood. The decrease in leptin secretion resulting from the decreased adipose mass results in a decrease in BMR and an increase in appetite. This may be the true evolutionary significance of leptin, namely that its disappearance from the blood lowers the BMR, consequently prolonging life during periods of starvation (**Figure 46.9b**).

Leptin was not discovered until 1994, but its existence was postulated decades before that by the pioneering work of Douglas Coleman, who investigated the nature of mutations in mice that result in obesity.

FEATURE INVESTIGATION

Coleman Revealed a Satiety Factor in Mammals

For many years, scientists wondered how most animals appeared to regulate their body mass around a predetermined set point, despite fluctuations in the food supply. They postulated that other parts of the body somehow communicated with the brain to signal when energy stores were above or below normal. In the 1970s, Canadian-American researcher Douglas Coleman tested this hypothesis in an experiment involving parabiosis, the surgical connection of the abdominal walls of two animals, such that the blood supply from one animal intermixes with that of the other.

Coleman used two strains of mice called ob and db mice, carrying different mutations that resulted in inherited forms of obesity, the excessive accumulation of body fat (an example of an ob mouse

Figure 46.10 **Coleman's parabiosis experiments revealed a satiety factor in wild-type mice that was absent in genetically obese mice.**

HYPOTHESIS Body weight is controlled by a factor that circulates in the blood. This factor is absent in strains of mice that have an inherited form of obesity.

KEY MATERIALS Two different strains of genetically obese mice, normal (wild-type) mice.

Experimental level | **Conceptual level**

1 Surgically connect the abdominal walls of an obese and normal (wt; wild-type) mouse. After a few days, blood vessels from each mouse cross to the other mouse. Monitor changes in body weight. Note: 2 different strains of obese mice were tested, called ob and db mice.

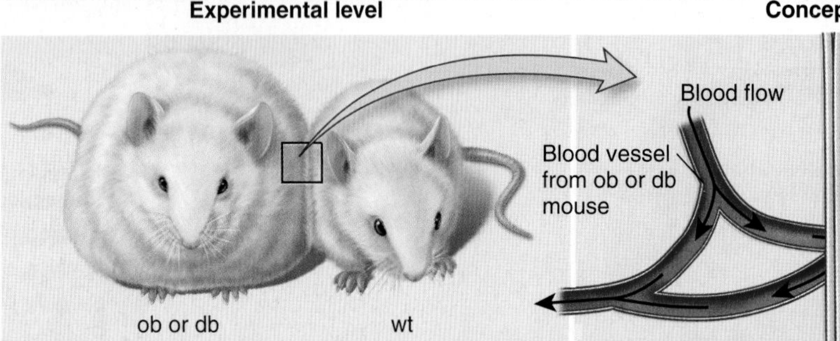

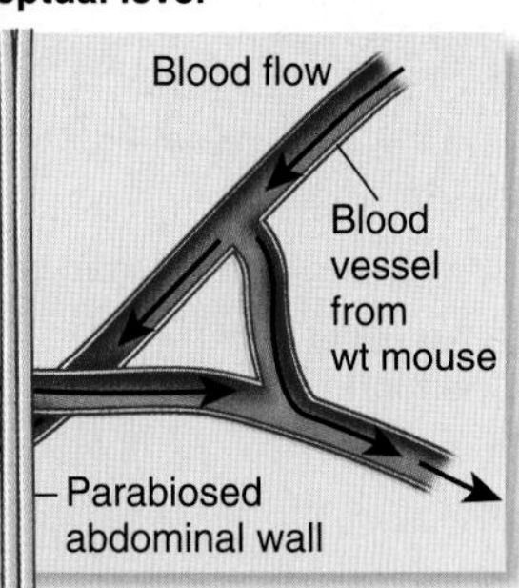

2 Feed mice a normal diet for several weeks, then visually inspect and weigh each pair.

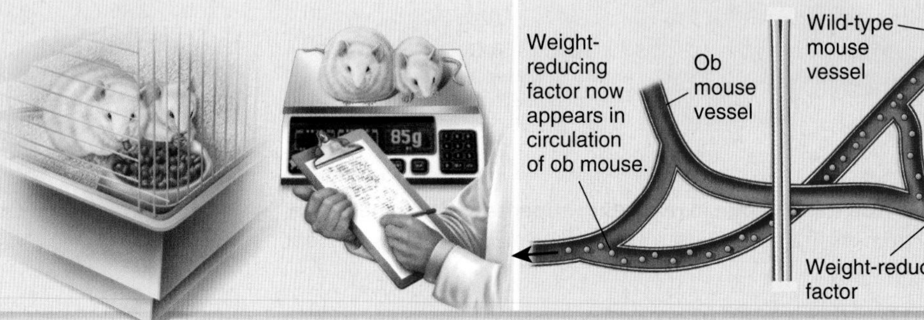

3 **THE DATA**

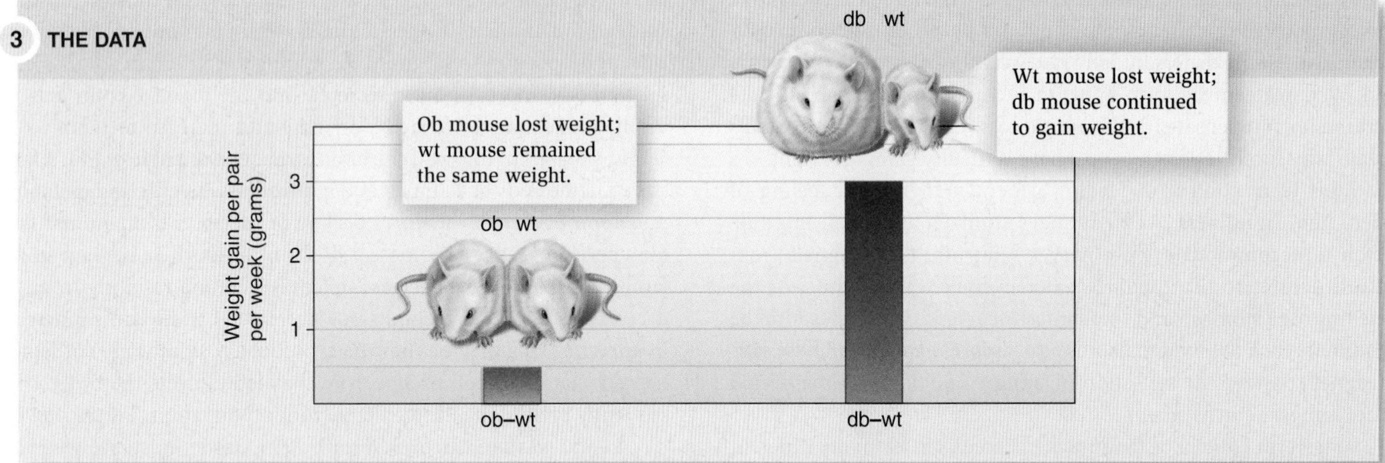

4 **CONCLUSION** Wild-type mice secrete a blood-borne factor that reduces body weight. The factor is absent from ob mice but present in db mice. Ob mice retain the ability to respond to the factor, unlike db mice, which cannot respond to it.

5 **SOURCE** Coleman, D.L. 1973. Effects of parabiosis of obese with diabetes and normal mice. *Diabetologia* 9:294–298.

is shown in the chapter-opening photo). Coleman first connected a wild-type (wt) mouse, one that lacked these mutations, with either an ob mouse or a db mouse, as shown in **Figure 46.10**. He discovered that when the circulatory system of the ob mouse was in contact with that of the wild-type mouse, the ob mouse ate less and gained less weight than usual. This suggested that the blood of the wild-type mouse contained a circulating factor that signals the brain when an animal has sufficient fat stored in its body and adjusts appetite accordingly. The ob animal apparently lacked this factor, but when exposed to it through the wild-type mouse's circulation, it responded in the appropriate way. The wild-type mouse of the parabiosis pair apparently retained a sufficient amount of the factor in its blood, because it maintained its body weight at a normal level.

Coleman noticed, however, that a db mouse continued to gain weight at an abnormally high rate even when its circulatory system was in contact with that of a wild-type mouse. In this case, the wild-type animal actually lost weight while the db animal remained obese. Coleman concluded that the db mouse must produce the same factor as the wild-type mouse, but for some reason, it was unable to respond to it. The wild-type mouse that was parabiosed to the db mouse lost weight because it received the factor from the db mouse, in addition to having its own supply of the factor. Thus, whether the factor was absent as in the ob mouse, or present but unable to function as in the db mouse, the resulting phenotype was the same—obesity.

In 1994, American molecular biologist Jeffrey Friedman and coworkers at Rockefeller University identified this circulating factor as the hormone leptin. The ob mice were found to be homozygous for a mutation in the leptin gene, which produced an inactive leptin molecule, whereas db mice produced leptin but did not respond to it. In fact, the db mice were found to produce even greater amounts of leptin than wild-type mice, which explained why the wild-type mouse in Coleman's experiments lost weight when parabiosed with a db mouse. Friedman and others later showed that adipose cells produce leptin in direct proportion to the total fat mass of an animal, as stated earlier. Biologists now know that all classes of vertebrates produce leptin.

At first, the work of Coleman and Friedman generated considerable excitement that leptin might be useful to treat obesity in humans, but this has thus far proven difficult. Why? Recent research has revealed that most obese humans are more like the db mice than the ob mice. That is, they produce leptin but fail to respond adequately to it, and therefore, simply increasing blood concentrations of leptin may not have a significant effect on body weight. However, other studies have shown that leptin normally acts in nonobese humans in a manner much like it does in wild-type rodents. As noted in the chapter introduction, researchers have identified rare individuals in whom leptin is not produced due to a mutation in the leptin gene. These individuals are extremely obese and respond well to injections of leptin, losing considerable weight. The body weight disorders of such individuals, therefore, are reminiscent of ob mice.

Experimental Questions

1. What observation led to the experiments conducted by Coleman?
2. What was the hypothesis tested by Coleman, and how did he test it?
3. How did the experimental linking of the bloodstreams of the wild-type mice and the mutant mice affect the body weight of both strains?

46.4 Regulation of Body Temperature

Learning Outcomes:

1. Provide examples of how changes in temperature affect chemical reactions, other protein functions, and membrane structure.
2. List and define the four terms used to categorize organisms based on their source of heat and ability to maintain body temperature.
3. Identify the four main mechanisms animals use to exchange heat with the environment.
4. Define several mechanisms by which animals can alter the rate of heat gain or loss.
5. Describe the mechanisms and adaptations demonstrated by animals to adjust to temperature extremes.

As we have seen, metabolic rate is linked to body temperature. In this section, we discuss why body temperature is important for the health and survival of all animals, and consider mechanisms by which animals gain or lose heat.

Temperature Affects Chemical Reactions, Protein Function, and Plasma Membrane Structure

Most animals can survive only in a relatively narrow range of temperatures. Temperature has an effect on animals' bodies in three main areas: chemical reactions (and the enzymes that catalyze them), protein function, and membrane structure. First, chemical reactions depend on temperature. Heat accelerates the motion of molecules, so as an animal's temperature increases, the rates at which the molecules in its body move and contact each other also increase. Consequently, the rate of most chemical reactions in animals doubles or even triples for every 10°C increase in body temperature. In addition, enzymes, which catalyze many reactions in the body, including those involved in metabolism, have an optimal temperature range for their maximal catalytic function. Low temperatures slow down chemical reactions, making it harder for an animal to remain active and carry out internal functions such as digestion, reproduction, and immunity. The latter is particularly important, as many vertebrates become susceptible to disease when their body temperatures are reduced for long periods.

A second effect of unusually high temperature is that it causes many proteins to become denatured; that is, they lose the three-dimensional structure that is crucial to their ability to function properly. This occurs because the bonds that form tertiary and quaternary protein structures result from weak interactions, such as hydrogen bonds, and can be disrupted by heat (refer back to Figure 3.16). Denaturation of enzymes is especially serious because of the major role they play in metabolism. Most animals have an upper limit of body temperature at which they can survive. To start with, most mammals have a resting body temperature of 35–38°C (95–100°F). In humans, a body temperature of 41°C (106°F) causes loss of protein function and breakdown of the nervous system, and a body temperature of 42–43°C (107–109°F) is fatal. Birds, which have slightly higher resting body temperatures than mammals (approximately 40–41°C [104–106°F]), cannot survive at body temperatures above 46–47°C (115–117°F). At sustained environmental temperatures greater than 50°C (122°F), nearly all animals die.

A third effect of temperature is that heat alters the structures of plasma and intracellular membranes. At low temperatures, membranes become less fluid and more rigid, primarily due to changes in membrane phospholipids. Rigid membranes are less able to perform biological functions, such as transporting ions and binding extracellular molecules to receptors on the membrane surface. Alternatively, if the temperature becomes too high, membranes can become leaky.

Animals can tolerate extreme cold better than extreme heat. For example, some animals can survive after freezing and thawing. This is normally dangerous because ice crystals form inside cells and rupture membranes. Also, the ice forms from water in the cells' cytosol, which dehydrates the cells. However, many ectotherms, including certain insects such as the woolly caterpillar, a few species of amphibia such as several types of frogs, and a very small number of reptiles such as the painted turtle, can prevent crystal formation in their cells. They do this by responding to ice on their skin surfaces with an enormous outpouring of glucose from the liver. The glucose enters the blood and cells, acting like antifreeze and lowering the freezing point so that the cells do not completely freeze solid (see Chapter 2 for a discussion of the colligative properties of water). Other regions of the body that are less critical, such as the lumen of the stomach and bladder, do freeze. These animals can have 65% or more of their bodies completely frozen for long periods, only to thaw during warm periods without harmful effects. The glucose is reabsorbed by the liver at that time. An alternative strategy to cope with extremely cold temperatures is seen in certain fishes inhabiting Arctic or Antarctic waters (refer back to Figure 23.8c). Such fishes do not freeze like the terrestrial vertebrates just described, but instead contain large amounts of highly specialized antifreeze proteins in their blood that stop ice crystal formation.

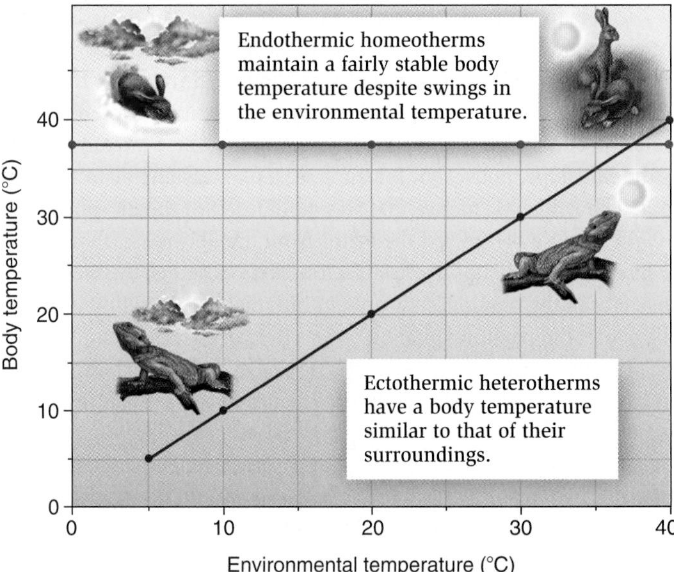

Figure 46.11 **Body temperature and environmental temperature in homeotherms and heterotherms.** Homeotherms maintain stable body temperatures across a wide range of environmental temperatures. Heterotherms, by contrast, have body temperatures that may vary widely, such as the ectotherm shown here.

Concept Check: *What thermoregulatory categories do humans fit into?*

Ectotherms and Endotherms May Have Fluctuating or Stable Body Temperatures

In the past, animals were classified as either "cold blooded" or "warm blooded." Cold-blooded animals were considered to be those that require an external heat source such as sunlight to warm themselves. By contrast, warm-blooded animals were considered to be those that used internally generated heat to maintain their body temperature. These terms are misleading and incorrect, however, because many cold-blooded animals generate considerable heat by exercising their skeletal muscles. Indeed, many have a body temperature during daylight hours that is at least as warm as that of "warm-blooded" animals like birds and mammals. Moreover, many so-called warm-blooded animals spend portions of their lives with greatly reduced body temperatures that are close to that of their environment.

Biologists now classify animals according to both their source of heat and their ability to maintain body temperature. Recall that ectotherms depend on external heat sources to warm their bodies, whereas endotherms use their own metabolically generated heat to warm themselves. **Homeotherms** maintain their body temperature within a narrow range, whereas **heterotherms** have body temperatures that vary with the environment (**Figure 46.11**). Most animals can be categorized as either endotherms or ectotherms *and* as either homeotherms or heterotherms. Generally, birds and mammals are endothermic and homeothermic. Other vertebrates and most invertebrates are ectothermic and heterothermic.

Not all animals, however, can be neatly classified into two categories at all times. Hibernating mammals, for example, are endotherms. They are homeothermic at times, but during the winter, their body temperature drops dramatically as their metabolism slows to conserve energy for the winter. Hibernators behave like heterotherms during the transition from fall to winter and again from winter to spring. During the winter, however, they are again homeothermic except for brief periods of arousal, but at a lower body temperature than at other times of year. Similarly, a fish swimming in deep ocean waters is an ectotherm but also homeothermic because the temperature of the water—and therefore of its body—remains essentially constant. Fishes that live in waters with fluctuating temperatures, by contrast, are ectothermic and heterothermic.

Even endothermic homeotherms do not have truly constant body temperatures. They have a narrow range of body temperatures that increases or decreases slightly in extreme climates, during exercise, or during sleep. The important feature is that birds and mammals can quickly adjust the body's mechanisms for retaining or releasing heat such that body temperature remains within the required narrow range. This provides the advantage that the body's chemical reactions are at optimal levels even when the environment imposes extreme challenges. The metabolic rate of a resting mammal, for example, is roughly six times greater than that of a comparably sized reptile. A suddenly awakened mammal is instantly capable of intense activity even on a winter day, but an icy-cold reptile could be at the mercy of a predator because of the time required to warm itself in order to flee.

Endothermy does have three major disadvantages, however. First, to produce sufficient heat by metabolic processes, endotherms must consume larger amounts of food to provide the nutrients used by cells in the formation of ATP, during which heat is generated. Small endotherms with high BMRs, such as shrews (see Figure 46.8), must eat almost continuously and may die if deprived of food for as little as a day. By contrast, many ectotherms, such as snakes, can go for weeks without eating. Second, endotherms have a greater risk of hyperthermia, or overheating, during periods of intense activity, even in cold weather. Third, because lowering body temperature requires the evaporation of bodily fluids (and thus a need for replenishment of fluid), endotherms are often restricted to environments where water is plentiful.

Animals Exchange Heat with the Environment in Four Ways

The surface of an animal's body can lose or gain heat from the external environment via four mechanisms: radiation, evaporation, convection, and conduction (Figure 46.12).

Radiation is the emission of electromagnetic waves by the surfaces of objects. The rate of emission is determined by the temperature of the radiating surface. Thus, if the body surface is warmer than the environment, the body loses heat at a rate that depends on the temperature difference. If the outside temperature is warmer than body temperature, the body gains heat, for instance from sunlight. We can observe radiated heat from an animal's body with imaging devices that detect infrared light (Figure 46.13).

Evaporation is the conversion of water from the liquid to the gaseous state. Animals can lose body heat through evaporation of water from the skin and membranes lining the respiratory tract, including the surface of the tongue. A large amount of energy in the form of heat is required to transform water from liquid to gas. Whenever water vaporizes from the body's surface, the heat required to drive the process is conducted from the surface, thereby cooling the animal.

Convection is the transfer of heat by the movement of air or fluid next to the body. For example, the air close to an endotherm's body is heated by conduction. Because warm air is less dense than cold air, the warm air near the body rises and carries away heat by convection. Convection is aided by creating currents of air around an animal's body. Humans do this by sitting near fans, but other animals can create cooling air currents by other means such as when an elephant waves its ears or a bat flaps its wings.

In **conduction**, the body surface loses or gains heat through direct contact with cooler or warmer substances. The greater the temperature difference, the greater is the rate of heat transfer. Different materials have different abilities to absorb heat, however. As we saw in Chapter 2, water has a higher specific heat than air, meaning that at any temperature, water will retain greater amounts of heat than will air. Consequently, aquatic animals in water that is 10°C lose considerably more heat in a short time than terrestrial animals lose in air that is 10°C. Indeed, on a hot day, terrestrial animals can lose heat efficiently by immersing themselves in cooler water. An animal's body surface area plays an important role in the rate of heat conduction across its body surface.

The four processes of heat transfer just described can be regulated in animals, such that heat is retained within the body at some times and lost at other times, as we see next.

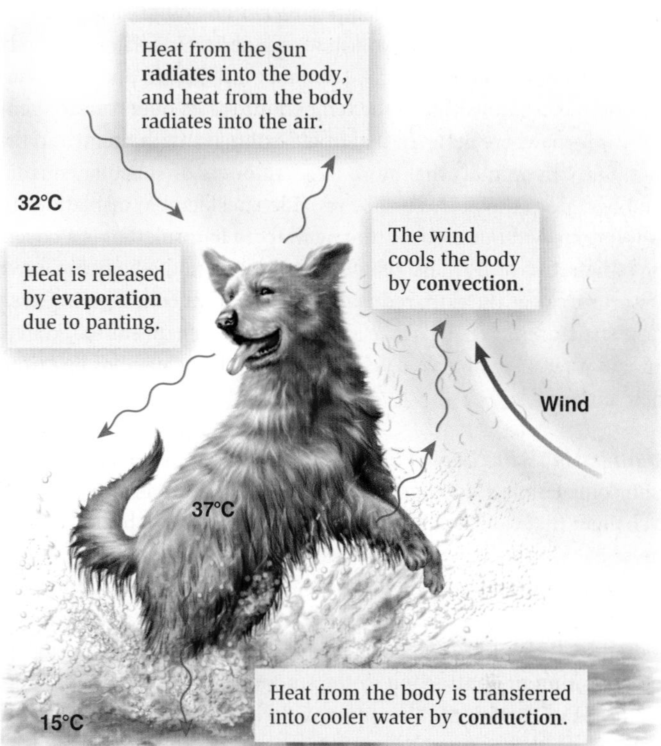

Figure 46.12 **Types of heat exchange.** The four ways in which animals exchange heat with the environment are radiation, evaporation, convection, and conduction.

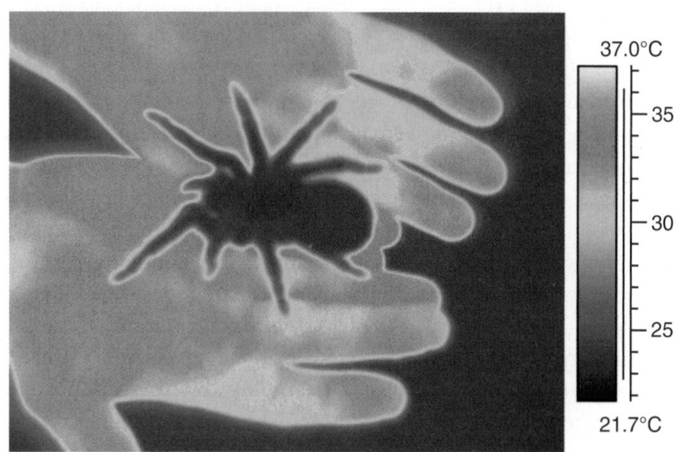

Figure 46.13 **Visualization of heat exchange in an ectotherm and an endotherm.** Thermal-imaging cameras can detect heat radiated from an animal's body. Note the warm skin of the endotherm (the human) and the cold skin of the ectotherm (the tarantula), even though both animals are at the same environmental temperature.

Several Mechanisms Can Alter Rates of Heat Gain or Loss in Endotherms

For purposes of temperature control, think of an endotherm's body as a central core surrounded by a shell consisting of skin and subcutaneous (just below the skin) tissue. Depending on the species, the temperature of the central core of endotherms is regulated at approximately

37–40°C (97–104°F), but the temperature of the outer surface of the skin varies considerably. If the skin were a perfect insulator, the body would never lose or gain heat by conduction. The skin does not insulate completely, however, so the temperature of its outer surface generally is somewhere between that of the external environment and the core. Only in animals that store large amounts of subcutaneous fat (blubber) does the body surface provide considerable insulation. In endotherms without blubber, the main form of insulation is a covering of hair, fur, or feathers, which traps heat from the body in a layer of warm air near the skin, reducing heat loss from conduction. Given these structures, then, let's take a look at four mechanisms that different endotherms use to regulate how much heat is gained or lost from their surface.

Changes in Skin Blood Flow Rather than acting as an insulator as in some animals, in many endotherms the skin functions as a heat exchanger that can be adjusted to increase or decrease heat loss from the body. Skin surface blood vessels dilate (widen to increase blood flow) on hot days to dissipate heat to the environment, and they constrict (get narrower to reduce blood flow) on cold days to retain body heat (Figure 46.14). Signals from the nervous system regulate the relaxation or contraction of the smooth muscles that control the opening and closing of these blood vessels. Diving birds and diving mammals are good examples of animals that use this mechanism. Ducks, seals, and walruses dramatically reduce the amount of blood flowing to the skin when they dive in cold waters. This allows them to retain body heat that would otherwise be conducted into the water. In many terrestrial endotherms, certain areas of skin play a more prominent role in heat exchange than others—such as an elephant's ears—so skin temperature varies with its location in the body.

Countercurrent Exchange Many endotherms and ectotherms regulate heat loss to the environment through **countercurrent exchange**, in which heat is transferred between fluids flowing in opposite directions. Countercurrent exchange regulates heat loss to the environment by returning heat to the body's core and keeping the core much warmer than the extremities. In endotherms, countercurrent exchange occurs primarily in the extremities—the flippers of dolphins, for example, or the legs of birds and certain other terrestrial animals (Figure 46.15a). As warm blood travels from the core through arteries down a bird's leg, heat moves by conduction from the artery to adjacent veins carrying cooler blood in the other direction (Figure 46.15b). By the time the arterial blood reaches the tip of the leg, its temperature has dropped considerably, reducing the amount of heat lost to the environment and returning the heat to the body's core.

Ectotherms such as many fishes use countercurrent exchange to warm their muscles. As the swimming muscles become active, they generate heat from metabolism that warms the blood in the veins leaving the muscles. The veins are in close contact with nearby arteries bringing fresh, oxygen-rich blood from the gills. In tuna, as in other ectotherms, arterial blood is cold. Blood entering the gills comes in contact with cold seawater, and its temperature rapidly adjusts to that of the seawater. As the cold blood reaches the muscles, however, heat from the warm veins *leaving* the muscles is conducted from the veins into the nearby arteries *entering* the muscle. In this way, the heat generated by the muscles is returned to the muscles rather than being lost from the gills. As warm temperatures stimulate the rate of chemical reactions, the muscles are able to operate more efficiently.

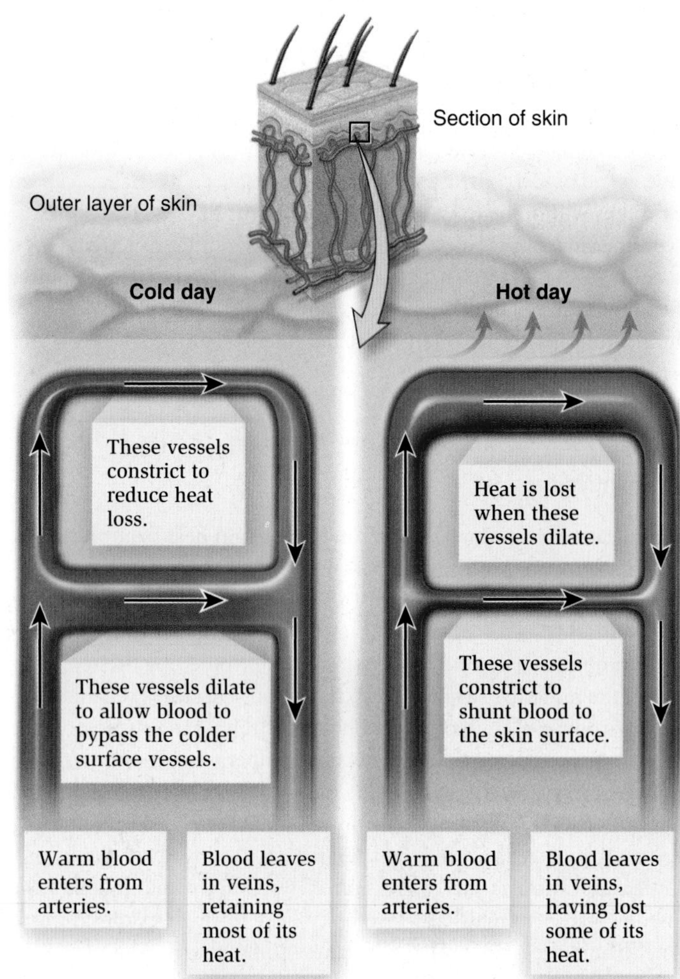

Figure 46.14 Regulation of heat exchange in the skin. As shown in this schematic illustration, the skin functions as a variable heat exchanger. The arrows in the blood vessels indicate direction of blood flow.

Evaporative Heat Loss Recall that some animals can lose body heat through evaporation of water from the skin and membranes lining the respiratory tract. Heat exchange in some mammals can be regulated by changing the rate of water evaporation through perspiration. Nerves to the sweat glands stimulate the production of sweat, a dilute solution containing sodium chloride. The most important factor determining evaporation rate—and therefore heat loss—is the water vapor concentration, or humidity, of the air. The discomfort you feel on a humid day is due to the failure of evaporation. Your sweat glands continue to secrete, but the sweat simply remains on your skin with body temperature remaining elevated, especially during exercise.

In endotherms that lack sweat glands, such as birds, or those that have very few glands, such as dogs and cats, panting (short, rapid breaths with the mouth open) promotes evaporation of water from the tongue surface. Panting has advantages over sweating, because no salt is lost, and panting provides the air current that promotes heat exchange by convection. However, the surface area of the mouth

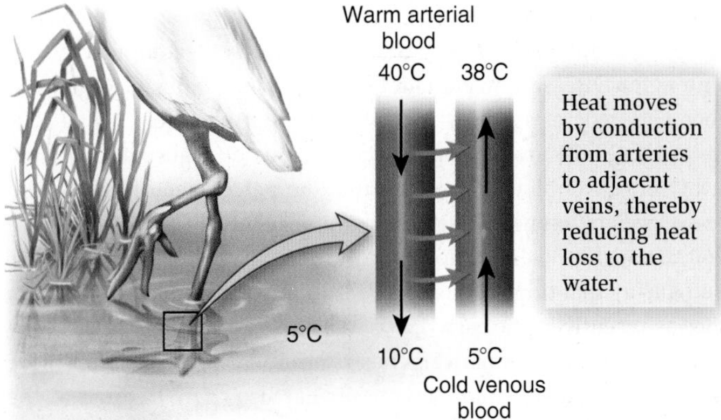

(a) Countercurrent heat exchange in the leg of an endotherm

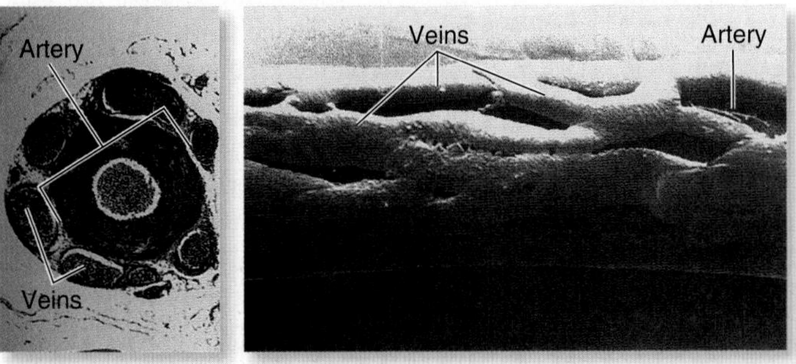

(b) Cross section and surface view of veins covering an artery

Figure 46.15 Countercurrent exchange. (a) Countercurrent exchange retains heat in the leg of an endotherm such as this bird. Black arrows in vessels indicate direction of blood flow. **(b)** An SEM of the arrangement of veins surrounding an artery in a bird's leg. The artery is almost completely covered by overlying veins, allowing efficient heat exchange between the vessels.

and tongue is relatively small, which limits the rate at which heat can be eliminated. Interestingly, many reptiles, though not endotherms, also pant on hot days, suggesting that panting evolved prior to endothermy.

Behavioral Adaptations Behavioral mechanisms can also alter heat loss by radiation, conduction, and evaporation. Two such behaviors involve changing exposed surface area and changing surroundings. On hot days, birds may ruffle their feathers and raise their wings, whereas many mammals will reduce their activity and spread their limbs; these postural changes increase the surface area available for heat transfer. Terrestrial endotherms seek shade, partially immerse themselves in water, or burrow into the ground when the sun is high. Pigs, which lack sweat glands, roll in the mud to cool down. Animals that neither sweat nor pant can still benefit from evaporative heat loss. The evaporation of fluids deposited on the body surface by licking the skin or splashing the skin with water also draws heat from the body.

Similarly, animals respond to cold temperatures with numerous behavioral adaptations. Huddling in groups, curling up into a ball, hunching the shoulders, burying the head and feet in feathers, and similar maneuvers reduce the surface area exposed to a cold environment and decrease heat loss by radiation and conduction. Changing environments is also a common strategy for coping with cold. Migration from cold to warmer regions occurs in numerous species of birds and mammals.

Muscle Activity and Brown Adipose Tissue Metabolism Increase Heat Production

We have discussed how heat is gained or lost to the environment and how heat can be retained by reducing blood flow to the skin on a cold day. Body temperature, however, is a balance between these factors and heat production. Changes in muscle activity constitute a major control of heat production for temperature regulation in endotherms.

When an endotherm is in its thermoneutral zone, no significant adjustments are necessary to maintain core body temperature. When exposed to conditions below the thermoneutral zone, however, core body temperature begins to decrease. The primary response to decreasing temperatures is to reduce the flow of blood to regions that permit conduction of heat. If this does not adequately reduce heat loss, skeletal muscle contraction is increased. This may lead to shivering, which consists of rapid muscle contractions without any locomotion. Virtually all of the energy liberated by the contracting muscles appears as internal heat, a process known as **shivering thermogenesis**. Many birds that remain in cold climates during the winter shiver almost continuously.

In many mammals, chronic cold exposure also induces **nonshivering thermogenesis**, an increase in the metabolic rate and therefore heat production that is not due to increased muscle activity. Nonshivering thermogenesis occurs primarily in **brown adipose tissue** (also called brown fat), a specialized tissue in small mammals such as hibernating bats, small rodents living in cold environments, and many newborn mammals, including humans. Brown adipose tissue is responsive to hormones and signals from the nervous system, which are activated when body temperature decreases. Unlike the adipose tissue discussed previously, which stores energy in the form of fat, brown adipose tissue metabolizes fat and generates heat as a by-product.

Endotherms Can Adapt to Chronic Changes in the Temperature of Their Environments

Although many animals—notably deep-sea-dwelling fishes and invertebrates—spend their lives in relatively unchanging environmental temperatures, most others do not. Such animals, therefore, must adapt to environmental changes using the mechanisms we have discussed in this section. In some cases, however, long-term exposure to a challenging environment, either very hot or very cold, results in a fine-tuning of the adaptive mechanisms that persist for as long as the animal lives in that environment. This process of **acclimatization** occurs particularly well in humans, who have more sweat glands than any known mammal. Increases in the amount and rate of sweat production and a decrease in the temperature threshold for initiating sweating, for example, can acclimatize an animal such as ourselves to chronic high temperatures. A person newly arrived in a hot climate initially has trouble coping. Body temperature increases, and too much activity can lead to a breakdown in the body's temperature-regulating systems. After several days to weeks, however, acclimatization occurs. Body temperature stabilizes, and the person finds it much easier to function at a normal activity level. Body temperature does not rise as much as it did initially upon moving to the hot climate because sweating begins sooner and the volume of sweat produced is greater. Long-term exposure to hot conditions also increases the dilation of skin blood vessels and blood flow to the skin, helping to dissipate heat by conduction.

Cold acclimatization has been less studied than heat acclimatization, but seasonal changes occur in many endotherms that live in variable climates. Birds grow an extra layer of insulating feathers, and mammals may grow additional fur in the winter. The additional insulation reduces heat loss up to 50% in some animals. Such feathers and fur are shed in the summer.

46.5 Impact on Public Health

Learning Outcomes:

1. Define body mass index (BMI) and explain how it is used to assess health risks associated with overweight and obesity.
2. Describe the impact of anorexia nervosa and bulimia nervosa on public health.

As we have seen, most animals, when provided adequate nutrients, maintain their body mass around a set point that is normal for their species. We rarely observe healthy animals in nature that are overweight. Generally, only domesticated animals become sufficiently sedentary that they gain excess, unnecessary weight (think of an overweight housecat). Humans, too, are prone to weight gain, particularly when living sedentary lives. Many people maintain a healthy body weight, but, as discussed in this section, an increasing number are unable to meet this goal.

Obesity Is a Global Health Issue

Excess body fat increases the risk of many diseases, including high blood pressure, cancer, heart disease, and diabetes. In diabetes, either insufficient insulin is available from the pancreas to control blood glucose concentrations, or the insulin that is present is no longer effective. In the U.S. alone, about 8% of the population—nearly 24,000,000 people—have diabetes mellitus. Of all diabetics in the U.S., more than 90% have the form called type 2 diabetes mellitus that results from insulin being ineffective. Compelling evidence has directly linked the incidence of this type of diabetes with being overweight. At what point does fat accumulation in humans start to pose a health risk? Historically, this question has been evaluated by research studies that investigate possible correlations between disease rates and some measure of body fat.

One of the currently preferred methods for assessing body fat and health risks is the **body mass index (BMI)**, a ratio of weight relative to height. A person's BMI is calculated by dividing his or her weight in kilograms by the square of his or her height in meters. For example, a 70-kg human with a height of 180 cm would have a BMI of 21.6 kg/m^2:

$$\begin{aligned} \text{BMI} &= 70 \text{ kg} / (1.8)^2 \text{ m} \\ &= 21.6 \text{ kg/m}^2 \end{aligned}$$

Current National Institutes of Health guidelines categorize BMIs of 25 or more as overweight, that is, as having increased health risk because of excess adipose tissue. BMIs of 30 or greater are considered obese, with a highly increased health risk. Data compiled by The Centers for Disease Control and Prevention in Atlanta, Georgia, and other U.S. federal agencies indicate that approximately 2/3 of U.S. adults age 20–74 are now overweight or obese (Figure 46.16). One of the more troubling statistics is that the percentage of adults who are overweight but not obese has remained relatively unchanged since 1960, at roughly 30–35%. However, the percentage of obese adults has risen during that time from about 13% to the current level of 33–34%. Since as recently as the early 1990s, the CDC estimates that the average body weight of Americans has risen by 10 pounds. Even more troubling, the rate of childhood obesity has also risen. In the U.S., the incidence of obesity in children age 6–11 has increased from 2–3% in 1960 to the current estimate of more than 15%.

The increase in overweight and obesity categories is not confined to the U.S., but has become a worldwide trend. According to the World Health Organization, more than 1 billion adults globally are overweight and 300 million are obese. One ray of hope is that the incidence of obesity in the U.S. population has leveled off since 2003, suggesting that efforts at educating the public about the health consequences of obesity may be starting to have a positive result.

Some studies indicate that genetic factors play an important role in obesity. Identical twins separated soon after birth and raised in different households have strikingly similar body weights as adults. Researchers hypothesize that natural selection favored the evolution of so-called **thrifty genes**, which boosted our ancestors' ability to store fat from each feast in order to sustain them through the next famine. Given today's abundance of high-fat foods in many countries, what was once a survival mechanism may now be a liability.

The methods and goals of treating obesity are undergoing extensive rethinking. An increase in body fat is generally due to an excess of energy intake over energy expenditure, and overweight people have traditionally been advised to follow a low-calorie diet. However, such diets alone have limited effectiveness, because over 90% of obese people regain most or all of their lost weight within 5 years. This disturbing phenomenon may be related to the observation that metabolic rate decreases as leptin levels decrease, sometimes decreasing sufficiently to prevent further weight loss on as little as 1,000 cal per day.

Research indicates that crash diets are not an effective long-term method for controlling weight. Instead, caloric intake should be set at a realistic level that can be maintained for the rest of one's life. This should lead to a slow, steady weight loss of no more than 1 pound per week until body weight stabilizes at a new, lower level. Most important, any program of weight loss should include increased physical activity. The exercise itself burns calories, but more importantly, it partially offsets the tendency for the metabolic rate to decrease. As a bonus, the combination of exercise and caloric restriction causes the person to lose more fat and less protein than with caloric restriction alone.

The impact of obesity on public health is enormous, accounting for many illnesses requiring hospitalization and chronic drug therapy and well over 100,000 premature deaths per year. Its impact is far-reaching on the economy as well. The economic toll of obesity-related illnesses is felt in the loss of worker-hours in the workplace, in the cost of hospital stays, physician office visits, nursing home care, and medication costs. In fact, current estimates are that as much as 10% of all U.S. health-care expenditures are directly or indirectly related to

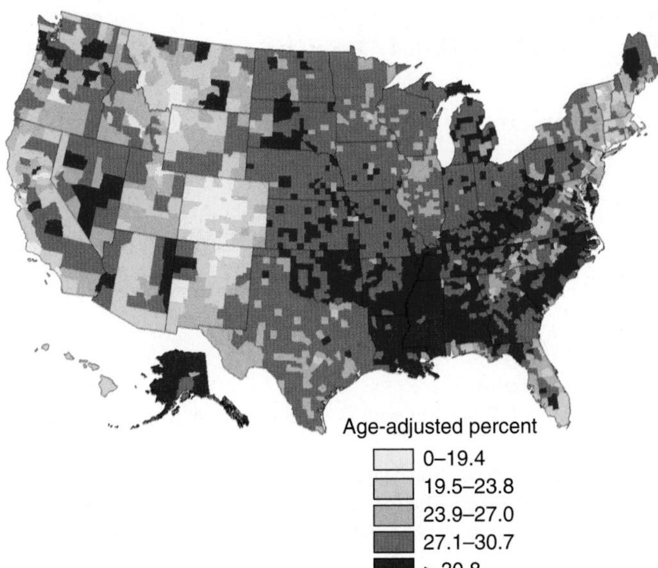

Figure 46.16 Obesity in U.S. adults. The Centers for Disease Control and Prevention estimate that many areas of the country experience obesity rates over 30% in adults (the darkest colored areas).

obesity! Obesity can affect society in unexpected ways. As an example, the increasing weight load of passengers forces airplanes to burn 350 million additional gallons of fuel each year, compared with just 15 years ago! This translates into nearly 4 million tons of additional pollution released into the atmosphere every year. In another recent example, for safety reasons, the United States Coast Guard has begun downgrading the allowable number of passengers on large state ferries due to the increasing average weight of U.S. adults.

Anorexia Nervosa and Bulimia Nervosa Can Have Serious Health Consequences

Being obese can be dangerous, but so can being extremely underweight. Two eating disorders that can leave people underweight are anorexia nervosa and bulimia nervosa.

Anorexia nervosa (often simply called anorexia) is characterized by weight loss and is found primarily in adolescent girls and young women. People with anorexia nervosa become pathologically obsessed with weight and body image, and they reduce their food intake to the point of starving. Anorexia nervosa may have psychological causes as well as a biological basis; recently, researchers have identified a possible genetic predisposition to anorexia. Many abnormalities are associated with the disorder, such as loss of menstrual periods, low blood pressure, low body temperature, and altered secretion of many hormones. These may be simply signs of starvation, although they may also result from malfunction of the parts of the brain that normally control appetite.

Bulimia nervosa (often simply called bulimia) involves recurrent episodes of binge eating followed by use of methods to prevent weight gain, including self-induced vomiting, laxatives, and diuretics, as well as strict dieting and overexercising. As with individuals with anorexia nervosa, people with bulimia are obsessed with body weight, although they are usually within 10% of their ideal weight. One of the most dangerous side effects of bulimia is dehydration due to purging. Vomiting, laxatives, and diuretics can also cause electrolyte imbalances in the body, most commonly low potassium levels, which can lead to symptoms ranging from lethargy to irregular heartbeat and death. Successful treatment of anorexia and bulimia is similar and relies primarily on counseling, nutritional education, and medications.

Summary of Key Concepts

46.1 Nutrient Use and Storage

- An animal's handling of nutrients has two phases: the absorptive state, during which ingested nutrients are entering the blood from the alimentary canal, and the postabsorptive state, during which the GI tract is empty of nutrients and the body's own stores must supply energy.
- Glucose and fats are the two major energy sources during the absorptive state. Much of the absorbed glucose immediately enters cells and is enzymatically broken down, providing energy required to synthesize ATP. Most absorbed triglycerides are stored in fat cells until the body requires additional energy. Amino acids are taken up by all body cells and used to synthesize proteins (Figure 46.1).
- The events that maintain blood glucose concentration in the postabsorptive state fall into two categories: (1) the reactions glycogenolysis and gluconeogenesis, which provide glucose to the blood, and (2) cellular use of fat for energy, which spares glucose for use by the nervous system (Figure 46.2, Table 46.1).

46.2 Regulation of the Absorptive and Postabsorptive States

- Tight control mechanisms, in the form of several hormones and the nervous system, maintain homeostatic levels of fuel in the blood. The hormone insulin acts on cells to facilitate the diffusion of glucose from blood into the cell cytosol via glucose transporters (GLUTs). All animal cells use GLUTs to transport glucose across their plasma membranes (Figure 46.3).
- In vertebrates, an increase in blood glucose concentration in the absorptive state stimulates the cells of the pancreas to secrete insulin; a decrease in glucose concentration removes the signal for secretion. In the postabsorptive state, when the blood glucose concentration decreases, glucose-monitoring regions in the hypothalamus stimulate production of glucose-elevating factors such as glucagon and norepinephrine (Figure 46.4).
- Exercise is any type of physical activity that increases an animal's metabolic rate. Exercise increases an animal's requirement for nutrients, including glucose and fatty acids, to provide energy.

46.3 Energy Balance

- An animal's metabolic rate refers to the amount of energy it uses in a given period of time to power all of its metabolic requirements.
- The most common measure for comparing metabolic rates of different species is the basal metabolic rate (BMR). Most of the basal

metabolism is due to the routine functions of the heart, liver, kidneys, and brain (Figures 46.5, 46.6).

- Many factors affect metabolism, including skeletal muscle activity, whether an animal has recently eaten, and body size (Figures 46.7, 46.8).
- When the daily amount of energy consumed equals the amount of energy expended, body weight remains stable. Tipping the balance in either direction causes weight gain or loss by increasing or decreasing total body energy content.
- Short-term control of feeding generally involves satiety signals that remove the sensation of hunger and set the time period before hunger returns again. Experiments by Coleman and Friedman investigated the hormone leptin as a satiety factor in mammals. Leptin has since been found in all classes of vertebrates (Figures 46.9, 46.10).

46.4 Regulation of Body Temperature

- Most animals can survive only in a relatively narrow temperature range that allows chemical reactions to proceed, maintains the structures of membranes, and avoids denaturing proteins. A few animals, however, have the ability to survive being partially frozen.
- Animals can be classified according to their source of heat and their ability to maintain body temperature. Ectotherms depend on external heat sources to warm their bodies, whereas endotherms use their own metabolically generated heat to warm themselves. Homeotherms maintain their body temperature within a narrow range, but heterotherms have body temperatures that vary with environmental conditions (Figure 46.11).
- The surface of an animal's body can lose or gain heat from the external environment via four mechanisms: radiation, evaporation, convection, and conduction (Figures 46.12, 46.13).
- The skin can function as a variable heat exchanger; blood vessels near the skin surface dilate to dissipate heat or constrict to retain it. Both endotherms and ectotherms regulate heat loss through countercurrent exchange, which retains heat by returning it to the body's core and keeping it warmer than the extremities. Heat exchange can also be regulated by changing the rate of water evaporation via perspiration. Behavioral mechanisms can alter heat loss by radiation, conduction, and convection (Figures 46.14, 46.15).
- Muscle activity (shivering thermogenesis) and brown adipose tissue metabolism (nonshivering thermogenesis) increase the production of heat.
- Acclimatization can fine-tune an animal's adaptive mechanisms to a changing environment.

46.5 Impact on Public Health

- Excess body fat increases the risk of many diseases. A body mass index (BMI) of 25 kg/m^2 or more is considered overweight; 30 kg/m^2 is considered obese (Figure 46.16).
- Being underweight is also unhealthy. People with anorexia nervosa become pathologically obsessed with weight and body image and reduce their food intake to the point of starving. Bulimia nervosa involves recurrent episodes of binge eating followed by use of methods to prevent weight gain, including self-induced vomiting, laxatives, and diuretics. Both eating disorders can have serious health risks and both are treated with counseling, nutritional education, and medications.

Assess and Discuss

Test Yourself

1. During the absorptive phase, an animal is
   a. fasting.
   b. relying entirely on stored molecules for energy.
   c. absorbing nutrients from a recently ingested meal.
   d. metabolizing lipids stored in adipose tissue to supply ATP to its cells.
   e. both a and b
2. Gluconeogenesis
   a. occurs when the liver synthesizes glucose from noncarbohydrate precursors.
   b. is the process by which glycogen is broken down to glucose.
   c. occurs primarily when an animal is in the absorptive phase.
   d. occurs when triglycerides are being formed and stored in adipose cells.
   e. none of the above
3. In the process of ______________, most tissues of the vertebrate body metabolize fat instead of glucose to ensure that ______________ tissue has an adequate supply of glucose.
   a. gluconeogenesis, muscle
   b. glucose sparing, epithelial
   c. glycogenolysis, nervous
   d. glucose sparing, nervous
   e. gluconeogenesis, epithelial
4. Ketones are compounds derived from
   a. glucose.
   b. glycogen.
   c. fatty acids.
   d. amino acids.
   e. proteins.
5. Insulin primarily regulates blood glucose concentrations by
   a. stimulating the recruitment of glucose transporter proteins from the cytosol to the plasma membrane for transport of glucose from extracellular to intracellular fluid.
   b. stimulating gluconeogenesis.
   c. suppressing glucose uptake by muscle tissue.
   d. stimulating the release of glucose from glycogen reserves in the liver.
   e. inhibiting the synthesis of new GLUT proteins.
6. The rate at which an animal uses energy is called
   a. body mass index.
   b. an animal's energy consumption.
   c. metabolic rate.
   d. an animal's energy expenditure.
   e. both c and d.
7. Which factor may increase metabolic rate?
   a. hibernation
   b. reduced muscle activity
   c. sleeping
   d. fasting
   e. consumption of a meal
8. Which molecule acts on brain centers to reduce appetite in mammals and other vertebrates?
   a. GLUT4
   b. glycogen
   c. leptin
   d. glucagon
   e. ketones
9. Animals that have body temperatures that are maintained within a narrow range are
   a. endotherms.
   b. ectotherms.
   c. homeotherms.
   d. heterotherms.
   e. both b and d.

10. The rate of heat loss in a mammal is regulated by
    a. the degree of blood flow at the surface of the skin.
    b. the amount of perspiration.
    c. behavioral adaptations.
    d. air currents near the animal's body.
    e. all of the above.

Conceptual Questions

1. Explain the functions of insulin. Why do you think a hormone such as insulin is required to carry out these functions?
2. Explain how appetite is controlled by the brain. What is the benefit of having a hormone released from adipose cells in proportion to total fat mass?
3.  A principle of biology is that *structure determines function*. How does this principle apply to countercurrent exchange?

Collaborative Questions

1. Discuss the differences between being ectothermic and endothermic and between being heterothermic and homeothermic.
2. Discuss four ways animals exchange heat with their environment.

Online Resource

www.brookerbiology.com

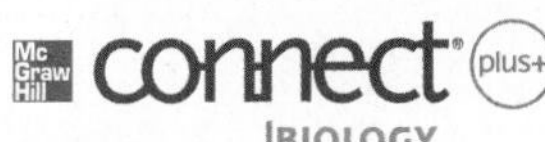

Stay a step ahead in your studies with animations that bring concepts to life and practice tests to assess your understanding. Your instructor may also recommend the interactive eBook, individualized learning tools, and more.

Circulatory Systems

47

Chapter Outline

47.1 Types of Circulatory Systems
47.2 Blood and Blood Components
47.3 The Vertebrate Heart and Its Function
47.4 Blood Vessels
47.5 Relationship Among Blood Pressure, Blood Flow, and Resistance
47.6 Adaptive Functions of Closed Circulatory Systems
47.7 Impact on Public Health
Summary of Key Concepts
Assess and Discuss

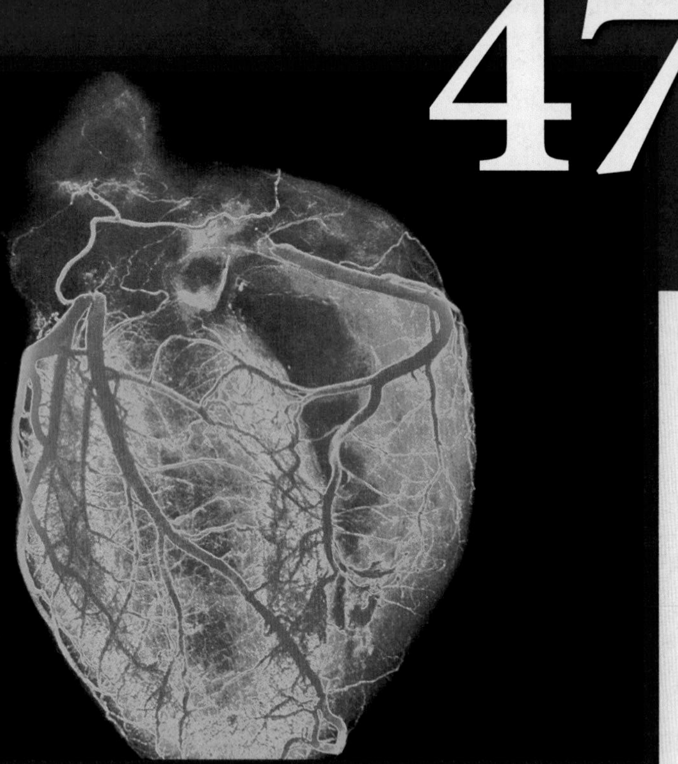

Image from cardiac angiography of a human heart (dye contrast injection to make the coronary blood vessels easier to visualize).

In the time it takes you to read this chapter introduction, three or four people in the United States will have suffered a heart attack, and one or more of those individuals will likely have died as a result. Cardiovascular disease—disease of the heart or blood vessels—is the leading cause of death in the U.S. and in much of the rest of the world. In the U.S. alone, an estimated 80,000,000 or more people have one or more diseases that affect the heart and blood vessels. About 1–1.5 million of those individuals will have a heart attack in the next year, between 400,000–600,000 of whom will die as a result.

The heart is a muscular pump that works ceaselessly. Because of this constant activity, heart muscle requires a steady and large amount of blood, bringing with it the nutrients and oxygen required to sustain that muscular effort. The chapter-opening photo shows the extensive network of blood vessels coursing through the human heart. Should any of those vessels become diseased, the regions of the heart supplied by those vessels can die. That is what happens during a heart attack. If a heart attack occurs, the damaged heart may not be able to pump blood forcefully enough to generate the pressure required for sufficient blood and oxygen to reach the cells of the body.

Circulatory systems transport necessary materials to all the cells of an animal's body, and transport waste products away from the cells so they can be released into the environment. In this chapter, we will examine how this occurs. Single-celled organisms are small enough that dissolved substances are able to diffuse into and out of the cell. In larger animals, however, dissolved substances must move greater distances between cells, as well as between the internal and external environments. The time required for osmosis of water and diffusion of solutes throughout the body and between environments would be too great to sustain life in such animals. In the circulatory systems of larger animals, the heart pumps blood—a fluid connective tissue—through vessels throughout the body. As we will see, they do this with either an open circulatory system or a closed circulatory system. An important feature of many circulatory systems is that they are capable of adjusting their activities to an animal's changing metabolic demands. For instance, we will see how physical activity increases the ability of the vertebrate heart to pump blood and directs blood to the skeletal muscles where it is most needed. Understanding the relationships among the heart, blood, and blood pressure is a major concept addressed in this chapter. We end by considering the major types of cardiovascular disease in humans and some of their causes, symptoms, and treatments.

47.1 Types of Circulatory Systems

Learning Outcomes:

1. Compare and contrast the different types of animal circulatory systems.
2. List the key structural and functional characteristics of open and closed circulatory systems.
3. Describe single, intermediate, and double circulations in vertebrates.

The two basic types of animal circulatory systems are open systems and closed systems. However, a simple type of circulation exists in certain invertebrates that have a gastrovascular cavity. Although this type of circulation does not possess the anatomic elements found in open and closed systems (that is, a heart, blood, and blood vessels), it nonetheless serves to move nutrients and wastes through the body.

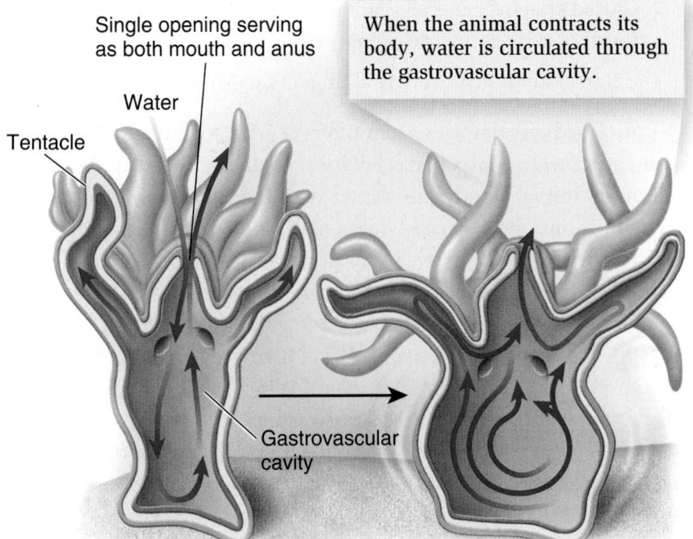
Single opening serving as both mouth and anus
When the animal contracts its body, water is circulated through the gastrovascular cavity.
Water
Tentacle
Gastrovascular cavity

Figure 47.1 **Circulation of water through the gastrovascular cavity of *Hydra*.** The animal's movements assist the flow of water throughout the cavity, which extends into the tentacles.

BioConnections: Look back at Figure 45.5. Can you see how the circulation of water in a gastrovascular cavity is related to the digestive functions of Hydra?

Gastrovascular Cavities

In cnidarians (jellyfish, hydras, sea anemones, and corals), ctenophores, and some flatworms, the surrounding water in which the animals live not only contains the nutrients and oxygen needed to sustain life, but also provides a sort of circulatory system. These animals rely on water currents to bring a steady supply of water into contact with their internal body surface. Water enters the **gastrovascular cavity**, a body cavity with a single opening to the outside that functions as both mouth and anus. As described in Chapter 45 (refer back to Figure 45.5), nutrients are digested within the cavity and distributed throughout the animal's body, and wastes are excreted into the cavity. Nutrients obtained from digested food reach all the body cells because water circulates throughout the entire cavity. Because the animals are only a few cells thick, all cells in their bodies are located close to the gastrovascular cavity or in slender extensions that branch from it.

In some cnidarians, cilia that line the opening of the gastrovascular cavity propel water through these extensions, increasing the circulation of water. In addition, muscular efforts of the body wall—as *Hydra* stretches and relaxes, for instance, or a jellyfish propels itself by contracting its bell—help circulate water throughout the cavity (Figure 47.1). The harder the muscles work, the more effectively they propel water, thus providing a rudimentary but effective circulation.

Open Circulatory Systems

Unlike gastrovascular cavities, all true circulatory systems generally have three basic components:

1. Blood or hemolymph
2. Vessels, a system of hollow tubes within the body through which blood or hemolymph travels
3. One or more **hearts**, muscular structures that pump blood or hemolymph through the vessels

Arthropods and some mollusks have an **open circulatory system**, with one or more hearts that pump fluid through vessels that open into the animal's body cavity, called the hemocoel (Figure 47.2a). Therefore, the fluid in the vessels and the interstitial fluid that surrounds cells mingle in one large, mixed compartment,

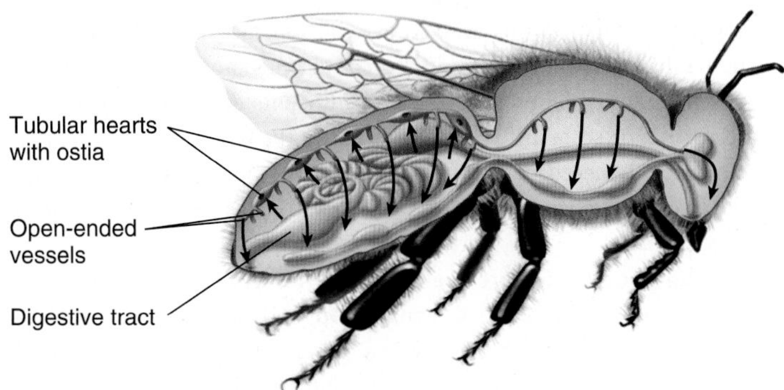
Tubular hearts with ostia
Open-ended vessels
Digestive tract

(a) An open circulatory system.

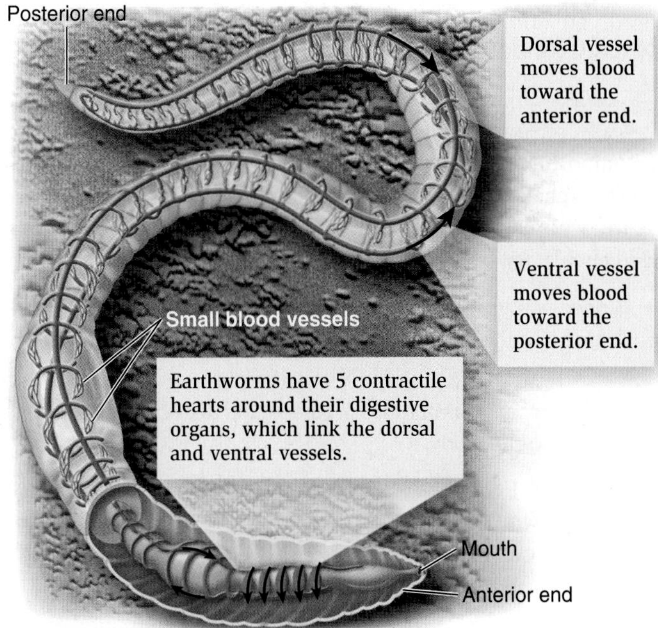
Posterior end
Dorsal vessel moves blood toward the anterior end.
Ventral vessel moves blood toward the posterior end.
Small blood vessels
Earthworms have 5 contractile hearts around their digestive organs, which link the dorsal and ventral vessels.
Mouth
Anterior end

(b) A closed circulatory system.

Figure 47.2 **Types of circulatory systems.** **(a)** In open circulatory systems, one or more muscular, tubelike hearts pump hemolymph through open-ended vessels, where it percolates through the body. In arthropods such as this honeybee, hemolymph reenters the heart through ostia. The arrows show the movement of hemolymph. **(b)** In closed circulatory systems, such as that of the earthworm, blood remains in vessels and hearts and recirculates without emptying into the body cavity. Arrows indicate direction of blood flow.

Concept Check: Why is it incorrect to think of open circulatory systems as "primitive" compared with closed systems?

rather than being located in separate body compartments. The mixed fluid is called **hemolymph**. Nutrients and wastes are exchanged by diffusion between the hemolymph and body cells, and the hemolymph eventually returns to the heart. In mollusks with open circulatory systems, hemolymph reenters the heart through vessels, whereas in arthropods, it returns directly to the heart through small openings called ostia.

In insects, hemolymph primarily transports nutrients and wastes; in mollusks, it also brings oxygen to body cells. Insects have a unique oxygen-delivery system that does not involve the circulatory system, as you will learn in Chapter 48. This is a major distinction between insect circulatory systems and those of other animals.

Open circulatory systems are metabolically inexpensive because they do not require much energy to pump hemolymph through an extensive network of blood vessels; recall that hemolymph simply enters the hemocoel through open vessels. In addition, open systems can adapt to changes in an animal's metabolic demands. As an insect takes flight, for example, its flight muscles contract more forcefully and rapidly, which acts to expand and compress the animal's thorax. This helps propel hemolymph more rapidly throughout the hemocoel and into and out of the hearts. In other words, as the animal's physical activity increases, its circulation becomes more efficient, recharging metabolically active cells with nutrients. The ability to adjust circulation to meet an animal's requirements is one of the most important features of any circulatory system.

Despite the phenomenal success of species with open systems, this type of circulation has certain limitations. For instance, because the hemolymph empties in bulk into the general body cavity, it cannot be selectively delivered to individual regions of the body whose metabolic activity may have increased relative to other regions. By contrast, during periods of increased physical activity in vertebrates, larger amounts of blood are delivered to skeletal muscles and away from less metabolically active structures, an extremely useful adaptation that results in greater endurance. This is possible because these animals have a closed circulatory system, as we see next.

Closed Circulatory Systems

In a **closed circulatory system**, blood and interstitial fluid are physically separated and differ in their components and chemical composition. Closed circulatory systems are found in earthworms, cephalopods (squids and octopuses), and all vertebrates (Figure 47.2b). Despite some differences in structure among taxa, closed circulatory systems share certain key features.

- **Blood**—a watery solution containing solutes to be transported throughout the body—is pumped under pressure by one or more contractile, muscular hearts.
- Blood remains within tubelike vessels that distribute it throughout the body.
- The solutes in blood can be exchanged with the environment and the body's cells.
- In most cases, blood contains disease-fighting cells and molecules.
- The activity of the closed circulatory system can be adjusted to match an animal's metabolic demands.
- Closed systems generally can heal themselves when wounded, by forming clots at the site of injury.
- These systems grow in size as an animal grows.

A closed circulatory system offers several advantages. First, animals can grow to a larger size, because blood can be directed to every cell of an animal's body, no matter how large. Nearly all body cells are within one- or two cell-widths of a blood vessel. Second, blood flow can be selectively increased or decreased to supply different parts of the body with the precise amount of blood needed at any given moment. After a meal, for example, more blood can be directed to the intestines to absorb nutrients, and on a hot day, additional blood can be routed to the skin to dissipate heat. Animals with an open circulatory system cannot make these adjustments.

The closed circulatory system of vertebrates can be divided into two major arrangements: single circulation and double circulation. In single circulation, blood is pumped under low pressure from the heart to the respiratory surface (for example, gills), where it picks up oxygen and drops off carbon dioxide. From there, blood circulates to the tissues of the body, where it releases oxygen and picks up carbon dioxide. The blood then circulates back to the heart. In double circulation, blood is pumped under low pressure from the heart to the lungs, and then back to the heart, to be pumped under high pressure to the tissues and finally returned again to the heart. Single circulations are seen in fishes, and double circulations are seen in crocodiles, birds, and mammals. Amphibians and most reptiles have an intermediate type of circulation that combines features of both.

Single Circulation: Fishes In the single circulation of fishes, the heart has a single filling chamber—an **atrium**—to collect blood from the tissues, and an exit chamber—a **ventricle**—to pump blood out of the heart (Figure 47.3a). Blood vessels called **arteries** carry blood away from the heart to the gills, which pick up oxygen from the water in which the fish swims and unload carbon dioxide into the water. The freshly oxygenated blood then circulates via other arteries to the rest of the body. There, oxygen and nutrients are delivered to cells, and carbon dioxide diffuses from cells into the blood. Finally, the partially deoxygenated blood is returned to the heart via blood vessels called **veins**, where it is pumped back to the gills for another load of oxygen.

An important feature of all respiratory surfaces is that they function best when the blood flowing through them is maintained at a low pressure. The fish heart does not generate high pressure when it pumps blood to the gills. This means that blood leaving the gills will also be under low pressure, therefore limiting the rate at which oxygenated blood can be delivered to the body's cells.

Intermediate Circulation: Amphibians and Most Reptiles Unlike fishes, most adult amphibians rely on lungs and their permeable skin to obtain oxygen and rid themselves of carbon dioxide. While amphibians are on land, their deoxygenated blood is pumped from the heart to the lungs and, to a lesser extent, the skin. Oxygen diffuses from the air into the blood vessels within the lungs and beneath the skin, and carbon dioxide diffuses in the opposite direction. While amphibians are under water, however, deoxygenated blood from the heart bypasses the lungs and is directed almost entirely to the skin

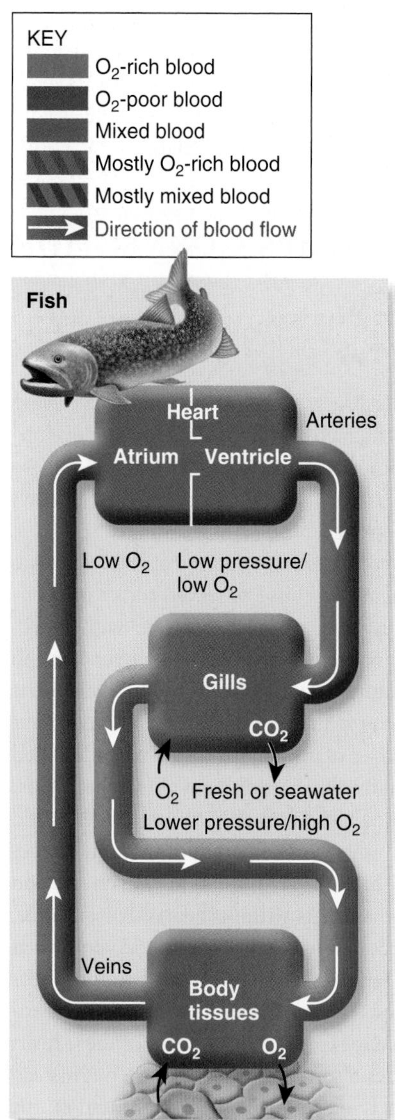

(a) Single circulation in fishes

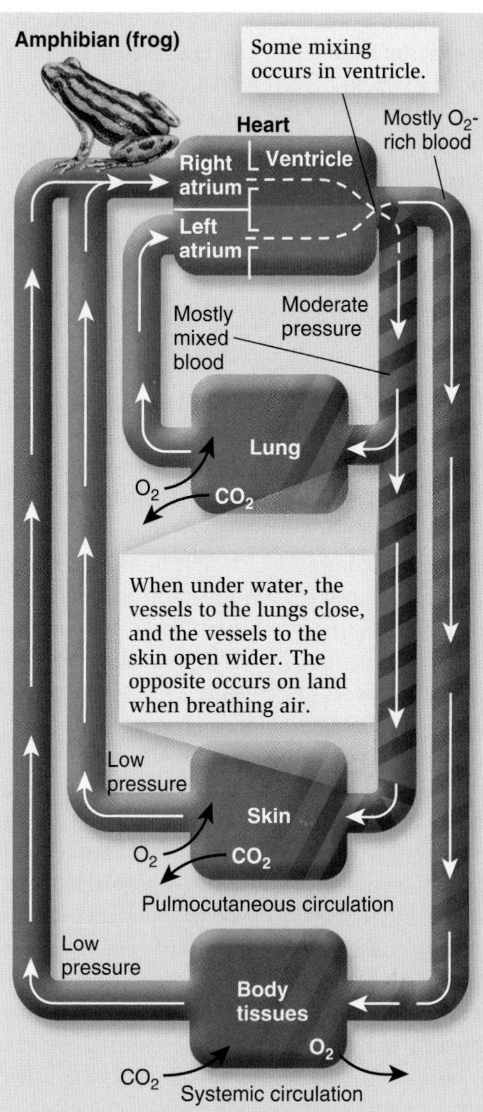

(b) Features of both single and double circulation in most amphibians

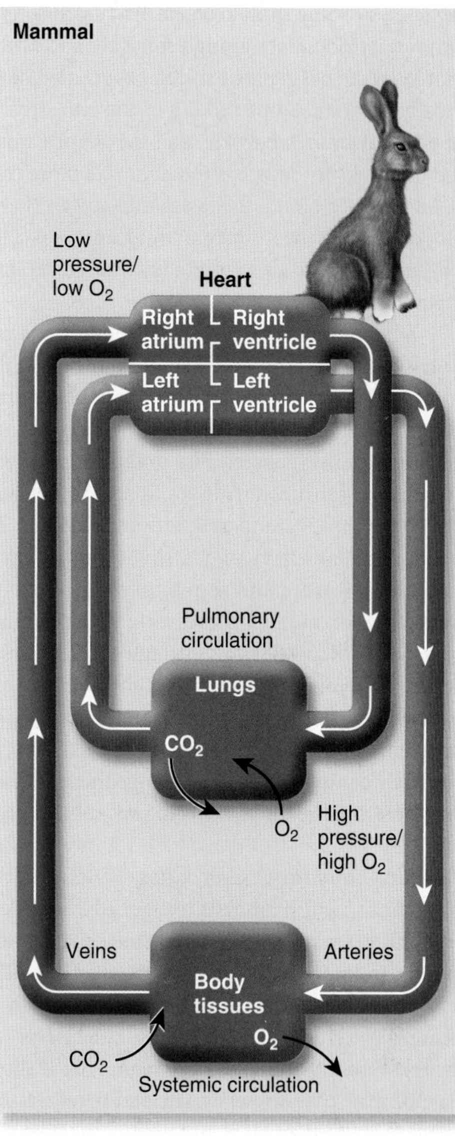

(c) Double circulation in crocodiles, birds, and mammals

Figure 47.3 **Representative vertebrate circulatory systems.** **(a)** Fishes have a single circulation in which blood is pumped from the heart to the gills, from where it circulates to the rest of the body tissues, and finally returns to the heart. **(b)** Amphibians, such as frogs, have a heart with a single ventricle that nevertheless has an internal structure that separates most of the oxygenated and deoxygenated blood entering it from the two atria. Some mixing does occur, however, as represented in this figure by the dashed lines. A significant advantage of this type of circulation is the ability to redirect blood to the skin instead of the lungs when under water. Because the skin is somewhat permeable to oxygen, this partly compensates for the lack of air-breathing. Note that blood returning from the skin has picked up oxygen, but is returned to the right atrium, not the left atrium as is the case with the lungs. **(c)** Birds and mammals have a double circulation, in which oxygenated blood is pumped under high pressure to the body's tissues, and deoxygenated blood is pumped under low pressure to the lungs. No mixing of the two types of blood occurs.

Concept Check: *What is the advantage of completely separating oxygenated and deoxygenated blood in a double circulation?*

so that oxygen in the water can diffuse across the skin and into the circulation.

Oxygenated blood from the lungs and skin does not travel directly to the rest of the body but instead returns to the heart, where it is pumped out to the tissues of the body. The amphibian heart, therefore, pumps oxygenated and deoxygenated blood to two separate locations. Blood is routed from the heart through different vessels to either the respiratory surfaces (the **pulmocutaneous circulation**) or the body tissues (the **systemic circulation**). This would not be possible if an amphibian's heart and circulation were the same as that of a fish, where blood travels through a single circulation. Instead, the amphibian heart has two separate atria for collecting blood (Figure 47.3b).

When an amphibian breathes air, oxygen-rich blood from the lungs is delivered to the left atrium. The right atrium receives blood that has passed through all of the body except the lungs and is therefore depleted of some of its oxygen. Each atrium of the amphibian heart empties its blood into a single large ventricle, which has flaps of tissue that keep the oxygenated and deoxygenated blood from the two atria mostly separate. Some mixing of the two streams of blood, however, does occur in the ventricle, and the slightly mixed blood traveling to the tissues is not fully oxygenated. The more oxygen that is dissolved in an animal's blood, the faster the oxygen is able to diffuse from the blood, through the interstitial fluid, and into cells where it is required. The mixed blood of amphibians, therefore, imposes limits on their metabolic activity.

Like amphibians, all of the reptiles except for crocodiles have hearts with two atria and a single ventricle, although the ventricle is somewhat more divided into two partially but incompletely separated chambers. These chambers allow more efficient separation of deoxygenated and oxygenated blood entering the ventricle from the two atria; this, in turn, contributes in part to the generally higher activity levels of reptiles compared with amphibians. Because these reptiles and amphibians have only a single ventricle that must pump blood to both the tissues and respiratory surfaces, blood must be pumped under low or moderate pressure in these animals to minimize the pressure of blood flowing through the lung tissue. This is important, because as you will learn later, fluid leaks out of capillaries under pressure in closed circulatory systems. Although not a significant problem outside the lungs, an accumulation of fluid in the lungs can seriously compromise gas exchange there. Having a lower blood pressure entering the lungs helps to prevent this occurrence.

Double Circulation: Crocodiles, Birds, and Mammals In crocodiles, birds, and mammals, oxygenated and deoxygenated blood are completely separated into two distinct circuits, the systemic circulation and the **pulmonary circulation** (Figure 47.3c). This is made possible by a heart that has two ventricles that are completely separated by a septum, or wall. The crocodilian circulatory system has evolved to be somewhat different from that of birds and mammals. It allows crocodiles to spend part of their lives under water, and it reflects that they are considerably less active than birds and mammals. When resting or under water, crocodiles can divert blood between the two circuits in ways that birds and mammals cannot. Nonetheless, all animals with double circulations have a left and right atrium and a left and right ventricle.

In a double circulation, the pulmonary circulation delivers oxygenated blood from the lungs to the left atrium, which then passes it on to the left ventricle. The left ventricle pumps blood to all the body's tissues via the systemic circulation. The systemic circulation delivers deoxygenated blood from the tissues to the right atrium, which then sends it to the right ventricle. The right ventricle pumps this blood to the lungs via the pulmonary circulation, and the cycle starts again.

An advantage of a double circulation is that the two ventricles can function as if they were, in effect, two hearts, each with its own ability to pump blood under different pressures. This means that blood from the right ventricle can be pumped under low pressure to the lungs, which, as we described earlier, is important for normal lung function. In a double circulation, however, once the blood picks up oxygen from the lungs, it can be returned to the left side of the heart. The left ventricle is more muscular than the right ventricle and therefore generates higher pressures when pumping blood. Thus, the oxygenated blood leaving the left side of the heart has sufficiently high pressure to reach all the cells of the animal's body and deliver oxygen and nutrients at a high rate, even to regions above the heart that must contend with gravity. This is no small feat, considering the distance blood must travel in some large animals. The left ventricle of a giraffe, for example, is particularly muscular compared with that of other animals.

Genomes & Proteomes Connection

A Four-Chambered Heart Evolved from Simple Contractile Tubes

The heart is the first vertebrate organ to become functional during embryonic development, and the factors responsible for its development have been extensively studied. Interestingly, at least one of these developmental factors has also recently shed light on how a heart develops four chambers in crocodiles, birds, and mammals.

The first heartlike organ appears to have evolved at least 500 mya. Ancestral hearts may have resembled those observed today in cephalochordates and urochordates—a simple linear tube with wavelike contractile properties but no valves or chambers (refer back to Figure 33.38). A valvelike structure first appears in the tubular hearts of some arthropods, such as *Drosophila*, but such hearts do not have clearly defined chambers. Analyses of animal genomes have identified numerous genes that are critical for the development of the chambered vertebrate heart, including a core set of five highly conserved genes and their proteins. Interestingly, however, one or more of these five genes are also expressed in linear tube hearts such as those just described, and even in contractile structures in animals without hearts. For example, as described earlier, cnidarians do not have a heart or blood vessels, but have muscle cells in their contractile bell that not only produce locomotion, but in the process circulate water through the animal's gastrovascular cavity. These muscle cells express at least three of these heart developmental genes. Nematodes also do not have a heart, but instead have a contractile pharynx that serves part of the same function; the pharyngeal muscles of these animals express four of the five key heart developmental genes that are found in all vertebrates. Thus, the genes required for the formation of a contractile, multichambered heart are ancient and conserved. This question arises, however: Why are some vertebrate hearts four-chambered, and others are not?

In a 2009 study, American researcher Benoit Bruneau and colleagues compared the expression of one of the major heart developmental genes, called *Tbx5*, in the hearts of amphibians, reptiles, birds, and mammals. They discovered that the expression of *Tbx5* became increasingly restricted in animals displaying greater separation of the ventricles into two chambers (**Figure 47.4**). In amphibians, which have a single ventricle, *Tbx5* was expressed throughout the developing ventricle. In reptiles in which an incomplete septum partially divides the ventricles, *Tbx5* expression was absent in a portion of the right half of the ventricle. In birds and mammals, *Tbx5* was completely absent from the developing right ventricle and highly expressed in the left ventricle. This suggests that a gradient of *Tbx5* is required for the ventricle to form two chambers and thereby a complete double circulation. The investigators also showed that if *Tbx5* expression was experimentally induced in the heart of an embryonic mouse in a pattern that mimicked that seen in lizards, the mouse

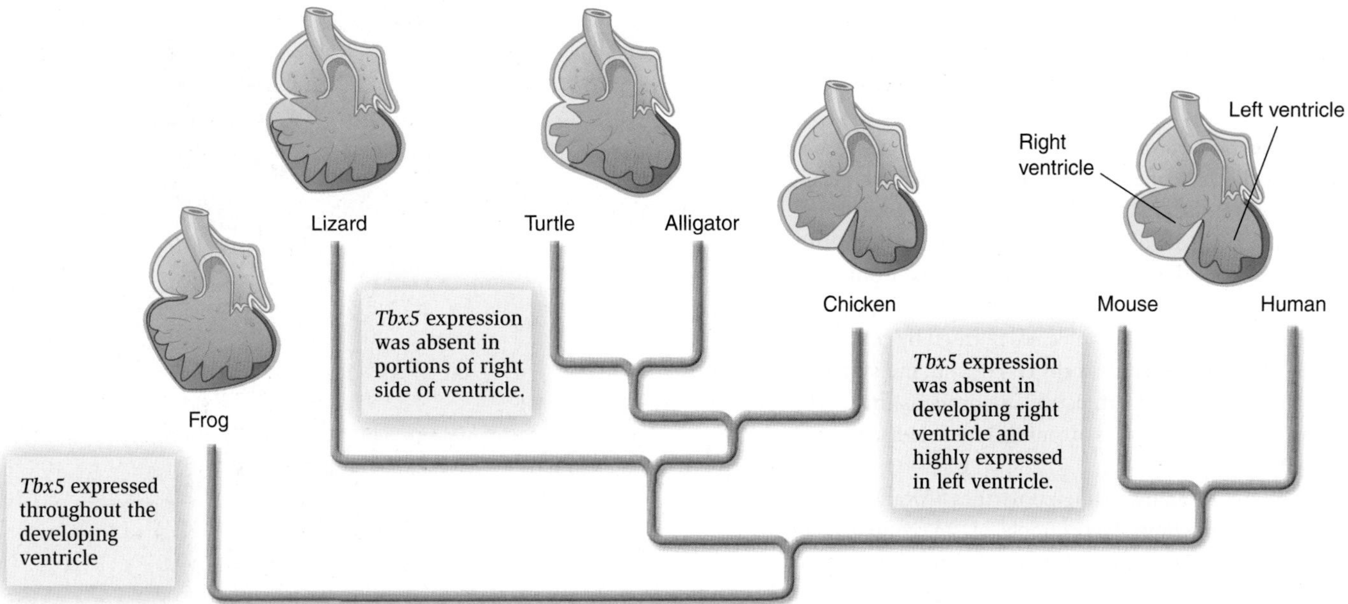

Figure 47.4 Expression of the *Tbx5* gene in embryonic vertebrate hearts. The blue color indicates where in the developing heart muscle the gene was expressed. Note: Structural details have been omitted for simplicity.

heart developed only three clearly defined chambers—there was no distinction between left and right ventricle!

From these and other studies, it appears that the expression of a core set of genes is vital for the formation of a heart or heartlike structure, and a smaller set of genes like *Tbx5* is critical for the division of the heart into chambers. The most common congenital organ defects in humans are those associated with the heart and its major vessels. The relatively high incidence of heart-related birth defects makes an understanding of the genetic control of the growth and development of the heart a great concern in medicine today.

47.2 Blood and Blood Components

Learning Outcomes:

1. List the four components of blood and the composition and functions of each.
2. Describe the process of blood clotting in mammals.

Blood is the transport medium of many animals' bodies. It moves necessary materials, including nutrients and gases such as oxygen, to all cells and takes away waste products, including carbon dioxide and other breakdown products of metabolism. What are the components of blood that allow it to perform its functions?

Blood Is Composed of Cells and Water with Dissolved Solutes

Blood is a fluid connective tissue consisting of cells and, in mammals, cell fragments suspended in a solution of water containing dissolved nutrients, proteins, gases, and other molecules. If we collect a blood sample and spin it in a centrifuge, the blood separates into three layers (Figure 47.5). These layers correspond to three of blood's four

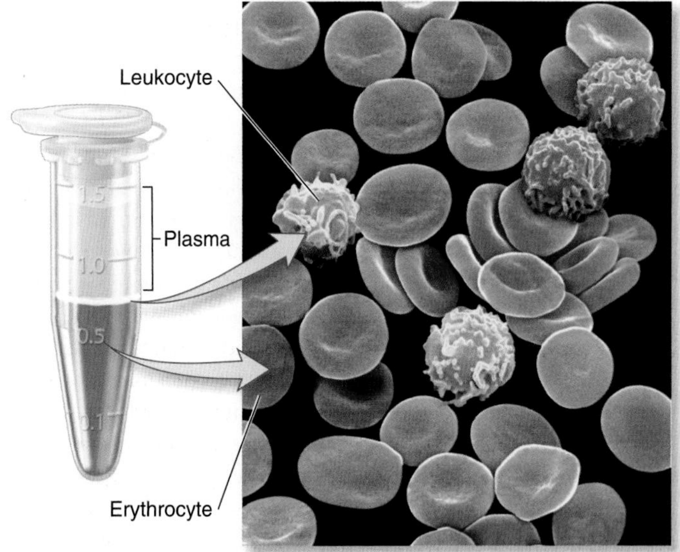

Figure 47.5 Components of blood. When a blood sample that has been prevented from clotting is centrifuged, it forms three visible layers. Leukocytes, shown in the scanning electron micrograph, are the white blood cells that make up part of the immune system. Erythrocytes are red blood cells and function to carry oxygen. An additional component of blood is not visible under these conditions because of small numbers and size. These are the platelets in mammals (or thrombocytes in other vertebrates) that participate in blood clotting reactions. *Note*: The leukocyte layer is enlarged and not to scale, for illustrative purposes.

BioConnections: Leukocytes are part of the immune system of animals. Are immune defenses unique to animals? (Look back at Figure 36.19 for a hint.)

components: plasma, leukocytes, and erythrocytes. (A fourth component, the much less abundant platelets or thrombocytes, is not readily visible in a centrifuged sample.) Let's take a look at each of these.

Plasma The top layer of the centrifuged blood sample is a yellowish solution called **plasma** (see Figure 47.5). Plasma typically makes up about 35–60% of the total volume of blood in vertebrates. It contains water and the dissolved organic and inorganic nutrients that were absorbed from the digestive tract or secreted from cells. Plasma also contains dissolved oxygen; waste products of metabolism, such as carbon dioxide; and other molecules released by cells, such as hormones. Plasma also transports cells of the immune system and cells involved in oxygen transport, as well as proteins that serve several important functions, such as forming blood clots, which seal off wounds to blood vessels.

In addition to transporting molecules throughout the body, plasma has other functions. For example, plasma contains buffers that help keep the body's pH within its normal range. It is also important in maintaining the fluid balance of cells. Changes in plasma salt and protein concentrations can affect the movement of water between intracellular and extracellular fluid compartments.

Leukocytes Beneath the plasma in our sample is a narrow white layer of **leukocytes**, also known as white blood cells (see Figure 47.5). Leukocytes develop from a specialized connective tissue (the marrow) of certain bones in vertebrates. Although there are several types—which we describe further in Chapter 53—all leukocytes perform vital functions that defend the body against infection and disease.

Erythrocytes The bottom visible layer of our blood sample consists of **erythrocytes**, also called red blood cells because of their color (see Figure 47.5). The term **hematocrit** refers to the volume of blood (expressed as percentage) that is composed of red blood cells, usually between 40 and 65% among vertebrates. Red blood cells serve the critical function of transporting oxygen throughout the body. There are approximately a thousand times more red blood cells than white blood cells in the circulation. Like leukocytes, red blood cells are derived from cells in the bone marrow. In most vertebrates, mature red blood cells retain their nuclei and other cellular organelles, but in all mammals (and a few species of fishes and amphibians), the nuclei are lost upon maturation. The lack of a nucleus and many other organelles in the mammalian red blood cell increases the cell's oxygen-carrying capacity and contributes to its characteristic biconcave shape (see Figure 47.5). The biconcave shape of the mammalian red blood cell increases its surface area relative to the flattened disc or oval shape seen in most other vertebrates. This is believed to make gas exchange between the red blood cell and the surrounding body fluids more efficient in mammals.

Oxygen is poorly soluble in plasma. Consequently, the amount of oxygen that dissolves in plasma usually cannot support a vertebrate's basal metabolic rate, let alone more strenuous activity. Within the cytosol of red blood cells, however, are large amounts of the protein **hemoglobin**. Each molecule of hemoglobin contains four protein subunits, each surrounding an atom of iron in a heme group. Each iron atom reversibly binds to a single molecule of oxygen (**Figure 47.6**). Thus, each hemoglobin molecule has four iron atoms and can bind four oxygen molecules. The oxygen attached to hemoglobin greatly increases the reservoir of oxygen in the blood and enables animals to be more active. Chapter 48 describes the mechanisms by which hemoglobin binds and releases oxygen.

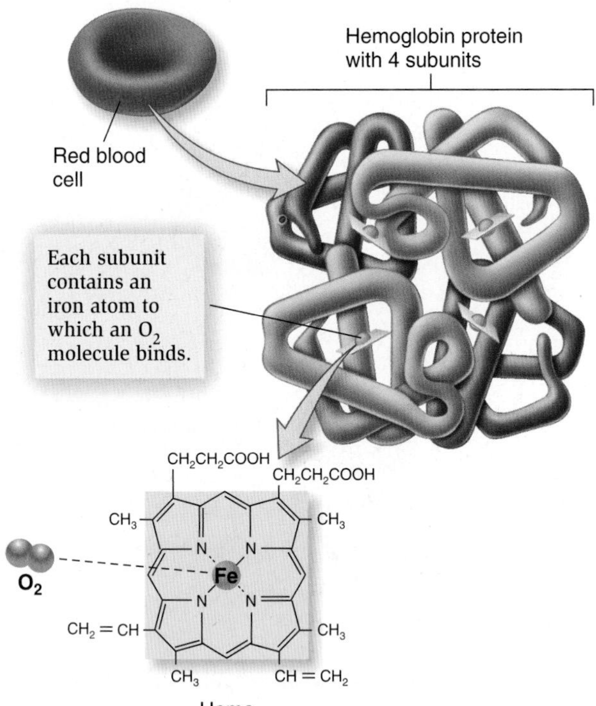

Figure 47.6 Hemoglobin. Erythrocytes contain large amounts of the protein hemoglobin. Oxygen binds reversibly to iron atoms in the heme portion of each subunit of hemoglobin.

BIOLOGY PRINCIPLE Structure determines function. The precise quaternary structure of hemoglobin permits its association with heme groups, which contain the iron atoms that bind oxygen molecules. Without a correct protein structure, the ability of hemoglobin to bind oxygen would be greatly compromised.

Concept Check: How many atoms of oxygen can be bound to a single molecule of hemoglobin?

Anemia refers to lower than normal amounts of hemoglobin, which reduces the amount of oxygen that can be stored in the blood. Symptoms of anemia vary, but can include fatigue, loss of energy, and shortness of breath. Among possible causes of anemia are loss of blood due to injury, decreased production of hemoglobin, impaired production of erythrocytes, or increased breakdown of erythrocytes, each of which reduces hematocrit.

Platelets Vertebrate blood has a fourth component called platelets in mammals and thrombocytes in other vertebrates. **Platelets** are cell fragments that lack a nucleus, whereas **thrombocytes** are intact cells; both play a crucial role in the formation of blood clots, which limit blood loss after injury. Like leukocytes and erythrocytes, platelets and thrombocytes are formed in the bone marrow.

Blood Clotting Is a Multistep Process

The formation of a blood clot in mammals requires several steps, two of which include platelets (**Figure 47.7a**). Upon injury to a blood vessel, platelets first secrete substances that cause platelets to clump

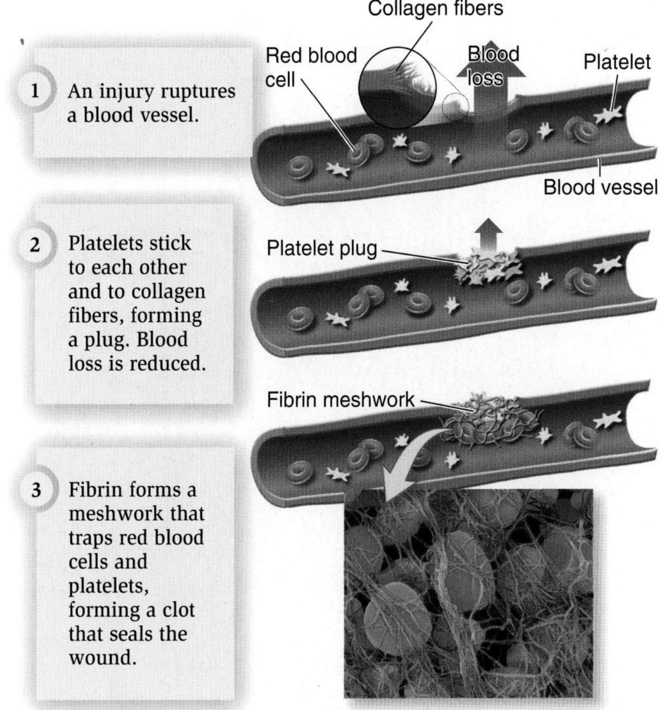

1 An injury ruptures a blood vessel.
Collagen fibers
Red blood cell
Blood loss
Platelet
Blood vessel
2 Platelets stick to each other and to collagen fibers, forming a plug. Blood loss is reduced.
Platelet plug
Fibrin meshwork
3 Fibrin forms a meshwork that traps red blood cells and platelets, forming a clot that seals the wound.

Figure 47.7 Platelets and the process of blood clot formation. A blood clot forms in two major steps: A platelet plug reduces initial blood loss, and a fibrin clot then seals the wound. An example of a fibrin clot is shown in the scanning electron micrograph.

BIOLOGY PRINCIPLE Living organisms maintain homeostasis. Because the blood carries the nutrients, hormones, proteins, and waste products that must circulate throughout an animal's body, a loss of blood due to injury can be catastrophic unless the wound is healed and the blood loss is stopped. In this way, blood clotting is a key feature that indirectly contributes to the ability of animals to maintain homeostasis.

together and bind to collagen fibers in the surrounding connective tissue at the wound site; this forms a plug that prevents continued blood loss. Second, other platelet secretions interact with plasma proteins to cause the precipitation from solution of the fibrous protein **fibrin**. Fibrin forms a meshwork of threadlike fibers that wrap around and between platelets and red blood cells, enlarging and thickening the plug to form a clot. Blood clotting begins within seconds and helps prevent injured animals from bleeding to death. Eventually, the body absorbs the clot as the injured vessel heals.

47.3 The Vertebrate Heart and Its Function

Learning Outcomes:

1. Explain the difference between neurogenic and myogenic hearts.
2. Describe the two phases of the excitation of the vertebrate heart.
3. Outline the events of the cardiac cycle, including the relationship between electrical and contractile events.
4. Analyze the meaning of the tracings on an electrocardiogram.

As described earlier, all vertebrate hearts have at least one upper (anterior) chamber called an atrium and at least one lower (posterior) chamber called a ventricle (Figure 47.8). In hearts with more than one atrium or ventricle, such as the mammalian heart, which has two of each chamber, the chambers are physically separated by a thick strip of connective tissue called a septum. Each atrium is fed by systemic or pulmonary veins; the systemic veins return blood from the body, and the pulmonary veins return blood from the lungs. Between the atria and ventricles there are one-way valves called **atrioventricular (AV) valves** that control the movement of blood between them. Each ventricle empties into the aorta or the pulmonary trunk. The **aorta** leads to the systemic circulation, and the **pulmonary trunk** divides into the pulmonary arteries that lead to the right and left lungs. Between each ventricle and the artery it empties into (either the aorta or the pulmonary trunk), there are one-way valves called **semilunar valves.**

The heart beats with a steady rhythm; each beat propels blood through the various chambers of the heart. Figure 47.8 (right) also depicts the route blood takes through the mammalian heart. Each ventricle in the mammalian heart functions as a separate pump. The right ventricle pumps deoxygenated blood returning from the body to the lungs, and the left ventricle pumps oxygenated blood coming from the lungs to the body. Blood enters the atria from systemic or pulmonary veins, moves down a pressure gradient through the AV valves into the ventricles, and is pumped out through the semilunar valves into the systemic and pulmonary arteries. Note that the blood can flow only one way through any of the valves.

Vertebrates Have Myogenic Hearts Capable of Beating on Their Own

What causes the heart to beat so steadily? Animals cannot consciously initiate heart contractions. The beating of the heart is initiated either by nerves or by intrinsic activity of the heart muscle cells themselves. Many arthropods and decapod crustaceans have a **neurogenic heart** that will not beat unless it receives regular electrical impulses from the nervous system. All vertebrates, however, have a **myogenic heart**; that is, the signaling mechanism that initiates contraction resides within the cardiac muscle itself. Cardiac muscle is distinguished by the interconnectedness between individual cardiac muscle cells, or myocytes. Each myocyte has membrane extensions that form interlocking networks with other myocytes. Within these networks, or intercalated discs, as they are called, are many gap junctions (refer back to Figure 10.11) that electrically couple the myocytes. The large number of gap junctions permits the rapid spread of electric current from cell to cell so that all parts of the heart are rapidly stimulated virtually as one. This electric current is the signal that increases the intracellular concentration of Ca^{2+} within the myocytes. The increase in Ca^{2+} triggers contraction (cell shortening) in a manner that is similar in many ways to skeletal muscle cells (see Chapter 44).

Myogenic hearts are electrically excitable and generate their own action potentials. The rate and forcefulness of the beating of myogenic hearts can, however, be regulated by the nervous system. Nonetheless,

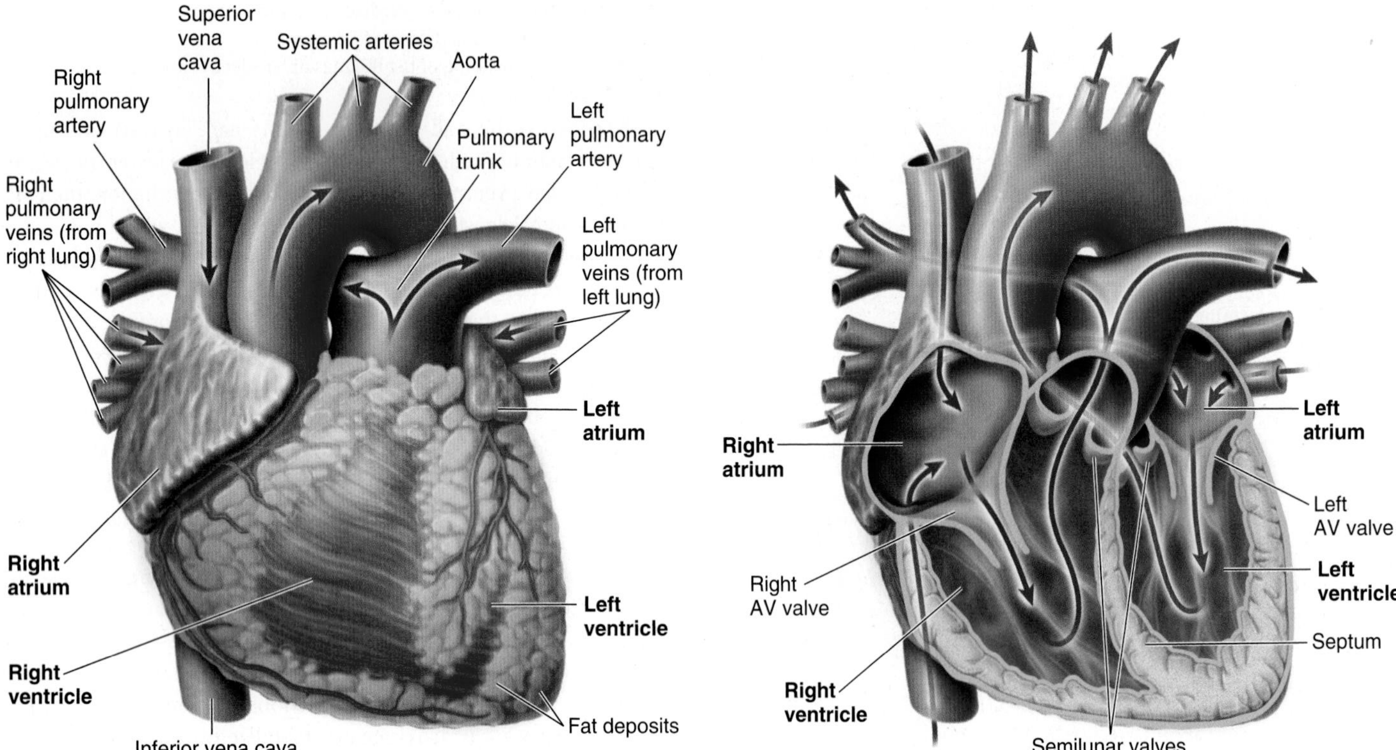

Figure 47.8 **The mammalian heart and circulation.** The figure shows the major blood vessels entering and leaving the heart, the locations of the valves, and the direction of blood flow through the chambers of the heart. Oxygenated blood is shown in red, deoxygenated in blue. Note that the pulmonary veins carry oxygenated blood because they return blood from the lungs, whereas veins from the systemic circulation carry deoxygenated blood.

myogenic hearts continue to beat on their own if dissected out of an animal and placed in a nutrient bath, even with no nerves present.

Excitation of the Vertebrate Heart Begins in the SA Node in the Atria and Spreads to the Ventricles via the AV Node

The electrical excitation of the vertebrate heart has two phases: atrial and ventricular. In atrial excitation, electrical signals are generated at the junction of the veins and single atrium in fishes, and within the wall of the right atrium in other vertebrates, at what is called the **sinoatrial (SA) node**, or **pacemaker** (Figure 47.9). The SA node is a collection of modified cardiac cells that have an inherently unstable resting membrane potential. Ion channels in the membranes of these cells are opened spontaneously and allow the influx of positively charged ions into the cytosol, thereby depolarizing the cell. These depolarizations produce action potentials in the SA node cells with a range of frequencies that are characteristic for a given species; the frequency determines an animal's heart rate. Once the SA node cells generate action potentials, the potentials quickly spread across one or both atria through the gap junctions described earlier. The action potentials trigger an influx of Ca^{2+} into the muscle cell cytosol, which activates contraction. Because the impulses spread very rapidly across the atria, both atria contract together almost as if they were one large muscle cell. Atrial contraction pumps blood through the AV valves into the ventricles.

To begin ventricular excitation, action potentials initiated in the SA node first reach another node of specialized cardiac cells, the **atrioventricular (AV) node**. The AV node is located near the junction of the atria and ventricles and conducts the electrical signals from the atria to the ventricles. This is important, because the atria and ventricles are separated by a connective tissue septum that does not contain gap junctions. Like the SA node, the AV node is electrically excitable, but its cells require a longer time to become excited than do the cells of the atria. This allows time for the atria to contract before the ventricles do. Fibers branching from the AV node spread electrical impulses along a conducting system of cardiac muscle cells called Purkinje fibers, which are specialized to rapidly conduct electricity. These fibers branch throughout the muscular walls of the ventricles, ensuring that all of the ventricular muscle gets depolarized quickly and nearly simultaneously (see Figure 47.9). In this way, the entire mass of both ventricles contracts nearly in unison in response to depolarization.

The Cardiac Cycle Has Two Phases: Diastole and Systole

Each beat of the vertebrate heart requires the coordinated activities of the atria and ventricles. The contraction and relaxation events that produce a single heartbeat are known as the **cardiac cycle**, which can be divided into two phases (see Figure 47.9). In the first phase, **diastole**, the ventricles are relaxed and fill with blood coming from the

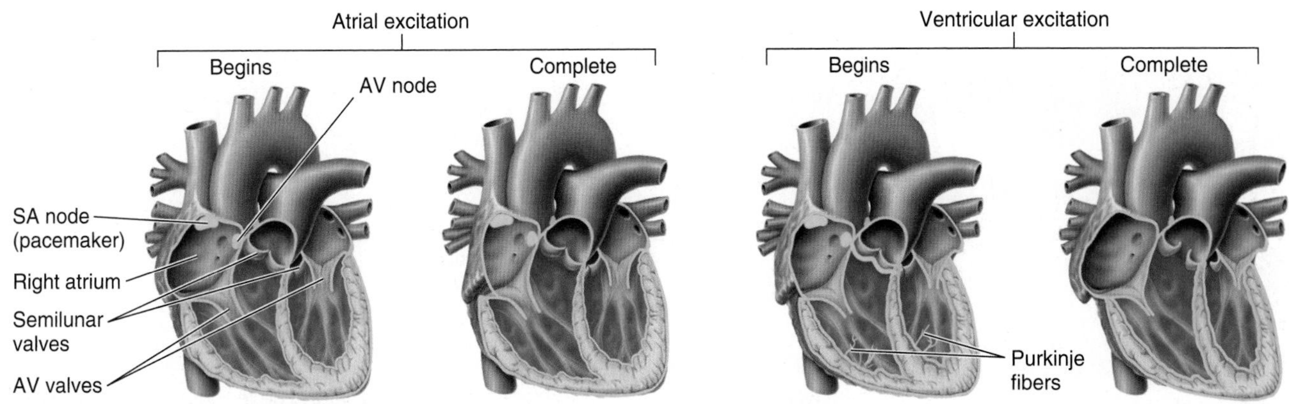

AV valves	OPEN	OPEN	Beginning to close	CLOSED
Semilunar valves	CLOSED	CLOSED	CLOSED	OPEN
Phase of cardiac cycle	Diastole	Diastole	Systole beginning	Systole
Atria	Relaxed	Contracted	Beginning to relax	Relaxed
Ventricles	Relaxed	Relaxed	Beginning to contract	Contracted
Chambers with highest pressure	All are low pressure	Atria (slightly)	Ventricles	Ventricles

Figure 47.9 Electrical and mechanical activity in the mammalian heart. Electrical activity, which is depicted in yellow, begins in the SA node and quickly spreads through the atria to the atrioventricular (AV) node. Branches from the AV node transmit electrical activity throughout the ventricles along the Purkinje fibers. The atria and ventricles do not contract until they have become electrically excited. Changes in fluid pressure gradients between the atria and ventricles, and between the ventricles and the aorta and pulmonary arteries, are the forces that open or close the four heart valves. *Note*: Most of the aortic valve is not visible in this drawing.

atria through the open AV valves. At the end of diastole, the atria contract and provide a boost of additional blood filling the ventricles. In the second phase, **systole**, the ventricles contract and eject the blood through the open semilunar valves.

The valves open and close as a result of pressure gradients established between the atria and ventricles, and the ventricles and arteries. During diastole, pressure in the ventricles is lower than in the atria and the arteries; therefore, the AV valves are open, but the semilunar valves are closed (Figure 47.10a). Once the ventricles are electrically excited, they begin contracting, and the pressure within the ventricles rapidly increases. This marks the beginning of systole. When the pressure in the ventricles exceeds that in the atria, the AV valves are forced closed. Closure of the AV valves makes the "lub" sound of the familiar "lub-dub" heard through a stethoscope. Ventricular pressure continues to rise; when it exceeds the pressure in the arteries, the semilunar valves open (Figure 47.10b). Blood is ejected through the open semilunar valves into the arteries.

Because the ventricular walls are thicker and stronger than the atrial walls, ventricular contractions generate much greater pressure. The pressure exerted by the ventricles on the blood within their cavities causes the AV valves to shut, preventing blood from flowing backward into the atria. This is why the electrical delay that is built into the AV node is important. If the atria and ventricles were excited and contracted at the same time, the AV valves would close too soon, and the ventricles would not receive their normal volume of blood.

The pressure in the arteries increases during systole. As this is happening, the ventricles begin to return to their resting, unexcited state, and the pressure in the ventricles decreases below that in the

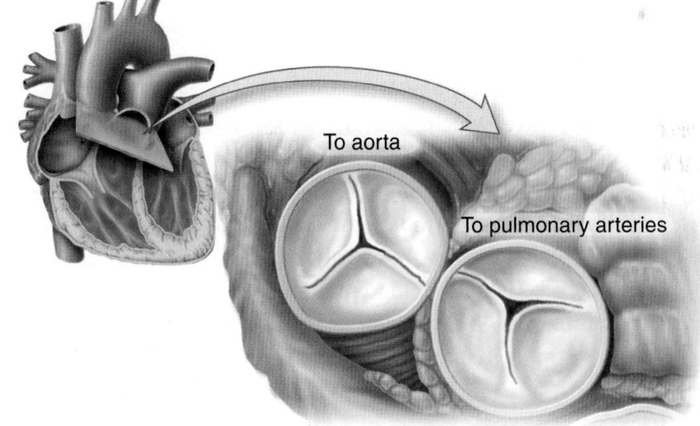

(a) Semilunar valves in nearly closed position

(b) Semilunar valves in opened position

Figure 47.10 Appearance and function of the semilunar valves. The four heart valves are sheathlike structures with flaps that open and close, as in this illustration of semilunar valves.

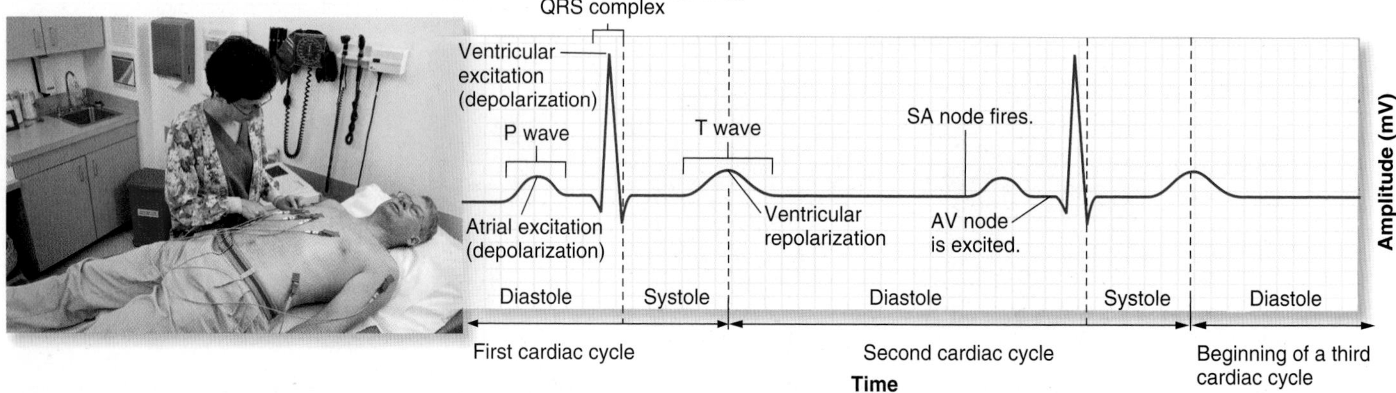

Figure 47.11 **An electrocardiogram (ECG).** Electrodes placed on the skin detect electrical impulses occurring in the heart, and an ECG visualizes them. The resultant waveform is a useful indicator of cardiac health. Note that the ventricular wave (QRS complex) is taller than the atrial wave (P wave). This is because the ventricles are larger than the atria and generate more electrical activity.

Concept Check: Why is it possible to detect electrical changes in the heart using electrodes placed on the surface of the skin? (Hint: Think about body fluids and their ability to conduct electricity.)

arteries. The higher pressure in the arteries closes the semilunar valves, which prevents blood from flowing back into the heart. Closure of the semilunar valves creates the second heart sound, "dub," heard through a stethoscope. Throughout systole, meanwhile, the atria continue to fill with blood, which raises the pressure in the atria. Soon the pressure in the atria exceeds that in the relaxed ventricles, and the AV valves open again, bringing a new volume of blood into the ventricles and starting a new diastole.

Blood pressure, which is defined as the force exerted by blood on the walls of blood vessels, is highest in the arteries during systole and lowest during diastole. For this reason, blood pressure is measured with two numbers, the systolic and diastolic pressures. For historical reasons, the units are usually given as mmHg (millimeters of mercury). Of vertebrates, blood pressures are highest in mammals and birds, and lowest in fishes. For example, a typical blood pressure in humans is around 120/80 mmHg (systolic/diastolic).

An ECG Tracks Electrical Events During the Cardiac Cycle

An **electrocardiogram** (**ECG** or **EKG**, the latter reflecting the original German spelling) is a medical test used to investigate the function of the heart. An ECG is a record of the electrical potentials between various points on the body. These potentials arise from the electrical signals generated during the cardiac cycle. Sensitive electrodes are placed on the surface of the body to monitor the wave of electricity initiated by the SA node, which travels through the atria, AV node, and ventricles. This procedure works because the body fluids that surround the heart conduct electricity, even the very weak impulses generated by a beating heart.

The trace on an ECG reveals several waves of electrical excitation (Figure 47.11). The first is the P wave, which begins when the SA node fires and ends when the two atria completely depolarize. In the cardiac cycle, the P wave is followed by atrial contraction, while the ventricles are in diastole. The next wave of excitation is a cluster of three waves, called the QRS complex. It begins when the branches from the AV node excite the ventricles and ends when both ventricles depolarize completely. In the cardiac cycle, the QRS complex is followed by ventricular contraction during systole. The final wave is the T wave, which results from the repolarization of the ventricles back to their resting state and is associated with ventricular relaxation as diastole begins again. No wave is visible for atrial repolarization because it occurs simultaneously with the large QRS complex.

The ECG monitor displays both the amplitude (strength) of the electrical signal and the direction that the signal is moving in the chest. From this information, physicians can determine whether a person's heart signals have a normal frequency, strength, duration, and pattern.

47.4 Blood Vessels

Learning Outcomes:

1. Explain the anatomic and functional distinctions among arteries, arterioles, capillaries, venules, and veins.
2. Describe the interaction of the lymphatic system with the circulatory system.

Now that we have discussed the vertebrate heart, let's examine the vessels that transport blood to and from the heart and throughout the body. Figure 47.12a illustrates the route that blood follows in a closed circulatory system. Blood is pumped by the heart to large arteries and then flows to small arteries and eventually to the smallest arteries, called arterioles. Arterioles bring blood to the smallest vessels, which are called capillaries. Gas and nutrient exchange occurs between the blood in the capillaries and the cells surrounding the capillaries. The blood then flows back to the heart, leaving the capillaries through the smallest veins, called venules, to small veins and, finally, to large veins.

Arteries Distribute Blood to Organs and Tissues

Arteries are thick-walled vessels that consist of layers of smooth muscle and connective tissue wrapped around a single-celled inner layer, the **endothelium**, which forms a smooth lining in contact with the blood (**Figure 47.12b**). Because thick layers of tissue surround the endothelium, most dissolved substances cannot diffuse across arteries. Instead, arteries act as conducting tubes that distribute blood leaving the heart to all the organs and tissues of an animal's body.

In vertebrates, the walls of the largest arteries, such as the aorta, also contain one or more layers of elastin, a protein with elastic properties (refer back to Figure 10.3). As the aorta stretches to accommodate blood arriving from the heart, the elastin layers also stretch. The thick layers of tissue in the aorta and other large arteries prevent them from stretching more than a small amount. When the heart relaxes as it readies for another beat, the elastin layers in the aorta and largest arteries recoil to their original state, something like releasing a stretched rubber band. The recoiling vessels generate a force on the blood within them; this helps prevent blood pressure from decreasing too much while the heart is relaxing during diastole.

Arterioles Distribute Blood to Capillaries

As arteries carry blood away from the heart, they branch repeatedly and become narrower to penetrate to the smallest reaches of an organ or tissue (**Figure 47.13**). Eventually, the vessels are little more than a single-celled layer of endothelium surrounded by one or two layers of smooth muscle and connective tissue (see Figure 47.12). These **arterioles** distribute blood to regions of the body in proportion to metabolic demands. This is accomplished by changing the diameter of arterioles, such that they widen, or dilate, in areas of high metabolic activity and narrow, or constrict, in inactive regions. Arterioles dilate when the smooth muscle cells around them relax, and constrict when these cells contract.

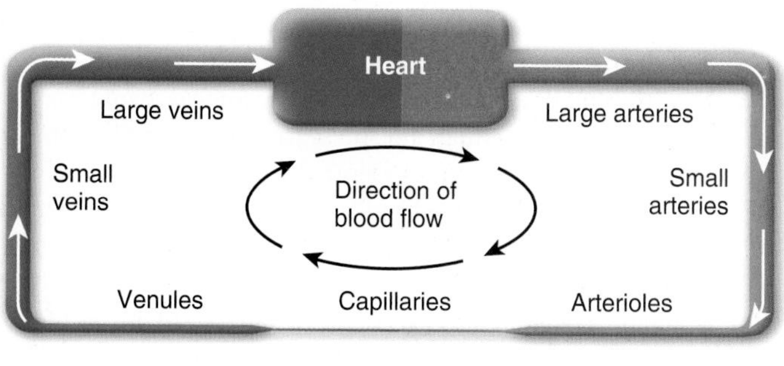

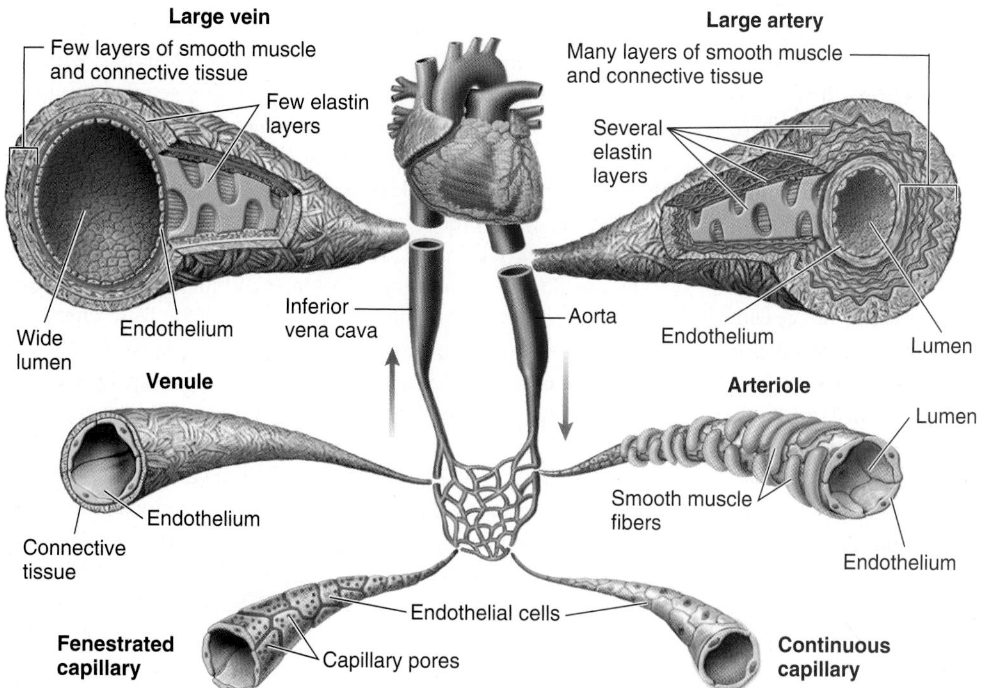

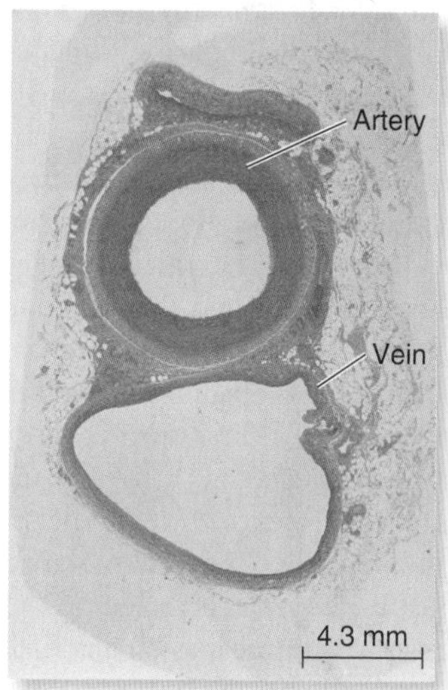

Figure 47.12 **Comparative features of blood vessels.** **(a)** Overview of blood flow through vessels in a closed circulatory system. Regions of gas exchange with the environment (e.g., lungs) have been omitted for simplicity. **(b)** Blood vessel anatomy. Sizes are not drawn to scale. Both types of capillaries are illustrated, but they do not actually appear together in this way. Fenestrated capillaries are typically found in secretory glands, such as the pancreas, or organs where blood is filtered, such as the kidney. Inset: Light micrograph (enlarged four times) of a medium-sized artery near a vein. Note the difference between the two vessels in wall thickness and lumen diameter.

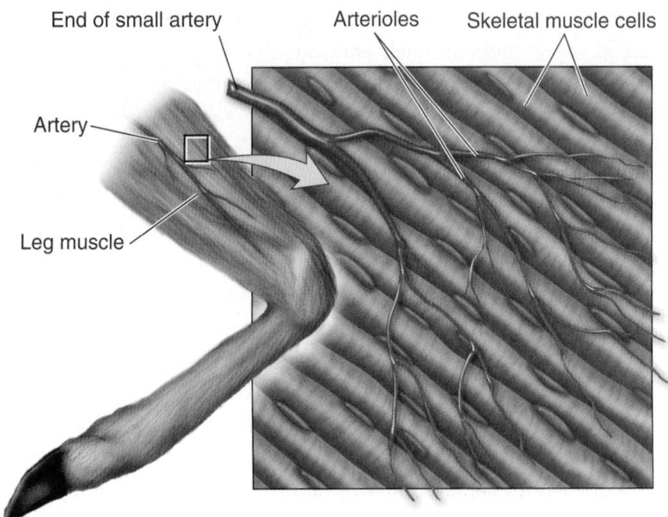

Figure 47.13 **Arteries and arterioles.** Arteries branch into smaller arteries, and into even smaller arterioles, in order to penetrate to all the cells of tissues and organs.

Concept Check: *What happens to the diameter of the arterioles of the animal's leg muscle when the animal is active?*

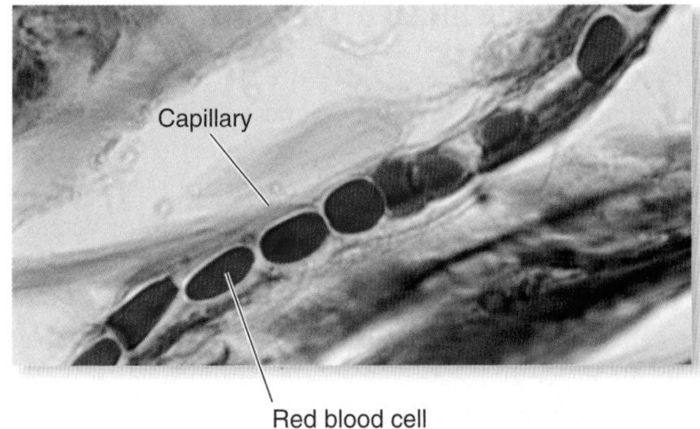

Figure 47.14 **Red blood cells moving through a capillary in single file, as seen through a light microscope.**

Capillaries Are the Site of Gas and Nutrient Exchange

Arterioles branch into the tiny thin-walled capillaries. At the capillaries, materials in the blood are delivered to the other cells of the body, and waste products and secretions from cells are delivered to the blood. **Capillaries** are tubes composed of a single-celled layer of endothelium resting on a layer of extracellular matrix called a basal lamina. Most capillaries have smooth walls and are the most common; fenestrated capillaries contain tiny openings called fenestrations (see Figure 47.12). Capillaries are the narrowest blood vessels in the body; essentially every cell in an animal's body is near one. The diameter of a capillary is about the same as the width of a red blood cell, so erythrocytes move through them in single file (**Figure 47.14**).

Blood enters capillaries under pressure that is created by the beating of the heart as blood is pumped into the arteries. This pressure forces some of the water in blood out through the fenestrations and other tiny openings in capillary walls into the interstitial fluid (**Figure 47.15**). These openings are wide enough to permit water with its solutes, including O_2, to leave the capillary, but not red blood cells and most proteins. Once in interstitial fluid, the solutes may diffuse into body cells. Conversely, body cells release secretions and metabolic waste products including CO_2 gas into interstitial fluid, from where they diffuse into capillaries to be carried away.

If the water that leaves a capillary were to remain in the interstitial fluid, the volume of plasma in the blood would decrease, and the interstitial fluid would swell. Most of the water that leaves at the beginning of a capillary, however, is recaptured at the capillary's end, due to two phenomena (see Figure 47.15). First, the blood pressure inside the capillary near the arteriole is higher than the fluid pressure of the surrounding interstitial fluid, which creates a force that moves water out of the capillary. However, by the time blood reaches the end of a capillary, its pressure has decreased considerably, partly as a result of the loss of water. Second, proteins are trapped within the capillary because they are too large to diffuse through the capillary walls or to leak out of the tiny openings. This creates an osmotic force that draws water back into the capillary from the interstitial fluid. Near the end of a capillary, the osmotic force drawing water into the capillary is greater than the pressure forcing water out; therefore, water and dissolved solutes reenter the capillary.

Despite the ability of capillaries to recapture their own fluid, the process is not 100% effective. Another set of vessels, those of the **lymphatic system**, collects excess fluid and returns it to the blood. The lymphatic system is also part of the defense mechanisms of the animal body (see Chapter 53).

Venules and Veins Return Blood to the Heart

Once blood travels through capillaries, picking up any substances secreted from the cells of the body, it enters the **venules**, which are small, thin-walled extensions of capillaries. The venules empty into the larger veins that return blood to the heart. The walls of veins are much thinner, less muscular, and more easily distended or stretched than those of arteries (see Figure 47.12).

By the time blood has traveled through the capillaries and reached the veins, the pressure of the blood is very low. Consequently, veins can fill with considerable volumes of blood, particularly veins in the lower parts of an animal's body, such as the legs, where gravity tends to cause blood to pool. Several factors assist the blood on its way toward the heart when flowing against gravity. One factor is stimulation of the contraction of smooth muscle in leg veins by the sympathetic nervous system, which compresses the veins and helps force blood back to the heart. Another factor is the activity of skeletal muscles in the limbs that assists the return of venous blood to the heart. For example, each time the leg skeletal muscles contract, they squeeze the veins passing through them (**Figure 47.16**). This alone would not move blood upward toward the heart, though, because each time a vein is squeezed, blood could be forced both up and down. This does not happen because one-way valves inside veins ensure that blood returning from below the heart moves in only one direction, toward the heart. By contrast, veins located above the heart,

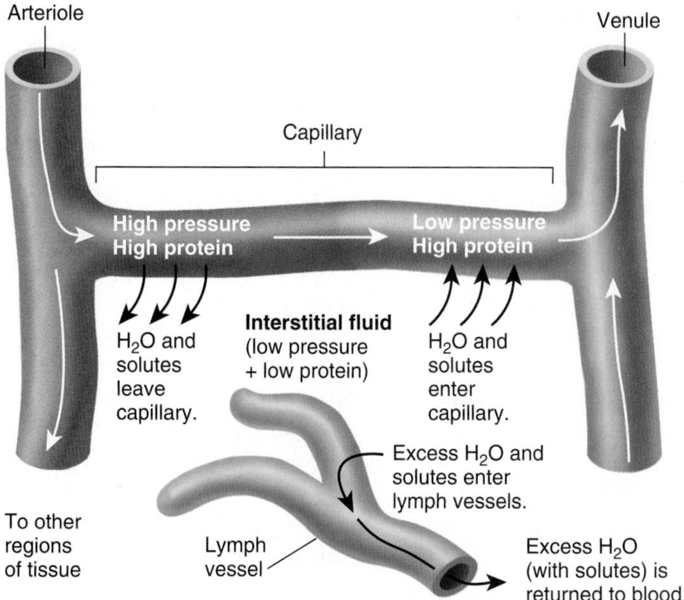

Figure 47.15 Water movement between capillaries and interstitial fluid. Water and dissolved solutes (other than proteins) exit capillaries near the arteriolar end because the capillary pressure is much greater than that of the interstitial fluid. As the volume of the water in the capillary decreases, however, the pressure within the capillary also decreases (but remains greater than interstitial fluid). Proteins remaining in the capillary contribute an osmotic force that tends to draw water back into the capillary. The combination of decreased pressure within the capillary and the osmotic force due to the proteins leads to the recapture of much of the water that left the capillary. Lymph vessels drain any excess fluid from the interstitial fluid. For clarity, the system shown here contains a single capillary connecting an arteriole and venule, instead of the typical situation where capillaries divide into numerous branching vessels.

like those in the necks of bipeds and some quadrupeds, lack valves because gravity pulls blood toward the heart.

Many people can easily observe the effects of gravity on venous blood flow. When the arms are held down by their side, the veins are visible on the backs of their hands. When the arms are raised above the head, the bulging veins quickly lose blood and become less visible. When blood from the veins returns to the heart, it must travel against gravity when your arms are at your side and with gravity when your arms are elevated. Blood drains from veins much more efficiently when gravity works in its favor.

47.5 Relationship Among Blood Pressure, Blood Flow, and Resistance

Learning Outcomes:

1. Describe the relationship among blood pressure, blood flow, and resistance to flow.
2. Distinguish between the effects of vasodilation and vasoconstriction on blood flow.
3. Explain how resistance and cardiac output determine blood pressure and flow.

Blood pressure is responsible for blood flow, the movement of blood through the vessels. It is not the same in all regions of an animal's body, however, because of resistance. **Resistance (R)** refers to the tendency of blood vessels to slow down the flow of blood through their lumens. As we saw earlier, the blood pressure in arteries is much higher than in veins. For this reason, we generally speak only of arterial pressure when discussing blood pressure. The relationship among blood pressure, blood flow, and resistance can be considered on two levels—local and systemic—as we see next.

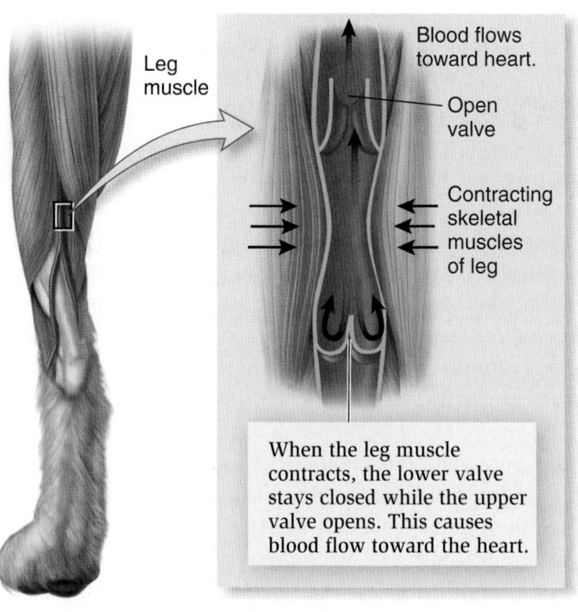

Figure 47.16 One-way valves in veins. Valves are typically present in the limbs, as shown in this dog's leg. **(a)** In some veins, one-way valves assist the return of blood to the heart against the force of gravity. **(b)** Blood moving in a vein without valves would flow in both directions.

Concept Check: Unlike most mammals, giraffes have one-way valves in the veins of their long necks. Which way do you think the valves open, toward the heart or toward the head, and why? (Hint: Picture the way a giraffe drinks from a body of water.)

Resistance Determines Local Blood Pressure and Flow

The physical characteristics of a blood vessel determine the amount of blood that flows through it and how fast that blood flows. This affects how much blood reaches particular areas of the body at any given time. For example, the diameter of different blood vessels (particularly arterioles) can be increased or decreased to change the amount of blood flowing through them. As stated earlier, the ability to distribute blood to different parts of the body in amounts proportional to that body part's metabolic requirements is a key adaptation of closed circulatory systems.

Resistance Resistance is a function of three variables: vessel radius, length, and blood viscosity. Wide, short tubes provide less resistance to flow than narrow, long tubes. Resistance is increased by the blood's viscosity, which is a measure of the blood's hematocrit. A high hematocrit increases the blood's viscosity—it makes the blood more sludgelike—and hinders its smooth flow through vessels.

The relationship among blood pressure, blood flow, and resistance is stated by Poiseuille's law, which was derived in the 1840s by Jean Marie Louis Poiseuille, a French physician and physiologist. His law is simplified here:

$$\text{Flow } (F) = \Delta \text{ Pressure } (P)/\text{Resistance } (R)$$

Stated mathematically, blood flow through a blood vessel is directly proportional to the difference (Δ) in pressure of the blood between the beginning and end of the vessel, and inversely proportional to the resistance created by that vessel. Changes in any of these three variables determine how much blood flows through different body regions at any moment. The equation can be rearranged as $\Delta P = F \times R$, which demonstrates that blood pressure depends on both blood flow and resistance. Poiseuille's law applies to blood flow through a single vessel, an organ, or the entire body.

Vasodilation and Vasocontriction Changes in arteriolar resistance are the major mechanism for increasing or decreasing blood flow to a region. In the short term, the length of an arteriole and the viscosity of blood do not normally change. Therefore, the radii of arterioles become the most important factor in determining minute-to-minute resistance.

The relationship between arteriolar radius and resistance is not linear. Resistance is inversely proportional to the radius of the vessel raised to the fourth power: $R \propto 1/r^4$, where $\propto$ means "proportional to," and r is the radius of the arteriolar lumen. Let's consider an arteriole with a radius that increases by a factor of 2. This would occur if the smooth muscles around the arteriolar wall relax sufficiently to allow the vessel to dilate and double its original radius. Because resistance is inversely proportional to the fourth power of vessel radius, an increase in radius by a factor of 2 will result in a decrease in resistance of 2^4 ($2 \times 2 \times 2 \times 2$), or 16-fold.

Vasodilation refers to an increase in blood vessel radius, and **vasoconstriction** refers to a decrease in blood vessel radius. The signals that control arteriolar radius come from three sources: locally produced substances, hormones, and nervous system inputs. Locally, metabolic by-products such as carbon dioxide, lactic acid, and other substances secreted by metabolically active tissues, cause nearby arterioles to vasodilate immediately. According to Poiseuille's law, this permits more blood flow to the active region, facilitating oxygen and nutrient delivery and waste removal. Hormones secreted by glands throughout the body can also regulate arteriolar radius. For example, some hormones cause arterioles that deliver blood to the small intestine to vasoconstrict during the fight-or-flight response, routing blood away from the intestine and to areas of more immediate need such as the heart and skeletal muscles. Smooth muscle cells that make up the outer wall of arterioles also receive inputs from nerves that can stimulate the muscles to contract or relax. One of the most important regulators of blood vessel diameters, however, is a gas, as we see next.

FEATURE INVESTIGATION

Furchgott Discovered a Vasodilatory Factor Produced by Endothelial Cells

As we have seen, the cells surrounding blood vessels can produce metabolic by-products that diffuse to arteriolar smooth muscle cells and cause these cells to relax and vasodilate. Beginning in the 1970s, American biochemist Robert Furchgott provided evidence that an unidentified substance released from within an artery—that is, from the endothelial lining of the vessel—also played a key role in regulating blood vessel radius.

For many years, Furchgott had used flattened strips of rabbit aorta in a culture bath to test whether various compounds stimulated or inhibited contraction of the smooth muscle in the artery. Although large arteries such as the aorta do not show the dramatic vasodilation and vasoconstriction that arterioles do, they were more useful in experiments because of their larger size.

One of the compounds he tested was the neurotransmitter acetylcholine (ACh). Scientists knew at the time that when ACh is injected into an animal, it causes vasodilation by relaxing smooth muscle in arterioles. However, in Furchgott's in vitro preparations, ACh caused the flattened strips of artery to contract. To determine if this apparent paradox could be the result of the method in which the tissues were prepared for in vitro tests, Furchgott compared the effects of ACh in different types of preparations. He discovered that ACh produced muscle relaxation when applied to circular rings of aorta, as described in step 1 in **Figure 47.17**.

How did Furchgott explain that the muscle contracted when ACh was applied to the flattened strips of aorta? After examining the two preparations of tissue, Furchgott realized that the lining of endothelial cells along the inner wall of the vessel had been scraped away during preparation of the strips, but the endothelial lining was still present in the rings of aorta. He hypothesized that the endothelial cells must produce a factor whose secretion was stimulated by ACh, and that not only prevented contraction, but even caused smooth muscles surrounding the blood vessel to relax. If so, this would explain the observation that when ACh was injected into rabbits, it resulted in vasodilation, because the intact vessels of the animals contained their endothelial lining and thus resembled Furchgott's aortic rings.

To test his hypothesis, Furchgott performed two experiments. As shown in step 2 of Figure 47.17, he used a wooden rod to scrape away the endothelium from circular rings of aorta and discovered that the relaxation effect of ACh was lost, as predicted. In step 3, he used a sandwich technique, in which he tested the effects of ACh on muscle contraction in a "denuded" strip of artery with the endothelium removed, and then again on the same strip, which was attached ("sandwiched") to another strip in which the endothelium remained intact. When ACh was added to the water bath before the sandwich was formed, the denuded strip contracted, as expected. When the denuded strip was attached to the intact strip in such a way that its muscle layer was exposed to the intact strip's endothelial surface, the denuded strip relaxed after ACh was added. This was consistent with the hypothesis that the endothelium released a vasodilatory factor that diffused to the denuded strip and relaxed its muscles.

Two other scientists, Americans Louis Ignarro and Ferid Murad, later determined that this vasodilatory factor was the gas nitric oxide (NO). The discovery of this function of NO revealed a new category of signaling molecules. Ignarro and Murad also revealed that nitroglycerin, which had for many years been used to treat patients with cardiovascular disease, acted by generating NO in the blood to produce vasodilation and lower blood pressure. Today NO is considered one of the most potent and important naturally occurring vasodilators in animals. Knowledge of the NO signaling pathway has led to new therapies for treating a wide variety of health disorders associated with blood vessels, including high blood pressure, glaucoma (an eye disease), and erectile dysfunction (see Chapter 51). For their efforts, Furchgott, Ignarro, and Murad shared the 1998 Nobel Prize in Physiology or Medicine.

Figure 47.17 **Furchgott's discovery that endothelial cells produce a vasodilatory substance.**

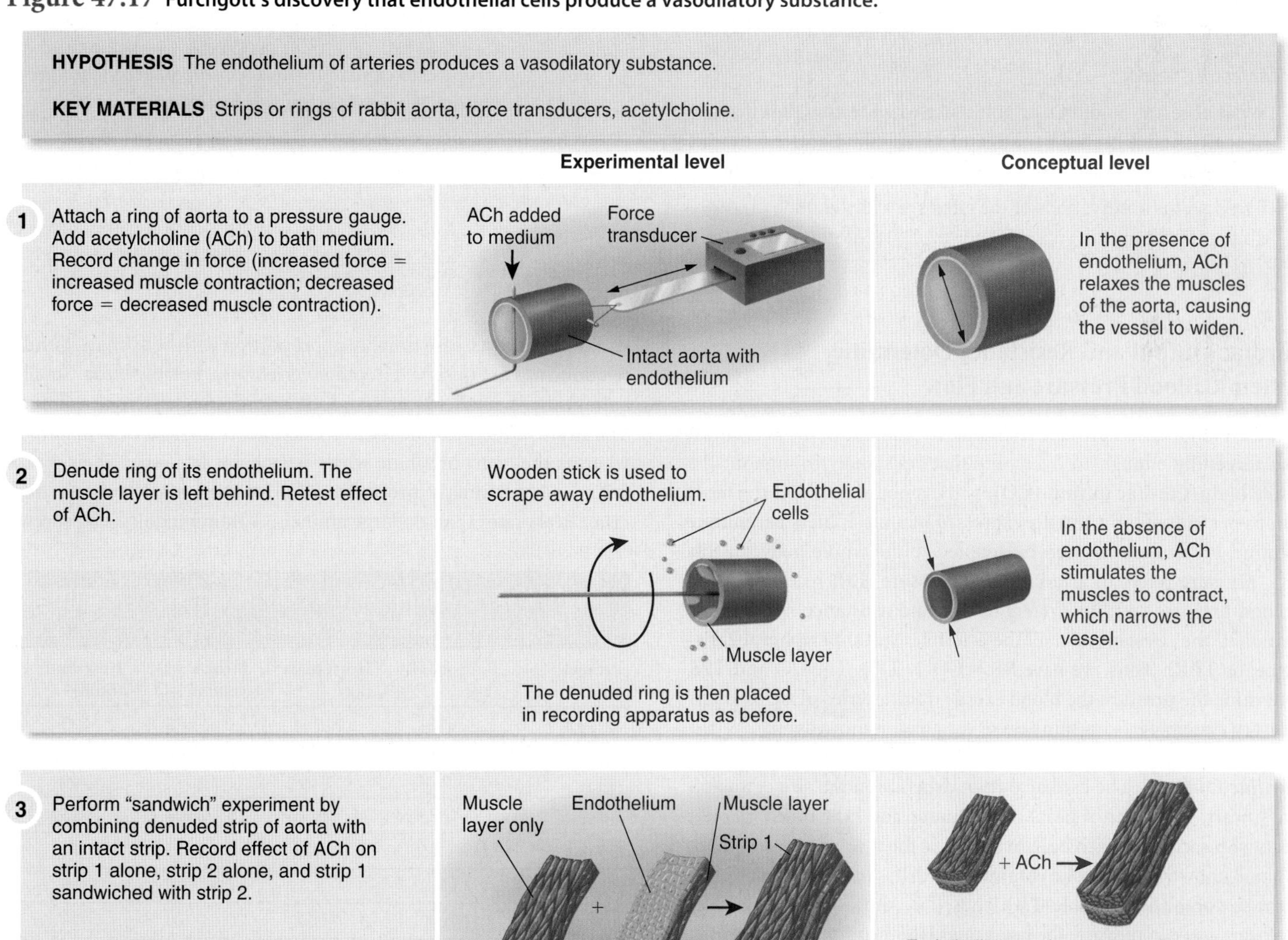
HYPOTHESIS The endothelium of arteries produces a vasodilatory substance.
KEY MATERIALS Strips or rings of rabbit aorta, force transducers, acetylcholine.
Experimental level
Conceptual level
1 Attach a ring of aorta to a pressure gauge. Add acetylcholine (ACh) to bath medium. Record change in force (increased force = increased muscle contraction; decreased force = decreased muscle contraction).
ACh added to medium
Force transducer
Intact aorta with endothelium
In the presence of endothelium, ACh relaxes the muscles of the aorta, causing the vessel to widen.
2 Denude ring of its endothelium. The muscle layer is left behind. Retest effect of ACh.
Wooden stick is used to scrape away endothelium.
Endothelial cells
Muscle layer
The denuded ring is then placed in recording apparatus as before.
In the absence of endothelium, ACh stimulates the muscles to contract, which narrows the vessel.
3 Perform "sandwich" experiment by combining denuded strip of aorta with an intact strip. Record effect of ACh on strip 1 alone, strip 2 alone, and strip 1 sandwiched with strip 2.
Muscle layer only
Endothelium
Muscle layer
Strip 1
Strip 2
Strip 1
Strip 2
Sandwich
+ ACh
Endothelium produces a vasodilating substance following ACh treatment that diffuses from strip 2 to strip 1, and causes strip 1 to relax.

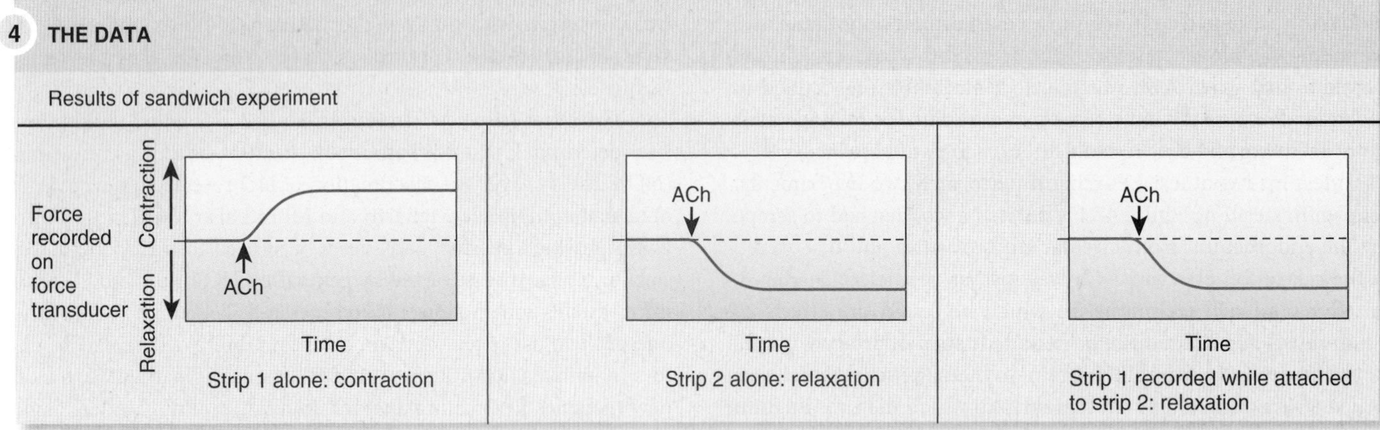

5 **CONCLUSION** Endothelial cells release a factor that diffuses to the surrounding smooth muscle, causing it to relax. Relaxation of the vascular smooth muscle causes dilation of the blood vessel.

6 **SOURCE** Furchgott, R.F., and Zawadzki, J.V. 1980. The obligatory role of endothelial cells in relaxation of arterial smooth muscle by acetylcholine. *Nature* 288:373–376.

Experimental Questions

1. What observation did Furchgott make when testing the effects of acetylcholine on different treatments of rabbit aorta? How did he explain the results?
2. Based on his observations of the effects of acetylcholine on different preparations of rabbit aorta, what hypothesis did Furchgott propose to explain the relationship of acetylcholine and vasodilation? How did he test this hypothesis?
3. What did Furchgott conclude based on his results?

Cardiac Output and Resistance Determine Systemic Blood Pressure and Flow

Because blood vessels provide resistance to blood flow, the heart must beat forcefully enough to overcome that resistance throughout the whole body. **Cardiac output (CO)** is the amount of blood the heart pumps per unit of time, usually expressed in units of liters per minute (L/min). Poiseuille's law can be adapted to the whole body. In this case, pressure refers to arterial blood pressure (BP) recorded using a blood pressure cuff, flow refers to CO, and resistance refers to the sum of all the resistances in all the arterioles (**total peripheral resistance**, or **TPR**). Thus, we have BP = CO × TPR. The CO and TPR determine the pressure the blood exerts in the arteries of a closed circulatory system. Let's examine these variables in more detail.

Cardiac Output The cardiac output depends on the size of an animal's heart, how often it beats each minute, and how much blood it ejects with each beat. Each beat, or stroke, of the heart ejects an amount of blood known as the **stroke volume (SV)** that is roughly proportional to the size of the heart (**Table 47.1**). Thus, if we know the stroke volume of a heart and can measure the heart rate (HR, the number of beats per minute, bpm), we can determine the CO. Simply put, CO = SV × HR.

Unsurprisingly, the CO of an elephant is far greater than that of a human, a dog, or a shrew. Note in Table 47.1 that heart mass and stroke volume increase roughly in proportion with body mass. Similar relationships are observed in other vertebrates, notably birds. Smaller animals have smaller hearts and, therefore, smaller stroke volumes. The heart of a typical shrew, for example, is the size of a small pea, whereas the heart of a blue whale is as large as a cow. Their stroke volumes are similarly proportioned. However, small animals have faster heart rates than do large animals. A hummingbird's or shrew's

Table 47.1 Comparative Features of Representative Mammalian Hearts

Animal	Body mass* (kg)	Heart mass* (kg)	Stroke volume* (L)	Heart rate* (bpm)†
Shrew‡	0.0024 (2.4 g)	0.000035 (35 mg)	0.000008 (8 μl)	835
Rat	0.20	0.001	0.0018	360
Dog	23	0.12	0.025	95
Human	75	0.38	0.075	70
Elephant	4,000	25	4.0	25
Blue whale	100,000	600	100	10

*Values are based on average body masses and resting conditions. In some cases, stroke volumes are estimates based on heart size. †bpm = beats per minutes. ‡The shrew reported here is the Etruscan shrew, one of the smallest known mammals. Its heart is somewhat larger than would be predicted for its body mass. Note its heart rate; at 835 bpm, the heart beats 14 times per second!

resting heart rate may be more than 800 bpm, whereas a blue whale's heart may beat only 10 times per minute (although the amount of blood ejected with each of those beats is enormous!). Recall from Chapter 46 that metabolic rate is relatively greater in smaller animals. The higher heart rates of small animals give them a greater cardiac output than would be predicted for the size of their hearts, which helps them meet the extraordinary oxygen and nutrient demands of such highly metabolic organisms.

Systemic Blood Pressure The greater the cardiac output and resistance to blood flow, the higher the blood pressure. Imagine that the circulatory system is like a faucet (the heart) connected to a garden hose (the arteries) (**Figure 47.18**). If the faucet is fully open, analogous to maximal cardiac output, and the hose is not blocked, analogous to low arteriolar resistance, the amount of water rushing into the hose will be high and so will the water pressure, representing blood pressure. If the faucet is only partially open, the water pressure will be lower. However, now imagine that the faucet is partially open but the end of the hose is constricted, representing a region of high arteriolar resistance. In that case, the pressure of the water in the hose will increase between the faucet and the point where the hose is squeezed, and will decrease beyond the constriction, as will the flow of water (representing blood).

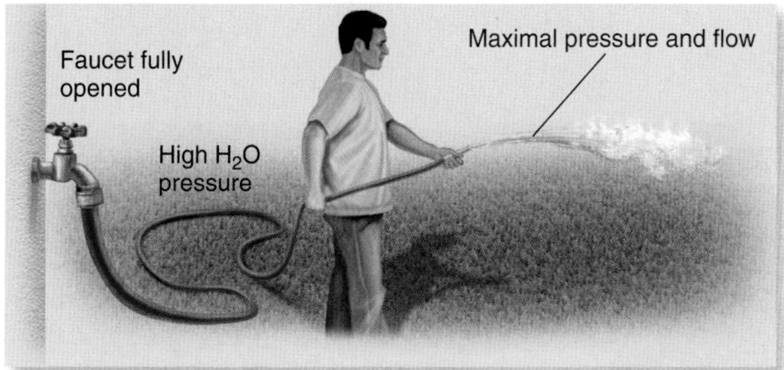

(a) Maximal cardiac output with low resistance

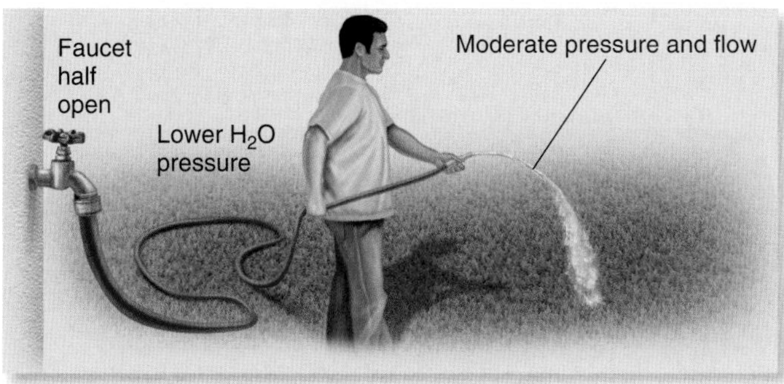

(b) Moderate cardiac output with low resistance

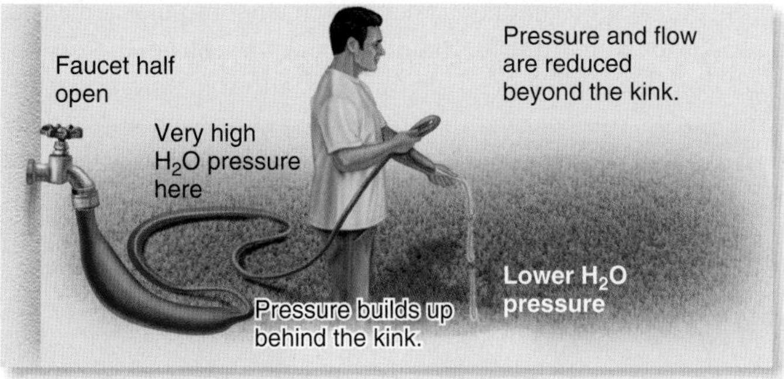

(c) Moderate cardiac output with high resistance

Figure 47.18 The relationship between cardiac output, resistance, and blood pressure. A hose analogy shows the way in which cardiac output (the faucet) and resistance (constriction of the hose) affect blood pressure.

Arterial blood pressure, therefore, is a function of how hard the heart is working and how constricted or dilated the various arterioles are. Blood pressure must be high enough for blood to reach all body tissues even at the farthest extremities but not high enough to damage blood vessels or force excess plasma out of capillaries. Blood pressure is roughly similar among mammals. Although species differences exist, they are not as diverse as, for example, species differences in heart rates. Blood pressure is relatively low in nonmammalian vertebrates and in invertebrates.

47.6 Adaptive Functions of Closed Circulatory Systems

Learning Outcomes:

1. Explain how the circulatory system adapts to exercise.
2. Describe what a baroreceptor is, and why such receptors are important features in the ability to respond to a change in blood pressure.

A circulatory system must not only pump blood and return it to the heart, it must also adapt to changes, including sleep, activity, and emergencies such as blood loss or dehydration. In this section, we examine how changes in the circulatory system help maintain cardiovascular activity during exercise and when an animal experiences a change in blood pressure.

Circulatory Function Adapts to Exercise

Nearly all animals have periods of rest and activity. Therefore, the activity of the heart must adjust to fluctuating metabolic demands. For this reason, blood must be sent to different areas in proportion to their requirements for oxygen and nutrients. Exercise provides a dramatic illustration of rapid changes in circulatory function.

We often think of exercise as voluntary. In nature, exercise means an increase in activity for any reason. Typically, an animal increases its physical activity when it needs something, such as food or shelter, or when it is escaping something, such as a predator. The increased activity may be momentary and moderate, as when a fish briefly swims faster to avoid the tentacles of an anemone, or it may last longer and be more intense, as when a cheetah sprints after a gazelle. Exercise may last for very long periods of time, as when a salmon migrates from the sea to its freshwater spawning grounds or a migratory goose flies thousands of miles over several weeks. The circulatory systems of animals must quickly adapt to these increased metabolic demands and adjust to the intensity, duration, and type of exercise.

We have already noted some adjustments, such as vasodilation and vasoconstriction. These are the result of changes in the radius of arterioles. During exercise, these vessels vasodilate in regions that require more energy and oxygen—for example, skeletal muscles and the heart—and vasoconstrict in areas where blood flow is less immediately essential, such as the gastrointestinal tract.

In addition, cardiac output increases during exercise to supply more blood to the tissues that require it. This happens by increasing stroke volume or heart rate, or both. In humans, for example, stroke volume increases from approximately 75 mL (milliliters) at rest to 110 mL during exercise, a change of 1.5-fold, and heart rate can almost triple from about 70 bpm to nearly 200 bpm. Thus, the combined effects of increasing stroke volume and heart rate raise the cardiac output by about 4.5-fold.

Whichever way cardiac output increases, one of the major mechanisms involved is an increase in blood concentrations of epinephrine, also known as adrenaline. **Epinephrine** is a hormone secreted by the adrenal glands, which are activated by the sympathetic nervous system during exercise. Epinephrine binds to receptors on heart ventricular muscle cells, making them contract more vigorously, thereby increasing stroke volume. In addition, epinephrine binds to receptors in the atrial SA node cells, stimulating them to initiate electrical signals at a faster rate, thereby increasing heart rate. The latter effect of epinephrine is enhanced by the release of the neurotransmitter **norepinephrine** (or noradrenaline) directly from sympathetic nerve endings onto the SA node cells.

At rest, the heart does not pump out its entire content of blood. A reservoir of blood remains in each ventricle at the end of a beat. During exercise, some of this reservoir is tapped when epinephrine stimulates the heart muscles. In addition, the mechanical process of exercise itself enhances the return of blood to the heart, as skeletal muscles squeeze the veins within them and open their one-way valves. The combination of more blood getting returned to the heart and the enhanced emptying of the built-in reservoir of blood from the ventricles provides increased stroke volume during exercise.

Because blood pressure equals cardiac output times total peripheral resistance (BP = CO × TPR), you might imagine that increased CO would raise BP to possibly harmful levels during exercise. Although BP does increase somewhat, it does not increase as much as you might predict. This is because the blood vessels of skeletal muscles dilate (that is, resistance is decreased) to allow more blood to enter; similarly, the vessels in the skin of terrestrial animals dilate as a means of dissipating heat. Therefore, even though CO increases in exercise, the TPR decreases, negating some of the effect on BP.

Regular, long-term exercise induces additional adaptive responses in the circulatory system. For example, the thickness of the muscle of the ventricles of the human heart is greater in athletes than in sedentary individuals, which makes the athletic heart a more efficient pump.

Baroreceptors Maintain Blood Pressure

How do vertebrate circulatory systems adapt to changes in blood pressure? One mechanism is by changing blood volume accordingly, which is largely the responsibility of the kidneys, as you will learn in Chapter 49. Another mechanism, however, involves a reflex that depends on special pressure-sensitive regions within the walls of certain arteries—notably the aorta and the carotid arteries—two large blood vessels that supply the brain with blood. These regions contain the endings of neurons, known as **baroreceptors** (from the Greek *baros*, weight or pressure), that are in constant electrical communication with the brain (**Figure 47.19**).

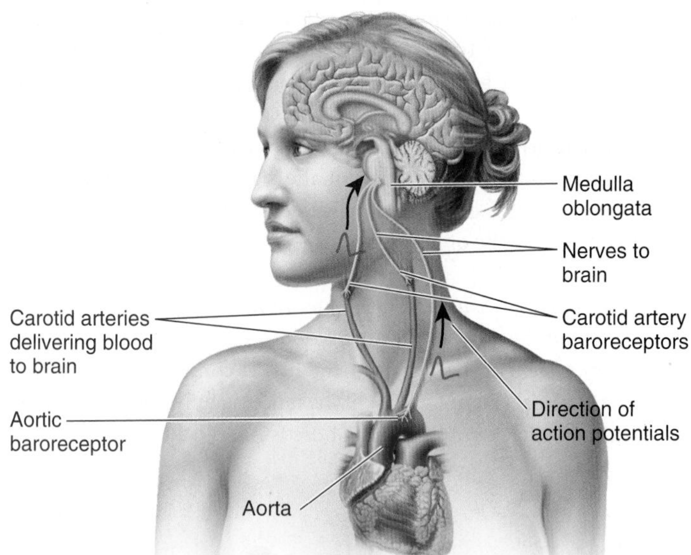

Figure 47.19 **Location of the major baroreceptors.** Increased blood pressure stretches arteries. This activates baroreceptors in the arteries, which send action potentials to the medulla oblongata in the brain. Nerve signals from the brain then initiate changes in heart function and total peripheral resistance that help restore blood pressure to normal.

BioConnections: In Chapters 41 to 43, we discussed the different types of ion channels found in neurons and the ways in which such channels open and close. What type of channels would you expect to be activated when baroreceptors are stimulated?

If blood pressure increases above the normal range, these large arteries stretch more than normal as a result, which makes the endings of the baroreceptors stretch too. The stretching of the baroreceptors opens ion channels in the cell membrane, resulting in the generation of action potentials. Nerves carry the signals to the brain, which interprets this as a signal that blood pressure is above normal. The response is a decrease in the amount of norepinephrine released from neurons of the autonomic nervous system onto the cardiac SA node and a decrease in the secretion of epinephrine from the adrenal glands. The result is a decreased heart rate and stroke volume, resulting in lower cardiac output. In addition, nerves throughout the body release less vasoconstricting neurotransmitters from their neurons onto the smooth muscle cells of arterioles, thereby promoting vasodilation and reducing total peripheral resistance. The combination of decreased cardiac output and decreased resistance lowers blood pressure toward normal.

By contrast, if blood pressure decreases below normal, the walls of these arteries would be less stretched than they normally are, and the baroreceptors would send fewer action potentials to the brain. This results in a sequence of events that are opposite to those just described. More norepinephrine and epinephrine are secreted, thereby increasing cardiac output and vasoconstriction and consequently increasing blood pressure toward normal. One situation in which blood pressure might decrease is **dehydration**, which is a reduction in the amount of water in the body. Because blood is mostly water, dehydration reduces blood volume, and arteries—including those that contain baroreceptors—are filled with less blood than normal. Blood pressure could also drop due to **hemorrhage**, loss of blood from a ruptured blood vessel. Dehydration is primarily a problem for terrestrial animals, but hemorrhage may occur in any animal that has been injured. In a hemorrhage, both plasma and blood cells are lost; therefore, hemorrhage may be more serious than dehydration. Like dehydration, though, hemorrhage results in reduced blood volume and therefore less baroreceptor activity.

47.7 Impact on Public Health

Learning Outcomes:

1. Describe the impact of hypertension and atherosclerosis on human health.
2. List the causes, symptoms, and current medical treatments for myocardial infarction (MI), or heart attack.

Cardiovascular disease—disease of the heart and blood vessels—accounts for more deaths each year in the U.S. than any other cause. Why is cardiovascular disease so devastating? One reason is that damage to these structures often occurs slowly, over many years, and without symptoms until the disease has reached late stages. Cardiovascular disease not only has a dramatic impact on the health of many Americans, it also has a significant impact on expenditures for health care services, medications, and lost productivity. In this section, we will consider several common cardiovascular disorders and their causes, symptoms, and treatments.

Hypertension and Atherosclerosis May Cause Heart and Blood Vessel Disease

Hypertension, or high blood pressure, refers to an arterial blood pressure that is chronically above normal. The normal range in humans varies from systolic/diastolic pressures of about 90/60 to 120/80 mmHg. A resting blood pressure above 140/90 mmHg defines hypertension, and values between 140/90 and 120/80 mmHg are considered borderline (sometimes called "prehypertension"). Hypertension can have many causes, including obesity, smoking, aging, kidney disease, excess male hormones, and genetic factors, although in many cases, the cause is unknown. It can often be treated with diet and exercise and with drugs that cause vasodilation, thereby reducing arterial resistance.

Hypertension rarely has any noticeable symptoms. For this reason, it is important to have your blood pressure checked regularly. Without treatment, hypertension can damage arteries, contributing to the formation of **plaques**—deposits of lipids, fibrous tissue, and smooth muscle cells—inside arterial walls that leads to a condition known as **atherosclerosis**, in which plaque causes the arteries to narrow and harden (Figure 47.20). Large plaques may occlude (block) the lumen of the artery entirely. In addition to hypertension, plaques are known to arise from a variety of factors, including calcium and fat deposits, and are correlated with obesity, high blood cholesterol concentrations, and smoking.

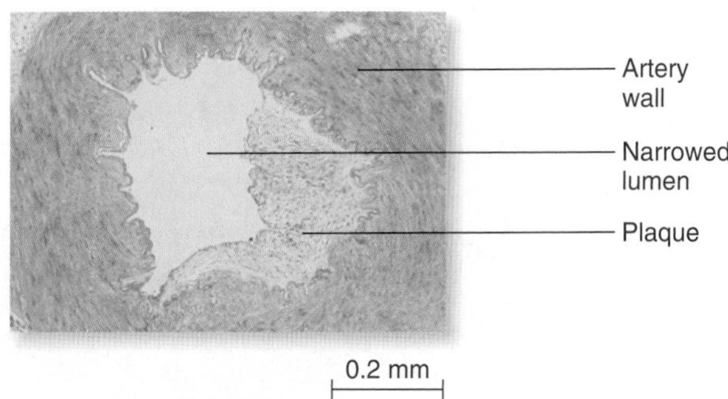

Figure 47.20 **An atherosclerotic plaque in an artery.** Compare this with the normal artery shown in the inset in Figure 47.12.

If plaques form in any artery, the regions of the body supplied with blood by that artery receive less oxygen and nutrients. Although atherosclerosis is dangerous anywhere, it is especially significant if it affects the **coronary arteries**, which carry oxygen and nutrients to the heart muscle. **Coronary artery disease** occurs when plaques form in the coronary vessels, and it can be life-threatening. One warning sign of coronary artery disease is **angina pectoris**, chest pain during exertion due to the heart being deprived of oxygen.

Myocardial Infarction Results in Death of Cardiac Muscle Cells

If a region of the heart is deprived of blood for an extended time, the result may be a **myocardial infarction (MI)**, or heart attack. This is usually caused by a blockage of one of the coronary arteries. Some heart attacks are relatively minor; in fact, the discomfort of a small heart attack may not even alarm someone enough to seek medical attention. A heart attack with no symptoms is called a silent heart attack. More serious heart attacks can lead to significant damage or death to the heart. Dead cardiac muscle tissue does not regenerate, diminishing the heart's ability to pump. Reduced pumping activity of the heart can result in congestive heart failure, in which the heart can't pump enough blood to meet the body's needs. As noted in the chapter introduction, each year in the U.S., more than 1 million people suffer a heart attack, many of which are fatal.

Preventing a heart attack in the first place is the best way to increase survival rates. Procedures are available that allow physicians to monitor the status of the coronary vessels in people thought to have heart disease. In the procedure called cardiac angiography, the coronary arteries can be visualized by injecting a dye into a person's veins and then taking an X-ray image of the chest, allowing a physician to determine if the vessels are narrowed by disease.

If a blockage is found, several common treatments can restore blood flow through a coronary artery. One is **balloon angioplasty**, in which a thin tube with a tiny, inflatable balloon at its tip is threaded through the artery to the diseased area. Inflating the balloon compresses the plaque against the arterial wall, widening the lumen. In most cases, a wire-mesh device called a stent is inserted into the artery after angioplasty has expanded it, providing a sort of lattice to hold the artery open (Figure 47.21). A treatment for more serious coronary artery disease is a **coronary artery bypass**, in which a small piece of healthy blood vessel is removed from one part of the body and surgically grafted onto the coronary circulation in such a way that blood bypasses the diseased artery.

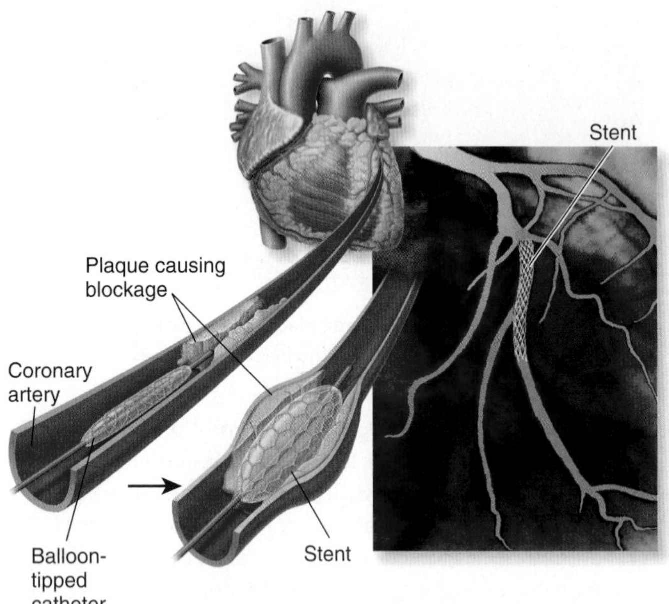

Figure 47.21 A treatment for blocked blood vessels. Balloon angioplasty can widen diseased arteries, followed by insertion of a stent. The inset shows a stent placed in a coronary artery of a human patient.

BIOLOGY PRINCIPLE Biology affects our society. Biologists investigating the mechanisms of blood flow, heart function, and blood pressure in animals can use that knowledge to develop treatments for common heart diseases in humans.

Summary of Key Concepts

47.1 Types of Circulatory Systems

- Circulatory systems transport necessary materials to all cells of an animal's body and transport waste products away from cells. The three basic types of circulatory systems are gastrovascular cavities, open circulatory systems, and closed circulatory systems.
- Cnidarians and some flatworms have a gastrovascular cavity that serves two functions: digestion of food and circulation of water containing the digested food (Figure 47.1).
- An open circulatory system contains three components: hemolymph, vessels, and one or more hearts. The vessels open into the animal's body cavity (Figure 47.2).
- In a closed circulatory system—found in earthworms, cephalopods, and all vertebrates—blood and interstitial fluid are physically separated by blood vessel walls and differ in their components and chemical composition (Figure 47.2).
- The closed circulatory systems of vertebrates can be divided into single or double circulations. In the double circulation of crocodiles, birds, and mammals, oxygenated and deoxygenated blood are completely separated into the systemic circulation and pulmonary circulation (Figures 47.3, 47.4).

47.2 Blood and Blood Components

- Blood is a fluid connective tissue consisting of cells and, in mammals, cell fragments suspended in a solution of water containing dissolved nutrients, proteins, gases, and other molecules. Blood has four components: plasma, leukocytes, erythrocytes, and platelets or thrombocytes (Figures 47.5, 47.6, 47.7).

47.3 The Vertebrate Heart and Its Function

- All vertebrate hearts have at least one atrium and one ventricle. In hearts with more than one atrium or ventricle, septae separate the atria and ventricles. Blood enters the atria from systemic or pulmonary veins, moves down a pressure gradient through the AV valves into the ventricles, and is pumped out through the semilunar valves into the systemic and pulmonary arteries (Figure 47.8).
- Many arthropods and decapod crustaceans have a neurogenic heart that cannot beat without regular electrical impulses from the nervous system; all vertebrates have a myogenic heart, which beats on its own without electrical stimulation, but which can be regulated by the nervous system.
- The electrical signals of the vertebrate heart are generated either at the junction of the veins and the single atrium (in fishes) or at the sinoatrial (SA) node (in other vertebrates). The electrical impulses spread through the atria and ventricles and stimulate the heart muscle to contract. The cardiac cycle has two phases: diastole and systole (Figures 47.9, 47.10, 47.11).

47.4 Blood Vessels

- Arteries are thick-walled vessels that carry blood away from the heart. Arterioles distribute blood to capillaries, and capillaries are the site of gas and nutrient exchange. Blood leaving the capillaries empties into venules and then into veins, which return it to the heart. One-way venous valves help to return blood to the heart (Figures 47.11, 47.12, 47.13, 47.14, 47.15, 47.16).

47.5 Relationship Among Blood Pressure, Blood Flow, and Resistance

- Blood pressure is the force exerted by blood on the walls of blood vessels, and it is responsible for moving blood through the vessels. Resistance refers to the tendency of blood vessels to slow down the flow of blood through their lumens.
- Local blood flow through a vessel is directly proportional to the pressure of the blood entering the vessel and inversely proportional to the resistance created by that vessel. Arteriole radius is the major factor that regulates resistance. The endothelium of arteries produces a vasodilatory substance (Figure 47.17).
- Cardiac output overcomes resistance to generate systemic blood pressure and depends on the size of an animal's heart, how often it beats each minute, and how strongly it contracts with each beat (Table 47.1).
- Cardiac output (CO) and total peripheral resistance (TPR) determine blood pressure (BP): $BP = CO \times TPR$. Blood pressure must be high enough for blood to adequately reach all body tissues and cells, but not so high as to damage blood vessels (Figure 47.18).

47.6 Adaptive Functions of Closed Circulatory Systems

- The circulatory system adjusts to fluctuating metabolic requirements through changes in cardiac output, vasodilation, and vasoconstriction. Baroreceptors maintain blood pressure by relaying information about circulatory system status to the brain, which initiates compensatory mechanisms (Figure 47.19).

47.7 Impact on Public Health

- Cardiovascular disease accounts for more deaths each year in the U.S. than any other cause. Cardiovascular disorders include hypertension, atherosclerosis, coronary artery disease, angina pectoris, and myocardial infarction (MI, or heart attack) (Figure 47.20).
- Cardiovascular diagnostic techniques and treatments include cardiac angiography, balloon angioplasty, and coronary artery bypass (Figure 47.21).

Assess and Discuss

Test Yourself

1. Hemolymph differs from blood in that it
   a. does not contain blood cells.
   b. is a mixture of fluid in blood vessels and the hemocoel.
   c. circulates through closed circulatory systems only.
   d. functions only in defense of the body and not transport.
   e. does not pass through a heart.
2. The heart chamber that receives oxygenated blood from the lungs of animals with a double circulation is
   a. the left atrium.
   b. the right ventricle.
   c. the left ventricle.
   d. the right atrium.
3. Amphibians are adapted for life in and out of water because
   a. they are the only vertebrates with an open circulatory system.
   b. oxygen diffuses out of their blood into the environment through their skin.
   c. oxygen diffuses into the blood from the lungs or through the skin.
   d. they have a four-chambered heart.
   e. they can direct blood away from the skin when they are submerged under water.
4. A major advantage of a double circulation is that
   a. blood can be pumped to the upper portions of the body by one circuit and to the lower portions of the body by the other circuit.
   b. each circuit can pump blood with differing pressures to optimize the function of each.
   c. the oxygenated blood can mix with the deoxygenated blood before being pumped to the tissues of the body.
   d. less energy is required to provide nutrients and oxygen to the tissues of the body.
   e. all of the above
5. The function of erythrocytes is to
   a. transport oxygen throughout the body.
   b. defend the body against infection and disease.
   c. transport chemical signals throughout the body.
   d. secrete the proteins that form blood clots.
   e. a and d are both correct.
6. The mammalian heart is an example of a myogenic heart, meaning
   a. it is composed of four chambers.
   b. it acts as two pumps for two different circulations.
   c. contraction is regulated solely by the nervous system.
   d. electrical activity is initiated by specialized cells of the heart.
   e. the nervous system plays no role in its function.
7. During systole of the cardiac cycle
   a. the ventricles of the heart are both relaxed.
   b. the ventricles of the heart are both filling.
   c. the ventricles of the heart are both contracting.
   d. the right ventricle is contracting while the left ventricle is at rest.
   e. the left ventricle is contracting while the right ventricle is at rest.
8. Considering blood flow through a closed circulation, which is the correct sequence of vessels beginning at the heart?
   a. arteriole, artery, capillary, vein, venule
   b. artery, capillary, arteriole, venule, vein
   c. vein, venule, capillary, arteriole, artery
   d. artery, arteriole, capillary, venule, vein
   e. artery, arteriole, capillary, vein, venule
9. Gas and nutrient exchange between the circulation and the tissues occurs primarily at
   a. the arteries.
   b. the arterioles.
   c. the capillaries.
   d. the venules.
   e. the veins.
10. Which of the following factors determines arterial blood pressure?
    a. cardiac output
    b. resistance
    c. arteriole diameter
    d. a, b, and c
    e. a and b only

Conceptual Questions

1. What are the three basic components of a true circulatory system?
2. Discuss the difference between closed and open circulatory systems. What advantages does a closed system provide?
3. A principle of biology is that *living organisms maintain homeostasis*. In what ways does the circulatory system contribute to homeostasis?

Collaborative Questions

1. Explain the cardiac cycle. Why must heart valves open in only one direction?
2. Discuss the types of closed circulatory systems found in vertebrates.

Online Resource

www.brookerbiology.com

Stay a step ahead in your studies with animations that bring concepts to life and practice tests to assess your understanding. Your instructor may also recommend the interactive eBook, individualized learning tools, and more.

Respiratory Systems

Chapter Outline

48.1 Physical Properties of Gases
48.2 Types of Respiratory Systems
48.3 Structure and Function of the Mammalian and Avian Respiratory Systems
48.4 Mechanisms of Oxygen Transport in Blood
48.5 Control of Ventilation in Mammalian Lungs
48.6 Adaptations to Extreme Conditions
48.7 Impact on Public Health
Summary of Key Concepts
Assess and Discuss

48

Llama (*Lama glama*) in the Andes Mountains, an example of an animal adapted to conditions of low oxygen pressure.

Have you ever wondered why it is easy for an animal like the llama in the chapter-opening photo to live at high altitudes, whereas most humans become ill when they ascend to such heights? What is it about the atmosphere in mountainous regions that makes it hard for us to function normally, and how do some people and animals adapt to those conditions if given sufficient time? Many such intriguing questions make the study of animal respiratory function fascinating. For example, how are you able to continue breathing while asleep? How does your breathing rate adjust to match your level of activity? If our respiratory system did not operate automatically, think how challenging, indeed impossible, life would be for us.

In Chapter 47, we saw how circulatory systems function to transport blood or hemolymph throughout an animal's body. These fluids also carry the oxygen (O_2) that animals require. The oxygen enters the body via a respiratory (breathing) organ. It then dissolves into bodily fluids and is distributed to cells. As described in Chapter 7, O_2 is used by mitochondria in the formation of ATP. One of the waste products of that reaction is carbon dioxide—CO_2—which diffuses out of cells. CO_2 is returned by the circulatory system to the respiratory organ and is then released into the environment, either the air or water. The process of moving O_2 and CO_2 in opposite directions between the environment, bodily fluids, and cells is called **gas exchange** (**Figure 48.1**). A **respiratory system** includes all of an animal's structures that contribute to gas exchange. This chapter deals with the mechanisms by which animals obtain and transport O_2, rid themselves of CO_2, and cope with the challenges imposed by their environments and their changing metabolic demands. We begin, however, with an overview of the properties of gases in air and water, which will help us understand how gases move between the environment and an animal's body fluids. We conclude the chapter by looking at several respiratory system disorders, some caused by lifestyle, others by genetics, including their typical causes, possible treatments, and impact on society.

48.1 Physical Properties of Gases

Learning Outcomes:

1. List the relative amounts of each gas that make up the majority of the air we breathe.
2. Describe the relationship between altitude and atmospheric pressure.
3. Describe what is meant by partial pressure and how it is calculated.
4. Explain the influence of pressure, temperature, and the presence of other solutes on the solubility of gases in water.

The exchange of O_2 and CO_2 between the environment and the fluids of an animal's body depends on the physical properties of these gases. These include such things as the solubility of the gas in water and the rate of diffusion of the gas. Air is composed of about 21% O_2, 78% nitrogen (N_2), and roughly 1% CO_2 and other gases. From a respiratory standpoint, N_2 can usually be ignored because it plays no role in energy production nor is it created as a waste product of metabolism. In this section, we begin by examining some of the basic properties of the two major gases that are important in respiration.

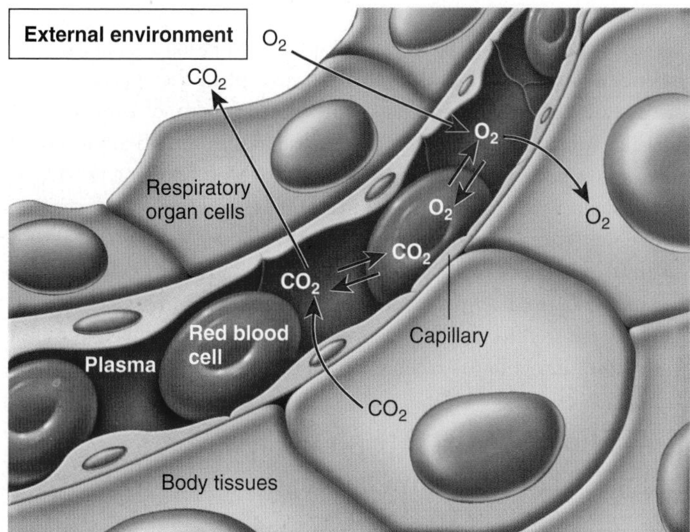

Figure 48.1 Overview of gas exchange between the environment, blood, and cells. In vertebrates, oxygen (O_2) diffuses from the environment across the cells of a respiratory organ into the blood (plasma and red blood cells). From there, O_2 diffuses into tissue cells, where it is used during the formation of ATP. Cells generate carbon dioxide (CO_2) as a waste product, which diffuses out of cells and into blood, and then across the respiratory organ into the environment. *Note:* Interstitial fluid is omitted for simplicity.

BIOLOGY PRINCIPLE Living organisms maintain homeostasis. Keep this principle in mind as you read this chapter. The exchange of O_2 and CO_2 illustrated here is absolutely essential for all homeostatic processes in animals, in part because oxygen is required for the production of much of an animal's ATP. The ATP, in turn, provides the energy required to sustain processes that contribute to homeostasis.

Gases Exert Pressure, Which Depends on Altitude

The gases in air exert pressure on the body surfaces of animals, although the pressure is not perceptible (unless it changes suddenly, like when your ears "pop" in a descending airplane). This pressure is called **atmospheric** (or **barometric**) **pressure**. It can be measured by noting how high a column of mercury is forced upward by the air pressure in a device called a mercury manometer. The traditional unit of gas pressure, therefore, is millimeters of mercury (mmHg)—the same as for blood pressure.

At sea level, atmospheric pressure is 760 mmHg. It decreases as we ascend to higher elevations because the gravitational effect of the Earth decreases and there are fewer gas molecules in a given volume of air (**Figure 48.2**). To visualize why gas pressure decreases at higher altitudes, think about how hydrostatic pressure decreases from the ocean floor to the ocean surface. Now, imagine that you are standing at the bottom of an "ocean" of air. The closer you get to the "ocean" surface (the top of the atmosphere), the lower the pressure exerted on your body. At an elevation of 1,700 m, as in Denver, the pressure is about 640 mmHg, but at the highest elevations continuously inhabited by humans—in the Andes Mountains at 5,500 m—the pressure falls to 375 mmHg.

Atmospheric pressure is the sum of the pressures exerted by each gas in air, in exact proportion to their amounts. The individual pressure of each gas is its **partial pressure**, symbolized by a capital P and a subscript depending on the gas. Thus, the partial pressure of oxygen (Po_2) in the air we breathe is 21% of the atmospheric pressure at sea level, or 160 mmHg (0.21×760 mmHg). The *percentage* of oxygen and other gases in air remains the same regardless of altitude, but the lower the atmospheric pressure, the lower the partial pressure of oxygen in air.

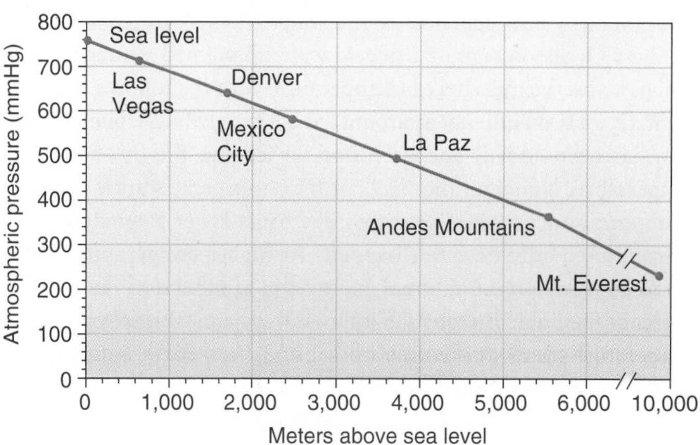

Figure 48.2 The relationship between atmospheric pressure and altitude. Atmospheric pressure decreases as altitude increases.

Concept Check: *Is there less percent oxygen in the air over Mt. Everest than there is at sea level?*

The partial pressure of oxygen in the environment provides the driving force for its diffusion from air or water across the respiratory surface and into the blood. All gases diffuse along pressure gradients, from regions of high pressure to regions of lower pressure. This is analogous to the diffusion of dissolved solutes from regions of high to low concentration, or the way heat moves from hotter to cooler regions. Consequently, the rate of oxygen diffusion into the bloodstream of a terrestrial animal decreases when the animal moves from sea level to a higher altitude, where the Po_2 is lower.

Pressure, Temperature, and Other Solutes Influence the Solubility of Gases

Gases dissolve in solution, including fresh water, seawater, and all body fluids. Gases such as O_2 exert their biological effects while in solution. However, most gases dissolve rather poorly in water. There is less O_2 in a given volume of water than in air, for example, which limits how active many aquatic organisms can be. Among the factors that influence the solubility of a gas in water, three are particularly important: the pressure of the gas, the temperature of the water, and the presence of other solutes.

The greater the pressure of a gas that comes into contact with water, the more of that gas that will dissolve, up to a limit that is specific for each gas at a given temperature. Not all gases dissolve equally. O_2 is less soluble in water than is CO_2, but more soluble than N_2. The partial pressure of gas in water is given in the same units as atmospheric pressure (mmHg). For example, if a solution of water is in contact with air that contains a Po_2 of 160 mmHg, the Po_2 of the water is also 160 mmHg.

Solubility also depends on the water's temperature, with more O_2 and CO_2 dissolving in 1 L of cold water than in 1 L of warm water. You can observe this effect of temperature with the following experiment. Open two bottles of a carbonated beverage (that is, one in which CO_2 has been added) and keep one on ice and the other at room temperature. Note that the "fizz" of the room-temperature beverage disappears sooner, as CO_2 escapes due to its lower solubility at that temperature. Why does this happen? At higher temperatures, gases in solution have more thermal energy and are therefore more likely to escape the liquid. In terms of biological systems, this effect of water temperature means that animals inhabiting very warm waters generally have less O_2 available to them than do animals living, for example, in the waters off Antarctica. Animals that live in shallow water such as ponds or tide pools, where water temperature can fluctuate widely over a single day, must be able to adapt to large swings in O_2.

Finally, the presence of other solutes, namely salts, reduces the amount of gas that dissolves in water. Thus, warm salty blood dissolves less O_2 than does cold water. Likewise, less O_2 dissolves in seawater than fresh water at any given temperature and pressure. Animals living in cold freshwater lakes, therefore, typically have more O_2 available to them than do animals living in warm salty seas (although other factors can also influence the amount of dissolved O_2, such as water depth and the amount of aquatic plant life).

With this understanding of the physical properties of O_2 and CO_2 in air and water, and their importance for biology, we turn now to a discussion of the variety of respiratory systems found in animals.

48.2 Types of Respiratory Systems

Learning Outcomes:

1. List features that are common to all gas-exchange surfaces in animals.
2. Describe the different challenges faced by water- and air-breathing animals.
3. Describe how an animal's body surface can exchange gases with the environment.
4. Diagram the process of countercurrent flow in gill ventilation.
5. Explain the unique aspects of the tracheal system of ventilation in insects.
6. Describe how lungs function in gas exchange in air-breathing vertebrates.

Despite the staggering array of animal shapes, sizes, and habitats, animals obtain oxygen from their surroundings with one of just four major types of gas-exchange organs: the body surface, gills, tracheae, or lungs. **Ventilation** is the process of bringing oxygenated water or air into contact with a gas-exchange organ. In this section, we will examine the mechanisms of ventilation used by animals with different respiratory systems.

The Structures of Respiratory Organs Are Adapted for Gas Exchange

All gas-exchange structures share certain common features. All of them have moist surfaces in which the gases can dissolve and diffuse. Further, all have adaptations that increase the amount of surface area available for gas exchange.

The extensive surface area of respiratory organs is coupled with an equally extensive blood circulation, except in the special case of insects, described later in this chapter. The density of capillaries in gills and lungs, for example, is among the highest found anywhere in an animal's body. The greater the amount of blood flowing to a respiratory surface, the more efficient the delivery and removal of gases.

Thick barriers greatly slow the diffusion of a gas. As a result, respiratory organs tend to be thin, delicate structures, which also means they are easily damaged. For instance, the inner structures of fish gills are so thin that they cannot support their own weight out of water, and they collapse when exposed to air.

Water-Breathing and Air-Breathing Animals Face Different Challenges

Exchanging gases in water or in air requires respiratory organs that can deal with the challenges of those different environments. Since water-breathing animals generally have less oxygen available to them than do air-breathing animals, an active water-breathing animal requires a very efficient means of extracting oxygen from water. In addition, unlike air-breathing animals, water-breathing animals must adapt to fluctuating oxygen availability when the temperature changes. By contrast, air-breathing animals must cope with the dryness of air. Passing dry air over the lung surfaces runs the risk of drying out the lungs, which would damage them, dehydrate the animal, and reduce the ability of gases to dissolve.

Water-breathing animals bring water, not air, over their respiratory surfaces and therefore do not usually risk dehydration in the way that air-breathing animals do. However, moving water over a gas-exchange surface has its own challenges. First, because water is denser than air, more energy (muscular work) is required to move water. Second, cold water moving over the richly vascular gills draws considerable heat from an animal's body, due to the high heat capacity of water. Third, moving either fresh or salty water over a gas-exchange surface can create osmotic movement of water across the surface, potentially resulting in water imbalances in an animal's body.

Some Animals Use Their Body Surface to Exchange Gases with the Environment

In those invertebrates that are only a few cell layers thick, such as platyhelminthes and cnidarians, O_2 and CO_2 can diffuse directly across the body surface. In this way, O_2 reaches all the interior cells, in some cases without any specialized transport mechanism or circulatory system.

Even in some large, complex animals, the body surface may be permeable to gases (**Figure 48.3**). Amphibians, and a few species of fishes including eels, have unusually permeable skin. On land, amphibians rely primarily on their lungs but also on their moist skin to obtain O_2 and release CO_2. Under water, however, the lungs are no longer useful because they are not suited for extracting O_2 from water. Although O_2 diffusion across the skin is less efficient than in the lungs, this secondary ability to exchange gas permits amphibians to spend prolonged times under water.

Some vertebrates such as seals spend part of their lives on land and part in water and can hold their breath under water for much

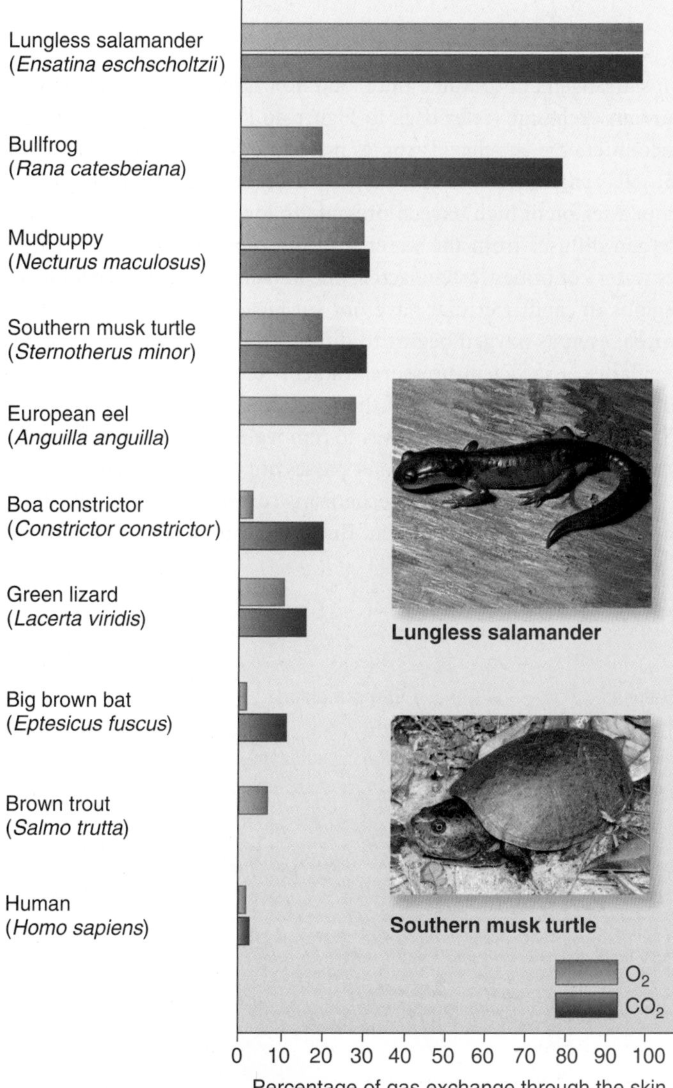

Figure 48.3 Gas exchange across body surfaces. This type of gas exchange occurs readily in amphibians, to a lesser extent in marine reptiles and some fishes, and to a negligible extent among mammals. *Note:* CO_2 determinations were not made in all species.

Concept Check: What would happen to gas exchange in lungless salamanders if the animals were allowed to dry out?

longer periods than can humans. Some reptiles, such as marine iguanas, can obtain O_2 from the water across the highly permeable mucous membranes of the nose, mouth, and anus. Although this provides less O_2 than does air-breathing, it nonetheless allows them to remain under water for extended periods, as when foraging for food. Other animals have highly specialized body regions with considerable surface area—like the wings of a bat—that permit some gas exchange with the environment (see Figure 48.3). Except for amphibians, however, no vertebrates can rely for long periods of time exclusively on their skin to obtain oxygen and remain active.

Water-Breathing Animals Use Gills for Gas Exchange

Most water-breathing animals use specialized respiratory structures called **gills**, which can be either external or internal. External gills are uncovered extensions from the body surface, as occur in many invertebrates and the larval forms of some amphibians. Internal gills, which occur in fishes, are enclosed in a protective cavity.

External Gills External gills vary widely in appearance, but all have a large surface area, often in the form of extensive projections (**Figure 48.4**). External gills may exist in one region of the body or be scattered over a large area. In many cases, they are ventilated by waving back and forth through the water. The ability to move external gills is particularly important for sessile invertebrates, which must otherwise rely on sporadic local water currents or muscular efforts of their bodies to create local currents for ventilation.

Despite the success of marine invertebrates, external gills have several limitations. First, they are unprotected and therefore are susceptible to damage from the environment. Second, because water is much denser than air, considerable energy is required to continually wave the gills back and forth through the water (think of the difference between waving your hands through air or water). Finally, their appearance and motion may draw the attention of predators.

(a) Mollusk (nudibranch) with clustered external gill tufts

(b) Larval salamander with external gills

Figure 48.4 Examples of animals with external gills. The gills of these animals have extensive projections to increase surface area, which facilitates oxygen diffusion from the surrounding water.

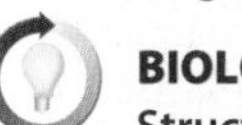

BIOLOGY PRINCIPLE Structure determines function. Note how the gills of the nudibranch and larval salamander have extensive surface modifications to increase the total area available for gas exchange. This is not unique to external gills; indeed, all gas-exchange surfaces have one or another type of surface area structural modification to increase their function.

Internal Gills By contrast, fishes have internal gills, which are covered by a bony plate called an operculum (**Figure 48.5a**). Fish gills have a more uniform appearance than external gills and are confined within the opercular cavity, the space beneath the operculum, which protects the gills and helps streamline the fish body.

The main support structure of gills are the gill arches, which contain gill filaments composed of numerous platelike structures called **lamellae** (**Figure 48.5b**). Blood vessels run the length of the filaments, with oxygen-poor blood traveling through the afferent vessel along one side of the filament, and oxygen-rich blood traveling through the efferent vessel along the other side. Within the lamellae are numerous capillaries, all oriented with blood flowing in the same direction, from the oxygen-poor vessel to the oxygen-rich one.

Water enters a fish's mouth and flows between the lamellae in the opposite direction to blood flowing through the lamellar capillaries. This arrangement of water and blood flow is an example of **countercurrent exchange** (refer back to Figure 46.15). As oxygenated water encounters the lamellae, it comes into close proximity with blood in the gill capillaries. Recall that oxygen diffuses along pressure gradients from a region of high oxygen pressure to low oxygen pressure. Thus, oxygen diffuses from the water into the capillaries of the lamellae. As water continues to flow across the lamellar surface, it encounters regions of capillaries that have not yet picked up oxygen—in other words, even as oxygen begins to diffuse from the water into the gill capillaries, a sufficient pressure gradient remains along the lamellae to permit diffusion of more of the remaining oxygen from the water. This is an extremely efficient way to remove as much oxygen from the water as possible before the water passes out of the operculum.

Fishes use one of two mechanisms to ventilate their gills: buccal pumping or ram ventilation. **Buccal pumping** (the buccal cavity

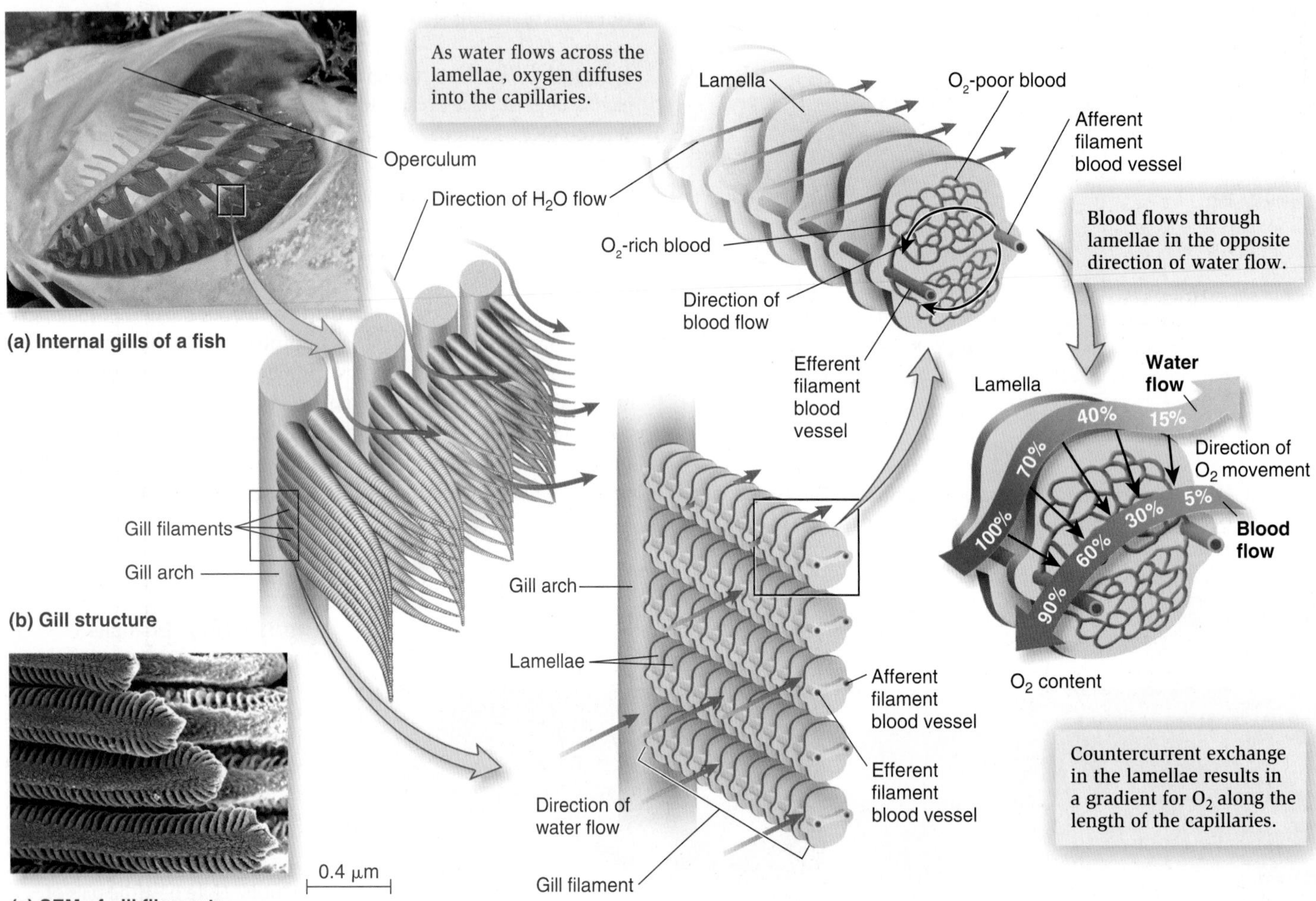

Figure 48.5 **Structure of fish gills.** **(a)** The operculum, which has been lifted open in this photo, protects the gills underneath. **(b)** The gills are composed of gill arches, from which numerous pairs of filaments arise. Thin, platelike lamellae are arrayed along the filaments. Blood flows through the capillaries of the gill filaments in the opposite direction of water flowing between lamellae, a process called countercurrent exchange. **(c)** Several filaments with their lamellae as revealed in a scanning electron micrograph.

Concept Check: *Why do fishes die when out of water? (Hint: Consider the thin, delicate appearance of the water-covered lamellae.)*

BioConnections: *You have learned about countercurrent exchange before; look back to Figure 46.15 for a description of how countercurrent exchange can regulate heat loss in a vertebrate.*

refers to the mouth) makes use of the muscles of the mouth and operculum to create a hydrostatic pressure gradient for water to flow in one direction (**Figure 48.6a**). First, the jaw is lowered, which enlarges the buccal cavity and lowers the pressure in the mouth, something like a suction pump. This draws water into the mouth, raising the water pressure again. At roughly the same time, the operculum begins to swing out from the body, enlarging the opercular cavity and lowering the water pressure there. This second suction pump is stronger than the buccal pump, so water flows down its hydrostatic pressure gradient from the outside into the mouth, and from there into the opercular cavity. Next, the mouth closes, and the buccal cavity is compressed. This creates positive pressure that forces water across the gills and out through the operculum, which is now open. A flap of tissue at the back of the mouth helps prevent accidental flow of water down the esophagus. Consequently, fishes do not swallow the water they inhale but instead send it on a one-way journey across their gills. In buccal pumping, a fish can remain stationary and still ventilate even in stagnant water by drawing water into its mouth.

The second way in which some fishes ventilate their gills, **ram ventilation**, consists of swimming with their mouths open, in essence using their swimming muscles to bring water into their buccal cavity and from there across their gills (**Figure 48.6b**). This method of ventilation is more energy-efficient than buccal pumping. Ram ventilation still requires energy expenditure, but in this case, the energy is used for swimming or for using muscles to remain stationary while facing upstream in moving water. Many fishes employ both methods of ventilation, using buccal pumping when swimming slowly or in stagnant water and switching to ram ventilation when swimming quickly or facing upstream. Some fishes—such as tuna—can only ram ventilate and therefore rarely stop swimming.

Both buccal pumping and ram ventilation are **flow-through systems**—water moves unidirectionally in such a way that the gills are constantly in contact with fresh, oxygenated water. As we will see, many air-breathing animals use a less efficient method of ventilating their lungs, called tidal ventilation, in which fresh air is breathed in and stale air is breathed out through the same route. Despite the efficiency of flow-through systems for maximizing oxygen diffusion from water, they are energetically costly because of the work required to overcome the density of water. Fishes may devote up to 10–20% of their resting metabolic rate simply to ventilating their gills, whereas a typical air-breather may allocate only 1–2% of its total energy usage to ventilating the lungs at rest.

Insects Use Tracheal Systems to Exchange Gases with the Air

Except for terrestrial isopods such as the pill bug (*Armadillidium vulgare*; see Figure 33.32b), which breathe air through very small, moist gills, the delicate nature of gill lamellae makes them unsuitable for gas exchange in air. Air-breathing probably evolved as an adaptation in aquatic animals inhabiting regions that were subject to periodic drought.

One of the major mechanisms that animals evolved to breathe air is the **tracheal system** found in insects. Running along the surface of both sides of an insect's body are tiny openings to the outside, called **spiracles**. Arising from the spiracles are **tracheae** (singular,

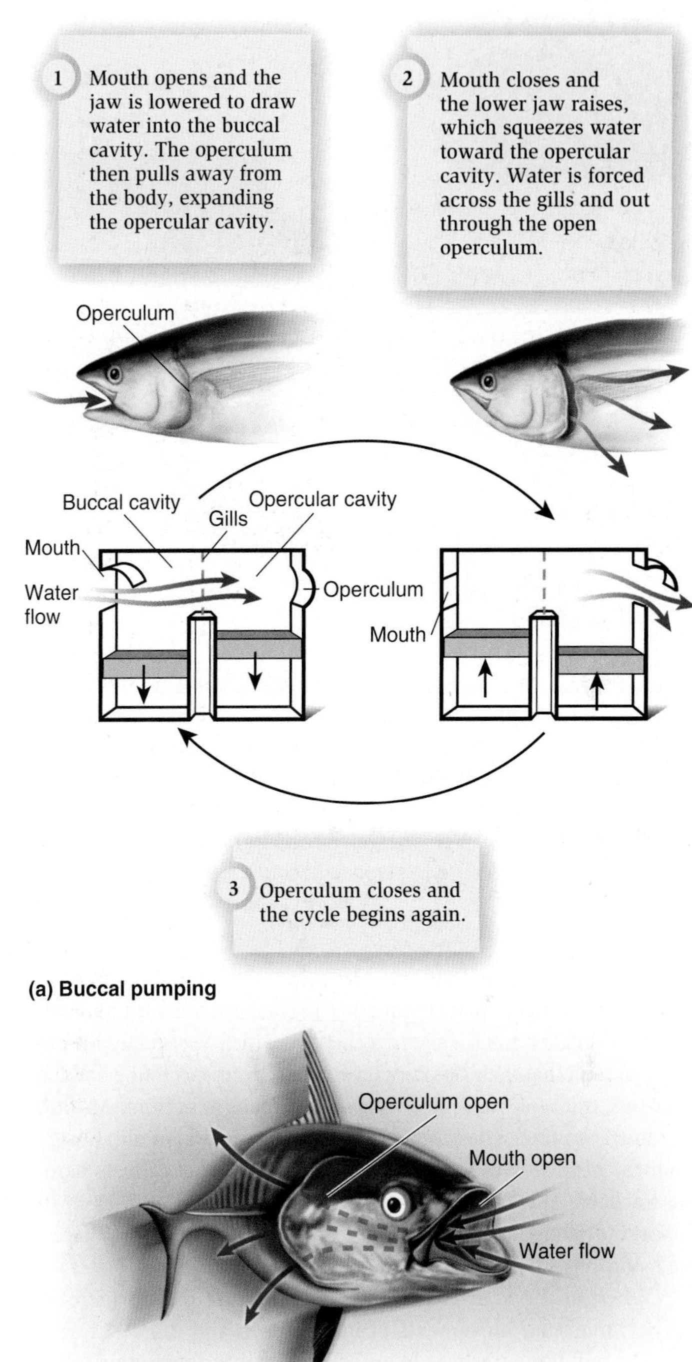
1 Mouth opens and the jaw is lowered to draw water into the buccal cavity. The operculum then pulls away from the body, expanding the opercular cavity.
2 Mouth closes and the lower jaw raises, which squeezes water toward the opercular cavity. Water is forced across the gills and out through the open operculum.
Operculum
Buccal cavity
Gills
Opercular cavity
Mouth
Water flow
Operculum
Mouth
3 Operculum closes and the cycle begins again.
(a) Buccal pumping
Operculum open
Mouth open
Water flow
(b) Ram ventilation

Figure 48.6 **Mechanisms of gill ventilation.** **(a)** Buccal pumping. The expansion of the buccal and opercular cavities acts like a suction pump, and closing these cavities acts like a pressure pump. **(b)** Ram ventilation. Many fishes ventilate their gills by swimming forward, such as this tuna, or by facing upstream in moving water with their mouths open. This causes the water to move through the mouth and operculum. Some species use both buccal pumping and ram ventilation, whereas some use only one. During both types of ventilation, flaps of tissue in the mouth, throat, and operculum help prevent water from moving in the reverse direction or down the esophagus.

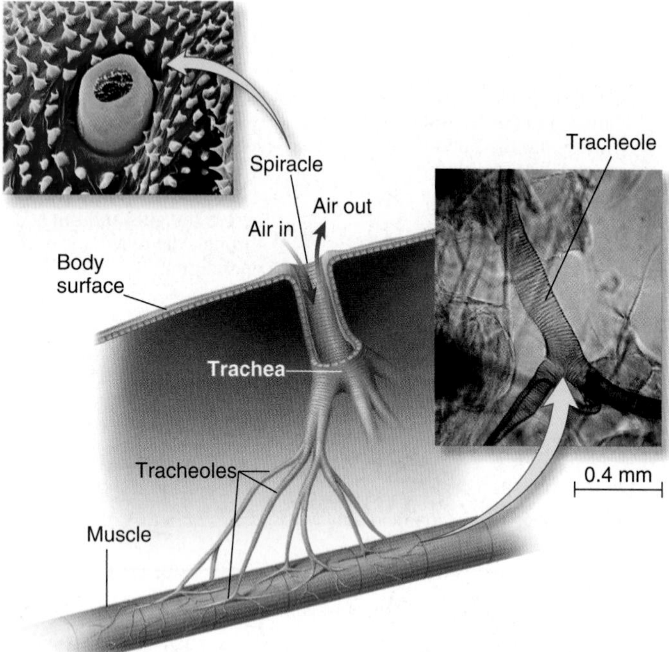

Figure 48.7 **The tracheal system of insects.** Air enters holes on the body surface called spiracles. Oxygen diffuses directly from the fluid-filled tracheole tips to cells that come into contact with the tips. The circulatory system plays no role in gas exchange. The micrographs illustrate a single spiracle and a few branching tracheoles of an insect.

Concept Check: How might the structure of the respiratory system of insects be related to why insects tend to be smaller than vertebrates?

trachea), sturdy tubes that are reinforced with the polysaccharide chitin to keep them open (Figure 48.7). Tracheae branch extensively into ever-smaller tubes called tracheoles, which eventually become small enough that their tips contact virtually every cell in the body. At their tips, tracheoles are filled with a small amount of fluid. Air flowing down the tracheoles comes into contact with this fluid. Oxygen from the air dissolves in the fluid, and from there it diffuses across the tracheole and into nearby cells. Carbon dioxide diffuses in the opposite direction, from cells into the tracheoles, and from there to the environment.

When an insect's oxygen demands increase due to increased activity, muscular movements of its abdomen and thorax draw air into and out of the tracheae like a bellows, a phenomenon first demonstrated in the early 20th century by Danish physiologist and Nobel laureate August Krogh, who studied the mechanics of ventilation in locusts. Krogh showed that an insect's muscles and tracheal system match ventilation with the animal's exercise intensity and oxygen requirements. This is particularly important in flying insects, whose metabolic demands are high. The faster an insect flies, the more oxygen it uses. Likewise, the more intensely and rapidly the body wall muscles move, the more air is moved in and out of the tracheae.

As discussed in Chapter 47, the open circulatory system of insects does not participate in gas exchange. Oxygen diffuses directly from air to trachea to tracheoles and finally to body cells. This mechanism of ventilation and oxygen delivery is very efficient. The metabolic rate of insect flight muscles is among the highest known of any tissue in any animal, and the tracheal system supplies enough oxygen to meet those enormous demands.

Air-Breathing Vertebrates Use Lungs to Exchange Gases

Except for some amphibians such as lungless salamanders (family Plethodontidae), all air-breathing terrestrial vertebrates use lungs to bring oxygen into the circulatory system and remove carbon dioxide from it. **Lungs** are internal paired structures that arise during embryonic life from the pharynx (throat). All lungs receive deoxygenated blood from the heart and return oxygenated blood back to the heart. Depending on the class of vertebrate, lungs may be filled using positive or negative pressure, and they may be ventilated by a tidal system or a flow-through system.

Most amphibians have lungs that are simple sacs with relatively little surface area, making them less-effective surfaces for oxygen diffusion than are the lungs of other vertebrates. A 30-g mouse lung, for example, has nearly 50 times more gas-exchange surface per cubic centimeter of lung tissue than a 30-g frog lung. The method by which amphibians ventilate their lungs is similar in some ways to the buccal pumping of fishes. A frog, for example, lowers its jaw, which decreases the air pressure inside the mouth cavity. According to Boyle's law, the pressure of a gas and its volume are inversely related (Figure 48.8). For example, when the volume in which a gas is contained increases, the pressure of the gas in that container decreases. By expanding its mouth cavity, therefore, a frog creates a pressure gradient for air to move from the atmosphere (higher pressure) into its mouth (lower pressure). The mouth and nostrils then close, and the mouth cavity is compressed by raising the bottom jaw. This raises the air pressure in the mouth, forcing air through a series of valves into the lungs. Thus, **positive pressure filling** means that frogs and most other amphibians gulp air and force it under pressure into the lungs, as if inflating a balloon. They may do this many times in a row before exhaling.

Except for a few species of lizards, all other terrestrial vertebrates use a different method of ventilation, called negative pressure filling. To understand how negative pressure filling works, we will first examine the anatomy of the respiratory tract in vertebrates, using the mammalian and avian systems as models.

48.3 Structure and Function of the Mammalian and Avian Respiratory Systems

Learning Outcomes:

1. Describe the components of the mammalian respiratory system and the structure of the mammalian lung.
2. Describe the process of ventilation in mammalian lungs.
3. Compare and contrast tidal ventilation in mammals and the flow-through breathing cycle of birds.

In vertebrates, the respiratory system includes all components of the body that contribute to the exchange of gas between the external

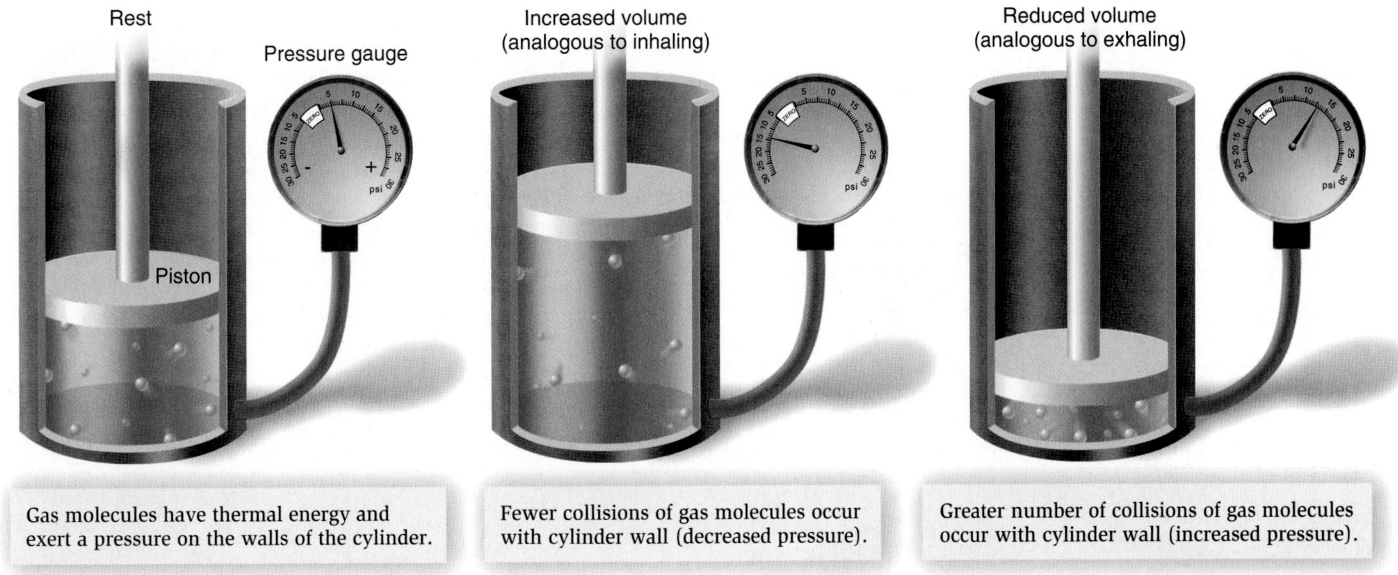
Rest
Pressure gauge
Piston
Increased volume
(analogous to inhaling)
Reduced volume
(analogous to exhaling)
psi
Gas molecules have thermal energy and exert a pressure on the walls of the cylinder.
Fewer collisions of gas molecules occur with cylinder wall (decreased pressure).
Greater number of collisions of gas molecules occur with cylinder wall (increased pressure).

Figure 48.8 **Boyle's law.** At a constant temperature, the volume and pressure of a gas are inversely related. This relationship explains the gas pressure gradients that ventilate the vertebrate lungs.

environment and the blood. In this section, we begin by examining the structures of the mammalian respiratory system and the mechanisms by which mammals ventilate their lungs. We then compare and contrast this with the highly specialized and unique way in which birds ventilate their lungs. Although both mammals and birds fill their lungs using negative pressure, mammals breathe by tidal ventilation, and birds use a flow-through system.

During Ventilation, Air Follows a Series of Branching Tubes in the Mammalian Respiratory System

In mammals, the respiratory system includes the nose, mouth, pharynx, larynx, trachea and branching tubes, lungs, and muscles and connective tissues that encase these structures within the thoracic (chest) cavity (Figure 48.9a). When humans and other mammals breathe, air first enters the nose and mouth, where it is warmed and humidified. These processes protect the lungs from drying out. While in the nose, the air is cleansed as it flows over a coating of sticky mucus in the nasal cavity. The mucus and hairs in the nasal cavity trap some of the larger dust and other particles that are inhaled with air. These are then removed by the body's defense cells or swallowed.

The inhaled air from the mouth and nose converges at the back of the throat, or **pharynx**, a common passageway for air and food. From there, air passes through the **larynx**. Within the larynx are the vocal cords, folds of tissue through which air passes to create sound. Air flows from the larynx into the **trachea**, a tube that leads to the lungs. At the opening to the larynx, a flap of tissue called the epiglottis prevents food from entering the trachea by closing when food is swallowed.

The trachea is ringed by C-shaped cartilage supports that provide rigidity and ensure that the trachea always remains open. The inner wall of the trachea is lined with cilia and cells called goblet cells (Figure 48.9b). These cells secrete mucus into the lumen of the trachea, coating the cilia. This mucous layer captures potentially harmful or irritating particles that escaped the nasal cleaning mechanism. The cilia continually beat the mucus and its trapped particles toward the mouth, where it can be expelled or swallowed. Without these active cilia, repeated coughing would be needed to force the mucus out of the trachea. This is why people who smoke develop a chronic "smoker's cough"—the smoke and its components reduce the number of cilia, causing mucus to pool in the trachea.

Inhaled air moves down the trachea as it branches into two smaller tubes, called **bronchi** (singular, bronchus), which lead to each lung. The bronchi branch repeatedly into smaller and smaller tubes, eventually becoming thin-walled **bronchioles** surrounded by circular rings of smooth muscle (Figure 48.9c). Bronchioles can dilate or constrict in a manner analogous to that of arterioles (see Chapter 47). They may partially constrict when a damaging particle—such as a small bit of inhaled pollutant or dust—gets past the mucous layers of the upper airways. Constriction of a bronchiole prevents foreign particles from reaching delicate lung tissue. Once the body's defense cells have removed the particles, the bronchiole reopens to its normal diameter.

The bronchioles empty into **alveoli** (singular, alveolus), the final, saclike regions of the lungs where gas exchange occurs (see Figure 48.9c). Until now, air has flowed through the air tubes without any gas exchange taking place. The alveoli are highly adapted for gas exchange and consist of two major types of cells. Type I cells are those across which gases diffuse, whereas type II cells are secretory cells (described later). The alveoli are only one-cell thick and resemble extremely thin sacs, appearing like bunches of grapes on a stem. Deoxygenated blood pumped from the right ventricle of the heart flows to the many capillaries surrounding the alveoli. Oxygen diffuses from the lumen of each alveolus across the alveolar cells, through the interstitial space outside the cells, and into the capillaries (Figure 48.9d). Carbon dioxide diffuses in the opposite direction. The

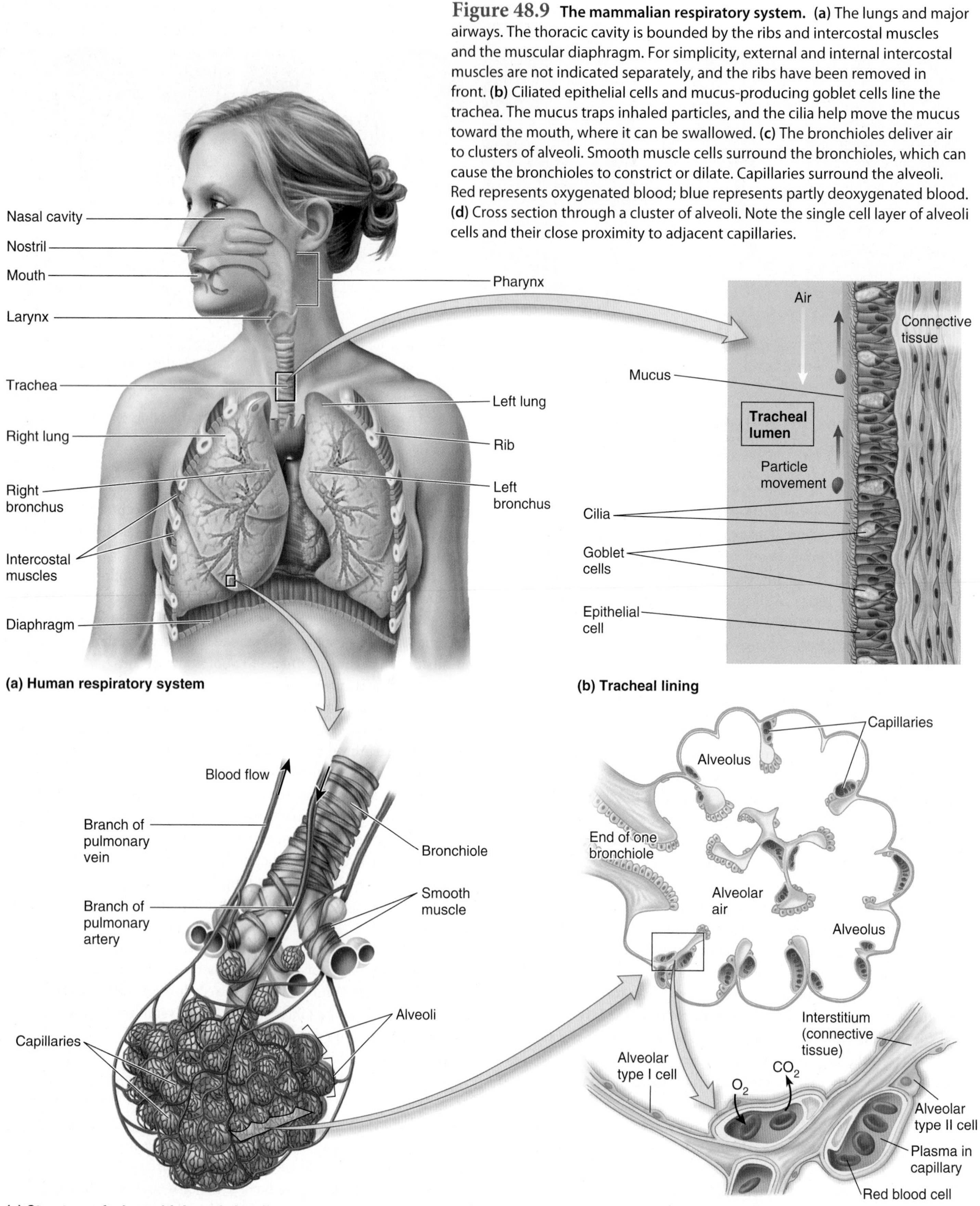

Figure 48.9 The mammalian respiratory system. **(a)** The lungs and major airways. The thoracic cavity is bounded by the ribs and intercostal muscles and the muscular diaphragm. For simplicity, external and internal intercostal muscles are not indicated separately, and the ribs have been removed in front. **(b)** Ciliated epithelial cells and mucus-producing goblet cells line the trachea. The mucus traps inhaled particles, and the cilia help move the mucus toward the mouth, where it can be swallowed. **(c)** The bronchioles deliver air to clusters of alveoli. Smooth muscle cells surround the bronchioles, which can cause the bronchioles to constrict or dilate. Capillaries surround the alveoli. Red represents oxygenated blood; blue represents partly deoxygenated blood. **(d)** Cross section through a cluster of alveoli. Note the single cell layer of alveoli cells and their close proximity to adjacent capillaries.

oxygenated blood from the lungs then flows to the left atrium of the heart and from there enters the left ventricle, where it is pumped out through the aorta to the rest of the body.

The Pleural Sacs Protect the Lungs

The lungs are soft, delicate tissues that could easily be damaged by the surrounding bone, muscle, and connective tissue of the thoracic cavity if not protected. Each lung is encased in a **pleural sac**, a double layer of thin, moist tissue. Between the two layers of tissue is a microscopically thin layer of water that acts as a lubricant and makes the two tissue layers adhere to each other.

In addition to protecting the lungs, the inner pleural sac adheres to its lung, and the outer pleural sac adheres to the chest wall. In this way, movements of the chest wall result in similar movements of the lungs. This is important because the lungs are not muscular and so cannot inflate themselves. Instead, as we will see, the lungs are inflated by the expansion of the thoracic cavity, which results from the contraction of muscles in the thorax.

Negative Pressure Is Used to Fill Both Mammalian and Avian Lungs

The way in which you inflate a balloon, by forcing air from your mouth into the balloon, is called positive pressure filling. It is reminiscent of how amphibians ventilate their lungs. **Negative pressure filling**, by contrast, is the mechanism by which reptiles, birds, and mammals ventilate their lungs. In this process, the volume of the lungs expands, creating a decreased pressure that draws air into the lungs. The process differs in some ways among classes of vertebrates, but in mammals, the work is provided by the **external intercostal muscles**, which surround and connect the ribs in the chest, and a large muscle called the **diaphragm** (see Figure 48.9a), which divides the thoracic cavity from the abdomen.

Let's follow the process when a mammal ventilates its lungs (**Figure 48.10**). At the start of a breath, the diaphragm contracts, pulling downward and enlarging the thoracic cavity. Simultaneously, the external intercostal muscles contract, moving the chest upward and outward, which also helps to enlarge the thoracic cavity. Recall that the pleural sacs adhere the lungs to the chest wall, so as the chest expands, the lungs expand with it. According to Boyle's law, as the volume of the lungs increases, the pressures of the gases within them must decrease. In other words, the pressure in the lungs becomes negative with respect to the outside air. Air, therefore, flows down its pressure gradient from outside the mouth and nose, into the lungs.

Once the lungs are inflated with air, the chest muscles and diaphragm relax and recoil back to their original positions as an animal exhales. This compresses the lungs and forces air out of the airways. Whereas inhaling requires the expenditure of significant amounts of energy, exhaling is mostly passive and does not require much energy. During exercise, however, exhalation is assisted by contraction of another set of rib muscles, called the internal intercostal muscles. The contraction of these muscles depresses the rib cage and provides a more forceful exhalation; this increases the rate at which ventilation occurs and helps empty the larger volume of air from the lungs.

Mammals Breathe by Tidal Ventilation

When mammals exhale, air leaves via the same route that it entered during inhalation, and no new oxygen is delivered to the airways at that time. As mentioned, this type of breathing is called **tidal ventilation** (think of air flowing in and out of the lungs like the ebb and flow of tides on the ocean). Tidal ventilation is less efficient than the unidirectional, flow-through system of fishes in which gills are always exposed to oxygenated water during all phases of the respiratory cycle.

As you likely know from experience, the lungs are neither fully inflated nor deflated at rest. For example, you could easily take a larger breath than normal if you wished, or exhale more than the usual amount of air. The volume of air that is normally breathed in and out at rest is the **tidal volume**, about 0.5 L in an average-sized human. Tidal volume is proportional to body size both among humans and between species. A 6-foot-tall adult, for example, has a larger tidal volume than a 4-foot-tall child because the adult has larger lungs. Similarly, horses have larger tidal volumes than humans, and humans have larger tidal volumes than dogs (**Table 48.1**). During exertion, the lungs can be inflated further than the resting tidal volume to provide

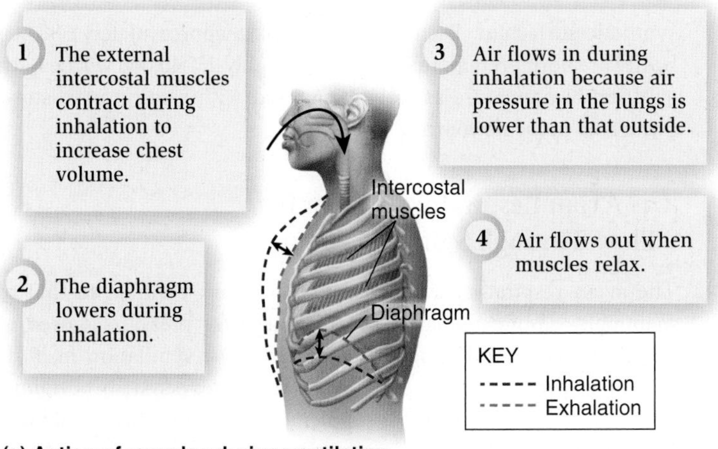

1 The external intercostal muscles contract during inhalation to increase chest volume.
2 The diaphragm lowers during inhalation.
3 Air flows in during inhalation because air pressure in the lungs is lower than that outside.
4 Air flows out when muscles relax.
Intercostal muscles
Diaphragm
KEY
Inhalation
Exhalation

(a) Action of muscles during ventilation

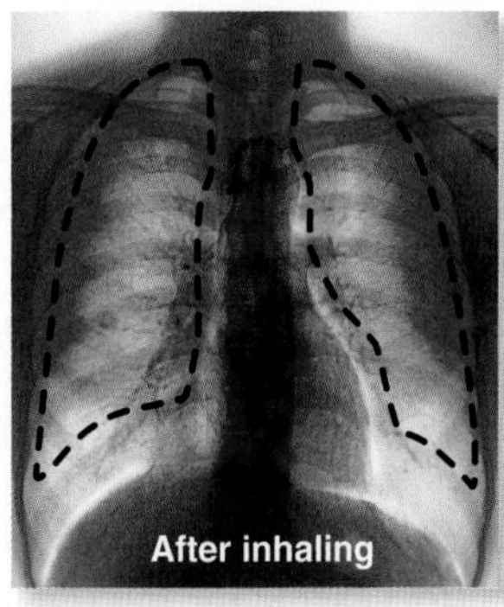

(b) Change in lung volume during ventilation

Figure 48.10 **Ventilation of the mammalian lung by negative pressure filling.** **(a)** The external intercostal muscles contract, which expands the chest cavity by moving the ribs up and out. The diaphragm also contracts, causing it to pull downwards, further expanding the cavity. The muscular efforts of ventilation require energy, whereas the return to the resting state by exhaling is primarily by recoil. **(b)** X-ray image of the chest of an adult man after inhaling. The volume of the lungs after exhaling is superimposed using dashed lines to illustrate the relative change in lung volume.

Table 48.1 Comparative Ventilatory Characteristics of Representative Mammals

Animal	Body mass (kg)	Breaths/min	Tidal volume
Shrew	0.0024	700	30 μL
Dog	25	20	0.27 L
Human	75	12	0.50 L
Horse	465	9	6.50 L

Note: All values are averages from resting animals. Tidal volume is the volume of air breathed in with each breath. Note that tidal volume increases as an animal's mass and lung size increase. By contrast, breathing rate is faster in smaller animals. Similar relationships occur in other vertebrates, notably birds.

additional oxygen. Likewise, the lungs can be deflated beyond their normal limits at rest, by exerting a strong effort during exhalation. The lungs never fully deflate, however, partly because they are held open by their adherence to the chest wall. This is important for a simple reason. Think again of our analogy of a balloon. It is much easier to fill a balloon that is already partly inflated than it is to inflate a completely empty balloon. The same is true of the lungs. The most difficult breath is the very first one that a newborn mammal takes—the only time its lungs are ever completely empty of air.

Surfactant Facilitates Lung Inflation by Reducing Surface Tension

Like all cells, those that make up the lining of the alveoli are surrounded by extracellular fluid. This fluid layer is where gases dissolve. Unlike other internal body cells, however, alveolar cells come into contact with air, creating an air/liquid interface along the inner surface of the alveoli. This results in surface tension (see Chapter 2, Figure 2.19g) within the alveoli. Surface tension results from the attractive forces between water molecules at the air/liquid interface and partly explains why droplets of water form beads. It also produces a force that makes alveoli tend to collapse as water molecules lining its surfaces are attracted to each other. If many or all of the alveoli collapsed, however, the amount of surface area available for gas exchange in the lungs would be greatly reduced. What prevents them from collapsing? The type II cells of the alveoli produce **surfactant**, a mixture of proteins and amphipathic lipids (that is, lipids with both polar and nonpolar regions), and secrete it into the alveolar lumen. There, it forms a barrier between the air and the fluid layer inside the alveoli. This barrier reduces surface tension in the alveolar walls, allowing them to remain open.

Surface tension is particularly important in the transition from fetal to postnatal life in mammals. Most mammalian fetuses are encased in fluid within the uterus. Consequently, their lungs do not have an air/liquid interface, and they do not start producing surfactant until the final stages of pregnancy. In humans, surfactant production begins around week 26 of gestation but does not increase to final levels until after week 33 (normal pregnancy length is about 40 weeks). If a human baby is born prematurely (defined as prior to week 37 of gestation), sufficient surfactant may not be available, and consequently, many alveoli may collapse after birth. This condition, known as **respiratory distress syndrome of the newborn**, can be partially alleviated by inserting a tube in the trachea and injecting synthetic surfactant. Each year in the U.S., approximately 500,000 babies are born prematurely; of those, roughly 25–40,000 will be born sufficiently premature as to be diagnosed with respiratory distress syndrome of the newborn.

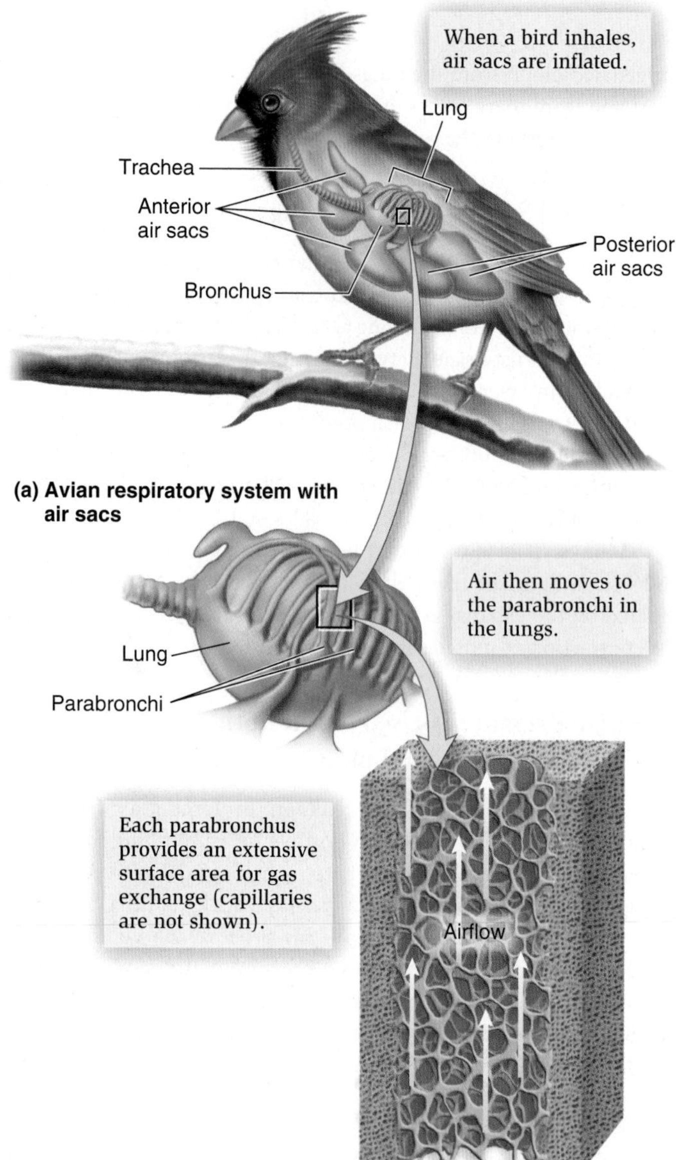
When a bird inhales, air sacs are inflated.
Lung
Trachea
Anterior air sacs
Posterior air sacs
Bronchus
(a) Avian respiratory system with air sacs
Air then moves to the parabronchi in the lungs.
Lung
Parabronchi
Each parabronchus provides an extensive surface area for gas exchange (capillaries are not shown).
Airflow
(b) Lung showing parabronchi and cross section of a parabronchus

Figure 48.11 **Respiratory system of a bird.** For simplicity, not all air sacs are shown.

The Avian Respiratory System Is a Flow-Through System

The avian respiratory system provides a stark contrast to that of mammals, despite the similarity that negative pressure is used to fill the lungs in both cases. Unlike mammals, ventilation in birds is a flow-through system. The avian respiratory system is unique among vertebrates in that it is supplemented with numerous **air sacs** (Figure 48.11a), which for simplicity we can group together as those at the anterior or posterior regions of the body. The air sacs—not the

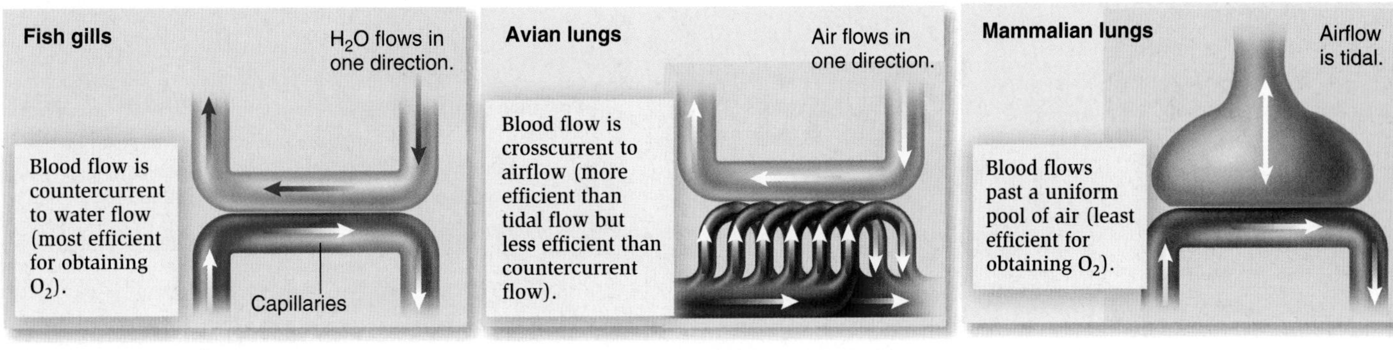

Figure 48.12 **Relationships between direction of flow of blood and water or air in different vertebrates.**

Concept Check: *What might you conclude about the relative amounts of oxygen available in water and air based on the differences described in this figure?*

lungs—expand when a bird inhales and shrink when it exhales. They expand by movements of the chest muscles, not by a diaphragm.

The air sacs, however, do not participate in gas exchange. When a bird inhales, air enters its trachea and from there moves into two bronchi. Instead of branching into bronchioles and then alveoli, however, avian bronchi branch into a series of parallel air tubes called **parabronchi** that make up the lungs (Figure 48.11b). The parabronchi are the regions of gas exchange. They have enormous surface area and an extensive network of blood capillaries. Blood flows through the lungs in a crosscurrent direction with respect to movement of air. This is less efficient in extracting oxygen than the countercurrent flow arrangement in fish gills, because air and blood do not move in exactly opposite directions along the length of the capillaries, but it is more efficient than the tidal ventilation system of mammalian lungs (Figure 48.12). Precisely how air moves through avian lungs and the function of the air sacs remained a mystery to physiologists for many years, however, until the ingenious work of Knut Schmidt-Nielsen, as described next.

Feature Investigation

Schmidt-Nielsen Mapped Airflow in the Avian Respiratory System

In 1967, American physiologist Knut Schmidt-Nielsen and coworkers conducted a series of experiments designed to determine the mechanisms of ventilation and gas exchange in avian lungs, using ostriches because of their large lung volumes. The researchers used O_2- and CO_2-sensitive probes to determine the concentrations of these gases in the birds' air sacs. In their first experiment, described in Figure 48.13, the investigators allowed the birds to breathe ordinary room air, which contained 21% O_2. They used special syringes to extract a gas sample from the large air sacs of the birds and analyzed the gas composition with the probes. They discovered that the O_2 content was high (21%) and the CO_2 content low in the posterior air sacs, with the reverse pattern in the anterior sacs. They concluded that fresh air must move through the trachea and into the posterior air sacs, while "stale" air, which had previously given up much of its O_2 and picked up CO_2 in the lungs, must have entered the anterior sacs.

Figure 48.13 **Schmidt-Nielsen and coworkers demonstrated the route that air travels in the avian lung.**

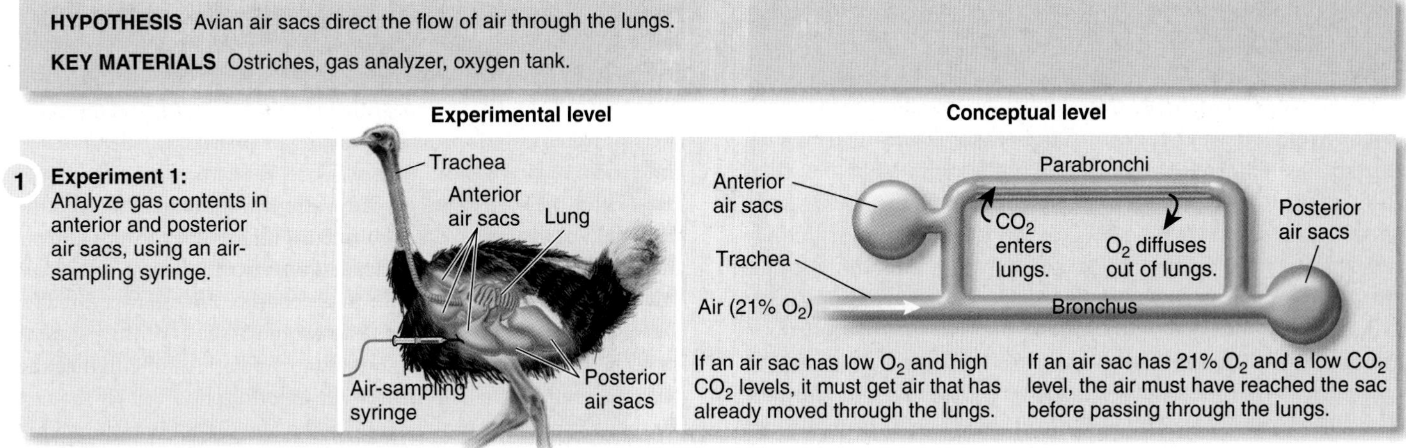

2 THE DATA

Results from Experiment 1:

Anterior air sacs

Low O_2; high CO_2

Posterior air sacs

21% O_2; low CO_2

The researchers proposed that air moves directly to posterior sacs, then through the lungs, and then to the anterior sacs.

How can this be demonstrated?

3 Experiment 2:
Give ostriches a single breath of pure O_2, which can be tracked as it moves through the respiratory system. Then remove the mask and allow the bird to breathe normal air.

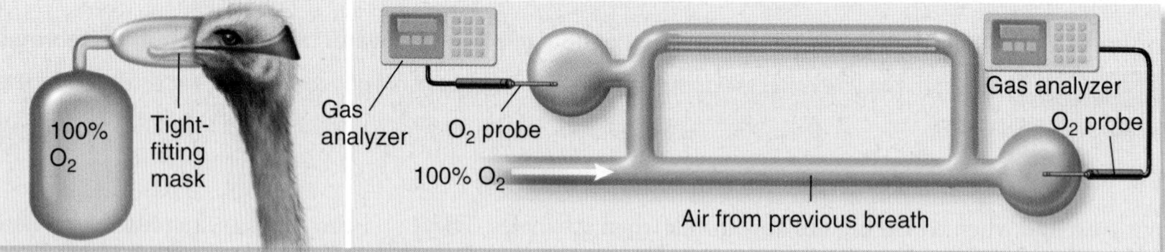

4 THE DATA

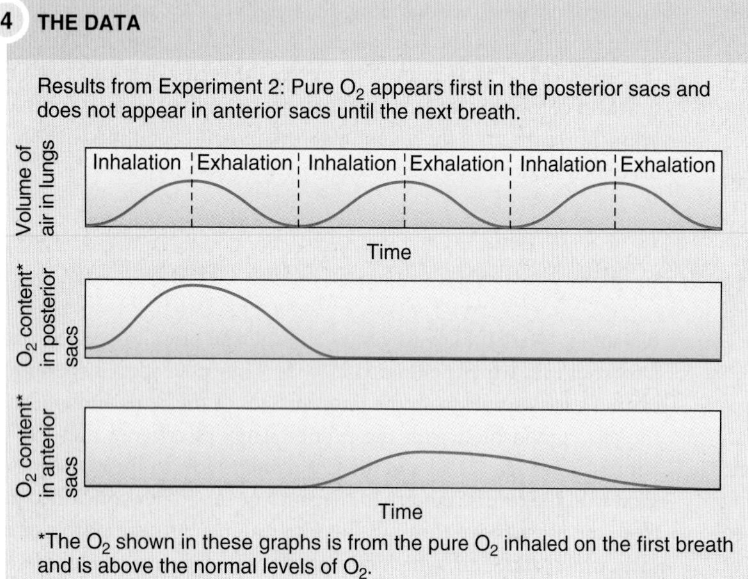

Results from Experiment 2: Pure O2 appears first in the posterior sacs and does not appear in anterior sacs until the next breath.
Volume of air in lungs
Inhalation
Exhalation
Inhalation
Exhalation
Inhalation
Exhalation
Time
O2 content* in posterior sacs
O2 content* in anterior sacs
Time
*The O2 shown in these graphs is from the pure O2 inhaled on the first breath and is above the normal levels of O2.

5 CONCLUSION

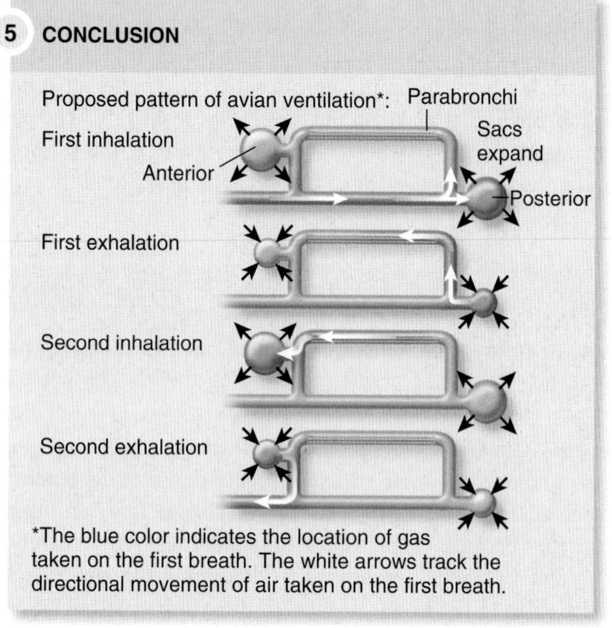

6 SOURCE Schmidt-Nielsen, K. et al. 1969. Temperature regulation and respiration in the ostrich. *The Condor* 71:341–352.

To test this hypothesis, they developed a method to track the movement of O_2 through the ostrich's respiratory system, as shown in the second experiment of Figure 48.13. In this experiment, the scientists fitted ostriches with face masks, through which the birds inhaled a single breath of pure (100%) O_2. O_2-sensitive probes were surgically inserted into the posterior and anterior air sacs to monitor when the pure O_2 appeared in each sac. The researchers discovered that during the first inhalation, high levels of O_2 appeared in the posterior air sac but not the anterior sac. As the ostrich exhaled, some O_2 remained in the posterior air sac, but still no O_2 appeared in the anterior one. When the animal inhaled its next breath (this one of plain air, not pure O_2), pure O_2 began appearing in the anterior sacs and disappearing from the posterior sacs. Finally, as the ostrich exhaled a second time, the pure O_2 in the anterior sacs decreased.

Schmidt-Nielsen concluded from this study that air flows through the trachea, down the bronchi, and into the posterior air sacs during inhalation. As a bird exhales, air exits the posterior sacs and flows through the parabronchi in a posterior to anterior direction. During its passage through the lungs, some of the O_2 diffuses into the capillaries, while CO_2 diffuses out of the capillaries and into the lungs. During the next inhalation, air at the anterior part of the lungs flows through the remainder of the lungs and then into the anterior air sacs, which serve as a sort of holding chamber. Finally, a second exhalation causes this air from the previous inhalation to exit the anterior sacs and leave the animal's body through the trachea.

Therefore, it takes two complete breaths for air to move from the environment into the lungs and back out again to the environment. Air flows through the lungs during inhalation, as it is drawn by negative pressure through the parabronchi and into the expanding anterior sacs. Air is also forced through the parabronchi by positive pressure during exhalation, as the posterior air sacs are compressed. The efficiency of the avian flow-through system, together with its crosscurrent blood supply, is a major reason why many birds are able to fly at altitudes with extremely low atmospheric pressures (even over Mt. Everest—see Figure 48.2).

Experimental Questions

1. What was the purpose of these studies?
2. Considering the first experiment only, what were the results, and what conclusions did the researchers make based on these results?
3. Explain the purpose and procedure of the second experiment conducted by Schmidt-Nielsen and the conclusions he drew from his results.

48.4 Mechanisms of Oxygen Transport in Blood

Learning Outcomes:

1. Describe the characteristics of respiratory pigments in the blood of vertebrates and the hemolymph of many invertebrates.
2. Analyze the oxygen-hemoglobin dissociation curve and explain how certain factors can modify the shape of the curve.
3. Write the bidirectional reaction between hemoglobin and oxygen.

Gases such as oxygen are soluble in the plasma component of blood, but as with all solutes, there are limits to how much of a gas can dissolve in a volume of water. The amount of gas dissolved in the body fluids is not sufficient to sustain life in most animals. In nearly all animals, therefore, the amount of oxygen in the body fluids must be increased above that which can be physically dissolved. This is made possible because of the widespread occurrence of oxygen-binding proteins, which increase the total amount of oxygen available to cells. In this section, we examine the nature, function, and evolution of these molecules.

Oxygen Binds to Respiratory Pigments

The oxygen-binding proteins that have evolved in animals are called **respiratory pigments**, because they have a color (blue or red). In vertebrates, the pigments are contained within red blood cells, whereas many invertebrates have these pigments in their hemolymph. Respiratory pigments are proteins containing one or more metal atoms in their cores; the metal atom binds to oxygen. In vertebrates and many marine invertebrates, the metal is typically iron (Fe^{2+}). **Hemoglobin** is the major iron-containing pigment and gives blood a red color when oxygen is bound. In decapod crustaceans, arachnids, and many mollusks (including cephalopods and some gastropods), the metal is copper (Cu^{2+}). The copper-containing pigment, **hemocyanin**, gives the blood or hemolymph a bluish tint.

Hemoglobin gets its name because it is a globular protein—which refers to the shape and water solubility of a protein—and because it contains a chemical group called a heme in its core. An atom of iron is bound within the heme group. In vertebrates, hemoglobin consists of four protein subunits, each with its own iron atom to which a molecule of oxygen can bind (refer back to Figure 47.6). Thus, a single hemoglobin protein can bind up to four molecules of oxygen.

Respiratory pigments share certain characteristics that make them ideal for transporting O_2. First, they all have a high affinity for binding O_2. Second, the binding between the pigment and O_2 is noncovalent and reversible. It would not be very useful for a protein to bind O_2 if it could not later unload the O_2 to cells that need it. The formula for the reversible oxygen binding to a respiratory pigment is:

$$Hb + O_2 \rightleftharpoons HbO_2$$

where Hb is hemoglobin, HbO_2 is hemoglobin with bound O_2 (called oxyhemoglobin), and the double arrows ($\rightleftharpoons$) indicate that the reaction is reversible.

The amount of pigment present in the blood is great enough to provide sufficient O_2 to meet all but the most strenuous exertion. In humans, for example, the presence of hemoglobin gives blood about 45 times more oxygen-carrying capacity than would plasma alone. Differences exist among respiratory pigments, however, at the molecular level. For example, the amino acid sequences of hemoglobins from different species vary, in some cases considerably. Human hemoglobin is similar but not identical to that of other mammals, and mammalian hemoglobins have numerous amino acid differences from other vertebrate hemoglobins. The differences in amino acid sequences cause slight changes in the overall three-dimensional shape of hemoglobin, which alters its affinity for O_2. As a general rule, animals with the highest metabolic rates tend to have hemoglobins with lower O_2 affinity, and animals with lower metabolic rates have higher affinities. That may sound paradoxical at first, but consider the preceding reactions again. Because the hemoglobin of small animals with high relative metabolic rates has lower O_2 affinity, the reverse reaction ($HbO_2 \rightarrow Hb + O_2$) will be more likely to occur when oxygen is diffusing from the blood to active cells. In other words, hemoglobin unloads its O_2 more readily in those animals whose cells use O_2 at the greatest rates.

The Amount of Oxygen Bound to Hemoglobin Depends on the Po_2 of Blood

Recall that the partial pressure of O_2 (Po_2) is a measure of its dissolved concentration. When Po_2 is high, more O_2 binds to hemoglobin, whereas fewer O_2 molecules will be bound when Po_2 is low. Figure 48.14 shows the relationship between O_2 binding and Po_2, known as an **oxygen-hemoglobin dissociation curve**, for humans. At a Po_2 of 100 mmHg, which is found in the blood leaving the lungs, each hemoglobin protein binds four O_2 molecules (far right of graph).

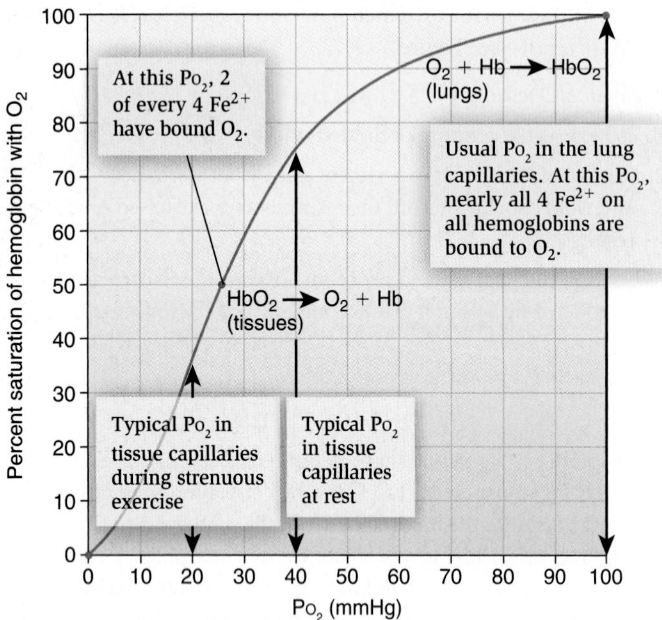

Figure 48.14 The human oxygen-hemoglobin dissociation curve. Depending on the partial pressure of oxygen (P_{O_2}), oxygen is either loaded onto hemoglobin, as in the lungs, or unloaded from hemoglobin, as in the rest of the body tissues. When P_{O_2} is high, more Fe^{2+} atoms are bound to O_2, and the hemoglobin is more saturated with O_2.

Concept Check: *Members of the Channichthyidae family of Antarctic icefish are the only vertebrates that do not have red blood cells and hemoglobin. Can you propose some mechanisms or adaptations that allow them to survive without the oxygen reservoir provided by hemoglobin? (Hint: Think about the effect of temperature on oxygen solubility and on an animal's metabolic rate.)*

It is 100% saturated with O_2 at that P_{O_2}. The P_{O_2} of blood leaving the tissue capillaries of other parts of the body is lower and depends on metabolic activity and exercise. At rest, the P_{O_2} of blood capillaries in these other parts of the body is typically around 40 mmHg. At this P_{O_2}, hemoglobin releases some O_2 molecules, and is about 75% saturated with O_2. At high levels of exercise, P_{O_2} in the capillaries drops even further (as low as 20 mmHg), and hemoglobin releases even more O_2. In this way, hemoglobin performs its role of O_2 delivery. In the lungs, it binds O_2, and elsewhere it releases some of the O_2 as needed.

You may have noticed that the curve in Figure 48.14 is not linear but S-shaped (sigmoidal). This is because the subunits of hemoglobin are said to cooperate with each other in binding O_2. Once a molecule of O_2 binds to one subunit's iron atom, the shape of the entire hemoglobin protein changes, making it easier for a second O_2 to bind to the next subunit, and so on. Thus, the relationship between O_2 pressure and the amount of O_2 bound to hemoglobin becomes very steep in the midrange of the curve, which represents the pressures that occur in the tissue capillaries throughout the body except in the lungs. This steepness allows O_2 release to be very sensitive to decreases in P_{O_2}.

One of the remarkable features of the oxygen-hemoglobin binding relationship is that it can be influenced by metabolic waste products such as CO_2 and H^+ and also by temperature. In addition to being transported as bicarbonate ions (HCO_3^-), about 25% of the CO_2 in blood is bound to hemoglobin. The remainder—about 7–10%—exists dissolved in solution in the plasma and within red blood cells. **Figure 48.15a** shows three curves, one obtained under normal resting conditions and the others in the presence of low or high levels of CO_2. Carbon dioxide binds to amino acids in the hemoglobin protein (not to the iron, like O_2), and when it does, the ability of hemoglobin to continue binding O_2 decreases. Note how an increase in CO_2 shifts the curve to the right, such that at any P_{O_2}, less O_2 is bound to hemoglobin. Another way of saying this is that at any P_{O_2}, more O_2 has been released from hemoglobin, thus becoming available to cells. A similar shift in the curve occurs with an increase in the concentration of H^+, which can also bind to hemoglobin; the effect of CO_2 and H^+on the oxygen-hemoglobin dissociation curve is known as the Bohr effect, after its discoverer, the Danish physiologist Christian Bohr. Elevated temperature also reduces the affinity of hemoglobin for O_2, resulting in a shifted curve.

Cells generate each of these products—CO_2, H^+, and heat—when they are actively metabolizing fuel. The metabolic products enter the surrounding blood vessels, where they disrupt the normal shape of hemoglobin, causing it to release more of its O_2 than would normally occur at that P_{O_2}. This phenomenon is a way in which individual body tissues can obtain more oxygen from the blood to match their changing metabolic demands. Thus, when an animal increases its physical activity, the skeletal muscles generate more waste products and heat than does, say, the skin. Therefore, the oxyhemoglobin in muscle blood vessels releases more oxygen to the cells where it is immediately required.

The shift in the oxygen-hemoglobin dissociation curve occurs in all classes of vertebrates (although not in all species), but it has different magnitudes in different species. Not surprisingly, perhaps, more metabolically active animals, such as mice, show a greater shift for a given increase in CO_2, H^+, or heat than do less-active animals. Recall from Chapter 47 that these same waste products of metabolism also cause local vasodilation of arterioles. The more metabolically active a tissue is, therefore, the more blood flow it receives, which means more oxygen-bound hemoglobin. Moreover, the oxygen is unloaded from hemoglobin more readily due to the shift in the curve. This is another example of how adaptive changes in circulatory and respiratory functions often occur in parallel.

The hemoglobins of metabolically active animals also have a lower affinity for oxygen (**Figure 48.15b**). In small, active animals the curves are displaced to the right of the human curve. In contrast, larger animals with slower metabolisms, such as the elephant, have curves shifted to the left of humans. At high oxygen pressures, such as those that occur in the lungs, these shifts have little relevance because nearly all the hemoglobin is bound to oxygen at those pressures. At lower P_{O_2}, however, such as those that would occur in the blood vessels of metabolically active tissues, the difference in the curves becomes significant. For example, look at the three curves at a P_{O_2} of 40 mmHg, a typical value found in tissues that are using oxygen at a resting rate. The mouse hemoglobin has less oxygen bound (its hemoglobin is less saturated) at that pressure than does the human hemoglobin, which, in turn, has less O_2 bound than does the elephant hemoglobin. In other words, the mouse hemoglobin has released more of its O_2 to the active tissues than have the other animal hemoglobins.

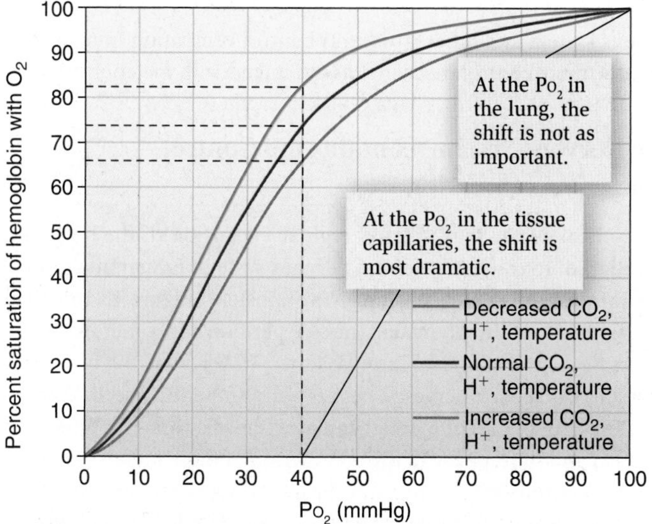

(a) Shifts in the oxygen-hemoglobin dissociation curve

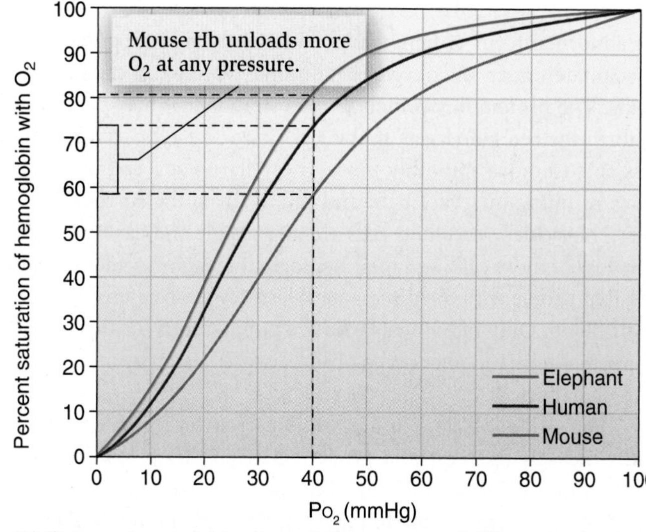

(b) Oxygen-hemoglobin dissociation curves of different animals

Figure 48.15 Differences in oxygen-hemoglobin dissociation curves under different physiological conditions and among different species. (a) Increasing or decreasing the amounts of CO_2 or H^+ (acid), or the temperature of the blood, shifts the oxygen-hemoglobin dissociation curve. Metabolically active tissues generate these products. The change in affinity of hemoglobin for oxygen (O_2) allows different tissues to obtain O_2 in proportion to their metabolic requirements. **(b)** Oxygen-hemoglobin dissociation curves for three mammals with low (elephant), moderate (human), or high (mouse) relative metabolic rates. For any Po_2, such as the one shown (40 mmHg), which is typical of the Po_2 of tissue capillaries, less O_2 is bound to mouse hemoglobin than to human or elephant hemoglobin, and less O_2 is bound to human hemoglobin than to elephant hemoglobin. Therefore, O_2 is unloaded from hemoglobin more readily in smaller animals.

Concept Check: *What would happen to the position of the middle curve in part (a) following infusion of bicarbonate ions (HCO_3^-) into the blood of a resting, healthy individual?*

Oxygen-carrying molecules are among the most ancient proteins found in animals. The globin gene family appears to have evolved from a single ancestral globin gene. Around 500 mya, the gene for this ancient molecule duplicated, eventually resulting in one gene that coded for myoglobin, a protein that stores oxygen in muscle cells, and the other for an early form of hemoglobin (refer back to Figure 21.8). Thus, animals gained the ability not only to store O_2 in tissues but also to transport it throughout the body via the circulation. Let's take a closer look at the molecular structure and evolution of hemoglobin and how a simple mutation in the sequence of one hemoglobin gene can have devastating consequences in humans.

Genomes & Proteomes Connection

Hemoglobin Evolved Over 500 Million Years Ago

Modern hemoglobin in adult mammals contains four subunits, designated α1, α2, β1, and β2. The two α and two β subunits are identical. Early in animal evolution, hemoglobin existed only as monomers (one subunit) or dimers (two subunits), which is still observed in lampreys. Later, gene duplication created the second set of subunits, leading to the four-subunit form, which first appeared in sharks and bony fishes. As we have seen in this chapter, the subunit structure of hemoglobin enables one hemoglobin protein to bind four oxygen molecules. Equally important, the subunits of hemoglobin cooperate with each other as shown by the sigmoidal shape of the oxygen-hemoglobin dissociation curve.

As noted earlier, modern hemoglobins differ among species within the same class of vertebrates and even more so between classes. This suggests that evolution of the hemoglobin genes continued even after the four-subunit structure appeared. The mutations that arose in the hemoglobin genes affected the structure of the hemoglobin protein. One consequence is that the affinity of hemoglobin for oxygen differs among animals. The differences in affinity may result, for example, in better oxygen-capturing ability of animals living in regions of low oxygen pressure, such as at high altitude. They also account, at least in part, for the ability of metabolically active animals to unload oxygen from hemoglobin more readily than less metabolically active organisms.

Despite the high frequency of mutations in the hemoglobin genes in vertebrates, certain regions of the molecule have remained well conserved. Not surprisingly, these are regions that are critically important for the function of the molecule. For instance, the amino acids that encase the heme group and iron atom of each subunit have changed little if at all during evolution. Similarly, the amino acid sequences associated with sites of subunit interactions have also resisted evolutionary change. Mutations that arise in these regions are typically not adaptive and may even be lethal.

Some mutations in the hemoglobin genes may appear to be detrimental because they negatively affect hemoglobin's function, and yet they provide a selective advantage under certain conditions. Consider the disease **sickle cell disease** (also called sickle cell anemia). In this case, substitution of a thymine for an adenine at one position in the hemoglobin gene results in a single amino acid substitution (from a glutamic acid to a valine; refer back to Figure 14.1). This produces an abnormal form of hemoglobin that tends to form polymers and precipitate under low oxygen conditions, typically in capillaries and veins. The protein forms long fibrous strands that may permanently deform the red blood cell, making it sickle-shaped. Sickled cells are less able to move smoothly through capillaries and can block blood flow, resulting in severe pain and cell death in the surrounding tissue. The sickled red blood cells are also fragile and easily destroyed. The loss of those cells and their hemoglobin results in anemia, meaning that people with the disease have less oxygen-carrying capacity in their blood. Only individuals who are homozygous for this mutation show the dramatic phenotype of the disease. Heterozygotes are relatively unaffected.

The sickle cell gene mutation is present in up to one-third of individuals living in malaria-prone regions of Africa, to lesser extents in the Middle East and Eastern Europe, and in roughly 8% of African Americans (refer back to Figure 24.6). Why should such an obviously harmful mutation have persisted in the human population? Recall from Chapter 24 that individuals who are heterozygous for the sickle cell allele are less likely to develop full-blown malaria than either those who are homozygous for the sickle cell allele or those who are homozygous for the normal allele, although the reason for this is uncertain. One favored hypothesis states that the parasite causing malaria lowers the oxygen pressure in red blood cells by consuming oxygen to support its own metabolism. This may render the red blood cell susceptible to sickling even in heterozygous individuals. The damaged cells are destroyed and removed from the circulation, which also removes the parasite and limits its ability to multiply. Another hypothesis states that the heterozygotic condition results in a more efficient immune response of the body to the presence of the parasite. Thus, although both malaria and sickle cell disease can be life-threatening, those who are heterozygous for the sickle cell allele develop neither pronounced anemia nor severe malaria and therefore have an advantage in areas where malaria is present. This phenomenon, an example of heterozygote advantage, explains why the sickle cell mutation persists among certain human populations.

48.5 Control of Ventilation in Mammalian Lungs

Learning Outcomes:

1. Describe the role of respiratory centers and chemoreceptors in the regulation of ventilation.
2. Explain the link between ventilation rate and H^+ concentration.
3. Describe how the activity of the respiratory centers in the brain of vertebrates is related to an animal's metabolic rate.

In the previous sections, you saw that different vertebrates have adapted to air-breathing in very different ways. Now let's look at how the mechanisms of breathing are controlled, using mammals as our example.

Lungs are neither muscles nor electrically excitable tissue. Therefore, they cannot initiate or regulate their own expansion. Nonetheless, lungs require a mechanism to rhythmically expand and recoil, because animals cannot always consciously control breathing, such as during sleep. In this section, we examine the ways in which the nervous system and other structures control ventilation in mammals and how these control mechanisms are linked with metabolism.

The Nervous System Contains the Control Center for Ventilation

The control center that initiates regular expansion of the lungs is a collection of nuclei in the central nervous system. In mammals, these **respiratory centers** are located in the brainstem (**Figure 48.16**). Neurons within certain brainstem nuclei periodically generate action potentials. These electrical impulses then travel from the brainstem through two sets of nerves. The first set stimulates the intercostal muscles, and the second set stimulates the diaphragm. When the lungs expand in response to the contraction of these muscles, stretch-sensitive neurons in the lungs and chest send signals to the respiratory centers, informing them that the lungs are inflated. This temporarily turns off the stimulating signal until the animal exhales, whereupon a new signal is sent to the breathing muscles.

Although the brainstem controls breathing automatically, it can be overridden. Animals that dive underwater—including humans when we go swimming—can hold their breath temporarily. We can

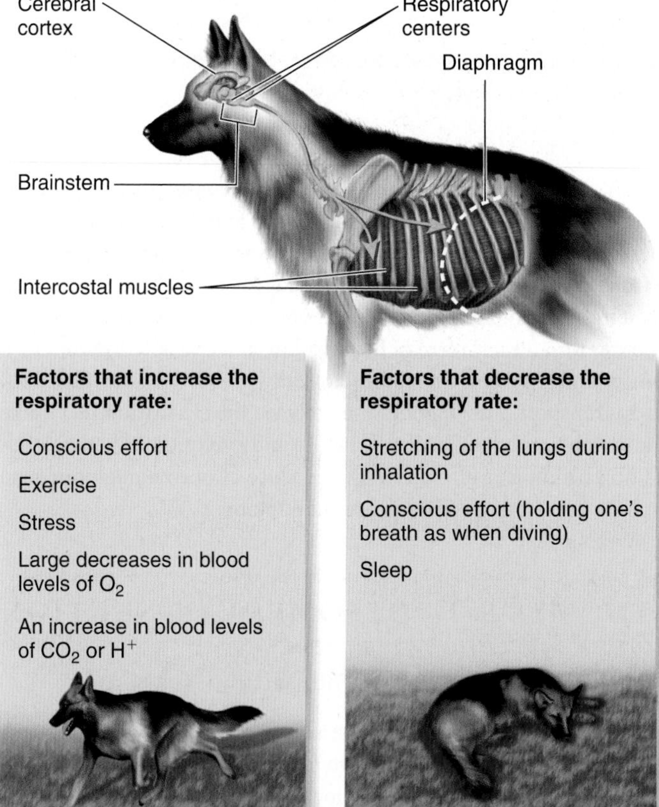

Figure 48.16 The control of breathing via respiratory centers in the mammalian brain. Neurons in the brainstem send action potentials along neurons in nerves that stimulate the intercostal muscles and diaphragm. The factors listed here can modulate the rate of action potential generation and therefore the respiratory rate.

BioConnections: The brainstem of vertebrates was described in Chapter 42. What parts of the brain are included in the brainstem?

also breathe faster than normal even while resting, at least for short periods of time. Normally, however, increased breathing occurs in response to physical activity. At such times, a variety of factors converge on the respiratory centers to increase the rate and strength of signals to the breathing muscles, resulting in faster and deeper breaths.

Chemoreceptors Modulate the Activity of the Respiratory Centers

The respiratory centers are influenced by the concentrations of certain substances in the blood, including the partial pressures of oxygen (P_{O_2}) and carbon dioxide (**P_{CO_2}**), and the concentration of hydrogen ions (H^+) (in other words, blood pH). **Chemoreceptors**—special cells located in the aorta, carotid arteries, and the brainstem—detect the circulating levels of these substances and relay that information through nerves or interneurons to the respiratory centers.

If the arterial P_{O_2} decreases well below normal, as might occur at high altitude or in certain respiratory diseases, the chemoreceptors signal the respiratory centers to increase the rate and depth of breathing to increase ventilation of the lungs. This brings in more oxygen. Similarly, a buildup of CO_2 in the blood, which would occur if an animal's ventilation were lower than normal (again, often the result of respiratory disease), signals the respiratory centers to stimulate breathing. The increased ventilation not only brings in more O_2, it also helps eliminate more CO_2. Finally, an increased concentration of H^+ in the blood activates chemoreceptors, which signal the brain that the blood is too acidic. This leads to an increase in the rate of breathing.

What is the link between ventilation rate and H^+ concentration? The concentrations of CO_2 and H^+ in the blood are related, because the concentration of H^+ in the fluid bathing the brainstem chemoreceptors reflects the amount of CO_2 produced by cells during metabolism. This is because roughly two-thirds of the CO_2 produced during metabolism is converted into a less toxic and more soluble form, bicarbonate ions (HCO_3^-). In the process, H^+ ions are formed according to the following reaction, where H_2CO_3 is a short-lived compound called carbonic acid that immediately dissociates to a H^+ and a HCO_3^-:

$$CO_2 + H_2O \rightleftharpoons H_2CO_3 \rightleftharpoons H^+ + HCO_3^-$$

This reaction is readily reversible. The first step is catalyzed in both directions by the enzyme carbonic anhydrase, which is present in high amounts in red blood cells. As you learned in Chapter 2, reactions of this type proceed according to the concentrations of the reactants and products. For example, if CO_2 concentrations were to increase for any reason, the forward reaction (from left to right as illustrated here) would be favored, and the pH of the blood would decrease because H^+ ions would increase. Conversely, if the concentrations of CO_2 decreased, the reactions would proceed from right to left, and the pH of the blood would increase as H^+ ions were removed from the blood by combining with HCO_3^-.

What happens when an animal increases its physical activity? During exercise, lactic acid—a by-product of metabolism—is released into the blood, and this decreases the pH of the blood. The increase in H^+ concentration that results from lactic acid activates chemoreceptors to stimulate increased ventilation. This, in turn, helps prevent CO_2 concentrations from increasing in the blood despite the increased metabolism that is occurring.

In body tissues, CO_2 concentrations are high because metabolism is generating the gas. As stated earlier, the reactions proceed largely to the right, and CO_2 and H_2O react to produce H^+ and HCO_3^-. The reaction is favored by the action of cell membrane transporters, that pump HCO_3^- out of the red blood cell into the plasma. As blood flows through the lungs, however, dissolved CO_2 diffuses into the alveoli and is exhaled. Therefore, CO_2 levels in the lung capillaries decrease; this changes the equilibrium of the reactions such that the reverse reactions (right to left) are favored. Consequently, HCO_3^- and H^+ are converted back to CO_2 and H_2O; the CO_2 is then exhaled. Additional HCO_3^- from the plasma is transported back into red blood cells, and generates more CO_2 that is then exhaled. These reactions are remarkably fast, occurring within the space of time required for blood to move through a capillary, typically less than 1 sec.

The chemoreceptors in the brain are very sensitive to certain drugs, such as ethanol (alcohol), opiates, and barbiturates. One action of these drugs is to decrease the sensitivity of chemoreceptors to CO_2, and thereby reduce ventilation to potentially life-threatening levels. Particularly when taken together, these drugs may be a lethal combination partly because of their inhibitory effects on ventilation.

Respiratory Center Activity Varies Among Mammals with Different Metabolic Rates

The respiratory centers are sensitive to a variety of other factors besides gases and hydrogen ions, such as sleep, stress, hormones, and body size. Recall from Chapter 46 that small animals tend to have higher relative metabolic rates than larger animals (refer back to Figure 46.8). Thus, small animals would be expected to require more oxygen per unit mass than would larger animals. This is not achieved by having disproportionately large lungs in small animals. Instead, small animals have faster breathing rates than do large animals. The respiratory centers are set at a higher rate in smaller, active animals. For example, a typical adult human takes about 12 to 14 breaths per minute at rest, whereas small rodents and shrews have astonishingly high rates of breathing, up to many hundreds of breaths per minute (see Table 48.1)! Thus, the respiratory activity of small, highly active animals is matched with their metabolism by increasing the resting breathing rate. This is similar to differences in heart rate among species (refer back to Table 47.1). Parallel changes in the heart and breathing rates are adaptations that allow small animals to have small hearts and lungs yet achieve high metabolic rates.

48.6 Adaptations to Extreme Conditions

Learning Outcomes:

1. List adaptations of the respiratory and circulatory systems of animals living at high altitudes.
2. Describe the adaptations of the respiratory and circulatory systems of diving animals.

Many animals including a wide range of vertebrates are able to live permanently or temporarily in low-oxygen environments. Many humans, for instance, live in mountainous regions, and llamas and mountain goats spend most of their lives at extremely high altitudes. Other animals may only transiently encounter periods of oxygen deprivation, such as some reptiles, birds, and mammals that dive under water to forage for food. As described in this section, several adaptations allow animals to exploit these environments, but all of them include changes in both cardiovascular and respiratory activities.

Life at High Altitudes Requires More Hemoglobin

Animals that live at high altitudes must have special adaptations that permit them to obtain the oxygen they require at such low atmospheric pressures. Llamas, such as the one shown in the chapter-opening photo, may live at altitudes up to 4,800 m, where the Po_2 is only about 85 mmHg (compared with 160 mmHg at sea level; see Figure 48.2). Llama hemoglobin is quite different in amino acid sequence from that of other mammals, giving it an extraordinarily high affinity for binding oxygen even at very low atmospheric pressure. In other words, their hemoglobin curves are shifted well to the left of a human's. In addition, llamas have larger hearts and lungs than would be predicted for their body size, and a higher number of red blood cells in a given volume of blood. These adaptations provide the oxygen-carrying capacity and cardiac output needed to deliver sufficient oxygen to the tissues, even at low partial pressures of oxygen.

When animals that normally inhabit lowland areas temporarily move to higher altitudes, they develop some of these same features. In humans, for example, the number of red blood cells increases from about 5.1×10^{12} cells/L to about 6.4×10^{12} cells/L. This is stimulated by the hormone erythropoietin, which is secreted by the kidneys when the arterial Po_2 is low. This hormone acts on bone marrow to stimulate maturation of new red blood cells. Moving to higher elevations also increases ventilation due to a higher rate of breathing. The number of capillaries in skeletal muscle increases at high altitudes, a modification that facilitates oxygen diffusion into muscle cells. Finally, myoglobin content increases in muscle cells, expanding the reservoir of oxygen in the cytosol of those cells. Consequently, after several days or weeks at high altitude, the cardiovascular and respiratory systems adapt together to maximize oxygen uptake, diffusion of oxygen into the blood, and the oxygen-carrying capacity of blood.

Diving Animals Have Adaptations That Permit Long Dives Under Water

Many birds, reptiles, and several species of mammals spend time under water, foraging for food or escaping predators. During that time, the animal cannot breathe. In short dives, this is not a problem. We are all familiar with our own ability to stay under water for a short time. After that time, the oxygen concentration in the blood decreases, the carbon dioxide concentration in the blood increases, and we must surface for air. Some marine mammals, however, have an astonishing ability to remain under water for very long periods—up to nearly 2 hours in beaked whales and elephant seals. During that time, they continue to be active and search for food. How do they do it?

Like high-altitude animals, many diving animals have unusually high numbers of red blood cells, allowing them to store more oxygen in their blood than nondiving animals. In some species, such as seals, the extra blood cells are stored in the spleen until they dive, at which time the spleen contracts like wringing a wet washcloth and ejects the blood cells into the circulation. When seals resurface, the blood cells are sequestered again in the spleen until needed. In addition to having more erythrocytes, diving mammals typically have larger blood volumes than comparably sized mammals that live exclusively on land. The muscles of diving mammals also usually contain large quantities of myoglobin and its bound oxygen. This means that the muscles do not need to consume the precious stores of oxygen circulating in blood. Instead, the blood and its oxygen can be routed to other critical structures that lack myoglobin, such as the eyes, certain glands, the brain, and the placenta if the animal is pregnant.

Eventually, even with these adaptations, the oxygen in muscles and blood becomes so depleted that the only way to prolong the dive is for cells to begin respiring anaerobically (see Chapter 7). This is a less efficient way of generating ATP than is aerobic respiration, but is sufficient to prolong the dives of such animals.

48.7 Impact on Public Health

Learning Outcomes:

1. Outline the causes, symptoms, and current treatments of several respiratory system diseases.
2. Describe the impact of smoking tobacco on respiratory health.

Respiratory diseases of all kinds (including lung cancer) afflict as many as 10% of the U.S. population and result in an estimated 3–400,000 deaths per year, making lung disease among the top three causes of death in the U.S. The economic impact of respiratory diseases on the U.S. economy is staggering, with recent estimates of \$40–150 billion per year in health-related costs and lost productivity. Many of these diseases are chronic—once they appear, they last for the rest of a person's lifetime. Lifestyle factors, such as smoking tobacco and exposure to air pollution, cause some respiratory disorders or make existing conditions worse. In this section, we examine a few of the most common respiratory disorders, as well as some of their causes and treatments.

Asthma Is a Disease of Hyperreactive Bronchioles

You learned earlier that the bronchioles deliver fresh air to the alveoli. Bronchioles are thin tubes surrounded by smooth muscle cells that can contract in the presence of airborne pollutants or other potentially damaging substances. In the disease **asthma**, however, the muscles around the bronchioles are hyperexcitable and contract more than usual. Contraction of these muscles narrows the bronchioles, a process called bronchoconstriction. This makes it difficult to move air in and out of the lungs, because resistance to airflow increases when the diameter of the airways decreases. Often, the resistance to airflow can be so great that the movement of air creates a characteristic wheezing sound.

Asthma tends to run in families and therefore has a genetic basis. Several known triggers can elicit wheezing, including exercise, cold air, and allergic reactions. The latter is of interest because asthma is believed to be partly the result of an imbalance in the immune system (see Chapter 53), which controls inflammation and other allergic responses. During flare-ups of asthma, a viscous, mucus-like fluid may inhibit the flow of air in and out of the airways and make symptoms worse.

The symptoms of asthma can be alleviated by inhaling an aerosol mist containing **bronchodilators**, compounds that bind to receptors located on the plasma membranes of smooth muscles that make up the outer part of bronchioles. These compounds, which are related to the neurotransmitter norepinephrine, cause bronchiolar smooth muscles to relax. This, in turn, allows the bronchioles to dilate (widen). To help reduce the inflammation of the lungs, patients may inhale a mist containing hormones with anti-inflammatory actions. Currently there is no cure for asthma, but with regular treatment and the avoidance of known triggers, most people with this disease can lead perfectly normal lives.

Tobacco Smoke Causes Respiratory Health Problems and Cancer

Smoking tobacco products is one of the leading global causes of death, contributing to about 430,000 deaths each year in the U.S. alone and over 5 million per year worldwide. According to the Centers for Disease Control and Prevention and the American Lung Association, people who smoke up to one pack of cigarettes each day live on average 7 years less than nonsmokers, and heavy smokers lose on average 15–25 years of life. Pregnant women who smoke run a high risk of their babies being born underweight, a potentially serious condition that may affect the newborn's long-term health.

Up to 85% of all new cases of lung cancer diagnosed each year are attributable to smoking, making lung cancer the leading cause of preventable death. Equally important, however, is that smoking is estimated to be responsible for nearly 30% of all cancers, including cancer of the mouth and throat, esophagus, bladder, pancreas, and ovaries. Smoking is also a leading cause of cardiovascular disease, high blood pressure, and stroke. Smoking as few as three to five cigarettes per day raises the risk of heart disease.

Because the products of tobacco smoke are inhaled directly into the lungs, the chemicals in smoke can do considerable damage to lung tissue. Even adolescents who have only recently started smoking have increased mucus (phlegm) production in their airways, shortness of breath, and reduced lung growth. Thousands of chemicals, including over 40 known cancer-causing compounds, have been identified in cigarette smoke. Some of these chemicals—such as formaldehyde—are toxic to all cells. Others, like the odorless gas carbon monoxide (CO), have harmful effects on lung function in particular. CO competes with oxygen for binding sites in hemoglobin, thereby reducing hemoglobin saturation. Heavy smokers who smoke more than a pack of cigarettes each day may have as much as 15% less oxygen-carrying capacity in their blood.

In addition to being a risk factor for lung cancer and cardiovascular disease, long-term smoking is the major cause of the serious and irreversible disease emphysema.

Emphysema Causes Permanent Lung Damage

Unlike asthma, in which the major problems are inflamed airways and hyperreactive bronchioles, **emphysema** involves extensive lung damage (Figure 48.17). The disease reduces the elastic quality of the lungs and the total surface area of the alveoli, which cuts the rate of oxygen diffusion from the lungs into the circulation. Consequently, one sign of emphysema is a lower than normal P_{O_2} in the arteries. It is also physically harder to exhale because of the loss of elasticity, and therefore, arterial CO_2 concentrations increase. Finally, the terminal ends of the bronchioles are often damaged, which increases resistance to airflow and creates asthma-like symptoms and shortness of breath.

Reduced blood oxygen and poor lung function limit the patient's ability to function, and in its late stages, emphysema results in a person being essentially bedridden. Oxygen therapy, in which the person breathes a mixture of air and pure oxygen from a portable gas tank, can provide some help. The extra oxygen increases the pressure gradient for oxygen from the alveoli to the lung capillaries, promoting oxygen diffusion into the blood.

In some cases, emphysema results from an enzyme deficiency in the lungs that destroys the protein that provides the recoil during exhalation, or it may result from chronic exposure to air pollution. However, the overwhelming majority of cases, 85%, are due to smoking. Toxins in cigarettes and other tobacco products damage the lungs by stimulating white blood cells to release proteolytic enzymes that degrade lung tissue. The likelihood of developing emphysema is strongly correlated with the quantity of cigarettes smoked during a person's lifetime.

Estimating how many people have emphysema is difficult because the symptoms appear gradually, but more than 3 million people have severe cases of the disease in the U.S., and at least 15,000 people die from it each year. Emphysema is a progressive disease. As the years go by, the disease worsens; although medical care can slow the rate at which this happens, it is not curable and does not get better.

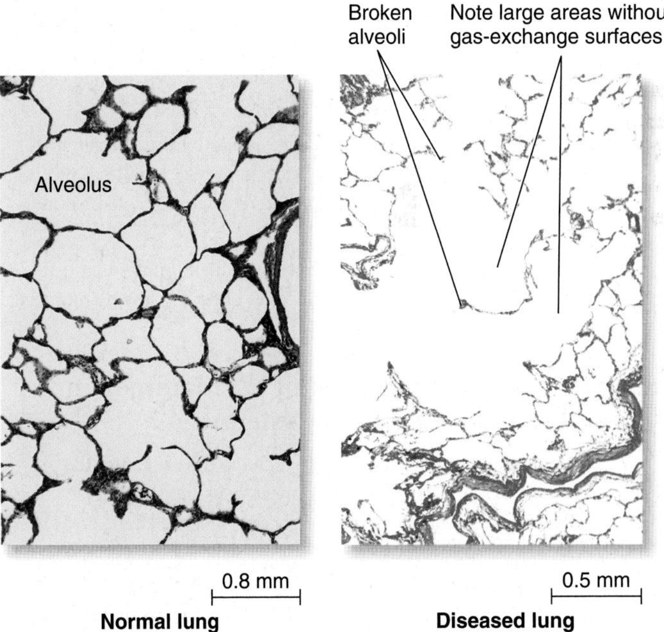

Figure 48.17 **The effects of emphysema.** These light micrographs compare a section of a normal lung (left) with that of a lung from a person who died of emphysema (right). The destruction of alveoli caused by this disease reduces the surface area for gas exchange in the lungs.

BIOLOGY PRINCIPLE Biology affects society. Lung disease affects millions of people. Biologists study the causes, mechanisms, and possible treatments of human diseases such as emphysema.

Summary of Key Concepts

48.1 Physical Properties of Gases

- Gas exchange is the process of moving oxygen and carbon dioxide in opposite directions between the environment, body fluids, and cells. The partial pressure of oxygen (Po_2) in the environment provides the driving force for its diffusion from air or water across a respiratory organ and into the blood. Atmospheric pressure decreases at higher elevations (Figures 48.1, 48.2).
- Three factors—the pressure of the gas, temperature of the water, and presence of any other solutes—are particularly important for affecting the solubility of a gas in water.

48.2 Types of Respiratory Systems

- Ventilation is the process of bringing oxygenated water or air into contact with a respiratory organ. All respiratory organs have moist surfaces in which gases can dissolve and diffuse, as well as other adaptations that increase the amount of surface area and blood flow. Water-breathing and air-breathing animals face different gas exchange challenges.
- The body surface is permeable to gases in some invertebrates and in amphibians, eels, and a few other species of fishes (Figure 48.3).
- Water-breathing animals use external or internal gills for gas exchange (Figures 48.4, 48.5, 48.6).
- Air-breathing animals have evolved two major mechanisms to exchange gas with the environment: the tracheal system in insects and lungs in terrestrial vertebrates. In insects, air moving down the tracheoles comes into contact with fluid at the tracheole tips. Oxygen from the air dissolves in this fluid and diffuses across the tracheole wall and into nearby cells (Figure 48.7).
- Air-breathing vertebrates generally use lungs to bring oxygen into the circulatory system and remove carbon dioxide. All lungs receive deoxygenated blood from the heart and return oxygenated blood to the heart.
- Frogs and most other amphibians ventilate their lungs with positive pressure filling, which uses the principles of Boyle's law (Figure 48.8).

48.3 Structure and Function of the Mammalian and Avian Respiratory Systems

- The mammalian respiratory system includes the nose, mouth, airways, lungs, and muscles and connective tissues that encase these structures within the thoracic (chest) cavity (Figure 48.9).
- Most reptiles and all birds and mammals ventilate their lungs by negative pressure filling. In mammals, the work is provided by the intercostal muscles and diaphragm. Mammals breathe by tidal ventilation. Tidal volume is proportional to body size within and between species (Figure 48.10, Table 48.1).
- The avian respiratory system, a flow-through system, is unique among vertebrates because it is supplemented with numerous air sacs. Inhaled air is stored in air sacs before passing through the lungs. Blood flows through the lungs in a crosscurrent direction with respect to the movement of oxygen (Figures 48.11, 48.12, 48.13).

48.4 Mechanisms of Oxygen Transport in Blood

- There are limits to how much of a gas can dissolve in water. These limits are overcome in nearly all animals by either transporting a gas reversibly bound to a protein carrier or by transforming the gas into a more soluble form.
- Animals have evolved a way to carry a reservoir of oxygen on respiratory pigments. Hemoglobin is the major iron-containing pigment in vertebrates and many invertebrates; hemocyanin is the copper-containing pigment in decapod crustaceans, arachnids, and many mollusks, including cephalopods and some gastropods.
- The amount of oxygen bound to hemoglobin depends on the Po_2 in the blood. Metabolic waste products can influence the oxygen-hemoglobin binding relationship (Figures 48.14, 48.15).
- The evolution of the globin gene family has resulted in several specialized hemoglobin proteins.

48.5 Control of Ventilation in Mammalian Lungs

- In mammals, respiratory centers in the brainstem initiate the rhythmic expansion of the lungs (Figure 48.16).
- Chemoreceptors detect blood levels of H^+ and the Pco_2 and Po_2. They relay this information through nerves to the respiratory centers, which, in turn, affect the breathing rate. The respiratory centers are also sensitive to factors such as sleep, stress, hormones, and body size.

48.6 Adaptations to Extreme Conditions

- Adaptations in respiratory and circulatory systems allow animals to exploit low-oxygen environments. Animals that inhabit high altitudes have larger hearts and lungs and have hemoglobin with a high affinity for binding oxygen. Many diving animals have unusually high numbers of red blood cells and also muscles with large quantities of the muscle protein myoglobin.

48.7 Impact on Public Health

- In asthma, the muscles around the bronchioles contract more than usual, increasing resistance to airflow.
- Smoking tobacco products is one of the leading global causes of death. Smoking is strongly linked to cancer, cardiovascular disease, stroke, and emphysema (Figure 48.17).

Assess and Discuss

Test Yourself

1. The driving force for diffusion of oxygen across the cells of a respiratory organ is
   a. the difference in the partial pressure of oxygen (Po_2) in the environment and in the blood.
   b. the humidity.
   c. the partial pressure of carbon dioxide (Pco_2) in the blood.
   d. the air temperature.
   e. the Pco_2 in the atmosphere.

2. Carbon dioxide is considered a harmful by-product of cellular respiration because it
   a. lowers the pH of the blood.
   b. lowers the H^+ concentration in the blood.
   c. competes with oxygen for transport in the blood.
   d. does all of the above.
   e. a and b only
3. The process of bringing oxygenated water or air into contact with a gas-exchange surface is
   a. respiration.
   b. gas exchange.
   c. ventilation.
   d. gas transport.
   e. exhalation.
4. The group of vertebrates with the greatest capacity for gas exchange across the skin is
   a. the fishes.
   b. the reptiles.
   c. the amphibians.
   d. the birds.
   e. the mammals.
5. The countercurrent exchange mechanism in fish gills
   a. maximizes oxygen diffusion into the bloodstream.
   b. is a less efficient mechanism for gas exchange compared with mammalian lungs.
   c. occurs because the flow of blood is in the same direction as water flowing across the gills.
   d. is the same phenomenon observed in birds' lungs.
   e. requires that the fish swallow water.
6. The tracheal system of insects
   a. consists of several tracheae that connect to multiple lungs within the different segments of the body.
   b. consists of extensively branching tubes that are in close contact with all the cells of the body.
   c. allows oxygen to diffuse across the thin exoskeleton of the insect to the bloodstream.
   d. cannot function without constant movement of the wings to move air into and out of the body.
   e. provides oxygen that is carried through the animal's body in hemolymph.
7. _____________ is secreted by type II alveolar cells in the mammalian lung to prevent the collapse of alveoli due to surface tension at the interface of air and extracellular fluid.
   a. Hemoglobin
   b. Myoglobin
   c. Mucus
   d. Water
   e. Surfactant
8. In negative pressure filling, air moves into the lungs when
   a. the volume of the thoracic cavity increases.
   b. the pressure in the thoracic cavity decreases.
   c. air is forced down the trachea by muscular contractions of the mouth and pharynx.
   d. all of the above
   e. a and b only
9. Which of the following factors does *not* alter the rate of breathing by influencing the chemoreceptors?
   a. P_{CO_2} in the blood
   b. P_{O_2} in the blood
   c. blood pH
   d. blood glucose levels
   e. H^+ concentration in the blood
10. With rare exceptions, the majority of oxygen is transported in the blood of vertebrates
    a. by binding to plasma proteins.
    b. by binding to hemoglobin in erythrocytes.
    c. as dissolved gas in the plasma.
    d. as dissolved gas in the cytoplasm in the erythrocytes.
    e. by binding to myoglobin.

Conceptual Questions

1. Define countercurrent exchange as it relates to gas exchange in fishes.
2. Explain some of the special adaptations for life at high altitudes; why are such adaptations necessary?
3. Two principles of biology are that *new properties emerge from complex interactions* and *structure determines function*. How are these principles related to what you have learned in this chapter about hemoglobin?

Collaborative Questions

1. Discuss two ways in which animals exchange gases in an aqueous environment. What special adaptations facilitate this exchange?
2. Discuss the components of the mammalian respiratory system.

Online Resource

www.brookerbiology.com

Stay a step ahead in your studies with animations that bring concepts to life and practice tests to assess your understanding. Your instructor may also recommend the interactive eBook, individualized learning tools, and more.

Excretory Systems and Salt and Water Balance

49

Chapter Outline

49.1 Principles of Homeostasis of Internal Fluids

49.2 Comparative Excretory Systems

49.3 Structure and Function of the Mammalian Kidney

49.4 Impact on Public Health

Summary of Key Concepts

Assess and Discuss

The human kidneys could filter the volume of water of this pool in a few months.

If you have ever noticed how quickly the water in an aquarium or a swimming pool becomes dirty if the filter is not functioning, you will have a good idea of the importance of filtering the wastes from an animal's body fluids. The human kidneys, for example, are remarkable filtration devices. Although each one is only about the size of a computer mouse, the kidneys are able to filter blood at a rate of 150–200 L/day. Considering that there are only 5 L or so of blood in a typical adult, that is an astonishingly effective filtration mechanism. Despite their small size, our kidneys could filter the entire contents of a medium-sized swimming pool in a few months. By the time a person reaches 50 years of age, his or her kidneys have filtered roughly 3,000,000 L of blood! Along the way, the kidneys not only remove waste products of metabolism from that filtered blood, but recapture useful substances such as sodium ions and water that form part of the liquid being filtered. In this way, the kidneys and other excretory organs found in animals contribute to homeostasis.

The ability of organisms to maintain homeostasis is one of the principles of biology and a common theme of the previous several chapters. Animals maintain a variety of physiological processes—including energy intake and usage, blood pressure, body temperature, and blood oxygen levels—within normal ranges. Homeostasis is also critical in the regulation of salt and water levels in body fluid compartments. As described in Chapter 2, the general term "salt" is used to refer to a compound formed from an attraction between a positively charged ion, such as sodium (Na^+), and a negatively charged ion, such as chloride (Cl^-). These ions are held together by ionic bonds, which are broken when the salt is dissolved in water. Changes in the concentrations of ions resulting from dissolved salts in the extracellular and intracellular fluids have the potential to disrupt proper cellular function; for example, they may alter the difference in electrical potential across plasma membranes in the cells of the heart and brain. Salt (ion) concentrations and water volumes in the different body fluid compartments are related to each other, because a major way in which water moves between compartments is by osmosis, which, in turn, depends on the numbers of dissolved solutes in water (refer back to Chapter 40 and Figure 40.8 for a discussion of osmolarity in animal body fluids). Consequently, changes in salt concentrations in body fluids may also affect cell volume, which then can cause cell disruption or death.

As we have learned, homeostasis is an energy-requiring process. A significant portion of most animals' daily energy expenditures goes toward maintaining salt and water homeostasis. The ability to do so is complicated by many factors, such as the environment and climate in which an animal lives and its access to sufficient supplies of drinking water.

In this chapter, we examine why salt and water balance is vital for survival, how it is affected by the requirement to eliminate metabolic wastes, and how different excretory organs participate in these processes. We then highlight some of the major features of the vertebrate and mammalian kidney, explore challenges posed by an animal's life history and environment, examine how the kidney eliminates wastes and regulates salt and water balance, and conclude by considering how kidney disease affects human health.

49.1 Principles of Homeostasis of Internal Fluids

Learning Outcomes:

1. Explain why a balance of water and electrolytes is critical for an animal's survival.
2. List ways in which water and ions (electrolytes) move across cell membranes and between body fluid compartments.

3. Describe the different kinds of nitrogenous wastes and their relative toxicities.
4. Compare and contrast osmotic adaptations of freshwater fish with those of marine fish.
5. List two ways of classifying animals according to how they adapt to osmotic challenges.

As we saw in Chapters 40 and 47, an animal's internal fluids exist in compartments within the body. In invertebrates, these fluids include the intracellular fluid and hemolymph, whereas in vertebrates, it includes the intracellular fluid and the extracellular fluid (the interstitial fluid surrounding cells and the plasma component of blood). Within the fluids are many salts, such as NaCl, KCl, and so on. As stated earlier, salts dissociate in solution into ions. Because ions from dissolved salts are electrically charged, they are also referred to as electrolytes; we will use these terms interchangeably. In this section, we will examine why water and salt homeostasis is so important and how an animal's electrolyte concentration and water volume are maintained within normal ranges.

The Balance of Water and Salts Is Critical for Survival

Maintenance of normal body water levels is of great importance for all animals. Not only is water the major portion of an animal's body mass, it is also the solvent that permits dissolved solutes to participate in chemical reactions. As described in Chapters 2 and 3, water also participates in important chemical reactions, notably hydrolysis reactions. In addition, water is the transport vehicle that brings oxygen and nutrients to cells and removes wastes generated by metabolism.

When an animal's water volume is reduced below the normal range, we say the animal is dehydrated. In terrestrial animals, dehydration may occur if sufficient drinking water is not available or when water is lost by evaporation (perspiring or panting). Dehydration can be a serious, potentially life-threatening condition. For example, because blood is roughly 50% plasma (water), blood volume may decrease in dehydrated animals. Decreased blood volume compromises the ability of the circulatory system to move nutrients and wastes throughout the body and to assist in the regulation of body temperature on hot days.

Salt balance is also very important for animals. A change of only a few percent in the extracellular fluid concentrations of potassium ions (K^+), for example, can trigger changes in nerve, heart, and skeletal muscle function by altering membrane potentials (see Chapter 41). Other electrolytes, such as calcium (Ca^{2+}), magnesium (Mg^{2+}), phosphate (PO_4^{3-}), and sulfate (SO_4^{2-}), also participate in various biological activities. These functions include serving as cofactors for enzyme activation, participating in bone formation, forming part of the extracellular matrices around cells, and activating cellular events such as exocytosis and muscle contraction. An imbalance in any of these ions can seriously disrupt cellular activities.

Water moves between adjacent body compartments by osmosis down an osmotic gradient (see Chapter 5; refer back to Figure 5.15). Changes in the salt concentration in one compartment will lead to changes in fluid distribution between the compartments. These changes can cause cells to shrink or swell. When, for example, the salt concentration of extracellular fluid increases, water moves by osmosis from inside cells to the extracellular fluid, causing the cells to shrink. Shrinking or swelling of cells in the brain, heart, and other vital organs can rupture plasma membranes, leading to cell death.

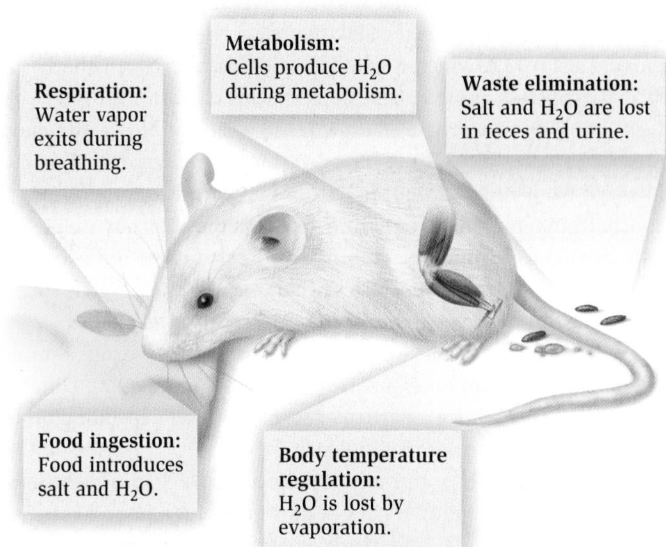
Metabolism:
Cells produce H_2O during metabolism.
Respiration:
Water vapor exits during breathing.
Waste elimination:
Salt and H_2O are lost in feces and urine.
Food ingestion:
Food introduces salt and H_2O.
Body temperature regulation:
H_2O is lost by evaporation.

Figure 49.1 **Types of obligatory salt and water exchanges in a terrestrial animal.** Obligatory exchanges with the environment occur as the result of necessary life processes.

Concept Check: Can animals completely avoid all the losses resulting from obligatory exchanges?

Whereas water moves between fluid compartments by osmosis, ions from dissolved salts such as NaCl move by different mechanisms. All ions have very limited ability to diffuse across plasma membranes, because of their high water solubility and low lipid solubility. As described in Chapter 5, ions may cross membranes by diffusion through channels formed by proteins that create a pore in the membrane (refer back to Figure 5.17). Alternatively, ions may be actively transported across epithelial cells such as those in the kidneys or in the gills. As this type of transport is an active process, it requires energy stored in the chemical bonds of ATP. Animals that face exceptional challenges to maintaining salt balance, such as marine fishes, must expend a considerable share of their daily energy budget to transport ions across epithelial cells.

Exchanges of Salt and Water with the Environment Are Obligatory

Many vital processes—eliminating nitrogenous wastes, obtaining oxygen and eliminating carbon dioxide, consuming and metabolizing food, and regulating body temperature—have the potential to disturb salt and water balance. Therefore, these processes require additional energy expenditure to minimize or reverse the disturbance. Exchanges of salt and water with the environment that occur as a consequence of such vital processes are called obligatory exchanges (because the animal is "obliged" to make them) (**Figure 49.1**).

Elimination of Nitrogenous Wastes When carbohydrates and fats are metabolized by animal cells, the major waste product is carbon dioxide gas (CO_2), which is exhaled or diffuses across the body surface. By contrast, proteins and nucleic acids contain nitrogen; when these molecules are broken down and metabolized, nitrogenous wastes are generated. **Nitrogenous wastes** are molecules that include nitrogen from amino groups ($—NH_2$). These wastes are toxic at high concentrations and must be eliminated from the body but, unlike carbon dioxide gas, cannot be eliminated by exhaling or diffusion. The elimination of nitrogenous wastes occurs via excretory organs such as the kidneys or other specialized structures, such as gills.

Nitrogenous wastes are usually found in three forms—ammonia (and ammonium ions), urea, or uric acid (**Figure 49.2**). Different animal groups produce a particular form of waste, depending on the species and the environment in which they live.

Ammonia (NH_3) and ammonium ions (NH_4^+) are the most toxic of the nitrogenous wastes because they disrupt pH, ion electrochemical gradients, and many chemical reactions that involve oxidations and reductions. Animals that excrete wastes in this form typically live in water. In marine invertebrates, NH_3 and NH_4^+ are continually excreted across the skin, whereas in freshwater and most saltwater fishes, these wastes are excreted via the gills and kidneys. Because NH_3 is so toxic, aquatic animals excrete it as quickly as it is formed. The chief advantage of excreting nitrogenous wastes in the forms of NH_3 or NH_4^+ is that energy is not required for their conversion to a less toxic product, as is the case for urea and uric acid. All mammals, most amphibians, some marine fishes, some reptiles, and some terrestrial invertebrates convert NH_3 into **urea**, which is then excreted. In addition to being less toxic than NH_3, urea does not require large volumes of water to be excreted. Animals can tolerate some accumulation of urea in their blood, tissues, and storage organs such as the urinary bladder. This conserves water, removes the necessity for constant excretion, and reduces the likelihood of toxicity. One drawback of producing urea is that the enzymatic conversion of NH_3 into urea requires a moderate expenditure of ATP and thus consumes part of an animal's total daily energy budget.

Birds, insects, and most reptiles produce **uric acid** or other nitrogenous compounds called purines. Like urea, these compounds are less toxic than ammonia, but they are even more energetically costly to synthesize from NH_3. However, because they are poorly soluble in water, they are not excreted in a watery urine but instead are packaged with other waste products and excess salts into a semisolid, partly dried precipitate that is excreted. The energy investment required to produce uric acid, therefore, is balanced against the water conserved by excreting nitrogenous wastes in this form.

Respiration-Related Water and Salt Exchanges The requirements for both respiration and water and electrolyte balance present different challenges to air- and water-breathing animals. To ventilate its lungs, an air-breathing animal moves air in and out of its airways. Water in the form of water vapor in the mouth, nasal cavity, and upper airways exits the body with each exhalation. As an animal becomes more active, it requires more O_2 and produces more CO_2. These changes are met by an increase in respiratory activity. Breathing becomes deeper and more rapid, which, in turn, increases the rate of water loss from the body. Therefore, respiration in animals with lungs is associated with significant water loss, as you can observe in cold weather when you can "see your breath."

As described in Chapter 48, small, active animals with high metabolic rates have faster breathing rates than do larger, less active animals. Consequently, the potential for water loss due to respiration is considerably greater in small animals, particularly in endotherms.

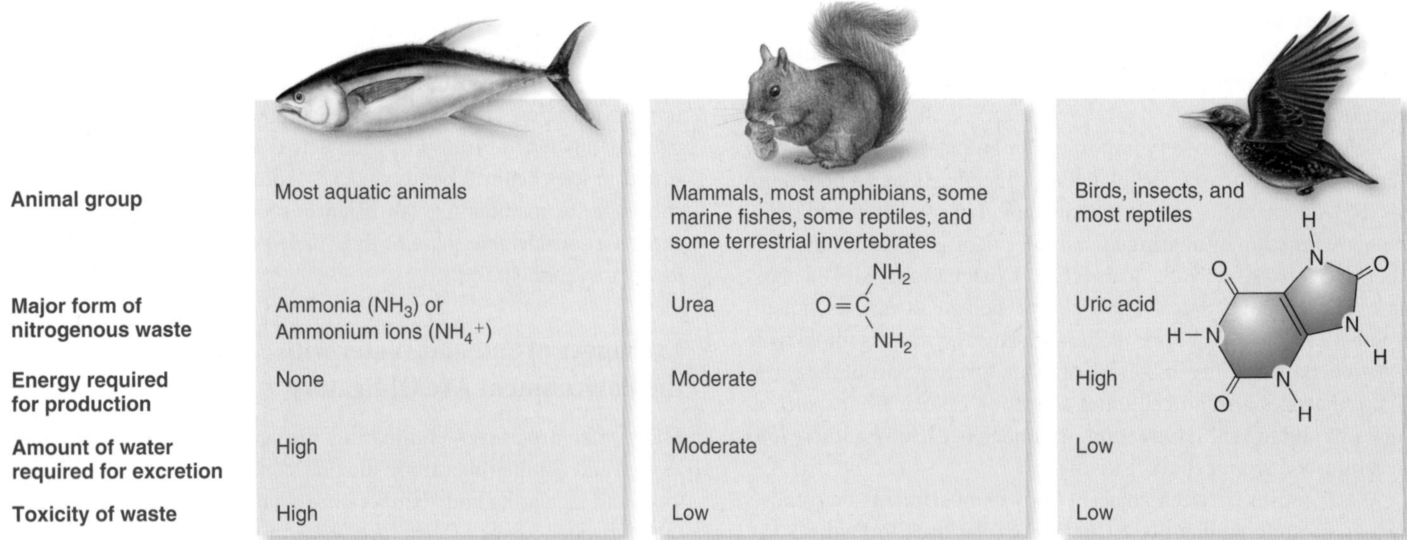

Animal group
Most aquatic animals
Mammals, most amphibians, some marine fishes, some reptiles, and some terrestrial invertebrates
Birds, insects, and most reptiles
Major form of nitrogenous waste
Ammonia (NH_3) or Ammonium ions (NH_4^+)
Urea
$O=C(NH_2)_2$
Uric acid
Energy required for production
None
Moderate
High
Amount of water required for excretion
High
Moderate
Low
Toxicity of waste
High
Low
Low

Figure 49.2 Nitrogenous wastes produced by different animal groups. The three forms of nitrogenous wastes, which are derived from the breakdown of proteins or nucleic acids, have different properties.

BIOLOGY PRINCIPLE Living organisms use energy. Energy from ATP is required to convert nitrogenous wastes into urea and uric acid. Different animals, therefore, must expend more or less energy each day to rid themselves of toxic nitrogenous wastes.

Hummingbirds, for example, may have 15 to 20 times the water loss per gram of body mass than would a large goose.

In water-breathing animals, the challenge of water and salt balance is more complex, because such animals move water, not air, over their respiratory organs (gills). Recall from Chapter 48 that gills, like all respiratory organs, are thin structures with large amounts of surface area and an extensive network of capillaries. Although these features make gills ideal for gas exchange by diffusion between the capillaries and the surrounding water, they also make them ideal for salt and water movement by diffusion and osmosis, respectively.

The solute concentration of a solution of water is known as the solution's **osmolarity**, expressed as milliosmoles/liter (mOsm/L). The number of dissolved solute particles determines a solution's osmolarity. For example, a 150 mM NaCl solution has an osmolarity of 300 mOsm/L, because each NaCl molecule dissociates into two ions, one Na^+ and one Cl^- ($2 \times 150 = 300$).

When differences occur in salt concentration between a water-breathing animal's body fluids and the surrounding water, respiration via the gills has the potential to disrupt salt and water balance. Fishes or other water-breathing animals that live in fresh water and those that live in salt water face opposite challenges in maintaining this balance (**Figure 49.3**). The internal fluid osmolarity of most fishes is usually within the range of 225–400 mOsm/L, similar to that of most other vertebrates. Because freshwater lakes and rivers have very little salt content (usually <25 mOsm/L), a high concentration gradient for salts could promote the loss of salts from a fish's body into the fresh water. Likewise, a high osmotic gradient favors the movement of water into a freshwater fish. Freshwater fishes, therefore, gain water and lose salt when ventilating their gills (Figure 49.3a). If left uncorrected, this would cause a dangerous decrease in blood salt concentrations.

Freshwater fishes maintain water and salt balance via two different mechanisms. First, their kidneys are adapted to producing copious amounts of dilute urine—up to 30% of their body mass per day (an amount that would be equivalent to about 25 L per day in an average-sized human!). Second, specialized gill epithelial cells actively transport Na^+ and Cl^- from the surrounding water into the fish's capillaries. Thus, these two important ions are recaptured from the water. Freshwater fishes rarely, if ever, drink water, except for any that might be swallowed with food.

Saltwater fishes have the opposite problem. They tend to gain salts and lose water across their gills, because seawater has a much higher osmolarity (about 1,000 mOsm/L) than that of their body fluids (Figure 49.3b). The gain of salts and the loss of water from the body are only partly offset by the kidneys, which in marine fishes produce very little urine so water can be retained in the body. The urine that is produced has a higher salt concentration than that of freshwater fishes. To prevent dehydration from occurring, marine fishes must drink. However, the only water available to them is seawater, which has a very high salt content. Paradoxically, therefore, marine fish drink seawater to replenish the water lost by osmosis through their gills. What does the fish do with all of the salt it ingested? The ingested salt must be eliminated, and this process is accomplished by gill epithelial cells. In contrast to the gills of freshwater fishes, which pump salt from the water into the fluids of the fish, the gills of marine fishes pump salt out of the fish and into the ocean. Thus, marine fishes drink seawater to replace the water lost through their gills by osmosis and then expend energy to transport the excess salt out of the body.

Ingestion-Related Fluid and Salt Balance Because foods contain salts and water, eating also involves an obligatory exchange of these substances. Some plant products are over 95% water by weight, and other foods may contain high amounts of sodium or other minerals. Therefore, the type of diet an animal consumes determines how much salt and water it ingests.

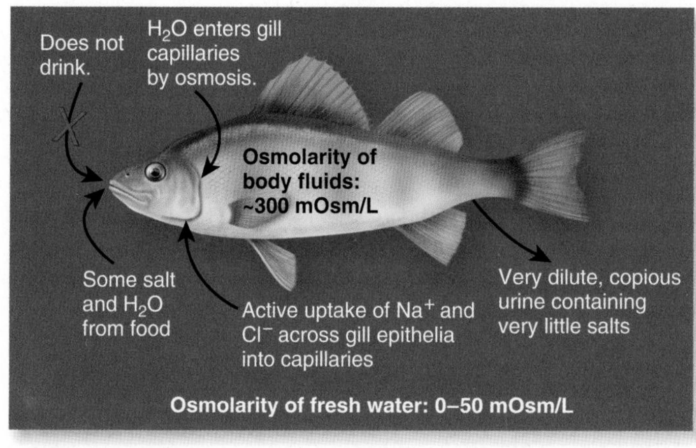

(a) Freshwater fish

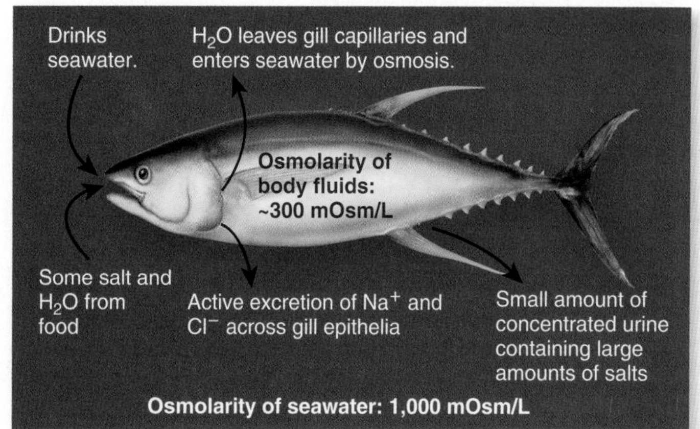

(b) Saltwater fish

Figure 49.3 **Salt and water balance in water-breathers.** Water-breathing creates osmoregulatory challenges due to diffusion of salts and osmosis of water across gills. These challenges differ between **(a)** freshwater and **(b)** saltwater fishes and are met by drinking or not drinking water, by active transport of salts across the gills, and by alterations in urine output.

BIOLOGY PRINCIPLE Living organisms use energy and living organisms maintain homeostasis. Energy from the hydrolysis of ATP is required for all of the active transport processes illustrated here. This energy is required to maintain homeostatic body fluid osmolarities despite the osmotic challenges imposed by the very different environments.

Once food has been digested and absorbed, the unusable parts of food are excreted as solid wastes. Some salt and water are lost by this route in most animals, but exceptions exist. Desert-dwelling kangaroo rats such as *Dipodomys panamintensis* produce fecal pellets that are almost completely dry, which helps these animals conserve water.

As noted earlier, marine fishes drink seawater. Other animals besides marine fishes may also drink seawater, either because fresh water is unavailable or because they ingest some with the food they eat. Many marine reptiles and birds, for example, ingest seawater when consuming prey or, in some cases, when they spend prolonged periods at sea and have no access to fresh water for drinking. These animals have specialized epithelial cells that line structures called salt glands, located in groups around the nostrils, mouth, and eyes (**Figure 49.4**). Salts move from the blood into the interstitial fluid, from which they are actively transported by the epithelial cells of the salt glands into the tubules of the gland. The salts and a small amount of fluid then collect into a central duct and are excreted as highly concentrated solutions. In general, vertebrates without salt glands cannot survive by drinking seawater, because they have no means of creating and excreting such a highly concentrated salt solution. Some marine mammals have been observed to occasionally drink small amounts of seawater, but most appear to never drink at all. These animals get their water from the food they eat.

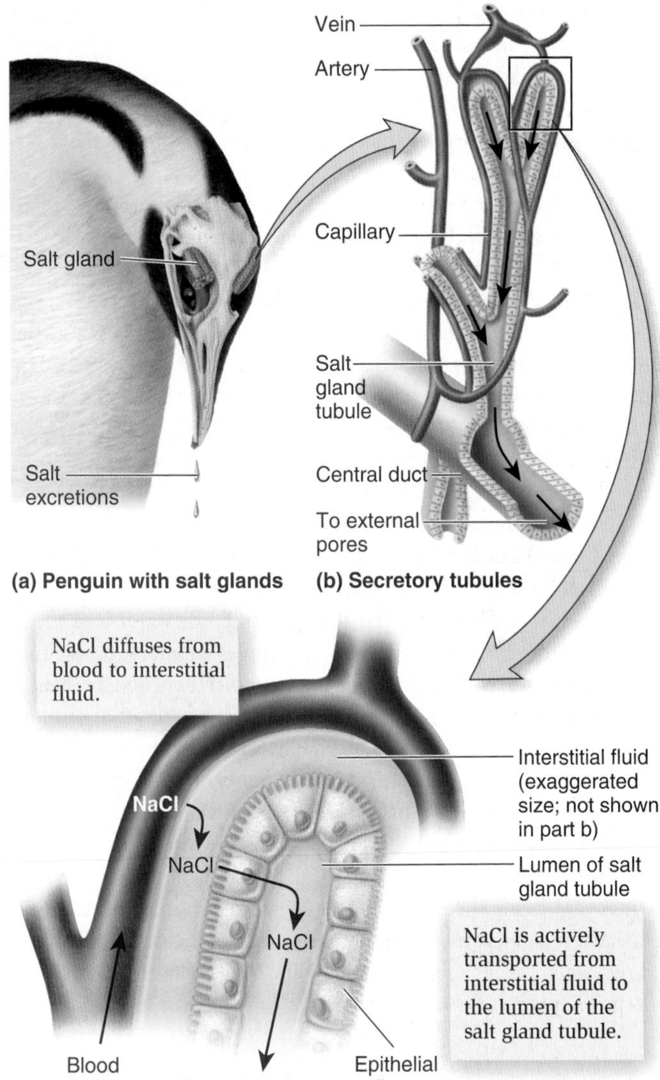

Figure 49.4 Salt glands as an adaptation for marine life. Many marine birds and reptiles have salt glands, which contain a network of secretory tubules that actively transport NaCl from the interstitial fluid into the tubule lumen. The viscous solution then moves through a central duct and to the outside environment through pores in the nose, around the eyes, and in other locations. The black arrows indicate direction of flow of blood or salt gland excretions.

Concept Check: *Why can't humans survive by drinking seawater?*

Regulation of Body Temperature with Water Endotherms use body water to cool off (refer back to Figure 46.12). For example, sweating and panting are used to cool the body. These behaviors use the evaporation of water to draw heat out of the body. In the process, however, the animal loses water and, in sweat, some salts. You know from tasting sweat that it is salty, but the saltiness of sweat and that of blood are not the same. Sweat is a hypoosmotic solution compared with blood; that is, it has a lower concentration of solutes, so the fluid left behind in the body after perspiration has both a lower volume and a higher salt concentration than normal.

Other than perspiration and panting, very little water is gained or lost directly across the body surface of most terrestrial vertebrates, because their skin is impermeable to water. Exceptions include amphibians (and also some invertebrates). In invertebrates, the rate of water loss across the body surface depends on whether the animal is soft-bodied, like worms, or covered in a waxy, water-impermeable cuticle, like most insects.

Metabolism and Water Balance When food molecules are metabolized to provide energy that will be stored in the chemical bonds of ATP, oxygen captures electrons and combines with hydrogen ions, thereby making water (refer back to Figure 7.8). This water is sometimes called "metabolic water" to indicate its origin. Some animals—especially desert dwellers, which lack ready access to drinking water—depend on this water to provide all or nearly all of their water requirements. In other animals, the production of metabolic water may sometimes result in more water than is required at that time. This excess water is eliminated by the excretory organs or through other routes. Because metabolism is always ongoing and is required for survival, the excretion or retention of metabolic water can be considered a type of obligatory exchange.

The significance of obligatory exchanges and their effects on homeostasis was dramatically illustrated by a long-term investigation by a research team at the University of Florida, as described next. Their discovery would lead to a revolution in our understanding of exercise physiology in humans.

FEATURE INVESTIGATION

Cade and Colleagues Discovered Why Athletes' Performances Wane on Hot Days

On a typically hot summer day in the mid-1960s in Gainesville, Florida, the University of Florida football team was practicing in full equipment. The players were rapidly becoming dehydrated and, unbeknownst to them, the osmolarity of their body fluids was increasing as their bodies produced copious amounts of dilute sweat in an effort to maintain body temperature. The athletes became aware of two things. First, they discovered that they did not need to urinate for long periods after a tough practice session, and second, their performance on the field suffered as they became increasingly fatigued and more susceptible to severe muscle cramps. Occasionally, players would need to receive medical treatment or even hospitalization for their symptoms. In extreme cases, athletes exercising in these conditions have been known to occasionally develop seizures—uncontrolled activity of neurons in the brain. This situation did not escape the notice of the team physicians and, notably, university faculty member and kidney specialist Robert Cade.

Many of the symptoms experienced by the players could be readily explained. The fatigue was directly related to loss of water from the body, which put a strain on the circulatory system and reduced blood flow to muscles and other organs. It was worsened by a decrease in blood glucose concentrations that were not being replenished during the long periods of strenuous activity. The muscle cramps and even occasional seizures arose from an imbalance in extracellular electrolytes—notably Na^+ and K^+—which are secreted outside the body by sweat glands in the process of perspiration. The resulting imbalance in extracellular electrolyte concentrations caused a change in the electrical potential across muscle and neuronal cell membranes, which triggered the spasms. Lastly, the decreased urine production is one of the body's mechanisms for reducing fluid loss when body water is decreasing.

The key question was how can these effects of extreme exercise best be reversed or prevented? The answer was simple and clever. Cade and his colleagues rejected the prevailing view that drinking any fluids during heavy exercise somehow contributed to cramps and other problems. Instead, they hypothesized that the best way to maintain salt and water homeostasis in a profusely sweating person is to restore to the body exactly what was lost; that is, the person should drink a solution that resembles sweat!

The first thing Cade needed to do was analyze precisely how much Na^+, K^+, and other ions are actually present in sweat. Fortunately, he had an abundance of human sweat at his disposal to analyze. Once the players left the field, their jerseys were wrung out, and the composition of the collected sweat was determined with an ion analyzer such as the flame spectrophotometer shown in Figure 49.5. The concentrations could then be compared with known values of

Figure 49.5 Cade and colleagues discovered a way to improve athletic performance and prevent salt and water imbalance during strenuous exercise.

HYPOTHESIS Athletic performance can be enhanced by maintaining the body's salt and H_2O balance during exercise.

KEY MATERIALS Supply of human sweat for analysis, ion analyzer, salt solution.

Experimental level
Conceptual level
1 Obtain human sweat from exercising athletes.
Sweat
Dilute solution of salts of unknown composition
2 Analyze composition of sweat using a flame spectrophotometer, which measures ion concentrations. Prepare artificial solution that mimics composition of sweat. Compare composition of both sweat and artificial solution to known ion concentration in human blood.
Flame spectrophotometer
Salts
Sweat
Artificial solution

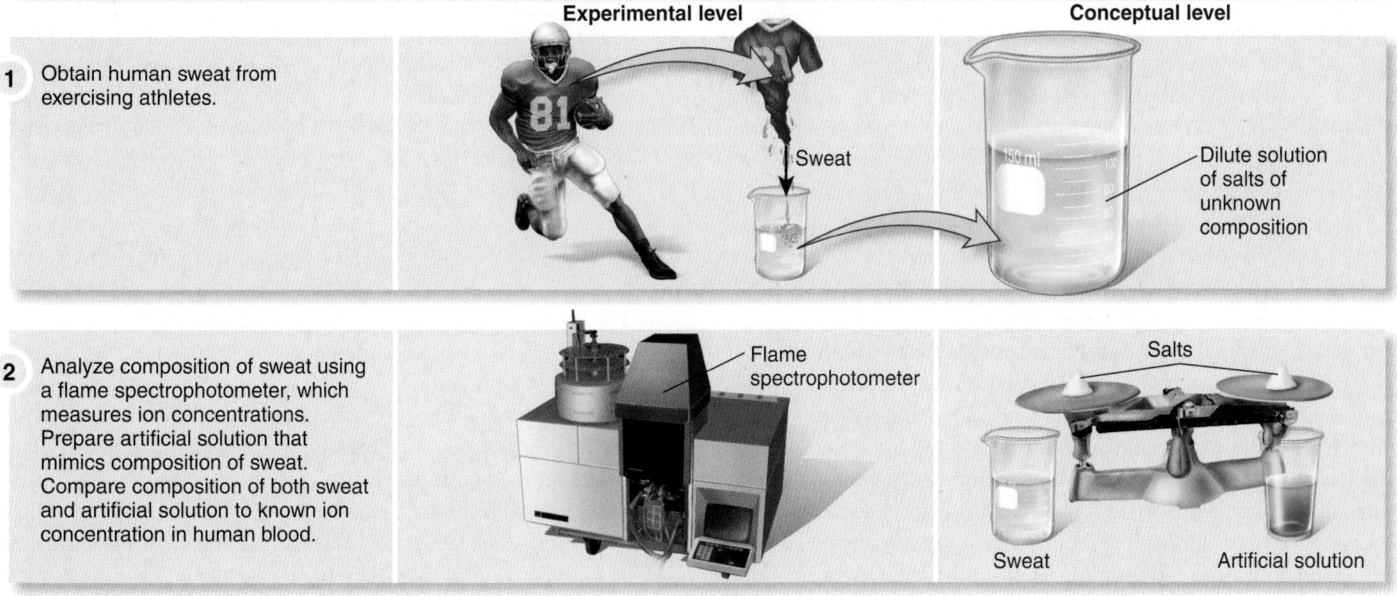

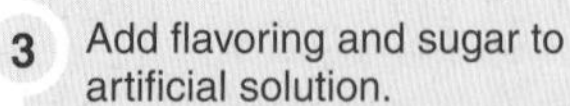

3 Add flavoring and sugar to artificial solution.

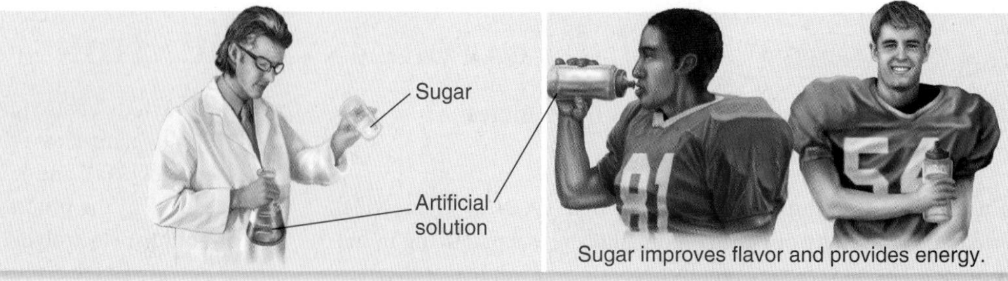

Sugar
Artificial solution
Sugar improves flavor and provides energy.

4 Provide freshman team with the artificial solution and varsity B-team with water. Hold scrimmage.

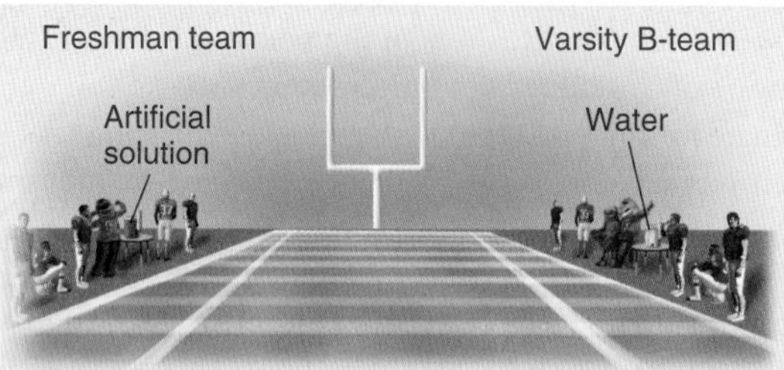

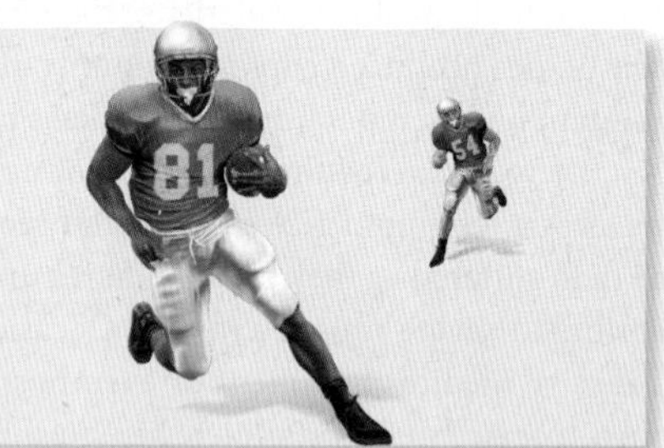

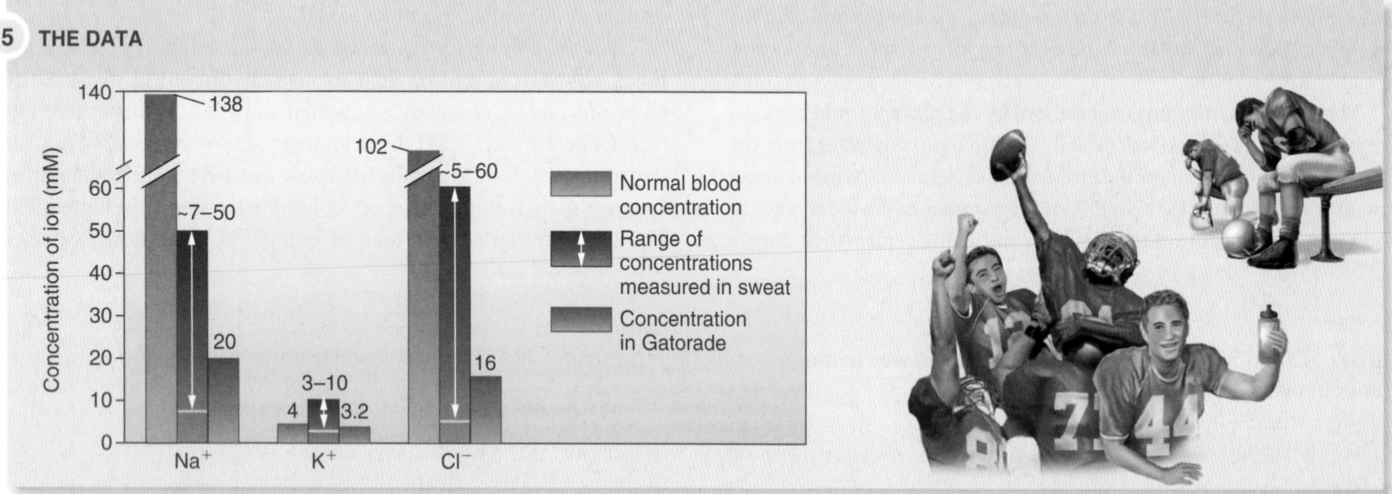

6 **CONCLUSION** Replacement of fluid with solute concentrations similar to those found in human sweat improves athletic performance compared to water replacement alone.

7 **SOURCE** Most of the original studies described here were published in expanded form in a later report. See Cade, R. et al. 1972. Effect of fluid, electrolyte and glucose replacement during exercise on performance, body temperature, rate of sweat loss, and compositional changes of extracellular fluid. *Journal of Sports Medicine and Physical Fitness* 12:150–156.

ion concentrations in human blood. Today, we know that the composition of human sweat can change under certain conditions and can vary among people, but Cade's results were typical. The athletes' sweat contained mostly Na^+, K^+, and Cl^- at concentrations that indicated the solution was dilute compared with blood. Once Cade completed this analysis, he simply made an artificial solution of a composition similar to human sweat. The next step was to have the players ingest the solution before and during the practice sessions and games. Improving its taste—adding some lemon flavoring and sugar—removed any inhibitions the players may have had about drinking it, while also providing an energy boost.

For the first trial, Cade gave the solution to the freshman players during an intrasquad scrimmage against the more experienced varsity B-team, whose members received only pure water to drink as a control. At first, the freshman team appeared overmatched by the B-team, as might be expected. In the second half of the scrimmage, however, the freshman team vastly outperformed the more experienced players and did not suffer the characteristic late-game fatigue the B-team experienced. Based on this test, the varsity A-team was given a similar solution to drink the next day during a game against a heavily favored opponent, whom they beat handily on a hot 39°C day. In subsequent years, Cade and other researchers would conduct carefully controlled experiments with humans and laboratory animals to confirm that a balanced solution of electrolytes similar to that present in sweat effectively improves exercise performance and reduces the possibility of dehydration and its consequences.

Because the solution was envisioned as "aid" for the team known as the University of Florida "Gators," the drink eventually came to be called Gatorade. The year after its introduction, the Gators enjoyed their most successful season ever. In 1965, the Kansas City Chiefs of the former American Football League became the first professional sports team to try the drink, and shortly thereafter, the team enjoyed its greatest success. Nowadays, Gatorade and certain similar "sports drinks" are used at sporting events around the world, and for good scientific reason.

The effectiveness of a solution like Gatorade is due to its ability to restore the correct amounts of both water and ions lost during exercise. Importantly, it is very rapidly absorbed because its osmolarity is close to that of body fluids. Many of the other sports drinks subsequently invented contain additional solutes, such as vitamins and other minerals, and many contain higher amounts of sugar. Because of the presence of these other solutes, these drinks may be very hyperosmotic (a greater concentration of solutes) relative to body fluids. Therefore, when ingested, they initially tend to draw water out of the interstitial fluid and into the gut lumen by osmosis. This slows down the rate at which the water from the drink gets absorbed into the blood.

The story of Gatorade is one of good common sense based on solid scientific principles of osmolarity and salt and water homeostasis. You can now understand why drinking a dilute salt solution during strenuous exercise is better than drinking water. Although drinking pure water prevents dehydration, if drunk in excess, it will actually reduce plasma salt concentrations to below normal. In other words, it will replace one type of salt imbalance with another.

Experimental Questions

1. What symptoms are sometimes seen in athletes after prolonged, strenuous exercise, particularly in hot weather? How are these symptoms related to water loss during exercise, and what did Cade and his colleagues hypothesize about this?
2. How did the researchers test their hypothesis?
3. What was the result of consuming the drink during exercise?

Animals Adapt to Osmotic Challenges by Regulating or Conforming

Animals adapt to osmotic challenges posed by the environment in one of two major ways. Some animals regulate their internal osmolarity at a very stable level, whereas others conform to the osmolarity of their environment (for example, the sea). Animals that maintain very stable internal salt concentrations and osmolarities, even when living in water with very different osmolarities than their body fluids or on land, are called **osmoregulators**. Such animals drink or excrete water and salt as necessary to maintain an internal osmolarity that is generally about 300 mOsm/L, or about one-third that of seawater and at least 10 times that of fresh water. All terrestrial animals are osmoregulators, as are all freshwater animals and many marine animals, including bony fishes and some crustaceans. Osmoregulators maintain stable cellular levels of ions and water, but this requires considerable expenditure of energy, primarily to pump ions into and out of epithelial cells.

Most marine invertebrates and some vertebrates—notably sharks—use a different means to control body fluid composition. In this case, the osmolarity of extracellular and intracellular fluids is matched with seawater. These animals are called **osmoconformers**, because their osmolarity conforms to that of their environment. The osmolarity of blood and other fluids of marine osmoconformers is like seawater, around 1,000 mOsm/L. An advantage of having body fluids conform to the osmolarity of the surrounding seawater is that there is much less tendency to gain or lose water by osmosis across the skin or gills. Thus, sharks and other osmoconformers expend less energy to compensate for water gain or loss than do other aquatic animals. However, osmoconformers are generally limited to the marine environment.

Vertebrate osmoconformers have a high concentration of uncharged molecules dissolved in their extracellular fluids. This allows the extracellular fluids and seawater to have similar osmolarities, but it prevents the excessive accumulation of ions in the body. The body fluids of sharks and other osmoconformers contain sugars, amino acids, and metabolic waste products—notably urea and an organic compound called trimethylamine oxide (TMAO). The total amount of salt and organic compounds in a shark's extracellular fluids produces an osmolarity very similar to that of seawater, even though the salt concentration is similar to that of osmoregulators.

Vertebrate osmoconformers cannot tolerate high ion concentrations in their body fluids any better than osmoregulators. One reason is because a proper ion balance is required for normal electrical signaling in their neurons and muscle cells. In addition, very high salt concentrations tend to disrupt the three-dimensional structure of many proteins, rendering them inactive. Consequently, the body fluids of vertebrate osmoconformers are less salty—that is, they have fewer ions—than seawater, as is also the case in all osmoregulators. Therefore, vertebrate osmoconformers tend to gain salt by diffusion across their gills. That excess salt is eliminated by the kidneys and a type of salt gland called the rectal gland.

49.2 Comparative Excretory Systems

Learning Outcomes:

1. Describe the general processes of filtration, reabsorption, secretion, and excretion.
2. Identify several invertebrate osmoregulatory organs, and compare and contrast the process of elimination in each.
3. List the general features of kidneys that are common to all vertebrates.
4. Relate kidney function to life history in vertebrates.

Although the mammalian kidney, and in particular that of humans, has been especially well studied, enough is known about other classes of animals to formulate general principles regarding the regulation of salt, water, and waste levels in the fluid compartments of an animal's body. This is an ancient and important function that arose early in the evolution of metazoans.

Animals make use of one or more different organs to rid themselves of metabolic wastes, excess water and salts, and toxins from their environment. Most excretory organs contain tubular structures lined with epithelial cells that have the capacity to actively transport ions across their membranes. Wastes are excreted out of the body by means of these tubes.

In some cases, animals may have considerable ability to regulate the rate at which waste is secreted and how much water is lost in the process. For example, even though a thirsty mammal on a hot, sunny day must continue to rid its body of soluble waste products, it must also conserve water. In this section, we consider the anatomy and physiology of invertebrate and vertebrate excretory organs.

Excretory Systems Use Four Basic Processes

Most excretory systems function by using one or more of the following processes: filtration, reabsorption, secretion, and excretion (Figure 49.6). In **filtration**, an organ acts like a sieve or filter, removing some of the water and its small solutes from the blood, interstitial fluid, or hemolymph, while excluding blood cells and large solutes such as proteins. A typical filtration system is that seen in the mammalian kidney, in which a portion of the plasma component of the blood is forced under pressure through leaky capillaries and into the kidney tubules. The material that passes through the filter and enters the excretory organ for either further processing or excretion is called a **filtrate**.

Some of the material in the filtrate can be recaptured and returned to the blood. This is an important feature of many excretory organs, because the formation of a filtrate is not selective, apart from the exclusion of proteins and blood cells. In other words, in order to filter the blood and remove soluble wastes, necessary molecules such as salts, sugars, and amino acids also get filtered in the process. Recapturing these useful solutes requires active transport pumps or other transport systems and is known as **reabsorption**. Much of the filtered water also gets reabsorbed along with useful solutes by osmosis.

In some cases, solutes may get excreted from the body in quantities greater than those found in the filtrate. How is this possible? Some solutes are actively transported from the interstitial fluid surrounding the epithelial cells of the tubules into the tubule lumens. This process, called **secretion**, supplements the amount of a solute that would normally be removed by filtration alone. This is often a way in which excretory organs eliminate particularly toxic compounds from an animal's body, and it can be very effective. Some marine fishes, for example, use secretion as the sole means of cleansing the blood. These animals do not form a filtrate at all.

Excretion is the process of expelling waste or harmful materials from the body. In animals that form a filtrate, the part of the filtrate that remains after reabsorption has been completed and that gets excreted is called **urine**.

Some Invertebrates Use a Filtration Mechanism to Cleanse Body Fluids

The simplest filtration mechanism in invertebrates is the protonephridia system of flatworms (Figure 49.7). **Protonephridia** (singular, protonephridium) are a series of branching tubules that filter fluids

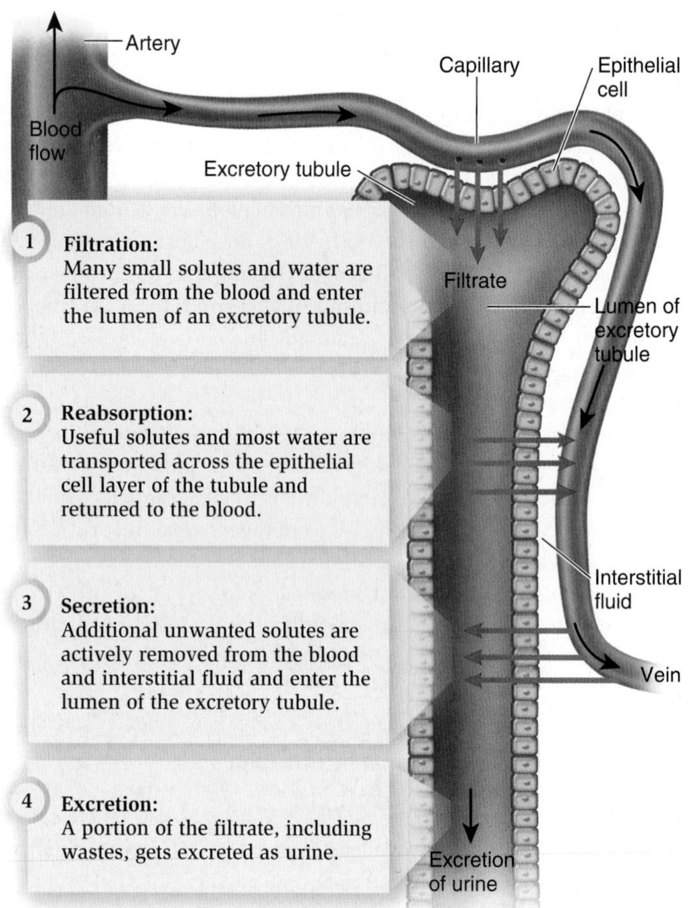

Figure 49.6 **Basic features of the function of many excretory systems.**

Concept Check: *What is the benefit of secreting substances into the tubule?*

from the body cavity by means of ciliated cells that cap the ends of the tubule branches. The beating of the cilia bears some resemblance to a flickering flame, which is why these cells are known as flame cells. As fluid is drawn through slitlike openings of the flame cells and into the lumen, it percolates through the tubule, where most solutes are reabsorbed back into the interstitial fluid. Excess water and some wastes travel through the tubules and exit the body through tiny openings in the body wall called nephridiopores. Much of the nitrogenous waste in flatworms actually diffuses across the body surface into the surrounding water; therefore, the protonephridia are primarily osmoregulatory organs. The urine is generally hypoosmotic compared with the rest of the body fluids, an adaptation for life in fresh water.

Annelids use a different filtration mechanism, called a metanephridial system (Figure 49.8). Pairs of **metanephridia** (singular, metanephridium) are located in each body segment and consist of a tubular network that begins with a funnel-like structure called a nephrostome. The nephrostomes collect coelomic fluid, which contains nitrogenous wastes, through tiny pores that exclude large solutes. Na^+, Cl^-, and other solutes are reabsorbed by active transport along the length of the tubules that extend from the nephrostomes, and from there diffuse into nearby capillaries. The nitrogenous wastes

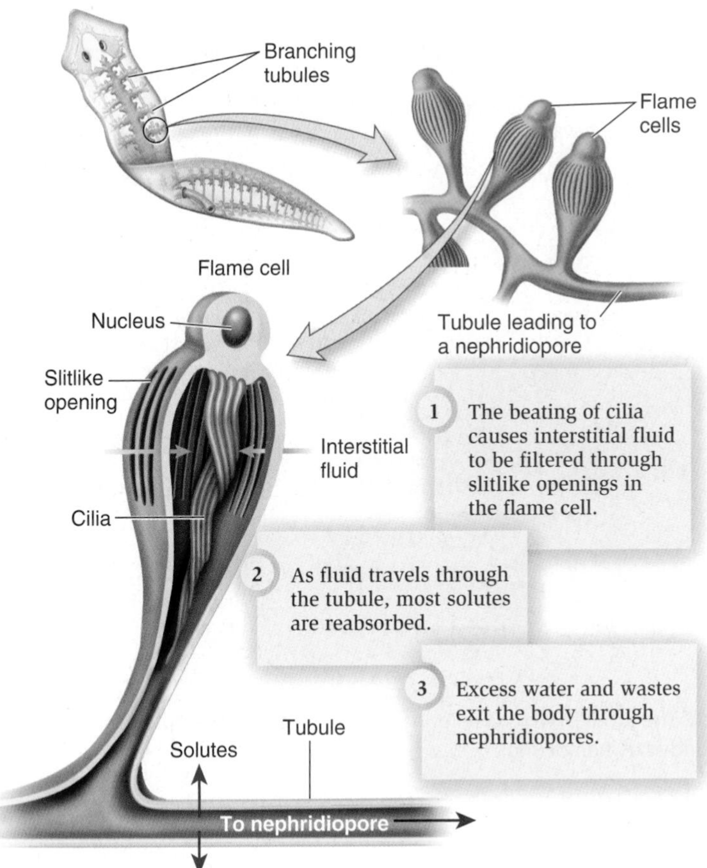

Figure 49.7 **The protonephridial filtration system of flatworms.** As the filtrate moves along the tubules, most solutes are reabsorbed. The final excreted fluid is typically hypoosmotic relative to body fluids.

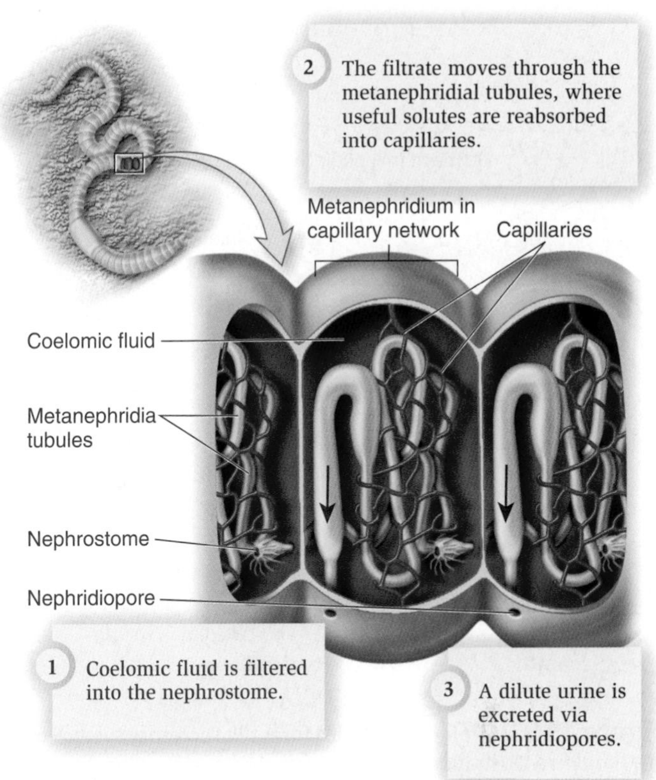

Figure 49.8 **The metanephridial filtration system of annelids.** Most internal body structures have been omitted for clarity. Only one of the two metanephridia in each segment is depicted; the other would be located behind the one shown.

remain behind in the tubules and are excreted through nephridiopores in the body wall. Many annelids live in watery environments and thus, like flatworms, excrete a hypoosmotic urine.

Insects Excrete Wastes by Means of Secretory Organs Rather Than Filtration

The insect excretory system is quite different from that of other invertebrates, because it involves secretion rather than filtration of body fluids. In insects, a series of narrow, extensive tubes called **Malpighian tubules** arises from the midgut and extends into the surrounding hemolymph (Figure 49.9). The cells lining the tubules actively transport salts and uric acid from the hemolymph into the tubule lumen. This secretion process creates an osmotic gradient that draws water into the tubules. The fluid moves from the tubules into the hindgut—the intestine and rectum—where much of the useful salt and water is reabsorbed. The nitrogenous wastes, any excess salts, and other waste compounds are excreted together with the feces through the anus.

Unlike other invertebrates, most terrestrial insects, apart from blood-sucking ones, excrete urine that is either isoosmotic (has the same total solute concentration) or hyperosmotic to body fluids. This is a testament to the efficiency with which the insect hindgut reabsorbs water, and reflects the general principle that life in dry environments is associated with a risk of dehydration.

The Kidney Is the Major Excretory and Filtration Organ in Vertebrates

The major excretory organ found in all vertebrates is the **kidney**. The kidneys of all vertebrates have many features in common. They typically contain specialized tubules composed of epithelial cells that participate in both salt and water homeostasis by promoting active transport of sodium, potassium, and other ions across their membranes. In addition, all kidneys participate in the excretion of wastes. In response to an animal's changing salt and water requirements, these processes can be controlled, that is, stimulated or inhibited, by the actions of nerves and hormones. Most vertebrate kidneys are filtration kidneys, with the exception of purely secretory kidneys found in some marine fishes. Finally, filtration in the kidneys is controlled by mechanical forces, such as the hydrostatic pressure exerted by blood entering the capillaries of the kidneys. The mammalian kidney has been especially well studied and is examined in detail later in this chapter.

The requirement to eliminate waste products while simultaneously maintaining salt and water homeostasis is common to all animals. How animals achieve this balance in highly disparate environments, and with different life histories, is the subject of our next discussion. In the following sections, note that when we refer to kidneys, we often use the adjective renal, meaning "pertaining to the kidneys." For example, we refer to renal physiology and renal functions.

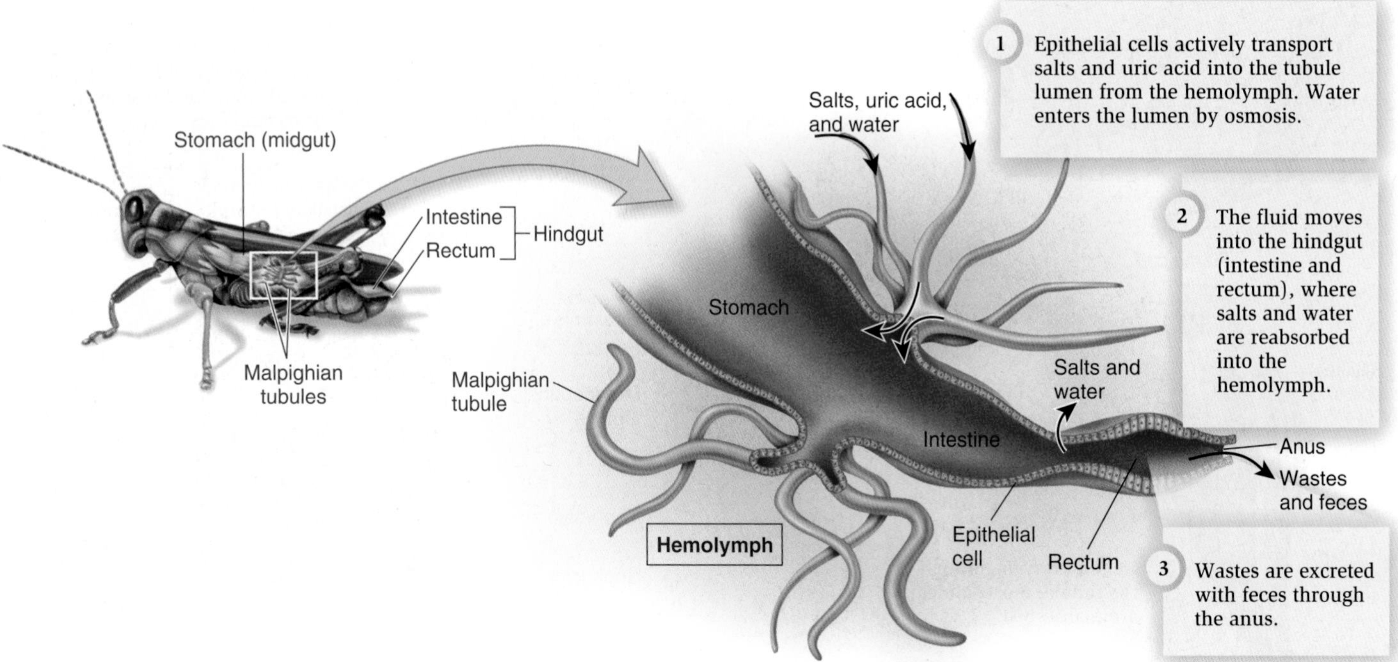
Stomach (midgut)
Intestine
Rectum
Hindgut
Malpighian tubules
Malpighian tubule
Salts, uric acid, and water
1 Epithelial cells actively transport salts and uric acid into the tubule lumen from the hemolymph. Water enters the lumen by osmosis.
2 The fluid moves into the hindgut (intestine and rectum), where salts and water are reabsorbed into the hemolymph.
Stomach
Salts and water
Intestine
Anus
Wastes and feces
Hemolymph
Epithelial cell
Rectum
3 Wastes are excreted with feces through the anus.

Figure 49.9 **Malpighian tubules form the secretory system of an insect.** The tubules, which are longer and more convoluted than shown in this simplified illustration, extend into the body cavity, where they are surrounded by hemolymph.

Kidney Function and Vertebrate Life History

As previously mentioned, not all animals have similar requirements for salt and water. **Table 49.1** summarizes some of the special requirements of different vertebrates and some major behavioral and physiological adaptations that allow them to maintain homeostasis.

The demands placed on an animal by its environment can often be predicted by examining the activities of its kidneys. For example, the kidneys of freshwater fishes are specially adapted for rapid, large-scale filtering of the blood and producing very dilute urine. Amphibians, because of their permeable skin, absorb fresh water from their environments, and consequently they have kidneys with an appearance and activity that resemble those of freshwater fishes. By contrast, the kidneys of marine fishes and desert mammals are adapted to produce urine that is more concentrated than that of freshwater fishes and amphibians. This is an important adaptation that helps conserve as much water as possible.

Some animals have a diet that is extreme in one way or another. This is often associated with differences in renal structure and function. Many mammals and insectivores, for example, subsist entirely or almost entirely on sporadic, high-protein meals. These animals consume so much protein at a single meal that they generate nitrogenous waste at a very high rate. Not surprisingly, perhaps, the blood also tends to be filtered through the kidneys of such animals at a very high rate, an adaptation that helps eliminate the toxic waste.

At the other extreme, polar bears may go 4 to 5 months without feeding. During the nonfeeding periods, they neither defecate nor urinate and subsist by breaking down stores of fat and protein in their bodies. Thus, even though they are not eating, they generate nitrogenous wastes such as urea. Despite not urinating during this time, the concentration of urea in their blood may actually decrease! The explanation for this surprising phenomenon appears to be that polar bears have the extraordinary ability to recycle nitrogenous waste into synthesis of new protein. This ability is of great interest to researchers who study the health consequences of kidney disease including the accumulation of urea in the bloodstream of humans.

49.3 Structure and Function of the Mammalian Kidney

Learning Outcomes:

1. Name the primary components of the urinary system in humans and the major anatomical features of the human kidney.
2. Describe the main parts of a nephron, and outline the steps in which each contributes to the formation of urine.
3. Explain how the actions of two hormones, aldosterone and antidiuretic hormone (ADH), mediate the final composition of urine.

In mammals, the two kidneys lie in the abdominal cavity (**Figure 49.10a**). The urine formed in each kidney collects in a central area called the renal pelvis. From there it flows through tubes called the **ureters** into the **urinary bladder**. Urine is eliminated via the **urethra**. Collectively, the kidneys, ureters, urinary bladder, and urethra constitute the **urinary system**.

Each kidney has an outer portion called the renal cortex and an inner portion called the renal medulla (**Figure 49.10b**). The cortex is the primary site of blood filtration. In the medulla, the filtrate becomes concentrated by the reabsorption of water back into the

Table 49.1 A Comparison of the Vertebrate Mechanisms of Osmoregulation in Different Environments

Animal group	Blood concentration relative to environment	Urine concentration relative to blood	Main nitrogenous waste	Osmoregulatory mechanisms and special features
Marine fish	Hypoosmotic*	Isoosmotic or hyperosmotic	Ammonia	Drinks seawater; secretes salt from gills
Freshwater fish	Hyperosmotic	Strongly hypoosmotic	Ammonia	Drinks no water; transports salt across gills
Amphibian (fresh water)	Hyperosmotic	Strongly hypoosmotic	Most secrete urea	Absorbs water through skin
Marine reptile	Hypoosmotic	Isoosmotic	Ammonia, urea, uric acid	Drinks seawater; hyperosmotic salt-gland secretion
Terrestrial reptile	—	Isoosmotic	Uric acid	Drinks fresh water
Marine bird	Hypoosmotic	Weakly hyperosmotic	Uric acid	Drinks seawater; hyperosmotic salt-gland secretion
Terrestrial bird	—	Weakly hyperosmotic	Uric acid	Drinks fresh water
Desert mammal	—	Strongly hyperosmotic	Urea	Drinks no water; very long loops of Henle; depends on water generated by cellular metabolism
Marine mammal	Hypoosmotic	Strongly hyperosmotic	Urea	Drinks no water
Other mammals	—	Strongly hyperosmotic	Urea	Drink fresh water

*The terms hypo-, iso-, and hyperosmotic indicate solutions with fewer, identical, or greater numbers of dissolved solutes compared with another fluid.

blood. In this section, we will examine the structural features of the kidney that allow it to function as a filtration system.

The Functional Units of the Kidney Are Called Nephrons

Depending on its size, the mammalian kidney contains as many as several million similar, single-cell-thick structures called **nephrons**. (We get the name of the medical specialty nephrology—the study of the kidney function and disease—from the word nephron.) As shown in **Figure 49.10c**, the filtering process begins at a region called the **renal corpuscle**. The renal corpuscle forms a filtrate from blood that is free of cells and proteins. This filtrate then leaves the renal corpuscle and passes through three different regions of the nephron. As the filtrate does so, substances are reabsorbed from it or secreted into it. Ultimately the filtrate remaining at the end of each nephron combines in the collecting duct. Let's look more closely at each part of the nephron and its associated structures.

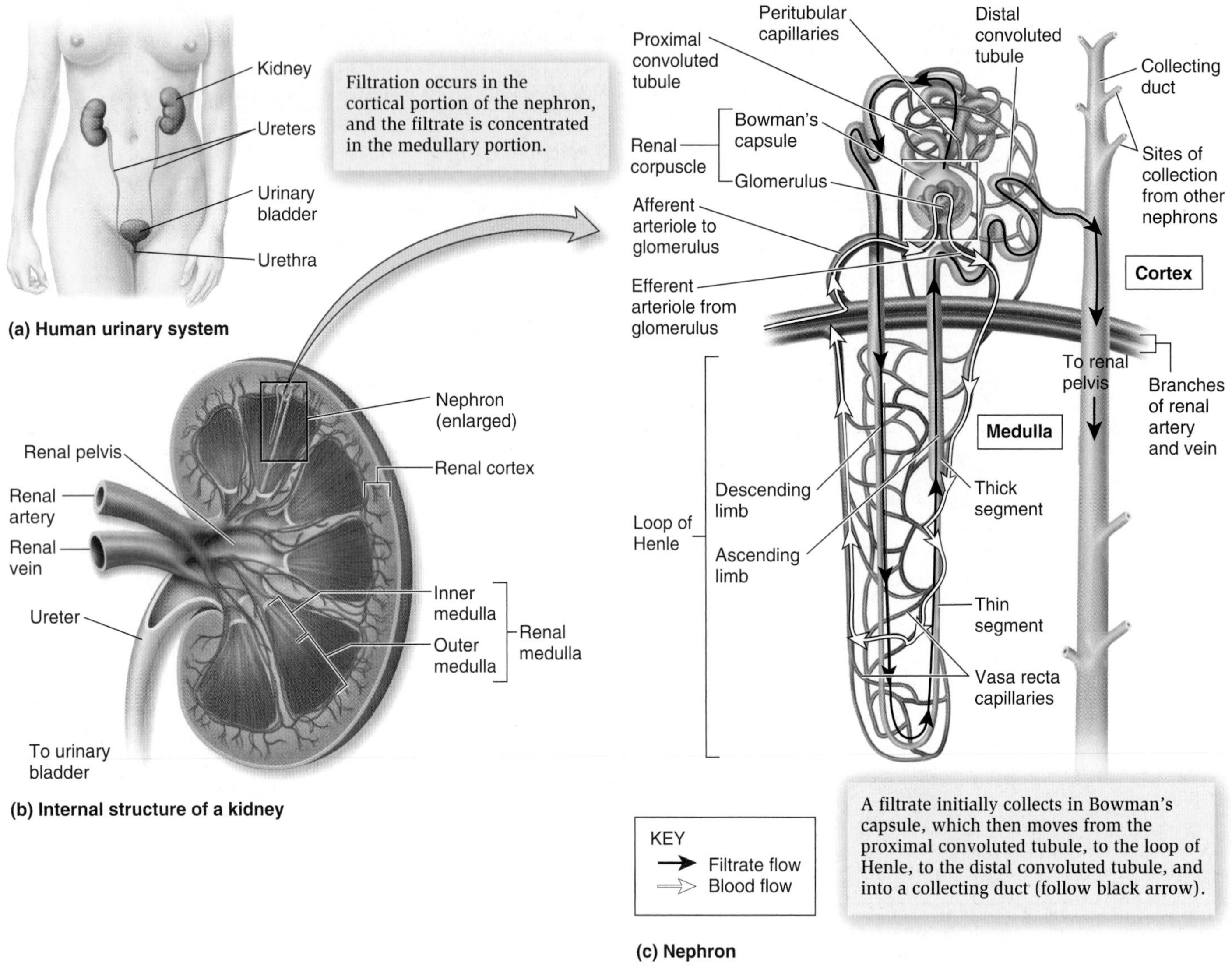

Figure 49.10 **The mammalian urinary system, including the basic functional unit of the nephron.** **(a)** The human urinary system in a woman. In the male, the urethra passes through the penis. **(b)** Enlarged view of a section through a kidney, showing the locations of the major internal structures and a single nephron (enlarged; at the scale of this illustration, nephrons would be microscopic). **(c)** Structure of a nephron. The nephron begins at Bowman's capsule and empties into a collecting duct. Many nephrons empty into a given collecting duct. Surrounding the nephron are capillaries, called peritubular capillaries in the cortex and vasa recta capillaries in the medulla.

The Renal Corpuscle Each renal corpuscle contains a cluster of capillaries called the **glomerulus** (plural, glomeruli) (Figure 49.11a). Each glomerulus is supplied with blood under pressure by an **afferent arteriole**, and blood exits the glomerulus via an **efferent arteriole**. The glomerulus protrudes into a fluid-filled space called **Bowman's capsule** (first identified by the English physiologist William Bowman in 1841). The combination of a glomerulus and a Bowman's capsule constitutes a renal corpuscle. The glomerular capillaries contain fenestrations, tiny holes that permit rapid flow of plasma out of the capillaries. This filtration is further modified by cells called podocytes that surround the capillaries and form filtration slits that allow the passage of small solutes, but are believed to help prevent proteins from entering the filtrate (Figure 49.11b).

The Tubular Part of the Nephron The tubule of the nephron, which is continuous with Bowman's capsule, is made of a single layer of epithelial cells joined by tight junctions and resting on a basement membrane. The basolateral surfaces of the epithelial cells have numerous Na^+/K^+-ATPase pumps in their membranes. The epithelial cells differ in structure and function along the tubule's length, and three distinct segments are recognized. The segment of the tubule that drains Bowman's capsule is the **proximal tubule** ("proximal" means adjacent to, with respect to the Bowman's capsule) (see Figure 49.10c). The next portion of the tubule is the **loop of Henle**, discovered by the German physician Friedrich Henle in 1861. It is a long, hairpin-shaped loop consisting of a descending limb coming from the proximal tubule and an ascending limb. The ascending limb

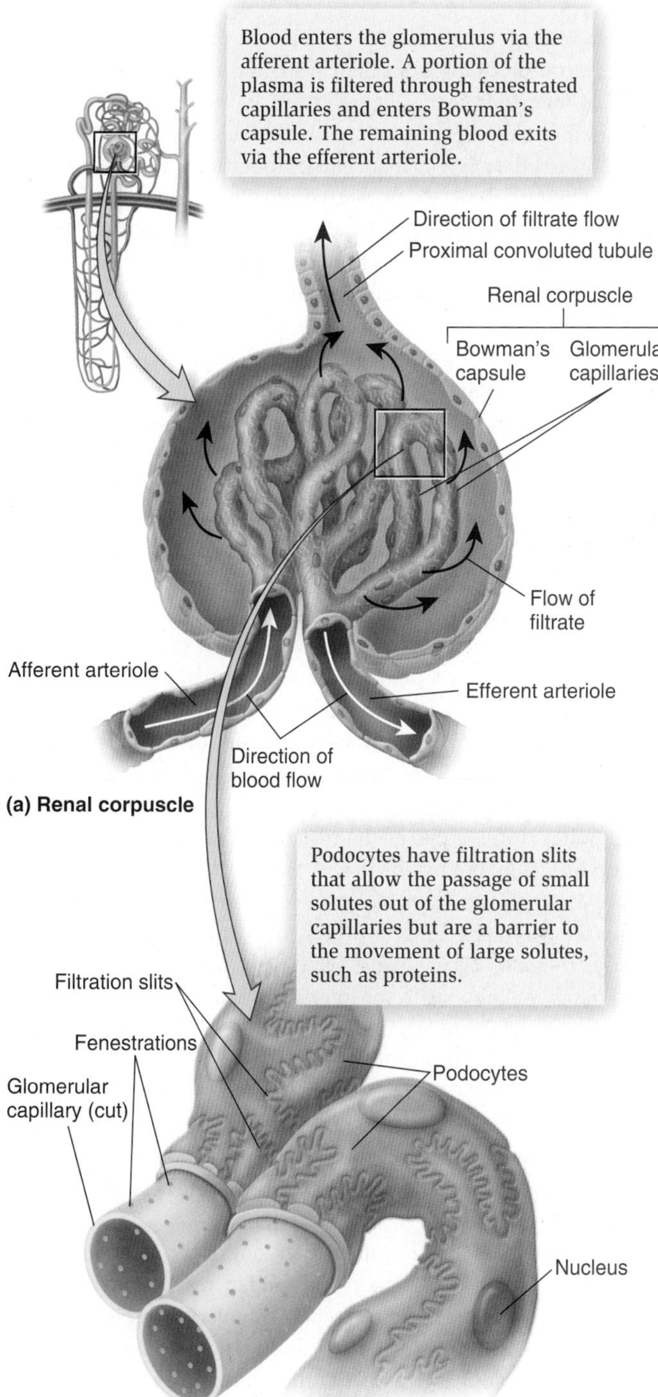

Blood enters the glomerulus via the afferent arteriole. A portion of the plasma is filtered through fenestrated capillaries and enters Bowman's capsule. The remaining blood exits via the efferent arteriole.
Direction of filtrate flow
Proximal convoluted tubule
Renal corpuscle
Bowman's capsule
Glomerular capillaries
Flow of filtrate
Afferent arteriole
Efferent arteriole
Direction of blood flow
(a) Renal corpuscle
Podocytes have filtration slits that allow the passage of small solutes out of the glomerular capillaries but are a barrier to the movement of large solutes, such as proteins.
Filtration slits
Fenestrations
Glomerular capillary (cut)
Podocytes
Nucleus
(b) Glomerular capillaries with podocytes

Figure 49.11 The structure and function of the renal corpuscle. A renal corpuscle is composed of Bowman's capsule and the capillaries that make up the glomerulus. It is here that the filtrate is first formed.

BIOLOGY PRINCIPLE Structure determines function. The structure of the capillaries in the renal glomerulus is suited to their function. Fenestrations permit the passage of plasma but not blood cells out of the capillaries. The structure of the podocytes is also suited to their function, in that filtration slits prevent the passage of plasma proteins into the filtrate. Damage to the structure of the podocytes, for example, might cause protein to leak into the filtrate and be lost in the urine.

has two segments: a thin and a thick segment. The thick segment leads to the next tubular segment, the **distal tubule** ("distal" means away from the Bowman's capsule). Fluid flows from the distal tubule into one of the many collecting ducts in the kidney.

Capillaries of the Nephron All along its length, each tubule is surrounded by capillaries. These include the **peritubular capillaries** in the cortex and the **vasa recta capillaries** in the medulla (see Figure 49.10c). Both sets of capillaries carry away reabsorbed solutes and water from the filtrate in the nephron and return them to the bloodstream. However, only the peritubular capillaries secrete solutes into the tubules.

A Filtrate Is Produced in Bowman's Capsule

Filtration begins as blood flows through the glomerulus and a portion of the plasma leaves the glomerular capillaries and filters into Bowman's capsule. Only about 15–20% of the plasma is filtered from the blood as it circulates through the glomerulus. Most of the blood, therefore, exits the glomerulus by an efferent arteriole (see Figure 49.11a). Proteins and blood cells are prevented from leaving the glomerular capillaries because of the small diameter of the fenestrations and the presumed actions of the filtration slits mentioned earlier. The fluid that enters Bowman's capsule is called the glomerular filtrate. The rate at which the filtrate is formed is called the **glomerular filtration rate (GFR)**. GFR can be increased by dilation (widening) of the afferent arteriole. When the afferent arteriole dilates, more blood enters the glomerulus, increasing the hydrostatic pressure in those capillaries and forcing more plasma through the fenestrations in the glomerular capillaries and into Bowman's capsule. This might happen, for example, when excess water in the body must be excreted in the urine.

By contrast, constriction of the afferent arteriole would reduce the amount of blood entering the glomerular capillaries and therefore would decrease GFR. This might occur following a loss of blood due to a severe injury. In such a scenario, reducing GFR results in less urine production, which, in turn, minimizes how much water is lost from the body and helps compensate for the blood lost due to the injury.

Useful Solutes Are Reabsorbed from the Filtrate in the Proximal Tubule

The filtrate flows from the renal corpuscle to the proximal tubule. Anywhere from two-thirds to 100% of a particular useful solute is reabsorbed from the filtrate in the proximal tubule. This includes Na^+, K^+, Cl^-, HCO_3^- (bicarbonate ion), and organic molecules such as glucose and amino acids. Some ions diffuse through channels in the membranes of the epithelial cells that form the proximal tubule. Others are actively transported across the tubule. Organic molecules generally are reabsorbed by being coupled to transport of ions such as Na^+. The reabsorption of solutes and water is enhanced by microvilli that extend from the luminal surface of the epithelial cells of the proximal tubule. This anatomical adaptation, called a **brush border**, creates an enormous surface area for transporters and channels to be localized (**Figure 49.12**).

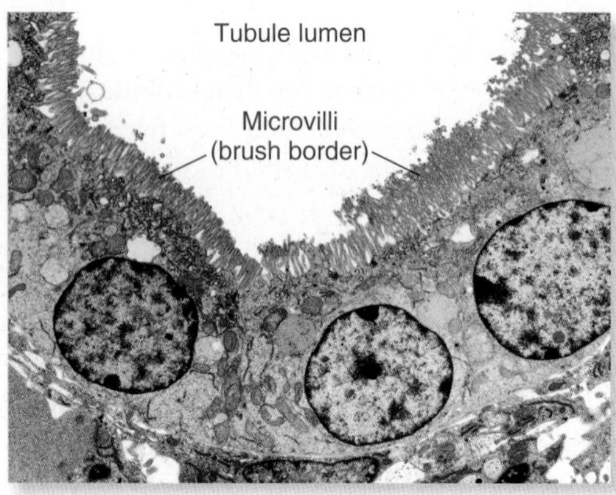

Figure 49.12 Electron micrograph of cuboidal epithelial cells of the proximal tubule of a rat nephron. Note the extensive brush border microvilli on the luminal surface of the cells.

BioConnections: In what other organ have you learned about a brush border (see Figure 45.9 for a hint)? What general conclusions can you draw about the function of such specialized epithelia?

Most of the water in the filtrate is reabsorbed by osmosis as the ions and organic molecules are transported from the lumen of the proximal tubule to the interstitial fluid. From the interstitial fluid, the solutes and water enter peritubular capillaries to return to the blood. Many of the reabsorptive mechanisms in the renal tubule have a limit to the amounts of material they can transport in a given amount of time—called the transport maximum (T_m). This occurs because the binding sites on the membrane-transport proteins become saturated with their ligands. Occasionally, certain molecules may reach such high concentrations in the blood that they exceed the T_m for that molecule in the renal tubule and consequently get excreted in the urine. For example, people who ingest very large quantities of water-soluble vitamins, such as vitamin C, have increased plasma concentrations of vitamin C. Eventually, the filtered load may exceed the tubular reabsorptive T_m for this vitamin, and any additional ingested vitamin C is excreted in the urine.

Other materials are actively secreted into the proximal tubule. Some solutes that are either not required by an animal or that are potentially toxic at high concentrations are removed by active transport mechanisms from the blood and secreted into the interstitial fluid and then into the proximal tubule. Examples of such substances include drugs (for example, penicillin), naturally occurring toxins, nucleoside metabolites, and ions such as K^+ and H^+. These solutes are excreted in the final urine.

By the time the filtrate leaves the proximal tubule, its volume and composition have changed considerably. The amount of salts and water it contains is much reduced, and the organic molecules have all been removed. However, the osmolarity of the filtrate at this stage is still about the same as that of blood.

Water and Salt Are Reabsorbed from the Filtrate Along the Loop of Henle

In the loop of Henle, filtrate moves down the descending limb of the loop, makes a U-turn at the bottom, and then moves back up the ascending limb of the loop. The permeabilities and transport characteristics of the epithelial cells lining the loop change over its length as it descends from the cortex into the medulla and then ascends to the cortex again (**Figure 49.13**).

The filtrate that leaves the proximal tubule has had solutes and water reabsorbed from it in about equal proportions, and the osmolarity of the filtrate is about 300 mOsm/L. This filtrate enters the descending limb of the loop of Henle, which is very permeable to water but not to Na^+ and Cl^-. Water leaves the filtrate by osmosis in this region because the surrounding interstitial fluid is hyperosmotic compared to the tubule contents.

The hyperosmolarity of the interstitial fluid originates from three sources. First, the initial upturn of the thin segment of the ascending limb of the loop of Henle is very permeable to Na^+ and Cl^- but not to water. Therefore, these ions diffuse at high rates out of the loop into the interstitial fluid, significantly increasing the osmolarity of the inner medulla. Second, the epithelial cells of the thick segment of the ascending limb of the loop of Henle actively transport some of the remaining Na^+ and Cl^- out of the filtrate and into the interstitial fluid of the outer medulla. Third, although urea is a waste product, some of the urea that is present in the filtrate does not get excreted in the urine, but instead diffuses out of the lower ends of the collecting ducts and into the inner medulla interstitial fluid. Collectively, these solutes create the osmotic force that draws water out of the filtrate in the descending limb. The water then enters local capillaries and rejoins the blood circulation.

As water diffuses out of the filtrate in the descending limb of the loop of Henle, the osmolarity of the filtrate increases from 300 to about 1,200 mOsm/L (or higher in some species). During its passage up the ascending limb, however, the osmolarity of the filtrate decreases to about 200 mOsm/L as ions diffuse out and are transported out of the tubule into the interstitial fluid. The ascending limb has two different segments, each with a slightly different structure. The thin segment of the ascending limb is permeable to Na^+ and Cl^-, but not water, so Na^+ and Cl^- diffuse out of the filtrate in the tubule. This begins to dilute the filtrate. The thick segment is similarly not permeable to water, but in this segment, epithelial cells actively transport Na^+ and Cl^- out of the filtrate, and the filtrate becomes increasingly more dilute.

As a consequence of ion movement out of the ascending limb, the osmolarity of the kidney interstitial fluid increases from renal cortex to inner medulla. This extracellular osmolarity gradient is what allows water to diffuse by osmosis from the descending limb of the loop of Henle all along its length. In other words, this is an example of a countercurrent exchange system, like those that operate in heat and gas exchange in some animals (refer back to Figures 46.15 and 48.5). A major difference between those countercurrent exchange systems and the one in the loop of Henle, however, is that the latter requires energy-dependent ion pumps to maintain the necessary concentration gradient. Because energy is used to increase—or multiply—the gradient, the loop of Henle system is also referred to as a **countercurrent multiplication system.**

The chief advantage to an animal provided by the loop of Henle is that the final volume of urine produced has been reduced and its contents concentrated by the recapture of water along the osmotic gradient. This is especially important in animals in which total body water stores are regularly in danger of being depleted. For example,

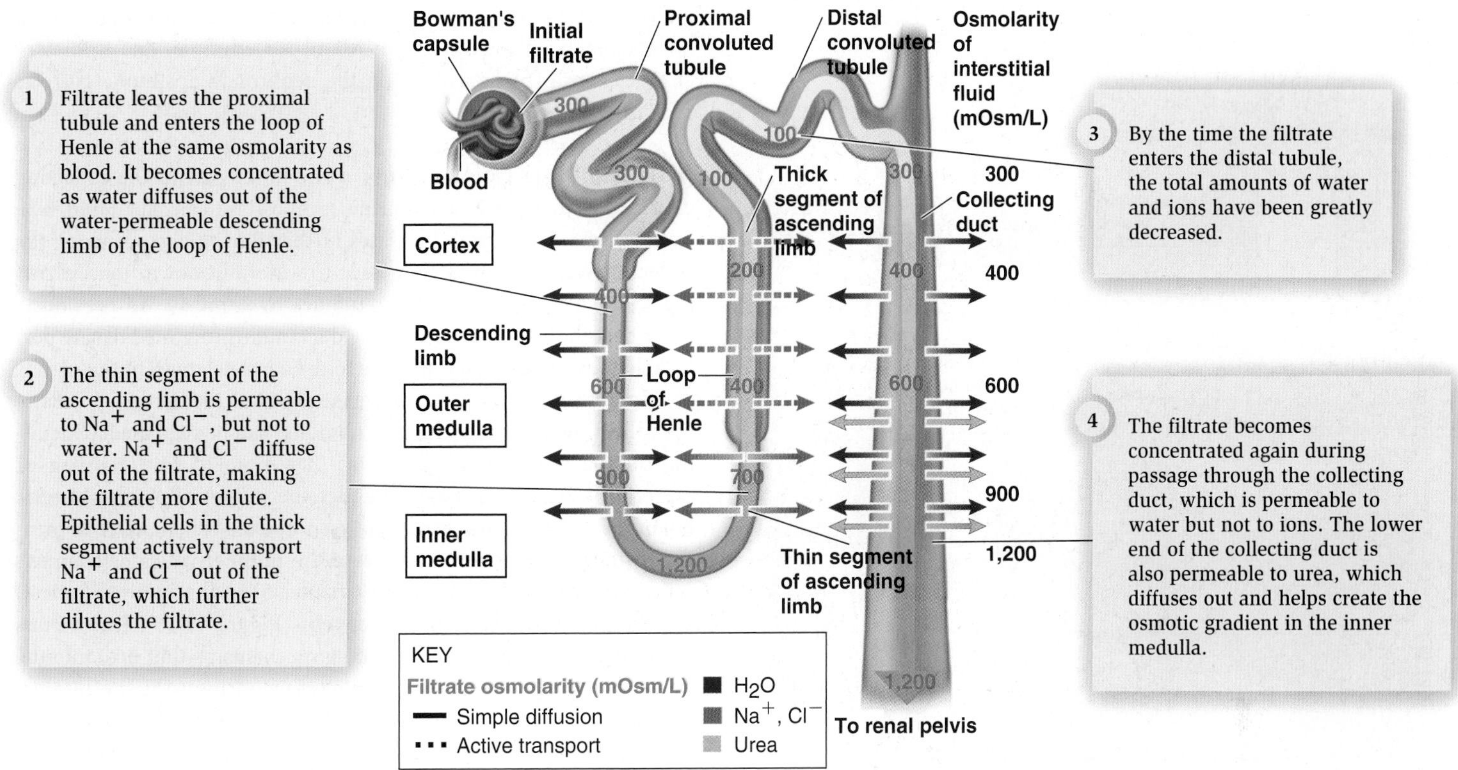

1 Filtrate leaves the proximal tubule and enters the loop of Henle at the same osmolarity as blood. It becomes concentrated as water diffuses out of the water-permeable descending limb of the loop of Henle.
2 The thin segment of the ascending limb is permeable to Na+ and Cl−, but not to water. Na+ and Cl− diffuse out of the filtrate, making the filtrate more dilute. Epithelial cells in the thick segment actively transport Na+ and Cl− out of the filtrate, which further dilutes the filtrate.
3 By the time the filtrate enters the distal tubule, the total amounts of water and ions have been greatly decreased.
4 The filtrate becomes concentrated again during passage through the collecting duct, which is permeable to water but not to ions. The lower end of the collecting duct is also permeable to urea, which diffuses out and helps create the osmotic gradient in the inner medulla.
Bowman's capsule
Initial filtrate
Proximal convoluted tubule
Distal convoluted tubule
Osmolarity of interstitial fluid (mOsm/L)
Blood
Cortex
Thick segment of ascending limb
Collecting duct
Descending limb
Loop of Henle
Outer medulla
Inner medulla
Thin segment of ascending limb
300
100
400
200
600
900
700
1,200
To renal pelvis
KEY
Filtrate osmolarity (mOsm/L)
Simple diffusion
Active transport
H2O
Na+, Cl−
Urea

Figure 49.13 **Tubule permeabilities and concentrations of the filtrate in the nephron and collecting duct.**

desert mammals such as the kangaroo rat tend to have longer loops of Henle than other mammals. The extra length of the loop provides for a very large osmotic gradient in the medulla and, therefore, more water-reabsorbing capacity. At the other extreme, animals that generally must eliminate excess water have reduced or absent loops of Henle. Freshwater fishes, for example, do not have loops of Henle in their nephrons. In their case, it is advantageous to excrete as much water as possible to compensate for the large amounts of water constantly entering the body by osmosis across the gills.

The Vasa Recta Maintain the Osmotic Gradient and Minimize the Loss of Useful Solutes

Why does the blood flowing through the vasa recta of the medulla not eliminate the countercurrent osmotic gradient set up by the loops of Henle? One would think that as plasma, having the usual osmolarity of ~300 mOsm/L, enters the highly concentrated environment of the medulla, two types of massive net diffusion or osmosis would occur: that of Na^+ and Cl^- into the capillaries and that of water out of them. Thus, the interstitial gradient would be "washed away." However, this does not happen because the vasa recta forms hairpin loops that run parallel to the loops of Henle and collecting ducts (see Figure 49.10c). Near the top of the loop of Henle, blood in the vasa recta has a normal osmolarity. As the blood flows down the loop deep into the inner medulla, Na^+ and Cl^- do diffuse into, and water moves out of, the vasa recta. However, after the bend in the loop is reached, the blood then flows up the ascending vessel loop, where the process is almost completely reversed. Thus, the hairpin-loop structure of the vasa recta minimizes excessive loss of solutes from the interstitial fluid by diffusion. At the same time, both the salt and water being reabsorbed from the loops of Henle and collecting ducts are carried away in equivalent amounts by the movement of water and solutes between fluid compartments. Therefore, the countercurrent gradient set up by the loops of Henle is maintained.

Salt and Water Concentrations Are Fine-Tuned in the Distal Tubule

By the time the filtrate reaches the distal tubule, most of the reabsorbed salt and water have already been restored to the blood, so the filtrate has been diluted to an osmolarity of about 100 mOsm/L (see Figure 49.13). However, the remaining concentrations of sodium, chloride, and potassium in the filtrate can still be fine-tuned to precisely match an animal's requirements for retaining or eliminating water and salts. This process is mediated by the actions of two hormones called aldosterone and antidiuretic hormone (ADH).

Aldosterone and Sodium Reabsorption and Potassium Secretion Aldosterone is a hormone produced by the adrenal glands. It acts on epithelial cells of the distal tubule and cortical collecting duct, stimulating the active transport of three molecules of Na^+ out of the filtrate in the tubule into the interstitial fluid (reabsorption) for every two molecules of K^+ it pumps into cells from the interstitial fluid and ultimately into the filtrate (secretion) (**Figure 49.14**). Water from

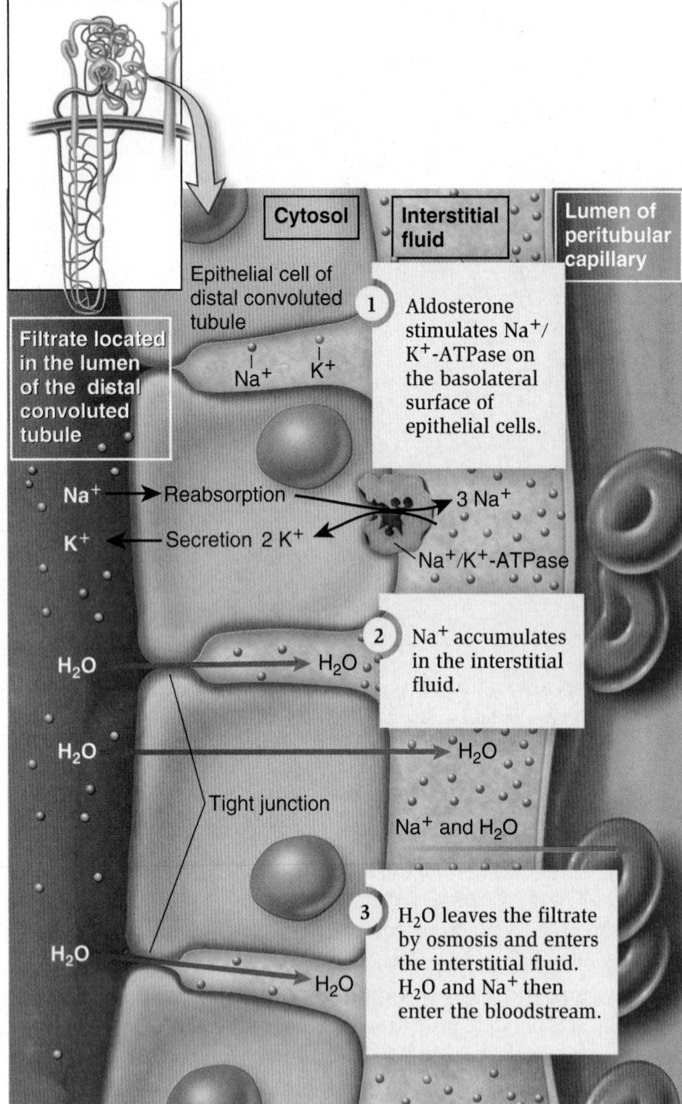

Figure 49.14 Action of aldosterone on distal tubule epithelial cells. Aldosterone stimulates Na^+/K^+-ATPase pumps in cells of the distal tubule; these pumps are expressed only on the side of the cells facing the interstitial fluid (the basolateral surface). This activity creates an osmotic gradient, as three Na^+ are reabsorbed from the filtrate for every two K^+ secreted into it. Water then leaves the tubule by osmosis. The net effect is reabsorption of Na^+ and water, and secretion of K^+, which gets excreted in the urine. Na^+ and water enter the peritubular capillaries and are carried away in the bloodstream.

BioConnections: What special property of epithelial cells is demonstrated by the action of aldosterone? Look back to Figures 10.17, 10.18, and 40.4 for help.

the filtrate follows the Na^+ by osmosis into the interstitial fluid and blood. This makes the filtrate that moves to the lower, medullary part of the collecting duct a bit more concentrated than it was before, with an osmolarity of about 400 mOsm/L. Aldosterone concentrations increase in the blood whenever the Na^+ concentration of the blood is lower than normal or the K^+ concentration is higher than normal; such imbalances might occur, for example, due to dietary changes. Through its actions on the nephron, aldosterone corrects such imbalances.

ADH and Water Reabsorption The lower part of the collecting duct is the final place where urine composition can be altered. The osmolarity of the filtrate increases further during passage along the collecting duct, which is permeable to water but not to ions, allowing water to diffuse out of the tubules by osmosis. As previously described, the cells of the collecting duct in the inner medulla are permeable to urea, which contributes to the osmotic gradient there.

In addition, however, the permeability of the epithelial cells of the collecting ducts to water (but not ions) can be regulated, depending on the body's requirement at that moment for retaining or excreting water. This happens under the influence of **antidiuretic hormone (ADH)**, also known as vasopressin. When ADH is present in the blood, it acts to increase the number of water channels called aquaporins (refer back to Chapter 5) in the luminal (apical) membranes of the collecting duct cells (Figure 49.15). It does this by stimulating intracellular signaling mechanisms that promote the fusion of storage vesicles containing aquaporins with the luminal membranes of the duct epithelial cells, resulting in the insertion of aquaporins into those membranes. Because the collecting ducts travel through the hyperosmotic medulla of the kidney, an osmotic gradient draws water from the filtrate in the duct into the duct epithelial cells. The water exits the cells on the other side (the basolateral surface), where another set of aquaporins is present. These aquaporins are always present in the basolateral membrane; unlike those on the luminal side of the cells, they do not require the presence of ADH for their insertion into the membrane. Once the water moves from the cells into the interstitial fluid, it enters the vasa recta capillaries and the blood. In this way, as the filtrate travels through the collecting ducts, it becomes greatly concentrated (hyperosmotic) compared to blood—as much as four to five times more concentrated in humans, for example.

ADH concentrations increase in the blood during situations where it is important to conserve water, such as when an animal is dehydrated. The ability of the kidneys to produce hyperosmotic urine is a major determinant of an animal's ability to survive in conditions where water availability is limited. By contrast, when the body's stores of water are plentiful, the concentration of ADH in the blood decreases. This results in endocytosis of portions of the luminal membranes, along with their aquaporins. This, in turn, decreases the water permeability of the collecting ducts. In such a case, urine increases in volume and becomes more dilute because less water diffuses out of the collecting ducts.

Biological membranes such as the plasma membranes of animal cells are composed of a lipid bilayer that inhibits the movement of water (refer back to Figure 5.1). The mysteries of how water moves rapidly across membranes like those of the collecting ducts, and how ADH controls the amount of water reabsorption, were solved by the discovery in the early 1990s of the aquaporins (from the Latin, meaning water pores), as described next.

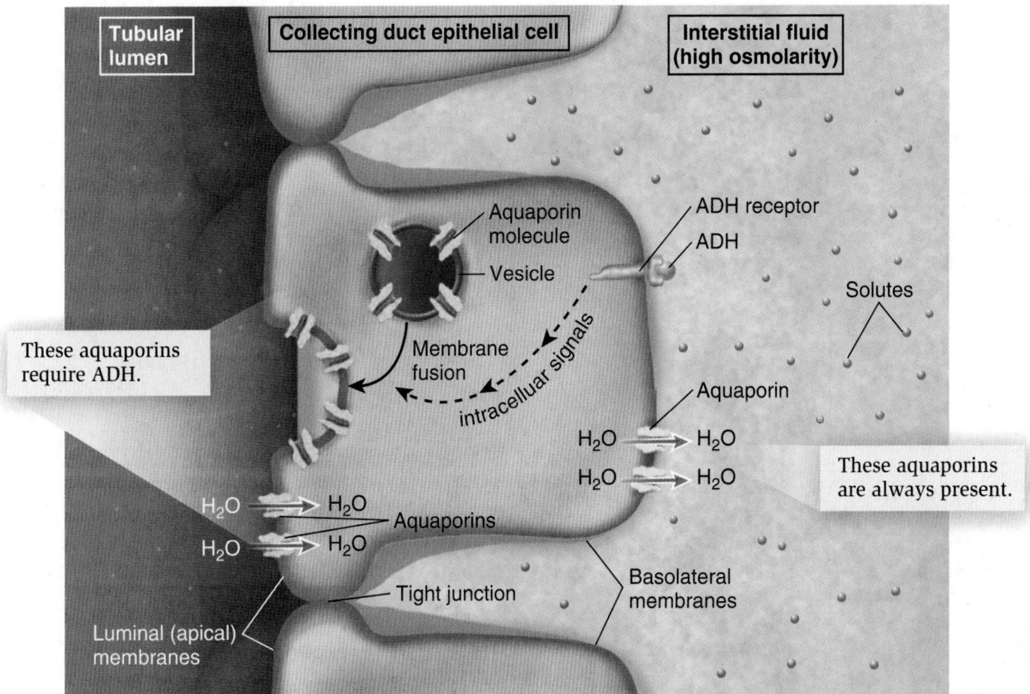

Figure 49.15 **The effect of antidiuretic hormone (ADH) on water reabsorption in the collecting ducts of the kidney.** Water molecules require a channel called an aquaporin to move through a membrane. The epithelial cells of the collecting duct express two sets of aquaporins. One set is always present on the basolateral side of the cells; the other set only inserts into the luminal membrane in the presence of ADH. ADH activates cell-signaling mechanisms that stimulate the fusion of intracellular storage vesicles containing aquaporins with the luminal membrane. Water moves by osmosis across the cell and into the interstitial fluid, and from there enters the vasa recta capillaries to be transported into the circulation (not shown). When ADH concentrations in the blood decrease (for example, when an animal is fully hydrated), the aquaporins on the luminal membrane are returned to the intracellular storage vesicles by the process of endocytosis.

GENOMES & PROTEOMES CONNECTION

Aquaporins Comprise a Large Family of Proteins That Are Ubiquitous in Nature

In the early 1990s, the American physiologist Peter Agre and colleagues discovered the first of what would eventually be recognized as a new family of proteins, called aquaporins, a discovery that earned Agre the Nobel Prize in Chemistry in 2003 (see Figure 5.18). Scientists now know that aquaporins are a subfamily of an even larger family of proteins with membrane transport capabilities, found in all kingdoms of living organisms. The functions of all the members of this family are not yet known, but some aquaporins can transport other small molecules, such as glycerol and urea.

Aquaporins are proteins with six transmembrane domains and two short loops in the membrane (**Figure 49.16**). In animals, the two short loops come together to form the three-dimensional core of the water channel. The importance of the loops is reflected in the observation that these portions are the most highly conserved sequences of the protein among different species. Water must pass through a zone of constriction, created by the loops, that reduces the channel opening to a pore that is about 30 picometers (30×10^{-12} m), or just about the width of a water molecule. Scattered along the inner part of the channel are arginine amino acids, which are positively charged. The charged arginines participate in hydrogen bonding with water molecules, facilitating their single-file movement through each channel at rates that have been estimated to be up to billions of water molecules per second!

Within the various extracellular and intracellular domains of aquaporin proteins are sites that can be modified by enzymes, such as protein kinases. This suggests that the opening and closing of these channels may be gated by stimuli, like the way ion channels are gated in neurons and other cells. In addition, the promoter region of certain aquaporin genes contains a site that is recognized by transcriptional activator proteins that are responsive to the presence of cAMP, a common intracellular signaling molecule and one that is generated by cells stimulated by ADH. Thus, in addition to its rapid effect on aquaporin insertion into cell membranes (see Figure 49.15), another mechanism by which ADH promotes osmosis of water out of the renal collecting ducts, reducing urine volume, is by stimulating the transcription of one or more aquaporin genes.

Our understanding of aquaporin function has allowed us to explain the molecular basis of one form of an inherited human disease called hereditary nephrogenic diabetes insipidus. People with this disease are unable to produce a concentrated urine and consequently lose large amounts of water. A mutation in an aquaporin

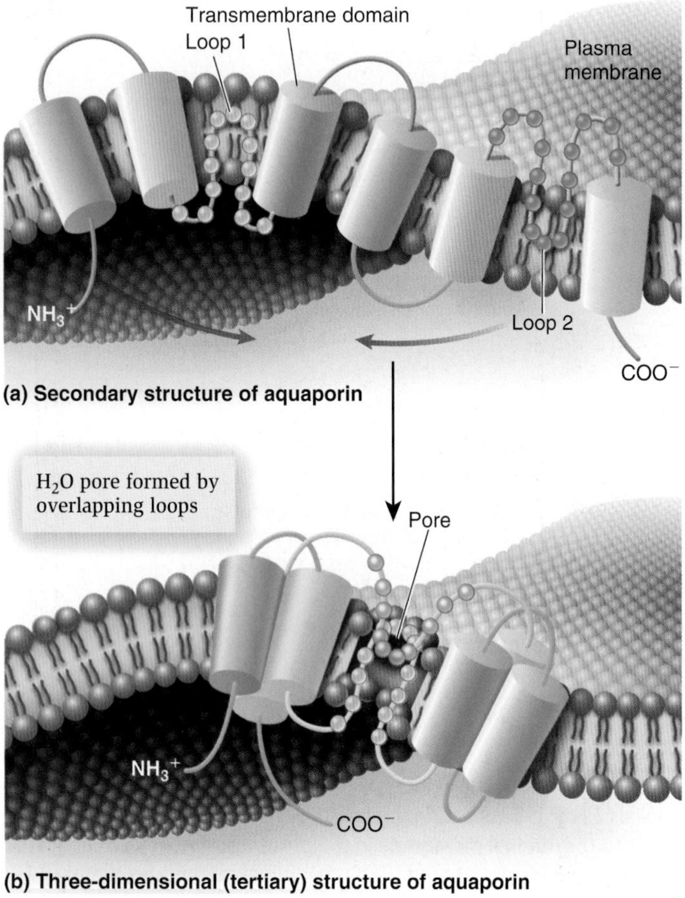

(a) Secondary structure of aquaporin

(b) Three-dimensional (tertiary) structure of aquaporin

Figure 49.16 Detailed aquaporin structure. All proteins of the aquaporin family share a similar structure, with six membrane-spanning domains (represented as cylinders) and two loops that come together to form a water channel. **(a)** Secondary structure of aquaporin. This highly schematic representation highlights the two separate regions of the molecule that come together to form the water pore. **(b)** Tertiary structure of aquaporin showing pore formation.

gene (*AQP2*) results in a form of the protein expressing any of several abnormalities: improper folding, impaired ability of the molecule to enter the plasma membrane, or impaired ability of the molecule to form a channel core. AQP2 is the aquaporin that is responsive to the presence of ADH. Because of this, water reabsorption from the kidneys is greatly impaired in individuals with mutations in the *AQP2* gene.

In addition, Agre's discovery may have widespread implications for other areas of biology and human health. For example, certain types of plant disease are associated with abnormal aquaporin expression in roots. In addition, some scientists are investigating the possibility that drugs that inhibit bacterial aquaporins may someday be useful antibiotics. The use of drugs to inhibit one class of aquaporins has even been suggested as a possible antiperspirant!

49.4 Impact on Public Health

Learning Outcomes:

1. Describe several common health issues related to the diseases of the kidneys.
2. Diagram the process of hemodialysis, and explain its strengths and limitations.

Diseases and disorders of the kidney are a major cause of illness in the human population. According to statistics released by the Centers for Disease Control and Prevention, up to 20 million people suffer from kidney disease in the U.S., with approximately 15–20,000 individuals receiving kidney transplants each year. This section gives an overview of kidney diseases and disorders in humans and also discusses some of the available treatments for these conditions.

Kidney Damage and Disease Result in Disruption of Homeostasis

Many diseases affect the kidneys. Diabetes, bacterial infections, allergies, congenital defects, kidney stones (accumulation of mineral deposits in nephron tubules), tumors, and toxic chemicals are some possible sources of kidney damage or disease. A buildup of pressure due to obstruction of the urethra or a ureter may damage one or both kidneys and increase the likelihood of a bacterial infection and eventual renal failure. The symptoms of renal failure (fatigue, weakness, swollen abdomen and other regions, changes in urine amount and composition, anemia, salt imbalances, nausea, muscle disturbances, among others) are similar regardless of the cause of the disease, and all stem from the condition known as **uremia**, the retention of urea and other waste products in the blood.

Assuming that a person with diseased kidneys continues to ingest a normal diet containing the usual quantities of nutrients and electrolytes, what problems might arise? Potentially toxic waste products that would normally enter the nephron tubules by filtration instead build up in the blood, because kidney damage significantly reduces the number of functioning nephrons. In addition, the excretion of K^+ is impaired because too few nephrons remain capable of normal tubular secretion of this ion. Increased K^+ in the blood is an extremely serious condition, because of the importance of stable extracellular concentrations of K^+ in the control of heart and nerve function.

The kidneys are still able to perform their homeostatic functions reasonably well as long as at least 20% or so of the nephrons are functioning normally. The remaining nephrons undergo alterations in function—filtration, reabsorption, and secretion—to compensate for the missing nephrons. For example, each remaining nephron increases its rate of K^+ secretion so that the total amount of K^+ excreted by the kidneys can be maintained at normal levels. The kidneys' regulatory abilities are limited, however. If, for example, someone with severe renal disease were to eat a diet high in K^+, the remaining nephrons might not be able to secrete enough K^+ to prevent its concentrations from increasing in the extracellular fluid.

Kidney Disease May Be Treated with Hemodialysis and Transplantation

Diseased kidneys may eventually reach a point where they can no longer excrete and reabsorb water and ions at rates that maintain salt and water homeostasis, nor excrete waste products as fast as they are produced. Adjusting a person's diet can help reduce the severity of

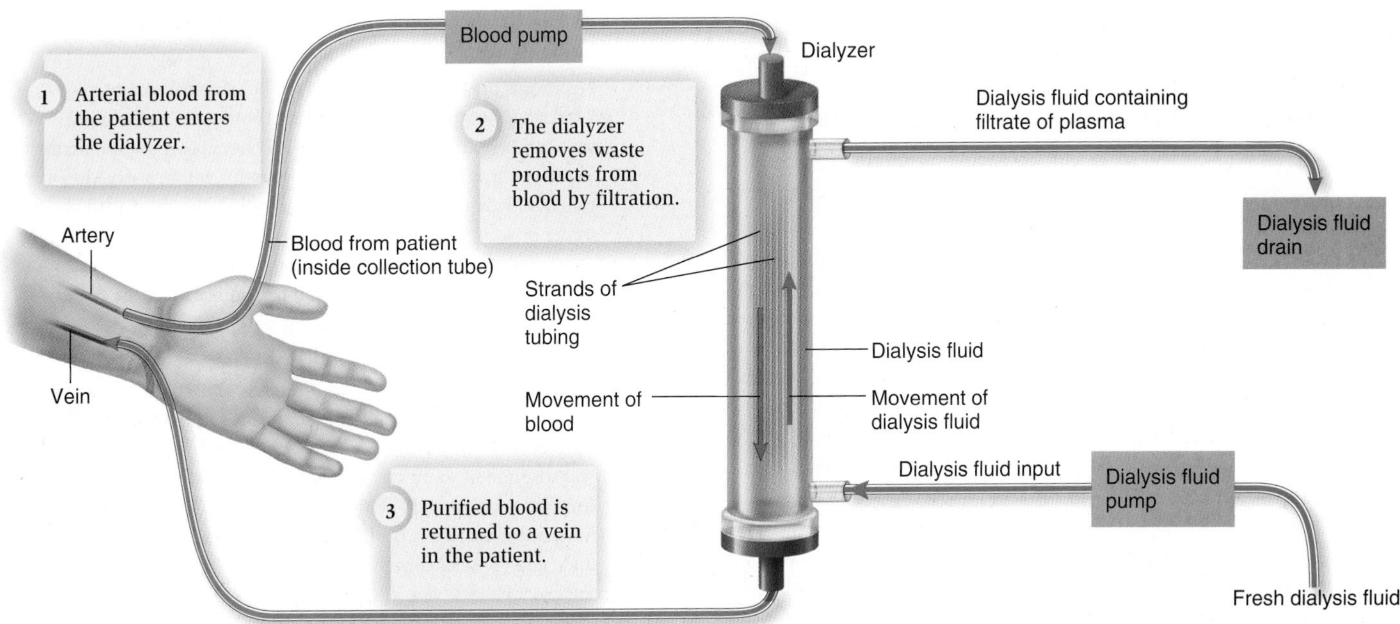

Figure 49.17 **Simplified diagram of hemodialysis.** The dialyzer is composed of many strands of very thin, sievelike tubing. In the dialyzer, blood within the dialysis tubing and the dialysis fluid bathing the tubing move in opposite directions (a countercurrent), which maximizes diffusion of substances out of the blood. The dialyzer provides a large surface area for diffusion of waste products out of the blood and into the dialysis fluid.

BioConnections: Note that modern medical technology is using a feature found in many animals—countercurrent exchange. Refer back to Figures 46.15 and 48.5 for examples of countercurrent exchange in animals, and to the text of this chapter for a discussion of a countercurrent multiplication system in the kidneys.

these problems; for example, lowering potassium intake reduces the amount of potassium to be excreted. However, such alterations may not eliminate the problems. In that case, doctors must use various procedures to artificially perform the kidneys' excretory functions.

The most important of these procedures is **hemodialysis**. The general term dialysis means to separate substances in solution using a porous membrane. In hemodialysis, blood from one of the patient's arteries is purified by redirecting it through a dialysis machine, which is called a dialyzer (Figure 49.17). Within the dialyzer, blood flows through cellophane tubing that is surrounded by a special dialysis fluid. The tubing is highly permeable to most solutes but relatively impermeable to protein and completely impermeable to blood cells. These characteristics are designed to be quite similar to those of the body's own capillaries. The dialysis fluid has ion concentrations similar to those in normal plasma but contains no urea or other substances that are to be completely removed from the plasma.

As blood flows through the tubing in the dialyzer, small solutes diffuse out into the dialysis fluid until an equilibrium is reached. If, for example, the patient's plasma potassium concentration is above normal, potassium diffuses out of the blood across the cellophane tubing and into the dialysis fluid. Similarly, waste products and excess amounts of other substances also diffuse into the dialysis fluid and thus are eliminated from the body. Note in Figure 49.17 that blood and dialysis fluid move in opposite directions through the dialyzer. This establishes an artificial countercurrent exchange system that increases the efficiency with which the blood is cleansed. The dialyzed, purified blood is then returned to one of the patient's veins through another type of tubing that leaves the dialyzer.

Some patients with reversible, temporary forms of kidney disease may require hemodialysis for only days or weeks. However, patients with chronic, irreversible kidney disease require treatment for the rest of their lives, unless they receive a kidney transplant. Such patients undergo hemodialysis several times a week. Each year nearly 400,000 Americans undergo some type of dialysis.

The treatment of choice for most patients with permanent kidney disease is kidney transplantation. Rejection of the transplanted kidney by the recipient's body is a potential problem with transplants, but great strides have been made in reducing the frequency of rejection. Many people who might benefit from a transplant, however, do not receive one, because the number of people needing a transplant far exceeds the number of donors. Currently, the major source of kidneys for transplanting is from recently deceased persons. Improved public understanding may lead many more individuals to give permission to have their kidneys and other organs used following their death. Recently, donation from a living, related donor has become more common, particularly with improved methods for preventing rejection. As noted earlier in this chapter, the mammalian kidney can perform its functions with only a fraction of its nephrons intact. Due to this large safety factor, a person who donates one of his or her kidneys can function quite normally with only one kidney.

Summary of Key Concepts

49.1 Principles of Homeostasis of Internal Fluids

- Exchanges of salt and water with the environment resulting from vital processes, such as respiration or the elimination of wastes, are called obligatory exchanges (Figure 49.1).
- Among the important products of the breakdown of proteins and nucleic acids are nitrogenous wastes, molecules that include nitrogen from amino groups (NH_2). Most aquatic animals produce ammonia (NH_3) and ammonium ions (NH_4^+), which are the most highly toxic nitrogenous wastes but require no energy to produce. Many animals, including all mammals, convert ammonia into urea, which is less toxic than ammonia and requires moderate expenditures of water and energy. Birds, insects, and most reptiles produce uric acid or purines. These nitrogenous wastes conserve water and are less toxic but energetically costlier than ammonia (Figure 49.2).
- The solute concentration of a solution of water is known as the solution's osmolarity. Fishes and other water-breathing animals that live in fresh water and those that live in salt water face opposite osmoregulatory challenges (Figures 49.3, 49.4).
- Robert Cade and coworkers discovered that fluid replacement during exercise is particularly beneficial if the fluid contains solutes at concentrations resembling those in sweat (Figure 49.5).
- Animals that maintain constant internal salt concentrations and osmolarities are called osmoregulators. Animals in which internal osmolarity conforms to the osmolarity of the environment are called osmoconformers.

49.2 Comparative Excretory Systems

- Most excretory organs operate by one or more of the following processes: (1) filtration, the removal of water and small solutes from the body fluids; (2) reabsorption, in which useful filtered solutes are returned to the body fluids via transport systems; (3) secretion, in which unnecessary solutes are actively transported from the blood or interstitial fluid and into the excretory tubule; and (4) excretion, in which waste is passed out of the body (Figure 49.6).
- In the protonephridial system of flatworms, a series of branching tubules filters fluids from the body cavity into the tubule lumens via the actions of ciliated flame cells (Figure 49.7).
- In the metanephridial system of annelids, pairs of metanephridia located in each body segment filter interstitial fluid and dilute urine is excreted via nephridiopores in the body wall (Figure 49.8).
- In insects, cells of Malpighian tubules secrete salts and uric acid from hemolymph into the tubule lumen; water follows by osmosis. After useful salts and water are reabsorbed into the hemolymph, wastes are excreted from the body (Figure 49.9).
- The mechanisms by which vertebrates osmoregulate vary depending on the animal's environment (Table 49.1).

49.3 Structure and Function of the Mammalian Kidney

- The urinary system in humans consists of the kidneys, ureters, urinary bladder, and urethra. Urine is excreted through the urethra. Each kidney is composed of an outer renal cortex and an inner renal medulla (Figure 49.10).
- Nephrons, the functional units of the kidney, are composed of a filtering component, called the renal corpuscle, and a tubule that empties into a collecting duct. Each renal corpuscle contains a cluster of capillaries called the glomerulus within a structure called Bowman's capsule. Each glomerulus is supplied with blood under pressure by an afferent arteriole. Each tubule of a nephron is composed of a proximal tubule, a loop of Henle, and a distal tubule. Tubules are surrounded by peritubular capillaries in the cortex and by the vasa recta in the medulla (Figure 49.11).
- Different portions of the tubule have different permeabilities to solutes and water. Most reabsorption of useful solutes occurs in the proximal tubule. Water and ion reabsorption continues along the loop of Henle, using a countercurrent exchange system. The reabsorption of solutes by the vasa recta minimizes the loss of solutes from the renal medulla. Fine-tuning of urine composition by the hormone aldosterone occurs in the distal tubule and upper collecting duct, and the final concentration of urine is determined by ADH in the lower collecting duct (Figures 49.12, 49.13, 49.14).
- Agre and coworkers discovered that water moves through plasma membranes through protein channels called aquaporins. Aquaporins regulate water reabsorption in the kidneys (Figures 49.15 and 49.16).

49.4 Impact on Public Health

- The symptoms of renal malfunction are similar and stem from uremia, the retention of urea and other waste products in the blood.
- One important treatment for kidney disease is hemodialysis, in which wastes in blood diffuse across a selectively permeable artificial membrane into a dialysis fluid (Figure 49.17).

Assess and Discuss

Test Yourself

1. A change in salt (electrolyte) concentrations in the body may result in
   a. altered membrane potentials that disrupt normal cell function.
   b. disruption of certain biochemical processes that occur in the cell.
   c. cell death.
   d. a and b only.
   e. a, b, and c.
2. Nitrogenous wastes are the by-products of the metabolism of
   a. carbohydrates. c. nucleic acids. e. both c and d.
   b. lipids. d. proteins.
3. Marine fishes avoid water balance problems by
   a. producing a large volume of dilute urine.
   b. not drinking water.
   c. having gill epithelial cells that recapture lost salts from the environment.
   d. producing small volumes of concentrated urine.
   e. a, b, and c.
4. Metabolic water is water
   a. necessary to stimulate the process of cellular respiration.
   b. found within cells.
   c. produced during cellular respiration.
   d. produced by sweat glands.
   e. used by cells during the uptake of glucose.

5. Animals that maintain a constant water balance despite changes in water concentrations in the environment are
   a. osmoregulators.
   b. osmoconformers.
6. The excretory system found in insects is composed of
   a. protonephridia.
   b. metanephridia.
   c. Malpighian tubules.
   d. a filtration kidney.
   e. a secretory kidney.
7. In the mammalian kidney, filtration is driven by
   a. solute concentration in the tubular filtrate.
   b. solute concentration in the blood.
   c. water concentration in the blood.
   d. water concentration in the tubular filtrate.
   e. hydrostatic pressure in the blood vessels of the glomerulus.
8. In the mammalian urinary system, the urine formed in the kidneys is carried to the urinary bladder by
   a. the collecting duct.
   b. the renal tubule.
   c. the renal pelvis.
   d. the ureters.
   e. the urethra.
9. Which of the following causes an increase in sodium reabsorption in the distal tubule?
   a. an increase in aldosterone concentrations
   b. an increase in antidiuretic hormone concentrations
   c. a decrease in aldosterone concentrations
   d. a decrease in antidiuretic hormone concentrations
   e. none of the above
10. Aquaporins are
    a. ion channels.
    b. water channels.
    c. receptors for aldosterone.
    d. small pores in the fenestrated capillaries of the glomerulus.
    e. both a and c.

Conceptual Questions

1. Define nitrogenous wastes, and list four types. What are some advantages and disadvantages of excreting different types of nitrogenous wastes?
2. List and define the three processes involved in urine production. Do all three processes occur for every substance that enters an excretory organ such as the kidney?
3. A principle of biology is that *living organisms maintain homeostasis.* Explain how salt glands are a homeostatic adaptation for marine life.

Collaborative Questions

1. Discuss two different types of filtration mechanisms found in invertebrates.
2. Briefly discuss the parts and functions of the nephron in the mammalian kidney.

Online Resource

www.brookerbiology.com

Stay a step ahead in your studies with animations that bring concepts to life and practice tests to assess your understanding. Your instructor may also recommend the interactive eBook, individualized learning tools, and more.

Endocrine Systems

50

Chapter Outline

50.1 Types of Hormones and Their Mechanisms of Action

50.2 Links Between the Endocrine and Nervous Systems

50.3 Hormonal Control of Metabolism and Energy Balance

50.4 Hormonal Control of Mineral Balance

50.5 Hormonal Control of Growth and Development

50.6 Hormonal Control of Reproduction

50.7 Hormonal Responses to Stress

50.8 Impact on Public Health

Summary of Key Concepts

Assess and Discuss

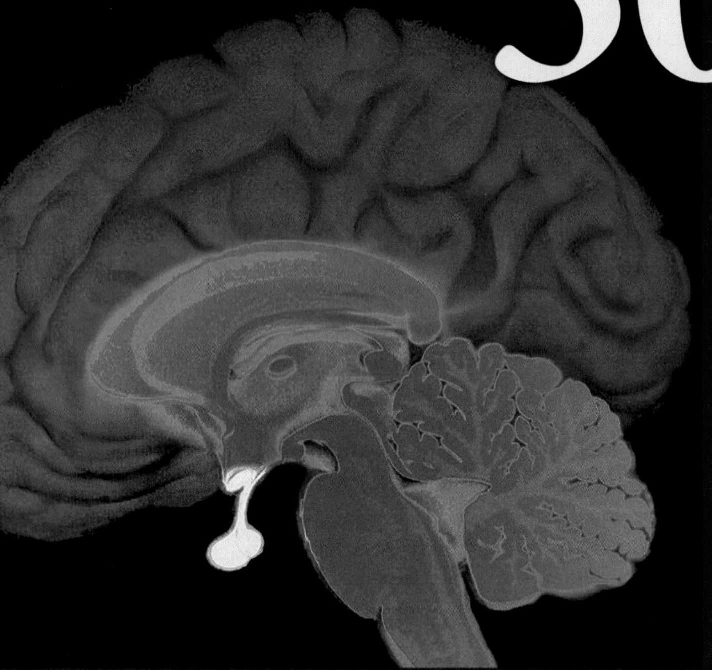

A section through a human brain, highlighting the pituitary gland and its connection to the hypothalamus (white). Both of these structures secrete numerous hormones. (Image is a three-dimensional MRI.)

A 22-year-old man was seen by his physician because of a complaint of hair loss and the appearance of severe acne over much of his face, neck, back, and shoulders; he also reported feelings of irritability and occasional aggression. Upon examination, the man was found to have additional symptoms, including hypertension (high blood pressure), an increased plasma cholesterol concentration, an increased hematocrit (red blood cell count; see Chapter 47), and, alarmingly, shrunken testes. When questioned, the patient admitted that for 6 months he had been taking an oral form of the illegal drug stanozolol in an effort to improve his physique by building more muscle mass. Stanozolol is a synthetic version of a hormone known as an androgen (from the Greek, *andros*, meaning man, and *genein*, meaning to produce). A **hormone** is a chemical produced by cells in the body, which circulates in the bloodstream and acts on one or more target tissues to produce one or more functions.

In males, androgens are normally produced by the testes, which are reproductive organs that contain endocrine glands. The most well-known androgen is testosterone. In normal amounts, androgens play a key role in the physiological events associated with puberty and reproduction, including the increase in skeletal muscle mass that accompanies male puberty. This young man hoped to increase his muscle mass by consuming large amounts of synthetic androgen. At high concentrations, however, androgens promote increased red blood cell production, which, in turn, makes the heart work harder and can lead to hypertension. Likewise, high concentrations of androgens increase the cholesterol concentration in the blood, promote fluid retention, damage the liver, cause hair loss, and increase the activity of the sebaceous glands (the oil-producing glands in the skin that make acne worse). At such concentrations, androgens also strongly inhibit the activity of the testes. In this last phenomenon, the body senses that sufficient androgens are already available in the blood; the testes do not make more androgens and thus shrink in size.

Fortunately, this person was educated by his doctor about the consequences of misuse of these powerful hormones, and he discontinued the practice. Had he continued to take androgens, all of his symptoms would have worsened, and he would have run the risk of serious and irreversible damage to his heart, liver, and other organs, while increasing his risk of cancer and other diseases.

Androgens and other hormones are found in all vertebrates and many invertebrates. Hormones are often produced in response to a homeostatic challenge, such as a change in an animal's blood pressure or body temperature. A chief function of hormones is to counter these challenges and maintain homeostasis. Not surprisingly, therefore, hormones affect a wide range of body functions, including gastrointestinal activity, blood pressure regulation, cholesterol balance, fluid and mineral balance, and reproduction, among others.

Hormones are made by cells in nearly all the body's organs. In addition, hormone-producing cells are often found in specialized glands, called **endocrine glands**, whose primary function is hormone synthesis and secretion. An example is the pituitary gland highlighted in the chapter-opening photo. Collectively, all the endocrine glands and other organs with hormone-secreting cells constitute the **endocrine system**.

In this chapter, we will first learn about the chemical nature of hormones and their mechanisms of action, how the endocrine and nervous systems interact, and the ways in which hormones influence such diverse functions as metabolism, growth, and reproduction. We conclude with a discussion of how hormones affect human health, including how synthetic hormones are misused, as in the example just described.

50.1 Types of Hormones and Their Mechanisms of Action

Learning Outcomes:

1. Define what a hormone is.
2. List and give examples of the three different chemical classes of hormones.
3. Describe the cellular location of receptors for lipid-soluble hormones and receptors for water-soluble hormones.
4. Describe the factors that regulate the concentration of a hormone circulating in the blood.

Figure 50.1 is an overview of the major endocrine glands and their hormones; you should take a moment and review this figure as preparation for the rest of the chapter. In this section, we will examine some of the general characteristics of hormones, how they act, and how they are controlled. Keep in mind that endocrine signaling is just one type of cell-to-cell communication. Refer back to Figure 9.3 for a review of the different types of cell-signaling mechanisms found in animals.

The Three Classes of Hormones Differ in Composition

Hormones fall into three broad classes: the amines, proteins/peptides, and steroid hormones (**Table 50.1**). The amines and the proteins/peptides generally share similar chemical properties and mechanisms of action, whereas the steroid hormones act very differently from the other two classes.

The amine hormones are derived from an amino acid, either tyrosine or tryptophan. As shown in **Figure 50.2**, tyrosine is the

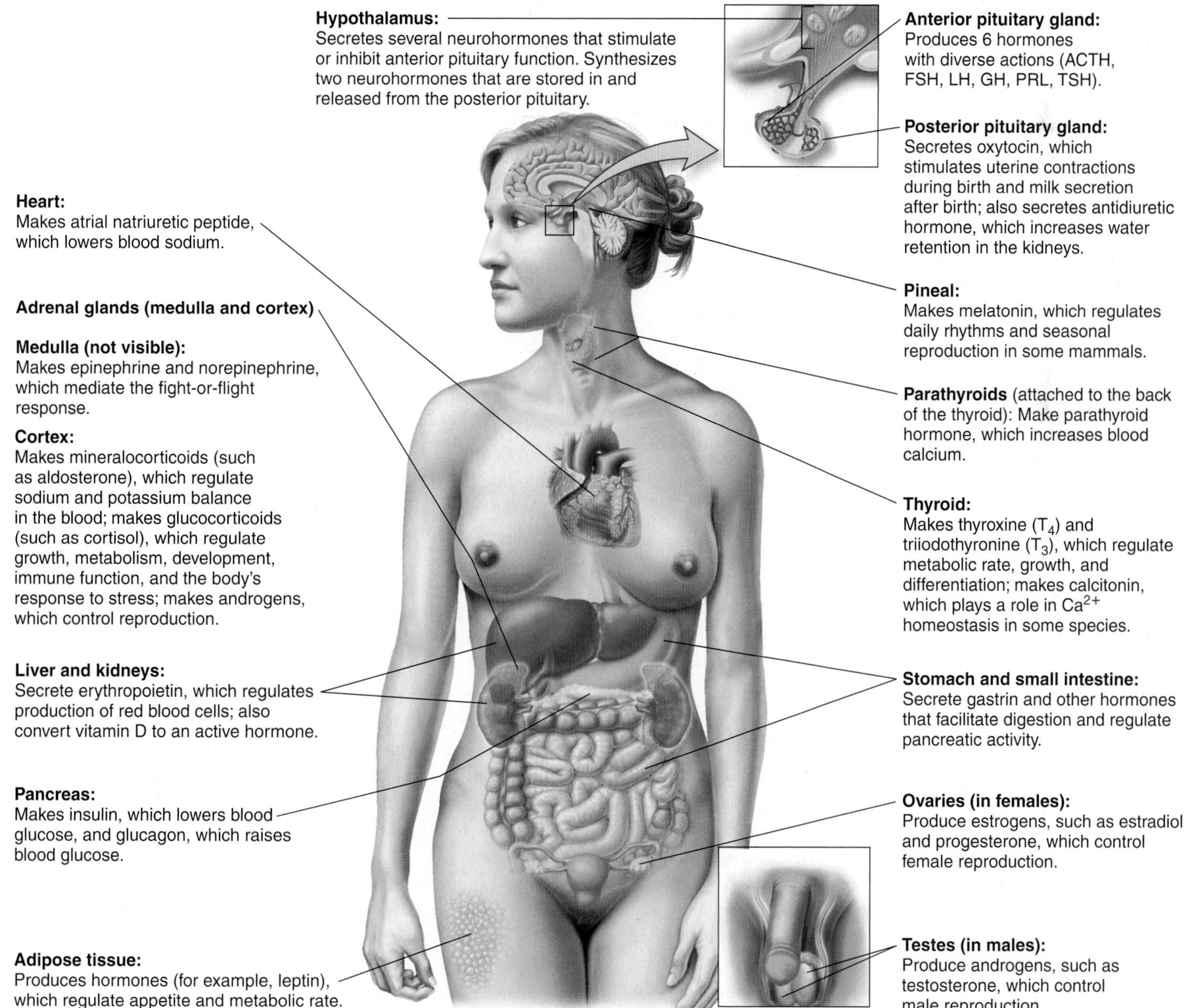

Hypothalamus:
Secretes several neurohormones that stimulate or inhibit anterior pituitary function. Synthesizes two neurohormones that are stored in and released from the posterior pituitary.
Anterior pituitary gland:
Produces 6 hormones with diverse actions (ACTH, FSH, LH, GH, PRL, TSH).
Posterior pituitary gland:
Secretes oxytocin, which stimulates uterine contractions during birth and milk secretion after birth; also secretes antidiuretic hormone, which increases water retention in the kidneys.
Heart:
Makes atrial natriuretic peptide, which lowers blood sodium.
Pineal:
Makes melatonin, which regulates daily rhythms and seasonal reproduction in some mammals.
Adrenal glands (medulla and cortex)
Medulla (not visible):
Makes epinephrine and norepinephrine, which mediate the fight-or-flight response.
Cortex:
Makes mineralocorticoids (such as aldosterone), which regulate sodium and potassium balance in the blood; makes glucocorticoids (such as cortisol), which regulate growth, metabolism, development, immune function, and the body's response to stress; makes androgens, which control reproduction.
Parathyroids (attached to the back of the thyroid): Make parathyroid hormone, which increases blood calcium.
Thyroid:
Makes thyroxine (T_4) and triiodothyronine (T_3), which regulate metabolic rate, growth, and differentiation; makes calcitonin, which plays a role in Ca^{2+} homeostasis in some species.
Liver and kidneys:
Secrete erythropoietin, which regulates production of red blood cells; also convert vitamin D to an active hormone.
Stomach and small intestine:
Secrete gastrin and other hormones that facilitate digestion and regulate pancreatic activity.
Pancreas:
Makes insulin, which lowers blood glucose, and glucagon, which raises blood glucose.
Ovaries (in females):
Produce estrogens, such as estradiol and progesterone, which control female reproduction.
Adipose tissue:
Produces hormones (for example, leptin), which regulate appetite and metabolic rate.
Testes (in males):
Produce androgens, such as testosterone, which control male reproduction.

Figure 50.1 **Overview of the endocrine system in humans.** This figure shows many of the major endocrine glands and other structures that constitute the human endocrine system, along with the major functions of the hormones produced by those glands.

Table 50.1 Chemical Classes of Hormones

Class	Chemical properties	Location of target cell receptor	Mechanism of action	Examples
Amines	Derived from tyrosine or tryptophan; small, water-soluble (except thyroid hormones, which are lipophilic)	Plasma membrane (except thyroid hormones, which act via intracellular receptors)	Stimulate second-messenger pathways (except thyroid hormones, which act via changes in gene transcription)	Epinephrine, norepinephrine, dopamine, thyroid hormones, melatonin
Proteins/Peptides	Water-soluble	Plasma membrane	Stimulate second-messenger pathways	Insulin, glucagon, leptin
Steroids	Derived from cholesterol, mostly lipid-soluble	Cytosol or nucleus	Usually stimulate gene transcription directly	Aldosterone, cortisol, testosterone, estradiol

precursor for the hormones epinephrine and norepinephrine, which are produced in the adrenal medulla and are important in the body's response to stress (the fight-or-flight response, discussed later), and for dopamine, a hormone made by the brain. Tyrosine is also the chemical backbone of the hormones made by the thyroid gland, which are important regulators of metabolic rate, growth, and development. The major hormone derived from tryptophan is melatonin, which is produced within an endocrine gland called the pineal gland located in the brain. Melatonin is important for controlling circadian rhythms, daily cycles of physiological or biochemical processes. In some mammals such as sheep and hamsters that breed only at certain times of year, melatonin controls seasonal cycles of reproduction.

Proteins and small polypeptides are the most abundant class of hormones. These hormones participate in numerous body functions such as metabolism, mineral balance, growth, and reproduction. Examples include insulin and glucagon made by the pancreas, and leptin made by adipose tissue.

Steroid hormones are synthesized from cholesterol, and thus all steroid hormones are lipids, unlike the other classes of hormones. Steroids are less soluble in water than are amines or protein/peptide hormones. Due to this limited solubility, steroids are usually bound to large, soluble proteins in the blood that serve as carriers. By combining with these proteins, steroids can reach high concentrations in the blood. The major steroid hormones found in vertebrates include aldosterone and cortisol made by the adrenal cortex, androgens made by the testes, and estrogens and progesterone made by the ovaries. In insects, 20-hydroxyecdysone is a major steroid hormone that has an important role in the processes of molting and metamorphosis.

Hormones Can Have Short-Term or Long-Term Effects

The effects of a given hormone may occur within seconds or require several hours to develop, and they may last for as short as a few minutes or as long as days. This is a key difference between the communication processes in the nervous and endocrine systems. In the nervous system, signals are transmitted from one cell to another within milliseconds, and the effect on the postsynaptic cell occurs immediately. In the endocrine system, one hormone may act on a cell quickly and for a very short time, whereas another may act very slowly and have a lingering action. The explanation for these differences lies in the mechanisms by which different hormones act.

Hormones Act Through Plasma Membrane or Intracellular Receptors

The amine and protein/peptide hormones are generally water-soluble, and the steroid hormones are lipid-soluble. Therefore, the amines and proteins/peptides are not able to cross plasma membranes and must bind to a receptor protein on the surface of a target cell. Steroid hormones, however, being lipids, can diffuse across the target cell's plasma membranes and bind to a receptor protein in either the cytosol or nucleus. Hormones that act through plasma membrane receptors tend to elicit fast responses, whereas those that act via intracellular receptors generally take longer.

Water-Soluble Hormones and Their Membrane Receptors All the amine and protein/peptide hormones, with one exception (thyroid hormones, discussed later), act by binding to a receptor protein located in the plasma membrane (refer back to Figure 9.5). Only cells having the proper receptors on their surfaces can respond to the hormone. Thus, although a hormone travels throughout the entire circulatory system, it activates only specific cells. The hormone interacts noncovalently and reversibly with the receptor. The reversibility of the binding between hormone and receptor is one way in which cells are prevented from being permanently stimulated.

Among cells throughout the body, different receptor proteins may recognize the same hormone. These different receptors, called subtypes or isoforms, may be the product of different genes or may be produced by alternative splicing, described in Chapter 13 (refer back to Figure 13.21). By binding to different receptor proteins, the same hormone is able to elicit differing, sometimes even opposite, responses, depending on where it binds in the body. These isoforms may have different affinities for the hormone, such that low concentrations of a hormone may cause one type of effect by binding to high-affinity receptors on one cell, and higher concentrations may cause entirely different actions by binding to low-affinity receptors on another cell. In this way, most hormones are able to serve more than one function.

The binding of a water-soluble hormone to a plasma membrane receptor initiates intracellular signaling pathways that often involve second messengers (refer back to Figures 9.10, 9.12 and 9.16). Three major signaling pathways that are activated by water-soluble hormones are those involving cyclic AMP, diacylglycerol and inositol trisphosphate, and receptor tyrosine kinases. These signal

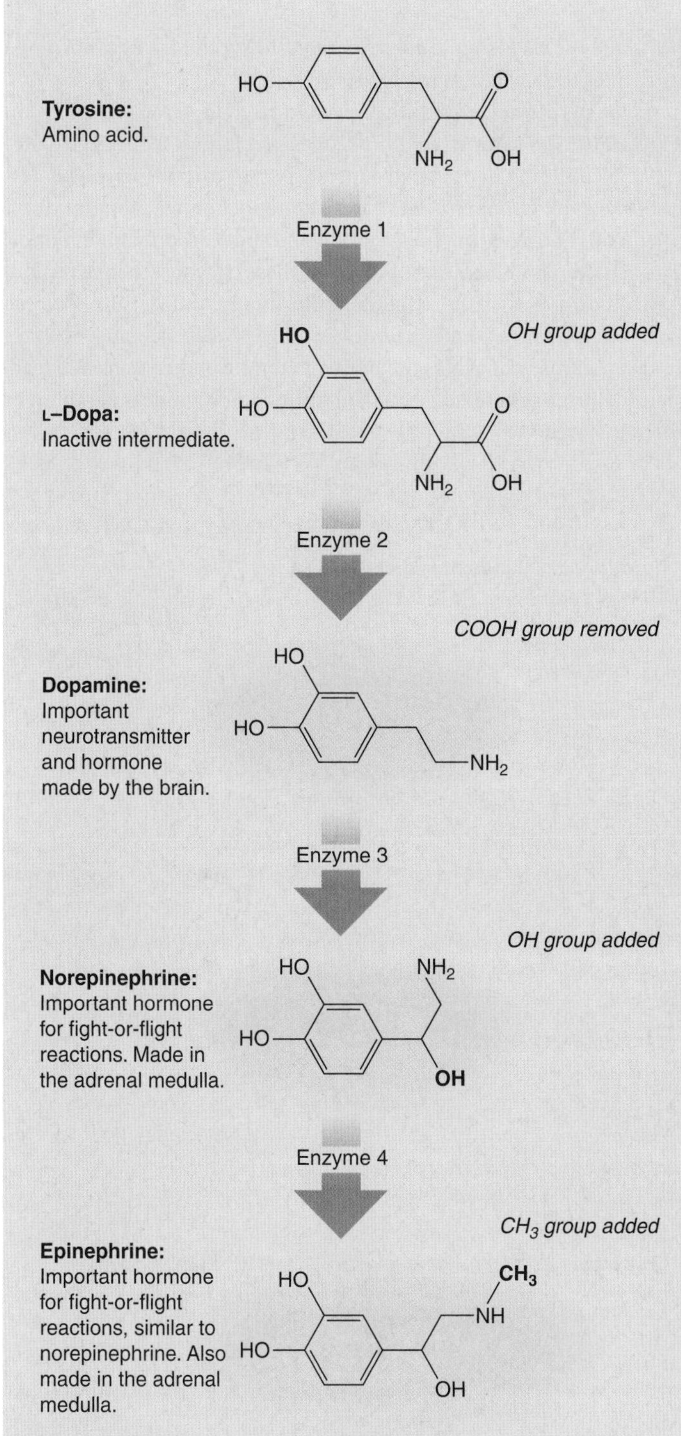

Figure 50.2 **Synthesis of the amine hormones dopamine, norepinephrine, and epinephrine.** The synthesis of thyroid hormones also begins with tyrosine but follows a different mechanism, as described in Figure 50.5b.

transduction processes may be rapid, occurring in some cases within seconds, and involve changing the activity of enzymes. Both of these features are important, because occasionally the rapidity of the cell response to a hormone can mean the difference between life and death, as may occur, for example, under fight-or-flight conditions.

The presence of enzymes in the signaling process ensures that the signal from a small amount of hormone is greatly amplified, because a single signaling molecule results in the production of many intracellular messengers (refer back to Figure 9.14). In this way, endocrine glands may synthesize and secrete relatively small amounts of hormone while still producing biological effects; this conserves energy and resources.

Activation of signaling pathways may also lead to changes in cellular activities that occur more slowly. These slower changes usually require activation or inhibition of genes in the nucleus, and are mediated by transcription factors (see Chapter 12) that are activated by the same signaling pathways described earlier. However, another means of altering gene transcription is through activation of receptors for lipid-soluble hormones.

Lipid-Soluble Hormones and Their Intracellular Receptors All steroid hormones and the thyroid hormones are lipophilic and bind to intracellular receptors. The complex of a steroid or thyroid hormone bound to its intracellular receptor functions as a transcriptional activator (or less frequently as an inhibitor) by binding to enhancers of particular genes (refer back to Figure 9.9). Once bound, transcription of a gene is increased, which increases the amount of that gene's protein product. These proteins may be important in a variety of cellular activities, such as regulating the number of ion pumps in membranes, and controlling cell differentiation, secretory activity, and growth.

Steroid and thyroid hormones can influence several genes within a single cell or different cells. In this way, one hormone can exert a variety of actions throughout the body. The physical changes that accompany puberty in mammals result from the actions of two steroid hormones—androgens in males and estrogens in females—and are among the most striking and commonly recognized examples of this widespread action.

Hormone Concentrations Depend on Rates of Synthesis and Removal

When hormones are produced by endocrine cells, they are released into the interstitial fluid between the cell and adjacent blood vessels. The hormones then diffuse into the blood. Although most hormones circulate in the blood at all times, their concentrations can increase or decrease dramatically when necessary. This can be accomplished in two major ways: by changing the rate of hormone production and release by an endocrine cell and by changing the rates at which hormones are inactivated or removed from the blood.

How hormones are synthesized and the rate of their synthesis differ for each of the three classes of hormones. Synthesis of amine hormones is mediated by enzymatic conversion of either tyrosine or tryptophan into their respective hormones. These reactions require several enzymes that add or remove chemical side groups to achieve the final product (see Figure 50.2). The synthesis of the hormone depends on the amounts and activities of the synthesizing enzymes. These enzymes are always present in the cell, but they can be greatly stimulated when additional hormone is required.

The protein and peptide hormones are generally synthesized at a steady rate or circadian rhythm in an unstimulated cell, until additional hormone is required. In that case, transcription factors within the cell direct the increased transcription of the gene coding for the hormone in question. Conversely, when less hormone is required, gene transcription is slowed or stopped. These hormones are too large and hydrophilic to diffuse across the plasma membrane and then into the extracellular fluid. Instead, they are packaged into secretory vesicles in much the same way as neurotransmitters are packaged in neuron axon terminals. This packaging provides a ready means of secreting the hormones by exocytosis and a reservoir of stored hormone available for immediate release when required. When a cell is stimulated to secrete its stored protein or peptide hormones, it also is typically stimulated to synthesize new hormone molecules to replace them.

The steroid hormones are derived from cholesterol, as depicted in Figure 50.3. To form steroid hormones, a cell must express enzymes capable of adding hydroxyl groups to specific carbons in the cholesterol skeleton. Other enzymes carry out reduction reactions, and still others can both hydroxylate a carbon and split carbon-carbon bonds. The activities and amounts of many of these enzymes change if more or less hormone is required. Steroid hormones, unlike water-soluble hormones, are not packaged into secretory vesicles because the steroid can diffuse out across the lipid membrane of the vesicle. Instead, steroid hormones are made on demand, and no significant amount of them is stored.

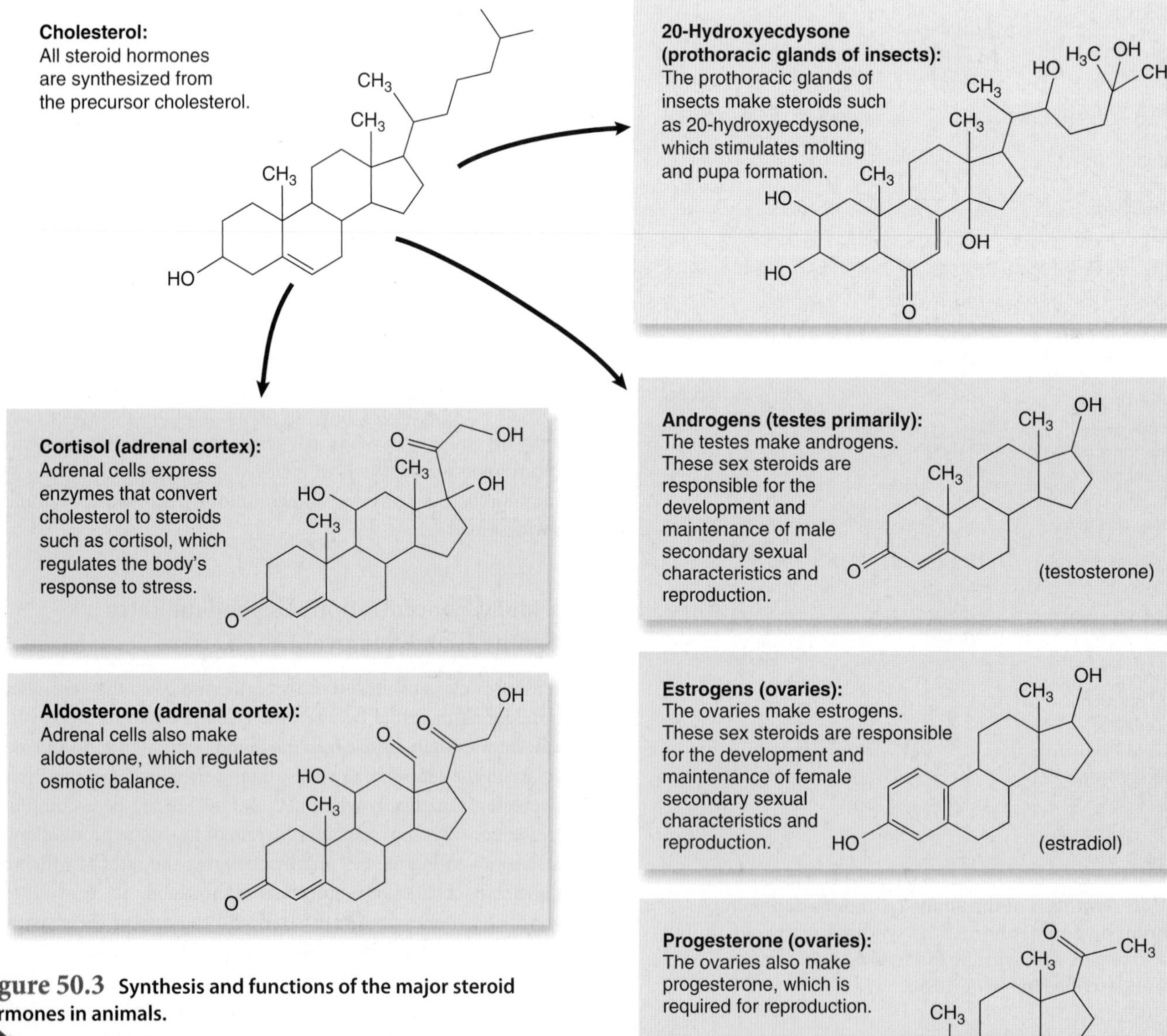

Figure 50.3 **Synthesis and functions of the major steroid hormones in animals.**

BIOLOGY PRINCIPLE Structure determines function. Notice how slight changes in the arrangements of side groups attached to the four-ring backbone create products with completely different functions. Keep in mind that the three-dimensional shape of a molecule may be changed much more by such chemical modifications than is visible in a two-dimensional depiction.

Although hormones carry out vital functions, excessive stimulation of cells by hormones may produce detrimental effects. For example, hormones that are activated if an animal loses blood typically help restore blood volume and pressure. This cannot continue indefinitely, however, because once blood volume and pressure are restored, it would be harmful if they continued to increase until their levels were above normal. Therefore, once a hormone enters the blood and performs its functions, it is usually prevented from exerting its effects indefinitely. This is accomplished in one or more ways:

- Hormones that bind to plasma membrane receptors may be engulfed by endocytosis into a cell, where lysosomal enzymes degrade the hormones.
- Small, water-soluble hormones are excreted in the urine.
- The liver chemically modifies many hormones to render them inactive and more easily excretable via the kidneys.
- Negative feedback processes (see Figure 40.13) turn off the signals that were responsible for stimulating the synthesis and secretion of the hormone.

Generally, these processes ensure that hormone concentrations in the blood remain within a normal range under most circumstances, but have the capacity to be increased or decreased beyond that range if required. One of the ways in which changes in hormone concentrations are initiated is through sensory input to an animal's brain. As we see next, the nervous system and endocrine system are functionally linked in many animals, including all vertebrates.

50.2 Links Between the Endocrine and Nervous Systems

Learning Outcomes:

1. Describe the anatomical connections among the hypothalamus, posterior pituitary, and anterior pituitary.
2. Describe the role of the hypothalamus and anterior pituitary gland in regulation of endocrine function.
3. List some of the major functions of the hormones of the posterior pituitary.

A key feature of the endocrine system in most animals is that the concentrations of many hormones in the extracellular fluid rise and fall with changes in an animal's environment. For this to happen, the hormone-producing cells of the endocrine system must somehow be informed of environmental changes. This often occurs when sensory input is received by an animal's nervous system, which, in turn, modulates the activity of one or more endocrine glands.

Sensory stimuli detected by the nervous system can activate the endocrine system. For example, when an antelope detects the presence of a nearby lioness, visual and olfactory sense information is relayed to the antelope's brain. The brain initiates responses in certain endocrine glands that release hormones to prepare the antelope for the possibility of an attack. As another example, the concentrations of several hormones fluctuate in the blood of certain fishes as they migrate back and forth between feeding grounds in the sea and freshwater spawning sites. These hormones are activated by different salinities in the environment, which are detected by sensory cells of the fish's nervous system. The hormones act to prepare the gills of the fish to handle the large changes in salinity of the water.

The common feature of these examples and many others is that a sensory cue, such as a predator or the salinity of water, must be perceived by a sensory receptor and converted into an endocrine response. Electrical signals are transmitted from the sensory receptors to different parts of the brain, including the hypothalamus. In this section, we explore how the hypothalamus and pituitary gland play the major roles in linking the nervous and endocrine systems.

The Hypothalamus and Pituitary Gland Are Physically Connected

As described in Chapter 42, the **hypothalamus** is a collection of several nuclei that are located at the bottom surface of the vertebrate brain (Figure 50.4a). Neurons in these nuclei are connected to an endocrine gland sitting directly below the hypothalamus, called the **pituitary gland** (see the bottom part of the highlighted structure in the chapter-opening photo). The pituitary gland in humans is made up of two lobes, the anterior and posterior lobes. Some vertebrate species have an intermediate lobe that is believed to be vestigial in primates.

The hypothalamus and pituitary are connected by a thin piece of tissue called the **infundibular stalk** and also by a system of blood vessels called portal veins. **Portal veins** differ from ordinary veins because not only do they collect blood from capillaries—like all veins do—but they then also form another set of capillaries, as opposed to returning the blood directly to the heart like other veins. The portal veins extend through the length of the infundibular stalk. Within the anterior lobe of the pituitary gland—often simply called the anterior pituitary gland—the portal veins empty into a second set of capillaries. This arrangement of blood vessels bypasses the general circulation and allows the hypothalamus to communicate directly with the anterior pituitary gland. Let's now explore the nature of this communication.

The Hypothalamus and Anterior Pituitary Gland Have Integrated Functions

As described in Chapter 42, the hypothalamic nuclei are vital for such diverse functions as reproduction, circadian rhythms, appetite, metabolism, and responses to stress. The hypothalamus has such wide-ranging effects in part because it acts as a master control, signaling the pituitary gland when to produce and secrete its many hormones. However, the hypothalamus communicates differently with the anterior and posterior lobes of the pituitary gland. Let's look at the interaction between the hypothalamus and anterior pituitary first.

Within the different nuclei of the hypothalamus are neurons that synthesize a class of hormones called neurohormones. A **neurohormone** is any hormone that is made in and secreted by neurons. All neurohormones are either amines or peptides. Although they are produced within neurons, these molecular signals are not referred to as neurotransmitters, because the endings of the neurons do not terminate in a synapse with another cell. Instead, the axon terminals from

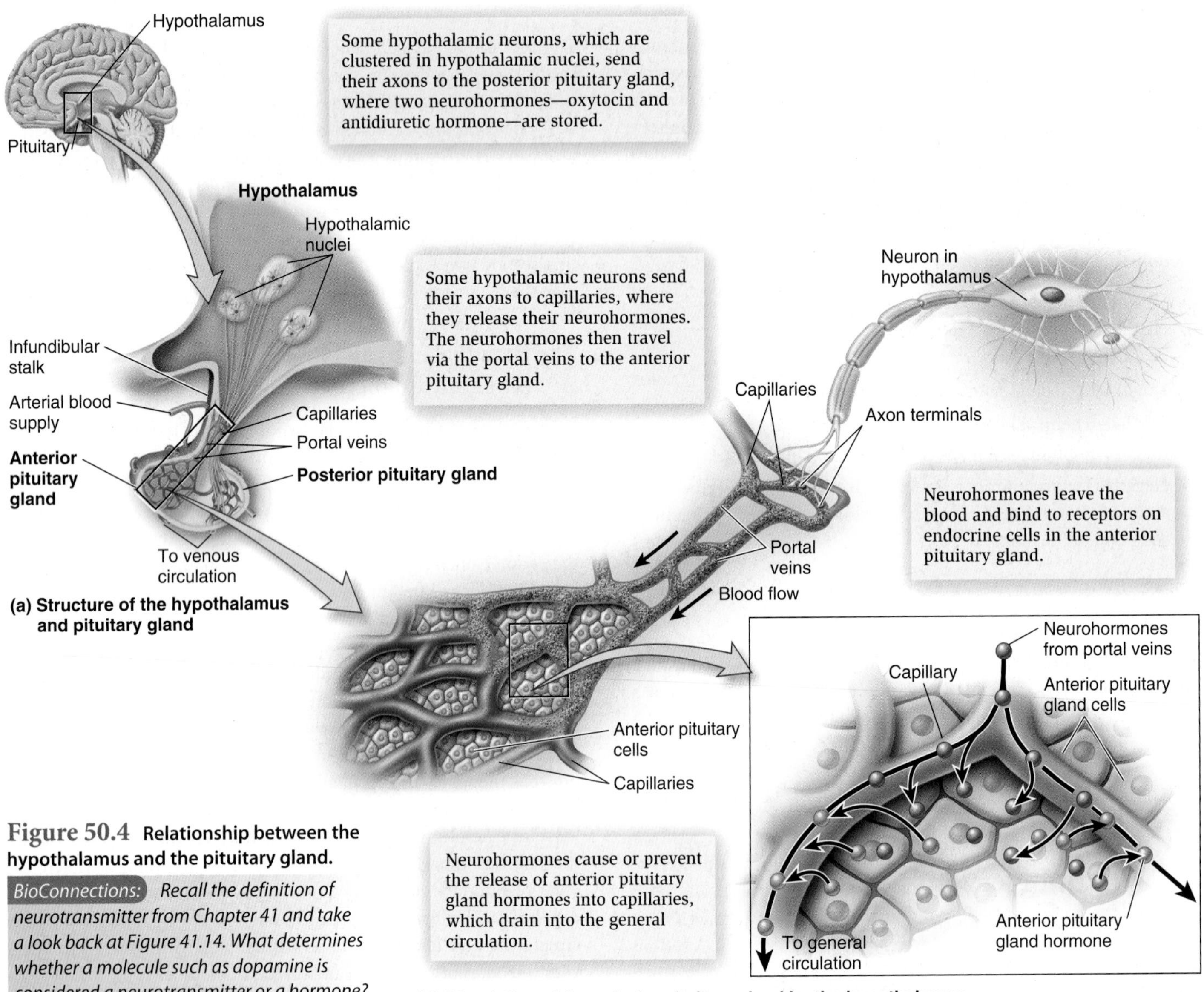
Hypothalamus
Pituitary
Some hypothalamic neurons, which are clustered in hypothalamic nuclei, send their axons to the posterior pituitary gland, where two neurohormones—oxytocin and antidiuretic hormone—are stored.
Hypothalamus
Hypothalamic nuclei
Infundibular stalk
Arterial blood supply
Anterior pituitary gland
Capillaries
Portal veins
Posterior pituitary gland
To venous circulation
(a) Structure of the hypothalamus and pituitary gland
Some hypothalamic neurons send their axons to capillaries, where they release their neurohormones. The neurohormones then travel via the portal veins to the anterior pituitary gland.
Neuron in hypothalamus
Capillaries
Axon terminals
Neurohormones leave the blood and bind to receptors on endocrine cells in the anterior pituitary gland.
Portal veins
Blood flow
Anterior pituitary cells
Capillaries
Neurohormones from portal veins
Capillary
Anterior pituitary gland cells
Neurohormones cause or prevent the release of anterior pituitary gland hormones into capillaries, which drain into the general circulation.
To general circulation
Anterior pituitary gland hormone
(b) Stimulation of the anterior pituitary gland by the hypothalamus

Figure 50.4 Relationship between the hypothalamus and the pituitary gland.

BioConnections: Recall the definition of neurotransmitter from Chapter 41 and take a look back at Figure 41.14. What determines whether a molecule such as dopamine is considered a neurotransmitter or a hormone?

the hypothalamus end next to capillaries. Here, the neurons secrete their neurohormones into the capillaries, which, in turn, collect into the portal veins. This allows neurohormones to be delivered directly from the hypothalamus to the cells of the anterior pituitary gland in a quick, efficient manner (Figure 50.4b).

In response to the presence of these hypothalamic neurohormones, the anterior pituitary gland synthesizes several different hormones. Table 50.2 lists the six major anterior pituitary hormones that have well-defined functions in vertebrates and the neurohormones that either stimulate or inhibit them. Recent research suggests the possibility for dual stimulatory and inhibitory control of several pituitary gland hormones, but this remains uncertain for most species. The stimulatory action of several of the hypothalamic neurohormones has historically led to them also being known as "hypothalamic-releasing hormones," because they cause the release, or secretion, of other hormones from the anterior pituitary. The six major hormones of the anterior pituitary gland are secreted into the general blood circulation, where they act on other endocrine glands or structures.

Understanding the function of the hormones of the anterior pituitary gland is an ongoing and active area of research. Several other peptides from the anterior pituitary have been characterized and suggested to be hormones, but their functions have not yet been clearly defined in any vertebrate species. One example is β-endorphin, which is released into the blood during times of stress. Because it is related in structure to the opiate-type molecules such as morphine, it has been suggested that β-endorphin may act in some species as a painkiller during times of stress, but this is uncertain.

The Posterior Pituitary Contains Axon Terminals from Hypothalamic Neurons That Store and Secrete Oxytocin and Antidiuretic Hormone

The posterior pituitary has a blood supply, but, in contrast to the anterior pituitary gland, it is not connected to the hypothalamus by portal veins and does not respond to neurohormones from the hypothalamus. Instead, the posterior pituitary is an extension of the

Table 50.2 Hormones of the Anterior Pituitary Gland and Hypothalamus

Anterior pituitary gland hormone	Stimulatory neurohormone from hypothalamus	Inhibitory neurohormone from hypothalamus	Major functions
Adrenocorticotropic hormone (ACTH)	Corticotropin-releasing hormone (CRH)	None known	Stimulates adrenal cortex to make glucocorticoids
Follicle-stimulating hormone (FSH)	Gonadotropin-releasing hormone (GnRH)	None known	Stimulates germ cell development and sex steroid production in gonads
Luteinizing hormone (LH)	GnRH	None known	Stimulates release of eggs in females; stimulates sex steroid production from gonads
Growth hormone (GH)	Growth hormone-releasing hormone (GHRH)	Somatostatin	Promotes linear growth; regulates glucose and fatty acid balance in blood
Prolactin (PRL)	TRH and other factors have been suggested as stimulators of PRL, but this is not unequivocal	Dopamine	Stimulates milk formation in mammals; participates in mineral balance in other vertebrates
Thyroid-stimulating hormone (TSH)	Thyrotropin-releasing hormone (TRH)	None known	Stimulates thyroid gland to make thyroid hormones

hypothalamus that lies in close contact with the anterior pituitary gland (see Figure 50.4). Axons from neurons extend from the hypothalamus into the posterior pituitary. In mammals, the axon terminals in the posterior pituitary store one of two hormones, oxytocin or antidiuretic hormone, that are produced by the cell bodies of those neurons. When the hypothalamus receives information that these hormones are required, they are released directly from the neuron axon terminals into the bloodstream.

Oxytocin concentrations increase in the blood of pregnant mammals just prior to birth. It stimulates contractions of the smooth muscles in the uterus, which facilitates the birth process and shortly afterward helps expel the placenta. Oxytocin stimulates the secretion of milk from the mammary glands of lactating females. When the mother's nipples are stimulated by the suckling of a newborn, neurons transmit a signal from there to the mother's hypothalamus, which stimulates the cells that produce oxytocin. Oxytocin is released from the posterior pituitary into the blood where it travels to smooth muscle cells surrounding the secretory components of the mammary glands. This stimulates the release of milk. Thus, oxytocin has two important and different functions, one during birth and one during lactation. In both cases, the hormone stimulates contraction of muscle cells. Recent evidence also supports a role of oxytocin in maternal/infant bonding and prosocial behavior in some species; whether this is true in humans, however, is unknown.

Antidiuretic hormone (ADH) gets its name because it acts on kidney cells to decrease urine production—a process known as antidiuresis. Diuresis is an increased loss of water in the urine, as happens, for example, when you drink large amounts of fluids. If the fluid content of the body is low, for example during dehydration or after a significant loss of blood, ADH is secreted into the blood from the posterior pituitary. It acts to increase the number of water-channel proteins called aquaporins (refer back to Figure 49.15) present in the membranes of kidney tubule cells; water is recaptured from the forming urine through these aquaporins. Minimizing the volume of water used to form urine is an adaptation that conserves body water when necessary.

At high concentrations, ADH also increases blood pressure by stimulating vasoconstriction of blood vessels, a function that accounts for the other common name of ADH, vasopressin. Like oxytocin, therefore, ADH has more than one function. Both of the major functions of ADH are related in that they contribute to maintaining blood pressure and fluid levels in the body. Oxytocin and ADH are well-studied examples of the evolution of hormones. Although these two hormones are found only in mammals, many invertebrates and all nonmammalian vertebrates secrete one or more peptides that are chemically similar to oxytocin and ADH but are not identical to the mammalian hormones. One of these is **vasotocin**, which combines some of the chemical structure of both oxytocin and ADH. Research indicates that an ancestral vasotocin gene duplicated at some point, and then the two genes evolved into the oxytocin and ADH genes found in mammals. Because only mammals lactate, the role of vasotocin must be different in birds, fishes, and other vertebrates than that of oxytocin in mammals. Research has shown that vasotocin is responsible for regulating salt and water balance in the blood of nonmammalian vertebrates. The observation that members of the vasotocin/oxytocin/ADH gene family arose early in animal evolution also suggests that oxytocin and ADH may have additional, unrecognized actions that have been retained in mammals. For example, human males have oxytocin in their blood, but its role cannot be the same as in females because, of course, only females give birth and lactate. The key roles of oxytocin in male humans and other animals have not been unequivocally identified.

50.3 Hormonal Control of Metabolism and Energy Balance

Learning Outcomes:

1. Describe the important anatomical features of the thyroid gland and the major function of thyroid hormones in adult animals.
2. List the hormones produced by the pancreas and adrenal glands, and provide a brief description of the major metabolic roles of each.
3. Explain the difference between type I and type II diabetes mellitus.

An important function of the endocrine system is to regulate the metabolic rate and energy balance. Hormones are partly responsible for regulating energy balance by modulating appetite, digestion, absorption of nutrients, and the concentrations of energy sources like glucose in the blood and its transport into cells. Although many hormones are involved in these processes, those from the thyroid gland, pancreas, adipose tissue, and the adrenal glands play particularly important roles, as described in this section.

Thyroid Hormones Contain Iodine and Regulate Metabolic Rate

The thyroid gland lies within the neck of vertebrates, straddling the trachea just below the larynx in mammals (Figure 50.5a). It consists of many small, spherical structures called follicles made of a shell of epithelial cells called follicular cells and a core of a gel-like substance called the **colloid**. The colloid consists primarily of large amounts of the protein **thyroglobulin (TG)**, which plays a major role in the synthesis of thyroid hormones.

Thyroid hormones are produced when the hypothalamic neurohormone thyrotropin-releasing hormone (TRH; see Table 50.2) stimulates the anterior pituitary gland to secrete thyroid-stimulating hormone (TSH) into the blood. TSH stimulates the follicular cells of the thyroid gland to begin the process of making thyroid hormones. The thyroid hormones have a negative feedback effect on the hypothalamus and anterior pituitary, preventing them from producing too much TRH and TSH, thereby keeping the circulating concentration of thyroid hormones in check.

Figure 50.5b shows the pathway leading to thyroid hormone synthesis. First, iodide—converted by the gut from dietary iodine—diffuses from the bloodstream into the interstitial fluid, from where it is transported across the basolateral membrane of the thyroid follicular cells. The iodide then diffuses through the apical membrane and enters the colloid, where it is oxidized and bonds to tyrosine side chains in thyroglobulin. When the thyroid follicular cells are stimulated by TSH, the apical membranes undergo endocytosis, bringing colloid with its iodinated thyroglobulin into the cell. The endocytotic vesicles fuse with lysosomes; there, lysosomal enzymes cleave the iodinated tyrosines from thyroglobulin to form two thyroid hormones, called thyroxine and triiodothyronine. These molecules are unique among hormones in that they contain iodine molecules: four in the case of **thyroxine**, also called **T_4**, and three in **triiodothyronine**, or **T_3**. Both T_4 and T_3 then diffuse out of the follicular cells across the basolateral membrane into the bloodstream. These hormones are carried throughout the body and diffuse into cells. Inside cells, most T_4 is converted by enzymes that remove one iodine, forming T_3. It is T_3 that actually binds to cellular receptor proteins. T_4 can be considered, therefore, as a circulating reservoir of the active hormone T_3.

A major action of thyroid hormones in adult animals is to stimulate energy consumption by many different cell types. This occurs in large part by increasing the number and activity of the Na^+/K^+-ATPase pumps in plasma membranes. As these pumps hydrolyze ATP, the cellular concentration of ATP decreases. This decrease is compensated for by increasing the cell's metabolism of glucose. Whenever metabolism is increased, heat production is increased. Consequently, a person with a hyperactive thyroid gland (hyperthyroidism; from the Greek *hyper*, meaning over or above) generally feels warm, whereas the opposite condition (hypothyroidism; from the Greek *hypo*, meaning under or below) results in a sensation of coldness. Biologists have estimated that up to 70% of the heat produced by some homeotherms is attributable solely to the actions of thyroid hormones on metabolic rate.

The observation that thyroid hormones cannot be made without iodine presents some interesting and unique features of these hormones. The availability of iodine in the diet of most animals varies. As a consequence, the ability to store large amounts of thyroglobulin in the colloid of the thyroid was an important evolutionary adaptation. In this way, during times when iodine ingestion is high, many thyroglobulin molecules have their tyrosine amino acids bound to iodines, one of the first steps in forming T_4 and T_3. During times of low iodine availability, this reservoir of iodinated tyrosines in thyroglobulin molecules can be tapped. Humans, for example, have at least a 2-month supply of thyroid hormones even if dietary iodine were to become unavailable. In most industrialized countries, iodine deficiency is rarely a problem since the introduction of iodized salt in the mid-20th century. However, in some regions of the world, this is still a major health problem.

The left side of Figure 50.6a shows the pattern of T_4 and T_3 production from a healthy thyroid gland when iodine ingestion is adequate. The right side of the figure shows the consequences when thyroid hormones are not produced in normal amounts, for example, due to a lack of iodine in the diet. In such a case, the decreased T_4 and T_3 concentrations provide less negative feedback on the hypothalamus and anterior pituitary gland, resulting in elevated TRH and, consequently, TSH concentrations. The thyroid gland responds to the increased TSH by increasing the cellular machinery needed to produce more and more thyroglobulin, even though in the absence of iodine, no additional T_4 or T_3 can be synthesized. What results is an overgrown gland that still lacks the resources to make thyroid hormone. This condition is known as an **iodine-deficient goiter** (Figure 50.6b). In humans, the problem can be alleviated either by adding iodine to the diet or by taking T_4 pills. Goiters are not unique to humans. Iodine deficiency is relatively common among vertebrates, and goiters are found frequently in reptiles and birds, particularly those that subsist on all-seed diets, which are generally low in iodine.

Hormones of the Pancreas and Adrenal Glands Regulate the Concentration of Energy-Yielding Molecules in the Blood

Thyroid hormones regulate an animal's metabolism. For metabolism to proceed normally, however, body cells must have adequate sources of energy available, usually in the form of glucose and fatty acids. The brain, in particular, must have a constant supply of glucose because brain cells have relatively limited storage capacity for fuel. Regulation of energy availability to cells is in large part accomplished by the hormones of the pancreas and the adrenal glands.

The pancreas is a complex organ that is both an exocrine and endocrine gland (Figure 50.7). An **exocrine gland** is one in which epithelial cells secrete chemicals into a duct, which carries those molecules directly to another structure or to the outside surface of the body. Familiar examples of exocrine glands include the sweat glands and salivary glands. Most of the mass of the pancreas consists of exocrine cells. The secretions of the exocrine pancreas empty into the small intestine, where they aid digestion.

Figure 50.5 The thyroid gland and synthesis of thyroid hormones. **(a)** Location and structure of the gland in a human. Vertebrates contain thyroid tissue in the neck region, but it is not always consolidated into a single structure as shown here. **(b)** The steps involved in production of thyroid hormones. Shown is a cross section through a small part of a single follicle in a thyroid gland, with a nearby capillary (not to scale). The blood delivers iodide to the gland and picks up thyroid hormones secreted by the gland.

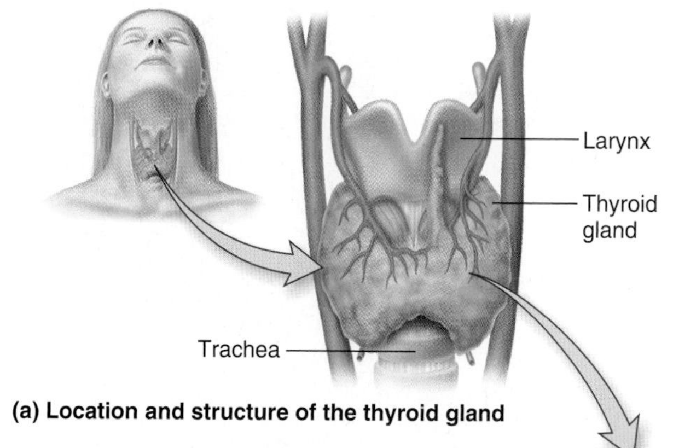
Larynx
Thyroid gland
Trachea

(a) Location and structure of the thyroid gland

Capillary
Interstitial fluid
Follicular cell of thyroid
Colloid
Iodide ion
Basolateral membrane
Apical membrane
2 Iodides bond to tyrosine residues in thyroglobulin.
1 Iodide is transported into the follicular cell.
Lysosome
4 The endocytotic vesicle fuses with lysosomes, which cleave off T_3 and T_4.
Iodinated TG
Thyroglobulin (TG)
Endocytotic vesicle
3 When the cell is stimulated by TSH, iodinated TG enters the cell by endocytosis.
5 T_3 and T_4 are secreted into the blood.
Triiodothyronine (T3)
Thyroxine (T4)

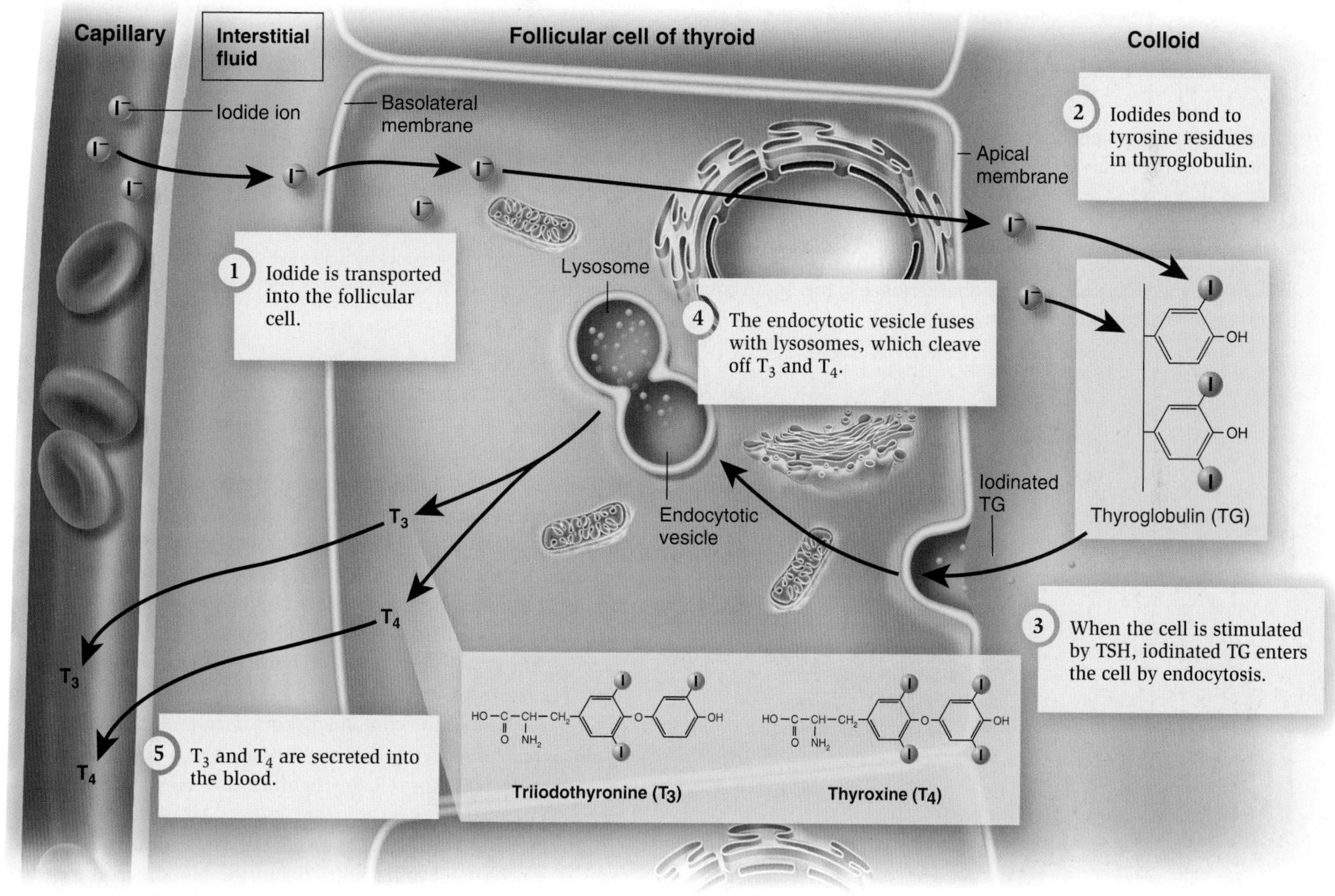

(b) Synthesis of thyroid hormones

The nonexocrine portion of the pancreas consists of endocrine cells that produce peptide hormones, notably insulin and glucagon. Spherical clusters of endocrine cells called **islets of Langerhans** are scattered in large numbers throughout the pancreas. Within the islets are alpha cells, which make **glucagon**, and beta cells, which make **insulin**. The actions of these two hormones are antagonistic with respect to each other—insulin decreases and glucagon increases blood glucose concentrations.

Maintaining normal glucose and other nutrient concentrations in the blood is a vital process that keeps cells functioning optimally. When an animal has not eaten for some time, its energy stores become depleted, and the blood glucose concentration decreases. Under these conditions, glucagon is secreted into the blood, where it acts on the liver (Figure 50.8, right side). The liver contains a limited supply of glucose in the form of stored glycogen. Glucagon stimulates the breakdown of glycogen into many molecules of glucose, which are then secreted into the blood. This process, known as glycogenolysis (refer back to Figure 46.2a), is stimulated within seconds by glucagon. A second action of glucagon on the liver is important for responses to prolonged fasting. In that case, glucagon stimulates the process

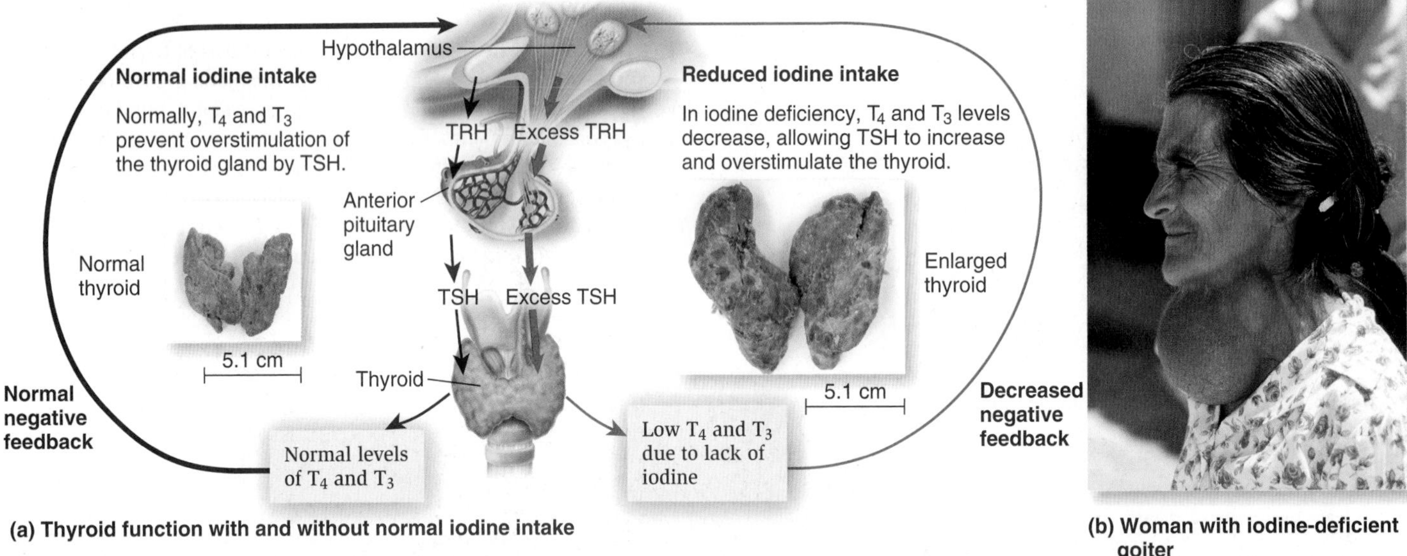

(a) Thyroid function with and without normal iodine intake

(b) Woman with iodine-deficient goiter

Figure 50.6 Consequences of normal and inadequate iodine in the diet. (a) With normal iodine intake, as shown on the left, T_4 and T_3 concentrations inhibit TSH secretion. Without enough iodine, as shown on the right, less T_4 and T_3 are synthesized, and the TSH concentration increases. This leads to the enlargement of the thyroid gland. **(b)** An extreme example of an enlarged thyroid gland, or goiter, due to iodine deficiency.

BIOLOGY PRINCIPLE Living organisms maintain homeostasis. Recall from Chapter 40 that a key way in which homeostasis is maintained is via negative feedback. Notice in this figure how a change in dietary intake of iodine results in decreased negative feedback and subsequent overgrowth of the thyroid gland.

of gluconeogenesis (refer back to Figure 46.2b), by which noncarbohydrates are converted into glucose, which is then released into the blood.

The adrenal glands, which sit atop the kidneys, also play a role in glucose metabolism by producing amine and steroid hormones. When blood glucose concentrations are lower than normal, neurons from the sympathetic nervous system activate the secretion of epinephrine by the adrenal medulla. Like glucagon, epinephrine also rapidly stimulates glycogenolysis and gluconeogenesis. In addition, **cortisol** is released from the adrenal cortex. After a delay, cortisol increases the expression of genes required for gluconeogenesis in the liver. These processes are vital to long-term survival in the absence of food. Because of the combined short-term and long-term actions of glucagon, epinephrine, and cortisol, the blood glucose concentration rarely decreases or remains significantly below normal except in extreme circumstances.

In contrast to fasting, after an animal eats a meal, the concentrations of glucose and other nutrients in the blood become elevated. Restoring the normal blood concentrations of glucose, fats, and amino acids is almost exclusively under the control of insulin, one of the few hormones that are absolutely essential for survival in animals.

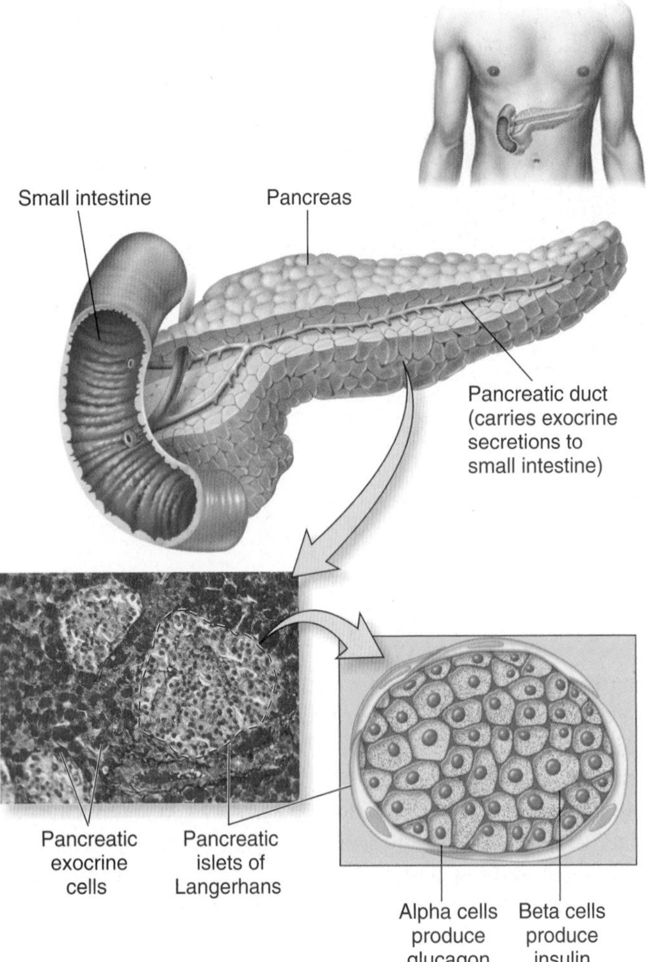

Figure 50.7 Location, appearance, and internal structure of the mammalian pancreas. Amidst the exocrine pancreas are scattered islets of Langerhans, which are endocrine tissue. Only the exocrine products are secreted into the intestine; the hormones from the islets of Langerhans are secreted into the blood (not shown).

BioConnections: The pancreas contains both exocrine and endocrine tissue. Is this property ever observed in other organs? (Hint: Think about the processes associated with digestion and absorption of food.)

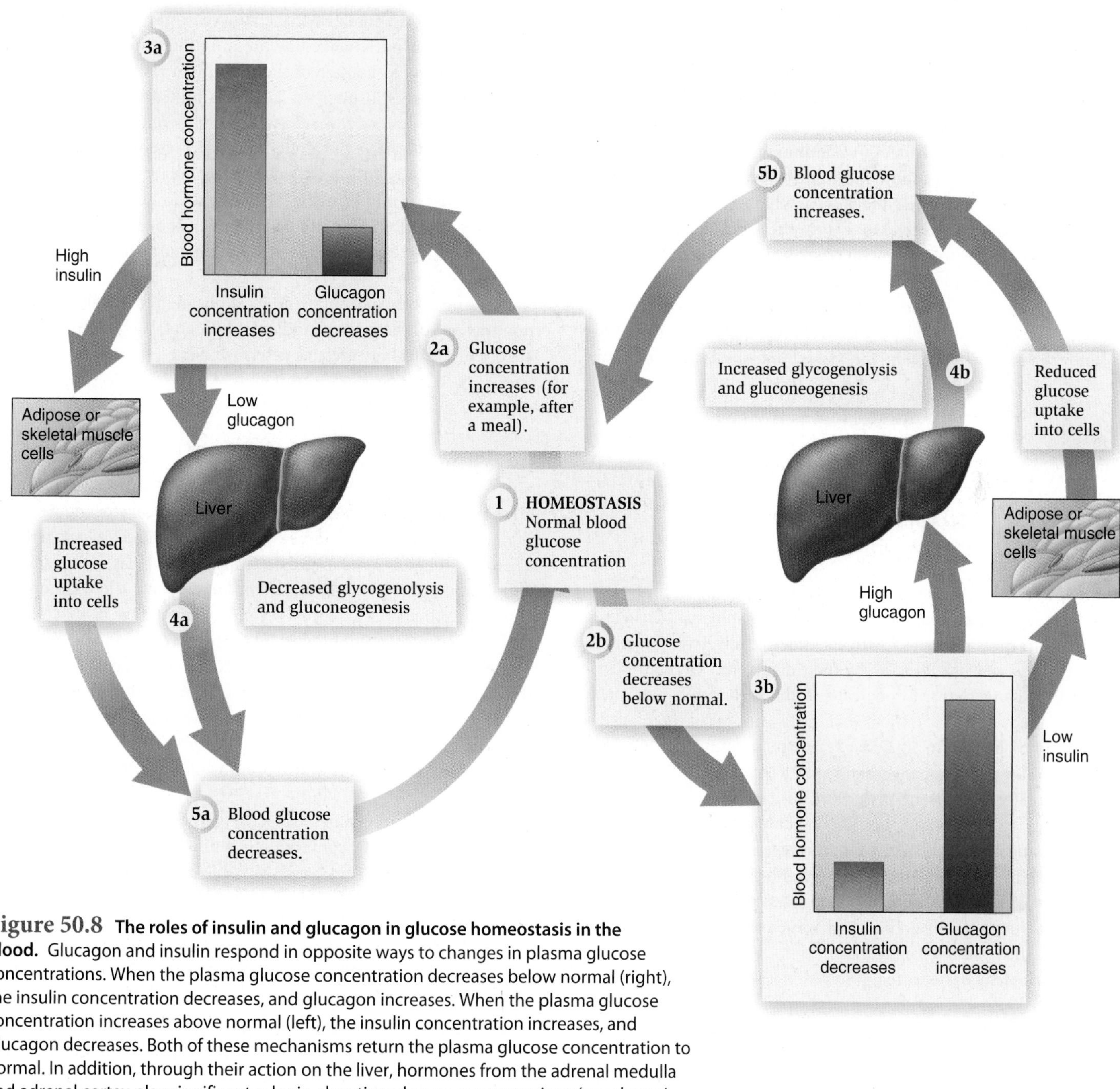

Figure 50.8 The roles of insulin and glucagon in glucose homeostasis in the blood. Glucagon and insulin respond in opposite ways to changes in plasma glucose concentrations. When the plasma glucose concentration decreases below normal (right), the insulin concentration decreases, and glucagon increases. When the plasma glucose concentration increases above normal (left), the insulin concentration increases, and glucagon decreases. Both of these mechanisms return the plasma glucose concentration to normal. In addition, through their action on the liver, hormones from the adrenal medulla and adrenal cortex play significant roles in elevating glucose concentrations (not shown).

The secretion of insulin is directly stimulated by an increased concentration of glucose in the blood (Figure 50.8, left side). Once in the blood, insulin acts on plasma membrane receptors located primarily in cells of adipose and skeletal and cardiac muscle tissues to facilitate the transport of glucose across the plasma membrane into the cytosol. Once in the cell, glucose is metabolized for cellular functions or converted to stored energy forms such as fat. As discussed in Chapter 46, the process of glucose transport involves the actions of proteins called glucose transporters (GLUTs; refer back to Figure 46.3). These proteins are located in cytosolic membrane-bound vesicles. The major function of insulin is to stimulate fusion of the vesicles with their GLUTs with the plasma membrane. Once the vesicle fuses with the plasma membrane, the GLUTs begin transporting glucose from the extracellular fluid into the cell.

When the blood glucose concentration returns to normal, the stimulus for insulin secretion disappears, and the insulin concentration in the blood decreases. This results in a decrease in the number of GLUTs in plasma membranes, because the GLUTs are subjected to endocytosis and thereby return to membrane vesicles in the cytosol. The actions of insulin are not limited to glucose transport, however, because the hormone also stimulates transport of amino acids into cells and promotes fat deposition in adipose tissue.

In the absence of sufficient insulin, as occurs in the disease **type 1 diabetes mellitus (T1DM)**, less extracellular glucose can cross plasma membranes, and consequently, glucose accumulates to a very high concentration in the blood. T1DM occurs in many mammals and other vertebrates. It is caused when the body's immune system mistakenly attacks and destroys the insulin-producing cells of the islets of Langerhans. One consequence of the disease is that muscle and fat cells cannot receive their normal amount of glucose to provide the ATP they require. In addition, the unusually large amount of glucose in the blood overwhelms the kidneys' ability to reabsorb it from the kidney filtrate, and glucose appears in the urine. Fortunately, this form of the disease is treatable with regular monitoring of blood glucose concentration and daily administration of insulin.

In humans, the more common form of diabetes, which accounts for roughly 90–95% of all cases (estimates range between 15 and 20 million people in the U.S. alone), is **type 2 diabetes mellitus (T2DM)**. In T2DM, the pancreas may function relatively normally and is not attacked by the immune system. However, the cells of the body lose much of their ability to respond to insulin for reasons that are still unclear. T2DM is linked to obesity and can often be prevented or reversed with weight control. Drugs are also available that improve the ability of cells to respond to insulin. Formerly, T2DM was known as "adult-onset" diabetes, because it usually appeared in middle age. This name is no longer used, because the rising tide of childhood obesity in developed countries like the U.S. has sparked a dramatic increase in the number of young people with the disease.

The discovery of insulin and its application to the human condition was one of the greatest and most influential achievements in the history of medical research, as described next.

FEATURE INVESTIGATION

Banting, Best, MacLeod, and Collip Were the First to Isolate Active Insulin

In the 19th century, scientists discovered that in addition to the exocrine part of the pancreas, the organ contains spherical clusters of cells that are not associated with its exocrine digestive functions. These clusters were called islets by their discoverer, German physiologist Paul Langerhans. By the early 20th century, German scientists had discovered that in dogs, complete removal of the pancreas resulted immediately in T1DM. Researchers assumed, therefore, that cells of the islets of Langerhans produced a factor of some kind that prevented diabetes—in other words, it helped maintain glucose homeostasis by preventing the blood glucose concentration from increasing uncontrollably. The factor, however, proved impossible to isolate using the chemical purification methods available at the time. Researchers assumed that during the process of grinding up a pancreas to produce an extract, the factor was destroyed by the digestive enzymes of the exocrine pancreas.

This problem was eventually solved in 1921 by a team of Canadian scientists working at the University of Toronto: surgeon Frederick Banting, medical student Charles Best, and biochemist James Bertram Collip. The work was performed in the laboratory of John MacLeod, a renowned expert in carbohydrate metabolism at the time. Banting had read a paper in a medical journal that described a deceased patient in whom the pancreatic duct, which carries digestive juices to the small intestine, had become clogged due to calcium deposits. The closed duct caused pressure to build up behind the blockage, which eventually caused the exocrine part of the pancreas to atrophy and die. The islets of Langerhans, however, survived intact. Banting hypothesized that if he were to tie off, or ligate, the pancreatic ducts of an animal, and then wait a sufficient time for the exocrine pancreas to die, he could more easily obtain an active glucose-lowering factor from the remaining islets without the problem of contamination by digestive enzymes.

Banting and Best proceeded to ligate the pancreatic ducts of several dogs as shown in Figure 50.9. After waiting 7 weeks, an amount of time they previously had determined was sufficient for the exocrine part of the pancreas to atrophy, they prepared extracts of the remaining parts of the pancreas, including its islets of Langerhans. This was done by removing the atrophied pancreas from each dog and grinding it up with a mortar and pestle in an acid solution. The extract was then injected into a second group of dogs in which diabetes had previously been induced by surgically removing the pancreas. The researchers discovered that the extract was capable of keeping the diabetic dogs alive for a brief time. However, they were unable to isolate sufficient quantities of the active factor in the extract to keep the experiments going longer than a day or so, nor were they able to obtain sufficiently pure factor to prevent side effects such as infection and fever.

This is where Collip's expertise came in. Collip developed a method to precipitate contaminating proteins from the extract by adding alcohol to the acid (step 3, Figure 50.9). Different proteins will precipitate from a solution in response to different concentrations of alcohol. Using a concentration too low to cause the active factor to precipitate, Collip was able to remove most of the contaminating proteins, leaving behind a much more purified and safer extract. The extract was then filtered to remove debris and further extracted to remove lipids. In this step, a hydrophobic solvent was added to the partially purified extract; the lipids in the extract dissolved in the solvent, which could then be decanted and removed. This was done because the researchers correctly believed that the factor was either a peptide or a small protein and by removing lipids, they would obtain a more purified preparation. Finally, the purified extract was concentrated by evaporating the alcohol, which increased its potency. When increasing amounts of this purified factor were injected into a second group of diabetic dogs, their blood glucose concentrations decreased and even returned to normal, and there was no longer any glucose in their urine. Diabetic dogs that received only a control solution that did not contain the factor continued to show a high glucose concentration in the blood and urine, compared with healthy controls.

In later experiments, two subsequent innovations enabled the researchers to obtain larger amounts of the factor and more accurately assess its potency without using the ligation procedure. First, the researchers chose the very large pancreases of cows, obtained at a local slaughterhouse, for the starting material from which to prepare the extracts. Second, Collip developed a highly sensitive assay to more precisely measure the concentration of glucose in the blood of an animal before and after injection of the purified factor. The combination of the improved chemical purification steps, large amounts of starting material, and improved assays for testing the extract proved to be the keys that enabled the team to test the factor, which they eventually named insulin, on human patients. The first successful test came in 1922 on a 14-year-old boy in Toronto, who was seriously ill from T1DM. The success of the team in rapidly isolating insulin and proving its effectiveness was so significant that Banting and MacLeod were awarded the 1923 Nobel Prize in Medicine or Physiology. The Nobel committee felt that Banting and MacLeod were the leaders of the project and that it was they who deserved the prize, but the two scientists disagreed. Banting shared the monetary portion of his award with Best, and MacLeod shared his with Collip.

Experimental Questions

1. How did Banting and Best propose to obtain the glucose-lowering factor produced by the pancreas?
2. What was Collip's contribution to the isolation of insulin?
3. What subsequent innovations led to large-scale production of insulin for human patients?

Figure 50.9 **The isolation of insulin by Banting, Best, MacLeod, and Collip.**

BIOLOGY PRINCIPLE Biology affects our society. The discovery of insulin and the methods to manufacture it in large amounts has improved (and in many cases, saved) the lives of many millions of people. Today, dozens of human diseases, including diabetes, are treatable with the use of hormones.

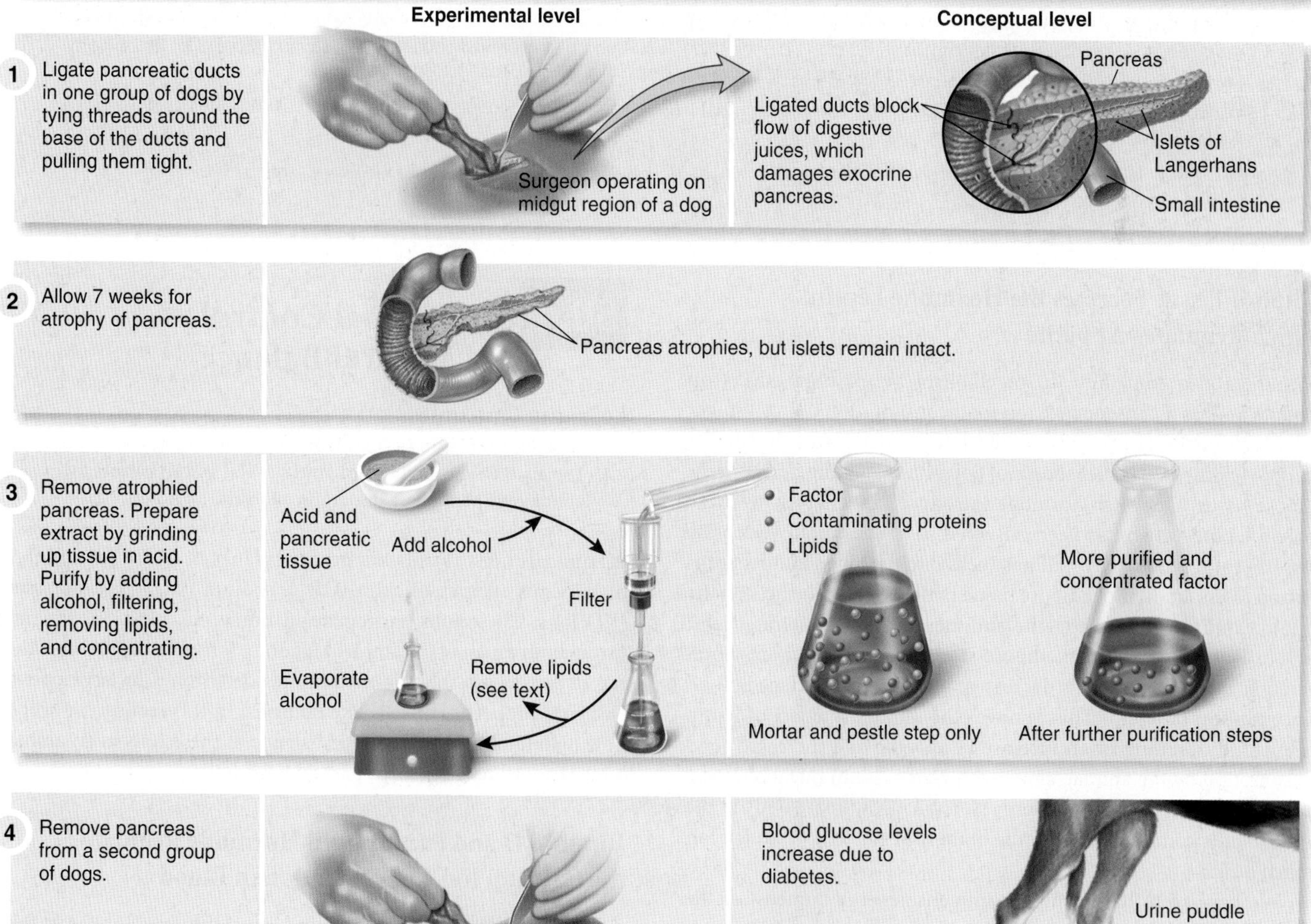

HYPOTHESIS Ligation of pancreatic ducts will cause atrophy of the exocrine pancreas, allowing extraction of an active glucose-lowering factor from the remaining portion of the pancreas.
KEY MATERIALS One group of dogs for ligation experiments; second group of dogs made diabetic by having pancreas removed.
Experimental level
Conceptual level
1 Ligate pancreatic ducts in one group of dogs by tying threads around the base of the ducts and pulling them tight.
Surgeon operating on midgut region of a dog
Ligated ducts block flow of digestive juices, which damages exocrine pancreas.
Pancreas
Islets of Langerhans
Small intestine
2 Allow 7 weeks for atrophy of pancreas.
Pancreas atrophies, but islets remain intact.
3 Remove atrophied pancreas. Prepare extract by grinding up tissue in acid. Purify by adding alcohol, filtering, removing lipids, and concentrating.
Acid and pancreatic tissue
Add alcohol
Filter
Evaporate alcohol
Remove lipids (see text)
Factor
Contaminating proteins
Lipids
More purified and concentrated factor
Mortar and pestle step only
After further purification steps
4 Remove pancreas from a second group of dogs.
Blood glucose levels increase due to diabetes.
Glucose appears in the urine, a sign of diabetes.
Urine puddle

5 Inject either purified extract or control solution into the diabetic dogs. Determine the levels of glucose in the blood and in the urine.

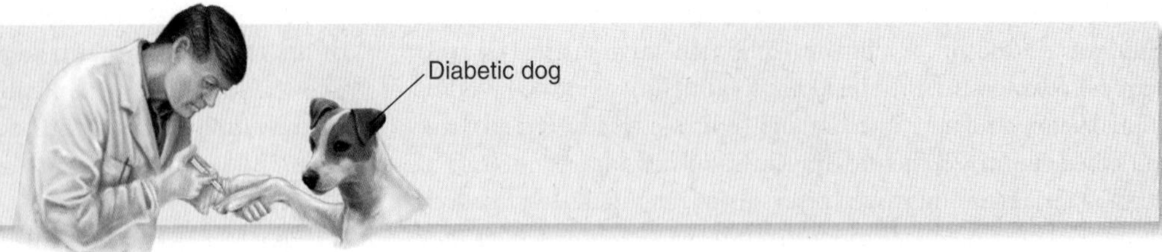

6 THE DATA

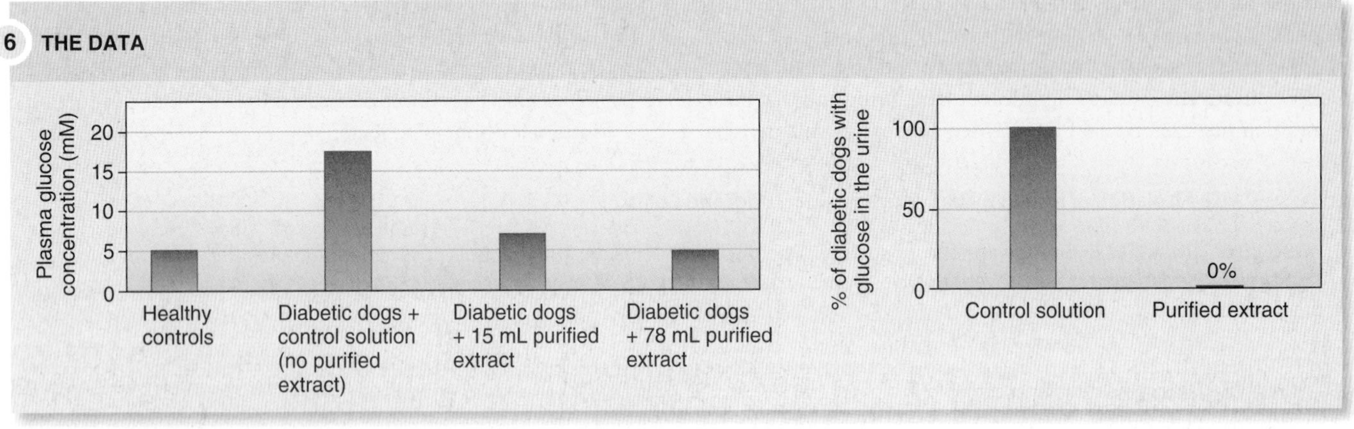

7 **CONCLUSION** Atrophy of the exocrine portion of the pancreas eliminated digestive enzymes that would have degraded the glucose-lowering factor (insulin) during its purification from the pancreas. Along with improved chemical purification procedures, this atrophy step allowed researchers to obtain a highly purified and fully active glucose-lowering factor.

8 **SOURCE** Banting, F.G., and Best, C.H. 1922. The internal secretion of the pancreas*Journal of Laboratory and Clinical Medicine* 7:256–271.

Adipose Tissue Secretes the Hormone Leptin, Which Regulates Appetite

Hormones also contribute to metabolism by exerting effects on appetite and, as a consequence, on food consumption. As you might imagine from its importance as the body's major storage site for energy-yielding fat, a chief source of appetite-regulating hormones is adipose tissue. One such hormone introduced in Chapter 46 is the protein leptin, which has been observed in species in all vertebrate classes. Leptin is released by adipose cells into the blood in direct proportion to the amount of adipose tissue in the body, and it acts on the hypothalamus to reduce appetite and increase metabolic rate (refer back to Figure 46.9). When adipose stores are low, however, leptin secretion from adipose cells decreases, resulting in a reduced blood concentration of leptin. This removes the inhibitory effect of leptin on appetite, resulting in an increase in appetite and increased food consumption by the animal, along with a decrease in metabolic rate. In these ways, the amount of energy stored in an animal's body in the form of fat is communicated to the brain to regulate how hungry an animal feels.

We have seen that hormones influence energy homeostasis by regulating nutrient concentrations in the blood, transport of nutrients into cells, appetite, and metabolic rate. In the next section, we consider how hormones control another feature of homeostasis, that of regulating concentrations of key minerals in the blood.

50.4 Hormonal Control of Mineral Balance

Learning Outcomes:

1. Describe the hormonal control of Ca^{2+} homeostasis.
2. Explain the regulation in the kidneys of Na^+ and K^+ balance by antidiuretic hormone, aldosterone, and atrial natriuretic factor.

All animals must maintain a proper balance in their cells and fluids of minerals such as calcium (Ca^{2+}), sodium (Na^+), and potassium (K^+) ions. These ions participate in numerous functions common throughout much of the animal kingdom. For example, the partitioning of ions across plasma membranes determines in large part the electrical properties of cells like neurons. In this section, we will see how maintaining a homeostatic balance of these ions in an animal's body is coordinated in large part by hormones.

Vitamin D and Parathyroid Hormone Regulate the Calcium Ion Concentration in Blood

Calcium ions play critical roles in neuronal transmission, heart function, muscle contraction, and numerous other events. Therefore, the concentration of Ca^{2+} in the blood is among the most tightly regulated variables in an animal's body.

Calcium is obtained from the diet and absorbed through the small intestine. Like all charged molecules, Ca^{2+} cannot readily cross plasma membranes, including those of the epithelial cells of the small intestine, and thus its transport must be regulated. This process is controlled by the action of a derivative of **vitamin D**. In most mammals that receive regular exposure to the Sun, skin cells produce vitamin D from a precursor called 7-dehydrocholesterol, a reaction that requires the energy of ultraviolet light (**Figure 50.10**). In addition, many animals obtain vitamin D from food or, in people today, from milk and other foods supplemented with the vitamin. This is especially important for people who live at latitudes that receive little sunlight for much or part of the year.

Before it can act, vitamin D must first be modified by two enzymes that both add one hydroxyl group to specific carbon atoms, first in the liver and then in the kidney. The final active product is called 1,25-dihydroxyvitamin D. This molecule is a hormone because it is secreted into the extracellular fluid by one organ—the kidneys—and then circulates in the blood and acts on a distant target tissue—the small intestine. The major function of 1,25-dihydroxyvitamin D in the intestine is to activate expression of a Ca^{2+} transporter in the intestinal epithelium, thereby stimulating the absorption of Ca^{2+}. The Ca^{2+} then enters the blood, from where the ions are delivered to tissues for such activities as building bone and maintaining nerve, muscle, and heart functions. If the active form of vitamin D is not present in the blood, the bones lose Ca^{2+} and become weakened. In children, this results in the weight-bearing bones of the legs becoming deformed, a condition called rickets.

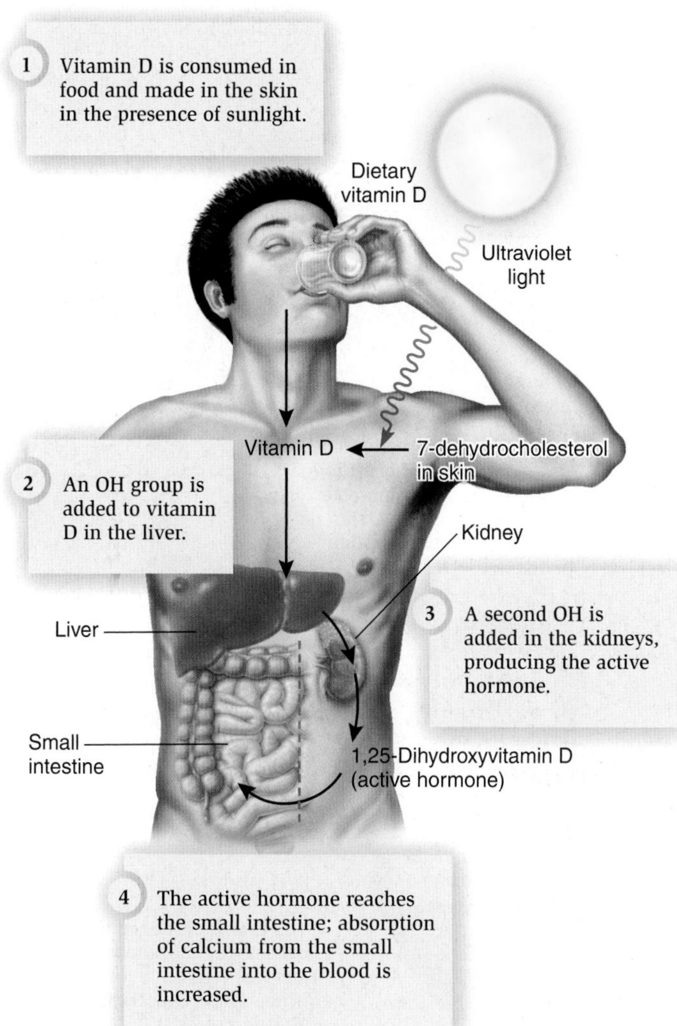

1 Vitamin D is consumed in food and made in the skin in the presence of sunlight.
Dietary vitamin D
Ultraviolet light
Vitamin D
7-dehydrocholesterol in skin
2 An OH group is added to vitamin D in the liver.
Kidney
Liver
3 A second OH is added in the kidneys, producing the active hormone.
Small intestine
1,25-Dihydroxyvitamin D (active hormone)
4 The active hormone reaches the small intestine; absorption of calcium from the small intestine into the blood is increased.

Figure 50.10 **Synthesis of the active hormone formed from vitamin D or its precursor in the skin.**

Concept Check: Would you predict that all mammals synthesize the active form of vitamin D using the energy of sunlight?

Even when Ca^{2+} is not present in the diet, or when 1,25-dihydroxyvitamin D is not formed in normal amounts because of insufficient exposure to sunlight, the blood concentration of Ca^{2+} does not normally decrease dramatically, because of a hormone called **parathyroid hormone (PTH)**. This hormone is secreted from several small glands called parathyroid glands, which in humans are attached to the back of the thyroid gland (**Figure 50.11a**). PTH acts on bone to stimulate the activity of cells that dissolve the mineral part of bone. This releases Ca^{2+}, which then enters the blood (**Figure 50.11b**). Typically, only a very small fraction of the total Ca^{2+} in bone is removed in this way. Therefore, besides providing a skeletal framework for the vertebrate body, bone also serves as an important reservoir of Ca^{2+}. PTH also acts to increase reabsorption of Ca^{2+} from the filtrate in the kidneys, such that less Ca^{2+} is excreted in the urine. Without PTH, Ca^{2+} homeostasis is not possible. The complete absence of PTH is fatal in humans and other mammals.

In addition to 1,25-dihydroxyvitamin D and PTH, another hormone called **calcitonin** plays a role in Ca^{2+} homeostasis in some vertebrates, notably fishes and possibly some mammals, including young children (but not adult humans). Calcitonin is produced in and secreted from cells in the thyroid gland. Its function is the opposite in many respects to that of PTH. Calcitonin promotes excretion of Ca^{2+} via the kidneys and deposition of Ca^{2+} into bone, thereby decreasing blood Ca^{2+} concentrations. This is especially important in marine fishes, because of the high Ca^{2+} content of seawater and the entry of this ion into the body fluids of the animals.

Several Hormones Regulate Sodium and Potassium Ions in Vertebrates

Like Ca^{2+}, concentrations of Na^+ and K^+ in the body fluids of most animals are tightly regulated, because these ions play crucial roles in membrane potential formation and action potential generation (see Chapter 41), among other functions. Like Ca^{2+}, Na^+ and K^+ are ingested in the diet and excreted in the urine in vertebrates at rates that maintain homeostatic concentrations in the body fluids. Unlike Ca^{2+}, however, no large reservoirs exist in the body for Na^+ and K^+. One of the key mechanisms that regulates the concentrations of these ions in the blood is altering the rate of Na^+, K^+, and water reabsorption from the urine as it is being formed in the kidneys. This is accomplished in large part by the actions of three hormones: ADH, aldosterone, and atrial natriuretic peptide (**Figure 50.12**). All three of these hormones exert their effects on the kidneys.

The vertebrate kidney normally reabsorbs most Na^+ and K^+ from the fluid filtered through the kidney glomerulus (refer back to

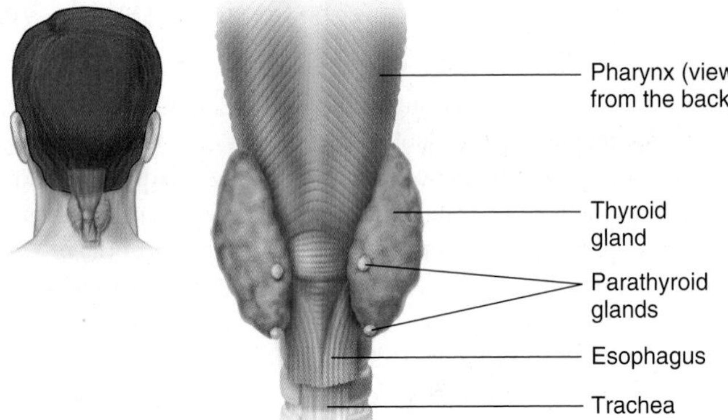

Pharynx (view from the back)

(a) Location of the parathyroid glands

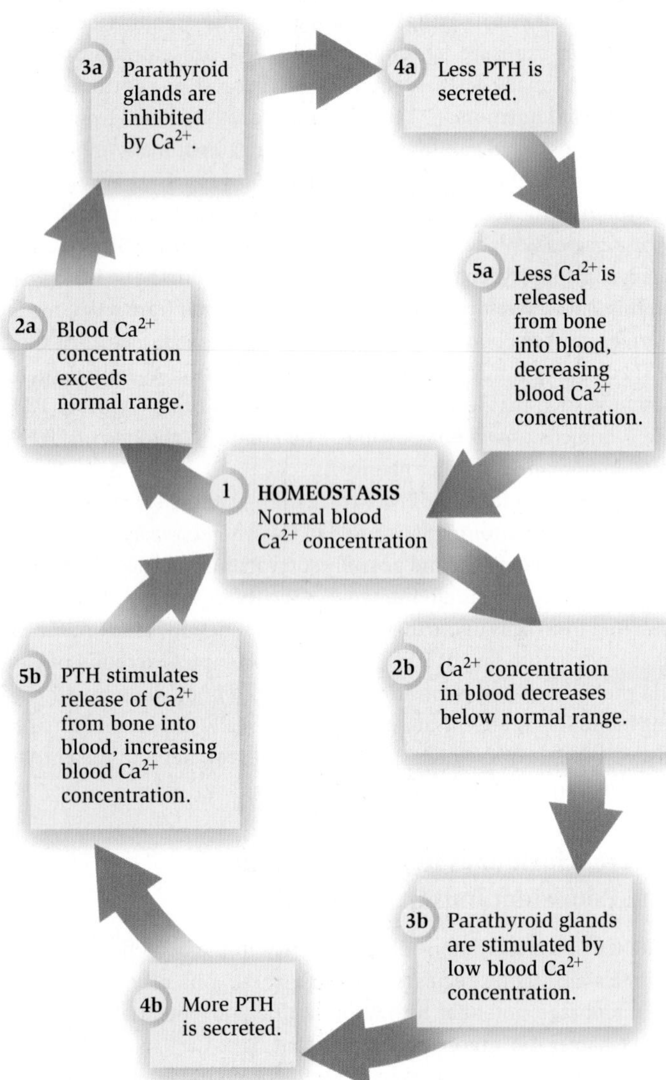

4a Less PTH is secreted.

4b More PTH is secreted.

(b) Homeostatic control of blood Ca^{2+} concentration

Figure 50.11 The role of parathyroid hormone in calcium homeostasis. (a) The four small parathyroid glands are located behind the thyroid. (b) The action of PTH: Steps 2a–5a occur when Ca^{2+} is in excess; steps 2b–5b occur when Ca^{2+} is below normal.

BIOLOGY PRINCIPLE Living organisms maintain homeostasis. Homeostatic control of Ca^{2+} concentrations is vitally important to life in most living organisms, and in all animals. Because of hormones like PTH, the Ca^{2+} concentration in the body fluids of humans rarely changes by more than a slight margin.

Figures 49.11 and 49.14). However, dietary and other changes can alter the concentrations of these ions in the blood. When this happens, the kidney works to restore the ions to their normal concentrations. For example, if the blood Na^+ concentration increases above normal, the osmolarity of the blood will increase. Osmoreceptors in the brain detect this and stimulate ADH secretion from the posterior pituitary. Recall from Chapter 49 that ADH acts on the kidney to reabsorb water from the forming urine. By increasing the amount of ADH available to act on the kidneys, less water is excreted in the urine. In addition, the synthesis and secretion of the adrenal steroid hormone **aldosterone** are directly inhibited in response to an increase in Na^+concentration. Normally, aldosterone increases Na^+ reabsorption in the kidney, and therefore, its absence results in more Na^+ excretion in the urine. Last, **atrial natriuretic peptide (ANP)** is

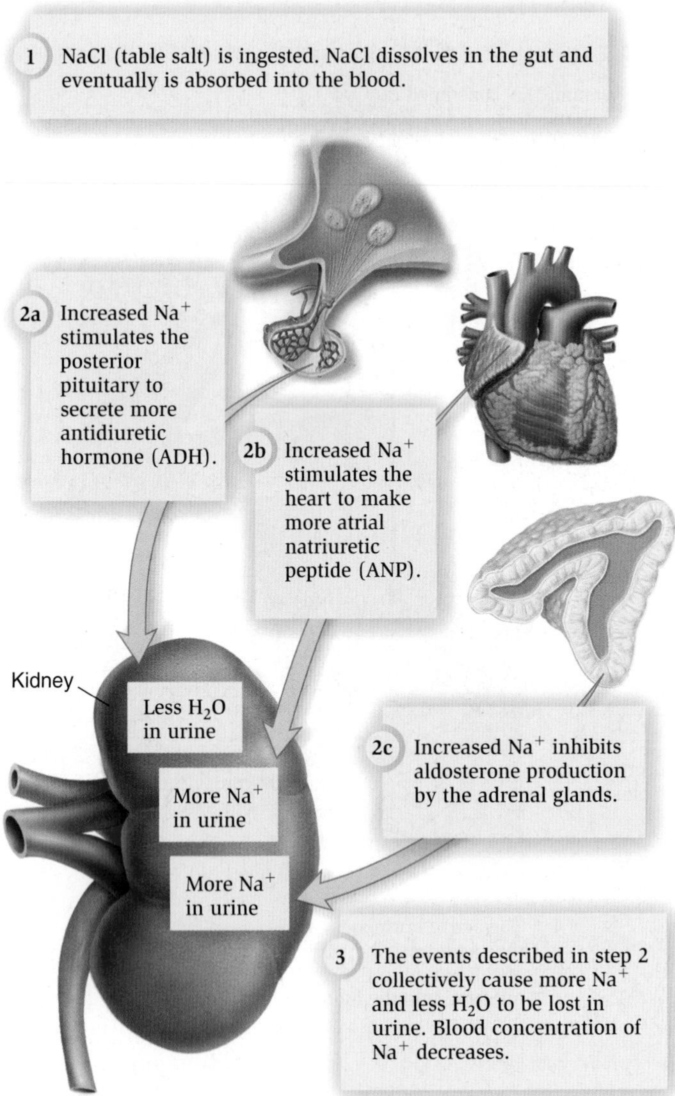

1 NaCl (table salt) is ingested. NaCl dissolves in the gut and eventually is absorbed into the blood.
2a Increased Na^+ stimulates the posterior pituitary to secrete more antidiuretic hormone (ADH).
2b Increased Na^+ stimulates the heart to make more atrial natriuretic peptide (ANP).
2c Increased Na^+ inhibits aldosterone production by the adrenal glands.
Kidney
Less H_2O in urine
More Na^+ in urine
More Na^+ in urine
3 The events described in step 2 collectively cause more Na^+ and less H_2O to be lost in urine. Blood concentration of Na^+ decreases.

Figure 50.12 An example of sodium balance achieved by the coordinated actions of three hormones.

Concept Check: Why is there more than one hormone that regulates Na^+ and K^+ balance?

secreted from the atria of the heart whenever the blood concentration of Na^+ increases. ANP causes a natriuresis (a loss of Na^+ in the urine; from the Latin *natrium*, meaning sodium) by decreasing Na^+ reabsorption. Thus, ANP and aldosterone have opposite effects on Na^+ balance in the body, which is why their concentrations in the blood tend to rise and fall in opposite directions. The combined effect of decreasing water loss in the urine and reabsorbing less Na^+ is to decrease the concentration of Na^+ in blood and other body fluids, returning its concentration to normal.

50.5 Hormonal Control of Growth and Development

Learning Outcomes:

1. List and describe the function of the peptide/protein hormones involved in the regulation of growth and development in vertebrates.
2. Identify the functions of the three major hormones that control growth and development in insects.

Hormones play a nearly universal role in controlling growth and development in animals. Growth can occur slowly over long periods or in brief spurts, with some animals exhibiting both types of growth. Many mammals, for example, grow slowly but steadily until puberty, then experience a period of rapid growth, followed by slower rates of growth, and finally cessation of growth. Many insects, however, grow and develop in spurts during molting periods. Although growth is determined by many factors, notably adequate nutrition, it is regulated in large part by the endocrine system.

Growth is distinguished from development, which is the process by which cells form tissues with specific functions, and tissues develop into larger and more complex structures. In this section, we will examine how both processes depend in part on the endocrine system.

Vertebrates Require a Balance of Several Hormones for Normal Growth

Several hormones stimulate growth in vertebrates. The anterior pituitary gland produces **growth hormone (GH)**, which is under the control of the hypothalamus (see Table 50.2). GH acts on the liver to produce another hormone, called **insulin-like growth factor-1 (IGF-1)**. In mammals, IGF-1 stimulates the elongation of bones, especially during puberty, when mammals become reproductively mature. This growth is further accelerated by the steroid hormones of the gonads, leading to the rapid pubertal growth spurt. Eventually, however, the gonadal hormones cause the growth regions of bone to seal, preventing any further bone elongation.

GH, IGF-1, and gonadal steroid hormones continue to be produced in adult vertebrates—including humans—even though growth has ceased. In adulthood, GH serves metabolic functions, such as helping to regulate the concentrations of glucose and fatty acids in the blood. The gonadal steroids are important for reproduction in adults (see Chapter 51).

In rare cases, a tumor of the GH-secreting cells of the anterior pituitary gland produces excess GH. If this occurs during childhood, a person with this disorder can grow exceedingly tall and is known as a **pituitary giant**. Soon after puberty, growth stops. If a tumor causes a high GH concentration after puberty, the excess GH causes many bones, such as those of the face, hands and feet, to thicken and enlarge, a condition known as **acromegaly** (Figure 50.13). People with acromegaly are generally treated with a synthetic form of somatostatin, which is the inhibitor of GH made by the hypothalamus (see Table 50.2). In many cases, however, the tumor must be surgically removed.

By contrast, if the pituitary fails to make adequate amounts of GH during childhood, the concentrations of GH and IGF-1 in the blood will be lower than normal. In such cases, growth is stunted, resulting in one of the many possible causes of **short stature**. Individuals with this condition can be treated with injections of recombinant human GH and will grow to relatively normal height, as long as treatment begins before puberty is completed, so that the bones can still elongate.

Hormones are also required during fetal life, not for growth but for development of the brain, lungs, and other organs into fully mature, functional structures. For example, cortisol from the fetal adrenal gland is vital for proper lung formation. Premature babies born before the adrenal glands have matured have lungs that are not capable of expanding normally; such babies require hospitalization to survive.

As another example, thyroid hormones influence development among all vertebrates. In amphibians, thyroid hormones play a critical role in metamorphosis, notably in tadpoles, where they promote the resorption of the tail and development of the legs (Figure 50.14a). This effect can be dramatically demonstrated by experimentally

Figure 50.13 Acromegaly in one individual from a pair of identical twins. This disease is caused by a high concentration of growth hormone, which leads to increased height and enlarged bones including those of the skull, hands, and feet.

Concept Check: *Did the individual on the left develop a growth hormone disorder before or after puberty?*

decreasing or increasing the tadpole's thyroid hormone concentration, which results respectively in a tadpole that does not transform into a frog or in a tadpole in which metamorphosis occurs sooner than normal, resulting in a tiny frog. In fishes, thyroid hormones play an equally critical role in development. For example, among species of flatfish such as flounder that live on the ocean floor, thyroid hormones are responsible for the characteristic change in appearance that occurs in these species as they settle into a sedentary existence on the ocean bottom. The fins and gill covers migrate to the dorsal surface facing the water; the dorsal body surface becomes pigmented; and, most remarkably, the eyes migrate to the same side of the head such that one eye is not unused on the side facing the ocean floor (Figure 50.14b). Each of these metamorphic events is under the direct control of thyroid hormones.

Invertebrates Grow and Develop in Spurts Under the Control of Three Major Hormones

Like vertebrate endocrine pathways, hormonal control systems in insects and other invertebrates often involve multiple glands and neural structures acting in concert. In insects, for example, the endocrine system is critical for the growth and development of larvae and their eventual development into pupae (Figure 50.15). In the case of larval growth and metamorphosis, specialized neurosecretory cells in the brain periodically secrete a hormone called **prothoracicotropic hormone (PTTH)**, which then stimulates a pair of endocrine glands called the prothoracic glands. These glands, located in the thorax, synthesize and secrete into the hemolymph a steroid hormone called **20-hydroxyecdysone**. This hormone is secreted

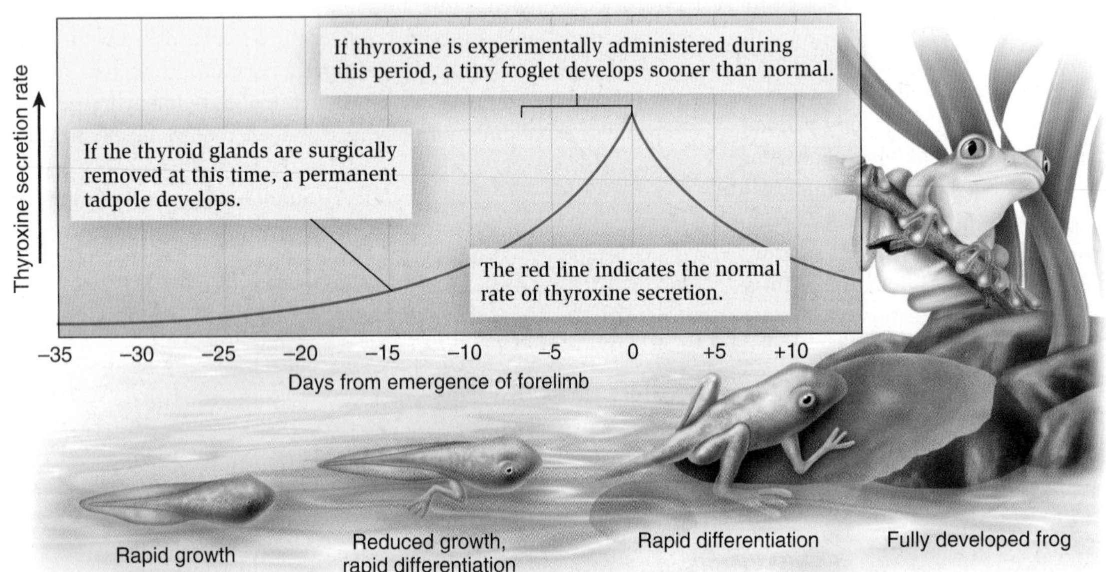
Thyroxine secretion rate
If thyroxine is experimentally administered during this period, a tiny froglet develops sooner than normal.
If the thyroid glands are surgically removed at this time, a permanent tadpole develops.
The red line indicates the normal rate of thyroxine secretion.
–35 –30 –25 –20 –15 –10 –5 0 +5 +10
Days from emergence of forelimb
Rapid growth
Reduced growth, rapid differentiation
Rapid differentiation
Fully developed frog
Stages of normal development

(a) The effect of thyroid hormones on tadpole development

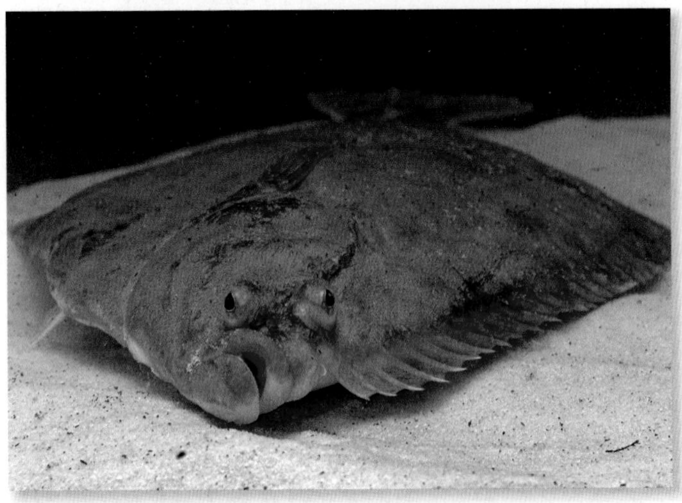

(b) Japanese flounder

Figure 50.14 Effects of thyroid hormones on animal development. **(a)** Experimental manipulation of thyroid hormone concentrations can slow down or accelerate development of tadpoles. **(b)** The eyes of a bottom-dwelling flounder (*Platichthys flesus*) migrate to one side of the head during development, partly in response to thyroid hormone.

only periodically. In response to each burst of secretion, the larva undergoes a rapid development and molts (sheds its cuticle). It then begins a new developmental period until it molts again in response to another episode of secretion. The molting process is known as ecdysis, from which the name of the hormone is derived (refer back to Figure 32.12).

Throughout larval development, paired neurosecretory structures behind the brain, called the corpus allata, secrete another hormone, a protein called **juvenile hormone (JH)**. JH determines the character of the molt that is induced by 20-hydroxyecdysone, by acting to prevent metamorphosis into an adult (hence the name "juvenile," which reflects the fact that JH fosters the larval stage). As the larva ages, however, the amount of JH it produces gradually declines until its concentration is nearly zero. During this time, 20-hydroxyecdysone continues to be periodically secreted. The decline in JH below a certain concentration results in the transition from larva to pupa in response to a burst of 20-hydroxyecdysone; the near absence of JH is a prerequisite for the final step of metamorphosis into an adult.

50.6 Hormonal Control of Reproduction

Learning Outcomes:

1. List the male and female sex steroids, their sites of production, and their roles in reproductive development.
2. Explain how nutrition and reproduction are linked through hormones.

The topic of reproduction is covered in detail in Chapter 51, but it is worth noting here that in all vertebrates and probably most invertebrates, reproduction is closely linked with endocrine function. In this section, we explore the most common reproductive hormones and their actions in male and female animals.

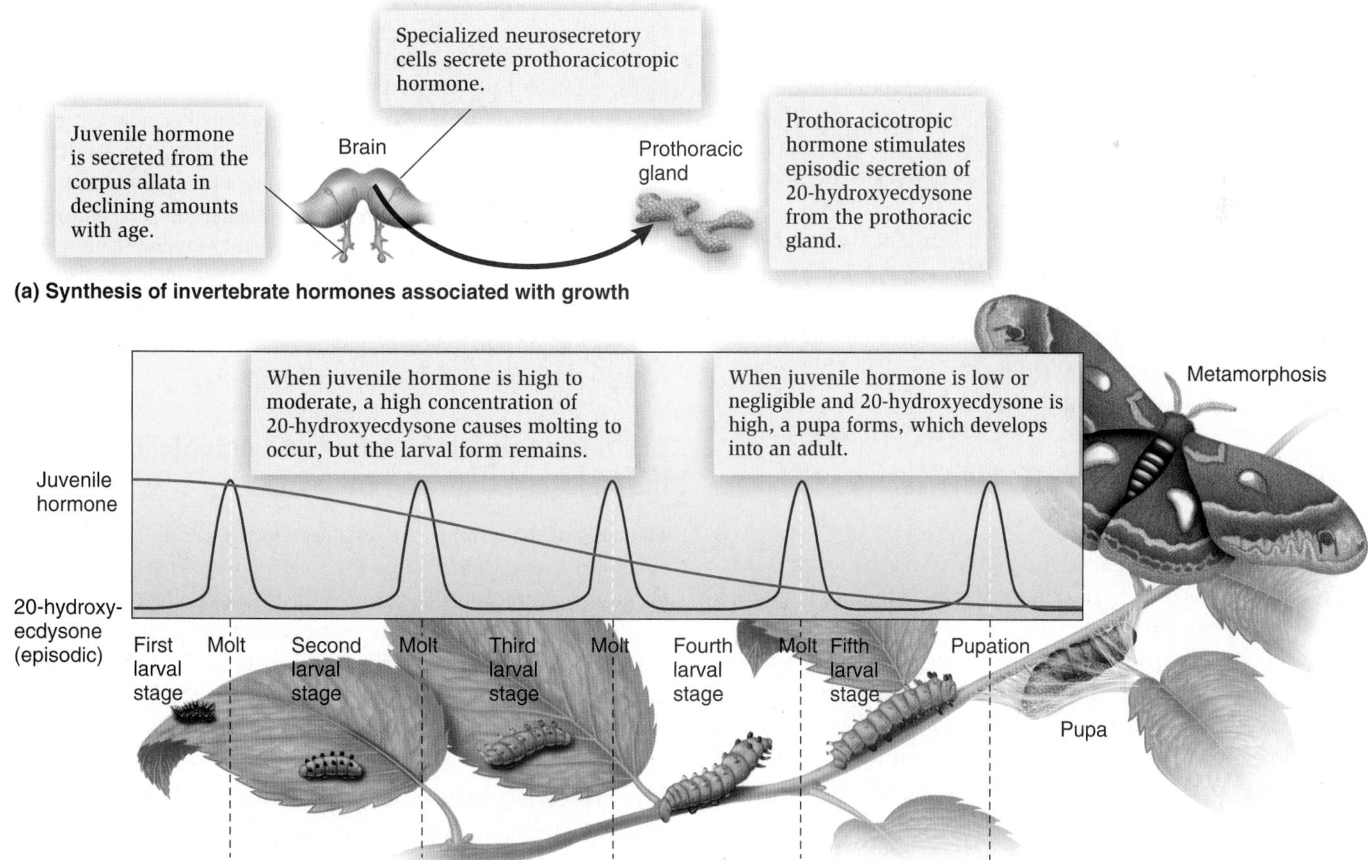

Specialized neurosecretory cells secrete prothoracicotropic hormone.
Juvenile hormone is secreted from the corpus allata in declining amounts with age.
Brain
Prothoracic gland
Prothoracicotropic hormone stimulates episodic secretion of 20-hydroxyecdysone from the prothoracic gland.
(a) Synthesis of invertebrate hormones associated with growth
When juvenile hormone is high to moderate, a high concentration of 20-hydroxyecdysone causes molting to occur, but the larval form remains.
When juvenile hormone is low or negligible and 20-hydroxyecdysone is high, a pupa forms, which develops into an adult.
Metamorphosis
Juvenile hormone
20-hydroxy-ecdysone (episodic)
First larval stage
Molt
Second larval stage
Molt
Third larval stage
Molt
Fourth larval stage
Molt
Fifth larval stage
Pupation
Pupa
(b) Effects of hormones on molting and pupation

Figure 50.15 **Hormonal control of insect development.** Development of insects requires the coordinated actions of three hormones. *Note:* The relative concentrations and patterns of 20-hydroxyecdysone and juvenile hormone in this figure are schematic and not representative of all insects.

Concept Check: *In what part of a cell do you predict the receptor for the steroid 20-hydroxyecdysone would be located?*

The Gonads Secrete Sex Steroids That Influence Most Aspects of Reproduction

Hormones produced by the gonads of animals play vital roles in nearly all aspects of reproduction, from reproductive behaviors to the ability to produce offspring. In males, the gonads are called testes. They produce and house the gametes, called sperm cells. In addition, the testes produce several related steroid hormones collectively termed **androgens**. In many vertebrates, including humans, the primary androgen is **testosterone**. The gonads of females are called the ovaries, which contain the female gametes, or egg cells. Other cells within the ovaries secrete two major steroid hormones, **progesterone** and a family of related hormones called **estrogens**. The major estrogen in many animals, including humans, is **estradiol**. Collectively, the male and female gonadal steroid hormones are referred to as the sex steroids.

The sex steroids are responsible for male- or female-specific reproductive changes associated with courtship and mating. In vertebrates, androgens from the developing testes are required for development of the male phenotype. Exposure of fetuses to abnormal concentrations of male or female hormones can lead to ambiguous phenotypes and sexual behavior after birth. Sex steroids are also chiefly responsible for development of secondary sex characteristics for each sex, which in humans include such things as growth of the appropriate external genitals, growth of the breasts in women, development of facial hair in men, and distribution and amount of fat and muscle in the body (look ahead to Chapter 51). Finally, sex steroids are required for maturation of gametes, the transition of young animals to reproductive maturity, and the ability of females to produce young.

The ability of the testes and ovaries to produce sex steroids depends on the presence of the gonadotropins secreted by the anterior pituitary, which are the same in both sexes and include **follicle-stimulating hormone (FSH)** and **luteinizing hormone (LH)** (see Table 50.2). Therefore, identical anterior pituitary gland hormones control the gonads in both male and female animals.

Nutrition and Reproduction Are Linked Through Hormones

An interesting feature of the endocrine control of reproduction is the observation that puberty is delayed in mammals, including humans, that are very undernourished. Similarly, fertility—the ability to produce offspring—is reduced in women and other adult female mammals under such conditions. This makes sense, because supporting the nutritional demands of a growing fetus would be difficult without good nutrition and energy stores. In such a case, it is more advantageous for an animal to delay reproduction until sufficient food is available.

How does the brain of a female mammal determine when sufficient energy is stored in her body to support pregnancy? The answer appears to be partly the result of hormonal signals. For example, if the amount of fat in a woman's body decreases, so does the concentration of leptin in her blood, as described earlier. Leptin has been demonstrated to stimulate synthesis and secretion of reproductive hormones such as FSH and LH. Consequently, undernourishment that causes a decrease in leptin results in reduced production of reproductive hormones, thereby contributing to a loss of fertility. Therefore, leptin acts as a link between adipose tissue and the reproductive system.

50.7 Hormonal Responses to Stress

Learning Outcomes:

1. Define stress in a biological context and how it is linked with homeostasis.
2. Name the hormones produced by the different regions of the adrenal glands in response to acute and long-term stress and describe the major functions of each.

Stress in animals is often defined as any real or perceived threat to survival or homeostasis. This can take many forms depending on the species. To a crab, a passing shadow may indicate a hungry shark passing overhead, but to a human, a shadow may simply mean that a cloud has passed in front of the Sun. Similarly, a severe storm can be extremely stressful for birds in exposed nests, but not to animals that live in sheltered environments. Extreme crowding is adaptive and nonstressful for penguins trying to keep warm on ice floes, but it is stressful to mice and rats.

In the 1920s, Hungarian-Canadian physiologist Hans Selye made the remarkable discovery that regardless of the nature of the stress in any given species, a mammal's adaptive responses to stress were highly similar. Later this was determined to be true because these responses depend on hormones from the adrenal glands. The adrenal glands, so named because they sit atop the kidneys (from the Latin *ad*, meaning toward, and *renis*, meaning kidney), are multifunctional glands containing an inner region called the adrenal medulla and an outer region called the adrenal cortex (**Figure 50.16a**). These two regions produce different hormones that affect the vertebrate response to stress. In this section, we will examine how these hormones function and how they act to prepare an animal to confront or escape a challenge—the fight-or-flight response.

The Inner Core of the Adrenal Glands Makes the Fight-or-Flight Hormones

The cells of the adrenal medulla secrete the amine hormones norepinephrine and epinephrine (**Figure 50.16b**). Together, these two hormones are responsible for most of the physiological reactions of the fight-or-flight response that were described in Chapter 42 and in Figure 9.17, and are summarized here in **Table 50.3**. These reactions include improved heart and lung function, increased production of energy sources by the liver, and increased alertness.

The Outer Regions of the Adrenal Glands Make Steroid Hormones in Response to Stress

The outer part of the adrenal gland, the cortex, is itself subdivided into three zones: the glomerulosa, the fasciculata, and the reticularis (see Figure 50.16b). The outer zone, the glomerulosa, is the region that makes the mineralocorticoid aldosterone, which acts to maintain Na^+, K^+, and water balance. The innermost cortical zone is known as the reticularis, which functions in humans to make certain

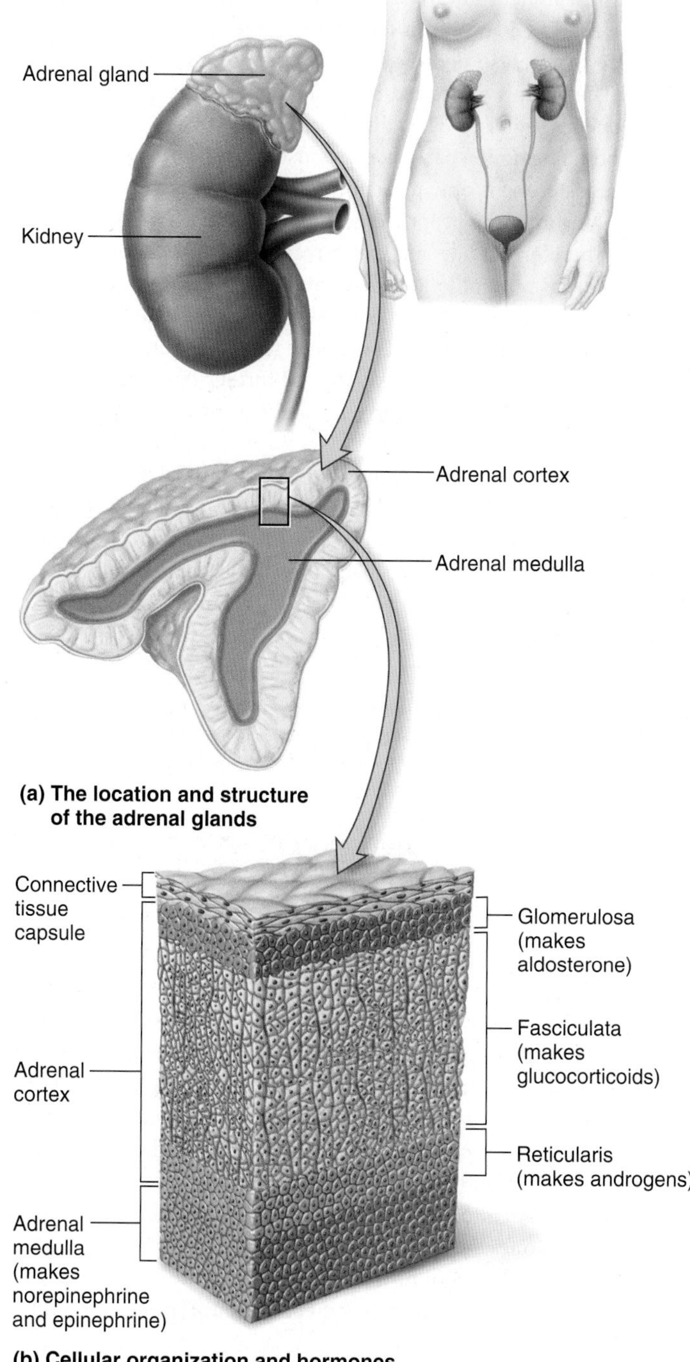

Figure 50.16 **Location, structure, and function of the adrenal glands.**

androgens, but whose function in other animals is not as clear. The bulk of the cortex is the middle zone, the fasciculata, which produces glucocorticoid hormones such as cortisol.

Table 50.4 summarizes some of the major actions of glucocorticoids in vertebrates. Glucocorticoids are catabolic hormones; that is, they promote the breakdown of molecules and macromolecules. For example, they act on bone, immune, muscle, and fat tissue to break down proteins and lipids to provide energy for the body's organs.

Table 50.3 The Fight-or-Flight Response

System	Reaction
Cardiovascular	Increases heart rate and strength of heart contractions to maximize pumping of blood to all parts of the body; dilates blood vessels entering tissues requiring more oxygen—such as skeletal muscle—and constricts blood vessels to regions of less immediate importance—such as the gut and kidney
Respiratory	Dilates small airways (bronchioles) to reduce resistance to airflow in mammals; increases rate and depth of breathing to maximize oxygen intake and carbon dioxide elimination
Metabolic	Increases glycogenolysis in muscle to provide glucose for muscle cells and in the liver to provide glucose to the blood, where it can reach all body cells; increases breakdown of adipose triglycerides into usable fuel (fatty acids) that can then enter the bloodstream; stimulates glucagon secretion, which acts on the liver to promote gluconeogenesis
Nervous	Increases arousal and alertness; inhibits nonessential functions such as appetite

Table 50.4 The Major Actions of Glucocorticoids

Site	Actions
Liver	Stimulate gluconeogenesis, thus providing glucose to the blood
Adipose tissue	Stimulate breakdown of triglycerides into fatty acids and glycerol for fuel
Muscle and adipose tissue	Inhibit sensitivity to insulin, making more glucose available to brain cells, which do not require insulin to move glucose across their plasma membranes
Bone	Inhibit bone growth and formation, because such processes require large amounts of nutrients that could be used to combat stress instead
Lungs	Stimulate lung maturation in the fetus
Immune	Suppress immune system function and reduce inflammation
Other	Regulate sodium and chloride balance in migratory fishes; stimulate nervous system development in most vertebrates; stimulate protein breakdown to provide amino acids to the liver for gluconeogenesis; inhibit reproduction

This is important for an animal that is facing an acute stress, because feeding and digestion both stop during the fight-or-flight response, leaving internal stores as the only source of energy. These adrenal steroids are called glucocorticoids because one of their major actions is to promote gluconeogenesis in the liver during times of stress, thus providing glucose to the blood.

Responding to acute stress is an important function of glucocorticoids, but excessive production of these steroids can create serious problems. Such a situation might arise if an animal were chronically stressed for some reason, for example, due to habitat destruction, intraspecies competition and aggression, or chronic illness. Glucocorticoids, together with amine hormones such as epinephrine, increase the activity of the heart and can raise blood pressure. Therefore, prolonged stress can lead to disorders of the heart and blood vessels. In addition, a chronic increase in glucocorticoid concentrations can suppress the immune system because of the catabolic actions of glucocorticoids on immune tissue. This may compromise an animal's ability to fight infection. For example, the prolonged stress a salmon experiences when it makes its exhausting swim upriver to its spawning grounds causes massive suppression of its immune system, and many mature salmon that die after spawning are found to be riddled with infections. Humans who are chronically stressed also tend to be susceptible to infections or to having dormant viruses suddenly flare up, such as the type of virus that causes cold sores.

Even an animal's ability to reproduce and grow is negatively affected by chronic stress, as noted most dramatically in medical situations involving humans. In children, for example, chronic exposure to a high concentration of glucocorticoids may stunt growth, at least temporarily. In women, chronic stress may result in the loss of monthly reproductive cycles. The explanation for this may be understood from an evolutionary perspective. Chronic stress generally suggests that an animal's fitness or survival is in jeopardy. Therefore, it is advantageous to have all energetically demanding activities—such as growth, reproduction, and even immune activity—come to a halt if these activities are not immediately required for staying alive.

Aldosterone and glucocorticoids such as cortisol are structurally similar, but distinct. These characteristics have enabled biologists to use these molecules as models for understanding how hormones and their receptors may have evolved, as described next.

Genomes & Proteomes Connection

Hormones and Receptors Evolved as Tightly Integrated Molecular Systems

We have learned that hormones act in all cases by binding to receptor proteins located on the cell surface or in the cytosol or nucleus. This raises an intriguing question: Which evolved first, a given hormone or its receptor? There would appear to be no selection pressure to evolve one without the other. Without a receptor, a hormone has no function, and without a hormone, the receptor has no function. Hormones and their receptors function as tightly integrated molecular systems. Each molecule depends on the other for biological activity. How such systems could have evolved, or whether they evolved together, has long been a major puzzle. However, recent research has generated intriguing new hypotheses regarding the evolution of these signaling systems.

American biologist and environmental scientist Joseph Thornton and colleagues recently addressed this puzzle by examining how two structurally related steroid hormones evolved separate activities and distinct receptors. Aldosterone, as noted, is a steroid hormone made by the adrenal cortex. Because it regulates the balance of two minerals (Na^+ and K^+) in the body, it is known as a mineralocorticoid. Cortisol is another steroid hormone made by the adrenal cortex and is known as a glucocorticoid because one of its major functions is to regulate glucose balance. The actions of these hormones are mediated by intracellular receptors known as the mineralocorticoid receptor (MR) and the glucocorticoid receptor (GR), respectively. Phylogenetic analysis of the gene sequences that code for these two receptors suggests that they arose at least 450 mya, after the evolution of the first vertebrates, by duplication of an ancient corticoid receptor (CR) gene.

The researchers analyzed the known sequences of the genes for the MR and GR of many vertebrate species, including the most ancient vertebrates, to deduce the theoretical sequence of an ancestral CR. They then synthesized this CR and tested its ability to bind aldosterone and cortisol. They discovered that the CR was capable of binding both hormones, particularly aldosterone. This was surprising, because aldosterone is only present in tetrapods, which arose around 300 mya, long after the proposed gene duplication event that created the MR. Therefore, it appears that a receptor with high affinity for aldosterone was present long before animals acquired the capacity to synthesize aldosterone. The receptor seems to have evolved to bind other steroids. Around 450 mya, the gene for the CR duplicated, and then the two resulting genes seem to have further mutated such that one receptor gained high affinity primarily for mineralocorticoids and the other for glucocorticoids.

These studies suggest that the ability of an animal to respond to aldosterone evolved long before aldosterone did, because a receptor with high affinity was already in place in animals. When aldosterone evolved in tetrapods, it was able to use the MR derivative of the ancestral CR. In this example, the answer to "which came first" appears to be that the receptor evolved first and the hormone later.

50.8 Impact on Public Health

Learning Outcomes:

1. Name several hormones used therapeutically and the disease or health problem for which they are prescribed.
2. Describe the risks associated with the misuse of hormones and with exposure to endocrine disruptors in the environment.

The list of endocrine-related diseases in the human population is lengthy, ranging from relatively common disorders such as diabetes and thyroid disease to rare ones that may affect only 1 in every 100,000 or more individuals. Hormones can be used to treat many of these diseases, but they can also be misused in ways that cause health problems. Also, the widespread industrial use of chemicals with hormone-like actions and their possible environmental consequences have become very much a part of the news. In this section, we consider how hormones are used therapeutically as well as misused, and then look at an example of how environmental factors that alter the function of the human endocrine system can have important public health implications.

Hormones Are Used to Treat Millions of People

Since the advent of synthetic hormone production, doctors have been able to provide individuals with hormones that their own bodies may fail to produce. For example, millions of people self-administer insulin each day to treat diabetes or take thyroxine to supplement an insufficient level of thyroid hormones. Many women take hormones to help induce pregnancy or control the symptoms of menopause. A growing number of men, too, are administered gonadal hormones if the levels produced by their bodies decline significantly in later life. Other examples include recombinant human growth hormone to treat abnormally short stature in prepubertal children, epinephrine inhalers to treat asthma, and glucocorticoids to treat inflammation, lung disease, and skin disorders, to name just a few.

Hormone Misuse Can Have Disastrous Consequences

As we saw at the beginning of this chapter, some individuals, typically those in competitive sports, self-administer hormones such as androgens. This practice may increase muscle mass and improve athletic performance. However, the price is steep. First, androgens exert negative feedback actions on FSH and LH secreted by the anterior pituitary gland. As a consequence, the anterior pituitary gland stops producing and secreting FSH and LH while the user is taking supplements containing androgens. In males, this causes the testes to shrink, as they no longer are making sperm, and the man becomes infertile; this is what happened to our case subject described at the beginning of the chapter. Androgen administration has also been linked to extreme aggressive behavior, cardiovascular disease and heart attacks, skin problems, and certain cancers. Women using androgens run similar health risks as men but, in addition, develop masculinizing traits, including increased growth of body hair and thinning scalp hair.

Another example of misuse is when a hormone is used to boost the number of red blood cells in the circulation in order to increase the oxygen-carrying capacity of the blood (often called blood-doping). This has become common among athletes participating in long-distance aerobic activities, such as competitive cycling and cross-country skiing. The hormone used is **erythropoietin (EPO)**, which acts by stimulating the maturation of red blood cells in the bone marrow and their release into the blood. EPO is normally made by the liver and kidneys in response to any situation where additional blood cells are required, such as following blood loss or when a person lives at high altitudes, where the oxygen pressure is low. When used abusively, however, the concentration of red blood cells can reach such a high level that the blood becomes much more viscous than normal. This puts a serious strain on the heart, which must work harder to pump the thickened blood. Since the 1990s, the international cycling community has been rocked by an alarming number of world-class European cyclists who died of heart attacks in the prime of their lives. These individuals had been using EPO to gain an unfair and, as it turns out, unwise advantage over their peers. Testing continues to detect injected EPO in the blood of some cyclists and other endurance athletes.

Synthetic Compounds May Act as Endocrine Disruptors

A recent and disturbing phenomenon is the growing amount of so-called **endocrine disruptors** found in lakes, streams, ocean water, and soil exposed to pollution runoff. These chemicals are derived in many cases from industrial waste and have molecular structures that in some cases resemble estrogen sufficiently to bind to estrogen receptors. If these compounds make their way into drinking water or food, they can exert estrogen-like actions or inhibit the actions of the body's own estrogen. This can lead to dramatic consequences on fertility and on the development of embryos and fetuses. The extent of the risk from endocrine disruptors is hotly debated, but the number of mature, functional germ cells produced in animals as diverse as mollusks and human males has declined dramatically during the past 50 years in the U.S. In addition, researchers throughout the world have noted feminization of freshwater fishes downstream of wastewater facilities. For example, male fishes that were exposed to such conditions during development show increased production of proteins normally made by females bearing eggs. They also show changes in gonadal structures that resemble the female appearance. Further research is urgently needed to provide more information on the consequences of these contaminants to animal endocrine systems.

Summary of Key Concepts

50.1 Mechanisms of Hormone Action and Control

- The endocrine glands and other organs with hormone-secreting cells constitute the endocrine system. Endocrine glands contain epithelial cells that secrete hormones into the bloodstream, where they circulate throughout the body. Although slower than the electrical signaling of nervous systems, chemical signaling complements nervous system regulation through its varying actions in multiple locations across widely ranging time frames (Figure 50.1).
- Hormones fall into three broad classes: the amines, proteins/peptides, and steroids. Water-soluble hormones (amines and proteins/peptides) act on receptor molecules located in the plasma membrane, whereas lipid-soluble hormones (steroids) act on intracellular receptors (Table 50.1).
- Synthesis of amine hormones is mediated by enzymatic conversion of either tyrosine or tryptophan into their respective hormones. The steroid hormones are derived from cholesterol (Figures 50.2, 50.3).

50.2 Links Between the Endocrine and Nervous Systems

- Sensory input from an animal's nervous system modulates the activity of certain endocrine glands and influences blood concentrations of many hormones.
- The hypothalamus is physically connected to the pituitary gland, which consists of the anterior and posterior lobes (Figure 50.4).
- Within the hypothalamus are numerous neurons that synthesize neurohormones and stimulate the anterior pituitary. In response, the anterior pituitary gland synthesizes six different hormones that respond to the presence of hypothalamic neurohormones. They are adrenocorticotropic hormone, follicle-stimulating hormone, luteinizing hormone, growth hormone, prolactin, and thyroid-stimulating hormone (Table 50.2).
- In mammals, the neuron terminals in the posterior pituitary gland store and secrete one of two hormones: oxytocin or antidiuretic hormone.

50.3 Hormonal Control of Metabolism and Energy Balance

- Hormones are partly responsible for regulating energy use by cells, that is, for modulating appetite, digestion, absorption of nutrients, and blood concentrations of energy sources such as glucose. Although many hormones are involved in these processes, two from the thyroid gland (thyroxine and triiodothyronine), two from the pancreas (insulin and glucagon), and one from adipose tissue (leptin) play especially important roles.

- The thyroid gland makes thyroid hormones, which contain iodine. A major action of thyroid hormones in adult animals is to stimulate energy consumption by many different cell types (Figures 50.5, 50.6).
- The endocrine pancreas produces the peptide hormones insulin and glucagon, which have opposite effects on the blood glucose concentration. The adrenal glands produce steroid hormones known as glucocorticoids, which increase the blood glucose concentration (Figure 50.7).
- Maintaining normal glucose and other nutrient concentrations in the blood is a vital process that keeps cells functioning optimally. The combined short-term and long-term actions of insulin, glucagon, epinephrine, and cortisol help maintain normal blood glucose concentrations during fasting (Figure 50.8).
- Groundbreaking research by Banting, Best, MacLeod, and Collip isolated insulin for therapeutic use in treating diabetes mellitus (Figure 50.9).
- Adipose tissue is an important source of appetite-regulating hormones, including the protein leptin, which acts on the hypothalamus to inhibit appetite.

50.4 Hormonal Control of Mineral Balance

- Because of the important roles that calcium plays in neuronal transmission, heart function, muscle contraction, and numerous other events, the concentration of Ca^{2+} in the blood is among the most tightly regulated variables in an animal's body. Vitamin D and parathyroid hormone, produced by the parathyroid glands, regulate the blood concentration of Ca^{2+} (Figures 50.10, 50.11).
- Na^+ and K^+ play crucial roles in membrane potential formation, action potential generation, and other functions. A key mechanism that regulates blood concentrations of these ions is to alter the rate of Na^+, K^+, and water reabsorption from the urine as it is being formed in the kidneys. This is accomplished in large part by the actions of ADH, aldosterone, and atrial natriuretic peptide (Figure 50.12).

50.5 Hormonal Control of Growth and Development

- Hormones play a crucial role in regulating growth and development. In vertebrates, normal growth depends on a balance between growth hormone, growth factor-1, and gonadal hormones. Thyroid hormones affect development among all vertebrates (Figures 50.13, 50.14).
- In insects, prothoracicotropic hormone, 20-hydroxyecdysone, and juvenile hormone control the growth of larvae and their development into pupae (Figure 50.15).

50.6 Hormonal Control of Reproduction

- Hormones produced by the gonads play vital roles in nearly all aspects of reproduction. The ability of the testes and ovaries to produce the sex steroids depends on the gonadotropins, which are the same in both sexes and include follicle-stimulating hormone and luteinizing hormone.

50.7 Hormonal Responses to Stress

- Regardless of the nature of the stress, animals' bodies respond to stress in similar ways. In vertebrates, hormones produced by the adrenal glands are the common denominator of the response to stress. Norepinephrine and epinephrine, produced by the adrenal medulla, are responsible for most physiological reactions of the fight-or-flight response (Figure 50.16, Table 50.3).
- Glucocorticoids promote breakdown of storage compounds to provide energy for cells during stress (Table 50.4).

50.8 Impact on Public Health

- Hormones are used therapeutically to treat a variety of human disorders, including diabetes, infertility, growth disorders, asthma, and inflammation.
- Androgen misuse can disrupt normal hormone concentrations and cause health risks such as cardiovascular disease; skin problems; cancer; infertility in men; and masculinizing traits in women. Blood-doping with erythropoietin can make blood dangerously thick.
- Endocrine disruptors, such as chemicals derived from industrial waste, may bind to estrogen receptors in animals' bodies. They may exert estrogen-like actions or inhibit the actions of the body's own estrogen.

Assess and Discuss

Test Yourself

1. Which is the defining feature of hormones?
   a. They are only produced in endocrine glands.
   b. They are secreted by one type of cell into the blood, where they may simultaneously reach many distant target cells, thereby altering cell function throughout the body.
   c. They are released only by neurons.
   d. They are never released by neurons.
   e. They are secreted into ducts, where they diffuse to another nearby gland or other structure.
2. Steroid hormones are synthesized from ______ and bind ______.
   a. proteins; membrane receptors
   b. fatty acids; membrane receptors
   c. tyrosine; intracellular receptors
   d. proteins; intracellular receptors
   e. cholesterol; intracellular receptors
3. Which of the following is *not* true about protein and peptide hormones?
   a. Most of them bind to receptors located on the cell membrane.
   b. Most of them are lipophilic.
   c. They are the most abundant class of hormones.
   d. They normally activate second messengers and intracellular signaling pathways.
   e. They bind noncovalently to receptors.
4. Which is *not* correct about the control of hormones?
   a. Many are regulated by negative feedback.
   b. In some cases, they may be controlled by changes in blood concentrations of certain nutrients such as glucose.
   c. In some cases, they may be controlled by changes in blood concentrations of certain minerals such as Ca^{2+}.

d. Their blood concentrations are controlled by their rates of synthesis and degradation.
e. The rate of degradation of a hormone may go up or down, but synthesis is always constant.

5. The hypothalamus and the pituitary gland are physically connected by
   a. arteries.
   b. the infundibular stalk and portal veins.
   c. the adrenal medulla.
   d. the spinal cord.
   e. the intermediate lobe.

6. Antidiuretic hormone
   a. increases water reabsorption in the kidneys.
   b. regulates blood pressure by constricting arterioles.
   c. decreases the volume of urine produced by the kidneys.
   d. increases blood pressure during times of high blood loss.
   e. does all of the above.

7. Which of the following pairs of hormones are involved in the regulation of blood Ca^{2+} concentration in vertebrates?
   a. aldosterone and ANP
   b. insulin and glucagon
   c. parathyroid hormone and 1,25-dihydroxyvitamin D
   d. prolactin and oxytocin
   e. thyroxine and TSH

8. In invertebrates, molting of larvae is stimulated by
   a. growth hormone.
   b. cortisol.
   c. juvenile hormone.
   d. 20-hydroxyecdysone.
   e. aldosterone.

9. Which of the following is *not* true of glucocorticoids?
   a. They stimulate maturation of the fetal lungs.
   b. They promote a decrease in the blood glucose concentration.
   c. They inhibit the sensitivity of cells to insulin.
   d. They are lipophilic (lipid soluble).
   e. They reduce inflammation.

10. Endocrine disruptors are
   a. chemicals released by the nervous system to override the endocrine system.
   b. chemicals released by the male of a species to decrease the fertility of other males.
   c. drugs used to treat overactive endocrine structures.
   d. chemicals derived from industrial waste that may alter endocrine function.
   e. all of the above.

Conceptual Questions

1. What is the function of leptin, and what is the benefit of an adipose-derived signaling molecule in this context? Why does an animal have an appetite?
2. Distinguish between type 1 and type 2 diabetes mellitus.
3. A principle of biology is that *living organisms maintain homeostasis.* How do the opposing actions of insulin and glucagon cooperate to maintain glucose homeostasis? When are they released into the blood? What might happen to a nonfasting mammal that was injected with a high dose of glucagon?

Collaborative Questions

1. Discuss the role of hormones in insect development.
2. Discuss the role of the different steroid hormones and where they are produced in the human body.

Online Resource

www.brookerbiology.com

Stay a step ahead in your studies with animations that bring concepts to life and practice tests to assess your understanding. Your instructor may also recommend the interactive eBook, individualized learning tools, and more.

Animal Reproduction

51

Chapter Outline

51.1 Asexual and Sexual Reproduction
51.2 Gametogenesis and Fertilization
51.3 Mammalian Reproductive Structure and Function
51.4 Pregnancy and Birth in Mammals
51.5 Impact on Public Health
Summary of Key Concepts
Assess and Discuss

These mud daubers, shown here mating, reproduce by sexual reproduction, which favors genetic variation in species.

Over a span of 40 years in the early 18th century, a woman known to history only as the "wife of Feodor Vassilyev," a Russian peasant, reportedly gave birth 27 times to a total of 69 children! If accurate, it is safe to say that this is about the uppermost realistic limit of human fertility, considering a pregnancy length of about 40 weeks, the rarity of a woman producing multiple births such as twins, and the reproductive life span of an adult woman. Impressive as such a number is, however, it pales in comparison with the numbers of offspring produced by many other animals. For example, many invertebrate species, such as the wasps known as mud daubers (shown in the chapter-opening photo), lay anywhere from several to dozens of eggs in 1 day. As an extreme example, a single female cod fish can lay several million eggs during one spawning season (although not all the eggs will result in offspring).

Indeed, when we examine all forms of life, we discover that individuals of every species are part of an unbroken cycle of life and death. Perpetuation of all life requires **reproduction**—the processes by which organisms replicate themselves and multiply. The biological mechanisms that favor successful reproduction in the animal kingdom are extraordinarily diverse. Many of the observable differences in animal behavior and anatomy are the result of adaptations that increase an animal's chances of reproducing. Both the behaviors and the anatomical specializations that promote reproduction are under the control of a variety of factors, particularly hormones.

In this chapter, we will begin by examining the diverse means of reproduction that occur throughout the animal kingdom, including asexual and sexual reproduction. Then we will highlight some of the anatomical, hormonal, and behavioral aspects of reproduction in mammals, including humans. We will conclude with a discussion of some key issues related to fertility (the ability to reproduce) in the human population today.

51.1 Asexual and Sexual Reproduction

Learning Outcomes:

1. Describe several differences between asexual and sexual reproduction.
2. Classify the three types of asexual reproduction.
3. List the advantages and disadvantages of sexual reproduction and asexual reproduction.

Asexual reproduction occurs when offspring are produced from a single parent, without the fusion of genetic material from another parent. The offspring are therefore clones of the single parent. **Sexual reproduction** is the production of a new individual by the joining of two haploid reproductive cells called **gametes**—one from each parent. This produces offspring that are genetically different from both parents. In this section, we will consider the processes of asexual and sexual reproduction, as well as their advantages and disadvantages.

Asexual Reproduction

Asexual reproduction occurs in some invertebrates and in a small number of vertebrate species. These animals use one of three major forms of asexual reproduction: budding, regeneration, or

(a) Budding

(b) Regeneration

Figure 51.1 **Examples of asexual reproduction.** **(a)** *Hydra* with a bud (on the left). **(b)** A starfish regenerating a complete new body from a single arm.

parthenogenesis. **Budding**, which is seen in cnidarians, occurs when a portion of the parent organism pinches off to form a complete new individual (Figure 51.1a). In this process, cells from the parent undergo mitosis and differentiate into specific types of structures before the new individual breaks away from the parent. At any one time, a parent organism may have one, two, or multiple buds forming simultaneously. Budding continues throughout such an animal's lifetime.

Some animals, including certain species of sponges, echinoderms, and worms, reproduce by the **regeneration** of a complete organism from small fragments of their body. In some sea stars and starfish, for example, an arm removed by injury or predation can grow into an entirely new individual (Figure 51.1b). Similarly, a flatworm bisected into two pieces can regenerate into two new individuals.

The asexual process called **parthenogenesis** (from the Greek, meaning virgin birth) is the development of offspring from an unfertilized egg. It occurs in several invertebrate classes and in a few species of fishes and reptiles. Animals produced by parthenogenesis are usually haploid. Some animals—such as rotifers, social insects such as ants and bees, and the freshwater crustacean *Daphnia*—may reproduce either parthenogenically or sexually, depending on the time of year or environmental conditions. In ants, honeybees, and wasps, for instance, haploid males (drones) are produced by parthenogenesis, and diploid females (workers and queens) are produced by sexual reproduction. Production of drones, whose major function is to mate with a queen, usually occurs in late spring. The queen stores sperm cells from several drones for up to 2 years, during which time she may lay hundreds of thousands of eggs, only some of which will combine with sperm. As stored sperm runs out, the remaining eggs become new drones.

Sexual Reproduction

Sexual reproduction occurs in most animal species, both invertebrate and vertebrate. As mentioned, sexual reproduction involves the joining of two gametes. The gametes are spermatozoa (usually shortened to **sperm**) from the male and **egg cells**, or **ova** (singular, **ovum**), from the female. When a sperm unites with an egg—a process called **fertilization**—each haploid gamete contributes its set of chromosomes to produce a diploid cell called a fertilized egg, or **zygote**. As the zygote undergoes cell divisions and begins to develop, it is called an **embryo**.

Advantages and Disadvantages of Asexual and Sexual Reproduction

Asexual reproduction provides a relatively simple way for an organism to produce many copies of itself, whereas sexual reproduction requires two individuals to produce offspring. What are the advantages and disadvantages of each method?

Asexual reproduction has numerous advantages over sexual reproduction. First, an animal can reproduce asexually even if it is isolated from others of its own species. Individuals can reproduce rapidly at any time because they need not seek out, attract, and mate with an opposite sex. Asexual reproduction, therefore, is an effective way of generating large numbers of offspring. Although many kinds of animals reproduce asexually, it is more prevalent in species that live in very stable environments, with little selection pressure for genetic diversity in a population.

Compared with asexual reproduction, sexual reproduction is associated with unique costs. Two types of gametes (sperm and eggs) must be made, males and females require specialized body parts to mate with each other, and the two sexes must be able to find each other. Yet given that the vast majority of eukaryotic species reproduce sexually, a question has intrigued biologists since the time of Darwin: What is the advantage of sexual reproduction for the perpetuation of a species?

In the context of species survival, the major difference between asexual and sexual reproduction is that sexual reproduction allows for greater genetic variation due to genetic recombination. Only certain alleles from each parent are passed on, and when a set of genes from one parent mixes with a different set from the other parent, the offspring are never exactly like either of their parents. Thus, a hallmark of sexual reproduction is increased genetic variation between successive generations. One prevalent hypothesis about why sex evolved is that sexual reproduction allows more rapid adaptation to environmental changes than does asexual reproduction.

In particular, sexual reproduction allows alleles within a species to be redistributed via crossing over and independent assortment across many generations. As a result, some offspring carry combinations of alleles that promote survival and reproduction, whereas other offspring may carry less favorable combinations. As described in Chapter 24, natural selection can favor those combinations that promote greater reproductive success while eliminating offspring with lower fitness. By comparison, the alleles of asexual organisms are not reassorted from generation to generation. As a result, it is more difficult to accumulate potentially beneficial alleles within individuals of these species. Also, as described next, sexual reproduction may facilitate elimination of harmful alleles from a population.

Feature Investigation

Paland and Lynch Provided Evidence That Sexual Reproduction May Promote the Elimination of Harmful Mutations in Populations

Evolutionary biologists have suggested that the inability of asexual species to reassort alleles may be a key disadvantage compared with sexually reproducing species. To investigate this question, American researchers Susanne Paland and Michael Lynch recently studied the persistence of mutations in populations of *Daphnia pulex*, a freshwater organism commonly known as the water flea. The researchers chose this organism because some natural populations reproduce asexually, and others reproduce sexually.

In their experiment, shown in **Figure 51.2**, Paland and Lynch studied the sequences of several mitochondrial genes in 14 sexually reproducing and 14 asexually reproducing populations of *D. pulex*. The researchers hypothesized that asexual populations would be less able to eliminate harmful mutations. As discussed in Chapter 14, random gene mutations that change the amino acid sequence of the encoded protein are much more likely to be harmful than beneficial. The alleles of sexually reproducing populations can be reassorted from generation to generation, thereby producing succeeding generations in which the detrimental alleles are lost from the population. As you can see in the data, the researchers discovered that both the sexual and asexual populations could eliminate highly deleterious mutations. Organisms harboring such mutations probably died rather easily. In addition, the sexual and asexual populations both retained mildly deleterious and neutral mutations. However, moderately deleterious mutations were eliminated from the sexual populations but not from the asexual ones. One interpretation of these data is that sexual reproduction allowed for the reassortment of beneficial and detrimental alleles, making it easier for sexually reproducing populations to eliminate those mutations that are moderately detrimental.

Figure 51.2 **Paland and Lynch demonstrated the importance of sexual reproduction in reducing the frequency of harmful genetic mutations.**

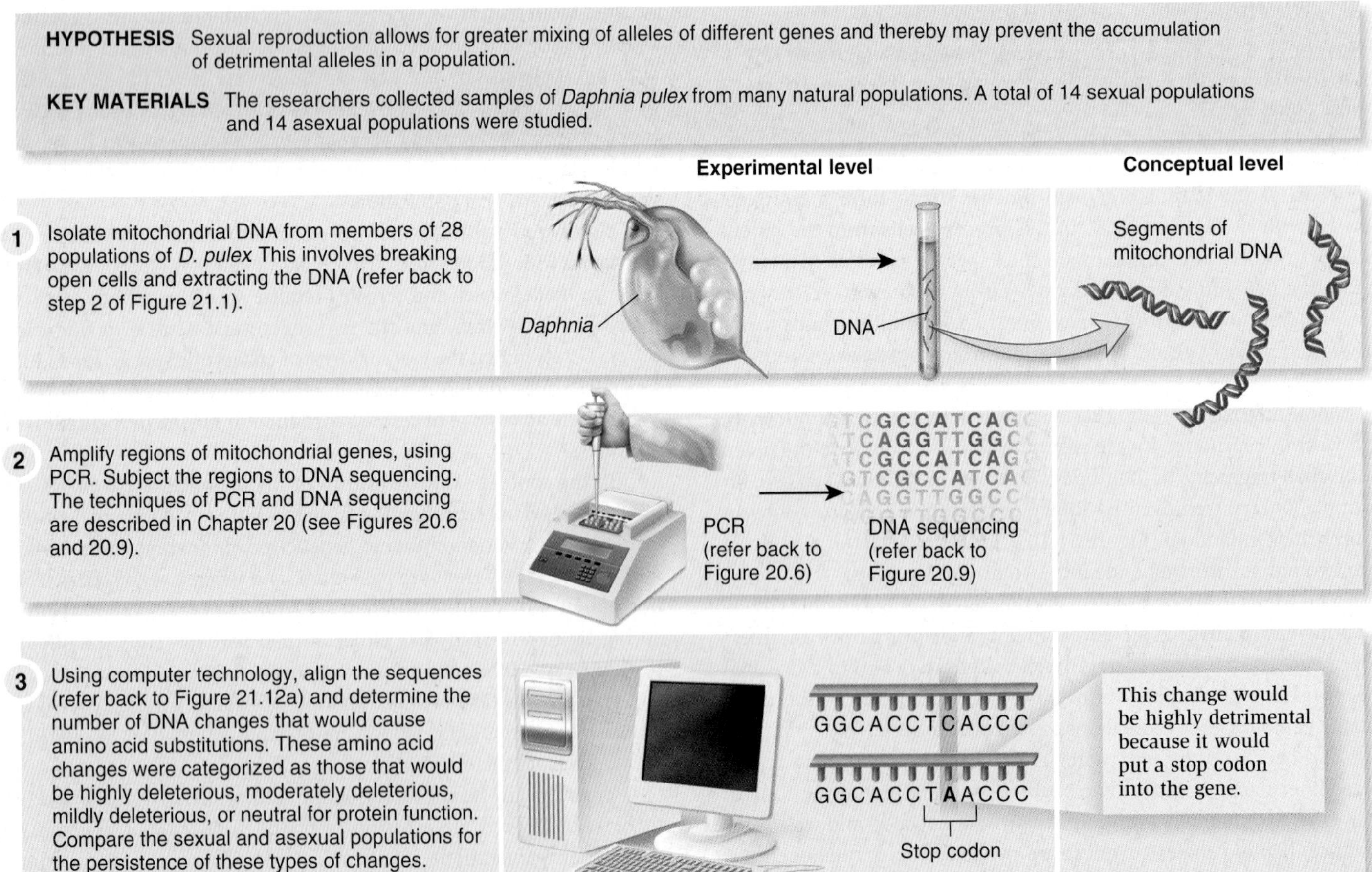

HYPOTHESIS Sexual reproduction allows for greater mixing of alleles of different genes and thereby may prevent the accumulation of detrimental alleles in a population.
KEY MATERIALS The researchers collected samples of Daphnia pulex from many natural populations. A total of 14 sexual populations and 14 asexual populations were studied.
Experimental level
Conceptual level
1 Isolate mitochondrial DNA from members of 28 populations of D. pulex This involves breaking open cells and extracting the DNA (refer back to step 2 of Figure 21.1).
Daphnia
DNA
Segments of mitochondrial DNA
2 Amplify regions of mitochondrial genes, using PCR. Subject the regions to DNA sequencing. The techniques of PCR and DNA sequencing are described in Chapter 20 (see Figures 20.6 and 20.9).
TCGCCATCAG
TCAGGTTGGC
TCGCCATCAG
GTCGCCATCA
AGGTTGGCC
PCR (refer back to Figure 20.6)
DNA sequencing (refer back to Figure 20.9)
3 Using computer technology, align the sequences (refer back to Figure 21.12a) and determine the number of DNA changes that would cause amino acid substitutions. These amino acid changes were categorized as those that would be highly deleterious, moderately deleterious, mildly deleterious, or neutral for protein function. Compare the sexual and asexual populations for the persistence of these types of changes.
GGCACCTCACCC
GGCACCTAACCC
Stop codon
This change would be highly detrimental because it would put a stop codon into the gene.

4 THE DATA

Results from step 3:

Types of amino acid substitutions (The amino acid substitutions were due to rare mutations that occurred in the natural populations of *D. pulex*.)	% of total amino acid substitutions	Allowed to persist in	
		Sexual populations	Asexual populations
Highly deleterious	73.2	No	No
Moderately deleterious	13.3	No	Yes
Mildly deleterious	4.4	Yes	Yes
Neutral	9.1	Yes	Yes

5 CONCLUSION Moderately deleterious mutations are less likely to persist in populations of animals that reproduce sexually.

6 SOURCE Paland, S., and Lynch, M. 2006. Transition to asexuality results in excess amino acid substitutions. *Science* 311:990–992.

Experimental Questions

1. How did Paland and Lynch test the hypothesis that sexual reproduction allowed for the reduction in deleterious mutations?
2. What did they discover?
3. What is the proposed evolutionary benefit of sexual reproduction?

51.2 Gametogenesis and Fertilization

Learning Outcomes:

1. Outline the processes of spermatogenesis and oogenesis.
2. List the advantages that internal fertilization provides.
3. Describe the two different types of hermaphroditism.
4. Compare viviparity, oviparity, and ovoviviparity.

This section covers how and where gametes are formed, and how two gametes join to form a new organism.

Sperm and Eggs Are Produced During the Process of Gametogenesis

Male and female gametes are formed within the **gonads**—the **testes** (singular, testis) in males and the **ovaries** (singular, ovary) in females. Some similarities are found in the ways gametes develop in the testes and the ovaries, as well as some differences.

Gametogenesis—the formation of gametes—begins with cells called germ cells, which multiply by mitosis, resulting in diploid cells (carrying two copies of each gene; denoted as $2n$) called **spermatogonia** (singular, spermatogonium) in males and **oogonia** (singular, oogonium) in females (**Figure 51.3**). Some of these cells become **primary spermatocytes** or **primary oocytes** that may begin the process of meiosis. Until this point, the development of sperm and eggs is similar. From then on, gametogenesis differs between the two types of gamete.

Spermatogenesis The formation of haploid sperm from the diploid germ cell is called **spermatogenesis**. As shown in Figure 51.3a, primary spermatocytes begin this process by undergoing the first of two meiotic divisions (meiosis I). (Refer back to Chapter 15 for a review of the processes of mitosis and meiosis.) In the primary spermatocyte, meiosis I produces two haploid (n) cells called **secondary spermatocytes**. These cells also undergo meiosis (meiosis II), producing four haploid **spermatids** that eventually differentiate into mature sperm cells. Gametogenesis in males, therefore, results in four gametes from each spermatogonium.

The most striking change in each spermatocyte as it differentiates into a sperm is the formation of a flagellum, also called the tail (Figure 51.3b). The movements of the tail require cellular energy and make the sperm motile. The sperm also has a head, which contains the nucleus that carries the chromosomes. At the tip of the head is a special structure called the **acrosome**, which contains proteolytic enzymes that help break down the protective outer layers surrounding the ovum. The head and tail are separated by a midpiece containing one or more mitochondria, depending on the species, that produce the ATP required for tail movements.

Oogenesis Whereas spermatogenesis produces four gametes from each primary spermatocyte, gametogenesis in the female, called **oogenesis**, results in the production of a single gamete from each primary oocyte (Figure 51.3c). The first meiotic division (meiosis I) in oogenesis results in one large cell—a **secondary oocyte**—plus a smaller cell, called a polar body, that eventually degenerates.

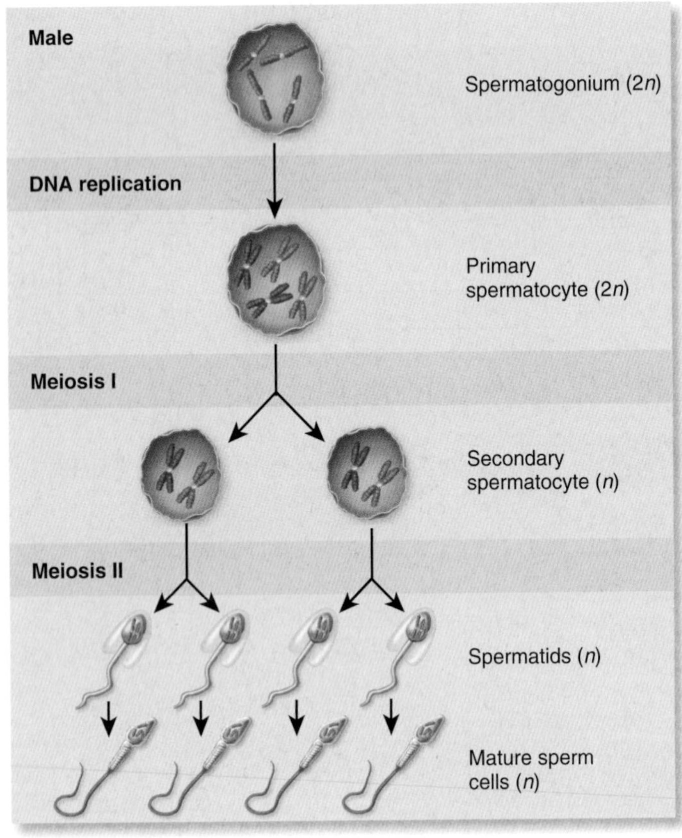

(a) Spermatogenesis

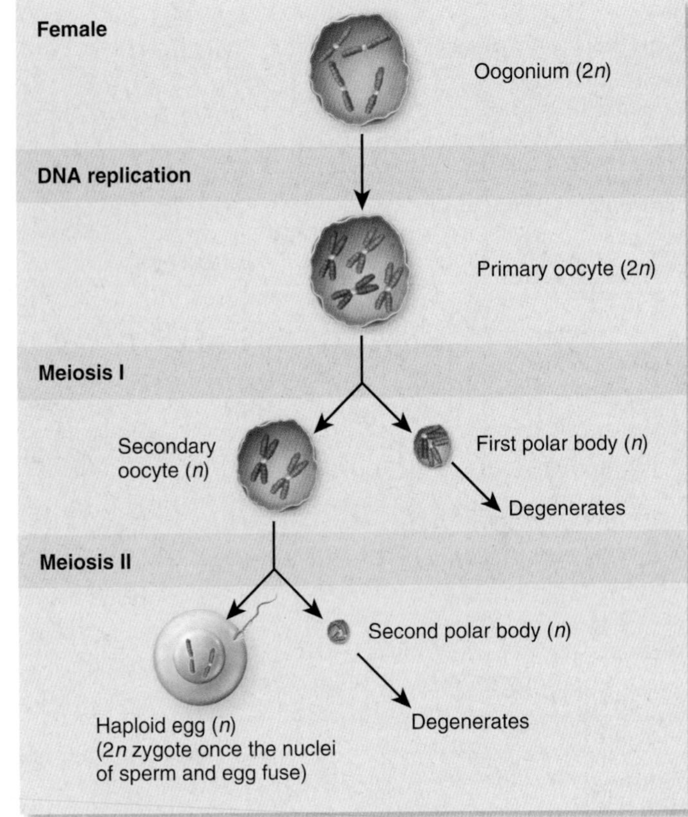

(c) Oogenesis

Acrosome
Head
Nucleus
Mitochondrion
Midpiece
Flagellum (tail)

(b) Mature human sperm

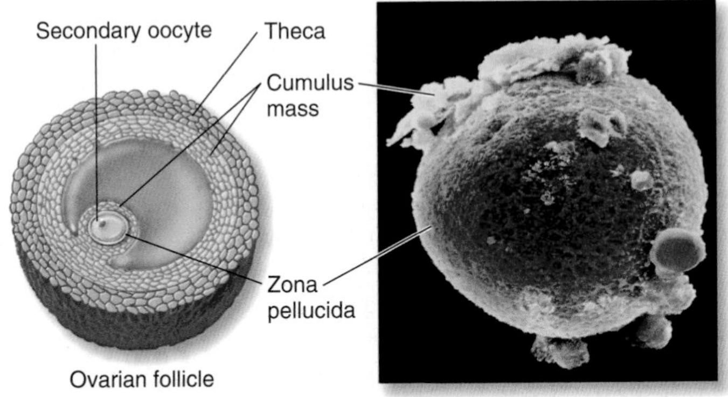

(d) Mature human follicle and oocyte

Figure 51.3 Gametogenesis and gametes in males and females. **(a)** In the process of spermatogenesis, male diploid (2*n*) germ cells undergo two meiotic divisions to produce mature haploid (*n*) sperm. **(b)** The characteristic head, midpiece, and flagellum (tail) of a mature human sperm, as seen in a drawing and accompanying scanning electron micrograph (SEM). **(c)** The process of oogenesis in females, which produces a haploid secondary oocyte that enters but does not complete meiosis II until it is fertilized. **(d)** Mature follicle and oocyte. The drawing depicts a secondary oocyte within its follicle; the SEM shows an isolated human oocyte covered by its zona pellucida and remnants of the cumulus mass.

BIOLOGY PRINCIPLE Structure determines function. The structure of a sperm cell is a good example of this principle of biology at the cellular level. Notice how each of the three major structural elements of the sperm cell is suited for a particular function. When all three elements function together, the cell is capable of fertilizing an egg.

Depending on the species, one or many oocytes can develop at a time. Within the ovaries, each oocyte undergoes growth and development within a structure called a **follicle** before it leaves the ovary in the process of **ovulation**. Once ovulated, a secondary oocyte can become fertilized if sperm are available. In mammals, the secondary oocyte is surrounded by three layers: an inner layer composed of glycoproteins called the **zona pellucida**, which surrounds the surface of the secondary oocyte, a second layer of cells called the cumulus mass which provides protection and nutritive support, and an outer layer called the theca which produces hormones that control oocyte growth

(Figure 51.3d). These layers play important roles in the fertilization of the secondary oocyte.

In mammals, oogenesis begins in the female embryo before birth: A cohort of germ cells develop into primary oocytes and enter meiosis I, which is arrested partway through the process. Meiosis I does not resume until puberty, the time when a mammal first becomes capable of reproducing. In selected primary oocytes, meiosis I is completed, producing haploid secondary oocytes. These cells then begin but do not complete meiosis II. A secondary oocyte will complete meiosis II and become a haploid egg if it is fertilized by a sperm. Once the haploid egg nucleus fuses with a haploid sperm nucleus, a diploid zygote is produced.

Fertilization Involves the Union of Sperm with Egg

Fertilization is a complex series of events by which the haploid male and female gametes unite and become a diploid zygote. Several important cellular and molecular processes must occur before the nuclei of the gametes can fuse. The mechanism by which the egg and sperm make contact has been studied extensively in sea urchins, and some evidence suggests that a similar process occurs in humans and other mammals. When chemical attractant molecules emitted by sea urchin eggs bind to nearby sperm, cellular respiration (that is, the breakdown of nutrients to synthesize ATP) within the sperm increases, which helps increase sperm motility. The sperm then swim toward the egg by following the attractant's concentration gradient. The attractants are species-specific; that is, sperm will respond to the attractants produced only by eggs of their own species.

For a sperm to physically contact the egg, it must first penetrate the layers surrounding the egg's plasma membrane. In mammals, when the head of a sperm contacts the layers surrounding a secondary oocyte, chemicals in the cumulus mass stimulate the breakdown of the membrane covering the acrosome of the sperm. Proteolytic enzymes released from the acrosome digest a local area of the zona pellucida, allowing the sperm to contact the plasma membrane of the secondary oocyte. The plasma membrane of the secondary oocyte then fuses with the sperm head, facilitating its movement into the secondary oocyte's cytoplasm.

Once a sperm fuses with the secondary oocyte, other sperm must be prevented from also fusing with the fertilized egg to avoid the formation of a polyploid zygote, a condition that few zygotes survive. In most mammals, the penetration of one sperm induces metabolic changes within the egg that prevent additional sperm from penetrating the zona pellucida and entering the egg (look ahead to Chapter 52).

In a Given Species, Fertilization Occurs Either Outside or Inside the Female

For sperm to fertilize eggs, the two gametes must physically come into contact. This can occur either outside or inside the female's body. When this occurs outside of the female, the process is called **external fertilization**. This type of fertilization occurs in aquatic environments, when eggs and sperm are released into the water close enough for fertilization to occur. The aqueous environment protects the gametes from drying out. Animals that reproduce by external fertilization show species-specific behaviors that bring the eggs and sperm together. For instance, very soon after a female fish lays her eggs, a male deposits his sperm in the water nearby and they spread over the clump of eggs. When frogs mate, the clasping behavior of the male stimulates the female to release her eggs into the water (**Figure 51.4**); the male then releases sperm onto the eggs. The fertilized eggs then develop outside the mother's body.

Figure 51.4 An example of external fertilization. The male frog (*Rana temporaria*) clasps the female, which stimulates her to release her eggs. He then releases his sperm (not visible here) over the eggs. This process occurs in aquatic environments, which protect gametes from drying out.

Concept Check: *What kinds of aquatic environments are most suitable for external fertilization?*

Although the aqueous environment protects against desiccation, eggs can be eaten by predators, washed away by currents, or subjected to potentially lethal changes in water temperature. Such environmental challenges have led to selection for some species, including many aquatic or amphibious animals, to release very large numbers of eggs at once, as noted previously for the female cod fish.

In contrast to external fertilization, most terrestrial animals and some aquatic animals reproduce by **internal fertilization**, in which sperm are deposited within the reproductive tract of the female during the act called **copulation**, as seen in the mud daubers in the chapter-opening photo. Internal fertilization protects the delicate gametes from environmental hazards and predation and also guarantees that sperm are placed and remain in very close proximity to eggs. Once fertilization occurs within the female, the zygotes then develop into offspring.

The behaviors and anatomical structures involved in achieving internal fertilization are extremely varied among species. Typically, mating involves accessory sex organs, which are reproductive structures other than the gonads. The external accessory sex organs involved in copulation are the genitalia (for example, the **penis** and the **vagina**), which are used to physically join the male and female so that sperm can be deposited directly into the female's reproductive tract. A penis or analogous structure is present in most insects, reptiles, some species of birds (ratites and waterfowl), and all mammals. However, males of other vertebrate species—including most birds—that reproduce by internal fertilization lack a structure that can be inserted into the female, so they deposit sperm in the female by cloacal contact. The cloaca is a common opening for the reproductive, digestive, and excretory systems in these animals (refer back to Figure 45.7).

Another form of internal fertilization involves an indirect means of depositing sperm. In this case, males produce small packets of sperm, called spermatophores, that are deposited externally and then inserted into the female's reproductive tract by either the male or the female. During copulation in cephalopods, the male uses a tentacle to transfer a spermatophore into the mantle cavity of the female. In spiders, the male places a droplet of sperm on a web and then uses a foreleg to insert the droplet into the female's reproductive tract. Subsequently the eggs are fertilized internally and deposited on the web or elsewhere in the environment.

Some Hermaphroditic Species Can Fertilize Themselves or Each Other

In some species, including many gastropods, worms, and some fish and reptiles, individuals have both male and female reproductive systems for part or all of their lives. This is called **hermaphroditism** (after the male and female Greek gods, Hermes and Aphrodite). In some hermaphroditic species, individuals can fertilize their own eggs with their own sperm, but in most hermaphroditic species, individuals exchange sperm with another individual. The latter situation has the selective advantage of creating additional genetic diversity in the population. An advantage of hermaphroditism is that all individuals of a sexually reproducing species can produce offspring.

There are two main types of hermaphroditism, in which animals are simultaneously both male and female or alternately male or female. This first type of hermaphroditism is known as synchronous hermaphroditism. Such hermaphrodites are often sessile animals or burrowing animals such as earthworms, which may live for long periods without encountering sexual partners. A single earthworm can fertilize its own eggs, or two earthworms can join together for several hours during which sperm from each worm fertilize the eggs of the other. In the latter case, individual worms act both as females (receiving sperm to fertilize their eggs) and males (giving sperm to fertilize another worm's eggs), so the offspring carry genes from both individuals.

The second type of hermaphroditism, called sequential hermaphroditism, involves sex reversal, in which a female may change into a male, or vice versa. In such animals, individuals express specific genes and therefore develop and maintain reproductive structures of either a male or a female, but not at the same time. This kind of sex reversal occurs in some species of animals with strong social hierarchies. In some reef-dwelling species of fishes, for example, a single dominant male defends a harem of several females within a specific territory. If that dominant male dies, the largest of the females reverses sex and becomes a male. Thus, these fishes are **protogynous**—that is, female first but capable of becoming males later during their life cycle. Oysters and clownfish are also sequential hermaphrodites, but they are **protandrous**; that is, they are males first and only later become females. An advantage of protandrous hermaphroditism is that males change into females when they are older and larger—and so are likely to be capable of producing a greater number of eggs.

Offspring Are Born Live or Hatch from Eggs That Are Laid or Retained Within the Mother

As we have seen, internal fertilization occurs within the female, and the resulting zygotes then continue their development into offspring. This occurs by three main processes. First, when most of embryonic development occurs within the mother and the animal is born alive, as in most mammals, the process is called **viviparity**. Second, if all or most of embryonic development occurs outside the mother and the embryo depends exclusively on yolk from an egg for nourishment, the process is called **oviparity**. Oviparity is the rule in avian species and is common in reptiles, fishes, amphibians, and insects. Among mammals, only the echidna and the platypus are oviparous. In some cases, embryos of oviparous animals grow within a protective covering, such as a shell, from which they hatch. Terrestrial animals lay eggs with shells that protect the future offspring from desiccation. Such eggs may have leathery shells, as in most reptiles and insects, or hard shells containing calcium carbonate, as in birds and many turtles. These eggs contain all the nutrients necessary for the development of the embryo. Oxygen enters, and carbon dioxide exits, through tiny pores in the shell. The eggs of aquatic species lack shells so that external fertilization may occur. Such eggs are protected from drying out by their environment.

Although oviparity reduces the female's metabolic investment in the young, it increases the incidence of predation. In many amphibian and reptile species, the large number of eggs laid increases the chances of some young surviving even if predation occurs. The energetic cost to the female of producing numerous eggs is high, but usually little or no parental care is involved, freeing the parent to devote energy to other activities. In birds, some fish, and some reptiles (crocodilians), however, incubation of the eggs after laying requires varying degrees of additional energy expenditure in the form of parental care, which typically continues after the eggs have hatched, and thus relatively few eggs are laid.

Lastly, some animals develop by a process called **ovoviviparity**, which features aspects of the first two modes of development. In this case, fertilized eggs covered with a thin shell hatch inside the mother's body, but the offspring receive no nourishment from the mother. Ovoviviparity occurs in sharks, lizards, some snakes, and some invertebrates.

51.3 Mammalian Reproductive Structure and Function

Learning Outcomes:

1. Describe the structure of the human male reproductive tract.
2. Outline the process of hormonal control of the male reproductive system.
3. Describe the structure of the human female reproductive tract.
4. Diagram the events of the ovarian cycle.
5. Outline the process of hormonal control of the reproductive cycle of the human female.
6. Explain how maternal hormones prepare the uterus to accept the embryo.

We turn now to a detailed look at the mammalian reproductive system with particular attention to the human system. For both sexes, we will begin with a description of the anatomy of the reproductive system, including the gonads and the accessory sex structures. We will then examine the hormones that control the production of the gametes and the preparation for and establishment of pregnancy.

The Human Male Reproductive Tract Is Specialized for Production and Ejaculation of Sperm

The external structures of the male reproductive tract—the genitalia—consist of the penis and the scrotum, the sac that contains the testes and holds them outside the body cavity (**Figure 51.5**). The testes develop within the body cavity, and just before birth in human males, they descend into the scrotum, where the temperature is approximately 2°C lower than core body temperature. The lower temperature is optimal for spermatogenesis.

Each testis is composed of tightly packed **seminiferous tubules** encased in connective tissue (**Figure 51.6**). Surrounding the tubules are Leydig cells—scattered endocrine cells that secrete the steroid hormone testosterone. Spermatogenesis begins at puberty and continues throughout life. It occurs all along the walls of the seminiferous tubules. Cells at the earliest stages of spermatogenesis, the spermatogonia, are located nearest the outer surface of the wall. Cells of more advanced stages are located progressively inward, such that the mature sperm are released into the tubule lumen. Cells within the seminiferous tubules are continuously developing from spermatogonia into spermatocytes and eventually to sperm, so at any one time, all types of cells are present along the seminiferous tubule. Support cells, called Sertoli cells, surround the developing spermatogonia and spermatocytes, providing them with nutrients and protection and playing a role in their maturation into sperm.

Sperm moving out of the seminiferous tubules are emptied into the **epididymis**, a coiled, tubular structure located on the surface of the testis (see Figure 51.6). The epididymis is very long—approximately 6 m in humans—and in humans, it takes about 2 to 12 days for new sperm to reach its end. Here the sperm complete their differentiation by becoming motile and gaining the capacity to fertilize ova.

Sperm leave the epididymis through the **vas deferens** (or ductus deferens), a muscular tube leading to the **ejaculatory duct**, which then connects to the urethra (see Figure 51.5). As noted in Chapter 49, the urethra originates at the bladder and extends to the end of the penis. In males, the urethra not only conducts urine but also carries **semen**, a mixture containing fluid and sperm, that is released during **ejaculation**—the movement of semen through the urethra by contraction of muscles at the base of the penis. These contractions during ejaculation contribute to the pleasurable sensation of orgasm.

The liquid components of semen are important for the survival and movement of sperm through the female reproductive tract. This liquid is formed by three accessory glands that secrete substances into the urethra to mix with the sperm. The **seminal vesicles** secrete the monosaccharide fructose, the main nutrient for sperm, as well as other factors that enhance sperm motility and survival. The

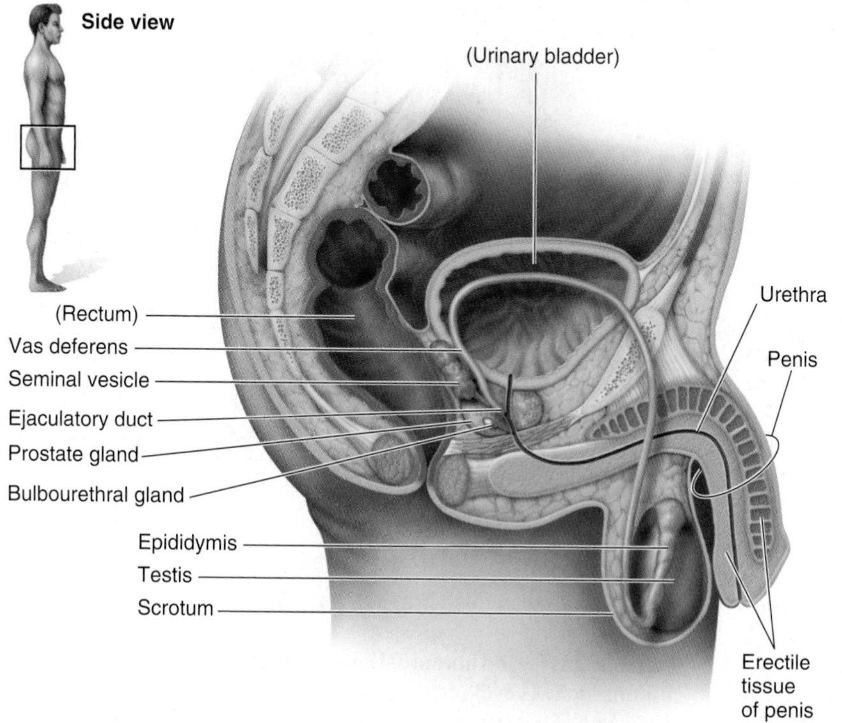

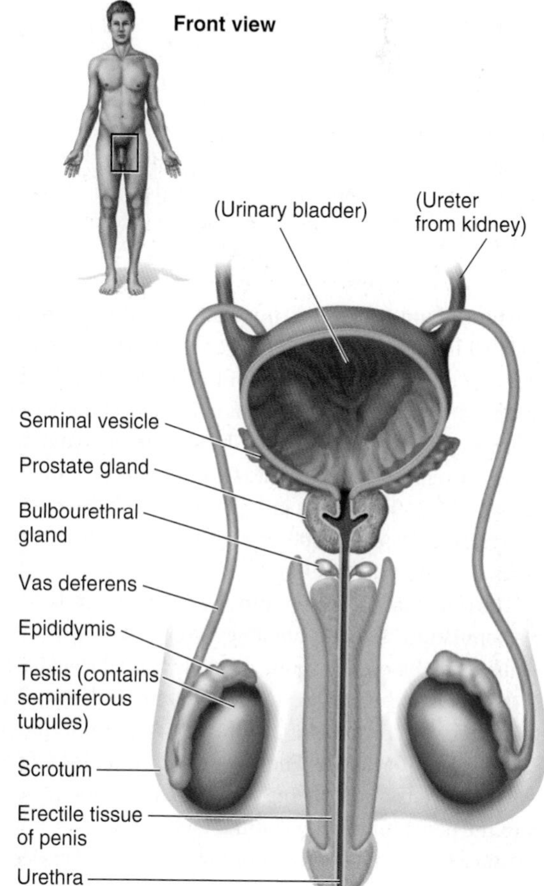

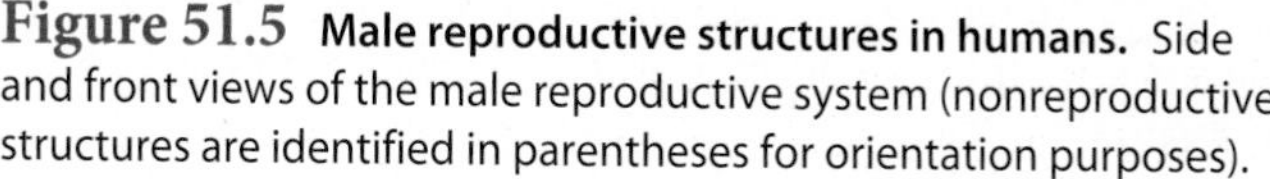
Figure 51.5 **Male reproductive structures in humans.** Side and front views of the male reproductive system (nonreproductive structures are identified in parentheses for orientation purposes).

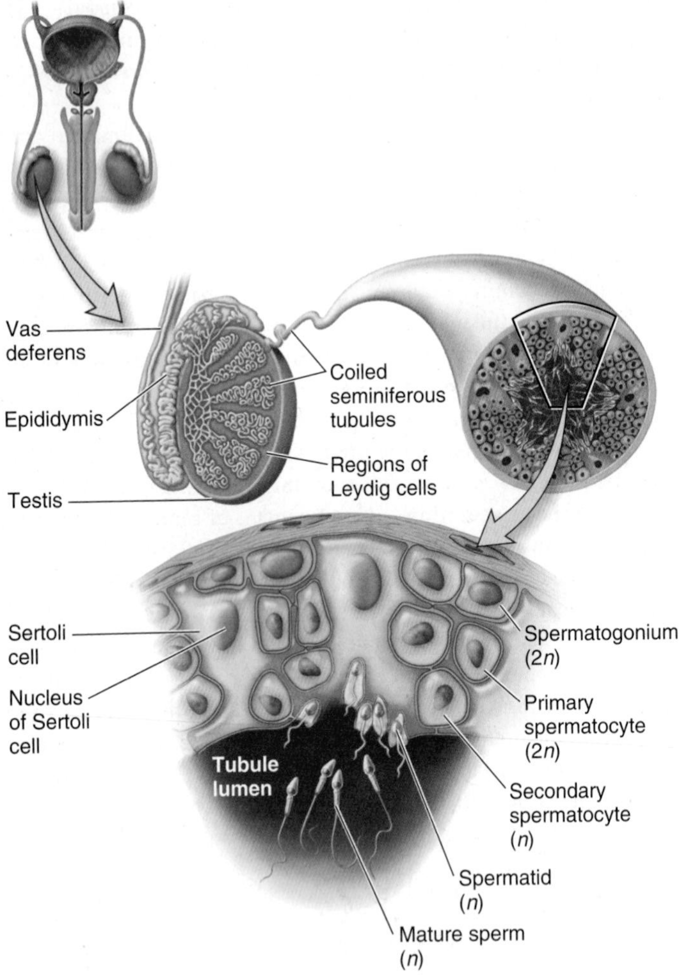

Figure 51.6 The internal structure of the human testis and associated structures. The top of this drawing shows the internal structure of a testis, epididymis, and vas deferens. At the bottom, the stages of spermatogenesis within the seminiferous tubule are shown.

prostate gland secretes a thin, alkaline fluid that protects sperm from acidic fluids in the urethra and within the female reproductive tract. An additional small amount of alkaline fluid is secreted by the **bulbourethral glands.** Secretions constitute about 95% of semen, and sperm make up only about 5% of total semen volume. The volume of semen released at ejaculation in humans is about 2 to 5 mL and contains 20–130 million sperm per milliliter. Although this seems like an excessive amount of sperm to fertilize a single ovum, relatively few sperm actually reach the egg.

Introduction of sperm into the female reproductive system during copulation is made possible by erection of the penis. Erection occurs when blood fills spongy erectile tissue located along the length of the penis (see Figure 51.5). Sexual arousal stimulates release of the gaseous neurotransmitter nitric oxide (NO) in the penis, causing vasodilation of arteries. The pressure of the blood flowing into the penis constricts nearby veins, causing a reduction in venous drainage from the penis, engorging it with blood. After ejaculation, NO release is reduced, causing a reversal of the vascular changes responsible for erection. Both physiological and psychological factors can result in an inability to achieve an erection, a medical condition known as erectile dysfunction (also called impotence). Orally administered drugs such as sildenafil and tadalafil (Viagra and Cialis) can increase the occurrence of erections by stimulating the same intracellular signaling events as NO in the smooth muscle cells of penile blood vessels.

Male Reproductive Function Requires the Actions of Testosterone

Recall from Chapter 50 that the hypothalamus is a structure at the base of the brain that synthesizes neurohormones, including gonadotropin-releasing hormone (GnRH) (see Table 50.2 and Figure 50.4). GnRH stimulates the anterior pituitary gland to release two gonadotropins: luteinizing hormone (LH) and follicle-stimulating hormone (FSH).

In males, LH stimulates the Leydig cells of the testes to produce androgens, particularly testosterone (**Figure 51.7**). Testosterone stimulates the growth of the male reproductive tract and the genitalia during development and puberty and the development of male secondary sex characteristics. In humans, these include facial hair growth, increased muscle size, and deepening of the voice. Examples

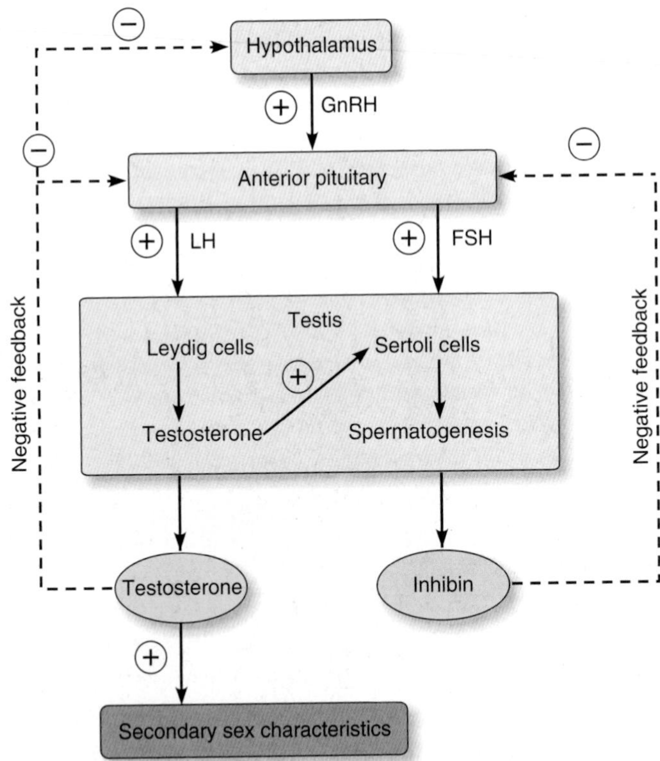

Figure 51.7 The hormonal control of male reproduction. In response to LH, Leydig cells in the testes secrete testosterone, which, along with FSH, acts on Sertoli cells to facilitate spermatogenesis. Negative signs indicate inhibitory effects via negative feedback, and plus signs indicate stimulatory effects.

Concept Check: *What would happen to spermatogenesis in a man taking testosterone to increase muscle mass or improve athletic performance?*

in other male mammals include growth of the horns of a bull, enlargement of the nose of proboscis monkeys, and formation of cheek pads in some apes. Testosterone-dependent secondary sex characteristics are not unique to mammals, however. One familiar example is the bright coloration of plumage in male birds such as the peacock.

The other pituitary gonadotropin, FSH, functions along with testosterone to stimulate spermatogenesis. FSH does this by stimulating the activity of the Sertoli cells within the seminiferous tubules. The Sertoli cells provide the nutritional and structural support necessary for development of the sperm. They also respond to testosterone produced by the Leydig cells, by stimulating mitosis and meiosis of the germ cells associated with them in the tubules.

Production of sperm and testosterone is kept in check by negative feedback mechanisms that control the amount of gonadotropins produced (see Figure 51.7). Testosterone in high concentrations inhibits the secretion of GnRH from the hypothalamus, so both LH and FSH are inhibited when the blood concentration of testosterone is high. Testosterone also directly inhibits LH secretion by the anterior pituitary gland. In addition, when Sertoli cells are activated by FSH, they secrete a protein hormone called inhibin, which enters the blood and inhibits further secretion of FSH. These feedback mechanisms maintain homeostatic concentrations of FSH and LH in the blood.

Before puberty, LH is not released in sufficient amounts to stimulate significant testicular production of testosterone, and the reproductive system is quiescent—spermatogenesis does not occur. Although the mechanisms that initiate puberty in mammals are still not completely understood, research has shown that increased GnRH production at that time initiates increased LH and FSH secretion from the pituitary. The testosterone induced by LH stimulates development of adult male characteristics. Testosterone is also responsible for an increased sex drive (libido) at this time.

The Female Reproductive Tract Is Specialized for Production and Fertilization of the Egg and Development of the Embryo

The female genitalia differentiate from the same embryonic tissues as the male genitalia. The female genitalia are composed of the larger, hair-covered outer folds called the **labia majora** (from the Latin, meaning major lips), which surround the external opening of the reproductive tract, plus the smaller, inner folds, called the **labia minora** (from the Latin, meaning minor lips; Figure 51.8). The labia majora originate from the same embryonic tissue as the scrotum of the male, whereas the labia minora originate from urethral primordial tissue. At the anterior part of the labia minora is the **clitoris**, which is erectile tissue of the same origin as the penis. Like the penis, the clitoris becomes engorged with blood during sexual arousal and is very sensitive to sexual stimulation. Unlike males, however, the openings of the reproductive tract and the urethra are separate in females. The opening of the urethra is located between the clitoris and the opening of the reproductive tract.

In mammals alone, the external opening of the reproductive tract leads to the vagina, a tubular, smooth muscle structure into which

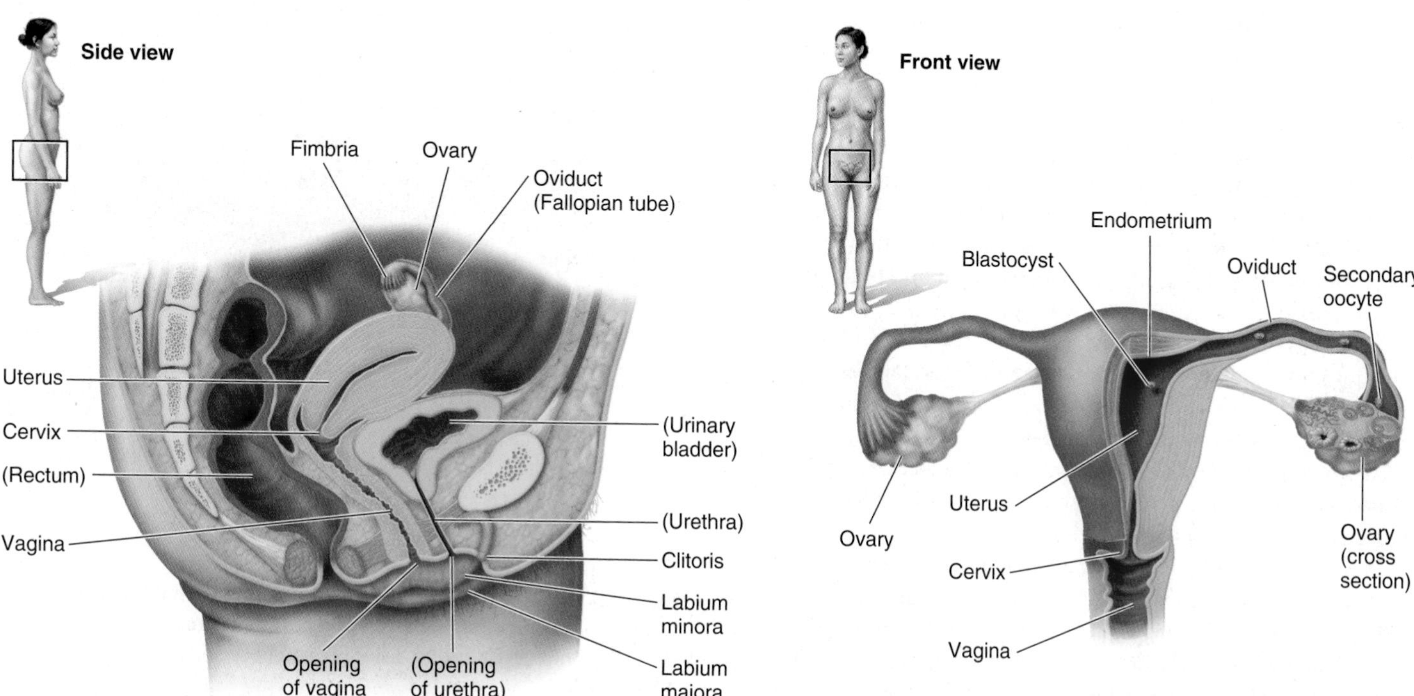

Figure 51.8 **Female reproductive structure and function in humans.** Side and front views of the female reproductive system (nonreproductive structures are identified in parentheses for orientation purposes). An oocyte moves from the ovary into the oviduct (Fallopian tube), where it may be fertilized and develop into a blastocyst. Subsequently, the blastocyst enters the uterus, where it may implant in the endometrium, the inner lining of the uterus.

sperm are deposited during copulation. At the end of the vagina is the **cervix**, the opening to the **uterus**, which in humans is about the size and shape of an inverted pear. Sperm pass through the cervix into the uterus, an organ specialized for carrying the developing embryo and fetus. It consists of an inner lining of glandular and secretory cells, called the endometrium, and a thick muscular layer, called the myometrium. We will discuss the functions of the uterus later in the chapter.

Oocytes develop within one of the two bilateral ovaries (see Figure 51.8), which are suspended within the abdominal cavity by connective tissue. In humans, each ovary is typically a little larger than an almond. Usually, a secondary oocyte leaves the ovary and is quickly drawn into a thin tube, the **oviduct** (also called the Fallopian tube), by the actions of undulating fimbriae (fingerlike projections) of the oviduct that extend out to the ovary.

The secondary oocyte is moved down the length of the oviduct by cilia on the oviduct's inner surface. For fertilization to take place, sperm must travel through the cervix and uterus and then into the oviduct, where fertilization typically occurs. Upon contact with a sperm, the secondary oocyte completes meiosis II, and the union of sperm and secondary oocyte creates a fertilized egg, or zygote. The zygote undergoes several cell divisions to become a **blastocyst**, a ball of approximately 32–150 cells that enters the uterus, where it will develop into an embryo (the details of this development are covered in Chapter 52).

Gametogenesis in Females Is a Cyclical Process Within the Ovaries

In contrast to spermatogenesis, which continues throughout postpubertal life in the testes in males, most female mammals appear to be born with all the primary oocytes they will ever have, although limited recent evidence in mice suggests that new oocytes may form later in life. At birth, each ovary in a human female has about 1 million primary oocytes, which are arrested in prophase of meiosis I. Most of these degenerate before the onset of puberty, when each ovary contains about 200,000 primary oocytes. Other than this degeneration, the ovaries are quiescent until puberty, when they begin to show cyclical activity.

The cells within the mammalian ovaries secrete a family of steroid hormones called estrogens. The most important estrogen is **estradiol**, which plays a critical role in ovulation and influences the secondary sex characteristics of females. The secondary sex characteristics, which begin to develop at puberty, include development of breasts, widening of the pelvis (an adaptation for giving birth), and a characteristic pattern of fat deposition.

The **ovarian cycle** involves the development of an ovarian follicle, the release of a secondary oocyte, and the formation and subsequent regression of a corpus luteum (Figure 51.9). During the first week of the ovarian cycle in humans, several primary oocytes that have been maturing for several months, each within their own follicle, now begin their final maturation steps. By the beginning of the second week, all but one of these growing follicles and its primary oocyte degenerate, and the single remaining follicle continues to develop and enlarge, and a new crop of immature follicles begins their slow growth period for future ovarian cycles. During that time, the primary oocyte of that follicle completes meiosis I, becomes a secondary oocyte, and begins meiosis II. The developing secondary oocyte is surrounded by cells of the cumulus mass and theca, which both protect and nurture it and which secrete estradiol. The estradiol is secreted into the blood, where it functions to control the secretion of LH and FSH from the anterior pituitary gland. Some estradiol is also secreted into the follicle, where it stimulates fluid secretion into the inner core of the follicle. As the follicle grows in response to continued stimulation by LH and FSH, the fluid pressure inside the follicle increases, until it begins to form a bulge. Eventually, ovulation occurs as the follicle ruptures, and the secondary oocyte, zona pellucida, and some surrounding supportive cells of the cumulus mass are released from the ovary.

Cells in the empty follicle subsequently undergo pronounced anatomical and physiological changes, developing into a structure called the **corpus luteum**. In humans, the corpus luteum is active for approximately the second half of the ovarian cycle. It is responsible for secreting hormones that stimulate the development of the uterus

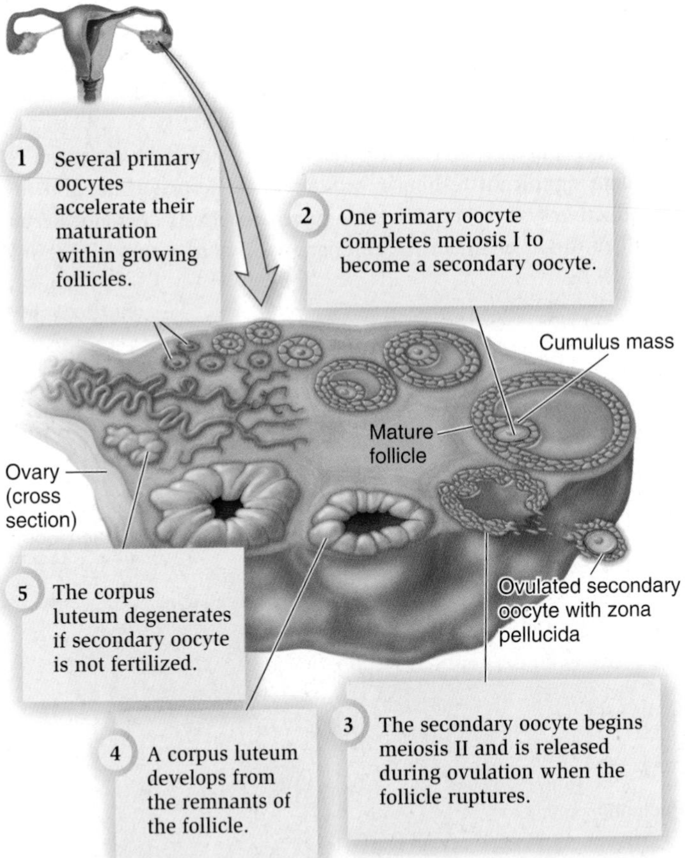

1 Several primary oocytes accelerate their maturation within growing follicles.
2 One primary oocyte completes meiosis I to become a secondary oocyte.
Cumulus mass
Mature follicle
Ovary (cross section)
Ovulated secondary oocyte with zona pellucida
5 The corpus luteum degenerates if secondary oocyte is not fertilized.
4 A corpus luteum develops from the remnants of the follicle.
3 The secondary oocyte begins meiosis II and is released during ovulation when the follicle ruptures.

Figure 51.9 **Follicle and oocyte development in the ovarian cycle.** Development of an oocyte and corpus luteum within the ovary are events that occur during a single ovarian cycle.

BIOLOGY PRINCIPLE Living organisms grow and develop. This figure illustrates that growth and development do not necessarily stop in animals once they have reached maturity. Organs such as the human ovary can continue to develop throughout much of adulthood, as shown in this figure. Likewise, male gametes continue to develop throughout life in humans, as shown in Figure 51.6.

required for sustaining the embryo in the event of a pregnancy. If pregnancy does not occur, the corpus luteum degenerates, and a new group of follicles with their primary oocytes develops.

In an adult female, the ovarian cycle may be as brief as a few days in small rodents or as lengthy as 15–16 weeks in elephants. In humans, a typical ovarian cycle lasts approximately 28 days. Ovulation and degeneration of additional primary oocytes continue throughout adulthood, with about 300 to 500 secondary oocytes ovulated over a woman's 30- to 40-year reproductive lifetime. Eventually, the oocytes become nearly depleted, and a woman stops having ovarian cycles, an event called **menopause**. The average onset of menopause in the U.S. is approximately 51 years of age. After menopause, a woman is no longer capable of becoming pregnant. At one time, scientists thought menopause was unique to humans, but it now appears that certain other female mammals, if they survive long enough, also become incapable of ovulation at some point.

The Ovarian Cycle Results from Changes in Hormone Secretion

We saw that in males, testosterone produced in the testes exerts a negative feedback on secretion of GnRH and LH. In females, however, the situation is more complicated. Although GnRH also stimulates release of LH and FSH in females, the resulting estradiol can have both negative and positive feedback effects on the gonadotropins. To understand this, let's examine the hormone changes that occur during the ovarian cycle in a human female (**Figure 51.10**).

The first half of the ovarian cycle is called the follicular phase of the cycle, because this is when the growth and differentiation of a cohort of follicles are occurring. The relatively low concentration of LH that exists during follicular development stimulates the cells of the follicle to make estradiol. The estradiol that is produced is important for enlargement and growth of the oocytes and it also is secreted into the blood, where it can influence the secretion of LH and FSH.

As these follicles develop, estradiol (and to a lesser extent, another steroid hormone called progesterone) production continues, and consequently, the concentration of estradiol in the blood slowly but steadily increases. Initially, estradiol exerts a negative feedback action on LH and FSH secretion, and all but the largest of follicles die. When the remaining follicle is fully developed and ready for ovulation, its production of estradiol increases, such that the blood concentration of estradiol increases sharply. At that time, the feedback action of estradiol on LH and FSH switches from negative to positive, by mechanisms that involve increased GnRH secretion from the hypothalamus. This results in a sudden, sharp surge in gonadotropin concentrations in the blood, particularly LH.

The LH released from the pituitary as a result of positive feedback by estradiol induces rupture of the follicle and ovulation. This type of ovulation is known as spontaneous ovulation, because it happens regularly on a cyclical basis without requiring any external stimulus. Some mammals, however, including rabbits, cats, and camels, undergo ovarian cycles that turn off unless mating occurs. If mating occurs at a time in the ovarian cycle when oocytes are ready, the mating act itself triggers hormonal events that result in ovulation. This mechanism, called induced ovulation, helps ensure that oocytes are not wasted by being ovulated when the female has not mated with a male.

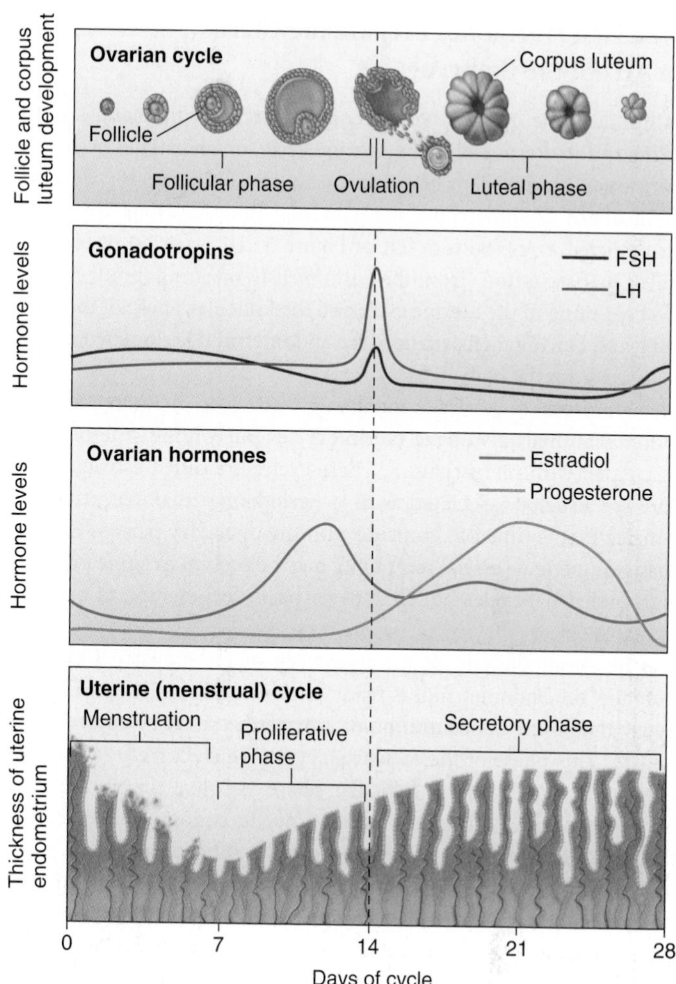

Figure 51.10 **The ovarian and uterine cycles in a human female.** The ovarian cycle is divided into the follicular and luteal phases. The uterine cycle is divided into menstruation and the proliferative and secretory phases.

Concept Check: *Would similar surges in the concentrations of FSH and LH be expected to occur in human males?*

Ovulation marks the end of the follicular phase and the beginning of the luteal phase of the ovarian cycle, named after the corpus luteum. Estradiol production decreases, and LH initiates development of the corpus luteum. The corpus luteum secretes progesterone, the dominant ovarian hormone of the luteal phase, plus some estradiol. Progesterone inhibits LH and FSH secretion, and it further prepares the uterus for receiving and nourishing the embryo. If fertilization of the secondary oocyte does not occur, the corpus luteum degenerates after 2 weeks, allowing LH and FSH to initiate development of a new set of oocytes and their follicles. However, if fertilization does occur, the blastocyst develops a surrounding layer of cells that secrete an LH-like hormone, called **chorionic gonadotropin**, which maintains the corpus luteum and its ability to secrete progesterone. Chorionic gonadotropin is the hormone that is tested for by home pregnancy tests; it is excreted in the urine, where its presence is readily detectable.

Maternal Hormones Prepare the Uterus to Accept the Embryo

In humans, the ovarian cycle occurs in parallel with changes in the lining of the uterus called the **uterine cycle**, or **menstrual cycle**. The hormones produced by the ovarian follicle influence the development of the endometrium, the glandular inner layer of the uterus. As depicted at the bottom left of Figure 51.10, a period of bleeding called **menstruation** (from the Latin *mensis*, meaning month) marks the beginning of the uterine cycle and the follicular phase of the ovarian cycle. During menstruation, the endometrium is sloughed off and released from the body.

Menstrual cycles are found in many primates, including humans. Other mammals also have uterine cycles but without the bleeding associated with menstruation. These cycles are called estrous cycles and are usually associated with a period of sexual receptivity in females that is timed to coincide with the ovulatory period. Cyclical changes in female sexual receptivity may be present in some primates with menstrual cycles but have never been documented to occur in humans.

By about the end of the first week of the menstrual cycle in humans, the endometrium is ready to grow again in response to the newly increasing concentration of estrogen secreted by a developing follicle. This phase of the menstrual cycle, which corresponds to the latter part of the ovarian follicular phase, is called the proliferative phase (see Figure 51.10). During this time, the endometrium becomes thicker and more vascularized. During the subsequent luteal phase of the ovarian cycle, progesterone from the corpus luteum initiates further endometrial growth, including the development of glands that secrete nutritive substances that sustain the embryo during its first 2 weeks in the uterus. This part of the menstrual cycle is called the secretory phase. If fertilization does not occur, degeneration of the corpus luteum and the associated decrease in progesterone and estrogen concentrations initiate menstruation and the beginning of the next uterine cycle. If fertilization does occur, however, and the blastocyst becomes embedded in the endometrium, pregnancy begins, as described next.

51.4 Pregnancy and Birth in Mammals

Learning Outcomes:

1. Describe the major developmental events in the first trimester of pregnancy in mammals.
2. Explain the relationship between the fetal and maternal structures of the placenta.
3. Outline the hormonal control of the birth process.
4. Describe several mechanisms by which animals synchronize the production of offspring with favorable environmental conditions.

Pregnancy, or gestation, is the time during which a developing embryo and fetus grows within the uterus of the mother. Physiologically, pregnancy is considered to begin not at fertilization but when the embryo is established in the uterine lining. This occurs within days of fertilization in animals with short gestation lengths but may take weeks in large animals with long gestations.

In mammals, gestation length varies widely and is roughly related to the size of adults in a particular species. Small animals such as hamsters and mice have gestation periods of 16 to 21 days, canines have longer pregnancies of about 60 to 65 days, humans average about 268 days, and the Asian elephant carries its fetus up to 660 days. The advantages of prolonging prenatal development are twofold: The embryo is protected while it is developing in the uterus, and the offspring can be more fully developed at birth. This is especially important for animals whose survival depends on mobility shortly after birth, such as horses and ruminants.

Gestation length is influenced not only by adult body size but also by the number of offspring in a single pregnancy. Rats, for example, which bear up to 12 or so offspring per litter, have a short gestation period and produce young that are relatively undeveloped at birth and are totally dependent on the mother. Horses, by contrast, have a long gestation period and typically give birth to a single, highly developed offspring. In this section, we will examine how mammals have evolved to retain their young in the uterus for extended periods, the structure and function of the nourishing placenta, and the role of hormones in pregnancy and birth.

Most Mammals Retain Their Young in the Uterus and Nourish Them via a Placenta

Three types of pregnancies are found in mammals, which correspond to the three clades of mammals (refer back to Chapter 34). Monotremes such as the platypus are the only mammals that lay fertilized eggs. In marsupials such as the kangaroo, the young are born while still extremely undeveloped. They then crawl up the mother's abdomen to her pouch, where they attach to a nipple to suckle and obtain nourishment. They remain there and mature within the pouch. Compared with marsupials, humans and other eutherian mammals retain their young within the uterus for a longer period of time, and nourish them via transfer of nutrients and gases through a structure called the **placenta**.

During pregnancy, many physiological changes occur in both the embryo and the mother. The first event of pregnancy in eutherian (placental) mammals is **implantation**, when the blastocyst embeds within the uterine endometrium, which typically occurs in humans around 8 to 10 days after fertilization. Initially, the implanted blastocyst receives nutrients directly from endometrial glands. However, shortly after implantation, newly developing embryonic tissues merge with the endometrium to form the placenta (Figure 51.11), which remains in place and grows larger as the embryo matures into a **fetus**. (In humans, an embryo is called a fetus after the eighth week of gestation.) The placenta, therefore, has a maternal portion and a fetal portion.

The placenta is rich in blood vessels from both the mother and the fetus. The maternal and fetal sets of vessels lie in close proximity. The fetal portion of the placenta, called the chorion, contains convoluted structures called chorionic villi that provide a large surface area containing capillaries for exchange of nutrients, gases, and other solutes. Nutrients and oxygen from the mother are carried through maternal arteries, where her blood pools in large areas of the fetal placenta surrounding the fetal capillaries. Solutes diffuse from the maternal blood into fetal capillaries, and from there flow into the umbilical

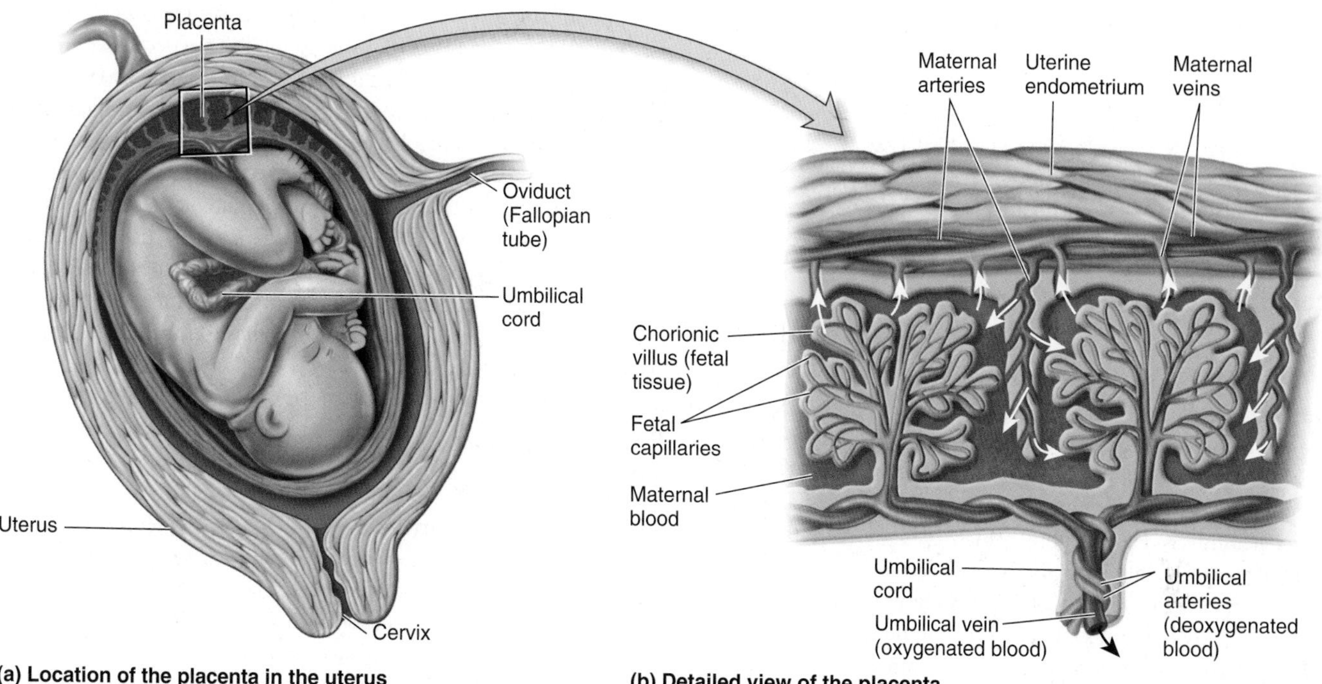

(a) Location of the placenta in the uterus **(b) Detailed view of the placenta**

Figure 51.11 **The structure of the placenta.** In all mammals, the placenta is composed of both fetal and maternal tissues. **(a)** Overview of placental structure in the human. **(b)** Enlarged view of the placenta showing the relationship between fetal and maternal structures. Note that in humans, blood in the fetal and maternal circulations does not mix.

BioConnections: Note that the umbilical arteries are shown in blue to signify that they carry deoxygenated blood, and the umbilical vein is shown in red to signify it carries oxygenated blood. Normally, arteries carry oxygenated blood, and veins carry deoxygenated blood, except in the pulmonary circulation (see Chapter 48). Why is the circulation through the placenta like that of the lungs?

vein, part of the fetal circulation. In turn, carbon dioxide and other waste products from the fetus are carried through the umbilical artery to the placenta, where they diffuse into the mother's circulation, from which they can be excreted. Because of this placental organization, the blood of the mother and the fetus do not mix.

Prenatal development in humans is generally described arbitrarily as having three trimesters, each of which lasts about 3 months (**Figure 51.12**). During the first 2 months of pregnancy, the organs of the embryo develop. At the end of the first trimester, the rudiments of the organs are present, and the developing fetus is about an inch long.

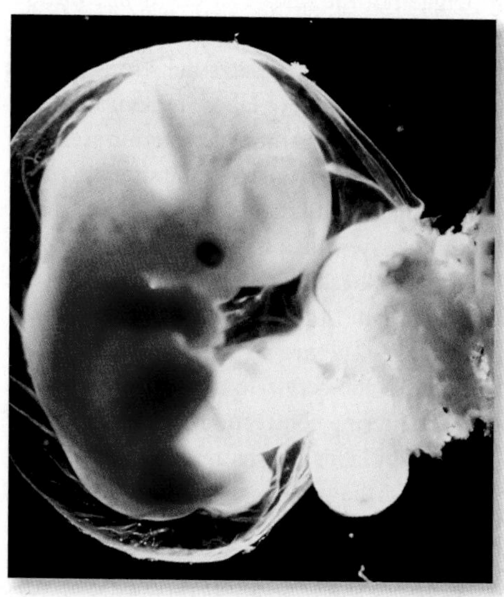

(a) First-trimester human embryo (6 weeks)

(b) Second-trimester human fetus (16 weeks)

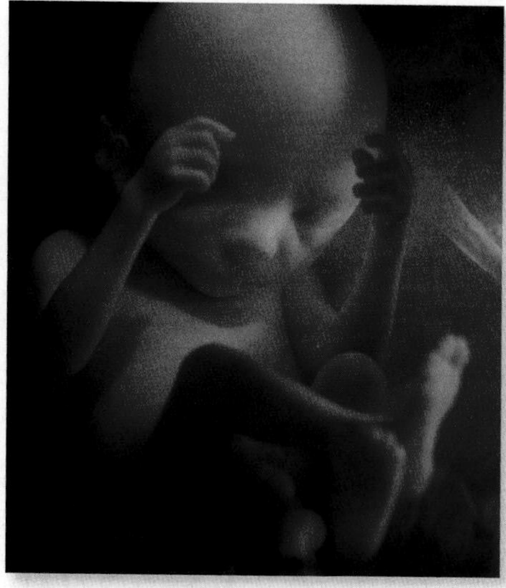

(c) Early third-trimester human fetus (about 24 weeks)

Figure 51.12 **Prenatal development in humans.**

The second trimester is an extremely rapid phase of growth. During the third trimester, the lungs of the fetus mature so that they are ready to function as gas-exchange organs.

Genomes & Proteomes Connection

The Evolution of the Globin Gene Family Has Been Important for Internal Gestation in Mammals

As discussed in Chapter 21, genes can become duplicated to create gene families. Gene families have been important in the evolution of complex traits because the various members of a gene family can enable the expression of complex, specialized forms and functions. An interesting example is the globin gene family in animals. Globin genes encode polypeptides that are subunits of proteins that function in oxygen binding. Hemoglobin, which is made in red blood cells, carries oxygen throughout the body in all vertebrates and many invertebrates, delivering oxygen to all of the body's cells. In humans, the globin gene family is composed of several homologous genes that were originally derived from a single ancestral globin gene (refer back to Figure 21.8).

All of the globin polypeptides are subunits of proteins that play a role in oxygen binding, but the various family members tend to have specialized functions. For example, certain globin genes are expressed only during particular stages of embryonic development. This has particular importance in placental mammals, because the oxygen demands of a growing embryo and fetus are quite different from the demands of its mother. These different demands are met by the differential expression of hemoglobin genes during prenatal development.

Altogether, five globin genes, designated α, β, γ, ε, and ζ, encode the major subunits that are found in hemoglobin proteins at different developmental stages. During embryonic development, the ε-globin and ζ-globin genes are turned on, resulting in embryonic hemoglobin with a very high affinity for oxygen (**Table 51.1**; refer also to Figure 13.3). At the fetal stage, these genes are turned off, and the α-globin and γ-globin genes are turned on, producing fetal hemoglobin with slightly less (but still high) affinity for oxygen. Finally, just before birth, expression of the γ-globin gene decreases, and the β-globin gene is turned on, resulting in adult hemoglobin, which has a lower affinity for oxygen than either the embryonic or fetal forms. The higher affinities of embryonic and fetal hemoglobins enable the embryo and fetus to remove oxygen from the mother's bloodstream and use that oxygen to meet their own metabolic demands. Therefore, the expression of different globin genes at particular stages of development enables placental mammals to develop in the uterus without either breathing on their own or being continually exposed to atmospheric oxygen (as occurs for vertebrate embryos that develop externally within shelled eggs).

Table 51.1 Globin Gene Expression During Mammalian Development

Stage of development	Globin genes expressed	Hemoglobin composition	Oxygen affinity (P_{50})*
Embryo	ε-globin and ζ-globin	Two ε-globin and two ζ-globin subunits	5–13.5 mmHg
Fetus	γ-globin and α-globin	Two γ-globin and two α-globin subunits	19.5 mmHg
Birth to adult	β-globin and α-globin	Two β-globin and two α-globin subunits	26.5 mmHg

*P_{50} values represent the partial pressure of oxygen required to half-saturate hemoglobin (see Chapter 48): A lower P_{50} indicates a higher affinity of hemoglobin for oxygen. The value for embryos is an estimate based on in vitro experiments. All values are for human hemoglobins.

Birth Is Dependent on Hormones That Elicit a Positive Feedback Loop

Birth—also called **parturition**—is initiated by the actions of several hormones and other factors secreted by the mother and the placenta (**Figure 51.13**). Toward the end of pregnancy, endocrine signals from the fetus stimulate the placenta to start secreting large amounts of estrogens such as estradiol into the maternal circulation. Estrogens have at least two major effects on uterine tissue at this time. First, they promote gap junction formation between uterine smooth muscle cells, which enables coordinated uterine contractions. Second, estrogens enhance uterine sensitivity to oxytocin.

Recall from Chapter 50 that oxytocin is a posterior pituitary hormone that stimulates contraction of uterine muscle. The high concentration of estradiol in the mother's blood near the end of pregnancy stimulates the production of oxytocin receptors within the cells of the smooth muscle layer of the uterus, making the uterus more sensitive to oxytocin. At the same time, the fetus usually positions itself with its head above the uterine cervix in preparation for birth. The pressure of the fetus's head pressing on the cervix stretches the smooth muscle of the uterus and cervix. This stretch is detected by neurons in these structures. Signals from the stretch-sensitive neurons are sent to the mother's hypothalamus, triggering the release of oxytocin from the posterior pituitary gland.

Binding of oxytocin to its receptors initiates the strong uterine muscle contractions that are the hallmark of **labor**. In addition to its direct action on uterine muscle, oxytocin stimulates uterine secretion of prostaglandins that act with oxytocin to increase the strength of the muscle contractions. The stronger contractions elicit more oxytocin release from the mother's pituitary, which causes yet stronger contractions, setting up a positive feedback loop that continues until the baby is born (refer back to Figure 40.14).

Labor occurs in three stages (**Figure 51.14**). The initial stage induces dilation and thinning of the cervix to allow passage of the fetus out of the uterus. As the uterine contractions get stronger and more frequent toward the end of labor, the fetus is pushed, usually headfirst, through the cervix and the vagina and out into the world; this is the second stage of labor. In the third and final stage, the contractions continue for a short while. Blood vessels within the placenta and umbilical cord contract and block further blood flow, making the newborn independent from the mother. As the oxytocin-induced contractions continue for a short period, the placenta detaches from the uterine wall and is delivered a few minutes after the birth of the baby.

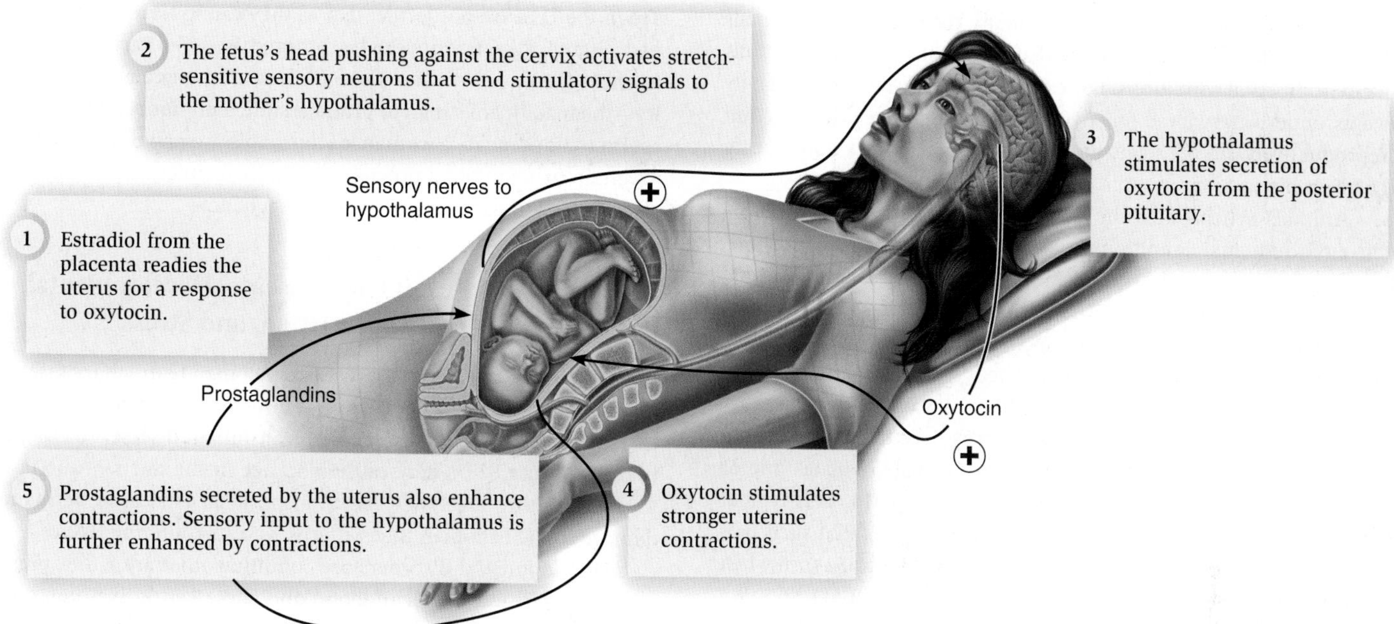

2 The fetus's head pushing against the cervix activates stretch-sensitive sensory neurons that send stimulatory signals to the mother's hypothalamus.
Sensory nerves to hypothalamus
+
3 The hypothalamus stimulates secretion of oxytocin from the posterior pituitary.
1 Estradiol from the placenta readies the uterus for a response to oxytocin.
Prostaglandins
Oxytocin
+
5 Prostaglandins secreted by the uterus also enhance contractions. Sensory input to the hypothalamus is further enhanced by contractions.
4 Oxytocin stimulates stronger uterine contractions.

Figure 51.13 **Hormonal control of parturition.** Birth relies on maternal hormones that act on the uterus, and neural signals from the uterus. In response to sensory neural input arising from the push of the fetus on the cervix, the maternal posterior pituitary gland releases oxytocin, which stimulates uterine smooth muscle contractions. The secretion of prostaglandins by the uterus also increases the strength of the contractions. Sensory receptors in the uterus detect the more forceful contractions and signal the mother's posterior pituitary gland to secrete more oxytocin, thus completing a positive feedback loop that further strengthens the contractions.

In mammals, the young are nurtured for a period after birth by milk produced within the mother's mammary glands and secreted via the nipples. The monotremes, in which the young of some species use their pliable bills to attach to the breast of the mother and draw milk directly through the skin, are an exception. The production of milk is called **lactation**. During pregnancy, an elevated blood progesterone concentration suppresses the secretion of the anterior pituitary hormone prolactin, which is required for milk production. In humans, lactation begins shortly after birth in response to a declining blood concentration of progesterone previously provided by the placenta. Whereas the formation of milk is dependent on prolactin, the actual release of milk from the breast depends on activation of smooth muscle cells surrounding secretory ducts in the breast. These cells are stimulated to contract by the presence of oxytocin. Oxytocin, therefore, plays a key role in two major and different processes: birth and lactation.

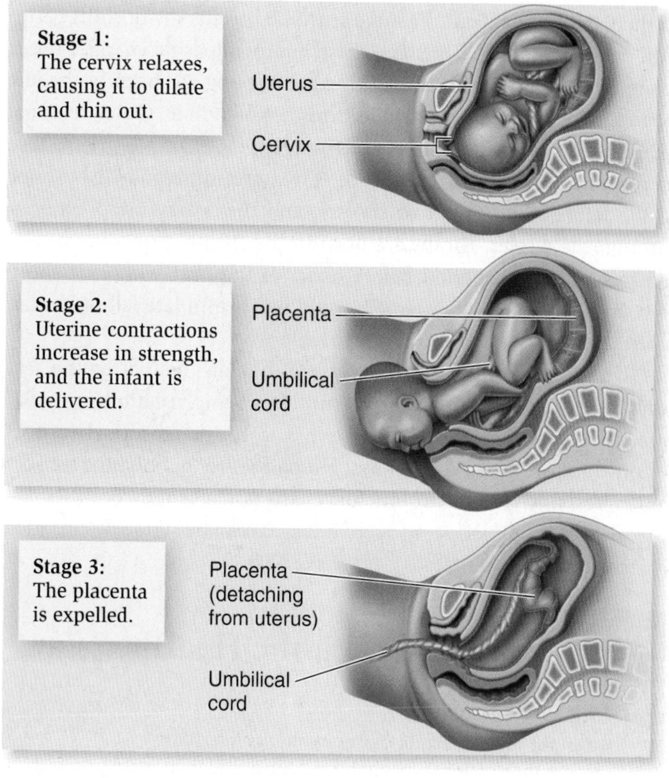

The cervix relaxes, causing it to dilate and thin out.

Figure 51.14 **The three stages of labor.**

Concept Check: *Many female mammals consume the placenta after giving birth. What is the benefit of this behavior?*

Timing of Reproduction with Favorable Times of Year

A species' survival depends on successful production of offspring, and the likelihood that those offspring will survive to adulthood and reproduce is influenced to a large degree by the environment into which they are born. For that reason, reproductive cycles often occur at times of the year when the likelihood of reproductive success is greatest—when the young will have sufficient nutrients to sustain them during their period of rapid growth. In general, animals that live under more uniformly favorable conditions—stable temperatures, ample rainfall, and abundant food—have less dramatic cycles of reproductive activity. Many tropical species, for example, reproduce several times each year. By contrast, many temperate zone animals have seasonal reproductive cycles that reflect the large fluctuations in environmental conditions. For example, insects such as the mayfly live as underwater larvae for a year before emerging into the adult (reproductively active) stage, but the emergence occurs only during the favorable summer months.

Animals synchronize the production of offspring with favorable environmental conditions by several means. In **sperm storage**, females store and nourish sperm in their reproductive tract for long periods of time, as long as 2 years in honeybees and 4 years in some turtles. Certain insectivorous bats such as the little brown bat (*Myotis lucifugus*) use a different strategy. These bats mate in the fall, but the ovarian cycle in females is halted before ovulation, and sperm are stored and nourished in the female's uterus over the winter. Upon arousal from hibernation in the spring, the female ovulates one or more oocytes, which are fertilized by the stored sperm. This type of reproductive cycle, called **delayed ovulation**, ensures that such bats mate when they are in prime condition and that young are born when temperatures and food supplies are optimal.

Other animals have **delayed implantation**, in which a fertilized egg reaches the uterus but does not implant until environmental conditions are more favorable for the newly produced young. This type of reproduction is common among carnivores, notably the mustelids (weasels) and some bears.

Finally, many animals (cats, some hamsters, sheep, and many birds) have seasonal periods of mating followed immediately by implantation and pregnancy. Such seasonal breeding results from neuroendocrine changes in the hypothalamus in response to changes in the length of daylight. In some seasonal breeders with short gestation periods, such as hamsters, these neuroendocrine changes occur when day length increases in spring. The young are born in the late spring or in summer when conditions are favorable. Other seasonal breeders with long gestation times—sheep, for example—reproduce in response to shorter day length. Ewes are impregnated by rams in the fall, as the days grow shorter, and they carry the pregnancy through the winter. Lambs are born in the spring, providing a relatively long growth period before onset of the next winter. If sheep were to mate in summer, lambs would be born in late fall or winter, when conditions are much less favorable.

Unlike the examples just described, humans neither have a seasonal breeding cycle nor alter the timing of implantation or ovulation. Moreover, sperm can survive in a woman's reproductive tract for only a short time—2 or 3 days. Human reproduction, therefore, is far less responsive to environmental changes, day length, or other factors. However, human reproductive success can be seriously curtailed by sickness and other factors, as we see next.

51.5 Impact on Public Health

Learning Outcomes:

1. Explain the most common causes of human infertility.
2. Compare the different types of birth control and their mechanisms of pregnancy prevention.

Human reproduction can be affected by many factors, some voluntary and some not. Approximately 5–10% of individuals of reproductive age in the U.S. are not fertile; that is, they cannot reproduce. In men and women alike, fertility can be compromised by a variety of factors. In this section, we discuss some of the common causes of **infertility**—the inability of a man to produce sufficient numbers or quality of sperm to impregnate a woman, or the inability of a woman to become pregnant or maintain a pregnancy. We then conclude by examining the methods in use today to prevent pregnancy.

Infertility May Result from Disease, Developmental Disorders, Inadequate Nutrition, and Stress

As many as 75% of infertility cases have some identifiable cause, and among the more prominent causes is disease. Primary among the diseases that affect fertility are sexually transmitted diseases (STDs). For example, some STDs may cause blockage in the ducts of the testes, thus preventing normal sperm transport, or can cause permanent damage to the Fallopian tubes, uterus, and surrounding tissues.

Developmental disorders are conditions that are either present at birth or arise during childhood and adolescence. In some developmental disorders that affect fertility, inherited mutations of genes that code for enzymes involved in the biosynthesis of reproductive hormones cause abnormal expression of those genes. The result is either too much or too little of one or more of these hormones, notably estradiol or testosterone. Other developmental disorders that compromise fertility include malformations of the cervix or oviducts.

Adequate nutrition is required for normal growth and development of all parts of the body, including the reproductive system. Because the reproductive system is not essential for an individual's survival, it often becomes inactive when nutrients are chronically scarce, such as during starvation. In this way, precious stores of energy in the body are preserved for vital functions, such as those of the brain and heart. Nutrition can also affect reproduction before adulthood. Undernourished children may enter puberty several years later than normal. The brains of mammals contain a center that monitors the body's fat stores. One of the triggers that initiate puberty in girls may be a signal—such as the hormone leptin—from adipose tissue to the brain. Very low fat stores in undernourished girls signal the brain that the body does not contain sufficient fuel to support the energetic demands of pregnancy; consequently, puberty is delayed.

Starvation or poor nutrition is considered a type of stress, defined as any real or perceived threat to an animal's homeostasis. Physical and psychological stress can and do affect fertility in humans. In the short term, stress can produce hormonal changes that are adaptive in meeting a crisis. However, long-term stress is damaging to many aspects of health, including reproductive health. Many nonessential functions, including the maintenance of menstrual cycles in women, can be suppressed by chronic stress. The reproductive consequences of stress, including starvation, appear to be much greater in females than in males, very likely because only females bear the energetic cost of pregnancy. Interestingly, from a reproduction viewpoint, the human body responds to long-term strenuous exercise in a way that is similar to its response to long-term stress. This is why many young ballerinas and gymnasts experience delayed puberty and why in female marathon runners, menstrual cycles may be abnormal or absent.

When the causes of infertility cannot be determined, a variety of factors come under suspicion. Among these possible causes are ingestion of toxins (for example, certain heavy metals such as cadmium), tobacco smoking, marijuana use, and injuries to the gonads. Recall also that as women age, they experience a loss of fertility, an event called menopause. Although reproductive function declines with age in men, they typically do not experience complete cessation of gamete production even at very advanced ages.

Among several currently available treatments for increasing the likelihood of pregnancy in infertile couples are hormone therapy for the woman to increase egg production and a collection of procedures known as **assisted reproductive technologies (ART)**. In the most common ART procedure, called in vitro fertilization, sperm and eggs collected from a man and a woman are placed together in culture dishes. Once the sperm have fertilized the eggs and the resulting zygotes have undergone several cell divisions, one or more embryos are inserted into a woman's uterus with the goal that one will implant. When this procedure was first used in 1978, the children born as a result came to be known as "test-tube babies." Since then, over 200,000 children have been born using this technology, which is typically effective about 30–35% of the time.

Contraception Usually Prevents Pregnancy

The use of methods to prevent fertilization or the implantation of a fertilized egg is termed **contraception**. Methods of contraception can be either permanent or temporary. The permanent forms of contraception surgically prevent the transport of gametes through the reproductive tract (Figure 51.15a). **Vasectomy** is a surgical procedure in men that severs the vas deferens, thereby preventing the release of sperm at ejaculation (however, semen is still released). In women, **tubal ligation** involves the cutting and sealing of the oviducts. This procedure prevents the movement of the egg from the oviduct into the uterus. These procedures are considered permanent, because it is difficult—sometimes impossible—to reverse the surgery.

Temporary methods of preventing fertilization include barrier methods, which prevent sperm from reaching an egg (Figure 51.15b). Barrier methods include **vaginal diaphragms**, which are placed in the upper part of the vagina just prior to intercourse and block movement of sperm to the cervix, and **condoms**, which are sheathlike membranes worn over the penis that collect the ejaculate. In addition to their contraceptive function, condoms significantly reduce the risk of STDs such as HIV infection, syphilis, gonorrhea, chlamydia, and herpes. Other types of contraception do not reduce this risk.

Another temporary form of contraception involves synthetic hormones. Oral contraceptives (birth control pills) are synthetic forms of estradiol and progesterone, taken by mouth, that prevent ovulation in women by inhibiting pituitary LH and FSH release. The hormones in these pills also affect the composition of cervical mucus such that sperm cannot easily pass through it into the uterus. In addition to the oral route, hormones can be administered by injections, skin patches, and vaginal rings.

The last temporary method of contraception involves placement in the uterus of an **intrauterine device (IUD)**, a small object that interferes with the endometrial preparation required for acceptance of the blastocyst. Unlike the other forms of contraception described here, an IUD works after fertilization—by preventing implantation (although some IUDs also inhibit sperm movement and survival in

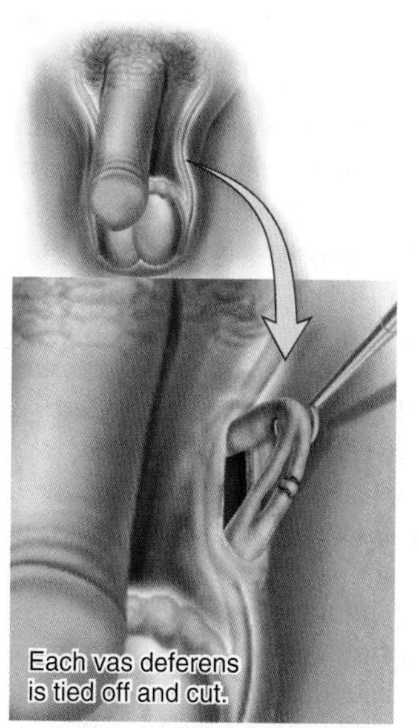

Each vas deferens is tied off and cut.

Vasectomy (<1.0%)

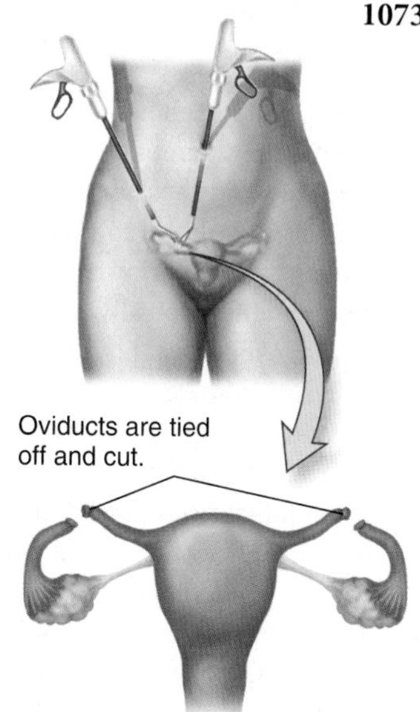

Oviducts are tied off and cut.

Tubal ligation (<1.0%)

(a) Permanent methods

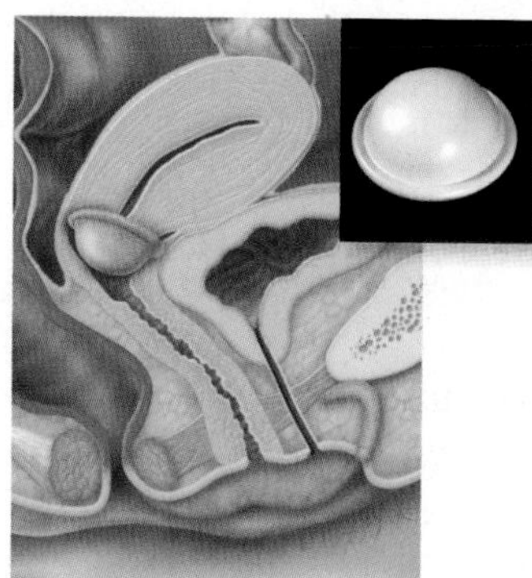

Diaphragm (5–20%)

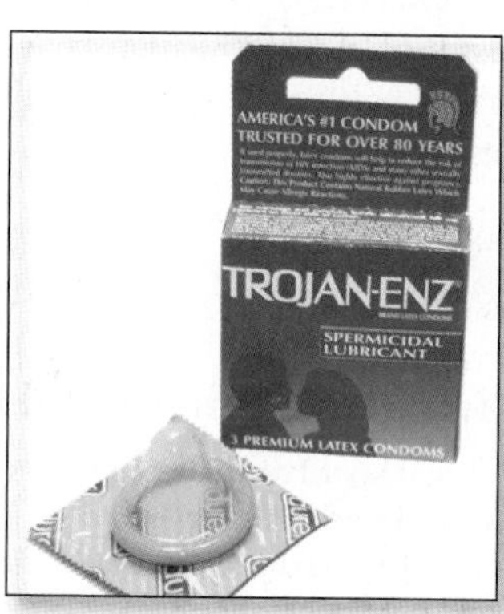

Condoms (male) (2–15%)

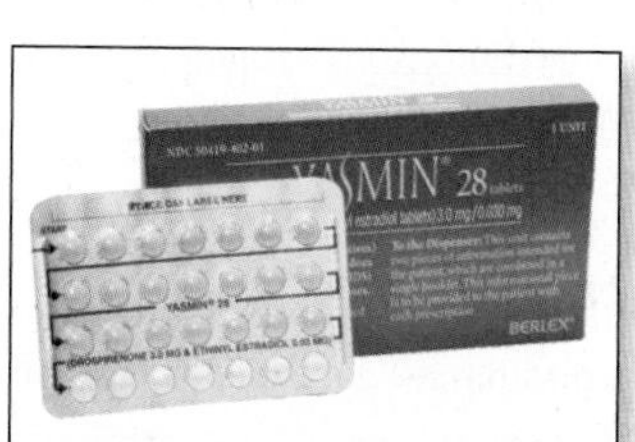

Oral contraceptive (1–2%)

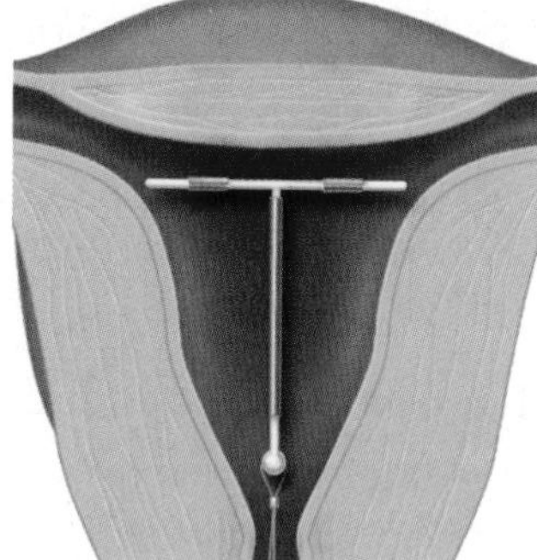

Intrauterine device (IUD) (1–2%)

(b) Temporary methods

Figure 51.15 Examples of contraceptive methods. These methods may be used by men or women to **(a)** permanently or **(b)** temporarily prevent pregnancy. The estimated first year failure rates for each method are given in parentheses (collected from data published by the U.S. Food and Drug Administration and other organizations). A failure rate of 10% means that 10 of every 100 women using that method of contraception will become pregnant in the first year of use. The large range for use of condoms and diaphragms is due to improper use of these devices by many people. Female condoms are also available and have a failure rate of approximately 20%.

the uterus). Although not as widely used as other contraceptives in the U.S., IUDs are the most commonly used means of contraception by women worldwide because of their effectiveness and simplicity of use.

In addition to the contraceptive methods used before or during intercourse, within 72 hours after intercourse, women can take a variety of drugs that typically interfere with ovulation or implantation. Approaches include a high dose of estrogen, or two large doses (12 hours apart) of a combined estrogen-progestin oral contraceptive. However, the drug RU486 (mifepristone) is more effective and produces fewer side effects; this drug antagonizes progesterone's effects on the endometrium, causing it to erode.

Used prior to the advent of modern contraception, and still in use today by individuals who prefer not to use contraceptives, is the rhythm method, which involves abstaining from sexual intercourse near the time of ovulation. Its main drawback is the difficulty in precisely pinpointing the time of ovulation. One problem with predicting the time of ovulation is that several of the detectable changes characteristic of the midpoint of the ovarian cycle—including a small rise in body temperature and changes in the cellular characteristics of the vaginal epithelium—occur only after ovulation. This problem, plus the fact that ovulation can occur any time between days 5 and 15 of the 28-day cycle, explains why the rhythm method has a relatively high failure rate.

Summary of Key Concepts

51.1 Asexual and Sexual Reproduction

- Asexual reproduction occurs when offspring are produced from a single parent, without the fusion of genetic material from two parents. Budding, regeneration, and parthenogenesis are mechanisms of asexual reproduction (Figure 51.1).
- Sexual reproduction is the production of a new individual by the joining of two haploid gametes: a sperm from the father and an egg from the mother. The union of a sperm and an egg—fertilization—produces a diploid zygote, which develops into an embryo.
- Sexual reproduction was shown by Paland and Lynch to be more effective in eliminating deleterious mutations from a population of *Daphnia* than asexual reproduction (Figure 51.2).

51.2 Gametogenesis and Fertilization

- Male and female gametes are formed within the gonads—the testes in males and the ovaries in females.
- Gametogenesis—the formation of gametes—begins with diploid primordial cells called germ cells. In spermatogenesis, one spermatogonium yields four sperm cells. In oogenesis, meiosis yields one egg. Within the ovaries, each oocyte undergoes growth and development within a structure called a follicle before it leaves the ovary in a process called ovulation (Figure 51.3).
- In external fertilization, sperm and eggs are released into an aquatic environment where they unite and avoid desiccation. Terrestrial animals use internal fertilization, in which sperm are deposited within the reproductive tract of the female during the act called copulation (Figure 51.4).
- In hermaphroditism, individuals can fertilize their own eggs with their own sperm, but in most hermaphroditic species, individuals exchange sperm with another individual.
- When an embryo develops within the mother and the mother gives birth to live young, the process is called viviparity. Development of an embryo that occurs primarily outside the mother, sometimes within a protective shell, is called oviparity. In ovoviviparity, eggs covered with a thin shell are produced and hatch inside the mother's body, but the offspring receive no nourishment from the mother.

51.3 Mammalian Reproductive Structure and Function

- Sperm produced within each testis move out of the seminiferous tubules and into the epididymis, which leads into the vas deferens, a muscular tube leading to the ejaculatory duct. The urethra conducts semen, a mixture containing fluid and sperm, during ejaculation. The fluid components of semen are produced in the seminal vesicles, the bulbourethral glands, and the prostate gland (Figures 51.5, 51.6, 51.7).
- The external female genitalia are composed of the labia majora, the labia minora, and the clitoris; the internal female genitalia consist of the vagina, uterus, oviducts, and each of two ovaries. In the ovary, primary oocytes develop into secondary oocytes (the ovarian cycle). Sperm typically fertilize secondary oocytes in the oviduct. The fertilized egg undergoes several cell divisions to become a blastocyst, a ball of cells that enters the uterus, where it will develop into an embryo (Figures 51.8, 51.9).
- In females, changes in hormone secretion produce the ovarian cycle and also control the uterine cycle or menstrual cycle (Figure 51.10).
- The cessation of ovarian cycles is called menopause.

51.4 Pregnancy and Birth in Mammals

- The time during which a developing embryo and fetus grow within the uterus of the mother is termed pregnancy, or gestation. Humans and other eutherian mammals retain and nourish their young within the uterus via transfer of nutrients and gases through a structure called the placenta (Figure 51.11).
- The first event of pregnancy is implantation, when the blastocyst imbeds within the uterine endometrium. Newly developing embryonic tissues merge with the endometrium to form the placenta, which remains in place and grows larger as the embryo matures into a fetus. There are three trimesters to human pregnancy (Figure 51.12).
- The evolution of the globin gene family contributed to the ability of placental mammals to develop inside the mother's uterus (Table 51.1).
- Birth, or parturition, is initiated by the actions of hormones produced by the placenta and by the mother's endocrine system. The hormone oxytocin stimulates the strong uterine muscle contractions that are the hallmark of the three-stage process called labor (Figures 51.13, 51.14).
- In mammals, the young are nurtured for a period after birth by milk produced in the process called lactation.
- Animals synchronize the production of offspring with favorable environmental conditions by several means, including sperm storage, delayed ovulation, delayed implantation, and seasonal mating.

51.5 Impact on Public Health

- Infertility is the inability of a man to produce sufficient numbers or quality of sperm to impregnate a woman, or the inability of a woman to become pregnant or maintain a pregnancy. A primary cause of infertility is STDs.
- The use of procedures to prevent fertilization or implantation of a fertilized egg is termed contraception. Methods of contraception include vasectomy and tubal ligation, vaginal diaphragms, condoms, oral contraceptives, and IUDs (Figure 51.15).

Assess and Discuss

Test Yourself

1. The development of offspring from unfertilized eggs is
   a. budding.
   b. cloning.
   c. fragmentation.
   d. parthenogenesis.
   e. implantation.
2. Which is considered an advantage of sexual reproduction?
   a. necessity to locate a mate
   b. increased energy expenditure in producing gametes that may not be used in reproduction
   c. increased genetic variation
   d. decreased genetic variation
   e. both a and b
3. Spermatogonia
   a. are germ cells.
   b. are diploid cells.
   c. are male gametes.
   d. have flagella.
   e. a and b only.
4. Compared with external fertilization, in internal fertilization,
   a. male gametes have a higher chance to come into close proximity to female gametes.
   b. gametes are less protected against predation or other harmful environmental factors.
   c. there is a decreased likelihood of desiccation of gametes.
   d. gametes come into contact only outside the mother's reproductive tract.
   e. b and c only.
5. Which of the following is an example of ovoviviparity?
   a. Honeybees lay soft eggs within the hive.
   b. Sharks hatch from shell-covered fertilized eggs within the female's body.
   c. Birds hatch from shell-covered fertilized eggs laid within a nest.
   d. Fetal mammals obtain nourishment from the mother through a placenta.
   e. Amphibian eggs are released into the water column, where they may be fertilized.
6. The fructose in semen is secreted by
   a. the epididymis.
   b. the seminiferous tubules.
   c. the seminal vesicles.
   d. the prostate gland.
   e. the bulbourethral glands.
7. A major function of FSH is to
   a. stimulate the development of the gonads during early development.
   b. stimulate spermatogenesis in males and oocyte maturation in females.
   c. increase the secretion of testosterone by the testes.
   d. regulate the secretion of the bulbourethral glands.
   e. inhibit the activity of Sertoli cells in the testes.
8. During the human ovarian cycle, ovulation is stimulated by
   a. a decrease in FSH secretion.
   b. an increase in progesterone secretion.
   c. an increase in LH secretion.
   d. the presence of semen in the vagina.
   e. a decrease in estradiol concentration in the bloodstream.
9. During the secretory phase of the menstrual cycle, endometrial glands secrete
   a. hormones that increase the likelihood of pregnancy.
   b. nutritive substances that sustain an embryo during the first 2 weeks of development.
   c. hormones that prevent ovulation.
   d. waste products into the lumen of the uterus.
   e. both a and c.
10. During the _________ stage of labor in mammals, the placenta is expelled from the uterus.
   a. first
   b. second
   c. third
   d. It is not expelled; the placenta is reabsorbed.

Conceptual Questions

1. Distinguish among viviparity, ovoparity, and ovoviviparity, and give examples of animals for each type. Which type characterizes humans? What is an advantage of viviparity?
2. How does the hypothalamus influence vertebrate reproduction?
3. A principle of biology is that *living organisms use energy*. What are some of the energy costs associated with sexual reproduction? What outweighs those costs and accounts for the observation that most animals reproduce sexually?

Collaborative Questions

1. Define asexual reproduction and give three examples.
2. Compare and contrast internal fertilization and external fertilization.

Online Resource

www.brookerbiology.com

Stay a step ahead in your studies with animations that bring concepts to life and practice tests to assess your understanding. Your instructor may also recommend the interactive eBook, individualized learning tools, and more.

Animal Development

52

Chapter Outline

52.1 Principles of Embryonic Development

52.2 General Events of Embryonic Development

52.3 Control of Cell Differentiation and Morphogenesis During Animal Development

52.4 Impact on Public Health

Summary of Key Concepts

Assess and Discuss

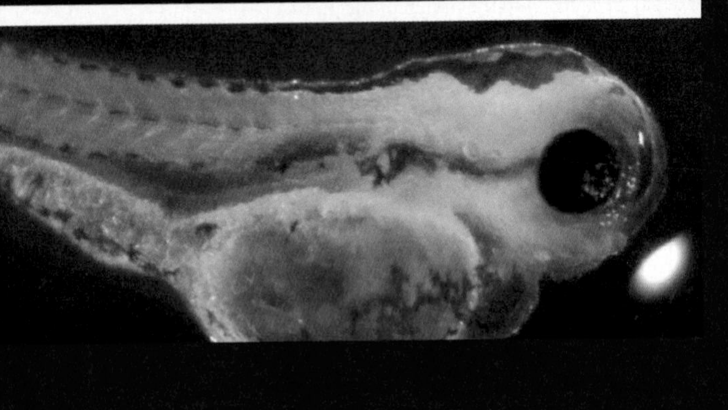

How embryonic development can go wrong—a normal zebrafish embryo (top) and one exposed to ethanol (bottom).

BIOLOGY PRINCIPLE Living organisms interact with their environment. As you read this chapter, consider the many environmental factors that animals such as zebrafish (*Danio rerio*) and humans are exposed to during embryonic development.

During the late 1950s and early 1960s in Europe, a drug called thalidomide was developed for use as an anticonvulsive treatment for epilepsy; it was also prescribed as a treatment for allergies. Although further testing revealed no beneficial effects on these conditions, it was noted that thalidomide seemed to calm nausea and help people sleep. In 1961, thalidomide was the most widely used sleeping medication in Europe, where it was also prescribed to combat morning sickness (nausea) in pregnant women.

Soon afterward, however, physicians in Europe and Australia noticed that babies born to mothers who had taken thalidomide sometimes had severely malformed limbs, with hands and feet often emanating directly from the body. It became clear that certain cells in human embryos, such as those that give rise to arms and legs, are selectively and adversely affected by this toxic compound. Ultimately, thousands of babies worldwide were deformed due to thalidomide. The drug was quickly eliminated from pharmaceutical markets, and the U.S. was largely spared from its effects due to prompt legislative action by the Food and Drug Administration that blocked the sale of the drug. Such tragedies underscore the importance of continued research into the mechanisms that regulate the normal development of an embryo. Environmentally or drug-induced developmental abnormalities occur throughout the animal kingdom. The chapter-opening photo, for example, shows the effect of ethanol (the alcohol in alcoholic beverages) on development of a zebrafish embryo. (An **embryo** is an early stage of a multicellular organism during which the organization of the organism is largely formed.) Such models are helping researchers understand the effects of alcohol consumption and other environmental factors on human embryo development.

In Chapter 19, we saw how the sequential actions of genes provide a program for the development of an organism from a fertilized egg to an adult. In this chapter, we will learn about the cellular and molecular processes that lead to the formation of an animal embryo. We will first briefly review some basic principles of development from Chapter 19. Then we will consider the five general events of embryonic development in animals. Next, we will examine the cellular and molecular mechanisms that control development. We conclude with an overview of how abnormal development impacts human health.

52.1 Principles of Embryonic Development

Learning Outcomes:

1. Define embryonic development and the establishment of a body plan.
2. Describe the process of cellular differentiation.

The process by which a fertilized egg (that is, a **zygote**) is transformed into an animal with distinct physiological systems and body parts is called **embryonic development**. The biological information that controls embryonic development resides in both the organism's genetic material—its DNA—and in the cytoplasm of the egg. A fertilized egg first becomes transformed into a cluster of cells without specialized functions and ultimately develops into a complex organism containing organs with specific and evolutionarily conserved functions. Embryonic development is often accompanied by growth, but they are different processes. Development produces organisms with a defined set of characteristics, whereas growth produces more or larger cells. Let's begin by reviewing some of the fundamental processes that underlie development, first introduced in Chapter 19.

As animals develop, cells arrange themselves in coordinated ways that lead to the establishment of a **body plan**. The final, adult body plan of most animals is organized along three axes: the **dorsoventral axis**, the **anteroposterior axis**, and the **left-right axis** (refer back to Figure 19.2a). Along these axes are often separate sections, or body segments, each containing specific body parts such as a wing or leg.

To establish the correct body plan, each cell in a developing animal must "know" where it is within the body, where it should move to, whether or not it should divide (or die), and what types of functions it will ultimately perform. This is possible because each cell receives positional information from its neighboring cells. This information is provided in a variety of ways, including intracellular and extracellular signaling molecules, and by cell-to-cell contacts (refer back to Figure 19.5).

In addition to being stimulated to divide or to die (by apoptosis), positional information may cause the migration of a cell or group of cells from one region of the embryo to another (refer back to Figure 19.3). It may also cause **cellular differentiation**—the process by which different cells within a developing organism acquire specialized forms and functions, due to the expression of cell-specific genes. (In the thalidomide cases we discussed in the introduction, for example, cells that should have differentiated into those capable of developing into limbs were destroyed or damaged.) Now let's look at the general events that occur during embryonic development in animals.

52.2 General Events of Embryonic Development

Learning Outcomes:

1. List the five major events that take place during embryonic development and describe the end result of each.
2. Compare and contrast the fast block to polyspermy and the slow block to polyspermy.
3. Describe the process of cleavage, beginning at fertilization and leading into gastrulation.
4. Trace the fates of cells in a gastrula as they differentiate into the three major germ layers.
5. Outline the early development of the nervous system during neurulation.
6. Explain the migration and roles of neural crest cells.

Even though the adult forms of animals vary immensely in size and morphology, embryonic development follows a similar pattern in most animals. As described in Chapter 32, most modern animals are triploblasts; that is, they develop from embryos with three germ cell layers. Such triploblasts include vertebrates, arthropods, echinoderms, and mollusks. Development in these animals can be categorized into five general events: fertilization, cleavage, gastrulation, neurulation, and organogenesis (Figure 52.1). In this section, we will examine the key aspects of each of the five general events of animal development. However, it should be noted that many species also go through an additional event called metamorphosis, which is a transition from a feeding larval form to an adult (refer back to Figures 19.7

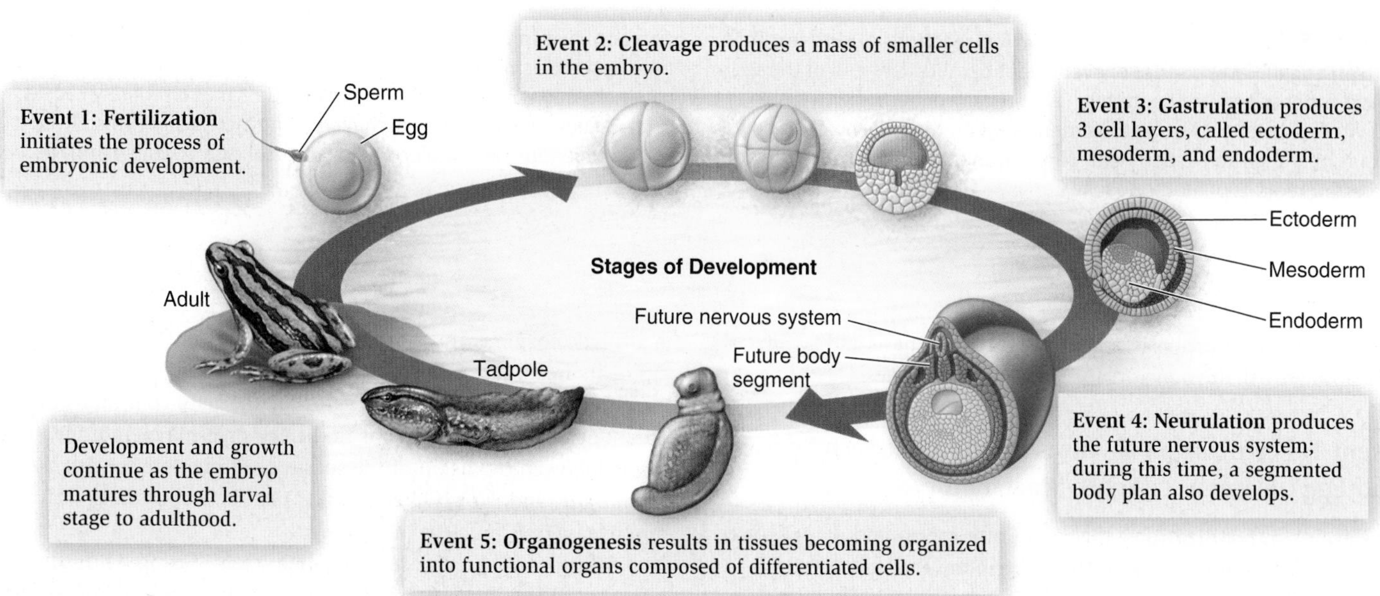

Figure 52.1 **Overview of events of embryonic development.** This figure shows the general events that all vertebrate embryos go through, using a frog as an example.

BioConnections: *What is the final process called by which a tadpole develops into an adult frog, and is this process unique to frogs? (Refer back to Figures 19.7 and 33.28 for a hint.)*

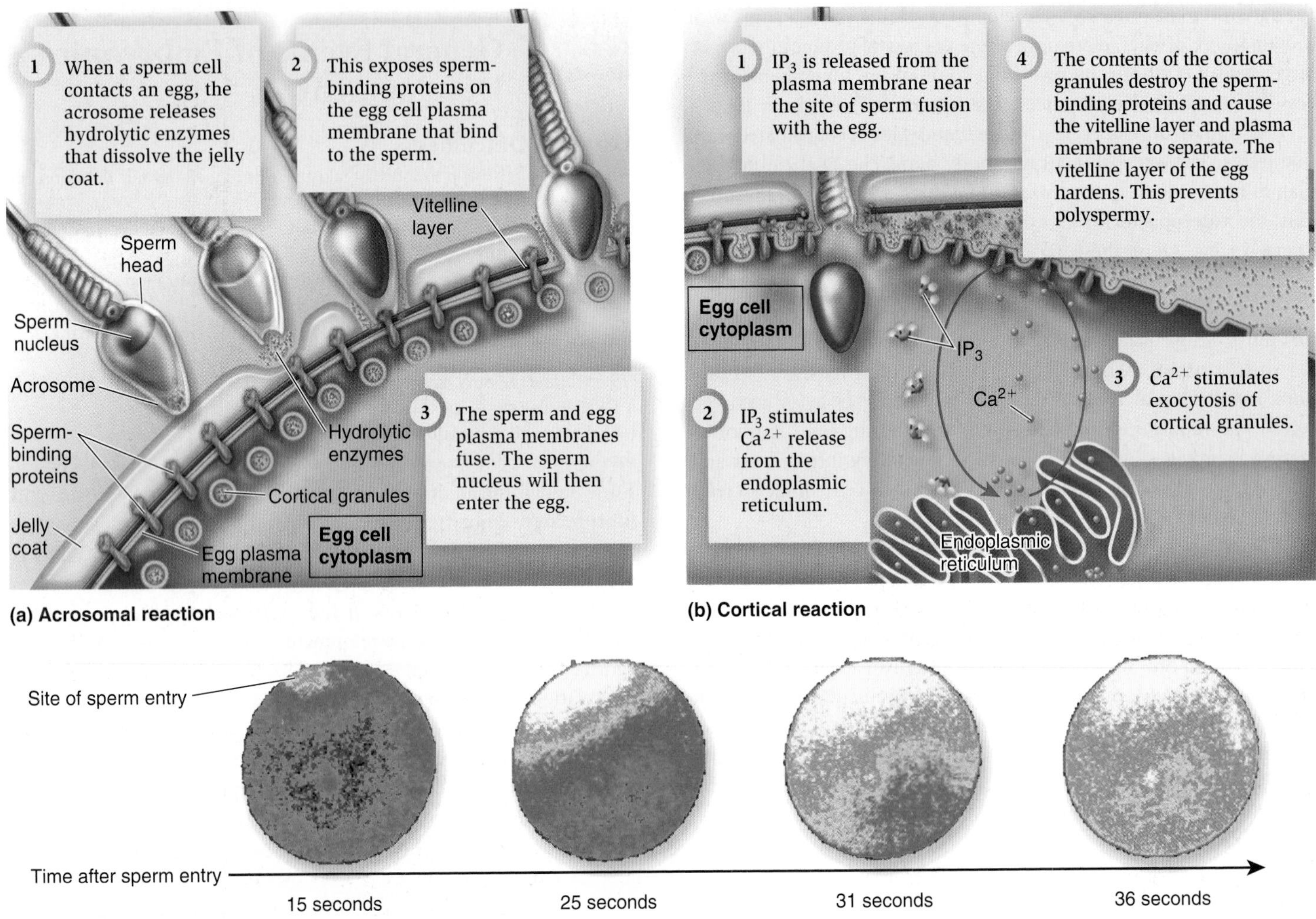

1 When a sperm cell contacts an egg, the acrosome releases hydrolytic enzymes that dissolve the jelly coat.
2 This exposes sperm-binding proteins on the egg cell plasma membrane that bind to the sperm.
3 The sperm and egg plasma membranes fuse. The sperm nucleus will then enter the egg.
Sperm head
Sperm nucleus
Acrosome
Sperm-binding proteins
Jelly coat
Vitelline layer
Hydrolytic enzymes
Cortical granules
Egg plasma membrane
Egg cell cytoplasm
(a) Acrosomal reaction
1 IP_3 is released from the plasma membrane near the site of sperm fusion with the egg.
2 IP_3 stimulates Ca^{2+} release from the endoplasmic reticulum.
3 Ca^{2+} stimulates exocytosis of cortical granules.
4 The contents of the cortical granules destroy the sperm-binding proteins and cause the vitelline layer and plasma membrane to separate. The vitelline layer of the egg hardens. This prevents polyspermy.
Egg cell cytoplasm
IP_3
Ca^{2+}
Endoplasmic reticulum
(b) Cortical reaction
Site of sperm entry
Time after sperm entry
15 seconds
25 seconds
31 seconds
36 seconds
(c) The Ca^{2+} wave of the cortical reaction in a sea urchin egg

Figure 52.2 The acrosomal and cortical reactions during fertilization of an egg by a sperm. **(a)** Acrosomal reaction. The contact of a sperm with an egg initiates a series of events that permits the head of the sperm to bind to the plasma membrane of the egg. This depolarizes the egg and blocks other sperm from entering. **(b)** Cortical reaction. Sperm fusion leads to an increased concentration of cytosolic Ca^{2+} that ultimately causes the vitelline layer of the egg to harden, creating the slow block to polyspermy. **(c)** Calcium wave in a sea urchin egg. The increase in cytosolic Ca^{2+} begins near the site of sperm entry and propagates throughout the egg. Green and blue represent regions of low cytosolic Ca^{2+}; yellow and red represent regions of high cytosolic Ca^{2+}.

and 33.28). Metamorphosis occurs after organogenesis and facilitates the rapid growth of young organisms into mature ones.

Event 1: Fertilization Involves a Union Between Sperm and Egg to Create a Zygote-Stage Embryo

The events in **fertilization**—the union of a sperm and an egg—are quite similar in all triploblasts. The description that follows summarizes some of the hallmark events following external fertilization in the sea urchin, a well-studied model organism.

In sea urchins, as in many animals, the sperm must penetrate a jelly-like layer consisting of glycoproteins and polysaccharides before contacting the plasma membrane of the egg. The sperm is able to do this because of the **acrosomal reaction**, in which proteases and other hydrolytic enzymes are released from the acrosome, a structure at the tip of its head, onto the jelly coat of the egg (**Figure 52.2a**). These enzymes dissolve a localized region of the jelly coat, allowing the sperm head to bind to proteins in the egg's plasma membrane. Binding is followed by fusion of the sperm head membrane with the egg membrane, and shortly thereafter by penetration of the sperm head and release of its nucleus into the egg.

Additional sperm are prevented from fusing with the egg because the fusion of sperm and egg plasma membranes depolarizes the egg. This depolarization blocks other sperm from binding to egg membrane proteins and is known as the **fast block to polyspermy**. Without this block, a single egg could receive chromosomes from two or more sperm, resulting in zygotes that fail to develop normally or at all. Polyspermy does occur in some animals, notably urodeles (newts and salamanders), but in such cases, the nucleus from only one sperm fuses with the nucleus of the egg, and the remaining nuclei are degraded inside the egg.

The acrosomal reaction is followed by the **cortical reaction** (Figure 52.2b). Normally, the cytosolic calcium ion (Ca^{2+}) concentration in eggs, as in most cells, is kept low by several mechanisms, including one in which Ca^{2+} is transported by an ATP-dependent pump out of the cytosol and into the endoplasmic reticulum. When the sperm binds to the egg, inositol trisphosphate (IP_3) (refer back to Figure 9.16) is released from the region of the plasma membrane nearest to the sperm entry point. IP_3 then binds to nearby sites on the endoplasmic reticulum (ER) and opens Ca^{2+} channels. Within 10 seconds after a sperm cell binds to an egg, Ca^{2+} is released from the lumen of the ER and into the cytosol. This signal is propagated across the entire ER, resulting in the transmission of a calcium wave across the egg over a period of about 30 seconds. This calcium wave can be visualized by injecting the cytosol of an unfertilized egg with a calcium-sensitive fluorescent dye that becomes highly fluorescent when Ca^{2+} is released from the ER (**Figure 52.2c**).

The release of Ca^{2+} in the cortical reaction has several important effects. First, membrane-bound vesicles in the egg's cytosol, called cortical granules, release enzymes and other substances that inactivate the sperm-binding proteins on the plasma membrane. In addition, the outer coating of the egg cell, known as the vitelline layer in sea urchins, or the zona pellucida in vertebrates, becomes hardened and begins to separate from the plasma membrane. These events create another barrier to more sperm fusing with the egg, a process called the **slow block to polyspermy**. Additionally, the burst of cytosolic Ca^{2+} leads to the activation of molecular signaling pathways that initiate the first cell cycle and triggers an increase in protein synthesis and metabolism within the egg cell.

Shortly afterward, the nucleus of the sperm fuses with the nucleus of the egg, creating a diploid zygote. The first cell division of the zygote occurs approximately 90 minutes after fertilization in sea urchins and amphibians, but it can take up to 24 hours in mammals.

Event 2: Cell Divisions Without Cell Growth Create a Cleavage-Stage Embryo

The initial cell cycles of embryos are unique because they involve repeated cell divisions without cell growth. The process by which these cell cycles occur is called **cleavage**. The embryonic cells repeatedly split in two, resulting in several generations of daughter cells that are roughly half the size of the cells that gave rise to them. These early cell cycles that lack cell growth are characterized as "biphasic," because they alternate only between the mitotic (M) phase and DNA synthesis (S) phase of the cell cycle—neither the G_1 nor G_2 phase occurs (see Chapter 15 and Figure 15.2 for a discussion of the cell cycle).

In most species in which development occurs outside the mother, where eggs can be eaten by predators, cell division during cleavage represents some of the fastest cell cycles found in nature. The cell cycle during cleavage in amphibians, for example, requires only 20 minutes. During each 20-minute cell cycle, complete genome replication, mitosis, and duplication of the nuclear envelope are followed by cytokinesis. In eutherian (placental) mammals, in which development occurs within the protective environment of the mother's body, biphasic cell divisions during cleavage are relatively slow, requiring about 12 hours to complete.

The two half-size daughter cells produced by each cell division during cleavage are known as **blastomeres**. Individual blastomeres are bound together, and the outer single-cell layer of blastomeres forms a sheet of epithelial cells that separates the embryo from its environment. After formation of the outer epithelial layer, the embryos of many animals take up water and form a cavity called a **blastocoel**. The embryo at this stage is called a **blastula**. The blastocoel provides a space into which cells will migrate to form the digestive tract and other structures of the embryo, as described in event 3.

Meroblastic Cleavage: Birds and Fishes Among triploblasts, cleavage-stage embryos can vary dramatically in size and appearance. This variation is in part related to whether or not the egg contains yolk and, if so, the location and amount of yolk that was deposited in the egg. Yolk is a nutrient-rich food store that is used by the developing embryo. The eggs of birds, some fishes, and some other vertebrates have large amounts of yolk. In the eggs of these species, yolk is most concentrated toward one end—or pole—of the egg, called the **vegetal pole**. Much less yolk, and much more cytoplasm, is concentrated near the opposite pole, called the **animal pole** (Figure 52.3). These poles form the apices of the vegetal and animal hemispheres, which determine in part the future anteroposterior (head-tail) and dorsoventral (back-front or top-bottom, depending on the species) axes of the embryo.

In some but not all species that exhibit animal and vegetal poles, cleavage of the zygote is called **meroblastic cleavage**, or incomplete cleavage, because only the region of the zygote and embryo containing the animal hemisphere undergoes cell division (Figure 52.4). Instead of forming a ball of cells (a blastula), in this type of cleavage, a flattened disc of blastomeres known as a **blastoderm** develops on top of the yolk mass.

Holoblastic Cleavage: Amphibians and Mammals In animals whose eggs have smaller amounts of yolk, cleavage during the first cell division is complete and bisects the entire zygote into two equal-sized blastomeres. Such **holoblastic cleavage** or complete cleavage occurs in amphibians and mammals (see Figure 52.4). In amphibians, cleavage-stage embryos form a blastula, as previously noted. In mammals, however, cleavage-stage embryos undergo a process called compaction, in which the amount of physical contact between cells is maximized. At this stage, the embryo in these species is called a

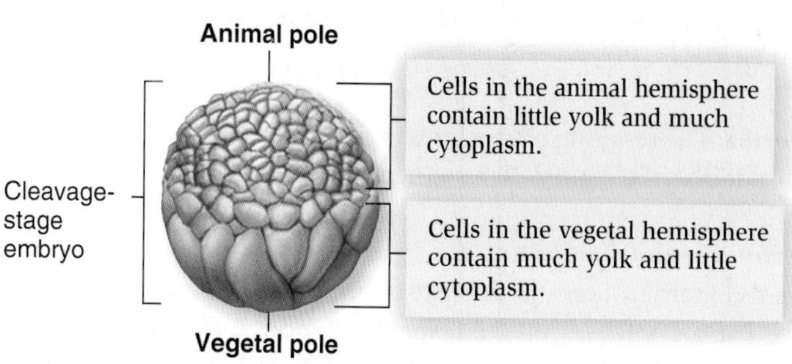

Animal pole
Cells in the animal hemisphere contain little yolk and much cytoplasm.
Cleavage-stage embryo
Cells in the vegetal hemisphere contain much yolk and little cytoplasm.
Vegetal pole

Figure 52.3 Polarity in an amphibian cleavage-stage embryo.

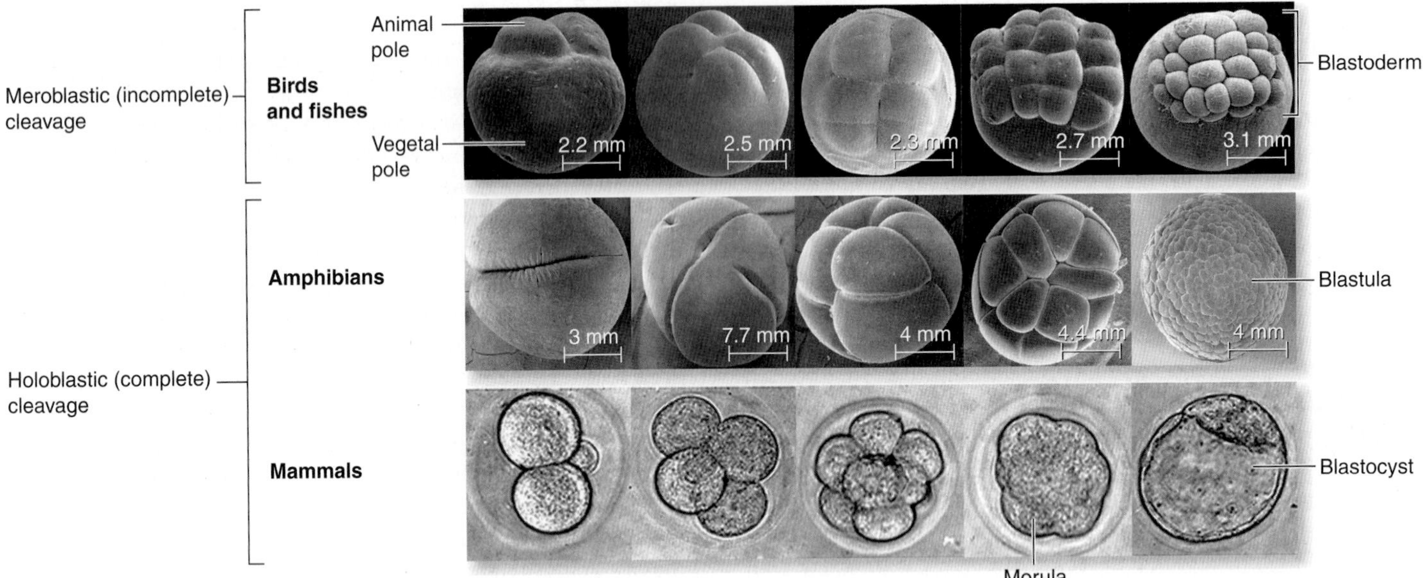

Figure 52.4 Meroblastic and holoblastic cleavage. As seen in these electron micrographs, early embryos of birds and many fishes undergo incomplete (meroblastic) cleavage, whereas most amphibian and mammalian embryos undergo complete (holoblastic) cleavage. The amount of yolk in the egg (not visible in these images) contributes to many of these morphological differences observed in various species. Source (top and middle row): © Dr. Richard Kessel & Dr. Gene Shih/Visuals Unlimited.

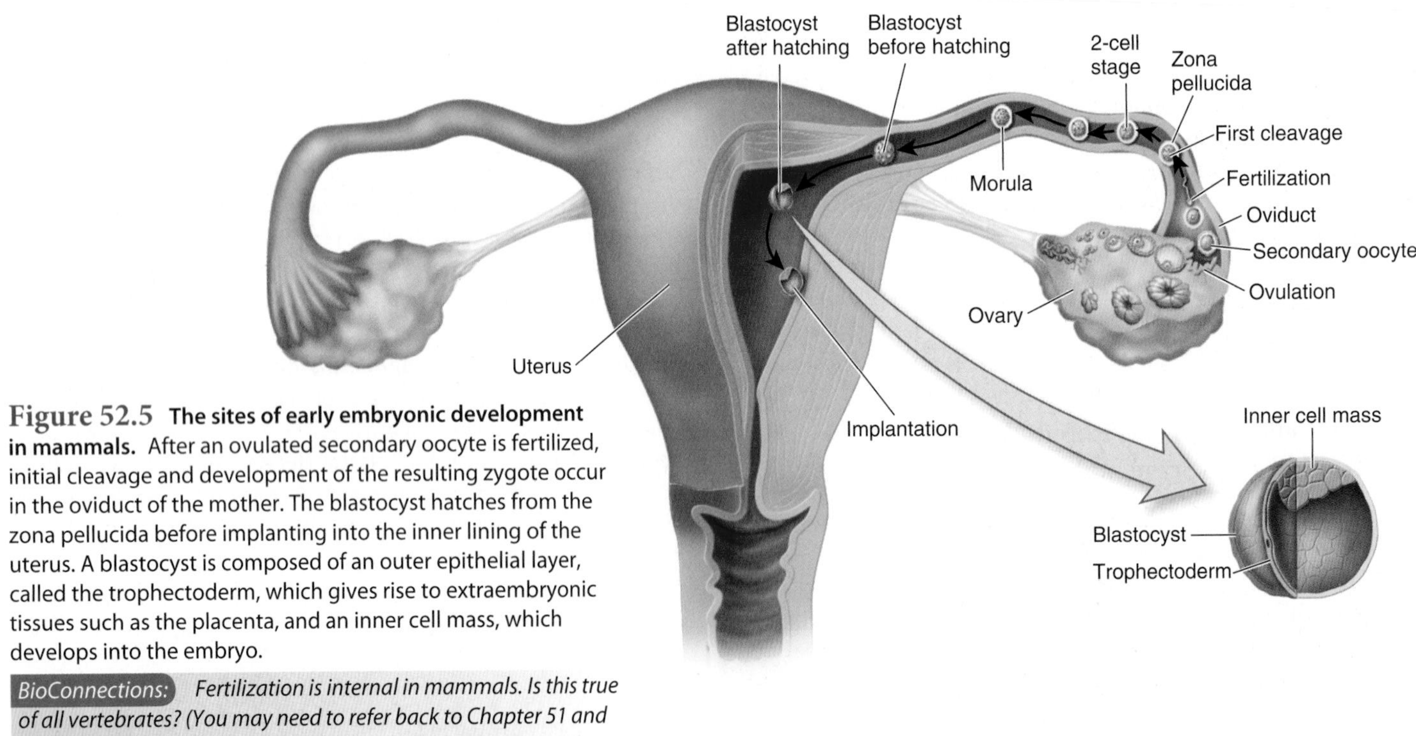

Figure 52.5 The sites of early embryonic development in mammals. After an ovulated secondary oocyte is fertilized, initial cleavage and development of the resulting zygote occur in the oviduct of the mother. The blastocyst hatches from the zona pellucida before implanting into the inner lining of the uterus. A blastocyst is composed of an outer epithelial layer, called the trophectoderm, which gives rise to extraembryonic tissues such as the placenta, and an inner cell mass, which develops into the embryo.

BioConnections: *Fertilization is internal in mammals. Is this true of all vertebrates? (You may need to refer back to Chapter 51 and Figure 51.4.)*

morula. The resulting blastomeres in mammals then proceed to form a **blastocyst**, the mammalian counterpart of a blastula.

Cleavage and Implantation in Mammals In mammals, the events of fertilization and cleavage occur in the oviduct (Figure 52.5). The blastocyst has a different morphological appearance than the blastula or blastoderm embryos in nonmammalian species, and no animal-vegetal polarity that is analogous to that of other chordates exists. The blastocyst consists of an outer epithelial layer called the **trophectoderm**, which gives rise to the placenta, and an inner layer called the inner cell mass, which develops into the embryo. Upon reaching the uterus, the embryo "hatches" from the zona pellucida, the layer of glycoproteins that surrounds the secondary oocyte and is retained up to this time preventing premature adhesion of the embryo to the oviduct. The embryo then becomes embedded in the endometrium of the mother's uterus, a process known as implantation (see Figure 52.5). This entire process takes 4 days in mice and about 8 to 10 days in humans.

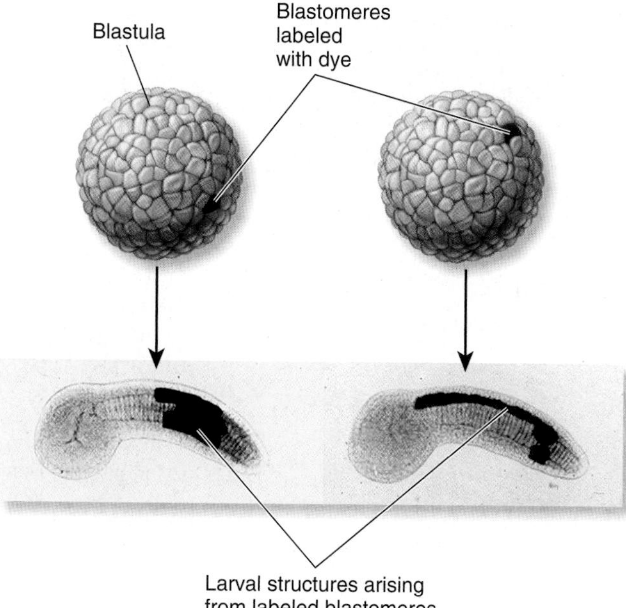

Figure 52.6 Fate mapping during embryonic development. In fate mapping, an individual embryonic cell (a blastomere) is labeled with a dye. The fate of that cell can then be followed, showing that different cells in the embryo (in this case, a urochordate blastula) form different structures in the tadpole-like larva, as shown by the dark regions.

Toward the end of cleavage, cell cycles become less synchronous, and the embryo begins to express its own genes. The embryo's shift from existing exclusively on maternal factors to developing in response to products derived from its own genome begins 6–24 hours after fertilization in vertebrates. This is followed by the next general event of development, called gastrulation.

Event 3: Gastrulation Establishes the Three Germ Layers in the Embryo

Gastrulation is one of the most dramatic events of embryonic development in animals because of the major cell movements that occur. During gastrulation, the hollow ball of cells that makes up a blastula or blastocyst is developed into a highly organized structure called a **gastrula** (refer back to Figure 32.6). In the gastrula-stage embryo, the three germ layers—**ectoderm**, **mesoderm**, and **endoderm**—become clearly established. These distinct germ layers are partially differentiated tissues that are easily recognized by their appearance under a light microscope. The layers occupy discrete regions of the embryo, with an outer ectoderm, a middle mesoderm, and an inner endoderm layer. Each type of germ layer eventually gives rise to different structures. The organization that emerges during gastrulation is most evident by the clear establishment of the digestive tube and body axes. Gastrulation is the first time when both the anteroposterior and dorsoventral body axes are clearly evident in the embryo.

Scientists have established the ultimate fate of the three germ layers with an experimental procedure called **fate mapping**. In this technique, a single cell or a small population of cells within an embryo is specifically labeled with a harmless dye, and the fate of these labeled cells is followed to a later stage of embryonic development (Figure 52.6). Use of this technique in a variety of vertebrates has shown that the ectoderm in the gastrula forms the epidermis and nervous system in the later embryo (Figure 52.7). The mesoderm gives rise to muscles, kidneys, blood, heart, limbs, connective tissues, and notochord, which is a key feature of all chordates, described later in this section of the chapter. The endoderm becomes the epithelial lining of the pancreas, thyroid, lungs, gut, liver, and urinary bladder.

Gastrulation occurs in all vertebrates. Many of the genes responsible for specifying gastrulation processes are conserved from fishes to humans. Although the developmental steps vary among vertebrate species, gastrulation usually is initiated by changes in a small number of cells within an external epithelial cell layer. This leads to a morphologically distinct structure that clearly defines the anteroposterior axis of the animal.

Some of our most detailed descriptions of the events in gastrulation come from the study of frog embryos. Three features of amphibians make them ideal for analyzing embryonic development using

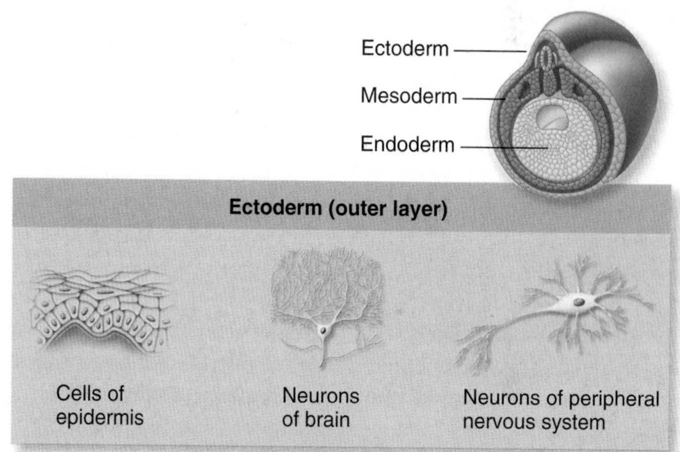

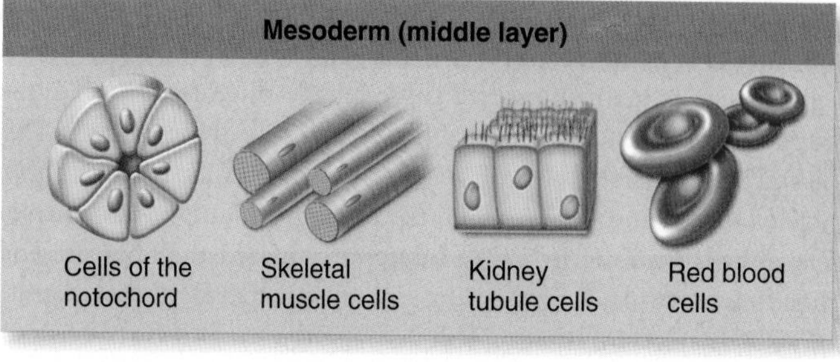

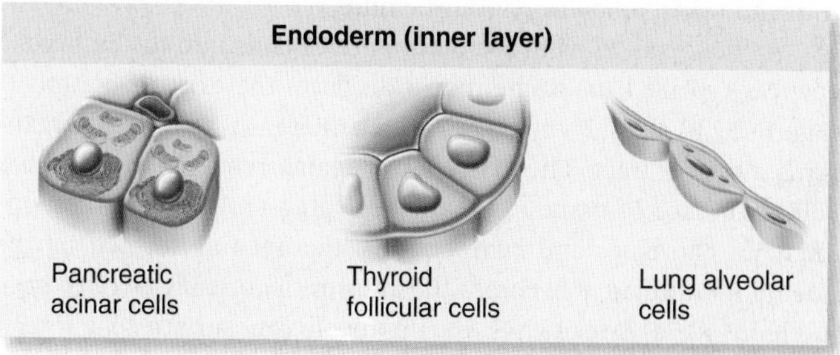

Figure 52.7 Examples of cell types derived from ectoderm, mesoderm, and endoderm.

BIOLOGY PRINCIPLE Living organisms grow and develop. The acquisition of specific cell types as shown here is an example of development; growth would entail an increase in the numbers of each of these cell types.

nongenetic mechanisms. First, fertilization and embryonic development occur outside the mother, making it easy to observe and manipulate embryos as they develop. Second, eggs are abundant—adult females carry thousands of eggs and can be induced to lay them by injecting a specific hormone—and are easily fertilized, usually by simply placing a male in an aquarium with an ovulating female. Third, amphibian embryos develop rapidly. In frogs, for example, development from fertilized egg to tadpole takes only about 35 hours. The major events in gastrulation as described in amphibians are depicted in Figure 52.8 and described next.

Invagination and Involution: Formation of Germ Layers and Archenteron Prior to gastrulation, the blastula is enclosed in a simple, spherical epithelial cell layer. Gastrulation begins when a band of tissue extending perpendicularly to the animal-vegetal axis at the widest part of the embryo **invaginates** (pinches in), pushing cells from the outside of the embryo to the inside (Figure 52.8, step 1). This process creates a small opening called the **blastopore**, which defines the anteroposterior axis of the animal. Invagination begins when a few epithelial cells located at the vegetal hemisphere of the blastula—called bottle cells—undergo a drastic change in their morphology, causing them to elongate toward their basal end and forcing the cells toward the interior of the embryo. The initiating site of invagination becomes what is called the dorsal lip of the blastopore. This change in morphology of only a few key cells in the embryo initiates the gastrulation process in amphibians.

Once the bottle cells change their shape and push into the interior of the embryo, other cell movements occur, and together these orchestrated movements establish the mesoderm and endoderm of the organism, including its future digestive tract. Just before invagination begins near the "equator" or midline of the embryo, cells of the animal hemisphere spread out and move downward. The bottle cells then form and move in at the blastopore. Animal hemisphere cells continue to migrate downward. When they arrive at the blastopore, they too enter the opening and subsequently migrate upward along the roof of the blastocoel, toward the animal pole of the embryo. This folding back of sheets of surface cells into the interior of the embryo is called **involution** (Figure 52.8, step 2). Both endodermal and mesodermal cells involute from the blastopore toward the opposite end of the embryo. After involution, dorsal mesodermal cells migrate toward the animal pole by crawling along the roof of the blastocoel, with endoderm following closely behind.

As the opening from the blastopore extends into the embryo, a new cavity called the **archenteron** displaces the existing blastocoel (Figure 52.8, steps 2 and 3). The archenteron becomes the organism's digestive tract. The blastopore opening remains sealed with a yolk-rich piece of tissue called the yolk plug until later in development. In chordates and echinoderms, the opening formed by the blastopore ultimately becomes the anus of the organism (refer back to Figure 32.7a). Meanwhile, during involution, surface cells spread from the animal hemisphere to surround the entire vegetal hemisphere to become the future ectoderm. The result of these cellular rearrangements is an embryo with three distinct germ layers.

Mechanisms for Changes in Cell Shape and Position How do cells change shape and position during the process of embryonic

1 **Formation of the blastopore by invagination of bottle cells.** Gastrulation is initiated by the invagination of bottle cells, which forms a blastopore. Invagination of bottle cells forces cells behind them to involute toward the future anterior end of the embryo. The curved arrows indicate directions of cell movements.

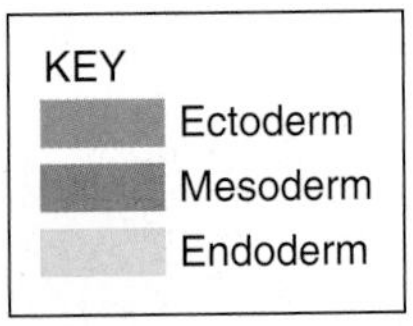

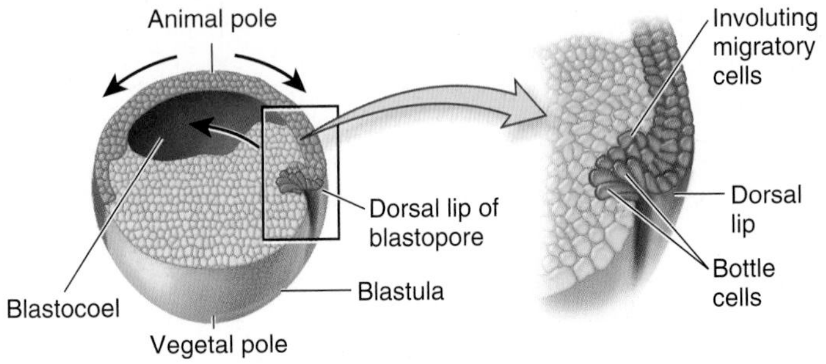

2 **Formation of the archenteron by invagination and involution.** The cavity that begins at the blastopore expands to form the archenteron (future digestive tract). Ectoderm spreads over the embryo.

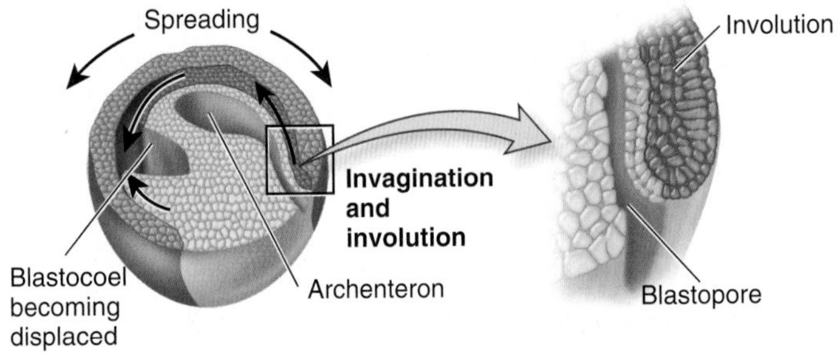

3 **Completion of gastrulation with the beginning of notochord formation.** By the end of gastrulation, the archenteron has displaced the blastocoel and becomes closed by a yolk plug. Involution continues; some of the involuting cells become the mesoderm layer of the gastrula. The dorsal surface of the gastrula begins to thicken, and the dorsal mesoderm begins to form the notochord.

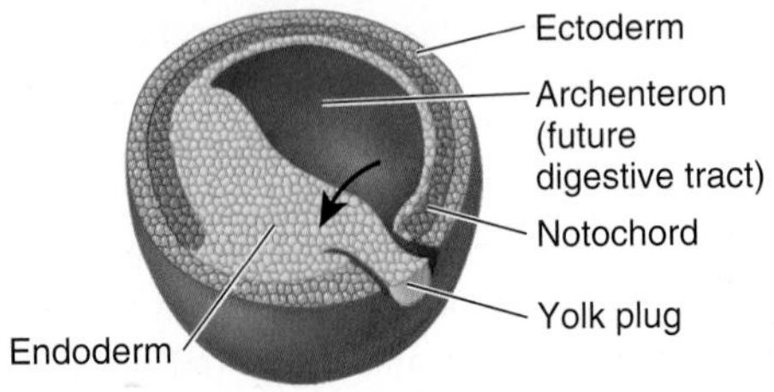

Figure 52.8 The events of gastrulation in amphibians.

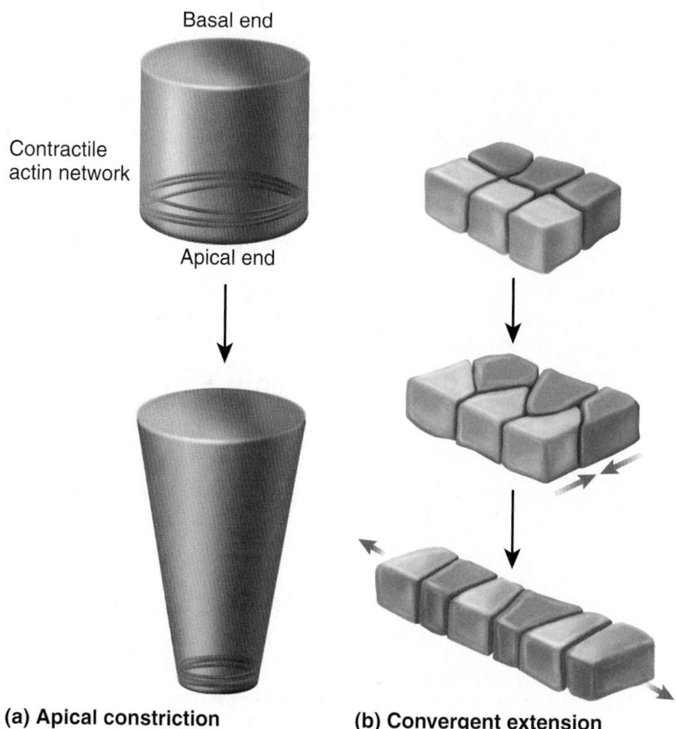

Figure 52.9 **Two mechanisms that affect cell shape and movement.**

development? Several processes are at work. For example, bottle cells initiate gastrulation through a process called **apical constriction** (**Figure 52.9a**). As discussed in Chapter 10, epithelial cells may have a ring of actin filaments that underlie anchoring junctions (adherens junctions) between neighboring cells (refer back to Figure 10.7). In apical constriction, the actin rings connected to the adherens junctions constrict, elongating the cells.

The spreading of ectoderm in the animal hemisphere toward the vegetal hemisphere is mediated by a cellular process called **convergent extension**. During this process, two rows of cells merge to form a single elongated layer (**Figure 52.9b**). Convergent extension produces the movement of sheets of cells.

Notochord Formation A distinguishing anatomical feature that begins forming at the end of gastrulation in all chordates is the **notochord**—a structure derived from mesoderm that provides rigidity along the anteroposterior axis in the dorsal side of the gastrula (see Figure 52.8, step 3). The presence of a notochord defines the phylum Chordates. The notochord persists in the trunk and tail of fishes and amphibians; in birds and mammals, the notochord disappears by the time vertebrae have formed.

In the amphibian embryo, after formation of the archenteron, the dorsal surface of the gastrula begins to thicken, and the dorsal mesoderm forms the notochord. The notochord elongates through convergent extension. By the time the notochord has formed, the dorsal ectoderm overlying the notochord begins to thicken, which initiates the next general event in development, called neurulation. Before we discuss that event, however, we consider another important event that occurs during gastrulation: the establishment of germ cells.

Primordial Germ Cells During gastrulation, a specialized group of cells arise called **primordial germ cells (PGCs)**. The PGCs often arise independently of the three germ layers in the embryo. This cell lineage has two primary functions: (1) to protect and propagate the genetic content of the species and (2) to undergo meiosis and differentiate into gametes—sperm or eggs—in the adult organism. PGCs are stem cells that can divide through mitosis to make copies of themselves. Some of the resulting daughter cells can later undergo meiosis and differentiate into gametes.

The eggs and resulting embryos of some species contain certain cytoplasmic determinants—called the **germ plasm** (**Figure 52.10a**)—that help define and specify the PGCs in the gastrula stage. In amphibian eggs, for example, the germ plasm occupies a small region of cytoplasm around the vegetal pole and contains a specific subset of maternal mRNAs. These cytoplasmic determinants are inherited by a subpopulation of blastomeres during cleavage. At the beginning of gastrulation, blastomeres that inherit these cytoplasmic determinants differentiate into PGCs. At a later stage of development, the PGCs migrate to primordial gonads that eventually will form testes or ovaries. In flies, by contrast, PGCs are the first cells to form at the posterior end of the embryo and are called pole cells (**Figure 52.10b**).

In mammals, neither germ plasm nor pole cells have been identified. Instead, a few mesoderm-like cells begin to express PGC-specific genes early during gastrulation. These cells migrate along the hindgut during gastrulation and, as in amphibians, eventually intermix with gonad primordial cells later in development. Thus, the establishment of PGCs occurs during the earliest phases of animal development and signifies the broad importance of this cell lineage for the propagation of species.

Event 4: Neurulation Involves Formation of the Central Nervous System and Segmentation of the Body

By studying development in several different vertebrate species, researchers are beginning to understand some of the fundamental steps in the formation of the central nervous system (CNS)—the brain and spinal cord—in vertebrates. The multistep embryological process responsible for initiating CNS formation is called **neurulation**

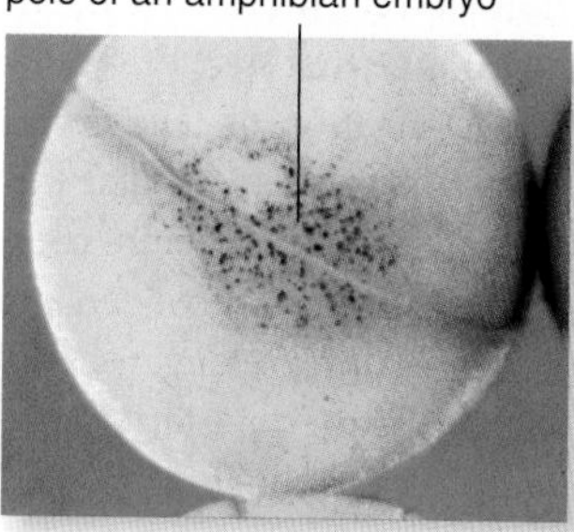

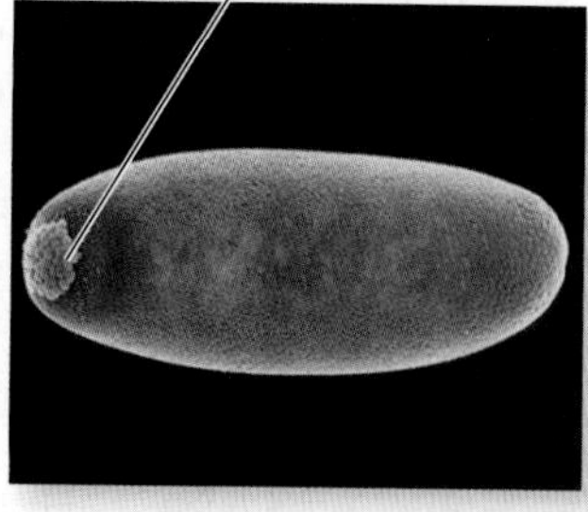

Figure 52.10 **Germ plasm and primordial germ cells (PGCs) during gastrulation.** (a) Germ plasm in an amphibian embryo visualized by labeling an mRNA that is specifically expressed in PGCs. (b) In fly embryos, the PGCs are called pole cells.

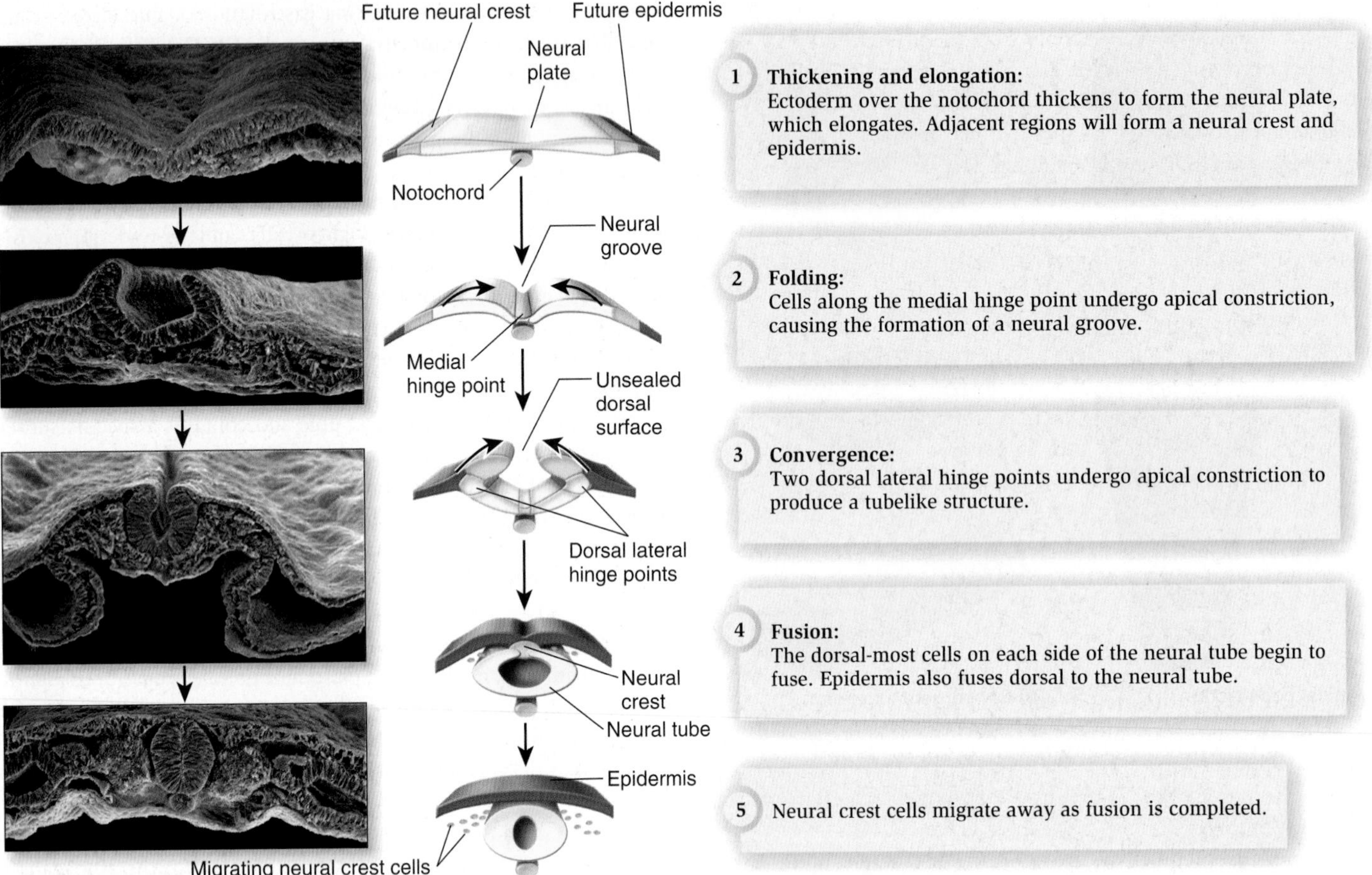

Figure 52.11 **Neurulation and the beginning of neural crest formation in vertebrates.** The four major steps of neurulation are (1) thickening and elongation; (2) folding, which creates the neural groove; (3) convergence, in which the neural tube begins to take shape; and (4) fusion, in which the neural tube is completed. In a later event (5), cells migrate away from the neural crest to form several other structures, including neurons of the peripheral nervous system.

(Figure 52.11). Neurulation occurs just after gastrulation and involves the formation of the **neural tube** from ectoderm located dorsal to the notochord. All neurons and their supporting cells in the CNS originate from neural precursor cells derived from the neural tube. During neurulation, the embryo also develops segmented structures. Next, we discuss the processes of neurulation and body segmentation.

Neural Tube Formation Neurulation in vertebrates occurs in four major steps, as shown in Figure 52.11. In the first step, ectoderm overlaying the notochord thickens by the elongation of cells in the dorsal region to form the neural plate, with adjacent regions that will eventually form the epidermis and a structure called the neural crest (discussed shortly). The neural plate then elongates by convergent extension, resulting in the formation of a single, dorsal elongated epithelial cell layer that is aligned with the animal's anteroposterior axis.

Next, the neural plate forms the neural tube through a series of apical constrictions (steps 2 and 3, Figure 52.11). In the second step, a column of cells along the midline of the neural plate—the medial hinge point—undergoes apical constriction. This initiates the folding phase of neurulation and leads to the formation of the neural groove. After folding, in the third step, bilateral columns of cells in the dorsal lateral hinge points then undergo apical constriction, leading to convergence of the two sides of the neural groove and generation of a tubelike structure that is not yet sealed on the dorsal surface.

In the fourth step of neurulation, called fusion, the dorsal-most cells on either side of the neural tube are released from adjacent ectoderm and fuse with each other, culminating in the closure of the neural tube. At the same time, ectoderm on either side of the neural tube moves toward the centerline, then up and over the neural tube, where it fuses and forms the dorsal epidermis of the embryo.

Neural Crest Formation Another important cell lineage that arises during neurulation is the **neural crest**, which is unique to vertebrates. It consists of cells that originate from the ectoderm overlaying the dorsal surface of the newly formed neural tube and that migrate to other regions of the embryo (Figure 52.11, step 5). Once these cells reach their final destination in the embryo, they differentiate into a variety of cell types different from those that arise from the rest of the ectoderm. All neurons and supporting cells of the peripheral nervous system in vertebrates are derived from neural crest cells. In addition, the neural crest gives rise to skeletal and cartilaginous structures in the head and face, melanocytes (specialized cells that provide pigmentation to the skin of vertebrates), the medulla (the inner region) of the adrenal glands, and connective tissue in numerous organs, notably the heart.

Segmentation and the Formation of Somites As embryos develop, distinct tissues acquire recognizable shapes and patterns, and the body often becomes segmented along the anteroposterior axis of the embryo. **Segmentation** allows individual body segments to have more specialized functions. In some animal taxa, such as insects, body segments are found in the adult animal. In vertebrates, the segmentation process helps define repeated structures such as vertebrae and ribs, which form later during development. Segmentation of the body plan along the anteroposterior axis becomes apparent during neurulation.

During neurulation, the mesoderm becomes segmented from the anterior end first, giving rise to blocklike structures of mesoderm called **somites** (Figure 52.12). The segmentation of somitic mesoderm continues toward the posterior end of the animal and into the tail, as can be seen in Figure 52.12b. The number of somite pairs that forms can vary considerably among species (50 in chicks, 65 in mice, and 500 in some snakes). Because the rate of embryonic development within a given species often varies with temperature and other environmental factors, stages of embryo development are often standardized according to the number of somites that have formed.

Somites arise from a group of loosely packed mesodermal cell aggregates that condense into epithelial somites, which consist of a hollow ball or vesicle enclosed by an epithelial layer. The scanning electron micrograph in Figure 52.12a shows a dorsal view of a chick embryo undergoing somite formation. These epithelial somites are located toward the anterior end, and unsegmented mesoderm can be seen extending toward the posterior end. The epithelial somites are transient structures, and the cells within them soon transform into two morphologically distinct structures that eventually form ribs/skeletal muscle and vertebrae, respectively.

Event 5: Organogenesis Is the Process of Organ Formation

As described in Chapter 40, organs are specialized structures that consist of arrangements of two or more tissue types. Most organs, such as the kidney, contain all four tissue types: nervous, muscle, epithelial, and connective tissue (refer back to Figures 10.15 and 40.1). The developmental event in which cells and tissues form organs is called **organogenesis**. Each germ layer gives rise to particular types of cells found within different organs (see Figure 52.7).

Many organs begin to form during or just after neurulation. However, these organs become functional at different times during development. For example, the lungs of mammals do not acquire the ability to function until shortly before birth. By contrast, the heart is the first functional organ to form in the vertebrate embryo. It begins to beat and pump blood before all the embryo's somites have formed (by 2.5 days after fertilization in chicks, 9 days in mice, and about 22 days in humans).

As we saw in Chapter 40, the development of different organs in animals is controlled by genes in the embryo, notably the *Hox* genes. *Hox* genes are important for establishing structures along the anteroposterior axis. Many of the genes controlling the processes of gastrulation, neurulation, and organogenesis encode secreted proteins or growth factors that induce cells in their local vicinity to differentiate along a specific developmental pathway. For example, the notochord

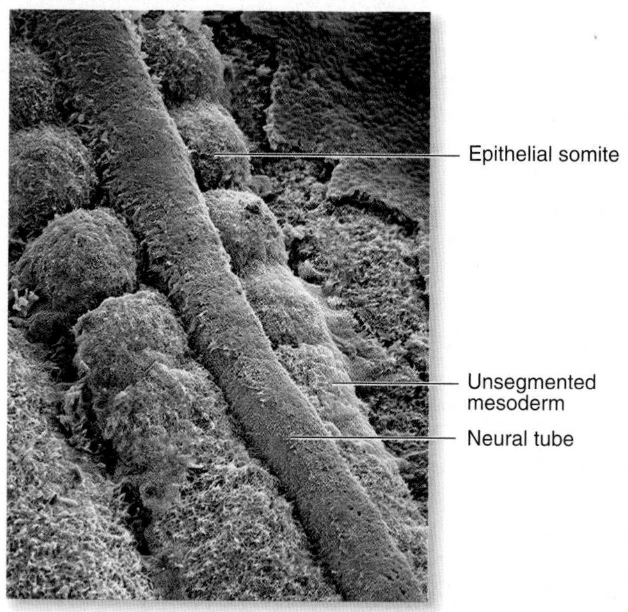

(a) Segmentation in an early chick embryo as seen with a scanning electron microscope

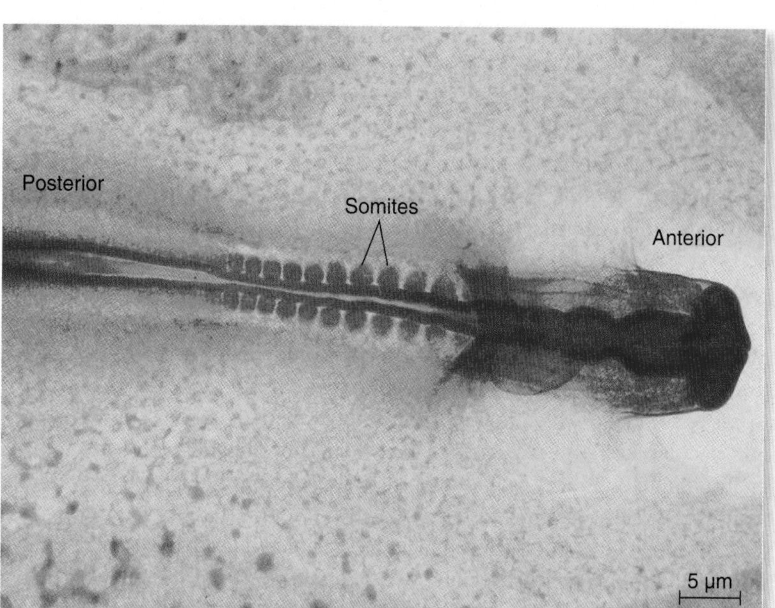

(b) Segmentation in an early chick embryo as seen with a light microscope

Figure 52.12 **Segmentation of mesoderm into somites during neurulation.** **(a)** This scanning electron micrograph shows early events in the segmentation process: the formation of epithelial somites toward the anterior end of a chick embryo. **(b)** Light-microscopic image of a chick embryo, showing body segments along the anteroposterior axis.

produces many signaling proteins that help establish tissue patterns in the embryo. Proteins produced within it induce segment-specific expression of the *Hox* genes in subsequent stages of development. Next, we will see how such growth factors can be studied during embryonic development.

52.3 Control of Cell Differentiation and Morphogenesis During Animal Development

Learning Outcomes:

1. Distinguish between autonomous and conditional specification.
2. Explain how the direct contact between the plasma membranes of neighboring cells affects embryonic development.
3. Describe how organizers can influence cells to differentiate.

Thus far, we have examined the five general events of embryonic development. In this section, we will turn our attention to certain molecular mechanisms that are crucial for cell differentiation and morphogenesis.

Positional Information Is Conveyed Internally and Externally to Each Cell

The process of embryonic development requires that cells divide, move to specific sites, and acquire distinct functional properties (that is, differentiate). Cells receive positional information that determines how they differentiate. There are two primary mechanisms of conveying such positional information. The first mechanism is an internal one, which involves the unequal acquisition by daughter cells of various cytoplasmic factors during cell division, a process known as **autonomous specification**. The second is a variety of external cell-to-cell signaling mechanisms, a process called **conditional specification**. These two mechanisms provide embryonic cells with a continuously changing internal and external environment, in which cells ultimately fulfill their unique spatial and functional fates. This differentiation process involves alterations of gene expression in which specific cells express a unique set of genes required for a particular function.

Two main molecular mechanisms mediate these forms of internal and external communication: morphogens and cell-to-cell contacts (refer back to Figure 19.5). **Morphogens** are molecules that impart positional information and promote developmental changes at the cellular level. Morphogens are the cytoplasmic factors involved in autonomous specification and can also be used in conditional specification as cell-to-cell signals.

The concentration of a morphogen determines its activity. For example, some morphogens are only effective above a certain threshold concentration. Others may direct cells along different developmental pathways at different concentrations. The concentrations of morphogens may vary within or between cells in an embryo. Some morphogenic gradients may be established in the cytoplasm of the oocyte. This is a form of autonomous specification (**Figure 52.13a**). In this process, the morphogen gradient results in daughter cells that have unequal amounts of the morphogen in their cytoplasm. The presence or lack of the morphogen at a threshold concentration will then determine how the cell differentiates. Second, a morphogenic gradient can be established in the embryo by secretion into extracellular fluids. This is a type of conditional specification (**Figure 52.13b**). For example, one or more cells may secrete a morphogen into the surrounding extracellular fluid at a specific stage of development. After secretion, the morphogen contacts neighboring cells, generating an intracellular signal such as a second messenger. This signal may influence the developmental fate of those cells. This process, whereby one or more cells governs the developmental fate of neighboring cells, is known as **induction**.

In contrast to morphogen secretion, the mechanism of direct cell-to-cell contacts used in conditional specification involves proteins that are present in the cell membrane and that can interact with other proteins in the cell membranes of other cells. This interaction generates intracellular signals. Specific cell-to-cell contacts play a major role in determining the final positioning of individual cells within different regions of an embryo.

Throughout development, mechanisms of autonomous and conditional specification function together so that distinct cells respond appropriately. In the rest of this section, we will look at some examples of how these mechanisms lead cells to differentiate into each of the numerous diverse cell types with unique functions in the developing organism. Such differentiation involves changes in both gene expression and subcellular organization. Identifying the precise autonomous and conditional signals that specify each cell lineage in the embryo is one of the biggest challenges facing developmental biologists in the 21st century.

Autonomous Specification: Asymmetric Distribution of Intracellular Morphogens in the Oocyte

An example of autonomous specification can be found during the cleavage events of embryonic development. During early cleavage, cell cycles can be extremely rapid, such that little or no gene transcription occurs during early cleavage. Consequently, nearly all cellular division and differentiation processes during cleavage are regulated by cytoplasmic factors that resided in the oocyte prior to fertilization. Subcellular components that were synthesized prior to fertilization are called maternal factors. These factors include many mRNAs and morphogens that are stockpiled in the maturing oocyte during oogenesis to facilitate cleavage in the absence of transcription.

Conditional Specification: Concentration Gradients of Extracellular Morphogens Control Cell Differentiation in Vertebrate Embryos

Conditional specification often involves cellular induction, the process by which a cell or group of cells induces a response in a neighboring group of cells in the embryo. The idea of cellular induction during embryonic development arose from experiments in which cells isolated from one part of an amphibian embryo were analyzed in the presence or absence of cells isolated from a different region of the embryo.

An example of cellular induction was observed in experiments performed by Dutch developmental biologist Pieter Nieuwkoop during the 1950s. Nieuwkoop isolated cells from a region near either the animal pole or the vegetal pole of late blastula stage (that is, during cleavage) amphibian embryos. When cells from around the animal pole, a region called the animal cap, were cultured in an appropriate

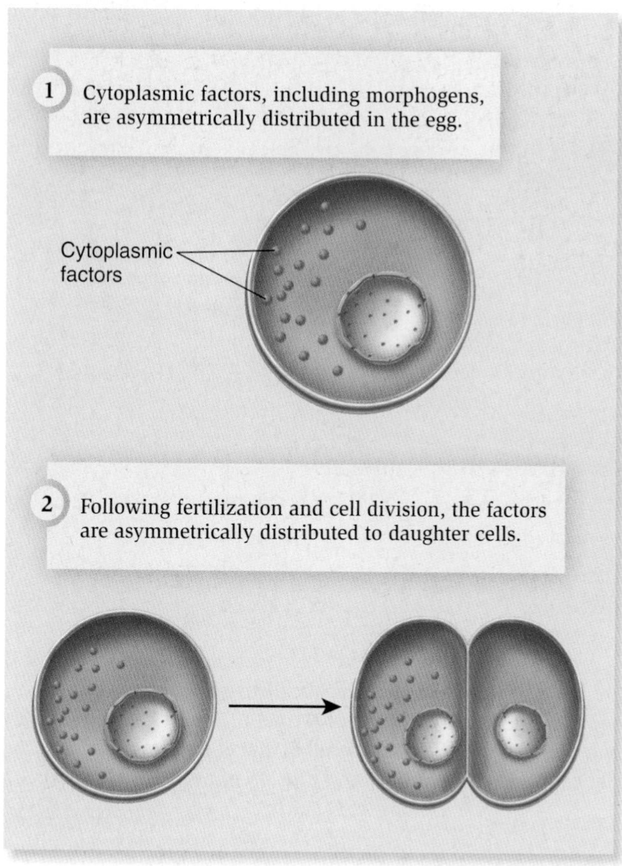

1 Cytoplasmic factors, including morphogens, are asymmetrically distributed in the egg.
Cytoplasmic factors
2 Following fertilization and cell division, the factors are asymmetrically distributed to daughter cells.

(a) Autonomous specification

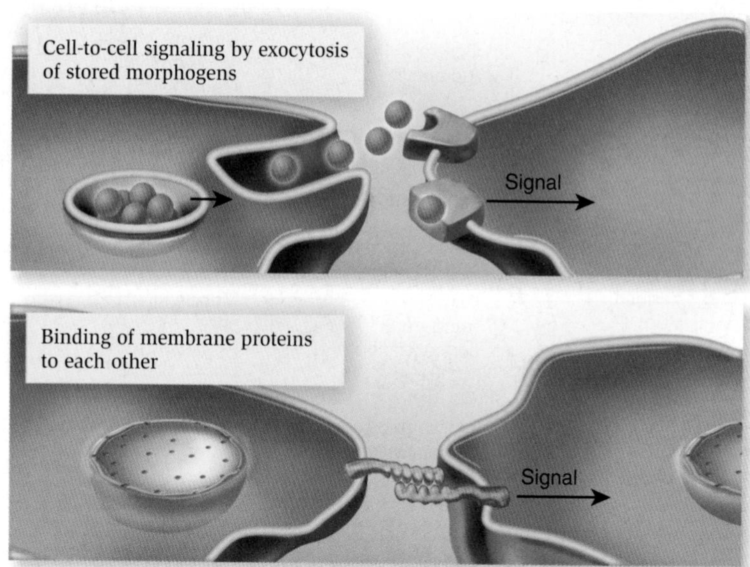

(b) Conditional specification

Figure 52.13 Mechanisms that convey positional information during embryonic development. **(a)** In autonomous specification, cell fate is determined through the unequal segregation of subcellular components during mitosis. **(b)** In conditional specification, cells respond to signals generated by neighboring embryonic cells or to membrane-bound proteins.

growth medium, they formed undifferentiated spherical structures composed primarily of ectodermal cells. When cells isolated from around the vegetal pole were cultured, undifferentiated clumps of endodermal cells resulted. However, when animal and vegetal cells were cultured together, Nieuwkoop observed the formation of mesodermal derivatives from the animal pole cells. This suggested that factors released by cells of the vegetal hemisphere could induce differentiation of animal hemisphere cells. This type of experiment, called an animal cap assay, has been used extensively to identify morphogens secreted by embryonic cells that induce cells in the animal hemisphere to differentiate into mesoderm. Nieuwkoop went on to show that different vegetal cells isolated from various positions along the dorsoventral axis induced specific types of mesoderm.

One family of proteins—named transforming growth factor betas (TGF-βs)—soon emerged as important morphogens involved in the induction of mesoderm. By purifying these proteins and then adding them to a culture medium containing animal caps, researchers demonstrated that TGF-β proteins have mesoderm-inducing activity. In the absence of TGF-βs, animal caps grow in culture to form clumps of ectodermal tissue that resemble skin cells. However, when purified TGF-βs are added to these cells, they differentiate into mesoderm. TGF-β proteins bind to a specific receptor molecule expressed on the surface of animal cap cells, inducing them to differentiate into mesoderm.

A key question was whether different TGF-β proteins specify different types of mesodermal cells, or whether a single protein at different concentrations can specify all types of mesodermal cells. The morphogenic activity of one TGF-β protein—known as activin—in an animal cap assay is shown in **Figure 52.14**. At low concentrations, activin induces the ectoderm to differentiate into mesodermal, blood-like cells. However, at incrementally higher concentrations, muscle cells, notochord, and fully differentiated heart cells are produced. These results show that certain TGF-β proteins exert different effects, depending on their concentrations.

Cell-to-Cell Contact Involves Binding of Membrane Proteins

During the major cell movements of embryonic development, how do cells receive information about where they are located? One way is that the different classes of cells are held together by cell-to-cell contact. Cells within ectoderm, mesoderm, and endoderm express different genes that encode distinct cadherin proteins, which are cell adhesion molecules (described in Figures 10.7 and 10.8). Only cadherins of the same type can bind to one another (**Figure 52.15**). As a result, for example, mesodermal cells tend to adhere to each other but not to endodermal or ectodermal cells.

This property of cells within distinct germ layers is best illustrated by cell dissociation and mixing experiments. When the cells of a gastrula are completely dissociated and then mixed with each other and allowed to reassociate, a fairly well-organized ball of cells forms. When the ectoderm, mesoderm, and endoderm cells within this ball are identified biochemically, the cells are observed to have sorted themselves out. The cells usually bind only to cells of the same type, and mesodermal cells tend to reside between the ectodermal and endodermal cells. This experiment indicates that cells within each germ layer can find each other and self-associate via cell-to-cell interactions.

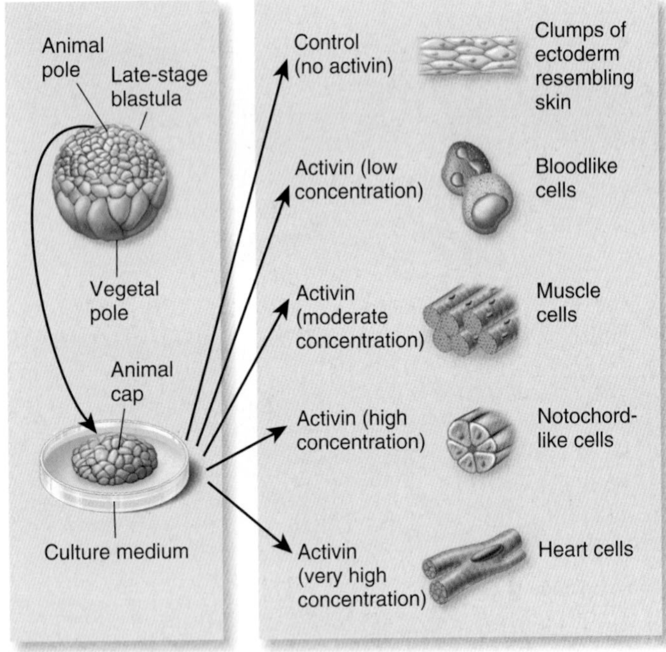

Figure 52.14 Use of an animal cap assay to demonstrate morphogen activity in conditional specification. Activin, a member of the TGF-β protein family, functions as a morphogen in an animal cap assay by inducing different types of mesoderm at different concentrations. In the absence of activin, mesodermal derivatives do not develop, and the animal caps form cells resembling skin cells. Low concentrations of activin induce the formation of mesodermal derivatives resembling blood cells, whereas moderate concentrations produce muscle cells and higher concentrations produce such mesodermal derivatives as notochord and heart cells.

Concept Check: Explain how a single type of molecule can exert different effects on cells at different concentrations.

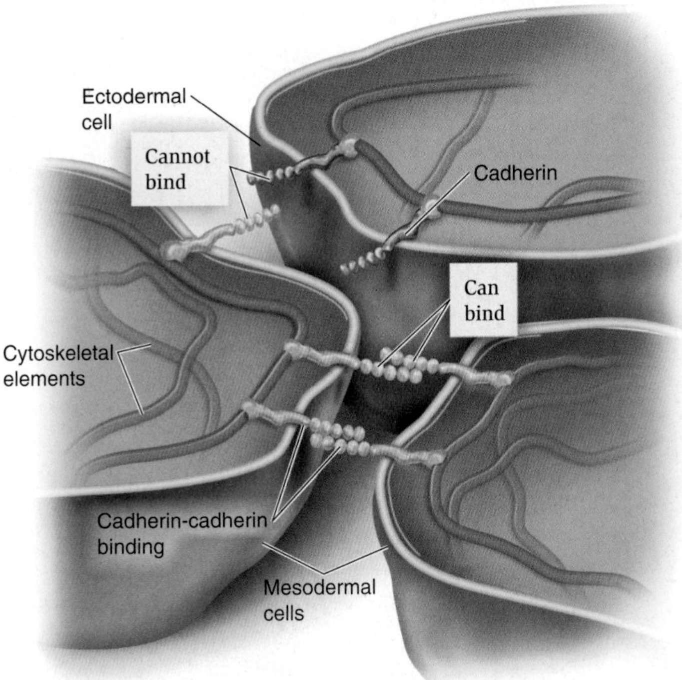

Figure 52.15 Role of cadherins in cell-to-cell contact. Developing cells express one type of cadherin. Similar cadherins bind with each other to promote cell-to-cell contacts. Cells expressing dissimilar cadherins cannot bind and thus do not form associations.

GENOMES & PROTEOMES CONNECTION

Groups of Embryonic Cells Can Produce Specific Body Structures Even When Transplanted into Different Animals

One of the early ideas to emerge from the study of developmental biology was the concept of the **morphogenetic field**, a group of embryonic cells that ultimately produce a specific body structure. Long before genes and their encoded proteins were identified, embryologists discovered that particular groups of embryonic cells in amphibians have a striking characteristic: They form complete body structures when transplanted to another site in the embryo. This has been observed for cells that form the eyes, limbs, and heart in vertebrates and the eyes, antennae, legs, and wings in insects.

One of the first morphogenetic fields postulated in vertebrates was the limb field. Between 1910 and 1930, two important observations were made concerning the limb field. First, removing this group of cells from either side of an early developing embryo led to an embryo that lacked a limb at the corresponding position. Second, transplanting a limb field to a new location within prelimb embryos led to the development of an additional limb at the new location. Interestingly, fate-mapping studies showed that certain cells within the field give rise to specific regions of the limb. These types of experiments suggested that cells within morphogenetic fields were uniquely specified to become particular embryonic structures before any physical evidence for the structure itself could be observed in the embryo.

In the 1920s, the German zoologists Hans Spemann and Hilde Mangold discovered an extremely important field with amazing properties in an early gastrula. This region of the gastrula is now known as **Spemann's organizer**. Their experiments involved dissecting a small piece of tissue from the dorsal lip of the blastopore in an early gastrula of a newt, and transplanting it to the opposite side of another gastrula (Figure 52.16). The result was the formation of a second notochord and neural tube during gastrulation and neurulation in the host embryo, and ultimately the formation of an entirely new body axis—the resulting embryo developed two bodies!

In their experiments, the researchers used two very closely related species of newts. *Triton taeniatus*, which is pigmented, served as the tissue donor, and *Triton cristatus*, which is nonpigmented, served as the host embryo. By transplanting a pigmented Spemann's organizer from the donor embryo into a nonpigmented host embryo, Spemann and Mangold could visually track the origin of the newly developed tissue. The results showed that the secondary notochord and neural tube were composed in large part of nonpigmented host cells, with some pigmented cells remaining from the transplanted tissue. This indicated that the transplanted tissue—which they named the "organizer"—had induced cells in the host to differentiate into neural tissue on the transplanted side of the embryo. More recent work has shown that the organizer secretes morphogens—unknown at the time of Spemann's experiments—responsible for inducing the formation of a new embryonic axis. In 1935, Spemann was acknowledged for his important discovery of cellular induction by receiving the first Nobel Prize (Physiology or Medicine) for studies in developmental biology.

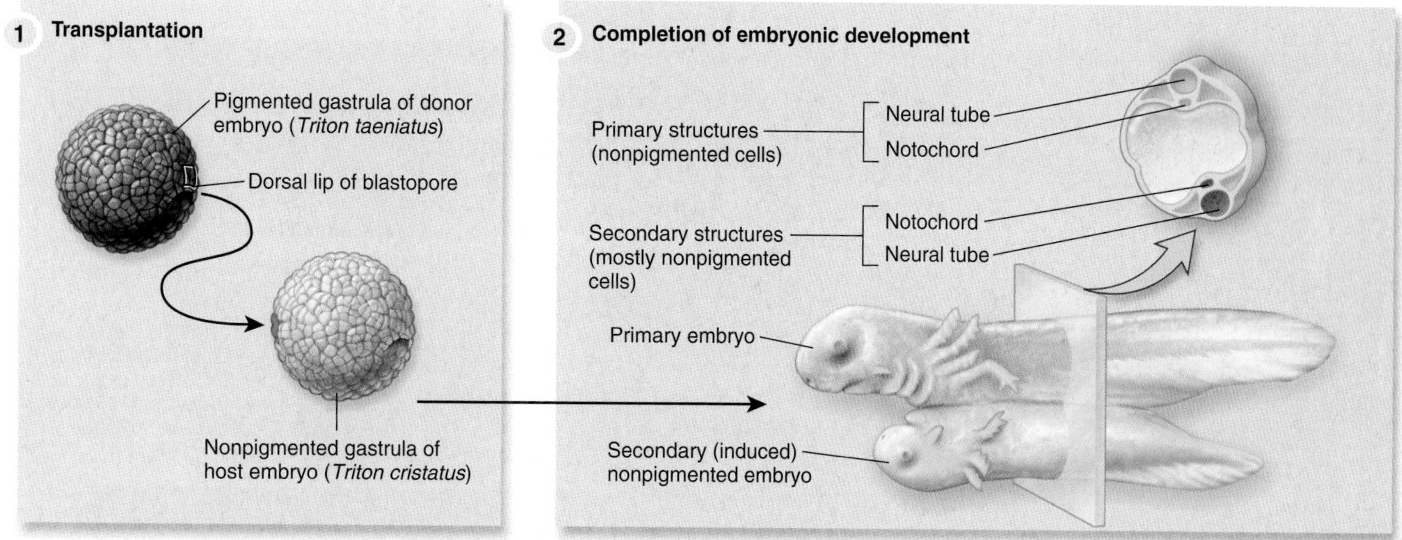

Figure 52.16 Experiment on newt embryos that led to the discovery of Spemann's organizer. Note that the secondary body axis formed by the transplanted organizer is composed of host (nonpigmented) tissue, indicating that the organizer (donor) tissue induced the host tissue to differentiate.

In the past two decades, scientists have identified several genes expressed specifically in Spemann's organizer. Strikingly, many of these genes, which were first discovered in amphibians, are conserved in all vertebrates. These organizer-specific genes are expressed in very small regions of early gastrula embryos, allowing researchers to identify the equivalent of Spemann's organizer in other species. The names given the organizer in different species vary. In chicks, it is called Hensen's node, whereas in mice, it is simply referred to as the node.

With advances in molecular genetics during the 1960s and 1970s, molecular biologists began to reexamine the concept of a morphogenetic field. An important goal of this new generation of molecular biologists was to identify the specific genes transcribed within Spemann's organizer that give this region the ability to form an entirely new body structure. They thought that identifying these genes would provide key insights into the genetic control systems that govern embryonic patterning during gastrulation in vertebrates. However, identifying unknown genes and their protein products was a daunting challenge. Eventually, in 1992, the first secreted morphogen expressed specifically in the organizer was isolated, as we see next.

FEATURE INVESTIGATION

Richard Harland and Coworkers Identified Genes Expressed Specifically in the Organizer

In the early 1990s, Richard Harland and his colleagues hypothesized that morphogens expressed in the organizer should promote the formation of dorsal structures such as those found in the head of embryos. To search for genes that promote the formation of dorsal structures and that are expressed in the organizer during gastrulation, Harland's group took advantage of a key property of unfertilized eggs of the African clawed frog (*Xenopus laevis*). When these eggs are exposed to UV light at their vegetal hemisphere, they fail to form tadpoles after the eggs are fertilized. However, embryos derived from UV-treated eggs do form primitive mesoderm, but only ventral mesoderm forms. Dorsal mesoderm, which normally gives rise to the notochord and somites, does not form in UV-treated eggs, and the resulting "ventralized" embryos soon die. The embryos appear to be lacking one or more morphogens needed for proper dorsal development.

To identify genes in the organizer that encode morphogens that promote the formation of structures such as the notochord, Harland and colleagues used a strategy called expression cloning, as shown in **Figure 52.17**. First, after dissecting a frog embryo, they isolated and purified mRNA from dorsal-lip tissue containing the organizer. Second, they constructed a cDNA library (see Figure 20.4) from the purified mRNA. This resulted in approximately 15,000 different cDNAs, each of which corresponded to a single gene that had been expressed in the organizer. Their next task was to transcribe each of these cDNAs back into mRNA in vitro. In a third step, the researchers injected these different mRNAs into thousands of unfertilized UV-treated eggs, which they then fertilized. Each egg was injected with a single type of mRNA. Their idea was that any mRNA that produced a protein with dorsal mesoderm-promoting activity might "rescue" the embryos and result in free-swimming tadpoles.

Using this procedure, the researchers identified one mRNA whose protein product rescued the embryo (step 3). They then injected increasing amounts of this mRNA into UV-treated eggs, fertilized the eggs, and examined the morphology of the resultant tadpoles. They discovered that the protein product from this mRNA acted as a morphogen because it induced different embryonic structures at different concentrations. At low concentrations, it partially rescued the embryos. At moderate concentrations, it resulted in the formation of normal embryos. However, at very high concentrations, the embryos developed too much dorsal mesoderm and not enough

Figure 52.17 Harland and colleagues discovered the morphogen noggin using expression cloning.

HYPOTHESIS Cells within Spemann's organizer express specific genes that encode proteins that regulate the development of dorsal structures during gastrulation.

KEY MATERIALS *Xenopus laevis* gastrulas and eggs.

	Experimental level	Conceptual level
1 Isolate mRNA from dorsal lip tissue, which contains the organizer.	Surgical scissors; Gastrula; Dorsal lip; Blastopore; Extract and purify mRNA.	mRNA. Contains many different mRNAs expressed in the organizer (~15,000 different mRNAs)
2 Create cDNA library. Use cDNA library to make 15,000 different mRNAs. (See Chapter 20 for a description of cDNA libraries.)	Plasmid; cDNA ligated into vector; + ; Bacterium; Each bacterium takes up 1 plasmid.	Clones of bacteria, each containing a different cDNA. DNA is isolated and transcribed into mRNA in vitro.
3 Expose unfertilized *Xenopus laevis* eggs to UV light, then inject with an mRNA. Fertilize the eggs to determine which, if any, develop. Note: Each egg was injected with a single type of mRNA.	UV; *Xenopus laevis* eggs; Vegetal pole; mRNA injected into egg	UV light inactivates a morphogen and prevents dorsal mesoderm formation; the embryos usually die. Most fertilized eggs formed only ventral mesoderm and died. One, however, developed into a tadpole (rescue).
4 Inject with UV-treated eggs increasing amounts of the mRNA that rescued the embryo in step 3.	Rescue mRNA from step 3; Egg; Fertilize	Fertilized egg; mRNA; Protein product of mRNA. Few copies of mRNA → Low number of proteins formed. Moderate number of copies of mRNA → Moderate number of proteins formed. Many copies of mRNA → Many proteins formed.

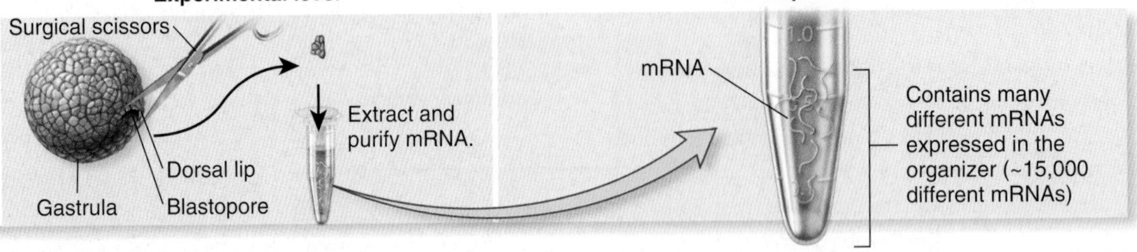

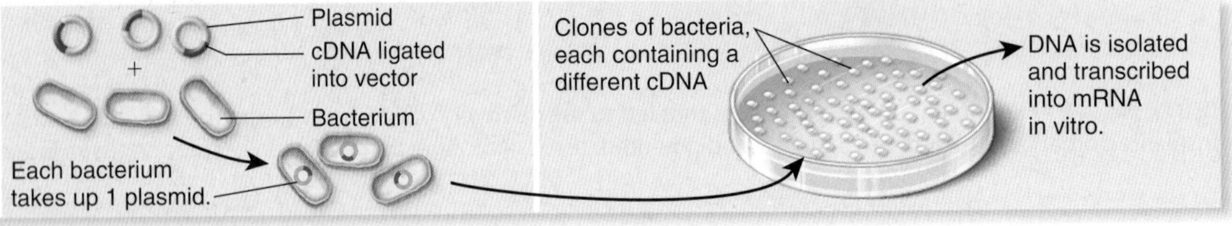

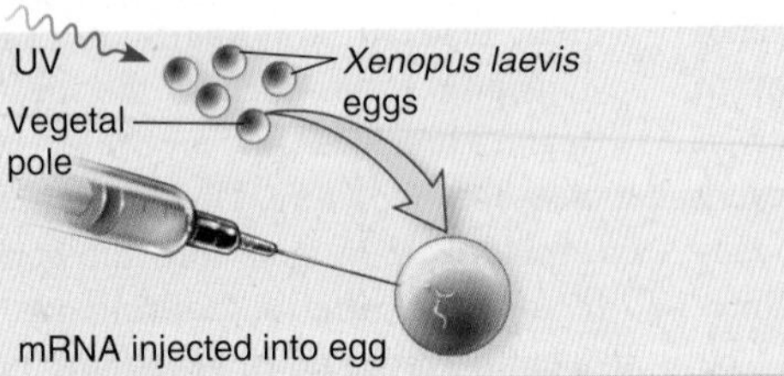

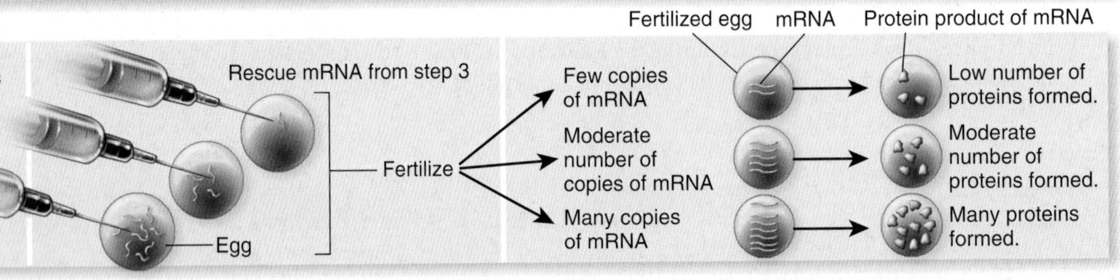

5 THE DATA

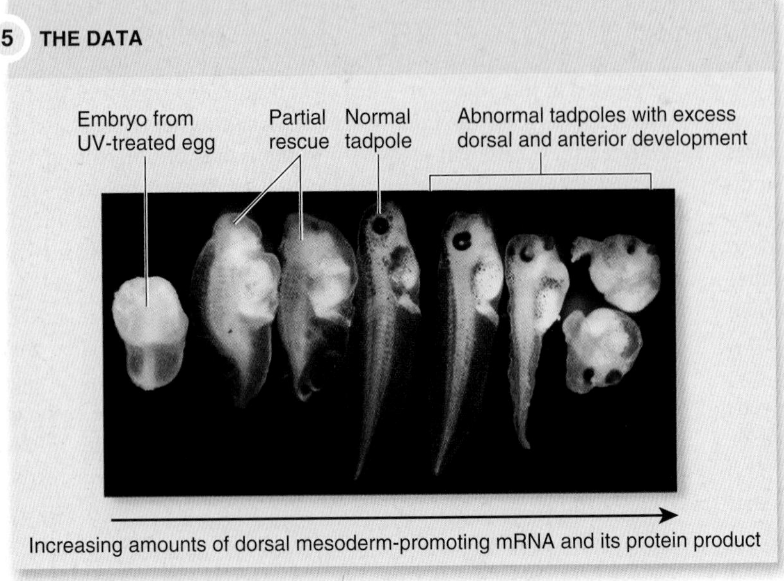

6 CONCLUSION Cells in Spemann's organizer secrete a morphogen—termed noggin—that directs normal dorsal development in *Xenopus laevis* embryos.

7 SOURCE Smith, W.C., and Harland, R.M. 1992. Expression cloning of noggin, a new dorsalizing factor localized to the Spemann organizer in *Xenopus* embryos. *Cell* 70:829–840.

ventral mesoderm. These abnormal embryos developed large heads and extremely small trunks. Further studies identified the gene that coded for this mRNA and verified that it was transcribed in Spemann's organizer. They named the gene *noggin* (a slang term for brain or head), because it caused embryos to develop large heads when present in abnormally high concentrations (see left side of the data).

Later work revealed that noggin protein promotes dorsal development by inhibiting ventral development. Noggin protein inhibits at least two other morphogens that are known to induce ventral, but not dorsal, mesoderm. These studies showed that antagonistically acting proteins expressed and secreted from certain cells in the organizer can specify precise structures in a concentration-dependent fashion. Further work revealed that *noggin* is expressed in the node of mice. Specific deletion of the *noggin* gene in mice leads to defects in dorsal structures. This indicates that *noggin* plays a fundamental role during gastrulation in mammals as well as amphibians.

The discovery of *noggin* and its mechanism of action revealed that the normal development of embryos requires a balance between stimulatory and inhibitory factors. The inhibition of developmental processes—such as the suppression of ventral mesoderm by *noggin* (and, as researchers have since learned, several other key gene products)—appears to be a normal part of the mechanism by which development of an organism occurs.

Experimental Questions

1. Why were scientists interested in identifying genes expressed exclusively in Spemann's organizer?
2. What hypothesis did Harland and colleagues test?
3. How did the scientists identify possible genes that are important for the development of dorsal structures, and how did they test the gene products to determine their activity?

52.4 Impact on Public Health

Learning Outcome:

1. Describe three human birth defects that are due to abnormalities in embryonic development.

Given the complexity and precision of the five general developmental events we have described in this chapter, it may not be surprising that on occasion these processes fail to properly occur. When this happens in humans, its impact can be devastating. Many health problems in humans are caused by genetic or environmental factors that disrupt embryonic development. As we have seen, the early development of an embryo is among the most complex processes found in nature, and a number of diseases and disorders in humans stem from problems that arise during this process.

An example of a condition involving defective embryonic development is **spina bifida**, caused by the failure of the neural tube to close at either the anterior or posterior end during neurulation. Signs and symptoms of spina bifida, which occurs in about 7 of every 10,000 births in the U.S., vary depending on the degree to which the neural tube fails to close. In minor forms, gaps may occur between a few vertebrae at the bottom of the back, and impairment of motor or sensory function may not be apparent at birth. However, neurological deterioration often becomes evident during childhood or adulthood. Surgical procedures soon after birth can improve the long-term quality of life for patients with spina bifida, but prevention of this condition is much more effective. Spina bifida appears to be related to deficient concentrations of folic acid, a common B vitamin needed during the rapid growth that occurs during embryonic development. Up to 75% of spina bifida cases may be prevented if women increase their dietary intake of folic acid before and in the first trimester of pregnancy.

Other embryonic defects are caused by foreign chemicals introduced during pregnancy. Given the deleterious effects of certain chemical compounds on specific phases of embryonic development in humans, it is not surprising that ingested or inhaled compounds such as tobacco, alcohol, and other drugs can also have severe and devastating consequences on a developing fetus. The leading overall cause of mental retardation worldwide is **fetal alcohol syndrome (FAS)**, which occurs in babies whose mothers drink large amounts of alcohol during pregnancy. In addition to cognitive disorders, children born with FAS also show malformed facial features and joints and altered overall growth characteristics (which are also seen in other vertebrate embryos exposed experimentally to ethanol; see the chapter-opening photo). These morphological features are likely caused by the generally detrimental effects that ethanol has on cell division. The cognitive deficits are thought to result from the death of developing CNS neurons, which are particularly susceptible to the toxic effects of ethanol.

Unlike the previous conditions, which are caused by environmental conditions surrounding the embryo, other embryonic defects are caused by genetic defects. The most common genetic disorder affecting humans is **Down syndrome**, which arises when a zygote receives three copies of chromosome 21. The effects in an embryo are complex, such that multiple physical and neurological disorders are associated with this syndrome.

Finally, cleft lip or palate (or both combined) are relatively common (roughly 1:1,000 births in the U.S.) developmental disorders that appear to share both a genetic and an environmental basis. During early development in humans, the tissues of the lip and palate exist as two structures that merge around weeks 5–7 of embryonic development. In cleft lip, the tissues of the upper lip fail to fuse, whereas in cleft palate, those of the palate fail to fuse; in some cases, these two deformities occur together. In the latter case, the result is a gap or groove that extends from the mouth into the nasal cavity. Cleft lip/palate is a disfiguring condition that leads to speech impediments, difficulty eating and drinking (for example, an affected infant cannot readily produce the suction needed to drink from a bottle or nipple), and repeated infections. The only treatment for cleft lip/palate is surgical repair of the affected tissues, which is often quite successful but which usually entails numerous surgical procedures over several years (**Figure 52.18**). Despite intensive research, the causes of cleft lip/palate are still uncertain. However, certain environmental agents have been linked with an increased risk of cleft lip/palate, notably maternal

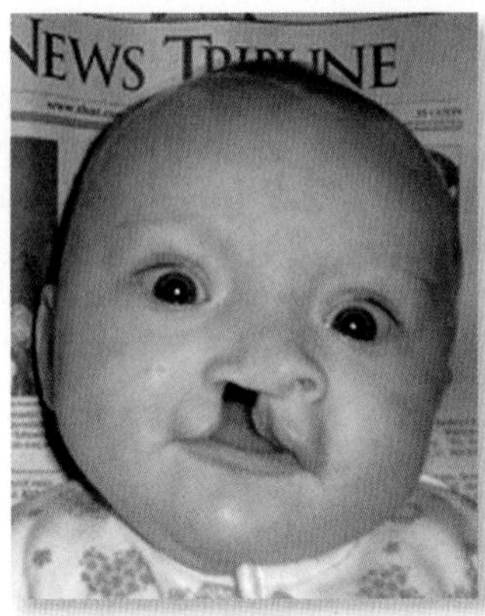
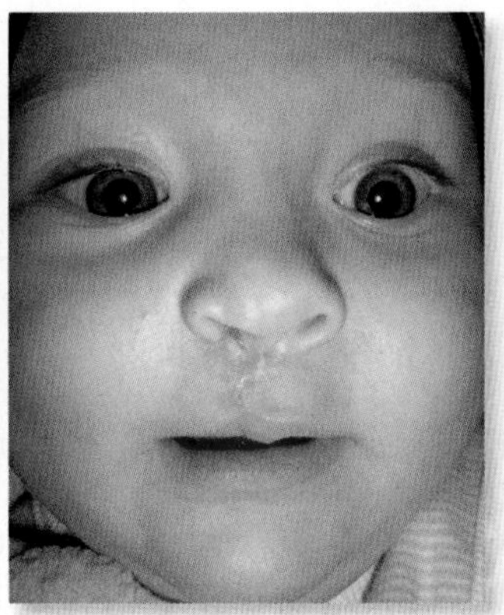

Figure 52.18 **Child with cleft lip/palate before and at two different ages after surgeries.**

ingestion of alcohol, lead, cocaine, or certain antiseizure drugs, or possibly vitamin deficiencies during pregnancy. Recently, researchers have identified several genes in which loss-of-function mutations affect the way in which the palate develops, contributing to a cleft.

The World Health Organization has compiled the number of birth defects around the world by geographic region in order to assess the extent of the impact of developmental disorders on public health. The resulting document, entitled *The World Atlas of Birth Defects*, indicates that regional differences in the incidences of a given birth defect can be dramatic. For example, the rate of spina bifida varies from 1 in every 10,000 births in France to about 15 to 30 in every 10,000 births in Mexico and Ireland. The occurrence of Down syndrome, by contrast, varies from about 7 cases per 10,000 births in Cuba to 22 per 10,000 births in Chile. Together, 27 different birth defects affect 2–3% of all babies born in the world today, an astonishing number when one considers the world's population of roughly 7 billion, with annual worldwide births of nearly 150 million.

Summary of Key Concepts

52.1 Principles of Embryonic Development

- The process by which a fertilized egg is transformed into an organism with distinct physiological systems and body parts is called embryonic development. The process by which different cells within a developing organism acquire specialized forms and functions, due to the expression of cell-specific genes, is called cellular differentiation. Four responses to positional information are cell division, cell migration, cell differentiation, and apoptosis.

52.2 General Events of Embryonic Development

- Development in many animals, including vertebrates, involves five general events: fertilization, cleavage, gastrulation, neurulation, and organogenesis (Figure 52.1).
- Major events in fertilization include the acrosomal reaction and the cortical reaction. During the acrosomal reaction, the binding of a sperm with the egg membrane triggers a series of events producing the fast block to polyspermy, which prevents other sperm from binding to the egg. During the cortical reaction, events produce a calcium wave that leads to additional barriers to sperm, a process called the slow block to polyspermy (Figure 52.2).
- Fertilization is followed by cleavage, which involves cell divisions without cell growth, and results in daughter cells called blastomeres. In many species, when the embryo forms an outer epithelial layer and an inner cavity, it is called a blastula. The mammalian counterpart of a blastula is a blastocyst.
- Cleavage-stage embryos in triploblast organisms have animal and vegetal hemispheres. The hemispheres determine in part the future anteroposterior and dorsoventral axes of the embryo (Figure 52.3).
- Incomplete or meroblastic cleavage occurs in birds, some fishes, and some other vertebrates whose eggs contain large amounts of yolk. Complete or holoblastic cleavage occurs in amphibians and mammals, whose eggs have smaller amounts of yolk. In mammals, cleavage occurs in the oviduct and the embryo implants in the uterine wall (Figures 52.4, 52.5).
- In fate mapping, a small population of embryonic cells is specifically labeled with a harmless dye, and the fate of these labeled cells is followed to a later stage of embryonic development (Figure 52.6).
- During gastrulation, the hollow ball of cells that makes up the blastula or blastocyst is converted into a gastrula, containing the three germ layers: endoderm, mesoderm, and ectoderm. Each germ layer gives rise to specific structures (Figure 52.7).
- Gastrulation begins when a band of tissue invaginates, creating a small opening called the blastopore. During gastrulation, the embryo forms a cavity called the archenteron, which will become the organism's digestive tract. By the end of gastrulation, the notochord has formed in chordates (Figure 52.8).
- During gastrulation, two cellular processes are crucial to development: apical constriction, in which a reduction in the diameter of the actin rings connected to the adherens junctions causes the cells to elongate toward their basal end, and convergent extension, in which two rows of cells merge to form a single elongated layer (Figure 52.9).
- Primordial germ cells (PGCs) become established during gastrulation. Amphibian eggs contain certain cytoplasmic determinants, called the germ plasm, that help define and specify the PGCs in the gastrula stage. In flies, PGCs are the first cells to form at the posterior end of the embryo and are called pole cells (Figure 52.10).
- Neurulation is the multistep embryological process responsible for initiating CNS formation. Neurulation occurs just after gastrulation and involves the formation of the neural tube from ectoderm located dorsal to the notochord. The neural crest, which gives rise to all neurons and supporting cells of the peripheral nervous system in vertebrates, arises during neurulation. The mesoderm becomes segmented from the anterior end first, giving rise to blocklike structures of mesoderm called somites (Figures 52.11 and 52.12).
- The developmental event during which cells and tissues form organs is called organogenesis.

52.3 Control of Cell Differentiation and Morphogenesis During Animal Development

- Two mechanisms convey positional information during embryonic development: autonomous specification, which involves internal factors, and conditional specification, which involves external signals. Morphogens and cell-to-cell contacts are the molecular mechanisms behind autonomous and conditional specification (Figure 52.13).
- Autonomous specification involves the asymmetric distribution of intracellular morphogens and other factors in the oocyte.
- A type of experiment called an animal cap assay has been used extensively to identify morphogens secreted by embryonic cells that induce cells in the animal hemisphere to differentiate into mesoderm. This is a type of conditional specification (Figure 52.14).
- In another type of cell-to-cell interaction, cells use membrane proteins called cadherins to bind to other like cells, ensuring that the cells will move together during development (Figure 52.15).
- An extremely important region in the early gastrula is known as Spemann's organizer. The organizer secretes morphogens responsible for inducing the formation of a new embryonic axis (Figure 52.16).
- Richard Harland and coworkers discovered the *noggin* gene in frog embryos and demonstrated that it acted to promote dorsal development by inhibiting ventral development (Figure 52.17).

52.4 Impact on Public Health

- Some important examples of conditions involving defective embryonic development are spina bifida, caused by the failure of the neural tube to close completely during neurulation; fetal alcohol syndrome (FAS), which occurs in babies whose mothers use alcohol during pregnancy; Down syndrome, the most common genetic disorder affecting humans, in which the embryo has three copies of chromosome 21; and cleft lip or palate, a developmental disorder with both genetic and environmental causes (Figure 52.18).

Assess and Discuss

Test Yourself

1. The acrosomal and cortical reactions occur during which event of development?
   a. fertilization
   b. cleavage
   c. gastrulation
   d. neurulation
   e. organogenesis
2. Cadherins are
   a. adhesion protein molecules that attach cells to extracellular material.
   b. genes necessary for proper germ layer formation.
   c. adhesion protein molecules that allow cells of a given germ layer to adhere to each other.
   d. germ layers.
   e. proteins that provide structural support to the cytoplasm of cells.
3. Cell differentiation that results from cell-to-cell signaling is
   a. autonomous specification.
   b. conditional specification.
   c. communicative specification.
   d. gastrulation.
   e. embryogenesis.
4. The three germ layers of triploblasts are established during
   a. fertilization.
   b. cleavage.
   c. blastula formation.
   d. gastrulation.
   e. neurulation.
5. The limbs of vertebrates are formed from
   a. the endoderm.
   b. the mesoderm.
   c. the ectoderm.
   d. all of the above.
6. In vertebrates, the digestive tract forms from
   a. the blastopore.
   b. the dorsal lip.
   c. the archenteron.
   d. the mesoderm.
   e. both a and d.
7. The cells that give rise to gametes
   a. are derived from the mesoderm.
   b. often arise independently from the three germ layers.
   c. originate directly from the tissue that will develop into the gonads.
   d. are some of the last cells of the embryo to differentiate.
   e. are formed the same way in all animals.
8. Cells of the neural crest
   a. give rise to the central nervous system.
   b. originate from the ectoderm.
   c. migrate to different areas of the body and differentiate into a variety of cells, including neurons of the peripheral nervous system.
   d. all of the above
   e. b and c only
9. Which is true of Spemann's organizer?
   a. It is unique to amphibian embryos.
   b. It arises early in cleavage-stage embryos.
   c. It secretes morphogens, including the protein called noggin.
   d. It is found in the dorsal lip region of early gastrula-stage embryos.
   e. c and d are both correct.
10. Morphogens
   a. are proteins that stimulate responses in the cells of an embryo.
   b. are important signaling proteins in the process of embryonic development.
   c. can elicit different cell responses at different concentrations.
   d. contribute to autonomous specification.
   e. all of the above

Conceptual Questions

1. What are the differences between autonomous and conditional specification?
2. During organogenesis in vertebrates, some organs develop and become functional sooner than others. Why is it important, for example, that the heart of a terrestrial vertebrate develop sooner than its lungs?
3. A principle of biology is that *living organisms grow and develop.* Distinguish among embryonic development, differentiation, and growth.

Collaborative Questions

1. Discuss the process of gastrulation.
2. Discuss the process of neurulation.

Online Resource

www.brookerbiology.com

Stay a step ahead in your studies with animations that bring concepts to life and practice tests to assess your understanding. Your instructor may also recommend the interactive eBook, individualized learning tools, and more.

Immune Systems

53

Chapter Outline

53.1 Types of Pathogens

53.2 Innate Immunity

53.3 Acquired Immunity

53.4 Impact on Public Health

Summary of Key Concepts

Assess and Discuss

In an immune system response, a macrophage engulfs numerous rod-shaped bacteria (false-color SEM).

All animals must defend themselves against environmental factors that threaten their survival, including predation, intraspecific conflict, accidents, and hazardous substances. In addition, animals must contend with various threats within their internal environment, including the invasion of potentially harmful microorganisms such as bacteria, the presence of foreign molecules such as the products of microorganisms, and the presence of abnormal cells such as cancer.

The ability of an animal to ward off these internal threats—an animal's **immunity**, or immune defenses—is the subject of this chapter. The cells and organs within an animal's body that contribute to immune defenses collectively constitute an animal's **immune system**. The study of immunity is called immunology. Immunologists examine the processes by which the immune system protects an animal from foreign matter, whether living or nonliving. In these processes, the body's immune defenses recognize the body's own molecules as "self" and attack anything that is foreign, or "nonself."

Immune defenses are often divided into two types: innate and acquired. **Innate immunity** refers to the body's defenses that are present at birth and that act against foreign materials in much the same way regardless of the specific identity of the invading material. Thus, the innate immunity of animals is also known as nonspecific immunity. Innate immunity includes the body's external barriers (skin and mucous membranes), plus a set of cellular and chemical defenses that oppose substances that breach those barriers. An example of an innate immune response is seen in the chapter-opening photo, in which a cell known as a macrophage is engulfing numerous bacteria. All animals have innate immune defenses.

In contrast, **acquired immunity** develops only after the body is exposed to foreign substances. This type of immunity is characterized by the ability of certain cells of the immune system to recognize a particular foreign substance and initiate a response that targets that substance specifically. For this reason, acquired immunity is also known as specific immunity. Another feature distinguishing acquired immunity is that repeated exposure to a foreign substance elicits greater and greater defense responses, unlike the situation for nonspecific immunity, in which each exposure to the foreign material elicits the same defense responses. Thus, acquired immunity is also called adaptive immunity. Acquired immune defenses have been identified in all vertebrates except for the jawless fishes, but they have not been unequivocally identified in invertebrates.

We begin the chapter with a brief overview of the different pathogens that cause disease in animals. We then consider the mechanisms that provide animals with innate and acquired defenses against harmful pathogens. We conclude with a discussion of the public health implications of some selected immunity-related conditions in humans.

53.1 Types of Pathogens

Learning Outcome:

1. List the three main types of pathogens that elicit immune responses and how each damages its host organism.

An animal's immune defenses must protect against a variety of foreign materials, but most important among them are disease-causing viruses and microorganisms, or **pathogens**. Pathogens exist in nearly every possible ecological niche on Earth. Both terrestrial and aquatic animals, including invertebrates and vertebrates, encounter each of the three major types of pathogens: certain bacteria, viruses, and eukaryotic parasites. A fourth type of infectious agent—proteins called prions—were considered in Chapter 18 (refer back to Figure 18.8) and will not be discussed here.

As discussed in Chapter 27, bacteria are single-celled prokaryotic organisms that lack a true nucleus. Bacteria can either damage tissues at open wound sites or release toxins that enter the bloodstream and disrupt functions in other parts of the body. Bacteria are responsible for many diseases and infections, including typhoid fever, strep throat, skin infections, middle ear infections, and food poisoning. The major ways in which bacteria gain entry into an animal's body are through direct bodily contact, open wounds, inhalation through the respiratory tract, and ingestion via fecal contamination of food or water. This last situation may arise because many infectious bacteria enter the intestines and are excreted in feces, which may be deposited near food or water sources used by some animals.

All viruses contain nucleic acid (DNA or RNA) within a protein coat (see Chapter 18). Unlike bacteria, viruses lack the metabolic machinery to synthesize the proteins they require to replicate themselves. Instead, they must infect a host cell and use its biochemical and genetic machinery, including nucleotides and energy sources, to make more viruses. The viral nucleic acid directs the host cell to synthesize the proteins required for viral replication. After entering a cell, some viruses, such as the common cold virus, multiply rapidly, kill the cell, and then infect other cells. Other viruses can lie dormant within host cells before suddenly undergoing rapid replication, which causes cell damage or death. Finally, certain viruses can transform their host cells into cancerous cells. Viruses are responsible for a great variety of illnesses, including some sexually transmitted diseases (refer back to Figure 18.2). Like bacterial infections, viral infections can spread rapidly among animals and can be lethal. Viruses typically enter an animal's body through the respiratory tract or through open wounds.

Eukaryotic parasites—whether protists, fungi, or worms—damage a host by using the host's nutrients for their own growth and reproduction or by secreting toxic chemicals. In humans, parasites account for an enormous number of cases of disease annually. For example, several hundred million people are infected each year with one of the mosquito-borne protists of the genus *Plasmodium* that cause malaria. Parasitic infections may enter a host through the bite of an infected insect, as in malaria; by ingestion of food or water containing parasitic organisms, such as roundworms; or in some cases, by penetrating the skin, as with blood flukes.

53.2 Innate Immunity

Learning Outcomes:

1. Identify the main innate defense mechanisms in animals, and explain why innate immunity is also called nonspecific immunity.
2. Explain how an animal's body surface provides protection from pathogens.
3. Describe the process of phagocytosis, and explain its importance in innate immune responses.
4. List the cell types involved in innate immunity and the functions of each.
5. Outline the sequence of events in the inflammatory response.
6. Describe the role of Toll proteins in innate immunity.

Innate (nonspecific) immune defenses protect against foreign cells or substances without having to recognize the invaders' specific identities. This type of defense mechanism is called innate because animals inherit the ability to perform these protective functions and because this type of immunity does not require prior exposure to invaders. Instead of distinguishing among foreign materials, nonspecific defenses recognize some general, conserved property marking the invader as foreign, such as a particular class of carbohydrate or lipid present in the cell walls of many different kinds of microbes.

In this section, we will consider the innate immune defenses. These include defenses at the body surfaces, the actions of phagocytic cells, the response to injury known as inflammation, and various proteins secreted by cells of the immune system that facilitate the destruction of pathogens.

The Body Surface Is an Initial Line of Defense

An animal's initial defense against pathogens is the barrier provided by a surface exposed to the external environment. Very few microorganisms can penetrate the intact skin or body surface of most animals, particularly the tough, thick, or scaly skin characteristic of many vertebrates or the rigid exoskeleton of many arthropods. In addition, glands in the body surfaces of both invertebrates and vertebrates secrete a variety of antimicrobial molecules, including mild acids and enzymes such as lysozyme that destroy bacterial cell walls.

The mucus secreted by cells in the mucous membranes lining the respiratory and upper gastrointestinal tracts of vertebrates also contains antimicrobial molecules. More importantly, mucus is sticky—microbes that become stuck in it are prevented from penetrating the mucous membrane barrier. They are either swept up by cilia into the pharynx and then swallowed, or they are engulfed by cells that are present in both tracts. Pathogens ingested with food are often destroyed by the acidic environment of an animal's midgut.

If a pathogen is able to penetrate a barrier and gain entry into an animal's internal tissues and fluids, other nonspecific defense mechanisms are activated. These mechanisms are mediated by several types of cells that reside in the body fluids and tissues, as described next.

Phagocytic Cells Defend Against Pathogens That Enter the Body

Several different types of cells in vertebrates play key roles in innate immunity (Figure 53.1). Many of these cells are **phagocytes**—cells capable of phagocytosis. **Phagocytosis** is a type of endocytosis in which the cell engulfs particulate matter, which usually is then destroyed by proteases or oxidizing compounds such as hydrogen peroxide. Phagocytes are found in the body fluids, such as hemolymph and blood, and also within various tissues and organs. They are present in all classes of animals and are among the most fundamental and ancestral forms of immune defenses.

In vertebrates, most phagocytes belong to the type of blood cells called white blood cells, or **leukocytes** (see Figure 53.1). All leukocytes are derived from a common type of stem cell (refer back to Figure 19.20), which in mammals and birds is found in the bone marrow. These stem cells give rise to several types of leukocytes and other cells that have specialized functions. The leukocytes involved in innate immunity include neutrophils, eosinophils, monocytes, macrophages, basophils, and natural killer cells.

Neutrophils are phagocytes and the most abundant leukocytes. They are found in blood and some may enter tissues during inflammation.

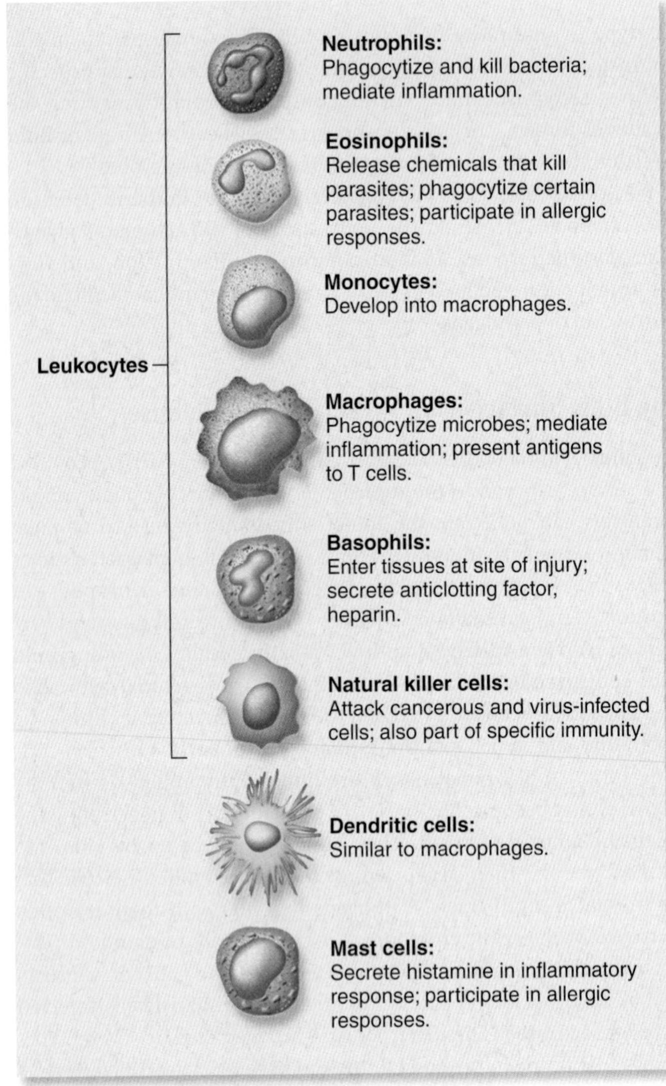

Leukocytes
Neutrophils:
Phagocytize and kill bacteria; mediate inflammation.
Eosinophils:
Release chemicals that kill parasites; phagocytize certain parasites; participate in allergic responses.
Monocytes:
Develop into macrophages.
Macrophages:
Phagocytize microbes; mediate inflammation; present antigens to T cells.
Basophils:
Enter tissues at site of injury; secrete anticlotting factor, heparin.
Natural killer cells:
Attack cancerous and virus-infected cells; also part of specific immunity.
Dendritic cells:
Similar to macrophages.
Mast cells:
Secrete histamine in inflammatory response; participate in allergic responses.

Figure 53.1 **Cells involved in innate immunity in vertebrates.** Note that six of these types of cells are leukocytes.

After neutrophils engulf bacteria by phagocytosis, the bacteria are destroyed within endocytotic vacuoles by proteases, oxidizing compounds, and antibacterial proteins called defensins. The production and release of neutrophils from bone marrow are greatly stimulated during the course of an infection. **Eosinophils** are found in the blood and in the mucosal surfaces lining the gastrointestinal, respiratory, and urinary tracts, where they fight off parasitic invasions. In some cases, eosinophils act by releasing toxic chemicals that kill parasites, and in other cases by phagocytosis. **Monocytes** are phagocytes that circulate in the blood for a short time, after which they migrate into tissues and organs and develop into **macrophages**. Macrophages are strategically located where they will encounter invaders, including epithelia in contact with the external environment, such as skin and the linings of respiratory and digestive tracts. Macrophages are large phagocytes capable of engulfing viruses and bacteria, as shown in the chapter-opening photo.

In contrast to these other leukocytes, **basophils** are secretory cells. They secrete an anticlotting factor called heparin at the site of an infection, which helps the circulation flush out the infected site. Basophils also secrete histamine, which attracts infection-fighting cells and proteins to the site.

Natural killer (NK) cells are another kind of leukocyte called lymphocytes, of which there are several types. We will learn more about the other types of lymphocytes later, because they play the major role in acquired immunity. NK cells, however, participate in both innate and acquired immunity. These cells are part of an animal's innate defenses because they recognize general features on the surface of cancerous cells or any virus-infected cells. NK cells arise in the bone marrow. They act by releasing chemicals into the vicinity of cancerous or virus-infected cells, thereby killing those cells.

In addition to leukocytes, two other types of cells derived from bone marrow stem cells play important roles in nonspecific immunity. **Dendritic cells** are scattered throughout most tissues, where they perform various macrophage-like functions. **Mast cells** are found throughout connective tissues, particularly beneath the epithelial surfaces of the body. Mast cells secrete many locally acting molecules, including histamine. Histamine and other substances are involved in inflammation, a fundamental component of the innate defense mechanism, which we now examine.

Inflammation Is an Innate Response to Infection or Injury

Inflammation is an innate local response to infection or injury. The functions of inflammation are to destroy or inactivate foreign invaders, to clear the infected region of dead cells and other debris, and to set the stage for tissue repair. The key cellular components of this process are neutrophils, macrophages, dendritic cells, and mast cells.

The events of inflammation are induced and regulated by chemical mediators. These include a family of proteins called **cytokines** that function in both innate and acquired immune defenses. Cytokines provide a chemical communication network that synchronizes the components of the immune response. Most cytokines are secreted by more than one type of immune system cell and also by certain nonimmune cells such as endothelial cells and fibroblasts.

The sequence of local events in a typical inflammatory response to a bacterial infection is summarized in Figure 53.2. A tissue injury such as that caused by a splinter begins the inflammation process, which results in the familiar signs and symptoms of local redness, swelling, heat, and pain.

Substances secreted into the extracellular fluid from injured tissue cells, mast cells, and neutrophils contribute to the inflammatory response. For example, histamine from mast cells and nitric oxide from endothelial cells (Figure 53.2, step 1) cause dilation of the small blood vessels in the infected and damaged area, inducing the vessels to leak (step 2).

These vascular changes provide two benefits. First, the increased blood flow to the inflamed area, which accounts for the redness and heat, speeds the delivery of beneficial proteins and leukocytes and increases local metabolism to facilitate healing. Second, the increased vascular permeability ensures that the plasma proteins that participate in inflammation can gain entry to the interstitial fluid. The swelling in an inflamed area also results from this increased leakiness of blood vessels.

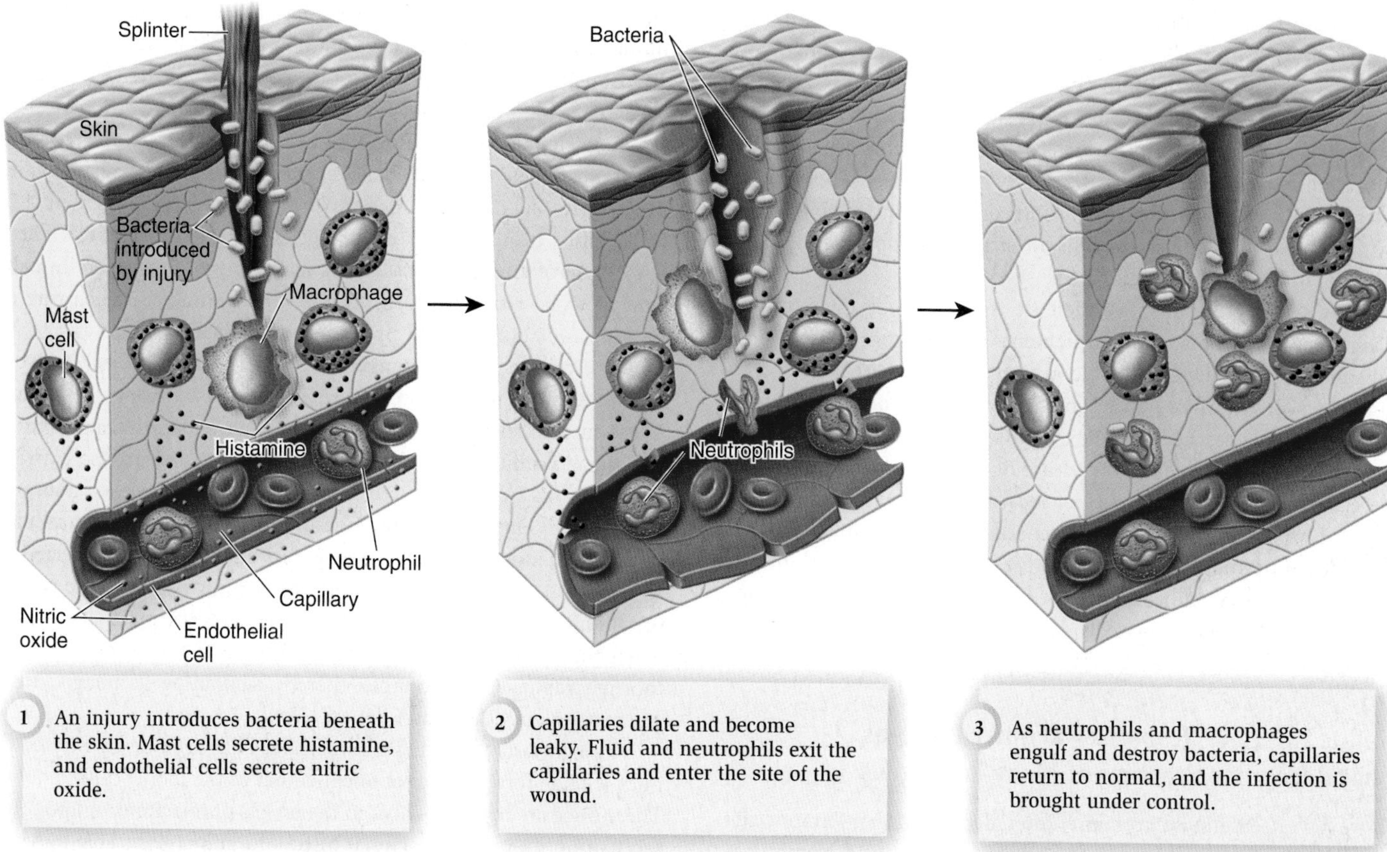
Splinter
Skin
Bacteria introduced by injury
Mast cell
Macrophage
Histamine
Neutrophil
Capillary
Nitric oxide
Endothelial cell
Bacteria
Neutrophils
1 An injury introduces bacteria beneath the skin. Mast cells secrete histamine, and endothelial cells secrete nitric oxide.
2 Capillaries dilate and become leaky. Fluid and neutrophils exit the capillaries and enter the site of the wound.
3 As neutrophils and macrophages engulf and destroy bacteria, capillaries return to normal, and the infection is brought under control.

Figure 53.2 **The events in inflammation.** Shown are the initial stages of inflammation in response to a penetrating wound that introduces bacteria beneath the skin.

Concept Check: *Inflammation is often associated with swelling of the inflamed area. Could this swelling have an adaptive value?*

BioConnections: *In Chapter 41, you learned about another role of histamine in animals, as a neurotransmitter that increases wakefulness. Why would antihistamines (drugs that block histamine receptors) used to treat allergies cause drowsiness?*

Once neutrophils and macrophages arrive at the site of the injury, they begin the process of phagocytizing and destroying invading microbes (Figure 53.2, step 3). The initial step in phagocytosis involves the interaction of phagocyte surface receptors with certain carbohydrates or lipids in the microbial cell walls. Subsequently, the neutrophils and macrophages also release antimicrobial substances into the extracellular fluid that can destroy microbes even before phagocytosis occurs. Other secreted substances such as nitric oxide function as inflammatory mediators. The result is positive feedback: Once phagocytes enter the area and encounter microbes, they release inflammatory mediators that bring even more phagocytes into the area.

Inflammation sets the stage for tissue repair. Nearby connective tissue cells called fibroblasts divide rapidly and secrete large quantities of collagen, the major component of extracellular matrices, and nearby blood vessel cells proliferate to restore blood supply. Ultimately, tissue repair may be imperfect, leaving a scar largely composed of fibroblasts, collagen, and other proteins.

Inflammation—indeed any stage of an illness—may be accompanied by **fever**, an increase in an animal's body temperature that results from an infection. This is distinguished from hyperthermia, an increase in body temperature resulting from any number of causes, such as overexertion. In humans, a temperature greater than 38°C (100.4°F) is considered to be a fever (normal is about 37°C [98.6°F]). The precise mechanisms by which a fever arises are not completely understood. It is clear, however, that in mammals, cytokines released into the circulation by activated macrophages act within the hypothalamus to raise the body's set point for temperature. The result is an increase in heat generation. Whereas a sustained high fever requires medical attention because of the damaging effects of high temperature on membrane function, enzyme activity, and other processes (see Chapter 46), it appears from animal models that a small or moderate increase in temperature stimulates leukocyte activity and proliferation and provides a less hospitable environment for at least some types of pathogens.

Defense Against Pathogens Includes Interferons and Complement Proteins

In addition to phagocytes and the inflammatory response, the body has at least two other types of innate defenses against invading pathogens. These defenses allow for extracellular destruction of pathogens without prior exposure to those pathogens. The first is used against viruses. **Interferons** are proteins that generally inhibit viral replication inside host cells. In response to viral infection, many cell types produce interferons and secrete them into the extracellular fluid. When the interferons bind to plasma membrane receptors on the secreting cell and on other cells, each cell synthesizes a variety of proteins that interfere with the ability of the viruses to replicate. Interferons are not specific. Many kinds of viruses induce interferon synthesis, and the same interferons, in turn, can inhibit the multiplication of many different kinds of viruses.

The second type of innate defense is provided by the family of plasma proteins known as **complement**. Inactive complement proteins normally circulate in the blood at all times and are activated by contact with the surface of microbes. Activation of complement proteins results in a cascade of events. Among their many actions, complement proteins stimulate the release of histamine from mast cells, thereby increasing permeability of local blood vessels, as described earlier. Five of the active proteins generated in the complement cascade form a multiunit protein called the **membrane attack complex (MAC)**, which, by embedding itself in the microbial plasma membrane, creates porelike channels. Water and ions enter the microbe through the channels, and the microbe bursts.

Innate Immune Responses Require Proteins That Recognize Features of Many Pathogens

As we have seen, innate immunity often depends on an immune cell recognizing some general molecular feature common to many types of pathogens. These features are called **pathogen-associated molecular patterns (PAMPs)**. Until recently, however, it was not known how that recognition was accomplished. In 1985, German biologist Christiane Nüsslein-Volhard and American biologist Eric Wieschaus were interested in how animal embryos differentiate into adults. In the course of their studies, they discovered a protein they named Toll (now called Toll-1) that was required for the proper dorsoventral orientation of the fruit fly *Drosophila melanogaster*. In a landmark study in 1996, however, it was discovered that Toll-1 also conferred on adult flies the ability to fight off fungal infections (see Feature Investigation that follows). It has since become clear that a family of Toll proteins exists in animals from nematodes to mammals, including humans, and these proteins are found in the plasma membranes and endosomal membranes of macrophages and dendritic cells, among others.

One function of the Toll proteins in vertebrates is to recognize and bind to ligands with PAMPs, such as lipopolysaccharide and other microbial lipids and carbohydrates; viral and bacteria nucleic acids; and a protein found in the flagellum of many bacteria. These PAMPs have conserved molecular features that are generally considered to be vital to the survival of that pathogen. When one of these ligands binds a Toll protein on the plasma membrane of an immune cell, second messengers are generated in the cell, triggering secretion of inflammatory mediators such as various cytokines. These in turn stimulate the activity of other leukocytes involved in the innate immune response. Some of these mediators also activate cells involved in the acquired immune response. Because many of the Toll proteins are plasma-membrane-bound, bind extracellular ligands, and induce second-messenger formation, they are referred to as receptors, and the family of proteins is known as **Toll-like receptors (TLRs)**. Despite this, not all TLRs generate intracellular signals when bound to a ligand; some TLRs induce attachment of a microbe to a macrophage, for example, thereby facilitating its phagocytosis and subsequent destruction.

The importance of TLRs in mammals has been demonstrated in mice with a mutated form of one member of the family called Toll-4. These mice are hypersensitive to the effects of injections of lipopolysaccharide (to mimic a bacterial infection) and are less able to ward off bacterial infection. In humans, recent studies suggest that certain naturally occurring variants in a specific TLR are associated with increased risk of certain diseases.

TLRs are currently an active area of investigation among biologists because of their immune significance. Certain domains of these receptors have also been identified in plants, where they may also be involved in disease resistance. Therefore, TLRs may be among the first immune defense mechanisms to ever evolve in living organisms. For their investigations of embryonic development and discovery of toll proteins, Nüsslein-Volhard and Wieschaus were awarded the 1995 Nobel Prize in Physiology or Medicine.

FEATURE INVESTIGATION

Lemaitre and Colleagues Identify an Immune Function for Toll Protein in *Drosophila*

As you've just learned, in 1985 Toll protein was identified as a critical molecule directing the proper dorsoventral development of *Drosophila* embryos. Interestingly, however, certain aspects of the sequence and function of Toll shared similarities with at least one important immune protein found in vertebrates. For example, both Toll and the receptor for the cytokine interleukin-1 were found to be transmembrane proteins with similar cytosolic sequences; moreover, both proteins activated a similar intracellular signaling pathway once they had bound an extracellular ligand. This led to the hypothesis proposed by French researcher Bruno Lemaitre and coworkers working in the laboratory of Jules Hoffman, that Toll protein may play a role in immunity in *Drosophila*.

To test this hypothesis, the researchers compared wild-type to mutant flies in which *Toll* was underexpressed due to a gene mutation (**Figure 53.3**). They then exposed the flies to a variety of microbial infections, by pricking them with a needle that had been dipped into water (as a control) or a concentrated solution of fungal spores. Survival of the flies was monitored over the next several days, and compared with the survival of wild-type flies with normal Toll protein. What they found was remarkable: within days of being infected with the fungus *Aspergillus fumigatus*, 100% of the mutant flies died, but only about 30% of the wild-type flies died (see the data for Figure 53.3). The mutant flies were observed to be covered in germinating hyphae of the fungus, clearly demonstrating that the flies were unable to fight the infection. Moreover, analysis of mRNA from the treated and untreated flies revealed that the mutant flies were unable to increase expression of a key antifungal gene after infection with *A. fumigatus*, unlike the situation in the wild-type animals.

Since then, these and other researchers have identified much of the mechanism by which Toll promotes immunity to certain microbes in *Drosophila* and, presumably, other insects. Activation of Toll protein induces expression of several antimicrobial genes. Unlike TLRs, however, in *Drosophila* Toll does not directly bind PAMPs associated with microbial membranes. Rather, microbial infection induces activation of extracellular signaling molecules in *Drosophila* that, in turn,

Figure 53.3 **Lemaitre and colleagues identified a role for Toll protein in immunity in *Drosophila*.**

HYPOTHESIS The Toll signaling pathway is required for the immune response in *Drosophila*.

KEY MATERIALS Wild-type *Drosophila*; mutant strain of *Drosophila* that under-expresses the receptor for Toll protein; spores of the fungus *A. fumigatus*; cDNA probe to detect mRNA of antifungal peptide drosomycin.

1 Prick flies with a clean needle that was previously dipped in water or a solution of *A. fumigatus* spores.

A. fumigatus is a common pathogen that *Drosophila* is normally able to fend off.

2 Follow survival rates of flies for 6 days.

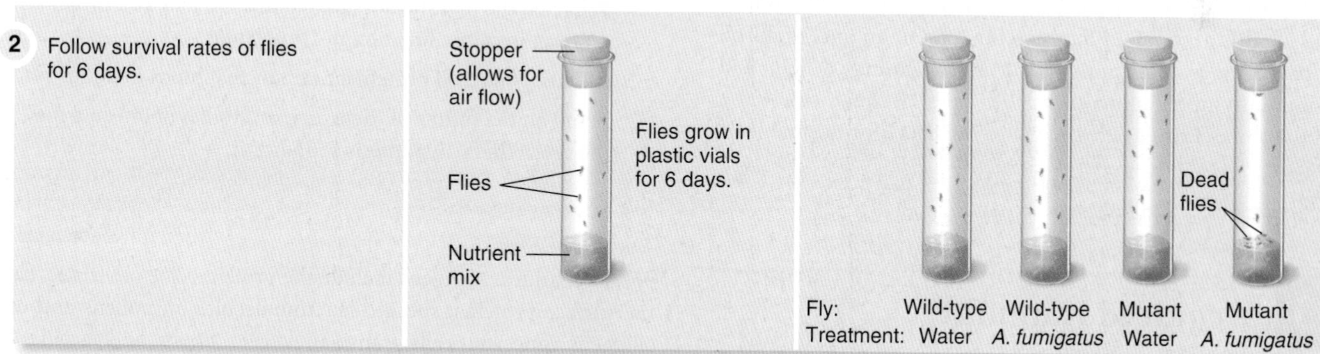

Experimental level

Conceptual level

3 Examine expression of antipathogen genes in wild-type and mutant flies, using Northern blot analysis:

1. Extract mRNA from flies.
2. Separate individual mRNAs by gel electrophoresis.
3. Transfer mRNAs from the gel to a nylon membrane.
4. Probe membrane with a radioactively labeled fragment of cDNA complementary to a region of the mRNA of interst.
5. Expose membrane to X-ray film.

Agarose gel

Nylon membrane

mRNA corresponding to drosomycin

Labeled cDNA probes

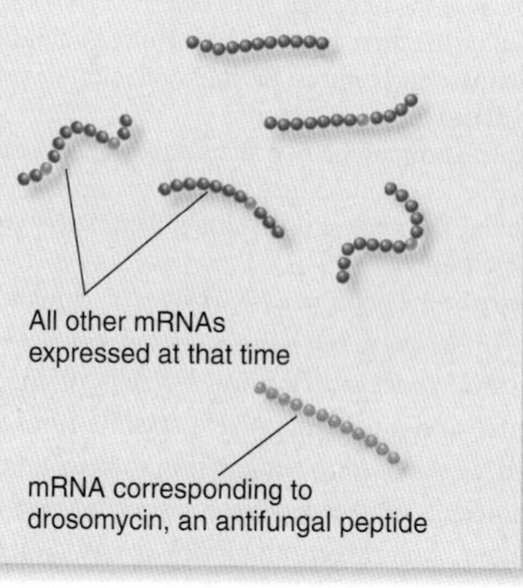

4 THE DATA

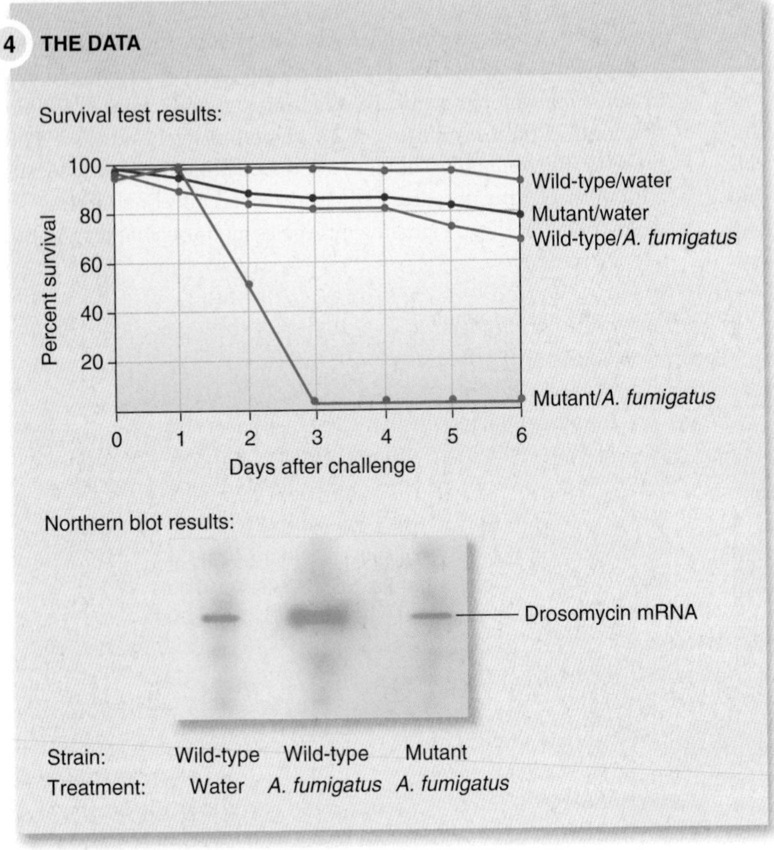

5 CONCLUSION Toll protein receptor is essential for survival in *Drosophila* exposed to a common fungal pathogen and for induction of antifungal genes.

6 SOURCE Lemaitre, B., Nicolas, E., Michaut, L., Reichhart, T., and Hoffmann, J.A. 1996. The dorsoventral regulatory gene cassette spätzle/toll/cactus controls the potent antifungal response in *Drosophila* adults. *Cell* 86:973–983.

bind to Toll that is located on cell membranes; this binding induces intracellular signals that activate gene expression. Despite this difference, the identification of an immune function of an ancestral Toll protein paved the way for our current understanding of TLRs and innate immunity in all animals; in 2011, Jules Hoffman received a share of the Nobel Prize in Physiology or Medicine for this pioneering research.

Experimental Questions

1. What led the investigators to hypothesize that Toll protein may serve an immune function in *Drosophila*?
2. Is *Drosophila* Toll protein a receptor that binds PAMPs?
3. Based on the results of their survival studies of infected flies, was the researchers' hypothesis supported?

53.3 Acquired Immunity

Learning Outcomes:

1. Describe the components of the immune system including the major classes of lymphocytes.
2. Outline the three stages of the acquired immune response.
3. Compare and contrast the humoral immune response with the cell-mediated immune response.
4. Explain how antibody structure and genetic rearrangement lead to the diverse set of antibodies found in mammals.
5. Outline the steps of the typical inflammatory and humoral immune responses.
6. Describe the importance of secondary immune responses.

In acquired (specific) immunity, cells of the immune system first encounter and later recognize a specific foreign cell or protein to be attacked, as opposed to recognizing some general feature of pathogens. Any molecule that can trigger a specific immune response is called an **antigen**. An antigen, therefore, is any molecule that the host does not recognize as self. Most antigens are either proteins or very large polysaccharides. Antigens include the protein coats of viruses, bacterial surface proteins, specific macromolecules on pollens and other allergens, cancerous cells, transplanted cells, toxins, and vaccines.

This type of immunity is found in all classes of vertebrates except jawless fishes and was once thought to be absent in invertebrates. Recently, however, scientists who study the evolution of immune systems have uncovered several interesting features of invertebrate immune function that suggest some invertebrates have a limited ability to adapt immune activity to a specific invader. Despite this intriguing finding, however, we will consider acquired immune responses in the context in which they are best understood, the jawed vertebrates.

The Immune System Consists of Lymphoid Organs, Tissues, and Cells

The cells of the immune system that are responsible for acquired immunity are a type of leukocyte mentioned earlier called **lymphocytes**. Like all leukocytes, lymphocytes circulate in the blood, but most of them reside in a group of organs and tissues that constitute the **lymphatic system** (Figure 53.4a). The system is composed primarily

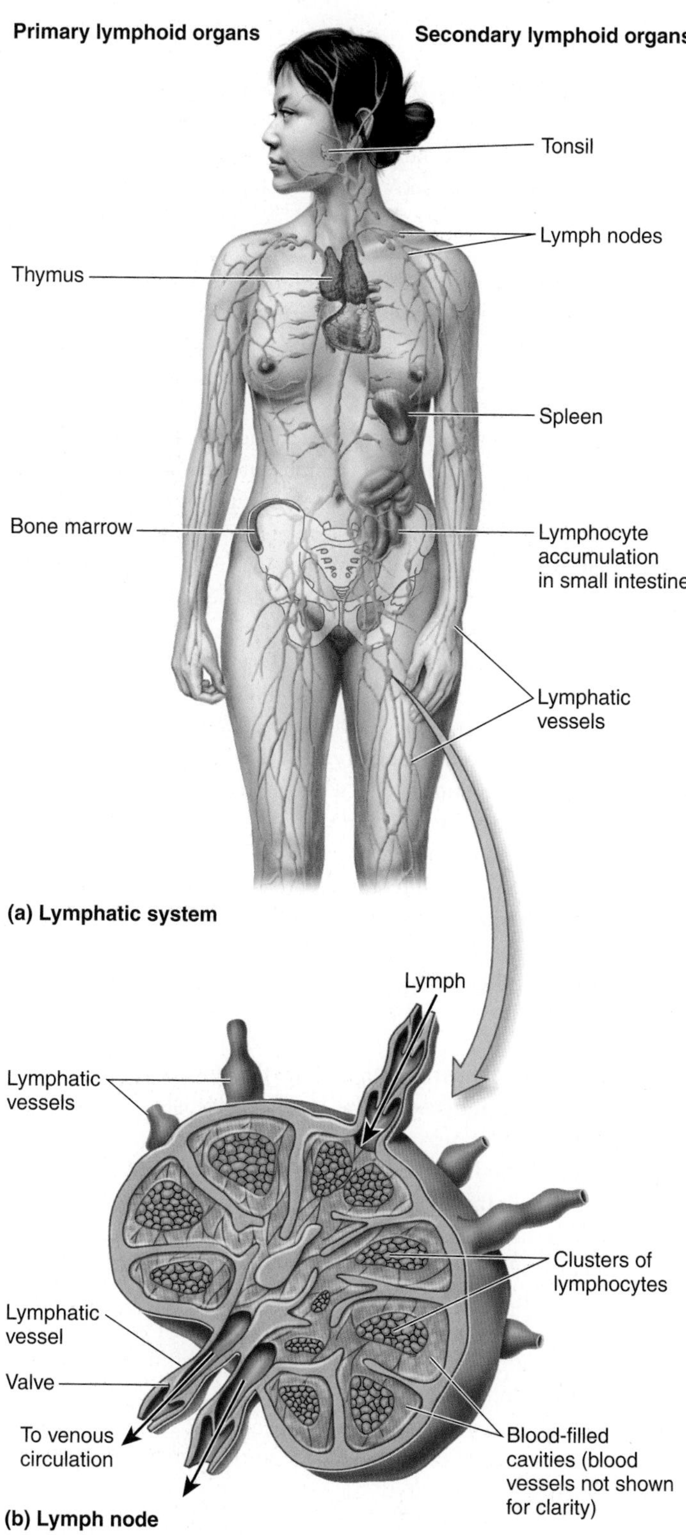

Figure 53.4 **The lymphatic system in humans.** **(a)** The major components of the human lymphatic system. Primary lymphoid organs are shown in red, and secondary lymphoid organs are shown in green. In adult humans, the primary lymphoid organs in bone are found in the sternum, ribs, parts of the skull, small regions of the femur and humerus, and, as shown here, the hip bones. **(b)** The structure of a lymph node. Lymph nodes occur along the course of lymphatic vessels, which drain interstitial fluid from tissues and return it to the venous circulation. Within a lymph node, lymph percolates through open cavities containing clusters of lymphocytes.

Concept Check: *From where does lymph arise?*

of a network of lymphatic vessels. These vessels drain the fluid known as lymph from the interstitial fluid. The lymph is transported through lymph vessels to lymph nodes, before eventually returning to the blood and circulatory system. Various lymphoid organs and tissues are located throughout the lymphatic system. They are grouped into primary and secondary lymphoid organs.

The primary lymphoid organs are the structures in which lymphocytes differentiate into mature immune cells. These are the bone marrow in birds and mammals and the thymus gland in all vertebrates. In animals without extensive bone marrow, specialized regions of other organs such as the kidney and liver serve as primary lymphoid organs.

The primary lymphoid organs supply mature lymphocytes to secondary lymphoid tissues and organs, where the lymphocytes multiply and function. These include the lymph nodes of mammals (Figure 53.4b), the spleen (found in all jawed vertebrates and the largest secondary lymphoid structure), the tonsils (small, rounded lymphoid organs in the pharyngeal region of mammals), and scattered lymphocyte accumulations in the linings of the intestinal, respiratory, genital, and urinary tracts. With some exceptions (for example, adult humans, in which the thymus gland is no longer very active), destruction of or damage to a primary lymphoid organ results in a severe inability to fight off infections. The loss of any of the secondary lymphoid organs, although not as serious, nevertheless increases the risk of local or systemic infections throughout an animal's life. For example, in humans, the spleen must occasionally be surgically removed due to injury or disease. Such individuals must be monitored carefully for the rest of their lives because of their increased vulnerability to infection.

After leaving the bone marrow or thymus gland, lymphocytes circulate between the secondary lymphoid organs, blood, lymph, and all the tissues of the body. Lymphocytes from all the secondary lymphoid structures continually leave those structures and are carried to the bloodstream. Simultaneously, some circulating lymphocytes leave venules all over the body to enter the interstitial fluid. From there, they reenter lymphatic vessels and are carried back to secondary lymphoid organs. This constant recirculation of lymphocytes increases the likelihood that any given lymphocyte will encounter an antigen it is specifically programmed to recognize.

Lymphocytes Provide Specific Immunity Against Antigens in Humoral and Cell-Mediated Immunity

Different kinds of lymphocytes participate in coordinated specific immune system responses (Figure 53.5). In addition to NK cells described earlier, the two major types of lymphocytes are **B cells** and **T cells.** B cells were first observed to mature in an avian organ called the bursa of Fabricius—thus the name B cells. In mammals, B cells mature within bone marrow. If stimulated by antigen, some B cells differentiate further into **plasma cells**, which synthesize and secrete **antibodies**, proteins that bind to and help destroy foreign molecules, as described later.

T cells may directly kill infected, mutated, or transplanted cells. They are called T cells because they mature within the thymus gland. T cells include two distinct types of lymphocytes: cytotoxic T cells and helper T cells. **Cytotoxic T cells** travel to the location of their targets, bind to these targets by recognizing an antigen, and kill those targets via secreted chemicals. In addition, responses mediated by cytotoxic

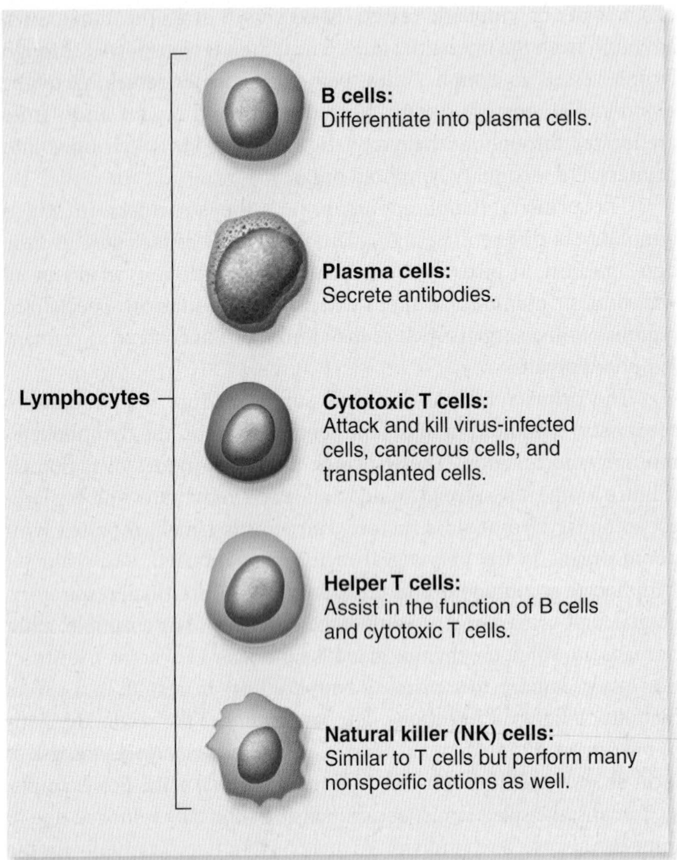

Figure 53.5 **Cells involved in acquired immunity in vertebrates.** Some important functions of each type of lymphocyte are included. The shape and color conventions shown in this figure will be used throughout this chapter.

T cells are directed against body cells that have become cancerous or infected by pathogens. NK cells also destroy such cells by secreting toxic chemicals.

As their name implies, **helper T cells** do not themselves function as "attack" cells. Instead, they assist in the activation and function of B cells and cytotoxic T cells. With only a few exceptions, B cells and cytotoxic T cells cannot function adequately unless they are stimulated by cytokines secreted from helper T cells.

Acquired Immune Responses Occur in Three Stages

Lymphocytes carry out their role by recognizing antigens such as those found on viruses, bacteria, and the surface of cancerous cells. The ability of lymphocytes to distinguish one antigen from another plays a central role in acquired immunity. Immunologists recognize two types of acquired immunity. In **humoral immunity**, plasma cells secrete antibodies that bind to antigens. In **cell-mediated immunity**, cytotoxic T cells directly encounter and destroy infected body cells, cancerous cells, or transplanted cells.

An acquired immune response can usually be divided into three stages (**Figure 53.6**): recognition of antigen, activation and proliferation of lymphocytes, and attack against a recognized antigen.

Stage 1: Recognition of Antigen During its development, each lymphocyte synthesizes a type of membrane receptor that can bind to a specific antigen. If subsequently the lymphocyte encounters that antigen, the antigen becomes bound to the receptor. This specific binding is the meaning of the word "recognize" in immunology. Antigens that bind to a lymphocyte receptor are said to be recognized by the lymphocyte. The ability of lymphocytes to distinguish one antigen from another, therefore, is determined by the nature of their plasma membrane receptors. Each lymphocyte is specific for just one type of antigen.

Stage 2: Activation of Lymphocytes In the second stage, the binding of an antigen to a receptor on a lymphocyte activates that lymphocyte. Upon activation, the lymphocyte undergoes multiple cycles of cell division. The result is the formation of many identical cells called clones that express the same receptor as the receptor that first recognized the antigen. The chemical nature of the antigen determines which individual lymphocytes will be activated to form clones. This process requires the function of helper T cells, which divide when activated and then secrete the cytokines that promote cell division. Some of the cloned lymphocytes become **effector cells**, the plasma cells and cytotoxic T cells that carry out the attack response, whereas others function as **memory cells**, which remain poised to recognize the antigen if it returns in the future.

In a typical person, the size of the lymphocyte population is staggering. Over 100 million different lymphocytes, each with the ability to recognize a unique antigen, are found in a person's immune system. This vast population explains why our bodies are able to recognize so many different antigens as foreign and eventually destroy them.

Stage 3: Attack Against Antigen In the third stage, the effector cells attack all antigens of the kind that initiated the immune response. Plasma cells carry out a humoral response by secreting antibodies into the blood. These antibodies then recruit and guide other molecules and cells that perform the actual attack. Activated cytotoxic T cells, by contrast, carry out cell-mediated immunity. They directly attack and kill the cells bearing the antigens.

Once the attack is successfully completed, the great majority of plasma cells and cytotoxic T cells that participated in it die by apoptosis. However, memory cells persist even after the immune response has been successfully completed, so they can recognize and fight off any future infection with the same type of antigen. Let's look at each of these three stages in more detail, first for humoral immunity and then cell-mediated immunity.

In Humoral Immunity, B Cells Produce Immunoglobulins That Serve as Receptors or Antibodies

In stage 1 of humoral immunity (see Figure 53.6, left side), B cells recognize antigens with the help of B-cell receptors. When B cells are activated in stage 2, they proliferate and differentiate into plasma cells, which secrete antibodies. These are proteins that travel all over the body to reach antigens identical to those that stimulated their

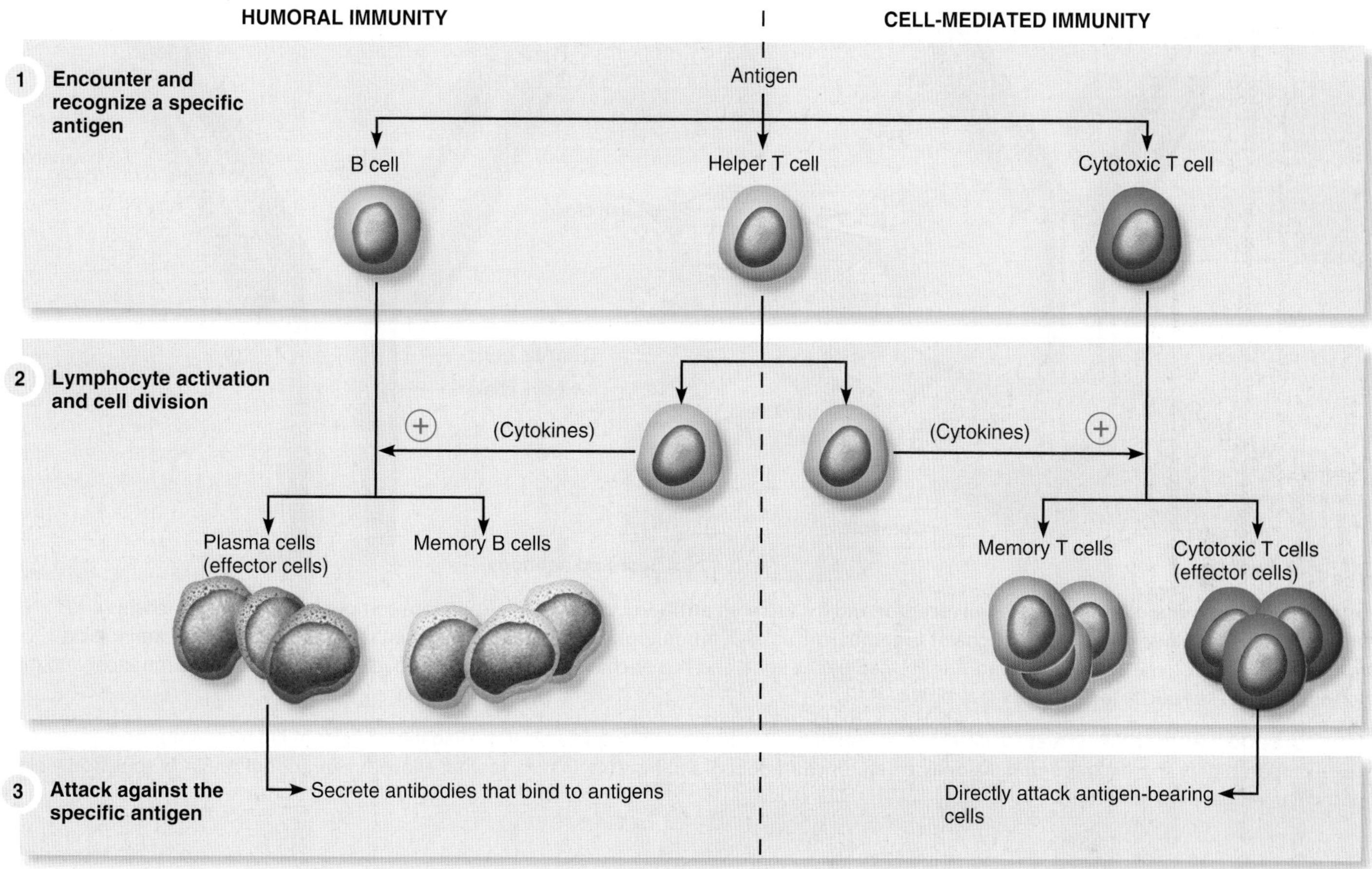

Figure 53.6 **The three stages of a specific (acquired) immune response.** All three cell types in step 1 recognize the same antigen. Helper T cells secrete cytokines that activate B cells and cytotoxic T cells, as indicated by the + symbols. Both B cells and T cells undergo cell division to form clones when activated, and in both cases, a portion of the cells are set aside as memory cells to fight off a future infection of the same type.

production. In the extracellular body fluids, antibodies combine with these antigens and guide an attack that eliminates the antigens or the cells bearing them, a process we will discuss in more detail later. Such antibody-mediated responses are also called humoral immune responses, the adjective "humoral" denoting communication by way of soluble chemical messengers (the old word "humors" was once used to refer to bodily fluids). Antibody-mediated responses are the major defense against bacteria, viruses, and other pathogens in the extracellular fluid, and against toxin molecules.

B-Cell Receptors and Antibodies B-cell receptors and the antibodies secreted by plasma cells share many structural and functional similarities (**Figure 53.7**). They are both members of a family of proteins called **immunoglobulins (Ig)**. However, there are some differences between them. B-cell receptors have a transmembrane domain that anchors them in the plasma membrane of the B cell (Figure 53.7a). Antibodies are soluble proteins that are secreted from plasma cells (Figure 53.7b). Interestingly, B-cell receptors and the antibodies made by plasma cells are encoded by the same genes. In plasma cells, the pre-mRNA is alternatively spliced, a phenomenon described in Chapter 13, so the transmembrane domain is not present in the protein. For this reason, the B-cell receptor in a particular B cell and secreted antibodies from the resulting plasma cells recognize the exact same antigen.

Immunoglobulin Structure Each immunoglobulin molecule is composed of four interlinked polypeptides: two long heavy chains and two short light chains (see Figure 53.7b). A hinge region that provides the molecule with flexibility separates the light chains and upper parts of the heavy chains from the lower parts of the heavy chains. One portion of an immunoglobulin is called the **constant region**: the amino acid sequence of the constant region is identical for all immunoglobulins of a given class and is what distinguishes the classes from each other. The constant regions are important for the binding of antibodies to immune cells and to complement proteins, and thus contribute to the eventual destruction of antigen. A defining feature of immunoglobulins, however, is their **variable region**, which gets its name because its sequence varies among different B cells. The variable region is the site that specifically recognizes a particular antigen.

Mammals have five classes of immunoglobulins, designated IgM, IgG, IgA, IgE, and IgD. All vertebrates have IgM molecules. These pentamers (made of five Ig molecules connected by disulfide bridges

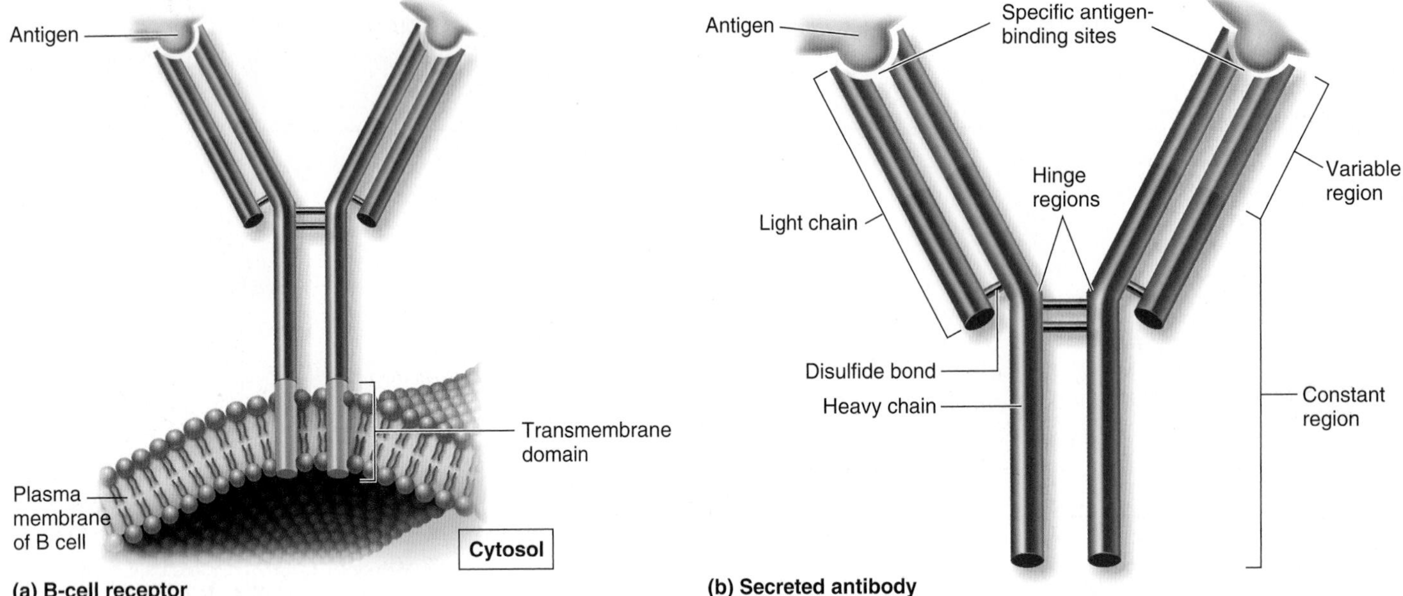

Figure 53.7 **Immunoglobulins. (a) B-cell receptor and (b) secreted antibody.** Immunoglobulins are composed of two heavy chains and two light chains. Disulfide bonds hold the chains together. Within each immunoglobulin class, the constant regions have identical amino acid sequences. In contrast, the antigen-binding sites formed by the light- and heavy-chain variable regions have unique amino acid sequences and give each antibody its specificity for a particular antigen.

BIOLOGY PRINCIPLE Structure determines function. The amino acid sequences of the variable regions of different immunoglobulins impart highly specific three-dimensional structures of immunoglobulins. This, in turn, is what allows an immunoglobulin to specifically bind a particular antigen and not others.

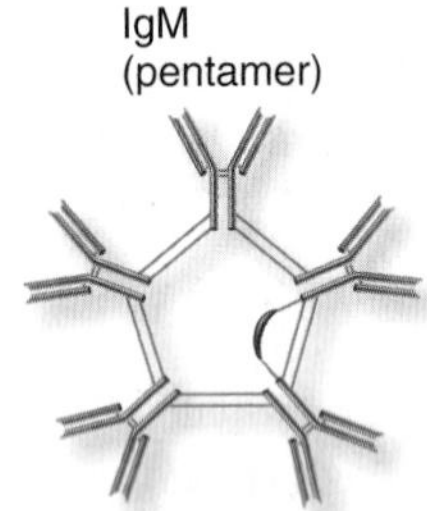

and other linkages) are the first Ig class produced after antigen exposure, but their blood concentration declines afterward. Some vertebrates have only some of the other classes and also express unique immunoglobulins not found in mammals. By contrast, invertebrates lack immunoglobulins. However, they have proteins containing regions called Ig domains with sequences that are similar to those found in immunoglobulins. These proteins may be ancestral to immunoglobulins and in some cases have been shown to carry out immune activities.

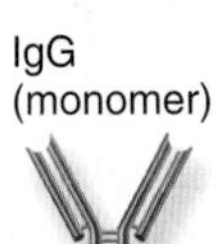

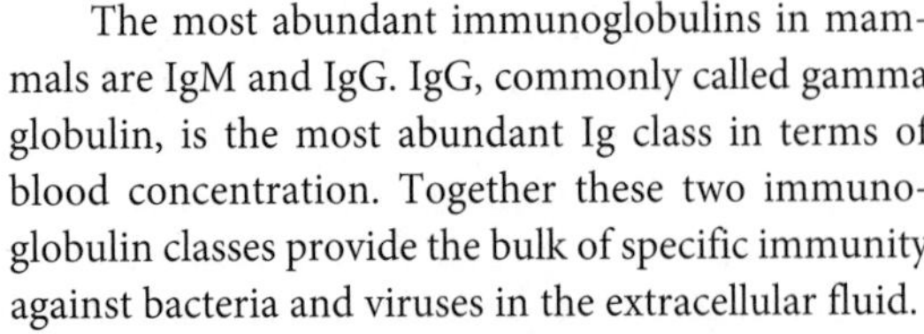

The most abundant immunoglobulins in mammals are IgM and IgG. IgG, commonly called gamma globulin, is the most abundant Ig class in terms of blood concentration. Together these two immunoglobulin classes provide the bulk of specific immunity against bacteria and viruses in the extracellular fluid.

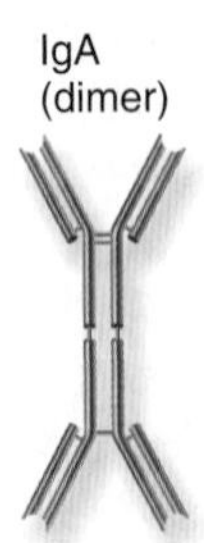

IgA antibodies exist as dimers and are secreted by plasma cells in the linings of the gastrointestinal, respiratory, and genitourinary tracts and in tear ducts and salivary glands. They act locally in the linings of these structures or on their surfaces, as they are present in their secretions. For example, IgA molecules secreted into saliva help keep animals' mouths relatively free of pathogens. IgA molecules are also secreted by the mammary glands of female mammals shortly after birth of their young and therefore are the major antibodies in milk.

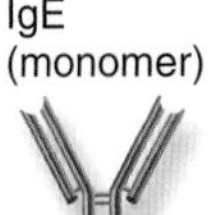

IgE antibodies are monomers that participate in defenses against multicellular eukaryotic parasites and also mediate allergic responses. Although present in blood at low concentrations, they also attach to mast cell membranes. When mast cell IgE molecules bind antigen, the mast cell secretes its histamine into the extracellular fluid, causing vasodilation and contributing to the allergic response. In people who are particularly sensitive to allergens, this response is easily demonstrated by a pinprick injection of an antigen, such as proteins associated with hay fever, into a small region under the skin, or by applying the antigens topically for an extended period of time—called patch testing. The resultant local inflammation and reddening of the skin are mediated in large part by IgE molecules.

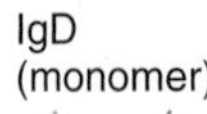
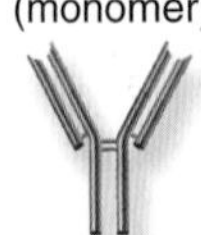

The functions of IgD are still unclear. However, IgD molecules are present both in blood and on the surface of B cells, and they are known to bind antigen on B cells, thus possibly contributing to B-cell activation.

The amino acid sequences of the variable regions vary widely from immunoglobulin to immunoglobulin in a given Ig class. The enormous number of variable sequences results in countless unique structures of immunoglobulins within each class. Thus, each of the five classes of antibodies contains up to millions of unique immunoglobulins, each capable of combining with only one specific antigen or, in some cases, with several antigens whose structures are very similar. The genetic explanation for this remarkable array of immunoglobulins was first identified in the 1970s, as we see next.

GENOMES & PROTEOMES CONNECTION

Recombination and Hypermutation Produce an Enormous Number of Different Immunoglobulin Proteins

The human genome contains about 200 genes that encode immunoglobulins. This raises an intriguing question. How can the body produce millions of different immunoglobulin proteins if there are only 200 immunoglobulin genes? The answer is that the 200 genes undergo a unique process involving gene rearrangements. This phenomenon was discovered by Japanese scientist Susumu Tonegawa and others in the 1970s.

Along the length of a typical human immunoglobulin gene are numerous gene segments that code for a piece of the final immunoglobulin protein (Figure 53.8). In light chains, these gene segments are of three types: variable, joining, and constant segments. A total of 40 variable (V) segments code for the antigen-binding site. These are next to four joining (J) segments and a single constant (C) segment. Each segment along the length of the gene is associated with recognition sequences that bind two enzymes, called RAG-1 and RAG-2 (for recombination-activating gene). These enzymes, which are expressed only in developing lymphocytes, cut randomly at the end of a V segment and at the beginning of a J segment. The intervening region is lost, and then other enzymes paste the V and J segments together. The result is a new, permanent immunoglobulin gene for that B cell. Because any V segment can be linked with any J segment, the number of possible final genes among different B cells is huge. Additionally, heavy chains have multiple segments that are spliced together in this way, except that they have yet another segment (designated D) and more V segments, yielding an even greater number of possible heavy-chain genes. Any heavy chain can combine with any light chain in a given B cell, which results in an immense number of possible combinations of immunoglobulins.

The number of possible immunoglobulins is increased even further in two more important ways within B cells. First, the joining process of the V, D, and J segments is not always precise. Occasionally, a few nucleotides may be lost at a joining end, resulting in a different amino acid sequence in the immunoglobulin protein. Second, in a subset of activated B cells, the DNA coding for the variable antigen-binding sites of immunoglobulins undergoes a unique process known as hypermutation, which primarily produces point mutations. The result is a hypervariable region of the light and heavy chains of all immunoglobulins.

The three processes of gene recombination, imprecise joining of gene segments, and hypermutation cause each lymphocyte within an individual's body to produce a unique type of immunoglobulin. The immune system can produce an incredibly diverse array of antibodies capable of recognizing many different antigens because the body makes hundreds of millions of different lymphocytes. Nearly any foreign antigen that is taken into the body is recognized by some lymphocytes in this large population.

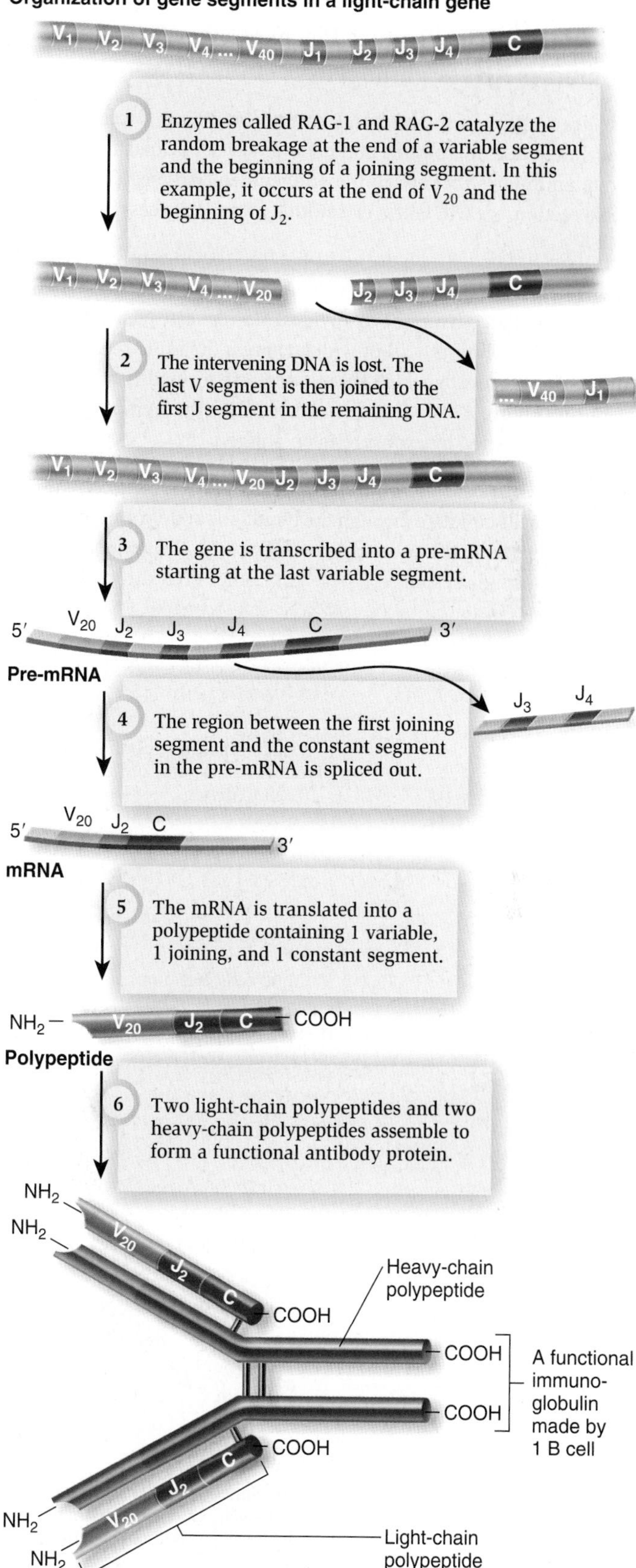

Organization of gene segments in a light-chain gene
V1 V2 V3 V4 ... V40 J1 J2 J3 J4 C
1 Enzymes called RAG-1 and RAG-2 catalyze the random breakage at the end of a variable segment and the beginning of a joining segment. In this example, it occurs at the end of V_{20} and the beginning of J_2.
V1 V2 V3 V4 ... V20 J2 J3 J4 C
2 The intervening DNA is lost. The last V segment is then joined to the first J segment in the remaining DNA.
... V40 J1
V1 V2 V3 V4 ... V20 J2 J3 J4 C
3 The gene is transcribed into a pre-mRNA starting at the last variable segment.
5′ V20 J2 J3 J4 C 3′
Pre-mRNA
4 The region between the first joining segment and the constant segment in the pre-mRNA is spliced out.
J3 J4
5′ V20 J2 C 3′
mRNA
5 The mRNA is translated into a polypeptide containing 1 variable, 1 joining, and 1 constant segment.
NH2 — V20 J2 C — COOH
Polypeptide
6 Two light-chain polypeptides and two heavy-chain polypeptides assemble to form a functional antibody protein.
NH2
NH2
V20
J2
C
COOH
Heavy-chain polypeptide
COOH
COOH
A functional immuno-globulin made by 1 B cell
COOH
NH2
NH2
Light-chain polypeptide

Figure 53.8 **The mechanism of immunoglobulin diversity.** While this figure shows events in a light-chain gene, events similar to those depicted here also occur in the heavy chains, creating even more structural diversity.

Activated B Cells Produce Plasma Cells Whose Antibodies Attack Pathogens

In stages 2 and 3 of the humoral immune response, B cells are activated and divide into plasma and memory cells. The plasma cells then secrete antibodies that attack the antigen detected (see Figure 53.6). In this section, we will take a closer look at these processes.

Clonal Selection B cells are activated by a specific antigen, with the aid of a helper T cell (a process which we will discuss later). When an antigen-stimulated lymphocyte divides and replicates itself, the progeny of this lymphocyte—all of which express the same receptor—are clones. The process by which these clones are formed is called **clonal selection** (Figure 53.9). This term emphasizes that lymphocyte proliferation is selected by exposure to an antigen.

Antibody Attack via Opsonization The antibodies secreted from the plasma cells circulate through the lymphatic system and the bloodstream. Eventually, the antibodies combine with the antigen that initiated the immune response. These antibodies then direct the attack against the pathogen to which they are now bound. Thus, immunoglobulins play two distinct roles in humoral immune responses. First, during antigen recognition, immunoglobulins (B-cell receptors) on the surface of B cells bind to antigen brought to them. Second, immunoglobulins (antibodies) secreted by the resulting plasma cells bind to pathogens bearing the same antigens, marking them as the targets to be attacked.

Instead of directly destroying the pathogens, antibodies bound to antigen on the pathogen surface inactivate the pathogens in various ways. Antibodies may physically link the pathogens to phagocytes (neutrophils and macrophages), complement proteins, or NK cells. This linkage—called **opsonization**—triggers the attack mechanism and ensures that only the pathogens, and not nearby body cells, are destroyed.

In a second mechanism, antibodies directed against toxins produced by bacterial pathogens in the extracellular fluid bind to the toxins, thereby preventing them from harming susceptible body cells. The antibody-antigen complexes that are formed are then destroyed by phagocytes.

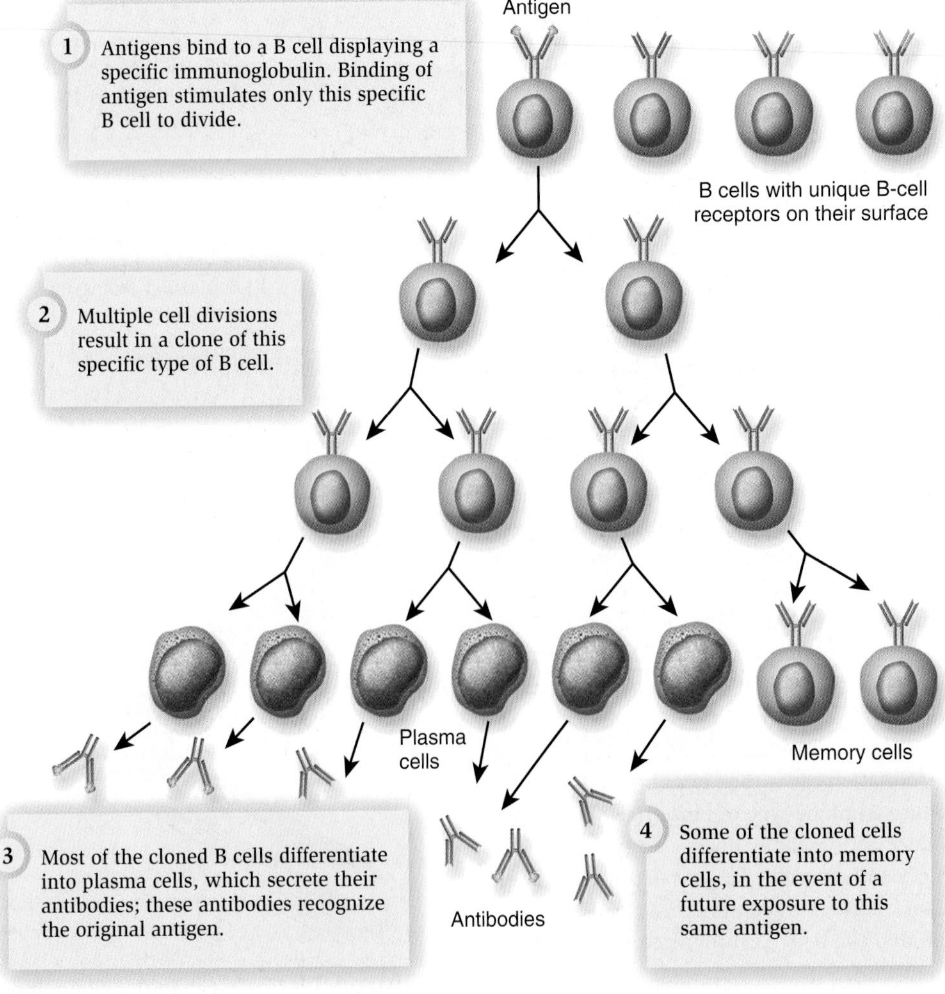

Antigen
1 Antigens bind to a B cell displaying a specific immunoglobulin. Binding of antigen stimulates only this specific B cell to divide.
B cells with unique B-cell receptors on their surface
2 Multiple cell divisions result in a clone of this specific type of B cell.
Plasma cells
Memory cells
3 Most of the cloned B cells differentiate into plasma cells, which secrete their antibodies; these antibodies recognize the original antigen.
Antibodies
4 Some of the cloned cells differentiate into memory cells, in the event of a future exposure to this same antigen.

Figure 53.9 **Clonal selection.** In this example, a B cell with a specific surface immunoglobulin recognizes an antigen and is stimulated to divide into a clone of identical cells.

In a similar way, antibodies produced against certain viral surface proteins bind to the viruses in the extracellular fluid, preventing them from attaching to the plasma membranes of potential host cells. As with bacterial toxins, the antibody-virus complexes that are formed are subsequently phagocytized.

In Cell-Mediated Immunity, T Cells Recognize Antigens Complexed with Self Proteins

In cell-mediated immunity, T cells recognize and are activated by antigen (see Figure 53.6). T-cell receptors for antigens have specific regions that differ from one T cell to another. As shown in **Figure 53.10**, they are composed of two polypeptides, each with a variable and constant region, along with a transmembrane domain. The variable regions recognize an antigen. As in B-cell development, multiple DNA rearrangements occur during T-cell maturation, leading to millions of distinct types of T cells, each with a receptor of unique specificity. For T cells, this maturation occurs as they develop in the thymus.

A T-cell receptor cannot bind to an antigen unless the antigen is already complexed with a receptor that is found on the surface of another cell, such as a macrophage. Such receptor proteins are encoded by a gene family known as the **major histocompatibility complex (MHC)**, and thus the proteins are called MHC proteins. Two major classes of MHC proteins are known. Class I MHC proteins are found on the surface of all human body cells except erythrocytes (that is, all nucleated cells). Class II MHC proteins are found primarily on the surface of macrophages, B cells, and dendritic cells.

The two different types of T cells have different MHC requirements. Cytotoxic T cells require antigen to be associated with class I MHC proteins, whereas helper T cells require class II MHC proteins. One reason for this difference stems from the presence of different proteins on their surfaces; helper T cells can be identified by a unique membrane protein called CD4, and cytotoxic T cells are identified by a membrane protein known as CD8. CD4 binds to class II MHC proteins, whereas CD8 binds to class I MHC proteins.

Antigen Presentation to Helper T Cells How do antigens, which are foreign, end up complexed with MHC proteins on the surface of the body's own cells? The answer involves the mechanism known as antigen presentation. As previously noted, helper T cells can bind antigen only when the antigen appears on the plasma membrane of a host cell complexed with the cell's class II MHC proteins. Cells bearing these complexes, therefore, function as **antigen-presenting cells (APCs)**. Because only macrophages, B cells, and dendritic cells express class II MHC proteins, only these cells can function as APCs for helper T cells.

Let's consider the function of macrophages as APCs for helper T cells (**Figure 53.11**). After a microbe or noncellular antigen has been phagocytized by a macrophage in a nonspecific (innate) response, antigens, such as proteins, are partially broken down into smaller peptide fragments by the macrophage's proteolytic enzymes within intracellular vesicles called endosomes. The resulting digested fragments then bind in the endosome to class II MHC proteins synthesized by the macrophage. Each fragment-MHC complex is then transported

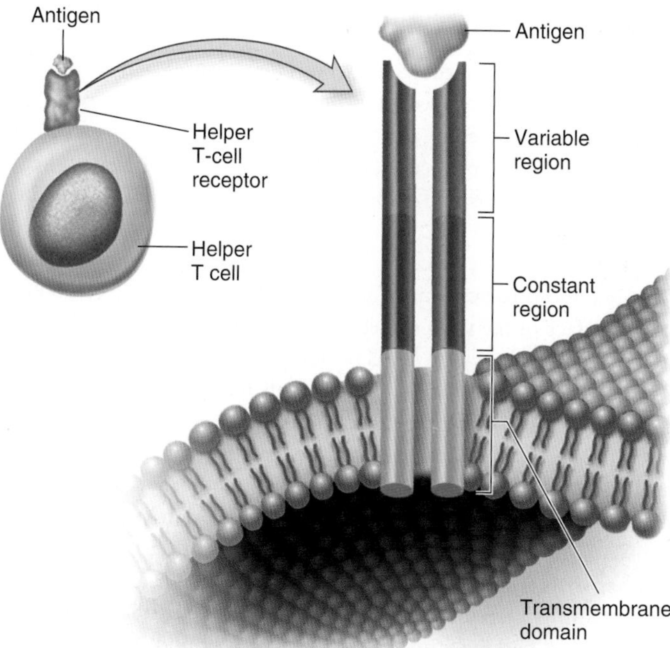

Figure 53.10 **Structure of a T-cell receptor in the plasma membrane.**

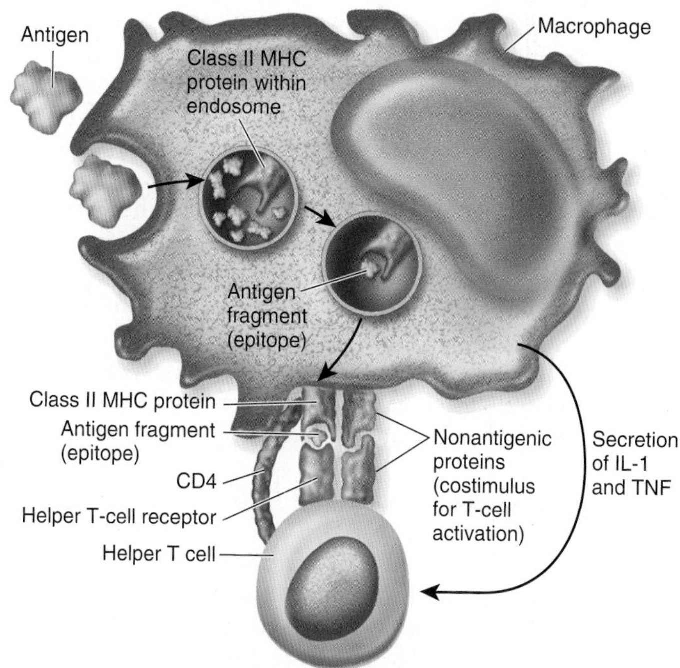

Figure 53.11 **Antigen presentation and helper T-cell activation.** In the initial events in helper T-cell activation, antigen fragments are complexed with a class II MHC protein within an antigen-presenting cell such as a macrophage. The complex is then displayed on the cell surface and binds to a helper T-cell receptor. Also required for T-cell activation are the binding of nonantigenic proteins between the APC and the attached helper T cell, and the actions of the cytokines interleukin 1 (IL-1) and tumor necrosis factor (TNF).

to the cell surface, where it is displayed in the plasma membrane. A specific helper T-cell receptor then binds this entire complex on the cell surface of the macrophage. The CD4 protein helps link the two cells. What is complexed to MHC proteins and presented to the helper T cells is not the intact antigen but instead a peptide fragment of the antigen—called an antigenic determinant, or **epitope**. Even so, it is customary to call this antigen presentation rather than epitope presentation.

B cells process antigen and present it to helper T cells in essentially the same way as macrophages do. The ability of B cells to present antigen to helper T cells is a second function of B cells in response to antigenic stimulation, in addition to their differentiation into antibody-secreting plasma cells.

The binding between the helper T-cell receptor and antigen bound to class II MHC proteins on an APC is the essential antigen-specific event in helper T-cell activation. However, by itself this specific binding does not result in T-cell activation. In addition, interactions occur between nonantigenic proteins on the surfaces of the attached helper T cell and the APC. These interactions provide a necessary costimulus for T-cell activation (see Figure 53.11).

Finally, the antigenic binding of the APC to the T cell plus the costimulus induces the APC to secrete large amounts of two cytokines—interleukin 1 (IL-1) and tumor necrosis factor (TNF). These molecules also stimulate the attached helper T cell.

Thus, the APC participates in the activation of a helper T cell in three ways: (1) antigen presentation, (2) provision of a costimulus, and (3) secretion of cytokines. Activated helper T cells then secrete cytokines that stimulate B cells and cytotoxic T cells.

Helper T Cells and B-Cell Activation Now we can go back to B cells and understand how they are activated by the actions of helper T cells. This process begins when a helper T cell specific for a particular antigen binds to a complex of that antigen and a class II MHC protein on an APC, activating the helper T cell. Along with other signals, this binding induces the activated helper T cell to divide. Some of the resulting activated helper T cells then bind to B cells that display the same antigen on their surfaces. This binding, along with additional cytokines, stimulates the B cell to go through the process of clonal selection. Thus, helper T cells are so named because their secretions help activate B cells that have bound antigen, in addition to their participation in activation of cytotoxic T cells and antigen presentation.

Antigen Presentation to Cytotoxic T Cells Unlike helper T cells, cytotoxic T cells require class I MHC proteins for activation. This distinction helps explain the major function of cytotoxic T cells—destruction of any of the body's own altered cells that have become cancerous or infected with viruses. The crucial point is that the antigens that complex with class I MHC proteins typically arise within body cells and are thus endogenous antigens.

How do such antigens arise? In viral infections, once a virus has entered a host cell, the expression of viral genes results in the synthesis of viral proteins, which are foreign to the cell. In cancerous cells, one or more of the cell's genes have become altered by chemicals, radiation, or other factors. The altered genes, called oncogenes, code for proteins that are not normally found in the body. Such abnormal proteins act as antigens.

In both virus-infected and cancerous cells, cytosolic enzymes hydrolyze some of the endogenously produced antigenic proteins into peptide fragments, which are transported into the endoplasmic reticulum. There the fragments are complexed with the host cell's class I MHC proteins and then shuttled by exocytosis to the plasma membrane, where a cytotoxic T cell specific for the antigen/MHC protein complex can bind to it. Once binding occurs, cytotoxic T cells release chemicals that kill the infected or cancerous cell, as discussed next.

Activated Cytotoxic T Cells Kill Infected or Cancerous Cells

The previous sections described how immune responses provide long-term defenses against bacteria, viruses, and individual foreign molecules that enter the body's extracellular fluid. We now examine how the body's own cells that have become infected by viruses or transformed into cancerous cells are destroyed by stage 3 cell-mediated immune responses (see Figure 53.6).

What is the value of destroying virus-infected host cells? First, and most importantly, such destruction prevents cells from making more viruses. Second, for cells that already are making mature viruses, it results in the release of the viruses into the extracellular fluid, where they can be neutralized by circulating antibody.

Role of Cytotoxic T Cells A typical cytotoxic T-cell response triggered by viral infection of a vertebrate's body cells is summarized in Figure 53.12. The response triggered by a cancerous cell would be similar. A virus-infected cell produces foreign proteins, viral antigens that are processed and presented on the plasma membrane of the cell complexed with class I MHC proteins. Cytotoxic T cells specific for the particular antigen bind to the complex (Figure 53.12, step 1). As with B cells, binding to antigen alone does not cause activation of the cytotoxic T cell. Cytokines from nearby activated helper T cells are also required.

Macrophages phagocytize extracellular viruses (or, in the case of cancer, antigens released from the surface of cancerous cells) and then process and present antigen, in association with class II MHC proteins, to the helper T cells (step 2). In addition, the macrophages provide a costimulus and also secrete IL-1 and TNF. The activated helper T cell releases IL-2 and other cytokines, which stimulate proliferation of the helper T cell.

IL-2 and other cytokines also act on the cytotoxic T cell bound to the surface of the virus-infected or cancerous cell, stimulating this attack cell to proliferate. Why is proliferation important if a cytotoxic T cell has already located and bound to its target? The answer is that there is rarely just one virus-infected or cancerous cell. By expanding the population of cytotoxic T cells capable of recognizing the particular antigen, the likelihood is greater that the other virus-infected or cancerous cells will be encountered by an appropriate cytotoxic T cell.

The cytotoxic T cells specific for that virus then find and bind to other virus-infected cells (step 3). The cytotoxic T cell releases the contents of its secretory vesicles directly into the extracellular space between itself and the target cell to which it is bound (thereby ensuring that other nearby host cells will not be killed). These vesicles contain proteases, and a protein called perforin, which is similar in structure to the proteins of the complement system's membrane

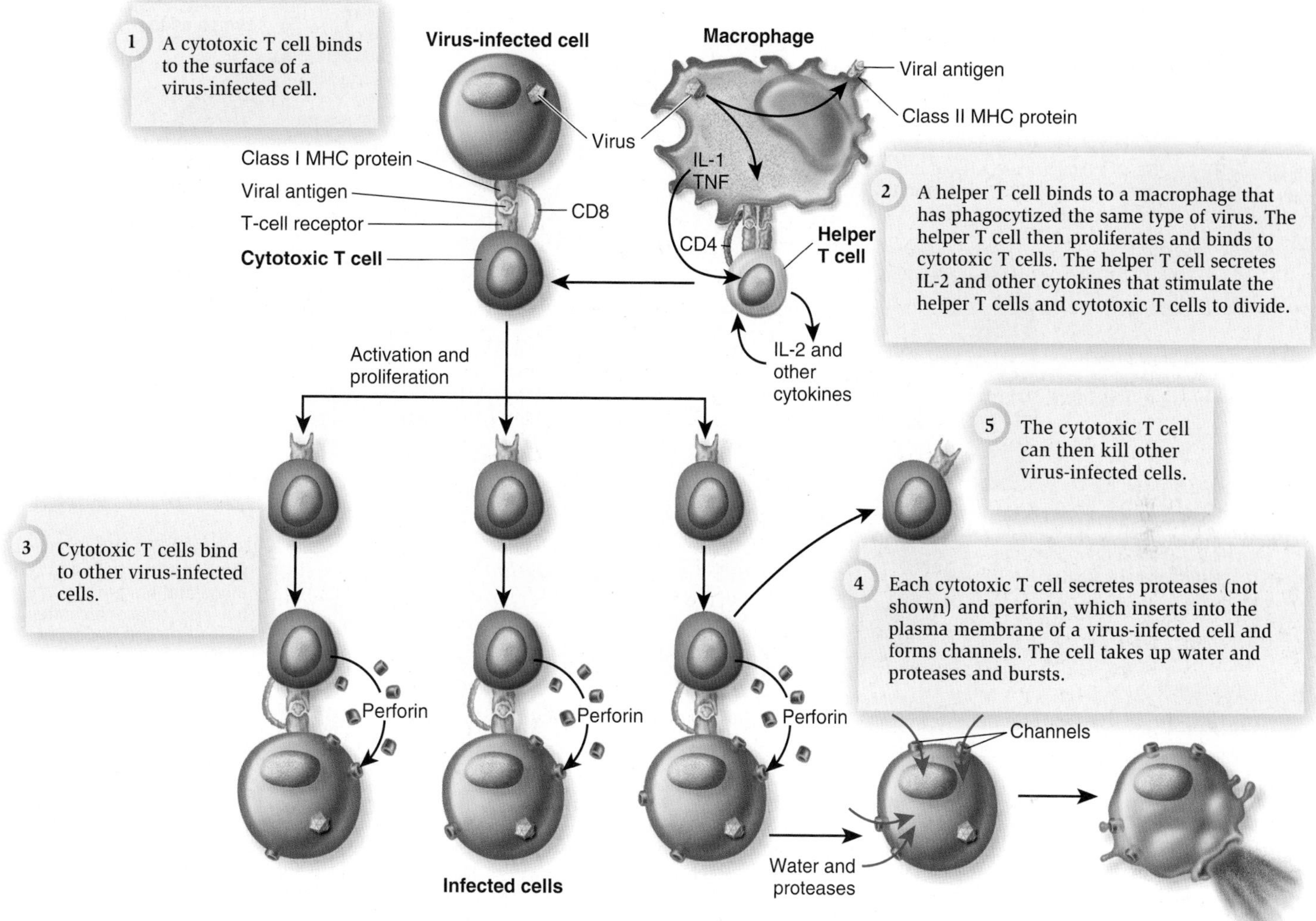

Figure 53.12 **Summary of events in the killing of virus-infected cells by cytotoxic T cells.** The sequence is similar for cancerous cells attacked by a cytotoxic T cell.

attack complex. Perforin is believed to insert into the target cell's membrane and form channels ("perforations") through the membrane (step 4). In this manner, it causes the attacked cell to take in the proteases secreted by the cytotoxic T cell, which is believed to induce apoptosis. The cell also takes in water through the pores, which causes it to burst. The cytotoxic T cell is not harmed by this process and can then continue to kill other virus-infected cells (step 5). Target cell killing by activated cytotoxic T cells occurs by several mechanisms, but this is one of the most important.

The body can eliminate viruses in two ways: through the humoral actions of antibodies in body fluids and through the cell-mediated killing of virus-infected cells by cytotoxic T cells. Although cytotoxic T cells play an important role in the attack against such cells, they are not the only mechanisms. NK cells also destroy virus-infected and cancerous cells by secreting toxic chemicals. As mentioned earlier, NK cells can recognize general features on the surface of such cells and participate in innate immunity. In addition, in a cell-mediated immune response, NK cells can be linked to such target cells by antibodies and then can destroy them by release of toxic molecules.

Summary: Example of an Acquired Immune Response

Let's bring together our discussion of the humoral immune system by looking in detail at one example. One classic humoral immune response is that which results in the destruction of bacteria. The sequence of events, which is quite similar to the humoral response to a virus in the extracellular fluid, is summarized in Figure 53.13. For this example, we consider the response in mammals, in which lymph nodes are present. Many features of the response, however, are similar in other vertebrates.

This process starts the same way as for nonspecific responses, with the bacteria penetrating one of the body's linings through an injury and entering the interstitial fluid (Figure 53.13, step 1). The bacteria then move with lymph into the lymphatic system and are carried to lymph nodes (step 2). Within the lymph node, a macrophage and a B cell recognize one of the bacteria as a foreign substance and bind to it.

As we have discussed, the process of B-cell activation usually requires the activation of helper T cells. The helper T cell binds to a complex of processed antigen and class II MHC protein on an APC

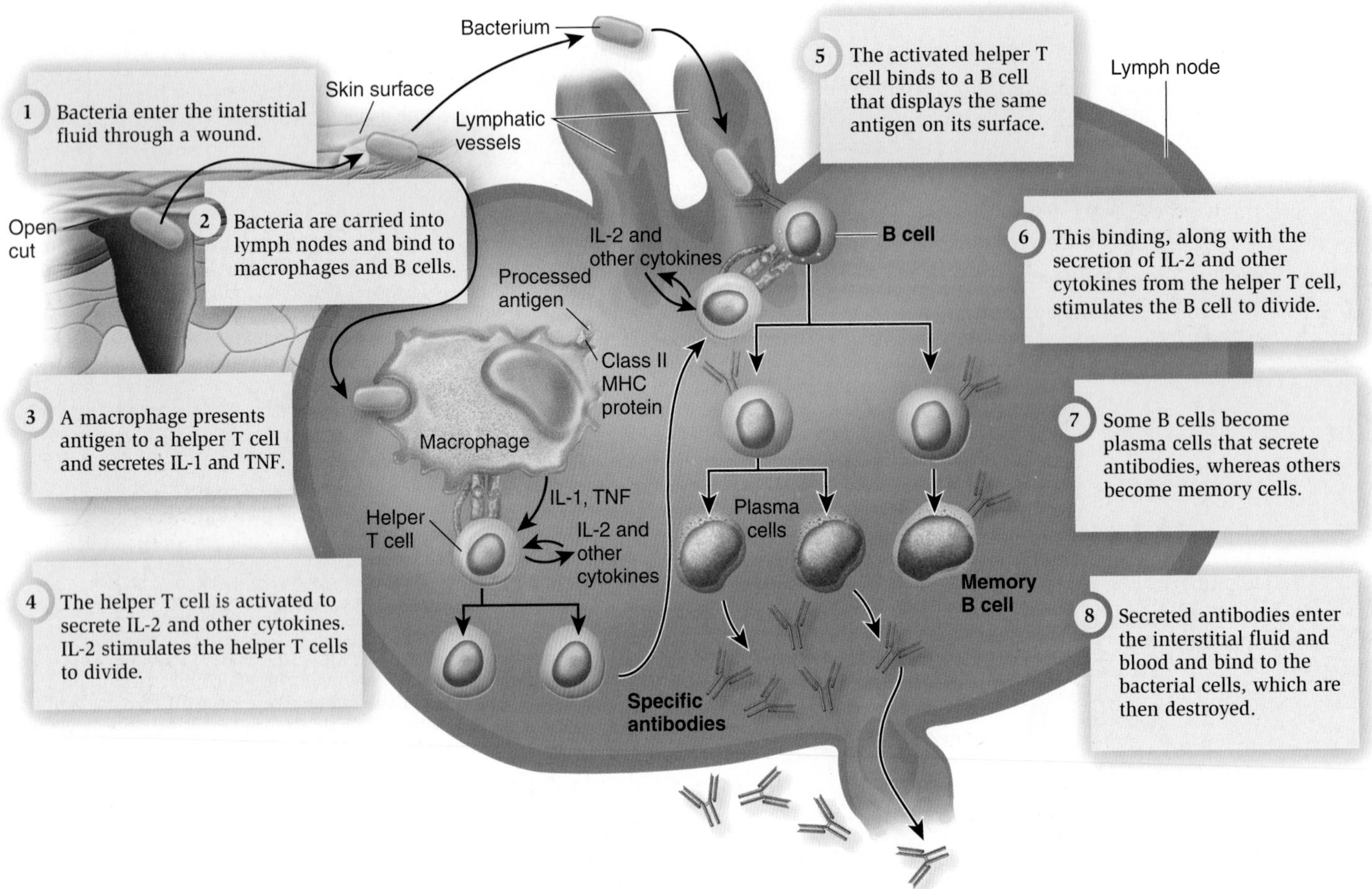

Figure 53.13 **Summary of events in a typical humoral immune response.** Most of the events depicted occur within a lymph node.

(step 3). In this case, the APC is a macrophage that has phagocytized the bacterium, hydrolyzed its proteins into peptide fragments, complexed the fragments with class II MHC proteins, and displayed the complexes on its surface. Once a helper T cell specific for the complex binds to it, the helper T cell becomes activated. The macrophage helps this process in two other ways: It provides a costimulus (not shown in step 3), and it secretes the cytokines IL-1 and TNF.

IL-1 and TNF stimulate the helper T cell to secrete another cytokine, IL-2. IL-2 stimulates the activated helper T cell to divide, which leads eventually to the formation of a clone of activated helper T cells (step 4). The activated helper T cells bind to B cells and also secrete IL-2 and other cytokines (step 5). Certain of these cytokines provide the additional signals that are usually required to activate nearby antigen-bound B cells to proliferate (step 6). These cells differentiate into memory cells, which help ward off possible future attacks by the same antigen, and plasma cells, which then secrete specific antibodies (step 7). The antibodies enter the bloodstream and bind to bacterial cells, which are then destroyed (step 8).

B Cells and T Cells That Recognize Self Molecules Must Be Killed or Inhibited

As we have seen, the lymphocytes responsible for the specific immune response in vertebrates are very capable killers of pathogens—so capable, in fact, that it raises a question: Why don't these cells attack and kill normal self cells? In other words, how does the body distinguish between self and nonself components and develop what is called **immune tolerance**, or tolerance of its own proteins and other molecules?

Recall that the huge diversity of lymphocyte receptors is ultimately the result of multiple random DNA-cutting/-recombination processes. It is virtually certain, therefore, that every animal possessing specific immune defenses would have clones of lymphocytes with receptors that could bind to that individual's own proteins. The continued existence and functioning of such lymphocytes would be disastrous, because such binding would launch an immune attack against all body cells expressing these proteins.

At least two mechanisms explain why individuals normally lack active lymphocytes that respond to self components. First, during early development in vertebrates, T cells are exposed to a wide mix of self proteins in the thymus. Those T cells with receptors capable of binding self proteins are destroyed by apoptosis in a process termed **clonal deletion**. The second process, termed **clonal inactivation**, occurs outside the thymus and causes potentially self-reacting T cells to become nonresponsive. B cells undergo similar processes. The mechanisms by which these two events occur are still under investigation.

Occasionally, however, these mechanisms fail, and the body's immune cells attack the body's own cells. When this happens in humans, it produces autoimmune disease. **Autoimmune diseases** are conditions in which the body's normal state of immune tolerance somehow breaks down, with the result that both humoral and cell-mediated attacks are directed against the body's own cells and tissues. A growing number of human diseases are being recognized as autoimmune in origin. Examples are multiple sclerosis, in which myelin around neurons is attacked; myasthenia gravis, in which the receptors for acetylcholine on skeletal muscle cells are the targets; rheumatoid arthritis, in which joints are damaged; systemic lupus erythematosus, in which numerous organs are damaged; and type 1 diabetes mellitus, in which the insulin-producing cells of the pancreas are destroyed. Treatments for autoimmune disease range from treating the symptoms (for example, administering insulin to individuals with diabetes) to suppressing the immune system with drugs.

Immunological Memory Is an Important Feature of Acquired Immunity

As we have learned, the acquired immune response to a given antigen depends on whether or not the body has previously been exposed to that antigen. Consider, for example, the humoral immune response. In mammals, antibody production in response to the first contact with an antigen occurs slowly, over a few weeks. This response to an initial antigen exposure is termed a **primary immune response** (**Figure 53.14**). Any subsequent infection by the same pathogen elicits an immediate and heightened production of additional specific antibodies against that particular antigen, a reaction termed a **secondary immune response**.

In the case of humoral immunity, this secondary response occurs more quickly, is stronger, and lasts longer because memory B cells that were produced in response to the initial antigen exposure are quickly stimulated to multiply and differentiate into thousands of plasma cells. These cells then produce large amounts of specific antibodies. The immune system's ability to produce this secondary response is called **immunological memory**.

Immunological memory explains why we and other animals are able to fight off many illnesses to which we have been previously exposed, such as many common childhood diseases. The acquired response to exposure to any type of antigen is known as **active immunity**. Active immunity not only results from natural exposure to antigens, it is also the basis for the artificial exposures to antigen that occur in vaccinations. In **vaccinations**, small quantities of living, dead, or altered microbes, small quantities of toxins, or harmless antigenic molecules derived from a microorganism or its toxin are injected into the body, resulting in a primary immune response, including the production of memory cells. Subsequent natural exposure to the immunizing antigen results in a rapid, effective response that can prevent or reduce the severity of disease.

In contrast to active immunity, another type of acquired immunity, called **passive immunity**, confers protection against disease through the direct transfer of antibodies from one individual to another. Passive immunity can occur naturally, as when IgG molecules cross the mammalian placenta to protect a fetus from various pathogens, or when a newborn mammal receives antibodies from breast milk. It can also occur artificially, as when a person is given an injection of IgG molecules shortly after exposure to hepatitis viruses. Recent advances in the creation of highly specific and pure antibodies, called **monoclonal antibodies** because they are derived from a single clone of cells prepared in a laboratory, have paved the way for the use of passive immunity to combat certain types of cancer. Because antibodies are proteins with a limited life span, the protection afforded by the transfer of antibodies in passive immunity is relatively short-lived, usually lasting only a few weeks or months.

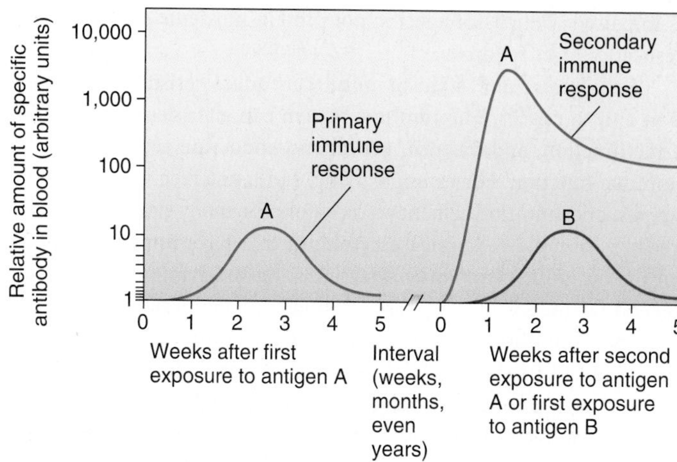

Figure 53.14 **Primary and secondary immune responses.** In a primary response, as shown on the left of this graph, an initial exposure to an antigen produces modest levels of specific antibody over a period of weeks. In a secondary response, subsequent exposure to the same antigen results in greater antibody production that occurs more rapidly and lasts longer than a primary response. (Note that the scale of the *y*-axis on the graph is logarithmic.) The secondary response is specific for that antigen. Exposure at that time to another antigen for the first time produces the usual primary response.

Concept Check: *What is the advantage of a secondary immune response?*

53.4 Impact on Public Health

Learning Outcomes:

1. Describe the influence of lifestyle on immunity.
2. Explain the role of the immune system in organ transplant rejection and allergic reactions.
3. Describe the effects of HIV on a human immune system and list current methods of treating HIV infection.

In this section, we will consider a few ways in which the functioning of the immune system can be affected by lifestyle, medical interventions, allergies, and destruction of immune cells. Collectively, the effects of disorders of the immune system have an almost immeasurable impact on public health in terms of worker productivity, health-care resources, and the economy.

Lifestyle Has an Important Influence on Immunity

Protein-calorie malnutrition is the single greatest contributor to decreased resistance to infection worldwide. When inadequate amino acids are available to synthesize essential proteins, immune function is impaired. Deficits of specific nonprotein nutrients can also lower resistance to infection.

Both stress and state of mind can affect resistance to infection and to cancer. The immune system can alter neural and endocrine function, and, in turn, neural and endocrine activity modifies immune function. For example, lymphoid tissue receives input from nerves, and immune cells have receptors for many hormones. Conversely, immune cells release cytokines that have important effects on the brain and endocrine system. Moreover, lymphocytes secrete several hormones that are also produced by endocrine glands. The multiple "mind-body" interactions that affect disease resistance are the subject of a field called psychoneuroimmunology.

Of the hormones associated with stress, the adrenal hormone cortisol has received the most attention due to its powerful suppressive activity on inflammation and specific immunity. Among other things, cortisol inhibits production of inflammatory mediators, reduces capillary permeability in injured areas, and suppresses the growth and activity of certain types of leukocytes. In this way, it acts as a sort of brake on the immune system. During chronic stress or when cortisol is used to treat certain illnesses for long periods of time, it may cause immunosuppression. This is a key link between stress and health. Chronic stress may lead to high cortisol concentrations that, by suppressing the body's immune responses, lowers resistance to infection.

Another feature of a person's lifestyle that appears to affect immune function is exercise. The influence of physical exercise on the body's resistance to infection and cancer has been debated for decades. Present evidence indicates that the intensity, duration, regularity, and psychological stress of the exercise all have important influences, both negative and positive, on a variety of immune functions, such as the numbers of circulating NK cells. Although evidence suggests that too much intense exercise can reduce immunity, most experts currently believe that moderate exercise and physical conditioning have net beneficial effects on the immune system and on disease resistance. A 2005 study suggested that exercise may be particularly beneficial in warding off the onset of breast cancer, one of the most common types of cancer in women.

Organ Transplants Are Medical Procedures That Can Cause Serious Immune Reactions

Organ transplants have saved numerous lives. However, they carry the possibility of provoking immune reactions that can threaten the life of the recipient. Since the mid-20th century, organ transplants from a healthy or recently deceased donor to a recipient have become widespread. The United Network for Organ Sharing reports that approximately 28,500 organs are transplanted in the U.S. each year, with kidney (17,000), liver (6,000), heart (2,000), and lung (1,800) accounting for most of the transplants. The major obstacle to successful transplantation of tissues and organs is a reaction called graft rejection, in which the immune system recognizes the transplant (also called a graft) as foreign and attacks it as it would any other foreign cells. Although B cells and macrophages play some role in graft rejection, cytotoxic T cells and helper T cells are mainly responsible. To minimize this possibility, transplant patients are given drugs that suppress immune function.

Except for grafts from identical twins, the class I MHC proteins on graft cells differ from those on the recipient's cells, as do the class II MHC proteins present on macrophages in the graft cells. Consequently, the recipient's T cells recognize the MHC proteins in the graft as foreign, and cytotoxic T cells (with the aid of helper T cells) destroy the graft cells.

Allergies Affect the Quality of Life of Millions of People

An **allergy** (also known as hypersensitivity) is a condition in which immune responses to environmental antigens cause inflammation and damage to body cells. Antigens that induce allergic reactions are called allergens. Common examples of allergens include ragweed pollen and animal dander. Most allergens themselves are relatively or completely harmless. It is the immune responses to them that cause the damage. In essence, then, allergy is immunity gone awry, for the response is of inappropriate strength and duration for the stimulus. In the U.S. alone, as many as 40 million people (about 13% of the population) suffer from allergies.

For any allergy to develop, a genetically predisposed person must first be exposed to the allergen—a process called sensitization. Subsequent exposures elicit the damaging immune responses we recognize as an allergy. Hypersensitivities can be broadly classified according to the speed of the response. Allergies that take up to several days to develop are considered delayed hypersensitivities. The skin rash that appears after contact with poison ivy is an example. More common are reactions considered immediate hypersensitivities, which can develop in minutes or up to a few hours. These allergies are also called IgE-mediated hypersensitivities because they involve IgE antibodies.

In immediate hypersensitivity, sensitization to the allergen leads to the production of specific antibodies and a clone of memory B cells. In individuals who are genetically susceptible to allergies, antigens that elicit immediate hypersensitivity reactions stimulate the production of IgE antibodies. Upon their release from plasma cells, these IgE molecules circulate throughout the body and become attached to mast cells in connective tissue. When the same antigen subsequently enters the body at some future time and binds with IgE that is bound to mast cells, the mast cell is stimulated to secrete many inflammatory mediators, including histamine, that then initiate an inflammatory response.

The signs and symptoms of IgE-mediated hypersensitivity reflect both the effects of inflammatory mediators and the body site in which the antigen–IgE–mast cell binding occurs. When, for example, a previously sensitized person inhales ragweed pollen, the antigen combines with the variable region of IgE, and the constant region of IgE binds to mast cells in the airways. The mast cells release their contents, which induce increased mucous secretion, increased blood flow, swelling of the epithelial lining, and contraction of the smooth muscle surrounding airways. These effects produce the congestion, runny nose, sneezing, and difficulty in breathing characteristic of hay fever. Antihistamines are drugs taken by people to block the action of histamine that is released during allergic responses. These drugs prevent histamine from binding to its receptor protein on its target cells, thereby preventing or relieving some of the symptoms of allergy.

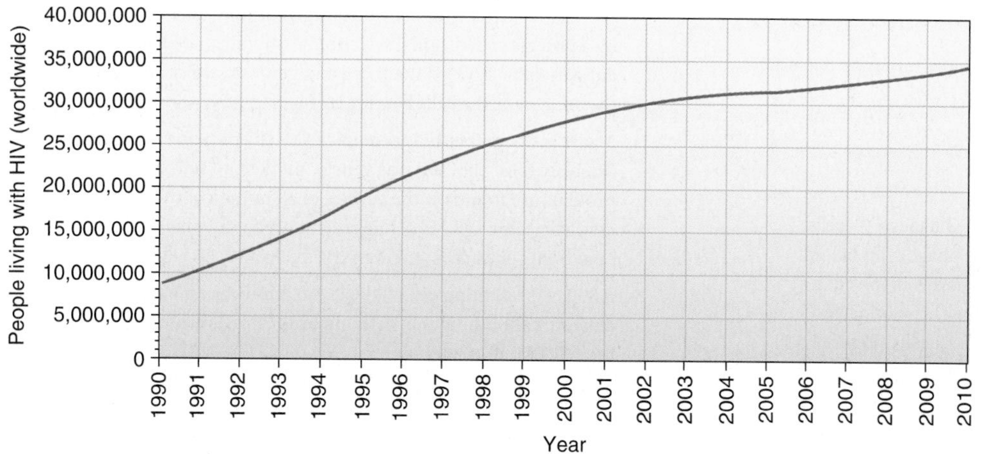

Figure 53.15 Worldwide incidence of people living with HIV/AIDS. Data are from the 2011 World Aids Day Report published by UNAIDS (The Joint United Nations Programme on HIV/AIDS).

Acquired Immune Deficiency Syndrome (AIDS) Is a Growing Pandemic

Acquired immune deficiency syndrome (AIDS) is caused by the **human immunodeficiency virus (HIV)**, which incapacitates the immune response by preferentially infecting helper T cells. HIV is a retrovirus, a virus that contains RNA as its genetic material. Once inside a helper T cell, HIV uses the enzyme reverse transcriptase to transcribe its RNA into DNA, which is then integrated into the chromosomal DNA of the host's T cells (refer back to Figure 18.4). Later, viral replication within the T cell results in the death of the cell.

HIV infects helper T cells because the CD4 protein in their plasma membranes acts as a receptor for an HIV capsid protein. However, binding to CD4 is not sufficient to enable HIV to enter the helper T cell. Another T-cell surface protein, which normally acts as a receptor for certain cytokines, must serve as a coreceptor. Interestingly, individuals possessing a mutation in this cytokine receptor are highly resistant to HIV infection, so much research is now focused on the possible therapeutic use of chemicals that can bind to and block this coreceptor.

HIV not only directly kills helper T cells, but it also indirectly causes additional helper T-cell death by inducing cytotoxic T cells to kill HIV-infected helper T cells. In addition, by still poorly understood mechanisms, HIV causes the death of many uninfected helper T cells by apoptosis. Without adequate numbers of helper T cells, neither B cells nor cytotoxic T cells can function normally. Both humoral and cell-mediated immunity are compromised. Many individuals with AIDS die from infections and cancers that ordinarily would be readily handled by a fully functional immune system.

AIDS, first described in 1981, has since reached pandemic proportions. About 33 to 35 million people worldwide are currently living with HIV infection, and an estimated 5,000–10,000 new infections occur in the world each day (Figure 53.15). The major routes of HIV transmission are (1) unprotected sexual intercourse with an infected partner; (2) transfer of contaminated blood or blood products between individuals, such as the sharing of needles among intravenous drug users, or, less commonly, as a result of a blood transfusion; (3) transfer from an infected mother to her child across the placenta or during delivery; or (4) transfer via breast milk during nursing.

The great majority of individuals now infected with HIV show no signs of AIDS. Their infections are diagnosed by the presence of anti-HIV antibodies or HIV RNA in the blood. However, if left untreated, HIV infection commonly develops into AIDS in about 10 years. During the first 5 years, killed helper T cells are typically replaced by new cells, so T-cell concentrations remain normal, and the individual remains asymptomatic. Over the next 5 years, T-cell concentrations begin to decline, until at some point, AIDS reveals itself in the form of opportunistic viral, bacterial, and fungal infections. Certain unusual cancers, such as Kaposi sarcoma, also occur with high frequency. In untreated individuals, death usually occurs within 2 years after the onset of AIDS symptoms.

Treatment for HIV-infected individuals has two components: one directed against the virus itself to delay progression of the disease, and one to prevent or treat the opportunistic infections and cancers that ultimately cause death. One current antiviral approach involves administering a combination of four drugs, known as HAART (highly active antiretroviral therapy). Two of the drugs inhibit the action of reverse transcriptase in converting viral RNA into DNA within the host cell, a third drug inhibits an HIV enzyme required for assembling new viruses, and a recently developed fourth class of drugs called fusion inhibitors prevents the virus from entering T cells. These treatments have been demonstrated to be effective in slowing the rate at which infection with HIV leads to AIDS. Unfortunately, however, the HAART regimen is associated with numerous side effects, including nausea, vomiting, diarrhea, metabolic disturbances resulting in part from insulin resistance, and liver damage. Much research is under way to find better treatments and ultimately to cure this disease.

Summary of Key Concepts

- An animal's cells and organs that collectively contribute to its immunity constitute the animal's immune system.
- In innate (nonspecific) immunity, the body's defenses are present at birth and act against foreign materials in much the same way regardless of the specific identity of the invading material. Acquired (specific) immunity develops only after the body is exposed to foreign substances and targets those foreign substances specifically.

53.1 Types of Pathogens

- Three major types of pathogens elicit an immune response: bacteria, viruses, and eukaryotic parasites.

53.2 Innate Immunity

- An important innate defense is composed of phagocytes—cells capable of phagocytosis. In vertebrates, most phagocytes belong to the blood cells called leukocytes. The leukocytes involved in innate immunity include neutrophils, eosinophils, monocytes, macrophages, basophils, and natural killer (NK) cells (Figure 53.1).
- Inflammation is an innate local response to infection or injury characterized by local redness, swelling, heat, and pain. The events of inflammation are induced and regulated by chemical mediators called cytokines (Figure 53.2).
- Antimicrobial proteins include interferons, which inhibit viral replication, and complement proteins, which kill microbes without prior phagocytosis. Activation of the complement proteins results in the formation of a membrane attack complex (MAC), which creates water channels in the microbial plasma membrane and causes the microbe to swell and burst.
- Toll-like receptors are evolutionarily ancient proteins that recognize conserved features of many pathogens (Figure 53.3).

53.3 Acquired Immunity

- A foreign molecule that the host does not recognize as self and that triggers an acquired immune response is an antigen.
- Leukocytes called lymphocytes are responsible for acquired immune responses. Most lymphocytes reside in a group of organs and tissues that constitute the lymphatic system (Figure 53.4).
- Lymphocytes responsible for acquired immunity are B and T cells. B cells differentiate into antibody-producing cells called plasma cells. T cells include cytotoxic T cells, which directly kill target cells, and helper T cells, which assist in the activation and function of B cells and cytotoxic T cells (Figure 53.5).
- Immunologists recognize two types of acquired immunity. In humoral immunity, plasma cells secrete antibodies that bind to antigens. In cell-mediated immunity, cytotoxic T cells directly attack and destroy abnormal body cells.
- Acquired immune responses occur in three stages. The first stage is recognition of antigen; the second is activation of lymphocytes; and the third is attack against antigen (Figure 53.6).
- In humoral immunity, B cells recognize antigens with B-cell receptors. When B cells are activated, they proliferate and differentiate into plasma cells, which secrete antibodies.
- Both B-cell receptors and antibodies belong to a family of proteins called immunoglobulins. Immunoglobulins contain a constant region, which is identical for all immunoglobulins of a given Ig class, and a variable region that serves as the antigen binding site (Figures 53.7, 53.8).
- The process known as hypermutation, which primarily involves numerous C to T point mutations, is crucial to enabling plasma cells to produce a diverse array of antibodies capable of recognizing many different antigens.
- B cells that are activated by an antigen differentiate into plasma cells by a process called clonal selection. Antibodies combine with the antigen that activated the B cell and guide an attack that eliminates the antigen or the cells bearing it (Figure 53.9).
- Major histocompatibility complex (MHC) proteins are cellular "identity tags" that serve as genetic markers of self. Class I MHC proteins are found on the surface of all human body cells except erythrocytes. Class II MHC proteins are found only on the surface of macrophages, B cells, and dendritic cells.
- Antigen-presenting cells (APCs) are cells bearing fragments of antigen, called antigenic determinants or epitopes, complexed with the cell's MHC proteins.
- The binding between a helper T-cell receptor and an antigen bound to class II MHC proteins on an APC is essential to helper T-cell activation. Once activated, helper T cells can help to activate both B cells and cytotoxic T cells (Figures 53.10, 53.11).
- Cell-mediated immune responses are mediated by cytotoxic T cells, which directly kill virus-infected and cancerous cells via secreted chemicals. Humoral immune responses are mediated by B cells and plasma cells. In both types of responses, helper T cells are required (Figures 53.12, 53.13).
- The process by which the body distinguishes between self and nonself components is called immune tolerance. Individuals normally lack active lymphocytes that respond to self components because of two mechanisms. T cells with receptors capable of binding self proteins are destroyed by apoptosis in a process termed clonal deletion. Clonal inactivation causes potentially self-reacting lymphocytes to become nonresponsive. When the body's immune cells attack the body's own cells, the result is autoimmune disease.
- Upon initial exposure to an antigen, the body produces a primary immune response. Any subsequent exposure to the same antigen elicits an immediate and heightened response termed a secondary immune response. The immune system's ability to produce this secondary response is called immunological memory (Figure 53.14).
- The acquired response to exposure to any type of antigen is known as active immunity. The artificial exposures to antigen that occur in vaccinations and immunizations also induce active immunity. In contrast, passive immunity confers protection against disease through the direct transfer of antibodies from one individual to another.

53.4 Impact on Public Health

- Factors that cause malfunction of the immune system include lifestyle; organ transplants; allergies; and acquired immune deficiency syndrome (AIDS), caused by the human immunodeficiency virus (HIV). AIDS reduces the body's immunity by killing helper T cells (Figure 53.15).

Assess and Discuss

Test Yourself

1. Which of the following is *not* an example of a barrier defense in animals?
   a. skin
   b. secretions from skin glands
   c. exoskeleton
   d. mucus
   e. antibodies

2. The leukocytes that are found in mucosal surfaces and that play a role in defending the body against parasitic infections are
   a. neutrophils.
   b. eosinophils.
   c. basophils.
   d. monocytes.
   e. NK cells.

3. The vascular changes of inflammation
   a. lead to an increase in bacterial cells at the injury site.
   b. decrease the number of leukocytes at the injury site.
   c. allow plasma proteins to move easily from the bloodstream to the injury site.
   d. increase the number of antibodies at the injury site.
   e. activate lymphocytes.

4. Which is correct regarding acquired immunity?
   a. Acquired immunity only requires the presence of helper T cells to function properly.
   b. Acquired immunity does not require exposure to a foreign substance.
   c. Acquired immunity is triggered by contact with a particular antigen.
   d. Acquired immunity includes inflammation.
   e. All of the above are correct.

5. Memory B cells are
   a. cloned lymphocytes that are active in subsequent infections.
   b. cloned lymphocytes that are active during a primary infection.
   c. NK cells that recognize cancer cells and destroy them.
   d. cells that produce antibodies.
   e. macrophages that have recognized self antigens.

6. The immunoglobulin that is passed from mother to fetus across the placenta is
   a. IgA.
   b. IgD.
   c. IgE.
   d. IgG.
   e. IgM.

7. The region of an antibody that is the antigen binding site is
   a. the constant region.
   b. the variable region.
   c. the heavy chain.
   d. the light chain.
   e. the hinge region.

8. A major difference between the activation of B cells and T cells is that
   a. T cells must interact with antigens bound to plasma membranes.
   b. B cells interact only with free antigens.
   c. B cells are not regulated by helper T cells.
   d. T cells produce antibodies.
   e. none of the above

9. Cells that process foreign proteins and complex them with their MHC proteins are called
   a. cytotoxic T cells.
   b. plasma cells.
   c. NK cells.
   d. antigen presenting cells.
   e. helper T cells.

10. HIV causes immune deficiency because the virus
   a. destroys all the cytotoxic T cells.
   b. preferentially destroys helper T cells that regulate the immune system.
   c. directly inactivates plasma cells.
   d. causes mutations that lead to autoimmune diseases.
   e. does all of the above.

Conceptual Questions

1. Distinguish between innate and acquired immunity.
2. Explain the function of cytotoxic T cells.
3. A principle of biology is that *living organisms interact with their environment.* Such interactions include potentially threatening environmental factors such as pathogens. Discuss three types of pathogens that affect the health of animals.

Collaborative Questions

1. Describe the basic structure of an immunoglobulin.
2. Describe the functions of helper T cells.

Online Resource

www.brookerbiology.com

Stay a step ahead in your studies with animations that bring concepts to life and practice tests to assess your understanding. Your instructor may also recommend the interactive eBook, individualized learning tools, and more.

UNIT VIII

ECOLOGY

Ecology is the study of interactions among organisms and between organisms and their environment. These interactions govern the number of species in an area and their population densities. Ecologists work at the largest scales of any biologists.

In Chapter 54, we introduce the field of ecology and discuss the effects of physical variables such as temperature and moisture. At the largest scales, variation in temperature and moisture create distinct large-scale habitats, called biomes. Chapter 55 discusses behavioral ecology and how behavior contributes to the fitness of organisms. We begin the chapter by investigating how different behaviors are achieved and end by examining group behavior and mating systems. The next two chapters examine population growth and the constraints to growth provided by competitors and natural enemies. In Chapter 56, we introduce the demographic tools needed to study population growth, provide simple mathematical models of growth, and examine the special case of human population growth. In Chapter 57, we discuss the effects of competition, mutualism, predation, herbivory, and parasitism on populations. Chapters 58 and 59 focus on communities and ecosystems. In Chapter 58, we consider the factors that influence the number of species in a community, and we examine different measures of diversity. Chapter 59 addresses the flow of energy and nutrients through the living and nonliving components of the environment. Finally, in Chapter 60 we address the conservation of life on Earth and the various strategies used to protect genetic, species, and ecosystem diversity. Throughout the unit, we'll examine the effects of humans on the environment, including pollution, global climate change, and the introduction of exotic species of plants and animals.

The following biology principles will be emphasized in this unit:

- ***Living organisms use energy:*** *In Chapter 59 we discuss the cycle of nutrients and energy flow in ecosystems.*
- ***Living organisms interact with their environment:*** *Chapter 54 provides examples of the influence of temperature, water, pH, salt concentration, and light on the distribution and abundance of organisms.*
- ***New properties of life emerge from complex interactions:*** *In Chapter 57, we see how the effects of natural enemies and abiotic factors can cascade through natural communities.*
- ***Biology is an experimental science:*** *Throughout the unit we provide numerous examples of experiments that ecologists have used to investigate how ecological systems function.*
- ***Biology affects our society:*** *In the last chapter of the unit we discuss some of the conservation efforts currently under way to save and secure life on Earth.*

Chapter Outline

54.1 The Scale of Ecology
54.2 Ecological Methods
54.3 The Environment's Effect on the Distribution of Organisms
54.4 Climate and Its Relationship to Biological Communities
54.5 Major Biomes
54.6 Continental Drift and Biogeography
Summary of Key Concepts
Assess and Discuss

An Introduction to Ecology and Biomes

54

Diminishing and disappearing populations. Population sizes of the Panamanian golden frog (*Atelopus zeteki*) have decreased greatly over the past 20 years, and populations of many other species of harlequin frogs have disappeared entirely. Ecologists are investigating the reasons for this decline.

BioConnections: *Look ahead to Section 60.3. What are the main threats to species? Are they natural or human-induced?*

In 2006, a study led by J. Alan Pounds of the Monteverde Cloud Forest Preserve in Costa Rica reported that two-thirds of the 110 species of harlequin frogs in mountainous areas of Central and South America had become extinct over the previous 20 years. The researchers noted that populations of other species, such as the Panamanian golden frog (*Atelopus zeteki*), had been greatly reduced (see chapter-opening photo). The question was why. The culprit was identified as a disease-causing fungus, *Batrachochytrium dendrobatidis*, but Pounds's study implicated global warming—a gradual increase in the average temperature of the Earth's atmosphere—as the agent causing outbreaks of the fungus. One effect of global warming is to increase the cloud cover, which reduces daytime temperatures and raises nighttime temperatures. Researchers believe that this combination has created favorable conditions for the spread of *B. dendrobatidis* and other diseases, which thrive in cooler daytime temperatures. Pounds, the team's lead researcher and an ecologist, was quoted as saying, "Disease is the bullet killing frogs, but climate change is pulling the trigger."

Ecology is the study of interactions among organisms and between organisms and their environments. Interactions among organisms are called **biotic** interactions, and those between organisms and their nonliving environment are termed **abiotic** interactions. These interactions, in turn, govern the numbers of species in an area and their population densities. In this first chapter of the ecology unit, we will introduce the four broad areas of ecology: organismal, population, community, and ecosystems ecology. Next, we will explore how ecologists approach and conduct their work. We will then turn our focus to abiotic interactions and examine the effects of factors such as temperature, water, light, pH, and salt concentrations on the distributions of organisms. We conclude with a consideration of climate and its large influence on **biomes**, the major types of habitats where organisms are found.

Before 1960, the field of ecology was dominated by taxonomy, natural history, and speculation about observed patterns. An ecologist's tools of the trade might have included sweep nets, quadrats (small, measured plots of land used to sample living things), and specimen jars. Since that time, the number of ecological studies has exploded, and ecologists have become active in investigating environmental change on local, regional, and global scales. Ecologists have embraced experimentation and adapted concepts and methods derived from agriculture, physiology, biochemistry, genetics, physics, chemistry, and mathematics. Their tools have kept pace with technological innovations. Now an ecologist's equipment is just as likely to include laptops, satellite-generated images, and chemical autoanalyzers.

Ecological studies have important implications in the real world, as will be amply illustrated by examples discussed throughout the unit. However, there is a distinction between ecology and **environmental science**, the application of ecology to real-world problems. To use an analogy, ecology is to environmental science as physics is to engineering. Both physics and ecology provide the theoretical framework on which to pursue more applied studies. Engineers rely on the principles of physics to build bridges. Environmental scientists rely on the principles of ecology to solve environmental problems.

(a) A single organism

(b) A population of zebras

(c) An African grassland community

(d) Nutrient flow in an African grassland community

Figure 54.1 **The scale of ecology.** **(a)** Organismal ecology. What is the temperature tolerance of this zebra? **(b)** Population ecology. What factors influence the growth of zebra populations in Africa? **(c)** Community ecology. What factors influence the number of species in African grassland communities? **(d)** Ecosystems ecology. How do water, energy, and nutrients flow among plants, zebras, and other herbivores and carnivores in African grassland communities?

54.1 The Scale of Ecology

Learning Outcome:

1. Describe and differentiate between the different scales at which ecologists work.

Ecology ranges in scale from the study of an individual organism through the study of populations to the study of communities and ecosystems (**Figure 54.1**). In this section, we introduce each of the broad areas of organismal, population, community, and ecosystem ecology and provide an investigation that helps illuminate the field of population ecology.

Organismal Ecology Investigates How Adaptations and Choices by Individuals Affect Their Reproduction and Survival

Organismal ecology is the study of the ways in which individual organisms meet the challenges of the abiotic and biotic environments. It can be divided into two subdisciplines. The first, **physiological ecology**, investigates how organisms are physiologically adapted to their environment and how the environment impacts the distribution of species. Much of this chapter discusses physiological ecology. The second area, **behavioral ecology**, focuses on how the behavior of individual organisms contributes to their survival and reproductive success, which, in turn, eventually affects the population density of the species. This is the topic of Chapter 55.

Population Ecology Describes How Populations Grow and Interact with Other Species

Population ecology focuses on groups of interbreeding individuals, called populations. A primary goal of population ecology is to understand the factors that affect a population's growth and determine its size and density. Although the attention of a population ecologist may be aimed at studying the population of a particular species, the relative abundance of that species is often influenced by its interactions with other species. Thus, population ecology includes the study of **species interactions**, such as predation, competition, and parasitism. Knowing what factors affect populations can help us lessen species endangerment, stop extinctions, and control invasive species.

Feature Investigation

Callaway and Aschehoug's Experiments Showed That the Secretion of Chemicals Gives Invasive Plants a Competitive Edge Over Native Species

One important topic in the area of population ecology concerns **introduced species** (also called exotic species), species that are moved from a native location to another location, usually by humans. Such species sometimes spread so aggressively that they crowd out native organisms, in which case they are considered **invasive species**. Of the 300 most invasive plants in the U.S., over half were brought in for gardening, horticulture, or landscaping purposes. Invasive species have traditionally been thought to succeed because they have escaped their natural enemies, primarily insects that are in the country of origin and not in the new locale. One way of controlling these species, therefore, has been to import the plant's natural enemies. This is known as **biological control**. However, an investigation of the population ecology of diffuse knapweed (*Centaurea diffusa*), a Eurasian plant that has established itself in many areas of North America, suggests a different reason for the success of invasive species.

American researchers Ragan Callaway and Erik Aschehoug hypothesized that the roots of this particular species secrete powerful toxins, called **allelochemicals**, that kill the roots of other species, allowing *Centaurea* to proliferate. To test their hypothesis, Callaway and Aschehoug collected seeds of three native Montana grasses, *Koeleria cristata*, *Festuca idahoensis*, and *Agropyron spicata*, and grew each of them with or without the exotic *Centaurea* species (**Figure 54.2**). As hypothesized, *Centaurea* depressed the biomass of the native grasses. When the experiments were repeated with grasses

Figure 54.2 **Experimental evidence of the effect of allelochemicals on plant production.**

HYPOTHESIS Exotic plants from Eurasia outcompete native Montana grasses by secreting allelochemicals from their roots.

KEY MATERIALS Seeds of *Centaurea diffusa* from Eurasia plus seeds of native Montana grasses.

	Experimental level	Conceptual level
1 Collect seeds of native Montana grasses and plant with and without seeds of invasive *C. diffusa* from Eurasia. Three months after sowing seeds, the plants are harvested, dried, and weighed.		*C. diffusa* significantly reduces biomass of native Montana grasses.
2 Collect seeds of grasses from Eurasia of the same three genera as the Montana grasses and plant with and without *C. diffusa*. Three months after sowing seeds, the plants are harvested, dried, and weighed.	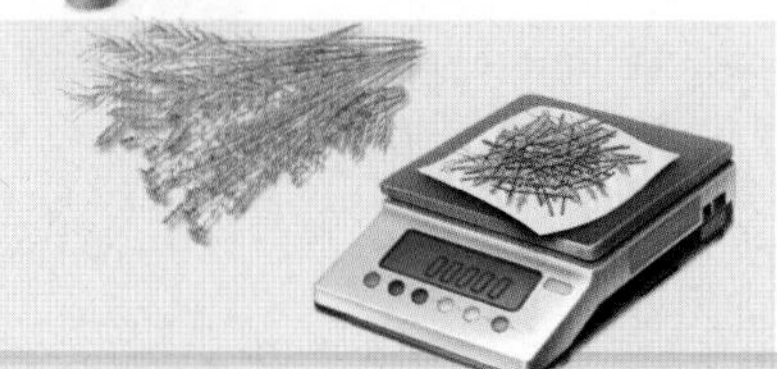	*C. diffusa* doesn't depress the biomass of grasses native to Eurasia as much.

3 **THE DATA***

*The biomass is that of the genus noted at the top of each graph.

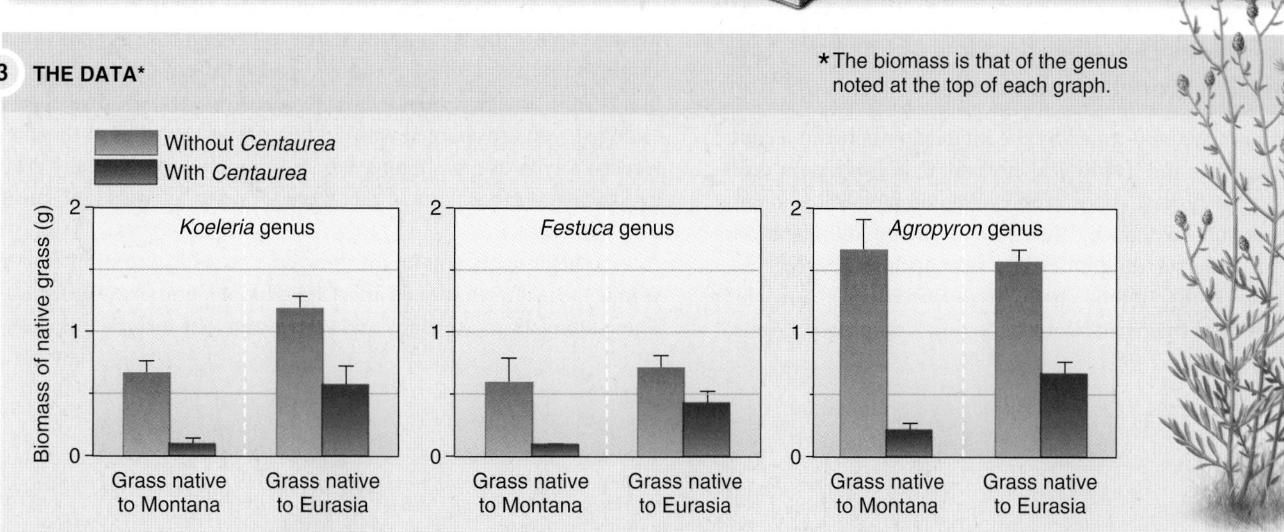

4 **CONCLUSION** *Centaurea diffusa*, a Eurasian grass, is invasive in the U.S. because it secretes allelochemicals, which inhibit the growth of native plants.

5 **SOURCE** Callaway, R.M., and Aschehoug, E.T. 2000. Invasive plants versus their old and new neighbors. *Science* 290:521–523.

native to Eurasia, *Koeleria laerssenii*, *Festuca ovina*, and *Agropyron cristatum*, the growth of each species was inhibited, but to a significantly lesser degree than the growth of the Montana species.

In other experiments not described in Figure 54.2, Callaway and Aschehoug added activated carbon to the soil, which absorbs the chemical excreted by the *Centaurea* roots. With activated carbon added, the Montana grass species increased in biomass compared with the previous experiments. The researchers concluded that *C. diffusa* outcompetes Montana grasses by secreting an allelochemical and that Eurasian grasses are not as susceptible to the chemical's effect because they coevolved with it. If the reason for the success of invasive plants can be attributed to the chemicals they secrete, this calls into question the effectiveness of biological control of invasive plants by importation of their natural enemies. This study on the population biology of an invasive plant has changed the way we think about why such species succeed and could affect the way we attempt to control them in the future.

Experimental Questions

1. Prior to Callaway and Aschehoug's study, what was the prevailing hypothesis of why invasive species succeed in new environments?
2. Briefly describe the evidence collected to support the allelochemical hypothesis.
3. What was the function of the activated carbon used in a subsequent test of the hypothesis?

Community Ecology Focuses on What Factors Influence the Number of Species in a Given Area

Community ecology studies how populations of species interact and form functional communities. For example, in a forest, there is a community of trees, herbs, shrubs, grasses, the herbivores that eat them, and the carnivores that prey on the herbivores.

Community ecology focuses on why certain areas have high numbers of species (that is, are species-rich), but other areas have low numbers of species (that is, are species-poor). Although ecologists are interested in species richness for its own sake, a link also exists between species richness and community function. Ecologists generally believe that species-rich communities perform better than species-poor communities. It has also been proposed that more species make a community more stable, that is, more resistant to disturbances such as introduced species. Community ecology also considers how species composition and community structure change over time and, in particular, after a disturbance, a process called succession.

Ecosystem Ecology Describes the Flow of Energy and Chemicals Through Communities

An **ecosystem** is a system formed by the interaction between a community of organisms and its physical environment. **Ecosystem ecology** deals with the flow of energy and cycling of chemical elements within an ecosystem. Following this flow of energy and chemicals necessitates an understanding of feeding relationships between species, called food chains. In food chains, each level is called a trophic level, and many food chains interconnect to form complex food webs. As we learned in Chapter 6, the second law of thermodynamics states that in every energy transformation, free energy is reduced because heat energy is lost in the process, and the entropy of the system increases. Therefore, a unidirectional flow of energy occurs through an ecosystem, with energy dissipated at every step. An ecosystem needs a recurring input of energy from an external source—in most cases, the Sun—to sustain itself. In contrast, chemicals such as nitrogen do not dissipate and constantly cycle between abiotic and biotic components of the environment.

54.2 Ecological Methods

Learning Outcome:

1. Explain how the five steps of hypothesis testing can be applied to an ecological research project.

How do ecologists go about studying their subject? In this section, we will explore the methods used by ecologists. Let's suppose you are employed by the United Nations' Food and Agriculture Organization (FAO), an agency that works to defeat hunger worldwide. As an ecologist, you are charged with finding out what causes outbreaks of locusts, a type of grasshopper whose population periodically erupts in Africa and other parts of the world, destroying crops and causing food shortages.

To begin with, you might draw up a possible web of interaction among the factors that could affect locust population size (Figure 54.3). These interactions are many and varied, and they include

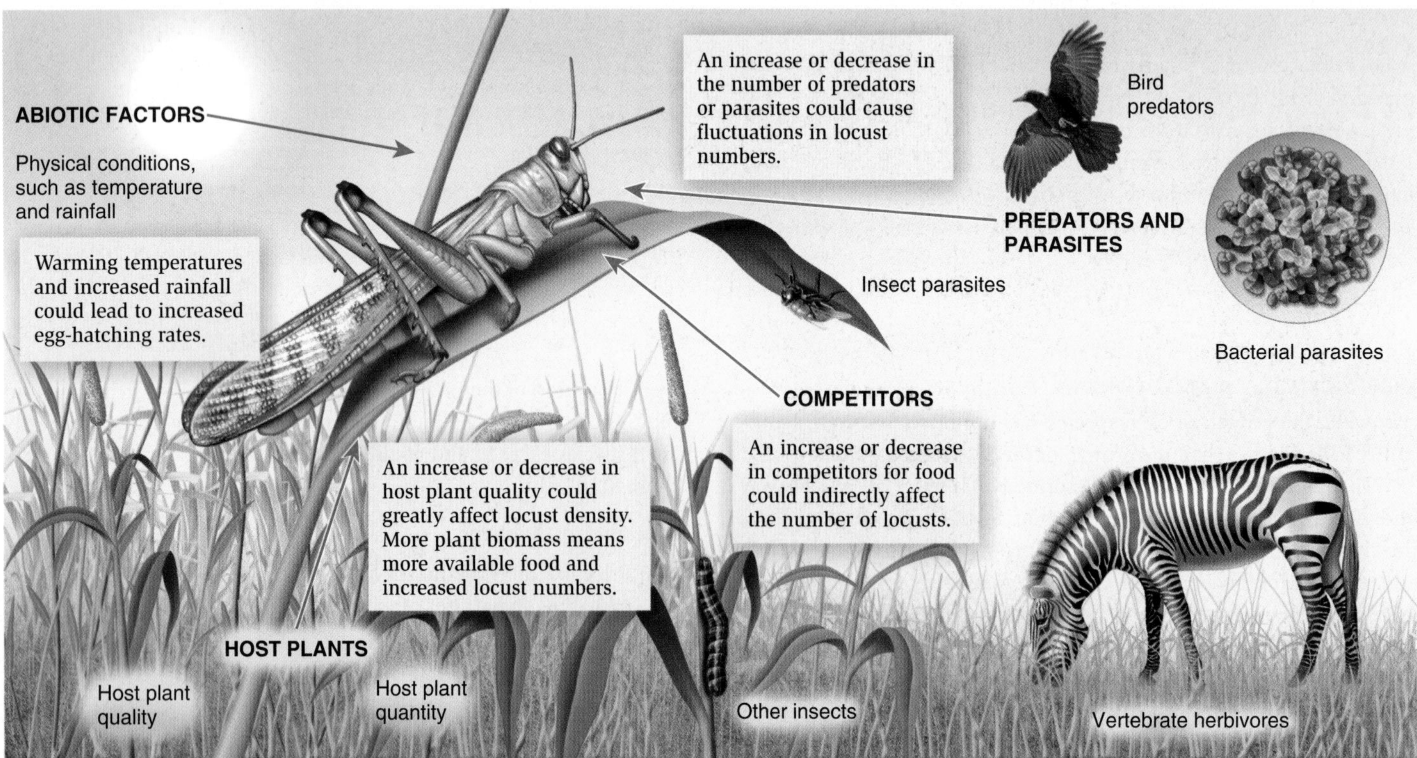

Figure 54.3 Depiction of factors that might influence locust population size.

- abiotic factors, such as temperature, rainfall, wind, and soil pH;
- natural enemies, including bird predators, insect parasites, and bacterial parasites;
- competitors, including other insects and larger vertebrate grazers;
- host plants, including increases or decreases in either the quality or quantity of the plants.

With such a vast array of factors to be investigated, where is the best place to start? As discussed in Chapter 1, hypothesis testing involves a five-stage process: (1) observations, (2) hypothesis formation, (3) experimentation, (4) data analysis, and (5) acceptance or rejection of the hypothesis.

Observations Are Made to Develop Hypotheses

In our study of locusts, we begin by carefully observing the organism in its native environment. We can analyze the fluctuations of locusts and determine if the populations vary with changes in the other phenomena, such as levels of parasitism, numbers of predators, or food supply. Let's say we observed that an inverse relationship exists between predation levels and locust numbers. As predation levels increase, locust numbers decrease. If we plotted this relationship graphically, the resulting graph would look like that depicted in Figure 54.4a. This result would give us some confidence that predation levels determined locust numbers, and this would be our hypothesis. In fact, we would have so much confidence that we could create a statistically determined line of best fit to represent a summary of the relationship between these two variables, which is shown in the graph.

However, if the points were not highly clustered, as in Figure 54.4b, we would have little confidence that predation affects locust density. Many statistical tests are used to determine whether or not two variables are significantly correlated. In the studies in this unit, unless otherwise stated, most graphs like Figure 54.5a imply that a meaningful relationship exists between the two variables. We call this type of relationship a significant **correlation**. In this graph, locust density shows a negative linear relationship with predation; therefore, we say that locust density is negatively correlated with predation.

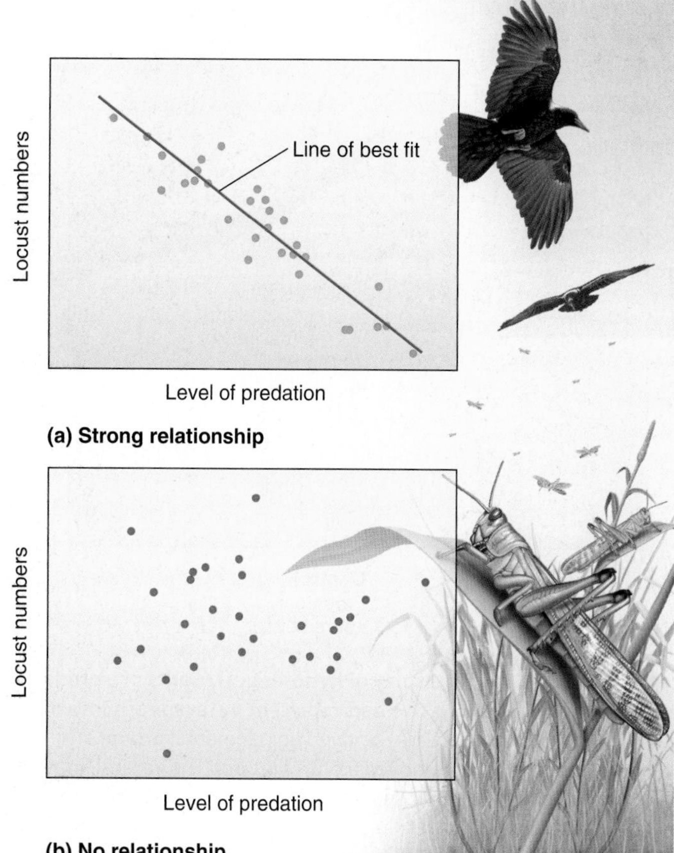

Figure 54.4 **Correlation of locust numbers with predation.** In this case, higher locust numbers are found in nature where predation levels are lowest. We can draw a line of best fit **(a)** to represent this relationship. In **(b)**, the relationship between locust numbers and predation levels might be so weak that we would not have much confidence in a linear relationship between the variables.

Concept Check: What would it mean if the line of best fit sloped in the opposite direction?

Experiments Are Used to Test Hypotheses

We have to be cautious when forming conclusions based on correlations. For example, large numbers of locusts could be associated with large, dense plants. We might conclude from this that food availability controls locust density. However, an alternative conclusion would be that large plants provide locusts refuge from bird predators, which cannot attack them in the dense interior. Although it would appear that biomass affects locust density by providing abundant food, in actuality, predation would still be the most important factor affecting locust density. Thus, correlation does not always mean causation. For this reason, after conducting observations, ecologists usually turn to experiments to test their hypotheses.

In our example, an experiment might involve removing predators from an area inhabited by locusts. If predators are having a significant effect, then removing them should cause an increase in locust numbers. Reduced predation might be achieved by putting a cage made of chicken wire over and around bushes containing locusts, so that birds are denied access. We would have two groups: a group of locusts with predators removed (the experimental group) and a group of locusts with predators still present (the control group), with equal numbers of locusts in both groups at the start of the experiment. Any differences in locust population density would be due solely to differences in predation. Experiments often have a defined time frame, and in our example, we could look at locust survivorship over the course of one generation of locusts.

Data Analysis Permits Rejection or Acceptance of a Hypothesis

Performing the experiments several times is called **replication**. We might replicate the experiment 5 times, 10 times, or even more. We would add up the total number of surviving locusts from each replicate and calculate the mean, which is the sum divided by the number of values. In the experimental group, let's suppose that the numbers

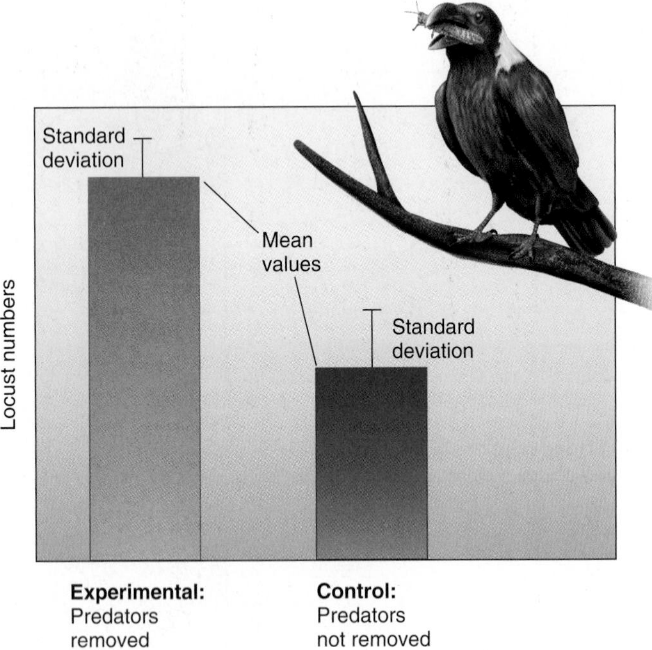

Figure 54.5 Graphic display of hypothetical results of a predator removal experiment. The two bars represent the average number of locusts where predators are removed (experimental group) and where predators are not removed (control group). The vertical lines (the standard deviations or standard errors) give an indication of how tightly the individual results are clustered around the mean. The shorter the lines, the tighter the cluster, and the more confidence we have in the result.

of surviving locusts in each replicate are 5, 4, 7, 8, 12, 15, 13, 6, 8, and 10; thus, the mean number surviving would be 8.8. In the control group, which still allows predator access, the numbers surviving might be 2, 4, 7, 5, 3, 6, 11, 4, 1, and 3, with a mean of 4.6. Without predators, the mean number of surviving locusts would therefore be almost double the average number surviving with predators. Our data analysis would give us confidence that predators were indeed a cause of our changes in locust numbers. The results of such experiments can be illustrated graphically by a bar graph (**Figure 54.5**). Ecologists can use a variety of tests to see if the differences between the control data and the experimental data are statistically significant, which means that the differences are not likely to have occurred as a result of random chance. We won't look at the mechanics of these tests, but in this unit, when experimental and control data are presented as differing, these are considered to be statistically significant differences unless stated otherwise.

By the way, it turns out that predation is not the primary factor that controls locust populations. The results we have been discussing are hypothetical. Weather, in particular, rain, is the most important feature governing locust population size. Moist soil allows eggs to hatch and provides water for germinating plants, allowing a ready source of food for the hatchling locusts. In general, physical or abiotic factors such as the availability of water usually have powerful effects in most ecological systems. In the next part of the chapter, we turn our attention to an examination of the effects of the physical environment on the distribution patterns of organisms.

54.3 The Environment's Effect on the Distribution of Organisms

Learning Outcomes:

1. Give examples of how extremes of temperature, both low and high, drastically affect the distribution and abundance of life on Earth.
2. Describe how global warming is gradually increasing the Earth's temperature and will likely affect species distributions.
3. Explain how other environmental factors such as wind, water availability, light availability, salt concentration, and pH of soil and water can affect the distributions of organisms.

Both the distribution patterns of organisms and their abundance are limited by physical features of the environment such as temperature, wind, availability of water and light, salinity, and pH (**Table 54.1**). In this section, we will examine these features of the environment.

Temperature Has an Important Effect on the Distribution of Plants and Animals

Temperature is perhaps the most important factor in the distribution of organisms because of its effect on biological processes and because of the inability of most organisms to regulate their body temperature precisely. For example, the organisms that form coral reefs secrete a calcium carbonate shell. Shell formation and coral deposition are accelerated at high temperatures but are suppressed in cold water. Coral reefs are therefore abundant only in warm water, and a close correspondence is observed between the 20°C isotherm for the average daily temperature during the coldest month of the year and the limits of the distribution of coral reefs (**Figure 54.6**). An isotherm is a line on a map connecting points of equal temperature. Coral reefs are located between the two 20°C isotherm lines that are formed above and below the equator.

Table 54.1 Selected Abiotic Factors and Their Effects on Organisms

Factor	Effect
Temperature	Low temperatures freeze many plants; high temperatures denature proteins. Some plants require fire for germination.
Wind	Wind amplifies effects of cool temperatures (wind chill) and water loss; creates pounding waves.
Water	Insufficient water limits plant growth and animal abundance; excess water drowns plants and other organisms.
Light	Insufficient light limits plant growth, particularly in aquatic environments.
Salinity	High salinity generally reduces plant growth in terrestrial habitats; affects osmosis in marine and freshwater environments.
pH	Variations in pH affect decomposition and nutrient availability in terrestrial systems; directly influences mortality in both aquatic and terrestrial habitats.

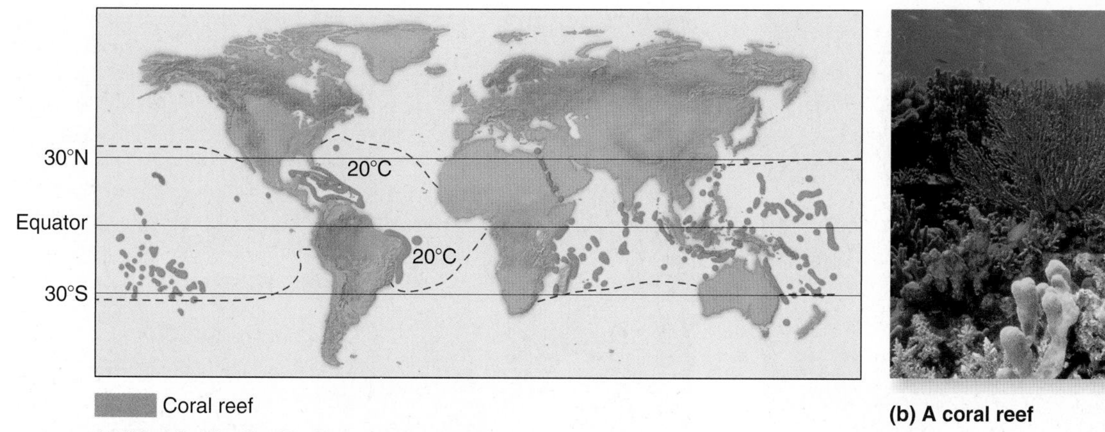

(a) Worldwide distribution of coral reefs

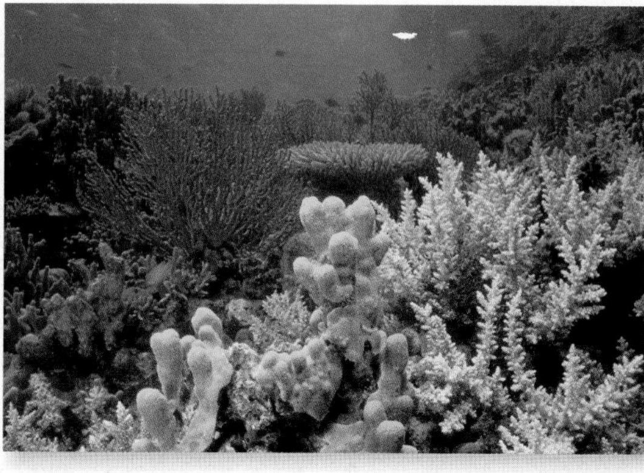
(b) A coral reef

Figure 54.6 **Worldwide locations of coral reefs.** **(a)** Coral reef formation is limited to waters bounded by the 20°C isotherm (dashed line), a line where the average daily temperature is 20°C during the coldest month of the year. **(b)** Coral reef from the Pacific Ocean.

Concept Check: Why are coral reefs limited to warm water?

Low Temperatures Frost is probably the single most important factor limiting the geographic distribution of tropical and subtropical plants. In plants, cold temperature can be lethal because cells may rupture if the water they contain freezes. In the Sonoran Desert in Arizona, saguaro cacti can easily withstand frost for one night as long as temperatures rise above freezing the following day, but they are killed when temperatures remain below freezing for more than 36 hours. This means that the cactus's distribution is limited to places where the temperature does not remain below freezing for more than one night (**Figure 54.7**).

The geographic range limits of endothermc animals are also affected by temperature. For example, the eastern phoebe (*Sayornis phoebe*), a small bird, has a northern winter range that coincides with an average minimum January temperature above 4°C. Such limits are probably related to the energy demands associated with cold temperatures. Cold temperatures mean higher metabolic costs, which are, in turn, dependent on high feeding rates. Below 4°C, the eastern phoebe cannot feed fast enough or, more likely, find enough food to keep warm.

High Temperatures High temperatures are also limiting for many plants and animals because relatively few species can survive internal temperatures more than a few degrees above their metabolic optimum. We have discussed how corals are sensitive to low temperatures; however, they are sensitive to very high temperatures as well. When temperatures are too high, the symbiotic algae that live within coral die and are expelled, causing a phenomenon known as coral bleaching. Once bleaching occurs, the coral tissue loses its color and turns a pale white (**Figure 54.8**). El Niño is a weather phenomenon characterized by a major increase in the water temperature of the equatorial Pacific Ocean. In the winter of 1982–1983, an influx of warm water from the eastern Pacific raised temperatures just 2–3°C for 6 months, which was enough to kill many of the reef-building corals on the coast of Panama. By May 1983, just a few individuals of one species, *Millepora intricata*, were alive.

The ultimate high temperatures that many terrestrial organisms face are the result of fire. However, some species depend on frequent low-intensity fires for their reproductive success. The longleaf pine (*Pinus palustris*) of the southeast U.S. produces serotinous cones, which

Figure 54.7 **Saguaro cacti in freeze-free zones.** A close correspondence is seen between the range of the saguaro cactus (dark green area) and the area in which temperatures do not drop below freezing (0°C) for more than 0.5 day.

BIOLOGY PRINCIPLE Living organisms interact with their environment. In many cases the distributional limits of organisms are set by the physical environment.

Figure 54.8 **Coral bleaching.** Mantanani Island, Malaysia.

remain sealed by pine resin until the heat of a fire melts them open and releases the seeds. In the west, giant sequoia trees are similarly dependent on periodic low-intensity fires for germination of their seeds. Such fires both enhance the release of seeds and clear out competing vegetation at the base of the tree so that seeds can germinate and grow. Fire-suppression practices that attempt to protect forests from fires can actually have undesirable results by preventing the regeneration of fire-dependent species. Furthermore, fire prevention can result in an accumulation of vegetation beneath the canopy (the understory) that may later fuel hotter and more damaging fires. The U.S. Forest Service uses controlled human-made fires to mimic the natural disturbance of periodic fires and maintain fire-dependent forest species (Figure 54.9).

Figure 54.9 **Giant sequoia.** A park ranger uses a drip torch to ignite a fire at Sequoia National Park in California. Periodic, controlled human-made fires mimic the sporadic wildfires that normally burn natural areas. Such fires are vital to the health of giant sequoia populations, because they serve to open the pine cones and release the seeds.

Concept Check: *Why are some fires very destructive to natural systems?*

The Greenhouse Effect The Earth is warmed by the **greenhouse effect.** In a greenhouse, sunlight penetrates the glass and raises temperatures, with the glass acting to trap the resultant heat inside. Similarly, solar radiation in the form of short-wave energy passes through the atmosphere to heat the surface of the Earth. At night, this energy is radiated from the Earth's warmed surface back into the atmosphere, but in the form of long-wave infrared radiation. Instead of letting it escape back into space, however, atmospheric gases absorb much of this infrared energy and radiate it back to the Earth's surface, causing its temperature to rise further (Figure 54.10). The greenhouse effect is a naturally occurring process that is responsible for keeping the Earth warm enough to sustain life. Without some type of greenhouse effect, global temperatures would be much lower than they are, perhaps averaging only −17°C compared with the existing average of +15°C.

The greenhouse effect is caused by a group of atmospheric gases that together make up less than 1% of the total volume of the atmosphere. These gases—primarily water vapor, carbon dioxide, methane, nitrous oxide, and chlorofluorocarbons—are referred to as greenhouse gases (Table 54.2).

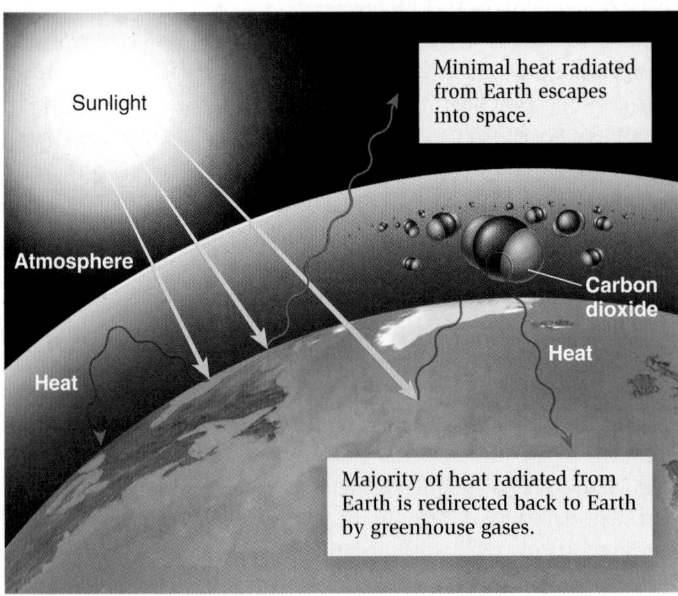

Figure 54.10 **The greenhouse effect.** Solar radiation, in the form of short-wave energy, passes through the atmosphere to heat the Earth's surface. Long-wave infrared energy is radiated back into the atmosphere. Much infrared energy is absorbed by atmospheric gases, including carbon dioxide molecules, and reflected back to Earth, causing global temperatures to rise.

Global Warming Ecologists are concerned that human activities are increasing the greenhouse effect and causing **global warming** (also called global climate change), a gradual elevation of the Earth's surface temperature. According to the Intergovernmental Panel on Climate Change 2007 report, warming of the climate is unequivocal, as is now evident from observations of increases in global average air

Table 54.2 Selected Greenhouse Gases and Their Contribution to Global Warming*

	Carbon dioxide (CO_2)	Methane (CH_4)	Nitrous oxide (N_2O)	Chlorofluorocarbons (CFCs)
Relative absorption in ppm of increase†	1	21	310	10,000
Atmospheric concentration (ppm‡)	385	1.75	0.315	0.0005
Contribution to global warming	73%	7%	19%	1%
Percent from natural sources; type of source	20–30%; volcanoes	70–90%; swamps, gas from termites and ruminants	90–100%; soils	0%
Major human-made sources	Fossil fuel use, deforestation	Rice paddies, landfills, biomass burning, coal and gas exploitation	Cultivated soil, fossil-fuel use, automobiles, industry	Previously manufactured products (for example, aerosol propellants) but now banned in the U.S. and the E.U.

*Water vapor is not included in this table.
†Relative absorption is the warming potential per unit of gas.
‡ppm = parts per million

and ocean temperatures, widespread melting of snow and ice, and rising average global sea level. Most greenhouse gases have increased in atmospheric concentration since industrial times. Of those increasing, the most important is carbon dioxide (CO_2). As Table 54.2 shows, although CO_2 has a lower global warming potential per unit of gas (relative absorption) than any of the other major greenhouse gases, its concentration in the atmosphere is much higher. Atmospheric levels of CO_2 have increased by about 24% between 1957 and 2011.

To predict the effect of global warming, most scientists focus on a future point, about 2100, when the concentration of atmospheric CO_2 will have doubled—that is, increased to about 700 ppm compared with the late-20th-century level of 350 ppm. Scientists estimate that at that time, average global temperatures will be somewhere in the range of 1–6°C (about 2–10°F) warmer than present and will increase an additional 0.5°C each decade. This increase in heat might not seem like much, but it is comparable to the warming that ended the last Ice Age. Future consequences would include a further contraction of snow cover and a decrease in sea ice extent, heat waves and drought in dry areas, heavier precipitation in moister areas, and an increase in hurricane and tornado intensity.

Assuming this scenario of gradual global warming is accurate, we need to consider what the consequences might be for plant and animal life. At the beginning of the chapter, we saw how global warming is believed to be contributing to the decline and extinction of some amphibian species. Although many species can adapt to slight changes in their environment, the anticipated changes in global climate are expected to occur too rapidly to be compensated for by normal evolutionary processes such as natural selection. Plant species in particular cannot simply disperse and move north or south into the newly created climatic regions that will be suitable for them. Many tree species take hundreds, even thousands, of years for seed dispersal. American paleobotanist Margaret Davis predicted that in the event of a CO_2 doubling, the sugar maple (*Acer saccharum*), which is presently distributed throughout the midwestern and northeastern U.S. and southeastern Canada, would die back in all areas except in northern Maine, northern New Brunswick, and southern Quebec (**Figure 54.11**). Of course, this contraction in the tree's distribution could be offset by the creation of new favorable habitats in central

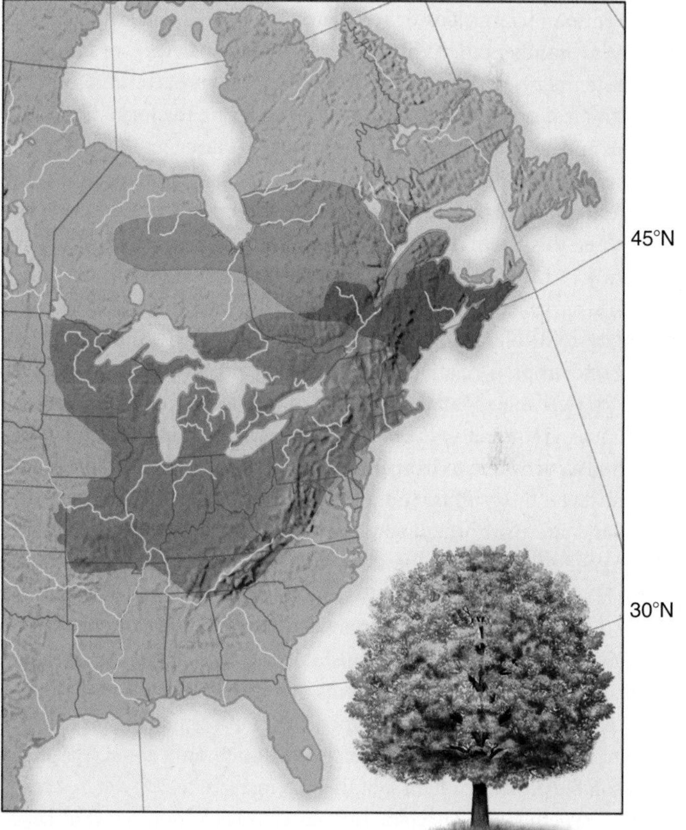

Figure 54.11 Possible changes in the range of sugar maples due to global warming. The present geographic range of the sugar maple (blue shading) and its potential range with doubled CO_2 levels (red shading) in North America. Purple shading indicates the region of overlap, which is the only area where the sugar maple would be found before it spread into its new potential range.

BIOLOGY PRINCIPLE Biology affects our society. Global warming will change the distributions of many species familiar to us.

Concept Check: What abiotic factors that influence sugar maple distribution might change in a world with elevated levels of CO_2?

Quebec. However, most scientists believe that the climatic zones would shift toward the poles faster than trees could migrate via seed dispersal; therefore, extinctions would occur. Interestingly, scientists are beginning to be able to genetically modify organisms to change their temperature tolerances, as described next.

Genomes & Proteomes Connection

Temperature Tolerance May Be Manipulated by Genetic Engineering

Below-freezing temperatures can be very damaging to plant tissue, either killing the plant or greatly reducing its productivity. Frost injury causes losses to agriculture of more than $1 billion annually in the U.S. Frost has been considered an unavoidable result of subfreezing temperatures, but genetic engineering is beginning to change this view.

Between 0°C and –40°C, pure water will be a liquid unless provided with an ice nucleus or template on which an ice crystal can be built. Any number of ions or molecules within an organism's cells can act as a template. Researchers discovered that some bacteria commonly found on leaf surfaces act as ice nuclei, triggering the formation of ice crystals and eventually causing frost damage. The genes that confer ice nucleation resistance have been identified, isolated, and deactivated in a genetically engineered strain of the bacteria *Pseudomonas syringae*. When this strain is allowed to colonize strawberries, frost damage is greatly reduced, and plants can withstand an additional 5°C drop in temperature before frost forms. Other techniques have been used to create cold-tolerant transgenic plants, but to date, the number of transgenic crops is relatively small and includes varieties of tomato, tobacco, rice, maize, alfalfa, cotton, canola and flax. Moreover, the techniques are not always perfect. For example, transgenic tomatoes that are less sensitive to frost may exhibit dwarfism and reduced fruit set. However, the promise of this technique for increasing agricultural yields and altering normal plant-distribution patterns is staggering.

At the other end of the temperature spectrum, heat shock proteins (HSPs) help organisms cope with the stress of high temperatures. At high temperatures, proteins may denature, that is, either unfold or bind to other proteins to form misfolded protein aggregations. HSPs act as molecular chaperones, proteins that help in the proper folding of other proteins, to prevent these types of events from taking place (refer back to Figure 4.31). HSPs normally constitute only about 2% of the cell's soluble protein content, but this can increase to 20% when a cell is stressed, whether by heat, cold, drought, or other conditions. The genes that encode HSPs are extremely common and are found in the genomes of all organisms, from bacteria to plants and animals. In 2006, several genes responsible for inducing the synthesis of HSPs were found in tomato and maize. As a result, transgenic tobacco plants have been produced that grow better than normal plants under higher temperatures.

In the tropics, high temperatures can substantially decrease the growth rates and productivity of many crop species. There is now substantial interest in identifying crop strains with naturally high HSP levels for use in crop-breeding programs. Given the projected continuation of global warming, such research seems particularly timely.

Wind Can Amplify the Effects of Temperature

Wind is created by temperature gradients. As air heats up, it becomes less dense and rises. As hot air rises, cooler air rushes in to take its place. For example, hot air rising in the tropics is replaced by cooler air flowing in from more temperate regions, thereby creating northerly or southerly winds.

Wind affects living organisms in a variety of ways. It increases the rate of heat loss by convection, the transfer of heat by the movement of air next to the body (the wind chill factor). Wind also contributes to water loss in organisms by increasing the rate of evaporation in animals and transpiration in plants. For example, the tree line in alpine areas is often determined by a combination of low temperatures and high winds, an environmental condition in which transpiration exceeds water uptake.

Winds can also intensify oceanic wave action, with resulting effects for aquatic organisms. On the ocean's rocky shore, seaweeds survive heavy surf by a combination of holdfasts and flexible structures. The animals of this zone have powerful organic glues and muscular feet to hold them in place (**Figure 54.12**).

The Availability of Water Has Important Effects on the Abundance of Organisms

Water has an important effect on the distribution of organisms. Cytoplasm is 85–90% water, and without moisture, there can be no life. As noted in Chapter 2, water performs crucial functions in all living organisms. It acts as a solvent for chemical reactions, takes part in hydrolysis and dehydration reactions, is the means by which animals eliminate wastes, and is used for support in plants and in some invertebrates as part of a hydrostatic skeleton.

The distribution patterns of many plants are limited by available water. Some plants, such as the water tupelo tree (*Nyssa aquatica*) in the southeast U.S., do best when completely flooded and are thus found predominantly in swamps. In contrast, coastal plants that grow on sand dunes have access to very little fresh water. Their roots penetrate deep into the sand to extract moisture. In cold climates, water can be present but locked up as permafrost and, therefore, be unavailable. Alpine trees stop growing at a point on the mountainside where they cannot take up enough moisture to offset transpiration losses. This point, known as the timberline, is readily apparent on many mountainsides. Not surprisingly, the density of many plants is limited by the availability of water.

Animals face problems of water balance, too, and their distribution and population density can be strongly affected by water availability. Because most animals depend ultimately on plants for food, their distribution is intrinsically linked to those of their food sources. Such a phenomenon regulates the number of buffalo (*Syncerus caffer*) in the Serengeti area of Africa. In this area, grass productivity is related to the amount of rainfall in the previous month. Buffalo density is governed by grass availability, so a significant correlation is found between buffalo density and rainfall (**Figure 54.13**). The only exception occurs in the vicinity of Lake Manyara, where groundwater promotes plant growth.

(a) Brown alga with a holdfast

(b) A mussel with byssal threads

Figure 54.12 Animals and plants of the intertidal zone adhering to their rocky surface. **(a)** The brown alga (*Laminaria digitata*) has a holdfast that enables it to cling to the rock surface. **(b)** The mussel (*Mytilus edulis*) attaches to the surface of a rock by proteinaceous threads (byssal threads) that extend from the animal's muscular foot.

Light Can Be a Limiting Resource for Plants and Algae

Because light is necessary for photosynthesis, it can be a limiting resource for plants. However, what may be sufficient light to support the growth of one plant species may be insufficient for another. Many plant species grow best in shady conditions, such as eastern hemlock (*Tsuga canadensis*). Its saplings grow in the understory below the forest canopy, reaching maximal photosynthesis at one-quarter of full sunlight. Other plants, such as sugarcane (*Saccharum officinarum*), continue to increase their photosynthetic rate as light intensity increases.

In aquatic environments, light may be an even more limiting factor because water absorbs light, preventing photosynthesis at depths greater than 100 m. Most aquatic plants and algae are limited to a fairly narrow zone close to the surface, where light is sufficient to allow photosynthesis to occur. This zone is known as the **photic zone**.

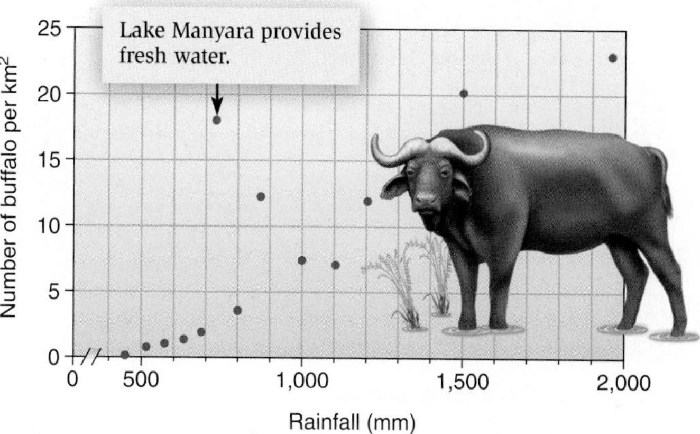

Figure 54.13 The relationship between the amount of rainfall and the density of buffalo. In the Serengeti area of Africa, buffalo density is very much dependent on grass availability, which itself depends on annual rainfall. The main exception is where there is permanent water, such as Lake Manyara. Greater water availability leads to greater grass growth and buffalo densities.

In marine environments, seaweeds at greater depths have wider thalli (leaflike light-gathering structures) than those nearer the surface, because wide thalli can collect more light. In addition, in aquatic environments, plant color changes with depth. At the surface, plants and algae appear green, as they are in terrestrial conditions, because they absorb red and blue light, but not green (**Figure 54.14a**). At greater depths, red light is mostly absorbed by water, leaving predominantly blue-green light. Red algae occur in deeper water because they possess pigments that enable them to utilize blue-green light efficiently, which reflect red light when we see them at the surface (**Figure 54.14b**).

The Concentration of Salts in Soil or Water Can Be Critical

Salt concentrations vary widely in aquatic environments and have a great effect on osmotic balance in animals. Oceans contain considerably more dissolved minerals than rivers because oceans continually receive the nutrient-rich waters of rivers, and the sun evaporates pure water from ocean surfaces, making concentrations of minerals such as salts even higher.

The phenomenon of osmosis influences how living organisms cope with different environments. Freshwater fishes cannot live in salt water, and saltwater fishes cannot live in fresh water. Each employs different mechanisms to maintain an osmotic balance with their environment (refer back to Figure 49.3). Freshwater fishes are hyperosmotic (having a greater concentration of solutes) to their environment and tend to gain water by osmosis as it diffuses through the thin tissue of the gills and mouth. To counter this, the fish continually eliminate water in the urine. However, to avoid losing all dissolved ions, many ions are reabsorbed into the bloodstream at the kidneys. Many marine fishes are hypoosmotic (having a lower concentration of solutes) to their environment and tend to lose water as seawater passes over the mouth and gills. They drink water to compensate for this loss, but the water contains a higher concentration of salt, which must then be excreted at the gills and kidneys.

(a) Green algae at the ocean surface

(b) Red algae at a greater depth

Figure 54.14 Algae growing at different ocean depths. **(a)** In the eastern Pacific Ocean, off the coast of California, these giant kelp floating at the ocean surface are green, just like terrestrial plants. **(b)** In contrast, at 75-m depth, in the McGrail Bank off of the Gulf of Mexico, most seaweeds are pink and red because the pigments can absorb the blue-green light that reaches such depths.

Figure 54.15 Plant adaptations for salty conditions. Special salt glands in the leaves of *Spartina* exude salt, enabling this grass to exist in saline intertidal conditions.

BioConnections: Look back to Section 38.2. Why can't most plants grow in salty habitats?

Salt in the soil also affects the growth of plants. In arid terrestrial regions, salt accumulates in soil where water settles and then evaporates. This can also be of great significance in agriculture, where continued watering in arid environments, together with the addition of salt-based fertilizers, greatly increases salt concentration in soil and reduces crop yields. A few terrestrial plants are adapted to live in saline soil along seacoasts. Here the vegetation consists largely of **halophytes**, species that can tolerate higher salt concentrations in their cell sap than regular plants. Species such as mangroves and *Spartina* grasses have salt glands that excrete salt to the surface of the leaves, where it forms tiny white salt crystals (Figure 54.15).

The pH of Soil or Water Can Limit the Distribution of Organisms

As discussed in Chapter 2, the pH of water can be acidic, alkaline, or neutral. Variation in pH can have a major effect on the distribution of organisms. Normal rainwater has a pH of about 5.6, which is slightly acidic because the absorption of atmospheric carbon dioxide (CO_2) and sulfur dioxide (SO_2) into rain droplets forms carbonic and sulfuric acids. However, most plants grow best at a soil water pH of about 6.5, a value at which soil nutrients are most readily available to plants. Only a few genera, such as rhododendrons and azaleas (*Rhododendron*), can live in soils with a pH of 4.0 or less. Furthermore, at a pH of 5.2 or less, nitrifying bacteria do not function properly, which prevents organic matter from decomposing. In general, alkaline soils containing chalk and limestone have a higher pH and sustain a much richer flora (and associated fauna) than do acidic soils (Figure 54.16).

Generally, the number of fishes and other species also decreases in acidic waters. The optimal pH for most freshwater fishes and bottom-dwelling invertebrates is between 6.0 and 9.0. Acidity in lakes increases the amount of toxic metals, such as mercury, aluminum, and lead, which can leach into the water from surrounding soil and rock. Both too much mercury and too much aluminum can interfere with gill function, causing fishes to suffocate.

Acid Rain The susceptibility of both aquatic and terrestrial organisms to changes in pH explains why ecologists are so concerned about **acid rain**, precipitation with a pH of less than 5.6. Acid rain results from the burning of fossil fuels such as coal, oil, and natural gas, which releases SO_2 and nitrogen oxide (NO_2) into the atmosphere. These react with oxygen in the air to form sulfuric acid and nitric acid, which falls to the Earth's surface in rain or snow. When this precipitation falls on rivers and especially lakes, it can turn them more acidic, and they lose their ability to sustain fishes and other aquatic life. For example, lake trout disappear from lakes in Ontario and the eastern U.S. when the pH dips below about 5.2. Although this low pH does not affect survival of the adult fish, it affects the survival of juveniles.

Acid rain is important in terrestrial systems, too. For example, acid rain can directly affect forests by killing leaves or pine needles, as has happened on some of the higher mountaintops in the Great Smoky Mountains. It can also greatly lower soil pH, which can result in a loss of essential nutrients such as calcium and nitrogen. Low soil calcium results in calcium deficiencies in plants, in the snails that consume the plants, and in the birds that eat the snails, ultimately causing weak eggshells that break before hatching. Decreased soil pH also kills certain soil microorganisms, preventing decomposition and recycling of nitrogen in the soil. Decreases in soil calcium and nitrogen weaken trees and other plants and may make them more susceptible to insect attack.

Acid rain is a common problem in the northeastern U.S. and Scandinavia, where sulfur-rich air drifts over from the Midwest and

(a) Rich flora on alkaline soil **(b) Sparse flora on acidic soil**

Figure 54.16 Species-rich flora of chalk grassland compared with species-poor flora of acid soils. **(a)** At Mount Caburn, in the lime-rich chalk hills of Sussex County, England, there is a much greater variety of plant and animal species than at **(b)** a heathland site elsewhere in England. Heathlands are a product of thousands of years of human clearance of natural forest areas and are characterized by acidic, nutrient-poor soils.

Concept Check: *Why do acidic soils support fewer species of plants and animals than lime-rich soils?*

the industrial areas of Britain, respectively, causing the deposition of highly acidic rain. The problem was particularly acute during the 1960s and 1970s, but decreased manufacturing and the use of low-sulfur coal and the introduction of sulfur-absorbing scrubbers on the smokestacks of coal-burning power plants have somewhat reduced the problem in recent years. Acid rain is clearly a problem with a wide-ranging effect on ecological systems.

54.4 Climate and Its Relationship to Biological Communities

Learning Outcomes:

1. Explain how global temperature differentials drive atmospheric circulation.
2. Explain how both mountains and large bodies of water can change local temperature and precipitation patterns.

Temperature, wind, precipitation, and light are components of **climate**, the prevailing weather pattern in a given region. As we have seen, the distribution and abundance of organisms are influenced by these factors. Therefore, to understand the patterns of abundance of life on Earth, ecologists need to study the global climate. In this section, we examine global climate patterns, focusing on how temperature variation drives atmospheric circulation and how features such as elevation and landmass can alter these patterns.

Atmospheric Circulation Is Driven by Global Temperature Differentials

Substantial differences in temperature occur over the Earth, mainly due to latitudinal variations in the incoming solar radiation. In higher latitudes, such as northern Canada and Russia, the Sun's rays hit the Earth obliquely and are spread out over more of the planet's surface than they are in equatorial areas (**Figure 54.17**). More heat is also lost in the atmosphere of higher latitudes because the Sun's rays travel a greater distance through the atmosphere, allowing more heat to be dissipated by cloud cover. The result is that 40% less solar energy strikes polar latitudes than equatorial areas. Generally, temperatures increase as the amount of solar radiation increases (**Figure 54.18**). However, at the tropics, both cloudiness and rain reduce average temperature, so temperatures do not continue to increase toward the equator.

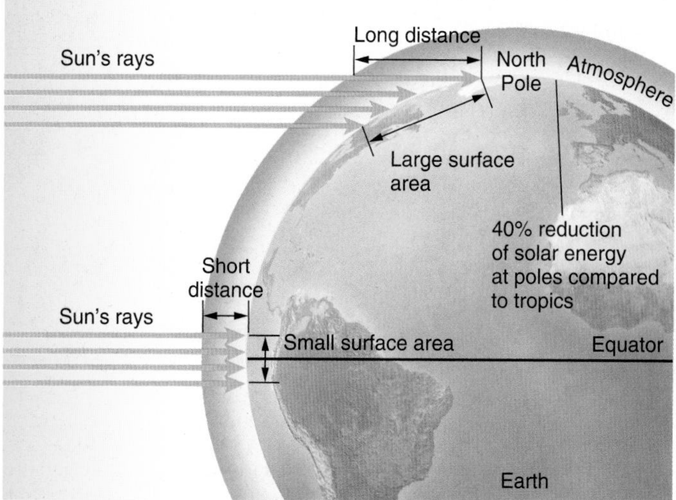

Figure 54.17 The intensity of solar radiation at different latitudes. In polar areas, the Sun's rays strike the Earth at an oblique angle and deliver less energy than at tropical locations. In tropical areas, the energy is concentrated over a smaller surface and travels a shorter distance through the atmosphere.

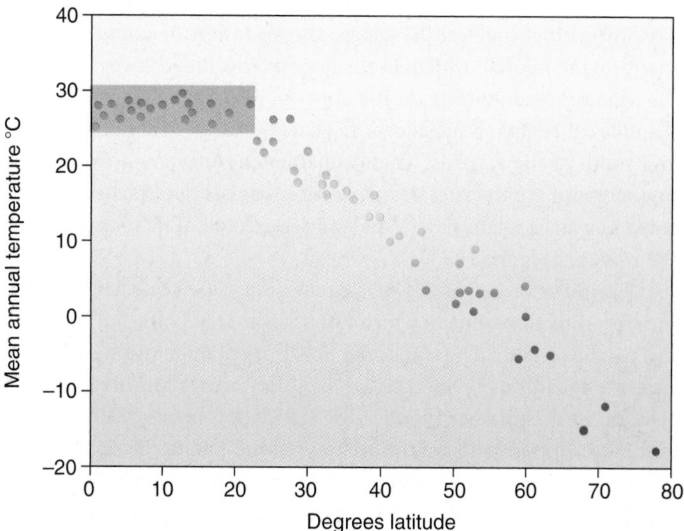

Figure 54.18 Variation of the Earth's temperature. The temperatures shown in this figure were measured at moderately moist continental locations of low elevation.

Concept Check: *Why is there a wide band of similar temperatures at the tropics?*

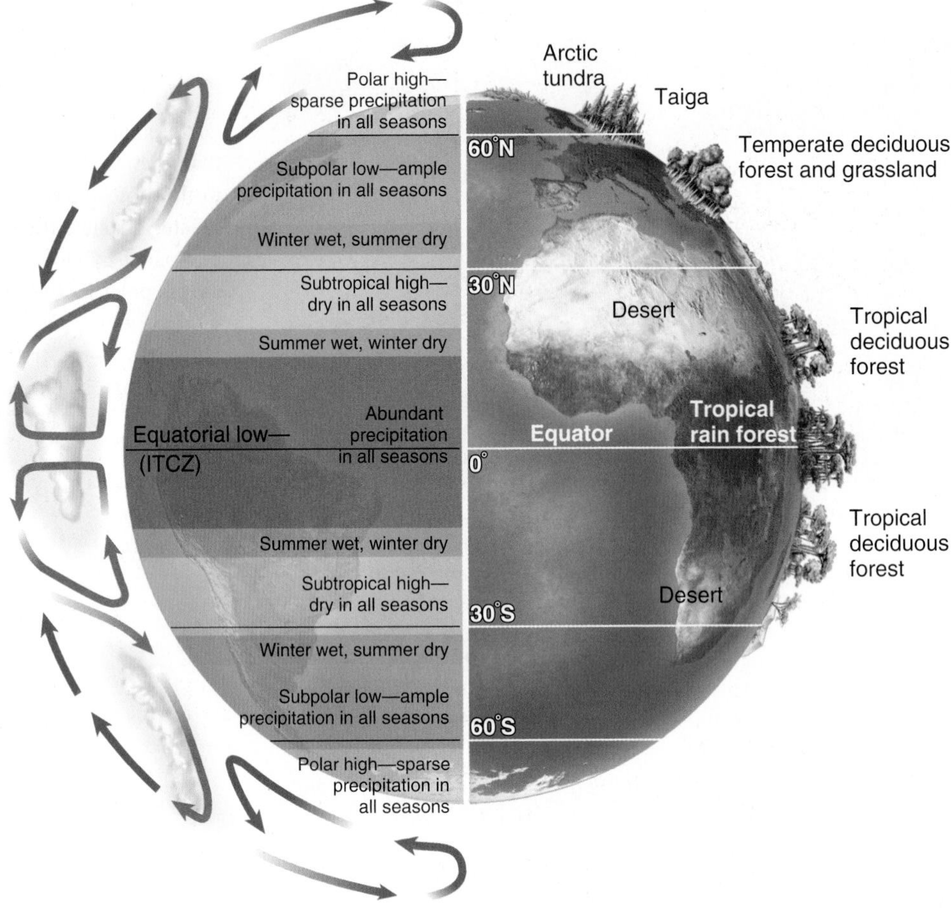

Figure 54.19 Global circulation based on a modified three-cell model. Tropical forests exist mainly in a band around the equator, where it is hot and rainy. At around 30° north and south, the air is hot and dry, and deserts exist. A secondary zone of precipitation exists at around 45° to 55° north and south, where temperate forests are located. The polar regions are generally cold and dry. The term high refers to areas of high pressure resulting from falling air. Lows refer to areas of low pressure resulting from rising air.

Global patterns of atmospheric circulation and precipitation are influenced by solar energy. In 1735, English meteorologist George Hadley made the initial contribution to a model of general atmospheric circulation. In his model, high temperatures at the equator cause the surface equatorial air to heat up and rise vertically into the atmosphere. The vertical rising of the hot air cools the land by convection (look back to Figure 46.12). As the warm air rises away from its source of heat, it cools and becomes less buoyant, but the cool air does not sink back to the surface because of the warm air behind it. The warm air rising near the equator forms towers of cumulus clouds that provide rainfall, which, in turn, maintains the lush vegetation of the equatorial rain forests. As the upper flow in this cell moves toward the poles, it begins to subside, or fall back to Earth, at about 30° north and south of the equator. These **subsidence zones** are areas of high pressure and are the sites of the world's tropical deserts, because the subsiding air is relatively dry, having released all of its moisture over the equator (**Figure 54.19**).

From the center of the subsidence zones, the surface flow splits into two directions, one of which flows toward the poles and the other toward the equator. The equatorial flow from both hemispheres meets near the equator in a region called the intertropical convergence zone (ITCZ). In the three-cell model, the circulation between 30° and 60° latitude, is opposite that of the cell nearest the equator because the net surface flow is poleward. Additional zones of high precipitation occur in this cell, usually between 45° and 55°. In the final circulation cell, at the poles, the air has cooled and descends, but it has little moisture left, explaining why many high-latitude regions are actually desert-like in condition. The distributions of the major biomes discussed in Section 54.5 are largely determined by temperature differences and the wind patterns they generate. Hot, tropical forest blankets the tropics, where rainfall is high. At about 30° latitude, the air cools and descends, but it is without moisture, so the hot deserts occur around that latitude. The middle cell of the circulation model shows us that at about 45° to 55° latitude, the air has warmed and gained moisture, so it ascends, dropping rainfall over the wet, temperate forests of the Pacific Northwest and Western Europe in the Northern Hemisphere and New Zealand and Chile in the Southern Hemisphere.

Elevation and Other Features of a Landmass Can Also Affect Climate

Thus far, we have considered how global temperatures and wind patterns affect climate. The geographic features of a landmass can also have an important effect. For example, the elevation of a region greatly influences its temperature range. On mountains, temperatures decrease with increasing elevation. This decrease is a result of a process known as **adiabatic cooling**, in which increasing elevation leads to a decrease in air pressure. When air is blown across the Earth's surface and up over mountains, it expands because of the reduced pressure. As it expands, it cools at a rate of about 10°C for every 1,000 m in elevation, as long as no water vapor or cloud formation occurs. (Adiabatic cooling is also the principle behind the function of a refrigerator, in which refrigerant gas cools as it expands coming out of the compressor.) A vertical ascent of 600 m produces a temperature change roughly equivalent to that brought about by an increase in latitude of 1,000 km. This explains why mountaintop vegetation, even in tropical areas, can have the characteristics of a colder biome.

Mountains can also influence patterns of precipitation. For example, when warm, moist air encounters the windward side of a mountain, it flows upward and cools, releasing precipitation in the form of rain or snow. On the side of the mountain sheltered from the wind (the leeward side), drier air descends, producing what is

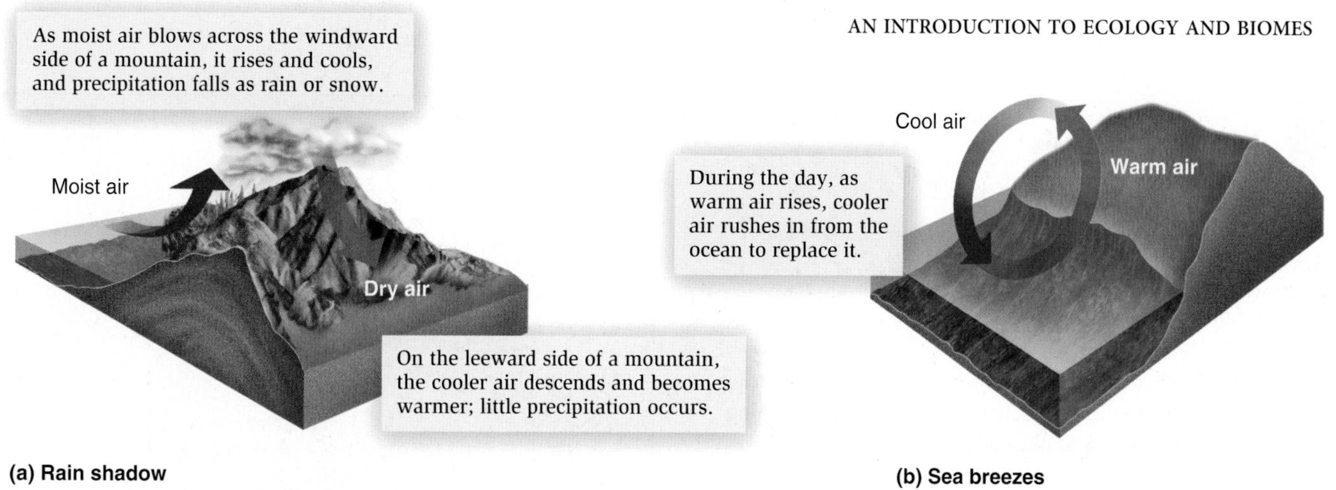

As moist air blows across the windward side of a mountain, it rises and cools, and precipitation falls as rain or snow.
Moist air
Dry air
On the leeward side of a mountain, the cooler air descends and becomes warmer; little precipitation occurs.
Cool air
Warm air
During the day, as warm air rises, cooler air rushes in from the ocean to replace it.
(a) Rain shadow
(b) Sea breezes

Figure 54.20 **The influence of elevation and proximity to water on climate.**

called a **rain shadow**, an area where precipitation is noticeably less (**Figure 54.20a**). In this way, the western side of the Cascade Range in Washington State receives more than 500 cm of annual precipitation, whereas the eastern side receives only 50 cm.

The proximity of a landmass to a large body of water can affect climate because land heats and cools more quickly than the sea does. Recall from Chapter 2 that water has a very high specific heat—the amount of energy required to raise the temperature of 1 gram of a substance by 1°C. The specific heat of the land is much lower than that of the water, allowing the land to warm quicker than water. During the day, the warmed air rises and cooler air flows in to replace it. This pattern creates the familiar onshore sea breezes in coastal areas (**Figure 54.20b**). At night, the land cools quicker than the sea, and so the pattern is reversed, creating offshore breezes. The sea, therefore, has a moderating effect on the temperatures of coastal regions and especially islands. The climates of coastal regions may differ markedly from those of their climatic zones. Many never experience frost, and fog is often evident. Thus, along coastal areas, different vegetation patterns may occur from those in areas farther inland. In fact, some areas of the U.S. would be deserts were it not for the warm water of the sea and the moisture-laden clouds that form above them.

Together with the rotation of the Earth, winds also create ocean currents. The major ocean currents act as "pinwheels" between continents, running clockwise in the ocean basins of the Northern Hemisphere and counterclockwise in those of the Southern Hemisphere (**Figure 54.21**). The Gulf Stream, equivalent in flow to 50 times the world's major rivers combined, brings warm water from the Caribbean and the U.S. coasts across the Atlantic Ocean, where it combines with the North Atlantic Drift to moderate the climate of Europe. The Humboldt Current brings cool conditions to the western coast of South America and almost to the equator, and the California Current brings cooler climate to the Hawaiian Islands.

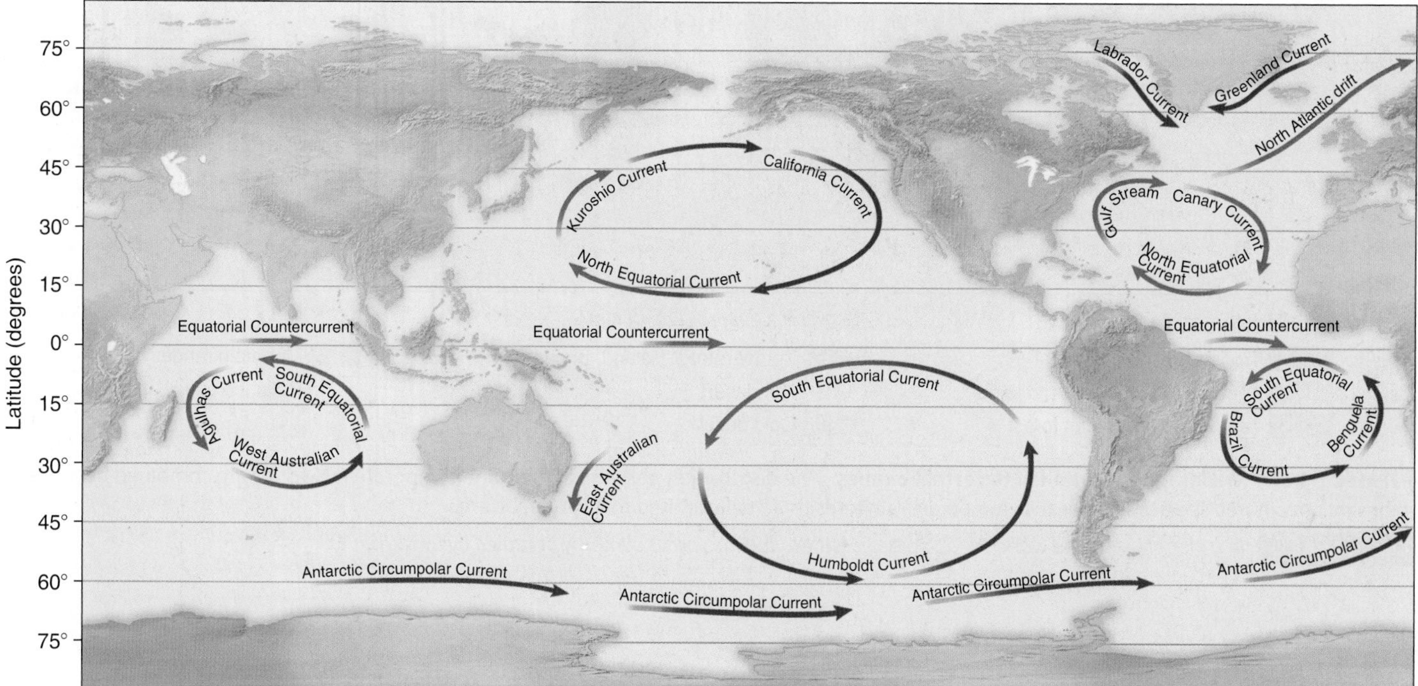

Figure 54.21 **Ocean currents of the world.** The red arrows represent warm water; the blue arrows, cold water.

54.5 Major Biomes

Learning Outcomes:

1. List 10 major terrestrial biomes, noting the temperature and rainfall pattern of each.
2. Discuss how changes in water salinity, oxygen content, depth, and current affect aquatic biomes.

Differences in climate on Earth help to define its different terrestrial biomes. Many types of classification schemes are used for mapping the geographic extent of terrestrial biomes, but one of the most useful was developed by the American ecologist Robert Whittaker, who classified biomes according to the physical factors of average annual precipitation and temperature (Figure 54.22). In this scheme, we recognize 10 terrestrial biomes (Figure 54.23). Aquatic biomes are generally differentiated by water salinity, current strength, water depth, oxygen content, and light availability. In this section, we explore the main characteristics of Earth's major terrestrial and aquatic biomes.

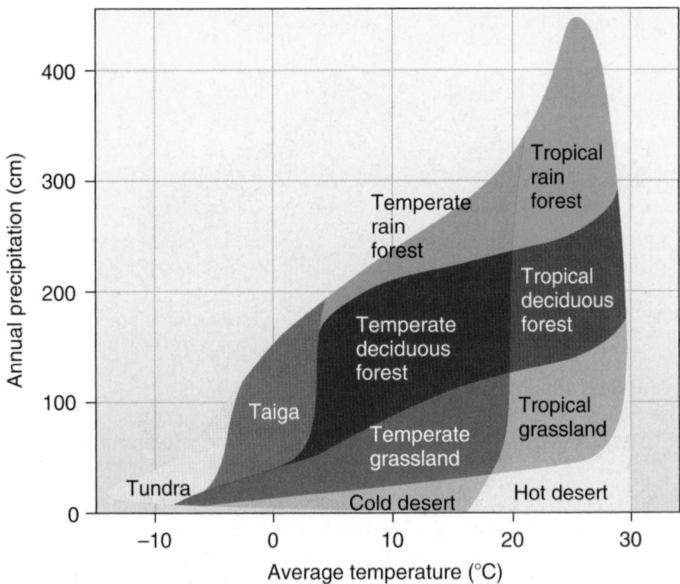

Figure 54.22 **The relationship between the world's terrestrial biome types and temperature and precipitation patterns.**

Concept Check: *What other factors may influence biome types?*

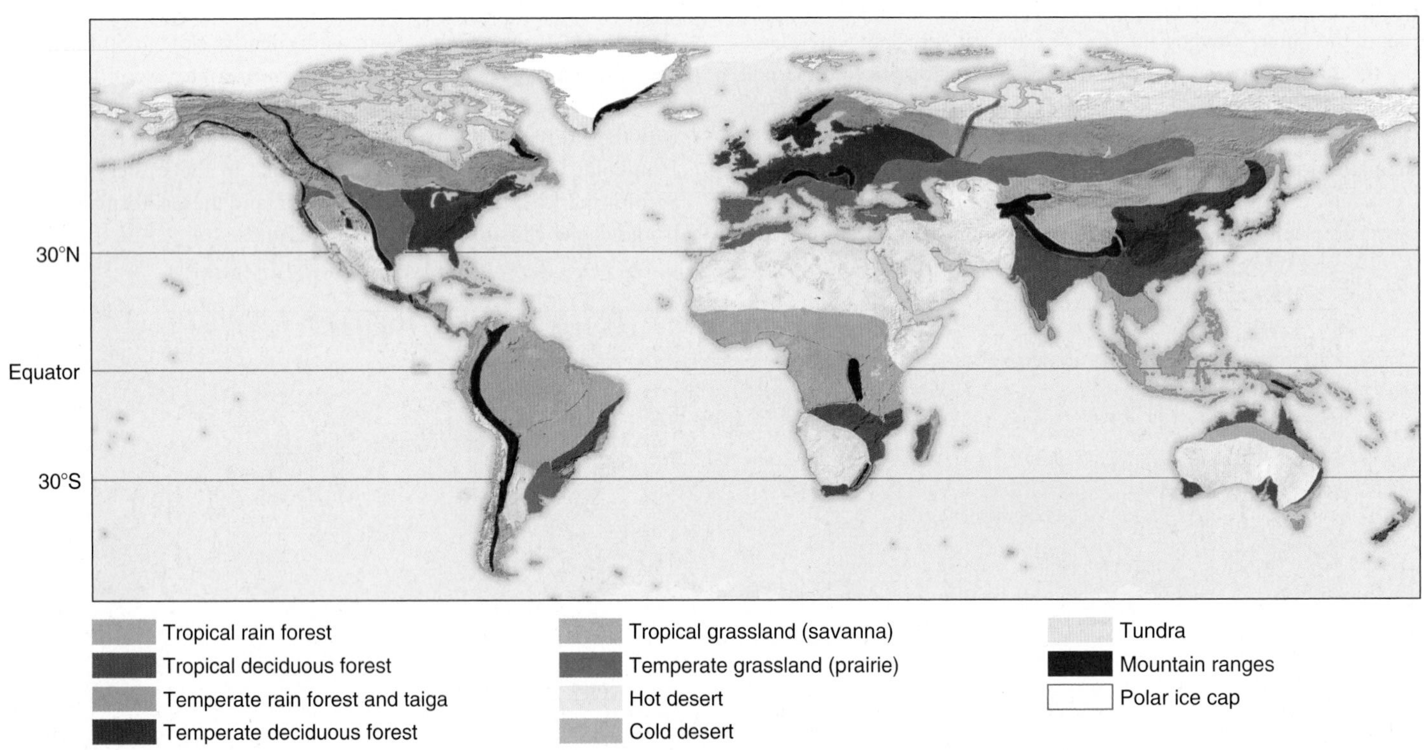

Figure 54.23 **Geographic location of terrestrial biomes.** The distribution patterns of taiga and temperate rain forest are combined because of their similarity in tree species and because temperate rain forest is actually limited to a very small area.

Concept Check: *In a globally warmed world, what biome might expand into areas currently occupied by tundra?*

Figure 54.24a–j illustrates the 10 terrestrial biomes and identifies their main characteristics. Although broad terrestrial biomes are a useful way of defining the main types of communities on Earth, ecologists acknowledge that not all communities fit neatly into 1 of these 10 major biome types. Also, one biome type often grades into another, as seen on mountain ranges (Figure 54.24k). Soil conditions can also influence biome type. In California, serpentine soils, which are dry and nutrient-poor, support only sparse vegetation. In the eastern U.S., most of New Jersey's coastal plain, called the Pine Barrens, consists of sandy, nutrient-poor soil that cannot support the surrounding deciduous forest and instead contains grasses and low shrubs growing among open stands of pygmy pitch pine and oak trees.

Tropical Rain Forest — Figure 54.24a

Tropical rain forest in Fiji

Physical Environment: Rainfall exceeds 230 cm per year, and the temperature is hot year round, averaging 25–29°C. Soils are often shallow and nutrient-poor.

Location: This biome is found in equatorial regions. Tropical forests cover much of northern South America, Central America, western and central Africa, Southeast Asia, and various islands in the Indian and Pacific oceans.

Plant Life: The numbers of plant species found in tropical forests can be staggering, often reaching as many as 100 tree species per square kilometer. Leaves often narrow to "drip-tips" at the apex so that rainwater drains quickly. Many trees have large buttresses that help support their shallow root systems. Little light penetrates the **canopy**, the uppermost layer of tree foliage, and the ground cover is often sparse. Vines and epiphytes, plants that live perched on trees and are not rooted in the ground, are common.

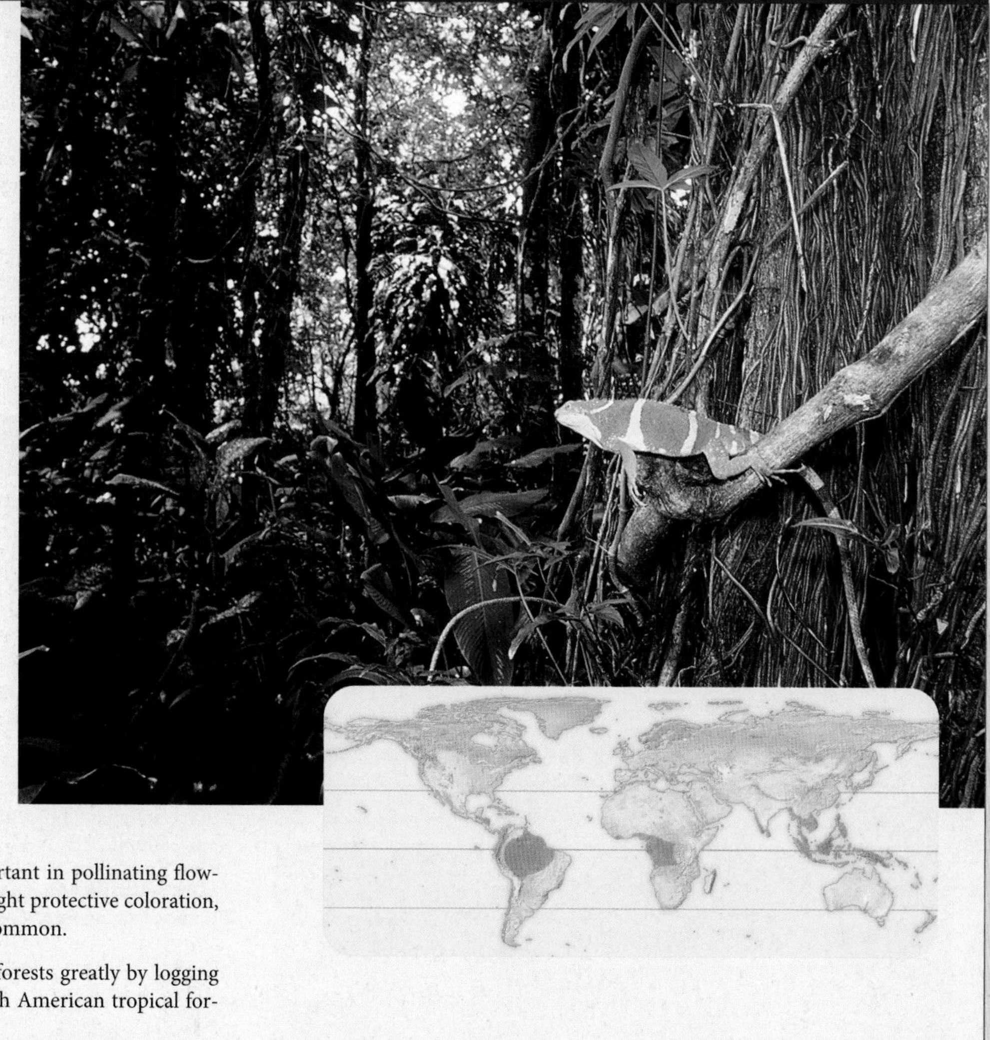

Animal Life: Animal life in the tropical rain forests is diverse; insects, reptiles, amphibians, and mammals are well represented. Large mammals, however, are not common. Because many of the plant species are widely scattered in tropical forests, plants do not typically rely on wind for pollination or to disperse their seed. Instead, animals are important in pollinating flowers and dispersing fruits and seeds. Mimicry and bright protective coloration, warning of bad taste or the existence of toxins, are common.

Effects of Humans: Humans are affecting tropical forests greatly by logging and by clearing the land for agriculture. Many South American tropical forests are cleared to create grasslands for cattle.

Terrestrial Biomes (continued)

Tropical Deciduous Forest

Figure 54.24b

Tropical deciduous forest in Bandhavgarh National Park, India

Physical Environment: Rainfall is substantial, at around 130–280 cm a year, and temperatures are hot year round, averaging 25–39°C. This biome experiences a distinct dry season that is often 2 to 3 months or longer. Soil water shortages can occur in the dry season.

Location: This biome exists in equatorial regions where rainfall is more seasonal than in tropical rain forests. Much of India consists of tropical deciduous forest, containing teak trees. Brazil, Thailand, and Mexico also contain tropical deciduous forest. At the wet edges of this biome, it may grade into tropical rain forests; at the dry edges, it may grade into tropical grasslands or savannas.

Plant Life: Because of the biome's distinct dry season, many of the trees in tropical deciduous forests shed their leaves, just as they do in temperate forests, and an understory of herbs and grasses may grow during this time. Indeed, because the canopy is often more open than in the tropical rain forest and more sunlight reaches the ground, a denser closed forest—what we might think of as a "tropical jungle"—exists at the forest floor. Where the dry season is 6 to 7 months long, tropical deciduous forests may contain shorter, thorny plants such as acacia trees, whose thorns deter moisture-seeking animals, and the forest is referred to as a tropical thorn forest.

Animal Life: The diversity of animal life is high, and species such as monkeys, antelopes, wild pigs, and tigers are present. However, as with plant diversity, animal diversity is less than that of tropical rain forests. Tropical thorn forests may contain more browsing mammals; hence, the development of plant thorns as a defense.

Effects of Humans: The soil of tropical deciduous forests is more fertile than that of tropical rain forests. Land is increasingly being logged and cleared for agriculture and a growing human population.

Temperate Rain Forest

Figure 54.24c

Hoh Rain Forest in Olympic National Park, Washington

Physical Environment: Rainfall is abundant, usually exceeding 200 cm a year. The condensation of water from dense coastal fogs augments the normal rainfall. Temperatures seldom drop below freezing in the winter, and summer temperatures rarely exceed 27°C.

Location: The area of this biome type is small, consisting of a thin strip along the northwest coast of North America from northern California through Washington State, British Columbia, and into southeastern Alaska (where it is called tongass). It also exists in southwestern South America along the Chilean coast. Indeed, it is found only in coastal locales because of the moderating influence of the ocean on air temperature.

Plant Life: The dominant vegetation type, especially in North America, consists of large evergreen trees such as western hemlock, Douglas fir, and Sitka spruce. The high moisture content allows epiphytes to thrive. Cool temperatures slow the activity of decomposers, so the litter layer is thick and spongy.

Animal Life: In North America, the temperate rain forest is rich in species such as mule deer, elk, squirrels, and numerous birds such as jays and nuthatches. Because of the abundant moisture and moderate temperatures, reptiles and amphibians are also common.

Effects of Humans: This biome is a prolific producer of wood and supplies much timber; logging threatens the survival of the forest in some areas.

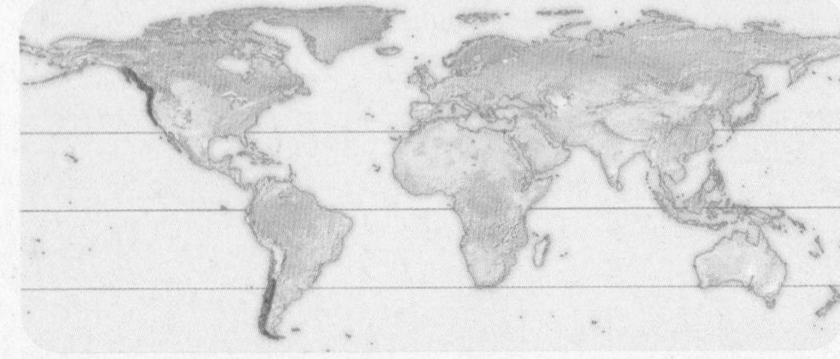

Temperate Deciduous Forest

Figure 54.24d

Temperate deciduous forest in Minnesota

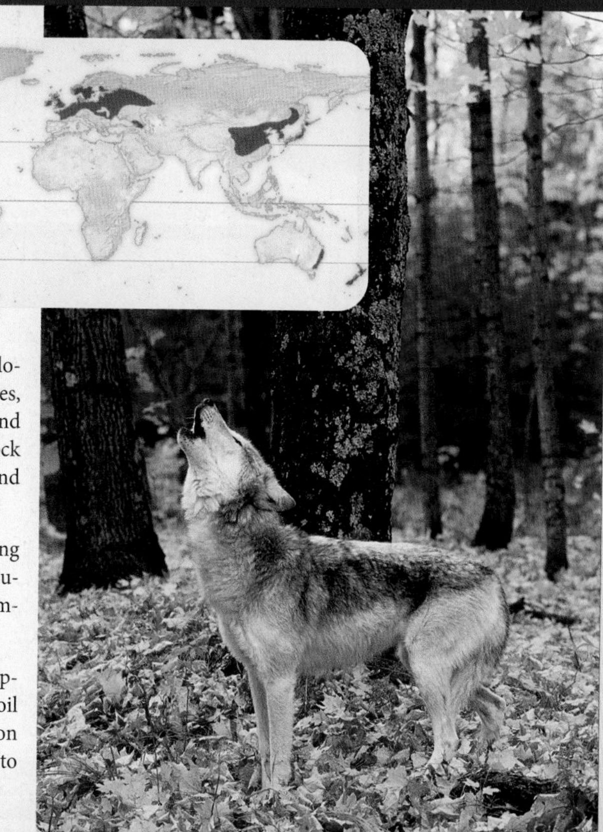

Physical Environment: Annual rainfall is generally between 75 and 200 cm. Temperatures fall below freezing each winter but not usually below −12°C.

Location: Large tracts of temperate deciduous forest are evident in the eastern U.S., Western Europe, and eastern Asia. In the Southern Hemisphere, eucalyptus forests occur in Australia, and stands of southern beech are found in southern South America, New Zealand, and Australia.

Plant Life: Species diversity is much lower in temperate deciduous forests than in the tropical forests, with about only three to four tree species per square kilometer, and several tree genera may be dominant in a given locality—for example, oaks, hickories, and maples are usually dominant in the eastern U.S. Commonly, leaves are shed in the fall and reappear in the spring. Many herbaceous plants flower in spring before the trees leaf out and block the light. Even in the summer, though, the forest is not as dense as in tropical forests, so ground cover is abundant.

Animal Life: Animals are adapted to the vagaries of the climate; many mammals hibernate during the cold months, birds migrate, and insects enter diapause, a condition of dormancy passed usually as a pupa. Reptiles, which depend on solar radiation for heat, are relatively uncommon. Mammals include squirrels, wolves, bobcats, foxes, bears, and mountain lions.

Effects of Humans: Logging has eliminated much of the temperate deciduous forest from populated portions of Europe and North America. Because the annual leaf drop promotes high soil nutrient levels, soils are rich and easily converted to agriculture. Much of the human population lives in the temperate deciduous forest, and both agriculture and development are the threats to the biome.

Temperate Coniferous Forest (Taiga)

Figure 54.24e

Temperate coniferous forest in Canada

Physical Environment: Precipitation is generally between 30 and 100 cm and often occurs in the form of snow. Temperatures are very cold, often below freezing for long periods of time.

Location: The biome of coniferous forests, known commonly by its Russian name, taiga, lies north of the temperate-zone forests and grasslands. Vast tracts of taiga exist in North America and Russia. In the Southern Hemisphere, little land area occurs at latitudes at which one would expect extensive taiga to exist.

Plant Life: Most of the trees are evergreens or conifers with tough needles, hence the similarity of taiga to temperate rain forest. In this biome, spruces, firs, and pines generally dominate, and the number of tree species is relatively low. Many of the conifers have conical shapes to reduce bough breakage from heavy loads of snow. As in tropical forests, the understory is sparse because the dense year-round canopies prevent sunlight from penetrating. Soils are poor because the fallen needles decay so slowly in the cold temperatures that a layer of needles builds up and acidifies the soil, reducing the numbers of understory species.

Animal Life: Reptiles and amphibians are rare because of the low temperatures. Insects are strongly periodic but may often reach outbreak proportions in times of warm temperatures. Mammals that inhabit this biome, such as bears, lynxes, moose, beavers, and squirrels, are heavily furred.

Effects of Humans: Humans have not extensively settled these areas, but they have been quite heavily logged. Exploration and development of oil and natural gas reserves are also a threat.

Tropical Grassland (Savanna) — Figure 54.24f

Tropical grassland of the Masai Mara Game Reserve in Kenya

Physical Environment: This biome includes hot, tropical areas, with a low or seasonal rainfall between 50 and 130 cm per year. There is often an extensive dry season. Temperatures average 24–29°C.

Location: Extensive savannas occur in Africa, South America, and northern Australia.

Plant Life: Wide expanses of grasses dominate savannas, but occasional thorny trees, such as acacias, may occur. Fire is prevalent in this biome, so most plants have well-developed root systems that enable them to resprout quickly after a fire.

Animal Life: The world's greatest assemblages of large mammals occur in the savanna biome. Herds of antelope, zebra, and wildebeest are found, together with their associated predators: cheetah, lion, leopard, and hyena. Termite mounds dot the landscape in some areas. The extensive herbivory of large grazers, together with frequent fires, may help maintain savannas and prevent their development into forests.

Effects of Humans: Savanna soils are often poor because the occasional rain leaches nutrients out. Nevertheless, conversion of this biome to agricultural land is rampant, especially in Africa. Overstocking of land for pasturage of domestic animals can greatly reduce grass coverage through overgrazing, turning the area desert-like. This process is known as **desertification**.

Temperate Grassland (Prairie) — Figure 54.24g

Temperate grassland in Wyoming State

Physical Environment: Annual rainfall is generally between 25 and 100 cm, too low to support a forest but higher than that in deserts. Temperatures in the winter sometimes fall below −10°C, whereas summers may be very hot, approaching 30°C.

Location: Temperate grasslands include the prairies of North America, the steppes of Russia, the pampas of Argentina, and the veldt of South Africa. In addition to the limiting amounts of rain, fire and grazing animals may also prevent the establishment of trees in the temperate grasslands. Where temperatures rarely fall below freezing and most of the rain falls in the winter, chaparral, a fire-adapted community featuring shrubs and small trees, occurs. Chaparral is seen at around 30° latitude, where cool ocean waters moderate the climate, as along the coasts of California, South Africa, Chile, and southwest Australia and in countries surrounding the Mediterranean Sea. Some ecologists recognize chaparral as a distinct biome type.

Plant Life: From east to west in North America and from north to south in Asia, grasslands show differentiation along moisture gradients. In Illinois, with an annual rainfall of 80 cm, tall prairie grasses such as big bluestem and switchgrass grow to about 2 m high. Along the eastern base of the Rockies, 1,300 km to the west, where rainfall is only 40 cm, prairie grasses such as buffalo grass and blue grama rarely exceed 0.5 m in height. Similar gradients occur in South Africa and Argentina.

Animal Life: Where the grasslands remain, large mammals are the most prominent members of the fauna: bison and pronghorn in North America, wild horses in Eurasia, and large kangaroos in Australia. Burrowing animals such as North American gophers and African mole rats are also common.

Effects of Humans: Prairie soil is among the richest in the world, having 12 times the humus layer of a typical forest soil. Worldwide, most prairies have been converted to agriculture, and original grassland habitats are among the rarest biomes in the world.

Hot Desert

Figure 54.24h

The Namib Desert, Namibia

Physical Environment: Rainfall is generally less than 30 cm per year. Temperatures are variable, from below freezing at night to as much as 50°C in the day.

Location: Hot deserts are found around latitudes of 30° north and south. Prominent deserts include the Sahara of North Africa, the Kalahari and Namib of southern Africa, the Atacama of Chile, the Sonoran of northern Mexico and the southwest U.S., and the Simpson of Australia.

Plant Life: Three forms of plant life are adapted to deserts: annuals, succulents, and desert shrubs. Annuals circumvent drought by growing only when there is rain. Succulents, such as the saguaro cactus and other barrel cacti of the southwestern deserts, store water. Desert shrubs, such as the spraylike ocotillo, have short trunks, numerous branches, and small, thick leaves that can be shed in prolonged dry periods. In many plants, spines or volatile chemical compounds serve as a defense against water-seeking herbivores.

Animal Life: To conserve water, desert plants produce many small seeds, and animals that eat those seeds, such as ants, birds, and rodents, are common. Reptiles are numerous, because high temperatures permit these ectothermic animals to maintain a warm body temperature. Lizards and snakes are important predators of seed-eating mammals.

Effects of Humans: Ambitious irrigation schemes and the prolific use of underground water have allowed humans to develop deserts and grow crops there. Salinization, a buildup in the salt content of the soil that results from irrigation in areas of low rainfall, is prevalent. Off-road vehicles can disturb the fragile desert communities.

Cold Desert

Figure 54.24i

The Gobi Desert of Mongolia

Physical Environment: Precipitation is less than 25 cm a year and is often in the form of snow. Rainfall usually comes in the spring. In the daytime, temperatures can be high in the summer, 21–26°C, but average around freezing, –2 to 4°C, in the winter.

Location: Cold deserts are found in dry regions at middle to high latitudes, especially in the interiors of continents and in the rain shadows of mountains. Cold deserts are found in North America (the Great Basin Desert), in eastern Argentina (the Patagonian Desert), and in central Asia (the Gobi Desert).

Plant Life: Cold deserts are relatively poor in terms of numbers of plant species. Most plants are small in stature, being only between 15 and 120 cm tall. Many species are deciduous and spiny. The Great Basin Desert in Nevada, Utah, and bordering states is a cold desert dominated by sagebrush.

Animal Life: As in hot deserts, large numbers of plants produce small seeds on which numerous ants, birds, and rodents feed. Many species live in burrows to escape the cold. In the Great Basin Desert, pocket mice, jackrabbits, kit foxes, and coyotes are common.

Effects of Humans: Agriculture is hampered because of low temperatures and low rainfall, and human populations are not extensive. If the top layer of soil is disturbed by human intrusions such as by off-road vehicles, erosion occurs rapidly and even less vegetation can exist.

Tundra

Figure 54.24j

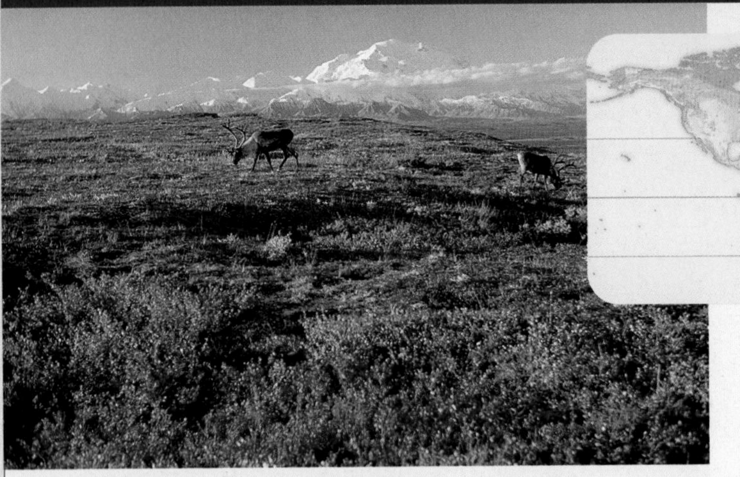

Denali National Park in Alaska

Physical Environment: Precipitation is generally less than 25 cm per year and is often locked up as snow and unavailable for plants. Deeper water can be locked away for a large part of the year in **permafrost**, a layer of permanently frozen soil. The growing season is short, only 50–60 days. Summer temperatures are only 3–12°C, and even during the long summer days, the ground thaws to less than 1 m in depth. Midwinter temperatures average −32°C.

Location: Tundra (from the Finnish *tunturia*, meaning treeless plain) exists mainly in the Northern Hemisphere, north of temperate coniferous forest, because there is very little land area in the Southern Hemisphere at the latitude where tundra would occur.

Plant Life: With so little available water, trees cannot grow. Vegetation occurs in the form of fragile, slow-growing lichens, mosses, grasses, sedges, and occasional shrubs, which grow close to the ground. Plant diversity is very low. In some places, desert conditions prevail because so little moisture falls.

Animal Life: Animals of the arctic tundra have adapted to the cold by having good insulation. Many birds, especially shorebirds and waterfowl, migrate. The fauna is much richer in summer than in winter. Many insects spend the winter at immature stages of growth, which are more resistant to cold than the adult forms. Larger animals include such herbivores as musk oxen and caribou in North America, called reindeer in Europe and Asia. Smaller animals include hares and lemmings. Common predators include arctic fox, wolves, and snowy owls, and polar bears near the coast.

Effects of Humans: Though this area is sparsely populated, mineral extraction, especially of oil, has the potential to significantly affect this biome. Ecosystem recovery from such damage would be very slow.

Mountain Ranges

Figure 54.24k

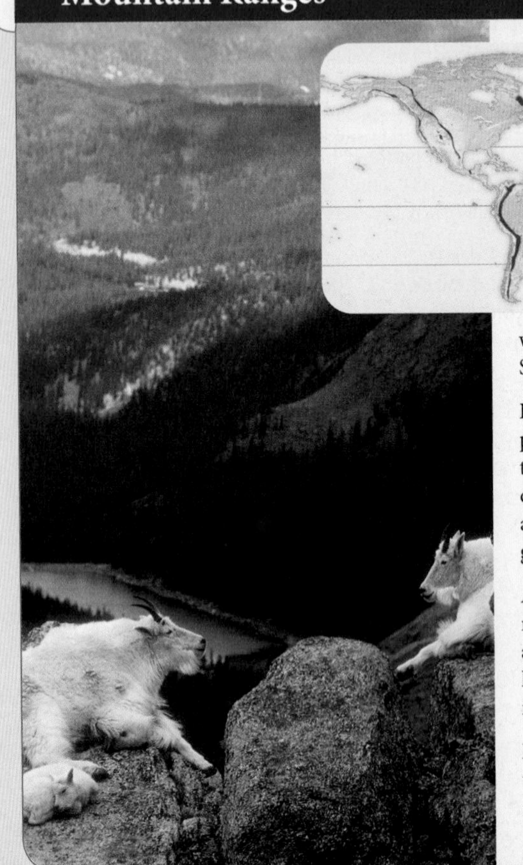

Rocky Mountains of Colorado

Physical Environment: Mountain ranges must be viewed differently from other biomes. On mountains, temperature decreases with increasing elevation through adiabatic cooling, as discussed previously. Thus, precipitation and temperature may change dramatically, depending on elevation and whether the mountainside is on the windward or leeward side.

Location: Mountain ranges exist in many areas of the world, but among the largest are the Himalayas in Asia, the Rockies in North America, and the Andes in South America.

Plant Life: A variety of biomes can be found on a single mountain range. Biome type may change from temperate forest through taiga and into tundra on an elevation gradient in the Rocky Mountains, and even from tropical forest to tundra on the highest peaks of the Andes in tropical South America. In tropical regions, daylight averages 12 hours per day throughout the year. Instead of a period of intense productivity, seen in arctic tundra, vegetation in the tropical alpine tundra exhibits slow but steady rates of photosynthesis and growth all year.

Animal Life: The animals of this biome are as varied as the number of habitats they contain. Generally, more species of plants and animals are found at lower elevations than at higher ones. At higher elevations, animals such as bighorn sheep and mountain goats climb the craggy slopes and have skidproof pads on their hooves. Birds of prey, such as eagles, are frequent predators of the furry rodents found at higher elevations, including guinea pigs and marmots.

Effects of Humans: Logging and agriculture at lower elevations can cause habitat degradation. Because of the steep slopes, mountain soils are often well drained, thin, and especially susceptible to erosion following agriculture.

Aquatic Biomes Consist of Marine and Freshwater Regions

Within aquatic environments, several different biome types are also recognized, including marine aquatic biomes (intertidal zone, coral reef, and open ocean) and freshwater habitats (lakes, rivers, and wetlands). These biomes are distinguished primarily by differences in salinity, oxygen content, depth, current strength, and availability of light (**Figures 54.25a–f**). Freshwater habitats are traditionally divided into **lentic**, or standing-water habitats (from the Latin *lenis*, meaning calm), and **lotic**, or running-water habitats (from the Latin *lotus*, meaning washed).

Intertidal Zone

Figure 54.25a

Olympic Coast National Marine Sanctuary in Washington State

Physical Environment: The **intertidal zone**, the area where the land meets the sea, is alternately submerged and exposed by the daily cycle of tides. The resident organisms are subject to huge daily variations in temperature, light intensity, and availability of seawater.

Location: Throughout the world, the area where the land meets the sea consists of sandy shore, mudflats, or rocky shore.

Plant Life: Plant life may be quite limited because the sand or mud is constantly shifted by the tide. Mangroves may colonize mudflats in tropical areas, and salt marsh grasses may colonize mudflats in temperate locations. On the rocky shore, green algae and seaweeds predominate.

Animal Life: Animal life may be quite diverse. On the rocky shore, sea anemones, snails, hermit crabs, and small fishes live in tide pools. On the rock face, there may be a variety of limpets, mussels, sea stars, sea urchins, snails, sponges, tube worms, whelks, isopods, and chitons. At low tides, organisms may be dry and vulnerable to predation by a variety of animals, including birds and mammals. High tides bring predatory fishes. Sandy or muddy shores may contain burrowing marine worms, crabs, and small isopods.

Effects of Humans: Human development has greatly reduced the beach area available to shorebirds and breeding turtles. Oil spills have greatly affected some rocky intertidal areas.

Coral Reef

Figure 54.25b

Caribbean coral reef

Physical Environment: Corals need warm water of at least 20°C but less than 30°C (refer back to Figure 54.6). They are also limited to the photic zone, where light penetrates and allows photosynthesis to occur. Sunlight is important because many corals harbor symbiotic algae, or dinoflagellates, that contribute nutrients to the animals and that require light to live.

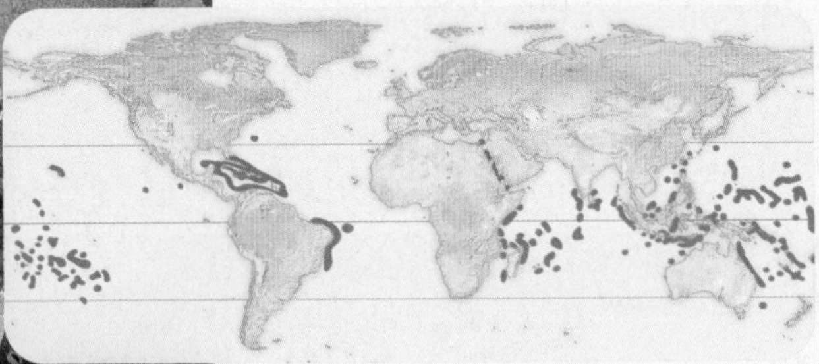

Location: **Coral reefs** exist in warm tropical waters where there are solid substrates for attachment and water clarity is good. The largest coral reef in the world is the Great Barrier Reef off the Australian coastline, but other coral reefs are found in the Atlantic Ocean, the Red Sea, and the Pacific and Indian Oceans.

Plant Life: Dinoflagellate algae live within the coral tissue, and a variety of red and green algae live on the coral reef surface.

Animal Life: An immense variety of microorganisms, invertebrates, and fishes live among the coral, making the coral reef one of the most interesting and species-rich biomes on Earth. Probably 30–40% of all fish species on Earth are found on coral reefs. Prominent herbivores include snails, sea urchins, and fishes. These are consumed by octopuses, sea stars, and carnivorous fishes. Many species are brightly colored, warning predators of their toxic nature.

Effects of Humans: Collectors have removed many corals and fishes for the aquarium trade, and marine pollution threatens water clarity in some areas. Perhaps the greatest threat to coral reefs is from global warming. Water temperatures that are too high (over 30°C) and high pH caused by elevated CO_2 levels both contribute to coral bleaching.

The Open Ocean

Figure 54.25c

Manta ray in the open ocean

Physical Environment: In the open ocean, sometimes called the **pelagic zone**, water depth averages 4,000 m. Nutrient concentrations are typically low, though the waters may be periodically enriched by ocean **upwelling**, the circulation of cold, mineral-rich nutrients from deeper water to the surface. Pelagic waters are mostly cold, only warming near the surface.

Location: Across the globe, covering 70% of the Earth's surface.

Plant Life: In the photic zone, many microscopic photosynthetic organisms **(phytoplankton)** grow and reproduce while drifting in ocean currents. Phytoplankton account for nearly half the photosynthetic activity on Earth and produce much of the world's oxygen.

Animal Life: Open-ocean organisms include **zooplankton**, minute drifting animal organisms consisting of some worms, copepods (tiny shrimplike creatures), small jellyfish, and small invertebrate and fish larvae that graze on the phytoplankton. The open ocean also includes free-swimming animals collectively called **nekton**, which can swim against the currents to locate food. Nekton includes large squids, fishes, sea turtles, and marine mammals. Only a few of these organisms live at any great depth. In some areas, a unique assemblage of animals is associated with deep-sea hydrothermal vents that spew hot (350°C) water rich in hydrogen sulfide. Large worms and other chemoautotrophic organisms exist together in this dark, oxygen-poor environment (refer back to Figure 22.3).

Effects of Humans: Oil spills and a long history of garbage disposal have polluted the ocean floors of many areas. Overfishing has caused many fish populations to crash, and the whaling industry has greatly reduced the numbers of most species of whales.

Lentic Habitats

Figure 54.25d

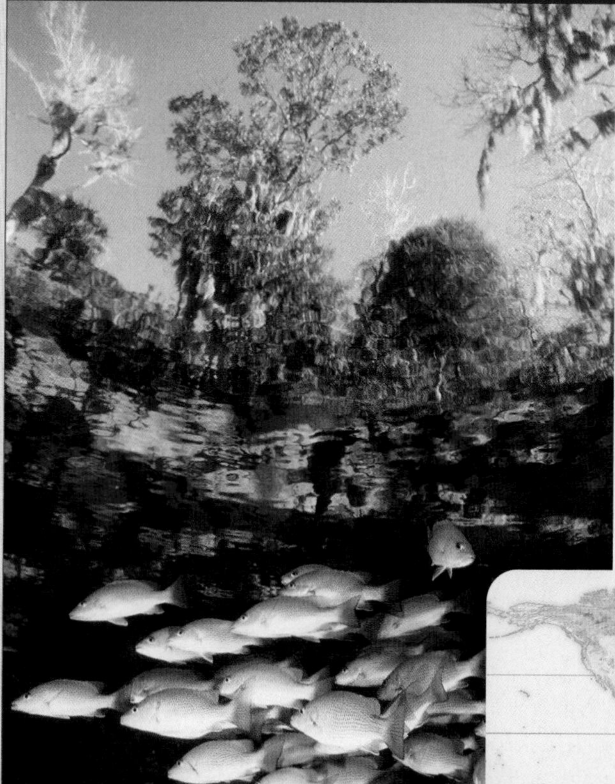

Everglades National Park, Florida

Physical Environment: The lentic habitat consists of still, often deep water. Its physical characteristics depend greatly on the surrounding land, which dictates what nutrients collect in the lake. Young lakes often start off clear and with little plant life. Such lakes are called **oligotrophic**. With age, the lake becomes richer in dissolved nutrients from erosion and runoff from surrounding land, with the result that cyanobacteria and algae spread, reducing the water clarity. Such lakes are termed **eutrophic**. The process of eutrophication occurs naturally but can be sped up by human activities (see Chapter 59).

Location: Throughout all the continents of the world.

Plant Life: In addition to phytoplankton, lentic habitats may have rooted vegetation, which often extends above the water surface (emergent vegetation), such as cattails, plus deeper-dwelling aquatic plants and algae.

Animal Life: Animals include fishes, frogs, turtles, crayfish, insect larvae, and many species of insects. In tropical and subtropical lakes, alligators and crocodiles are common.

Effects of Humans: Agricultural runoff, including fertilizers and sewage, can greatly increase lake nutrient levels and speed up the process of eutrophication, resulting in phytoplankton blooms and fish kills. In some areas, invasive species of invertebrates and fishes are out-competing native species.

Lotic Habitats

Figure 54.25e

Fast-flowing river in the Pacific Northwest

Physical Environment: In lotic habitats, flowing water prevents nutrient accumulations and phytoplankton blooms. The current also mixes water thoroughly, providing a well-aerated habitat of relatively uniform temperature. The current, oxygen level, and clarity are greater at the source of a stream (its headwaters) than in the lower reaches of rivers. Nutrient levels are generally less in headwaters.

Location: On all continents except Antarctica.

Plant Life: In slow-moving streams and rivers, algae and rooted plants may be present; in swifter-moving rivers, leaves from surrounding forests are the primary food source for animals.

Animal Life: Lotic habitats have a fauna completely different from that of lentic waters. Animals are adapted to stay in place despite an often-strong current. Many of the smaller organisms are flat and attach themselves to rocks to avoid being swept away. Others live on the underside of large boulders, where the current is much reduced. Fish such as trout may be present in rivers with cool temperatures, high oxygen, and clear water. In warmer, murkier waters, catfish and carp may be abundant.

Effects of Humans: Animals of lotic systems are not well adapted for low-oxygen environments and thus are particularly susceptible to oxygen-reducing pollutants such as sewage. Dams across rivers have prevented the passage of migratory species such as salmon.

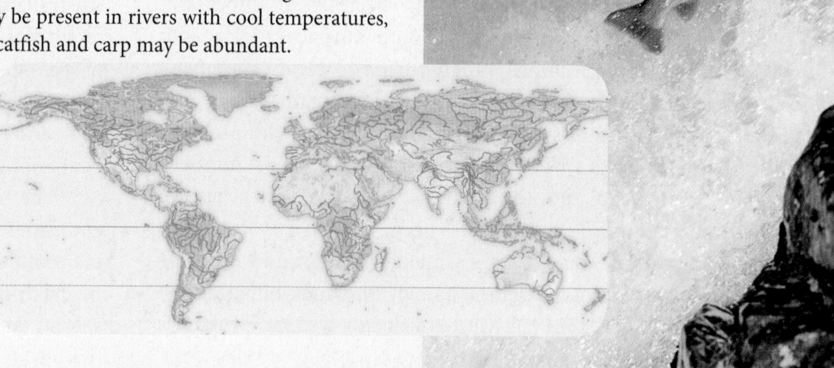

Wetlands

Figure 54.25f

Yellow Waters River, Kakadu National Park, Northern Territory, Australia

Physical Environment: At the margins of both lentic and lotic habitats, wetlands may develop. Wetlands are areas regularly saturated by surface water or groundwater. They range from marshes (treeless areas where herbaceous species predominate), to swamps (wet areas dominated by trees), and bogs (depressions dominated by marshes). Many wetlands are seasonally flooded when rivers overflow their banks or lake levels rise. Some wetlands also develop along estuaries, areas where river water merges with ocean water, and high tides can flood the land. Because of generally high nutrient levels, oxygen levels are fairly low. Temperatures vary substantially with location.

Location: Worldwide, except in Antarctica.

Plant Life: Wetlands are among the most productive, species-rich areas in the world. In North America, floating plants such as lilies and rooted species such as sedges, cattails, cypress, and gum trees predominate.

Animal Life: Most wetlands are rich in animal species. Wetlands are a prime habitat for wading and diving birds. In addition, they are home to a profusion of insects, from mosquitoes to dragonflies. Vertebrate predators include many amphibians, reptiles, otters, and alligators.

Effects of Humans: Long mistakenly regarded as wasteland by humans, many wetlands have been drained and developed for housing and industry. Wetlands play a valuable role in protecting coastal communities from hurricanes, and the loss of wetlands in Louisiana contributed to the severity of effects of hurricane Katrina in 2005.

54.6 Continental Drift and Biogeography

Learning Outcomes:

1. Describe how the distribution of species on Earth may result from continental drift.
2. Explain the concept of biogeographical regions.

A knowledge of biomes does not tell us everything about the distribution of plant and animal life on Earth. An understanding of evolution and geological change over large scales and time periods helps explain some of the patterns we see. **Biogeography**, the study of the geographic distribution of extinct and living species, is an important part of the field of ecology. For example, South America, Africa, and Australia all have similar biomes, ranging from tropical to temperate, yet each continent has distinctive animal life. South America is inhabited by sloths, anteaters, armadillos, and monkeys with prehensile tails. Africa possesses a wide variety of antelopes, zebras, giraffes, lions, baboons, the okapi, and the aardvark. Australia, which has no native placental mammals except bats, is home to a variety of marsupials such as kangaroos, koala bears, Tasmanian devils, and wombats, as well as the egg-laying monotremes, namely, the duck-billed platypus and four species of echidnas. Most continents also have distinct species of plants; for example, eucalyptus trees are native only in Australia. In South American deserts, succulent plants belong to the family Cactaceae, the cacti. In Africa, they belong to the genus *Euphorbia*, the spurges. In North America, the pines, *Pinus* spp., and firs, *Abies* spp., are common, but they do not occur south of the mountains of central Mexico. In contrast, palms are common in South America and do not generally occur north of the mountains of central Mexico, except for several genera in southern California and Florida.

A plausible explanation for these species distributions is that abiotic factors are of paramount importance and that each region supports the fauna best adapted to it. However, the spread of introduced species has proved this explanation incorrect: European rabbits introduced into Australia proliferated rapidly, and eucalyptus from Australia grows well in California. The best explanation is that different floras and faunas are the result of the independent evolution of separate, unconnected populations, which have generated different species in different places. A knowledge of biogeography, continental drift, and evolution is therefore of great importance in understanding contemporary distributions of species.

The relative location of the landmasses on Earth has changed enormously over time as a result of **continental drift**, the slow movement of the Earth's surface plates (refer back to Figure 22.10). Continental drift explains the occurrences of similar living plant and animal species, and fossils, in South America, Africa, India, Antarctica, and

Australia. Many of these fossils were of large land animals, such as the Triassic reptiles *Lystrosaurus* and *Cynognathus*, that could not have easily dispersed among continents, or of plants whose seeds were not likely to be dispersed far by wind, such as the fossil fern *Glossopteris*. Also, the discovery of abundant fossils in Antarctica was proof that this presently frozen land must have been situated much closer to temperate areas in earlier geological times. The current distribution of the essentially flightless bird family, the ratites, in the Southern Hemisphere is also the result of continental drift. The common ancestor of these birds occurred in Gondwana, a supercontinent that included South America, Africa, and Australia. As Gondwana split apart, genera evolved separately in each continent so that today we have ostriches in Africa, emus in Australia, and rheas in South America (refer back to Figure 26.15).

Continental drift is not the only mechanism that creates widely separate populations of closely related species, called disjunct distributions. The distributions of many present day species are relics of once much broader distributions. For example, there are currently four living species of tapir: three in Central and South America and one in Malaysia (**Figure 54.26**). Fossil records reveal a much more widespread distribution over much of Europe, Asia, and North America. The oldest fossils of the ancestral *Paleotapirus* come from Europe, making it likely that this was the center of origin of tapirs. Dispersal of later evolving *Protapirus* resulted in a more widespread distribution. Cooling resulted in the demise of tapirs in all areas except the tropical locations.

Another well-known example of a disjunct distribution is the restricted distribution of monotremes and marsupials. These animals were once plentiful all over North America and Europe. They spread into the rest of the world, including South America and Australia, at the end of the Cretaceous period when, although the continents were separated, land bridges existed between them. Later, placental

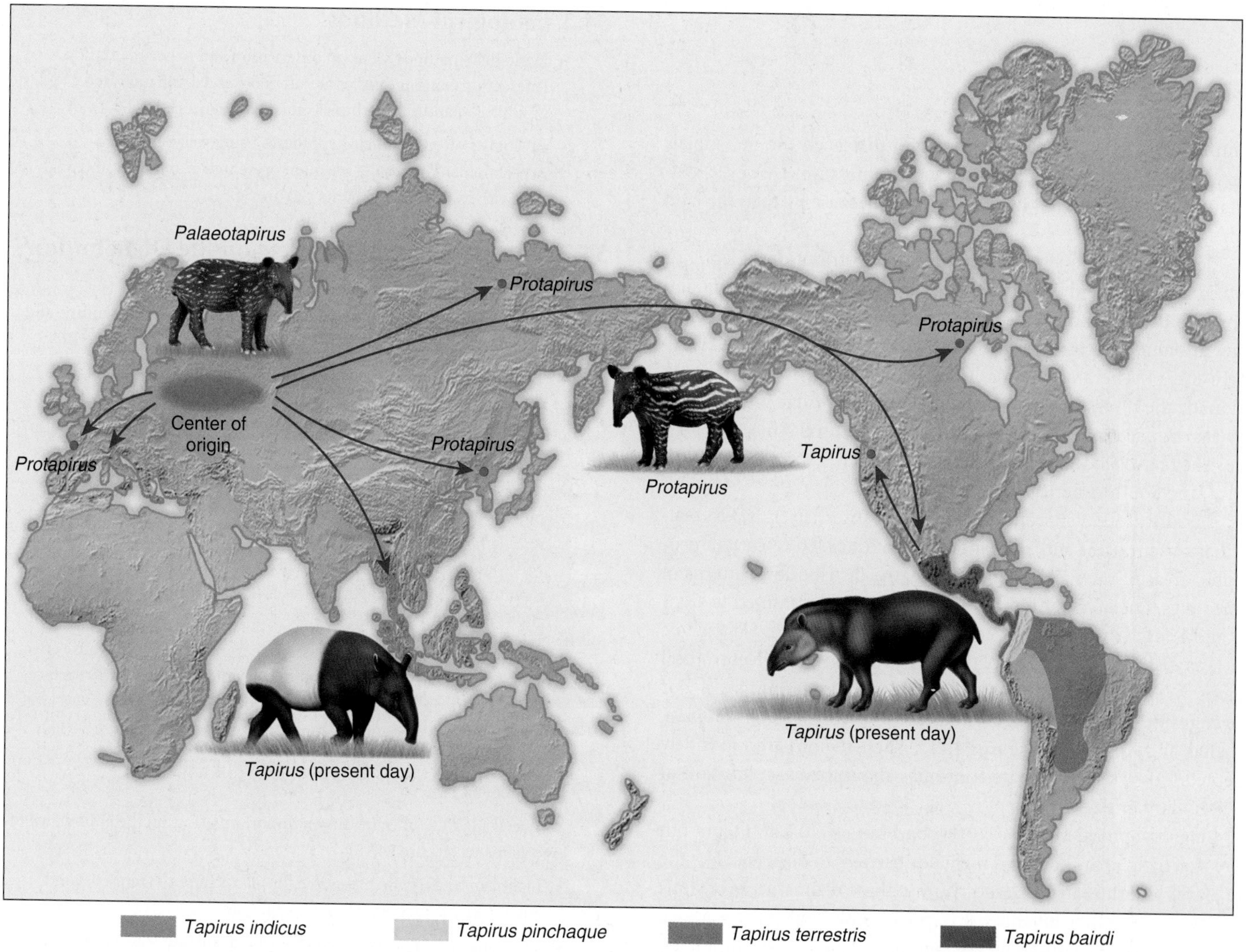

Figure 54.26 Tapir distribution. There are four living tapir species, three in Central and South America and one in Malaysia. Fossil evidence suggests a European origin of the ancestral *Paleotapirus* and a dispersal of later evolving *Protapirus*. A more widespread distribution followed, with tapirs dying out in other regions (marked with a red dot) possibly due to climate change.

BIOLOGY PRINCIPLE All life is related by an evolutionary history. Knowing that all tapir species share a common European ancestor makes it easier to explain their current distribution pattern.

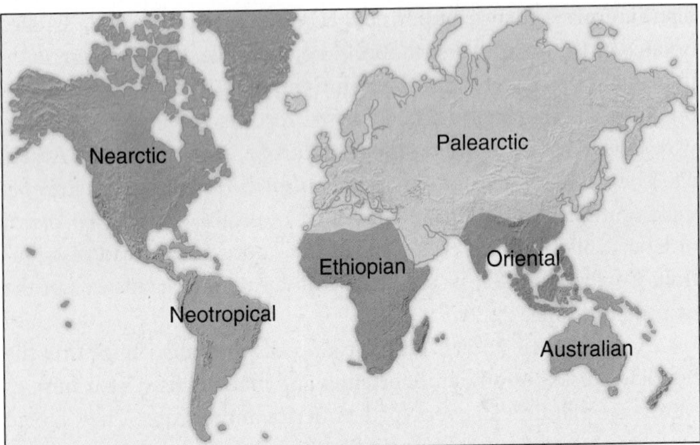

Figure 54.27 **The biogeographic regions proposed by A. R. Wallace.** Note that the borders do not always demarcate continents.

mammals evolved in North America and displaced the marsupials there, apart from a few species such as the opossum. However, placental mammals could not invade Australia because by then the land bridge was broken.

Elephants and camels also have disjunct distributions. Elephants evolved in Africa and subsequently dispersed through Eurasia and across the Bering land bridge from Siberia to North America, where many are found as fossils. They subsequently became extinct everywhere except Africa and India. Camels evolved in North America and made the reverse trek across the Bering land bridge into Eurasia; they also crossed into South America via the Central American isthmus. They have since become extinct everywhere except Asia, North Africa, and South America.

Alfred Russel Wallace was one of the earliest scientists to realize that certain plant and animal taxa were restricted to certain geographic areas of the Earth. For example, the distribution patterns of guinea pigs, anteaters, and many other groups are confined to Central and South America, from central Mexico southward. The whole area was distinct enough for Wallace to proclaim it the Neotropical regions. Wallace went on to divide the world's biota into six major **biogeographic regions**: Nearctic, Palearctic, Neotropical, Ethiopian, Oriental, and Australian (Figure 54.27). These regions are still widely accepted today, though debate continues about the exact location of the boundary lines.

Biogeographical regions correspond largely to continents but more exactly to areas bounded by major barriers to dispersal, like the Himalayas and the Sahara Desert. Within these realms, areas of similar climates are often inhabited by species with similar appearance and habits but from different taxonomic groups. For example, the kangaroo rats of North American deserts, the jerboas of central Asian deserts, and the hopping mice of Australian deserts look similar and occupy similar hot, arid environments, but they arose from different lineages, belonging to the families Heteromyidae, Dipodidae, and Muridae, respectively. As noted in Chapter 23, this phenomenon, called convergent evolution, has led to the emergence of similar species that have evolved from different taxonomic ancestors.

Summary of Key Concepts

54.1 The Scale of Ecology

- Ecologists study the interactions among organisms and between organisms and their environments. The field of ecology can be subdivided into broad areas of organismal, population, community, and ecosystem ecology (Figure 54.1).
- Organismal ecology considers how individuals are adapted to their environment and how the behavior of an individual organism contributes to its survival and reproductive success and the population density of the species. Population ecology explores those factors that influence a population's growth, size, and density. Community ecology studies how populations of species interact and form functional communities. Ecosystem ecology examines the flow of energy and cycling of nutrients among organisms within a community and between organisms and the environment (Figure 54.2).

54.2 Ecological Methods

- Ecological methods focus on observation and experimentation. Interactions among species are often observed and analyzed graphically, and a hypothesis is formed (Figures 54.3, 54.4).
- Ecologists often test their hypotheses using well-replicated experiments. The results are often presented graphically and analyzed via a variety of statistical tests (Figure 54.5).

54.3 The Environment's Effect on the Distribution of Organisms

- Abiotic factors such as temperature, wind, water, light, salinity, and pH can have powerful effects on ecological systems (Table 54.1).
- Temperature exerts important effects on the distribution of organisms because of its effect on biological processes and the inability of many organisms to regulate their body temperature (Figures 54.6, 54.7, 54.8, 54.9).
- The greenhouse effect is the process in which short-wave solar radiation passes through the atmosphere to warm the Earth and is radiated back into the atmosphere as long-wave infrared radiation. Much of this radiation is absorbed by atmospheric gases and radiated back to the Earth's surface, causing its temperature to rise (Figure 54.10, Table 54.2).
- An increase in atmospheric gases is increasing the greenhouse effect, causing global warming—a gradual elevation of the Earth's surface temperature. Ecologists expect that global warming will have a large effect on the distribution of the world's organisms (Figure 54.11).
- Wind can amplify the effects of temperature and modify wave action (Figure 54.12).
- The availability of water has an important effect on the abundance of organisms (Figure 54.13).
- Light can be a limiting resource for plants in both terrestrial and aquatic environments (Figure 54.14).
- The concentration of salts and the pH of soil and water can limit the distribution of organisms (Figures 54.15, 54.16).

54.4 Climate and Its Relationship to Biological Communities

- Global temperature differentials are caused by variations in incoming solar radiation and patterns of atmospheric circulation (Figures 54.17, 54.18, 54.19).

- Elevation and the proximity between a landmass and large bodies of water can similarly affect climate (Figures 54.20, 54.21).

54.5 Major Biomes

- Climate has a large effect on biomes, major types of habitats characterized by distinctive plant and animal life (Figures 54.22, 54.23).
- Terrestrial biomes are generally named for their climate and vegetation type and include tropical rain forest, tropical deciduous forest, temperate rain forest, temperate deciduous forest, temperate coniferous forest (taiga), tropical grassland (savanna), temperate grassland (prairie), hot and cold deserts, and tundra. In mountain ranges, biome type may change on an elevation gradient (Figure 54.24).
- Within aquatic environments, biomes include marine aquatic biomes (the intertidal zone, coral reef, and open ocean) and freshwater lakes, rivers, and wetlands. These are distinguished by differences in salinity, oxygen content, depth, current strength (lentic versus lotic), and availability of light (Figure 54.25).

54.6 Continental Drift and Biogeography

- The location of many fossils and living organisms can be explained by evolution in one supercontinent followed by subsequent continental drift.
- Other distinct distribution patterns are relics of once much broader distributions (Figure 54.26).
- Six major biogeographical regions—Nearctic, Palearctic, Neotropical, Ethiopian, Oriental and Australian—each of which contains a distinct fauna and flora, are recognized today (Figure 54.27).

Assess and Discuss

Test Yourself

1. Which of the following is probably the most important factor in the distribution of organisms in the environment?
   a. light
   b. temperature
   c. salinity
   d. water availability
   e. pH
2. The greenhouse effect is
   a. a new phenomenon resulting from industrialization.
   b. due to the absorption of solar radiation by atmospheric gases.
   c. responsible for the natural warming of the Earth.
   d. all of the above.
   e. b and c only.
3. An examination of the temperature tolerances of locusts would best be described by which ecological subdiscipline?
   a. organismal ecology
   b. population ecology
   c. community ecology
   d. ecosystem ecology
   e. both a and b
4. Physics is to engineering as ecology is to
   a. biology.
   b. environmental science.
   c. chemistry.
   d. mathematics.
   e. statistics.
5. The most common biome type, by area occupied, is the
   a. open ocean.
   b. tropical rainforest.
   c. tundra.
   d. hot desert.
   e. lentic habitats.
6. What is the driving force that determines the circulation of the atmospheric air?
   a. temperature differences of the Earth
   b. winds
   c. ocean currents
   d. mountain ridges
   e. all of the above
7. In this biome, rainfall is between 25 cm and 100 cm and temperatures vary between −10°C in winter and 30°C in summer. Where are you?
   a. tropical rainforest
   b. tropical deciduous forest
   c. savanna
   d. prairie
   e. temperate deciduous forest
8. What characteristics are commonly used to identify the biomes of the Earth?
   a. temperature
   b. precipitation
   c. vegetation
   d. all of the above
   e. a and b only
9. Young lakes are often clear and with little plant life. Such lakes are called
   a. oligotrophic.
   b. eutrophic.
   c. lotic.
   d. lentic.
   e. pelagic.
10. Which gas contributes most to human-caused global warming?
   a. carbon dioxide
   b. nitrous oxide
   c. sulfur dioxide
   d. methane
   e. chlorofluorocarbons

Conceptual Questions

1. If mountains are closer to the Sun than valleys, why aren't they hotter?
2. Why are fires generally more frequent in prairies than in hotter, drier deserts?
3. A principle of biology is that *living organisms interact with their environment.* In most locations on Earth, at about 30° latitude, air cools and descends, and hot deserts occur. Florida is situated between 31°N and 24°N. Why does it not support a desert biome?

Collaborative Questions

1. The so-called Telegraph fire, near Yosemite National Park, in 2008, was one of the worst in California that year, burning more than 46 square miles covered by timber that has not burned in over 100 years. What could be done to prevent such a catastrophic fire in the park itself?
2. Based on your knowledge of biomes, identify the biome in which you live. In your discussion, list and describe the organisms that you have observed in your biome. Why might your observations not fit the biome predicted to occur from temperature/precipitation profiles?

Online Resource

www.brookerbiology.com

Stay a step ahead in your studies with animations that bring concepts to life and practice tests to assess your understanding. Your instructor may also recommend the interactive eBook, individualized learning tools, and more.

Behavioral Ecology

55

Chapter Outline

55.1 The Influence of Genetics and Learning on Behavior
55.2 Local Movement and Long-Range Migration
55.3 Foraging Behavior
55.4 Communication
55.5 Living in Groups
55.6 Altruism
55.7 Mating Systems
Summary of Key Concepts
Assess and Discuss

Killdeer (*Charadrius vociferus*) removing an eggshell from its nest. What is the selective advantage of this behavior?

After their young hatch, nesting birds often pick up the empty eggshells and carry them away from the nest. One might think that they are being neat and tidy or are minimizing the risk of bacterial infection to the chicks, but there is more to the behavior than this. The chicks and unhatched eggs are well camouflaged in the nest, but the white color of the empty eggshell quickly attracts the attention of predators such as crows that would kill and eat the chicks or remaining eggs. By removing the old eggshells, the parents are increasing the chances that their offspring—and thus their genes—will survive. Although this behavior is likely to be an instinctive activity, it is promoted because birds that performed this activity likely have a higher rate of chick survival than those that don't, ensuring that the genes that code for this behavior are passed on.

Behavior is the observable response of organisms to external or internal stimuli. In this chapter, we focus our attention on the field of **behavioral ecology**, the study of how behavior contributes to the differential survival and reproduction of organisms. Contemporary behavioral ecology builds on earlier work that focused primarily on how organisms behave. In the early 20th century, scientific studies of animal behavior, termed **ethology** (from the Greek *ethos*, meaning habit or manner), focused on the specific genetic and physiological mechanisms of behavior. These factors are called **proximate causes**. For example, we could hypothesize that male deer rut or fight with other males in the fall because a change in day length stimulates the eyes, brain, and pituitary gland and triggers hormonal changes in their bodies. The founders of ethology, ethologists Karl von Frisch, Konrad Lorenz, and Niko Tinbergen, shared the 1973 Nobel Prize in Physiology or Medicine for their pioneering discoveries concerning the proximate causes of behavior.

However, we could also hypothesize that male deer fight to determine which deer get to mate with the most female deer and pass on their genes. This hypothesis leads to a different answer than the one that is concerned with changes in day length. This answer focuses on the adaptive significance of fighting to the deer, that is, on the effect of a particular behavior on reproductive success. These factors are called **ultimate causes** of behavior. Since the 1970s, behavioral ecologists have focused more on understanding the ultimate causes of behavior.

In this chapter, we will explore the role of both proximate and ultimate causes of behavior. We begin the chapter by investigating how behavior is achieved, examining the roles of both genetics and the environment. In doing so, we will examine the important contributions of ethologists von Frisch, Lorenz, and Tinbergen. We consider how different behaviors are involved in movement, gathering food, and communication. Later, we investigate how organisms interact in groups, whether an organism can truly behave in a way that benefits others at a cost to itself, and how behavior shapes different mating systems. The chapter focuses on animal behavior, because the behavior of other organisms is more limited and less well understood.

55.1 The Influence of Genetics and Learning on Behavior

Learning Outcomes:

1. Describe the differences among innate behavior, conditioning, and learning.
2. Distinguish between classical conditioning and operant conditioning.
3. Give examples of how genetics and learning influence most behaviors.

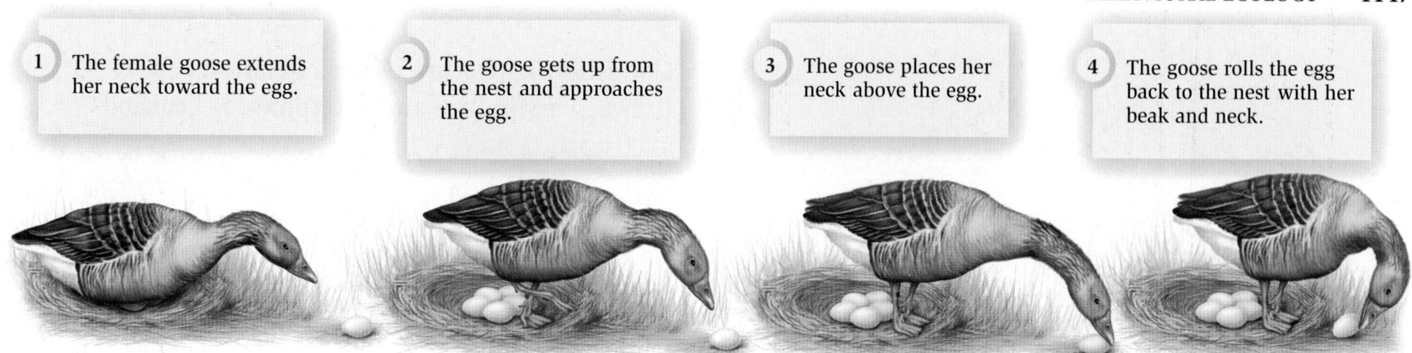

1 The female goose extends her neck toward the egg.
2 The goose gets up from the nest and approaches the egg.
3 The goose places her neck above the egg.
4 The goose rolls the egg back to the nest with her beak and neck.

Figure 55.1 **A fixed action pattern as an example of innate behavior.** Female geese retrieve eggs that have rolled outside the nest through a set sequence of movements. The goose completes this entire sequence even if a researcher takes the egg away before the goose has rolled it back to the nest.

Behavior is controlled by both genetics and the environment, and in this chapter, we will discuss the influence of both. Determining to what degree a behavior is influenced by genes versus the environment will depend on the particular genes and environment examined. However, in a few cases, changes in behavior may be caused by variation in just one gene. Even if a given behavior is influenced by many genes, if one gene is altered, it is possible that the entire behavior can change. To use the analogy of baking a cake, a change in one ingredient of the recipe may change the taste of the cake, but that does not mean that the one ingredient is responsible for the entire cake.

In this section, we begin by examining how genes can affect behavior and consider several examples of simple genetically programmed behaviors. Later, we explore several types of learned behavior, including classical and operant conditioning and cognitive learning, and conclude the section by exploring an example of the interaction of genetics and learning on behavior.

Genes for behavior act on the development of the nervous system and musculature—physical traits that evolve through natural selection. Many genes are needed for the proper development and function of the nervous system and musculature. Even so, as described by Rothenbuhler's and Brown's work (see Genomes & Proteomes Connection), variation in a single gene can have a dramatic influence on behavior.

Fixed Action Patterns Are Genetically Programmed

Behaviors that seem to be genetically programmed are referred to as **innate** (also called instinctual). Although we recognize that the expression of genes varies, often in response to environmental stimuli, some behavior patterns evidently are genetically quite fixed. Most individuals will exhibit the same behavior regardless of the environment. A spider will spin a specific web without ever seeing a member of its own species build one. The courtship behaviors of many bird species are so stereotyped as to be virtually identical.

A classic example of innate behavior is the egg-rolling response in geese (Figure 55.1). If an incubating goose notices an egg out of the nest, she will extend her neck toward the egg, get up, and then roll the egg back to the nest using her beak. Such behavior functions to improve fitness because it increases the survival of offspring. Eggs that roll out of the nest get cold and fail to hatch. Geese that fail to exhibit the egg-rolling response would pass on fewer of their genes to future generations.

Egg-rolling behavior is an example of what ethologists term a **fixed action pattern (FAP)**, a behavior that, once initiated, continues until completed. For example, if the egg is removed while the goose is in the process of rolling it back toward the nest, the goose still completes the FAP, as though she were rolling back the now-absent egg to the nest. The stimulus to initiate this behavior is obviously a strong one, which ethologists term a **sign stimulus**. The sign stimulus for the goose is that an egg had rolled out of the nest. According to ethologists, this stimulus acts on the goose's central nervous system, which provides a neural stimulus to initiate the motor program or FAP. Interestingly, any round object, from a wooden egg to a volleyball, can elicit the egg-rolling response. Although sign stimuli usually have certain key components, they are not necessarily very specific.

Niko Tinbergen's study of male stickleback fish provides another classic example of an FAP. Male sticklebacks, which have a characteristic red belly, will attack other male sticklebacks that invade their territory. Tinbergen found that sticklebacks attacked small, unrealistic model fish having a red ventral surface (the sign stimulus), while ignoring a realistic male stickleback model that lacked a red underside (Figure 55.2).

GENOMES & PROTEOMES CONNECTION

Some Behavior Results from Simple Genetic Influences

An example of the effect of genes on behavior was demonstrated in biologist W. C. Rothenbuhler's 1964 work on honeybees. Some strains of bees are termed hygienic; that is, they detect and remove diseased larvae from the nest. This behavior involves two distinct maneuvers: uncapping the wax cells and then discarding the dead larvae. Other strains are not hygienic and do not exhibit such behavior. Using genetic crosses, Rothenbuhler demonstrated that one recessive gene (*u*) controls cell uncapping and another recessive gene (*r*) controls larval removal. Double recessives (*uurr*) are hygienic strains, and double dominants (*UURR*) are nonhygienic strains. When the two strains were crossed, all the F_1 hybrids were nonhygienic (*UuRr*). When the F_1 hybrids were crossed with the pure hygienic strain (*uurr*), four different genotypes were produced: one-quarter of the offspring were hygienic (*uurr*), one-quarter were nonhygienic and showed neither behavior (*UuRr*), one-quarter uncapped the cells but

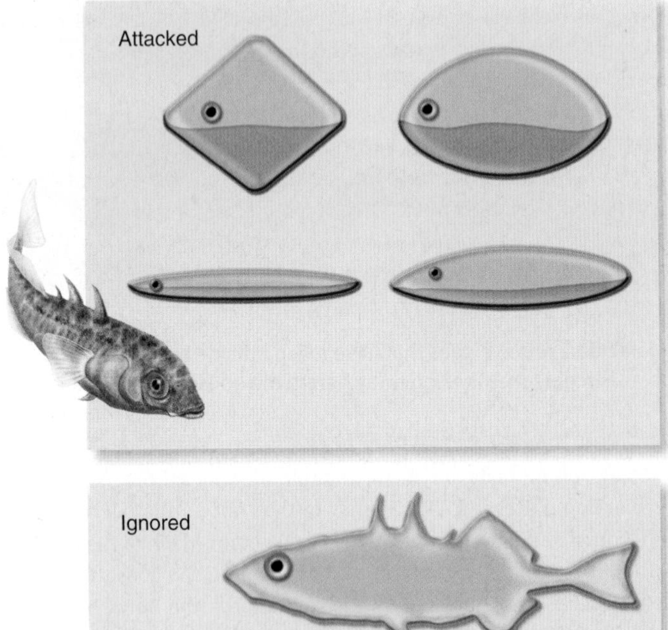

Figure 55.2 **A fixed action pattern elicited by a sign stimulus.** The sign stimulus for male sticklebacks to attack other males entering their territory is a red ventral surface. In experiments, male sticklebacks attacked all models that had a red underside, while ignoring a realistic model of a stickleback that lacked the red belly.

failed to remove the larvae (*uuRr*), and one-quarter removed the larvae but only if the cells were uncapped for them (*Uurr*).

More recently, in 2004, American neuroscientist Barry Richmond and colleagues showed how the work ethic of monkeys is affected by a gene expressed in a region of the brain called the rhinal cortex. Most primates, humans and monkeys included, tend to work harder when a deadline looms. Richmond's team trained four monkeys to release a lever at the exact moment a spot on a computer screen changed color from red to green. The monkeys had to complete this task three times and only on the third trial did they receive a food reward, regardless of how they performed on the first two trials. As an indication of how many trials were left, the monkeys could see a gray bar on the screen. As the bar became brighter the monkeys knew they were reaching the last trial and they worked more diligently for the reward. In the first two trials, the monkeys made more errors than in the last trial. Next, the team switched off the gene known to be involved in processing reward signals. To do this, the researchers injected a short strand of DNA into the monkey's brain. The effects were only temporary, 10–12 weeks, but during that time the monkeys were unable to determine how many trials were left before the reward was given and they worked vigilantly to receive the reward on every trial, making few errors even on trials one and two. Could such studies be performed with humans? Sufferers of obsessive-compulsive disorders and people with bipolar disease (manic-depression) also work for little personal reward.

Conditioning Occurs When a Relationship Between a Stimulus and a Response Is Learned

Although many of the behavioral patterns exhibited by animals are largely innate, sometimes animals can make modifications to their behavior based on previous experience, a process that involves learning. Perhaps the simplest form of learning is **habituation**, in which an organism learns to ignore a repeated stimulus. For example, animals in African safari parks become habituated to the presence of vehicles containing tourists; these vehicles are neither a threat nor a benefit to them. Birds can become habituated to the presence of a scarecrow, resulting in damage to crops. Habituation can be a problem at airports, where birds eventually ignore the alarm calls designed to scare them away from the runways.

Habituation is a form of nonassociative learning, a change in response to a repeated stimulus without association with a positive or negative reinforcement. Alternatively, an association may gradually develop between a stimulus and a response. Such a change in behavior is termed **associative learning**. In associative learning, a behavior is changed or conditioned through the association. The two main types of associative learning are termed classical conditioning and operant conditioning.

In **classical conditioning**, an involuntary response comes to be associated positively or negatively with a stimulus that did not originally elicit the response. This type of learning is generally associated with the Russian psychologist Ivan Pavlov. In his original experiments in the 1920s, Pavlov restrained a hungry dog in a harness and presented small portions of food at regular intervals (refer back to Figure 40.16). The dog would salivate whenever it smelled the food. Pavlov then began to sound a metronome when presenting the food. Eventually the dog would salivate at the sound of the metronome, whether or not the food was present. Classical conditioning is widely observed in animals. For example, many insects quickly learn to associate certain flower odors with nectar rewards and other flower odors with no rewards. In humans, the sound of a dentist's drill is enough to produce a feeling of uneasiness, tension, and sweaty palms.

In **operant conditioning**, an animal's behavior is reinforced by a consequence, either a reward or a punishment. The classic example of operant conditioning is associated with the American psychologist B. F. Skinner, who placed laboratory animals, usually rats, in a specially devised cage with a lever that came to be known as a Skinner box. If the rat pressed on the lever, a small amount of food would be dispensed. At the beginning of the experiment, the rat would often bump into the lever by accident, eat the food, and continue exploring its cage. Later, it would learn to associate the lever with obtaining food. Eventually, if it was hungry, the rat would almost continually press the lever. Operant conditioning, also called trial-and-error learning, is common in animals. Often it is associated with negative rather than positive reinforcement. For example, toads eventually refuse to strike at insects that sting, such as wasps and bees, and birds will learn to avoid bad-tasting butterflies (Figure 55.3). In humans, giving children a reward for completing homework is a positive reinforcer.

(a) Blue jay eating monarch

(b) Vomiting reaction

Figure 55.3 Operant conditioning, also known as trial-and-error learning. **(a)** A young blue jay will eat a monarch butterfly, not knowing that it is noxious. **(b)** After the first experience of vomiting after eating a monarch, a blue jay will avoid the insects in the future.

Concept Check: What's the difference between operant conditioning and classical conditioning?

BioConnections: Look forward to Figure 57.9d. How might operant conditioning be related to the similar appearance of king snakes and coral snakes?

Cognitive Learning Involves Conscious Thought

Cognitive learning refers to the ability to solve problems with conscious thought and includes activities such as perception, analysis, judgment, recollection, and imagining. In the 1920s, German psychologist Wolfgang Köhler conducted a series of classic experiments with chimpanzees that suggested animals could exhibit cognitive learning. In the experiments, a chimpanzee was left in a room with bananas hanging from the ceiling and out of reach (Figure 55.4). Also present in the room were several wooden boxes. At first, the chimp tried in vain to jump up and grab the bananas. After a while, however, it began to arrange the boxes one on top of another underneath the fruit. Eventually, the chimp climbed the boxes and retrieved the fruit.

Many other examples of such behavior have been observed. Chimps strip leaves off twigs and use the twigs to poke into ant nests, withdrawing the twig and licking the ants off. Captive ravens have been shown to retrieve meat suspended from a branch by a string, even though they have never encountered the problem before. They pull up on the string, step on it, and then pull up on the string again, repeating the process until the meat is within reach.

Both Genetics and Learning Influence Most Behaviors

Much of the behavior we have discussed so far has been presented as either innate or learned, but the behavior we observe in nature is usually a mixture of both. Bird songs present a good example. Many birds learn their songs as juveniles, when they hear their parents sing. If juvenile white-crowned sparrows are raised in isolation, their adult songs do not resemble the typical species-specific song (Figure 55.5). If they hear only the song of a different species, such as the song sparrow, they again sing a poorly developed adult song. However, if they hear the song of the white-crowned sparrow, they will learn to sing a fully developed white-crowned sparrow song. The birds are genetically programmed to learn, but they will sing the correct song only if the appropriate instructive program is in place to guide learning.

Another example of how innate behavior interacts with learning can occur during a limited time period of development, called a **critical period**. At this time, many animals develop irreversible species-specific behavior patterns. This process is called **imprinting**. One of the best examples of imprinting was demonstrated by the Austrian ethologist Konrad Lorenz in the 1930s. Lorenz noted that young birds

Figure 55.4 Cognitive behavior involving problem-solving ability. This chimp has devised a solution to the problem of retrieving bananas that were initially out of its reach.

BioConnections: Look back at Figure 42.17. Does learning affect brain structure?

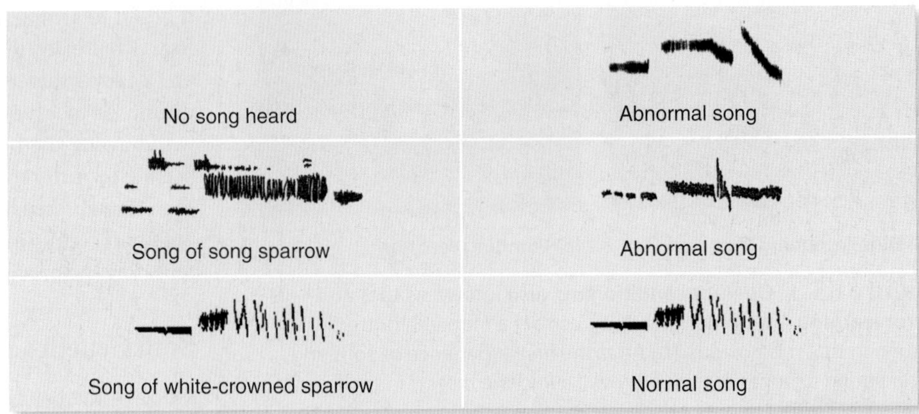

Figure 55.5 The interaction between genetics and learning. The lines represent the different sound frequencies produced by the birds over a short time interval. The juvenile white-crowned sparrow will sing an abnormal song if it is kept in isolation or hears only the song of a different species. However, the juvenile will sing the normal white-crowned sparrow song if exposed to it.

Concept Check: Cuckoos lay their eggs in other birds' nests, so their young are reared by parent birds of a different species. However, unlike the white-crowned sparrow, adult cuckoos always sing their own distinctive song, not that of the host species they hear as juveniles. How is this possible?

of some species imprint on their mother during a critical period that is usually within a few hours after hatching. This behavior serves them well, because in many species of ducks and geese, it would be hard for the mother to keep track of all her offspring as they walk or swim. After imprinting takes place, the offspring keep track of the mother.

The survival of the young ducks requires that they quickly learn to follow their mother's movements. Lorenz raised greylag geese from eggs, and soon after they hatched, he used himself as the model for imprinting. As a result, the young goslings imprinted on Lorenz and followed him around (**Figure 55.6**). For the rest of their life, they preferred the company of Lorenz and other humans to geese. Studies have shown that even an object as foreign as a black box, watering can, or flashing light can be imprinted on if it is the first moving object the chick sees during the critical period. In nature, if young geese are not provided with any stimulus during the critical period, they will fail to imprint on anything, and without parental care, they will almost certainly die.

Other animals imprint in different ways. Newborn shrews imprint on the scent of their mother. Mothers also can imprint on their own young within a few hours. For example, a relatively common trick used in sheep farming is to disguise a lamb whose mother has died or abandoned it by wrapping it in the fleece of another ewe's stillborn lamb. That second ewe will then care for the abandoned lamb because it smells like her own. In these situations, the innate behavior is the ability to imprint soon after birth, and the factors in the environment are the stimulus to which the imprinting is directed.

Innate behavior can interact with learning during animal migration. Inexperienced juvenile birds migrate in a particular direction but fail to correct for deviations if they are blown off course. Experienced adult birds, on the other hand, can often correct for storm-induced displacement, indicating they have developed more complex navigational skills. Many complex behaviors are involved in movement and migration, as we will explore in the next section.

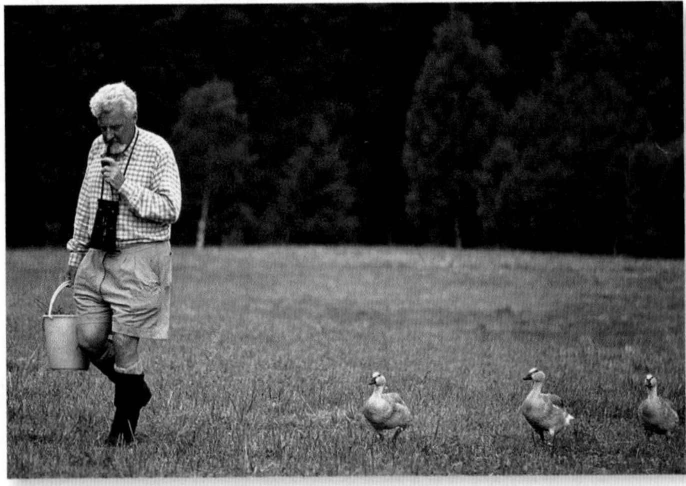

Figure 55.6 Konrad Lorenz being followed by his imprinted geese. Newborn geese follow the first object they see after hatching and later will follow that particular object only. They normally follow their mother but can be induced to imprint on humans. The first thing these young geese saw after hatching was ethologist Konrad Lorenz.

55.2 Local Movement and Long-Range Migration

Learning Outcomes:

1. Distinguish between kinesis and taxis, two different types of local movement.
2. Describe the three mechanisms animals use during migration.

Organisms need to find their way, both locally and over what can be extremely long distances. Locally, organisms continually need to locate sources of food, water, mates, and perhaps nesting sites. Migration involves the longer-distance seasonal movement of animals, usually between overwintering areas and summer breeding sites; these are often hundreds or even thousands of kilometers apart. Several different types of behavior may be involved in these movements.

In this section, we begin by exploring local movement and how animals can use landmarks to guide their movements. We then consider migration and examine the possible mechanisms used by migrating animals to find their way.

Local Movement Can Involve Kinesis, Taxis, and Memory

The simplest forms of movement are mere responses to stimuli. A **kinesis** is a movement in response to a stimulus, but one that is not directed toward or away from the source of the stimulus. A simple experiment often done in classrooms is to observe the activity levels of woodlice, sometimes called sow bugs or pill bugs, in dry areas and moist areas. The woodlice move faster in drier areas, and they slow down when they reach moist environments. This behavior tends to keep them in damper areas, which they prefer in order to avoid desiccation.

A **taxis** is a more directed type of response either toward (positive taxis) or away from (negative taxis) an external stimulus. Cockroaches exhibit negative phototaxis, meaning they tend to move away from light. Under low-light conditions, the photosynthetic unicellular flagellate *Euglena gracilis* shows positive phototaxis and moves toward a light source. Sea turtle hatchings are also strongly attracted to light. On emerging from their nests, they crawl toward the brightest location, traditionally the reflected moonlight on the ocean's surface. Lighted houses on the shore can disorient the hatchlings, however, and lead them to wander away from the ocean and succumb to dehydration, exhaustion, and predation. This is why beachfront property owners are requested to turn their lights down in turtle-hatching season. Male silk moths orient themselves in relation to wind direction (anemotaxis). If the air current carries the scent of a female moth, they will move upwind to locate it. Some freshwater fishes orient themselves to the currents of streams. Many fishes exhibit positive rheotaxis (from the Greek *rheos*, meaning current), in that they swim against the water current to prevent being washed downstream.

Sometimes memory and landmarks may be used to aid in local movements. Dutch-born ethologist Niko Tinbergen showed how the female digger wasp uses landmarks to relocate her nests, as described next.

FEATURE INVESTIGATION

Tinbergen's Experiments Show That Digger Wasps Use Landmarks to Find Their Nests

In the sandy, dry soils of Europe, the solitary female digger wasp (*Philanthus triangulum*) digs four to five nests in which to lay her eggs. Each nest stretches obliquely down into the ground for 40–80 cm. The wasp follows this by performing a sequence of apparently genetically programmed events. She catches and stings a honeybee, which paralyzes it; returns to the nest; drags the bee into the nest; and lays an egg on it. The egg hatches into a larva, which feeds on the paralyzed bee. However, the larva needs to ingest five to six bees before it is fully developed. This means the wasp must catch and sting four to five more bees for each larva. She can carry only one bee at a time. After each visit, the wasp must seal the nest with soil, find a new bee, relocate the nest, open it, and add the bee. How does the wasp relocate the nest after spending considerable time away? Niko Tinbergen observed the wasps hover and fly around the nest each time they took off. He hypothesized that they were learning the nest position by creating a mental map of the landmarks in the area.

To test his hypothesis, Tinbergen experimentally adjusted the landmarks around the burrow that the wasps might be using as cues (**Figure 55.7**). First, he put a ring of pinecones around the nest entrance to train the wasp to associate the pinecones with the nest. Then, when the wasp was out hunting, he moved the circle of pinecones a distance from the real nest and constructed a sham nest, making a slight depression in the sand and mimicking the covered entrance of the burrow. On returning, the wasp flew straight to the sham nest and tried to locate the entrance. Tinbergen chased it away. When it returned, it again flew to the sham nest. Tinbergen repeated this nine times, and every time the wasp chose the sham nest. Tinbergen got the same result with 16 other wasps, and not once did they choose the real nest.

Figure 55.7 How Niko Tinbergen discovered the digger wasp's nest-locating behavior.

Concept Check: *How would you test what type of spatial landmarks are used by female digger wasps?*

HYPOTHESIS Digger wasps (*Philanthus triangulum*) use visual landmarks to locate their nests.

STARTING LOCATION The female digger wasp excavates an underground nest, to which she returns daily, bringing food to the larvae located inside.

Experimental level

1 Place a ring of pinecones around the nest to train the wasp to associate pinecones with the nest.

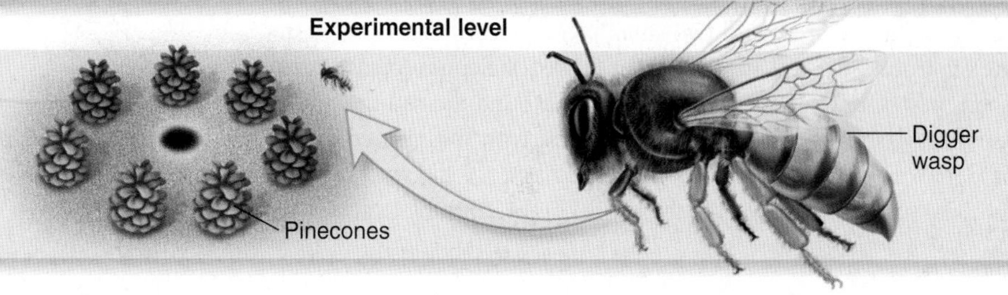

2 After the wasp leaves the nest to hunt, move the pinecones 30 cm from the real nest. The wasp returns and flies to the center of the pinecone circle instead of the real nest. Repeated experiments yield similar results (see data), indicating that the wasp uses landmarks as visual cues.

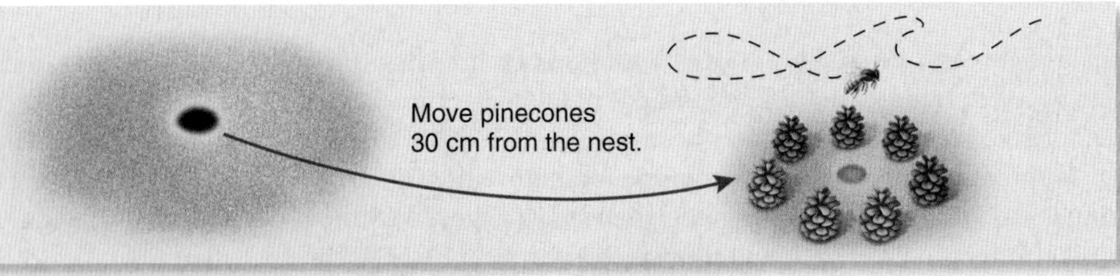
Move pinecones
30 cm from the nest.

3 To test whether it is the shape or the smell of the pinecones that elicits the response, perform the same experiment as above, except use pinecones with no scent and add 2 small pieces of cardboard coated with pine oil.

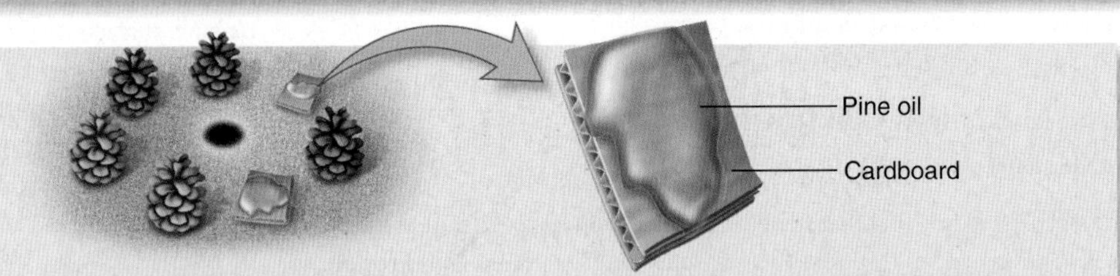
Pine oil
Cardboard

4 After the wasp leaves the nest, move the pinecones 30 cm from the nest, but leave the scented cardboard at the nest. The wasps again fly to the pinecone nest (see data), indicating that it is the arrangement of cones, not their smell, that elicits the learning.

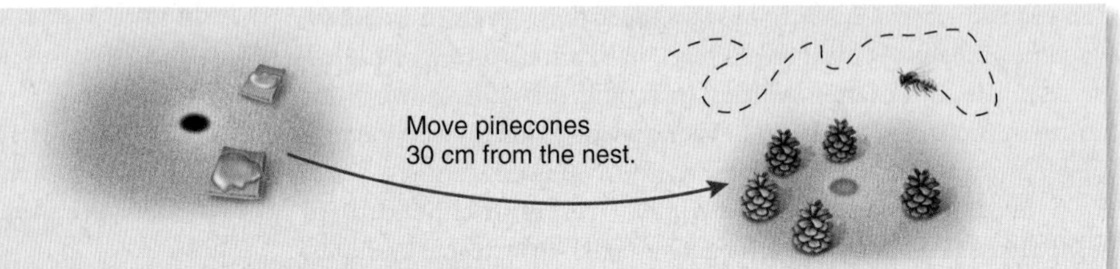
Move pinecones
30 cm from the nest.

5 **THE DATA***

Results from steps 1 and 2:

Wasp #	Number of return visits per wasp to real nest without pinecones	Number of return visits per wasp to sham nest with pinecones
1–17	0	~9

Results from steps 3 and 4:

Wasp #	Number of return visits per wasp to real nest with scented cardboard	Number of return visits per wasp to sham nest with pinecones
18–22	0	~6

*Seventeen wasps, numbered 1–17, were studied as described in steps 1 and 2. Five wasps, numbered 18–22, were studied as described in steps 3 and 4.

6 **CONCLUSION** Digger wasps remember the positions of visual landmarks and use them as aids in local movements.

7 **SOURCE** Tinbergen, N. 1951. The study of instinct. Clarendon Press, Oxford.

Next Tinbergen experimented with the type of stimulus that might be eliciting the learning. He hypothesized that the wasps could be responding to the distinctive scent of the pinecones rather than their appearance. He trained the wasps by placing a circle of pinecones that had no scent and two small pieces of cardboard coated in pine oil around the real nest. He then moved the cones to surround a sham nest and left the scented cardboard around the real nest. The returning wasps again ignored the real nest with the scented cardboard and flew to the sham. He concluded that for the wasps, sight was apparently more important than smell in determining landmarks.

Experimental Questions

1. What observations were important for the development of Niko Tinbergen's hypothesis explaining how digger wasps located their nests?
2. How did Tinbergen test the hypothesis that the wasps were using landmarks to relocate the nest? What were the results?
3. Did the Tinbergen experiment rule out any other cue the wasps may have been using besides the sight of pinecones?

Migration Involves Long-Range Movement and More Complex Spatial Navigation

As well as navigating over short-range distances, many animal species undergo **migration**, long-range seasonal movement. Migrations usually involve a movement away from a birth area to feed and a return to the birth area to breed, with the movement generally being linked to seasonal availability of food. For example, nearly half the birds of North America migrate to South America to escape the cold winters and feed, returning to North America in the spring to breed. Arctic terns that breed in Arctic Canada and Asia in summer migrate to the Antarctic to feed in the winter and then return to breed. This staggering journey involves up to a 40,000-km (25,000-mile) round-trip, most of it over the open ocean, during which the birds must stay airborne for days at a time!

Many mammals, including wildebeest and caribou, make migrations that track the appearance of new vegetation on which they feed. The monarch butterfly of North America migrates to overwinter in California, Mexico, and possibly south Florida and Cuba (Figure 55.8). An interesting point about the northward journey of the monarch is that it involves several generations of butterflies to complete. On their way back to the northern U.S. and Canada, the butterflies lay eggs and die. The caterpillars develop on milkweed plants, and the resultant adults continue to journey farther north. This cycle happens several times in the course of the return journey. The northward and southward migrations are unique in that none of the individuals has ever been to the destinations before; therefore, the ability to migrate must be an innate behavior.

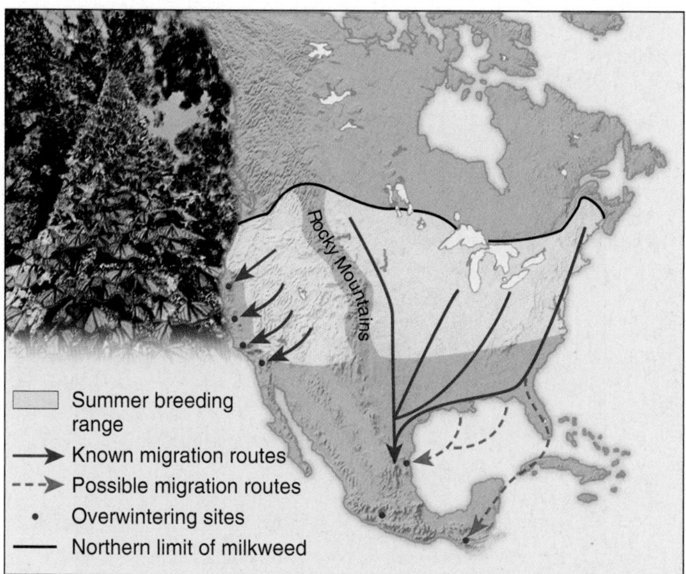

Figure 55.8 Monarch butterfly migration. Many monarch butterflies east of the Rocky Mountains migrate to a small area in Mexico to avoid the cold northern weather. Here they roost together in large numbers in fir trees (inset). Some butterflies may stay in Florida and Cuba. Butterflies west of the Rockies overwinter in mild coastal California locations.

Concept Check: *Why is this an unusual example of migration?*

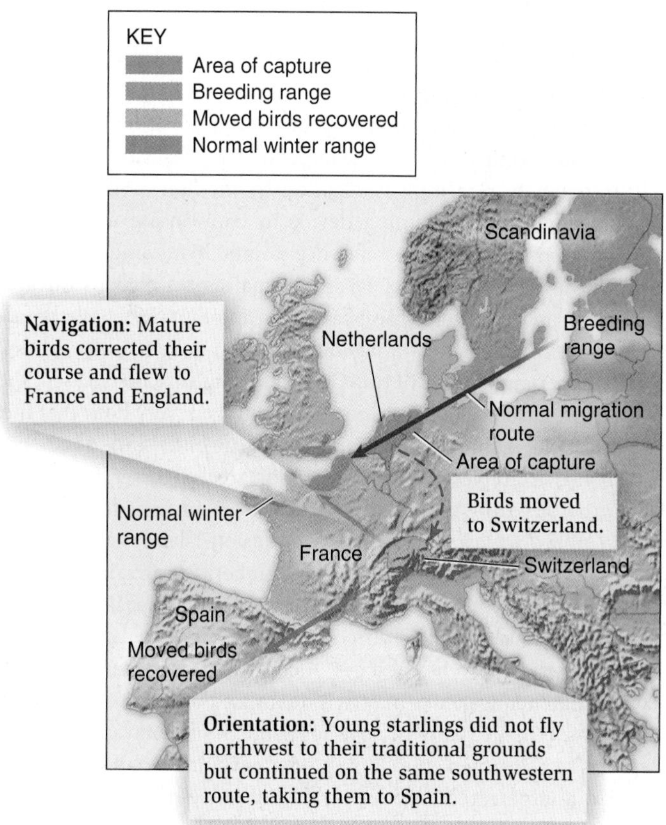

Figure 55.9 Orientation versus navigation. Starlings normally migrate from breeding grounds in Scandinavia and northeastern Europe through the Netherlands and northern Germany to overwintering sites in France and England. This involves a southwest flight. When juveniles were captured in the Netherlands and moved to Switzerland, they continued on in a southwestern direction and ended up in Spain. When adult birds were captured and moved, they changed course and flew to the normal overwintering areas.

How do migrating animals find their way? Three mechanisms may be involved: piloting, orientation, and navigation. In **piloting**, an animal moves from one familiar landmark to the next. For example, many whale species migrate between summer feeding areas and winter calving grounds. Gray whales migrate between the Bering Sea near Alaska to coastal areas of Mexico. Features of the coastline, including mountain ranges, and rivers, may aid in navigation. In **orientation**, animals have the ability to follow a compass bearing and travel in a straight line. **Navigation** involves the ability not only to follow a compass bearing but also to set or adjust it.

An experiment with starlings helps illuminate the difference between orientation and navigation (Figure 55.9). European starlings breed in Scandinavia and northeastern Europe and migrate in a southwest direction toward coastal France and southern England to spend the winter. Migrating starlings were captured and tagged in the Netherlands and then transported south to Switzerland and released. Juvenile birds, which had never made the trip before, flew southwest in their migration and were later recaptured in Spain. Adult birds, with more experience, returned to their normal wintering range by adjusting their course by approximately 90°. This implies that the adult birds can actually navigate, whereas the juveniles rely on orientation.

Many species use a combination of navigational reference points, including the position of the Sun, the stars (for nighttime travel), and Earth's magnetic field. Homing pigeons have magnetite in their beaks that acts as a compass to indicate direction (refer back to Section 43.4, Electromagnetic Sensing). Navigation by the Sun or the stars also requires the use of a timing device to compensate for the ever-changing position of these reference points. Many migrants, therefore, possess the equivalent of an internal clock. Pigeons integrate their internal clock with the position of the Sun. Researchers have altered the internal clock of pigeons by keeping them under artificial lights for certain periods of time. When the pigeons are released, they display predictable deviations in their flight. For every hour that their internal clock is shifted, the orientation of the birds shifts about 15°.

Not all examples of animal migration are well understood. Green sea turtles feed off the coast of Brazil yet swim east for 2,300 km (1,429 miles) to lay their eggs on Ascension Island, an 8-km-wide island in the center of the Atlantic Ocean between Brazil and Africa. It is not known why the turtles lay their eggs on this speck of an island or how they succeed in finding it. Perhaps fewer predators exist on Ascension than on other beaches. A combination of magnetic orientation and chemical cues may help them find it. Thus, although scientists have made many discoveries about animal navigation, much remains to be learned about how animals acquire a map sense.

To a large extent, local and long-distance movement involves searching for food. In the next section, we will investigate how such foraging decisions are made.

55.3 Foraging Behavior

Learning Outcomes:

1. Describe and give examples of optimal foraging.
2. Outline the costs and benefits of defending a territory.

Food gathering, or foraging, often involves decisions about whether to remain at a resource patch and look for more food or look for a completely new patch. The analysis of these decisions is often performed in terms of **optimality theory**, which predicts that an animal should behave in a way that maximizes the benefits of a behavior minus its costs. In this case, the benefits are the nutritional or caloric value of the food items, and the costs are the energetic or caloric costs of movement. When the difference between the energetic benefits of food gathering and the energetic costs of food gathering is maximized, an organism is said to be optimizing its foraging behavior. Optimality theory can also be used to investigate other behavioral issues such as how large a territory to defend. Too small a territory would contain insufficient resources, such as food and mates, and too large a territory would be too energetically costly to defend. Theoretically, then, there is an optimal territory size for a given individual.

Optimal Foraging Entails Maximizing the Benefits and Minimizing the Costs of Food Gathering

Optimal foraging proposes that in a given circumstance, an animal seeks to obtain the most energy possible with the least expenditure of energy. The underlying assumption of optimal foraging is that natural selection favors animals that are maximally efficient at propagating their genes and at performing all other functions that serve this purpose. In this model, the more net energy an individual gains in a limited time, the greater the reproductive success.

Shore crabs (*Carcinus maenas*) eat many different-sized mussels but tend to feed preferentially on intermediate-sized mussels, which give them the highest rate of energy return (**Figure 55.10**). Very large mussels yield more energy, but they take so long for the crab to open that they are actually less profitable, in terms of energy yield per unit time spent, than smaller sizes. Very small mussels are easy to crack open but contain so little flesh that they are not worth the effort. This leaves intermediate-sized mussels as the preferred size. Of course, the intermediate-sized mussels may take a longer time to locate, because more crabs are looking for them, so crabs eat some less profitable but more frequently encountered sizes of mussels. The result is that the diet consists of mussels in a range of sizes around the preferred optimal size.

In some cases, animals do not forage optimally. For example, animals seek not only to maximize food intake but also to minimize the risk of predation. Some species may only dart out to take food from time to time. The risk of predation thus has an influence on foraging behavior. Many animals also maintain territories to minimize competition with other individuals and control resources, whether food, mates, or nesting sites. As we will see, defending these territories also has an energetic cost.

Defending Territories Has Costs and Benefits

Many animals or groups of animals, such as a pride of lions, actively defend a **territory**, a fixed area in which an individual or group excludes other members of its own species, and sometimes other

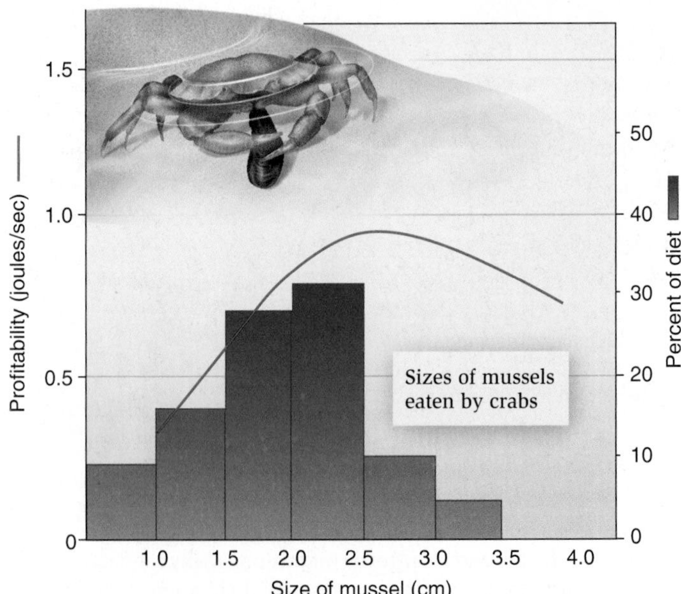

Figure 55.10 Optimal foraging behavior in shore crabs. When offered a choice of equal numbers of each size mussel, shore crabs (*Carcinus maenas*) prefer intermediate-sized mussels that provide the highest rate of energy return. Profitability is the energy yield (joules) per second of time used in breaking open the shell.

(a) Golden-winged sunbird

(b) Cheetah

(c) Nesting gannets

Figure 55.11 Differing territory sizes among animals. (a) The golden-winged sunbird of East Africa (*Nectarinia reichenowi*) has a medium territory size that depends on the number of flowers it can obtain resources from and defend. (b) Cheetahs (*Acinonyx jubatus*) hunt over large areas and can have extensive territories. This male is urine-marking part of his territory in the southern Serengeti, near Ndutu, Tanzania. (c) Nesting gannets (*Morus bassanus*) have much smaller territories, in which each bird is just beyond the pecking range of its neighbor.

species, by aggressive behavior or territory marking. Optimality theory predicts that territory owners tend to optimize territory size according to the costs and benefits involved. The primary benefit of a territory is that it provides exclusive access to a particular resource whether it be food, mates, or sheltered nesting sites. Large territories provide more of a resource but may be costly to defend, while small territories that are less costly to defend may not provide enough.

In studies of the territorial behavior of the golden-winged sunbird (*Nectarinia reichenowi*) in East Africa, American ornithologists Frank Gill and Larry Wolf measured the energy content of nectar as the benefit of maintaining a territory and compared it to the energy costs of activities such as perching, flying, and fighting (**Figure 55.11a**). Defending the territory ensured that other sunbirds did not take nectar from available flowers, thus increasing the amount of nectar in each flower. In defending a territory, the sunbird gained 780 Calories (kilocalories) a day in extra nectar content. However, the sunbird also spent 728 Calories in defense of the territory, yielding a net gain of 52 Calories a day and making territorial defense advantageous.

Territory size differs considerably among species. Because cheetahs need large areas to be able to hunt successfully, they establish large territories relative to their size (**Figure 55.11b**). In contrast, territories set up solely to defend areas for mating or nesting are often relatively small. For example, male sea lions defend small areas of beach. The preferred areas contain the largest amount of females and are controlled by the largest breeding bulls. The size of the territory of some nesting birds, such as gannets, is determined by how far the bird can reach to peck its neighbor without leaving its nest (**Figure 55.11c**).

Territories may be held for a season, a year, or the entire lifetime of the individual. Ownership of a territory needs to be periodically proclaimed; thus, communication between individuals is necessary for territory owners. This may involve various types of signaling, which we discuss next.

55.4 Communication

Learning Outcome:

1. Give examples of how animals use chemical, auditory, visual, and tactile communication.

Communication is the use of specially designed signals or displays to modify the behavior of others. It may be used for many purposes, including defining territories, maintaining contact with offspring, courtship, and contests between males. The use of different forms of communication between organisms depends on the environment in which they live. For example, visual communication plays little role in the signals of nocturnal animals. Similarly, for animals in dense forests, sounds are of prime importance. Sound, however, is a temporary signal. Scent can last longer and is often used to mark the large territories of some mammals. In this section, we outline the various types of communication—chemical, auditory, visual, and tactile—that occur among animals.

Chemical Communication Is Often Used to Mark Territories or Attract Mates

The chemical marking of territories is common among animals, especially among members of the canine and feline families (see Figure 55.11b). Scent trails are often used by social insects to recruit workers to help bring prey to the nest. Fire ants (genus *Solenopsis*) attack large, living prey, and many ants are needed to drag the prey back to the nest. The scout that finds the prey lays down a scent trail from the prey back to the nest. The scent excites other workers, which follow the trail to the prey. The scent marker is very volatile, and the trail effectively disappears in a few minutes to avoid mass confusion over old trails.

Animals frequently use chemicals to attract mates. Female moths attract males by powerful chemical attractants called **pheromones**.

Male moths have receptors that can detect as little as a single molecule. Among social organisms, some individuals use pheromones to manipulate the behavior of others. For example, a queen bee releases pheromones that suppress the reproductive system of workers, which ensures that she is the only reproductive female in the hive.

Auditory Communication Is Often Used to Attract Mates and to Deter Competitors

Many organisms communicate by making sound. Because the ground can absorb sound waves, sound travels farther in the air, which is why many birds and insects perch on branches or leaves when singing. Air is on average 14 times less turbulent at dawn and dusk than during the rest of the day, so sound carries farther then, which helps explain the preference of most animals for calling at these times. Some insects utilize the very plants on which they feed as a medium of song transmission. Many male leafhopper and planthopper insects vibrate their abdomens on leaves and create species-specific courtship songs that are transmitted by adjacent vegetation and are picked up by nearby females of the same species.

Although many males use auditory communication to attract females, some females use calls to attract the attention of males. Female elephant seals scream loudly when approached by a nondominant male. This attracts the attention of the dominant male, which drives the nondominant male away. In this way, the female is guaranteed a mating with the strongest male. Sound production can attract predators as well as mates. Some bats listen for the mating calls of male frogs to find their prey. Parasitic flies detect and locate chirping male crickets and then deposit larvae on or near them. The larvae latch onto and penetrate the cricket and eventually kill it. Sound may also be used by males during competition over females. In many animals, lower-pitched sounds come from larger males, so by calling to one another, males can gauge the size of their opponents and decide whether it is worth fighting.

Visual Communication Is Often Used in Courtship and Aggressive Displays

In courtship, animals use a vast number of visual signals to identify and select potential mates. Competition among males for the most impressive displays to attract females has led to elaborate coloration and extensive ornamentation in some species. For example, peacocks and males of many bird species have developed elaborate plumage to attract females.

Male fireflies have developed light flashes that are species specific with regard to number and duration of flashes (**Figure 55.12a**). Females respond with a flash of their own. Such bright flashes are also bound to attract predators. Some female fireflies use mimicry to their advantage. Female *Photuris versicolor* fireflies mimic the flashing responses normally given by females of other species, such as *Photinus tanytoxus*, in order to lure the males of those species close enough to eat them.

Visual signals are also used to resolve disputes over territories or mates. Deer and antelope have antlers or horns that they use to display and spar over territory and females. Most of these matches never develop into outright fights, because the males gauge their opponent's strength by the size of these ornaments (look ahead to Figure 55.21b). Among insects, the "horns" of rhinoceros beetles and the eye stems of stalk-eyed flies send similar signals (**Figure 55.12b**).

Tactile Communication Is Used to Strengthen Social Bonds and to Convey Information About Food

Animals often use tactile communication to establish bonds between group members. Primates frequently groom one another, and canines and felines may nuzzle and lick each other. Many insects use tactile communication to convey information on the whereabouts of food. Members of the ant genus *Leptothorax* feed on immobile prey such as dead insects. When a scouting ant encounters such prey, it usually needs an additional worker to help bring it back to the nest. Rather than laying a scent trail, which is energetically costly, the scout ant recruits a helper and physically leads it to the food source. The helper runs in tandem with the scout, its antennae touching the scout's abdomen.

Perhaps the most fascinating example of tactile communication among animals is the dance of the honeybee, elegantly studied by German ethologist Karl von Frisch in the 1940s. Bees commonly live in large hives; in the case of the European honeybee (*Apis mellifera*), the hive consists of 30,000–40,000 individuals. The flowering plants on which the bees forage can be located miles from the hive and are distributed in a patchy manner, with any given patch usually containing many flowers that store more nectar and pollen than an individual bee can carry back to the nest. The scout bee that locates the resource patch returns to the hive and recruits more workers to join it (**Figure 55.13a**). Because it is dark inside the hive, the bee uses

(a) Firefly flashing

(b) Male rhinoceros beetles fighting

Figure 55.12 **Visual communication.** **(a)** Communication between fireflies is conducted by species-specific light flashes emitted by organs located on the underside of the abdomen. **(b)** The horns of these rhinoceros beetles provide a signal about the strength of their owners.

BIOLOGY PRINCIPLE Structure determines function. In both these cases morphological features influence animal behavior.

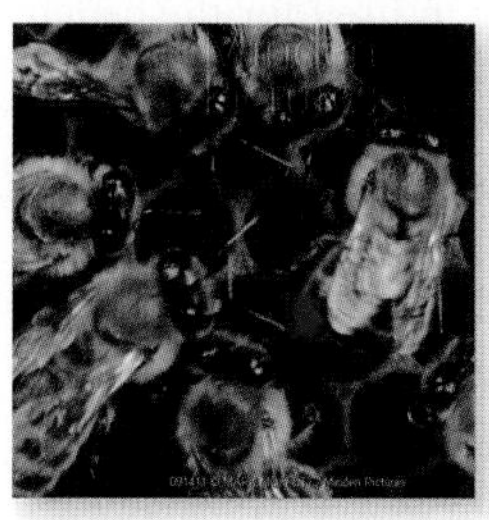

(a) Bees clustering around a recently returned scout, shown on the right

(b) Round dance

(c) Waggle dance: The angle of the waggle to the vertical orientation of the honeycomb corresponds to the angle of the food source from the Sun.

Figure 55.13 **Tactile communication among honeybees regarding food sources.** **(a)** Bees gather around a newly returned scout to receive information about nearby food sources. **(b)** If the food is less than 50 m away, the scout performs a round dance. **(c)** If the food is more than 50 m away, the scout performs a waggle dance, which conveys information about its location. If the dance is performed at a 30° angle to the right of the hive's vertical plane, then the food source is located at a 30° angle to the right of the Sun.

a tactile signal. The scout dances on the vertical side of a honeycomb, and the dance is monitored by other bees, which follow and touch her to interpret the message. If the food is relatively close to the hive, less than 50 m away, the scout performs a round dance, rapidly moving in a circle, first in one direction and then the other. The other bees know the food is relatively close at hand, and the smell of the scout tells them what flower species to look for (Figure 55.13b).

If the food is more than 50 m away, the scout will perform a different type of dance, called a "waggle dance." In this dance, the scout traces a figure 8, in the middle of which she waggles her abdomen and produces bursts of sound. Again, the other bees maintain contact with her. Occasionally, the scout regurgitates a small sample of nectar so the bees know the type of food source they are looking for. The truly amazing part of the waggle dance is that the angle at which the central part of the figure 8 deviates from the vertical direction of the comb represents the same angle at which the food source deviates from the point at which the sun hits the horizon (Figure 55.13c). The direction is always up-to-date, because the bee adjusts the dance as the Sun moves across the sky.

Much communication occurs not only to defend territories but also to communicate information to other individuals in the population, including potential mates. Although living on your own and maintaining a territory has advantages, living in a group also has its benefits, including ready availability of mates and increased protection from predators. In the following section, we examine group living and the behavior it engenders.

55.5 Living in Groups

Learning Outcome:

1. Detail the costs and benefits of living in groups.

As we have seen, much of animal behavior is directed at other animals. Some of the more complex behavior occurs when animals live together in groups such as flocks or herds. If a central concern of ecology is to explain the distribution patterns of organisms, then one of our most important tasks is to understand the reason for such variation in the degree of group living. One way to approach this question is to assess the costs and benefits involved. Although group living increases competition for food and the spread of disease, it also has benefits that compensate for the costs involved. Many of these benefits relate to locating food sources, assistance in rearing offspring, access to mates, and group defense against predators. Group living can reduce predator success in at least two ways: through increased vigilance and through protection in numbers.

Living in Large Groups May Reduce the Risk of Predation Because of Increased Vigilance

For many predators, success depends on the element of surprise. If an individual is alerted to an attack, the predator's chance of success is lowered. A woodpigeon (*Columba palumbus*) in a flock takes to the air when it spots a goshawk (*Accipiter gentilis*). Once one pigeon takes flight, the other members of the flock are alerted and follow suit. If each individual in a group occasionally scans the environment for predators, the larger the group, the less time an individual forager needs to devote to vigilance and the more time it can spend feeding. This is referred to as the **many-eyes hypothesis** (Figure 55.14). Of course, cheating is a possibility, because some birds might never look up, relying on others to keep watch while they keep feeding. However, the individual that happens to be scanning when a predator approaches is most likely to escape, a fact that tends to discourage cheating.

Living in Groups Offers Protection by the Selfish Herd

Group living also provides protection in sheer numbers. Typically, predators take one prey animal per attack. In any given attack, an individual antelope in a herd of 100 has a 1 in 100 chance of being

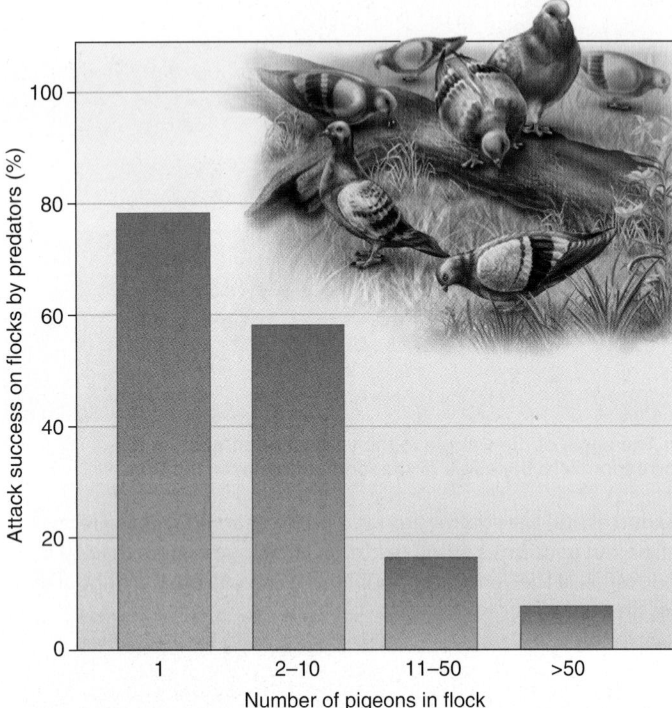

Figure 55.14 **Living in groups and the many-eyes hypothesis.** The larger the number of woodpigeons, the less likely an attack will be successful.

Concept Check: *What other advantages are there to large groups of individuals when being attacked by a predator?*

selected, whereas a single individual has a 1 in 1 chance. Large herds may be attacked more frequently than a solitary individual, but a herd is unlikely to attract 100 times more attacks than an individual, often because of the territorial nature of predators. Furthermore, large numbers of prey are able to defend themselves better than single individuals, which usually choose to flee. For example, groups of nesting black-headed gulls mob a crow, thereby reducing the crow's ability to steal the gulls' eggs.

Research has shown that within a group, each individual can minimize the danger to itself by choosing the location that is as close to the center of the group as possible. This was the subject of a famous paper, "The Geometry of the Selfish Herd," by the British evolutionary biologist W. D. Hamilton. The explanation of this type of defense is that predators are likely to attack prey on the periphery because they are easier to isolate visually. Many animals in herds tend to bunch close together when they are under attack, making it physically difficult for the predator to get to the center of the herd.

Overall, group size may be the result of a trade-off between the costs and benefits of group living. Although much group behavior serves to reduce predation, other complex behavior occurs in groups, including grooming behavior and behavior that appears to benefit the group at the expense of the individual. For example, a honeybee stings a potential hive predator to discourage it. The bee's stinger is barbed, and once it has penetrated the predator's skin, the bee cannot withdraw it. The bee's only means of escape is to tear away part of its abdomen, leaving the stinger behind and dying in the process. In the next section, we explore the reasons for such apparent altruistic behavior, in which an individual incurs costs to itself for the benefit of others.

55.6 Altruism

Learning Outcomes:

1. List the arguments in favor and against the concept of group selection.
2. Describe how the concept of kin selection can explain altruistic behavior.
3. Explain eusociality as an example of altruism.

In Chapter 23, we learned that a primary goal of an organism is to pass on its genes, yet we see many instances in which some individuals forego reproducing altogether, apparently to benefit the group. How do ecologists explain **altruism**, a behavior that appears to benefit others at a cost to oneself? In this section, we begin by discussing whether such behavior evolved for the good of the group or for the good of the individual. As we will see, most altruistic acts serve to benefit the individual's close relatives. We explore the concept of kin selection, which argues that acts of self-sacrifice indirectly promote the spread of an organism's genes, and see how this plays out in an extreme form in the genetics of social insect colonies. Last, we examine reciprocal altruism, instances of altruism among nonkin.

In Nature, Individual Selfish Behavior Is More Likely Than Altruism

One of the first attempts to explain the existence of altruism was called **group selection**, the premise that natural selection produces outcomes beneficial for the whole group or species. In 1962, the British ecologist V. C. Wynne-Edwards argued that a group containing altruists, each willing to subordinate its interests for the good of the group, would have a survival advantage over a group composed of selfish individuals. In concept, the idea of group selection seemed straightforward and logical: a group that consisted of selfish individuals would overexploit its resources and die out, but the fitness of a group with altruists would be enhanced.

In the late 1960s, the idea of group selection came under severe attack. Leading the charge was the American evolutionary biologist George C. Williams, who argued that evolution acts through the individual; that is, adaptive traits generally are selected for because they benefit the survival and reproduction of the individual rather than the group. Some of Williams' arguments against group selection follow.

Mutation Mutant individuals that readily use resources for themselves or their offspring have an advantage in a population in which individuals limit their resource use. Consider a species of bird in which a pair lays only two eggs; that is, it has a clutch size of two, and the resources are not overexploited for the good of the group. Laying two eggs ensures a replacement of the parent birds but prevents a population explosion. Imagine a mutant bird arises that lays three eggs. If the population is not overexploiting its resources, sufficient food may be available for all three young to survive. If this happens, the three-egg genotype eventually becomes more common than the two-egg genotype.

Immigration Even in a population in which all pairs laid two eggs and no mutations occurred to increase clutch size, selfish individuals that laid more could still immigrate from other areas. In nature, populations are rarely sufficiently isolated to prevent immigration of selfish mutants from other populations.

Resource Prediction Group selection assumes that individuals are able to assess and predict future food availability and population density within their own habitat. There is little evidence that they can. For example, it is difficult to imagine that songbirds would be able to predict the future supply of the caterpillars that they feed to their young and adjust their clutch size accordingly.

Most ecologists accept individual gain as a more plausible result of natural selection than group selection. Population size is more often controlled by competition in which individuals strive to command as much of a resource as they can. Such selfishness can cause some seemingly surprising behaviors. For example, male Hanuman langurs (*Semnopithecus entellus*) kill infants when they take over groups of females from other males (**Figure 55.15**). The reason for the behavior is that when they are not nursing their young, females become sexually receptive much sooner, hastening the day when the male can father his own offspring. Infanticide ensures that the male can father more offspring, and the genes governing this tendency spread by natural selection.

Apparent Altruistic Behavior in Nature Is Often Associated with Kin Selection

If individual selfishness is more common than group selection, how do we account for what appear to be examples of altruism in nature? Some propose that the answer lies in a concept known as **kin selection**, selection for behavior that lowers an individual's own fitness but enhances the reproductive success of a relative. Because all offspring have copies of their parents' genes, parents taking care of their young are actually caring for copies of their own genes. Genes for altruism toward one's young are favored by natural selection and become more numerous in the next generation, because offspring have copies of those same genes.

The probability that any two individuals will share a copy of a particular gene is a quantity, r, called the **coefficient of relatedness**. During meiosis in a diploid species, any given copy of a gene has a 50% chance of segregating into an egg or sperm. A mother and father are on average related to their children by an amount $r = 0.5$, because half of a child's genes come from its mother and half from its father. By similar reasoning, brothers or sisters are related by an amount $r = 0.5$ (they share half their mother's genes and half their father's); grandchildren and grandparents, by 0.25; and cousins, by 0.125 (**Figure 55.16**). In 1964, ecologist W. D. Hamilton realized the implication of the coefficient of relatedness for the evolution of altruism. An organism not only can pass on its genes through having offspring, but also can pass them on through ensuring the survival of siblings, nieces, nephews, and cousins. This means an organism has a vested interest in protecting its brothers and sisters, and even their offspring.

The term **inclusive fitness** is used to designate the total number of copies of genes passed on through one's relatives, as well as one's own reproductive output. Hamilton proposed that an altruistic gene is favored by natural selection when

$$rB > C$$

where r is the coefficient of relatedness of donor (the altruist) to the recipient, B is the benefit received by the recipient of the altruism, and C is the cost incurred by the donor. This is known as **Hamilton's rule**.

Imagine two sisters who are not yet mothers. One has a rare kidney disease and needs a transplant from her sister. Let's assume both sisters will have two children of their own. The risk of the transplant to the donor involves a 1% chance of dying, but the benefit to the recipient involves a 90% chance of living and having children. In this example, $r = 0.5$, $B = 0.9 \times 2 = 1.8$, and $C = 0.01 \times 2 = 0.02$. Because the genetic benefit (rB) of 0.9 is much greater than the genetic cost (C) of 0.02, it makes evolutionary sense to proceed with the transplant. Although humans are unlikely to do this type of

Figure 55.15 Infanticide as selfish behavior. Male Hanuman langurs (*Semnopithecus entellus*) can act aggressively toward the young of another male, even killing them, hastening the day the females become sexually receptive and thus the time when the males can father their own offspring. Note that the mother is running with the infant.

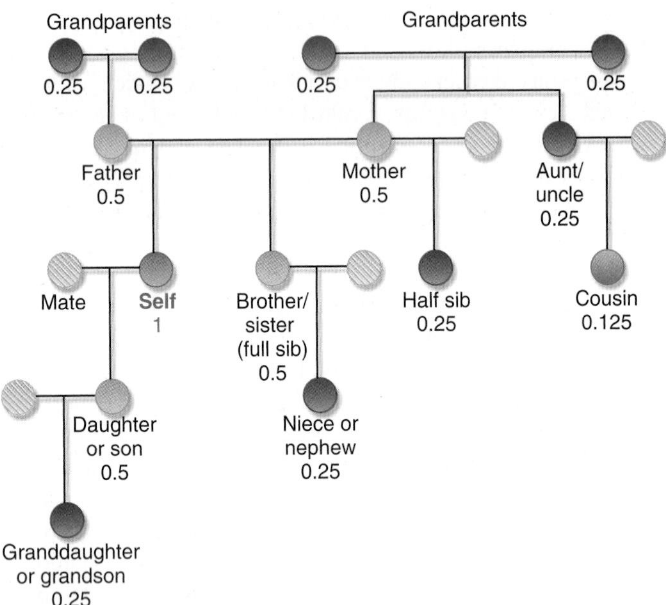

Figure 55.16 Degree of genetic relatedness to self in a diploid organism. Pink hatched circles represent completely unrelated individuals.

Concept Check: In theory, should you sacrifice your life to save two sisters or nine cousins?

calculation before deciding whether to risk their lives to save their siblings from a life-threatening event, this example shows how such behavior could arise and spread in nature.

Let's examine a situation involving altruism within a group of animals. Many insect larvae, especially caterpillars, are soft-bodied creatures. They rely on possessing a bad taste or toxin to deter predators and advertise this condition with bright warning colors. For example, noxious *Datana ministra* caterpillars, which feed on oaks and other trees, have bright red and yellow stripes and adopt a specific posture with head and tail ends upturned when threatened (**Figure 55.17**). Unless it is born with an innate avoidance of this prey type, a predator has to kill and eat one of the caterpillars in order to learn to avoid similar individuals in the future. It is of no personal use to the unlucky caterpillar to be killed. However, animals with warning colors often aggregate in kin groups because they hatch from the same egg mass. In this case, the death of one individual is likely to benefit its siblings, which are less likely to be attacked by the same bird in the future, and thus its genes will be preserved. This explains why the genes for bright color and a warning posture are successfully passed on from generation to generation. In a case where $r = 0.5$, B might be 50, and $C = 1$, the benefit of 25 is greater than 1, so the genes for this behavior are favored by natural selection.

A common example of altruism in social animals occurs when a sentry raises an alarm call in the presence of a predator. This behavior has been observed in Belding's ground squirrels (*Spermophilus beldingi*). The squirrels feed in groups, with certain individuals acting as sentries and watching for predators. As a predator approaches, the sentry typically gives an alarm call, and the group members retreat into their burrows. Similar behavior occurs in prairie dogs (*Cynomys* spp.) (**Figure 55.18**). In drawing attention to itself, the caller is at a higher risk of being attacked by the predator. However, in many groups, those closest to the sentry are most likely to be offspring or brothers or sisters; thus, the altruistic act of alarm

Figure 55.17 Altruistic behavior or kin selection? *Datana ministra* caterpillars exhibit a bright, striped warning pattern to advertise their bad taste to predators.

Concept Check: Why do these caterpillars congregate in clusters?

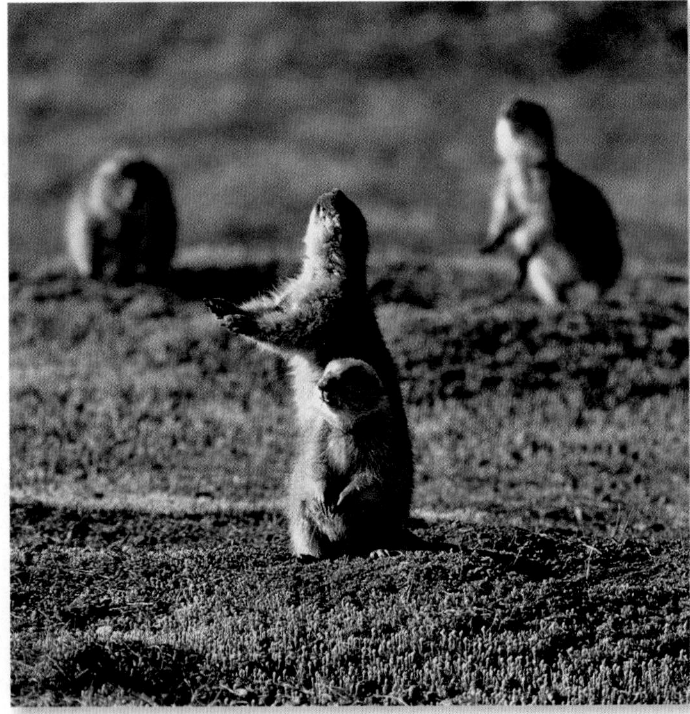

Figure 55.18 Alarm calling, a possible example of kin selection. This prairie dog sentry is emitting an alarm call to warn other individuals, which are often close kin, of the presence of a predator. It is believed that by doing so, the sentry draws attention away from the others but becomes an easier target itself.

BIOLOGY PRINCIPLE The genetic material provides a blueprint for sustaining life. The similarities in DNA between kin promote behavior whereby some animals act to save the lives of their close relatives.

calling is reasoned to be favored by kin selection. Supporting this is the observation that most alarm calling is done by females, because they are more likely to stay in the colony where they were born and have kin nearby, whereas the males are more apt to disperse far from the colony.

Altruism in Eusocial Animals Arises Partly from Genetics and Partly from Lifestyle

Perhaps the most extreme form of altruism is the evolution of sterile castes in social animals, in which the vast majority of females, known as workers, rarely reproduce themselves but instead help one reproductive female (the queen) to raise offspring, a phenomenon called **eusociality**. In insects, the explanation of eusociality lies partly in the particular genetics of most social insect reproduction. Females develop from fertilized eggs and are diploid, the product of fertilization of an egg by a sperm. Males develop from unfertilized eggs and are haploid.

Such a system of sex determination is called the **haplodiploid system** (refer back to Figure 16.14d). If they have the same parents, each daughter receives an identical set of genes from her haploid father. The other half of a female's genes comes from her diploid mother, so the coefficient of relatedness (r) of sisters is 0.50 (from father) + 0.25 (from mother) = 0.75. The result is that females are more related to their sisters (0.75) than they would be to their own offspring (0.50). This suggests it is evolutionarily advantageous for females to stay in the nest or hive and care for other female offspring of the queen, which are their full sisters.

Elegant though these types of explanations are, they do not explain the whole picture. Large eusocial colonies of termites exist, but termites are diploid, not haplodiploid. In this case, how do we account for the existence of eusociality? In the 1970s, American evolutionary biologist Richard Alexander suggested it was the particular lifestyle of these animals, rather than genetics, that promoted eusociality. He argued that in a normal diploid organism, females are related to their daughters by 0.50 and to their sisters by 0.50, so it should matter little to them whether they rear siblings or daughters of their own. He predicted, well before eusociality was discovered to occur in mammals, that a eusocial mammalian species could exist when certain conditions were met, including that the nests or burrows be enclosed and subterranean, in order to house a large colony, and that the colony have a food supply such as large tubers and roots. In addition, the soil would need to be hard, dry clay to keep the colony safe from digging predators. He proposed that the colony would be defended by a few members of the colony willing to give their lives in defense of others, and he posited the existence of mechanisms by which a queen could manipulate other individuals.

At the time, Alexander had no idea that a mammal with such characteristics existed. Surprisingly, subsequent discoveries confirmed the existence of a eusocial mammal that satisfied all of the predictions of Alexander's model: the naked mole rat (*Heterocephalus glaber*). Naked mole rats are diploid species that live in arid areas of Africa in large underground colonies where only one female, the queen, produces offspring (**Figure 55.19**). A renewable food supply is present in the form of tubers of the plant *Pyrenacantha kaurabassana*.

Figure 55.19 **A naked mole rat colony (*Heterocephalus glaber*).** In this mammalian species, most females do not reproduce; only the queen (shown resting on workers) has offspring.

These weigh up to 50 kg and can provide food for a whole colony. Because the burrows are hard packed, there are few ways to attack them, and a heroic effort by a mole rat blocking the entrance can effectively stop a predator (commonly a rufous-beaked snake). The queen mole rat does indeed manipulate the colony members; she suppresses reproduction in other females by producing a pheromone in her urine that is passed around the colony by grooming. Hence, the mole rats seem to have evolved the appropriate behavior to exploit this ecological niche. As Alexander argued, lifestyle characteristics can provide an explanation for the evolution of eusociality in species such as termites and naked mole rats, in which both sexes are diploid.

Unrelated Individuals May Engage in Altruistic Acts If the Altruism Is Likely to Be Reciprocated

Even though we have argued that kin selection can explain instances of apparent altruism, cases of altruism are known to exist between unrelated individuals. What drives this type of behavior appears to be a "You scratch my back, I'll scratch yours" type of reciprocal altruism, in which the cost to the animal of behaving altruistically is offset by the likelihood of a return benefit. This occurs in nature, for example, when unrelated chimps groom each other.

American biologist Gerald Wilkinson has noted that female vampire bats exhibit reciprocal altruism via food sharing. Vampire bats can die after 60 hours without a blood meal, because they can no longer maintain their correct body temperature. Adult females share their food with their young, the young of other females, and other unrelated females that have not fed. The females and their dependent young roost together in groups of 8 to 12. A hungry female will solicit food from another female by approaching and grooming her. The female being groomed then regurgitates part of her blood meal for the other. The roles of blood donor and recipient are often reversed, and Wilkinson showed that unrelated females are more likely to share with those that had recently shared with them. The probability of a female getting a free lunch is decreased because the roost consists of individuals that remain associated with each other for long periods of time.

55.7 Mating Systems

Learning Outcomes:

1. Compare and contrast promiscuous, monogamous, polygynous, and polyandrous mating systems.
2. List and describe the two forms of sexual selection.

In nature, males produce millions of sperm, but females produce far fewer eggs. It would seem that the majority of males are superfluous because one male could easily fertilize all the females in a local area. If one male can mate with many females, why in most species does the sex ratio remain at approximately 1 to 1? The answer lies with natural selection. Let's consider a hypothetical population that contains 10 females to every male; each male mates, on average, with 10 females. A parent whose children were exclusively sons could expect to have 10 times the number of grandchildren of a parent with the same number of daughters. Under such conditions, natural selection would favor the spread of genes for male-producing tendencies, and males would become prevalent in the population. If the population were mainly males, females would be at a premium, and natural selection would favor the spread of genes for female-producing tendencies. Such constraints operate on the numbers of both male and female offspring, keeping the sex ratio at about 1:1. This idea was developed in 1930 by the British geneticist Ronald Fisher and has come to be known as Fisher's principle.

Even though the sex ratio is fairly even in most species, that doesn't mean that one female always mates with one male or vice versa. Four different types of mating systems occur in nature (**Figure 55.20**). In some species, mating is promiscuous, with each female and each male mating with multiple partners within a breeding season. In monogamy, each individual mates exclusively with one partner over at least a single breeding cycle and sometimes for longer.

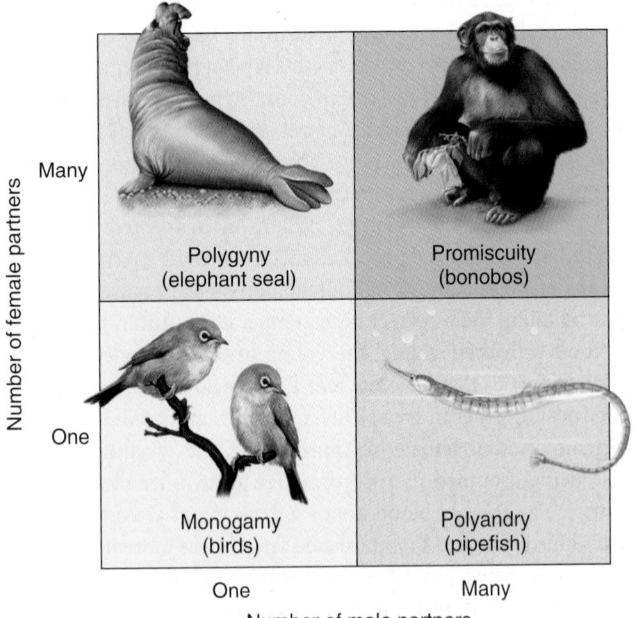

Figure 55.20 **The four different animal mating strategies.**

In contrast, polygamy is a system in which either males or females mate with more than one partner in a breeding season. There are two types of polygamy. In polygyny (Greek for many females), one male mates with more than one female, but females mate only with one male. In polyandry (Greek for many males), one female mates with several males, but males mate with only one female.

In Promiscuous Mating Systems, Each Male or Female Mates with Multiple Partners

Chimpanzees and bonobos are somewhat **promiscuous**; each male mates with many females and vice versa. Here sex alleviates conflict within the social group, but sometimes promiscuity is favored in unpredictable environments. Females that maximize the genetic diversity of their offspring are more likely to have at least some offspring that will survive in a changing world. Intertidal and terrestrial mollusks are also usually promiscuous. Individuals copulate with several partners and eggs are fertilized with sperm from several different individuals. These mollusks are slow moving and they risk desiccation when searching for a mate. The risk of not finding a mate is believed to promote promiscuous mating.

In Monogamous Mating Systems, Males and Females Are Paired for at Least One Reproductive Season

In **monogamy**, each individual mates exclusively with one partner over at least a single breeding cycle and sometimes for longer. Males and females do not exhibit much **sexual dimorphism**, a pronounced difference in the morphologies of the two sexes within a species, and are generally similar in body size and structure (**Figure 55.21a**). Several hypotheses explain the existence of monogamy. The first is the **mate-guarding hypothesis**, which suggests that males stay with a female to protect her from being fertilized by other males. Such a strategy may be advantageous when receptive females are widely scattered and difficult to find.

The **male-assistance hypothesis** maintains that males remain with females to help them rear their offspring. Monogamy is common among birds, about 70% of which are socially monogamous; that is, the pairings remain intact during at least one breeding season. According to the male-assistance hypothesis, monogamy is prevalent in birds because eggs and chicks take a considerable amount of parental care. Most eggs need to be incubated continuously if they are to hatch, and chicks require almost continual feeding. It is therefore in the male's best interest to help raise his young, because he would have few surviving offspring if he did not.

The **female-enforced monogamy hypothesis** suggests that females stop their male partners from being polygynous. Male and female burying beetles (*Nicrophorus defodiens*) work together to bury small, dead animals, which provide a food resource for their developing offspring. Males release pheromones to attract other females to the site. However, while an additional female might increase the male's fitness, the additional developing offspring might compete with the offspring of the first female, decreasing her fitness. As a result, on smelling these pheromones, the first female interferes with the male's attempts at signaling, preserving the monogamous relationship.

(a) Monogamous species

(b) Polygynous species

(c) Polyandrous species

Figure 55.21 **Sexual dimorphism in body size and mating system.** **(a)** In monogamous species, such as these Manchurian cranes, *Grus japonensis*, males and females do not exhibit pronounced sexual dimorphism and appear very similar. **(b)** In polygynous species, such as elk, *Cervus canadensis*, males are bigger than females and have large horns with which they engage in combat over females. **(c)** In polyandrous species, females are usually bigger, as with these golden silk spiders, *Nephila clavipes*.

Recent research by American neuroscientists Larry Young and Elizabeth Hammock has shown that social behavior such as fidelity may have a genetic basis. These researchers found that fidelity of male voles depends on the length of a short tandem repeat sequence (STR) in a gene that codes for a key hormone receptor. Adult male voles with the long version of the STR were more apt to form pair bonds with female partners and nurture their offspring than were voles with the short version.

In Polygynous Mating Systems, One Male Mates with Many Females

In **polygyny** (Greek, meaning many females), one male mates with more than one female in a single breeding season. Physiological constraints often dictate that female organisms must care for the young. Because of these constraints, at least in many organisms with internal fertilization, such as mammals and some fishes, males are able to mate with and then desert several females. Polygynous systems are therefore associated with uniparental care of young, with males contributing little. Sexual dimorphism is typical in polygynous mating systems, with males developing a larger body size to boost success in competition over mates (**Figure 55.21b**). Sexual maturity is often delayed in males that fight because of the considerable time it takes to reach a sufficiently large size to compete for females.

Polygyny is influenced by the temporal or spatial distribution of breeding females and by the availability of resources. In cases when all females are sexually receptive within the same narrow period of time, little opportunity exists for a male to garner all the females for himself. When female reproductive receptivity is spread out over weeks or months, there is much more opportunity for males to mate with more than one female. Where some critical resource is patchily distributed and in short supply, certain males may dominate the resource and breed with more than one visiting female. The major source of nestling death in the lark bunting (*Calamospiza melanocorys*), which lives in North American grasslands, is overheating from too much exposure to the Sun. Prime territories are therefore those with abundant shade, and some males with shaded territories attract two females, even though the second female can expect no help from the male in the process of rearing young. Males in some exposed territories remain bachelors for the season. From the dominant male's point of view, polygyny is advantageous; from the female's point of view, there may be costs. Although by choosing dominant males, a female may be gaining access to good resources, she will have to share these resources with other females.

Sometimes males defend a group of females without commanding a resource-based territory. This pattern is more common when females naturally congregate in groups or herds, perhaps to avoid predation, as in horses, zebras, and some deer, and where space is limited, as with southern elephant seals. Usually the largest and strongest males command most of the matings, but being a dominant male is usually so exhausting that males may only manage to remain the strongest male for a year or two.

Polygynous mating can occur where neither resources nor groups of females are defended. In some instances, particularly in birds and mammals, males display in designated communal courting areas called **leks** (**Figure 55.22**). Females come to these areas specifically to find a mate, and they choose a prospective mate after the males have performed elaborate displays. Most females seek to mate with the best male, so a few of the flashiest males perform the vast majority of the matings. At a lek of the white-bearded manakin (*Manacus manacus*) of South America, one male accounted for 75% of the 438 matings when there were as many as 10 males. A second male mated 56 times (13% of matings), but six others mated only a total of 10 times.

In Polyandrous Mating Systems, One Female Mates with Many Males

Polyandry (Greek, meaning many males), in which one female mates with several males, is more rare than polygyny. Nevertheless, it occurs in some species of birds, fishes, and insects. Sexual dimorphism is present, with the females being the larger of the sexes (see **Figure 55.21c**). In the Arctic tundra, the summer season is short but

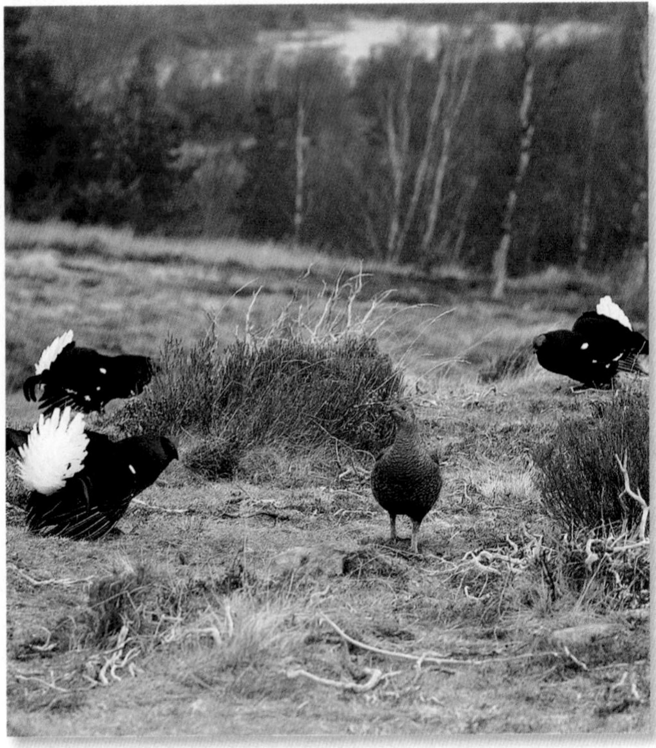

Figure 55.22 Male birds at a lek. Black grouse (*Tetrao tetrix*) congregate at a moorland lek in Scotland in April. Females visit the leks, and males display to them.

very productive, providing a bonanza of insect food for 2 months. The productivity of the breeding grounds of the spotted sandpiper (*Actitis macularia*) is so high that the female becomes rather like an egg factory, laying up to five clutches of four eggs each in 40 days. Her reproductive success is limited not by food but by the number of males she can find to incubate the eggs, and females compete for males, defending territories where the males sit.

Polyandry is also seen in some species where egg predation is high, and males are needed to guard the nests. For example, in the pipefish (*Syngnathus typhle*), males have brood pouches that provide eggs with safety and a supply of oxygen- and nutrient-rich water. Females produce enough eggs to fill the brood pouches of two males and may mate with more than one male.

Sexual Selection Involves Mate Choice and Mate Competition

As we learned in Chapter 24, **sexual selection** is a form of natural selection that promotes traits that improve an individual's mating success. Recall that sexual selection can take two forms. In intersexual selection, members of one sex, usually females, choose mates based on particular characteristics, such as the color of plumage or the sound of a courtship song (refer back to Figure 24.7b). In intrasexual selection, members of one sex, usually males, compete over partners, and the winner performs most of the matings (refer back to Figure 24.7a).

Intersexual Selection Females have many different ways to choose their prospective mates. Female hangingflies (genus *Hylobittacus*) demand a nuptial gift of a food package, an insect prey item that the male has caught (Figure 55.23). Such a nutrient-rich gift may permit females to produce more eggs. The bigger the gift, the longer it takes the female to eat it and the longer the male can copulate with her. Females do not mate with males that do not offer such a package. Female spiders and mantids sometimes eat their mate during or after copulation, with the male's body constituting the ultimate nuptial gift.

Figure 55.23 Female choice of males based on nuptial gifts. A male hangingfly, on the left, has presented a nuptial gift, a small moth, to a female, and now mates with her while she consumes the meal.

Males may also have parenting skills that females desire. Among 15-spined sticklebacks (*Spinachia spinachia*), males perform cleaning, guarding, and fanning the offspring. Males display their parental skills through body shakes during courtship, and females prefer to mate with males that shake their bodies the most energetically, apparently using this cue to assess the quality of the male as a potential father.

Often, females choose mates without the offering of obvious material benefits and make their choices based on plumage color or courtship display. The male African long-tailed widowbird (*Euplectes progne*) has long tail feathers that he displays to females via aerial flights. Swedish researcher Malte Andersson experimentally shortened the tails of some birds by clipping their tail feathers, and lengthened the tails of others by taking the clippings and sticking them onto other birds with superglue. Males with experimentally lengthened tails attracted four times as many females as males with shortened tails, and they fathered more clutches of eggs (Figure 55.24).

Some researchers have suggested ornaments such as excessively long tail feathers function as a sign of an individual's genetic quality, in that the bearer must be very healthy in order to afford this energetically costly trait. This hypothesis is called the handicap principle. However, in some species of birds, other important benefits may be associated with plumage quality. Bright colors are often caused by pigments called carotenoids that help stimulate the immune system to fight diseases. In zebra finches and red jungle fowl, colorful plumage has been associated with heightened resistance to disease, suggesting that females that choose such males are choosing genetically healthier mates.

On rare occasions, a sex role reversal occurs, and the male discriminates among females. In the Mormon cricket (*Anabrus simplex*), males mate only once because they provide a nutrient-rich nuptial gift of a spermatophore to females, which is energetically costly

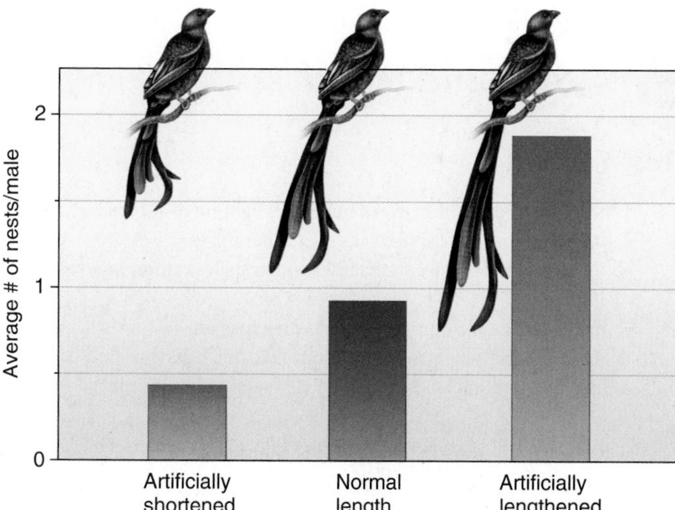

Figure 55.24 Female choice based on male appearance. Males with artificially lengthened tails mate with more females and, therefore, have more nests than males with a normal or artificially shortened tail.

Concept Check: Why do you think it is rarer for female birds or mammals to have more colorful plumage or elaborate adornments than males of the same species?

to produce. In this case, males choose heavier females to mate with because these females have more eggs, and the males can father more offspring.

Intrasexual Selection In many species, females do not actively choose their preferred mate; instead, they mate with competitively superior males. In such cases, dominance is determined by fighting or by ritualized sparring (Figure 55.25). Outcomes may be dictated by the size of weapons, such as antlers or horns, or by body size. In the southern elephant seal (*Mirounga leonina*), females haul up onto the beach to give birth and gain safe haven for their pups from marine predators. Following birth, they are ready to mate. In this situation, dominant males are able to command a substantial group of females and constantly lumber across the beach to fight other males. Over the course of many generations, such competition results in an increased male body size.

Large body size does not always guarantee access to females. Smaller male elephant seals may intercept females in the ocean and attempt to mate with them there, rather than on the beach, where the competitively dominant males patrol. Such "satellite" males, which are unable to acquire and defend territories, move around the edge of the mating arena. For example, small male frogs hang around ponds waiting to intercept females headed toward the call of dominant males. Thus, even though competitively dominant males father most offspring, smaller males can have some reproductive success.

Ultimately, as we have seen, most behaviors have evolved to maximize an individual's reproductive output. In a successful group of individuals, this leads to population growth. But we are not knee-deep in sandpipers or elephant seals, so there must be some constraints on reproductive output. In Chapter 56, we turn to the realm of population ecology to explore how populations grow and what factors limit their growth.

Figure 55.25 Intrasexual selection between elephant seals. These male elephant seals are fighting to maintain control of a group of females.

Concept Check: During fights between males, some pups are crushed as the males lumber across the beach. Why aren't the males careful to avoid the pups?

BioConnections: Look back to Figure 24.7. What anatomical feature of male fiddler crabs is enlarged as a result of intrasexual selection?

Summary of Key Concepts

55.1 The Influence of Genetics and Learning on Behavior

- Behavior is usually due to the interaction of an organism's genes and the environment.
- Genetically programmed behaviors are termed innate and often involve a sign stimulus that initiates a fixed action pattern (Figures 55.1, 55.2).
- Organisms can often make modifications to their behavior based on previous experience, a process called learning. Some forms of learning include habituation, classical conditioning, operant conditioning, and cognitive learning (Figures 55.3, 55.4).
- Much behavior is a mixture of innate and learned behaviors. A good example of this occurs in a process called imprinting, in which animals develop strong attachments that influence subsequent behavior (Figures 55.5, 55.6).

55.2 Local Movement and Long-Range Migration

- The simplest forms of local movement involve kinesis, taxis, and memory (Figure 55.7).
- Many animals undergo long-range seasonal movement called migration in order to feed or breed. They do this using three mechanisms: piloting (the ability to move from one landmark to the next), orientation (the ability to follow a compass bearing), and navigation (the ability to set, follow, and adjust a compass bearing) (Figures 55.8, 55.9).

55.3 Foraging Behavior

- Animals use complex behavior in food gathering or foraging. Optimality theory views foraging behavior as a compromise between the costs and benefits involved. The theory of optimal foraging assumes that animals modify their behavior to keep the ratio of their energy uptake to energy expenditure high (Figure 55.10).
- The size of a territory, a fixed area in which an individual or group excludes other members of its own species, tends to be optimized according to the costs and benefits involved (Figure 55.11).

55.4 Communication

- Communication is a form of behavior. The use of different forms of communication between organisms depends on the environment in which they live.
- Chemical communication often involves marking territories; auditory and visual forms of communication are often used to attract mates. A fascinating form of tactile communication involves the dance of the honeybee (Figures 55.12, 55.13).

55.5 Living in Groups

- Many benefits of group living relate to defense against predators, offering protection through sheer numbers and through what is called the many-eyes hypothesis or the geometry of the selfish herd (Figure 55.14).

55.6 Altruism

- Altruism is behavior that benefits others at a cost to oneself. One of the first hypotheses to explain altruism, called group selection, suggested that natural selection produced outcomes beneficial for the group. Biologists now believe that most apparently altruistic acts are often associated with outcomes beneficial to those most closely related to the individual, a concept termed kin selection (Figures 55.15, 55.16, 55.17, 55.18).
- Altruism among eusocial animals may arise partly from the unique genetics of the animals and partly from lifestyle (Figure 55.19).
- Altruism is known to exist among nonrelated individuals that live in close proximity for long periods of time.

55.7 Mating Systems

- Four types of mating systems are found among animals: promiscuity, monogamy, polygyny, and polyandry. Relative body size of males and females depends on mating system (Figures 55.20, 55.21).
- Polygynous mating can often occur in situations when males dominate a resource, defend groups of females, or display in common courting areas called leks (Figure 55.22).
- Sexual selection takes two forms: intersexual selection, in which the female chooses a mate based on particular characteristics, or intrasexual selection, in which males compete with one another for the opportunity to mate with a female (Figures 55.23, 55.24, 55.25).

Assess and Discuss

Test Yourself

1. What is the proximate cause of male deer fighting over females?
   a. to determine their supremacy over other males
   b. to injure other males so that these other males cannot mate with females
   c. to maximize the number of genes they pass on
   d. because changes in day length stimulate this behavior
   e. because fighting helps rid the herd of weaker individuals
2. Geotaxis is a response to the force of gravity. Fruit flies placed in a vial will move to the top of the vial. This is an example of _______ geotaxis.
   a. positive
   b. neutral
   c. innate
   d. negative
   e. learned
3. Certain behaviors seem to have very little environmental influence. Such behaviors are the same in all individuals regardless of the environment and are referred to as ________ behaviors.
   a. genetically programmed
   b. instinctual
   c. innate
   d. all of the above
   e. b and c only
4. Patrick has decided to teach his puppy a few new tricks. Each time the puppy responds correctly to Patrick's command, the puppy is given a treat. This is an example of
   a. habituation.
   b. classical conditioning.
   c. operant conditioning.
   d. imprinting.
   e. orientation.
5. Whales have magnetite in their retinas, which aids in navigation during migration by
   a. piloting.
   b. locating the position of the Sun.
   c. use of the Earth's magnetic fields.
   d. locating the positions of the stars.
   e. none of the above.
6. For group living to evolve, the benefits of living in a group must be greater than the costs of group living. Which of the following is an example of a benefit of living in a group?
   a. reduced spread of disease and/or parasites
   b. increased food availability
   c. reduced competition for mates
   d. decreased risk of predation
   e. all of the above
7. The modification of behavior based on prior experience is called
   a. a fixed action pattern.
   b. learning.
   c. navigation.
   d. adjustment behavior.
   e. innate.
8. When an individual behaves in a way that reduces its own fitness but increases the fitness of others, the organism is exhibiting
   a. kin selection.
   b. group selection.
   c. altruism.
   d. selfishness.
   e. ignorance.
9. In ants, which employ a haplodiploid mating system, fathers are related to sons by $r =$
   a. 0
   b. 0.125
   c. 0.25
   d. 0.5
   e. 0.75

10. In a polygynous mating system,
    a. one male mates with one female.
    b. one female mates with many different males.
    c. one male mates with many different females.
    d. many different females mate with many different males.

Conceptual Questions

1. Some male spiders are eaten by the females after copulation. How can this act be seen to benefit the males?
2. Male parental care occurs in only 7% of fishes and amphibian families with internal fertilization but in 69% of families with external fertilization. Propose an explanation for why this is so.
3. A principle of biology is that *new properties emerge from complex interactions*. Male brown bears (*Ursus arctos*) can be infanticidal, killing cubs when they move into a new territory. Explain why bear hunting may have severe consequences for bear populations.

Collaborative Questions

1. Whooping cranes (*Grus americana*) are an endangered species that are bred in captivity to increase their numbers. One problem with this approach is that these cranes are migratory. In the absence of other cranes, can you think of an innovative way human researchers might have used crane behavior to ensure their safe passage to overwintering sites?
2. Discuss several ways in which organisms communicate with each other.

Online Resource

www.brookerbiology.com

Stay a step ahead in your studies with animations that bring concepts to life and practice tests to assess your understanding. Your instructor may also recommend the interactive eBook, individualized learning tools, and more.

Population Ecology

Chapter Outline

56.1 Understanding Populations
56.2 Demography
56.3 How Populations Grow
56.4 Human Population Growth
Summary of Key Concepts
Assess and Discuss

56

A population of black-footed ferrets in Meeteetse, Wyoming.

The last known population of black-footed ferrets, *Mustela nigripes*, was discovered in 1981 near Meeteetse, Wyoming. Shortly thereafter, all but 18 of the 100 known ferrets in Meeteetse died of canine distemper. The remainder were captured between 1985 and 1987, inoculated against distemper, and bred in captivity, with the intent of reestablishing the population in the wild later on. Since then, populations have been established in Arizona, Colorado, Montana, South Dakota, Utah, Wyoming, and Chihuahua, Mexico.

In Wyoming, an area called Shirley Basin was one of those targeted for reintroductions of captive-born animals. During 1991 to 1994, Shirley Basin received 228 ferrets, but distemper again triggered a decline in the population size. By 1997, only 5 ferrets were found. Extinction seemed imminent. Monitoring efforts, which might disturb the animals, decreased. Surprisingly, by 2003, a total of 52 animals were found, and by 2006, 223 were present. How did the number of ferrets increase this fast?

A **population** can be defined as a group of interbreeding individuals occupying the same area at the same time. In this way, we can think of a population of water lilies in a particular lake, the lion population in the Ngorongoro crater in Africa, or the human population of New York City. The boundaries of a population can be a little difficult to define, though they may correspond to geographic features such as the boundaries of a lake or forest or be contained within a mountain valley or a certain island. Individuals may enter or leave a population, such as the human population of New York City or the deer population in North Carolina. Thus, populations are often fluid entities, with individuals moving into (immigrating) or out of (emigrating) an area.

This chapter explores **population ecology**, the study of what factors affect population size and how these factors change over space and time. To study populations, we need to employ some of the tools of **demography**, the study of birth rates, death rates, age distributions, and the sizes of populations. We begin our discussion by examining the ways that ecologists measure and categorize populations. We will explore characteristics of populations and how growth rates are determined by the number of reproductive individuals in the population and their fertility rate. These data are used to construct simple mathematical models that allow us to analyze population growth, such as that of the black-footed ferrets, and predict future growth. We will also look at the factors that limit the growth of populations and conclude the chapter by using the population concepts and models to explore the growth of human populations.

56.1 Understanding Populations

Learning Outcomes:

1. List the different techniques ecologists use to measure population density.
2. Identify the three main patterns of dispersion observed in nature.
3. Describe the difference between semelparity and iteroparity—two different reproductive strategies.

Within their areas of distribution, organisms occur in varying numbers. We recognize this pattern by saying a plant or animal is "rare" in one place and "common" in another. For more precision, ecologists quantify distribution further and talk in terms of population **density**—the numbers of organisms in a given unit area or volume. Population growth affects population density, and knowledge of both can help us make decisions about the management of species. How long will it take for a population of an endangered species to recover to a healthy level if we protect it from its most serious threats? For example, how quickly will the black-footed ferret populations increase in Wyoming? A knowledge of population growth rates and population densities would allow us to predict future ferret population sizes. Since 1994, several large parts of Georges Bank, an area of the sea floor in the North Atlantic that was once one of the world's richest fishing grounds, have been closed to commercial fishing because of overfishing. How long will it take for populations to recover? How many fishes can we reasonably trawl from the sea and still ensure that an adequate population will exist for future use? Such information is vital in making determinations of size limits, catch quotas, and length of season for fisheries to ensure an adequate future population size.

In this section, we will examine density and other characteristics of populations within their habitats. We will also discuss the different reproductive strategies organisms use and how ecologists assign individuals to different groups called age classes.

Ecologists Use Many Different Methods to Quantify Population Density

The simplest method for measuring population density is to visually count the number of organisms in a given area. We can reasonably do this only if the area is small and the organisms are relatively large. For example, we can readily determine the number of gumbo limbo trees (*Bursera simaruba*) on a small island in the Florida Keys. Normally, however, population ecologists calculate the density of plants or animals in a small area and use this figure to estimate the total abundance over a larger area.

For plants, algae, or other sessile organisms such as intertidal animals, it is fairly easy to count numbers of individuals per square meter or, for larger organisms such as trees, numbers per hectare (an area of land equivalent to 2.471 acres). However, many plant individuals are clonal; that is, they grow in patches of genetically identical individuals, so that rather than count individuals, we can also use the amount of ground covered by plants as an estimate of vegetation density.

Plant ecologists use a sampling device called a **quadrat**, a square frame that often, but not always, measures 50 × 50 cm and encloses an area of 0.25 m^2 (**Figure 56.1a**). They then count the numbers of plants of a given species inside the quadrat to obtain a density estimate per square meter. For example, if you counted densities of 20, 35, 30, and 15 plants in four quadrats, you could reliably say that the density of this species was 25 individuals per 0.25 m^2, or 100/m^2. For larger plants, such as trees, a quadrat would be ineffective. To count such organisms, many ecologists perform a **line transect**, in which a long piece of string is stretched out and any tree along its length is counted. For example, to count tree species on larger islands in the Florida Keys, we could lay out a 100-m line transect and count all the trees within 1 m on either side of the transect. In effect, this transect is little more than a long, thin quadrat encompassing 200 m^2. By performing five such transects, we could obtain estimates of tree density per 1,000 m^2 and then extrapolate that to a number per hectare or per island.

Several different sampling methods exist for quantifying the density of animals, which are more mobile than plants. Suction traps, like giant aerial vacuum cleaners, can suck flying insects from the sky. Pitfall traps set into the ground can catch species such as spiders, lizards, or beetles wandering over the surface (**Figure 56.1b**). Sweep nets can be passed over vegetation to dislodge and capture the insects feeding there. Mist nets—very fine netting spread between trees—can entangle flying birds and bats (**Figure 56.1c**). Baited snap traps, such as mouse traps, or live traps can snare terrestrial animals (**Figure 56.1d**). Population density can thus be estimated as the number of animals caught per trap or per unit area where a given number of traps are set, for example, 10 traps per 100 m^2 of habitat.

Sometimes population biologists capture animals and then tag and release them (**Figure 56.2**). The rationale behind the **mark-recapture technique** is that after the tagged animals are released, they mix freely with unmarked individuals and within a short time are randomly mixed within the population. The population is resampled, and the numbers of marked and unmarked individuals are recorded.

(a) Quadrat

(b) Pitfall trap

(c) Mist net

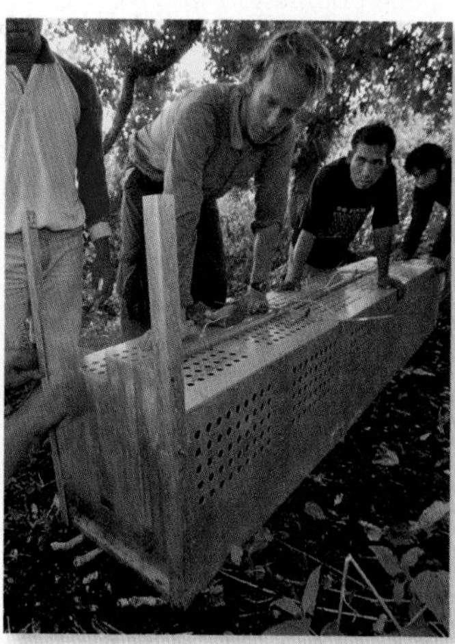
(d) Live mammal trap

Figure 56.1 Sampling techniques. **(a)** Quadrats are frequently used to count the number of plants per unit area. **(b)** Pitfall traps set into the ground catch wandering species such as beetles and spiders. **(c)** Mist nets consist of very fine mesh to entangle birds or bats. **(d)** Baited live traps catch terrestrial animals, including Komodo dragons, as here on Rinca Island, Indonesia.

Figure 56.2 The mark-recapture technique for estimating population size. An ear tag identifies this Rocky Mountain goat (*Oreamnos americanus*) in Olympic National Park, Washington. Recapture of such marked animals permits estimates of population size.

Concept Check: If we mark 110 Rocky Mountain goats and recapture 100 goats, 20 of which have ear tags, what is the estimate of the total population size?

We assume that the ratio of marked to unmarked individuals in the second sample is the same as the ratio of marked individuals in the first sample to the total population size. Thus,

$$\frac{\text{Number of individuals marked in first catch}}{\text{Total population size, } N} = \frac{\text{Number of marked recaptures in second catch}}{\text{Total number of second catch}}$$

Let's say we catch 50 largemouth bass in a lake and mark them with colored fin tags. A week later, we return to the lake and catch 40 fish and 5 of them were previously tagged fish. If we assume no immigration or emigration has occurred, which is quite likely in a closed system like a lake, and we assume there have been no births or deaths of fish, then the total population size is given by rearranging the equation:

$$\text{Total population size, } N = \frac{\text{Number of marked individuals in first catch} \times \text{Total number of second catch}}{\text{Number of marked recaptures in second catch}}$$

Using our data,

$$N = \frac{50 \times 40}{5} = \frac{2{,}000}{5} = 400$$

From this equation, we estimate that the lake has a total population size of 400 largemouth bass. This could be useful information for game and fish personnel who wish to know the total size of a fish population in order to set catch limits.

However, the mark-recapture technique can have drawbacks. Some animals that have been marked may learn to avoid the traps. Recapture rates will then be low, resulting in an overestimate of population size. Imagine that instead of 5 tagged fish out of 40 recaptured fish, we get only 2 tagged fish. Now our population size estimate is 2,000/2 = 1,000, a dramatic increase in our population size estimate. On the other hand, some animals can become "trap-happy," particularly if the traps are baited with food. This would result in an underestimate of the population size.

Because of the limitations of the mark-recapture technique, ecologists also use other, more novel methods to estimate population density. For some larger terrestrial or marine species, captured animals can be fitted with radio collars and followed remotely, using an antennal tracking device. Their home ranges can be determined and population estimates developed based on the area of available habitat. For many species with valuable pelts, we can track population densities through time by examining pelt records taken from trading stations. We can also estimate relative population density by examining catch per unit effort, which is especially valuable in commercial fisheries. We can't easily expect to count the number of fishes in an area of ocean, but we can count the number caught, say, per 100 hours of trawling. For some species that leave easily recognizable fecal pellets, like rabbits or deer, we can count pellet numbers, and, if we know the pellet production per individual and how long pellets last in the environment, we can estimate population size. For frogs or birds, we can count chorusing or singing individuals. We can also count leaf scars or chewed leaves and, if we know the responsible herbivores and the rates of herbivory, use these to estimate the density of the animals that damage them.

Populations Show Different Degrees of Spacing Among Individuals

Individuals within a population show different patterns of **dispersion**; that is, they can be clustered together or spread out to varying degrees. The three basic kinds of dispersion patterns are clumped, uniform, and random.

The type of dispersion observed in nature can tell us a lot about what processes shape group structure. The most common dispersion pattern is **clumped**, because resources in nature tend to be clustered. For example, certain plants may do better in moist conditions, and moisture is greater in low-lying areas (**Figure 56.3a**). Social behavior among animals that aggregate into flocks or herds reflects a clumped pattern.

On the other hand, competition may cause a **uniform** dispersion pattern among individuals, as among trees in a forest. At first, the pattern of trees and seedlings may appear random as seedlings develop from seeds dropped at random, but competition among roots may cause some trees to be outcompeted by others, causing a thinning out and resulting in a relatively uniform distribution. Thus, the dispersion pattern starts out random but ends up uniform. Uniform dispersions may also result from social interactions, as among some nesting birds, which tend to keep an even distance from one other (**Figure 56.3b**).

Perhaps the rarest dispersion pattern is **random**, in which the probability of finding an individual at any point in an area is equal,

Figure 56.3 Three types of dispersion. (a) A clumped distribution pattern, as in these plants clustered around an oasis, often results from the uneven distribution of a resource, in this case, water. (b) A uniform distribution pattern, as in these nesting black-browed albatrosses (*Diomedea melanophris*) on the Falkland Islands, may be a result of competition or social interactions. (c) A random distribution pattern, as in these bushes at Leirhnjukur Volcano in Iceland, is the least common form of spacing.

Concept Check: *What is the distribution pattern of students in a half-empty classroom?*

BioConnections: *Look back to Figure 55.11b. What is the likely dispersion pattern of cheetahs in the wild?*

because resources in nature are rarely randomly spaced. Where resources are common and abundant, as in moist, fertile soil, the dispersion patterns of plants may lack a pattern as plants germinate from randomly dispersed wind-blown seeds (**Figure 56.3c**).

Reproductive Strategies May Differ Among Species

To better understand how populations grow in size, let's consider their reproductive strategies. Some organisms produce all of their offspring in a single reproductive event. This pattern, called **semelparity** (from the Latin *semel*, meaning once, and *parere*, meaning to bear), is common in insects and invertebrates and also occurs in organisms such as salmon, bamboo grasses, and agave plants (**Figure 56.4a**). These individuals reproduce once only and die. Semelparous organisms, like agaves, may live for many years before reproducing, or they may be annual plants that develop from seed, flower, and drop their own seed within a year.

Other organisms reproduce in successive years or breeding seasons. The pattern of repeated reproduction at intervals throughout the life cycle is called **iteroparity** (from the Latin *itero*, meaning to

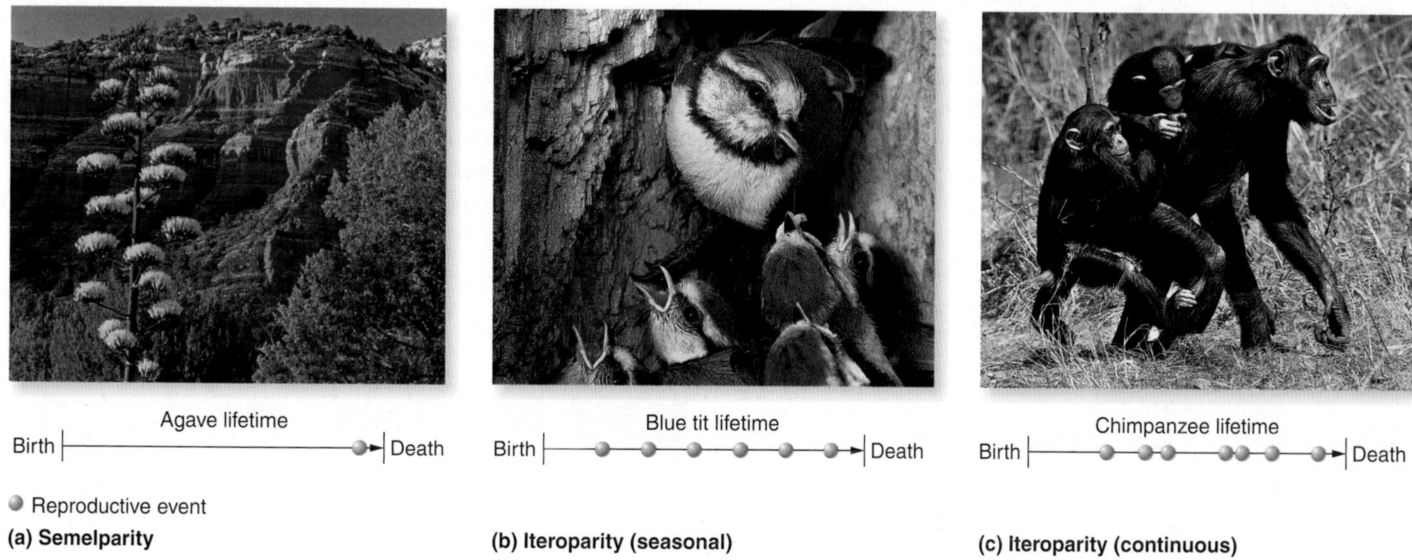

Figure 56.4 Differences in reproductive strategies. Species such as (a) agave plants (*Agave shawii*) are semelparous, meaning they breed once in their lifetime and then die. This contrasts with (b) blue tits (*Parus caeruleus*) and (c) chimpanzees (*Pan troglodytes*), which are iteroparous and breed more than once in their lifetime.

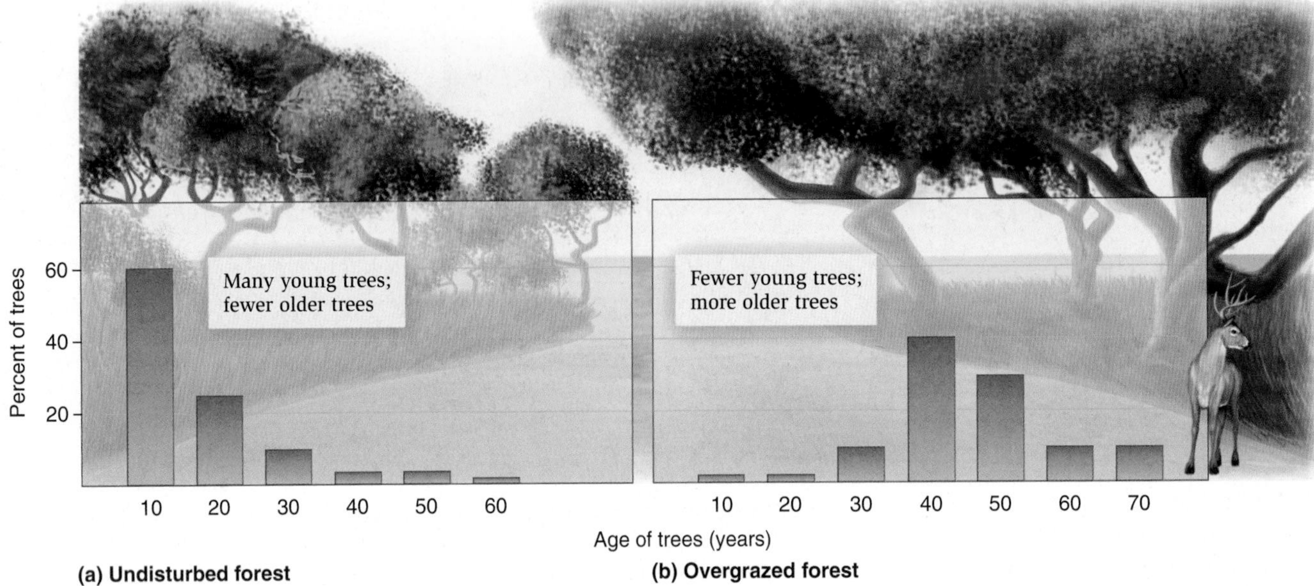

Figure 56.5 **Theoretical age distribution of two populations of gumbo limbo trees in the Florida Keys.** **(a)** Age distribution of an island with no Key deer with numerous young trees, many of which die as the trees age and compete with one another for resources, leaving relatively few big, older trees. **(b)** Age distribution of a forest where overgrazing by Key deer has reduced the abundance of young trees, leaving only trees in the older age classes.

repeat). It is common in most vertebrates, perennial plants, and trees. Among iteroparous organisms, much variation occurs in the number of reproductive events and in the number of offspring per event. Many species, such as birds or trees in temperate areas, have distinct breeding seasons (seasonal iteroparity) that lead to distinct generations (Figure 56.4b). For a few species, individuals reproduce repeatedly and at any time of the year. This is termed continuous iteroparity and is exhibited by some tropical species, many parasites, and many primates (Figure 56.4c).

Why do species reproduce in a semelparous or iteroparous mode? The answer may lie in part in environmental uncertainty. If survival of juveniles is very poor and unpredictable, then selection favors repeated reproduction and a long reproductive life to increase the chance that juveniles will survive in at least some years. If the environment is stable, then selection favors a single act of reproduction, because the organism can devote all its energy to making offspring, not maintaining its own body. Under favorable circumstances, annual plants produce more seeds per unit biomass than trees, which have to invest a lot of energy in maintenance. However, when the environment becomes stressful, annuals run the risk of their seeds not germinating. They must rely on some seeds successfully lying dormant and germinating after the environmental stress has ended.

The Size of an Age Class Can Indicate How Quickly a Population Might Grow

The reproductive strategy employed by an organism has a strong effect on the subsequent age classes of a population. Semelparous organisms often produce groups of same-aged young called **cohorts** that grow at similar rates. Iteroparous organisms generally have many young of different ages because the parents reproduce frequently. The age classes of populations can be characterized by specific categories, such as years in mammals, stages (eggs, larvae, or pupae) in insects, or size in plants.

We expect that a population that is increasing should have a large number of young, whereas a decreasing population should have few young. An imbalance in age classes can have a profound influence on a population's future. For example, in an overexploited fish population, the larger, older reproductive age classes are often removed. If the population is overfished for several years, there will be no young fish to move into the reproductive age class to replace the removed fish, and the population may collapse. Other populations experience removal of younger age classes. In the Florida Keys, populations of Key deer overgraze young gumbo limbo trees, leaving older trees, whose foliage is too tall for them to reach (Figure 56.5). This can have disastrous effects on the future population of trees, for although the forest might consist of healthy mature trees, when these die, there will be no replacements. To accurately examine how populations grow, we need to examine and understand the demography of the population.

56.2 Demography

Learning Outcomes:

1. Describe the difference between information summarized in a life table and in a survivorship curve.
2. Differentiate among type I, II, and III survivorship curves and give examples of organisms that exhibit those survivorship curves.
3. Analyze age-specific fertility data and predict a population's growth.

One way to determine how a population will change is to examine a cohort of individuals from birth to death. For most animals and plants, this involves marking a group of individuals in a population as soon as they are born or germinate and following their fate through their lifetime. For some long-lived organisms, such as tortoises, elephants, or trees, this is impractical, so a snapshot approach is used, in which researchers examine the age structure of a population at one point in time. Recording the presence of juveniles and mature individuals, researchers use this information to construct a **life table**—a table that provides data on the number of individuals alive in each particular age class. Age classes can be created for any time period, but they often represent 1 year. Only females are included in these tables because only females produce offspring. In this section, we will determine how to construct life tables and plot survivorship curves, which show at a glance the general pattern of population survival over time.

Life Tables and Survivorship Curves Summarize Survival Patterns

Let's examine a life table for the North American beaver (*Castor canadensis*). Prized for their pelts, by the mid-19th century, these animals had been hunted and trapped to near extinction. Beavers began to be protected by laws in the 20th century, and populations recovered in many areas, often growing to what some considered to be nuisance status. In Newfoundland, Canada, legislation supported trapping as a management technique. From 1964 to 1971, trappers provided mandibles from which teeth were extracted for age classification. If many mandibles were obtained from, say, 1-year-old beavers, then such animals were probably common in the population. If the number of mandibles from 2-year-old beavers was low, then we know there was high mortality for the 1-year-old age class. From the mandible data, researchers constructed a life table (**Table 56.1**). The number of individuals alive at the start of the time period (in this case, a year) is referred to as n_x, where n is the number, and x refers to the particular age class. By subtracting the value of n_x from the number alive at the start of the previous year, we can calculate the number dying in a given age class or year, d_x. Thus $d_x = n_x - n_x$+1. For example, in Table 56.1, 273 beavers were alive at the start of their sixth year (n_5), and only 205 were alive at the start of the seventh year (n_6); thus, 68 died during the sixth year: $d_5 = n_5 - n_6$, or $d_5 = 273 - 205 = 68$.

A simple but informative exercise is to plot numbers of surviving individuals at each age, creating a **survivorship curve** (**Figure 56.6**). The value of n_x, the number of individuals, is typically expressed on a log scale. Ecologists use a log scale to examine rates of change with time, not change in absolute numbers. Although we could accomplish the same thing with a linear scale, the use of logs makes it easier to examine a wide range of population sizes. For example, if we start with 1,000 individuals and 500 are lost in year 1, the log of the decrease is

$$\log_{10} 1{,}000 - \log_{10} 500 = 3.0 - 2.7 = 0.3 \text{ per year}$$

If we start with 100 individuals and 50 are lost, the log of the decrease is similarly

$$\log_{10} 100 - \log_{10} 50 = 2.0 - 1.7 = 0.3 \text{ per year}$$

In both cases, the rates of change are identical, even though the absolute numbers are different. Plotting the n_x data on a log scale ensures that regardless of the size of the starting population, the rate of change of one survivorship curve can easily be compared with that of another species.

Survivorship curves generally fall into one of three patterns (**Figure 56.7**). In a type I curve, the rate of loss for juveniles is relatively low, and most individuals are lost later in life, as they become older

Table 56.1 Life Table for the Beaver (*Castor canadensis*) in Newfoundland, Canada

Age (years), x	Number alive at start of year, n_x	Number dying during year, d_x	Proportion alive at start of year, l_x	Age-specific fertility, m_x	$l_x m_x$
0–1	3,695	1,995	1.000	0.000	0
1–2	1,700	684	0.460	0.315	0.145
2–3	1,016	359	0.275	0.400	0.110
3–4	657	286	0.178	0.895	0.159
4–5	371	98	0.100	1.244	0.124
5–6	273	68	0.074	1.440	0.107
6–7	205	40	0.055	1.282	0.071
7–8	165	38	0.045	1.280	0.058
8–9	127	14	0.034	1.387	0.047
9–10	113	26	0.031	1.080	0.033
10–11	87	37	0.024	1.800	0.043
11–12	50	4	0.014	1.080	0.015
12–13	46	17	0.012	1.440	0.017
13–14	29	7	0.007	0.720	0.005
14+	22	22	0.006	0.720	0.004

Net reproductive rate, $\Sigma l_x m_x = 0.938$

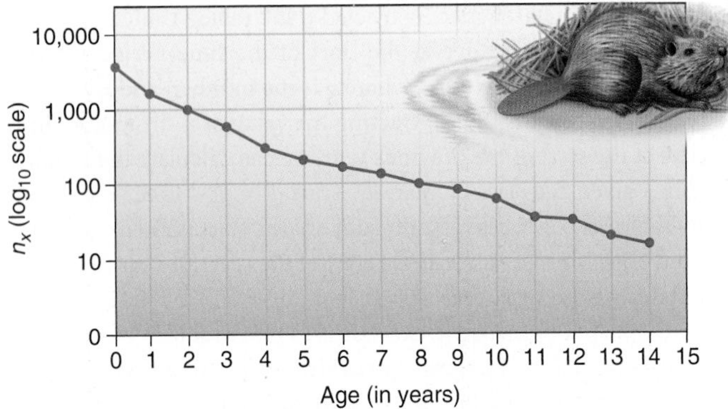

Figure 56.6 **Survivorship curve for the North American beaver.** The survivorship curve is generated by plotting the number of surviving individuals, n_x, from any given cohort of young, usually measured on a log scale, against age. This survivorship curve shows a fairly uniform rate of decline through time.

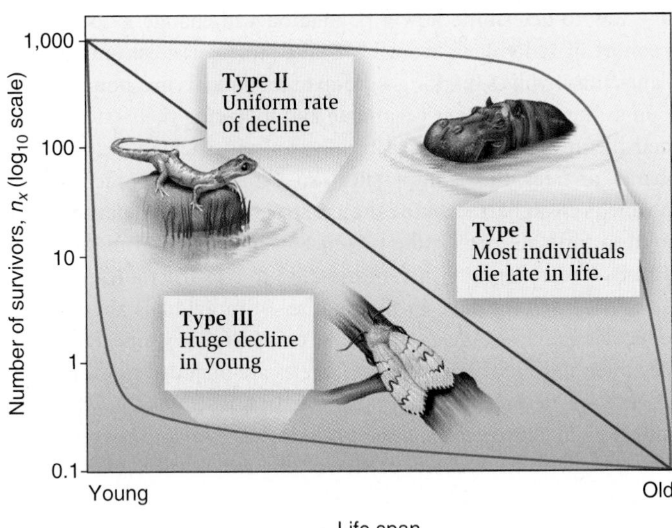

Figure 56.7 **Idealized survivorship curves.**

Concept Check: *Which type of survivorship curve would you expect in (a) mussels and (b) turtles?*

and more prone to sickness and predators (see the Feature Investigation that follows). Organisms that exhibit type I survivorship have relatively few offspring but invest much time and resources in raising their young. Many large mammals, including humans, exhibit type I curves. At the other end of the scale is a type III curve, in which the rate of loss for juveniles is relatively high, and the survivorship curve flattens out for those organisms that have avoided early death. Many fishes and marine invertebrates fit this pattern. Most of the juveniles die or are eaten, but a few reach a favorable habitat and thrive. For example, once they find a suitable rock face on which to attach themselves, barnacles grow and survive very well. Many insects and plants also fit the type III survivorship curve, because they lay many eggs or release hundreds of seeds, respectively. Type II curves represent a middle ground, with fairly uniform death rates over time. Species with type II survivorship curves include many birds, small mammals, reptiles, and some annual plants. The North American beaver population exhibits this survivorship curve. Keep in mind, however, that these are generalized curves and that few populations fit them exactly.

Feature Investigation

Murie's Collections of Dall Mountain Sheep Skulls Permitted the Accurate Construction of Life Tables

The Dall mountain sheep (*Ovis dalli*) lives in mountainous regions, including the Arctic and sub-Arctic regions of Alaska. In the late 1930s, the U.S. National Park Service was bombarded with public concerns that wolves were responsible for a sharp decline in the population of Dall mountain sheep in Denali National Park (then Mt. McKinley National Park). Shooting the wolves was advocated as a way of increasing the number of sheep. Because meaningful data on sheep mortality were nonexistent, the Park Service enlisted American biologist Adolph Murie to determine whether the wolves killed enough sheep to justify controlling the wolf population. In addition to spending many hours observing interactions between wolves and sheep, Murie also collected sheep skulls, determining the sheep's age at death by counting annual growth rings on the horns.

In 1947, American ecologist Edward Deevey put Murie's data in the form of a life table that listed each age class and the number of skulls in it (**Figure 56.8**). Although Murie had collected 608 skulls, Deevey expressed the data per 1,000 individuals to allow for comparison with other life tables. From the data, Deevey constructed a survivorship curve. For the Dall mountain sheep in Denali National Park, there was a slight initial decline in survivorship as young lambs were lost; then the survivorship curve flattened out, indicating that the sheep survived well through about age 7 or 8. Then the number of sheep declined rapidly as they aged. These data underlined what Murie had previously observed, which was that wolves preyed primarily on the most vulnerable members of the sheep population—the youngest and the oldest. Such predation would not be expected to dramatically reduce the sheep population. The Park Service ultimately ended their limited wolf-control program.

Experimental Questions

1. What problem led to the study conducted by Murie on the Dall mountain sheep population of Denali National Park?
2. Describe the survivorship curve developed by Deevey based on Murie's data.
3. How did the Murie and Deevey data affect the decision of the Park Service on the control of the wolf population?

Figure 56.8 **Examining the survivorship curve of a Dall mountain sheep population reveals information on the cause of death.**

HYPOTHESIS Culling the wolf population would protect reproductively active adults in the Dall mountain sheep population.

STARTING LOCATION Denali National Park (formerly known as Mt. McKinley National Park) in Alaska, where wolf predation of sheep is common.

	Experimental level	Conceptual level
1 Collect sheep skulls lying on the ground.		Only skulls with horns are collected in this sampling technique.
2 Determine the age of the skulls by counting their growth rings.		Annuli are the annual growth rings used to estimate a horned animal's age.
3 Organize the data into a life table (see step 4) and construct a survivorship curve using the data.		Survivorship curve for the Dall mountain sheep shows the number of sheep alive in each age class on a log scale, plotted against age in years.

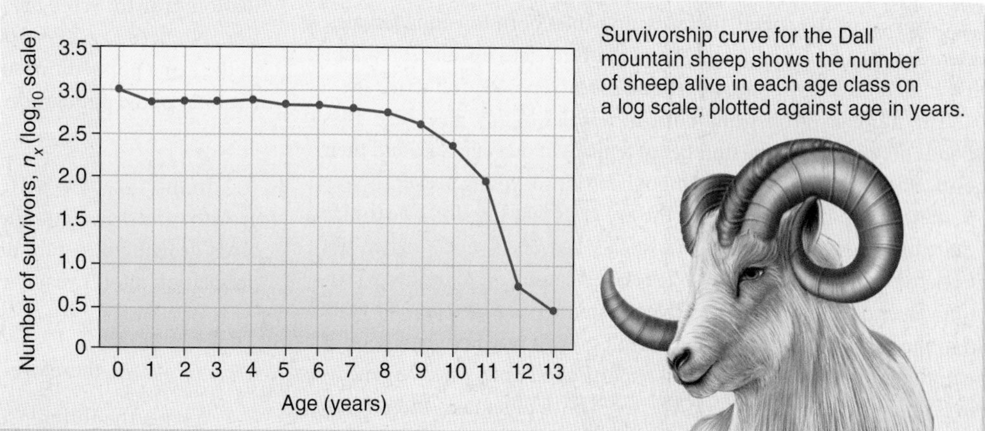

4 THE DATA

Results used in step 3:

Age class	Number alive, n_x	$\log_{10} n_x$
0–1	1,000	3.00
1–2	801	2.90
2–3	789	2.90
3–4	776	2.89
4–5	764	2.88
5–6	734	2.86
6–7	688	2.84
7–8	640	2.81
8–9	571	2.76
9–10	439	2.64
10–11	252	2.40
11–12	96	1.98
12–13	6	0.78
13–14	3	0.48

5 CONCLUSION Most Dall mountain sheep die when very young or very old. Culling the wolf population would not greatly increase sheep survival.

6 SOURCE Deevey, E.S. Jr. 1947. Life tables for natural populations of animals. *Quarterly Review of Biology* 22:283–314.

Age-Specific Fertility Data Can Help to Predict Population Growth

To calculate how a population grows, we need information on birth rates as well as mortality and survivorship rates. For any given age, we can determine the proportion of female offspring that are born to females of reproductive age. Using these data, we can determine an **age-specific fertility rate**, called m_x. For example, if 100 females of a given age produce 75 female offspring, $m_x = 0.75$. With this additional information, we can calculate the growth rate of a population.

First, we use the survivorship data to find the proportion of individuals alive at the start of any given age class. This age-specific survivorship rate, termed l_x, equals n_x/n_0, where n_0 is the number alive at time 0, the start of the study, and n_x is the number alive at the beginning of age class *x*. Let's return to the beaver life table in Table 56.1. The proportion of the original beaver population still alive at the start of the sixth age class, l_5, equals $n_5/n_0 = 273/3{,}695$, or 0.074. This means that 7.4% of the original beaver population survived to age 5. Next we multiply the data in the two columns, l_x and m_x, for each row, to give us a column $l_x m_x$, an average number of offspring per female. This column represents the contribution of each age class to the overall population growth rate. An examination of the beaver age-specific fertility rates illustrates a couple of general points. First, for this beaver population in particular, and for many organisms in general, there are no babies born to young females. As females mature sexually, age-specific fertility goes up, and it remains fairly high until later in life, when females reach postreproductive age.

The number of offspring born to females of any given age class depends on two things: the number of females in that age class and their age-specific fertility rate. Thus, although fertility of young beavers is very low, there are so many females in the age class that $l_x m_x$ for 1-year-olds is quite high. Age-specific fertility for older beavers is much higher, but the relatively few females in these age classes cause $l_x m_x$ to be low. Maximum values of $l_x m_x$ occur for females of an intermediate age, 3–4 years old in the case of the beaver. The overall growth rate per generation is the number of offspring born to all females of all ages, where a generation is defined as the mean period between birth of females and birth of their offspring. Therefore, to calculate the generational growth rate, we sum all the values of $l_x m_x$, that is, $\Sigma l_x m_x$, where the Σ symbol means "sum of." This summed value, R_0, is called the **net reproductive rate**.

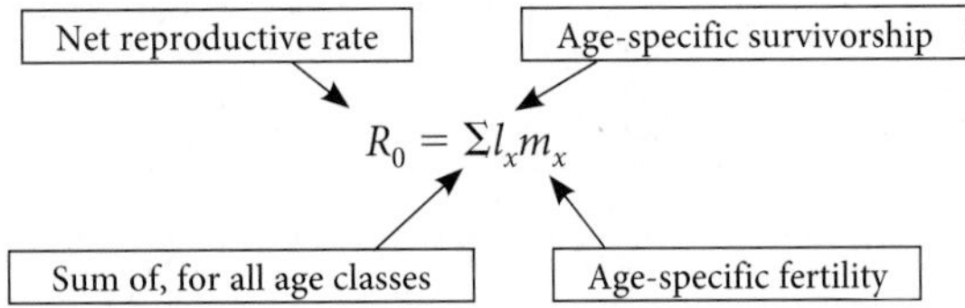

To calculate the future size of a population, we simply multiply the number of individuals in the population by the net reproductive rate. Thus, the population size in the next generation, N_{t+1}, is determined by the number in the population now, at time *t*, which is given by N_t, multiplied by R_0.

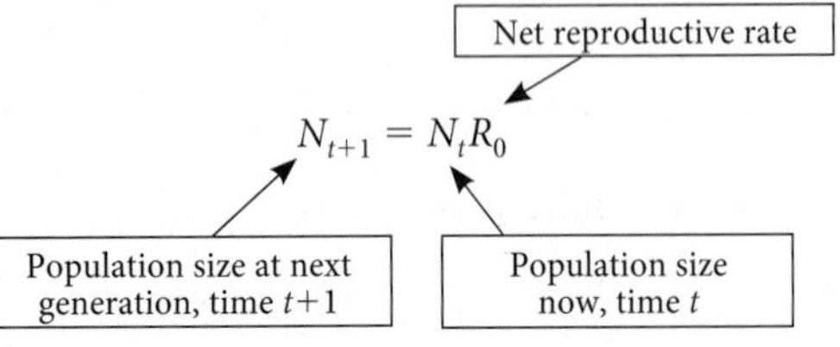

Let's consider an example in which the number of beavers alive now, N_t, is 1,000, and $R_0 = 1.1$. This means the beaver population is reproducing at a rate that is 10% greater than simply replacing itself. The size of the population next generation, N_{t+1}, is given by

$$N_{t+1} = N_t R_0$$
$$N_{t+1} = 1{,}000 \times 1.1$$
$$= 1{,}100$$

Therefore, the number of beavers in the next generation is 1,100, and the population will have grown larger.

In determining population growth, much depends on the value of R_0. If $R_0 > 1$, then the population will grow. If $R_0 < 1$, the population is in decline. If $R_0 = 1$, the population size stays the same, and we say it is at **equilibrium**. In the case of the beavers, Table 56.1 reveals that $R_0 = 0.938$, which is less than 1, and, therefore, the population is declining. This is valuable information, because it tells us that at that time, the beaver population in Newfoundland needed some form of protection (perhaps bans on trapping and hunting) in order to attain a population level at equilibrium.

Because of the effort involved in calculating R_0, the net reproductive rate, ecologists sometimes use a shortcut to predict population growth. Imagine a bird species that breeds annually. To measure population growth, ecologists count the number of birds in the population, N_0. Let's say $N_0 = 100$. The next year, ecologists count 110 birds in the same population, so $N_1 = 110$. The **finite rate of increase**, λ, is the ratio of the population size from one year to the next, calculated as

$$\lambda = N_1/N_0$$

In this case, $\lambda = 1.10$, or 10%. λ is often given as percent annual growth, and *t* is a number of years. Let's consider a population of birds growing at a rate of 5% per year. To calculate the size of the population after 5 years, we substitute λ for R_0:

$$N_t = 100, \lambda = 1.05, \text{and } t = 5$$
$$\text{therefore, } N_{t+5} = 100\,(1.05)^5 = 127.6$$

What's the difference between R_0 and λ? R_0 represents the net reproductive rate per generation. λ represents the finite rate of population change over some time interval, often a year. When species are annual breeders that live 1 year, such as annual plants, $R_0 = \lambda$. For species that breed for multiple years, $R_0 \neq \lambda$. Just as

$$N_t = N_0 R_0{}^t, \text{ where } t = \text{a number of generations}$$
$$\text{so } N_t = N_0 \lambda^t, \text{ where } t = \text{a number of time intervals}$$

Populations grow when R_0 or $\lambda > 1$; populations decline when R_0 or $\lambda < 1$; and they are at equilibrium when R_0 or $\lambda = 1$.

56.3 How Populations Grow

Learning Outcomes:

1. Predict the population growth of continuously breeding organisms using the per capita growth rate (*r*).
2. Distinguish between exponential growth and logistic growth.
3. Explain how density-dependent factors and density-independent factors regulate population size.
4. Compare and contrast life history strategies of *r*-selected and *K*-selected species.

Life tables can provide accurate information about how populations can grow from generation to generation. However, other population growth models can provide valuable insights into how populations grow over shorter time periods. The simplest of these assumes that populations grow if, for any given time interval, the number of births is greater than the number of deaths. In this section, we will examine two different types of these simple models. The first assumes resources are not limiting, and it results in prodigious growth. The second, and perhaps more biologically realistic, assumes resources are limiting, and it results in limits to growth and eventual stable population sizes. We then consider how other factors might limit population growth, such as natural enemies, and discuss the overall life history strategies exhibited by different species.

Knowing the Per Capita Growth Rate Helps Predict How Populations Will Grow

The change in population size over any time period can be written as the number of births per unit time interval minus the number of deaths per unit time interval.

For example, if in a population of 1,000 rabbits, there were 100 births and 50 deaths over the course of 1 year, then the population would grow in size to 1,050 the next year. We can write this formula mathematically as

$$\frac{\text{Change in numbers}}{\text{Change in time}} = \text{Births} - \text{Deaths}$$

or

$$\frac{\Delta N}{\Delta t} = B - D$$

The Greek letter delta, Δ, indicates change, so that ΔN is the change in number, and Δt is the change in time; B is the number of births per time unit; and D is the number of deaths per time unit.

Often, the numbers of births and deaths are expressed per individual in the population, so the birth of 100 rabbits to a population of 1,000 would represent a per capita birth rate, b, of 100/1,000, or 0.10. Similarly, the death of 50 rabbits in a population of 1,000 would be a per capita death rate, d, of 50/1,000, or 0.05. Now we can rewrite our equation giving the rate of change in a population.

$$\frac{\Delta N}{\Delta t} = bN - dN$$

For our rabbit example,

$$\frac{\Delta N}{\Delta t} = 0.10 \times 1{,}000 - 0.05 \times 1{,}000 = 50$$

so if $\Delta t = 1$ year, the rabbit population would increase by 50 individuals in a year.

Ecologists often simplify this formula by representing $b - d$ as r, the **per capita growth rate**. Thus, $bN - dN$ can be written as rN. Because ecologists are also interested in population growth rates over very short time intervals, so-called instantaneous growth rates, instead of writing

$$\frac{\Delta N}{\Delta t}$$

they write

$$\frac{dN}{dt}$$

which is the notation of differential calculus. The equations essentially mean the same thing, except that dN/dt reflects very short time intervals. Thus,

$$\frac{dN}{dt} = rN = (0.10 - 0.05)N = 50$$

Exponential Growth Occurs When the Per Capita Growth Rate Remains Above Zero

How do populations grow? Clearly, much depends on the value of the per capita growth rate, r. When $r < 0$, the population decreases; when $r = 0$, the population remains constant; and when $r > 0$, the population increases. When $r = 0$, the population is often referred to as being at equilibrium, where no changes in population size will occur and there is **zero population growth**.

Even if r is only fractionally above 0, population increase is rapid, and when plotted graphically, a characteristic J-shaped curve results (Figure 56.9). We refer to this type of population growth as **exponential growth**, also known as geometric growth. When conditions are optimal for the population, r is at its maximum rate and is called the **intrinsic rate of increase** (denoted r_{max}). Thus, the rate of population growth under optimal conditions is $dN/dt = r_{max}N$. The larger the value of r_{max}, the steeper the slope of the curve. Because population growth depends on the value of N as well as the value of r, the population increase is even greater as time passes.

How do field data fit this simple model for exponential growth? Population growth cannot go on forever, as envisioned under

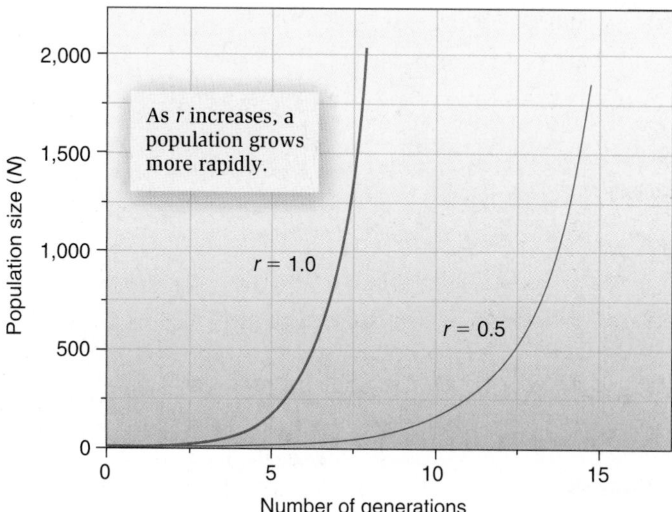

Figure 56.9 **Exponential population growth.** As the value of r increases, the slope of the curve gets steeper. In theory, a population with unlimited resources could grow indefinitely.

exponential growth. But initially at least, in a new and expanding population when resources are not limited, exponential growth is often observed. Let's look at a few examples. Tule elk (*Cervus elaphus nannodes*) is a subspecies of elk that is native to California. Hunted nearly to extinction in the 19th century, less than a dozen individuals survived on a private ranch. In the 20th century, reintroductions resulted in the recovery of tule elk to around 3,500 individuals. One reintroduction was made in March 1978 at Point Reyes National Seashore in California, where 10 animals—2 males and 8 females—were released. By 1993, the herd had reached 214 individuals, and it continued to grow in an exponential fashion until 1998, when the herd size stood at 549 (**Figure 56.10a**). This was deemed an excessive number for the size of the available habitat, and animals were removed to begin herds in other locations. Since then, herd size at Point Reyes has been maintained at around 350.

The growth of the recovering black-foot ferret population in Wyoming that we mentioned at the beginning of the chapter also fits the exponential growth pattern (**Figure 56.10b**). The ferrets had been reintroduced in 1991, but the population declined and languished for many years, so that by 1997, only five were living. However, from 2000 to 2006, the population grew in an exponential fashion. A value of $r = 0.47$ was calculated for the increase in population size of the ferrets during those years. Ecologists have noted that in some cases, populations seem to languish at low levels before conditions become favorable for population growth.

The growth of some introduced species also seems to fit the pattern of exponential growth. The rapid expansion of rabbits after their introduction into southern Australia in the late 19th century is a case in point. In 1859, British immigrant Thomas Austin received two dozen European rabbits from England. Rabbit gestation lasts a mere 31 days, and in southern Australia, each female rabbit could produce up to 10 litters of at least six young each year. The rabbits had essentially no enemies and ate the grass used by sheep and other grazing animals. Even when two-thirds of the population was shot for sport, which was the purpose of the initial introduction, the population grew into the millions within a few short years. By 1875, rabbits were reported on the west coast of Australia, having moved over 1,760 km across the continent despite the deployment of huge, thousand-kilometer-long fences ("rabbit-proof fences") meant to contain them.

Finally, one of the most prominent examples of exponential growth is the growth of the global human population, which, because of its great importance, we will examine separately in Section 56.4.

Logistic Growth Occurs in Populations in Which Resources Are Limited

Despite its applicability to rapidly growing populations, the exponential growth model is not appropriate in many situations. The model assumes unlimited resources, which is not typically the case in the real world. For most species, resources become limiting as populations grow, and the per capita growth rate decreases. The upper boundary for the population size is known as the **carrying capacity** (*K*). A more realistic equation to explain population growth, one that takes into account the amount of available resources, is

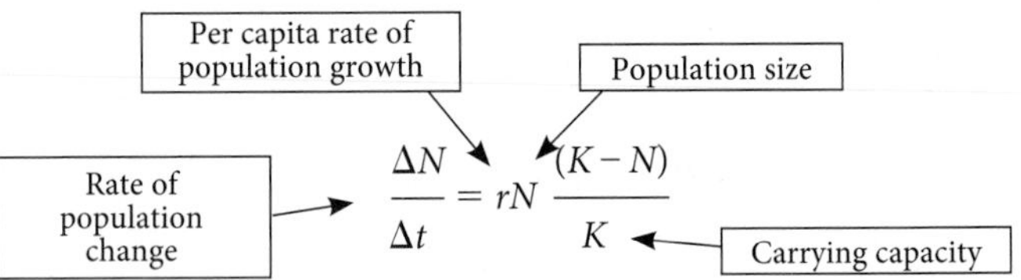

where $(K - N)/K$ represents the proportion of the carrying capacity that is unused by the population. This equation is called the **logistic equation**.

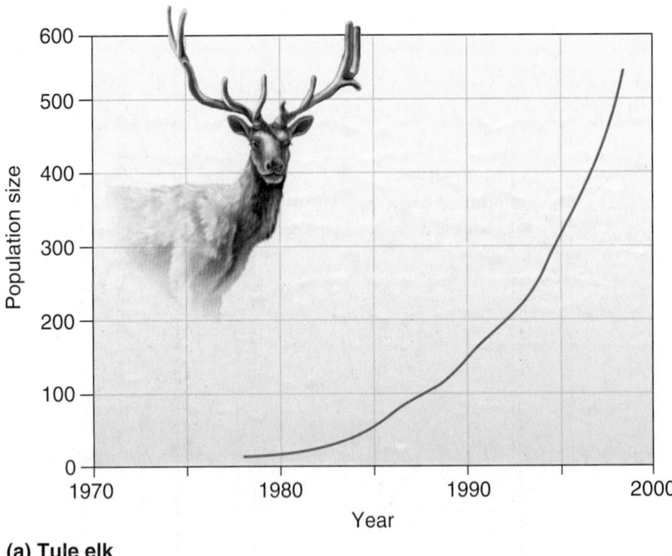

(a) Tule elk

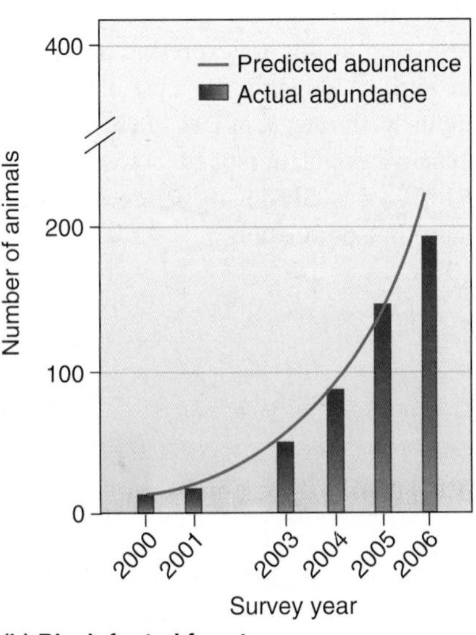

(b) Black-footed ferrets

Figure 56.10 Exponential growth following reintroduction of a population into a habitat. **(a)** A population of tule elk (*Cervus elaphus nannodes*) reintroduced to Point Reyes National Seashore in 1978 fits a pattern of exponential growth. **(b)** Black-footed ferrets (*Mustela nigripes*) reintroduced to Shirley Basin, Wyoming, since 2000. No survey was conducted in 2002.

As the population size, N, grows, it moves closer to the carrying capacity, K, with fewer available resources for population growth. At large values of N, the value of $(K - N)/K$ becomes small, and population growth is small. If $K = 1{,}000$, $N = 900$, and $r = 0.1$, then

$$\frac{dN}{dt} = (0.1)(900) \times \frac{(1{,}000 - 900)}{1{,}000}$$

$$\frac{dN}{dt} = 9$$

In this instance, population growth is 9 individuals per unit of time.

Let's consider how an ecologist would use the logistic equation. First, the value of K would come from intense field and laboratory work from which researchers would determine the amount of resources, such as food, needed by each individual and then determine the amount of available food in the wild. Field censuses determine N, and field censuses of births and deaths per unit time provide r. When this type of population growth is plotted over time, an S-shaped growth curve results (**Figure 56.11**). This pattern, in which the growth of a population slows down as it approaches K, is called **logistic growth**.

Does the logistic growth model provide a better fit to growth patterns of plants and animals in the wild than the exponential model? In some instances, such as laboratory cultures of bacteria and yeasts, the logistic growth model provides a very good fit (**Figure 56.12**). In nature, however, variations in temperature, rainfall, or resources can cause changes in carrying capacity and thus in population size. The uniform conditions of temperature, moisture, and resource levels of the laboratory do not usually exist. In addition, time lags may occur between changes in carrying capacity and changes in reproduction. For instance, pregnant females are still likely to give birth even when resources are declining. This can lead to temporary overshoots of population density beyond the carrying capacity. Therefore, there are relatively few exact fits of the logistic growth model to population growth in the field. Instead, populations tend to fluctuate around the limits suggested by the logistic, with frequent overshoots and undershoots.

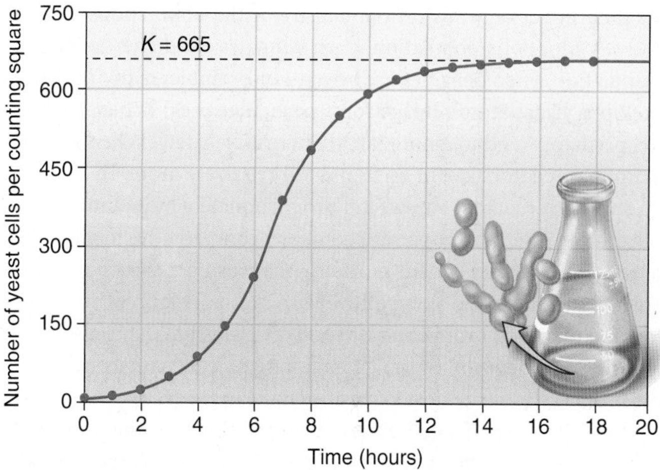

Figure 56.12 **Logistic growth of yeast cells in culture.** Early tests of the logistic growth curve were validated by growth of yeast cells in laboratory cultures. These populations showed the typical S-shaped growth curve.

BIOLOGY PRINCIPLE Biology is an experimental science. Tests of population growth models are more accurately performed using manipulative experiments than by simple field observations.

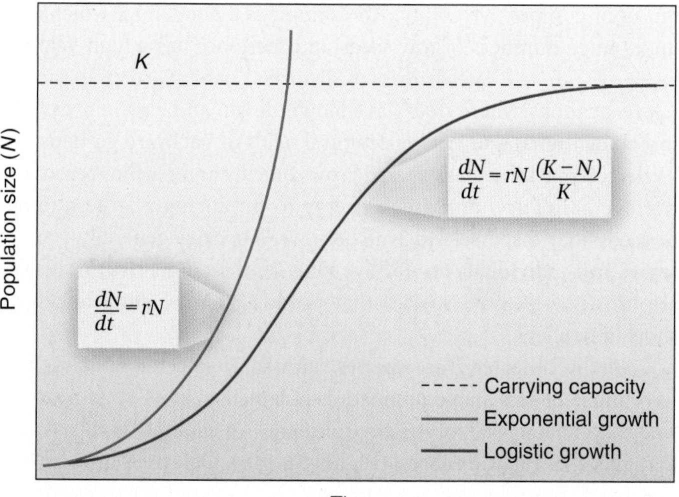

Figure 56.11 **Exponential versus logistic growth.** Exponential (J-shaped) growth occurs in an environment with unlimited resources, whereas logistic (S-shaped) growth occurs in an environment with limited resources.

Concept Check: *What is the population growth per unit of time when $r = 0.1$, $N = 500$, and $K = 1{,}000$?*

Is the logistic model of little value because it fails to describe population growth accurately? Not really. It is a useful starting point for thinking about how populations grow, and it seems intuitively correct. However, the carrying capacity is a difficult feature of the environment to identify for most species, and it also varies with time and according to local climate patterns. For these reasons, logistic growth is difficult to measure accurately.

Also, as we will discover, populations are affected by interactions with other species. In Chapter 57, we will examine how predators, parasites, and competitors affect population densities and explore situations in which species interactions commonly limit population growth. As described next, such population limitations are often influenced by a process known as density dependence.

Density-Dependent Factors May Regulate Population Sizes

A **density-dependent factor** is a mortality factor whose influence increases with the density of the population. Parasitism, predation, and competition are some of the many density-dependent factors that may reduce the population densities of living organisms and stabilize them at equilibrium levels. Such factors can be density-dependent in that their effect depends on the density of the population; they kill relatively more of a population when densities are higher and less of a population when densities are lower. For example, many predators develop a visual search image for a particular prey. When a prey is rare, predators tend to ignore it and kill relatively few. When a prey is common, predators key in on it and kill relatively more. In England, for example, predatory shrews kill proportionately more moth pupae in leaf litter when the pupae are common compared with when they are rare. Density-dependent mortality may also occur as population densities increase and competition for scarce resources increases, reducing offspring production or survival. Parasitism may also act in a density-dependent manner. Parasites are able to pass from host to host more easily as the host's densities increase.

Density dependence can be detected by plotting mortality, expressed as a percentage, against population density (**Figure 56.13**). If a positive slope results and mortality increases with density, the factor tends to have a greater effect on dense populations than on sparse ones and is clearly acting in a density-dependent manner.

A **density-independent factor** is a mortality factor whose influence is not affected by changes in population size or density. When mortality is plotted against density, a flat line results. In general, density-independent factors are physical factors, including weather, drought, freezes, floods, and disturbances such as fire. For example, in hard freezes, the same proportion of organisms such as birds or plants are usually killed, no matter how large the population size.

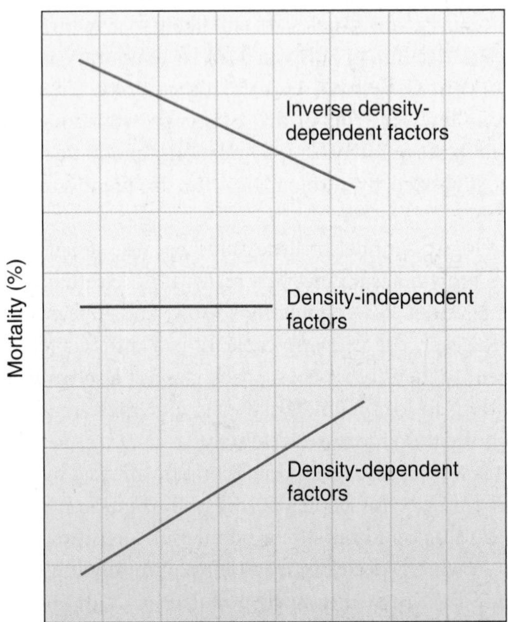

Figure 56.13 **Three ways that factors affect mortality in response to changes in population density.** For a density-dependent factor, mortality increases with population density; for a density-independent factor, mortality remains unchanged. For an inverse density-dependent factor, mortality decreases as a population increases in size.

Concept Check: Which types of factors tend to stabilize populations at equilibrium levels?

Finally, a mortality factor that decreases with increasing population size is considered an **inverse density-dependent factor**. In this case, a negative slope results when mortality is plotted against density. For example, if a territorial predator such as a lion always killed the same number of wildebeest prey, regardless of wildebeest density, it is acting in an inverse density-dependent manner, because it is taking a smaller proportion of the population at higher density. Some mammalian predators, being highly territorial, often act in this manner on herbivore density.

Determining which factors act in a density-dependent or density-independent fashion has large practical implications. Foresters, game managers, and conservation biologists alike are interested in learning how to maintain populations. For example, if a specific disease were to act in a density-dependent manner on white-tailed deer, there wouldn't be much point in game managers attempting to kill off predators such as mountain lions to increase herd sizes for hunters, because proportionately more deer would be killed by disease.

Life History Strategies Incorporate Traits Relating to Survival and Competitive Ability

The population parameters we have discussed—including iteroparity versus semelparity, exponential versus logistic growth, and density-dependent versus density-independent factors—have important implications for how populations grow and indeed for the reproductive success of populations and species. These reproductive strategies can be viewed in the context of a much larger picture of life history strategies, sets of physiological and behavioral features that incorporate not only reproductive traits but also survivorship and length of life characteristics, habitat type, and competitive ability.

When comparing many different species, life history strategies follow a continuum. At the one end are species, termed ***r*-selected species**, that have a high rate of per capita population growth (r), but poor competitive ability. An example is a dandelion, which produces huge numbers of tiny seeds and therefore has a high value of r (**Figure 56.14a**). Weeds exist in disturbed habitats such as gaps in a forest canopy where trees have blown down, and also in areas disturbed by humans such as agricultural fields or backyard gardens. An r-selected species such as a weed grows quickly and reaches reproductive age early, devoting much energy to producing a large number of seeds that disperse widely. These weed species generally remain small, and individuals do not live long. In the animal world, insects are mostly r-selected species that produce many young and have short life cycles.

At the other end are species, termed ***K*-selected species**, that have more or less stable populations adapted to exist at or near the carrying capacity (K), of the environment. An example is an oak tree that exists in a mature forest (**Figure 56.14b**). Oak trees grow slowly and reach reproductive age late, having to devote much energy to growth and maintenance. A K-selected species like a tree grows large and shades out r-selected species like weeds, eventually outcompeting them. Such trees live a long time and produce seeds repeatedly every year when mature. These seeds are bigger than those of r-selected species, but do not disperse widely. Acorns contain a large food reserve that helps them grow, whereas dandelion seeds must rely on whatever nutrients they can gather from the soil where they land. Mammals,

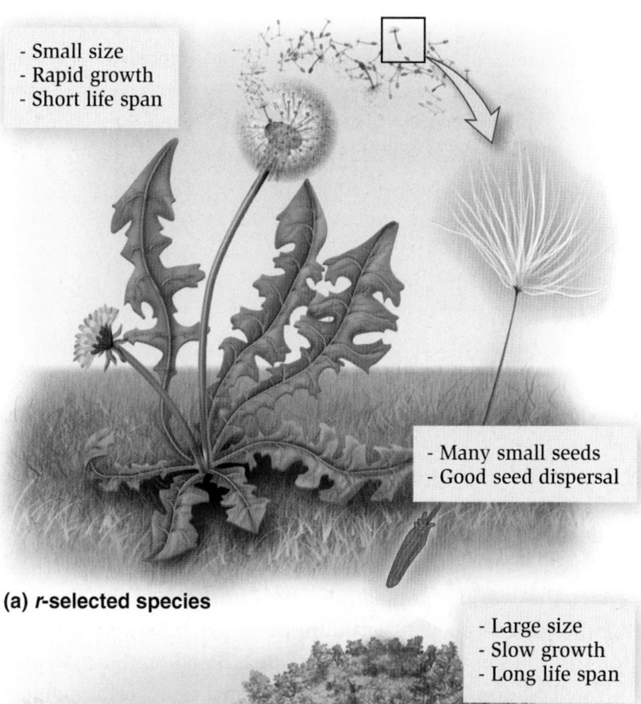

(a) *r*-selected species

(b) *K*-selected species

Figure 56.14 Life history strategies. Differences in traits of a dandelion **(a)** and an oak tree **(b)** illustrate some of the differences between *r*- and *K*-selected species.

BioConnections: Look back at Figure 39.21. What is unusual about the reproductive strategy of the mother of thousands plant?

such as elephants, that grow slowly, have few young, and reach large sizes are typical of *K*-selected animal species. **Table 56.2** compares the general characteristics of *r*- and *K*-selected species.

In a human-dominated world, almost every life history attribute of a *K*-selected species sets it at risk of extinction. First, *K*-selected species tend to be larger, so they need more habitat in which to live. For example, Florida panthers need huge tracts of land to establish their territories and hunt for deer (look ahead to Figure 60.15c). *K*-selected species tend to have fewer offspring, so their populations cannot recover as fast from disturbances such as fire or overhunting. California condors, for example, produce only a single chick every other year. *K*-selected species breed at a later age, and their time to grow from a small population to a larger population is long. Gestation time in elephants is 22 months, and elephants take at least 7 years to become sexually mature. Large trees, such as the giant sequoia; large terrestrial mammals, such as elephants, rhinoceroses, and grizzly bears; and large marine mammals, such as blue whales and sperm whales, all run the risk of extinction. Interestingly, the coast redwood seems to be an exception, a fact perhaps attributable to its unusual genome (see the following Genomes & Proteomes Connection).

Table 56.2 Characteristics of *r*- and *K*-Selected Species

Life history feature	*r*-selected species	*K*-selected species
Development	Rapid	Slow
Reproductive rate	High	Low
Reproductive age	Early	Late
Body size	Small	Large
Length of life	Short	Long
Competitive ability	Weak	Strong
Survivorship	High mortality of young	Low mortality of young
Population size	Variable	Fairly constant
Dispersal ability	Good	Poor
Habitat type	Disturbed	Not disturbed
Parental care	Low	High

What are the advantages to being a *K*-selected species? In a world not disturbed by humans, *K*-selected species would fare well. However, in a human-dominated world, many *K*-selected species are selectively logged or hunted, or their habitat is altered, and the resulting small population sizes make extinction a real possibility.

Genomes & Proteomes Connection

Hexaploidy Increases the Growth of Coastal Redwood Trees

Besides being home to the world's most massive tree, the giant sequoia (*Sequoiadendron giganteum*), California is also the location of the world's tallest tree, the coast redwood (*Sequoia sempervirens*), a towering giant that can grow to over 90 m and can live for up to 2,000 years (**Figure 56.15**). These trees are currently confined to a relatively small 700-km strip along the Pacific coast from California to southern Oregon, an area characterized by year-long moderate temperatures, heavy winter rains, and dense summer fog. Interestingly, because this climate was far more common in an earlier era, these trees were once dispersed throughout the Northern Hemisphere.

How is this huge species different from other tree species? In 1948, researchers made the startling discovery that the tree is a hexaploid; that is, each of its cells contains six sets of chromosomes, with 66 chromosomes in total. (Keep in mind that humans have two sets of chromosomes in every cell.) Although hexaploidy is not unknown in grasses and shrubs, it is unusual in trees and particularly gymnosperms. The

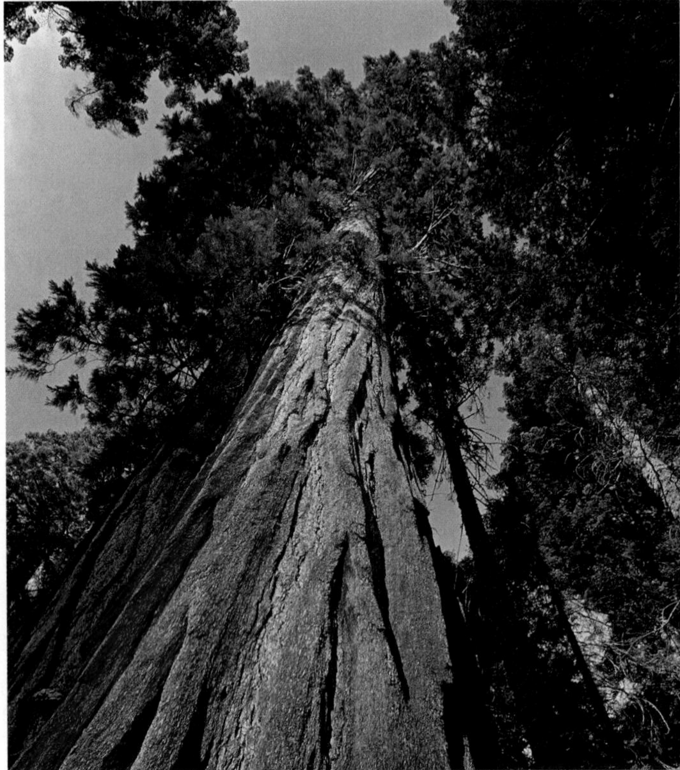

Figure 56.15 The coastal redwood (*Sequoia sempervirens*)—a hexaploid conifer. The coast redwood can grow to over 90 m, and the oldest living trees are over 2,000 years old. Their great genetic variation may help explain their incredible growth and longevity.

BIOLOGY PRINCIPLE The genetic material provides a blueprint for reproduction. Despite being large, *K*-selected species, coastal redwoods grow faster than any other known conifer and this is due to their unusual hexaploid genome.

coast redwood is the only known hexaploid conifer. Having this quality means each tree may have several different alleles for any given gene, which leads to a very genetically diverse population. American molecular biologist Chris Brinegar has found that hardly any two trees have exactly the same genetic constitution. Such genetic diversity allows greater adaptation to environmental conditions and more adaptations against insect or fungal pests. Indeed, living redwoods have no known lethal diseases, and pests do not cause significant damage. What's more, with six sets of genes, trees also have the potential for great variety in their gene products, the proteins, which may help explain their prodigious growth. It grows faster than any conifer on Earth, and this is why it is an exception to most *K*-selected species.

56.4 Human Population Growth

Learning Outcomes:

1. Describe the pattern of human population growth.
2. Graph the demographic transition.
3. Detail the differences in age structure and human fertility across different countries.
4. Explain the concept of an ecological footprint.

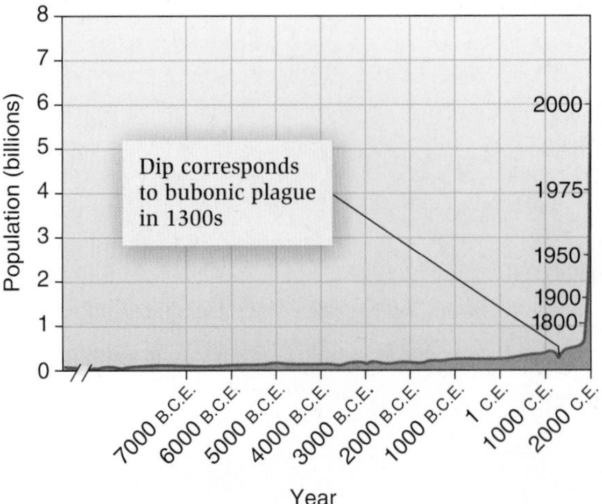

Figure 56.16 The growth pattern of the human population through history. If, and when, human population growth will level off are issues of considerable debate.

In 2011, the world's population was estimated to be increasing at the rate of 145 people every minute: 2 per minute in developed nations and 143 in less-developed nations. Based on this rate, one of the United Nations' 2010 projections pointed to a world population reaching 10 billion near the year 2100. In this section, we examine human population growth trends in more detail and discuss how knowledge of the human population's age structure can help predict its future growth. We then investigate the carrying capacity of the Earth for humans and explore how the concept of an ecological footprint, which measures human resource use, can help us determine this carrying capacity.

Human Population Growth Fits an Exponential Pattern

Until the beginning of agriculture and the domestication of animals, about 10,000 B.C.E., the average rate of population growth was very low. With the establishment of agriculture, the world's population grew to about 300 million by 1 C.E. and to 800 million by the year 1750. Between 1750 and 1998, a relatively tiny period of human history, the world's human population surged from 800 million to 6 billion (**Figure 56.16**). In 2012, the number of humans was estimated at 7 billion. If the population reaches 10 billion by 2100, as the U.N. projects, when and at what level will the human population level off?

Human populations can exist at equilibrium densities in one of two ways:

1. *High birth and high death rates.* Before 1750, this was often the case, with high birth rates offset by deaths from wars, famines, and epidemics.
2. *Low birth and low death rates.* In Western Europe, beginning in the 18th century, better health and living conditions reduced the death rate. Eventually, social changes such as increasing education for women and marriage at a later age reduced the birth rate.

The shift in birth and death rates that accompanies development is known as the **demographic transition** (**Figure 56.17**). In the first stage of the transition, birth and death rates are both high, and the population is in equilibrium. In the second stage of this transition, the

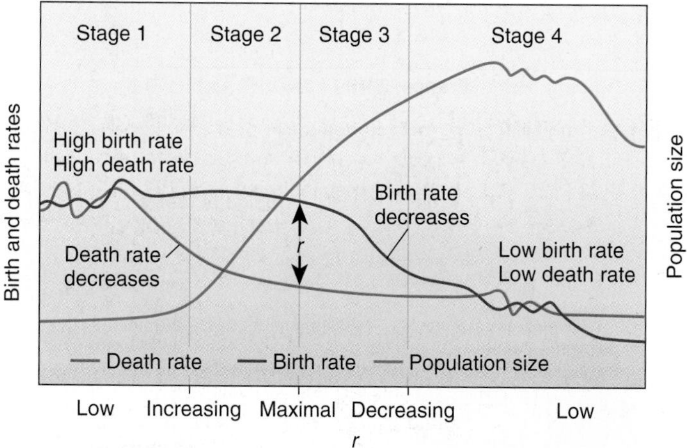

Figure 56.17 **The classic stages of the demographic transition.** The difference between the birth rate and the death rate determines the rate of population increase or decrease.

death rate declines first, but the birth rate remains high. High rates of population growth result. In the third stage, the birth rates drop and death rates stabilize, so population growth rates become lower. In the fourth stage, both birth and death rates are low, and the population is again at equilibrium.

The pace of the demographic transition between countries differs, depending on culture, economics, politics, and religion. This is illustrated by comparing the demographic transition in Sweden and Mexico (**Figure 56.18**). In Mexico, the demographic transition occurred more recently and was typified by a faster decline in the death rate, reflecting rapid improvements in public health. A relatively longer lag occurred between the decline in the death rate and the decline in the birth rate, however, with the result that Mexico's population growth rate is still well above Sweden's, perhaps reflecting differences in culture or the fact that in Mexico, the demographic transition is not yet complete.

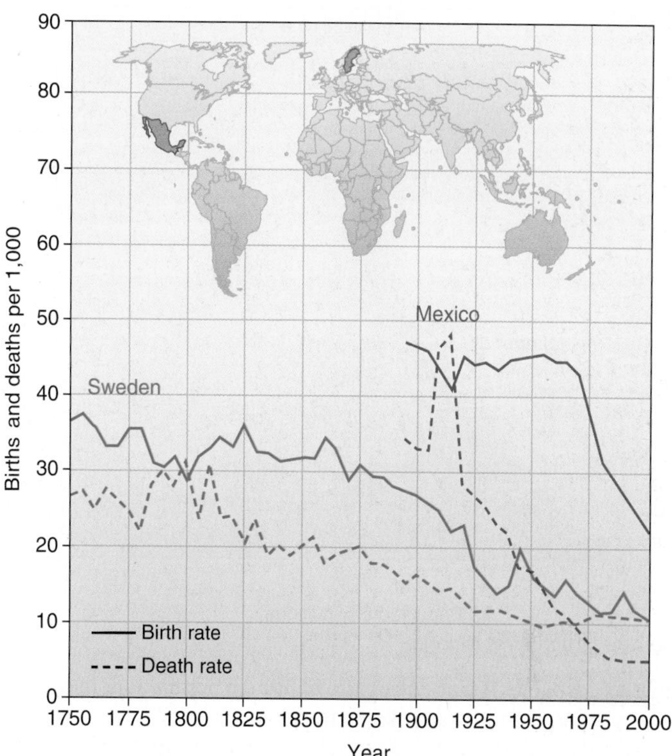

Figure 56.18 **The demographic transition in Sweden and Mexico.** Although the demographic transition began earlier in Sweden than it did in Mexico, the transition was more rapid in Mexico, and the overall rate of population increase remains higher. (The spike in the death rate in Mexico prior to 1920 is attributed to the turbulence surrounding the Mexican Revolution.)

Knowledge of a Population's Age Structure Can Help Predict Its Future Growth

Changes in the age structure of a population also characterize the demographic transition. In all populations, **age structure** refers to the relative numbers of individuals of each defined age group. This information is commonly displayed as a population pyramid (**Figure 56.19**). In West Africa, for example, children younger than age 15 make up nearly half of the population, creating a pyramid with a wide base and narrow top. Even if fertility rates decline, there will still be a huge increase in the population as these young people move into childbearing age. The age structure of Western Europe is much more balanced. Even if the fertility rate of young women in Western Europe increased to a level higher than that of their mothers, the annual numbers of births would still be relatively low because of the low number of women of childbearing age.

Human Population Fertility Rates Vary Widely Around the World

Most estimates propose that the human population will grow to between 10 and 11 billion people by the middle of the 22nd century. Global population growth can be examined by looking at the **total fertility rate (TFR)**, the average number of live births a woman has during her lifetime (**Figure 56.20**). The total fertility rate differs considerably between geographic areas. In Africa, the total fertility rate of 4.6 in 2010 has declined substantially since the 1970s, when it was around 6.7 children per woman. In Latin America and Southeast Asia, the rates have declined considerably from the 1970s and are now at around 2.3. Canada and most countries in Europe have a TFR of less than 2.0 (it is slightly above in the U.S.); in Russia, fertility rates have dropped to 1.34. In China, although the TFR is only 1.7, the population there will still continue to increase until at least 2025 because of the large number of women of reproductive age. Although the global TFR has declined from 4.47 in the 1970s to 2.52 in 2010, this is still greater than the average of 2.3 needed for zero population growth. The replacement rate is slightly higher than 2.0, to replace mother and father, due to natural mortality prior to reproduction. The replacement rate varies globally, from 2.1 in developed countries to between 2.5 and 3.3 in developing countries.

The wide variation in fertility rates makes it difficult to predict future population growth. The 2010 United Nations report shows world population projections to the year 2100 for three different growth scenarios: low, medium, and high (**Figure 56.21**). The three scenarios are based on three different assumptions about fertility rate. Using a low fertility rate estimate of only 1.5 children per woman, the population would reach a maximum of about 8 billion people by 2050. A more realistic assumption may be to use the fertility rate estimate

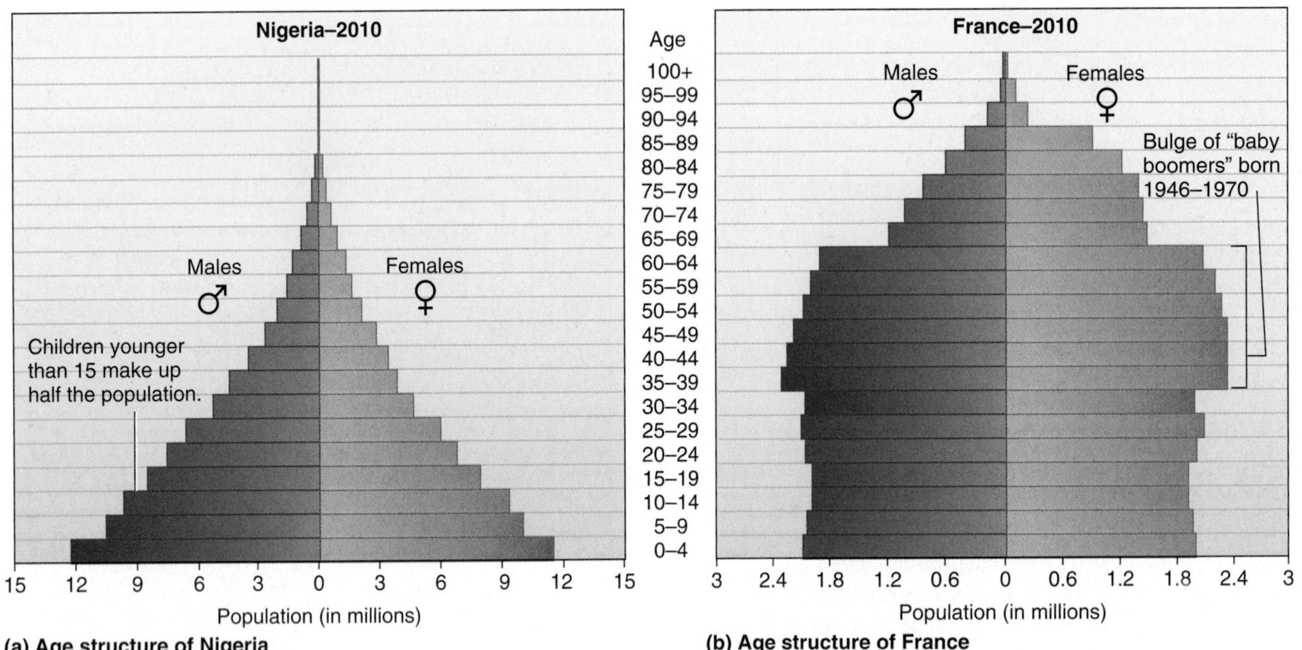

Figure 56.19 **The age structure of human populations in Nigeria and France, as of 2010.** **(a)** In developing areas of the world such as Nigeria, there are far more children than any other age group. Population growth is rapid. **(b)** In the developed countries of Western Europe, the age structure is more evenly distributed. The bulge represents those born in the post-World War II "baby boom," when birth rates climbed due to stabilization of political and economic conditions. Population growth is close to zero.

Concept Check: *If the population pyramid in (a) was inverted, what would you conclude about the age structure of the population?*

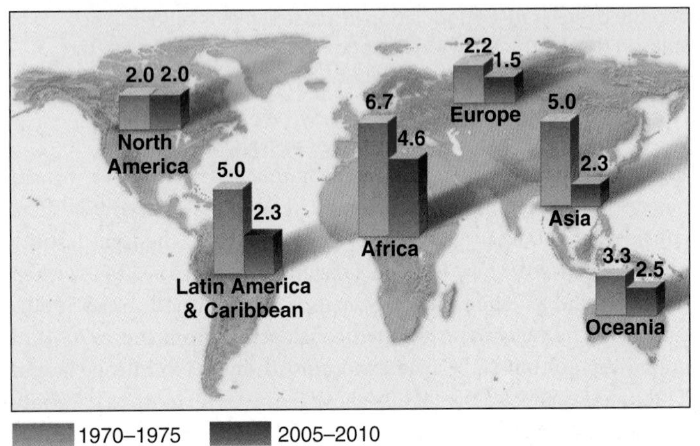

Figure 56.20 **Total fertility rates (TFRs) among major regions of the world.** Data refer to the average number of children born to a woman during her lifetime.

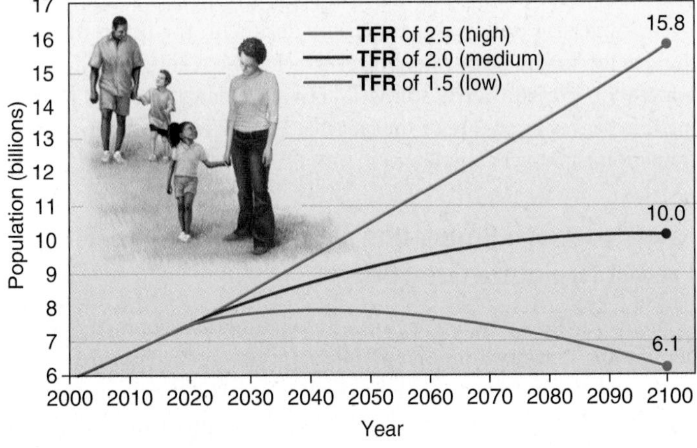

Figure 56.21 **Population predictions for 2000–2100, using three different total fertility rates (TFRs).**

BIOLOGY PRINCIPLE **Biology affects our society.** How TFR is calculated has a great influence on assumptions about how human global population size will change over the next 90 years.

of 2.0 or even 2.5, in which case the population would continue to rise to 10 or 16 billion, respectively.

The Concept of an Ecological Footprint Helps Estimate Carrying Capacity

What is the Earth's carrying capacity for the human population and when will it be reached? Estimates vary widely. Much of the speculation on the upper boundary of the world's population size centers on lifestyle. To use a simplistic example, if everyone on the planet ate meat extensively and drove large cars, then the carrying capacity would be a lot less than if people were vegetarians and used bicycles as their main means of transportation.

In the 1990s, Swiss researcher Mathis Wackernagel and his coworkers calculated how much land is needed for the support of each person on Earth. Everybody has an effect on the Earth, because they consume the land's resources, including crops, wood, fossil fuels, minerals, and so on. Thus, each person has an **ecological footprint**, the aggregate total of productive land needed for survival in a sustainable world. The average footprint size for everyone on the planet is about

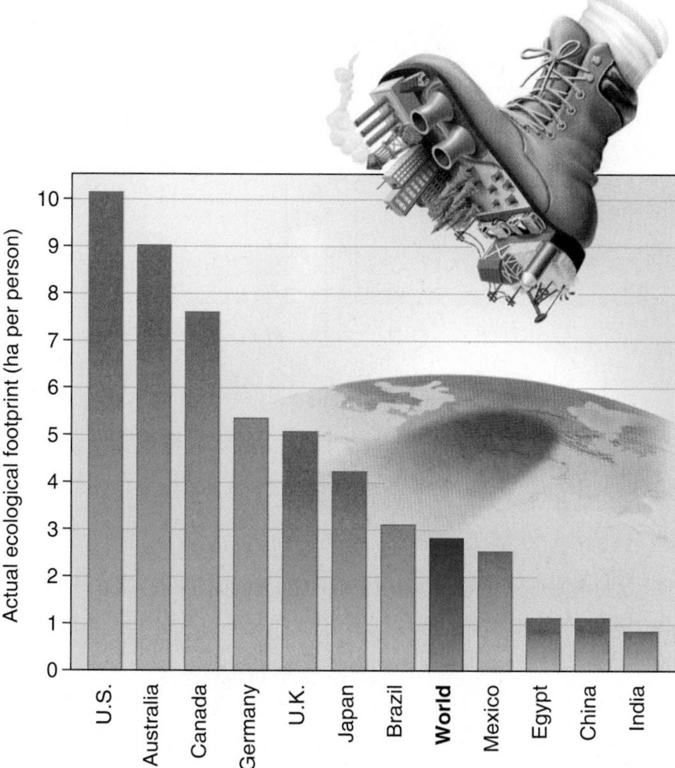

Figure 56.22 **Ecological footprints of different countries.** The term ecological footprint refers to the amount of productive land needed to support the average individual of that country.

Concept Check: *What is your ecological footprint?*

3 hectares (1 ha = 10,000 m^2), but a wide variation is found around the globe (Figure 56.22). The ecological footprint of the average Canadian is 7.5 hectares versus about 10 hectares for the average American.

In most developed countries, the largest component of land is for energy, followed by food and then forestry. Much of the land needed for energy serves to absorb the CO_2 emitted by the use of fossil fuels. If everyone required 10 hectares, as the average American does, we would need three Earths to provide us with the needed resources. Many people in less-developed countries are much more frugal in their use of resources. However, globally we are already beyond the Earth's carrying capacity for humans if we were to live in a sustainable manner. This has happened because many people currently live in an unsustainable manner, using more resources than can be regenerated in any given year.

What's your personal ecological footprint? Several different calculations are available on the Internet that you can use to find out. A rapidly growing human population combined with an increasingly large per capita ecological footprint makes it increasingly difficult to preserve other species on the planet, a subject we will examine further in our discussion of conservation biology (Chapter 60).

Summary of Key Concepts

56.1 Understanding Populations

- Population ecology studies how populations grow and what factors promote and limit growth. Ecologists measure population density, the numbers of organisms in a given unit area, in many ways, including the mark-capture technique (Figures 56.1, 56.2).
- Individuals within populations show different patterns of dispersion, including clumped (the most common), uniform, and random. Individuals also exhibit different reproductive strategies, and populations have different age classes (Figures 56.3, 56.4, 56.5).

56.2 Demography

- Life tables summarize the survival pattern of a population. Survivorship curves illustrate life tables by plotting the numbers of surviving individuals at different ages. Age-specific fertility and survivorship data help determine the overall growth rate per generation, or the net reproductive rate (R_0) (Figures 56.6, 56.7, 56.8, Table 56.1).

56.3 How Populations Grow

- The per capita growth rate (r) helps determine how populations grow over any time period. When r is > 0, exponential (J-shaped) growth occurs. Exponential growth can be observed in an environment where resources are not limited. Logistic (S-shaped) growth takes into account the upper boundary for a population, called carrying capacity, and occurs in an environment where resources are limited (Figures 56.9, 56.10, 56.11, 56.12). Density-dependent factors are mortality factors whose influence varies with population density. Density-independent factors are those whose influence does not vary with density (Figure 56.13).
- Life history strategies are a set of features including reproductive traits, survivorship and length of life characteristics, and competitive ability. Life history strategies can be viewed as a continuum, with *r*-selected species (those with a high rate of population growth but poor competitive ability) at one end and *K*-selected species (those with a lower rate of population growth but better competitive ability) at the other (Figures 56.14, 56.15, Table 56.2).

56.4 Human Population Growth

- Up to the present, human population growth has fit an exponential growth pattern. Human populations have been moving from states of high birth and death rates to low birth and death rates, a shift called the demographic transition (Figures 56.16, 56.17, 56.18).
- Differences in the age structure of a population, the numbers of individuals in each age group, are also characteristic of the demographic transition (Figure 56.19).
- Although they have been declining worldwide, total fertility rates (TFRs) differ markedly between less-developed and more-developed countries. Predicting the growth of the human population depends on the total fertility rate that is projected (Figures 56.20, 56.21).
- The ecological footprint refers to the amount of productive land needed to support each person on Earth. Because people in many countries live in a nonsustainable manner, globally we are already in an ecological deficit (Figure 56.22).

Assess and Discuss

Test Yourself

1. A student decides to conduct a mark-recapture experiment to estimate the population size of mosquitofish in a small pond near his home. In the first catch, he marked 45 individuals. Two weeks later, he captured 62 individuals, of which 8 were marked. What is the estimated size of the population based on these data?
   a. 134
   b. 349
   c. 558
   d. 1,016
   e. 22,320

Questions 2–4 refer to the following table:

Age	n_x	d_x	l_x	m_x	$l_x m_x$
0	100	35	1.00	0	0
1	65	?	0.65	0	0
2	45	15	?	3	1.35
3	30	20	0.30	1	?
4	10	10	0.10	1	0.10
5	0	0	0.00	1	0.0

2. How many individuals die between their first and second birthday?
   a. 65
   b. 45
   c. 35
   d. 25
   e. 20
3. What proportion of newborns survive to age 2?
   a. 0.55
   b. 0.45
   c. 0.35
   d. 0.20
   e. 0.15
4. What is the net reproductive rate?
   a. 5
   b. 2.5
   c. 1.75
   d. 1.45
   e. 0.80
5. _______________ survivorship curves are usually associated with organisms that have high mortality rates in the early stages of life.
   a. Type I
   b. Type II
   c. Type III
   d. Types I and II
   e. Types II and III
6. If the net reproductive rate (R_0) is equal to 0.5, what assumptions can we make about the population?
   a. This population is essentially not changing in numbers.
   b. This population is in decline.
   c. This population is growing.
   d. This population is in equilibrium.
   e. none of the above
7. The maximum number of individuals a certain area can sustain is known as
   a. the intrinsic rate of growth.
   b. the resource limit.
   c. the carrying capacity.
   d. the logistic equation.
   e. the equilibrium size.

Questions 8 and 9 refer to the following generalized growth patterns as plotted on arithmetic scales. Match the following descriptions with the patterns indicated below.

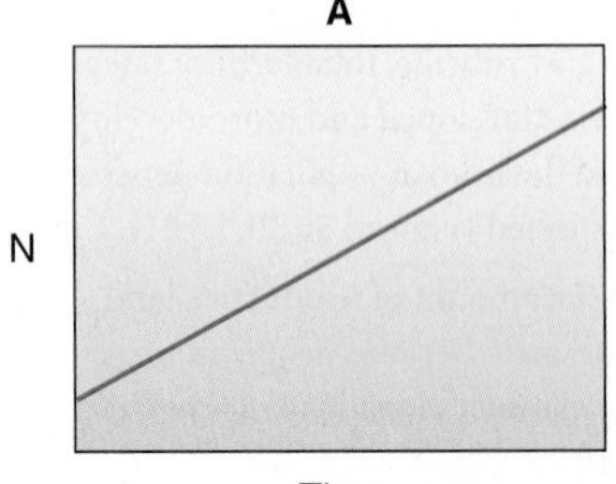

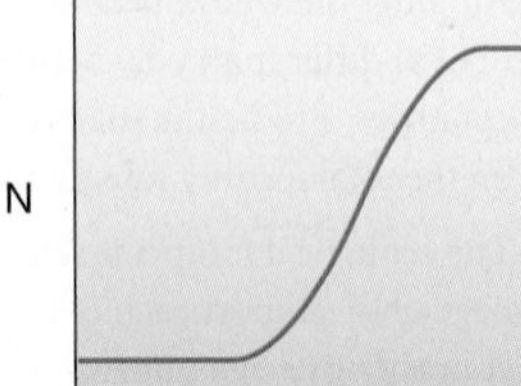

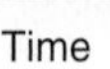

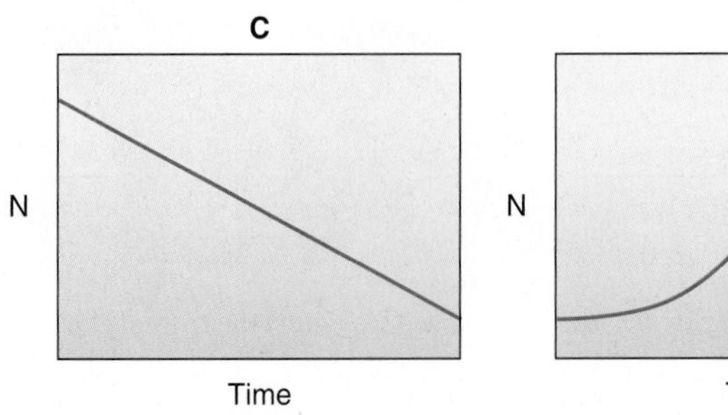

Each pattern may be used once, more than once, or not at all.

8. Which pattern is found when a population exhibits a constant per capita growth rate?
   a. A
   b. B
   c. C
   d. D
   e. none of the above
9. Which pattern is found when a population is heading toward extinction?
   a. A
   b. B
   c. C
   d. D
   e. none of the above
10. The amount of land necessary for survival for each person in a sustainable world is known as
    a. the sustainability level.
    b. an ecological impact.
    c. an ecological footprint.
    d. survival needs.
    e. all of the above.

Conceptual Questions

1. As a researcher, you are using the mark-recapture procedure (see Figure 56.2) with large mouth bass. Say you do a poor job tagging the fish, and 20% of the tags fall off. How does this influence your estimate of population size? Does it increase or decrease?
2. Using the logistic equation, calculate population growth when K = 1,000, N = 500, and r = 0.1 and when K = 1,000, N = 100, and r = 0.1. Compare the results with those shown in Section 56.3, where K = 1,000, N = 900, and r = 0.1.
3. A principle of biology is that *living organisms interact with their environment*. Imagine two types of ponds. In one, the pond dries out when there is little rain, but in the other the water levels fluctuate but there is always some water present. Contrast the reproductive strategies of organisms that might live in each pond.

Collaborative Questions

1. Discuss what might limit human population growth in the future.
2. Describe where students on campus might show each type of dispersion pattern, and explain why this might occur.

Online Resource

www.brookerbiology.com

Stay a step ahead in your studies with animations that bring concepts to life and practice tests to assess your understanding. Your instructor may also recommend the interactive eBook, individualized learning tools, and more.

Chapter Outline

57.1 Competition

57.2 Predation, Herbivory, and Parasitism

57.3 Mutualism and Commensalism

57.4 Bottom-Up and Top-Down Control

Summary of Key Concepts

Assess and Discuss

Species Interactions

57

In this species interaction, a shark is feeding on a ray, which in turn feeds on bay scallops.

In 2007, marine biologist Ransom Myers and his colleagues showed that overfishing severely depleted the numbers of 11 shark species that occurred along the eastern seaboard of the U.S. Several shark species had declined by over 99% since the 1950s. Because of strong interactions between the sharks and other marine species, this drastic reduction of the shark population had at least two other effects. First, there was a large increase in the main prey species of the sharks, rays and skates. Second, the increase in rays and skates reduced the densities of their prey—bay scallops (*Argopecten irradians*). Such losses contributed to the closure of the bay scallop industry in North Carolina.

In this chapter, we turn from considering populations on their own to investigating how they interact with populations of other species that live in the same locality. Such species interactions can take a variety of forms (**Table 57.1**). **Competition** can be defined as an interaction that affects both species negatively (–/–), as both species compete over food or other resources. Sometimes this interaction is quite one-sided, being detrimental to one species and neutral to the other, an interaction called **amensalism** (–/0). **Predation, herbivory,** and **parasitism** all have a positive effect on one species and a negative effect on the other (+/–). **Mutualism** is an interaction in which both species benefit (+/+), whereas **commensalism** benefits one species and leaves the other unaffected (+/0). To illustrate how species interact in nature, let's consider a rabbit population in a woodland community (**Figure 57.1**). To determine what factors influence the size and density of the rabbit population, we need to understand each of its possible species interactions. For example, the rabbit population could be limited by the quality of available food. It is also likely that other species, such as deer, use the same resource and thus compete with the rabbits. The rabbit population could be limited by predation from foxes or by the virus that causes the disease myxomatosis, which is usually spread by fleas and mosquitoes. It is also possible that other associations, such as mutualism or commensalism, may occur.

This chapter examines each of these species interactions in turn, beginning with competition, an important interaction among species. We conclude with a discussion of bottom-up and top-down effects, both of which are influential in controlling population densities within ecological systems.

Table 57.1 Summary of the Types of Species Interactions

Nature of interaction	Species 1*	Species 2*
Competition	–	–
Amensalism	–	0
Predation, herbivory, parasitism	+	–
Mutualism	+	+
Commensalism	+	0

*+ = positive effect; – = negative effect; 0 = no effect.

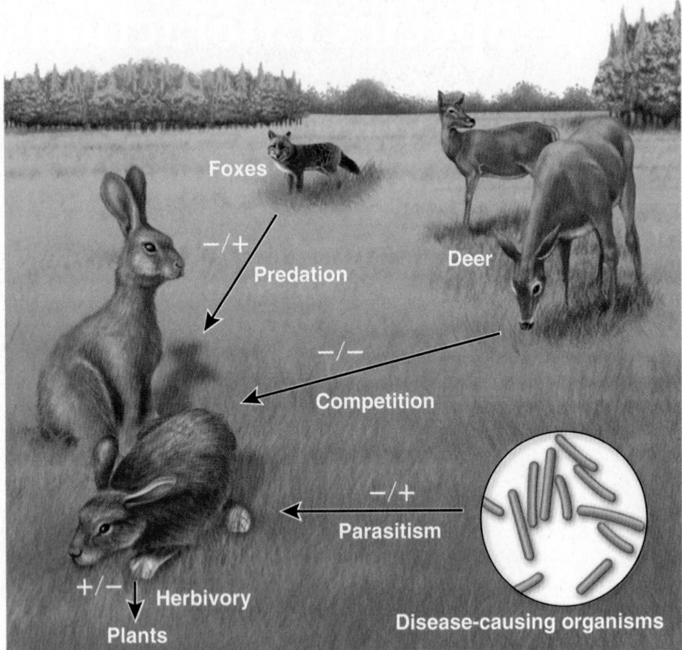

Figure 57.1 **Species interactions.** These rabbits can interact with a variety of species, experiencing predation by foxes, competition with deer for food, and parasitism from various disease-causing organisms. Herbivory occurs when rabbits feed on various plants. The effects of each species on the other are shown by the terms assigned to the arrows, as discussed in the text.

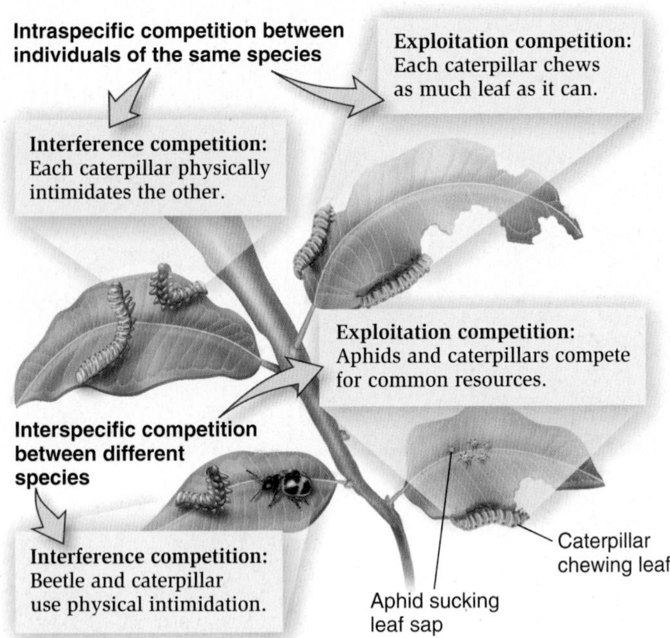

Figure 57.2 **The different types of competition in nature.**

Concept Check: How would you classify competition between vultures feeding on roadkill?

57.1 Competition

Learning Outcomes:

1. Describe the different types of competition that occur in nature.
2. Explain how the competitive exclusion principle can lead to resource partitioning among species.
3. Give examples of how morphological differences may allow species to coexist.

In this section, we will see how ecologists have studied different types of competition and how they have shown that the competitive effects of one species on another can change as the environment changes. Although species may compete for resources, we will also learn how sufficient differences in lifestyle or morphology can reduce the overlap in their habitat use, thus allowing them to coexist.

Several Different Types of Competition Occur in Nature

Several different types of competition are found in nature (Figure 57.2). Competition may be **intraspecific** (between individuals of the same species) or **interspecific** (between individuals of different species). Competition can also be characterized by the mechanism by which it occurs. In **exploitation competition**, organisms compete indirectly through the consumption of a limited resource, with each obtaining as much as it can. For example, when fly maggots compete in a mouse carcass, not all the individuals can command enough of the resource to survive and become adult flies. In **interference competition**, individuals interact directly with one another by physical force or intimidation. Often this force is ritualized into aggressive behavior associated with territoriality (refer back to Figure 55.25). In these cases, strong individuals survive and take the bulk of the resources, and weaker ones perish or, at best, survive under suboptimal conditions.

Competition between species is not always equal. An extreme asymmetric competition can be observed between plants, in which one species secretes and produces chemicals from its roots that inhibit the growth of another species. In Chapter 54's Feature Investigation, we saw how diffuse knapweed, an introduced species, secretes root chemicals called allelochemicals into the surrounding environment that kill the roots of native grass species—a phenomenon called **allelopathy**.

Field Studies Show Competition Occurs Frequently in Nature

By reviewing studies that have investigated competition in nature, we can see how frequently it occurs and in what particular circumstances it is most important. In a 1983 review of field studies by American ecologist Joseph Connell, competition was found in 55% of 215 species surveyed, demonstrating that it is indeed frequent in nature. Generally in studies of single pairs of species utilizing the same resource, competition is almost always reported

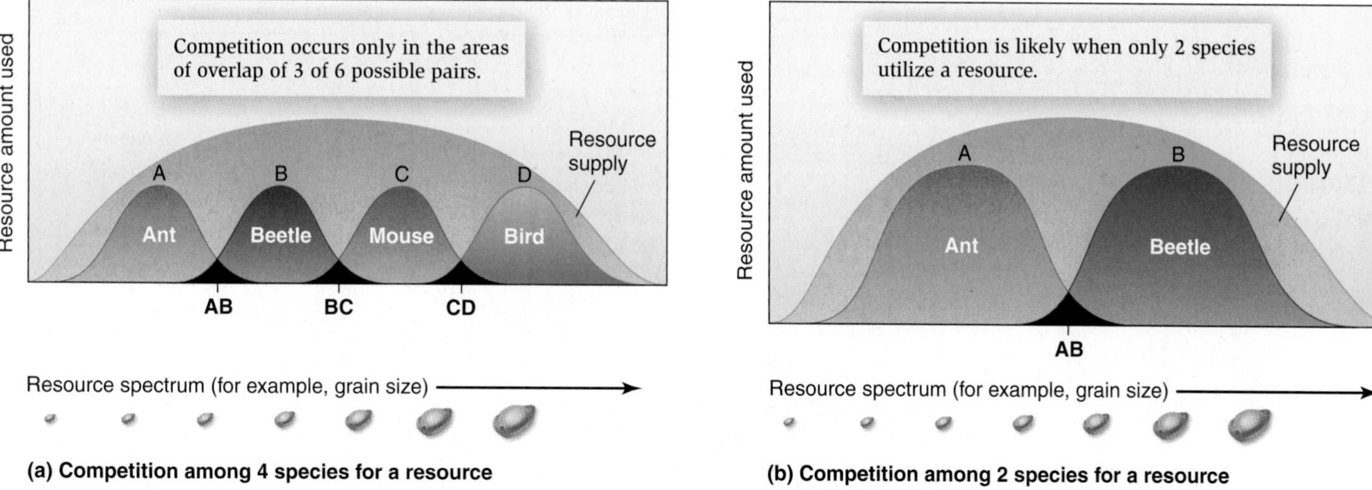

Figure 57.3 The frequency of competition according to the number of species involved. **(a)** Resource supply and utilization curves of four species, A, B, C, and D, along the spectrum of a hypothetical resource such as grain size. If competition occurs only between species with adjacent resource utilization curves, competition would be expected between three of the six possible pairings: A and B; B and C; and C and D. **(b)** When only two species utilize a resource set, competition would nearly always be expected between them.

Concept Check: *If five species utilized the resource set in part (a), what percent of the interactions would be competitive?*

(90%), whereas in studies involving more species, the frequency of competition drops to 50%. Why should this be the case? Imagine a resource such as a series of different-sized grains with four species—ants, beetles, mice, and birds—feeding on it (Figure 57.3a). The ants feed on the smallest grain, the beetles and mice on the intermediate sizes, and the birds, on the largest. If only adjacent species competed with each other, competition would be expected only between the ant–beetle, beetle–mouse, and mouse–bird. Thus, competition would be found in only three out of the six possible species pairs (50%). Naturally, the percentage would vary according to the number of species on the resource spectrum. If only three species occur along the spectrum, we would expect competition in two of the three pairs (67%). If just two species utilized the resource spectrum, however, we would expect competition in almost 100% of the cases (Figure 57.3b).

Some other general patterns were evident from Connell's review. Plants showed a high degree of competition, perhaps because they are rooted in the ground and cannot easily escape or perhaps because they are competing for the same set of limiting nutrients—water, light, and minerals. Marine organisms tended to compete more than terrestrial ones, perhaps because many of the species studied lived in the intertidal zone and were attached to the rock face, in a manner similar to that of plants. Because the area of the rock face is limited, competition for space is quite important.

Species May Coexist If They Do Not Occupy Identical Niches

Although competition is common, researchers have proposed several mechanisms by which two competing species can coexist. One states that similar species can coexist if they occupy different niches. A **niche** is often thought of as the area where an organism can be found, but it also can convey what an organism does in a community, including how it feeds. In 1934, the Russian microbiologist Georgyi Gause began to study competition between three protist species, *Paramecium aurelia*, *Paramecium bursaria*, and *Paramecium caudatum*, all of which fed on bacteria and yeast, which, in turn, fed on an oatmeal medium in a culture tube in the laboratory. The bacteria occurred more in the oxygen-rich upper part of the culture tube, and the yeast in the oxygen-poor lower part of the tube. Because each species was a slightly different size, Gause calculated population growth as a combination of numbers of individuals per milliliter of solution multiplied by their unit volume to give a population volume for each species. When grown separately, population volume of all three *Paramecium* species followed a logistic growth pattern (Figure 57.4a; refer back to Figure 56.11). When Gause cultured *P. caudatum* and *P. aurelia* together, *P. caudatum* went extinct (Figure 57.4b). Both species utilized bacteria as food, but *P. aurelia* grew at a rate six times faster than *P. caudatum*.

However, when Gause cultured *P. caudatum* and *P. bursaria* together, neither went extinct (Figure 57.4c). The population volume of each was much less than when they were grown alone, because some competition occurred between them. Gause discovered, however, that *P. bursaria* was better able to utilize the yeast in the lower part of the culture tubes. *P. bursaria* have tiny green algae inside them, which produce oxygen and allow *P. bursaria* to thrive in the lower oxygen levels at the bottom of the tubes. From these experiments, Gause concluded that two species with exactly the same requirements cannot live together in the same place and use the same resources; that is, occupy the same niche. His conclusion was later termed the **competitive exclusion principle**.

If complete competitors drive one species to local extinction, at least in the laboratory, how different must two species be to coexist, and in what features do they usually differ? To address such questions,

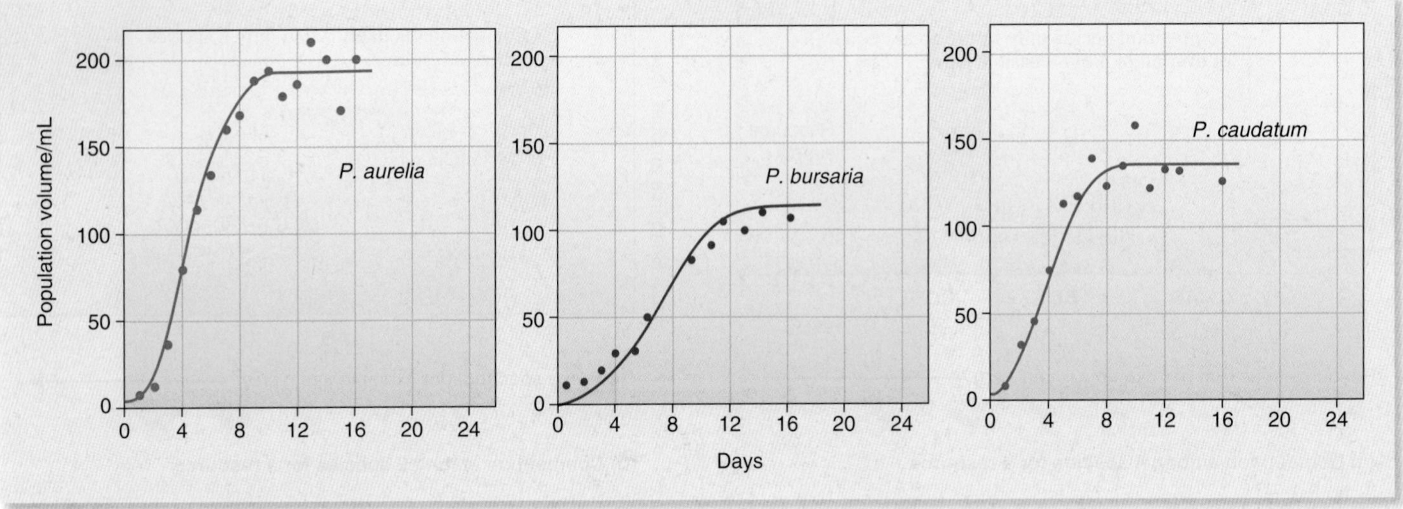

(a) Each *Paramecium* species grown alone

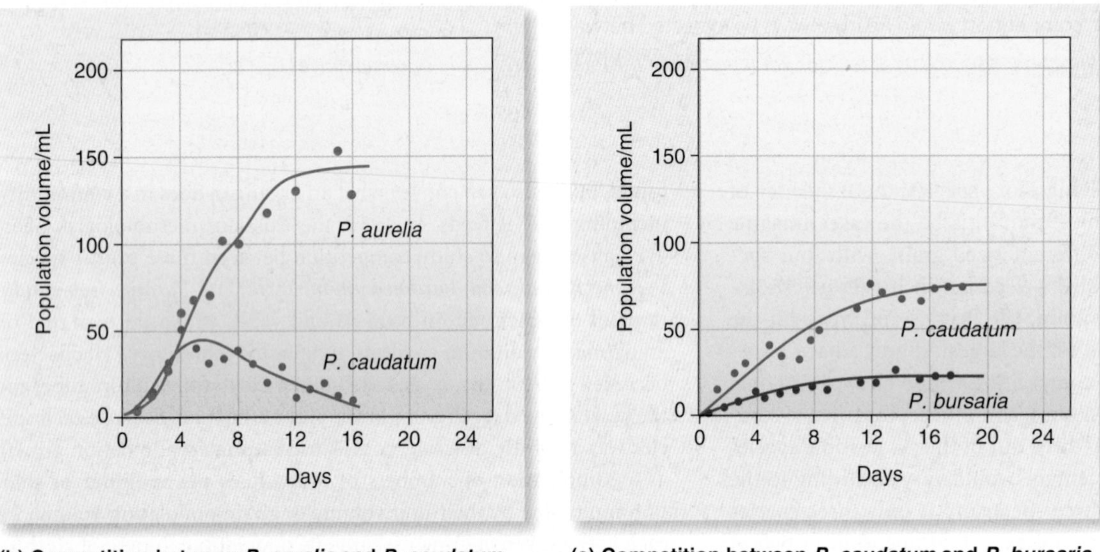

(b) Competition between *P. aurelia* and *P. caudatum* (c) Competition between *P. caudatum* and *P. bursaria*

Figure 57.4 **Competition among *Paramecium* species.** **(a)** When grown alone, each of three species, *Paramecium aurelia*, *Paramecium bursaria*, and *Paramecium caudatum*, grows according to the logistic model. **(b)** When *P. aurelia* is grown with *P. caudatum*, the density of *P. aurelia* is lower than when grown alone, and *P. caudatum* goes extinct. **(c)** When *P. caudatum* is grown with *P. bursaria*, the population densities of both are lowered, but they coexist.

in 1958, American ecologist Robert MacArthur examined coexistence between five species of warblers feeding within spruce trees in New England. All belonged to the genus *Dendroica*, so these closely related bird species would be expected to compete strongly, possibly sufficiently strongly to cause extinctions. MacArthur found that the species occupied different heights and portions in the tree, and therefore, each probably fed on a different range of insects (**Figure 57.5**). In addition, the Cape May warbler fed on flying insects and tended to remain on the outside of the trees.

The term **resource partitioning** describes the differentiation of niches, both in space and time, that enables similar species to coexist in a community, just as the five species of warblers feeding in different parts of a spruce tree. We can think of resource partitioning as reflecting the results of past competition, in which competition leads the inferior competitor to eventually occupy a different niche. British ornithologist David Lack examined competition and coexistence among about 40 species of British passerines, or perching birds (**Figure 57.6**). As a group, these birds had fairly similar lifestyles. Most segregated according to some resource factor, with habitat being the most common one. For example, although all the passerines fed on insects, some would feed exclusively in grasslands, others in forests, some low to the ground, and others high in trees, where the insects present would likely be different. Birds also segregated by size—so bigger species would take different-sized food from that of smaller species—and by feeding habit—with some feeding on insects on foliage, others on tree trunks, and so on. Some species also fed

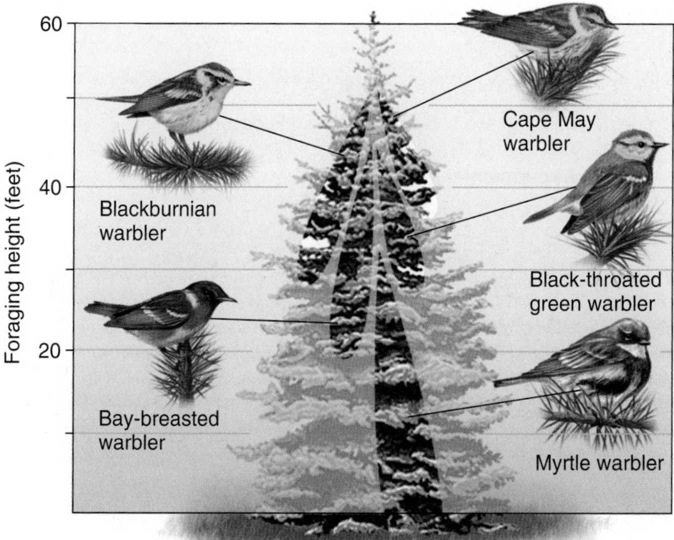

Figure 57.5 Resource partitioning. Among five species of warblers feeding in North American spruce trees, each species prefers to feed at a different height and portion of the tree, thus reducing competition.

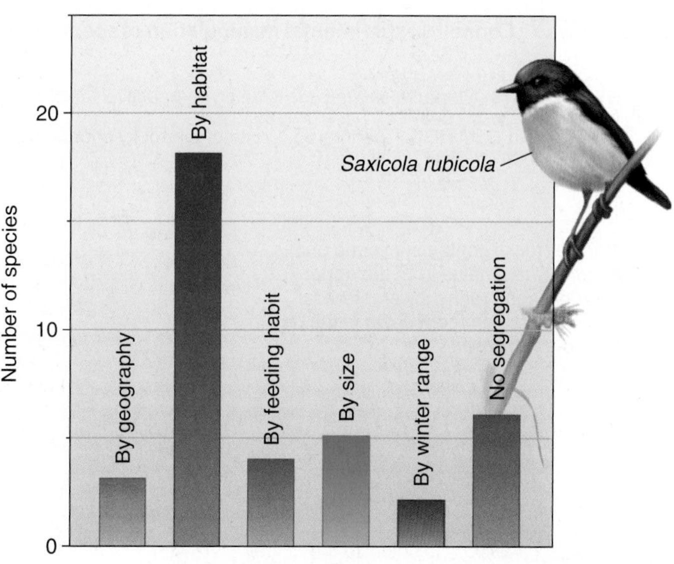

Figure 57.6 Segregation of 40 bird species according to resource factor. Among 40 species of passerine birds, most segregation is by habitat, followed by size, feeding habit, geography, and type of winter range they forage in. In about 15% of cases, no obvious segregation was observed. More than half of all bird species, including *Saxicola rubicola*, are passerines, also known as perching birds.

Concept Check: *Do you think these results for passerine birds are typical for most other species? For example, do most other species segregate by habitat?*

in different winter ranges, whereas others occurred in different parts of the country (separation by geography). About 15% of bird species showed no segregation at all.

Most species perform best over a physiologically optimal range of conditions called the **fundamental niche**. However, if some part of the fundamental niche is occupied by competitors, the range of an organism may be limited to an area known as the **realized niche**, where the competitor is absent. Researchers have established that one of the best methods of determining an organism's fundamental niche is to temporarily remove one of the competing species and examine the effect on the other species. A now-classic example of this method involved a study of the interactions between two species of barnacles conducted on the west coast of Scotland, as described next.

Feature Investigation

Connell's Experiments with Barnacle Species Revealed Each Species' Fundamental and Realized Niches

Chthamalus stellatus and *Semibalanus balanoides* (formerly known as *Balanus balanoides*) are two species of barnacles that dominate the Scottish coastline. Each organism's realized niche on the intertidal zone is well defined. *Chthamalus* occurs in the upper intertidal zone, and *Semibalanus* is restricted to the lower intertidal zone. Connell sought to determine what the range of *Chthamalus* adults might be in the absence of competition from *Semibalanus* (**Figure 57.7**).

To do this, Connell obtained rocks from high on the rock face, just below the high-tide level, where only *Chthamalus* grew. These rocks already contained young and mature *Chthamalus*. He then moved the rocks into the *Semibalanus* zone, fastened them down with screws, and allowed *Semibalanus* to also colonize them. Once *Semibalanus* had colonized these rocks, he took the rocks out, removed all the *Semibalanus* organisms from one side of the rocks with a needle, and then returned the rocks to the lower intertidal zone, screwing them down once again. As seen in the data, the mortality of *Chthamalus* on rock halves with *Semibalanus* was fairly high. On the *Semibalanus*-free halves, however, *Chthamalus* survived well.

In other studies, Connell also monitored survival of natural patches of both barnacle species where both occurred on the intertidal zone at the upper margin of the *Semibalanus* distribution. In a period of unusually low tides and warm weather, when no water reached any barnacles for several days, desiccation became a real threat to both species' survival. During this time, young *Semibalanus* suffered a 92% mortality rate, and older individuals, a 51% mortality rate. At the same time, young *Chthamalus* experienced a 62% mortality rate compared with a rate of only 2% for more-resistant older individuals. Clearly, *Semibalanus* is not as resistant to desiccation as *Chthamalus* and could not survive in the upper intertidal zone where *Chthamalus* occurs. *Chthamalus* is more resistant to desiccation than *Semibalanus* and can be found higher in the intertidal zone. Thus, whereas the lower limit of *Chthamalus* was set by competition with *Semibalanus*, the upper limit was controlled by desiccation. Although the potential

Figure 57.7 **Connell's experimental manipulation of species indicated the presence of competition.**

HYPOTHESIS *Chthamalus stellatus* is being competitively excluded from the lower intertidal zone by the species *Semibalanus balanoides*.

STARTING LOCATION The intertidal zone of the rocky shores of the Scottish coast, where the two species of barnacles occur.

Experimental level

1 Transfer rocks containing young and mature *Chthamalus* from the upper intertidal zone to the lower intertidal zone, and fasten them down in the new location with screws.

High-tide level
Upper intertidal zone
Mean-tide level
Lower intertidal zone
Low-tide level
Chthamalus stellatus
Semibalanus balanoides

2 Allow *Semibalanus* to colonize the rocks.

3 After the colonization period is over, remove *Semibalanus* from half of each rock with a needle (leaving the other half undisturbed). Return the rocks to the lower intertidal zone, and fasten them down once again.

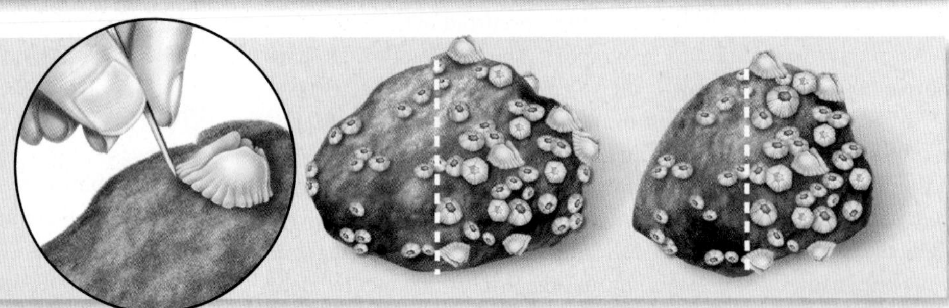

4 Monitor the survival of *Chthamalus* on both sides of the rocks.

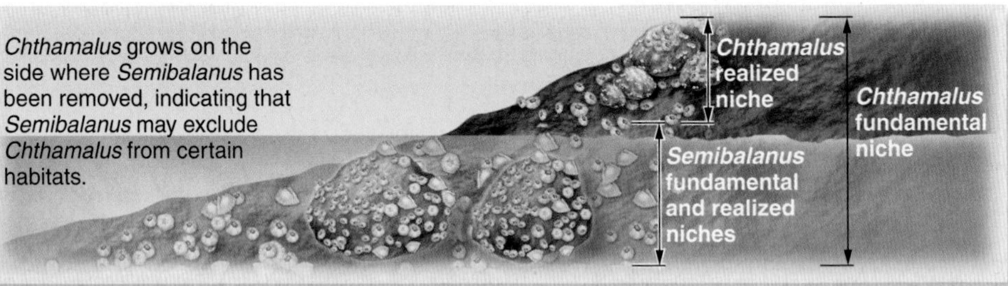
Chthamalus grows on the side where Semibalanus has been removed, indicating that Semibalanus may exclude Chthamalus from certain habitats.
Chthamalus realized niche
Chthamalus fundamental niche
Semibalanus fundamental and realized niches

5 **THE DATA**

Rock No.	Side of rock	% *Chthamalus* mortality over 1 year	
		Young barnacles	**Mature barnacles**
13b	*Semibalanus* removed	35	0
	Semibalanus not removed	90	31
12a	*Semibalanus* removed	44	37
	Semibalanus not removed	95	71
14a	*Semibalanus* removed	40	36
	Semibalanus not removed	86	75

6 **CONCLUSION** The data from this study indicate that *Chthamalus* is not found on the lower rock face because of competition with *Semibalanus.* Other studies indicate that *Chthamalus* occupies the upper rock face because it is more resistant to desiccation.

7 **SOURCE** Connell, J.H. 1961. The influence of interspecific competition and other factors on the distribution of the barnacle *Chthamalus stellatus. Ecology* 42:710–732.

distribution, the fundamental niche, of *Chthamalus* extends over the entire intertidal zone, its actual distribution, the realized niche, is restricted to the upper zone.

Experimental Questions

1. Describe the realized niches for the two species of barnacles used in Connell's experiment.
2. Outline the procedure Connell used in the experiments.
3. How did Connell explain the presence of *Chthamalus* in the upper intertidal zone if *Semibalanus* was shown to outcompete the species in the first experiment?

Morphological Differences May Allow Species to Coexist

Although the competitive exclusion principle acknowledges that competitors with exactly the same requirements cannot coexist, some partial level of competition may exist that is not severe enough to drive one of the competitors to extinction or to a different niche. In 1959, British-born American biologist G. Evelyn Hutchinson examined the sizes of mouthparts or other body parts important in feeding and compared their sizes across species when they were **sympatric** (occurring in the same geographic area) and **allopatric** (occurring in different geographic areas). Hutchinson's hypothesis was that when species were sympatric, each species tended to specialize on different types of food. This was reflected by differences in the size of body parts associated with feeding, also called feeding characters. The tendency for two species to diverge in morphology and thus resource use because of competition is called **character displacement**. Alternatively, in areas where species were allopatric, there was no need to specialize on a particular prey type, so the size of the feeding character did not evolve to become larger or smaller; rather it retained a "middle of the road" size that allowed species to exploit the largest range of prey size distribution.

One of the classic cases of character displacement involves a study of Galápagos finches, several closely related species of finches Charles Darwin discovered on the Galápagos Islands (refer back to Figure 23.4). When two species, *Geospiza fortis* and *Geospiza fuliginosa*, are sympatric, their beak sizes (bill depths) are different: *G. fortis* has a larger bill depth, which enables it to feed on bigger seeds, whereas *G. fuliginosa* has a smaller bill depth, which enables it to crack small seeds more efficiently. However, when both species are allopatric, that is, existing on different islands, their bills are more similar in depth. Researchers studying *Geospiza* concluded that the bill depth differences evolved in ways that minimized competition.

How great must differences between characters be in order to permit coexistence? Hutchinson noted that the ratio between feeding characters when species were sympatric (and thus competed) averaged about 1.3 (**Table 57.2**). In contrast, the ratio between feeding characters when species were allopatric (and did not compete) was closer to 1.0. Hutchinson proposed that the value of 1.3, a roughly 30% difference, could be used as an indication of the amount of

Table 57.2 Comparison of Feeding Characters of Sympatric and Allopatric Species

		Measurement (mm) when		Ratio* when	
Animal (character)	**Species**	**Sympatric**	**Allopatric**	**Sympatric**	**Allopatric**
Weasels (skull)	*Mustela erminea*	50.4	46.0	1.28	1.07
	Mustela ivalis	39.3	42.9		
Mice (skull)	*Apodemus flavicollis*	27.0	26.7	1.09	1.04
	Apodemus sylvaticus	24.8	25.6		
Nuthatches (beak)	*Sitta tephronota*	29.0	25.5	1.23	1.02
	Sitta neumayer	23.5	26.0		
Galápagos finches (beak)	*Geospiza fortis*	12.0	10.5	1.43	1.13
	Geospiza fuliginosa	8.4	9.3		
Average ratio				1.26	1.06

*Ratio of the larger to smaller character.

difference necessary to permit two species to coexist. One problem with Hutchinson's ratio is that some differences of 1.3 between similar species might have evolved for reasons other than competition. Some ecologists have argued that we should not conclude that a 30% difference is a strong indicator of coexistence. Nevertheless, although the actual ratio may be disputed, Hutchinson's findings show that competition in nature can cause character displacement.

57.2 Predation, Herbivory, and Parasitism

Learning Outcomes:

1. List and describe strategies animals use to avoid predation.
2. Give examples of the effects of predators on prey populations.
3. Explain why plants and herbivores are said to be in an "evolutionary arms race."
4. Describe variations on the parasitic lifestyle.

Predation, herbivory, and parasitism are interactions that have a positive effect for one species and a negative effect for the other. These categories of species interactions can be classified according to how lethal they are for the prey and the length of association between the consumer and prey (**Figure 57.8**). Each has particular characteristics that set it apart. Herbivory usually involves nonlethal predation on plants, whereas predation generally results in the death of the prey. Parasitism, like herbivory, is typically nonlethal and differs from predation in that the adult parasite typically lives and reproduces for long periods in or on the living host (refer back to Figure 33.9).

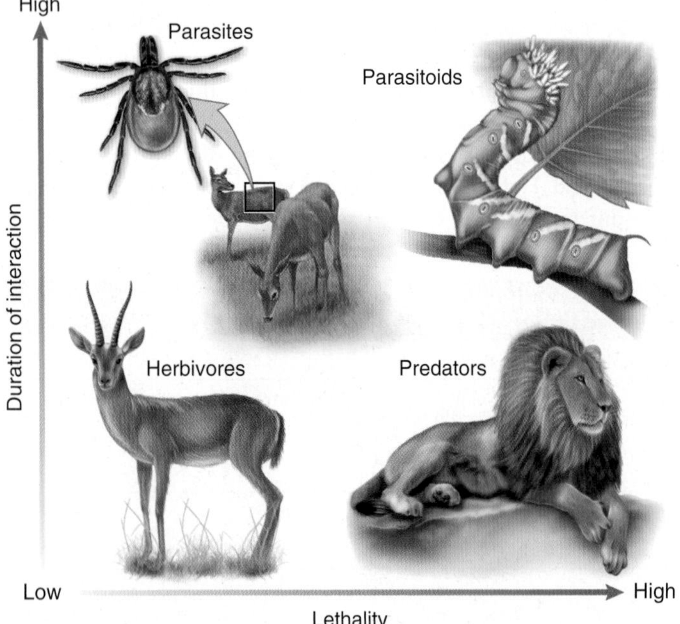

Figure 57.8 Possible interactions between populations. Lethality represents the probability that an interaction results in the death of the prey. Duration represents the length of the interaction between the consumer and the prey.

Concept Check: *Where might omnivores fit in this figure?*

Parasitoids, insects that lay eggs in living hosts, have features in common with both predators and parasites. They always kill their prey, as predators do, but unlike predators, which immediately kill their prey, parasitoids kill the host more slowly. Parasitoids are common in the insect world and include parasitic wasps and flies that feed on many other insects.

In this section, we begin by looking at antipredator strategies and how, despite such strategies, predation remains a factor affecting the density of prey. We survey the strategies plants use to deter herbivores and how, in turn, herbivores overcome host plant defenses. Finally, we investigate parasitism, which may be the predominant lifestyle on Earth, and explore the growing role of genomics in the fight against parasites.

Animals Have Evolved Many Antipredator Strategies

The variety of strategies that animals have evolved to avoid being eaten suggests that predation is a strong selective force. Common strategies include chemical defense; forms of camouflage and mimicry; displays of intimidation; agility; armor; and altering of reproductive patterns, as in masting.

Chemical Defense A great many species have evolved chemical defenses against predation. One of the classic examples of a chemical defense involves the bombardier beetle (*Stenaptinus insignis*), which has been studied by German-born American entomologist Tom Eisner and coworkers. These beetles possess a reservoir of hydroquinone and hydrogen peroxide in their abdomen. When threatened, they eject the chemicals into an "explosion chamber," where the subsequent release of oxygen causes the whole mixture to be violently ejected as a hot spray (about 88°C, or 190°F) that can be directed at the beetle's attackers (**Figure 57.9a**). Many other arthropods, such as millipedes, also have chemical sprays, and the phenomenon is also found in vertebrates, as anyone who has had a close encounter with a skunk can testify.

Often associated with a chemical defense is an **aposematic coloration**, or warning coloration, which advertises an organism's unpalatable taste. For instance, the ladybird beetle's bright red color warns of the toxic defensive chemicals it exudes when threatened, and many tropical frogs have bright warning coloration that calls attention to their skin's lethality (**Figure 57.9b**). Monarch butterfly caterpillars feed exclusively on milkweed, which contains toxic chemicals called cardiac glycosides that pass into the caterpillars. In the 1960s, American entomologist Lincoln Brower and coworkers showed that after inexperienced blue jays ate a monarch butterfly and suffered a violent vomiting reaction, they learned to associate the striking orange-and-black appearance of the butterfly with a noxious reaction (refer back to Figure 55.3).

Cryptic Coloration **Cryptic coloration** is an aspect of camouflage, the blending of an organism with the background of its habitat. Cryptic coloration is a common method of avoiding detection by predators. For example, many grasshoppers are green and blend in with the foliage on which they feed. Stick insects mimic branches and twigs with their long, slender bodies. In most cases, these animals stay perfectly still when threatened, because movement alerts a predator.

(a) As it is held by a tether attached to its back, this bombardier beetle (*Stenaptinus insignis*) directs its hot, stinging spray at a forceps "attacker."

(b) Aposematic coloration advertises the poisonous nature of this blue poison arrow frog (*Dendrobates azureus*) from South America.

(c) Cryptic coloration allows this Pygmy sea horse (*Hippocampus bargibanti*) from Bali to blend in with its background.

(d) In this example of Batesian mimicry, an innocuous scarlet king snake (*Lampropeltis elapsoides*) (left) mimics the poisonous coral snake (*Micrurus fulvius*) (right).

(e) In a display of intimidation, this porcupine fish (*Diodon hystrix*) puffs itself up to look threatening to its predators.

Figure 57.9 Antipredator adaptations.

Concept Check: *According to the classification of species interactions in Table 57.1, how would you classify Batesian and Müllerian mimicry?*

BioConnections: *Refer back to Figure 33.13. Which types of antipredator adaptations are possessed by mollusks?*

Maintenance of a fixed body posture is referred to as catalepsis. Cryptic coloration is prevalent in the vertebrate world, too. Many sea horses adopt a body shape and color pattern similar to the environment in which they are found (**Figure 57.9c**).

Mimicry **Mimicry**, the resemblance of a species (the mimic) to another species (the model), also secures protection from predators. There are two major types of mimicry. In **Müllerian mimicry**, two or more toxic species converge to look the same, thus reinforcing the basic distasteful design. One example is the black-and-yellow-striped bands of several different types of bees and wasps. Müllerian mimicry is also found among noxious Amazonian butterflies. The viceroy butterfly (*Limenitis archippus*) and the monarch butterfly (*Danaus plexippus*) are examples of Müllerian mimicry. Both species are unpalatable and look similar, but the viceroy can be distinguished from the monarch by a black line that crosses its wings.

Batesian mimicry is the mimicry of an unpalatable species (the model) by a palatable one (the mimic). Some of the best examples involve flies, especially hoverflies of the family Syrphidae, which are striped black and yellow and resemble stinging bees and wasps but are themselves harmless. Among vertebrates, the nonvenomous scarlet king snake (*Lampropeltis elapsoides*) mimics the venomous coral snake (*Micrurus fulvius*), thereby gaining protection from would-be predators (**Figure 57.9d**).

Displays of Intimidation Some animals put on displays of intimidation in an attempt to discourage predators. For example, a cat arches its back, a frilled lizard extends it collar, and a porcupine fish inflates itself when threatened in order to appear larger (**Figure 57.9e**). All of these animals use displays to deceive potential predators about the ease with which they can be eaten.

Fighting Though many animals developed horns and antlers for sexual selection, they can also be used in defense against predators (refer back to Figure 34.25). Invertebrate species often have powerful claws, pincers, or, in the case of scorpions, venomous stingers that can be used in defense as well as offense.

Agility Some groups of insects, such as grasshoppers, have a powerful jumping ability to escape the clutches of predators. Many frogs are prodigious jumpers, and flying fish can glide above the water to escape their pursuers.

Armor The shells of tortoises and turtles are a strong means of defense against most predators, as are the quills of porcupines (refer back to Figure 34.23c). Many beetles have a tough exoskeleton that protects them from attack from other arthropod predators such as spiders.

Masting **Masting** is the synchronous production of many progeny by all individuals in a population to satiate predators and thereby allow some progeny to survive. Masting is more commonly discussed as a strategy of trees, which tend to have years of unusually high seed production that reduces predation. However, a similar phenomenon is exhibited by the emergence of 13-year and 17-year periodical cicadas (genus *Magicicada*). These insects are termed periodical because the emergence of adults is highly synchronized to occur once every 13 or 17 years. Adult cicadas live for only a few weeks, during which time females mate and deposit eggs on the twigs of trees. The eggs hatch 6 to 10 weeks later, and the nymphs drop to the ground, burrow beneath the soil, and begin a long subterranean development, feeding on the contents of the xylem of roots. Because the xylem is low in nutrients, it takes many years for nymphs to develop, though there appears to be no physiological reason why some cicadas couldn't emerge after, say, 12 years of feeding, and others after 14. Their synchrony of emergence is thought to maximize predator satiation. Worth noting in this context is the fact that both 13 and 17 are prime numbers, and thus predators on a shorter multiannual cycle cannot repeatedly utilize this resource. For example, a predator that bred every 3 years could not rely on cicadas always being present as a food supply.

How common is each of these defense types? No one has done an extensive survey of the entire animal kingdom. However, in 1989, American biologist Brian Witz surveyed studies that documented antipredator mechanisms in arthropods, mainly insects. By far the most common antipredator mechanisms were chemical defenses and associated aposematic coloration, noted in 51% of the examples. Other types of defense mechanisms occurred with considerably less frequency.

Despite the Impressive Array of Defenses, Predators Can Still Affect Prey Densities

The importance of predation on prey populations may depend on whether the system is donor-controlled or predator-controlled. In a donor-controlled system, prey supply is determined by factors other than predation, such as food supply, so that removal of predators has no significant effect on prey density. Examples include predators that feed on intertidal communities in which space is the limiting factor that controls prey populations.

In a predator-controlled system, the action of predator feeding eventually reduces the supply of prey. Therefore, the removal of predators would probably result in large increases in prey abundance. Research studies have shown that predators can have a significant effect on prey populations. Considerable data exist on the interaction of the Canada lynx (*Lynx canadensis*) and its prey, snowshoe hares (*Lepus americanus*), because of the value of the pelts of both animals. In 1942, British ecologist Charles Elton analyzed the records of furs traded by trappers to the Hudson's Bay Company in Canada over a 100-year period. Analysis of the records showed that a dramatic 9- to 11-year cycle existed for as long as records had been kept (**Figure 57.10**). As hare density increases, there is an increase in density of the lynx, which then depresses hare numbers. This is followed by a decline in the number of lynx, and the cycle begins again. Using radio collars to track individual hares, researchers were able to determine that 90% of individuals died of predation. However, more recent research has shown that hare densities may increase in times of food surpluses. When researchers added food supplements to large experimental areas containing both hares and lynxes, the hare densities increased threefold but still continued to cycle.

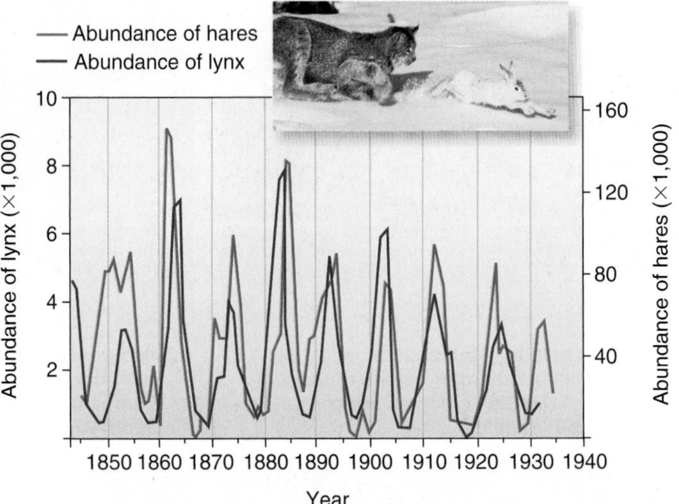

Figure 57.10 **Effect of predator on prey populations.** The 9- to 11-year oscillation in the abundance of the snowshoe hare (*Lepus americanus*) and the Canada lynx (*Lynx canadensis*) was revealed from pelt trading records of the Hudson's Bay Company.

Invasive species provide striking examples of the effects of predators. The brown tree snake (*Boiga irregularis*) was inadvertently introduced by humans to the island of Guam, in Micronesia, shortly after World War II. The growth and spread of its population over the next 40 years closely coincided with a precipitous decline in the island's forest birds. On Guam, the snake had no natural predators to control it. Because the birds on Guam did not evolve with the snake, they had no defenses against it. Eight of the island's 11 native species of forest birds went extinct by the 1980s, such as the Guam rail and Micronesian kingfisher (**Figure 57.11**).

One of the biggest reductions in prey in response to predation has been the systematic decline of various whale species as a result of the human whaling industry. The history of whaling has been characterized by a progression from larger, more valuable or easily caught species to smaller, less valuable or easily caught ones, as numbers of the original targets have been depleted (**Figure 57.12**). In 1986, the International Whaling Commission belatedly enacted a moratorium on all commercial whaling. Following the moratorium, the populations of some whales have increased. Blue whales are thought to have quadrupled their numbers off the California coast during the 1980s, and numbers of the California gray whales have recovered to prewhaling levels, showing the effect an absence of a predator can have.

Ecologists have found that in nearly 1,500 predator-prey studies, over two-thirds (72%) showed a large depression of prey density by predators. Thus, we can conclude that in the majority of cases,

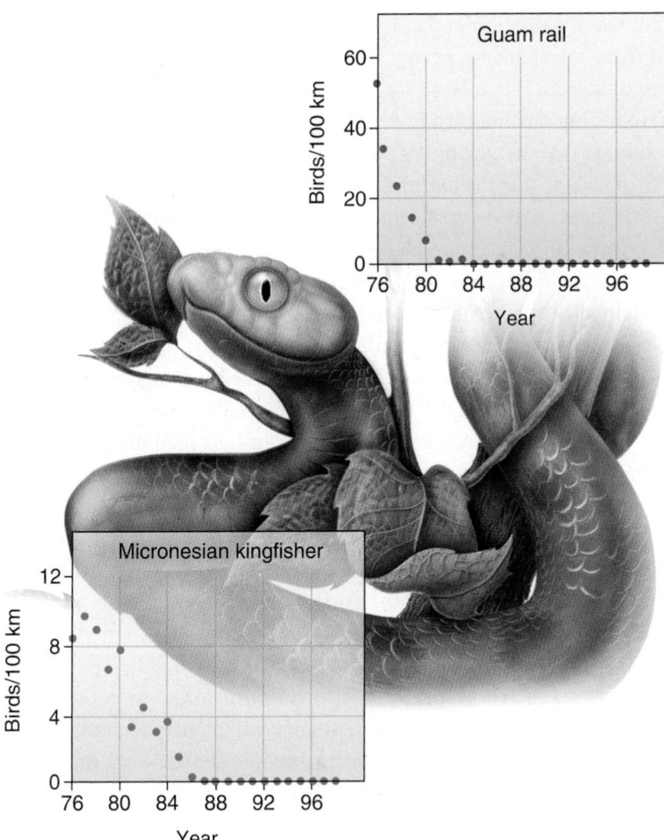

Figure 57.11 Predation by an invasive species. Population trends for two native Guam birds, as indicated by 100-km roadside surveys conducted from 1976 to 1998, show a precipitous decline because of predation by the introduced brown tree snake.

Concept Check: *Why can invasive predators have such strong effects on native prey?*

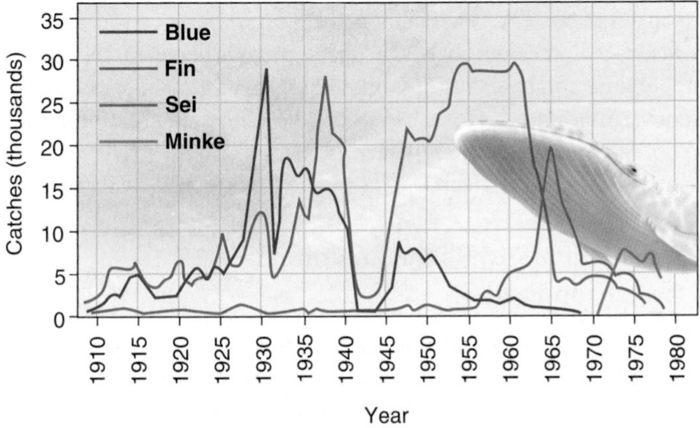

Figure 57.12 Sequential decline of whale catches in the Antarctic due to human predation. Whale catches are believed to be directly related to whale population sizes. The catches of the blue whale, the first species to be strongly affected by human predation, started a precipitous decline in the 1940s, as the whale was hunted to very low levels. Humans then began hunting more-abundant fin whales, and then sei and minke whales, as each species became depleted.

predators influence the abundance of their prey in their native environment. The variety of antipredator mechanisms discussed earlier also shows how predation is important enough to select for the evolution of chemical defenses, camouflage, and mimicry in prey. Taken together, these data indicate that predation is a powerful force in nature.

Plants and Herbivores May Be Engaged in an Evolutionary Arms Race

Herbivory involves the predation of plants or similar life forms such as algae. Such predation can be lethal to the plants, especially for small species, but often it is nonlethal, because many plant species, particularly larger ones, can regrow. We can distinguish two types of herbivores: generalist herbivores and specialist herbivores. Generalist herbivores, which are usually mammals, can feed on many different plant species. Specialist herbivores, which are typically insects, are often restricted to one or two species of host plants. There are, however, exceptions. Pandas are specialists because they feed only on bamboo, and koalas specialize on eucalyptus trees. On the other hand, grasshoppers are generalists, because they feed on a wide variety of plant species, including agricultural crops.

Plants present a luscious green world to any organism versatile enough to use it, so why don't herbivores eat more of the food available to them? After all, unlike most animals, plants cannot move to escape being eaten. Two hypotheses have been proposed to answer the question of why more plant material is not eaten. First, predators and parasites may keep herbivore numbers low, thereby sparing the plants. The many examples of the strength of predation provide evidence for this view. Second, the plant world is not as helpless as it appears. The sea of green is armed with defensive spines, tough cuticles, noxious chemicals, and more. Let's take a closer look at plant defenses against herbivores and the ways that herbivores attempt to overcome them.

Plant Defenses Against Herbivores An array of unusual and powerful chemicals is present in plants, including alkaloids (nicotine in tobacco, morphine in poppies, cocaine in coca, and caffeine in coffee), phenolics (lignin in wood and tannin in leaves), and terpenoids (in peppermint) (Figure 57.13a–c). Such compounds are not part of the primary metabolic pathway that plants use to obtain energy and are therefore referred to as **secondary metabolites**. Most of these chemicals are bitter tasting or toxic, thereby deterring herbivores from feeding. The staggering variety of secondary metabolites in plants, over 25,000, may be testament to the large number of organisms that feed on plants. In an interesting twist, many of these compounds have medicinal properties that have proved to be beneficial to humans. In addition to containing chemical compounds, many plants have an array of mechanical defenses, such as thorns and spines (Figure 57.13d).

An understanding of plant defenses is of great use to agriculturalists. The more that crops can defend themselves against pests, the higher the crop yield. The ability of plants to prevent herbivory via either chemical or mechanical defenses is also known as **host plant resistance**. One serious problem associated with commercial development of host plant resistance is that it may take a long time to breed into plants—between 10 and 15 years. This is because of the long time it takes to identify the responsible chemicals and develop

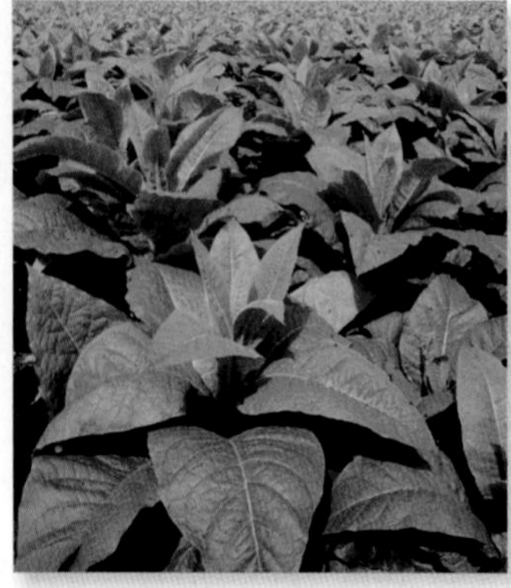

(a) Alkaloids in tobacco

(b) Phenolics in tea

(c) Terpenoids in peppermint

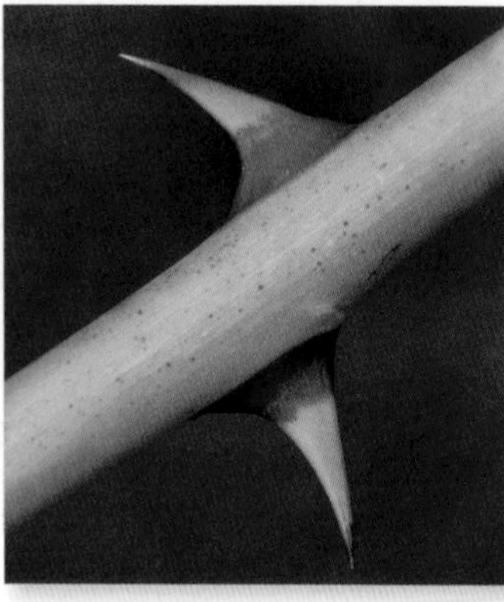

(d) Thorns on rose stems

Figure 57.13 Defenses against herbivory. Plants possess an array of unusual and powerful chemicals, including **(a)** alkaloids, such as nicotine in tobacco, **(b)** phenolics, such as tannins in tea leaves (near Mount Fuji, Japan), and **(c)** terpenoids in peppermint leaves. **(d)** Mechanical defenses include plant spines and thorns, as on this shrub, *Rosa multiflora.*

Concept Check: Of the defenses shown here, which type would be most effective in deterring invertebrate herbivores?

BioConnections: Capsaicin is an alkaloid produced by which species of plant? Refer back to Figure 30.22.

the resistant genetic lines. Also, resistance to one pest may come at the cost of increasing susceptibility to other pests. Finally, some pest strains can overcome the plant's mechanisms of resistance.

Despite these problems, host plant resistance is a good tactic for the farmer. Host plant resistance reduces the need for chemical insecticides and is less environmentally harmful, generally having few side effects on other species in the community. About 75% of cropland in the U.S. utilizes pest-resistant plant varieties, most of these being resistant to plant pathogens. Bt corn is a variety of corn that has been genetically modified to incorporate a gene from the soil bacterium *Bacillus thuringiensis* (Bt) that encodes a protein, Bt toxin, that is toxic to some insects (refer back to Figure 20.14). Genetic engineers have also produced Bt cotton, Bt tomato, and genetically modified varieties of many other crop species.

Overcoming Plant Resistance Herbivores can often overcome plant defenses. They can detoxify many poisons, mainly by two chemical pathways: oxidation and conjugation. Oxidation, the most important of these mechanisms, occurs in the liver of mammals and in the midgut of insects. It involves catalysis of the secondary metabolite to a corresponding alcohol by a group of enzymes known as mixed-function oxidases (MFOs). Conjugation, often the next step in detoxification, occurs by uniting the harmful compound or its oxidation product with another molecule to create an inactive and readily excreted product.

In addition, certain chemicals that are toxic to generalist herbivores actually increase the growth rates of adapted specialist species, which put the chemicals to good use in their own metabolic pathways. The Brassicaceae, the plant family that includes mustard, cabbage, and other species, contains acrid-smelling mustard oils called glucosinolates, the most important one of which is sinigrin. Large white butterflies (*Pieris brassicae*) preferentially feed on cabbage over other plants. They are able to detoxify even high levels of sinigrin. If newly hatched larvae are fed an artificial diet, they do much better when sinigrin is added to it. When larvae are fed cabbage leaves on hatching from eggs and are later switched to an artificial diet without sinigrin, they starve rather than eat. In this case, the secondary metabolite has become an essential feeding stimulant.

The Effects of Herbivores on Plant Populations A good method for estimating the effects of herbivory on plant populations is to remove the herbivores and examine subsequent growth and reproductive output. Analyses of hundreds of such experiments have yielded several interesting generalizations. First, herbivory in aquatic systems is more extensive than in terrestrial ones. Aquatic systems contain species, such as algae, that are especially susceptible to herbivory, presumably because these organisms are the least sophisticated in terms of their ability to manufacture complex secondary metabolites. In terms of terrestrial systems, grasses and shrubs are significantly affected by herbivores, but woody plants such as trees are less so. Large and long-lived trees can draw on substantial resource reserves to buffer the effect of herbivores.

Second, invertebrate herbivores such as insects have a stronger effect on plants than vertebrate herbivores such as mammals, at least in terrestrial systems. Thus, although large grazers like bison in North America or antelopes in Africa might be considered to be of huge importance in grasslands, it is more likely that grasshoppers are the more significant herbivores because of their sheer weight of numbers. In forests, invertebrate grazers such as caterpillars have greater access to canopy leaves than vertebrates and are also likely to have a greater effect.

Parasitism Might Be the Predominant Lifestyle on Earth

When one organism feeds on another but does not normally kill it outright, the organism is termed a **parasite**, and the prey, a **host**. Some parasites remain attached to their hosts for most of their life. For example, tapeworms spend their entire adult life inside the host's alimentary canal and even reproduce within their host. Others, such as the lancet fluke, have more complex life cycles that require multiple hosts (**Figure 57.14**). To facilitate transmission, many parasites induce changes in the behavior of one host, making that host more susceptible to being eaten by a second host. Notice that in Figure 57.14, fluke parasites cause a change in the ant's behavior, so

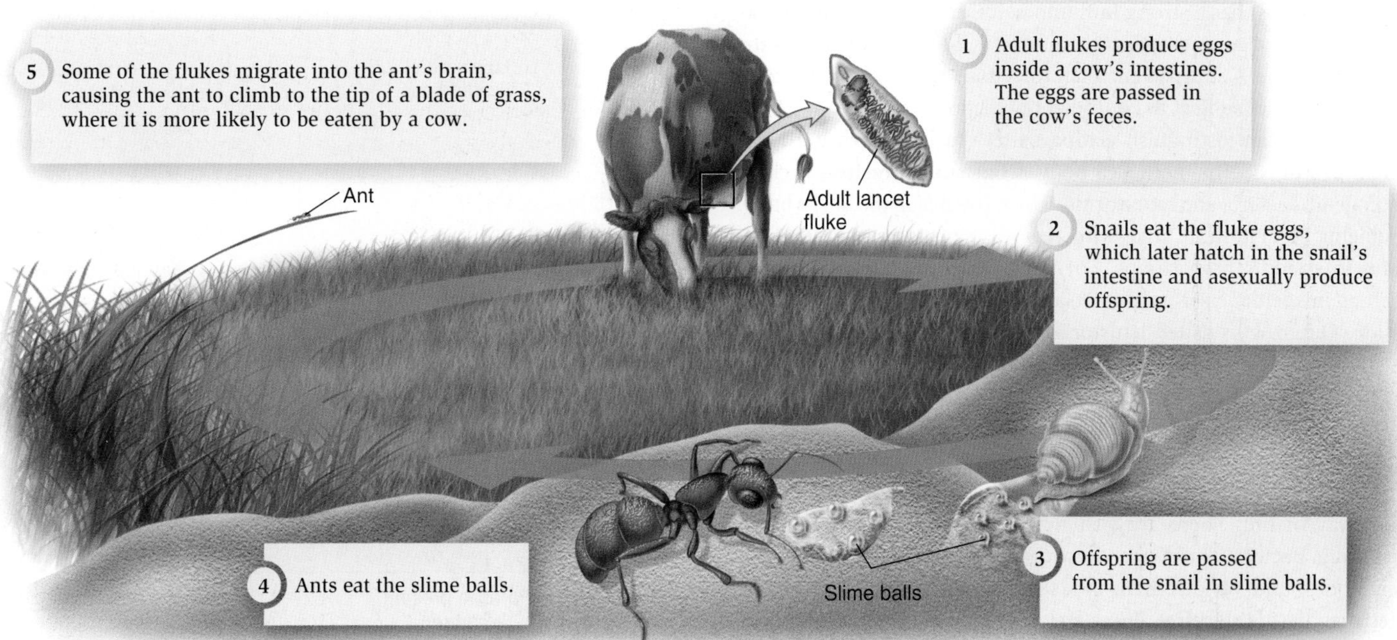

5 Some of the flukes migrate into the ant's brain, causing the ant to climb to the tip of a blade of grass, where it is more likely to be eaten by a cow.
1 Adult flukes produce eggs inside a cow's intestines. The eggs are passed in the cow's feces.
Ant
Adult lancet fluke
2 Snails eat the fluke eggs, which later hatch in the snail's intestine and asexually produce offspring.
4 Ants eat the slime balls.
Slime balls
3 Offspring are passed from the snail in slime balls.

Figure 57.14 The life cycle of the lancet fluke. The lancet fluke (*Dicrocoelium dendriticum*) causes behavioral changes in ants (one of its three hosts) that increase its transmission rate.

that it climbs to the tip of the blade of grass, where it is exposed and likely to be ingested by a cow.

Great variation exists in parasites and their lifestyle. Some parasites, such as ticks and leeches, drop off their hosts after prolonged periods of feeding. Others, like mosquitoes, remain attached for relatively short periods. Even parasites with a short attachment period can manipulate the feeding behavior of their hosts. The malaria parasite, a single-celled species in the genus *Plasmodium*, has a complex life cycle involving two hosts, mosquitoes and vertebrates (refer back to Figure 28.29). *Plasmodium* interferes with the ability of the mosquito to draw up blood from its vertebrate hosts. This increases the number of attacks the infected mosquitoes make in order to try and obtain enough blood. Increased attack rates maximize the transmission rates of the *Plasmodium* itself. British epidemiologist Jacob Koella and colleagues showed that most uninfected mosquitoes generally fed on just one human host at night, and only 10% bit more than one person. Multiple biting of different hosts increased to 22% in malaria-infected mosquitoes. In addition, the saliva of the infected mosquitoes was changed, making the host's blood flow less freely into the mouthparts. Similar behavior is exhibited by leishmaniasis parasites in sand flies and bubonic plague parasites in fleas.

Some flowering plants are parasitic on other plants. **Holoparasites** lack chlorophyll and are totally dependent on the host plant for their water and nutrients. One famous holoparasite is the tropical *Rafflesia arnoldii*, which lives most of its life within the body of its host, a *Tetrastigma* vine, which grows in rain forests (Figure 57.15). Only the *Rafflesia* flower develops externally. It is a massive flower, 1 m in diameter, and the largest known in the world. **Hemiparasites** generally photosynthesize, but depend on their hosts for water and mineral nutrients. Mistletoe (*Viscum album*) is a hemiparasite that grows on the stems of trees. Hemiparasites usually have a broader range of hosts than do holoparasites, which may be confined to a single or a few host species.

Figure 57.15 A holoparasite. *Rafflesia arnoldii*, the world's biggest flower, lives as a holoparasite in Indonesian rain forests.

BioConnections: More than 4,500 species of plants live as complete or partial parasites. What is an example of another important parasitic plant we learned about in Chapter 37? Refer back to Figure 37.20.

Parasites that feed on one species or just a few closely related hosts are termed **monophagous**. By contrast, **polyphagous** species can feed on many different host species, often from more than one family. We can also distinguish parasites as **microparasites** (for example, pathogenic bacteria), which multiply within their hosts, sometimes within the cells, and **macroparasites** (such as schistosomes), which live in the host but release infective juvenile stages outside the host's body.

Usually, the host has a strong immunological response to microparasitic infections. For macroparasitic infections, however, the immunological response is short-lived. Such infections tend to be persistent, and the hosts are subject to continual reinfection.

Last, we can distinguish **ectoparasites**, such as ticks and fleas, which live outside of the host's body, from **endoparasites**, such as pathogenic bacteria and tapeworms, which live inside the host's body. Problems of definition arise with regard to plant parasites, which seem to straddle both camps. For example, some parasitic plants, such as mistletoe, exist partly outside of the host's body and partly inside. Outgrowths called haustoria penetrate inside the host plant to tap into nutrients. Being endoparasitic on a host seems to require greater specialization than ectoparasitism. Therefore, ectoparasitic animals such as leeches feed on a wider variety of hosts than do endoparasites such as liver flukes.

As we have seen throughout this textbook, parasitism is a common way of life. There are vast numbers of species of parasites, including bacteria, protozoa, flatworms (flukes and tapeworms), nematodes, and various arthropods (ticks, mites, and fleas). Parasites may outnumber free-living species by four to one. Most plant and animal species harbor many parasites. For example, on average, each mammal species hosts two cestode species, two trematodes, four nematodes, and one acanthocephalan. For birds, the figures are even higher. A free-living organism that does not harbor parasitic individuals of a number of species is a rarity.

As with studies of other species interactions, a direct method for determining the effect of parasites on their host population is to remove the parasites and to reexamine the population. However, this is difficult to do, primarily because of the small size and unusual life histories of many parasites, which make them difficult to remove from a host completely. The few cases of experimental removal confirm that parasites can reduce host population densities. The nests of birds such as blue tits are often infested with parasitic blowfly larvae that feed on the blood of nestlings. In 1997, French biologist Sylvie Hurtrez-Bousses and colleagues experimentally reduced blowfly larval parasites of young blue tits in nests in Corsica. Parasite removal was cleverly achieved by taking the nests from 145 nest boxes, removing the young, microwaving the nests to kill the parasites, and then returning the nests and chicks to the wild. The success of chicks in microwaved nests was compared with that in nonmicrowaved (control) nests. The parasite-free blue tit chicks had greater body mass at fledging, the time when feathers first grow (Figure 57.16). Perhaps more important was the fact that complete nest failure, that is, death of all chicks, was much higher in control nests than in treated nests.

Because parasite removal studies are difficult to do, ecologists have also examined the strength of parasitism as a mortality factor by studying introduced parasite species. Evidence from natural populations suggests that introduced parasites have substantial effects on their hosts. For example, chestnut blight (*Cryphonectria parasitica*), a fungus from Asia, was accidentally introduced to New York around 1904. At that time, the American chestnut tree (*Castanea dentata*), was one of the most common trees in the eastern United States. It was said that a squirrel could jump from one chestnut to another all the way from Maine to Georgia without touching the ground. By the 1950s, the airborne fungus had significantly reduced the density of American chestnut trees in North Carolina (Figure 57.17).

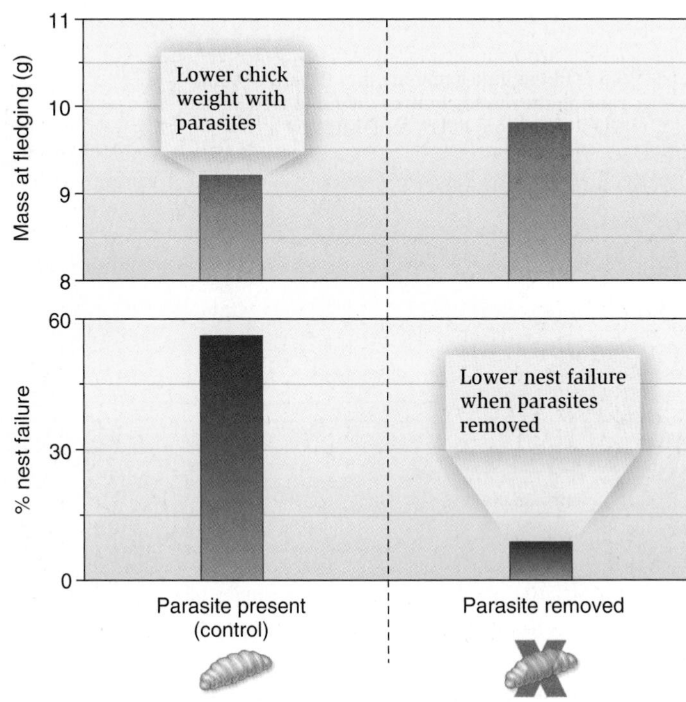

Figure 57.16 Parasite removal experiments. The left side shows the results when blowfly larvae were present in the nests of young blue tits. The right side shows the results when these parasites were removed.

BIOLOGY PRINCIPLE Biology is an experimental science. In ecology the best tests of hypotheses are usually provided by experiments, in this case performed in the field rather than in the laboratory.

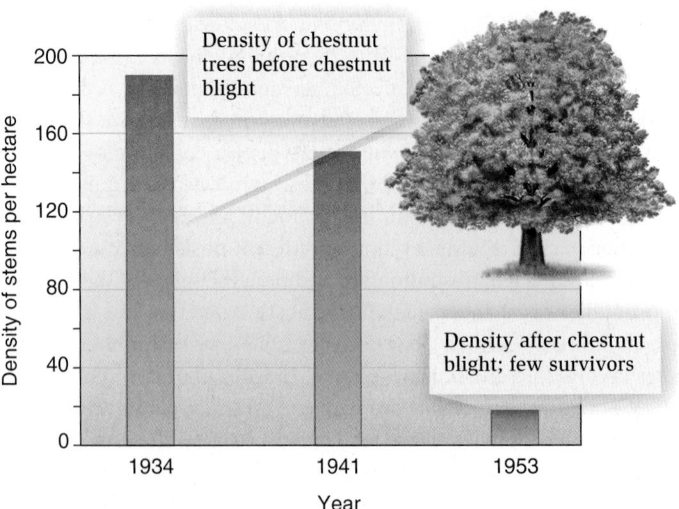

Figure 57.17 Effects of introduced parasites on American chestnut trees. The reduction in density of American chestnut trees in North Carolina following the 1904 introduction of chestnut blight disease from Asia shows the severe effect that parasites can have on their hosts. By the 1950s, this once-widespread species was virtually eliminated.

Eventually, it eliminated nearly all chestnut trees across North America. In Europe and North America, Dutch elm disease has similarly devastated populations of elms (*Ulmus*). The disease wiped out 25 million of Britain's original 30 million elm trees between the 1960s and the 1990s. The creation of transgenic plants through recombinant DNA methods is a recent development in the fight against plant diseases.

Genomes & Proteomes Connection

Transgenic Plants May Be Used in the Fight Against Plant Diseases

Many important native forest trees, which are also grown in urban landscapes, have been almost entirely wiped out by diseases spread by the importation of exotic plants. Sudden oak death is a recently recognized disease that is killing tens of thousands of oak trees and other plant species in California. The symptoms vary between species but include leaf spots, oozing of a dark sap through the bark, and twig dieback. Although sudden oak death is a forest disease, the organism causing this disease is known to infect many woody ornamental plants, such as rhododendrons, that are commonly sold by nurseries. In March 2004, a California nursery was found to have unknowingly shipped plants infected with sudden oak death to all 50 states. Following this discovery, California nurseries halted shipments of trees to other states in an attempt to stop the spread of the disease, originally thought to have been imported on rhododendrons. In 2004, scientists mapped out the genome sequence of the disease-carrying protist, *Phytophthora ramorum*. Identifying the genes and their proteins may help scientists develop specific diagnostic tests to quickly detect the presence of sudden oak death in trees, which is currently impossible to detect until a year or more after the tree is infected.

Scientists hope for much from the field of genomics in their fight against disease-causing plant parasites. Many scientists have suggested limited, cautious transfer of resistance genes from the original host species in the source regions of the disease to newly threatened species. Original host species have usually evolved over millions of years of exposure to these diseases and have acquired genes that provide resistance. In the regions of recent introduction of parasites, there has been no selection for resistance, so the host plants are often killed en masse. Transgenic trees that have received pathogen resistance-enhancing genes could be produced and then be replanted in forests or urban areas. An advantage of this technique over traditional cross-breeding strategies involving two different species is that transgenic methods involve the introduction of fewer genes to the native species. Also, fewer tree generations would be required to develop resistance. For example, using traditional breeding technology, Asian chestnut trees (*Castanea mollissima*) are being bred with American chestnut trees (*C. dentata*) to reduce the susceptibility of the latter to chestnut blight. The resultant hybrid is often significantly altered in appearance from the traditional American chestnut, and the process takes more than a decade to produce trees that are ready to plant. Transgenic technology could minimize these drawbacks. American biologist William Powell and colleagues are working to enhance the American chestnut's resistance by inserting a gene taken from wheat. The gene, which encodes an enzyme called oxalate oxidase, destroys a toxin produced by the fungus that causes chestnut blight. The whole process takes up to 2 years to produce plants ready to be transplanted back to the wild. The first two transgenic chestnuts were planted in New York in 2006 and 17 more followed in 2007 (**Figure 57.18**). In 2009, 500 blight-resistant trees were planted in Virginia, Tennessee, and North Carolina, but it will take many years to determine how successful these transplants will be.

Figure 57.18 **Newly planted transgenic American chestnut trees in New York.**

BIOLOGY PRINCIPLE The genetic material provides a blueprint for sustaining life. Even a relatively small amount of new genetic material may make the difference between life and death for American chestnut trees.

57.3 Mutualism and Commensalism

Learning Outcomes:

1. Distinguish between the different types of mutualisms in nature.
2. Give examples of the species interaction called commensalism.

In this section, we will examine interactions that are beneficial to at least one of the species involved. In mutualism, both species gain from the interaction. For example, in mutualistic pollination systems, the plant benefits by the transfer of pollen, and the pollinator typically gains a nectar meal. In commensalism, one species benefits, and the other remains unaffected. For example, in some forms of seed dispersal, barbed seeds are transported to new germination sites in the fur of mammals. The seeds benefit, but the mammals are generally unaffected.

It is interesting to note that humans have entered into mutualistic relationships with many species. For example, the association of humans with plants has resulted in some of the most far-reaching ecological changes on Earth. Humans have planted huge areas of the Earth with crops, allowing these plant populations to reach densities they never would attain on their own. In return, the crops have led to expanded human populations because of the increased amounts of food they provide.

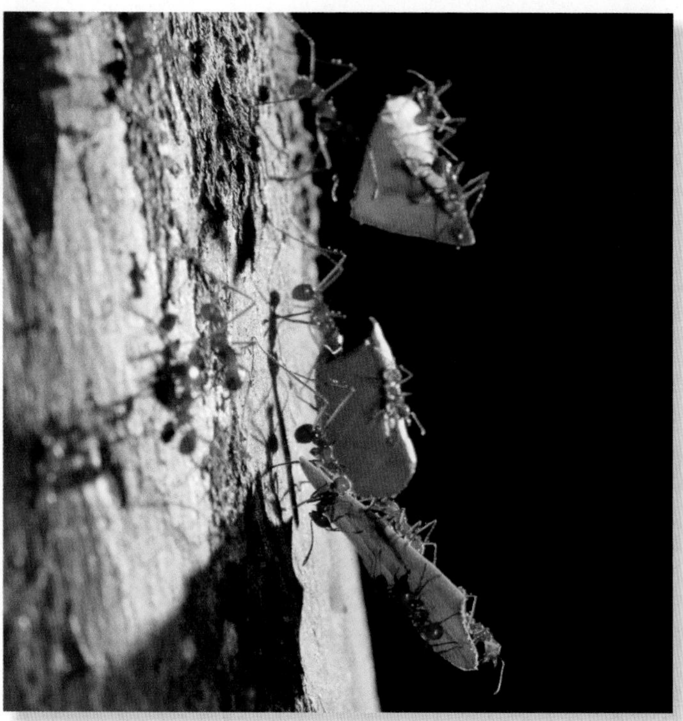

Figure 57.19 **Trophic mutualism.** Leaf-cutter ants, *Atta cephalotes*, cut leaves and chew them to a pulp underground, where fungi develop in the pulp.

Mutualism Is an Association Between Two Species That Benefits Both

Different types of mutualisms occur in nature. In **trophic mutualisms**, both species receive a benefit in the form of resources transfer energy and nutrients. In **defensive mutualisms**, one species receives food or shelter in return for defending another species; they often involve an animal defending a plant or an herbivore. **Dispersive mutualisms** are interactions in which a species receives food in return for transporting the pollen or seeds of its partner.

Trophic Mutualism Leaf-cutter ants of the group Attini, of which there are about 210 species, enter into a mutualistic relationship with a fungus. A typical colony of about 9 million ants has the collective biomass of a cow and harvests the equivalent of a cow's daily requirement of fresh vegetation. Instead of consuming the leaves directly, however, the ants chew them into a pulp, which they store underground as a substrate on which the fungus grows (**Figure 57.19**). The ants shelter and tend the fungus, helping it reproduce and grow and weeding out competing fungi. In turn, the fungus produces specialized structures known as gongylidia, which serve as food for the ants. In this way, the ants circumvent the chemical defenses of the leaves, which are digested by the fungus.

Defensive Mutualism One of the most commonly observed mutualisms occurs between ants and aphids. Aphids are fairly defenseless creatures and are easy prey for most predators. The aphids feed on plant sap and have to process a significant amount of it to get their required nutrients. In doing so, they excrete a lot of fluid, and some of the sugars still remain in the excreted fluid, which is called "honeydew." The ants drink the honeydew and, in return, protect the aphids from an array of predators, such as ladybird beetle larvae, by driving the predators away. In some cases, the ants herd the aphids like cattle, moving them from one area to another.

In other cases, ants enter into a mutualistic relationship with a plant itself. One of the most famous cases involves acacia trees in Central America, whose large thorns provide food and nesting sites for ants (**Figure 57.20**). In return, the ants bite and discourage both insect and vertebrate herbivores from feeding on the trees. They also trim away foliage from competing plants and kill neighboring plant shoots, ensuring more light, water, and nutrient supplies for the acacias. In this case, neither species can live without the other, a concept called **obligatory mutualism**. This contrasts with **facultative mutualism**, in which the interaction is beneficial but not essential to the survival and reproduction of either species. For example, ant-aphid mutualisms are generally facultative. Both species benefit from the association, but each could live without the other.

Dispersive Mutualism Many examples of plant-animal mutualisms involve pollination and seed dispersal. From the plant's perspective, an ideal pollinator would be a specialist, moving quickly among individuals but retaining a high fidelity to a plant species. Two ways that plant species in an area promote the pollinator's species fidelity is by synchronized flowering within a species and by sequential flowering of different species through the year. The plant should provide just enough nectar to attract a pollinator's visit. From the pollinator's perspective, it would be best to be a generalist and obtain nectar and pollen from as many flowers as possible in a small area, thus minimizing the energy spent on flight between patches. This suggests that although mutualisms are beneficial to both species, their optimal needs are quite different.

Ants defending an acacia plant in exchange for food and shelter

Figure 57.20 **Defensive mutualism.** Ants, usually *Pseudomyrmex ferruginea*, make nests inside the large, hornlike thorns of the bull's horn acacia and defend the plant against insects and mammals. In return, the acacia (*Acacia collinsii*) provides two forms of food to the ants: protein-rich granules called Beltian bodies and nectar from extrafloral nectaries (nectar-producing glands that are physically apart from the flower).

Concept Check: Is the relationship between ants and bull's horn acacia an example of facultative or obligatory mutualism?

BioConnections: Refer back to Figure 31.26. What is the name given to the association between the hyphae of certain fungi and the roots of most plants which exist together in a trophic mutualism?

Figure 57.21 **Dispersive mutalism.** This blackbird (*Turdus merula*) is an effective seed disperser.

BioConnections: Refer back to Table 30.1. Birds have excellent color vision and are also involved in the pollination of flowers. What is a common color of bird-pollinated plants?

Mutualistic interactions are also highly prevalent in the seed-dispersal systems of plants. Fruits provide a balanced diet of proteins, fats, and vitamins. In return for this juicy meal, animals unwittingly disperse the enclosed seeds, which pass through the digestive tract unharmed. Fruits eaten by birds and mammals often have attractive colors (**Figure 57.21**); those that attract nocturnal bats are not brightly colored but instead give off a pungent odor.

In Commensalism, One Partner Receives a Benefit While the Other Is Unaffected

Commensalism is an interaction between species in which one benefits and the other is neither helped nor harmed. Such is the case when orchids or other epiphytes grow in forks of tropical trees. The tree is unaffected, but the orchid gains support and increased exposure to sunlight and rain. Cattle egrets feed in pastures and fields among cattle, whose movements stir up insect prey for the birds. The egrets benefit from the association, but the cattle generally do not. One of the best examples of commensalism involves **phoresy**, in which one organism uses a second organism for transportation. Hummingbird flower mites feed on the pollen of flowers and travel between flowers in the nostrils (nares) of hummingbirds. The flowers the mites inhabit live only a short while before dying, so the mites relocate by scuttling into the nares of visiting hummingbirds and hitching a ride to the next flower. When the hummingbird visits a new flower, the mites disembark. Presumably, the hummingbirds are unaffected.

Some commensalisms involve one species "cheating" on the other without harming it. In the bogs of Maine, the grass-pink orchid (*Calopogon pulchellus*) produces no nectar, but it mimics the nectar-producing rose pogonia (*Pogonia ophioglossoides*) and is therefore still visited by bees. Another example involves bee orchids (*Ophrys apifera*) that mimic the appearance and scent of female bees. Males try to copulate with the flowers and in the process pick up and transfer pollen (**Figure 57.22a**). The stimuli of the bee orchid flowers are so effective that male bees prefer to mate with them even in the presence of actual female bees! Many plants have essentially cheated their potential mutualistic seed-dispersal agents out of a meal by developing seeds with barbs or hooks that lodge in the animals' fur or feathers rather than their stomachs (**Figure 57.22b**). In these cases, the plants receive free seed dispersal, and the animals receive nothing, except perhaps minor annoyance. This type of relationship is fairly common; most hikers and dogs have at some time gathered spiny or sticky seeds as they wandered through woods or fields.

(a) An orchid without nectar mimicking a female bee

(b) Seed dispersal via hooked seeds

Figure 57.22 **Commensalisms.** **(a)** Bee orchids (*Ophrys apifera*) mimic the shape of a female bee. Male bees copulate with the flowers, transferring pollen but getting no nectar reward. **(b)** Hooked seeds of burdock (*Arctium minus*) have lodged in the fur of a white-footed mouse (*Peromyscus leucopus*). The plant benefits from the relationship by the dispersal of its seeds, and the animal is not affected.

57.4 Bottom-Up and Top-Down Control

Learning Outcome:

1. Describe bottom-up control and top-down control as conceptual models of how species interactions limit population size.

In this chapter, we have seen that interactions between species, such as competition, predation, and parasitism, are important in nature. Let's return to a question we posed in the consideration of population ecology in Chapter 56: How can we determine which factors, along with abiotic factors such as temperature and moisture, are the most important in affecting population size? The question is one asked by many applied biologists, such as foresters, marine biologists, and conservation biologists, who are interested in managing a population's size, as well as ecologists who in general are interested in determining a population's size.

Some ecologists stress the importance of so-called bottom-up factors, such as plant quality and abundance in controlling herbivores and the predators that feed on them. Others stress the importance of top-down factors, such as predators and parasites, acting to control herbivore or plant prey (Figure 57.23). In the beginning of the chapter, we noted how a decline in the size of shark populations along the east coast had led to an increase in their main prey, rays and skates, and thereby a decrease in bay scallops, the prey of rays and skates. This is a top-down effect known as a trophic cascade, because its effects cascade down to all feeding levels of the system. In this section, we will briefly discuss some of the evidence for the existence of bottom-up versus top-down control.

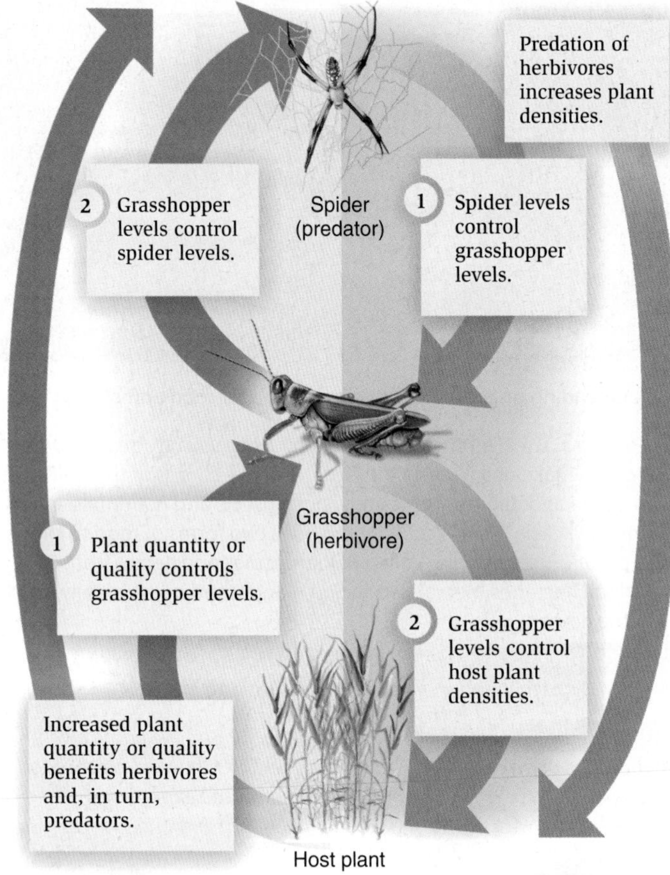

Figure 57.23 Bottom-up control versus top-down control. **(a)** Bottom-up control proposes that host plant quantity or quality limits the density of herbivores, such as grasshoppers, which, in turn, sets limits on the abundance of predators, such as spiders. Taken together, this means that high quantity and quality of host plants would result in increased numbers of predators because of higher densities of the herbivores they prey on. **(b)** Top-down control proposes that predators limit the number of herbivores, which, in turn, increases host plant density. Taken together, this means that high levels of predation would result in high densities of host plants because there would be fewer herbivores.

Concept Check: You add fertilizer to a bush, and this increases spider density on the bush. What type of control is this an example of?

Bottom-Up Control Suggests Food Limitation Influences Population Densities

At least two lines of evidence suggest that bottom-up effects are important in limiting population sizes. First, we know there is a progressive lessening of available energy passing from plants through herbivores to carnivores and to secondary carnivores (carnivores that eat other carnivores). This line of evidence, based on the thermodynamic properties of energy transfer, suggests that the quantity and quality of plants regulates the population size of all other species that rely on them.

Second, much evidence supports the **nitrogen-limitation hypothesis** that organisms select food in terms of the nitrogen content of the tissue. This is largely due to the different proportions of nitrogen in plants and animals. Animal tissue generally contains about 10 times as much nitrogen as plant tissue. For this reason, animals favor high-nitrogen plants. Fertilization has repeatedly been shown to benefit herbivores. Nearly 60% of 186 studies investigating the effects of fertilization on herbivores reported that increasing a plant's tissue nitrogen concentration through fertilization had strong positive effects on herbivore population sizes, survivorship, growth, and fecundity.

Top-Down Control Suggests Natural Enemies Influence Population Densities

Top-down models suggest that predators control populations of their prey (ultimately, herbivores) and that these herbivores control plant populations. Supporting evidence comes from the world of biological control, where natural enemies are released in order to control agricultural pests such as weeds. Many weeds are invaders that were accidentally introduced to an area from a different country, as seeds that lodge in ships' ballasts or in agricultural shipments. Over 50% of the

(a) Before biological control

(b) After biological control

Figure 57.24 **Successful biological control of prickly pear cactus.** The prickly pear cactus (*Opuntia stricta*) in Chinchilla, Australia, **(a)** before and **(b)** after control by the cactus moth (*Cactoblastis cactorum*).

BIOLOGY PRINCIPLE **Biology affects our society.** Control of prickly pear by the cactus moth cleared hundreds of thousands of hectares of cacti, allowing sheep to graze the area and farming to thrive.

190 major weeds in the U.S. are invasive species. Many of these weeds have become separated from their native natural enemies, which is one reason the weeds become so prolific. Because chemical control is expensive and may have unwanted environmental side effects, many land managers have reverted to biological control, in which the invasive species is reunited with its native natural enemy.

Ecologists have seen many successes in the biological control of weeds. St. John's wort (*Hypericum perforatum*), a pest in California pastures, was controlled by two beetles from its homeland in Europe. Likewise, alligator weed has been controlled in Florida's rivers by the alligatorweed flea beetle (*Agasicles hygrophila*) from South America. The prickly pear cactus (*Opuntia stricta*) provides a prime example of effective biological control of a weed. The cactus was imported into Australia in the 19th century and quickly established itself as a major pest of rangeland. The small cactus moth (*Cactoblastis cactorum*) was introduced in the 1920s and, within a short time, successfully saved hundreds of thousands of acres of valuable rangeland from being overrun by the cacti (**Figure 57.24**). The numerous examples showing that pest populations are controlled when reunited with their natural enemies provide strong evidence of top-down control in nature.

Current thinking is that both bottom-up and top-down control are important in affecting population size, with communities varying in their degrees of importance. Species interactions can clearly be very important in influencing both the growth of individual populations and the structure of communities—groups of species living in a particular area. What factors determine the numbers of species in a given area? What factors influence the stability of a community, and what are the effects of disturbances on community structure? In the next chapter, on community ecology, we will explore these and other questions.

Summary of Key Concepts

57.1 Competition

- Species interactions can take a variety of forms that differ based on their effect on the species involved (Figure 57.1, Table 57.1).
- Competition can be categorized as intraspecific (between individuals of the same species) or interspecific (between individuals of different species), and as exploitation competition or interference competition (Figure 57.2).
- Laboratory and field experiments show that competition occurs frequently in nature (Figures 57.3, 57.4).
- The competitive exclusion hypothesis states that two species with the same resource requirements cannot occupy the same niche. Resource partitioning and morphological differences between species allow them to coexist in a community (Figures 57.5, 57.6, 57.7, Table 57.2).

57.2 Predation, Herbivory, and Parasitism

- The most common antipredator strategies are chemical defense and aposematic coloration (Figures 57.8, 57.9).
- Despite these defenses, oscillations in predator-prey cycles, the effect of introduced species, and examples of human predation illustrate that predators can have a large effect on prey densities (Figures 57.10, 57.11, 57.12).
- Plants have also evolved an array of defenses against herbivores, including chemical defenses, such as secondary metabolites, and mechanical defenses, such as thorns and spines (Figure 57.13).
- Parasitism is a common lifestyle on Earth, and some parasites have complex life cycles involving multiple hosts (Figures 57.14, 57.15).
- Evidence from experimental removal of parasites and from the study of introduced plant and animal parasites confirms that parasites can greatly reduce prey densities (Figures 57.16, 57.17, 57.18).

57.3 Mutualism and Commensalism

- Mutualism is an association between two species that benefits both. In trophic mutualisms, both species receive a benefit in the form of resources; defensive mutualisms typically involve an animal defending either a plant or herbivore; and dispersive mutualisms involve animals that disperse a plant's pollen or seeds (Figures 57.19, 57.20, 57.21).
- In commensal relationships, one partner receives a benefit while the other is not affected (Figure 57.22).

57.4 Bottom-Up and Top-Down Control

- Bottom-up models propose that plant quality or quantity regulates the abundance of all herbivore and predator species; top-down models propose that the abundance of predators controls herbivore and plant densities (Figures 57.23, 57.24).

Assess and Discuss

Test Yourself

1. A species interaction in which one species benefits but the other species is unharmed is called
   a. mutualism.
   b. amensalism.
   c. parasitism.
   d. commensalism.
   e. mimicry.
2. Two species of birds feed on similar types of insects and nest in the same tree species. This is an example of
   a. intraspecific competition.
   b. interference competition.
   c. exploitation competition.
   d. mutualism.
   e. none of the above.
3. According to the competitive exclusion hypothesis,
   a. two species that use the exact same resource show very little competition.
   b. two species with the same niche cannot coexist.
   c. one species that competes with several different species for resources will be excluded from the community.
   d. all competition between species results in the extinction of at least one of the species.
   e. none of the above is correct.
4. In Lack's study of British passerine birds, different species seem to segregate based on resource factors, such as location of prey items. This differentiation among the niches of these passerine birds is known as
   a. competitive exclusion.
   b. intraspecific competition.
   c. character displacement.
   d. resource partitioning.
   e. allelopathy.
5. Divergence in morphology that is a result of competition is termed
   a. competitive exclusion.
   b. resource partitioning.
   c. character displacement.
   d. amensalism.
   e. mutualism.
6. Tapeworms have
   a. low lethality and low duration of interaction.
   b. low lethality and high duration of interaction.
   c. high lethality and low duration of interaction.
   d. high lethality and high duration of interaction.
   e. none of the above.
7. Ticks are regarded as
   a. monophagous endoparasites.
   b. monophagous ectoparasites.
   c. polyphagous endoparasites.
   d. polyphagous ectoparasites.
   e. none of the above.
8. Batesian mimicry differs from Müllerian mimicry in that
   a. in Batesian mimicry, both species possess the chemical defense.
   b. in Batesian mimicry, one species possesses the chemical defense.
   c. in Müllerian mimicry, one species has several different mimics.
   d. in Müllerian mimicry, one species has several different chemical defenses.
   e. in Batesian mimicry, cryptic coloration is always found.
9. Deadly nightshade is protected from herbivores by
   a. the alkaloid capsaicin.
   b. the alkaloid atropine.
   c. the phenolic anthocyanin.
   d. the phenolic tannin.
   e. the terpenoid β-carotene.
10. Parasitic plants that rely solely on their host for nutrients are called
    a. hemiparasites.
    b. fungi.
    c. holoparasites.
    d. monophagous.
    e. polyphagous.

Conceptual Questions

1. Can the removal of ectoparasites from the coat of one primate by another primate (grooming) be viewed in terms of selfish behavior we discussed in Chapter 55? Why or why not?
2. A principle of biology is that *biology affects our society*. Crop pests cost millions of dollars to control annually. What factors do you think might limit such losses?

Collaborative Questions

1. Explain how the reintroduction of wolves in Yellowstone National Park might be beneficial.
2. Detail several antipredator strategies that animals have evolved.
3. Can you think of examples of mimicry used by predators to catch prey rather than used by prey to avoid being eaten? Look back to Figure 55.12 and the associated text.

Online Resource

www.brookerbiology.com

Stay a step ahead in your studies with animations that bring concepts to life and practice tests to assess your understanding. Your instructor may also recommend the interactive eBook, individualized learning tools, and more.

Chapter Outline

58.1 Differing Views of Communities
58.2 Patterns of Species Richness
58.3 Calculating Species Diversity
58.4 Species Diversity and Community Stability
58.5 Succession: Community Change
58.6 Island Biogeography
Summary of Key Concepts
Assess and Discuss

58 Community Ecology

A massive volcanic explosion in 1883 on the island of Krakatau in Indonesia destroyed two-thirds of the island, originally 11 km long and covered in tropical rain forest. Life on the remaining part was eradicated, suffocated by tens of meters of red-hot ash. Nine months after the eruption, however, the first reported sign of life was a spider spinning its web. By 1896, there were 11 species of ferns and 15 species of flowering plants, mainly grasses. Most plant species at that time had been wind- or sea-dispersed. By the 1920s, 40 years after the original eruption, birds had become abundant and had dispersed additional plant species by excreting their seeds. Ecologists have been examining the return of different species to the island ever since.

So far in this unit, we have examined ecology in terms of abiotic factors such as temperature and moisture, the behavior of individual organisms, the growth of populations, and interactions between species. Most populations, however, exist not on their own, but together with populations of many other species. This assemblage of many populations that live in the same place at the same time is known as a **community**. For example, a tropical forest community consists of not only tree species, vines, and other vegetation, but also the insects that pollinate them, the herbivores that feed upon the plants, and the predators and parasites of the herbivores. Communities occur on a wide range of scales, and one community can be nested within another. For example, the tropical forest community also encompasses smaller communities, such as the water-filled recesses of bromeliads, which form a microhabitat for different species of insects and their larvae. Both of these entities—the tropical forest and the bromeliad tank—are viable communities, depending on one's frame of reference with regard to scale.

Community ecology is the study of how groups of species interact and form functional communities. In Chapter 57, we considered the interactions between individual species. In this chapter, we widen our focus to explore the factors that influence the number and abundance of species in a community. We begin by examining the nature of ecological communities. Are communities loose assemblages of species that happen to live in the same place at the same time, or are they more tightly organized groups of mutually dependent species? Community ecology also addresses what factors influence the number of species in a community.

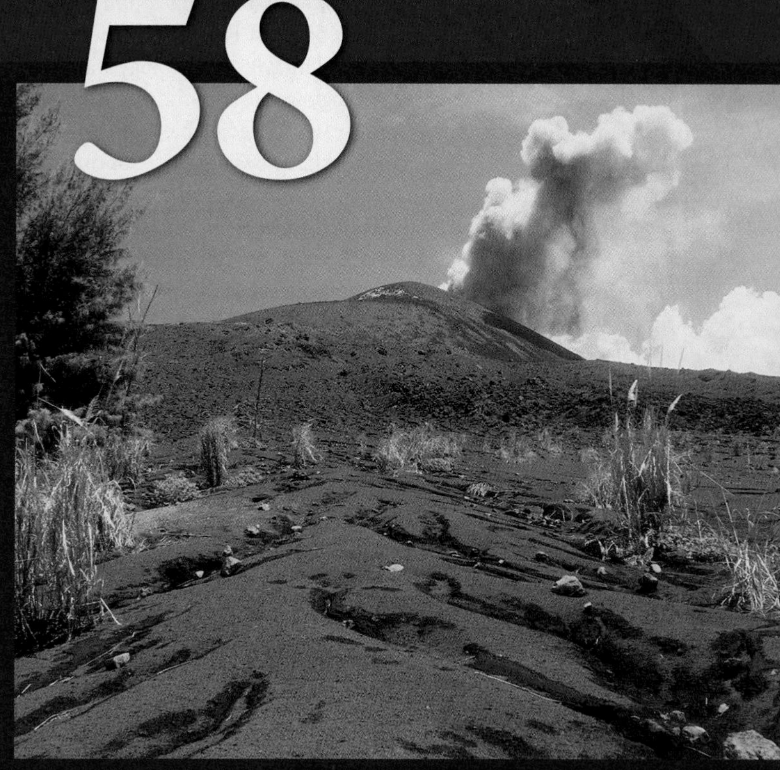

Krakatau. Many of the species formerly present on Krakatau have returned following the 1883 eruption that covered the island in volcanic dust.

We explore why, on a global scale, the number of species is usually greatest in the tropics and declines toward the poles. However, ecologists recognize that communities may change, for example, following a disturbance such as a fire or a volcanic eruption. This recovery tends to occur in a predictable way, which ecologists have termed succession. In certain situations—for example, on islands recovering from physical disturbance—the structure of the community tends toward an equilibrium determined by the balance between the rates of immigration and extinction.

58.1 Differing Views of Communities

Learning Outcomes:

1. Distinguish between Clements's organismic model and Gleason's individualistic model to explain the nature of an ecological community.
2. Describe the principle of species individuality.

Ecologists have long held differing views on the nature of an ecological community and its structure and functions. Some of the initial work in the field of community ecology considered a community to be equivalent to a superorganism, in much the same way that the body of an animal is more than just a collection of organs. In this view, individuals, populations, and communities have a stable relationship with one another that resembles the associations found between cells, tissues, and organs. American botanist Frederic Clements, the champion of this viewpoint, suggested in 1905 that ecology was to the study of communities what physiology was to the study of individual organisms. This view of community, with predictable and integrated associations of species separated by sharp boundaries, is termed the **organismic model**.

Clements's ideas were challenged in 1926 by American botanist Henry Allan Gleason. Gleason proposed an **individualistic model**, which described a community as an assemblage of species coexisting primarily because of similarities in their physiological requirements and tolerances. Although acknowledging that some assemblages of species were fairly uniform and stable over a given region, Gleason suggested that distinctly structured ecological communities usually do not exist. Instead, communities are loose assemblages of species distributed independently along an environmental gradient. Viewed in this way, communities do not necessarily have sharp boundaries, and associations of species are much less predictable and integrated than in Clements's organismic model.

By the 1950s, many ecologists had abandoned Clements's view in favor of Gleason's. In particular, American plant ecologist Robert Whittaker's studies proposed the **principle of species individuality**, which states that each species is distributed according to its physiological needs and population dynamics and that most communities intergrade, or merge into one another gradually. For example, let's consider an environmental gradient such as a moisture gradient on an uninterrupted slope of a mountain. Whittaker proposed that four hypotheses could explain the distribution patterns of plants and animals on the gradient (Figure 58.1):

1. Competing species, including dominant plants, exclude one another along sharp boundaries. Other species evolve toward a close, perhaps mutually beneficial association with the dominant species. Communities thus develop along the gradient, each zone containing its own group of interacting species giving way at a sharp boundary to another assemblage of species. This corresponds to Clements's organismic model.
2. Competing species exclude one another along sharp boundaries but do not become organized into groups of species with parallel distributions.
3. Competition does not usually result in sharp boundaries between species. However, the adaptation of species to similar physical variables results in the appearance of groups of species with similar distributions.
4. Competition does not usually produce sharp boundaries between species, and the adaptation of species to similar physical variables does not produce well-defined groups of species with similar distributions. The centers and boundaries of species populations are scattered along the environmental gradient. This corresponds to Gleason's individualistic model.

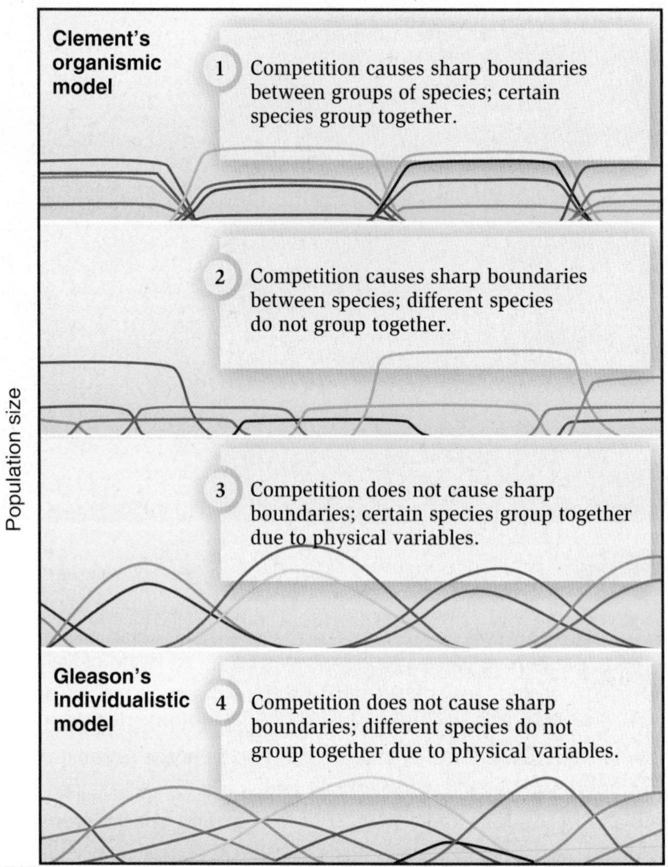

Figure 58.1 **Four hypotheses for the distribution patterns of plants and animals along an environmental gradient.** Each curve in each part of the figure represents one species and the way its population might be distributed along an environmental gradient.

BioConnections: Look back to Section 1.3. What's the difference between a hypothesis and a theory?

To test these possibilities, Whittaker examined the vegetation on various mountain ranges in the western U.S. He sampled plant populations along an elevation gradient from the tops of the mountains to the bases and collected data on physical variables, such as soil moisture.

The results supported the fourth hypothesis, that competition does not produce sharp boundaries between species and that adaptation to physical variables does not result in defined groups of species. Whittaker concluded that his observations agreed with Gleason's predictions that (1) each species is distributed in its own way, according to its genetic, physiological, and life cycle characteristics; and (2) most communities grade into each other continuously rather than form distinct, clearly separated groups. The composition of species at any one point in an environmental gradient is largely determined by abiotic factors such as temperature, water, light, pH, and salt concentrations (refer back to Chapter 54).

Even though most communities intergrade along environmental gradients such as a mountain slope, ecologists recognize distinct differences between communities. The community at the top of a mountain is quite different from that at the bottom, so distinguishing between communities on a broad scale is useful. Also, some sharp boundaries between groups of species sometimes do exist, especially related to

Figure 58.2 An example of a sharp boundary between two communities. In New Zealand's Dun Mountain area, the sparse vegetation on the serpentine soil on the left contrasts with that of the beech forest on the nonserpentine soil on the right.

BIOLOGY PRINCIPLE Living organisms interact with their environment. Serpentine soil is nutrient-poor and relatively few plants species are adapted to live in it.

physical differences such as water quality and soil type that cause distinct communities to develop. For example, serpentine soils are rich in metals, including magnesium, iron, and nickel, but poor in plant nutrients. The species that have adapted to these harsh conditions form a unique community restricted to this area (Figure 58.2). Such sharp boundaries are not common between neighboring communities.

58.2 Patterns of Species Richness

Learning Outcomes:

1. Identify the latitudinal gradient of species richness.
2. List and describe four hypotheses for observed patterns of species richness.

Community ecology addresses what factors influence the number of species in a community, or **species richness**. Globally, the number of species of most taxa varies according to latitudinal gradient, generally increasing from polar to temperate areas and reaching a maximum in the tropics. For example, the species richness of North American birds increases from Arctic Canada to Panama (Figure 58.3). A similar pattern exists for mammals, amphibians, reptiles, and plants. Although the latitudinal gradient of species richness is an important pattern, species richness is also influenced by topographical variation. More mountains mean more hilltops, valleys, and differing habitats; thus, the number of birds is greater in the mountainous western U.S. Species richness is also reduced by the peninsular effect, in which the number of species decreases as a function of distance from the main body of land. Ecologists have noted that species richness also depends on the degree to which an environment is disturbed, with more species richness observed in environments with an intermediate disturbance level.

Many hypotheses for the variation in species richness have been advanced. We will consider several hypotheses for patterns of species richness. Although they are treated separately here, these hypotheses are not mutually exclusive. All can contribute to patterns of species richness.

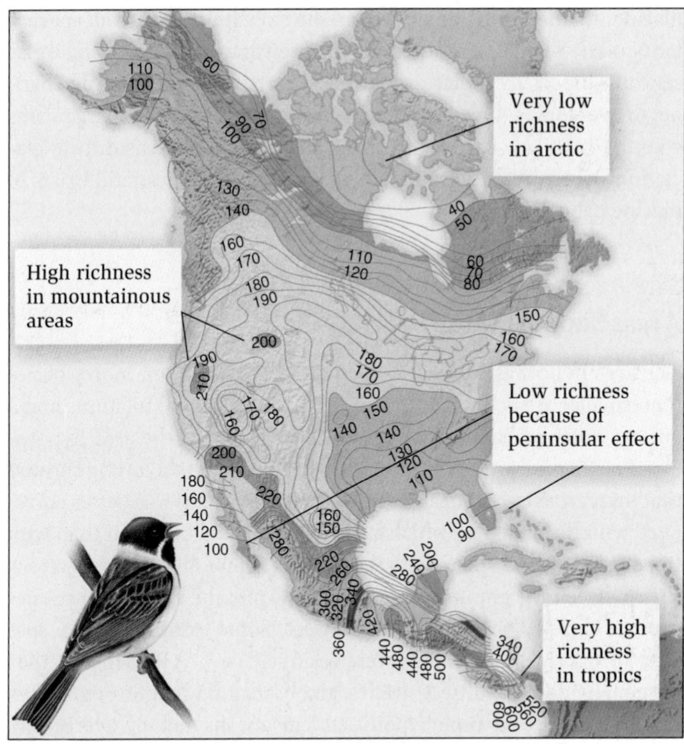

Figure 58.3 Species richness of birds in North America. The values indicate the numbers of different species in a given area. Contour lines show equal numbers of bird species, with colors indicating incremental changes. Note the pronounced latitudinal gradient toward the tropics and the high diversity in California and northern Mexico, regions of considerable topographical variation and habitat diversity.

The Time Hypothesis Suggests Communities Diversify with Age

Many ecologists argue that communities diversify, or gain species, with time. Therefore temperate regions have less rich communities than tropical ones because they are younger and have only more recently (relatively speaking) recovered from glaciations and severe climatic disruptions. The time hypothesis proposes that resident species of the temperate zone have not yet evolved new forms to exploit vacant niches. In addition, it suggests that species that could possibly live in temperate regions have not migrated back from the unglaciated areas into which the Ice Ages drove them.

In support of the time hypothesis, ecologists compared the species richness of bottom-dwelling invertebrates, such as worms, in historically glaciated (covered with ice) and unglaciated lakes in the Northern Hemisphere that occur at similar latitudes. Lake Baikal in Siberia is an ancient, unglaciated temperate lake and contains a very diverse fauna. For example, 580 species of invertebrates are found in the bottom zone. Great Slave Lake, a comparably sized lake that was once glaciated at the same latitude in northern Canada, contains only four species in the same zone.

However, ecologists recognize drawbacks to the time hypothesis. For example, this hypothesis may help explain variations in the species richness of terrestrial organisms, but it has limited applicability to marine organisms. Although we might not expect terrestrial species, particularly plants, to redistribute themselves quickly following a glaciation—especially if there is a physical barrier like the English Channel to overcome—there seems to be no reason that marine organisms couldn't relatively easily shift their distribution patterns during glaciations, yet the latitudinal gradient of species richness still exists in marine habitats.

The Area Hypothesis Suggests Large Areas Support More Species

The **area hypothesis** proposes that larger areas contain more species than smaller areas because they can support larger populations and a greater range of habitats. Much evidence supports the area hypothesis. For example, in 1974, American ecologist Donald Strong showed that insect species richness on tree species in Britain was better correlated with the area over which a tree species could be found than with time of habitation since the last Ice Age (**Figure 58.4**). The relationship between the amount of available area and the number of species present is called the **species-area effect**. Some introduced tree species, such as apple and lime, were relatively new to Britain, but they bore many different insect species, an observation that Strong argued did not support the time hypothesis. This means that, on average, an individual willow tree standing next to an individual maple tree in the same location would support an order of magnitude more insect species because of its greater abundance in that habitat.

The large, climatically similar area of the tropics has been proposed as a reason why the tropics have high species richness. However, the area hypothesis seems unable to explain why, if increased richness is linked to increased area, more species are not found in certain regions such as the vast contiguous landmass of Asia. Furthermore, although tundra may be the world's largest biome in terms of land mass, it has low species richness. Finally, the largest marine system, the open ocean, which has the greatest volume of any habitat, has fewer species than tropical nearshore waters, which have a relatively small volume.

The Productivity Hypothesis Suggests That More Energy Permits the Existence of More Species

The **productivity hypothesis** proposes that greater production by plants results in greater overall species richness. An increase in plant productivity, the total weight of plant material produced over time, leads to an increase in the number of herbivores and hence an increase in the number of predator, parasite, and scavenger species. Production itself is influenced by factors such as temperature and rainfall, because many plants grow better where it is warm and wet. For example, in 1987, Canadian biologist David Currie and colleagues showed that the species richness of trees in North America is best predicted by the **evapotranspiration rate**, the rate at which water moves into the atmosphere through the processes of evaporation from the soil and transpiration of plants, both of which are influenced by the amount of solar energy (**Figure 58.5**).

Once again, however, there are exceptions to this rule. In 1993, American researchers Robert Latham and Robert Ricklefs showed that although patterns of tree richness in North America support the productivity hypothesis, the pattern does not hold for broad comparisons between continents. For example, the temperate forests of eastern Asia support substantially higher numbers of tree species (729) than do climatically similar areas of North America (253) or Europe (124). These three areas have different evolutionary histories and different neighboring areas from which species might have invaded.

Some tropical seas, such as the southeast Pacific off of Colombia and Ecuador, have low productivity but high species richness. On the other hand, the sub-Antarctic Ocean has a high productivity but low species richness. Estuarine areas, where rivers empty into the sea, are similarly very productive yet low in species, presumably because they represent stressful environments for many organisms that are

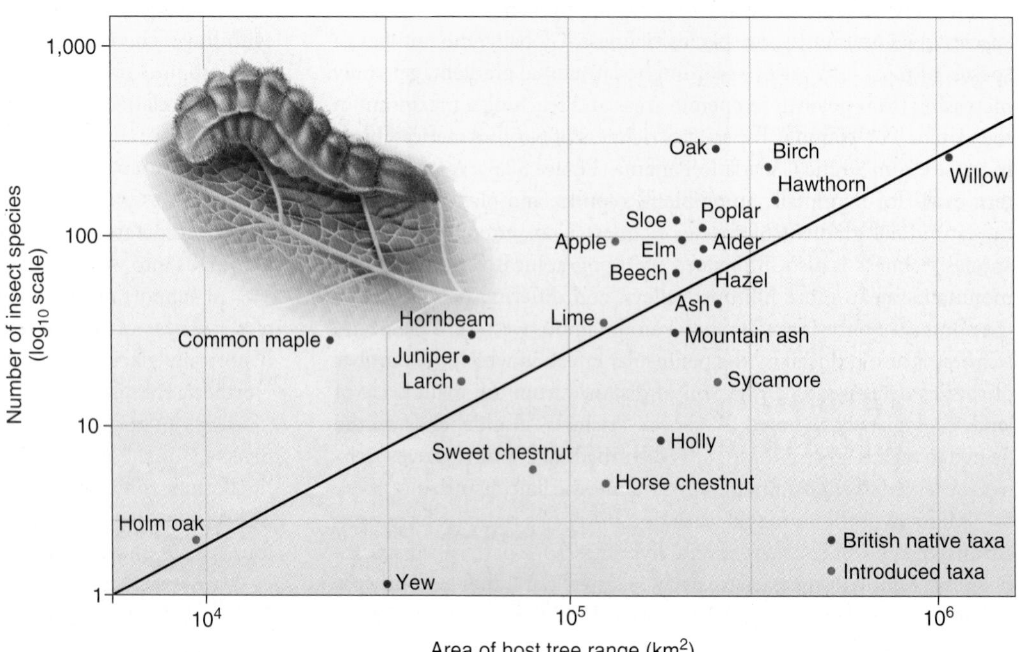

Figure 58.4 Relationship between species richness on British host trees and area. A positive correlation is found between insect species richness and the host tree's present range, in square kilometers (km^2).

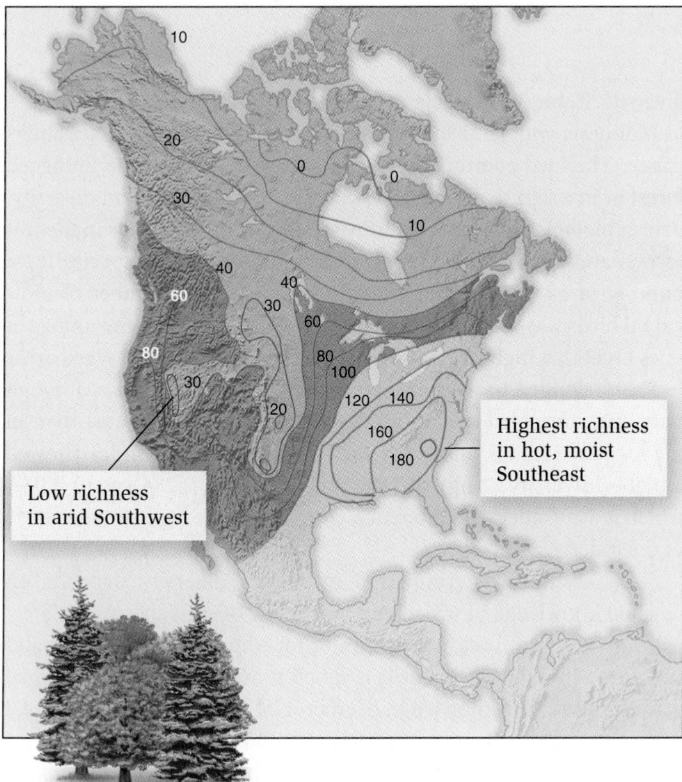

Figure 58.5 **Tree species richness in North America.** Contour lines show equal numbers of tree species, with colors indicating incremental changes. Tree species richness and evapotranspiration rates are highest in the southeast.

Concept Check: Why doesn't the species richness of trees increase in mountainous areas of the West, as it does for birds?

alternately inundated by fresh water and salt water with daily changes in the tide. Some lakes that are polluted with fertilizers also have high productivity but low species richness.

The Intermediate-Disturbance Hypothesis Proposes That Moderately Disturbed Communities Contain More Species

American ecologist Joseph Connell has argued that the highest numbers of species are maintained in communities with intermediate levels of environmental disturbance, a concept called the **intermediate-disturbance hypothesis** (Figure 58.6a). Disturbance in communities may be brought about by many different phenomena such as droughts, fires, floods, and hurricanes or by species interactions such as herbivory, predation, or parasitism. Recall from Chapter 56 that some species, termed *r*-selected species, are better dispersers than other species, and that *K*-selected species are better competitors (refer back to Figure 56.14). Connell reasoned that at high levels of disturbance, only colonists that were *r*-selected species would survive, giving rise to low species richness. This is because these species would be the only ones able to disperse quickly to a highly disturbed area. At low rates of disturbance, competitively dominant *K*-selected species would outcompete all other species, which would also yield low species richness. The most species-rich communities would lie somewhere in between.

Connell argued that natural communities fit into this model fairly well. Tropical rain forests and coral reefs are both examples of communities with high species richness. Coral reefs exhibit highest species richness in areas disturbed by hurricanes, and the richest tropical forests occur where disturbance by storms causes landslides and tree falls. The fall of a tree creates a hole in the rain forest canopy known as a light gap, where direct sunlight is able to reach the rain forest floor. The light gap is rapidly colonized by *r*-selected species, such as small herbaceous plants, which are well adapted for rapid growth. Although these pioneering species grow rapidly, they are eventually overtaken by *K*-selected species, such as mature trees, which fill in the gap in the canopy (**Figure 58.6b**). Although environmental events such as hurricanes and tree falls are fairly frequent events in these communities, their occurrence in any one area is usually of intermediate frequency.

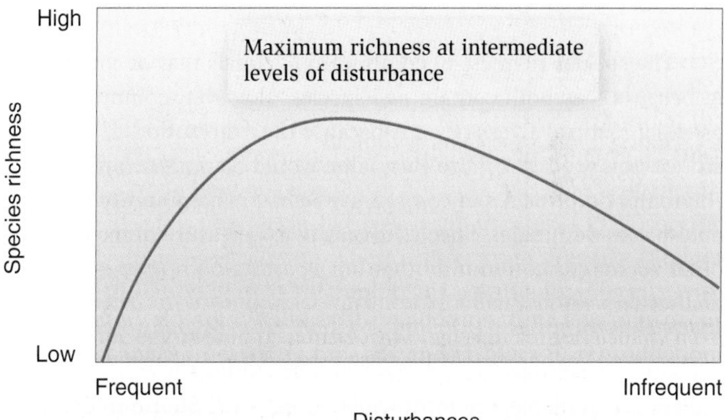

(a) Relationship between species richness and disturbances

(b) Light gap in a tropical rain forest

Figure 58.6 **The intermediate-disturbance hypothesis of community organization.** **(a)** This hypothesis proposes that species richness is highest at intermediate levels of disturbances caused by events such as fires or windstorms. **(b)** A light gap in a tropical rain forest in Costa Rica promotes the growth of small herbaceous species until trees colonize the light gap and gradually grow over and outcompete the smaller species.

Concept Check: According to the intermediate-disturbance hypothesis, why are there so many species in the tropics?

58.3 Calculating Species Diversity

Learning Outcomes:

1. Define species diversity.
2. Calculate the Shannon diversity index.

So far, we have discussed communities in terms of variations in species richness. However, ecologists need to take into account not only the number of species in a community but also their frequency of occurrence, or **relative abundance**. For example, consider two hypothetical communities, A and B, both with two species and 100 total individuals.

	Number of individuals of species 1	Number of individuals of species 2
Community A	99	1
Community B	50	50

The species richness of community B equals that of community A, because they both contain two species. However, community B is considered more diverse than A because the distribution of individuals between species is more even. One would be much more likely to encounter both species in community B than in community A, where one species dominates. **Species diversity** is a measure of the diversity of an ecological community that incorporates both species number and relative abundance.

To measure the species diversity of a community, ecologists calculate what is known as a diversity index. Although many different indices are available, the most widely used is the **Shannon diversity index (H_S)**, which is calculated as

$$H_S = -\Sigma p_i \ln p_i$$

where p_i is the proportion of individuals belonging to species i in a community, ln is the natural logarithm, and Σ indicates summation. For example, for a species in which there are 50 individuals out of a total of 100 in the community, p_i is 50/100, or 0.5. The natural log of 0.5 is −0.693. For this species, $p_i \ln p_i$ is then $0.5 \times -0.693 = -0.347$. For a hypothetical community with 5 species and 100 total individuals, the Shannon diversity index is calculated as follows:

	Species	Abundance	p_i		$p_i \ln p_i$
	1	50	0.5		−0.347
	2	30	0.3		−0.361
	3	10	0.1		−0.230
	4	9	0.09		−0.217
	5	1	0.01		−0.046
Total	5	100	1.00	$\Sigma p_i \ln p_i$	−1.201

In this example, even the rarest species, species 5, contributes some value to the index. If a community had many rare species, their contributions would accumulate. This makes the Shannon diversity index very valuable to conservation biologists, who often study rare species and their importance to the community. Remember, too, that in the equation, the negative sign in front of the summation changes these values to positive, so the index actually becomes 1.201, not −1.201.

Values of the Shannon diversity index for real communities often fall between 1.5 and 3.5, with the higher the value, the greater the diversity. **Table 58.1** calculates the diversity of two bird communities in Indonesia with similar species richness but differing species abundance. The bird communities were surveyed in a pristine unlogged forest or in a selectively logged lowland forest. To document diversity, British biologist Stuart Marsden established census stations in the two forests and recorded the type and number of all bird species for a number of 10-minute periods. Although a greater number of individual birds was seen in the logged areas (2,358) than in the unlogged ones (1,824), a high proportion of the individuals in the logged areas (0.386) belonged to just one species, *Nectarinia jugularis*. Although only one more bird species was found in the unlogged area than in the logged area, calculation of the Shannon diversity index showed a higher diversity of birds in the unlogged area, 2.284 versus 2.037, which is a considerable difference, considering the logarithmic nature of the index.

An accurate determination of species diversity depends on detailed knowledge of which and how many of each species are present. This is relatively easy to determine for communities of vertebrates and some invertebrates, but it is much more difficult for microbial communities. Yet knowledge of microbial communities is of great importance, because microbes carry out vital functions such as nitrogen fixation and decomposition. As described next, with the advent of modern molecular tools, our knowledge of the species diversity of microbial communities is beginning to expand.

Genomes & Proteomes Connection

Metagenomics May Be Used to Measure Species Diversity

Bacteria are abundant members of all communities and are vital to their functioning. They serve as food sources for other organisms and participate in the decomposition process. However, most microorganisms are taxonomically unknown, mainly because they cannot be cultivated on known culture media. The field of **metagenomics** seeks to identify and analyze the collective genetic material contained in a community of organisms, including those that are not easily cultured in the laboratory. Metagenomics techniques have been in existence only since the early 1990s, but significant progress has already been made in providing data that have advanced our understanding of which bacteria are present in various communities and how they function.

The process involves four main steps (**Figure 58.7**). First, an environmental sample containing an unknown number of bacterial species is collected, and its DNA is isolated from the cells using chemical or physical methods. Because the genomic DNA of each species is relatively large, it is cut up into fragments with restriction enzymes (refer back to Figures 20.1 to 20.4, which detail the steps used to clone genes and create a DNA library). Second, the fragments are combined with vectors, small units of DNA that can be inserted into a model laboratory organism, usually a bacterium. The third step begins with transformation in which the DNA from step 2 is taken up into bacterial cells. Individual bacteria are then grown on a selective medium so that only the transformed cells survive. Each cell grows into a colony of cloned cells. A collection of thousands of clones, each containing a

Table 58.1 Shannon Diversity Index of Bird Species on Logged and Unlogged Sites in Indonesia

	Unlogged			*Logged*		
Species	*N*	p_i	$p_i \ln p_i$	*N*	p_i	$p_i \ln p_i$
Nectarinia jugularis, olive-backed sunbird	410	0.225	−0.336	910	0.386	−0.367
Ducula bicolor, pied imperial pigeon	230	0.126	−0.261	220	0.093	−0.221
Philemon subcorniculatus, grey-necked friarbird	210	0.115	−0.249	240	0.102	−0.233
Nectarinia aspasia, black sunbird	190	0.104	−0.235	120	0.051	−0.152
Dicaeum vulneratum, ashy flowerpecker	185	0.101	−0.232	280	0.119	−0.253
Ducula perspicillata, white-eyed imperial pigeon	170	0.093	−0.221	180	0.076	−0.196
Phylloscopus borealis, arctic warbler	160	0.088	−0.214	140	0.059	−0.167
Eos bornea, red lory	88	0.048	−0.146	73	0.031	−0.108
Ixos affinis, golden bulbul	76	0.042	−0.133	31	0.013	−0.056
Geoffroyus geoffroyi, red-cheeked parrot	44	0.024	−0.089	54	0.023	−0.087
Rhyticeros plicatus, Papuan hornbill	24	0.013	−0.056	27	0.011	−0.050
Cacatua moluccensis, Moluccan cockatoo	12	0.007	−0.035	1	0.001	−0.007
Tanygnathus megalorynchos, great-billed parrot	9	0.005	−0.026	11	0.005	−0.026
Electus roratus, electus parrot	7	0.004	−0.022	0	0	0
Macropygia amboinensis, brown cuckoo-dove	6	0.003	−0.017	7	0.003	−0.017
Cacomantis sepulcralis, ruby-breasted cuckoo	3	0.002	−0.012	0	0	0
Trichoglossus haematodus, rainbow lorikeet	0	0	0	64	0.027	−0.097
Total	**1,824**	**1.0**		**2,358**	**1.0**	
Shannon diversity index			**2.284**			**2.037**

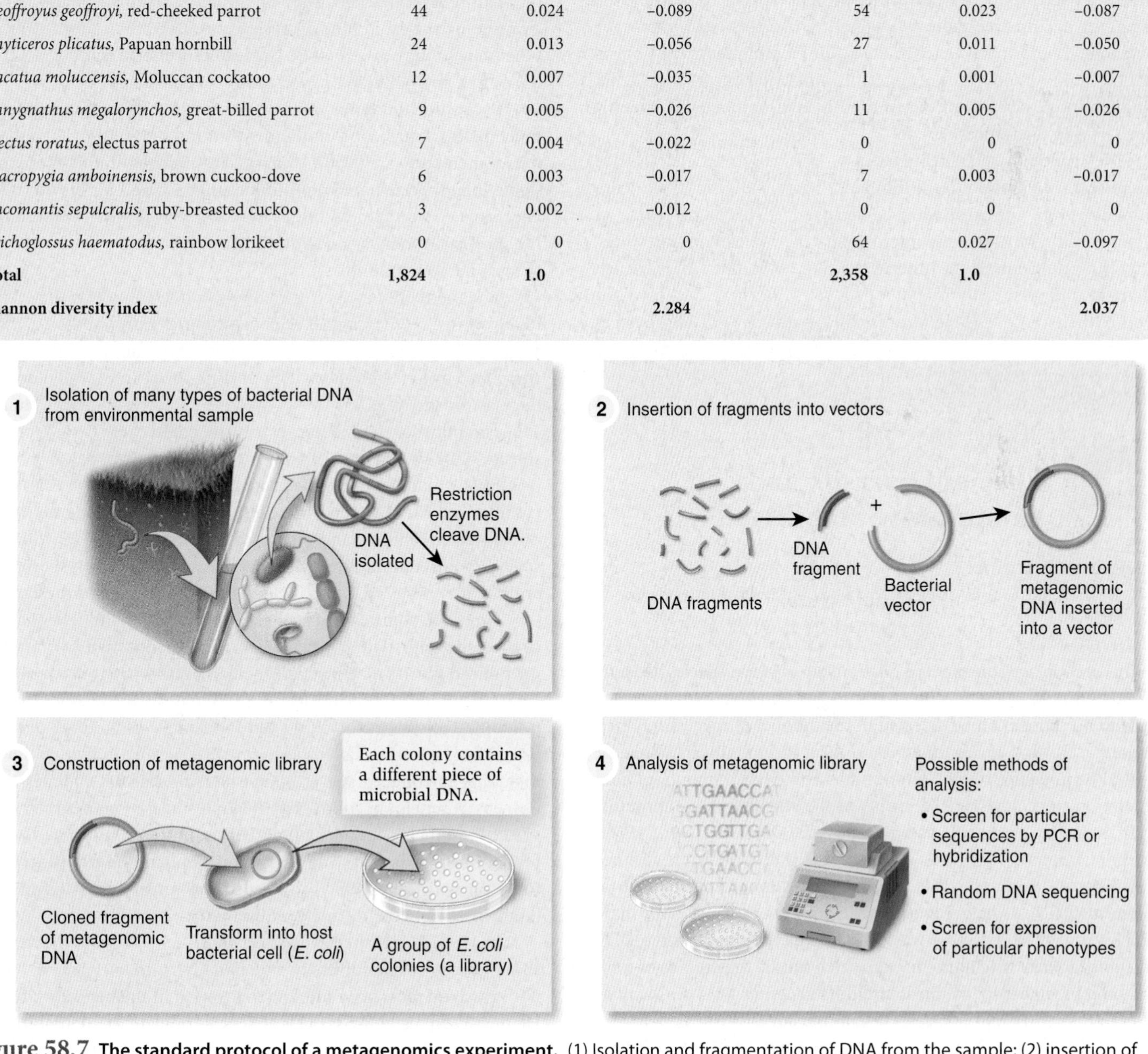

Figure 58.7 The standard protocol of a metagenomics experiment. (1) Isolation and fragmentation of DNA from the sample; (2) insertion of fragments into bacterial vectors; (3) insertion of cloned DNA into host bacterial cell (here, *Escherichia coli*), and culturing in selective growth media to create a DNA library; and (4) analysis of DNA sequences and protein expression.

different piece of microbial DNA, is called a DNA library. Lastly, the DNA from the DNA library is analyzed. In some cases, expression of the new DNA results in the synthesis of a new protein that changes the phenotype of the host, for example, a new enzyme that is detected by a chemical technique or an unusual color or shape in the model organism.

In 2004, Australian Earth scientist Jill Banfield and colleagues used metagenomics techniques to identify the five dominant species of bacteria living at temperatures of 42°C (107°F) and pH 0.8 in the acidic wastewater (the same pH as battery acid) from a mine in California. They detected 2,033 proteins from these species. This represented the first large-scale proteomics-level expression of a natural microbial community. One of the proteins, a cytochrome, oxidizes iron and probably influences the rate of breakdown of acid mine drainage products. Many other proteins appear responsible for defending against free radicals, suggesting that this is an important metabolic trait for persistence in the acidic environment. The hope is that the team can now identify enzymes and metabolic pathways that can help in the cleanup of this and other environmentally contaminated sites in the future.

Metagenomic sequencing is also being used to characterize the microbial communities of humans. There is a belief that a core human microbiome exists which, when changed, may have effects on human health. Most of these microbial partners live in our intestines, extracting nutrients from otherwise indigestible parts of our diet. Many of these also detoxify potentially harmful chemicals from our food. Other microbes defend us against pathogens. Using metagenomics to gain a better understanding of our microbial community could be of immense medical value.

58.4 Species Diversity and Community Stability

Learning Outcome:

1. Describe the diversity-stability hypothesis and evaluate the evidence supporting it.

In this section, we consider the relationship between species diversity and community stability. A community is often seen as stable when little to no change can be detected in the number of species and their abundance over a given time period. The community may then be said to be in equilibrium. Community stability is an important consideration to ecologists. A decrease in the stability of a community over time may alert ecologists to a possible problem. In the 1950s, the populations of many bird species in the U.S. and Europe declined precipitously (look ahead to Figure 59.8). Raptor species such as peregrine falcons, bald eagles, and osprey were particularly hard hit. Eventually, the decline was traced to use of the pesticide DDT (dichlorodiphenyltrichloroethane), which caused eggshells to become thin and break before the birds could hatch. After DDT was banned later in the 1970s, raptor species began to recover.

We begin our discussion by exploring the question of whether communities with more species are more stable than communities with fewer species. We then examine the link between diversity and stability, using evidence from the field. Finally, we look at the relationship from a different angle and consider whether or not stable communities are more species-rich than communities that have been disturbed.

The Diversity-Stability Hypothesis States That Species-Rich Communities Are More Stable Than Those with Fewer Species

Community stability may be viewed in several different ways. Some communities, such as extreme deserts, are considered stable because they are resistant to change by anything other than water. Other communities, such as river communities, are considered stable because they can recover quickly after a disturbance, such as pollution, being cleansed by the rapid flow of fresh water. Lake communities, on the other hand, may be seen as less stable because there is often no drainage outlet, and pollutants can accumulate quickly.

Because maintaining community stability is seen as important, much research has gone into understanding the factors that enhance it. In general, research shows that species-rich communities are more stable than species-poor communities. Even so, ecologists debate the effects of species richness. For example, are species-rich communities more resistant to invasion by introduced species, such as weeds, than species-poor communities?

The link between species richness and stability was first explicitly proposed by the English ecologist Charles Elton in the 1950s. He suggested that a disturbance in a species-rich community would be cushioned by large numbers of interacting species and would not produce as drastic an effect as it would on a species-poor community. Thus, an introduced predator or parasite could cause extinctions in a species-poor system but possibly not in a more diverse system, where its effects would be buffered by interactions with more species in the community. Elton argued that outbreaks of pests are often found on cultivated land or land disturbed by humans, both of which are species-poor communities with few naturally occurring species. His argument became known as the **diversity-stability hypothesis**.

However, some ecologists began to challenge Elton's association of diversity with stability. Ecologists pointed out many examples of introduced species that have assumed pest proportions in species-rich areas, including rabbits in Australia and pigs in North America. They noted that disturbed or cultivated land may suffer from pest outbreaks not because of its simple nature but because individual species, including introduced species, often have no natural enemies in the new environment, in contrast to the long associations between native species and their natural enemies. For example, in Europe, coevolved predators such as foxes prevent rabbit populations from increasing to pest proportions. What was needed was research to determine if a link existed between diversity and stability.

In 1996, American ecologist David Tilman reported the relationship between species diversity and stability from an 11-year study of 207 grassland plots in Minnesota that varied in their species richness. He measured the biomass of every species of plant, in each plot, at the end of every year and obtained the average species biomass. He then calculated how much this biomass varied from year to year through

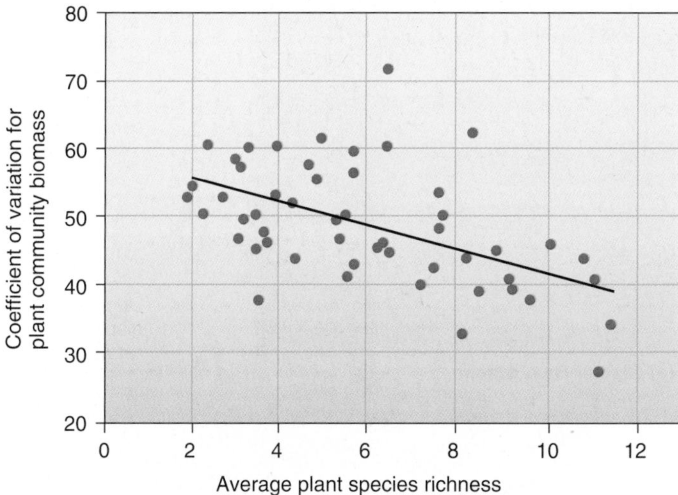

Figure 58.8 **Biomass variation and species richness.** Tilman's 11-year study of grassland plots in Minnesota revealed that year-to-year variability in community biomass was lower in species-rich plots. Each dot represents an individual plot. Only the plots from one field are graphed.

a statistical measure called the coefficient of variation. Less variation in biomass signified community stability. Year-to-year variation in plant community biomass was significantly lower in plots with greater plant species richness (**Figure 58.8**). The results showed that greater diversity enhances community stability.

Tilman suggested that diversity stabilizes communities because they are more likely to contain disturbance-resistant species that, in the event of a disturbance, could grow and compensate for the loss of disturbance-sensitive species. For example, when a change in climate such as drought decreased the abundance of competitively dominant species that thrived in normal conditions, unharmed drought-resistant species increased in mass and replaced them. Such declines in the number of susceptible species and compensatory increases in other species acted to stabilize total community biomass. Although ecologists recognize a link between species diversity and community stability, they are also aware that, over long periods of time, communities may experience severe disturbances. The change in composition and structure of communities that follows occurs in a predictable way termed succession, which is described next.

58.5 Succession: Community Change

Learning Outcomes:

1. Distinguish between primary and secondary succession.
2. Compare and contrast facilitation, inhibition, and tolerance as mechanisms of succession.

At 8:32 a.m. on May 18, 1980, Mount St. Helens, a previously little-studied peak in the Washington Cascades, erupted. The blast felled trees over a 600-km^2 area, and the landslide that followed—the largest in recorded history—destroyed everything in its path, killing nearly 60 people and millions of animals. However, since the 1980 eruption, much of the area has experienced a relatively rapid recovery of plant and animal communities (**Figure 58.9**).

Ecologists have developed several terms to describe how community change occurs. The term **succession** describes the gradual and continuous change in species composition of a community following a disturbance. **Primary succession** refers to succession on a newly exposed site that has no biological legacy in terms of plants, animals, or microbes, such as bare ground caused by a volcanic eruption or the sediment created by the retreat of glaciers. In primary succession on land, the plants must often build up the soil, and thus a long time—even hundreds of years—may be required for the process. Only a tiny proportion of the Earth's surface is currently undergoing primary succession, for example, around Mount St. Helens and the volcanoes in Hawaii and off the coast of Iceland, and behind retreating glaciers in Alaska and Canada.

(a) 1980

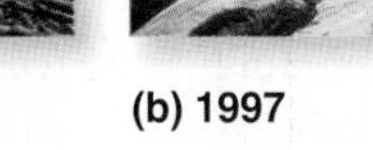

(b) 1997

Figure 58.9 **Succession on Mount St. Helens.** **(a)** The initial blast occurred on May 18, 1980. **(b)** By 1997, many of the areas initially flattened by the blast and covered in ash developed low-lying vegetation, and new trees sprouted up between the old dead tree trunks.

Secondary succession refers to succession on a site that has already supported life but has undergone a disturbance such as a fire, tornado, hurricane, or flood (as in the 2004 tsunami in Indonesia). In terrestrial areas, soil is already present. Clearing a natural forest and farming the land for several years is an example of a severe forest disturbance that does not kill all native species. Some plants and many soil bacteria, nematodes, and insects are still present. Secondary succession occurs if farming is ended. The secondary succession in abandoned farmlands (also called old fields) can lead to a pattern of vegetation quite different from one that develops after primary succession following glacial retreat. For example, the plowing and added fertilizers, herbicides, and pesticides may have caused substantial changes in the soil of an old field, allowing species that require a lot of nitrogen to colonize. These species would not be present for many years in newly created glacial soils.

Frederic Clements is often viewed as the founder of successional theory. His work in the early 20th century emphasized succession as proceeding through several stages to a distinct end point or **climax community**. Although disturbance can return a community from a later stage to an earlier stage, generally the community progresses in one direction. Clements's depiction of succession focused on a process termed facilitation, but two other mechanisms of succession—inhibition and tolerance—have since been described. Let's examine the evidence for each of them.

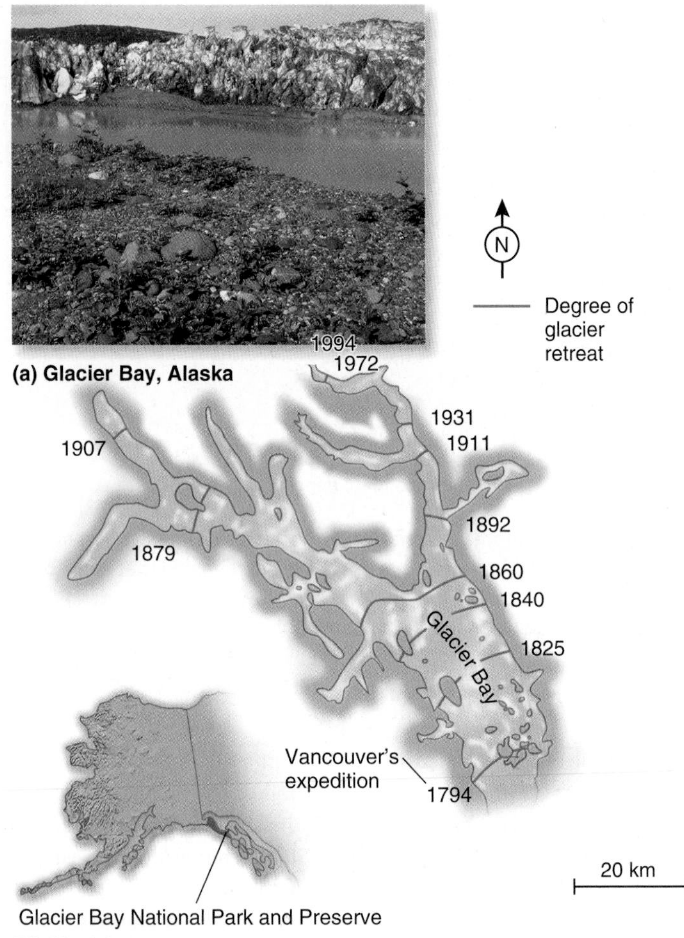

Figure 58.10 The degree of glacier retreat at Glacier Bay, Alaska, since 1794. **(a)** Primary succession begins on the bare rock and soil evident at the edges of the retreating glacier. **(b)** The lines reflect the position of the glacier in 1794 and its subsequent retreat northward.

Concept Check: Why do ecologists sometimes view walking through Glacier Bay as the equivalent of being in a time machine?

Facilitation Assumes Each Invading Species Creates a More Favorable Habitat for Succeeding Species

A key assumption of Clements is that each colonizing species makes the environment a little different—a little shadier or a little richer in soil nitrogen—so that it becomes more suitable for other species, which then invade and outcompete the earlier residents. This process, known as **facilitation**, continues until the most competitively dominant species has colonized, when the community is at climax. The composition of the climax community for any given region is thought to be determined by climate and soil conditions.

Succession following the gradual retreat of Alaskan glaciers is often used as a specific example of facilitation as a mechanism of succession. Over the past 200 years, the glaciers in Glacier Bay have undergone a dramatic retreat of nearly 100 km (**Figure 58.10**). Succession in Glacier Bay follows a distinct pattern of vegetation. As glaciers retreat, they leave moraines—deposits of stones, pulverized rock, and debris that serve as soil. In Alaska, the bare soil has a low nitrogen content and scant organic matter. In the pioneer stage, the soil is first colonized by a black crust of cyanobacteria, mosses, lichens, horsetails (*Equisetum variegatum*), and the occasional river beauty (*Epilobium latifolium*) (**Figure 58.11a**). Because the cyanobacteria are nitrogen fixers (refer back to Figure 37.17), the soil nitrogen increases a little, but soil depth and litterfall (fallen leaves, twigs, and other plant material) are still minimal. At this stage, there may be a few seeds and seedlings of dwarf shrubs of the rose family commonly called mountain avens (*Dryas drummondii*), alders (*Alnus sinuata*), and spruce, but they are rare. After about 40 years, mountain avens dominates the landscape (**Figure 58.11b**). Soil nitrogen increases, as does soil depth and litterfall, and alder trees begin to invade.

At about 60 years, alders form dense, close thickets (**Figure 58.11c**). Alders have nitrogen-fixing bacteria that live mutualistically in their roots and convert nitrogen from the air into a biologically useful form. Soil nitrogen dramatically increases, as does litterfall. Spruce trees (*Picea sitchensis*) begin to invade at about this time. After about 75 to 100 years, the spruce trees begin to overtop the alders, shading them out. The litterfall is still high, and the large volume of needles turns the soil acidic. The shade causes competitive exclusion of many of the original understory species, including alder, and only mosses carpet the ground. At this stage, seedlings of western hemlock (*Tsuga heterophylla*) and mountain hemlock (*Tsuga mertensiana*) may also occur. After 200 years, a mixed spruce-hemlock climax forest results (**Figure 58.11d**).

What other evidence is there of facilitation? Experimental studies of early primary succession on Mount St. Helens, which show that decomposition of fungi allows mosses and other fungi to colonize the

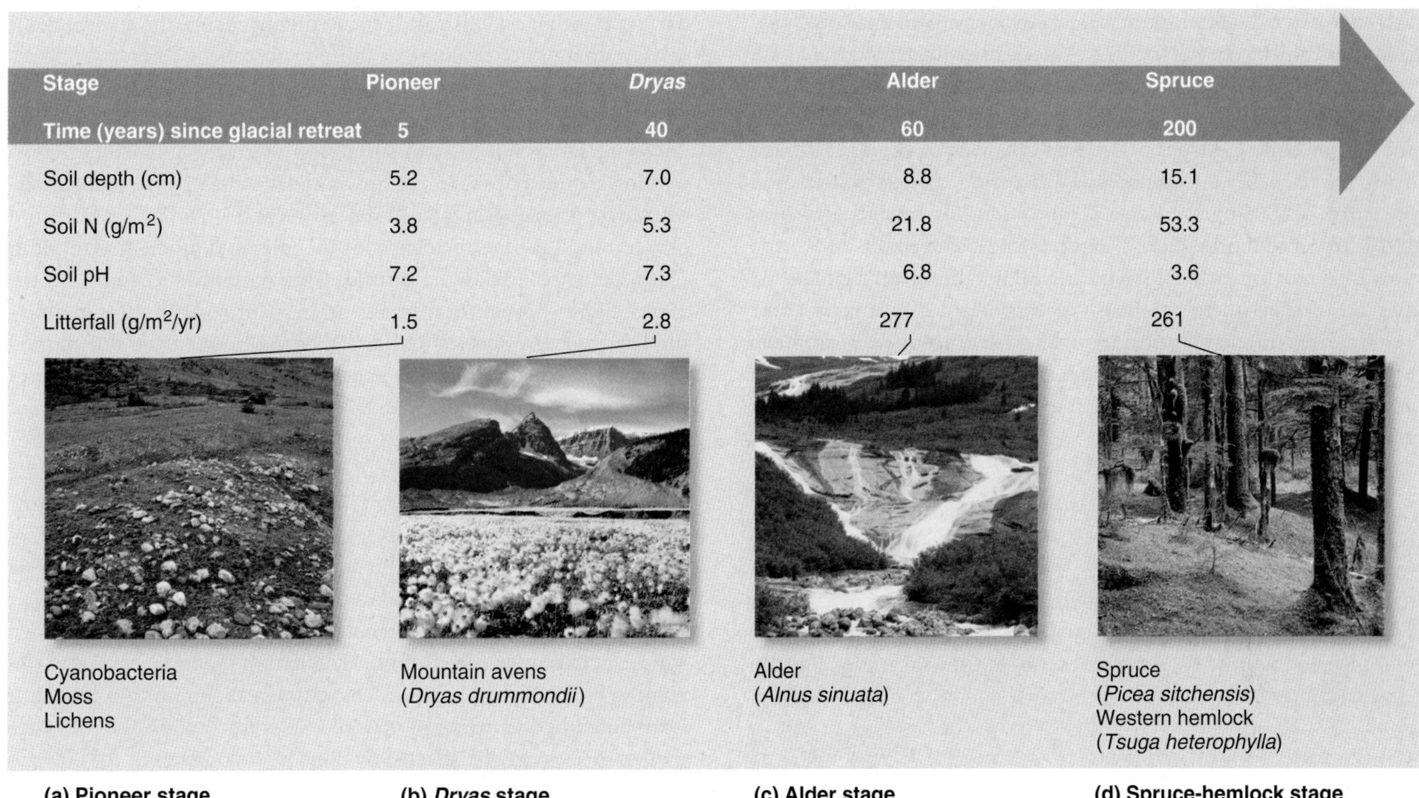

Stage
Pioneer
Dryas
Alder
Spruce
Time (years) since glacial retreat
5
40
60
200
Soil depth (cm)
5.2
7.0
8.8
15.1
Soil N (g/m²)
3.8
5.3
21.8
53.3
Soil pH
7.2
7.3
6.8
3.6
Litterfall (g/m²/yr)
1.5
2.8
277
261
Cyanobacteria
Moss
Lichens
Mountain avens
(Dryas drummondii)
Alder
(Alnus sinuata)
Spruce
(Picea sitchensis)
Western hemlock
(Tsuga heterophylla)
(a) Pioneer stage
(b) Dryas stage
(c) Alder stage
(d) Spruce-hemlock stage

Figure 58.11 **The pattern of primary succession at Glacier Bay, Alaska.** **(a)** The first species to colonize the bare ground following retreat of the glaciers are small species such as cyanobacteria, moss, and lichens. **(b)** Mountain avens (*Dryas drummondii*) is a flower common in the *Dryas* stage. **(c)** Soil nitrogen and litterfall increase rapidly as alder (*Alnus sinuata*) invade. Note also the appearance of a few spruce trees higher up the valley. **(d)** Spruce (*Picea sitchensis*) and hemlock (*Tsuga heterophylla*) trees comprise a climax spruce-hemlock forest at Glacier Bay, with moss carpeting the ground. Two hundred years ago, glaciers occupied this spot.

Concept Check: *Is facilitation the only mechanism fueling succession at Glacier Bay?*

BioConnections: *Refer back to Section 37.3. In what genus are the nitrogen-fixing bacteria that occur in nodules on the roots of alder trees?*

soil, provide evidence of facilitation. In New England salt marshes, *Spartina* grass facilitates the establishment of beach plant communities by stabilizing the rocky substrate and reducing water velocity, which enables other seedlings to emerge. Succession on sand dunes also supports the facilitation model, in that pioneer plant species stabilize the sand dunes and facilitate the establishment of subsequent plant species. The foredunes, those nearest the shoreline, are the most frequently disturbed and are maintained in a state of early succession, whereas more stable communities develop farther away from the shoreline.

Succession also occurs in aquatic communities. Although soils do not develop in marine environments, facilitation may still be encountered when one species enhances the quality of settling and establishment sites for another species. When experimental test plates used to measure settling rates of marine organisms were placed in the Delaware Bay, researchers discovered that certain cnidarians enhanced the attachment of tunicates, and both facilitated the attachment of mussels, the dominant species in the community. In this experiment, the smooth surface of the test plates prevented many species from colonizing, but once the surface became rougher, because of the presence of the cnidarians, many other species were able to colonize. In a similar fashion, early colonizing bacteria, which create biofilms on rock surfaces, can facilitate succession of other organisms.

Inhibition Implies That Early Colonists Prevent Later Arrivals from Replacing Them

Although data on succession in some communities fit the facilitation model, researchers have proposed alternative hypotheses of how succession may operate. In the process known as **inhibition**, early colonists prevent colonization by other species. For example, removing the litter of *Setaria faberi*, an early successional plant species in New Jersey old fields, causes an increase in the biomass of a later species, *Erigeron annuus*. The release of toxic compounds from decomposing *Setaria* litter or physical obstruction by the litter itself blocks the establishment of *Erigeron*. Without the litter present, however, *Erigeron* dominates and reduces the biomass of *Setaria*. Plant species, such as some grasses, ferns, vines, pine trees, and bamboo, that grow in dense thickets can inhibit succession, as can many introduced plant species.

Inhibition has been seen as the primary method of succession in the marine intertidal zone, where space is limited. In this habitat, early successional species are at a great advantage in maintaining possession of valuable space. In 1974, American ecologist Wayne Sousa created an environment for testing how succession works in the intertidal zone by scraping rock faces clean of all algae or putting out fresh boulders or concrete blocks. The first colonists of these areas were the green algae *Ulva*. By removing *Ulva* from the substrate, Sousa showed that the large red alga *Chondracanthus canaliculatus* was able to colonize more quickly (**Figure 58.12**). The results of Sousa's study indicate that early colonists can inhibit rather than facilitate the invasion of subsequent colonists. Succession may eventually occur because early colonizing species, such as *Ulva*, are more susceptible than later successional species, such as *Chondracanthus*, to the rigors of the physical environment and to attacks by herbivores, such as crabs (*Pachygrapsus crassipes*).

Tolerance Suggests That Early Colonists Neither Facilitate nor Inhibit Later Colonists

In 1977, researchers Joseph Connell and Australian ecologist Ralph Slatyer proposed a third mechanism of succession, which they termed **tolerance**. In this process, any species can start the succession, but the eventual climax community is reached in a somewhat orderly fashion. The species that establish and remain do not change the environment in ways that either facilitate or inhibit subsequent colonists. Species have differing tolerances to the intensity of competition that results as more species accumulate. Relatively competition-intolerant species are more successful early in succession when the intensity of competition is low and resources are abundant. Relatively competition-tolerant species appear later in succession and at climax. Connell and Slatyer found the best evidence for the tolerance model in American plant ecologist Frank Egler's earlier work on floral succession. In the 1950s, Egler showed that succession in plant communities is determined largely by species that already exist in the ground as buried seeds or old roots. Whichever species germinates first or regenerates from roots initiates the succession sequence. Germination or root regeneration, in turn, depends on the timing of a disturbance. For example, an early-season tree fall would promote early-germinating species to grow in the subsequent light gap, whereas a late-season tree fall would promote the growth of late-germinating species. As succession proceeds, earlier germinating or regenerating species may be outcompeted by different species.

The key distinction between the three models is in the manner in which succession proceeds. In the facilitation model, species replacement is facilitated by previous colonists; in the inhibition model, it is inhibited by the action of previous colonists; and in the tolerance model, species may be affected by previous colonists, but they do not require them (**Figure 58.13**).

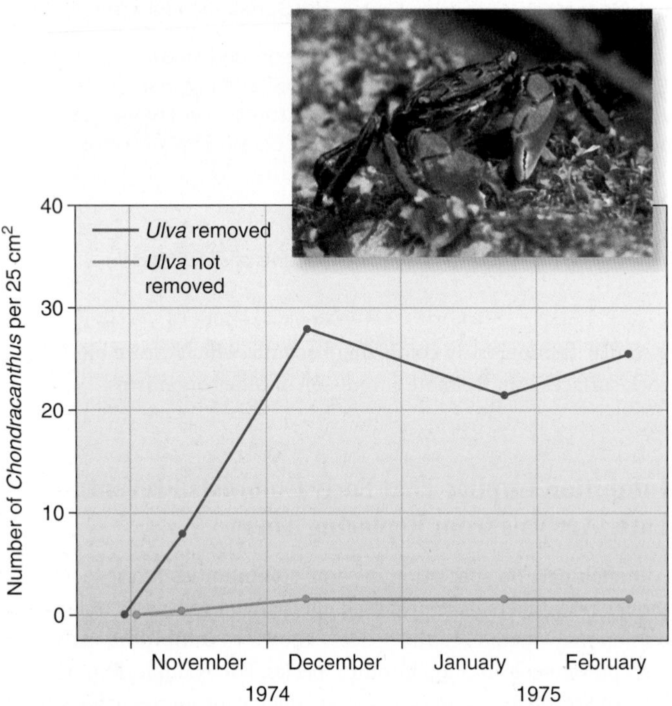

Figure 58.12 Inhibition as a primary method of succession in the marine intertidal zone. Removing *Ulva* from intertidal rock faces allowed colonization by *Chondracanthus*. The inset shows *Ulva* on a rock face with the striped shore crab *Pachygrapsus crassipes*, a herbivore.

BIOLOGY PRINCIPLE Biology is an experimental science. Waiting for ecological disturbance and following succession is unpredictable and time-consuming. Experimentally creating disturbances provides a better starting point for such studies.

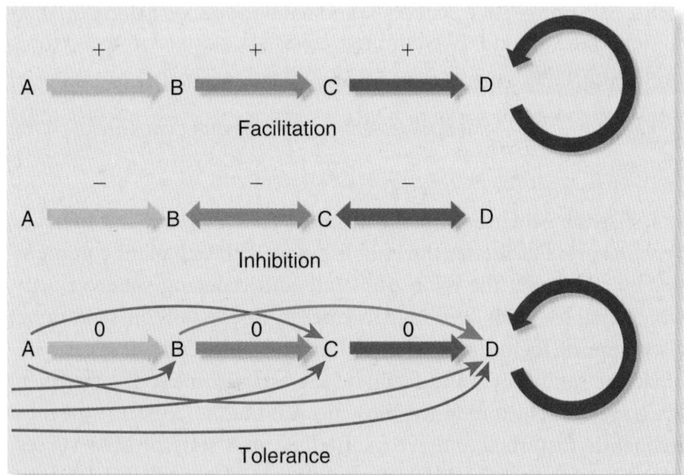

Figure 58.13 Three models of succession. A, B, C, and D represent four stages or seres. D represents the climax community. An arrow indicates "is replaced by," and + = facilitation, – = inhibition, and 0 = no effect. The facilitation model is the classic model of succession. In the inhibition model, early arriving species outcompete later arriving species. The tolerance model much depends on which species gets there first. The colored arrows show that succession may bypass some stages in the tolerance model.

Concept Check: *Inhibition implies competition exists between species, with early-arriving species tending to outcompete later arrivals, at least for a while. Does competition or mutualism feature more prominently in facilitation?*

58.6 Island Biogeography

Learning Outcomes:

1. Describe the equilibrium model of island biogeography.
2. List the predictions of the model and evaluate whether the evidence supports all the predictions.

Research has suggested that succession on islands differs from that on mainlands. In the 1960s, two eminent American ecologists, Robert MacArthur and E. O. Wilson, developed a comprehensive model to explain the process of succession on new islands, where a gradual buildup of species proceeds from a sterile beginning. Their model, termed the **equilibrium model of island biogeography**, holds that the number of species on an island tends toward an equilibrium number that is determined by the balance between two factors: immigration rates and extinction rates. In this section, we explore island biogeography and how well the model's predictions are supported by experimental data.

The Island Biogeography Model Suggests That During Succession, Gains in Immigration Are Balanced by Losses from Extinction

MacArthur and Wilson's model of island biogeography suggests that species repeatedly arrive on an island and either thrive or become extinct. The rate of immigration of new species is highest when no species are present on the island. As the number of species accumulates, the immigration rate decreases, since subsequent immigrants are more likely to represent species already present on the island. The rate of extinction is low at the time of first colonization, because few species are present and many have large populations. With the addition of new species, the populations of some species diminish, so the probability of extinction by chance alone increases. Over time, the number of species tends toward an equilibrium, $\hat{S}$, in which the rates of immigration and extinction are equal. Species may continue to arrive and go extinct, but the number of species on the island remains approximately the same.

MacArthur and Wilson reasoned that when plotted graphically, both the immigration and extinction lines would be curved, for several reasons (**Figure 58.14a**). First, species arrive on islands at different rates. Some organisms, including plants with seed-dispersal mechanisms and winged animals, are more mobile than others and arrive quickly. Other organisms arrive more slowly. This pattern causes the immigration curve to start off steep but get progressively shallower. On the other hand, extinctions rise at accelerating rates, because as later species arrive, competition increases and more species are likely to go extinct. As noted previously, earlier-arriving species tend to be *r*-selected species, which are better dispersers, whereas later-arriving species are generally *K*-selected species, which are better competitors. Later-arriving species usually outcompete earlier-arriving ones, causing an increase in extinctions.

The strength of the island biogeography model was that it generated several testable predictions:

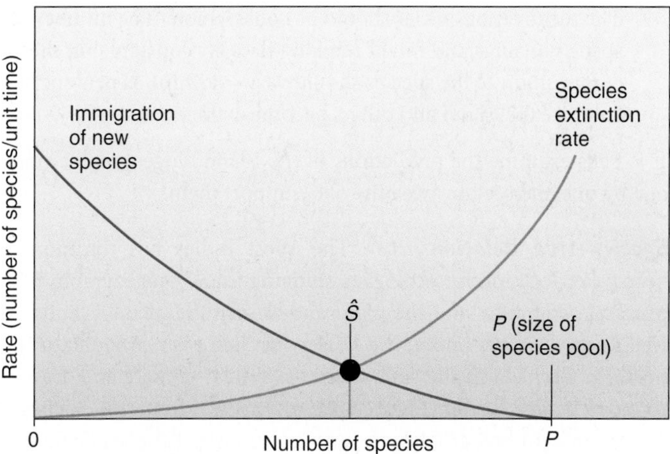

(a) Effects of immigration and extinction on species number

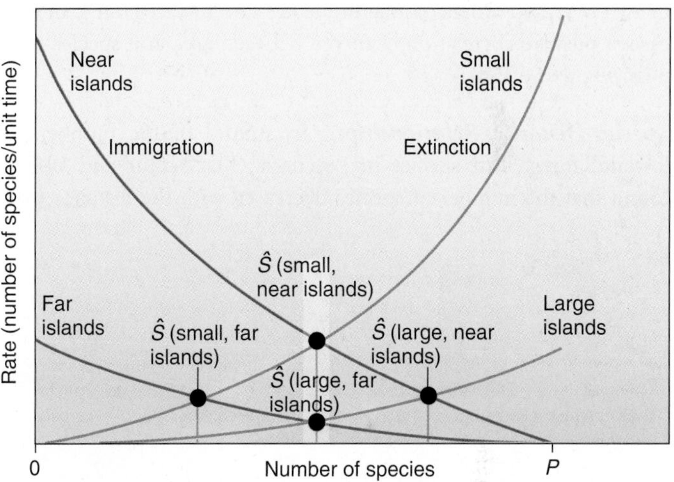

(b) Added effects of island size and proximity to the mainland on species number

Figure 58.14 MacArthur and Wilson's equilibrium model of island biogeography. **(a)** The interaction of immigration rate and extinction rate produces an equilibrium number of species on an island, $\hat{S}$, can vary from 0 species to *P* species, the total number of species available to colonize. **(b)** $\hat{S}$ varies according to the island's size and distance from the mainland. An increase in distance (near to far) lowers the immigration rate. An increase in island area (small to large) lowers the extinction rate.

Concept Check: Can you think of a scenario where there would be large numbers of species on a small island?

BioConnections: Look forward to Figure 60.12. How might the model of island biogeography be useful in the design of nature reserves?

1. The number of species should increase with increasing island size (area), a concept known as the species-area effect (see Figure 58.4). Extinction rates would be greater on smaller islands because population sizes would be smaller and more susceptible to extinction (**Figure 58.14b**).
2. The number of species should decrease with increasing distance of the island from the mainland, or the **source pool**, the pool of potential species available to colonize the island. Immigration rates would be greater on islands near the source pool because species do not have as far to travel (see Figure 58.14b).

3. The turnover of species should be considerable. The number of species on an island might remain relatively constant, but the composition of the species should vary over time as new species colonize the island and others become extinct.

Let's examine the predictions of the island biogeography model one by one and see how well the data support them.

Species-Area Relationships The West Indies has traditionally been a key location for ecologists studying island biogeography. The physical geography and the plant and animal life of the islands are well known. Furthermore, the Lesser Antilles, from Anguilla in the north to Grenada in the south, enjoy a similar climate and are surrounded by deep water (**Figure 58.15a**). In 1999, Robert Ricklefs and American ornithologist Irby Lovette summarized the available data on the richness of species of four groups of animals—birds, bats, reptiles and amphibians, and butterflies—across 19 islands that varied in area over two orders of magnitude (13 km^2 to 1,510 km^2). In each case, a positive correlation occurred between area and species richness (**Figure 58.15b**).

Species-Distance Relationships In studies of the numbers of lowland forest bird species in Polynesia, MacArthur and Wilson found that the number of species decreased with the distance from the source pool of New Guinea (**Figure 58.16**). They expressed the richness of bird species on the islands as a percentage of the number of bird species found on New Guinea. A significant decline in this percentage was observed with increasing distance. More-distant islands contained lower numbers of species than nearer islands. This research substantiated the prediction of species richness declining with increasing distance from the source pool.

Species Turnover Studies involving species turnover on islands are difficult to perform because detailed and complete species lists are needed over long periods of time, usually many years and often decades. The lists that do exist are often compiled in a casual way and are not usually suitable for comparison with more modern data. In 1980, British researcher Francis Gilbert reviewed 25 investigations carried out to demonstrate turnover and found a lack of this type of rigor in nearly all of them. Furthermore, most of the observed turnover in these studies, usually less than 1% per year, or less than one species per year, appeared to be due to immigrants that never became established rather than to the extinction of well-established species. More recent studies have revealed similar findings, suggesting that the rates of turnover are low rather than high, giving little conclusive support to the third prediction of the equilibrium model of island biogeography.

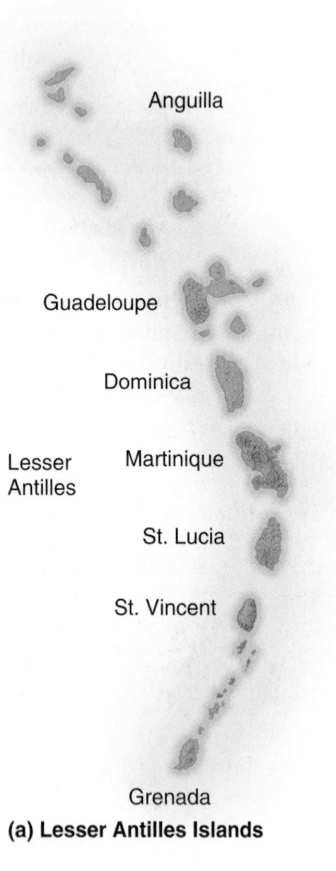

(a) Lesser Antilles Islands

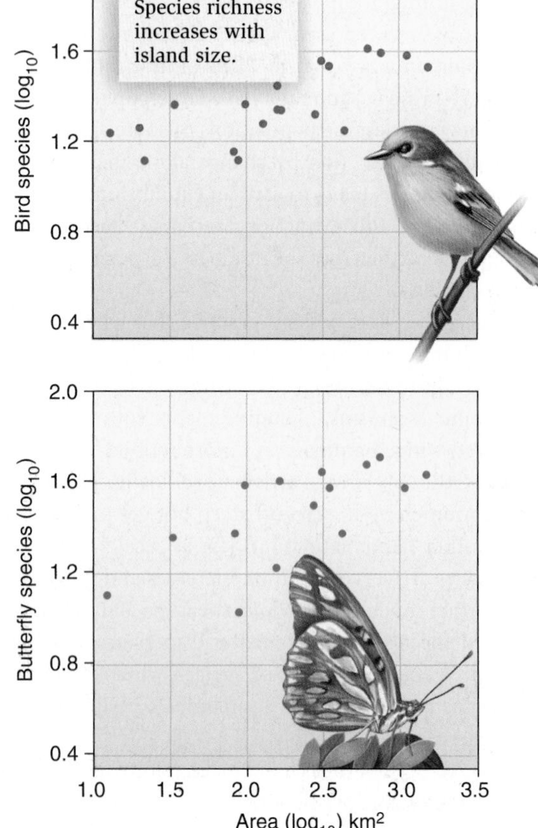

(b) Relationship between species richness and island size

Figure 58.15 Species richness and island size. **(a)** The Lesser Antilles extend from Anguilla in the north to Grenada in the south. **(b)** On these islands, the number of bird and butterfly species increases with the area of an island. Note that these relationships are traditionally plotted on a double logarithmic scale, a so-called log-log plot, in which the horizontal axis is the logarithm to the base 10 of the area and the vertical axis is the logarithm to the base 10 of the number of species. A linear plot of the area versus the number of species would be difficult to produce, because of the wide range of area and richness of species involved. Logarithmic scales condense this variation to manageable limits.

Concept Check: *How large is the change in bird species richness across islands in the Lesser Antilles?*

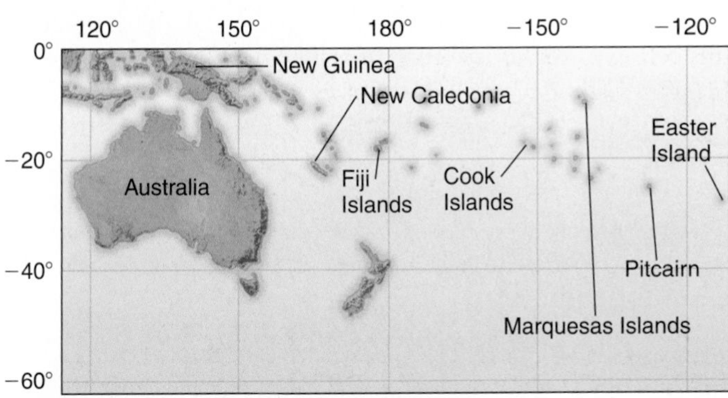

(a) New Guinea and neighboring islands

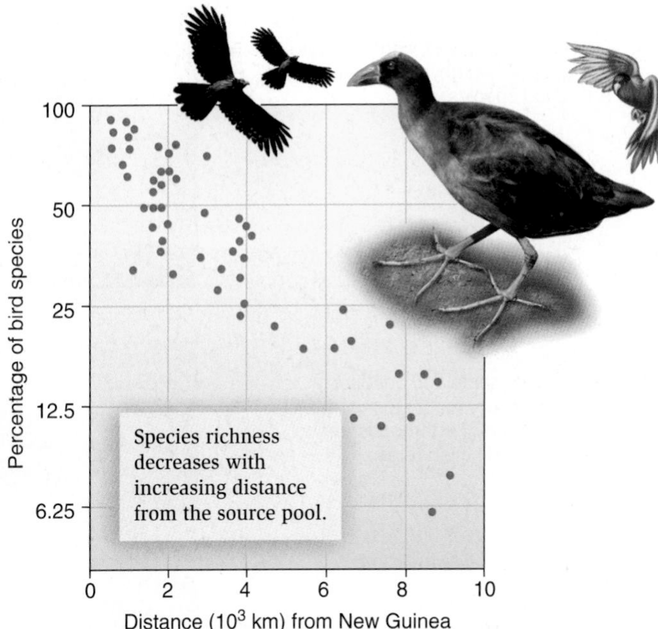

(b) Relationship between species richness and distance from source

Figure 58.16 **Species richness and distance from the source pool.** **(a)** Map of Australia, New Guinea, and these Polynesian Islands: New Caledonia, Fiji Islands, Cook Islands, Marquesas Islands, Pitcairn, and Easter Island. **(b)** The numbers of bird species on the islands decreases with increasing distance from the source pool, New Guinea. The species richness is expressed as the percentage of bird species on New Guinea.

FEATURE INVESTIGATION

Simberloff and Wilson's Experiments Tested the Predictions of the Equilibrium Model of Island Biogeography

In the 1960s, American ecologists Daniel Simberloff and E. O. Wilson conducted possibly the best test of the equilibrium model of island biogeography ever performed, using islands in the Florida Keys. They surveyed small red mangrove (*Rhizophora mangle*) islands, 11–25 m in diameter, for all terrestrial arthropods. They then enclosed each island with a plastic tent and had the islands fumigated with methyl bromide, a short-acting insecticide, to remove all arthropods on them. The tents were removed, and periodically thereafter Wilson and Simberloff surveyed the islands to examine recolonization rates. At each survey, they counted all the arthropod species present, noting any species not there at the previous census and the absence of others that were previously there but had presumably gone extinct (results for four of the islands are shown in **Figure 58.17**). In this way, they estimated turnover of species on islands.

After 250 days, all but one of the islands had a similar number of arthropod species to that before fumigation, even though population densities were still low. The data indicated that recolonization rates were higher on islands nearer to the mainland than on far islands—as the island biogeography model predicts. However, the data, which consisted of lists of species on islands before and after extinctions, provided little support for the prediction of substantial turnover. Rates of turnover were low, only 1.5 extinctions per year, compared

Figure 58.17 **Simberloff and Wilson's experiments on the equilibrium model of biogeography.**

HYPOTHESIS Island biogeography model predicts higher species richness for islands closer to the mainland and significant turnover of species on islands.

STARTING LOCATION Mangrove islands in the Florida Keys.

Experimental level | **Conceptual level**

1 Take initial census of all terrestrial arthropods on 4 mangrove islands. Erect a framework over each mangrove island.

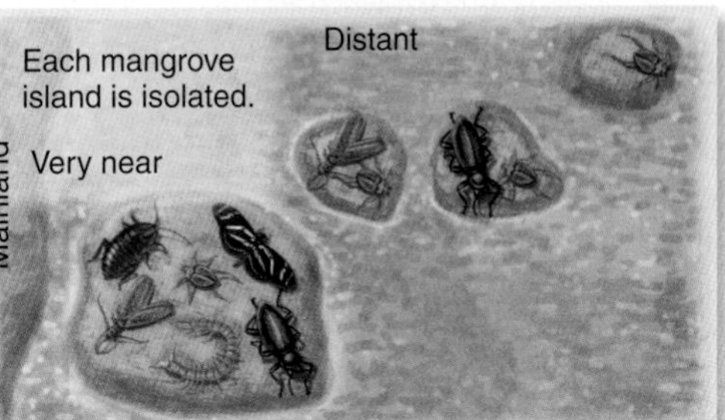

2 Cover the framework with tents and fumigate with methyl bromide to kill all arthropod species.

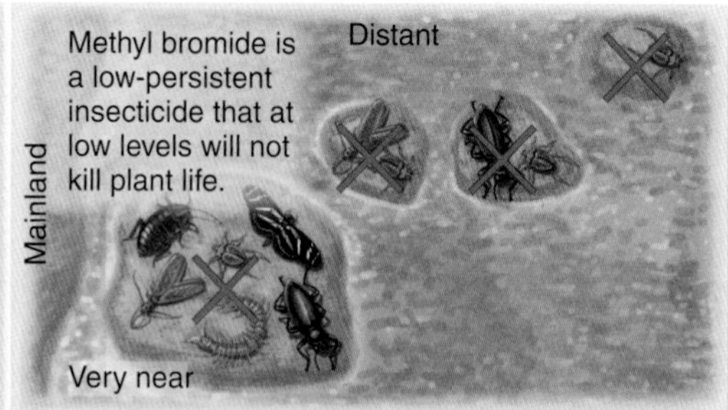

Methyl bromide is a low-persistent insecticide that at low levels will not kill plant life.
Distant
Mainland
Very near

3 Remove the tents and conduct censuses every month to monitor recolonization of arthropods and to determine extinction rates.

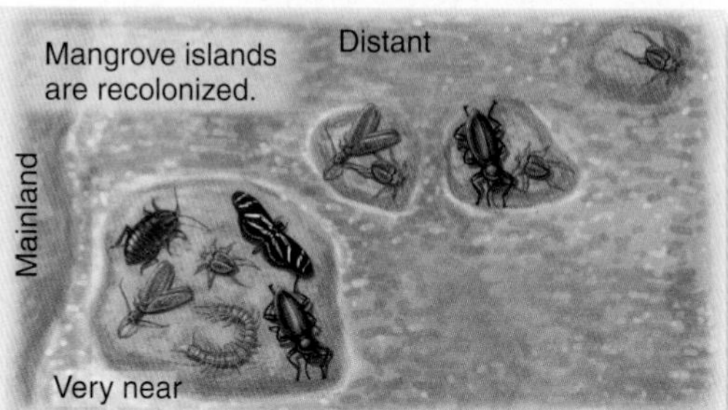

4 **THE DATA** Island E2 was closest to the mainland and supported the highest number of species both before and after fumigation. E3 and ST2 were at an intermediate distance from the mainland, and E1 was the most distant.

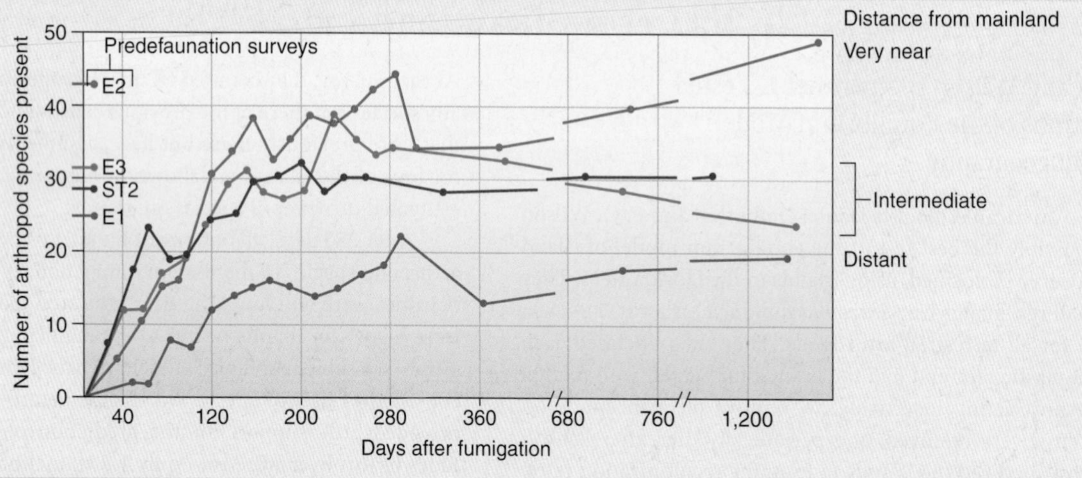

5 **CONCLUSION** Island distance from the mainland influences species richness on mangrove islands in the Florida Keys. However, species turnover is minimal, and species richness changes little following initial recolonization.

6 **SOURCE** Simberloff, D.S. 1978. Colonization of islands by insects: immigration, extinction and diversity. pp. 139–153 in L.A. Mound and N. Waloff (eds.). Diversity of insect faunas. *Blackwell Scientific Publications*, Oxford, U.K.

with the 15 to 40 species found on the islands within a year. Simberloff and Wilson concluded that turnover probably involves only a small subset of transient or less important species, with the more important species remaining permanent after colonization.

Experimental Questions

1. What was the purpose of Simberloff and Wilson's study?
2. Why did the researchers conduct a thorough species survey of arthropods before experimental removal of all the arthropod species?
3. What did the researchers conclude about the relationship between island proximity to the mainland and species richness and turnover?

The equilibrium model of island biogeography has stimulated much research confirming the strong effects of area and distance on species richness. However, species turnover appears to be low rather than considerable, which suggests that succession on most islands is a fairly orderly process. This means that colonization is not a random process and that the same species seem to colonize first and other species gradually appear in the same order.

It is also important to note that the principles of island biogeography have been applied to wildlife preserves, which are essentially islands in a sea of developed land consisting of agricultural fields or urban sprawl. Conservationists have therefore utilized the model of island biogeography in the design of nature preserves, a topic we will return to in Chapter 60.

Summary of Key Concepts

58.1 Differing Views of Communities

- Community ecology studies how groups of species interact and form functional communities. Ecologists have differing views on the nature of a community. In one view, communities are tightly organized groups of mutually dependent species; in another, they are loose assemblies of species that happen to live in the same place at the same time (Figure 58.1).
- Although many observations support the idea that communities are loose assemblages of species, sharp boundaries between groups of species do exist, especially related to physical differences that cause distinct communities to develop (Figure 58.2).

58.2 Patterns of Species Richness

- The number of species of most taxa varies according to geographic location, generally increasing from polar areas to tropical areas (Figure 58.3).
- Different hypotheses for the variations in species richness have been advanced, including the time hypothesis, the area hypothesis, the productivity hypothesis, and the intermediate disturbance hypothesis (Figures 58.4, 58.5, 58.6).

58.3 Calculating Species Diversity

- The most widely used measure of the species diversity of a community, called the Shannon diversity index, takes into account both species richness and species abundance (Table 58.1).
- The field of metagenomics seeks to identify and analyze the genomes contained in a community of microorganisms (Figure 58.7).

58.4 Species Diversity and Community Stability

- Community stability is an important consideration in ecology. The diversity-stability hypothesis maintains that species-rich communities are more stable than communities with fewer species. Tilman's field experiments, which showed that year-to-year variation in plant biomass decreased with increasing species diversity, established a link between diversity and stability (Figure 58.8).

58.5 Succession: Community Change

- Succession describes the gradual and continuous change in community structure over time. Primary succession refers to succession on a newly exposed site with no prior biological legacy; secondary succession refers to succession on a site that has already supported life but has undergone a disturbance (Figures 58.9, 58.10).
- Three mechanisms have been proposed for succession. In facilitation, each species facilitates or makes the environment more suitable for subsequent species. In inhibition, initial species inhibit later colonists. In tolerance, any species can start the succession, and species replacement is unaffected by previous colonists (Figures 58.11, 58.12, 58.13).

58.6 Island Biogeography

- In the equilibrium model of island biogeography, the number of species on an island tends toward an equilibrium number determined by the balance between immigration rates and extinction rates (Figure 58.14).
- The model predicts that the number of species increases with increasing island size; that the number of species decreases with distance from the source pool; and that turnover is high (Figures 58.15, 58.16).
- Simberloff and Wilson's experiments on mangrove islands in the Florida Keys provided support for the first tenet of the island biogeography model but refuted the third tenet (Figure 58.17).

Assess and Discuss

Test Yourself

1. A community with many individuals but few different species would exhibit
   a. low abundance and high species complexity.
   b. high stability.
   c. low species richness and high abundance.
   d. high species diversity.
   e. high abundance and high species richness.
2. Which of the following statements best represents the productivity hypothesis regarding species richness?
   a. The larger the area, the greater the number of species that will be found there.
   b. Temperate regions have a lower species richness due to the lack of time available for migration after the last ice age.
   c. The number of species in a particular community is directly related to the amount of available energy.
   d. As invertebrate productivity increases, species richness will increase.
   e. Species richness is not related to primary productivity.
3. Ecologists began to question Elton's link of increased stability to increased diversity because
   a. mathematical models showed communities with high diversity had high stability.
   b. cultivated land undergoes few outbreaks of pests.
   c. highly disturbed areas have high numbers of species.
   d. pest outbreaks are caused by lack of long associations with natural enemies, not because they occur in simple systems.
   e. all of the above.

4. Metagenomics is a field of study that
   a. is the analysis of a collection of genome sequences obtained from an environmental site.
   b. focuses on the microbial genomes contained in a community.
   c. compares the genomes of similar species in different communities.
   d. none of the above.
   e. both a and b.
5. Extreme fluctuations in species abundance
   a. lead to more diverse communities.
   b. are usually seen in early stages of community development.
   c. may increase the likelihood of extinction.
   d. have very little effect on species richness.
   e. are characteristic of stable communities.
6. Which of the following statements best represents the relationship between species diversity and community disturbance?
   a. Species diversity and community stability have no relationship.
   b. Communities with high levels of disturbance are more diverse.
   c. Communities with low levels of disturbance are more diverse.
   d. Communities with intermediate levels of disturbance are more diverse.
   e. Communities with intermediate levels of disturbance are less diverse.
7. The process of primary succession occurs
   a. around a recently erupted volcano.
   b. on a newly plowed field.
   c. on a hillside that has suffered a mudslide.
   d. on a recently flooded riverbank.
   e. on none of the above.
8. Early colonizers excluding subsequent colonists from moving into a community is referred to as
   a. facilitation.
   b. competitive exclusion.
   c. secondary succession.
   d. inhibition.
   e. natural selection.
9. A tree falls in a forest in spring and flowers germinate in the light gap. Following a tree fall in autumn, different species of flowers germinate in the light gaps. This illustrates the principle of
   a. facilitation.
   b. tolerance.
   c. inhibition.
   d. primary succession.
   e. climax communities.
10. On which types of island would you expect species richness to be greatest?
   a. small, near mainland
   b. small, distant from mainland
   c. large, near mainland
   d. large, distant from mainland
   e. Species richness is equal on all these types of islands.

Conceptual Questions

1. Re-examine Figure 58.4. How does the relationship shown relate to what we learned in Chapter 57 about the influence of secondary metabolites on herbivores?
2. List some possible ecological disturbances, their likely frequency in natural communities, and the severity of their effects.
3. A principle of biology is that *biology is an experimental science.* In the nutrient-poor heathlands of Europe, scotch heather (*Calluna vulgaris*) and cross-leaved heath (*Erica tetralix*) are gradually replaced by variegated purple moor grass (*Molinia caerulea*) and wavy hair grass (*Deschampsia flexuosa*). Adding *Calluna* litter or nitrogen fertilizer speeds up this process. Explain this phenomenon and which mechanism of succession is supported.

Collaborative Questions

1. Distinguish between the time hypothesis, area hypothesis, and productivity hypothesis as explanations for the latitudinal gradient in species richness.
2. Calculate the species diversity of the following four communities. Which community has the highest diversity? What is the maximum diversity each community could have?

	Relative abundance of species				
Community	Species 1	Species 2	Species 3	H_S	Maximum possible diversity
1	90	10	—		
2	50	50	—		
3	80	10	10		
4	33.3	33.3	33.3		

Online Resource

www.brookerbiology.com

Stay a step ahead in your studies with animations that bring concepts to life and practice tests to assess your understanding. Your instructor may also recommend the interactive eBook, individualized learning tools, and more.

Chapter Outline

59.1 Food Webs and Energy Flow

59.2 Biomass Production in Ecosystems

59.3 Biogeochemical Cycles

Summary of Key Concepts

Assess and Discuss

59 Ecosystem Ecology

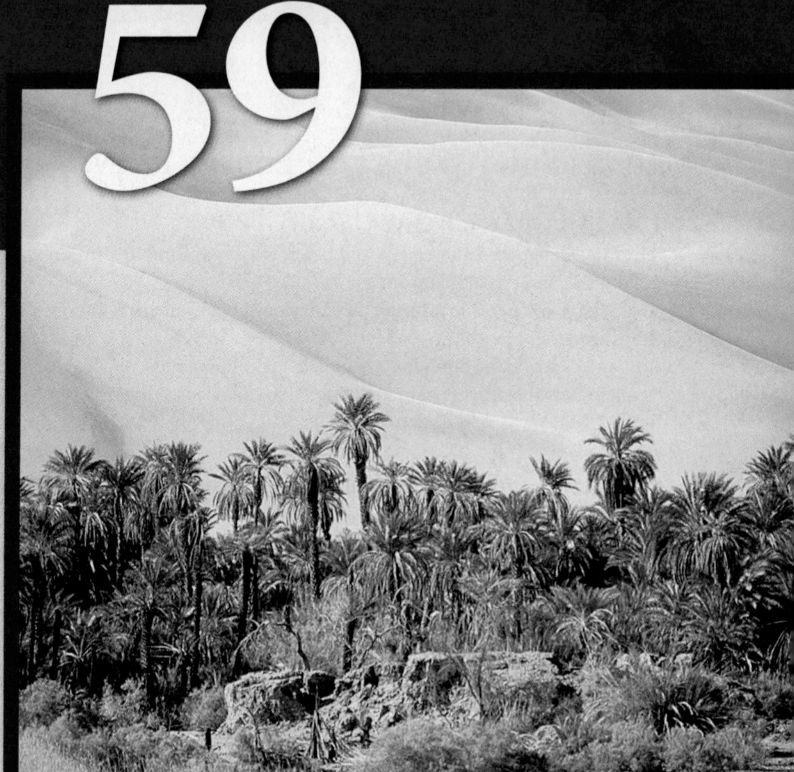

Mandara Lake Oasis, Libya, an example of a large-scale ecosystem.

Familiar backyard earthworms are known for their ability to convert organic matter such as dead leaves into rich humus, improving soil fertility. However, earthworms are not native everywhere. North American glaciations exterminated earthworms from hardwood forests in Wisconsin and Minnesota some 11,000 to 14,000 years ago. In the absence of earthworms to break up litter into small pieces, slower acting fungi and bacteria were the main decomposers. A thick forest floor formed, and carbon built up in the soil. Earthworms from Europe and Asia initially were introduced into Wisconsin and Minnesota by European settlers and have continued to be transported to the area through a range of human activities, including the dumping of fishing bait. Organisms that are beneficial in one location can be destructive when introduced to another, however, and earthworms are no exception. In northern forests, the worms accelerate the cycling of nutrients through the soil, drastically altering the structure of the forest floor soil and releasing soil carbon into the atmosphere. According to Cindy Hale, an American biologist who has been studying the effect of earthworms on northern hardwood forests, "They have a cascading effect on plants, animals, and soil organisms. And we know they're causing significant damage to some forests. Their effect could be really profound."

The term **ecosystem** was coined in 1935 by the British plant ecologist A. G. Tansley to describe the system formed by the interaction between a community of organisms and its physical environment. **Ecosystem ecology** deals with the flow of energy and cycling of chemical elements within an ecosystem. As with the concept of a community, the ecosystem concept can be applied at any scale. A small pond inhabited by protozoa and insect larvae is an ecosystem, and an oasis with its plants, frogs, fishes, and birds constitutes another. Most ecosystems cannot be regarded as having definite boundaries. Even in a clearly defined pond ecosystem, species may be moving in and out (**Figure 59.1**). Nevertheless, studying ecosystem ecology allows us to use the common currency of energy and chemicals to compare the functions between and within ecosystems.

In investigating the dynamics of an ecosystem, at least three major constituents can be measured: the flow of energy, the production of biomass, and cycling of elements through ecosystems. We begin the chapter by exploring **energy flow**, the movement of energy through

Figure 59.1 **A small ecosystem.** Even in this pond ecosystem, frogs or other species such as birds may move in and out, importing or exporting nutrients and energy with them.

an ecosystem. In examining energy flow, our main task will be to document the complex networks of feeding relationships and to measure the efficiency of energy transfer between organisms in an ecosystem. Next, we will focus on the measurement of **biomass**, the total mass of living matter in a given area, usually measured in grams or kilograms per square meter. We will examine the amount of biomass produced through photosynthesis, termed primary production, and the amount of biomass produced by the organisms that are the consumers of primary production. In the last section, we will examine **biogeochemical cycles**, the movement of chemicals through ecosystems, and explore the cycling of elements, such as phosphorus, carbon, and nitrogen, and the effects that human activities are having on these ecosystem-wide processes.

59.1 Food Webs and Energy Flow

Learning Outcomes:

1. Distinguish between autotrophs and heterotrophs and among primary, secondary, and tertiary consumers.
2. Describe two ways of measuring the efficiency of consumers as energy transformers.
3. List and describe the different types of ecological pyramids.
4. Explain how the process of biomagnification can occur at higher trophic levels.

Most organisms either make their own food using energy from sunlight or feed on other organisms. Simple feeding relationships between organisms can be characterized by an unbranched **food chain**, a linear depiction of energy flow, with each organism feeding on and deriving energy from the preceding organism. Each feeding level in the chain is called a **trophic level** (from the Greek *trophos*, meaning feeder), and different species feed at different levels. In a food-chain diagram, an arrow connects each trophic level with the one above it (**Figure 59.2**).

In this section, we will consider the flow of energy in a food chain and examine a food web, a more complex model of interconnected food chains. We will then explore two of the most important features of food webs—chain length and the pyramid of numbers—and learn how the passage of nutrients through food webs can result in the accumulation of harmful chemicals in the tissues of organisms at higher trophic levels.

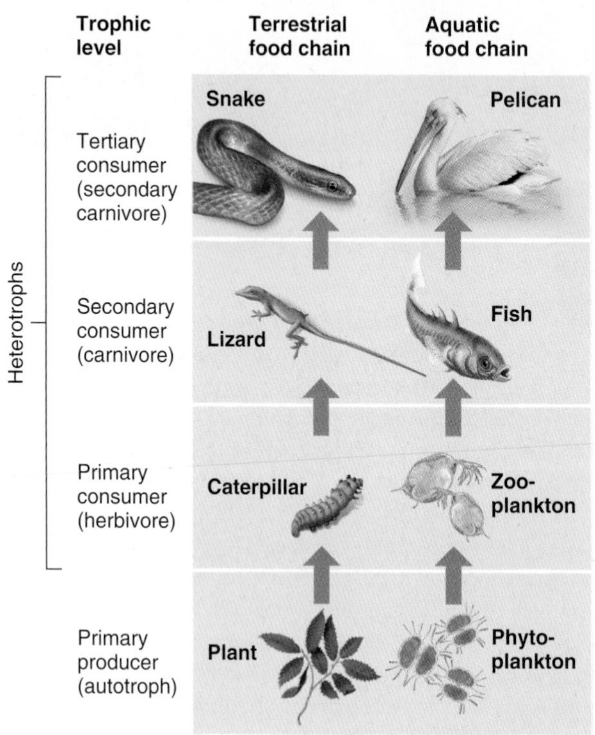

Figure 59.2 **Food chains.** Two examples of the flow of food energy up the trophic levels: a terrestrial food chain and an aquatic food chain.

BioConnections: In these two food chains, plants and protists (phytoplankton) are the producers. Look back at Section 27.5. What other organisms are producers and could also support food chains?

The Main Trophic Levels Within Food Chains Consist of Primary Producers, Primary Consumers, and Secondary Consumers

Food chains typically consist of organisms that obtain energy in different ways. **Autotrophs** harvest light or chemical energy and store that energy in carbon compounds. Most autotrophs, including plants, algae, and photosynthetic bacteria, use sunlight for this process. These organisms, called **producers**, form the base of the food chain. They produce the energy-rich organic molecules upon which nearly all other organisms depend.

Organisms in trophic levels above the primary producers are termed **heterotrophs**. These organisms must consume organic molecules from their environment to sustain life and thus receive their nutrition by eating other organisms. Organisms that obtain their food by consuming primary producers are termed **primary consumers** and include most protists, most animals, and even some plants such as mistletoe, which is parasitic on other plants. Animals that eat plants are also called **herbivores**. Organisms that eat primary consumers are **secondary consumers**. Animals that eat other animals are also called **carnivores** (from the Latin *carn*, meaning flesh). Organisms that feed on secondary consumers are **tertiary consumers**, and so on. Thus, energy enters a food chain through producers, via photosynthesis, and is passed up the food chain to primary, secondary, and tertiary consumers (see Figure 59.2).

At each trophic level, many organisms die before they are eaten. Much energy from the first trophic level, such as the plants, goes unconsumed by herbivores. Instead, unconsumed plants die and decompose in place. This material, along with dead remains of animals and waste products, is called **detritus**. Consumers that get their energy from detritus, called **detritivores**, or **decomposers**, break down dead organisms from all trophic levels (**Figure 59.3**). In terrestrial systems, detritivores probably carry out 80–90% of the

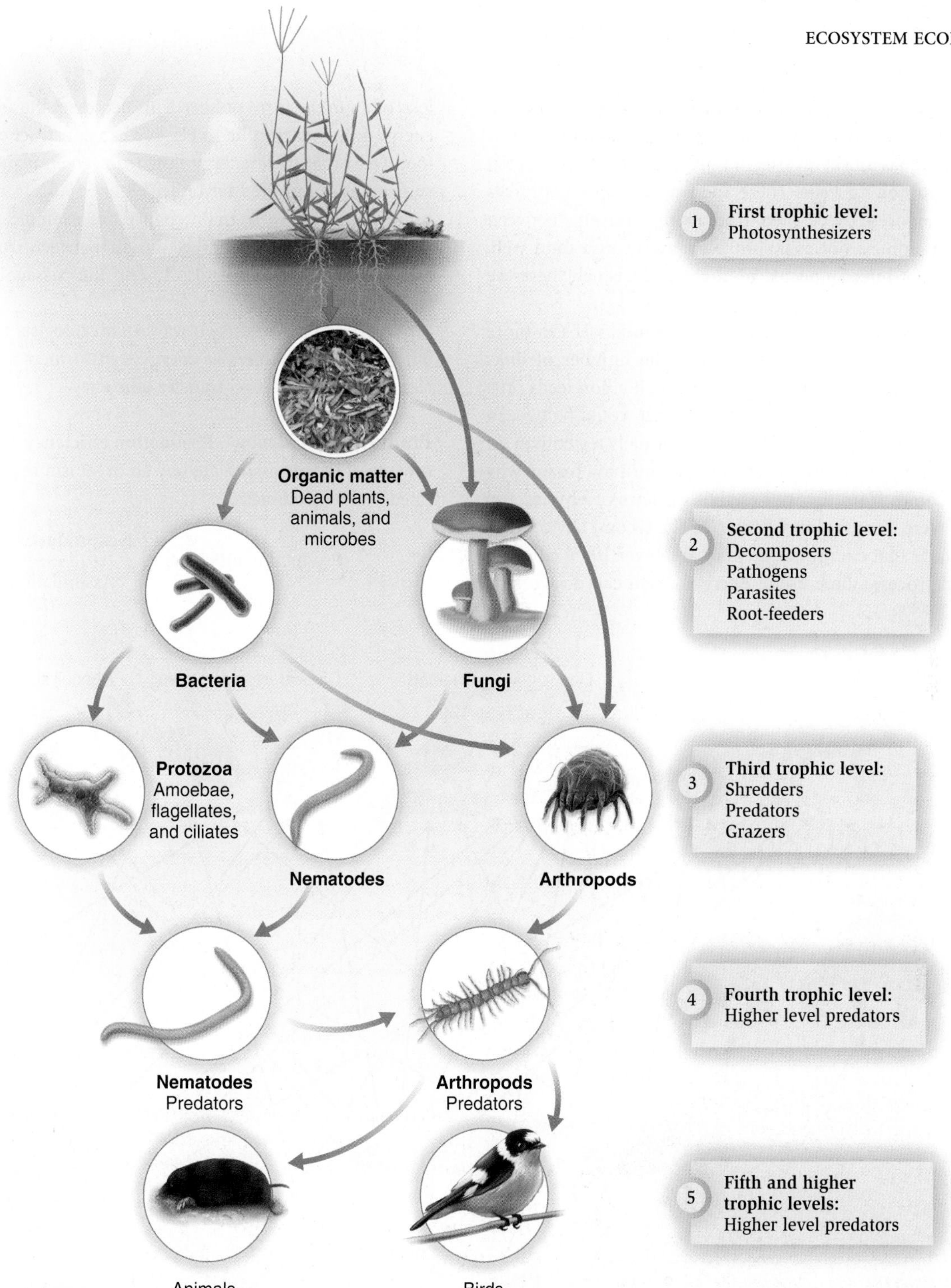

Figure 59.3 **Decomposers (detritivores) feeding on dead plant and animal matter.** Many dead plants and animals are eaten by a variety of decomposers. Here, bacteria and fungi feed on rotting plant material. These may, in turn, support a variety of predators, including centipedes or larger predators such as mammals and birds, which will also feed on the animal carcass.

Concept Check: *At which trophic level do decomposers feed?*

BioConnections: *Name two fungal phyla that are active in decomposition. (Hint: Refer back to Table 31.1.)*

consumption of plant matter, with different species working in concert to extract most of the energy. Detritivores may, in turn, support a community of predators that feed on them. As we noted at the beginning of the chapter, changes in the decomposer community can lead to changes in nutrient cycling.

In Most Food Webs, Chain Lengths Are Short

The consumption of species between trophic levels varies widely. For example, many different herbivore species may feed on the same plant species. Also, each species of herbivore may feed on several different plant species. Such branching of food chains also occurs at other trophic levels. For instance, on the African savanna, cheetahs, lions, and hyenas all eat a variety of prey, including wildebeest, impala, and Thompson's gazelle. These, in turn, eat a variety of trees and grasses. It is more correct, then, to draw relationships between these plants and animals not as a simple chain but as a **food web**, a complex model of interconnected food chains in which there are multiple links among species (Figure 59.4).

Let's examine some of the characteristics of food webs in more detail. The concept of chain length refers to the number of links between the trophic levels involved. For example, if a lion feeds on a zebra, and a zebra feeds on grass, the chain length would be two. In many food webs, chain lengths tend to be short, usually fewer than six levels, even including parasites and detritivores. The main reason why they are short comes from the well-established laws of physics and chemistry that were discussed in Chapter 2. The second law of thermodynamics states that energy conversions are never 100% efficient. In any transfer process, some useful energy, which can do work, is lost, often in the form of heat. This decreases the amount of available energy at higher trophic levels. We can construct energy budgets for food webs that trace energy flow from green plants to tertiary consumers (and if needed beyond) (Figure 59.5). In each trophic level, some energy is lost to maintenance, for example, to maintain body temperature. Because energy transfer between trophic levels is not 100% efficient, energy is also lost in the passage from one trophic level to another.

As described next, two ways that ecologists use to evaluate the efficiency of consumers as energy transformers are production efficiency and trophic-level transfer efficiency.

Production Efficiency **Production efficiency** is defined as the percentage of energy assimilated by an organism that becomes incorporated into new biomass.

$$\text{Production efficiency} = \frac{\text{Net productivity}}{\text{Assimilation}} \times 100$$

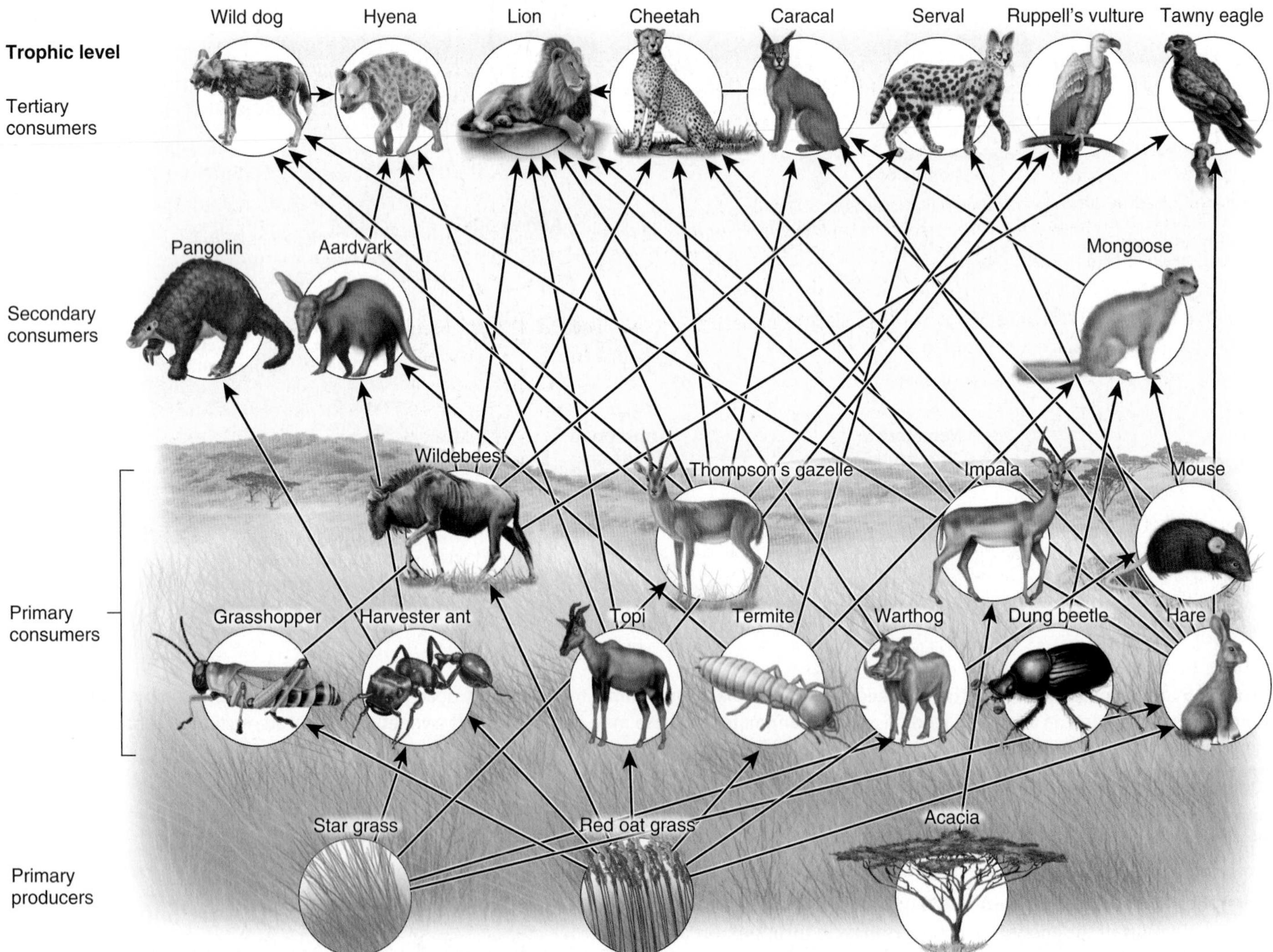

Figure 59.4 A food web from an African savanna ecosystem. Each trophic level is occupied by different species. Generally, each species feeds on, or is fed upon by, more than one species.

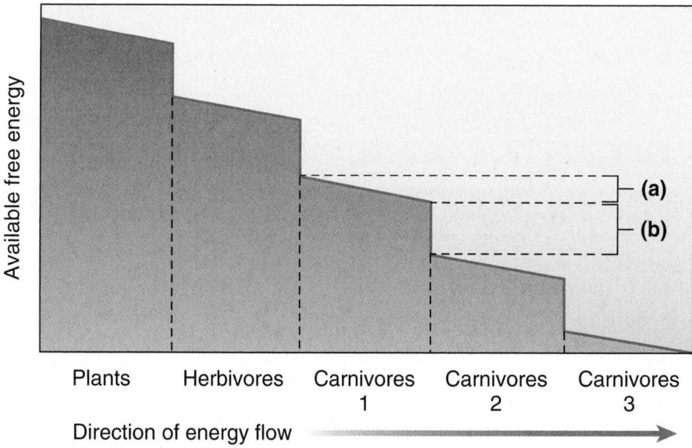

Figure 59.5 **Energy flow through a food web.** In this food web there are five trophic levels and four links between the trophic levels. **(a)** Energy lost as heat in a single trophic level. **(b)** Energy lost in the conversion from one trophic level to another.

BIOLOGY PRINCIPLE Living organisms use energy. Within trophic levels, energy is lost to maintenance, and between trophic levels, energy is lost to imperfect efficiency of transfer.

Here, net productivity is the energy, stored in biomass, that has accumulated over a given time span, and assimilation is the total amount of energy taken in by an organism over the same time span. Invertebrates generally have high production efficiencies that average about 10–40% (Figure 59.6a). Microorganisms also have relatively high production efficiencies. Vertebrates tend to have lower production efficiencies than invertebrates, because they devote more energy to sustaining their metabolism than to new biomass production. Even within vertebrates, much variation occurs. Fishes, which are ectotherms, typically have production efficiencies of around 10%, and birds and mammals, which are endotherms, have production efficiencies in the range of 1–2% (Figure 59.6b). In large part, the difference reflects the energy cost of maintaining a constant body temperature. Production efficiencies are higher in young animals, which are rapidly accruing biomass, than in older animals, which are not. This is the main reason behind the practice of harvesting young animals for meat, at about the time when they first attain adult mass.

One consequence of differing production efficiencies is that sparsely vegetated deserts can support populations of ectotherms such as snakes and lizards, whereas mammals might easily starve. The largest living lizard known, the Komodo dragon, eats the equivalent of its own weight every 2 months, whereas a cheetah consumes approximately four times its own weight in the same period.

Trophic-Level Transfer Efficiency The second measure of efficiency of consumers as energy transformers is **trophic-level transfer efficiency**, which is the amount of energy at one trophic level that is acquired by the trophic level above and incorporated into biomass. This provides a way of examining energy flow between trophic levels, not just in an individual species. Trophic level transfer efficiency is calculated as follows:

$$\text{Trophic-level transfer efficiency} = \frac{\text{Production at trophic level } n}{\text{Production at trophic level } n-1} \times 100$$

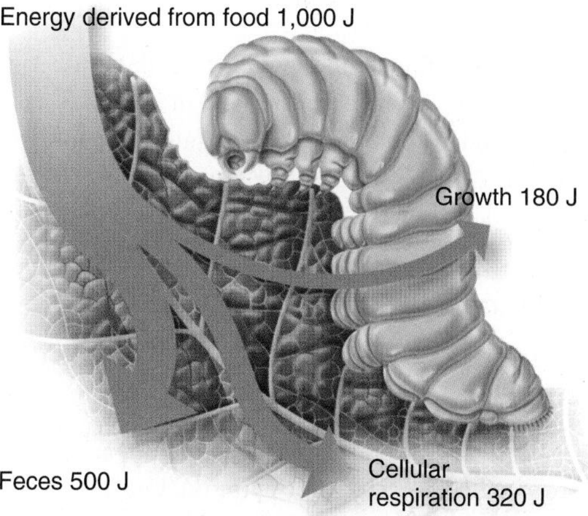

(a) High production efficiency of an invertebrate

(b) Low production efficiency of a vertebrate

Figure 59.6 **Production efficiencies.** **(a)** This caterpillar, an invertebrate, chews leaves to obtain its energy. If a mouthful of food contains 1,000 joules (J) of energy, about 320 J is used in cellular respiration to fuel metabolic processes (32%), and 500 J (50%) is lost in feces. As a result about 180 J of the 500 g is assimilated to be converted into insect biomass, a production efficiency of 36%. **(b)** The production efficiency of this squirrel, a mammal, is much lower.

BIOLOGY PRINCIPLE Living organisms maintain homeostasis. For the squirrel, maintaining a constant body temperature reduces its production efficiency.

Concept Check: What is the production efficiency of the squirrel, using the numbers in the figure?

For example, recall from Chapter 54 that zooplankton are minute drifting animal organisms that graze on microscopic photosynthetic organisms called phytoplankton. If there were 14 g/m^2 of zooplankton in a lake (trophic level n) and 100 g/m^2 of phytoplankton production (trophic level $n - 1$), the trophic level efficiency would be 14%. Trophic-level transfer efficiency appears to average around 10%, though there is much variation. Trophic-level transfer efficiency is generally low for two reasons. First, many organisms cannot digest all their prey. They take only the easily digestible plant leaves or animal tissue such as muscles and guts, leaving the hard wood or energy-rich bones behind. Second, much of the energy assimilated by animals is used in maintenance, so most energy is lost from the system as heat. The 10% average transfer rate of energy from one trophic level to another also necessitates short food webs of no more than four or five levels. Relatively little energy is available for the higher levels.

Ecological Pyramids Describe the Distribution of Numbers, Biomass, or Energy Between Trophic Levels

Trophic-level transfer efficiencies can be expressed in a graphical form called an ecological pyramid. One of the best-known pyramids, described by British ecologist Charles Elton in 1927, is the **pyramid of numbers**, in which the number of individuals decreases at each trophic level, with a large number of individuals at the base and fewer individuals at the top. Elton used a small pond as an example, in which the numbers of protozoa may run into the millions and those of *Daphnia*, their predators, number in the hundreds of thousands. Hundreds of beetle larvae may feed on *Daphnia*, and tens of fishes feed on the beetles. Many other examples of this type of pyramid are known. For example, in a grassland, there may be hundreds of individual plants per square meter, dozens of insects that feed on the plants, a few spiders feeding on the insects, and birds that feed on the spiders (**Figure 59.7a**).

Ecologists have, however, discovered many exceptions to this pyramid. One single producer such as an oak tree can support hundreds of herbivorous beetles, caterpillars, and other primary consumers, which, in turn, may support thousands of predators. This is called an inverted pyramid of numbers (**Figure 59.7b**).

One way to reconcile this apparent exception is to weigh the organisms in each trophic level, creating a **pyramid of biomass**. For example, an oak tree weighs more than all its herbivores and predators combined. At the bottom of the pyramid is the **standing crop**, the total autotroph biomass in an ecosystem present at any one point in time. Looking at the biomass at each trophic level rather than at numbers of organisms shows an upright pyramid. In 1957, American ecologist

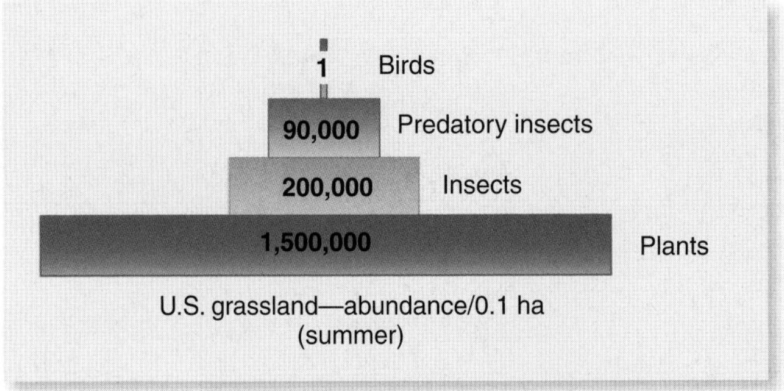

(a) Pyramid of numbers

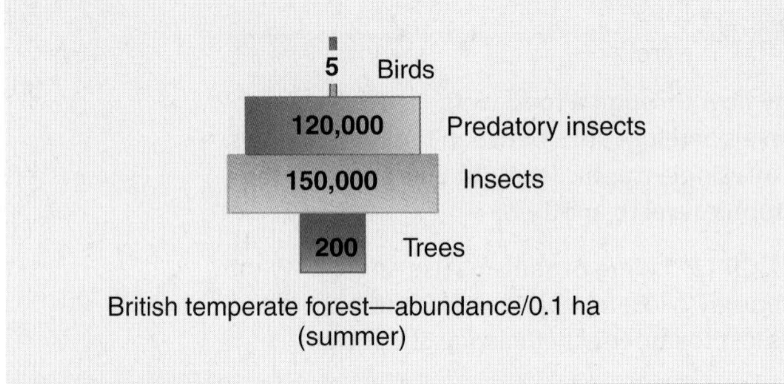

(b) Inverted pyramid of numbers

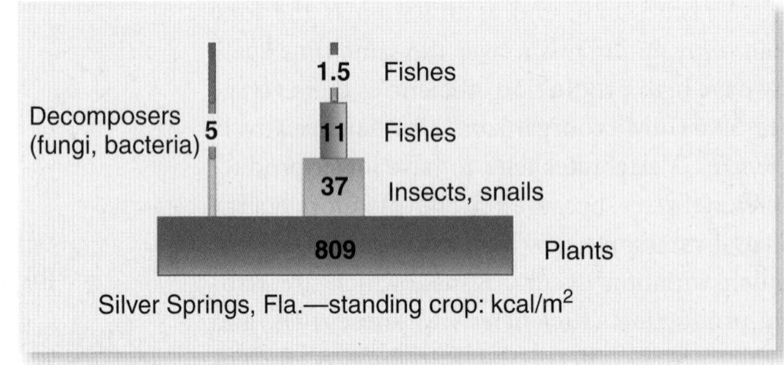

(c) Pyramid of biomass

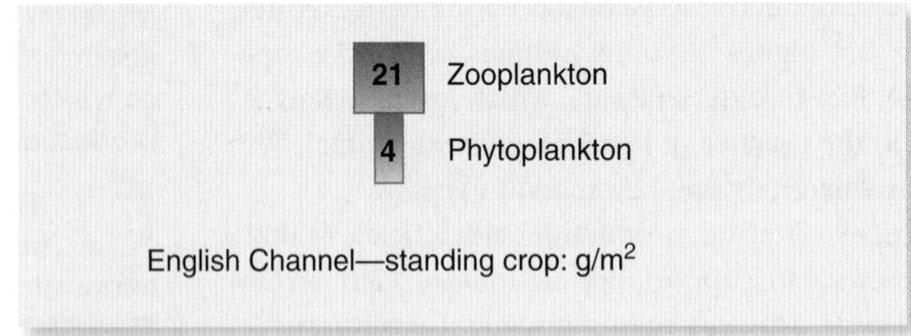

(d) Inverted pyramid of biomass

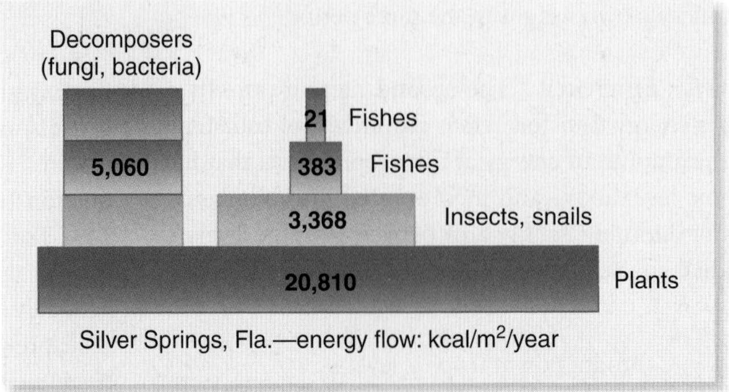

(e) Pyramid of energy

Figure 59.7 Ecological pyramids in food webs. **(a)** In this pyramid of numbers, the abundance of species in an American grassland decreases with increasing trophic level. **(b)** An inverted pyramid of numbers based on organisms living in a British temperate forest. **(c)** When the amount of biological material is used instead of numbers of individuals, the pyramid is termed a pyramid of biomass. Note the presence of decomposers that decompose material at all trophic levels. **(d)** An inverted pyramid of biomass in the English Channel. **(e)** A pyramid of energy for Silver Springs, Florida. Note the large energy production of decomposers, despite their small biomass.

Howard Odum measured the pyramid of biomass for a freshwater ecosystem, Silver Springs, Florida (**Figure 59.7c**). Beds of eelgrass (genus *Sagittaria*) and attached algae make up most of the producers. Insects, snails, herbivorous fishes, and turtles eat the producers. Other fishes form the secondary and tertiary consumers. Odum also noted the presence of fungi and bacteria, which were involved in decomposition on all trophic levels.

Even when biomass is used as the measure, inverted pyramids can still occur, albeit rarely. In some marine and lake systems, the biomass of phytoplankton supports a higher biomass of zooplankton (**Figure 59.7d**). This is possible because the rate of production of phytoplankton biomass is much higher than that of zooplankton, and the small phytoplankton standing crop processes large amounts of energy.

However, by expressing the pyramid in terms of production rate, it is no longer inverted. The **pyramid of energy**, which shows the rate of energy production rather than standing crop, is never inverted (**Figure 59.7e**). The laws of thermodynamics ensure that the highest amounts of free energy are found at the lowest trophic levels. The energy pyramid for Silver Springs also shows that large amounts of energy pass through decomposers, despite their relatively small biomass.

Biomagnification Can Occur in Higher Trophic Levels

The tendency of certain chemicals to concentrate in higher trophic levels in food chains, a process called **biomagnification**, presents a problem for certain organisms. The passage of DDT in food chains provides a startling example.

Dichlorodiphenyltrichloroethane (DDT) was first synthesized by chemists in 1874. In 1939, its insecticidal properties were recognized by Paul Müller, a Swiss scientist who won the 1948 Nobel Prize in Physiology or Medicine for his discovery and subsequent research on the uses of the chemical. The first important application of DDT was in human health programs during and after World War II, particularly as a means of controlling mosquito-borne malaria; at that time, its use in agriculture also began. The global production of DDT peaked in 1970, when 175 million kg of the insecticide was manufactured.

DDT has several chemical and physical properties that profoundly influence the nature of its ecological effect. First, DDT is persistent in the environment. It is not rapidly degraded to other, less toxic chemicals by microorganisms or by physical agents such as light and heat. The typical persistence in soil of DDT is about 10 years, which is two to three times longer than the persistence of most other insecticides. Another important characteristic of DDT is its low solubility in water and its high solubility in fats or lipids. In the environment, most lipids are present in living tissue. Therefore, because of its high lipid solubility, DDT tends to concentrate in biological tissues.

Because biomagnification occurs at each step of the food chain, organisms at higher trophic levels can amass especially high concentrations of DDT in their lipids. A typical pattern of biomagnification is illustrated in **Figure 59.8**, which shows the relative amounts of DDT found in a Lake Michigan food chain. The highest concentration of the insecticide was found in gulls, tertiary consumers that feed on fishes, which are the secondary consumers that eat small insects. An unanticipated effect of DDT on bird species was its interference with the metabolic process of eggshell formation. The result was thin-shelled eggs that often broke under the weight of incubating birds (**Figure 59.9**). DDT was responsible for a dramatic decrease in the populations of many birds due to failed reproduction. Relatively high levels of the chemical were also found to be present in some game fishes, which became unfit for human consumption.

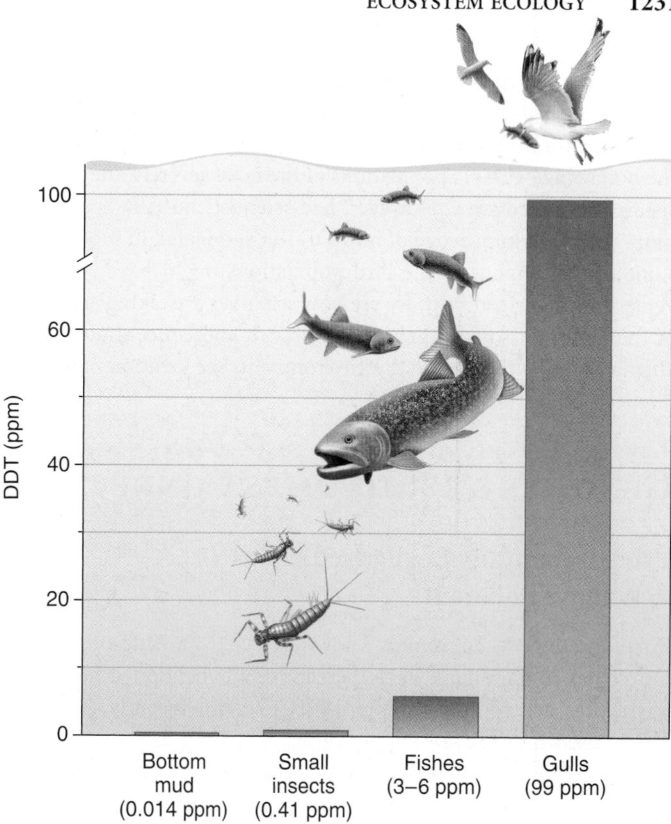

Figure 59.8 **Biomagnification in a Lake Michigan food chain.** The DDT tissue concentration in gulls, a tertiary consumer, was about 240 times that in the small insects sharing the same environment. The biomagnification of DDT in lipids causes its concentration to increase at each successive link in the food chain.

Cl
Cl
H
C
CCl_3
DDT (dichlorodiphenyltrichloroethane)
• Persists in environment
• High solubility in lipids
• Found in high concentrations at higher trophic levels

Figure 59.9 **Thinning of eggshells caused by DDT.** These ibis eggs are thin-shelled and have been crushed by the incubating adult.

Because of growing awareness of the adverse effects of DDT, most industrialized countries, including the U.S., had banned the use of the chemical by the early 1970s. The good news is that following the outlawing of DDT, populations of the most severely affected bird species have recovered. However, had scientists initially possessed a more thorough knowledge of how DDT accumulated in food chains, some of the damage to the bird populations might have been prevented. As described next, we see how scientists are taking advantage of the ability of some organisms to absorb and concentrate certain chemicals to help clean up the environment (see Genomes and Proteomes Connection).

GENOMES & PROTEOMES CONNECTION

Using Genetically Engineered Plants to Remove Pollutants

Trichloroethylene, commonly known as TCE, is a widespread environmental contaminant. Forty percent of all abandoned hazardous waste sites are contaminated with TCE. Until recently, TCE was widely used in a range of applications, including as a dry cleaning solvent and anesthetic, and continues to be used as a degreasing agent for metal parts. It is a human carcinogen and causes a range of neurological effects.

Scientists have been searching for safe ways to remove TCE and other toxic chemicals from hazardous waste sites. One approach, termed bioremediation, involves the use of plants that take up pollutants. After taking up the chemicals, the plants are harvested and the site replanted until pollutants have been reduced to environmentally safe levels. Recently, scientists have genetically engineered poplar trees to efficiently remove TCE from the environment. Gene products of mammalian cytochrome genes oxidize a range of compounds, including TCE. In a 2007 study, American ecologist Sharon Doty and colleagues incorporated the gene that produces cytochrome in rabbit livers into poplar trees. These genetically engineered trees were found to metabolize TCE and other pollutants nearly one hundred times faster than unaltered poplars. If used in the field, the trees would be cut down before flowering so there would be no chance of pollination with wild relatives.

59.2 Biomass Production in Ecosystems

Learning Outcomes:

1. Explain how biomass production is calculated.
2. Explain the different ways in which primary production is influenced in terrestrial and aquatic ecosystems.
3. Describe the factors that limit secondary production in ecosystems.

In this section, we will take a closer look at biomass production in ecosystems. Because the bulk of the Earth's biosphere, 99.9% by mass, consists of producers, when we measure ecosystem biomass production, we are primarily interested in plants, algae, or cyanobacteria. Because these photosynthetic organisms represent the first, or primary, trophic level, their production is called **gross primary production (GPP)**. Gross primary production is equivalent to the carbon fixed during photosynthesis. **Net primary production (NPP)** is GPP minus the energy used during cellular respiration (R) of photosynthetic organisms.

$$NPP = GPP - R$$

NPP is thus the amount of energy available to primary consumers. Unless otherwise noted, the term **primary production** refers to NPP.

Primary Production Is Influenced in Terrestrial Ecosystems by Water, Temperature, and Nutrient Availability

In terrestrial systems, water is a major determinant of primary production, and primary production shows an almost linear increase with annual precipitation, at least in arid regions. Likewise, temperature, which affects production primarily by slowing or accelerating plant metabolic rates, is also important. American ecologist Michael Rosenzweig noted that, on a logarithmic scale, the evapotranspiration rate could predict the aboveground primary production with good accuracy in North America (Figure 59.10). Recall from Chapter 58 that the evapotranspiration rate measures the amount of water entering the atmosphere through the processes of evaporation from the soil and transpiration of plants, so it is a measure of both temperature and available water. For example, a desert will have a low

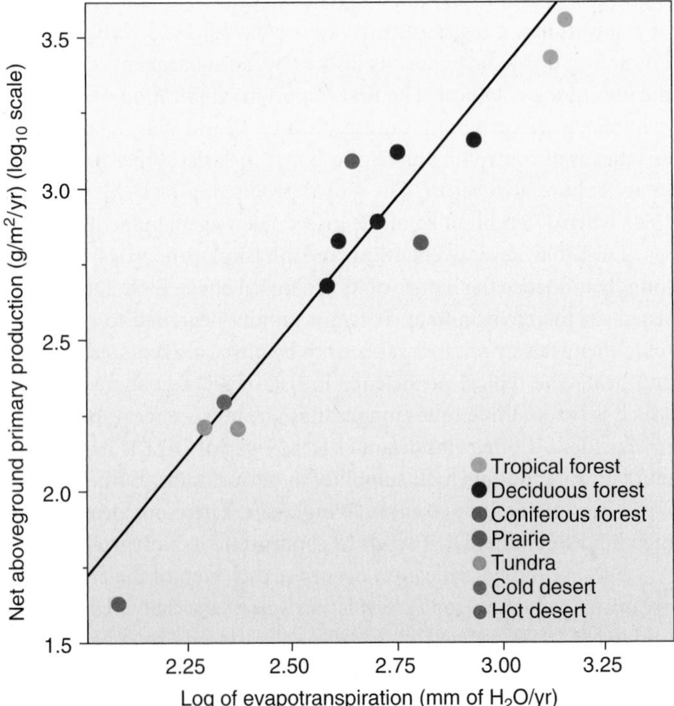

Figure 59.10 **Positive correlation between the evapotranspiration rate and primary production.** Warm, humid environments are ideal for plant growth. Dots represent different ecosystems.

evapotranspiration rate because water availability is low despite high temperature. The evapotranspiration rate is maximized when both temperature and moisture are at high levels, as in tropical rain forests.

A lack of **nutrients**, key elements in usable form, particularly nitrogen and phosphorus, can also limit primary production in terrestrial ecosystems, as farmers know only too well. Fertilizers are commonly used to boost the production of annual crops. In 1984, Susan Cargill and Robert Jefferies showed how a lack of both nitrogen and phosphorus was limiting to salt marsh sedges and grasses in subarctic conditions in Hudson Bay, Canada (**Figure 59.11**). Of the two nutrients, nitrogen was the **limiting factor**, the one in the shortest supply for growth; without it, the addition of phosphorus did not increase production. However, once nitrogen was added and was no longer limiting, phosphorus became the limiting factor. The addition of nitrogen and phosphorus together increased production the most. This result supports a principle known as **Liebig's law of the minimum**, named for Justus von Liebig, a 19th-century German chemist, which states that species biomass or abundance is limited by the scarcest factor. This factor can change, as the Hudson Bay experiment showed. When sufficient nitrogen is available, phosphorus becomes the limiting factor. Once phosphorus becomes abundant, then productivity will be limited by another nutrient.

Primary Production in Aquatic Ecosystems Is Limited Mainly by Light and Nutrient Availability

Of the factors limiting primary production in aquatic ecosystems, the most important are available light and available nutrients. Light is particularly likely to be in short supply because water readily absorbs light. At a depth of 1 m, more than half the solar radiation has been absorbed. By 20 m, only 5–10% of the radiation remains. The decrease in light is what limits the depth of algal growth.

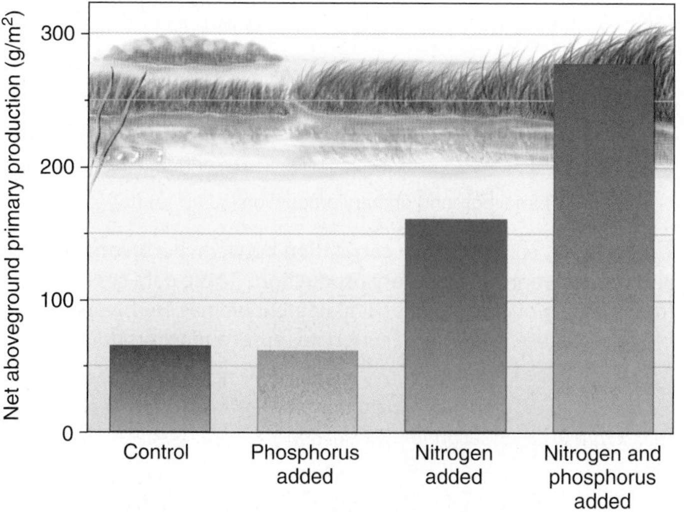

Figure 59.11 Limitation of primary production by nitrogen and phosphorus. Net aboveground primary production of a salt marsh sedge (*Carex subspathacea*) in response to nutrient addition. Nitrogen is the limiting factor. After nitrogen is added, phosphorus becomes the limiting factor.

The most important nutrients affecting primary production in aquatic systems are nitrogen and phosphorus, because they occur in very low concentrations. Whereas soil contains about 0.5% nitrogen, seawater contains only 0.00005% nitrogen. Enrichment of the aquatic environment by the addition of nitrogen and phosphorus occurs naturally in areas of upwellings, where cold, deep, nutrient-rich water containing sediment from the ocean floor is brought to the surface by strong currents, resulting in very productive ecosystems and plentiful fishes. Some of the largest areas of upwelling occur in the Antarctic and along the coasts of Peru and California. However, as noted in Section 37.2, too much nutrient supply can be harmful to aquatic systems, resulting in large, unchecked growths of algae called algal blooms. When the algae die, they are consumed by bacteria that, as they respire, deplete the surrounding water of oxygen, causing dead zones with little oxygen to support other aquatic life. Such dead zones are prominent along coastal areas where fertilizer-rich rivers discharge into the oceans.

Primary Production Is Greatest in Areas of Abundant Warmth and Moisture

Knowing which factors limit primary production helps ecologists understand why the mean net primary production varies across the different biomes on Earth. Modern methods of estimating productivity use orbiting satellites to measure differences in the electromagnetic radiation reflected back from the vegetation of different ecosystems on Earth (**Figure 59.12**). When we look at the oceans, bright greens, yellows, and reds indicate high chlorophyll concentrations. Some of the highest marine chlorophyll concentrations occur at continental margins, where river nutrients pour into the oceans. Upwellings along coasts also bring nutrient-rich water to the surface. Northern oceans, and to a lesser extent southern oceans, are also very productive because seasonal temperature variations allow vertical mixing of all layers, bringing nutrient-rich water to the surface. In the spring, the high light and nutrients permit rapid phytoplankton growth until the nutrients are all used up. Many other marine areas, including tropical oceans, are highly unproductive.

Over land, productivity is measured as the Normalized Difference Vegetation Index or NDVI, which is an estimate of the photosynthetically absorbed radiation over land surfaces. Plants absorb much visible light but reflect light at near-infrared wavelengths. This difference in absorption allows scientists to estimate photosynthetic rates and hence primary production. The productivity of forests from all parts of the world, from the tropics to northern temperate areas, is similar, but production is often higher in temperate than tropical habitats. This matches the pattern of productivity observed in the oceans. Although tropical forests enjoy warm temperatures and abundant rainfall, such conditions weather soils rapidly. Tropical soils are low in available forms of most plant nutrients because of loss through leaching. In contrast, temperate soils tend to have much greater concentrations of essential nutrients because of lower rates of nutrient loss and more frequent grinding of fresh minerals by the cycles of continental glaciations over the past 3 million years. Prairies and savannas are also highly productive because their plant biomass usually dies and decomposes each year, returning a portion of the nutrients to the soil, and temperatures and rainfall are not limiting.

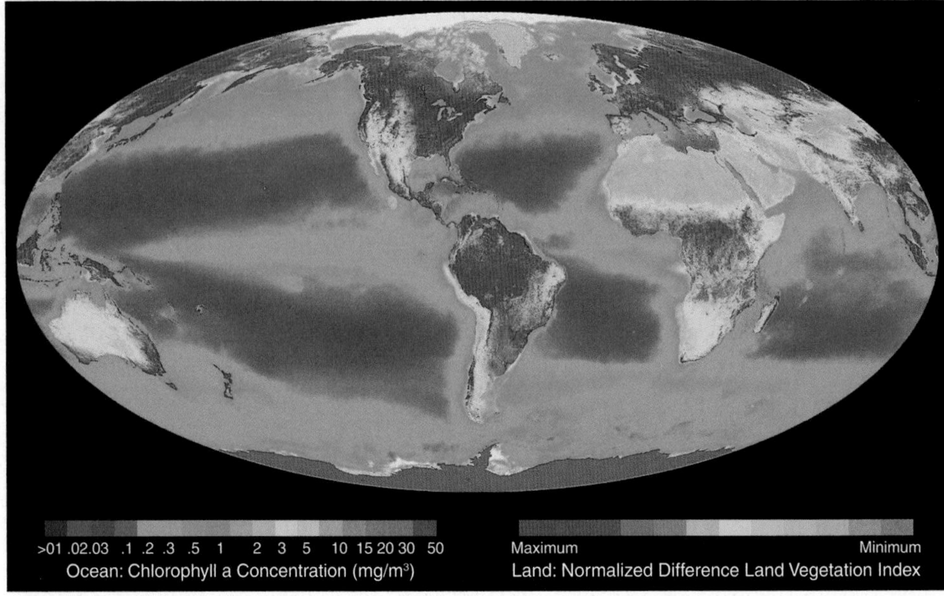

Figure 59.12 Primary productivity measured by satellite imagery. Ocean chlorophyll concentrations and the Normalized Difference Vegetation Index (NDVI) on land provide good data on marine and terrestrial productivity, respectively.

Deserts and tundra have low productivity because of a lack of water and low temperatures, respectively. Wetlands tend to be extremely productive, primarily because water is not limiting and nutrient levels are high.

Secondary Production Is Generally Limited by Available Primary Production

What factors control **secondary production**—the productivity of herbivores, carnivores, and decomposers? This is a complex question, but it is generally thought to be limited largely by available primary production. A strong relationship exists between primary production in a variety of biomes and the biomass of herbivores (**Figure 59.13**). This means that more plant biomass, and thus more primary production, leads to an increased biomass of consumers. This is not such a trivial answer as might be assumed; for example, secondary production could be limited by the availability of a particular nutrient or by the presence of natural enemies.

As we have noted before, trophic-level transfer efficiency averages about 10%. Thus, after one link in the food web, only 1/10 of the energy captured by plants is transferred to herbivores, and after two links in the food web, only 1/100 of the energy fixed by plants goes to carnivores. Thus, secondary production is generally much smaller than that of primary production. In 1962, American ecologist John Teal examined energy flow in a Georgia salt marsh (**Figure 59.14**). In salt marshes, most of the energy from the Sun goes to two types of organisms: *Spartina* plants and marine algae. The *Spartina* plants are rooted in the ground, whereas the algae float on the water surface or live on the mud or on *Spartina* leaves at low tide. These photosynthetic organisms absorb about 6% of the sunlight. Most of the plant energy, 77.6%, is used in plant and algal cellular respiration. Of the energy that is accumulated in plant biomass, 22.4%, most dies in place

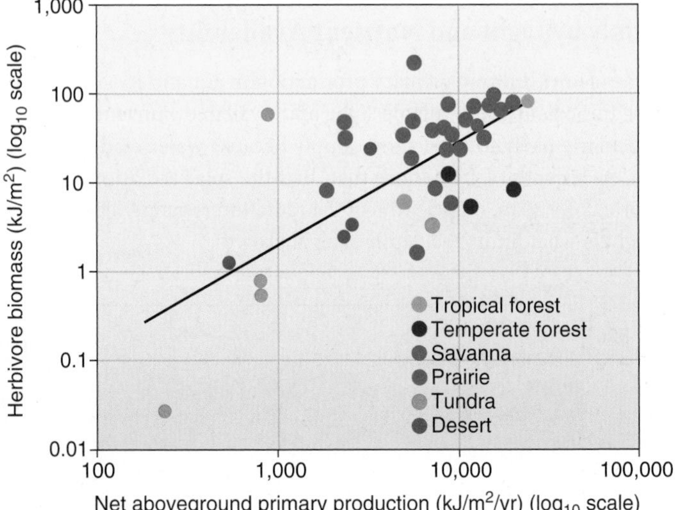

Figure 59.13 A positive correlation between herbivore biomass and net aboveground primary production. These data are taken from a variety of case studies from different biomes. Herbivore biomass can be considered a surrogate for secondary production.

Concept Check: *What does this relationship imply about the effects of plant secondary metabolites, many of which taste bad and some of which are toxic, on secondary production?*

and rots on the muddy ground, to be consumed by bacteria. Bacteria are the major decomposers in this system, followed distantly by nematodes and crabs, which feed on tiny food particles as they sift through the mud. Some of this dead material is also removed from the system (exported) by the tide. The herbivores take very little of the plant production, around 0.6%, eating only a small proportion of the *Spartina* and none of the algae. A fraction of herbivore biomass

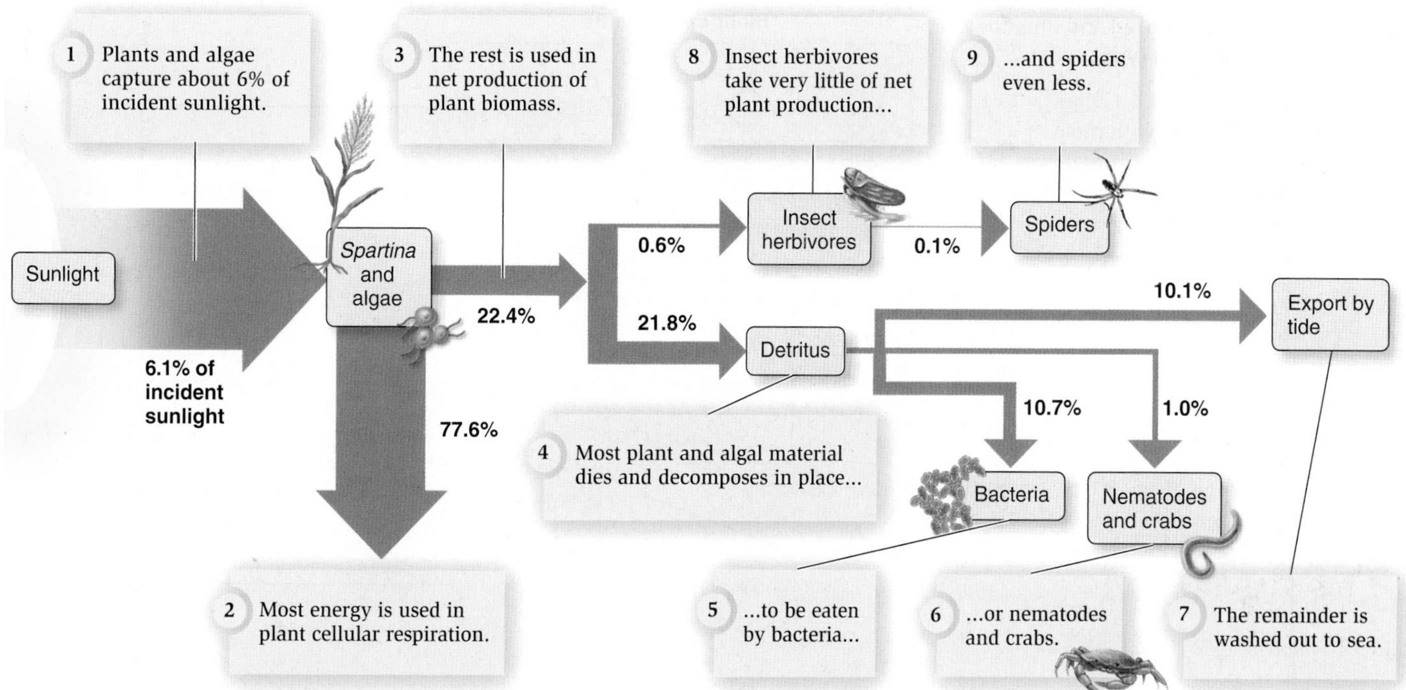

Figure 59.14 **Energy-flow diagram for a Georgia salt marsh.** Numbers represent the percentage of gross primary production that flows into different trophic levels or is used in plant respiration.

is then consumed by spiders. Overall, if we view the species in ecosystems as transformers of energy, then plants and algae are by far the most important organisms on the planet, bacteria are next, and animals are a distant third.

59.3 Biogeochemical Cycles

Learning Outcomes:

1. Describe the phosphorus cycle and the causes of eutrophication.
2. Outline the steps of the carbon cycle and the environmental effects of elevated atmospheric concentrations of CO_2.
3. List the five main steps of the nitrogen cycle and the human influences on it.
4. Describe the processes of the water cycle and how they are affected by humans.

As we have seen, a unit of energy moves through an ecosystem only once, passing through the trophic levels of a food web from producer to consumer and dissipating as heat. In contrast, chemical elements such as carbon or nitrogen are recycled, moving from the physical environment to organisms and back to the environment, where the cycle begins again. Whereas an ecosystem constantly receives energy in the form of light, chemical elements are available in limited amounts and are recycled. Because the movements of chemicals through ecosystems involve biological, geological, and chemical transport mechanisms, they are termed **biogeochemical cycles.** Biological mechanisms involve the absorption of chemicals by living organisms and their subsequent release back into the environment. Geological mechanisms include weathering and erosion of rocks, and elements transported by surface and subsurface drainage. Chemical transport mechanisms include dissolved matter in rain and snow, atmospheric gases, and dust blown by the wind.

In addition to the basic building blocks of hydrogen, oxygen, and carbon, the elements required in the greatest amounts by living organisms are phosphorus and nitrogen. In this section, we take a detailed look at the cycles of these nutrients. These cycles can be divided into two broad types: (1) local cycles, such as the phosphorus cycle, which involve elements with no atmospheric mechanism for long-distance transfer; and (2) global cycles, which involve an interchange between the atmosphere and the ecosystem. Global nutrient cycles, such as the carbon and nitrogen cycles, unite the Earth and its living organisms into one giant interconnected ecosystem called the **biosphere**. Most biogeochemical cycles involve assimilation of nutrients from the soil by plants and animals, and decomposition of plants and animals, which releases nutrients back into the soil. A generalized and simplified biogeochemical cycle involves both biotic and abiotic components (**Figure 59.15**). In our discussion of biogeochemical cycles, we will take a particular interest in the alteration of these cycles through human activities, such as the burning of fossil fuels, that increase nutrient inputs.

Phosphorus Cycles Locally Between Geological and Biological Components of Ecosystems

All living organisms require phosphorus, which becomes incorporated into ATP, the compound that provides energy for most metabolic processes. Phosphorus is a key component of other biological molecules such as DNA and RNA, and it is also an essential mineral that in many animals helps maintain a strong, healthy skeleton.

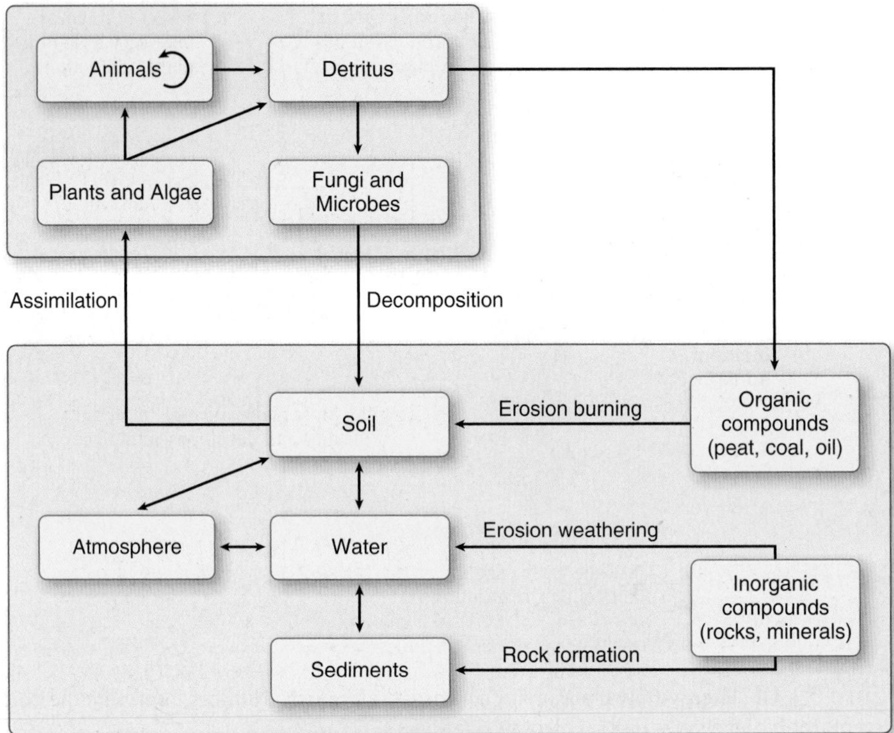

Figure 59.15 A generalized and simplified model of a biogeochemical cycle.

BIOLOGY PRINCIPLE Living organisms interact with their environments. The biomass of organisms is affected by the availability of nutrients in the soil.

The phosphorus cycle is a relatively simple cycle (Figure 59.16). Phosphorus has no gaseous phase and thus no atmospheric component; that is, it is not moved by wind or rain. As a result, phosphorus tends to cycle only locally. The Earth's crust is the main storehouse for this element. Weathering and erosion of rocks release phosphorus into the soil. Plants have the metabolic means to absorb dissolved ionized forms of phosphorus, the most important of which occurs as phosphate (HPO_4^{2-} or $H_2PO_4^-$). Herbivores obtain their phosphorus only from eating plants, and carnivores obtain it by eating herbivores. When plants and animals excrete wastes or die, the phosphorus becomes available to decomposers, which release it back to the soil.

Leaching and runoff eventually wash much phosphate into aquatic systems, where plants and algae utilize it. Phosphate that is not taken up into the food chain settles to the ocean floor or lake bottom, forming sedimentary rock. Phosphorus can remain locked in sedimentary rock for millions of years, becoming available again through the geological process of uplift.

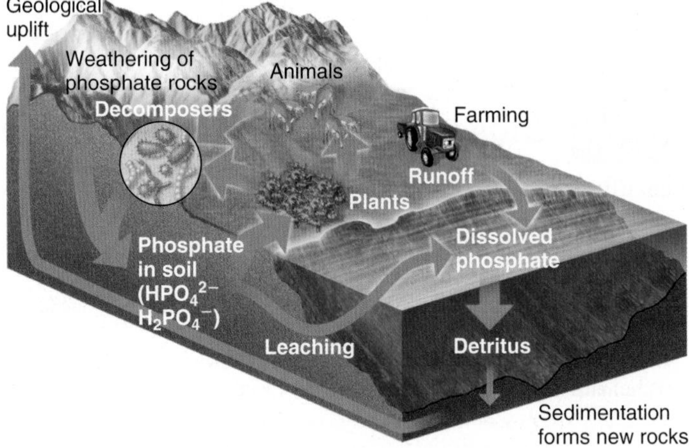

Figure 59.16 The phosphorus cycle. Unlike other major biogeochemical cycles, the phosphorus cycle does not have an atmospheric component and thus cycles only locally. The widths of the lines indicate the relative contribution of each process to the cycle.

Human Influences on the Phosphorus Cycle Plants can take up phosphate so rapidly and efficiently that they often reduce soil concentrations of phosphorus to extremely low levels, so phosphorus becomes a limiting factor, as noted previously (see Figure 59.11). As more phosphorus is added to an aquatic ecosystem, the production of algae and aquatic plants increases. In a pivotal 1974 study, Canadian biologist David Schindler showed that an overabundance of phosphorus caused the rapid growth of algae and plants in an experimental lake in Canada (Figure 59.17). What is the consequence of the rapid growth of algae and plants? When the algae and plants die, they sink to the bottom, where bacteria decompose them and consume the dissolved oxygen in the water. Dissolved oxygen concentrations can drop too low for fishes to breathe, killing them. The process by which elevated nutrient levels lead to an overgrowth of algae and the subsequent depletion of water oxygen concentrations is known as **eutrophication**. Cultural eutrophication refers to the enrichment of water with nutrients derived from human activities, such as fertilizer use and sewage dumping.

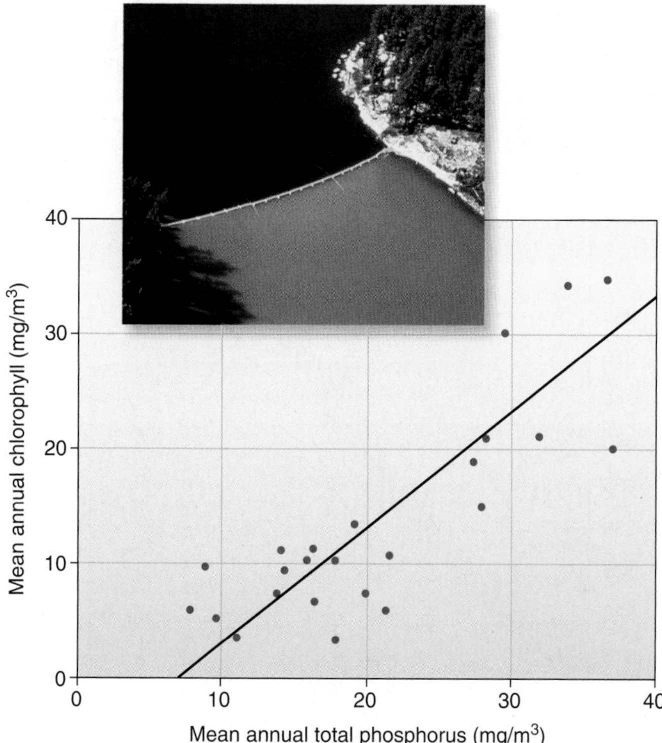

Figure 59.17 The relationship between primary production and total phosphorus concentration. As shown in this graph, primary production (measured by chlorophyll concentration) increases linearly with an increase in phosphorus. Each dot represents a different lake. The aerial photograph shows the contrast in water quality of two basins of an experimental lake in Canada separated by a plastic curtain. Carbon and nitrogen were added to the upper basin, and carbon, nitrogen, and phosphorus were added to the lower basin. The bright green color is from a surface film of algae that resulted from the added phosphorus.

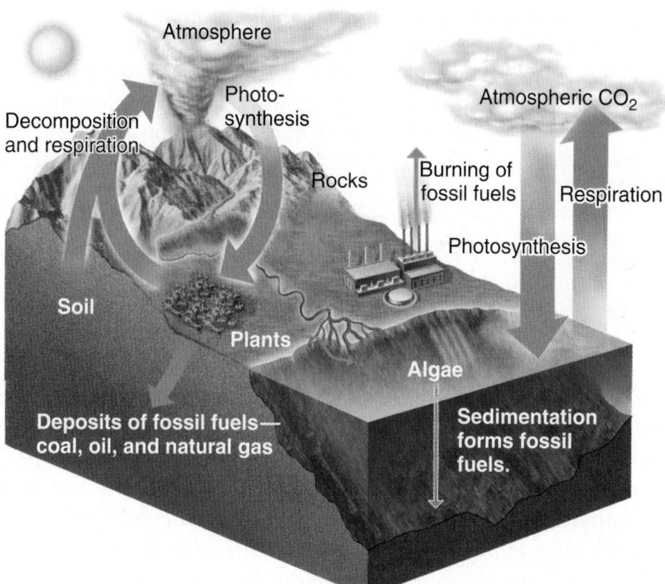

Figure 59.18 The carbon cycle. Each year, plants and algae remove about one-seventh of the CO_2 in the atmosphere. Animal respiration is so small it is not represented. The width of the arrows indicates the relative contribution of each process to the cycle.

Concept Check: Where are the greatest stores of global carbon?

BioConnections: Refer back to Table 2.2. Carbon is one of just four elements that account for the vast majority of atoms in living organisms. What are the other three and, therefore, what biogeochemical cycles might be the most important to us?

Lake Erie became eutrophic in the 1960s due to the runoff of fertilizer rich in phosphorus from farms and to the industrial and domestic pollutants released from the many cities along its shores. Fish species such as white fish and lake trout became severely depleted. Based on research such as Schindler's that showed the dramatic effect of phosphorus on a lake system, the U.S. and Canada teamed together to reduce the levels of discharge by 80%, primarily through eliminating phosphorus in laundry detergents and maintaining strict controls on the phosphorus content of wastewater from sewage treatment plants. Fortunately, lake systems have great potential for recovery after phosphorous inputs are reduced, and Lake Erie has experienced fewer algal blooms, clearer water, and a restoration of fish populations.

Carbon Cycles Among Biological, Geological, and Atmospheric Pools

The movement of carbon from the atmosphere into organisms and back again is known as the carbon cycle (**Figure 59.18**). Carbon dioxide (CO_2) is present in the atmosphere at a level of about 395 parts per million (ppm), or about 0.04%. Autotrophs, primarily plants, algae, and cyanobacteria, acquire CO_2 from the atmosphere or water and incorporate it into the organic matter of their own biomass via photosynthesis. Each year, plants, algae, and cyanobacteria remove approximately one-seventh of the CO_2 from the atmosphere. At the same time, respiration and the decomposition of plants recycle a similar amount of carbon back into the atmosphere as CO_2. Much material from primary producers is also transformed into deposits of coal, gas, and oil, which are collectively known as **fossil fuels**. Herbivores can return some CO_2 to the atmosphere, eating plants and breathing out CO_2, but the amount flowing through this part of the cycle is minimal. Chemical processes such as diffusion and absorption of CO_2 into and out of oceans also contribute to changes in atmospheric CO_2.

Over time, much carbon is also incorporated into the shells of marine organisms, which eventually form huge limestone deposits on the ocean floor or in terrestrial rocks, where turnover is extremely low. As a result, rocks and fossil fuels contain the largest reserves of carbon. Natural sources of CO_2 such as volcanoes, hot springs, and fires release large amounts of CO_2 into the atmosphere. In addition, human activities, primarily deforestation and the burning of fossil fuels, are increasingly causing large amounts of CO_2 to enter the atmosphere together with large volumes of particulate matter.

Human Influences on the Carbon Cycle Direct measurements over the past five decades show a steady rise in atmospheric CO_2

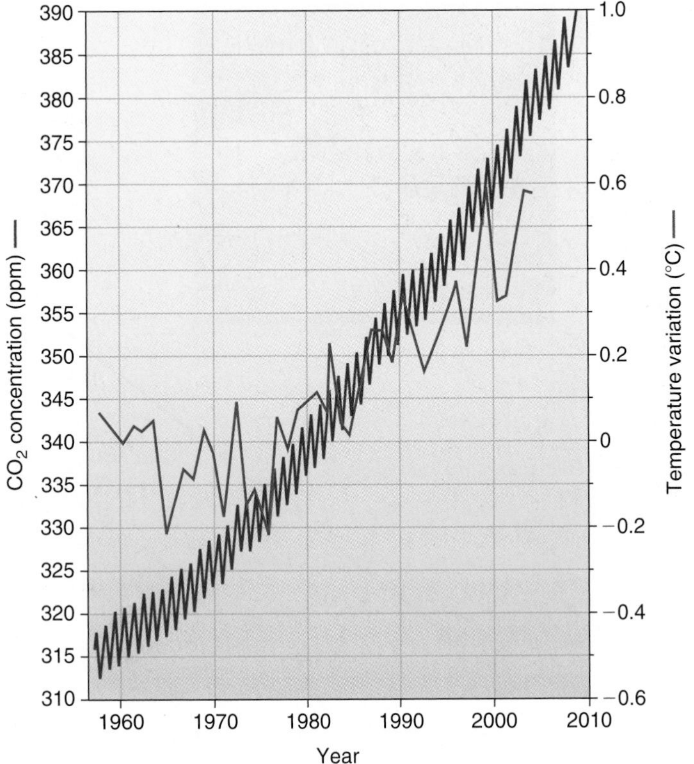

Figure 59.19 **The increase in atmospheric CO_2 levels and temperatures due to the burning of fossil fuels.** From 1958 to 2008, atmospheric CO_2 shows an increase of nearly 20%. In addition, the graph shows a seasonal variation in CO_2 (shown in blue). Temperatures are annual deviations from the 1961–1990 average (shown in red). Measurements were recorded at Mauna Loa Observatory in Hawaii.

Concept Check: *Why does the amount of CO_2 fluctuate seasonally in the graph?*

BioConnections: *Refer back to Figures 29.17 and 22.9. The Earth's atmospheric CO_2 concentration has fluctuated dramatically over time. At what point, and in which period, was it highest?*

(**Figure 59.19**), a pattern that shows no sign of slowing. Because of its increasing concentration in the atmosphere, CO_2 is the most troubling of the greenhouse gases, which are a primary cause of global warming (refer back to Table 54.2). Elevated atmospheric CO_2 has other dramatic environmental effects, boosting plant growth but lowering the amount of herbivory (see Feature Investigation).

The amount of CO_2 in the atmosphere shows a seasonal variation, as can be seen in Figure 59.19 (blue line). Concentrations of atmospheric CO_2 are lowest during the Northern Hemisphere's summer and highest during the winter, when photosynthesis is minimal. This phenomenon occurs in both of the Earth's hemispheres. Because there is more land in the Northern than in the Southern Hemisphere, and therefore more vegetation, concentrations of atmospheric CO_2 are lowest during the Northern Hemisphere's summer. The vegetation has a maximum photosynthetic activity during the summer, reducing the global amount of CO_2. During the Northern Hemisphere's winter, photosynthesis is low, and decomposition is relatively high, causing a global increase in the gas.

Feature Investigation

Stiling and Drake's Experiments with Elevated CO_2 Showed an Increase in Plant Growth but a Decrease in Herbivory

How will forests of the future respond to elevated CO_2? To begin to answer such a question, ecologists ideally would enclose large areas of forests with chambers, increase the CO_2 content within the chambers, and measure the responses. This has proven to be difficult for two reasons. First, it is hard to enclose large trees in chambers, and second, it is expensive to increase CO_2 levels over such a large area. However, in a discovery-based investigation, ecologists Peter Stiling and Bert Drake were able to increase CO_2 levels around small patches of forest at the Kennedy Space Center in Cape Canaveral, Florida. In much of Florida's forests, trees are small, only 3–5 m when mature, because frequent lightning-initiated fires prevent the growth of larger trees. In the 1990s, Stiling and Drake teamed up with NASA engineers to create 16 circular, open-topped chambers (**Figure 59.20**). In eight of these they increased atmospheric CO_2 to double their ambient levels, from 360 ppm to 720 ppm, the latter of which is the atmospheric concentration predicted by the end of the 21st century. The experiments commenced in 1996 and lasted until 2007. Plants produced more biomass in elevated CO_2, because CO_2 is limiting to plant growth, but the data revealed much more.

Because the chambers were open-topped, insect herbivores could come and go. Insect herbivores cause the largest amount of herbivory in North American forests, because most vertebrate herbivores cannot access the high foliage. Censuses were conducted of all damaged leaves, but focused on leaves damaged by leaf miners, the most common type of herbivore at this site. Leaf miners are small moths whose larvae are small enough to burrow between the surfaces of plant leaves and mine tunnels through the leaves.

Densities of damaged leaves, including those damaged by leaf miners, were lower in elevated CO_2 in every year studied. Part of the reason for the decline was that even though plants increased in mass, the existing soil nitrogen was diluted over a greater volume of plant material, so the nitrogen level in leaves decreased. This increased insect mortality by two means. First, poorer leaf quality directly increased insect death because leaf nitrogen levels may have been too low to support the normal development of the leaf miners. Second, lower leaf quality increased the amount of time insects needed to feed to gain sufficient nitrogen. Increased feeding times, in turn, led to increased exposure to natural enemies, such as predatory spiders and ants, and parasitoids, so top-down mortality also increased (see the data of Figure 59.20). Thus, in a world of elevated CO_2, plant growth may increase, and herbivory could decrease.

Figure 59.20 **The effects of elevated atmospheric CO_2 on insect herbivory.**

GOAL To determine the effects of elevated CO_2 on a forest ecosystem; effects on herbivory are highlighted here.

STUDY LOCATION Patches of forest at the Kennedy Space Center in Cape Canaveral, Florida.

	Experimental level	Conceptual level
1 Erect 16 open-top chambers around native vegetation. Increase CO_2 levels from 360 ppm to 720 ppm in half of them.		Expected atmospheric CO_2 level is 720 ppm by end of the 21st century. Open-top chambers allow movement of herbivores in and out of chambers.
2 Conduct a yearly count of numbers of insect herbivores per 200 leaves in each chamber.	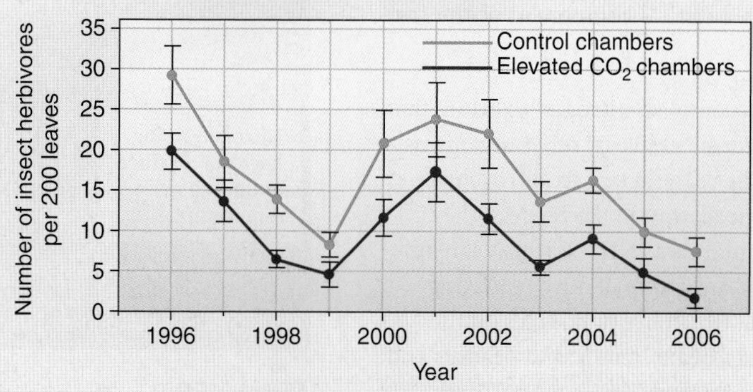	
3 Count number of herbivores that died due to nutritional inadequacy. Monitor attack rates on insect herbivores by natural enemies such as predators and parasitoids.	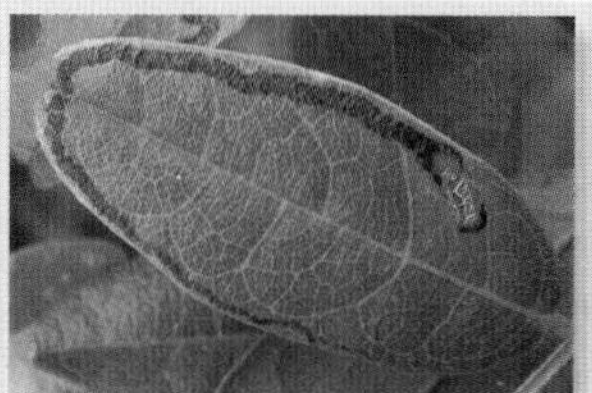	Elevated CO_2 reduces foliar nitrogen, inhibits normal insect development, and prolongs the feeding time of herbivores, allowing natural enemies greater opportunities to attack them.

4 THE DATA

Source of mortality*	Elevated CO_2 (% mortality)	Control (% mortality)
Nutritional inadequacy	10.2	5.0
Predators	2.4	2.0
Parasitoids	10.0	3.2

*Data refer only to mortality of larvae within leaves and do not sum to 100%. Mortality of eggs on leaves, pupae in the soil, and flying adults is unknown.

5 CONCLUSION Elevated CO_2 decreases insect herbivory in a Florida forest.

6 SOURCE Stiling, P., and Cornelissen, T. 2007. How does elevated carbon dioxide (CO_2) affect plant-herbivore interactions? A field experiment and meta-analysis of CO_2-mediated changes on plant chemistry and herbivore performance. *Global Change Biology* 13:1823–1842.

Experimental Questions

1. What was the hypothesis of Stiling and Drake's experiment?
2. What was the purpose of increasing the CO_2 levels in only half of the chambers in the experiment and not all of the chambers?
3. What were the results of the experiment?

The Nitrogen Cycle Is Strongly Influenced by Biological Processes That Transform Nitrogen into Usable Forms

Nitrogen is an essential component of proteins, nucleic acids, and chlorophyll. Because 78% of the Earth's atmosphere consists of nitrogen gas (N_2), it may seem that nitrogen should not be in short supply for organisms. However, nitrogen is often a limiting factor in ecosystems because N_2 molecules must be broken apart before the individual nitrogen atoms are available to combine with other elements. Because of its triple bond, N_2 is very stable, and only certain bacteria can break it apart into usable forms such as ammonia (NH_3). This process, called nitrogen fixation, is a critical component of the five-part nitrogen cycle (Figure 59.21):

1. A few species of bacteria can accomplish **nitrogen fixation**, that is, convert atmospheric N_2 to forms usable by other organisms (refer back to Figure 37.18). The bacteria that fix nitrogen are fulfilling their own metabolic needs, but in the process, they release ammonia (NH_3) or ammonium (NH_4^+), which can be used by some plants. An important group of nitrogen-fixing bacteria are known as rhizobia, which live in nodules on the roots of legumes, including peas, beans, lentils, and peanuts and some woody plants. In more natural systems, such as forests and savannas, nitrogen-fixing bacteria such as *Frankia* form a symbiosis with actinorhizal plants. Cyanobacteria are important nitrogen fixers in aquatic systems.
2. In the process of **nitrification**, soil bacteria convert NH_3 or NH_4^+ to nitrate (NO_3^-), a form of nitrogen commonly used by plants. The bacteria *Nitrosomonas* and *Nitrococcus* first oxidize the forms of ammonia to nitrite (NO_2^-), after which the bacteria *Nitrobacter* converts NO_2^- to NO_3^-.
3. **Assimilation** is the process by which inorganic substances are incorporated into organic molecules. In the nitrogen cycle, organisms assimilate nitrogen by taking up NH_3, NH_4^+, and NO_3^- formed through nitrogen fixation and nitrification and incorporating them into other molecules. Plant roots take up these forms of nitrogen through their roots, and animals assimilate nitrogen from plant tissue.
4. Ammonia can also be formed in the soil through the decomposition of plants and animals and the release of animal waste. **Ammonification** is the conversion of organic nitrogen to NH_3 and NH_4^+. This process is carried out by bacteria and fungi. Most soils are slightly acidic and, because of an excess of H^+, the NH_3 rapidly gains an additional H^+ to form NH_4^+. Because many soils lack nitrifying bacteria, ammonification is the most common pathway for nitrogen to enter the soil.
5. **Denitrification** is the reduction of NO_3^- to N_2. Denitrifying bacteria, which are anaerobic and use NO_3^- in their metabolism instead of O_2, perform the reverse of their nitrogen-fixing counterparts by delivering N_2 to the atmosphere. This process delivers only a relatively small amount of nitrogen to the atmosphere.

Human Influences on the Nitrogen Cycle Human alterations of the nitrogen cycle have approximately doubled the rate of nitrogen input to the cycle. Industrial fixation of nitrogen for the production of fertilizer makes a significant contribution to the pool of nitrogen-containing material in the soils and waters of agricultural regions. Fertilizer runoff can cause eutrophication of rivers and lakes, and, as the resultant algae die, decomposition by bacteria depletes the oxygen level of the water, resulting in fish kills. Excess NO_3^- in surface or groundwater systems used for drinking water are also a health hazard, particularly for infants. In the body, NO_3^- is converted to NO_2^-, which then combines with hemoglobin to form methemoglobin, a type of hemoglobin that does not carry oxygen. In infants, the production of large amounts of NO_2^- can cause methemoglobinemia, a dangerous condition in which the level of O_2 carried through the body decreases. Finally, burning fossil fuels releases not only carbon but also nitrogen in the form of nitrous oxide (N_2O), which contributes to air pollution. N_2O can also react with rainwater to form nitric acid (HNO_3), a component of acid rain, which decreases the pH of lakes and streams and increases fish mortality (see Chapter 54).

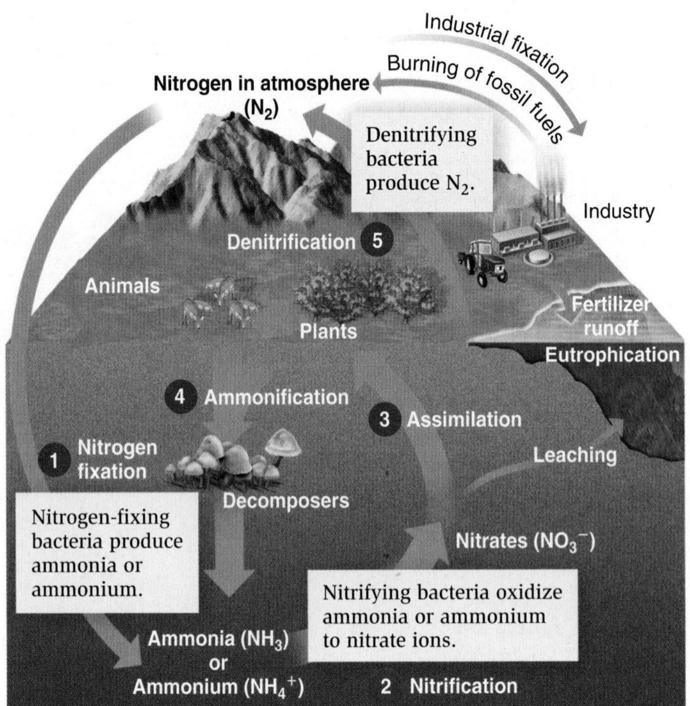

Figure 59.21 **The nitrogen cycle.** The five main parts of the nitrogen cycle are (1) nitrogen fixation, (2) nitrification, (3) assimilation, (4) ammonification, and (5) denitrification. The recycling of nitrogen from dead plants and animals into the soil and then back into plants is of paramount importance because this is the main pathway for nitrogen to enter the soil. The width of the arrows indicates the relative contribution of each process to the cycle.

The dramatic effects of human activities on nutrient cycles in general and the nitrogen cycle in particular were illustrated by a famous long-term study by American ecosystem ecologists Gene Likens, Herbert Bormann, and their colleagues at Hubbard Brook Experimental Forest in New Hampshire in the 1960s. Hubbard Brook is a 3,160-hectare reserve that consists of six catchments along a mountain ridge. A catchment is an area of land where all water eventually drains to a single outlet. In Hubbard Brook, each outlet is fitted with a

(a) Hubbard Brook dam and weir

(b) Hubbard Brook Experimental Forest, New Hampshire

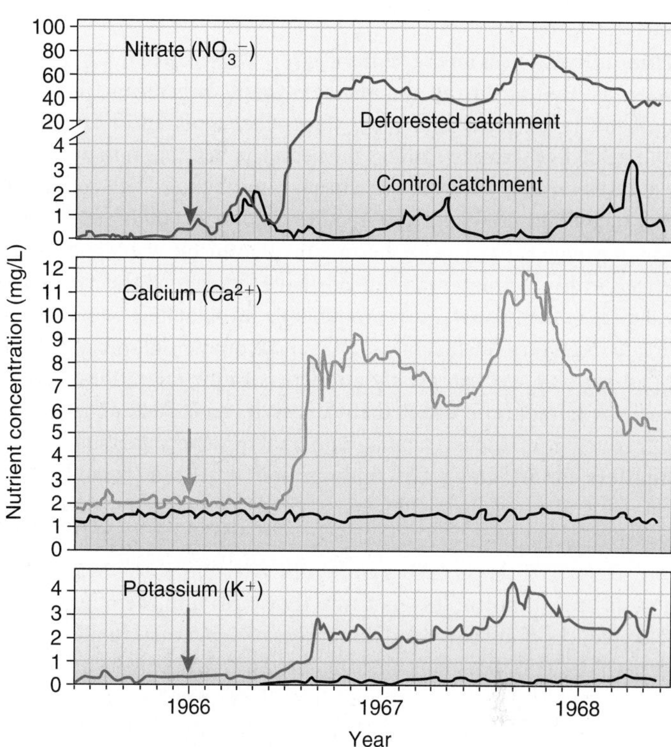

(c) Nutrient concentrations in deforested and control catchments

Figure 59.22 The effects of deforestation on nutrient concentrations. **(a)** Concrete dam, which concentrates water flow and permits accurate measurement of discharge rate, used to monitor nutrient flow from a Hubbard Brook catchment. **(b)** Deforested catchment at Hubbard Brook. **(c)** Nutrient concentrations in stream water from the experimentally deforested catchment and a control catchment at Hubbard Brook. The timing of deforestation is indicated by arrows.

BIOLOGY PRINCIPLE Biology is an experimental science. Although many experiments are performed in the laboratory, some ecological experiments are done on a large scale in the field.

permanent concrete dam that enables researchers to monitor the outflow of water and nutrients (Figure 59.22a). In this large-scale experiment, researchers felled all of the trees in one of the Hubbard Brook catchments (Figure 59.22b). The catchment was then sprayed with herbicides for 3 years to prevent regrowth of vegetation. An untreated catchment was used as a control.

Researchers monitored the concentrations of key nutrients in the flow of water exiting the two catchments for over 3 years. Their results revealed that the overall export of nutrients from the disturbed catchment rose to many times the normal rate (Figure 59.22c). The researchers determined that two phenomena were responsible. First, the enormous reduction in plants reduced water uptake by vegetation and led to 40% more runoff discharged to the streams. This increased outflow caused greater rates of chemical leaching and rock and soil weathering. Second, and more significantly, in the absence of nutrient uptake in spring, when the deciduous trees would have resumed photosynthesis, the inorganic nutrients released by decomposer activity were simply leached in the drainage water. Similar processes operate in terrestrial ecosystems where clearance of forests is significant.

The Water Cycle Is Largely a Physical Process of Evaporation and Precipitation

The water cycle, also called the hydrological cycle, differs from the cycles of other nutrients in that very little of the water that cycles through ecosystems is chemically changed by any of the cycle's components (Figure 59.23). It is a physical process, fueled by the Sun's energy, rather than a chemical one, because it consists of essentially two phenomena: evaporation and precipitation. Even so, the water cycle has important biological components. Over land, 90% of the water that reaches the atmosphere is moisture that has passed through plants and exited from the leaves via evapotranspiration. Only about 2% of the total volume of Earth's water is found in the bodies of organisms or is held frozen or in the soil. The rest cycles between bodies of water, the atmosphere, and the land.

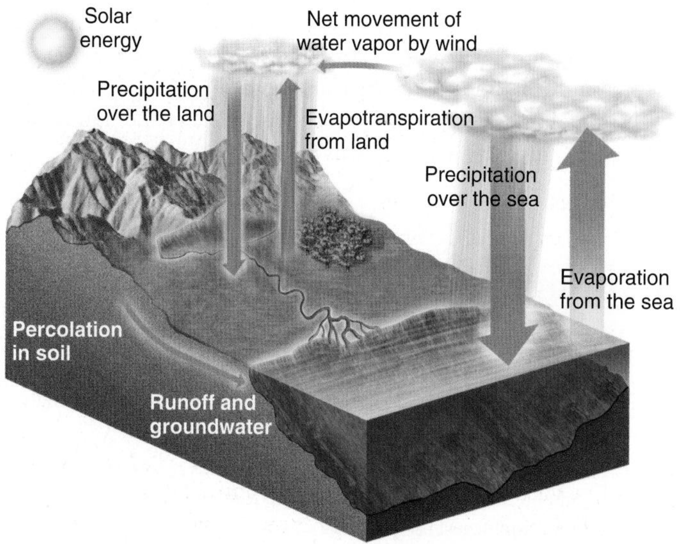

Figure 59.23 The water cycle. This cycle is primarily a physical process, not a chemical one. Solar energy drives the water cycle, causing evaporation of water from the ocean and evapotranspiration from the land. This is followed by condensation of water vapor into clouds and precipitation. The width of the arrows indicates the relative contribution of each step to the cycle.

Human Influences on the Water Cycle As we noted in Chapter 54, water is limiting to the abundance of many organisms, including humans. It takes 228 L of water to produce a pound of dry wheat, and 9,500 L of water to support the necessary vegetation to produce a pound of meat. Industry is also a heavy user of water, with commodities such as oil, iron, and steel requiring up to 20,000 L of water per ton of product. To increase the amount of available water and also to create hydroelectric power, humans have interrupted the hydrological cycle in many ways, most prominently through the use of dams to create reservoirs. Such dams, such as those on the Columbia River in Washington State, can greatly interfere with the migration of fishes such as salmon and affect their ability to reproduce and survive. Other activities, such as tapping into underground water supplies, or **aquifers**, for drinking water removes more water than is put back by rainfall and can cause shallow ponds and lakes to dry up and sinkholes to develop.

Deforestation can also significantly alter the water cycle. When forests are cut down, less moisture transpires into the atmosphere. This reduces cloud cover and diminishes precipitation, subjecting the area to drought. Reestablishing the forests, which calls for increased water, then becomes nearly impossible. Such a problem has occurred on the island of Madagascar, located off the east coast of Africa. In this country, clearing of the forests for cash crops such as cotton, coffee, and tobacco has been so rapid and extensive that areas have become devoid of vegetation (**Figure 59.24**). In Madagascar, as in so many other areas on Earth, deforestation and other environmental degradations are having a deleterious effect on much of the natural habitat. Appropriately, in the following chapter, Chapter 60, we finish our study of ecology in particular, and biology in general, with a discussion of what we gain from ecosystem diversity and how best to conserve the diversity of ecosystems for future generations.

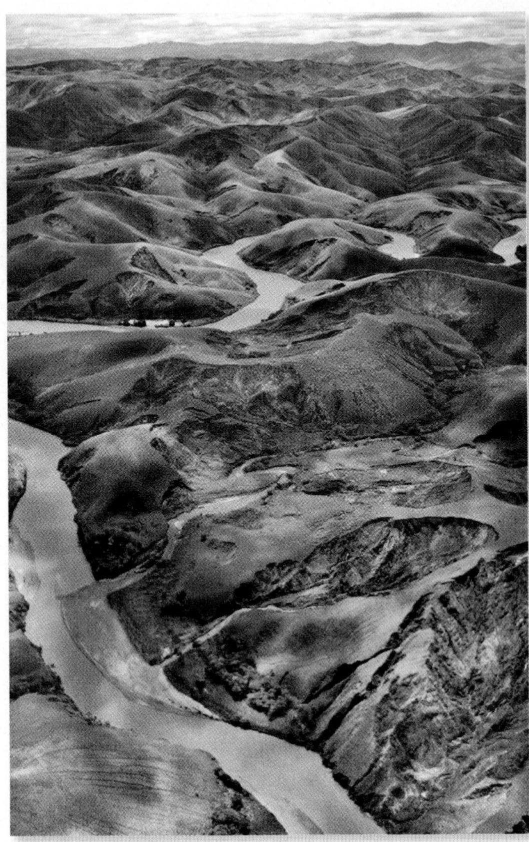

Figure 59.24 **Severe erosion following deforestation in Madagascar.** After clearing and a few years of farming, the shallow soil can no longer support crops and is susceptible to erosion by rainfall.

BIOLOGY PRINCIPLE Biology affects our society. Interruptions in biogeochemical cycles, such as changes in the water cycle, can have severe repercussions for human societies.

Summary of Key Concepts

59.1 Food Webs and Energy Flow

- Ecosystem ecology concerns the movement of energy and materials through organisms and their communities (Figure 59.1).
- Simple feeding relationships between organisms can be characterized by an unbranched food chain, and each feeding level in the chain is called a trophic level (Figure 59.2).
- Organisms that obtain energy from light or chemicals are called autotrophs and are primary producers. Organisms that feed on other organisms are called heterotrophs. Those organisms that feed on primary producers are called primary consumers. Animals that eat plants are also called herbivores. Organisms that feed on primary consumers are called secondary consumers. Animals that eat other animals are called carnivores. Consumers that get their energy from the remains and waste products of organisms are called detritivores or decomposers (Figure 59.3).
- Food webs are a more complex model of interconnected food chains in which multiple links occur between species. Food webs tend to have five or fewer levels. Energy conversions are not 100% efficient, and usable energy is lost within each trophic level and from one trophic level to the next (Figures 59.4, 59.5).
- Production efficiency measures the percentage of energy assimilated that becomes incorporated into new biomass. Trophic-level transfer efficiency measures the energy available at one trophic level that is acquired by the level above. These efficiencies can be expressed in the form of ecological pyramids, the best known being the pyramid of numbers (Figures 59.6, 59.7).
- The increase in the concentration of a substance in living organisms, called biomagnification, can occur at each trophic level of the food web (Figures 59.8, 59.9).

59.2 Biomass Production in Ecosystems

- Plant production can be measured as gross primary production, whereas net primary production is gross primary production minus the energy released during respiration via photosynthetic organisms.
- Net primary production in terrestrial ecosystems is limited primarily by temperature and the availability of water and nutrients. Net primary production in aquatic ecosystems is limited mainly by availability of light and nutrients (Figures 59.10, 59.11).
- Ecosystems differ in their net primary production. Secondary production is limited by available primary production (Figures 59.12, 59.13, 59.14).

59.3 Biogeochemical Cycles

- Elements such as phosphorus, carbon, nitrogen, and sulfur recycle from the physical environment to organisms and back in what are called biogeochemical cycles (Figure 59.15).
- The phosphorus cycle lacks an atmospheric component and thus is a local cycle. An overabundance of phosphorus can cause the overgrowth of algae and subsequent depletion of oxygen levels, called eutrophication (Figures 59.16, 59.17).
- In the carbon cycle, autotrophs incorporate CO_2 from the atmosphere into their biomass; decomposition of plants and respiration recycles most of this CO_2 back to the atmosphere. Human activities, primarily the burning of fossil fuels, are causing increased amounts of CO_2 to enter the atmosphere. Experiments have shown that elevated levels of CO_2 result in an increase in plant growth but a decrease in herbivory (Figures 59.18, 59.19, 59.20).
- The nitrogen cycle has five parts: nitrogen fixation, nitrification, assimilation, ammonification, and denitrification. In the nitrogen cycle, atmospheric nitrogen is unavailable for use by most organisms and must be converted to usable forms by certain bacteria. The activities of humans, including fertilizer use, fossil fuel use, and deforestation, have dramatically altered the nitrogen cycle (Figures 59.21, 59.22).
- The water cycle is a physical rather than a chemical process because it consists of essentially two phenomena: evaporation and precipitation. Alteration of the water cycle by deforestation can result in regional climatic changes because a reduction in transpiration causes a decrease in cloud cover and precipitation (Figures 59.23, 59.24).

Assess and Discuss

Test Yourself

1. The amount of energy that is fixed during photosynthesis is known as
   a. net primary production.
   b. biomagnification.
   c. trophic-level transfer efficiency.
   d. gross primary production.
   e. production efficiency.
2. Chemoautotrophic bacteria are
   a. primary consumers.
   b. secondary consumers.
   c. tertiary consumers.
   d. primary producers.
   e. decomposers.
3. When considering the average food chain, which of the following statements is true?
   a. Secondary consumers are the most abundant organisms in an ecosystem.
   b. The more lengths in the food chain, the more stable the ecosystem.
   c. Biomass decreases as you move up the food chain.
   d. The trophic level with the highest species abundance is usually the primary producers.
   e. All of the above are true.
4. Which organisms are the most important consumers of energy in a Georgia salt marsh?
   a. *Spartina* grass and algae
   b. insects
   c. spiders
   d. crabs
   e. bacteria
5. The amount of energy that is fixed during photosynthesis is the
   a. net primary production.
   b. trophic level transfer efficiency.
   c. gross primary production.
   d. net secondary production.
   e. production efficiency.
6. Primary production in aquatic systems is limited mainly by
   a. temperature and moisture.
   b. temperature and light.
   c. temperature and nutrients.
   d. light and nutrients.
   e. light and moisture.
7. The evapotranspiration rate
   a. can be used as a predictor for primary production.
   b. is increased when temperature decreases.
   c. is not affected by temperature.
   d. is highest in deserts.
   e. can be predicted by measuring only the water content of the soil.
8. Eutrophication is
   a. caused by an overabundance of nitrogen, which leads to an increase in bacteria populations.
   b. caused by an overabundance of nutrients, which leads to an increase in algal populations.
   c. the normal breakdown of algal plants following a pollution event.
   d. normally seen in dry, hot regions of the world.
   e. none of the above.
9. Terrestrial primary producers acquire the carbon necessary for photosynthesis from
   a. decomposing plant material.
   b. carbon monoxide released from the burning of fossil fuels.
   c. carbon dioxide in the atmosphere.
   d. carbon sources in the soil.
   e. both a and d.
10. Nitrogen fixation is the process
    a. that converts organic nitrogen to NH_3.
    b. by which plants and animals take up NO_3^-.
    c. by which bacteria convert NO_3^- to N_2.
    d. by which N_2 is converted to NH_3 or NH_4^+.
    e. all of the above.

Conceptual Questions

1. At what trophic level does a carrion beetle feed?
2. Explain why chain lengths are short in food webs.
3. A principle of biology is that *living organisms use energy*. What is a fundamental difference between the passage of energy and the passage of nutrients through ecosystems?

Collaborative Questions

1. What might the atmospheric concentration of CO_2 be in 2100? Discuss what effects this might have on the environment.
2. The Earth's atmosphere consists of 78% nitrogen. Why is nitrogen a limiting nutrient and how can we increase the supply of nitrogen to plants?

Online Resource

www.brookerbiology.com

Stay a step ahead in your studies with animations that bring concepts to life and practice tests to assess your understanding. Your instructor may also recommend the interactive eBook, individualized learning tools, and more.

Biodiversity and Conservation Biology

60

Chapter Outline

60.1 What Is Biodiversity?
60.2 Why Conserve Biodiversity?
60.3 The Causes of Extinction and Loss of Biodiversity
60.4 Conservation Strategies
Summary of Key Concepts
Assess and Discuss

The gastric brooding frog of Queensland, Australia. Discovered only in 1973, the species became extinct by the mid-1980s.

In 2009, Jeff Corwin, an American conservationist and host for programs on *Animal Planet* and other television networks, published a book entitled *100 Heartbeats: The Race to Save the Earth's Most Endangered Species*. The Hundred Heartbeat Club was created earlier by biologist E. O. Wilson to highlight the plight of animal species, such as Spix's macaw (*Cyanopsitta spixii*) in Brazil, the Chinese river dolphin (*Lipotes vexillifer*), and the Philippine eagle (*Pithecophaga jefferyi*), that have 100 or fewer individuals left alive (and hence that number of heartbeats away from extinction). Corwin's book was a result of 2 year's work traveling and researching such issues as the amphibian-killing chytrid fungus (refer back to the chapter opening photo of Chapter 54), the killing of elephants for their tusks, and the destruction of Indonesia's rainforests. Sadly, many of the species in the Hundred Heartbeat Club are still headed toward extinction or have recently become extinct.

The café marron (*Ramosmania rodriguesii*), a wild relative of the coffee plant that is native to a tiny island of the coast of Mauritius, was assumed to be extinct until 1979, when one surviving tree was identified. Today, cuttings from the tree are being cultured in London's Kew Gardens. The plant may contain genes that would allow coffee to be grown in a wider range of soils and elevations. The gastric brooding frog of Australia (*Rheobatrachus* spp.) incubated its young in its stomach and gave birth to froglets through its mouth. In the brooding period females were thought to be able to switch off digestive acid production. Scientists were interested in how this may have been accomplished because it could have had a bearing in the treatment of gastric ulcers in humans. In the 1980s, however, not long after its discovery, the frog became extinct.

Biological diversity, or **biodiversity**, encompasses the genetic diversity of species, the variety of different species, and the different ecosystems they form. The field of **conservation biology** uses principles and knowledge from molecular biology, genetics, and ecology to protect and sustain the biological diversity of life. Because it draws from nearly all chapters of this textbook, a discussion of conservation biology is an apt way to conclude our study of biology. In this chapter, we begin by examining the question of why biodiversity should be conserved and explore how much diversity is needed for ecosystems to function properly. We then survey the main threats to the world's biodiversity. For many species, multiple threats result from human activities, ranging from habitat loss, overexploitation, and the effects of introduced species to climate change and pollution. Even if species are not exterminated, many may exist only in very small population sizes. We will see how small populations face special problems such as inbreeding, genetic drift, and limited mating, emphasizing the importance of genetics in conservation biology.

Last, we consider what is being done to help conserve the world's endangered plant and animal life. This includes identifying global areas rich in species and establishing parks and refuges of the appropriate size, number, and connectivity. We also discuss conservation of particularly important types of species and outline how ecologists have been active in restoring damaged habitats to a more natural condition. We then examine how captive-breeding programs have been useful in building up populations of rare species prior to their release back into the wild. Some programs have also used modern genetic techniques such as cloning to help breed and perhaps eventually increase populations of endangered species.

60.1 What Is Biodiversity?

Learning Outcome:

1. List and describe the three levels of biodiversity.

Biodiversity can be examined on three levels: genetic diversity, species diversity, and ecosystem diversity. Each level of biodiversity provides valuable benefits to humanity.

Genetic diversity consists of the amount of genetic variation occurring within and between populations. Maintaining genetic variation in the wild relatives of crops may be vital to the continued success of crop-breeding programs. For example, in 1977, Rafael Guzman, a Mexican biologist, discovered a previously unknown wild relative of corn, *Zea diploperennis*, that is resistant to many of the viral diseases that infect domestic corn, *Zea mays*. Genetic engineers believe that this relative has valuable genes that can improve current corn crops. Because corn is the third-largest crop on Earth, the discovery of *Z. diploperennis* may well turn out to be critical to the global food supply.

The second level of biodiversity concerns species diversity, the number and relative abundance of species in a community (refer back to Chapter 58). Species diversity is an area on which much public attention is focused. In 1973, the U.S. Endangered Species Act (ESA) was enacted, which was designed to protect both endangered and threatened species. **Endangered species** are those species that are in danger of extinction throughout all or a significant portion of their range. **Threatened species** are those likely to become endangered in the future. Many species are currently threatened. According to the International Union for Conservation of Nature and Natural Resources (IUCN), more than 25% of the fish species that live on coral reefs and 22% of all mammals, 12% of birds, and 31% of amphibians are threatened with extinction.

The last level of biodiversity is ecosystem diversity, the diversity of structure and function within an ecosystem. Conservation at the level of species diversity has largely focused attention on species-rich ecosystems such as tropical rain forests. Some scientists have argued that other relatively species-poor ecosystems are also highly threatened and need to be conserved. In North America, many of the native prairies have been converted to agricultural use, especially in Midwestern states such as Illinois and Iowa. In some counties, remnants of prairie exist only inside cemetery plots, which have been spared from the plow (**Figure 60.1**).

Figure 60.1 **Ecosystem biodiversity.** This small cemetery in Bureau County, Illinois, contains the remains of a natural prairie ecosystem. Most of the prairie has been plowed under for agriculture.

60.2 Why Conserve Biodiversity?

Learning Outcomes:

1. Detail the benefits of biological diversity to human welfare.
2. Provide graphical representations of possible relationships between biodiversity level and ecosystem function.
3. Describe experimental evidence that shows how species diversity and ecosystem function are linked.

Why should biodiversity be a concern? American biologists Paul Ehrlich and E. O. Wilson have suggested that the loss of biodiversity should be an area of great concern for at least three reasons. First, humans depend on plants, animals, and microorganisms for a wide range of food, medicine, and industrial products. The second reason focuses on preserving the array of essential services provided by ecosystems, such as clean air and water. Finally, Ehrlich and Wilson propose that we have an ethical responsibility to protect what are our only known living companions in the universe. In this section, we examine some of the primary reasons why preserving biodiversity matters and explore the link between biodiversity and ecosystem function.

Human Society Benefits Economically from Increased Biodiversity

During the latter half of the 20th century, the reduction of the Earth's biological diversity emerged as a critical issue, one with implications for public policy. A major concern was that loss of plant and animal resources would impair future development of important products and processes in agriculture, medicine, and industry. For example, as previously noted, *Z. diploperennis*, the wild relative of corn discovered in Mexico, is resistant to many corn viruses. Its genes are currently being used to develop virus-resistant types of corn. However, *Z. diploperennis* occurs naturally in only a few small areas of Mexico and could easily have been destroyed by development or cultivation of the land. If we allow such species to go extinct, we may unknowingly threaten the health of the food supply on which much of the world depends.

The pharmaceutical industry is heavily dependent on plant products. An estimated 50,000–70,000 plant species are used in traditional and modern medicine. About 25% of the prescription drugs in the U.S. alone are derived from plants, and the 2009 market value of such drugs was estimated to be $300 billion, accounting for a little less than half the global pharmaceutical market. Many medicines come from plants found only in tropical rain forests. These include quinine, a drug from the bark of the Cinchona tree (*Cinchona officinalis*) (**Figure 60.2**), which is used for treating malaria, and vincristine, a drug derived from rosy periwinkle (*Catharanthus roseus*), which is a treatment for leukemia and Hodgkin disease. Many chemicals of therapeutic importance are likely to be found in the numerous rain forest plant species that have not yet been fully analyzed. The continued destruction of rain forests thus could mean the loss of potential lifesaving medical treatments.

Individual species are valuable for research purposes. The blood of the horseshoe crab (*Limulus polyphemus*) clots when exposed to toxins produced by some bacteria. Pharmaceutical industries use the blood enzyme responsible for this clotting to ensure that their products are free of bacterial contamination. Desert pupfishes, in the genus *Cyprinodon*, found in isolated desert springs in the U.S. Southwest,

Figure 60.2 The value of biodiversity. Bark of the Cinchona tree (*Cinchona officinalis*), found only in tropical rain forests, is used to produce quinine, an effective treatment for malaria.

BioConnections: *Is malaria caused by a bacterium, a protist, or an insect? Refer back to Figure 28.29.*

tolerate salinity twice that of seawater and are valuable models for research on human kidney diseases.

Natural Ecosystems Provide Essential Services to Humans

Beyond the direct economic gains from biodiversity, humans benefit enormously from the essential services that natural ecosystems provide (**Table 60.1**). Forests soak up carbon dioxide, maintain soil fertility, and retain water, preventing floods; estuaries provide water filtration and protect rivers and coastal shores from excessive erosion. The loss of biodiversity can disrupt an ecosystem's ability to carry out such functions. Other ecosystem functions include the maintenance of populations of natural predators to regulate pest outbreaks and reservoirs of pollinators to pollinate crops and other plants.

A 1997 paper in the journal *Nature* by economist Robert Costanza and colleagues made an attempt to calculate the monetary value of ecosystems to various economies. They came to the conclusion that, at the time, the world's ecosystems were worth more than $33 trillion a year, nearly twice the gross national product of the world's economies combined ($19 trillion) (**Table 60.2**). Due to its massive size, open ocean has the greatest total global value of all ecosystems. Another way to view ecosystem value is its dollar value per hectare. From this perspective, shallow aquatic ecosystems, such as estuaries and swamps, are extremely valuable because of their role in nutrient cycling, water supply, and disturbance regulation. They also serve as

Table 60.1 Examples of the World's Ecosystem Services

Service	Example
Atmospheric gas supply	Regulation of carbon dioxide, ozone, and oxygen levels
Climate regulation	Regulation of carbon dioxide, nitrogen dioxide, and methane levels
Water supply	Irrigation; water for industry
Pollination	Pollination of crops
Biological control	Pest population regulation
Wilderness and refuges	Habitat for wildlife
Food production	Crops; livestock
Raw materials	Fossil fuels; timber
Genetic resources	Medicines; genes for plant resistance
Recreation	Ecotourism; outdoor recreation
Cultural	Aesthetic and educational value
Disturbance regulation	Storm protection; flood control
Waste treatment	Sewage purification
Soil erosion control	Retention of topsoil; reduction of accumulation of sediments in lakes
Nutrient cycling	Nitrogen, phosphorus, carbon, and sulfur cycles

Table 60.2 Valuation of the World's Ecosystem Services

Ecosystem type	Total global value* ($ trillion)	Total value (per ha) ($)	Main ecosystem service
Open ocean	8,381	252	Nutrient cycling
Coastal shelf	4,283	1,610	Nutrient cycling
Estuaries	4,100	22,832	Nutrient cycling
Tropical forest	3,813	2,007	Nutrient cycling/raw materials
Seagrass and algal beds	3,801	19,004	Nutrient cycling
Swamps	3,231	19,580	Water supply/disturbance regulation
Lakes and rivers	1,700	8,498	Water supply
Tidal marsh	1,648	9,990	Waste treatment/disturbance regulation
Grasslands	906	232	Waste treatment/food production
Temperate forest	894	302	Climate regulation/waste treatment/lumber
Coral reefs	375	6,075	Recreational/disturbance regulation
Cropland	128	92	Food production
Desert	0	0	
Ice and rock	0	0	
Tundra	0	0	
Urban	0	0	
Total	33,260		

*In 1997 values

nurseries for aquatic life. These habitats, once thought of as useless wastelands, are among the ecosystems most endangered by pollution and development. A 2009 European Union Cost of Inaction Report suggested there could be a loss of $20 trillion in the value of ecosystem services by 2050 if no additional policy action is taken. This would be equivalent to a loss of 7% of the gross domestic product of 2050.

There Are Ethical Reasons for Conservation of Biodiversity

Arguments can also be made against the loss of biodiversity on ethical grounds. As only one of many species, it has been argued that humans have no right to destroy other species and the environment around us. American philosopher Tom Regan suggests that animals should be treated with respect because they have a life of their own and therefore have value apart from anyone else's interests. American law professor Christopher Stone, in an influential 1972 article titled "Should Trees Have Standing?" has argued that entities such as non-human natural objects like trees or lakes should be given legal rights just as corporations are treated as individuals for certain purposes. As E. O. Wilson proposed in a 1984 concept known as biophilia, humans have innate attachments with species and natural habitats because of our close association over millions of years.

Although these benefits of species diversity may seem obvious to us now, it was only relatively recently that the link between ecosystem services and biodiversity was uncovered.

FEATURE INVESTIGATION

Ecotron Experiments Showed the Relationship Between Biodiversity and Ecosystem Function

In the early 1990s, American ecologist Shahid Naeem and colleagues used a series of 14 environmental chambers in a facility termed the Ecotron, at Silwood Park, England, to determine how biodiversity affects ecosystem functioning. These chambers contained terrestrial communities that differed only in their level of biodiversity (**Figure 60.3**). The number of species in each chamber was manipulated to create high-, medium-, and low-diversity ecosystems, each with four trophic levels. The trophic levels consisted of primary producers (annual plants), primary consumers (insects, snails, and slugs), secondary consumers (parasitoids that fed on the herbivores), and decomposers (earthworms and soil insects). The experiment ran for just over 6 months, and species were added only after the trophic level below them was established. For example, parasitoids were not added until herbivores were abundant.

Researchers monitored and analyzed a range of measures of ecosystem function, including community respiration, decomposition, nutrient retention rates, and community productivity. The data of Figure 60.3 focuses only on community productivity. The result was that community productivity, expressed as percentage change in vegetation cover (the amount of ground covered by leaves of plants), increased as species richness increased. This occurred because of a greater variety of plant growth forms that could utilize light at different levels of the plant canopy. A larger ground cover also meant a larger plant biomass and greater community productivity, and increased decomposition and nutrient uptake rates. For the first time, ecologists had provided an experimental demonstration that the loss of biodiversity can alter or impair the functioning of an ecosystem.

Experimental Questions

1. What was the goal of Shahid Naeem and colleagues in their experiment at Silwood Park, England?
2. What was the hypothesis tested by the researchers?
3. How did the researchers test for ecosystem functioning?

Figure 60.3 **Ecotron experiments comparing species diversity and ecosystem function.**

Concept Check: *What is one of the dangers in interpreting these results?*

HYPOTHESIS Reduced biodiversity can lead to reduced ecosystem functioning.

STARTING LOCATION Ecotron, a controlled environment facility at the Natural Environment Research Council (NERC) Centre for Population Biology, in Silwood Park, England.

	Experimental level	Conceptual level
1 Construct 14 identical experimental chambers.	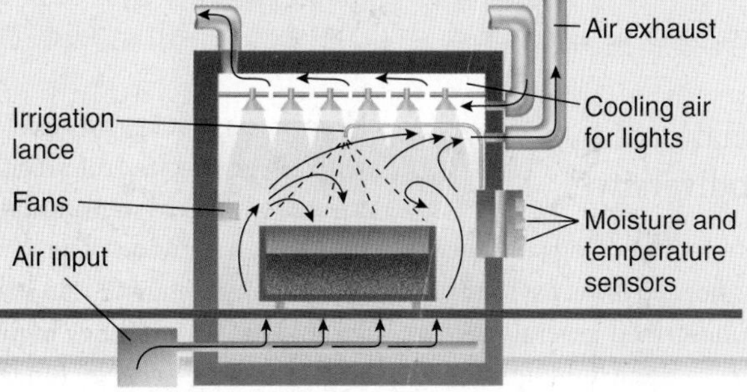 Irrigation lance; Fans; Air input; Air exhaust; Cooling air for lights; Moisture and temperature sensors	Temperature- and humidity-controlled chambers are used to control environmental conditions and allow identical starting conditions in all chambers.

2 Add different combinations of species to the 14 chambers. The species added were based on 3 types of model communities (food webs), each with 4 trophic levels but with varying degrees of species richness.

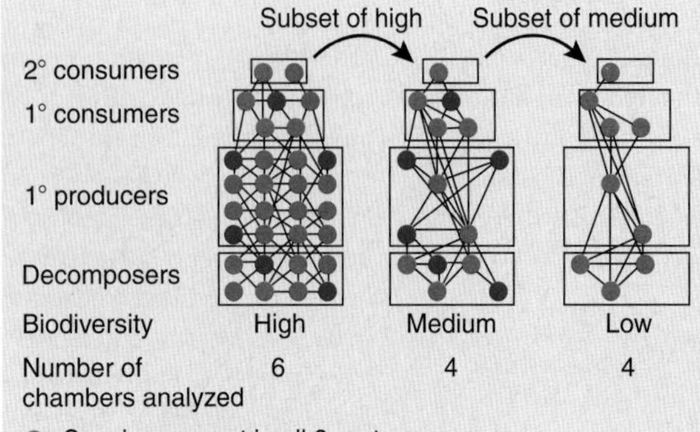

The three diagrams to the left each represent a single chamber. Circles represent species, and lines connecting them represent biotic interactions among the species. Note that each lower-biodiversity community is a subset of its higher-diversity counterpart and that all community types have 4 trophic levels.

3 Measure and analyze a range of processes, including vegetation cover and nutrient uptake.

Measurements help determine how each different type of community functions.

4 THE DATA

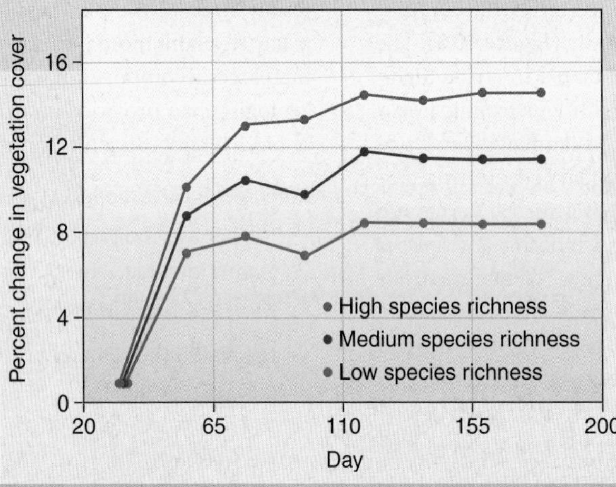

Plant productivity is linked to community diversity as measured by the percent change in vegetation cover from initial conditions.

Data reveal that low-diversity communities have lower vegetation cover and thus are less productive than high-diversity communities.

5 **CONCLUSION** Increases in biodiversity lead to increases in ecosystem function. In this case, increased plant diversity results in greater vegetation cover.

6 **SOURCE** Naeem, S. et al. 1994. Declining biodiversity can alter the performance of ecosystems. *Nature* 368:734–737.

Ecologists Have Described Several Relationships Between Ecosystem Function and Biodiversity

Because biodiversity affects the health of ecosystems, ecologists have explored the question of how much diversity is needed for ecosystems to function properly. In doing so, they have described several possible relationships between biodiversity and ecosystem function. In the 1950s, ecologist Charles Elton proposed in the **diversity-stability hypothesis** that species-rich communities are more stable than those with fewer species (refer back to Chapter 58). If we use stability as a measure of ecosystem function, Elton's hypothesis suggests a linear correlation between diversity and ecosystem function (**Figure 60.4a**). Australian ecologist Brian Walker proposed an alternative to this idea, termed the **redundancy hypothesis** (**Figure 60.4b**). According to this hypothesis, ecosystem function asymptotes at extremely low levels of diversity so most additional species are functionally redundant.

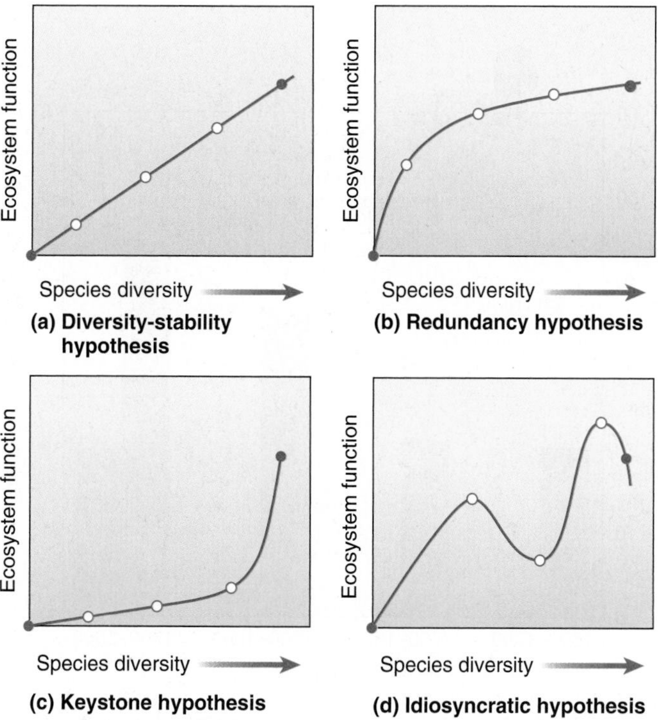

Figure 60.4 Graphical representations of possible relationships between ecosystem function and biodiversity. The two solid dots represent the end points of a continuum of species richness. The first dot is at the origin, where there are no species and no community services. The second dot represents natural levels of species diversity. The relationship is strongest in (a) and weakest in (d).

Two other alternatives relating species richness and ecosystem services have been proposed. The first, termed the **keystone hypothesis** (Figure 60.4c), supposes ecosystem function plummets as soon as biodiversity declines from its natural levels. Lastly, the **idiosyncratic hypothesis** addresses the possibility that ecosystem function changes as the number of species increases or decreases but that the amount and direction of change are unpredictable (Figure 60.4d).

Determining which model is most correct is very important, as our understanding of the effect of species loss on ecosystem function can greatly affect the way we manage our environment.

Field Experiments Have Been Used to Determine How Much Diversity Is Needed for Normal Ecosystem Function

In the mid-1990s, David Tilman and colleagues performed experiments in the field to determine how much biodiversity was necessary for proper ecosystem functioning. Tilman's previous experiments had suggested that species-rich grasslands were more stable; that is, they were more resistant to the ravages of drought and recovered from drought more quickly than species-poor grasslands (refer back to Figure 58.8). In the newer experiments, Tilman's group sowed plots, each 3 m × 3 m and on comparable soils, with seeds of 1, 2, 4, 6, 8, 12, or 24 species of prairie plants. Exactly which species were sown into each plot was determined randomly from a pool of 24 native species. The treatments were replicated 21 times, for a total of 147 plots. The results showed that more-diverse plots had increased productivity and used more nutrients, such as nitrate (NO_3^-), than less-diverse plots (Figure 60.5a,b). Furthermore, the frequency of invasive plant species (species not originally planted in the plots) decreased with increased plant species richness (Figure 60.5c). In a separate experiment, where plots were planted with 1, 2, 4, 8, or 16 species, Tilman and colleagues showed that increased diversity also reduced the severity of attack by foliar fungal diseases (Figure 60.5d).

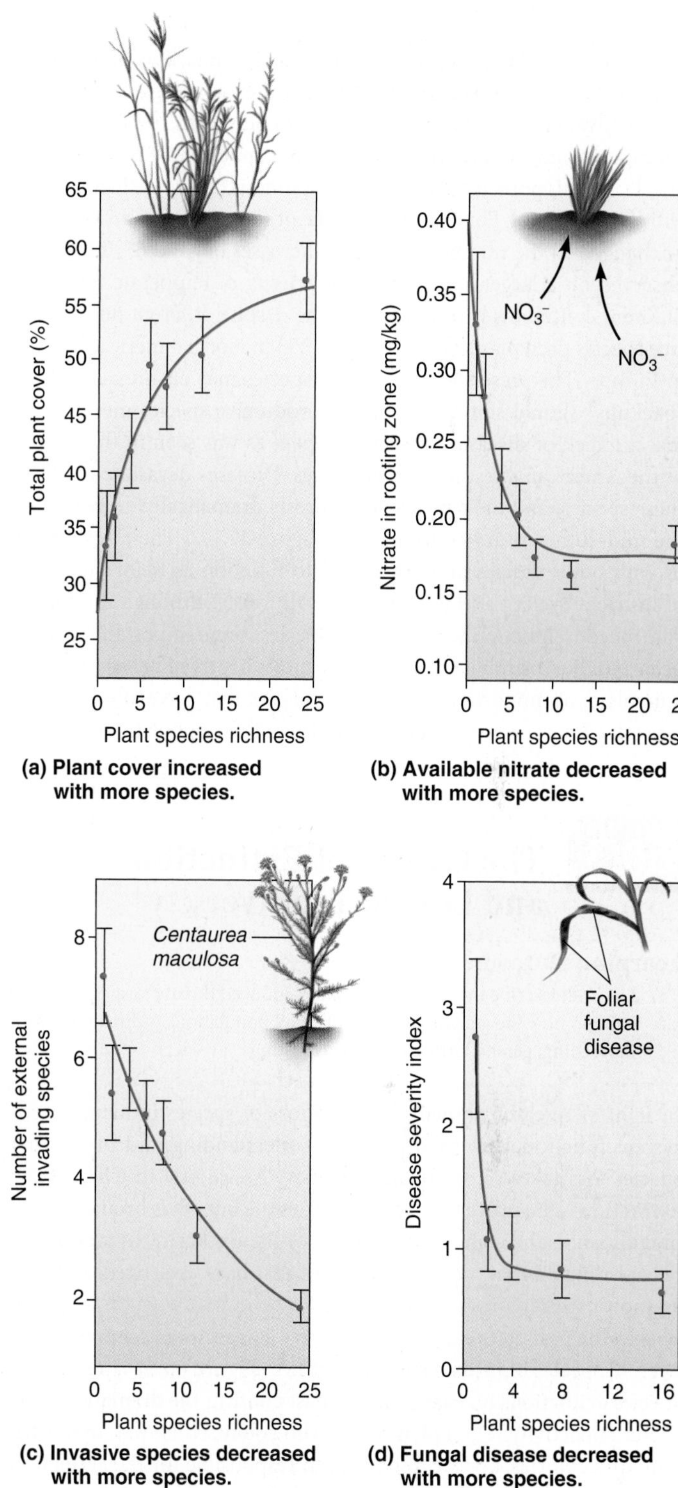

Figure 60.5 **Increased species richness improves community function.**

Although Tilman's experiments show a relationship between species diversity and ecosystem function, they also suggest that most of the advantages of increasing diversity come with the first 5 to 10 species, beyond which adding more species appears to have little to no effect. This supports the redundancy hypothesis (compare Figure 60.5a with Figure 60.4c). For example, uptake of nitrogen remains relatively unchanged as the number of species increases beyond 6. This is also observed on a larger scale. The productivity of temperate forests in different continents is roughly the same despite different numbers of tree species present—729 in East Asia, 253 in North America, and 124 in Europe. The presence of more tree species may ensure a supply of "backups" should some of the most-productive species die off from insect attack or disease. This can happen, as was seen in the demise of the American chestnut and elm trees. Diseases devastated both of these species, and their presence in forests dramatically decreased by the mid-20th century (refer back to Figure 57.17). The forests filled in with other species and continued to function as before in terms of nutrient cycling and gas exchange. However, although the forests continued to function without these species, some important changes occurred. For example, the loss of chestnuts deprived bears and other animals of an important source of food and may have affected their reproductive capacity and hence the size of their populations.

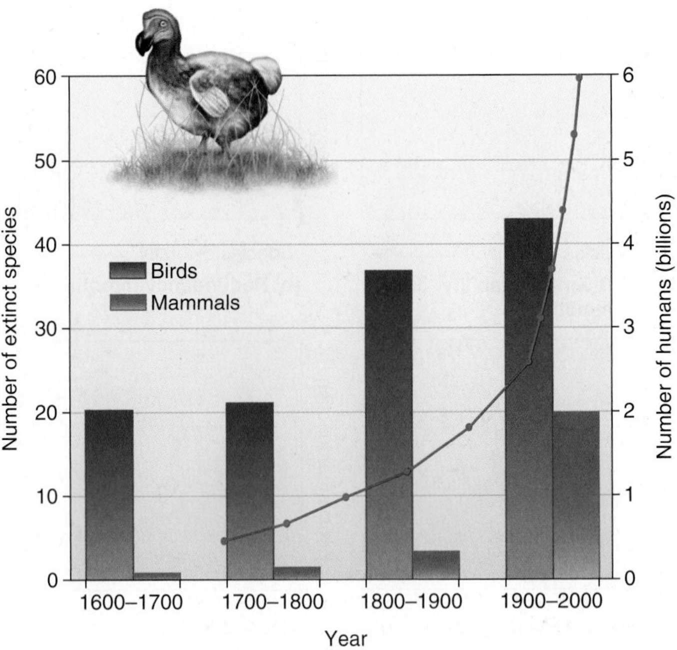

Figure 60.6 **Animal extinctions since the 17th century in relation to human population growth.** Increasing numbers of known extinctions in birds and mammals are concurrent with the exponential increase in the global human population. These data suggest that as the human population increases, more and more species will go extinct. The inset depicts a dodo bird, which went extinct within 100 years after contact with humans.

Concept Check: Why might the increasing human population result in an increase in the extinction rate of other species?

60.3 The Causes of Extinction and Loss of Biodiversity

Learning Outcomes:

1. List and describe the four main human-induced threats to species.
2. Explain how the genetic diversity of small populations is threatened by inbreeding, genetic drift, and limited mating.

In light of research showing that the loss of species influences ecosystem function, the importance of understanding and preventing species loss takes on particular urgency. As we saw in Chapter 22, **extinction**—the process by which species die out—has been a natural phenomenon throughout the history of life on Earth. In fact, it has been estimated that most of all species that have ever lived on Earth are now extinct.

In the past 100 years, approximately 20 species of mammals and over 40 species of birds have gone extinct (**Figure 60.6**). The rates of species extinctions on islands in the past confirm the dramatic effects of human activity. The Polynesians, who colonized Hawaii in the 4th and 5th centuries, appear to have been responsible for the extinction of half of the 100 or so species of land birds in the period between their arrival and that of the Europeans in the late 18th century. A similar effect was felt in New Zealand, which was colonized by settlers some 500 years later than Hawaii. In New Zealand, an entire avian megafauna, consisting of huge land birds, was exterminated over the course of a century, probably through a combination of hunting and large-scale habitat destruction through burning.

The term **biodiversity crisis** is often used to describe this elevated loss of species. Many scientists believe that the rate of loss is higher now than during most of geological history, and most suggest that the growth in the human population has led to the increase in the number of extinctions of other species. In fact, most scientists believe that we are in the middle of a sixth mass extinction.

To understand the process of extinction in more modern times, ecologists need to examine the role of human activities and their environmental consequences. In this section, we examine why species have gone extinct in the past and look at the factors that are currently threatening species with extinction.

The Main Threats to Species Are Human-Induced

Although all causes of extinctions are not known, introduced species, direct exploitation, and habitat destruction have been identified as the most important human-induced threats. In addition, climate change is increasingly being viewed as a significant human-induced threat to species.

Introduced Species **Introduced species** are those species moved by humans from a native location to another location. Most often the species are transported for agricultural purposes or as sources of timber, meat, or wool. Others, such as plants, insects, or marine organisms, are sometimes unintentionally transported in the cargo of ships or planes. Regardless of their method of introduction, some introduced species become **invasive species**, spreading and outcompeting native species for space and resources.

(a) Introduced species

(b) Direct exploitation

(c) Habitat destruction

Figure 60.7 **Causes of extinction.** **(a)** Many Hawaiian honeycreepers such as this Ou (*Psittirostra psittacea*) were exterminated by avian malaria from introduced mosquito species. **(b)** The passenger pigeon (*Ectopistes migratorius*), which may have once been among the most abundant bird species on Earth, was hunted to extinction for its meat. **(c)** The ivory-billed woodpecker (*Campephilus principalis*), the third-largest woodpecker in the world, was long thought to be extinct in the southeastern U.S. because of habitat destruction, but a possible sighting occurred in 2004. This nestling was photographed in Louisiana in 1938.

We can categorize the interactions between introduced and native species into competition, predation, and disease. Competition can eliminate local populations and cause huge reductions in the densities of native species, but it has not yet been clearly shown to exterminate entire species. On the other hand, many recorded cases of extinction have been due to predation. Introduced predators such as rats, cats, and mongooses have accounted for at least 43% of recorded extinctions of birds on islands. Lighthouse keepers' cats have annihilated populations of ground-nesting birds on small islands around the world. The brown tree snake, which was accidentally introduced onto the island of Guam, has decimated the country's native bird populations (refer back to Figure 57.11). Parasitism and disease carried by introduced organisms have also been important in causing extinctions as we saw in the case of the American chestnut. Avian malaria in Hawaii, spread by introduced mosquito species, is believed to have contributed to the demise of up to 50% of native Hawaiian birds (**Figure 60.7a**).

Direct Exploitation Direct exploitation, particularly the hunting of animals, has been the cause of many extinctions in the past. Two remarkable North American bird species, the passenger pigeon (*Ectopistes migratorius*) and the Carolina parakeet (*Conuropsis carolinensis*), were hunted to extinction by the early 20th century. *E. migratorius* was once the most common bird in North America, probably accounting for over 40% of the entire bird population (**Figure 60.7b**). Their total population size was estimated to be over 3 billion birds. Habitat loss because of deforestation was a contributing factor in their demise, but hunting as a cheap source of meat was the main factor. The flocking behavior of the birds made them relatively easy targets for hunters, who used special firearms to harvest the birds in quantity. In 1876, in Michigan alone, over 1.6 million birds were killed and sent to markets in the eastern U.S. *C. carolinensis*, the only species of parrot native to the eastern U.S., was similarly hunted to extinction by the early 1900s.

Many whale species were driven to the brink of extinction prior to the 1988 moratorium on commercial whaling (refer back to Figure 57.12). Steller's sea cow (*Hydrodamalis gigas*), a 9-m-long manatee-like mammal, was hunted to extinction in the Bering Strait only 27 years after its discovery by humans in 1740. A poignant example of human excess in hunting was the dodo (*Raphus cucullatus*), a flightless bird native only to the island of Mauritius that had no known predators (see Figure 60.6). A combination of overexploitation, habitat destruction, and introduced species led to its extinction within 100 years of the arrival of humans. Sailors hunted it for its meat; rats, pigs, and monkeys brought to the island by humans destroyed the dodo's eggs and chicks in their ground nests; and forest destruction reduced its habitat.

Habitat Destruction Habitat destruction through **deforestation**, the conversion of forested areas to nonforested land, has historically been a prime cause of the extinction of species. About one-third of the world's land surface is covered with forests, and much of this area is at risk of deforestation through human activities such as development, farming, animal grazing, or logging. Although tropical forests are probably the most threatened forest type, with rates of deforestation in Africa, South America, and Asia varying between 0.6% and 0.9% per year, the destruction of forests is a global phenomenon. The ivory-billed woodpecker (*Campephilus principalis*), the largest woodpecker in North America and an inhabitant of wetlands and forests of the southeastern U.S., was widely assumed to have gone extinct in the 1950s due to destruction of its habitat by heavy logging (**Figure 60.7c**). In 2004, the woodpecker was purportedly sighted in the Big Woods area of eastern Arkansas, though this has not been confirmed despite concerted efforts.

Deforestation is not the only form of habitat destruction. Prairies are often replaced by agricultural crops. The average area of land under cultivation worldwide averages about 11%, with an additional 24% given over to rangeland; however, this amount varies tremendously between regions. For example, Europe uses 28% of its land for crops and pasturelands, with the result that many of its native species went extinct long ago. Wetlands also have been drained for agricultural purposes. Others have been filled in for urban or industrial development. In the U.S., as much as 90% of the freshwater marshes and 50% of the estuarine marshes have disappeared. Urbanization, the development of cities on previously natural or agricultural areas, is the most human-dominated and fastest-growing type of land use worldwide and devastates the land more severely than practically any other form of habitat degradation.

Climate Change As mentioned at the beginning of Chapter 54, human-induced climate change, or global warming, has been implicated in the dramatic decrease in the population sizes of frog species in Central and South America. We also noted that the distribution of trees would also change as climate zones shifted, with such shifts occurring faster than many plants could migrate via seed dispersal (refer back to Figure 54.11). Indeed, a recent study of six biodiversity-rich regions employed computer models to simulate the movement of species' ranges in response to changing climate conditions. The models predicted that unless greenhouse gas emissions are cut drastically, climate change will cause 15–37% of the species in those regions to go extinct by the year 2050.

Other ecological properties of species, not just range limits, may change with global warming, including population densities and phenology, the timing of biological events such as flowering, egg laying, or migration in relation to climate. In 2003, American ecologist Terry Root and colleagues examined the phenologies of 694 species over the past 30 years. Over an average decade, the estimated mean number of days changed in spring phenology was 5.1 days earlier. For example, the North American common murre (*Uria aalge*) bred, on average, 24 days earlier per decade, and Fowler's toad (*Bufo fowleri*) bred 6.3 days earlier. Potentially more critical than the absolute change in phenology is the possible disruption of timing between associated species such as herbivores and host plants or predators and prey. If the phenologies of all species are sped up by global warming at the same rate, there is no problem. However, limited evidence suggests that this is not the case. Only 11 systems have been examined in detail, but in 7 of them, interacting species responded differently enough to temperature changes to put them more out of synchrony than they were earlier. For example, in the Colorado Rockies, yellow-bellied marmots (*Marmota flaviventris*) now emerge from hibernation 23 days earlier than they did in 1975, changing the relative phenology of the marmots with the emergence of their food plants and causing food shortages.

Small Populations Are Threatened by the Loss of Genetic Diversity

Even if habitats are not destroyed, many become fragmented, leading to the development of small, isolated populations. Such populations are more vulnerable to the loss of genetic diversity resulting from three factors: inbreeding, genetic drift, and limited mating.

Inbreeding **Inbreeding**, which is a form of mating among genetically related relatives, is more likely to take place in nature when population size becomes very small and the number of potential mates shrinks drastically (see Chapter 24). In many species, the health and survival of offspring decline as populations become more inbred. This is because inbreeding produces homozygotes, which because of an increase in recessive genes are less fit, thereby decreasing the reproductive success of the population.

One of the most striking examples of the effects of inbreeding in conservation biology involves the greater prairie chicken (*Tympanuchus cupido*). The male birds have a spectacular mating display that involves inflating the bright orange air sacs on their throat, stomping their feet, and spreading their tail feathers. The prairies of the Midwest were once home to millions of these birds, but as the prairies were converted to farmland, the population sizes of the bird shrank dramatically. The population of prairie chickens in Illinois decreased from 25,000 in 1933 to less than 50 in 1989. At that point, according to studies by Ronald Westemeier and colleagues, only 10 to 12 males existed. Because of the decreasing numbers of males, inbreeding in the population had increased. This was reflected in the steady reduction in the hatching success of eggs (**Figure 60.8**). The prairie chicken population had entered a downward spiral toward extinction from which it could not naturally recover. In the early 1990s, conservation biologists began trapping prairie chickens in Kansas and Nebraska, where populations remained larger and more genetically diverse, and moved them to Illinois, bringing an infusion of new genetic material into the population. This transfer resulted in a rebounding of the egg-hatching success rate to over 90% by 1993.

Genetic Drift In small populations, the chance is greater that some individuals will fail to mate successfully purely by chance. For example, finding a mate may be increasingly difficult as population size decreases. If an individual that fails to mate possesses a rare gene, that genetic information will not be passed on to the next generation, resulting in a loss of genetic diversity from the population. **Genetic drift** refers to the random change in allele frequencies attributable to chance (refer back to Figure 24.10). Because the likelihood of an allele being represented in just one or a few individuals is higher in small populations than in large populations, small, isolated populations are particularly vulnerable to this type of reduction in genetic diversity. Such isolated populations will lose a percentage of their original diversity over time, approximately at the rate of $1/(2N)$ per generation, where N = population size. This has a greater effect in smaller versus larger populations:

If $N = 500$, then $\frac{1}{2N} = 1/1{,}000 = 0.001$, or 0.1% genetic diversity lost per generation

If $N = 50$, then $\frac{1}{2N} = 1/100 = 0.01$, or 1.0% genetic diversity lost per generation

Due to genetic drift, a population of 500 will lose only 0.1% of its genetic diversity in a generation, whereas a population of 50 will lose 1%. Such losses become magnified over many generations. After 20 generations, the population of 500 will lose 2% of its original genetic variation, but the population of 50 will lose about 20%! For organisms

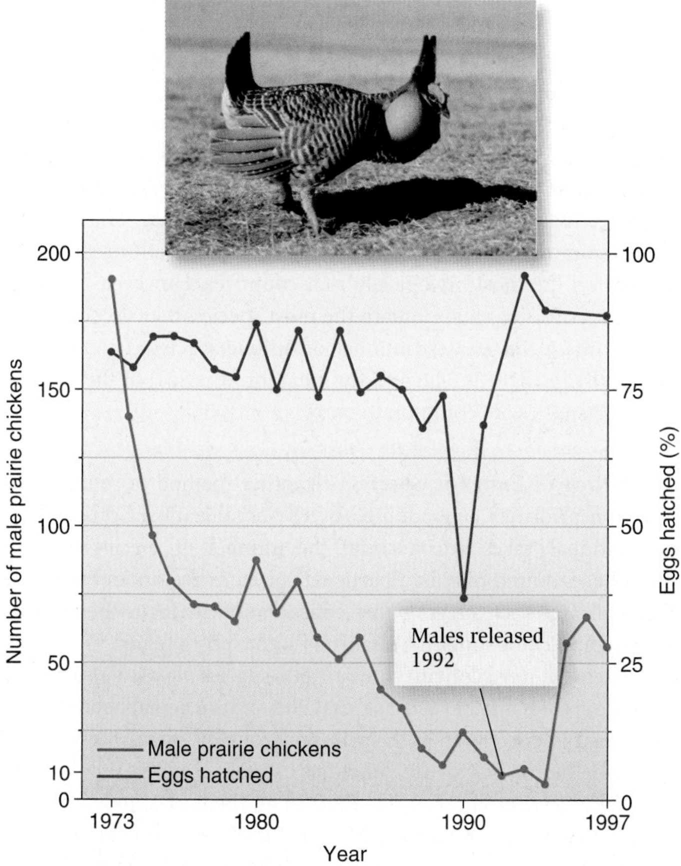

Figure 60.8 Changes in the abundance and egg-hatching success rate of prairie chickens. As the number of males decreased, inbreeding increased, resulting in a decrease in fertility, as indicated by a reduced egg-hatching rate. An influx of males in the early 1990s increased the egg-hatching success rate dramatically.

BIOLOGY PRINCIPLE New properties emerge from complex interactions. The conversion of prairies to farmland reduced the available prairie chicken habitat in Illinois, and populations of the bird decreased dramatically. The consequent increase in inbreeding led to a path towards extinction that was difficult to reverse.

Concept Check: *Is the fitness of all organisms decreased by inbreeding?*

that breed annually, this would mean a substantial loss in genetic variation over 20 years. Once again, this effect becomes more severe as the population size decreases.

As with inbreeding, the effects of genetic drift can be countered by immigration of individuals into a population. Even relatively low immigration rates of about one immigrant per generation (or one individual moved from one population to another) can be sufficient to counter genetic drift in a population of 100 individuals.

Limited Mating In many populations, the **effective population size**, the number of individuals that contribute genes to future populations, may be smaller than the number of individuals in the population, particularly in animals with a harem mating structure in which only a few dominant males breed. For example, dominant elephant seal bulls control harems of females, and a few males command all the matings (refer back to Figure 55.25). If a population consists of breeding males and breeding females, the effective population size is given by

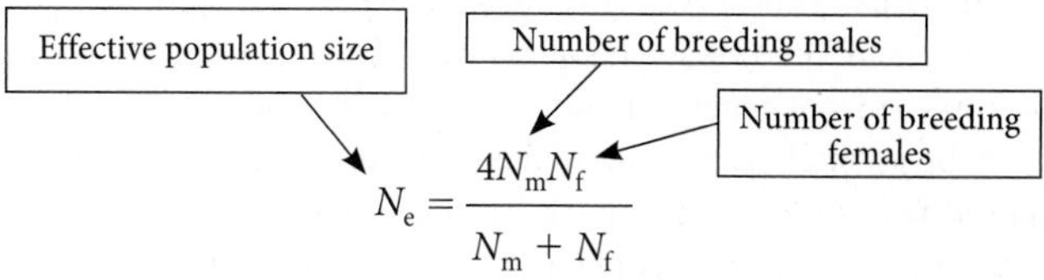

$$N_e = \frac{4N_m N_f}{N_m + N_f}$$

In a population of 500, a 50:50 sex ratio, and all individuals breeding, the effective population size (N_e) = (4 × 250 × 250)/(250 + 250) = 500, or 100% of the actual population size. However, if 10 males breed with 250 females, N_e = (4 × 10 × 250)/(10 + 250) = 38.5, or 8% of the actual population size.

Knowledge of effective population size is vital to ensuring the success of conservation projects. One notable project in the U.S. involved planning the sizes of reserves designed to protect grizzly bear populations in the contiguous 48 states. The grizzly bear (*Ursus arctos*) has declined in numbers from an estimated 100,000 in 1800 to less than 1,000 at present. The range of the species is now less than 1% of its historical range and is restricted to six separate populations in four states (Figure 60.9). Research by American biologist Fred Allendorf has

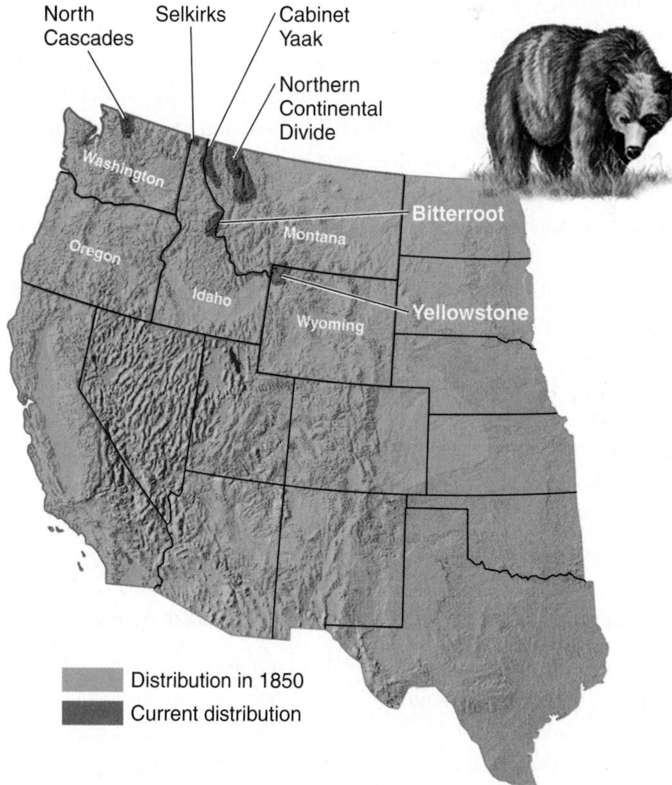

Figure 60.9 Past and current ranges of the grizzly bear. The range of the grizzly bear is currently less than 1% of its historical range. The current range in the continental U.S. has contracted to just six populations in four states, as the population size has shrunk from 100,000 before the West was settled to about 1,000 today.

Concept Check: *If 500 male and 500 female grizzlies exist today, but only 25% of the males breed, what is the effective population size?*

indicated that the effective population size of grizzly populations is generally only about 25% of the actual population size because not all bears breed. Thus, even fairly large, isolated populations, such as the 200 bears in Yellowstone National Park, are vulnerable to the harmful effects of loss of genetic variation because the effective population size may be as small as 50 individuals. Allendorf and his colleagues proposed that an exchange of grizzly bears between populations or zoo collections would help tremendously in promoting genetic variation. Even an exchange of two bears per generation between populations would greatly reduce the loss of genetic variation.

60.4 Conservation Strategies

Learning Outcomes:

1. Detail the different criteria that conservation biologists use to target areas for protection.
2. Explain how the principles of the model of island biogeography and landscape ecology are used to create nature preserves.
3. Describe different approaches conservation biologists use to protect individual species.
4. Define restoration ecology and the approaches used to restore degraded ecosystems and populations of species.

In their efforts to maintain the diversity of life on Earth, conservation biologists are currently active on many fronts. How do they decide on which areas or species to focus and which strategies to employ to protect diversity? We begin this section by discussing how conservation biologists identify the global habitats richest in species. Next, we explore the concept of nature reserves and consider questions such as how large conservation areas should be and how far apart they should be situated. These questions are within the realm of landscape ecology, which studies the spatial arrangement of communities and ecosystems in a geographic area. Next we discuss how conservation efforts often focus on certain species that can have a disproportionate influence on their ecosystem. We will also examine the field of restoration ecology, focusing on how wildlife habitats can be established from degraded areas and how captive breeding programs have been used to reestablish populations of threatened species in the wild. We conclude by returning to the theme of genomes and proteomes to show how modern molecular techniques of cloning can contribute to the fight to save critically endangered species.

Conservation Seeks to Establish Protected Areas

Currently, about 12.85% of the global land area is under some form of protection. There are more than 160,000 separate protected areas, with more added daily. Conservation biologists often must make decisions regarding which habitats should be protected. Many conservation efforts have focused on saving habitats in so-called megadiversity countries, because they often have the greatest number of species. However, more recent strategies have promoted preservation of certain key areas with the highest levels of unique species or the preservation of representative areas of all types of habitat, even relatively species-poor areas.

Megadiversity Countries One method of targeting areas for conservation is to identify those countries with the greatest numbers of species, the **megadiversity countries**. Using the number of plants, vertebrates, and selected groups of insects as criteria, American biologist Russell Mittermeier and colleagues have determined that just 17 countries are home to nearly 70% of all known species. Brazil, Indonesia, and Colombia top the list, followed by Australia, Peru, Mexico, Madagascar, China, and nine other countries. The megadiversity country approach suggests that conservation efforts should be focused on the most biologically rich countries. However, although megadiversity areas may contain the most species, they do not necessarily contain the greatest number of unique species. The mammal species list for Peru is 344, and for Ecuador, it is 271; of these, however, 208 species are common to both.

Areas Rich in Endemic Species Another method of setting conservation priorities, one adopted by the organization Conservation International, takes into account the number of species that are **endemic**, or found only in a particular place or region and nowhere else. This approach suggests that conservationists focus their efforts on geographic **hot spots**. To qualify as a hot spot, a region must meet two criteria: It must contain at least 1,500 species of vascular plants as endemic species and have lost at least 70% of its original habitat. Vascular plants were chosen as the primary group of organisms to determine whether or not an area qualifies as a hot spot, mainly because most other terrestrial organisms depend on them to some extent.

Conservationists Norman Myers, Russell Mittermeier, and colleagues identified 34 hot spots that together occupy a mere 2.3% of the Earth's surface but contain 150,000 endemic plant species, or 50% of the world's total (**Figure 60.10**). Of these areas, the Tropical Andes and Sundaland (the region including Malaysia, Indonesia, and surrounding islands) have the most endemic plant species (**Table 60.3**). This approach proposes that protecting geographic hot spots will prevent the extinction of a larger number of endemic species than would protecting areas of a similar size elsewhere. The main argument against using hot spots as the criterion for targeting conservation efforts is that the areas richest in endemic species—tropical rain forests—would receive the majority of attention and funding, perhaps at the expense of protecting other areas.

Representative Habitats In a third approach to prioritizing areas for conservation, scientists have recently argued that we need to conserve representatives of all major habitats. Prairies, such as some of those in the U.S., are a case in point. An example is the Pampas region of South America, which is arguably the most threatened habitat on the continent because of conversion of its natural grasslands to ranch land and agriculture. The Pampas does not compare well in richness or endemics with the rain forests, but it is a unique area that without preservation could disappear (**Figure 60.11**). By selecting habitats that are most distinct from those already preserved, many areas that are threatened but not biologically rich may be preserved in addition to the less immediately threatened, but richer, tropical forests.

Ecologists are divided as to which is the best way to identify areas for habitat conservation. Some ecologists feel that the best approach might be one that creates a "portfolio" of areas to conserve, containing some areas of high species richness, others with large numbers of endemic species, and some with various habitat types.

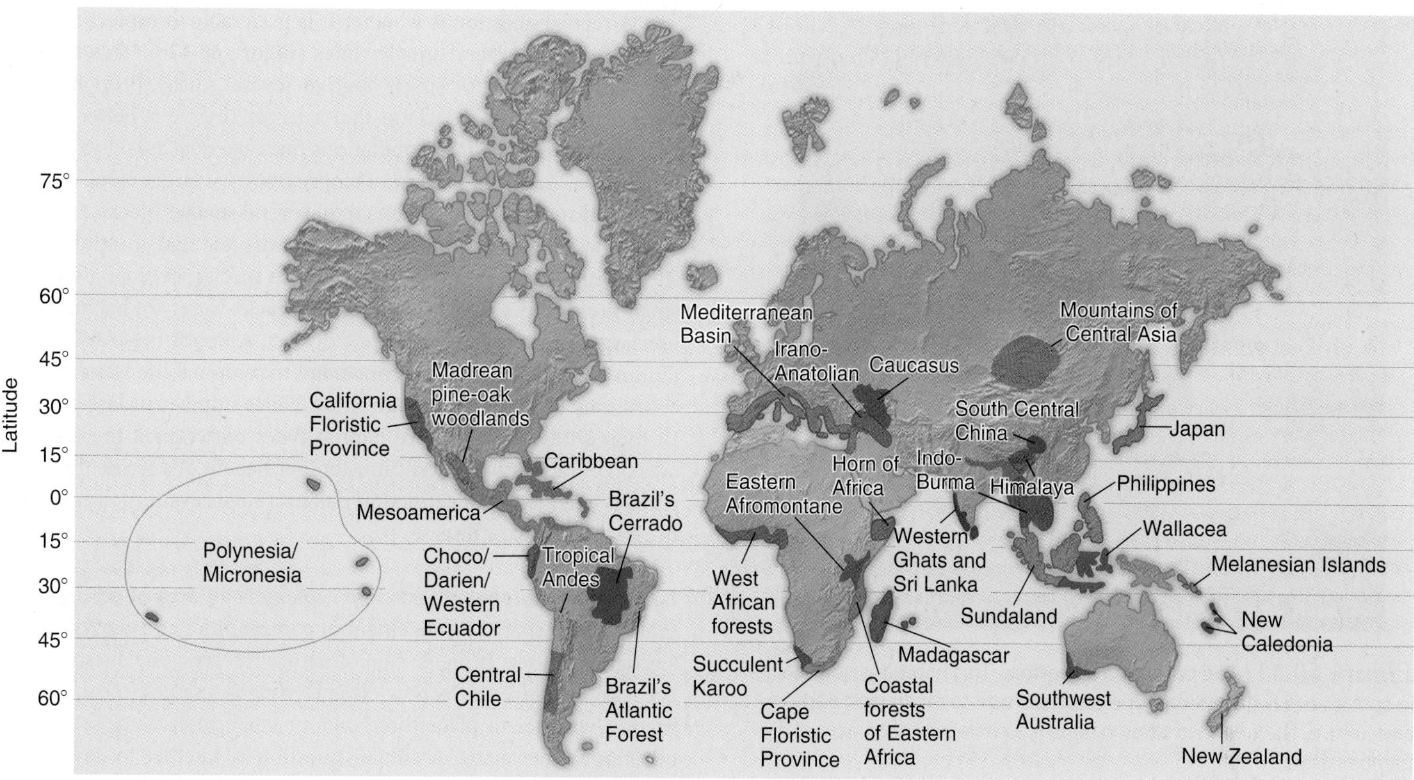

Figure 60.10 **Location of major biodiversity hot spots around the world.** Hot spots have high numbers of endemic species.

The Theory and Practice of Reserve Design Incorporate Principles of Island Biogeography and Landscape Ecology

After identifying areas to preserve, conservationists must determine the size, arrangement, and management of the protected land. Among the questions conservationists ask is whether one large reserve is preferable to an equivalent area composed of smaller reserves. Ecologists also need to determine whether parks should be close together or far apart and whether or not they should be connected by strips of suitable habitat to allow the movement of plants and animals between them. Conservationists also need to consider that park design is often contingent on economic factors. Let's examine some of the many issues that conservationists address in the creation and management of protected land.

The Role of Island Biogeography In exploring the equilibrium model of island biogeography (refer back to Chapter 58), we noted that nature reserves and sanctuaries are, in essence, islands in a sea of human-altered habitat. Seen this way, the tenets of the equilibrium

Table 60.3 Numbers of Endemic Species Present in the Top 10 Hot Spots of the World, Ranked by the Numbers of Endemic Plants

Rank	Hot spot	Plants	Birds	Mammals	Reptiles	Amphibians	Freshwater fishes
1	Tropical Andes	15,000	584	75	275	664	131
2	Sundaland	15,000	146	173	244	172	350
3	Mediterranean Basin	11,700	32	25	77	27	63
4	Madagascar	11,600	183	144	367	226	97
5	Brazil's Atlantic Forest	8,000	148	71	94	286	133
6	Indo-Burma	7,000	73	73	204	139	553
7	Caribbean	6,550	167	41	468	164	65
8	Cape Floristic Province	6,210	6	4	22	16	14
9	Philippines	6,091	185	102	160	74	67
10	Brazil's Cerrado	4,400	16	14	33	26	200

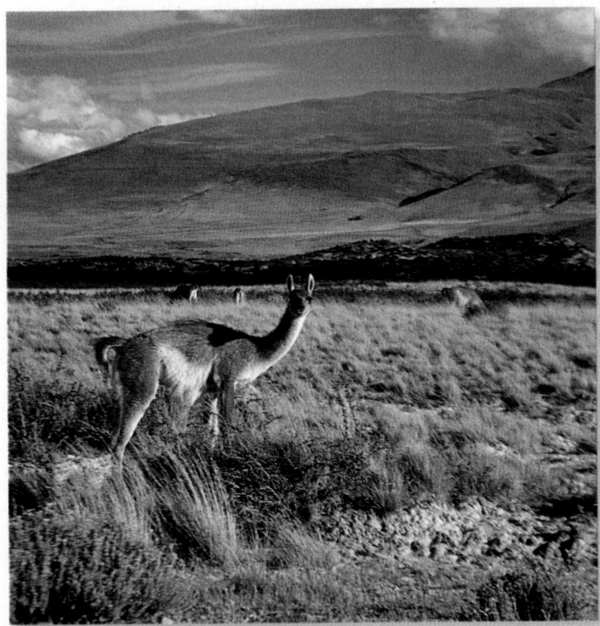

Figure 60.11 The pampas, Argentina. This habitat is not rich in species but is threatened due to conversion to ranch land and agriculture. The guanaco, shown here, is a characteristic grazer of pampas grass.

model of island biogeography can be applied not only to a body of land surrounded by water but also to nature reserves. One question for conservationists is how large a protected area should be (**Figure 60.12a**). According to island biogeography, the number of species should increase with increasing area (the species-area effect). Thus, the larger the area, the greater the number of species protected. In addition, larger parks have other benefits. For example, they are beneficial for organisms that require large spaces, including migrating species and species with extensive territories, such as lions and tigers.

A related question is whether it is preferable to protect a single, large reserve or several smaller ones (**Figure 60.12b**). This is called the **SLOSS debate** (for single large or several small). Proponents of the single, large reserve claim that a larger reserve is better able to preserve more and larger populations than an equal area divided into small areas. According to island biogeography, a larger block of habitat should support more species than several smaller blocks.

However, many empirical studies suggest that multiple small sites of equivalent area will contain more species, because a series of small sites is more likely to contain a broader variety of habitats than one large site. Looking at a variety of sites, American researchers Jim Quinn and Susan Harrison concluded that animal life was richer in collections of small parks than in a smaller number of larger parks. In their study, having more habitat types outweighed the effect of area on biodiversity. In addition, another benefit of a series of smaller parks is a reduction of extinction risk by a single event such as a wildfire or the spread of disease.

Landscape Ecology Landscape ecology is an area of ecology that examines the spatial arrangement of communities and ecosystems in a geographic area. In the design of nature reserves, one question that needs to be addressed is how close to situate reserves to each other, such as whether to place three or four small reserves close to each other or farther apart. A similar question is whether to have a linear or a cluster arrangement of small reserves. Island biogeography suggests that if an area must be fragmented, the sites should be as close as possible to permit dispersal (**Figure 60.12c,d**). In practice, however, having small sites far apart may preserve more species than having them close together, because once again, distant sites are likely to incorporate slightly different habitats and species.

Landscape ecologists have also suggested that small reserves should be linked together by **movement corridors**, thin strips of land that may permit the movement of species between patches (**Figure 60.12e**). Such corridors ideally facilitate movements of organisms that are vulnerable to predation outside of their natural habitat

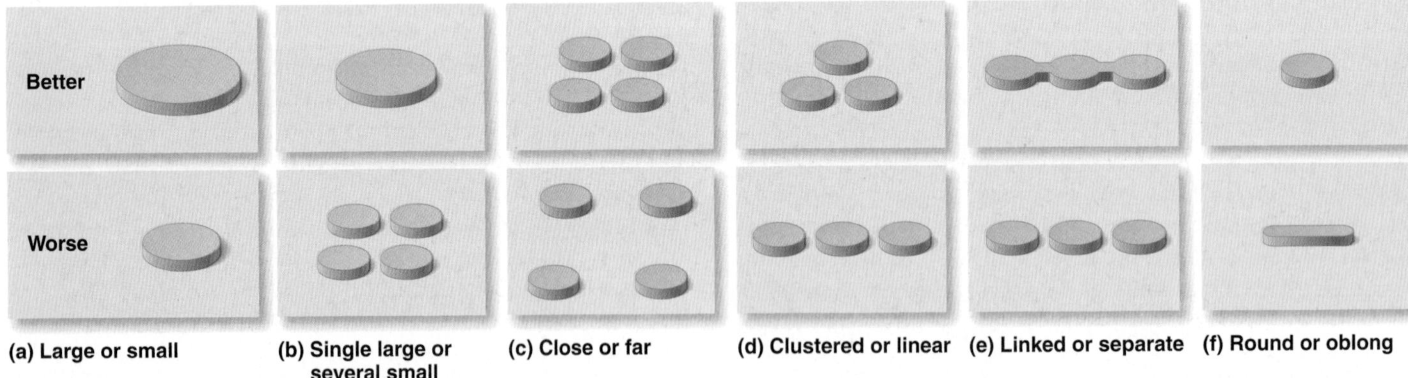

Figure 60.12 The theoretical design of nature reserves. **(a)** A larger reserve will hold more species and have low extinction rates. **(b)** A given area should be fragmented into as few pieces as possible. **(c)** If an area must be fragmented, the pieces should be as close as possible to permit dispersal. **(d)** To enhance dispersal, a cluster of fragments is preferable to a linear arrangement. **(e)** Maintaining or creating corridors between fragments may also enhance dispersal. **(f)** Circular-shaped areas minimize the amount of edge effects. The labels "better" and "worse" refer to theoretical principles generated by the equilibrium model of island biogeography, but empirical data have not supported all the predictions.

Concept Check: *What are some of the potential risks in connecting areas via movement corridors?*

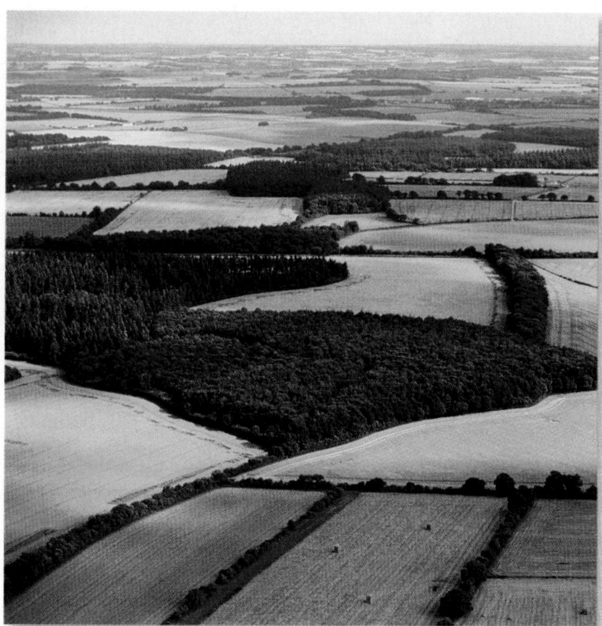

Figure 60.13 **Movement corridors.**

Concept Check: *Why would these European hedgerows act as movement corridors?*

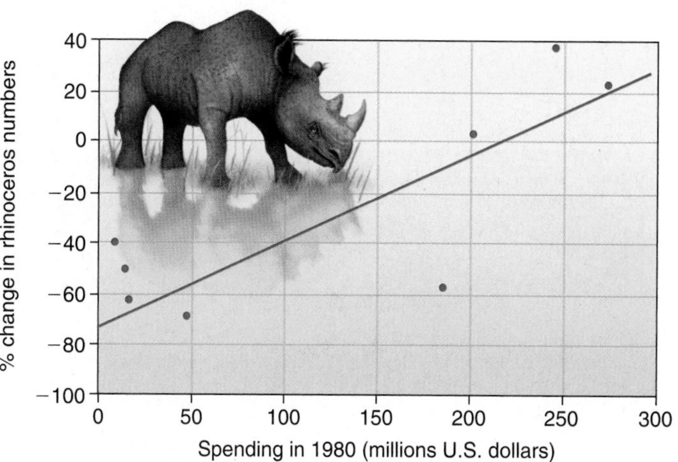

Figure 60.14 **The economics of conservation.** A positive relationship is seen between the percent change in the number of black rhinoceros and conservation spending in various African countries between 1980 and 1984.

or that have poor powers of dispersal between habitat patches. In this way, if a disaster befalls a population in one small reserve, immigrants from neighboring populations can more easily recolonize it. This avoids the need for humans to physically move new plants or animals into an area.

Several types of habitat may function as movement corridors, including hedgerows in Europe, which facilitate movement and dispersal of species between forest fragments (**Figure 60.13**). In China, corridors of habitat have been established to link small, adjacent populations of giant pandas. Riparian habitats, vegetated corridors bordering watercourses, are thought to help facilitate movement of species between habitats. In Florida, debate continues about whether to establish a movement corridor to allow bears to move between the Ocala and Osceola National Forests. However, disadvantages are associated with movement corridors. First, corridors also can facilitate the spread of disease, invasive species, and fire between small reserves. Second, it is not yet clear if species would actually use such corridors.

Finally, parks are often designed to minimize **edge effects**, the special physical conditions that exist at the boundaries or edges of ecosystems. Habitat edges, particularly those between natural habitats such as forests and developed land, are often different in physical characteristics from the habitat core. For example, the center of a forest is shaded by trees and has less wind and light than the forest edge, which is unprotected. Many forest-adapted species therefore shy away from forest edges and prefer forest centers. Because the amount of edge is minimized, circular parks are generally preferable to oblong parks (**Figure 60.12f**).

Economic Considerations in Conservation Although the principles of the model of island biogeography and landscape ecology are useful in illuminating conservation issues, in reality, there is often little choice as to the location, size, shape, and extent of nature reserves. Management practicalities, such as costs of acquisition and maintenance, and politics often override ecological considerations, especially in developing countries, where costs for large reserves may be relatively high. Economic considerations often enter into the choice of which areas to preserve. Many countries protect areas in those regions that are the least economically valuable rather than choosing areas to ensure a balanced representation of the country's biota. In the U.S., most national parks were historically chosen for their scenic beauty, not because they preserve the richest habitat for wildlife.

When designing nature reserves, countries need to consider how to finance their management. Interestingly, the amount of money spent to protect nature reserves may better determine species extinction rates than reserve size. Theoretically, large areas minimize the risk of extinctions because they contain sizable populations. In Africa, several parks, such as Serengeti and Selous in Tanzania, Tsavo in Kenya, and Luangwa in Zambia, are large enough to fulfill this theoretical ideal. However, in the 1980s, populations of black rhinoceroses and elephants declined dramatically within these areas because of poaching—the illegal killing of animals or removing of plants. A larger park is also more difficult to patrol. Research has shown that the rates of decline of rhinos and elephants, largely a result of poaching, have been related directly to conservation efforts and spending (**Figure 60.14**). Populations of the remaining black rhinos, lowland gorillas, and pygmy chimpanzees in Africa, and the vicuna, a llama-like animal in South America, have all shown the greatest stabilization in areas that have been heavily patrolled and where economic resources have been concentrated.

The Single-Species Approach Focuses Conservation Efforts on Particular Types of Species

Much public awareness of the biodiversity crisis results from efforts to preserve individual species that are at risk of extinction. The single-species approach to conservation focuses on saving species that are

(a) Indicator species: Polar bear

(b) Umbrella species: Northern spotted owl

(c) Flagship species: Florida panther

Figure 60.15 **Indicator, umbrella, and flagship species.** **(a)** Polar bears have been called an indicator species of global climate change. **(b)** The Northern spotted owl is considered an umbrella species for the old-growth forest in the Pacific Northwest. **(c)** The Florida panther has become a flagship species for Florida.

deemed particularly important. As with habitat conservation, there are different approaches to identifying which ecologically important species to focus effort on.

Indicator Species Some conservation biologists have suggested that certain organisms can be used as **indicator species**, those species whose status provides information on the overall health of an ecosystem. Corals are good indicators of marine processes such as siltation, the accumulation of sediments transported by water. Because siltation reduces the availability of light, the abundance of many marine organisms decreases in such situations, with corals among the first to display a decline in health. Coral bleaching is also an indicator of climate change (refer back to Figure 54.8). A proliferation of the dark variety of the peppered moth (*Biston betularia*) has been shown to be a good indicator of air pollution. The darker-colored moths flourish because predators are less able to detect them on trees darkened by soot. Polar bears (*Ursus maritimus*) are thought to be a mammalian indicator species for global climate change (**Figure 60.15a**). Scientists believe that global warming is causing the ice in the Arctic to melt earlier in the spring than in the past. Because polar bears rely on the ice to hunt for seals, the earlier breakup of the ice is leaving the bears less time to feed and build the fat that enables them to sustain themselves and their young. A U.S. Geological Survey concluded that future reduction of arctic ice could result in a loss of two-thirds of the world's polar bear population within 50 years. In May 2008, the polar bear was listed as a threatened species under the U.S. Endangered Species Act (ESA).

Umbrella Species **Umbrella species** are species whose habitat requirements are so large that protecting them would protect many other species existing in the same habitat. The Northern spotted owl (*Strix occidentalis*) of the Pacific Northwest is considered to be an important umbrella species (**Figure 60.15b**). A pair of birds needs at least 800 hectares of old-growth forest for survival and reproduction, so maintaining healthy owl populations is thought to help ensure survival of many other forest-dwelling species. In the southeast area of the U.S., the red-cockaded woodpecker (*Picoides borealis*) is often seen as the equivalent of the spotted owl, because it requires large tracts of old-growth long-leaf pine (*Pinus palustris*), including old diseased trees in which it can excavate its nests.

Flagship Species In the past, conservation resources were often allocated to a **flagship species**, a single large or instantly recognizable species. Such species were typically chosen because they were attractive and thus more readily engendered support from the public for their conservation. The concept of the flagship species, typically a charismatic vertebrate such as the American buffalo (*Bison bison*), has often been used to raise awareness for conservation in general. The giant panda (*Ailuropoda melanoleuca*) is the World Wildlife Fund's emblem for endangered species, and the Florida panther (*Puma concolor*) has become a symbol of the state's conservation campaign (**Figure 60.15c**).

Keystone Species A different conservation strategy focuses on **keystone species**, species within a community that have a role out of proportion to their abundance or biomass. The beaver, a relatively small animal, can completely alter the composition of a community by building a dam and flooding an entire river valley (**Figure 60.16**). The resultant lake may become a home to fish species, wildfowl, and aquatic vegetation. A decline in the number of beavers could have serious ramifications for the remaining community members, promoting fish die-offs, waterfowl loss, and the death of vegetation adapted to waterlogged soil. In the southeastern U.S., gopher tortoises (*Gopherus polyphemus*) can be regarded as keystone species because the burrows they create provide homes for an array of other animals, including mice, opossums, frogs, snakes, and insects. Many of these creatures depend on the gopher tortoise burrows and would be unable to survive without them.

American tropical ecologist John Terborgh considers palm nuts and figs to be keystone species because they produce fruit during otherwise fruitless times of the year and are thus critical resources for tropical forest fruit-eating animals, including primates, rodents, and many birds. Together, these fruit eaters account for as much as three-quarters of the tropical forest animal biomass. Without the fruit

Figure 60.16 **Keystone species.** The American beaver creates large dams across streams, and the resultant lakes provide habitats for a great diversity of species.

trees, wholesale extinction of these animals could occur. Note that a keystone species is not the same as a **dominant species**, one that has a large effect in a community because of its abundance or high biomass. For example, *Spartina* cordgrass is a dominant species in a salt marsh because of its large biomass, but it is not a keystone species. The American chestnut was another dominant species before it was greatly reduced in abundance by an introduced parasite (refer back to Figure 57.17).

Restoration Ecology Attempts to Rehabilitate Degraded Ecosystems and Populations

Although the preservation of umbrella, flagship, or keystone species is a valuable conservation strategy, another approach is to rehabilitate previously degraded habitat. **Restoration ecology** is the full or partial repair or replacement of biological habitats and/or their populations that have been degraded or destroyed. It can focus on restoring or rehabilitating a habitat, or it can involve reintroducing species or returning species to the wild following captive breeding. Following open-pit mining for coal or phosphate, huge tracts of disturbed land must be replenished with topsoil, and a large number of species such as grasses, shrubs, and trees must be replanted. Aquatic habitats can be restored by reducing human impacts and replanting vegetation. In Florida, where seagrass beds are vulnerable to damage by motor-driven boat propellers, efforts have focused on closing off areas to motorboats and replanting previously damaged beds.

Habitat Restoration The three basic approaches to habitat restoration are complete restoration, rehabilitation, and ecosystem replacement. In complete restoration, conservationists attempt to return a habitat to its condition prior to the disturbance. Under the leadership of American ecologist Aldo Leopold, the University of Wisconsin pioneered the restoration of prairie habitats as early as 1935, converting agricultural land back to species-rich prairies (**Figure 60.17a**). The second approach aims to return the habitat to something similar to, but a little less than, full restoration, a goal called rehabilitation. In Florida, phosphate mining involves removing a layer of topsoil or "overburden," mining the phosphate-rich layers, returning the overburden, and replanting the area. Exotic species such as cogongrass (*Imperata cylindrica*), an invasive Southeast Asian species, often invade these disturbed areas, and the biodiversity of the restored habitat is usually not comparable to that of unmined areas (**Figure 60.17b**). The third approach, termed replacement, makes no attempt to restore what was originally present but instead replaces the original ecosystem with a different one. The replacement could be an ecosystem that is simpler but more productive, as when deciduous forest is replaced after mining by grassland to be used for public recreation.

Although any of these approaches can be employed in the habitat restoration process, complete restoration is not always the desired endpoint. In some cases, it is appropriate, but in many cases complete restoration is so difficult or expensive as to be impractical. Ecosystem replacement is particularly useful for land that has been significantly damaged by past activities. It would be nearly impossible to re-create

(a) Complete restoration

(b) Rehabilitation

(c) Ecosystem replacement

Figure 60.17 **Habitat restoration.** **(a)** The University of Wisconsin pioneered the practice of complete restoration of agricultural land to native prairies. **(b)** In Florida, phosphate mines are so degraded that complete restoration is not possible. After topsoil is replaced, some exotic species such as cogongrass often invade, allowing only habitat rehabilitation. **(c)** These old open-pit mines in Middlesex, England, have been converted to valuable freshwater habitats, replacing the wooded area that was originally present.

BIOLOGY PRINCIPLE **Biology affects our society.** Restoration of human-degraded habitats can lead to recovery of habitat and biodiversity.

the original landscape of an area that was mined for stone or gravel. In these situations, however, wetlands or lakes may be created in the open pits (**Figure 60.17c**).

Bioremediation Restoration can also involve **bioremediation**, the use of living organisms, usually microorganisms or plants, to degrade sewage or detoxify polluted habitats such as dump sites or oil spills.

In 1975, a leak from a military fuel storage facility released 80,000 gallons of jet fuel into the sandy soil at Hanahan, South Carolina. Soon the groundwater contained toxic chemicals such as benzene. By the 1980s, it was found that naturally occurring microorganisms in the soil were actively consuming many of these toxic compounds and converting them into carbon dioxide. In 1992, nutrients were delivered in pipes to the contaminated soils to speed up the action of the natural microbial community. By the end of 1993, contamination had been reduced by 75%. The increasing interest in bacterial genomes is providing opportunities for understanding the genetic and molecular bases of degradation of organic pollutants. Many novel biochemical reactions have been discovered.

Heavy metals such as cadmium or lead are not readily absorbed by microorganisms. Phytoremediation, a form of bioremediation that involves the use of plants, is valuable in these cases. Plants absorb contaminants in the root system and store them in root biomass or transport them to stems and leaves. After plants are removed from the area, a lower level of soil contamination will remain. Several growth/harvest cycles may be needed to achieve cleanup. Sunflower (*Helianthus annuus*) has been used to extract arsenic and uranium from soils. Pennycress (*Thalsphi caenilescens*) is an accumulator of zinc and cadmium, and lead may be removed by Indian mustard (*Brassica juncea*) and ragweed (*Ambrosia artemisifolia*). Some polychlorinated biphenyls (PCBs) have been removed by transgenic plants containing genes for bacterial enzymes. As discussed in Chapter 59, poplar trees are being genetically engineered to better degrade the chemical trichloroethylene (TCE).

Reintroductions and Captive Breeding Reintroducing species to areas where they previously existed is a valuable conservation strategy. As noted in Chapter 56, both black-footed ferrets and tule elk have been successfully reintroduced into areas where they once occurred (refer back to Figure 56.10). Reintroductions may increase genetic diversity and reduce the effects of inbreeding (see Figure 60.8). In some cases, organisms are bred in captivity before being released back into the wild. Captive breeding, the propagation of animals and plants outside their natural habitat to produce stock for subsequent release into the wild, has proved valuable in reestablishing breeding populations following extinction or near extinction. Zoos, aquariums, and botanical gardens often play a key role in captive breeding, propagating species that are highly threatened in the wild. They also play an important role in public education about the loss of biodiversity and the use of restoration programs.

Several classic programs illustrate the value of captive breeding and reintroduction. The peregrine falcon (*Falco peregrinus*) became extinct in nearly all of the eastern U.S. by the mid-1960s, a decline that was linked to the effects of DDT (refer back to Figure 59.9). In 1970, American biologist Tom Cade gathered falcons from other parts of the country to start a captive breeding program at Cornell University. Since then, the program has released thousands of birds into the wild, and in 1999, the peregrine falcon was removed from the list of endangered species. A captive breeding program is also helping save the California condor (*Gymnogyps californicus*) from extinction. At a cost of $35 million, this is the most expensive species conservation project ever undertaken in the U.S. In the 1980s, there were only 22 known condors, some in captivity and some in the wild. Scientists made the decision to capture the remaining wild birds in order to protect and breed them (**Figure 60.18a**). By 2011, the captive population numbered 203 individuals, and 181 birds were living in the Grand Canyon area of Arizona; Zion National Park, Utah; the western coastal mountains of California; and northern Baja California, Mexico (**Figure 60.18b**). A milestone was reached in 2003, when a pair of captive-reared California condors bred in the wild.

Because the number of individuals in any captive breeding program is initially small, care must be exercised to avoid inbreeding. Matings are usually carefully arranged to maximize resultant genetic variation in offspring. The use of genetic engineering to clone endangered species is a new area that may eventually help bolster populations of captive-bred species.

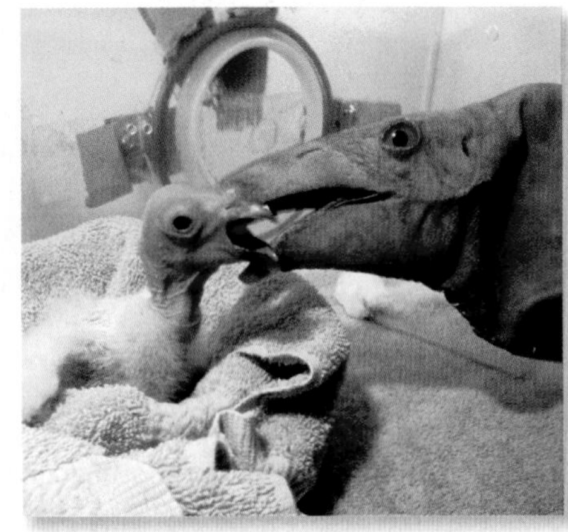

(a) A condor chick being fed using a puppet

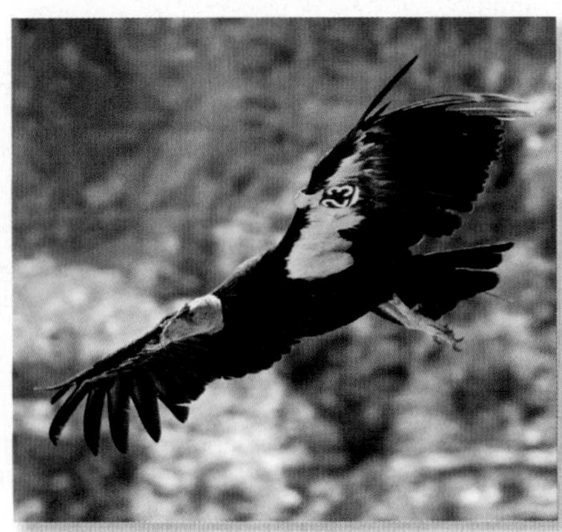

(b) A released captive-bred condor

Figure 60.18 Captive breeding programs. The California condor (*Gymnogyps californicus*), the largest bird in the U.S., with a wingspan of nearly 3 m, has been bred in captivity in California. **(a)** A researcher at the San Diego Wild Animal Park feeds a chick with a puppet so that the birds will not become habituated to the presence of humans. **(b)** This captive-bred condor soars over the Grand Canyon. Note the tag on the underside of its wing.

Genomes & Proteomes Connection

Can Cloning Save Endangered Species?

In 1997, geneticist Ian Wilmut and colleagues at Scotland's Roslin Institute announced to the world that they had cloned a now-famous sheep, Dolly, from mammary cells of an adult ewe (refer back to Figure 20.15). Since then, interest has arisen among conservation biologists about whether the same technology might be used to save species on the verge of extinction. Scientists were encouraged that in January 2001, an Iowa farm cow called Bessie gave birth to a cloned Asian gaur (*Bos gaurus*), an endangered species. The gaur, an oxlike animal native to the jungles of India and Burma, was cloned from a single skin cell taken from a dead animal. To clone the gaur, scientists removed the nucleus from a cow's egg and replaced it with a nucleus from the gaur's cell. The treated egg was then placed into the cow's womb. Unfortunately, the gaur died from dysentery two days after birth, although scientists believe this was unrelated to the cloning procedure. In 2003, another type of endangered wild cattle, the Javan banteng (*Bos javanicus*), was successfully cloned (**Figure 60.19**). In 2005, clones of the African wildcat (*Felis libyca*) successfully produced wildcat kittens. This is the first time that clones of a wild species have bred. In 2009 a cloned Pyrenean ibex (*Capra pyrenaica*) was born but lived only 7 minutes due to physical defects in the lungs. The last wild Pyrenean ibex had died in Spain in 2000. Other candidates for cloning include the Sumatran tiger (*Panthera tigris*) and the giant panda. Cloning extinct animals such as the woolly mammoth (*Mammuthus primigenius*) or Tasmanian tiger (*Thylacinus cynocephalus*) would be more difficult due to a lack of preserved DNA.

Figure 60.19 **Cloning an endangered species.** In 2004, this 8-month-old cloned Javan banteng (*Bos javanicus*) made its public debut at the San Diego Zoo.

BIOLOGY PRINCIPLE The genetic material provides a blueprint for sustaining life. Genetic cloning utilizes a somatic cell, which contains the complete DNA or genetic blueprint of the animal that is to be cloned.

BioConnections: Livestock and even pets have been cloned as well as endangered species (refer back to Figure 20.15 and the chapter-opening photograph of Chapter 20). What are some of the arguments in favor and against genetic cloning of endangered species?

Despite the promise of cloning, a number of issues remain unresolved:

1. Scientists would have to develop an intimate knowledge of different species' reproductive cycles. For sheep and cows, this was routine, based on the vast experience in breeding these species, but eggs of different species, even if they could be harvested, often require different nutritive media in laboratory cultures.
2. Because it is desirable to leave natural mothers available for breeding, scientists will have to identify surrogate females of similar but more common species that can carry the fetus to term.
3. Some argue that cloning does not address the root causes of species loss, such as habitat fragmentation or poaching, and that resources would be better spent elsewhere, for example, in preserving the remaining habitat of endangered species.
4. Cloning might not be able to increase the genetic variability of the population. However, if it were possible to use cells from deceased animals, for example, from their skin, these clones could theoretically reintroduce lost genes back into the population.

Many biologists believe that while cloning may have a role in conservation, it is only part of the solution and that we should address what made the species go extinct before attempting to restore it.

Conservation is clearly a matter of great importance, and a failure to value and protect our natural resources adequately could be a grave mistake. Some authors, most recently the American ecologist and geographer Jared Diamond, have investigated why many societies of the past—including Angkor Wat, Easter Island, and the Mayans—collapsed or vanished, leaving behind monumental ruins. Diamond has concluded that the collapse of these societies occurred partly because people inadvertently destroyed the ecological resources on which their societies depended. Modern nations such as Rwanda face similar issues. The country's population density is the highest in Africa, and it has a limited amount of land that can be used for growing crops. By the late 1980s, the need to feed a growing population led to the wholesale clearing of Rwanda's forests and wetlands, with the result that little additional land was available to farm. Increased population pressure, along with food shortages fueled by environmental scarcity, were likely contributing factors in igniting the genocide of 1994.

As we hope you have seen throughout this textbook, an understanding of biology is vital to comprehending and helping to solve many of society's problems. Within this large field, genomics and proteomics may have a huge potential for improving people's lives and society at large. These disciplines offer the opportunity to unlock new diagnoses and treatments for diseases, to improve nutrition and food production, and even to help restore biological diversity.

Summary of Key Concepts

60.1 What Is Biodiversity?

- Biodiversity represents diversity at three levels: genetic diversity, species diversity, and ecosystem diversity. Conservation biology uses knowledge from molecular biology, genetics, and ecology to protect the biological diversity of life (Figure 60.1).

60.2 Why Conserve Biodiversity?

- The preservation of biodiversity has been justified because of its economic value, because of the value of ecosystem services, and on ethical grounds (Tables 60.1, 60.2, Figure 60.2).
- Four models exist that describe the relationship between biodiversity and ecosystem function: diversity-stability, redundancy, keystone, and idiosyncratic (Figure 60.3).
- Experiments both in the laboratory and in the field have shown that increased biodiversity results in increased ecosystem function (Figures 60.4, 60.5).

60.3 The Causes of Extinction and Loss of Biodiversity

- Extinction—the process by which species die out—has been a natural phenomenon throughout the history of life on Earth. Extinction rates in recent times, however, have been much higher than in the past, a phenomenon called the biodiversity crisis (Figure 60.6).
- The main causes of extinctions have been and continue to be introduced species, direct exploitation, and habitat destruction (Figure 60.7).
- Reduced population size can lead to a reduction of genetic diversity through inbreeding, genetic drift, and limited mating, which reduces effective population size. Inbreeding, mating among genetically related relatives, can lead to a reduction in fertility (Figure 60.8).
- Knowledge of a species' effective population size is vital to ensure the success of conservation projects (Figure 60.9).

60.4 Conservation Strategies

- Habitat conservation strategies commonly target megadiversity countries, countries with the largest number of species; biodiversity hot spots, areas with the largest number of endemic species, those unique to the area; and representative habitats, areas that represent the major habitats (Figures 60.10, 60.11, Table 60.3).
- Conservation biologists employ many strategies in protecting biodiversity. Principles of the equilibrium model of island biogeography and landscape ecology are used in the theory and practice of park reserve design to determine, for example, whether the park should take the form of one single or several small reserves (Figures 60.12, 60.13).
- Economic considerations also play an important role in reserve creation, and it has been shown that conservation spending is positively related to population size (Figure 60.14).
- The single-species approach focuses conservation efforts on indicator species, umbrella species, flagship species, and keystone species (Figures 60.15, 60.16).
- Restoration ecology seeks to repair or replace populations and their habitats. Three basic approaches to habitat restoration are complete restoration, rehabilitation, and ecosystem replacement (Figure 60.17).
- Captive breeding is the propagation of animals outside their natural habitat and reintroducing them to the wild (Figure 60.18).
- Cloning of endangered species has been accomplished on a very small scale and despite its limitations may have a role in conservation biology (Figure 60.19).

Assess and Discuss

Test Yourself

1. Which of the following statements best describes an endangered species?
   a. a species that is likely to become extinct in a portion of its range
   b. a species that has disappeared in a particular community but is present in other natural environments
   c. a species that is extinct
   d. a species that is in danger of becoming extinct throughout all or a significant portion of its range
   e. both b and d
2. Biological diversity is important and should be preserved because
   a. food, medicines, and industrial products are all benefits of biodiversity.
   b. ecosystems provide valuable services to us in many ways.
   c. many species can be used as valuable research tools.
   d. we have an ethical responsibility to protect our environment.
   e. all of the above are correct.
3. The research conducted by Tilman and colleagues demonstrated that
   a. as diversity increases, productivity increases.
   b. as diversity decreases, productivity increases.
   c. areas with higher diversity demonstrate less efficient use of nutrients.
   d. species-richness increases lead to an increase in invasive species.
   e. increased diversity results in increased susceptibility to disease.
4. Approximately what percentage of genetic variation remains in a population of 25 individuals after three generations?
   a. 98 c. 94 e. 84
   b. 96 d. 92
5. What is the effective population size of an island population of parrots of 30 males and 30 females, in which only 10 of the males breed?
   a. 10 c. 30 e. 60
   b. 20 d. 40
6. Saving endangered habitats, such as the Argentine pampas, focuses on
   a. saving genetic diversity.
   b. saving keystone species.
   c. conservation in a megadiversity country.
   d. preserving an area rich in endemic species.
   e. preservation of a representative habitat.
7. Geographic hotspots are those areas rich in
   a. species. d. biodiversity.
   b. habitats. e. endemic species.
   c. rare species.

8. A new canine distemper pathogen that decimates a population of black-footed ferrets is known as a(n):
   a. keystone species.
   b. dominant species.
   c. indicator species.
   d. umbrella species.
   e. flagship species.
9. Small strips of land that connect and allow organisms to move between small patches of natural habitat are called
   a. biological conduits.
   b. edge effects.
   c. movement corridors.
   d. migration pathways.
   e. landscape breaks.
10. Bioremediation is
    a. a process that restores a disturbed habitat to its original state.
    b. a process that uses microbes or plants to detoxify contaminated habitats.
    c. the legislation requiring rehabilitation of a disturbed habitat.
    d. a process of capturing all of the living individuals of a species for breeding purposes.
    e. the process of removing tissue from a dead organism in the hopes of cloning it.

Conceptual Questions

1. Why do managers go to the expense of keeping stud books and moving males and females between zoos to produce offspring?
2. Which types of species are most vulnerable to extinction?
3. A principle of biology is that *biology affects our society*. What is the value of increased biodiversity for human society?

Collaborative Questions

1. Discuss several causes of species extinction.
2. You are called upon to design a park to maximize biodiversity in a tropical country. What are your recommendations?

Online Resource

www.brookerbiology.com

Stay a step ahead in your studies with animations that bring concepts to life and practice tests to assess your understanding. Your instructor may also recommend the interactive eBook, individualized learning tools, and more.

Appendix A

Periodic Table of the Elements

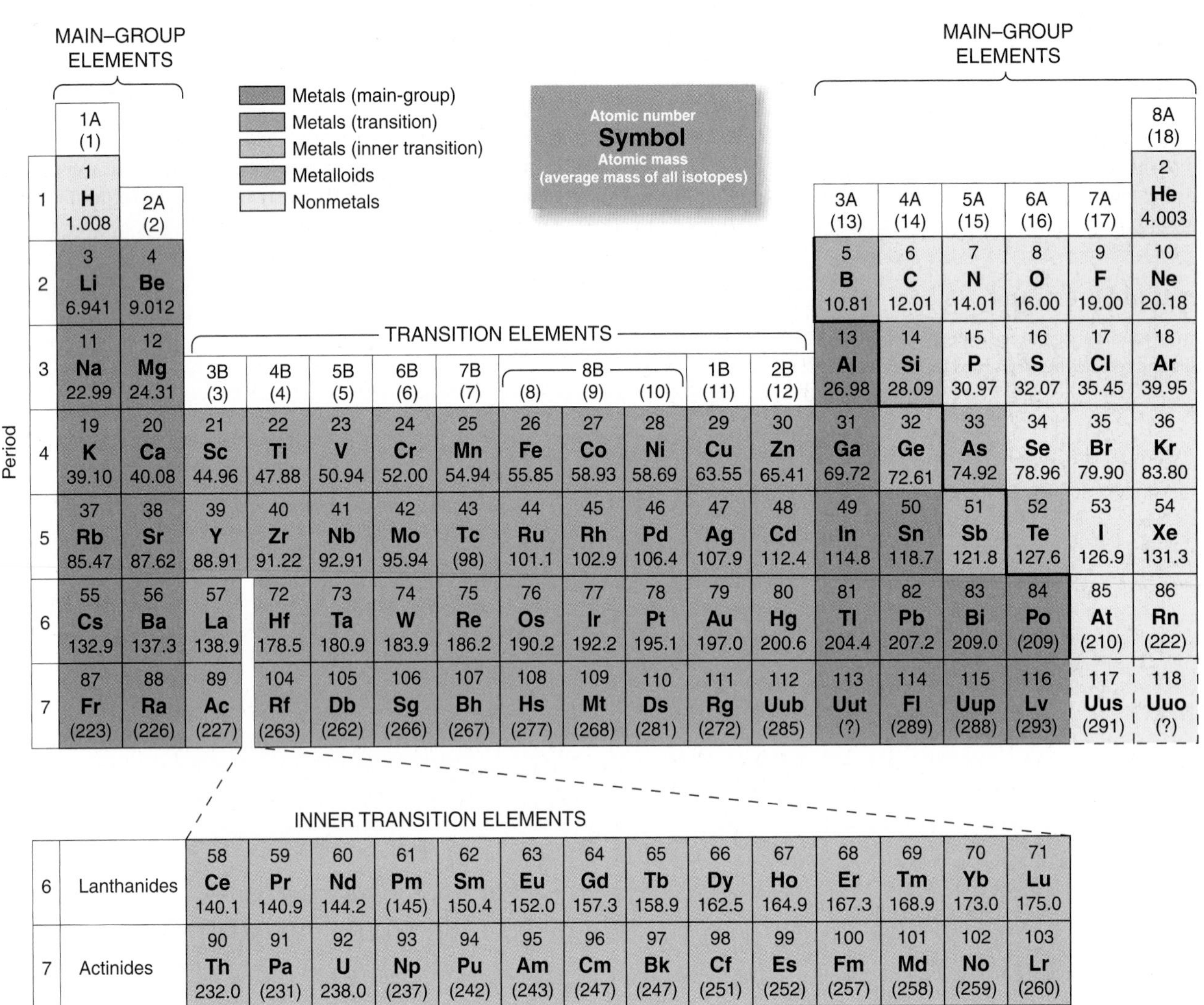

Period	1A (1)	2A (2)	3B (3)	4B (4)	5B (5)	6B (6)	7B (7)	8B (8)	8B (9)	8B (10)	1B (11)	2B (12)	3A (13)	4A (14)	5A (15)	6A (16)	7A (17)	8A (18)
1	1 **H** 1.008																	2 **He** 4.003
2	3 **Li** 6.941	4 **Be** 9.012											5 **B** 10.81	6 **C** 12.01	7 **N** 14.01	8 **O** 16.00	9 **F** 19.00	10 **Ne** 20.18
3	11 **Na** 22.99	12 **Mg** 24.31											13 **Al** 26.98	14 **Si** 28.09	15 **P** 30.97	16 **S** 32.07	17 **Cl** 35.45	18 **Ar** 39.95
4	19 **K** 39.10	20 **Ca** 40.08	21 **Sc** 44.96	22 **Ti** 47.88	23 **V** 50.94	24 **Cr** 52.00	25 **Mn** 54.94	26 **Fe** 55.85	27 **Co** 58.93	28 **Ni** 58.69	29 **Cu** 63.55	30 **Zn** 65.41	31 **Ga** 69.72	32 **Ge** 72.61	33 **As** 74.92	34 **Se** 78.96	35 **Br** 79.90	36 **Kr** 83.80
5	37 **Rb** 85.47	38 **Sr** 87.62	39 **Y** 88.91	40 **Zr** 91.22	41 **Nb** 92.91	42 **Mo** 95.94	43 **Tc** (98)	44 **Ru** 101.1	45 **Rh** 102.9	46 **Pd** 106.4	47 **Ag** 107.9	48 **Cd** 112.4	49 **In** 114.8	50 **Sn** 118.7	51 **Sb** 121.8	52 **Te** 127.6	53 **I** 126.9	54 **Xe** 131.3
6	55 **Cs** 132.9	56 **Ba** 137.3	57 **La** 138.9	72 **Hf** 178.5	73 **Ta** 180.9	74 **W** 183.9	75 **Re** 186.2	76 **Os** 190.2	77 **Ir** 192.2	78 **Pt** 195.1	79 **Au** 197.0	80 **Hg** 200.6	81 **Tl** 204.4	82 **Pb** 207.2	83 **Bi** 209.0	84 **Po** (209)	85 **At** (210)	86 **Rn** (222)
7	87 **Fr** (223)	88 **Ra** (226)	89 **Ac** (227)	104 **Rf** (263)	105 **Db** (262)	106 **Sg** (266)	107 **Bh** (267)	108 **Hs** (277)	109 **Mt** (268)	110 **Ds** (281)	111 **Rg** (272)	112 **Uub** (285)	113 **Uut** (?)	114 **Fl** (289)	115 **Uup** (288)	116 **Lv** (293)	117 **Uus** (291)	118 **Uuo** (?)

INNER TRANSITION ELEMENTS

Period	Series														
6	Lanthanides	58 **Ce** 140.1	59 **Pr** 140.9	60 **Nd** 144.2	61 **Pm** (145)	62 **Sm** 150.4	63 **Eu** 152.0	64 **Gd** 157.3	65 **Tb** 158.9	66 **Dy** 162.5	67 **Ho** 164.9	68 **Er** 167.3	69 **Tm** 168.9	70 **Yb** 173.0	71 **Lu** 175.0
7	Actinides	90 **Th** 232.0	91 **Pa** (231)	92 **U** 238.0	93 **Np** (237)	94 **Pu** (242)	95 **Am** (243)	96 **Cm** (247)	97 **Bk** (247)	98 **Cf** (251)	99 **Es** (252)	100 **Fm** (257)	101 **Md** (258)	102 **No** (259)	103 **Lr** (260)

The complete Periodic Table of the Elements. Group numbers are different in some cases from those presented in Figure 2.5, because of the inclusion of transition elements. In some cases, the average atomic mass has been rounded to one or two decimal places, and in others only an estimate is given in parentheses due to the short-lived nature or rarity of those elements. The symbols and names of some of the elements between 112–118 are temporary until the chemical characteristics of these elements become better defined. Element 117 is currently not confirmed as a true element, and little is known about element 118. The International Union of Pure and Applied Chemistry (IUPAC) has recently proposed adopting the name copernicium (Cp) for element 112 in honor of scientist and astronomer Nicolaus Copernicus.

Appendix B

Answer Key

Answers to Collaborative Questions can be found on the website.

Chapter 1

Concept Checks

Figure 1.5 It would be at the population level.

Figure 1.6 In monkeys, the tail has been modified to grasp onto things, such as tree branches. In skunks, the tail is modified with a bright stripe; the tail can stick up and act as a warning signal to potential predators. In cattle, the tail has long hairs and is used to swat insects. Many more examples are possible.

Figure 1.7 Natural selection is a process that causes evolution to happen.

Figure 1.9 A tree of life suggests that all living organisms evolved from a single ancestor by vertical evolution with mutation. A web of life assumes that both vertical evolution with mutation and horizontal gene transfer were important mechanisms in the evolution of new species.

Figure 1.11 Taxonomy helps us appreciate the unity and diversity of life. Organisms that are closely related evolutionarily are placed in smaller groups.

Figure 1.12 The genome stores the information to make an organism's proteins. In and of itself, the genome is merely DNA. The traits of cells and organisms are largely determined by the structures and functions of the hundreds or thousands of different proteins they make.

Figure 1.13 Male and female clownfish have the same genomes. Hormones affect the expression of their genes, thereby altering their proteomes. Differences in proteomes determine the morphological differences between males and females.

Figure 1.15 A researcher can compare the results in the experimental group and control group to determine if a single variable is causing a particular outcome in the experimental group.

Figure 1.16 After the *CF* gene was identified by discovery-based science, researchers realized that the *CF* gene was similar to other genes that encoded proteins that were already known to be transport proteins. This provided an important clue that the *CF* gene also encodes a transport protein.

BioConnections

Figure 1.4 This figure is emphasizing that living organisms grow and develop.

Figure 1.10 Fungi are more closely related to animals.

Feature Investigation Questions

1. In discovery-based science, a researcher does not need to have a preconceived hypothesis. Experimentation is conducted in the hope that it may have practical applications or may provide new information that will lead to a hypothesis. By comparison, hypothesis testing occurs when a researcher forms a hypothesis that makes certain predictions. Experiments are conducted to see if those predictions are correct. In this way, the hypothesis may be accepted or rejected.
2. This strategy may be described as a five-stage process:
   1. Observations are made regarding natural phenomena.
   2. These observations lead to a hypothesis that tries to explain the phenomena. A useful hypothesis is one that is testable because it makes specific predictions.
   3. Experimentation is conducted to determine if the predictions are correct.
   4. The data from the experiment are analyzed.
   5. The hypothesis is accepted or rejected.
3. In an ideal experiment, the control and experimental groups differ by only one factor. Biologists apply statistical analyses to their data to determine if the control and experimental groups are likely to be different from each other because of the single variable that is different between the two groups. This provides an objective way to accept or reject a hypothesis.

Test Yourself

1. d 2. a 3. c 4. c 5. d 6. b 7. d 8. d 9. a 10. b

Conceptual Questions

1. Principles (a) through (f) apply to individuals whereas (g) and (h) apply to populations.
2. The unity among different species occurs because modern species have evolved from a group of related ancestors. Some of the traits in those ancestors are also found in modern species, which thereby unites them. The diversity is due to the variety of environments on the Earth. Each species has evolved to occupy its own unique environment. For every species, many traits are evolutionary adaptations to survival in a specific environment. For this reason, evolution also promotes diversity.
3. The principles are outlined in Figure 1.4. Students can rephrase these principles in their own words.

APPENDIX B

Chapter 2

Concept Checks

Figure 2.4 An energy shell is a region outside the nucleus of an atom occupied by electrons of a given energy level. More than one orbital can be found within an electron shell. An orbital may be spherical or dumbbell-shaped and contains up to two electrons.

Figure 2.9 The octet rule states that atoms are stable when they have eight electrons in their outermost shell. Oxygen has six electrons in its outer shell. When two oxygen atoms share two pairs of electrons, each atom has eight electrons in its outer shell, at least part of the time.

Figure 2.11 Strand separation requires energy, because the DNA strands are held together by a large number of hydrogen bonds. Although each hydrogen bond is weak, collectively the vast number of such bonds in a molecule of DNA adds up to a considerable strength.

Figure 2.17 The oil would be in the center of the soap micelles.

Figure 2.20 It is 10^{-6} M. Since $[H^+][OH^-]$ always equals 10^{-14} M, if $[H^+] = 10^{-8}$ M (that is, pH 8.0), then $[OH^-]$ must be 10^{-6} M.

BioConnections

Figure 2.20 At a pH of 5.0, the H^+ concentration would be 10^{-5}M, as can be seen from the figure but which can also be calculated by the equation $pH = -\log_{10}[H^+]$. (From this information, you can also determine that the OH^- concentration must be 10^{-9} M, because the product of the H^+ and OH^- concentrations must be equal to 10^{-14} M).

Feature Investigation Questions

1. Scientists were aware that atoms contained charged particles. Many believed that the positive charges and mass were evenly distributed throughout the atom.
2. Rutherford was testing the hypothesis that atoms are composed of positive charges evenly distributed throughout the atom. Based on this model of the structure of the atom, alpha particles, which are positively charged nuclei of helium atoms, should be deflected as they pass through the foil, due to the presence of positive charges spread throughout the gold foil.
3. Instead of detecting slight deflection of most alpha particles as they passed through the gold foil, the majority, 98%, of the alpha particles passed directly through the gold foil without deflection. A much smaller percentage either

deflected or bounced back from the gold foil. Rutherford suggested that since most of the alpha particles passed unimpeded through the gold foil, most of the volume of atoms is empty space. Rutherford also proposed that the bouncing back of some of the alpha particles indicated that most of the positively charged particles were concentrated in a compact area. These results ran counter to the hypothesized model.

Test Yourself

1. b 2. b 3. b 4. d 5. e 6. e 7. e 8. c 9. e 10. b

Conceptual Questions

1. Covalent bonds are bonds in which atoms share electrons. A hydrogen bond is a weak polar covalent bond that forms when a hydrogen atom from one polar molecule becomes electrically attracted to an electronegative atom. A nonpolar covalent bond is one between two atoms of similar electronegativities, such as two carbon atoms. The van der Waal forces are temporary, weak bonds, resulting from random electrical forces generated by the changing distributions of electrons in the outer shells of nearby atoms. The strong attraction between two oppositely charged atoms forms an ionic bond.
2. Within limits, bonds within molecules can rotate and thereby change the shape of a molecule. This is important because it is the shape of a molecule that determines, in part, the ability of that molecule to interact with other molecules. Also, when two molecules do interact through such forces as hydrogen bonds, the shape of one or both molecules may change as a consequence. The change in shape is often part of the mechanism by which signals are sent within and between cells.
3. A good example of emergent properties at the molecular level is that of the formation of sodium chloride (NaCl), a solid white crystalline compound that is very important for most living organisms. In their elemental states, sodium is a soft, highly reactive metal and chlorine is a toxic gas. When they combine through ionic bonds, the two elements produce a completely new and harmless substance found in all the world's oceans and soils. Another example described in this chapter is water, a liquid that is vital for all life but which is formed from two gases, hydrogen and oxygen, with very different properties.

APPENDIX B

Chapter 3

Concept Checks

Figure 3.1 Due to the fact that he had earlier purified urea from urine and then formed urea crystals, he already knew what urea crystals looked like. As seen in this figure, they are quite distinctive looking. Therefore, when he reacted ammonia and cyanic acid and got a compound that formed crystals, the distinctive look of the crystals made him realize that he had synthesized urea.

Figure 3.6 One reason is that the binding of a molecule to an enzyme depends on the spatial arrangements of the atoms in that molecule. Enantiomers have different spatial relationships that are mirror images of each other. Therefore, one may bind very tightly to an enzyme and the other may not be recognized at all.

Figure 3.7 Recall from Figure 3.5 that the reverse of a dehydration reaction is called a hydrolysis reaction, in which a molecule of water is added to the molecule being broken down, resulting in the formation of monomers.

Figure 3.11 Hydrogenation is the addition of hydrogens to double-bonded carbon atoms, changing them from unsaturated to saturated. This causes them to be solid at room temperature.

Figure 3.12 The phospholipids would be oriented such that their polar regions dissolved in the water layer and the nonpolar regions dissolved in the oil. Thus, the phospholipids would form a layer at the interface between the water and oil.

Figure 3.15 71; one less than the number of amino acids in the polypeptide.

Figure 3.19 If the primary structure of protein 1 were altered in some way, this would, in turn, most likely alter the secondary and tertiary structures of protein 1. Therefore, it is possible that the precise fit between proteins 1 and 2 would be lost and that the two proteins would lose the ability to interact.

Figure 3.24 Yes. The opposite strand must be the mirror image of the first strand, because pairs can form only between A and T, and G and C. For instance, if a portion of the first strand is AATGCA, the opposite strand along that region would be TTACGT.

BioConnections

Figure 3.8 Cellulose is believed to be the most abundant organic molecule on earth. In addition to being part of plant cells, it is also found in many other organisms, including many protists.

Figure 3.21 Many intracellular signaling molecules, such as those described in Chapter 9, are not proteins and therefore do not have domain structure. Some are very small molecules with a relatively simple structure; some are actually ions like Ca^{2+}.

Feature Investigation Questions

1. Many scientists assumed that protein folding was directed by some cellular factor, meaning some other molecule in the cytoplasm, and therefore, protein folding could not occur spontaneously. Others assumed that protein folding was determined somehow by the ribosome, because this organelle is primarily responsible for synthesizing proteins.
2. Anfinsen was testing the hypothesis that the information necessary for determining the three-dimensional shape of a protein is contained within the protein itself. In other words, the chemical characteristics of the amino acids that make up a protein determines the three-dimensional shape.
3. The urea disrupts hydrogen bonds and ionic interactions that are necessary for protein folding. The mercaptoethanol disrupted the S—S bonds that also form between certain amino acids of the same polypeptide chains. Both substances essentially allow the polypeptide chain to unfold, disrupting the three-dimensional shape. Anfinsen removed the urea and mercaptoethanol from the protein solution by size-exclusion chromatography. After removing the urea and mercaptoethanol, Anfinsen discovered that the protein refolded into its proper three-dimensional shape and became functional again. This was important because the solution contained only the protein and lacked any other cellular material that could possibly assist in protein folding. This demonstrated that the protein could refold into the functional conformation.

Test Yourself

1. b 2. b 3. e 4. b 5. c 6. b 7. b 8. d 9. b 10. b

Conceptual Questions

1. Isomers are two structures with an identical molecular formula but with different structures and arrangements of atoms within the molecule. There are two major types of isomers: structural and stereoisomers. Because many chemical reactions in biology depend on the actions of enzymes, which are often highly specific for the spatial arrangement of atoms in a molecule, one isomer of a pair may have biological functions, and the other may not.
2. Saturated fatty acids are saturated with hydrogen and have only single (C—C) bonds, whereas unsaturated fatty acids have one or more double (C=C) bonds. The double bonds in unsaturated fatty acids alters their shape, resulting in a kink in the structure. Saturated fatty acids are unkinked and are better able to stack tightly together. Fats containing saturated fatty acids have a higher melting point than those containing mostly unsaturated fatty acids; consequently, saturated fats tend to be solids at room temperatures, and unsaturated fatty acids are usually liquids at room temperature.
3. The structures of macromolecules in all cases determines their function. For example, the structure of a protein determines its three-dimensional shape. This, in turn, allows a protein to interact specifically with certain other molecules. Certain intracellular signaling molecules, for example, have shapes that are determined by the structural arrangement of various protein domains; by themselves, these domains may have no function, but when combined in a precise way they create a functional protein. Likewise, the structure of different lipids determines such functional characteristics as male/female differences, cellular membrane formation, and energy storage. The different structures of polysaccharides determine their usefulness as energy stores, or as components of plant cell walls.

Chapter 4

Concept Checks

Figure 4.1 You would use transmission electron microscopy. The other methods do not have good enough resolution.

Figure 4.3 The primary advantage is that it gives an image of the 3-D surface of a material.

Figure 4.6 They have different proteomes.

Figure 4.7 Centrioles: Not found in plant cells; their role is not entirely clear, but they are found in the centrosome, which is where microtubules are anchored.

Chloroplasts: Not found in animal cells; function in photosynthesis.

Cell wall: Not found in animal cells; important in cell shape.

Figure 4.12 Both dynein and microtubules are anchored in place. Using ATP as a source of energy, dynein tugs on microtubules. Because the microtubules are anchored, they bend in response to the force exerted by dynein.

Figure 4.15 The nuclear lamina organizes the nuclear envelope and also helps to organize/anchor the chromosomes. The nuclear matrix is inside the nucleus and helps to organize the chromosomes into chromosome territories.

Figure 4.18 The protein begins its synthesis in the cytosol and then is further synthesized into the ER. It travels via vesicles to the *cis*-, *medial*-, and *trans*-Golgi and then is secreted.

Figure 4.22 Of these three choices, membrane transport is probably the most important because it regulates which molecules can enter the cell and participate in metabolism and which products of metabolism are exported from the cell. Cell signaling may affect metabolism, but its overall effect is probably less important than membrane transport.

Figure 4.24 It increases the surface area where ATP synthesis takes place, thereby making it possible to increase the amount of ATP synthesis.

Figure 4.27 Bacteria and mitochondria are similar in size; they both have circular chromosomes; they both divide by binary fission; and they both make ATP. Bacterial chromosomes are larger, and they make all of their own cellular proteins. Mitochondria chromosomes are smaller, and they import most of their proteins from the cytosol.

Figure 4.29 The signal sequence of an ER protein is recognized by SRP, which halts translation. The emerging protein and its ribosome are then transferred to the ER membrane, where translation resumes.

Figure 4.31 If chaperone proteins were not found in the cytosol, the mitochondrial matrix protein would start to fold, which might prevent it from being able to pass through the channel in the outer and inner mitochondrial membrane. Normally, a protein is threaded through this channel in an unfolded state.

BioConnections

Figure 4.6 Alternative splicing produces proteins with slightly different structures, because they have certain regions that have different amino acid sequences. The functions of such proteins are often similar, but specialized for the cell type in which they are expressed.

Figure 4.8 The surfaces of cells involved with gas exchange are highly convoluted. This provides a much greater surface area, thereby facilitating the movement of gasses across the membrane.

Figure 4.11 The type of movement shown in part (b) occurs during muscle contraction.

Figure 4.13 Cilia carry out a variety of functions, including motility, moving food particles into a feeding groove, and dispersing across moist surfaces.

Figure 4.16 During cell division, the chromosomes condense and form more compact structures.

Figure 4.26 These processes are very similar in that DNA replication occurs and then mitochondria or bacterial cells split in two. They are different in that bacteria have cell walls and septa must form between the two daughter cells, which does not occur when mitochondria divide.

Feature Investigation Questions

1. In a pulse-chase experiment, radioactive material is provided to cells. This is referred to as the pulse, or single administration of the radioactive material to the cells. After a few minutes, a large amount of nonradioactive material is provided to the cells to remove or "chase away" any of the remaining radioactive material.

   The researchers were attempting to determine the movement of proteins through the different compartments of a cell. Radioactive amino acids were used to label the proteins and enable the researchers to visualize where the proteins were at different times.
2. Pancreatic cells produce large numbers of proteins that are secreted from the cell. The number and final location of the proteins would allow the researchers an ideal system for studying protein movement through the cell.
3. Using electron microscopy, the researchers found that the proteins, indicated by radioactivity, were first found in the ER of the cells. Later the radioactivity moved to the Golgi and then into vesicles near the plasma membrane.

   The researchers concluded that secreted proteins move through several cellular compartments before they are secreted from the cell. Also, the movement of proteins through these compartments is not random but follows a particular pathway: ER, Golgi, secretory vesicles, plasma membrane, and, finally, secreted.

Test Yourself

1. d 2. d 3. b 4. c 5. e 6. a 7. e 8. e 9. a It would go there first, because targeting to the ER occurs cotranslationally. 10. c It is true they carry out metabolism, but so do eukaryotic cells.

Conceptual Questions

1. There are a lot of possibilities. The interactions between a motor protein (dynein) and cytoskeletal filaments (microtubules) cause a flagellum to bend. The interaction between V-snares and T-snares causes a vesicle to fuse with the correct target membrane.
2. If the motor is bound to a cargo and the motor can walk along a filament that is fixed in place, this will cause the movement of the cargo when the motor is activated. If the motor is fixed in place and the filament is free to move, this will cause the filament to move when the motor is activated. If both the motor and filament are fixed in place, the activation of the motor will cause the filament to bend.
3. ATP synthesis occurs along the inner mitochondrial membrane. The invaginations of this membrane greatly increase its surface area, thereby allowing for a greater amount of ATP synthesis.

Chapter 5

Concept Checks

Figure 5.3 Probably not. The hydrophobic tails of both leaflets touch each other, so the heavy metal would probably show a single, thick dark line. Osmium tetroxide shows two parallel lines because it labels the polar head groups, which are separated by the hydrophobic interior of the membrane.

Figure 5.4 More double bonds and shorter fatty acyl tails make the membrane more fluid. Changing the cholesterol concentration can also affect fluidity, but that depends on the starting level of cholesterol. If cholesterol was at a level that maximized stability, increasing the cholesterol concentration would probably increase fluidity.

Figure 5.5 The low temperature prevents lateral diffusion of membrane proteins. Therefore, after fusion, all of the mouse proteins would stay on one side of the fused cell, and all of the human proteins would remain on the other.

Figure 5.7 Lipids are transferred to the other leaflet of the ER via enzymes called flippases.

Figure 5.8 The most common way for a transmembrane segment to form is that it contains a stretch (about 20) of amino acids that mostly have hydrophobic (nonpolar) side chains.

Figure 5.12 Although both of these molecules penetrate the bilayer fairly quickly, methanol has a polar $-OH$ group and therefore crosses a bilayer more slowly than methane.

Figure 5.15 Water will move from outside to inside, from the hypotonic medium into the hypertonic medium.

Figure 5.17 The purpose of gating is to regulate the function of channels.

Figure 5.22 The Na^+/K^+-ATPase could reach the point where the protein was covalently phosphorylated and Na^+ was released on the outside. At that stage, the reaction would stop, because it needs K^+ to proceed through the rest of the cycle.

Figure 5.23 The protein coat is needed for the membrane to bud from its site and form a vesicle.

BioConnections

Figure 5.6 Transmembrane proteins called cell adhesion molecules bind to each other to promote cell-to-cell adhesion. In addition, they can bind to molecules in the extracellular matrix, such as collagen fibers, thereby causing a cell to adhere to the extracellular matrix.

Figure 5.11 Leucine would cross an artificial membrane more easily because it is more hydrophobic than lysine.

Figure 5.13 Gradients of sodium and potassium ions are important for the conduction of action potentials.

Feature Investigation Questions

1. Most cells allow movement of water across the cell membrane by passive diffusion. However, it was noted that certain cell types had a much higher rate of water movement, indicating that something different was occurring in these cells.
2. The researchers identified water channels by characterizing proteins that are present in red blood cells and kidney cells but not other types of cells. Red blood cells and kidney cells have a faster rate of water movement across the membrane than other cell types. These cells are more likely to have water channels. By identifying proteins that are found in both of these types of cells but not in other cells, the researchers were identifying possible candidate proteins that function as water channels. In addition, CHIP28 had a structure that resembled other known channel proteins.

Agre and his associates experimentally created multiple copies of the gene that produces the CHIP28 protein and then artificially transcribed the genes to produce many mRNAs. The mRNAs were injected into frog oocytes where they could be translated to make the CHIP28 proteins. After altering the frog oocytes by introducing the CHIP28 mRNAs, they compared the rate of water transport in the altered oocytes versus normal frog oocytes. This procedure allowed them to introduce the candidate protein to a cell type that normally does not have the protein present.

3. After artificially introducing the candidate protein into the frog oocytes, the researchers found that the experimental oocytes took up water at a much faster rate in a hypotonic solution as compared to the control oocytes. The results indicated that the presence of the CHIP28 protein did increase water transport into cells.

Test Yourself

1. c 2. c 3. b 4. d 5. b 6. e 7. d 8. e 9. e 10. c

Conceptual Questions

1. See Figure 5.1 for the type of drawing you should have made. The membrane is considered a mosaic of lipid, protein, and carbohydrate molecules. The membrane exhibits properties that resemble a fluid because lipids and proteins can move relative to each other within the membrane.
2. Integral membrane proteins can contain transmembrane segments that cross the membrane, or they may contain lipid anchors. Peripheral membrane proteins are noncovalently bound to integral membrane proteins or to the polar heads of phospholipids.
3. Lipid bilayers, channels, and transporters cause the plasma membrane to be selectively permeable. This allows a cell to take up needed nutrients from its extracellular environment and to export waste products into the environment.

Chapter 6

Concept Checks

Figure 6.2 The solution of dissolved Na^+ and Cl^- has more entropy. A salt crystal is very ordered, whereas the ions in solution are much more disordered.

Figure 6.4 If a large amount of ADP was broken down, it would be more difficult for the cell to make ATP, which is made by the attachment of a phosphate group to ADP. The ATP cycle would be inhibited.

Figure 6.5 It speeds up the rate. When the activation energy is lower, it takes less time for reactants to reach a transition state where a chemical reaction can occur. It does not affect the direction of a reaction.

Figure 6.6 The activation energy is lowered during the second step when the substrates undergo induced fit.

Figure 6.7 At a substrate concentration of 0.5 mM, enzyme A would have a higher velocity. Enzyme A would be very near its V_{max}, whereas enzyme B would be well below its V_{max}.

Figure 6.12 The oxidized form is NAD^+.

Figure 6.15 Protein degradation eliminates proteins that are worn out, misfolded, or no longer needed by the cell. Such proteins could interfere with normal cell function. In addition, the recycling of amino acids saves the cell energy.

BioConnections

Figure 6.8 Thermophilic bacteria have enzymes that function at a very high temperature, perhaps with a temperature optimum of 50 to 70°C.

Figure 6.9 Without RNase P, the tRNAs could not be converted to their mature forms. The pre-tRNAs would be too large to fit into the ribosome. Therefore, translation would be inhibited.

Figure 6.13 Feedback inhibition prevents the excessive breakdown of carbohydrates and thereby prevents the excessive synthesis of ATP. Cells don't waste energy making ATP if they don't need it.

Feature Investigation Questions

1. RNase P has both a protein and RNA subunit. To determine which subunit has catalytic function, it was necessary to purify them individually and then see which one is able to cleave ptRNA.
2. The experimental strategy was to incubate RNase P or subunits of RNase P with ptRNA and then run a gel to determine if ptRNA had been cleaved to a mature tRNA and a 5′ fragment. The control without protein was to determine if the RNA alone could catalyze the cleavage. The control without RNA was to determine if some other factor in the experiment (for instance, Mg^{2+} or protein) was able to cleave the ptRNA.
3. The critical results occurred when the researchers incubated the purified RNA subunit at high Mg^{2+} concentrations with the ptRNA. Under these conditions, the ptRNA was cleaved. These results indicate that the RNA subunit has catalytic activity. A high Mg^{2+} concentration is needed to keep it catalytically active in the absence of a protein subunit.

Test Yourself

1. d 2. e 3. b 4. d 5. a 6. c 7. b 8. c 9. e 10. a

Conceptual Questions

1. Exergonic reactions are spontaneous. They proceed in a particular direction. An exergonic reaction could be slow or fast. By comparison, an endergonic reaction is not spontaneous. It will not proceed in a particular direction unless free energy is supplied. An endergonic reaction can be fast or slow.
2. During feedback inhibition, the product of a metabolic pathway binds to an allosteric site on an enzyme that acts earlier in the pathway. The product inhibits this enzyme, thereby preventing the overaccumulation of the product.
3. Recycling of amino acids and nucleotides is important because it conserves a great deal of energy. Cells don't have to remake these building blocks, which would require a large amount of energy. Eukaryotes primarily use the proteasome to recycle proteins.

Chapter 7

Concept Checks

Figure 7.2 The first phase is named the energy investment phase because some ATP is used up. The second phase is called the cleavage phase because a 6-carbon molecule is broken down into two 3-carbon molecules. The energy liberation phase is so named because NADH and ATP are made.

Figure 7.3 The molecules that donate phosphates are 1,3-bisphosphoglycerate and phosphoenolpyruvate.

Figure 7.6 For each acetyl group that is oxidized, the main products are 2 CO_2, 3 NADH, 1 $FADH_2$, and 1 GTP.

Figure 7.8 It is called cytochrome oxidase because it removes electrons from (oxidizes) cytochrome *c*. Another possible name would be oxygen reductase because it reduces oxygen.

Figure 7.10 No. The role of the electron transport chain is to make an H^+ electrochemical gradient. It is the H^+ electrochemical gradient that drives ATP synthase. If the H^+ electrochemical is made another way, such as by bacteriorhodopsin, the ATP synthase still makes ATP.

Figure 7.11 The γ subunit turns clockwise, when viewed from the intermembrane space. The β subunit in the back right is in conformation 3, and the one on the left is in conformation 1.

Figure 7.14 The advantage is that the cell can use the same enzymes to metabolize different kinds of organic molecules. This saves the cell energy because it is costly to make a lot of different enzymes, which are composed of proteins.

BioConnections

Figure 7.2 Glycolytic muscle fibers rely on glycolysis for their ATP needs. Because glycolysis does not require oxygen, such muscle fibers can function without oxygen.

Figure 7.4 FDG is radiolabeled so it can be specifically detected by a PET scan.

Feature Investigation Questions

1. The researchers attached an actin filament to the γ subunit of ATP synthase. The actin filament was fluorescently labeled so the researchers could determine if the actin filament moved when viewed under the fluorescence microscope.
2. When functioning in the hydrolysis of ATP, the actin filament was seen to rotate. The actin filament was attached to the γ subunit of ATP synthase. The rotational movement of the filament was the result of the rotational movement of the enzyme. In the control experiment, no ATP was added to stimulate enzyme activity. In the absence of ATP, no movement was observed.
3. No, the observation of counterclockwise rotation is the opposite of what would be expected inside the mitochondria. During the experiment, the enzyme was not functioning in ATP synthesis but instead was running backwards and hydrolyzing ATP.

Test Yourself

1. a 2. d 3. b 4. b 5. a 6. d 7. b 8. d 9. d 10. b

Conceptual Questions

1. The purpose of the electron transport chain is to pump H^+ across the inner mitochondrial membrane to establish a H^+ electrochemical gradient. When the H^+ flows back across the membrane through ATP synthase, ATP is synthesized.
2. The movement of H^+ through the *c* subunits causes the γ subunit to rotate. As it rotates, it sequentially alters the conformation of the subunits, where ATP is made. This causes (1) ADP and P_i to bind with moderate affinity, (2) ADP and P_i to bind very tightly such that ATP is made, and (3) ATP to be released.
3. As discussed in this chapter, the phases of glucose metabolism are regulated in a variety of ways. For example, key enzymes in glycolysis and the citric acid cycle are regulated by the availability of substrates and by feedback inhibition. The electron transport chain is regulated by the ATP/ADP ratio. Such regulation ensures that a cell does not waste energy making ATP when it is in sufficient supply. Also, the production of too much NADH is potentially harmful because at high levels it has the potential to haphazardly donate its electrons to other molecules and promote the formation of free radicals, highly reactive chemicals that damage DNA and cellular proteins.

Chapter 8

Concept Checks

Figure 8.1 Both heterotrophs and autotrophs carry out cellular respiration.

Figure 8.3 The Calvin cycle can occur in the dark as long as there is sufficient CO_2, ATP, and NADPH.

Figure 8.4 Gamma rays have higher energy than radio waves.

Figure 8.5 To drop down to a lower orbital at a lower energy level, an electron could release energy in the form of heat, release energy in the form of light, or transfer energy to another electron by resonance energy transfer.

Figure 8.7 By having different pigment molecules, plants can absorb a wider range of wavelengths of light.

Figure 8.8 ATP and NADPH are made in the stroma. O_2 is made in the thylakoid lumen.

Figure 8.9 Noncyclic electron flow produces equal amounts of ATP and NADPH. However, plants usually need more ATP than NADPH. Cyclic photophosphorylation allows plants to make just ATP, thereby increasing the relative amount of ATP.

Figure 8.10 Because these two proteins are homologous, this means that the genes that encode them were derived from the same ancestral gene. Therefore, the amino acid sequences of these two proteins are expected to be very similar, though not identical. Because the amino acid sequence of a protein determines its structure, two proteins with similar amino acid sequences would be expected to have similar structures.

Figure 8.13 No. The enhancement effect occurs because the flashes activate both photosystem II and photosystem I. Light at 700 nm is needed to activate P700 in photosystem I.

Figure 8.14 An electron has its highest amount of energy just after it has been boosted by light in PSI.

Figure 8.15 NADPH reduces organic molecules and makes them more able to form C—C and C—H bonds.

Figure 8.18 The arrangement of cells in C_4 plants makes the level of CO_2 high and the level of O_2 low in the bundle sheath cells.

Figure 8.19 When there is plenty of moisture and it is not too hot, C_3 plants are more efficient. However, under hot and dry conditions, C_4 and CAM plants have the advantage because they lose less water and avoid photorespiration.

BioConnections

Figure 8.2 Two guard cells make up one stoma.

Figure 8.18 Water is taken up by the roots of plants and moves via the vascular system to the leaves.

Feature Investigation Questions

1. The researchers were attempting to determine the biochemical pathway of the process of carbohydrate synthesis in plants. The researchers wanted to identify different molecules produced in plants over time to determine the steps of the biochemical pathway.
2. The purpose for using ^{14}C was to label the different carbon molecules produced during the biochemical pathway. The researchers could "follow" the carbon molecules from CO_2 that were incorporated into the organic molecules during photosynthesis. The radioactive isotope provided the researchers with a method of labeling the different molecules.

   The purpose of the experiment was to determine the steps in the biochemical pathway of photosynthesis. By examining samples from different times after the introduction of the labeled carbon source, the researchers would be able to determine which molecules were produced first and, thus, products of the earlier steps of the pathway versus products of later steps of the pathway.

   The researchers used two-dimensional paper chromatography to separate the different molecules from each other. Afterward, the different molecules were identified by different chemical methods. The text describes the method of comparing two-dimensional paper chromatography results of unknown molecules to known molecules and identifying the unknown with the known molecule it matched.
3. The researchers were able to determine the biochemical process that plants use to incorporate CO_2 into organic molecules. The researchers were able to identify the biochemical steps and the molecules produced at these steps in what is now called the Calvin cycle.

Test Yourself

1. c 2. c 3. c 4. a 5. b 6. b 7. c 8. b 9. e 10. c

Conceptual Questions

1. The two stages of photosynthesis are the light reactions and the Calvin cycle. The key products of the light reactions are ATP, NADPH, and O_2. The key product of the Calvin cycle is carbohydrate. The initial product is G3P, which is used to make sugars and other organic molecules.
2. NADPH is used during the reduction phase of the Calvin cycle. It donates its electrons to 1,3-BPG.
3. At the level of the biosphere, the role of photosynthesis is to incorporate carbon dioxide into organic molecules. These organic molecules can then be broken down, by autotrophs and by heterotrophs, to make ATP. The organic molecules made during photosynthesis are also used as starting materials to synthesize a wide variety of organic molecules and macromolecules that are made by cells.

Chapter 9

Concept Checks

Figure 9.1 It is glucose.

Figure 9.3 Endocrine signals are more likely to exist for a longer period of time. This is necessary because endocrine signals called hormones travel relatively long distances to reach their target cells. Therefore, the hormone must exist long enough to reach its target cells.

Figure 9.4 The effect of a signaling molecule is to cause a cellular response. Most signaling molecules do not enter the cell. Therefore, to exert an effect, they must alter the conformation of a receptor protein, which, in turn, stimulates an intracellular signal transduction pathway that leads to a cellular response.

Figure 9.6 Phosphorylation of a protein via a kinase involves ATP hydrolysis, which is an exergonic reaction. The energy from this reaction usually alters the conformation of the phosphorylated protein, thereby influencing its function. Phosphorylation is used to regulate protein function.

Figure 9.7 The α subunit has to hydrolyze its GTP to GDP. This changes the conformation of the α subunit so that it can reassociate with the β and γ subunits.

Figure 9.12 The signal transduction pathway begins with the G protein and ends with protein kinase A being activated. The cellular response involves the phosphorylation of target proteins. The phosphorylation of target proteins will change their function in some way, which is how the cell is responding.

Figure 9.13 Depending on the protein involved, phosphorylation can activate or inhibit protein function. Phosphorylation of phosphorylase kinase and glycogen phosphorylase activates their function, whereas it inhibits glycogen synthase.

Figure 9.14 Signal amplification allows a single signaling molecule to affect many proteins within a cell, thereby amplifying a cellular response.

Figure 9.20 The initiator caspase is part of the death-inducing signaling complex. It is directly activated when a cell receives a death signal. The initiator caspase then activates the executioner caspases, which degrade various cellular proteins and thereby cause the destruction of the cell.

BioConnections

Figure 9.2 Auxin causes cells to elongate. The cells on the nonilluminated side accumulate more auxin, which causes this side to grow faster and bend toward the light.

Figure 9.5 Most receptors and enzymes bind their ligands noncovalently and with high specificity. Enzymes, however, convert their ligands (which are reactants) into products, whereas receptors undergo a conformation change after a ligand binds.

Figure 9.10 The GTP-bound form of Ras is active and promotes cell division. To turn the pathway off, Ras hydrolyzes GTP to GDP. If this cannot occur due to a mutation, the pathway will be continuously on, and uncontrolled cell division will result.

Feature Investigation Questions

1. Compared with control rats, those injected with prednisolone alone would be expected to have a decrease in the number of cells because it suppresses ACTH synthesis. Therefore, apoptosis would be higher. By comparison, prednisolone + ACTH would have a normal number of cells because the addition of ACTH would compensate for effects of prednisolone. ACTH alone would be expected to have a greater number of cells; apoptosis would be inhibited.
2. Yes, when injected with ACTH, prednisolone probably inhibited the ability of the rats to make their own ACTH. Even so, they were given ACTH by injection, so they didn't need to make their own ACTH to prevent apoptosis.
3. The lowest level of apoptosis would occur in the ACTH alone group, because they could make their own ACTH plus they were given ACTH. With such high levels of ACTH, they probably had the lowest level of apoptosis; it was already known that ACTH promotes cell division.

Test Yourself

1. d 2. c 3. d 4. e 5. a 6. d 7. e 8. e 9. e 10. b

Conceptual Questions

1. Cells need to respond to a changing environment, and cells need to communicate with each other.
2. In the first stage, a signaling molecule binds to a receptor, causing receptor activation. In the second stage, one type of signal is transduced or converted to a different signal inside the cell. In the third stage, the cell responds in some way to the signal, possibly by altering the activity of enzymes, structural proteins, or transcription factors. When the estrogen receptor is activated, the second stage, signal transduction, is not needed because the estrogen receptor is an intracellular receptor that directly activates the transcription of genes to elicit a cellular response.
3. Cell signaling allows cells to respond to environmental changes. For example, if a yeast cell is exposed to glucose, cell signaling will allow it to adapt to that change and utilize glucose more readily. Likewise, cell signaling allows plants to grow toward light. In addition, cells in a multicellular organism respond to changes in signaling molecules, such as hormones, and thereby coordinate their activities.

Chapter 10

Concept Checks

Figure 10.1 The four functions of the ECM in animals are strength, structural support, organization, and cell signaling.

Figure 10.2 The extension sequences of procollagen prevent fibers from forming intracellularly.

Figure 10.3 The proteins would become more linear, and the fiber would come apart.

Figure 10.4 GAGs are highly negatively charged molecules that tend to attract positively charged ions and water. Their high water content gives GAGs a gel-like character, which makes them difficult to compress.

Figure 10.5 Because the secondary cell wall is usually rigid, it prevents cell growth. If it were made too soon, it might prevent a cell from attaining its proper size.

Figure 10.7 Adherens junctions and desmosomes are cell-to-cell junctions, whereas hemidesomosomes and focal adhesions are cell-to-ECM junctions.

Figure 10.9 Tight junctions in your skin prevent harmful things like toxins and viruses from entering your body. They also prevent materials like nutrients from leaking out of your body.

Figure 10.10 As opposed to the results shown in Figure 10.10, the dye would be on the side of the cell layer facing the intestinal tract. You would see dye up to the tight junction on this side of the cells, but not on the side of the tight junction facing the blood.

Figure 10.13 Middle lamellae are similar to anchoring junctions and desmosomes in that they all function in cell-to-cell adhesion. However, their structures are quite different. Middle lamellae are composed primarily of carbohydrates that involve linkages between negatively charged carbohydrates and divalent cations. By comparison, anchoring junctions and desmosomes hold cells together via proteins such as cadherins and integrins.

Figure 10.15 Connective tissue would have the most extensive ECM.

Figure 10.16 Dermal tissue would be found on the surfaces of leaves, stems, and roots.

Figure 10.19 Simple epithelium and epidermis are one cell layer thick, whereas stratified epithelium and periderm are several cell layers thick. All of these types of tissues form coverings on the surfaces of animals and plants and (in animals) on the surfaces that line internal organs.

Figure 10.21 Both ground tissue in plants and connective tissue in animals are important in supporting the organism. These tissues have a large amount of ECM that provides structural support.

BioConnections

Figure 10.9 If tight junctions did not exist, substances in the lumen of your intestine might directly enter your blood. This could be potentially harmful if you consumed something with a toxic molecule in it. Likewise, materials from blood could be lost by diffusing into the lumen of your small intestine.

Figure 10.14 Plasmodesmata facilitate the movement of nutrients in a cell-to-cell manner. This is called symplastic transport.

Feature Investigation Questions

1. The purpose of this study was to determine the sizes of molecules that can move through gap junctions from one cell to another.
2. The researchers used fluorescent dyes to visibly monitor the movement of material from one cell to an adjacent cell through the gap junctions. First, single layers of rat liver cells were cultured. Next, fluorescent dyes with molecules of various masses were injected into particular cells. The researchers then used fluorescence microscopy to determine whether or not the dyes were transferred from one cell to the next.
3. The researchers found that molecules of masses less than 1,000 daltons (Da) could pass through the gap junction channels. Molecules of masses larger than 1,000 Da could not pass through the gap junctions. Further experimentation revealed variation in gap junction channel size of different cell types. However, the upper limit of the gap junction channel size was determined to usually be around 1,000 Da.

Test Yourself

1. e 2. c 3. b 4. e 5. e 6. d 7. e 8. d 9. e 10. a

Conceptual Questions

1. The primary cell wall is synthesized first between the two newly made daughter cells. It is relatively thin and allows cells to expand and grow. The secondary cell wall is made in layers by the deposition of cellulose fibrils and other components. In many cell types, it is relatively thick.
2. Cadherins and integrins are both membrane proteins that function as cell adhesion molecules. They also can function in cell signaling. Cadherins bind one cell to another cell, whereas integrins bind a cell to the extracellular matrix. Cadherins require calcium ions to function, but integrins do not.
3. Cell junctions are important in the proper arrangement of cells in a multicellular organism. In animals, for example, cells junctions allow cells to recognize and bind to each other. This is very important during embryonic development. In addition, cell junctions adhere cells to the extracellular matrix. Likewise, in plants, the cell wall and middle lamella are important in forming connections between plant cells that gives plants their correct morphology and function.

Chapter 11

Concept Checks

Figure 11.4 ^{35}S was used to label phage proteins, whereas ^{32}P was used to label phage DNA.

Figure 11.6 Cytosine is found in both DNA and RNA.

Figure 11.7 The phosphate is attached to the number 5′ carbon in a single nucleotide. In a DNA strand, it is attached to both the 5′ carbon and 3′ carbon.

Figure 11.8 A phosphoester bond is a single covalent bond between a phosphorus atom and an oxygen atom. A phosphodiester linkage involves two phosphoester bonds. This linkage occurs along the backbone of DNA and RNA strands.

Figure 11.10 Because it is antiparallel and obeys the AT/GC rule, it would be 3′–CTAAGCAAG–5′.

Figure 11.13 It would be 1/8 half-heavy and 7/8 light.

Figure 11.17 The oxygen in a new phosphoester bond comes from the sugar.

Figure 11.19 The lagging strand is made discontinuously in the direction opposite to the movement of the replication fork.

Figure 11.20 When primase is synthesizing a primer in the lagging strand, it moves from left to right in this figure. After it is done making a primer, it needs to hop to the opening of the replication fork to make a new primer. This movement is from right to left in this figure.

Figure 11.22 Telomerase uses a short strand of RNA as template to make the DNA repeat sequence.

Figure 11.25 Proteins hold the bottoms of the loops in place.

Figure 11.26 Proteins that compact the radial loop domains are primarily responsible for the X shape.

BioConnections

Figure 11.1 When a bacterium dies, it may release some of its DNA into the environment. Such DNA can be taken up via transformation by living bacteria, even bacteria of other species. If the DNA that is taken up encodes an antibiotic resistance gene, such a gene may be incorporated into the genome of the living bacterium and make it resistant to an antibiotic.

Figure 11.26 If chromosomes did not become compact, they might get tangled up with others during cell division, which would prevent the even segregation of chromosomes into the two daughter cells.

Feature Investigation Questions

1. Previous studies had indicated that mixing different strains could lead to transformation or the changing of a strain into a different one. Griffith had shown that mixing heat-killed type S with living type R would result in the transformation of the type R to type S. Though mutations could cause the changing of the identity of certain strains, the type R to type S transformation was not due to mutation but was more likely due to the transmission of a biochemical substance between the two strains. Griffith recognized this and referred to the biochemical substance as the "transformation principle." If Avery, MacCleod, and McCarty could determine the biochemical identity of this "transformation principle," they could identify the genetic material for this organism.
2. A DNA extract contains DNA that has been purified from a sample of cells.
3. The researchers could not verify that the DNA extract was completely pure and did not have small amounts of contaminating molecules, such as proteins and RNA. The researchers were able to treat the extract with enzymes to remove proteins (using protease), RNA (using RNase), or DNA (using DNase). Removing the proteins or RNA did not alter the transformation of the type R to type S strains. Only the enzymatic removal of DNA disrupted the transformation, indicating that DNA is the genetic material.

Test Yourself

1. a 2. b 3. d 4. d 5. b 6. c 7. b 8. d 9. d 10. c

Conceptual Questions

1. The genetic material must contain the information necessary to construct an entire organism. The genetic material must be accurately copied and transmitted from parent to offspring and from cell to cell during cell division in multicellular organisms. The genetic material must contain variation that can account for the known variation within each species and among different species.

   Griffith discovered something called the transformation principle, and his experiments showed the existence of biochemical genetic information. In addition, he showed that this genetic information can move from one individual to another of the same species. In his experiments, Griffith took heat-killed type S bacteria and mixed them with living type R bacteria and injected them into a live mouse, which died after the injection. By themselves, these two strains would not kill the mouse, but when they were put together, the genetic information from the heat-killed type S bacteria was transferred into the living type R bacteria, thus transforming the type R bacteria into type S.
2. In the case of the Hershey and Chase experiment, a radioactive isotope of sulfur was used to label the protein in the viral protein coat. The DNA was labeled using a radioactive isotope of phosphorus. This was an ideal way of labeling the different components, because sulfur is found in proteins but not DNA, and phosphorus is found in DNA but not proteins. By labeling the two candidate molecules with the radioactive isotopes, Hershey and Chase could determine the genetic material by seeing which isotope entered the bacterial cells.
3. In a DNA double helix, the two strands hydrogen-bond with each other according to the AT/GC rule. This provides the basis for DNA replication. In addition, as described in later chapters, hydrogen bonding between complementary bases is the basis for the transcription of RNA, which is needed for gene expression.

Chapter 12

Concept Checks

Figure 12.1 A person with two defective copies of phenylalanine hydroxylase would have phenylketonuria.

Figure 12.2 The ability to convert ornithine into citrulline is missing.

Figure 12.3 The usual direction of flow of genetic information is from DNA to RNA to protein, though exceptions occur.

Figure 12.4 If a terminator was removed, transcription would occur beyond the normal stopping point. Eventually, RNA polymerase would encounter a terminator from an adjacent gene, and transcription would end.

Figure 12.9 The ends of structural genes do not have a poly T region that acts as a template for the synthesis of a poly A tail. Instead, the poly A tail is added after the pre-mRNA is made by an enzyme that attaches many adenine nucleotides in a row.

Figure 12.10 A structural gene would still be transcribed into RNA if the start codon was missing. However, it would not be translated properly into a polypeptide.

Figure 12.11 It would bind to a 5′–UGG–3′codon, and it would carry tryptophan.

Figure 12.13 The function of the anticodon in tRNA is to recognize a codon in an mRNA.

Figure 12.16 Each mammal is closely related to the other mammals, and *E. coli* and *Serratia marcescens* are also closely related. The mammals are relatively distantly related to the bacterial species.

Figure 12.18 A region near the 5′ end of the mRNA is complementary to a region of rRNA in the small subunit. These complementary regions hydrogen-bond with each other to promote the binding of the mRNA to the small ribosomal subunit.

BioConnections

Figure 12.5 Both DNA and RNA polymerase use DNA strands as a template and connect nucleotides to each other in a 5′ to 3′ direction based on the complementarity of base pairing. One difference is that DNA polymerase needs a pre-existing strand, such as a RNA primer, to begin DNA replication, whereas RNA polymerase can begin the synthesis of RNA on a bare template strand. Another key difference is that DNA polymerase connects deoxyribonucleotides, whereas RNA polymerase connects ribonucleotides.

Figure 12.14 The attachment of an amino acid to a tRNA is an endergonic reaction. ATP provides the energy to catalyze this reaction.

Feature Investigation Questions

1. A triplet mimics mRNA because it can cause a specific tRNA to bind to the ribosome. This was useful to Nirenberg and Leder because it allowed them to correlate the binding of a tRNA carrying a specific amino acid with a triplet sequence.
2. The researchers were attempting to match codons with appropriate amino acids. By labeling one amino acid in each of the 20 tubes for each codon, the researchers were able to identify the correct relationship by detecting which tube resulted in radioactivity on the filter.
3. The AUG triplet would have shown radioactivity in the methionine test tube. Even though AUG acts as the start codon, it also codes for the amino acid methionine. The other three codons act as stop codons and do not code for an amino acid. In these cases, the researchers would not have found radioactivity trapped on filters.

Test Yourself

1. b 2. d 3. d 4. b 5. c 6. e 7. d 8. d 9. d 10. b

Conceptual Questions

1. Beadle and Tatum had the insight from their studies that a single gene controlled the synthesis of a single enzyme. In later years, it became apparent that genes code for all proteins and that some proteins consist of more than

one polypeptide chain. So the modern statement is one gene codes for each polypeptide.

Confirmation of their hypothesis came from studies involving arginine biosynthesis. Biochemists had already established that particular enzymes are involved in a pathway to produce arginine. Intermediates in this pathway are ornithine and citrulline. Mutants in single genes disrupted the ability of cells to catalyze just one reaction in this pathway, thereby suggesting that a single gene encodes a single enzyme.

2. Each of these 20 enzymes catalyzes the attachment of a specific amino acid to a specific tRNA molecule.
3. During transcription, a DNA strand is used as a template for the synthesis of RNA. Most genes encode mRNAs, which contain the information to make polypeptides. During translation, an mRNA binds to a ribosome and a polypeptide is made, which becomes a unit within a functional protein.

Chapter 13

Concept Checks

Figure 13.2 Gene regulation causes each cell type to express its own unique set of proteins, which, in turn, are largely responsible for the morphologies and functions of cells.

Figure 13.6 The *lacZ*, *lacY*, and *lacA* genes are under the control of the *lac* promoter.

Figure 13.7 Negative control refers to the action of a repressor protein, which inhibits transcription when it binds to the DNA. Inducible refers to the action of a small effector molecule. When it is present, it promotes transcription.

Figure 13.11 In this case, the repressor keeps the *lac* operon turned off unless lactose is present in the environment. The activator allows the bacterium to choose between glucose and lactose.

Figure 13.12 Both proteins are similar in that they repress transcription. They prevent RNA polymerase from transcribing the operons. They are different with regard to the effects of their small effector molecules. For the lac repressor, the binding of allolactose causes a conformational change that prevents the repressor from binding to its operator site. In contrast, the binding of tryptophan to the trp repressor allows it to bind to its operator site. Another difference is that the lac repressor binds to the DNA sequence found in the *lac* operator site, whereas the trp repressor recognizes a different DNA sequence that is found in the *trp* operator site.

Figure 13.16 When an activator interacts with mediator, it causes RNA polymerase to proceed to the elongation phase of transcription.

Figure 13.18 Some histone modifications may promote a loosening of chromatin structure, whereas others cause the chromatin to become more compact.

Figure 13.21 The advantage of alternative splicing is that it allows a single gene to encode two or more polypeptides. This enables organisms to have smaller genomes, which is more efficient and easier to package into a cell.

Figure 13.23 When iron levels rise in the cell, the iron binds to IRP and removes it from the mRNA that encodes ferritin. This results in the rapid translation of ferritin protein, which can store excess iron. Unfortunately, ferritin storage does have limits, so iron poisoning can occur if too much is ingested.

BioConnections

Figure 13.10 In eukaryotic cells, cAMP acts as a second messenger in signal transduction pathways.

Figure 13.19 A nucleosome is composed of DNA wrapped around an octamer of histone proteins.

Feature Investigation Questions

1. The first observation was the identification of rare bacterial strains that had constitutive expression of the *lac* operon. Normally, the genes are expressed only when lactose is present. These mutant strains expressed the genes all the time. The researchers also observed that some of these strains had mutations in the *lacI* gene. These two observations were key to the development of hypotheses explaining the relationship between the *lacI* gene and the regulation of the *lac* operon.
2. The correct hypothesis is that the *lacI* gene encodes a repressor protein that inhibits the operon.
3. The researchers used an F′ factor to introduce the wild-type *lacI* gene into the cell. In this case, the cells that contained the F′ factor had both a mutant copy of the gene and a normal copy of the gene. By creating a merozygote with an F′ factor with a normal copy of the *lacI* gene, regulation of the *lac* operon was restored. The researchers concluded that the normal *lacI* gene produced adequate amounts of a diffusible protein that could interact with the operator on the chromosomal DNA as well as the F′ factor DNA and regulate transcription.

Test Yourself

1. d 2. b 3. c 4. c 5. c 6. d 7. c 8. c 9. d 10. c

Conceptual Questions

1. In an inducible operon, the presence of a small effector molecule causes transcription to occur. In repressible operons, a small effector molecule inhibits transcription. The effects of these small molecules are mediated through regulatory proteins that bind to the DNA. Repressible operons usually encode anabolic enzymes, and inducible operons encode catabolic enzymes.
2. a. regulatory protein; b. small effector molecule; c. segment of DNA; d. small effector molecule; and e. regulatory protein.
3. Gene regulation offers key advantages such as (1) proteins are made only when they are needed; (2) proteins are made in the correct cell type; and (3) proteins are made at the correct stage of development. These advantages are important for reproduction and sustaining life.

Chapter 14

Concept Checks

Figure 14.1 At neutral pH, glutamic acid is negatively charged. Perhaps the negative charges repel each other and prevent hemoglobin proteins from aggregating into fiber-like structures.

Figure 14.3 Only germ-line cells give rise to gametes (sperm or egg cells). A somatic cell cannot give rise to a gamete and therefore cannot be passed to offspring.

Figure 14.4 This is a mutation that occurred in a somatic cell, so it cannot be transmitted to an offspring.

Figure 14.6 A thymine dimer is harmful because it can cause errors in DNA replication.

Figure 14.7 If we divide 44 by 2 million, the rate is 2.2×10^{-5}.

Figure 14.8 UvrC and UvrD are responsible for removing the damaged DNA. UvrC makes cuts on both sides of the damage, and then UvrD removes the damaged region.

Figure 14.9 The Sun has UV rays and other harmful radiation that could damage the DNA. This person has a defect in the nucleotide excision repair pathway. Therefore, his DNA is more likely to suffer mutations, which cause growths on the skin.

Figure 14.11 Growth factors turn on a signaling pathway that ultimately leads to cell division.

Figure 14.14 The type of cancer associated with this fusion is leukemia, which is a cancer of blood cells. The gene fusion produces a chimeric gene that is expressed in blood cells because it has the *bcr* promoter. The abnormal fusion protein promotes cancer in these cells.

Figure 14.15 Checkpoints prevent cell division if a genetic abnormality is detected. This helps to properly maintain the genome, thereby minimizing the possibility that a cell harboring a mutation will divide to produce two daughter cells.

Figure 14.16 Cancer would not occur if both copies of the *Rb* gene and both copies of the *E2F* gene were rendered inactive due to mutations. An active copy of the *E2F* gene is needed to promote cell division.

BioConnections

Figure 14.10 Epithelial cells form sheets that separate certain tissues from each other. When the integrity of the sheet is compromised, metastasis may occur.

Figure 14.11 Drugs that inhibit protein kinases may be used to combat cancer if they target the protein kinases that are overactive in certain forms of cancer.

Feature Investigation Questions

1. Some biologists believed that heritable traits may be altered by physiological events. This suggests that mutations may be stimulated by certain needs of the organism. Others believed that mutations are random. If a mutation had a beneficial effect that improved survival and/or reproductive success, these mutations would be maintained in the population through natural selection.
2. The Lederbergs were testing the hypothesis that mutations are random events. By subjecting the bacteria to some type of environmental stress, the bacteriophage, the researchers would be able to see if the stress induced mutations or if mutations occurred randomly.

3. When looking at the number and location of colonies that were resistant to viral infection, the pattern was consistent among the secondary plates. This indicates that the mutation that allowed the colonies to be resistant to viral infection occurred on the master plate. The secondary plates introduced the selective agent that allowed the resistant bacteria colonies to survive and reproduce while the other colonies were destroyed. Thus, mutations occurred randomly in the absence of any selective agent.

Test Yourself

1. c 2. d 3. d 4. e 5. d 6. b 7. b 8. b 9. c 10. e

Conceptual Questions

1. Random mutations are more likely to be harmful than beneficial. The genes within each species have evolved to work properly. They have functional promoters, coding sequences, terminators, and so on, that allow the genes to be expressed. Mutations are more likely to disrupt these sequences. For example, mutations within the coding sequence may produce early stop codons, frameshift mutations, and missense mutations that result in a nonfunctional polypeptide. On rare occasions, however, mutations are beneficial; they may produce a gene that is expressed better than the original gene or produce a polypeptide that functions better.
2. A spontaneous mutation originates within a living cell. It may be due to spontaneous changes in nucleotide structure, errors in DNA replication, or products of normal metabolism that may alter the structure of DNA. The causes of induced mutations originate from outside the cell. They may be physical agents, such as UV light or X-rays, or chemicals that act as mutagens. Both spontaneous and induced mutations may cause a harmful phenotype such as a cancer. In many cases, induced mutations are avoidable if the individual can prevent exposure to the environmental agent that acts as a mutagen.
3. Mutations may alter the expression of a gene and/or alter the function of a protein encoded by a gene. In many cases, such changes are harmful because a gene may not be expressed at the correct level, or the protein may not function as well as the normal (nonmutant) protein.

Chapter 15

Concept Checks

Figure 15.1 Chromosomes are readily seen when they are compacted in a dividing cell. By adding such a drug, you increase the percentage of cells that are actively dividing.

Figure 15.2 Interphase consists of the G_1, S, and G_2 phases of the cell cycle.

Figure 15.6 As shown in the inset, each object is a pair of sister chromatids.

Figure 15.7 The astral microtubules, which extend away from the chromosomes, are important for positioning the spindle apparatus within the cell. The polar microtubules project into the region between the two poles. Polar microtubules that overlap with each other play a role in the separation of the two poles. Kinetochore microtubules are attached to kinetochores at the centromeres and are needed to sort the chromosomes.

Figure 15.8 The mother cell (in G_1 phase) and the daughter cells have the same chromosome composition. They are genetically identical.

Figure 15.9 Cytokinesis in both animal and plant cells separates a mother cell into two daughter cells. In animal cells, cytokinesis involves the formation of a cleavage furrow, which constricts like a drawstring to separate the cells. In plants, the two daughter cells are separated by the formation of a cell plate, which forms a cell wall between the two daughter cells.

Figure 15.13 The mother cell is diploid with two sets of chromosomes, whereas the four resulting cells are haploid with one set of chromosomes.

Figure 15.14 The reason for meiosis in animals is to produce gametes. These gametes combine during fertilization to produce a diploid organism. Following fertilization, the purpose of mitosis is to produce a multicellular organism.

Figure 15.16 Inversions and the translocations shown here do not affect the total amount of genetic material.

BioConnections

Figure 15.3 Checkpoints prevent cancer by checking the integrity of the genome. If abnormalities in DNA structure are detected or if a chromosome is not properly attached to the spindle, the checkpoint will delay cell division until the problem is fixed. If it cannot be fixed, the checkpoint will initiate the process of apoptosis, thereby killing a cell that may harbor mutations. This prevents the proliferation of cells that have the potential to be cancerous.

Figure 15.10 Though the process is not entirely understood, myosin motor proteins tug on the actin filaments. Actin monomers are gradually released during this process, which makes the contractile ring get smaller and smaller.

Feature Investigation Questions

1. Researchers had demonstrated that the binding of progesterone to receptors in oocytes caused the cells to progress from the G_2 phase of the cell cycle to mitosis. It appeared that progesterone acted as a signaling molecule for the progression through the cell cycle.
2. The researchers proposed that progesterone acted as a signaling molecule that led to the synthesis of molecules that cause the cell to progress through the cell cycle. These changes led to the maturation of the oocyte.
   To test their hypothesis, donor eggs were exposed to progesterone for either 2 or 12 hours. Control donor oocytes were not exposed to progesterone. Cytosol from each treatment was then transferred to recipient oocytes. The researchers recorded whether or not the recipient oocytes underwent maturation.
3. The oocytes that were exposed to the progesterone for only 2 hours did not induce maturation in the recipient oocytes, whereas the oocytes that were exposed to progesterone for 12 hours did induce maturation in the recipient oocytes. The researchers suggested that a time span greater than 2 hours is needed to accumulate the proteins that are necessary to promote maturation.

Test Yourself

1. b 2. e 3. b 4. e 5. c 6. a 7. c 8. d 9. b 10. c

Conceptual Questions

1. In diploid species, chromosomes are present in pairs, one from each parent, and contain similar gene arrangements. Such chromosomes are homologous. When DNA is replicated, two identical copies are created, and these are sister chromatids.
2. There are four copies. A karyotype shows homologous chromosomes that come in pairs. Each member of the pair has replicated to form a pair of sister chromatids. Therefore, four copies of each gene are present. See the inset to Figure 15.1.
3. Mitosis is a process that produces two daughter cells with the same genetic material as the original daughter cell. In the case of plants and animals, this allows a fertilized egg to develop into a multicellular organism composed of many, genetically identical cells.

Chapter 16

Concept Checks

Figure 16.2 Having blue eyes is a variant (also called a trait). A character is a more general term, which in this case would refer to eye color.

Figure 16.4 In this procedure, stamens are removed from the purple flower to prevent self-fertilization.

Figure 16.5 The reason why offspring of the F_1 generation exhibit only one variant of each character is because one trait is dominant over the other.

Figure 16.6 The ratio of alleles (*T* to *t*) is 1:1. The reason why the phenotypic ratio is 3:1 is because *T* is dominant to *t*.

Figure 16.7 It was *Pp*. To produce white offspring, which are *pp*, the original plant had to have at least one copy of the *p* allele. Because it had purple flowers, it also had to have one copy of the *P* allele. So, its genotype must be *Pp*.

Figure 16.8 If the linked hypothesis had been correct, the ratio would have been 3 round, yellow to 1 wrinkled, green.

Figure 16.10 The word segregate means that alleles are separated into different places. In this case, the alleles are segregated into different cells during the process of meiosis. Alleles are located on chromosomes. A diploid cell has two copies of each allele. During meiosis, a diploid cell divides twice to produce four haploid cells that each have only one copy of an allele.

Figure 16.11 There would be four possible ways of aligning the chromosomes, and eight different types of gametes (*ABC*, *abc*, *ABc*, *abC*, *Abc*, *aBC*, *AbC*, *aBc*) could be produced.

Figure 16.12 No. If two parents are affected with the disease, they must be homozygous for the mutant allele if it's recessive. Two homozygous parents would have to produce all affected offspring. If they don't, then the inheritance pattern is not recessive.

Figure 16.13 All affected offspring having at least one affected parent suggests a dominant pattern of inheritance.

Figure 16.14 The person would be a female. In mammals, the presence of the Y chromosome causes maleness. Therefore, without a Y chromosome, a person with a single X chromosome would develop into a female.

Figure 16.18 No. You need a genetically homogenous population to study the norm of reaction. A wild population of squirrels is not genetically homogenous, so it could not be used.

BioConnections

Figure 16.9 Sexual reproduction is the process in which two haploid gametes (for example, sperm and egg) combine with each other to begin the life of a new individual. Each gamete contributes one set of chromosomes. The resulting zygote has chromosomes that occur in pairs (one from each parent). The members of each pair are called homologs of each other; they carry the same types of genes.

Figure 16.10 Alleles segregate from each other during the process of meiosis. Meiosis begins with a diploid mother cell that has pairs of genes, which may be found in different alleles. During meiosis, these pairs of genes separate and end up in different haploid cells. Therefore, each haploid cell has only one copy of each gene. In other words, each haploid cell has only one allele of a given gene.

Feature Investigation Questions

1. Morgan was testing the hypothesis of use and disuse. This hypothesis suggests that if a structure is not used, over time, it will diminish and/or disappear. In Morgan's experiments, originally he was testing to see if flies reared in the dark would lose some level of eye development.
2. When the F_1 individuals were crossed, only male F_2 offspring expressed the white eye color. At this time, Morgan was aware of sex chromosome differences between male and female flies. He realized that since males only possess one copy of X-linked genes, this would explain why only F_2 males exhibited the recessive trait.
3. In a cross between a white-eyed male and a female that is heterozygous for the white and red alleles, 1/2 of the female offspring would have white eyes. Also, a cross between a white-eyed male and a white-eyed female would yield all offspring with white eyes.

Test Yourself

1. c 2. b Mendel's law of segregation refers to the separation of the two alleles into separate cells. Meiosis is the cellular division process that produces haploid cells. During the first meiotic division, a diploid cell divides to produce haploid cells. This is the phase in which the two alleles segregate, or separate, from each other. 3. d 4. c 5. e 6. d 7. d 8. d 9. d 10. c

Conceptual Questions

1. Two affected parents having an unaffected offspring would rule out recessive inheritance. If two unaffected parents had an affected offspring, dominant inheritance is ruled out. However, it should be noted that this answer assumes that no new mutations are happening. In rare cases, a new mutation could cause or alter these results. For recessive inheritance, two affected parents could have an unaffected offspring if the offspring had a new mutation that converted the recessive allele to the dominant allele. Similarly for dominant inheritance, two unaffected parents could have an affected offspring if the offspring inherited a new mutation that was dominant. Note: New mutations are expected to be relatively rare.
2. The individual probabilities are as follows: $AA = 0.25$; $bb = 0.5$; $CC = 0.5$; and $Dd = 0.5$. These are determined by making small Punnett squares. We use the product rule to calculate the probability of $AAbbCCDd = (0.25)(0.5)(0.5)(0.5) = 0.03125$, or 3.125%.
3. The environment is needed so that genes can be expressed. For example, organic molecules and energy are needed for transcription and translation. In addition, environmental factors influence the outcome of traits. For example, sunlight can cause a tanning response, thereby affecting the darkness of the skin.

Chapter 17

Concept Checks

Figure 17.1 The recessive allele is the result of a loss-of-function mutation. In a *Ccpp* individual, the enzyme encoded by the *P* gene is defective.

Figure 17.4 Crossing over occurred during oogenesis in the heterozygous female of the F_1 generation to produce the recombinant offspring of the F_2 generation.

Figure 17.5 One strategy would be to begin with two true-breeding parental strains: *alal dpdp* and $al^+al^+\ dp^+dp^+$ and cross them together to get F_1 heterozygotes $al^+al\ dp^+dp$. Then testcross female F_1 heterozygotes to male *alal dpdp* homozygotes. In the F_2 generation, the recombinant offspring would be $al^+al\ dpdp$ and $alal\ dp^+dp$, and the nonrecombinants would be $al^+al\ dp^+dp$, and *alal dpdp*.

Figure 17.7 The gene is located in the chloroplast DNA. In this species, chloroplasts are transmitted from parent to offspring via eggs but not via sperm.

Figure 17.9 The Barr body is much more compact than the other X chromosome in the cell. This compaction prevents most of the genes on the Barr body from being expressed.

Figure 17.11 Only the male genes are transcriptionally active in the offspring. In this case, half the offspring would be normal, and half would be dwarf. The dwarf offspring would have inherited the *Igf-2* allele from their father.

BioConnections

Figure 17.4 Crossing over occurs during prophase of meiosis I.

Figure 17.6 The evolutionary origin of these organelles is an ancient endosymbiotic relationship. Mitochondria are derived from purple bacteria, and chloroplasts are derived from cyanobacteria.

Feature Investigation Questions

1. Bateson and Punnett were testing the hypothesis that the gene pairs that influence flower color and pollen shape would assort independently of each other. The two traits were expected to show a pattern consistent with Mendel's law of independent assortment.
2. The expected results were a phenotypic ratio of 9:3:3:1. The researchers expected 9/16 of the offspring would have purple flowers and long pollen, 3/16 of the offspring would have purple flowers and round pollen, 3/16 of the offspring would have red flowers and long pollen, and 1/16 of the offspring would have red flowers and round pollen.
3. Though all four of the expected phenotype groups were seen, they were not in the predicted ratio of 9:3:3:1. The number of individuals with the phenotypes found in the parental generation (purple flowers and long pollen or red flowers and round pollen) was much higher than expected. Bateson and Punnett suggested that the gene controlling flower color was somehow coupled with the gene that controls pollen shape. This would explain why these traits did not always assort independently.

Test Yourself

1. d 2. c 3. d 4. c 5. d 6. b 7. a 8. d 9. a 10. a

Conceptual Questions

1. The correct answer is 6%. Individuals that are *Aabb* are recombinants that occurred as a result of crossing over. Because the genes are 12 map units apart, we expect that 12% will be recombinants. However, there are two types of recombinants: *Aabb* and *aaBb*, which would occur in equal amounts. Therefore, we expect 6% to be *Aabb*.
2. This may happen due to X inactivation. As a matter of bad luck, a female embryo may preferentially inactivate the X chromosome carrying the normal allele in the embryonic cells that will give rise to the eyes. If the X chromosome carrying the allele for color blindness is preferentially expressed, one or both eyes may show color blindness to some degree.
3. When genes are linked along the same chromosome, researchers can set up testcrosses and determine the percentage of recombinant offspring that results from those testcrosses. This percentage can be used as a map distance between genes.

Chapter 18

Concept Checks

Figure 18.3 Viruses vary with regard to their structure and their genomes. Genome variation is described in Table 18.1.

Figure 18.5 The advantage of the lytic cycle is that the virus can make many copies of itself and proliferate. However, sometimes the growth conditions may not be favorable to make new viruses. The advantage of the lysogenic cycle is that the virus can remain latent until conditions become favorable to make new viruses.

Figure 18.9 There appears to be three nucleoids in the bacterial cell to the far right.

Figure 18.11 The loop domains are held in place by proteins that bind to the DNA at the bases of the loops. The proteins also bind to each other.

Figure 18.12 Bacterial chromosomes and plasmids are similar in that they typically contain circular DNA molecules. However, bacterial chromosomes are usually much longer than plasmids and carry many more genes. Also, bacterial chromosomes tend to be more compacted due to the formation of loop domains and supercoiling.

Figure 18.13 16 hours is the same as 32 doublings. So, $2^{32} = 4,294,967,296$. (The actual number would be much less because the cells would deplete the growth media and grow more slowly than the maximal rate.)

Figure 18.16 Yes. The two strains would have mixed together, allowing them to conjugate. Therefore, there would have been colonies on the plates.

Figure 18.17 During conjugation, only one strand of the DNA from an F factor is transferred from the donor to the recipient cell. The single-stranded DNA in both cells is then used as a template to create double-stranded F factor DNA in both cells.

Figure 18.19 Transduction is not a normal part of the phage life cycle. It is a mistake in which a piece of the bacterial chromosome is packaged into a phage coat and is then transferred to another bacterial cell.

BioConnections

Figure 18.4 Viral release occurs as a budding process in which a membrane vesicle is formed that surrounds the capsid. Similarly, exocytosis involves the formation of a membrane vesicle that encloses some type of cargo.

Figure 18.9 A nucleoid is not a membrane-bound organelle. It is simply a region where a bacterial chromosome is found. A cell nucleus in a eukaryotic cell has an envelope with a double membrane.

Figure 18.18 Griffiths was able to show that genetic material was transferred to type R bacteria, which converted them to type S. This occurred via transformation. Later, Avery, MacLeod, and McCarty determined that DNA was the material that was being transferred.

Feature Investigation Questions

1. Lederberg and Tatum were testing the hypothesis that genetic material could be transferred from one bacterial strain to another.
2. The experimental growth medium lacked particular amino acids and biotin. The mutant strains were unable to synthesize these particular amino acids or biotin. Therefore, they were unable to grow due to the lack of the necessary nutrients. The two strains used in the experiment each lacked the ability to make two essential nutrients necessary for growth. The appearance of colonies growing on the experimental growth medium indicated that some bacterial cells had acquired the normal genes for the two mutations they carried. By acquiring these normal genes, the ability to synthesize the essential nutrients was restored.
3. Bernard Davis placed samples of the two bacterial strains in different arms of a U-tube. A filter allowed the free movement of the liquid in which the bacterial cells were suspended, but prevented the actual contact between the bacterial cells. After incubating the strains in this environment, Davis found that genetic transfer did not take place. He concluded that physical contact between cells of the two strains was required for genetic transfer.

Test Yourself

1. c 2. e 3. c 4. b 5. e 6. a 7. d 8. d 9. b 10. c

Conceptual Questions

1. Viruses are similar to living cells in that they contain a genetic material that provides a blueprint to make new viruses. However, viruses are not composed of cells, and by themselves, they do not carry out metabolism, use energy, maintain homeostasis, or even reproduce. A virus or its genetic material must be taken up by a living cell to replicate.
2. Conjugation—The process involves a direct physical contact between two bacterial cells in which a donor cell transfers a strand of DNA to a recipient cell.
   Transformation—This occurs when a living bacteria takes up genetic information that has been released from a dead bacteria.
   Transduction—When a virus infects a donor cell, it incorporates a fragment of bacterial chromosomal DNA into a newly made virus particle. The virus then transfers this fragment of DNA to a recipient cell.
3. Horizontal gene transfer is the transfer of genes from another organism without being the offspring of that organism. These acquired genes sometimes increase survival and therefore may have an evolutionary advantage. Such genes may even promote the formation of new species. From a medical perspective, an important example of horizontal gene transfer is when one bacterium acquires antibiotic resistance from another bacterium and then itself becomes resistant to that antibiotic. This phenomenon is making it increasingly difficult to treat a wide variety of bacterial diseases.

Chapter 19

Concept Checks

Figure 19.3 Cell division and cell migration are common in the earliest stages of development, whereas cell differentiation and apoptosis are more common as tissues and organs start to form.

Figure 19.4 If apoptosis did not occur, the fingers would be webbed.

Figure 19.5 During development, positional information may a cause a cell to respond in one of four ways: cell division, cell migration, cell differentiation, and cell death.

Figure 19.6 Cell division and migration would be the most prevalent in the early phases of development, such as phase 1 and 2.

Figure 19.8 The larva would have anterior structures at both ends and would lack posterior structures such as a spiracle.

Figure 19.9 The Bicoid protein functions as a transcription factor. Its function is highest in the anterior end of the embryo.

Figure 19.11 This embryo has 15 pink stripes, each of which corresponds to a portion of the 15 segments in the embryo.

Figure 19.14 The last abdominal segment would have legs!

Figure 19.18 Stem cells can divide, and they can differentiate into specific cell types.

Figure 19.20 Hematopoietic stem cells are multipotent.

Figure 19.23 Most stem cells in plants are found in meristems, which are located at the tips of roots and shoots.

Figure 19.24 The pattern would be sepal, petal, stamen, stamen.

BioConnections

Figure 19.3 First, the nucleus and cell shrink. Next, multiple extensions (blebs) are formed from the plasma membrane. Finally, this blebbing continues to occur until the cell is broken up into small fragments.

Figure 19.5 Some receptors recognize signals that convey positional information. After the signal binds to this type of receptor, a signal transduction pathway is activated that may cause a cell to divide, migrate, differentiate, or undergo apoptosis.

Figure 19.17 As the number of *Hox* genes increases, the body plans of animals become more complex.

Feature Investigation Questions

1. The researchers were interested in the factors that cause cells to differentiate. For this particular study, the researchers were attempting to identify genes involved in the differentiation of muscle cells.
2. Using genetic technology, the researcher compared the gene expression in cells that could differentiate into muscle cells with the gene expression in cells that could not differentiate into muscle cells. Though many genes were expressed in both, the researchers were able to isolate three genes that were expressed in muscle cell lines that were not expressed in the nonmuscle cell lines.
3. Again, using genetic technology, each of the candidate genes was introduced into a cell that normally did not give rise to skeletal muscle. This procedure was used to test whether or not these genes played a key role in muscle cell differentiation. If the genetically engineered cell gave rise to muscle cells, the researchers would have evidence that a particular candidate gene was involved in muscle cell differentiation. Of the three candidate genes, only one was shown to be involved in muscle cell differentiation. When the *MyoD* gene was expressed in fibroblasts, these cells differentiated into skeletal muscle cells.

Test Yourself

1. c 2. d 3. e 4. b 5. c 6. e 7. a 8. a 9. b 10. d

Conceptual Questions

1. a. This would be consistent with a defect in a segmentation gene, such as a gap gene.

   b. This would be consistent with a mutation in a homeotic gene because the characteristics of a particular segment have been changed.
2. Both types of genes encode transcription factors that bind to the DNA and regulate the expression of other genes. The effects of *Hox* genes are to determine the characteristics of certain regions of the body, whereas the *myoD* gene is cell specific—it causes a cell to become a skeletal muscle cell.

3. Maternal effect genes control the formation of body axes, such as the anteroposterior and dorsoventral axes. Next, the segmentation genes divide the embryo into segments, though visible segments are lost in many animal species at later stages of development. Finally, the homeotic genes determine the characteristics of each segment.

Chapter 20

Concept Checks

Figure 20.2 No. A recombinant vector has been made, but it has not been cloned. In other words, many copies of the recombinant vector have not been made yet.

Figure 20.3 The insertion of chromosomal DNA into the vector disrupts the *lacZ* gene, thereby preventing the expression of β-galactosidase. The functionality of *lacZ* can be determined by providing the growth medium with a colorless compound, X-Gal, which is cleaved by β-galactosidase into a blue dye. Bacterial colonies containing recircularized vectors form blue colonies, whereas colonies containing recombinant vectors carrying a segment of chromosomal DNA will be white.

Figure 20.5 The 600-bp piece would be closer to the bottom. Smaller pieces travel faster through the gel.

Figure 20.6 The primers are complementary to sequences at each end of the DNA region to be amplified.

Figure 20.7 It means that part of their inserts are exactly the same, but other regions are not.

Figure 20.9 If a dideoxynucleotide ddNTP is added to a growing DNA strand, the strand can no longer grow because the 3′—OH group, the site of attachment for the next nucleotide, is missing.

Figure 20.10 A fluorescent spot identifies a cDNA that is complementary to a particular DNA sequence. Because the cDNA was generated from mRNA, this technique identifies a gene that has been transcribed in a particular cell type under a given set of conditions.

Figure 20.11 The reason why the A and B chains are made as fusion proteins is because the A and B chains are rapidly degraded when expressed in bacterial cells by themselves. The fusion proteins, however, are not.

Figure 20.13 Only the T DNA within the Ti plasmid is transferred to a plant cell.

Figure 20.15 Not all of Dolly's DNA came from a mammary cell. Her mitochondrial DNA came from the oocyte donor.

Figure 20.16 The bands match suspect 2.

BioConnections

Figure 20.2 Plasmids are small, circular molecules of DNA that exist independently of the bacterial chromosome. They have their own origin of replication. Many plasmids carry genes that convey some type of selective advantage to the host cell, such as antibiotic resistance.

Figure 20.6 Primers are needed in a PCR experiment because DNA polymerase cannot begin DNA replication on a bare template strand.

Feature Investigation Questions

1. Gene therapy is the introduction of cloned genes into living cells to correct genetic mutations. The hope is that the cloned genes will correct or restore the normal gene function, thereby eliminating the clinical effects of the disease.

   ADA deficiency is a recessive genetic disorder in which an enzyme, adenosine deaminase, is not functional. The absence of this enzyme causes a buildup of deoxyadenosine, which is toxic to lymphocytes. When lymphocytes are destroyed, a person's immune system begins to fail, leading to a severe combined immunodeficiency disease (SCID).
2. The researchers introduced normal copies of the ADA gene into lymphocytes, restoring normal cell metabolism. The researchers isolated lymphocytes from the patient and used a viral vector to introduce the gene into the lymphocytes. These lymphocytes were then reintroduced back into the patient.
3. Following several rounds of treatment with gene therapy, researchers were able to document continued production of the correct enzyme by the lymphocytes over the course of 4 years. However, because the patients were also receiving other forms of treatment, it was not possible to determine if the gene therapy reduced the negative effects of the genetic disease.

Test Yourself

1. e 2. d 3. b 4. b 5. b 6. c 7. d 8. c 9. e 10. e

Conceptual Questions

1. The restriction enzyme cuts the plasmid at a specific site, leaving sticky ends. The gene of interest, cut with the same enzyme, has complementary sticky ends that allow hydrogen bonding between the gene of interest and the plasmid. The connections are then made permanent, using DNA ligase that connects the DNA backbones.
2. A ddNTP is missing an oxygen at the 3′ position. This prevents the further growth of a DNA strand, thereby causing chain termination.
3. One example is Bt corn in which corn plants carry genes that make it herbicide- and pest-resistant. These types of strains have become very popular among farmers. A similar approach was used to create Bt cotton. You may find other examples by searching the web.

Chapter 21

Concept Checks

Figure 21.2 One reason is that more complex species tend to have more genes. A second reason is that species vary with regard to the amount of repetitive DNA present in their genome.

Figure 21.5 For DNA transposons, inverted repeats are recognized by transposase, which cleaves the DNA and inserts the transposon into a new location.

Figure 21.6 Retroelements. A single retroelement can be transcribed into multiple copies of RNA, which can be converted to DNA by reverse transcriptase, and inserted into multiple sites in the genome.

Figure 21.9 Differential gene regulation. Genes that encode metabolic enzymes are highly expressed in liver cells, whereas those same genes are expressed in lower amounts in muscle cells. Conversely, genes that encode cytoskeletal and motor proteins are highly expressed in muscle cells, but less so in liver cells.

Figure 21.11 The two main advantages of having a computer program translate a genetic sequence are that it's faster and probably more accurate.

Figure 21.12 It is possible for orthologs to have exactly the same DNA sequence if neither of them has accumulated any new mutations that would cause their sequences to become different. This is likely only for closely related species that have diverged relatively recently from each other.

BioConnections

Figure 21.8 The family members that are expressed at early stages of development (embryonic and fetal stages) have a higher affinity for oxygen than the adult form. This allows the embryo and fetus to obtain oxygen from its mother's bloodstream.

Figure 21.10 Reversible post-translation covalent modifications provide a way to modulate protein function. Certain types can turn off protein function, whereas others can turn on protein function. These modifications provide a rapid way for a cell to control protein function.

Feature Investigation Questions

1. The goal of the experiment was to sequence the entire genome of *Haemophilus influenzae*. By conducting this experiment, the researchers would have information about genome size and the types of genes the bacterium has.
2. One strategy requires mapping the genome prior to sequencing. After mapping is completed, each region of the genome is then sequenced. The shotgun approach does not require mapping of the genome prior to sequencing. Instead, many fragments are randomly sequenced.

   The advantage of the shotgun approach is the speed at which the sequencing can be conducted because the researchers do not have to spend time mapping the genome first. The disadvantage is that because the researchers are sequencing random fragments, some fragments may be sequenced more than necessary.
3. The researchers were successful in sequencing the entire genome of the bacterium. The genome size was determined to be 1,830,137 base pairs, with a predicted 1,743 structural genes. The researchers were also able to predict the function of many of these genes. More importantly, the results were the first complete genomic sequence of a living organism.

Test Yourself

1. c 2. e 3. a 4. c 5. e 6. b 7. b 8. d 9. c 10. c

Conceptual Questions

1. a. yes

   b. No, it's only one chromosome in the nuclear genome.

c. yes

d. yes

2. The two main reasons why the proteomes of eukaryote species are usually much larger than their genomes are alternative splicing and post-translational covalent modifications. During alternative splicing, a pre-mRNA is spliced in two or more different ways to yield two or more different polypeptides. Post-translational covalent modifications can affect protein structure in a variety of ways, including proteolytic processing; disulfide bond formation; the attachment of prosthetic group, sugars, or lipids; phosphorylation; acetylation; and methylation.
3. The genome contains the information for the production of cellular proteins; it is a blueprint. The production of proteins is largely responsible for determining cellular characteristics, which, in turn, determine an organism's traits.

Chapter 22

Concept Checks

Figure 22.2 Organic molecules form the chemical foundation for the structure and function of living organisms. Modern organisms can synthesize organic molecules. However, to explain how life got started, biologists need to explain how organic molecules were made prior to the existence of living cells.

Figure 22.3 These vents release hot gaseous substances from the interior of the Earth. Organic molecules can form in the temperature gradient between the extremely hot vent water and the cold water that surrounds the vent.

Figure 22.4 A liposome is more similar to real cells, which are surrounded by a membrane that is composed of a phospholipid bilayer.

Figure 22.5 Certain chemicals, such as RNA molecules, have properties that provide advantages and therefore cause them to increase in number relative to other molecules.

Figure 22.7 In a sedimentary rock formation, the layer at the bottom is usually the oldest.

Figure 22.8 For this time frame, you would analyze the relative amounts of the rubidium-87 and strontium-87 isotopes.

Figure 22.14 Most animal species, including fruit flies, fishes, and humans, exhibit bilateral symmetry.

BioConnections

Figure 22.4 Phospholipids are amphipathic molecules; they have a polar end (the head groups) and a nonpolar end (the two fatty acyl tails). Phospholipids form a bilayer such that the heads interact with water, whereas the tails are shielded from the water. This is an energetically favorable structure.

Figure 22.12 First, the process of membrane invagination created the nuclear envelope. Second, endocytosis may have enabled an ancient archaeon to take up a bacterial cell. Over time, bacterial genes were transferred to the nucleus, which gave rise to the eukaryotic nuclear genome. An engulfed bacterial cell eventually became a mitochondrion, and an engulfed cyanobacterial cell became a chloroplast in algae and plants.

Figure 22.18 Two key features are mammary glands and hair. They also have specialized teeth, external ears, and enlarged skulls that harbor highly developed brains. Mammals are typically endothermic.

Feature Investigation Questions

1. Chemical selection occurs when a particular chemical in a mixture has advantageous properties that allow it to increase in number compared to the other chemicals in the mixture. Bartel and Szostak hypothesized that variation in the catalytic abilities of RNA molecules would allow for chemical selection in the laboratory. Bartel and Szostak proposed to select for RNA molecules with higher catalytic abilities.
2. The short RNA molecules allowed the researchers to physically separate the mixture of longer RNA molecules based on catalytic properties. Long RNA molecules with catalytic abilities would covalently bond with the short RNA molecules. The short RNA molecules had a specific region that caused them to be attracted to column beads in the experimental apparatus. The long RNA molecules that did not have catalytic abilities passed through the column and therefore could be separated from the ones that had catalytic activity and became bound to the column beads.
3. The researchers found that with each round of selection, the enzymatic activity of the selected pool of RNA molecules increased. These results provided evidence that chemical selection could improve the functional characteristics of a group of molecules. Much of the explanation of the evolution of life on Earth is theoretical, meaning it is based on scientific principles but has not been experimentally verified. Researchers are attempting to develop laboratory experiments that test the explanations of the evolution of life. The experiment conducted by Bartel and Szostak provided experimental data to support the hypothesis of chemical selection as a possible mechanism for the early evolutionary process that led to living cells.

Test Yourself

1. b 2. e 3. b 4. c 5. e 6. a 7. d 8. c 9. b 10. d

Conceptual Questions

1. Nucleotides and amino acids were produced prior to the existence of cells.

   Nucleotides and amino acids became polymerized to form DNA, RNA, and proteins.

   Polymers became enclosed in membranes.

   Polymers enclosed in membranes evolved cellular properties.
2. The relative ages of fossils can be determined by the locations in sedimentary rock formation. Older fossils are found in lower layers. A common way to determine the ages of fossils is via radioisotope dating, which is often conducted using a piece of igneous rock from the vicinity of the fossil. A radioisotope is an unstable isotope of an element that decays spontaneously, releasing radiation at a constant rate. The half-life is the length of time required for a radioisotope to decay to exactly one-half of its initial value. To determine the age of a rock (and that of a nearby fossil), scientists can measure the amount of a given radioisotope as well as the amount of the decay product.
3. Several examples are described in this chapter. In some cases, catastrophic events like volcanic eruptions and glaciers caused mass extinctions, which allowed new species to evolve and flourish. In other cases, changing environmental conditions (for example, changes in temperature and moisture) played key roles. One interesting example is adaptation to terrestrial environments. Plant species evolved seeds that are dessication resistant, whereas animal species evolved eggs. Mammalian species evolved internal gestation.

Chapter 23

Concept Checks

Figure 23.2 A single organism does not evolve. Populations may evolve from one generation to the next.

Figure 23.7 Due to a changing global climate, the island fox became isolated from the mainland species. Over time, natural selection resulted in adaptations for the population on the island and eventually resulted in a new species with characteristics that are somewhat different from the mainland species.

Figure 23.8 Many answers are possible. One example is the wing of a bird and the wing of a bat.

Figure 23.11 The relative sizes of traits are changing. For example, in dogs, the lengths of legs, body size, and so on, are quite different. Artificial selection is often aimed at changing the relative sizes of body parts.

Figure 23.13 Rhesus and green monkeys = 0, Congo puffer fish and European flounder = 2, and Rhesus monkey and Congo puffer fish = 10. Pairs that are closely related evolutionarily have fewer differences than do pairs that are more distantly related.

Figure 23.14 Orthologs have similar gene sequences because they are derived from the same ancestral gene. The sequences are not identical because after the species diverged, each one accumulated different random mutations that changed their sequences.

Figure 23.15 It creates multifunctional proteins that may have new properties that can be acted upon by natural selection.

Figure 23.17 Humans have one large chromosome 2, but this chromosome is divided into two separate chromosomes in the other three species. In chromosome 3, the banding patterns among humans, chimpanzees, and gorillas are very similar, but the orangutan has a large inversion that flips the arrangement of bands in the centromeric region.

BioConnections

Figure 23.2 Both natural selection and chemical selection involve processes in which the relative proportions of something in a population increases compared with something else. In natural selection, it is the relative proportions of individuals with certain traits that increases. In chemical selection, molecules with certain characteristics increase their relative numbers compared with other molecules.

Figure 23.13 When comparing homologous genes or proteins, species that are closely related evolutionarily have more similar sequences than do more distantly related species.

Figure 23.16 The three mechanisms of horizontal gene transfer between bacterial species are conjugation, transformation and transduction.

Feature Investigation Questions

1. The island has a moderate level of isolation but is located near enough to the mainland to have some migrants. The island is an undisturbed habitat, so the researchers would not have to consider the effects of human activity on the study. Finally, the island had an existing population of ground finches that would serve as the study organism over many generations.
2. First, the researchers were able to show that beak depth is a genetic trait that has variation in the population. Second, the depth of the beak is an indicator of the types of seeds the birds can eat. The birds with larger beaks can eat larger and drier seeds; therefore, changes in the types of seeds available could act as a selective force on the bird population.

   During the study period, annual changes in rainfall occurred, which affected the seed sizes produced by the plants on the island. In the drier year, fewer small seeds were produced, so the birds would have to eat larger, drier seeds.
3. The researchers found that following the drought in 1978, the average beak depth in the finch population increased. This indicated that birds with larger beaks were better able to adapt to the environmental changes due to the drought and produce more offspring. This is direct evidence of the phenomenon of natural selection.

Test Yourself

1. d 2. d 3. b 4. b 5. b 6. d 7. c 8. b 9. d 10. e

Conceptual Questions

1. Some random mutations result in a phenotype with greater reproductive success. If so, natural selection results in a greater proportion of such individuals in succeeding generations. These individuals are more likely to survive and reproduce, which means they have evolved to be better adapted to their environment.
2. The process of convergent evolution produces two different species from different lineages that show similar characteristics because they occupy similar environments. An example is the long snout and tongue of both the giant anteater, found in South America, and the echidna, found in Australia. This enables these animals to feed on ants, but the two structures evolved independently. These observations support the idea that evolution results in adaptations to particular environments.
3. Homologous structures are two or more structures that are similar because they are derived from a common ancestor. An example is the same set of bones that is found in the human arm, turtle arm, bat wing, and whale flipper. The forearms in these species have been modified to perform different functions. This supports the idea that all of these animals evolved from a common ancestor by descent with modification.

Chapter 24

Concept Checks

Figure 24.2 If C^R is 0.4, then C^W must be 0.6, because the allele frequencies add up to 1.0. The heterozygote ($2pq$) equals 2(0.4)(0.6), which equals 0.48, or 48%.

Figure 24.3 Over the short run, alleles that confer better fitness would be favored and increase in frequency, perhaps enhancing diversity. Over the long run, however, an allele that confers high fitness in the homozygous state may become monomorphic, thereby reducing genetic diversity.

Figure 24.4 Stabilizing selection eliminates alleles that give phenotypes that deviate significantly from the average phenotype. For this reason, it tends to decrease genetic diversity.

Figure 24.6 If malaria was eradicated, there would be no selective advantage for the heterozygote. The H^S allele would eventually be eliminated because the H^SH^S homozygote has a lower fitness. Directional selection would occur.

Figure 24.7 This is likely to be a form of intersexual selection. Such traits are likely to be involved in mate choice.

Figure 24.11 The bottleneck effect decreases genetic diversity. This may eliminate adaptations that promote survival and reproductive success. Therefore, the bottleneck effect makes it more difficult for a population to survive.

Figure 24.13 Gene flow tends to make the allele frequencies in neighboring populations more similar to each other. It also promotes genetic diversity by introducing new alleles into populations.

BioConnections

Figure 24.12 There are lots of possibilities. The idea is that you are changing one codon to another codon that specifies the same amino acid. For example, changing a codon from GGA to GGG is likely to be neutral because both codons specify glycine.

Figure 24.14 Inbreeding favors homozygotes. Initially, inbreeding would result in more homozygotes in a population. Over the long run, however, if a homozygote has a lower fitness, inbreeding would accelerate the elimination of the allele from the population.

Feature Investigation Questions

1. The two species of cichlids used in the experiment are distinguishable by coloration, and the researchers were testing the hypothesis that the females make mate choices based on this variable.
2. Individual females were placed in tanks that contained one male from each species. The males were held in small glass tanks to limit their movement but allowed the female to see each of the males. The researchers recorded the courtship behavior between the female and males and the number of positive encounters between the female and each of the different males. This procedure was conducted under normal lighting and under monochromatic lighting that obscured the coloration differences between the two species. Comparing the behavior of the females under normal light conditions and monochromatic light conditions allowed the researchers to determine the importance of coloration in mate choice.
3. The researchers found that the female was more likely to select a mate from her own species in normal light conditions. However, under monochromatic light conditions, the species-specific mate choice was not observed. Females were as likely to choose males of the other species as they were males of their own species. This indicated that coloration is an important factor in mate choice in these species of fish.

Test Yourself

1. d 2. c 3. c 4. e 5. b 6. c 7. b 8. d 9. b 10. a

Conceptual Questions

1. The frequency of the disease is a genotype frequency because it represents individuals with the disease. If we let q^2 represent the genotype frequency, then q equals the square root of 0.04, which is 0.2. If $q = 0.2$, then $p = 1 - q$, which is 0.8. The frequency of heterozygous carriers is $2pq$, which is $2(0.8)(0.2) = 0.32$, or 32%.
2. Directional selection—This is when natural selection favors an extreme phenotype that makes the organism better suited to survive and reproduce in its environment. As a result, the extreme phenotype will become predominant in the population. This can occur either through new mutation or through a prolonged environmental change. In addition to selecting for a certain phenotype, the opposite end of the extreme is removed from the gene pool.
   - Stabilizing selection—In this type of selection, natural selection favors individuals with intermediate phenotypes, whereas organisms with extreme phenotypes are less likely to reproduce. This selection tends to prevent major changes in the phenotypes of populations.
   - Disruptive selection—This type of selection favors both extremes and removes the intermediate phenotype. It is also known as diversifying selection.
   - Balancing selection—This type of selection results in a balanced polymorphism in which two or more alleles are stably maintained in a population. Examples include heterozygote advantage, as in the sickle cell allele, and negative frequency-dependent selection, as in certain prey.
   - Sexual selection—This is a type of natural selection that is directly aimed at reproductive success. It can occur by any of the previous four mechanisms. Male coloration in African cichlids is an example.
3. Genetic drift involves random changes in the genetic composition of a population from one generation to the next. Neutral changes in DNA sequences may happen randomly, and these are most likely to accumulate in a population due to genetic drift. This is evolution at the level of DNA, but it does not affect phenotype.

Chapter 25

Concept Checks

Figure 25.1 There are a lot of possibilities. Certain grass species look quite similar. Elephant species look very similar. And so on.

Figure 25.3 The temporal isolation is an example of a prezygotic isolating mechanism. Because the species breed at different times of the year, hybrid zygotes are not formed between the two species.

Figure 25.5 Hybrid sterility is a type of postzygotic isolating mechanism. A hybrid forms between the two species, but it is sterile.

Figure 25.11 The offspring would inherit 16 chromosomes from *Galeopsis tetrahit*, and from the hybrid, it would inherit anywhere from 8 to 16. So the answer is 24 to 32. The hybrid parent would always pass the 8 chromosomes that are found in pairs. With regard to the 8 chromosomes not found in pairs, it could pass 0 to 8 of them.

Figure 25.12 The insects on different host plants would tend to breed with each other, and natural selection would favor the development of traits that are an advantage for feeding on that host. Over time, the accumulation of genetic changes may lead to reproductive isolation between the populations of insects.

Figure 25.14 If the *Gremlin* gene was underexpressed, less Gremlin protein would be produced. Because Gremlin protein inhibits apoptosis, more cell death would occur, and the result would probably be smaller feet, and maybe they would not be webbed.

Figure 25.15 By comparing the number of *Hox* genes in many different animal species, a general trend is observed that animals with more complex body structures have a greater number of *Hox* genes.

Figure 25.18 The tip of the mouse's tail might have a mouse eye!

BioConnections

Figure 25.2 Female choice is a prezygotic isolating mechanism.

Figure 25.7 The Hawaiian Islands have many different ecological niches that can be occupied by birds. The first founding bird inhabitants evolved to occupy those niches, thereby evolving into many different species.

Figure 25.15 The *Hox* genes expressed along the anteroposterior axis during early embryonic development are homeotic genes. In insects that contain discrete body segments, each *Hox* gene determines the structures that will ultimately form in those segments. Although more complex animals such as mammals do not display discrete segments, the expression of the *Hox* genes controls what structures will form along the anteroposterior axis.

Feature Investigation Questions

1. Podos hypothesized that the morphological changes in the beak would also affect the birds' songs. A bird's song is an important component for mate choice. If changes in the beak alter the song of the bird, reproductive ability would be affected. Podos suggested that changes in the beak morphology could thus lead to reproductive isolation among the birds.
2. Podos first caught male birds in the field and collected data on beak size. The birds were banded for identification and released. Later, the banded birds' songs were recorded and analyzed for range of frequencies and trill rates. The results were then compared with similar data from other species of birds to determine if beak size constrained the frequency range and trill rate of the song.
3. The results of the study did indicate that natural selection on beak size due to changes in diet could lead to changes in song. Considering the importance of bird song to mate choice, the changes in the song could also lead to reproductive isolation.

   The phrase "by-product of adaptation" refers to changes in the phenotype that are not directly acted on by natural selection. In the case of the Galápagos finches, the changes in beak size were directly related to diet; however, as a consequence of that selection, the song pattern was also altered. The change in song pattern was a by-product.

Test Yourself

1. b 2. b 3. e 4. d 5. c 6. a 7. b 8. d 9. c 10. c

Conceptual Questions

1. Prezygotic isolating mechanisms prevent the formation of the zygote. An example is mechanical isolation, the incompatibility of genitalia. Postzygotic isolating mechanisms act after the formation of the zygote. An example is inviability of the hybrid that is formed. (Other examples shown in Figure 25.2 would also be correct.) Postzygotic mechanisms are more costly because some energy is spent in the formation of a zygote and its subsequent growth.
2. The concept of gradualism suggests that each new species evolves continuously over long spans of time (Figure 25.13a). The principal idea is that large phenotypic differences that produce new species are due to the accumulation of many small genetic changes. According to the punctuated equilibrium model, species exist relatively unchanged for many generations. During this period, the species is in equilibrium with its environment. These long periods of equilibrium are punctuated by relatively short periods during which evolution occurs at a far more rapid rate. This rapid evolution is caused by relatively few genetic changes.
3. One example involves the *Hox* genes, which control morphological features along the anteroposterior axis in animals. An increase in the number of *Hox* genes during evolution is associated with an increase in body complexity and may have spawned many different animal species.

Chapter 26

Concept Checks

Figure 26.2 A phylum is broader than a family.

Figure 26.3 Yes. They can have many common ancestors, depending on how far back you go in the tree. For example, dogs and cats have a common ancestor that gave rise to mammals, and an older common ancestor that gave rise to vertebrates. The most recent common ancestor is the point at which two species diverged from each other.

Figure 26.4 An order is a smaller taxon that would have a more recent common ancestor.

Figure 26.9 A hinged jaw is the character common to the salmon, lizard, and rabbit, but not to the lamprey.

Figure 26.10 Changing the second G to an A is common to species A, B, and C, but not to species G.

Figure 26.13 The kiwis are found in New Zealand. Even so, the kiwis are more closely related to Australian and African flightless birds than they are to the moas, which were found in New Zealand.

Figure 26.15 Gorillas and humans would be expected to have fewer genetic differences because their common ancestor (named C) is more recent than that of orangutans and gorillas, which is ancestor B.

Figure 26.16 Monophyletic groups are based on the concept that a particular group of species descended from a common ancestor. When horizontal gene transfer occurs, not all of the genes in a species were inherited from the common ancestor, so this muddles the concept of monophyletic groups.

BioConnections

Figure 26.1 The domains Bacteria and Archaea have organisms with prokaryotic cells.

Figure 26.14 There are lots of possibilities. The idea is that you are changing one codon to another codon that specifies the same amino acid. For example, changing a codon from GGA to GGG is likely to be neutral because both codons specify glycine.

Feature Investigation Questions

1. Molecular paleontology is the sequencing and analysis of DNA obtained from extinct species. Tissue samples from specimens of extinct species may contain DNA molecules that can be extracted, amplified, and sequenced. The DNA sequences can then be compared with living species to study evolutionary relationships between modern and extinct species.

   The researchers extracted DNA from tissue samples of moas, extinct flightless birds that lived in New Zealand. The DNA sequences from the moas were compared with the DNA sequences of modern species of flightless birds to determine the evolutionary relationships of this particular group of organisms.
2. The researchers compared the DNA sequences of the extinct moas and modern kiwis of New Zealand to the emu and cassowary of Australia and New Guinea, the ostrich of Africa, and rheas of South America. All of the birds are flightless. With the birds selected, the researchers could look for similarities between birds over a large geographic area.
3. The sequences were very similar among the different species of flightless birds. Interestingly, the sequences of the kiwis of New Zealand were more similar to those of the modern species of flightless birds found on other land masses than they were to those of the moas found in New Zealand.

   The researchers constructed a new evolutionary tree that suggests that kiwis are more closely related to the emu, cassowary, and ostrich. Also, based on the results of this study, the researchers suggested that New Zealand was colonized twice by ancestors of flightless birds. The first ancestor gave rise to the now-extinct moas. The second ancestor gave rise to the kiwis.

Test Yourself

1. c 2. d 3. e 4. d 5. b 6. d 7. b 8. b 9. c 10. e

Conceptual Questions

1. The scientific name of every species has two parts, which are the genus name and the species epithet. The genus name is always capitalized, but the species name is not. Both names are italicized. An example is *Canis lupus*.
2. If neutral mutations occur at a relatively constant rate, they act as a molecular clock on which to measure evolutionary time. Genetic diversity between species that is due to neutral mutation gives an estimate of the time elapsed since the last common ancestor. A molecular clock can provide a timescale to a phylogenetic tree.
3. Morphological analysis focuses on morphological features of extinct and modern species. Many traits are analyzed to obtain a comprehensive picture of two species' relatedness. Convergent evolution leads to similar traits that arise independently in different species as they adapt to similar environments. Convergent evolution can, therefore, cause errors if a researcher assumes that a particular trait arose only once and that all species having the trait have the same common ancestor.

Chapter 27

Concept Checks

Figure 27.5 The cell will tend to float because it is full of intact gas vesicles.

Figure 27.11 The motion of the stiff filament of a prokaryotic flagellum is more like that of a propeller shaft than the flexible arms of a human swimmer.

Figure 27.13 Cells having pili tend to move with a twitching or gliding motion.

Figure 27.14 When DNA sequencing studies show that samples contain many uncultured bacterial species, the fluorescence method is preferred, though it requires the use of a fluorescence microscope. Under such conditions, the culture method will underestimate the bacterial numbers. But when the goal is to estimate numbers of bacteria whose culture preferences are known, the culture method may provide good estimates.

Figure 27.15 Endospores allow bacterial cells to survive treatments and environmental conditions that would kill ordinary cells.

Figure 27.18 Structural similarities to bacterial flagella and pili indicate that these types of attack systems evolved from these structures.

BioConnections

Figure 27.6 Like the bacterium *Magnetospirillum magnetotacticum*, birds such as homing pigeons and migratory fishes such as rainbow trout have the capacity to sense and respond to magnetic fields.

Figure 27.2 The microscopic protist *Giardia intestinalis* likewise uses flagella to move within the human small intestine.

Figure 27.14 Two.

Feature Investigation Questions

1. Many bacteria are known to produce organic compounds that function as antibiotics, and are potential food sources for chemoheterotrophic bacteria.
2. Researchers isolated and cultivated bacteria from different types of soils, then grew the cultured bacteria on media that contained one of several common types of antibiotics as the only source of organic food.
3. It was important to know if soil bacteria are a source of antibiotic resistance that can be medically significant.

Test Yourself

1. c 2. b 3. c 4. d 5. a 6. a 7. b 8. e 9. d 10. d

Conceptual Questions

1. Small cell size and simple division processes allow many bacteria to divide much more rapidly than eukaryotes. This helps to explain why food can spoil so quickly and why infections can spread very rapidly within the body. Other factors also influence these rates.
2. Pathogen populations naturally display genetic variation in their susceptibility to antibiotics. When such populations are exposed to antibiotics, even if initially only a few cells are resistant, the numbers of resistant cells will eventually increase and could come to dominate natural populations.
3. Humans. When humans pollute natural waters with high levels of fertilizers originating from sewage effluent or crop field runoff, cyanobacterial populations are able to grow large enough to produce harmful blooms.

Chapter 28

Concept Checks

Figure 28.7 After particles are ingested via feeding grooves, particles are enclosed by membrane vesicles and then digested by enzymes.

Figure 28.8 The intestinal parasite *Giardia intestinalis* is transmitted from one person to another via fecal wastes, whereas the urogenital parasite *Trichomonas vaginalis* can be transmitted by sexual activity.

Figure 28.17 Flagellar hairs function like oars, helping to pull cells through the water.

Figure 28.18 Kelps are harvested for the production of industrially useful materials. In addition, they nurture fishes and other wildlife of economic importance.

Figure 28.21 Genes that encode cell adhesion and extracellular matrix proteins are likely essential to modern choanoflagellates' ability to attach to surfaces, where they feed. Similar proteins are involved in the formation of multicellular tissues in animals. Evolutionary biologists would say that ancient choanoflagellates were preadapted for the later evolution of multicellular tissues in early animals.

Figure 28.24 Cysts allow protists to survive conditions that are not suitable for growth. One such condition would be the dry or cold environment outside a parasitic protist's warm, moist host tissues.

Figure 28.29 Gametes of *Plasmodium falciparum* undergo fusion to produce zygotes while in the mosquito host.

BioConnections

Figure 28.5 The amoebocytes of sponges, which carry food to other cells, move similarly to amoeboid protists.

Figure 28.25 Because the only cell in the *Chlamydomonas* life cycle that is diploid is the zygote, other phases of the life cycle are haploid, and therefore homologous copies do not affect gene expression.

Figure 28.27 The life cycle of diatoms is similar to that of animals.

Feature Investigation Questions

1. One strain had earlier been reported to be toxic to fishes, whereas the other had been reported to be nontoxic, a difference that could be attributed to differing experimental conditions. The investigators wanted to determine the degree of toxicity of the two strains when grown under the same conditions.
2. Producing toxins requires considerable ATP and other resources, so many organisms produce such compounds only when needed. In the case of *Pfiesteria shumwayae*, this might be when a major food source, fish, was present, but not when they fed primarily upon algal cells. The investigators needed to know if this dinoflagellate produces toxin even when feeding on algae alone (which would not require toxin production) or only when exposed to fishes.
3. The team knew that fishermen and scientists had suffered amnesia and other neurological impairments when they were near water containing large populations of the genus *Pfiesteria*. These observations suggested that the toxin was volatile or suspended in water droplets that people could inhale. As a precaution, they used the biohazard containment system to avoid personal harm. The use of biohazard containment systems is generally recommended for scientists who work with hazardous or potentially hazardous biological materials.

Test Yourself

1. c 2. a 3. b 4. b 5. e 6. b 7. e 8. d 9. b 10. c

Conceptual Questions

1. Protists are amazingly diverse, reflecting the occurrence of extensive adaptive radiation after the origin of eukaryotic cells, widespread occurrence of endosymbiosis, and adaptation to many types of moist habitats, including the tissues of animals and plants. As a result of this extensive diversity, protists cannot be classified into a single kingdom or phylum.
2. Several protists, including the apicomplexans *Cryptosporidium parvum* and *Plasmodium falciparum* and the kinetoplastids *Leishmania major* and *Trypanosoma brucei*, cause many cases of illness around the world, but few treatments are available, and organisms often evolve drug resistance. Genomic data allow researchers to identify metabolic features of these parasites that are not present in humans and are therefore good targets for development of new drugs. An

example is provided by metabolic pathways of the apicoplast, a reduced plastid that is present in cells of the genus *Plasmodium*. Because the apicoplast plays essential metabolic roles in the protist but is absent from humans, drugs that disable apicoplast metabolism would kill the parasite without harming the human host.

3. Most protist cells cannot survive outside moist environments, but cysts have tough walls and dormant cytoplasm that allow them to persist in habitats that are unfavorable for growth. While cysts play important roles in the asexual reproduction and survival of many protists, they also allow protist parasites such as *Entamoeba histolytica* (the cause of amoebic dysentery) to spread to human hosts who consume food or water that have been contaminated with cysts. Widespread contamination can sicken thousands of people at a time.

Chapter 29

Concept Checks

Figure 29.3 Liverworts grow very close to surfaces such as soil or tree trunks. Raising their sporophytes off the surface helps to disperse spores into air currents.

Figure 29.4 Wind speed varies, so if the moss released all the spores at the same time into a weak air current, the spores would not travel very far and might have to compete with the parent plant for scarce resources. By releasing spores gradually, some spores may enter strong gusts of wind that carry them long distances, reducing competition with the parent.

Figure 29.8 Larger sporophytes are able to capture more resources for use in producing larger numbers of progeny and therefore have greater fitness than do smaller sporophytes.

Figure 29.13 The capacity to produce both wood and seeds are key features of lignophytes.

Figure 29.15 The polyester cutin found in cuticle, sporopollenin on spore walls, and lignin on water-conducting tracheids of vascular tissues are resistant to decay and thus help plants fossilize.

Figure 29.16 During the Carboniferous period (Coal Age), atmospheric oxygen levels reached historic high levels that were able to supply the large needs of giant insects, which obtain oxygen by diffusion.

Figure 29.21 Because the veins of fern leaves reflect the vascular systems of branched-stem systems, you might infer that leaves evolved from more highly branched stem systems would be more densely veined, that is, have more veins per unit area than fern leaves. This is actually the case for leaves of seed plants.

Figure 29.22 Although some angiosperm seeds, such as those of corn and coconut, contain abundant endosperm, many angiosperm embryos consume most or all of the nutritive endosperm during their development.

Figure 29.24 Because the lacy integument of *Runcaria* does not completely enclose the megasporangium, it probably did not function to protect the megasporangium before fertilization nor as an effective seed coat after fertilization, as do the integuments of modern seed plants. However, the lacy integument of *Runcaria* might have retained the megasporangium on the parent sporophyte during the period of time when nutrients flowed from parent to developing ovule and seed. That function would prevent megasporangia from dropping off the parent plant before fertilization occurred, allow the parent plant to provide nutrients needed during embryo development, and allow seeds time to absorb and store more nutrients from the parent. Such a function would illustrate how one mutation having a positive reproductive benefit can lay the foundation for subsequent mutations that confer additional fitness. *Runcaria* illustrates a first step in the multistage evolutionary process that gave rise to modern seeds.

BioConnections

Figure 29.19 Microvilli characteristic of the animal placenta and small intestine, like the transfer cell-wall protrusions that occur in the plant placenta, vastly increase cell membrane area, thereby providing space for many transport proteins, resulting in relatively high flux of materials across the cell membrane.

Figure 29.23 The amniotic egg characteristic of many animals, like the seeds of plants, provides protection and nutrients to the developing embryo.

Feature Investigation Questions

1. The experimental goals were to determine the rate at which organic molecules produced by gametophyte photosynthesis were able to move into sporophytes and to investigate the effect of sporophyte size on the amount of organic molecules transferred from the gametophyte.
2. The investigators shaded sporophytes with black glass covers to ensure that all of the radioactive organic molecules detected in sporophytes at the end of the experiment came originally from the gametophyte.
3. The investigators measured the amount of radioactivity in gametophytes and sporophytes, and in sporophytes of different sizes. These measurements indicated the relative amounts of labeled organic compounds that were present in different plant tissues.

Test Yourself

1. c 2. d 3. d 4. e 5. b 6. a 7. c 8. e 9. c 10. b

Conceptual Questions

1. Charophycean algae, particularly the complex genera *Chara* and *Coleochaete*, share many features of structure, reproduction, and biochemistry with land plants. Examples include cell division similarities and plasmodesmata and sexual reproduction by means of flagellate sperm and eggs.
2. Bryophytes are well adapted for sexual reproduction when water is available for fertilization. Their green gametophytes efficiently transfer nutrients to developing embryos, enhancing their growth into sporophytes. Their sporophytes are able to produce many genetically diverse spores as the result of meiosis and effectively disperse these spores by means of wind.
3. Vascular tissues allow tracheophytes to effectively conduct water from roots to stems and to leaves. Waxy cuticle helps prevent loss of water by evaporation through plant surfaces. Stomata allow plants to achieve gas exchange under moist conditions and help them avoid losing excess water under arid conditions.

Chapter 30

Concept Checks

Figure 30.4 The nitrogen-fixing cyanobacteria that often occur within the coralloid roots of cycads are photosynthetic organisms that require light. If coralloid roots occurred underground, symbiotic cyanobacteria would not receive enough light to survive.

Figure 30.10 Ways in which conifer leaves are adapted to resist water loss include low surface area/volume, needle- or scale-shape, thick surface coating of waxy cuticle, and stomata that are sunken into the leaf and are therefore less exposed to drying winds.

Figure 30.12 Wide vessels are commonly present in the water transport tissues of angiosperms and much less commonly in other plants. The vessels occasionally found in nonangiosperms are thought to have evolved independently from those of angiosperms.

Figure 30.20 A large, showy perianth would not be useful to grass plants because they are wind pollinated; such a perianth would interfere with pollination in grasses. By not producing a showy perianth, grasses increase the chances of successful pollination and save resources that would otherwise be consumed during perianth development.

Figure 30.24 The flower characteristics of *Brighamia insignis* shown in this figure (white color and deep, narrow nectar tubes) are consistent with pollination by a moth (see Table 30.1).

Figure 30.26 Importantly, ears of modern *Zea mays* do not readily shatter when the fruits are mature, as do those of teosinte. This feature enables human harvesting of the fruits.

BioConnections

Figure 30.2 Modern forests are dominated by seed plants, gymnosperms and angiosperms, whereas nonseed plants dominated *Archaeopteris* forests.

Figure 30.8 The wind-dispersed seeds of the gymnosperm pine resemble the wind-dispersed seeds of the flowering plant maple in bearing winglike structures that enhance transport in air.

Feature Investigation Questions

1. The investigators obtained many samples from around the world because they wanted to increase their chances of finding as many species as possible.
2. The researchers grew plants in a greenhouse under consistent environmental conditions because they wanted to reduce the possible effect of environmental variation on the ratio of cannabinoids produced.
3. Although cannabinoids are produced in glandular hairs that cover the plant surface, these compounds are most abundant on leaves near the flowers. Collecting such leaves reduces the chances that compounds might be missed by the analysis.

Test Yourself

1. d 2. a 3. e 4. e 5. b 6. d 7. e 8. c 9. d 10. e

Conceptual Questions

1. Consult Figure 30.15 to see how plant biologists think stamens and pistils might have evolved from leaves that bore sporangia. Then consider how green leaves surrounding stamens and pistils might have been transformed into petals, sepals, or tepals.
2. Apple, strawberry, and cherry plants coevolved with animals that use the fleshy, sweet portion of the fruits as food and excrete the seeds, thereby dispersing them. Humans have sensory systems similar to those of the target animals and likewise are attracted by the same colors, odors, and tastes.
3. A sunflower is not a single flower, but rather is an inflorescence, a group of flowers.

Chapter 31

Concept Checks

Figure 31.4 Fungal hyphae growing into a substrate having a much higher solute concentration will tend to lose cell water to the substrate, a process that could inhibit fungal growth. This process explains how salting or drying foods helps to protect them from fungal degradation and thus are common preservation techniques.

Figure 31.6 You might filter the air entering the patient's room and limit the entry of visitors and materials that could introduce fungal spores from the outside environment.

Figure 31.27 Modern AM (arbuscular mycorrhizal fungi), also known as Glomeromycota, do not occur separately from plant hosts, as far as is known.

Figure 31.28 Ectomycorrhizal fungi provide their plant partners with water and minerals absorbed from a much larger area of soil than plant roots can exploit on their own.

BioConnections

Figure 31.7 The *Saccharomyces cerevisiae* genome is only 12 million base pairs in size, relatively small for a eukaryote.

Figure 31.10 The amanitin toxin, by interfering with the function of RNA polymerase II, inhibits transcription in eukaryotic cells.

Feature Investigation Questions

1. Plants growing on soils up to 65°C would be expected to have fungal endophytes that aid in heat stress tolerance.
2. The investigators cured some of their *Curvularia protuberata* cultures of an associated virus; then they compared the survival of plants infected with fungal endophytes that had virus versus endophytes lacking virus under conditions of heat stress. Only plants having fungal endophytes that possessed the virus were able to survive growth on soils of high temperature.
3. The fungus *C. protuberata* might be used to confer heat stress tolerance to crop plants, as the investigators demonstrated in tomato.

Test Yourself

1. c 2. b 3. e 4. b 5. a 6. d 7. e 8. b 9. e 10. a

Conceptual Questions

1. Fungi are like animals in being heterotrophic, having absorptive nutrition, and storing surplus organic compounds in their cells as glycogen. Fungi are like plants in having rigid cell walls and reproducing by means of walled spores that are dispersed by wind, water, or animals.
2. Toxic or hallucinogenic compounds likely help to protect the fungi from organisms that would consume them.
3. Some fungi partner with algae or cyanobacteria to form lichens. Some fungi associate with plant roots to form mycorrhizae. Some fungi grow as endophytes within the bodies of plants. In all cases, the heterotrophic fungi receive photosynthetic products from the autotrophic partner.

Chapter 32

Concept Checks

Figure 32.4 Simple choanoflagellates are single-celled organisms. Only later, when such organisms became colonial and groups of cells acquired specialized functions, as in sponges, can we consider them early animals.

Figure 32.8 The coelom functions as a hydrostatic skeleton, which aids in movement. This feature permitted increased burrowing activity and contributed to the development of a profusion of wormlike body shapes.

Figure 32.12 The main members of the Ecdysozoa are the arthropods (insects, spiders, and crustaceans) and the nematodes.

BioConnections

Figure 32.6 A shared derived character.

Figure 32.11 Yellow

Feature Investigation Questions

1. The researchers sequenced the complete gene that encodes small subunit rRNA from a variety of representative taxa of animals to determine their phylogenetic relationships, particularly the relationships of arthropods to other animal taxa.
2. The results indicated a monophyletic clade containing arthropods and nematodes, plus several other smaller phyla. This clade was called the Ecdysozoa. The results of this study indicated that nematodes were more closely related to the arthropods than previously believed.
3. The fruit fly, *Drosophila melanogaster*, and the nematode, *Caenorhabditis elegans*, have been widely studied to understand early development. Under the traditional phylogeny, these two species were not considered to be closely related, so similarities in development were assumed to have arisen early in animal evolution. With the closer relationship indicated by this study, these similarities may have evolved after the divergence of the Ecdysozoan clade. This puts into question the applicability of studies of these organisms to the understanding of human biology.

Test Yourself

1. b 2. c 3. e 4. c 5. c 6. d 7. d 8. d 9. b 10. e

Conceptual Questions

1. (1) Absence or existence of different tissue types. (2) Type of body symmetry. (3) Patterns of embryonic development.
2. The evolution of a coelom cushioned the internal organs in fluid, preventing injury from external forces. In addition, the coelom enabled the internal organs to grow and move independently of the outer body wall. Finally, in some invertebrates, the coelom acts as a hydrostatic skeleton that supports the body and permits movement.
3. Polyphyletic.

Chapter 33

Concept Checks

Figure 33.2 Sponges aren't eaten by other organisms because they produce toxic chemicals and contain needle-like silica spicules that are hard to digest.

Figure 33.4 The dominant life stages are jellyfish: medusa; sea anemone: polyp; Portuguese man-of-war: polyp (in a large floating colony).

Figure 33.5 Cnidocytes are not reused. New ones form to replace the old used ones.

Figure 33.7 Having no specialized respiratory or circulatory system, flatworms obtain oxygen by diffusion. A flattened shape ensures no cells are too far from the body surface.

Figure 33.11 (1) A ciliary feeding device, and (2) a respiratory device are the two main functions of the lophophore.

Figure 33.12 Technically, most mollusks pump hemolymph into vessels and then into tissues. The hemolymph collects in open, fluid-filled cavities called sinuses, which flow into the gills and then back to the heart. This is known as an open circulatory system. Only closed circulatory systems pump blood, as occurs in the cephalopods.

Figure 33.17 Some advantages of segmentation are organ duplication, minimization of body distortion during movement, and specialization of some segments.

Figure 33.19 An annelid is segmented and possesses a true coelom, whereas a nematode is unsegmented and has a pseudocoelom. In addition, nematodes molt, but annelids do not.

Figure 33.20 Other parasitic nematodes in humans are roundworms, *Ascaris lumbricoides*; hookworms, *Necator americanus*; and pinworms, *Enterobius vermicularis*.

Figure 33.25 All arachnids have a body consisting of two tagmata: a cephalothorax and an abdomen. Insects have three tagmata: a head, thorax, and abdomen.

Figure 33.27 Two key insect adaptations are the development of wings and an exoskeleton that reduced water loss and aided in the colonization of land.

Figure 33.33 In embryonic development, deuterostomes have radial cleavage, indeterminate cleavage, and the blastopore becomes the anus. (In protostomes, cleavage is spiral and determinate, and the blastopore becomes the mouth.)

Figure 33.34 Two unique features of an echinoderm are an internal skeleton of calcified plates and a water vascular system.

BioConnections

Figure 33.14 Mollusks arose in the Cambrian period, 543–490 mya. Three hundred million years later ammonites flourished, yet none are alive today.

Figure 33.21 Because most species can excrete urine that is isoosmotic or hyperosmotic to the body fluids.

Figure 33.27 Some insects, such as flies, have chemoreceptors on their feet, whereas others, such as moths, smell through their antennae.

Figure 33.30 These organs, called statocysts, are located at the base of the antennules.

Feature Investigation Questions

1. The researchers tested the hypothesis that an octopus can learn by observing the behavior of another octopus.
2. The results indicated that the observer learned by watching the training of the other octopus. The observer was much more likely to choose the same color ball that the demonstrator was trained to attack. These results seem to support the hypothesis that octopuses can learn by observing the behavior of others.
3. The untrained octopuses had no prior exposure to the demonstrators. The results indicated that these octopuses were as likely to attack the white ball as the red ball. No preference for either color was indicated. The untrained octopuses acted as a control. This is an important factor to ensure the results from the trials using observers indicate response to learning and not an existing preference for a certain color.

Test Yourself

1. b 2. d 3. d 4. d 5. b 6. c 7. b 8. a 9. c 10. a

Conceptual Questions

1. The five main feeding methods used by animals are (1) suspension feeding, (2) decomposition, (3) herbivory, (4) predation, and (5) parasitism. Suspension feeding is usually used to filter out food particles from the water column. A great many phyla, including sponges, rotifers, lophophorates, some mollusks and echinoderms and tunicates, are filter feeders. Decomposers usually feed on dead material such as animal carcasses or dead leaves. For example, many fly and beetle larvae feed on dead animals, and earthworms consume dead leaves from the surface of the Earth. Earthworms and crabs also sift through soil or mud, eating the substrate and digesting the soil-dwelling bacteria, protists, and dead organic material. Herbivores eat plants or algae and are especially common in the arthropoda. Adult moths and butterflies also consume nectar. Snails are also common plant feeders. Predators feed on other animals, killing their prey, and may be active hunters or sit-and-wait predators. Many scorpions and spiders actively pursue their prey, whereas web-spinning spiders ambush their prey using webs. Parasites also feed on other animals but do not normally kill their hosts. Endoparasites, which includes flukes, tapeworms, and nematodes, live inside their hosts. Ectoparasites (ticks and lice) live on the outside of their hosts.
2. Gametes dry out on land, and internal fertilization prevents this from happening. Also, water facilitates movement of gametes, reducing the need for internal fertilization.
3. Complete metamorphosis has four stages: egg, larva, pupa, and adult. The larval stage is often spent in an entirely different habitat from that of the adult, and larval and adult forms utilize different food sources. Incomplete metamorphosis has only three stages: egg, nymph, and adult. Young insects, called nymphs, look like miniature adults when they hatch from their eggs.

Chapter 34

Concept Checks

Figure 34.1 Vertebrates (but not invertebrates) usually possess a (1) notochord; (2) dorsal hollow nerve chord; (3) pharyngeal slits; (4) postanal tail, exhibited by all chordates; (5) vertebral column; (6) cranium; (7) endoskeleton of cartilage or bone; (8) neural crest; and (9) a diversity of internal organs.

Figure 34.7 Ray-finned fishes (but not sharks) have a (1) bony skeleton; (2) mucus-covered skin; (3) swim bladder; and (4) operculum covering the gills.

Figure 34.9 Both lungfishes and coelocanths are Sarcopterygians, having lobe fins.

Figure 34.11 The advantages to animals that moved onto land included an oxygen-rich environment and a bonanza of food in the form of terrestrial plants and the insects that fed on them.

Figure 34.14 No. Caecilians and some salamanders give birth to live young.

Figure 34.15 Besides the amniotic egg, other critical innovations in amniotes are thoracic breathing; internal fertilization; a thicker, less permeable skin; and more efficient kidneys.

Figure 34.17 Snakes evolved from tetrapod ancestors but subsequently lost their limbs. Some species have tiny vestigial limbs.

Figure 34.21 Adaptations in birds to reduce body weight for flight include a lightweight skull; reduction of organ size; and a reduction of organs outside of breeding season. Also female birds have one ovary and relatively few eggs, and no urinary bladder.

Figure 34.29 Defining features of primates are grasping hands; eyes situated on the front of the head to facilitate binocular vision; a large brain; and digits with flat nails instead of claws.

BioConnections

Figure 34.5 Collagen-secreting cells. Cartilage is not mineralized and is softer and more flexible than bone.

Figure 34.13 Yes, the blood vessels to the lungs close and those to skin open wider. The opposite occurs when on land and frogs breath air.

Figure 34.18 Both classes have four-chambered hearts and care for their young.

Figure 34.21 Air is constantly being moved across the lungs, both in inhalation and exhalation. Also, birds employ a cross-current blood supply to the lungs.

Figure 34.27 None. The bloodstreams of fetus and mother are brought into close contact in the placenta, but they do not mix.

Feature Investigation Questions

1. The researchers were interested in determining the method in which *Hox* genes controlled limb development.
2. The researchers bred mice that were homozygous for certain mutations in specific *Hox* genes. This allowed the researchers to determine the function of individual genes.
3. The researchers found that homozygous mutants would develop limbs of shorter lengths compared to the wild-type mice. The reduced length was due to the lack of development of particular bones in the limb, specifically, the radius, ulna, and some carpels. These results indicated that simple mutations in a few genes could lead to dramatic changes in limb development.

Test Yourself

1. c 2. d 3. a 4. d 5. d 6. c 7. c 8. a 9. c 10. d

Conceptual Questions

1. Both taxa have external limbs that move when the attached muscles contract or relax. The difference is that arthropods have external skeletons with the muscles attached internally, whereas vertebrates have internal skeletons with the muscles attached externally.
2. Endothermy (warm-bloodedness) probably evolved independently in both birds and mammals. If the common ancestor of reptiles and birds were endothermic, the chances are that all reptiles would be endothermic.
3. Possibly. Both birds and reptiles lay amniotic eggs and possess scales, though these only cover the legs in birds. Birds and crocodilians also share a four-chambered heart. Finally, birds share many skeletal similarities with certain dinosaurs.

Chapter 35

Concept Checks

Figure 35.6 As in the case of shoots, the capacity to divide the root into two equal pieces by means of a line drawn from the circular edges through the center would indicate that a root has superficial radial symmetry. In order to determine that an organ has radial symmetry at the cellular level, you would have to compare the microscopic views of randomly chosen, wedge-shaped pieces of cross slices. If the structure of the wedges is similar, the organ has radial symmetry at the microscopic level.

Figure 35.8 Locating stomata on the darker and cooler lower leaf surface helps reduce water loss from the leaf.

Figure 35.12 A twig having five sets of bud scale scars is likely to be approximately 6 years old.

Figure 35.17 Cactus stems are green and photosynthetic, playing the role served by the leaves of most plants.

Figure 35.21 A woody stem builds up a thicker layer of wood than inner bark in part because older tracheids and vessel element walls are not lost during shedding of bark, which is the case for secondary phloem. In addition, plants typically produce a greater volume of xylem than phloem tissue per year, in part because vessel elements are relatively wide. A large volume of water-conducting tissue helps plants maintain a large amount of internal water.

Figure 35.25 Lateral roots are produced from internal meristematic tissue because roots do not produce axillary buds like those from which shoot branches develop. Internal production of branch roots helps to prevent them from shearing off as the root tip grows through abrasive soil.

BioConnections

Figure 35.5 Apical-basal polarity of the plant body resembles anterior-posterior polarity in the animal body.

Figure 35.9 Plant cells often possess a large vacuole, whereas animal cell vacuoles are relatively small.

Figure 35.11 There is no difference in microtubule structure among eukaryotes.

Feature Investigation Questions

1. The advantages of using natural plants include the opportunity to avoid influencing plants with unnatural environmental factors, such as artificial light, and the exposure of all experimental plants to similar growth conditions. In addition, the investigators studied the leaves of some large trees, which would be hard to accommodate in a greenhouse.
2. Pinnately veined leaves were splinted to prevent their breaking, since they were cut at the single main vein, which has both support and conducting functions.
3. Sack and associates measured leaf water conduction at two or more places on each leaf because the effect of cutting a vein might have affected some portions of leaves more than others.

Test Yourself

1. d 2. c 3. b 4. a 5. c 6. a 7. d 8. e 9. b 10. d

Conceptual Questions

1. If overall plant architecture were bilaterally symmetrical, plants would be shaped like higher animals, with a distinct front (ventral surface) and back (dorsal surface). By comparison with radially symmetrical organisms, bilaterally symmetrical plants would have reduced ability to deploy branches and leaves in a way that would fill available lighted space and would thus be unable to take optimal photosynthetic advantage of their habitats.
2. If leaves were generally radially symmetrical (shaped like spheres or cylinders), leaves would not have maximal ability to absorb sunlight, and they would not be able to optimally disperse excess heat from their surfaces.
3. Although tall herbaceous plants exist (palms and bamboo are examples), the additional support and water-conducting capacity that are provided by secondary xylem allow woody plants to grow tall.

Chapter 36

Concept Checks

Figure 36.4 Auxin efflux carriers could be located on the upper sides of root cells, thereby allowing auxin to move upward in roots.

Figure 36.6 Once a callus has been established from a single plant having desirable characteristics using plant tissue culture, the callus can be divided into many small calluses. A grower could transfer these to separate containers having the appropriate hormone mixtures to induce root and shoot growth, then transplant the young plant clones to soil. This would allow the grower to produce many identical plants.

Figure 36.9 The triple response of dicot seedlings to internally produced ethylene allows them to protect the delicate apical meristem from damage as the seedling emerges through the soil.

Figure 36.10 The active conformation of phytochrome absorbs far-red light. Such absorption causes the active conformation of phytochrome to change to the inactive conformation and to move out of the nucleus and into the cytosol.

Figure 36.11 The inactive conformation of phytochrome would absorb the red portion of sunlight, thereby converting phytochrome into the active conformation.

Figure 36.12 Exposing plants to brief periods of darkness during the daytime will have no effect on flowering because flowering is determined by night length.

Figure 36.14 Yes, just as shoots exhibit negative gravitropism in upward growth, roots are capable of using negative phototropism to grow downward, because light decreases with depth in the soil.

Figure 36.16 In some plants, aerenchyma development is genetically determined and occurs even in the absence of flooding. In other cases, aerenchyma develops only under flooding conditions as a result of controlled cell death.

Figure 36.17 Predators are more likely to be able to find their prey if the latter are concentrated and exposed while feeding on plants. Plants benefit when predation removes herbivores, a process that lessens damage to plants.

Figure 36.19 Similar suites of protective plant hormones, such as jasmonic acid, are used in both types of defenses.

BioConnections

Figure 36.3 The defense hormone/plasma membrane receptor is the plant signaling system most similar to the general diagram shown in Figure 9.4.

Figure 36.18 The lipopolysaccharide envelope characteristic of Gram-negative bacteria is a feature by which plants recognize pathogens.

Feature Investigation Questions

1. The experimental procedure that Briggs would use to test the hypothesis that light destroys auxin would follow a similar protocol as his other experiment. The figure on the next page shows this experimental protocol.
2. Hypothetically, auxin enhances the rate at which cell membrane proton pumps acidify the plant cell wall, thereby allowing cells to extend. Although the evidence for acid effects on cell-wall extension is strong, the molecular basis of possible auxin effects on proton pumps is not as yet clear.
3. A small number of seedling tips could display atypical responses for a variety of reasons. The investigators actually performed the experiment with many replicate seedling tips (coleoptiles), in order to gain confidence that the responses are general.

Test Yourself

1. c 2. c 3. a 4. e 5. d 6. d 7. c 8. d 9. d 10. d

Conceptual Questions

1. Behavior is defined as the responses of living things to a stimulus. Therefore, because plants display many kinds of responses to diverse stimuli, they display behavior.
2. Many kinds of disease-causing bacteria and fungi occur in nature, and these organisms evolve very quickly, producing diverse elicitors. Thus, plants must maintain a stock of resistance genes, each having many alleles.
3. Talking implies a conversation with "listeners" who detect a message and respond to it. Thus, plants that exude volatile compounds that attract enemies of herbivores could be interpreted as "talking" to those enemies. The message is "Hey, you guys, there's food for you over here." In addition, research has revealed that some plants near those under attack respond to volatile compounds by building up defenses. "Talking" to other plants does not enhance the "talker's" fitness. But the ability to "listen" enhances the "listener's" fitness, because it can take preemptive actions to prevent attack.

(b) Briggs experiment 1

HYPOTHESIS Light destroys auxin on lit side of shoot tips, causing unequal auxin distribution. Unlit side should grow more than lit side.

STARTING MATERIALS Corn seedlings.

Experimental level | **Conceptual level**

1 Collect auxin into agar blocks from:
A dark-grown tips
B tips grown with directional light

Dark-grown tip A; Auxin diffusion; Directional light-grown tip B

Light. If light destroys auxin on one side, less auxin will enter the block.

2 Place agar blocks on right side of decapitated shoots.

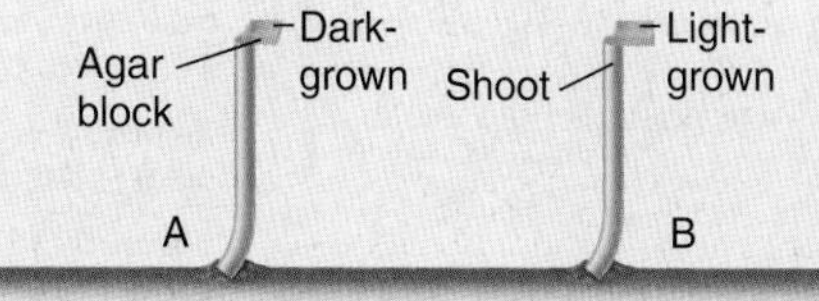

If the block on the right side has less auxin, it will cause less bending.

3 **THE DATA**

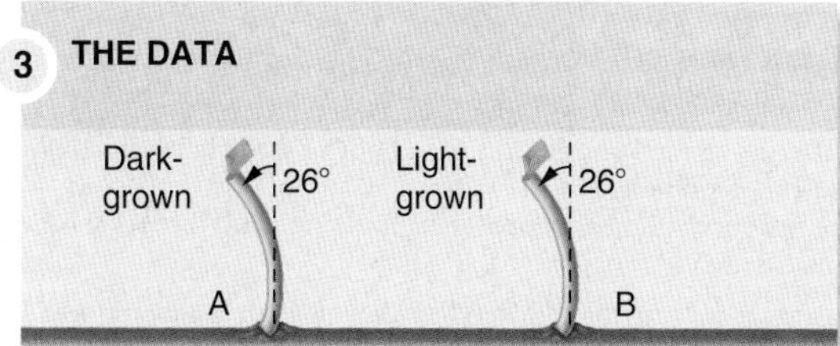

4 **CONCLUSION**

Similar bending demonstrates that light did not destroy auxin in the directionally lit shoot tip. If it had, less auxin would have been present in the agar block in B, and the degree of bending would have been less. The hypothesis described under conceptual level (above) is incorrect.

Chapter 37

Concept Checks

Figure 37.5 Plastids that occur in a cluster near the nucleus would have more rubisco than plastids at the periphery.

Figure 37.6 Chlorosis is not always a sign of iron deficiency; it can be a deficiency symptom for several mineral nutrients, including zinc in corn.

Figure 37.10 Mineral leaching occurs more readily from sandy soils than from clay soils.

Figure 37.12 Soil crusts containing nitrogen-fixing cyanobacteria increase soil fertility, fostering the growth of larger plants that stabilize soils against erosion and provide forage for animals.

Figure 37.13 Oxygen, which makes up 21% of Earth's present atmosphere, can bind to the active site of nitrogenase, thereby inactivating it.

BioConnections

Figure 37.7 Land whose topsoil has been damaged will likely not function normally to support a wild community or agriculture; rehabilitation might be accomplished by restoring topsoil and planting vegetation that holds the soils in place and increases its fertility.

Figure 37.20 Both the tapeworm and the dodder obtain organic food from a host, an animal or plant, respectively.

Feature Investigation Questions

1. During this period of time, the amount of phosphorus in plant tissues had significantly decreased, but plant growth had not yet been affected. Thus, a monitoring system based on gene expression changes that occur during this time would allow farmers time to apply fertilizer in order to prevent crop losses resulting from nutrient deficiency.
2. *SDQ1* expression is induced by phosphorus deficiency. This gene fosters replacement of plastid phospholipids with sulfur-containing lipids, thereby reducing the plant's phosphorus requirement.
3. They used genetic engineering techniques to place a reporter gene under the control of the *SQD1* promoter, so that when *SQD1* was expressed, the reporter gene was expressed also. After growing plants in nutrient solutions containing various levels of phosphorus, they removed sample leaves and treated them with a compound that turns blue when the reporter gene is expressed. When they saw blue leaves, the investigators could infer (1) that the plants from which those leaves had been taken were beginning to experience phosphorus deficiency, and (2) that application of fertilizer at this point could prevent damage to the plants.

Test Yourself

1. e 2. a 3. c 4. b 5. e 6. c 7. d 8. d 9. a 10. b

Conceptual Questions

1. Agricultural experts are concerned that adding excess fertilizer to crop fields increases the costs of crop production. Ecologists are concerned that excess fertilizers will wash from crop fields into natural waters and cause harmful overgrowths of cyanobacteria, algae, and aquatic plants. Methods for closely monitoring crop nutrient needs so that only the appropriate amount of fertilizer is applied would help to allay both groups' concerns.
2. Use Figure 37.10 as a reference. A first arrow could be drawn from a root to rhizobia in the soil, and the arrow labeled "flavonoids." A second arrow could be drawn from rhizobia to roots and labeled "Nod factors." A third arrow from rhizobia to roots could be labeled "infection proteins." A fourth arrow from roots to rhizobia could be labeled "nodulins" and the resulting nodule environmental conditions, which influence the formation of bacterioids. A fifth arrow could represent the flow of fixed nitrogen from bacterioids to plant. A sixth arrow could represent the flow of organic compounds from plant to bacteroids.

Chapter 38

Concept Checks

Figure 38.5 When placed in pure water, a turgid cell having a water potential of 1.0 will lose water, because 1 is greater than 0. When placed in pure water, a plasmolyzed cell having a water potential of –1.0 MPa will gain water. When placed in pure water, a flaccid cell having a water potential of –0.5 MPa will gain water. This is because water moves from a region of higher water potential to a region of lower water potential, and 0 is greater than –0.5.

Figure 38.11 You would likely see stained rings or helical ribbons extending up the insides of the long walls of extensible tracheids. You would not see staining at the ends of tracheids, where they connect to form cell files.

Figure 38.12 The large perforations in vessel element end walls allow an air bubble to extend from one element to another, thereby clogging vessels and preventing water flow through them. In contrast, the much smaller pores in the end walls of tracheids do not allow water to flow as efficiently as it does through vessels, but these smaller pores also retard the movement of air bubbles. As a result, air bubbles are confined to a single tracheid where they do little harm.

Figure 38.13 Root pressure can help to reverse embolism, thereby aiding water flow through xylem.

Figure 38.16 The evaporation of water has a powerful cooling effect because it disperses heat so effectively. Water has the highest heat of vaporization of any known liquid.

Figure 38.17 You could model a stomatal guard cell with an elongate balloon by partially inflating it, then attaching thick tape along one side to represent thickened inner walls and circles of string or thin tape to represent radial cellulose, then adding more air to the balloon. The balloon should curve as it expands, just as a guard cell does when the stomatal pore opens. Two such balloons could be used to model both guard cells and the stomatal pore.

Figure 38.19 In its desert habitat, times of drought and contrasting availability of water sufficient to support the development and photosynthetic function of leaves do not occur at predictable times, as is the case for temperate forests. For this reason, ocotillo leaf abscission is not amenable to the evolution of genetic mechanisms that allow leaf drop to be precisely timed in anticipation of the onset of drought.

Figure 38.25 You could note the relatively few genes that are plotted along the middle left side of the triangle, then try to localize the encoded proteins within the tissues of very young stem tissue, using microscopy.

BioConnections

Figure 38.8 Tight junctions of intestinal epithelium and Casparian strips of endodermal cells of plant roots both incorporate materials that form a tight seal, preventing movement of materials from one location to another.

Figure 38.21 Stomatal guard cell and phloem companion cell development both begin with an unequal cell division.

Feature Investigation Questions

1. This design allowed investigators to compensate for variation among plants, which might have influenced the results had they used separate plants for experiments and controls.
2. Transpiration! Water evaporating from the surfaces of leaves exerted a tension on the water column of the xylem, pulling sap and water through it.
3. The effects of ions on sap flow rates did not directly depend on a biological process, so xylem sap of the same ionic concentration moved through dead plants at the same rate as in living plants.

Test Yourself

1. c 2. b 3. e 4. d 5. a 6. e 7. d 8. e 9. a 10. c

Conceptual Questions

1. In the case of plant fertilizers, more is not better, because the ion concentration of overfertilized soil may become so high as to draw water from plant cells. In this case, the cells would be bathed in a hypertonic solution and would likely lose water to the solution. If plant cells lose too much water, they will die.
2. When the natural vegetation is removed, transpiration stops, so water is not transported from the ground to the atmosphere, where it may be an important contributor to local rainfall. Extensive removal of plants actually changes local climates in ways that reduce agricultural productivity and human survival.
3. You cannot assume that an ocotillo plant lacking leaves is dead, because this plant responds to drought by shedding its leaves, and living plants can produce new leaves when the drought stress is relieved. However, if the ocotillo plants do not produce new leaves after normal rainstorms, you might suspect that they have died.

Chapter 39

Concept Checks

Figure 39.2 Because gametophytes are haploid, they lack the potential for allele variation at each gene locus that is present in diploid sporophytes. Hence, gametophytes are more vulnerable to environmental stresses. By living within the diploid tissues of flowers, flowering plant gametophytes are protected to some extent, and the plant does not lose its gamete-producing life cycle stage.

Figure 39.3 Some flowers lack some of the major flower parts.

Figure 39.4 By clustering its stamens around the pistil, the hibiscus flower increases the chance that a pollinator will both pick up pollen and deliver pollen from another hibiscus flower on the same trip.

Figure 39.7 The absence of showy petals often correlates with wind pollination, because large petals would interfere with the shedding of pollen in the wind.

Figure 39.10 The rim flowers of *Gerbera* inflorescences have bilateral symmetry, conferred by expression of a *CYCLOIDEA*-like gene. Rim flowers also possess showy petals that attract pollinators, but lack pollen-producing stamens. By contrast, central flowers display radial symmetry, lack showy petals, and possess pollen-producing stamens.

Figure 39.12 The maximum number of cells in a mature male gametophyte of a flowering plant is three: a tube cell and two sperm cells.

Figure 39.13 Female gametophytes are not photosynthetic and cannot produce their own food. Enclosed within ovules, female gametophytes lack direct access to the outside environment. Carpels contain veins of vascular tissue that bring nutrients from sporophytic tissue to ovules.

Figure 39.16 An embryo in which the TOPLESS genes were nonfunctional would have two roots and no shoots.

Figure 39.17 During their maturation, the cotyledons of eudicot seeds absorb the nutrients originally present in endosperm.

BioConnections

Figure 39. 15 The plant pollen tube is analogous in function to a human penis in that both structures accomplish internal fertilization. The plant pollen tube grows long enough to deposit sperm at the micropyle within the body of the female gametophyte, much as an animal penis deposits sperm within the female's body. In both cases, sperm are more likely to survive and accomplish fertilization than if they were deposited outside the female body.

Figure 39.21 The plant is similar to the animal *Obelia* because both can reproduce asexually by means of multicellular structures that when released can grow into an adult.

Feature Investigation Questions

1. The large flowers of this lily enabled investigators to more easily mark petals and record the positions of marks over time.
2. Time-lapse video reduced the amount of time investigators would have to spend recording changes in the positions of petal marks.
3. Results obtained by using mathematical models can be compared with actual measurements to assess the accuracy of the models. An accurate mathematical model indicates a relatively full understanding of the physical processes involved. If models are sufficiently accurate, they can be utilized in other situations.

Test Yourself

1. a 2. b 3. d 4. b 5. d 6. e 7. b 8. c 9. d 10. c

Conceptual Questions

1. Pollen grains are vulnerable to mechanical damage and microbial attack during the journey through the air from the anthers of a flower to a stigma. Sporopollenin is an extremely tough polymer that helps to protect pollen cells from these dangers. The function of the beautiful sculptured patterns of sporopollenin on pollen surfaces is unclear.
2. The embryos within seeds are vulnerable to mechanical damage and microbial attack after they are dispersed. Seed coats protect embryos from these dangers and also help to prevent seeds from germinating until conditions are favorable for seedling survival and growth.
3. Flower diversity is an evolutionary response to diverse pollination circumstances. For example, plants such as oak and corn that are wind-pollinated produce flowers having a poorly developed perianth. If such wind-pollinated flowers had large, showy perianths, they would get in the way of pollen dispersal or acquisition. On the other hand, flowers that are pollinated by animals often have diverse shapes and attractive petals of differing colors or fragrances that have coevolved with different types of animal pollinators.

Chapter 40

Concept Checks

Figure 40.2 Locomotion is the movement of an animal's body from one place to another. This is achieved by the actions of skeletal muscle. However, smooth muscle contraction promotes movement of internal structures, like those of the digestive system, and contraction of cardiac muscle causes movement (beating) of the heart.

Figure 40.6 No. The brain, for example, does not contain muscle tissue (although the blood vessels supplying the brain do contain smooth muscle).

Figure 40.7 Blood, including plasma and blood cells, would leak out of the blood vessel into the interstitial space. The fluid level of the bloodstream would decrease, and that of the interstitial space near the site of the injury would increase. Eventually the blood that entered the interstitial space would be degraded by enzymes, resulting in the characteristic skin appearance of a bruise. If the injury were very severe, the fluid level in the blood could decrease to a point where the various tissues and organs of the body would not receive sufficient nutrients and oxygen to function normally.

Figure 40.8 A decrease in intracellular fluid volume, like that shown in the cell in this figure, would result in an increase in intracellular solute concentration (likewise, an increase in intracellular fluid volume would decrease intracellular solute concentrations). This may have drastic consequences on cell function. For example, some solutes, like Ca^{2+} and certain other ions, are toxic to cells at high concentrations.

Figure 40.10 Surface area is important to any living organism that needs to exchange materials with the environment. A good example of a high surface area/volume ratio is that of most tree leaves. This makes leaves ideally suited for such processes as light absorption (required for photosynthesis; see Chapter 8) and the exchange of gases and water with the environment.

Figure 40.12 No, not necessarily. Body temperature, for example, is maintained at different set points in birds and mammals. Other vertebrates and most invertebrates do not have temperature set points; their body temperature simply conforms close to that of the environment. As another example, a giraffe has a set point for blood pressure that is higher than that of a human being, because a giraffe's circulatory system must generate enough pressure to pump blood up its long neck to its brain.

Figure 40.15 In nature, an animal such as a horse would have the same type of responses shown here if threatened by a predator. Upon sensing the presence of the predator, the horse's respiratory and circulatory systems would begin increasing their activities in preparation for the possibility that the horse might have to flee or defend itself. This would occur even before the horse began to flee.

BioConnections

Figure 40.6 Note that both the stomach and the intestine depicted in Figures 40.6 and 45.9 contain layers of muscle wrapped around the lumen. Although you will learn later that the stomach and intestine have many different functions, this similarity in anatomy suggests that both of these organs may perform the similar activity of mechanically breaking apart chunks of food, and propelling the contents from one region to another.

Feature Investigation Questions

1. Pavlov studied feedforward regulation of saliva production that occurs in hungry dogs even before they receive food. He hypothesized that the feedforward response could be conditioned to other, nonrelevant stimuli such as sounds, as long as the sounds were presented simultaneously with food.
2. Pavlov remained outside the room where the dog was housed when the conditioning stimulus—a metronome—was started. In addition, the room was carefully sealed to prevent any other stimuli, including smells, sights, and sounds, from interfering with the conditioning response.
3. He measured the amount of saliva secreted by salivary glands in the dog's mouth by collecting the saliva through a tube and funnel, and then recording the number of drops. He discovered that once a dog had become conditioned to hearing the sound of the metronome whenever presented with food, the sound itself was sufficient to stimulate the feedforward response of salivation. This experiment revealed that feedforward processes could be modulated by experience and learning.

Test Yourself

1. d 2. c 3. d 4. b 5. c 6. c 7. d 8. b 9. d 10. e

Conceptual Questions

1. Structure and function are related in that the function of a given organ, for example, depends in part on the organ's size, shape, and cellular and tissue arrangement. Clues about a physical structure's function can often be obtained by examining the structure's form. For example, the extensive surface area of a moth's antennae suggests that the antennae are important in detecting the presence of airborne chemicals. Likewise, any structure that contains a large surface area for its volume is likely involved in some aspect of signal detection, cell-cell communication, or transport of materials within the animal or between the animal and the environment. Surface area increases by a power of 2, and volume increases by a power of 3 as an object enlarges; this means that in order to greatly increase surface area of a structure such as an antenna, without occupying enormous volumes, specializations must be present (such as folds) to package the structure in a small space.
2. Homeostasis is the ability of animals to maintain a stable internal environment by adjusting physiological processes, despite changes in the external environment. Examples include maintenance of salt and water balance, pH of body fluids, and body temperature. Some animals conform to their external environment to achieve homeostasis, but others regulate their internal environment themselves.
3. Maintaining homeostasis requires continual supplies of energy. Animals consume food, and the energy from that food helps sustain activities that maintain variables such as body temperature and pH, and synthesis of complex molecules. Without this energy, it would be difficult or impossible for animals to maintain many important biological processes within a narrow range despite changes in the environment.

Chapter 41

Concept Checks

Figure 41.4 Many reflexes, such as the knee-jerk reflex, cannot be prevented once started. Others, however, can be controlled to an extent. Open your eyes widely and gently touch your eyelashes. A reflex that protects your eye will tend to make you close your eyelid. However, you can overcome this reflex with a bit of difficulty if you need to, for example, when you are putting in contact lenses.

Figure 41.5 The squid axon is not coated in myelin sheaths. This is another feature of the squid giant axon that makes it a convenient model for conducting in vitro experiments such as the one depicted in this figure.

Figure 41.7 Yes, the flow of K^+ down its chemical gradient does create an electrical gradient because K^+ is electrically charged. The net flow of K^+ will stop when the chemical gradient balances the electrical gradient. This occurs at the equilibrium potential.

Figure 41.10 When the K^+ channels open (at 1 msec), the Na^+ channels would still be opened, so the part of the curve that slopes downward would not occur as rapidly, and perhaps the cell would not be able to restore its resting potential.

Figure 41.12 The action potential can move faster down an axon. This is especially important for long axons, such as those that carry signals from the spinal cord to distant muscles.

Figure 41.14 In the absence of such enzymes, neurotransmitters would remain in the synapse for too long, and the postsynaptic cell could become overstimulated. In addition, the ability of the postsynaptic cell to respond to multiple, discrete inputs from the presynaptic cell would be compromised.

BioConnections

Figure 41.1 This is an example of a feedforward response, most famously demonstrated by the conditioning experiments of Ivan Pavlov. In this case, the peripheral and central nervous systems interact to prepare the hyena for feeding.

Figure 41.8 Water molecules move through membrane channels called aquaporins.

Feature Investigation Questions

1. Loewi was aware that electrical stimulation of the vagus nerve associated with heart muscle would slow down the rate of heart contractions in a frog. Also, he knew that electrical stimulation of other nerves associated with the frog heart produced opposite results. If the effects of the different nerves on heart muscle were mediated directly by electrical activity only, the heart muscle cells would have no way to distinguish between stimulatory and inhibitory signals. Loewi hypothesized that nerves released chemicals onto heart muscle cells and that it was these different chemicals that produced the varied effects on the heart.
2. Loewi placed two hearts in separate chambers, one heart with its vagus nerve intact and the other with its vagus nerve removed. He electrically stimulated the vagus nerve of the first heart, then removed some of the saline solution surrounding the heart and transferred it to the second heart. He then observed whether or not the second heart responded as if its vagus nerve had been intact and had been stimulated.
3. When fluid from the saline solution around the stimulated heart was added to the saline solution of the second, unstimulated heart, the rate of contraction in the second heart was decreased just as if its own vagus nerve had been intact

and was stimulated. This suggested that chemicals were released into the saline solution of the first heart following the electrical stimulation of its vagus nerve, and that it was these chemicals that caused the cardiac muscle to slow its rate of contraction. The results did support Loewi's hypothesis.

Test Yourself

1. c 2. d 3. e 4. e 5. b 6. d 7. b 8. e 9. a 10. e

Conceptual Questions

1. In a graded potential, a weak stimulus causes a small change in the membrane potential, whereas a strong stimulus produces a greater change. Graded potentials occur along the dendrites and cell body. If a graded potential reaches the threshold potential at the axon hillock, an action potential results. This is a change in the membrane potential that is of a constant value and is propagated from the axon hillock to the axon terminal.
2. An increase in extracellular Na^+ concentration would slightly depolarize neurons, thereby changing the resting membrane potential. This effect would be minimal, however, because the resting membrane is not very permeable to Na^+. However, the shape of the action potentials in such neurons would be a little steeper, and the peak a little higher, because the electrochemical gradient favoring Na^+ entry into the cell through voltage-gated channels would be greater.
3. Neurons are among the most highly complex cells in an animal's body, with numerous extensions of the cell body. These extensions provide considerable surface area that allows for an extraordinary number of cell-to-cell contacts with other neurons, making them ideally suited for intercellular communication. In addition, myelin sheaths provide a structure that speeds up electric signaling along the axon, facilitating communication even more.

Chapter 42

Concept Checks

Figure 42.4 Not necessarily. Brain mass is not the sole determinant of intelligence or the ability to perform complex tasks. The degree of folding of the cerebral cortex is also important.

Figure 42.6 A spinal nerve is composed of both afferent and efferent neurons.

Figure 42.7 The major symptom experienced by patients undergoing a lumbar puncture is headache, in part because the brain is no longer cushioned adequately by CSF. Within 24–48 hours, however, the CSF is replenished to normal levels.

Figure 42.9 Damage to the cerebellum would result in loss of balance and a lack of fine motor control, such as picking up small objects or making graceful, smooth movements.

Figure 42.11 It was in her right hand.

Figure 42.16 Thinking requires energy! Even daydreaming requires energy; imagine how much energy the brain uses when you concentrate for 60 minutes on a difficult exam. In fact, you just expended energy thinking about this question!

BioConnections

Figure 42.1 As defined in Chapter 32, animals are multicellular heterotrophs (cannot make their own food) whose cells lack a cell wall. Most animals have a nervous system, muscles, the ability to move about during at least some phase of their life cycle, and to reproduce sexually.

Feature Investigation Questions

1. Gaser and Schlaug hypothesized that repeated exposure to musical training would increase the size of certain areas of the brain associated with motor, auditory, and visual skills. All three skills are commonly used in reading and performing musical pieces.
2. The researchers used MRI to examine the areas of the brain associated with motor, auditory, and visual skills in three groups of individuals: professional musicians, amateur musicians, and nonmusicians. The researchers found that certain areas of the brain were larger in the professional musicians compared to the other groups, and larger in the amateur musicians compared to the nonmusicians.
3. Schmithorst and Holland found that, when exposed to music, certain regions of the brains of musicians were activated differently compared with the brains of nonmusicians. This study supports the hypothesis that there is a difference in the brains of musicians versus nonmusicians.

The experiment conducted by Gaser and Schlaug compared the size of certain regions of the brain among professional musicians, amateur musicians, and nonmusicians. Schmithorst and Holland, however, were also able to detect functional differences between musicians and nonmusicians.

Test Yourself

1. b 2. a 3. e 4. c 5. c 6. b 7. e 8. b 9. a 10. d

Conceptual Questions

1. All animals with nervous systems have reflexes, which allow rapid behavioral responses to changes in the environment. When a cnidarian senses a tactile stimulus, its nerve net responds immediately and the animal reflexively contracts nearly all of its muscles, making the animal a smaller target. This behavior protects the animal from predators. When you hear a loud, unexpected, and frightening sound (such as a firecracker), you hunch your shoulders and slightly lower your head; this reflex protects you from danger by minimizing exposure of your neck and head to danger. Dilation of the pupils of the eyes in darkness, and constriction of the pupils in bright light, are reflexes that help us see in the dark and protect our retinas in bright light. Reflexes are particularly adaptive because they occur rapidly, typically with very few synapses involved, and without the need for conscious thought.
2. White matter consists of the myelinated axons that are bundled together in large tracts in the central nervous system and which connect different CNS regions. The lipid-rich myelin gives the tracts a whitish appearance. It is distinguished from gray matter, which are the cell bodies, dendrites, and some unmyelinated axons of neurons in the CNS.
3. The activities of animal nervous systems are replete with examples of new properties emerging from complex interactions. For example, you learned about reflexes in this chapter, which are behaviors that emerge from interactions between individual neurons that form communication circuits between the peripheral and central nervous systems. You also learned about such "higher" properties such as conscious thought, which also emerges from the interactions between many individual cells, each of which is in communication with up to hundreds of thousands of other cells. Individually, the cells cannot "think," but networked together in elaborate ways, a person like yourself can think, remember, plan ahead, and interpret your environment.

Chapter 43

Concept Checks

Figure 43.3 To think about what types of touch you are aware of, let's take the example of sitting in a chair reading this textbook while holding it on your lap. You are aware of the constant weight of the book, the brush of the pages on your fingertips as you turn a page, a gentle breeze that may be circulating in your environment, the deep pressure from regularly adjusting your posture in your chair, an itch you may have on your skin, and the heat or cold of the room. Even a simple exercise such as this one is filled with stimuli of numerous types and durations.

Figure 43.10 This orientation permits animals to detect circular or angular movement of the head in three different planes. The fluid in a canal that is oriented in the same plane as the plane of movement will respond maximally to the movement. For example, the canal that is oriented horizontally would respond greatest to horizontal movements, while the other two canals would not. Overall, by comparing the signals from the three canals, the brain can interpret the motion in three dimensions.

Figure 43.19 Because red-green color blindness is a sex-linked recessive gene, males require only a single defective allele on an X chromosome, whereas females require two defective alleles, one on each X chromosome.

Figure 43.27 Salt is a vital nutrient needed to maintain plasma membrane potentials and fluid balance in animals' bodies. Sugar provides glucose and other monosaccharides, important energy-yielding compounds. Sour (acidic) foods, like citrus fruits, provide nutrients and important antioxidants (vitamin C, for example) that protect against disease. Bitter substances are often toxic, and their bad taste discourages animals from eating them.

BioConnections

Figure 43.1 The term sensory receptor refers to a type of cell that can respond to a particular type of stimulus. The term membrane receptor refers to a protein within a cell membrane that binds a ligand, thereby generating signals that initiate a cellular response.

Figure 43.4 Cilia are cell extensions that contain in their internal structure microtubules and motor proteins that cause the cilia to beat, or move, in a coordinated fashion. Stereocilia are membrane projections that are not motile, but instead are deformed by the movements of surrounding fluids.

Figure 43.9 Statoliths are also found in the roots and shoots of plants. They serve as a gravity-detection mechanism that results in roots growing downward, and shoots upward.

Feature Investigation Questions

1. One possibility is that many different types of odor molecules might bind to one or just a few types of receptor proteins, with the brain responding differently depending on the number or distribution of the activated receptors. The second hypothesis is that organisms can make a large number of receptor proteins, each type binding a particular odor molecule or group of odor molecules. According to this hypothesis, it is the *type* of receptor protein, and not the number or distribution of receptors, that is important for olfactory sensing.

   The researchers extracted RNA molecules from the olfactory receptor cells of the nasal epithelium. They then used this RNA to identify genes that encoded G-protein-coupled receptor proteins.
2. In their study, they identified 18 different genes that encoded different G-protein-coupled receptor proteins.
3. The results of the experiment conducted by Buck and Axel support the hypothesis that animals discriminate between different odors based on having a variety of receptor proteins that recognize different odor molecules. Current research suggests that each olfactory receptor cell has a single type of receptor protein that is specific to particular odor molecules. Because most odors are due to multiple chemicals that activate many different types of odor receptor proteins, the brain detects odors based on the combination of the activated receptor proteins. Odor seems to be discriminated by many olfactory receptor proteins, which are in the membrane of separate olfactory receptor cells.

Test Yourself

1. d 2. d 3. a 4. c 5. e 6. d 7. b 8. b 9. d 10. b

Conceptual Questions

1. Sensory transduction—The process by which incoming stimuli are converted into neural signals. An example would be the signals generated in the retina when a photon of light strikes a photoreceptor.

   Perception—An awareness of the sensations that are experienced. An example would be an awareness of what a particular visual image is.
2. The organ of Corti contains the hair cells and sensory neurons that initiate signaling. The hair cells sit on top of the basilar membrane, and their stereocilia are embedded in the tectorial membrane at the top of the organ of Corti. Pressure waves of different frequencies cause the basilar membrane to vibrate at particular sites. This bends the stereocilia of hair cells back and forth, sending oscillating signals to the sensory neurons. Consequently, the sensory neurons send intermittent action potentials to the CNS via the auditory nerve. Hair cells at the end of the basilar membrane closest to the oval window respond to high-pitched sounds, and lower-pitched sounds trigger hair cell movement further along the basilar membrane.
3. Of the various senses, the sense of olfaction (smell) is least important for the survival of humans. As diurnal animals, we rely largely on our visual sense. Sounds are a critical way to learn about impending danger, such as a car horn, but also is our major means of communication. Other senses, such as the ability to sense pain, have acutely important functions from time to time. Olfaction, though often a pleasurable sense and at times a protective one (think of the smell of spoiled food), nonetheless provides little survival advantage to us. In fact, many people spend much of their lives with greatly diminished olfactory abilities, whether from chronic allergies or other problems, and are not hindered in any significant way. The story is very different for animals such as nocturnal mammals, which rely very heavily on olfaction to find food, locate mates, and avoid predators.

Chapter 44

Concept Checks

Figure 44.1 Yes. In addition to not having a requirement to shed their skeletons periodically, animals with endoskeletons can use their skin as an efficient means of heat transfer (and, to an extent in amphibians, water transfer). In addition, the body surface of such animals is often a highly sensitive sensory organ.

Figure 44.3 If a tendon is torn, its ability to link a muscle to bone is reduced or lost. Therefore, when a muscle such as the one shown in this illustration contracts, it will not be able to move the bone from which the tendon has become dislodged.

Figure 44.8 The ATP concentration in cells becomes depleted after death, because oxygen and nutrients are not being provided to cells. Consequently, the cross-bridge cycle becomes locked before step 3. Without ATP, the cross-bridges cannot dissociate until many hours later, when the muscle tissue sufficiently decomposes.

Figure 44.11 Na^+ enters the muscle cell because all cells have an electrochemical gradient for Na^+ that favors diffusion of Na^+ from extracellular to intracellular fluid (see Chapter 41). This is because cells have a negative membrane potential and because Na^+ concentrations are higher in the extracellular fluid. The acetylcholine receptor on skeletal muscle cells is also a ligand-gated ion channel; when acetylcholine binds the receptor, it induces a shape change that opens the channel. This allows the entry of Na^+ into the cell.

Figure 44.14 No. The data are expressed "per kg"; this means that when normalized to a standard body mass (1 kg), the amount of energy expended for any type of locomotion by a small animal tends to be greater than that of a larger animal. However, these are *relative* values. For example, the *absolute* amount of energy expended by a tiny minnow is much less than that of a large tuna over any given distance.

BioConnections

Figure 44.10 Voltage-gated Ca^{2+} channels exist in the terminals of all axons that communicate by chemical signaling (neurotransmitter release). In those cases, depolarization of the axon terminal opens Ca^{2+} channels, allowing Ca^{2+} to enter the terminal and trigger exocytosis of stored vesicles containing neurotransmitter molecules.

Figure 44.11 As described in Figures 41.2 and 41.12, myelin is a lipid-rich membrane sheath that speeds up conduction of action potentials along an axon. Action potentials are regenerated at discrete lengths along the axon wherever the myelin sheath is interrupted by a node of Ranvier. This is known as saltatory conduction.

Feature Investigation Questions

1. PPAR-δ is a nuclear receptor that regulates the expression of genes that enable cells to more efficiently burn fat instead of glucose for energy.
2. Evans suggested that if PPAR-δ were highly activated in mice, the mice would lose weight because of the high level of fat metabolism.
3. They developed transgenic mice with highly activated PPAR-δ. Then they fed the transgenic mice and a strain of normal mice high-fat diets. They then compared the weights of the two strains of mice to determine if the change in PPAR-δ activity affected weight. The weights of the transgenic mice were considerably lower than those of the normal mice. These results supported the hypothesis that highly activated PPAR-δ would lead to lower weight gain due to fat metabolism. Interestingly, the researchers also discovered that the transgenic mice could perform prolonged exercise for a much longer time than the normal mice. The muscle tissue of the transgenic mice was more specialized for long-term exercise.

Test Yourself

1. c 2. e 3. d 4. c 5. e 6. c 7. a 8. e 9. c 10. e

Conceptual Questions

1. Exoskeletons are on the outside of an animal's body, and endoskeletons are inside the body. Both function in support and protection, but only exoskeletons protect an animal's outer surface. Exoskeletons must be shed when an animal grows, whereas endoskeletons grow with an animal.
2. a. The cycle begins with the binding of an energized myosin cross-bridge to an actin molecule on a thin filament.

   b. The cross-bridge moves, and the thin filaments slide past the thick filaments.

   c. The ATP binds to myosin, causing the cross-bridge to detach.

   d. The ATP bound to myosin is hydrolyzed by ATPase, re-forming the energized state of myosin.
3. The use of energy released by the hydrolysis of ATP is fundamental to muscle function and locomotion. Recall that ATP must be hydrolyzed during the cross-bridge cycle for skeletal muscle cells to shorten. Energy is also used to maintain calcium ion balance in the sarcoplasmic reticulum and is used in all forms of locomotion. The amount of energy expended by an animal during locomotion reflects how well they are adapted to the environment in which they must move.

Chapter 45

Concept Checks

Figure 45.4 After a large blood meal, the body mass of a flying blood-sucking animal increases sufficiently as to make it nearly impossible to fly. The problem is solved, however, by a unique adaptation that allows such animals to concentrate the nutrients from blood and excrete most of the water portion of blood as soon as they begin eating. By the time the meal is finished, much of the water they consumed has already been excreted.

Figure 45.7 Sauropod dinosaurs were herbivores that probably contained a gizzard-type stomach in which stones helped to grind coarse vegetation. Such stones would have become smooth after months or even years of rumbling around in the gizzard. Some of these sauropods are known to have lacked the sort of grinding teeth characteristic of modern mammalian herbivores, and thus a gizzard would have aided in their digestion much as it does in modern birds.

Figure 45.10 By having bile stored in a gallbladder, bile can be released precisely when needed in response to a meal, which is particularly useful for animals that consume large or infrequent meals. In the absence of a gallbladder, bile flows into the intestine continuously and cannot be increased to match the amount or timing of food intake.

Figure 45.11 Secondary active transport requires energy provided by ATP. Thus, absorption of nutrients by this mechanism is an energy-requiring event, and some portion of an animal's regular nutrient consumption is used to provide the energy required to absorb the nutrients.

BioConnections

Figure 45.11 Transmembrane transport processes are not unique to animals, and one or more types are found in virtually all cells.

Figure 45.13 CCK inhibits stomach activity. This is an example of negative feedback. The arrival of chyme in the small intestine stimulates CCK, which promotes digestion as shown in the figure. At the same time, CCK inhibits contraction of the smooth muscles of the stomach so that the entry of chyme into the small intestine is slowed down. This allows time for controlled digestion and absorption of nutrients in the intestine, without the intestine becoming overfilled with chyme. Simultaneously, CCK inhibits acid production by the stomach so that the pH of the intestine does not become dangerously low before bicarbonate ions are able to neutralize it.

Feature Investigation Questions

1. The researchers severed the nerves that connected to the small intestine in a dog. Following the removal of the nerves, the researchers introduced an acidic solution directly into the intestine of the dog. The introduction of the acid into the intestine caused pancreatic secretion. This suggested that non-neural factors must have mediated communication between the digestive tract and the secretory cells of the pancreas.
2. Other researchers were not convinced that all the nerves were dissected from the intestine, because of the technical difficulty in performing such a procedure. To provide more conclusive evidence of other regulatory factors that were produced by the intestine, the researchers conducted a second experiment. First, they dissected a portion of a small intestine from a dog, treated it with acid, ground it up to produce a mash, and then filtered the mash to obtain an extract. The extract—which was expected to contain any secretions of the intestine that occurred following acid exposure—was then injected into the circulatory system of a second dog. The results indicated that the second dog had pancreatic secretion following the injection.
3. The results suggested that factors were secreted by the small intestine following exposure of the intestine to acid, as would occur when chyme enters the intestine from the stomach. These factors probably reached the pancreas through the bloodstream. The researchers called these factors hormones. Thus, the digestive system was regulated not only by the nervous system, but also by chemical secretions, and different parts of the digestive system were able to communicate with each other via hormones.

Test Yourself

1. d 2. e 3. e 4. b 5. c 6. b 7. d 8. c 9. a 10. e

Conceptual Questions

1. Digestion is the breakdown of large molecules into smaller ones by the action of enzymes and acid. Absorption is the transport of digested molecules and small molecules that do not require digestion, across the epithelial cells of the alimentary canal and from there into the extracellular fluid of an animal.
2. The crop is a dilation of the esophagus, which stores and softens food. The gizzard contains swallowed pebbles that help pulverize food. Both of these functions are adaptations that assist digestion in birds, which do not have teeth and therefore do not chew food. Humans, like many animals, can chew food before swallowing.
3. Carnivores eat live animal flesh and/or fluids or may scavenge dead animals. Carnivores' teeth are adapted for seizing, grasping, piercing, biting, slicing, tearing, or holding prey; they generally do not chew their food extensively, but may chew to facilitate swallowing. Herbivores have powerful jaw muscles and large, broad molars for grinding tough, fibrous plant material; they may also have incisors adapted for nipping grass or other vegetation. Simply examining the type of teeth an animal has is often sufficient to determine whether that animal eats vegetation, animals, or both.

Chapter 46

Concept Checks

Figure 46.3 The time required for the vesicles to move to the plasma membrane and fuse with it is much shorter than the time required for new GLUTs to be synthesized by activation of GLUT genes. Thus, the action of insulin on cells is very quick, because the GLUTs are already synthesized.

Figure 46.4 The glycerol and fatty acids used to make glucose are the breakdown products of triglycerides that were stored in adipose tissue during the absorptive period. The amino acids used to make glucose are derived from the breakdown of protein in muscle and other tissue.

Figure 46.6 Even though the goose was resting, sampling the air from the mask would be only a rough estimate of BMR. That is because the artificial setting and the placement of the mask would be enough of a stimulus to affect the activity and behavior of the goose, thereby increasing its metabolism.

Figure 46.7 As shown in Figure 46.7, for humans exercise is a voluntary activity. In nature, however, "exercise" is often a component of the fight-or-flight reaction, such as when an animal attempts to escape danger. During such times, digestion and absorption of food are less important than providing as much blood flow, oxygen, and nutrients as possible to skeletal muscle. The gut, therefore, temporarily reduces its activity and requires less blood flow.

Figure 46.8 Nearly all animals today show a similar relationship between body mass and metabolic rate, and there is no reason why it should not always have been true. Thus, the tiny 1-foot-tall ancestral horse *Eohippus* most likely had a higher BMR than do today's larger horses.

Figure 46.11 Humans are homeothermic endotherms. We maintain our body temperature within a very narrow range, and we supply our own body heat.

BioConnections

Figure 46.3 Exocytosis, a feature characteristic of animal cells, involves the fusion of intracellular vesicles with the plasma membrane, resulting in the release of the vesicle contents into the extracellular fluid. See Figure 5.26 for a general description, and Figure 41.14 for a specific example unique to animal cells.

Feature Investigation Questions

1. Scientists were interested in knowing why animals seemed to regulate their body mass around a particular level, even though many animals experience changes in food supply throughout the year. This seemed to indicate that a mechanism existed within the body that monitored when fuel stores were higher or lower than normal, and that initiated changes in behavior and metabolism to compensate.
2. Coleman hypothesized that communication regarding energy status must take place between the brain and the rest of the body. He suggested that chemical signals were transported through the blood from outside the brain to feeding or satiety centers within the brain, where they regulated appetite and thus body weight. He tested this by linking the blood circulations of normal mice and genetically obese mice and then monitoring the mice for changes in body weight.
3. In most cases, the obese mice lost weight and ate less during the experimental procedure. This confirmed that something in the bloodstream of the wild-type mice was regulating body weight but was missing in the obese mice. When the unknown factor crossed into the bloodstream of the obese mice, it caused them to lose weight. In another group of parabiosed mice, however, the wild-type

mice lost weight, but the obese mice did not. Coleman concluded that these obese mice were not able to respond to the chemical signal that regulates body weight, even though they made the signal themselves and it was active in their parabiosed wild-type partners.

Test Yourself

1. c 2. a 3. d 4. c 5. a 6. e 7. e 8. c 9. c 10. e

Conceptual Questions

1. Insulin acts on adipose and skeletal muscle cells to facilitate the diffusion of glucose from extracellular fluid into the cell cytosol. This is accomplished by increasing the translocation of glucose-transporter (GLUT) proteins from the cytosol to sites within the plasma membrane of insulin-sensitive cells. Insulin also inhibits glycogenolysis and gluconeogenesis in the liver, which decreases the amount of glucose secreted into the blood by the liver. Insulin is required for glucose transport because like many other polar molecules, glucose cannot move across the lipid bilayer of a plasma membrane by simple diffusion. The inhibitory effects of insulin on liver function help to ensure that liver glycogen stores will be spared for the postabsorptive period.
2. Appetite is controlled by a satiety center in the brain that receives signals from the stretched stomach and intestines after a meal. When digestion and absorption are complete, the stomach and intestines return to their original size, and the brain no longer senses that an animal feels "full." In addition, appetite is controlled by leptin, a hormone secreted by adipose cells in direct proportion to the amount of fat stored in an animal's body. When leptin levels in the blood are high, appetite is suppressed. When leptin levels are low, as occurs when an animal is losing weight, appetite is increased. The presence of a hormone that is released into the blood in proportion to fat mass in the body allows the brain to monitor the amount of energy stored in the body. A decrease in the concentration of leptin in the blood, for example, is the mechanism that communicates to the brain that fat stores are lower than normal. This initiates the sensation of hunger, which encourages an animal to seek food.
3. Countercurrent heat exchange is a mechanism for retaining body heat. The physical arrangement (structure) of arteries and veins in an animal's body can contribute to the very important function of thermoregulation. As warm blood travels through arteries down a bird's leg, for example, heat moves by conduction from the artery to adjacent veins carrying cooler blood in the other direction, toward the heart. By the time the arterial blood reaches the tip of the leg, its temperature has dropped considerably, reducing the amount of heat loss to the environment, while the heat is returned to the body's core via the warmed veins.

Chapter 47

Concept Checks

Figure 47.2 Open circulatory systems evolved prior to closed systems. However, this does not mean that open systems are in some way inferior to closed circulatory systems. It is better to think of open systems as being ideally suited to the needs of those animals that have them. Arthropods are an incredibly successful order of animals, with the greatest number of species, and inhabiting virtually every ecological niche on the planet. Clearly, their type of circulatory system has not prevented arthropods from achieving their great success.

Figure 47.3 Keeping oxygenated and deoxygenated blood fully separate allows the arterial blood of birds and mammals to provide the maximum amount of oxygen to tissues. This means that those tissues can achieve higher metabolisms and be more active at all times.

Figure 47.6 Each hemoglobin molecule contains four subunits, each of which has an iron atom at its core. Each iron binds one oxygen molecule (O_2); therefore, a total of eight oxygen atoms can bind to one hemoglobin molecule.

Figure 47.11 Body fluids, both extracellular and intracellular, contain large amounts of charged ions, which are capable of conducting electricity. The slight electric currents generated by the beating heart muscle cells are conducted through the surrounding body fluids by the movements of ions in those fluids. This is recorded by the surface electrodes and amplified by the recording machine.

Figure 47.13 When the animal is active, the arterioles of its leg muscle would dilate, bringing more blood and, consequently, nutrients and oxygen to the active muscle tissue.

Figure 47.16 The valves open toward the heart. When the head is upright, the valves are open, and blood drains from the head to the right atrium by gravity. When the giraffe lowers its head to drink, however, gravity would prevent the venous blood from reaching the heart; instead, blood would pool in the head and could raise pressure in the head and brain. The valves in the neck veins work the same way as those in the legs of other animals, helping to propel blood against gravity to the heart.

BioConnections

Figure 47.1 Water circulation mixes food with enzymes and brings digested food into close contact with all interior cells.

Figure 47.5 Immune defenses are found in most living organisms. Many bacteria produce antibacterial secretions that kill other bacteria. Plants, as shown in Figure 36.19, have a wide array of pathogen-fighting mechanisms.

Figure 47.19 Baroreceptors are mechanoreceptors. Like all mechanoreceptors (for example, those in distensible or deformable structures such as the urinary bladder and stomach), their ion channels are opened by physical deformation or stretching of the plasma membrane. They are, therefore, mechanically gated ion channels.

Feature Investigation Questions

1. Furchgott noted that acetylcholine had different effects on the rabbit aorta depending on the manner in which the aorta was isolated and prepared. When applied to flattened strips of the aorta, acetylcholine caused contraction of the aorta smooth muscle; however, when applied to circular rings of the aorta, acetylcholine caused relaxation. Furchgott suggested that the difference was due to the absence of the endothelial layer of tissue in the flattened strips of aorta.
2. Furchgott hypothesized that acetylcholine stimulated the endothelial cells to secrete a substance that functioned as a vasodilator, causing the muscle layer to relax. Furchgott performed several experiments to test his hypothesis. He compared the effects of acetylcholine on circular rings of aorta that either had the endothelial layer intact or experimentally removed. The results of this experiment demonstrated that when the endothelial layer was present, relaxation occurred in the presence of acetylcholine. Removal of the endothelial layer, however, resulted in contraction of muscle in the presence of acetylcholine. In a second experiment, a strip of the aorta with the endothelial layer removed was put in contact with a strip of aorta with an intact endothelial layer. When this "sandwiched" treatment was exposed to acetylcholine, both muscle layers relaxed.
3. Furchgott concluded that the endothelial layer produced a vasodilator in the presence of acetylcholine. The vasodilator diffused from the intact strip of muscle to the denuded strip and caused the muscle layer to relax.

Test Yourself

1. b 2. a 3. c 4. b 5. a 6. d 7. c 8. d 9. c 10. d

Conceptual Questions

1. The three main components of a circulatory system are (1) blood or hemolymph, an internal body fluid containing dissolved solutes; (2) blood vessels, a system of hollow tubes within the body through which blood travels; and (3) one or more hearts, muscular structures that pump blood through the blood vessels.
2. ***Closed circulatory system***—In a closed circulatory system, the blood and interstitial fluid are contained within tubes called blood vessels and are transported by a pump called the heart. All of the nutrients and oxygen that tissues require are delivered directly to them by the blood vessels. Advantages of closed circulatory systems are that different parts of an animal's body can receive blood flow in proportion to that body part's metabolic requirements at any given time. Due to its efficiency, a closed circulatory system allows organisms to become larger.

   Open circulatory system—In an open circulatory system, the organs are bathed in hemolymph that ebbs and flows into and out of the heart(s) and body cavity, rather than blood being directed to all cells. Like a closed circulatory system, there are a pump and blood vessels, but these two structures are less developed and less complex compared to a closed circulatory system. Partly as a result, organisms such as mollusks and arthropods are generally limited to being relatively small, although exceptions do exist.
3. A circulatory system permits delivery of the nutrients and oxygen required by cells to maintain energy-demanding processes, such as pumping ions across cellular membranes, contracting muscle cells including those of the heart, cell division, protein synthesis, and many others. In addition, the circulatory system removes soluble waste products, which, if allowed to accumulate, would be toxic to cells. Many circulatory systems are capable of adapting to changing metabolic requirements, thus ensuring that homeostasis is maintained whether an animal is resting or active.

Chapter 48

Concept Checks

Figure 48.2 Regardless of whether the atmosphere is measured on Mt. Everest or at sea level, the percentage of gas molecules that are oxygen remains close to 21%. However, the pressure exerted by those gas molecules decreases as one ascends in elevation.

Figure 48.3 If a lungless salamander were to dry out, its capacity for gas exchange would be greatly reduced. Gases diffuse into and out of the body of the salamander by dissolving in the moist fluid layer on the skin.

Figure 48.5 Imagine holding several thin sheets of a wet substance, such as paper. If you wave them in the air, what happens? The sheets stick to one another because of surface tension and other properties of moist surfaces. This is what happens to the lamellae in gills when they are in air. When the lamellae stick to each other, the surface area available for gas exchange is reduced and the fish suffocates.

Figure 48.7 Several factors probably limit insect body size, but the respiratory system most likely is one such factor. If an insect grew to the size of a human, for example, the trachea and tracheoles would be so large and extensive that there would be little room for any other internal organs in the body! Also, the mass of the animal's body and the forces generated during locomotion would probably collapse the tracheoles. Finally, diffusion of oxygen from the surface of the body to the deepest regions of a human-sized insect would take far too long to support the metabolic demands of internal structures.

Figure 48.12 Because fishes have the most efficient means of extracting oxygen from their environment, one might conclude that this is an adaptation to cope with low environmental oxygen. Based on that logic, you would conclude that the oxygen content of water was less than that of air, which is indeed correct.

Figure 48.14 The waters off the coast of Antarctica are extremely cold, rarely warmer than 0.30°C. As we saw earlier in this chapter, more oxygen dissolves in cold water than in warm water, and therefore icefish have the potential to obtain more oxygen across their gills. Cold temperatures also decrease the metabolic rate of the animals, because all chemical reactions slow down at low temperature. Thus, the oxygen demands of icefish are lower than those of warm-water fish. Several other adaptations have evolved to enable these animals to live without hemoglobin. Large gills with exceptionally high surface area facilitate diffusion of oxygen into the animal's blood. In addition, cardiovascular adaptations evolved to help increase the total amount of oxygen in the blood and its ability to be pumped to all body tissues. For example, icefish have larger blood volumes and a larger heart than warm-water fish of a similar size. Also, the absence of red blood cells makes the blood less viscous (makes it "thinner") and therefore easier to pump through the body.

Figure 48.15 An increase in the blood concentration of HCO_3^- would favor the reaction $HCO_3^- + H^+ \rightarrow H_2CO_3 \rightarrow CO_2 + H_2O$. This would reduce the H^+ concentration of the blood, thereby raising the pH; the CO_2 formed as a result would be exhaled. These changes would shift the hemoglobin curve to the left of the usual position.

BioConnections

Figure 48.16 The brainstem includes the midbrain, pons and medulla oblongata. See Figure 42.9 for an illustration of the major parts of the human brain.

Feature Investigation Questions

1. The study conducted by Schmidt-Neilsen intended to determine the route of air through the avian respiratory system. This would provide a better understanding of the functions of the air sacs and the process of gas exchange in birds.
2. The first experiment by Schmidt-Neilsen compared the composition of air between the posterior and anterior air sacs. Oxygen content was high in the posterior sacs but low in the anterior sacs. Carbon dioxide levels, however, were low in the posterior sacs but high in the anterior sacs. The researchers concluded that when inhaled, the air moves first to the posterior sacs; then to the lungs where oxygen diffuses into the blood and carbon dioxide diffuses into the lungs; and, finally, to the anterior sacs before being exhaled.
3. The second experiment by Schmidt-Neilsen was conducted to verify the pathway of air through the respiratory system of the bird. In this experiment, the researcher monitored oxygen levels by surgically implanting oxygen probes in the anterior and posterior air sacs. The bird was fitted with a face mask and allowed to take one breath of pure oxygen. The researcher was then able to track the movement of this oxygen through the respiratory tract. Schmidt-Neilsen concluded that it takes two complete breaths for air to move from the environment through the lungs and back out again to the environment. The two breaths are required to move the air from the posterior air sacs through the lungs and, finally, to the anterior air sacs before exiting the body.

Test Yourself

1. a 2. e 3. c 4. c 5. a 6. b 7. e 8. e 9. d 10. b

Conceptual Questions

1. Countercurrent exchange maximizes the amount of oxygen that can be obtained from the water in fishes. Oxygenated water flows across the lamellae of a fish gill in the opposite direction in which deoxygenated blood flows through the capillaries of the lamellae. In this way, a diffusion gradient for oxygen is maintained along the entire length of the lamellae, facilitating diffusion of oxygen even when much of it has already entered the blood.
2. Animals that live at high altitudes face the special challenge of obtaining oxygen where the atmospheric pressure is low. When atmospheric pressure is low, the partial pressure of oxygen in the air is also low. This means that there is less of a driving force for the diffusion of oxygen from the air into the body of the animal. Several adaptations have arisen that help animals cope with such habitats. For example, many high-altitude animals have more red blood cells and have hemoglobin with a higher affinity for oxygen than that of sea-level animals. This means their hemoglobin can bind oxygen even at the low partial pressures of high altitudes, thereby saturating their blood with oxygen. In addition, such animals generally have larger hearts and lungs for their body size than animals that live at lower altitudes. Animals that move to high altitude show increases in the number of red blood cells in their circulation and in respiratory rates. The number of capillaries in skeletal muscle increases to facilitate oxygen diffusion into the muscle cells. Myoglobin content of muscle cells also increases, expanding the reservoir of oxygen in the cytosol.
3. Hemoglobin is a protein with quaternary structure (see Chapter 3) in which the different subunits cooperate to bind up to a total of four oxygen molecules. It is the structure of the subunits and their relationship to each other that contributes to their ability to bind O_2 and to the nonlinear relationship of the oxygen-hemoglobin dissociation curve. In addition, however, interactions of hemoglobin with other molecules, such as CO_2, change the structure of hemoglobin in such a way that its properties change. Under such conditions, hemoglobin is less able to bind O_2 and consequently it releases the gas. Any molecule that binds to hemoglobin will alter its structure and change its properties; these revert to the original state once the bound molecules are released. A particularly dramatic example of the structure and function of hemoglobin being related is that which occurs in sickle cell disease, due to a mutation.

Chapter 49

Concept Checks

Figure 49.1 No, obligatory exchanges must always occur, but animals can minimize obligatory losses through modifications in behavior. For example, terrestrial animals that seek shade on a hot, sunny day reduce evaporative water loss. As another example, reducing activity minimizes water loss due to respiration.

Figure 49.4 Humans cannot survive by drinking seawater because we do not possess specialized salt glands to rid ourselves of the excess sodium and other ions ingested with seawater. The human kidneys cannot eliminate that much salt. The high blood levels of sodium and other ions would cause changes in cellular membrane potentials, disrupting vital functions of electrically excitable tissue such as cardiac muscle and nerve tissue.

Figure 49.6 Secretion of substances into excretory organ tubules is advantageous because it increases the amount of a substance that gets removed from the body by the excretory organs. This is important, because many substances that get secreted are potentially toxic. Filtration, though efficient, is limited by the volume of fluid that can leave the capillaries and enter the excretory tubule.

BioConnections

Figure 49.12 A brush border composed of microvilli is also present along the epithelial cell layer of the vertebrate small intestine (see Chapter 45). In the intestine, the brush border serves to increase the absorption of nutrients. In both the intestine and the proximal tubule of nephrons, therefore, a brush border provides extensive surface area for the transport of substances between a lumen and the epithelial cells (and from there to extracellular fluid).

Figure 49.14 Epithelial cells like those in the kidney tubules can distribute proteins between the luminal and basolateral sides of the plasma membrane. In this way, the Na^+/K^+-ATPase pumps that are stimulated by aldosterone are present and active only on one side of the cell, the basolateral surfaces. If the pumps were activated on the luminal surface of the cell, aldosterone would not be able to promote reabsorption of

APPENDIX B

Na^+ and water, because Na^+ would also be transported from the cell into the tubule lumen.

Feature Investigation Questions

1. Symptoms of prolonged, heavy exercise include fatigue, muscle cramps, and even occasionally seizures. Fatigue results from the reduction in blood flow to muscles and other organs. Muscle cramps and seizures are the results of imbalances in plasma electrolyte levels. Cade and his colleagues hypothesized that maintaining proper water and electrolyte levels would prevent these problems, and that if water and electrolyte levels were maintained, athletic performance should not decrease as rapidly with prolonged exercise.
2. To test their hypothesis, the researchers created a drink that would restore the correct proportions of lost water and electrolytes within the athletes. If the athletes consumed the drink during exercise, they should not experience as much fatigue or muscle cramping, and thus their performance should be enhanced compared with a control group of athletes that drank only water.
3. The performance of a group of exercising athletes given the electrolyte-containing drink was better than that of the control group that drank only water during exercise. This could be attributed to the replacement of normal electrolyte levels by the drink.

Test Yourself

1. e 2. e 3. d 4. c 5. a 6. c 7. e 8. d 9. a 10. b

Conceptual Questions

1. Nitrogenous wastes are the breakdown products of the metabolism of proteins and nucleic acids. They consist of ammonia, ammonium ions, urea, and uric acid. The predominant type of waste excreted depends in part on an animal's environment. For example, aquatic animals typically excrete ammonia and ammonium ions, whereas many terrestrial animals excrete primarily urea and uric acid. Urea and uric acid are less toxic than the other types but require energy to be synthesized. Urea and uric acid also result in less water excreted, an adaptation that is especially useful for organisms that must conserve water, such as many terrestrial species.
2. During filtration, an organ acts like a sieve or filter, removing some of the water and its small solutes from the blood, interstitial fluid, or hemolymph, while retaining blood cells and large solutes such as proteins. Reabsorption is the process whereby epithelial cells of an excretory organ recapture useful solutes that were filtered. Secretion is the process whereby epithelial cells of an excretory organ transport unneeded or harmful solutes from the blood to the excretory tubules for elimination. Some substances such as glucose and amino acids are reabsorbed but not secreted, while some other substances such as toxic compounds are not reabsorbed and are secreted. Still other substances, namely proteins, are not filtered at all.
3. Salt glands contain a network of secretory tubules that actively transport NaCl from the extracellular fluid into the tubule lumen. This solution then moves through a central duct and to the outside environment through pores in the nose, around the eyes, and in other locations. The ability to remove salt from body fluids is an adaptation for many marine reptiles and birds, which do not have ready access to fresh water and would otherwise run the risk of having very high levels of salts in their blood and other body fluids.

Chapter 50

Concept Checks

Figure 50.10 Not all mammals use the energy of sunlight to synthesize vitamin D. Many animals, such as those that inhabit caves or that are strictly nocturnal, rarely are exposed to sunlight. Some of these animals get their vitamin D from dietary sources. How others maintain calcium balance without dietary or sunlight-derived active vitamin D remains uncertain.

Figure 50.12 Sodium and potassium ion balance is of vital importance for most animals because of the critical role these ions play in nervous system and muscle function. It is more the rule than the exception that such important physiological variables are under multiple layers of control. This grants a high degree of fine-tuning capability such that these ions—and other similarly important molecules—rarely exceed or fall below the normal range of concentrations for a given animal.

Figure 50.13 The great height of the twin on the left in Figure 50.13 clearly indicates that his condition arose prior to puberty. The enlarged bones further suggest that the disease continued for a time after puberty, when further linear bone growth was no longer possible.

Figure 50.15 Because 20-hydroxyecdysone is a steroid hormone, you would predict that its receptor would be intracellular. All steroid hormones interact with receptors located either in the cytosol or, more commonly, in the nucleus. The hormone-receptor complex then acts to promote or inhibit transcription of one or more genes. The receptor for 20-hydroxyecdysone is indeed found in cell nuclei.

BioConnections

Figure 50.4 When dopamine is secreted from an axon terminal into a synapse where it diffuses to a postsynaptic cell, it is considered a neurotransmitter. When it is secreted from an axon terminal into the extracellular fluid, from where it diffuses into the blood, it is considered a hormone.

Figure 50.7 In addition to the pancreas, certain other organs in an animal's body may contain both exocrine and endocrine tissue or cells. For example, you learned in Chapters 45 and 46 that the vertebrate alimentary canal is composed of several types of secretory cells. Some of these cells release hormones into the blood that regulate the activities of the pancreas and other structures, such as the gallbladder. Other cells of the alimentary canal secrete exocrine products such as acids or mucus into the gut lumen that directly aid in digestion or act as a protective coating, respectively.

Feature Investigation Questions

1. Banting and Best based their procedure on a medical condition that results when pancreatic ducts are blocked. The exocrine cells will deteriorate in a pancreas that has obstructed ducts; however, the islet cells are not affected. The researchers proposed to experimentally replicate the condition to isolate the cells suspected of secreting the glucose-lowering factor. From these cells, they assumed they would be able to extract the substance of interest without contamination or degradation due to exocrine products.
2. The extracts obtained by Banting and Best did contain insulin, the glucose-lowering factor, but were of low strength and purity. Collip developed a procedure to obtain a more purified extract with higher concentrations of insulin.
3. The researchers chose to use bovine pancreases as their starting material for preparing the extracts. Because of the large size of these animals and their availability at local slaughterhouses, the researchers were able to obtain great yields of insulin. Second, Collip developed a highly sensitive assay for monitoring changes in blood glucose levels after injection of insulin. This allowed the researchers to better estimate how much insulin was in a preparation and how much was necessary to give to a patient.

Test Yourself

1. b 2. e 3. b 4. e 5. b 6. e 7. c 8. d 9. b 10. d

Conceptual Questions

1. Leptin acts in the hypothalamus to reduce appetite and increase metabolic rate. Because adipose tissue is typically the most important and abundant source of stored energy in an animal's body, the ability to relay information to the appetite and metabolism centers of the brain about the amount of available adipose tissue is a major benefit. In this way, the brain's centers can indirectly monitor the minute-to-minute energy status in the body. A decrease in leptin, for example, would indicate that a decrease in adipose tissue existed—as might occur during a fast. Removal of the leptin signal would cause appetite to increase and metabolism to decrease, thereby conserving energy. The presence of an appetite and the subjective sensations associated with hunger is a motivation that drives an animal to seek food at the expense of other activities, such as seeking shelter, finding a mate, and so on.
2. Type 1 DM is characterized by insufficient production of insulin due to the immune system destroying the insulin-producing cells of the pancreas. In type 2 DM, insulin is still produced by the pancreas, but adipose and muscle cells do not respond normally to insulin.
3. Insulin acts to lower blood glucose concentrations, for example, after a meal, whereas glucagon elevates blood glucose, for example, during fasting. Insulin acts by stimulating the insertion of glucose transporter proteins into the cell membrane of muscle and fat cells. Glucagon acts by stimulating glycogenolysis in the liver. If a high dose of glucagon were injected into an animal, including humans, the blood concentration of glucagon would increase rapidly. This would stimulate increased glycogenolysis, resulting in blood glucose concentrations that were above normal.

Chapter 51

Concept Checks

Figure 51.4 Aquatic environments in which the water is stagnant or only gently moving, as shown in this figure, are generally best for external fertilization. Fast-moving bodies of water reduce the likelihood of a sperm contacting an egg and increase the chances that gametes will be washed away in the current. Many river-dwelling fishes lay eggs in gently moving streams, and many marine fishes do so in relatively shallow waters.

Figure 51.7 The elevated testosterone levels would inhibit LH and FSH production through negative feedback. This would result in reduced spermatogenesis and possibly even infertility (an inability to produce sufficient sperm to cause a pregnancy).

Figure 51.10 FSH and LH concentrations do not surge in males, but instead remain fairly steady, because the testes do not show cyclical activity. Sperm production in males is constant throughout life after puberty.

Figure 51.14 Pregnancy and subsequent lactation require considerable energy and, therefore, nutrient ingestion. Consuming the placenta provides the female with a rich source of protein and other important nutrients.

BioConnections

Figure 51.11 In addition to its other functions, the placenta must serve the function of the lungs for the fetus, because the fetus' lungs are not breathing air during this time. Arteries always carry blood away from the heart; veins carry blood to the heart. Consequently, blood leaving the heart of the fetus and traveling through arteries to the placenta is deoxygenated. As blood leaves the placenta and returns to the heart, the blood has become oxygenated as oxygen diffuses from the maternal blood into fetal blood. That oxygenated blood then gets pumped from the fetal heart through other arteries to the rest of the fetus' body.

Feature Investigation Questions

1. Using *Daphnia*, Paland and Lynch compared the accumulation of mitochondrial mutations between sexually reproducing populations and asexually reproducing populations.
2. The results—that sexually reproducing populations had a lower rate of deleterious mutations compared with asexually reproducing populations—indicate that sexual reproduction does decrease the accumulation of deleterious mutations, at least in this species.
3. Sexual reproduction allows for mixing of the different alleles of genes with each generation, thereby increasing genetic variation within the population. This could prevent the accumulation of deleterious alleles in the population.

Test Yourself

1. d 2. c 3. e 4. a 5. b 6. c 7. b 8. c 9. b 10. c

Conceptual Questions

1. In viviparity, most of embryonic development occurs within the mother, and the animal is born alive, as occurs in most mammals. If all or most of embryonic development occurs outside the mother and the embryo depends exclusively on yolk from an egg for nourishment, the process is called oviparity; this occurs in most vertebrates and in insects. In ovoviviparity, which occurs in some reptiles, sharks, and some invertebrates, fertilized eggs covered with a very thin shell hatch inside the mother's body, but the offspring receive no nourishment from the mother. Humans are viviparous. An advantage of viviparity is that the embryo and fetus develop in a protected environment.
2. Cells of the hypothalamus produce two important hormones that regulate reproduction. GnRH stimulates the anterior pituitary gland to release two gonadotropic hormones, LH and FSH. These two hormones regulate the production of gonadal hormones and development of gametes in both sexes. In addition, increased secretion of GnRH contributes to the initiation of puberty. The hypothalamus also produces oxytocin, a hormone that is stored in the posterior pituitary gland and that acts to stimulate milk release during lactation. Finally, changes in neuroendocrine activity in the hypothalamus are linked to seasonal changes in day length and therefore contribute to seasonal breeding in certain mammals.
3. Sexual reproduction requires that males and females of a species produce different gametes and that these gametes come into contact with each other. This requires males and females to expend energy to locate mates. It also may require specialized organs for copulation and in some cases requires the production of very large numbers of gametes to increase the likelihood that the eggs are fertilized. These costs are outweighed by the genetic diversity afforded by sexual reproduction.

Chapter 52

Concept Checks

Figure 52.14 Different concentrations of a signaling protein can exert different effects on cells when, for example, different cells express different isoforms of a plasma membrane receptor for the protein. If one cell expresses a high-affinity receptor and another cell a low-affinity receptor, the two cells would respond to the signaling protein at different concentrations. Likewise, the different receptors may be linked with different second-messenger molecules generated within the cell. These messengers, such as cAMP and Ca^{2+}, may have different effects on cell function.

BioConnections

Figure 52.1 The process by which a tadpole develops into an adult frog is called metamorphosis. This process is widespread in animals and occurs in many arthropods, certain fishes, numerous marine invertebrates such as gastropods, and amphibia.

Figure 52.5 No, all vertebrates do not use internal fertilization. External fertilization is common in fishes and amphibia; these animals lay unfertilized eggs, over which males deposit sperm (see Chapter 51).

Feature Investigation Questions

1. Knowing the genes expressed in this region of a developing embryo would provide important information about the control of the patterning of embryonic tissues and structures.
2. Harland and colleagues tested the hypothesis that cells within the Spemann organizer expressed certain genes important in the development of dorsal structures, such as the notochord.
3. The scientists used a procedure called expression cloning. In this process, they isolated the various mRNAs that were present in the tissue of the dorsal lip of the embryo. After purifying these mRNAs, they produced a cDNA library. This library contained all the genes expressed in the particular tissue at that particular time of development. The scientists then transcribed the different genes in the cDNA library into mRNAs and injected these into UV-damaged eggs, which were subsequently fertilized. UV-damaged fertilized eggs fail to develop dorsal structures. The scientists were interested in any mRNA that "rescued" the developing embryo and restored some level of normal development. One protein, noggin, was found to rescue the embryos and acted as a morphogen.

Test Yourself

1. a 2. c 3. c 4. d 5. b 6. c 7. b 8. e 9. e 10. e

Conceptual Questions

1. Autonomous specification results from the asymmetrical distribution of intracellular proteins and mRNAs during the cleavage events of embryonic development. The resulting daughter cells will contain different amounts of these cytoplasmic determinants, and this will direct these cells into different developmental fates. Conditional specification results from the interactions of proteins on the extracellular surface of the cell membranes of different cells or from proteins secreted from one cell and acting on another cell. This type of specification determines where a cell ends up within the embryo and what type of cell develops.
2. The timing of the final development of an embryo's organs is typically linked with the requirement for that organ's function. In mammals, for example, the heart is required early in development to pump blood through the embryonic and fetal circulation, thereby delivering nutrients and removing wastes. Fully functional lungs, however, are not required until the animal is born and begins breathing air for the first time.
3. Embryonic development is the process by which a fertilized egg is transformed into an organism with distinct physiological systems and body parts. Cell differentiation is the process by which different cells within a developing organism acquire specialized forms and functions, due to the expression of cell-specific genes. Growth is the enlargement of an embryo, as cells divide and/or enlarge.

Chapter 53

Concept Checks

Figure 53.2 Although swelling is one of the most obvious manifestations of inflammation, it has no significant adaptive value of its own. It is a consequence of fluid leaking out of blood vessels into the interstitial space. It can, however, contribute to pain sensations, because the buildup of fluid may cause distortion of connective tissue structures such as tendons and ligaments. Pain, while obviously unpleasant, is an important signal that alerts many animals to the injury and serves as a reminder to protect the injured site.

Figure 53.4 Recall from Chapter 47 that as blood circulates, a portion of the plasma—the fluid part of blood—exits venules and capillaries and enters the interstitial fluid. Most of the plasma is reabsorbed back into the capillaries, but a portion gets left behind. That excess fluid is drained away by lymph vessels and becomes lymph. Without lymph vessels, fluid would accumulate outside of the blood, in the interstitial fluid.

Figure 53.14 Because an animal may encounter the same type of pathogen many times during its life, having a secondary immune response means that future infections will be fought off much more efficiently.

BioConnections

Figure 53.2 When histamine receptors are blocked by antihistamines, histamine cannot promote wakefulness and thus drowsiness ensues. Some antihistamines are designed such that they cannot get into the extracellular fluid of the brain; these drugs are still effective in inhibiting allergic reactions but do not cause drowsiness.

Feature Investigation Questions

1. The amino acid sequence of Toll protein shared similarities with a portion of a protein known to be involved in immune responses in vertebrates. In addition, activation of Toll protein and the vertebrate immune protein (a cytokine receptor) resulted in the generation of some of the same intracellular signals. This suggested that in addition to its characterized role in embryonic development, Toll may also be important in immune functions in flies.
2. No, Toll protein is not a receptor that recognizes pathogen-associated molecular patterns (PAMPs) expressed on microbial surfaces, and thus it is distinguishable from Toll-like receptors in vertebrates. Toll is, however, a transmembrane protein that binds to extracellular signals; these signals arise, however, not from the microbes themselves but rather from proteins that are endogenous to flies and that are generated during infections.
3. Yes, the results of the survival study clearly implicated Toll as a protein required for the induction of antimicrobial proteins and the ability to withstand fungal infection. Thus, the investigators' hypothesis was supported.

Test Yourself

1. e 2. b 3. c 4. c 5. a 6. d 7. b 8. a 9. d 10. b

Conceptual Questions

1. Innate immunity is present at birth and is found in all animals. These defenses recognize general, conserved features common to a wide array of pathogens and include external barriers, such as the skin, and internal defenses involving phagocytes and other cells. Acquired immunity develops *after* an animal has been exposed to a *particular* antigen. The responses include humoral and cell-mediated defenses. Acquired immunity appears to be largely restricted to vertebrates. Unlike innate immunity, in acquired immunity, the response to an antigen is greatly increased if an animal is exposed to that antigen again at some future time.
2. Cytotoxic T cells are "attack" cells that are responsible for cell-mediated immunity. Once activated, they migrate to the location of their targets, bind to the targets by combining with an antigen on them, and directly kill the targets via secreted chemicals.
3. Pathogens are disease-causing microorganisms and include bacteria, viruses, and eukaryotic parasites such as certain protists, fungi and small worms. Bacteria are single-celled prokaryotes that lack a true nucleus but are capable of reproducing on their own, whereas viruses are nucleic acids packaged in a protein coat that require a host cell to reproduce.

Chapter 54

Concept Checks

Figure 54.4 Higher predation would occur where locust numbers are highest. This means that predators would be responding to an increase in prey density by eating more individuals.

Figure 54.6 Cold water suppresses the ability of the coral-building organisms to secrete their calcium carbonate shell.

Figure 54.9 In some areas when fire is prevented, fuel, in the form of old leaves and branches, can accumulate. When a fire eventually occurs, it can be so large and hot that it destroys everything in its path, even reaching high into the tree canopy.

Figure 54.11 Temperature and rainfall.

Figure 54.16 Acid soils are low in essential plant and animal nutrients such as calcium and nitrogen and are lethal to some soil microorganisms that are important in decomposition and nutrient cycling.

Figure 54.18 This occurs because increasing cloudiness and rain at the tropics maintain fairly constant temperatures across a wide latitudinal range.

Figure 54.22 Soil conditions can also influence biome type. Nutrient-poor soils, for example, may support vegetation different from that of the surrounding area.

Figure 54.23 Taiga.

BioConnections

Chapter Opener The main causes of extinctions are introduced species, direct exploitation, habitat destruction, and climate change. All are human-induced.

Figure 54.15 Plants cannot readily absorb salty water because of its highly negative water potential.

Feature Investigation Questions

1. Most believe that invasive species succeed in new environments due to the lack of natural enemies and that diseases and predators present in the original environment controlled the growth of the population. When these organisms are introduced into a novel environment, the natural enemies are usually absent. This allows for an unchecked increase in the population of the invasive species.
2. Callaway and Aschehoug were able to demonstrate through a controlled experiment that the presence of *Centaurea*, an invasive species, reduced the biomass of three other native species of grasses by releasing allelochemicals. Similar experiments using species of grasses that are found in the native region of *Centaurea* indicate that these species have evolved defenses against the allelochemicals.
3. The activated charcoal helps to remove the allelochemical from the soil. The researchers conducted this experiment to provide further evidence that the chemical released by the *Centaurea* was reducing the biomass of the native Montana grasses. With the removal of the chemicals by the addition of the charcoal, the researchers showed an increase in biomass of the native Montana grasses compared with the experiments lacking the charcoal.

Test Yourself

1. b 2. e 3. a 4. b 5. a 6. a 7. d 8. d 9. a 10. a

Conceptual Questions

1. Mountains are cooler than valleys because of adiabatic cooling. Air at higher altitudes expands because of decreased pressure. As it expands, air cools, at a rate of 10°C for every 1,000 m in elevation. As a result, mountain tops can be much cooler than the plains or valleys that surround them.
2. For several reasons. First, lightning strikes from electrical storms are usually more frequent in prairies than in deserts. Second, the vegetation in a prairie is more continuous and the biomass more extensive than in a desert, so fires burn more frequently and for longer.
3. Florida is a peninsula that is surrounded by the Atlantic Ocean and the Gulf of Mexico. Differential heating between the land and the sea creates onshore sea breezes on both the east and west coasts. These breezes often drift across the whole peninsula, bringing heavy rain.

Chapter 55

Concept Checks

Figure 55.3 In classical conditioning, an involuntary response comes to be associated with a stimulus that did not originally elicit the response, as with Pavlov's dogs salivating at the sound of a metronome.

Figure 55.5 The ability to sing the same distinctive song must be considered innate behavior because the cuckoo has had no opportunity to learn its song from its parents.

Figure 55.7 Tinbergen manipulated pinecones, but not all digger wasp nests are surrounded by pinecones. You could manipulate branches, twigs, stones, and leaves to determine the necessary size and dimensions of objects that digger wasps use as landmarks.

Figure 55.8 This is an unusual example because the return trip involves several different generations to complete: One generation overwinters in Mexico, but these individuals lay eggs and die on the return journey, and their offspring continue the return trip.

Figure 55.14 The individuals in the center of the group are less likely to be attacked than those on the edge of the group. This is referred to as the geometry of the selfish herd.

Figure 55.16 Because of the genetic benefit, the answer is nine cousins. Consider Hamilton's rule, expressed in the formula $rB > C$. Using cousins, $B = 9$, $r = 0.125$, and $C = 1$, and $1.125 > 1$. Using sisters, $B = 2$, $r = 0.5$, and $C = 1$. Because rB would not be greater than C, there would be no net genetic benefit in self-sacrifice.

Figure 55.17 All the larvae in the group are likely to be the progeny of one egg mass from one adult female moth. The death of the one caterpillar teaches a predator to avoid the pattern and benefits the caterpillar's close kin.

Figure 55.24 Because sperm are cheaper to produce than eggs, males try to maximize their fitness through attracting multiple females, whereas female fitness is maximized by choosing a mate with good genetic quality and parenting skills. Colorful plumage and elaborate adornments may be signals of the male's overall health.

Figure 55.25 The males aren't careful because it is likely the pups were fathered in the previous year by a different male. Being a harem master is demanding, and males may often only perform this role for a year or two.

BioConnections

Figure 55.3 Toxic or bad-tasting prey species converge on the same color patterns to reinforce the basic distasteful design.

Figure 55.4 According to studies of humans and other animals, learning a task increases the size of the brain regions that are associated with learning and memory.

Figure 55.25 One claw is enlarged and used in fights over females and to block burrows containing females so that other males cannot enter.

Feature Investigation Questions

1. Tinbergen observed the activity of digger wasps as they prepared to leave the nest. Each time, the wasp hovered and flew around the nest for a period of time before leaving. Tinbergen suggested that during this time, the wasp was making a mental map of the nest site. He hypothesized that the wasp was using characteristics of the nest site, particularly landmarks, to help relocate it.
2. Tinbergen placed pinecones around the nest of the wasps. When the wasps left the nest, he removed the pinecones from the nest site and set them up in the same pattern a distance away, constructing a sham nest. For each trial, the wasps would go directly to the sham nest, which had the pinecones around it. This indicated to Tinbergen that the wasps identified the nest based on the pinecone landmarks.
3. No. Tinbergen also conducted an experiment to determine if the wasps were responding to the visual cue of the pinecones or the chemical cue of the pinecone scent. The results of this experiment indicated that the wasps responded to the visual cue of the pinecones and not their scent.

Test Yourself

1. d 2. d 3. d 4. c 5. c 6. d 7. b 8. c 9. a 10. c

Conceptual Questions

1. The donation of the male's body to the female is the ultimate nuptial gift. It is possible that this meal enables the females to produce more eggs. In this way, the male's genes will be passed on to future generations.
2. Certainty of paternity influences degree of parental care. With internal fertilization, certainty of paternity is relatively low. With external fertilization, eggs and sperm are deposited together, and paternity is more certain. This explains why males of some species, such as mouth-breeding cichlid fish, are more likely to engage in parental care.
3. As male bears are killed by hunters, new males move into a territory and kill existing cubs. Thus, not only are bears killed directly by hunters, but population growth is also slowed as cubs are killed and population recovery is prolonged.

Chapter 56

Concept Checks

Figure 56.2 The total population size, N, would be estimated to be $110 \times 100/20$, or 550.

Figure 56.3 In a half-empty classroom, the distribution is often clumped because friends sit together.

Figure 56.7 (a) type III, (b) type II

Figure 56.11 $dN/dt = 0.1 \times 500\ (1000 - 500)/1000 = 25$.

Figure 56.13 Only density-dependent factors operate in this way.

Figure 56.19 There were very few juveniles in the population and many mature adults. The population would be in decline.

Figure 56.22 Many different ecological footprint calculators are available on the Internet. Does altering inputs such as type of transportation, amount of meat eaten, or amount of waste generated make a difference?

BioConnections

Figure 56.3 Uniform. Territorial marking is likely to keep cheetahs well separated from each other.

Figure 56.14 It has lost the ability to produce viable seeds but it makes thousands of fully formed plantlets, borne on its leaves.

Feature Investigation Questions

1. It became apparent that the sheep population was declining. Some individuals felt that the decline in the population was due to increased wolf predation having a negative effect on population growth. This led to the suggestion of culling the wolf population to reduce the level of predation on the sheep population.
2. The survivorship curve is very similar to a typical type I survivorship curve. This suggests that survival is high among young and reproductively active members of the population and that mortality rates are higher for older members of the population. One difference between the actual survivorship curve and a typical type I curve is that the mortality rate of very young sheep was higher in the actual curve, and then it leveled off after the second year. This suggests that very young and older sheep are more at risk for predation.
3. It was concluded that wolf predation was not the primary reason for the drop in the sheep population. It appeared that wolves prey on the vulnerable members of the population and not on the healthy, reproductively active members. The Park Service determined that several cold winters may have had a more important effect on the sheep population than wolf predation did. Based on these conclusions, the Park Service ended a wolf population control program.

Test Yourself

1. b 2. e 3. b 4. c 5. c 6. b 7. c 8. d 9. c 10. c

Conceptual Questions

1. Increase. Instead of recapturing 5 tagged fish, we only recapture 4. Population size is now estimated as $50 \times 40/4 = 2000/4 = 500$. Our population estimate has increased to 500 when in fact it is more likely that 400 fish occur in the lake.
2. At medium values of N, $(K - N)/K$ is closer to a value of 1, and population growth is relatively large. If $K = 1{,}000$, $N = 500$, and $r = 0.1$, then

$$\frac{dN}{dt} = (0.1)(500) \times \frac{(1{,}000 - 100)}{1{,}000}$$

$$\frac{dN}{dt} = 25$$

However, if population sizes are low ($N = 100$), $(K - N)/K$ is so small that growth is low.

$$\frac{dN}{dt} = (0.1)(100) \times \frac{(1{,}000 - 100)}{1{,}000}$$

$$\frac{dN}{dt} = 9$$

By comparing these two examples with that shown in Section 56.3, we see that growth is small at high and low values of N and is greatest at immediate values of N. Growth is greatest when $N = K/2$. However, when expressed as a percentage, growth is greatest at low population sizes. Where $N = 100$, percentage growth = 9/100 = 9%. Where $N = 500$, percentage growth = 25/500 = 5%, and where $N = 900$, percentage growth = 9/100 = 1%.

3. In the ponds that dry out, species would tend to be semelparous, producing all their offspring in a single reproductive rate while water is present. In the permanently wet ponds, species would be iteroparous, reproducing repeatedly over the course of the year.

Chapter 57

Concept Checks

Figure 57.2 Individual vultures often fight one another over small carcasses. These interactions would constitute intraspecific interference competition.

Figure 57.3 There would be 10 possible pairings (AB, AC, AD, AE, BC, BD, BC, CD, CE, DE), of which only neighboring species (AB, BC, CD, DE) competed. Therefore, competition would be expected in 4/10 pairings, or 40% of the cases.

Figure 57.7 A 1974 review by Tom Schoener examined segregation in a more wide-ranging literature review of over 80 species, including slime molds, mollusks, and insects, as well as birds. He found segregation by habitat occurred in the majority of the examples, 55%. The other most common form of segregation was by food type, 40%.

Figure 57.8 Omnivores, such as bears, can feed on both plant material, such as berries, and animals, such as salmon. As such, omnivores may act as both predators and herbivores depending on what they are feeding on.

Figure 57.9 Batesian mimicry has a positive effect for the mimic, and the model is unaffected, so it is a +/0 relationship, like commensalism. Müllerian mimicry has a positive effect on both species, so it is a +/+ relationship, like mutualism.

Figure 57.11 Because there is no evolutionary history between invasive predators and native prey, the native prey often have no defenses against these predators and are very easily caught and eaten.

Figure 57.13 Invertebrate herbivores can eat around mechanical defenses; therefore, chemical defenses are probably most effective against invertebrate herbivores.

Figure 57.20 It's an example of facultative mutualism, because in this case, both species can live without the other.

Figure 57.23 Fertilizer increases plant quality and hence herbivore density, which, in turn, increases the density of spiders. This is a bottom-up effect.

BioConnections

Figure 57.10 Most mollusks are heavily armored. However, sea slugs have lost their shells. These species are aposematically colored, advertising a poisonous body. In addition, some octopuses are poisonous, and most can eject an inky chemical smoke screen.

Figure 57.13 Red hot chili peppers.

Figure 57.20 Mycorrhizae.

Figure 57.21 Red.

Feature Investigation Questions

1. The two species of barnacles can be found in the same intertidal zone, but there is a distinct difference in niche of each species. *Chthamalus stellatus* is found only in the upper intertidal zone. *Semibalanus balanoides* is found only in the lower tidal zone.
2. Connell moved rocks with young *Chthamalus* from the upper intertidal zone into the lower intertidal zone to allow *Semibalanus* to colonize the rocks. After the rocks were colonized by *Semibalanus*, he removed *Semibalanus* from one side of each rock and returned the rocks to the lower intertidal zone. This allowed Connell to observe the growth of *Chthamalus* in the presence and the absence of *Semibalanus*.
3. Connell observed that *Chthamalus* was more resistant to desiccation than *Semibalanus*. Though *Semibalanus* was the better competitor in the lower intertidal zone, the species was at a disadvantage in the upper intertidal zone when water levels were low. This allowed *Chthamalus* to flourish and outcompete *Semibalanus* in a different region of the intertidal zone.

Test Yourself

1. d 2. c 3. b 4. d 5. c 6. b 7. d 8. b 9. b 10. c

Conceptual Questions

1. Interspecific and interference competition.
2. Yes, it is possible that by removing parasites from a neighbor, an individual may be reducing the likelihood of the parasite spreading. You scratch my back, I'll scratch yours, and together we will both be better off.
3. There are at least three reasons why we don't see more herbivory in nature. First, plants possess an array of defensive chemicals, including alkaloids, phenolics, and terpenes. Second, many herbivore populations are reduced by the action of natural enemies. We see evidence for this in the world of biological control. Third, the low nutritive value of plants ensures herbivore populations remain low and unlikely to affect plant populations.

Chapter 58

Concept Checks

Figure 58.5 Species richness of trees doesn't increase because rainfall in the western United States is low compared with that in the east.

Figure 58.6 Hurricanes, tropical storms, heavy rainfall, and mudslides are disturbances that maintain a mosaic of disturbed and undisturbed habitats, favoring high species richness in the tropics.

Figure 58.10 As we walk forward from the edge of the glacier to the mouth of the inlet, we are walking backward in ecological time to communities that originated hundreds of years ago.

Figure 58.11 No, competition is also important. For example, the shade from later-arriving species, such as spruce trees, causes competitive exclusion of some of the original understory species.

Figure 58.13 Competition features more prominently. Although early colonists tend to make the habitat more favorable for later colonists, it is the later colonists who outcompete the earlier ones, and this fuels species change.

Figure 58.14 If a small island was extremely close to the mainland, it could continually receive migrating species from the source pool. Even though these species could not complete their life cycle on such a small island, extinctions would rarely be recorded because of this continual immigration.

Figure 58.15 At first glance, the change looks small, but the data are plotted on a log scale. On this scale, an increase in bird richness from 1.2 to 1.6 equals an increase from 16 to 40 species, a change of over 100%.

BioConnections

Figure 58.1 A hypothesis is a proposed idea, whereas a theory is a broad explanation backed by extensive evidence.

Figure 58.11 *Frankia*.

Figure 58.14 The model helps conservationists design the best shaped and optimally placed nature reserves.

Feature Investigation Questions

1. Simberloff and Wilson were testing the three predictions of the theory of island biogeography. One prediction suggested that the number of species should increase with increasing island size. Another prediction suggested that the number of species should decrease with increasing distance of the island from the source pool. Finally, the researchers were testing the prediction that the turnover of species on islands should be considerable.
2. Simberloff and Wilson used the information gathered from the species survey to determine whether the same types of species recolonized the islands or if colonizing species were random.
3. The data suggested that species richness did increase with island size. Also, the researchers found that in all but one of the islands, the number of species was similar to the number of species before fumigation.

Test Yourself

1. c 2. c 3. d 4. b 5. c 6. d 7. a 8. d 9. b 10. c

Conceptual Questions

1. Much of what we learned about secondary metabolites in Chapter 57 related to how these chemicals reduced herbivory. This graph shows that a tree's range influences species richness of herbivores. As such it suggests a reduced role for secondary metabolites in influencing herbivore species richness, although abundance of individual herbivore species may still be influenced by the presence of secondary metabolites.

Disturbance	Frequency	Severity of Effects
Forest fire	Low to high, depending on lightning frequency	High to low, depending on frequency
Hurricane	Low	Severe
Tornado	Very low	Severe
Floods	Medium to high in riparian areas	Fairly low; many communities can recover quickly
Disease epidemics	Low	High; may cause catastrophic losses of species
Droughts	Low	Potentially severe
High winds	High	May kill large trees and create light gaps
Hard freezes	Low	May cause deaths to tropical species, such as mangroves

2. Facilitation. *Calluna* litter enriches the soil with nitrogen, facilitating the growth of the grasses. Adding fertilizer also increases soil nitrogen.

Chapter 59

Concept Checks

Figure 59.3 It depends on the trophic level of their food, whether dead vegetation or dead animals. Many decomposers feed at multiple trophic levels.

Figure 59.6 The production efficiency is (16/823) × 100, or 1.9%.

Figure 59.13 On a population level, plant secondary metabolites can deter herbivores from feeding. However, on an ecosystem level, these effects are not as important because higher primary production tends to result in higher secondary production.

Figure 59.18 The greatest stores are in rocks and fossil fuels.

Figure 59.19 It fluctuates because less CO_2 is emitted from vegetation in the summer and more is emitted in the winter. This pattern is driven by the large land masses of the Northern Hemisphere relative to the smaller land masses of the Southern Hemisphere.

BioConnections

Figure 59.2 Cyanobacteria.

Figure 59.3 Basidiomycetes and Zygomycetes.

Figure 59.9 In organic fertilizers, most minerals are bound to organic molecules and are released relatively slowly. In inorganic fertilizers, the minerals are not bound up in this way and are immediately available to plants. However, they are also more easily leached out by heavy rainfall.

Figure 59.18 Oxygen, hydrogen, and nitrogen.

Figure 59.19 About 450 mya, in the Ordovician period.

Feature Investigation Questions

1. The researchers were testing the effects of increased carbon dioxide levels on the forest ecosystem. The researchers were testing the effects of increased CO_2 levels on primary production as well as other trophic levels in the ecosystem.
2. By increasing the CO_2 levels in only half of the chambers, the researchers were maintaining the control treatments necessary for all scientific studies. By maintaining equal numbers of control and experimental treatments, the researchers could compare data to determine what effects the experimental treatment had on the ecosystem.
3. The increased CO_2 levels led to an increase in primary productivity, as expected. Since photosynthetic rate is limited by CO_2 levels, increases in the available CO_2 should increase photosynthetic rates. Interestingly, though, the increase in primary productivity did not lead to an increase in herbivory. The results indicated that herbivory actually decreased with increased CO_2 levels.

Test Yourself

1. d 2. d 3. d 4. a 5. a 6. d 7. a 8. b 9. c 10. d

Conceptual Questions

1. Carrion beetles are decomposers. They feed on dead animals such as mice, at trophic level 3 or 4. Mice generally feed on vegetative material (trophic level 1) or crawling arthropods (trophic level 2), so mice themselves feed at trophic level 2 or 3.
2. Chain lengths are short in food webs because there is low production efficiency and only a 10% rate of energy transfer from one level to another, so only a few links can be supported.
3. A unit of energy passes through a food web only once and energy is lost at each transfer between trophic levels. In contrast, chemicals cycle repeatedly through food webs and may become more concentrated at higher trophic levels.

Chapter 60

Concept Checks

Figure 60.3 It is possible that the results are driven by what is known as a sampling effect. As the numbers of species in the community increase, so does the likelihood of including a "superspecies," a species with exceptionally large individuals that would use up resources. In communities with higher diversity, care has to be taken that increased diversity is driving the results, not the increased likelihood of including a superspecies.

Figure 60.6 The extinction rate could increase because an increasing human population requires more space to live, work, and grow food, resulting in less available habitat and resources for other species.

Figure 60.8 No, some species, such as self-fertilizing flowers, appear to be less affected by inbreeding.

Figure 60.9 The effective population size (N_e) would be = (4 × 125 × 500) / (125 + 500), or 400.

Figure 60.12 Corridors might also promote the movement of invasive species or the spread of fire between areas.

Figure 60.13 They act as habitat corridors because they permit movement of species between forest fragments.

BioConnections

Figure 60.2 A protist. Mosquitoes are the vectors.

Figure 60.19 Genetic cloning could be used to save threatened species or even to resurrect recently extinct species. Cloning may theoretically be able to increase genetic variability of populations if it were possible to use cells from deceased animals. However, cloning is not a panacea because habitat loss, poaching, or invasive species may still prevent reintroductions of species back into the wild.

Feature Investigation Questions

1. The researchers hoped to replicate terrestrial communities that differed only in their level of biodiversity. This would allow the researchers to determine the relationship between biodiversity and ecological function.
2. The hypothesis was that ecological function was directly related to biodiversity. If biodiversity increased, the hypothesis suggested that ecological function should increase.
3. The researchers tested for ecosystem function by monitoring community respiration, decomposition, nutrient retention rates, and productivity. All of these indicate the efficiency of nutrient production and use in the ecosystem.

Test Yourself

1. d 2. e 3. a 4. c 5. c 6. e 7. e 8. a 9. c 10. b

Conceptual Questions

1. To reduce the risks associated with inbreeding in especially small populations.
2. The most vulnerable are those with small population sizes, low rates of population growth, *K*-selected (Chapter 56), with inbreeding and possible harem mating structure, tame and unafraid of humans, possibly limited to islands, flightless, possibly valuable to humans as timber, a source of meat or fur, or desirable by collectors (Chapter 60).
3. Increased species diversity increases ecosystem function. Ecosystem functions such as nutrient cycling, regulation of atmospheric gasses, pollination of crops, pest regulation, water purity, storm protection, and sewage purification are all likely to be increased by increased species diversity. In addition, increased plant species diversity increases likely availability of new medicines for humans.

Glossary

A

A band A wide, dark band in a myofibril produced by the orderly parallel arrangement of the thick filaments in the middle of each sarcomere.
abiotic The term used to describe interactions between organisms and their nonliving environment.
abortion A procedure or circumstance that causes the death of an embryo or fetus after implantation.
abscisic acid One of several plant hormones that help a plant cope with environmental stress.
abscission The process by which plants drop their leaves.
absolute refractory period The period during an action potential when the inactivation gate of the voltage-gated sodium channel is closed; during this time, it is impossible to generate another action potential.
absorption The process in which digested nutrients are transported from the digestive cavity into an animal's circulatory system.
absorption spectrum A diagram that depicts the wavelengths of electromagnetic radiation that are absorbed by a pigment.
absorptive nutrition The process whereby an organism uses enzymes to digest organic materials and absorbs the resulting small food molecules into its cells.
absorptive state One of two alternating phases in the utilization of nutrients; occurs when ingested nutrients enter the blood from the gastrointestinal tract. The other phase is the postabsorptive state.
acclimatization A long-term and persistent physiological adaptation to an extreme environment.
accommodation In the vertebrate eye, the process in which contraction and relaxation of the ciliary muscles adjust the lens according to the angle at which light enters the eye.
acetylcholinesterase An enzyme located on membranes of postsynaptic cells that respond to the neurotransmitter acetylcholine, such as in muscle fibers in a neuromuscular junction; breaks down excess acetylcholine released into the synaptic cleft.
acid A molecule that releases hydrogen ions (H^+) in solution.
acid hydrolase A hydrolytic enzyme found in lysosomes that functions at acidic pH and uses a molecule of water to break a covalent bond.
acid rain Precipitation with a pH of less than 5.6; results from the burning of fossil fuels.
acidic A solution that has a pH below 7.
acoelomate An animal that lacks a fluid-filled body cavity.
acquired antibiotic resistance The common phenomenon of a previously susceptible strain of bacteria becoming resistant to a specific antibiotic.
acquired immunity A specific immune defense that develops only after an animal is exposed to a foreign substance; believed to be unique to vertebrates.
acquired immunodeficiency syndrome (AIDS) A disease caused by the human immunodeficiency virus (HIV) that leads to a defect in the immune system of infected individuals.
acrocentric A chromosome in which the centromere is near one end.
acromegaly A condition in which a person's growth hormone level is abnormally elevated, causing many bones to thicken and enlarge.
acrosomal reaction An event in fertilization in which enzymes released from a sperm's acrosome break down the outer layers of an egg cell, allowing the entry of the sperm cell's nucleus into the egg cell.
acrosome A special structure at the tip of a sperm's head containing proteolytic enzymes that break down the protective outer layers of the egg cell at fertilization.
actin A cytoskeletal protein.
actin filament A thin type of protein filament composed of actin proteins that forms part of the cytoskeleton and supports the plasma membrane; plays a key role in cell strength, shape, and movement.
action potential An electrical signal along a cell's plasma membrane; occurs in animal neuron axons and muscle cells and in some plant cells.
action spectrum The rate of photosynthesis plotted as a function of different wavelengths of light.
activation energy An initial input of energy in a chemical reaction that allows the molecules to get close enough to cause a rearrangement of bonds.
activator A transcription factor that binds to DNA and increases the rate of transcription.
active immunity An animal's ability to fight off a pathogen to which it has been previously exposed. Active immunity can develop as a result of natural infection or artificial immunization.
active site The location in an enzyme where a chemical reaction takes place.
active transport The transport of a solute across a membrane against its gradient (from a region of low concentration to a region of higher concentration). Active transport requires an input of energy.
adaptations The processes and structures by which organisms adjust to changes in their environment.
adaptive radiation The process whereby a single ancestral species evolves into a wide array of descendant species that differ greatly in their habitat, form, or behavior.
adenine (A) A purine base found in DNA and RNA.
adenosine triphosphate (ATP) A molecule that is a common energy source for all cells.
adenylyl cyclase An enzyme in the plasma membrane that synthesizes cAMP from ATP.
adherens junction A mechanically strong cell junction between animal cells that typically occurs in bands. The cells are connected to each other via cadherins, and the cadherins are linked to actin filaments on the inside of the cells.
adhesion The ability of two different substances to bind to each other; the ability of water to be attracted to, and thereby adhere to, a surface that is not electrically neutral.
adiabatic cooling The process in which increasing elevation leads to a decrease in air temperature.
adventitious root A root that is produced on the surfaces of stems (and sometimes leaves) of vascular plants; also, roots that develop at the bases of stem cuttings.
aerenchyma Spongy plant tissue with large air spaces.
aerobe An organism that requires oxygen to survive.
aerobic respiration A type of cellular respiration in which O_2 is consumed and CO_2 is released.
aerotolerant anaerobe A microorganism that does not use oxygen but is not poisoned by it either.
afferent arterioles Blood vessels that provide a pathway for blood into the glomeruli of the vertebrate kidney.
affinity The degree of attraction between an enzyme and its substrate.
aflatoxins Fungal toxins that cause liver cancer and are a major health concern worldwide.
age-specific fertility rate The rate of offspring production for females of a certain age; used to calculate how a population grows.
age structure The relative numbers of individuals of each defined age group in a population.
AIDS *See* acquired immunodeficiency syndrome.
air sac A component of the avian respiratory system; air sacs—not lungs—expand when a bird inhales and shrink when it exhales. They do not participate in gas exchange, but help direct air through the lungs.
akinete A thick-walled, food-filled cell produced by certain bacteria or protists that enables them to survive unfavorable conditions in a dormant state.
aldosterone A steroid hormone made by the adrenal glands that regulates salt and water balance in vertebrates.
algae (singular, **alga**) A term that applies to about 10 phyla of protists, including both photosynthetic and nonphotosynthetic species; often also includes cyanobacteria.
alimentary canal In animals, the single elongated tube of a digestive system, with an opening at either end through which food and eventually wastes pass from one end to the other.
alkaline A solution with a pH above 7.
alkaloids A group of secondary metabolites that contain nitrogen and usually have a cyclic, ringlike structure. Examples include caffeine and nicotine.
allantois One of the four extraembryonic membranes in the amniotic egg. It serves as a disposal sac for metabolic wastes.
Allee effect The phenomenon that some individuals will fail to mate successfully purely by chance, for example, because of the failure to find a mate.
allele A variant form of a gene.
allele frequency The number of copies of a particular allele in a population divided by the total number of alleles in that population.
allelochemical A powerful plant chemical, often a root exudate, that kills other plant species.
allelopathy The suppressed growth of one species due to the release of toxic chemicals by another species.
allergy Hypersensitivity reaction to a harmless substance.
allopatric The term used to describe species occurring in different geographic areas.
allopatric speciation A form of speciation that occurs when a population becomes geographically isolated from other populations and evolves into one or more new species.
alloploid An organism having at least one set of chromosomes from two or more different species.
alloploidy *See* alloploid.
allosteric site A site on an enzyme where a molecule can bind noncovalently and affect the function of the active site.
alpha (α) helix A type of protein secondary structure in which a polypeptide forms a repeating helical structure stabilized by hydrogen bonds.
alternation of generations The phenomenon that occurs in plants and some protists in which the life cycle alternates between multicellular diploid organisms, called sporophytes, and multicellular haploid organisms, called gametophytes.
alternative splicing The splicing of pre-mRNA in more than one way to create two or more different polypeptides.
altruism Behavior that appears to benefit others at a cost to oneself.
alveolus (plural, **alveoli**) 1. Saclike structures in the lungs where gas exchange occurs. 2. Saclike cellular features of the protists known as alveolates.

Alzheimer disease (AD) The leading worldwide cause of dementia; characterized by a loss of memory and intellectual and emotional function (formerly called Alzheimer's disease).

AM fungi A phylum of fungi that forms mycorrhizal associations with plants.

amensalism One-sided competition between species, in which the interaction is detrimental to one species but not to the other.

Ames test A test that helps ascertain whether or not an agent is a mutagen by using a strain of a bacterium, *Salmonella typhimurium*.

amino acid The building blocks of proteins. Amino acids have a common structure in which a carbon atom, called the α-carbon, is linked to an amino group (NH_2) and a carboxyl group (COOH). The α-carbon also is linked to a hydrogen atom and a particular side chain.

aminoacyl site (A site) One of three sites for tRNA binding in the ribosome during translation; the other two are the peptidyl site (P site) and the exit site (E site). The A site is where incoming tRNA molecules bind to the mRNA (except for the initiator tRNA).

aminoacyl tRNA *See* charged tRNA.

aminoacyl-tRNA synthetase An enzyme that catalyzes the attachment of amino acids to tRNA molecules.

amino terminus *See* N-terminus.

ammonia (NH_3) A highly toxic nitrogenous waste typically produced by many aquatic animal species.

ammonification The conversion of organic nitrogen to NH_3 and NH_4^+ during the nitrogen cycle.

amnion The innermost of the four extraembryonic membranes in the amniotic egg. It protects the developing embryo in a fluid-filled sac called the amniotic cavity.

amniotes A group of tetrapods with amniotic eggs that includes turtles, lizards, snakes, crocodiles, birds, and mammals.

amniotic egg A type of egg produced by amniotic animals that contains the developing embryo and the four separate extraembryonic membranes that it produces: the amnion, the yolk sac, the allantois, and the chorion.

amoeba (plural, **amoebae**) A protist that moves by pseudopodia, which involves extending cytoplasm into filaments or lobes.

amoebocyte A mobile cell within a sponge's mesophyl that absorbs food from choanocytes, digests it, and carries the nutrients to other cells.

amphibian An ectothermic, vertebrate animal that metamorphoses from a water-breathing to an air-breathing form but must return to the water to reproduce.

amphipathic Molecules containing a hydrophobic (water-fearing) region and a hydrophilic (water-loving) region.

ampulla (plural, **ampullae**) 1. A muscular sac at the base of each tube foot of an echinoderm; used to store water. 2. A bulge in the walls of the semicircular canals of the mammalian inner ear; important for sensing circular motions of the head.

amygdala An area of the vertebrate forebrain known to be critical for understanding and remembering emotional situations.

amylase A digestive enzyme in saliva and the pancreas involved in the digestion of starch.

anabolic reaction A metabolic pathway that involves the synthesis of larger molecules from smaller precursor molecules. Such reactions usually require an input of energy.

anabolism A metabolic pathway that results in the synthesis of cellular molecules and macromolecules; requires an input of energy.

anaerobic Refers to a process that occurs in the absence of oxygen; a form of metabolism that does not require oxygen.

anaerobic respiration The breakdown of organic molecules in the absence of oxygen.

anagenesis The pattern of speciation in which a single species is transformed into a different species over the course of many generations.

analogous structure A structure that is the result of convergent evolution. Such structures have arisen independently, two or more times, because species have occupied similar types of environments on Earth.

anaphase The phase of mitosis during which the sister chromatids separate from each other and move to opposite poles; the poles themselves also move farther apart.

anchoring junction A type of junction between animal cells that attaches cells to each other and to the extracellular matrix (ECM).

androecium The aggregate of stamens that forms the third whorl of a flower.

androgens Steroid hormones produced by the male testes (and, to a lesser extent, the adrenal glands) that affect most aspects of male reproduction.

anemia A condition characterized by lower than normal levels of hemoglobin, which reduces the amount of oxygen that can be stored in the blood.

aneuploidy An alteration in the number of particular chromosomes so that the total number of chromosomes is not an exact multiple of a set.

angina pectoris Chest pain during exertion due to the heart being deprived of oxygen.

angiosperm A flowering plant. The term means enclosed seed, which reflects the presence of seeds within fruits.

animal cap assay A type of experiment used to identify proteins secreted by embryonic cells that induce cells in the animal pole to differentiate into mesoderm.

animal pole In triploblast organisms, the pole of the egg with less yolk and more cytoplasm.

Animalia A eukaryotic kingdom of the domain Eukarya.

animals Multicellular heterotrophs with cells that lack cell walls. Most animals have nerves, muscles, the capacity to move at some point in their life cycle, and the ability to reproduce sexually, with sperm fusing directly with eggs.

anion An ion that has a net negative charge.

annual A plant that dies after producing seed during its first year of life.

anorexia nervosa An eating disorder characterized by aversion to food, starvation, and illness secondary to malnourishment.

antagonist A muscle or group of muscles that produces oppositely directed movements at a joint.

antenna complex *See* light-harvesting complex.

anterior Refers to the end of an animal where the head is found.

anteroposterior axis In bilateral animals, one of the three axes along which the adult body pattern is organized; the others are the dorsoventral axis and the right-left axis.

anther The uppermost part of a flower stamen, consisting of a cluster of microsporangia that produce and release pollen.

antheridia Round or elongate gametangia that produce sperm in plants.

anthropoidea A member of a group of primates that includes the monkeys and the hominoidea; these species are larger-brained and diurnal.

antibiotic A chemical, usually made by microorganisms, that inhibits the growth of certain other microorganisms.

antibody A protein secreted by plasma cells that is part of the immune response; antibodies travel all over the body to reach antigens identical to those that stimulated their production, combine with these antigens, then guide an attack that eliminates the antigens or the cells bearing them.

anticodon A three-nucleotide sequence in tRNA that is complementary to a codon in mRNA.

antidiuretic hormone (ADH) A hormone secreted by the posterior pituitary gland that acts on kidney cells to decrease urine production.

antigen Any foreign molecule that the host does not recognize as self and that triggers a specific immune response.

antigen-presenting cell (APC) Cells bearing fragments of antigen, called antigenic determinants or epitopes, complexed with the cell's major histocompatibility complex (MHC) proteins.

antiparallel The arrangement in DNA where one strand runs in the 5′ to 3′ direction while the other strand is oriented in the 3′ to 5′ direction.

antiporter A type of transporter that binds two or more ions or molecules and transports them in opposite directions across a membrane.

anus The final portion of the alimentary canal through which solid wastes are expelled.

aorta In vertebrates, a large blood vessel that exits a ventricle of the heart and leads to the systemic circulation.

apical-basal-patterning genes A category of genes that are important in early stages of plant development during which the apical and basal axes are formed.

apical-basal polarity An architectural feature of plants in which they display an upper, apical pole and a lower, basal pole; shoot apical meristem occurs at the apical pole, and root apical meristem occurs at the basal pole.

apical constriction A cellular process during gastrulation that occurs in bottle cells, where a reduction in the diameter of the actin rings connected to the adherens junctions causes the cells to elongate toward their basal end.

apical meristem In plants, a group of actively dividing cells at a growing tip.

apical region The region of a plant seedling that produces the leaves and flowers.

apomixis A natural asexual reproductive process in which plant fruits and seeds are produced in the absence of fertilization.

apoplast The continuum of water-soaked cell walls and intercellular spaces in a plant.

apoplastic transport The movement of solutes through cell walls and the spaces between cells.

apoptosis Programmed cell death.

aposematic coloration Warning coloration that advertises an organism's unpalatable taste.

aquaporin A transport protein in the form of a channel that allows the rapid diffusion of water across the cell membrane.

aqueous humor A thin liquid in the anterior cavity behind the cornea of the vertebrate eye.

aqueous solution A solution made with water.

aquifer An underground water supply.

arbuscular mycorrhizae Symbiotic associations between AM fungi and the roots of vascular plants.

Archaea One of the three domains of life; the other two are Bacteria and Eukarya.

archaea When not capitalized, refers to a cell or species within the domain Archaea.

archegonia Flask-shaped plant gametangia that enclose an egg cell.

archenteron A cavity formed in an animal embryo during gastrulation that will become the organism's digestive tract.

area hypothesis The proposal that larger areas contain more species than smaller areas because they can support larger populations and a greater range of habitats.

arteriole A single-celled layer of endothelium surrounded by one or two layers of smooth muscle and connective tissue that delivers blood to the capillaries and distributes blood to regions of the body in proportion to metabolic demands.

artery A blood vessel that carries blood away from the heart.

artificial selection *See* selective breeding.

asci (singular, **ascus**) Fungal sporangia shaped like sacs that produce and release sexual ascospores.

ascocarp The type of fruiting body produced by ascomycete fungi.

ascomycetes A phylum of fungi that produce sexual spores in saclike asci located at the surfaces of fruiting bodies known as ascocarps.

ascospore The type of sexual spore produced by the ascomycete fungi.

aseptate The condition of not being partitioned into smaller cells; usually refers to fungal cells.

asexual reproduction A reproductive strategy that occurs when offspring are produced from a single parent, without the fusion of gametes from two parents. The offspring are therefore clones of the parent.

A site *See* aminoacyl site.

assimilation During the nitrogen cycle, the process by which plants and animals incorporate the ammonia and NO_3^- formed through nitrogen fixation and nitrification.

assisted reproductive technologies (ART) Any of a group of methods used to produce a pregnancy by artificial mechanisms.

associative learning A change in behavior due to an association between a stimulus and a response.

asthma A disease in which the smooth muscles around the bronchioles contract more than usual, decreasing airflow in the lungs.

AT/GC rule Refers to the phenomenon that an A in one DNA strand always hydrogen-bonds with a T in the opposite strand, and a G in one strand always bonds with a C.

atherosclerosis The condition in which large plaques may occlude (block) the lumen of an artery.

atmospheric (barometric) pressure The pressure exerted by the gases in air on the body surfaces of animals.

atom The smallest functional unit of matter that forms all chemical substances and cannot be further broken down into other substances by ordinary chemical or physical means.

atomic mass An atom's mass relative to the mass of other atoms. By convention, the most common form of carbon, which has six protons and six neutrons, is assigned an atomic mass of exactly 12.

atomic nucleus The center of an atom; contains protons and neutrons.

atomic number The number of protons in an atom.

ATP *See* adenosine triphosphate.

ATP-dependent chromatin remodeling enzyme An enzyme that catalyzes a change in the positions of nucleosomes.

ATP synthase An enzyme that utilizes the energy stored in a H^+ electrochemical gradient for the synthesis of ATP via chemiosmosis.

atrial natriuretic peptide (ANP) A peptide secreted from the atria of the heart whenever blood levels of sodium increase; ANP causes a natriuresis by decreasing sodium reabsorption in the kidney tubules.

atrioventricular (AV) node Specialized cardiac cells in most vertebrates that sit near the junction of the atria and ventricles and conduct the electrical events from the atria to the ventricles.

atrioventricular (AV) valve A one-way valve into the ventricles of the vertebrate heart through which blood moves from the atria.

atrium In the heart, a chamber to collect blood from the tissues.

atrophy A reduction in the size of a structure, such as a muscle.

audition The ability to detect and interpret sound waves; present in vertebrates and arthropods.

autoimmune disease In humans and many other vertebrates, a disorder in which the body's normal state of immune tolerance breaks down, with the result that attacks are directed against the body's own cells and tissues.

autonomic nervous system The division of the peripheral nervous system that regulates homeostasis and organ function.

autonomous specification The unequal acquisition of cytoplasmic factors during cell division in a developing vertebrate embryo.

autophagosome A double-membrane structure enclosing cellular material destined to be degraded; produced by the process of autophagy.

autophagy A process whereby cellular material, such as a worn-out organelle, becomes enclosed in a double membrane and is degraded.

autopolyploidy The condition of having more than two copies of the entire nuclear genome, which may result when all homologous chromosome pairs do not separate during meiosis.

autosomes All of the chromosomes found in the cell nucleus of eukaryotes except for the sex chromosomes.

autotomy In echinoderms, the ability to detach a body part, such as a limb, that will later regenerate.

autotroph An organism that has metabolic pathways that use energy from either inorganic molecules or light to make organic molecules.

auxin One of several types of hormones considered to be "master" plant hormones because they influence plant structure, development, and behavior in many ways.

auxin efflux carrier One of several types of PIN proteins, which transport auxin out of plant cells.

auxin influx carrier A plasma membrane protein that transports auxin into plant cells.

auxin-response genes Plant genes that are regulated by the hormone auxin.

avirulence gene (*Avr* gene) A gene in a plant pathogen that encodes a virulence-enhancing elicitor, which causes plant disease.

Avogadro's number As first described by Italian physicist Amedeo Avogadro, 1 mole of any element contains the same number of atoms—6.022×10^{23}.

axillary bud A bud that occurs in the axil, the upper angle where a twig or leaf emerges from a stem.

axillary meristem A meristem produced in the axil, the upper angle where a twig or leaf emerges from a stem. Axillary meristems generate axillary buds, which can produce flowers or branches.

axon An extension of the plasma membrane of a neuron that is involved in sending signals to neighboring cells.

axon hillock The part of the axon closest to the cell body; typically where an action potential begins.

axon terminal The end of the axon that sends electrical or chemical messages to other cells.

axoneme The internal structure of eukaryotic flagella and cilia consisting of microtubules, the motor protein dynein, and linking proteins.

B

bacilli (singular, **bacillus**) Rod-shaped prokaryotic cells.

backbone The linear arrangement of phosphates and sugar molecules in a DNA or RNA strand.

Bacteria One of the three domains of life; the other two are Archaea and Eukarya.

bacteria (singular, **bacterium**) When not capitalized, refers to a cell or species within the domain Bacteria.

bacterial artificial chromosome (BAC) A cloning vector derived from F factors that can contain large DNA inserts.

bacterial colony A clone of genetically identical cells formed from a single cell.

bacteriophage A virus that infects bacteria.

bacteroid A modified bacterial cell of the type known as rhizobia present in mature root nodules of some plants.

balanced polymorphism The phenomenon in which two or more alleles are kept in balance and maintained in a population over the course of many generations.

balancing selection A type of natural selection that maintains genetic diversity in a population.

balloon angioplasty A common treatment to restore blood flow through a blood vessel. A thin tube with a tiny, inflatable balloon at its tip is threaded through the artery to the diseased area; inflating the balloon compresses the plaque against the arterial wall, widening the lumen.

barometric pressure *See* atmospheric pressure.

baroreceptor A pressure-sensitive region within the walls of certain arteries that contains the endings of nerve cells; these regions sense and help to maintain blood pressure in the normal range for an animal.

Barr body A highly condensed X chromosome present in female mammals.

basal body A site at the base of flagella or cilia from which microtubules grow. Basal bodies are anchored on the cytosolic side of the plasma membrane.

basal metabolic rate (BMR) The metabolic rate of an animal under resting conditions, in a postabsorptive state, and at a standard temperature.

basal nuclei Clusters of neuronal cell bodies in the vertebrate forebrain that surround the thalamus and lie beneath the cerebral cortex; involved in planning and learning movements.

basal region The region of a plant seedling that produces the roots.

basal transcription A low level of transcription resulting from just the core promoter.

basal transcription apparatus In a eukaryotic structural gene, refers to the complex of RNA polymerase II, general transcription factors (GTFs), and a DNA sequence containing a TATA box.

base 1. A molecule that when dissolved in water lowers the H^+ concentration. 2. A component of nucleotides that is a single or double ring of carbon and nitrogen atoms.

base pair The structure in which two bases in opposite strands of DNA hydrogen-bond with each other.

base substitution A mutation that involves the substitution of a single base in the DNA for another base.

basic local alignment search tool (BLAST) *See* BLAST.

basidia Club-shaped cells that produce sexual spores in basidiomycete fungi.

basidiocarp The type of fruiting body produced by basidiomycete fungi.

basidiomycetes A phylum of fungi whose sexual spores are produced on the surfaces of club-shaped structures (basidia).

basidiospore A sexual spore of the basidiomycete fungi.

basilar membrane A component of the mammalian ear that vibrates back and forth in response to sound and bends the stereocilia in one direction and then the other.

basophil A type of leukocyte that secretes the anticlotting factor heparin at the site of an infection, which helps flush out the infected site; basophils also secrete histamine, which attracts infection-fighting cells and proteins.

Batesian mimicry The mimicry of an unpalatable species (the model) by a palatable one (the mimic).

Bayesian method One method used to evaluate a phylogenetic tree based on an evolutionary model.

B cell A type of lymphocyte responsible for specific immunity.

behavior The observable response of organisms to external or internal stimuli.

behavioral ecology A subdiscipline of organismal ecology that focuses on how the behavior of an individual organism contributes to its survival and reproductive success, which, in turn, eventually affects the population density of the species.

benign tumor A precancerous mass of abnormal cells.

beta (β) pleated sheet A type of protein secondary structure in which regions of a polypeptide lie parallel to each other and are held together by hydrogen bonds to form a repeating zigzag shape.

bidirectional replication The process in which DNA replication proceeds outward from the origin in opposite directions.

biennial A plant that does not reproduce during the first year of life but may reproduce within the following year.

bilateral symmetry An architectural feature in which the body or organ of an organism can be divided along a vertical plane at the midline to create two halves.

Bilateria Bilaterally symmetric animals.

bile A substance produced by the liver that contains bicarbonate ions, cholesterol, phospholipids, a number of organic wastes, and a group of substances collectively termed bile salts. Bile emulsifies fats so they can be absorbed by the small intestine.

bile salts A group of substances produced in the liver that solubilize dietary fat and increase its accessibility to digestive enzymes.

binary fission The process of cell division in bacteria and archaea in which one cell divides into two cells.

binocular (or stereoscopic) vision A type of vision in animals having two eyes located at the front of the head; the overlapping images coming into both eyes are processed together in the brain to form one perception. Binocular vision enables depth perception.

binomial nomenclature The standard method for naming species. Each species has a genus name and species epithet.

biochemistry The study of the chemistry of living organisms.

biodiversity The diversity of life forms in a given location.

biodiversity crisis The idea that there is currently an elevated loss of species on Earth, far beyond the normal historical extinction rate of species.

biofilm An aggregation of microorganisms that secrete adhesive mucilage, thereby gluing themselves to surfaces.

biogeochemical cycle The continuous movement of nutrients such as nitrogen, carbon, sulfur, and phosphorus from the physical environment to organisms and back.

biogeographic region One of six geographic regions which divide up the world's biota: Nearctic, Palearctic, Neotropical, Ethopian, Oriental, and Australian.

biogeography The study of the geographic distribution of extinct and modern species.

bioinformatics A field of study that uses computers to study biological information.

biological control The use of an introduced species' natural enemies to control its proliferation.

biological diversity *See* biodiversity.

biological nitrogen fixation Nitrogen fixation that is performed in nature by certain prokaryotes.

biological species concept An approach used to distinguish species, which states that a species is a group of individuals whose members have the potential to interbreed with one another in nature to produce viable, fertile offspring but cannot successfully interbreed with members of other species.

biology The study of life.

bioluminescence A phenomenon in living organisms in which chemical reactions give off light rather than heat.

biomagnification The increase in the concentration of a substance in living organisms from lower to higher trophic levels in a food web.

biomass A quantitative estimate of the total mass of living matter in a given area, usually measured in grams or grams per square meter.

biome A major type of habitat characterized by distinctive plant and animal life.

bioremediation The use of living organisms, usually microbes or plants, to detoxify polluted habitats such as dump sites or oil spills.

biosphere The regions on the surface of the Earth and in the atmosphere where living organisms exist.

biosynthetic reaction Also called an anabolic reaction; a chemical reaction in which small molecules are used to synthesize larger molecules.

biotechnology The use of living organisms or the products of living organisms for human benefit.

biotic The term used to describe interactions among organisms.

biparental inheritance An inheritance pattern in which both the male and female gametes contribute organellar genes to the offspring.

bipedal Having the ability to walk on two feet.

bipolar cells Cells in the vertebrate eye that make synapses with photoreceptors and relay responses to the ganglion cells.

bivalent Homologous pairs of sister chromatids associated with each other, lying side by side.

blade The flattened portion of a leaf.

BLAST (basic local alignment search tool) A computer program that can identify homologous genes that are found in a database.

blastocoel A cavity formed in a cleavage-stage vertebrate embryo (blastula); provides a space into which cells of the future digestive tract will migrate.

blastocyst The mammalian counterpart of a blastula.

blastoderm A flattened disc of dividing cells in the embryo of animals that undergo incomplete cleavage; occurs in birds and some fishes.

blastomere The two half-size daughter cells produced by each cell division during cleavage.

blastopore A small opening created when a band of tissue invaginates during gastrulation. It forms the primary opening of the archenteron to the outside.

blastula An animal embryo at the stage when it forms an outer epithelial layer and an inner cavity.

blending inheritance An early hypothesis of inheritance that stated that the genetic material that dictates hereditary traits blends together from generation to generation, and the blended traits are then passed to the next generation.

blood A fluid connective tissue in animals consisting of cells and (in mammals) cell fragments suspended in a solution of water containing dissolved nutrients, proteins, gases, and other molecules.

blood-doping An example of hormone misuse in which the number of red blood cells in the circulation is boosted to increase the oxygen-carrying capacity of the blood.

blood pressure The force exerted by blood on the walls of blood vessels; blood pressure is responsible for moving blood through the vessels.

body mass index (BMI) A method of assessing body fat and health risk that involves calculating the ratio of weight compared with height; weight in kilograms is divided by the square of the height in meters.

body plan The organization of cells, tissues, and organs within a multicellular organism; also known as a body pattern.

bone A relatively hard component of the vertebrate skeleton; a living, dynamic tissue composed of organic molecules and minerals.

bottleneck effect A situation in which a population size is dramatically reduced and then rebounds. While the population is small, genetic drift may rapidly reduce the genetic diversity of the population.

Bowman's capsule A saclike structure that houses the glomerulus at the beginning of the tubular component of a nephron in the mammalian kidney.

brain Organ of the central nervous system of animals that functions to process and integrate information.

brainstem The part of the vertebrate brain composed of the medulla oblongata, the pons, and the midbrain.

brassinosteroid One of several plant hormones that help a plant to cope with environmental stress.

bronchi (singular, **bronchus**) Tubes branching from the trachea and leading into the lungs.

bronchiole A thin-walled, small tube branching from the bronchi and leading to the alveoli in mammalian lungs.

bronchodilator A compound that binds to the muscles of the bronchioles of the lung and causes them to relax, thereby widening the bronchioles and easing breathing.

brown adipose tissue A specialized tissue in small mammals such as hibernating bats, small rodents living in cold environments, and many newborn mammals, including humans, that can help to generate heat and maintain body temperature.

brush border The collective name for the microvilli in the vertebrate small intestine.

bryophytes Liverworts, mosses, and hornworts, the modern nonvascular land plants.

buccal pumping A form of breathing in which animals take in water or air into their mouths, then raise the floor of the mouth, creating a positive pressure that pumps water or air across the gills or into the lungs; found in fishes and amphibians.

bud A miniature plant shoot having a dormant shoot apical meristem.

budding A form of asexual reproduction in which a portion of the parent organism pinches off to form a complete new individual.

buffer A compound that acts to minimize pH fluctuations in the fluids of living organisms. Buffer systems can raise or lower pH as needed.

bulbourethral gland Paired accessory glands in the human male reproductive system that secrete an alkaline mucus that protects sperm by neutralizing the acidity in the urethra.

bulimia nervosa An eating disorder characterized by cycles of binge eating followed by purging and vomiting to reduce body weight.

bulk feeders Animals that eat food in large pieces.

bulk flow The mass movement of liquid in a plant caused by pressure, gravity, or both.

C

C_3 plant A plant that incorporates CO_2 into organic molecules via RuBP to make 3PG, a three-carbon molecule.

C_4 plant A plant that uses PEP carboxylase to initially fix CO_2 into a four-carbon molecule and later uses rubisco to fix CO_2 into simple sugars; an adaptation to hot, dry environments.

cadherin A cell adhesion molecule found in animal cells that promotes cell-to-cell adhesion.

calcitonin A hormone that plays a role in Ca^{2+} homeostasis in some vertebrates.

calcium wave A brief increase in cytosolic Ca^{2+} concentrations in an egg that has been penetrated by a sperm cell; the change in Ca^{2+} moves through the cell and contributes to the slow block to polyspermy.

callose A carbohydrate that plays crucial roles in plant development and plugging wounds in plant phloem.

calorie The amount of heat required to raise the temperature of 1 gram of water 1°C. The Calorie (dietary unit) is equivalent to a kilocalorie, or 1,000 calories.

Calvin cycle The second stage in the process of photosynthesis. During this cycle, ATP is used as a source of energy, and NADPH is used as a source of high-energy electrons so that CO_2 can be incorporated into carbohydrate.

calyx The sepals that form the outermost whorl of a flower.

Cambrian explosion An event during the Cambrian period (543–490 mya) in which there was an abrupt increase (on a geological scale) in the diversity of animal species.

cAMP *See* cyclic adenosine monophosphate.

CAM (crassulacean acid metabolism) plants C_4 plants that open their stomata at night to take up CO_2.

cancer A disease caused by gene mutations that lead to uncontrolled cell growth.

canopy The uppermost layer of tree foliage.

capillary A tiny thin-walled vessel that is the site of gas and nutrient exchange between the blood and interstitial fluid.

capping The process in which a 7-methylguanosine is covalently attached at the 5′ end of mature mRNAs of eukaryotes.

capsid A protein coat enclosing a virus's genome.

CAP site One of two regulatory sites near the *lac* promoter; this site is a DNA sequence recognized by the catabolite activator protein (CAP).

capsule A very thick, gelatinous glycocalyx produced by certain strains of bacteria that may help them avoid being destroyed by an animal's immune (defense) system.

carapace The hard protective cuticle covering the cephalothorax of a crustacean.

carbohydrate An organic molecule often with the general formula, $C(H_2O)$; a carbon-containing compound that includes starches, sugars, and cellulose.

carbon fixation A process in which carbon from inorganic CO_2 is incorporated into an organic molecule such as a carbohydrate.

carboxyl terminus *See* C-terminus.

carcinogen An agent that increases the likelihood of developing cancer, usually a mutagen.

carcinoma A cancer of epithelial cells.

cardiac cycle The events that produce a single heartbeat, which can be divided into two phases, diastole and systole.

cardiac muscle A type of muscle tissue found only in hearts in which physical and electrical connections between individual cells enable many of the cells to contract simultaneously.

cardiac output (CO) The amount of blood the heart pumps per unit time, usually expressed in units of L/min.

cardiovascular disease Diseases affecting the heart and blood vessels.

carnivore An animal that consumes animal flesh or fluids.

carotenoid A type of photosynthetic or protective pigment found in plastids that imparts a color that ranges from yellow to orange to red.

carpel A flower shoot organ that produces ovules that contain female gametophytes.

carrier *See* transporter.

carrying capacity (*K*) The upper boundary for a population size.

Casparian strips Suberin ribbons on the walls of endodermal cells of plant roots; prevent apoplastic transport of ions into vascular tissues.

caspase An enzyme that is activated during apoptosis.

catabolic reaction A metabolic pathway in which a molecule is broken down into smaller components, usually releasing energy.

catabolism A metabolic pathway that results in the breakdown of larger molecules into smaller molecules. Such reactions are often exergonic.

catabolite activator protein (CAP) An activator protein for the *lac* operon.

catabolite repression In bacteria, a process whereby transcriptional regulation is influenced by the presence of glucose.

catalase An enzyme within peroxisomes that breaks down hydrogen peroxide to water and oxygen gas.

catalyst An agent that speeds up the rate of a chemical reaction without being consumed during the reaction.

cataract An accumulation of protein in the lens of the eye; causes blurring and poor night vision.

cation An ion that has a net positive charge.

cation exchange With regard to soil, the process in which hydrogen ions are able to replace mineral cations on the surfaces of humus or clay particles.

cDNA *See* complementary DNA.

cDNA library A type of DNA library in which the inserts are derived from cDNA.

cecum The first portion of a vertebrate's large intestine.

cell The simplest unit of a living organism.

cell adhesion A vital function of the cell membrane that allows cells to bind to each other. Cell adhesion is critical in the formation of multicellular organisms and provides a way to convey positional information between neighboring cells.

cell adhesion molecule (CAM) A membrane protein found in animal cells that promotes cell adhesion.

cell biology The study of individual cells and their interactions with each other.

cell body A part of a neuron that contains the cell nucleus and other organelles.

cell coat Also called the glycocalyx, the carbohydrate-rich zone on the surface of animal cells that shields the cell from mechanical and physical damage.

cell communication The process through which cells can detect and respond to signals in their extracellular environment. In multicellular organisms, cell communication is also needed to coordinate cellular activities within the whole organism.

cell cycle The series of phases a eukaryotic cell progresses through from its origin until it divides by mitosis.

cell differentiation The phenomenon by which cells become specialized into particular cell types.

cell division The process in which one cell divides into two cells.

cell doctrine *See* cell theory.

cell junctions Specialized structures that adhere cells to each other and to the ECM.

cell-mediated immunity A type of specific immunity in which cytotoxic T cells directly attack and destroy infected body cells, cancer cells, or transplanted cells.

cell nucleus The membrane-bound area of a eukaryotic cell in which the genetic material is found.

cell plate In plant cells, a structure that forms a cell wall between the two daughter cells during cytokinesis.

cell signaling A vital function of the plasma membrane that involves cells sensing changes in their environment and communicating with each other.

cell surface receptor A receptor found in the plasma membrane that enables a cell to respond to different kinds of signaling molecules.

cell theory A theory that states that all organisms are made of cells, cells are the smallest units of living organisms, and new cells come from pre-existing cells by cell division.

cell-to-cell communication A form of cell communication that occurs between two different cells.

cellular differentiation The process by which different cells within a developing organism acquire specialized forms and functions due to the expression of cell-specific genes.

cellular respiration A process by which living cells obtain energy from organic molecules and release waste products.

cellular response Adaptation at the cellular level that involves a cell responding to signals in its environment.

cellulose The main macromolecule of the primary cell wall of plants and many algae; a polymer made of repeating molecules of glucose attached end to end.

cell wall A relatively rigid, porous structure located outside the plasma membrane of prokaryotic, plant, fungal, and certain protist cells; provides support and protection.

centiMorgan (cM) *See* map unit (mu).

central cell In the female gametophyte of a flowering plant, a large cell that contains two nuclei; after double fertilization, it forms the first cell of the nutritive endosperm tissue.

central dogma Refers to the steps of gene expression at the molecular level. DNA is transcribed into mRNA, and mRNA is translated into a polypeptide.

central nervous system (CNS) In vertebrates, the brain and spinal cord.

central region The region of a plant seedling that produces stem tissue.

central vacuole An organelle that often occupies 80% or more of the cell volume of plant cells and stores a large amount of water, enzymes, and inorganic ions.

central zone The area of a plant shoot meristem where undifferentiated stem cells are maintained.

centrioles A pair of structures within the centrosome of animal cells. Most plant cells and many protists lack centrioles.

centromere The region where the two sister chromatids are tightly associated; the centromere is an attachment site for kinetochore proteins.

centrosome A single structure often near the cell nucleus of eukaryotic cells that forms a nucleating site for the growth of microtubules; also called the microtubule-organizing center.

cephalization The localization of a brain and sensory structures at the anterior end of the body of animals.

cephalothorax The fused head and thorax structure in species of the class Arachnida and Crustacea.

cerebellum The part of the vertebrate hindbrain, along with the pons, responsible for monitoring and coordinating body movements.

cerebral cortex The surface layer of gray matter that forms the outer part of the cerebrum of the vertebrate brain.

cerebral ganglia A paired structure in the head of invertebrates that receives input from sensory cells and controls motor output.

cerebrospinal fluid Fluid that exists in ventricles within the central nervous system and surrounds the exterior of the brain and spinal cord; it absorbs physical shocks to the brain resulting from sudden movements or blows to the head.

cerebrum A region of the vertebrate forebrain that is responsible for the higher functions of conscious thought, planning, and emotion, as well as control of motor function.

cervix A fibrous structure at the end of the female vagina that forms the opening to the uterus.

channel A transmembrane protein that forms an open passageway for the direct diffusion of ions or molecules across a membrane.

chaperone A protein that keeps other proteins in an unfolded state during the process of post-translational sorting.

character A characteristic of an organism, such as the appearance of seeds, pods, flowers, or stems.

character displacement The tendency for two species to diverge in morphology and thus resource use because of competition.

character state A particular variant of a given character.

charged tRNA A tRNA with its attached amino acid; also called aminoacyl tRNA.

charophyceans The lineages of freshwater green algae that are most closely related to the land plants.

checkpoint One of three critical regulatory points found in the cell cycle of eukaryotic cells. At these checkpoints, a variety of proteins act as sensors to determine if a cell is in the proper condition to divide.

checkpoint protein A protein that senses if a cell is in the proper condition to divide and prevents a cell from progressing through the cell cycle if it is not.

chemical energy The potential energy contained within covalent bonds in molecules.

chemical equilibrium A state in a chemical reaction in which the rate of formation of products equals the rate of formation of reactants.

chemical mutagen A chemical that causes mutations.

chemical reaction The formation and breaking of chemical bonds, resulting in a change in the composition of substances.

chemical selection Occurs when a chemical within a mixture has special properties or advantages that cause it to increase in amount. May have played a key role in the formation of an RNA world.

chemical synapse A synapse in which a chemical called a neurotransmitter is released from the axon terminal of a neuron and acts as a signal from the presynaptic to the postsynaptic cell.

chemiosmosis A process for making ATP in which energy stored in an ion electrochemical gradient is used to make ATP from ADP and P_i.

chemoautotroph An organism able to use energy obtained by chemical modifications of inorganic compounds to synthesize organic compounds.

chemoheterotroph An organism that must obtain organic molecules both for energy and as a carbon source.

chemoreceptor A sensory receptor in animals that responds to specific chemical compounds.

chiasma The connection at a crossover site of two chromosomes.

chimeric gene A gene formed from the fusion of two gene fragments to each other.

chitin A tough, nitrogen-containing polysaccharide that forms the external skeleton of many insects and the cell walls of fungi.

chlorophyll A photosynthetic green pigment found in the chloroplasts of plants, algae, and some bacteria.

chlorophyll *a* A type of chlorophyll pigment found in plants, algae, and cyanobacteria.

chlorophyll *b* A type of chlorophyll pigment found in plants, green algae, and some other photosynthetic organisms.

chloroplast A semiautonomous organelle found in plant and algal cells that carries out photosynthesis.

chloroplast genome The chromosome found in chloroplasts.

chlorosis The yellowing of plant leaves caused by various types of mineral deficiencies.

choanocyte A specialized cell of sponges that functions to trap and eat small particles.

chondrichthyans Members of the class Chondrichthyes, including sharks, skates, and rays.

chordate An organism that has or at some point in its life has had a notochord and a hollow dorsal nerve cord; includes all vertebrates and some invertebrates.

chorion One of the four extraembryonic membranes in the amniotic egg. It exchanges gases between the embryo and the surrounding air.

chorionic gonadotropin A luteinizing hormone (LH)-like hormone made by the blastocyst and placenta that maintains the corpus luteum.

chromatin Refers to the biochemical composition of chromosomes, which contain DNA and many types of proteins.

chromosome A discrete unit of genetic material composed of DNA and associated proteins. Eukaryotes have chromosomes in their cell nuclei and in plastids and mitochondria.

chromosome territory A distinct, nonoverlapping area where each chromosome is located within the cell nucleus of eukaryotic cells.

chromosome theory of inheritance An explanation of how the steps of meiosis account for the inheritance patterns observed by Mendel.

chylomicron Large fat droplet coated with amphipathic proteins that perform an emulsifying function similar to that of bile salts; chylomicrons are formed in intestinal epithelial cells from absorbed fats in the diet.

chyme A solution of water and partially digested food particles in the stomach and small intestine.

chymotrypsin A protease involved in the breakdown of proteins in the small intestine.

chytrids Simple, early-diverging phyla of fungi; commonly found in aquatic habitats and moist soil, where they produce flagellate reproductive cells.

cilia (singular, **cilium**) Cell appendages that have the same internal structure as flagella and function like flagella to facilitate cell movement; cilia are shorter and more numerous on cells than are flagella.

ciliate A protist that moves by means of cilia, which are tiny hairlike extensions that occur on the outside of cells and have the same internal structure as flagella.

circadian rhythm Internal biological clock system that can be found in plants, animals, and other organisms.

circulatory system A system that transports necessary materials to all cells of an animal's body and transports waste products away from cells. Three basic types are gastrovascular cavities, open systems, and closed systems.

***cis*-acting element** *See cis*-effect.

***cis*-effect** A DNA segment that must be adjacent to the gene(s) that it regulates. The *lac* operator site is an example of a *cis*-acting element.

cisternae Flattened, fluid-filled tubules of the endoplasmic reticulum.

***cis/trans* isomers** Organic molecules with the same chemical composition but existing in two different configurations determined by the positions of hydrogen atoms on the two carbons of a CC double bond. When the hydrogen atoms are on the same side of the double bond, it is called a *cis* isomer; when on the opposite sides of the double bond, it is a *trans* isomer.

citric acid cycle A cycle that results in the breakdown of carbohydrates to CO_2; also known as the Krebs cycle.

clade A group of species derived from a single common ancestor.

cladistic approach An approach used to construct a phylogenetic tree by comparing primitive and shared derived characters.

cladogenesis A pattern of speciation in which a species is divided into two or more species.

cladogram A phylogenetic tree constructed by using a cladistic approach.

clamp connection In basidiomycete fungi, a structure that helps distribute nuclei during cell division.

clasper An extension of the pelvic fin of a chondrichthyan, used by the male to transfer sperm to the female.

class In taxonomy, a subdivision of a phylum.

classical conditioning A type of associative learning in which an involuntary response comes to be associated positively or negatively with a stimulus that did not originally elicit the response.

cleavage A succession of rapid cell divisions with no significant growth that produces a hollow sphere of cells called a blastula.

cleavage furrow In animal cells, an area that constricts like a drawstring to separate the cells during cytokinesis.

climate The prevailing weather pattern of a given region.

climax community A distinct end point of succession.

clitoris Located at the anterior part of the female labia minora, erectile tissue that becomes engorged with blood during sexual arousal and is very sensitive to sexual stimulation.

clonal deletion One of two mechanisms that explain why normal individuals lack active lymphocytes that respond to self components; T cells with receptors capable of binding self proteins are destroyed by apoptosis.

clonal inactivation One of two mechanisms that explain why normal individuals lack active lymphocytes that respond to self components; the process occurs outside the thymus and causes potentially self-reacting T cells to become nonresponsive.

clonal selection The process by which an antigen-stimulated lymphocyte divides and forms a clone of cells, each of which recognizes that particular antigen.

cloning Making many copies of something such as a DNA molecule.

closed circulatory system A circulatory system in which blood flows throughout an animal entirely within a series of vessels and is kept separate from the interstitial fluid.

closed conformation Tightly packed chromatin that cannot be transcribed into RNA.

clumped The most common pattern of dispersion within a population, in which individuals are gathered in small groups.

cnidocil On the surface of a cnidocyte, a hairlike trigger that detects stimuli.

cnidocyte A characteristic feature of cnidarians; a stinging cell that functions in defense or the capture of prey.

coacervates Droplets that form spontaneously from the association of charged polymers such as proteins, carbohydrates, or nucleic acids surrounded by water.

coactivator A protein that increases the rate of transcription but does not directly bind to the DNA itself.

coat protein A protein that surrounds a membrane vesicle and facilitates vesicle formation.

cocci Sphere-shaped prokaryotic cells.

cochlea A coiled structure in the inner ear of mammals that contains the auditory receptors (organ of Corti).
coding sequence The region of a gene or a DNA molecule that encodes the information for the amino acid sequence of a polypeptide.
coding strand The DNA strand opposite to the template (or noncoding strand).
codominance The phenomenon in which a single individual expresses two alleles.
codon A sequence of three nucleotide bases that specifies a particular amino acid or a stop codon; codons function during translation.
coefficient of relatedness (*r*) The probability that any two individuals will share a copy of a particular gene.
coelom A fluid-filled body cavity in an animal.
coelomate An animal with a true coelom.
coenzyme An organic molecule that participates in a chemical reaction with an enzyme but is left unchanged after the reaction is completed.
coevolution The process by which two or more species of organisms influence each other's evolutionary pathway.
cofactor Usually an inorganic ion that temporarily binds to the surface of an enzyme and promotes a chemical reaction.
cognitive learning The ability to solve problems with conscious thought and without direct environmental feedback.
cohesion The ability of like molecules to noncovalently bind to each other; the attraction of water molecules for each other.
cohesion-tension theory The explanation for long-distance water transport as the combined effect of the cohesive forces of water and evaporative tension.
cohort A group of organisms of the same age.
coleoptile A protective sheath that encloses the first bud of the epicotyl in a mature monocot embryo.
coleorhiza A protective envelope that encloses the young root of a monocot.
colinearity rule The phenomenon whereby the order of homeotic genes along the chromosome correlates with their expression along the anteroposterior axis of the body.
collagen A protein secreted from animal cells that forms large fibers in the extracellular matrix.
collecting duct A tubule in the mammalian kidney that collects urine from nephrons.
collenchyma cells Flexible cells that make up collenchyma tissue.
collenchyma tissue A plant ground tissue that provides support to plant organs.
colligative property A property of a solution that depends only on the concentration of solute molecules.
colloid A gel-like substance in the follicles of the thyroid gland.
colon A part of a vertebrate's large intestine consisting of three relatively straight segments—the ascending, transverse, and descending portions. The terminal portion of the descending colon is S-shaped, forming the sigmoid colon, which empties into the rectum.
colony hybridization A method that uses a labeled probe to identify bacterial colonies that contain a desired gene.
combinatorial control The phenomenon whereby a combination of many factors determines the expression of any given gene.
commensalism An interaction that benefits one species and leaves the other unaffected.
communication The use of specially designed visual, chemical, auditory, or tactile signals to modify the behavior of others.
community An assemblage of populations of different species that live in the same place at the same time.
community ecology The study of how populations of species interact and form functional communities.
compartmentalization A characteristic of eukaryotic cells in which many organelles separate the cell into different regions. Cellular compartmentalization allows a cell to carry out specialized chemical reactions in different places.
competent The term used to describe bacterial strains that have the ability to take up DNA from the environment.
competition An interaction that affects two or more species negatively, as they compete over food or other resources.
competitive exclusion principle The proposal that two species with the same resource requirements cannot occupy the same niche.
competitive inhibitor A molecule that binds to the active site of an enzyme and inhibits the ability of the substrate to bind.
complement The family of plasma proteins that provides a means for extracellular killing of microbes without prior phagocytosis.
complementary Describes the specific base pairing that occurs between strands of nucleic acids; A pairs only with T (in DNA) or U (in RNA), and G pairs only with C.
complementary DNA (cDNA) DNA molecules that are made from mRNA as a starting material.
complete flower A flower that possesses all four types of flower organs.
complete metamorphosis During development in the majority of insects, a dramatic change in body form from larva to a very different looking adult.
compound A molecule composed of two or more different elements.
compound eye A type of image-forming eye in arthropods and some annelids consisting of several hundred to several thousand light detectors called ommatidia.
computational molecular biology An area of study that uses computers to characterize the molecular components of living things.
concentration The amount of a solute dissolved in a unit volume of solution.
condensation reaction A chemical reaction in which two or more molecules are combined into one larger molecule by covalent bonding, with the loss of a small molecule.
conditional specification The acquisition by cells of specific properties through a variety of cell-to-cell signaling mechanisms in a developing vertebrate embryo.
conditioned response The learned response that is elicited by a newly conditioned stimulus.
conditioned stimulus A new stimulus that is delivered at the same time as an old stimulus, and that over time, is sufficient to elicit the same response.
condom A sheathlike membrane worn over the penis; in addition to their contraceptive function, condoms significantly reduce the risk of contracting and transmitting sexually transmitted diseases.
conduction The process in which the body surface loses or gains heat through direct contact with cooler or warmer substances.
cone pigment Any of several types of visual pigments found in the cones of the vertebrate eye.
cones 1. Photoreceptors found in the vertebrate eye; they are less sensitive to low levels of light but can detect color. 2. The reproductive structures of coniferous plants.
congenital hypothyroidism A condition characterized by poor differentiation of the central nervous system due to a failure of neurons to become myelinated in fetal development; results in profound mental defects.
congestive heart failure The condition resulting from the failure of the heart to pump blood normally; results in fluid buildup in the lungs (congestion).
conidia A type of asexual reproductive cell produced by many fungi.
conifers A phylum of gymnosperm plants, Coniferophyta.
conjugation A type of genetic transfer between bacteria that involves a direct physical interaction between two bacterial cells.
connective tissue Clusters of cells that connect, anchor, and support the structures of an animal's body; includes blood, adipose (fat-storing) tissue, bone, cartilage, loose connective tissue, and dense connective tissue.
connexon A channel that forms gap junctions consisting of six connexin proteins in one cell aligned with six connexin proteins in an adjacent cell.
conservation biology The study that uses principles and knowledge from molecular biology, genetics, and ecology to protect the biological diversity of life.
conservative mechanism In this incorrect model for DNA replication, both parental strands of DNA remain together (are conserved) following DNA replication. The two newly made daughter strands also occur together.
consortia A community of many microbial species.
constant region The portions of amino acid sequences in the heavy and light chains that are identical for all immunoglobulins of a given class.
constitutive gene An unregulated gene that has constant levels of expression in all conditions over time.
contig A series of clones that contain overlapping pieces of chromosomal DNA.
continental drift The process by which, over the course of billions of years, the major landmasses, known as the continents, have shifted their positions, changed their shapes, and, in some cases, have become separated from each other.
contraception The use of birth control procedures to prevent fertilization or implantation of a fertilized egg.
contractile vacuole A small, membrane enclosed, water-filled compartment that eliminates excess liquid from the cells of certain protists.
contrast In microscopy, relative differences in the lightness, darkness, or color between adjacent regions in a sample.
control group The sample in an experiment that is treated just like an experimental group except that it is not subjected to one particular variable.
convection The transfer of heat by the movement of air or water next to the body.
convergent evolution The process whereby two different species from different lineages show similar characteristics because they occupy similar environments.
convergent extension A cellular process during gastrulation that is crucial to development; two rows of cells merge to form a single elongated layer.
convergent trait *See* analogous structure.
coprophagy The practice of certain birds and mammals in which feces are consumed to maximize absorption of water and nutrients.
copulation The process of sperm being deposited within the reproductive tract of the female.
corepressor A small effector molecule that binds to a repressor protein to inhibit transcription.
core promoter Refers to the TATA box and the transcriptional start site of a eukaryotic structural gene.
Coriolis effect The effect of the Earth's rotation on the surface flow of wind.
cork cambium A secondary meristem in a plant that produces cork tissue.

cornea A thin, clear layer on the front of the vertebrate eye.
corolla The petals of a flower, which occur in the whorl to the inside of the calyx and the outside of the stamens.
corona The ciliated crown of members of the phylum Rotifera.
coronary artery An artery that carries oxygen and nutrients to the heart muscle.
coronary artery bypass A common treatment to restore blood flow through a coronary artery. A small piece of healthy blood vessel is removed from one part of the body and surgically grafted onto the coronary circulation in order to bypass the diseased artery.
coronary artery disease A condition that occurs when plaques form in the coronary arteries.
corpus callosum The major tract that connects the two hemispheres of the cerebrum.
corpus luteum A structure that develops from a ruptured follicle following ovulation; it is responsible for secreting hormones that stimulate the development of the uterus during pregnancy.
correlation A meaningful relationship between two variables.
cortex The area of a plant stem or root beneath the epidermis that is largely composed of parenchyma tissue.
cortical reaction An event in fertilization in which IP_3 and calcium signaling produces barriers to more than one sperm cell binding to and uniting with an egg; called the slow block to polyspermy.
cortisol A steroid hormone made by the adrenal cortex.
cotranslational sorting The sorting process in which the synthesis of certain eukaryotic proteins begins in the cytosol and then halts temporarily until the ribosome has become bound to the ER membrane.
cotransporter *See* symporter.
cotyledon An embryonic seed leaf.
countercurrent exchange 1. An arrangement of water and blood flow in which water enters a fish's mouth and flows between the lamellae of the gills in the opposite direction to blood flowing through the lamellar capillaries. 2. An arrangement of blood vessels under the skin of vertebrates that contributes to heat retention.
countercurrent multiplication system The mechanism by which the vertebrate loop of Henle reabsorbs salts and water along it length.
covalent bond A chemical bond in which two atoms share a pair of electrons.
CpG island A cluster of CpG sites. CG refers to the nucleotides of C and G in DNA, and p refers to a phosphodiester linkage.
cranial nerve A nerve in the peripheral nervous system that is directly connected to the brain.
cranium A protective bony or cartilaginous housing that encases the brain of a craniate.
crenation The process of cell shrinkage that occurs if animal cells are placed in a hypertonic medium.
cristae Projections of the highly invaginated inner membrane of a mitochondrion.
critical innovations New features that foster the diversification of phyla.
critical period A limited period of time in which many animals develop species-specific patterns of behavior.
crop A storage organ that is a dilation of the lower esophagus; found in most birds and many invertebrates, including insects and some worms.
cross-bridge A region of myosin molecules that extend from the surface of the thick filaments toward the thin filaments in skeletal muscle.
cross-bridge cycle During muscle contraction, the sequence of events that occurs between the time when a cross-bridge binds to a thin filament and when it is set to repeat the process.
cross-fertilization Fertilization that involves the union of a female gamete and a male gamete from different individuals.
crossing over The exchange of genetic material between homologous chromosomes during meiosis; allows for increased variation in the genetic information that each parent may pass to the offspring.
cross-pollination The process in which a stigma receives pollen from a different plant of the same species.
cryptic coloration The blending of an organism with the background color of its habitat; also known as camouflage.
cryptochrome A type of blue-light receptor in plants and protists.
cryptomycota A newly discovered group of eukaryotes proposed to be an early-diverging lineage of fungi.
C-terminus The location of the last amino acid in a polypeptide; also known as the carboxyl terminus.
CT scan Computerized tomography, which is an X-ray technique used to examine the structure of bones and soft tissues, including the brain.
cupula A gelatinous structure within the lateral line organ of fishes that detects changes in water movement.
cuticle A coating of wax and cutin that helps to reduce water loss from plant surfaces. Also, a nonliving covering that serves to both support and protect an animal.
cycads A phylum of gymnosperm plants, Cycadophyta.
cyclic adenosine monophosphate (cAMP) A small effector molecule that acts as a second messenger and is produced from ATP.
cyclic AMP (cAMP) *See* cyclic adenosine monophosphate.
cyclic electron flow *See* cyclic photophosphorylation.
cyclic photophosphorylation During photosynthesis, a pattern of electron flow in the thylakoid membrane that is cyclic and generates ATP alone.
cyclin A protein responsible for advancing a cell through the phases of the cell cycle by binding to a cyclin-dependent kinase.
cyclin-dependent kinase (cdk) A protein responsible for advancing a cell through the phases of the cell cycle. Its function is dependent on the binding of a cyclin.
cyst A one-to-few celled structure that often has a thick, protective wall and can remain dormant through periods of unfavorable climate or low food availability.
cytogenetics The field of genetics that involves the microscopic examination of chromosomes.
cytokines A family of proteins that function in both nonspecific and specific immune defenses by providing a chemical communication network that synchronizes the components of the immune response.
cytokinesis The division of the cytoplasm to produce two distinct daughter cells.
cytokinin A type of plant hormone that promotes cell division.
cytoplasm The region of the cell that is contained within the plasma membrane.
cytoplasmic inheritance *See* extranuclear inheritance.
cytosine (C) A pyrimidine base found in DNA and RNA.
cytoskeleton In eukaryotes, a network of three different types of protein filaments in the cytosol called microtubules, intermediate filaments, and actin filaments.
cytosol The region of a eukaryotic cell that is inside the plasma membrane and outside the organelles.
cytotoxic T cell A type of lymphocyte that travels to the location of its target, binds to the target by combining with an antigen on it, and directly kills the target via secreted chemicals.

D

dalton (Da) A measure of atomic mass. One dalton equals one-twelfth the mass of a carbon atom.
data mining The extraction of useful information and often previously unknown relationships from sequence files and large databases.
database A large number of computer data files that are collected, stored in a single location, and organized for rapid search and retrieval.
daughter strand The newly made strand in DNA replication.
day-neutral plant A plant that flowers regardless of the night length, as long as day length meets the minimal requirements for plant growth.
deafness Hearing loss, usually caused by damage to the hair cells within the cochlea.
death-inducing signaling complex (DISC) A complex consisting of death receptors, adaptor proteins, and procaspase that initiates apoptosis via the extrinsic pathway.
death receptor A type of receptor found in the plasma membrane of eukaryotic cells that can promote apoptosis when it becomes activated.
decomposer A consumer that gets its energy from the remains and waste products of other organisms.
defecation The expulsion of feces that occurs through the anus of an animal's digestive canal.
defensive mutualism A mutually beneficial interaction often involving an animal defending a plant or herbivore in return for food or shelter.
deforestation The conversion of forested areas by humans to nonforested land.
degenerate In the genetic code, the observation that more than one codon can specify the same amino acid.
dehydration A reduction in the amount of water in the body.
dehydration reaction A type of condensation reaction in which a molecule of water is lost.
delayed implantation A reproductive cycle in which a fertilized egg reaches the uterus but does not implant until later, when environmental conditions are more favorable for the newly produced young.
delayed ovulation A reproductive cycle in which the ovarian cycle in females is halted before ovulation and sperm are stored and nourished in the female's uterus over the winter. Upon arousal from hibernation in the spring, the female ovulates one or more eggs, which are fertilized by the stored sperm.
deletion A type of mutation in which a segment of genetic material is missing.
demographic transition The shift in birth and death rates accompanying human societal development.
demography The study of birth rates, death rates, age distributions, and the sizes of populations.
dendrite A treelike extension of the plasma membrane of a neuron that receives electrical signals from other neurons.
dendritic cell A type of cell derived from bone marrow stem cells that plays an important role in nonspecific immunity; these cells are scattered throughout most tissues, where they perform various macrophage functions.
denitrification The reduction of nitrate (NO_3^-) to gaseous nitrogen (N_2).
density In the context of populations, the numbers of organisms in a given unit area.

density-dependent factor A mortality factor whose influence increases with the density of the population.
density-independent factor A mortality factor whose influence is not affected by changes in population density.
deoxynucleoside triphosphates Individual nucleotides with three phosphate groups.
deoxyribonucleic acid (DNA) One of two classes of nucleic acids; the other is ribonucleic acid (RNA). A DNA molecule consists of two strands of nucleotides coiled around each other to form a double helix, held together by hydrogen bonds according to the AT/GC rule.
deoxyribose A five-carbon sugar found in DNA.
depolarization The change in the membrane potential that occurs when a cell becomes less polarized, that is, less negative relative to the surrounding fluid.
dermal tissue The covering on various parts of a plant.
descent with modification Darwin's theory that existing life-forms on our planet are the product of the modification of pre-existing life-forms.
desertification The overstocking of land with domestic animals that can greatly reduce grass coverage through overgrazing, turning the area more desert-like.
desmosome A mechanically strong cell junction between animal cells that typically occurs in spotlike rivets.
determinate cleavage In animals, a characteristic of protostome development in which the fate of each embryonic cell is determined very early.
determinate growth A type of growth in plants that is of limited duration, such as the growth of flowers.
determined The term used to describe a cell that is destined to differentiate into a particular cell type.
detritivore *See* decomposer.
detritus Unconsumed plants that die and decompose, along with the dead remains of animals and animal waste products.
deuterostome An animal whose development exhibits radial, indeterminate cleavage and in which the blastopore becomes the anus; includes echinoderms and vertebrates.
development In biology, a series of changes in the state of a cell, tissue, organ, or organism; the underlying process that gives rise to the structure and function of living organisms.
developmental genetics A field of study aimed at understanding how gene expression controls the process of development.
diaphragm A large muscle that subdivides the thoracic cavity from the abdomen in mammals; contraction of the diaphragm enlarges the thoracic cavity during inhalation.
diarrhea A common intestinal disorder arising from ingested microbes or other causes; usually runs its course within one or two days but, in serious cases, can require hospitalization.
diastole The phase of the cardiac cycle in which the ventricles fill with blood coming from the atria through the open AV valves.
diazotroph A bacterium that fixes nitrogen.
dideoxy chain-termination method The most common method of DNA sequencing; utilizes dideoxynucleotides as a reagent.
differential gene regulation The phenomenon in which the expression of genes differs under various environmental conditions and in specialized cell types.
diffusion In a solution, the process that occurs when a solute moves from a region of high concentration to a region of lower concentration.
digestion The process of breaking down nutrients in food into smaller molecules that can be absorbed across the intestinal epithelia and directly used by cells.
digestive system In animals, the long tube through which food is processed. In a vertebrate, this system consists of the alimentary canal plus several associated structures.
dihybrid An offspring that is a hybrid with respect to two traits.
dihybrid cross A cross in which the inheritance of two different traits is followed.
dikaryotic The occurrence of two genetically distinct nuclei in the cells of fungal hyphae after mating has occurred.
dikaryotic mycelium A fungal body that is made of cells that each possess two genetically distinct nuclei.
dimorphic fungi Fungi that can exist in two different morphological forms.
dinosaur A term, meaning "terrible lizard," used to describe some of the extinct fossil reptiles.
dioecious The term to describe plants that produce staminate and carpellate flowers on separate plants.
diploblastic Having two distinct germ layers—ectoderm and endoderm—but not mesoderm.
diploid Refers to cells containing two sets of chromosomes; designated as $2n$.
diploid-dominant species Species in which the diploid organism is the prevalent organism in the life cycle. Animals are an example.
direct calorimetry A method of determining basal metabolic rate that involves quantifying the amount of heat generated by the animal.
direct repair Refers to a DNA repair system in which an enzyme finds an incorrect structure in the DNA and directly converts it back to the correct structure.
directionality In a DNA or RNA strand, refers to the orientation of the sugar molecules within that strand. Can be 5′ to 3′ or 3′ to 5′.
directional selection A pattern of natural selection that favors individuals at one extreme of a phenotypic distribution.
disaccharide A carbohydrate composed of two monosaccharides.
discovery-based science The collection and analysis of data without the need for a preconceived hypothesis; also called discovery science.
discrete trait A trait with clearly defined phenotypic variants.
dispersion A pattern of spacing in which individuals in a population are clustered together or spread out to varying degrees.
dispersive mechanism In this incorrect model for DNA replication, segments of parental DNA and newly made DNA are interspersed in both strands following the replication process.
dispersive mutualism A mutually beneficial interaction often involving plants and pollinators that disperse their pollen, and plants and fruit eaters that disperse the plant's seeds.
dissociation constant An equilibrium constant between a ligand and a protein, such as a receptor or an enzyme.
distal tubule The segment of the tubule of the nephron through which fluid flows into one of the many collecting ducts in the kidney.
disulfide bridge Covalent chemical bond formed between two cysteine residues in a protein; important in the tertiary structure of proteins.
diversifying selection A pattern of natural selection that favors the survival of two or more different genotypes that produce different phenotypes.
diversity-stability hypothesis The proposal that species-rich communities are more stable than those with fewer species.
DNA (deoxyribonucleic acid) The genetic material that provides a blueprint for the organization, development, and function of living things.
DNA fingerprinting A technology that identifies particular individuals using properties of their DNA.
DNA helicase An enzyme that uses ATP to separate DNA strands during DNA replication.
DNA library A collection of recombinant vectors, each containing a particular fragment of chromosomal DNA (cDNA).
DNA ligase An enzyme that catalyzes the formation of a covalent bond between nucleotides in adjacent DNA fragments to complete the replication process.
DNA methylase An enzyme that attaches methyl groups to bases in DNA.
DNA methylation A process in which methyl groups are attached to bases in DNA.
DNA microarray A technology used to monitor the expression of thousands of genes simultaneously.
DNA polymerase An enzyme responsible for covalently linking nucleotides together during DNA replication.
DNA primase An enzyme that synthesizes a primer for DNA replication.
DNA repair systems One of several systems to reverse DNA damage before a permanent mutation can occur.
DNA replication The process by which DNA is copied.
DNase An enzyme that digests DNA.
DNA sequencing A method to determine the base sequence of DNA.
DNA supercoiling A method of compacting chromosomes through the formation of additional coils around the long, thin DNA molecule.
DNA topoisomerase An enzyme that alleviates DNA supercoiling during DNA replication.
DNA transposon A type of transposable element that moves as a DNA molecule.
domain 1. A defined region of a protein with a distinct structure and function. 2. One of the three major categories of life: Bacteria, Archaea, and Eukarya.
domestication A process that involves artificial selection of plants or animals for traits desirable to humans.
dominant A term that describes the displayed trait in a heterozygote.
dominant species A species that has a large effect in a community because of its high abundance or high biomass.
dormancy A phase of metabolic slowdown in a plant.
dorsal Refers to the upper side of an animal.
dorsoventral axis In bilateral animals, one of the three axes along which the adult body pattern is organized; the others are the anteroposterior axis and the right-left axis.
dosage compensation The phenomenon that gene dosage is compensated between males and females. In mammals, the inactivation of one X chromosome in the female reduces the number of expressed copies (doses) of X-linked genes from two to one.
double bond A bond that occurs when the atoms of a molecule share two pairs of electrons.
double fertilization In angiosperms, the process in which two different fertilization events occur, producing both a zygote and the first cell of a nutritive endosperm tissue.
double helix Two strands of DNA hydrogen-bonded with each other. In a DNA double helix, two DNA strands are twisted together to form a structure that resembles a spiral staircase.
Down syndrome A human disorder caused by the inheritance of three copies of chromosome 21.
Duchenne muscular dystrophy An inherited, X-linked disorder of humans causing muscle weakness and muscle degeneration.

duodenum The first part of the vertebrate small intestine arising from the stomach.
duplication A type of mutation in which a section of a chromosome occurs two or more times.
dynamic instability The oscillation of a single microtubule between growing and shortening phases; important in many cellular activities, including the sorting of chromosomes during cell division.

E

Ecdysis The process by which an animal molts, or breaks out of its old exoskeleton, and secretes a newer, larger one.
Ecdysozoa A clade of molting animals that encompasses primarily the arthropods and nematodes.
echolocation The phenomenon in which certain species listen for echoes of high-frequency sound waves in order to determine the distance and location of an object.
ECM *See* extracellular matrix.
ecological footprint The amount of productive land needed to support each person on Earth.
ecological species concept An approach used to distinguish species; considers a species within its native environment and states that each species occupies its own ecological niche.
ecology The study of interactions among organisms and between organisms and their environments.
ecosystem The biotic community of organisms in an area as well as the abiotic environment affecting that community.
ecosystem engineer A keystone species that creates, modifies, and maintains habitats.
ecosystems ecology The study of the flow of energy and cycling of nutrients among organisms within a community and between organisms and the environment.
ecotypes Genetically distinct populations adapted to their local environments.
ectoderm In animals, the outermost layer of cells formed during gastrulation that covers the surface of the embryo and differentiates into the epidermis and nervous system.
ectomycorrhizae Beneficial interactions between temperate forest trees and soil fungi that coat their roots.
ectoparasite A parasite that lives on the outside of the host's body.
ectotherm An animal that largely depends on the environment to warm its body.
ectothermic Referring to an animal that is an ectotherm.
edge effect A special physical condition that exists at the boundary or edge of an area.
effective population size The number of individuals that contribute genes to future populations, often smaller than the actual population size.
effector A molecule that directly influences cellular responses.
effector cell A cloned lymphocyte that carries out the attack response during specific immunity.
efferent arteriole A blood vessel that carries blood away from a glomerulus of the vertebrate kidney.
egestion In animals, the process of undigested material passing out of the body.
egg The mature female gamete; also called an ovum.
ejaculation The movement of semen through the urethra by contraction of muscles at the base of the penis.
ejaculatory duct The structure in the male reproductive system through which sperm leave the vas deferens and enter the urethra.
elastin A protein that makes up elastic fibers in the extracellular matrix of animals.
electrical synapse A synapse that directly passes electric current from the presynaptic to the postsynaptic cell via gap junctions.
electrocardiogram (ECG) A record of the electrical impulses generated by the cells of the heart during the cardiac cycle.
electrochemical gradient The combined effect of both an electrical and chemical gradient across a membrane; determines the direction that an ion will move.
electrogenic pump A pump that generates an electrical gradient across a membrane.
electromagnetic receptor A sensory receptor in animals that detects radiation within a wide range of the electromagnetic spectrum, including visible, ultraviolet, and infrared light, as well as electrical and magnetic fields in some animals.
electromagnetic spectrum All possible wavelengths of electromagnetic radiation, from relatively short wavelengths (gamma rays) to much longer wavelengths (radio waves).
electron A negatively charged particle found in orbitals around an atomic nucleus.
electron microscope A microscope that uses an electron beam for illumination.
electron shell The region around an atom's nucleus where electrons reside; larger atoms have more electron shells than smaller atoms.
electron transport chain (ETC) A group of protein complexes and small organic molecules within the inner membranes of mitochondria and chloroplasts and the plasma membrane of prokaryotes. The components accept and donate electrons to each other in a linear manner and produce a H^+ electrochemical gradient.
electronegativity A measure of an atom's ability to attract electrons to its outer shell from another atom.
element A substance composed of specific types of atoms that cannot be further broken down by ordinary chemical or physical means.
elicitor A compound produced by bacterial and fungal pathogens that promotes virulence.
elongation factor A protein that is needed for the growth of a polypeptide during translation.
elongation stage The second step in transcription or translation where RNA strands or polypeptides are made, respectively.
embryo The early stages of development in a multicellular organism during which the organization of the organism is largely formed.
embryogenesis The process by which embryos develop from single-celled zygotes by mitotic divisions.
embryonic development The process by which a fertilized egg is transformed into an organism with distinct physiological systems and body parts.
embryonic germ cell (EG cell) A cell in the early mammalian embryo that later gives rise to sperm or egg cells. These cells are pluripotent.
embryonic stem cell (ES cell) A cell in the early mammalian embryo that can differentiate into almost every cell type of the body. These cells are pluripotent.
embryophyte A synonym for the land plants.
emerging virus A newly arising virus.
emphysema A progressive disease characterized by a loss of elastic recoil ability of the lungs, usually resulting from chronic tobacco smoking.
empirical thought Thought that relies on observation to form an idea or hypothesis, rather than trying to understand life from a nonphysical or spiritual point of view.
emulsification A process during digestion that disrupts large lipid droplets into many tiny droplets, thereby increasing their total surface area and exposure to lipase action.
enantiomer One of a pair of stereoisomers that exist as mirror images.
endangered species Those species that are in danger of extinction throughout all or a significant portion of their range.
endemic The term to describe organisms that are naturally found only in a particular location.
endergonic Refers to chemical reactions that require an addition of free energy and do not proceed spontaneously.
endocrine disruptor A chemical found in polluted water and soil that resembles a natural hormone; a common example are chemicals that resemble estrogen and can bind to estrogen receptors in animals.
endocrine glands Structures that contain epithelial cells that secrete hormones into the bloodstream, where they circulate throughout the body.
endocrine system All the endocrine glands and other organs containing hormone-secreting cells.
endocytosis A process in which the plasma membrane invaginates, or folds inward, to form a vesicle that brings substances into the cell.
endoderm In animals, the innermost layer of cells formed during gastrulation; lines the gut and gives rise to many internal organs.
endodermis In vascular plants, a thin cylinder of root tissue that forms a barrier between the root cortex and the central core of vascular tissue.
endomembrane system A network of membranes that includes the nuclear envelope, the endoplasmic reticulum, Golgi apparatus, lysosomes, vacuoles, and plasma membrane.
endomycorrhizae Partnerships between plants and fungi in which the fungal hyphae grow into the spaces between root cell walls and plasma membranes.
endoparasite A parasite that lives inside the host's body.
endophyte A mutualistic fungus that lives compatibly within the tissues of various types of plants.
endoplasmic reticulum (ER) A convoluted network of membranes in a cell's cytoplasm that forms flattened, fluid-filled tubules or cisternae.
endoskeleton An internal hard skeleton covered by soft tissue; present in echinoderms and vertebrates.
endosperm A nutritive tissue that increases the efficiency with which food is stored and used in the seeds of flowering plants.
endospore A cell with a tough coat that is produced in certain bacteria and then released when the enclosing bacterial cell dies and breaks down.
endosporic gametophyte A plant gametophyte that grows within the confines of microspore or megaspore walls.
endosymbiont A smaller species that lives within a larger species in a symbiotic relationship.
endosymbiosis A symbiotic relationship in which the smaller species—the symbiont—lives inside the larger species.
endosymbiosis theory A theory that mitochondria and chloroplasts originated from bacteria that took up residence within a primordial eukaryotic cell.
endosymbiotic Describes a relationship in which one organism lives inside the other.
endothelium The single-celled inner layer of a blood vessel; forms a smooth lining in contact with the blood.
endotherm An animal that generates most of its body heat by metabolic processes.
endothermic Referring to an animal that is an endotherm.
energy The ability to promote change or to do work.
energy flow The movement of energy through an ecosystem.

energy intermediate A molecule such as ATP or NADH that stores energy and is used to drive endergonic reactions in cells.
energy shell In an atom, an energy level of electrons occupied by one or more orbitals; each energy level is a characteristic distance from the nucleus, with outer shells having more energy than inner shells.
enhancement effect The phenomenon in which maximal activation of the pigments in photosystems I and II is achieved when organisms are exposed to two wavelengths of light.
enhancer A response element in eukaryotes that increases the rate of transcription.
enthalpy (*H*) The total energy of a system.
entomology The study of insects.
entropy The degree of disorder of a system.
environmental science The application of ecology to real-world problems.
enzyme A protein that acts as a catalyst to speed up a chemical reaction in a cell.
enzyme-linked receptor A receptor found in all living species that typically has two important domains: an extracellular domain, which binds a signaling molecule, and an intracellular domain, which has a catalytic function.
enzyme-substrate complex The binding between an enzyme and its substrate.
eosinophil A type of phagocyte found in large numbers in mucosal surfaces lining the gastrointestinal, respiratory, and urinary tracts, where they fight off parasitic infections.
epicotyl The portion of an embryonic plant stem with two tiny leaves in a first bud; located above the point of attachment of the cotyledons.
epidermis A layer of dermal tissue that helps protect a plant from damage.
epididymis A coiled, tubular structure located on the surface of the testis in which sperm complete their differentiation.
epigenetic inheritance An inheritance pattern in which modification of a gene or chromosome during egg formation, sperm formation, or early stages of embryonic growth alters gene expression in a way that is fixed during an individual's lifetime.
epinephrine A hormone secreted by the adrenal glands; also known as adrenaline.
episome A plasmid that can integrate into a bacterial chromosome.
epistasis A gene interaction in which the alleles of one gene mask the expression of the alleles of another gene.
epithalamus A region of the vertebrate forebrain that includes the pineal gland.
epithelial tissue In animals, a sheet of densely packed cells that covers the body, covers individual organs, and lines the walls of various cavities inside the body.
epitope Antigenic determinant; the peptide fragments of an antigen that are complexed to MHC proteins and presented to a helper T cell.
equilibrium 1. In a chemical reaction, occurs when the rate of the forward reaction is balanced by the rate of the reverse reaction. 2. In a population, the situation in which the population size stays the same.
equilibrium model of island biogeography A model to explain the process of succession on new islands; states that the number of species on an island tends toward an equilibrium number that is determined by the balance between immigration rates and extinction rates.
equilibrium potential In membrane physiology, the membrane potential at which the flow of an ion is at equilibrium, with no net movement in either direction.
ER lumen A single compartment enclosed by the ER membrane.
ER signal sequence A sorting signal in a polypeptide usually located near the amino terminus that is recognized by SRP (signal recognition particle) and directs the polypeptide to the ER membrane.
erythrocyte A cell that serves the critical function of transporting oxygen throughout an animal's body; also known as a red blood cell.
erythropoietin (EPO) A hormone made by the liver and kidneys in response to any situation where additional red blood cells are required.
E site *See* exit site.
esophagus In animals, the tubular structure that forms a pathway from the pharynx to the stomach.
essential amino acid An amino acid that is required in the diet of particular organisms.
essential fatty acid A polyunsaturated fatty acid, such as linoleic acid, that cannot be synthesized by animal cells and must therefore be consumed in the diet.
essential element In plants, a chemical element that is required for metabolism, sometimes by functioning as an enzyme cofactor.
essential nutrient In animals, a compound that cannot be synthesized from any ingested or stored precursor molecule and so must be obtained in the diet in its complete form. In plants, those substances needed to complete reproduction while avoiding the symptoms of nutrient deficiency.
estradiol The major estrogen in many vertebrates, including humans.
estrogens Steroid hormones produced by the ovaries that affect most aspects of female reproduction.
ethology Scientific studies of animal behavior.
ethylene A plant hormone that is particularly important in coordinating plant developmental and stress responses.
euchromatin The less condensed regions of a chromosome; areas that are capable of gene transcription.
eudicots One of the two largest lineages of flowering plants in which the embryo possesses two seed leaves.
Eukarya One of the three domains of life; the other two are Bacteria and Archaea.
eukaryote One of the two categories into which all forms of life can be placed. The distinguishing feature of eukaryotes is cell compartmentalization, including a cell nucleus; includes protists, fungi, plants, and animals.
eukaryotic Refers to organisms having cells with internal compartments that serve various functions; includes all members of the domain Eukarya.
Eumetazoa A subgroup of animals having more than one type of tissue and, for the most part, different types of organs.
euphyll A leaf with branched veins.
euphyllophytes The clade that includes pteridophytes and seed plants.
euploid An organism that has a chromosome number that is a multiple of a chromosome set ($1n$, $2n$, $3n$, etc.).
eusociality An extreme form of altruism in social insects in which the vast majority of females, known as workers, do not reproduce. Instead, they help one reproductive female (the queen) raise offspring.
Eustachian tube In mammals, a connection from the middle ear to the pharynx; maintains the pressure in the middle ear at atmospheric pressure.
eustele In plants, a ring of vascular tissue arranged around a central pith of nonvascular tissue; typical of progymnosperms, gymnosperms, and angiosperms.
eutherian A placental mammal and member of the subclass Eutheria.
eutrophic Waters that contain relatively high levels of nutrients such as phosphate or nitrogen and typically exhibit high levels of primary productivity and low levels of biodiversity.
eutrophication The process by which elevated nutrient levels in a body of water lead to an overgrowth of algae or aquatic plants and a subsequent depletion of water oxygen levels when these photosynthesizers decay.
evaporation The transformation of water from the liquid to the gaseous state at normal temperatures. Animals use evaporation as a means of losing excess body heat.
evapotranspiration rate The rate at which water moves into the atmosphere through the processes of evaporation from the soil and transpiration of plants.
evolution The phenomenon that populations of organisms change from one generation to the next. As a result, some organisms become more successful at survival and reproduction.
evolutionarily conserved The term used to describe homologous DNA sequences that are very similar or identical between different species.
evolutionary developmental biology (evo-devo) A field of biology that compares the development of different organisms in an attempt to understand ancestral relationships between organisms and the developmental mechanisms that bring about evolutionary change.
evolutionary lineage concept An approach used to distinguish species; states that a species is derived from a single distinct lineage and has its own evolutionary tendencies and historical fate.
excitable cell The term used to describe neurons and muscle cells because they have the capacity to generate electrical signals.
excitation-contraction coupling The sequence of events by which an action potential in the plasma membrane of a muscle fiber leads to cross-bridge activity.
excitatory postsynaptic potential (EPSP) The response from an excitatory neurotransmitter that depolarizes the postsynaptic membrane; the depolarization brings the membrane potential closer to the threshold potential that would trigger an action potential.
excretion In animals, the process of expelling waste or harmful materials from the body.
exercise Any physical activity that increases an animal's metabolic rate.
exergonic Refers to chemical reactions that release free energy and occur spontaneously.
exit site (E site) One of three sites for tRNA binding in the ribosome during translation; the other two are the peptidyl site (P site) and the aminoacyl site (A site). The uncharged tRNA exits from the E site.
exocrine gland A gland in which epithelial cells secrete chemicals into a duct, which carries those molecules directly to another structure or to the outside surface of the body.
exocytosis A process in which material inside a cell is packaged into vesicles and excreted into the extracellular medium.
exon A portion of RNA that is found in the mature mRNA molecule after splicing is finished.
exon shuffling A form of mutation in which exons and their flanking introns are inserted into genes and thereby create proteins with additional functional domains.
exonuclease An enzyme that cleaves off nucleotides, one at a time, from the end of a DNA or RNA molecule.
exoskeleton An external skeleton made of chitin and protein that surrounds and protects most of the body surface of animals such as insects.
exosome A multiprotein complex that degrades mRNA.

expansin A protein that occurs in the plant cell wall and fosters cell enlargement.
experimental group The sample in an experiment that is subjected to some type of variation that does not occur for the control group.
exploitation competition Competition in which organisms compete indirectly through the consumption of a limited resource.
exponential growth Rapid population growth that occurs when the per capita growth rate remains above zero.
extensor A muscle that straightens a limb at a joint.
external fertilization Fertilization that occurs in aquatic environments, when eggs and sperm are released into the water in close enough proximity for fertilization to occur.
external intercostal muscles Muscles of the rib cage that contract during inhalation, thereby expanding the chest.
extinction The end of the existence of a species or a group of species.
extinction vortex A downward spiral toward extinction from which a species cannot naturally recover.
extracellular fluid The fluid in an organism's body that is outside of the cells.
extracellular matrix (ECM) A network of material that is secreted from animal cells and forms a complex meshwork outside of cells. The ECM provides strength, support, and organization.
extranuclear inheritance In eukaryotes, the transmission of genes that are located outside the cell nucleus.
extremophile An organism that occurs primarily in extreme habitats.
eye The visual organ in animals that detects light and sends signals to the brain.
eyecup An eye in planaria that detects light and its direction but which does not form an image.

F

facilitated diffusion A method of passive transport of solutes down a concentration gradient that involves the aid of a transport protein.
facilitation A mechanism for succession in which a species facilitates or makes the environment more suitable for subsequent species.
facultative anaerobe A microorganism that can use oxygen in aerobic respiration, obtain energy via anaerobic fermentation, or use inorganic chemical reactions to obtain energy.
facultative mutualism An interaction between mutualistic species that is beneficial but not essential to the survival and reproduction of either species.
family In taxonomy, a subdivision of an order.
fast block to polyspermy A depolarization of the egg that blocks other sperm from binding to the egg membrane proteins.
fast fiber A skeletal muscle fiber containing myosin with a high rate of ATP hydrolysis.
fast-glycolytic fiber A skeletal muscle fiber that has high myosin ATPase activity but cannot make as much ATP as oxidative fibers because its source of ATP is glycolysis; best suited for rapid, intense actions.
fast-oxidative fiber A skeletal muscle fiber that has high myosin ATPase activity and can make large amounts of ATP; used for long-term activities.
fate The ultimate morphological features that a cell or a group of cells will adopt.
fate mapping A technique in which a small population of cells within an embryo is specifically labeled with a harmless dye, and the fate of these labeled cells is followed to a later stage of embryonic development.
feedback inhibition A form of regulation in which the product of a metabolic pathway inhibits an enzyme that acts early in the pathway, thus preventing the overaccumulation of the product.
feedforward regulation The process by which an animal's body begins preparing for a change in some variable before it even occurs.
female-enforced monogamy hypothesis The hypothesis that a male is monogamous due to various actions employed by his female mate.
female gametophyte A haploid multicellular plant generation that produces one or more eggs but does not produce sperm cells.
fermentation The breakdown of organic molecules to produce energy without any net oxidation of an organic molecule.
fertilization The union of two gametes, such as an egg cell with a sperm cell, to form a zygote.
fertilizer A soil addition that enhances plant growth by providing essential elements.
fetal alcohol syndrome The physical and mental defects that may arise in a child born to a woman who consumes alcohol during pregnancy.
fetus The maturing embryo after the eighth week of gestation in humans.
fever An increase in an animal's temperature, typically due to infection.
F factor A type of bacterial plasmid called a fertility factor that plays a role in bacterial conjugation.
F_1 generation The first generation in a genetic cross.
F_2 generation The second generation in a genetic cross.
fiber A type of tough-walled plant cell that provides support.
fibrin A protein that forms a meshwork of threadlike fibers that wrap around and between platelets and blood cells, enlarging and thickening a blood clot.
fibrous root system The root system of monocots, which consists of multiple adventitious roots that grow from the stem base.
fight-or-flight The response of vertebrates to real or perceived danger; associated with increased activity of the sympathetic branch of the autonomic nervous system.
filament 1. The elongate portion of a flower's stamen; contains vascular tissue that delivers nutrients from parental sporophytes to anthers. 2. In fishes, a part of the gills.
filtrate In the process of filtration in an excretory system, the material that passes through the filter and enters the excretory organ for either further processing or excretion.
filtration The passive removal of water and small solutes from the blood during the production of urine.
finite rate of increase In ecology, the ratio of a population size from one year to the next.
first law of thermodynamics States that energy cannot be created or destroyed; also called the law of conservation of energy.
fitness The relative likelihood that a genotype will contribute to the gene pool of the next generation as compared with other genotypes.
5′ cap The 7-methylguanosine cap structure at the 5′ end of most mature mRNAs in eukaryotes.
fixed action pattern (FAP) An animal behavior that, once initiated, will continue until completed.
fixed nitrogen Atmospheric nitrogen that has been combined with other elements into a form of nitrogen that can be used by plants. An example is ammonia, NH_3.
flaccid A plant cell in which the concentration of solutes is the same as that in the external fluid environment. A flaccid cell has a water content higher than a plasmolyzed cell, but lower than a turgid cell.
flagella (singular, **flagellum**) Relatively long cell appendages that facilitate cellular movement or the movement of extracellular fluids.
flagellate A protist that uses one or more flagella to move in water or cause water motions useful in feeding.
flagship species A single large or instantly recognizable species.
flame cell A cell that exists primarily to maintain osmotic balance between an organism's body and surrounding fluids; present in flatworms.
flavonoid A type of phenolic secondary metabolite that provides plants with protection from ultraviolet damage or imparts color to flowers.
flexor A muscle that bends a limb at a joint.
florigen The hypothesized flowering hormone, now identified as the FT (flowering time) protein that moves from leaves, where it is produced, into the shoot apex.
flower A reproductive shoot; a short stem that produces reproductive organs instead of leaves.
flowering plants The angiosperms, which produce ovules within the protective ovaries of flowers. The ovules develop into seeds, and the ovaries develop into fruits, which function in seed dispersal.
flow-through system The method of ventilation in fishes in which water moves unidirectionally such that the gills are constantly in contact with fresh, oxygenated water. Buccal pumping and ram ventilation are examples.
fluid-feeder An animal that licks or sucks fluid from plants or animals and does not need teeth except to puncture an animal's skin.
fluidity A property of biomembranes in which individual molecules remain in close association yet have the ability to move rotationally or laterally within the plane of the membrane. Membranes are semifluid.
fluid-mosaic model The accepted model of the plasma membrane; its basic framework is the semifluid phospholipid bilayer with a mosaic of proteins. Carbohydrates may be attached to the lipids or proteins.
fMRI *See* functional magnetic resonance imaging.
focal adhesion A mechanically strong cell junction that connects an animal cell to the extracellular matrix (ECM).
follicle A structure within an animal ovary where each ovum undergoes growth and development before it is released.
follicle-stimulating hormone (FSH) A gonadotropin that stimulates follicle development.
food chain A linear depiction of energy flow between organisms, with each organism feeding on and deriving energy from the preceding organism.
food-induced thermogenesis A rise in metabolic rate for a few hours after eating that produces heat.
food vacuole *See* phagocytic vacuole.
food web A complex model of interconnected food chains in which there are multiple links between species.
foot In mollusks, a muscular structure usually used for movement.
forebrain One of three major divisions of the vertebrate brain; the other two divisions are the midbrain and hindbrain.
fossil Recognizable preserved remains of past life on Earth.
fossil fuel A fuel formed in the Earth from protist, plant, or animal remains, such as coal, petroleum, and natural gas.
founder effect Genetic drift that occurs when a small group of individuals separates from a larger population and establishes a colony in a new location.
fovea A small area on the retina directly behind the lens, where an image is most sharply focused.

frameshift mutation A mutation that involves the addition or deletion of a number of nucleotides that are not in multiples of three.
free energy (*G*) In living organisms, the amount of available energy that can be used to do work.
free radical A molecule containing an atom with a single, unpaired electron in its outer shell. A free radical is unstable and interacts with other molecules by removing electrons from their atoms.
frequency In regard to sound, the number of complete wavelengths that occur in 1 second, measured in hertz (Hz).
frontal lobe One of four lobes of the cerebral cortex of the human brain; important in a variety of functions, including judgment and conscious thought.
fruit A structure that develops from flower organs, encloses seeds, and fosters seed dispersal in the environment.
fruiting bodies The visible fungal reproductive structures that are composed of densely packed hyphae that typically grow out of the substrate.
functional genomics Genomic methods aimed at studying the expression of a genome.
functional group A group of atoms with chemical features that are functionally important. Each functional group exhibits the same properties in all molecules in which it occurs.
functional magnetic resonance imaging (fMRI) A technique used to determine changes in brain activity while a person is performing specific tasks.
fundamental niche The optimal range in which a particular species best functions.
fungi A eukaryotic kingdom of the domain Eukarya.
fungus-like protists Heterotrophic protists that often resemble true fungi in having threadlike, filamentous bodies and absorbing nutrients from their environment.

G

G_0 A phase in which cells exit the cell cycle and postpone making the decision to divide.
G_1 The first gap phase of the cell cycle.
G_2 The second gap phase of the cell cycle.
gallbladder In many vertebrates, a small sac underneath the liver that is a storage site for bile; allows the release of large amounts of bile to be precisely timed to the consumption of fats.
gametangia Specialized structures produced by many land plants in which developing gametes are protected by a jacket of tissue.
gamete A haploid cell that is involved with sexual reproduction, such as a sperm or egg cell.
gametic life cycle A type of life cycle where all cells except the gametes are diploid, and gametes are produced by meiosis.
gametogenesis The formation of gametes.
gametophyte In plants and many multicellular protists, the haploid stage that produces gametes by mitosis.
ganglion A group of neuronal cell bodies in the peripheral nervous system that is involved in a similar function.
ganglion cells Cells in the vertebrate eye that send their axons into the optic nerve.
gap gene A type of segmentation gene; a mutation in this type of gene may cause several adjacent segments to be missing in the larva.
gap junction A type of junction between animal cells that provides a passageway for intercellular transport.
gas exchange The process of moving oxygen and carbon dioxide in opposite directions between the environment and blood and between blood and cells.
gas vesicle A cytoplasmic structure used to adjust buoyancy in cyanobacteria and certain other bacteria that live in aquatic habitats.
gastrin A hormone secreted by cells of the vertebrate stomach that stimulates acid production by stomach cells.
gastrovascular cavity In certain invertebrates such as cnidarians, a body cavity with a single opening to the outside; it functions as both a digestive system and circulatory system.
gastrula A stage of an animal embryo that is the result of gastrulation and has three cellular layers: the ectoderm, endoderm, and mesoderm.
gastrulation In animals, a process in which an area in the blastula invaginates and folds inward, creating different embryonic cell layers called germ layers.
gated A property of many channels that allows them to open and close to control the diffusion of solutes through a membrane.
gel electrophoresis A technique used to separate macromolecules by using an electric field that causes them to pass through a gel matrix.
gene A unit of heredity that contributes to the characteristics or traits of an organism. At the molecular level, a gene is composed of organized sequences of DNA.
gene addition The insertion of a cloned gene into the genome of an organism.
gene amplification An increase in the copy number of a gene.
gene cloning The process of making multiple copies of a gene of interest.
gene expression Gene function both at the level of traits and at the molecular level.
gene family A group of homologous genes within a single species.
gene flow Occurs when individuals migrate between different populations and results in changes in the genetic composition of the resulting populations.
gene interaction A situation in which a single trait is controlled by two or more genes.
gene knockout An organism in which both copies of a functional gene have been replaced with nonfunctional copies. Experimentally, this can occur via gene replacement.
gene mutation A relatively small change in DNA structure that alters a particular gene.
gene pool All of the genes in a population.
genera (singular, **genus**) In taxonomy, a subdivision of a family.
general lineage concept A widely accepted approach used to distinguish species; states that each species is a population of an independently evolving lineage.
general transcription factors (GTFs) Five different proteins that play a role in initiating transcription at the core promoter of structural genes in eukaryotes.
generative cell In a seed plant, one of the cells resulting from the division of a microspore; a generative cell divides to produce two sperm cells.
gene regulation The ability of cells to control their level of gene expression.
gene replacement The phenomenon in which a cloned gene recombines with the normal gene on a chromosome and replaces it.
gene therapy The introduction of cloned genes into living cells in an attempt to cure disease.
genetic code A code that specifies the relationship between the sequence of nucleotides in the codons found in mRNA and the sequence of amino acids in a polypeptide.
genetic drift The random change in a population's allele frequencies from one generation to the next that is attributable to chance. It occurs more quickly in small populations.
genetic engineering The direct manipulation of genes for practical purposes.
genetic map A chart that shows the linear arrangement of genes along a chromosome.
genetic mapping The use of genetic crosses to determine the linear order of genes that are linked to each other along the same chromosome.
genetically modified organisms (GMOs) *See* transgenic.
gene transfer The process by which genetic material is transferred from one bacterial cell to another.
genome The complete genetic composition of a cell or a species.
genomic imprinting A phenomenon in which a segment of DNA is imprinted, or marked, in a way that affects gene expression throughout the life of the individual who inherits that DNA.
genomic library A type of DNA library in which the inserts are derived from chromosomal DNA.
genomics Techniques that are used in the molecular analysis of the entire genome of a species.
genotype The genetic composition of an individual.
genotype frequency In a population, the number of individuals with a given genotype divided by the total number of individuals.
geological timescale A time line of the Earth's history from its origin about 4.55 bya to the present.
germination In plants, the process in which an embryo absorbs water, becomes metabolically active, and grows out of the seed coat, producing a seedling.
germ layer An embryonic cell layer such as ectoderm, mesoderm, or endoderm.
germ line Cells that give rise to gametes such as egg and sperm cells.
germ plasm Cytoplasmic determinants that help define and specify the primordial germ cells in the gastrula stage of animal development.
gestation *See* pregnancy.
giant axon A very large axon in certain species such as squids that facilitates high-speed neuronal conduction and rapid responses to stimuli.
gibberellic acid A type of gibberellin.
gibberellin A plant hormone that stimulates both cell division and cell elongation.
gills Specialized filamentous organs in aquatic animals that are used to obtain oxygen and eliminate carbon dioxide.
ginkgos A phylum of gymnosperms; Ginkgophyta.
gizzard The muscular portion of the stomach of birds and some reptiles that is capable of grinding food into smaller fragments.
glaucoma A condition in which drainage of aqueous humor in the eye becomes blocked and the pressure inside the eye increases. If untreated, this pressure damages cells in the retina and leads to irreversible loss of vision.
glia Cells that surround the neurons; a major class of cells in nervous systems that perform various functions.
global warming A gradual elevation of the Earth's surface temperature caused by an increasing greenhouse effect.
glomerular filtration rate (GFR) The rate at which a filtrate of plasma is formed in all the glomeruli of the vertebrate kidneys.
glomerulus A cluster of interconnected, fenestrated capillaries in the renal corpuscle of the kidney; the site of filtration in the kidney.
glucagon A hormone found in animals that stimulates the processes of glycogenolysis, gluconeogenesis, and the synthesis of ketones in the liver.
glucocorticoid A steroid hormone that regulates glucose balance and helps prepare the body for stress situations.

gluconeogenesis A mechanism for maintaining blood glucose level; enzymes in the liver convert noncarbohydrate precursors into glucose, which is then secreted into the blood.

glucose sparing A metabolic adjustment that reserves the glucose produced by the liver for use by the nervous system.

glycocalyx 1. An outer viscous covering surrounding a bacterium that traps water and helps protect bacteria from drying out. 2. A carbohydrate-rich zone on the surface of animal cells; also called a cell coat.

glycogen A polysaccharide found in animal cells (especially liver and skeletal muscle) and sometimes called animal starch; also, the major carbohydrate storage of fungi.

glycogenolysis A mechanism for maintaining blood glucose level; stored glycogen can be broken down into molecules of glucose which are then secreted into the blood.

glycolipid A lipid that has carbohydrate attached to it.

glycolysis A metabolic pathway that breaks down glucose to pyruvate.

glycolytic fiber A skeletal muscle fiber that has few mitochondria but possesses both a high concentration of glycolytic enzymes and large stores of glycogen.

glycoprotein A protein that has carbohydrate attached to it.

glycosaminoglycan The most abundant type of polysaccharide in the extracellular matrix (ECM) of animals, consisting of repeating disaccharide units that give a gel-like character to the ECM of animals.

glycosidic bond A bond formed between two sugar molecules.

glycosylation The attachment of carbohydrate to a protein or lipid, producing a glycoprotein or glycolipid.

glyoxysome A specialized organelle within plant seeds that contains enzymes needed to convert fats to sugars.

gnathostomes All vertebrate species that possess jaws.

gnetophytes A phylum of gymnosperms; Gnetophyta.

Golgi apparatus A stack of flattened, membrane-bound compartments that performs three overlapping functions: secretion, processing, and protein sorting.

gonadotropins Hormones secreted by the anterior pituitary gland that are the same in both sexes; gonadotropins influence the ability of the testes and ovaries to produce the sex steroids.

gonads The testes in males and the ovaries in females, where the gametes are formed.

G protein An intracellular protein that binds guanosine triphosphate (GTP) and guanosine diphosphate (GDP) and participates in intracellular signaling pathways.

G-protein-coupled receptors (GPCRs) A common type of receptor found in the cells of eukaryotic species that interacts with G proteins to initiate a cellular response.

graded potential A depolarization or hyperpolarization in a neuron that varies with the strength of a stimulus.

gradualism A concept suggesting that species evolve continuously over long spans of time.

grain The characteristic single-seeded fruit of cereal grasses such as rice, corn, barley, and wheat.

Gram stain A staining process that can help to identify bacteria and predict their responses to antibiotics.

granum A structure composed of stacked membrane-bound thylakoids within a chloroplast.

gravitropism Plant growth in response to the force of gravity.

gray matter Brain tissue that consists of neuronal cell bodies, dendrites, and some unmyelinated axons.

greenhouse effect The process in which short-wave solar radiation passes through the atmosphere to warm the Earth but is radiated back to space as long-wave infrared radiation. Much of this radiation is reflected by atmospheric gases back to Earth's surface, causing its temperature to rise.

groove In the DNA double helix, an indentation where the atoms of the bases make contact with the surrounding water.

gross primary production (GPP) The measure of biomass production by photosynthetic organisms; equivalent to the carbon fixed during photosynthesis.

ground meristem In plants, a type of primary plant tissue meristem that gives rise to ground tissue.

ground tissue Most of the body of a plant, which has a variety of functions, including photosynthesis, storage of carbohydrates, and support. Ground tissue can be subdivided into three types: parenchyma, collenchyma, and sclerenchyma.

group selection A premise that attempts to explain altruism. States that natural selection produces outcomes beneficial for the whole group or species rather than for individuals.

growth An increase in weight or size.

growth factors Proteins in animals that stimulate certain cells to grow and divide.

growth hormone (GH) A hormone produced in vertebrates by the anterior pituitary gland; GH acts on the liver to produce insulin-like growth factor-1 (IGF-1).

guanine (G) A purine base found in DNA and RNA.

guard cell A specialized plant cell that allows epidermal pores (stomata) to close when conditions are too dry and to open under moist conditions, allowing the entry of CO_2 needed for photosynthesis.

gustation The sense of taste.

gut The gastrointestinal (GI) tract of an animal.

guttation Droplets of water at the edges of leaves that are the result of root pressure.

gymnosperm A plant that produces seeds that are exposed rather than seeds enclosed in fruits.

gynoecium The aggregate of carpels that forms the innermost whorl of a flower.

H

habituation The form of nonassociative learning in which an organism learns to ignore a repeated stimulus.

hair cell A mechanoreceptor in animals that is a specialized epithelial cell with deformable stereocilia.

half-life 1. In the case of organic molecules in a cell, refers to the time it takes for 50% of the molecules to be broken down and recycled. 2. In the case of radioisotopes, the time it takes for 50% of the molecules to decay and emit radiation.

halophile A bacterium or archaeon that can live in an extremely salty environment.

halophyte A plant that can tolerate higher than normal salt concentrations and can occupy coastal salt marshes or saline deserts.

Hamilton's rule The proposal that an altruistic gene will be favored by natural selection when $rB > C$, where r is the coefficient of relatedness of the donor (the altruist) to the recipient, B is the benefit received by the recipient, and C is the cost incurred by the donor.

haplodiploid system A genetic system in which females develop from fertilized eggs and are diploid but males develop from unfertilized eggs and are haploid.

haploid Containing one set of chromosomes; designated as $1n$.

haploid-dominant species Species in which the haploid organism is the prevalent organism in the life cycle. Examples include fungi and some protists.

haplorrhini Larger-brained diurnal species of primates; includes monkeys, gibbons, orangutans, gorillas, chimpanzees, and humans.

Hardy-Weinberg equation An equation ($p^2 + 2pq + q^2 = 1$) that relates allele and genotype frequencies; the equation predicts an equilibrium if no new mutations are formed, no natural selection occurs, the population size is very large, the population does not migrate, and mating is random.

heart A muscular structure that pumps blood through blood vessels.

heart attack *See* myocardial infarction (MI).

heartburn More properly called gastroesophageal reflux; the movement of acidic stomach contents upward into the esophagus, typically causing pain.

heat of fusion The amount of heat energy that must be withdrawn or released from a substance to cause it to change from the liquid to the solid state.

heat of vaporization The heat required to vaporize 1 mole of any substance at its boiling point under standard pressure.

heavy chain A part of an immunoglobulin molecule.

H^+ electrochemical gradient A transmembrane gradient for H^+ composed of both a membrane potential and a concentration difference for H^+ across a membrane.

helper T cell A type of lymphocyte that assists in the activation and function of B cells and cytotoxic T cells.

hematocrit The volume of blood that is composed of red blood cells, usually between 40 and 65% in vertebrates.

hemidesmosome A mechanically strong cell junction that connects an animal cell to the extracellular matrix (ECM).

hemiparasite A parasitic organism that photosynthesizes, but lacks a root system to draw water and thus depends on its host for that function.

hemispheres The two halves of the cerebrum.

hemizygous The term used to describe the single copy of an X-linked gene in a male.

hemocyanin A copper-containing pigment that binds oxygen and gives blood or hemolymph a bluish tint.

hemodialysis A medical procedure used to artificially perform the kidneys' function.

hemoglobin An iron-containing protein that binds oxygen and is found within the cytosol of red blood cells.

hemolymph Blood and interstitial fluid combined in one fluid compartment; present in many invertebrates.

hemophilia An inherited disorder characterized by the deficiency of a specific blood clotting factor.

hemorrhage A loss of blood from a ruptured blood vessel.

herbaceous plant A plant that produces little or no wood and is composed mostly of primary vascular tissues.

herbivore An animal that eats only plants.

herbivory Refers to herbivores feeding on plants.

hermaphrodite In animals, an individual that can produce both sperm and eggs.

hermaphroditism A form of sexual reproduction in which individuals have both male and female reproductive systems.

heterochromatin The highly compacted regions of chromosomes that are usually transcriptionally inactive because of their tight conformation.

heterochrony Evolutionary changes in the rate or timing of developmental events.

heterocyst A specialized cell of some cyanobacteria in which nitrogen fixation occurs.

heterospory In plants, the formation of two different types of spores: microspores and megaspores; microspores produce male gametophytes, and megaspores produce female gametophytes.

heterotherm An animal that has a body temperature that is not constant; both ectotherms and endotherms may be heterotherms.

heterotroph Organisms that cannot produce their own organic molecules and thus must obtain organic food from other organisms.
heterotrophic Requiring organic food from the environment.
heterozygote advantage A phenomenon in which a heterozygote has a higher Darwinian fitness than either corresponding homozygote.
heterozygous An individual with two different alleles of the same gene.
hibernation The state of torpor in an animal that can last for months.
highly repetitive sequence A DNA sequence found tens of thousands or even millions of times throughout a genome.
hindbrain One of three major divisions of the vertebrate brain; the other two divisions are the midbrain and forebrain.
hippocampus The area of the vertebrate forebrain that functions in establishing memories for spatial locations, facts, and the sequence of events.
histone acetyltransferase An enzyme that loosens the compaction of chromatin by attaching acetyl groups to histone proteins.
histone code hypothesis Refers to the pattern of histone modification recognized by particular proteins. The pattern of covalent modifications of amino terminus tails provides binding sites for proteins that subsequently affect the degree of chromatin compaction.
histones A group of proteins involved in the formation of nucleosomes that aid in the compaction of eukaryotic DNA.
HIV *See* human immunodeficiency virus.
holoblastic cleavage A complete type of cell cleavage in certain animals in which the entire zygote is bisected into two equal-sized blastomeres.
holoparasite A parasitic organism that lacks chlorophyll and is totally dependent on a host plant for its water and nutrients.
homeobox A 180-bp sequence within the coding sequence of homeotic genes.
homeodomain A region of a homeotic protein that functions in binding to the DNA.
homeostasis The process whereby living organisms regulate their cells and bodies to maintain relatively stable internal conditions.
homeostatic control system A system designed to regulate particular variables in an animal's body, such as body temperature; consists of a set point, sensor, integrator, and effectors.
homeotherm An animal that maintains its body temperature within a narrow range.
homeotic gene A gene that controls the developmental fate of particular segments or regions of an animal's body.
hominin Either an extinct or modern species of humans.
hominoidea (hominoid) A member of a group of primates that includes gibbons, orangutans, gorillas, chimpanzees, and humans.
homologous genes Genes derived from the same ancestral gene that have accumulated random mutations that make their sequences slightly different.
homologous structures Structures that are similar to each other because they are derived from the same ancestral structure.
homolog A member of a pair of chromosomes in a diploid organism.
homology A fundamental similarity that occurs due to descent from a common ancestor.
homozygous An individual with two identical copies of an allele.
horizontal gene transfer A process in which an organism incorporates genetic material from another organism without being the offspring of that organism.
hormone A chemical messenger that is produced in a gland or other structure and acts on distant target cells in one or more parts of an animal or plant.
hornworts A phylum of bryophytes; Anthocerophyta.
host The prey organism in a parasitic association.
host cell 1. A cell that is infected by a virus, fungus, or a bacterium. 2. A eukaryotic cell that contains photosynthetic or nonphotosynthetic endosymbionts.
host plant resistance The ability of plants to prevent herbivory.
host range The number of species and cell types that a virus or bacterium can infect.
hot spot A human-impacted geographic area with a large number of endemic species. To qualify as a hot spot, a region must contain at least 1,500 species of endemic vascular plants and have lost at least 70% of its original habitat.
***Hox* genes** In animals, a class of genes involved in pattern formation in early embryos.
Human Genome Project A 13-year international effort coordinated by the U.S. Department of Energy and the National Institutes of Health that characterized and sequenced the entire human genome.
human immunodeficiency virus (HIV) A retrovirus that is the causative agent of acquired immune deficiency syndrome (AIDS).
humoral immunity A type of specific immunity in which plasma cells secrete antibodies that bind to antigens.
humus A collective term for the organic constituents of soils.
hybridization A situation in which two individuals with different characteristics are mated or crossed to each other; the offspring are referred to as hybrids.
hybrid zone An area where two populations can interbreed.
hydrocarbon Molecules with predominantly hydrogen–carbon bonds.
hydrogen bond A weak chemical attraction between a partially positive hydrogen atom of a polar molecule and a partially negative atom of another polar molecule.
hydrolysis reaction A chemical reaction that utilizes water to break apart molecules.
hydrophilic Refers to ions and molecules that contain polar covalent bonds and will dissolve in water.
hydrophobic Refers to molecules that do not have partial charges and therefore are not attracted to water molecules. Such molecules are composed predominantly of carbon and hydrogen and are relatively insoluble in water.
hydrostatic skeleton A fluid-filled body cavity in certain soft-bodied invertebrates that is surrounded by muscles and provides support and shape.
hydroxide ion An anion with the formula, OH^-.
hypermutation A process that primarily involves numerous C to T point mutations that are crucial to enabling lymphocytes to produce a diverse array of immunoglobulins capable of recognizing many different antigens.
hyperpolarization The change in the membrane potential that occurs when the cell becomes more polarized.
hypersensitive response (HR) A plant's local defensive response to pathogen attack.
hypertension High blood pressure.
hyperthermophile An organism that thrives in extremely hot temperatures.
hyperthyroidism A medical condition resulting from a hyperactive thyroid gland.
hypertonic Any solution that causes a cell to shrink due to osmosis of water out of the cell.
hypha A microscopic, branched filament of the body of a fungus.
hypocotyl The portion of an embryonic plant stem located below the point of attachment of the cotyledons.
hypothalamus A part of the vertebrate brain located below the thalamus; it controls functions of the gastrointestinal and reproductive systems, among others, and regulates many basic behaviors such as eating and drinking.
hypothesis In biology, a proposed explanation for a natural phenomenon based on previous observations or experimental studies.
hypothesis testing Also known as the scientific method, a strategy for testing the validity of a hypothesis.
hypothyroidism A medical condition resulting from an underactive thyroid gland.
hypotonic Any solution that causes a cell to swell when placed in that solution.
H zone In a myofibril, a narrow, light region in the center of the A band that corresponds to the space between the two sets of thin filaments in each sarcomere.

I

I band In a myofibril, a light band that lies between the A bands of two adjacent sarcomeres.
idiosyncratic hypothesis The possibility that ecosystem function changes as the number of species increases or decreases but that the amount and direction of change is unpredictable.
immune system The cells and organs within an animal's body that contribute to immune defenses.
immune tolerance The process by which the body distinguishes between self and nonself components.
immunity The ability of an animal to ward off internal threats, including harmful microorganisms, foreign molecules, and abnormal cells such as cancer cells.
immunoglobulin A Y-shaped protein with two heavy chains and two light chains that provides immunity to foreign substances; antibodies are a type of immunoglobulin.
immunological memory The immune system's ability to produce a secondary immune response.
imperfect flower A flower that lacks either stamens or carpels.
implantation The first event of pregnancy, when the blastocyst embeds within the uterine endometrium.
imprinting 1. The development of a species-specific pattern of behavior that occurs during a critical period; a form of learning, with a large innate component. 2. In genetics, the marking of DNA that occurs differently between males and females.
inactivation gate A string of amino acids that juts out from a channel protein into the cytosol and blocks the movement of ions through the channel.
inborn error of metabolism A genetic defect in the ability to metabolize certain compounds.
inbreeding Mating among genetically related relatives.
inbreeding depression The phenomenon whereby inbreeding produces homozygotes that are less fit, thereby decreasing the reproductive success of a population.
inclusive fitness The term used to designate the total number of copies of genes passed on through one's relatives, as well as one's own reproductive output.
incomplete dominance The phenomenon in which a heterozygote that carries two different alleles exhibits a phenotype that is intermediate between the corresponding homozygous individuals.
incomplete flower A flower that lacks one or more of the four flower organ types.
incomplete metamorphosis During development in some insects, a gradual change in body form from a nymph into an adult.

incurrent siphon A structure in a tunicate used to draw water through the mouth.

indeterminate cleavage In animals, a characteristic of deuterostome development in which each cell produced by early cleavage retains the ability to develop into a complete embryo.

indeterminate growth Growth in which plant shoot apical meristems continuously produce new stem tissues and leaves, as long as conditions remain favorable.

indicator species A species whose status provides information on the overall health of an ecosystem.

indirect calorimetry A method of determining basal metabolic rate in which the rate at which an animal uses oxygen is measured.

individualistic model A view of the nature of a community that considers it to be an assemblage of species coexisting primarily because of similarities in their physiological requirements and tolerances.

individual selection The proposal that adaptive traits generally are selected for because they benefit the survival and reproduction of the individual rather than the group.

induced fit Occurs when a substrate(s) binds to an enzyme and the enzyme undergoes a conformational change that causes the substrate(s) to bind more tightly to the enzyme.

induced mutation A mutation brought about by environmental agents that enter the cell and then alter the structure of DNA.

inducer In transcription, a small effector molecule that increases the rate of transcription.

inducible operon In this type of operon, the presence of a small effector molecule causes transcription to occur.

induction 1. In development, the process by which a cell or group of cells governs the developmental fate of neighboring cells. 2. In molecular genetics, refers to the process by which transcription has been turned on by the presence of a small effector molecule.

industrial nitrogen fixation The human activity of producing nitrogen fertilizers.

infertility The inability to produce viable offspring.

inflammation An innate local response to infection or injury characterized by local redness, swelling, heat, and pain.

inflorescence A cluster of flowers on a plant.

infundibular stalk The structure that physically connects the hypothalamus to the pituitary gland.

ingestion In animals, the act of taking food into the body.

ingroup In a cladogram, a group of interest.

inheritance The acquisition of traits by their transmission from parent to offspring.

inheritance of acquired characteristics Jean-Baptiste Lamarck's incorrect hypothesis that species change over the course of many generations by adapting to new environments.

inhibition A mechanism for succession in which early colonists exclude subsequent colonists.

inhibitory postsynaptic potential (IPSP) The response from an inhibitory neurotransmitter that hyperpolarizes the postsynaptic membrane; this hyperpolarization reduces the likelihood of an action potential.

initiation factor A protein that facilitates the interactions between mRNA, the first tRNA, and the ribosomal subunits during the initiation stage of translation.

initiation stage The first step in the process of transcription or translation.

initiator tRNA A specific tRNA that recognizes the start codon AUG in mRNA and binds to it, initiating translation.

innate The term used to describe behaviors that seem to be genetically programmed.

innate immunity The body's defenses that are present at birth and act against foreign materials in much the same way regardless of their specific identity; includes the skin and mucous membranes, plus various cellular and chemical defenses.

inner bark The thin layer of secondary phloem that carries out most of the sugar transport in a woody stem.

inner ear One of the three main compartments of the mammalian ear. The inner ear is composed of the bony cochlea and the vestibular system, which plays a role in balance.

inner segment The part of the vertebrate photoreceptors (rods and cones) that contains the cell nucleus and cytoplasmic organelles.

inorganic chemistry The study of the nature of atoms and molecules, with the exception of those that contain rings or chains of carbon.

insulin A hormone found in animals that regulates metabolism in several ways, primarily by regulating the blood glucose concentration.

insulin-like growth factor-1 (IGF-1) A hormone in mammals that stimulates the elongation of bones, especially during puberty.

integral membrane protein A protein that cannot be released from the membrane unless it is dissolved with an organic solvent or detergent. Includes transmembrane proteins and lipid-anchored proteins.

integrase An enzyme, sometimes encoded by viruses, that catalyzes the integration of the viral genome into a host-cell chromosome.

integrator The part of a homeostatic control system in which a variable is compared to a set point; typically a nucleus in the brain.

integrin A cell adhesion molecule found in animal cells that connects cells to the extracellular matrix.

integument In plants, a structure that encloses the megasporangium to form an ovule.

interference competition Competition in which organisms interact directly with one another by physical force or intimidation.

interferon A protein that generally inhibits viral replication inside host cells.

intermediate-disturbance hypothesis The proposal that moderately disturbed communities are more diverse than undisturbed or highly disturbed communities.

intermediate filament A type of protein filament of the cytoskeleton that helps maintain cell shape and rigidity.

internal fertilization Fertilization that occurs in terrestrial animals in which sperm are deposited within the reproductive tract of the female during copulation.

interneuron A type of neuron that forms interconnections between other neurons.

internode The region of a plant stem between adjacent nodes.

interphase The G_1, S, and G_2 phases of the cell cycle. It is the portion of the cell cycle during which the chromosomes are decondensed and found in the nucleus.

intersexual selection Sexual selection between members of the opposite sex.

interspecies hybrid The offspring resulting from the mating of two different species.

interspecific competition Competition between individuals of different species.

interstitial fluid The fluid that surrounds cells.

intertidal zone The area where the land meets the sea, which is alternately submerged and exposed by the daily cycle of tides.

intracellular fluid The fluid inside cells.

intranuclear spindle A spindle that forms within an intact nuclear envelope during nuclear division in fungi and some protists.

intrasexual selection Sexual selection between members of the same sex.

intraspecific competition Competition between individuals of the same species.

intrauterine device (IUD) A small object that is placed in the uterus and interferes with the endometrial preparation required for acceptance of the blastocyst; used as a form of contraception.

intrinsic rate of increase The situation in which conditions are optimal for a population and the per capita growth rate is at its maximum rate.

introduced species A species moved by humans from a native location to another location.

intron Intervening DNA sequences that are found in between the coding sequences of genes.

invaginate The act of pinching inward, as during early embryonic development in animals.

invasive cell A cancer cell that can invade healthy tissues.

invasive species Introduced species that spread on their own, often outcompeting native species for space and resources.

inverse density-dependent factor A mortality factor whose influence decreases as population size or density increases.

inversion A type of mutation that involves a change in the direction of the genetic material along a single chromosome.

invertebrate An animal that lacks vertebrae.

in vitro Meaning, "in glass" as in a test tube. An alternative to studying a process in a living organism that involves studying components outside of their natural locations. This can involve studying cells or components of cells in the laboratory.

in vivo Meaning, "in life." Studying a process in living organisms.

involution During embryogenesis, the folding back of sheets of surface cells into the interior of an embryo.

iodine-deficient goiter An overgrown thyroid gland that is incapable of making thyroid hormone due to a lack of dietary iodine.

ion An atom or molecule that gains or loses one or more electrons and acquires a net electric charge.

ion electrochemical gradient A dual gradient for an ion that is composed of both an electrical gradient and a chemical gradient for that ion.

ionic bond The bond that occurs when a cation binds to an anion.

ionotropic receptor One of two types of postsynaptic receptors, the other being a metabotropic receptor. Consists of a ligand-gated ion channel that opens in response to binding of a neurotransmitter.

iris The circle of pigmented smooth muscle and connective tissue that is responsible for eye color.

iron regulatory element (IRE) A response element within the ferritin mRNA to which the iron regulatory protein binds.

iron regulatory protein (IRP) An RNA-binding protein that regulates the translation of the mRNA that encodes ferritin.

islets of Langerhans Spherical clusters of endocrine cells that are scattered throughout the pancreas; the cells secrete insulin or glucagon, among other hormones.

isomers Two structures with an identical molecular formula but different structures and characteristics.

isotonic Condition in which the solute concentrations on both sides of a plasma membrane are equal, which does not cause a cell to shrink or swell.

isotope An element that exists in multiple forms that differ in the number of neutrons they contain.

iteroparity The pattern of repeated reproduction at intervals throughout an organism's life cycle.

J

joint The juncture where two or more bones of a vertebrate endoskeleton come together.

juvenile hormone A hormone made in arthropods that inhibits maturation from a larva into a pupa.

K

karyogamy The process of nuclear fusion.

karyotype A photographic representation of the chromosomes in an actively dividing cell.

K_d The dissociation constant between a ligand and its receptor.

ketones Small compounds generated from fatty acids. Ketones are made in the liver and released into the blood to provide an important energy source during prolonged fasting for many tissues, including the brain.

keystone hypothesis The idea that ecosystem function plummets as soon as biodiversity declines from its natural levels.

keystone species A species within a community that has a role out of proportion to its abundance.

kidney The major excretory organ found in all vertebrates.

kcal (kilocalorie) One thousand calories; the amount of heat energy required to raise the temperature of 1 kg of water by 1°C.

kinesis A movement in response to a stimulus, but one that is not directed toward or away from the source of the stimulus.

kinetic energy Energy associated with movement.

kinetic skull A characteristic of lizards and snakes in which the joints between various parts of the skull are extremely mobile.

kinetochore A group of proteins that bind to a centromere and are necessary for sorting each chromosome.

kingdom A taxonomic group; the second largest division after domain.

kin selection Selection for behavior that lowers an individual's own fitness but enhances the reproductive success of a relative.

K_M The substrate concentration at which an enzyme-catalyzed reaction is half of its maximal value.

knowledge The awareness and understanding of information.

Koch's postulates A series of steps used to determine whether a particular organism causes a specific disease.

***K*-selected species** A type of life history strategy where species have a low rate of per capita population growth but good competitive ability.

K/T event An ancient cataclysm that involved at least one large meteorite or comet that crashed into the Earth near the present-day Yucatán Peninsula in Mexico about 65 mya.

L

labia majora In the female mammalian genitalia, large outer folds that surround the external opening of the reproductive tract.

labia minora In the female mammalian genitalia, smaller, inner folds near the external opening of the reproductive tract.

labor The strong rhythmic contractions of the uterus that serve to deliver a fetus during childbirth.

***lac* operon** An operon in the genome of *E. coli* that contains the genes for the enzymes that allow it to metabolize lactose.

lac repressor A repressor protein that regulates the *lac* operon.

lactation In mammals, a period after birth in which the young are nurtured by milk produced by the mother.

lacteal A lymphatic vessel in the center of each intestinal villus; lipids are absorbed by the lacteals, which eventually empty into the circulatory system.

lagging strand During DNA replication, a DNA strand made as a series of small Okazaki fragments that are eventually connected to each other to form a continuous strand.

lamellae (singular, **lamella**) Platelike structures in the internal gills of fishes that branch from structures called filaments; gas exchange occurs here.

larva A free-living organism that is morphologically very different from the embryo and adult.

larynx The segment of the respiratory tract that contains the vocal cords.

latent The term used to describe a prophage or provirus that remains inactive for a long time.

lateral line system Microscopic sensory organs in fishes and some toads that allows them to detect movement in surrounding water.

lateral meristem *See* secondary meristem.

law of independent assortment States that the alleles of different genes assort independently of each other during gamete formation.

law of segregation States that two copies of a gene segregate from each other during gamete formation and during transmission from parent to offspring.

leaching The dissolution and removal of inorganic ions as water percolates through materials such as soil.

leading strand During DNA replication, a DNA strand made in the same direction that the replication fork is moving. The strand is synthesized as one long continuous molecule.

leaf abscission The process by which a leaf drops after the formation of an abscission zone.

leaflet 1. Half of a phospholipid bilayer. 2. A portion of a compound leaf.

leaf primordia Small outgrowths that occur at the sides of a shoot apical meristem and develop into young leaves.

leaf vein In plants, a bundle of vascular tissue in a leaf.

learning The ability of an animal to make modifications to a behavior based on previous experience; the process by which new information is acquired.

leaves Flattened plant organs that emerge from stems and function in photosynthesis.

left-right axis In bilaterally symmetric animals, the left and right sides of the body.

leghemoglobin A protein found in legume plants that helps to regulate local oxygen concentrations around rhizobial bacteroids in root nodules.

legume A member of the pea (bean) family; also their distinctive fruits.

lek A designated communal courting area in certain species of birds.

lens 1. A structure of the eye that focuses light. 2. The glass components of a light microscope or the electromagnetic parts of an electron microscope that allow the production of magnified images of microscopic structures.

lentic Refers to a freshwater habitat characterized by standing water.

lenticels Passages in the outer bark of a woody plant stem that allow inner stem tissues to accomplish gas exchange.

leptin A hormone produced by adipose cells in proportion to fat mass; controls appetite and metabolic rate.

leukocyte A cell that develops from the marrow of certain bones of vertebrates; all leukocytes (also known as white blood cells) perform vital functions that defend the body against infection and disease.

lichens The mutualistic association between particular fungi and certain photosynthetic green algae or cyanobacteria. This association results in a body form distinctive from that of either partner alone.

Liebig's law of the minimum States that species' biomass or abundance is limited by the scarcest factor.

life cycle The sequence of events that characterize the steps of development of the individuals of a given species.

life table A table that provides data on the number of living individuals in a population in particular age classes.

ligand An ion or molecule that binds to a protein, such as an enzyme or a receptor.

ligand-gated ion channel A type of cell surface receptor that binds a ligand and functions as an ion channel. Ligand binding either opens or closes a channel.

ligand-receptor complex The structure formed when a ligand and its receptor noncovalently bind.

light chain 1. A part of an immunoglobulin molecule. 2. Two of the polypeptides that make up each myosin molecule.

light-harvesting complex A component of photosystem II and photosystem I composed of several dozen pigment molecules that are anchored to proteins in the thylakoid membranes of a chloroplast. The role of these complexes is to absorb photons of light.

light microscope A microscope that utilizes light for illumination.

light reactions The first of two stages in the process of photosynthesis. During the light reactions, photosystem II and photosystem I absorb light energy and produce ATP, NADPH, and O_2.

lignin A tough polymer that adds strength and decay resistance to cell walls of tracheids, vessel elements, and other cells of plants.

lignophytes Modern and fossil seed plants and seedless ancestors that produced wood.

limbic system In the vertebrate forebrain, the areas involved in the formation and expression of emotions; also plays a role in learning, memory, and the perception of smells.

limiting factor A factor whose amount or concentration limits the rate of a biological process or a chemical reaction.

lineage A progression of changes in a series of ancestors.

line transect A sampling technique used by plant ecologists in which the number of plants located along a length of string are counted.

linkage The phenomenon that two genes close together on the same chromosome are transmitted as a unit.

linkage group A group of genes that usually stay together during meiosis.

lipase The major fat-digesting enzyme from the pancreas.

lipid A molecule composed predominantly of hydrogen and carbon atoms. Lipids are nonpolar and therefore very insoluble in water. They include fats (triglycerides), phospholipids, waxes, and steroids.

lipid-anchored protein A type of integral membrane protein that is attached to the membrane via a lipid molecule.

lipid-exchange protein A protein that extracts a lipid from one membrane, diffuses through the cell, and inserts the lipid into another membrane.

lipid raft In a membrane, a group of lipids and proteins that float together as a unit in a larger sea of lipids.

lipolysis The enzymatic breakdown of triglycerides into fatty acids and either monoglycerides or glycerol.

lipopolysaccharides Lipids with covalently bound carbohydrates; prevalent in the thin, outer envelope that encloses the cell walls of Gram-negative bacteria.

liposome A vesicle surrounded by a lipid bilayer.

liver An organ in vertebrates that performs diverse metabolic functions and is the site of bile production.

liverworts A phylum of bryophytes; formally called Hepatophyta.
loam A type of soil that contains a mixture of sand, silt, and clay and is ideal for plant cultivation.
lobe fins The Actinistia (coelacanths), Dipnoi (lungfishes), and tetrapods; also called Sarcopterygii.
lobe-finned fishes Fishes in which the fins are part of the body; the fins are supported by skeletal extensions of the pectoral and pelvic areas.
locomotion The movement of an animal from place to place.
locus The physical location of a gene on a chromosome.
logistic equation An equation that relates the growth of populations to their carrying capacity, *K*.
logistic growth The pattern in which the growth of a population typically slows down as it approaches the carrying capacity.
long-day plant A plant that flowers in spring or early summer, when the night period is shorter (and thus the day length is longer) than a defined period.
long-term potentiation (LTP) The long-lasting strengthening of the connection between neurons that is believed to be part of the mechanism of learning and memory.
loop domain In bacteria, a chromosomal segment that is folded into loops by the attachment to proteins; a method of compacting bacterial chromosomes.
loop of Henle A segment of the tubule of the nephron of the kidney containing a sharp hairpin-like loop that contributes to reabsorption of ions and water. It consists of a descending limb coming from the proximal tubule and an ascending limb leading to the distal tubule.
lophophore A horseshoe-shaped crown of tentacles used for feeding in several invertebrate species.
Lophotrochozoa A clade of animals that encompasses the mollusks, annelids, and several other phyla; they are distinguished by two morphological features: the lophophore, a crown of tentacles used for feeding, and the trochophore larva, a distinct larval stage.
lotic Refers to a freshwater habitat characterized by running water.
lumen The internal space or hollow cavity of an organelle or an organ, such as the stomach or a blood vessel.
lungfishes The Dipnoi; fish with primitive lungs that live in oxygen-poor freshwater swamps and ponds.
lung In terrestrial vertebrates, internal paired structures used to bring O_2 into the circulatory system and remove CO_2.
luteinizing hormone (LH) A gonadotropin that controls the production of sex steroids in both males and females.
lycophyll A relatively small leaf having a single unbranched vein; produced by lycophytes.
lycophytes Members of a phylum of vascular land plants whose leaves are lycophylls, Lycopodiophyta.
lymphatic system A system of vessels along with a group of organs and tissues where most leukocytes reside. The lymphatic vessels collect excess interstitial fluid and return it to the blood.
lymphocyte A type of leukocyte that is responsible for specific immunity; the two types are B cells and T cells.
lysogenic cycle The growth cycle of a bacteriophage consisting of integration, prophage replication, and excision.
lysosome A small organelle found in animal cells that contains acid hydrolases that degrade macromolecules.
lytic cycle The growth cycle of a bacteriophage in which the production and release of new viruses lyses the host cell.

M

macroalgae Photosynthetic protists that can be seen with the unaided eye; also known as seaweeds.
macroevolution Evolutionary changes that create new species and groups of species.
macromolecule Many molecules bonded together to form a polymer. Carbohydrates, proteins, and nucleic acids (for example, DNA and RNA) are important macromolecules found in living organisms.
macronutrient An element required by plants in amounts of at least 1 g/kg of plant dry matter.
macroparasite A parasite that lives in a host but releases infective juvenile stages outside the host's body.
macrophage A type of phagocyte capable of engulfing viruses and bacteria.
macular degeneration An eye condition in which photoreceptor cells in and around the fovea of the retina are lost; one of the leading causes of blindness in the U.S.
madreporite A sievelike plate on the surface of an echinoderm where water enters the water vascular system.
magnetic resonance imaging (MRI) An imaging method that relies on the use of magnetic fields and radio waves to visualize the internal structure of an organism's body.
magnification The ratio between the size of an image produced by a microscope and a sample's actual size.
major depressive disorder A neurological disorder characterized by feelings of despair and sadness, resulting from an imbalance in neurotransmitter levels in the brain.
major groove A groove that spirals around the DNA double helix; provides a location where a protein can bind to a particular sequence of bases and affect the expression of a gene.
major histocompatibility complex (MHC) A gene family that encodes the plasma membrane self proteins that must be complexed with an antigen for T-cell recognition to occur.
male-assistance hypothesis A hypothesis to explain the existence of monogamy that maintains that males remain with females to help them rear their offspring.
male gametophyte A haploid multicellular plant life cycle stage that produces sperm.
malignant tumor A growth of cells that has progressed to the cancerous stage.
Malpighian tubules Delicate projections from the digestive tract of insects and some other taxa that function as an excretory organ.
mammal A vertebrate that is a member of the class Mammalia that nourishes its young with milk secreted by the female's mammary glands. Another distinguishing feature is hair.
mammary gland A gland in female mammals that secretes milk.
manganese cluster A site where the oxidation of water occurs in photosystem II during photosynthesis.
mantle In mollusks, a fold of skin draped over the visceral mass that secretes a shell in those species that form shells.
mantle cavity The chamber in a mollusk mantle that houses delicate gills.
many-eyes hypothesis The idea that increased group size decreases predators' success because of increased detection of predators.
map distance The distance between genes along chromosomes, which is calculated as the number of recombinant offspring divided by the total number of offspring times 100.
mapping The process of determining the relative locations of genes or other DNA segments along a chromosome.
map unit (mu) A unit of distance on a chromosome equivalent to a 1% recombination frequency.
mark-recapture technique The capture and tagging of animals so they can be released and recaptured, allowing an estimate of population size.
marsupial A member of a group of seven mammalian orders and about 280 species found in the subclass Metatheria.
mass extinction When many species become extinct at the same time.
mass-specific BMR The amount of energy expended per gram of body mass.
mastax The circular muscular pharynx in the mouth of rotifers.
mast cell A type of cell derived from bone marrow stem cells that plays an important role in nonspecific immunity.
mate-guarding hypothesis The hypothesis that a male is monogamous to prevent his mate from being fertilized by other males.
maternal effect An inheritance pattern in which the genotype of the mother determines the phenotype of her offspring.
maternal effect gene A gene that follows a maternal effect inheritance pattern.
maternal inheritance A phenomenon in which offspring inherit particular genes only from the female parent (through the egg).
matrotrophy In plants, the phenomenon in which zygotes remain enclosed within gametophyte tissues, where they are sheltered and fed.
matter Anything that has mass and takes up space.
maturation-promoting factor (MPF) A factor, now known to be a complex of cyclin and cyclin-dependent kinase, important in the division of all types of eukaryotic cells.
mature mRNA In eukaryotes, transcription produces a long RNA, pre-mRNA, which undergoes certain processing events before it exits the nucleus; mature mRNA is the final functional product.
maximum likelihood One method used to evaluate a phylogenetic tree based on an evolutionary model.
mean fitness of the population The average reproductive success of members of a population.
mechanoreceptor A sensory receptor in animals that transduces mechanical energy such as pressure, touch, stretch, movement, and sound.
mediator A large protein complex that plays a role in initiating transcription at the core promoter of structural genes in eukaryotes.
medulla oblongata The part of the vertebrate hindbrain that coordinates many basic reflexes and bodily functions, such as breathing.
medusa A type of cnidarian body form that is motile and usually floats mouth down.
megadiversity country Those countries with the greatest numbers of species; used in targeting areas for conservation.
megaspore In seed plants and some seedless plants, a large spore that produces a female gametophyte within the spore wall.
meiosis The process by which haploid cells are produced from a cell that was originally diploid.
meiosis I The first division of meiosis in which the homologs are separated into different cells.
meiosis II The second division of meiosis in which sister chromatids are separated into different cells.
Meissner's corpuscle Structures that sense touch and light pressure and lie just beneath the skin surface of an animal.
melatonin A hormone produced by the pineal gland of vertebrates; plays a role in light-dependent behaviors such as seasonal reproduction and daily rhythms.

membrane attack complex (MAC) A multiunit protein formed by the activation of complement proteins; the complex creates water channels in the microbial plasma membrane and causes the microbe to swell and burst.

membrane potential The difference between the electric charges outside and inside a cell; also called a potential difference (or voltage).

membrane transport The movement of ions or molecules across a cell membrane.

memory The retention of information over time.

memory cell A cloned lymphocyte that remains poised to recognize a returning antigen; a component of specific immunity.

Mendelian inheritance The inheritance patterns of genes that segregate and assort independently.

meninges Three layers of sheathlike membranes that cover and protect the brain and spinal cord.

meningitis A potentially life-threatening infectious disease in which the meninges become inflamed.

menopause The event during which a woman permanently stops having ovarian cycles.

menstrual cycle The cyclical changes that occur in the uterus in parallel with the ovarian cycle in a female mammal. Also called the uterine cycle.

menstruation A period of bleeding at the beginning of the menstrual cycle in a female mammal.

meristem In plants, an organized tissue that includes actively dividing cells and a reservoir of stem cells.

meroblastic cleavage An incomplete type of cell cleavage in which only the region of the egg containing cytoplasm at the animal pole undergoes cell division. Occurs in birds and some fishes.

merozygote A strain of bacteria containing an F′ factor.

mesoderm In animals, a layer of cells formed during gastrulation that develops between the ectoderm and endoderm; gives rise to the skeleton, muscles, and much of the circulatory system.

mesoglea A gelatinous substance between the epidermis and the gastrodermis in the Radiata.

mesohyl A gelatinous, protein-rich matrix in between the choanocytes and the epithelial cells of a sponge.

mesophyll The internal tissue of a plant leaf; the site of photosynthesis.

messenger RNA (mRNA) RNA that contains the information to specify a polypeptide with a particular amino acid sequence.

metabolic cycle A biochemical cycle in which particular molecules enter while others leave; the process is cyclical because it involves a series of organic molecules that are regenerated with each turn of the cycle.

metabolic pathway In living cells, a series of chemical reactions in which each step is catalyzed by a specific enzyme.

metabolic rate The total energy expenditure of an organism per unit of time.

metabolism The sum total of all chemical reactions that occur within an organism. Also, a specific set of chemical reactions occurring at the cellular level.

metabotropic receptor A G-protein-coupled receptor that initiates a signaling pathway in response to a neurotransmitter. One of two types of postsynaptic receptors, the other being an ionotropic receptor.

metacentric A chromosome in which the centromere is near the middle.

metagenomics A field of study that seeks to identify and analyze the collective microbial genomes contained in a community of organisms, including those not easily cultured in the laboratory.

metamorphosis The process in which a pupal or juvenile organism changes into a mature adult with very different characteristics.

metanephridia Excretory filtration organs found in a variety of invertebrates.

metaphase The phase of mitosis during which the chromosomes are aligned along the metaphase plate.

metaphase plate A plane halfway between the poles of the spindle apparatus on which the sister chromatids align during the metaphase stage of mitosis.

metastasis The process by which cancer cells spread from their original location to distant parts of the body.

Metazoa The collective term for animals.

methanogens Several groups of anaerobic archaea that convert CO_2, methyl groups, or acetate to methane, and release it from their cells.

methanotroph An aerobic bacterium that consumes methane.

methyl-CpG-binding protein A protein that binds methylated sequences and inhibits transcription.

micelle The sphere formed by long amphipathic molecules when they are mixed with water. In animals, micelles aid in the absorption of poorly soluble products during digestion.

microbiome All of the microorganisms in a particular environment.

microclimate Local variations of the climate within a given area.

microevolution Changes in a population's gene pool from generation to generation.

microfilament *See* actin filament.

micrograph An image taken with the aid of a microscope.

micronutrient An element required by plants in amounts at, or less than, 0.1 g/kg of plant dry matter; also known as a trace element.

microparasite A parasite that multiplies within its host, usually within the cells.

micropyle A small opening in the integument of a seed plant ovule through which a pollen tube grows.

microRNAs (miRNAs) Small RNA molecules, typically 22 nucleotides in length, that silence the expression of specific mRNAs by inhibiting translation.

microscope A magnification tool that enables researchers to study very small structures such as cells.

microspore In seed plants and some seedless plants, a relatively small spore that produces a male gametophyte within the spore wall.

microsporidia Single-celled fungi that parasitize animal cells.

microtubule A type of hollow protein filament composed of tubulin proteins that is part of the cytoskeleton and is important for cell shape, organization, and movement.

microtubule-organizing center *See* centrosome.

microvillus (plural **microvilli**) Small projections in the surface membranes of epithelial cells in the small intestine and many other absorptive cells.

midbrain One of three major divisions of the vertebrate brain; the other two divisions are the hindbrain and the forebrain.

middle ear One of the three main compartments of the mammalian ear; contains three small bones called ossicles that connect the eardrum with the oval window.

middle lamella An extracellular layer in plants composed primarily of carbohydrate; cements adjacent plant cell walls together.

migration Long-range seasonal movement among animals in order to feed or breed.

mimicry The resemblance of an organism (the mimic) to another organism (the model).

mineral An inorganic ion or inorganic molecule required by a living organism.

mineralization The general process by which phosphorus, nitrogen, CO_2, and other minerals are released from organic compounds.

mineralocorticoid A steroid hormone such as aldosterone that regulates the balance of sodium and potassium ions in the body.

minor groove A smaller groove that spirals around the DNA double helix.

miRNA *See* microRNAs.

missense mutation A base substitution that changes a single amino acid in a polypeptide sequence.

mitochondrial genome The chromosome found in mitochondria.

mitochondrial matrix A compartment inside the inner membrane of a mitochondrion.

mitochondrion A semiautonomous organelle found in eukaryotic cells that supplies most of a cell's ATP.

mitogen-activated protein kinase (MAP kinase) A type of protein kinase that is involved with promoting cell division.

mitosis In eukaryotes, the process in which nuclear division results in two nuclei, each of which receives the same complement of chromosomes.

mitotic cell division A process whereby a eukaryotic cell divides to produce two new cells that are genetically identical to the original cell.

mitotic spindle The structure responsible for organizing and sorting the chromosomes during mitosis; also called the mitotic spindle apparatus.

mixotroph An organism that is able to use autotrophy as well as phagotrophy or osmotrophy to obtain organic nutrients.

M line In a myofibril, a narrow, dark band in the center of the H zone where proteins link the central regions of adjacent thick filaments.

model organism An organism studied by many different researchers so they can compare their results and determine scientific principles that apply more broadly to other species.

moderately repetitive sequence A DNA sequence found a few hundred to several thousand times in a genome.

molar A term used to describe a solution's molarity; a 1 molar solution contains 1 mole of a solute in 1 L of water.

molarity The number of moles of a solute dissolved in 1 L of water.

mole The amount of any substance that contains the same number of particles as there are atoms in exactly 12 g of carbon.

molecular biology A field of study spawned largely by genetic technology that looks at the structure and function of the molecules of life.

molecular clock A method for estimating evolutionary time; based on the observation that neutral mutations occur at a relatively constant rate.

molecular evolution The molecular changes in genetic material that underlie the phenotypic changes associated with evolution.

molecular formula A representation of a molecule that consists of the chemical symbols for all of the atoms present and subscripts that indicate how many of those atoms are present.

molecular homologies Similarities at the molecular level that indicate that living species evolved from a common ancestor or interrelated group of common ancestors.

molecular mass The sum of the atomic masses of all the atoms in a molecule.

molecular pharming An avenue of research that involves the production of medically important proteins in agricultural crops or animals.

molecular systematics A field of study that involves the analysis of genetic data, such as DNA sequences, to

identify and study genetic homology and construct phylogenetic trees.
molecule Two or more atoms that are connected by chemical bonds.
monoclonal antibodies Antibodies of a specific type that are derived from a single clone of cells.
monocots One of the two largest lineages of flowering plants in which the embryo produces a single seed leaf.
monocular vision A type of vision in animals that have eyes on the sides of the head; the animal sees a wide area at one time, though depth perception is reduced.
monocyte A type of phagocyte that circulates in the blood for only a few days, after which it takes up permanent residence in various organs as a macrophage.
monoecious The term to describe plants that produce carpellate and staminate flowers on the same plant.
monogamy A mating system in which one male mates with one female, and most individuals have mates.
monohybrid The F_1 offspring, also called single-trait hybrids, of true-breeding parents that differ with regard to a single trait.
monohybrid cross A cross in which the inheritance of only one trait is followed.
monomer An organic molecule that can be used to form larger molecules (polymers) consisting of many repeating units of the monomer.
monomorphic gene A gene that exists predominantly as a single allele in a population.
monophagous The term used to describe parasites that feed on one or a few closely related species.
monophyletic group A group of species, a taxon, consisting of the most recent common ancestor and all of its descendants.
monosaccharide A simple sugar.
monosomic An aneuploid organism that has one too few chromosomes.
monotreme A member of the mammalian order Monotremata, which consists of three species found in Australia and New Guinea: the duck-billed platypus and two species of echidna.
morphogen A molecule that imparts positional information and promotes developmental changes at the cellular level.
morphogenetic field A group of cells believed to differentiate into a single body structure.
morphology The structure or form of a body part or an entire organism.
morula An early stage in a mammalian embryo in which physical contact between cells is maximized by compaction.
mosaic An individual with somatic cells that are genetically different from each other.
mosses A phylum of bryophytes; Bryophyta.
motor end plate The region of a skeletal muscle cell that lies beneath an axon terminal at the neuromuscular junction.
motor neuron A neuron that sends signals away from the central nervous system and elicits some type of response from a gland, muscle or other structure.
motor protein A category of cellular proteins that uses ATP as a source of energy to promote movement; consists of three domains called the head, hinge, and tail.
movement corridor Thin strips of habitat that may permit the movement of individuals between larger habitat patches.
M phase The sequential events of mitosis and cytokinesis.
mRNA *See* messenger RNA.
Müllerian mimicry A type of mimicry in which many noxious species converge to look the same, thus reinforcing the basic distasteful design.
multicellular Describes an organism consisting of more than one cell, particularly when cell-to-cell adherence and signaling processes and cellular specialization can be demonstrated.
multimeric protein A protein with more than one polypeptide chain; also said to have a quarternary structure.
multiple alleles Refers to the occurrence of a gene that exists as three or more alleles in a population.
multiple sclerosis (MS) A disease in which the patient's own body attacks and destroys myelin as if it were a foreign substance; impairs the function of myelinated neurons that control movement, speech, memory, and emotion.
multipotent A term used to describe a stem cell that can differentiate into several cell types, but far fewer than pluripotent cells.
muscle A grouping of muscle cells (fibers) bound together by a succession of connective tissue layers.
muscle fiber Individual cell within a muscle.
muscle tissue Bundles of muscle fibers that are specialized to contract when stimulated.
muscular dystrophy A group of diseases associated with progressive degeneration of skeletal and cardiac muscle fibers.
mutagen An agent known to cause mutation.
mutant allele An allele that has been altered by mutation.
mutation A heritable change in the genetic material of an organism.
mutualism A symbiotic interaction in which both species benefit.
myasthenia gravis A disease characterized by loss of acetylcholine receptors on skeletal muscle, due to the body's own immune system destroying the receptors.
mycelium A fungal body composed of microscopic branched filaments known as hyphae.
mycorrhizae (singular, **mycorrhiza**) Associations between the hyphae of certain fungi and the roots of plants.
myelin sheath In the nervous system, an insulating layer made up of specialized glial cells wrapped around the axons.
myocardial infarction (MI) The death of cardiac muscle cells, which can occur if a region of the heart is deprived of blood for an extended time.
myofibril Rodlike collection of myofilaments within a muscle fiber (cell); contains thick and thin filaments.
myogenic heart A heart in which the signaling mechanism that initiates contraction resides within the cardiac muscle itself.
myoglobin An oxygen-binding protein that provides an intracellular reservoir of oxygen for muscle fibers.
myosin A motor protein found abundantly in muscle cells and also in other cell types.

N

NAD^+ Nicotinamide adenine dinucleotide; a dinucleotide that functions as an energy intermediate molecule. It combines with two electrons and H^+ to form NADH.
NADPH Nicotinamide adenine dinucleotide phosphate; an energy intermediate that provides the energy and electrons to drive the Calvin cycle during photosynthesis.
natural killer (NK) cell A type of leukocyte that participates in both nonspecific and specific immunity; recognizes general features on the surface of cancer cells or any virus-infected cells.
natural selection The process that eliminates those individuals that are less likely to survive and reproduce in a particular environment, while allowing other individuals with traits that confer greater reproductive success to increase in numbers.
nauplius The first larval stage in a crustacean.
navigation A mechanism of migration that involves the ability not only to follow a compass bearing but also to set or adjust it.
negative control Transcriptional regulation by repressor proteins.
negative feedback loop A homeostatic system in animals in which a change in the variable being regulated brings about responses that move the variable in the opposite direction.
negative frequency-dependent selection A pattern of natural selection in which the fitness of a genotype decreases when its frequency becomes higher; the result is a balanced polymorphism.
negative pressure filling The mechanism by which reptiles, birds, and mammals ventilate their lungs.
nekton Free-swimming animals in the open ocean that can swim against currents to locate food.
nematocyst In a cnidarian, a powerful capsule with an inverted coiled and barbed thread that functions to immobilize small prey.
neocortex The layer of the brain that evolved most recently in mammals.
nephron One of several million single-cell-thick tubules that are the functional units of the mammalian kidney.
Nernst equation The formula that gives the equilibrium potential for an ion at any given concentration gradient: $E = 60 \text{ mV} \log_{10}([X_{extracellular}]/[X_{intracellular}])$.
nerve A structure found in the peripheral nervous system that is composed of multiple myelinated neurons bound by connective tissue; carries information to or from the central nervous system.
nerve cord In many invertebrates, a ventral structure that extends from the anterior end of the animal to the tail; a dorsal nerve cord is found in chordates.
nerve net Interconnected neurons with no central control organ.
nervous system Groups of cells that sense internal and environmental changes and transmit signals that enable an animal to respond in an appropriate way.
nervous tissue Clusters of cells that initiate and conduct electrical signals from one part of an animal's body to another part.
net primary production (NPP) Gross primary production minus the energy lost in plant cellular respiration.
net reproductive rate The population growth rate per generation.
neural crest In vertebrates, a group of embryonic cells derived from ectoderm that disperse throughout the embryo and contribute to the development of the skeleton and other structures, including peripheral nerves.
neural tube In chordates, a structure formed from ectoderm located dorsal to the notochord; all neurons and their supporting cells in the central nervous system originate from neural precursor cells derived from the neural tube.
neurogenesis The production of new neurons by cell division.
neurogenic heart A heart that will not beat unless it receives regular electrical impulses from the nervous system.
neurohormone A hormone made in and secreted by neurons whose cell bodies are in the hypothalamus.
neuromodulator Another term for a neuropeptide, which is a neurotransmitter that can alter or modulate the response of a postsynaptic neuron to other neurotransmitters.
neuromuscular junction The junction between a motor neuron's axon and a skeletal or cardiac muscle fiber.
neuron A highly specialized cell found in nervous systems of animals that communicates with other cells by electrical or chemical signals.

neuroscience The scientific study of nervous systems.

neurotransmitter A small signaling molecule that is released from an axon terminal and diffuses to a postsynaptic cell where it elicits a response.

neurulation The embryological process responsible for initiating central nervous system formation.

neutral theory of evolution States that most genetic variation is due to the accumulation of neutral mutations that have attained high frequencies in a population via genetic drift.

neutral variation Genetic variation in which natural selection does not favor any particular genotype.

neutron A neutral particle found in the center of an atom.

neutrophil A type of phagocyte and the most abundant type of leukocyte. Neutrophils engulf bacteria by endocytosis.

niche The unique set of habitat resources a species requires as well as its effect on the ecosystem.

nitrification The conversion by soil bacteria of ammonia (NH_3) or ammonium (NH_4^+) to nitrate (NO_3^-), a form of nitrogen commonly used by plants.

nitrogen fixation A specialized metabolic process in which certain prokaryotes use the enzyme nitrogenase to convert inert atmospheric nitrogen gas (N_2) into ammonia (NH_3); also, the industrial process by which humans produce NH_3 fertilizer from N_2.

nitrogenase An enzyme used in the biological process of fixing nitrogen.

nitrogen-limitation hypothesis The proposal that organisms select food based on its nitrogen content.

nitrogenous waste Degradation product of proteins and nucleic acids that are toxic at high concentrations and must be eliminated from the body.

nociceptor A sensory receptor in animals that responds to extreme heat, cold, and pressure, as well as to certain molecules such as acids; also known as a pain receptor.

node The region of a plant stem from which one or more leaves, branches, or buds emerge.

nodes of Ranvier Exposed areas in the axons of myelinated neurons that contain many voltage-gated Na^+ channels and are the sites of regeneration of action potentials.

Nod factor Nodulation factor; a substance produced by nitrogen-fixing bacteria in response to flavonoids secreted from the roots of potential host plants. Nod factors bind to receptors in plant root membranes, starting a process that allows the bacteria to invade roots.

nodule A small swelling on a plant root that contains nitrogen-fixing bacteria.

nodulin One of several plant proteins that foster root nodule development.

noncoding strand *See* template strand.

noncompetitive inhibitor A molecule that binds to an enzyme at a location that is outside the active site and inhibits the enzyme's function.

noncyclic electron flow The combined action of photosystem II and photosystem I in which electrons flow in a linear manner to produce NADPH.

non-Darwinian evolution The idea that much of the modern variation in gene sequences is explained by neutral variation rather than adaptive variation.

nondisjunction An event in which the chromosomes do not sort properly during cell division.

nonpolar covalent bond A strong bond formed between two atoms of similar electronegativities in which electrons are shared between the atoms.

nonpolar molecule A molecule composed predominantly of nonpolar bonds.

nonrandom mating The phenomenon that individuals choose their mates based on their genotypes or phenotypes.

nonrecombinant An offspring whose combination of traits has not changed from the parental generation.

nonsense codon *See* stop codon.

nonsense mutation A mutation that changes a normal codon into a stop codon; this causes translation to be terminated earlier than expected, producing a truncated polypeptide.

nonshivering thermogenesis An increase in an animal's metabolic rate that is not due to increased muscle activity; occurs primarily in brown adipose tissue.

nonspecific immunity *See* innate immunity.

nonvascular plant A plant that does not produce lignified vascular tissue; includes the bryophytes.

norepinephrine A type of neurotransmitter; also known as noradrenaline.

norm of reaction A description of how a trait may change depending on environmental conditions.

notochord A defining characteristic of all chordate embryos; consists of a flexible rod that lies between the digestive tract and the nerve cord.

N-terminus The location of the first amino acid in a polypeptide; also known as the amino terminus.

nuclear envelope A double-membrane structure that encloses the cell's nucleus.

nuclear genome The chromosomes found in the nucleus of a eukaryotic cell.

nuclear lamina A collection of filamentous proteins that line the inner nuclear membrane; part of the nuclear matrix.

nuclear matrix A filamentous network of proteins that is found inside the nucleus and lines the inner nuclear membrane. The nuclear matrix serves to organize the chromosomes.

nuclear pore A passageway for the movement of molecules and macromolecules into and out of the nucleus; formed where the inner and outer nuclear membranes make contact with each other.

nucleic acid An organic molecule composed of nucleotides. The two types of nucleic acids are deoxyribonucleic acid (DNA) and ribonucleic acid (RNA).

nucleoid region A site in a bacterial cell where the genetic material (DNA) is located.

nucleolus A prominent region in the nucleus of nondividing cells where ribosome assembly occurs.

nucleosome A structural unit of eukaryotic chromosomes composed of an octamer of histones (eight histone proteins) wrapped with DNA.

nucleotide An organic molecule having three components: one or more phosphate groups, a five-carbon sugar (either deoxyribose or ribose), and a single or double ring of carbon and nitrogen atoms known as a base.

nucleotide excision repair (NER) A common type of DNA repair system that removes (excises) and repairs a region of the DNA where damage has occurred.

nucleus (plural, **nuclei**) 1. In cell biology, an organelle found in eukaryotic cells that contains most of the cell's genetic material. 2. In chemistry, the region of an atom that contains protons and neutrons. 3. In neurobiology, a group of neuronal cell bodies in the brain that are devoted to a particular function.

nutrient Any substance taken up by a living organism that is needed for survival, growth, development, repair, or reproduction.

nutrition The means of providing nutrients for cell survival.

O

obese According to current National Institutes of Health guidelines, a person having a body mass index (BMI) of 30 kg/m^2 or more.

obligate aerobes Microorganisms that require oxygen.

obligate anaerobes Microorganisms that are poisoned by oxygen.

obligatory mutualism An interaction in which two mutualistic species cannot live without each other.

occipital lobe One of four lobes of the cerebral cortex of the human brain; controls aspects of vision and color recognition.

ocelli Photosensitive organs in some animal species.

octet rule The phenomenon that some atoms are most stable when their outer shell is full with eight electrons.

Okazaki fragments Short segments of DNA synthesized in the lagging strand during DNA replication.

olfaction The sense of smell.

olfactory bulbs Part of the limbic system of the forebrain of vertebrates; the olfactory bulbs carry information about odors to the brain.

oligodendrocytes Glial cells that produce the myelin sheath around neurons in the central nervous system.

oligotrophic The term used to describe aquatic systems that are low in nutrients such as phosphate and combined nitrogen and are consequently low in primary productivity and biomass, but typically high in species diversity.

ommatidium (plural, **ommatidia**) An independent visual unit in the eye of insects that functions as a separate photoreceptor capable of forming an independent image.

omnivore An animal that can survive on plants and animals for food and which typically consumes both.

oncogene A type of mutant gene derived from a proto-oncogene. An oncogene is overactive, thus contributing to uncontrolled cell growth and promoting cancer.

one-gene/one-enzyme hypothesis An early hypothesis by Beadle and Tatum that suggested that one gene encodes one enzyme. It was later modified to the one-gene/one-polypeptide theory.

one-gene/one-polypeptide theory The concept that one structural gene codes for one polypeptide.

oogenesis Gametogenesis in a female animal resulting in the production of an egg cell.

oogonium (plural, **oogonia**) In animals, diploid germ cells that give rise to the female gametes, the eggs.

open circulatory system In animals, a circulatory system in which hemolymph, which is not different than the interstitial fluid, flows throughout the body and is not confined to special vessels.

open complex Also called the transcription bubble; a small bubble-like structure between two DNA strands that occurs during transcription.

open conformation Loosely packed chromatin that can be transcribed into RNA.

operant conditioning A form of behavior modification; a type of associative learning in which an animal's behavior is reinforced by a consequence, either a reward or a punishment.

operator A DNA sequence in bacteria that is recognized by activator or repressor proteins that regulate the level of gene transcription.

operculum A protective flap that covers the gills of a bony fish.

operon An arrangement of two or more genes in bacteria that are under the transcriptional control of a single promoter.

opsin A protein that is a component of visual pigments in the vertebrate eye.

opsonization The process by which an antibody binds to a pathogen and provides a means to link the pathogen with a phagocyte.

optic disc In vertebrates, the point on the retina where the optic nerve leaves the eye.

optic nerve A structure of the vertebrate eye that carries electrical signals to the brain.

optimal foraging The concept that in a given circumstance, an animal seeks to obtain the most energy possible with the least expenditure of energy.

optimality theory The theory that predicts an animal should behave in a way that maximizes the benefits of a behavior minus its costs.

orbital The region surrounding the nucleus of an atom where the probability is high of finding a particular electron.

order In taxonomy, a subdivision of a class.

organ Two or more types of tissue combined to perform a common function.

organelle A subcellular structure or membrane-bound compartment with its own unique structure and function.

organic chemistry The study of carbon-containing molecules.

organic farming The production of crops without the use of commercial inorganic fertilizers, growth substances, and pesticides.

organic molecule A carbon-containing molecule, so named because they are found in living organisms.

organism A living thing that maintains an internal order that is separated from the environment.

organismal ecology The investigation of how adaptations and choices by individuals affect their reproduction and survival.

organismic model A view of the nature of a community that considers it to be equivalent to a superorganism; individuals, populations, and communities have a relationship to each other that resembles the associations found between cells, tissues, and organs.

organizing center A group of cells in a plant shoot meristem that ensures the proper organization of the meristem and preserves the correct number of actively dividing stem cells.

organ of Corti Coiled structure in the vertebrate ear responsible for detecting sound.

organogenesis The developmental stage during which cells and tissues form organs in animal embryos.

organ system Different organs that work together to perform an overall function in an organism.

orientation A mechanism of migration in which animals have the ability to follow a compass bearing and travel in a straight line.

origin of replication A site within a chromosome that serves as a starting point for DNA replication.

ortholog A homologous gene in different species.

osmoconformer An animal whose osmolarity conforms to that of its environment.

osmolarity The solute concentration of a solution of water, expressed as milliosmoles/liter (mOsm/L).

osmoregulator An animal that maintains stable internal salt concentrations and osmolarities, even when living in water with very different osmolarities than its body fluids.

osmosis The movement of water across membranes to balance solute concentrations. Water diffuses from a solution that is hypotonic (lower solute concentration) into a solution that is hypertonic (higher solute concentration).

osmotic adjustment The process by which a plant cell modifies the solute concentration of its cytosol.

osmotic lysis Occurs when a cell in a hypotonic environment takes up so much water that it ruptures.

osmotic pressure The hydrostatic pressure required to stop the net flow of water across a membrane due to osmosis.

osmotroph An organism that relies on osmotrophy (uptake of small organic molecules) as a form of nutrition.

osteichythan A clade that includes all vertebrates with a bony skeleton.

osteomalacia Bone deformation in adults due to inadequate mineral intake or absorption from the intestines.

osteoporosis A disease in which the mineral and organic components of bone are reduced.

otolith Granules of calcium carbonate found in the gelatinous substance that embeds hair cells in the vertebrate ear.

outer bark Protective layers of mostly dead cork cells that cover the outside of woody stems and roots.

outer ear One of the three main compartments of the mammalian ear; consists of the external ear, or pinna, and the auditory canal.

outer segment The highly convoluted plasma membranes found in the rods and cones of the eye.

outgroup In a cladogram, a species or group of species that does not exhibit one or more shared derived characters found in the ingroup.

ovarian cycle The events beginning with the development of an ovarian follicle, followed by release of a secondary oocyte, and concluding with formation and subsequent degeneration of a corpus luteum.

ovary 1. In animals, the female gonad where eggs are formed. 2. In plants, the lowermost portion of the pistil that encloses and protects the ovules.

oviduct A thin tube with undulating fimbriae (finger-like structures) that is connected to the uterus and extends out to the ovary; also called the Fallopian tube.

oviparity Development of an embryo outside the mother, usually in a protective shell or other structure from which the young hatch.

oviparous An animal whose young hatch from eggs laid outside the mother's body.

ovoviparous An animal that retains fertilized eggs covered by a protective sheath or other structure within the body, where the young hatch.

ovoviviparity Development of an embryo involving aspects of both viviparity and oviparity; fertilized eggs covered with a protective sheath are produced and hatch inside the mother's body, but the offspring receive no nourishment from the mother.

ovulation The process by which a mature oocyte is released from an ovary.

ovule In a seed plant, a megaspore-producing megasporangium and enclosing tissues known as integuments.

ovum (plural, **ova**) *See* egg.

oxidation A process that involves the removal of electrons; occurs during the breakdown of small organic molecules.

oxidative fiber A skeletal muscle fiber that contains numerous mitochondria and has a high capacity for oxidative phosphorylation.

oxidative phosphorylation A process during which NADH and $FADH_2$ are oxidized to make more ATP via the phosphorylation of ADP.

oxygen-hemoglobin dissociation curve A curve that represents the relationship between the partial pressure of oxygen and the binding of oxygen to hemoglobin proteins.

oxytocin A hormone secreted by the posterior pituitary gland that stimulates contractions of the smooth muscles in the uterus of a pregnant mammal, facilitating the birth process; after birth, it is important in milk secretion.

P

pacemaker *See* sinoatrial (SA) node.

Pacinian corpuscle Structure located deep beneath the surface of an animal's skin that responds to deep pressure or vibration.

paedomorphosis The retention of juvenile traits in an adult organism.

pair-rule gene A type of segmentation gene; a mutation in this gene may cause alternating segments or parts of segments to be deleted.

paleontologist A scientist who studies fossils.

palisade parenchyma Photosynthetic ground tissue of the plant leaf mesophyll that consists of closely packed, elongate cells adapted to efficiently absorb sunlight.

palmate A type of leaf vein pattern in which veins radiate outward, resembling an open hand.

pancreas In vertebrates, an elongated gland located behind the stomach that secretes digestive enzymes and a fluid rich in bicarbonate ions.

parabronchi In birds, a series of parallel air tubes that make up the lungs and are the regions of gas exchange.

paracrine Refers to a type of cellular communication in which molecules are released into the interstitial fluid and act on nearby cells.

paralogs Homologous genes within a single species.

paraphyletic group A group of organisms that contains a common ancestor and some, but not all, of its descendants.

parapodia Fleshy, footlike structures in the polychaetes that are pushed into the substrate to provide traction during movement.

parasite A predatory organism that feeds off another organism but does not normally kill it.

parasitism A symbiotic association in which one organism feeds off another but does not normally kill it.

parasympathetic division The division of the autonomic nervous system that is involved in maintaining and restoring body functions.

parathyroid hormone (PTH) A hormone that acts on bone to stimulate the activity of cells that dissolve the mineral part of bone.

Parazoa A subgroup of animals lacking specialized tissue types or organs, although they may have several distinct types of cells; the one phylum in this group is the Porifera (sponges).

parenchyma cell A type of plant cell that is thin-walled and alive at maturity.

parenchyma tissue A plant ground tissue that is composed of parenchyma cells.

parental strand The original strand in DNA replication.

parietal lobe One of four lobes of the cerebral cortex of the human brain; receives and interprets sensory input from visual and somatic pathways.

parthenogenesis An asexual process in which an offspring develops from an unfertilized egg.

partial pressure The individual pressure of each gas in the air; the sum of these pressures is known as atmospheric pressure.

particulate inheritance The idea that the determinants of hereditary traits are transmitted intact from one generation to the next.

parturition The birth of an organism.

passive diffusion Diffusion through a membrane without the aid of a transport protein.

passive immunity A type of acquired immunity that confers protection against disease through the direct transfer of antibodies from one individual to another.

passive transport The diffusion of a solute across a membrane in a process that is energetically favorable and does not require an input of energy.

paternal inheritance A pattern in which only the male gamete contributes particular genes to the offspring.

pathogen A microorganism that causes disease symptoms in its host.

pathogen associated molecular pattern (PAMP) Common molecules found in many pathogens that trigger an innate immune response.
pattern formation The process that gives rise to a plant or animal with a particular body structure.
pedal glands Glands in the foot of a rotifer that secrete a sticky substance that aids in attachment to the substrate.
pedicel 1. A flower stalk. 2. A narrow, waistlike point of attachment between the body parts of spiders and some insects.
pedigree analysis An examination of the inheritance of human traits in families.
pedipalps In spiders, a pair of appendages that have various sensory, predatory, or reproductive functions.
peer-review process A procedure in which experts in a particular area evaluate papers submitted to scientific journals.
pelagic zone The open ocean, where the water depth averages 4,000 m and nutrient concentrations are typically low.
penis A male external accessory sex organ found in many animals that is involved in copulation.
pentadactyl limb A limb ending in five digits.
PEP carboxylase An enzyme in C_4 plants that adds CO_2 to phosphoenolpyruvate (PEP) to produce the four-carbon compound oxaloacetate.
pepsin An active enzyme in the stomach that begins the digestion of protein.
peptide bond The covalent bond that links amino acids in a polypeptide.
peptidoglycan A polymer composed of carbohydrates crosslinked with peptides that is an important component of the cell walls of most bacteria.
peptidyl site (P site) One of three sites for tRNA binding in the ribosome during translation; the other two are the aminoacyl site (A site) and the exit site (E site). The P site holds the tRNA carrying the growing polypeptide chain.
peptidyl transfer reaction During translation, the transfer of the polypeptide from the tRNA in the P site to the amino acid at the A site.
per capita growth rate The per capita birth rate minus the per capita death rate; the rate that determines how populations grow over any time period.
perception An awareness of the sensations that are experienced.
perennial A plant that lives for more than 2 years, often producing seeds each year after it reaches reproductive maturity.
perfect flower A flower that has both stamens and carpels.
perianth The term that refers to flower petals and sepals collectively.
pericarp The wall of a plant's fruit.
pericycle A cylinder of plant tissue having cell division (meristematic) capacity that encloses the root vascular tissue.
periderm The outer layers of a woody stem, composed of cork cambium, layers of cork tissue produced by the cork cambium, and associated parenchyma cells, together forming outer bark.
peripheral membrane protein A protein that is noncovalently bound to regions of integral membrane proteins that project out from the membrane, or they are noncovalently bound to the polar head groups of phospholipids.
peripheral nervous system (PNS) In vertebrates, all nerves and ganglia outside the brain and spinal cord.
peripheral zone The area of a plant shoot meristem that contains dividing cells that will eventually differentiate into plant structures.
periphyton Communities of microorganisms that are attached by mucilage to underwater surfaces such as rocks, sand, and plants.
peristalsis In animals, the rhythmic, spontaneous waves of muscle contractions that propel food through the digestive system.
peritubular capillary A capillary near the junction of the cortex and medulla that surrounds the nephron of the mammalian kidney.
permafrost A layer of permanently frozen soil found in tundra.
peroxisome A relatively small organelle found in all eukaryotic cells that catalyzes detoxifying reactions.
personalized medicine A medical practice in which information about a patient's genotype is used to individualize their medical care.
petal A flower organ that usually serves to attract insects or other animals for pollen transport.
petiole A stalk that connects a leaf to the stem of a plant.
P generation The parental generation in a genetic cross.
pH The mathematical expression of a solution's hydrogen ion (H^+) concentration, defined as the negative logarithm to the base 10 of the H^+ concentration.
phage *See* bacteriophage.
phagocyte A cell capable of phagocytosis; phagocytes provide nonspecific defense against pathogens that enter the body.
phagocytic vacuole A vacuole that functions in the degradation of food particles or bacteria; also called a food vacuole.
phagocytosis A form of endocytosis that involves the formation of a membrane vesicle, called a phagocytic vacuole, which engulfs a particle such as a bacterium.
phagotroph An organism that specializes in phagotrophy (particle feeding) by means of phagocytosis as a form of nutrition.
pharyngeal slit A defining characteristic of all chordate embryos. In early-diverging chordates, pharyngeal slits develop into a filter-feeding device, and in some advanced chordates, they form gills.
pharynx A portion of the vertebrate alimentary canal; also known as the throat.
phenolics A group of secondary metabolites that contain a benzene ring covalently linked to a single hydroxyl group. Includes tannins, lignins, and flavonoids.
phenotype The characteristics of an organism that are the result of the expression of its genes.
pheromone A powerful chemical attractant used to manipulate the behavior of others.
phloem A specialized conducting tissue in a plant's stem.
phloem loading The process of conveying sugars to sieve-tube elements for long-distance transport.
phoresy A form of commensalism in which individuals of one species use individuals of a second species for transportation.
phosphodiesterase An enzyme that breaks down cAMP into AMP.
phosphodiester linkage Refers to a double linkage (two phosphoester bonds) that holds together adjacent nucleotides in DNA and RNA strands.
phospholipid A class of lipids that are similar in structure to triglycerides, but the third hydroxyl group of glycerol is linked to a phosphate group instead of a fatty acid; a key component of biological membranes.
phospholipid bilayer The basic framework of the cellular membrane, consisting of two layers of lipids.
phosphorylation The attachment of a phosphate to a molecule.
photic zone A fairly narrow zone close to the surface of an aquatic environment, where light is sufficient to allow photosynthesis to exceed respiration.
photoautotroph An organism that uses the energy from light to make organic molecules from inorganic sources.
photoheterotroph An organism that is able to use light energy to generate ATP but must take in organic compounds from the environment.
photon A discrete particle that makes up light. A photon is massless and travels in a wavelike pattern.
photoperiodism A plant's ability to measure and respond to amounts of light; used as a way of detecting seasonal change.
photopsin A protein in the cone cell of an animal's eye; detects color vision.
photoreceptor A specialized cell in an animal that responds to visible light energy; in plants, molecules that respond to light.
photorespiration The metabolic process occurring in C_3 plants that occurs when the enzyme rubisco combines with O_2 instead of CO_2 and produces only one molecule of 3PG instead of two, thereby reducing photosynthetic efficiency.
photosynthesis The process whereby light energy is captured by plant, algal, or bacterial cells and is used to synthesize organic molecules from CO_2 and H_2O (or H_2S).
photosystem I (PSI) A distinct complex of proteins and pigment molecules in chloroplasts that absorbs light during the light reactions of photosynthesis.
photosystem II (PSII) A distinct complex of proteins and pigment molecules in chloroplasts that generates oxygen from water during the light reactions of photosynthesis.
phototropin The main blue-light sensor involved in phototropism in plants.
phototropism The tendency of a plant to grow toward a light source.
phylogenetic tree A diagram that describes a phylogeny; such a tree is a hypothesis of the evolutionary relationships among various species, based on the information available to and gathered by systematists.
phylogeny The evolutionary history of a species or group of species.
phylum (plural, **phyla**) In taxonomy, a subdivision of a kingdom.
physical mutagen A physical agent, such as ultraviolet light, that causes mutations.
physiological ecology A subdiscipline of organismal ecology that investigates how organisms are physiologically adapted to their environment and how the environment impacts the distribution of species.
phytochrome A red and far-red-light receptor in plants.
phytoplankton Microscopic photosynthetic protists that float in the water column or actively move through water.
phytoremediation The process of removing harmful metals from soils by growing hyperaccumulator plants on metal-contaminated soils, then harvesting and burning the plants to ashes for disposal and/or metal recovery.
pigment A molecule that can absorb light energy.
pili (singular, **pilus**) Threadlike surface appendages that allow bacteria to attach to each other during conjugation or to move across surfaces.
piloting A mechanism of migration in which an animal moves from one familiar landmark to the next.
pinnate A type of leaf vein pattern in which veins appear feather-like.
pinocytosis A form of endocytosis that involves the formation of membrane vesicles from the plasma membrane as a way for cells to internalize the extracellular fluid.
pistil A flower structure that may consist of a single carpel or multiple, fused carpels and is differentiated into stigma, style, and ovary.

pit A thin-walled circular area in a plant cell wall where secondary wall materials such as lignin are absent and through which water moves.
pitch The tone of a sound wave that depends on its length and frequency.
pituitary giant A person who has a tumor of the GH-secreting cells of the anterior pituitary gland and thus produces excess GH during childhood and, if untreated, during adulthood; the person can grow very tall before growth ceases after puberty.
pituitary gland A multilobed endocrine gland sitting directly below the hypothalamus of the brain.
placenta A structure through which humans and other eutherian mammals retain and nourish their young within the uterus via the transfer of nutrients and gases.
placental transfer tissue In plants, a nutritive tissue that aids in the transfer of nutrients from maternal parent to embryo.
plant A multicellular eukaryotic organism that is photosynthetic, generally lives on land, and is adapted in many ways to cope with the environmental stresses of life on land.
Plantae A eukaryotic kingdom of the domain Eukarya.
plant tissue culture A laboratory process to produce thousands of identical plants having the same desirable characteristics.
plaques 1. Deposits of lipids, fibrous tissue, and smooth muscle cells that may develop inside arterial walls. 2. A bacterial biofilm that may form on the surfaces of teeth.
plasma The fluid part of blood that contains water and dissolved solutes.
plasma cell A cell that synthesizes and secretes antibodies.
plasma membrane The biomembrane that separates the internal contents of a cell from its external environment.
plasmid A small circular piece of DNA found naturally in many strains of bacteria and occasionally in eukaryotic cells; can be used as a vector in cloning experiments.
plasmodesma (plural, **plasmodesmata**) A membrane-lined, ER-containing channel that connects the cytoplasm of adjacent plant cells.
plasmogamy The fusion of the cytoplasm between two gametes.
plasmolysis The shrinkage of algal or plant cytoplasm that occurs when water leaves the cell by osmosis, with the result that the plasma membrane no longer presses on the cell wall.
plastid A general name given to organelles found in plant and algal cells that are bound by two membranes and contain DNA and large amounts of either chlorophyll (in chloroplasts), carotenoids (in chromoplasts), or starch (in amyloplasts).
platelets Cell fragments in the blood of mammals that play a crucial role in the formation of blood clots.
pleiotropy The phenomenon in which a mutation in a single gene can have multiple effects on an individual's phenotype.
pleural sac A double layer of moist sheathlike membranes that encases each lung.
pluripotent Refers to the ability of embryonic stem cells to differentiate into almost every cell type of the body.
point mutation A mutation that affects only a single base pair within DNA or that involves the addition or deletion of a single base pair to a DNA sequence.
polar cell The highest latitude cell in the three-cell model of atmospheric circulation.
polar covalent bond A covalent bond between two atoms that have different electronegativities; the shared electrons are closer to the atom of higher electronegativity than the atom of lower electronegativity. This distribution of electrons around the atoms creates a polarity, or difference in electric charge, across the molecule.
polarized 1. In cell biology, refers to cells that have different sides, such as the apical and basal sides of epithelial cells. 2. In neuroscience, refers to the electrical gradient across a neuron's plasma membrane.
polar molecule A molecule containing significant numbers of polar bonds.
polar transport The process whereby auxin flows primarily downward in shoots.
pole A structure of the spindle apparatus defined by each centrosome.
pollen In seed plants, tiny male gametophytes enclosed by sporopollenin-containing microspore walls.
pollen coat A layer of material that covers the sporopollenin-rich pollen wall.
pollen grain The immature male gametophyte of a seed plant.
pollen tube In seed plants, a long, thin tube produced by a pollen grain that delivers sperm to the ovule.
pollen wall A tough, sporopollenin wall at the surface of a pollen grain.
pollination The process in which pollen grains are transported to an angiosperm flower or a gymnosperm cone primarily by means of wind or animal pollinators.
pollination syndromes The pattern of coevolved traits between particular types of flowers and their specific pollinators.
pollinator An animal that carries pollen between angiosperm flowers or cones of gymnosperms.
polyandry A mating system in which one female mates with several males, but males mate with only one female.
poly A tail A string of adenine nucleotides at the 3′ end of most mature mRNAs in eukaryotes.
polycistronic mRNA An mRNA that contains the coding sequences for two or more structural genes.
polycythemia A condition of increased hemoglobin due to increased hematocrit.
polygenic A trait in which several or many genes contribute to the outcome of the trait.
polygyny A mating system in which one male mates with several females in a single breeding season, but females mate with only one male.
polyketides A group of secondary metabolites produced by diverse organisms. Examples include streptomycin, erythromycin, and tetracycline.
polymer A large molecule formed by linking many smaller molecules called monomers.
polymerase chain reaction (PCR) A technique to make many copies of a gene in vitro; primers are used that flank the region of DNA to be amplified.
polymorphic gene A gene that commonly exists as two or more alleles in a population.
polymorphism The phenomenon that many traits or genes may display variation within a population.
polyp A type of cnidarian body form that is sessile and occurs mouth up.
polypeptide A linear sequence of amino acids; the term denotes structure.
polyphagous Parasites that feed on many host species.
polyphyletic group A group of organisms that consists of members of several evolutionary lines and does not include the most recent common ancestor of the included lineages.
polyploid An organism that has three or more sets of chromosomes.
polyploidy In an organism, the state of having three or more sets of chromosomes.
polysaccharide Many monosaccharides linked to form long polymers.
pons The part of the vertebrate hindbrain, along with the cerebellum, responsible for monitoring and coordinating body movements.
population A group of individuals of the same species that occupy the same environment and can interbreed with one another.
population ecology The study of how populations grow and what factors promote or limit growth.
population genetics The study of genes and genotypes in a population.
portal vein A vein that not only collects blood from capillaries—like all veins—but also forms another set of capillaries, as opposed to returning the blood directly to the heart.
positional information Molecules that are provided to a cell that allow it to determine its position relative to other cells.
positive control Transcriptional regulation by activator proteins.
positive feedback loop In animals, the acceleration of a process, leading to what is sometimes called an explosive system.
positive pressure filling The method by which amphibians ventilate their lungs. The animals gulp air and force it under pressure into the lungs, as if inflating a balloon.
postabsorptive state One of two alternating phases in the utilization of nutrients; occurs when the gastrointestinal tract is empty of nutrients and the body's own stores must supply energy. The other phase is the absorptive state.
postanal tail A defining characteristic of all chordate embryos; consists of a tail of variable length that extends posterior to the anal opening.
posterior Refers to the rear (tail-end) of an animal.
postsynaptic cell The cell that receives the electrical or chemical signal sent from a neuron.
post-translational covalent modification A process of changing the structure of a protein, usually by covalently attaching functional groups; this process greatly increases the diversity of the proteome.
post-translational sorting The uptake of proteins into the nucleus, mitochondria, chloroplasts, or peroxisomes that occurs after the protein is completely made in the cytosol (that is, completely translated).
postzygotic isolating mechanism A mechanism that prevents interbreeding by blocking the development of a viable and fertile individual after fertilization has taken place.
potential energy The stored energy that a substance possesses due to its structure or location.
power stroke In muscle, a conformation change in the myosin cross-bridge that results in binding between myosin and actin and the movement of the actin filament.
P protein Phloem protein; the proteinaceous material used by plant phloem as a response to wounding.
prebiotic soup The medium formed by the slow accumulation of organic molecules in the early oceans over a long period of time prior to the existence of life.
predation An interaction in which the action of a predator results in the death of its prey.
predator An animal that kills its prey.
prediction An expected outcome based on a hypothesis that can be shown to be correct or incorrect through observation or experimentation.
pregnancy The time during which a developing embryo and fetus grows within the uterus of the mother. The period of pregnancy is also known as gestation.
preinitiation complex The structure of the completed assembly of RNA polymerase II and GTFs at the TATA box prior to transcription of eukaryotic structural genes.

pre-mRNA In eukaryotes, the mRNA transcript prior to any processing.
pressure-flow hypothesis Explains sugar translocation in plants as a process driven by differences in turgor pressure between cells of a sugar source, where sugar is produced, and cells of a sugar sink, where sugar is consumed.
pressure potential (P) The component of water potential due to hydrostatic pressure.
presynaptic cell The neuron that sends an electrical or chemical signal to another cell.
prezygotic isolating mechanism A mechanism that stops interbreeding by preventing the formation of a zygote.
primary active transport A type of transport that involves pumps that directly use energy to transport a solute against a gradient.
primary cell wall In plants, a relatively thin and flexible cell wall that is synthesized first between two newly made daughter cells.
primary consumer An organism that obtains its food by eating primary producers; also called a herbivore.
primary electron acceptor The molecule to which a high-energy electron from an excited pigment molecule such as P680* is transferred during photosynthesis.
primary endosymbiosis The process by which a eukaryotic host cell acquires prokaryotic endosymbionts. Mitochondria and the plastids of green and red algae are examples of organelles that originated with primary endosymbiosis.
primary growth Plant growth that occurs from primary meristems and produces primary tissues and organs of diverse types.
primary immune response The response to an initial exposure to an antigen.
primary meristem A meristematic tissue that increases plant length and produces new organs.
primary metabolism The synthesis and breakdown of molecules and macromolecules that are found in all forms of life and are essential for cell structure and function.
primary oocyte In animals, a cell that undergoes meiosis to begin the process of egg production.
primary plastid A plastid that originated from a prokaryote as the result of primary endosymbiosis.
primary production Production by autotrophs, normally green plants.
primary spermatocyte In animals, a cell that undergoes meiosis to begin the process of sperm production.
primary structure The linear sequence of amino acids of a polypeptide; one of four levels of protein structure.
primary succession Succession on newly exposed sites that were not previously occupied by soil and vegetation.
primary vascular tissue Plant tissue composed of primary xylem and phloem, which is the conducting tissue of nonwoody plants.
primer A short segment of RNA, typically 10 to 12 nucleotides in length, that is needed to begin DNA replication.
primordial germ cells (PGCs) In animals, the embryonic cells that eventually give rise to gametes.
principle of parsimony The concept that the preferred hypothesis is the one that is the simplest.
principle of species individuality A view of the nature of a community in which each species is distributed according to its physiological needs and population dynamics; most communities intergrade continuously, and competition does not create distinct vegetational zones.
prion An infectious protein that causes disease by inducing the abnormal folding of other protein molecules.
probability The chance that an event will have a particular outcome.
proboscis The coiled tongue of a butterfly or moth, which can be uncoiled, enabling it to drink nectar from flowers.
procambium In plants, a type of primary tissue meristem that produces vascular tissue.
producer An organism that synthesizes the organic compounds used by other organisms for food.
product The end result of a chemical reaction.
production efficiency The percentage of energy assimilated by an organism that becomes incorporated into new biomass.
productivity hypothesis The proposal that greater production by plants results in greater overall species richness.
product rule The probability that two or more independent events will occur is equal to the product of their individual probabilities.
progesterone A hormone secreted by the female ovaries that plays a key role in pregnancy.
progymnosperms An extinct group of plants having wood but not seeds, which evolved before the gymnosperms.
prokaryote One of the two categories into which all forms of life can be placed. Prokaryotes lack a nucleus and include bacteria and archaea.
prokaryotic Refers to organisms having cells lacking a membrane-enclosed nucleus and cell compartmentalization; includes all members of the domains Bacteria and Archaea.
prometaphase The phase of mitosis during which the mitotic spindle is completely formed.
promiscuous In ecology, a term for animals that have different sexual mates every year or breeding season.
promoter The site in the DNA where transcription begins.
proofreading The ability of DNA polymerase to identify a mismatched nucleotide and remove it from the daughter strand.
prophage Refers to the DNA of a phage that has become integrated into a bacterial chromosome.
prophase The phase of mitosis during which the chromosomes condense and the nuclear membrane begins to vesiculate.
proplastid Unspecialized structures that form plastids.
prostate gland A structure in the male reproductive system that secretes a thin fluid that protects sperm once they are deposited within the female reproductive tract.
prosthetic group Small molecules that are permanently attached to the surface of an enzyme and aid in catalysis.
protandrous Describes an animal that is born male but that may become phenotypically female under certain conditions.
protease An enzyme that cuts proteins into smaller polypeptides.
proteasome A molecular machine that is the primary pathway for protein degradation in archaea and eukaryotic cells.
protein A functional unit composed of one or more polypeptides. Each polypeptide is composed of a linear sequence of amino acids.
protein kinase An enzyme that transfers phosphate groups from ATP to a protein.
protein kinase cascade The sequential activation of multiple protein kinases.
protein phosphatase An enzyme responsible for removing phosphate groups from proteins.
protein-protein interactions The specific interactions between proteins that occur during many critical cellular processes.
protein subunit An individual polypeptide within a functional protein; most functional proteins are composed of two or more polypeptides.
proteoglycan A glycosaminoglycan in the extracellular matrix linked to a core protein.
proteolysis A processing event within a cell in which enzymes called proteases cut proteins into smaller polypeptides.
proteome The complete complement of proteins that a cell or organism can make.
proteomics Techniques used to identify and study groups of proteins.
prothoracicotropic hormone (PTTH) A hormone produced in certain invertebrates that stimulates a pair of endocrine glands called the prothoracic glands.
protist A eukaryotic organism that is not a member of the animal, plant, or fungal kingdoms; lives in moist habitats and is typically microscopic in size.
Protista Formerly a eukaryotic kingdom. Protists are now placed into seven eukaryotic supergroups.
protobiont The term used to describe the first nonliving structures that evolved into living cells.
protoderm In plants, a type of primary tissue meristem that generates the outermost dermal tissue.
protogynous Describes an hermaphroditic animal born female but which is capable of assuming a male phenotype at some point in its life history.
proton A positively charged particle found in the nucleus of an atom. The number of protons in an atom is called the atomic number and defines each type of element.
protonephridia Simple excretory organs found in flatworms that are used to filter out wastes and excess water.
proton-motive force *See* H^+ electrochemical gradient.
proto-oncogene A normal gene that, if mutated, can become an oncogene.
protostome An animal whose development exhibits spiral determinate cleavage and in which the blastopore becomes the mouth; includes mollusks, annelid worms, and arthropods.
protozoa A term commonly used to describe diverse heterotrophic protists.
proventriculus The glandular portion of the stomach of a bird.
provirus Refers to viral DNA that has become incorporated into a eukaryotic chromosome.
proximal tubule The segment of the tubule of the nephron in the kidney that drains Bowman's capsule.
proximate cause A specific genetic and physiological mechanism of behavior.
pseudocoelomate An animal with a pseudocoelom.
P site *See* peptidyl site.
pteridophytes A phylum of vascular plants having euphylls, but not seeds; Pteridophyta.
pulmocutaneous circulation The routing of blood from the heart to the gas exchange organs (lungs and skin) of frogs and some other amphibians.
pulmonary circulation The pumping of blood from the right side of the heart to the lungs to pick up oxygen from the atmosphere and release carbon dioxide.
pulmonary hypertension A condition that usually results from a diseased or damaged left ventricle that fails to pump out the usual amount of blood with each beat of the heart. This causes blood to back up in the pulmonary vessels, raising their pressure.
pulmonary trunk A major artery leaving the heart of some air-breathing vertebrates including humans that splits into the left and right pulmonary arteries and brings blood to the lungs.
pulse-chase experiment A procedure in which researchers administer a pulse of radioactively labeled materials to cells so that they make radioactive products. This is followed by the addition of nonlabeled materials called a chase.
pump A transporter that directly couples its conformational changes to an energy source, such as ATP hydrolysis.

punctuated equilibrium A concept that suggests that the tempo of evolution is more sporadic than gradual. Species rapidly evolve into new species followed by long periods of equilibrium with little evolutionary change.
Punnett square A common method for predicting the outcome of simple genetic crosses.
pupa A developmental stage in some insects that undergo metamorphosis; occurs between the larval and adult stages.
pupil A small opening in the eye of a vertebrate that transmits different patterns of light emitted from images in the animal's field of view.
purine The bases adenine (A) and guanine (G), with double rings of carbon and nitrogen atoms.
pyramid of biomass A measure of trophic-level transfer efficiency in which the organisms at each trophic level are weighed.
pyramid of energy A measure of trophic-level transfer efficiency in which rates of energy production are used rather than biomass.
pyramid of numbers An expression of trophic-level transfer efficiency in which the number of individuals decreases at each trophic level, with a huge number of individuals at the base and fewer individuals at the top.
pyrimidine The bases thymine (T), cytosine (C), and uracil (U) with a single ring of carbon and nitrogen atoms.

Q

quadrat A sampling device used by plant ecologists consisting of a square frame that often encloses 0.25 m^2.
quantitative trait A trait that shows continuous variation over a range of phenotypes.
quaternary structure The association of two or more polypeptides to form a protein; one of four levels of protein structure.
quorum sensing A mechanism by which prokaryotic cells are able to communicate by chemical means when they reach a critical population size.

R

radial cleavage A mechanism of animal development in which the cleavage planes are either parallel or perpendicular to the vertical axis of the embryo.
radial loop domain A loop of chromatin, often 25,000 to 200,000 base pairs in size, that is anchored to the nuclear matrix.
radial pattern A characteristic of the body pattern of plants.
radial symmetry 1. In plants, an architectural feature in which embryos display a cylindrical shape, which is retained in the stems and roots of seedlings and mature plants. In addition, new leaves or flower parts are produced in circular whorls, or spirals, around shoot tips. 2. In animals, an architectural feature in which the body can be divided into symmetrical halves by many different longitudinal planes along a central axis.
Radiata Radially symmetric animals; includes cnidarians and ctenophores.
radiation The emission of electromagnetic waves by the surfaces of objects; a method of heat exchange in animals.
radicle An embryonic root, which extends from the plant hypocotyl.
radioisotope An isotope found in nature that is inherently unstable and usually does not exist for long periods of time. Such isotopes decay and emit energy in the form of radiation.
radioisotope dating A common way to estimate the age of a fossil by analyzing the elemental isotopes within the accompanying rock.
radula A unique, protrusible, tonguelike organ in a mollusk that has many teeth and is used to eat plants, scrape food particles off of rocks, or bore into shells of other species.
rain shadow An area on the side of a mountain that is sheltered from the wind and experiences less precipitation.
ram ventilation A mechanism used by fishes to ventilate their gills; fishes swim or face upstream with their mouths open, allowing water to enter into their buccal cavity and across their gills.
random The rarest pattern of dispersion within a population, in which the location of individuals lacks a pattern.
random sampling error The deviation between the observed and the expected outcomes.
rate-limiting step The slowest step in a pathway.
ray-finned fishes The Actinopterygii, which includes all bony fishes except the coelacanths and lungfishes.
reabsorption In the production of urine, the process in which useful solutes in the filtrate are recaptured and transported back into the body fluids of an animal.
reactant A substance that participates in a chemical reaction and becomes changed by that reaction.
realized niche The actual range of an organism in nature.
reading frame Refers to the way in which codons are read during translation, in groups of three bases beginning with the start codon.
receptacle The enlarged region at the tip of a flower peduncle to which flower parts are attached.
receptor 1. A cellular protein that recognizes a signaling molecule. 2. A structure capable of detecting changes in the environment of an animal, such as a touch receptor.
receptor-mediated endocytosis A common form of endocytosis in which a receptor is specific for a given cargo.
receptor potential The membrane potential in a sensory receptor cell of an animal.
receptor tyrosine kinase A type of enzyme-linked receptor found in animal cells that can attach phosphate groups onto tyrosines that are found in the receptor itself or in other cellular proteins.
recessive A term that describes a trait that is masked by the presence of a dominant trait in a heterozygote.
reciprocal translocation A type of mutation in which two different types of chromosomes exchange pieces, thereby producing two abnormal chromosomes carrying translocations.
recombinant An offspring that has a different combination of traits from the parental generation.
recombinant DNA technology The use of laboratory techniques to isolate and manipulate fragments of DNA.
recombinant vector A vector containing a piece of chromosomal DNA.
recombination frequency The frequency of crossing over between two genes.
rectum The last segment of the large intestine of animals that empties into the anus, the posterior opening of the alimentary canal to the external environment.
red blood cell *See* erythrocyte.
redox reaction A type of reaction in which an electron that is removed during the oxidation of an atom or molecule is transferred to another atom or molecule, which becomes reduced; short for a reduction-oxidation reaction.
reduction A process that involves the addition of electrons to an atom or molecule.
reductionism An approach that involves reducing complex systems to simpler components as a way to understand how the system works. In biology, reductionists study the parts of a cell or organism as individual units.
redundancy hypothesis A biodiversity proposal that is an alternative to the rivet hypothesis. In this model, most species are said to be redundant because they could simply be eliminated or replaced by others with no effect.
reflex arc A simple circuit that allows an organism to respond rapidly to inputs from sensory neurons and consists of only a few neurons.
regeneration A form of asexual reproduction in which a complete organism forms from small fragments of its body.
regulatory element In eukaryotes, a DNA sequence that is recognized by regulatory transcription factors and regulates the expression of genes.
regulatory gene A gene whose function is to regulate the expression of other genes.
regulatory sequence In the regulation of transcription, a DNA sequence that functions as a binding site for genetic regulatory proteins. Regulatory sequences control whether a gene is turned on or off.
regulatory transcription factor A protein that binds to DNA in the vicinity of a promoter and affects the rate of transcription of one or more nearby genes.
relative abundance The frequency of occurrence of species in a community.
relative refractory period The period near the end of an action potential when voltage-gated potassium channels are still open; during this time a new action potential can be generated if a stimulus is sufficiently strong to raise the membrane potential to threshold.
relative water content (RWC) The property often used to gauge the water content of a plant organ or entire plant; RWC integrates the water potential of all cells within an organ or plant and is thus a measure of relative turgidity.
release factor A protein that recognizes a stop codon in the termination stage of translation and promotes the termination of translation.
renal corpuscle A filtering component in the nephron of the kidney.
repetitive sequence Short DNA sequences that are present in many copies in a genome.
replica plating A technique in which a replica of bacterial colonies is transferred from one petri plate to a new petri plate.
replication 1. The copying of DNA strands. 2. The performing of experiments several or many times.
replication fork The area where two DNA strands have separated and new strands are being synthesized.
repressible operon In this type of operon, a small effector molecule inhibits transcription.
repressor A transcription factor that binds to DNA and inhibits transcription.
reproduction The generation of offspring by sexual or asexual means.
reproductive cloning The cloning of a multicellular organism, such as a plant or animal.
reproductive isolating mechanisms Mechanisms that prevent interbreeding between different species.
reproductive isolation Refers to the concept that a species cannot successfully interbreed with other species.
reproductive success The likelihood of contributing fertile offspring to the next generation.
resistance (R) The tendency of blood vessels to slow down the flow of blood through their lumens.
resistance gene (*R* gene) A plant gene that has evolved as part of a defense system in response to pathogen attack.

resolution In microscopy, the ability to observe two adjacent objects as distinct from one another; a measure of the clarity of an image.
resonance energy transfer The process by which energy (not an electron itself) can be transferred to adjacent pigment molecules during photosynthesis.
resource partitioning The differentiation of niches, both in space and time, that enables similar species to coexist in a community.
respiration Metabolic reactions that a cell uses to get energy from food molecules and release waste products.
respiratory centers Several regions of the brainstem in vertebrates that initiate expansion of the lungs.
respiratory chain *See* electron transport chain.
respiratory distress syndrome of the newborn The situation in which a human baby is born prematurely, before sufficient surfactant is produced in the lungs, causing the collapse of many alveoli.
respiratory pigment A large protein that contains one or more metal atoms that bind to oxygen.
respiratory system All components of the body that contribute to the exchange of gas between the external environment and the blood; in mammals, includes the nose, mouth, airways, and lungs and the muscles and connective tissues that encase these structures within the thoracic (chest) cavity.
resting potential The difference in charges across the plasma membrane in an unstimulated neuron.
restoration ecology The full or partial repair or replacement of biological habitats and/or their populations that have been damaged.
rest-or-digest The response of vertebrates to situations associated with nonstressful states, such as feeding; mediated by the parasympathetic branch of the autonomic nervous system.
restriction enzyme An enzyme that recognizes particular DNA sequences and cleaves the DNA backbone at two sites.
restriction point A point in the cell cycle in which a cell has become committed to divide.
restriction sites The base sequences recognized by restriction enzymes.
reticular formation An array of nuclei in the brainstem of vertebrates that plays a major role in controlling states such as sleep and arousal.
retina A sheetlike layer of photoreceptors at the back of the vertebrate eye.
retinal A derivative of vitamin A that is capable of absorbing light energy; a component of visual pigments in the vertebrate eye.
retroelement A type of transposable element that moves via an RNA intermediate.
retrovirus An RNA virus that utilizes reverse transcription to produce viral DNA that can be integrated into the host cell genome.
reverse transcriptase A viral enzyme that catalyzes the synthesis of viral DNA starting with viral RNA as a template.
rhizobia The collective term for proteobacteria involved in nitrogen-fixation symbioses with plants.
rhodopsin The visual pigment in the rods of the vertebrate eye.
ribonucleic acid (RNA) One of two classes of nucleic acids; the other is deoxyribonucleic acid (DNA). RNA consists of a single strand of nucleotides.
ribonucleoprotein A complex between an RNA molecule and a protein.
ribose A five-carbon sugar found in RNA.
ribosomal RNA (rRNA) An RNA that forms part of ribosomes, which provide the site where translation occurs.
ribosome A structure composed of proteins and rRNA that provides the site where polypeptide synthesis occurs.
ribozyme A biological catalyst that is an RNA molecule.
rickets A condition in children characterized by bone deformations due to inadequate mineral intake or malabsorption in the intestines.
right-left axis In bilateral animals, one of the three axes along which the adult body pattern is organized; the others are the dorsoventral axis and the anteroposterior axis.
ring canal A central disc in the water vascular system of echinoderms.
RNA *See* ribonucleic acid.
RNA-induced silencing complex (RISC) A complex consisting of miRNA or siRNA and proteins; mediates RNA interference.
RNA interference (RNAi) Refers to a type of mRNA silencing; miRNA or siRNA interferes with the proper expression of an mRNA.
RNA polymerase The enzyme that synthesizes strands of RNA during gene transcription.
RNA processing A step in gene expression between transcription and translation in eukaryotes; the RNA transcript, termed pre-mRNA, is modified in ways that make it a functionally active mRNA.
RNase An enzyme that digests RNA.
RNA world A hypothetical period on primitive Earth when both the information needed for life and the enzymatic activity of living cells were contained solely in RNA molecules.
rod Type of photoreceptor found in the vertebrate eye; they are very sensitive to low-intensity light but do not readily discriminate different colors. Rods are utilized mostly at night, and they send signals to the brain that generate a black-and-white visual image.
root A plant organ that provides anchorage in the soil and also fosters efficient uptake of water and minerals.
root apical meristem (RAM) The region of rapidly dividing cells at plant root tips.
root hair A specialized, long, thin root epidermal cell that functions to absorb water and minerals, usually from soil.
root meristem The collection of cells at the root tip that generate all of the tissues of a plant root.
root pressure Osmotic pressure within roots that causes water to rise for some distance through a plant stem, under conditions of high soil moisture or low transpiration.
root-shoot axis The general body pattern of plants in which the root grows downward and the shoot grows upward.
root system The collection of roots and root branches produced by root apical meristems.
rough endoplasmic reticulum (rough ER) The part of the ER that is studded with ribosomes; this region plays a key role in the initial synthesis and sorting of proteins that are destined for the ER, Golgi apparatus, lysosomes, vacuoles, plasma membrane, or outside of the cell.
rRNA *See* ribosomal RNA.
***r*-selected species** A type of life history strategy, where species have a high rate of per capita population growth but poor competitive ability.
rubisco The enzyme that catalyzes the first step in the Calvin cycle in which CO_2 is incorporated into an organic molecule.
Ruffini corpuscle Tactile (touch) receptors in the skin of mammals that respond to deep pressure and vibration.
ruminants Animals such as sheep, goats, llamas, and cows that have complex stomachs consisting of several chambers.

S

saltatory conduction The conduction of an action potential along an axon in which the action potential is regenerated at each node of Ranvier instead of along the entire length of the axon.
sarcoma A tumor of connective tissue such as bone or cartilage.
sarcomere One complete unit of the repeating pattern of thick and thin filaments within a myofibril.
sarcoplasmic reticulum A cellular organelle that provides a muscle fiber's source of the calcium involved in muscle contraction; a specialized form of the endoplasmic reticulum.
satiety A feeling of fullness.
saturated fatty acid A fatty acid in which all the carbons are linked by single covalent bonds.
scanning electron microscopy (SEM) A type of microscopy that utilizes an electron beam to produce an image of the three-dimensional surface of biological samples.
scavenger An animal that eats the remains of dead animals.
Schwann cells The glial cells that form myelin on axons that travel outside the brain and spinal cord.
science In biology, the observation, identification, experimental investigation, and theoretical explanation of natural phenomena.
scientific method A series of steps to test the validity of a hypothesis. This approach often involves a comparison between control and experimental groups.
sclera The white of the vertebrate eye; a strong outer sheath that in the front is continuous with a thin, clear layer known as the cornea.
sclereid Star- or stone-shaped plant cells having tough, lignified cell walls.
sclerenchyma tissue A rigid plant ground tissue composed of tough-walled fibers and sclereids.
secondary active transport A type of membrane transport that involves the utilization of a pre-existing gradient to drive the active transport of another solute.
secondary cell wall A thick rigid plant cell wall that is synthesized and deposited between the plasma membrane and the primary cell wall after a plant cell matures and has stopped increasing in size.
secondary consumer An organism that eats primary consumers; also called a carnivore.
secondary endosymbiosis A process that occurs when a eukaryotic host cell acquires a eukaryotic endosymbiont having a primary plastid.
secondary growth Plant growth that occurs from secondary meristems and increases the girth of woody plant stems and roots.
secondary immune response An immediate and heightened production of additional specific antibodies against the particular antigen that previously elicited a primary immune response.
secondary meristem A meristem in woody plants forming a ring of actively dividing cells that encircle the stem.
secondary metabolism The synthesis of chemicals that are not essential for cell structure and growth and are usually not required for cell survival but are advantageous to the organism.
secondary metabolite Molecules that are produced by secondary metabolism.
secondary oocyte In animals, the large haploid cell that is produced when a primary oocyte undergoes meiosis I during oogenesis.
secondary phloem The inner bark of a woody plant.
secondary plastid A plastid that has originated by the endosymbiotic incorporation of a eukaryotic cell containing a primary plastid into a eukaryotic host cell.
secondary production The measure of production of heterotrophs and decomposers.

secondary spermatocytes In animals, the haploid cells produced when a primary spermatocyte undergoes meiosis I during spermatogenesis.
secondary structure The bending or twisting of proteins into α helices or β sheets; one of four levels of protein structure.
secondary succession Succession on a site that has previously supported life but has undergone a disturbance.
secondary xylem In plants, a type of secondary vascular tissue that is also known as wood.
second law of thermodynamics States that the transfer of energy or the transformation of energy from one form to another increases the entropy, or degree of disorder, of a system.
second messengers Small molecules or ions that relay signals inside the cell.
secretion 1. The export of a substance from a cell. 2. In the production of urine, the process in which some solutes are actively transported into the tubules of the excretory organ; this supplements the amount of a solute that would normally be removed by filtration alone.
secretory pathway A pathway for the movement of larger substances, such as carbohydrates and proteins, out of a cell.
secretory vesicle A membrane vesicle carrying different types of materials that fuses with the cell's plasma membrane to release the contents extracellularly.
seed A reproductive structure having specialized tissues that enclose plant embryos; produced by gymnosperms and flowering plants, usually as the result of sexual reproduction.
seed coat A hard and tough covering that develops from the ovule's integuments and protects a plant embryo.
seed plant The informal name for gymnosperms and angiosperms.
segmentation The division of an animal's body into clearly defined regions.
segmentation gene A gene that controls the segmentation pattern of an animal embryo.
segment-polarity gene A type of segmentation gene; a mutation in this gene causes portions of segments to be missing either an anterior or a posterior region and for adjacent regions to become mirror images of each other.
segregate To separate, as in chromosomes during mitosis.
selectable marker A gene whose presence can allow organisms (such as bacteria) to grow under a certain set of conditions. For example, an antibiotic-resistance gene is a selectable marker that allows bacteria to grow in the presence of the antibiotic.
selective breeding Programs and procedures designed to modify traits in domesticated species.
selectively permeable The property of membranes that allows the passage of certain ions or molecules but not others.
selective serotonin reuptake inhibitors Drugs used to treat major depressive disorder that act by increasing concentrations of serotonin in the brain.
self-compatible The reproductive state of plants that can serve as both mother and father to their progeny.
self-fertilization Fertilization that involves the union of a female gamete and male gamete from the same individual.
self-incompatibility (SI) Rejection of pollen that is genetically too similar to the pistil of a plant.
selfish DNA hypothesis The hypothesis that transposable elements exist because they have the characteristics that allow them to insert themselves into the host cell DNA but do not provide any advantage.
self-pollination The process in which pollen from the anthers of a flower is transferred to the stigma of the same flower or between flowers of the same plant.
self-splicing The phenomenon that RNA itself can catalyze the removal of its own intron(s); occurs in rRNA and tRNA.
SEM *See* scanning electron microscopy.
semelparity A reproductive pattern in which organisms produce all of their offspring in a single reproductive event.
semen A mixture containing fluid and sperm that is released during ejaculation.
semicircular canal Structures of the vertebrate ear that can detect a range of motions of the head.
semiconservative mechanism The correct model for DNA replication; double-stranded DNA is half conserved following replication, resulting in new double-stranded DNA containing one parental strand and one daughter strand.
semifluid A type of motion within biomembranes; considered two-dimensional because movement occurs only within the plane of the membrane.
semilunar valves One-way valves into the systemic and pulmonary arteries through which blood is pumped from the ventricles.
seminal vesicles Paired accessory glands in the male reproductive system that secrete fructose, the main nutrient for sperm, into the urethra.
seminiferous tubule A tightly packed tubule in the testis, where spermatogenesis takes place.
senescent Cells that have doubled many times and have reached a point where they have lost the capacity to divide any further.
sense A system in an animal that consists of sensory cells that respond to a specific type of chemical or physical stimulus and send signals to the central nervous system, where the signals are received and interpreted.
sensor A structure such as a sensory receptor or a nucleus in the brain that detects a signal in a homeostatic control system.
sensory neuron A neuron that detects or senses information from the outside world, such as light, sound, touch, and heat; sensory neurons also detect internal body conditions such as blood pressure and body temperature.
sensory receptor In animals, a specialized cell whose function is to receive sensory inputs.
sensory transduction The process by which incoming stimuli are converted into neural signals.
sepal A flower organ that occurs in a whorl located outside whorls of petals of eudicot plants.
septum (plural, **septa**) A cross wall; examples include the cross walls that divide the hyphae of most fungi into many small cells and the structure that separates the old and new chambers of a nautilus.
sere Each phase of succession in a community; also called a seral stage.
setae Chitinous bristles in the integument of many invertebrates.
set point The normal value for a controlled variable, such as blood pressure, in an animal.
sex chromosomes A distinctive pair of chromosomes that are different in males and females.
sex-influenced inheritance The phenomenon in which an allele is dominant in one sex but recessive in the other.
sex linked Refers to genes that are found on one sex chromosome but not on the other.
sex pili Hairlike structures made by bacterial F^+ cells that bind specifically to other F^- cells.
sexual dimorphism A pronounced difference in the morphologies of the two sexes within a species.
sexual reproduction A process that requires a fertilization event in which two gametes unite to produce a cell called a zygote.
sexual selection A type of natural selection that is directed at certain traits of sexually reproducing species that make it more likely for individuals to find or choose a mate and/or engage in successful mating.
Shannon diversity index (H_S) A means of measuring the diversity of a community; $H_S = -\Sigma p_i \ln p_i$.
shared derived character A trait that is shared by a group of organisms but not by a distant common ancestor.
shared primitive character A trait shared with a distant ancestor.
shattering The process by which ears of wild grain crops break apart and disperse seeds.
shell A tough, protective covering on an amniotic egg that is impermeable to water and prevents the embryo from drying out.
shivering thermogenesis Rapid muscle contractions in an animal, without any locomotion, in order to raise body temperature.
shoot The portion of a plant comprised of stems and leaves.
shoot apical meristem (SAM) The region of rapidly dividing plant cells at plant shoot apices.
shoot meristem The tissue that produces all aerial parts of the plant, which include the stems as well as lateral structures such as leaves and flowers.
shoot system The collection of plant organs produced by shoot apical meristems.
short-day plant A plant that flowers only when the night length is longer than a defined period.
short stature A condition characterized by stunted growth; formerly called pituitary dwarfism.
short tandem repeat sequences (STRs) Short sequences repeated many times in a row and found in multiple sites in the genome of humans and other species; often vary in length among different individuals.
shotgun DNA sequencing A strategy for sequencing an entire genome by randomly sequencing many different DNA fragments.
sickle cell disease A disease due to a genetic mutation in a hemoglobin gene in which sickle-shaped red blood cells are less able to move smoothly through capillaries and can block blood flow, resulting in severe pain and cell death of the surrounding tissue.
sieve plate The perforated end wall of a mature sieve-tube element.
sieve plate pore One of many perforations in a plant's sieve plate.
sieve-tube elements A component of the phloem tissues of flowering plants; thin-walled cells arranged end to end to form transport pipes.
sigma factor A protein that plays a key role in bacterial promoter recognition and recruits RNA polymerase to the promoter.
signal Regarding cell communication, an incoming or outgoing agent that influences the properties of cells.
signal recognition particle (SRP) A protein/RNA complex that recognizes the ER signal sequence of a polypeptide, pauses translation, and directs the ribosome to the ER to complete translation.
signal transduction pathway A group of proteins that convert an initial signal to a different signal inside a cell.
sign stimulus In animals, a trigger that initiates a fixed-action pattern of behavior.
silencer A regulatory element in eukaryotes that prevents transcription of a given gene.
silencing RNAs (siRNAs) Small RNA molecules, typically 22 nucleotides in length, that silence the expression of specific mRNAs by promoting their degradation.

silent mutation A gene mutation that does not alter the amino acid sequence of the polypeptide, even though the nucleotide sequence has changed.

simple Mendelian inheritance The inheritance pattern of traits affected by a single gene that is found in two variants, one of which is completely dominant over the other.

simple translocation A type of mutation in which a single piece of chromosome is attached to another chromosome.

single-factor cross *See* monohybrid cross.

single lens eye Type of eye found in vertebrates and some invertebrates with only one lens, as opposed compound eyes of insects with many lenses.

single nucleotide polymorphism (SNP) A type of genetic variation in a population in which a particular gene sequence varies at a single nucleotide.

single-strand binding protein A protein that binds to both of the single strands of parental DNA and prevents them from re-forming a double helix during DNA replication.

sinoatrial (SA) node A collection of modified cardiac cells in the right atrium of most vertebrates that spontaneously and rhythmically generates action potentials that spread across the entire atria; also known as the pacemaker of the heart.

siRNAS *See* silencing RNAs.

sister chromatids The two duplicated chromatids that are still joined to each other after DNA replication.

skeletal muscle A type of muscle tissue that is attached by tendons to bones in vertebrates and to the exoskeleton of invertebrates.

skeleton A structure or structures that serve one or more functions related to support, protection, and locomotion.

sliding filament mechanism The way in which a muscle fiber shortens during muscle contraction.

SLOSS debate In conservation biology, the debate over whether it is preferable to protect one single, large reserve or several smaller ones.

slow block to polyspermy Events initiated by the release of Ca^{2+} that produce barriers to more sperm penetrating an already fertilized egg.

slow fiber A skeletal muscle fiber containing myosin with a low rate of ATP hydrolysis.

slow-oxidative fiber A skeletal muscle fiber that has a low rate of myosin ATP hydrolysis but has the ability to make large amounts of ATP; used for prolonged, regular activity.

small effector molecule With regard to transcription, refers to a molecule that exerts its effects by binding to a regulatory transcription factor, causing a conformational change in the protein.

small intestine In vertebrates, a tube that leads from the stomach to the large intestine where nearly all digestion of food and absorption of food nutrients and water occur.

smooth endoplasmic reticulum (smooth ER) The part of the ER that is not studded with ribosomes. This region is continuous with the rough ER and functions in diverse metabolic processes such as detoxification, carbohydrate metabolism, accumulation of calcium ions (Ca^{2+}), and synthesis and modification of lipids.

smooth muscle A type of muscle tissue that surrounds hollow tubes and cavities inside the body's organs; it is not under conscious control.

soil horizon Layers of soil, ranging from topsoil to bedrock.

solute A substance dissolved in a liquid.

solute potential (S) The component of water potential due to the presence of solute molecules.

solution A liquid that contains one or more dissolved solutes.

solvent The liquid in which a solute is dissolved.

somatic cell The type of cell that constitutes all cells of an animal or plant body except those that give rise to gametes.

somatic embryogenesis The production of plant embryos from body (somatic) cells.

somatic nervous system The division of the peripheral nervous system that senses the external environmental conditions and controls skeletal muscles.

somites Blocklike structures resulting from the segmentation of mesoderm during neurulation.

soredia An asexual reproductive structure produced by lichens consisting of small clumps of hyphae surrounding a few algal cells that can disperse in wind currents.

sorting signal A short amino acid sequence in a protein that directs the protein to its correct location; also known as a traffic signal.

source pool The pool of species on the mainland that is available to colonize an island.

spatial summation Occurs when two or more postsynaptic potentials are generated at one time along different regions of the dendrites and their depolarizations and hyperpolarizations sum together.

speciation The formation of new species.

species A group of related organisms that share a distinctive form in nature and (for sexually reproducing species) are capable of interbreeding.

species-area effect The relationship between the amount of available area and the number of species present.

species concepts Different approaches for distinguishing species.

species diversity A measure of biological diversity that incorporates both the number of species in an area and the relative distribution of individuals among species.

species interactions A part of the study of population ecology that focuses on interactions such as predation, competition, parasitism, mutualism, and commensalism.

species richness The numbers of species in a community.

specific heat The amount of energy required to raise the temperature of 1 gram of a substance by 1°C.

specific immunity *See* acquired immunity.

specificity Refers to the concept that enzymes recognize specific substrates.

Spemann's organizer An extremely important morphogenetic field in the early gastrula; the organizer secretes morphogens responsible for inducing the formation of a new embryonic axis.

sperm Refers to a male gamete that is generally smaller than the female gamete (egg).

spermatid In animals, a haploid cell produced when the secondary spermatocytes undergo meiosis II; these cells eventually differentiate into sperm cells.

spermatogenesis Gametogenesis in a male animal resulting in the production of sperm.

spermatogonium In animals, a diploid germ cell that gives rise to the male gametes, the sperm.

spermatophytes All of the living and fossil seed plant phyla.

sperm storage A method of synchronizing the production of offspring with favorable environmental conditions in which female animals store and nourish sperm in their reproductive tract for long periods of time.

S phase The DNA synthesis phase of the cell cycle.

spicules Needle-like structures that are usually made of silica and form lattice-like skeletons in sponges, possibly helping to reduce predation.

spina bifida A developmental abnormality of the neural tube, which fails to close.

spinal cord In chordates, the structure that connects the brain to all areas of the body and together with the brain constitutes the central nervous system.

spinal nerve A nerve that connects the peripheral nervous system and the spinal cord.

spiracle One of several pairs of pores on the body surface of insects through which air enters and exits the body.

spiral cleavage A mechanism of animal development in which the planes of cell cleavage are oblique to the axis of the embryo.

spirilli Rigid, spiral-shaped prokaryotic cells.

spirochaetes Flexible, spiral-shaped prokaryotic cells.

spliceosome A complex of several subunits known as snRNPs that removes introns from eukaryotic pre-mRNA.

splicing The process whereby introns are removed from RNA and the remaining exons are connected to each other.

spongin A tough protein that lends skeletal support to a sponge.

spongocoel A central cavity in the body of a sponge.

spongy parenchyma Photosynthetic ground tissue of the plant leaf mesophyll that contains round cells separated by abundant air spaces.

spontaneous mutation A mutation resulting from abnormalities in biological processes.

sporangia Structures that produce and disperse the spores of plants, fungi, or protists.

sporangium Singular of sporangia.

spore A haploid, typically single-celled reproductive structure of fungi and plants that is dispersed into the environment and is able to grow into a new fungal mycelium or plant gametophyte in a suitable habitat.

sporic life cycle *See* alternation of generations.

sporophyte The diploid generation of plants or multicellular protists that have a sporic life cycle; this generation produces haploid spores by the process of meiosis.

sporopollenin The tough material that composes much of the walls of plant spores and helps to prevent cellular damage during transport in air.

stabilizing selection A pattern of natural selection that favors the survival of individuals with intermediate phenotypes.

stamen A flower organ that produces the male gametophyte, pollen.

standard metabolic rate (SMR) The metabolic rate of ectotherms measured at a standard temperature for each species—one that approximates the average temperature that a species normally encounters.

standing crop The total biomass in an ecosystem at any one point in time.

starch A polysaccharide composed of repeating glucose units that is produced by the cells of plants and some algal protists.

start codon A three-base sequence—usually AUG—that specifies the first amino acid in a polypeptide.

statocyst An organ of equilibrium found in many invertebrate species.

statolith 1. Tiny granules of sand or other dense objects located in a statocyst that aid equilibrium in many invertebrates. 2. In plants, a starch-heavy plastid that allows both roots and shoots to detect gravity.

stem A plant organ that produces buds, leaves, branches, and reproductive structures.

stem cell A cell that divides so that one daughter cell remains a stem cell and the other can differentiate into a specialized cell type. Stem cells construct the bodies of all animals and plants.

stereocilia Deformable projections from epithelial cells called hair cells that are bent by movements of fluid or other stimuli.

stereoisomers Isomers with identical bonding relationships, but different spatial positioning of their atoms.

sternum The breastbone of a vertebrate.

steroid A lipid containing four interconnected rings of carbon atoms; functions as a hormone in animals and plants.

steroid receptor A transcription factor that recognizes a steroid hormone and usually functions as a transcriptional activator.

sticky ends Single-stranded ends of DNA fragments that will hydrogen-bond to each other due to their complementary sequences.

stigma In a flower, the topmost portion of the pistil, which receives and recognizes pollen of the appropriate species or genotype.

stomach A saclike organ in some animals that most likely evolved as a means of storing food; it partially digests some of the macromolecules in food and regulates the rate at which the contents empty into the small intestine.

stomata Surface pores on plant surfaces that can be closed to retain water or open to allow the entry of CO_2 needed for photosynthesis and the exit of O_2 and water vapor.

stop codon One of three three-base sequences—UAA, UAG, and UGA—that signals the end of translation; also called termination codon or nonsense codon.

strain Within a given species, a lineage that has genetic differences compared to another lineage.

strand A structure of DNA (or RNA) formed by the covalent linkage of nucleotides in a linear manner.

strepsirrhini Smaller species of primates; includes bush babies, lemurs, and pottos.

streptophyte Land pants (embryophytes) and close relatives among the green algae.

streptophyte algae The green algae that are closely related to land plants (embryophytes).

stretch receptor A type of mechanoreceptor found widely in an animal's organs and muscle tissues that can be distended.

striated muscle Skeletal and cardiac muscle with a series of light and dark bands perpendicular to the muscle's long axis.

stroke The condition that occurs when blood flow to part of the brain is disrupted.

stroke volume (SV) The amount of blood ejected with each beat, or stroke, of the heart.

stroma The fluid-filled region of the chloroplast between the thylakoid membrane and the inner membrane.

stromatolite A layered calcium carbonate structure in an aquatic environment generally produced by cyanobacteria.

strong acid An acid that completely ionizes in solution.

structural gene Refers to most genes, which produce an mRNA molecule that contains the information to specify a polypeptide with a particular amino acid sequence.

structural isomers Isomers that contain the same atoms but in different bonding relationships.

style In a flower, the elongate portion of the pistil through which the pollen tube grows.

stylet A sharp, piercing organ in the mouth of nematodes and some insects.

submetacentric A chromosome in which the centromere is off center.

subsidence zones Areas of high pressure that are the sites of the world's tropical deserts because the subsiding air is relatively dry, having released all of its moisture over the equator.

subspecies A subdivision of a species; this designation is used when two or more geographically restricted groups of the same species differ, but not enough to warrant their placement into separate species.

substrate 1. The reactant molecules and/or ions that bind to an enzyme at the active site and participate in a chemical reaction. 2. The organic compounds such as soil or rotting wood that fungi use as food.

substrate-level phosphorylation A method of synthesizing ATP that occurs when an enzyme directly transfers a phosphate from an organic molecule to ADP.

succession The gradual and continuous change in species composition and community structure over time.

sugar sink The plant tissues or organs in which more sugar is consumed than is produced by photosynthesis.

sugar source The plant tissues or organs that produce more sugar than they consume in respiration.

sum rule The probability that one of two or more mutually exclusive outcomes will occur is the sum of the probabilities of the possible outcomes.

supergroup One of the seven subdivisions of the domain Eukarya.

surface area-to-volume (SA/V) ratio The ratio between a structure's surface area and the volume in which the structure is contained.

surface tension A measure of how difficult it is to break the interface between a liquid and air.

surfactant A mixture of proteins and amphipathic lipids produced in certain alveolar cells that prevents the collapse of alveoli by reducing surface tension in the lungs.

survivorship curve A graphical plot of the numbers of surviving individuals at each age in a population.

suspension feeder An aquatic animal that sifts water, filtering out the organic matter and expelling the rest.

suspensor A short chain of cells at the base of an early angiosperm embryo that provides anchorage and nutrients.

swim bladder A gas-filled, balloon-like structure that helps a fish to remain buoyant in the water even when the fish is completely stationary.

symbiosis An intimate association between two or more organisms of different species.

symbiotic Describes a relationship in which two or more different species live in direct contact with each other.

sympathetic division The division of the autonomic nervous system that is responsible for rapidly activating body systems to provide immediate energy in response to danger or stress.

sympatric The term used to describe species occurring in the same geographic area.

sympatric speciation A form of speciation that occurs when members of a species that initially occupy the same habitat within the same range diverge into two or more different species.

symplast All of a plant's protoplasts (the cell contents without the cell walls) and plasmodesmata.

symplastic transport The movement of a substance from the cytosol of one cell to the cytosol of an adjacent cell via membrane-lined channels called plasmodesmata.

symplesiomorphy *See* shared primitive character.

symporter A type of transporter that binds two or more ions or molecules and transports them in the same direction across a membrane; also called a cotransporter.

synapomorphy *See* shared derived character.

synapse A junction where a nerve terminal meets a target neuron, muscle cell, or gland and through which an electrical or chemical signal passes.

synapsis The process of forming a bivalent.

synaptic cleft The extracellular space between a neuron and its target cell.

synaptic plasticity The formation of additional synaptic connections that occurs as a result of learning.

synaptic signaling A specialized form of paracrine signaling that occurs in the nervous system of animals.

synergids In the female gametophyte of a flowering plant, the two cells adjacent to the egg cell that help to import nutrients from maternal sporophyte tissues.

syntrophy The phenomenon in which one species lives off the products of another species.

systematics The study of biological diversity and evolutionary relationships among organisms, both extinct and modern.

systemic acquired resistance (SAR) A whole-plant defensive response to pathogenic microorganisms.

systemic circulation The pumping of blood from the left side of an animal's heart to the body to drop off O_2 and nutrients and pick up CO_2 and wastes. The blood then returns to the right side of the heart.

systemic hypertension An arterial blood pressure above normal; in humans, normal blood pressure ranges from systolic/diastolic pressures of about 90/60 to 120/80 mmHg; often called hypertension or high blood pressure.

systems biology A field of study in which researchers investigate living organisms in terms of their underlying networks—groups of structural and functional connections—rather than their individual molecular components.

systole The second phase of the cardiac cycle, in which the ventricles contract and eject the blood through the open semilunar valves.

T

tagmata The fusion of body segments into functional units.

taproot system The root system of eudicots, consisting of one main root with many branch roots.

taste buds Structures located in the mouth and tongue of vertebrates that contain the sensory cells, supporting cells, and associated neuronal endings that contribute to taste sensation.

TATA box One of three features found in most eukaryotic promoters; the others are the transcriptional start site and regulatory elements.

taxis A directed type of response to a stimulus that is either toward or away from the stimulus.

taxon A group of species that are evolutionarily related to each other. In taxonomy, each species is placed into several taxons that form a hierarchy from large (domain) to small (genus).

taxonomy The field of biology that is concerned with the theory, practice, and rules of classifying living and extinct organisms and viruses.

T cells A type of lymphocyte that directly kills infected, mutated, or transplanted cells.

telocentric A chromosome in which the centromere is at the end.

telomerase An enzyme that catalyzes the replication of the telomere.

telomere A region at the ends of eukaryotic chromosomes where a specialized form of DNA replication occurs.

telophase The phase of mitosis during which the chromosomes decondense and the nuclear membrane re-forms.

TEM *See* transmission electron microscopy.

temperate phage A bacteriophage that may spend some of its time in the lysogenic cycle.

template strand The DNA strand that is used as a template for RNA synthesis or DNA replication.

temporal lobe One of four lobes of the cerebral cortex of human brain; necessary for language, hearing, and some types of memory.

temporal summation Occurs when two or more postsynaptic potentials arrive at the same location in a dendrite in quick succession and their depolarizations and hyperpolarizations sum together.

tepal A flower perianth part that cannot be distinguished by appearance as a petal or a sepal.

termination codon *See* stop codon.

termination stage The final stage of transcription or translation in which the process ends.

terminator A sequence that specifies the end of transcription.

terpenoids A group of secondary metabolites synthesized from five-carbon isoprene units. An example is β-carotene, which gives carrots their orange color.

territory A fixed area in which an individual or group excludes other members of its own species, and sometimes other species, by aggressive behavior or territory marking.

tertiary consumer An organism that feeds on secondary consumers.

tertiary endosymbiosis The acquisition by eukaryotic protist host cells of plastids from cells that possess secondary plastids.

tertiary plastid A plastid acquired by the incorporation into a host cell of an endosymbiont having a secondary plastid.

tertiary structure The three-dimensional shape of a single polypeptide; one of four levels of protein structure.

testcross A cross to determine if an individual with a dominant phenotype is a homozygote or a heterozygote. Also, a cross to determine if two different genes are linked.

testis (plural, **testes**) In animals, the male gonad, where sperm are produced.

testosterone The primary androgen in many vertebrates, including humans.

tetrad *See* bivalent.

tetraploid An organism or cell that has four sets of chromosomes.

tetrapod A vertebrate animal having four legs or leglike appendages.

thalamus A region of the vertebrate forebrain that plays a major role in relaying sensory information to appropriate parts of the cerebrum and, in turn, sending outputs from the cerebrum to other parts of the brain.

theory In biology, a broad explanation of some aspect of the natural world that is substantiated by a large body of evidence. Biological theories incorporate observations, hypothesis testing, and the laws of other disciplines such as chemistry and physics. A theory makes valid predictions.

thermodynamics The study of energy interconversions.

thermoreceptor A sensory receptor in animals that responds to cold and heat.

theropods A group of bipedal saurischian dinosaurs.

thick filament A section of the repeating pattern in a myofibril composed almost entirely of the motor protein myosin.

thigmotropism Touch responses in plants.

thin filament A section of the repeating pattern in a myofibril that contains the cytoskeletal protein actin, as well as two other proteins—troponin and tropomyosin—that play important roles in regulating contraction.

30-nm fiber Nucleosome units organized into a more compact structure that is 30 nm in diameter.

thoracic breathing Breathing in which coordinated contractions of muscles expand the rib cage, creating a negative pressure to suck air in and then forcing it out later; found in amniotes.

threatened species Those species that are likely to become endangered in the future.

threshold concentration The concentration above which a morphogen will exert its effects but below which it is ineffective.

threshold potential The membrane potential, typically around −50 to −55 mV, which is sufficient to trigger an action potential in an electrically excitable cell such as a neuron.

thrifty genes Genes that boosted our ancestors' ability to store fat from each feast in order to sustain them through the next famine.

thrombocyte Intact cell in the blood of vertebrates other than mammals that plays a crucial role in the formation of blood clots; in mammals, cell fragments called platelets serve this function.

thylakoid A flattened, platelike membranous region found in cyanobacterial cells and the chloroplasts of photosynthetic protists and plants; the location of the light reactions of photosynthesis.

thylakoid lumen The fluid-filled compartment within the thylakoid.

thylakoid membrane A membrane within the chloroplast that forms many flattened, fluid-filled tubules that enclose a single, convoluted compartment. It contains chlorophyll and is the site where the light-dependent reactions of photosynthesis occurs.

thymine (T) A pyrimidine base found in DNA.

thymine dimer In DNA, a type of pyrimidine dimer that can cause a mutation; a site where two adjacent thymine bases become covalently crosslinked to each other.

thyroglobulin A protein found in the colloid of the thyroid gland that is involved in the formation of thyroid hormones.

thyroxine (T_4) A weakly active thyroid hormone that contains iodine and helps regulate metabolic rate; it is converted by cells into the more active triiodothyronine (T_3).

tidal ventilation A type of breathing in mammals in which the lungs are inflated with air and then the chest muscles and diaphragm relax and recoil back to their original positions as an animal exhales. During exhalation, air leaves via the same route that it entered during inhalation, and no new oxygen is delivered to the airways at that time.

tidal volume The volume of air that is normally breathed in and out at rest.

tight junction A type of junction between animal cells that forms a tight seal between adjacent epithelial cells and thereby prevents molecules from leaking between cells; also called an occluding junction.

Ti plasmid Tumor-inducing plasmid found in *Agrobacterium tumefaciens*; it is used as a cloning vector to transfer genes into plant cells.

tissue The association of many cells of the same type, for example, muscle tissue.

tolerance A mechanism for succession in which any species can start the succession, but the eventual climax community is reached in a somewhat orderly fashion; early species neither facilitate nor inhibit subsequent colonists.

Toll-like receptor Receptor proteins that recognize nonspecific antigens in microbes; key part of the innate immune system.

tonoplast The membrane of the central vacuole in a plant or algal cell.

topsoil The uppermost layer of a soil.

torpor The strategy in endotherms of lowering internal body temperature to just a few degrees above that of the environment in order to conserve energy.

torus The nonporous, flexible central region of a conifer pit that functions like a valve.

total fertility rate The average number of live births a female has during her lifetime.

total peripheral resistance (TPR) The sum of all the resistance in all arterioles.

totipotent The ability of a fertilized egg to produce all of the cell types in the adult organism; also the ability of unspecialized plant cells to regenerate an adult plant.

toxins Compounds that have adverse effects in living organisms; often produced by various protist and plant species.

trace element An element that is essential for normal function in living organisms but is required in extremely small quantities.

trachea 1. A sturdy tube arising from the spiracles of an insect's body; involved in respiration. 2. The name of the tube leading to the lungs of air-breathing vertebrates.

tracheal system The respiratory system of insects consisting of a series of finely branched air tubes called tracheae; air enters and exits the tracheae through spiracles, which are pores on the body surface.

tracheary elements Water-conducting cells in plants that, when mature, are always dead and empty of cytosol; include tracheids and vessel elements.

tracheid A type of dead, lignified plant cell in xylem that conducts water, along with dissolved minerals; also provides structural support.

tracheophytes A term used to describe vascular plants.

tract A parallel bundle of myelinated axons in the central nervous system.

traffic signal *See* sorting signal.

trait An identifiable characteristic; usually refers to a variant.

transcription The use of a gene sequence to make a copy of RNA.

transcriptional start site The site in a eukaryotic promoter where transcription begins.

transcription factor A protein that influences the ability of RNA polymerase to transcribe genes.

transduction A type of genetic transfer between bacteria in which a virus infects a bacterial cell and then subsequently transfers some of that cell's DNA to another bacterium.

***trans*-effect** In both prokaryotes and eukaryotes, a form of genetic regulation that can occur even though two DNA segments are not physically adjacent. The action of the lac repressor on the *lac* operon is a *trans*-effect.

transepithelial transport The process of moving solutes across an epithelium, such as in the gut of animals.

transfer RNA (tRNA) An RNA that carries amino acids and is used to translate mRNA into polypeptides.

transformation A type of genetic transfer between bacteria in which a segment of DNA from the environment is taken up by a competent cell and incorporated into the bacterial chromosome.

transgenic The term used to describe an organism that carries genes that were introduced using molecular techniques such as gene cloning.

transitional form An organism that provides a link between earlier and later forms in evolution.

transition state In a chemical reaction, a state in which the original bonds have stretched to their limit; once this state is reached, the reaction can proceed to the formation of products.

translation The process of synthesizing a specific polypeptide on a ribosome.

translocation 1. A type of mutation in which one segment of a chromosome becomes attached to a different chromosome. 2. A process in plants in which phloem transports substances from a source to a sink.

transmembrane gradient A situation in which the concentration of a solute is higher on one side of a membrane than on the other.

transmembrane protein A protein that has one or more regions that are physically embedded in the hydrophobic region of a cell membrane's phospholipid bilayer.

transmembrane segment A region of a membrane protein that is a stretch of nonpolar amino acids that spans or traverses the membrane from one leaflet to the other.

transmembrane transport The export of material from one cell into the intercellular space and then into an adjacent cell.

transmission electron microscopy (TEM) A type of microscopy in which a beam of electrons is transmitted through a biological sample to form an image on a photographic plate or screen.

transpiration The evaporative loss of water from plant surfaces into sun-heated air.

transporter A membrane protein that binds a solute and undergoes a conformational change to allow the movement of the solute across a membrane; also called a carrier.

transport protein Proteins embedded within the phospholipid bilayer that allow plasma membranes to be selectively permeable by providing a passageway for the movement of some but not all substances across the membrane.

transposable element (TE) A segment of DNA that can move from one site to another.

transposase An enzyme that facilitates transposition.

transposition The process in which a short segment of DNA moves within a cell from its original site to a new site in the genome.

transverse tubules (T-tubules) Invaginations of the plasma membrane of skeletal muscle cells that open to the extracellular fluid and conduct action potentials from the outer surface to the myofibrils.

triacylglycerol *See* triglyceride.

trichome A projection, often hairlike, from the epidermal tissue of a plant that offers protection from excessive light, ultraviolet radiation, extreme air temperature, or attack by herbivores.

triglyceride A molecule composed of three fatty acids linked by ester bonds to a molecule of glycerol; also known as a triacylglycerol.

triiodothyronine (T_3) A thyroid hormone that contains iodine and helps regulate metabolic rate.

triplet A group of three bases that function as a codon.

triploblastic Having three distinct germ layers: endoderm, ectoderm, and mesoderm.

triploid An organism or cell that has three sets of chromosomes.

trisomic An aneuploid organism that has one too many chromosomes.

tRNA *See* transfer RNA.

trochophore larva A distinct larval stage of many invertebrate phyla.

trophectoderm The outer layer of cells in a developing mammalian blastocyst; continuous with the ectoderm layer.

trophic level Each feeding level in a food chain.

trophic-level transfer efficiency The amount of energy at a trophic level that is acquired by the trophic level above and incorporated into biomass.

trophic mutualism A mutually beneficial interaction between two species in which both species receive the benefit of resources.

tropism In plants, a growth response that is dependent on a stimulus that occurs in a particular direction.

tropomyosin A rod-shaped protein that plays an important role in regulating muscle contraction.

troponin A small globular-shaped protein that plays an important role in regulating muscle contraction through its ability to bind Ca^{2+}.

***trp* operon** An operon of *E. coli* that encodes enzymes required to make the amino acid tryptophan, a building block of cellular proteins.

true-breeding line A strain that continues to exhibit the same trait after several generations of self-fertilization or inbreeding.

trypsin A protease involved in the breakdown of proteins in the small intestine.

T-snare A protein in a target membrane that recognizes a V-snare in a membrane vesicle.

tubal ligation A means of contraception that involves the cutting and sealing of the fallopian tubes in a woman, thereby preventing movement of a fertilized egg into the uterus.

tube cell In a seed plant, one of the cells resulting from the division of a microspore; stores proteins and forms the pollen tube.

tube feet Echinoderm structures that function in movement, gas exchange, feeding, and excretion.

tumor An abnormal overgrowth of cells.

tumor-suppressor gene A gene that when normal (that is, not mutant) encodes a protein that prevents cancer; however, when a mutation eliminates its function, cancer may occur.

tunic A nonliving structure that encloses a tunicate, made of protein and a cellulose-like material called tunicin.

turgid The term used to describe a plant cell whose cytosol is so full of water that the plasma membrane presses right up against the cell wall; as a result, turgid cells are firm or swollen.

turgor pressure *See* osmotic pressure.

20-hydroxyecdysone A hormone produced by the prothoracic glands of arthropods that stimulates molting.

two-factor cross *See* dihybrid cross.

type 1 diabetes mellitus (T1DM) A disease in which the pancreas does not produce sufficient insulin; as a result, extracellular glucose cannot cross plasma membranes, and glucose accumulates to very high concentrations in the blood.

type 2 diabetes mellitus (T2DM) A disease in which the pancreas produces sufficient insulin, but the cells of the body lose much of their ability to respond to insulin.

U

ubiquitin A small protein in eukaryotic cells that directs unwanted proteins to a proteasome by its covalent attachment.

ulcer An erosion of the mucosal surface of the alimentary canal; typically occurs in the lower esophagus, stomach, or duodenum.

ultimate cause The reason a particular behavior evolved, in terms of its effect on reproductive success.

umbrella species A species whose habitat requirements are so large that protecting them would protect many other species existing in the same habitat.

unconditioned response An action that is elicited by an unconditioned stimulus.

unconditioned stimulus A trigger that elicits an original response.

uniform A pattern of dispersion within a population in which individuals maintain a certain minimum distance between themselves to produce an evenly spaced distribution.

uniporter A type of transporter that binds a single ion or molecule and transports it across a membrane.

unipotent A term used to describe a stem cell found in the adult that can produce daughter cells that differentiate into only one cell type.

unsaturated The quality of certain lipids containing one or more CwC double bonds.

unsaturated fatty acid A fatty acid that contains one or more CwC double bonds.

upwelling In the ocean, a process that carries mineral nutrients from the bottom waters to the surface.

uracil (U) A pyrimidine base found in RNA.

urea A nitrogenous waste commonly produced in many terrestrial species, including mammals.

uremia A condition characterized by the presence of nitrogenous wastes, such as urea, in the blood; typically results from kidney disease.

ureter A structure in the mammalian urinary system through which urine flows from the kidney into the urinary bladder.

urethra The structure in the mammalian urinary system through which urine is eliminated from the body.

uric acid A nitrogenous waste produced by birds, insects, and reptiles.

urinary bladder The structure in the mammalian urinary system that collects urine before it is eliminated.

urinary system The structures that collectively act to filter blood or hemolymph and excrete wastes, while recapturing useful compounds.

urine The part of the filtrate formed in the kidney that remains after all reabsorption of solutes and water is complete.

uterine cycle *See* menstrual cycle.

uterus A small, pear-shaped organ capable of enlarging and specialized for carrying a developing fetus in female mammals.

V

vaccination The injection into the body of small quantities of weakened or dead pathogens, resulting in the development of immunity to those pathogens without causing disease.

vacuole Specialized compartments found in eukaryotic cells that function in storage, the regulation of cell volume, and degradation.

vagina The birth canal of female mammals; also functions to receive sperm during copulation.

vaginal diaphragm A barrier method of preventing fertilization in which a diaphragm is placed in the upper part of the vagina just prior to intercourse; blocks movement of sperm to the cervix.

valence electron An electron in the outer shell of an atom that is available to combine with other atoms. Such electrons allow atoms to form chemical bonds with each other.

van der Waals forces Attractive forces between molecules in close proximity to each other, caused by the variations in the distribution of electron density around individual atoms.

variable region A unique domain within an immunoglobulin that serves as the antigen-binding site.

vasa recta capillaries Capillaries in the medulla in the nephron of the kidney.

vascular bundle Primary plant vascular tissues that occur in a cluster.

vascular cambium A secondary meristematic tissue of plants that produces both wood and inner bark.

vascular plant A plant that contains vascular tissue. Includes all modern plant species except liverworts, hornworts, and mosses.

vascular tissue Plant tissue that provides both structural support and conduction of water, minerals, and organic compounds.

vas deferens A muscular tube through which sperm leave the epididymis.

vasectomy A surgical procedure in men that severs the vas deferens, thereby preventing the release of sperm at ejaculation.
vasoconstriction A decrease in blood vessel radius; an important mechanism for directing blood flow away from specific regions of the body.
vasodilation An increase in blood vessel radius; an important mechanism for directing blood flow to specific regions of the body.
vasotocin A peptide hormone that is responsible for regulating salt and water balance in the blood of nonmammalian vertebrates.
vector A type of DNA that acts as a carrier of a DNA segment that is to be cloned.
vegetal pole In triploblast organisms, the pole of the egg where the yolk is most concentrated.
vegetative growth The production of new nonreproductive tissues by the shoot apical meristem and root apical meristem during seedling development and growth of mature plants.
vein 1. In animals, a blood vessel that returns blood to the heart. 2. In plants, a bundle of vascular tissue in a leaf.
veliger In mollusks, a free-swimming larva that has a rudimentary foot, shell, and mantle.
ventilation The process of bringing oxygenated water or air into contact with a respiratory surface such as gills or lungs.
ventral Refers to the lower side of an animal.
ventricle In the heart, a chamber that pumps blood out of the heart.
venule A small, thin-walled extension of a capillary that empties into larger vessels called veins that return blood to the heart for another trip around the circulation.
vertebrae A bony or cartilaginous column of interlocking structures that provides support and also protects the nerve cord, which lies within its tubelike structure.
vertebrate An organism with a backbone.
vertical evolution A process in which species evolve from pre-existing species by the accumulation of mutations.
vesicle A small membrane-enclosed sac within a cell.
vessel In a plant, a pipeline-like file of dead, water-conducting vessel elements.
vessel element A type of plant cell in xylem that conducts water, along with dissolved minerals and certain organic compounds.
vestibular system The organ of balance in vertebrates, located in the inner ear next to the cochlea.
vestigial structure An anatomical feature that has no apparent function but resembles a structure of a presumed ancestor.
vibrios Comma-shaped prokaryotic cells.
villus (plural, **villi**) Finger-like projections extending from the luminal surface into the lumen of the small intestine; these are specializations that aid in digestion and absorption.
viral envelope A structure enclosing a viral capsid that consists of a membrane derived from the plasma membrane of the host cell; is embedded with virally encoded spike glycoproteins.
viral genome The genetic material of a virus.
viral reproductive cycle The series of steps that result in the production of new viruses during a viral infection.
viral vector A type of vector used in cloning experiments that is derived from a virus.
viroid An RNA particle that infects plant cells.
virulence The ability of a microorganism to cause disease.
virulent phage A phage that follows only the lytic cycle.
virus A small infectious particle that consists of nucleic acid enclosed in a protein coat.
visceral mass In mollusks, a structure that rests atop the foot and contains the internal organs.
vitamin An organic nutrient that serves as a coenzyme for metabolic and biosynthetic reactions.
vitamin D A vitamin that is converted into a hormone in the body; regulates the calcium level in the blood through an effect on intestinal transport of calcium ions.
vitreous humor A thick liquid in the large posterior cavity of the vertebrate eye, which helps maintain the shape of the eye.
viviparity Development of an embryo within the mother, resulting in a live birth.
viviparous The term used to describe an animal whose embryos develop within the uterus, receiving nourishment from the mother via a placenta.
V_{max} The maximal velocity of an enzyme-catalyzed reaction.
volt A unit of measurement of potential difference in charge (electrical force) such as the difference between the interior and exterior of a cell.
voltage-gated ion channels Ion channels that open and close in response to changes in the amount of electric charge across a membrane.
V-snare A protein incorporated into a vesicle membrane during vesicle formation that is recognized by a T-snare in a target membrane.

W

water potential The potential energy of water.
water vascular system A network of canals powered by water pressure generated by the contraction of muscles; enables extension and contraction of the tube feet, allowing echinoderms to move slowly.
wavelength The distance from the peak of one sound wave or light wave to the next.
waxy cuticle A protective, waterproof layer of polyester and wax present on most surfaces of vascular plant sporophytes.
weak acid An acid that only partially ionizes in solution.
weathering The physical and chemical breakdown of rock.
white blood cell *See* leukocyte.
white matter Brain tissue that consists of myelinated axons that are bundled together in large numbers to form tracts.
whorls In a flower, concentric rings of sepals and petals (or tepals), stamens, and carpels.
wild-type allele One or more prevalent alleles in a population.
wood A secondary plant tissue composed of numerous pipelike arrays of dead, empty, water-conducting cells whose walls are strengthened by an exceptionally tough secondary metabolite known as lignin.
woody plant A type of plant that produces both primary and secondary vascular tissues.

X

X inactivation The phenomenon in which one X chromosome in the somatic cells of female mammals is inactivated, meaning that its genes are not expressed.
X inactivation center (Xic) A short region on the X chromosome known to play a critical role in X inactivation.
X-linked gene A gene found on the X chromosome but not on the Y.
X-linked inheritance The pattern displayed by pairs of dominant and recessive alleles located on X chromosomes.
X-ray crystallography A technique in which researchers purify molecules and cause them to form a crystal. When a crystal is exposed to X-rays, the resulting pattern can be analyzed mathematically to determine the three-dimensional structure of the crystal's components.
xylem A specialized conducting tissue in plants that transports water, minerals, and some organic compounds.
xylem loading The process by which root xylem parenchyma cells transport ions and water across their membranes into the long-distance conducting cells of the xylem, which include the vessel elements and tracheids.

Y

yeast A fungus that can occur as a single cell and that reproduces by budding.
yolk sac One of the four extraembryonic membranes in the amniotic egg. The yolk sac encloses a stockpile of nutrients, in the form of yolk, for the developing embryo.

Z

zero population growth The situation in which no changes in population size occur.
Z line A network of proteins in a myofibril that anchors thin filaments at the ends of each sarcomere.
zona pellucida The glycoprotein covering that surrounds a mature oocyte.
zone of elongation The area above the root apical meristem of a plant where cells extend by water uptake, thereby dramatically increasing root length.
zone of maturation The area above the zone of elongation in a plant root where root cell differentiation and tissue specialization occur.
zooplankton Aquatic organisms drifting in the open ocean or fresh water; includes minute animals consisting of some worms, copepods, tiny jellyfish, and the small larvae of invertebrates and fishes.
Z scheme A model depicting the series of energy changes of an electron during the light reactions of photosynthesis. The electron absorbs light energy twice, resulting in an energy curve with a zigzag shape.
zygomycete A phylum of fungi that produces distinctive, large zygospores as the result of sexual reproduction.
zygospore A dark-pigmented, thick-walled spore that matures within the zygosporangium of zygomycete fungi during sexual reproduction.
zygote A diploid cell formed by the fusion of two haploid gametes.
zygotic life cycle The type of life cycle of most unicellular protists in which haploid cells develop into gametes. Two gametes then fuse to produce a diploid zygote.

Photo Credits

Front Matter: Page v: © Ian J. Quitadamo, Ph.D.

Contents: I: © Dr. Parvinder Sethi; II: © PCN Photography; III: © Daniel Gage, University of Connecticut. Appeared in Schultz, S. C., Shields, G. C. and Steitz, T. A. (1991) "Crystal structure of a CAP-DNA complex: the DNA is bent by 90 degrees," *Science*, 253:1001; IV: © George Bernard/SPL/Photo Researchers, Inc.; V: © Morales/Getty Images; VI: © Gerald & Buff Corsi/Visuals Unlimited; VII: © John Rowley/Getty Images RF; VIII: © Mike Lockhart.

Chapter 1: Opener: © Georgette Douwma/Photo Researchers, Inc.; 1.1: © Photo W. Wüster, courtesy Instituto Butantan; 1.2: © blickwinkel/Alamy; 1.3: © SciMAT/Photo Researchers, Inc.; 1.4a: © David Scharf/Photo Researchers, Inc.; 1.4b: © Alexis Rosenfeld/Photo Researchers, Inc.; 1.4c: © Cathlyn Melloan/Stone/Getty Images; 1.4d: © Adam Jones/Visuals Unlimited; 1.4e: © Patti Murray/Animals Animals; 1.4f: © Paul Hanna/Reuters/Corbis; 1.4g: © Mehgan Murphy, National Zoo/AP Photo; 1.4h: © HPH Publishing/Getty Images; 1.4i: © Louise Pemberton; 1.4j: © Maria Teijeiro/Getty Images RF; 1.4k: © Corbis/SuperStock RF; 1.4l: © Bill Barksdale/agefotostock; 1.10a: © Dr. David M. Phillips/Visuals Unlimited; 1.10b: © B. Boonyaratanakornkit & D.S. Clark, G. Vrdoljak/EM Lab, U of C Berkeley/Visuals Unlimited; 1.10c (protists): © Dr. Dennis Kunkel Microscopy/Visuals Unlimited; 1.10c (plants): © Kent Foster/Photo Researchers, Inc.; 1.10c (fungi): © Carl Schmidt-Luchs/Photo Researchers, Inc.; 1.10c (animals): © Fritz Polking/Visuals Unlimited; 1.13: © Georgette Douwma/Photo Researchers, Inc.; 1.14a: © Fred Bavendam/Minden Pictures; 1.14b: © Eastcott/Momatiuk/Animals Animals; 1.14c: © Ton Koene/Visuals Unlimited; 1.14d: © Northwestern, Shu-Ling Zhou/AP Photo; 1.14e: © Andrew Brookes/Corbis; 1.14e (inset): © Alfred Pasieka/Photo Researchers, Inc.; 1.17: © Dita Alangkara/AP Photo.

Unit I: 2: © Dr. Parvinder Sethi; 3: © de Vos, A. M., Ultsch, M., Steitz, A.A., (1992), "Human growth hormone and extracellular domain of its receptor: crystal structure of the complex," *Science*, 255(306). Image by Daniel Gage, University of Connecticut.

Chapter 2: Opener: © Dr. Parvinder Sethi; 2.6: © The McGraw-Hill Companies, Inc./Al Telser, photographer; 2.12b: © Charles D. Winters/Photo Researchers, Inc.; 2.17 (top right): © Jeremy Burgess/Photo Researchers, Inc.; 2.19b: © Aaron Haupt/Photo Researchers, Inc.; 2.19d: © Chris McGrath/Getty Images; 2.19e: © Dana Tezarr/Getty Images; 2.19f: © Anthony Bannister/Gallo Images/Corbis; 2.19g: © Hermann Eisenbeiss/Photo Researchers, Inc.

PHOTO CREDITS

Chapter 3: Opener: © de Vos, A. M., Ultsch, M., Steitz, A.A., (1992), "Human growth hormone and extracellular domain of its receptor: crystal structure of the complex," *Science*, 255(306). Image by Daniel Gage, University of Connecticut; 3.1: © The McGraw-Hill Companies, Inc./Al Telser, photographer; 3.11a: © Tom Pantages; 3.11b: © Felicia Martinez/PhotoEdit; 3.13: © Adam Jones/Photo Researchers, Inc.

Unit II: 4: © Biophoto Associates/Photo Researchers, Inc.; 5: © Tom Pantages; 6: © 3660 Group Inc./Custom Medical Stock Photo; 7: © PCN Photography; 8: © Travel Pix Ltd/Getty Images; 9: © David McCarthy/SPL/Photo Researchers, Inc.; 10: © altrendo panoramic/Getty Images.

Chapter 4: Opener: © Biophoto Associates/Photo Researchers, Inc.; 4.2a-b: © Images courtesy of Molecular Expressions; 4.3a: © Dr. Donald Fawcett & L. Zamboni/Visuals Unlimited; 4.3b, 4.4b: © Dr. Dennis Kunkel Microscopy/Visuals Unlimited; 4.6a: © Ed Reschke/Getty Images; 4.6b: © Eye of Science/Photo Researchers, Inc.; Table 4.1(1): © Thomas Deerinck/Visuals Unlimited; Table 4.1(2–3): © Dr. Gopal Murti/Visuals Unlimited; 4.12a: © Courtesy Charles Brokaw/California Institute of Technology, 1991. "Microtubule sliding in swimming sperm flagella, direct and indirect measurements on sea urchin and tunicate spermatozoa," *Journal of Cell Biology*, 114:1201–15, issue cover image; 4.12b: Courtesy of Dr. Barbara Surek, Culture Collection of Algae at the University of Cologne (CCAC); 4.12c: © SPL/Photo Researchers, Inc.; 4.13 (top left): © Aaron J. Bell/Photo Researchers, Inc.; 4.13 (middle): © Dr. William Dentler/University of Kansas; 4.15 (top right): © Dr. Donald Fawcett/Visuals Unlimited; 4.15 (middle right): © Dr. Richard Kessel & Dr. Gene Shih/Visuals Unlimited; 4.16: Courtesy of Felix A. Habermann; 4.17 (right): © Dennis Kunkel Microscopy, Inc./Phototake; 4.19: © Lucien G. Caro, Ph.D., and George E. Palade, M.D., 1964. "Protein synthesis, storage, and discharge in the pancreatic exocrine: an autoradiographic study," *Journal of Cell Biology*, 20:473–95, Fig. 3; 4.20a: © E.H. Newcomb & S.E. Frederick/Biological Photo Service; 4.20b: Courtesy Dr. Peter Luykx, Biology, University of Miami; 4.20c: © Dr. David Patterson/Photo Researchers, Inc.; 4.21 (inset): © The McGraw-Hill Companies, Inc./Al Telser, photographer; 4.24: © Dr. Donald Fawcett/Visuals Unlimited; 4.25: © Dr. Jeremy Burgess/Photo Researchers, Inc.; 4.26: © T. Kanaseki and Dr. Donald Fawcett/Visuals Unlimited.

Chapter 5: Opener: © Tom Pantages; 5.3a-b: © The McGraw-Hill Companies, Inc./Al Telser, photographer; 5.16(1–2): © Carolina Biological Supply/Visuals Unlimited; 5.18: Courtesy Dr. Peter Agre. From GM Preston, TP Carroll, WP Guggino, P Agre (1992), "Appearance of water channels in *Xenopus* oocytes expressing red cell CHIP28 protein," *Science*, 256(5055):385–7.

Chapter 6: Opener: © 3660 Group Inc./Custom Medical Stock Photo; 6.1a: © moodboard/Corbis RF; 6.1b: © amanaimages/Corbis RF; 6.10(4–5): From Altman, S., (1990). Nobel Lecture, "Enzymatic Cleavage of RNA by RNA," *Bioscience Reports*, 10:317–37, Fig. 7. © The Nobel Foundation; 6.14: © Liu, Q., Greimann, J.C., and Lima, C.D., (2006), "Reconstitution, activities, and structure of the eukaryotic exosome," *Cell*, 127:1223–37. Graphic generated using DeLano, W.L. (2002), The PyMOL Molecular Graphics System (San Carlos, CA, USA, DeLano Scientific). With permission from Elsevier.

Chapter 7: Opener: © PCN Photography; 7.4: © Custom Medical Stock Photo; 7.13: From Noji, H., Yoshida, M. (2001), "The rotary machine in the cell, ATP Synthase," *Journal of Biological Chemistry*, 276(3):1665–8. © 2001 The American Society for Biochemistry and Molecular Biology; 7.14: © Ernie Friedlander/Cole Group/Getty Images RF; 7.16a: © Homer W Sykes/Alamy; 7.16b: © Jeff Greenberg/The Image Works.

Chapter 8: Opener: © Travel Pix Ltd/Getty Images; 8.2(1): © Norman Owen Tomalin/Bruce Coleman Inc./Photoshot; 8.2(2): © J. Michael Eichelberger/Visuals Unlimited; 8.2(3): © Dr. George Chapman/Visuals Unlimited; 8.12b: © Reproduction of Fig. 1A from Ferreira, K.N., Iverson, T.M., Maghlaoui, K., Barber, J. and Iwata, S., "Architecture of the photosynthetic oxygen evolving center," *Science*, 303(5665):1831–8 © 2004. With permission from AAAS; 8.16(6): Calvin, M., "The path of carbon in photosynthesis," *Nobel Lecture, December 11, 1961*, pp. 618–44, Fig. 4. © The Nobel Foundation; 8.17a: © David Noton Photography/Alamy; 8.17b: © David Sieren/Visuals Unlimited; 8.19(1): © Wesley Hitt/Getty Images; 8.19(2): © John Foxx/Getty Images RF.

Chapter 9: Opener: © David McCarthy/SPL/Photo Researchers, Inc.; 9.2 (inset): © Robert J. Erwin/Photo Researchers, Inc.; 9.15(1–2): Courtesy of Brian J. Bacskai, from Bacskai et al., *Science*, 260:222–6, 1993. With permission from AAAS; 9.18(1–4): © Prof. Guy Whitley/Reproductive and Cardiovascular Disease Research Group at St. George's University of London; 9.19(4): © Dr. Thomas Caceci, Virginia–Maryland Regional College of Veterinary Medicine.

Chapter 10: Opener: © altrendo panoramic/Getty Images; 10.1 (left): © Dr. Dennis Kunkel Microscopy/Visuals Unlimited/Corbis; 10.1 (right): Courtesy of Dr. Joseph Buckwalter/University of Iowa; 10.6: © Dr. Dennis Kunkel Microscopy/Visuals Unlimited; 10.10: © Dr. Daniel Friend; 10.11: Courtesy Dr. Dan Goodenough/Harvard Medical School; 10.13: © Purbasha Sarkar; 10.14: © E.H. Newcomb & W.P. Wergin/Biological Photo Service; 10.19a: © Ed Reschke/Getty Images; 10.19b: © Biodisc/Visuals Unlimited; 10.20: © Robert Brons/Biological Photo Service; 10.21: © J.N.A. Lott/Biological Photo Service.

Unit III: 11: © Jean Claude Revy/ISM/Phototake; 12: © Kiseleva and Dr. Donald Fawcett/Visuals Unlimited; 13: © Daniel Gage, University of Connecticut. Appeared in Schultz, S. C., Shields, G. C., and Steitz, T. A. (1991) "Crystal structure of a CAP-DNA complex: the DNA is bent by 90 degrees," *Science*, 253:1001; 14: © Yvette Cardozo/Workbook Stock/Getty Images; 15: © Biophoto Associates/Photo Researchers, Inc.; 16: © Radu Sigheti/Reuters; 17: © Dave King/Dorling Kindersley/Getty Images; 18: © CAMR/A. Barry Dowsett/Photo Researchers, Inc.; 19: © Medical-on-Line/Alamy; 20: © Corbis; 21: © Coston Stock/Alamy.

Chapter 11: Opener: © Jean Claude Revy/ISM/Phototake; 11.3b: © Eye of Science/Photo Researchers, Inc.; 11.13: © Meselson, M., Stahl, F., (1958) "The replication of DNA in *Escherichia coli*," *PNAS*, 44(7):671–82, Fig. 4a; 11.24a: Photo courtesy of Dr. Barbara Hamkalo; 11.26a: © Dr. Gopal Murti/Visuals Unlimited; 11.26b: © Ada L. Olins and Donald E. Olins/Biological Photo Service; 11.26c: Courtesy Dr. Jerome B. Rattner, Cell Biology and Anatomy, University of Calgary; 11.26d: Courtesy of Paulson, J.R. & Laemmli, U.K. James R. Paulson, U.K. Laemmli, "The structure of histone-depleted metaphase chromosomes," *Cell*, 12:817–28, Copyright Elsevier 1977; 11.26e-f: © Peter Engelhardt/Department of Virology, Haartman Institute.

Chapter 12: Opener: © Kiseleva and Dr. Donald Fawcett/Visuals Unlimited; 12.15a: Reprinted from Seth A. Darst, "Bacterial RNA polymerase," *Current Opinion in Structural Biology*, 11(2):155–62, © 2001, with permission from Elsevier.

Chapter 13: Opener: © Daniel Gage, University of Connecticut. Appeared in Schultz, S. C., Shields, G. C. and Steitz, T. A. (1991) "Crystal structure of a CAP-

DNA complex: the DNA is bent by 90 degrees," *Science*, 253:1001; 13.2a: © Ed Reschke/Getty Images; 13.2b: © Triarch/Visuals Unlimited; 13.2c: © SIU BioMed/Custom Medical Stock Photo; 13.10: © Thomas Steitz, Howard Hughes Medical Institution, Yale University.

Chapter 14: Opener: © Yvette Cardozo/Workbook Stock/Getty Images; 14.1a-b: © Stan Flegler/Visuals Unlimited; 14.1c: Courtesy of Thomas Wellems and Robert Josephs. Electron Microscopy and Image Processing Laboratory, University of Chicago; 14.4: © Otero/gtphoto; 14.9: © Dr. Kenneth Greer/Visuals Unlimited; 14.10b: © St. Bartholomew's Hospital/Photo Researchers, Inc.; 14.17(1–5): © Dr. Oscar Auerbach, reproduced with permission.

Chapter 15: Opener: © Biophoto Associates/Photo Researchers, Inc.; 15.1(4): © Burger/Photo Researchers, Inc.; 15.1(5): Courtesy of the Genomic Centre for Cancer Research and Diagnosis, CancerCare Manitoba, University of Manitoba, Winnipeg, Manitoba, Canada; 15.8a-f: Photographs by Conly L. Rieder, East Greenbush, New York, 12061; 15.9a: © Dr. David M. Phillips/Visuals Unlimited; 15.9b: © Carolina Biological Supply Company/Phototake; 15.19a-b: © A. B. Sheldon; 15.20a: © Sylvan Wittwer/Visuals Unlimited; 15.20b: © B. Runk/S. Schoenberger/Grant Heilman Photography.

Chapter 16: Opener: © Radu Sigheti/Reuters; 16.1: © SPL/Photo Researchers, Inc.; 16.3: © Nigel Cattlin/Photo Researchers, Inc.; 16.19: © March of Dimes.

Chapter 17: Opener: © Dave King/Dorling Kindersley/Getty Images; 17.9a-b: Courtesy of I. Solovei, University of Munich (LMU); 17.11: Courtesy of Dr. Argiris Efstratiadis.

Chapter 18: Opener: © CAMR/A. Barry Dowsett/Photo Researchers, Inc.; 18.1: © Norm Thomas/Photo Researchers, Inc.; 18.3a: © Robley C. Williams/Biological Photo Service; 18.3b: Reprinted from R.C. Valentine and H.G. Pereira, "Antigens and structure of the adenovirus," *Journal of Molecular Biology*, 13(2):71–83, © 1965, with permission from Elsevier; 18.3c: © Chris Bjornberg/Photo Researchers, Inc.; 18.3d: © Omikron/Photo Researchers, Inc.; 18.6: Cynthia Goldsmith/CDC; 18.7: © Photograph by J. W. Randles, at Albay Research Center, Philippines; 18.8: © Eye of Science/Photo Researchers, Inc.; 18.9: Carl Robinow & Eduard Kellengerger, "The bacterial nucleoid revisited," *Microbiological Reviews*, 1994 June, 58(2):211–32. Fig. 4 © American Society for Microbiology; 18.12 (right): © Stanley Cohen/Photo Researchers, Inc.; 18.13 (bottom): © Dr. Jeremy Burgess/SPL/Photo Researchers, Inc.; 18.17: © Dr. L. Caro/SPL/Photo Researchers, Inc.

Chapter 19: Opener: © Medical-on-Line/Alamy; 19.1a: © Herman Eisenbeiss/Photo Researchers, Inc.; 19.1b: © Sinclair Stammers/Photo Researchers, Inc.; 19.1c: © J-M. Labat/Photo Researchers, Inc.; 19.1d: © Mark Smith/Photo Researchers, Inc.; 19.1e: © Nigel Cattlin/Photo Researchers, Inc.; 19.4a–19.6(1): Courtesy of the National Museum of Health and Medicine, Washington, D.C.; 19.6(2): © Congenital Anomaly Research Center of Kyoto University; 19.6(3–4): Courtesy of the National Museum of Health and Medicine, Washington, D.C.; 19.9b–19.11(1): Christiane Nüsslein-Volhard, "Determination of the embryonic axes of *Drosophila*," *Development*, Supplement 1, 1991, pp. 1–10, Fig. 5, © The Company of Biologists Limited 1991; 19.11(2–4): © Jim Langeland, Steve Paddock and Sean Carroll/University of Wisconsin-Madison; 19.12a-b: Courtesy of the Archives, California Institute of Technology; 19.14a: © Juergen Berger/Photo Researchers, Inc.; 19.14b: © F. R. Turner, Indiana University; 19.24a: © Darwin Dale/Photo Researchers, Inc.; 19.24b-c: © John Bowman.

Chapter 20: Opener: © Corbis; 20.10 (bottom): © Alfred Pasieka/Photo Researchers, Inc.; 20.14a: © Bill Barksdale/Agstockusa/agefotostock; 20.15(6): © Roslin Institute/Phototake; 20.16a: © Leonard Lessin.

Chapter 21: Opener: © Coston Stock/Alamy; 21.2b: © The Picture Store/SPL/Photo Researchers, Inc.; 21.2c: © Photo by Michael Beckmann, Institute of Geobotany and Botanical Garden, Halle, Germany; 21.4a: © Topham/The Image Works; 21.4b: © Jerome Wexler/Visuals Unlimited.

Unit IV: 22: © George Bernard/SPL/Photo Researchers, Inc.; 23: © Mark Dadswell/Getty Images; 24: © Dynamic Graphics Group/IT StockFree/Alamy RF; 25: © Frederic B. Siskind; 26: © Peter Weimann/Animals Animals.

Chapter 22: Opener: © George Bernard/SPL/Photo Researchers, Inc.; 22.1a: © Stanley M. Awramik/Biological Photo Service; 22.1b: © Michael Abbey/Visuals Unlimited; 22.3b: © CSSF/NEPTUNE Canada; 22.4a: A.I. Oparin. From *The Origin of Life*, New York: Dover, 1952; 22.4b: © Mary Kraft; 22.7: © Simon Fraser/SPL/Photo Researchers, Inc.; 22.11a: © Dirk Wiersma/SPL/Photo Researchers, Inc.; 22.11b: © Roger Garwood & Trish Ainslie/Corbis; 22.13a: Courtesy of Dr. Barbara Surek, Culture Collection of Algae at the University of Cologne (CCAC); 22.13b: © Bill Bourland/micro*scope; 22.13c-d: © Dr. Cristian A. Solari, Department of Ecology and Evolutionary Biology, University of Arizona; 22.14: We thank Prof. Jun-yuan Chen for permission to use this image; 22.15a: © Newman & Flowers/Photo Researchers, Inc.; 22.15b: Courtesy of the Smithsonian Institution, Photograph by Chip Clark. *Marrella splendens*, most commonly occurring fossil from the Burgess Shale. Collected by Charles D. Walcott for the Smithsonian Research Collections; 22.16a: © Francois Gohier/Photo Researchers, Inc.; 22.16b: © DK Limited/Corbis; 22.17: © Ken Lucas/Visuals Unlimited; 22.19: © Jason Edwards/Getty Images RF.

Chapter 23: Opener: © Mark Dadswell/Getty Images; 23.1a: © PoodlesRock/Corbis; 23.3a: © Worldwide Picture Library/Alamy; 23.3b: © Gerald & Buff Corsi/Visuals Unlimited; 23.5 (middle): © T. Daeschler/VIREO; 23.7a: © Wm. Grenfell/Visuals Unlimited; 23.7b: © Prisma Bildagentur AG/Alamy; 23.8a (left): © Tom Brakefield/Corbis; 23.8a (right): © Tom McHugh/Photo Researchers, Inc.; 23.8b (left): © David Sieren/Visuals Unlimited; 23.8b (right): © 2003 Steve Baskauf/bioimages.vanderbilt.edu; 23.8c (left): © Jonathan Bird/Getty Images; 23.8c (right): © David Wrobel/SeaPics.com; 23.9a: © Mark Raycroft/Minden Pictures; 23.9b: © Martin Rugner/agefotostock; 23.9c: © Juniors Bildarchiv/agefotostock.

Chapter 24: Opener: © Dynamic Graphics Group/IT StockFree/Alamy RF; 24.1: © Paul Harcourt Davies/SPL/Photo Researchers, Inc.; 24.5a: Courtesy Mark McNair, University of Exeter; 24.7a: © Gerald Cubitt; 24.7b: © Topham/The Image Works; 24.7c: © Photo by Darren P. Croft, University of Exeter UK; 24.8a-b: © Ole Seehausen.

Chapter 25: Opener: © Frederic B. Siskind; 25.1a (left): © Mark Smith/Photo Researchers, Inc.; 25.1a (right): © Pascal Goetgheluck/ardea.com; 25.1b (left): © Gary Meszaros/Visuals Unlimited; 25.1b (right): © robin chittenden/Alamy; 25.3a: © C. Allan Morgan/Getty Images; 25.3b: © Bryan E. Reynolds; 25.4a: © Rod Planck/Photo Researchers, Inc.; 25.4b: © Ron Austing/Photo Researchers, Inc.; 25.5 (top left): © Mark Boulton/Photo Researchers, Inc.; 25.5 (top right): © Carolina Biological Society/Visuals Unlimited; 25.5 (bottom): © Stephen L. Saks/Photo Researchers, Inc.; 25.6 (left): © Hal Beral/V&W/imagequestmarine.com; 25.6 (right): © Amar and Isabelle Guillen/Guillen Photography/Alamy; 25.7 (top right): © FLPA/Alamy; 25.7b(1–3, 6): © Jack Jeffrey Photography; 25.7b(4–5): © Jim Denny; 25.12: © Dr. Sara Via, Department of Biology and Department of Entomology, University of Maryland; 25.14a: Courtesy Ed Laufer; 25.14b-c: Courtesy of Dr. J.M. Hurle. Originally published in *Development*. 1999 Dec. 126(23):5515–22; 25.17: © Gary Nafis; 25.18a-b: © Prof. Walter J. Gehring, University of Basel.

Chapter 26: Opener: © Peter Weimann/Animals Animals.

Unit V: 27: © Dr. Jeremy Burgess/SPL/Photo Researchers, Inc.; 28: © Photograph by H. Cantor-Lund reproduced with permission of the copyright holder J. W. G. Lund; 29: © Craig Tuttle/Corbis; 30: © Gallo Images/Corbis; 31: © Brian Lightfoot/naturepl.com; 32: © Morales/Getty Images; 33: © Georgie Holland/agefotostock; 34: © Ken Catania/Visuals Unlimited.

Chapter 27: Opener: © Dr. Jeremy Burgess/SPL/Photo Researchers, Inc.; 27.2a: © Linda Graham: 27.2b: © Eye of Science/Photo Researchers, Inc.; Table 27.1(1): © Dr. David M. Phillips/Visuals Unlimited; Table 27.1(2): © Michael J. Daly, Uniformed Services University of the Health Sciences, Bethesda, MD, U.S.A.; Table 27.1(3): © Wim van Egmond/Visuals Unlimited; Table 27.1(4): © Prof. Dr. Hans Reichenbach/GBF; 27.3a: © Linda Graham; 27.3b: © Wim van Egmond/Visuals Unlimited; 27.3c: © Michael Abbey/Visuals Unlimited; 27.3d: © Lee W. Wilcox; 27.4: © Nigel Cattlin/Visuals Unlimited; 27.5: © H. Stuart Pankratz/Biological Photo Service; 27.6: © Dr. Richard P. Blakemore, University of New Hampshire; 27.7a: © SciMAT/Photo Researchers, Inc.; 27.7b-d: © Dennis Kunkel Microscopy; 27.8: © Science Photo Library/Alamy RF; 27.9a: © CNRI/Photo Researchers, Inc.; 27.9b: © Lee W. Wilcox; 27.12a: © Dennis Kunkel Microscopy/Phototake; 27.12b: © Scientifica/Visuals Unlimited; 27.13: © Dr. Terry Beveridge/Visuals Unlimited; 27.14a: © Scientifica/RML/Visuals Unlimited; 27.14b: © Michael Gabridge/Visuals Unlimited; 27.14c–27.15a: © Lee W. Wilcox; 27.15b: © Dr. Kari Lounatmaa/Photo Researchers, Inc.; 27.17: © Peter Durben.

Chapter 28: Opener & 28.1: © Photographs by H. Cantor-Lund reproduced with permission of the copyright holder J. W. G. Lund; 28.2: © Andrew Syred/Photo Researchers, Inc.; 28.3a: © Brian P. Piasecki; 28.3b: © Roland Birke/Phototake; 28.3c-e: © Linda Graham; 28.4: © Dr. Dennis Kunkel Microscopy/Visuals Unlimited; 28.5: © Dr. Stanley Flegler/Visuals Unlimited; 28.8a: © Dr. David M. Phillips/Visuals Unlimited; 28.8b: © Jerome Paulin/Visuals Unlimited; 28.9a: © Wim van Egmond/Visuals Unlimited; 28.9b: © Ross Waller, University of Melbourne, Australia; 28.9c: © Eye of Science/Photo Researchers, Inc.; 28.11a: © Doug Sokell/Visuals Unlimited; 28.11b: © Andrew J. Martinez/Photo Researchers, Inc.; 28.12: © Joe Scott, Department of Biology, College of William and Mary, Williamsburg, VA 23187; 28.14a: © Dennis Kunkel Microscopy; 28.14b: © Isao Inouye; 28.14c: © Stockbyte/Getty Images RF; 28.16a: © Lee W. Wilcox; 28.16b: Cover photograph of *Eukaryotic Cell*, March 2007, 6(3). Related article: Alvin C. M. Kwok, Carmen C. M. Mak, Francis T. W. Wong, Joseph T. Y. Wong, "Novel method for preparing spheroplasts from cells with an internal cellulosic cell wall," pp. 563–7. doi: 10.1128/EC.00301-06. Copyright © 2007 American Society for Microbiology. Reproduced with permission from American Society for Microbiology;

28.18a: © Linda Graham; 28.18b: © Jeff Rotman/Photo Researchers, Inc.; 28.19a: © Claude Nuridsany & Marie Perennou/SPL/Photo Researchers, Inc.; 28.19b: © O. Roger Anderson, Columbia University, Lamont-Doherty Earth Observatory; 28.21: © Stephen Fairclough, King Lab, University of California at Berkeley; 28.22: © Lee W. Wilcox; 28.23 (inset): © NCSU Center for Applied Aquatic Ecology; 28.24: © Linda Graham; 28.29 (inset): © Angelika Strum, Rogerio Amino, Claudia van de Sand, Tommy Regen, Silke Retzlaff, Annika Rennenberg, Andreas Krueger, Jorg-Matthias Pollok, Robert Menard, Volker T. Heussler, "Manipulation of host hepatocytes by the malaria parasite for delivery into liver sinusoids," *Science*, September 2006, 313(5791):1287–90. Fig. 1c. Reproduced with permission from AAAS.

Chapter 29: Opener: © Craig Tuttle/Corbis; 29.1 (inset 1): © Roland Birke/Phototake; 29.1 (inset 2): © the CAUP image database, http://botany.natur.cuni.cz/algo/database; 29.1 (inset 3–6): © Lee W. Wilcox; 29.1 (inset 7): © Ed Reschke/Getty Images; 29.1 (inset 8): © Patrick Johns/Corbis; 29.1 (inset 9): © Philippe Psaila/Photo Researchers, Inc.; 29.1 (inset 10): © Fancy Photography/Veer RF; 29.1 (inset 11): © Fred Bruemmer/Getty Images; 29.1 (inset 12): © Gallo Images/Corbis; 29.2a (left): © Lee W. Wilcox; 29.2a (right): © Linda Graham; 29.2b (left): © the CAUP image database, http://botany.natur.cuni.cz/algo/database; 29.2b (right): © Lee W. Wilcox; 29.3a: © Dr. Jeremy Burgess/SPL/Photo Researchers, Inc.; 29.3b–29.4: © Lee W. Wilcox; 29.4 (inset): © Eye of Science/Photo Researchers, Inc.; 29.5: © Lee W. Wilcox; 29.7 (top right inset): © Larry West/Photo Researchers, Inc.; 29.7 (bottom insets): © Linda Graham; 29.9–29.10a: © Lee W. Wilcox; 29.10b: © S. Solum/PhotoLink/Getty Images RF; 29.10c: © Patrick Johns/Corbis; 29.10d: © Rich Reid/Animals Animals; 29.11a: © Linda Graham; 29.11b: © Martha Cook; 29.12 (foreground inset): © Carolina Biological Supply Company/Phototake; 29.12 (inset 1): © Ernst Kucklich/Getty Images; 29.12 (inset 2–3): © Linda Graham; 29.12 (inset 4–7): © Lee W. Wilcox; 29.12 (inset 8): © Dr. Richard Kessel & Dr. Gene Shih/Visuals Unlimited; 29.13: © Stephen P. Parker/Photo Researchers, Inc.; 29.14: © Brand X Pictures/PunchStock RF; 29.15: © Photo by Steven R. Manchester, University of Florida Courtesy Botanical Society of America, St. Louis, MO., www.botany.org; 29.16–29.17 (left): © Lee W. Wilcox; 29.17 (middle): © Charles McRae/Visuals Unlimited; 29.17 (right): © David R. Frazier/The Image Works; 29.18: © Marjorie C. Leggitt; 29.19: Courtesy Prof. Roberto Ligrone. Fig. 6 in Ligrone et al., *Protoplasma* (1982) 154:414–25; 29.22c: © Lee W. Wilcox.

Chapter 30: Opener: © Gallo Images/Corbis; 30.3a: © Philippe Psaila/Photo Researchers, Inc.; 30.3b: © Ed Reschke/Getty Images; 30.4a-b: © Lee W. Wilcox; 30.5a: © Karlene V. Schwartz; 30.5b: © Fancy Photography/Veer RF; 30.5c: © TOPIC PHOTO AGENCY IN/agefotostock; 30.6a: © Lee W. Wilcox; 30.6b: © Bryan Pickering/Eye Ubiquitous/Corbis; 30.8a: © Zach Holmes Photography; 30.8b: © Duncan McEwan/naturepl.com; 30.8c: © Ed Reschke/Getty Images; 30.10a: © Steven P. Lynch; 30.10b: © Ken Wagner/Phototake; 30.10c: © Lee W. Wilcox; 30.11a: © Robert & Linda Mitchell; 30.11b: © 2004 James M. Andre; 30.11c: © Michael & Patricia Fogden/Corbis; 30.12: © Bill Ross/Corbis; 30.17: © Sangtae Kim, Ph.D.; 30.18a: © Medioimages/PunchStock RF; 30.18b: © Ed Reschke/Getty Images; 30.20a: © Neil Joy/Photo Researchers, Inc.; 30.20b: © Royalty-Free/Corbis; 30.20c: Image released under GFDL license. Photographer Florence Devouard, France 2003; 30.21a-e: © Lee W. Wilcox; 30.21f: © Dr. James Richardson/Visuals Unlimited; 30.21g: © foodanddrinkphotos co/agefotostock; 30.21h: © Jerome Wexler/Visuals Unlimited; 30.22a: © Eddi Boehnke/zefa/Corbis; 30.22b: © Jonathan Buckley/GAP Photo/Getty Images; 30.22c: © Science Photo Library/Alamy; 30.23 (top inset): © Phil Schermeister/Getty Images; 30.23 (middle inset): © Joao Luiz Bulcao; 30.24: © Jack Jeffrey/Photo Resource Hawaii; 30.25: © Beng & Lundberg/naturepl.com.

Chapter 31: Opener: © Brian Lightfoot/naturepl.com; 31.4a: Kaminskyj, S.G.W., and Heath, I.B. (1996), "Studies on *Saprolegnia ferax* suggest the general importance of the cytoplasm in determining hyphal morphology," *Mycologia*, 88:20–37, Fig 16. Mycological Society of America. Allen Press, Lawrence Kansas; 31.5a: © Agriculture and Agri-Food Canada, Southern Crop Protection and Food Research Centre, London ON; 31.5b: CDC; 31.6: © Dr. Dennis Kunkel Microscopy/Visuals Unlimited; 31.7: © Medical-on-Line/Alamy: 31.8a: © Felix Labhardt/Taxi/Getty Images; 31.8b: © Bob Gibbons/ardea.com; 31.9: © Rob Casey/Alamy RF; 31.10: © Gary Meszaros/Visuals Unlimited; 31.11: © David Q. Cavagnaro/Getty Images; 31.13: © Photograph by H. Cantor-Lund reproduced with permission of the copyright holder J. W. G. Lund; 31.14: © Dr. Raquel Martín and collaborators; 31.15a (top right): © Mike Peres/Custom Medical Stock Photo; 31.15b (bottom right): © William E. Schadel/Biological Photo Service; 31.16: © Yolande Dalpé, Agriculture and Agri-Food Canada; 31.17a: Micrograph courtesy of Timothy M. Bourett, DuPont Crop Genetics, Wilmington, DE USA; 31.17b: © Charles Mims; 31.18b (middle inset): © Ed Reschke/Getty Images; 31.19: © Nacivet/Getty Images; 31.20 (right inset): © Dr. Jeremy Burgess/Photo Researchers, Inc.; 31.20 (left inset): © Biophoto Associates/Photo Researchers, Inc.; 31.21a: © Dayton Wild/Visuals Unlimited; 31.21b: © Mark Turner/Botanica/Getty Images; 31.22: © N. Allin & G.L. Barron/Biological Photo Service; 31.23: © Dr. Eric Kemen and Dr. Kurt W. Mendgen; 31.24 (left): © Nigel Cattlin/Photo Researchers, Inc.; 31.24 (right): © Herve Conge/ISM/Phototake; 31.25a: Courtesy Bruce Klein. Reprinted with permission; 31.25b: Cover photograph of *The Journal of Experimental Medicine*, April 19, 1999, 189(8). Copyright © 1999 by The Rockefeller University Press; 31.26: © Dr. D.P. Donelley and Prof. J.R. Leake, University of Sheffield, Department of Animal & Plant Sciences; 31.27a: © Mark Brundrett; 31.28a: © Jacques Landry, Mycoquebec.org; 31.28b: Courtesy of Larry Peterson and Hugues Massicotte; 31.30a: © Joe McDonald/Corbis; 31.30b: © Lee W. Wilcox; 31.30c: © Ed Reschke/Getty Images; 31.30d: © Lee W. Wilcox.

Chapter 32: Opener: © Morales/Getty Images; 32.1a: © waldhaeusl.com/agefotostock; 32.1b: © Enrique R. Aguirre Aves/Getty Images; 32.1c: © Bartomeu Borrell/agefotostock; 32.2: © Publiphoto/Photo Researchers, Inc.; 32.5a: © E. Teister/agefotostock; 32.5b: © Gavin Parsons/Getty Images; 32.5c: © Tui de Roy/Minden Pictures; 32.12: © Dwight Kuhn.

Chapter 33: Opener: © Georgie Holland/agefotostock; 33.2a: © Norbert Probst/agefotostock; 33.5b: © Minden Pictures/Masterfile; 33.6: © Matthew J. D'Avella/SeaPics.com; 33.8a: © Wolfgang Poelzer/Wa./agefotostock; 33.8b: © Biophoto Associates/Photo Researchers, Inc.; 33.11a: © Wim van Egmond/Visuals Unlimited; 33.11b: © Fred Bavendam/Minden Pictures; 33.13a: © Andrew J. Martinez/Photo Researchers, Inc.; 33.13b: © Kjell Sandved/Visuals Unlimited; 33.13c: © Dr. William Weber/Visuals Unlimited; 33.13d: © Hal Beral/Corbis; 33.13e: © Alex Kerstitch/Visuals Unlimited; 33.14: © Frank Boxler/AP Photo; 33.15b: © Jonathan Blair/Corbis; 33.18a: © WaterFrame/Alamy; 33.18b: © J. W. Alker/agefotostock; 33.18c: © Colin Varndell/Getty Images; 33.18d: © St. Bartholomew's Hospital/Photo Researchers, Inc.; 33.19: © Biophoto Associates/Photo Researchers, Inc.; 33.20: © Johnathan Smith; 33.22: © James L. Amos/Photo Researchers, Inc.; 33.24a-c: © NASA/SPL/Photo Researchers, Inc.; 33.25a: © Duncan Usher/ardea.com; 33.25b: © Paul Freed/Animals Animals; 33.25c: © Dr. Dennis Kunkel Microscopy/Visuals Unlimited; 33.25d: © Roger De LaHarpe/Gallo Images/Corbis; 33.26a: © David Aubrey/Corbis; 33.26b: © Larry Miller/Photo Researchers, Inc.; 33.29a: © Alex Wild/myrmecos.net; 33.29b: © Christian Ziegler/Minden Pictures/National Geographic Stock; 33.31: © Wim van Egmond/Visuals Unlimited; 33.32a: © Kjell Sandved/Visuals Unlimited; 33.32b: © Richard Walters/Visuals Unlimited; 33.32c: © Franklin Viola/Animals Animals; 33.34: © Leslie Newman & Andrew Flowers/Photo Researchers, Inc.; 33.37a: © Natural Visions/Alamy; 33.38c: © Reinhard Dirscherl/Visuals Unlimited.

Chapter 34: Opener: © Ken Catania/Visuals Unlimited; 34.2: © Pat Morris/ardea.com; 34.3a: © Breck P. Kent/Animals Animals; 34.3b: © Jacana/Photo Researchers, Inc.; 34.5a: © Valerie & Ron Taylor/ardea.com; 34.5b: © Jeff Rotman/naturepl.com; 34.5c: © Oxford Scientific/Getty Images; 34.5d: © Bill Curtsinger/National Geographic/Getty Images; 34.7a: © Reinhard Dirscherl/Visuals Unlimited; 34.7b: © Andrew Dawson/agefotostock; 34.7c: © Luc Novovitch/Getty Images; 34.8: © Peter Scoones/SPL/Photo Researchers, Inc.; 34.9: © D. R. Schrichte/SeaPics.com; 34.13a: © Don Vail/Alamy; 34.13b-c: © Dwight Kuhn; 34.14a: © Gregory G. Dimijian/Photo Researchers, Inc.; 34.14b: © Juan-Manuel Renjifo/agefotostock; 34.14c: © Gary Meszaros/Photo Researchers, Inc.; 34.16a: © Pat Morris/ardea.com; 34.16b: © Fabio Pupin/Visuals Unlimited/Corbis; 34.16c: © Jim Merli/Visuals Unlimited; 34.17: © Michael & Patricia Fogden/Minden Pictures; 34.18a: © Warren Jacobi/Corbis RF; 34.18b: © J. & C. Sohns/Animals Animals; 34.21c: © Gilbert S. Grant/Photo Researchers, Inc.; 34.22a: © B. G. Thomson/Photo Researchers, Inc.; 34.22b: © Jean-Claude Canton/Bruce Coleman Inc./Photoshot; 34.22c: © Morales/agefotostock; 34.22d: © Brand X Pictures/PunchStock RF; 34.22e: © Rick & Nora Bowers/Visuals Unlimited; 34.22f: © Mervyn Rees/Alamy; 34.23a: © Eric Baccega/agefotostock; 34.23b: © Charles Krebs/Corbis; 34.23c: © Anthony Bannister/Photo Researchers, Inc.; 34.24a: © Image Source/Corbis RF; 34.24b: © Joe McDonald/Corbis; 34.24c: © mauritius images GmbH/Alamy; 34.24d: © DLILLC/Corbis RF; 34.24e: © Ken Lucas/Visuals Unlimited; 34.25a: © Martin Harvey/Getty Images; 34.25b: © John Shaw/Photo Researchers, Inc.; 34.25c: © Paul A. Souders/Corbis; 34.27a: © Dave Watts/naturepl.com; 34.27b: © Theo Allofs/Visuals Unlimited; 34.27c: © Jeffrey Oonk/Foto Natura/Minden Pictures; 34.29a: © David Haring/DUPC/Getty Images; 34.29b: © Gerard Lacz/Animals Animals; 34.29c: © Martin Harvey/Corbis; 34.30a: © Joe McDonald/Corbis; 34.30b: © Creatas/PunchStock RF; 34.30c: © Tetra Images RF/Getty Images.

Unit VI: 35: © Frans Lanting/Corbis; 36: © Gerald & Buff Corsi/Visuals Unlimited; 37: © Dwight Kuhn; 38: © Barry Mason/Alamy RF; 39: © E.R. Degginger/Animals Animals.

Chapter 35: Opener: © Frans Lanting/Corbis; 35.4a: © Linda Graham; 35.4b: © Howard Rice/Getty Images; 35.5a: © James Mann/ABRC/CAPS; 35.5b: © Prof. Dr. Gerd Jürgens/Universität Tübingen. Image Courtesy Hanno Wolters; 35.6: Figure adapted from Jackson, D. and Hake, S. (1999), "Control of phylotaxy in maize by the ABPHYL1 gene," *Development*, 126:315–23, © The Company of Biologists Limited 1999; 35.7a-b: © Lee W. Wilcox; 35.7c: © Dr. Dennis Drenner/Visuals Unlimited; 35.7d-35.13d: © Lee W. Wilcox; 35.15: © Eye of Science/Photo Researchers, Inc.; 35.16 (left): Figure adapted from Jackson, D. and Hake, S. (1999), "Control of phylotaxy in maize by the ABPHYL1 gene," *Development*, 126:315–23,

© The Company of Biologists Limited 1999; 35.17a: © Ed Reschke/Getty Images; 35.17b: © John Farmar; 35.17c: © Steve Terrill/Corbis; 35.17d: © John Gerlach/Animals Animals; 35.18–35.19: © Lee W. Wilcox; 35.21: © Bruce Iverson; 35.22a: © Stephen Ingram/Animals Animals; 35.22b: © Peter E. Smith, Natural Sciences Image Library; 35.25: © Lee W. Wilcox.

Chapter 36: Opener: © Gerald & Buff Corsi/Visuals Unlimited; 36.1a: © Digital Photography by Ash Kaushesh, University of Central Arkansas, Conway, Arkansas 72035 USA/Image courtesy Botanical Society of America, www.botany.org, St. Louis, MO 63110; 36.1b: © Lee W. Wilcox; 36.11(1–5): © Prof. and Mrs. M. B. Wilkins/University of Glasgow; 36.12 (top left): © Pat O'Hara/Corbis; 36.12 (top right): © Henry Schleichkorn/Custom Medical Stock Photo; 36.12 (middle left): © Robert Maier/Animals Animals; 36.12 (middle right): © Michael Gadomski/Animals Animals; 36.12 (bottom left): © Pat O'Hara/Corbis; 36.12 (bottom right): © Henry Schleichkorn/Custom Medical Stock Photo; 36.13: © Lee W. Wilcox; 36.16: Courtesy Dr. Malcolm Drew, Texas A&M University; 36.18 (inset): © G. R. "Dick" Roberts/Natural Sciences Image Library.

Chapter 37: Opener: © Dwight Kuhn; 37.2: © Robert Ziemba/agefotostock; 37.3a-b: © Raymon Donahue; 37.4: © Royal Botanical Gardens Kew/Wellcome Trust Medical Photographic Library; 37.5: Confocal fluorescence micrograph courtesy Simon D.X. Chuong; 37.6: © Geoff Kidd/SPL/Photo Researchers, Inc.; 37.7b: Courtesy of C.A. Stiles, University of Wisconsin; 37.7c: © Ruddy Gold/agefotostock: 37.8: USDA Photo; 37.12: Photo courtesy of Jayne Belnap, U.S. Geological Survey; 37.16: © J. Hyvönen; 37.16 (inset): Photo by Birgitta Bergman, Department of Botany, Stockholm University (Sweden); 37.17: © M. Kalab/Custom Medical Stock Photo; 37.19a: © Dani/Jeske/Animals Animals; 37.19b: © Breck P. Kent/agefotostock; 37.20: © Charles T. Bryson, USDA Agricultural Research Service, Bugwood.org.

Chapter 38: Opener: © Barry Mason/Alamy RF; 38.8 (right): © James S. Busse; 38.10: © Jerry Cooke, Inc./Animals Animals; 38.11b: © Astrid & Hanns/Frieder Michler/Photo Researchers, Inc.; 38.12b-c: © John N.A. Lott/Biological Photo Service; 38.13: © David Q. Cavagnaro/Getty Images; 38.16: © Adalberto Rios Szalay/Sexto Sol/Stockbyte/Getty Images RF; 38.17a: © Andrew Syred/Photo Researchers, Inc.; 38.19a: © Dan Suzio/Photo Researchers, Inc.; 38.19b: © Dr. William J. Weber/Visuals Unlimited; 38.20a–38.22: © Lee W. Wilcox.

Chapter 39: Opener: © E.R. Degginger/Animals Animals; 39.3b: © Lee W. Wilcox; 39.5a: © Natural Sciences Image Library; 39.5b: © Lee W. Wilcox; 39.7a: © Scott Sinklier/agefotostock; 39.7b: © Chris Russell, 11eggs.com; 39.8a: © Richard Shiell/Animals Animals; 39.8b: © Richard Jorgensen; 39.9a-b: Courtesy of John Innes Centre; 39.10: © Burke/Triolo Productions/Getty Images RF; 39.11(6): Haiyi Liang and L. Mahadevana, "Growth, geometry, and mechanics of a blooming lily," *PNAS*, April 5, 2011, 108(14): 5516–21, Fig. 2c; 39.12b: © RMF/Scientifica/Visuals Unlimited; 39.14 (top left): © Biodisc/Visuals Unlimited; 39.15: Courtesy J.M. Escobar-Restrepo and A.J. Johnston, University of Zurich, Institute of Plant Biology. From Bernasconi et al., "Evolutionary ecology of the prezygotic stage," *Science*, 303(5660):971–5, Fig. 2, © 2004. Reprinted with permission from AAAS; 39.18–39.21: © Lee W. Wilcox.

Unit VII: 40: © John Rowley/Getty Images RF; 41: © James Cavallini/Photo Researchers, Inc.; 42: Courtesy Ann-Shyn Chiang, Tsing Hua Chair Professor/Brain Research Center & Institute of Biotechnology/National Tsing Hua University; 43: © Jane Burton/Bruce Coleman Inc./Photoshot; 44: © Stephen Dalton/Photo Researchers, Inc.; 45: © Richard Hutchings/PhotoEdit; 46: © The Rockefeller University/AP Photo; 47: © SPL/Photo Researchers, Inc.; 48: © brianlatino/Alamy RF; 49: © Guido Alberto Rossi/agefotostock; 50: © Sovereign/ISM/Phototake; 51: © David Liebman/Pink Guppy; 52: Matsui et al., "Effects of ethanol on photoreceptors and visual function in developing zebrafish," *Investigative Ophthalmology & Visual Science*, October 2006, 47(10):4589–97, Fig. 1. © Association for Research in Vision and Ophthalmology. Images Courtesy of Jonathan Matsui; 53: © SPL/Photo Researchers, Inc.

Chapter 40: Opener: © John Rowley/Getty Images RF; 40.2 (left): © Michael Abbey/Photo Researchers, Inc.; 40.2 (middle): © Sinclair Stammers/Photo Researchers, Inc.; 40.2 (right): © Dr. Richard Kessel/Visuals Unlimited; 40.5 (blood): © Dennis Kunkel Microscopy, Inc./Phototake; 40.5 (adipose tissue): © Ed Reschke/Getty Images; 40.5 (bone): © Innerspace Imaging/Photo Researchers, Inc.; 40.5 (cartilage): © Dr. John D. Cunningham/Visuals Unlimited; 40.5 (loose connective tissue, dense connective tissue): © The McGraw-Hill Companies, Inc./Al Telser, photographer; 40.8 (left, middle): © Dr. Stanley Flegler/Visuals Unlimited; 40.8 (right): © Dr. David M. Phillips/Visuals Unlimited; 40.10a: © Biophoto Associates/Photo Researchers, Inc.; 40.10b: © Pete Oxford/Minden Pictures; 40.10c: © Thomas Deerinck/Visuals Unlimited; 40.10d: © Anthony Bannister/Photo Researchers, Inc.; 40.15 (top left): © Mitch Wojnarowicz/The Image Works.

Chapter 41: Opener: © James Cavallini/Photo Researchers, Inc.; 41.2a: © James Cavallini/BSIP/Phototake; 41.5a: Courtesy of Dr. R. F. Rakowski, Ohio University, Athens, OH; 41.14b: © The McGraw-Hill Companies, Inc./Al Telser, photographer.

Chapter 42: Opener: Courtesy Ann-Shyn Chiang, Tsing Hua Chair Professor/Brain Research Center & Institute of Biotechnology/National Tsing Hua University; 42.6b: © Jan Leestma, M.D./Custom Medical Stock Photo; 42.14a: © Premaphotos/Alamy; 42.16a-b: © Dr. Scott T. Grafton/Visuals Unlimited; 42.18 (left): © The McGraw-Hill Companies, Inc./Al Telser, photographer.

Chapter 43: Opener: © Jane Burton/Bruce Coleman Inc./Photoshot; 43.8: © Gary Meszaros/Photo Researchers, Inc.; 43.12: © Daniel Heuclin/Photo Researchers, Inc.; 43.14a: © Eye of Science/Photo Researchers, Inc.; 43.16: © Omikron/Photo Researchers, Inc.; 43.17 (right): © Dr. Donald Fawcett/Visuals Unlimited; 43.19b: © Steve Allen/Getty Images RF; 43.23a: © Corbis/SuperStock RF; 43.23b: © Cal Vornberger/Getty Images; 43.24b: Courtesy of Russell Jurenka, Iowa State University; 43.24c (left): © Anthony Bannister/Photo Researchers, Inc.; 43.24c (right): Courtesy of Louisa Howard, Dartmouth College; 43.28a-c: © The Royal Society for the Blind of South Australia Inc.

Chapter 44: Opener: © Stephen Dalton/Photo Researchers, Inc.; 44.1a (right): © Dwight Kuhn; 44.1b: © Michele Hall/SeaPics.com; 44.1c (left): © Georgette Douwma/Photo Researchers, Inc.; 44.1c (right): © The Natural History Museum, London; 44.5b: © Dr. H.E. Huxley; 44.7: © Hansell H. Stedman, M.D.; 44.11a: © Dr. Donald Fawcett/Photo Researchers, Inc.; 44.12 (right): © Mark Richards/PhotoEdit; 44.15a: © Dr. LR/Photo Researchers, Inc.; 44.15b: © Tim Arnett, University College London.

Chapter 45: Opener: © Richard Hutchings/PhotoEdit; 45.2: © Daniel Hornbach and Tony Deneka, Courtesy of Daniel Hornbach, Macalester College; 45.2b: © Armin May forest/Getty Images; 45.3a: © David A. Northcott/Corbis; 45.3b: © John Giustina/Getty Images RF; 45.3c: © Kim Karpeles/agefotostock; 45.3d: © Reinhard Dirscherl/Getty Images; 45.3e: © CFranke/F1Online RF/Photolibrary; 45.4a: © Royalty-Free/Corbis; 45.4b: © Sarah Klockars-Clauser; 45.15c: © Javier Domingo/Phototake.

Chapter 46: Opener: © The Rockefeller University/AP Photo; 46.9a: © Paul McCormick/Getty Images; 46.9b: © William S. Clark/Frank Lane Picture Agency/Corbis; 46.13: © Nutscode/T Service/Photo Researchers, Inc.; 46.15b: Courtesy of Uffe Midtgård, University of Copenhagen; 46.16: CDC.

Chapter 47: Opener: © SPL/Photo Researchers, Inc.; 47.5: © Dr. David M. Phillips/Visuals Unlimited; 47.7: © Dennis Kunkel Microscopy; 47.11: © The McGraw-Hill Companies, Inc./Rick Brady, photographer; 47.12b: © Carolina Biological Supply/Visuals Unlimited; 47.14: © Ed Reschke/Getty Images; 47.20: © Biophoto Associates/SPL/Photo Researchers, Inc.; 47.21: © Sovereign/TSM/Phototake.

Chapter 48: Opener: © brianlatino/Alamy RF; 48.3 (top): © Ken Lucas/Visuals Unlimited; 48.3 (bottom): © Todd Pierson; 48.4a: © Hal Beral/V&W/The Image Works; 48.4b: © Jane Burton/Bruce Coleman Inc./Photoshot; 48.5a: © Jane Burton/naturepl.com; 48.5c: © Electron Microscopy Unit, Royal Holloway, University of London; 48.7 (top left): © Microfield Scientific Ltd/Photo Researchers, Inc.; 48.7 (right): © Ed Reschke/Getty Images; 48.10b: © Pr. M. Brauner/Photo Researchers, Inc.; 48.17 (left): © Astrid & Hanns-Frieder Michler/SPL/Photo Researchers, Inc.; 48.17 (right): © Dr. John D. Cunningham/Visuals Unlimited.

Chapter 49: Opener: © Guido Alberto Rossi/agefotostock; 49.12: © Steve Gschmeissner/SPL/Photo Researchers, Inc.

Chapter 50: Opener: © Sovereign/ISM/Phototake; 50.6a (left, right insets): © S. Goodwin & Dr. Max Hincke, Division of Clinical and Functional Anatomy, University of Ottawa; 50.6b: © Bob Daemmrich/The Image Works; 50.7 (bottom left): © Carolina Biological Supply/Visuals Unlimited; 50.13: From Robert F. Gagel, M.D., and Ian E. McCutcheon, M.D., "Pituitary Gigantism," *New England Journal of Medicine*, 1999, 340:524, Fig. 1A. © 2007 Massachusetts Medical Society. All rights reserved; 50.14b: © blickwinkel/Alamy.

Chapter 51: Opener: © David Liebman/Pink Guppy; 51.1a: © Clouds Hill Imaging Ltd./Corbis; 51.1b: © WaterFrame/Alamy; 51.3b: © Dr. Stanley Flegler/Visuals Unlimited; 51.3d: © P.M. Motta & G. Familiari/Univ. La Sapienza/Photo Researchers, Inc.; 51.4: © FLPA/Alamy; 51.12a: © Claude Edelmann/Photo Reseachers, Inc.; 51.12b: © Nestle/Petit Format/Photo Researchers, Inc.; 51.12c: © Photo Lennart Nilsson/Scanpix; 51.15 (diaphragm, condoms, oral contraceptive): © The McGraw-Hill Companies, Inc./Jill Braaten, photographer.

Chapter 52: Opener: Matsui et al., "Effects of ethanol on photoreceptors and visual function in developing zebrafish," *Investigative Ophthalmology & Visual Science*, October 2006, 47(10):4589–97, Fig. 1. © Association for Research in Vision and Ophthalmology. Images Courtesy of Jonathan Matsui; 52.2: Courtesy Dr. M. Whitaker, University of Newcastle upon Tyne; 52.4 (top row, middle row): © Dr. Richard Kessel & Dr. Gene Shih/Visuals Unlimited; 52.4 (bottom row): © Tom Fleming; 52.6 (bottom): Courtesy Hiroki Nishida, Biological Sciences, Osaka University; 52.10a: Courtesy Dr. Laurence D. Etkin, University of

Texas M.D. Anderson Cancer Center. With Permission of Malgorzata Kloc. This article was published in *Mechanisms of Development*, 75(1–2), Malgorzata Kloc, Carolyn Larabell, Agnes Pui-Yee Chan and Laurence D. Etkin, "Contribution of METRO pathway localized molecules to the organization of the germ cell lineage," pp. 81–93. Elsevier Science Ireland Ltd. July 1998; 52.10b: © F.R. Turner/Indiana University; 52.11–52.12a: Courtesy Kathryn Tosney; 52.12b: © Ed Reschke/Getty Images; 52.17(5): © Richard Harland, U.C. Berkeley; 52.18: © Courtesy Edward M. King.

Chapter 53: Opener: © SPL/Photo Researchers, Inc.

Unit VIII: 54: © David M. Dennis/Animals Animals; 55: © J. Heidecker/VIREO; 56: © Mike Lockhart; 57: © Stephen Wong & Takako Uno; 58: © DEA/C. DANI-I.JESKE/Getty Images; 59: © Juan Carlos Muñoz/agefotostock; 60: © D. Parer & E. Parer-Cook/Auscape/The Image Works.

Chapter 54: Opener: © David M. Dennis/Animals Animals; 54.1a: © Brand X Pictures/PunchStock RF; 54.1b: © Alain Pons/Biosphoto; 54.1c: © Paul Springett/Alamy RF; 54.1d: © Art Wolfe/Photo Researchers, Inc.; 54.6b: © Michael McCoy/Photo Researchers, Inc.; 54.8: © Jonathan Bird/Getty Images; 54.9: © Raymond Gehman/Corbis; 54.12a: © FLPA/D P Wilson/agefotostock; 54.12b: © Biophoto Associates/Photo Researchers, Inc.; 54.14a: © Gregory Ochocki/Photo Researchers, Inc.; 54.14b: Image courtesy of FGBNMS/UNCW-NURC/NOAA; 54.15: © Virginia P. Weinland/Photo Researchers, Inc.; 54.16a: © Peter Wakely/English Nature; 54.16b: © G.A. Matthews/SPL/Photo Researchers, Inc.; 54.24a: © Jean-Paul Ferrero/ardea.com; 54.24b: © Theo Allofs/theoallofs.com; 54.24c: © ARCO/K Wothe/agefotostock; 54.24d: © altrendo nature/Altrendo/Getty Images; 54.24e: © Tom & Pat Leeson; 54.24f: © Joe McDonald/Visuals Unlimited; 54.24g: © D. Robert & Lorri Franz/Corbis; 54.24h: © Joe McDonald/Visuals Unlimited; 54.24i: © Art Wolfe/Getty Images; 54.24j: © Michio Hoshino/Minden Pictures; 54.24k: © Howie Garber/Animals Animals; 54.25a: © Nature Picture Library/Alamy; 54.25b: © Stephen Frink/Corbis; 54.25c: © Jeffrey L. Rotman/Corbis; 54.25d: © Phillip Colla/OceanLight/Bruce Coleman Inc./Photoshot; 54.25e: © Tom & Pat Leeson; 54.25f: © Larry Mulvehill/Photo Researchers, Inc.

Chapter 55: Opener: © J. Heidecker/VIREO; 55.3a-b: © L.P. Brower, Sweet Briar College; 55.4: © Lilo Hess/Time Life Pictures/Getty Images; 55.5 (left): © Joe McDonald/Corbis; 55.6: © Nina Leen/Time Life Pictures/Getty Images; 55.8: © Frans Lanting/Corbis; 55.11a: © Tony Camacho/Photo Researchers, Inc.; 55.11b: © Gregory G. Dimijian/Photo Researchers, Inc.; 55.11c: © Getty Images RF; 55.12a: © Phil Degginger/Alamy; 55.12b: © Sachiko Kono/amanaimages/agefotostock RF; 55.13a: © Mark Moffett/Minden Pictures; 55.15: © Cyril Ruoso/Biosphoto; 55.17: © Peter Stiling; 55.18: © Danita Delimont/Alamy; 55.19: © Raymond Mendez/Animals Animals; 55.21a: © Masahiro Iijima/ardea.com; 55.21b: © WILDLIFE GmbH/Alamy; 55.21c: © Millard H. Sharp/Photo Researchers, Inc.; 55.22: © Chris Knights/ardea.com; 55.23: © Stephen A. Marshall; 55.25: © Minden Pictures/Masterfile.

Chapter 56: Opener: © Mike Lockhart; 56.1a: © Paul Glendell/Alamy; 56.1b: © Nigel Cattlin/Photo Researchers, Inc.; 56.1c: © The Baxter Bulletin, Kevin Pieper/AP Photo; 56.1d: © Cyril Ruoso/JH Editorial/Minden Pictures; 56.2: © W. Wayne Lockwood, M.D./Corbis; 56.3a: © Phil Banko/Corbis; 56.3b: © Fritz Polking/Frank Lane Picture Agency/Corbis; 56.3c: © Bob Krist/Corbis; 56.4a: © Doug Sherman/Geofile; 56.4b: © John Foxx/ImageState RF/agefotostock; 56.4c: © M. Watson/ardea.com; 56.15: © Royalty-Free/Corbis.

Chapter 57: Opener: © Stephen Wong & Takako Uno; 57.9a: © Dr. Thomas Eisner/Visuals Unlimited; 57.9b: © Hans D. Dossenbach/ardea.com; 57.9c: © Thomas Aichinger/V&W/The Image Works; 57.9d (left): © Suzanne L. & Joseph T. Collins/Photo Researchers, Inc.; 57.9d (right): © Michael Fogden/OSF/Animals Animals; 57.9e: © Paul Springett/Alamy RF; 57.10 (inset): © Tom & Pat Leeson/Photo Researchers, Inc.; 57.13a: © Glen Allison/Getty Images; 57.13b: © Toyofumi Mori/The Image Bank/Getty Images; 57.13c: © Gilbert S. Grant/Photo Researchers, Inc.; 57.13d: © John Dudak/Phototake; 57.15: © A & J Visage/Alamy; 57.18: © Dr. William A. Powell, SUNY-ESF; 57.19: © Konrad Wothe/Minden Pictures/Corbis; 57.20: © Michael & Patricia Fogden/Corbis; 57.21: © Mike Wilkes/naturepl.com; 57.22a: © E.A. Janes/agefotostock; 57.22b: © Dwight Kuhn; 57.24a-b: From Alan P. Dodds, *The Biological Campaign Against Prickly Pear*. Published under the Authority of the Commonwealth Prickly Pear Board. Brisbane, Queensland. 30th October, 1940.

Chapter 58: Opener: © DEA/C.DANI-I.JESKE/Getty Images; 58.2: © G.R. "Dick" Roberts/Natural Sciences Image Library; 58.6b: © Plowes Proteapix; 58.9a: © Gary Stewart/AP Photo; 58.9b: © David M. Dennis/Animals Animals; 58.10a: © Charles D. Winters/Photo Researchers, Inc.; 58.11a: © Tom Bean; 58.11b: © James Hager/agefotostock; 58.11c: © Howie Garber/AccentAlaska.com; 58.11d: © Tom Bean; 58.12 (inset): © Wayne Sousa/University of California, Berkeley; 58.17: Courtesy Dr. D. Simberloff, University of Tennessee.

Chapter 59: Opener: © Juan Carlos Muñoz/agefotostock; 59.1: © Stephen Dalton/Photo Researchers, Inc.; 59.9: © Robert T. Smith/ardea.com; 59.12: Courtesy of NASA and GeoEye Inc. Copyright 2012. All rights reserved; 59.17: © E. Debruyn, Courtesy of Experimental Lakes Area, Fisheries and Oceans Canada. Reproduced with the permission of the Minister of Public Works and Government Services Canada, 2010; 59.20: © Peter Stiling; 59.22a: © Science VU/Visuals Unlimited; 59.22b: Provided by the Northern Research Station, Forest Service, USDA; 59.24: © Frans Lanting/Corbis.

Chapter 60: Opener: © D. Parer & E. Parer-Cook/Auscape/The Image Works; 60.1: © Willard Clay Photography; 60.2: © Heather Angel/Natural Visions; 60.3: © Pete Manning, Ecotron Facility, NERC Centre for Population Biology; 60.7a: © The Bridgeman Art Library; 60.7b: © Topham/The Image Works; 60.7c: © James T. Tanner/Photo Researchers, Inc.; 60.8: © Bruce Coleman Inc./Photoshot; 60.11: © Vicki Fisher/Alamy; 60.13: © Andrew Parker/Alamy; 60.15a: © Robert E. Barber/Alamy; 60.15b: © Rick A. Brown/rick@moosephoto.com; 60.15c: © T. Kitchin & V Hurst/NHPA/Photoshot; 60.16: © Wildlife/Alamy; 60.17a: © University of Wisconsin-Madison Arboretum; 60.17b: Courtesy of DL Rockwood, School of Forest Resources and Conservation, University of Florida, Gainesville, FL; 60.17c: © Sally A. Morgan/Ecoscene/Corbis; 60.18a: © Corbis; 60.18b: © Simpson/Photri Images/Alamy; 60.19: © Ken Bohn/UPI/Newscom.

Index

Page numbers followed by *f* denote figures; those followed by *t* denote tables.

A

A band, 905*f*, 906
ABC model of flower development, 397
Abelson leukemia virus, 292*t*
abiotic interactions, 1117, 1122–29
abiotic synthesis, 439
abl gene, 290, 291*f*
A blood type, 338
ABO blood types, 337–38
abomasum, 931
abortion, 393
abscisic acid
  chemical structure, 740*t*
  major functions in plants, 740*t*, 746
  response to water stress, 785
  seed maturation role, 806
absolute refractory period, 840, 840*f*
absorption of food, 920, 921*f*, 928
absorption spectra, 158, 159*f*
absorptive nutrition of fungi, 619
absorptive state, 944–46, 949
acacia (*Acacia collinsii*), 1134, 1202, 1203*f*
Acanthodii, 688
Acanthoplegma, 562*f*
Accipiter gentilis, 1157
Accipitriformes, 702*t*
acclimatization, 959–60
accommodation, 887, 887*f*
accuracy of replication, 228
ACE inhibitors, 1*f*
Acer saccharum, 1125, 1125*f*
Acetabularia, 555*f*
acetaldehyde, 151
acetylcholine
  discovery, 846–47
  diseases hindering actions of, 917
  major effects, 845*t*, 846
  at neuromuscular junction, 911
  receptor types, 188
  vasodilation effects on aorta, 978–79
acetylcholinesterase, 911, 917
acetyl CoA, 140, 141
N-acetylgalactosamine, 338
acetyl groups, 137–38, 140, 142
acetylsalicylic acid, 1*f*, 118
acid hydrolases, 83–84
acidic solutions, 38
acid rain, 762, 1128–29, 1240
acids
  defined, 38
  in innate immune systems, 1095
  pH scale, 38–39
  in soils, 1128–29
Acinonyx jubatus, 1155, 1155*f*
acne, 544
acoelomates, 644, 645*f*, 657
acquired antibiotic resistance, 378
acquired immune deficiency syndrome (AIDS), 361*f*, 367, 1113
acquired immunity
  cell-mediated response, 1107–9
  defined, 1094
  humoral response, 1102–7
  immune system cells, 1100
  organs, tissues, and cells of, 1100–1101
  stages, 1102–11, 1103*f*
  summary example, 1109–10
acrocentric chromosomes, 314, 315*f*
acromegaly, 1047
acrosomal reaction, 1078, 1078*f*
acrosomes, 1059, 1060*f*
ACTH. *See* adrenocorticotropic hormone (ACTH)
actin
  role in mitosis, 307
  role in muscle contraction, 906–7, 907*f*, 908–9
  in skeletal muscle structure, 905*f*, 906
  structure and function, 4, 75–76, 75*t*
actinobacteria, 536*t*, 548, 767
Actinopterygii, 9, 11*f*, 690–91
action potentials
  conduction, 841–42, 841*f*
  generation, 834, 839–41
  relation to stimulus intensity, 876, 877*f*
  SA node, 972
  in skeletal muscle, 904, 910–11, 910*f*
  in Venus flytrap, 769
action spectra, 159, 159*f*
Actitis macularia, 1164
activation
  B cells, 1106, 1108
  complement proteins, 1098
  as second stage of acquired immunity, 1102, 1103*f*
activation energy, 122–23
activators
  defined, 260, 261*f*
  functions in eukaryotes, 269, 270, 270*f*, 272–73, 273*f*
  positive control of *lac* operon, 266, 267*f*
active immunity, 1111
active sites, 123
active transport
  between body fluid compartments, 820
  defined, 105
  gradient formation by, 106
  illustrated, 105*f*
  nutrient absorption via, 928, 933, 934*f*
  in plant cells, 773–74, 774*f*
  process overview, 111–13, 112*f*, 113*f*
active trapping mechanisms, 769
activin, 1087, 1088*f*
acyl transferase, 104*f*
adaptations, 482. *See also* evolution
adaptive immunity, 1094
adaptive radiation, 501–2
adenine
  base pairing in DNA, 61–62, 62*f*, 220, 220*f*, 221
  bonding in DNA, 220, 220*f*, 221
  molecular structure, 61, 217, 218*f*
adenosine deaminase (ADA) deficiency, 416–17
adenosine diphosphate (ADP), 76, 142
adenosine monophosphate (AMP), 248, 249*f*. *See also* cyclic adenosine monophosphate (cAMP)
adenosine triphosphate (ATP)
  activation energy example, 122–23
  breakdown in nitrogen fixation, 763, 764*f*
  as citric acid cycle regulator, 142
  as energy source for endergonic reactions, 121–22, 122*f*
  energy storage, 47
  formation by aerobic respiration, 136–38, 137*f*
  formation in anaerobic conditions, 150–51, 150*f*
  formation in mitochondria, 86–87
  in glycolysis, 138–39
  homeostatic role, 140, 823*t*
  hydrolysis as exergonic reaction, 120, 120*f*
  hydrolysis in muscle fibers, 911–12
  hydrolysis-synthesis cycle, 121–22, 122*f*
  ion pump energy from, 112–13, 773–74
  regulation of production, 825
  role in muscle contraction, 908, 909
  synthesis in chloroplasts, 159
  synthesis in oxidative phosphorylation, 145–47
  synthesis reactions, 129
  use by motor proteins, 76
adenoviruses, 360, 361*t*, 362*f*
adenylyl cyclase, 183, 184*f*, 266
adherens junctions, 199, 199*f*
adhesion, 38, 86
adhesive proteins, 194–95, 816
adiabatic cooling, 1130
adipose tissue. *See also* fats
  appetite-regulating hormones, 1044
  basic features, 817*f*
  endocrine functions, 1031*f*
  puberty and, 1072
ADP. *See* adenosine diphosphate (ADP)
adrenal cortex, 189–90, 1031*f*, 1050–51
adrenal glands, 189, 1031*f*, 1040
adrenal medulla, 1031*f*, 1050, 1084
adrenocorticotropic hormone (ACTH), 189, 1037*t*
adult stem cells, 393
adventitious roots, 723
aerenchyma, 750–51, 751*f*
aerial rootlets, 467, 468*f*
aerobic exercise, 912–13. *See also* exercise
aerobic respiration, 136. *See also* respiration
aerotolerant anaerobes, 544
afferent arterioles, 1020, 1020*f*, 1021, 1021*f*
afferent neurons, 833
affinity, enzyme-substrate, 123–24
aflatoxins, 623–24
African clawed frog (*Xenopus laevis*), 381, 1089, 1090*f*
African forest elephant (*Loxodonta cyclotis*), 514, 514*f*
African long-tailed widowbird (*Euplectes progne*), 1164
African savanna elephant (*Loxodonta africana*), 514
African wildcat (*Felis libyca*), 1261
Afrotheria, 706
agar, 559, 742–43
agarose, 559
Agasicles hygrophila, 1205
age classes, 1172, 1173
Age of Fishes, 455
age-specific fertility rate, 1176
age structure, 1183, 1184*f*
aggregate fruits, 609–11, 610*f*
agility, 1195
Agre, Peter, 109, 110*f*, 1025
agriculture
  biotechnology applications, 412–15, 1197–98
  domestication of angiosperms for, 615
  legume-rhizobia symbioses, 767
  organic, 762
  parasitic plants, 770
  prairie destruction, 1245, 1252
  secondary succession following, 1216
  selective breeding, 469, 470*f*, 471*f*
  smart plants for, 764–66
  use of bacteria, 550
Agrobacterium tumefaciens
  chromosomes, 421
  gall formation by, 536, 537*f*
  producing transgenic plants with, 412–13, 413*f*
  type IV secretion system, 549, 549*f*
Agrostis capillaris, 483, 485*f*
Aguinaldo, Anna Marie, 649

A-horizon, 760, 761*f*
Ailuropoda melanoleuca, 1258
Aintegumenta gene, 396, 396*t*
air bubbles in plant vessels, 780–81
air pollution, 635, 1258. *See also* pollution
air sacs, 701, 996–99, 996*f*
Ajaia ajaja, 701*f*
akinetes, 543–44, 543*f*
alanine, 54*f*
alarm calling, 1160–61
albinism, 321
alcoholism, 80, 1091
aldehydes, 44*t*
alder (*Alnus sinuata*), 767, 1216, 1217*f*
aldosterone
  molecular structure, 1034*f*
  sodium regulation, 1023–24, 1046
aleurone, 807
Alexander, Richard, 1161
Alexandrium catenella, 561*f*
algae. *See also* seaweeds
  basic features, 553
  evolution examples, 453, 453*f*
  importance to life, 553
  industrial products from, 561–62
  in lichens, 634–35
  nutrition, 564
  overfertilization, 763, 1233, 1236
  photosynthesis studies, 164–65, 167, 168–69*f*
  similarity to plants, 558–60
algal blooms, 1233
alimentary canals, 928–33
alkaline solutions, 38
alkaloids, 611, 611*f*, 1197, 1198*f*
alkaptonuria, 236
alkylation, 284, 285*t*
alkyl groups, 287
allantois, 696, 696*f*
allele frequency, 480–81
alleles
  defined, 324
  frequency in populations studied, 480–81, 492
  genetic drift, 489–91
  homozygous vs. heterozygous individuals, 325
  independent assortment, 326–27, 329–30, 329*f*
  interactions between, 344–46
  linked, 346–50
  multiple, 337–38
  pedigree analysis, 330–31
  polymorphic genes, 479
  recording in Punnett square, 325–26
  redistribution via sexual reproduction, 1057–58
  segregation, 324–25, 328–29, 328*f*
allelochemicals, 1118–19, 1188
allelopathy, 1188
Allendorf, Fred, 1253–54
allergies, 1112
Alligator mississippiensis, 332, 698
alligators, 698, 698*f*
alligatorweed flea beetle (*Agasicles hygrophila*), 1205
Allis, David, 272
allolactose, 261, 262, 263*f*
allopatric speciation, 500–502, 506–7, 1193
alloploid organisms, 504
allopolyploids, 504–5, 608
allosteric sites, 125, 130–31, 909
Alnus sinuata, 1216, 1217*f*
Alper, Tikvah, 368
α -antitrypsin deficiency, 336*t*
α^1-carbons, 53
alpha cells, 1039, 1040*f*
α chains, 195, 196
α-globin genes, 429, 1070
α-glucose, 47, 47*f*
α helices
  Pauling's discoveries, 219–20
  as secondary structure of proteins, 55, 56*f*
α-ketoglutarate, 129, 143*f*
α-ketoglutarate dehydrogenase, 141, 143*f*
α-latrotoxin, 843
alpha particles, 22–23
α-proteobacteria, 452
α subunits, 1001
α-tropomyosin, 274–75
alternation of generations, 313*f*, 314, 578, 717, 793–94, 794*f*
alternative exons, 274*f*, 275
alternative splicing, 72, 242, 274–75, 431, 431*f*
altitude, 987, 987*f*, 1004
Altman, Sidney, 126
altruism, 1158–61
Altschul, Stephen, 434
Alu sequence, 425, 426, 427
Alveolata, 560–61, 561*f*
alveoli
  in dinoflagellates, 561, 561*f*
  in mammals, 821, 821*f*, 993, 994*f*
  surfactant in, 996
Alzheimer, Alois, 872
Alzheimer disease, 872–73
amacrine cells, 891
Amanita muscaria, 623*f*
Amanita virosa, 624
Amblyomma hebraeum, 671*f*
Amborella trichopoda, 606, 608*f*, 609
Ambrosia artemisifolia, 1260
amensalism, 1187
American alligator (*Alligator mississippiensis*), 698
American avocet (*Recurvirostra americana*), 701*f*
American buffalo (*Bison bison*), 1258
American chestnut (*Castanea dentata*), 1200–1201, 1200*f*, 1250, 1259
American crocodile (*Crocodylus acutus*), 698
American vetch plant, 730*f*
American woodcock (*Scolopax minor*), 892, 892*f*
Ames, Bruce, 285
Ames test, 285–86, 286*f*
amines, 1031–32, 1032*t*, 1033
amino acids
  absorption in digestive tract, 935
  codons, 243–44
  as essential animal nutrients, 922
  identifying triplets with, 246
  immunoglobulin regions, 1103, 1104
  metabolism, 149
  mutations causing substitutions, 280–81
  as neurotransmitters, 845*t*, 846
  origins on Earth, 439–40
  as protein building blocks, 53
  ranking for hydrophobicity, 99–100
  reabsorption in kidney, 1021
  recycling, 132
  sequenced in genetic code, 243–44
  storage and use, 945–46, 945*f*
amino acid sequence databases, 433*t*
amino acid sequences
  computer analysis, 432–33, 432*f*
  evolution evidence from, 471–72, 472*f*
aminoacyl (A) site, 250, 250*f*, 254
aminoacyl tRNA, 248
aminoacyl-tRNA synthetases, 248–49, 249*f*
amino ends of polypeptides, 245
amino groups
  in amino acids, 53, 55*f*
  basic features, 43, 44*t*
  in polypeptides, 245
2-aminopurine, 284, 285*t*
Amish, 491
ammonia
  at deep-sea vents, 440
  in Earth's early atmosphere, 439
  as fertilizer, 763
  as nitrogenous metabolic waste, 1010, 1010*f*
  produced during nitrogen fixation, 545, 763, 1240
  reactions with water, 38
  release in nitrogen fixation, 763
ammonification, 1240
ammonites, 662–63, 663*f*
ammonium, 129
ammonium cyanate, 43
amnion, 696
amniotes
  birds, 699–702
  key characteristics, 696–97
  mammals, 702–12
  reptiles, 697–99
amniotic egg, 640, 696–97, 696*f*. *See also* eggs
amoebae, 555, 555*f*
amoebic dysentery, 567
amoebocytes, 653
amoeboid movement, 555
Amoebozoa, 562
amphibians
  appearance during Devonian period, 455
  circulatory systems, 966–68
  embryonic development, 1081–83, 1082*f*
  evolution, 465*f*, 692, 694–95
  gas exchange through skin, 988
  hearing sense, 881, 881*f*
  holoblastic cleavage, 1079–80, 1080*f*
  kidneys, 1018
  locomotion in water, 915
  lungs, 992
  modern species, 695–96
  nervous systems, 859*f*
amphipathic molecules, 35, 51, 98
Amphiprion, 9, 11*f*
ampicillin, 403
amp^R gene, 402, 403
ampullae, 678, 882, 883*f*
ampullae of Lorenzini, 689
amygdala, 866
amylase, 929, 933
amyloplasts, 87
anabolic reactions, 128, 129, 130
anabolism, 74
Anabrus simplex, 1164–65
anaerobic respiration, 150, 450
anagenesis, 517
analgesia, congenital, 340
analogous structures, 468–69, 520
anaphase (mitosis), 305, 307*f*
anaphase I (meiosis), 311*f*, 312
anaphase II (meiosis), 312
anatomical homologies, 470, 471*f*. *See also* homologies
anatomy, defined, 14
anchoring junctions
  basic features, 198, 198*t*
  in epithelial tissue, 207
  types, 199–200, 199*f*
ancient DNA analysis, 524–25
Andersen, Dorothy, 17
Anderson, W. French, 416
Andersson, Malte, 1164
androgens
  abuse of synthetic versions, 1030, 1053
  effects on human male, 1064–65
  functions, 1030, 1034*f*, 1050
  site of production, 1031*f*, 1050
anemia
  due to nutrient deficiency, 923*t*, 924*t*

with sickle cell disease, 1002
symptoms and causes, 970
anemones
Hox genes, 509*f*
interactions with clownfish, 1*f*, 12–13
key characteristics, 648*t*, 681*t*
anemotaxis, 1151
aneuploidy, 316, 318
Anfinsen, Christian, 58–60
angina pectoris, 983
angiography, 983
angiosperms. *See also* flowering plants
coevolution, 613–15
development during Cenozoic, 457
diversity, 609–13
early ecological impact, 587–88
evolution, 604–9
first appearance, 456
fossils, 586*f*
human uses, 605, 615, 715
seeds, 592
angiotensin-converting enzyme (ACE) inhibitors, 1*f*
Anguilla anguilla, 989*f*
Anguis fragilis, 697*f*
animal cap assay, 1087, 1088*f*
animal cells
basic types, 283
cell junctions, 198, 199–203
cleavage during mitosis, 305, 306*f*
extracellular matrix functions, 194–97
locations of genetic material, 351*f*, 423–24
major structures, 70, 71*f*, 74*f*
proteins involved in cytokinesis, 307–8
animal pole, 1079, 1079*f*
animals
in aquatic biomes, 1139–42
basic features, 10*f*
body organization, 813–21
classification, 640–45, 646*f*
cloning, 400*f*
in coniferous forest biomes, 1135
in deciduous forest biomes, 1134, 1135
in desert biomes, 1137
developmental stages, 258–59, 385–95
effects of chromosome number changes, 316–17, 318*t*
extracellular matrix functions, 194–97
feeding behaviors, 922–26, 927*f*
genetic engineering, 411–12, 414–15
on geological timescale, 449*f*
in grassland biomes, 1136
homeostatic mechanisms, 822–29
introns in genes, 242
key characteristics, 638–40
life cycle, 313*f*, 314
locomotion, 915–16
model organisms, 381
molecular diversity, 646–50
in mountain biomes, 1138
naming, 517
pattern formation, 382–84, 385*f*
plant predation on, 756, 769–70
as pollinators, 613–14
in rain forest biomes, 1133, 1134
relation of forms to functions, 821–22
rRNA gene sequences compared, 251*f*
skeleton types, 901–3
tissue types, 205, 814–16, 814–18*f*
in tundra, 1138
water in, 27, 34
anions, 31
aniridia, 380, 511
Anisotremus taeniatus, 500, 501*f*
Anisotremus virginicus, 500, 501*f*
annealing, 405*f*, 406
annelids
classification, 666–67
excretory systems, 665, 665*f*, 1016–17, 1017*f*
genome comparison to arthropods, 650
hydrostatic skeleton, 902, 902*f*
key characteristics, 648*t*, 665–66, 681*t*
nervous systems, 855*f*, 856
segmentation, 646*f*
annuals, 720, 1137
Anodorhynchus hyacinthinus, 701*f*
Anopheles mosquito, 569
anorexia nervosa, 961
Anseriformes, 702*t*
antacids, 940
antagonists, 915
anteaters, 467, 468*f*
antenna complex, 162
antennae, 822*f*, 893*f*, 894
Antennapedia complex, 389, 389*f*, 390
anterior cavity, in single-lens eye, 886, 887*f*
anterior chamber, in single-lens eye, 886, 887*f*
anterior end, 642
anterior pituitary gland, 1031*f*, 1035–36, 1036*f*
anteroposterior axis, 382, 382*f*, 1077, 1082
antheridia, 579
anthers, 605, 606*f*, 795
Anthozoa, 656, 656*f*
anthrax, 69, 544
anthropoidea, 707, 708*f*
antibiotic-resistant bacteria, 7–8, 8*f*, 377*f*, 378, 403, 547
antibiotics
actions on bacterial ribosomes, 250
bacterial breakdown, 545–47
bacterial resistance, 7–8, 8*f*, 377*f*, 378, 403, 547
discoveries, 1
saliva's function as, 929
susceptibility of different bacteria, 541
using bacteria to produce, 550
antibodies
B-cell receptors versus, 1103
functions, 1102–3, 1106
from plasma cells, 1101
anticoagulants, 926
anticodons, 245, 248, 248*f*
antidiuretic hormone
major functions, 1037
mediation of salt-water balance, 1023, 1024, 1025, 1025*f*
site of production, 1031*f*, 1037
sodium regulation, 1046
antifreeze proteins, 467–68, 956
antigen-presenting cells, 1107–8
antigens
acquired immune response, 1102–9
defined, 1100
presentation to helper T cells, 1107–8
on surface of red blood cells, 338, 338*t*
antihistamines, 1112
anti-inflammatory drugs, 884
antioxidants, 33, 611
antiparallel strands in DNA, 221
antiporters, 111, 111*f*
Antirrhinum majus, 397, 799, 800*f*
antiviral drugs, 367–68
antlers, 704, 705*f*
ants
communication, 1156
as fluke hosts, 1198–99
key characteristics, 673*t*
mutualistic relationships, 1202, 1202*f*, 1203*f*
reproduction, 1057
anurans, 695. *See also* frogs
anus, 933
aorta, 971, 972*f*
apes, monkeys versus, 707
aphids, 639*f*, 1202
apical-basal-patterning genes, 396, 396*t*
apical-basal polarity, 720, 720*f*
apical cells, 395, 396*f*
apical constriction, 1083, 1083*f*
apical dominance, 742
apical membrane, 815, 816*f*
apical meristems, 577, 718
apical region, 395, 396*f*
Apicomplexa, 560, 572
apicoplast, 572
Apis mellifera, 1156–57, 1157*f*
Aplysia californica, 867
Aplysia punctata, 867*f*
Apoda, 695
Apodiformes, 702*t*
apomixis, 809
apoplastic transport, 777, 777*f*
apoplasts, 777
apoptosis
hormonal control, 189–90
importance, 174, 188
nematode studies, 668
pattern formation and, 382*f*, 383
process overview, 174*f*, 188, 188*f*
signal transduction pathways, 190, 191*f*
as tissue-forming process, 205
tumor suppression by, 292–93
apoptosomes, 190
aposematic coloration, 1194, 1195*f*
appendicular skeleton, 903
appendix, 933
appetite, 952–55, 1044
apples, 797*f*
apple scab, 627
AQP2 gene, 1026
aquaporins
in collecting ducts, 1024, 1025*f*
discovery, 109, 110*f*, 1025–26
effects of ADH on, 1025, 1025*f*, 1037
molecular structure, 1026*f*
in plant cells, 773
aquatic biomes, 1132, 1139–42
aquatic birds' foot structure, 4
aquatic communities, succession in, 1217
aquatic ecosystems, 1233
aquatic plants, photic zone, 1127, 1128*f*
aqueous humor, 886, 887*f*, 897
aqueous solutions, 34
aquifers, 1242
Aquila verreauxii, 701*f*
ARABIDILLO gene, 734
Arabidopsis thaliana
alternative splicing, 275*t*
embryogenesis, 805*f*
gene regulation, 257
genome size, 275*t*, 424*t*
GNOM mutants, 720*f*
guard cell development, 729
membrane protein genes, 100*t*
as model organism, 13, 381–82, 381*f*, 729
mutations, 397
phytochrome genes, 748
root development, 734
sclerenchyma cells, 208, 208*f*
in smart plant study, 765–66
whole-genome duplications, 608
arachidonic acid, 922
Arachnida, 670
arachnids, 648*t*, 681*t*
arachnoid mater, 859, 860*f*
Arber, Werner, 401
arbuscular mycorrhizae, 632, 632*f*
arbuscular mycorrhizal (AM) fungi, 619, 624*t*, 626–27

arbuscules, 632, 632*f*
archaea
  bacteria versus, 450–51
  basic features, 9, 10*f*
  cell structure, 69
  diversity, 534
  in eukaryotic cell origins, 451–52
  genomes, 420–23
  horizontal gene transfer, 537–38
  membrane protein genes, 100*t*
  motility, 541–42
  role in carbon cycle, 545
Archaea (domain)
  ancestral to Eukarya, 535
  basic features, 9, 10*f*, 533–34
  in taxonomy, 515
Archaean eon, 450–52
Archaeoglobus fulgidus, 100*t*
Archaeopteris, 598, 599*f*
Archaeopteryx, 456, 457*f*
Archaeopteryx lithographica, 699, 700*f*
archegonia, 579
archenteron, 642, 1082, 1082*f*
Arctium minus, 1203*f*
area hypothesis, 1210
arginine
  molecular structure, 54*f*
  synthesis, 237–38, 237*f*
arginines, 1025
Argopecten irradians, 1187
arid conditions, plant responses to, 776
Armadillidium vulgare, 991
Armillaria ostoyae, 193
Armillaria ostoyaef, 618
armor, 1196
army ants (Eciton burchelli), 675*f*
Arnold, William, 161
arteries, 966, 975, 975*f*, 983
arterioles
  of kidneys, 1020, 1020*f*, 1021*f*
  structure and function, 975, 975*f*, 976*f*
Arthrobotrys anchonia, 630*f*
Arthropoda, 648*t*, 681*t*
arthropods
  body plan, 454, 668–69
  establishment on land, 450
  excretory systems, 669, 1017, 1018*f*
  exoskeletons, 902–3, 902*f*
  genome comparison to annelids, 650
  key characteristics, 648*t*, 681*t*
  locomotion, 916
  nervous systems, 855*f*
  photoreception in, 886, 886*f*
  relation to nematodes, 649–50
  segmentation, 646*f*
  sound sensitivity, 879
  subphyla, 669–74, 670*t*
artificial selection, 469
Asarum canadensis, 632*f*
Ascaris lumbricoides, 668
ascending limb, loop of Henle, 1020–21, 1020*f*, 1022
Aschehoug, Erik, 1118–19
asci, 627, 628*f*
ascocarps, 627, 628*f*
ascomycetes
  basic features, 619, 624*t*, 627–28
  life cycle, 628*f*
  septal pores, 627*f*
ascospores, 627, 628*f*
aseptate hyphae, 620
asexual reproduction
  advantages and disadvantages, 1057
  in annelids, 666
  basic forms, 1056–57
  binary fission, 542–43
  in dandelions, 793
  defined, 303, 1056
  in flatworms, 657
  in flowering plants, 793, 808–9
  of fungi, 622, 626*f*, 628*f*
  in lichens, 635
  of protists, 567
  in sponges, 654
Asian chestnut (*Castanea mollissima*), 1201
Asian elephant (*Elephas maximus*), 514
asparagine, 54*f*, 104
aspartate, 845*t*
aspartic acid, 54*f*
Aspergillus nidulans, 621*f*
Aspergillus versicolor, 622*f*
aspirin, 1*f*, 118
ASPM gene, 866
assassin bug, 673*t*
assimilation, in nitrogen cycle, 1240
assisted reproductive technologies, 1073
associative learning, 1148
assortative mating, 492
Asteroidea, 678, 678*t*
asthma, 1004–5
astral microtubules, 304, 305*f*
astrocytes, 833
Atelopus zeteki, 1117
AT/GC rule, 221, 223–24, 223*f*
atherosclerosis, 983, 983*f*
atmosphere
  circulation patterns, 1129–30, 1130*f*
  early Earth, 439, 440, 448–50
  plants' impact, 586–87, 587*f*
atmospheric pressure, 987, 987*f*
atomic mass, 26
atomic number, 25–26
atoms
  basic structure, 21–23
  defined, 6, 21
  electron orbitals, 24–25
  isotopes, 26–27
  as level of biological organization, 5*f*, 6
  mass, 26
  types of chemical bonds, 28–31
ATP. *See* adenosine triphosphate (ATP)
ATP-dependent chromatin-remodeling complexes, 271, 272
ATP synthase
  experimental verification, 146
  role in oxidative phosphorylation, 145, 146–47, 146*f*, 148–49*f*
atria
  amphibian heart, 967–68
  in double circulation systems, 968
  fish heart, 966
  vertebrate heart, 971, 972*f*
atrial excitation, 972
atrial natriuretic peptide, 1031*f*, 1046–47
atrioventricular node, 972, 974, 974*f*
atrioventricular valves, 971, 972*f*, 973
atrophy, muscular, 913
attachment, in viral reproductive cycle, 363, 364*f*
audition, 879–82, 898
auditory communication, 1156
auricles, 657
Austin, Thomas, 1178
Australian lungfish (*Neoceratodus forsteri*), 691*f*
Australopithecus, 457, 710
Australopithecus afarensis, 710
Australopithecus africanus, 710
Australopithecus garhi, 710
autocrine signaling, 176, 176*f*
autoimmune diseases, 851, 917–18, 1110–11
autonomic nervous system, 860–62, 861*f*
autonomous specification, 1086, 1087*f*
autophagosomes, 133, 133*f*
autophagy, 132–33, 133*f*
autophosphorylation, 182
autopolyploidy, 504, 608
autoradiography, 167, 168*f*
autosomes, 299, 331
autotomy, 678
autotrophs
  among prokaryotes, 544
  defined, 155, 451
  as early life-forms, 451
  in food chains, 1226
Aux/IAA repressors, 741
auxin
  in carnivorous plants, 770
  imbalance in seedling stems, 746
  major functions in plants, 740*t*, 741–43
  phototropism and, 175, 739, 742–43
  response to statolith position, 750
  role in leaf development, 725
  transmembrane transport, 776
auxin efflux carriers, 741
auxin influx carrier (AUX1), 741, 741*f*
auxin-response genes, 741
Averof, Michalis, 645
Avery, Oswald, 214–15
Aves, 699. *See also* birds
avian malaria, 1251
avian respiratory system, 996–99, 996*f*
avirulence genes, 752, 753*f*
avocets, 701*f*
Avogadro's number, 26
Axel, Richard, 895–96
axes, 382, 1077
axial skeleton, 903
axillary buds, 725
axillary meristems, 725
axonemes, 77–78
axon hillock, 832, 833*f*, 841
axons
  action potentials, 839–41
  conduction disorders, 851
  defined, 832, 833*f*
  effect of diameter on action potential speed, 842
axon terminals, 832, 833*f*
azidothymidine, 367

B

Bacillariophyceae, 561
bacilli, 539, 539*f*
Bacillus anthracis, 69, 421*t*, 544
Bacillus subtilis, 100*t*, 251*f*, 370*f*
Bacillus thuringiensis, 413, 414*f*, 550, 1198
backbone of nucleic acids, 218, 219*f*
Bacskai, Brian, 185
bacteria. *See also* infectious diseases
  anaerobic respiration, 150, 150*f*
  antibiotic resistance, 7–8, 8*f*, 377*f*, 378, 403, 547
  archaea versus, 450–51
  biotechnology applications, 410–11, 549–50, 1232
  cell structure, 69
  cellulose digesting, 704, 931
  challenges to distinguishing species, 496
  discovery of genetic material from, 213–17
  diversity, 534, 535–37
  drugs derived from, 1, 2*f*
  endosymbiosis theory, 88
  in eukaryotic cell origins, 451–52
  eukaryotic cells versus, 9, 93
  gene regulation, 258, 259, 260–67
  genetic structures, 370–72, 371*f*, 372*f*, 420–23
  gene transfer between, 373–78, 474–75, 537–38
  infections causing diarrhea, 940

INDEX

  introns in genes, 242
  in large intestine, 548, 933, 1214
  messenger RNA, 244, 244f
  model organisms, 13
  motility, 541–42
  mutation studies, 282–83
  nitrogen fixing, 763, 767–69, 1216, 1240
  nutrition and metabolism, 544–45
  as pathogens, 69, 359, 548–49, 872, 1095
  photosynthesis, 155
  plant defenses, 752
  rate of DNA synthesis, 226
  reproduction, 372, 373f
  ribosomes, 249–50, 249t, 250f
  RNA polymerase, 241
  role in carbon cycle, 545–47
  rRNA gene sequences compared, 251f
  sizes, 360, 538
  in symbiotic associations, 547–49, 767–69
  translation vs. eukaryotes, 254t
  as ulcer cause, 940
  use in Ames test, 285–86
  use in gene cloning, 401–3
  use in genetic engineering, 412–13, 413f
Bacteria (domain)
  basic features, 9, 10f, 533–34
  diversity, 534, 535–37
  membrane protein genes, 100t
  in taxonomy, 515
bacterial artificial chromosomes (BACs), 406
bacterial colonies, 372
bacterial infections, 1096–97, 1097f
bacterial meningitis, 359
bacteriomes, 452
bacteriophages
  DNA discoveries from, 215–17, 216f
  fd, 361t
  hosts and characteristics, 361t
  λ (lambda), 361t, 363–66, 364–65f
  latency, 366
  protein coats, 361, 362f
  Qβ, 361t
  release, 365f, 366
  T1, 283
  T2, 215–17, 216f, 366
  T4, 361t, 362f
  transduction by, 377, 378f
bacteriorhodopsin, 99f, 146, 146f
Bacteroidetes, 536t, 547
bacteroids, 769
baculoviruses, 361t
Baikal, Lake, 1209
balance, sensory systems, 882–83, 883f
balanced polymorphism, 484, 485
balancing selection, 483–86
bald cypress (*Taxodium distichum*), 603
baleen whales, 924, 925f
ball-and-socket joints, 904f
balloon angioplasty, 983, 984f
bamboo, 723
banding patterns of chromosomes, 315, 475, 475f
Banfield, Jill, 1214
Banks, Jody, 583
Banting, Frederick, 1042, 1043
Barber, James, 163
barbules, 701
barcoding, 675
barnacles, 676, 676f, 1191–93
barometric pressure, 987
baroreceptors, 860, 982–83
Barr, Murray, 353
Barr bodies, 353, 353f, 354–55
barrier methods of contraception, 1073, 1073f
bar-tailed godwit (*Limosa lapponica*), 916
Bartel, David, 443, 444f, 446
basal bodies, 78
basal cells
  olfactory epithelium, 894, 894f
  in plants, 395, 396f
basal membrane, 815, 816f
basal metabolic rate
  major influences, 951–52, 952f
  measuring, 950–51, 951f
basal nuclei, 865t, 866
basal region, plant embryos, 395, 396f
basal transcription, 269
base analogues, 284
basement membrane, 294
base pairs, 221, 280
bases
  defined, 38
  directionality in nucleic acids, 218
  evolutionarily conserved sequences, 251
  in nucleotides, 60, 61f, 217–18
  pH scale, 38–39
  spontaneous mutations, 284
  in structural genes, 239, 239f
base substitutions, 280, 280t
basic helix-loop-helix (bHLH) proteins, 729
basidia, 628–29, 629f
basidiocarps, 629, 629f
basidiomycetes
  basic features, 619, 624t, 628–29
  life cycle, 629f
  septal pores, 627f
basidiospores, 628–29, 629f
basilar membrane, 880–81, 880f, 881f
basophils, 1096, 1096f
Batesian mimicry, 1195, 1195f
Bateson, William, 344–45, 346, 348, 388, 397
Batrachochytrium dendrobatidis, 625, 1117
bats
  echolocation, 882
  evolutionary changes, 6f
  fungal diseases, 630
  as pollinators, 609f
  wing structure, 916
Bayer, Martin, 395
Bayliss, William, 937, 939
bay scallops (*Argopecten irradians*), 1187
B blood type, 338
B-cell receptors, 1103, 1104f
B cells
  activation, 1106, 1108
  antigen presentation by, 1108
  autoimmune responses, 1110–11
  basic features, 1101, 1102f
  immunoglobulin production, 1102–5
bcr gene, 290, 291f
bdellovibrios, 537
Beadle, George, 237
Beagle voyage, 460–61
beaks
  diversity in modern birds, 701
  electromagnetic receptors in, 884–85
  evolution of morphology, 462f, 463–64
  relation of morphology to bird's song, 502, 503–4f
  type versus diet in finches, 462t
bean seedlings, 807, 808f
beavers
  age-specific fertility rates, 1176
  as keystone species, 1258, 1259f
  survivorship curves, 1173–74, 1174f
  teeth, 704f
bedbug, 673t
bedrock, 760
bee orchids (*Ophrys apifera*), 1203, 1203f
beer, malting, 807
Beers, Eric, 790
bees
  altruism, 1158
  colony collapse disorder, 625
  color vision, 889
  communication, 1156–57
  innate behaviors, 1147–48
  key characteristics, 673t
  males as haploid organisms, 316, 332
  mutualistic relationships, 1203
  open circulatory systems, 965f
  reproduction, 1057
  sex determination, 332
behavior
  defined, 737, 1146
  hormonal effects in plants, 740–46
  major influences on, 1147–50
  plant potential for, 737–39
  plant responses to environment, 747–54
behavioral ecology. *See also* ecology
  altruism, 1158–61
  communication, 1155–57
  defined, 1118, 1146
  foraging behavior, 1154–55
  group living, 1157–58
  local movement and long-range migration, 1150–54
  mating systems, 1162–65
behavioral isolation, 498, 498f, 499f
Beijerinck, Martinus, 360
Belding's ground squirrels (*Spermophilus beldingi*), 1160
Beltian bodies, 1203f
beluga whales, 279f
Bendall, Fay, 165
benign growths, 288
Benson, Andrew Adam, 165
benzo(*a*)pyrene, 285, 285t
Berg, Paul, 400
beriberi, 923t
Bernal, John, 441
Bernard, Claude, 823
Bertram, Ewart, 353
Best, Charles, 1042, 1043
β-amyloid, 872, 873
beta cells, 1039, 1040f
β-endorphin, 1036
β-galactosidase
  bacterial regulation, 258, 261, 262f, 264–65f
  in recombinant DNA technology, 403, 410, 411f
β-globin gene
  functions, 429
  heterozygote advantage, 485
  mutations, 280, 281f
  polymorphic, 479
  similarity between species, 433–34, 1070
β-glucose, 47, 47f
β-lactamase, 403
β-lactoglobulin, 412
β-mercaptoethanol, 58, 59, 59f
β pleated sheets, 55, 56f
β subunits, 1001
Betta splendens, 690
bicarbonate ions
  as buffer, 39
  carbon dioxide breakdown producing, 1003
  from pancreas, 932, 937
  reabsorption in kidney, 1021
bicoid gene product, 386, 387f
Bicoid protein, 386
bidirectional replication, 224, 224f
Bienertia cycloptera, 759f
biennials, 720
biflagellates, 453
Big Bang, 438
big brown bat (*Eptesicus fuscus*), 989f

bilateral symmetry
advantages, 642
body cavity types with, 645*f*
of early animals, 453, *454*
of flowers, 799
of leaves, 722, 722*f*
Bilateria, 642–43, 642*f*
bilayers, 35, 51, 98. *See also* plasma membranes
bile, 932
bile salts, 932, 935
bimodal breathers, 690
binary fission
bacterial reproduction via, 372, 373*f*
mitochondria and chloroplasts, 88, 88*f*
by most prokaryotic organisms, 542–43
bindin, 499
binocular vision, 707, 892, 892*f*
binomial nomenclature, 9, 516–17
biochemical regulation, 130–31
biochemistry, 138
biodiversity
defined, 1244
importance, 1245–50
levels of, 1245
biodiversity crisis, 1250
biofilms, 540
biogenic amines, 845*t*, 846
biogeochemical cycles
carbon, 1237–38, 1237*f*
defined, 1226
nitrogen, 1240–41, 1240*f*
overview, 1235, 1236*f*
phosphorus, 1235–37, 1237*f*
water, 1241–42, 1241*f*
biogeographic regions, 1144, 1144*f*
biogeography
continental drift and, 1142–44
evolution evidence from, 465*t*, 466–67
bioinformatics, 431–35
biological control, 1118, 1204–5
biological diversity, 532
biological evolution. *See* evolution
biological mechanisms in ecosystems, 1235
biological membranes. *See* plasma membranes
biological nitrogen fixation, 763
biological species concept, 497
biology
defined, 1
major principles, 2–5
as science, 13–18
unity and diversity, 6–13
bioluminescence
bacteria, 547, 548*f*
in ctenophores, 656
protists, 565
biomagnification, 1231–32, 1231*f*
biomass, 1226, 1232–35
biomes
aquatic, 1132, 1139–42
defined, 1117
relation to temperature and precipitation, 1132, 1132*f*
terrestrial, 1132*f*, 1133–38
biophilia, 1247
bioremediation, 410–11, 550, 1232, 1260
biosphere
biogeochemical cycles, 1235
defined, 6, 155
as level of biological organization, 5*f*, 6
percentage of producers, 1232
biosynthetic reactions, 129
biotechnology. *See also* genetic engineering
defined, 400
selected applications, 410–17
use of fungi, 635–36
biotic interactions, 1117
biotin, 923*t*
biparental inheritance, 352
bipedalism, 709–10
biphasic cell cycles, 1079
bipolar cells, 890
birds
alimentary canal, 930, 930*f*
basic features, 699, 700–701
captive breeding, 1260, 1260*f*
circulatory systems, 968
clutch size, 483
competition and coexistence between, 1190–91, 1191*f*
countercurrent exchange mechanism, 958, 959*f*
DDT impact on populations, 1214
as dinosaur descendants, 456
electromagnetic sensing, 884–85
evolution, 699
extinctions, 1196, 1250
foot structure, 4
on geological timescale, 449*f*
meroblastic cleavage, 1079, 1080*f*
monogamy among, 1162
natural selection, 461, 462*t*, 463–64
navigation, 1153–54
orders, 702, 702*t*
relation to crocodiles, 698
in reptile clade, 519
respiratory systems, 996–99
song studies, 502, 503–4*f*, 1149, 1150*f*
species diversity example, 1212, 1213*t*
species richness in North America, 1209, 1209*f*
vestigial structures, 471*t*
wing structure, 916
birds' nest fungi, 623
Birgus latro, 638, 638*f*
birth. *See also* reproduction
in mammals, 1070–72
as positive feedback example, 825, 826*f*
birth defects, 1091–92
birth rates, human, 1182–83
Bishop, Michael, 292
Bison bison, 1258
1,3-bisphosphoglycerate, 141*f*
Biston betularia, 1258
bithorax complex, 388, 388*f*, 389, 390
bivalents, 309, 309*f*, 310*f*
bivalves
feeding behaviors, 924, 925*f*
sea star predation of, 678
shell structure, 660, 661, 662, 662*f*
Bivalvia, 661
blackberries, 806, 807*f*
blackbird (*Turdus merula*), 1203*f*
Blackburn, Elizabeth, 229, 230
black-footed ferrets (*Mustela nigripes*), 1168, 1168*f*, 1178, 1178*f*
black grouse (*Tetrao tetrix*), 1164*f*
black rhinoceros (*Diceros bicornis*), 705*f*, 1257
black widow (*Latrodectus mactans*), 671, 843
blade of leaf, 726
Blaese, R. Michael, 416
blastocoels, 1079
blastocysts, 1066, 1080, 1080*f*
blastoderm, 1079, 1080*f*
blastomeres, 1079, 1081*f*
Blastomyces dermatitidis, 630
blastomycosis, 630
blastopores, 643, 644*f*, 1082, 1082*f*
BLAST program, 434–35
blastula, 642, 1079, 1081*f*
blebs, 188, 188*f*
Blechnum capense, 582*f*
blending inheritance, 321, 324
blind spot, 887, 887*f*
Bliss, Timothy, 867
Blobel, Günter, 89
blood
adaptations for feeding on, 926, 927*f*
characteristic of closed circulatory systems, 966
clotting, 970–71
components, 969–70, 969*f*
formation in bone marrow, 903
oxygen transport mechanisms, 999–1002
phagocytic cells, 1095–96
stem cells, 392, 392*f*
tissue features, 817*f*
viruses infecting, 361*f*
blood clotting disorders, 332–33, 353–54
blood doping, 1053
blood flukes, 658
blood glucose concentration, 947–50
blood pressure
adaptations to changes, 981–83
effects of ADH on, 1037
feedback loops, 825, 825*f*
hypertension, 983
measuring, 974
relation to flow and resistance, 977–81
in veins and venules, 976
blood types, 337–38
blood vessels. *See also* capillaries
flow and resistance, 977–81
main types, 974–77
responses during inflammation, 1096
blooming process, 800–802
blowfly, 893*f*, 894
blowfly larvae, 1200, 1200*f*
blubber, 958
blue-light receptors, 747
blue poison arrow frog (*Dendrobates azureus*), 1195*f*
blue-ringed octopus (*Hapalochlaena lunulata*), 662*f*
blue tits, 1200, 1200*f*
blue whales, 1196, 1197*f*
BMP4 protein, 508, 508*f*
boa constrictor (*Constrictor constrictor*), 471*t*, 989*f*
bobcat (*Lynx rufus*), 703*f*
body fluid compartmentalization, 819–21
body fluid homeostasis, 1008–15
bodyguard wasp (*Cotesia rubecula*), 751
body mass, relation to brain size, 857–58
body mass index, 960
body plans
animals, 642–43, 1077
annelids, 665*f*, 666
arthropods, 668–69
crustaceans, 676*f*
echinoderms, 677–78, 677*f*
embryonic formation in animals, 817
flatworms, 657*f*
Hox genes' role, 509–10, 646–47
lancelets, 680*f*
mollusks, 661, 661*f*
nematodes, 668
rotifers, 659*f*
sharks, 689
sponges, 653*f*
body size, metabolic rate and, 951–52, 952*f*
body surfaces, gas exchange across, 988–89, 989*f*
body temperatures
changes to reduce metabolic demands, 952
of modern birds, 701
regulation, 824, 824*f*, 955–60, 1012
upper limits, 955
body weight
maintaining, 952–55
obesity and public health, 960–61
Boehmeria nivea, 730
bogs, 1142
Bohr, Christian, 1000
Bohr effect, 1000

Boiga irregularis, 1196, 1251
bombardier beetle (*Stenaptinus insignis*), 1194, 1195*f*
bond rotation, 32*f*
bonds. *See* chemical bonds
bone marrow, 1101
bone marrow transplants, 392
bone morphogenetic protein 4 (BMP4), 508
bones
  animal skeleton types, 901–3
  as calcium reservoir, 1045
  components, 903
  disorders, 916–17
  of modern birds, 700*f*, 701
  tissue features, 817*f*
bonobos, 1162
bont tick (*Amblyomma hebraeum*), 671*f*
bony fishes, 690–92, 690*f*. *See also* fishes
book lungs, 669
Boore, Jeffrey, 669–70
Bormann, Herbert, 1240
boron, 757*t*
Borrelia burgdorferi, 421
Bosch, Carl, 763
Bos gaurus, 1261
Bos javanicus, 1261, 1261*f*
Botox, 544, 843
Botrychium lunaria, 582*f*
Bottjer, David, 453
bottle cells, 1082, 1082*f*
bottleneck effect, 490, 490*f*
bottom-up control, 1204
botulism, 544, 843
Boveri, Theodor, 328
bovine spongiform encephalitis, 370
Bowman, William, 1020
Bowman's capsule, 1020, 1020*f*, 1021, 1021*f*
Boyer, Paul, 147
Boyle's law, 992, 993*f*
Boysen-Jensen, Peter, 742
brachiation, 707
Brachiopoda, 648*t*, 660, 681*t*
brachiopods
  appearance in fossil record, 454, 455*f*
  key characteristics, 648*t*, 660, 681*t*
Brachiosaurus, 456
bracts, 729, 730*f*
brain
  of arthropods, 669
  defined, 831
  evolution, 856–58, 907–8
  glucose transport in, 948, 950
  nutrition spurring evolution, 710–11
  of primates, 707
  structure in humans, 862–66, 862*f*
  structure in vertebrates, 856
  viruses infecting, 361*f*
brainstem, 863, 1002–3
Branchiostoma, 679
branch site, 242, 242*f*
Branton, Daniel, 99*t*
Brassicaceae, 1198
Brassica juncea, 1260
Brassica oleracea, 469, 470*f*
brassinolide, 740*t*
brassinosteroids, 740*t*, 746
Brazilian arrowhead viper, 1*f*
BRCA-1 gene, 292*t*
BRCA-2 gene, 292*t*
bread mold fungi, 625–26, 626*f*
bread wheat, 615
breathing. *See* respiratory systems; ventilation
Brenner, Sydney, 668
Briggs, Winslow, 742
Brighamia insignis, 614, 614*f*
Brinegar, Chris, 1182
brittle stars, 678
bromodeoxyuridine, 869
5-bromouracil, 284, 285*t*
bronchi, 993, 994*f*
bronchioles, 993, 994*f*, 1004–5
bronchoconstriction, 1004
bronchodilators, 1005
brooding, 701
Brooker, Robert, 111
Brower, Lincoln, 1194
brown alga (*Laminaria digitata*), 561–62, 1127*f*
brown fat cells, 87, 959
Brownian motion, 33
Browning, Adrian, 589
brown recluse (*Loxosceles reclusa*), 671
brown rot fungi, 633
brown tree snake (*Boiga irregularis*), 1196, 1197*f*, 1251
brown trout (*Salmo trutta*), 989*f*
Bruneau, Benoit, 968
brush border, 931, 1021, 1022*f*
bryophytes
  distinguishing features, 577–78, 580–81
  early ecological impact, 586
  lack of gibberellin growth response, 745, 745*f*
  life cycle, 578, 579–80
Bryozoa, 648*t*, 660, 681*t*
Bt crops, 413, 414*f*, 1198
Bubo scandiacus, 892, 892*f*
buccal pumping, 694, 990–91, 991*f*
Buck, Linda, 895–96
budding
  defined, 1057, 1057*f*
  viral release via, 366
  of yeasts, 622, 622*f*
buds, 718, 725
bud scales, 730*f*
buffalo (*Syncerus caffer*), 1126, 1127*f*
buffers, 39
Buffon, George, 460
Bufo fowleri, 1252
bulbnose unicorn fish (*Naso tonganus*), 538
bulbourethral glands, 1063*f*, 1064
bulimia nervosa, 961
bulk feeders, 639*f*, 924, 925*f*
bulk flow, 778–79
bulldog, 469*f*
bullfrog (*Rana catesbeiana*), 989*f*
burdock (*Arctium minus*), 1203*f*
Burgess Shale, 454, 455*f*
Burkholder, JoAnn, 565
Burkholderia pseudomallei, 548
Bursera simaruba, 1169
Burton, Harold, 869
burying beetles (*Nicrophorus defodiens*), 1162
Bush, Guy, 506
butterflies
  color vision, 889
  feeding stimulants, 1198
  key characteristics, 673*t*
  metamorphosis, 675*f*
  migration, 1153, 1153*f*
  as pollinators, 609*f*
  predation defenses, 1149*f*, 1194, 1195
  taste sense, 896
buttress roots, 732, 732*f*
bypass surgery, 983

C

C_3 plants, 169–70
C_4 photosynthesis, 759
C_4 plants, 170–71, 170*f*, 171*f*, 759
Cacops, 692, 692*f*
Cactoblastis cactorum, 1205, 1205*f*
cactus finches, 461, 462*t*
cactus moth (*Cactoblastis cactorum*), 1205, 1205*f*
cactus spines, 730*f*
Cade, Robert, 1013–15
Cade, Tom, 1260
cadherins, 199, 200*f*, 1087, 1088*f*
caecilians, 695, 695*f*
Caenorhabditis elegans
  alternative splicing, 275*t*
  genome size, 275*t*, 424*t*
  as model organism, 13, 381, 381*f*, 668
  nervous system, 832
  relation to arthropods, 649–50
café marron (*Ramosmania rodriguesii*), 1244
caffeine, 187, 585, 611*f*
Calamospiza melanocorys, 1163
calcitonin, 1031*f*, 1045
calcium
  as essential animal nutrient, 922, 924*t*
  as plant macronutrient, 757*t*
  regulation by bone tissues, 903
calcium binding, 739
calcium carbonate, 451
calcium-dependent protein kinases (CDPKs), 572, 739
calcium ions
  at cell anchoring junctions, 199
  in cortical reaction, 1078*f*, 1079
  homeostatic role, 823*t*
  regulation by bone tissues, 903
  regulation by hormones, 1044–45, 1046*f*
  role in muscle contraction, 908, 909–11, 909*f*
  as second messengers, 186–87, 186*f*, 739
  in smooth ER, 80
calico cats, 343, 353, 354*f*
California condor (*Gymnogyps californicus*), 1181, 1260, 1260*f*
California Current, 1131, 1131*f*
California gray whales, 1196
California sea slug (*Aplysia californica*), 867
Callaway, Ragan, 1118–19
Calliarthron, 559*f*
callose, 788, 802, 805
calluses, 744, 744*f*
calmodulin, 187
Calopogon pulchellus, 1203
calories, 950
Calothorax lucifer, 701*f*
Calvin, Melvin, 165, 167
Calvin cycle
  discovery, 167, 168–69*f*
  location, 156*f*
  overview, 154, 156, 165–67, 166*f*
Cambrian explosion, 454, 455*f*, 640
Cambrian period, 448, 454, 510
camellia, 760*f*
camels, 1144
cAMP, 183–85
Campephilus principalis, 1251, 1251*f*
CAM plants, 171, 171*f*
Canada lynx (*Lynx canadensis*), 1196, 1196*f*
cancer. *See also* mutations
  causes, 288–95
  cervical, 361*f*
  DNA microarray applications, 410*t*
  immune system response, 1108–9, 1109*f*
  inherited susceptibility, 279
  overview, 288
  smoking related, 294, 295, 1005
  use of cannabis with, 612
cancer cells, 139, 230
Candida albicans, 622
Canidae, 516
canine teeth, 924, 925
Canis lupus, 469, 516, 516*f*
cannabidiol, 612
cannabinoids, 612–13
Cannabis genus, 612–13

Cannabis sativa, 730
Cannon, Walter, 823
canopy, 1133
5′ cap, 243, 243*f*, 251
cap-binding proteins, 243, 251
Capecchi, Mario, 693
Cape May warbler, 1190
Cape thick-tailed scorpion (*Parabuthus capensis*), 671*f*
capillaries
  in alveoli, 993, 994*f*
  in fishes' gills, 990
  of nephrons, 1020, 1020*f*, 1021*f*
  in placenta, 1068
  responses during inflammation, 1097*f*
  structure and function, 975*f*, 976, 976*f*
  water movement in, 976, 977*f*
capping, 243, 243*f*
Capra pyrenaica, 1261
Capsaicin, 611
capsids, 360–61, 362*f*, 366
CAP site, 261
capsules of bacterial cells, 69–70, 213, 539–40
captive breeding, 1260
carapaces, 676
carbenicillin, 413
carbohydrates
  as animal nutrients, 921, 921*t*
  basic types, 46–49
  digestion and absorption, 933–35, 934*f*
  in membranes, 98*f*, 103–4
  storage and use, 945, 945*f*
  synthesis in Calvin cycle, 165–67
carbon. *See also* organic molecules
  in animals and plants, 27, 27*t*, 42
  atomic mass, 26
  chemical features, 43–45
  double bonds, 28–29
  in nucleotides, 218
  as plant macronutrient, 757*t*
carbonaceous chondrites, 440
carbon cycle, 545–47, 1237–38
carbon dioxide
  atmospheric circulation, 1237–38
  blood buildup triggering ventilation, 1003
  Calvin cycle, 165–67
  diffusion into blood, 987*f*
  ecological impact of increases, 1238, 1239*f*
  as greenhouse gas, 1125, 1125*t*
  in hemoglobin, 1000
  homeostatic role, 823*t*
  impact on plant growth, 759
  plants as reservoir, 586–87
  from reaction of methane and oxygen, 33
  reduction in photosynthesis, 155
  release in aerobic respiration, 136
  release in citric acid cycle, 141
  release in pyruvate breakdown, 140
  site of uptake in plants, 155
carbon fixation, 166, 166*f*
carbon-hydrogen bonds. *See* hydrocarbons
carbonic acid, 38, 39, 1003
carbonic anhydrase, 1003
Carboniferous period, 455–56, 586–87, 688
carbon monoxide, 845*t*, 846, 1005
carbonyl groups, 44*t*
carbonyl sulfide, 441
carboxyl ends of polypeptides, 245
carboxyl groups
  in amino acids, 53, 55*f*
  basic features, 44*t*
  in fatty acids, 49
  in polypeptides, 245
Carcharhinus albimarginatus, 689*f*
Carcharias taurus, 689*f*
Carcharodon carcharias, 688
carcinogens, 288
carcinomas, 294–95, 294*f*
Carcinus maenas, 1154, 1154*f*
cardiac angiography, 983
cardiac cycle, 972–74
cardiac glycosides, 1194
cardiac muscle
  basic features, 814, 815*f*
  electrical properties, 971
  functions, 904
cardiac output, 980–81, 982
cardiomyopathy, 352*t*
cardiovascular diseases, 964, 983, 984*f*
Cargill, Susan, 1233
caribou (*Rangifer tarandus*), 705*f*
Carnegiea gigantea, 609*f*
Carnivora, 516
carnivores
  defined, 921, 1226
  feeding behaviors, 924–25, 926*f*
  frogs and toads as, 695
  hominins as, 710–11
carnivorous plants, 756, 756*f*, 769–70, 769*f*
Carolina parakeet (*Conuropsis carolinensis*), 1251
β-carotene, 158*f*
carotene, 564
carotenoids, 87, 158, 158*f*, 1164
carpellate flowers, 798, 798*f*
carpels
  in *Arabidopsis*, 397
  evolution, 606, 607*f*
  functions, 605, 606, 607*f*, 795
carrageenan, 559
Carroll, Sean, 645
carrying capacity, 1178–79, 1184–85
cartilage, 208, 817*f*, 993
cartilaginous skeletons, 688–89
Casparian strips, 777, 778*f*
caspase-12 gene, 709
caspases, 190, 191*f*, 292
Castanea dentata, 1200–1201, 1200*f*, 1250
Castanea mollissima, 1201
castings, worm, 666
Castor canadensis, 1173–74
catabolic reactions, 128–29, 130, 130*f*
catabolism, defined, 74
catabolite activator protein (CAP), 257*f*, 266, 267*f*
catabolite repression, 266
catalase, 85, 122
catalepsis, 1195
catalysts
  defined, 33, 122
  early polymers as, 441
  emergence of proteins as, 446
  RNA as, 442–45, 443*f*, 444–45*f*
cataracts, 897, 898*f*
catastrophism, 460
catchments, 1240–41
catch per unit effort, 1170
catecholamines, 845*t*, 846
catfish, 896
Catharanthus roseus, 597*f*, 1245
cation exchange, 762, 762*f*
cations, 31, 762
cats, 400*f*, 934
cattle egrets, 1203
Cattleya orchids, 610*f*
Caudipteryx zoui, 699, 700*f*
Cavalli-Sforza, Luca, 375
CD4 proteins, 1107, 1113
CD8 proteins, 1107
cDNA libraries, 404
Ceanothus, 767
Cech, Thomas, 128
cecum, 933
cell adhesion, 86, 86*f*, 384, 384*f*
cell adhesion molecules, 199, 384, 384*f*
cell biology, 14, 65, 66
cell bodies, 832, 833*f*
cell communication
  cell adhesion molecules' role, 200
  chemical receptor activation, 178–81
  at gap junctions, 176, 201
  hormonal signaling, 187–88
  overview, 174–77
  signal transduction pathways, 85, 177, 177*f*, 181–87
  through plasmodesmata, 203
cell cycle
  checkpoint proteins, 292, 301, 301*f*
  overview in eukaryotes, 299–301, 300*f*
  protein degradation during, 132
cell differentiation
  during animal development, 814, 1077
  animal stem cells, 390–95
  as major feature of development, 381, 1077
  pattern formation and, 382*f*, 383
  role of gene regulation, 258
  skeletal muscle, 393–95
  as tissue-forming process, 205
cell division
  bacterial, 372, 373*f*
  mitotic, 303–8
  overactivity linked to cancer, 288–90
  overview, 297
  pattern formation and, 382, 382*f*, 383
  regulation by tumor-suppressor genes, 292–94
  studies, 302–3
  as tissue-forming process, 205
cell expansion
  ethylene's effects, 746
  in leaf development, 725
  in plant growth, 723–24, 723*f*, 724*f*
cell-free translation systems, 245
cell growth, 205
cell junctions
  direct signaling via, 176, 201
  types in animals and plants, 198–204
cell-mediated immunity, 1102, 1107–9
cell membranes. *See* plasma membranes
cell migration
  pattern formation and, 382, 382*f*, 383
  as tissue-forming process, 205
cell movement, 1081, 1082–83
cell plate, 305, 306*f*
cells
  as basic units of life, 2, 3*f*
  defined, 6
  discoveries, 66, 67
  estimated number in human body, 297
  evolution, 441–43
  formation by proteins, 12*f*
  as level of biological organization, 5*f*, 6
  overview of structures, 69–73
  prokaryotic vs. eukaryotic organisms, 9, 93
  systems biology, 92–95
cell signaling
  apoptosis, 188–90
  at gap junctions, 176, 201
  general principles, 174–77
  hormone responses in multicellular organisms, 187–88
  illustrated, 12*f*, 86*f*
  overview, 85
  in plants, 739, 739*f*
  positional information, 1086
  receptor activation, 178–81
  role of extracellular matrix, 194, 816
  signal transduction pathways, 85, 177, 177*f*, 181–87
  by water-soluble hormones, 1032–33
cell-signaling proteins, 430*t*

cell structure
- cytosol, 73–78
- endomembrane system, 79–86
- nuclei, 78–79
- overview, 69–73
- protein sorting to organelles, 89–92
- semiautonomous organelles, 86–89
- systems biology, 92–95

cell surface receptors, 178–80
cell theory, 2, 66
cell-to-cell communication, 175, 176*f*. *See also* cell communication; cell signaling
cell-to-cell contacts, 1086–87, 1087*f*, 1088*f*
cellular induction, 1086–87
cellular regulation, 130
cellular respiration
- in anaerobic conditions, 150–51
- carbohydrate, protein, and fat metabolism, 149
- citric acid cycle, 141–42
- defined, 136
- glycolysis, 138–39, 138*f*, 140–41*f*
- overview of metabolic pathways, 136–38, 137*f*
- oxidative phosphorylation, 142–47
- during photosynthesis, 155, 155*f*
- pyruvate breakdown, 139–40, 140*f*

cellular responses, 175, 177, 177*f*
cellular water potential, 775, 776*f*
cellulose
- fungal breakdown, 633
- mammalian nutrition from, 704, 931
- microfibril direction in plant cell walls, 724, 724*f*
- molecular structure, 48*f*, 49
- structure and function, 197–98, 198*f*
- in wood, 730

cell walls
- basic functions in prokaryotes, 69–70, 540
- defense mechanisms in plants, 752, 753*f*
- expansion in angiosperms, 723–24, 723*f*, 724*f*, 725, 746
- of fungi, 619, 625
- mycorrhizal interactions, 632, 632*f*
- for prokaryotes in harsh conditions, 543–44
- structure and function in plants, 72*f*, 197–98, 197*f*

Cenozoic era, 457
Centaurea diffusa, 1118–19
centiMorgans, 350
centipedes, 672, 672*f*
central cells, 804
central dogma of gene expression, 238–39
central nervous system
- brain structure, 862–66
- elements of, 832, 855*f*, 856, 859*f*
- embryonic formation, 1083–84
- interactions with PNS, 858
- neural pathways in, 876–77
- stem cells, 868
- structure, 858–59
- viruses infecting, 361*f*

central region, 395, 396*f*
central spindle, 307, 308, 308*f*
central vacuoles, 72*f*, 84, 84*f*
central zone, 396
centrioles, 71*f*, 74, 304
centromeres
- classifying chromosomes by location, 314, 315*f*
- defined, 304
- lacking in bacterial chromosomes, 421

centrosomes
- in animal cells, 71*f*, 74
- functions, 304, 305*f*
- during mitosis, 305, 306–7*f*

Centruroides, 672
cephalization, 642, 657, 856
cephalochordates, 679–80, 680*f*
Cephalopoda, 661, 662
cephalopods, 662–65, 915
cephalothoraces, 670, 676
Ceratium hirundinella, 625*f*
cercariae, 658
cerebellum, 862, 865*t*
cerebral cortex
- evolution, 856–57, 866
- structure and functions, 856, 864–65, 864–65*f*, 865*t*

cerebral ganglia, 657, 856
cerebrospinal fluid, 859, 860*f*, 872
cerebrum, 856–57, 863–66
cervical cancer, 361*f*
cervix, 1065*f*, 1066
Cervus elaphus, 486, 1163*f*
Cervus elaphus nannodes, 1178, 1178*f*
cestodes, 657*t*, 658
cetaceans, 915
CFTR gene, 17, 330, 336–37
Chadradriiformes, 702*t*
chain length, in food webs, 1227
chain termination, 406
chambered heart, evolution, 968–69
chambered nautilus (*Nautilus pompilius*), 663*f*
Chambers, Geoffrey, 524
channels
- functions, 109, 109*f*, 110*f*
- ligand-gated, 180, 180*f*
- in plant cells, 773, 774*f*

chaparral, 1136
chaperone proteins, 92, 92*f*
character displacement, 1193
characters. *See also* inheritance; traits
- in cladistics, 521
- defined, 322
- findings producing Mendel's laws, 323–27
- gene interactions, 344–46
- in Mendel's studies, 323*f*

character states, 521
Charadrius vociferus, 1146*f*
Chara zeylanica, 577*f*
Chargaff, Erwin, 219
charged tRNA, 248
Chase, Martha, 215–17
checkpoint proteins, 292, 301, 301*f*
checkpoints, 301, 301*f*
cheetah (*Acinonyx jubatus*), 1155, 1155*f*, 1229
chelicerae, 671
Chelicerata, 670–72
Chelonia mydas, 697*f*
chemical bonds, 28–31
chemical communication, 1155–56
chemical defenses, 1194, 1197
chemical energy, 119, 119*t*
chemical equilibrium, 33, 121
chemical evolution, 442–46
chemical gradients, 837–38, 837*f*
chemical mutagens, 284–85, 284*t*
chemical reactions
- defined, 33, 118
- enzymes' role, 122–23
- importance of water, 37
- major principles, 118–22
- temperature dependence, 955

chemicals, 1091, 1235–42
chemical selection, 442–46
chemical signals
- plant responses to, 738
- role in homeostasis, 828–29, 947–50

chemical synapses, 842–43, 844*f*
chemical weathering, 761
chemiosmosis
- defined, 129, 138
- experimental verification, 146
- in oxidative phosphorylation, 145

chemoautotrophs, 544
chemoheterotrophs, 544
chemoreceptors
- defined, 877
- in insects, 893–94
- in mammals, 894–97, 1003
- respiratory center control via, 1003

Chen, Jun-Yuan, 453
cherry, 610*f*
chestnut blight (*Cryphonectria parasitica*), 627, 1200–1201, 1200*f*
Chetverikov, Sergei, 481
chewing, 924–25
Chiba, Yasutane, 87
chick embryo, 1085*f*
chickenpox, 367
chickens, 381, 508
chigger mite (*Trombicula alfreddugesi*), 671*f*, 672
Chilopoda, 672
chimeric genes, 290–91, 291*f*
chimpanzee (*Pan troglodytes*)
- chromosomes, 475, 475*f*
- classification, 707–8, 708*f*
- human genome versus, 709
- learning ability, 1149
- mating behavior, 1162
- nucleotide differences from humans, 528
- skull development, 510, 511*f*

Chimpanzee Sequencing and Analysis Consortium, 708, 709
chinch bug, 673*t*
Chinese liver fluke (*Clonorchis sinensis*), 658, 659*f*
Chinese river dolphin (*Lipotes vexillifer*), 1244
CHIP28 protein, 109, 110*f*
chitin
- basic functions, 49
- in cell walls of most fungi, 619, 625
- in exoskeletons, 902
- as extracellular matrix component, 197
- in tracheae, 992

chitons, 661
Chlamydiae, 536*t*, 537
Chlamydomonas, 77*f*, 555*f*, 568*f*
Chlamydomonas reinhardtii, 84, 453, 453*f*
chloramphenicol, 250
Chlorarachniophyta, 562
Chlorella pyrenoidosa, 167, 168–69*f*
chloride ions
- channel opening causing IPSPs, 843
- cystic fibrosis impact on transport, 17, 336–37
- formation, 31
- inhibitory effects, 848–49
- movement in kidney, 1021, 1022, 1023
- resting potential role, 836

chlorine, 757*t*, 924*t*
Chloroflexi, 536*t*
chlorofluorocarbons, 1125*t*
Chlorokybus atmophyticus, 577*f*
chlorophyll
- in aquatic ecosystems, 1233
- genetic influences on synthesis, 352
- green color, 154, 155

chlorophyll *a*, 158, 158*f*
chlorophyll *b*, 158, 158*f*
Chlorophyta, 558
chlorophytes, 576*f*, 577
chloroplast genome
- importance, 350
- location in cells, 351*f*
- maternal inheritance, 351–52
- nuclear genome compared, 87–88, 424
- origins, 452

chloroplasts
- chlorophyll in, 158
- evolution, 88, 452
- formation, 87–88
- homologous genes, 161, 162*f*
- mechanism of action, 159–60, 160*f*

INDEX

chloroplasts, *continued*
in plant cells, 72*f*
as site of photosynthesis, 155
structure and function, 86*f*, 87, 87*f*, 95, 155, 156*f*
chlorosis, 760, 760*f*
choanocytes, 641, 653, 653*f*
choanoflagellates, 563, 563*f*, 640–41, 642*f*
cholecystokinin, 937, 937*f*
cholera (*Vibrio cholerae*), 69, 421, 537, 940
cholesterol
in membranes, 98*f*, 102
molecular structure, 52, 52*f*, 1034*f*
as precursor molecule, 1034*f*
solubility, 34–35
Cholodny, N. O., 742
Chondracanthus canaliculatus, 1218, 1218*f*
chondrichthyans, 688–89
chondroitin sulfate, 197*f*
Chondrus crispus, 559*f*
Chordata
distinguishing features, 516
invertebrate subphyla, 679–80, 681*f*
key characteristics, 648*t*, 678–79, 681*t*
segmentation, 646*f*
chordates
distinguishing features, 516
as early body plan group, 454
key characteristics, 648*t*, 678–79, 681*t*
nervous systems, 855*f*
segmentation, 646*f*
chorion, 696, 696*f*
chorionic gonadotropin, 1067
chorionic villi, 1068, 1069*f*
choroid, 886, 887*f*
chromatin
in animal cells, 71*f*
defined, 78, 230, 271
distribution in nucleus, 79*f*
in nuclear matrix, 231
in plant cells, 72*f*
structural changes, 271–73
chromatin-remodeling complexes, 271, 272–73
chromium, 924*t*
chromophores, 748
chromoplasts, 87
chromosomal translocations, 290–91, 291*f*, 315, 315*f*
chromosomes
bacterial, 370–71, 371*f*, 421
composition, 78
early studies, 213
evolution, 475, 475*f*
examining microscopically, 298–99
hexaploidy, 1181–82
hybrid, 499–500
linked genes, 346–50
location, 12*f*
loss leading to cancer, 294
meiosis overview, 308–14
Mendel's inheritance laws and, 327–30
mitotic cell division, 304–7, 304–7*f*
molecular structure in eukaryotes, 230–31, 232*f*
origins of replication, 224, 229–30
polyploidy, 504–6
sets and homologous pairs, 299
variations in eukaryotes, 314–18
viral integration, 363, 364*f*
chromosome territories, 79, 79*f*
chronic myelogenous leukemia (CML), 290–91, 291*f*
chronic stress, 1052
chronic wasting disease, 370*t*
Chthamalus stellatus, 1191–93, 1192*f*
Church, George, 545
chylomicrons, 935, 936*f*, 945
chyme, 930, 932
chymotrypsin, 935
chytrids, 619, 624*t*, 625, 625*f*
cicada, 673*t*, 1196
cichlids, 478*f*, 487–89
cilia
basic features, 76–78, 77*f*
in ctenophores, 656, 656*f*
in gastrovascular cavities, 965
of inner tracheal wall, 993, 994*f*
of olfactory receptors, 894
of oviducts, 1066
of protists, 555
in protonephridia, 1016, 1017*f*
of trochophore larvae, 648, 648*f*
ciliary muscles, 887, 887*f*
ciliated cells, 295
ciliates, 555, 568, 571*f*
Ciliophora, 560
Cimbex axillaris, 388
Cinchona (*Cinchona officinalis*), 1245, 1246*f*
circadian rhythms, 738, 1032
circulatory systems
adaptations to changes, 981–83
amphibian, 694
in annelids, 666
of arthropods, 669
basic functions, 964
blood clotting, 970–71
blood components, 969–70
blood pressure, flow, and resistance, 977–81
blood vessels, 974–77
of cephalopods, 663
disorders, 983
heart functions, 971–74
major organs in animals, 818*t*
of modern birds, 701
in mollusks, 661
types, 964–69
cirri, 652
Cirripedia, 676
cis-acting element, 264
cis-effect, 264
cis-retinal, 890, 891*f*, 892
cisternae, 79, 80*f*
cis-trans isomers, 45, 45*f*
citrate, 143*f*
citrate synthase, 141, 143*f*
citric acid cycle
in cellular respiration, 137–38, 141–42
overview, 142*f*, 143*f*
citronella, 611
citrulline, 237–38
clades, 517, 647
cladistics, 521–24
cladogenesis, 500
cladograms, 521
Cladophora, 555*f*
clamp connections, 629, 629*f*
clamp proteins, 227
clams, 648*t*, 681*t*
claspers, 689
classes, in taxonomy, 516
classical conditioning, 663, 1148
classification of organisms, 9–11
claudin, 201
Claviceps purpurea, 624, 624*f*
clay, 441, 442, 761–62, 761*f*
cleavage furrow, 305, 306*f*, 307
cleavage of zygotes, 642
cleavage phase
in animal development, 643, 1077*f*, 1079–81, 1079*f*, 1080*f*
autonomous specification during, 1086
glycolysis, 138, 138*f*
cleft palate, 1091–92
Clements, Frederic, 1208, 1216
climate change, 1117, 1252, 1258. *See also* global warming
climates, 7, 1129–31
climax communities, 1216–18
clitoris, 1065, 1065*f*
cloaca, 1062
clonal deletion, 1110
clonal inactivation, 1110
clonal selection, 1106, 1106*f*
clones, lymphocyte, 1102
cloning
animals, 400*f*, 414–15
endangered species, 1261
genes, 400–406, 413
of plants by somatic embryogenesis, 809
Clonorchis sinensis, 658, 659*f*
closed circulatory systems
adaptive functions, 981–83
basic features, 965*f*, 966–68
of cephalopods, 663
closed conformations, 271
Clostridium acetylbutylicum, 472–73, 473*f*
Clostridium botulinum, 544, 843
Clostridium genus, 544
Clostridium perfringens, 544
Clostridium tetani, 544
clotting disorders, 332–33
clotting process, 970–71, 971*f*
clownfish
characteristics, 1*f*, 12–13
harems, 13*f*
taxonomy, 9, 11*f*
club mosses, 581*f*
clumped dispersion, 1170, 1171*f*
clutch size, 483
Cnidaria
facilitation of succession by, 1217
gastrovascular cavity, 965
key characteristics, 648*t*, 654–56, 681*t*
nervous systems, 855–56
cnidocils, 655, 655*f*
cnidocytes, 655, 655*f*
CO1 gene, 675
coacervates, 441–42, 442*f*
coactivators, 270
coal, 455, 586
coastal climates, 1131
coast redwood (*Sequoia sempervirens*), 1181–82, 1182*f*
coat proteins, 91, 91*f*
Cobbania corrugata, 588*f*
cocaine, 851
cocci, 539, 539*f*
Coccidioides immitis, 630
coccidiomycosis, 630
coccolithophorids, 560, 560*f*
coccoliths, 560, 560*f*
cochlea
damage to, 898
mammals, 879, 879*f*, 880, 880*f*
cochlear duct, 880, 880*f*
Cockayne syndrome, 288
cockroaches, 1151
coconut, 610*f*, 611, 797, 797*f*
coconut cadang-cadang viroid, 368, 369*f*
coding sequences
defined, 244, 244*f*
discovery of introns within, 241
mutations, 280–81
coding strands (DNA), 240, 240*f*
codominance, 335*t*, 338
codons
defined, 243
Nirenberg and Leder's identification, 246–47

INDEX

positions in genetic code, 244*t*, 432–33
silent mutations, 280
coefficient of relatedness, 1159
coelacanth (*Latimeria chalumnae*), 691, 691*f*
coelomates, 644, 645*f*
coeloms
in annelids, 665–66
defined, 643
echinoderms, 678
in mollusks, 661
types, 644, 645*f*
Coen, Enrico, 397
coenzyme A, 103, 104*f*, 140
coenzyme Q, 142
coenzymes, defined, 125
coevolution, 613–15, 633
cofactors, 125
Coffea arabica, 611*f*
cognitive learning, 1149. *See also* learning
cogongrass (*Imperata cylindrica*), 1259
cohesin, 304
cohesion, 38, 783
cohesion-tension theory, 783–84
cohorts, 1172
coiled coil domains, 61*f*
cold conditions
animal adaptations, 952, 956, 958
behavioral changes for, 959
biomes featuring, 1135, 1137, 1138
effects on distribution of organisms, 1123, 1123*f*
fishes' resistance to, 467–68
genetically engineered tolerance, 1126
seasonal acclimatization, 960
cold desert biomes, 1137
cold sores, 361*f*
Coleman, Douglas, 953–55
Coleochaete pulvinata, 577*f*
Coleoptera, 673*t*
coleoptiles, 742, 806, 806*f*, 808
coleorhiza, 806, 806*f*
colicins, 372
colinearity rule, 389
collagen
in bone, 817*f*, 903
discovery, 194
formation, 195*f*, 196
micrographs, 194*f*
in nematode cuticle, 668
platelet binding to, 971, 971*f*
secretion in response to inflammation, 1097
strength, 55, 195
as structural protein, 195, 195*t*, 816
types, 195–96
collared flycatchers (*Ficedula albicollis*), 483, 484*f*
collecting ducts, 1020*f*, 1024, 1025*f*
collenchyma, 206
collenchyma cells, 721
collenchyma tissue, 721, 722*f*
colligative properties, 36
Collins, Francis, 17
Collip, James Bertram, 1042, 1043
colloblasts, 655
colloid, 1038
colon, 548, 933
colonial bentgrass (*Agrostis capillaris*), 483, 485*f*
colonization, 490–91
coloration
to attract pollinators, 613–14
cephalopods' ability to change, 663
cyanobacteria, 536
flower variations, 799
predation defenses, 1194–95
sexual selection and, 486, 487–89, 487*f*
color blindness, 889, 890*f*
color vision, 889, 892–93
col-plasmids, 372
Columba palumbus, 1157
Columbiformes, 702*t*
columella cells, 733
columnar epithelial cells, 295
combinatorial control, 267–68
comb jellies, 656, 656*f*
comma-shaped prokaryotes, 539
commensalism, 1187
common bile duct, 932*f*, 933
common cold, 361*f*
common hemp nettle (*Galeopsis tetrahit*), 504–5
communication, animal, 1155–57. *See also* cell communication
communities
defined, 6, 1207
diversity-stability hypothesis, 1214–15
ecological models, 1207–9
as level of biological organization, 5*f*, 6
species diversity, 1212–14
species richness, 1209–11
succession in, 1215–23
community ecology, 1118*f*, 1120, 1207
companion cells, 730, 787, 788–89
compartmentalization
animal body fluids, 819–21
cellular regulation by, 130
in eukaryotic cells, 70, 74*f*
competence factors, 377
competent bacteria, 377, 403
competition
defined, 1187
between ecological communities, 1208
from introduced species, 1118–19, 1251
species richness and, 1211
in succession, 1216–18
types, 1188–94
competitive exclusion principle, 1189
competitive inhibitors, 124–25, 125*f*
complement, in innate immunity, 1098
complementary base sequences, 221
complementary DNA (cDNA), 404–5, 409
complete flowers, 605, 798
complete gut, 656
complete metamorphosis, 674, 675*f*
compost, 760, 761*f*
compound eyes, 886, 886*f*
compound leaves, 726, 726*f*
compound microscopes, 66–67
compounds, defined, 28
computational molecular biology, 432–35
computerized tomography, 869
concentration, defined, 35
concentration gradient, 106, 107*f*
condensation reactions, 37, 45
conditional specification, 1086–87, 1087*f*
conditioning, 826–28, 1148
condoms, 1073, 1073*f*
condors, 1181, 1260, 1260*f*
conduction, 957
conduction disorders, 851
Condylura cristata, 684, 684*f*
cone pigments, 888
cones, 887–88, 888*f*, 889, 890, 891*f*
cone shells, 661
confocal fluorescence microscopy, 68*f*
conformational changes
ATP synthase, 146–47, 147*f*
caused by prions, 369–70
ligand • receptor complexes, 178, 178*f*
regulation, 260, 261*f*
conformers, 823
Confuciusornis sanctus, 699, 700*f*
congenital analgesia, 340
congenital hypothyroidism, 851
congestive heart failure, 983
conidia, 622, 622*f*
conifers
diversity, 600–603
first appearance, 456
hexaploid, 1181–82
as major biome feature, 1134, 1135
as plant phylum, 577
survival to modern times, 599
conjugation
cellular mechanisms, 375–76, 376*f*
ciliate reproduction via, 568, 571*f*
defined, 373, 373*t*
discovery, 374–75
of toxins, 1198
connective tissue
blood, 969–71
general features in animals, 205, 205*f*, 816, 817*f*
ground tissue compared, 208
Connell, Joseph, 1188, 1189, 1218
connexin, 201
connexons, 201, 201*f*
conservation biology, defined, 1244
conservation strategies
protected areas, 1254–57
restoration ecology, 1259–60
single-species approach, 1257–59
conservative mechanism, 221, 222*f*, 223
conserved regions, 896
constant regions, immunoglobulin, 1103
constant segments, immunoglobulin, 1105
constitutive exons, 274*f*, 275
constitutive genes, 257
Constrictor constrictor, 989*f*
contact-dependent signaling, 176, 176*f*
contigs, 406, 407*f*
continental drift, 450, 450*f*, 467, 1142–44
continuous iteroparity, 1172
continuous variation, 343*t*, 345
contour feathers, 700–701
contraception, 1073–74
contractile fibers, in cnidarians, 655
contractile ring, 307, 308, 308*f*
contractile vacuoles, 84, 84*f*, 107, 108*f*
contraction, skeletal muscle, 906–11
contrast, in microscopes, 67
control groups, 15–16
Conuropsis carolinensis, 1251
convection, 957
Convention on International Trade in Endangered Species of Wild Fauna and Flora (CITES), 599
convergent evolution, 465*t*, 467–69, 520, 1144
convergent extension, 1083, 1083*f*
Cooper, Alan, 524
Cope's giant salamander (*Dicamptodon copei*), 511, 511*f*, 696
copepods, 676
copper
as essential animal nutrient, 922, 924*t*
as plant micronutrient, 757*t*
in respiratory pigments, 999
as trace element, 27
Coprinus disseminatus, 629*f*
copulation, 1061–62, 1064
Coraciiformes, 702*t*
coral bleaching, 1123, 1124*f*, 1140, 1258
coral crab, 676*f*
coralloid roots, 599, 600*f*
coral reefs
as biome, 1140
formation, 654–55
species richness, 1211
temperature dependence, 1122, 1123*f*
corals, 654, 1258

coral snake (*Micrurus fulvius*), 1195, 1195*f*
Corchorus capsularis, 730
corepressors, 267
core promoters, 269, 269*f*
cork cambium, 731*f*, 732
cork cells, 730–31, 732
cork oak tree (*Quercus suber*), 732
corn (*Zea mays*)
  discovery of wild relative, 1245
  domestication, 615
  flower development, 798, 798*f*
  germination, 808, 808*f*
  prop roots, 732
  selective breeding, 469, 471*f*
  transpiration in, 784
cornea, 886, 887, 887*f*
corn smut (*Ustilago maydis*), 629*f*
corona, of rotifers, 658, 659*f*
coronary artery bypass, 983
coronary artery disease
  causes and symptoms, 983
  trans fatty acids and, 51
  treatments, 983, 984*f*
corpus allata, 1049
corpus callosum, 863, 864*f*
corpus luteum, 1066–67
correlations, 1121
Correns, Carl, 337, 351
corridors, between reserves, 1256–57
cortex
  adrenal, 189–90, 1031*f*, 1050–51
  cerebral, 856–57, 864–66
  of flowering plant root, 734, 734*f*
  of flowering plant stem, 721
  renal, 1018, 1020*f*
  rhinal, 1148
Corti, Alfonso, 880
cortical granules, 1078*f*, 1079
cortical reaction, 1078*f*, 1079
cortisol, 1034*f*, 1040, 1112
Corwin, Jeff, 1244
Costanza, Robert, 1246*t*
Cotesia rubecula, 751
cotranslational sorting, 89–92, 90*f*, 91*f*
cotyledons
  in flowering plant seeds, 717*f*, 719, 805, 805*f*
  in monocots versus eudicots, 607, 805
  in plant development, 395, 396*f*
countercurrent exchange
  body heat regulation, 958, 959*f*
  in fishes' gills, 990, 990*f*
  in loop of Henle, 1022
countercurrent multiplication systems, 1022
courtship displays, 1164
covalent bonds
  in amino acid formation, 53
  basic features, 28–30, 28*f*, 29*f*
  energy storage, 128
  in fatty acids, 50
  molecular shapes and, 32–33
  in organic molecules, 43
CpG islands, 273, 294
crabs
  basic features, 638, 676*f*, 677
  optimal foraging behavior, 1154, 1154*f*
cracking beaks, 701*f*
cranial nerves, 858
cranium, 684–85
crassulacean acid metabolism, 171
Crenarchaeota, 535
crenation, 821
creosote bushes, 808
crested porcupine (*Hystrix africaeaustralis*), 703*f*
Cretaceous period, 456–57
Crick, Francis, 220, 238
crickets, 498, 499*f*, 673*t*
Crinoidea, 678, 678*t*
cristae, 86, 87*f*, 138
critical innovations of algae, 576, 576*f*
critical period, 1149–50
crocodiles, 698, 698*f*, 968
Crocodylus acutus, 698
crop, 930
cross-bridge cycling, 908–9, 908*f*
cross-bridges, 905*f*, 906–7, 907*f*
cross-fertilization, 322–23, 593
crossing over
  linkage and, 348–50
  misalignments in, 427, 428*f*
  overview, 309, 309*f*
cross-pollination, 795
crown gall tumors, 412, 413, 421
crustaceans
  exoskeletons, 668
  key characteristics, 648*t*, 676–77, 681*t*
  stretch receptors, 878
crustose lichens, 634, 635*f*
Cryphonectria parasitica, 627, 1200–1201
cryptic coloration, 1194–95
cryptic female choice, 486
cryptochromes, 747
Cryptococcus neoformans, 630
cryptomonads, 560
cryptomycota, 619, 624*t*, 625
Cryptosporidium parvum, 553, 567
crystals, urea, 43*f*
Ctenophora, 648*t*, 656, 681*t*
C-terminus, 53, 55*f*, 245
CTR1 protein kinase, 745
CT scans, 869
Cubozoa, 656, 656*f*
cuckoos, 471*t*
cud, 931
cultural eutrophication, 1236
Culver, Kenneth, 416
cumulus mass, 1060, 1061
cup fungi, 627
cupula, 882, 883*f*, 884
Currie, Alastair, 189
Currie, David, 1210
Curvularia protuberata, 633–34
Curvularia thermal tolerance virus, 633–34
cuticles
  of arthropods, 668–69
  ecdysozoans, 667
  as evaporation barrier in plants, 784
  of flowering plant stems, 721
  of leaves, 728
  nematodes, 668
cuttings, 808
cuttlefish, 662
Cuvier, Georges, 460
Cyanea arctica, 655
cyanide, 145
cyanobacteria
  as autotrophs, 544
  basic features, 536, 536*t*
  blooms, 533, 533*f*
  in cycad roots, 599, 600*f*
  diversity, 536
  early emergence, 451
  fossils, 438, 438*f*, 449*f*, 451*f*
  in lichens, 634, 635
  nitrogen fixing, 763, 763*f*, 767, 1216, 1240
  site of Calvin cycle, 165
cyanogen bromide, 410
Cyanopsitta spixii, 1244
cycads, 577, 599
cyclic adenosine monophosphate (cAMP)
  CAP control by, 266
  phosphodiesterase's effects, 187
  as second messenger, 183–85, 739
  serotonin's effects, 185, 186*f*
  synthesis and breakdown, 183*f*
cyclic electron flow, 160, 161*f*
cyclic guanosine monophosphate (cGMP), 889, 890
cyclic photophosphorylation, 160, 161*f*
cyclin-dependent kinases, 293, 301, 301*f*
cyclins, 301, 301*f*
CYCLOIDEA gene, 799, 800*f*
cyclooxygenase, 99, 118
cyclosporine, 635
cyclostomes, 686–87
Cynomys spp., 1160–61, 1160*f*
cysteine, 54*f*, 58, 468
cystic fibrosis
  gene replacement research, 411–12
  as recessive disease, 330, 336–37
  scientific discoveries, 16–17, 434
cysts, of protists, 567
cytochrome b_6-*f*, 161, 162*f*
cytochrome b-c_1, 143, 144*f*, 145, 161, 162*f*
cytochrome *c*, 143, 144*f*
cytochrome oxidase, 143, 144*f*, 145
cytogenetics, 298
cytokines, 1096, 1110
cytokinesis
  in eukaryotic cell cycle, 300
  mitosis, 305, 306*f*, 307*f*
  proteins' role in animal cells, 307–8
cytokinins, 740*t*, 744
cytoplasm, 69, 93
cytoplasmic factors, 1086, 1087*f*
cytosine
  base pairing in DNA, 61–62, 62*f*, 220, 220*f*, 221
  bonding in DNA, 220, 220*f*, 221
  molecular structure, 61, 217, 218*f*
cytoskeletons
  in animal cells, 71*f*
  basic functions, 74–76
  formation by proteins, 12*f*
  in plant cells, 72*f*
  role in movement, 76–78, 93
cytosol
  in animal cells, 71*f*
  optimal pH environment for enzymes, 126
  in plant cells, 72*f*
  plant defense mechanisms in, 752
  protein synthesis, 89
  structures and functions, 73–78, 93, 94*f*
cytosolic leaflets, 98*f*, 103
cytotoxic T cells
  actions of, 1101–2, 1102*f*, 1108–9, 1109*f*
  activation, 1107, 1108
  HIV effects on, 1113

D

D1 protein, 758*f*
dachshund, 469*f*
Dactylorhiza sambucina, 479, 479*f*
Daeschler, Ted, 465
dairy products, 933–35
Dall mountain sheep (*Ovis dalli*), 1174, 1175*f*
daltons, 26
D-amino acids, 53
dams, 1242
damselflies, 673*t*
Danaus plexippus, 1149*f*, 1153, 1194
dandelions, 793, 797, 809, 1180, 1181*f*
Danio rerio, 381, 381*f*, 1076*f*
Dantas, Gautam, 545

INDEX

Daphne Major, 463–64, 463*f*
Daphnia pulex, 1058
dark adaptation, 892
Dart, Raymond, 710
Darwin, Charles
  carnivorous plant studies, 769
  on complex organ development, 511
  on defining species, 497
  earthworm studies, 666
  evolutionary theory overview, 460–63
  finch observations, 461, 466
  as founder of evolutionary theory, 459, 460, 463, 495
  phototropism studies, 175, 737, 742
Darwin, Erasmus, 459, 460
Darwin, Francis, 175, 737, 742
data analysis, 16
databases, 433, 433*t*, 434–35
data mining, 433
Datana ministra caterpillars, 1160, 1160*f*
daughter cells
  in bacterial cell division, 372, 373*f*
  in meiosis, 311*f*, 312
  in mitosis, 297, 300, 300*f*, 303, 305, 307
  of stem cells, 391, 391*f*, 392
daughter strands (DNA), 221
Davis, Allen, 693
Davis, Bernard, 375
Davis, Margaret, 1125
Davis, Robert, 393
dawn redwood (*Metasequoia glyptostroboides*), 601*f*, 603
daylilies, 318*f*
day-neutral plants, 748
DDBJ database, 433*t*
DDT, 1214, 1231–32
deacetylated histones, 273
dead zones, 763, 1233
deafness, 898
deamination, 284, 285*f*
death-inducing signaling complex, 190, 191*f*
death rates, human, 1182–83
death receptors, 190, 191*f*
decapods, 676*f*, 677
decibels, 898
deciduous plants, 785
decomposers
  bacteria as, 545, 1234
  in food chains, 1226–27, 1227*f*, 1234
  fungi as, 630
  protists as, 563
de Duve, Christian, 84
deep-sea vent hypothesis, 440, 441*f*
deep-sea vent organisms, 535
deer, 704, 704*f*, 705*f*
Deevey, Edward, 1174
defecation, 933
defense and protective proteins, 430*t*
defenses
  cactus spines, 730*f*
  of cnidarians, 655, 655*f*
  common types among animals, 1194–96
  as function of secondary metabolites, 611
  plant responses to herbivores and pathogens, 751–54
  of protists, 565–66
  quills in mammals, 703, 703*f*
  of sea cucumbers, 678
  of sponges, 653–54
defensive mutualism, 1202, 1203*f*
definitive hosts, 658
deforestation, 1242, 1242*f*, 1251
degeneracy of genetic code, 243–44
degradative plasmids, 372
degradosomes, 132
dehydration
  as air breather risk, 988, 1010–11
  with bulimia, 961
  defined, 1009
  effect on blood pressure, 983
  physical symptoms, 1013
  risks, 1009
dehydration reactions
  in amino acid formation, 53, 55*f*
  defined, 37, 45
  illustrated, 46*f*
  in triglyceride formation, 49, 49*f*
7-dehydrocholesterol, 1045
Deinococcus, 536*t*
delayed implantation, 1072
delayed ovulation, 1072
deletions, 315, 315*f*
DELLA proteins, 744–45, 745*f*
delocalized electrons, 158
Delsuc, Frédéric, 680
dementia, 872–73
Demodex brevis, 672
Demodex canis, 672
demographic transition, 1182–83, 1183*f*
demography, 1168, 1172–76
denaturation
  by high temperatures, 955
  in PCR technique, 405*f*, 406
  of ribonuclease, 58–59
dendrites, 832, 833*f*
dendritic cells, 1096, 1096*f*
Dendrobates azureus, 1195*f*
Dendrobates tinctorius, 496, 496*f*
denervation atrophy, 913
denitrification, 544, 1240
dense connective tissue, 817*f*
density-dependent factors in population growth, 1179–80
density-independent factors in population growth, 1180
dental plaque, 540, 540*f*
deoxyadenosine, 416
deoxynucleoside monophosphates, 225*f*
deoxynucleoside triphosphates (dNTPs), 225*f*, 226, 406
deoxyribonucleic acid. *See* DNA (deoxyribonucleic acid)
deoxyribonucleotides, 60, 61*f*
deoxyribose, 61, 217, 218*f*
depolarization, 838, 840*f*
depression, 850
depth perception, 892
de Queiroz, Kevin, 497
derived characters, 521–22
dermal tissue, 205–6, 206–7
dermatophytes, 630
descending limb, loop of Henle, 1020, 1020*f*
desert biomes, 1137
desertification, 1136
desert pupfishes, 1245–46
desert shrubs, 1137
Desmidium, 555*f*
Desmodus rotundus, 926
desmosomes, 199, 199*f*
dessication resistance in barnacles, 1191
detergents, 35
determinate cleavage, 643, 644*f*
determinate growth, 718
determined cells, 381
detritivores, 1226–27, 1227*f*
detritus, 1226
deuterostomes
  chordate overview, 678–80
  echinoderm overview, 677–78
  embryonic development, 643, 644*f*
development
  as characteristic of life, 3*f*
  defined, 4, 380
  embryonic stage, 1076–86
  evolutionary changes and, 507–12
  of flowering plants, 718–19, 798–802
  functions of gene regulation, 258–59
  growth versus, 1077
  major stages, 380–81
  model organisms, 381–82
  pattern formation, 382–84, 385*f*, 1077
developmental disorders, 1072
developmental genetics, 380
developmental homologies, 470
Devonian period, 455, 510, 692
D-glucose, 47, 47*f*
diabetes mellitus. *See* type 1 diabetes; type 2 diabetes
diacylglycerol, 186, 186*f*
dialysis, 1027
Diamond, Jared, 1261
Diamond v. Chakrabarty, 411
diaphragm, 995
diarrhea, 940
diastole, 972–73
diatoms
  consumption by protists, 554, 554*f*
  reproduction, 568, 570*f*
  structure, 561
Dicamptodon copei, 511, 511*f*, 696
dicer, 275, 276*f*
Diceros bicornis, 705*f*
Dichanthelium lanuginosum, 633
dichlorodiphenyltrichloroethane (DDT), 1214, 1231–32
dichromatic animals, 889
Dictyostelium discoideum, 562
dideoxy chain-termination method (dideoxy sequencing), 406–9
dideoxyguanosine triphosphate, 407*f*
dideoxynucleoside triphosphates (ddNTPs), 406, 407
dideoxynucleotides, 407*f*
diencephalon, 856, 863
Diener, Theodor, 368
diethylurea, 106, 106*f*
dieting, 960
differential gene regulation, 188, 196, 384
differential interference contrast microscopy, 68*f*
differentiation. *See* cell differentiation; specialization
diffraction patterns, 218–19*f*
diffuse knapweed (*Centaurea diffusa*), 1118–19, 1188
diffusion
  between body fluid compartments, 820
  defined, 105
  illustrated, 105*f*
  membrane barriers, 106
  nutrient absorption via, 928
  osmosis, 107
  in plant cells, 773
digestion
  cystic fibrosis impact, 16
  defined, 920, 921*f*
  impact on metabolic rate, 951
digestive cecae, 669
digestive systems
  animal mechanisms, 933–36
  of arthropods, 669
  disorders, 940, 941*f*
  major structures in animals, 818*t*, 928–33
  parasites, 668
  Pavlov's conditioning studies, 826–28
  principles, 927–28
  regulation in animals, 936–39
  viruses infecting, 361*f*
digger wasps (*Philanthus triangulum*), 1151–52
digits, of primates, 707
dihybrid crosses, 326–27, 327*f*, 330
dihydroxyacetone phosphate, 140*f*
1,25-dihydroxyvitamin D, 1045
diisopropyl phosphorofluoridate, 125
dikaryotic mycelia, 627, 628*f*

dimers, 163
DiMichele, William, 586
dimorphic fungi, 630
Dinobryon, 564, 564*f*
dinoflagellates, 560–61, 565, 567
dinosaurs
  basic features, 699
  bird ancestors, 699, 700*f*
  extinction, 588
  food sources, 599
  on geological timescale, 449*f*, 456
Dinozoa, 560
Diodon hystrix, 1195*f*
dioecious plants, 798
Dionaea muscipula, 206
diploblastic organisms, 643
diploid cells, 308, 312, 313–14
diploid-dominant species, 313*f*, 314
diploid organisms, 299, 316
Diplopoda, 672
Dipnoi, 691
Dipodomys panamintensis, 1012
Diptera, 673*t*
direct calorimetry, 950, 951*f*
direct intercellular signaling, 176, 176*f*
directionality of DNA strands, 218
directional selection, 483, 484*f*
direct repair, 287, 287*t*
disaccharides, 47, 47*f*, 196, 197*f*
discovery-based science, 15
discrete traits, 345
diseases. *See also* bacteria; genetic disorders; viruses
  autoimmune, 851, 917–18, 1110–11
  cardiovascular, 964, 983, 984*f*
  caused by mutations, 196, 336–37, 352–53, 352*t*
  caused by viroids and prions, 368–70
  cyst-borne, 567
  extinctions from, 1251
  fungal, 625, 630–31
  genetically engineered treatments, 411–12, 416–17, 1197–98, 1201
  human deaths from, 359, 367
  mineral deficiency, 924*t*
  mitochondrial, 352*t*
  muscular-skeletal, 916–18
  nervous system, 850–51, 866, 871–73
  obesity as risk factor, 960
  parasitic, 556, 557, 563, 568–72, 625, 658, 659*f*, 672
  proteome changes with, 73
  resistance, 478
  respiratory, 1004–5
  selected viruses, 361*f*, 367
  stem cell potential applications, 392–93
  vitamin deficiency, 923*t*
disjunct distributions, 1143–44
dispersion patterns, 1170–71
dispersive mechanism, 221, 222*f*, 223
dispersive mutualism, 1202–3, 1203*f*
dissociation constant, 178
distal convoluted tubules, 1020*f*, 1021, 1021*f*, 1023, 1023*f*
distance from source pool, 1220, 1221*f*
disturbances, 1211, 1211*f*, 1214–15
disulfide bridges, 57*f*, 58, 60
diuresis, 1037
diurnal animals, 888
diversifying selection, 483, 485*f*
diversity of life, 6–13
diversity-stability hypothesis, 1214–15, 1248, 1249*f*
diving animals, 988–89, 1004
DNA (deoxyribonucleic acid). *See also* genomes
  amount in genomes of different species, 424–25
  in bacteria, 370–78
  basic functions, 60
  as characteristic of life, 4
  in chloroplasts, 87–88
  compaction around histones, 230
  discovery as genetic material, 212–17
  evolution, 446
  hydrogen bonds, 30, 30*f*
  molecular structure, 60, 61–62, 62*f*, 217–21
  mutations, 280–86
  recombinant technology, 400
  relation to mRNA and polypeptide of gene, 244–45, 245*f*
  repair systems, 286–88
  replication overview, 221–24, 304
  sequences on homologous chromosomes, 299
  as successor to RNA world, 446
  template and coding strands, 223, 223*f*, 240–41, 240*f*, 241*f*
  theory of genetic material, 15
  transcription overview, 239–41, 240*f*, 241*f*
  viral integration, 363, 364*f*
DNA-binding domains, 60, 61*f*
DNA fingerprinting, 415–16, 415*f*
DNA helicase, 225, 225*f*
DNA libraries, 404–5, 404*f*, 422, 1214
DNA ligase, 228, 228*t*, 402, 402*f*
DNA methylation
  genomic imprinting via, 356, 805
  links to cancer, 294
  regulation via, 271, 273–74
DNA microarrays, 409, 409*f*, 410*t*
DNA polymerase
  chromosome shortening and, 229–30
  role in replication, 225–26, 225*f*, 226*f*, 227, 228*t*
  specialized functions, 228–29
  spontaneous mutations involving, 284
DNA primase, 226, 227, 228*t*
DNA replication
  basic features, 223–24
  molecular mechanism, 224–30
  proposed mechanisms, 221–23, 222*f*
DNase, 190, 215
DNA sequences. *See also* gene sequences
  computer analysis, 432–33
  as evidence of evolutionary relationships, 520
  molecular clocks based on, 528, 529*f*
  rules for change over time, 523–24
  shared derived characters, 522, 523*f*
DNA sequencing
  dideoxy and pyrosequencing methods, 406–9
  from extinct organisms, 524–25
  shotgun method, 421–23
DNA supercoiling, 371, 371*f*
DNA testing, 212
DNA topoisomerase, 225, 225*f*
DNA transposons, 426, 426*f*
Dobzhansky, Theodosius, 459, 470, 496
dodder (*Cuscuta pentagona*), 770, 770*f*
dodo (*Raphus cucullatus*), 1251
dogfish shark (*Scyliorhinus canicula*), 689*f*
dogs
  Pavlov's conditioning studies, 826–28, 1148
  selective breeding for, 469
dolichol, 104
Dolly the sheep, 414–15, 1261
dolphins, 6*f*, 704*f*, 915
domains
  protein structure, 60, 61*f*
  of repressors and activators, 260
  taxonomic, 9, 10*f*, 11*f*, 515
domestication of flowering plants, 615
dominant species, 1259
dominant traits, 324, 331, 335–36
donkey (*Equus asinus*), 499, 500*f*
donor-controlled systems, 1196
dopamine, 1032, 1033*f*
dormancy of seeds, 797, 805–6, 807
dorsal grooves in trilobites, 670
dorsal hollow nerve cord, 679
dorsal lip of blastopore, 1082, 1082*f*
dorsal side, 642
dorsal structures, morphogens for, 1089
dorsoventral axis, 382, 382*f*, 1077
dosage compensation, 354
Doty, Sharon, 1232
double bonds, 28–29, 43, 50
double circulation, 966, 967*f*, 968
double crossovers, 474*f*
double fertilization, 592, 605, 796, 804–5
double helix (DNA), 217, 219–21
Down, John Langdon, 317
Down syndrome, 317, 318*t*, 1091, 1092
downy feathers, 700
dragonflies, 455–56, 647*f*, 673*t*
Drake, Bert, 1238
Dreissena polymorpha, 660
drones, 1057
Drosera intermedia, 756*f*
Drosera rotundifolia, 756*f*
Drosophila melanogaster
  alternative splicing, 275*t*
  brain structure, 856
  compound eyes, 886*f*
  developmental stages, 385–90, 385*f*
  eye mutations, 511, 512*f*
  genetic map, 350*f*
  genome size, 275*t*, 424*t*
  highly repetitive DNA sequences, 425
  linked genes, 348, 349*f*
  membrane protein genes, 100*t*
  as model organism, 13, 381, 381*f*
  normal chromosome numbers, 316
  relation to nematodes, 649–50
  sex determination, 331–32
  sex-linked traits, 333–34
  species in Hawaiian Islands, 501
  Toll proteins, 1098–1100
drought, plant responses to, 751
drugs
  anti-inflammatory, 884
  antiviral, 367–68
  biodiversity's importance for, 1245–46
  biotechnology applications, 410, 635
  chemoreceptor sensitivity, 1003
  discoveries, 1*f*
  effects on embryos, 1076, 1091
  effects on spider webs, 671*f*
  for HIV infections, 1113
  membrane protein binding, 99
  nervous system disorders and, 850–51
  plant sources, 597, 1245
  resistant bacteria, 7–8, 8*f*, 377*f*, 378, 403
  use of bacteria to produce, 550
Dryas drummondii, 1216, 1217*f*
dry seasons, 1134
Dubois, Eugene, 711
Duchenne muscular dystrophy, 918
duck-billed platypus (*Ornithorhynchus anatinus*), 704, 706*f*, 884
duckweed, 718, 719, 719*f*
duodenum, 932, 932*f*, 940, 941*f*
duplications, 315, 315*f*
dura mater, 859, 860*f*
Dutch elm disease, 627, 1201
dwarf plants, 744, 746
dyeing poison frog (*Dendrobates tinctorius*), 496, 496*f*
dynamic instability, 74
dynein, 76, 76*f*, 78
dysplasia, 294*f*, 295
dystrophin, 918

E

E1 and E2 conformations, 112
E2F protein, 293–94, 293*f*
eagles, 701*f*
eardrum, 879, 879*f*, 880
early-diverging ferns, 582*f*
ears, 470, 879–81
Earth
  environmental changes, 448–50
  formation, 438
  major periods for life-forms, 450–57
  origins of life, 439–46
earthworms
  benefits of segmentation, 665–66
  body plan, 665*f*, 666
  closed circulatory systems, 965*f*
  as coelomates, 645*f*
  ecological impact, 666, 1225
  hydrostatic skeleton, 644, 902, 902*f*
  reproduction, 1062
eastern hemlock (*Tsuga canadensis*), 1127
eastern meadowlark (*Sturnella magna*), 498, 499*f*
eastern phoebe (*Sayornis phoebe*), 1123
Eastwood, Daniel, 633
Ebola virus, 361*f*
E-cadherin, 199
ecdysis, 647–48, 647*f*, 667, 902
Ecdysozoa, 647–48, 649–50
echidna, 467, 468*f*, 704
Echinodermata, 648*t*, 677–78, 681*t*
echinoderms
  as early body plan group, 454
  endoskeletons, 677–79, 903
  key characteristics, 648*t*, 677–78, 681*t*
  nervous systems, 855*f*, 856
Echinoidea, 678, 678*t*
Echinops bannaticus, 424–25, 424*f*
Echinops nanus, 424–25, 424*f*
echolocation, 882
Eciton burchelli, 675*f*
ecological factors, distinguishing species by, 497
ecological footprint, 1184–85, 1185*f*
ecological pyramids, 1230–31
ecological species concept, 497
ecology. *See also* behavioral ecology; ecosystems; population ecology
  biogeography and, 1142–44
  climates and, 1129–31
  defined, 14, 1116, 1117
  influences on organism distribution, 1122–29
  levels of study, 1118–20
  major biomes, 1132–42
  study methods, 1120–22
economic benefits of biodiversity, 1245–46
economic considerations in conservation, 1257
*Eco*RI enzyme, 402, 402*f*, 402*t*
ecosystem diversity, 1245
ecosystem ecology, 1225
ecosystems
  biodiversity requirements, 1248–50
  biogeochemical cycles, 1235–42
  biomass production, 1232–35
  defined, 6, 1225
  food webs, 1226–32
  as level of biological organization, 5*f*, 6
  natural benefits, 1246–47, 1246*t*
ecosystems ecology, 1118*f*, 1120
Ecotron experiments, 1247, 1247–48*f*
ecotypes, 496
ectoderm, 642–43, 1081, 1081*f*
ectomycorrhizae, 632, 632*f*
ectoparasites, 1200
Ectopistes migratorius, 1251, 1251*f*
ectothermic species
  classification, 956
  countercurrent exchange mechanisms, 958
  dinosaurs as, 699
  resistance to freezing, 956
  standard metabolic rate, 950
  thermoreceptor importance, 884
edge effects, 1257
Edidin, Michael, 102
Edward syndrome, 318*t*
effective population size, 1253–54
effector caspases, 190, 191*f*
effector cells, 1102
effectors, 739, 824
efferent arterioles, 1020, 1020*f*, 1021, 1021*f*
efficiency of energy transformation, 1228–30
egestion, 920, 921*f*
egg cells
  fertilization, 1061, 1066, 1078–79, 1078*f*
  of flowering plants, 322, 323*f*, 795
  formation, 1059–61
  maternal inheritance in, 352
  mutations, 283
  as stem cells, 391–92
egg-rolling response, 1147, 1147*f*
eggs
  amniotic, 640, 696–97, 696*f*
  DDT effects, 1231
  impact of inbreeding on hatching success, 1252, 1253*f*
  mammalian, 1068
  variations, 1062
Egler, Frank, 1218
egrets, 1203
Ehlers-Danlos syndrome, 196
Ehrlich, Paul, 1245
Eimer's organs, 684
Einstein, Albert, 157
Eisner, Tom, 1194
ejaculation, 1063
ejaculatory duct, 1063, 1063*f*
elastin
  of arteries, 975, 975*f*
  as structural protein, 195, 195*f*, 816
  structure and function, 195, 196*f*
elder-flowered orchid, 479
Eldredge, Niles, 506
electrical gradients, 837–38, 837*f*
electrical stimulation in muscle, 910–11
electrical synapses, 842
electric charge
  ions, 31
  neuron functions and, 834–38
  of subatomic particles, 22
  water molecules, 34
electrocardiograms, 974, 974*f*
electrochemical gradient
  defined, 106
  of electrogenic pump, 112–13
  as energy form, 119*t*
  illustrated, 107*f*
  in oxidative phosphorylation, 144*f*, 145, 146
  resting potential role, 837–38, 837*f*
electrogenic pump, 112–13, 113*f*
electrolytes, 1009
electromagnetic receptors, 877, 884–85
electromagnetic spectrum, 157, 157*f*
electronegativity, 29–30, 43
electron microscopes, 67–69, 99*t*, 100–101
electrons
  basic features, 22
  in chemical bonds, 28–31
  light absorption, 157–58, 158*f*, 162
  orbitals, 24–25
  role in oxidative phosphorylation, 142–47
electron shells
  basic principles, 24–25
  carbon atom, 43, 43*f*
  covalent bond formation, 28–29, 28*f*, 29*f*
  ionic bond formation, 31
electron transport chains
  in anaerobic conditions, 150–51, 150*f*
  in oxidative phosphorylation, 142–45, 144*f*
  in photosynthesis, 159, 161
electrophoresis, 405
electrostatic attraction, 762, 762*f*
elements, 21, 25–26, 27
elephantiasis, 668, 668*f*
elephants
  basal metabolic rate, 951
  classification, 514
  disjunct distribution, 1144
  oxygen-hemoglobin dissociation curve, 1000, 1001*f*
  poaching, 1257
  tusks, 514, 704*f*
elephant seals, 1156, 1165, 1165*f*
Elephas maximus, 514
elevation, climate and, 1130–31
elicitors, 752, 753*f*
elk (*Cervus elaphus*), 486, 1163*f*, 1178
Ellis-van Creveld syndrome, 491
El Niño, 1123
elongation factors, 248*t*, 252
elongation stage, 240, 240*f*, 252–54, 252*f*, 253*f*
elongation zone, 733
Elton, Charles, 1196, 1214, 1230, 1248
Embden, Gustav, 138
EMBL database, 433*t*
embolisms, in plant vessels, 780–81
embryogenesis, 805, 805*f*
embryonic development
  basic principles, 1076–77
  cell differentiation, 1086–91
  defects in, 1091–92
  sequence of events, 1077–86
embryonic germ cells, 392
embryonic stage
  amniote, 696–97
  in animal development, 643, 1076–86
  defined, 380
  developmental homologies, 470
  duration in humans, 258–59
  pattern formation, 383, 383*f*, 384, 395–96, 396*f*
  reproductive cloning, 414–15
  segmentation during, 384, 385*f*, 386–87
embryonic stem cells, 392
embryophytes, 579
embryos (animal)
  cleavage, 643, 644*f*, 1077*f*, 1079–81, 1079*f*, 1080*f*
  defined, 1057, 1076
  development concepts, 1076–77
  development events, 1077–86
  human, 1069*f*
embryos (plant)
  defined, 1057
  development in flowering plants, 717–18, 796–97, 805
  origins in plants, 588–91
  production from somatic cells, 808–9
emergent properties, 4
emerging viruses, 367
Emerson, Robert, 161, 163–64
emotional stress, impact on digestion, 937
emotions, brain centers for, 866
emphysema, 1005, 1005*f*
empirical thought, 460
emulsification, 935
enantiomers, 45, 45*f*, 47*f*
Encephalartos laurentianus, 599
endangered species, 1261

Endangered Species Act, 1245
endemic species, 467, 1254, 1255*f*
endergonic reactions, 120, 121–22, 122*f*
endocannabinoids, 612
endocrine disruptors, 1053
endocrine glands, 1030, 1031*f*
endocrine signaling, 176–77, 176*f*
endocrine systems. *See also* hormones
  control of growth and development, 1047–49
  control of metabolism and energy balance, 1037–44
  defined, 1030
  functions in animals, 818*t*
  hormone types, 1031–35
  major glands, 1031*f*
  mineral balance, 1044–47
  nervous system links, 1035–37
  public health impact, 1052–53
  role in homeostasis, 947–50
  role in reproduction, 1049–50
endocytosis, 113–14, 115*f*
endoderm, 642, 1081, 1081*f*
endodermis
  plant root, 733, 734, 734*f*, 777
  statocytes in, 750
  transport across, 777–78, 778*f*
endomembrane system
  defined, 78
  major structures and functions, 79–86, 93–94, 94*f*
endometrium, 1066, 1067*f*, 1068
endomycorrhizae, 632, 632*f*
endoparasites, 1200
endophytes, 633–34
endoplasmic reticulum
  calcium ion release in fertilized egg cell, 1078*f*, 1079
  phospholipid synthesis, 103–5
  structure and functions, 79–80, 80*f*
  synthesis of proteins for, 89–92
endorphin, 845*t*
endoskeletons
  chordate versus echinoderm, 678–79
  echinoderms, 677–78, 902*f*, 903
  variations, 903
  of vertebrates, 685
endosomes, 1107–8
endosperm
  as defining feature of angiosperms, 604
  development, 605
  double fertilization role, 592, 796, 804
  nutrients in, 804
endospores, 543*f*, 544
endosporic gametophytes, 593
endosymbiosis
  horizontal gene transfer via, 537–38
  primary, secondary, and tertiary, 559–60, 559*f*
  in protist evolution, 88, 452, 556
endosymbiosis theory, 88, 452
endothelium of arteries, 975, 975*f*, 978–79
endothermic species
  acclimatization, 959–60
  birds as, 701
  classification, 956
  dinosaurs as, 699
  disadvantages, 956–57
  with high metabolic rates, 952
  mechanisms of heat gain and loss, 957–59
energy. *See also* metabolism
  activation, 122–23
  for chemical reactions, 33
  defined, 24, 119
  efficiency of transformation, 1228–30
  electron shells, 24–25
  from fats, 51
  forms, 119
  from glucose, 47
  impact on emergence of species, 451
  from light, 157
  major principles, 118–22
  metabolism overview, 128–31
  from polysaccharides, 48–49
  of state changes in water, 36
  use as characteristic of life, 2–3, 3*f*
energy balance
  hormonal control, 952–55, 1038–44
  measuring, 950–51
  metabolic rate variables, 951–52
energy budgets, 1227
energy flow, 1225–32
energy intermediates, 128–29
energy investment phase (glycolysis), 138, 138*f*
energy liberation phase (glycolysis), 138, 138*f*
English ivy, 467, 468*f*
enhancement effect, 164–65, 164*f*
enhancers, 269, 389
enolase, 139
Ensatina eschscholtzii, 989*f*
Entamoeba histolytica, 567
Enterobius vermicularis, 668
Enterococcus faecalis, 378
enteropeptidase, 935
enthalpy, 120
entomology, 672
entropy, 119, 120*f*
entry stage, in viral reproductive cycle, 363, 364*f*
environmental conditions. *See also* natural selection
  animal homeostasis and, 823
  body temperatures and, 467–68, 952, 956, 959–60
  cell communication and, 174, 175, 299, 301, 430*t*
  cell structures responding to, 88, 93
  changes in Earth's history, 448–50
  distinguishing species by, 497
  effects on distribution of organisms, 1122–29
  effects on photosynthesis, 169, 170
  effects on seed germination, 806–8
  evolutionary role, 467, 507
  gene regulation for, 266, 268*f*, 410*t*
  hormones affecting plant responses, 746
  impact on speciation, 506
  influence on phenotypes, 338–39, 345
  mutations due to, 284
  phenotypic variation based on, 338–39, 345
  polyploid plant advantages, 318
  reproductive cycles and, 1071–72
  sex determination effects, 332
  viral reproduction and, 366
environmental interactions
  as characteristic of life, 3–4, 3*f*
  influence on phenotypes, 338–39, 345
  sex determination by, 332
environmental science, ecology versus, 1117
environmental stimuli
  for flower production, 797–98
  for plants, 738–39, 738*f*, 747–54
enzyme adaptation, 261
enzyme-linked receptors, 179, 179*f*, 181–83, 182*f*
enzymes
  actions on cellulose, 49
  as catalysts, 33, 122
  early association with genes, 236, 237–38
  enantiomer recognition, 45
  hydrogen bonds, 30
  in metabolic pathways, 74, 128–31
  overview of major functions, 122–28
  in peroxisomes, 85
  proteins as, 12*f*
  of proteolysis, 81
  in small intestine, 931
enzyme-substrate complexes, 123–25
eons, 448
eosinophils, 1096, 1096*f*
Ephedra, 603, 604*f*
ephedrine, 603
epicotyls, 806, 806*f*, 807
epidermal growth factor, 181–83, 182*f*
epidermis
  defined, 205–6
  of leaves, 728–29
  plant root, 733–34
  structure in plants, 207–8, 207*f*, 721
Epidermophyton floccosum, 622
epididymis, 1063, 1063*f*, 1064*f*
epigenetic inheritance, 274, 353–56
epiglottis, 993
epilepsy, 863
epinephrine
  cellular regulation by, 130
  effect on glucose metabolism, 1040
  exercise effects, 982
  molecular structure, 1033*f*
  signal transduction pathway, 183, 184*f*
  site of production, 1031*f*, 1032
  varied cellular responses to, 187
epiphytes, 1133, 1134, 1203
episomes, 367, 372
epistasis, 343*t*, 344–45, 344*f*
epithalamus, 863, 865*t*
epithelial cells, as sensory receptors, 876*f*
epithelial tissue
  alimentary canals, 928–29
  dermal tissue compared, 206–7
  general features in animals, 205, 205*f*, 815, 816*f*
  in vertebrate stomach, 818*f*
epitopes, 1108
epsilon-globin genes, 429, 1070
Epstein-Barr virus, 361*f*, 361*t*
Eptesicus fuscus, 989*f*
Epulopiscium, 538
equilibrium, 33, 481, 506–7
equilibrium constants, 121
equilibrium model of island biogeography, 1219–23
equilibrium populations, 1176
equilibrium potential, 838
equilibrium sense, 882–83, 883*f*
Equus, 7*f*, 466*f*
Equus asinus, 499, 500*f*
Equus grevyi, 495*f*
Equus quagga boehmi, 495*f*
erbB gene, 290*t*
erectile dysfunction, 1064
erection, 846, 1064
ergot, 624, 624*f*
Erie, Lake, 1237
Erigeron annuus, 1217
Eriksson, Peter, 868
Erk, 182
ER lumen, 79
Errantia, 666
errors, replication, 228
ER signal sequence, 89–91, 90*f*, 91*f*
erythrocytes
  intracellular fluid volumes, 820–21, 820*f*
  as major component of blood, 969*f*, 970
  malaria's effects, 570
  surface antigens, 338, 338*t*
erythromycin, 250
erythropoietin, 1004, 1031*f*, 1053
Escherichia coli
  alternative splicing, 275*t*
  anaerobic respiration, 150, 150*f*
  bacteriophages, 215–17, 216*f*, 361*t*, 366
  cell size, 371
  chromosomes, 370
  DNA polymerase in, 228

DNA replication studies, 222
evolution, 549
gene regulation of lactose utilization, 258, 258f, 260–67
gene transfer studies, 374–76
genome size, 275t, 421t
homologous genes, 472–73, 473f
membrane protein genes, 100t
as model organism, 13
mutation studies, 282–83
nucleotide excision repair, 287–88, 287f
as proteobacterium, 537
reproduction rate, 372
ribosomal subunits, 249
rRNA gene sequences, 251f
as source of insulin, 400, 410, 411f
esophagus, 929–30, 940
Essay on the Principle of Population (Malthus), 460
essential elements, 756–57, 757t
essential fatty acids, 50
essential nutrients for animals, 922
ester bonds, 49
estradiol
as major estrogen, 1050, 1066
molecular structure, 1034f
role in gametogenesis, 1066, 1067
role in parturition, 1070, 1071f
estrogens
functions, 1034f
impact on bone density, 917
molecular structure, 52, 52f
receptors, 180–81, 181f
role in parturition, 1070
site of production, 1031f, 1050
estrous cycles, 1068
estuaries, 1142
ethanol
effects on embryos, 1091
formation in anaerobic conditions, 151, 151f
response to GABA, 849
ethical issues
biodiversity, 1247
cloning, 415
raised by Human Genome Project, 429
stem cells, 392–93
ethology, 1146
ethylene
chemical structure, 740t
effect on aerenchyma formation, 751
fruit ripening effects, 806
major functions in plants, 740t, 745–46
as plant hormone, 177
role in leaf abscission, 786
ethyl methanesulfonate, 284, 285t
Euarchontoglires, 706
eucalyptus trees, 1142
euchromatin, 231
eudicots
defined, 607
distinguishing features, 719t
examples, 719
flower structure, 607, 608f
flower variations, 799
leaf attachment points, 726
seed structure, 805, 806
taproot system, 723
Euglena, 558, 558f
Euglena gracilis, 1151
euglenoids, 557–58
Eukarya
basic features, 9, 10f
membrane protein genes, 100t
subdivisions, 515
eukaryotic cells
basic features, 70, 71f, 72f, 94f
cell cycle overview, 297–301
chromosome variations, 314–18
cytosol structures and functions, 73–78
on geological timescale, 449f
meiosis overview, 308–14
mitotic cell division, 303–7
nucleus and endomembrane system structures, 78–86, 93–94
oocyte maturation, 302–3
origins, 451–52
protein sorting to organelles, 89–92
ribosomes, 249, 249t, 250
semiautonomous organelles, 86–89, 94–95
systems biology, 92–95
transcription and translation, 238–39, 238f
typical size, 360
eukaryotic chromosomes. *See also* chromosomes
linked genes, 346–50
locations, 12f, 423–24
meiosis overview, 308–14
mitotic cell division, 304–7, 304–7f
molecular structure, 230–31, 232f
telomerase actions, 229–30, 229f
variations in structure and number, 314–18
eukaryotic organisms
defined, 9
gene regulation, 258–60, 267–76
genome features, 423–29
on geological timescale, 449f
prokaryotic versus, 515t
RNA processing, 241–43
transcription and translation, 238–39, 238f, 241, 249, 250, 251–52, 254
eukaryotic parasites, 1095
Eumetazoa, 642, 642f
Euonymus fortunei, 467
Euphausiacea, 677
euphylls, 591, 592f
Euplectes progne, 1164
euploid organisms, 316
Eurasian rosefinch, 501f, 502
European eel (*Anguilla anguilla*), 989f
European honeybee (*Apis mellifera*), 1156–57, 1157f
European medicinal leech (*Hirudo medicinalis*), 926
European starlings, 1153
European tree frog (*Hyla arborea*), 901f
Euryarchaeota, 535
eusociality, 1161
Eustachian tube, 879f, 880, 881
eustele, 598
eutherians, 706
eutrophication, 1141, 1236–37
eutrophic lakes, 1141, 1237
Evans, Ron, 913
evaporation
body heat losses via, 957, 958–59
defined, 38
functions in plants, 772
water transport role, 783–84
wind's effects, 1126
evapotranspiration rate, 1210, 1211f, 1232–33, 1232f
evergreen conifers, 603. *See also* conifers
evisceration, 678
evolution. *See also* plant evolution
amphibians, 465f, 692, 694–95
birds, 699
as characteristic of life, 3f
cladistics overview, 521–24
coevolution in angiosperms, 613–15
convergent, 465t, 467–69, 520, 1144
defined, 459
early concepts, 460
hemoglobin, 1001–2
human, 707–12
jaws, 687–88, 688f
limbs, 692, 693–94
mammals, 702–3
mechanisms of change, 6–8, 474–75
molecular processes, 472–75
nervous systems, 854–58
origins of life on Earth, 439–46
overview of Darwin's theory, 460–63
photoreception, 891–93
phylogeny overview, 517–20
rate of change, 506–7, 507f
taxonomy overview, 9–11, 514–17
timescales, 527–28
types of evidence, 465–72
of viruses, 368
evolutionarily conserved base sequences, 251
evolutionary developmental biology (evo-devo), 507–12
evolutionary histories, 4
evolutionary lineage concept, 497
evolutionary relationships, 250–51, 251f, 497
Excavata, 556–58
excisionase, 363
excitable cells, 838
excitation-contraction coupling, 910–11
excitatory neurotransmitters, 846
excitatory postsynaptic potentials, 843, 844, 845f
excited state, 158f
excretion, defined, 1016
excretory systems
in annelids, 665, 665f, 1016–17, 1017f
of arthropods, 669, 1017
in crustaceans, 676
disorders, 1026–27
in flatworms, 657, 1016, 1017f
kidney structures and functions, 1018–26
major organs in animals, 818t
in mollusks, 661
in rotifers, 659
types compared, 1015–18
executioner caspases, 190, 191f
exercise
circulatory adaptations, 981–82
glycolysis in anaerobic conditions, 151
impact on immune system, 1112
impact on partial pressure of oxygen, 999–1000
lactic acid release, 1003
maintaining water and salt balance, 1013–15
muscle adaptation, 912–13
nutrient sources, 949–50
exergonic reactions, 120, 121
exhaling, 995
exit (E) site, 250, 250f
exocrine glands, 1038
exocytosis, 113, 114f, 843
exons
defined, 241
as percentage of human genome, 425
splicing, 242–43, 274–75, 274f
exon shuffling
mechanisms of action, 473, 474f
role in microevolution, 481, 482t
exonucleases, 132, 132f
exoskeletons
of arthropods, 668–69
as defense against pathogens, 1095
major features, 902–3, 902f
molting, 647–48, 647f, 667
exosomes, 132, 132f
expansins, 723, 724f, 725
experimental groups, 15–16
experiments, 4–5, 1121
exploitation competition, 1188
exponential population growth, 1177–78, 1179f, 1182–83
expression cloning, 1089, 1090f

extant species, 514
extension sequences, 195, 195*f*
extensors, 915
external fertilization, 1061, 1061*f*
external gills, 989, 989*f*
external intercostal muscles, 995
extinctions
  causes, 1250–52
  defined, 448
  global warming effects, 1117
  island biogeography model, 1219–20, 1221–22
  K-selected species risk, 1181
  mass scale, 450, 456, 588
extinct species, 514
extracellular digestion, 654, 927
extracellular fluid, 819, 996
extracellular leaflets, 98*f*
extracellular matrix
  characterizing animal cells, 639
  in connective tissue, 208, 208*f*, 816
  epithelial tissue, 815
  structure and function, 194–97, 194*f*
extracellular proteins, 12*f*
extranuclear inheritance, 343*t*, 350–53
extraterrestrial hypothesis, 440
extremophiles, 535
extrinsic proteins, 99
extrusomes, 565
eye color, 299, 333–34
eyecups, 885–86, 885*f*
eyeless gene, 511, 512*f*
eyes
  of arthropods, 669
  evolution, 511–12, 512*f*, 891–93
  genetic disorders, 380
  of invertebrates, 885–86
  processing signals from, 890–91
  receptors, 887–90
  single-lens structures, 886–87, 887*f*
eyespots, 558, 885

F

F_1 generation, 324
F_2 generation, 324
facilitated diffusion
  between body fluid compartments, 820
  defined, 105, 105*f*
  nutrient absorption via, 928, 933, 934*f*
  in plant cells, 773, 774*f*
facilitation of succession, 1216–17
factor VIII, 353, 354, 410
facultative anaerobes, 544
facultative mutualism, 1202
$FADH_2$
  in electron transport chain, 143–45
  formation in cellular respiration, 137–38, 141, 144*f*
failure rates for contraception, 1073*f*
fairy rings, 629
Falco peregrinus, 1260
fall field cricket (*Gryllus pennsylvanicus*), 498, 499*f*
falsifiability, 15
FAMA gene, 729
families, in taxonomy, 516
Farquhar, Marilyn, 199
fascicles, 905, 905*f*
fasciculata, 1051
FASS gene, 724
fast block to polyspermy, 1078
fast fibers, 912
fast-glycolytic fibers, 912
fast-oxidative fibers, 912
fate (developmental), 387–88
fate mapping, 1081, 1081*f*
fats
  absorption in digestive tract, 931, 935, 936*f*
  basic features, 49–51
  brown cells, 87
  homeostatic role, 823*t*
  metabolism, 149
  relation to blood leptin levels, 953*f*, 1044
  storage and use, 945, 945*f*
fat-soluble vitamins, 922, 923*t*, 935
fatty acids
  as essential animal nutrients, 922
  increased blood concentration, 949, 950
  in triglycerides, 49–51, 49*f*
fatty acyl tails
  carbon acetyl unit removal, 149
  in electron micrographs, 100*f*
  location in membranes, 98, 99
  membrane fluidity and, 102
fd phage, 361*t*
feathers, 699, 700–701, 700*f*
feather stars, 652, 652*f*, 678
Feder, Jeffrey, 506
feedback, role in homeostasis, 825
feedback inhibition, 130–31, 131*f*
feedforward regulation, 825–28
feeding behaviors. *See also* nutrition (animal)
  animals, 922–26, 927*f*
  foraging, 1154–55
  metabolic influences, 952–55
feeding characters, 1193
feeding groove, 556
feet
  of cephalopods, 663
  echinoderms, 677*f*, 678
  in mollusks, 661
felines, 515, 922, 934
Felis libyca, 1261
female-enforced monogamy hypothesis, 1162
female gametophytes, 795, 804
fenestrated capillaries, 975*f*, 976, 1020
fermentation, 150–51
ferns
  diversity, 582*f*
  gametophytes and sporophytes, 313*f*, 314
  leaf evolution, 591, 592*f*
  life cycle, 584*f*
ferredoxin, 159, 160
ferrets, 1168
ferritin, 276
fertility of soils, 760, 762–63
fertility plasmids, 372
fertility rates, 1176, 1183–84, 1184*f*
fertilization
  defined, 1057
  double, 592, 605
  as embryonic development stage, 1077*f*, 1078–79, 1078*f*
  of flowering plants, 795–97
  in humans, 1064, 1066, 1067
  internal, 697, 1061–62
  in seed plants, 592, 593
  variations in animals, 1061–62
fertilizers
  impact on primary production, 1233
  runoff, 763, 1233, 1236–37, 1240
  soil, 762–63, 764
fetal alcohol syndrome, 1091
fetal stage, 259, 996
fetus, human images, 1069*f*
fever, 1097
F factors, 264, 372, 376, 376*f*
fiber, 49
fibers
  collagen, 196
  phloem, 730
  sclerenchyma tissue, 721
fibrin, 971, 971*f*
fibroblasts, 393, 1097
fibronectin, 194–95, 195*t*, 199, 200*f*, 816
fibrous root system, 723
Ficedula albicollis, 483, 484*f*
Fick diffusion equation, 820
fiddler crab (*Uca paradussumieri*), 486, 487*f*
fight-or-flight response
  hormonal control, 183, 187, 1050, 1051*t*
  sympathetic division control, 862
  vasoconstriction in, 978
figs, 1258
filaments
  of cyanobacteria, 536, 537*f*
  in flowering plants, 605, 795
filose pseudopodia, 562, 562*f*
filter feeders
  crustaceans, 676
  lancelets, 679
  mollusks, 661, 662*t*
  tunicates, 680
filtrate, 1016, 1018–24
filtration
  as basic excretory system function, 1016
  in invertebrate excretory systems, 1016–17
  in mammalian kidney, 1018–24
fimbriae, 1066
finches
  beak types and diets compared, 462*t*
  character displacement, 1193
  Darwin's observations, 461
  evolutionary adaptations, 462*f*, 463–64
  reproductive isolation, 502, 503–4*f*
fingerprinting, DNA, 415–16, 415*f*
finite rate of increase, 1176
Fiorito, Graziano, 663
fir (*Abies spp.*), 1142
Fire, Andrew, 275
fire, 1123–24, 1136
fire ants (*genus Solenopsis*), 1155
fireflies, 1156, 1156*f*
Firmicutes, 536*t*, 544
first law of thermodynamics, 119
Fischer, Emil, 123
Fischer, Hans, 158
fishapods, 465–66, 692
Fisher, Ronald, 478
fishes
  appearance during Devonian period, 455
  circulatory systems, 966
  feeding behaviors, 925
  locomotion, 915
  meroblastic cleavage, 1079, 1080*f*
  sexual selection, 487–89
  taxonomy, 9, 11*f*
  vision, 892–93
  water and salt balance, 1011, 1011*f*, 1127
fitness, 482–83
fixed action patterns, 1147, 1148*f*
fixed nitrogen sources, 763–64
flaccid cells, 775, 775*f*
flagella
  of algae, 453
  basic features, 76–78, 77*f*
  as common feature of bacteria, 93
  of cryptomycota, 619, 625, 625*f*
  defined, 70
  prokaryote diversity, 541–42, 542*f*
  of protists, 555, 561–62, 563, 563*f*
  of sperm, 1059, 1060*f*
flagellates, 555
flagship species, 1258

flame cells
  in flatworms, 657, 1016, 1017*f*
  in rotifers, 659
flame spectrophotometer, 1013, 1013*f*
flatus, 933
flatworms
  as acoelomates, 645*f*, 657
  excretory system, 657, 1016, 1017*f*
  Hox genes, 509*f*
  key characteristics, 648*t*, 657, 657*f*, 681*t*
  main classes, 657–58, 657*t*
  photoreception in, 885–86, 885*f*
  reproduction, 1057
flavins, 747
flavivirus, 361*f*
flavonoids, 613–14, 767–68
flax (*Linum usitatissimum*), 730
FLC, 798
fleas, 673*t*
Flemming, Walther, 305
fleshy fruits, 610*f*
flexors, 915
flies, 673*t*
flight, 699, 916, 992
flightless birds
  body plans, 702
  evolutionary study, 524–25, 525–26*f*, 527*f*
flippases, 103, 104*f*
floods, 450, 750–51
floral tubes, 609
Florida panther (*Puma concolor*), 493, 1181, 1258, 1258*f*
flounder (*Platichthys flesus*), 1048, 1048*f*
flowering plant reproduction
  asexual, 808–9
  double fertilization, 592, 605, 796, 804–5
  embryo development, 717–18, 796–97, 805
  flower production, 718–19, 797–802
  fruit formation, 797, 806
  gametophytes, 794, 795, 802–4
  overview, 605, 606*f*, 717–19, 793–97
  seed formation, 797, 805–6
  seed germination, 797, 806–8
flowering plants. *See also* plants
  coevolution, 613–15
  diversity, 609–13
  epistasis, 344–45
  evolution, 604–9
  flower color, 335–36, 336*f*
  generalized structure, 605*f*
  on geological timescale, 449*f*
  growth and development, 395–97, 720–24
  hormones, 740–46
  human uses, 605, 615, 715
  incomplete dominance, 337, 337*f*
  introns in genes, 242
  life cycle, 716–19, 717*f*
  nutrient sources, 766–70
  nutritional needs, 756–60
  organ systems, 718, 719
  parasitic, 1199, 1199*f*
  as plant phylum, 577
  polymorphism, 479, 479*f*
  polyploid, 318, 318*f*
  potential behaviors, 737–39
  responses to environmental stimuli, 747–54
  root systems, 732–34
  sex determination, 332
  shoot systems, 724–32
  soil's importance, 760–66
  transport across cell membranes, 773–76
  transport across tissues, 776–78
  transport over long distances, 778–90
flowers
  defined, 794
  development, 718–19, 797–802
  evolution, 605–6
  reproductive role, 605, 606*f*
  structure, 794–95, 795*f*
flow-through systems, 991, 996–97
flu, 361*f*
fluid feeders, 639*f*, 925–26, 927*f*
fluidity of plasma membranes, 101–2, 103*f*
fluid-mosaic model, 98, 98*f*
flukes, 658, 659*f*, 1198–99
fluorescence, 158
fluorescence detectors, 407, 408*f*
fluorescence microscopy, 68*f*, 542
fluorine isotopes, 26
fluorodeoxyglucose, 26
[18F]-fluorodeoxyglucose, 139
flying, 699, 916, 992
focal adhesions, 199, 199*f*
folic acid, 923*t*, 1091
foliose lichens, 634, 635*f*
follicles
  of ovaries, 1060, 1060*f*, 1066–67, 1066*f*
  of thyroid gland, 1038, 1039*f*
follicle-stimulating hormone
  from anterior pituitary, 1037*t*, 1050
  effects on human female, 1067
  effects on human male, 1065
  synthetic androgens' effect, 1053
follicular cells, 1038, 1039*f*
follicular phase, ovarian cycle, 1067
food chains, 1120, 1226–28, 1226*f*
food gathering behavior, 1154–55
food-induced thermogenesis, 951
food industry, 549, 635
food intake. *See* digestive systems; feeding behaviors; nutrition
foot-and-mouth disease, 360
foraging behavior, 1154–55
Foraminifera, 562, 562*f*
forebrain, 856, 857*f*, 863–66, 865*t*
forensics, 416
forest biomes, 1132*f*, 1133–35, 1250. *See also* deforestation; tropical rain forests
formaldehyde, 439, 1005
form and function in animal organs, 821–22
forward primers, 405*f*, 406
fos gene, 290*t*
fossil fuels
  acid rain and, 1128–29
  carbon dioxide release, 1237, 1238, 1238*f*
fossil record, 448
fossils
  analysis, 447–48
  arthropods, 670*f*
  bird ancestors, 699, 700*f*
  continental drift evidence from, 1142–43
  cyanobacteria, 438, 438*f*
  evolution evidence from, 465–66, 465*t*
  formation, 446–47
  hominoid, 457, 710–12
  plant evolution clues from, 586
  support for punctuated equilibrium, 507
founder effect, 490–91
four-chambered heart, evolution, 968–69
four-o'clock plant (*Mirabilis jalapa*), 337, 337*f*, 351–52
fovea, 887, 893
Fowler's toad (*Bufo fowleri*), 1252
foxes, 467, 467*f*
FOXP2 gene, 709
fragrances of flowers, 799
frameshift mutations, 281
France, age structure, 1184*f*
Frankia, 767
Franklin, Rosalind, 219, 220
free energy, 120, 145
free neuronal endings, 878*f*, 884
free radicals, 33, 150, 898
freeze fracture electron microscopy, 99*t*, 100*f*, 101
freezing
  animal adaptations, 956
  effects on distribution of organisms, 1123, 1123*f*
  fishes' resistance to, 467–68, 956
  genetically engineered tolerance, 1126
frequency, of sound waves, 879, 880, 881
freshwater biomes, 1141–42
freshwater fishes, 1011, 1011*f*, 1127. *See also* fishes
Friedman, Jeffrey, 955
Friend, Daniel, 201
Frisch, Karl von, 1146, 1156
frogs
  basic features, 695
  diet, 923
  diploid and tetraploid, 317*f*
  distinguishing species, 496, 496*f*
  embryonic development, 1081–83
  external fertilization, 1061, 1061*f*
  jumping abilities, 901, 901*f*
  lungs, 992
  nervous systems, 859*f*
  rain forest extinctions, 1117
  skin surface area, 822*f*
frontal lobe, 864, 864*f*
fructose
  absorption in digestive tract, 933, 934*f*
  molecular structure, 47*f*
  in semen, 1063
fructose-1,6-bisphosphate, 138, 140*f*
fructose-6-phosphate, 140*f*
fruit flies, 13. *See also Drosophila melanogaster*
fruiting bodies, 620–21, 620*f*, 623–24
fruits
  development, 806
  ethylene's effects, 746
  functions and diversity, 609–11, 610*f*, 614–15, 797
  as organ systems, 719
  ripening process, 203
fruticose lichens, 634–35, 635*f*
Frye, Larry, 102
FT protein, 798
FtsZ protein, 542
fucoxanthin, 564
fumarate, 143*f*
Funaria hygrometrica, 589
function, relation to structure, 4
functional genomics, 406
functional groups, 43–44, 44*t*
functional MRI, 869–71, 869*f*
functional sites (translation), 250
fundamental niche, 1191
fungi
  basic features, 10*f*, 620–22
  biotechnology applications, 635–36
  as decomposers and pathogens, 630–31
  diversity, 624–29
  evolution, 618–19
  giant species, 193
  life cycle, 313*f*, 314
  reproduction, 622–24
  in symbiotic relationships, 631–35, 766, 1202
fungus-like protists, 554, 554*f*
fur, 703
Furchgott, Robert, 978–79
FUS3 gene, 809
fused joints, 904*f*
fusion, 36, 1084, 1084*f*
fusion inhibitors, 1113
fusion proteins, 410

G

G_1 phase, 299, 300, 301, 301*f*
G_2 phase, 299, 300, 301, 301*f*
Gage, Fred, 868–69
galactose, 47, 47*f*, 933, 934*f*
galactoside transacetylase, 261, 262*f*
Galápagos finches. *See* finches
Galápagos Islands, 461, 463–64
Galeopsis tetrahit, 504–5
gallbladder, 932*f*, 933
Galliformes, 702*t*
galls, 412, 413, 536, 537*f*
Gallus gallus, 381
Galvani, Luigi, 834
gametangia, 579, 584*f*, 626
gametes. *See also* sexual reproduction
    of bryophytes, 579
    defined, 1056
    differences among eukaryotes, 314
    DNA methylation, 356
    epigenetic inheritance and, 355, 356
    flowering plants, 717
    formation, 1059–61
    fungi, 623, 626, 627, 628*f*
    haploid nature, 299
    hormonal control of development, 1050
    linkage and crossing over effects, 348–50
    maternal inheritance and, 352
    mutations, 283, 283*f*
    of protists, 567, 568*f*
    recording in Punnett square, 325–26
    self-fertilization, 322
gametic isolation, 498*f*, 499
gametic life cycles of protists, 568, 570*f*
gametogenesis, 1059–61, 1066–67
gametophytes
    of bryophytes, 578, 578*f*, 579, 580
    defined, 313*f*, 314
    development in flower tissues, 719
    in flowering plant life cycle, 717, 794
    lycophyte and pteridophyte, 583, 584*f*
    production in flowers, 795
    structure and function in flowering plants, 802–4
gametophytic SI, 803, 803*f*
γ-aminobutyric acid, 845*t*, 846, 848–49
γ_c cytokine receptor, 416
γ-globin genes, 429, 1070
gamma globulin, 1104
gamma rays, 157, 285
ganglia, 858
ganglion cells of retina, 890
gannets (*Morus bassanus*), 1155, 1155*f*
gap genes, 387, 388*f*
gap junctions
    basic features, 198, 198*t*
    cardiac muscle, 176, 201, 971
    channel size, 202, 202*f*
    mechanisms of action, 201, 201*f*
Garces, Helena, 809
garden pea (*Pisum sativum*), 322–23, 807, 808*f*
Garrod, Archibald, 236–37
Garstang, Walter, 680
gaseous neurotransmitters, 846
Gaser, Christian, 870, 871
gases, 845*t*, 986–88
gas exchange, 986, 987*f*
gas gangrene, 544
gastric brooding frog (*Rheobatrachus* spp.), 1244
gastric ulcers, 940, 941*f*
gastrin, 937, 937*f*, 1031*f*
gastroenteritis, 361*f*
gastroesophageal reflux, 940
Gastropoda, 661, 662*f*
gastropod locomotion, 916
gastrovascular cavity
    as circulatory system, 965, 965*f*
    digestion via, 927, 927*f*
    in flatworms, 657
    in Radiata, 654
gastrula, 642, 1088
gastrulation, 642, 643*f*, 1081–83
gated channels, 109, 109*f*, 110*f*, 838–42
Gatorade, 1015
gaur (*Bos gaurus*), 1261
Gause, Georgyi, 1189
Gaut, Brandon, 615
GDP. *See* guanosine diphosphate (GDP)
geese
    egg-rolling response, 1147, 1147*f*
    imprinted behavior, 1150, 1150*f*
Gehring, Walter, 511
gel electrophoresis, 126, 127*f*, 404*f*, 405, 407
gemmae, 578*f*
Gemmata obscuriglobus, 539
GenBank database, 433*t*
gene addition, 411
gene amplification, 290, 291*f*
gene cloning
    DNA libraries, 404–5
    gel electrophoresis and PCR, 405–6
    steps, 400–404, 401*f*
gene duplication, 427–29, 428*f*, 481, 482*t*
gene expression. *See also* inheritance
    defined, 235, 257
    early studies, 236–38
    genetic code, 243–47, 244*t*, 245*f*
    genotype vs. phenotype, 325
    hormone effects, 13
    molecular-level steps, 238–39
    in organizers, 1089–91
    RNA processing in eukaryotes, 241–43, 242*f*
    transcription, 239–41, 240*f*, 241*f*
    translation basics, 244–45, 247–51
    translation stages, 251–54, 252*f*, 253*f*, 254*f*
    xylem, phloem, and nonvascular tissue, 789–90
gene expression and regulatory proteins, 430*t*
gene families, 427–29, 473, 1070
gene flow, 492
gene interactions, 344–46
gene knockout, 411
gene pools, 479
genera, 516
generalist herbivores, 1197
general lineage concept, 497
General Sherman tree, 193, 193*f*
general transcription factors, 269, 270, 270*f*
generation times, 507
generative cells, 802
gene regulation
    defined, 257
    microarray data, 410*t*
    occurrence in eukaryotes and bacteria, 259–60
    overview, 130
    purpose, 258–59
    RNA processing in eukaryotes, 274–75
    transcription in bacteria, 259, 260–67
    transcription in eukaryotes, 267–74
    translation in eukaryotes, 275–76
gene replacement, 411–12
gene repressors, 740
genes
    altruism, 1159–61
    Alzheimer disease related, 872
    avirulence, 752, 753*f*
    basic functions, 4, 69, 239
    control of flower structures, 798–802
    current theory, 15
    defined, 4, 239
    differential regulation, 188, 196
    duplications, 427–29, 473
    early association with enzymes, 236, 237–38
    early discoveries, 327
    evolution, 472–75
    extranuclear, 350–53
    homologous, 161, 162*f*, 471–72
    interactions between, 344–46
    linked, 346–50
    loci, 328
    Mendel's observations, 324
    for muscle cell differentiation, 393–95
    mutations, 280–86, 294, 295, 336–37, 380, 389, 482
    percentage encoding transmembrane proteins, 99–100
    polymorphic, 479
    resistance, 752, 753*f*
    role in behavior, 1147–48, 1149–50
    structural vs. regulatory, 238
    transfer between species, 7–8, 377–78, 474–75, 667
gene sequences. *See also* DNA sequences
    evolution evidence from, 471–72
    from extinct organisms, 524–25
    homologous, 520
    mechanisms producing, 472–75
    neutral variation, 491
    shared derived characters, 522, 523*f*
gene silencing, 269, 799
gene therapy, 416–17
genetically modified organisms, 411
genetic code, 243–47
genetic disorders
    adenosine deaminase (ADA) deficiency, 416–17
    aniridia, 380
    biotechnology efforts against, 411–12, 416–17
    cleft palate, 1091–92
    congenital analgesia, 340
    cystic fibrosis, 16–17, 330, 336–37, 411–12
    Down syndrome, 317, 1091, 1092
    Ehlers-Danlos syndrome, 196
    Huntington disease, 331
    Klinefelter syndrome, 318*t*, 354, 354*t*
    nucleotide excision repair systems, 288
    pedigree analysis, 330–31
    phenylketonuria, 336*t*, 337, 338–39, 339*f*
    recessive alleles causing, 330, 336–37, 336*t*
    sickle cell disease, 280, 281*f*, 478
    triple X syndrome, 317, 318*t*, 354, 354*t*
    Turner syndrome, 317, 318*t*, 354, 354*t*
    xeroderma pigmentosum, 279, 288, 288*f*
    X-linked, 332–33, 353–54
genetic diversity
    chromosome structure and, 314
    impact on small populations, 1252–54
    importance, 1245
    random chromosome alignment and, 312
genetic drift
    effect on populations, 489–91
    role in microevolution, 481, 482*t*
    small populations and, 1252–53
genetic engineering. *See also* biotechnology
    of animals, 411–12, 414–15
    defined, 400
    of plants, 412–13, 764–66, 1198, 1232
genetic mapping, 349–50, 350*f*
genetic material
    as characteristic of life, 3*f*, 4
    current theory, 15
    DNA discoveries, 212–17
    early discoveries, 327
    locations in eukaryotic cells, 351*f*, 423–24
    in mitochondria and chloroplasts, 88–89
    mutations, 280–86

genetic variation
  in Darwin's natural selection hypothesis, 463
  microarray data, 410*t*
  selective breeding and, 469
  as sexual reproduction advantage, 1057–58
gene transfer, 7–8, 373–78
genital herpes, 361*f*
genitalia, human, 1063–66, 1063*f*, 1065*f*
genome doubling, 607
genomes. *See also* DNA (deoxyribonucleic acid)
  alternative splicing's impact, 242, 274, 275, 431
  aquaporins, 1025–26
  ATP-using proteins in, 121–22
  bacterial, 370–72, 420–23
  basic functions, 9–11
  cancer clues in, 139, 295
  cerebral cortex evolution, 866
  collagen synthesis, 195–96
  for color vision, 889
  control of guard cell development, 729
  cytokinesis proteins, 307–8
  defined, 4, 12*f*, 69, 217, 420
  development and, 381, 390
  distinguishing species by, 497
  DNA replication support, 228–29
  eukaryotic cell origins, 451–52
  eukaryotic organisms, 423–29
  as evolutionary evidence, 9, 161, 250–51, 472–75
  eye evolution and, 511–12, 512*f*
  flowering plant evolution and, 607–9
  fungal, 633
  gibberellin functions, 744–45
  globin, 1070
  for glucose transport proteins, 948
  heart, 968–69
  hemoglobin, 1001–2
  hexaploid species, 1181–82
  homologous genes, 161
  hormones and receptors, 1052
  human versus chimpanzee, 709
  immunoglobulin, 1105
  for lactose intolerance, 934–35
  land plant evolution clues from, 583–85
  location in eukaryotic cells, 93
  maintenance of integrity, 292–93, 301
  mapping, 349
  metagenomics, 1212–14, 1213*f*
  of mitochondria and chloroplasts, 87–88, 350–53, 424
  morphological variety, 258, 259*f*
  of multicellular organisms, 193
  myosin proteins, 907–8
  neuronal regulation, 848–49
  parasitic excavates, 557
  percent devoted to transmembrane proteins, 99–100
  pleiotropy and, 336–37
  polymorphism, 479–80
  proteomes versus, 4, 12*f*, 70–73, 430–31
  role in organism's characteristics, 12–13
  sequencing, 406–9, 421–23, 432–33
  size compared between species, 275*t*
  techniques for studying, 406–9, 431–35
  for temperature tolerance, 1126
  transgenic organisms, 1201
  of transplanted cells, 1088–89
  viral, 291, 292, 360, 361–63, 367, 368
  xylem and phloem development, 789–90
genomic imprinting, 343*t*, 355–56, 355*f*
genomic libraries, 404
genomics, 9, 400, 406–9
genotype
  defined, 325
  determining with testcross, 326, 326*f*
  frequency in populations studied, 480–81
  recording in Punnett square, 325–26
genotype frequency, 480–81
Genpept database, 433*t*
genus name, 516–17
genus Solenopsis, 1155
geographic isolation. *See* isolated populations
geological mechanisms in ecosystems, 1235
geological timescale, 448, 449*f*
Geomyces destructans, 630
Georges Bank, 1169
Geospiza fortis, 463, 463*f*, 1193
Geospiza fuliginosa, 1193
geraniol, 799
Gerbera hybrida, 799–800, 800*f*
germ cells, 1059, 1083
germination
  pollen in flowers, 795–96, 803–4
  of seeds, 797, 806–8
germ layers, 642, 1081, 1081*f*
germ-line cells, 283
germ-line mutations, 283, 283*f*
germ plasm, 1083, 1083*f*
gestation periods in mammals, 1068
Ghadiri, M. Reza, 441
giant anteater, 467, 468*f*
giant axons, 666
giant horsetail (*Equisetum telmateia*), 582*f*
giant kelps, 562, 562*f*, 567
giant panda (*Ailuropoda melanoleuca*), 1258
giant rhubarb, 767
giant saguaro, 730*f*
giant sequoia (*Sequoiadendron giganteum*), 193, 597–98, 1124, 1124*f*, 1181
giant species, 193, 618
Giardia intestinalis, 556, 557, 557*f*
giardiasis, 556
gibberellic acid, 725, 740*t*, 807, 807*f*
gibberellins, 740*t*, 744
Gibbs, J. Willard, 120
GID1 proteins, 744–45, 745*f*
Giemsa stain, 315
Gila monster (*Heloderma suspectum*), 697*f*, 698
Gilbert, Francis, 1220
Gill, Frank, 1155
gill arches, 688
gill filaments, 990, 990*f*
gill ridges, 470
gills
  in arthropods, 669
  in fishes, 966
  in mollusks, 661
  surface area, 822
  water and salt balance mechanisms, 1011, 1011*f*, 1127
  water-breathing types, 989–91
gill-withdrawal reflex, 867–68
Gilman, Alfred, 179
Gilula, Norton, 201
Ginkgo biloba, 599–600
ginkgos, 577
girdling, 732
gizzards, 930, 930*f*
Glacier Bay, 1216, 1216*f*
glaciers
  impact on life-forms, 450, 454, 1209
  mass extinctions and, 456
  succession following, 1216, 1216*f*
glass sponges, 653
glaucoma, 897, 898*f*
Gleason, Henry Allan, 1208
glia (glial cells), 833, 833*f*
gliding, 916
global warming
  current studies, 1124–26
  extinctions resulting from, 1117, 1252
  indicator species, 1258
  methane's potential, 545
globe thistle, 424–25
globin
  gene family, 427–29, 428*f*, 473, 1001, 1070
  gene regulation, 259, 259*f*
glomerular filtration rate, 1021
glomerulosa, 1050
glomerulus, 1020, 1020*f*, 1021*f*
Glomus, 627*f*
glucagon
  functions, 949, 1039–40
  site of production, 1031*f*, 1039, 1040*f*
glucocorticoid receptors, 1052
glucocorticoids
  functions, 1051–52, 1051*t*
  site of production, 1031*f*
gluconeogenesis, 946, 1040
glucose
  absorption in digestive tract, 933, 934*f*
  antifreeze properties, 956
  basic properties, 47, 47*f*
  breakdown from glycogen, 80, 183
  breakdown into pyruvate, 129, 138–39, 138*f*
  as cellular respiration energy source, 137–38
  in cellulose molecule, 49, 197, 198*f*
  electrochemical gradient, 107*f*
  homeostatic role, 823*t*, 824, 824*f*, 947–49
  lac operon inhibition, 266, 268*f*
  maintaining blood levels, 946–47, 948–49, 949*f*, 1039–42
  reabsorption in kidney, 1021
  reaction with ATP, 122–23
  storage and use, 945, 945*f*
  yeasts' response to, 175, 175*f*
glucose-6-phosphatase, 80
glucose-6-phosphate, 140*f*
glucose sparing, 947
glucose transport proteins, 947–48, 948*f*, 1041
glucosinolates, 1198
glutamate, 129, 845*t*, 846
glutamic acid, 54*f*, 280, 281*f*
glutamine, 54*f*, 129, 468
GLUT proteins, 947–48, 948*f*, 1041
glycans, 197*f*, 198
glyceraldehyde-3-phosphate
  formation in Calvin cycle, 166–67
  formation in glycolysis, 138, 140*f*, 149
glyceraldehyde-3-phosphate dehydrogenase, 139
glycerol
  metabolism, 149
  release in lipolysis, 946, 947
  in triglyceride formation, 49, 49*f*
glycine
  codons, 243, 244
  molecular structure, 53, 54*f*
  as neurotransmitter, 845*t*
glycocalyces, 69
glycogen
  breakdown into glucose, 80, 183, 946
  in liver, 1039
  molecular structure, 48, 48*f*
  in skeletal muscle, 912
glycogenolysis, 946, 1039
glycogen phosphorylase, 183, 442
glycogen synthase, 183
glycolipids, 98, 98*f*, 103
glycolysis
  in anaerobic conditions, 150–51, 151*f*
  in cancer cells, 139
  in cellular respiration, 137
  overview, 129
  steps and phases, 138–39, 138*f*, 140–41*f*
glycolytic fibers, 912, 913–14

- glycoproteins
  - formation, 103
  - location in membranes, 98f
  - of viruses, 361, 362f
- glycosaminoglycans (GAGs)
  - basic functions, 49
  - in extracellular matrix, 196–97, 197f, 816
- glycosidic bonds, 47
- glycosylation, 80, 103–5
- glycosyl transferase, 338
- glyoxysomes, 85
- gnathostomes
  - basic features, 687–88
  - bony fishes, 690–92
  - sharks and rays, 688–89
  - tetrapod origins and evolution, 692–96
- Gnetales, 603, 604f
- *Gnetum*, 603, 604f
- goblet cells, 993, 994f
- Goethe, Johann Wolfgang von, 725
- goiter, 1038, 1040f
- golden silk spiders (*Nephila clavipes*), 1163f
- golden-winged sunbird (*Nectarinia reichenowi*), 1155, 1155f
- Golgi, Camillo, 80
- Golgi apparatus
  - in animal cells, 71f
  - exocytosis, 113, 114f
  - glycosylation in, 104–5
  - in plant cells, 72f
  - protein sorting to, 91–92
  - structure and functions, 80–81, 81f
- gonadotropin-releasing hormone, 1064, 1067
- gonads, 1050, 1063, 1063f
- Gondwana, 1143
- gongylidia, 1202
- *Gonium pectorale*, 453, 453f
- goose barnacles, 676f
- gopher tortoises (*Gopherus polyphemus*), 1258
- *Gopherus polyphemus*, 1258
- gorilla (*Gorilla gorilla*)
  - chromosomes, 475, 475f
  - classification, 707–8, 708f
- Gorter, Evert, 99t
- goshawk (*Accipiter gentilis*), 1157
- Gould, Elizabeth, 868
- Gould, John, 461
- Gould, Stephen Jay, 506
- *GPR56* gene, 866
- G-protein-coupled receptors (GPCRs)
  - for acetylcholine, 188
  - chemoreceptors as, 895–96
  - opsins, 888
  - signal transduction pathways, 183–87
  - structure and functions, 179–80, 180f
- G proteins, 179–80, 180f
- graded membrane potentials, 839, 839f
- gradualism, 506–7
- graft rejection, 1112
- grains, as fruits, 610f, 611
- Gram, Hans Christian, 540
- Gram-negative bacteria, 536t, 540–41, 540f
- Gram-positive bacteria, 540–41, 540f
- grana, 87, 87f, 155, 156f
- Grant, Peter, 463
- Grant, Rosemary, 463
- Grant's zebra (*Equus quagga boehmi*), 495f
- grape, 610f
- grasping hands, 707
- grasses, rhizomes, 732
- grass flowers, 610f
- grasshoppers, 673t, 1197
- grassland biomes, 1136
- grass-pink orchid (*Calopogon pulchellus*), 1203
- grassquit finch, 461
- grass seeds, 807, 807f
- gravity
  - atmospheric pressure and, 987
  - effect on venous blood flow, 977
  - plant responses to, 749–50
- gray fox (*Urocyon cinereoargenteus*), 467, 467f
- gray matter, 859, 860f
- gray wolf (*Canis lupus*), 516, 516f
- Grb, 182
- Great Barrier Reef, 654, 1140
- Great Basin Desert, 1137
- greater apes, 707
- greater prairie chicken (*Tympanuchus cupido*), 1252, 1253f
- Great Slave Lake, 1209
- great white shark (*Carcharodon carcharias*), 688
- green algae, 558, 564, 575–77
- greenhouse effect, 1124
- greenhouse gases
  - contribution to global warming, 1125, 1125t
  - main types, 1124, 1125t
  - methane as, 545
- green lizard (*Lacerta viridis*), 989f
- green sea turtles, 1154
- green turtle (*Chelonia mydas*), 697f
- Greider, Carol, 229, 230
- *Gremlin* gene, 508
- Grendel, F., 99t
- Grevy's zebra (*Equus grevyi*), 495f
- greyhound, 469f
- Griffith, Frederick, 213–14, 376
- Griffith, John Stanley, 368
- grizzly bear (*Ursus arctos*), 639f, 1253–54
- gross primary production, 1232
- Groudine, Mark, 290
- ground finches, 461, 462t
- ground state, 158f
- ground tissue, 206, 208
- group living, 1157–58
- group selection, 1158
- growth
  - as characteristic of life, 3f
  - defined, 4
  - development versus, 1077
  - in flowering plants, 718
  - hormonal control, 1047–49
  - of populations, 1176–85
  - as tissue-forming process, 205
- growth factors
  - defined, 181
  - overactivity linked to cancer, 288–90
  - signal transduction pathways, 181–83, 182f, 289f
- growth hormone, 1037t, 1047
- growth rates, 510–11
- growth rings, 731, 731f
- *Grus japonensis*, 1163f
- *Gryllus pennsylvanicus*, 498, 499f
- *Gryllus veletis*, 498, 499f
- GTP. *See* guanosine triphosphate (GTP)
- Guam rail, 1196, 1197f
- guanine
  - base pairing in DNA, 61–62, 62f, 220, 220f, 221
  - bonding in DNA, 220, 220f, 221
  - molecular structure, 61, 217, 218f
- guanine triphosphate, 141, 143f
- guanosine diphosphate (GDP), 179–80, 290
- guanosine monophosphate (cGMP), 890, 891f
- guanosine triphosphate (GTP), 179–80, 290
- guard cells
  - development, 729, 729f
  - as epidermal cells, 206
  - functions, 729, 785, 785f, 786f
- Guerrier-Takada, Cecilia, 126
- Gulf Stream, 1131, 1131f
- gumbo limbo trees (*Bursera simaruba*), 1169
- *Gunnera*, 767, 767f
- Gunning, Brian, 589
- guppy (*Poecilia reticulata*), 486, 487f
- *GUS* gene, 765
- Gustafsson, Lars, 483
- guttation, 781, 781f
- Guzman, Rafael, 1245
- *Gymnogyps californicus*, 1260, 1260f
- gymnosperms, 592, 597–603
- *Gymnothorax meleagris*, 691f
- gyri, 856, 866

H

- H1N1 influenza, 367
- HAART (highly active antiretroviral therapy), 1113
- Haber, Fritz, 763
- habitats
  - destruction, 1251–52
  - distinguishing species by, 497, 498
  - isolation, 498, 498f
  - labeling protists by, 554, 555f
  - restoration, 1259–60
- habituation, 1148
- Hadley, George, 1130
- *Haemophilus influenzae*
  - chromosomes, 370
  - colorized micrograph, 359f
  - connection to meningitis, 872
  - as first genome to be sequenced, 421, 422
  - genome size, 421t
  - membrane protein genes, 100t
- hagfish, 686–87, 687f
- hair, as characteristic of mammals, 703, 703f
- hair cells
  - equilibrium sense, 882, 882f, 883f
  - hearing sense, 879, 880–81, 881f, 898
  - lateral line systems, 883–84
  - mechanisms of action, 878–79, 878f
- hairs, root. *See* root hairs
- Haldane, J. B. S., 439, 478
- Hale, Cindy, 1225
- half-lives, 26, 131, 447
- Halloween snake (*Pliocercus euryzonus*), 698f
- hallucinogenic substances, 624
- halophytes, 1128
- halteres, 388
- Hamilton, W. D., 1158, 1159
- Hamilton's rule, 1159
- Hammock, Elizabeth, 1163
- Hammond, John, 765
- handicap principle, 1164
- hands of primates, 707
- hangingflies, 1164, 1164f
- hantavirus, 361f
- Hanuman langurs (*Semnopithecus entellus*), 1159, 1159f
- *Hapalochlaena lunulata*, 662f
- haplodiploid systems, 332, 1161
- haploid cells, 308, 312, 313–14
- haploid-dominant species, 313f, 314
- haplorrhini, 707, 708f
- haptophytes, 560
- Harberd, Nicholas, 744
- Hardy, Godfrey Harold, 480
- Hardy-Weinberg equation, 480–81, 481f
- Hardy-Zuckerman 4 feline sarcoma virus, 292t
- Harland, Richard, 1089
- harlequin frogs, 1117
- Harrison, Susan, 1256
- Hatch, Marshall, 170
- haustoria, 630, 630f, 752, 770, 1200
- Hawaiian honeycreepers, 1251, 1251f
- Hawaiian Islands, speciation in, 501
- Hayes, Marianne, 303

Hayes, William, 375
hazardous waste. *See* pollution
hearing, 879–82, 898
heart
  amphibian, 694
  cardiac output, 980–81
  diseases, 983, 984*f*
  embryonic formation, 1085
  endocrine functions, 1031*f*
  epinephrine's effects, 187
  evolution, 968–69
  exercise effects, 981–82
  in fishes, 689, 966, 967*f*
  neurotransmitter actions on, 846–47
  in open circulatory systems, 965–66
  structures and functions in vertebrates, 971–74, 972*f*
  tissues, 6
  types in closed circulatory systems, 966–69
heart attacks, 983
heartburn, 940
heart rate, 826, 980–81, 982
heart sounds, 973, 974
heart stage, 395, 396*f*
heat. *See also* body temperatures; temperature
  denaturation by, 955
  effect on chemical reactions, 123, 125, 955
  effects on distribution of organisms, 1123–24
  as energy form, 119*t*
  genetically engineered tolerance, 1126
  impact on plasma membranes, 955–56
  photorespiration and, 169–71
heat capacity, 36
heat of fusion, 36
heat of vaporization, 36
heat shock proteins, 1126
heat stress, 633–34
heavy metals, 483, 485*f*, 1260
Hebert, Paul, 675
Hedera helix, 467
hedgerows, 1257
Heitz, Emil, 231
Helianthus, 608, 610*f*
Helianthus annuus, 1260
Helicobacter pylori, 537, 940
helium, 22*f*
Heloderma horridum, 698
Heloderma suspectum, 697*f*, 698
helper T cells
  actions of, 1102, 1102*f*, 1107–8
  AIDS destruction, 367, 367*f*, 1113
hematocrit, 970, 978
hematomas, 667
hematopoietic stem cells, 392, 392*f*
hemicellulose, 197*f*, 198, 730
hemidesmosomes, 199, 199*f*
hemiparasites, 1199
Hemiptera, 673*t*
hemispheres (cerebral), 863–65, 864*f*
Hemitripterus americanus, 468
hemizygous individuals, 332
hemochromatosis, 97
hemocoel, 661, 669
hemocyanin, 663, 999
hemodialysis, 1027, 1027*f*
hemoglobin
  component polypeptides, 238
  evolution, 1001–2
  functions, 970
  gene family, 427–29, 1070
  gene regulation of oxygen uptake, 259, 259*f*
  high altitude adaptations, 1004
  malaria's effects, 570
  as multimeric protein, 56
  mutations, 280, 281*f*
  oxygen binding, 999–1001
hemolymph
  in arthropods, 669, 966
  characteristic of open circulatory systems, 819, 966
  in mollusks, 661, 966
hemolysis, 821
hemophilia A, 332–33, 336*t*, 353–54
hemorrhage, 983
hemorrhagic fever, 361*f*
hemp (*Cannabis sativa*), 730
Henle, Friedrich, 1020
Hennig, Willi, 514
Hensen's node, 1089
heparin, 1096
hepatitis B virus, 292*t*, 361*f*
herbaceous plants, vascular tissues, 729. *See also* plants
herbivores
  defined, 921, 1226
  digestive systems, 931
  feeding behaviors, 925, 926*f*
  plant defenses, 751–52, 1197–98
herds, 1157–58
hereditary nephrogenic diabetes insipidus, 1025–26
heritability, 4. *See also* inheritance
hermaphrodites
  clownfish, 12–13
  ctenophores as, 656
  main types, 1062
  sponges as, 654
herpes simplex viruses, 361*f*, 361*t*, 367
herpesviruses, episomes, 367
Hershey, Alfred, 215–17
Hertz, Heinrich, 879
Heterocephalus glaber, 1161, 1161*f*
heterochromatin, 231, 232*f*
heterocytes, 545
heterodont mammals, 704
heterokonts, 561
heterospory, 591, 593
heterotherms, 956, 956*f*
heterotrophs
  among prokaryotes, 544
  animals as, 639–40
  defined, 155, 451
  as early life-forms, 451
  in food chains, 1226
  fungi as, 619
heterozygote advantage, 484–85
heterozygous individuals
  defined, 325
  dominance and, 325, 335–38
  identifying, 326
  inbreeding and, 492–93
  law of segregation and, 328–29
Hevea brasiliensis, 611
hexaploidy, 1181–82
hexapods, 672
hexokinase, 123, 124*f*
hexoses, 46–47
hibernation, 952, 956
hibiscus, 609*f*
Hibiscus cannabinus, 730
Hib vaccine, 872
hierarchies, 53–56, 514
high altitude adaptations, 1004
high blood pressure treatments, 1*f*
highly active antiretroviral therapy (HAART), 1113
highly conserved proteins, 307
highly repetitive sequences, 425
high temperatures, effects on distribution of organisms, 1123–24. *See also* heat; temperature
Hill, Robin, 165
Hillig, Karl, 612
hindbrain, 856, 857*f*, 862, 865*t*
hinge joints, 904*f*
hippocampus, 866, 867, 868–69
Hippocampus bargibanti, 1195*f*
Hirudo medicinalis, 666*f*, 926
histamine
  in allergic response, 1112
  from basophils, 1096
  major effects, 845*t*, 846
  from mast cells, 1096, 1097*f*
histidine
  Ames test for, 285–86
  as essential animal nutrient, 922
  molecular structure, 54*f*
histone acetyltransferase, 271, 273
histone code hypothesis, 272
histones
  in archaea and eukaryotes, 535
  DNA compaction around, 230
  modifications affecting transcription, 271–73, 271*f*
histone variants, 271, 273
Histoplasma capsulatum, 630–31
histoplasmosis, 631
hoatzin, 699
hobbits, 712
Hoffman, Jules, 1098, 1100
Holbrook, N. Michelle, 781
holdfasts, 1126, 1127*f*
Holland, Scott, 870–71
Holley, Robert, 248
holoblastic cleavage, 1079–80
holoparasites, 1199, 1199*f*
Holothuria edulis, 678*f*
Holothuroidea, 678, 678*t*
homeobox, 389, 389*f*
homeodomain, 389, 389*f*
homeostasis
  in animals, 822–29
  ATP regulation, 140
  as characteristic of life, 3*f*, 4
  defined, 4, 813, 823
  hormone regulation, 829, 947–50, 1030
  of internal body fluids, 1008–15
  nervous system regulation, 860–62, 947–50
  role of hypothalamus, 863
  role of vertebrate skeleton, 903
homeostatic control systems, 824–25
homeotherms, 956, 956*f*
homeotic genes, 387–90, 397
homing pigeons, 884–85, 1154
hominins, early species, 457, 709–12
hominoids
  on geological timescale, 449*f*, 457
  key characteristics, 707, 708*f*
homodont mammals, 704
Homo erectus, 711, 712
Homo ergaster, 711, 712
Homo florensiensis, 712
homogentisic acid, 236, 236*f*
Homo habilis, 457, 710
Homo heidelbergensis, 711
homologies
  defined, 299, 470
  evolution evidence from, 465*t*, 470–72
  in systematics, 519–20
homologous chromosomes
  crossing over, 309, 309*f*, 310*f*
  law of segregation and, 328–29
  separation in meiosis I, 309–12
  similarity, 299
homologous genes
  computer identification, 433–35
  defined, 161, 472
  evolution evidence from, 471–72

homologous genes, *continued*
- formation, 427, 428*f*, 472–75
- homeotic, 390

homologous structures, 470
homologs, 299
Homo neanderthalensis, 711
homophilic binding, 199
Homo sapiens
- alternative splicing, 275*t*
- genome size, 275*t*, 424*t*
- membrane protein genes, 100*t*
- origins and spread, 457, 712, 712*f*

homozygous individuals
- defined, 325
- dominance and, 325, 335–38
- effects of genomic imprinting, 355–56
- gene interactions, 344*f*, 345
- identifying, 326
- inbreeding and, 492–93

honeybee colony collapse disorder, 625
honeybees. *See also* bees
- altruism, 1158
- color vision, 889
- communication, 1156–57
- innate behaviors, 1147–48
- open circulatory systems, 965*f*
- reproduction, 1057

honeycreepers (*Drepanidinae*), 501–2, 501*f*
honeysuckle vine (*Lonicera japonica*), 738*f*
Hooke, Robert, 67
hookworms (*Necator americanus*), 668
hooves, 704
horizons, soil, 760, 761*f*
horizontal cells, 891
horizontal gene transfer
- in angiosperms, 608–9
- in archaea and bacteria, 537–38
- mechanisms of action, 377–78, 378*f*, 474–75
- in parasitic excavates, 557
- role in microevolution, 7–8, 481, 482*t*
- role in species evolution, 528–29, 530*f*

hormones
- affecting adrenal glands, 189
- clownfish, 13
- contraceptive, 1073, 1074
- control of growth and development, 1047–49
- control of metabolism and energy balance, 1037–44
- control of mineral balance, 1044–47
- control of reproduction, 1049–50 (*see also* reproduction)
- defined, 176–77, 1030
- digestive system regulation, 937, 937*f*
- discovery, 939
- ending signals from, 1035
- fruit development, 806
- health issues, 1030, 1052–53
- of hypothalamus, 863
- impact on bone density, 917
- impact on gene expression, 13
- intracellular receptors, 180–81, 181*f*
- major types in plants, 740–46, 740*t*
- misuse, 1053
- molecular structure, 52*f*
- plant defenses using, 746, 751–52, 753, 754
- plant responses to, 738, 739, 739*f*
- role in homeostasis, 829, 947–50, 1030
- role in leaf development, 725–26
- stress responses, 1050–52
- types and functions in animals, 1031–35
- varied cellular responses to, 187–88
- of vasoconstriction, 978

hormone therapy, 1073
hornbirds, 471*t*
horns, 704, 705*f*
hornworts, 577, 578*f*
Horowitz-Scherer, Rachel, 230
horses
- evolutionary changes, 7, 7*f*, 466, 466*f*
- gestation period, 1068
- mating with donkey, 499, 500*f*
- phylogenetic tree, 519–20, 520*f*

horseshoe crab (*Limulus polyphemus*), 1245
Horvitz, Robert, 668
host cells, 360, 363–67, 364–65*f*
host-derived cell membrane, 570, 572*f*
host plant resistance, 1197–98
host ranges, 360
hosts, defined, 1198
hot desert biomes, 1137
Hotopp, Julie Dunning, 667
hot spots, geographic, 1254, 1255*f*
Hox genes
- body segment specialization, 645, 646*f*
- clusters in vertebrates, 686
- control of body patterns, 509–10, 646–47, 1085–86
- control of limb development, 692, 692*f*, 693–94
- control of organ development, 819
- expression in mouse, 390*f*
- organization in animals, 390, 819

Hubbard Brook Experimental Forest, 1240–41, 1241*f*
human genome
- signal transduction proteins, 181, 187–88
- size, 275*t*, 420

Human Genome Project, 429
human immunodeficiency virus (HIV)
- hosts and characteristics, 361*t*, 367
- means of infection, 367
- reproductive cycle, 363–67, 364–65*f*
- target tissues, 361*f*, 1113

human impacts. *See also* pollution
- on aquatic biomes, 1139–42
- on biogeochemical cycles, 1236–37, 1240–41, 1242
- bioremediation efforts, 410–11, 550, 1232
- deforestation, 1242, 1242*f*, 1251
- extinctions, 1250–52
- lichen sensitivity to, 635
- on terrestrial biomes, 1133–38

human microbiome, 548
Human Microbiome Project, 548
humans
- chromosome abnormalities, 317, 318*t*
- developmental stages, 258–59
- early species, 457
- ear structures, 879*f*
- estimated number of cells in, 297
- evolution, 707–12
- fertility limits, 1056
- gas exchange across skin, 989*f*
- gastrointestinal tract, 928, 929*f*
- genome size, 275*t*, 424*t*
- introns in genes, 242
- limb development, 383, 383*f*
- nervous system overview, 858–66
- normal chromosome numbers, 314
- nucleotide differences from chimpanzees, 528
- nucleotide excision repair systems, 288
- oxygen-hemoglobin dissociation curve, 1000, 1000*f*, 1001*f*
- pedigree analysis, 330–31
- population growth, 1182–85
- reproduction, 1062–68
- respiratory system, 821, 821*f*, 992–96, 994*f*
- skull development, 510, 511*f*
- vestigial structures, 470, 471*t*
- viral latency, 367

Humboldt Current, 1131, 1131*f*
humidity, 779, 958
hummingbirds
- beak type, 701*f*
- mites, 1203
- muscle mass, 916
- as pollinators, 609*f*
- torpor, 952
- water loss, 1011

humoral immunity
- defined, 1102
- elements of, 1102–7
- secondary immune response, 1111
- summary example, 1109–10, 1110*f*

humus, 760, 761*f*
hunchback gene, 386
Hundred Heartbeat Club, 1244
Huntington disease, 331, 331*f*
Huntonia huntonesis, 670*f*
Hurtrez-Bousses, Sylvie, 1200
Hutchinson, G. Evelyn, 1193–94
Hutton, James, 460
hyacinthe macaw (*Anodorhynchus hyacinthinus*), 701*f*
hyaluronic acid, 197
hybrid breakdown, 498*f*, 499–500
hybrid inviability, 498*f*, 499
hybridization, 322, 498
hybrid sterility, 498*f*, 499, 505
hybrid zones, 502, 502*f*
hydras
- gastrovascular cavity, 965*f*
- hydrostatic skeleton, 901–2
- key characteristics, 648*t*, 681*t*
- reproduction, 1057*f*

hydrocarbons, 43, 51, 52
hydrochloric acid, 38, 930
Hydrodamalis gigas, 1251
hydrogen
- in animals and plants, 27, 27*t*
- atomic mass, 26
- atomic number, 25
- covalent bonds, 28
- in Earth's early atmosphere, 439
- as plant macronutrient, 757*t*, 759
- structure, 22*f*

hydrogen bonds
- basic features, 30–31, 30*f*
- in DNA, 61, 62*f*, 220, 220*f*, 221
- in protein structures, 55, 57, 57*f*
- of water, 35*f*, 36

hydrogen cyanide, 439
hydrogen fluoride, 28, 28*f*
hydrogen ions
- active transport, 112*f*, 773–74
- blood buildup triggering ventilation, 1003
- movement in anaerobic respiration, 150, 150*f*
- in oxidative phosphorylation, 144*f*, 145–46, 146*f*
- pH and, 38–39
- in photosynthesis, 159–60, 160*f*
- in pure water, 38

hydrogen peroxide
- beneficial uses, 33
- in bombardier beetles, 1194
- breakdown by peroxisomes, 84–85, 122
- as plant defense signal, 752, 753

hydrogen sulfide, 155
hydrological cycle, 1241–42, 1241*f*
hydrolysis, 120*f*
hydrolysis reactions
- defined, 37, 37*f*, 46
- as exergonic reaction, 120
- in lysosomes, 83–84
- polymer breakdown by, 46*f*
- as protein energy source, 121–22, 122*f*

hydrolytic enzymes, 931
hydrophilic molecules, 34
hydrophobic molecules. *See also* lipids
- amino acids, 99–100
- defined, 34

INDEX

in phospholipid bilayers, 106
in protein structures, 57, 57*f*
in transmembrane proteins, 103, 104*f*
hydroponic studies, 760, 765
hydroquinone, 1194
hydrostatic pressure, 775, 901–2
hydrostatic skeletons
in annelids, 644, 665–66, 902, 902*f*
in cnidarians, 656, 901–2
defined, 644, 901
major features, 901–2
water's incompressibility and, 37
hydroxide ions, 38
20-hydroxyecdysone, 1034*f*, 1048–49, 1049*f*
hydroxyl groups
basic features, 44*t*
of glucose, 47*f*
in glycerol, 49, 49*f*
p-hydroxyphenylpyruvic acid, 236
Hydrozoa, 656, 656*t*
hyenas, 832
Hyla chrysoscelis, 317*f*
Hyla versicolor, 317*f*
Hymenoptera, 673*t*
hyoid bone, 711–12
Hypericum perforatum, 1205
hypermutation, 1105
hyperosmolarity, 1022
hyperosmotic organisms, 1127
hyperparasites, 674
hyperplasia, 294*f*, 295
hyperpolarization
defined, 838
in photoreceptors, 889, 890, 890*f*, 891*f*
hypersensitive response, 753, 753*f*
hypersensitivity, 1112
hypertension, 1*f*, 983
hyperthermia, 957, 1097
hyperthermophiles, 535
hyperthyroidism, 1038
hypertonic solutions, 107, 820*f*, 821
hyphae
of chytrids, 625*f*
growth, 621–22
structure and function, 620, 620*f*
hypocotyls, 806, 806*f*, 807
hypoosmotic organisms, 1127
hypothalamus
appetite control, 953
blood glucose regulation, 949
functions, 863, 865*t*, 1031*f*
relation to pituitary, 1035–37, 1036*f*
hypotheses
defined, 5
formation, 14–15
testing, 15–16, 16*f*, 17, 1121–22
hypothyroidism, 1038
hypotonic solutions, 107, 820*f*, 821
Hyracotherium, 7, 466, 466*f*
Hystrix africaeaustralis, 703*f*
H zone, 905*f*, 906

I

I band, 905*f*, 906
ibuprofen, 118
ice, molecular structure, 36, 36*f*
Ice Ages, 457, 467
ichthyosis, 288
identical twins, 414
idiosyncratic hypothesis, 1249, 1249*f*
Ig domains, 1104
IgE-mediated hypersensitivities, 1112
Igf1 gene, 469
Igf2 gene, 355
Ignarro, Louis, 979
igneous rock, 447
ileum, 933
immigration
island biogeography model, 1219, 1219*f*, 1220
selfish behavior and, 1159
immune systems
acquired defenses, 1100–1111
disorders, 416, 851
of flowering plants, 753–54
glucocorticoid effects, 1051–52
health issues, 1111–13
imbalance leading to asthma, 1004
innate defenses, 1094, 1095–1100
major organs in animals, 818*t*
trimeric signaling molecules, 190
viruses infecting, 361*f*, 367–68
immune tolerance, 1110
immunity
acquired, 1094, 1100–1111
health issues, 1111–13
innate, 1095–1100
major types, 1094
immunoglobulins, 1103–4, 1104*f*, 1105
immunological memory, 1111
immunology, 1094
Imperata cylindrica, 1259
imperfect flowers, 605, 798
implantation
delayed, 1072
as embryonic development stage, 1080, 1080*f*
as pregnancy stage, 1068
impotence, 1064
imprinted genes, 355–56
imprinting, behavioral, 1149–50
inactivation gate, 840
inborn errors of metabolism, 236–37
inbreeding, 492–93, 493*f*, 1252
inbreeding depression, 493
incisors, 704, 925
inclusive fitness, 1159
incomplete dominance, 335*t*, 337, 337*f*
incomplete flowers, 605, 798
incomplete metamorphosis, 674, 675*f*
incubation of eggs, 1062
incurrent siphon, 680
incus, 879, 879*f*, 880, 880*f*
independent assortment
law of, 326–27, 329–30, 329*f*
linkage versus, 346–50
indeterminate cleavage, 643, 644*f*
indeterminate growth, 718, 719*f*
Indian mustard (*Brassica juncea*), 1260
Indian peafowl (*Pavo cristatus*), 486, 487*f*
Indian pipe (*Monotropa uniflora*), 757, 758*f*
indicator species, 1258
indirect calorimetry, 950–51, 951*f*
individualistic model of communities, 1208
indoleacetic acid, 740*t*
induced fit, 123, 124*f*
induced mutations, 284, 284*f*
induced ovulation, 1067
inducers, 262
inducible operons, 262
induction, 384, 1086–87
industrial nitrogen fixation, 763–64
industries
use of algae, 561–62
use of bacteria, 549
use of fungi, 635
infanticide, 1159, 1159*f*
infectious diseases
cyst-borne, 567
human deaths from, 359, 367
malaria, 553, 568–72
resistance, 478
selected viruses, 361*f*, 367
infertility, human, 1072–73
inflammation, 871–72, 1096–97, 1097*f*
inflorescences, 609, 610*f*, 799–800
influenza virus
emerging strains, 367
hosts and characteristics, 361*t*
structure, 362*f*
target tissues, 361*f*
infrared radiation
greenhouse effect, 1124, 1124*f*
sensory systems, 885, 885*f*
infundibular stalk, 1035
Ingenhousz, Jan, 154
ingestion, food, 920, 921*f*
ingroups, 521
inhaling, 995
inheritance. *See also* gene expression
chromosome theory, 327–30
defined, 211, 321
epigenetic, 274, 353–56
extranuclear, 350–53
Mendel's laws, 322–27
pedigree analysis in humans, 330–31
probability in, 339–41
sex chromosomes, 331–34
varying patterns, 335–39
inheritance of acquired characteristics, 460
inherited disorders. *See* genetic disorders
inhibition of succession, 1217–18
inhibitors, enzyme, 124–25, 125*f*
inhibitory neurotransmitters, 846
inhibitory postsynaptic potentials, 843, 844–45
initiation factors, 248*t*, 251
initiation stage
transcription, 240, 240*f*, 241
translation, 251–52, 252*f*, 253*f*
initiator caspases, 190, 191*f*
injectisomes, 548
ink sacs, 663
innate behaviors, 1147
innate immunity, 1094, 1095–1100
inner bark, 598, 730–31
inner cell mass, 1080, 1080*f*
inner ear, 879, 879*f*
inner segment of photoreceptors, 888, 888*f*
inorganic chemistry, 20
inorganic fertilizers, 762–63, 764
inorganic nutrients, 920
inorganic soil constituents, 760–62, 761*f*
inositol trisphosphate, 186, 186*f*, 739, 1079
insecticides, 550
insects
appearance during Devonian period, 455
chemoreception, 893–94
circulatory systems, 966
color vision, 889
compound eyes, 886, 886*f*
diversity, 672–74
excretory systems, 669, 1017, 1018*f*
growth regulation, 1048–49
Hox genes, 509, 509*f*
key characteristics, 648*t*, 681*t*
plant predation on, 756, 756*f*, 769–70, 769*f*
predation by fungi, 630
respiratory systems, 821, 821*f*, 991–92, 992*f*
sex determination, 331–32
viruses infecting, 361*t*
instantaneous growth rates, 1177

insulin
  discovery, 1042–43, 1043–44*f*
  enzyme-linked receptors for, 179
  formation, 81
  ineffectiveness in type 2 diabetes, 960
  laboratory production technology, 400, 410, 411*f*
  major functions, 410, 411*f*, 1040–41, 1041*f*
  regulation of metabolism, 947, 948–49, 948*f*, 949*f*
  signal transduction pathway, 85
  site of production, 1031*f*, 1039, 1040*f*
insulin-like growth factor 1, 469, 1047
insulin-like growth factor 2, 355
integral membrane proteins, 99
integrase, 363, 364*f*, 426, 427*f*
integration stage, viral reproductive cycle, 363, 364*f*
integrators, 824
integrins, 199–200, 200*f*
integumentary systems, 818*t*
integuments, 592, 593*f*, 594, 795
interbreeding barriers, 496–500
intercalated discs, 971
interference competition, 1188
interferons, 1098
intergenic regions, 281*t*
interleukin 1, 1108
intermediate circulation, 966–68, 967*f*
intermediate-disturbance hypothesis, 1211, 1211*f*
intermediate filaments, 74–75, 75*t*
intermediate hosts, 658
intermembrane space, 86, 87*f*
internal fertilization, 697, 1061–62
internal gills, 990–91
internal intercostal muscles, 995
internal organs. *See* organs
internal stimuli for plants, 738
International Barcoding of Life project, 675
International Code of Botanical Nomenclature, 517
International Commission on Zoological Nomenclature, 517
interneurons, 834, 834*f*
internodes, 725, 725*f*
interphase, 299, 305, 306*f*
intersexual selection, 486–89, 1164
interspecies hybrids, 498
interspecific competition, 1188
interstitial fluid
  active transport of solutes from, 1016
  capillary exchanges with, 976, 977*f*
  exchanges in nephrons, 1022
  as extracellular fluid, 819
  filtration in protonephridia, 1016, 1017*f*
  hormone release into, 1033
  hyperosmolarity, 1022
intertidal zone, 1139, 1191–93
intertropical convergence zone, 1130
intestines
  internal surface area, 822*f*
  tight cell junctions, 200–201, 200*f*, 207
intimidation, 1195
intracellular digestion, 927
intracellular fluid, 819
intracellular receptors, 180–81
intrasexual selection, 486, 1165
intraspecific competition, 1188
intrauterine device, 1073–74, 1073*f*
intrinsic pathway, apoptosis, 190
intrinsic rate of increase, 1177
introduced species
  challenges to diversity-stability hypothesis, 1214
  chemical secretions, 1118–19
  extinctions and, 1250–51
  mortality from, 1200–1201
  success in new territories, 1142
introns
  absence from cDNA, 404
  defined, 241
  as percentage of human genome, 425
  splicing, 242–43, 274–75, 274*f*
invagination, 1082, 1082*f*
invasive species, 1118–19, 1196, 1205, 1250
inverse density-dependent factors, 1180
inversions, 315, 315*f*, 709
invertebrate grazers, 1198
invertebrates
  acquired immunity, 1100
  diversity, 652
  early animals as, 453, 454
  Ecdysozoa, 667–77
  energy production efficiency, 1229
  growth regulation, 1048–49
  Lophotrochozoa, 656–67
  Parazoa features, 653–54
  Radiata features, 654–56
in vitro fertilization, 1073
involution, 1082, 1082*f*
iodide, 1038, 1039*f*
iodine
  as essential animal nutrient, 924*t*
  required by thyroid, 1038, 1040*f*
  thyroid disorders and, 851
iodine-deficient goiter, 1038, 1040*f*
ion channels, 180, 180*f*
ionic bonds
  basic features, 31, 31*f*
  in protein structures, 57, 57*f*
  in salts, 1008
  water solubility, 34
ionizing radiation, 285
ionotropic receptors, 848, 849*f*
ion pumps, 112–13, 113*f*, 773–74
ions
  defined, 31
  movement across neuron membranes, 835–38
  passage at gap junctions, 201, 201*f*, 202
  soil nutrients, 762
  of water, 38
i promoter, 261
iris (flowering plant), 749*f*
iris (of eye), 380, 886, 887*f*
iron
  in bacteria, 538–39
  disorders involving, 97
  as essential animal nutrient, 922, 924*t*
  in hemoglobin, 970, 999
  homeostatic role, 823*t*
  as plant micronutrient, 757*t*, 760*f*
  regulation of absorption, 276, 276*f*
  as trace element, 27
iron regulatory element, 276
iron regulatory protein, 276
irreversible inhibitors, 125
island biogeography model, 1219–23, 1255–56
island dwarfing, 467
island fox (*Urocyon littoralis*), 467, 467*f*
islands, evolutionary role, 467, 490–91
islets of Langerhans, 1039, 1040*f*, 1042
isocitrate, 143*f*
isocitrate dehydrogenase, 141, 142, 143*f*
isoforms, 1032
isolated populations, 467, 490–91, 498, 500–502
isoleucine, 54*f*, 922
isomers
  of amino acids, 53
  basic features, 44–45, 45*f*
  of glucose, 47*f*
Isopoda, 677
isopropyl alcohol, 45*f*
Isoptera, 673*t*
isotherms, 1122
isotonic solutions, 107
isotopes, 26–27
Isthmus of Panama, 500, 501*f*
Isurus oxyrinchus, 688
iteroparity, 1171–72
Ivanovski, Dmitri, 360
ivory-billed woodpecker (*Campephilus principalis*), 1251, 1251*f*
Iwata, So, 163

J

Jackson, David, 400
Jacob, François, 261, 264
Jacobs syndrome, 318*t*
Jakoba libera, 556, 557*f*
Jansen, Zacharias, 67
jasmonic acid
  plant defenses using, 752, 752*f*, 754
  as plant hormone, 738
Javan banteng (*Bos javanicus*), 1261, 1261*f*
jaws
  characteristic of most vertebrates, 686
  evolution, 687–88, 688*f*
  of insects, 674
Jefferies, Robert, 1233
jejunum, 933
jellyfish, 648*t*, 654–56, 681*t*
Jenkins, Farish, 465
Johannsen, Wilhelm, 324
Johanson, Donald, 709*f*, 710
joining segments, 1105
joints, 903, 904*f*
Jorgensen, Rich, 799
jumping, 901, 916
junctional folds, 911
jun gene, 290*t*
juniper (*Juniperus scopularum*), 602*f*
Juniperus scopularum, 602*f*
Jurassic period, 456
jute (*Corchorus capsularis*), 730
juvenile hormone, 1049

K

Kaiser, A. Dale, 400
Kalanchoë daigremontiana, 809, 809*f*
kanamycin, 412
Kandel, Eric, 867–68
kangaroo rats, 1012, 1023
Kan^R gene, 412
karyogamy, 627, 628*f*
karyotypes, 298–99, 298*f*
kelps, 562
kenaf (*Hibiscus cannabinus*), 730
keratins, 55, 75, 924
Kerr, John, 189
Keshan disease, 924*t*
ketones, 947
Kew Botanical Garden, 716
Key deer, 1172, 1172*f*
keystone hypothesis, 1249, 1249*f*
keystone species, 1258–59, 1259*f*
kidneys
  disorders, 1026–27
  endocrine functions, 1031*f*
  filtering capacity, 1008
  major functions, 1017–18
  sodium and potassium regulation in, 1045–47, 1046*f*
  structure and function in mammals, 1018–24, 1025*f*
  water conservation by, 697
kidney transplantation, 1027
killdeer (*Charadrius vociferus*), 1146*f*
kilocalories, 950
Kimura, Motoo, 491

INDEX

kinases, 293, 301, 301*f*
kinesin, 76, 76*f*
kinesis, 1151
kinetic energy, 119
kinetic skull, 698, 698*f*
kinetochore, 304, 304*f*, 305, 306–7*f*
kinetochore microtubules, 304, 305*f*
kinetoplastids, 558*f*
King, Nicole, 563
kingdoms, 9, 515
Kinosita, Kazuhiko, 147, 148*f*
kin selection, 1159–61
Kirschvink, Joseph, 450
kiwis, 525
Klinefelter syndrome, 318*t*, 354, 354*t*
Knoll, Max, 67
knowledge, theories as, 15
KNOX protein, 725, 726
Knudson, Alfred, 293
koalas, 1197
Koch, Robert, 548
Koch's postulates, 548
Koella, Jacob, 1199
Köhler, Wolfgang, 1149
Kolbe, Hermann, 43
Kolreuter, Joseph, 321
Komodo dragon, 1229
Korarchaeota, 535
Kornberg, Arthur, 225
Kortschak, Hugo, 170
Koshland, Daniel, 123
Kozak, Marilyn, 252
Kraatz, Ernest G., 387–88
Krakatau, 1207
Krebs, Hans, 142
krill, 677
Krogh, August, 992
K-selected species, 1180–81, 1181*t*, 1211
K/T event, 588
kudu (*Tragelaphus strepsiceros*), 705*f*
Kühne, Wilhelm, 122
Kühne, Willi, 907
Kusky, Timothy, 440

L

labia majora, 1065, 1065*f*
labia minora, 1065, 1065*f*
labor, 1070, 1071*f*
Laccaria bicolor, 632*f*
Lacerta viridis, 989*f*
Lack, David, 483, 1190
lac operon, 260–67, 262*f*, 281
lac repressor, 261–64
lactase, 934–35
lactate, 151, 151*f*
lactation, 1071
lacteals, 931, 932*f*, 935
lactic acid
  formation in anaerobic conditions, 151, 151*f*
  NADH role in formation, 146
  release into blood, 1003
  transport across membranes, 111
Lactobacillus plantarum, 421*t*, 539*f*
Lactococcus lactis, 539*f*
lactose
  as common disaccharide, 47
  digestion, 933–35
  lac operon, 260–67, 268*f*
  utilization by *E. coli*, 258, 258*f*
lactose permease, 111, 258, 261, 262*f*
lacunae, 208
lacZ gene, 402, 403
Laetiporus sulphureus, 629*f*
lagging strands, 226–28, 226*f*
Lake, James, 649
lakes
  acid rain impact, 1128
  as biomes, 1141
  cyanobacteria overwintering in, 543–44
  eutrophic, 1141, 1237
  invasive mollusks, 660
  oxygen in, 988
  protists in, 564
  species richness, 1209, 1211
  stability of communities, 1214
Lama glama, 986
Lamarck, Jean Baptiste, 282, 321, 460
λ phage
  hosts and characteristics, 361*t*
  reproductive cycle, 363–66, 364–65*f*
lamellae, 990, 990*f*
Laminaria digitata, 1127*f*
laminin, 194–95, 195*t*, 816
L-amino acids, 53
lampreys, 687, 687*f*
Lampropeltis elapsoides, 1195, 1195*f*
lancelets, 679–80, 680*f*
lancet fluke, 1198–99, 1199*f*
landmarks, wasps' use of, 1151–52
landmass formation, 450, 467, 1142–43
landscape ecology, 1256–57
Langerhans, Paul, 1042
Langmuir, Irving, 99*t*
language, brain control, 864
lanthanum, 201, 201*f*
Lap-Chee Tsui, 17
large intestine, 933
large ribosomal subunits, 249–50, 251, 253*f*
Larix laricina, 603
lark bunting (*Calamospiza melanocorys*), 1163
larvae
  crustacean, 676, 676*f*, 677
  Drosophila melanogaster, 385*f*, 386, 386*f*
  flatworms, 658
  insect, 672, 674
  lamprey, 687
  mollusks, 661
  sponges, 654
  trochophore, 648, 648*f*
  tunicate, 680, 681*f*
larval salamander, 989*f*
larynx, 993, 994*f*
La Sima de Los Huesos, 711
Lassar, Andrew, 393
late blight, 561
latency, 366–67
lateral gene transfer, 667
lateral line systems, 689, 883–84, 883*f*
lateral meristems, 731
Latham, Robert, 1210
Lathyrus odoratus, 344
Latimeria chalumnae, 691*f*
Latrodectus mactans, 671
Laurasiatheria, 706–7
Lauterbur, Paul, 869
Lavoisier, Antoine, 950
law of independent assortment, 326–27, 329–30, 329*f*
law of segregation
  chromosome theory and, 328–29
  exceptions to, 350–53
  Mendel's studies producing, 323–25
laws of thermodynamics, 119
L-dopa, 866
leaching, 761, 762, 778
leading strands, 226–27, 226*f*
leaf abscission, 785–87
leaf-cutter ants, 1202, 1202*f*
leaf cuttings, 808
leaflets, 98, 98*f*, 103, 726
leaf miners, 1238
leaf primordia, 722, 722*f*, 725–26
leaf veins, 722–23
leafy sea dragon (*Phycodurus eques*), 691*f*
Leakey, Louis, 710
learning
  brain centers for, 866
  cellular processes, 867–71
  cognitive, 1149
  conditioning, 826–28, 1148
  defined, 867
  by octopuses, 663–65
  role in homeostasis, 826–28
leaves. *See also* photosynthesis
  abscission, 785–87
  adaptations to environmental stress, 726–29
  as angiosperm organ, 718
  chlorosis, 760, 760*f*
  of conifers, 603, 603*f*
  of cycads, 599
  evolutionary importance, 591
  as feature of vascular plants, 582
  folding in sensitive plant, 737, 738*f*, 750, 751*f*
  modifications, 729, 730*f*
  shade adaptations, 758, 758*f*
  sites of photosynthesis, 155
  structure and development, 722–23, 725–26
Leber's hereditary optic neuropathy, 352*t*, 353
LEC1 gene, 809
Leder, Philip, 245, 246
Lederberg, Esther, 282–83, 375
Lederberg, Joshua, 282–83, 374–75
leeches, 666–67, 666*f*, 926, 927*f*
left-right axis, 382, 382*f*, 1077
legal rights of species, 1247
leghemoglobin, 769
legumes, 610*f*, 611, 767–69
Leishmania, 558, 558*f*
leks, 1163, 1164*f*
Lemaitre, Bruno, 1098, 1099*f*
Leman, Luke, 441
Lemna gibba, 799
Lemna species, 718, 719*f*
lemon, 610*f*
lenses
  cataracts, 897, 898*f*
  in compound eyes, 886, 886*f*
  refraction in, 887
  in single-lens eyes, 886, 887*f*
lenticels, 732
lentic habitats, 1141
Leopold, Aldo, 1259
Lepidoptera, 673*t*
leptin
  appetite regulation, 953*f*, 955, 1044
  fertility and, 1050
  links to obesity, 944, 955
Leptospira, 539*f*
Leptothorax ants, 1156
Lepus americanus, 1196, 1196*f*
Lesser Antilles, 1220, 1220*f*
lesser apes, 707
leucine, 54*f*, 81–83, 922
leucoplasts, 87
leukemia
  in gene therapy patients, 417
  potential causes, 290–91, 291*f*
  therapeutic agents from plants, 597
leukocytes
  immune functions, 1095–96, 1097
  as major component of blood, 969*f*, 970
  types, 1096*f*

INDEX

levels of organization, 5–6, 5*f*, 14
Levine, Arnold, 292
Lewis, Edward, 388
Leydig cells, 1063, 1064, 1064*f*
L-glucose, 47, 47*f*
Liang, Haiyi, 800
lice, 673*t*
lichens, 634–35
Liebig, Justus von, 1233
Liebig's law of the minimum, 1233
life, 448–57
life cycle events, sexual reproduction, 313–14
life history strategies, 1180–81
life stage, impact on animal diet, 923
lifestyle and immunity, 1111–12
life tables, 1173, 1173*t*, 1174, 1175*f*, 1176
ligand • receptor complex, 178
ligand-gated ion channels, 180, 180*f*, 838, 839*f*
ligands, 109, 178, 838
light
  effect of availability on organism distribution, 1127, 1128*f*
  effect on sexual selection, 488–89
  as energy form, 119*t*
  essential to plant growth, 757–58
  impact on primary production in aquatic ecosystems, 1233
  movement toward or away from, 1151
  photosynthesis, 2–3
  plant detection, 747–49
  sensory systems for, 885–93
  stomata response to, 785
light gaps, 1211
light-harvesting complexes, 159, 161–62, 163*f*
light microscopes, 67, 68*f*
light reactions
  location, 156*f*
  molecular features, 161–65
  overview, 154, 156, 157–61
lignins
  as feature of vascular plants, 582, 780
  functions in plant stems, 730
  fungal breakdown, 633
  in secondary cell wall, 198, 780
  in wood, 598, 730
Liguus fasciatus, 662*f*
Likens, Gene, 1240
Lilium casablanca, 800–802
limb field, 1088
limbic system, 865*t*, 866
limbs, 692, 693–94
Limenitis archippus, 1195
limestone, 654, 1237
limited mating, 1253–54
limiting factors, 757, 1233
Limosa lapponica, 916
Limulus polyphemus, 1245
lineages, 7, 497
linear relationships, 527–28
linear structure of glucose, 47*f*
lines of best fit, 1121, 1121*f*
line transect, 1169
linkage
  crossing over and, 348–50
  defined, 343*t*, 346
  discovery, 346–48
linkage groups, 346
linker domains, 61*f*
linker proteins, 148*f*, 199
Linnaeus, Carolus, 460, 514
linoleic acid, 50*f*, 51, 922
Linum usitatissimum, 730
lionfish (*Pterois volitans*), 691*f*
lipase, 935
lipid-anchored proteins, 99, 99*f*
lipid exchange proteins, 103
lipid rafts, 101–2
lipids
  absorption in digestive tract, 935, 936*f*
  as animal nutrients, 921, 921*t*
  basic types, 49–52
  synthesis at ER membrane, 103, 104*f*
  synthesis in smooth ER, 80
lipid-soluble hormones, 1033
Lipman, David, 434
lipolysis, 946, 947
lipoprotein lipase, 945
liposomes, 442, 442*f*
Lipotes vexillifer, 1244
little brown bat (*Myotis lucifugus*), 1072
Littoraria irrorata, 631
liver
  cystic fibrosis impact, 16
  endocrine functions, 1031*f*
  glucagon's effects, 1039–40
  glucose supplies to blood for exercise, 950
  glycogen storage, 80, 950
  role in digestion, 932–33, 932*f*
liver cells, protein abundance, 430, 430*f*
liver fluke, 658, 659*f*
liverworts, 577, 578*f*, 589*f*
livestock, transgenic, 412
living organisms
  importance of water, 37–38
  levels of organization, 5–6, 5*f*
  major principles, 2–5
  model organisms, 13
  unity and diversity, 6–13
lizards, 698
llama (*Lama glama*), 986, 1004
loam, 761, 762
Lobban, Peter, 400
lobe-finned fishes, 691–92
lobes, cerebral cortex, 864–65
lobsters, 677, 896, 902–3
local movement, 1151–52
loci of genes, 328, 328*f*
lock-and-key model, 32, 123
locomotion, 901, 915–16
locusts, 1120–22
Loewenstein, Werner, 202
Loewi, Otto, 846
logarithmic scales, 1173, 1220*f*
logistic equation, 1178–79
logistic population growth, 1178–79, 1179*f*
log-log plots, 1220*f*
Lohka, Manfred, 303
long-day plants, 748
long-distance transport in plants, 778–90
longhorn sculpin (*Trematomus nicolai*), 468, 468*f*
longleaf pine (*Pinus palustris*), 1123–24, 1258
long RNA molecules, 443–45, 444*f*
long-term memory, 868, 868*f*
long-term potentiation, 867
Lonicera japonica, 738*f*
loop domains, 371, 371*f*
loop of Henle, 1020–21, 1020*f*, 1022–23, 1023*f*
loose connective tissue, 817*f*
lophophores, 648, 648*f*, 660, 660*f*
Lophotrochozoa
  Annelida, 665–67
  Bryozoa and Brachiopoda, 658–59
  key characteristics, 648
  as major protostome clade, 647
  Mollusca, 659–63
  overview, 656–57
  Platyhelminthes, 657–58
  Rotifera, 658–59
Lorenz, Konrad, 1146, 1149–50
loss-of-function alleles, 335–36
lotic habitats, 1141
lotus (*Nelumbo nucifera*), 807
Lovette, Irby, 1220
Loxodonta africana, 514
Loxodonta cyclotis, 514
Loxosceles reclusa, 671
Lucifer hummingbird (*Calothorax lucifer*), 701*f*
Lucy skeleton, 710
lumen, 79, 928–29
lung cancer
  cellular changes leading to, 294–95, 294*f*
  normal tissues compared, 289*f*
  PET scan, 139*f*
  as smoking risk, 294, 295, 1005
lungfishes, 691, 691*f*, 692
lungless salamander (*Ensatina eschscholtzii*), 989*f*
lungs
  alveoli structure, 821, 993, 994*f*
  in circulatory systems, 966–68, 967*f*
  disorders, 289*f*, 294–95, 1004–5
  embryonic formation, 1085
  epinephrine's effects, 187
  fungal diseases, 631
  pleural sacs, 995
  ventilation in mammals, 995–96, 1002–3
  vertebrate variations, 992
Lupinus alba, 764
luteal phase, 1067, 1067*f*, 1068
lutein, 564
luteinizing hormone
  from anterior pituitary, 1037*t*, 1050
  effects on human female, 1067
  effects on human male, 1064–65
  synthetic androgens' effect, 1053
Lycoperdon perlatum, 623*f*
lycophylls, 591, 592*f*
lycophytes
  lack of gibberellin growth response, 745, 745*f*
  leaves, 591
  as plant phylum, 577
  secondary metabolites, 585
  spores, 593
  structures and life cycle, 581–83
Lycopodium obscurum, 581*f*
Lycosa tarantula, 671*f*
Lyell, Charles, 460, 461
Lyme disease, 421, 672
lymphatic systems
  basic features, 1100–1101
  major organs in animals, 818*t*, 1101
  parasites, 668
  return of fluids to blood, 976
lymph nodes, 1101, 1101*f*
lymphocytes
  activation, 1102, 1106–9
  autoimmune responses, 1110–11
  cell-mediated immune response, 1107–9
  genetic disorders attacking, 416
  humoral immune response, 1102–7
  immunoglobulin production, 1102–5
  in lymphatic system, 1100–1101
  main types, 1101–2, 1102*f*
  natural killer cells, 1096
Lynch, Michael, 1058
lynx, 1196, 1196*f*
Lynx canadensis, 1196, 1196*f*
Lynx rufus, 703*f*
Lyon, Mary, 353
lysine, 54*f*, 922
lysis, osmotic, 107, 108*f*
lysogenic cycle, 363, 364*f*, 366*f*
lysogeny, 366
lysosomes
  in animal cells, 71*f*

autophagy, 132–33, 133*f*
structure and function, 83–84, 94
lysozyme, 365*f*, 366
lytic cycle, 366, 366*f*
Lømo, Terje, 867

M

MacAlister, Cora, 729
MacArthur, Robert, 1190, 1219, 1220
macaws, 701*f*
MacLeod, Colin, 214–15
MacLeod, John, 1042, 1043
macroalgae, 554
macroevolution, 459, 495
macromolecules, 5*f*, 6, 42
macronuclei, 568, 571*f*
macronutrients, 757, 757*f*, 757*t*
macroparasites, 1199–1200
macrophages
as antigen-presenting cells, 1107–8
immune functions, 1094*f*, 1096, 1096*f*, 1097
phagocytosis, 84, 114
macular degeneration, 897, 898*f*
Madagascar deforestation, 1242, 1242*f*
Madagascar periwinkle (*Catharanthus roseus*), 597*f*
mad cow disease, 370*t*
madreporite, 678
MADS box genes, 397
MADS domains, 397
maggots, 672
magnesium
atomic mass, 26
atomic number, 25
as essential animal nutrient, 924*t*
as plant macronutrient, 757*t*
magnesium ions, 126, 158
magnetic fields, 884–85
magnetic resonance imaging, 869–71, 869*f*
magnetite, 538, 884–85, 1154
magnetosomes, 538–39
Magnetospirillum magnetotacticum, 538–39, 539*f*
magnification of microscopes, 67
Magnoliids, 607
Mahadevan, Lakshminarayanan, 800
Mahlberg, Paul, 612
major depressive disorder, 850
major grooves in DNA, 221, 221*f*
major histocompatibility complex (MHC) proteins, 1107–8, 1112
Mako shark (*Isurus oxyrinchus*), 688
malacostracans, 677
malaria
avian, 1251, 1251*f*
human costs, 553, 569
parasite life cycle, 1199
resistance, 478, 485, 1002
transmission, 568–72
malate, 143*f*, 171
male-assistance hypothesis, 1162
male gametophytes, 795, 802–3
maleylacetoacetic acid, 236*f*
malignant growths, 288
Maller, James, 303
malleus, 879, 879*f*, 880, 880*f*
malnutrition, 1111–12
Malpighian tubules, 669, 1017, 1018*f*
Malthus, Thomas, 460
malting, 807
maltose, 47, 933
Mammalia, 516
mammals
Australian species, 467
chemoreception, 894–96
circulatory systems, 968
developmental stages, 258–59
development during Cenozoic, 457
diversity, 704–7
ears and hearing, 879–81
egg-laying, 467
evolution, 702–3
on geological timescale, 449*f*
holoblastic cleavage, 1079–80, 1080*f*
Hox genes, 509, 509*f*
pregnancy and birth, 1068–72
primate characteristics, 707
primate evolution, 707–12
reproductive cloning, 414–15
respiratory systems, 821, 821*f*, 992–96, 994*f*, 1002–3
stem cell differentiation, 390–95
mammary cells, 414–15
mammary glands, 703
Mammuthus primigenius, 1261
Manacus manacus, 1163
manatees, 471*t*
Manchurian cranes (*Grus japonensis*), 1163*f*
Mandara Lake Oasis, 1225*f*
manganese, 757*t*, 924*t*
manganese cluster, 163
mange, 672
Mangold, Hilde, 1088
mangrove finches (*Cactospiza heliobates*), 461, 462*t*
mangroves, 732*f*
Mansfield, Peter, 869
mantle cavities, 661
mantles, 661
Manyara, Lake, 1126, 1127*f*
many-eyes hypothesis, 1157, 1158*f*
map distance, 350
MAP-kinases, 182
maple leaves, 16*f*
maple seeds, 610*f*, 611
maple syrup, 779
mapping (DNA), 406
mapping (human genome), 429
map units, 350
marathon running, 136
Marchantia polymorpha, 578*f*
Margulis, Lynn, 88
marijuana, 851
marine biomes, 1139–40
marine mammals, 988–89, 1004
Markert, Clement, 302–3
mark-recapture technique, 1169–70
Marmota flaviventris, 1252
Márquez, Luis, 633, 634
Marrella, 455*f*
Marsden, Stuart, 1212
Marshall, Barry, 940
marshes, 1142
marsupials
classification, 706, 706*f*
geographic distribution, 467, 1142, 1143–44
undeveloped young, 1068
mass, 26, 35
mass extinctions, 450, 456, 588
mass-specific BMR, 951–52, 952*f*
mastax, 658, 659*f*
mast cells, 1096, 1096*f*, 1104
master plate, 282*f*, 283
masting, 1196
Masui, Yoshio, 302–3
mate choice, 486–89, 506
mate-guarding hypothesis, 1162
maternal effect genes, 386, 387, 388*f*
maternal factors, 1086
maternal inheritance, 351–53, 351*f*
maternal myopathy, 352*t*
mating behavior
limited mating, 1253–54
nonrandom, 481, 482*t*, 492–93
sexual selection and, 486–87, 488–89
variations, 1162–65
mating systems, 1162–65
matrotrophy, 579
matter, defined, 21
maturation, in cell cycle, 302
maturation-promoting factor, 302*f*, 303
maturation zone, 733
mature mRNA, 241
maxillipeds, 645
Maximow, Alexander, 723
maximum likelihood approach, 523–24
Mayer, Adolf, 360
Mayr, Ernst, 496, 497, 511
McCarty, Maclyn, 214–15
McClintock, Barbara, 425, 426*f*
McClung, C. E., 331
McEwen, Bruce, 868
McKusick, Victor, 491
mealybugs, 452
mean fitness of the population, 483
measles, 361*f*
mechanical defenses, 1197, 1198*f*
mechanical energy, 119*t*
mechanical isolation, 498–99, 498*f*
mechanoreception
audition, 879–82
balance, 882–83
lateral line systems, 883–84
receptor types, 877–79
mechanoreceptors, 877–79
medial hinge point, 1084, 1084*f*
mediator, 269–70, 270*f*
medications, 99. *See also* drugs
medicine
biotechnology applications, 400, 410, 411–12, 416–17
importance of SNPs, 479–80
stem cell potential applications, 392–93
medium ground finch *Geospiza fortis*, 463, 463*f*
medulla oblongata, 862, 865*t*
medusae, 654*f*, 655, 655*f*
meetings, scientific, 18
megabase pairs, 421
megadiversity countries, 1254
megapascals, 775
megasporangia, 591, 594, *594*, 606*f*
megaspores
female gametophytes from, 591
heterospory and, 593, 594*f*
production in plant life cycle, 606*f*, 795
megasporophylls, 606
Megazostrodon, 456, 456*f*
meiosis
defined, 308
in flowering plants, 717, 802
in gametogenesis, 1059–61
mitosis versus, 312–13, 312*t*
overview of cellular events, 309–12
in protist reproduction, 567, 568–72*f*, 572
relation to sexual reproduction, 313–14
meiosis I
in gametogenesis, 1059, 1060*f*, 1061
nondisjunction during, 316, 317*f*
segregation in, 328–29
stages, 308*f*, 309–12, 310–11*f*
meiosis II
following fertilization in mammals, 1066
in gametogenesis, 1059, 1060*f*, 1061
stages, 308*f*, 312
Meissner corpuscles, 877, 878*f*
Mek, 182

INDEX

melanin, 174
melanocytes, 1084
melatonin
  functions, 1032
  site of production, 863, 1031*f*, 1032
melioidosis, 548
Mello, Craig, 275
membrane attack complex (MAC), 1098
membrane lipids in archaea, 535, 535*f*
membrane potential
  changes conducting neuron signals, 838–42
  creation in plant cells, 773–74
  defined, 835
  measuring, 835, 836*f*
  in photoreceptors, 889–90, 890*f*
  in retina cells, 890
membrane transport
  active transport overview, 111–13
  defined, 85
  exocytosis and endocytosis, 113–14, 114*f*, 115*f*
  illustrated, 86*f*
  major types, 105–7, 108*f*
  processes in plant cells, 773–76
  transport protein types, 108–11
memory, 866, 867–71
memory cells, 1102
Mendel, Gregor, 321, 322–27, 344
Mendelian inheritance
  chromosome theory and, 327–30
  defined, 335, 343*t*
  Mendel's laws, 322–27
  patterns, 335*t*
  pedigree analysis in humans, 330–31
  phenotype variations with, 335–39
  probability in, 339–41
  sex chromosomes and X-linked traits, 331–34
meninges, 859, 860*f*, 872
meningitis, 359, 871–72
menisci, 783, 784*f*
menopause, 1067
menstrual cycle, 1067*f*, 1068
menstruation, 1067*f*, 1068
mental retardation, 1091
Menten, Maud, 124
meristemoids, 729
meristems
  axillary, 725
  defined, 718
  development in plant embryo, 395, 396*f*, 717*f*
  secondary, 731
mermaid's purse, 689*f*
meroblastic cleavage, 1079, 1080*f*
Merostomata, 670
merozoites, 570, 572*f*
merozygotes, 264
Merychippus, 466*f*
Meselson, Matthew, 221–23
mesencephalon, 856
mesoderm
  as Bilateria characteristic, 643
  in flatworms, 657
  formation in animal embryo, 1081, 1081*f*
  induction, 1087
mesoglea, 654, 654*f*
Mesohippus, 466*f*
mesohyl, 653
Mesonychoteuthis hamiltoni, 662
mesophyll, 155, 208, 722
Mesostigma viridae, 577*f*
Mesozoic era, 456, 599
messenger RNA
  from alternative splicing, 242, 274–75, 431, 431*f*
  basic functions, 4, 131
  nucleotide recycling, 131–32, 132*f*
  polycistronic, 260
  processing in eukaryotes, 239, 241–43
  regulation in bacteria, 259
  regulation in eukaryotes, 260
  role in transcription and translation, 238–40, 244–45, 247, 248*t*, 251–52, 253–54
  in viral reproductive cycle, 363
metabolic cycles, 142
metabolic enzymes, 430*t*
metabolic pathways
  basic types, 74, 128–31
  cellular respiration, 136–38
  microarray data, 410*t*
metabolic rate
  defined, 944, 950
  food intake and, 952–55
  measuring, 950–51
  relation to hemoglobin's oxygen affinity, 999, 1000
  respiratory center activity and, 1003
  variables controlling, 951–52
metabolic water, 1012
metabolism. *See also* energy
  blood glucose concentration, 947–50
  of carbohydrates, proteins, and fats, 149
  defined, 2, 74, 118, 944
  energy balance, 950–55
  hormonal control, 1037–38
  organic molecule recycling, 131–33
  overview, 128–31
  role of cytosol, 74
  role of endoplasmic reticulum, 80
metabotropic receptors, 848, 849*f*
metacentric chromosomes, 314, 315*f*
metacercariae, 658
metagenomics, 1212–14, 1213*f*
metamorphosis
  advantages, 640, 667
  amphibian, 695
  dietary changes and, 923
  hormone regulation, 1047–48
  in insect development, 386, 674
metanephridia, 661, 1016–17, 1017*f*
metaphase
  chromosome compaction, 231, 232*f*
  mitosis, 305, 307*f*
metaphase checkpoint, 301, 301*f*
metaphase I, 310*f*, 312
metaphase plate
  meiosis, 310*f*, 312
  mitosis, 305, 307*f*
Metasequoia glyptostroboides, 601*f*, 603
metastasis
  defined, 288, 289*f*
  in lung cancer, 294*f*, 295
Metatheria, 706
Metazoa, 642
metencephalon, 856
meteorites
  impact on life-forms, 450
  mass extinctions and, 456–57, 588
  as origin of organic molecules, 440
methane
  in Earth's early atmosphere, 439
  as greenhouse gas, 1125*t*
  prokaryote consumption, 545
  reaction with oxygen, 33
Methanobacterium thermoautotrophicum, 421*t*
Methanococcus jannaschii, 100*t*
methanogens, 545
Methanopyrus, 535
Methanosarcina mazei, 537
methanotrophs, 545
methemoglobin, 1240
methicillin-resistant *Staphylococcus aureus*, 378, 541
methionine, 54*f*, 922
methylation, 294, 356
methyl-CpG-binding proteins, 273–74
methyl-directed mismatch repair, 287*t*
methyl groups
  attachment to DNA, 273–74, 294, 356
  basic features, 44*t*
7-methylguanosine cap, 243, 243*f*
methyl salicylate, 754
Mexican beaded lizard (*Heloderma horridum*), 698
Mexico, demographic transition, 1183, 1183*f*
Meyerhof, Otto, 138
Meyerowitz, Elliot, 397
Meyers, Ronald, 863–64
MHC proteins, 1107–8
micelles, 35, 35*f*, 935, 936*f*
Michaelis, Leonor, 124
Michigan, Lake, 1231, 1231*f*
microarrays, 409, 409*f*, 410*t*
microbiome, human, 548
microcephalin gene, 866
microelectrodes, 835, 836*f*
microevolution, 459, 481, 482*t*
microfibrils, 197, 198*f*, 724, 724*f*
microfilaments, 75–76, 292–93
microfilariae, 668
microglia, 833
micrographs, 66
Micronesian kingfisher, 1196, 1197*f*
micronuclei, 568, 571*f*
micronutrients, 757, 757*f*, 757*t*
microorganisms. *See also* bacteria
  anaerobic respiration, 150, 150*f*
  bioremediation by, 410–11, 550, 1232, 1260
  cell structure, 69
  cellulose digesting, 704, 931
microparasites, 1199–1200
micropyle, 592, 795–96
microRNA 165/6, 734
microRNAs (miRNAs), 275, 276*f*, 726, 752
microscopic examination of chromosomes, 298–99
microscopy, 66–69
microsporangia, 591, *594*, 606*f*
microspores
  in flowering plant life cycle, 606*f*
  heterospory and, 593, 594*f*
  male gametophytes from, 591, 795
  mitosis in, 802
  size, 795
microsporidia, 619, 624*t*, 625
microsporophylls, 605
microtubules
  in cilia and flagella, 77–78
  in mitotic spindle, 304, 305, 305*f*, 306–7*f*
  in plant cell walls, 724, 724*f*
  structure and function, 74, 75*t*
microvilli
  of intestines, 931, 932*f*
  of proximal tubules, 1021, 1022*f*
  of taste cells, 896, 897*f*
Micrurus fulvius, 1195, 1195*f*
midbrain, 856, 857*f*, 862–63, 865*t*
middle ear, 879, 879*f*
middle lamellae, 198, 198*t*, 203, 203*f*
Miescher, Friedrich, 217
mifepristone, 1074
migration
  of cells, 205
  challenges of, 1150–51
  effect on alleles in populations, 492
  electromagnetic sensing and, 884–85
  by flying, 916
  navigation for, 1153–54
  role in microevolution, 481, 482*t*

milk
  as characteristic of mammals, 703
  genetically engineered products, 412
  immunoglobulins in, 1104
  lactose intolerance, 933–35
Millepora intricata, 1123
Miller, Stanley, 439–40
millipedes, 672, 672*f*
mimicry, 1195, 1195*f*
Mimosa pudica, 737, 738*f*, 750, 751*f*
mineral elements, 27, 27*t*, 903
mineralization, 764
mineralocorticoid receptors, 1052
mineralocorticoids, 1031*f*, 1052
minerals
  absorption in digestive tract, 928, 936
  as essential animal nutrients, 922, 924*t*
  hormone regulation, 1044–47
mining, restoration after, 1259–60
minor grooves, in DNA, 221, 221*f*
Mirabilis jalapa, 337, 351
miracidia, 658
Mirounga angustirostris, 490
Mirounga leonina, 1165
misaligned crossovers, 427, 428*f*
missense mutations, 280, 281*f*, 290, 291*f*
mist, 784, 784*f*
mistletoe (*Viscum album*), 1199
mist nets, 1169, 1169*f*
Mitchell, Peter, 145
mites, 672
mitochondria
  in animal cells, 71*f*
  electron transport chain in, 142, 144*f*
  evolution, 88, 452, 556
  formation, 87–88
  homologous genes, 161, 162*f*
  oxidative phosphorylation in, 138
  in plant cells, 72*f*
  protein sorting to, 92, 92*f*
  structure and function, 86–87, 86*f*, 87*f*, 94–95
mitochondrial Eve, 712
mitochondrial genome
  human ancestry data based on, 712
  importance, 350
  location in cells, 351*f*
  maternal inheritance, 352–53
  nuclear genome compared, 87–88, 424
  origins, 452
  replication, 88*f*
mitochondrial matrix, 86, 87*f*
mitochondrial pathway (apoptosis), 190
mitogen-activated protein kinases, 182
mitosis
  in bryophyte reproduction, 579*f*
  meiosis versus, 312–13, 312*t*
  occurrence in eukaryotic cell cycle, 300
  in protist reproduction, 567, 568, 568–72*f*
  stages, 305, 306–7*f*
mitotic cell division, 303–8
mitotic cyclin, 301
mitotic spindle, 304, 305*f*
Mittermeier, Russell, 1254
mixed-function oxidases, 1198
mixotrophs, 564
M line, 905*f*, 906
Mnemiopsis leidyi, 656
Mnium, 578*f*
moas, 525
model organisms, 13, 381–82
moderately repetitive sequences, 425
modular structures, shoot systems, 725, 725*f*
molarity, 35
molars, 704*f*, 924, 925
molar solutions, 35
molds, 237–38
molecular biology, 14
molecular clocks, 527–28
molecular evolution, 459
molecular formulas, 28
molecular homologies, 470–75
molecular mass, 35
molecular paleontology, 524–25
molecular pharming, 412, 412*f*
molecular systematics, 520
molecules
  defined, 21
  as level of biological organization, 5*f*, 6
  reactive, 33
  shape changes, 32–33
  types of bonds within, 28–31
moles, 684, 684*f*
Mollusca, 648*t*, 660–63, 681*t*
mollusks
  classification, 661–63
  as early body plan group, 454
  feeding behaviors, 924, 925*f*
  key characteristics, 648*t*, 660–61, 681*t*
  mating behavior, 1162
  nervous systems, 855*f*, 856
molting
  evolutionary history, 667
  exoskeletons, 647–48, 647*f*, 667, 903
  hormonal control, 1048–49
molybdenum, 757*t*
Monactinus, 555*f*
monarch butterfly (*Danaus plexippus*), 1149*f*, 1153, 1194
monetary value of ecosystems, 1246–47, 1246*t*
monkeys, 707, 1148
Monoclea, 589*f*
monoclonal antibodies, 1111
monocots
  defined, 607
  distinguishing features, 719*t*
  examples, 719
  fibrous root system, 723
  flower structure, 607, 608*f*
  flower variations, 799
  leaf attachment points, 726
  seed structure, 805, 806
monocular vision, 892, 892*f*
monocytes, 1096, 1096*f*
Monod, Jacques, 261, 264
monoecious plants, 798, 798*f*
monogamy, 1162–63, 1163*f*
monogeneans, 657*t*, 658
monohybrid crosses, 323–25
monohybrids, 324
monomers
  in cellulose, 49
  defined, 45
  in nucleic acids, 60
  in proteins, 53
  of simple sugars, 46–47
monomorphic genes, 479
mononucleosis, 361*f*
monophagous parasites, 1199
monophyletic groups, 517–19, 519*f*
Monopterous gene, 396*t*
monosaccharides
  absorption in digestive tract, 933, 934*f*
  basic features, 46–47, 47*f*
Monosiga brevicollis, 563
monosodium glutamate, 897
monosomic organisms, 316
monotremes
  classification, 704
  echidna, 467
  geographic distribution, 1142, 1143
  nursing in, 1071
Monotropa uniflora, 757, 758*f*
monounsaturated fatty acids, 50
Monstera deliciosa, 758*f*
Moore, Peter, 253
moraines, 1216
Morgan, Thomas Hunt, 333–34, 346, 348, 350
Mormon cricket (*Anabrus simplex*), 1164–65
morphine, 851
morphogenetic field, 1088, 1089
morphogens
  concentration effects, 1086
  distribution in egg, 386
  major functions, 383–84, 1086
  from organizer cells, 1088, 1089–91
morphological analysis, 519–20
morphology
  animal classification by, 643–45
  defined, 380
  distinguishing species by, 496
  impact on species coexistence, 1193–94
  molecular classification versus, 647–50
mortality factors, density-dependent, 1179–80
morula, 1080, 1080*f*
Morus bassanus, 1155, 1155*f*
mosaics, 283, 353
mosquitos, 569, 673*t*
mosses
  basic features, 578*f*
  gametophytes and sporophytes, 314, 794
  as plant phylum, 577
mother-of-pearl, 661
moths
  antennae, 822*f*, 893*f*, 894
  as indicator species, 1258
  key characteristics, 673*t*
  orientation, 1151
motility, 541–42, 554–55
motion sickness, 882–83
motor end plates, 911, 911*f*
motor neurons, 833–34, 834*f*, 860
motor proteins, 76–78, 76*f*, 430*t*
mountain avens (*Dryas drummondii*), 1216, 1217*f*
mountain hemlock (*Tsuga mertensiana*), 1216
mountains, 1130–31, 1138, 1209
Mount St. Helens, 1215, 1215*f*, 1216–17
Mourer-Chauviré, Cécile, 524
mouse (*Mus musculus*)
  β-globin genes, 433–34, 433*f*
  basal metabolic rate, 951
  feeding characters, 1193*t*
  genome size, 424*t*
  genomic imprinting, 355–56, 355*f*
  Hox genes, 390, 390*f*
  as model organism, 13, 381, 411–12
  oxygen-hemoglobin dissociation curve, 1000, 1001*f*
mouth, 668, 929
mouthparts of insects, 674, 674*f*
movement
  as animal characteristic, 639
  bacteria, 541–42
  by plants, 737, 738*f*, 750, 751*f*
  skeletal muscle adaptations for, 911–14
  structures producing, 76–78, 93
movement corridors, 1256–57
M phase, 299, 300
mucigel, 733
mucilage
  of carnivorous plants, 769–70
  cyanobacteria colonies, 536, 537*f*
  functions for cells, 539–40, 540*f*
  as protist defense, 565
mucosa, alimentary canal, 928

mucus
  of cystic fibrosis patients, 16
  as defense against pathogens, 1095
  in mammalian respiratory system, 993, 994*f*
mud daubers, 1056*f*
mudpuppy (*Necturus maculosus*), 989*f*
mules, 499, 500*f*
Müller, Paul, 1231
Müllerian mimicry, 1195
Mullis, Kary, 405
multicellular organisms
  advantages, 193
  animals as, 639–40
  cell junctions, 198–204
  cell walls and cell matrices, 194–98
  defined, 193
  functions of gene regulation, 258–59
  on geological timescale, 449*f*, 453
  reproductive cloning, 414–15
  somatic cell mutations, 283–84
  tissues, 204–8
multimeric proteins, 56
multiple alleles, 337–38
multiple fruits, 610*f*, 611
multiple sclerosis, 851, 1111
multipotent stem cells, 392
multiregional hypothesis, 712
Murad, Ferid, 979
Murie, Adolph, 1174
muscle cells, 393–95, 430, 430*f*
muscle fibers, 905, 905*f*
muscles. *See also* skeletal muscle
muscle tissue
  in annelids, 902, 902*f*
  general features in animals, 205, 205*f*
  glycolysis in anaerobic conditions, 151
  main types in animals, 814, 815*f*, 904
muscular dystrophy, 917, 918
muscular-skeletal systems, 818*t*
mushrooms, 620–21, 623–24
musicians, 870–71
Mus musculus. *See* mouse (*Mus musculus*)
mussel (*Mytilus edulis*), 1127*f*
Mustela nigripes, 1168, 1168*f*, 1178, 1178*f*
mutagens, 284–86
mutant alleles, 335
mutations. *See also* cancer
  in chemical selection of RNA, 442–43, 443*f*
  of chromosome structures, 315, 315*f*
  defined, 236
  diseases caused by, 196, 336–37, 352–53, 352*t*
  DNA repair systems, 286–88
  early studies, 236–38
  elimination from populations, 1058
  in flowering plants, 397
  of genes, 280–86, 294, 295, 336–37, 380, 389, 482
  in hemoglobin, 1001–2
  of HIV, 367
  inherited susceptibility, 279
  limb development, 693–94
  with neutral effects, 491, 523, 527
  from radiation exposure, 157
  role in microevolution, 481, 482*t*
  selfish behavior and, 1158
  vertical descent with, 7
MUTE gene, 729
mutualism
  bacteria with eukaryotes, 547–48
  defined, 547, 1187
  between fungi and other species, 631–35
  types, 1202–3
myasthenia gravis, 917–18, 1111
mycelia
  basic features, 620, 620*f*
  growth, 621–22, 621*f*, 622*f*
  reproduction, 623
myc gene, 290
Mycobacterium tuberculosis, 421*t*
mycoheterotrophy, 766
Mycoplasma pneumoniae, 549, 550*f*
mycorrhizae, 631–32, 631*f*, 766
myelencephalon, 856
myelin, 859
myelin sheath, 833, 833*f*, 842
Myers, Norman, 1254
Myers, Ransom, 1187
MYH16 gene, 907–8
myocardial infarction, 983
myoclonic epilepsy, 352*t*
myocytes, 971
myofibrils, 905–6, 905*f*
myogenic genes, 393–95
myogenic hearts, 971–72
myoglobin
  functions, 429, 912
  high altitude adaptations, 1004
  in oxidative fibers, 912
myometrium, 1066
myosin
  ancient mutation, 907–8
  discovery, 907
  role in mitosis, 307
  role in muscle contraction, 906–7, 907*f*, 908–9
  in skeletal muscle structure, 905*f*, 906
  structure, 76*f*
Myotis lucifugus, 1072
myriapods, 672
Myrmecophaga tridactyla, 467
myrtle (*Myrica*), 767
myxobacteria, 537
Myxococcus xanthus, 536*t*, 542
myxomatosis, 1187

N

Na^+/K^+-ATPase, 123
Na^+/K^+-ATPase pump, 112–13, 113*f*, 836–37
nacre, 661
NAD^+
  as citric acid cycle regulator, 141, 142
  consumption in anaerobic conditions, 150–51
  formation in cellular respiration, 137, 144*f*
  reduction, 129, 130*f*
NADH
  as citric acid cycle regulator, 142
  in electron transport chain, 143–45
  formation in cellular respiration, 137–38, 140, 141, 145
  redox reaction, 129, 130*f*
  removal in anaerobic conditions, 150–51
  use in organic molecule synthesis, 145–46
NADH dehydrogenase, 144*f*, 145, 150
$NADP^+$, 160
$NADP^+$ reductase, 159
NADPH, 156, 159, 160
Naeem, Shahid, 1247
Nägeli, Karl, 213, 327
naked mole rat (*Heterocephalus glaber*), 1161, 1161*f*
Nanoarchaeota, 535
Naso rectifrons, 438*f*
Naso tonganus, 538
Nathans, Daniel, 401
natriuresis, 1047
natural killer cells, 1096, 1096*f*, 1109
natural selection
  artificial selection versus, 469
  Darwin's hypothesis, 462*f*, 463
  defined, 7
  differing patterns, 483–86
  evidence, 466, 467
  modern studies, 463–64
  reproductive success and, 482–83
  role in microevolution, 481, 482*t*
  sexual selection, 486–89
nature reserves, 1255–57, 1256*f*
nauplius, 676, 676*f*
nautiluses, 662, 663
navigation, 1153–54, 1153*f*
N-cadherin, 199
Neanderthals, 711–12
Necator americanus, 668
neck length, 645, 646*f*
necrotic spots in plants, 753, 753*f*
nectar, 609, 609*f*
nectar-feeding beaks, 701*f*
Nectarinia jugularis, 1212
Nectarinia reichenowi, 1155, 1155*f*
Necturus maculosus, 989*f*
Needleman, Saul, 434
negative control
  defined, 260, 261*f*
  lac operon, 261–64
  trp operon, 267, 268*f*
  tumor suppression by, 293–94
negative feedback loops, 825, 1065
negative frequency-dependent selection, 485–86
negative pressure filling, 992, 995
negative reinforcement, 1148
Neisseria gonorrhoeae, 536–37
nekton, 1140
Neljubov, Dimitry, 745
nematocysts, 655, 655*f*
Nematoda, 648*t*, 681*t*
nematodes
  key characteristics, 648*t*, 667–68, 681*t*
  as model organisms, 13
  predation by fungi, 630, 630*f*
  as pseudocoelomates, 645*f*
  relation to arthropods, 649–50
Neoceratodus forsteri, 691*f*
Neotropical regions, 1144
Nephila clavipes, 1163*f*
nephridiopores, 1016, 1017*f*
nephrons, 1019–21
nephrostomes, 1016, 1017*f*
Nernst, Walther, 838
Nernst equation, 838
nerve cells. *See also* neurons
  morphology, 71*f*
  neuron types, 833–34
  proteomes, 72–73
  specialization, 832–33
nerve cord, 679
nerve nets, 654, 855–56
nerve rings, 855*f*, 856
nerves, defined, 831, 858
nervous systems. *See also* sensory systems
  alkaloid effects, 611
  in annelids, 666
  cellular components, 831–34
  digestive system regulation, 936–37
  disorders, 850–51, 866, 871–73, 871*t*
  electrical properties, 834–38
  endocrine system links, 1035–37
  evolution and development, 854–58
  in flatworms, 657
  learning and memory, 867–71
  major organs in animals, 818*t*
  in mollusks, 661, 663
  role in homeostasis, 947–50
  in rotifers, 659
  signal conduction, 838–42
  structure and functions in vertebrates, 858–66

synapses, 842–49
ventilation controls, 1002–3
nervous tissue
general features in animals, 205, 205*f*, 814–15
prion diseases, 368–70
viruses infecting, 361*f*
net primary production, 1232
net reproductive rate, 1176
neural crest, 685, 1084, 1084*f*
neural pathways, 876–77
neural plate, 1084, 1084*f*
neural tube, 856, 1084, 1084*f*, 1091
neurofibrillary tangles, 872
neurogenesis, 868–69
neurogenic hearts, 971
neurogenic muscle weakness, 352*t*
neurohormones, 1035–36, 1036*f*
neuromodulators, 846
neuromuscular junctions, 846, 911, 911*f*
neurons
basic features, 832, 833*f*
defined, 831
electrical properties, 834–38
functional groups in CNS, 858
gastrointestinal tract, 929, 936–37
morphology, 71*f*, 72
production in adult brain, 868–69
as sensory receptors, 876*f*
signal conduction, 838–42
surface area, 822*f*
synapses, 842–49
variations in animals, 814–15, 815*f*
viral infections, 361*f*
neuropeptides, 845*t*, 846
neuroscience, 831
Neurospora crassa, 237–38
neurotransmitters
classification, 845–46
functions, 842, 844*f*
in hair cells, 878–79
movement during synaptic transmission, 843
paracrine signaling, 176, 828
neurulation, 1083–85
neutral theory of evolution, 527
neutral variation
molecular clocks based on, 527
prevalence in populations, 491, 491*f*, 523
neutrons, 22, 26
neutrophils, 1095–96, 1096*f*, 1097
New Guinea, 1220, 1221*f*
newts, 306–7*f*, 1088, 1089*f*
NF1 gene, 292*t*
niacin, 923*t*
niches, competition and, 1189–93
nickel, 757*t*
Nicolson, Garth, 98, 99*t*
nicotinamide adenine dinucleotide. *See* NAD^+
nicotinamide adenine dinucleotide phosphate. *See* NADPH
Nicrophorus defodiens, 1162
Nierhaus, Knud, 250
Nieuwkoop, Pieter, 1086–87
Nigeria, age structure, 1184*f*
night blindness, 923*t*
night vision, 888
Nirenberg, Marshall, 245, 246
nitrate reductase, 150
nitric oxide
involvement in male erection, 1064
as neurotransmitter, 845*t*, 846
as plant hormone, 738
release during inflammation, 1097*f*
vasodilatory effects, 979
nitrification, 1240
Nitrobacter, 1240
Nitrococcus, 1240
nitrogen
in animals and plants, 27, 27*t*
atomic structure, 24–25, 24*f*
in inorganic fertilizers, 763
from lichens, 635
plant acquisition from animals, 769–70
as plant macronutrient, 757*t*, 1233
nitrogenase, 763, 764*f*
nitrogen cycle, 1240–41, 1240*f*
nitrogen fixation
biological and industrial mechanisms, 763–64
defined, 763
in nitrogen cycle, 1240
plant-bacterial symbioses, 767–69
process requirements, 545
role in succession, 1216
nitrogen-limitation hypothesis, 1204
nitrogen mustards, 284, 285*t*
nitrogenous waste elimination, 1010
nitroglycerin, 979
Nitrosomonas, 536–37, 1240
nitrous acid, 284, 285*f*
nitrous oxide
from fossil fuel burning, 1240
as greenhouse gas, 1125*t*
as plant defense signal, 752, 753
NK-like gene, 390, 509*f*
N-linked glycosylation, 104
nociceptors, 877, 884
nocturnal animals, 888, 893
nodes
of phylogenetic trees, 517
in shoot systems, 725, 725*f*
nodes of Ranvier, 833, 833*f*, 842
Nod factors, 768, 768*f*
nodules, legume root, 767–69, 768*f*
nodulins, 769
noggin gene, 1091
noise-induced hearing loss, 898
Nomarski microscopy, 68*f*
nonassociative learning, 1148
noncoding sequences, 281, 425
noncompetitive inhibitors, 125, 125*f*
noncontroversial outgroups, 522–23
noncyclic electron flow, 159, 160
non-Darwinian evolution, 491
nondisjunction, 316, 317, 317*f*, 608
nonpolar covalent bonds
in amino acids, 53, 54*f*
with carbon atoms, 30, 43, 44*f*
defined, 30
nonpolar molecules, 30, 34–35
nonrandom mating, 481, 482*t*, 492–93
nonrecombinants, 348
nonsense mutations, 280–81
nonshivering thermogenesis, 959
nonspecific immunity, 1094
nonvascular plants, 581
nonwebbed feet, 508–9, 508*f*
norepinephrine
heart rate effects, 982
molecular structure, 1033*f*
site of production, 1031*f*, 1032
Normalized Difference Vegetation Index, 1233, 1234*f*
norm of reaction, 338, 339*f*
North American apple maggot fly (*Rhagoletis pomenella*), 506
North American beaver (*Castor canadensis*), 1173–74
North American common murre (*Uria aalge*), 1252
Northern elephant seal (*Mirounga angustirostris*), 490
Northern leopard frog (*Rana pipiens*), 302, 496, 496*f*
Northern spotted owl (*Strix occidentalis*), 1258, 1258*f*
Norwalk virus, 361*f*
nose (human), 894*f*
Nosema ceranae, 625, 625*f*
notochord
distinguishing feature of Chordata, 516, 679
formation, 1082*f*, 1083
structure and function, 679
Nottebohm, Fernando, 868
Nowell, Peter, 290
N-terminal protein-binding domains, 61*f*
N-terminus, 53, 55*f*, 245
nuclear envelopes
in animal cells, 71*f*
during meiosis, 309, 310*f*, 311*f*
during mitosis, 305
in plant cells, 72*f*
structure and function, 78, 79*f*
nuclear genome
bacterial and archaeal contributions, 452
mitochondrial and chloroplast genomes compared, 87–88
structural variety, 423–25
Nuclearia, 563, 618, 619*f*
nuclear lamina, 78, 231
nuclear matrix, 78–79, 231
nuclear pores
in animal cells, 71*f*
in plant cells, 72*f*
structure and function, 78, 79*f*
nuclei (atomic), 22, 23
nuclei (cell)
absent from erythrocytes, 970
in animal cells, 71*f*
of ciliates, 568
defined, 78
discoveries, 66
eukaryotic cells distinguished by, 9
intermediate filaments, 75
mitosis, 305, 306–7*f*
in plant cells, 72*f*
structures and functions, 78–79, 79*f*, 93, 94*f*
nuclei (central nervous system), 858, 1035
nucleic acids
as animal nutrients, 921, 921*t*
basic types, 60–62
molecular structure, 217–21
in viruses, 360, 362*f*, 1095
nucleoid regions, 69, 370, 370*f*
nucleoli
in animal cells, 71*f*
functions, 79, 79*f*
in plant cells, 72*f*
rRNA synthesis, 250
nucleosome-free regions, 272, 272*f*
nucleosomes
changes in position, 271, 271*f*, 272–73
structure, 230, 230*f*
nucleotide excision repair, 287–88, 287*f*
nucleotides
in bacterial chromosomes, 370, 371*f*
components, 60
defined, 217
in DNA, 61–62, 217–18, 218*f*
molecular structure, 217–18
origins on Earth, 439–40
recycling, 131–32
spontaneous mutations, 284
nucleotide sequence databases, 433*t*
nudibranchs, 662*f*, 989*f*
nuptial gifts, 1164, 1164*f*
Nüsslein-Volhard, Christiane, 386, 1098
nutation, 737
nuthatches, 1193*t*

- nutrients
  - animals' major requirements, 920, 921–22, 921*t*, 923*t*, 924*t*
  - defined, 756, 920
  - impact on primary production in ecosystems, 1233
  - use and storage, 944–47
- nutrition
  - defined, 920
  - of fungi, 619
  - plants' needs, 756–60
  - plants' sources, 154, 766–70
  - prokaryotes' sources, 544
  - of protists, 563–64
- nutrition (animal)
  - digestion mechanisms, 933–36
  - digestion principles, 927–28
  - digestion regulation, 936–39
  - digestive systems, 928–33
  - as energy source, 921–22
  - impact on immune system, 1111–12
  - links to reproduction, 1050, 1072
  - modes, 639, 922–26, 927*f*
  - variations, 920–21
  - water and salt balance mechanisms, 1011–12, 1127
- *Nyssa aquatica*, 1126

O

- oak trees, 1180, 1181*f*, 1201
- *Obelia*, 655*f*
- obesity
  - Coleman's leptin studies, 953–55
  - diabetes and, 1042
  - genetic influences, 235
  - leptin deficiency and, 944
  - public health impact, 960–61
- obligate aerobes, 544
- obligate anaerobes, 544
- obligatory exchanges, 1009–112
- obligatory mutualism, 1202
- O blood type, 338
- observational learning, by octopuses, 663–65
- observations, in hypothesis testing, 1121
- *OCA2* gene, 299
- occipital lobe, 864*f*, 865, 869
- occludin, 201
- ocean currents, 1131, 1131*f*
- oceans
  - biomes, 1139–40
  - effects on climate, 1131
  - natural benefits, 1246–47
  - primary production, 1233
- ocean waves, 1126
- Ocellaris clownfish, 9, 11*f*, 12–13
- ocelli, 655, 657
- Ochoa, Severo, 245
- ocotillo (*Fouquieria splendens*), 785, 786*f*
- octamers, 230
- octet rule, 28
- octopuses
  - key characteristics, 662–63, 662*f*
  - learned behavior, 663–65
  - nervous system, 832
- *Octopus vulgaris*, 663
- *Odobenus rosmarus*, 703*f*
- Odonata, 673*t*
- odors, perception, 893–96
- Odum, Howard, 1231
- Ohno, Susumu, 353
- oils, 34–35, 50–51
- Okazaki, Reiji, 227
- Okazaki, Tsuneko, 227
- Okazaki fragments, 227
- *Olavius algarvensis*, 547
- oleic acid, 50
- olfactory bulbs, 866, 894
- oligodendrocytes, 833
- oligotrophic lakes, 1141
- O-linked glycosylation, 105
- omasum, 931, 931*f*
- ommatidia, 669, 886, 886*f*
- omnivores, 921, 925, 926*f*
- onabotulinumtoxinA, 843
- oncogenes
  - defined, 288
  - formation, 290–91, 290*t*, 1108
  - mechanisms of action, 288–90
  - viral, 291–92
- one-gene/one-enzyme hypothesis, 238
- one-way valves in veins, 976, 977*f*
- onion seedlings, 807, 808*f*
- oocytes
  - cell division studies, 302–3
  - development, 1066–67, 1066*f*
  - positional information, 384, 386
- oogenesis, 1059–61, 1060*f*
- oogonia, 1059, 1060*f*
- Oparin, Alexander, 439, 441
- open circulatory systems
  - of arthropods, 669, 966
  - basic features, 965–66, 965*f*
  - in mollusks, 661, 966
- open complexes, 240, 240*f*
- open conformations, 271
- open ocean biomes, 1140
- operant conditioning, 1148, 1149*f*
- operator, 261, 281
- operculum, 690, 990, 990*f*, 991
- operons, 260–67, 262*f*
- *Ophiostoma*, 627
- Ophiuroidea, 678, 678*t*
- *Ophrys apifera*, 1203, 1203*f*
- opiate peptides, 846
- Opisthokonta, 515, 556, 563, 563*f*
- opossum, 706
- opportunistic feeders, 923
- opportunistic infections, 367
- opposable thumb, 707
- opsins, 888, 892–93
- opsonization, 1106
- optic nerve, 886, 887, 887*f*
- optimal foraging, 1154
- optimality theory, 1154–55
- *Opuntia stricta*, 1205, 1205*f*
- oral contraceptives, 1073, 1073*f*
- orangutan (*Pongo pygmaeus*)
  - chromosomes, 475, 475*f*
  - classification, 706, 706*f*
- orbitals, 24–25, 157–58
- orchids
  - commensalism, 1203, 1203*f*
  - structures, 609, 610*f*
- orders, in taxonomy, 516
- Ordovician period, 454, 455*f*
- organelles
  - defined, 70
  - endomembrane system, 79–86
  - genomes, 350–53
  - nuclei, 78–79
  - protein sorting to, 89–92
  - recycling, 132–33
  - semiautonomous, 86–89, 94–95
- organic chemistry, 20, 42
- organic farming, 762
- organic fertilizers, 762
- organic matter in soils, 760, 761*f*
- organic molecules
  - in bone, 903
  - carbohydrates, 46–49
  - chemical bonds, 43–45
  - defined, 42
  - early studies, 42–43
  - formation, 45–46
  - lipids, 49–52
  - nucleic acids, 60–62
  - origins on Earth, 439–40
  - prokaryote production and consumption, 545–47
  - proteins, 52–58
  - recycling, 131–33
- organic nutrients, 920, 921–22
- organic polymers, 440–42
- organismal ecology, 1118, 1118*f*
- organismic model of communities, 1208
- organisms, 2, 5*f*, 6
- organization, levels of, 5–6, 5*f*
- organizers, 1088–89
- organizing center, 396
- organ of Corti, 880, 881*f*
- organogenesis, 1085–86
- organs
  - characteristic of vertebrates, 686
  - defined, 6, 204
  - embryonic formation, 1085–86
  - gene control of development, 819
  - as level of biological organization, 5*f*, 6
  - of modern birds, 701
  - plant reproduction from, 808
  - systems in animals, 817, 818*t*
  - systems in flowering plants, 718, 718*f*
  - tissues, 816, 818*f*
- organ transplants, 1112
- Orgel, Leslie, 441
- orientation, 1153, 1153*f*
- *On the Origin of Species* (Darwin), 461
- origins of replication
  - bacterial, 370, 371, 371*f*
  - defined, 224
  - multiple copies in eukaryotes, 425
- ornithine, 237–38
- ornithischians, 699
- *Ornithomimus*, 588, 588*f*
- *Ornithorhynchus anatinus*, 704, 706*f*
- orthologs, 434, 472, 473*f*. *See also* homologous genes
- Orthoptera, 673*t*
- *Oryza sativa*, 424*t*, 615
- osmium tetroxide, 82*f*, 83, 100
- osmoconformers, 1015
- osmolarity, 1011, 1022
- osmoregulators, 1015, 1019*t*
- osmosis
  - Agre's findings, 109, 110*f*
  - cell expansion and, 723–24, 724*f*
  - in hyphae, 621
  - in photosynthesis, 773
  - process overview, 106–7, 108*f*
  - water absorption via, 928, 931
- osmotic adjustment in plant cells, 776
- osmotic lysis, 107, 108*f*
- osmotrophs, 563, 619
- ossicles, 879, 879*f*, 880, 880*f*
- osteichthyans, 690–92
- osteoblasts, 903
- osteoclasts, 903
- osteocytes, 903
- osteomalacia, 917
- osteoporosis, 917, 917*f*
- ostia, 669
- ostracods, 676
- ostriches, 997–98
- otoliths, 882, 883*f*
- Ou (*Psittirostra psittacea*), 1251*f*
- outer bark, 730–31
- outer ear, 879, 879*f*

outer segment of photoreceptors, 887–88, 888*f*
outgroups, 521
Out of Africa hypothesis, 712
ova, 1057. *See also* egg cells
oval window, 879, 879*f*, 880, 880*f*
ovarian cycle, 1066–67, 1066*f*
ovaries
  endocrine functions, 1031*f*
  of flowering plants, 322, 323*f*, 605, 795
  gametogenesis, 1066–67
  structure, 1065*f*, 1066, 1066*f*
overeating, 940
overfishing, 1172, 1187
overgrazing, 1172, 1172*f*
oviducts, 1065*f*
oviparous species, 689, 1062
Ovis dalli, 1174, 1175*f*
ovoviviparous species, 689, 1062
ovulation, 1060, 1066, 1072
ovules
  in conifers, 600
  egg cell formation, 322
  evolution, 593–94
  in flowering plants, 719, 795
  general structure, 592, 593*f*
  structure in pea plants, 323*f*
Owen, Richard, 699
owls
  binocular vision, 892, 892*f*
  sound location, 881
  as umbrella species, 1258
oxalate oxidase, 1201
oxaloacetate
  in C_4 photosynthesis, 759
  in citric acid cycle, 141, 142, 143*f*
  discovery, 170
  role in limiting photorespiration, 170, 171, 171*f*
oxidation, 129, 1198
oxidative fibers, 912, 913–14
oxidative phosphorylation, 138, 142–47, 144*f*
oxygen. *See also* circulatory systems; respiratory systems
  in animals and plants, 27, 27*t*
  atmospheric changes over time, 448–50, 454
  blood transport mechanisms, 999–1002
  diffusion out of blood, 987*f*
  double bonds, 28, 29*f*
  emission by leaf stomata, 155
  extracting from water versus air, 988
  formation by photosynthesis, 154–56
  homeostatic role, 823*t*
  measuring animals' consumption, 950–51, 951*f*
  nitrogenase binding, 763
  partial pressure, 987
  as plant macronutrient, 757*t*, 759
  production by rain forests, 154
  prokaryote responses, 544
  reaction with methane, 33
oxygen delivery, regulation of, 259
oxygen-hemoglobin dissociation curve, 1000*f*, 1001*f*
oxygen therapy, 1005
oxyhemoglobin, 999
oxytocin
  functions, 1037
  role in lactation, 1071
  role in parturition, 1070, 1071*f*
  site of production, 1031*f*
ozone layer, 157, 454

P

p16 gene, 292*t*
p53 gene, 292–93, 292*t*, 293*f*, 471–72
P680, 159, 162
P700, 159–60
Pääbo, Svante, 524, 709
pacemaker, 972*f*, 973*f*
Pachygrapsus crassipes, 1218, 1218*f*
Pacific yew (*Taxus brevifolia*), 597
Pacinian corpuscles, 877, 878*f*
paedomorphosis, 510–11, 511*f*, 696
painkillers, 99, 118
pain receptors, 877, 884
pair-rule genes, 387, 388*f*
Palade, George, 81–83, 199
Paland, Susanne, 1058
paleontologists, 446
Paleotapirus, 1143, 1143*f*
palindromic sequences, 402
palisade parenchyma, 722, 722*f*
palmate venation, 726–27, 726*f*
palm nuts, 1258
palms, 1142
pampas biomes, 1136
Pampas region, 1254, 1256*f*
Panamanian golden frog (*Atelopus zeteki*), 1117
Panamic porkfish (*Anisotremus taeniatus*), 500, 501*f*
pancreas
  cystic fibrosis impact, 16
  endocrine functions, 1031*f*, 1038–39, 1042
  insulin secretion, 949
  Palade's studies, 81–83
  role in digestion, 932–33, 932*f*, 935, 937–39
Pancrustacea, 670
pandas, 1197
Pangaea, 456
Panthera tigris, 1261
panthers, 493, 1181, 1258, 1258*f*
panting, 958–59
pantothenic acid, 923*t*
Papilio xuthus, 889
papillae, 896, 897*f*
papillomavirus, 361*f*
parabiosis, 953, 954*f*
parabronchi, 996*f*, 997, 999
Parabuthus capensis, 671*f*
paracrine signaling, 176, 176*f*, 828
paralogous genes, 819
paralogs, 427, 473
paramecia
  animals versus, 647, 647*f*
  cilia, 77*f*
  competition and coexistence, 1189, 1190*f*
  contractile vacuoles, 108*f*
  horizontal gene transfer, 474, 474*f*
  reproduction, 571*f*
Paramecium, 77*f*
Paramecium aurelia, 1189, 1190*f*
Paramecium bursaria, 1189, 1190*f*
Paramecium caudatum, 1189, 1190*f*
Paranthropus, 710
paraphyletic groups, 519, 519*f*, 534
parapodia, 665, 666, 867
parasites
  annelid, 666–67
  defined, 547, 1198
  flatworm, 658
  fungi, 625
  lamprey, 687
  as major pathogen type, 1095
  measuring effects, 1200–1201
  plants as, 770, 1199, 1200
  population density factors, 1180
  protist, 556, 557, 563, 568–72
  roundworm, 668
  ticks and mites, 672
  variations, 1198–1200
parasitoids, 1194
parasympathetic division, 861*f*, 862
parathyroid glands, 1031*f*, 1045, 1046*f*
parathyroid hormone
  impact on bone density, 917
  major functions, 1031*f*
  vitamin D regulation, 1045, 1046*f*
Parazoa, 642, 642*f*, 653–54
Pardee, Arthur, 264
parenchyma cells, 206, 721, 730
parenchyma tissue, 206, 721, 722–23, 722*f*, 779
parental strands (DNA), 221, 223–24, 223*f*
parietal lobe, 864–65, 864*f*
Parkinson disease, 866
parks, 1255–57, 1256*f*
Parnas, Jacob, 138
parrotfish, 925, 926*f*
parsimony principle, 523, 524*f*
Parthenocissus quinquefolia, 615
parthenogenesis, 659, 1057
partial pressure, 987
partial pressure of oxygen, 999–1001, 1003
particulate inheritance, 321, 324
partly apoplastic phloem loading, 788*f*, 789
parturition, in mammals, 1070–72
parvovirus, 361*t*
passenger pigeon (*Ectopistes migratorius*), 1251, 1251*f*
Passeriformes, 702*t*
passerines, 1191, 1191*f*
passive immunity, 1111
passive transport
  defined, 105
  gradient dissipation by, 106
  illustrated, 105*f*
  in plant cells, 773
passive trapping mechanisms, 769
Patau, Klaus, 354
Patau syndrome, 318*t*
patch testing, 1104
patents, 411, 429
paternal inheritance, 352
paternity cases, 416
pathogen-associated molecular patterns (PAMPs), 1098
pathogenic protists, 563
pathogens. *See also* diseases; infectious diseases
  defined, 548
  diarrhea causing, 940
  discovery, 548
  evolution, 549, 550*f*
  fungal, 630–31
  major types, 1094–95
  mechanisms of action, 548–49
  plant defenses against, 752–54
pattern formation
  in evolution, 508–9
  phases for *Drosophila*, 386–90
  role in animal and plant development, 382–84, 385*f*, 395–96, 1077
  stem cell differentiation, 390–95
Pauling, Linus, 219, 520
Pavlov, Ivan, 826, 937, 1148
Pavo cristatus, 486
Pax6 gene, 380, 511–12, 512*f*
Payen, Anselme, 198
pea aphid (*Acyrthosiphon pisum*), 506
peafowl (*Pavo cristatus*), 486, 487*f*
pea plants, 322–23, 610*f*. *See also* garden pea; *Pisum sativum*
pearls, 661
peat mosses, 586
pectinases, 203
pectins, 197*f*, 198, 203
pectoral muscles, in flying vertebrates, 916
pedal glands, 658–59, 659*f*
pedicellariae, 678
pedicels, 605, 671
pedigree analysis in humans, 330–31

pedipalps, 671
peer-review process, 18
pelagic zone, 1140
Pelecaniformes, 702*t*
Pelecanus onocrotalus, 701*f*
pelican (*Pelicanus occidentalis*), 700*f*, 701*f*
pelt records, 1170
penicillins, 378, 541
peninsular effect, 1209
penis, 1061, 1063, 1063*f*, 1064
pennycress (*Thalsphi caenilescens*), 1260
pentadactyl limbs, 692
pentoses, 46–47, 60, 61*f*
PEP carboxylase, 170–71
peppered moth (*Biston betularia*), 1258
pepsin, 125–26, 930
pepsinogen, 930
peptide bonds, 53, 245
peptide hormones, 1034
peptidoglycan, 535*f*, 540, 541*f*
peptidyl (P) site, 250, 250*f*, 254
peptidyltransferase center, 253
peptidyl transfer reaction, 253, 253*f*
peptidyl tRNA, 252
per capita growth rate, 1177–78
perception, 876
Perciformes, 9, 11*f*
peregrine falcon (*Falco peregrinus*), 1260
perennials, 720
perfect flowers, 605, 798
perforin, 1108–9
perianth, 605, 606, 608*f*, 798
pericarp, 797
pericycle, 733, 734, 734*f*
periderm, 207–8, 207*f*, 732
Peridinium limbatum, 567*f*
perilymph, 880
periodical cicadas, 1196
periodic table of elements, 25–26, 25*f*
peripheral membrane proteins, 99
peripheral nervous system
  divisions, 859–62
  elements of, 832, 855*f*, 856
  interactions with CNS, 858
peripheral zone, 396
periphyton, 554
peristalsis, 929–30
peritubular capillaries, 1020*f*, 1021
permafrost, 1138
permanent protein modifications, 431, 431*f*
permeability, 106, 106*f*
Permian period, 456
Peromyscus leucopus, 1203*f*
peroxisomes
  in animal cells, 71*f*
  catalase formation, 85, 122
  formation, 85*f*
  in plant cells, 72*f*
  structure and function, 84–85, 94
perspiration, 813
petals
  in *Arabidopsis*, 397
  evolution, 606
  floral tube development from, 609
  functions, 605, 795
petioles, 726, 786–87, 787*f*
Petrogale assimilis, 706, 706*f*
Petromyzon marinus, 687*f*
PET scanning. *See* positron-emission tomography
petunia (*Petunia hybrida*), 799, 799*f*
Pfiesteria shumwayae, 565–66
P generation, 324
pH
  effects on organism distribution, 1128–29
  homeostatic role, 823*t*
  organisms adapted to extremes, 535
  overview of scale, 38–39, 39*f*
Phaeophyceae, 561–62
phages. *See* bacteriophages
phagocytosis
  defined, 114, 556
  digestion via, 927
  by leukocytes, 1095–96, 1097
  protists characterized by, 563
  vacuoles in, 84, 84*f*
phagotrophs, 563, 618
Phakopsora pachyrhizi, 630
Phallus impudicus, 623*f*
Phanerochaete chrysosporium, 633
Phanerozoic eon, 454–57
pharmaceuticals. *See* drugs
pharyngeal slits, 679, 680, 681*f*
pharynx
  connection to inner ear, 880
  in flatworms, 657
  functions in vertebrates, 929–30
  in mammals, 993, 994*f*
phase contrast microscopy, 68*f*
phenolics, 611, 611*f*, 1197, 1198*f*
phenology, 1252
phenotypes
  defined, 325
  gene interactions, 344–46
  intermediate, 337
  selective breeding for, 469, 470*f*
pH environment, effect on enzymes, 125–26
phenylalanine
  breakdown, 236, 236*f*
  codon and anticodon, 245
  as essential animal nutrient, 922
  inability to metabolize, 336*t*, 337, 338–39
  molecular structure, 54*f*
phenylalanine hydroxylase, 337, 338, 434–35, 435*t*
phenylketonuria, 336*t*, 337, 338–39, 339*f*
pheophytin, 159, 163
pheromones, 893*f*, 894, 1155–56, 1162
Philanthus triangulum, 1151–52
Philippine eagle (*Pithecophaga jefferyi*), 1244
Philodina, 659*f*
phloem
  arrangement in stems, 722*f*
  basic functions, 582
  bulk flow forces, 779, 788–89
  development, 721, 789–90
  in leaves, 723
  primary, 729–30, 731*f*
  in roots, 734
  secondary, 730–32
  structure, 787–88, 788*f*
phloem loading, 788–89
phoresy, 1203
phosphatase enzymes, 764
phosphates
  basic features, 44*t*
  mining, 1259
  in nucleotides, 60, 61*f*, 217, 218, 218*f*, 223
  plant uptake, 764–66, 1236
  regulation by bone tissues, 903
  release in muscle contraction, 908
  role in DNA replication, 225*f*, 226
phosphatidylcholine, 51*f*
phosphodiesterase
  caffeine's effects on, 187
  in photoreceptors, 890, 891*f*
  role in signal transduction, 185
phosphodiester linkages, 218
phosphoenolpyruvate, 129, 141*f*, 170–71
phosphoenolpyruvate (PEP) carboxylase, 759
phosphoester bonds, 218
phosphofructokinase, 139
2-phosphoglycerate, 141*f*
3-phosphoglycerate
  formation in Calvin cycle, 166–67, 169–70
  formation in glycolysis, 141*f*
phosphoglycolate, 170
phospholipase C, 186
phospholipid bilayer, 98, 106. *See also* plasma membranes
phospholipids
  basic features, 51–52
  molecular structure, 51*f*
  movement within membranes, 101, 101*f*
  role in membrane structures, 98
  synthesis at ER membrane, 103, 103–5, 104*f*
phosphorus
  biogeochemical cycle, 1235–37, 1237*f*
  as essential animal nutrient, 924*t*
  in inorganic fertilizers, 763
  as plant macronutrient, 757*t*, 1233
  plant uptake, 764–66
phosphorylase kinase, 183
phosphorylation
  activation energy, 122–23
  defined, 60, 121
  in ion pump mechanism, 112
  of receptors, 179, 181–83
photic zone, 1127, 1128*f*, 1140
photoautotrophs, 155, 544, 564
photoheterotrophs, 544
photons, 157
photoperiodism, 747, 748–49
photophosphorylation, 159
photopsins, 888
photoreception
  evolution, 891–93
  eyecups and compound eyes, 885–86
  retina's functions, 890–91
  rods and cones, 887–89, 888*f*, 889*f*, 890*f*, 891*f*
  signal transduction pathway, 889–90
  single-lens eyes, 886–87
photoreceptors
  Darwin's hypothesis, 511
  defined, 877, 885
  type in plants, 747–49
photorespiration, 169–71
photosensitivity, 288
photosynthesis
  by algae, 553
  Calvin cycle, 156, 165–67, 166*f*
  carbon dioxide requirements, 759
  chloroplasts' role, 87
  by cyanobacteria, 533, 536, 536*t*
  defined, 2–3
  environmental conditions limiting, 1127, 1128*f*
  importance to life, 154
  leaf size and, 591
  leaf structure and, 722–23
  light reactions, 156, 157–65
  nutrient requirements for, 756
  overview, 154–56
  variations, 169–71
photosystem I
  combined action with photosystem II, 159–60, 160*f*
  discovery, 163–65
  molecular features, 161
photosystem II
  combined action with photosystem I, 159–60, 160*f*
  discovery, 163–65
  molecular features, 161–63, 164*f*
phototaxis, 1151
phototropin, 747
phototropism
  auxin's role, 175, 739, 742–43
  defined, 175, 175*f*, 739
Photuris versicolor fireflies, 1156
Phthiraptera, 673*t*

phycobilins, 536
Phycodurus eques, 691*f*
phycoerythrin, 564
phyla, 516
Phyllidia ocellata, 662*f*
phylogenetic trees
basic principles, 517–20
producing, 521–24
phylogeny, defined, 517
Physcomitrella, 583–85
Physcomitrella patens, 586
physical mapping, 406
physical mutagens, 284*t*, 285, 285*t*
physiological ecology, 1118
physiology, defined, 14
phytochrome, 747–48, 747*f*, 748*f*, 749
phytol tail, 158
Phytophthora infestans, 554, 554*f*, 561
Phytophthora ramorum, 1201
phytoplankton, 554, 1140, 1229, 1231
phytoremediation, 1260
phytosterols, 102
pia mater, 859, 860*f*
PIBIDS, 288
Picea sitchensis, 1216, 1217*f*
Piciformes, 702*t*
Picoides borealis, 1258
Pieris brassicae, 1198
PIF3, 748
pigeons, 1154
pigments, 157–59. *See also* coloration
pili
conjugation via, 376, 376*f*
defined, 70
prokaryote diversity, 542
pill bug (*Armadillidium vulgare*), 676*f*, 677, 991
Pillitteri, Lynn, 729
piloting, 1153
pine (*Pinus* spp.), 600, 601*f*, 1142
pineal gland, 863, 1031*f*
pineapple, 610*f*, 611
Pine Barrens, 1133
PIN genes, 741
pinnate venation, 726–27, 726*f*
pinnipeds, 915
pinocytosis, 114
Pinus palustris, 1123–24
Pinus resinosa, 632*f*
pinworms (*Enterobius vermicularis*), 668
pioneer stage, 1216, 1217*f*
pipefish (*Syngnathus typhle*), 1164
PIR database, 433*t*
pistils
evolution, 607*f*
of flowers, 605, 795
role in pollen germination, 803–4
Pisum sativum, 322–23
pitch, wavelength and, 879
pitcher plant (genus *Nepenthes*), 769, 769*f*
pitfall traps, 1169, 1169*f*
pith, 731, 731*f*
Pithecophaga jefferyi, 1244
pits, in tracheids, 602, 602*f*, 780, 780*f*
pituitary giants, 1047
pituitary gland
functions, 863, 1031*f*
relation to hypothalamus, 1035–37, 1036*f*
pit vipers, 885, 885*f*
pivot joints, 904*f*
placenta
delivery, 1070, 1071*f*
structure and functions, 1068–69, 1069*f*
placental mammals, 467, 1068, 1143–44
placental transfer tissues, 588–91
Placobdelloides jaegerskioeldi, 667
Placodermi, 688
planarians
nervous systems, 855*f*, 856
photoreception in, 885–86, 885*f*
Planctomycetes, 536*t*
plankton, 554, 676, 677, 1140
Plantae, 558
plant behavior, 737–39
plant cells
cell plate formation, 305, 306*f*
locations of genetic material, 351*f*, 423–24
water absorption, 723–24, 723*f*, 724*f*, 725
plant diseases, 368
plant evolution
ancient evidence, 586–88
angiosperms, 604–9
embryo development, 588–91
fungi, 618–19
genome evidence, 583–85
leaves and seeds, 591–94
plants. *See also* flowering plants
ancestry, 575–77
in aquatic biomes, 1139–42
as autotrophs, 451
basic features, 10*f*, 575, 577
bryophytes, 577–81
cell junctions, 198, 203–4
cell structures, 70, 72*f*, 74*f*
cell wall composition, 197–98
classification, 558*f*, 577–86
in coniferous forest biomes, 1135
deciduous forest biomes, 1134, 1135
in desert biomes, 1137
development, 395–97
effects of chromosome number changes, 318, 318*f*
embryos, 588–91
in grassland biomes, 1136
herbivore defenses, 751–52, 1197–98
incidence of polyploidy, 505
innovations of seed plants, 598*t*
life cycle, 313*f*, 314
mammalian nutrition from, 704
measuring population density, 1169
model organisms, 13, 381–82, 381*f*
in mountain biomes, 1138
naming, 517
nutrient sources, 154, 766–70
parasitic, 770, 1199, 1200
pattern formation, 382
phototropism, 175, 175*f*
in rain forest biomes, 1133, 1134
reproduction, 577
selective breeding, 469, 470*f*, 471*f*
similarity to algae, 558–60
tundra, 1138
water in, 27, 34
plant tissue culture, 744, 744*f*
plaques, 872, 983, 983*f*
plasma
as extracellular fluid, 819
filtered in glomerulus, 1020, 1021
as major component of blood, 817*f*, 969*f*, 970
plasma cells, 1101
plasma membranes. *See also* receptors
active transport overview, 111–13
of animal cells, 71*f*
of bacterial cells, 69, 70*f*, 538–39
cystic fibrosis impact, 17, 336–37
defense mechanisms in plants, 752, 753*f*
electrical charge in neurons, 835–38
endodermal, 777–78
as endomembrane system elements, 78, 94
exocytosis and endocytosis, 113–14, 115*f*
fluidity, 101–2, 103*f*
glucose transport, 47
GLUT proteins, 947, 948*f*
heat's effects on, 955–56
hormone transport, 1032–33
major functions, 85–86, 86*f*
phospholipid arrangement, 51–52, 51*f*
in plant cells, 72*f*
protein sorting across, 89–92
secretory vesicle fusion with, 81
sperm-egg fusion, 1078, 1078*f*
structure, 85, 98–101, 98*f*
surface area vs. cell volumes, 73
synthesis of components, 103–5
transport in plant cells, 773–76
transport process overview, 105–7, 108*f*
transport protein types, 108–11
plasmapheresis, 918
plasmids
general features, 371, 421
occurrence in bacteria, 371–72, 372*f*
in transgenic plants, 412–13, 413*f*
as vectors, 401, 402–3*f*, 406, 413
as virus precursors, 368
plasmodesmata
basic features, 198, 198*t*, 204, 204*f*
discovery, 203
of sieve-tube elements, 787
tissue-level transport via, 776–77, 777*f*, 788
Plasmodium falciparum
effects on host mosquitos, 1199
human costs, 553
life cycle, 568–72, 1199
resistance to, 485
plasmogamy, 627, 628*f*
plasmolysis, 107, 775
plasmolyzed cells, 775, 775*f*
plasticity, 868, 869–71
plastids
chloroplasts as, 87
in euglenoids, 558
origins, 452
in *Plasmodium*, 572
primary, 559–60, 559*f*
plastocyanin, 159, 160
plastoquinone, 163
platelets
clotting functions, 970–71, 971*f*
as major component of blood, 817*f*, 969*f*, 970
Platichthys flesus, 1048, 1048*f*
Platyhelminthes, 648*t*
platyhelminths
key characteristics, 648*t*, 657, 681*t*
nervous systems, 855*f*, 856
platypus, 704, 706*f*, 884
pleiotropy, 336–37
Pleodorina californica, 453, 453*f*
Pliocercus euryzonus, 698*f*
Pliohippus, 466*f*
plumage, 1164
pluripotent stem cells, 392, 393
pneumatophores, 732, 732*f*
Pneumocystis jiroveci, 367, 630
pneumonia, 367
poaching, 1257
podocytes, 1020, 1021*f*
Podos, Jeffrey, 502
Poecilia reticulata, 486
Pogonia ophioglossoides, 1203
poinsettia, 730*f*
point mutations, 280–81, 280*t*
Poiseuille, Jean Marie Louis, 978
Poiseuille's law, 978, 980
poisons. *See* toxins
polar bear (*Ursus maritimus*), 635, 1018
polar bodies, 1059, 1060*f*

polar covalent bonds
in amino acids, 53, 54*f*
with carbon atoms, 43, 44*f*
defined, 29
in water, 29*f*, 34
polarity in epithelial tissue, 207, 815
polarity of character states, 522
polarization of neurons, 835
polar microtubules, 304, 305*f*
polar molecules, 30
pole cells, 1083, 1083*f*
pole of mitotic spindle, 304
poliomyelitis, 917
pollen
development in flowering plants, 719
gametic isolation, 499
germination in flowering plants, 795–96
Mendel's studies, 323
in seed plants, defined, 592
sperm cells in, 322
pollen grains, 795, 802–3
pollen tubes
of flowering plants, 795–96, 803*f*
in *Ginkgo biloba*, 600
growth into ovule, 605, 804, 804*f*
pollination
defined, 592
of flowers, 605, 609, 795
mutualistic relationships, 1202–3
pollination syndromes, 614, 614*t*
pollinators
mutualistic relationships, 1202–3
structures attracting, 609, 609*f*, 613–14
pollution
acid rain, 762, 1128–29
bioremediation, 410–11, 550, 1232, 1260
endocrine disruptors in, 1053
indicator species, 635, 1258
lichen sensitivity, 635
phosphorus, 1236–37
resistant plants, 483, 485*f*
polyandry, 1162, 1163–64, 1163*f*
poly A tail, 243, 243*f*
polychlorinated biphenyls (PCBs), 1260
polycistronic mRNA, 260
polyethylene glycol, 416
polygamy, 1162
polygenic traits, 345, 346*f*
Polygonum, 606*f*
polygyny, 1162, 1163, 1163*f*
polymerase chain reaction (PCR), 405–6, 405*f*, 415
polymers
defined, 45
evolution, 440–42
formation and breakdown, 46, 46*f*
proteins as, 53
polymicrogyria, 866
polymorphic genes, 479
polymorphism
balanced, 484, 485
causes, 479–80
defined, 479
lactose intolerance as, 934
polypeptides
amino acid sequences, 243–44, 280–81
basic structure, 55*f*
common sizes, 54
control of synthesis, 1034
defined, 53, 238
direction of synthesis, 245
effects of mutations, 280–81
as hormone class, 1032
RNA translation to, 4, 243–45
synthesis during elongation stage of translation, 252–54, 252*f*, 253*f*
synthesis on ribosomes, 89
polyphagous parasites, 1199
polyphyletic groups, 519
Polyplacophora, 661, 662*f*
polyploid organisms, 316, 316*f*, 317*f*, 318
polyploidy, 504–6, 607–8
polyploid zygote, 1061
polyps, cnidarian, 654–55, 654*f*, 655*f*
polysaccharides
basic features, 48–49
in extracellular matrix, 194, 196–97
in plant cell walls, 197–98
Polysiphonia life cycle, 569*f*
polysomes, 254
polyspermy, 1078–79
polyunsaturated fatty acids, 50
Pomacentridae, 9, 11*f*
Pongo pygmaeus, 706, 706*f*
pons, 862, 865*t*
poplars, genetically engineered, 1232
population density, 1168–71
population ecology
bottom-up versus top-down control, 1204–5
competition between species, 1188–94
defined, 1168
demography, 1172–76
dispersion patterns, 1170–71
goals of, 1118, 1118*f*
growth factors, 1176–82
herbivory, 1197–98
human population growth, 1182–85
mutualism and commensalism, 1202–3
parasitism, 1180, 1198–1201 (*see also* parasites)
predation, 1174, 1179–80, 1194–97 (*see also* predation)
reproductive strategies, 1171–72
sampling methods, 1168–70
population genetics
allele and genotype frequencies, 480–81
defined, 478
genetic drift, 489–91
natural selection, 482–86
sexual selection, 486–89
population growth
exponential, 1177–78
factors regulating, 1178–82
human, 1182–85
measuring, 1177
populations
changes in, 479
defined, 6, 459, 478, 1168
as level of biological organization, 5*f*, 6
measuring, 1168–72
population size
effective, 1253–54
genetic drift and, 489–91, 1252–53
p orbitals, 24, 24*f*
porcupine fish (*Diodon hystrix*), 1195, 1195*f*
Porifera, 648*t*, 653–54, 653*f*, 681*t*
porkfish (*Anisotremus virginicus*), 500, 501*f*
Porphyra, 558
porphyrin ring, 158
portal veins, 1035
Portuguese man-of-war, 655, 655*f*
Posidonia oceanica, 193
positional information
pattern formation and, 382–84, 382*f*, 386, 1077
signaling, 1086
positive controls
defined, 260, 261*f*
lac operon, 266, 267*f*
positive feedback loops
birth as example, 826*f*, 1070
defined, 825
in inflammatory response, 1097
positive pressure filling, 992
positive reinforcement, 1148
positron-emission tomography (PET) scans, 26–27, 27*f*, 139
postabsorptive state, 945, 946–47, 949
postanal tail, 679
posterior cavity, in single-lens eye, 886, 887*f*
posterior chamber, in single-lens eye, 886
posterior end, 642
posterior pituitary gland, 1031*f*, 1036–37, 1036*f*
postsynaptic cell, 842, 843*f*
postsynaptic membrane receptors, 848
post-translational covalent modification, 431, 431*f*
post-translational sorting, 89, 90*f*, 92, 92*f*
postzygotic isolating mechanisms, 498, 498*f*, 499–500
potassium, 757*t*, 763, 924*t*
potassium ions
disorders involving, 1026
electrochemical gradient, 837
homeostatic role, 823*t*, 1009
hormone regulation, 1045–46
ion pump, 112–13
movement during action potentials, 840, 841
movement in kidney, 1021, 1023, 1024*f*
resting potential role, 836–38
potato spindle tuber disease, 368
potato tubers, 732
potential energy, 119
Pounds, J. Alan, 1117
powdery mildews, 627
Powell, Mike, 901
Powell, William, 1201
power stroke, 908
PPAR-δ gene, 913–14
P protein, 788, 788*f*
prairie biomes, 1136
prairie chickens, 1252, 1253*f*
prairie dogs (*Cynomys* spp.), 1160–61, 1160*f*
prairies
destruction, 1245, 1252, 1254, 1256*f*
remnants, 1245
restoration, 1259
prebiotic soup, 439
prebiotic synthesis, 439–40
Precambrian, 448
precipitation
acidic, 1128–29
in biome classification, 1132*f*
in coniferous forest biomes, 1135
in deciduous forest biomes, 1134, 1135
in desert biomes, 1137
in grassland biomes, 1136
mountains' influence, 1130–31, 1131*f*
in rain forest biomes, 1133, 1134
tundra, 1138
predation
by annelids, 666
defensive strategies, 1194–96
on eggs, 1062
extinctions from, 1251
feeding behaviors, 924–25, 926*f*
by frogs and toads, 695
by fungi, 630, 630*f*
group living protections, 1157–58
herbivory, 1197–98
historical development, 640
impact of coloring, 486, 487*f*
influence on prey animals' foraging, 1154
natural selection and, 485–86
by plants, 756*f*, 769–70
as population control, 1196–97, 1204–5
population density factors in, 1180
by sea stars, 678

sensory systems for, 881, 884, 885, 892, 893
sentries against, 1160–61
suggested impact on species diversity, 454
use of signals in, 1156
predator-controlled systems, 1196
predictions, in scientific method, 15
prednisolone, 189–90
pregnancy
effects on mammals, 1068–70
heartburn, 940
smoking's impact, 1005
tests for, 1067
preinitiation complex, 269, 270*f*, 273
pre-miRNA, 275, 276*f*
pre-mRNA, 241, 242, 242*f*, 274–75, 431
pre-RNA, 239
pre-siRNA, 275
pressure
effect on gas solubility, 987
relation to gas volume, 992, 993*f*
skin receptors for, 877, 878*f*
pressure-flow hypothesis, 789, 789*f*
pressure potential, 775
presynaptic cell, 842, 843*f*
prezygotic isolating mechanisms, 498–99
prickly pear cactus (*Opuntia stricta*), 1205, 1205*f*
prickly rhubarb, 767
primary active transport, 111, 112*f*
primary cell wall, 197–98, 197*f*
primary consumers, 1226
primary electron acceptor, 162, 163*f*
primary endosymbiosis, 559–60, 559*f*
primary growth, 721, 721*t*
primary immune response, 1111, 1111*f*
primary lymphoid organs, 1101, 1101*f*
primary meristems, 721
primary oocytes, 1059, 1060*f*, 1066
primary plastids, 559–60
primary production, 1232–34
primary spermatocytes, 1059, 1060*f*
primary stem tissues, 721–22
primary structure of proteins, 54, 56*f*
primary succession, 1215
primary vascular tissues
major features, 729–30
in roots, 734
structure and development, 721–22
primates
classification, 707
color vision, 889
evolution, 707–12
key characteristics, 707
primer extension, 406
primers, 405–6, 405*f*
primitive characters, 521
primordial germ cells, 1083
principle of parsimony, 523, 524*f*
principle of species individuality, 1208
prions, 368–70, 369*f*
probability, in genetics, 339–41
probing beaks, 701*f*
proboscis of insects, 674, 893, 893*f*
procaspases, 190, 191*f*
processivity, 227
procollagen, 195, 195*f*
producers, 545, 1226
production efficiency, 1228–29
productivity hypothesis, 1210–11
product rule, 340
products, defined, 33
progesterone
effects on human female, 1067
in frogs, 302
molecular structure, 1034*f*
prolactin suppression, 1071
site of production, 1050
proglottids, 658, 658*f*
programmed cell death. *See also* apoptosis
importance, 174
nematode studies, 668
as plant defense, 752, 753
in primary vascular tissues of plants, 730
process overview, 174*f*
progymnosperms, 598
proinsulin, 81
prokaryotic cells
basic features, 69–70
capsules, 539–40
cell wall structures, 540–41, 543–44
complex adaptations, 538–39
division, 542–43
on geological timescale, 449*f*, 450–51, 450–52
transcription and translation, 238*f*, 239
prokaryotic organisms. *See also* archaea; bacteria
biotechnology applications, 549–50
defined, 9
diversity and evolution, 533–38
ecological role, 545–47
eukaryotic versus, 515*t*
fossils, 438*f*
nutrition and metabolism, 544–45
as pathogens, 548–49
reproduction, 542–44
structure and movement, 538–42
symbiotic relationships, 547–48
transcription and translation, 238*f*, 239
prolactin, 1037*t*, 1071
proliferative phase, 1067*f*, 1068
proline, 54*f*, 243, 245
prometaphase, 305, 306*f*
prometaphase I, 309–12, 310*f*
promiscuous mating systems, 1162
promoters
lac operon, 261
mutations, 281, 281*t*
role in transcription, 239, 239*f*
proofreading, 228, 367
prophages, 363, 366
prophase, 305, 306*f*
prophase I, 309, 310*f*, 313
Propionibacterium acnes, 544
proplastids, 87, 351*f*, 352, 352*f*
propyl alcohol, 45*f*
prostaglandins
effect on pain perception, 884
nutritional requirements for production, 922
role in parturition, 1070, 1071*f*
prostate gland, 1063*f*, 1064
prosthetic groups, 125
protandrous hermaphrodites, 12–13, 1062
Protapirus, 1143, 1143*f*
protease inhibitors, 368
proteases
antiviral drug actions on, 367–68
in Avery, MacLeod, and McCarty's study, 215
functions, 81, 128, 132
proteasomes, 132, 133*f*
protected areas, 1254–57
protective attributes. *See* defenses
protein-calorie malnutrition, 1111–12
protein kinase A, 183–85, 184*f*, 868, 868*f*
protein kinase C, 187
protein kinase cascades, 182, 182*f*
protein kinases, 179
protein phosphatases, 185
protein-protein interactions, 58, 58*f*, 69
proteins. *See also* proteomes
as animal nutrients, 921, 921*t*
associations with membranes, 98–100, 98*f*, 99*f*, 102, 103–5
breakdown and recycling, 132
categories, 12*f*
as cause of dominant traits, 335–36
control of synthesis, 1034
digestion and absorption, 935
in DNA, 217
DNA replication, 224–26
effects of genetic mutations on, 280–81
epistatic interactions, 345
evolution, 446
in extracellular matrix, 194–96, 816
formed by essential amino acids, 922
gene expression producing, 238, 239
general categories, 429–30, 430*t*
as glucose source, 946–47
heat shock, 1126
as hormone class, 1032
infectious, 368–70
involved in cytokinesis in animals, 307–8
lipid exchange, 103
major types, 53*t*
metabolism, 149
molecular structure, 52–58
overactivity linked to cancer, 288–90
pathogen recognition, 1098
proteomics and, 9–11
relay, 182
role in cell differentiation, 258, 259*f*
role in reproduction, 4
secretory pathway, 81–83, 81*f*
of signal transduction pathway, 177
sorting to organelles, 89–92
synthesis for long-term memory, 868
from transgenic organisms, 412
transport functions in membranes, 12*f*, 108–13
use of ATP as energy source, 121–22, 122*f*
viewed as genetic material in early research, 213
in viruses, 360–61, 362*f*
protein subunits
ATP synthase, 146–47
defined, 56
photosystem II, 163, 164*f*
Proteobacteria, 536–37, 536*t*
proteoglycans, 105, 194*f*, 197, 197*f*
proteolysis, 81, 431
proteomes. *See also* proteins
alternative splicing's impact, 242, 274, 275, 431
analysis for ATP binding, 122
ATP-using proteins in, 121–22
defined, 4, 9–11, 12*f*, 70, 420
of multicellular versus unicellular species, 193
protein categories, 429–30, 430*t*
role in cell differentiation, 258, 259*f*
role in organism's characteristics, 12–13, 70–73
proteomics, 11
Proterozoic eon, 451, 453
Proteus mirabilis, 542*f*
prothoracic glands, 1048
prothoracicotropic hormone, 1048
protists
animals versus, 647, 647*f*
basic features, 10*f*
cilia and flagella, 77*f*
competition and coexistence, 1189
defenses, 565–66
defined, 553
evolution and relationships, 556–63, 640–41
informal classification, 553–55
nutrition, 563–64
reproduction, 567–72
use of vacuoles, 84*f*
protobionts, 441–43, 442*f*
protogynous hermaphrodites, 1062

protonephridia
  in flatworms, 657, 1016, 1017*f*
  in rotifers, 659
proton-motive force, 145
protons, 22, 25–26
proto-oncogenes, 290–91
protostomes, 643, 644*f*
Prototheria, 704
protozoa, 553–54, 554*f*
proventriculus, 930, 930*f*
proviruses, 363
proximal convoluted tubules, 1020, 1020*f*, 1021*f*, 1023*f*
proximate causes of behavior, 1146
Prusiner, Stanley, 369
Psaronius, 455, 456*f*
Pseudoceros bifurcus, 658*f*
pseudocoelomates, 644, 645*f*, 659
pseudoephedrine, 603
pseudogenes, 429, 896
Pseudomonas aeruginosa, 549
Pseudomonas syringae, 754, 1126
Pseudomyrmex ferruginea, 1203*f*
pseudopodia, 555, 562
Pseudosalix handleyi, 586*f*
pseudostratified ciliated columnar epithelium, 816*f*
Psilocybe, 624
Psilotum nudum, 583*f*
Psittaciformes, 702*t*
Psittirostra psittacea, 1251*f*
psychoactive substances in fungi, 624
psychoneuroimmunology, 1112
pteridophytes
  diversity, 582*f*
  as plant phylum, 577
  spores, 593
  structures and life cycle, 581–83
Pterocarpus hayesii, 732*f*
Pterois volitans, 691*f*
pterosaurs, 916
ptRNAs, 126
puberty, nutrition and, 1072
Puccinia graminis, 631*f*
puffballs, 623, 623*f*
pufferfish, 841
pulmocutaneous circulation, amphibian, 967
pulmonary circulation, birds and mammals, 967*f*, 968
pulmonary trunk, 971, 972*f*
pulmonary veins, 971, 972*f*
pulse-chase experiments, 81–83, 589
pulvini, 750, 751*f*
Puma concolor, 1258, 1258*f*
Puma concolor coryi, 493
pumps, active transport, 111
punctuated equilibrium, 506–7
Pundamilia nyererei, 487–89
Pundamilia pundamilia, 487–89
Punnett, Reginald, 325, 344–45, 346, 348
Punnett squares
  Hardy-Weinberg equation illustration, 480–81, 481*f*
  predictions with, 325–26, 339–40
pupae, 385*f*
pupils, 886, 887*f*
purines
  in DNA, 61
  molecular structure, 217, 218*f*
  as nitrogenous metabolic waste, 1010
Purkinje fibers, 972
purple bacteria, 88
P wave, 974*f*
Pycnogonida, 670
Pygmy sea horse (*Hippocampus bargibanti*), 1195*f*
pyramid of biomass, 1230–31, 1230*f*
pyramid of energy, 1230*f*, 1231
pyramid of numbers, 1230, 1230*f*
Pyrenean ibex (*Capra pyrenaica*), 1261
pyrimidines, 61, 217, 218*f*
Pyrococcus horikoshii, 100*t*
pyrophosphate
  DNA sequencing using, 409
  in DNA synthesis, 225*f*, 226
  release during translation, 248, 249*f*
pyrosequencing, 409
pyruvate
  breakdown in cellular respiration, 137, 139–40, 140*f*
  glucose breakdown into, 129, 138–39, 140–41*f*
pyruvate dehydrogenase, 140
pyruvate kinase, 139

Q

Qβ phage, 361*t*
QRS complex, 974*f*
quadrats, 1169, 1169*f*
Quahog clam (*Mercenaria mercenaria*), 662*f*
quantitative traits, 345
Quaternary period, 457
quaternary structure of proteins, 56, 56*f*
queens, 1161
quiescent center, 733
quills, 703
quinine, 1245, 1246*f*
Quinn, Jim, 1256
quorum sensing, 540

R

rabbits, 1178, 1187, 1188*f*
rabies, 361*f*
Racker, Efraim, 146
radial cleavage, 643, 644*f*
radial loop domains, 231, 231*f*
radial patterns
  echinoderms, 677
  in plants, 382, 382*f*, 395
radial symmetry, 720, 799
Radiata, 642, 642*f*, 654–56
radiation
  effects on DNA, 285
  heat exchange via, 957
  solar, 1124, 1124*f*, 1129–30
radicles, 806, 806*f*
radio collars, 1170
radioisotopes, 26, 447
Radiolaria, 562, 562*f*
radiometric dating, 447
radula, 661
Raf, 182
Rafflesia arnoldii, 1199, 1199*f*
raf gene, 290*t*
RAG-1 and RAG-2 enzymes, 1105
ragweed (*Ambrosia artemisifolia*), 1260
ragweed (genus *Ambrosia*), 803
rain forests. *See* temperate rain forests; tropical rain forests
rain shadows, 1131, 1131*f*
rainwater, pH, 1128
ramie (*Boehmeria nivea*), 730
Ramosmania rodriguesii, 1244
ram ventilation, 991, 991*f*
Rana catesbeiana, 989*f*
Rana pipiens, 302, 496, 496*f*
Rana utricularia, 496, 496*f*
random coiled regions, 55
random dispersion, 1170–71
random genetic drift, 489–91
random mating, 481
random sampling error, 340, 490
Rangifer tarandus, 705*f*
Ranunculus adoneus, 737, 737*f*
Raphus cucullatus, 1251
ras gene, 290*t*
Ras protein, 182, 290, 290*f*
rate-limiting steps, 131
ratites, 1143
rats, 433–34, 433*f*, 1068
Ray, John, 460, 514
ray-finned fishes, 690–91
rays, 689, 689*f*, 884
Rb gene, 292*t*, 293, 723
Rb protein, 293–94, 723
reabsorption, 1016, 1021–24, 1025*f*
reactants, 33
reaction centers, 161, 162, 163*f*
reading frames, 244, 432–33, 432*f*
realized niche, 1191
receptacles of flowers, 605
receptor-mediated endocytosis, 114, 115*f*
receptors. *See also* sensory systems
  activation, 177, 177*f*, 178–81, 182*f*, 739
  differential regulation, 188
  for light in plants, 747–49
  lock-and-key model, 32
  in major signal transduction pathways, 181–87
  role in cell communication, 174
  sensory types, 875–77
  types responding to signaling molecules, 178–81
  for water-soluble hormones, 1032–33
receptor tyrosine kinases, 181–83, 182*f*
recessive traits
  defined, 324
  due to allele loss of function, 335–36
  for genetic disorders, 330, 336–37, 336*t*
  inbreeding and, 492–93
  protein functions and, 335–37
reciprocal altruism, 1161
reciprocal translocations, 315, 315*f*
recognition of antigens, 1102
recombinant DNA technology, 400, 410, 411*f*
recombinants, 348, 349–50
recombinant vectors, 403–4, 403*f*
recombination-activating genes, 1105
recombination frequency, 350
rectum, 933
Recurvirostra americana, 701*f*
recycling organic molecules, 131–33
red algae
  adaptations for depth, 1127, 1128*f*
  basic features, 558–59
  nutrition, 564
  reproduction, 567, 569*f*
red blood cells. *See* erythrocytes
red-cockaded woodpecker (*Picoides borealis*), 1258
red-green color blindness, 889, 890*f*
rediae, 658
red-light receptors, 747–48
red mangrove (*Rhizophora mangle*), 1221–22
red muscle fibers, 912
redox reactions
  basic features, 129
  electron transport chain, 142, 143
  in photosynthesis, 155, 162–63
reducing atmosphere hypothesis, 439–40
reduction
  in Calvin cycle, 166–67, 166*f*
  carbon dioxide, 155
  defined, 129
reduction division, 312
reductionism, 14
redundancy hypothesis, 1248, 1249*f*, 1250
reflex arcs, 834, 835*f*, 867–68
refractory periods, 840–41
Regan, Tom, 1247
regeneration, reproduction by, 1057, 1057*f*

regressive evolution, 368
regulation. *See also* gene regulation
citric acid cycle, 141–42
collagen synthesis, 196
digestion in vertebrates, 936–39
glycolysis, 139
homeostasis based on, 823–29
metabolic pathways, 130–31
oxidative phosphorylation, 145
regulatory elements in eukaryotes, 269
regulatory genes, 238
regulatory sequences, 240, 260, 281
regulatory transcription factors, 260, 269, 270
rehabilitation, 1259
Reich, David, 709
reindeer, 705*f*
reintroducing species, 1260
relative abundance of species, 1212–14
relative permeability, 106, 106*f*
relative refractory period, 840–41, 840*f*
relative water content, 775–76
relay proteins, 182
release factors, 248*t*
renal corpuscles, 1019–20
renal cortex, 1018, 1020*f*
renal failure, 1026
renal medulla, 1018, 1020*f*
renal pelvis, 1018, 1020*f*
repeatability, in scientific method, 16
Repenomamus, 703
repetitive sequences, 425
replacement, of habitats, 1259
replacement rate, 1183
replica plating, 282*f*, 283
replication forks, 223, 224, 224*f*, 226–27, 226*f*, 227*f*
replication of experiments, 1121–22
repolarization, 840*f*
representative habitats, 1254
repressible operons, 267
repressors
defined, 260, 261*f*
functions in eukaryotes, 269, 270
lac operon, 261–64
reproduction
amphibian, 695
animal characteristics, 639–40
in annelids, 666
bacteria, 372, 373*f*
bony fishes, 690
bryophytes, 579–80
as characteristic of life, 3*f*, 4
in crustaceans, 676
cycads, 599
defined, 1056
echinoderms, 678
in flatworms, 657
flowering plants, 605, 606*f*, 717–19, 793–97
of fungi, 622–24, 625–27, 626*f*, 628*f*, 629*f*
gametogenesis and fertilization, 1059–62
hormonal control, 1049–50
impact of stress on, 1052
land plant features, 577
in lichens, 635
lycophyte and pteridophyte, 583, 584*f*
mammalian, 1062–68
mating systems, 1162–65
mollusks, 661
nematode, 668
placental transfer tissues in plants, 588–91
protists, 567–72
public health issues, 1072–74
in rotifers, 659
sexual selection and, 486–89
sexual versus asexual, 1056–58 (*see also* sexual reproduction)
in sharks, 689
in sponges, 654
success and natural selection, 482–83
viruses, 359, 363–67, 364–65*f*
reproductive cells, 453
reproductive cloning, 414–15
reproductive cycles, 1071–72
reproductive isolating mechanisms, 497–500
reproductive isolation, 496–97
reproductive rates, 1176
reproductive strategies, 1171–72
reproductive success, 482–83, 486–89. *See also* natural selection
reproductive systems, 361*f*, 818*t*
reptiles
circulatory systems, 968
classification, 697–99
dominant periods in Earth's history, 456, 640
on geological timescale, 449*f*
reserves, 1255–57, 1256*f*
resin ducts, 602–3
resistance exercise, 912, 913
resistance in blood vessels, 977–81
resistance plasmids, 372
resolution of microscopes, 67
resonance energy transfer, 161–62, 163*f*
resource partitioning, 1190
resources
ecological footprint, 1184–85, 1185*f*
as restraint on population growth, 1178–79, 1188–89
selfish behavior and, 1159
respiration. *See also* cellular respiration
amniotes, 697
amphibians, 694
in bony fishes, 690
chondrichthyans, 689
defined, 2
effects on osmolarity in water breathers, 1011
feedforward regulation in mammals, 826
water loss via, 988, 1010–11
respiratory centers, 1002–3
respiratory chain, 143. *See also* electron transport chain
respiratory distress syndrome of the newborn, 996
respiratory pigments, 999
respiratory systems
adaptations to environmental extremes, 1003–4
in birds, 996–99
disorders, 1004–5
insects and mammals compared, 821, 821*f*
major organs in animals, 818*t*
in mammals, 992–96, 994*f*, 1002–3
oxygen transport mechanisms, 999–1002
types, 988–92
respiratory tracts, viruses infecting, 361*f*
respiratory trees, 678
resting potential, 835–37
restoration ecology, 1259–60
rest-or-digest response, 862
restriction enzymes, 401–2, 402*f*
restriction point, 301
restriction sites, 402
reticular formation, 863
reticularis, 1050–51
reticulum, 931, 931*f*
retina
image formation, 887, 890–91
in single-lens eyes, 886, 887*f*
retinal, 45, 888, 890
retinoblastoma, 293–94, 723
retroelements (retrotransposons), 426–27, 427*f*
retroviral insertion, 291, 291*f*
retroviruses, 363, 416, 1113
reuptake, neurotransmitter, 843
reverse primers, 405*f*, 406
reverse transcriptase
absence of proofreading function, 367
role in transposition, 426, 427*f*
use in gene cloning, 404
in viral reproductive cycle, 363, 364*f*, 1113
reverse transcription, 363
reversible inhibitors, 124–25
reversible protein modifications, 431, 431*f*
rewards, 1148
R factors, 372
rhabdoms, 886, 886*f*
rhabdovirus, 361*f*
Rhagoletis pomenella, 506
Rheinberger, Hans-Jörg, 250
Rheobatrachus, 1244
rheotaxis, 1151
rheumatoid arthritis, 1111
rhinal cortex, 1148
rhinoceros, 704, 705*f*, 1257
rhinoceros beetles, 1156, 1156*f*
rhinoviruses, 361*f*
Rhizaria, 560, 561*f*, 562, 562*f*
rhizobia, 767–69, 1240
Rhizobium, 536, 545
rhizoids, 584*f*
rhizomes, 732
Rhizophora mangle, 1221–22
Rhizopus stolonifer, 625–26, 626*f*
Rhodophyta, 558–59, 559*f*
rhodopsin, 888, 889*f*
Rhopalaea crassa, 681*f*
rhythm method, 1074
ribbon model of DNA, 221
ribonuclease, 54, 56*f*, 58–60
ribonucleic acid. *See* RNA (ribonucleic acid)
ribonucleotides, 60, 61*f*, 443
ribose, 62, 217, 218*f*
ribosomal RNA
defined, 239
evolutionary relationships, 250–51, 251*f*
repetitive sequences, 425
ribosome assembly from, 249–50
ribosomal subunits, 249–50, 249*t*, 250*f*
ribosomes
in animal cells, 71*f*
basic functions, 69
formation, 79
in plant cells, 72*f*
role in translation, 248*t*, 249–50
on rough ER, 80*f*
ribosome subunits, 79
ribozymes
defined, 243
ribosomes as, 253
RNA molecules as, 442, 443, 446
ribulose bisphosphate (RuBP), 166, 166*f*, 167
rice (*Oryza sativa*), 615
Richmond, Barry, 1148
richness of species, 1209–11
rickets, 917, 917*f*, 923*t*, 1045
Ricklefs, Robert, 1210, 1220
ring canal, 678
ring structure of glucose, 47*f*
Riordan, John, 17
riparian habitats, 1257
ripening fruit, 203, 746
RNA (ribonucleic acid)
basic functions, 4, 60
enzyme functions, 126–28, 442, 443–46
enzymes degrading, 54, 56*f*
evolution, 442–46
molecular structure, 60, 62
processing in eukaryotes, 241–43, 274–75
small ribosomal subunit, 646–47

RNA (ribonucleic acid), *continued*
synthesis during transcription, 240–41
synthetic, 245
viral integration, 363, 364*f*
of viroids, 368
RNA interference, 275
RNA polymerase
forms in eukaryotes, 241, 269
regulation, 261–62, 263*f*, 269–70
role in transcription, 240–41, 240*f*
RNA primers, 226*f*, 227, 227*f*
RNA processing, 239, 241–43, 274–75
RNase, 215
RNase P, 126, 127*f*, 128*t*
RNA splicing
alternative, 72, 242, 274–75
common in eukaryotes, 260
molecular mechanisms, 241–43, 242*f*
RNA transcripts, 238, 239–41. *See also* transcription
RNA world, 368, 443, 446
robber crab (*Birgus latro*), 638, 638*f*
Robertson, J. D., 99*t*
rock wallaby (*Petrogale assimilis*), 706, 706*f*
Rocky Mountain goat (*Oreamnos americanus*), 1170*f*
Rodbell, Martin, 179
rods, 887–88, 888*f*, 889*f*, 890, 891*f*
rod-shaped prokaryotes, 539
Root, Terry, 1252
root apical meristems
formation, 395, 396*f*
structure, 733, 733*f*
tissues arising from, 720, 723
root cap, 733, 733*f*, 750, 750*f*
root cortex, 734, 734*f*
root hairs
ion uptake, 774, 774*f*
production from epidermal cells, 733, 734*f*
responses to low soil phosphorus, 764
rhizobia interactions with, 768–69
rooting phylogenetic trees, 522–23
root pressure, 781
roots
as angiosperm organ, 718
of cycads, 599
external variations, 732
as feature of vascular plants, 582
internal variations, 732–34
mycorrhizal partners, 631–32, 631*f*, 632*f*, 766
nitrogen-fixing bacteria, 767–69
phosphate uptake adaptations, 764–66
responses to gravity and touch, 749–50, 750*f*
structure and development, 723
symplastic and apoplastic transport, 777, 777*f*
root-shoot axis, 382, 382*f*, 395
root sprouts, 808
root systems, 718, 718*f*, 732–34
roseate spoonbill (*Ajaia ajaja*), 701*f*
Rosenzweig, Michael, 1232
rose pogonia (*Pogonia ophioglossoides*), 1203
rosy periwinkle (*Catharanthus roseus*), 1245
rotary machine, 146–47, 148–49*f*
rotavirus, 361*f*
Rothenbuhler, W. C., 1147
Rotifera, 648*t*, 658–59, 659*f*, 681*t*
rough endoplasmic reticulum, 71*f*, 72*f*, 80, 80*f*
round dance, 1157, 1157*f*
round window, 880, 880*f*
roundworms, 648*t*, 667–68, 681*t*. *See also* nematodes
Rous, Peyton, 291
Rous sarcoma virus (RSV), 292
Rozella allomycis, 625
r-selected species, 1180–81, 1181*t*, 1211
RU486, 1074
rubber, 611
rubella virus, 361*f*
rubisco, 166, 169–70
Ruffini corpuscles, 877, 878*f*
rumen, 931, 931*f*
ruminants, 931, 931*f*
Runcaria heinzelinii, 594, 594*f*
running, 915–16
Ruska, Ernst, 67
rusts, 630, 631*f*
Rutherford, Ernest, 22–23
Rwanda, 1261
rybozymes, 126–28

S

Saccharomyces cerevisiae
alternative splicing, 275*t*
budding in, 622*f*
genome size, 275*t*, 424*t*
membrane protein genes, 100*t*
as model organism, 13, 636
use in food industry, 635
Saccharum officinarum, 1127
saccule, 882, 883*f*
*Sac*I enzyme, 402*t*
Sack, Lawren, 726
saguaro cactus (*Carnegiea gigantea*), 609*f*, 1123, 1123*f*
Sahelanthropus tchadensis, 457, 710
salamanders
basic features, 695–96
nervous system, 832
paedomorphosis, 511, 511*f*
salicylic acid, 1, 118, 746
salinity changes, 1035
salinization, 1137
saliva
anticoagulants in, 926
functions, 826, 929
immunoglobulins in, 1104
salivation, 826–28, 1148
Salmonella enterica, 537, 548
Salmonella typhimurium, 285–86, 286*f*, 377
Salmo trutta, 989*f*
saltatory conduction, 842
salt glands, 1012, 1012*f*, 1128, 1128*f*
salt marshes, 1234–35, 1235*f*
salt marsh snail (*Littoraria irrorata*), 631
salts. *See also* sodium chloride
defined, 31, 1008
internal balance, 1009–12
reabsorption in kidney, 1022
removal from marine animals, 1012, 1012*f*, 1127
in soils, 1128, 1137
saltwater fishes, 1011, 1011*f*, 1127. *See also* fishes
Salvini-Plawen, Luitfried von, 511
sample sizes, 340
sampling methods, 1169
sand, 761, 761*f*, 762*f*
sand dollars, 678
sand tiger shark (*Carcharias taurus*), 689*f*
sarcomas, 291
sarcomeres, 905*f*, 906
sarcoplasmic reticulum, 910, 910*f*
Sarcopterygii, 691–92
satellite imagery, 1233, 1234*f*
satiety, 953–55
saturated fatty acids, 50, 50*f*
saurischians, 699
savanna biomes, 1136, 1228*f*
sawfly (*Cimbex axillaris*), 388
Saxicola rubicola, 1191*f*
Sayornis phoebe, 1123
scales, 729, 730*f*
scanning electron microscopy, 69, 69*f*
Scarecrow gene, 396*t*
scarlet king snake (*Lampropeltis elapsoides*), 1195, 1195*f*
scavengers, 924
scent trails, 1155
Schimper, Andreas, 88
Schistosoma, 658
schistosomiasis, 658
Schlaug, Gottfried, 870, 871
Schleiden, Matthias, 66
Schmidt-Nielsen, Knut, 997
Schmithorst, Vincent, 870–71
Schwann, Theodor, 66
Schwann cells, 833, 833*f*
SCID-X1, 416–17
science, 13–18
scientific method, 13–17
sclera, 886, 887*f*
sclereids, 721
sclerenchyma tissue
connective tissue compared, 208
in stem structure, 721, 722*f*
structural support from, 206, 208, 208*f*
scolex, 658, 658*f*
Scolopax minor, 892, 892*f*
Scolopendra heros, 672, 672*f*
scooping beaks, 701*f*
scorpions, 671–72, 671*f*
Scotto, Pietro, 663
scrapie, 369, 370*t*
scrotum, 1063, 1063*f*
scurvy, 923*t*
Scyliorhinus canicula, 689*f*
Scyphozoa, 656, 656*t*
sea breezes, 1131, 1131*f*
sea cucumbers, 678, 678*f*
sea horses, 1195, 1195*f*
sea lamprey (*Petromyzon marinus*), 687*f*
sea level atmospheric pressure, 987
sea lilies, 678
seals, 988, 1004
sea raven (*Hemitripterus americanus*), 468, 468*f*
sea slugs, 867–68
seasonal breeding, 1072
sea squirts, 680, 924
sea stars
body plan, 677*f*
endoskeletons, 902*f*
key characteristics, 648*t*, 681*t*
nervous systems, 855*f*, 856
reproduction, 1057
sea turtles, 1151, 1154
sea urchin
classification, 678
fertilization, 1078–79, 1078*f*
gametic isolation, 499
key characteristics, 648*t*, 681*t*
seawater consumption, 1011, 1012, 1127
seaweeds. *See also* algae
adaptations for depth, 1127, 1128*f*
protists as, 554
resistance to surf, 1126, 1127*f*
sporic life cycle, 567, 569*f*
seborrhea, 923*t*
secondary active transport, 111–12, 112*f*
secondary cell wall, 197*f*, 198
secondary consumers, 1226
secondary endosymbiosis, 559*f*, 560
secondary growth, 721, 721*t*
secondary immune response, 1111, 1111*f*
secondary lymphoid tissues, 1101, 1101*f*
secondary meristems, 731, 731*f*
secondary metabolites
as defenses, 751, 1197
of flowering plants, 611–13

importance to plants, 585
plant sources, 597
secondary oocytes, 1059, 1060*f*, 1066
secondary phloem, 730–32
secondary plastids, 560
secondary production, 1234–35
secondary sex characteristics
in females, 1066
hormonal control, 1050
in males, 1064–65
for reproductive success, 486
transcription factor activation and, 177
secondary spermatocytes, 1059, 1060*f*
secondary structure of proteins, 54–55
secondary succession, 1216
secondary vascular tissues, 730–32
secondary xylem, 730–32. *See also* wood
second law of thermodynamics, 119
second messengers
advantages, 185
basic functions, 183–85
types, 186–87
types in plants, 739
secretin, 937, 937*f*
secretion, 1016, 1017, 1018*f*
secretory pathway, 81–83, 81*f*
secretory phase, ovarian cycle, 1067*f*, 1068
secretory vesicles, 81, 81*f*
Sedentaria, 666
sedimentary rocks, 446–47, 447*f*
sedimentation coefficient, 249–50
seed banks, 716
seed coat
formation, 592, 593*f*, 797
major functions, 806
seedless plants, 586–87
seedlings
apical-basal polarity, 720*f*
of flowering plants, 717*f*, 718
germination, 807–8, 808*f*
triple response to ethylene, 746, 746*f*
seeds
of conifers, 600, 602*f*
dispersal in fruits, 609–11, 610*f*, 614–15, 806, 1203
ecological advantages, 593
evolutionary importance, 591–94
flowering plant diversity, 716*f*
formation in flowers, 797
germination, 797, 806–8
mutualistic dispersal, 1203
Seehausen, Ole, 487–89, 506
segmentation
in animals, 644–45, 646*f*
benefits, 665–66
during development, 384, 385*f*, 386–87
embryonic formation, 1085
in trilobites, 670
segmentation genes, 386–87
segmented worms. *See also* annelids
classification, 666–67
hydrostatic skeleton, 902, 902*f*
key characteristics, 648*t*, 665–66, 681*t*
nervous systems, 855*f*, 856
segment-polarity genes, 387, 388*f*
segregation, Mendel's law, 323–25, 328–29
Selaginella moellendorffii, 583–85
selectable markers, 403
selective attention, 863
selective breeding, 465*t*, 469, 469*f*, 470*f*
selective permeability, 105
selective serotonin reuptake inhibitors (SSRIs), 851
selenium, 924*t*
self-assembling viruses, 363
self-compatible plants, 803
self-fertilization, 322, 330, 635
self-incompatible plants, 803
selfish DNA hypothesis, 427
selfish herd, 1158
self-pollination, 795, 803
self-replication, 441, 442
self-splicing, 243
Selye, Hans, 1050
semelparity, 1171, 1172
semen, 1063–64
semiautonomous organelles, 86–89, 94–95
Semibalanus balanoides, 1191–93, 1192*f*
semicircular canals, 882, 883*f*
semiconservative mechanism, 221, 222*f*, 223
semifluidity of plasma membranes, 101–2, 101*f*
semilunar valves, 971, 972*f*, 973, 973*f*
seminal vesicles, 1063–64
seminiferous tubules, 1063, 1063*f*
Senebier, Jean, 154
senses, defined, 875
sensitive plant (*Mimosa pudica*), 737, 738*f*, 750, 751*f*
sensitization, 1112
sensors, 824
sensory inputs, 865, 865*f*
sensory neurons
graded membrane potentials, 839
structure and function, 833, 834*f*, 860
sensory receptors, 875–77
sensory systems. *See also* nervous systems
audition, 879–82
balance, 882–83
electromagnetic, 884–85
lateral line, 689, 883–84
mechanoreceptor types, 877–79
photoreception, 885–93
receptors, 875–77
sensory transduction, 876
sentries, 1160–61
sepals
appearance, 605
in *Arabidopsis*, 397
evolution, 606
functions, 795
septa
ascomycetes and basidiomycetes, 627, 627*f*
between atria, 971, 972*f*
of nautilus shells, 663
septate hyphae, 620
sequential hermaphroditism, 1062
Sequoiadendron giganteum, 193, 597–98, 1124, 1124*f*
Sequoia sempervirens, 1181–82, 1182*f*
sequoia trees, 193
serine, 54*f*
serine/threonine kinases, 179
serotinous cones, 1123–24
serotonin
depression and, 851
impact on cAMP production, 185, 186*f*
major effects, 845*t*, 846
serpentine soils, 1209, 1209*f*
Serpula lacrymans, 633
Serratia marcescens, 251*f*
Sertoli cells, 1063, 1064*f*, 1065
sessile species, 639
setae, 665, 666, 902
Setaria faberi, 1217
set points, 824, 952, 953
severe combined immunodeficiency (SCID), 416
sex chromosomes
in *Ginkgo biloba*, 600
inheritance patterns, 331–34
in karyotypes, 299
X inactivation, 353–55
sex differences, 331–32
sex hormones, 13, 177
sex-linked genes, 332–34
sex pili, 376, 376*f*
sex ratio, 1162
sex reversal, 1062
sex steroids, 1050
sexual dimorphism, 486, 710, 1162, 1163
sexually transmitted diseases
bacteria, 536–37
HIV/AIDS, 367–68, 1113
infertility caused by, 1072
Trichomonas, 556, 557*f*
sexual reproduction
advantages and disadvantages, 1057–58
animal characteristics, 639–40
in annelids, 666
barriers between species, 496–500
defined, 1056
echinoderms, 678
elements of, 1057
fertilization, 1061–62
in flatworms, 657
in flowering plants, 605, 606*f*, 717–19, 793–97
of fungi, 623–24, 626*f*, 628*f*
gametogenesis, 1059–61
in lichens, 635
life cycle events, 313–14
meiosis overview, 308–14
nematode, 668
of protists, 567–68
sexual selection and, 486–89
in sponges, 654
sexual selection, 486–89, 506, 1164–65
SH2 domain, 60, 61*f*
shade, 749, 758, 758*f*
Shannon diversity index, 1212, 1213*t*
shape, prokaryotic cells, 539
shared derived characters, 521–22, 521*f*, 522*f*
shared primitive characters, 521, 521*f*, 522*f*
Shark Bay, Australia, 451
sharks
electromagnetic receptors, 884
impact of overfishing on, 1187
major characteristics, 688–89, 689*f*
as osmoconformers, 1015
shattering, 615
sheep, 414, 1072
shelf fungi, 629*f*
shells
of amniotic eggs, 696–97, 1062
of cephalopods, 662–63
of mollusks, 660, 661, 662*f*
turtles, 697
shingles, 367
Shirley Basin, 1168
shivering thermogenesis, 959
shoot apical meristems
detailed structure, 396*f*
formation, 395, 396*f*
leaf primordia, 722, 722*f*
tissues arising from, 720–21, 721*f*
shoots, 718, 749–50
shoot systems
defined, 718, 718*f*
leaf development, 725–29
stem tissues, 729–32
shore crabs (*Carcinus maenas*), 1154, 1154*f*
short-day plants, 748
short-interfering RNAs (siRNAs), 275, 276*f*, 799
short RNA molecules, 444*f*, 445
short stature, 1047
short tandem repeat sequences, 415, 1163
short-term memory, 868, 868*f*
shotgun DNA sequencing, 421–22, 422–23*f*

shrews, 1150
shrimp, 676*f*
Shubin, Neil, 465
sialic acid, 709
Siamese fighting fish (*Betta splendens*), 690
sickle cell disease
  areas of high occurrence, 478, 485*f*, 1002
  as example of heterozygote advantage, 485
  mutations causing, 280, 281*f*, 1001–2
  polymorphic gene, 479
side chains, 53, 54*f*
sieve plate pores, 787
sieve plates, 787
sieve-tube elements
  formation and structure, 787, 788*f*
  phloem loading, 788–89
  structure, 730, 731*f*
sieving beaks, 701*f*
sigma factor, 240, 240*f*
sigmoid colon, 933
signal amplification, 185, 185*f*
signal recognition particles, 89, 91*f*
signals, 174, 178–81. *See also* cell signaling
signal transduction pathways
  anchoring junctions in, 200
  defined, 85, 177, 177*f*
  enzyme-linked, 181–83, 182*f*
  G-protein-coupled, 183–87
  in photoreceptors, 890, 891*f*
  for water-soluble hormones, 1032–33
sign stimuli, 1147
sildenafil, 1064
silencers, 269, 799
silent mutations, 280
silk glands, 671
silk moths, 1151
silt, 761–62, 761*f*
siltation, 1258
Silurian period, 454–55
silvertip shark (*Carcharhinus albimarginatus*), 689*f*
Simberloff, Daniel, 1221–22
simian sarcoma virus, 292*t*
simple chordates, 509*f*
simple columnar epithelium, 816*f*, 818*f*
simple cuboidal epithelium, 816*f*
simple diffusion, 820
simple epithelium, 207, 207*f*, 816*f*
simple leaves, 726, 726*f*
simple Mendelian inheritance, 335
simple squamous epithelium, 816*f*, 818*f*
simple translocations, 315, 315*f*
Simpson, George Gaylord, 497
Singer, S. Jonathan, 98, 99*t*
single circulation, 966, 967*f*
single-lens eyes, 886–87, 887*f*
single-nucleotide polymorphisms (SNPs), 479–80
single-species approach, 1257–59
single-strand binding proteins, 225
Sinha, Neelima, 809
sinigrin, 1198
sinoatrial node, 972, 973*f*, 974, 974*f*
sinuses, in mollusks, 661
Siphonaptera, 673*t*
siphons, 663
siphuncle, 663
sister chromatids
  bivalent formation, 309, 309*f*
  compaction, 304
  defined, 298*f*, 299
  law of segregation and, 328–29
  during meiosis, 309–12, 310–11*f*, 313
  during mitosis, 305, 306–7*f*
size-exclusion chromatography, 59
skates, 689
skeletal muscle
  basic features, 814, 815*f*, 904
  cell differentiation genes, 393–95
  contraction mechanisms, 906–11
  contraction moving blood through veins, 976, 977*f*
  disorders, 917–18
  functions, 911–15
  impact of activity on body heat, 959
  impact of activity on metabolic rate, 951, 952*f*
  locomotion, 915–16
  structure, 905–6, 905*f*
skeletons, 901–3, 916–17
skin
  of amniotes, 697
  as defense against pathogens, 1095
  heat exchange, 958
  mechanoreceptors in, 877, 878*f*
  pigmentation, 345, 346*f*
  viruses infecting, 361*f*
  wrinkling, 196
skin cancer, 279
skin cells, 71*f*, 72–73
Skinner, B. F., 1148
Skinner box, 1148
Skou, Jens, 112
skull
  human versus chimpanzee development, 510, 511*f*
  mammalian, 704
  myosin mutation's effects, 907–8
Slack, Roger, 170
Slatyer, Ralph, 1218
sleeping sickness, 675
sleep movements, 750
sliding filament mechanism, 906–7, 907*f*
slime molds, 562
SLOSS debate, 1256
slow block to polyspermy, 1079
slow fibers, 912
slow-oxidative fibers, 912
slow worm (*Anguis fragilis*), 697*f*
7SL RNA gene, 425
slugs, 661
small effector molecules, 260, 261*f*
small intestine
  digestion and absorption in, 931, 933–35, 934*f*
  endocrine functions, 1031*f*
  parasites, 668
  tissue arrangement, 931, 932*f*
smallpox, 361*f*
small ribosomal subunit (SSU) rRNA gene
  advantages for analysis, 528, 646–47
  animal versus protist, 647*f*
  bacterial versus mammalian, 251, 251*f*
  chordate versus nonchordate, 679, 680*f*
  Ecdysozoa discovery from, 649–50
  use in Cooper et al. flightless bird study, 525, 525–26*f*
small ribosomal subunits
  role in translation, 251, 253*f*
  varieties in different species, 249–50
smart plants, 764–66
smell senses, 893–97
Smith, Hamilton, 401, 422
smoker's cough, 993
smoking
  damage to tracheal cilia, 993
  lung cancer and, 294, 295, 1005
  respiratory disorders from, 1005
smooth endoplasmic reticulum
  in animal cells, 71*f*
  functions, 80, 80*f*
  in plant cells, 72*f*
  role in lipid digestion, 935, 936*f*
smooth muscle
  in alimentary canals, 928
  basic features, 814, 815*f*
  functions, 904
snails, 648*t*, 661–62, 681*t*
snakes
  basic features, 698
  electromagnetic sensing, 885, 885*f*
  locomotion, 916
  venom, 1*f*
snapdragon (*Antirrhinum majus*), 799, 800*f*
snap traps, 1169, 1169*f*
Snowball Earth hypothesis, 450
snow buttercup (*Ranunculus adoneus*), 737, 737*f*
snowshoe hares (*Lepus americanus*), 1196, 1196*f*
snowy owl (*Bubo scandiacus*), 892, 892*f*
snRNP subunits, 242, 242*f*
social behavior of primates, 707
sodium, 28, 31, 757*t*
sodium chloride
  crystals, 21*f*
  ionic bonds, 31, 31*f*
  physical properties, 28
  water solubility, 34, 34*f*
sodium hydroxide, 38
sodium ions
  electrochemical gradient, 107*f*
  homeostatic role, 823*t*
  hormone regulation, 1045–47
  ion pump, 112–13
  movement during action potentials, 839–40, 841
  movement in kidney, 1021, 1022, 1023–24, 1024*f*
  resting potential role, 836–38
sodium-potassium pump
  overview, 112–13, 113*f*
  resting potential role, 836–37, 840
  stimulation by thyroid hormones, 1038
soil bacteria, 1, 2*f*, 545–47, 767–69
soil base, 760
soils
  fixed nitrogen, 763–64
  influence on biome types, 1133
  pH, 1128–29
  phosphorus, 764–66
  physical structure, 760–63
  salt concentrations, 1128, 1137
  serpentine, 1209, 1209*f*
  tropical versus temperate, 1233
solar radiation, 1124, 1124*f*, 1129–30. *See also* sunlight
Soltis, Douglas, 608
Soltis, Pamela, 608
solubility of gases, 987–88
solute potential, 775
solutes
  concentration, 35
  defined, 34, 106
  in intracellular versus extracellular fluids, 819–20
  movement between compartments, 820
solutions, 34, 36
solvents, 34
somatic cells
  cloning, 414–15
  defined, 283
  inheritance patterns, 353, 355, 356
  mutations, 283–84, 283*f*, 284*f*
  plant embryos from, 395, 808–9
  role in multicellular organisms, 453
  sex chromosomes, 331
somatic embryogenesis, 808–9
somatic nervous system, 860
somites, 1085, 1085*f*
song patterns of birds
  learning, 1149, 1150*f*
  reproductive isolation from, 502, 503–4*f*
sonicators, 265*f*
s orbitals, 24, 24*f*
soredia, 635

sori, 584*f*
sorting signals
cotranslational, 89–92, 90*f*, 91*f*
defined, 89
post-translational, 92, 92*f*
Sos, 182
sound communication, 1156
sound location, 881
sound waves, 879–82
source pools, 1219, 1220
Sousa, Wayne, 1218
Southern leopard frog (*Rana utricularia*), 496, 496*f*
Southern musk turtle (*Sternotherus minor*), 989*f*
Southern Right Whale, 925*f*
soybean pathogens, 630
space-filling models, 30*f*
Spartina grasses, 1128, 1128*f*, 1234, 1259
spatial summation, 844, 845*f*
specialist herbivores, 1197
specialization
of animal tissues, 814–16
of body segments, 645, 666
influences in plants, 721
of mammalian teeth, 703–4, 704*f*
nerve cells, 832–33
speciation, 495, 500–507
species
classification, 9–11, 460, 515–16 (*see also* taxonomy)
community diversity, 1212–14
community richness, 1209–11
defined, 6, 459, 495
diversity-stability hypothesis, 1214–15
endangered and threatened, 1245, 1261
evolutionary histories, 4
gene transfer between, 7–8, 8*f*, 377–78, 474–75
identifying, 495–500
invasive, 1118–19
mechanisms of development, 500–506
pace of development, 506–7
strains within, 373
total numbers estimated, 495–96
species-area effect, 1210, 1219, 1220, 1256
species concepts, 497
species diversity, 1245
species epithets, 517
species individuality principle, 1208
species interactions
bottom-up versus top-down control in, 1204–5
competition, 1188–94
herbivory, 1197–98
mutualism and commensalism, 1202–3
parasitism, 1180, 1198–1201 (*see also* parasites)
as population ecology subtopic, 1118
predation, 1174, 1179–80, 1194–97 (*see also* predation)
species richness, 1209–11, 1220, 1249, 1249*f*
species turnover, 1220, 1221–22, 1223
specific heat of water, 36, 1131
specific immunity, 1094
speech, 709, 711–12, 864
SPEECHLESS gene, 729
Spemann, Hans, 1088
Spemann's organizer, 1088–89, 1089*f*
spermatids, 1059, 1060*f*
spermatogenesis, 1059, 1060*f*, 1063, 1065
spermatogonia, 1059, 1060*f*, 1063
spermatophores, 1062
sperm cells
of bryophytes, 579
defined, 1057
ejaculation, 1063–64
fertilization, 1066, 1078–79, 1078*f*
flagella, 77*f*
of flowering plants, 322, 323*f*, 795, 804
formation, 1059
intersexual selection for, 486
mutations, 283, 283*f*
Spermophilus beldingi, 1160
sperm storage, 1072
Sperry, Roger, 863–64
Sphagnum mosses, 580*f*, 586
S phase, 299, 300
Sphenisciformes, 702*t*
spherical prokaryotes, 539
sphincter muscles, 940
spicules, 653
spiders, 670*f*, 671
spike glycoproteins, 361, 362*f*
spike mosses, 581*f*
spina bifida, 1091, 1092
Spinachia spinachia, 1164
spinal cord, 516, 832, 856
spinal nerves, 859
spines, 729, 730*f*
spinnerets, 671
spiracles, 669, 821, 821*f*, 991, 992*f*
spiral cleavage, 643, 644*f*
spiral prokaryotes, 539
spirilli, 539, 539*f*
spirochaetes, 536*t*, 539, 539*f*
Spix's macaw (*Cyanopsitta spixii*), 1244
spleen, removal, 1101
spliceosomes, 242, 242*f*
3′ splice site, 242, 242*f*
5′ splice site, 242, 242*f*
splice site mutations, 281*t*
splicing. *See* RNA splicing
split-brain surgery, 863–64
sponges
cell adhesion molecules, 384
choanocytes, 641, 642*f*, 653
Hox genes and, 390, 509, 509*f*
key characteristics, 648*t*, 653–54, 653*f*, 681*t*
lack of nervous system, 855
spongin, 654
spongocoel, 653, 653*f*
spongy parenchyma, 722, 722*f*
spontaneous abortions, 316
spontaneous mutations, 284, 284*f*
spontaneous ovulation, 1067
spontaneous reactions, 120
sporangia
of bryophytes, 578*f*, 579
of ferns, 584*f*
in flowering plants, 795, 802
of fungi, 625–26
of seed plants, 591, 594
spores
of bryophytes, 579–80
as common feature of land plants, 577
of flowering plants, 795
of fungi, 621, 623*f*, 626*f*, 627*f*, 628*f*
as haploid cells, 314
lycophytes and pteridophytes, 593
of protists, 567, 569*f*
of seed plants, 591
sporic life cycles
of bryophytes, 578
of protists, 567, 569*f*
sporocysts, 658
sporophytes
of bryophytes, 578, 578*f*, 579–81
defined, 313*f*, 314
development from seedlings, 718–19, 718*f*
of ferns, 584*f*
in flowering plant life cycle, 717, 794
sex determination, 332
sporophytic SI, 803–4, 803*f*
sporopollenin, 579, 802–3
sporozoites, 569–70, 572*f*
sports drinks, 1015
spotted sandpiper (*Actitis macularia*), 1164
spring field cricket (*Gryllus veletis*), 498, 499*f*
S proteins, 803*f*, 804
spruce (*Picea sitchensis*), 1216, 1217*f*
SQD1 gene, 765
Squamata, 698
squids, 648*t*, 662–63, 681*t*
src gene, 290*t*, 292
SRY gene, 331
SSU rRNA. *See* small ribosomal subunit (SSU) rRNA gene
St. John's wort (*Hypericum perforatum*), 1205
stability of communities, 1214–15
stabilizing selection, 483, 484*f*
Stahl, Franklin, 221–23
stamens
in *Arabidopsis*, 397
evolution, 605, 607*f*
functions, 322, 605, 795
nonfunctional, 608*f*
structure, 323*f*
staminate flowers, 798, 798*f*
Stammzelle, 723
standard fluorescence microscopy, 68*f*
standard metabolic rate, 950
standing crop, 1230
stanozolol, 1030
stapes, 879, 879*f*, 880, 880*f*
Staphylococcus aureus, 378
starch
digestion in vertebrates, 933
formation in Calvin cycle, 166
molecular structure, 48, 48*f*
Starling, Ernest, 937, 939
starlings, 1153, 1153*f*
star-nosed mole (*Condylura cristata*), 684, 684*f*
start codon, 244, 252
states of water, 35–36
statistical analysis, 16
statocysts, 655, 882, 882*f*
statocytes, 750
statoliths, 750, 750*f*, 882*f*
STAT proteins, 60, 61*f*
stearic acid, 50*f*
Stedman, Hansell, 907, 908
Steere, Russell, 101
Stegosaurus, 699*f*
Steitz, Thomas, 253
Steller's sea cow (*Hydrodamalis gigas*), 1251
stem cells
in central nervous system, 868
defined, 391
differentiation in animals, 390–95
in plant meristems, 723, 733
stems
as angiosperm organ, 718
basic features in land plants, 582
modifications, 732
vascular tissues, 729–32
stents, 983, 984*f*
steppe biomes, 1136
stereocilia
in audition, 880, 881*f*
of lateral line systems, 884
mechanisms of action, 878–79, 878*f*
stereoisomers, 45, 45*f*, 47*f*
Sternotherus minor, 989*f*
sternum, 701
steroids
control of synthesis, 1034, 1034*f*
as hormone class, 1032
molecular structure, 52, 52*f*
receptors, 180–81, 181*f*
sterols, 52
Steward, F. C., 809

stickleback fishes, 1147, 1148*f*, 1164
sticky ends, 402, 402*f*
stigmas
  animal pollination, 613
  defined, 605, 795
  evolution, 607*f*
  genotype, 803*f*, 804
  pollen tube formation, 606*f*, 795, 804, 804*f*
  in typical flower structure, 795*f*, 796*f*
Stiling, Peter, 1238
stimulus intensity, 876, 877*f*
stinging cells, 655, 655*f*
stingrays, 689*f*
stinkhorn fungi, 623, 623*f*
STM gene, 809
Stoeckenius, Walther, 146
stolons, 732
stomach
  endocrine functions, 1031*f*
  functions in animals, 928, 930–31
  smooth muscle, 814
  tissues in vertebrates, 816, 818*f*
stomata
  carbon dioxide absorption, 759
  defined, 155
  distribution, 722
  as epidermal cells, 206
  photorespiration and, 170–71, 583
  structure, 583*f*
  water vapor loss through, 784, 785
Stone, Christopher, 1247
stop codons, 244, 254, 432*f*, 433
Strahl, Brian, 272
Stramenopila, 560, 561–62, 561*f*, 562*f*
strands of DNA, 217, 218, 219*f*
stratified epithelium, 207, 207*f*, 816*f*
stratified squamous epithelium, 816*f*
strawberries, 609–11, 610*f*
strepsirrhini, 707, 708*f*
Streptococcus pneumoniae, 213–15, 539–40
Streptomyces griseus, 2*f*
streptomycin, 1, 2*f*
Streptophyta, 558–60
streptophyte algae, 575–77, 576*f*, 577*f*
streptophytes, 575
stress
  hormonal responses, 1050–52
  impact on digestion, 937
  impact on immune system, 1112
  reproductive effects, 1072
stretch receptors, 878
striated muscle, 906. *See also* skeletal muscle
Strigiformes, 702*t*
strigolactones, 742
Strix occidentalis, 1258
stroke volume, 980, 982
stroma
  basic features, 87, 87*f*, 155, 156*f*
  as site of Calvin cycle, 165
stromatolites, 451, 451*f*
Strong, Donald, 1210
strong acids, 38
Struck, Torsten, 666
structural genes
  defined, 238
  point mutations, 280*t*
  transcription, 239, 269–70, 272*f*, 273*f*
structural isomers, 44, 45*f*
structural proteins, 194–95, 430*t*, 816
structure, relation to function, 4
structures, evolutionary changes, 6
Sturnella magna, 498, 499*f*
Sturnella neglecta, 498, 499*f*
Sturtevant, Alfred, 349
styles, of flowers, 605, 795
stylets, 668
subarachnoid space, 859
subatomic particles
  discoveries, 22–23
  electron orbitals, 24–25
  isotopes, 26–27
  mass, 26
  types described, 21–22
suberin, 732, 777, 787
submetacentric chromosomes, 314, 315*f*
subsidence zones, 1130
subsoil, 760
subspecies, 496
substantia nigra, 866
substrate, of fungi, 619
substrate-level phosphorylation, 129, 137
substrates (enzyme), 123–25, 125*f*
subtypes, 1032
subunits
  ATP synthase, 146–47
  hemoglobin, 1001
  photosystem II, 163
succession
  basic mechanisms, 1215–18
  island biogeography model, 1219–23
succinate reductase, 144*f*, 145
succinyl CoA, 143*f*
succulants, 1137, 1142
sucker shoots, 808
sucrose
  active transport, 111–12, 112*f*
  bulk flow in plants, 788
  formation in Calvin cycle, 166
  molecular structure, 47, 47*f*
suction traps, 1169
sudden oak death, 1201
sugarcane (*Saccharum officinarum*), 1127
sugar maple (*Acer saccharum*), 779*f*, 785, 1125, 1125*f*
sugars. *See also* glucose
  basic features, 46–47
  increases in plant cells under stress, 776
  in nucleotides, 217, 218
  in phloem sap, 776, 779, 788–89
  pressure-flow hypothesis, 789
  in xylem sap, 779
sugar sinks, 789
sugar sources, 789
sulci, 856
sulfates, 44*t*
sulfhydryl groups, 44*t*, 58
Sulfolobus, 535, 544
Sulfolobus solfataricus, 421*t*
sulfur, 757*t*, 924*t*
sulfur shelf fungus (*Laetiporus sulphureus*), 629*f*
Sulston, John, 668
Sumatran tiger (*Panthera tigris*), 1261
sum rule, 340–41
sundews, 756*f*, 769–70
sunflower (*Helianthus annuus*), 609, 610*f*, 1260
Sun leaves, 758, 758*f*
sunlight. *See also* light
  effect of availability on organism distribution, 1127, 1128*f*
  greenhouse effect, 1124, 1124*f*
  impact on primary production in aquatic ecosystems, 1233
  inherited sensitivity, 279
Sun tracking, 737, 750
supergroups, 9, 515, 556
supernovas, 438
supporting cells, olfactory epithelium, 894, 894*f*
suprachiasmatic nuclei, 863
surface antigens, 338, 338*t*
surface area
  behavioral changes for heat loss, 959
  of cells, 73
  within mitochondria, 86
  relation to volume, 822
  of respiratory surfaces, 821–22, 988
  small intestine, 931
surface area-to-volume (SA/V) ratio, 822
surface tension, 37*f*, 38, 996
surfactant, 996
survival advantage, 482–83
survivorship curves, 1173–74, 1174*f*
suspension feeders, 639*f*, 924, 925*f*
suspensor, 395, 396*f*, 805, 805*f*
Sussex, Ian, 725
Sutherland, Earl, 185
Sutter, Nathan, 469
Sutton, Walter, 328
Svedberg, Theodor, 250
Svedberg units, 249–50
swamps, 1142
sweating
  acclimatization, 959–60
  evaporation rates, 958
  maintaining water and salt balance, 1013–15
  water loss via, 1012
Sweden, demographic transition, 1183, 1183*f*
sweep nets, 1169
sweet pea (*Lathyrus odoratus*), 344–45, 344*f*, 346–48, 347*f*
swim bladder, 690
swimmerets, 676
swimming mechanics, 915
swine flu, 367
Swiss-Prot database, 433*t*
symbiotic relationships
  bacteria in, 547–49, 767–69
  clownfish and anemones, 12
  fungi in, 631–35, 766
  mitochondria and chloroplasts, 88
symmetry in body plans, 642–43
symmetry in flowers, 799–800
Symons, Robert, 400
sympathetic division, 861*f*, 862
sympatric speciation, 504–6, 1193
symplastic phloem loading, 788, 788*f*
symplastic transport, 776–77, 777*f*
symplasts, 777
symplesiomorphy, 521
symporters, 111, 111*f*, 774
synapomorphy, 521
synapses
  additions for long-term memory, 868
  defined, 176, 842
  multiple inputs, 843–45
  types, 842–43
synapsis of bivalents, 309, 309*f*
synaptic cleft, 842
synaptic plasticity, 868
synaptonemal complex, 309, 309*f*
Syncerus caffer, 1126, 1127*f*
synchronous hermaphroditism, 1062
synergids, 804
Syngnathus typhle, 1164
synthesis stage, in viral reproductive cycle, 363, 365*f*
synthetic hormones, 1052–53, 1073
synthetic RNA, 245
Syrphidae, 1195
systematics, 514, 517–20
systemic acquired resistance, 753–54
systemic circulation, 967, 968
systemic lupus erythematosus, 1111
systemic veins, 971
systemin, 746, 751–52, 754
systems biology
  bioinformatics techniques, 431–35
  cell systems, 92–95
  defined, 4, 14
systole, 973–74
Szostak, Jack, 230, 443, 444*f*, 446

T

T1 bacteriophages, 283
T2 phage, 215–17, 216*f*, 366
T4 bacteriophage, 361*t*, 362*f*
Tachyglossus aculeatus, 467
tactile communication, 1156–57, 1157*f*
tadalafil, 1064
tadpoles, 695, 695*f*, 923, 1047–48
Taenia pisiformis, 658*f*
tagged animals, 1169–70
tagmata, 669, 670
taiga biomes, 1135
tails, 470, 679
tamarack (*Larix laricina*), 603
tannins, 732
Tansley, A. G., 1225
tapetum lucidum, 875*f*, 893
tapeworms, 658, 658*f*, 1198
tapirs, 1143, 1143*f*
taproot system, 723
Taq polymerase, 406
Tasmanian tiger (*Thylacinus cynocephalus*), 1261
taste buds, 896–97, 897*f*
taste pores, 896, 897*f*
taste senses, 893–97
TATA box, 269, 269*f*
Tatum, Edward, 237, 374–75
Taung child, 710
taxis, 1151
Taxodium distichum, 603
taxol, 597, 611
taxonomy
  cladistics, 521–24
  defined, 514
  elements of, 9–11, 514–17
  phylogenetic trees and, 517–20
taxons, 515, 517
Taxus baccata, 602*f*
Taxus brevifolia, 597
Tay-Sachs disease, 336*t*
Tbx5 gene, 968–69, 969*f*
T cells
  AIDS destruction, 367, 367*f*, 1113
  autoimmune responses, 1110–11
  basic features, 1101–2, 1102*f*
  functions, 1107–9
T DNA, 412–13, 413*f*
Teal, John, 1234
tearing beaks, 701*f*
tectorial membrane, 880
teeth
  adaptations for varied feeding behaviors, 924–25, 926*f*
  differentiation in mammals, 703–4, 704*f*
  evolutionary changes in horses, 7
  in sharks, 688–89
telencephalon, 856
telocentric chromosomes, 314, 315*f*
telomerase, 229–30, 229*f*
telomeres, 229–30, 229*f*, 421
telophase, 305, 307*f*
telophase I, 311*f*, 312
temperate coniferous forests, 1135
temperate deciduous forests, 1135
temperate grasslands, 1136
temperate phages, 366
temperate rain forests, 1134
temperature
  in biome classification, 1132*f*
  changes in Earth's history, 448
  in coniferous forest biomes, 1135
  in desert biomes, 1137
  effect on chemical reactions, 123, 125
  effect on enzymes, 125, 126*f*
  effect on fats, 50*f*
  effect on gas solubility, 988
  effects on distribution of organisms, 1122–26
  grassland biomes, 1136
  homeostatic control systems, 824, 824*f*, 955–60
  homeostatic role, 823*t*
  organic molecule stability, 43
  organisms adapted to extremes, 535, 956
  role in atmospheric circulation, 1129–30
  for seed germination, 807
  sensory systems, 884
  sex determination by, 332, 697
  states of water and, 36
  in temperate biomes, 1134–35, 1136
  in tropical biomes, 1133–34, 1136
  tundra, 1138
temperature tolerance, 1126
template strands (DNA), 223, 223*f*, 240–41, 240*f*, 241*f*, 406
temporal isolation, 498, 498*f*, 499*f*
temporal lobe, 864*f*, 865
temporal summation, 844
tendons, 905, 905*f*
tendrils, 729, 730*f*
tensile strength of collagen, 195
tentacles
  in bryozoans and brachiopods, 660
  in cephalopods, 662
  in cnidarians, 655
  in ctenophores, 656
teosinte, 615
tepals, 605, 608*f*
Terborgh, John, 1258
termination codons, 244
termination stage
  transcription, 240*f*, 241
  translation, 254, 254*f*
terminators, 239, 239*f*
termites, 673*t*, 1161
terpenes, 611, 611*f*, 799
terpenoids, 611, 751, 1197, 1198*f*
terrestrial biomes, 1132*f*, 1133–38
territories, defending, 1154–55
tertiary consumers, 1226
tertiary endosymbiosis, 559*f*, 561
Tertiary period, 457
tertiary plastids, 561
tertiary structure of proteins, 55–56, 56*f*
testcrosses, 326, 326*f*, 348, 349–50
testes
  endocrine functions, 1031*f*, 1050
  human, 1063, 1063*f*, 1064*f*
testosterone
  cells producing, 1063
  effects on human male, 1064–65
  molecular structure, 52, 52*f*, 1034*f*
  as primary vertebrate androgen, 1050
tetanus, 544
tetracycline, 377, 377*f*
tetrahydrocannabinol (THC), 612–13
Tetrahymena thermophila, 77*f*
Tetrao tetrix, 1164*f*
tetraploid organisms, 316, 608
tetrapods
  appearance during Devonian period, 455, 510
  locomotion, 916
  origins and evolution, 692–96
  transitional forms, 465–66
tetrodotoxin, 841
thalamus, 863, 865*t*
thalidomide, 1076
thalli, 1127
Thalsphi caenilescens, 1260
Thaumarchaeota, 535
theca, 1060
theories, 15
Theria, 706
Therman, Eeva, 354
thermocyclers, 406
thermodynamics, 119
thermogenesis, 951
thermoneutral zone, 950, 959
thermoreceptors, 877, 884
Thermus, 536*t*
Thermus aquaticus, 550
theropods, 456, 699
thick filaments, 905*f*, 906–7
thigmotropism, 750
thin filaments, 905*f*, 906–7
Thiomargarita namibiensis, 538, 544
30-nm fiber, 230, 231*f*
thoracic breathing, 697
thorns, 1198*f*
Thornton, Joseph, 1052
threatened species, 1245
threonine, 54*f*, 922
threshold concentration, 384
threshold potential, 839
thrifty genes, 235, 960
thrombocytes, 970
thumb, primate, 707
Thylacinus cynocephalus, 1261
thylakoid lumen, 155, 156*f*
thylakoid membrane
  basic features, 87, 87*f*, 155, 156*f*
  chlorophyll bound to, 158
  light-harvesting complex, 161–62, 163*f*
thylakoids
  in cyanobacteria, 538, 538*f*
  structure, 155, 156*f*
thymine
  base pairing in DNA, 61–62, 62*f*, 220, 220*f*, 221
  bonding in DNA, 220, 220*f*, 221
  dimer formation, 285, 285*f*
  molecular structure, 61, 217, 218*f*
thymine dimers, 285, 285*f*
thymus gland, 1101, 1101*f*
thyroglobulin, 1038, 1039*f*
thyroid gland, 1031*f*, 1038, 1039*f*
thyroid hormones
  disorders involving, 851
  effects on metabolic rate, 1038
  role in metamorphosis, 1047–48
thyroid-stimulating hormone, 1037*t*, 1038
thyrotropin-releasing hormone, 1038
thyroxine, 1031*f*, 1038, 1039*f*
ticks, 671*f*, 672
tidal ventilation, 991, 995–96
tidal volume, 995
tight junctions
  basic features, 198, 198*t*
  in epithelial tissue, 207
  structure and function, 200–201, 200*f*, 201*f*
Tiktaalik roseae, 465–66, 465*f*, 692
Tilman, David, 1214–15, 1249–50
timberline, 1126
time hypothesis, species richness, 1209–10
Tinbergen, Niko, 1146, 1147, 1151–52
Ti plasmid, 412–13, 413*f*
TIR1 complex, 741
tissue plasminogen activator, 410
tissues
  defined, 6, 814
  formation processes, 204–5
  as level of biological organization, 5*f*, 6
  plant transport between, 776–78
  selected viruses affecting, 361*f*
  types in animals, 205–8, 642, 814–16, 814–18*f*
  types in plants, 205–8, 577
  varying metabolic rates, 951, 952*f*
toads, 695
tobacco mosaic virus, 360, 360*f*, 361*t*, 362*f*, 363

tobacco smoking. *See* smoking
tolerance of succession, 1218
Toll-like receptors, 1098
Toll proteins, 1098–1100
Tonegawa, Susumu, 269, 1105
tongass, 1134
Tonicella lineata, 662*f*
tools, 710–11
top-down control, 1204–5
topsoil, 760, 761*f*
tori, 602, 602*f*
torpor, 952
total fertility rate, 1183–84, 1184*f*
total peripheral resistance, 980
totipotent plant tissues, 395, 412
totipotent stem cells, 391–92
touch
  plant responses to, 737, 738*f*, 750, 751*f*
  skin receptors for, 877, 878*f*
toxins
  affecting neurotransmitters, 843
  antibody binding to, 1106
  biomagnification, 1231–32
  birds' susceptibility, 701
  blocking action potentials, 841
  of centipedes, 672
  from cyanobacteria, 533
  detoxifying, 1198, 1232
  effects on embryos, 1076, 1091
  in endospores, 544
  in fungi, *623*, 623–24
  insecticides, 550
  as protist defense, 565–66
  removal in kidneys, 1022
  secretion, 1016
  in spider venom, 671
trace elements, 27, 27*t*
tracheae, 991–92, 993, 994*f*
tracheal systems, 669, 991–92, 992*f*
tracheary elements, 779–81
tracheids
  of conifers, 600–602, 602*f*
  formation by apoptosis, 383
  functions in plant stems, 730, 730*f*
  plant phyla containing, 581
  in pteridophyte stem, 583*f*
  structure, 779–80, 780*f*
tracheoles, 821, 821*f*
tracheophytes, 581, 583*f*
trachioles, 992, 992*f*
tracts, of central nervous system, 858
traffic signals, 89
Tragelaphus strepsiceros, 705*f*
traits. *See also* inheritance
  defined, 322
  effects of linkage, 346–50
  gene interactions, 344–46
  Mendel's observations, 324–27
  pedigree analysis, 330–31
  regressive evolution, 368
  selective breeding for, 465*t*, 469, 470*f*, 471*f*
  variations within species, 463
transactivation domains, 61*f*
transcription
  overview, 239–40
  regulation in bacteria, 259, 260–67
  regulation in eukaryotes, 259–60, 267–76
  steps in, 238–39, 240–41, 240*f*
  translation compared, 247
transcriptional start site, 269, 269*f*
transcription factor II D, 270, 270*f*
transcription factors
  defined, 241
  effects of cellular responses on, 177, 181
  gene regulation, 269, 270, 270*f*, 1034
  gibberellin's effects, 744–45, 745*f*
  KNOX protein, 725, 726
  phosphorylation, 182, 183
  role in pattern formation, 384, 385*f*
transducin, 890, 891*f*
transduction
  cellular mechanisms, 377, 378*f*
  defined, 373, 373*f*
  sensory, 876
trans-effects, 264
transepithelial transport, 207, 933
trans fatty acids, 51
transfer RNA
  basic functions, 239, 244–45
  binding to ribosomes, 246, 250
  molecular structure, 248, 248*f*
  role in translation, 248–49, 248*t*, 251, 252–53, 253*f*
transformation
  cellular mechanisms, 376–77
  defined, 373, 373*t*
  Griffith's discoveries, 213–14
transforming growth factor betas, 1087
transgenic organisms, 411–13, 414*f*, 1126, 1201
transitional epithelium, 816*f*
transitional forms
  as evidence of evolutionary relationships, 465–66, 465*f*
  feathered dinosaurs, 699, 700*f*
  fishapods, 465–66, 465*f*, 692
  in human evolution, 710–12
transition state, glucose-ATP, 123
translation
  defined, 238
  genetic code and, 243–47
  machinery of, 247–51
  regulation in eukaryotes, 275–76
  stages, 251–54, 252*f*, 253*f*, 254*f*
translation factors, 248*t*
translocation, of sugar in plants, 789
translocations, chromosomal, 290–91, 315, 315*f*
transmembrane gradient, 106, 107*f*
transmembrane proteins
  defined, 99, 99*f*
  insertion in ER membrane, 103, 104*f*
  movement, 102, 102*f*, 103*f*
transmembrane transport, 776, 777*f*
transmissible spongiform encephalopathies, 370
transmission electron microscopy
  overview, 68–69
  SEM compared, 69*f*
  viewing membranes with, 100–101, 100*f*
transpiration
  adaptations to reduce, 785–87
  functions in plants, 772
  water transport role, 783–84, 784*f*
  wind's effects, 1126
transplantation, 1027, 1088–89, 1112
transport (membrane)
  active transport overview, 111–13
  exocytosis and endocytosis, 113–14, 114*f*, 115*f*
  major types, 105–7, 108*f*
  processes in plant cells, 773–76
  protein types, 108–11
transporters, 111, 111*f*, 773
transport maximum, 1022
transport proteins
  basic functions, 12*f*, 108
  major types, 108–11, 109*f*, 110*f*, 111*f*, 773, 774*f*
  as proteome category, 430*t*
transposable elements, 425–27
transposase, 426
transposition, 425–27
transposons, 284*t*, 426, 426*f*
trans-retinal, 890, 891*f*, 892
transverse tubules, 910, 910*f*
tree diseases, 368, 369*f*
tree finches, 461, 462*t*
trees
  appearance during Devonian period, 455
  dermal tissue, 207–8, 207*f*
  trunk structure, 731–32, 731*f*
tree snail (*Liguus fasciatus*), 662*f*
trematodes, 657*t*, 658
Trematomus nicolai, 468
TrEMBL database, 433*t*
Trembley, Abraham, 654
trial-and-error learning, 1148, 1149*f*
Triassic period, 456
trichloroethylene, 1232, 1260
trichomes, 206, 728–29, 728*f*
Trichomonas vaginalis, 556, 557, 557*f*
triglycerides, 49–51. *See also* fats
triiodothyronine, 1031*f*, 1038, 1039*f*
trilobites, 454, 455*f*, 670, 670*f*
trimeric proteins, 190
trimesters, prenatal development, 1069–70
trimethylamine oxide, 1015
triple bonds, 29, 43
triplets, 244, 246
triple X syndrome, 317, 318*t*, 354, 354*t*
triploblastic organisms, 643, 1077
triploid organisms, 316
trisomic organisms, 316, 317
Triticum, 610*f*
Triticum aestivum, 318, 318*f*, 615
Triton cristatus, 1088, 1089*f*
Triton taeniatus, 1088, 1089*f*
tRNA, 126
trochophore larvae, 648, 648*f*
Trombicula alfreddugesi, 671*f*, 672
trophectoderm, 1080, 1080*f*
trophic cascade, 1204
trophic levels, 1120, 1226–32, 1226*f*, 1227*f*
trophic-level transfer efficiency, 1229–30
trophic mutualism, 1202
tropical deciduous forests, 1134
tropical grasslands, 1136
tropical rain forests
  biome characteristics, 1133
  mycorrhizal associations in, 766, 766*f*
  pharmaceutical potential, 1245
  photosynthesis, 154*f*
  soils, 760, 761*f*
  species richness, 1211
  transpiration in, 784, 784*f*
tropics, species richness, 1210
tropism, 739
tropomyosin, 906, 907*f*, 909–10, 909*f*
troponin, 906, 907*f*, 909–10, 909*f*
trp operon, 267, 268*f*
true-breeding lines, 322, 344
true fungi, 618
truffles, 623, 628, 628*f*
Trypanosoma, 558
Trypanosoma brucei, 558*f*
trypsin, 122, 935
tryptophan
  as essential animal nutrient, 922
  molecular structure, 54*f*
  as precursor molecule, 1032
  regulation, 267, 268*f*
tryptosomiasis, 675
tsetse flies, 675
T-snares, 91*f*, 92
Tsuga canadensis, 1127
Tsuga heterophylla, 1216, 1217*f*
Tsuga mertensiana, 1216
T-tubules, 910, 910*f*

tubal ligation, 1073, 1073*f*
tube cells, 802
tube feet, 677*f*, 678
tube worms, 639*f*, 666
tubulin, 74
tule elk (*Cervus elaphus nannodes*), 1178, 1178*f*
tumor necrosis factor, 1108
tumors. *See also* cancer
  defined, 288
  lung cancer, 294*f*, 295
  microarray data, 410*t*
  mutant genes in, 295
tumor-suppressor genes, 288, 292–95
tundra, 1138, 1210
tunicates, 680, 681*f*, 924
tunicin, 680
turbellarians, 657–58, 657*t*, 658*f*
Turdus merula, 1203*f*
turgid cells, 774, 775*f*
turgor pressure, 84, 774–76
Turkana boy, 711
Turner syndrome, 317, 318*t*, 354, 354*t*
turnover of species, 1220, 1221–22, 1223
turtles, 697–98, 1154
tusks, 514, 704*f*
T wave, 974, 974*f*
twins, 414
two-dimensional paper chromatography, 167
two-hit hypothesis for retinoblastoma, 293
tympanic canal, 880, 880*f*
tympanic membrane
  amphibians, 881, 881*f*
  mammals, 879, 879*f*, 880
Tympanuchus cupido, 1252
type 1 diabetes
  as autoimmune disease, 1111
  insulin discovery, 1042–43
  lab-produced insulin, 400, 410, 411*f*
  symptoms and causes, 1041–42
type 2 diabetes, 960, 1042
type I cells, in alveoli, 993
type II cells, in alveoli, 993, 996
type III secretion systems, 548, 549*f*
type IV secretion systems, 548–49
Tyrannosaurus rex, 456, 699*f*
tyrosine
  molecular structure, 54*f*, 1033*f*
  phenylalanine breakdown to, 236, 236*f*
  as precursor molecule, 1031–32
tyrosine kinases, 179, 181–83, 182*f*

U

ubiquinone, 142, 143, 144*f*, 145, 150
ubiquitin, 132, 133*f*, 431
Uca paradussumieri, 486
ulcers, 940, 941*f*
ultimate causes of behavior, 1146
ultracentrifuge, 250
ultraviolet radiation
  high energy, 157
  mutations by, 285, 285*f*
  phenolic absorption, 611
  plant protections, 758
  required for vitamin D production, 1045
  signals from, 174
Ulva, 547
umami, 897
umbilical arteries, 1069, 1069*f*
umbilical veins, 1068–69, 1069*f*
umbrella species, 1258
uncoating, 363, 364*f*
ungated channels, 837
unicells, 536, 537*f*
unicorn fish (*Naso rectifrons*), 438*f*
uniform dispersion, 1170, 1171*f*
uniformitarianism, 460
uniporters, 111, 111*f*
unipotent stem cells, 392
unity of life, 6–13
unsaturated fatty acids, 50, 50*f*, 102
unusable energy, 120
upwelling, 1140
uracil, 62, 217, 218*f*
urbanization, 1252
urea
  in Anfinsen's ribonuclease studies, 58, 59, 59*f*
  in kidney filtrate, 1022
  membrane permeability to, 106
  molecular structure, 106*f*
  as nitrogenous metabolic waste, 1010, 1010*f*
  Wöhler's studies, 42–43
uremia, 1026
ureters, 1018
urethra
  excretory function, 1018
  in human female, 1065
  in human male, 1063, 1063*f*
Urey, Harold, 439–40
Uria aalge, 1252
uric acid, 1010, 1010*f*
urinary bladder, 1018
urinary system, 1018
urine
  defined, 1016
  of flatworms, 1016
  of insects, 1017
  mutagen levels, 286
  sodium and potassium concentrations, 1045–47, 1046*f*
urochordates, 680, 681*f*
Urocyon cinereoargenteus, 467, 467*f*
Urocyon littoralis, 467, 467*f*
Urodela, 695–96
usable energy, 120
Ustilago maydis, 629*f*
uterine cycle, 1067*f*, 1068
uterine milk, 695
uterus, structure, 1065*f*, 1066
utricle, 882, 883*f*
U-tubes, 375, 375*f*
Uvr proteins, 287–88, 287*f*

V

vaccinations, 368, 872, 1111
vacuoles
  contractile, 107, 108*f*
  structure and function, 84, 84*f*, 94
vagina, 1061, 1065–66, 1065*f*
vaginal diaphragms, 1073, 1073*f*
vagus nerve, 847
valence electrons, 25
valine
  as essential animal nutrient, 922
  molecular structure, 54*f*
  mutations involving, 280, 281*f*
valves
  atrioventricular, 971, 972*f*, 973
  semilunar, 971, 972*f*, 973, 973*f*
  in veins, 976–77, 977*f*
vampire bat (*Desmodus rotundus*), 926, 1161
van Alphen, Jacques, 487–89, 506
van der Waals forces, 31, 57, 57*f*
Van Helmont, Jan Baptista, 154
van Leeuwenhoek, Anton, 67
van Niel, Cornelis, 155
Van Valen, Leigh, 497
vaporization of water, 36
variable regions, immunoglobulin, 1103, 1104
variable segments, immunoglobulin gene, 1105
varicella-zoster, 367
variegated leaves, 352, 352*f*
variola virus, 361*f*
Varki, Ajit, 709
Varmus, Harold, 292
vasa recta, 1023
vasa recta capillaries, 1020*f*, 1021
vascular bundles, 721–22, 722*f*
vascular cambium, 598, 731, 731*f*, 790
vascular plants, 581, 586–87. *See also* plants
vascular tissue
  appearance during Silurian, 455
  general features in plants, 206
  lacking in bryophytes, 581
  primary and secondary in angiosperms, 729–32
vas deferens, 337, 1063, 1063*f*
vasectomy, 1073, 1073*f*
vasoconstriction, 978, 981–82
vasodilation, 978–79, 981–82
vasopressin, 1024
vasotocin, 1037
vector DNA, 401
vegetal pole, 1079, 1079*f*
vegetarian finches (*Platyspiza crassirostris*), 461, 462*t*
vegetative growth, 718
veins
  in angiosperm leaves, 722–23
  evolution in leaves, 591
  in fishes, 966
  one-way valves, 976, 977*f*
  patterns in leaves, 726
  structure and function, 975*f*, 976–77
veldt biomes, 1136
veligers, 661
velocity of chemical reactions, 124, 125*f*
venation, 726, 727–28*f*
venom
  lizards with, 698
  of scorpions, 672
  of spiders, 671
  treatments derived from, 1*f*
Venter, Craig, 422
ventilation
  avian system, 996–99
  defined, 988
  mammalian system, 992–96, 1002–3
  types of systems, 988–92
  water loss via, 988, 1010–11
ventral side, 642
ventricles
  amphibian heart, 968
  in double circulation systems, 968
  evolution, 968–69
  fish heart, 966
  human brain, 859, 860*f*
  vertebrate heart, 971, 972–74, 972*f*
ventricular excitation, 972
Venturia inaequalis, 627
venules, 975*f*, 976
Venus flytrap (*Dionaea muscipula*), 206, 769, 769*f*
vernalization, 798
Vernanimalcula guizhouena, 454
Verreaux's eagle (*Aquila verreauxii*), 701*f*
vertebral column, 645, 684
Vertebrata, 684
vertebrates
  amniote features, 696–702
  brain structure, 856
  classification, 678–80
  comparative anatomy, 470
  cyclostomes, 686–87
  digestive systems, 928–37, 929*f*

vertebrates, *continued*
diversity, 684
as early body plan group, 454
energy production efficiency, 1229
excretory systems, 1017–18
on geological timescale, 449*f*
growth regulation, 1047–48
heart structures and functions, 971–74, 972*f*
homeostasis in, 823–24
key characteristics, 648*t*, 681*t*, 684–86
respiratory systems, 821, 821*f*, 992–99
single-lens eyes, 886–87
skeletal structures, 903, 904*f*
tetrapod origins and evolution, 692–96
vertical descent with mutation, 7
vertical evolution, 7, 474, 528
vertical gene transfer, 377
vertigo, 883
vesicles
in axon terminals, 843
defined, 78
exocytosis and endocytosis, 113–14, 114*f*, 115*f*
formation during meiosis, 309, 310*f*
formation during mitosis, 305, 306*f*
in hyphal tips, 621, 621*f*
protein sorting via, 91–92
secretory, 81, 81*f*
vessel elements
functions in plant stems, 730, 730*f*
structure, 780–81, 781*f*
vessels, in angiosperms, 604, 780–81, 781*f*
vestibular canal, 880, 880*f*
vestibular system, 882–83, 883*f*
vestigial structures, 470, 471*t*
Via, Sara, 506
Vibrio cholerae, 69, 421, 537, 539*f*, 940
Vibrio fischeri, 548*f*
vibrios, 539, 539*f*
vibrissae, 703
viceroy butterfly (*Limenitis archippus*), 1195
vigilance in groups, 1157
villi
chorionic, 1068, 1069*f*
of intestines, 931, 932*f*
vinblastine, 597
vincristine, 597, 1245
viral envelope, 360–61
viral gastroenteritis, 361*f*
viral genomes, 360, 361–63
viral hepatitis, 361*f*
viral reproductive cycles, 363–67, 364–65*f*
viral vectors, 401, 406
Virchow, Rudolf, 66, 194, 297
Virginia creeper (*Parthenocissus quinquefolia*), 615
viroids, 368
virulence, 752
virulence plasmids, 372
virulent phages, 366
viruses
active immune responses to, 1108–9, 1109*f*
cancer causing, 291–92
discovery, 360
emerging types, 367–68
genomes, 360, 361–63, 367, 368
innate defenses against, 1098
meningitis, 872
origins, 368
as pathogens, 1095
plant defenses, 752
reproduction, 359, 363–67, 364–65*f*
structure, 360–61
symbiotic relationships with, 633–34
transduction by, 377, 378*f*
as vectors in gene cloning, 401, 406
visceral mass, 661
Viscum album, 1199
visible light, 157
vision, 885–93, 897, 898*f*
visual communication, 1156, 1156*f*
visual pigments, 888, 889–90, 889*f*
vitalism, 42
vitamin A, 923*t*
vitamin B_1, 923*t*
vitamin B_2, 923*t*
vitamin B_6, 923*t*
vitamin B_{12}, 923*t*
vitamin C, 923*t*
vitamin D, 917, 923*t*, 1045
vitamin E, 923*t*
vitamin K, 923*t*
vitamins
absorption in digestive tract, 928, 935
antioxidants, 33
as coenzymes, 125
role in animal nutrition, 921, 922, 923*t*
vitelline layer, 1078*f*, 1079
vitreous humor, 886, 887*f*
in vitro fertilization, 392–93
in vitro translation systems, 245, 246
viviparous species, 689, 1062
vocal cords, 993
Vogt, Peter, 292
volcanic eruptions
community recovery from, 1207, 1215
impact on life-forms, 450, 1207
mass extinctions and, 456
Volta, Alessandro, 834, 835
voltage-gated ion channels
during action potentials, 839–40, 843
opening and closing forces, 838, 839*f*
volts, 835
volume
of cells, 73
relation to gas pressure, 992, 993*f*
relation to surface area, 822
volvocine green algae, 453
Volvox aureus, 453, 453*f*
von Haeseler, Arndt, 524
von Mayer, Julius, 154
V-snares, 91–92, 91*f*

W

Wächtershäuser, Günter, 440
Wackernagel, Mathis, 1184
waggle dance, 1157, 1157*f*
Walcott, Charles, 454, 455*f*
Wald, George, 888
Walker, Brian, 1248
Walker, John, 147
walking, 915–16
walking legs, 676
Wallace, Alfred Russel, 461, 482, 1144
Wallin, Ivan, 88
walrus (*Odobenus rosmarus*), 703*f*
warbler finches (*Certhidea olivacea*), 461, 462*t*
Warburg, Otto, 139
Warburg effect, 139
Warren, J. Robin, 940
warts, 361*f*
wasps
key characteristics, 673*t*
nervous system, 832
reproduction, 1056*f*, 1057
use of landmarks, 1151–52
water
absorption by plant cells, 723–24, 725, 774–76
absorption in digestive tract, 928, 931, 935–36
animal locomotion in, 915
in animals and plants, 27, 34, 819
audition in, 881
basic properties, 33–39, 957
covalent bonds, 28, 29
in dormant seeds, 806
effect of availability on organism distribution, 1126, 1127*f*
effects on climate, 1131
as essential plant nutrient, 759
in extracellular fluid, 819, 1009
gas solubility in, 987–88
homeostatic role, 823*t*, 1009–15
impact on primary production in ecosystems, 1232–33
internal balance with salts, 1009–12
in intracellular fluid, 819, 1009, 1025*f*
long-distance transport in plants, 778–84
movement across membranes, 106–7, 108*f*, 109
movement between compartments, 820–21, 1009–15
movement in and out of capillaries, 976, 977*f*
movement in kidney, 1022, 1024, 1025*f*
plant conservation, 582–83, 785–87
from reaction of methane and oxygen, 33
required in photosynthesis, 723
seed absorption, 807
site of uptake in plants, 155
systems for extracting oxygen from, 988
water cycle, 1241–42, 1241*f*
water flea (*Daphnia pulex*), 1058
water potential, 775, 776*f*
water-soluble hormones, 1032–33
water-soluble vitamins, 922, 923*t*
water stress, 785
water tupelo tree (*Nyssa aquatica*), 1126
water vapor, 439. *See also* transpiration
water vascular systems, 677*f*, 678
Watkinson, Sarah, 633
Watson, James, 220, 250, 429
wavelength, 157, 157*f*, 159*f*, 879
waxes, 52
weak acids, 38
weasels, 1193*t*
weathering, 760–61
webbed feet, 4, 508–9, 508*f*
web construction, 671*f*
web of life, 8*f*, 529, 530*f*
weed species, 1180, 1204–5
weevils, 673*t*
weight, mass versus, 26
weight control, 235, 952–55
weightlifting, 257, 912, 913
Weinberg, Wilhelm, 480
Weintraub, Harold, 393
Weismann, August, 213, 327
Welwitschia mirabilis, 603–4
Went, Frits, 742
Westemeier, Ronald, 1252
Western hemlock (*Tsuga heterophylla*), 1216, 1217*f*
western meadowlark (*Sturnella neglecta*), 498, 499*f*
wetlands biomes, 1142, 1252
whales
feeding behaviors, 924, 925*f*
locomotion, 915
population declines, 1196, 1197*f*, 1251
vestigial structures, 471*t*
wheat, 169*f*, 615
wheat rust, 630, 631*f*
whisk fern (*Psilotum nudum*), 582*f*
White, Philip, 765
white-bearded manakin (*Manacus manacus*), 1163
white butterflies (*Pieris brassicae*), 1198
white cliffs of Dover, 560*f*
white-crowned sparrows, 1149, 1150*f*
white-footed mouse (*Peromyscus leucopus*), 1203*f*
white-lipped pit viper (*Cryptelytrops albolabris*), 885*f*

white matter, 859, 860*f*
whitemouth moray eel (*Gymnothorax meleagris*), 691*f*
white muscle fibers, 912
white nose syndrome, 630
white oak, 169*f*
white pelican (*Pelecanus onocrotalus*), 701*f*
white rot fungi, 633
white willow, 1*f*
Whittaker, Robert, 515, 1132, 1208
whorls, 798, 799
wide-field fluorescence microscopy, 68*f*
Wieschaus, Eric, 386, 1098
wild mustard plant (*Brassica oleracea*), 469, 470*f*
wild-type alleles, 335
wild-type strains, 237
Wilkins, Maurice, 219, 220
Wilkinson, Gerald, 1161
Williams, George C., 1158
Wilmut, Ian, 414, 1261
Wilson, Allan, 524
Wilson, E. O., 495, 1219, 1220, 1221–22, 1244, 1245, 1247
Wilson, Henry V., 384
wind
  effects on distribution of organisms, 1126
  plant responses to, 750
  role in ocean currents, 1131
winged fruits, 610*f*, 611
wings
  insect evolution, 672
  of insect orders, 673*t*
  of modern birds, 700*f*
  mutations, 388, 388*f*
wintercreeper, 467, 468*f*
witchweeds (genus *Striga*), 770
Witz, Brian, 1196
Woese, Carl, 515
Wöhler, Friedrich, 42–43
Wolbachia pipientis, 667
Wolf, Larry, 1155
wolf spider (*Lycosa tarantula*), 671, 671*f*
Wollemia nobilis, 603
wolves, 704*f*, 1174
wood
  cellulose discovery, 198
  composition, 730, 731*f*
  evolution, 598
Woodcock, Christopher, 230
woodlice, 1151
woodpecker finches (*Camarhynchus pallidus*), 461, 462*t*
woodpeckers, 1251, 1258
woodpigeon (*Columba palumbus*), 1157
woody plants. *See also* plants
  dermal tissue, 207–8, 207*f*
  evolution, 598
  symbiotic relationships, 767
  vascular tissues, 729, 730–32
woody roots, 734
woolly mammoth (*Mammuthus primigenius*), 1261
work, defined, 950
World Atlas of Birth Defects, 1092
worms. *See* annelids; flatworms; nematodes
wounds, plant responses to, 751–52, 788
Wright, Sewall, 478, 489
Wright, Stephen, 615
wrinkling, 196
Wuchereria bancrofti, 668
Wunsch, Christian, 434
Wyllie, Andrew, 189
Wynne-Edwards, V. C., 1158

X

X chromosomes
  inactivation, 353–55
  sex determination, 331–32
  traits linked to, 332–34
Xenartha, 706
Xenopus laevis, 381, 1089, 1090*f*
xeroderma pigmentosum, 279, 288, 288*f*
X-Gal, 403
X inactivation, 343*t*, 353–55
X inactivation center (Xic), 354–55
Xist gene, 355
X-linked genes, 332–34, 335*t*
X-O system, 331–32
XPD gene, 292*t*
X-ray crystallography, 163, 164*f*
X-ray diffraction, 218–19, 219*f*
X-rays, 157, 285
xylem
  arrangement in stems, 722*f*
  basic functions, 582
  bulk flow forces, 779, 783–84
  development, 721
  genes for developing, 789–90
  in leaves, 723
  primary, 729–30, 730*f*
  in roots, 734
  secondary, 730–32
  structure, 779–83
X-Y system, 331, 332*f*

Y

Yamanaka, Shinya, 393
Yasumura, Yuki, 744
Y chromosome, 331
yeast artificial chromosomes (YACs), 406
yeasts
  glycolysis in anaerobic conditions, 151
  as model organisms, 13, 636
  reproduction, 622, 622*f*
  response to glucose, 175, 175*f*
yellow-bellied marmots (*Marmota flaviventris*), 1252
yellow fever, 361*f*
Yersinia pestis, 548
yew (*Taxus baccata*), 602*f*
yolk, 1079
yolk plug, 1082, 1082*f*
yolk sac, 696, 696*f*
Yoshida, Masasuke, 147, 148*f*
Young, Larry, 1163
yucca species, 497

Z

Zea diploperennis, 1245
Zea mays, 615
zeatin, 740*t*
zebrafish (*Danio rerio*), 381, 1076*f*
zebra mussel (*Dreissena polymorpha*), 660
zebras, 495*f*
zero population growth, 1177, 1183
zeta-globin genes, 429, 1070
Zhao, Chengsong, 790
zinc, 757*t*, 924*t*
zinnias, 609*f*
Z line, 905*f*, 906
zoecium, 660, 660*f*
zona pellucida, 1060, 1079
zone of elongation, 733, 733*f*
zone of maturation, 733, 733*f*
zooplankton, 1140, 1229, 1231
Z scheme, 165, 165*f*
Zuckerkandl, Emile, 520
Z-W system, 332
zygomycetes, 619, 624*t*, 625–26, 626*f*
zygosporangia, 626
zygospores, 626, 626*f*
zygotes
  defined, 308, 380, 1057
  in flowering plants, 719, 796, 804, 805
  in mammals, 1066
  of rotifers, 659
  of sponges, 654
zygotic life cycle
  of bryophytes, 578, 579*f*
  of protists, 567, 568*f*